SME
MINERAL PROCESSING & EXTRACTIVE METALLURGY HANDBOOK

VOLUME TWO

SME MINERAL PROCESSING & EXTRACTIVE METALLURGY HANDBOOK

VOLUME TWO

MANAGING EDITOR: ROBERT C. DUNNE
EDITORS: S. KOMAR KAWATRA & COURTNEY A. YOUNG

PUBLISHED BY THE
SOCIETY FOR MINING, METALLURGY & EXPLORATION

Society for Mining, Metallurgy & Exploration (SME)
12999 E. Adam Aircraft Circle
Englewood, Colorado, USA 80112
(303) 948-4200 / (800) 763-3132
www.smenet.org

The Society for Mining, Metallurgy & Exploration (SME) is a professional society whose more than 15,000 members represent all professionals serving the minerals industry in more than 100 countries. SME members include engineers, geologists, metallurgists, educators, students, and researchers. SME advances the worldwide mining and underground construction community through information exchange and professional development.

ISBN 978-0-87335-385-4
Ebook 978-0-87335-386-1

Library of Congress Cataloging in Publication Control Number: 2018054404

Contents

PART 4: CLASSIFICATION AND WASHING

PART 5: TRANSPORT AND STORAGE

PART 6: PHYSICAL SEPARATIONS

PART 7: FLOTATION

PART 8: SOLID AND LIQUID SEPARATION

PART 9: DISPOSAL

VOLUME TWO

PART 10: HYDROMETALLURGY

Foreword

Mineral processing and extractive metallurgy are atypical disciplines, requiring a combination of knowledge, experience, and art. This was recognized by Peter Ritter von Rittinger in the aptly titled *Lehrbuch der Aufbereitungskunde*, or *Textbook of Processing Art*, that was published in Berlin in 1867. Subsequent handbooks continued the pattern, combining technical analyses of processes with numerous examples and detailed descriptions of successfully operating process plants. Arthur F. Taggart's 1927 *Handbook of Ore Dressing* was, in the author's words, intended to "serve as a reference handbook for engineers ... and also as a text-book for students." Its compact size made it a true handbook—one that could be easily carried in an engineer's knapsack to a mill site in Chuquicamata, Johannesburg, Butte, or Kalgoorlie. With a second edition appearing in 1945, it was the only book on the shelf in many mills.

The *SME Mineral Processing Handbook* published in 1985 rapidly became a standard reference, providing comprehensive coverage of all the unit operations in mineral processing, descriptions of process plants for more than 26 minerals and materials, and small sections on hydrometallurgy and pyrometallurgy. The authors of its many chapters contributed their collective expertise unselfishly to provide a handbook that was truly useful to all the practitioners of mineral processing—students, engineers, mill managers, and operators.

This new *SME Mineral Processing and Extractive Metallurgy Handbook* provides up-to-date coverage of all mineral processing unit operations, but it also has much larger sections on hydrometallurgy and pyrometallurgy as well as a section on management and reporting, which discusses such topics as health and safety, community and social issues, and project management. Once again, each chapter is authored by an acknowledged expert. These selfless experts, recruited by the editors, have each made an invaluable contribution. The chapters were skillfully organized and refined by the managing editor, with the incomparable assistance of SME's book publishing team. The new SME handbook is truly a timely document, addressing the new technologies and important cultural and social issues that are important to today's minerals community.

With the release of the *SME Mineral Processing and Extractive Metallurgy Handbook*, SME continues to fulfill its mission as the world's premier publisher and distributor of technical information related to the mineral industries. I am proud to have had some association with the society's continuing efforts and ongoing success in that endeavor.

Michael G. Nelson
Professor and Chair, Mining Engineering Department
University of Utah, Salt Lake City, Utah

Preface

A *handbook* is a concise reference book intended to provide ready reference material such as facts on a particular subject; details regarding how equipment works, how it operates, and how it is controlled; and guidance for problem solving. Such a text should be easy to consult and should provide quick answers where possible. This new *SME Mineral Processing and Extractive Metallurgy Handbook* has been designed and written to meet these criteria and is intended for those interested about, and working in, these industries in the 21st century.

The handbook contains the most current information on unit operations used in mineral processing, hydrometallurgy, and pyrometallurgy. Integrated optimization and value-chain processes that have gained importance during the last 20 years, such as mine-to mill, mine-to-market, and mine-to-metal, are explained, and examples of mine-to-market and mine-to-metal applications are inherent in the commodity chapters. An important input to these processes depends on proper and quantitative ore characterization. The latest available analytical methods and automated mineral identification systems that are used for ore characterization are described in separate chapters and will help metallurgists select the most appropriate procedures for their particular situations. The important design and equipment requirements for processes such as carbon-in-pulp, carbon-in-leach, bacterial tank leaching, heap and in situ leaching, and pressure oxidation will help both those looking at these options for a new project and operators of these processes. How best to model and control these and other methods is an important aspect when optimizing metallurgical operation performance. Chapters on modeling and simulation as well as process control and operational intelligence provide excellent overviews of what techniques are available and what works. The importance of proper and appropriate health and safety programs, environmental stewardship, and attention to community and social engagement are also covered in these pages. Those interested in recycling, graphene, and "battery" minerals will also find chapters appropriate to these topics. Overall, the handbook contains a wealth of valuable information.

The search for an editor to take on the task of producing an updated handbook began in 2012, as the previous handbook went out of print in 2003 and there was a steady and undeniable demand for a new version. Professors S. Komar Kawatra and Courtney A. Young from Michigan Technological University and Montana Tech, respectively, submitted a joint handbook proposal that incorporated both mineral processing and extractive metallurgy. The proposal was unanimously approved by the SME Information Publishing Committee in 2013. After six years of intense work by the authors, reviewers, editors, and SME book publishing team, the new *SME Mineral Processing and Extractive Metallurgy Handbook* is now here, comprising two volumes of 2,312 pages, with 128 chapters.

A project of this magnitude owes much appreciation to the commitment of the many people who brought it to fruition, including 192 authors and co-authors, and 102 technical reviewers, to realize a product of technical and practical worth. I applaud Kawatra and Young for taking on the challenge of such a large, complex, and diverse project. An unexpected obstacle that hindered the long-term progress of the task was the turbulent time in the mining industry during 2014–2017, created by the global recession in the mining industry. It was an extremely stressful and unsettling stretch for all those associated with the handbook. Many of the authors who agreed to contribute were retrenched or relocated, which meant that many were unable to fulfil their commitment and new authors had to be found. During this period, I was asked to join the technical team and assist with locating new contributors and provide technical management of the project.

Another issue that reared its head and is aptly described by Arthur F. Taggart is that "No one who has not tried to write a technical chapter in a handbook, or a part thereof, realizes how little he or any other one person knows about the subject that he considers his specialty." Consequently, many authors needed to spend much more of their personal and precious time either late at night and/or over weekends to source the appropriate information and complete the development of their chapters. On behalf of Kawatra, Young, and I, we extend our appreciation and sincere thanks to all the authors, co-authors, and technical reviewers for their dedication and commitment to make the handbook a reality. My special and heartfelt thanks goes to Michael G. Nelson and Thomas Battle who were stalwarts willing to give a lot more of their valuable time to take on additional authorship and reviewing tasks at short notice. Undoubtably, owners of the new handbook will greatly appreciate everyone's efforts in delivering these two volumes of technical and practical excellence.

The five professional and dedicated individuals who supported SME efforts and were intimately associated with the production of the handbook require special mention and a sincere thank-you for a job well done: Jane Olivier, manager of book publishing, who initiated the project; Diane Serafin, senior technical editor; Terese Platten, technical editor; Karen Ehrmann, permissions editor; and Stephen Adams who composed the pages. Readers of the handbook will appreciate the enormity of the overall editing task that included obtaining thousands of permissions and redrawing or reworking almost 3,000 figures to produce this high-quality handbook.

Finally, SME's management and Mineral & Metallurgical Processing Division members are thanked for their support and encouragement throughout the project.

Robert C. Dunne
Managing Editor
Gooseberry Hill, Western Australia

About the Editors

Robert C. Dunne

Robert C. Dunne, now retired, has worked in the mining industry for 41 years and is currently an adjunct professor at both Curtin University (Gold Technology Group) and the University of Queensland (Julius Kruttschnitt Mineral Research Centre [JKMRC]) in Australia. He graduated from the University of Witwatersrand in Johannesburg, South Africa, with a BScEng degree in metallurgy.

During his career, Dunne held group executive positions in metallurgical development and technology and also in innovation, and he spent time at Newcrest Mining, Anglovaal, and Anglo American. He was the assistant director of ore dressing at Mintek for 14 years, and during this time, he visited most of the gold and base metal mines in South Africa and initiated mineral processing research and development at the chemical engineering departments of Cape Town and Stellenbosch universities. He immigrated to Australia in 1986 and spent three years at the Western Australian School of Mines as senior lecturer in the Minerals Engineering and Extractive Metallurgy Department and oversaw the new Research and Development Department. At the time of his retirement in 2013, he was the Fellow Metallurgy at Newmont Mining Corporation in Denver, Colorado.

In 2016, Dunne received the Robert H. Richards Award from the American Institute of Mining, Metallurgical, and Petroleum Engineers for pioneering work in the application of semiautogenous milling, flotation, and gravity concentration for gold and copper recovery. He was a Society for Mining, Metallurgy & Exploration (SME) Henry Krumb Lecturer in 2012 with his presentation titled Water in Mining. In 2017, Dunne was nominated as the Australasian Institute of Mining and Metallurgy's (AusIMM's) Metallurgical Society G.D. Delprat Distinguished Lecturer, and his presentation titled Water: Facts, Perceptions and Conflicts was delivered to local AusIMM branches throughout Australia. As an invited conference speaker, Dunne has given discourses on refractory gold, gravity separation, gold flotation, and mine-to-mill application. He has also been a contributor to numerous short courses on gold processing and flotation. Dunne has authored and co-authored more than 80 technical papers in conference proceedings and professional journals and has contributed to numerous chapters on gold and precious metal flotation.

Dunne is a member of SME, AusIMM, and the Southern African Institute of Mining and Metallurgy. He has been a reviewer for *Minerals Engineering*; *Mineral Processing and Extractive Metallurgy: Transactions of the Institutions of Mining and Metallurgy*; and for research applications for the Natural Sciences and Engineering Research Council of Canada. He has also represented the mining industry on several committees, including the Chamber of Minerals and Energy of Western Australia, and as a board member of JKMRC and a panel member of the Gold Advisory Group at the A.J. Parker Cooperative Research Centre for Hydrometallurgy at Murdoch University in Western Australia.

Although retired, Dunne provides consulting services through his metallurgical consulting company and enjoys mentoring young engineers and advancing new ideas and technologies within the mining industry. Spending more time with family, sons Kevin and James and daughter Tracy, and his five grandchildren, is a favorite pastime, as is traveling the world with his wife Deidre.

S. Komar Kawatra

S. Komar Kawatra is a professor of chemical engineering at Michigan Technological University in the United States. Kawatra's primary research focuses on analysis, control, and optimization of mineral processing operations and the treatment and utilization of industrial waste materials. He obtained an MS degree in physics from the former University of Poona in India and a PhD in metallurgy from the University of Queensland in Brisbane, Queensland, Australia. His doctoral work was carried out at Mount Isa Mines in Mount Isa, Australia.

Kawatra's research experience includes work at the Bhabha Atomic Research Centre in Trombay, Bombay, India; the Julius Kruttschnitt Mineral Research Centre in Brisbane; and the Canada Centre for Mineral and Energy Technology in Ottawa, Ontario, Canada. He was also a research associate at the University of Alberta in Edmonton, Canada. He moved to Michigan Technological University in 1977 and became chair of the Department of Chemical Engineering from 2007 to 2017 and, during this time, established several endowments. Kawatra has been continuously active in research projects of academic, industrial, and governmental interest and has received several patents for technologies stemming from his research.

Since 1997, Kawatra has been editor in chief of the Society for Mining, Metallurgy & Exploration's (SME's) *Minerals & Metallurgical Processing* journal. Since 2001, he has also been editor in chief of *Mineral Processing and Extractive Metallurgy Review*, an international journal published by Taylor and Francis. He has been invited to give many keynote addresses and plenary lectures on the future of mineral processing and extraction, and has also authored or edited nine books and more than 200 peer-reviewed publications.

Kawatra has received several SME awards, including the Distinguished Member Award, the Arthur F. Taggart Award for advances in coal flotation technology, and the Antoine M. Gaudin Award for sustainable by-product management treatment and utilization. He has also received an SME Presidential Citation in honor of his outstanding contributions to SME.

Kawatra has received prestigious awards from several other organizations, including the Robert H. Richards Award and the Frank F. Aplan Award from the American Institute of Mining, Metallurgical, and Petroleum Engineers; the Michigan Technological University Research Award and Graduate Student Mentor Award; and the IEEE Certificate of Service from the Institute of Electrical and Electronics Engineers. He has also received the Michigan Association of Governing Boards Distinguished Faculty Member Award for extraordinary contribution to Michigan higher education. Kawatra is a Fulbright Scholar, having completed the Fulbright Scholarship Program in 2012.

Kawatra is grateful to his graduate students and Leeyana Gupta for their assistance during the course of his career.

Courtney A. Young

Courtney A. Young is the Department Head and Lewis S. Prater Distinguished Professor of Metallurgical and Materials Engineering at Montana Technological University (Montana Tech). As a faculty member there for the past 25 years, Young has taught various topics centered around mineral processing and extractive metallurgy to more than 2,000 students. He has also worked with numerous local and international companies doing research in these areas, focusing on mining sustainability issues such as water remediation, slag recycling, tailings repurposing, and waste treatment.

Young is a graduate of three premier mineral processing and extractive metallurgy institutions, having obtained his BS degree in mineral processing engineering from the former Montana College of Mineral Science and Technology in 1984, his MS degree in mining and minerals engineering from Virginia Polytechnic Institute and State University in 1987, and his PhD degree in metallurgical engineering from the University of Utah's College of Mines and Earth Sciences in 1994.

Young is extremely active in the Society for Mining, Metallurgy & Exploration (SME), having served as a co-advisor to its local student chapter since 1995 and on at least 30 committees, including 18 as chair or co-chair. He has organized and edited seven symposia and their corresponding proceedings, chaired and organized 15 sessions at SME meetings, and presented more than 30 papers, with about half being published in SME proceedings or the *Minerals & Metallurgical Processing* journal. He has co-taught an extremely popular short course in mineral processing and extractive metallurgy with his good friend, Corby Anderson. The course will continue to be offered annually at the SME Annual Conference and Expo.

Young has authored and co-authored 213 publications and presentations. Because of his devoted efforts to this handbook, SME, and the mining industry in general, Young has been accordingly recognized with plenary lectures and has received many awards, including the Outstanding Young Engineer Award and Millman of Distinction Award from the Mineral & Metallurgical Processing Division of SME; the Frank F. Aplan Award and Mineral Industry Education Award from the American Institute of Mining, Metallurgical, and Petroleum Engineers; and two Presidential Citations and the Distinguished Member Award from SME. His and others' research at Montana Tech have led to the establishment of a Materials Science PhD program, a first for Montana Tech, with materials being specifically defined to include minerals and metals.

Young acknowledges that none of this would have happened without the support of his wife, Miriam, of 30 years and their daughters, Jessica and Jamie.

Contributing Authors

Russell D. Alley
President
Minerals Advisory Group LLC
Tucson, Arizona, USA

Peter Amelunxen
President
Aminpro
Lima, Peru

Basak Anameric
Chief Scientist
Basak Anameric Consulting
Grand Rapids, Minnesota, USA

Corby G. Anderson
Harrison Western Professor
Dept. of Metallurgical & Materials Engineering
Colorado School of Mines
Golden, Colorado, USA

Josh Andres
Senior Metallurgist
Freeport-McMoRan Inc.
Morenci, Arizona, USA

Ian Arbuthnot
Retired
Outotec Pty Ltd.
Perth, Western Australia, Australia

Esau Arinaitwe
Group Leader
Solvay Technology Solutions
Stamford, Connecticut, USA

Jari Aromaa
Senior Lecturer
Aalto University
Espoo, Finland

Kevin Ausburn
Chief Mineralogist
FLSmidth USA Inc.
Midvale, Utah, USA

Mark G. Aylmore
Senior Research Fellow
John de Laeter Centre, Faculty of Science & Engineering
Curtin University
Perth, Western Australia, Australia

Manfred Bach
Senior Sales Manager Alumina
FLSmidth Wiesbaden GmbH
Walluf, Hessen, Germany

Andrew Bamber
Principal and Managing Director
Bara Consulting Ltd.
London, United Kingdom

Arthur Barnes
President
Metallurgical Process Consultants Ltd.
Kamloops, British Columbia, Canada

Osvaldo Bascur
Principal, Academia–Industry Innovation
OSIsoft LLC
Houston, Texas, USA

Thomas P. Battle
Extractive Metallurgy Consultant
Self-Employed
Charlotte, North Carolina, USA

Wolfgang Baum
Managing Director
Ore & Plant Mineralogy LLC
San Diego, California, USA

Richard (Ted) Bearman
Director
Bear Rock Solutions Pty Ltd.
Attadale, Western Australia, Australia

Megan Becker
Associate Professor
Dept. of Chemical Engineering, University of Cape Town
Cape Town, South Africa

Jose A. Botin
Retired Professor
Universidad Politecnica de Madrid
Madrid, Spain

Sylvie C. Bouffard
Manager, Studies Optimization
BHP
Saskatoon, Saskatchewan, Canada

David J. Bowman
Principal Engineer
Bear Rock Solutions Pty Ltd.
Attadale, Western Australia, Australia

Dee Bradshaw
Emeritus Professor
Dept. of Chemical Engineering, University of Cape Town
Cape Town, South Africa

Corale L. Brierley
President/Principal
Brierley Consultancy LLC
Highlands Ranch, Colorado, USA

James A. Brierley
Biohydrometallurgy Consultant
Brierley Consultancy LLC
Highlands Ranch, Colorado, USA

Bill Burton
Chief Engineer, Precious Metals Recovery
FLSmidth Salt Lake City Inc.
Salt Lake City, Utah, USA

Frank Cappuccitti
President
Flottec LLC
Boonton, New Jersey, USA

Jannette L. Chorney
Adjunct Professor
Montana Tech
Butte, Montana, USA

William M. Cross
Professor
Materials & Metallurgical Engineering
South Dakota School of Mines & Technology
Rapid City, South Dakota, USA

Frank Crundwell
Director
CM Solutions Pty Ltd.
Johannesburg, South Africa

Jan Czarnecki
Adjunct Professor
University of Alberta
Edmonton, Alberta, Canada

Michael Daniel
Consulting Metallurgist & Director
CMD Consulting Pty Ltd.
Brisbane, Queensland, Australia

Avimanyu Das
Associate Professor, Metallurgical & Materials Engineering
Montana Tech
Butte, Montana, USA

Dean M. David
Technical Director–Process
Mining & Minerals Australia Division, Wood
Perth, Western Australia, Australia

Alex de Andrade
General Manager Equipment & Technology
Mineral Technologies Pty Ltd.
Gold Coast, Queensland, Australia

Jerome P. Downey
Professor of Extractive Metallurgy
Montana Tech
Butte, Montana, USA

Michael J. Dry
Owner
Arithmetek Inc.
Peterborough, Ontario, Canada

Martin du Plessis
Lead Mechanical Engineer
Fluor
Perth, Western Australia, Australia

Kristy-Ann Duffy
Process Consultant
Consulting & Technology, Mining & Minerals Processing
Hatch
Brisbane, Queensland, Australia

Robert C. Dunne
Principal
Robert Dunne Consulting
Gooseberry Hill, Western Australia, Australia

Timothy C. Eisele
Assistant Professor, Department of Chemical Engineering
Michigan Technological University
Houghton, Michigan, USA

Cathy Evans
Senior Research Fellow
Julius Kruttschnitt Mineral Research Centre
Brisbane, Queensland, Australia

Daryl Evans
Director of Metallurgy
IMO Pty Ltd.
Perth, Western Australia, Australia

Ken Evans
Formerly with Rio Tinto Alcan (Retired)
Evans Technical Consulting Limited
Chalfont St. Peter, United Kingdom

Steven C. Evans
Principal Engineer (Retired)
Westinghouse Electric Company, Western Zirconium Plant
Ogden, Utah, USA

Raymond S. Farinato
Senior Research Fellow
Solvay Technology Solutions
Stamford, Connecticut, USA

Brian Flintoff
Consultant
Sigmation Inc.
Kelowna, British Columbia, Canada

Olof Forsén
Professor Emeritus
Aalto University
Espoo, Finland

Kevin S. Fraser
Co-Director & Principal Metallurgist
High-Pressure Metallurgy, Metals, Autoclave Technology
Hatch Ltd.
Mississauga, Ontario, Canada

Michael L. Free
Professor
University of Utah
Salt Lake City, Utah, USA

Rudi Frischmuth
Senior Metallurgist
High-Pressure Metallurgy, Metals, Autoclave Technology
Hatch Ltd.
Mississauga, Ontario, Canada

Francis H. (Sam) Froes
Consultant
Tacoma, Washington, USA

Xu Gao
Assistant Professor
Institute of Multidisciplinary Research for Adv. Materials
Tohoku University
Sendai, Miyagi, Japan

Mike Garska
Senior Process Engineer
Hudson Ranch Energy Service
Calipatria, California, USA

David B. George
Principal Consultant
David B. George and Associates LLC
Heber City, Utah, USA

David George-Kennedy
Chief Advisor, Smelting & Refining
Rio Tinto Copper & Diamonds
Salt Lake City, Utah, USA

Mike Germain
Process Metallurgist
Mineral Technologies Pty Ltd.
Gold Coast, Queensland, Australia

Andrea R. Gerson
Professor & Managing Director
Blue Minerals Consultancy
Middleton, South Australia, Australia

Aidan Giblett
Senior Technical Advisor—Mineral Processing
Newmont Mining Services
Subiaco, Western Australia, Australia

J.R. Goode
Principal & Consulting Metallurgist
J.R. Goode & Associates
Toronto, Ontario, Canada

Nolan Goodweiler
Senior Metallurgist
Climax Molybdenum Company
Parshall, Colorado, USA

Muxing Guo
Senior Researcher
Department of Materials Engineering
Katholieke Universiteit Leuven
Leuven, Belgium

Kenneth N. Han
Professor Emeritus
South Dakota School of Mines & Technology
Rapid City, South Dakota, USA

Greg Harbort
Technical Director—Process
Wood plc
Brisbane, Queensland, Australia

David Harbottle
Associate Professor
School of Chemical & Process Engineering
University of Leeds
West Yorkshire, United Kingdom

Daniel Harris
Consultant & Professional Board Director
EnergySource LLC
Reno, Nevada, USA

Brian R. Hart
Adjunct Professor & Senior Research Scientist
Univ. of Western Ontario & Surface Science Western
London, Ontario, Canada

Howard Haselhuhn
Applications Engineer–Industrial Minerals
Solvay Technology Solutions
Stamford, Connecticut, USA

Nick Hazen
President
Hazen Research Inc.
Golden, Colorado, USA

Jason Heath
Senior Metallurgist
Talison Lithium
Greenbushes, Western Australia, Australia

Paul J. Henkels
Senior Technical Manager
United States Gypsum Company
Chicago, Illinois, USA

Brandt Henriksson
Vice President Thickeners & Clarifiers
Outotec OY
Espoo, Finland

Leonard Hill
Technical Services Director
Freeport-McMoRan Inc.
Phoenix, Arizona, USA

J. Brent Hiskey
Professor Emeritus, Mining & Geological Engineering
University of Arizona
Tucson, Arizona, USA

Lauri Holappa
Emeritus Professor
Aalto University
Espoo, Finland

John Michael Holden
Materials Handling Engineer
Retired
Littleton, Colorado, USA

John Hollow
Owner
CRCW Consulting & Management Services
Castle Rock, Colorado, USA

Rick Q. Honaker
Professor
Mining Engineering, University of Kentucky
Lexington, Kentucky, USA

Hsin-Hsiung Huang
Professor
Department of Metallurgical & Materials Engineering
Montana Tech
Butte, Montana, USA

Peter Huxtable
Principal
Huxtable Associates
Calver, Derbyshire, United Kingdom

M. Ashraf Imam
Faculty/Research Professor
George Washington University
Washington, D.C., USA

Dusty Jacobson
Senior Cushing & Screening Specialist
Metso Minerals Industries Inc.
Waukesha, Wisconsin, USA

Alex Jankovic
Principal Consultant
Consulting & Technology, Mining & Minerals Processing
Hatch
Brisbane, Queensland, Australia

Matthew Jeffrey
Process Manager
Newmont USA Limited
Englewood, Colorado, USA

Bill Johnson
Principal Consultant & Adjunct Professor
Mineralis Consultants Pty Ltd. & University of Queensland
Brisbane, Queensland, Australia

Adam Johnston
Chief Metallurgist
Transmin Metallurgical Consultants
Lima, Peru

Myungwon Jung
Postdoctoral Researcher
Metal Processing Institute, Worcester Polytechnic Institute
Worcester, Massachusetts, USA

Sarma S. Kanchibotla
Professor
JKMRC, Sustainable Minerals Institute
University of Queensland
Brisbane, Queensland, Australia

S. Komar Kawatra
Professor & Chair, Department of Chemical Engineering
Michigan Technological University
Houghton, Michigan, USA

Jon J. Kellar
Nucor Professor & Douglas W. Fuerstenau Professor
South Dakota School of Mines & Technology
Rapid City, South Dakota, USA

Harold E. Kelley
Director of Technical Services
TungsteMet
Oak Hill, West Virginia, USA

Brandon Kern
Technical Service Specialist, Ion Exchange
The Dow Chemical Company
Midland, Michigan, USA

Bern Klein
Professor, Norman B. Keevil Institute of Mining Engineering
University of British Columbia
Vancouver, British Columbia, Canada

Brian Knorr
Director, Research & Development
Metso Minerals Industries Inc.
York, Pennsylvania, USA

Jaisen N. Kohmuench
Managing Director
Eriez Manufacturing Pty Ltd.
Melbourne, Victoria, Australia

Stacy Kramer
Vice President Global H&S
Freeport-McMoRan Inc.
Phoenix, Arizona, USA

Tom Krumins
Process Engineer
High-Pressure Metallurgy, Metals, Autoclave Technology
Hatch Ltd.
Mississauga, Ontario, Canada

Martin C. Kuhn
Chief Executive Officer
Minerals Advisory Group LLC
Tucson, Arizona, USA

Halvor Kvande
Professor Emeritus
Norwegian University of Science & Technology
Trondheim, Norway

Richard LaDouceur
Postdoctoral Research Scholar
Montana Tech
Butte, Montana, USA

Sai Wei Lam
Senior Process Engineer
Wood
Perth, Western Australia, Australia

Tim Laros
Owner
Filtration Technologies LLC
Park City, Utah, USA

Mike Larson
Applications Engineering Manager
Moly-Cop USA
Ewen, Michigan, USA

Marko Latva-Kokko
R&D Manager—Hydromet Equipment
Outotec
Pori, Finland

Dariusz Lelinski
Global Product Manager of Flotation
FLSmidth Minerals
Salt Lake City, Utah, USA

Meg Dietrich LeVier
Consulting Analytical Chemist
K. Marc LeVier & Associates Inc.
Highlands Ranch, Colorado, USA

Alison Lewis
Dean, Faculty of Engineering & Built Environment
Dept. of Chemical Engineering, University of Cape Town
Cape Town, South Africa

Leendert (Leon) Lorenzen
Managing Director (Professor)
Lorenzen Consultants, Mintrex (Stellenbosch Univ.)
Perth, Australia (Stellenbosch, South Africa)

Norman O. Lotter
President
Flowsheets Metallurgical Consulting Inc.
Sudbury, Ontario, Canada

David Love
Principal Consultant & Director
Water Logics
Adelaide, Australia

Mari Lundström
Professor
Aalto University
Espoo, Finland

John Lupo
Senior Director Geotechnical and Hydrology
Newport Mining Corporation
Denver, Colorado, USA

Gerald H. Luttrell
E. Morgan Massey Professor
Mining & Minerals Engineering, Virginia Tech
Blacksburg, Virginia, USA

David C. Lynch
Professor Emeritus
University of Arizona
Tucson, Arizona, USA

Phillip J. Mackey
President
P.J. Mackey Technology Inc.
Kirkland, Quebec, Canada

Deepak Malhotra
President
Resource Development Inc.
Wheat Ridge, Colorado, USA

Michael J. Mankosa
Executive Vice President of Global Technology
Eriez Manufacturing
Erie, Pennsylvania, USA

John G. Mansanti
President and CEO
Crystal Peak Minerals Inc.
Salt Lake City, Utah, USA

John O. Marsden
President & Consultant
Metallurgium
Phoenix, Arizona, USA

Jacob Masliyah
Professor Emeritus
University of Alberta
Edmonton, Alberta, Canada

Hiroyuki Matsuura
Associate Professor
Department of Materials Engineering, University of Tokyo
Tokyo, Japan

Edward F. McCarthy
Manager
Performance Minerals LLC
Morgan Hill, California, USA

Ed McGowan
Safety Manager
Kinross Gold Mining (Bald Mountain) Inc.
Elko, Nevada, USA

Terry McNulty
President
T.P. McNulty and Associates Inc.
Tucson, Arizona, USA

David G. Meadows
Manager of Global Process Technology
Bechtel Mining and Metals
Phoenix, Arizona, USA

Glenn C. Miller
Professor
Dept. of Natural Resources & Environmental Science
University of Nevada–Reno
Reno, Nevada, USA

Paul Miller
Managing Director
Sulphide Resource Processing Pty Ltd.
Perth, Western Australia, Australia

Brajendra Mishra
Director of Metal Processing Institute
Mechanical Engineering, Worcester Polytechnic Institute
Worcester, Massachusetts, USA

Michael S. Moats
Associate Professor
Missouri University of Science & Technology
Rolla, Missouri, USA

Stephen Morrell
President
SMC Testing Pty Ltd.
Brisbane, Queensland, Australia

Robert D. Morrison
Former Chief Technologist
Julius Kruttschnitt Mineral Research Centre
University of Queensland
Brisbane, Queensland, Australia

Trond Muri
Senior Project Engineer
PASSER
Vear, Vestfold, Norway

D.R. Nagaraj
Principal Research Fellow
Solvay Technology Solutions
Stamford, Connecticut, USA

Michael G. Nelson
Professor & Chair, Mining Engineering Department
University of Utah
Salt Lake City, Utah, USA

Khosrow Nikkhah
Senior Process Engineer
AMEC Foster Wheeler
Vancouver, British Columbia, Canada

Aaron Noble
Associate Professor
Mining & Minerals Engineering, Virginia Tech
Blacksburg, Virginia, USA

Stefan Norgaard
Operations Manager
Australian Laboratory Services
Australasian Institute of Mining and Metallurgy
Perth, Western Australia, Australia

Daniel A. Norrgran
Former Manager Minerals & Materials Processing Division
Eriez Manufacturing
Erie, Pennsylvania, USA

Ricardo Maerschner Ogawa
Product Manager, Mining Screens
Metso Brasil Indústria e Comércio
Sorocaba, Brazil

Joohyun Park
Professor
Department of Materials & Chemical Engineering
Hanyang University
Ansan, Korea

Bruce Peachey
President & Senior Consultant
New Paradigm Engineering Ltd.
Edmonton, Alberta, Canada

Murray Pearson
Director, Technology Development
High-Pressure Metallurgy, Metals, Autoclave Technology
Hatch Ltd.
Mississauga, Ontario, Canada

P. Chris Pistorius
POSCO Professor of Materials Science & Engineering
Carnegie Mellon University
Pittsburgh, Pennsylvania, USA

Graham Popplewell
Technical Director & Senior Fellow, Diamond Processing
Fluor Corporation
Perth, Western Australia, Australia

Joe Poveromo
Raw Materials & Ironmaking Global Consulting
Self-Employed
Bethlehem, Pennsylvania, USA

Robert J. Pruett
Formerly with Imersys Oilfield Solutions
Sandersville, Georgia, USA

Brian Putland
Principal Metallurgist
Orway Mineral Consultants
Perth, Western Australia, Australia

Randall Pyper
General Manager
Kappes, Cassiday & Associates Australia Pty Ltd.
Perth, Western Australia, Australia

Benny E. Raahauge
General Manager, Alumina & Pyro Technology
FLSmidth
Copenhagen, Denmark

Shashi Rao
Metallurgical Engineer
Natural Resources Research Institute, Coleraine Labs
University of Minnesota
Coleraine, Minnesota, USA

Amy J. Richins
Research Associate, Mining Engineering Department
University of Utah
Salt Lake City, Utah, USA

Tiago Ramos Riebeiro
Researcher
Laboratory of Metallurgical Processes
IPT (Institute for Technological Research)
São Paulo, Brazil

S. Jayson Ripke
Manager–R&D
Midrex Technologies Inc.
Pineville, North Carolina, USA

Joanna Robertson
Director of Metal Recovery
Freeport-McMoRan Technology Center
Tucson, Arizona, USA

David Rohaus
(Retired)
United States Steel Corporation
Pittsburgh, Pennsylvania, USA

Kym Runge
Program Leader, Separation
JKMRC, Sustainable Minerals Institute
University of Queensland
Brisbane, Queensland, Australia

Rackel San Nicolas
Lecturer
Department of Infrastructure Engineering
The University of Melbourne
Parkville, Victoria, Australia

Vicki J. Scharnhorst
Operations Manager
Tetra Tech Inc.
Denver, Colorado, USA

Mark E. Schlesinger
Professor of Metallurgical Engineering
Missouri University of Science & Technology
Rolla, Missouri, USA

Christopher Schmitz
Chief Mine Engineer
Climax Molybdenum Company
Leadville, Colorado, USA

Henry Schnell
Retired
Formerly with AREVA (Paris, France)
Eagle Bay, British Columbia, Canada

Fred Schoenbrunn
Director for Thickeners
FLSmidth
Salt Lake City, Utah, USA

Rebecca Sciberras
Lead Metallurgist
Orway Mineral Consultants
Perth, Western Australia, Australia

Andrew Scogings
Principal Consultant
CSA Global Pty Ltd.
Perth, Western Australia, Australia

Thom Seal
Director of the Institute for Mineral Resource Studies
Mining & Metallurgical Eng. Dept.
University of Nevada–Reno
Reno, Nevada, USA

Roger D. Sharpe
Director, GeoTechnical & Mining Services
United States Gypsum Company
Chicago, Illinois, USA

Andreas Siegmund
Principal
LanMetCon LLC
Lantana, Texas, USA

Roger St.C. Smart
Emeritus Professor & Senior Consultant
Univ. of South Australia & Blue Minerals Consultancy
Mawson Lakes, South Australia, Australia

Stuart Smith
Technical Director
Metifex Pty Ltd.
Brisbane, Queensland, Australia

Matthew Soderstrom
Business Director Metal Extraction Products
Solvay S.A.
Tempe, Arizona, USA

Peter-Hans ter Weer
Director
TWS Services and Advice Bauxite & Alumina Consultancy
Huizen, The Netherlands

Daniel Thomas
Principal Consultant, Alumina
Advisian, WorleyParsons Group
Brisbane, Queensland, Australia

Philip Thompson
Senior Minerals Processing Advisor
FLSmidth
Salt Lake City, Utah, USA

Larry G. Twidwell
(Retired) Emeritus Professor of Metallurgical Engineering
Metallurgy/Materials Engineering Dept., Montana Tech
Butte, Montana, USA

John L. Uhrie
President Consulting Services–Americas
RPM Global USA Inc.
Greenwood Village, Colorado, USA

Walter Valery
Global Director
Consulting & Technology, Mining & Minerals Processing
Hatch
Brisbane, Queensland, Australia

Jannie S. J. van Deventer
Chief Executive Officer
Zeobond Pty Ltd.
Melbourne, Victoria, Australia

Edgar E. Vidal
Director of Business Dev. for Defense, Nuclear & Science
Materion Beryllium & Composites
Mayfield Heights, Ohio, USA

Judith C. Vidal
Building Energy Science Group Manager
National Renewable Energy Laboratory
Golden, Colorado, USA

Brant Walkley
Research Associate
The University of Sheffield
Sheffield, United Kingdom

Henry Walqui
Director, Field Pilot Plant Programs
CiDRA Minerals Processing
Windsor, Connecticut, USA

Chengtie Wang
PhD Candidate
University of British Columbia
Vancouver, British Columbia, Canada

Barry Welch
Professor Emeritus
University of Auckland
Auckland, New Zealand

Nicholas J. Welham
Principal
Welham Metallurgical Services
Perth, Western Australia, Australia

Jessica M. Wempen
Assistant Professor of Mining Engineering
University of Utah
Salt Lake City, Utah, USA

Greg Wilkie
Program Coordinator
CRC Ore
Brisbane, Queensland, Australia

Richard Williams
Regional Manager Western Australia
McLanahan Corporation
Perth, Western Australia, Australia

Gary R. Wilson
Graduate Student
Department of Metallurgical & Materials Engineering
Montana Tech
Butte, Montana, USA

David Wiseman
Principal
David Wiseman Pty Ltd.
Blackwood, South Australia, Australia

Zhenghe Xu
Tech Professor (Dean)
Univ. of Alberta (Southern Univ. of Science & Technology)
Edmonton, Alberta, Canada (Shenzhen, China)

Jorge L. Yordan Hernandez
Retired
Formerly with Rio Tinto Minerals
Parker, Colorado, USA

Courtney A. Young
Department Head & Professor
Metallurgical & Materials Engineering
Montana Tech
Butte, Montana, USA

Patrick Zhang
Research Director
Florida Industrial & Phosphate Research Institute
Bartow, Florida, USA

Technical Reviewers

Jeffrey F. Adams
Specialist Process Engineer, Hydrometallurgy
Hatch Ltd.
Mississauga, Ontario, Canada

John E. Angove
Principal
AFT Metallurgy
North Beach, Western Australia, Australia

Dan Apai
Senior Civil Engineer
Fluor Canada
Vancouver, British Columbia, Canada

Tony Bagshaw
Retired
Formerly with Minerals Research Institute of WA
Nedlands, Western Australia, Australia

Amanda S. Barnard
Chief Research Scientist, Data61
CSIRO
Docklands, Victoria, Australia

Thomas Battle
Ext. Metallurgy Consultant & Senior Consulting Engineer
Kingston Process Metallurgy
Charlotte, North Carolina, USA

Richard (Ted) Bearman
Director
Bear Rock Solutions Pty Ltd.
Attadale, Western Australia, Australia

Gavin Becker
Principal
Gavin Becker & Associates
Ormiston, Queensland, Australia

Sylvie C. Bouffard
Manager, Studies Optimization
BHP
Saskatoon, Saskatchewan, Canada

Quentin Campbell
Professor, School of Chemical and Minerals Engineering
North-West University
Potchefstroom, South Africa

John A. Cole
Senior Director of Process Development
Newmont Mining Corporation
Englewood, Colorado, USA

Rob Coleman
Account Director
Outotec
Brisbane, Queensland, Australia

Avimanyu Das
Associate Professor, Metallurgical & Materials Engineering
Montana Tech
Butte, Montana, USA

Dean M. David
Technical Director–Process
Mining and Minerals Australia Division, Wood
Perth, Western Australia, Australia

Emmanuel De Moor
Assistant Professor
Department of Metallurgical & Materials Engineering
Colorado School of Mines
Golden, Colorado, USA

Carl E. Defilippi
Engineering Manager
Kappes, Cassiday & Associates
Reno, Nevada, USA

Jaroslaw W. Drelich
Professor, Department of Materials Science & Engineering
Michigan Technological University
Houghton, Michigan, USA

David C. Duffy
Mining Engineer
ACG Materials
Norman, Oklahoma, USA

Robert C. Dunne
Principal
Robert Dunne Consulting
Gooseberry Hill, Western Australia, Australia

Akbar Farzanegan
Professor, School of Mining, College of Engineering
University of Tehran
Tehran, Iran

Brian Flinthoff
Consultant
Sigmation Inc.
Kelowna, British Columbia, Canada

Robert J. Fraser
Global Director, Hydrometallurgy
Hatch Ltd.
Mississauga, Ontario, Canada

Mike Garska
Senior Process Engineer
Hudson Ranch Energy Service
Calipatria, California, USA

Andrea R. Gerson
Professor & Managing Director
Blue Minerals Consultancy
Middleton, South Australia, Australia

Aidan Giblett
Senior Technical Advisor—Mineral Processing
Newmont Mining Services
Subiaco, Western Australia, Australia

Rick S. Gilbert
Vice President Process Technology
Freeport-McMoRan Inc.
Phoenix, Arizona, USA

Eric J. Grimsey
Emeritus Professor, Minerals Engineering
Curtin University
Perth, Western Australia, Australia

Fathi Habashi
Professor Emeritus of Extractive Metallurgy
Laval University
Quebec City, Quebec, Canada

Ian Harmsworth
Engineering Manager
Nexus Engineering Services Pty Ltd.
Brisbane, Queensland, Australia

Steven D. Hart
Principal Advisor Processing
Newmont Mining Corporation
Perth, Western Australia, Australia

Steve Hearn
Senior Staff Engineer
Huntsman
Centennial, Colorado, USA

J. Brent Hiskey
Professor Emeritus
Department of Mining & Geological Engineering
University of Arizona
Tucson, Arizona, USA

Kirsty Hollis
Process Manager
Phu Bia Mining
Vientiane, Laos

Ralph Holmes
Honorary Fellow
CSIRO Mineral Resources Flagship
Melbourne, Victoria, Australia

Matthew Jeffrey
Process Manager
Newmont USA Limited
Englewood, Colorado, USA

Bill Johnson
Principal Consultant & Adjunct Professor
Mineralis Consultants Pty Ltd. & University of Queensland
Brisbane, Queensland, Australia

Adam Johnston
Chief Metallurgist
Transmin Metallurgical Consultants
Lima, Peru

Richard H. Johnson
Principal
JohnsonMCS LLC
Bellaire, Texas, USA

Ronel Kappes
Process Manager
Newmont Mining Corporation
Englewood, Colorado, USA

S. Komar Kawatra
Professor & Chair, Department of Chemical Engineering
Michigan Technological University
Houghton, Michigan, USA

Jon J. Kellar
Nucor Professor & Douglas W. Fuerstenau Professor
South Dakota School of Mines & Technology
Rapid City, South Dakota, USA

Harold E. Kelley
Director of Technical Services
TungsteMet
Oak Hill, West Virginia, USA

Bern Klein
Professor, Norman B. Keevil Institute of Mining Engineering
University of British Columbia
Vancouver, British Columbia, Canada

Gary Kordosky
(Retired) Consultant in Hydrometallurgy & Solvent Extraction
Formerly with Cognis
Tucson, Arizona, USA

Halvor Kvande
Professor Emeritus
Norwegian University of Science & Technology
Trondheim, Norway

James Kyle
Professor
Murdoch University
Murdoch, Western Australia, Australia

Luis M. La Torre
Principal Metallurgist
Transmin Metallurgical Consultants
Lima, Peru

Jaeheon Lee
Associate Professor
Department of Mining & Geological Engineering
University of Arizona
Tucson, Arizona, USA

Alison Lewis
Dean, Faculty of Engineering & Built Environment
Dept. of Chemical Engineering, University of Cape Town
Cape Town, South Africa

Jens Lichter
Head of Comminution
Anglo American
Denver, Colorado, USA

David McCallum
Senior Experimental Scientist
CSIRO Mineral Resources
Clayton, Victoria, Australia

Robbie McDonald
Sr. Research Scientist, Pressure Hydrometallurgy
Precious & Base Metal Hydrometallurgy
CSIRO Mineral Resources
Karawara, Western Australia, Australia

Thomas F. McIntyre
Consulting Engineer
REC Silicon
Butte, Montana, USA

Terry McNulty
President
T.P. McNulty and Associates Inc.
Tucson, Arizona, USA

James P. Metsa
Director of Sales & Technical Support
Weir Minerals
Madison, Wisconsin, USA

Paul Miller
Managing Director
Sulphide Resource Processing Pty Ltd.
Perth, Western Australia, Australia

Michael S. Moats
Associate Professor
Missouri University of Science & Technology
Rolla, Missouri, USA

Robert D. Morrison
Former Chief Technologist
Julius Kruttschnitt Mineral Research Centre
University of Queensland
Brisbane, Queensland, Australia

Brij Moudgil
Distinguished Professor & Alumni Professor
Materials Science & Engineering
University of Florida
Gainesville, Florida, USA

D.R. Nagaraj
Principal Research Fellow
Solvay Technology Solutions
Stamford, Connecticut, USA

Andrew Neale
Executive Director–Technology
PT Merdeka Copper and Gold
Jakarta, Indonesia

Michael G. Nelson
Professor & Chair, Mining Engineering Department
University of Utah
Salt Lake City, Utah, USA

Aaron Noble
Associate Professor
Mining & Minerals Engineering, Virginia Tech
Blacksburg, Virginia, USA

Jim Orlich
Senior Director Metallurgical Technology
Newmont Mining Corporation
Englewood, Colorado, USA

Jaye Pickarts
Consultant
Rare Element Resources Inc.
Littleton, Colorado, USA

P. Chris Pistorius
POSCO Professor of Materials Science & Engineering
Carnegie Mellon University
Pittsburgh, Pennsylvania, USA

Holger Plath
Vice President, Minerals
ThyssenKrupp Industrial Solutions (USA)
Atlanta, Georgia, USA

Malcolm Powell
Professorial Research Fellow
JKMRC, Sustainable Minerals Institute
University of Queensland
Brisbane, Queensland, Australia

Randall Pyper
General Manager
Kappes, Cassiday & Associates Australia Pty Ltd.
Perth, Western Australia, Australia

W. John Rankin
Honorary Fellow
CSIRO Mineral Resources
Clayton, Victoria, Australia

Samira Rashidi
Process Engineer
ThyssenKrupp Industrial Solutions (USA)
Atlanta, Georgia, USA

Timothy Reeves
Principal Hydrologist
Tetra Tech
Golden, Colorado, USA

S. Jayson Ripke
Senior Applications Engineer
Solvay
York, South Carolina, USA

Andrew Robertson
Principal
Andrew Robertson & Associates
East Freemantle, Western Australia, Australia

Dave Robinson
Technology Leader, In Situ Recovery & Processing
Mining3
Perth, Western Australia, Australia

Angus A. Rockett
Professor
Department of Metallurgical & Materials Engineering
Colorado School of Mines
Golden, Colorado, USA

Greg Roset
VP & General Manager of Recycling Operations
Stillwater Mining Company
South Columbus, Montana, USA

Kym Runge
Program Leader, Separation
JKMRC, Sustainable Minerals Institute
University of Queensland
Brisbane, Queensland, Australia

Axel Schippers
Professor, Resource Geochemistry
Federal Institute for Geosciences & Natural Resources
Hannover, Germany

Mark E. Schleisinger
Professor of Metallurgical Engineering
Missouri University of Science & Technology
Rolla, Missouri, USA

Henry Schnell
Retired
Formerly with AREVA (Paris, France)
Eagle Bay, British Columbia, Canada

Fred Schoenbrunn
Director for Thickeners
FLSmidth
Salt Lake City, Utah, USA

Scott Schuey
Process Manager, Sr. Technical Advisor
Refractory Hydrometallurgical Technology
Newmont Mining Corporation
Englewood, Colorado, USA

Robert A. Seitz
Consultant
Seitz Solutions LLC
Phoenix, Arizona, USA

Nicholaos Serpentzis
Project Director
Fluor Australia Pty Ltd.
Perth, Western Australia, Australia

Roger St.C. Smart
Emeritus Professor & Senior Consultant
Univ. of South Australia & Blue Minerals Consultancy
Mawson Lakes, South Australia, Australia

Stuart Smith
Technical Director
Metifex Pty Ltd.
Brisbane, Queensland, Australia

James G. Speight
Consultant
CD&W Inc.
Laramie, Wyoming, USA

D. Erik Spiller
Research Professor, Kroll Institute for Extractive Metallurgy
Colorado School of Mines
Golden, Colorado, USA

Regis R. Stana
President
R Squared S Inc.
Lakeland, Florida, USA

Patrick R. Taylor
Director, Kroll Institute for Extractive Metallurgy
Colorado School of Mines
Golden, Colorado, USA

Frank Trask
Technical Director
Mining & Process Solutions Pty Ltd.
Darch, Western Australia, Australia

Jack Tryon
Health, Safety & Compliance Specialist
Newmont Mining Corporation
Englewood, Colorado, USA

Larry G. Twidwell
(Retired) Professor Emeritus of Metallurgical Engineering
Metallurgy/Materials Engineering Dept., Montana Tech
Butte, Montana, USA

Maureen T. Upton
Principal
Maureen Upton Ltd.
Denver, Colorado, USA

André Vien
Senior Manager Metallurgy & Process Development
Freeport-McMoRan Mining Company
Phoenix, Arizona, USA

David Wiseman
Principal
David Wiseman Pty Ltd.
Blackwood, South Australia, Australia

Zhenghe Xu
Tech Professor (Dean)
Univ. of Alberta (Southern Univ. of Science & Technology)
Edmonton, Alberta, Canada (Shenzhen, China)

Jorge L. Yordan Hernandez
Retired
Formerly with Rio Tinto Minerals
Parker, Colorado, USA

Courtney A. Young
Department Head & Professor
Metallurgical & Materials Engineering
Montana Tech
Butte, Montana, USA

Patrick Zhang
Research Director
Florida Industrial & Phosphate Research Institute
Bartow, Florida, USA

Xia Zhang
Metallurgist
Freeport-McMoRan Copper & Gold Inc.
Tucson, Arizona, USA

PART 10

Hydrometallurgy

CHAPTER 10.1

Agglomeration Pretreatment

Sylvie C. Bouffard

Pretreatment for hydrometallurgy is a very broad subject, but within this chapter, it is limited to ore agglomeration for the benefit of heap leaching. The reason for narrowing the scope stems from the injustice that this chapter would do to other forms of pretreatment, complex and widespread, that merit their own coverage. Such is the case with crushing, grinding, classification, screening, separation of all types, flotation, and pyrometallurgical pretreatments, which are all covered in other chapters of this handbook.

Other pretreatments for hydrometallurgy are worth noting here but will not be covered in more detail, mostly because of their small-scale experimental nature. Such pretreatments include biooxidation, ammonia leaching (e.g., Feng and Van Deventer 2011), alkaline leaching (e.g., Celep et al. 2011), mycohydrometallurgy (e.g., Ofori-Sarpong et al. 2011, 2013), and microwave use (e.g., Olubambi et al. 2007; Olubambi 2009).

A further restriction to the scope of this chapter is the type of ores discussed; specifically the exclusion of coal and iron ore, both good candidates for agglomeration. Iron and coal are not discussed in this chapter because their post-agglomeration treatment is pyrometallurgical in nature.

From 2005 to 2015, four review-type articles were published on the subject of agglomeration as a pretreatment for hydrometallurgy (Bouffard 2005; Lewandowski and Kawatra 2009a; Moats and Janwong 2008; and Dhawan et al. 2013). Yet, during this decade, there were few research-type articles published on this subject, and those published were predominantly academic in nature, reporting on small-scale testing. In previous decades, more publications of large-scale testing, at times in operations themselves, were reported. This chapter reviews published works on agglomeration for hydrometallurgy, beginning with the benefits of agglomeration, followed by equipment design, quality control, and concluding with commodity-specific observations and recommendations.

BENEFITS OF AGGLOMERATION

The work of T.C. Scrutton in 1905 (as cited in Dorr and Bosqui 1950) is one of the earliest references to agglomeration. Scrutton agglomerated ore by rolling it down a chute inclined at 60° and stacked the agglomerates in a vat. The initial work on iron ore pelletization began a few years later in 1911 and quickly expanded to manganese, fluorspars, and phosphate. The greatest advances in agglomeration for hydrometallurgy, more precisely for heap leaching, occurred in the late 1970s at the U.S. Bureau of Mines (USBM) in Reno, Nevada (Potter 1983). In the 1960s, this organization developed cyanide heap leaching for recovering precious metals from low-grade ores. Some commercial heaps commissioned prior to the USBM agglomeration studies had poor solution permeability, resulting in low gold and silver recovery over the leach cycle. The wide range of particle sizes in the heap, the size segregation during stacking, and the ore mineralogy (e.g., clays) were the ultimate culprits of low permeabilities. The agglomeration work of pioneers, such as Gene McClelland and Paul Chamberlin, contributed to the rapid commercialization of cyanide heap leaching. The USBM determined the optimum practical requirements (moisture level, binder type, binder dosage, curing time, agglomeration equipment, residence time) for agglomeration of precious metal ores and tailings.

In 1979, the first pilot-scale heap leach test was performed on crushed and agglomerated ore in eastern Nevada. The first commercial agglomeration heap leach began operation in 1980. According to Gomes (1983), there were 36 commercial operations in the western United States that agglomerated ore in 1983. Gold or silver extraction from agglomerated crushed ore was as high as 90% in sometimes as little as 10 days. Agglomeration at an Arizona silver heap leach increased recovery from 37%–90%, and the leaching time dropped from 90 to only 7 days. Significant improvements were also obtained at a northern Nevada gold heap leach operation where the gold extraction increased by 60%, and the leach time was reduced by half from 50 to 20–30 days. The Alligator Ridge mine in northeastern Nevada reported a reduction of the cyanide heap leach time from 60–90 days down to 30–40 days, with gold recoveries of 70% (DeMull and Womack 1984; Strachan and Van Zyl 1987). Agglomeration was a technical and economical breakthrough technology for heap leaching of clayey ore and ores containing high fines content. Up to 80% of the metal value of tailings could be extracted in 20–70 days.

Sylvie C. Bouffard, Manager, Studies Optimization, BHP, Saskatoon, Saskatchewan, Canada

In the copper industry, the thin layer leaching concept developed in the 1980s by Sociedad Minera Pudahuel, originally for copper oxides and then applied to copper sulfides, contributed to the phenomenal expansion of heap leaching, solvent extraction, and electrowinning in South America. In general, the ore, crushed to <12–16 mm, was agglomerated with concentrated sulfuric acid, stacked in heaps typically 6–8 m tall, and cured for a couple of days prior to irrigation with raffinate or intermediate pregnant leach solution.

The ultimate benefits of agglomeration are higher and/or faster metal extraction and recovery from heaps, notwithstanding lower reagent consumption and faster rinsing of the heap with potentially lower volumes of wash water used at closure. The benefits of agglomeration are summarized as follows:

- From the perspective of the leach chemistry, agglomeration reduces the travel time of reagents and increases the initial recovery rate. These benefits arise because of the faster contact between the sulfide minerals and the reagents contained in the agglomeration solution, although undesirable chemical reactions between gangue minerals and reagents are also possible (e.g., sulfuric acid and carbonates in copper sulfide ores).
- From the perspective of stacking, agglomeration helps minimize or avoid size segregation (typically larger particles at the bottom) and fines migration, and generally helps form a more structurally stable heap (less slumping), capable of bearing its weight and those of the lifts above the base level. Agglomeration does not prevent the expansion of swelling clays on contact with water but avoids the formation of impermeable zones by distributing clay particles more evenly into the heap. According to Kappes (2002), higher, slower or staged water additions in agglomeration allow clays to swell in the drum.
- From the perspective of hydrodynamics, agglomeration increases the heap permeability, which translates into less solution retained overall, less stagnant solution, and a higher volume of pores filled with air for greater oxygen transfer, with the expectation that such parameters remain relatively constant over the long leach cycle. Agglomeration helps minimize solution ponding, channeling, and potentially slope failure.

AGGLOMERATION EQUIPMENT

Agglomeration begins with wetting ore particles and evolves through agglomerate nucleation, growth, consolidation, and breakage. Growth by layering does not change the number of agglomerates but increases their mass. The coalescence process (two agglomerates forming one) has the opposite effect. Breakage produces more uniformly sized agglomerates by splitting large, poorly consolidated agglomerates into smaller ones. During and after agglomeration, solid bridges (such as crystals or interlocking fibers) and liquid bridges are formed.

According to 24 of the 43 responding mining companies that crushed ore before heap leaching (Kappes 2002), 8 did not agglomerate, 5 turned to belt agglomeration, and 11 used rotating drums. Belt conveyor and drum agglomeration are the two principal pieces of industrial-scale agglomeration equipment employed. Some references to rotating disc agglomerators and pug mills are included.

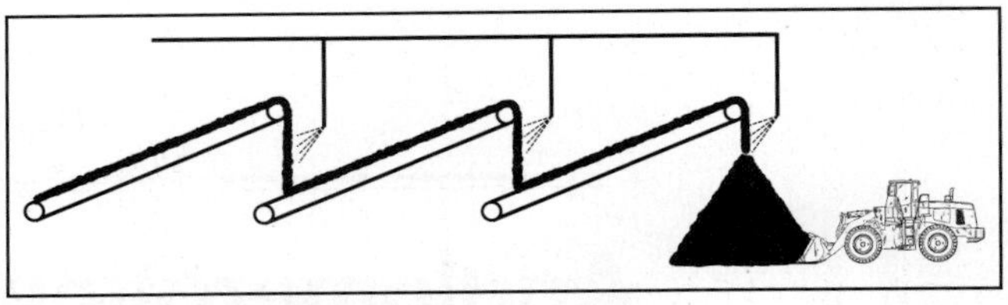

Adapted from Chamberlin 1986

Figure 1 Belt conveyor agglomeration

Belt Agglomeration

Conveyor belts are well suited to agglomerate ore typically containing less than 15% of –104 µm (150 mesh size) fines (Figure 1). Belt conveyors can be operated in two ways:

1. When all belt conveyors are inclined at approximately the same angle (about 15°) and are moving in the same direction, agglomeration occurs when particles touch each other at the transfer point between belts or when they bounce on the belt upon landing. Dispersion bars hanging at the discharge of a belt improve the mixing of the ore in its free fall of 1.2–1.8 m.
2. When the ore falls from a low-angle conveyor moving relatively slowly onto a high-angle (35°–55°) conveyor moving rapidly in the opposite direction, agglomeration occurs primarily on the high-angle belt. Too high of a belt angle can cause the agglomerates to slide down the belt rather than roll.

A solution can be sprayed at the transfer points or along the belts. LeHoux (1997) recommends staging the points of solution addition at the end of the conveyor trains to avoid buildup of slimes underneath the equipment and solution running down the belt. Too little solution results in excessive dusting at the transfer points. Too much solution results in spillage at the transfer points, more frequent shutdown for cleanup, solution running down the belt, as well as compaction of agglomerates on landing on the pile (rather than rolling down the pile).

Drum Agglomeration

An agglomeration drum (also called a rotary drum, balling drum, or agglomerator) is well suited for ores containing high clays or a large fines content. Chamberlin (1986) prefers using an agglomeration drum rather than a belt conveyor when a binder must be added or when an application combines a chemical reaction during agglomeration. Drum agglomeration consists of injecting ore into a cylindrical, inclined drum that rotates to impart rolling, cascading, and tumbling of the ore particles (Figure 2). Two manufacturers of agglomeration drums are FEECO International and Sepro Mineral Systems, to name a few. According to industrial drum design data, the length of the drum is typically three times longer than the drum diameter (Bouffard 2005). The drum length rarely exceeds 15 m. Sepro Mineral Systems offer drums from 1.8 m in diameter by 5 m long with one 45-kW motor to 3.6 m in diameter by 10 m long with four 60-kW motors. The motors power either chain and sprocket drives (low power), friction drives (low power), gear and pinion drives (high power), pneumatic tire drives, or direct drive assemblies (specialty applications). Drums can be equipped with spray systems;

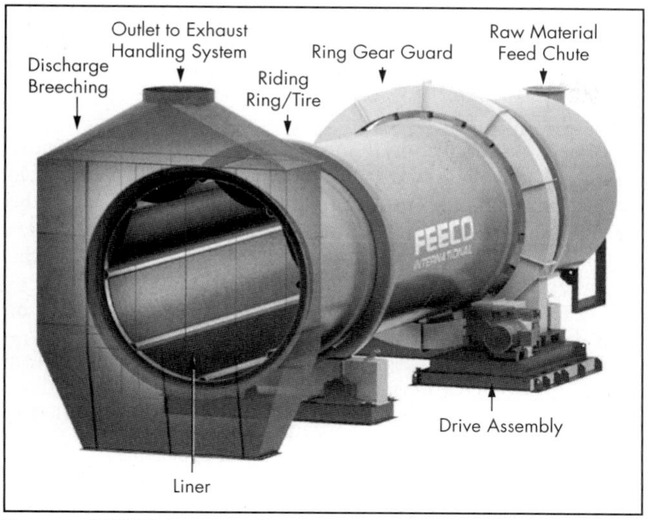

Courtesy of FEECO International
Figure 2 Drum agglomeration equipment

various liners; screw conveyor feeders; scrapers; gear lubrication systems; variable speeds (9–17 rpm), slopes (5°–8°), and frequency drives; rotary control consoles; tire pressure monitoring systems; and safety limit switches.

Regular adjustment of drum bearings prevents serious wear and damage to the drum. Normal wear and tear causes the drum to fall out of alignment, resulting in excessive chatter and drive component vibration, formation of grooves and gouges on the face of the tires, and wear and damage to the thrust rollers, pinions, and girth gear. Regular inspection and realignment of a drum are important preventive maintenance tasks having the greatest impact on costs, downtime, and longevity.

The interior of the drum may be rubber-lined to prevent corrosion and equipped with loose chains or rubber strips to prevent ore from sticking. The solution is pumped through nozzles or perforated pipes preferentially located along the first two-thirds of the drum length. In rare applications, a drum agglomerator may be equipped with screens on the discharge end to separate oversize from undersize agglomerates. Recycle ratios between 2:1 and 5:1 are common in iron ore pelletization and fertilizer granulation circuits. Recycling affects the moisture content and the agglomerate size distribution.

Miller (2010) published a comprehensive study on drum agglomeration design. He developed a mathematical model for predicting power draw and optimum fill as a function of the ore feed rate. The model considers drum geometry, total static and operating volumes, solids residence time based on operating volume, and critical speed. Miller verified the model predictions with the performance of industrial drums.

Bouffard (2005) developed an empirical relationship calculating the drum's ore throughput as a function of drum diameter, drum length, and rotation speed. This author demonstrated that the rotation speed of laboratory and commercial drum agglomerators used in heap leaching, in the fertilizer and iron ore industries, lies between 30% and 50% of the critical speed. The rotation speed translates into a residence time that Chamberlin (1986) suggested to be at least 60 seconds for coarse ore and 240 seconds for fines.

Abbaszadeh et al. (2013) built a 1-m-wide by 3-m-long drum agglomerator to agglomerate a gold/copper ore. The position of the spray nozzles in the drum, the load in the drum, the uniformity of the ore on entry into the drum, and the residence time were important factors.

Other Agglomeration Equipment

Ores containing low clay content and fairly coarse particles can also be agglomerated by simpler means, such as spraying the agglomeration solution on the sides of the heap or onto the ore as it discharges the stacking conveyor as stacking progresses. As the ore is rolling down the slopes, fines will stick to coarse particles. This is referred to as stockpile agglomeration.

Pug mills or paddle mixers (Figure 3) and disc agglomerators (Figure 4) are rarely used in heap leaching (Kappes 2002). A pug mill is a device equipped with a horizontal trough in which a central shaft slowly rotates, to which is attached mixing blades, bars, rods, or paddles. This creates a kneading and folding-over motion that thoroughly blends the

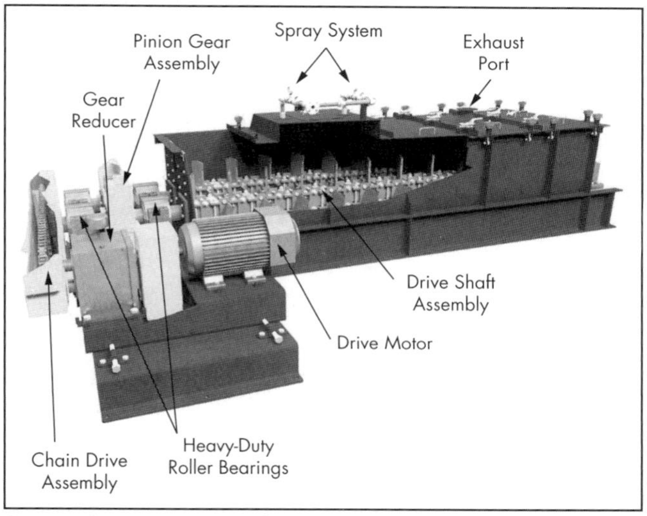

Courtesy of FEECO International
Figure 3 Pug mill

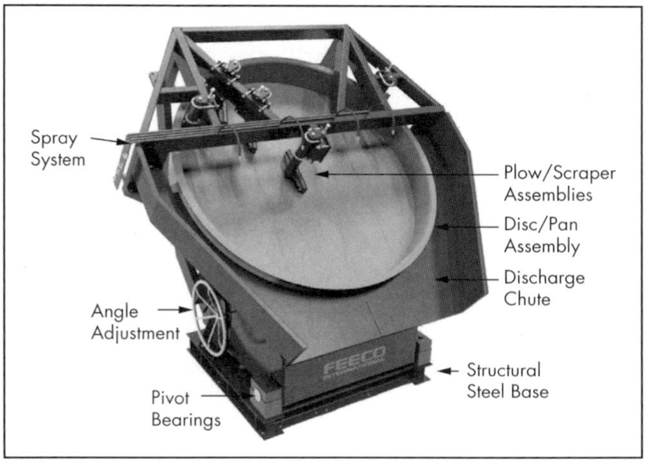

Courtesy of FEECO International
Figure 4 Disc agglomerator

materials. A disc agglomerator is a rotating, tilted disc or pan with a rim, primarily used in the iron ore, agricultural, and chemical industries. Disc agglomerators have lower throughputs than drum agglomerators but produce a product in a narrower range size and have fewer maintenance requirements (FEECO International 2017).

OPERATING CONDITIONS AND QUALITY CONTROL

Heap Stacking, Irrigation, and Operation

The beginning of this chapter cited the benefits of greater permeability possible with agglomeration. To achieve good permeability, ore handling and irrigation practices are just as important as agglomeration.

There are at least three types of stacking systems: (1) a haul truck for dump leaching, (2) a mobile conveyor unit (or grasshopper) combined with radial stacker for heaps, and (3) a spreader conveyor employed primarily for dynamic heap leach pads of constant width and height. Because spreader conveyors travel across the entire width of the pile, there is less segregation across the length of the pile. Radial stackers tend to create discontinuities at the intersection between ridges and fingers (Scheffel 2002) and were the cause for the size gradation measured across the 9-m height of a heap (Kinard and Schweizer 1987).

Miller (2003) demonstrated a proportional relationship between heap height and its bulk density. The heap bulk density increases with heap height, almost in a linear fashion. A lower bulk density should help minimize the retention of solution and keep more pore spaces filled with air.

Scheffel (2002) recommends ripping the heap surface two to four times in crisscross direction and to rip the lower lift before stacking the next lift. Uhrie and Koons (2001) have shown that truck traffic areas should be ripped to a 2.4-m depth before irrigation.

There are three common methods of irrigation: drip emitters, wobble sprinklers, and reciprocating sprinklers. Drip emitters have the gentlest impact on the heap surface but are susceptible to hole plugging. Rainbird sprinklers are undesirable for heaps stacked with agglomerates. The jets cause disintegration of the agglomerates, unless the heap surface is covered by coarse particles to dampen the impact of the droplets, as practiced by the Cerro Rico operation of Compañia Minera del Sur S.A., Bolivia. Heavy rainfall in tropical climates and stalled sprinklers are responsible for agglomerate disintegration, which can be prevented with laying shadecloth or similar material on the heap.

Like ripping, remining the heap (ore turnover by hydraulic excavators), pioneered at the Australian Bullabulling gold heap leach operation and then the Girilambone Copper Company, may increase recovery (e.g., copper recovery up by 2%–10% during the same leach time).

The following subsections review methods for determining agglomerate quality: moisture content, size distribution, attrition, flowability, crush strength, and green/wet strength. Many of these direct methods can be automated and used in operations. Other methods indirectly determine the quality of agglomerates by testing for bulk density, slump, compression, porosity, permeability, and leachability of agglomerates stacked in a bed (typically a column). A lower bulk density, a lower slump, a higher porosity, and so forth, are the benefits of proper agglomeration. These indirect methods are not specifically reviewed in this section but are cited elsewhere in this chapter. Care should be exercised when scaling up parameters from columns to heaps, as the column walls are a significant influence on bulk density and permeability (Bouffard 2008).

Moisture Quality Control

There is no specific guideline for the appropriate moisture addition during agglomeration; it depends on the particle size distribution, the initial moisture content of the ore, the ore mineralogy, and the use of binders. Here are some rules of thumb for determining the moisture content before the agglomeration begins or during the operation.

1. Use this empirical equation relating the density of solid and liquid (before agglomeration):

$$\omega = \frac{1}{1 + 2.17\frac{\rho_s}{\rho_l}} \text{ for particle size} > 30\ \mu\text{m} \qquad \textbf{(EQ 1)}$$

where ω is the moisture content in kg liquid/kg dry ore, ρ_s is the solid density, and ρ_l is the density of the liquid having similar properties as water. For a solid density of 2.8 g/cm^3 and a water density of 1.0 g/cm^3, the moisture content is approximated to be 14 wt %. This equation would be unsuitable for tailings, which contain a greater proportion of fines, thus typically requiring twice as much moisture as crushed ore (15–30 wt % vs. 5–15 wt %).

2. Perform a soak test (before agglomeration). Vethosodsakda et al. (2013) suggested that the optimum moisture content of agglomerates prepared in batch can be determined from the volume of water soaked up in 1 hour by the non-agglomerated ore placed in a column whose bottom is submerged in water. The amount of ore, and more precisely the height of the ore in the column, was found to be very influential in this soak test. A height of 10–18 cm for a column diameter of 15 cm was preferred.

3. Use the filter cake test (before agglomeration). Chamberlin (1986) estimated that the moisture content of the agglomerates is 1%–3% less than the moisture content of a dewatered filter cake.

4: Perform a sheen test (during agglomeration). According to the sheen test, there should be no moisture glistening on the surface of agglomerates.

5. Use squeeze and drop tests (during agglomeration). Squeeze or drop shatter tests have been used to control the amount of acid and moisture added in batch or continuous agglomeration.

6. Check for electrical conductivity (during agglomeration). Fernández (2003) and Velarde (2005) introduced the use of electrical conductivity to evaluate the proper amount of moisture for agglomerating a copper ore. The electrical conductivity increases exponentially with the moisture added, with the largest signal detected when a liquid film forms around the agglomerates. Fernández (2003) related conductivity readings to the optimum moisture content of 4% (no clay in ore), 6.5% (medium clayey ore), and 10% (high clayey ore).

Some additional recommendations concerning moisture addition are discussed in the following paragraphs.

If the material to be agglomerated is too wet, it should be dried, because excessive moisture does not produce individual agglomerates but rather clumps of agglomerates that do not roll down the slopes of the heap. A starting material that is too wet also limits the quantity of soluble reagents that can be mixed with the ore during agglomeration.

Tibbals (1987) suggested that the amount of input energy for agglomeration of crushed ore was more important than the method of solution addition. However, when the ore particle size is smaller than the droplet size, as for tailings, some authors (Tibbals 1987; Eisele and Pool 1987) agree that the moisture should be added as droplets rather than as an atomizing spray. Fines adhere to the droplet to create a nucleus.

Some researchers incorrectly equate agglomerate drying and agglomerate curing. Agglomerates should remain moist at all times to avoid disintegration and other negative effects on leaching (Connelly and West 2009). Cement-based agglomerates that were allowed to dry out immediately after agglomeration disintegrated when less than 50 kg/t of cement was added.

Size Distribution Quality Control

There is no denying that agglomerating particles, particularly fines smaller than 104 µm, has been beneficial; the question is rather how wide the agglomerate size distribution should be. Views are mixed on this subject. On the one hand, Dhawan et al. (2012) postulate that a wider distribution is preferable to mono-sized agglomerates. On the other hand, Lipiec and Bautista (1998) recommend uniformly sized agglomerates. Bouffard (2008) also recommends a narrow size distribution, hypothesizing that mono-sized agglomerates have fewer points of contact between them when stacked in a bed. Bouffard and West-Sells (2009) later proved that the agglomerate size distribution affects the bed moisture content. With fewer points of contact, there is less of a tendency for the solution to accumulate between surfaces, thereby reducing the length of diffusion channels and increasing the gas–liquid surfaces.

Bouffard (2008) controlled the size distribution of agglomerates using the following methods:

- Reduce the amount of fines in the ore size distribution, thereby avoiding the production of agglomerates that are too small. Screening the ore can reduce the amount of fines.
- Ensure a narrow initial size distribution of the ore to produce a narrow agglomerate size distribution. This could mean crushing run-of-mine ore to eliminate boulders or, at other times, adding coarse particles (Serrano 2003).
- Recycle undersized agglomerates to skew the feed size distribution and avoid too wide an agglomerate size distribution.
- Control the moisture added at just the right level to produce a narrow size distribution of agglomerates (too little or too much moisture added produces a wide size distribution).
- Extend the time of agglomeration to enable growth and breakage phenomena to form better-shaped agglomerates. This can be done with more conveyor transfer points, a longer drum operated at lower speed and set at a lower angle, or with recycling agglomerates.

The agglomerate size distribution can be measured with online cameras and digital imaging, or screening freeze-dried, air-dried, or wet agglomerates in a less automated way. To understand the growth and breakage phenomena occurring during agglomeration, agglomerates can be wet-screened to literally split the agglomerates into their constituent particles.

Strength Quality Control

Here are a range of methods for measuring agglomerate strength. Some are simple but subjective, suitable for in-field measurement, and others require laboratory testing.

- **Squeeze test.** The interpretation of this very subjective method is mixed. Some argue that after squeezing a bunch of agglomerates in one's hand, the clump should hold if it is poked; others believe the clump should fall apart.
- **Drop test.** Good agglomerates should roll on the ground after they have been thrown up in the air. A variant on this test is the measurement of the highest height of the drop that leads to complete disintegration. Another variant is the number of falls that an agglomerate can sustain when dropped from a fixed height.
- **Attrition test.** Herkenhoff (1987) proposed to tumble dry agglomerates and measure the proportion of abraded fines.
- **Soak test.** There are multiple variations on this theme. Chamberlin (1986) suggested that good agglomerates submerged in water should not disintegrate for many hours. Rather than dipping pellets, the author placed them in a burette and covered the top with glass wool. He then applied water at increasing flow rates and measured the fines content in the discharge solution. Milligan and Engelhardt (1984) measured the amount of fines produced when dipping pellets in water 10 times. The pellets must be previously cured for 6 hours at 90°C and cooled before dipping. Milligan and Engelhardt (1984) proposed a more general soak test in which a known amount of agglomerates between 9.5 mm and 12.7 mm in size are placed onto a 10-mesh screen, which is then submerged in water for 24 hours. After this period, the ore remaining on the screen is weighed.
- **Drainage test.** The USBM (Heinen et al. 1979; McClelland et al. 1985; Eisele and Pool 1987) rated agglomerates to be of high strength if a column filled with agglomerates and flooded with water drained rapidly.
- **Percolation test.** Known informally as the *Kappes test* (Pyper et al. 2015), this percolation method determines the optimum binder requirements for clayey ores. The test involves soaking cured agglomerates in a column for a set time and then tapping the column walls. The test records bed slump, flow rate, agglomerate disintegration, bubble formation, and fines release as indicators of agglomerate strength.
- **Chemical resistance test.** Yijun et al. (2002) measured the number of intact agglomerates remaining in a column subjected to sequential irrigation with water, 20–30 g/L sulfuric acid, and finally 50–100 g/L sulfuric acid. The number of unbroken agglomerates relative to the original number of agglomerates was a proxy for agglomerate strength.

CAPITAL AND OPERATING COSTS

Estimates proposed by Rose et al. (1990) and Kappes (2002) indicate that agglomeration and stacking account for 6%–10% of the total capital costs (Table 1). For cyanide heap leach operations, agglomeration and stacking account for 1%–2% of total operating costs; cement, in a dose of 10 kg/t, for 9%–15%; and other reagents (e.g., cyanide, lime) for 3%–5% (Kappes

Table 1 Heap leaching operation scale influences agglomeration/stacking unit capital, operating, and total costs

Production	400 t/d (Rose et al. 1990)	1,000 t/d (Rose et al. 1990)	3,000 t/d (Kappes 2002)	15,000 t/d (Kappes 2002)
Capital costs				
Agglomeration/stacking, thousand $	460	N/A*	1,000	3,500
Total capital cost, thousand $	4,600	N/A	14,000	53,600
Percentage for agglomeration	10%	N/A	7%	6.5%
Operating costs				
Agglomeration/stacking, $/t	N/A	1.15	0.20	0.10
Reagent for agglomeration, $/t	N/A	N/A	1.00	1.00
Total operating costs, $/t	N/A	7.20	11.50	8.30
Percentage for agglomeration	N/A	16%	10%	13%

Note: All figures are in U.S. dollars.
*N/A = not available

2002). By comparison, a typical Nevada cyanide heap leach operation of 30,000 t/d of coarse-crushed, non-agglomerated ore would have operating costs about 30% less than a similar cyanide heap leach operation agglomerating 15,000 t/d of fine-crushed ore (Kappes 2002).

CONDITIONS SPECIFIC TO CERTAIN COMMODITIES

The following subsections describe agglomeration conditions and results specific to gold, copper, and nickel.

Gold

Agglomeration of gold ores commonly utilizes cement and lime as binders and typically a cyanide solution as an agglomeration solution.

A dosage of 1 kg of sodium cyanide (NaCN)/t of ore is fairly typical for agglomeration. The use of a strong cyanide solution can be a safety hazard for hydrogen cyanide (HCN) emanation at too low solution pH. To control alkalinity and hence minimize cyanide losses as HCN at pH less than 9, 1.5–25 kg lime/t ore is added during agglomeration. Agglomeration with a cyanide solution does not affect the chemistry of the agglomerates. The benefits of adding cyanide during agglomeration on the rate of metal recovery from a heap are not conclusive (Chamberlin 1983 vs. O'Brien 1982). According to Kappes (2002), some operations have stopped adding cyanide during agglomeration because the cyanide code requires containment of spillage of agglomerates in transport from the agglomeration equipment to the heap.

Cement is comprised of 50%–70% tricalcium silicate ($3CaO \times SiO_2$), 15%–30% dicalcium silicate ($2CaO \times SiO_2$), 5%–10% tricalcium aluminate ($3CaO \times Al_2O_3$), 5%–15% tetracalcium aluminoferrite ($4CaO \times Al_2O_3 \times Fe_2O_3$), and various hydrated forms of gypsum ($CaO \times SO_3 \times H_2O$) (Kosmatka et al. 2002). Curing refers to hydration reactions between calcium silicates and water to form calcium hydroxide and calcium silicate hydrate. The chemical composition of calcium silicate hydrate varies, but it typically contains three parts lime to two parts silicate. The calcium silicate hydrate forms dense bonds between particles. Electron microscopy showed calcium to be uniformly distributed throughout agglomerates of clayey particles, even though it typically accounts for less than 1% of the material agglomerated (Heinen et al. 1979).

Cement does not cure by drying. If the moisture is too low, cement stops gaining strength. Hydration resumes after resaturation; the strength increases again. Approximately 40 kg of water per 100 kg of cement is necessary for curing. The strength continues to increase provided that unhydrated cement is still present and the temperature remains favorable. Although 28 days of curing is the standard in the concrete industry, 8–24 hours sufficed in previous agglomeration studies of crushed ore (Chamberlin 1986; McClelland 1986a, 1986b; Eisele and Pool 1987; Zárate and Guzmán 1987). Seventy-two hours was preferable for tailings (Eisele and Pool 1987). The installation of solution header lines and emitters on top of the heap typically takes more than 3 days, and thus agglomerates have sufficient time to cure before the cyanide solution is applied.

Some recommendations on the dosage of cement to use are as follows:

- There are five types of cement under the Canadian Standards Council Standard A5 (CAN/CSA-A5-98), and eight types of cement under the ASTM C150 standard. Cement types I, III, and I/II performed similarly at equivalent dosages added. The handling of cement-based agglomerates after agglomeration is more important to the agglomerate strength than the types of cement used.
- Eisele and Pool (1987) recommend a dosage of 7.5 kg cement/t ore and the same dosage of lime for crushed ore or tailings. Zárate and Guzmán (1987) recommend as much of each binder, that is, 8–12 kg cement/t crushed ore and 8–12 kg lime/t crushed ore. McClelland (1986a, 1986b) and Heinen et al. (1979), however, recommend 2.8–5.5 kg cement/t ore, without or with 3.9 kg of lime/t ore. In operations, the cement dosage ranges from 2.5 to 5 kg/t for crushed ore and 5 to 17 kg/t for tailings.
- Zárate and Guzmán (1987) found that the quality of cement/lime-based agglomerates was optimum at a moisture of 17 wt % but declined at higher water dosages. All three methods employed in these authors' work to measure the agglomerate quality (dip, flooding, compaction) recommended the same moisture content.
- Amaratunga and Hmidi (1997) used gypsum alone and in combination with cement to agglomerate tailings. Agglomerates were stronger with the combination of cement and gypsum.

Following are some positive and negative experiences using cement-agglomerated ore:

- At Little Bald Mountain in Nevada, cement agglomeration reduced slumping from 24% to 8% (Tibbals 1987).

- At the Alligator Ridge mine in northeastern Nevada, agglomerating gold ores with cyanide reduced the overall cyanide consumption and increased the initial recovery (DeMull and Womack 1984). However, adding more than 50 g NaCN/t of ore did not improve the gold extraction and increased the overall cyanide consumption.
- Although a very clayey ore had been well agglomerated with 5 kg lime/t ore and 4 kg cement/t ore, trenches dug across the 9-m height of the heap showed uneven material size gradation (Kinard and Schweizer 1987).
- Keuhey and Coughlin (1984) reported a rare instance of lower gold extraction from cement-based agglomerates. Eight percent less gold was recovered after 32 days from a column containing agglomerates prepared with 10 kg cement type II/t ore.

Some recommendations about lime-agglomerated ore and heap leaches are as follows:

- At the Alligator Ridge mine in northeastern Nevada, agglomeration with 50 g NaCN/t ore and 1.5–5 kg lime/t ore increased gold recovery and shortened the leach cycle by half (DeMull and Womack 1984; Strachan and Van Zyl 1987).
- Litz (1993) reported on the development and patent of a modified lime (Walker and Oliphant 1992) for use as a binder.
- Walker and Oliphant (1992) also patented a mixture comprised of 10%–80% calcareous component, such as quick lime, 5%–50% siliceous-calcareous component, and 10%–80% sulfated component, such as gypsum. The mixture should be added in the amount of up to 2 wt % to precious metal ores for cyanide heap leaching.

Here are some experiences with other polymeric binders for agglomeration of gold ores:

- Milligan was unsuccessful with both wood fibers and molasses but was slightly successful with 1–2 kg/t carboxymethylcellulose/t ore.
- At the USBM, Heinen et al. (1979) observed faster drainage with 0.05 kg polyethylene oxide/t ore mixed with lime and added to the agglomerates.
- Nivens and Given (1993) used polymer aids for agglomeration of gold ores. Newmont gold mines (Carlin, Nevada) and Philex Mining's Sibutad gold project in the Philippines also employed Nalco binders and cement to agglomerate gold ores.
- Brewer Gold found that column gold recovery was superior with the use of 125 g/t Nalco Extract-Ore 9760/t ore and 2.5 kg cement/t ore compared to an ore agglomerated with 7.5 kg cement/t ore (Pautler et al. 1990). In the field, the dual mixture of cement and Nalco 9760 reduced channeling. In addition, the bulk density of the heap with the dual mixture was 3% lower than the heap with cement alone.
- The Betz Dearborn Company developed binder HL 9121 for agglomeration of gold ores. HL 9121 is a cross-linked, borated polyvinyl alcohol added in a dose of 50 g/t ore to 3 kg lime/t ore and 3 kg cement/t ore. In columns, it improved the gold extraction by 5% under simulated heavy rainfall conditions (Polizzotti et al. 1997) and increased the drainage rate by five times (Polizzotti 1993).

Copper

Key issues for the agglomeration of copper ores include the amount of sulfuric acid added, the binder added, and the curing time.

Copper heap leaching operations agglomerate with water and sulfuric acid. On average, 15–25 kg sulfuric acid/t ore is added to 60–100 kg water/t ore. The final moisture content ranges from 7.5 to 12.5 wt %, which is a typical moisture content for agglomerated crushed ore. Sulfuric acid is thought to react with gangue minerals (e.g., kaolinite), to render them amorphous and to inhibit silica dissolution (Cruz et al. 1980; Farias et al. 1995). Curing times with sulfuric acid have been reported from 14 to 336 hours (Lu et al. 2007). Holle (1996) found that more acid added during agglomeration increased the copper recovery by 30%. Morenci mine-for-leach operation also recovered 15% more copper with the addition of 5 kg sulfuric acid/t ore. The increase in copper recovery was proportional to the acid dosage (Uhrie et al. 2003).

Cement is not a suitable binder for ores leached with sulfuric acid: Calcium, sodium, and magnesium sulfates attack calcium aluminate hydrates and calcium hydroxide. A search for a suitable binder to agglomerate copper ores has been underway, with the following findings to date:

- The Cuban Research Center for the Mining-Metallurgical Industry has developed a proprietary binder termed *Additive 1* to agglomerate clayey copper ores (Serrano 2003). The author claims that Additive 1 is low cost, resistant to acid solution, and forms porous agglomerates with good mechanical resistance.
- Nalco binders were utilized by the Toquepala/Caujone copper mine in Chile and the Ray mines in Arizona to agglomerate copper ores. Nifty Copper operation in Washington (United States) has used two other products commercialized by Nalco, Extract-Ore 9560 being one of them, used in a dose of 1 kg/t ore (Efthymiou et al. 1998). Extract-Ore is a medium-molecular-weight latex copolymer of moderate anionic charge.
- Kodali (2010) and Kodali et al. (2011) used stucco (calcium sulfate hemihydrate) in combination with a sulfuric acid solution to agglomerate copper ore prior to leaching the agglomerates in columns. Greater bed permeability and larger copper leaching after 110 days were obtained when using the combination of stucco and acid for agglomeration.
- Hill (2013) tested the effects of polymeric binders and sulfuric acid for agglomeration of a copper ore. The author measured the hydraulic conductivity, the permeability in columns, the agglomerate size distribution, and the column bulk density.
- Kawatra and Lewandowski published extensively on the testing of binders for copper ores. Lewandowski et al. (2006) identified five nonionic polymers as the most promising binders for agglomerating copper ores. The screening was performed with an adapted soak-test method. Agglomerates prepared with these five binders were then subjected to an adapted percolation test. Lewandowski and Kawatra (2009b) preferred cationic polyacrylamide binders over nonionic polyacrylamide binders for agglomeration of copper ore using a sulfuric acid solution of pH 2. Cationic binders created more stable

agglomerates. Using surface charge analysis, the authors found the cationic binders neutralized the negative surface charge of the ore. Hydrogen bonding occurred when the hydrogen bond donor sites on the polyacrylamide chains were attracted to the silanol group hydrogen bond acceptor sites in the ore.

- Following from the earlier works of West-Sells et al. (2007) and Bouffard (2009) on the agglomeration of copper ores with sulfuric acid, elemental sulfur, and lignosulfonate, De Oliveira et al. (2014) mixed a copper ore with fine elemental sulfur (up to 13.3 kg sulfur/t ore) and an inoculum before placing the agglomerates in columns for sulfuric acid leaching. Compared to the control column, more copper was leached from the ore agglomerated with sulfur and microorganisms.
- Barriga Vilca (2013) found it unnecessary to add ferric sulfate to the sulfuric acid solution used for agglomeration of a secondary copper ore: There was no difference in copper recovery after curing.

Nickel

Several researchers, based at the University of Queensland, at the Ian Wark Research Institute, and at the Commonwealth Scientific and Industrial Research Organisation (CSIRO), have published 11 articles on the agglomeration of nickel laterite ores from 2011 to 2014. Following is a summary of their published findings, from the earliest to the latest:

- By increasing the water-to-sulfuric-acid ratio, Nosrati et al. (2011) observed faster nucleation and granule growth of nickel laterite ore particles crushed to less than 2 mm, forming agglomerates 5–40 mm in size. The faster growth of agglomerates was not necessarily ideal to achieve high compressive strength.
- Liu (2012) used a pellet press to agglomerate nickel laterite fines with water or sulfuric acid, and later allowed some to dry and others to remain moist. The pellets were stronger if agglomerated with water and if dried. In these experiments, the conditions tested were not typical of heap leaching.
- Liu et al. (2012) postulated that the agglomeration of nickel laterite ore begins with a kernel that subsequently grows by coalescence. The population balance model that was developed clearly predicted the agglomerate size distribution under the one condition tested.
- Nosrati et al. (2012a) found that an acidic solution of high concentration hindered the growth of agglomerates, producing smaller agglomerates.
- Nosrati et al. (2012b) used micro-tomography to reveal the agglomerate structure and pore volume.
- Addai-Mensah et al. (2013) found that the compressive strength of agglomerates was higher with the increasing ratio of fines to coarse particles. It took as long as 8–14 minutes in batch mode to agglomerate the ore into agglomerates 5–40 mm in size.
- Nosrati et al. (2013) hypothesized about the formation of solid bridges between particles agglomerated with acid. Solid bridges provide structural stability to the dry agglomerates, but on contact with an acidic solution, the acid-soluble bridges dissolve and the structural stability is greatly diminished.
- Quast et al. (2013) crushed nickel laterite ore to two different particle sizes (–38 μm and 2–15 mm) and agglomerated each batch with a sulfuric acid solution before leaching in columns. After 100 days of leaching, agglomerates made of –38 μm particles had leached only 10% more nickel than those made of coarser particles, and it is difficult to envision heaps made entirely of agglomerates made of –38 μm particles.
- Xu et al. (2013) recommended mixing goethitic laterite ore particles with clays during agglomeration and recommended allowing agglomerates to dry out moderately to accelerate leaching.
- Nosrati et al. (2014) measured the agglomerate properties using agglomerate size, compressive strength, three-dimensional microstructure analysis, and laboratory columns leaching tests. Of interest in this study were the effects of ore mineralogy and particle size on agglomeration.
- Quaicoe et al. (2014) performed the soak test (dipping agglomerates in a sulfuric acid solution) as a proxy for the compressive strength of the agglomerates, which were prepared by mixing ore particles smaller than 2 mm with 30 wt % sulfuric acid for 15 minutes. In this work, the authors compared the compressive strength of agglomerates made of saprolitic and geothitic ores. The latter had higher compressive strength.

Other works on the pretreatment of laterite ores are assigned to Guo et al. (2011) for their use of sodium carbonate (Na_2CO_3) roasting to break the mineral lattice and Janwong (2012) for his observation of two types of laterite agglomerates (layered and coalesced) using scanning electron microscopy and X-ray spectroscopy.

Other Commodities

There are few published works on the agglomeration of other types of oxide- or sulfide-bearing ores. Four articles refer to the agglomeration of uranium ores:

1. Scheffel (1982) reported on a 544,000-t heap stacked with agglomerates made of uranium ore mixed with 45–68 kg sulfuric acid/t ore. Some prior success had been obtained using Polyox WSR301 and sulfuric acid, but its addition was considered unnecessary.
2. Videau and Roche (1990) reported on the uniform leaching of a 1,000-t heap stacked with agglomerates made of uranium ore mixed with sodium silicate, concentrated sulfuric acid, and water. The success of the leach was attributed to the silicates added during agglomeration, as solution ponding and channeling were evident in another 1,000-t heap stacked with ore agglomerated without silicates.
3. Jianhua et al. (2004) used N_6O_2 and sulfuric acid to agglomerate uranium ore prior to acid leaching in columns.
4. Sinclair (2010) performed small and large column tests with agglomerated uranium ore.

Following are a few references for sulfide ores:

- Lastra and Chase (1984) used gypsum to agglomerate sulfide ores leached with acid.
- Perez et al. (1990) agglomerated with a sodium or calcium hypochlorite solution and cement to destroy, modify, or passivate sulfide minerals. After curing, the agglomerates were rinsed with water and then cyanide-leached.

Ahmadiantehrani et al. (1991) also used hypochlorite and cement to agglomerate sulfidic gold ores and achieved 80% gold recovery.

- Misra et al. (1996) found the combination of fly ash and cement to be superior to either material alone with regard to the strength of agglomerates made of pyrite tailings.
- According to the sulfide bioheap model developed at the University of British Columbia, sulfide heaps can also benefit from agglomerating the ore with the leaching solution (Bouffard 2003).
- Nalco Holding Company has developed polymeric binders (9704, CX-2131, CX-2134, CX-2185, CX-2194, 98DF108, and 97DF125) that would not inhibit the microorganisms present and would not be consumed by microorganisms (Nalco Water 2017).
- Saririchi et al. (2012) tested the bioleaching in small columns of low-grade zinc sulfide ores agglomerated with an inoculum.

CONCLUSION

Agglomeration for gold and silver ores is a mature process with no recent significant development. For copper, researchers have concluded that there is a need for an agglomeration binder to be added to the sulfuric acid solution; research in the last decade has focused on polymeric binders. For nickel, many recent articles have been published, limited to laboratory tests under conditions, which are at times atypical of heap leach environments.

While agglomeration was a breakthrough technology for heap leaching gold and silver ores of high fines and clay content, the benefit of agglomeration for copper ores is not as well substantiated. The slow (>300 days), difficult-to-control, and ever larger heap may overwrite the imperfections of agglomerates. The benefits of increased recovery and shorter leach cycles must be weighed against the 5%–10% extra capital costs for the agglomeration equipment and instrumentation, as well as the higher operating costs for labor, energy, and agglomeration medium and binder. In-line instrumentation is displacing the rather subjective manual methods for measuring agglomerate moisture content and size distribution, but no in-line method is yet available for measuring agglomerate strength. Agglomeration may also require adapting the crushing and screening processes to produce the right ore size distribution, configuring stacking equipment to place the agglomerates more gently, and modifying irrigation practices to distribute the leaching solution more evenly.

With the broader adoption of high-pressure grinding rolls (HPGRs), one must question whether this technology could be a substitute to agglomeration for non-clayey ores, delivering a similar gain in recovery without the additional cost of agglomeration, and whether the combination of agglomeration and HPGR, known to generate large amounts of fines, could deliver greater recovery than would agglomeration alone.

REFERENCES

Abbaszadeh, S., Haws, N., Henley, M., Taylor, S., Swanepoel, F., and Eichhorn, R. 2013. Pilot-scale drum agglomeration of an oxide ore with high fines and variable moisture. Presented at the Heap Leach Conference, Vancouver, BC.

Addai-Mensah, J., Quaicoe, I., Nosrati, A., and Robinson, D.J. 2013. Understanding lateritic ore agglomeration behaviour as a precursor to enhanced heap leaching. *Ghana Min. J.* 14:41–50.

Ahmadiantehrani, M., Hendrix, J.L., and Nelson, J.H. 1991. Hypochlorite pretreatment of a low grade carbonaceous gold ore, Part II: Effects of agglomeration, temperature and bed height. In *EPD Congress*. Edited by D.R. Gaskell and J.P. Hager. Pittsburgh: TMS. pp. 63–79.

Amaratunga, L.M., and Hmidi, N. 1997. Cold-bond agglomeration of gold mill tailings for backfill using gypsum betahemihydrate and cement as low cost binders. *Can. Metall. Q.* 36(4):283–288.

ASTM C150/C150M-17. 2017. *Standard Specification for Portland Cement.* West Conshohocken, PA: ASTM International.

Barriga Vilca, A. 2013. Studies on the curing and leaching kinetics of mixed copper ores. M.A.Sc., University of British Columbia, Vancouver.

Bouffard, S.C. 2003. Understanding the heap biooxidation of sulfidic refractory gold ores. Ph.D. thesis, University of British Columbia, Vancouver.

Bouffard, S.C. 2005. Agglomeration processes in the mining industry. *Miner. Proc. Extr. Met. Rev.* 26(3/4):233–294.

Bouffard, S.C. 2008. Agglomeration for heap leaching: Equipment design, agglomerate quality control, and impact on the heap leach process. *Min. Eng.* 21:1115–1125.

Bouffard, S.C. 2009. Use of lignosulfonate for sulfur biooxidation and copper leaching. *Min. Eng.* 22:100–103.

Bouffard, S.C., and West-Sells, P.G. 2009. Hydrodynamic behavior of heap leach piles: Influence of testing scale and material properties. *Hydrometallurgy* 98(1/2):136–142.

CAN/CSA-A5-98. 1998. *Portland Cement.* Ottawa, ON: Standards Council of Canada.

Celep, O., Alp, I., Paktunc, D., and Thibault, Y. 2011. Implementation of sodium hydroxide pretreatment for refractory antimonial gold and silver ores. *Hydrometallurgy* 108:109–114.

Chamberlin, P.D. 1983. Heap leaching and pilot testing of gold and silver ores. In *Papers Given at the Precious-Metals Symposium.* Nevada Bureau of Mines and Geology Report 36. Edited by V.E. Kral, J.A. Hall, R.B. Blakestad, H.F. Bonham Jr., G.B. Hartley Jr., G.E. McClelland, J.A. McGlasson, and P. Mousette-Jones. Reno, NV: Nevada Bureau of Mines and Geology. pp. 77–83.

Chamberlin, P.D. 1986. Agglomeration: Cheap insurance for good recovery when heap leaching gold and silver ores. *Min. Eng.* 38(12):1105–1109.

Connelly, D., and West, J. 2009. Trips and traps for copper heap leaching. In *ALTA Copper 2009.* Melbourne, Victoria: ALTA Metallurgical Services.

Cruz, A., Lastra, M., and Menacho, J. 1980. Sulfuric acid leaching of copper silicate ores. In *Leaching and Recovering Copper from As-Mined Materials*. Edited by W.J. Schlitt. Littleton, CO: SME-AIME. pp. 61–70.

De Oliveira, D.M., Sobral, L.G.S., Olson, G.J., and Olson, S.B. 2014. Acid leaching of a copper ore by sulfur-oxidizing microorganisms. *Hydrometallurgy* 147-148:223–227.

DeMull, T.J., and Womack, R.A. 1984. Heap leaching practice at Alligator Ridge. In *Au and Ag Heap and Dump Leaching Practice*. Edited by J.B. Hiskey. New York: SME-AIME. pp. 9–21.

Dhawan, N., Sadegh Safarzadeh, M., Miller, J.D., Moats, M.S., Rajamani, R.K., and Lin, C.-L. 2012. Recent advances in the application of X-ray computed tomography in the analysis of heap leaching systems. *Miner. Eng.* 35(2):75–86.

Dhawan, N., Sadegh Safarzadeh, M., Miller, J.D., Moats, M.S., and Rajamani, R.K. 2013. Crushed ore agglomeration and its control for heap leach operations. *Miner. Eng.* 41:53–70.

Dorr, J.V.N., and Bosqui, F.L. 1950. *Cyanidation and Concentration of Gold and Silver Ores*, 2nd ed. New York: McGraw-Hill.

Efthymiou, M., Henderson, L., and Pyper, R. 1998. Practical aspects of the application of Extract-Ore 9560 for heap leach agglomeration at Nifty Copper operation. Presented at the ALTA 1998 Copper Hydrometallurgy Forum.

Eisele, J.A., and Pool, D.L. 1987. Agglomeration heap leaching of precious metals. *CIM Bulletin* 80(902):31–34.

Farias, L.L, Reghezza, A.I., Cruz, A.R., Menacho, J.L., and Zivkovic, Y.D. 1995. Acid leaching of copper ores. In *Copper Hydrometallurgy Short Course, Copper 95—Cobre 95*. Edited by D.B. Dreisinger and J.D. Vasquez. Montreal, QC: Canadian Institute of Mining, Metallurgy and Petroleum. pp. 13–17.

FEECO International. 2017. *The Agglomeration Handbook*. Green Bay, WI: FEECO. http://go.feeco.com/acton/attachment/12345/f-00b2/1/-/-/-/-/Agglomeration-Handbook.pdf.

Feng, D., and Van Deventer, J.S.J. 2011. Oxidative pretreatment in thiosulphate leaching of sulphide gold ores. *Int. J. Miner. Process.* 94(1-2):28–34.

Fernández, G.V. 2003. The use of electrical conductivity in agglomeration and leaching. In *Proceedings of the Copper 2003–Cobre 2003 5th International Conference*, Vol. 6, Book 1: *Hydrometallurgy of Copper*. Edited by P.A. Riveros, D.G. Dixon, D.B. Dreisinger, and J.M. Menacho. Montreal, QC: Canadian Institute of Mining, Metallurgy and Petroleum. pp. 161–175.

Gomes, J. 1983. Heap leaching of gold ores. In *Fifth Annual Conference on Alaskan Placer Mining*. Edited by B.W. Campbell, J.A. Madonna, and S.M. Husted. Fairbanks, AK: University of Alaska Mineral Industry Research Laboratory. pp. 80–83.

Guo, Q., Qu, J., Qi, T., Wei, G., and Han, B. 2011. Activation pretreatment of limonitic laterite ores by alkali-roasting method using sodium carbonate. *Miner. Eng.* 24(8):825–832.

Heinen, H.J., McClelland, G.E., and Lindstrom, R.E. 1979. *Enhancing Percolation Rates in Heap Leaching of Gold-Silver Ores*. Report of Investigation 8388. Washington, DC: U.S. Bureau of Mines.

Herkenhoff, E.C. 1987. Heap leaching: agglomerate or deslime? *Eng. Min. J.* 188(6):32–39.

Hill, G.R. 2013. Agglomeration and leaching of a crushed secondary sulphide copper ore. M.S. thesis, University of Utah, Salt Lake City.

Holle, H. 1996. Process innovations at Girilambone Copper Company's heap leach/SX/EW operation. In *Randol at Vancouver '96: 5th Annual Global Mining Opportunities and 2nd Annual Copper Hydromet Roundtable*. Edited by N. McPherson Nilles. Golden, CO: Randol International. pp. 289–299.

Janwong, A. 2012. The agglomeration of nickel laterite ore. Ph.D. dissertation, University of Utah, Salt Lake City.

Jianhua, L., Qiulin, X., and Mao, Y. 2004. Acidic heap leaching of cracked granite type uranium ore by agglomeration. *Uranium Min. Metall.* 23(3):134–137.

Kappes, D.W. 2002. Precious metal heap leach design and practice. In *Mineral Processing Plant Design, Practice, and Control*. Edited by A.L. Mular, D.N. Halbe, and D.J. Barratt. Littleton, CO: SME. pp. 1606–1630.

Keuhey, K.Y., and Coughlin, W.E. 1984. Getty Mining Company's approach to heap leaching at the Mercur mine. SME Preprint No. 83-419. Littleton, CO: SME.

Kinard, D.T., and Schweizer, A.A. 1987. Engineering properties of agglomerated ore in a heap leach pile. In *Geotechnical Aspects of Heap Leach Design*. Edited by D. Van Zyl. Littleton, CO: SME. pp. 55–64.

Kodali, P. 2010. Pretreatment of copper ore prior to heap leaching. M.S. thesis, University of Utah, Salt Lake City.

Kodali, P., Depci, T., Dhawan, N., Wang, W., Lin, C.L., and Miller, J.D. 2011. Evaluation of stucco binder for agglomeration in the heap leaching of copper ore. *Miner. Eng.* 24(8):886–893.

Kosmatka, S.H., Kerkhoff, B., Panarese, W.C., MacLeod, N.F., and McGrath, R.J. 2002. *Design and Control of Concrete Mixtures*, 7th ed. Engineering Bulletin 101. Ottawa, ON: Cement Association of Canada.

Lastra, M.R., and Chase, C.K. 1984. Permeability, solution delivery, and solution recovery: Critical factors in dump and heap leaching of gold. *Miner. Eng.* 36(11):1537–1539.

LeHoux, P.L. 1997. Agglomeration practice at Kennecott Barneys Canyon Mining Co. In *Global Exploitation of Heap Leachable Gold Deposits*. Edited by D.M. Hausen, W. Petruck, and R.D. Hagni. Warrendale, PA: The Minerals, Metals & Materials Society. pp. 243–249.

Lewandowski, K.A., and Kawatra, S.K. 2009a. Binders for heap leaching agglomeration. *Miner. Metall. Process.* 26(1):1–24.

Lewandowski, K.A., and Kawatra, S.K. 2009b. Polyacrylamide as an agglomeration additive for copper heap leaching. *Int. J. Miner. Process.* 91(3-4):88–93.

Lewandowski, K.A., Gurtler, J.A., Eisele, T.C., and Kawatra, S.K. 2006. Determination of the acid resistance of copper ore agglomerates in heap leaching. In *Sohn International Symposium: Advanced Processing of Metals and Materials*, Vol. 6. *New, Improved and Existing Technologies*. Edited by F. Kongoli and R.G. Reddy. Warrendale, PA: The Minerals, Metals & Materials Society.

Lipiec, A.I., and Bautista, R.G. 1998. Considerations in the extraction of metal from heap leaches. *Trans. Indian Inst. Met.* 51(1):7–16.

Litz, J.E. 1993. Hydrometallurgy: A review of 1992 activities. *Min. Eng.* 46:609.

Liu, L.X. 2012. Effect of binder properties on the strength, porosity and leaching behaviour of single nickel laterite pellet. *Adv. Powder Technol.* 23(4):472–477.

Liu, L.X., Robinson, D.J., and Addai-Mensah, J. 2012. Population balance based modelling of nickel laterite agglomeration behaviour. *Powder Technol. Lausanne* 223:92–97.

Lu, J., Dreisinger, D.B., and West-Sells, P. 2007. Acid curing and agglomeration. In *Cu 2007: John E. Dutrizac International Symposium on Copper Hydrometallurgy*. Edited by P.A. Riveros, D.G. Dixon, D.B. Dreisinger, and M.J. Collins. Montreal, QC: Canadian Institute of Mining, Metallurgy and Petroleum. pp. 453–464.

McClelland, G.E. 1986a. Agglomerated and unagglomerated heap leaching behavior is compared in production heaps. *Min. Eng.* 38(7):500–503.

McClelland, G.E. 1986b. Heap leaching and agglomeration: Heap leaching practice in the United States. In *Gold 100: Proceedings of the International Conference on Gold*, Vol. 2. *Extractive Metallurgy of Gold*. Edited by H. Wagner. Johannesburg: Southern African Institute of Mining and Metallurgy. pp. 57–69.

McClelland, G.E., Pool, D.L, Hunt, A.H., and Eisele, J.A. 1985. *Agglomeration and Heap Leaching of Finely Ground Precious-Metal-Bearing Tailings*. Information Circular 9034. Washington, DC: U.S. Bureau of Mines.

Miller, G. 2003. Ore geotechnical effects on copper heap leach kinetics. In *Hydrometallurgy 2003: Proceedings of the 5th International Symposium Honoring Professor Ian M. Ritchie: Leaching and Solution Purification*. Edited by C. Young, A. Alfantaxi, C. Anderson, D.B. Dreisinger, B. Harris, and A. James. Warrendale, PA: The Minerals, Metals & Materials Society. pp. 329–342.

Miller, G. 2010. Agglomeration drum selection and design process. In *Proceedings of the XXV International Mineral Processing Congress*. Melbourne, Victoria: Australasian Institute of Mining and Metallurgy. pp. 193–202.

Milligan, D.A., and Engelhardt, P.R. 1984. Agglomerated heap leaching at Anaconda's Darwin silver recovery project. SME Preprint No. 183-430. Littleton, CO: SME.

Misra, M., Yang, K., and Mehta, R.K. 1996. Application of fly ash in the agglomeration of reactive mine tailings. *J. Hazard. Mater.* 51:181–192.

Moats, M.S., and Janwong, A. 2008. The art and science of crushed ore agglomeration for heap leaching. In *Hydrometallurgy 2008: Proceedings of the Sixth International Symposium*. Edited by C.A. Young, P.R. Taylor, C.G. Anderson, and Y. Choi. Littleton, CO: SME. pp. 912–917.

Nalco Water. 2017. Agglomeration aids. Naperville, IL: Nalco Water. www.ecolab.com/nalco-water/solutions/agglomeration-aids#f:@websolutions=[Agglomeration%2Aids]&f:@webapplications=[Mineral%20Processing%20Plant,Granulation]. Accessed Dec. 18, 2017.

Nivens, R.E., Given, K.M. 1993. Benefits of polymer agglomeration at Newmont gold company. SME Preprint No. 93-189. Littleton, CO: SME.

Nosrati, A., Addai-Mensah, J., Robinson, D.J., and Farrow, J. 2011. Investigation of the fundamentals of nickel laterite ore agglomeration process. Presented at Chemeca 2011: Engineering a Better World, Sydney, Australia, September 18–21.

Nosrati, A., Addai-Mensah, J., and Robinson, D.J. 2012a. Drum agglomeration behaviour of nickel laterite ore: Effect of process variables. *Hydrometallurgy* 125-126:90–99.

Nosrati, A., Skinner, W., Robinson, D.J., and Addai-Mensah, J. 2012b. Microstructure analysis of mineral ore agglomerates for enhanced heap leaching. *Powder Technol.* 232:106–112.

Nosrati, A., Robinson, D.J., Addai-Mensah, J. 2013. Establishing nickel laterite agglomerate structure and properties for enhanced heap leaching. *Hydrometallurgy* 134-135:66–73.

Nosrati, A., Quast, K., Xu, D., Skinner, W., Robinson, D.J., and Addai-Mensah, J. 2014. Agglomeration and column leaching behaviour of nickel laterite ores: Effect of ore mineralogy and particle size distribution. *Hydrometallurgy* 146:29–39.

O'Brien, R.T. 1982. Agglomeration pre-treatment in the heap leaching of gold and silver. In *Seminar on the Carbon-In-Pulp Technology for the Extraction of Gold*. Melbourne, Victoria: Australasian Institute of Mining and Metallurgy. pp. 297–311.

Ofori-Sarpong, G., Osseo-Asare, K., and Tien, M. 2011. Fungal pretreatment of sulphides in refractory gold ores. *Miner. Eng.* 24:499–504.

Ofori-Sarpong, G., Osseo-Asare, K., and Tien, M. 2013. Mycohydrometallurgy: Biotransformation of double refractory gold ores by the fungus, *Phanerochaete chrysosporium*. *Hydrometallurgy* 137:38–44.

Olubambi, P.A. 2009. Influence of microwave pretreatment on the bioleaching of low-grade complex sulphide ores. *Hydrometallurgy* 95:159–165.

Olubambi, P.A., Potgieter, J.H., Hwang, J.Y., and Ndlovu, S. 2007. Influence of microwave heating on the processing and dissolution behaviour of low-grade complex sulphide ores. *Hydrometallurgy* 89:127–135.

Pautler, J.B., Gross, A.E., and Strominger, M.G. 1990. New polymeric agglomeration aid improves heap leach efficiency at Brewer gold. In *Advances in Gold and Silver Processing: Proceedings of the Symposium at GOLDTech 4*. Edited by M.C. Fuerstenau and J.L. Hendrix. Littleton, CO: SME. pp. 15–21.

Perez, J.W., Barrois, M.J., McCord, T.H., and O'Neil, G.R. 1990. Pretreatment/agglomeration as a vehicle for refractory gold ore treatment. U.S. Patent 4,961,777.

Polizzotti, D.M. 1993. Agglomerating agents for clay containing ores. U.S. Patent 5,211,920.

Polizzotti, D.M., Liao, W.P., and Roe, D.C. 1997. Cationic block polymer agglomeration agents for mineral bearing ores. U.S. Patent 5,668,219.

Potter, G.M. 1983. Some factors in the design of heap leaching operations. In *Papers Given at the Precious-Metals Symposium*. Nevada Bureau of Mines and Geology Report 36. Edited by V.E. Kral, J.A. Hall, R.B. Blakestad, H.F. Bonham Jr., G.B. Hartley Jr., G.E. McClelland, J.A. McGlasson, and P. Mousette-Jones. Reno, NV: Nevada Bureau of Mines and Geology. pp. 69–76.

Pyper, R., Kappes, D.W., and Albert, T. 2015. Evaluation of agglomerates using the Kappes percolation test. In *Proceedings of the 3rd International Conference on Heap Leach Solutions*. Edited by M. Evatz, M.E. Smith, and D. Van Zyl. Vancouver, BC: InfoMine. pp. 319–328.

Quaicoe, I., Nosrati, A., Skinner, W., and Addai-Mensah, J. 2014. Agglomeration and column leaching behaviour of geothitic and saprolitic nickel laterite ores. *Miner. Eng.* 65:1–8.

Quast, K., Xu, D., Skinner, W., Hilder, T., Addai-Mensah, J., and Robinson, D.J. 2013. Column leaching of nickel laterite agglomerates: Effect of feed size. *Hydrometallurgy* 134-135:144–149.

Rose, W. R., Pyper, R., Eguivar, J., and Mirabal, C. 1990. Heap leaching with fine grinding and agglomeration at Potosi, Bolivia. In *Advances in Gold and Silver Processing: Proceedings of the Symposium at GOLDTech 4*. Edited by M.C. Fuerstenau and J.L. Hendrix. Littleton, CO: SME. pp. 23–32.

Saririchi, T., Roosta Azad, R., Arabian, D., Molaie, A., and Nemati, F. 2012. On the optimization of sphalerite bioleaching: The inspection of intermittent irrigation, type of agglomeration, feed formulation and their interactions on the bioleaching of low-grade zinc sulfide ores. *Chem. Eng. J.* 187:217–221.

Scheffel, R.E. 1982. Retreatment and stabilization of the Naturita tailing pile using heap leaching techniques. In *Interfacing Technologies in Solution Mining: Proceedings of the Second SME-SPE International Solution Mining Symposium*. Edited by W.J. Schlitt and J.B. Hiskey. Littleton, CO: SME-AIME. pp. 311–321.

Scheffel, R.E. 2002. Copper heap leach design and practice. In *Mineral Processing Plant Design, Practice, and Control*. Edited by A.L. Mular, D.N. Halbe, and D.J. Barratt. Littleton, CO: SME. pp. 1571–1605.

Serrano, E.M. 2003. New agglomeration techniques for the acid percolation leaching of clay. In *Proceedings of the Copper 2003–Cobre 2003 5th International Conference, Vol. 6, Book 1: Hydrometallurgy of Copper*. Edited by P.A. Riveros, D.G. Dixon, D.B. Dreisinger, and J.M. Menacho. Montreal, QC: Canadian Institute of Mining, Metallurgy and Petroleum. pp. 189–201.

Sinclair, G. 2010. Applying heap leach technology to the treatment of Ranger ores. IAEA Consultancy on Low-Grade Uranium Ore. www.iaea.org/OurWork/ST/NE/NEFW/documents/RawMaterials/TM_LGUO/4f%20Sinclair%20Ranger%20Heap%20Leach.pdf.

Strachan, C., and Van Zyl, D. 1987. Feasibility assessment for increasing heap thicknesses at the Alligator Ridge mine. In *Geotechnical Aspects of Heap Leach Design*. Edited by D. Van Zyl. Littleton, CO: SME. pp. 65–76.

Tibbals, R.L. 1987. Agglomeration practice in the treatment of precious metal ores. In *Proceedings of the International Symposium on Gold Metallurgy*. Edited by R.S. Salter, D.M. Wyslouzil, and G.W. McDonald. New York: Pergamon Press. pp. 77–86.

Uhrie, J.L., and Koons, G.J. 2001. The evaluation of deeply ripping truck-dumped copper leach stockpiles. *Min. Eng.* 53(12):54–56.

Uhrie, J.L., Wilton, L.E., Rood, E.A., Parker, D.B., Griffin, J.B., and Lamanna, J.R. 2003. The metallurgical development of the Morenci MFL project. In *Proceedings of the Copper 2003–Cobre 2003 5th International Conference, Vol. 6, Book 1: Hydrometallurgy of Copper*. Edited by P.A. Riveros, D.G. Dixon, D.B. Dreisinger, and J.M. Menacho. Montreal, QC: Canadian Institute of Mining, Metallurgy and Petroleum. pp. 29–37.

Velarde, G. 2005. Agglomeration control for heap leaching processes. *Miner. Process. Extr. Metall. Rev.* 26(3-4):219–231.

Vethosodsakda, T., Free, M.L., Janwong, A., and Moats, M.S. 2013. Evaluation of liquid retention capacity measurements as a tool for estimating optimal ore agglomeration moisture content. *Int. J. Miner. Process.* 119:58–64.

Videau, G., and Roche, M. 1990. Développement industriel d'un nouveau procédé d'agglomération de mineral uranifere par boulettage avant traitement par lixiviation en tas (voie acide): Application au mineral très argileux de Nord-Aquitaine Industrie Minérale Mines et Carrières. *Les Techniques* 2-3(March-April):135–142.

Walker, D.D., Jr., and Oliphant, J. 1992. Composition and method for agglomerating ore. U.S. Patent 5,116,417.

West-Sells, P.G., Bouffard, S.C., Tshilombo, F.A., and Bruysteneyn, A. 2007. Acid generation by in-situ sulfur biooxidation for copper heap leaching. In *Cu 2007: John E. Dutrizac International Symposium on Copper Hydrometallurgy*, Vol. 4, Book 1. Edited by P.A. Riveros, D.G. Dixon, D.B. Dreisinger, and M.J. Collins. Montreal, QC: Canadian Institute of Mining, Metallurgy and Petroleum. pp. 323–334.

Xu, D., Liu, L.X., Quast, Q., Addai-Mensah, J., and Robinson, D.J. 2013. Effect of nickel laterite agglomerate properties on their leaching performance. *Adv. Powder Technol.* 24(4):750–756.

Yijun, Z., Jianhua, L., Tieqiu, Li., and Pingru, Z. 2002. Application of agglomerated acid heap leaching of clay-bearing uranium ore in China. In *Proceedings of the Recent Developments in Uranium Resources and Production with Emphasis on In-Situ Leach Mining*. Beijing, China: IAEA-OECD Nuclear Energy Agency, the Bureau of Geology, and China National Nuclear Corporation. pp. 243–249.

Zárate, G.E., and Guzmán, J.C. 1987. Gold tailings processing by heap leaching. In *Small Mines Development in Precious Metals*. Littleton, CO: SME. pp. 151–156.

CHAPTER 10.2

Solution Mining and In Situ Leaching

J. Brent Hiskey, John G. Mansanti, and Terry McNulty

Solution mining and in situ leaching of copper, water-soluble salts, and uranium are discussed in this chapter. Among the water-soluble salts are boron compounds, lithium compounds, magnesium chloride, potassium chloride, and various sodium compounds including table salt.

DEFINITIONS

The terminology used in this section follows industry conventions but is intended to be generalized and may differ from the definitions used by specialists such as those in the field of hydrology.

- *Solution mining* includes all methods for dissolving or liquefying minerals or materials and capturing the resulting fluid for commercial recovery of contained values. "Solution mining is an interdisciplinary field involving geology, chemistry, hydrology, extractive metallurgy, mining engineering, process engineering, and economics" (Bartlett 1992).
- *In situ leaching* (ISL) is a form of solution mining that relates to extraction of metals, elements, or compounds from an undisturbed location or mineral deposit.
- An *aquifer* is a formation beneath the ground surface that not only contains water but also is capable of producing water, for example, by way of extraction through a well.
- An *aquitard* is a formation either above or below an aquifer that resists the flow of water.
- The *permeability* of a geologic structure, usually a sedimentary unit for the purposes of this chapter, is a measure of the capability for transmitting a fluid. The rate at which a fluid can be transmitted over a specific distance in a specific time interval is defined experimentally in darcies. One darcy equals a flow rate of 1 $cm^3/s/cm^2$ of flow cross section for a fluid having a viscosity of 1 cP under a pressure gradient of 1 atm/cm.
- *Porosity* in a rock or sedimentary unit is quantified by the ratio of pore volume to the volume of the rock. This can be approximated by one minus the ratio of bulk density of a core section to the expected specific gravity of a massive fragment of the same material, both expressed as weight per unit of volume.

COPPER MINERALS

In the broadest sense of the term, *copper solution mining* refers to the extracting and recovery of copper by dump and stockpile leaching, heap leaching, and ISL methods. This section will focus on the ISL of copper. Regardless of the method, solution mining is an extremely site-specific endeavor and very dependent on the geologic setting and copper mineralogy. Bateman (1951) summarized important copper minerals associated with the three major zones of a typical copper deposit. This information is listed in Table 1.

Common copper minerals in the uppermost portion (oxidized zone) are classified for the most part as oxyanion

Table 1 Zones of mineralization for copper

Mineralized Zone	Mineral	Simplified Formula
Oxidized zone (secondary)	Native copper	Cu
	Malachite*	$Cu_2(OH)_2CO_3$
	Brochantite*	$Cu_4(OH)_6SO_4$
	Antlerite*	$Cu_3(OH)_4SO_4$
	Atacamite*	$Cu_2(OH)_3Cl$
	Azurite*	$Cu_3(OH)_2(CO_3)_2$
	Chrysocolla*	$CuSiO_3{\cdot}2H_2O$
	Cuprite*	Cu_2O
	Tenorite*	CuO
	Chalcanthite*	$CuSO_4{\cdot}5H_2O$
Supergene enrichment zone (secondary)	Chalcocite*	Cu_2S
	Covellite	CuS
	Native copper	Cu
Hypogene zone (primary)	Chalcopyrite	$CuFeS_2$
	Bornite	Cu_5FeS_4
	Enargite*	Cu_3AsS_4
	Tetrahedrite*	$Cu_{12}Sb_4S_{13}$
	Tennantite*	$Cu_{12}As_4S_{13}$
	Covellite	CuS

Source: Forrester 1946

*Always in relative descending position indicated.

J. Brent Hiskey, Professor Emeritus, Mining & Geological Engineering, University of Arizona, Tucson, Arizona, USA
John G. Mansanti, President and CEO, Crystal Peak Minerals Inc., Salt Lake City, Utah, USA
Terry McNulty, President, T.P. McNulty and Associates Inc., Tucson, Arizona, USA

Table 2 Dissolution reactions for selected copper minerals

Mineralized Zone	Mineral	Leaching Reaction	
Oxidized zone (secondary)	Native copper	$Cu + 2H^+ + ½O_2 \rightarrow Cu^{2+} + H_2O$	
	Malachite	$CuCO_3{\cdot}Cu(OH)_2 + 4H^+ \rightarrow 2Cu^{2+} + CO_2 + 3H_2O$	
	Brochantite	$CuSO_4{\cdot}3Cu(OH)_2 + 6H^+ \rightarrow 4Cu^{2+} + SO_4^{2-} + 6H_2O$	
	Antlerite	$CuSO_4{\cdot}2Cu(OH)_2 + 4H^+ \rightarrow 3Cu^{2+} + SO_4^{2-} + 4H_2O$	
	Atacamite	$3CuO{\cdot}CuCl_2{\cdot}3H_2O + 3H^+ \rightarrow 4Cu^{2+} + 2Cl^- + 6H_2O$	
	Azurite	$2CuCO_3{\cdot}Cu(OH)_2 + 6H^+ \rightarrow 3Cu^{2+} + 2CO_2 + 4H_2O$	
	Chrysocolla	$CuO{\cdot}SiO_2{\cdot}2H_2O + 2H^+ \rightarrow Cu^{2+} + SiO_2{\cdot}xH_2O + (3-x)H_2O$	
	Cuprite	$Cu_2O + 2H^+ \rightarrow Cu^{2+} + Cu + H_2O$	
	Tenorite	$CuO + 2H^+ \rightarrow Cu^{2+} + H_2O$	
Supergene enrichment zone (secondary)	Chalcocite	$Cu_2S + 2H^+ + ½O_2 \rightarrow Cu^{2+} + CuS + H_2O$	First stage
	Covellite	$CuS + 2H^+ + ½O_2 \rightarrow Cu^{2+} + H_2SO_4$	Second stage
	Native copper	$Cu + 2H^+ + ½O_2 \rightarrow Cu^{2+} + H_2O$	
Hypogene zone (primary)	Chalcopyrite	$CuFeS_2 + 16Fe^{3+} + 8H_2O \rightarrow Cu^{2+} + 17Fe^{2+} + 2SO_4^{2-} + 16H^+$	
	Bornite	$Cu_5FeS_4 + 36Fe^{3+} + 16H_2O \rightarrow 5Cu^{2+} + 37Fe^{2+} + 4SO_4^{2-} + 32H^+$	
	Enargite	$Cu_3AsS_4 + 35Fe^{3+} + 20H_2O \rightarrow 3Cu^{2+} + 35Fe^{2+} + H_3AsO_4 + 4SO_4^{2-} + 37H^+$	
	Covellite	$CuS + 8Fe^{3+} + 4H_2O \rightarrow Cu^{2+} + 8Fe^{2+} + SO_4^{2-} + 8H^+$	

minerals (i.e., hydroxides, carbonates, sulfates, chlorides). Naturally, these minerals, along with tenorite, are of special importance because they dissolve rapidly in acid solution. Chrysocolla dissolves in acid solution with the formation of a remnant layer of $SiO_2{\cdot}xH_2O$. The dissolution kinetics follows a shrinking core model involving ion diffusion through the hydrated silica product layer, whereas acidic dissolution of cuprite involves a disproportionation reaction that yields only half of the copper to solution. Native copper is the only mineral that requires an oxidant for leaching. In general, the dissolution kinetics of the oxide copper minerals is relatively fast as compared with the sulfides. Table 2 shows the leaching reactions for the copper minerals listed in the various mineral zones.

Under normal conditions, secondary and primary copper sulfide minerals require the presence of an oxidant for dissolution. In aqueous solution, dissolved oxygen serves the role. Pyrite commonly found in the supergene enrichment and hypogene zones can undergo aqueous oxidation to yield an acid ferric sulfate solution that can effectively dissolve copper sulfide minerals. The acid ferric sulfate dissolution of copper sulfide appears to have a reasonably straightforward leaching reaction. For example, the leaching of chalcocite can be represented by the following reactions:

$$Cu_2S + 2Fe^{3+} \rightarrow Cu^{2+} + CuS + 2Fe^{2+}$$
First stage **(EQ 1)**

$$CuS + 2Fe^{3+} \rightarrow Cu^{2+} + S + 2Fe^{2+}$$
Second stage **(EQ 2)**

However, the fundamental leaching of chalcocite is much more complex than characterized by the general reactions shown in Equations 1 and 2. Compositionally intermediate phases between chalcocite and covellite include several nonstoichiometric copper sulfides that comprise a progressive dissolution series. In order of decreasing Cu:S ratio, these minerals include djurleite ($Cu_{1.94}S$), digenite ($Cu_{1.8}S$), anilite ($Cu_{1.75}S$), geerite ($Cu_{1.6}S$), spionkopite ($Cu_{1.4}S$), and yarrowite ($Cu_{1.12}S$). Sulfide minerals have excellent semiconducting properties, and chalcocite and covellite are no exception. In fact, covellite is characterized by metallic-like conduction. Therefore, copper sulfide dissolves in aqueous solution according to an electrochemical mechanism. The anodic dissolution step for the first stage of chalcocite dissolution is illustrated by the following general reaction:

$$Cu_2S \rightarrow Cu_{2-x} + xCu^{2+} + 2xe^- \quad \textbf{(EQ 3)}$$

Acid ferric sulfate dissolution of chalcocite up to about $x = 0.9$, corresponding to 45% copper recovery, exhibits relatively fast kinetics. The dissolution of covellite is considerably slower.

Primary copper minerals are sparingly soluble in acid ferric sulfate solutions. Chalcopyrite, economically the most important copper sulfide mineral, has been recognized for many years (Sullivan 1933) as having extremely slow leach kinetics. Even under relatively aggressive conditions, leaching is slow. Numerous kinetic investigations have focused on chalcopyrite to better understand the leaching fundamentals. Surface passivation is believed to play a major role in retarding the leaching rate of chalcopyrite.

Historical Examples

The historical fabric of copper solution mining is very rich. Passive hydrometallurgical systems for the extraction and recovery of copper have been used for millennia. These systems refer to the recovery of copper-bearing solutions from natural seeps and springs, and from drainages associated with copper mining activities. Early attempts in China to recover copper by passive methods have been described by Pu (1982) and Lung (1986) in considerable detail. Pharmacologists in the East Han Dynasty (25–220 AD) clearly understood the formation of gall (copper sulfate) springs and described in exquisite terms the precipitation of copper on iron from its solution. Engineering leaching systems for in situ copper extraction are more recent. These modern examples clearly demonstrate the application and integration of various scientific and technological disciplines, including hydrometallurgy, hydrology, geology, geochemistry, rock mechanics, environmental chemistry, and process engineering.

Passive Systems

An interesting early example of passive solution mining is the recovery of copper-rich solution from the ancient mines on Cyprus by the Romans. Galen (see Angelatou 2004), a naturalist and physician, reported in 166 AD the operation of ISL of copper on Cyprus.

Surface water was allowed to percolate through the permeable rock, and was collected in amphorae. In the process of percolation through the rock, copper minerals dissolved so that the concentration of copper sulfate in solution was high. The solution was allowed to evaporate until copper sulfate crystallized.

Pliny (23–79 AD), in *Historia Naturalis XXXIII–XXXV* (Rackham 1952), reported that a similar practice for the extraction of copper in the form of copper sulfate was widely practiced in Spain. The recovery of copper from mine drainages at Rio Tinto, Spain predates the large-scale commercial heap leaching operations (1876) described by Taylor and Whelan (1942). In 1556, Diego Delgado observed and reported copper cementation at Rio Tinto, and Don Alvaro Alonso de Garfias was granted a patent to recover copper by iron cementation from mine water in 1661.

Bushnell (1911) described crude attempts by Morgan as early as 1888 to recover copper from mine drainage in Butte, Montana, United States. Improved attempts by Miller to recover copper by cementation from water originating from the St. Lawrence mine were marginally better. In 1901, Ledford secured a three-year lease on the water from the St. Lawrence and Anaconda mines at a 25% royalty. This was a very successful enterprise and, when the lease expired, the companies decided to build their own precipitation plants. By 1906, similar passive solution mining was extremely well developed at Bisbee, Arizona, United States.

Engineered Systems

Copper ISL can be conveniently divided into two primary activities. One method involves the ISL of copper values associated with old mining remnants (i.e., pit slopes, block caving subsidence zones, underground stopes). The other comprises the extraction of copper from the confines of a deposit in its original geologic setting (i.e., undisturbed ore). This is commonly referred to as "true" ISL. This ISL method can be performed with or without matrix modification (permeability enhancement). Blasting, hydraulic fracturing ("hydrofracing"), and chemical stimulation have been suggested as ways to improve deposit permeability.

Greenwood (1926) reviewed the ISL of copper at Cananea Consolidated Copper Co. in Sonora, Mexico, in the 1920s. Underground leaching of remnant copper ore in old shrinkage stopes was accomplished in a controlled fashion. Solutions were allowed to infiltrate naturally through old workings, or in some cases, solution was delivered to the surface of broken ore via sprays in perforated pipe and allowed to percolate through the stope. Bulkheads were constructed in certain drifts to collect and store leach solution. In-mine cementation launders were installed in select locations to recover copper. This avoided pumping solutions to a surface recovery plant.

Miami mine. The Miami mine located near Miami, Arizona, is the oldest continuously operated ISL facility in the world (Kyle et al. 2011). Block caving of the ore body created a huge surface subsidence area containing broken and highly fractured low-grade mixed copper oxide–sulfide material. The first attempts to leach this material in place were in 1942 with only the application of water. This proved unsuccessful since there was not enough pyrite in the rock to maintain the autogenous generation of acid. Application of a solution containing about 5 g/L H_2SO_4 was effective in leaching the copper. Solution has been applied with sprays, via ponds,

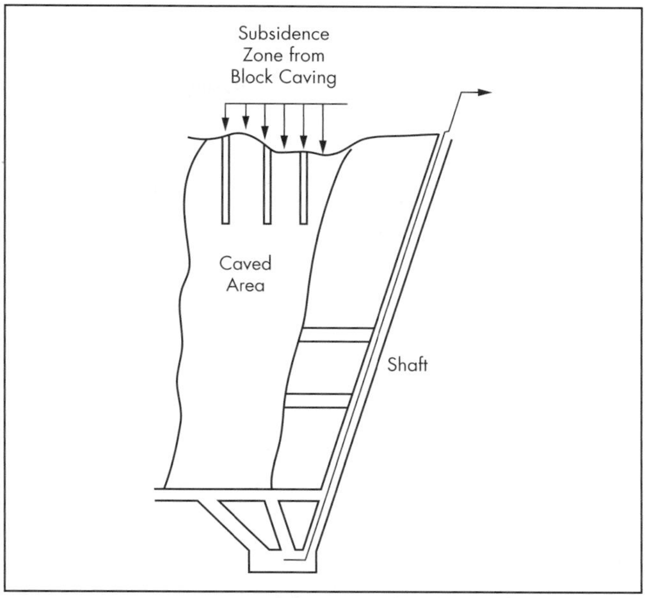

Figure 1 Representation of the in situ leaching of a block caved subsidence zone

and by injection holes. Ponding proved to be ineffective and was discontinued. Cased injection holes, with a perforated bottom section, reach an average depth of 59.5 m (195 ft). Leach solution percolates by gravity from the surface and the injection points through the caved ore and then through raises and drawpoints until it reaches the final haulage level, about 305 m (~1,000 ft). Solution flow is controlled by weir boxes in the underground workings. Solutions are pumped to the surface and a solvent extraction electrowinning (SX-EW) plant. A general pictorial representation of the Miami system is shown in Figure 1.

Old Reliable. The Old Reliable project carried out by Ranchers Exploration and Development Corporation is a special example of an in situ operation incorporating matrix modification. Longwell (1974) reported that the ore body was a nearly vertical breccia pipe containing 3.6 Mt (4×10^6 st) of 0.80% Cu. The most abundant copper minerals were chalcocite, chalcopyrite, malachite, chalcanthite, and chrysocolla. Historic mining operations had removed approximately 2.72 kt (30,000 st) of ore from two main adits at the 30.5-m and 61-m (100-ft and 200-ft) levels and associated workings. Development mining was done to prepare the ore body for blasting (i.e., rubblization). This consisted of an additional adit located 51 m (167 ft) above the 30.5-m (100-ft) level and extra crosscuts, which resulted in a total of 2,591 m (8,500 ft) of tunnels and crosscuts. Ranchers in partnership with E.I. du Pont de Nemours and Company designed a 3.63-Mt (4-million-lb) ANFO (ammonium nitrate and fuel oil) blast using 1,829 m (6,000 ft) of coyote tunnels for emplacement of the explosives and stemming. The blast was detonated on March 9, 1972, and it was estimated that fragmentation resulted in a mean particle diameter of 22.9–25.4 cm (9–10 in.) in the ore zone. The hillside was terraced with 6.1-m (20-ft) benches varying from 4.6 to 54.9 m (15–180 ft) wide. Leach solutions were applied by a sprinkler system at a total solution flow rate of 4.17 m^3/min (100 gpm), equivalent to an application rate of 0.183 L min^{-1} m^{-2} (0.0045 gpm/ft^2). Leach solution percolated through the rubblized ore, and pregnant leach solution (PLS) flowed from the hillside at about the position of the existing water table to a catchment basin. Copper was recovered by cementation on scrap iron in a simple launder plant.

ISL was conducted at the Old Reliable mine during two periods: initially from August 1972 to September 1974, and resumption of operation from August 1979 to January 1981. A total of 5.898 kt (13 million lb) of copper was recovered during these two phases of leaching. This represented a total copper recovery of approximately 20%. Even at this low level of copper extraction, the project was an economic success; however, it was in many ways a technical disappointment. Operational problems included solution channeling and the formation of a perched water table.

White Pine. Conventional mining and milling operations at the White Pine mine located in the Keweenaw Peninsula region of northern Michigan, United States, were terminated in 1995. Prior to this, Copper Range Company, the operating entity, had initiated a test program to evaluate the potential of ISL at the White Pine mine. The Michigan Department of Environmental Quality issued a permit in 1993 to perform the necessary test work. The plan was to conduct in-stope leaching of rubblized ore with acid ferric sulfate solutions and collect the PLS for copper recovery by SX-EW. The resource available for ISL was estimated at ~189 Mt, or million metric tons (208,207,000 st), averaging 1.04% Cu. Copper minerals in the White Pine deposit occur principally in two forms: as a very fine-grained chalcocite (Cu_2S) and as native copper (Cu). The chalcocite accounted for an estimated 85%–90% of the contained copper value.

Room-and-pillar mining was the principal ore extraction method used at White Pine. Remnant pillars were to be rubblized by a controlled blast with a resulting collapse of the mine. Certain portions of the mine had already collapsed by natural processes. Leaching was to be accomplished by circulation of acid ferric sulfate solution through rubble piles created from the blasted pillars. Pregnant leach solution would be collected and pumped to a surface SX-EW plant. In May 1996, Copper Range Company announced approval of permits to in situ leach the remnant ore in the White Pine mine; however, opposition from environmental groups and regional Native American tribes was extremely strong and the project was postponed. One contentious issue was the delivery of sulfuric acid across tribal lands in Wisconsin. On May 30, 1997, Copper Range Company declared an end to the project.

Laboratory Simulation Techniques

Metallurgical testing is a critical step in the evaluation and development of a copper ore body for ISL. An ore testing program relies on the receipt of representative samples either as individual samples or as composites. Ore samples are usually obtained from an exploration drilling program. They are received as small- or large-diameter diamond core or possibly as cuttings and chips from percussion drilling. These samples should be accompanied by a comprehensive geologic description (core log). In terms of ore characterization, the first task is to obtain a total copper assay by standard analytical techniques. This usually is followed by a determination of the acid soluble (oxide) copper content. Early in an ore characterization program, a carefully planned and executed process mineralogy study should be carried out. This will identify and quantify the key mineralogical attributes of the ore, including ore mineralogy, gangue mineralogy, the presence of deleterious minerals and elements, grain size and shape, mineral assemblages, surface coatings, and texture. Process mineralogy studies can be carried out by a combination of optical and electron microscopic techniques. Automated systems such as mineral liberation analysis (MLA) instruments are an extremely valuable tool for quantitative process mineralogy information. Sequential diagnostic leach tests are another useful procedure in evaluating the ore mineralogy.

Leaching tests usually follow a sequence from bottle roll experiments to small and large column tests, and whole core leaching tests. These tests are preliminary and help provide important data for the design of a pilot field experiment.

Bottle Rolls

Cyanide bottle-roll tests are an industry-standard initial step in evaluating the leaching characteristics of gold ores. These tests have been successfully adapted to assessing the leaching performance of copper ores and to determining lixiviant requirements. Bottle-roll tests are performed at ambient temperature and pressure and are usually conducted for relatively short times (~24–72 hours) using a fairly coarse feed size, equivalent to a top size of approximately 4 mesh (4.7 mm [~0.2 in.]). Ore slurries are prepared by pulping the coarse ore with leaching solution to reach 40–50 wt % solids. Solution samples are withdrawn from the bottle at prescribed time intervals and analyzed by standard techniques for dissolved copper, pH, free acid, and certain elements and species as required. When the test is terminated, the slurry is filtered to separate the PLS from the solid residue. The PLS volume is measured and the solution analyzed. The leached residue is carefully washed, dried, and analyzed; alternatively, it is screened and each size fraction analyzed to determine metal extraction as a function of particle size. Bottle-roll tests represent a rapid metallurgical test that provides vital information regarding ultimate copper recovery as a function of screen size, copper recovery rate, and maximum sulfuric acid consumption.

Column Leach Tests

The next phase of a metallurgical test program typically consists of conducting leaching experiments in columns. These tests provide more detailed information about the copper recovery, kinetic data, and reagent consumption. In general, column leaching tests provide a systematic study of mineralized ore types using larger quantities of sample and a larger particle size. Columns are normally operated for extended periods of time. There are several benefits to column leaching: (1) it can serve to optimize the lixiviant chemistry; (2) it can provide a more accurate picture of acid consumption; (3) it can define cation loading; and (4) it can be operated under locked-cycle conditions. All of these benefits help establish the criteria for larger tests.

Columns can be operated under both percolation and inundation (flooded) conditions. Furthermore, they can be leached in sequential order to simulate solution "stacking," where reagents are added to low-grade PLS, which is then returned to the circuit in an effort to increase PLS grade. Frequently, columns are run in a locked-cycle mode to more accurately simulate actual processing conditions. In a locked-cycle test, after the PLS reaches a certain copper concentration, it is directed to a copper solvent extraction (SX) circuit. Dissolved copper is extracted and the raffinate is recycled to the leaching circuit after addition of makeup water and reagents. If the PLS is below the set copper concentration, the weak PLS is returned directly to leaching after makeup water and reagent adjustments. It is critical that steady state be achieved during locked-cycle experiments. Under locked-cycle leaching, the system reaction chemistry dictates that all PLS will eventually be treated by solvent extraction.

After a thorough analysis of all leach solutions, the leach residues are carefully rinsed and dried, and the residue metal content determined. Physical examination of the residues can yield valuable information regarding factors such as physical degradation. In addition, MLA examination of residues can quantitatively determine factors such as preferential copper mineral dissolution, the generation of fines and degradation, the precipitation of second phases, the degree of ion exchange by clays, and copper mineral alteration, such as replacement of copper by aluminum in chrysocolla, expressed as $(Cu,Al)_2H_2Si_2O_5(OH)_4{\cdot}nH_2O$.

Whole Core Experiments

Whole core leaching experiments more accurately simulate the chemical and hydrological interactions at the solution/rock interface. This is especially true for copper mineralization that is distributed along fractures and within pore spaces of the rock. Therefore, leach fluids will more likely approximate the solution chemistry produced during in situ mining. Leach parameters such as copper concentration in the PLS, copper recovery, acid consumption, and gangue mineral reactions can be determined more precisely. Special apparatuses can be designed and utilized to leach the core under pressure. These experimental systems allow the study of pressure as a variable and the real-time monitoring of permeability changes during leaching. An increased permeability may result when substantial amounts of copper dissolve from fracture surfaces and pores. If significant enough leaching occurs, channels may form and solution short-circuiting may result. Permeability may actually decrease if second-phase solids such as gypsum precipitate along fluid channels or within pores. Core-leaching experiments also furnish important information regarding the physical nature of the leached sample when postmortem examinations are performed.

Paulson and Kuhlman (1989) have documented in detail various laboratory core-leaching apparatus and experiments. Three systems were designed: (1) a low-pressure system with an acrylic reaction cell; (2) a medium-pressure system with a polyvinyl chloride (PVC) reaction cell; and (3) an elevated-pressure system with stainless-steel components. The low, medium, and elevated pressure systems could handle pressures up to 345 kPa, 1,034 kPa, and 6,895 kPa (50 psi, 150 psi, and 1,000 psi), respectively. All three systems were designed to operate under axial flow conditions. A cross-sectional schematic of the elevated-pressure core-leaching cell is depicted in Figure 2. The main components are constructed of Type 316 stainless steel. The entire experimental setup for the elevated-pressure cell is shown in Figure 3.

Hiskey and Oner (1992) discussed the use of a laboratory system for leaching of large whole-core samples that was designed to operate at moderate pressures (<3,450 kPa [<500 psi]). This system was designed and fabricated to operate under radial flow conditions. Leach fluids flowed from the outside perimeter of the core to a center collection well. Copper-bearing PLS moved from the center well by displacement to a collection vessel that was sampled on a regular basis. Experiments were conducted with a 103-mm (4-in.)-diameter core sample and a core section length of 103 mm (4 in.). A 25.4-mm (1-in.) center well was drilled in the core to serve as a production well. The system was designed for convenient removal of the core during the course of the test. The partially

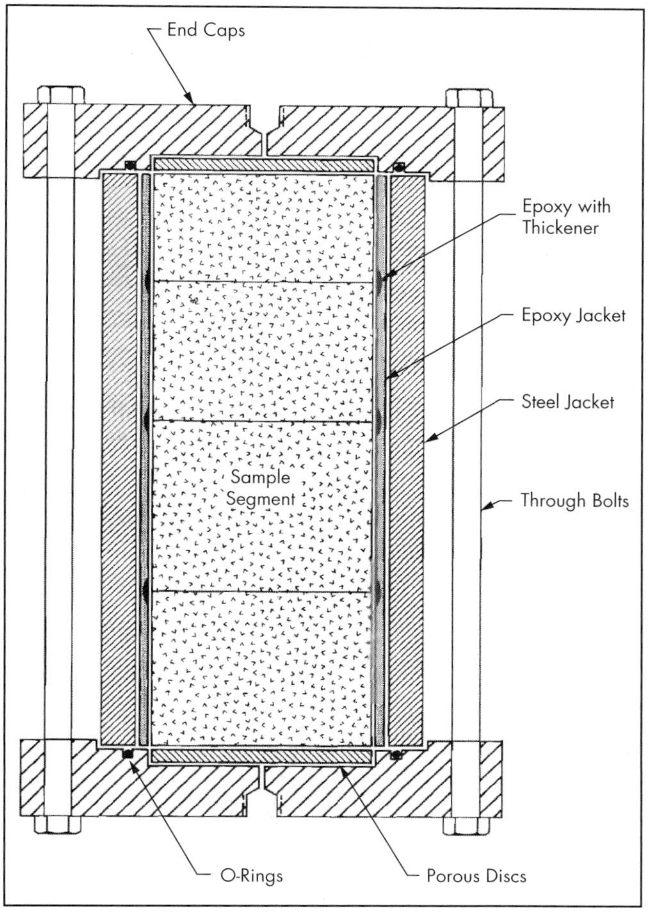

Source: Paulson and Kuhlman 1989

Figure 2 Elevated-pressure core-leaching cell

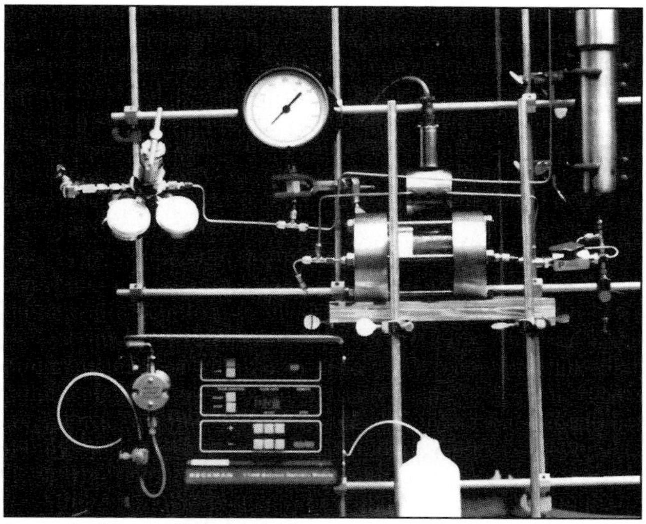

Source: Paulson and Kuhlman 1989

Figure 3 Experimental setup for whole core leaching

leached core could be examined by macro computed tomography (CT) imaging to determine the development of leach channels and solution pathways.

Box Experiments

The large whole core-leaching systems described earlier work well with competent intact core samples. Core samples that are highly fractured and fragmented would require another system. HDI Curis, at its Florence Copper Project (M3 Engineering and Technology Corporation 2013), developed a unique box method for laboratory leaching of natural highly fractured and unconsolidated core. In these tests, rectangular boxes made of transparent welded plastic sheet were loaded with multiple sections of fractured core, undisturbed as much as possible, and contacted with leaching solution flowing normal to the long axis of the core. Each core section was wrapped in plastic screen mesh to preserve the size, shape, position, and orientation of the core fragments. The stabilized core was then loaded horizontally into the boxes on a layer of paraffin wax to support the weight of the core interval and to prevent solution channeling around the sample. The spaces between the ends of the core and the box were also filled with paraffin wax to restrict fluid flow to the central portions of the core. In addition, interstitial spaces and large gaps between core fragments were filled with acid-washed silica sand to encourage fluid flow through the core intervals. Dilute aqueous sulfuric acid, typically 10 g/L free H_2SO_4, was introduced to one end of a sealed box at a hydraulic head of about 0.305 m (12 in.), as prescribed by the aquifer injection permit, and drained by gravity from the other end. Subsequent tests were conducted in vertical stainless-steel cylinders that enabled leaching at a pressure of 689.5–827.4 kPa (100–120 psi) to match the approximate hydraulic head over the mineralized formation. Another description of the Florence Copper Project is included in a comprehensive analysis of copper ISL by Sinclair and Thompson (2015).

The laboratory simulation technique used by Excelsior Mining Corp. for its Gunnison ISL copper project is also worthy of note (M3 Engineering and Technology Corporation 2017). Its "core tray" method involved exposing the core only at major fractures, using epoxy resin to coat the unfractured rock, minimizing consumption of acid by barren gangue. Leaching solution flowed along the top surfaces of the horizontal core samples, and solution drained downward through exposed mineralized fractures. This method was expected to provide more reliable indications of copper extraction from exposed mineralization and acid consumption per unit of copper dissolved than could be determined from a field trial. Leaching solution was initially applied at a concentration of 1 g/L free H_2SO_4 that was gradually increased to 15 g/L.

With box experiments, it is difficult to assess permeability effects; however, other leach data can be fully evaluated, including data from lock-cycle tests. As with the large whole-core leaching experiments, the box tests can be run for extended times, often more than 200 days. They can also include rinse cycles and postleach studies of the core.

Lixiviant Chemistry and Gangue Mineral Interactions

The main criteria determining the choice of leaching agent (lixiviant selection) depend primarily on the physical and chemical properties of the material to be leached. Other factors that are important in the selection process are reagent cost, selectivity toward the particular material to be leached, ability to regenerate the leaching agent, and the chemical stability of the leaching agent. Three important conditions are taken into account when determining the appropriate lixiviant for a given mineral phase: (1) solution pH, (2) type of oxidant, and (3) the role of complexing ligands. All three factors can be suitably evaluated using oxidation potential–pH relationships and the graphically effective Eh–pH diagram.

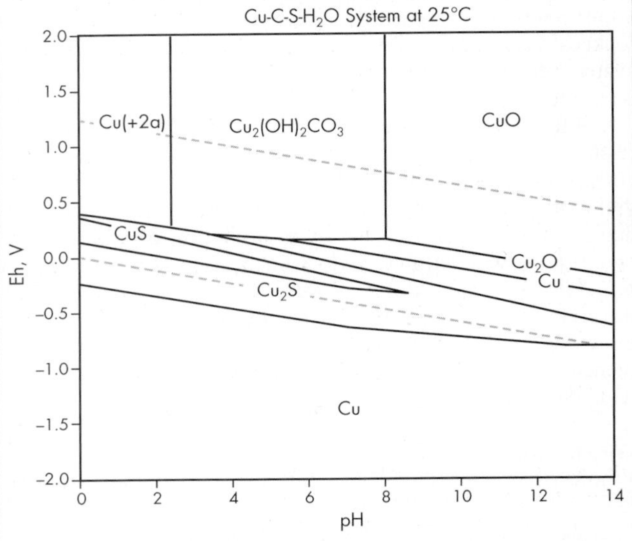

Figure 4 Eh–pH diagram for the Cu-S-O-H_2O system at 298 K

Eh–pH Diagrams

The Eh–pH (Pourbaix) diagrams are thermodynamic stability diagrams showing equilibrium relationships between solid (mineral) phases, as well as dissolved species. These drawings are an excellent means of treating geochemical, mineral, and hydrometallurgical equilibria. They represent a convenient and efficient means of graphically displaying a large amount of thermodynamic data. Constructing the Eh–pH diagram is quite simple for a single element system; however, it becomes extremely complex for multicomponent systems. An excellent source of information on the application of Eh–pH relationships to hydrometallurgy is by Pourbaix (1966). Garrels and Christ (1965) have developed the details of constructing Eh–pH diagrams and have discussed various diagrams of interest to copper geochemistry and extractive metallurgy. Several software packages model reaction chemistry and equilibria. The Outotec HSC Chemistry software has a comprehensive and fairly robust Eh–pH feature. The equilibria for the Cu-S-C-H_2O system at 298 K, dissolved species activity 10^{-3} M, and pressure of 101.3 kPa (1 atm) are depicted in Figure 4. This diagram shows that for economic copper leaching to occur, the Eh must be relatively high and the pH relatively low (i.e., the upper left-hand corner of the diagram).

Gangue Mineral Interactions

Dreier (1999) has provided an excellent discussion of the interactions between gangue minerals and the lixiviant and the controlling role of Eh–pH microenvironments. He states that "the principal obstacle to successful leaching is the persistence within the leaching system of low Eh–high pH microenvironments." Low-grade copper porphyry ores consist mostly

Table 3 Different categories of gangue mineral reactivity

Gangue Reactivity	Minerals	Example Mineral Formulas
Highly reactive	Calcite Dolomite Siderite	$CaCO_3$ $CaMg(CO_3)_2$ $FeCO_3$
Reactive	Hornblende Pyroxene Ca plagioclase	$Ca_2(Mg,Fe,Al)_5(Si,Al)_8O_{12}(OH)_2$ $(Ca,Na)(Mg,Fe,Al)(Si, Al)_2O_6$ $Ca(Mg,Fe)_3Si_4O_{12}$
Moderately reactive	Orthoclase Biotite Albite	$KAlSi_3O_8$ $KFe_3(AlSi_3O_{10})(OH)_2$ $NaAlSi_3O_8$
Slightly reactive	Quartz Sericite Kaolinite	SiO_2 $KAl_2(AlSi_3O_{10})(OH)_2$ $Al_4Si_4O_{10}(OH)_8$
Nonreactive	Quartz	SiO_2

Adapted from Dreier 1999

of rock-forming minerals and alteration products. These silicate components are the main source of acid consumption and to a large degree control the leaching behavior of copper ores. If the ore contains significant amounts of carbonate minerals, they also consume large amounts of acid. Dreier has divided silicate and carbonate minerals into five categories according to the rate at which they react with acid. Table 3 shows some representative silicate minerals for these five categories.

Dissolution rates and degree of acid consumption for rock-forming minerals are shown in Figure 5. This figure was adapted from Dreier. The dissolution rates of K-feldspar and Ca-plagioclase at pH 2 are 2×10^{-15} and 2×10^{-13} moles cm^{-2} s^{-1}, respectively. By comparison, the leaching rate for chrysocolla at pH 2 is approximately 1×10^{-4} moles cm^{-2} s^{-1}.

The overall dissolution reactions for several of the silicates shown in Figure 5 are given as follows. Reactions 4 and 5 illustrate the general stoichiometry for the acid-consuming nature of Ca-plagioclase and biotite, respectively.

Ca-plagioclase:

$$CaAl_2Si_2O_8 + 8H^+ \rightarrow Ca^{2+} + 2Al^{3+} + 2SiO_2 + 4H_2O \qquad \textbf{(EQ 4)}$$

Biotite:

$$H_2KMg_{(3x)}(Fe)_{6-3x}Al(SiO_4)_3 + 10H^+ \rightarrow K^+ + 3xMg^{2+} + (6-3x)Fe^{2+} + Al^{3+} + 3SiO_2 + 6H_2O \qquad \textbf{(EQ 5)}$$

Acid consumption, on a molar basis, is very high for these silicate minerals. Even though the actual rate of dissolution is extremely slow, the huge amount of silicate surface available for reaction in the microenvironment results in an enormous macro effect on solution pH. The following sequence of reactions illustrates the consumption of acid associated with the dissolution and alteration of K-feldspar.

K-feldspar alteration to K-mica:

$$3KAlSi_3O_8 + 2H^+ + 12H_2O \rightarrow KAl_3Si_3O_{10}(OH)_2 + 2K^+ + 6H_4SiO_4 \qquad \textbf{(EQ 6)}$$

K-mica alteration to kaolinite:

$$2KAl_3Si_3O_{10}(OH)_2 + 2H^+ \rightarrow 3Al_2Si_2O_5(OH)_4 + 2K^+ \qquad \textbf{(EQ 7)}$$

Kaolinite dissolution:

$$Al_2Si_2O_5(OH)_4 + 6H^+ \rightarrow 2Al^{3+} + H_2O + 2H_4SiO_4 \qquad \textbf{(EQ 8)}$$

Overall reaction:

$$KAlSi_3O_8 + 4H^+ + 4H_2O \rightarrow K^+ + Al^{3+} + 3H_4SiO_4 \qquad \textbf{(EQ 9)}$$

The main acid-generating reaction in copper-leaching systems is the aqueous oxidation of pyrite. Pyrite (FeS_2) is an extremely common sulfide mineral and is found in a wide variety of geologic settings. It is present in porphyry copper deposits ranging from 2% to 6% by volume. Pyrite oxidizes very slowly in the presence of oxygen and ferric ion; however, its aqueous oxidation is aided by the presence of autotrophic bacteria. Acid generation by pyrite from direct oxidation with dissolved oxygen occurs according to the following reaction:

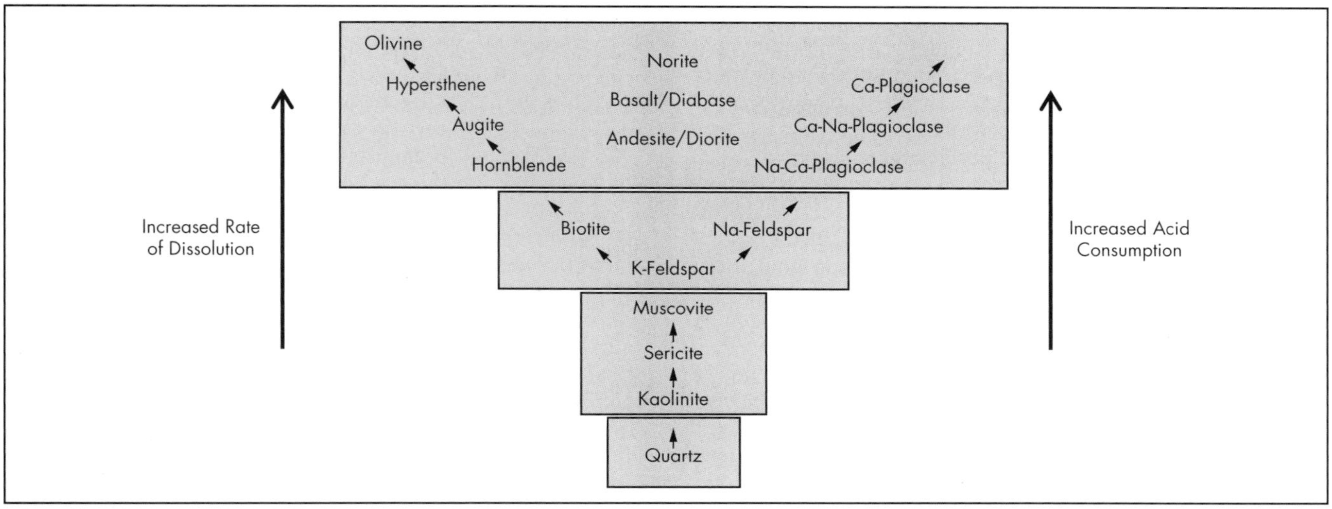

Adapted from Dreier 1999

Figure 5 Relative acid dissolution rates for some silicate minerals and the resulting acid consumption

$$FeS_2 + 15/4O_2 + 7/2H_2O \rightarrow 2Fe(OH)_{3(solid)} + 2SO_4^{2-} + 4H^+ \quad \textbf{(EQ 10)}$$

The corresponding reaction with ferric ion as the oxidant is

$$FeS_2 + 15Fe^{3+} + 8H_2O \rightarrow 16Fe^{2+} + 2SO_4^{2-} + 16H^+ \quad \textbf{(EQ 11)}$$

Acidithiobacillus microbes, which are indigenous with sulfide minerals, serve to catalyze the oxidation of ferrous ion to ferric and the oxidation of elemental sulfur to sulfate, both of which facilitate the acid-generating capacity of pyrite. Ferric ion hydrolysis to form hydrated hematite, hydronium jarosite, or jarosite salts results in the generation of acid. The molar ratio of pyrite to copper is important in maintaining the autogenous generation of acid in copper deposits. If it is too low, acid needs to be added to the system to maintain stable leaching conditions.

Pilot-Scale In Situ Leaching Tests

Pilot-scale tests were extensions of batch agitated (beaker) leaches and static small-diameter column tests and were intended to provide scale-up information needed for design of a commercial operation.

Lawrence Livermore Laboratory

Project Plowshare was a U.S. Atomic Energy Commission (AEC) program to develop the use of nuclear explosives for peaceful construction purposes. One application was the use of a contained nuclear explosive to break ore in a deep-seated disseminated primary copper deposit prior to in situ chemical mining. In essence, a rubblized chimney of broken ore would be constructed well below the water table. The plan was to introduce oxygen or air into the bottom of the chimney through drill holes and recover copper PLS through a solution recovery well. Lewis et al. (1972) described the results of a pilot-scale experiment conducted at the Lawrence Livermore Laboratory to simulate and model the chemistry of copper extraction from a chalcopyrite ore.

A pilot-scale leaching experiment was conducted in a large, specially constructed autoclave with an internal capacity of 4.4 m^3 (43 ft^3). The autoclave was hand loaded with 5.8 t (6.4 st) of primary porphyry copper ore from the San Manuel mine, Arizona. The ore was mainly quartz and sericite with minor amounts of feldspar, chlorite, biotite, and calcite. The predominant copper mineral was chalcopyrite, and the copper content was 0.7% Cu with a pyrite/chalcopyrite mole ratio of 2.0. The particle size distribution was 0.1–160 mm (0.004–6.3 in.), and approximately 50% of the ore had a particle size greater than 80 mm (3.15 in.). A total of 1.4 m^3 (370 gal) of distilled water was added to the vessel. After it was sealed, the autoclave was operated at 90°C and 2.76 MPa (400 psi) oxygen pressure. Oxygen flow rate was 2.4×10^4 cm^3 (standard temperature and pressure, STP) min^{-1} (6.34 gpm) (first 90 days) and 1.2×10^4 cm^3 (STP) min^{-1} (3.17 gpm) (remainder of the test) to account for the consumption of oxygen by the leaching reactions. Autoclave leaching was allowed to proceed for more than 717 days. Fractional copper dissolution at this point was 0.43. Acid-generating reactions produced a rapid decrease in solution pH to about 1.7 in the first 60 days. This was followed by a slight increase in pH to 1.9, which stayed relatively constant for the remainder of the test.

A mixed kinetic model for leaching the chalcopyrite ore was proposed by Braun and co-workers (1974). The rate of reaction was controlled by the rate of oxygen diffusion through the reacted shell of the ore fragment to unreacted sulfide particles. The reaction zone is only a few sulfide particle diameters in thickness. Overall kinetics are directly associated with physical aspects such as particle porosity, ore and mineral density, surface roughness, and tortuosity. This was acceptable for the particulate system employed, but the chemical reaction model would require integration with a hydrologic model to create a predictive reactive fluid flow model for actual ISL and scale-up to a field pilot test.

Magma Copper Company

The Florence Copper Project is a copper in situ leach target located in Pinal County, Arizona approximately 5.6 km (3.5 mi) from the town of Florence. The HDI Curis laboratory core-box program described earlier was a sequel to the Magma investigation. The ore body is a porphyry copper deposit containing approximately 390 Mt (430 million st) of copper oxide ore that lie 137–366 m (450–1,200 ft) below the surface. Mineralization in the oxide zone comprises principally chrysocolla with lesser amounts of "copper wad," tenorite, cuprite, native copper, and only traces of azurite and brochantite. The copper grade averages 0.33% Cu. The oxide zone is located below the water table at a depth of roughly 137 m (450 ft) below ground level. Ambient groundwater elevation is between 61 and 70 m (200 and 230 ft) above the top of the oxide zone. There is a 6.1–12.2 m (20–40 ft) clay layer approximately 18–33 m (60–100 ft) above bedrock, which acts as an upper aquitard, and there is a low-hydraulic conductivity layer beneath the bedrock oxide unit. These confining aquitards serve to control the vertical movement of solutions.

Magma Copper Company acquired the Florence property from Continental Oil Company (Conoco) in July 1992. In January 1993, Magma commenced a prefeasibility study and concluded that ISL and SX-EW was the preferred technique for developing the Florence deposit. Magma was purchased in January 1996 by Broken Hill Proprietary Company Limited, creating BHP Copper Inc. Work at Florence continued with planning for a field optimization test facility to demonstrate hydrologic control and to validate predicted copper recovery. The BHP test facility consisted of a pilot five-spot well pattern (see the description later in this chapter) with adjacent observation wells and a solution storage tank and evaporation pond. A total of 20 wells were drilled for the test facility. Four injection wells were drilled on a spacing of 21.6 m (71 ft) with a center recovery well. During the test, an injection rate of 0.15 m^3 min^{-1} (40 gpm) was applied to each well with an aggregate solution recovery rate of 0.70 m^3 min^{-1} (190 gpm). The test operated for approximately 90 days and successfully demonstrated that the ore zone had sufficient hydraulic conductivity to support fluid flow in the pattern and that hydraulic control of injected fluids could be maintained. Robertson and Hall (2011) have discussed in detail the hydrologic parameters associated with the Florence Copper Project. Current copper recovery estimates project approximately 70% Cu recovery in four years of commercial well-field production.

Kennecott Copper Corporation

In 1967, Kennecott Copper Corporation announced plans to conduct a nuclear mining experiment at its Safford deposit located approximately 14.5 km (9 mi) northeast of Safford, Arizona. The project, called "Sloop," was part of the AEC's

Plowshare Program. The proposed nuclear blast would produce a rubblized chimney of fragmented ore approximately 134 m (440 ft) high and 61 m (200 ft) in diameter, containing about 1.18 Mt (1.3 million st) of copper ore. Fortunately, a nuclear device was never detonated at Safford; however, Kennecott during the late 1960s and most of the 1970s conducted an extensive investigation of "true" ISL of the Safford deposit.

The Safford test program involved many scientists and technologists and, for the time, a huge expenditure of research and development funds. It was estimated that Kennecott expended more than $20 million for this project over a three-year period. The Safford ore body was a deep-seated target with primary sulfide mineralization and extremely low permeability. Deep wells were placed in two-spot and five-spot patterns. The ore horizon was between 975 and 1,000 m (3,200 and 3,280 ft) at depth, and the formation had 3% porosity. Laboratory core data in five wells indicated permeabilities ranging from 0.001 to 0.4 mD. Conventional acid ferric sulfate leaching was unsuccessful. A field test at 981–1016 m (3,218–3,332 ft) deep using 0.1 molar NH_3 and 2 molar NH_4NO_3 solution injected at 18.9 L min^{-1} (5 gpm) showed significant permeability enhancement under these conditions. Little has been published regarding the Kennecott in situ project at Safford.

USBM/Asarco/Freeport

In 1988, the U.S. Bureau of Mines initiated a cooperative agreement with Asarco to investigate in situ recovery of copper at the Santa Cruz joint venture (Asarco and Phelps Dodge, later Freeport-McMoRan). The Santa Cruz deposit is located approximately 11.3 km (7 mi) west of Casa Grande, Arizona. A field experiment was planned and executed during the early 1990s. The deposit is moderately deep oxide mineralization, lying between 366 and 549 m (1,200 and 1,800 ft) deep. The mineralized zone contains approximately 88 Mt (97 million st) of ore averaging 0.7% acid soluble copper. At Santa Cruz, atacamite ($Cu_2(OH)_3Cl$) and chrysocolla ($CuO{\cdot}SiO_2{\cdot}2H_2O$) are the predominant copper oxide minerals occurring as fracture-fillings in quartz monzonite and granodiorite porphyry. The deposit was considered too deep to mine by conventional surface methods and too low-grade for underground mining.

Considerable laboratory research was performed by the USBM as part of this project. Physical, mineralogical, and chemical features of the ore body were characterized by this research. Field research involved a five-spot well pattern with additional monitoring wells drilled to a targeted depth. Initial measurements collected hydrologic data that included salt-water tracer tests in the five-spot pattern. The tracer experiments started on March 14, 1991, and continued through July. Tracer solution was injected at a rate of 0.087 m^3 min^{-1} (23 gpm) at a surface pump pressure of 1,585–1,998 kPa (230–290 psi). All but one of the recovery wells delivered solution with slowly increasing salt concentrations. Near the end of the salt tracer phase, a bromide slug experiment was tried. Environmental permitting was required for acid leaching at the pilot test facility. During the permitting process, a small SX-EW plant with a capacity of 0.19 m^3 min^{-1} (50 gpm) was constructed. Acid leaching of the ore body from the pilot five-spot commenced in January 1996 at an injection rate of 0.095 m^3 min^{-1} (25 gpm). Regrettably, the USBM was closed at about the same time and documentation of the results from this test is sketchy at best. One valuable outcome from the Santa Cruz project was the publication of the *Generic In Situ Copper Mine Design Manual* (Davidson 1988), which is available to the public.

WATER-SOLUBLE MINERALS

Because of the nature of their solubility, water-soluble minerals present somewhat unique opportunities for their extraction and recovery. Water-soluble minerals are extracted from seas, lakes, salars, brine reservoirs, abandoned underground mines, and bedded deposits for the concentration of salts and mineral products. Brines containing soluble minerals are withdrawn directly from natural bodies of brine using pumps or brine elevators and from shallow aquifers using surface trenches or wells. Naturally occurring brines are also pumped from deeper brine aquifers for the recovery of soluble minerals. Solution mining or ISL of soluble mineral beds using ambient or high-temperature water or engineered brines is utilized to selectively extract targeted minerals.

Solution mining of bedded deposits may be preferred to conventional methods when deposits are too deep or too geologically complex for conventional underground mining. Avoiding hazardous mine gases, such as methane or hydrogen sulfide; eliminating the need to place personnel at increased depths; leaving insoluble impurities in the solution caverns; and minimizing the surface tailings storage requirement all favor a solution mining option. Solution mining is more energy-sensitive than conventional mining and consumes more water per short ton of product; therefore, the availability of water and access to inexpensive energy or solar evaporation are important differentiators.

Halite (NaCl), referred to herein as salt; potash (KCl); and trona ($Na_2CO_3{\cdot}NaHCO_3{\cdot}H_2O$) represent the largest product volumes produced by solution mining. Approximately 50% of the United States' 2014 salt production, 44.1 Mt (49 million st) was extracted by solution mining (Bolen 2015). Magnesium chloride, sodium sulfate, borax ($Na_2B_4O_7{\cdot}10H_2O$), lithium, bromine, and iodine are also produced from soluble sources. In addition to the mineral value, the storage capacity of the resulting void may have commercial value in the storage of hydrocarbons and compressed gases. These solution cavities are also used for the disposal of wastes.

The recovery of soluble minerals from surface expressions such as seas, lakes, and shallow aquifers is presented elsewhere in this handbook, specific to the individual commodity. Therefore, this section focuses on the solution mining of nonbrine, subsurface deposits.

The extraction of subsurface soluble minerals by dissolution of the mineral represents some of the earlier methods of mining. Chinese miners used bamboo drills and bamboo pipes to extract salt brines from deep brine reservoirs or from salt beds as early as 300 BC (Warren 2006). In the 1920s, German attempts at solution mining of carnallite ($KCl{\cdot}MgCl_2{\cdot}6H_2O$) were unsuccessful and were abandoned. Success was finally achieved for carnallite in the 1980s by DEUSA at Bleicherode/Kehmstedt in Germany. Following multiple unsuccessful attempts, solution mining has only had a broader commercial application in the last 50 years. In the 1960s, solution mining was expanded to include the extraction of potash from sylvite ore (KCl) at Kalium's (now Mosaic's) Belle Plaine operation in Saskatchewan, Canada. In the 1980s, Royal Dutch Shell (now Nedmag) started solution mining bischofite ($MgCl_2{\cdot}6H_2O$) and carnallite at the Veendam mine in the Netherlands, and in the 1990s, FMC (now Tronox) started solution mining of trona ($Na_2CO_3 \cdot NaHCO_3 \cdot 2H_2O$) in Wyoming, United States. In addition, borax ($Na_2B_4O_7 \cdot 10H_2O$) is solution mined in California, United States, and nahcolite ($NaHCO_3$) was solution mined from bedded deposits in western Colorado, United States.

The brines extracted from solution mines may be delivered to evaporative pans, solar ponds, evaporators, and crystallizers to recover the soluble mineral. Some salt operations are captive to chlor-alkali plants (Kyle et al. 2011). With rock salt, total evaporation is often required. For other salts, such as potash, cooling the brines recovers the mineral and mother liquor is produced, heated, and returned to the mine for restoration of the brine.

Principles and Key Parameters

A multitude of variables must be considered in the design and decision to use solution mining. Key considerations include the following:

- Thickness of the deposit and depth of the deposit
- Grade of the deposit and nature and quantity of associated impurities
- Depth and temperature of the deposit
- Uniformity of the deposit bedding and the absence of anomalies
- Geomechanics of the deposit and overlying rock members
- Homogeneity and extent of the deposit
- Dip of the deposit
- Proximity of overlying aquifers

Soluble minerals do not have the strength of hard rock minerals; they plastically deform when stressed, and creep is exacerbated by elevated temperatures. Impurities will influence the phase chemistry of the extraction brines and, along with temperature, will influence the rate of dissolution. Stability relationships in complex brines are defined with the aid of classical phase chemistry.

Dissolution

The main dissolution mechanism is free convection or density-driven movement and mixing at the cavern walls. The boundary layer at the cavern side walls is influenced by cavern flow patterns, wall roughness, and impurities. Denser saturated brine flows downward along the cavern walls creating convection currents, flow along the walls, and brine density stratification in the cavern. Unsaturated brines are convected upward. This results in a brine concentration gradient that determines the rate of dissolution. Dissolution rates are a direct function of temperature and the concentration gradient or an inverse function with brine saturation. As a result of the concentration gradient, increased dissolution rates occur higher on the cavern wall and at the roof of the cavern and, if unmanaged, result in caverns with the shape of an inverted cone. The dissolution rate at the roof of the cavern is 1.5–2 times the dissolution rate of the side walls. Therefore, immiscible liquids or compressed gases, such as oil, diesel fuel, nitrogen, and air, are used to create an inert blanket at the top of the cavern. In addition to the inert blanket, the injection flow rate, the progressive relocation of injection and extraction strings, and the routine reversal of injection and extraction flows can be managed to determine the shape of the cavern. The Solution Mining Research Institute supported a significant volume of research work on salt dissolution and the development and measurement of solution-mined cavern shapes; Nigbor (1982) provides a good summary of the work by Snow, Durie, Jessen, Saberian, and others.

There is a contrast in dissolution methods for different soluble minerals that is best described by comparing solution mining of salt and potash. In the case of salt, solution caverns are mined in relatively high-purity deposits resulting in brine-filled voids with the relatively small accumulation of insoluble materials at the base of the cavern. Dissolution by convection is relatively unobstructed. For potash that is less pure, 30%–40% KCl, there are two methods of dissolution: one dissolves both the salt and the potash, and another selectively dissolves the potash, taking advantage of sylvite's relatively higher solubility at elevated temperatures, and leaving the salt in the cavern. The latter approach results in caverns containing a honeycomb of mineralization as the target mineral is removed and the insoluble minerals remain behind (Husband 1971). This honeycomb obstructs the natural convection patterns and decreases the relative dissolution rates. In the case of potash, after the initial cavern development is completed to create a void and a functioning level of surface area, approximately 60% of the deposit is not potash and, depending on the mining method, may not be solubilized and thus remains underground.

Cavern production is the product of the dissolution rate and surface area. During cavern development, additional surface area is created as the cavern size increases, thereby increasing the overall mass production from the cavern.

Solution Mining Methods

Solution mining methods can be divided into pressurized and unpressurized systems, with the majority of minerals extracted from pressurized systems. In a pressurized system, high-pressure pumps inject water or brine into the mining cavern, displacing brines as extraction brines. Depending on the pressure drop of the system, extraction brine is directly displaced or assisted by lift pumps. In a nonpressurized system, brine is pumped into an open cavity, such as mine workings and flows, to a lower point in the mine where the saturated extraction brine is pumped to the surface via vertical wells.

Production wells in a pressurized system are similar to oil and gas wells. After casings are established for the protection of local aquifers, typically there are three concentric production strings. The inert blanket is managed through the outer annulus, while the injection and extraction brines are introduced or removed through the remaining annulus or the final string, depending on preference for direct or reverse cavern flow.

The pressurized systems can be configured as single-, dual-, or multi-well caverns. Single-well configurations are common for mining in salt domes or for the development of storage caverns where long, cylindrical Thermos-shaped caverns are preferred. A dual-well system is common in bedded salts where two wells are paired to provide lateral cavern growth and overlap of the caverns. When dual-well systems establish hydraulic connectivity, they have the advantage of being much more productive than a single-well cavern and produce brine with increased saturation. Salt mining since about 1955 has used hydraulic fracturing to initially connect the two wells. Paired well spacing can vary depending on bed thickness and designed cavern roof span. This configuration is used in potash solution mines in Saskatchewan and Europe. Multi-well systems are usually the product of mature single- or dual-cavern systems where caverns are allowed to overlap or wells are required in later development to remove stranded brine from subsurface topographically low points or sumps.

An alternative to vertical wells is the use of horizontal well technology adopted from the oil and gas industry. Directional drilling is used to develop horizontal wells in the seam or just below thin mineralized seams. A second horizontal well and sometimes a series of lateral wells are developed to intercept the first well, establishing a dual-well system. White River Nahcolite Minerals in Rifle, Colorado, and Intrepid Potash in Moab, Utah (United States), use this method.

Applications

Salt is solution mined at depths from 400 to 2,900 m (1,312–9,512 ft), with the deepest cavern at Frisia Salt in the Netherlands. Single-well caverns are generally constructed in salt domes, and caverns range from 500 to 800 m (1,640–2,624 ft) high. An inert blanket is used to manage the diameter of the cavern, typically 50–100 m (164–328 ft) in diameter, and is used in the case of direct circulation. After the development of a sump for the accumulation of insoluble materials is complete, the extraction tubing is progressively raised to form a near-cylindrical cavern. The injection string may need to be raised as the cavern matures and insolubles accumulate at the bottom. Flow rates are initially very low to provide high brine concentrations and increase as the cavern develops and increased surface area is created. High flows can be maintained after about a year of development, and the life of successful solution caverns may be as long as 30–40 years (Warren 2006).

Since the 1930s, the Trump method, identified by an inert blanket, has been used to control vertical development. Since the 1950s, dual- and multi-well caverns are generally used for salt in bedded deposits, using the Trump method (Nigbor 1982). Wells located between 60 and 183 m (197 and 600 ft) apart are operated as single wells until the caverns connect. At that time, one well is converted to the injection well and the other well is converted to the extraction well.

In Saskatchewan, Canada, the rule of thumb is to use conventional mining to a depth of 1,100 m (3,608 ft) and to use solution mining at depths greater than 1,450 m (4,756 ft) (Halabura and Hardy 2007). Mosaic's Belle Plaine operation is the largest potash solution mine and extracts potash from three ore members at a depth of 1,600 m (5,248 ft); bed thicknesses vary from 9 to 15 m (30 to 49 ft). The new K+S Bethune Project is in the final stages of construction and will extract potash from the same members at similar depths. A dual-well configuration is used with 9–14 caverns directionally drilled from one pad or cluster. Thirty or more clusters may be in various stages of development to support the level of potash production. The spacing between the paired wells determines the areal extent and the roof span of the cavern. The developed cavern will have a near-oval shape with cylindrical-type walls. Cavern spacing within the cluster is predetermined to retain boundary pillars between the caverns in each cluster. Cavern spacing is designed to provide a 30%–40% extraction ratio. At deposit depth, the spacing between the paired wells ranges from 80 to 100 m (262 to 328 ft) and salt temperatures are approximately 57°C (135°F). Mosaic's Hershey operations solution mined potash in a similar manner from 1990 to 2013 in Michigan at bed depths of 2,300 m (7,544 ft) (Plosz 2005).

The solution mining method employs five stages of development and operation, summarized as follows.

1. **Sump development:** A sump is solution mined with fresh water at the base of each well in the salt bed just below the potash bed. The inert blanket is managed to prevent dissolution of potash.
2. **Undercut:** Solution mining with fresh water continues in the salt bed until connection is established between the two wells. The inert blanket is used to limit vertical growth in the caverns
3. **Roof development:** Solution mining with fresh water continues with injection into one well and extraction from the paired well. The flow may be alternated from well to well, and the roof is developed to about 50%–60% of the designed cavern roof area.
4. **Primary mining:** Hot fresh water is used to solution mine both the salt and the potash, creating the mining cavern. Progressive horizontal cuts from 1.5 to 3.0 m (5–10 ft) and cyclic injection are used to increase the height of the cavern until the cavern reaches the top of the potash beds
5. **Secondary mining:** Solution mining switches from hot fresh water to hot recycle brine from the plant to selectively dissolve sylvite from the walls and the roof.

In some cases, a forced roof collapse is used in preparation for mining with hot recycle brine. Cavity development lasts four to five years, and the cavities can produce for more than 20 years. About 70% of production originates as primary mining and 30% as secondary mining. For potash mining, injection brine temperatures of 10°–20°C (50°–68°F) greater than the salt temperature balance dissolution rates and energy losses to wall rock (Asch et al. 2013).

DEUSA solution-mines potash from carnallite at Bleicherode in Germany. The company uses a dual-well system with hot dissolution. This is a complex system, and magnesium and potassium levels must be closely controlled to avoid blinding beds or increasing potash losses by precipitating KCl in the caverns. Mining occurs in a 50°C (122°F), 33-m (108-ft) bed at 400 m (1,312 ft) deep with a well spacing of 60 m (197 ft) and uses a compressed air blanket. Hot magnesium chloride mother liquor from the process, which is saturated in NaCl but undersaturated in KCl, is heated to 80°C (176°F) and injected into the cavern. Carnallite is leached and NaCl precipitates out in the cavern, whereas KCl and $MgCl_2$ report to the brine. Because of the high solubility of magnesium chloride, halite and kieserite ($MgSO_4 \cdot H_2O$) have limited solubility and remain behind in the cavern. Eight percent of the cavern volume is filled with insoluble residues. The extraction brine reports to the process plant where carnallite and halite are crystallized and the resulting mother liquor is used to produce bischofite or returned as injection brine. The carnallite is decomposed to produce magnesium chloride brine, leaving sylvite and halite crystals. A hot leach and crystallization step recovers the sylvite from the halite.

Even more complex phase chemistry and geomechanics are encountered at the Veendam mine, which solution mines magnesium from carnallite and bischofite beds at depths of 1,400–1,800 m (459–5,904 ft) and in situ temperatures of 80°C (176°F). A double-well system is used and, with experience, well spacing increased progressively from 100 to 140 to 200 m (328 to 460 to 656 ft). Bischofite is very plastic and paste-like at these depths and squeezes into the cavern. Cavern pressures must be managed to control the convergence of the cavern, and the cavern diameter remains almost constant as dissolution and convergence are balanced. An estimated 30%–40% of the production is generated from salts squeezed into the cavern. Salt is "squeeze mined" in a similar fashion at Frisia, where 100°C (212°F) salt squeezes into the cavern; managed convergence and dissolution result in constant cavern sizes.

Solution mining of potash also occurs from unintentionally or intentionally flooded underground workings, such as the Potash Corporation of Saskatchewan's Patience Lake (Canada), and Intrepid's Moab and HB (Carlsbad, New Mexico, United States) operations. Salt-saturated brine is injected at a higher elevation in the workings, and sylvite is dissolved from pillars through convection as the brine migrates to the low spots in the mine. As the brine stratifies, the concentrated brine reports to the lower areas of the mine and is pumped to the surface using vertical wells. Sylvite is recovered either in ponds that are cooler relative to underground temperatures, or in solar ponds or crystallizers.

The development of horizontal caverns is also used at Moab and at White River Nahcolite. Directional drilling is used to intercept another well bore, either horizontal or vertical, maintaining bed contact just below the base of the target mineral bed. Hot water is used to undercut the target mineral bed and to increase the surface area of the developing cavern. Moab uses lateral wells and a pitchfork design to connect to the target well, whereas White River develops parallel pairs of horizontal and vertical wells to create a chamber and pillar arrangement (Day 1994). Hot-salt saturated brine or water are used, respectively, to mine the two deposits.

In 1996, FMC began introducing trona tailings into abandoned underground workings. The transport brine migrates to lower areas in the mine panels, leaching pillars and increasing the alkalinity of the brine. The brine is returned to the surface using vertical wells. The ability to process the brine was critical to the project's success, and an evaporation, lime neutralization, decahydrate crystallization, and monohydrate crystallization (ELDM) process plant was developed where evaporation and liming produce a concentrated brine. Sodium carbonate decahydrate crystals are recovered and converted to sodium carbonate monohydrate. The deposition of tailings and recovery of leach brines is practiced by other Wyoming operators as well.

Also similar to potash practice, the Granger trona mine in Wyoming was flooded in 2007 to solution mine the pillars and remaining ore. Mine water is pumped from low spots in the mine for processing.

Fort Cady Minerals solution mines borax at Hector, California. Weak hydrochloric acid is injected into colemanite beds 300 m (984 ft) deep to extract boron-rich brines to the surface.

Environmental Considerations

Salts deposits and beds form impervious zones and provide a barrier to solution releases from caverns. Leaching with saturated brines further minimizes the potential for solution migration beyond planned cavities. In North America, deep-well injection is used to dispose of bitterns or purged brines, whereas in Europe, ocean, sea, or riverine disposal is also used. Precise brine accounting and monitoring wells are used to ensure that all mine and disposal solutions are appropriately contained and managed.

Subsidence associated with solution mining can also be regulated, depending on local governmental agencies. It is a key permit limitation, especially for deeper solution mines, and regulatory subsidence limits are integral to development and operating strategies. Surface monitoring programs, including the use of light detection and ranging, are common.

URANIUM DEPOSITS

Refer to Chapter 12.41, "Uranium," for a discussion of uranium chemistry, conventional milling, solution treatment, precipitation, product specifications, and other considerations relevant to conventional uranium ore processing.

Historical Examples 1960–1990

Solution mining in its simplest form is typified by treatment of mine water that has been enriched through dissolution of uranium by oxygenated groundwater. This has probably been done in many mining districts, but it has been reported that in 1965, Kerr-McGee and United Nuclear-Homestake Partners jointly produced about 91 kg (200 lb) of U_3O_8 daily from approximately 11.4 m^3 min^{-1} (3,000 gpm) of mine water in the United States at Ambrosia Lake District, New Mexico (Merritt 1971). In this case, uranium probably was present as anionic uranyl tricarbonate complexes and water treatment was effected with transportable resin ion exchange columns that were eluted at the owners' mills.

Engineered solution mining, or in-place leaching, involves the intentional introduction of leaching solutions into an ore body after the ore has been rendered permeable and provisions have been made for solution collection and treatment. There have been many examples for various metals and minerals, but the following apply specifically to uranium.

During the 1960s, Canmet had a bacterial leaching program that led to a cooperative agreement with the Kerr-Addison and Agnew Lake mines. A 0.61-m (2-ft) diameter by 4.6-m (15-ft) column was loaded with 1.8 t (2 st) of Agnew Lake uranothorite ore and leached with solutions that were acidified by bacterial oxidation of pyrite and pyrrhotite in the ore. The test predicted 70% uranium extraction in one year, leading to a test in which a stope containing 45 t (50 st) of broken ore was leached for 30 weeks, dissolving 57% of the uranium (Ehrlich and Brierley 1990).

This encouraging information led Kerr-Addison in 1976 to invest $37 million in an underground bacterial leaching project at Agnew Lake. The anticipated production rate was 453,700 kg (1,000,000 lb)/yr of U_3O_8, but actual production in 1977 was only 31,760 kg (70,000 lb). Because of the steep dip of the ore zone, solution distribution in the inclined rubblized stopes was poor. This factor and other complications led to closure of the project in 1980.

In 1984, Denison Mines and Canmet began development of another underground bioleaching project for uranium recovery from the Denison ore body. An extensive study was made to optimize rock fragmentation, bacterial nutrients, spraying versus flooding, solution temperature, and other variables. Commercial production began in 1987 and 381,125 kg (840,000 lb) of U_3O_8 were recovered during that year. A major contributor to the success of the project was a very favorable ore body configuration that allowed blasting of the upper ore zone down into the cavities in the lower ore zone that had been mined by the room-and-pillar method (Ehrlich and Brierley 1990).

True ISL of uranium was begun in 1963 by Utah Construction and Mining in Wyoming's Shirley Basin. Leaching was done with a sulfuric acid solution, and annual production was about 90,744 kg (200,000 lb) of U_3O_8 through 1970. However, Utah determined that ISL costs were higher than those obtained by conventional milling at other company operations, so the Shirley Basin ISL project was replaced with a conventional mine and mill (Pool 2000). Several ISL operations began in Texas, United States, during the mid-1970s. In the United States, it is common to refer to ISL as ISR for "in situ recovery."

Physical and Chemical Constraints for In Situ Leaching

Although there are many geologic classifications of uranium deposits, the type that is most conducive to ISL of uranium is the so-called roll-front deposit characterized by deposition of uranium minerals under a reducing environment immediately down gradient of an oxidizing environment. The physical setting is usually an ancient stream channel or "paleochannel," and mobile hexavalent uranium has been reduced to relatively insoluble tetravalent uranium by agents ranging from sulfide ion to carbonaceous material such as decaying wood. Bhattacharyya et al. (2017) discussed the discovery of evidence revealing biogenic reduction of hexavalent uranium.

For ISL to be carried out with effective hydraulic control, the uranium-mineralized zone must be situated beneath or enveloped within an aquifer. Ideally, this zone will be essentially horizontal and will be bounded above and below by aquitards such as clay layers. During the development phase of the project, a thorough understanding of groundwater hydrology must be established. With the aid of core drilling, the combined grade and thickness of uranium mineralization is defined for each core intercept and a mine plan devised that will ensure access to an economically viable reserve expressed in pounds of recoverable U_3O_8.

Hydraulic control of an ISL well field requires that slightly (0.5%–2%) more water must be withdrawn from the aquifer than is injected into the mineralized formation. This excess water is balanced by bleeding an equivalent flow rate from the processing plant as a barren solution (either SX raffinate or ion exchange [IX] strip solution) that must be discarded. Ordinarily, this bleed stream is evaporated in a lined pond, but an injection well may be needed under unusual circumstances such as a site with high rainfall and negative net annual precipitation.

Technical and Economic Advantages Versus Conventional Milling

Compared with conventional milling using either an acidic or alkaline lixiviant, ISL occupies and disturbs less surface area, allowing less expensive closure and minimal audible and visible impact on residences and commercial entities. An in situ operation likely will create an evaporation pond that must eventually be closed and reclaimed, but it will leave neither waste rock nor tailings. Although each project is unique, it is fair to generalize that an in situ uranium operation will also require less capital investment and have lower operating and maintenance costs than conventional milling. It may also be less expensive than heap leaching; an instructive comparison for a specific collection of deposits has been published by Beahm (2007).

In contrast to conventional open pit mining, ISL requires fewer pieces of major equipment such as blasthole drills, excavators, haul trucks, crushers, grinding mills, leach tanks, and residue washing thickeners. Consequently, capital costs *may* be lower, depending on the cost of well field development and well drilling. Because of its relative simplicity, uranium ISL potentially offers quicker project development and engineering, faster ramp-up to design capacity, and possibly a higher return on invested capital. Furthermore, the lack of dependence on drilling, blasting, crushing, and grinding equates to less energy needed per unit of production. Likely, radiation exposure also will be minimized.

Economic Disadvantages

The ISL method typically dissolves less uranium than a conventional milling circuit and more slowly, for the following reasons:

- Because of channeling and short circuiting of the leaching solution, contact with mineralized grains is relatively poor.
- Oxidation of U^{4+} is the rate controlling step, and oxidant contact may be ineffective.
- Sulfide minerals will react with the oxidant, producing sulfuric acid and reducing lixiviant pH.

Domestic uranium production by ISL expanded during the 1970s, and by 1981 there were 11 operations, mostly in southern Texas. Initially, sulfuric acid solutions were injected, but in nearly all cases, alkaline leaching solutions were adopted later. Ammonium carbonate was tried, but aquifer restoration was difficult, so sodium carbonate/bicarbonate solutions and oxygenated groundwater with added carbon dioxide became the preferred lixiviants. Declining uranium prices led to closure of some of these operations, along with nearly all of the conventional mines and mills, but some ISL mines continue to produce uranium and new ones have been added in the United States in Wyoming, Texas, and Nebraska.

Well-Field Design and Operation

ISL wells for uranium extraction are generally adaptations of conventional water well designs. They are drilled with rotary rigs, and the drill hole is reamed to a diameter of 17.8–25.4 cm (7–10 in.) Usually, three casing centralizers are used, and the casings are typically 10–15 cm (4–6 in.) PVC or fiberglass-reinforced plastic, the latter being preferred for higher pressures and corrosive solutions. The casings are cemented up to the surface from the bottom ("point") of the casing, and the drill hole through the mineralized sandstone is uncased and may be under-reamed to increase formation contact area. If there are multiple economic zones, the casing is perforated to match depths of mineralized intervals, and the perforated sections are lined with stainless-steel screens.

The most common well pattern is a five-spot, which is a square configuration with a central injection well and a production well at each corner. Corner-to-corner spacing is typically 15–61 m (50–200 ft) with an average of about 23 m (75 ft). Well depths are typically 61–274 m (200–900 ft) but may reach 610 m (2,000 ft) in some overseas operations. Submersible stainless-steel pumps are used in the United States and Australia, but air lifts are preferred in the former Soviet Union because of operational simplicity and low cost. Figure 6 illustrates a typical well completion technique as practiced in the western United States. It is based on a drawing provided by Uranium Energy Corporation and typifies that company's well completion technique at the Palangana mine near Duval City, Texas.

A fence of monitor wells is installed around the well field to detect migration of leaching solutions outside the active patterns and to intercept and extract those fluids before they contaminate the aquifer. Typically, some monitor wells are completed in the aquifer overlying the production zone, as well as in the aquifer underlying the production zone. A ring of monitor wells encircles the wellfield. At the Palangana mine, the ring wells are spaced 122 m (400 ft) outside the outer production wells in the field. The monitor wells are cased to ensure that solutions are controlled.

The solution flow rate from a single well cased to a 10-cm (4-in.) nominal diameter and reamed to 20 cm (8 in.) diameter in the screened intervals will typically be 19–76 L min^{-1} (5–20 gpm). Flow rate can be estimated using the following equation (Huff 2002):

$$Q = (1.035 \times 10^{-4} kh(\Delta p))/\mu[\ln(d/r_w) - 0.619] \quad \textbf{(EQ 12)}$$

where
- Q = flow rate, gpm
- k = permeability, millidarcies, mD
- h = vertical ore interval open to flow, ft
- Δp = applied differential pressure, psig
- μ = fluid viscosity, cP
- $\ln$ = natural logarithm
- d = spacing between injection and production wells, ft
- r_w = well bore radius, not casing radius, ft

Conversion of this equation to metric units is potentially misleading, especially since formation permeability is universally expressed in millidarcies, and the darcy is a mixture of metric and conventional units. Commercially viable uranium roll front formations typically have permeabilities in the range of 300–1,000 mD. Pregnant and barren leach solutions ordinarily have viscosities of 0.8–1.0 cP.

Typical ore grades are in the range of 0.04%–0.15% U_3O_8, and well bores are logged electrically or radiometrically to define mineralized bed thicknesses and grades. The grade-thickness product (G×T) is used in reserve calculations, and a typical cutoff is in the range of 6–30.5 cm% (0.2–1.0 ft%) U_3O_8, depending on uranium price and expected production costs. In the United States, the most common ore minerals are uraninite and coffinite and both are readily oxidized. Accessory minerals include quartz, clays, feldspars, calcite, pyrite, and carbonaceous materials.

Injected leach solutions contain an oxidant, usually oxygen or hydrogen peroxide, at concentrations of 50–500 mg/L. In the United States, the lixiviant is typically oxygenated groundwater to which carbon dioxide is added to complex uranium and to adjust alkalinity to pH 7–9. The carbon dioxide concentration is about 500–1,000 mg/L. In this case, the dissolved uranium is complexed as uranyl tricarbonates. In Australia and Kazakhstan, the lixiviant is dilute sulfuric acid and the dissolved uranium is complexed as uranyl sulfate. Alkaline leaching is necessary when the calcium carbonate content of the ore renders sulfuric acid uneconomic. However, alkaline leach extractions are lower than those obtained with sulfuric acid, and alkaline leaching is roughly twice as

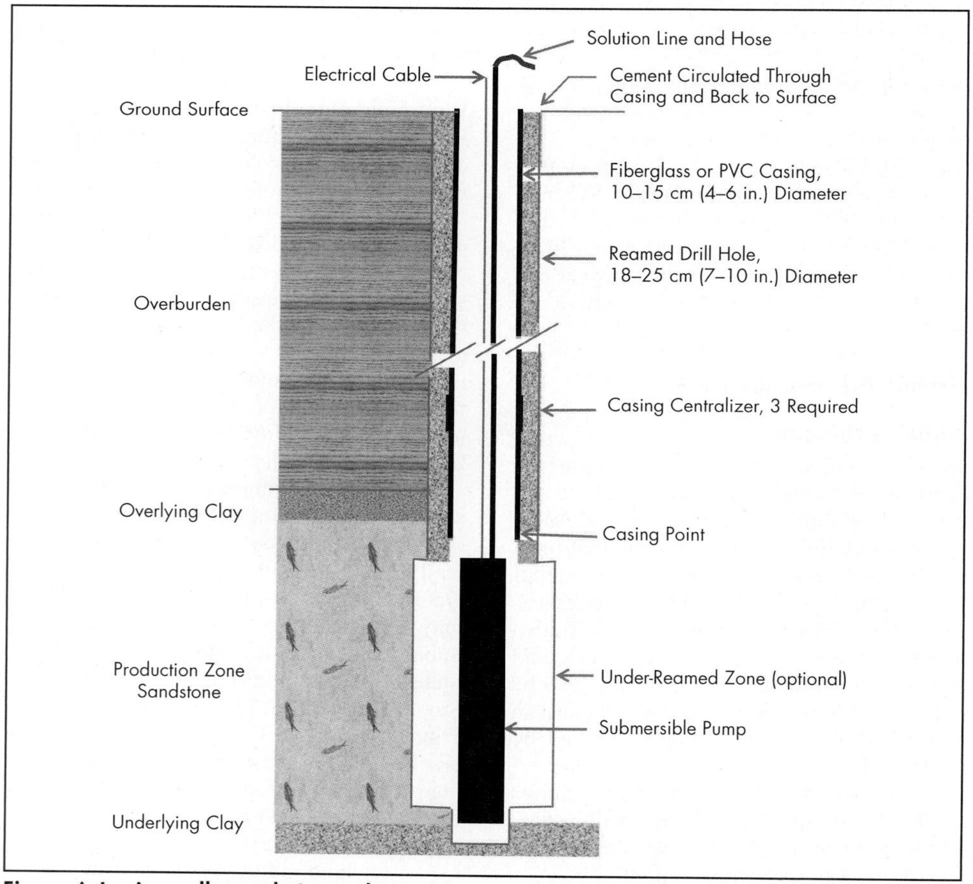

Figure 6 In situ well completion technique (not to scale)

expensive. The uranium concentration in the PLS is normally in the range of 35 to 200 mg/L U_3O_8 (Stano 2002; Pool 2000; WNA 2017a).

In practice, the PLS is pumped to a holding tank, clarified, and pumped through IX columns loaded with strong-base anionic resin beads. The resin is eluted (stripped) with alkaline or acidic chloride solution, concentrating the uranium by a factor of 5–20. The resulting eluate is neutralized and uranium is precipitated. Precipitation with ammonia produces ammonium diuranate crystals, whereas precipitation with hydrogen peroxide yields uranyl peroxide. The precipitated slurry is thickened, filtered or centrifuged, and dried. Most U.S. operations now use rotary vacuum dryers that have virtually no particulate emissions. The barren solution after IX is reconstituted and injected into the well field. It is common for PLS from small remote well fields to be treated with portable IX columns that can be eluted at a central processing plant.

Aquifer restoration requirements vary among countries and may be subject to local regulatory considerations. In Australia and in former Soviet Union countries, the groundwater may be so poor and the site so remote from habitation that restoration requirements are minimal. In the United States, restoration may be required on an individual element basis, which can be very expensive. Furthermore, groundwater quality in the United States ranges from good in Wyoming and New Mexico to relatively poor in Nebraska and Texas. Restoration of an aquifer is done by injecting clean water or by drawing aquifer water through the mined patterns and treating, for example, by reverse osmosis (RO). The clean RO permeate is injected into well patterns while the brine is either evaporated or injected into a disposal well. Radium is removed from waste solutions by resin ion exchange or precipitation with barium chloride.

Lixiviants and Gangue Interactions

As noted previously, sulfuric acid solutions react with calcium carbonate, and more than 2% $CaCO_3$ in a mineralized formation will usually consume an uneconomical amount of acid unless local acid costs are unusually low. Furthermore, the gypsum reaction product may gradually reduce permeability and impair solution flow rates. In alkaline leaching, the most likely gangue reaction is with naturally occurring gypsum, where dissolved sodium carbonate ions will react to form soluble sodium sulfate while precipitating calcium carbonate.

Laboratory Simulation Techniques

Representative core intervals from throughout the deposit can be subjected to standardized simulation procedures to establish the responsiveness of the uranium minerals to oxidation and complex formation and to subsequent recovery of a commercially acceptable product.

Pressurized bottle rolls are a common method for alkaline ISL, where a pulverized sample is added to five pore volumes of lixiviant containing low concentrations of hydrogen peroxide and sodium bicarbonate. The bottle is pressurized to expected formation hydrostatic pressure with gaseous carbon dioxide and is rolled for 24 hours. The depressurized slurry is filtered, the solids are repulped with fresh solution, and the bottle is again pressurized. This sequence is repeated for a total of 30 pore volumes and a metallurgical balance is calculated on the basis of solution volumes and assays and residue solids weight and assay. For sulfuric acid leaching, it is acceptable to simulate ISL using vertical plastic columns, as is done during simulation of heap leaching of oxide copper ores or nonrefractory gold ores. The laboratories at several ISL operations in the United States conduct leaching tests at ambient pressure. At least one commercial laboratory adds sodium carbonate to the lixiviant, rather than relying on the carbonate/bicarbonate equilibrium.

These laboratory techniques will predict maximum reagent consumption and maximum uranium dissolution, so a conservative factor should be applied to the results, based on the experience of the client and the laboratory manager. Also, laboratory tests on small pulverized samples should not be relied on to predict with any reliability the performance of a well pattern unless sampling and compositing have been done diligently. In uranium formations for which there is little or no ISL experience, it is essential to drill and leach a commercial-scale well pattern. As a general rule, overall recovery of uranium to yellowcake is in the range of 60% to 75% with alkaline leaching and 70% to 85% with acidic leaching.

Commercial In Situ Operations

In 2016, 48% of the world's uranium was produced by ISL operations in Kazakhstan, Uzbekistan, the United States, Australia, China, and Russia. Global ISL production in 2016 was 30.06 million kg (66.25 million lb) (WNA 2017a). Most ISL operations produce 454–2,269 million kg/yr (1–5 million lb/yr) of U_3O_8, but the largest consolidated ISL operation in Kazakhstan, a Katco/Areva joint venture, produced more than 4 million kg (8.8 million lb) in 2016, down slightly from 2014 and 2015. In contrast, total 2014 ISL production in the United States was 1.13 million kg (2.48 million lb), down from a peak of 1.92 million kg (4.23 million lb) in 2014 (WNA 2017b).

REFERENCES

Angelatou, V. 2004. The contribution of biotechnology in the development of mineral processing technology. In *Advances in Mineral Resources Management and Environmental Geotechnology*. Edited by Z. Agloutantis and K. Komnitsas. Santorini, Greece: Heliotopos Conferences.

Asch, R., Grüschow, N., Rembe, M., and Rejaee, M. 2013. Temperature effects on sylvinite solution mining. Presented at the Solution Mining Research Institute Spring 2013 Conference, Lafayette, LA, April 22–23.

Bartlett, R.W. 1992. *Solution Mining: Leaching and Fluid Recovery of Materials*. Philadelphia: Gordon and Breach Science Publishers. p. 1.

Bateman, A.M. 1951. *The Formation of Mineral Deposits*. New York: John Wiley and Sons.

Beahm, D. 2007. Antelope uranium project: Comparison of ISR and heap leach extraction. *Min. Eng.* (Sept.):17–23.

Bhattacharyya, A., Campbell, K.M., Kelly, S.D., Roebbert, Y., Weyer, S., Bernier-Latmani, R., and Borch, T. 2017. Biogenic non-crystalline U(IV) revealed as major component in uranium ore deposits. *Nature Comm.* 8(15538).

Bolen, W.P. 2015. Salt. In *2013 Minerals Yearbook*. Reston, VA: U.S. Geological Survey. pp. 63.1–63.25.

Braun, R.L., Lewis, A.E., and Wadsworth, M.E. 1974. In-place leaching of primary sulfide ores: Laboratory leaching data and kinetics model. *Met. Metall. Trans.* 5(1717–1727).

Bushnell, F. 1911. Precipitation of copper from mine waters. *Min. Sci. Press*. (Nov. 18): 649.

Davidson, D.H. 1988. *Generic In Situ Copper Design Manual.* Minneapolis: U.S. Bureau of Mines, U.S. Department of the Interior.

Day, R. 1994. White River Nahcolite solution mine. SME Preprint No. 94-210. Littleton, CO: SME.

Dreier, J.E. 1999. The chemistry of copper heap leaching. SME short course, copper heap leach. Presented at the Annual SME Meeting.

Ehrlich, H.L., and Brierley, C.L. 1990. Bioleaching of uranium. In *Microbial Mineral Recovery.* Edited by H.L. Ehrlich and C.L. Brierley. New York: McGraw-Hill.

Forrester, J.D. 1946. *Principles of Field and Mining Geology.* New York: John Wiley and Sons.

Garrels, R.M., and Christ, C.L. 1965. *Solutions, Minerals, and Equilibria.* New York: Harper and Row. p. 450.

Greenwood, C.C. 1926. Underground leaching at Cananea. *Eng. Min. J.* 121(13):518–521.

Halabura, S.P., and Hardy, M.P. 2007. An overview of solution mining of potash in Saskatchewan. Presented at the Solution Mining Research Institute Fall 2007 Conference, Halifax, Nova Scotia, October 8–9.

Hiskey, J.B., and Oner, G. 1992. Computer assisted in situ leaching experiments for copper oxide ores. *APCOM '92 Computer Applications in the Mineral Industries.* Edited by Y.C. Kim. Littleton, CO: SME. pp. 73–84.

Huff, R.V. 2002. In situ leaching. In *SME Mining Reference Handbook.* Edited by R.L. Lowrie. Littleton, CO: SME.

Husband, W.H.W. 1971. Application of solution mining to the recovery of potash. SME Preprint No. 71-AS-31. New York: SME.

Kyle, J.I., Maxwell, D.K., and Alexander, B.L. 2011. In-situ techniques of solution mining. In *SME Mining Engineering Handbook.* Edited by P. Darling. Englewood, CO: SME. pp. 1103–1119.

Lewis, A.E., Braun, R.L., and Higgins, G.H. 1972. Nuclear chemical mining of primary copper sulfides. Presented at the Third IAEA Panel on Peaceful Uses of Nuclear Explosions, Vienna, Austria, November 26–30.

Longwell, R.L. 1974. In place leaching of a mixed copper ore body. In *Solution Mining Symposium 1974.* Edited by F.F. Aplan, W.A. McKinney, and A.D. Pernichele. New York: AIME.

Lung, T.N. 1986. The history of copper cementation on iron: The world's first hydrometallurgical process from medieval China. *Hydrometallurgy* 17:113–129.

Merritt, R.C. 1971. The Extractive Metallurgy of Uranium. Germantown, MD: U.S. Atomic Energy Commission.

M3 Engineering and Technology Corporation. 2013. *Florence Copper Project: NI 43-101 Technical Report Pre-Feasibility Study, Florence, Pinal County, Arizona.* Tucson, AZ: M3 Engineering and Technology Corporation.

M3 Engineering and Technology Corporation. 2017. *Gunnison Copper Project: NI 43-101. Technical Report Feasibility Study, Cochise County, Arizona.* Tucson, AZ: M3 Engineering and Technology Corporation.

Nigbor, M.T. 1982. State of the Art of Solution Mining for Salt, Potash, and Soda Ash. Open-File Report 142-82. Washington, DC: U.S. Bureau of Mines.

Paulson, S.E., and Kuhlman, H.L. 1989. Laboratory core-leaching and petrologic studies to evaluate oxide copper ores for in situ mining. USBM IC 9216 In Situ Mining. Presented at the Bureau of Mines Technology Transfer Seminars, Phoenix, AZ, and Salt Lake City, UT.

Plosz, R.E. 2005. Solution mining of deep potash deposits in the Michigan basin. Presented at the Solution Mining Research Institute Spring 2005 Conference, Syracuse, NY, April 17–20.

Pool, T.C. 2000. In situ leaching of uranium: History, technology, and economics. SME Preprint No. 00-83. Littleton, CO: SME.

Pourbaix, M. 1966. *Atlas of Electrochemical Equilibria in Aqueous Solutions.* New York: Pergamon Press.

Pu, Y.D. 1982. The history and present status of practice and research work on solution mining in China. In *Interfacing Technologies in Solution Mining.* Edited by W.J. Schlitt and J.B. Hiskey. Littleton, CO: SME.

Rackham, H. 1952. *Pliny: Natural History, Books XXXIII–XXXV.* Cambridge, MA: Harvard University Press.

Robertson, M., and Hall, D. 2011. *Review of Curis Resources (Arizona) Inc. Applications PZC-32-11-MGPA and PZC-33-11-MGPA for general plan amendments for Florence Copper Project, Florence, Arizona.* Technical Memorandum. Phoenix: Arizona Dept. of Environmental Quality.

Sinclair, L., and Thompson, J. 2015. In situ leaching of copper: Challenges and future prospects. *Hydrometallurgy* 157:306–324.

Stano, J.R. 2002. In situ leaching of uranium. In *SME Mining Reference Handbook.* Edited by R.L. Lowrie. Littleton, CO: SME.

Sullivan, J.D. 1933. Chemical and physical features of copper leaching. *Trans. AIME* 106:515.

Taylor, J.H., and Whelan, P.F. 1942. The leaching of cuperous pyrites and the precipitation of copper at Rio Tinto, Spain. *Bull. Inst. Min. Metall.* 457:1–36.

Warren, J.K. 2006. Evaporites as exploited minerals. In *Evaporites: Sediments, Resources and Hydrocarbons.* Berlin: Springer Science and Business Media. pp. 893–909.

WNA (World Nuclear Association). 2017a. In Situ Leaching of Uranium. www.world-nuclear.org/information-library/nuclear-fuel-cycle/mining-of-uranium/in-situ-leach-mining-of-uranium.aspx. Accessed May 2018.

WNA (World Nuclear Association). 2017b. World Uranium Mining Production. www.world-nuclear.org/information-library/nuclear-fuel-cycle/mining-of-uranium/world-uranium-mining-production.aspx. Accessed May 2018.

CHAPTER 10.3

Heap and Dump Leaching

Randall Pyper, Thom Seal, John L. Uhrie, and Glenn C. Miller

Heap leaching is a hydrometallurgical recovery process where broken ore of appropriate characteristics is stacked upon an engineered liner, and then the surface of the heap is irrigated with a water-based lixiviant. Leach solution travels through the ore under unsaturated fluid flow conditions, contacting and leaching the metal or mineral of interest. Solution exits the bottom of the heap through slotted pipes and a drainage gravel layer placed above the liner. The metal-rich pregnant leach solution (often termed *PLS* or simply the *preg*) is then collected by gravity and sent to a process facility for recovery of the metal values. The solution exiting the metal recovery section, referred to as *raffinate* in copper leaching or the *barren* in gold leaching, is refortified with the lixiviant chemicals and pumped to the top of the heap.

Heap leaching is often broken into two principle distinctions, run-of-mine or *ROM* dump leaching and crushed ore heap leaching. Dump leaching consists of truck end-dumping of ore broken only by the drilling and blasting of mining activity. After completion of ore placement, the surface is ripped using bulldozers to break up compaction layers and improve solution distribution uniformity.

Crushed ore heap leaching involves reducing ROM ore to a predetermined optimal target size distribution, sometimes with topsizes as small as 12.5 mm (0.5 in.) or even finer through multiple stages of crushing. Crushing increases surface area and mineral liberation which, in turn, result in higher, more economic recovery. Crushed ores are often agglomerated to bind fines and clays with the larger rocks then placed on the heap using mobile stacker conveyors to improve ore wetting, leaching, and solution percolation through the heap.

Following stacking and surface preparation, ore is irrigated using drip emitters or sprinklers to evenly apply leach solutions to the heap surface. For successful heap leach operations, the ore must be of high porosity or must be fractured to ensure thorough contact between the metal or mineral in question and the lixiviant/leach solution (Heinen and Porter 1969).

A simplistic heap leach flow sheet is illustrated in Figure 1. A unit of ore under irrigation is called a *cell*. Solution discharges from the cells to the pregnant or PLS pond. This pond typically has several backup solution or storm ponds to handle temporary solution surges from weather events. The heap leach process is designed to contain all solutions from the environment and recover the dissolved metal or mineral values.

For copper and other metal sulfide minerals, the heap leach process has evolved to incorporate bacteria and nutrients into the barren solution promoting the bio-oxidation of sulfides. Aeration of the heap is often required in these cases. Similar applications to heap leaching include in situ leaching, rubble leaching, and vat leaching. Refer to Chapter 10.4, "Vat Leaching," and Chapter 10.2, "Solution Mining and In Situ Leaching," for more detailed information.

Heap leach operations are found on all continents and are used to recover a wide variety of metals, principally copper, gold, and silver. Less commonly, other metals such as cobalt, nickel, uranium, and zinc are recovered. In 2006, 10% of the world's gold production (Marsden and House 2006) and 18% of the world's copper production (ICSG 2015) were produced from heap leaching.

The primary advantages of heap or dump leaching as compared to milling ores are

- Lower capital costs,
- Lower operating costs,
- No liquid/solid separation step, and
- No tailings disposal.

The primary disadvantages of heap or dump leaching as compared to milling are

- Lower metal recovery,
- Longer process time,
- Slow response to process changes, and
- No by-product credits.

Randall Pyper, General Manager, Kappes, Cassiday & Associates Australia Pty Ltd., Perth, Western Australia, Australia
Thom Seal, Director of the Institute for Mineral Resource Studies, Mining & Metallurgical Eng. Dept., University of Nevada–Reno, Reno, Nevada, USA
John L. Uhrie, President Consulting Services – Americas, RPM Global USA Inc., Greenwood Village, Colorado, USA
Glenn C. Miller, Professor, Dept. of Natural Resources & Environmental Science, University of Nevada–Reno, Reno, Nevada, USA

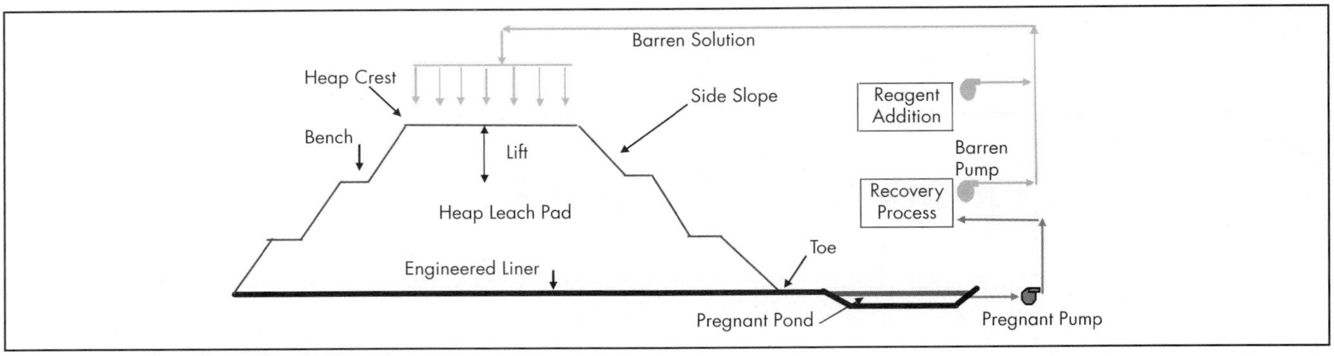

Figure 1 Simplistic heap leach flow sheet

FUNDAMENTALS

What appears to be a simple, empirical process has long been recognized as being very complex, involving many critical factors that encompass several scientific disciplines, such as physics, chemistry, geology, biology, and hydrology (Bhappu et al. 1969).

Dump leaching has mostly been applied to gold, silver, and copper ores, whereas heap leaching has been applied to many different minerals including copper, gold, silver, uranium, vanadium, zinc, nickel, and cobalt. However, as the principal applications are copper and precious metals, most of the following discussion will refer to these metals.

Five factors are essential to the successful heap leaching of any metal-bearing ore or valuable mineral:

1. A lixiviant must be selected to optimally dissolve the target valuable mineral and minimally dissolve gangue minerals.
2. The lixiviant diffuses to the site of the reaction.
3. A chemical reaction must occur.
4. The solution containing the dissolved metal or mineral must diffuse away from the reaction site and back into the main solution stream.
5. The dissolved target metal or mineral must be recoverable from the main pregnant solution.

In leaching a given ore, the rate of metal extraction can be rapid or slow depending on the mineralogy, lixiviant, and the chemical dissolution reaction. The factors of penetration, dissolution, and diffusion go on simultaneously and not in successive steps. The ore is wetted from the surface, and solution travels down through the heap due to gravity under variably saturated conditions as described by the Richards equation (Sullivan 1931) and characteristic curves (Van Genuchten 1980).

Percolation Theory

Leach solutions passing through a heap (and all flow regimes where unsaturated flow conditions exist) can be divided into three zones: the saturated zone (where all pores are filled with water), the unsaturated zone (where pores are only partially filled with water and that water is held in place by capillary forces), and the capillary fringe (where pores are filled with water being held in place by capillary forces). However, the flow regime is essentially composed of only the wide channels and macropores carrying large quantities of water by gravity flow (Bartlett 1993).

When rock is fragmented during mining and crushing, the bulk density of the material decreases with a resulting increase in void space. This *swell factor* can be 25%–40% by volume. On stacking and leaching, the ore settles/compacts with wetting and because of the weight of ore placed above it. Depending on the heap height and because of this compaction, the volume percent of voids in the bottom of the heap may be less than 5%. When the regional solution application rate is greater than the heap permeability, solution reaches a bottleneck and migrates horizontally until it reaches a region of better permeability to continue gravity flow downward.

If a solution pathway is not available, the solution volume tends to build up, leading to surface ponding and internal *perched* water tables. These can promote heap instability. If near the edge of a heap, solution can *blow out* the side of the heap, resulting in erosion damage to side slopes.

Solution flowing to a more permeable zone results in channeling, with the potential to cause fines migration and dry zones under the bottlenecks. The factor controlling the liquid volume in a heap is the particle size distribution (PSD). Solution fills all voids from rock sizes less than 0.3 mm (48 mesh) with the exclusion of air and rock sizes coarser than 1 mm (10–20 mesh) where solution drainage is almost complete. Capillary forces rather than gravity flow prevail for finer particles (fines).

As ore gangue minerals react with leach solutions, degradation or decrepitation of rocks can occur, resulting in more fines being produced. These fines will change the regional permeability of the heap (Bartlett 1998).

Normal heap leach operations have a wide distribution of ore particle sizes and uneven ore mixing, so there will be a range of localized hydraulic conductivities and percolation flow velocities (Bartlett 1993; O'Kane et al. 1999; Orr 2002). Figure 2 presents the ranges of water conductivity, water drainage, and air permeability for various particle sizes, ranging from clays to sand to gravel (Bartlett 1993). Thus, large variations in hydraulic conductivity (permeability), in combination with high solution application, will result in significant flow channeling ultimately leading to diluted pregnant solution grades and overall longer leach periods to target extraction (Bouffard and Dixon 2000).

Clays and significant fines content in ores pose the highest threat to suitable heap permeability and metal extraction efficiency. Poor permeability is the key factor in many failed heap leach operations. Agglomeration and appropriate stacking techniques obviate the clays/fines issues. Chapter 10.1,

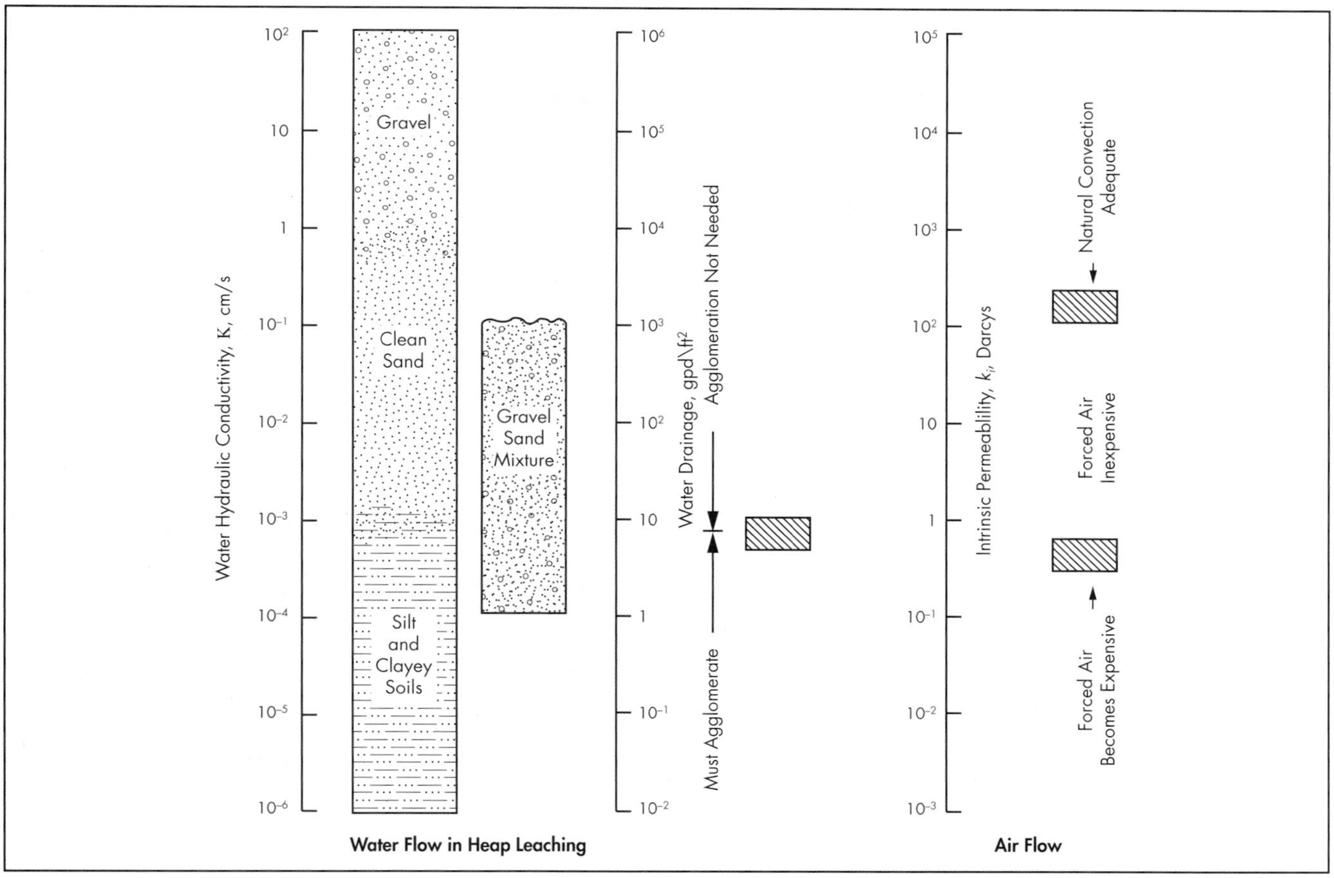

Source: Barlett 1993

Figure 2 Effect of grain size on saturated hydraulic conductivity

"Agglomeration Pretreatment," presents the issues related to this pretreatment step.

Leach Chemistry

Leach chemistry factors and issues that must be considered for successful heap leaching of gold and copper are provided in the chapters on gold, silver, and copper hydrometallurgy elsewhere in this handbook.

Other metals of interest in heap or dump leaching include uranium, nickel, cobalt, zinc, and recently, some rare earth elements. Of these, uranium heap leaching has had the most attention. Because of its many oxidation states, there are numerous complex uranium minerals (Merritt 1971). Some uranium minerals can be leached with sulfuric acid or in alkaline conditions employing sodium carbonate. More-complex/less-oxidized uranium minerals require stronger acid and sometimes higher temperatures.

Heap leaching of nickel laterite ores using sulfuric acid was advanced during the recent commodities boom period of 2007–2013. The Talvivaara nickel bioleach operation in Finland (Saari and Riekkola-Vanhanent 2012) is currently processing ores containing nickel sulfide (pentlandite). Cobalt is recovered when heap leaching nickel–cobalt or copper–cobalt ores.

LEACH PADS

The leach pad itself is a key element of the heap or dump leach process. Selection of the type of leach pad, detailed pad design, liner selection, construction techniques, and assessment of geotechnical risks are all part of the overall leach pad development process.

Types of Leach Pad

Three main types of leach pads are used in heap and dump leaching. These include flat or expanding pads, valley fill pads, and on/off leach pads. Selection of pad type is influenced by many variables, including available topography, leach cycles, closure issues, and so forth. The choice of pad type influences capital costs, stacking options, solution application methodology, solution collection, recovery plant sizing, and closure issues.

Flat or Expanding Leach Pad

The ideal leach pad is located within a sufficiently large area with a 2%–3% downslope and 1% across slope, clayey in situ material below the topsoil, nearby sources of clay or low-permeability soil for the pad base, and gravel for the protective/drainage layer. This allows construction of a regular-shaped (rectangular) *flat* leach pad that can be stacked easily with mobile conveyor systems and expanded in the upslope direction at a relatively low installation cost. Solution collection and storage systems that form key aspects of the overall project are also straightforward regarding design, construction, and utility.

In reality, very few mine sites cooperate to the extent that all these factors are favorable. Slopes can be as low as 1% downslope or quite high (up to 20%) with rocky ground, no available clays or gravels, and so forth. Expanding pads, however, offer the most overall cost benefits and ease of operation. Thus, most leach pads are built as expanding pads even though the "flat" identifier may not be accurate.

Collection systems incorporating individual leach cell outlets and collection ditches or pipes can likewise be expanded upslope with the pad. Process ponds can be located logically and sized for the life of the operation. Stormwater pond capacity may need to be increased as the leach pad expands.

Expanding pads with multiple cell outlets also offer greater flexibility in multistage leaching, with recycling of low tenor intermediate leach solution (ILS) onto newer ore to maintain upgraded PLS, which is treated at a constant flow rate over the project life.

Expanding pads offer relatively easy access for stacking of upper lifts. In this case, when the first lift of the initial leach pad is completed, the decision to go up or go out needs to be made. This is usually based on cost factors, which often indicate going up rather than building new pads. Operability issues, such as long leach cycles, increased metal inventories in multiple lifts, and risks of channeling may force an expansion of the leach pad prior to building upper lifts.

Valley Fill Pads

Valley fill pads are built in areas of steep terrain where large, flat areas are not available. A valley fill pad usually consists of constructing a dam across a suitable valley, with ore stacked behind the dam as shown in Figure 3. The dam wall and downslope leach pad area form an internal solution collection/storage pond, and the natural topography directs solutions to the pond. Steep walls of the valley and dam offer potential difficulties in pad preparation and liner installation.

As ore is stacked in multiple lifts, the lining is extended up the valley. Valley fill lifts are more appropriately stacked with trucks or with an advancing conveyor stacking system that operates on top of the lift rather than on the relatively steep base. Softer, more compressible ores should perhaps be avoided in a valley fill heap.

A valley fill pad can be designed to contain all process solutions as well as incident rainfall. Otherwise there can be external stormwater ponds connected by a spillway from the internal pond.

The main disadvantages of valley fill pads are higher cost of installation, reduced leaching flexibility with a single process pond, and often a requirement for staged increases in solution treatment facilities. The latter is caused by falling PLS grades as the pad expands and rainfall dilution cannot be diverted away from the internal pond.

On/Off Pads

The on/off leach pad concept employs a permanent, fixed-size leach pad. Ore is stacked, leached, rinsed, then removed either to a separate stockpile where it is abandoned or to a secondary leach pad where leaching continues. Several smaller

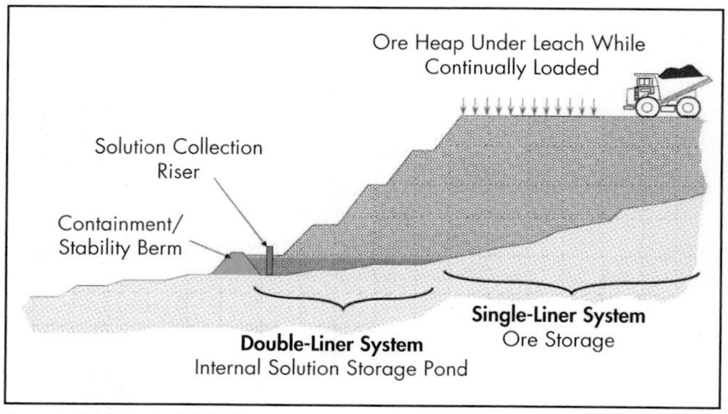

Source: Galea et al. 2010
Figure 3 Valley fill leach system

Courtesy of FLSmidth
Figure 4 Racetrack on/off leach pad

on/off gold heap leach operations have been built (Bernard 1995). Currently, several very large copper heap leach operations in Chile employ on/off leaching systems (Taylor 2017).

The on/off concept is most useful for relatively fast-leaching ores (generally 30 days or less), in areas with limited available leach pad area or in high-rainfall environments as the combined pad/pond collection area is minimized.

Some large heap leach operations employ a *racetrack* arrangement, employing mechanized stacking and unloading systems that can move large volumes of ore quickly and at reduced operating costs (although upfront capital costs are high for these systems). See Figure 4. The leach pad is constructed in two parallel sections with stacking equipment advancing across one section before rotating around to stack on the opposite section. The reclaiming system follows a similar pattern, removing leached/washed material.

Solution collection systems can be directed either to the center section between the two pads or to the outer sides of each pad. The pads are subdivided into cells with separate, dedicated collection systems directing off-flows to ILS or PLS ponds.

The leach pad base must be able to withstand vehicle traffic on a routine basis so is more robust than a typical flat pad. Pads are often made of layered asphalt with an impervious rubberized layer, concrete, or standard composite clay/high-density polyethylene (HDPE) but with thick protective gravel layers.

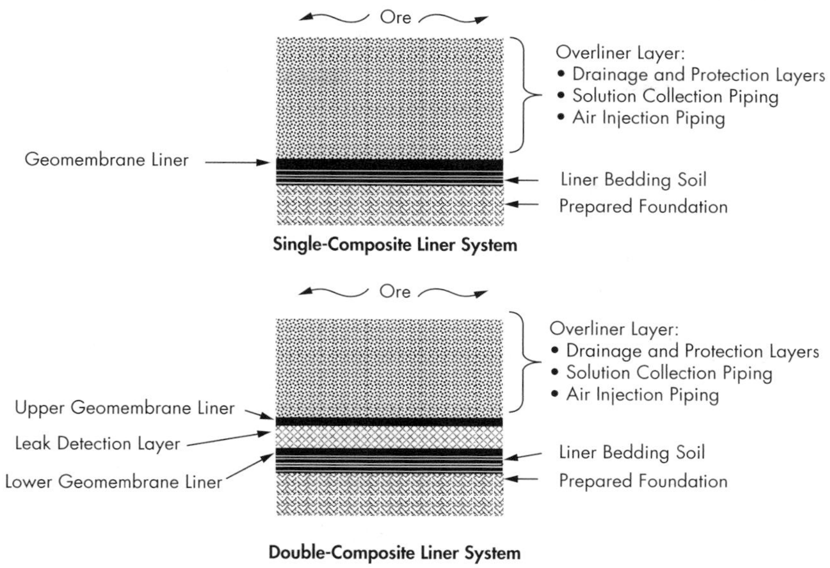

Source: Galea et al. 2010
Figure 5 Typical liner construction

One advantage for on/off operations is the capability to collect representative tailings (*ripios*) samples during pad unloading for a more reliable metallurgical balance and heap recovery assessment. Another advantage is the relatively small amount of unrecoverable gold-in-inventory (gold dissolved in solution remaining within the ripios) compared to multiple lift heaps.

The main disadvantages include high capital cost, higher operating costs caused by double handling of material, and a limit to leach times available. In the latter case, leach cycles required to achieve optimal recovery may not be available as the requirement for getting new ore under leach to maintain cash flow overrides the relatively minimal additional recovery.

Pad and Pond Design and Construction

Constructing a heap leach pad or pond system begins with obtaining the necessary federal, state, and local permits based on the development of a feasibility-level engineering design. These permits are numerous, cover areas addressing land access through to reclamation plans, and vary depending on location and land title. Prior to the permit application, the heap leach facility needs to be designed for size, location, and ore stacking/treatment rate, with detailed solution management and reclamation plans.

The heap leach pad design is conducted by a professional engineer. The design areas include crushed and/or ROM ore rock properties; seismic data for pad stability; water balance and solution management; pad, pond, and solution piping containment and monitoring; heap height; future expansion; and reclamation and closure plans. A site investigation by a geotechnical engineer is often required to assess construction issues, suitability of available in situ and borrow materials, surface runoff conditions, groundwater or seepages, and other issues that may impact construction costs and timing as well as the stability of the heap or dump during operations. Figure 5 presents the most typical liner systems in use in modern heap and dump leach design. The single liner system is generally acceptable in most jurisdictions.

Monitoring systems are incorporated to monitor for leaks in the pad or ponds. Additional leak detection methods may be required, which could involve a collection system under the geomembrane or between geomembrane layers daylighting in the downslope collection ditch for continuous monitoring.

The greatest cost in heap leach pad construction is the foundation earthworks necessary to provide the firm, stable loading surface. The pad must be sufficiently level so that it is possible to line with geomembrane and, if specified in the design criteria, be operable for a retreat-type mobile stacker (approximately 6% maximum gradient). The pad must also have surface area necessary to meet leach cycle requirements on the next to last lift.

Cutting and filling of the leach pad and pond area are performed by mine equipment and construction equipment alike, depending on fleet availability and any spatial constraints involved in foundation preparation usually dictated by subsurface geotechnical conditions, topography, and geomorphology. The base is then covered with approximately 300 mm (12 in.) of compacted soil of low permeability and lined with a geosynthetic liner such as HDPE or linear low-density polyethylene (LLDPE).

In areas where clay is unavailable, bentonite-impregnated geotextile known as *geosynthetic clay liner* or GCL has been used to form the low-permeability layer. GCL is also used on steeper slopes where compaction equipment cannot be safely operated. In all cases, a geotechnical evaluation of the composite liner system is necessary to ensure overall heap stability. The key aspects of leach pad design are discussed by Thiel and Smith (2004) and Breitenbach (2005).

Specifications for permeability of the clay base are normally 1×10^{-6} to 1×10^{-8} cm/s. Quality control testing of the clay layer is required during placement. More information on liner selection and installation is provided by Breitenbach (1994, 2004).

A protective cushion layer of ~300 mm (~12 in.) of compacted soil or clay is often installed on the geomembrane to provide protection from vehicular traffic. Then a network of perforated drainage pipe is laid on top of the liner or the cushion layer. The collection network is designed to provide sufficient flow-carrying capacity to a point of collection. This speeds collection of pregnant solutions and minimizes hydraulic head on the liner thereby limiting risk of any leaked solution from migrating beneath the pad and into the environment.

Finally, the network is covered with a free-draining overliner rock material, often ore crushed to a coarse gravel size distribution of –50 mm (–2 in.). If no bedding layer is installed over the geomembrane, then the drainage gravel topsize should be –12 mm (approximately –0.5 in.), or finer if the crushed rock is particularly angular. A layer of geotextile fabric over the geomembrane could be installed thus allowing coarser or more angular drain rock. Heap stability may be compromised when using geotextile over geomembrane for steeper leach pads.

Containment berms around the perimeter of the heap are necessary to keep leach solutions in the system and to prevent surface runoff from entering the system. These are usually constructed from compacted soil and are overlain with geomembrane as an integral part of the pad liner. Berm heights of 0.5–1.0 m (1.6–3.3 ft) are employed for internal dividing berms as well as for external berms along the top and ends of the pad. The main berm along the downslope edge where solution collection ditches and pipes are located can be much higher to ensure that any heap slumping does not impact on solution collection.

Generally, solution ponds containing process solutions (cyanide for gold/silver, acid for copper, etc.) are built using the same procedures as leach pads, but often with double geomembrane liners with leak detection monitoring systems incorporated under each liner layer. Stormwater ponds or raw water ponds not expected to routinely contain lixiviant chemicals are usually built with a single geomembrane liner.

Process ponds are usually built as rectangular designs with a low corner or sump to allow most of the solution to be pumped out. The leak detection system is placed under the sump. Pond wall slopes of 3H:1V are typical as steeper walls are difficult to adequately compact. An allowance is made in volume calculations for spillways and freeboard. Ponds are often fenced and netted or covered with bird balls to reduce wildlife exposure and evaporative losses.

A typical pond system will have interconnecting overflow spillways such that the entire pond system fills before any solution overflows to the environment. Emergency unlined catchments may also be part of the overall solution management system.

The geomembrane liners for berms, ditches, and ponds are often textured to reduce slipping hazards if significant foot traffic is expected in the area. Often, liners under the ore and in areas not continually exposed to sunlight are thinner at 1.0–1.5 mm for HDPE and LLDPE while exposed areas and areas with high foot traffic are thicker, usually at 1.5–2.0 mm.

A key part of the approach to pad and pond design is a geotechnical evaluation of the leach pad/pond sites for overall suitability. This includes assessment of available construction materials for preparation of the base foundation, the low-permeability layer, and any overliner materials. Evaluation of the ore stacking plan is required to ensure pad stability over the life of the operation and beyond.

The ultimate height of the heap is established by evaluation of several primary constraints:

- Liner and drainage system integrity with increasing load
- Intrinsic ore characteristics governing permeability reductions with height (geomechanical aspects)
- Seismic conditions in the area
- Heap stability based on the interfaces between ore and the liner system
- Slope stability under various conditions of saturation

Prior to construction, surface water diversion channels are cut around the pad/ponds to direct surface runoff away from the area under construction, protect process facilities during operations, and limit water balance impacts.

ORE PREPARATION

Prior to placement of ore onto the leach pad, some ore preparation is usually required. This can include addition of pH control chemicals (usually lime for gold/silver), crushing, agglomeration, screen upgrading, and so forth. The initial preparation step in the leaching process is mining, which is common to all approaches.

Blasting to increase fines generation, or sometimes to limit fines generation, followed by dozing, loading, hauling, and tipping onto the pad or into feed stockpiles can affect processing and leaching performance. Some degree of blending can be accomplished in the pit. The mining approach can also have a major impact on the moisture level of feed ores, which affects downstream handling properties.

Otherwise, the choice of ROM dump leaching versus crushed heap leaching is usually a question of economics where many variables such as extraction, reagent consumption, assumed metal price, project life, capital expense, and operating expense are considered.

Run of Mine

Dump leaching to process ROM ore is affected by the size distribution of the feed ore because reagent consumptions and metal extraction are often impacted by the exposed surface area of the material being leached. Thus, finer blasting may improve leach results while avoiding the additional capital and operating costs of a crushing operation.

For gold and silver operations, leaching pH is often controlled through addition of quicklime or hydrated lime to the ore prior to stacking, usually by dosing lime onto the ore in the back of the haul truck at a lime-dosing station. Other methods of lime dosing include dosing the empty truck as it returns from the pad, spraying a lime slurry onto the ore either in the haul truck or as it is dumped on the pad, or adding bagged lime onto the ore in the pit prior to/during mining. The latter approach is most suitable to very small operations.

There is no easy method of adding acid during stacking when dump leaching copper or other ores using sulfuric acid as lixiviant.

Crushing

Leach extraction generally increases as material is reduced in size, thus liberating the target mineral for leaching. Size reduction is generally accomplished through standard crushing techniques, although recently, more heap leach projects are considering finer crushing incorporating vertical shaft impactors (VSIs), high-pressure grinding rolls (HPGRs), or quad roll crushers. After primary crushing, ore is often screened to remove fines and avoid the recrushing of this size fraction.

VSI machines have been employed to crush heap leach ores as fine as 2.36 mm (8 mesh) (Rose et al. 1990). HPGR crushing to as fine as 4 mm (0.16 in.) has been evaluated at a laboratory scale. Gold Fields' Tarkwa mine employed an HPGR unit for heap leach feed at a crush size of P_{50} 3 mm (0.12 in.), treating some 4.4 Mt/yr (4.8 million stpy [short tons per year]). Golden Queen Mining's Soledad Mountain operation uses an HPGR to produce a product with P_{70} of 6 mm (0.24 in.). Improvements in gold recovery of up to 10% over standard crushing circuit product have been reported (McNab 2006; Von Michaelis 2009).

Although agglomeration, as discussed next, is not required for all crushed ores, addition of pH modifiers and/or lixiviants during crushing/stacking can give a head start to leaching of the desired metal. Lime or cement in dry form as well as cyanide solution can be added to gold/silver ores on the product conveyor, and sulfuric acid can be dosed to ore on a conveyor belt or at a transfer point. Care must be taken to prevent spills or leaks of hazardous solutions from conveyors onto unlined/uncontained areas.

Agglomeration

Agglomeration improves ore permeability to both air and liquids by eliminating segregation of fines, resulting in a more uniform heap. Well-operated agglomeration also allows for improved reagent and solution addition to ore. Controlled reagent addition has obvious cost benefits, while controlled solution addition also results in drier agglomerates, which are less likely to plug chutes.

Moisture addition can play a key role in optimizing agglomerate quality, and this is especially true with cement agglomeration (Pyper et al. 2015). Unfortunately the wetter, sticky agglomerates can impact downstream conveying operations.

Agglomeration quality control is often visually done. Techniques involving measurement of electrical conductivity have been successfully used for agglomeration process control (Velarde 2003; Bouffard 2005), and infrared and microwave moisture analyzers have been employed for controlling solution addition. Otherwise, many operations control agglomerating reagent feeders and solution additions through a ratio setpoint controller linked to the feed rate of material into the agglomerating drum.

Most operations employ rotating drums for agglomerating, as this is generally the most effective method of blending in reagents and forming strong agglomerates. However, belt and stockpile agglomeration can be considered (Chamberlin 1986).

The key factors in drum agglomeration are residence time and the volume of material in the drum as a percentage of total volume. A detailed discussion of drum agglomeration design and selection can be found in Miller (2005).

Blending

Blending strategies are incorporated for maintaining relatively uniform grade, ore characteristics, or moisture content of material fed to the heap leach. Clayey ores are often blended with rockier ores to assist in maintaining crushing plant throughput and possibly obviating the need for an agglomeration step.

Pulp Agglomeration

Pulp agglomeration (Zaebst 1994; Jones 2000) can be considered when there are relatively small volumes of higher grade ore in the ore reserve. The high-grade material is selectively mined and ground in a small grinding circuit. The ground slurry can then be leached in an agitated tank or blended in with lower-grade ore in the agglomeration system prior to stacking and heap leaching. If slurry volumes are low, they can be pumped directly to agglomeration; otherwise, they can be filtered with the cake added to the agglomeration feed.

STACKING

The three main methods for stacking heap leach ore include truck stacking, conveyor stacking, and excavator stacking. Truck and excavator stacking can be employed for dump leaching. In any case, the leach pad design must take into consideration the stacking approach.

Truck Stacking

Truck stacking is the most common method of stacking for dump leaching and is also employed at heap leach operations with competent, low clay/fines ores. Many operations employ dump construction techniques essentially identical to waste dump construction and maintenance, including watering of haul/access roadways on the dump to control dust. In these cases, compaction and breakdown of surface ore can be significant.

Alternate truck stacking scenarios can be employed where truck traffic is limited to dedicated roadways on the dump surface (Chamberlin 1981). Ripping is nearly always recommended after truck stacking. Studies have shown that compaction is usually limited to 1.0–2.0 m (3.3–6.6 ft) by normal mining truck traffic. Single pass or cross-ripping can be considered.

Conveyor Stacking

Conveyor stacking systems are common in heap leach operations and often include one or more overland conveyors connecting the crushing plant to the leach pad. Usually a tripper conveyor diverts ore from the overland running along the upslope edge of the leach pad onto a string of portable grasshopper conveyors that ultimately feed a radial stacker conveyor. Often a horizontal indexing conveyor feeds the stacker, and the stacker has a stinger conveyor extension on the end that can extend several meters. Both of these approaches reduce the number of relocations of the system (Figure 6).

Grasshopper stacker systems can operate in either retreat or advance mode. Most grasshopper-type stacking systems operate in retreat mode where the stacker operates on the leach pad base or on top of an older lift, stacking new ore in lifts from 6- to 12-m (20- to 40-ft) tall. In this case, the pad base/lower lift surface must be relatively flat and firm to allow the equipment to be relocated easily.

For upper lift stacking, compaction from the stacking equipment can be significant, and it is recommended that equipment have low ground pressure. This is accomplished either through minimizing the weight of the equipment or employing wide tires, tracks, or skids to spread the weight over a larger area.

Minimizing weight is difficult for stackers required to stack above 6 m (20 ft) in height and for high-tonnage

Figure 6 Grasshopper stacking system

Figure 7 Mobile stacker at on/off heap leach

operations where the equipment is large and must be robust and long lasting. Thus, radial stackers with tires often have high ground pressures and can sink into the operating surface, causing problems with both radial and towing movements. A grader or dozer is employed to rip the surface where the radial stacker has been operating.

For retreat stacking, the ore is stacked from the bottom (low end) of the leach pad with the stacking gear retreating upslope. As new lift surface is created, it is placed under leach with no impact on continued stacking operations.

Advance stacking essentially employs the same equipment as retreat stacking, but operates on top of the newly generated leach pad surface. The stacking system is extended through addition of grasshopper conveyors until the entire lift is completed. Advance stacking has the advantage of allowing a lighter stacker conveyor as it does not have to lift ore to any height. And it can build lifts over uneven or steep terrain, leaving a flat surface for the next lift. The main disadvantage is interference of the equipment (and related access roadways) with leaching activities. See Cook and Kappes (2013).

Track-mounted mobile stacking systems operating on a racetrack on/off leach pad or for large multiple lift operations are more automated and generate very flat surfaces (Figure 7). Mobile conveyor stacking systems offer low operating costs, which are offset by high capital expenditure.

Excavator Stacking

Excavator stacking can be employed for crushed or ROM ore and is considered when truck traffic on the surface of the new lift is expected to be problematic for permeability (Figure 8). The ore is delivered by haul trucks to the toe of the lift being stacked. Like truck stacking, it is a higher operating expenditure (OPEX) and lower capital expenditure (CAPEX) option compared to purchase of a conveyor system.

LEACHING

For theoretical aspects of the leaching processes for the relevant metals, refer to applicable chapters in this handbook. The practical aspects of solution application to heaps and dumps are covered in the following sections.

Figure 8 Excavator stacking a 5-m (16-ft) heap

Irrigation

Operational solution application (*leaching*) can be conducted employing a variety of sprinkler and drip emitter systems. Sprinkler systems are constructed principally from polyvinyl chloride or HDPE plastic because of light weight, low cost, ease of installation, and corrosion-resistant properties. Dripper lines are generally HDPE.

The goal of solution application is to have uniform surface coverage to allow for optimum wetting uniformity leading to maximal metal extraction. To minimize surface ponding, the application rate should match or be less than the hydraulic conductivity or acceptance rate of the ore. Solution application rates vary from 2.4 to 19.6 L/h/m^2 (0.001 to 0.008 gpm/ft^2) with typical applicate rates of 8–12 L/h/m^2 (0.003–0.005 gpm/ft^2).

The application method has been shown to significantly affect copper dump leaching effectiveness (Jackson and Ream 1980; Schlitt 1984; Uhrie 1998). Drip emitters are believed to be able to deliver solution more precisely than other application methods. They apply relatively large drops of leach

Source: Galea et al. 2010

Figure 9 Heap leaching solution application methods

solution directly to the heap/dump surface and also have the advantage of minimizing evaporation losses of both solution and reagents. Emitters also provide constant flow under a wide range of pressures (34–20 kPa, 5–30 psig). Emitter flow and spacing determine the application rate. The advantages of drip emitters are further delineated in Dixon et al. (2012).

Application methods are shown in Figure 9. Wobblers, D-ring sprinklers, rotating impact sprinklers, and misters are also used. Wind-borne losses from sprinklers and misters involve loss of reagents and potential environmental issues (Jackson et al. 1979). Pressure regulators employed in the sprinkler system can limit uneven coverage and wind loss issues related to pressure.

Solution Control

Leach pads are usually divided into cells or bays to separate leach solution off-flows from different blocks of ore for leach management, metallurgical accounting, and reconciliation purposes. For ores with long leach cycles or for large tonnage operations, use of multistage leaching is employed to provide additional leach time while limiting the size of the recovery plant capacity. Older ore is leached with barren solution from the recovery plant, and off-flow from this ore is recycled onto new ores, with the upgraded PLS then treated for metal removal.

Typically a two-stage, countercurrent leach regime is employed with ILS recycled to new ore. Three-stage leaching has been practiced at several operations where barren leach solution (BLS) is pumped to the oldest ore, ILS off-flow is then pumped to partially leached ore with off-flow recycle leach solution (RLS) ultimately pumped to the newest ore (Lawry et al. 1994).

One disadvantage of multistage leaching is an increase in the metal in inventory. Because of the delays in metal recovery (and related cash flow) from multiple lift heaps or dumps, considerations have been made for full or partial interlift lining systems (Echeverria et al. 2015). For a full interlift liner, the lower lift surface is graded and compacted, then a geomembrane is installed along with drainage pipe and drain rock. Solution is collected and directed down the face of the lower lifts to the cell outlet. If the lower lift ore is particularly clayey, compaction of the surface material to a low-permeability layer may obviate the use of the geomembrane.

As environmental standards become more stringent, water management procedures become critical to managing the heap leaching operation. Water management may include using different application methods to maximize or minimize evaporation, installing purpose-built evaporators and geomembrane "raincoats" on top or side slope surfaces to divert rainwater out of the collection system (Breitenbach and Smith 2007; Manning and Kappes 2016).

Side Slopes

Leaching side slopes of a heap or dump presents specific problems and is generally less effective, with metal extraction from side slope ore likely to be compromised. Neither sprinklers nor drip emitters present foolproof options for side slope leaching. Small sprinklers with a gentle sprinkling pattern offer a reasonable compromise for side slope leaching.

PONDS AND SOLUTION STORAGE

Solution containment is a key aspect of heap leaching. The vast majority of operations employ geomembrane-lined ponds to contain process solutions. A typical heap leach operation would involve PLS, BLS (or raffinate), and stormwater ponds. ILS/RLS ponds are required if multiple-stage leaching is employed.

For operations incorporating solvent extraction for metal recovery, settling ponds allow solids to settle out prior to treatment. Raw water for makeup is often stored in ponds, and there may be other lined process ponds involved with the metal recovery steps.

In areas of migratory bird flightpaths and where severe winters are encountered, solution tanks are often substituted for process ponds while lined stormwater or event ponds are employed for emergency containment. Several operations in Kazakhstan and Russia employ belowground process solution tanks for freeze protection. The tank option reduces flexibility in solution management but can force improved operational efficiencies.

Process pond sizing usually incorporates the following factors:

- Minimum leach solution pumping allowance (12 or 24 hours of operation)
- An allowance for draindown solution from active leach areas in the event of a power failure
- A minimum volume for pump operation/priming
- An allowance for nominal rainfall events that may occur a few times in a normal year
- Freeboard/spillway allowance

Stormwater pond sizing is based on a design rainfall event, such as a 1-in-100 year, 24-hour rainfall event over the entire catchment area (pad, ponds, and ditches). The design event is dictated by local conditions as well as regulatory guidelines. Experience shows that stormwater ponds that are designed for only the 1-in-100 year, 24-hour event will most likely fill and overflow at least once during a project life. Accumulation of longer-term extreme precipitation events should also be taken into consideration.

Given that leach pads are often built in stages, stormwater capacity has to match the rainfall requirements from the pad extensions. Often the initial stormwater pond capacity is sized for the initial installation plus the first expansion, with additional stormwater capacity added at the same time as the second pad expansion. A runoff coefficient is often included in pond sizing for the expected rainfall collected on active or inactive cells of the leach pad.

SAMPLE SELECTION AND TESTING

Ore samples for metallurgical evaluations are typically chosen in a collaborative process involving geologists, metallurgists, and mine modelers/planners. Geometallurgical domains and ore classification criteria are identified to reflect the deposit characteristics and variability (Olson Hoal et al. 2013). The ore body must be broken down into categories covering process-significant variables appropriate for the type of deposit being sampled. These include

- Oxidation state,
- Ore grade,
- Depth,
- Rock unit or lithology,
- Alteration/weathering, and
- Sulfide and gangue mineralogy.

Surface grab samples should be avoided. Near-surface bulk oxide samples can be of limited use for oxide deposits with respect to oxidation state, friability, and so forth, but should be used sparingly and in no instance be given any greater influence than a typical variability composite sample taken from drill core.

Variables to Evaluate

The cataloging of drill-core samples generates a table of composites representing various characteristics of the ores to be leached. Testing programs follow on from these composite selections but with these additional variables to be evaluated:

- Crush size/size distribution
- Spatial distribution within the ore zones (e.g., east end vs. west end)

Of key importance to the leach process is crush size/size distribution. Higher and faster extractions are expected at finer crush sizes, but the additional extraction may not be sufficient to offset CAPEX and OPEX for crushing, agglomeration, and conveyor stacking steps, in which case ROM ore dump leaching is the selected approach.

Types of Tests for Heap Leach Evaluation

The types of tests that are typically employed in assessing the amenability of an ore body to heap leach processing are

- Quick leach,
- Bottle roll on ground or pulverized ore,
- Bottle roll on coarse-crushed samples,
- Agglomeration,
- Screen analyses with fraction assays,
- Bench columns,
- Large columns and cribs, and
- Field trials.

The quick leach test (QLT) can be used for early identification of heap or dump leach potential. The main evaluation approach for heap leaching is the column test. However, shorter-term bottle roll leach tests have found a niche in the testing program to provide comparative data for variability samples and sometimes for crush size in a shorter time frame and at lower costs than column testing.

A ground ore bottle roll test (BRT) over a 24- or 48-hour leach period is usually conducted at a PSD with a P_{80} of ~75 µm (200 mesh) employing a pertinent leaching reagent regime. This basically gives the near-maximum metal extraction that could be expected from an ore in a milling situation. The results should mirror the quick leach results and are used to compare extractions from coarser size distributions. The BRTs can be conducted for kinetic evaluations with solution samples at 2, 4, 8, and 24 hours to assess the rate of leaching and to adjust leach chemistry.

Coarse ore BRTs are usually conducted on 2- to 5-kg (4.4- to 11-lb) portions of ore crushed to a specific size distribution or topsize, with rolling conducted on an intermittent basis of 30 to 60 seconds per hour. The intermittent rolling allows improved contact of the ore with leach solutions and air/oxygen but reduces the grinding effect of continuous rolling (and potential increased liberation). Intermittent bottle roll tests (IBRTs) are typically conducted for 4 to 10 days at 30%–50% solids with solution samples at logical intervals to determine leach kinetics and adjust leach chemistry.

One drawback of the IBRT is the relatively short leach time in comparison with column or field leaching. Thus for slow leaching gold/silver ores or some copper minerals, the IBRT may not be useful. Another drawback of the IBRT is the buildup of chemical species in the leach solution. This does not simulate heap leaching where ore is subjected to fresh leach solution on a continual basis.

Also part of the IBRT is a size analysis of the test residue, with metal assays on three to five fractions. This might indicate that all fractions leach to the same value, thus suggesting that a coarser crush size could be considered. Or it could show that there is a finer fraction size where extraction increases significantly. This break point can direct future test work at a finer crush size. IBRTs will give an indication of extraction as well as reagent consumptions/requirements but not geomechanical aspects.

Prior to column testing, an assessment of ore permeability should be conducted to determine whether agglomeration is required. A typical approach is a version of a falling head permeability test used in soils analysis. More involved geotechnical evaluation including triaxial testing can be used, but the simpler test is usually sufficient to determine the requirement for field agglomeration.

Otherwise, the clay/fines content in the material can be assessed, where ≥10% –75 µm generally indicates that agglomeration would be required (Chamberlin 1986). A detailed assessment of the hydrodynamics of stacked ore is presented in Robertson et al. (2013).

If agglomeration is deemed necessary, laboratory agglomeration in a cement mixer or on a rolling cloth is required to optimize conditions. A grid of three or more small (3–4 kg [6.6–8.8 lb]) batch tests at different binder levels and/or water additions should be set up for each ore composite, with agglomerated samples allowed to cure for some minimum time (24 hours for cement). The agglomerates can be evaluated in one of several ways to determine strength, including slump/percolation testing in columns (Pyper et al. 2015), dunk tests, and soak tests (Vethosodsakda et al. 2013).

Ore samples intended for column testing are typically crushed in jaw crushers to the size of interest; often near ROM, P_{80} 37 mm (1.5 in.), and P_{80} 12 mm (0.5 in.). These sizes typically represent ROM, secondary crushed, and tertiary crushed product. After passing a sample through the crusher, the product should be screened and oversize material returned to the crusher until the desired sized distribution is achieved.

For competent core, laboratory crushing often does not generate the amount of fines that blasting, mining, and field crushing do. In this case, the size distribution of the material to be tested should be adjusted to match *as-processed* size distributions to avoid a negative bias in the leach results caused by a lower degree of liberation. The product size distribution of a commercial cone crusher can be used as a target for PSD adjustment. Adjustment involves screening the entire composite at three of the coarser size fractions and comparing the fractions to the commercial product sizings. Splitting out and crushing a portion of the coarser fractions to finer sizes may be required to generate a PSD similar to the expected field operation.

Column testing at coarse crush sizes to represent ROM ore is limited to the size of the drill core available. PQ core (diameter 85 mm [3.3 in.]) is the most common of the larger core sizes used in exploration drilling, although core up to 300 mm (11.8 in.) can be obtained using specialty drill rigs.

Prior to column testing, splits are taken from the composite sample. One split is pulverized to less than 75 µm (200 mesh) and analyzed for head grade, multi-element analysis, and mineralogy. Mineralogy is often done with a combination of optical microscopy and scanning electron microscope techniques such as Quantitative Evaluation of Minerals by Scanning Electron Microscopy (QEMSCAN) or mineral liberation analysis.

Another split is screened into pertinent size fractions with each fraction pulverized and submitted for key metal assays. The results of this PSD with assays are later compared to the column residue sample PSD.

Copper head samples are typically further analyzed for acid-soluble copper and sulfide solubility using procedures such as the sequential copper leach procedure, QLT, or similar procedures modified for specific deposit characteristics.

Precious metal samples are typically analyzed for cyanide-soluble gold and for total precious metals by fire assay, as well as silver, organic carbon, carbonate, sulfide sulfur, sulfate sulfur, arsenic, antimony, mercury, base metals, and so forth. Larger pulverized or ground ore splits are generally employed in BRTs.

Two principal methods are used for preparation of column charges; if care is taken in sample preparation, either method should produce statistically reproducible results:

1. Samples are blended through statistically defensible techniques such as coning and quartering, then split through rotary or riffle splitters.
2. The entire sample can be screened into predetermined size fractions, then each size fraction is riffle split. Column charges are made up by recombining size fractions in appropriate percentages to rebuild representative samples.

If required, agglomeration of the column charge is typically performed in a cement mixer and done on a batch basis. Moisture must be added slowly until agglomerates are formed that are considered satisfactory or until the target moisture addition as per the agglomeration optimization test work is obtained. The quantity of moisture and reagents added must be recorded and included in the metallurgical balance of the test. Agglomeration of the sample will employ cement or lime addition (precious metals), acid addition (copper), or polymer binders (gold, copper, uranium, nickel laterites). Samples are then loaded into columns in a manner that reduces size segregation. Tapping the column walls during loading compacts the material to better simulate the consistency expected in field operations.

If possible, samples should be tested at the intended operational PSD and the operational lift height. To reduce wall effects and other preferential flow patterns, a rule of thumb is that column diameter should be roughly six times the largest ore particle diameter. A more or less standard minimum ore bed height is 1.8–2.0 m (~6 ft), but taller columns are more representative of field lift heights.

Column solution application practices vary between gold/silver and copper column leach testing. Typically for gold, solution is applied at or near the field application rate irrespective of column height. Leach testing is generally stopped when a solution-to-ore ratio of approximately 3 is reached; however, short columns (<3 m [9.8 ft]) often are run to much higher ratios, which then impact on scale-up assessments.

Copper leaching is more sensitive to ore diffusion and reaction kinetics; therefore, solution is generally applied at a ratio of column height to application rate and using a set cycle time. This results in solution being applied based on a solution volume-to-mass-of-ore ratio similar to gold testing, but within a set period. This has been shown to result in more representative estimates of acid consumption and extraction (Rood 2000).

Previous research has shown that small columns are generally well aerated, but occasionally are air starved. To avoid limitations in oxygen content, copper sulfide leach columns should be aerated at a flux similar to that intended for the field scale application. This is usually done using laboratory compressed air and controlled through flowmeters. Aeration rates are generally 100% of the air requirements necessary to achieve target extractions over the intended leach cycle time.

Columns can be irrigated on a continuous or intermittent basis using metering pumps. Pregnant solution should be collected on a regular basis (daily or on alternate days), weighed, and sampled. It is important that assays be carried out in a timely fashion to avoid changes in solution chemistry caused by oxidation, precipitation, evaporation, and so forth. Samples must be carefully measured for volume and other elements of interest. General analyses are as follows:

- Copper—copper, total iron, ferric iron, Eh, pH, and free acid
- Precious metals—gold, silver, (possibly copper and mercury), free cyanide, and pH

Often, the pregnant solution is passed through a metal recovery step (solvent extraction for copper or activated carbon for precious metals), then recycled after adjusting the solutions to the target leach conditions. This allows monitoring of the buildup of other species, which could affect downstream operations and ultimate process plant design.

Following completion of leaching, columns should be broken down and the tailings/residue assayed for grade and any other metal of interest as well as residual moisture content. A portion of the tailings should be submitted for a PSD with fraction assays to compare with the head sample. This can sometimes give an indication of expected extractions at coarser or finer crush sizes.

Metallurgical balance calculations will result in extraction curves based on the tailings assays and pregnant solution samples taken over the leach cycle. The extraction based on metal recovered in the pregnant solution treatment step can also be compared. Other data generated from each column test should include

- Days leaching,
- Solution-to-ore ratio of applied leach solution,
- Reagent consumptions (after allowing for reagents lost to samples and remaining in final solutions),
- Ore slump,
- Final ore bulk density,
- Head and tailings moisture content, and
- Percolation/permeability problems or other physical observations.

Column and bottle roll tests are often conducted in duplicate to provide a further level of comfort in leach performance. However, if the leaching behavior has previously been shown to be relatively consistent and if sample availability is limited, then random duplicates are often acceptable. If possible, column (and bottle roll) tests should be conducted using site water. This is especially important if project water quality is poor. A full water analysis and pH buffer curve are required to identify potential problems. A few operations employ seawater for processing due to unavailability of fresh water. These factors need to be taken into account when designing the column test program.

Scaling tendencies of makeup water and recirculating solutions are important for final design and during operations. Water treatment chemical suppliers often assist in assessing scaling tendencies of water and solution samples. Additional testing on physical aspects of the ore samples is required to assist in project design and cost estimating.

Field Trials

Full field trials of crushed ore at the proposed project stack height can be conducted as another option for evaluating heap leach potential and obtaining operating and cost data. Field trials were common in the early stages of the industry, including for uranium, copper, and gold, and most recently for nickel laterites. As experience was gained in extrapolating laboratory column test data to field operations, the prevalence of field trials has decreased.

For the most part, field trials are now deemed necessary only when there are significant unknowns in the scale-up process such as very high clay content, unusual leach behavior, or new process technology (Efthymiou et al. 1998). Financial institutions or company directors may insist on a field trial to reduce perceived project risk.

Run-of-Mine Ore Evaluations

As discussed, determination of crush size is normally conducted through one or more series of bottle roll or column tests at different crush sizes. However, when very coarse sizes show good leach behavior, it becomes difficult to verify expected field performance in bench-scale laboratory programs. Large-diameter whole core with minimal crushing (to generate some fines as per mining) can be employed in columns. Even so, the results still require extrapolation to a size distribution expected in field conditions—often with rock sizes of 1 m (3.3 ft) or larger. Bartlett (1998) presents a methodology for this type of extrapolation and discusses the risks involved while Bennett et al. (2003) discusses modeling of copper dump leaching from column leach data.

Otherwise, large columns or cribs can be set up with as-mined samples employed in a scaled-up version of a bench column test. Most commercial testing laboratories have facilities for large column tests or cribs to heights of 6 m (19.7 ft) or taller and crib dimensions up to 4 × 4 m (13 × 13 ft).

Another option for ROM ore evaluations is a full field trial where a very large sample is mined, dosed with reagent as needed, stacked, and leached. Field trials can range from a few thousand to 500,000 t (metric tons) (551,155.7 st [short tons]), with relevant facilities eventually incorporated into the full project design. However, the same drawbacks exist as for heap leach field trials, and even for large column or crib tests.

In several instances, heap leach operations have been set up based on a column test database of crushed ore, then once operations have begun, field testing of ROM ore is undertaken. Often, lower-grade ROM ore is placed as upper lifts on crushed ore heaps to save on capital for leach pads. See Gőkdere et al. (2015) for one gold mine's approach to ROM testing.

SCALE-UP TO FULL OPERATION

Scale-up to a full production operation is required when a successful column test program is translated into a final project design. For the scale-up process, taller columns (closer to full field lift height) provide a more reliable database. Column tests are generally considered an ideal leaching environment, so predictions of field extraction are based on final column test extractions but with some allowance (discounting) for field conditions that are seldom optimum or ideal.

Generally, final column extractions are discounted by 2%–10% for gold and copper ores and possibly by higher amounts for silver ores. The discounting is based on expected operational issues that may impact crush size or agglomerate quality, channeling within the heap, weather events, or other potential factors. One source of lost extraction is because of the difficulty in leaching side slopes effectively. Another is metal tied up in leach solutions in lower lift ores that may or may not be washed out over the course of the project life.

Another factor to consider when scaling up is the field leach time that will be available. Although the main advantage

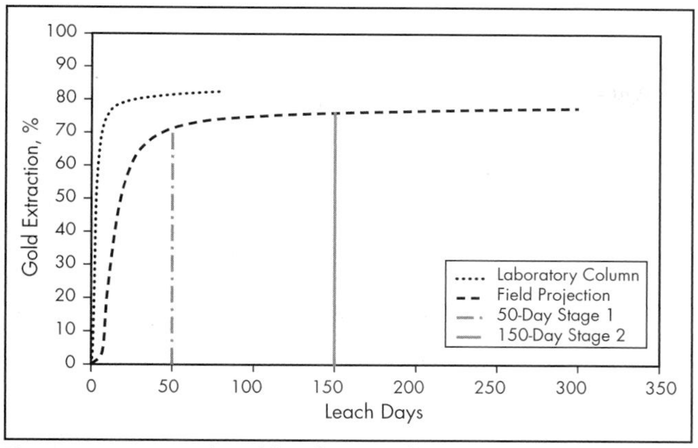

Figure 10 Extrapolated leach curve for a gold ore

of heap leaching over other process routes is time, leach pad and pumping design puts limits on the leach time available for ore in a given cell in a given lift. If leach time is limited compared to the scaled-up column time requirement, then further discounting may be required.

Typically, column extraction data used to develop production or field leach curves often take the shape of

$$F(t) = e^{-kt}$$

where

F(t) = metal extraction at time t
k = a fitting parameter constant
t = time

Other methods for predicting recovery from column data are described by Bartlett (1998). Bartlett uses the *shrinking core* model summed over multiple size fractions to allow for scale-up extraction predictions. The most common approach when larger (taller) column test data are available is based on the solution-to-ore ratio (sometimes called *flux*) that was required in the column to achieve a target extraction for a given leach time.

A typical scale-up situation is shown in Figure 10 where the laboratory extraction curve is shown along with the projected field leach curve including a 4% discount to the final laboratory extraction. Note where a 150-day leach cycle has been designated, which can be a single stage of leaching or a 50-day primary leach cycle (generating PLS) and a 100-day secondary leach cycle (generating ILS). When these time requirements are considered, another 2% discount is required from the final column recovery. It is possible that the discounted 2% of total gold will be extracted during leaching of upper lifts, but this not guaranteed and is thus considered upside potential when evaluating the economics of the project.

Another value to project for field operations is the bulk density of the stacked ores. Because of wall effects in leach columns, bulk density values are often lower in the columns than in actual heaps. This factor is employed in scaling of column test results as well as in crushing plant design and leach pad sizing.

When determining expected metal production from a heap or dump leach project, it is important to note that there will be losses in downstream operations that must be included in the final recovery values (as opposed to extraction values from the heap or dump).

LEVEL OF EVALUATION

As with any project evaluation, there are usually three stages prior to final design and construction of a heap or dump leach operation.

Initial Evaluations (Scoping Studies)

Testing can be limited to BRTs and possibly a few column tests.

Prefeasibility

Significantly more data are required involving IBRT and column tests on ore samples representing all the deposit lithologies. Additional tests are conducted to verify crush size and agglomeration requirements. For copper leaching, the leach regime (acid dosing, acid strength during leaching, aeration, bacteria, etc.) could encompass many column tests to assist in process design and confirm field requirements.

Final Feasibility Study and Final Design

The database must be completed to ensure appropriate ore treatment equipment selection, leach pad and pump sizing, recovery plant design, and other operational issues through decommissioning and closure. Confirmation testing is conducted, including reagent optimizations and additional geomechanical testing to verify permeability and heap stability at design lift heights and number of lifts.

Bench-scale or mini-pilot runs of solution treatment/metal recovery steps may be warranted (solvent extraction) employing solutions generated in column and bottle roll testing. Key column tests should include appropriate wash and draindown procedures at the end of the leach cycle to develop data for closure options. For cyanidation, water washing in batches with low-level cyanide speciation analyses will generate data for use in closure planning and submissions to appropriate regulatory agencies. Similarly, copper leach tests should include evaluations of residual solutions and their potential requirements for neutralization.

Production Models

Many models have been developed for predicting heap and dump leach performance, including several references in this chapter (Cathles and Apps 1975; Cathles and Schlitt 1980; Dixon and Hendrix 1993a, 1993b, 1993c; Keller et al. 2015; Marsden and Botz 2017; McBride et al. 2015; Roman et al. 1974) with each having additional references. Also refer to Chapter 2.5, "Modeling and Simulation."

Best practice for modeling extraction and production on an ongoing basis includes taking regular samples or monthly composites of leach ore for analysis and quality control (QC) leach testing and then trending extraction over time based on laboratory as well as production results. This historical data is then used to refine extraction formulas.

Whatever methodology is used in the leach production plan, it must be reviewed regularly and revised to match operational reality. Mines with historic production records and an extensive column leach database have an advantage.

Proper sample collection and evaluation are required for overall performance assessment and reconciliation of the ore

Table 1 Heap and dump key production indicators

• Head grade	• Solution-to-ore ratio–cumulative, m^3/t
• Tonnage stacked	• Lixiviant concentration, on
• Ore blend	• Lixiviant concentration, off
• Crush size, P_{100} and P_{80}	• Leach solution pH, on
• Moisture content	• Leach solution pH, off
• Agglomeration reagents	• Pregnant leach solution (PLS) grade
• Agglomeration water	• PLS treated
• Agglomerate quality	• Metal recovered
• New area under leach	• Efficiency of recovery
• Total area under leach	• Total metal in inventory
• Application rate, $L/h/m^2$	• Recoverable metal in inventory

body with the block/mine model (Chieregati and Pitard 2009). For ROM dump leaching, blasthole or grade control samples usually constitute the only source of assay data for ore to the dump. However, solution analyses must still be conducted.

Production Monitoring

Many operations now have sophisticated data collection and control systems. But with or without them, overall heap or dump leach production monitoring involves tracking a significant amount of data. These range from start and end dates for leach cells to rainfall to reagent additions to month-end metal inventories (heaps, ponds, recovery system).

The metal production model must be updated monthly to verify assumptions for short- to mid-term forecasting. QC bottle roll and column test results on actual stacked ore composite samples are critical to update the model results and projections.

Heap and dump surveys are required as part of the overall reconciliation with mine data. It is often the case that ore tonnage delivered to the ROM stockpile does not match weightometer readouts of ore processed through the crushing plant. In this case, ore bulk density based on crusher tonnage and heap survey volumes can be checked against mine department volumes and assumed swell factors.

Heap and dump surveys are also critical for monitoring slump. Excessive slump will lead to extraction/production losses through compression and elimination of void spaces in the heap. A 10% or greater slump is the value above which a fall-off in extraction will become noticeable (Pyper et al. 2015). Table 1 shows some key production indicators (KPIs) for heap and dump operations. There will be other KPIs for specific aspects of individual projects and depending on the metal(s) being leached.

DECOMMISSIONING

Decommissioning of spent heap and dump leaching operations cannot be adequately addressed in this chapter. Indeed, the approach to final closure of a heap or dump is highly site specific with several options available. The general consensus is that decommissioning is much more involved and time consuming than had been allowed for by many companies.

The steps include winding down of operations after final stacking of ore, efforts to increase extraction from older heaps and dumps, the appropriate rinsing approach, draindown issues, and final closure, including recontouring and capping.

Winding Down Operations

Even in well-operated heaps, there could be a recoverable 3%–10% metal left as inventory after completion of leaching caused by compaction, fines migration, heap settlement, clay zones, solution management, and zones with adverse chemistry. Attempts to recover these additional values can involve re-ripping the surface, *turning* (or re-mining) all or part of a lift with an excavator, or even drilling and blasting to generate new flow paths. These are followed by continuous or intermittent leaching. Targeted releaching is perhaps the most sophisticated approach to maximizing metal extraction from older heaps (Seal 2007, 2011, 2015).

Closure of Precious Metals Heaps

Because heap leaching of gold and silver ore is relatively young (approximately 50 years), closure of precious metals heaps was not considered a major issue until a sufficient number of heaps had been operated for many years and closure activities were initiated—beginning in the late 1980s. The procedures for closure of heaps are dependent on the climate of the area, particularly the rainfall, but also on the quality of the off-flow draining from the heaps. In very wet areas, heap leach operations will be closed differently than in arid regions. Ultimately, drainage from heaps in wet areas will likely be managed until the drainage water quality meets discharge standards (Parshley et al. 2012).

How long this takes depends on the amount of rainfall and the geochemistry of the heap, as well as the regulatory issues in the specific state or country. Ultimately, the industry and consultants have realized that fresh water rinsing would simply *not* allow a walk-away solution. In almost every case, the quality of the released water will be sufficiently poor that some sort of management will be required. And, particularly in arid regions, allowing meteoric water to rinse the heaps to remove the salts will require a time frame of several years to decades (Decker and Tyler 1999).

In arid environments, closure procedures generally involve store-and-release caps in which heaps are covered with sufficient growth medium to capture the water during the periods of precipitation, and allow vegetation to absorb the water for plant growth during the drier periods. The goals are to prevent or minimize water from penetrating into the closed heap and to create a diverse plant community that can support productive post-mine land uses.

Closure Chemistry of Cyanide Heaps and Dumps

Following active cyanide leaching of a heap, addition of cyanide is discontinued, and the solution is continuously recirculated to capture remaining precious metals. This begins the process of reducing the volume of the recirculating fluids. In arid regions, this is most commonly accomplished by sprinklers or misters to evaporate water, a process that can take months to years.

The chemistry of cyanide heaps must address the following issues:

- Cyanide removal (Johnson 2015)
- Salt accumulation
- Mercury releases (Flynn and Haslam 1995)
- Arsenic and other oxyanions (Miller et al. 1999)

Management of Closed Heaps: To Rinse or Not to Rinse

The initial approach to heap closure was to recirculate solution in a heap until weak acid dissociable (WAD) cyanide was down to 0.2 mg/L, followed by rinsing with fresh water. As observed in column studies (Miller et al. 1999), the mobile constituents (e.g., chloride and nitrate) are largely removed in less than one pore volume of water passing through the heap material under unsaturated flow conditions. Sulfate and selenium are 90% reduced in three to four pore volumes, but as the mobile constituents and acidity are removed, calcite retained in the heap material raises the pH to 8.3–8.5, and arsenic and vanadium concentrations in the column effluent (after six pore volumes) increase to greater than the applicable drinking water standard. Thus, management of the rinsed heap effluent will still be required, even if most of the salts have been removed. The rinsed heaps will likely still be discharging poor-quality water that will need to be managed, probably by evaporative processes. For heap closures in arid climates, rinsing does not solve the contaminant issue. Plus, rinsing is costly and consumes water.

The option that is now more commonly accepted (Parshley et al. 2012) is to establish a store-and-release cap for a heap. Complete elimination of drainage from a heap is unlikely and depends on the amount of rainfall as well as the closure methods used. Regulatory agencies in Nevada (United States), particularly the Nevada Division of Environmental Protection (NDEP), are increasingly convinced that closure of heaps will require caps that can eliminate (or substantially reduce) the amount of water allowed to penetrate the heaps each year. NDEP also requires evaporation basins that capture drainage from the heaps during the 100- to 500-year events. But nobody can predict how climate change will affect precipitation patterns during the upcoming decades into the future.

Heaps are most often composed of oxidized ore, and acidification is not common. However, when sufficient sulfidic rock is present in the spent ore to generate acids, the problems with closure can be substantially increased. Acidic water draining from the site will need to be managed, primarily by neutralization and evaporation.

In areas where meteoric water is substantial, using store-and-release caps may not be successful, and a combination of processes may be required to limit the amount of water that will be discharged. But ultimately, discharge of detoxified solution to surface water may be one of few options. The mobile constituents will be discharged relatively soon after closure (years) while the less mobile constituents will be discharged over a longer period. Permitting a heap in regions that receive large amounts of meteoric water should recognize that closure of the heaps may be problematic.

Closure of Acid Heaps

Although experience with closure of precious metals heaps has established methods for closure, much less experience is available in the literature on closure of copper or uranium heaps where sulfuric acid is employed as lixiviant (Borden et al. 2006; Smith 2002). Although heap leaching of copper ores has increased during the past 40 years because of improved recovery technology, closure of these heaps is generally more complicated than gold heaps.

First, the heaps are often much larger, owing to large copper ore bodies, where decades of acidic leaching has been conducted. Sulfuric acid requires neutralization with lime or some other alkaline material, which is generally unfeasible for large heaps. Concentrations of total contaminants being released from a closed copper heap can be large. In one case, the total dissolved solids in off-flow solution was 90,000 mg/L, with a pH of 2.9 (Borden et al. 2006).

Climate is perhaps the most critical factor for closing heaps. In extremely dry regions of the Atacama Desert in northern Chile, rainfall is often nonexistent. Consequently, the challenges for closing copper heaps are much less severe as the heaps will ultimately drain, and water from rainfall will be minor. In other parts of the world where higher levels of precipitation exist, however, closure will be a major challenge. Even in relatively arid areas, meteoric water is expected to result in long-term drainage when the heap is not properly closed (Smith 2002). Closure of copper heaps will require additional discussions and research.

Heap Capping and Covers

As discussed in previous sections, heap and dump capping and covers are key elements in successful closure. Heap capping has three main goals:

1. Reduce the potential for erosion.
2. Mitigate seepage from the heap into the natural groundwater.
3. Mitigate and/or reduce runoff and rainfall infiltration.

The general approach to heap and dump rehabilitation is broken down as follows:

1. Push down the side slopes to achieve a target overall angle of repose of 18–21 degrees, depending on regional environmental guidelines. Some jurisdictions may require "geomorphic reclamation" where the dumps are contoured to mimic the surrounding terrain (Stebbins 2015).
2. Place a containment berm around the entire leach pad/pond footprint to prevent ingress of surface runoff and release of potentially contaminated solutions.
3. Place a multilayered store-and-release cover on the top surface of the heap. These layers can consist of waste rock fill, sand, compacted clay, GCL, random soils, topsoil, and vegetation.
4. Place a containment bund around the top surface of the heap to reduce the potential for erosion of side slopes from surface runoff.
5. Place a thick cover of armoring rock and/or topsoil with vegetation on the side slopes.

A review of cover options is presented in Rykaart and Caldwell (2006), and more recent advances in acid rock drainage control are discussed in INAP's *Global Acid Rock Drainage Guide* (2009).

New Technology for Drying Heaps

Preventing meteoric water from infiltrating the heap is an established method for reducing drainage volume. Alternatively, additional methods for removal of water from the heap by evaporation can also potentially be utilized. Seal and Kiley (2015) have proposed such a system for using forced air to evaporate water from the interior of heaps. This technology has been demonstrated under laboratory conditions using actual heap material).

COSTS

Capital and operating costs for heap or dump leach operations can vary considerably depending on location, topography, ore type, downstream process requirements, project size, infrastructure requirements, and several other factors.

For a comprehensive analysis of heap leach costs for gold deposits, InfoMine has recently released a 200-page *Gold Heap Leach Cost Estimating Guide* (2016). For further accuracy, it would be necessary to prepare a preliminary feasibility study. Where the project economics are complex, it is often (but not always) necessary to follow this up with a full feasibility study.

REFERENCES

Bartlett, R.W. 1993. Soil decontamination by percolation leaching. In *Extraction and Processing for the Treatment and Minimization of Wastes*. Edited by J. Hager, B. Hansen, W. Imrie, J. Pusatori, and V. Rarnachandran. Warrendale, PA: The Minerals, Metals & Materials Society. pp. 411–424.

Bartlett, R.W. 1998. *Solution Mining Leaching and Fluid Recovery of Materials*, 2nd ed. Routledge.

Bennett, C.R., Cross, M., Croft, T.N., Uhrie, J.L., Green, C.R., and Gebhardt, J. 2003. A comprehensive copper stockpile leach model: Background and model sensitivity. In *Copper 2003–Cobre 2003*. Vol. 6, Hydrometallurgy of Copper (Book 2). Edited by P.A. Riveros, D. Dixon, D.B. Dreisinger, and J. Menacho. Montreal, QC: Canadian Institute of Mining, Metallurgy and Petroleum. pp. 563–579.

Bernard, G.M. 1995. The start-up of Cerro Colorado. SME Preprint No. 95–181. Littleton, CO: SME.

Bhappu, R.B., Johnson, P.H., and Brierley, J.A., and Reynolds, D.H. 1969.Theoretical and practical studies on dump leaching. *Trans. AIME* 244:307–320.

Borden, R.K., Peacey, V., and Vinton, B. 2006. Groundwater response to the end of forty years of copper heap leach operations, Bingham Canyon, Utah. Presented at the 7th International Conference on Acid Rock Drainage, St. Louis, MO, March 26–30. Lexington, KY: American Society of Mining and Reclamation.

Bouffard, S.C. 2005. Review of agglomeration practice and fundamentals in heap leaching. *Miner. Process. Extr. Metall. Rev*. 26:233–294.

Bouffard, S.C., and Dixon, D. 2000. Investigative study into the hydrodynamics of heap leach processes. *Metall. Mater. Trans. B* 32(5):763–776.

Breitenbach, A.J. 1994. Heap leach pad liner systems in North America. *Min. Eng*. 46(1):46–48.

Breitenbach, A.J. 2004. Improvement in slope stability performance of lined heap leach pads from operation to closure. *Geotech. Fabr. Rep. Mag*. 22(1):18–25.

Breitenbach, A.J. 2005. Heap leach pad design and construction practices in the 21st century. www.ore-max.com/pdfs/resources/heap_leach_pad_design_and_construction_practices_in_the_21st_century.pdf. Accessed September 2018.

Breitenbach, A.J., and Smith, M.E. 2007. Geomembrane raincoat liners in the mining heap leach industry. *Geosynthetics* 25(2):32–39.

Cathles, L.M., and Apps, J.A. 1975. A model of the dump leaching process that incorporates oxygen balance, heat balance, and air convection. *Metall. Mater. Trans. B* 6:617–624.

Cathles, L.M., and Schlitt, W.J. 1980. A model of the dump leaching process that incorporates oxygen balance, heat balance, and two dimensional air convection. In *Leaching and Recovering Copper from As-Mined Materials*. Edited by W.J. Schlitt. New York: SME-AIME.

Chamberlin, P.D. 1981. Heap leaching and pilot testing of gold and silver ores. *Min. Cong. J.* (April):47–52.

Chamberlin, P.D. 1986. Agglomeration for the heap leaching of gold and silver ores. SME Preprint No. 86–106. Littleton, CO: SME.

Chieregati, A.C., and Pitard, F.F. 2009. The challenge of sampling gold. Presented at the Fourth World Conference on Sampling and Blending, Cape Town, South Africa, October 19–23.

Cook, C., and Kappes, D. 2013. Conveying and stacking systems design for heap leach applications. In *Proceedings of the Heap Leach Solutions Conference 2013*. Vancouver, BC: InfoMine.

Decker, D.L., and Tyler, S.W. 1999. Hydrodynamics and solute transport in heap leach mining. In *Closure, Remediation and Management of Precious Metals Heap Leach Facilities*. Edited by D. Kosich and G. Miller. Reno, NV: University of Nevada Press.

Dixon, D.G., and Hendrix, J.L. 1993a. A general model for leaching one or more solid reactants from porous ore pellets. *Metall. Mater. Trans. B* 24(1):157–169.

Dixon, D.G., and Hendrix, J.L. 1993b. A mathematical model for heap leaching of one or more solid reactants from porous ore pellets. *Metall. Mater. Trans. B* 24(6):1087–1101.

Dixon, D.G., and Hendrix, J.L. 1993c. Theoretical basis for variable order assumption in the kinetics of leaching of discrete grains. *AIChE J.* 39(5):904–907.

Dixon, S.N., Shearer, M.N., and Steele, D.J. 2012. Comparison of spray and drip leaching systems. www.ore-max.com. Accessed June 2018.

Echeverria, N., Romo, D., and Suazo, I. 2015. Hydraulic performance of a permanent heap leach with an intermediate drainage system. In *Proceedings of the Heap Leach Solutions Conference 2015*. Vancouver, BC: InfoMine.

Efthymiou, M., Henderson, L., and Pyper, R. 1998. Practical aspects of the application of extract-ore 9560 for heap leach agglomeration at Nifty copper operation. Presented at the Alta Copper Hydrometallurgy Forum, Brisbane, Queensland.

Flynn, C.M., and Haslam, S.M. 1995. *Cyanide Chemistry—Precious Metals Processing and Waste Treatment*. Information Circular IC-9429. Washington, DC: U.S. Bureau of Mines.

Galea, W., Lupo, J., and Gutierrez, A. 2010. AMEC Minproc presentation of 20 July.

Gőkdere, V., Engin, A., and Suleyman, S.C. 2015. Gold heap leach testing using low-grade sulfide run-of-mine ores in a full-scale leach pad. In *Proceedings of the Heap Leach Solutions Conference 2015*. Vancouver, BC: InfoMine.

Heinen, H.J., and Porter, B. 1969. *Experimental Leaching of Gold from Mine Waste*. Information Circular IC-7250. Washington, DC: U.S. Bureau of Mines.

ICSG (International Copper Study Group). 2015. ICSG statistical database. www.icsg.org/index.php/external-database. Lisbon, Portugal.

INAP (International Network for Acid Prevention). 2009. *Global Acid Rock Drainage Guide*. www.gardguide.com. Accessed June 2018.

InfoMine. 2016. *Gold Heap Leach Cost Estimating Guide*. Vancouver, BC: InfoMine. http://costs.infomine.com/goldheapleachcosts/.

Jackson, J.S., and Ream, B.P. 1980. Solution management in dump leaching. In *Leaching and Recovering Copper from As-Mined Materials* Edited by W.J. Schlitt. New York: SME-AIME.

Jackson, J.S., Adamson, D.L., and Schlitt, W.J. 1979. *The Operating Characteristics of Small Sprinklers and Their Use in Dump Leaching*. Kennecott Copper Company Technical Report RTR 79–2, Project No. R6369. Bingham Canyon, UT: Kennecott Utah Copper.

Johnson, C.A. 2015. The fate of cyanide in leach wastes at gold mines: An environmental perspective. *Appl. Geochem.* 57:194–205.

Jones, A. 2000. Pulp agglomeration at Homestake Mining Company's Ruby Hill mine. Presented at the Randol Gold and Silver Forum 2000, Golden, CO.

Keller, J., Milczarek, M., Bilskie, J., Amponsah, I., and Zhan, G. 2015. Calibration and predictive performance of a two-domain empirical metal recovery model. In *Proceedings of the Heap Leach Solutions Conference 2015*. Vancouver, BC: InfoMine.

Lawry, A., Gelfi, P., Mitchell, I., Pyper, R., and Dunne, R. 1994. An overview of the Telfer 10 Mtpa dump leach operation. Presented at Recent Trends in Heap Leaching, Bendigo, Victoria, Australia.

Manning, T., and Kappes, D. 2016. Heap leaching of gold and silver ores. In *Gold Ore Processing*, 2nd ed. Vol. 15, Project Development and Operations. Edited by M.D. Adams. Amsterdam: Elsevier. pp. 413–428.

Marsden, J.O., and Botz, M.M. 2017. Heap leach modelling: A review of approaches to metal production forecasting. *Miner. Metall. Process*. 34(2):53–64.

Marsden, J.O., and House, C.I. 2006. *The Chemistry of Gold Extraction*. Littleton, CO: SME.

McBride, D., Gebhardt, J.E., Croft, T.N., and Cross, M. 2015. Capturing the hydrodynamics of heap leaching in freezing conditions. Presented at Computational Modelling '15, Minerals Engineering International, Falmouth, UK.

McNab, B. 2006. Exploring HPGR technology for heap leaching of fresh rock gold ores. Presented at the IIR Crushing and Grinding Conference, Townsville, Queensland, Australia, March 29–30.

Merritt, R.C. 1971. *The Extractive Metallurgy of Uranium*. Golden, CO: Colorado School of Mines Research Institute.

Miller, G. 2005. Agglomeration drum selection and design process. In *Alta 2005 Copper Conference*. Melbourne, Victoria: Alta Metallurgical Services.

Miller, G.C., Hoonhout, C., Miller, W.W., and Miller, M.M. 1999. Geochemistry of closed heaps: A rationale for drainage water quality. In *Closure, Remediation and Management of Precious Metals Heap Leach Facilities*. Edited by D. Kosich and G. Miller. Reno, NV: Center for Environmental Sciences and Engineering, University of Nevada.

O'Kane, M., Barbour, S.L., and Haug, M.D. 1999. A framework for improving the ability to understand and predict the performance of heap leach piles. In *Proceedings of the Copper 99—Cobre 99 International Conference*. Vol. 4, Hydrometallurgy of Copper. Edited by S.K. Young, D.B. Dreisinger, R.P. Hackl, and D.G. Dixon. Warrendale, PA: The Minerals, Metals & Materials Society. pp. 409–419.

Olson Hoal, K., Woodhead, J., and Smith, K.S. 2013. The importance of mineralogical input into geometallurgy programs. In *The Second AusIMM International Geometallurgy Conference: GeoMet 2013*. Melbourne, Victoria: Australasian Institute of Mining and Metallurgy. pp. 17–26.

Orr, S. 2002. Enhanced heap leaching—I. Insights. *Min. Eng.* (September):49–56.

Parshley, J.V., Willow, M.A., and Bowell, R.J. 2012. The evolution of cyanide heap leach closure methods. In *Mine Closure 2012*. Edited by A.B. Fourie and M. Tibbett. Perth: Australian Centre for Geomechanics.

Pyper, R.A., Kappes, D.W., and Albert, T. 2015. Evaluation of agglomerates using the Kappes percolation test. In *Proceedings of the Heap Leach Solutions Conference 2015*. Vancouver, BC: InfoMine.

Robertson, S., Guzman, A., and Miller, G. 2013. Implications of hydrodynamic testing for heap leaching design. In *Proceedings of the 5th International Seminar on Process Hydrometallurgy: Hydroprocess 2013*. Edited by F. Valenzuela and C. Young. Santiago, Chile: Gencamin.

Roman, R.J., Benner, B.R., and Becker, G.W. 1974. Diffusional model for heap leaching and its application to scale-up. *Trans. SME* 256:247–252.

Rood, E.A. 2000. A comparison of column testing techniques. In *Randol Copper Hydromet Roundtable 2000: Conference and Exhibition*. Golden, CO: Randol International.

Rose, W.R., Pyper, R.A., Eguívar, J., and Mirabal, C. 1990. Heap leaching with fine grinding and agglomeration at Potosí, Bolivia. In *Advances in Gold and Silver Processing: Proceedings of the Symposium at GoldTECH 4*. Edited by M.C. Fuerstenau and J.L. Hendrix. Littleton, CO: SME.

Rykaart, M., and Caldwell, J. 2006. Covers—State of the art review [online course]. www.infomine.com.

Saari, P., and Riekkola-Vanhanent, M. 2012. Talvivaara bioheapleaching process. *J. S. Afr. Inst. Min. Metall.* 112(12):1013–1020.

Schlitt, W.J. 1984. The role of solution management in heap and dump leaching. In *Au and Ag Heap and Dump Leaching Practice*. Edited by J.B. Hiskey. Littleton, CO: SME-AIME. pp. 69–83.

Seal, T. 2007. Hydro-Jex: Heap leach pad stimulation technology; ready for worldwide industrial adoption? SME Preprint No. 07-123. Littleton, CO: SME.

Seal, T. 2011. Hydro-Jex operations at Anglogold Ashanti's Cripple Creek and Victor Gold Mine. SME Preprint No. 11–093. Littleton, CO: SME.

Seal, T. 2015. Understanding the leaching efficiency of the Hydro-Jex technology. In *Proceedings of the Heap Leach Solutions 2015*. Vancouver, BC: InfoMine.

Seal, T., and Kiley, T. 2015. Drying heaps, dumps and piles for closure. In *Proceedings of the Heap Leach Solutions 2015*. Vancouver, BC: InfoMine.

Smith, M. 2002. Potential problems in dump leaching. *Min. Mag.* (July).

Stebbins, S. 2015. Estimating the costs of geomorphic reclamation practices upon closure of spent heaps and waste stockpiles. Presented at Heap Leach Solutions 2015, Reno, NV, September 12–16.

Sullivan, J.D. 1931. *Factors Involved in the Heap Leaching of Copper Ores*. Information Circular IC-6425. Washington, DC: U.S. Bureau of Mines.

Taylor, A. 2017. Heap leaching and its application to copper, gold, uranium and nickel ores. In *Alta Short Course*. Melbourne, Victoria: Alta Metallurgical Services.

Thiel, R., and Smith, M.E. 2004. State of the practice review of heap leach pad design issues. *Geotext. Geomembr.* 22(6):555–568.

Uhrie, J.L. 1998. Use of electrical resistivity geophysics in the stockpile comparison of raffinate application techniques: Drip emitters versus wobblers. In *Fourth Annual Randol Copper Hydromet Roundtable '98: Conference and Exhibition*. Golden, CO: Randol International.

Van Genuchten, M.T. 1980. A closed-form equation for predicting the hydraulic conductivity of unsaturated soils. *Soil Sci. Soc. Am. J.* 44(2):892–898.

Velarde, G. 2003. The use of electrical conductivity in agglomeration and leaching. In *Copper 2003–Cobre 2003*. Vol. 6, Hydrometallurgy of Copper (Book 2). Edited by P.A. Riveros, D. Dixon, D.B. Dreisinger, and J. Menacho. Montreal, QC: Canadian Institute of Mining, Metallurgy and Petroleum. pp. 563–579.

Vethosodsakda, T., Free, M.L., Janwong, A., and Moats, M.S. 2013. Evaluation of liquid retention capacity measurements as a tool for estimating optimal ore agglomeration moisture content. *Int. J. Miner. Process.* 119:58.

Von Michaelis, H. 2009. How energy efficient is HPGR? Presented at World Gold Conference 2009, Johannesburg, South Africa.

Zaebst, R.J. 1994. Pump agglomeration at the Castle Mountain Mine. Presented at the 100th Annual Northwest Mining Association Convention, Spokane, WA.

CHAPTER 10.4

Vat Leaching

Adam Johnston

BASIC CONCEPT

In vat-leaching processes, crushed and sometimes agglomerated ore is piled in a vessel. Leach solution is flooded into the vessel and usually irrigated upward through the ore charge. The emerging pregnant leach solution (PLS) is collected and sent for metal recovery, while the depleted ore is disposed of in residue dumps (Gray 1975; Knobler and Werner 1962; LaChapelle and Dyas 1993; Tapia and Kelley 1998).

A typical flow arrangement is shown in Figure 1 and a typical equipment arrangement is shown in Figure 2.

VAT-LEACHING FLOW SHEET

Ore is usually crushed to –6 to 12 mm. In some cases, the ore is deslimed or agglomerated. Reagents such as acid, lime, or cyanide may be added during the vat loading process. The vats are loaded with ore. Leaching solution is introduced from below and completely submerges the ore. Solution continues to pass through the vat charge for 24 to 168 hours. The solution is then drained from the vat, the waste is removed, and the vat is loaded with a new charge of ore to begin a new cycle. The ore loading is done in short lifts to avoid the concentration of fines or coarse material in one area of the vat, which may make the flow of leach solution uneven.

Because the vats are completely flooded shortly after the ore is loaded, agglomeration cure times are very short. Hard, discrete agglomerates that are formed for heap leaching are not typically achieved in a vat leach. Because the vat solution irrigation is upflow, the strength and durability of agglomerates in a vat are not as important as they are in a heap leach.

Solution flow rates are typically between 300 and 1,000 L/m^2/h. The solution may pass through vats several times in a recirculation, countercurrent, or intermediate-leach-solution (ILS) flow arrangement. Vat-leaching-solution flow rates are designed to control one or more of the following:

- **Reagents concentration.** If the ore consumes reagent quickly, then a faster solution flow rate is required.
- **PLS tenor.** If downstream processes require a certain tenor of PLS, then the flow rate through each vat may be manipulated to achieve it.
- **PLS turbidity.** If there are a lot of fines, then flow rates may be low to reduce entrainment of fines in the PLS.

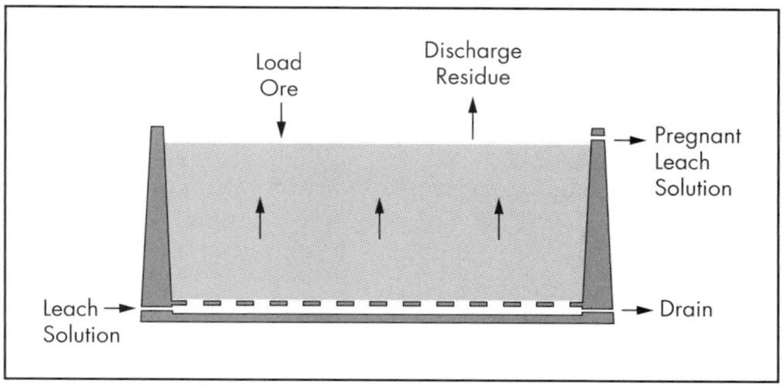

Figure 1 Typical solution flow arrangement

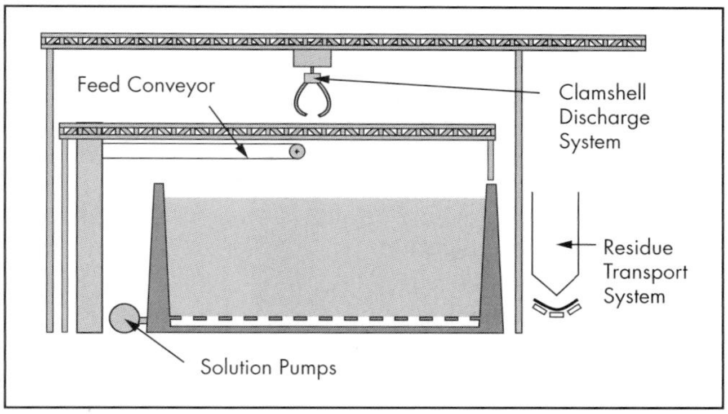

Figure 2 Typical equipment arrangement

- **Reagent penetration.** If reagents are consumed quickly, then a higher flow rate may be selected to achieve similar leaching conditions throughout the vat.

Carbon dioxide gas is generated in acid leaching of ores that contain a significant amount of carbonate minerals. This gas forms as pockets in the ore bed, which grow in size until they achieve enough buoyancy to break through the ore bed and form a channel to the surface. These channels become the paths of least resistance to the leach solutions and become progressively larger as the solution flow increases in the channel

Adam Johnston, Chief Metallurgist, Transmin Metallurgical Consultants, Lima, Peru and washes out the finer material. A typical vat-leaching circuit is shown in Figure 3.

On a larger scale, vats lost favor in the 1980s when activated carbon and solvent extraction were developed to recover gold and copper, respectively, from low-tenor PLSs. Since then, there have not been any major new vat-leaching facilities developed.

Although a heap leach solution typically approaches chemical equilibrium with the ore that it is in contact with, a leach solution passes through a vat quickly, so the entire height of a vat is in contact with a similar leaching solution.

Vat leaching has some technical advantages over heap leaching (Coburn et al. 1976), which may lead operations to select this technology (Cope 1991):

- Faster leaching times
  - Lower ore and metal inventory
  - Lower reagent consumption
- More homogeneous contact between ore and leaching solutions
- Less area required
- Less personnel required for operations
- Upflow irrigation, which is less prone to channeling or blockages
- Lower surface area affected by evaporation or rainfall
- Ability to have several stages of countercurrent flow between the ore and the leaching solution, which facilitates
  - The ability to generate higher PLS,
  - Less solution pumping, and
  - Better washing efficiency for lower soluble metal losses.

However, there are also several disadvantages that must be evaluated:

- Higher capital cost of installation
- Less flexibility for leaching times
- Typically a finer crush size required
- Unstable PLS tenor
- Possible undesirable chemical reactions
- Low leach time, which negates slower leaching mechanisms, such as bioleaching
- Anaerobic conditions, because it is a flooded system and air cannot be added

Vat leaching is typically slower and achieves lower recovery compared to agitated tank leaching, but it has the advantage of lower comminution costs, and the residue can be stacked rather than requiring a tailings dam (Murdoch 1990).

WASTE REMOVAL

The vats are drained by gravity before the solids are removed. The drain solution can have some suspended solids entrained and should be throttled to avoid high initial flow rates.

Overhead clamshell reclaimers, operated by an electric motor on a traveling crane arrangement have been the most successful approach to large vat unloading. The equipment is easy to maintain and would be easy to automate.

The following alternatives have been attempted:

- **Backhoe or mechanical excavator.** This type of machine has limited reach, requiring that the vat be narrow.
- **Front-end loader.** This requires that one of the walls of the vat be dismountable or that there is a ramp to drive the equipment in. This arrangement can be problematic because of operations and maintenance issues. The doors themselves can leak and stress, and the walls and subfloor irrigation can be damaged by the front-end loader.
- **Hydraulic removal by fluidizing the charge.** This has the advantage of allowing for a semicontinuous leaching process (Mackie and Trask 2009).

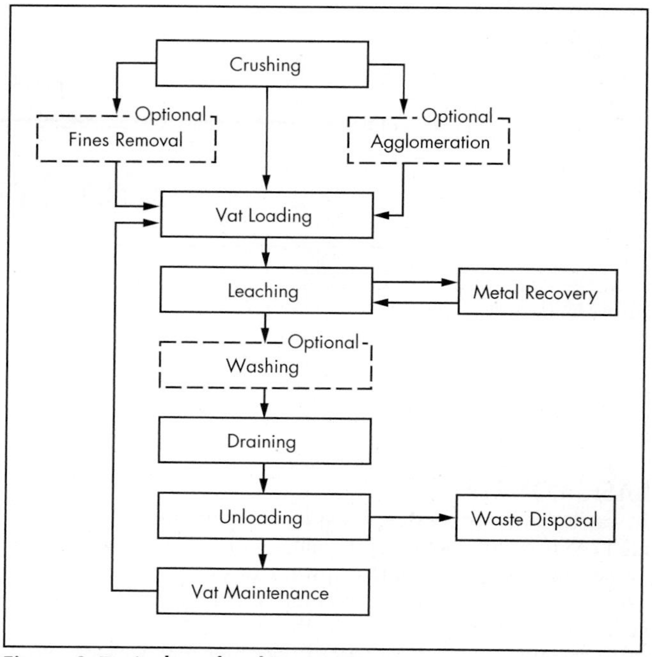

Figure 3 Typical vat-leaching circuit

VAT CONSTRUCTION

Circular steel vats have been used, but larger plants have used concrete rectangular boxes, reducing construction costs and allowing for bulk unloading methods.

If acid proofing the concrete is required, then high-density polyethylene sheeting, polymer panels, asphalt, and sulfur concrete have been used. The height of the vats is usually dependent on the optimization of the reinforced concrete structures. All of the major concrete vat facilities that have been built have had a live leaching height of approximately 6–7 m. The width of the vat is dependent on the ore loading and discharging technologies to be used. If a tripping conveyor and clamshell arrangement is to be used, then the vats would typically be 25–35 m wide. Vats may be either joined together in rows or stand separately. The inner walls have to be able to withstand a charge on one side but not the other. False floors are installed into the vat that must

- Provide structural support to the charge above,
- Allow for even irrigation of the leach solution,
- Be resistant to the unloading equipment, and
- Allow for drainage of the void solution before the vat is unloaded.

In some cases, wooden sleepers have been used to build the floor, then a polymer or canvas mesh has been installed over the top layer of sleepers, supported by a wooden lattice. A further layer of sleepers can be installed over the mesh to provide further support of the charge as well as protection from the discharge equipment.

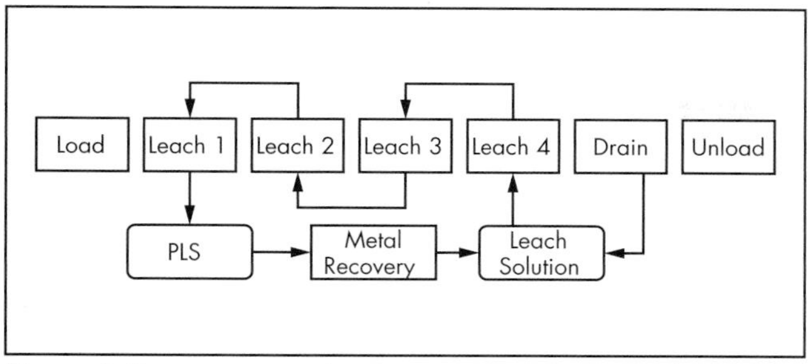

Figure 4 Countercurrent leach arrangement

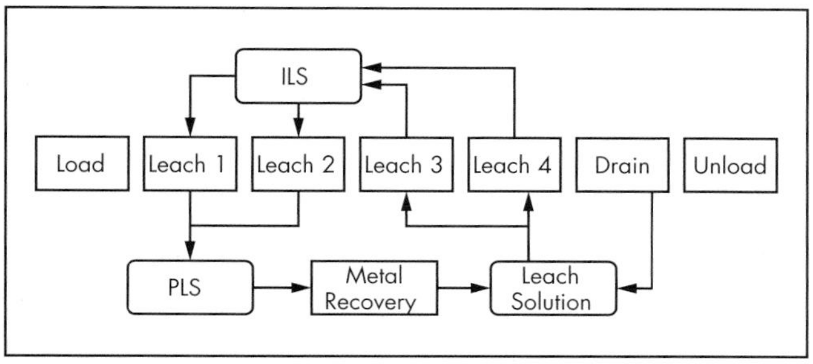

Figure 5 Common intermediate-leach-solution arrangement

Other alternatives such as formed concrete blocks with solution channels have been used but are not favored because of their high cost. In acid environments, they have been made from sulfur or polymer concrete to protect from the acid.

In gold operations, oxygen may be added to the incoming leach solution at this point as it is under pressure. The vat walls must also be protected from acid solutions. Solution pumps can be arranged at either the top or bottom of the vat.

SOLUTION MANAGEMENT

Existing vat operations have only used one or two ILSs, though the plants were designed for complete countercurrent operation. Countercurrent operation allows for more predictable performance with less operator interaction but produces a very high-grade PLS and does not allow for flexibility on cycle times and conditions for each vat. Figure 4 shows a countercurrent leach arrangement.

In some instances, solution has been recirculated in the same vat to continue leaching while achieving a higher PLS tenor. Figure 5 shows a common ILS arrangement. The first PLS to emerge from the vats will have high concentrations of suspended solids. Typically, this would be treated in a clarifier before recovering the metal from the solution.

TEST WORK

Conventional bottle-roll leaching tests do not typically give a good representation of the vat-leaching performance. In a bottle roll, there is attrition of the particles, breaking passivation layers of precipitates. Constant reagent concentrations are hard to control in a bottle-roll test, especially if consumption is high.

It is recommended that the test work be done with upflow column leaches, where the solution is either recirculated or has a single pass, depending on requirements for reagent control. The parameters that should be tested are

- Particle size, including crush size and deslimed sands;
- Solution flow rate;
- Reagent concentration; and
- Agglomeration and cure.

The tests should be controlled for temperature because the thermodynamics of a small test may be significantly different to an industrial-sized vat. The leach solution should be prepared from either a mature process solution or a synthetic solution that has the same chemical composition as the projected process solution. Some tests should be done at the full vat height and final projected conditions to ensure that the performance is stable and efficient throughout the entire height.

No information is available about scaling up recovery or leaching kinetics between batch laboratory tests and industrial tests. Empirical observation suggests that well-run batch tests in the laboratory will give similar extraction rates to an industrial vat.

ENVIRONMENTAL CONSIDERATIONS

The residue that is discharged from the vats may or may not have been washed with water. Typical industrial applications do not wash with water so that the solution that is lost in the residue acts as a bleed for unwanted species that accumulate in the process solution, such as iron, chlorides, or copper. Consequently, the residue that is discharged will have residual metals in solution, reagents, and will be saturated with solution that will continue to drain. Therefore, containment or treatment of the residues should be considered to avoid this drainage from contaminating ground and water. The residue also contains fine solids that could cause dusting and acid mine drainage.

REFERENCES

Coburn, J.L., Keane, J.M., Chase, C.K., and Bhappu, R.R. 1976. Analysis of submerged vat leaching versus trickle leaching. SME-AIME Preprint No. 75-AS-348. New York: SME-AIME.

Cope, L.W. 1991. Vat leaching—Historic gold processing technique again viable. *Min. Eng.* 43:1131–1132.

Gray, R.E. 1975. Vat leaching practices at Anaconda's Yerington mine operation. SME-AIME Preprint No. 75-AS-306. New York: SME-AIME.

Knobler, R.R., and Werner, J. 1962. The Mantos Blancos operation: Chile's new integrated copper producer. *Min. Eng.* 14:40–45.

LaChapelle, G.W., and Dyas, J.S. 1993. Cyprus Miami property overview and mine modernization. *Min. Eng.* 45(4):344–346.

Mackie, D., and Trask, F. 2009. Continuous vat leaching—First copper pilot trials. In *Proceedings of ALTA Copper 2009*. Melbourne, Victoria: ALTA Metallurgical Services.

Murdoch, R.B. 1990. Small scale vat leaching. In *Pacific Rim Congress 90: An International Congress on the Geology, Structure, Mineralisation, Economics and Feasibility of Mining Development in the Pacific Rim, Including Feasibility Studies of Mines in Remote, Island, Rugged and High Rainfall Locations*. Melbourne, Victoria: Australasian Institute of Mining and Metallurgy.

Tapia, G., and Kelley, R.J. 1998. Leaching practices at Mantos Blancos. In *Latin American Perspectives: Exploration, Mining, and Processing*. Edited by O.A. Bascur. Littleton, CO: SME. pp. 249–257.

CHAPTER 10.5

Tank Leaching

Marko Latva-Kokko

Tank leaching lies at the core of many hydrometallurgical processes. Leaching is usually performed in mechanically agitated tanks, but there are also alternatives such as pneumatic agitation and jet mixing. Generally, all of a hydrometallurgical plant's revenue is derived from its reactors, and thus even marginal underperformance matters to a plant's economy. The initial investment usually has a low impact on the lifetime costs of typical reactors. Instead, the process result and availability of the equipment make the difference. The hydrometallurgical reactor is not just a mixing tank, and a complete approach requires detailed knowledge of both the process and the equipment.

There is a great variation in tank leaching applications with very different process conditions. Temperatures range from room temperature to over 100°C. Two or three phases can be present, with all kinds of leaching agents and various gases. The required retention time in tank leaching extends from a few minutes to several days depending on the application. This sets some special requirements for reactor design. The process environment can be highly corrosive and abrasive, simultaneous solids suspension and gas dispersion is common, and scaling can be intensive. The most feasible reactor design is always application- and case-specific. There are significant case-by-case variations in the industry's raw materials, processes, and construction material requirements. The cornerstone of successful reactor design is the identification of case-specific critical agitation duties to overcome the factors limiting the reaction rate. In addition, the overall plant unit design, which fulfills the operation, safety, availability, and maintenance aspects, is equally important.

In many hydrometallurgical applications, declining ore grades have forced the use of larger throughputs, leading to higher solution and slurry flows and larger equipment. Larger equipment further emphasizes the importance of accurate reactor specifications to avoid excessive costs through over-dimensioning. With proper reactor design, significant improvement in process performance in terms of energy consumption, gas utilization efficiency, and equipment lifetime can be achieved (Latva-Kokko et al. 2014).

AGITATED TANK LEACHING

The vast majority of industrial leaching operations are performed in mechanically agitated reactors. These devices can be very different in terms of size and construction depending on the application. The smallest reactors are only some cubic meters in volume whereas the largest range up to a few thousand cubic meters. Figure 1 illustrates the different components of a mechanically agitated leaching reactor.

The cylindrical tank with a flat bottom is by far the most commonly used tank design in hydrometallurgy. This is due to its superior mechanical strength against hydrostatic pressure inside the tank that enables minimum wall thickness and use of construction material. In smaller size categories, square and rectangular tanks can also be found. Other tank bottom designs include the dished and cone bottom. In addition, conical fillets that mimic the shape of the dished bottom are being used. These bottom types ease drainage of the tank and can decrease the mixing energy required for solids suspension (Taca and Paunescu 2000). On the other hand, they increase the price of the reactor and decrease the effective slurry volume.

A common height to tank diameter (H/T) ratio is from 1 to 1.5, but much higher ratios can also be used, especially if there is a need to minimize the leaching plant surface area. A high slurry height also generates higher hydrostatic pressure at the bottom of the tank. This improves the leaching kinetics in applications using gas by increasing the solubility of gas in solution, increasing the content of effective molecules in dispersed gas bubbles, and decreasing their maximum stable bubble size (Latva-Kokko and Riihimäki 2012). At H/T ratios of up to 1.2, a single impeller is usually enough to keep the contents well mixed; with higher ratios, two or more impellers are required.

The purpose of baffle plates in leaching tanks is to prevent swirling and vortex formation. They also transform tangential flows into vertical flows, providing top-to-bottom mixing. Baffling also has an effect on the power draw of the impeller. The industrial standard for a wall baffle configuration consists of three or four vertical plates with a width of T/12 or T/10. Plates typically have T/60 spacing to the tank

Marko Latva-Kokko, R&D Manager—Hydromet Equipment, Outotec, Pori, Finland

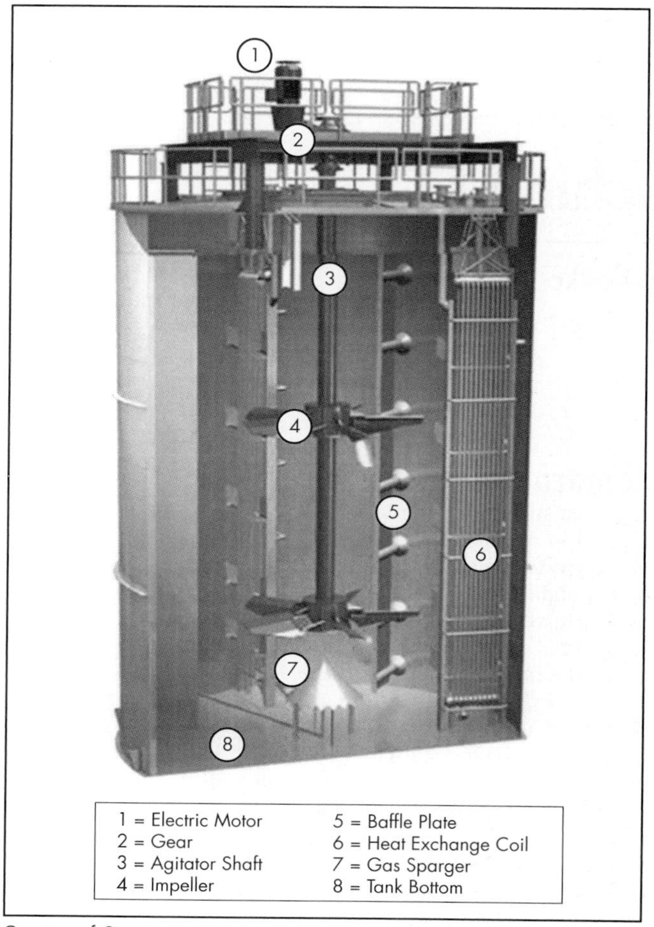

Courtesy of Outotec

Figure 1 Mechanically agitated leaching tank with its components

wall to minimize the dead zones, especially in solid–liquid systems (Hemrajani and Tatterson 2004).

Impellers

The agitator is naturally the most important part of a mechanically agitated tank. A vast variety of different impellers with specific characteristics are available. The selection of an impeller should be made based on the case-specific critical agitation function to optimize reactor performance for any given application. Generally, impellers can be divided into different categories based on their flow pattern, pumping, and shear capabilities. Two main flow patterns, axial and radial, are illustrated in Figure 2. Axial flow impellers have a single up-and-down flow pattern, whereas the radial flow impeller produces two circulating loops, one below and one above the impeller.

Conventional impeller types, such as the propeller, Rushton disc turbine (RDT), and pitched blade turbine (PBT), are still common, but more efficient impellers are available that are specially designed for hydrometallurgical applications. The impellers shown in Figure 3 are categorized according to their flow pattern, pumping, and shear properties. The impeller diameter (D) is typically 0.3 to 0.5 times the tank diameter and the impeller is located less than one D distance from the bottom. If multiple impellers are used, the distance between impellers is usually kept higher than one D.

Agitator Power Draw

The amount of power drawn by an agitator relies on several parameters and is given by:

$$P = N_p \rho N^3 D^5 \qquad \textbf{(EQ 1)}$$

where

P = power
N_p = power number
ρ = mixture density
N = impeller speed
D = impeller diameter

Courtesy of Outotec

Figure 2 Axial and radial flow patterns

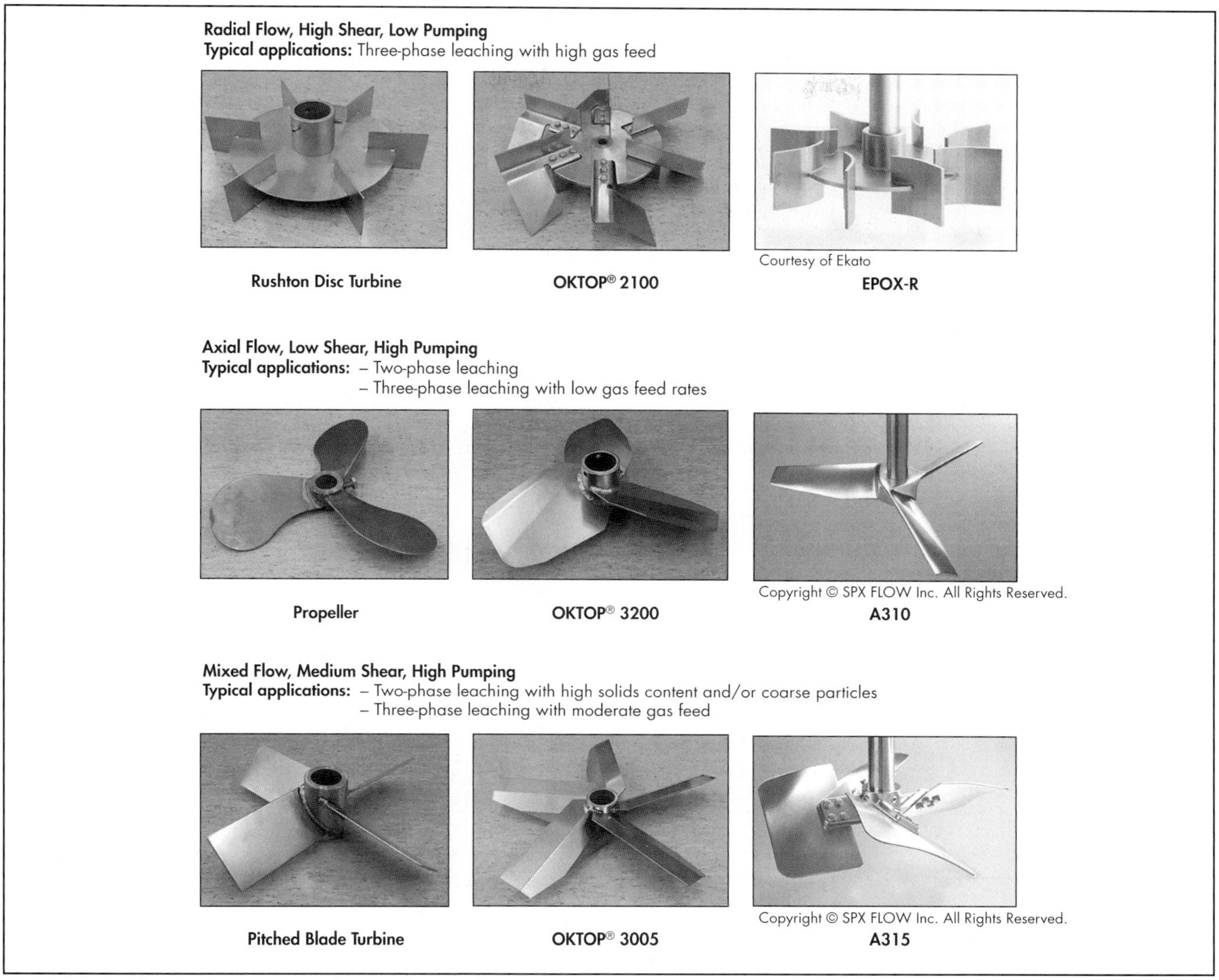

Figure 3 Impeller types categorized according to their properties

The power number depends on several factors including Reynolds number and impeller characteristics. This formula serves to demonstrate the impact impeller speed and, more particularly, impeller diameter have on an agitator's power draw.

Two-Phase Leaching

In several applications in hydrometallurgy, solids suspension is the main duty of agitation. Leaching processes that do not require gas feed typically belong to this category. Here the leaching rate is often controlled by the chemical reaction rate. In these applications, the process performance of the equipment can be improved by increasing the homogeneity of the suspension. Uniform suspension utilizes the equipment volume fully, providing a longer retention time for solids in a continuous process. Good overall mixing decreases the reagent gradients inside the slurry, which enhances leaching. If the process is known to induce scaling, mixing should be designed to generate higher flow velocities than those required for suspension of the solids alone.

Maximizing the degree of solids suspension with minimal energy consumption has been widely studied (Hosseini et al. 2010; Tahvildarian et al. 2011). The factors that have the largest influence on the required mixing power are the particle size distribution, specific density of solids, and solids content. A typical mixer arrangement in two-phase leaching applications is a downward pumping hydrofoil impeller, which generates a strong axial flow that can carry solid particles close to the surface. Figure 4 shows that off-bottom and uniform solids suspension can be systematically achieved with lower energy consumption by using a wide-blade hydrofoil impeller than by using a standard 45° pitched, four-bladed turbine. This is in good agreement with Wu et al.'s (2006) study on energy efficiency in axial flow impellers, which concludes that PBTs are approximately 7% less efficient than commercial hydrofoil impellers. In addition to impeller type, tank configurations, including baffle plates and the shape of the bottom, have an impact on solids suspension.

The degree of solids suspension has a major effect on the required mixing power, as shown in Figure 4. It is much easier to achieve a partial suspension where all the solids particles are moving than a homogeneous suspension. Liquid to solid mass transfer does not greatly improve as a function of

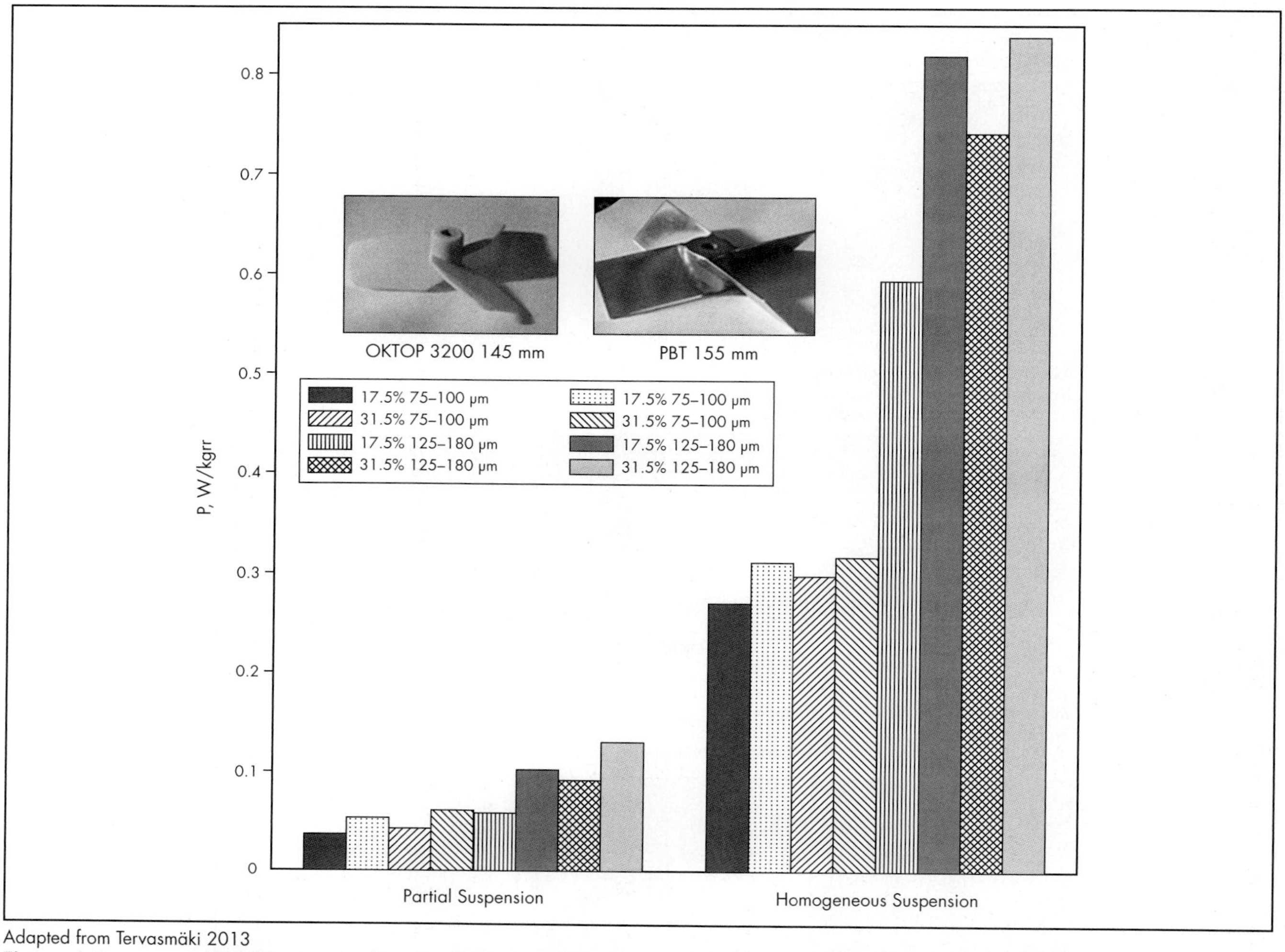

Adapted from Tervasmäki 2013

Figure 4 Power consumption per total mass of slurry required to achieve partial and uniform suspension with different impellers, solid concentrations (wt %), and particle size fractions. Tests were made with silica sand–water slurry in a flat-bottomed tank with a diameter of 362 mm.

homogeneity after complete off-bottom suspension. However, leaching reactors should always be designed for uniform suspension if they are operated in continuous mode. This is because the retention time of the solid particles can be as much as 40% lower in a reactor series that is being operated in a state of partial suspension compared to uniform suspension.

Three-Phase Leaching

Simultaneous solids suspension and gas dispersion commonly occur in the hydrometallurgical industry. Because of poor oxygen solubility and high oxygen demand, several hydrometallurgical operations are controlled by the rate of oxygen transfer from the gas to the aqueous phase. Among all the factors that affect the rate of oxygen mass transfer, the most important is the reactor configuration and geometry (Filippou et al. 2000). Ideal impellers for gas dispersion are considered to be those that induce radial flow and high shear, such as RDT. However, the presence of gas affects the performance of the impeller and its ability to suspend solids. Thus, three-phase leaching places some special demands on agitator design.

The volumetric mass transfer coefficient, k_La, can be used as a quantitative measure of the mass transfer capability of a reactor. The general equation for the estimation of k_La that is frequently found in the literature follows. It shows that k_La is a function of mixing power intensity (P/V) and superficial gas velocity, v_s. The letters A, B, and C refer to reactor- and application-specific coefficients.

$$k_La = A\,(P/V)^B\,v_s{}^C \qquad \textbf{(EQ 2)}$$

Impeller performance in a three-phase system is illustrated in Figure 5 with a silica sand, water, and air system. The sand particle size range was between 50 and 200 µm. The solids concentration in these tests was kept at 400 g/L, and the amount of air feed was varied between 0 and 26.5 Nm^3/h. The impeller rotation speed required for complete off-bottom suspension of all solid particles was determined visually.

The selection of an impeller for three-phase leaching applications strongly depends on the type and amount of the gas. If the gas feed rate per reactor volume is low, a downward pumping mixed-flow impeller, such as PBT or OKTOP

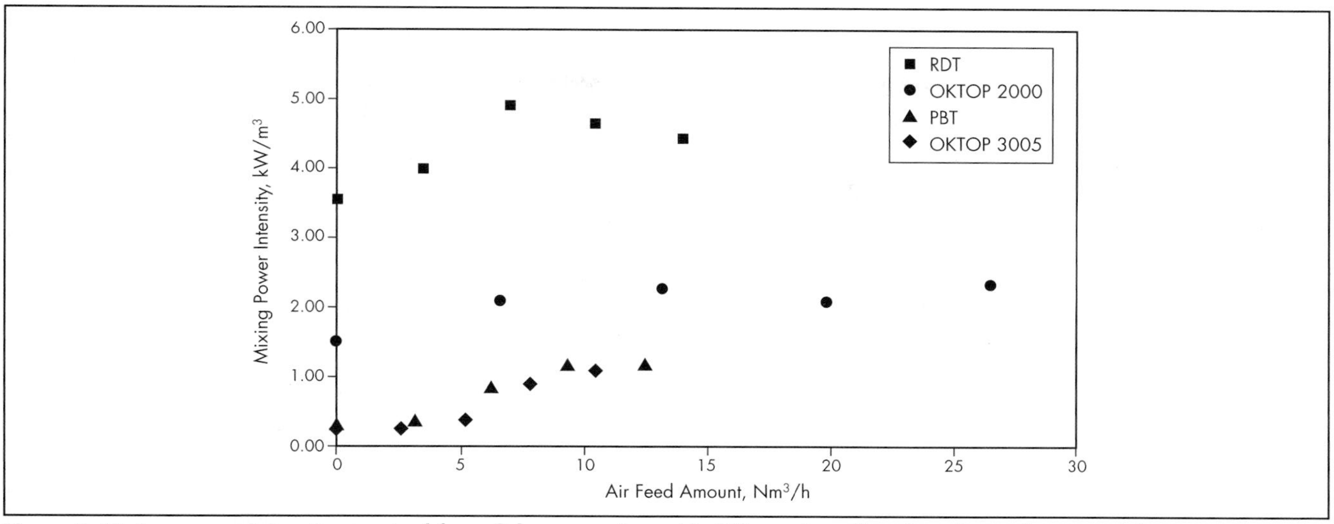

Figure 5 Mixing power intensity required for solids suspension with different impellers in a three-phase model system (400 g/L silica sand 50–200 μm)

3005, is a good choice since it can suspend solids with the lowest mixing power. Mixed-flow impellers offer reasonable pumping efficiency, similar to that of an axial flow impeller, combined with higher shear rates akin to a radial flow impeller. Thus good solids suspension can be achieved while providing reasonable levels of gas dispersion. However, gas feed strongly decreases the PBT's solids suspension performance, making it applicable only with low gas feed rates. These impellers might also be feasible when the gas used is inexpensive, such as air, and it is simply enough to avoid flooding of the impeller. Flooding of the impeller occurs when excessive amounts of gas are handled by the impeller; at this point gas dispersion drops drastically and small bubbles are no longer generated by the shearing action of the impeller.

With large industrial reactors, the speed of the impeller tips is usually the limiting factor, because of erosive wear. Thus, a high power number impeller is required to achieve decent gas dispersion and good gas utilization efficiency at high gas feed rates. As shown in Figure 5, impellers that have a pure radial flow pattern, such as the RDT, require significantly high mixing power to suspend all the particles in the presence of gas. The impeller tip speeds at industrial scale would be extremely high. The OKTOP 2000, an impeller that is specially designed for three-phase systems, can suspend all the solids with a tolerable impeller tip speed at an air feed rate twice as high as any other impeller in this comparison.

Instrumentation and Automation

There are several ways to improve the performance and process control of tank leaching by means of instrumentation and automation. Common instrumentation includes temperature and pH measurements and automated sampling with various metal analyses. It is also possible to monitor and control the solids suspension degree inside a slurry tank. The system includes two integrated measurement probes that are used to monitor the cloud height of the slurry and the thickness of the settled layer of solids at the bottom of the tank. Measurement is based on electrical resistance tomography (Latva-Kokko et al. 2016). For automated control, the mixer motor must be equipped with a variable-speed drive. Because of the rapid development of measurement technologies and data processing capabilities, this aspect is set to become increasingly more important in tank leaching.

OTHER TANK LEACHING SYSTEMS

The vast majority of hydrometallurgical leaching tanks are equipped with top-entry agitators, but there are also other alternatives, some of which are described briefly in this chapter.

Pachuca Tanks

The Pachuca tank is a large cylindrical vessel with a steep conical bottom. Pachuca tanks typically are 4–10 m in diameter and up to 15 m high. They may also include a central draft tube that can either be short or reach almost to the surface of the tank. Because there is no agitator, to agitate the slurry in the tank, air is injected at the base of the bottom cone as in airlift reactors. Air bubbles carry slurry particles from the bottom of the tank to the surface, as illustrated in Figure 6. From the surface, the slurry flows back through the annular space between the draft tube and tank wall (Mehrotra and Shekhar 2000).

Pachuca tanks have been used in various applications in the minerals industry, such as in the alumina industry and leaching of gold. However, they have been widely displaced by mechanically agitated reactors or draft tube circulators. The main disadvantage of the Pachuca tank is the large quantity of high-pressure air that is required for the agitation. With some ore types there can also be problems with the building up and peeling off of solid masses from the tank walls. The height of the Pachuca tank is another disadvantage and usually necessitates pumping of the pulp. Therefore, the overall energy consumption of a Pachuca tank train is much higher than of mechanically agitated reactors.

Draft Tube Reactors

The draft tube circulator is a reactor configuration where mechanical agitation is enhanced with a draft tube. These reactors typically have a high height-to-diameter ratio of 2:1 or more. The diameter of the draft tube usually ranges from 0.2 to 0.4 times the tank diameter. The impeller is located at

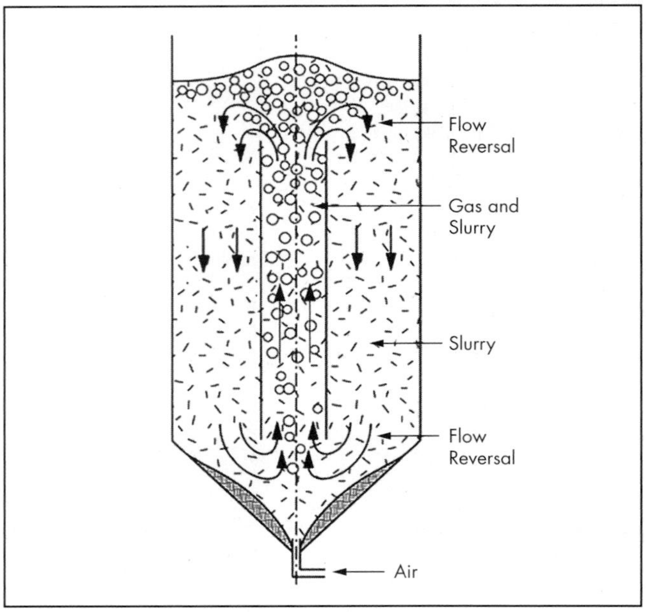

Source: Roy et al. 1998
Figure 6 Operating principle of Pachuca tank

least partly inside the draft tube. The most common design is a top-entry agitator with a downward pumping impeller located at the top of the draft tube. Here the slurry flow is down inside the draft tube and up in the reactor annulus. The flow velocity has to be high enough to avoid sedimentation and lift particles up to the surface. In general, with a draft tube, solids suspension can be achieved with lower mixing power per volume input than in a basic tank. The major limitation of most draft tube circulators is their gas feed capacity, and typically they are not suitable for leaching applications that require large amounts of gas. In addition, the resuspension of settled solids after shutdown can be challenging.

A draft tube circulator that is specially designed for three-phase leaching is shown in Figure 7. Here a bottom-entry agitator is used and the impeller is located at the bottom of the draft tube. The impeller disperses the gas that is introduced below it and induces recirculation through the draft tube so that the direction of flow is downward inside the tube and upward outside of it. These reactors are used for the atmospheric leaching of zinc concentrate. They have a high gas feed capacity and good oxygen utilization efficiency because of the high hydrostatic pressure and local mixing power intensity at the bottom part of the reactor, where most of the gas-to-liquid mass transfer takes place.

Jet Mixing

Mixing of solids requires the input of energy to achieve suspension and, in the case of three-phase leaching, the dispersion of gas. This energy can be implemented mechanically with a rotating impeller or pneumatically as in Pachuca tanks. The third option is to generate a high-velocity jet of fluid with a pump. These jet mixers are not frequently used in leaching applications, but they are common in large storage tanks, where the contents have to be homogenized but the mixing time is not an issue (Grenville and Nienow 2004). Jet mixers do not contain any moving parts inside the tank. They are driven by pumps located on the ground next to the tank. There

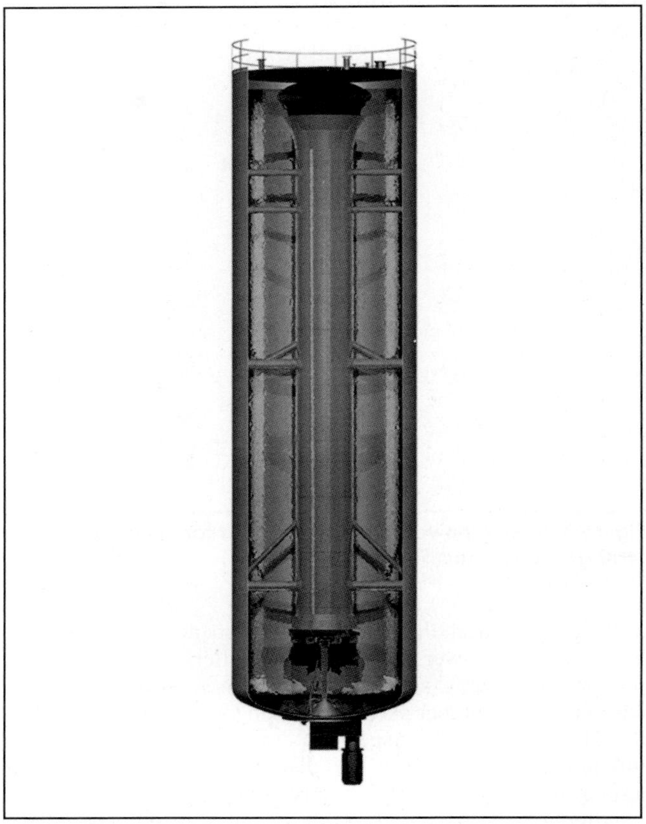

Courtesy of Outotec
Figure 7 Draft tube reactor with bottom-entry agitator

can be one or several jets in one vessel, and their orientation can be different. Usually, the jet enters from the side of the tank close to the bottom and is directed toward the opposite top corner, as illustrated in Figure 8. However, for leaching applications, a centrally located downward pushing axial jet is the most feasible choice since it suspends solids with the lowest amount of energy.

REACTOR DESIGN

Certain reactor-design-related issues repeat themselves in the hydrometallurgical industry. The faulty selection or improper dimensioning of an agitator leads to a decrease in process performance and availability. Inadequate mechanical design or the wrong material selection may have a considerable impact on the availability, lifetime, and operating cost of the whole reactor system. One of the factors leading to these errors is the cost savings made in the project implementation phase. Another factor may be failing to see the whole picture of the required functions that agitation needs to perform in the process and focusing only on certain aspects of agitation that are considered critical.

Suitable specification material has to be collected to start reactor design. The most notable factors defining reactor design are listed in Table 1. The starting point for reactor design is usually the identification of the reaction limiting factor. Other information collected in the table defines the reactor design leading to required vessel size, agitator type, and possible accessories such as a gas sparger and heating or cooling coils.

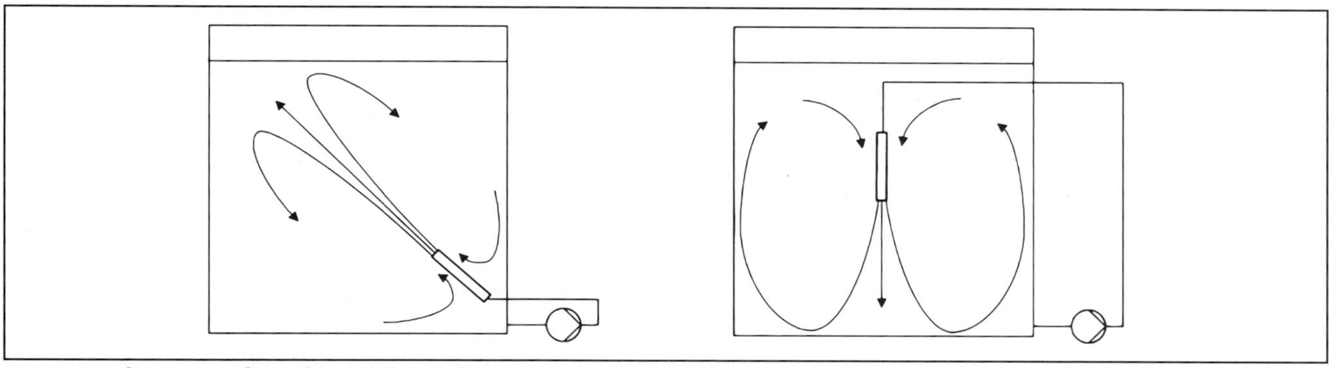

Figure 8 Side entry and axial jet mixing tanks

Table 1 Primary defining factors in stirred-tank reactor design

Factor	Parameters
Reactor duty	Reaction limiting factor, required mixing intensity, critical agitation duty
Solution properties	Mass flow, composition, temperature, density, etc.
Solids properties	Mass flow, density, hardness, particle size
Gas dispersion	Mass flow, density, solubility
Heat transfer	Endothermic or exothermic reaction

Source: Latva-Kokko et al. 2014

Experimental Testing

Adequately performed chemical batch tests with authentic and representative raw material samples create the foundation for successful leaching reactor design. Typically these tests are carried out in 2–5-L reactors in conditions where the reaction rate is not limited by the mixing intensity or the available reactive gas. Thus, an excessive amount of gas feed and oversized mixing intensity are used. These batch tests provide the most feasible process conditions and basic kinetic data for mass and heat balance calculations. Batch test data is converted for a continuous process based on the residence time distribution in a completely mixed stirred tank reactor series (Latva-Kokko et al. 2016).

In some cases, equipment capacity can also restrict the kinetic rate. Usually, limitations in gas feed capacity and solids suspension ability, for instance, are well known by equipment vendors, but their actual effect on the process kinetics should be tested experimentally. These tests must be done on a larger scale, ~20 L, so the industrial process conditions can be scaled down representatively. Ideally, the test equipment is geometrically similar to the industrial design. In addition, the reaction rate limiting factor must be known, since scale-down routines differ by phenomenon. For instance, it might not be possible or feasible to feed as much gas on an industrial scale as has been used in chemical batch tests.

Gas Sparging

Gas feed positioning has an effect on the gas dispersion properties of the tank. Gas is usually fed below the impeller through a sparger. The sparger can be a plain pipe, a ring with a smaller or larger diameter than the impeller, a cone, or a cartwheel-like structure. The purpose of the sparger is to distribute the gas evenly to the impeller. With radial flow impellers there is no need for a special sparger arrangement, since the flow pattern ensures good distribution of gas bubbles. However, a cone sparger can be installed to provide mechanical protection for the gas feed pipe and prevent blocking during shutdown situations. With mixed-flow impellers it is important to feed the gas directly below and rather close to the impeller so that gas bubbles do not bypass the high shear zone of the impeller in the rising fluid flow. Gas may also be introduced down the shaft of the agitator (using a rotating joint at the top of the shaft), exiting the bottom of the shaft at a point below the lower impeller. This method is commonly used in gold leaching (carbon-in-leach or carbon-in-pulp) applications where air (or oxygen) is used to enhance leaching kinetics.

The gas-to-liquid mass transfer can also be increased by using high-velocity gas feed spargers. These devices feed gas at so-called supersonic speed, that is, more than 400 m/s. A high-velocity gas jet creates high local shear rates that induce fine bubbles. Obtaining a high gas feed velocity requires a high gas pressure, and because use of a compressor creates more sources for energy losses, they are less economical than direct dispersion of gas with an agitator (Jung and Keller 2016). In addition, there are some nozzle wear and blocking issues in the use of these high-velocity gas feed spargers.

Mechanical Issues

Many things can go wrong in the mechanical design of an agitator. The shaft may be too narrow or too thin so that it bends in operation. On the other hand, a very hard and rigid shaft can break under heavy loads, such as those during start-up situations. Even a small dissymmetry in impeller blades or simply operating at resonance frequency may cause severe vibration of the whole reactor. Perhaps the most challenging aspect of mechanical design is the long-term endurance of the agitator. During operation, impeller blades are exposed to severe fatigue stress and wear by the combined effect of corrosion and abrasion. Usually, the corrosion resistance of a stainless steel weld is not as high as that of the base material.

In addition to corrosion, the welded joints of impellers are vulnerable to fatigue failure. Welded joints can thus be considered as the weakest link of an impeller. An agitator design that does not contain any welded joints of impeller blades or their fastenings is superior to welded structures in terms of corrosion resistance and mechanical strength. It also enables easy and quick replacement of worn-out blades with new ones. Figure 9 shows a series of impellers that do not contain any welded joints.

Figure 9 Impellers containing no welded joints and their welded duplicates

Construction Materials

Highly corrosive and abrasive conditions are often present in tank leaching. Therefore, the selection of materials is a crucial part of reactor design. The starting point for case-specific material selection is the corrosion resistance of the material under process conditions. The temperature and chemical composition determine the corrosivity of the solution. In hydrometallurgical applications, both uniform and localized corrosion must be taken into account. Uniform corrosion typically occurs in acids and hot alkaline solutions. Localized corrosion, such as pitting or crevice corrosion, is of concern in acidic solutions that contain chloride ions and oxidizing ions like Fe^{3+} (Latva-Kokko et al. 2014).

Stainless steels traditionally have been a common construction material for hydrometallurgical reactors. Austenitic steel grades such as 316L (1.4432) and 904L (1.4539) are typical choices thanks to their good fabrication characteristics. Of these grades, the higher alloyed 904L is more corrosion resistant. Duplex steel grades that have an austenitic-ferritic structure provide excellent corrosion resistance together with higher mechanical strength and surface hardness (Ekman and Berqvist 2008). Results from some corrosion tests are presented in Table 2.

As shown in Table 2, the different metal ion contents in solution greatly impact the corrosion resistance of different stainless-steel grades. In atmospheric operation, even higher corrosion resistance can be achieved by the use of fiberglass-reinforced plastic, which is able to withstand very high chloride content. In oxidizing conditions, titanium and its alloys can provide excellent corrosion resistance even in hot, acidic, and chloride-rich environments (Laihonen and Lindgren 2013).

Leaching reactors are also exposed to erosive wear because of the high solids content and abrasive components found in many ores and concentrates. Impeller blades in particular tend to wear because the velocities of the particles are highest next to the impeller. Impact and sliding wear are the two main wear mechanisms in impellers, and the erosion rate is strongly dependent on the impeller tip speed (Keller 2007). In addition to impeller tip speed, factors that most affect the intensity of erosive wear are solids content, particle size and shape, particle specific density, and hardness. Impeller blades, hubs, and shafts are often protected from wear through the use of polymeric coatings. Rubbers (either natural or synthetic) or polymers such as polyurethane are used to protect agitator components from wear.

Tank Leaching Plant

When developing a leaching plant, the correct specification of the reactor and agitator is only a part of the work required for successful operation. When multiple reactors form a plant unit and plant units form a complete plant, a comprehensive design is required to achieve a solution that has the required operation and maintenance design factors implemented.

In hydrometallurgical reactor design, it is common practice to comply with industry-specific standards, so process-specific requirements are not the main focus. Reactors in a certain process area should be seen as part of a larger reactor plant unit that includes, in addition to the reactor solution, transfer systems, access platforms, piping, and electrification. In addition to metallurgical performance, plant design should secure high availability with safe and easy operation and maintenance.

One of the key features in leaching plant design is connecting reactors in series. Inlet or outlet pipes must be designed

Table 2 Results from corrosion tests conducted at 90°C with different stainless-steel grades

		Austenitic 316L		LDX 2101		Duplex 2205	
Solution	**Cl^-, mg/L**	**mm/yr***	**Loc.†**	**mm/yr**	**Loc.**	**mm/yr**	**Loc.**
10 g/L H_2SO_4	200	0.80	Yes	<0.01	No	<0.01	No
10 g/L H_2SO_4, 0.1 g/L Cu^{2+}	200	<0.01	No	<0.01	No	<0.01	No
10 g/L H_2SO_4, 1.0 g/L Cu^{2+}	200	<0.01	No	<0.01	No	<0.01	No
10 g/L H_2SO_4, 10 g/L Cu^{2+}	200	<0.01	No	<0.01	No	<0.01	No
10 g/L H_2SO_4, 0.1 g/L Fe^{2+}	200	<0.01	No	<0.01	No	<0.01	No
10 g/L H_2SO_4, 1.0 g/L Fe^{2+}	200	0.03	No	<0.01	No	<0.01	No
10 g/L H_2SO_4, 10 g/L Fe^{2+}	200	<0.01	No	<0.01	No	<0.01	No
10 g/L H_2SO_4, 0.1 g/L Fe^{3+}	200	0.01	No	0.01	No	<0.01	No
10 g/L H_2SO_4, 1.0 g/L Fe^{3+}	200	0.17	Yes	0.21	No	<0.01	No
10 g/L H_2SO_4, 10 g/L Fe^{3+}	200	0.84	Yes	<0.01	No	<0.01	No

Source: Latva-Kokko et al. 2014

*mm/yr = Uniform corrosion rate in millimeters per year (if <0.1 mm/yr, material is corrosion resistant).

†Loc. = Localized pitting or crevice corrosion occurs (yes/no).

Courtesy of Outotec

Figure 10 Gravity flow launder connecting reactors in series

to avoid short circuiting flows. With a traditional overflow channel between the tanks, there is a high probability that part of the slurry will bypass the main reactor volume. Especially with sulfidic particles that are hydrophobic by nature and tend to form a froth layer at the tank surface, there can be a significant decrease in leaching yield if an overflow connection is used. Pipes that connect reactors clearly below the surface are a better option, but usually not very flexible. A gravity flow launder with inlet and outlet pipes, as shown in Figure 10, provides easy bypass of any or even several reactors at the same time for maintenance.

REFERENCES

Ekman, S., and Berqvist, A. 2008. Suitable steel grades for hydrometallurgical applications. In *Hydrometallurgy 2008: Proceedings of the Sixth International Symposium*. Edited by C.A. Young, P.R. Taylor, C.G. Anderson, and Y. Choi. Littleton, CO: SME.

Filippou, D., Cheng, T., and Demopoulos, G.P. 2000. Gas–liquid oxygen mass-transfer; from fundamentals to applications in hydrometallurgical systems. *Miner. Process. Extr. Metall. Rev.* 20:447–502.

Grenville, R., and Nienow, A. 2004. Jet mixing in tanks. In *Handbook of Industrial Mixing*. Edited by E. Paul, V. Atiemo-Obeng, and S. Kresta. Hoboken, NJ: John Wiley and Sons.

Hemrajani, R., and Tatterson, G. 2004. Mechanically stirred vessels. In *Handbook of Industrial Mixing*. Edited by E. Paul, V. Atiemo-Obeng, and S. Kresta. Hoboken, NJ: John Wiley and Sons.

Hosseini, S., Patel, D., Ein-Mozaffari, F., and Mehrvar, M. 2010. Study of solid–liquid mixing in agitated tanks through electrical resistance tomography. *Chem. Eng. Sci.* 65(4):1374–1384.

Jung, J., and Keller, W. 2016. Process and cost optimized agitator solutions for hydrometallurgical base metals processing. *World Metall.–ERZMETALL.* 69(2):108–118.

Keller, W. 2007. POX autoclaves: New advantages in impeller design for highly abrasive ores. Presented at the ALTA Copper 2007 International Conference, Castlemaine, Victoria, Australia.

Laihonen, P., and Lindgren, M. 2013. The combined effect of fluorides and ferric ions on the uniform corrosion of titanium and titanium alloys in sulfuric acid. In *Proceedings of Material Science and Technology 2013*. Montreal, QC: Materials Science and Technology.

Latva-Kokko, M., and Riihimäki, T. 2012. Effect of temperature in leaching of low grade sulphide ore at ambient temperature: Development of hydrostatic pressure reactor. In *Pressure Hydrometallurgy 2012*. Westmount, QC: Canadian Institute of Mining, Metallurgy and Petroleum.

Latva-Kokko, M., Hirsi, T., Lindgren, M., and Ritasalo, T. 2014. Influence of reactor design to process performance in hydrometallurgical applications. In *Hydrometallurgy 2014: Proceedings of the 7th International Symposium on Hydrometallurgy*. Vol. II. Westmount, QC: Canadian Institute of Mining, Metallurgy and Petroleum.

Latva-Kokko, M., Hirsi, T., and Ritasalo, T. 2016. Sustainable agitator and reactor design for demanding applications in hydrometallurgy. In *Hydroprocess 2016: 8th International Seminar on Process Hydrometallurgy*. Edited by F. Valenzula. Santiago, Chile: Gecamin.

Mehrotra, S., and Shekhar, R. 2000. Studies on particle suspension in air-agitated Pachuca tanks. In *Processing of Fines (2)*. Edited by P. Bhattacharyya, R. Singh, and N. Goswami. Jamshedpur, India: NML Jamshedpur.

Roy, G., Shekhar, R., and Mehrotra, S. 1998. Particle suspension in (air-agitated) Pachuca tanks. *Metall. Mater. Trans. B* 29B(2):339–349.

Taca, C., and Paunescu, M. 2000. Suspension of solid particles in spherical stirred vessels. *Chem. Eng. Sci.* 55:2989–2993.

Tahvildarian, P., Ng, H., D'Amato, M., Drappel, S., Ein-Mozaffari, F., and Upreti, S.R. 2011. Using electrical resistance tomography images to characterize the mixing of micron-sized polymeric particles in a slurry reactor. *Chem. Eng. J.* 172(1):517–525.

Tervasmäki, P. 2013. *Comparison of Solids Suspension Criteria by Electrical Impedance Tomography and Visual Measurements*. Department of Process and Environmental Engineering, University of Oulu, Finland.

Wu, J., Graham, L., Nguyen, B., and Mehidi, M. 2006. Energy efficiency study on axial flow impellers. *Chem. Eng. Process.* 45:631.

CHAPTER 10.6

Pressure Leaching and Oxidation

Rudi Frischmuth, Tom Krumins, Murray Pearson, and Kevin S. Fraser

Pressure leaching and pressure oxidation are hydrometallurgical processes applied to the extraction of many metals. The processes occur above atmospheric boiling temperature, require a sealed reactor vessel, and often operate in highly corrosive and oxidizing environments. Pressure leaching and pressure oxidation have been applied to a broad range of processes over a wide range of conditions and include similar features such as

- Slurry feed: Chemical conditioning, preheat, and high-pressure pumps;
- Reactor vessel: Digester and leach or oxidation autoclave;
- Pressure letdown: Flash vessel;
- Off-gas handling: Atmospheric condenser and scrubber; and
- Ancillary systems: Acid, oxygen, air, steam, and water injection systems, and an agitator seal system.

Pressure leaching is primarily applied when the goal is to solubilize desirable elements, recover the solution fraction, and reject the residual solids. The process can operate under alkaline or acidic conditions and may include the use of oxygen gas (or an alternative oxidant such as hydrogen peroxide) to oxidize and leach base metal sulfide minerals. The most common alkaline pressure leach processes are applied in the extraction of alumina (e.g., Bayer process), tungsten, uranium, and rare earth elements, as well as nickel, cobalt, and copper sulfide minerals. The most common acidic processes are applied in the extraction of platinum group metals (PGMs), nickel–cobalt laterites, and uranium, as well as copper, nickel, cobalt, molybdenum, and zinc sulfides. Acidic pressure leaching of iron from an ilmenite slag to produce synthetic rutile is a process in which the solid fraction contains the product and the impurities leached into solution are rejected (solution is recycled).

Pressure oxidation is applied when the goal is to decompose sulfide minerals for the recovery of desirable elements from the solids fraction and the rejection of the solution fraction. Pressure oxidation is mostly applied to refractory gold-bearing ore or concentrates where the gold is locked in the sulfide minerals and recovery using conventional cyanide leaching is not effective. The process is mostly operated under acidic conditions; however, alkaline pressure oxidation is also practiced.

Variations of pressure leaching and pressure oxidation processes have been developed over time and have become known by different terms. For the sake of simplicity and clarity, the aforementioned fundamental process definitions have been applied consistently in this chapter unless otherwise identified.

This chapter provides a review of pressure leaching and pressure oxidation unit process systems, including discussions of process design, metallurgical test work, key economic drivers for operating and capital costs, and typical practices for equipment selection and design. Readers are directed to the respective commodity sections elsewhere in this handbook for more detailed discussions of process metallurgy fundamentals.

HISTORICAL DEVELOPMENT

Pressure leaching and pressure oxidation processes have developed sporadically over the last 120 years, driven by mineralogical complexity, fluctuation in metal demand, commercial opportunity, and the technological progression of process equipment. These key drivers continue to influence development and application of new technologies in the industry today.

Pressurized extractive metallurgy began with the alkaline leaching of alumina from bauxite ores using the Bayer process, named after its inventor, Karl Josef Bayer. The process, which was patented in 1888 and first applied in St. Petersburg, Russia, in 1892, continues to be applied in aluminum production to this day (Habashi 1995).

Many leaching and recovery processes were postulated at the turn of the 20th century (Habashi 2004). Those that progressed into commercial application at the time were supported

Rudi Frischmuth, Senior Metallurgist, High-Pressure Metallurgy, Metals, Autoclave Technology, Hatch Ltd., Mississauga, Ontario, Canada
Tom Krumins, Process Engineer, High-Pressure Metallurgy, Metals, Autoclave Technology, Hatch Ltd., Mississauga, Ontario, Canada
Murray Pearson, Director, Technology Development, High-Pressure Metallurgy, Metals, Autoclave Technology, Hatch Ltd., Mississauga, Ontario, Canada
Kevin S. Fraser, Co-Director & Principal Metallurgist, High-Pressure Metallurgy, Metals, Autoclave Technology, Hatch Ltd., Mississauga, Ontario, Canada

by testing and understanding of the process chemistry for the complete extraction and recovery process, the economical availability of reagents, and suitable process equipment technology. The 1919 U.S. patent for the alkaline digestion and extraction of tungsten (Giles and Giles 1919) provides insight into the pressure leaching equipment applied at the time. As shown in Figure 1, the patent illustrates the use of batchwise process equipment that combines pressure leaching and grinding to liberate minerals and promote the alkaline leaching of tungsten minerals.

The mid-20th century was a period when pressure leaching technology expanded to a range of metals. This century also witnessed the development of high-capacity process equipment. Pressure leaching was applied to alkaline and acidic uranium extraction (Fraser and Thomas 2010); the extraction of cobalt from an arsenic cobalt sulfide concentrate in acidic oxidizing conditions (Mitchell 1956); the extraction of cobalt, nickel, and copper from a mixed sulfide concentrate in alkaline conditions using ammonia; and the acidic pressure leaching of nickel–cobalt laterites (Chalkley et al. 2004). The extraction of tungsten also evolved to produce a high-purity ammonium paratungstate for industrial consumption (Kurtak 1998).

Pressure hydrometallurgy was first extended to uranium at Eldorado Nuclear's Beaverlodge Mill (Saskatchewan, Canada) in 1953 (Fraser and Thomas 2010). The oxidative process included a 96-hour alkali–carbonate Pachuca leach circuit consisting of twenty-four 255-m^3 vessels sparged with steam for temperature control. Pachuca technology was later extended to the first application of pressure hydrometallurgy for nickel laterite at Moa Bay, Cuba, which included five trains of four vessels 3 m in diameter and 15 m tall, agitated by the injection of superheated steam to an internal draft tube (Mason and Gulyas 1999).

Importantly, during the mid-20th century, the technology for pressure leach equipment evolved from small-scale batch processes to the use of continuous multicompartment horizontal autoclaves. This evolution of autoclave technology supported an increase in treatment rate and improvement in economies of scale. One of the first horizontal autoclaves was introduced for pressure oxidative leaching of an arsenic-rich cobalt sulfide concentrate at the Garfield Refinery in Utah (United States). The process, designed to operate at 190°C and 3,400 kPa, experienced many corrosion and erosion challenges during the first few years of operation (Mitchell 1956). The challenges and subsequent development resulted in the introduction of specialized materials, including titanium for agitator components and slurry discharge and ceramics for pressure letdown control. Undoubtedly, the lessons learned from the Garfield Refinery and subsequent commercial production of titanium and ceramic components supported the design of pressure oxidation and pressure leach circuits in the years to follow.

After many years of testing and development, pressure leaching was applied to the extraction of zinc from sulfide concentrates. The first commercial zinc pressure leach plant was introduced in 1981 as an expansion of the Cominco Trail operations in Canada (Weir 1985) and was later introduced to other plants in Canada and Germany. The process, operating at 150°C, was the first to target partial oxidation of sulfide to elemental sulfur (instead of complete oxidation to sulfate) and to overcome challenges with the formation and agglomeration of molten elemental sulfur in the autoclave. The goal of promoting elemental sulfur formation rather than sulfate is to lower the operating costs associated with oxygen consumption and acid neutralization while still achieving high zinc extraction.

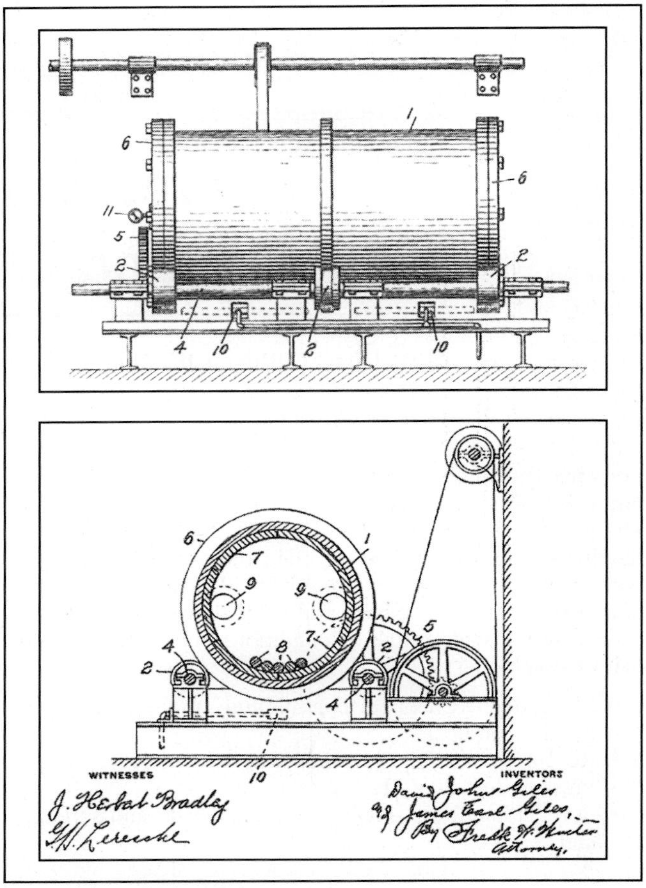

Source: Giles and Giles 1919

Figure 1 Historical tungsten extraction "grinding" autoclave

In March 1985, gold was produced from the McLaughlin pressure oxidation plant in California (United States). The project was the first to pretreat sulfidic refractory gold ore and built upon lessons learned from Anglo American Corporation's 1977–1981 Western Deeps/Vaal Reefs uranium pressure leach demonstration plant in South Africa. The São Bento pressure oxidation plant in Brazil followed in 1986, and another seven operations were constructed between 1986 and 1993, including the Goldstrike and Twin Creeks operations in Nevada (United States), the Campbell mine in Canada, and the Porgera mine in Papua New Guinea. The first commercial alkaline pressure oxidation plant, Mercur (Utah), was commissioned in 1988 (St. Louis and Edgecombe 1990). Subsequent refractory gold projects have included the Lihir operation in Papua New Guinea, Macraes in New Zealand, and the Pueblo Viejo operation in the Dominican Republic.

The 1990s witnessed rapid development in the acidic pressure leaching of nickel and cobalt from laterites, with a series of plants constructed in Western Australia. The projects were initiated as a result of an increase in nickel and cobalt prices; advantages in project location; acid availability; low mining complexity; and advances in the design of solvent extraction, electrowinning, thickening, and autoclave processes (Kyle 1996). Although aspects from early pressure leach plants were incorporated into the designs, the project start-ups were prolonged and had many technical challenges (Nice 2004). Of the three projects constructed—Bulong, Cawse, and Murrin Murrin—only Murrin Murrin has continued to operate and achieve nameplate throughput. Lessons learned from these first-generation projects have since supported the development of a second generation of laterite projects, including Coral Bay in the Philippines; Ravensthorpe in Western Australia; and Goro Nickel in New Caledonia; and, more recently, a third generation, which includes the Ambatovy project in Madagascar and Ramu in Papua New Guinea.

In the mid- to late 1990s, various interests turned to the use of pressure leaching of copper concentrates as an alternative to conventional smelting (King et al. 1993; Defreyne et al. 2006; Marsden et al. 2007). The CESL (Cominco Engineering Services Ltd.) process was one of the first methods developed and progressed to a demonstration plant (Defreyne et al. 2008). The process, which was developed by operators of the first zinc leach plant at Trail, applies similar process conditions to the zinc pressure leach, but with the addition of chloride to promote copper sulfide oxidation and elemental sulfur formation.

Medium- and high-temperature pressure leaching of copper concentrate was also under development in the United States around the same time. The processes, which are well documented by Marsden et al. (2007), include fine grinding of copper concentrate to promote oxidation to elemental sulfur under medium-temperature conditions and direct electrowinning of copper from the leach solution. Under high-temperature conditions, the oxidation reaction progresses to sulfate and sulfuric acid, and the acid can be recovered for use elsewhere in the process (e.g., heap leach operations, as in the case for the demonstration plant at the Bagdad mine and the commercial operation at the Morenci mine in Arizona, United States).

The expansion of pressure leaching and oxidation from the 1950s and the associated development in process equipment have firmly cemented the processes as mature metallurgical technologies. Many of the challenges encountered during the evolution of the processes have led to new techniques or processes that provide the modern metallurgical engineer with extensive research data and project references to work with. Key developments in pressure leaching and pressure oxidation are illustrated in Figure 2.

APPLICATION AND SELECTION

The selection of pressure leaching or pressure oxidation for the recovery of metals depends on many factors particular to the material to be treated (e.g., mineralogy and grade, deposit size, geographic location). In some cases, the selection may be subject to evaluation against other treatment technologies. Pressure leaching and oxidation processes maintain the key advantages of occurring in aqueous process conditions, which often have advantages for metal recovery and environmental emissions control while providing fast reaction rates. These advantages become more apparent when the material or ore source has specific challenges and other processes, such as atmospheric leaching, are not as effective or appropriate.

For example, the increase in the oxidation kinetics associated with the pressure oxidation of gold-bearing refractory sulfides provides a significant advantage over biological oxidation for high-sulfide throughput. A typical metric used for biological oxidation is 7 kg of sulfide sulfur oxidized per cubic meter of reactor process volume per day (7 kg $S^{2-}/m^3/d$), whereas a typical metric for pressure oxidation is 700 kg $S^{2-}/m^3/d$. Similarly, biological oxidation typically occurs over 4 to 5 days (96–120 hours), whereas a 60- to 90-minute retention time is typical for pressure oxidation. This order-of-magnitude unit capacity difference provides an advantage in equipment size and plant layout for pressure oxidation when a comparatively high-sulfide-sulfur throughput is required. An early economic assessment of applying pressure oxidation for the pretreatment of a sulfidic refractory gold ore body is described in the study by Singh et al. (2013).

Select commercial pressure leach and oxidation process applications are summarized in Table 1 and illustrated relative to a temperature–pressure-saturated steam curve in Figure 3. In general, the processes can be categorized into low to medium temperature (130°–160°C) and high temperature (200°–260°C). In low- to medium-temperature conditions, the saturated steam pressure is less than 1,000 kPa(g), whereas at high-temperature conditions, the pressure is significantly higher at 2,000–4,500 kPa(g). The high-temperature processes require special consideration to the circuit design and materials of construction, especially those operating in oxidizing or acidic process conditions.

A key differentiator between sulfide and nonsulfide processes is the energy balance. The oxidation of sulfide sulfur is exothermic (i.e., heat-generating); therefore, the pressure oxidation process does not require energy input as long as the feed grade is sufficiently high (approximately >5 wt % sulfide sulfur). The process is said to be autothermal when no heat energy input is required to sustain the operation. Nonsulfide leaching processes often require significant heat energy input to sustain the target operating temperature. Samples of pressure leaching and pressure oxidation chemical reactions and heats of reaction are provided in Table 2 (Outotec 2016).

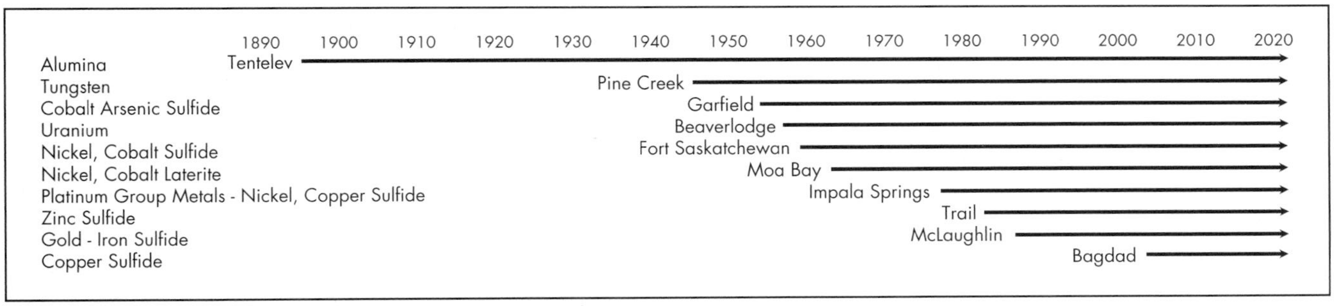

Figure 2 History of pressure leaching and pressure oxidation

Table 1 Pressure leach and oxidation process comparison

Aspect	Pressure Leach									Pressure Oxidation
Application	Aluminum (Bayer)	Tungsten	Nickel, cobalt, copper (sulfide)	Uranium	Nickel–cobalt (laterites)	Zinc (sulfides)	Copper/molybdenum (sulfides)	Platinum group metals	Titanium	Gold/silver (refractory in sulfides)
Process objective	Solubilization	Solubilization	Solubilization	Solubilization	Solubilization	Solubilization	Solubilization	Solubilization/ solids enrichment	Solids enrichment	Liberation
Feed material	Ore (bauxite)	Flotation concentrate	Flotation concentrate	Ore	Ore (upgraded)	Flotation concentrate	Flotation concentrate	Matte	Slag	Ore/flotation concentrate
Common minerals	Gibbsite, boehmite	Scheelite, wolframite	Pentlandite	Uraninite, brannerite, coffinite	Goethite, garnierite, asbolite	Sphalerite	Chalcopyrite/ molybdenite	Digenite, covellite, polydymite, millerite	Ilmenite	Pyrite, arsenopyrite, realgar, orpiment
pH	Alkaline	Alkaline	Alkaline	Alkaline/acid	Acid	Acid	Acid	Acid	Acid	Acid/alkaline
Operating temperature, °C	140–270	175–250	110–120	70–180	230–270	130–150	140–225	140	145–165	190–230
Operating pressure, kPa	400–5,500	1,000–4,000	800–900	300–1,700	3,000–5,700	1,100–1,400	1,000–3,200	600	350–500	1,600–3,350
Reagent	Sodium hydroxide	Sodium carbonate/ sodium hydroxide	Ammonia	Sodium hydroxide/sulfuric air/oxygen	Sulfuric acid	Oxygen	Oxygen/acid	Oxygen	Hydrochloric acid	Oxygen
Vessel configuration	Multiple vertical	Vertical/ horizontal	Horizontal	Multiple vertical/ horizontal	Horizontal	Horizontal	Horizontal	Horizontal	Vertical	Horizontal
Batch/ continuous	Continuous	Batch/ continuous	Continuous	Continuous	Continuous	Continuous	Continuous	Continuous	Batch	Continuous
Heat recovery	3 to 12 Stages	—	1 Stage	0 to 3 Stages	3 to 4 Stages	—	—	—	—	0 to 3 Stages
Project examples	Pinjarra, Australia		Fort Saskatchewan, Canada	Key Lake, Canada	Murrin Murrin, Australia	Trail, Canada	Morenci, United States	Impala Platinum, South Africa	Quebec Iron and Titanium, Canada	Goldstrike, United States
	Alumar, Belem, Brazil	Pine Creek, United States	Kwinana, Australia	McClean Lake, Canada	Ravensthorpe, Australia Goro Nickel, New Caledonia	Kidd Creek, Canada	Bagdad, United States	Lonmin, South Africa		Twin Creeks, United States
	Weiqiao, China			Dominion Uranium, South Africa	Ambatovy, Madagascar	Ruhr Zinc, Germany				Pueblo Viejo, Dominican Republic

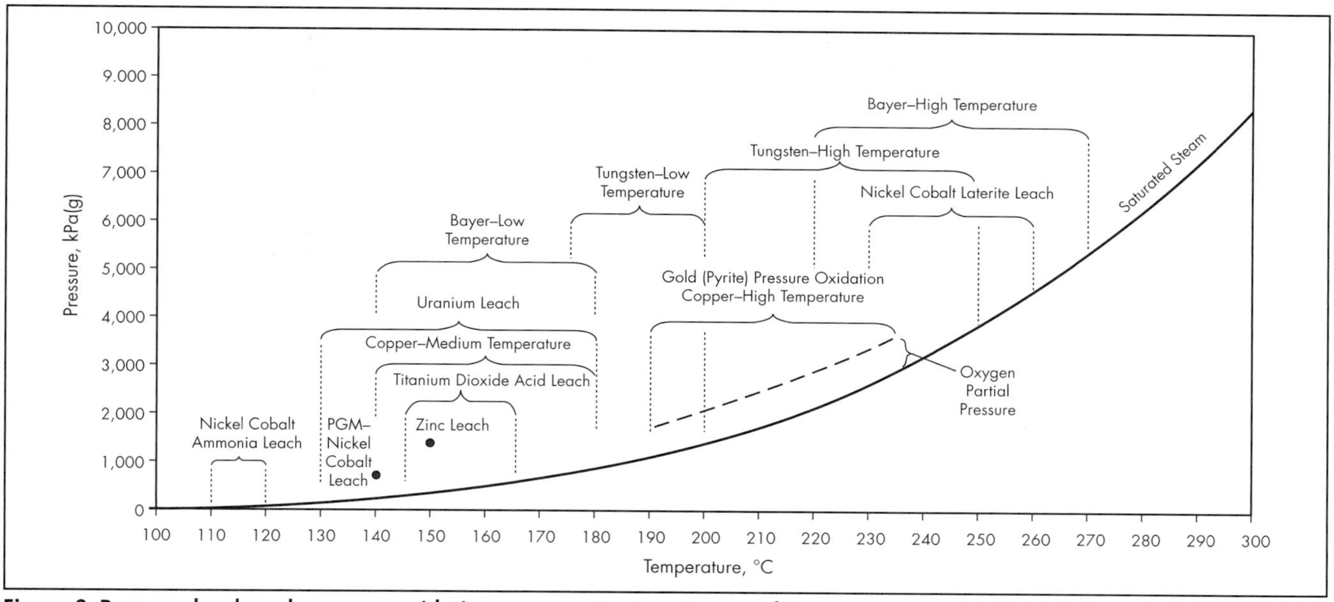

Figure 3 Pressure leach and pressure oxidation pressure–temperature conditions

Table 2 Pressure leach and pressure oxidation heats of reaction

Mineral	Chemical Formula	Heat of Reaction, mJ/kg Mineral
Uraninite (acid)	$UO_2 + Fe_2(SO_4)_3 \rightarrow UO_2SO_4 + 2FeSO_4$	–0.5
Beohmite (Bayer)	$AlO(OH) + NaOH + H_2O \rightarrow NaAl(OH)_4$	–0.3
Wolframite	$(Fe, Mn)WO_4 + 2NaOH \rightarrow NaWO_4 + (Fe, Mn)(OH)_2$	–0.1
Uraninite (alkaline)	$UO_3 + Na_2CO_3 + H_2O \rightarrow Na_2UO_2(CO_3)_3 + 2NaOH$	0.0
Goethite (Ni bearing)	$Ni(OH)_2 + H_2SO_4 \rightarrow NiSO_4 + 2H_2O$	1.0
Geothite (Co bearing)	$Co(OH)_2 + H_2SO_4 \rightarrow CoSO_4 + 2H_2O$	2.0
Goethite (leach)	$2FeOOH + 3H_2SO_4 \rightarrow Fe_2(SO_4)_3 + 4H_2O$	2.3
Arsenopyrite	$2FeAsS + 7O_2 + 2H_2O \rightarrow 2FeAsO_4 + 2H_2SO_4$	8.9
Sphalerite	$ZnS + O_2 \rightarrow ZnSO_4$	8.9
Chalcopyrite	$CuFeS_2 + 4O_2 \rightarrow CuSO_4 + FeSO_4$	9.0
Metal sulfide (NH_3 leach)	$MS + 2O_2 + 4NH_3 \rightarrow M(NH_3)_2 + (NH_3)_2SO_4$ (M = Ni, Co, Cu)	10.0 (Ni)
Pyrite	$2FeS_2 + 7O_2 + 2H_2O \rightarrow 2FeSO_4 + 2H_2SO_4$	12.1

Source: Outotec 2016

The need for heat input to the process strongly influences the application of heat recovery and recycle from the reactor discharge to the reactor feed. Nonsulfide pressure leach processes that require significant heat input use multiple stages of pressure letdown and heat transfer to maximize energy recovery. Acid leach processes may also receive 30°–40°C of supplementary heating from acid dilution. Pressure leach or whole-ore pressure oxidation projects with low sulfide sulfur grades (3.0%–4.5%) typically include two stages of letdown and heat recovery and the ability to directly add steam to the reactor to maintain the operating temperature. Ores with sulfide sulfur grades between 4.5% and 6% typically include one stage of letdown and heat recovery to ensure that autothermal operation is maintained. The Mercur alkaline process was designed for 1.7% sulfide sulfur and uses three stages of preheating to achieve an operating temperature of 225°C (Mason 1990). Whole-ore or concentrate feeds with sulfide sulfur feed grades greater than 6% usually have excess heat, requiring energy dissipation instead of energy input.

General flow-sheet configurations for the Bayer pressure leach process, the nickel laterite pressure leach process, and pressure oxidation are shown in Figures 4–6. Key points of variation between the three example flow sheets include the following:

- **Bayer leach:** Vertical plug flow digesters; extensive multistage letdown and indirect-contact heat recovery system
- **Nickel laterite leach:** Horizontal multicompartment autoclave with mixing agitators and steam, acid, and air addition; multistage letdown and direct-contact heat recovery system
- **Pressure oxidation:** Horizontal multicompartment autoclave with gas dispersion agitators and oxygen and cooling water addition; one stage of letdown and potential direct-contact heat recovery to allow operation at lower sulfide sulfur grades

The selected process conditions for pressure leach or pressure oxidation processes (e.g., operating temperature, pressure, retention time, and reagent addition) are highly

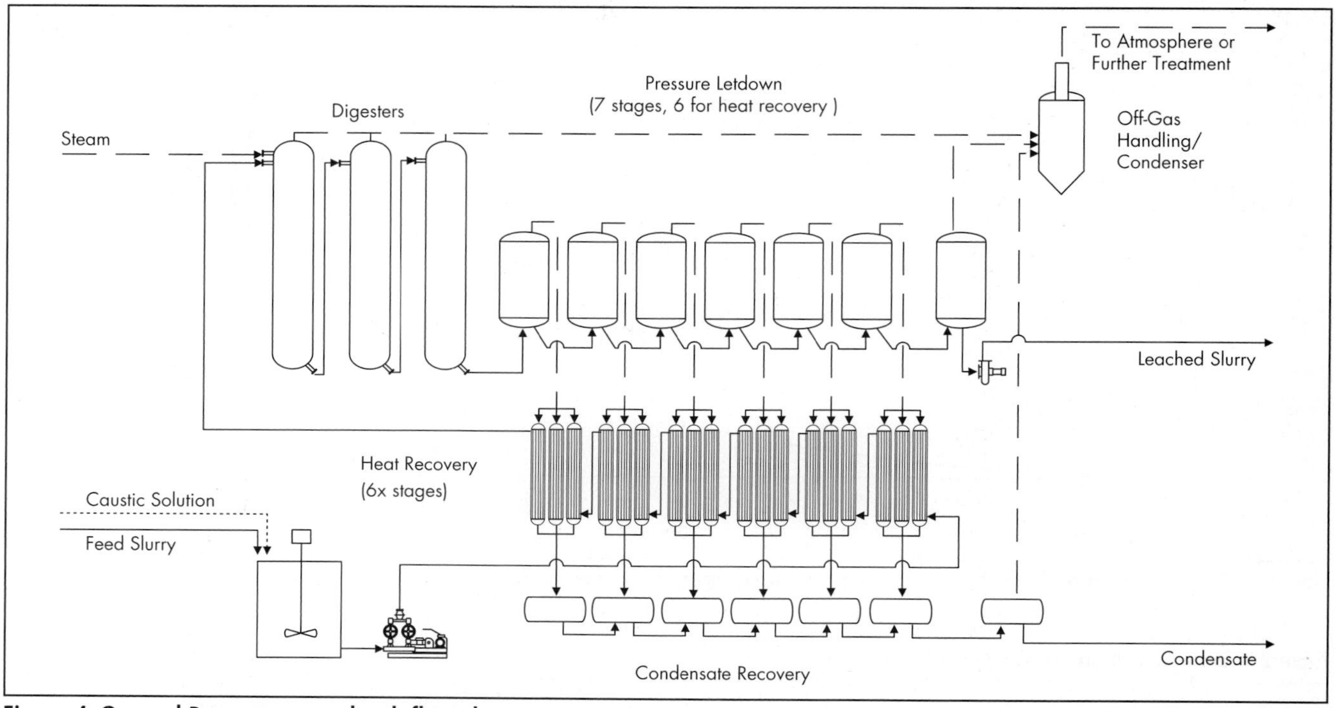

Figure 4 General Bayer pressure leach flow sheet

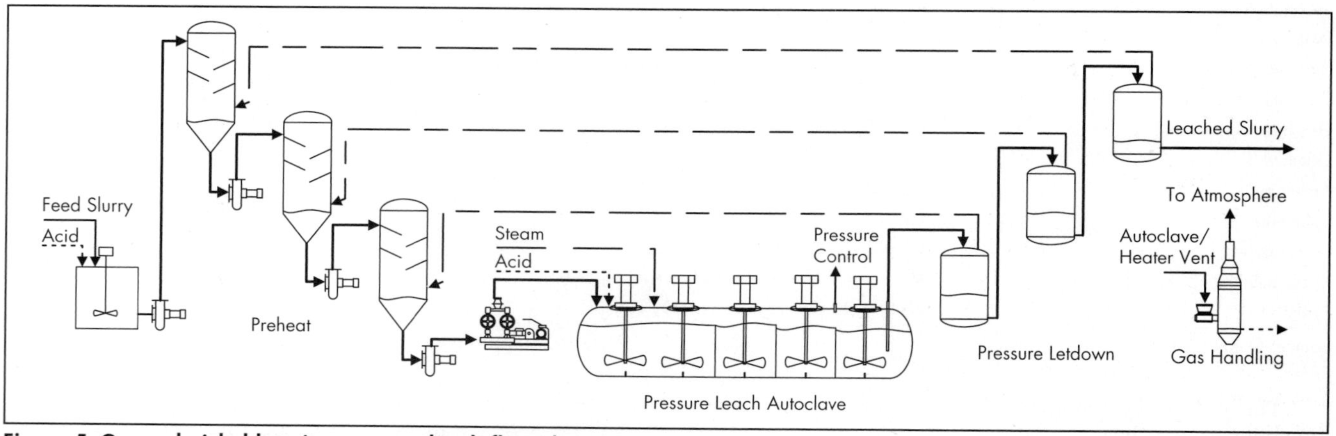

Figure 5 General nickel laterite pressure leach flow sheet

dependent on the specific mineralogical composition of the ore or concentrate and the goal of the process. As a result, the pressure leach or oxidation circuit is designed specifically for a type of feed material at an economical mass throughput. The treatment of substantially different feed material or throughput is often not possible without significant modification to the circuit.

Where applicable, concentrating the ore using flotation may be an effective way to improve the process so that additional equipment or continuous energy addition may not be required. The benefit of concentrating an ore is dependent on the sulfide sulfur and pay metal recovery, the achievable concentrate sulfide sulfur grade, and the rejection of undesirable minerals that affect the process (e.g., carbonates). Concentration is often selected under the belief that the autoclave size will be smaller due to the decrease in the mass to be processed; however, this is often not the case, as the autoclave size is more dependent on sulfide sulfur throughput than mass throughput. A comparison of whole-ore and concentrate refractory gold pressure oxidation facility configurations is provided in Table 3. Some pressure oxidation circuits actively treat a combination of whole ore and concentrate to achieve autothermal operation or to maximize pay metal production.

An assessment of process conditions and the potential benefit of alternate circuit configurations, such as mineral concentration, are important to process flow-sheet development. Completing the necessary supporting metallurgical test work is also a critical phase of project development.

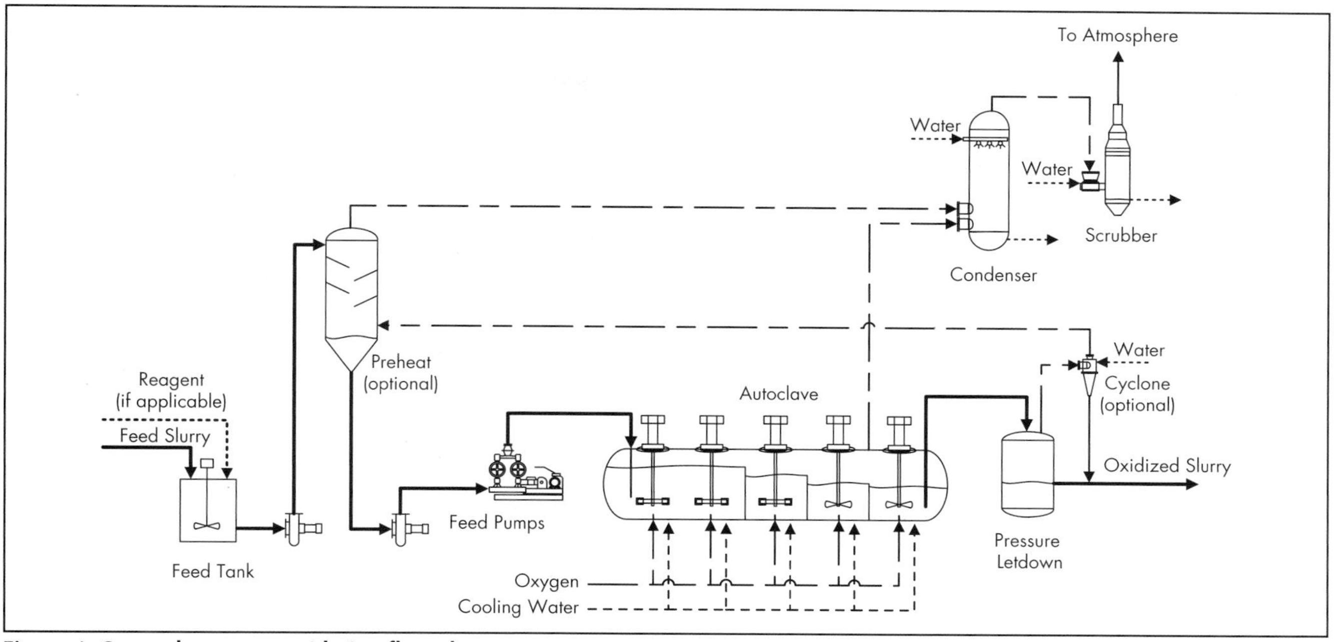

Figure 6 General pressure oxidation flow sheet

Table 3 Refractory gold pressure oxidation facility process configurations

Whole Ore	Concentrate
Getchell	Amursk
Goldstrike	Campbell
Lihir*	Con mine
Lone Tree	Kittila
McLaughlin	Macraes
Mercur	Porgera
Pueblo Viejo	São Bento
Twin Creeks*	

*Operations that treat whole ore and concentrate.

Table 4 Sample economic comparison of test program impacts

Description	Limited Test Program	Extensive Test Program
Total process test-work cost	$1 Million	$5 Million
Process commissioning delay	0 Months	3 Months
Process capacity after 36 months	81%	94%
Revenue at 100% process capacity*	$10 Million per month	
Net present value of first 36 months†	$85 Million	$85 Million

*Revenue is assumed to scale proportionally to process capacity.
†Includes cost of test work and impact of commissioning delay at 5% discount rate.

Metallurgical Test Work and Process Flow-Sheet Development

The success of a commercial operation is directly affected by the degree of test work performed during its development, and this is especially true for more complex high-pressure processes (McNulty 1998). Therefore, the correct balance must be obtained between the extent of test work performed and the risk tolerance in reaching commercial design capacity in a timely and cost-effective manner following commissioning. The goals of a test-work program should be to

- Define feed characteristics, such as ore mineralogy, including potential variability;
- Confirm process parameters and operating conditions;
- Understand the complete process chemistry, including materials of construction requirements;
- Pilot-test any process operations with higher degrees of risk/unknowns; and/or
- Establish criteria for engineering design and equipment sizing.

For process design, it is imperative that test work be performed on samples that are representative of the overall material. This is particularly important to an economic assessment of the process in terms of operating costs and engineering design requirements. When a high degree of variability is expected in the feed, additional test work may be required to establish appropriate design boundaries. The presence of potentially troublesome components (e.g., As, Cl, CO_3, F, Hg, Sb) should be determined in the initial analysis. If present, the test program should then be expanded to assess their impact.

An example of balancing test work with the timeliness of implementation is provided in Table 4. The example is based on an analysis of plant data for complex processes with different degrees of test work and the process start-up production curves provided in Figure 7 (McNulty 1998). The example shows that additional test work at the incremental cost of $4 million with a 3-month delay has a negligible bearing on the economics for the first 36 months of operation, and that additional test work establishes better performance (and greater value) for the following years. Increased unit operating costs that would be expected at lower production capacities have been ignored, which would further favor the extensive test program scenario.

Test work is performed batchwise and continuously (pilot), depending on a project's stage of development. Batch work is typically performed in 2-L vessels to initially indicate whether the material is amenable to the proposed process and can be used as a comparison to alternative processes. Further batch tests can later be performed to establish operating conditions and assess ore variability to form a basis for scoping or prefeasibility studies. Typically, mass and energy balances are carried out with the tests to support the understanding of process chemistry and process cooling or heating requirements.

Continuous pilot tests support feasibility studies with a greater level of detail and provide proof-of-concept assurance to the previously completed batch work. Shorter duration campaigns (e.g., 12–24 hours, compared to 60–100 hours for a full feasibility campaign) may be completed during the prefeasibility stages of a project to validate process configurations. Continuous pilot tests represent commercial operation more closely than batch tests, provide more detailed inputs to engineering design (e.g., scale formation, corrosion, off-gas composition), and demonstrate the impact of recycle streams within the process. Continuous test work (and expected commercial operation performance) can differ in performance from batch test work (e.g., required retention time, process chemistry, extraction extents) and is therefore critical to a project's success. Pilot autoclaves usually operate at a capacity of 20–70 L, resulting in a scale-up ratio for commercial operations ranging between 8,000 and 20,000 (Adams et al. 2004).

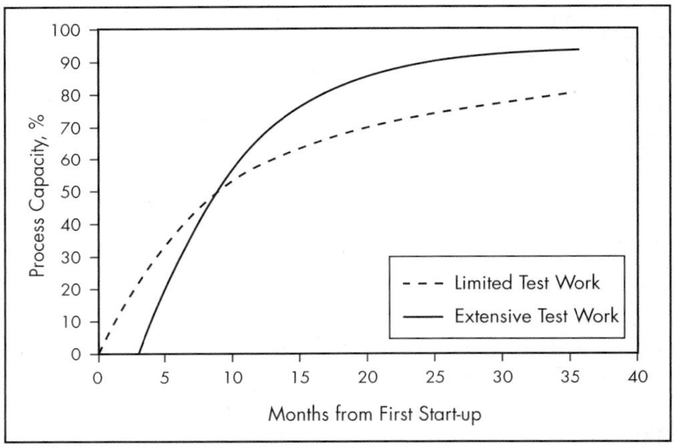

Adapted from McNulty 1998

Figure 7 Sample comparison of test program impact on start-up process capacity

A general process flow for test work is demonstrated in Figure 8. A basic sample test program specific to a pressure oxidation process is provided in Table 5, which excludes all feed analysis (e.g., mineralogy), downstream processing (e.g., solid–liquid separation), or supplementary work (e.g., environmental). Material estimates are provided on a whole-ore basis; that is, the total mass required would be significantly greater if a concentrate is required for the test program (e.g., a concentrate with a mass recovery of 20% would require 5 times the mass input to the program). The sample pilot program only considers a single ore feed and does not include additional run time for campaigns of differing feed material. The overall costs are indicative for a similar test program for an acidic pressure leach or pressure oxidative leach process. The total cost would be higher for additional feed material campaigns and variability testing or for testing various process configurations.

EQUIPMENT SIZING, SELECTION, AND MATERIALS OF CONSTRUCTION

An early understanding of the overall plant throughput and the key drivers determining the selected throughput needs to be developed from the perspective of the leach or oxidation process. It is not uncommon during early project development for mine planning to limit its focus to the pay metals and neglect the variability of minerals that may affect the leaching or oxidation process. For example, the mine and ore delivery plan for a refractory gold deposit may be initially optimized to gold production and neglect the impact on the sulfide sulfur feedrate variability to the pressure oxidation circuit, potentially resulting in an incorrectly sized autoclave vessel and oxygen plant. Depending on the complexity of the ore and the number of pay metals, it is typical for the mine plan, ore blending, and delivery strategy, as well as the plant production profile, to undergo several iterations before the optimum throughput is selected. At this stage, attention should be paid to understanding the ore mineralogy, metal deportment, and gangue mineralogy that may interfere with the pressure leach or pressure oxidation process.

In many cases, the pressure leach or oxidation circuit and its ancillary equipment are the most capital-intensive process areas in the plant, and maximizing their utilization is important to optimizing operating economics. To support this philosophy, it is prudent to ensure that the upstream ore preparation circuit is designed to maintain the required feed to the leach or

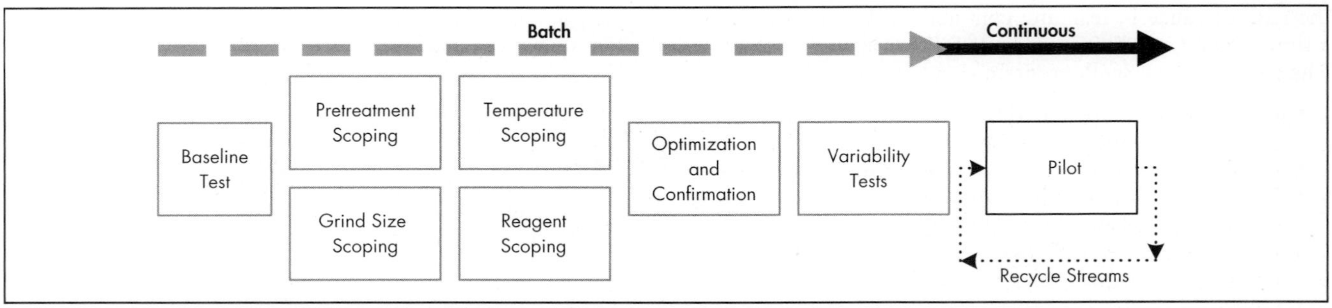

Figure 8 Sample process flow for test program

Table 5 Sample pressure oxidation test-work program

Description	Quantity	Total Mass Required	Total Approximate Cost*, US$
Batch autoclave tests, 2 L			
Baseline gold recovery test	2	1 kg	1,000–3,000
Process conditions scoping (four variables at three levels)	4 × 3 = 12	6 kg	30,000–40,000
Process conditions optimization (two variables at three levels)	2 × 3 = 6	3 kg	15,000–25,000
Ore variability tests	12	6 kg	30,000–40,000
Gold recovery (one per test)	30	NA†	25,000–35,000
Batch test total	—	**16 kg**	**100,000–150,000**
Continuous autoclave pilot, 72 hours		5–15 kg/h‡	
Setup, management, takedown	1	—	55,000–65,000
Operator labor	72 hours	—	110,000–140,000
Shift (12-hour) assays (feed and discharge for seven shifts)	7 × 2 = 14	—	5,000–15,000
Autoclave profile assays (8 six-compartment assays)	8 × 6 = 48	—	25,000–35,000
Gold recovery (one per assay)	14 + 48 = 62	—	60,000–90,000
Discharge mineralogy (X-ray diffraction, QEMSCAN)	2	—	5,000–15,000
Corrosion testing	1	—	5,000–15,000
Scale characterization	1	—	3,000–6,000
Gas sampling	1	—	10,000–20,000
Continuous autoclave pilot total		**350–1,000 kg‡**	**275,000–400,000**
Total for pressure oxidation test-work program		**355–1,015 kg‡**	**375,000–550,000**

*Quarter 4, 2015.
†NA = Not applicable; tests performed on material allocated to other tests.
‡Mass dependent on ore sulfur content, apparatus oxidation capacity, and retention time.

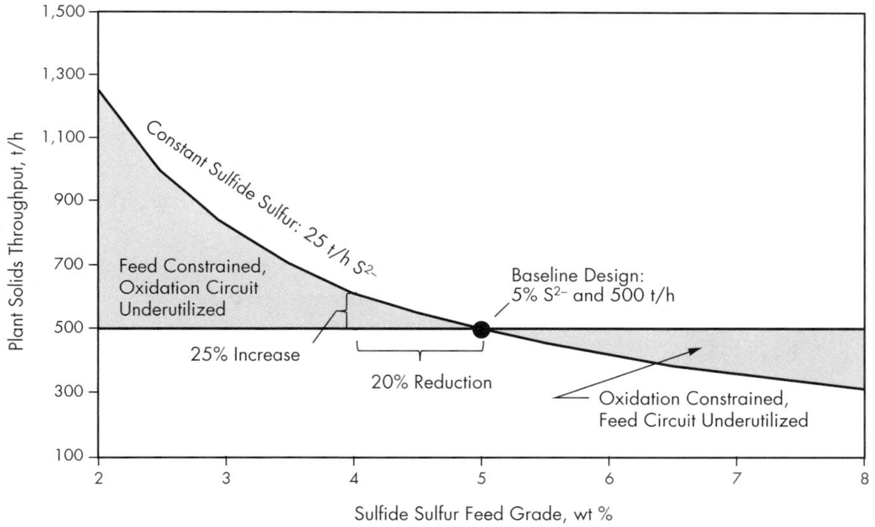

Figure 9 Impact of sulfur feed grade deviation on plant throughput and equipment utilization

oxidation circuit, with consideration of the anticipated range in feed grades, ore hardness, or other physical characteristics. For example, the capacity of a sulfide flotation concentrator for a pressure oxidation circuit should be determined by the lowest anticipated sulfide feed grade and sulfide recovery to ensure that the sulfide sulfur throughput is maintained to the pressure leach or oxidation circuit. In this way, by design, the plant should mostly be limited by the pressure leach circuit and not by the upstream ore preparation circuit. The design of a concentrator for a pressure leach circuit may be constrained by other factors, including transport or pregnant solution concentrations; however, a similar philosophy should be considered where applicable.

Figure 9 demonstrates the required concentrator throughput for changes in the sulfide sulfur feed grade from a baseline case. As shown in the figure, a 20% decrease in the sulfide sulfur feed grade from the baseline case of 5 wt % sulfide sulfur at 500 t/h requires a 25% increase in the ore preparation circuit capacity to maintain a constant feed sulfide tonnage to a pressure oxidation circuit. The increase is independent of any design requirements associated with variation in the feed, such as hardness. Sizing the ore preparation circuit from the perspective of the pressure leach or oxidation circuit often results in a revision of the ore stockpiling and plant feed blending strategy to minimize sulfide feed grade variation and the potential need for a larger circuit.

Feed Systems—Heating

Heat recovery is essential to the operation of a process in which the net exothermic heat of reaction is insufficient to raise the feed slurry temperature to the desired operating temperature. In pressure leaching operations in which the net heat of reaction is endothermic, efficient recovery of heat from the autoclave discharge stream is essential to minimize the usage of boiler steam. In pressure oxidation circuits in which the autoclave feed sulfide sulfur grade is insufficient to achieve the desired operating temperature, the operating temperature can be achieved by using single- or multistage heat recovery stages to preheat the feed slurry. The exact sulfide grade required to achieve a selected operating temperature is dependent on several factors, including mineralogy, feed density, oxidation reaction rate, retention time in the first autoclave compartment, rate of gas vent from the autoclave, and heat loss from the vessel (Mason 1990). An example is provided in Figure 10, where the autothermal sulfide grade is lowered for each additional stage of preheating (i.e., higher feed slurry temperature).

As of 2017, feed slurry pumping temperature is limited by the feed pump check valve elastomers to approximately 200°C. For applications in which the incoming slurry temperature is higher than the elastomeric material's rated working temperature (e.g., after preheating), water-cooled "dead legs" are used to protect the elastomeric materials. The water-cooled slurry legs allow the diaphragms to displace hotter process slurry at a distance, thus allowing for a higher operating temperature. The temperature of steam to the first stage of preheating aligns with the coincident pressure letdown vessel from which steam is generated and is usually slightly above the ambient boiling temperature. Intermediate heating stage set points are typically determined through process modeling and balancing volumetric steam flows from each source, with consideration of process requirements.

Direct-Contact Condensation

One of the most widely used and successful unit operations for preheating of slurries is a form of baffled direct-contact condenser (or slurry preheater vessel), as shown in Figure 11. Feed slurry enters the vessel through a distributor pipe at the top and flows downward by gravity while flash steam enters at the bottom of the vessel and flows countercurrently upward. Multiple angled baffles within the vessel redirect the slurry back and forth across the heater, decelerating the steam and forcing it to contact curtains of slurry. Slurry preheaters are thermally efficient and can achieve approach temperatures of <10°C—and sometimes <5°C—provided that sufficient energy is available in the steam to heat the slurry and the presence of noncondensable gas is negligible (where the *approach temperature* is defined as the difference in temperature between the incoming steam and the final heated slurry).

A rule of thumb for heating slurries is that every 10 kg of steam condensed will raise the temperature of 1 t of slurry by

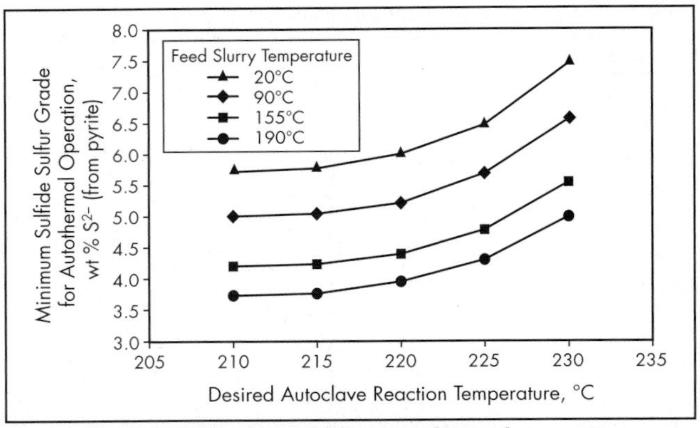

Figure 10 Sulfide feed grade for autothermal operation

External View: Courtesy of Newcrest Mining Limited

Figure 11 Direct-contact slurry heater with segmented baffles

approximately 6.5°C, based on the latent heat of steam and a slurry specific heat capacity equal to 70% that of water.

The mechanical design of slurry heaters generally follows international codes and standards for the design of unfired pressure vessels, particularly the American Society of Mechanical Engineers' standard, *ASME Boiler and Pressure Vessel Code* (BPVC), Section VIII, Division 1 (ASME 2015a); Australian National Standard AS 1210; or British Standard BS PD 5500. The vessel diameter and baffle spacing are dictated by the volumetric flow rate of the flash steam, whereas the number of baffles and baffle geometry are determined by the slurry properties. The performance of a baffled heat exchanger is highly dependent on slurry fluid properties (e.g., yield stress, apparent viscosity) as well as its thermal properties (e.g., conductivity, specific heat capacity).

Segmented baffle design has been applied successfully in many refractory gold slurry heaters since its original development by Anglo American Corporation in 1977 (Fraser and Thomas 2010). Heater vessel sizing generally follows the design principles established by James Fair (1993) for the design of baffle tray fractionation columns for the petroleum

Courtesy of Ma'aden Bauxite and Alumina Company
Figure 12 Jacketed pipe indirect heat exchangers for slurry heating in tube digester

refining industry. These principles include the number of baffles, gas or vapor rise velocity, and expected heat/mass transfer rate (in nephelometric turbidity units) as a function of the liquid-to-gas ratio. Several other baffle designs have also been tried with varying degrees of success, such as lathe grids, disks and donuts, and cups and cones.

A disadvantage of direct-contact heating is the dilution of the process stream. For multistage heating from 40° to 250°C, direct-contact heating adds approximately 370 kg of condensate per metric ton of autoclave feed slurry. If feed dilution affects the process (e.g., increased heating requirements), there may be an economic advantage to apply indirect slurry heating.

Indirect Slurry Heating

Indirect heating is accomplished without dilution of the process slurry, resulting in a lower volumetric flow rate through the circuit, and hence, smaller equipment and piping for the same ore treatment capacity can be selected. The Bayer process requires a high degree of heat recovery and minimal process liquor dilution for economic treatment of low-grade bauxite. Indirect heating has been used extensively in alumina plants since the mid-1960s. Modern alumina tube digester plants use jacketed pipe heat exchangers to preheat bauxite slurry prior to entering the tube digester. Examples include Rio Tinto's Yarwun alumina refinery in Australia; Ma'aden alumina refinery in Ras Al-Khair, Saudi Arabia; and the Emirates alumina refinery in the United Arab Emirates. Figure 12 shows a photo of a high-pressure tube digester plant constructed at the Ma'aden Bauxite and Alumina Company refinery.

Over the past 30 years, there has been substantial research conducted on indirect heat exchange, including the construction and operation of a 1/1,000-scale integrated pilot plant that used indirect heating of nickel laterite slurry from 100° to 250°C using shell-and-tube heat exchangers by AMAX and Inco Technical Services Ltd. in 1999, as well as research by Anglo American and Anaconda Nickel. The authors designed, constructed, and operated an indirect slurry heating demonstration plant in 1998 for the Murrin Murrin nickel laterite facility, as shown in Figure 13.

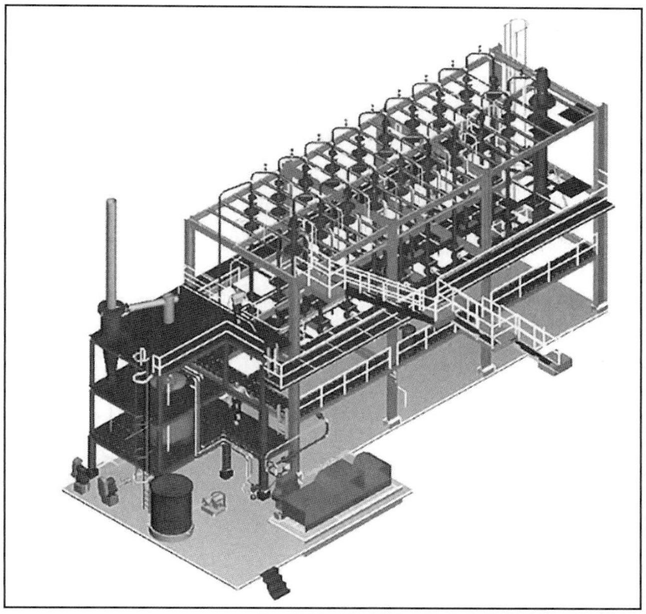

Figure 13 Murrin Murrin indirect heating demonstration plant

The challenges associated with the indirect heating of slurries are primarily associated with non-Newtonian rheology and hindered settling, where the interaction of fine particles results in rheopectic (shear thinning) behavior coupled with a high yield stress. A high apparent viscosity makes it difficult to achieve turbulent flow (i.e., a Reynolds number greater than 2,000), and heat transfer is dominated by conduction and diffusion rather than convection. Under a high apparent viscosity scenario, the effective heat transfer surface area needed to achieve the specified heating duty is substantial, and multiple heat exchanger units are required for each stage of heat recovery. For example, in nickel–cobalt laterite pressure leaching, heat transfer coefficients up to 600 $W/m^2 \cdot °C$ have been observed for a 1:1 blend of saprolite and limonite.

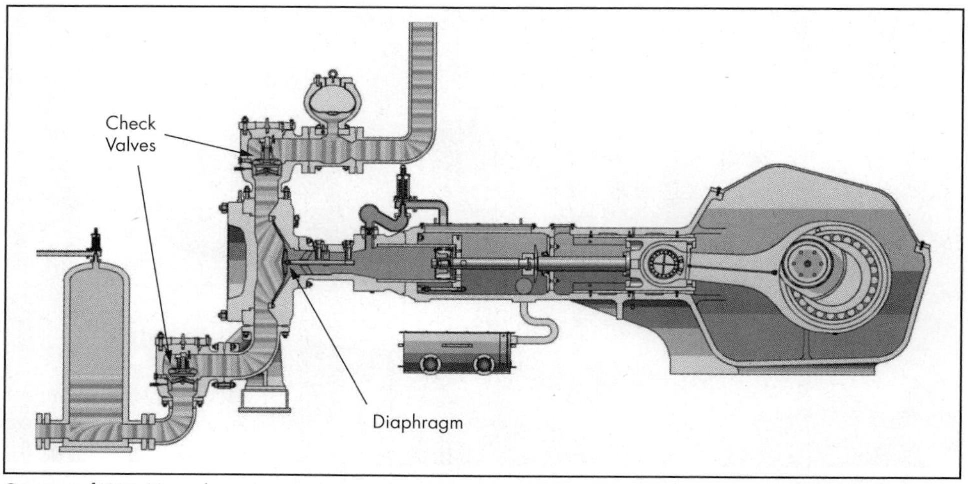

Courtesy of Weir Minerals

Figure 14 Positive-displacement slurry feed pump configuration

However, it is not uncommon to see values as low as 300 W/$m^2 \cdot °C$ for very fine limonite slurries.

An additional challenge of indirect heat exchange is maintaining the heat transfer rate in the presence of heat exchanger tube fouling, which is often a result of precipitating solids on hot surfaces. For example, the insulating effect of 1-mm-thick hematite scale on the inner surface of a 25-mm-diameter heat exchanger tube can reduce the overall heat transfer coefficient by 50%. For commercial operations, a means of tube cleaning has been developed using automated water-jet lancing to address this issue; however, the heat exchanger unit operation still needs to be taken off-line for cleaning duty.

Indirect heat exchangers are more susceptible to erosion and channeling when applied to ore slurries, leading to issues such as cross-contamination of exchange media due to wear-through of heat exchange boundaries or plugging as a result of fouling or solids accumulation. Therefore, more maintenance is usually required for this configuration, and redundant equipment may be required to maintain production rates during maintenance activities.

Pumping

The reactor vessel feed pump is typically a positive-displacement pump, most commonly a piston diaphragm configuration, where one or more hydraulically operated elastomeric diaphragms are situated between two check valves. The rated service temperature for feed pump diaphragm materials (Buna-N, HBN, EPDM) is 60°–90°C, but in a high-temperature service (greater than 100°C), the diaphragms are protected by a water-cooled "dead leg" or static section of slurry that allows the diaphragms to displace the higher-temperature slurry at a distance of 1.5–3.0 m from the diaphragm housing. A feed accumulator and discharge dampener are typically installed on the suction side and discharge side of the pump, respectively, to reduce pressure variation from the pulsating nature of the pump and to reduce energy losses associated with slurry acceleration. The pump is usually equipped with a variable-speed drive to control the pump stroke rate, and the associated mass flow is determined using in-line volumetric flow rate and density measurements. Figure 14 demonstrates a crankshaft-driven, single-acting, high-pressure piston diaphragm pump with feed accumulator and discharge dampener (Weir Group PLC 2016).

The reactor vessel feed pumps generally require a thorough preventive maintenance program to ensure reliable operating time. The highest frequency of wear occurs on the check valves and diaphragms, which account for about 30%–50% of the routine maintenance downtime. It is common practice to operate multiple feed pumps in parallel at low stroke rates to reduce the wear rate of wetted parts and to minimize process feed interruption from unplanned downtime. Figure 15 shows an example of the impact on stroke rate for typical check valve operating life. Parallel feed pumps are often sized to operate at 66%–75% of their design throughput to reduce the overall impact when one pump is off-line. Depending on project economics, it may be advantageous to select pumps that are each capable of delivering the full process design throughput for complete redundancy. At the Twin Creeks pressure oxidation facility, a third feed line was eventually added to each autoclave not only to expand capacity but also to improve pump run time and reduce the impact of pump maintenance (Krumins et al. 2014). It is common to have a centrifugal "charge" pump preceding the autoclave feed pump to provide adequate net positive suction head, but this is not always required.

A basket-style slurry strainer is typically used in advance of the reactor vessel feed pump to prevent oversize material from entering the pump and causing premature failure of the check valves or diaphragm. Quick change-out strainers or self-cleaning strainers should be used where excess oversize material is anticipated. More recent refractory gold plants such as Pueblo Viejo and alumina tube digesters at Yarwun, Ma'aden, and Emirates Alumina use self-cleaning strainers where oversize material can be discharged while the process remains online.

Multistage centrifugal pumps have been proposed as an alternative to positive-displacement pumps; however, they pose the risk of catastrophic failure of the pump casing due to reverse impeller rotation, as centrifugal pumps permit reverse flow when their motors are de-energized. Pump failure can be minimized with the installation of antireverse devices such as dynamic clutches; however, reverse process flow is still possible. For most pressure vessel operators, the safety risk and

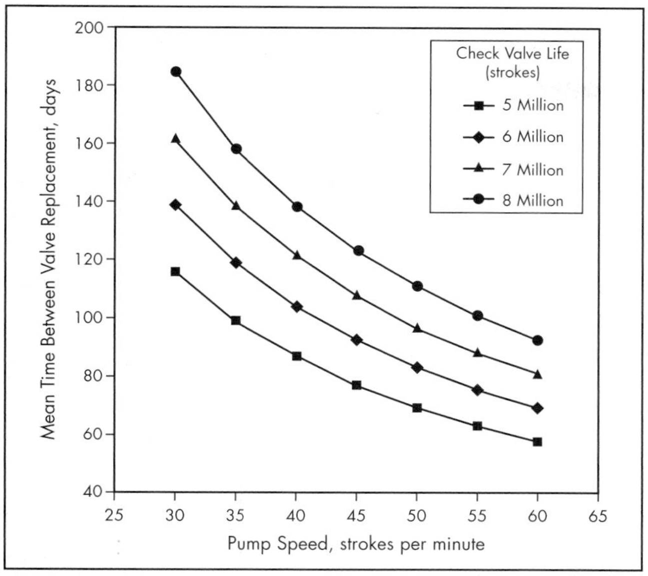

Figure 15 Impact of pump stroke rate on valve replacement requirement

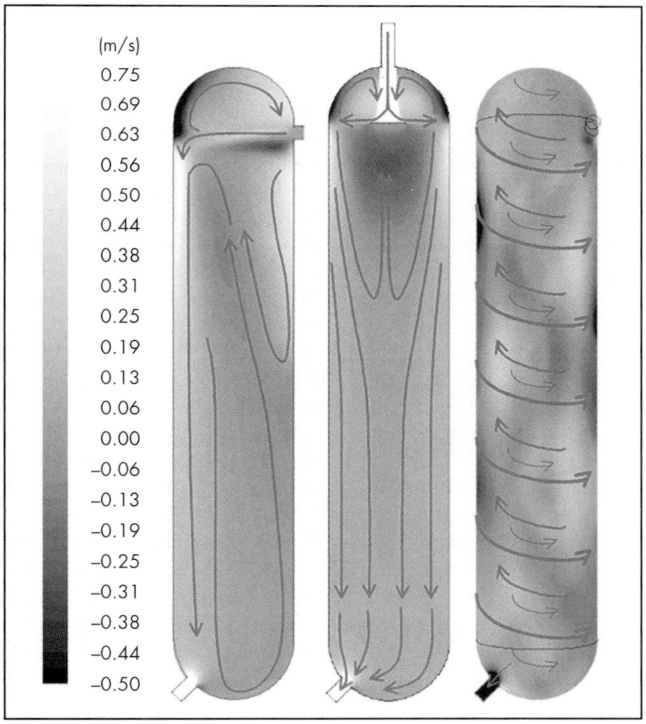

Source: Woloshyn et al. 2006. Copyright 2006 by The Minerals, Metals & Materials Society. Used with permission.

Figure 16 Comparison of axial velocity and the characteristic flow patterns in digesters with normal, central, and tangential slurry inlets

the cost of pump seal maintenance outweigh any potential savings in capital cost; therefore, positive-displacement piston diaphragm pumps are selected for most feed applications to autoclaves and high-pressure reactors.

Reactor Vessel

Vessel type and configuration. Reactor vessels are most often one of three configurations, depending on the application and project constraints:

1. Vertical digesters
2. Horizontal compartmentalized and mixed vessels
3. One or more vertically oriented vessels, mixed or lightly agitated by draft tubes

Vertical digesters, as used in alumina processing, are typically used as a series of agitated or unagitated vessels that either use direct steam injection to maintain temperature or are indirectly heated using internal steam coils or external heaters. The vessels are commonly 3–4.5 m in diameter and up to 30 m tall to promote an ideal residence time distribution for the chemical reaction (Hill and Sehnke 2006). Digester agitators may divide the vessel into a series of multiple compartments with staged baffle plates to prevent short-circuiting within the vessel.

Developments in digester technology have led to the use of plug flow tube digesters, which have been implemented in recently constructed facilities and have been able to reduce plant capital and operating costs (Gorst et al. 2013). A simulation of fluid flow through various tube digester configurations is provided in Figure 16 (Woloshyn et al. 2006).

Pressure acid leach and pressure oxidation autoclaves are typically configured as horizontal vessels with four to eight stages of agitation (process dependent), separated by internal compartment walls. The compartments act as a series of continuous stirred-tank reactors and provide a residence time distribution that ensures that the feed material is given sufficient opportunity to react. In oxidative processes, the initial stages (typically the first 25%–50% of the vessel) contain the most intensive oxidation, driven by oxygen mass transfer rates into the slurry, whereas the latter stages are required to satisfy kinetic limitations. A typical arrangement is shown in Figure 17.

Many autoclaves are designed with combined initial stages (i.e., an enlarged first compartment), which benefits the process in three ways:

1. For rapid reactions, the concentration of reaction products is distributed over a greater volume, reducing the potential for local precipitation and scale deposition, which can result in increased maintenance requirements.
2. Back-mixing of slurry allows overall compartment temperature moderation. Heat generated in the second stage can back-mix and heat the first stage, or cooling water introduced in the second stage can moderate the temperature in the first and/or third stages.
3. Acid generated by pressure oxidation has a greater concentration in the first stage (i.e., the acid produced in the second stage can mix back into the first stage), promoting soluble iron and increased oxidation rates or overcoming carbonate neutralization, if present.

Slurry cascades over the internal compartment walls, similar to a series of overflow-configured tanks. Alternatively, each wall may have an underflow port, allowing slurry flow through the bottom of the wall. An underflow configuration facilitates movement of coarse reaction product solids to the discharge end of the vessel with the intent of improved slurry level control and improved ease of cleanout.

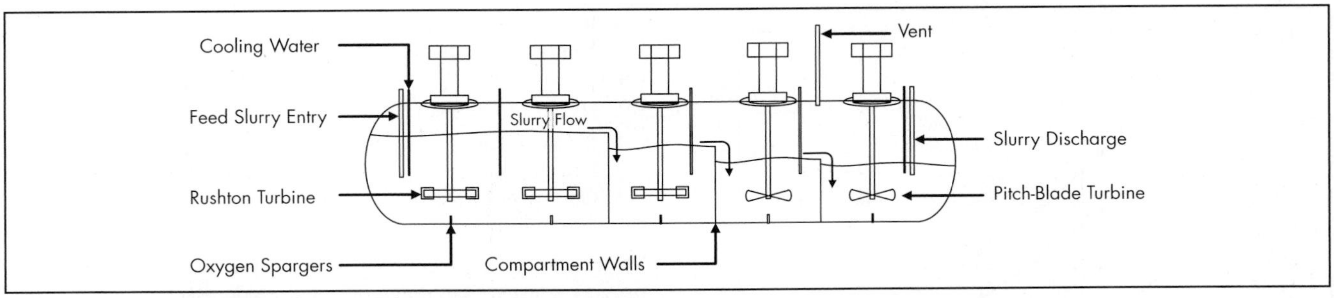

Figure 17 Typical oxidative horizontal autoclave arrangement

In oxidative processes, oxygen is delivered to each stage, usually below the agitator impeller and proportional to the expected oxidation profile through the vessel. High-pressure cooling water is directly injected into the slurry at a well-mixed location in each stage to achieve the target process temperature. The water addition rate is based on local temperature measurement and is directly indicative of the sulfur oxidation in each stage.

In some cases, multiple horizontal autoclaves can be used in series at staged conditions to optimize operations. Such is the case at Hudbay Minerals' zinc pressure leach plant in Canada, which was designed to leach 75% of the zinc in a low-acid leach autoclave, followed by pressure letdown, thickening, and re-leaching in a high-acid leach autoclave to recover the remaining zinc (Krysa 1995).

Autoclave vessel process diameters range from 1.0 to 6.0 m, and tan–tan lengths range from 6 to 45 m for demonstration-size vessels and commercial-size vessels, respectively. The corresponding process volumes can be as low as 4 m^3 for demonstration units and as high as 865 m^3 for commercial units. The largest autoclave (by volume) currently in operation is at Lihir Gold, Papua New Guinea, and is approximately 47.7 m long and 5.6 m in diameter (Collins et al. 2011). A comparison of vessel sizes for select operations is provided in Figure 18.

It may not be economically feasible to fabricate and transport large equipment to certain mine sites; therefore, alternative options such as on-site fabrication—or a series of vessels (e.g., vertical pots)—need to be considered. A series of vertically oriented vessels performs similarly to a combined horizontal vessel, with each stage segregated to its own piece of equipment. In this configuration, there is a loss of synergy with respect to the reduction of construction materials, piping, and instrumentation. However, this configuration may be advantageous if there are major restrictions on vessel fabrication and transportation and if the process is not prone to scale formation. The comparative cost implications are lessened for processes that do not require expensive materials of construction. Furthermore, a series of vessels may allow for bypasses around a vessel to continue operation during maintenance or descaling if this is not prohibited by piping and valve costs. An example of this configuration is the five-stage pressure oxidative leach for the recovery of cobalt from slag and copper production in Zambia (Munnik et al. 2003).

A vertical vessel is also appropriate if only one reaction stage is required. This is more often the case for processes that operate batchwise. For example, two 15-m^3 batch brick-lined autoclaves were commissioned at Xstrata Copper Canada's Canadian Copper Refinery in 2006 (now Glencore) and were designed to operate at a rate of 14 batches per week. The autoclaves operate in a two-stage batch process, first leaching nickel at 160°C in the absence of oxygen and then, in the same vessel, dissolving copper and tellurium at 110°–140°C through pressure oxidation (Doucet and Stafiej 2007).

Initial reactor vessel sizing. Sizing and selection for reactor vessels require careful consideration of many aspects of a project. One important aspect is the overall plant throughput and anticipated range in feed composition and circuit throughput, as discussed in the introduction to this section. Once the circuit feed throughput and variability are defined, the reactor vessel sizing can progress with consideration of typical process condition parameters: temperature, pressure, particle size distribution, retention time, reagent addition, and a leach recovery or oxidation extent target.

The approach to reactor vessel sizing depends on the phase of project development. In the preliminary phases, vessel sizing may be based on benchmarking or theoretical process modeling. As the project develops, the vessel sizing is revised with data obtained from a batch metallurgical testwork program and is updated during the feasibility and design project phases with pilot metallurgical test work, depending on the process. Projects with challenging mineralogy and oxidation processes require pilot test work to support the vessel design, whereas a noncomplex ore and widely used technology may use batch test work and benchmarking to support the vessel design (e.g., alumina digesters).

Initial pressure leach or pressure oxidation vessel sizing can be established with theoretical process modeling and industry benchmarking using operations with similar expected feed composition and operating conditions. Theoretical process modeling, as described by Papangelakis and Demopolous (1992) and Crundwell and Bryson (1992), can be used for initial vessel sizing and assessment of initial compartment configurations; however, the models do require fundamental assumptions of the oxidation or leach rate and an initial particle size distribution. Use of the methodology is iterative with adjustment to the vessel configuration and size until the target reaction extent is achieved. Industry benchmarking can be used to support the theoretical model assumptions or used in a simpler way with the comparison of leach rate or oxidation capacity metrics. A value of 0.7 sulfide sulfur metric tons oxidized per cubic meter of process volume per day (0.7 t S^{2-}/$m^3 \cdot d$) is a typical vessel oxidation capacity for a whole-ore pressure oxidation operation with a pyritic feed sulfide sulfur grade between 6% and 8%.

Consideration of transportation restrictions between the vessel fabrication facility and the project site is important for the sizing, selection, and design of pressure vessel equipment.

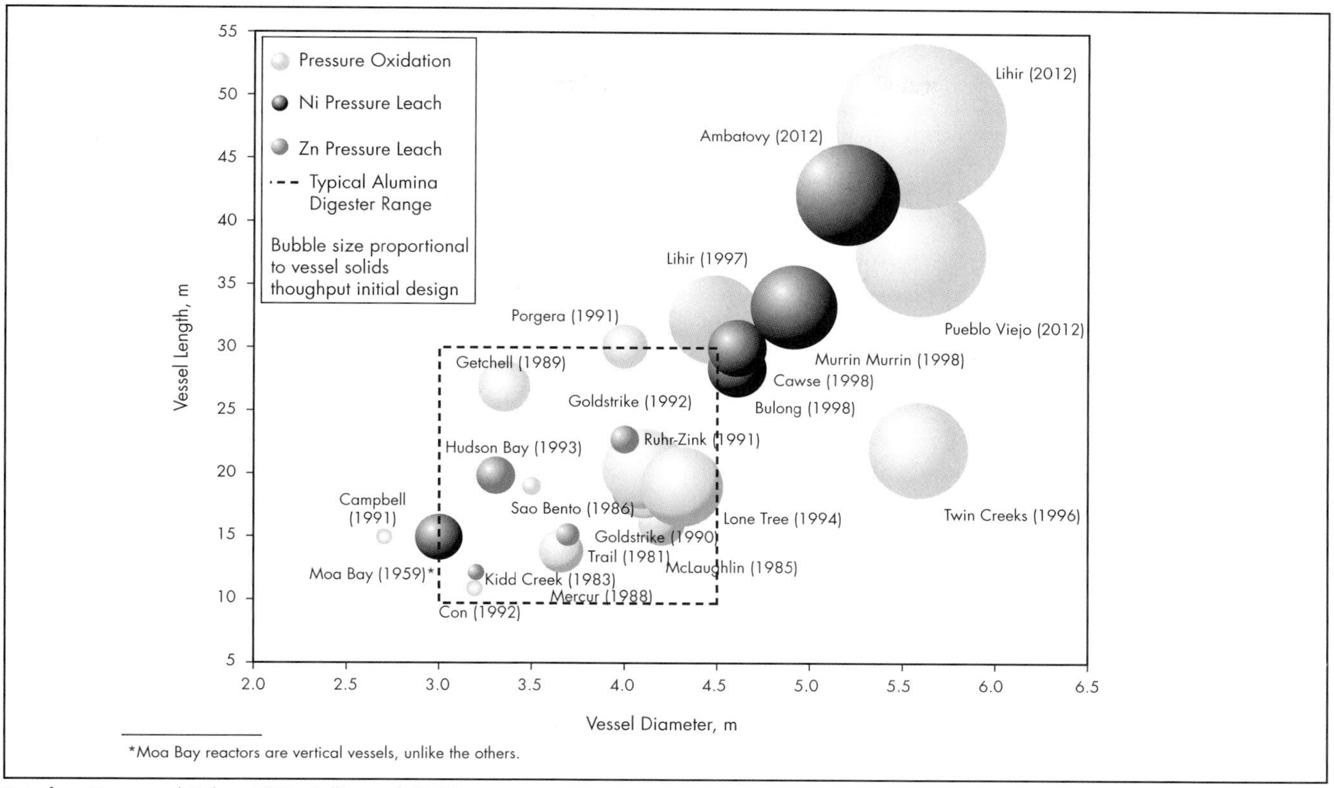

Data from Mason and Gulyas 1999; Collins et al. 2011

Figure 18 Comparison of reactor vessel sizes for various pressure hydrometallurgy processes

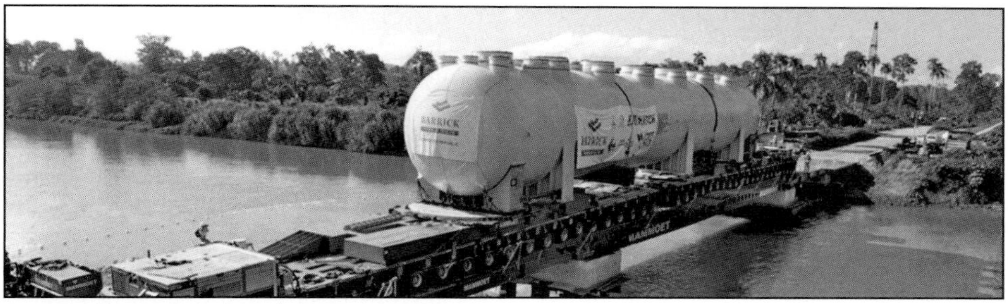

Courtesy of Mammoet USA North Inc.

Figure 19 Autoclave vessel transportation for the Pueblo Viejo Project, Dominican Republic

Transport restrictions include axle loading, load width and height, clearance requirements for overpasses and signage, road gradients, and turning radii. The cost of specialized transport for vessels over 100 t can be as much as the cost of fabrication, and delays in the delivery of key equipment can result in costly route traffic disruption and construction delay claims. Figure 19 shows the transport of a 780-t autoclave that is 37.6 m long and 5.6 m in diameter (Hatch Ltd. 2016).

Alternatively, the use of prefabricated assemblies enables conventional transport equipment but requires more assembly/fabrication at the site. A transport and associated site assembly/fabrication study should be conducted in the early stages of a project to determine which approach will be most cost-effective and least disruptive to communities along the transport route.

Preliminary reactor vessel sizing. Key process conditions are tested and selected during the batch test-work phase. The results from batch test work can be used to develop a preliminary mass and energy balance of the proposed commercial process to estimate the volumetric flow rates by compartment. These flow rates are then used to estimate a preliminary process volume and vessel size based on retention time. The mass and energy balance can be developed from first principles or by using process modeling software such as METSIM and supplementing batch test-work data with simulated values, where data are limited in the early stages of project development. The preliminary size can then provide information for an assessment of the mechanical constraints (primarily the vessel transportation requirements and fabrication).

Reactor vessel scale-up from pilot test work. Although batch test work is important to establish process conditions, pilot test work is required to verify the conditions and obtain data to support the vessel design, particularly for an oxidation process. Oxidation kinetics change from batch operation (static) to continuous (dynamic) operation, the latter of which emulates a commercial process and allows compartment profile sampling and composition analysis. Continuous piloting also provides a valuable opportunity to test different autoclave compartment configurations and ore blends at the target conditions and to identify challenges that may not have been apparent in the batch test work (Simmons 1995).

The key information obtained during pilot-plant test work is a leach or oxidation profile consistent with the selected vessel compartment configuration (e.g., one-, two-, or three-stage first compartment). Figure 20 provides an example of cumulative sulfide sulfur oxidation profiles for both a single- and dual-stage first compartment of the same ore type. The significant difference in the oxidation profile between the one- and two-stage first compartments is mostly associated with the acidification of carbonate minerals from the recycling of acid generated during the oxidation reaction. In this case, the change in compartment configuration enables a reduction in the cost to pre-acidify the autoclave feed.

It is not unusual for the oxidation profile information to indicate a range of oxidation rates with different ore blends or conditions. A design oxidation rate needs to be established based on a statistical analysis of all data and the selection of a defining ore blend. Once established, the oxidation profile information sourced from the pilot test work is used to update the mass and energy balance, and the vessel process volume is recalculated based on the cumulative retention time of all reactor compartments. Some important considerations for final vessel sizing include the following:

1. The density of water decreases as the water temperature increases; however, dissolved metals and acid concentration increase the solution density. The use of process modeling software accounts for both of these effects.
2. An allowance must be made for gas holdup for reactor designs with gas addition. The allowance needs to be tempered with consideration to any gas generated during the process, introduced inert gases, and gas solubility.
3. Final consideration must be given for any mechanical or physical constraints associated with fabrication and transportation to a site.

Design standards. Pressure vessels such as autoclaves impose a potential risk to health and safety if they are not designed, fabricated, installed, operated, and/or maintained in accordance with pressure vessel code requirements. The ASME BPVC is globally recognized and widely accepted for the design and fabrication of unfired pressure vessels and satisfies the regulatory requirements of authorities in many countries.

Within the ASME BPVC, a vessel can be designed using Section VIII Division 1 or Division 2 (ASME 2015a, 2015b). Division 1 is commonly referred to as "design by rule" in which the component sizes and material thicknesses are governed by preset rules, equations, and conservative values of allowable material stresses that have been developed by materials researchers and adopted by the ASME Code Committee. The most commonly used pressure vessel code tends to be Division 1, where a design-by-rule method is implemented to achieve a code-compliant design.

Within Division 2, a vessel (or components of a vessel) may be designed using either *Part 4: Design by Rule* or *Part 5: Design by Analysis*. Typically, design by analysis is used for nonstandard components that are not addressed through design by rule. Division 2 is considered a more sophisticated approach because it permits increased utilization of material strength (i.e., higher allowable stress) and generally results in thinner, lighter vessel wall components. The greater reliance on the material properties of Division 2 is accompanied by additional—and more stringent—quality requirements for materials and fabrication compared to those of Division 1. Division 2 is typically only used for a vessel when the additional design effort is compensated by material cost savings or weight restrictions for transport or when this is required by the end user.

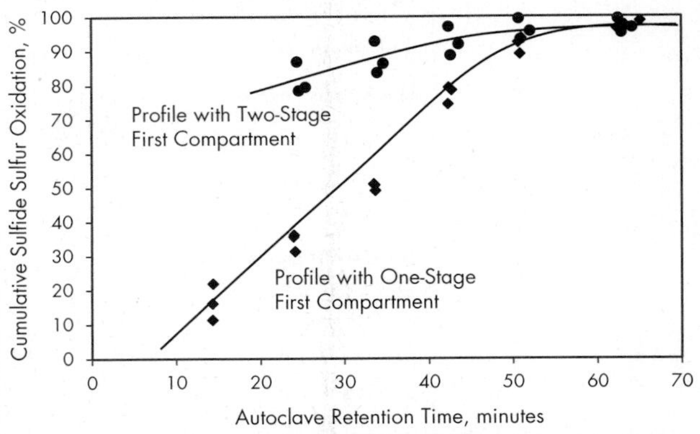

Figure 20 Cumulative sulfide sulfur oxidation as a function of cumulative retention time

The use of Section VIII Division 1 is sufficient for most autoclave applications. Where design rules are not provided to address a particular aspect of the design, then the use of Division 2 is considered acceptable to qualify that aspect. Examples of this provision are the assessment of fatigue in agitator nozzles and thermal displacement stresses at support locations for metal-clad autoclaves. However, in this case the design is still conducted using Division 1 allowable stresses in such applications because the basic vessel design is still based on Division 1 requirements.

Pressure vessels conforming to ASME requirements receive a U stamp in addition to other markings that may be required to identify registration or compliance with a local regulatory body. In the United States, some jurisdictions require National Board registration identified with an NB mark. In Canada, a pressure vessel must be registered in the province or territory of its end user and be stamped with a Canadian Registration Number certifying federal approval of the vessel for legal operation. If the vessel is built outside of Canada for Canadian use, it must also be registered with the National Board. For an ASME pressure vessel to be used within the European Union (EU), the vessel must demonstrate full compliance with the Pressure Equipment Directive (PED; EU Directive 97/23/EC). Within the EU, the PED is a legal requirement in which compliance is identified by CE marking, indicating that the marked vessel is permitted for use anywhere in the EU. The PED dictates essential safety requirements governing the design and fabrication of pressure vessels. Instead of providing a design methodology for unfired pressure vessels, the PED references the harmonized standard BS EN 13445 that has demonstrated PED conformity. The use of BS EN 13445 is not mandatory, and the ASME BPVC can be used pending completion of a PED conformity

Courtesy of Patrick Lauzon, Hatch

Figure 21 Pressure vessel refractory lining installation in progress

assessment. ASME Section VIII Division 2 is comparable to BS EN 13445, making compliance with PED easier than with ASME Section VIII Division 1. PED compliance may be achieved by following a guide published by ASME that outlines the additional requirements to meet the PED based on ASME Section VIII Division 1. Australian vessels typically comply with Australian National Standard AS 1210.

For autoclaves, internal refractory lining design requirements are not addressed by pressure vessel codes. Within the design codes, an internal refractory lining is considered a dead load. A vessel designed in accordance with a pressure vessel code will have adequate strength, allowing the vessel to deform under specified operating or design conditions without concern. However, an internal refractory lining responds poorly to vessel deformation, which can induce excessive compressive stresses in the refractory face course and tensile stresses in the refractory back course. Both are detrimental to lining stability, potentially resulting in brick spalling and/or complete lining failure. Because a refractory lining has a very low resistance to induced bending stress, additional calculations and finite element analyses (FEAs) are commonly completed to investigate vessel deformation and the interaction between the refractory lining and the vessel. Vessel deformations are investigated and reinforcement options are developed to minimize deformation gradients in areas that could affect lining stability. Because there is no specific standard addressing these design practices, the combination of experience and the criteria of BS EN 14879, Annex D (formerly DIN 28060) are potential resources to develop a basis for determining acceptable deformation gradients in a refractory-lined vessel (Donohue et al. 2012).

Areas of concern for excessive deformation gradients tend to be discontinuities inherent to vessel designs (e.g., nozzles, re-pads, supports, stiffeners, and heads). A hemispherical head would be preferred for lining stability; however, a semielliptical 2:1 head is common in autoclave applications to suit process requirements. The transition from vessel cylinder end to the dished portion of the head is typically a region of concern. The suggested best practice to reduce vessel deformation gradients in an autoclave is simply increasing the shell thickness because the addition of stiffeners or re-pads creates additional discontinuities with the potential to induce bending stresses in the refractory lining. Depending on the design pressure of the autoclave and the shell thickness required by code, it may be determined that the additional shell thickness for lining stability is minimal in comparison. As additional shell thickness benefits refractory lining stability, it is advantageous to design the vessel using ASME BPVC Section VIII Division 1 (ASME 2015a) because the code generally results in a thicker vessel shell compared to using Section VIII Division 2 (ASME 2015b) or BS EN 13445.

Lining systems. Many reactor vessels are made of C-Mn-Si steel and are fire clay refractory lined, with an acid-resistant protective membrane between the shell and thermally insulating refractory lining. Figure 21 shows the installation of a four-course refractory lining within an autoclave. Designing the lining system with respect to stresses is critical to its performance, maintenance requirements, and longevity. A balance must be obtained between keeping the brick under compression (due to low tensile strengths) without imposing so much compression that bricks are spalled or crushed. Therefore, many considerations must be made with respect to the vessel size, operating conditions, and selection of materials. Furthermore, operating procedures should be established to limit the rate of pressurization to approximately 350 kPa/h and rate of temperature change to approximately 10°C/h (or as recommended by the brick lining supplier) to minimize transient stresses on the lining. Sufficient brick must be installed to maintain a membrane temperature less than the maximum working temperature of the selected membrane material, which is usually between 85° and 120°C.

Historically, chemically pure (>99.95% w/w Pb) lead has been used as the protective membrane for pressure oxidation and zinc pressure leach processes; however, many lining suppliers have developed alternatives (e.g., elastomeric, polymeric, bituminous mastic) to meet the needs of various vessel designs and user preferences. As of 2010, chemical lead membranes were still in service after 54 years in the Moa Bay leach reactors. Similarly, a bituminous mastic membrane had provided more than 15 years of service with no failures or repairs at the Twin Creeks pressure oxidation facility (Wei et al. 2010).

One of the greatest challenges is the design of the lining system around vessel nozzles. Nozzles are troublesome, as their geometry and proximity to one another promote hotter surface temperatures (McMullen 2014), which can lead to membrane creep and failure (Donohue et al. 2013). In such cases, the membrane may be replaced with high-nickel alloy weld overlay (e.g., Inconel 625, Hastelloy C2000, Alloy 59), which has a greater resistance to temperature creep, although at a higher cost (Bristowe et al. 2004). Alloy nozzle seal rings are typically required, with a spiral-wound gasket seal to ensure that the autoclave process fluids cannot contact the steel shell where the lining ends. In smaller-diameter nozzles, access does not permit the installation of refractory brick for thermal protection. In these cases, either fire clay or thermoplastic insulating inserts are used (Fraser et al. 2008). Detailed finite element analysis is typically performed during detailed engineering to determine the optimal nozzle lining design.

For processes operating above 235°C, such as pressure acid leaching (250°–270°C), the design and installation of structurally stable refractory linings become more difficult and costly with additional lining thickness (Donohue et al. 2012). For such applications, the industry has adopted

Table 6 Comparison of refractory and titanium-clad autoclave linings

Argument	Refractory Lining	Titanium-Clad Lining
Advantages	Excellent corrosion resistance in both oxidizing and reducing sulfuric acid environments Excellent abrasion resistance Excellent resistance to oxidation and ignition (pyrophoricity)	Excellent corrosion resistance in oxidizing environments Higher service temperature (up to 315°C) is permitted by pressure vessel codes Can be in direct contact with process media, resulting in a smaller and lighter vessel for a given process volume
Disadvantages	Lower service temperature limits (<250°C) due to stability limitations in lining thickness Higher maintenance costs associated with face course refractory relines Requires larger vessel to accommodate lining thickness for a given process volume	Susceptible to pitting and/or crevice corrosion in reducing environments Reduced abrasion resistance, especially to siliceous ore slurries Potential for ignition in enriched oxygen environments (pyrophoricity) Weld repair of corroded cladding is difficult and expensive

Source: Pearson et al. 2010

metal-clad construction—specifically explosion welding—as a means of bonding corrosion-resistant metal such as titanium to a C-Mn-Si steel substrate. This has been shown to be a cost-effective alternative to refractory lining at vessel sizes of 300 t or more (Pearson et al. 2010). Although it varies with project specifics (e.g., vessel sizes, materials, fabricator shop loading), a typical delivery period for a large, horizontal, titanium-clad autoclave is in the range of 80 weeks, whereas that of an equivalently sized brick-lined vessel is approximately 60 weeks, including 8 weeks for on-site brick installation (Zunti and Pearson 2004).

The titanium layer is considered a corrosive barrier and is not included in the design for pressure containment (e.g., wall thickness). The explosion-bonded material is created in flat sheets, rolled or formed into the sections that make up the pressure vessel, and sealed using a batten strap welding technique. A comparison of the two autoclave lining options is provided in Table 6.

Mixing and agitation. Mixing and solids suspension are typically required for most slurry leaching applications in which reagents (e.g., acid) are added to produce a chemical reaction, and maintaining solids suspension is important. Depending on slurry solid content and viscosity, mixing and solids suspension duty can be performed using hydrofoil or pitch-blade, turbine-style axial flow impellers. For low agitation requirements, pneumatic or steam-based agitation is possible in specifically designed vessels (e.g., Pachucas) to reduce wear components and has been successful for uranium and alumina leaching applications.

Agitator impeller tip speeds are typically maintained below 6 m/s during normal operation to limit abrasive wear from the slurry on both the agitator and the vessel lining. Some operations use protective coatings (e.g., titanium dioxide, chromium oxide) to reduce agitator wear and extend the service life; however, regular maintenance, refurbishment, or replacement is still expected. Agitator materials are typically selected to be similar to vessel internals (e.g., titanium), according to process conditions. In the case of titanium agitators in high-oxygen concentration environments, the upper shaft is fabricated from an ignition-resistant material (e.g., nickel alloy) as a safety measure against breach of containment, should the lower titanium agitator shaft ignite. Many agitators are designed with impellers that must be assembled within the vessel itself to reduce requirements for larger vessel nozzles, which can be problematic in terms of thermal lining design, gasket seal tightness, and long-term maintenance of the nozzle.

The primary design intent of pressure oxidation autoclave agitators is oxygen gas dispersion, typically achieved through the use of high-shear radial turbines, such as Rushton turbines. These agitators typically achieve power-per-unit volume inputs of 1 to 3 kW/m^3 (Pieterse 2004). The theoretical unit-power requirement is based on several variables, including oxygen partial pressure, oxygen solubility, and operating temperature, which determine the system oxygen mass transfer coefficient ($k_L a$) to suit the oxidation reaction. A simulation of radial fluid flow for gas dispersion is provided in Figure 22.

For the latter stages of pressure oxidation, when the oxidation reaction is typically kinetics limited, the agitation duty changes from gas dispersion to solids suspension. At this point, oxygen addition and dispersion are typically minimal, as most of the oxidation reaction has already occurred in the initial high-power mixing stages. In this regime, pitch-blade or hydrofoil impellers can be used and typically require substantially lower power input (approximately 0.5–1.0 kW/m^3) to achieve the required solids suspension.

Positive isolation requirements. Pressure leaching and pressure oxidation processes have many inherent hazards; therefore, effective isolation or disconnection of equipment from every energy and process source is essential for safe operational and maintenance practices. Table 7 lists common hazards of pressure hydrometallurgical processes that require positive isolation.

The autoclave piping design needs to consider the appropriate isolation requirement for the service and isolation under different scenarios (e.g., full shutdown vs. hot standby). Many of the risks from hazardous conditions cannot be mitigated using single-block (on/off) valves; rather, single-block valves with bleed streams can only be considered sufficient until positive isolation is provided by other acceptable means such as a process shutdown. Double-block-and-bleed valves (i.e., a pair of block valves fitted into a line with a bleed valve in the space between the block valves) are advisable for all systems containing hazardous fluids to provide effective isolation, as long as the bleed line can be verified as being clear and routed to a safe destination.

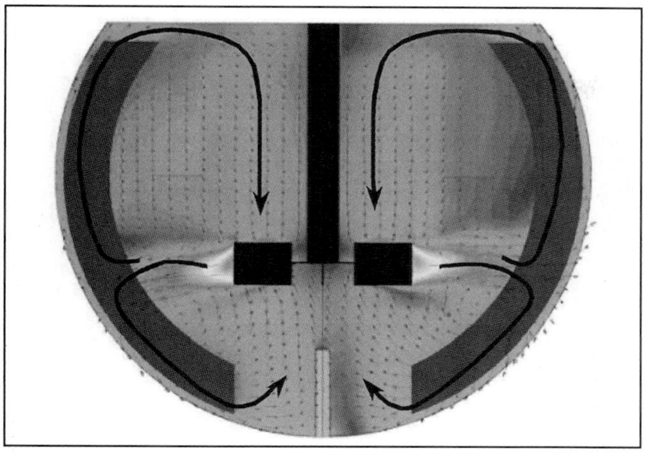

Source: Nicolle et al. 2009

Figure 22 Radial slurry flow within an autoclave compartment (axial vessel view)

Table 7 Hazardous conditions typical for pressure hydrometallurgy

Hazard	Description
Elevated temperature	Above 100°C
Elevated pressure	Above 1.0 MPa
Corrosive materials	Highly acidic or alkaline slurries
Oxygen-enriched environment	Oxygen concentration greater than 35 vol %
Asphyxiation	Abnormal oxygen levels
Flammable materials	Flash point <60°C
Pyrophoric materials	Potential for spontaneous ignition in air
Explosive	Potential to exceed lower explosive limits
Radiation	>1 mSv per annum dosage rates
Reduced temperature	Less than –5°C
Toxic materials	For example, lead mortar

Isolation valves for hazardous fluids typically subjected to physically demanding conditions, such as corrosive and abrasive slurry flow, are designated as "severe service." Examples of severe service valves include autoclave inlet and outlet (slurry and vent) valves, acid injection block valves, oxygen block valves, and steam injection block valves. Only 50–100 valve cycles are typically expected for slurry duty prior to failure (leakage) due to the demanding nature of the service. Severe service valve performance has a direct impact on the mechanical availability of the processes, as failure to achieve positive isolation results in protracted, unscheduled shutdowns (Lauzon et al. 2004) if the valves fail on demand. Key features of severe service valves include the use of bidirectional seating and ceramic coatings in the design of balls and seats. Both the design and selection of severe service valves are critical to extend valve life, as their sizes, construction materials, and specialized design typically result in significant maintenance or replacement costs. Valve repair or replacement costs can represent up to 85% of the annual maintenance cost of the piping system (Caruana et al. 2010).

Pressure Letdown

Dedicated process equipment is required to return the pressurized reactor products to ambient or near-ambient conditions. This is typically accomplished through the use of a refractory-lined phase separator (flash vessel) in which slurry is subjected to a controlled pressure letdown and the slurry and steam phases are separated. In this process, steam rapidly evolves from the solution phase (flashing) as the slurry temperature and enthalpy decrease to maintain thermodynamic equilibrium through the pressure reduction. The result is a highly abrasive, high-velocity flow field that requires the use of ceramic materials to achieve equipment service life.

The letdown arrangement can be a single- or two-stage configuration. In a single-stage configuration, a partial pressure reduction occurs through the autoclave level control/letdown valve and the final pressure reduction occurs as the combined steam and slurry enter the flash vessel through a ceramic-lined blast tube. In a two-stage configuration, the pressure reduction through the level control valve is maintained above the saturated steam pressure (to avoid flashing), and the final pressure reduction occurs across a fixed orifice (choke) located within the vessel at approximately midheight. Bottom-entry flash vessel arrangements have been used in alumina digester processes to dissipate energy into the slurry volume. In this design, a recirculating draft tube prevents slurry from flashing as it enters the vessel.

Water may be added to the incoming slurry to modify the flash point of the slurry and defer flashing from the level control valve to the choke outlet and to reduce the total amount of flash steam produced. This can be an effective technique for increasing slurry flow through a capacity-limited system to the physical limits of the piping and choke size.

Energy dissipation is critical to flash vessel design. Slurry typically enters the vessel at or above sonic velocity as a result of choked flow (Blackmore et al. 2014). Therefore, consideration of the overall slurry momentum balance is critical to understanding how the high-energy stream is dissipated within the vessel. Furthermore, the depth to which the slurry jet penetrates the vessel operating volume (slurry pool) should be determined, such that the risk of excessive wear and potential breach of containment can be prevented through selection of adequate slurry pool depth. The inclusion of a sacrificial silicon–carbide impingement block is often installed at the base of the flash vessel to minimize the risk of vessel wear. Figure 23 demonstrates the modeled variation of top-entry slurry penetration depth during start-up (Blackmore et al. 2014).

The steam evolved from the flash vessel is typically near the ambient boiling point of water (100°C) and has the potential to be used as a low-grade heat source in other parts of the process, such as preheating slurry to the reactor vessel. Should higher-temperature steam be required for process heating, the opportunity exists to let down the system pressure in stages (i.e., a series of flash vessels) to generate steam at an intermediate temperature and pressure. An example of such practice is the Twin Creeks pressure oxidation facility, where two stages of pressure letdown supply flash steam to two stages of slurry preheating, allowing for operation at lower sulfide grades (Yernberg 1996).

For alumina processes, the number of flash vessels is an important economic consideration. Low-temperature plants that process largely gibbsitic bauxite may require three of four stages of pressure letdown, whereas high-temperature digestion for boehmitic or diasporic bauxite may require 8–12 stages (Haneman 2012).

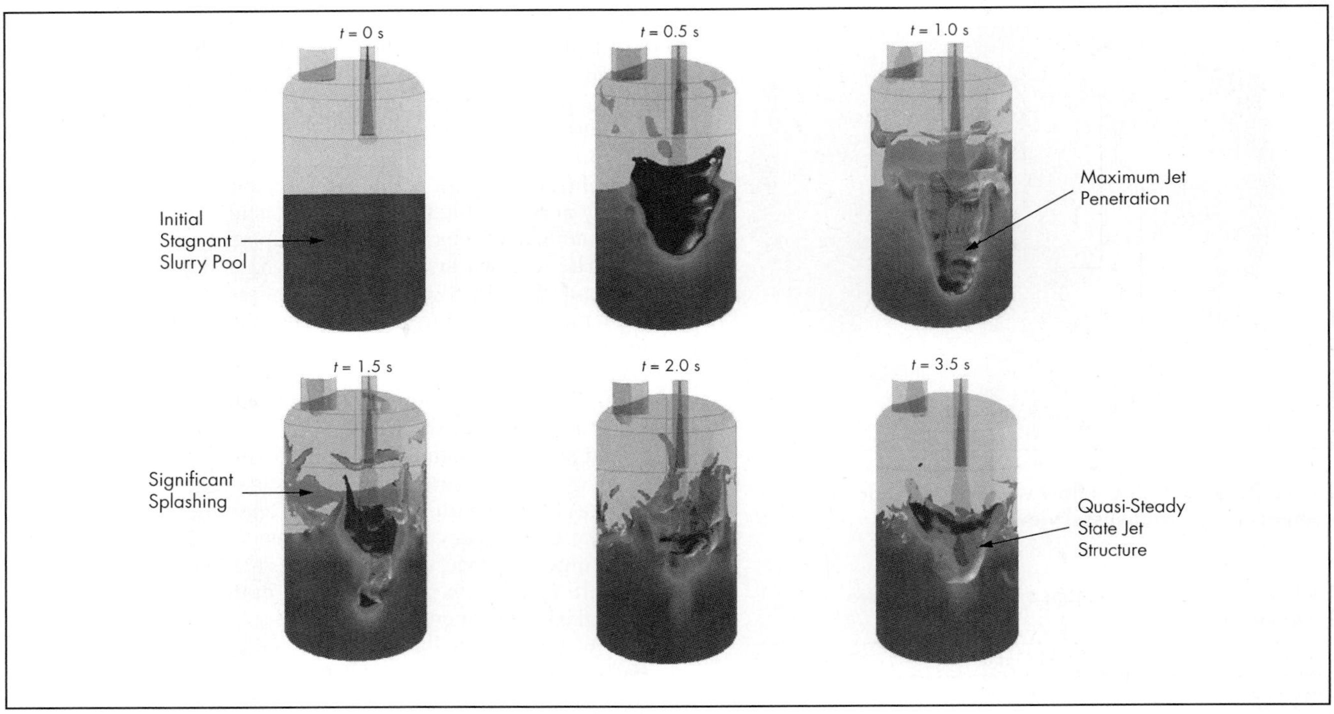

Source: Blackmore et al. 2014

Figure 23 Slurry jet penetration for top-entry flash vessel operation

The diameter of the flash vessel is typically governed by a steam mass flux (kg/s·m^2) to allow for sufficient separation from the slurry and minimization of carryover. For large-capacity processes, the vessel diameter may be limited by the stability of the refractory lining in the domed upper head of the vessel or by transport restrictions. In this case, the duty may be split between two or more vessels in parallel. The mechanical design of pressure letdown vessels generally follows international codes and standards for the design of unfired pressure vessels, particularly ASME BPVC Section VIII, Division 1 (ASME 2015a); Australian National Standard AS 1210; or British Standard BS PD 5500.

Off-Gas Handling

Cyclone systems. A cyclone may be used as an initial measure to remove entrained slurry and recover solids material from the flash vessel steam stream, particularly where there could be an impact on the downstream process or if it is important to recover the solids or solution streams. As shown in Figure 24, the cyclone feed gas is fed tangentially into the top of the cyclone, where the rotational inertia of the gas and gravity promote separation of entrained material. The gas exits the cyclone through a vertical dip pipe (vortex finder) in the top center of the vessel. The cyclone geometry defines the recovery efficiency and size of the entrained slurry or solution droplets that are removed. A differential pressure reduction of 10–40 kPa is typical for cyclone operation; however, the differential will vary depending on the operation of the downstream off-gas system.

The cyclone typically experiences high velocity with variation in solids content and therefore abrasion-resistant materials are required to reduce the maintenance frequency and to achieve the required equipment service life. Refractory or engineered ceramic linings allow for resurfacing of wear areas within the cyclone. The internal surface of the cyclone is commonly wetted with a small flow of water to promote the transport of captured material and to reduce solids buildup on internal surfaces.

Condenser systems. Water-based atmospheric condenser (quench) systems are frequently used to reduce the volume and temperature of waste steam from the process, to lessen the load on downstream gas cleaning processes, and to reduce environmental emissions. The condenser system also provides an opportunity to recover low-grade waste heat from the process. As shown in Figure 25, water is typically sprayed in direct contact with the combined process vent gas (flash and autoclave vents) to promote the condensation of steam. Spray nozzles are selected to atomize the water and increase surface contact area between it and the gas for improved performance. The condenser does not typically include any packing if fouling and scale formation (e.g., the condensation of elemental sulfur) will lead to significant maintenance requirements and increased process downtime. Sufficient gas contact time of 2–5 seconds must be included to achieve sufficient condensation. Water should be added at a rate such that the operation is condensation limited rather than heat transfer limited, if possible. Depending on its temperature, the water addition rate typically varies between 6 and 12 times the mass flow rate of the gas (i.e., the liquid-to-gas mass ratio).

Alternative quench media, such as process or tailings slurry, may be considered where the removal of elemental sulfur is desired prior to gas scrubbing. If slurry is used, cascading baffles are typically used as opposed to spray nozzles (similar to slurry heater vessels). In all cases, any chemical interaction between both streams must be considered, as well

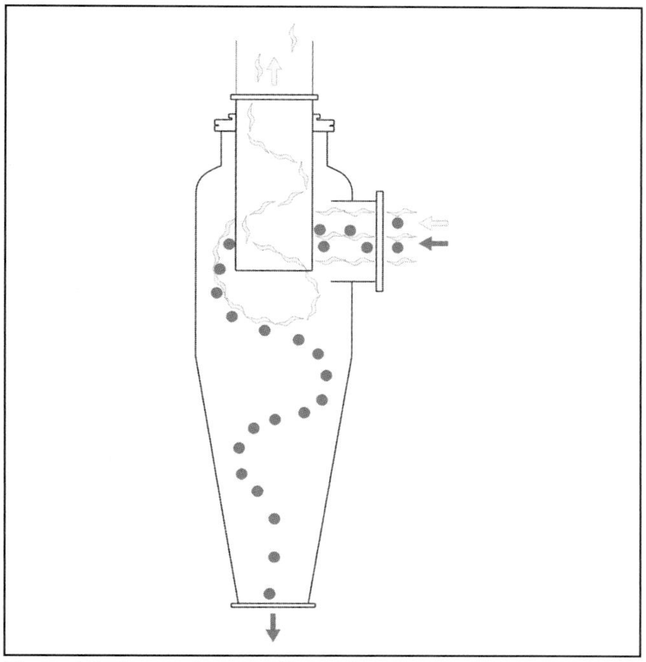

Figure 24 Gas cyclone configuration

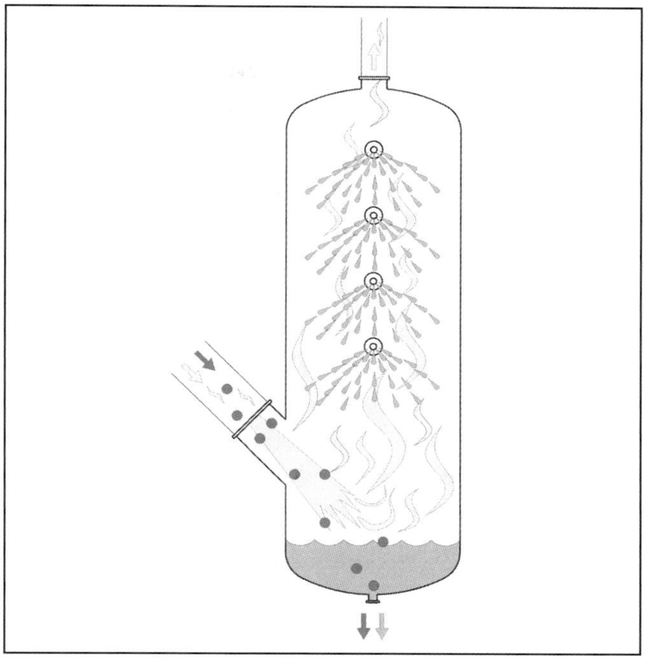

Figure 25 Condenser vessel example

as the potential for any component of the quench media to be stripped by the gas and emitted from the process.

Scrubber systems. High-energy wet scrubbers typically provide a final stage of vent gas cleaning (>99% particulate removal) prior to the gas being discharged to the atmosphere. In the wet scrubber, the remaining gas from the condenser system comes into contact with water, typically at a rate of 1.3 L/m^3 of gas, and is accelerated through a venturi and a 90° bend to promote the removal of particulate matter (see Figure 26). A pressure differential reduction of approximately 11 kPa is frequently required to provide adequate solids removal. A cyclonic separator body induces final separation of droplets from the vent gas. If required, a mist eliminator can be installed prior to gas discharge to recover residual condensed droplets; however, fouling of the mist eliminator and the necessity for cleaning should be considered.

Ideally, scrubber water should be relatively free of particulate matter and should be used in a single-pass system for best performance (i.e., not recycled). Recirculating the scrubber solution can be problematic with respect to fouling and scaling, similar to the condenser system. Any chemical interaction between the water and gas must be considered, as well as the potential for any component of the scrubber water to be stripped by the gas and emitted from the process.

Special systems. In some cases, process off-gas may contain contaminants that require additional treatment, as they are not effectively removed by conventional process equipment. Examples of such contaminants are elemental sulfur and mercury, which can be volatilized under high temperatures and oxidative conditions (Krumins et al. 2013). Elemental sulfur can be removed using condenser systems with slurry to promote agglomeration of sulfur particles (Fraser and McCombe 2011). Pressure oxidation processes have used carbon-bed technology for removal of mercury to meet regulatory emissions targets.

The necessity of any special treatment requirements should be assessed on a case-by-case basis from an analysis of the process feed material, pilot test off-gas analysis, or stack tests during operation.

Oxygen safety. Pressure oxidation and pressure oxidative leach process vents typically contain unused oxygen. Residual oxygen can concentrate as a result of condensing steam, increasing the risk of oxygen-enriched metal ignition (e.g., titanium). Therefore, where applicable, the process vent and downstream equipment design needs to consider selection of ignition-resistant materials of construction, appropriate instrumentation, and process controls such that the risk of oxygen-induced ignition is mitigated. An example of ignition risk mitigation is controlling the autoclave overpressure to an oxygen concentration of less than 35% using the measured autoclave vapor temperature and pressure (Frischmuth et al. 2014). Readers should refer to the best practices for oxygen safety and design principles from organizations such as the Compressed Gas Association and the European Industrial Gases Association.

Ancillary Systems and Services

Oxygen. Pressure oxidation and the pressure oxidative leach process typically require a supply of high-purity (>90%) oxygen. Depending on the quantity required, oxygen is either supplied as delivered liquid oxygen (for small plants and pilot work) or produced on-site from an energy-intensive (450–550 kW·h/t O_2) air separation plant. On-site production is most commonly from vacuum pressure swing adsorption (VPSA) or a cryogenic separation process, where the former is typically used for capacities less than 500 t/d. The VPSA process uses molecular sieves to filter out nitrogen, water (humidity), and hydrocarbons and typically produces a product at approximately 93 vol % O_2. The cryogenic process distills oxygen and nitrogen from air based on differences in boiling points and can produce oxygen at greater than 99% purity (McMullen et al. 2014).

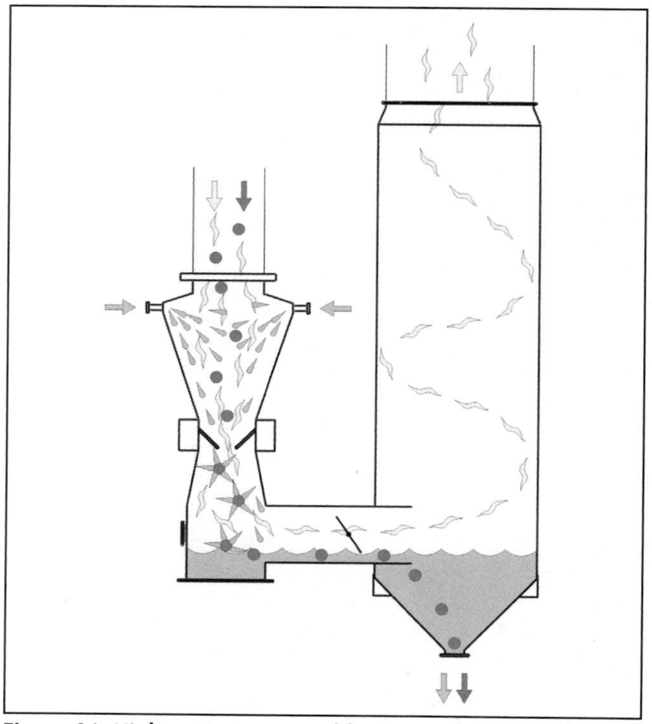

Figure 26 High-energy wet scrubber

Although the oxygen purities of both processes are relatively high, they are an important consideration for processes with sensitivity to oxygen partial pressure. The lower the oxygen purity, the higher the system operating pressure (and cost) must be to achieve the same oxygen partial pressure in the autoclave. Furthermore, additional impurities result in greater venting requirements to maintain oxygen partial pressure and increase the size of gas handling equipment.

Acid. Sulfuric acid is a principal reagent for acidic pressure leaching of laterites and can either be imported or produced by burning elemental sulfur in an on-site acid plant. On-site production has the added advantage of producing steam and power for the process. Acid injection to the reactor vessel is accomplished using tantalum-lined titanium dip pipes (lances) located to introduce the acid into well-mixed zones. The exothermic heat of dilution results in a temperature increase that affects material corrosion resistance; therefore, the lances are typically water cooled.

For pressure oxidation, most processes are self-sufficient in generating acid through the oxidation of sulfide minerals. However, in the presence of carbonates, pre-acidification of the feed may be necessary. This can be achieved through direct acid contact prior to autoclave feed, the recovery of acid from the autoclave discharge in a countercurrent wash circuit, or a combination thereof, depending on the sulfur-to-carbonate ratio of the feed material.

Services and utilities. High-pressure systems typically require a supply of water, air, and steam (continuous or intermittent) at a pressure exceeding that of the process. It is imperative that these systems be designed to avoid backflow of corrosive material from the high-pressure process.

Seal systems (e.g., for agitation) must be carefully designed to contain the process pressure and operated such that pressure cycling, wear, and corrosion of the seal components are minimized. Barrier fluids often use viscosity modifiers such as diethylene glycol or glycerin to maintain adequate lubrication at the seal mating faces.

Safety instrumented system. The ideal approach to reduce risk in a metallurgical process plant is to design safe and stable processes. However, because of the large quantity of potential energy inherent to pressurized processes, such as pressure leaching or pressure oxidation, reducing risk through elimination or substitution is limited, and engineering or administrative controls are often required. The identification and control of hazards is typically undertaken during the early stages of project execution using a hazard and operability study and layer-of-protection analysis based on the equipment design and process control developed during prior studies. If the residual risk from either activity exceeds acceptable levels, the additional protection of a safety instrumented system (SIS) may be required.

The SIS is a highly reliable instrumentation and control system with the sole purpose of monitoring critical process variables and initiating a controlled shutdown in the event of unsafe process deviations. An SIS, separate and redundant to the process control system, contains one or more safety instrumented functions, typically consisting of a field device (sensor) to detect an unsafe condition, a logical operation (logic solver) to initiate one or more specified actions, and a final control device (final element) to execute the specified actions (Pearson et al. 2014). The design, implementation, operation, and maintenance of the SIS follow internationally recognized standards such as IEC 61508 and IEC 61511/ANSI/ISA-84.00.

Materials of Construction

Table 8 highlights the typical differences in materials use for constructing the different types of processes. Pressure acid leaching has the highest relative cost per ton of material feed to the circuit due to the quantity of exotic (expensive) materials of construction needed for corrosion resistance at elevated temperatures and free acid concentrations of 25–50 g/L. This is followed by chloride-assisted oxidative leaching, which requires exotic materials that are resistant to chloride-induced stress corrosion cracking and pitting, although at a lower temperature and pressure. Acidic pressure oxidation (without chlorides) is the second-lowest relative cost, followed by alkaline pressure oxidation or caustic digestion, which use conventional carbon steels and 316 stainless steel.

General Cost Considerations

Capital costs. Capital expenditure costs for pressure hydrometallurgical facilities vary greatly with process conditions (pressure and temperature), materials of construction (equipment, piping, valves), and site location (site access, availability of skilled trades, supporting infrastructure, and site conditions). Much of the process equipment is designed for the specified plant throughput, with materials of construction that are specifically selected to be cost-effective yet suitable for unique process conditions. Such considerations include temperature, acidity/alkalinity, oxidation–reduction potential, solids abrasion, settling velocity, presence of multiphase flow, presence of flashing flow, presence of supersonic flow (such as vent gas depressurization), and mass transfer requirements.

Table 9 compares the direct capital costs of four pressure oxidation projects/studies completed by the authors between

Table 8 Typical materials of construction of pressure hydrometallurgical process facilities

Process	Pressure Acid Leach	Cl^--Assisted Oxidative Leach	Acidic Oxidative Leach	Alkaline/Caustic Leach
Relative Cost	**Highest→**			**Lowest**
Slurry Preheaters				
High-pressure vessel shell/head	C-Mn-Si steel	NA	C-Mn-Si steel	C-Mn-Si steel
Low-pressure vessel shell/head	C-Mn-Si steel	NA	Alloy 2507	C-Mn-Si steel
Membrane	Halogenated butyl rubber	NA	Vinyl ester resin	None
Refractory brick	Acid-proof fire clay	NA	Acid-proof fire clay	Acid-proof fire clay
High-pressure steam duct	Ti Grade 12	NA	Ti Grade 12	316 stainless steel
Low-pressure steam duct	Alloy 2507	Ti Grade 16	Alloy 2507	316 stainless steel
Interstage heater feed pumps	High-Cr iron	NA	CD4MCu	High-Cr iron
Autoclave Feed Pumps				
Diaphragm housing	Cast Ti Grade 3/5	CD4MCu	CD4MCu	Nodular cast iron
Check valves	Cast Ti Grade 3/5	CD4MCu	CD4MCu	Stellite-coated steel
Diaphragm	EPDM	Butyl rubber	EPDM	EPDM
Drop legs	Ti Grade 12	NA	Alloy 255	C-Mn-Si steel
Suction accumulator	Ti Grade 12	Alloy 2507	Alloy 2205	C-Mn-Si steel
Discharge dampener	Ti Grade 12	Alloy 2507	Alloy 2507	C-Mn-Si steel
Slurry strainers	Ti Grade 2/12	Alloy 2507	Alloy 2205/2507	C-Mn-Si steel
Reactor Vessels				
Vessel shell/heads	C-Mn-Si steel	C-Mn-Si steel	C-Mn-Si steel	C-Mn-Si steel
Membrane	Ti Grade 17—explosion welding cladding	Bituminous mastic	Homogeneously bonded lead or bituminous mastic	—
Refractory brick	NA	Silica/carbon/fire clay	Acid-proof fire clay	Acid-proof fire clay
Mortar	NA	Silicate/litharge-glycerol	Silicate/litharge-glycerol	Calcium salt
Oxygen/acid lances	Ta-clad Ti Grade 12	Alloy 255 or Alloy 2507	Ta-2.5W or Ta-clad Cu alloy	Inconel 625
Dip pipes/internals	Ti Grade 12	Ti Grade 16	Ti Grade 12	316 stainless steel
Vent piping/valves	Ti Grade 12	Alloy 59	Ti Grade 12/Ti-45Nb	316 stainless steel
Agitators	Ti Grade 5/12	Alloy 59/Ti Grade 12	Alloy 625/Ti Grade 12	316 stainless steel
Pressure Letdown				
Vessel shell/heads	C-Mn-Si steel	C-Mn-Si steel	C-Mn-Si steel	C-Mn-Si steel
Membrane	Halogenated butyl rubber	Bituminous mastic	Vinyl ester resin	—
Refractory brick	Acid-proof fire clay	Acid-proof fire clay	Acid-proof fire clay	Acid-proof fire clay
Mortar	AR-20-C vinyl ester	—	VET94 vinyl ester	—
Letdown valve body	Ti Grade 12	Ti Grade 12	Ti Grade 12	Hastelloy C
Letdown valve trim	Ceramic SiC	Ceramic SiC	Ceramic SiC	Ceramic SiC

NA = not applicable

2012 and 2016 that are representative of various types of processes, plant sizes, and site locations. All costs exclude indirect project costs such as engineering, procurement, and construction management services; construction equipment and services; camp costs; freight, transportation; and owner's costs and exclude any normalization for escalation, currency exchange, bulk material unit rates, labor rates, or labor productivity. As shown in the table, unit capital costs per metric ton of feed vary widely from project to project. An alternative metric is the unit capital cost per ton per day of sulfide sulfur oxidized by the facility (US\$ per t/d S^{-2}). However, these unit costs also vary considerably depending on the infrastructure, buildings, utilities, and electrical services needed to support the process facility, and Table 9 illustrates this variability.

Building costs (local excavation, fill, concrete, architectural, structural steel) for pressure hydrometallurgical facilities vary according to the area and volume of process buildings. For prefeasibility level studies, these can be estimated parametrically using a standard unit cost per building area (US\$/m^2) for building foundations and containment areas and a standard unit cost per building volume (US\$/m^3) for architectural and structural steel in specific locales, based on wind and snow loads, seismic zone, and ambient temperature range.

The direct cost of piping and valves for pressure acid leach and pressure oxidation facilities is generally 45%–55% of the installed mechanical equipment cost. A significant proportion of this cost is for severe service ball valves used for autoclave vessel and service isolation. The cost ratio is more valid for a range of design pressures from ASME Class 300 to 900, as the cost of piping, valves, and pressure equipment is equally affected by design pressure.

The cost of low-voltage (<600 V) and medium-voltage (4.16–13.8 kV) three-phase electrical systems varies as a function of connected load from US\$800–US\$1,000 per kilowatt. High-voltage systems (>38 kV) with redundant power supplies (substations, transformers, and switchgear) vary from US\$1,500–US\$2,000 per kilovolt-ampere of connected load; therefore, it is important to define the available supply voltage

Table 9 Comparison of direct capital costs of pressure oxidation facilities

Design Consideration	Project 1	Project 2	Project 3	Project 4
Environment	Tropical, highlands	Temperate, mountains	Arctic Circle	Arid, high elevation
Design oxidation capacity	High	Medium	Medium	Low
Feed to pressure oxidation plant	Whole ore	Concentrate	Concentrate	Concentrate
Unit capital expenditure, US$ per t/d run-of-mine	13,000	6,000	6,000	11,000
Unit capital expenditure, US$ per t/d S^{-2}	190,000	130,000	580,000	330,000
Distribution of Costs by Category, % of Total Direct Cost of Pressure Oxidation Facilities Only				
Earthworks, mass excavation, %	Excluded	0.2	0.6	0.1
Local excavation, fill, concrete, %	1.7	2.9	9.5	7.1
Architectural, %	3.5	2.5	0.7	0.5
Structural steel, %	5.7	7.9	13.4	5.9
Mechanical equipment, %	51.6	55.1	42.2	36.9
Piping and valves, %	24.4	28.6	17.3	28.5
Electrical systems, %	8.6	2.5	12.1	18.4
Instrumentation and control, %	4.5	0.4	4.0	2.6

and capacity when preparing the scope of work for an electrical estimate.

Instrumentation and control system costs vary with the type of control system (distributed control system or programmable logic controller), field communications network (analog, digital, HART, Fieldbus Foundation), and input/output (I/O) count. A typical pressure hydrometallurgy circuit has 100 control loops and 1,500 discrete I/O points per train. The average installed unit costs in the second quarter of 2016 were $1,500 per control loop and $1,000 per discrete I/O point, including control hardware, cable and tray, termination, continuity check, calibration, and software programming.

A 2016 study of construction cost overruns on mining and metallurgical projects identified a strong direct correlation between cost overruns and project duration and weaker indirect relationships between cost overruns and project size (plant capacity), the presence of local skilled trades, and existing infrastructure (Pearson et al. 2016). The study examined the ratio of actual project completion costs to estimated project development costs for a sample of 98 mining and metallurgical projects completed between 1997 and 2015. The frequency distribution of project overruns generated from this study is shown in Figure 27. The outcome of this study reaffirmed the historical norm, showing a shift in the mean of the sample distribution (1.35) from that of a normally distributed population (1.0), implying a systematic bias toward underestimating project development costs. One possible explanation for this is that although the number of unit operations in these metallurgical facilities is comparable to that of the chemical process industry, they can be more complex because of the natural variation in the plant feed. These facilities require a great deal more custom engineering and more detailed execution planning due to the remoteness of the host mine sites. Hence, they are more susceptible to cost overruns associated with procurement and construction.

Oftentimes, the cost overrun can be attributed to the underestimation of bulk quantities such as excavation, engineered fill, concrete, steel and cable, and the labor hours associated with their movement and installation. This growth in bulk quantities can occur at several stages of a project: between basic engineering and detailed engineering due to better definition of vendor-supplied equipment; between completion of the project estimate and completion of purchasing activities due to revised take-offs from detailed engineering drawings; and between completion of bulk material purchasing and construction completion due to material loss, overpour, and wastage. Estimators typically include a growth allowance of 5%–20% for bulk materials in their project cost estimates, depending on the level of engineering and procurement definition at the time the estimate is prepared to account for growth in the bulk quantities.

Operating costs. Pressure leach and oxidation circuits are complex, and the operating costs are often notably higher than those of other mineral processing circuits. The comparatively higher cost is primarily associated with labor, maintenance, and reagents. Nonsulfide pressure leach circuits also require significant energy input to sustain the operating temperature, which is why large facilities are often integrated with a power plant or an acid plant to generate steam at a substantially lower cost than using fuel sources (e.g., diesel). The energy costs for a Bayer process typically represent approximately 35% of overall operating costs (Gorst et al. 2013).

Operations and maintenance labor costs often account for a significant proportion (>20%) of the operating costs for a pressure leach or oxidation facility. The costs are mostly due to the additional labor count required to manage, operate, and maintain the circuit compared to other mineral processing or hydrometallurgy facilities (e.g., a flotation concentrator or free-milling gold leach plant). Specifically, dedicated personnel include technical staff (e.g., metallurgists), area operations supervisors, control room operators, and a mechanical maintenance team. Electrical, instrumentation, process control, and reliability maintenance support are usually spread throughout multiple areas of the facility; however, additional personnel may be required to adequately support the pressure leach or oxidation circuit. Labor costs for a Bayer process are typically a lower proportion of overall costs—between 5% and 10% (Gorst et al. 2013). A project-specific labor plan and estimate of the labor rates are required to establish requirements and develop a cost estimate.

Pressure oxidation and pressure leach circuits typically include specialized process valves with exotic construction materials that increase regular maintenance consumable costs. A factor of 3.5%–4% of the total direct cost for the circuit is

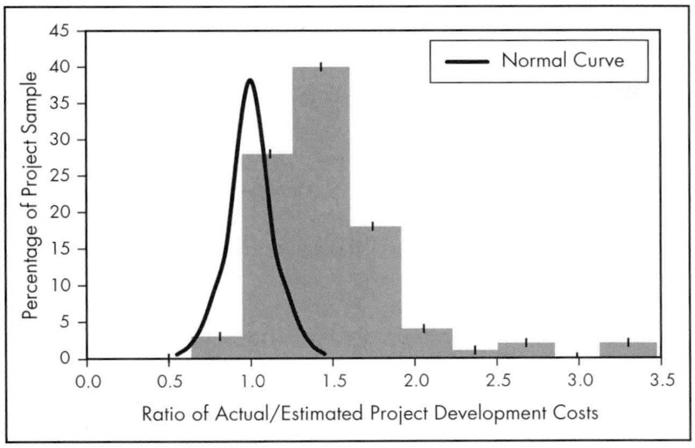

Source: Pearson et al. 2016

Figure 27 Distribution of capital cost overruns for 98 mining and metallurgical projects (1997–2015)

often applied when estimating the maintenance consumable costs. The material cost is independent of contractor costs that are mostly associated with the annual or biannual shutdowns and an influx of personnel to complete the necessary vessel cleanout, refractory repairs, inspections, rebuilds, and other routine maintenance activities.

The reagent consumption and costs for pressure leaching and pressure oxidation circuits are mostly associated with the use of sulfuric acid, sodium hydroxide, or oxygen. Minor reagent consumption is associated with mechanical seal lubricant and demineralized water. Higher water treatment costs will be incurred if a boiler is required for continuous operation.

Sulfuric acid is consumed in acidic pressure leaching of nickel laterites, slurry conditioning for pressure oxidation, and pressure leaching of select base metals. An acid unit consumption of 200–500 kg/t of dry ore is a typical range for nickel laterite leaching (Kyle 1996), 25 t/t of nickel product (Crundwell et al. 2011), and up to approximately 75 kg/t of sulfuric acid (100%) for acidification ahead of pressure oxidation (Thomas 1994). The unit price of sulfuric acid is linked to global sulfur and fertilizer demand and can be quite volatile, ranging from US$50/t to more than US$150/t (excluding shipping). Operations typically arrange long-term supply contracts from sulfur and sulfuric acid producers such as copper smelters and oil refineries to ensure supply and reduce supply cost volatility.

Sodium hydroxide is predominantly used in the digestion of bauxite ores via the Bayer process and represents approximately 15% of operating costs (Gorst et al. 2013). The sodium hydroxide unit consumption is highly dependent on the reactive silica content of the ore but is ~100 kg/t of ore based on current minimization and recovery practices (Smith 2009).

High-purity oxygen (>90 vol %) is predominantly used for pressure oxidation and pressure oxidative leaching. The oxygen unit consumption is based on the oxidation reaction stoichiometry and efficiency. For a pyrite (FeS_2) feed material, the stoichiometric addition rate is ~1.9 kg O_2/kg S. The stoichiometry changes with the presence of minerals such as arsenopyrite (FeAsS) and stibnite (Sb_2S_3) or other sulfides. The oxygen usage efficiency (utilization) is the approach to the stoichiometric requirement and is dependent on oxygen mass transfer, sulfide reactivity, and the presence of noncondensable gases that increase the reactor vent rate. The use of high-purity oxygen reduces the noncondensable nitrogen content and improves oxygen utilization, although less significantly than reducing the carbonate content in the reactor feed. Typical oxygen utilization ranges from 85% to 95%; however, it can be less than 80% in the presence of carbonates.

The production of high-purity oxygen for use in oxidation processes typically requires 450–550 kW·h/t of oxygen, which corresponds to a unit cost of US$45–US$50 per metric ton at a power price of US$0.10/kW·h. This equates to US$94–US$104 per metric ton of sulfide sulfur in pyrite, assuming 90% oxygen utilization. In a sale-of-gas arrangement (i.e., the oxygen plant is owned and operated by others—sometimes referred to as "over the fence"), the unit cost of oxygen is often higher due to a monthly facilities charge, which is often a primary means for the vendor to obtain a return on capital and limit financial risk. The ability of a vendor to recover and sell nitrogen and argon may reduce the monthly operating fee; however, most mineral deposits are not located near other industries, so opportunities are limited, and supply costs are often elevated for this reason.

REFERENCES

Adams, M.D., van der Meulen, D.R., Lunt, D.J., and Anderson, P. 2004. Selection of piloting parameters in pressure hydrometallurgy. In *Proceedings of the MetPlant 2004 Conference: Metallurgical Plant Design and Operating Strategies*. Melbourne, Victoria: Australasian Institute of Mining and Metallurgy.

AS 1210:2010. *Australian Standard Pressure Vessels*. Sydney, New South Wales: Standards Australia International.

ASME (American Society of Mechanical Engineers). 2015a. *ASME Boiler and Pressure Vessel Code Section VIII: Rules for Construction of Pressure Vessels—Division 1*. New York: ASME.

ASME (American Society of Mechanical Engineers). 2015b. *ASME Boiler and Pressure Vessel Code Section VIII: Rules for Construction of Pressure Vessels—Division 2: Alternate Rules*. New York: ASME.

Blackmore, A., Shah, U., Pearson, M., Plikas, A., and Sim, B. 2014. Flashing of pressurized slurry in metallurgical processing: A review of the state of the art, operational experiences, and recommendations for future research. Presented at the ASME 2014 4th Joint US-European Fluids Engineering Division Summer Meeting collocated with the ASME 2014 12th International Conference on Nanochannels, Microchannels, and Minichannels, Chicago, IL, August 3–7.

Bristowe, W., Hanson, A., and Pearson, M. 2004. Quality assurance programs for fabrications of specialized vessels and exotic alloy piping. In *Proceedings of the 34th Annual Hydrometallurgy Meeting of the CIM*. Montreal, QC: Canadian Institute of Mining, Metallurgy and Petroleum.

BS EN 13445-3:2009. *Unfired Pressure Vessels—Part 3: Design*. London: British Standards Institute.

BS EN 14879-6:2009. *Organic Coating Systems and Linings for Protection of Industrial Apparatus and Plants Against Corrosion Caused by Aggressive Media—Part 6: Combined Linings with Tile and Brick Layers. Annex D: Test Methods for Tolerance and Limit Deviations*. Dublin: National Standards Authority of Ireland.

BS PD 5500:2009. *Specification for Unfired Fusion Welded Pressure Vessels*. London: British Standards Institute.

Caruana, T., Clappison, J., and Mercer, L.N. 2010. Design and fabrication of titanium piping for pressure hydrometallurgy service. In *Proceedings of Titanium 2010: 26th Annual Conference*. Broomfield, CO: International Titanium Association.

Chalkley, M.E., Cordingley, P., Freeman, G., Budac, J., Krentz, R., and Scheie, H. 2004. Fifty years of pressure hydrometallurgy at Fort Saskatchewan. In *Proceedings of the 34th Annual Hydrometallurgy Meeting of the CIM*. Montreal, QC: Canadian Institute of Mining, Metallurgy and Petroleum.

Collins, M.J., Hasenbank, A., Parekh, B., and Hewitt, B. 2011. Design of the new Lihir gold pressure oxidation autoclave. In *Proceedings of the 50th Conference of Metallurgists: COM 2011*. Montreal, QC: MetSoc.

Crundwell, F.K., and Bryson, A.W. 1992. The modelling of particulate leaching reactors—The population balance approach. *Hydrometallurgy* 29(1-3):275–295.

Crundwell, F.K., Moats, M.S., Ramachandran, V., Robinson, T.G., and Davenport, W.G. 2011. *Extractive Metallurgy of Nickel, Cobalt and Platinum Group Metals*. Oxford: Elsevier. p. 131.

Defreyne, J., Grieve, W., Jones, D.L., and Mayhew, K. 2006. The role of iron in the CESL process. In *Proceedings of the Third International Symposium on Iron Control in Hydrometallurgy*. Montreal, QC: Canadian Institute of Mining, Metallurgy and Petroleum.

Defreyne, J., Brace, T., Miller, C., Omena, A., Matos, M., and Cobral, T. 2008. Commissioning UHC: A Vale copper refinery based on CESL technology. In *Hydrometallurgy 2008: Proceedings of the Sixth International Symposium*. Edited by C.A. Young, P.R. Taylor, C.G. Anderson, and Y. Choi. Littleton, CO: SME.

Donohue, I., Barrette, E., and Pearson, M. 2012. Refractory lining design for pressure oxidation autoclaves. In *Proceedings of the 51st Conference of Metallurgists: COM 2012*3. Westmount, QC: Canadian Institute of Mining, Metallurgy and Petroleum.

Donohue, I., Lauzon, P., and Pearson, M. 2013. Thermal design of refractory lined pressure oxidation autoclaves. In *Proceedings of the 52nd Conference of Metallurgists: COM 2013*. Westmount, QC: Canadian Institute of Mining, Metallurgy and Petroleum.

Doucet, M., and Stafiej, J. 2007. Processing of high-nickel slimes at the CCR refinery. In *Copper 2007, Vol. V: Copper Electrorefining and Electrowinning*. Edited by G.E. Houlachi, J.D. Edwards, and T.G. Robinson. Montreal, QC: Canadian Institute of Mining, Metallurgy and Petroleum. pp. 173–182.

EU Directive 97/23/EC. 1997. Directive of the European Parliament and of the Council of 29 May 1997 on the approximation of the laws of the Member States concerning pressure equipment. In *Official Journal of the European Communities*, L 181/1. Luxembourg: Office for Official Publications of the European Communities.

Fair, J.R. 1993. How to design baffle tray columns. *Hydrocarb. Process*. 72(5):75–78.

Fraser, K.S., and McCombe, W.E. 2011. Method for removing sulfur from a gas stream. U.S. Patent 20,110,174,155.

Fraser, K.S., and Thomas, K.G. 2010. Uranium extraction history using pressure leaching. In *Proceedings of the 3rd Annual International Conference on Uranium and the 40th Annual Hydrometallurgical Meeting*. Montreal, QC: Canadian Institute of Mining, Metallurgy and Petroleum.

Fraser, K.S., Lauzon, P.H., Cooper, A.W., and Koning, A.J. 2008. Insulating inserts for elevated temperature process vessels. U.S. Patent 7,381,384.

Frischmuth, R., Krumins, T., and Zunti, L. 2014. Reducing titanium risk with overpressure control in pressure oxidation. In *Proceedings of the 53rd Conference of Metallurgists: COM 2014*. Westmount, QC: Canadian Institute of Mining, Metallurgy and Petroleum.

Giles, D.J., and Giles, J.E. 1919. Process of extracting tungsten and similar metals from their ores. U.S. Patent 1,293,402.

Gorst, J., O'Connor, M., and Haneman, B. 2013. High capacity tube digestion technology reduces capital and operating costs for refineries. Presented at ICSOBA 2013, Krasnoyarsk, Russia.

Habashi, F. 1995. Bayer's process for alumina production: A historical perspective. *Bull. Hist. Chem.* 17:15–19.

Habashi, F. 2004. The origins of pressure hydrometallurgy. In *Proceedings of the 34th Annual Hydrometallurgy Meeting of the CIM*. Montreal, QC: Canadian Institute of Mining, Metallurgy and Petroleum.

Haneman, B. 2012. Optimising digestion flash tank design for the alumina industry. Presented at the Ninth AQW Conference (International Alumina Quality Workshop), Perth, Australia, March 18–22.

Hatch Ltd. 2016. Autoclave technology. www.hatch.com/Expertise/Services-and-Technologies/Autoclave-Technology. Accessed July 11, 2016.

Hill, V.G., and Sehnke, E.D. 2006. Bauxite. In *Industrial Minerals and Rocks*, 7th ed. Edited by J.E. Kogel, N.C. Trivedi, J.M. Barker, and S.T. Krukowski. Littleton, CO: SME. pp. 227–238.

IEC 61508. 2010. *Functional Safety of Electrical/Electronic/Programmable Electronic Safety-Related Systems*. Geneva: International Electrotechnical Commission.

IEC 61511/ANSI/ISA-84.00. 2004. *Functional Safety: Safety Instrumented Systems for the Process Industry*. Geneva: International Electrotechnical Commission.

King, J.A., Dreisinger, D.B., and Knight, D.A. 1993. The total pressure oxidation of copper concentrates. In *Proceedings of the Paul E. Queneau International Symposium, Extractive Metallurgy of Copper, Nickel, and Cobalt—Volume 1: Fundamentals Aspects*. Warrendale, PA: The Minerals, Metals & Materials Society.

Krumins, T., Stunguris, C., Zunti, L., and Blaskovich, S. 2013. Mercury removal from pressure oxidation vent gas at Newmont Mining Corporation's Twin Creeks facility. In *Proceedings of the 2013 International Conference on Advanced Materials Science and Technology (ICAMST 2013)*. Zurich, Switzerland: Trans Tech Publications.

Krumins, T., Eichhorn, M., Zunti, L., and Ruff, F.C. 2014. Capacity enhancement at Newmont Mining Corporation's Twin Creeks whole ore pressure oxidation facility. In *Proceedings of the 7th International Symposium on Hydrometallurgy 2014*. Westmount, QC: Canadian Institute of Mining, Metallurgy and Petroleum.

Krysa, B. 1995. Zinc pressure leaching at HBMS. *Hydrometallurgy* 39:71–77.

Kurtak, J. 1998. History of Pine Creek: A world class tungsten deposit. *Min. Eng.* (December):42–47.

Kyle, J.H. 1996. Pressure acid leaching of Australian nickel/cobalt laterites. In *Proceedings of Nickel '96, Mineral to Market*. Melbourne, Victoria: Australasian Institute of Mining and Metallurgy.

Lauzon, P., Mercer, L., Pearson, M., and Irvine, J. 2004. Severe service ball valve design analysis utilizing a modified Kepner-Tregoe method. In *Proceedings of the 34th Annual Hydrometallurgy Meeting of the CIM*. Montreal, QC: Canadian Institute of Mining, Metallurgy and Petroleum.

Marsden, J.O., Wilmot, J.C., and Hazen, N. 2007. Medium-temperature pressure leaching of copper concentrates—Part 1: Chemistry and initial process development. *Miner. Metall. Proc.* 24:193–204.

Mason, P.G. 1990. Energy requirements for the pressure oxidation of gold-bearing sulfides. *JOM* 42(9):15–18.

Mason, P.G., and Gulyas, J.W. 1999. Pressure hydrometallurgy: No longer regarded with trepidation for the treatment of gold and base metal ores and concentrates. In *1999 EPD Congress*. Edited by B. Mishra. Warrendale, PA: The Minerals, Metals & Materials Society. pp. 585–616.

McMullen, D. 2014. Incorporating radiant heat exchange into finite element models of hydrometallurgical process equipment. In *Proceedings of the 7th International Symposium on Hydrometallurgy 2014*. Westmount, QC: Canadian Institute of Mining, Metallurgy and Petroleum.

McMullen, D., Price, J., Pearson, M., Ugarenko, C., van Warmerdam, K., Kember, C., and Thayer, M. 2014. Design, construction, and commissioning of air separation units in remote locations. In *Proceedings of the 53rd Conference of Metallurgists: COM 2014*. Westmount, QC: Canadian Institute of Mining, Metallurgy and Petroleum.

McNulty, T. 1998. Developing innovative technology. *Min. Eng.* (October):50–55.

Mitchell, J.S. 1956. Pressure leaching and reduction at the Garfield Refinery. *Min. Eng.* (November):1093–1095.

Munnik, E., Singh, H., Uys, T., Bellino, M., Harris, B., Fraser, K.S., and du Plessis, J. 2003. Development and implementation of a novel pressure leach process for the recovery of cobalt and copper at Chambishi, Zambia. In *Hydrometallurgy 2003—Fifth International Conference in Honor of Professor Ian Ritchie*. Vol. 1, Leaching and Solution Purification. Warrendale, PA: The Minerals, Metals & Materials Society.

Nice, R.W. 2004. The Western Australian nickel laterite projects—What have we learned? In *Proceedings of the 34th Annual Hydrometallurgy Meeting of the CIM*. Montreal, QC: Canadian Institute of Mining, Metallurgy and Petroleum.

Nicolle, M., Nel, G., Plikas, T., Shah, U., Zunti, L., Bellino, M., and Pieterse, H.J.H. 2009. Mixing system design for the Tati Activox® autoclave. In *SAIMM Southern African Hydrometallurgy Conference 2009*. Johannesburg: Southern African Institute of Mining and Metallurgy.

Outotec. 2016. HSC Chemistry version 6.12 (software). www.hsc-chemistry.com.

Papangelakis, V.G., and Demopolous, G.P. 1992. Reactor models for a series of continuous stirred tank reactors with a gas-liquid-solid leaching system: Part I. Surface reaction control. *Metall. Mater. Trans. B* 23(6):847–856.

Pearson, M., Bristowe, W., Stunguris, C., and Gothard, S. 2010. A comparison of refractory lined carbon steel and titanium EXW clad pressure vessels for specific operating conditions. In *Proceedings of Titanium 2010: 26th Annual Conference*. Broomfield, CO: International Titanium Association.

Pearson, M., McCombe, W., and Gonzalez-Sardi, L. 2014. Implementation of safety instrumented systems in pressure oxidation facilities. In *Proceedings of the 53rd Conference of Metallurgists: COM 2014*. Westmount, QC: Canadian Institute of Mining, Metallurgy and Petroleum.

Pearson, M., Oughtred, C., and Wong-Cameron, K. 2016. Statistical analysis of parameters influencing capital overruns on mining projects. In *Proceedings of 2016 AACE International Transactions*. Red Hook, NY: AACE.

Pieterse, H.J.H. 2004. Oxidation autoclave agitation review. In *Proceedings of the 34th Annual Hydrometallurgy Meeting of the CIM*. Montreal, QC: Canadian Institute of Mining, Metallurgy and Petroleum.

Simmons, G.L. 1995. Pressure oxidation process development for treating refractory carbonaceous ores at Twin Creeks. SME Preprint No. 95-65. Littleton, CO: SME.

Singh, S., Eichhorn, M., and Fraser, K.S. 2013. Approach for early evaluation of pressure oxidation for a refractory gold ore deposit. In *Proceedings of the 52nd Conference of Metallurgists: COM 2013*. Westmount, QC: Canadian Institute of Mining, Metallurgy and Petroleum.

Smith, P. 2009. The processing of high silica bauxites—Review of existing and potential processes. *Hydrometallurgy* 98(1-2):162–176.

St. Louis, R.M., and Edgecombe, J.M. 1990. Recovery enhancement in the Mercur autoclave circuit. In *Proceedings of the Gold 1990 Symposium*. Littleton, CO: SME. pp. 443–450.

Thomas, K.G. 1994. *Research, Engineering Design and Operation of a Pressure Hydrometallurgy Facility for Gold Extraction*. Den Haag, Netherlands: CIP-Gegevens Koninkluke Bibliotheek. p. 130.

Wei, J., Moe, G., and Pearson, M. 2010. Quality assurance activities for corrosion resistant membranes in pressure vessels. In *Proceedings of the 2010 ASME Pressure Vessels and Piping Conference*. New York: American Society of Mechanical Engineers.

Weir Group PLC. 2016. GEHO® TZPM. www.global.weir/products/product-catalogue/geho-tzpm. Accessed July 11, 2016.

Weir, R.D. 1985. Advances in pressure hydrometallurgy. In *Frontier Technology in Mineral Processing*. Edited by J.F. Spisak and G.V. Jergensen II. Littleton, CO: SME-AIME. pp. 99–112.

Woloshyn, J., Oshinowo, L., and Rosten, J. 2006. Digester design using CFD. In *Light Metals*. Edited by D. Donaldson and B. Raahauage. Warrendale, PA: The Minerals, Metals & Materials Society. pp. 939–944.

Yernberg, W.R. 1996. Santa Fe Pacific Gold targets million ounces/year production. *Min. Eng.* 48(9):39–43.

Zunti, L., and Pearson, M. 2004. A comparison of refractory lined and metal-clad process vessels for specific operating conditions. In *Proceedings of the 34th Annual Hydrometallurgy Meeting of the CIM*. Montreal, QC: Canadian Institute of Mining, Metallurgy and Petroleum.

CHAPTER 10.7

Bioleaching

James A. Brierley and Corale L. Brierley

Bioleaching, also termed *biohydrometallurgy*, *biomining*, *minerals biooxidation*, and *biooxidation pretreatment*—the latter two terms when employed for precious metals ores and concentrates in which the gold is embedded in a sulfide mineral—is a hydrometallurgical process that uses certain types of microorganisms as catalysts to produce the chemical lixiviant for leaching sulfide minerals. Bioleaching is one of several commercially employed technologies for processing sulfidic ores and concentrates. The decision as to which technology is best suited for a mine site is based on many factors, including test results, mineralogy, ore grade, mine-site topography, and economics.

Bioleaching, which is a naturally occurring process, has inadvertently been used for more than 2,000 years to leach copper (Rossi 1990), but not until the late 1940s and 1950s was it recognized that microorganisms were involved (Colmer and Hinkle 1947; Colmer et al. 1950; Temple and Colmer 1951; Bryner et al. 1954). The first bioleach patent, issued in 1958 to Kennecott Copper Corporation (Zimmerley et al. 1958), claimed the use of bacteria for a cyclic process of ferric sulfate oxidation of sulfide minerals wherein the ferric sulfate served as the lixiviant for oxidation of the sulfides, and bacteria regenerated the ferric sulfate by oxidizing ferrous sulfate. Since these discoveries more than half a century ago, much has been learned about the microorganisms involved and the interaction between the microorganisms and minerals. Bioleaching is now commercially applied for leaching base and precious metals in heaps and stirred-tank reactors with approximately 20% of the world's mined copper and about 3% of the world's mined gold produced by bioleaching technology (Schippers et al. 2014).

This chapter summarizes fundamental microbial and chemical aspects of bioleaching and biooxidation pretreatment, describes test work and scale-up essentials, and reviews commercial operating factors for heaps and stirred-tank reactors.

MICROBIOLOGY AND CHEMISTRY OF BIOLEACHING

Extensive studies of the microorganisms involved in bioleaching began in the late 1960s, followed by recognition of their catalytic role coupled with the process chemistry. The discovery and characterization of robust suites of microorganisms tolerating high concentrations of dissolved metals (30–50 g/L), low pH values (pH <1), high ionic strengths (7.6 M), and populating temperature habitats from less than 15°C to about 90°C contributed significantly to the design of bioleach systems and amplified the types of minerals that could be processed by bioleaching (Watling 2015).

Bioleaching Microorganisms

Early studies by Beck (1967) identified the bacterium *Thiobacillus ferrooxidans*, renamed *Acidithiobacillus ferrooxidans* by Kelly and Wood (2000), as a microbial catalyst for the oxidation of ferrous iron (Fe^{2+}) to ferric iron (Fe^{3+}). Sulfur oxidation by *Acidithiobacillus thiooxidans* was first reported in the early 1920s; by the 1930s this organism's function in acid rock drainage formation was known (Ehrlich 2004). The occurrence of *A. thiooxidans* in copper leaching environments was reported in the 1950s by Bryner and colleagues (1954). The *Acidithiobacillus* species described in these publications are mesophilic bacteria—bacteria growing best between 25° and 40°C—that oxidize ferrous iron and sulfur.

Thermophilic microorganisms involved in bioleaching were discovered in the 1960s and 1970s. The moderately thermophilic *Sulfobacillus* species were discovered in base metal heap leach operations (J.A. Brierley and LeRoux 1977; J.A. Brierley and Lockwood 1977; Golovacheva and Karavaiko 1978). The extremely thermophilic and acidophilic archaea, first found to flourish in acid hot springs (J.A. Brierley 1966; Brock et al. 1972), were demonstrated to have a role in bioleaching by J.A. Brierley and Brierley (1986); these archaea, commonly referred to as "extremophiles" or "extreme thermoacidophiles," are genetically distinct from bacteria and classified as Archaebacteria (Woese 1982). The extreme thermoacidophiles are of specific interest for bioleaching primary sulfide minerals such as chalcopyrite, molybdenite (C.L. Brierley 1974; J.A. Brierley and Brierley 1978), and enargite (Lee et al. 2011). Mesophilic and

James A. Brierley, Biohydrometallurgy Consultant, Brierley Consultancy LLC, Highlands Ranch, Colorado, USA
Corale L. Brierley, President/Principal, Brierley Consultancy LLC, Highlands Ranch, Colorado, USA

Table 1 Bioleaching microorganisms and their temperature ranges for growth

Microorganism	Temperature Range, °C
Mesophilic and Moderately Thermophilic Bacteria	
Acidimicrobium ferrooxidans	<30–55; optimum 45–50
Acidithiobacillus caldus	32–52; optimum 45
Acidithiobacillus ferrooxidans	10–37; optimum 30–35
Acidithiobacillus thiooxidans	10–37; optimum 28–30
Leptospirillum ferrooxidans	Not available; optimum 28–30
Sulfobacillus acidophilus	<30–55; optimum 45–50
Sulfobacillus thermosulfidooxidans	20–60; optimum 45–48
Sulfobacillus thermotolerans	20–60; optimum 40
Mesophilic and Moderately Thermophilic Archaea	
Ferroplasma acidiphilum	15–45; optimum 35
Ferroplasma cupricumulans	22–63; optimum 54
Extremely Thermophilic Archaea	
Acidianus brierleyi	45–75; optimum ≈70
Acidianus infernus	65–96; optimum ≈90
Metallosphaera sedula	50–80; optimum 75
Sulfolobus metallicus	50–75; optimum 65
Sulfurococcus yellowstonensis	40–80; optimum 60

moderately thermophilic archaea have also been isolated and identified (Golyshina et al. 2000) since the discovery of the extremely thermophilic and acidophilic archaea.

Bioleaching microbes are customarily grouped according to their temperature ranges of growth: mesophilic bacteria and archaea, 5° to ≈50°C; moderately thermophilic bacteria and archaea, 45° to ≈60°C; and extremely thermophilic archaea, 60° to 80°C. Table 1 shows examples of bacteria and archaea involved with metal sulfide bioleaching (Schippers 2007; Schippers et al. 2014). The overlapping temperature ranges for the different groups of microorganisms illustrate their capacity to share bioleach environments as a group or consortium. Research and development in the late 1990s and first decade of the 2000s demonstrated that bioleaching of sulfide minerals was accomplished not by just a few genera and/or species of bacteria, as earlier believed, but rather by consortia of different genera of bacteria and archaea spanning a range of temperatures and physical conditions. Optimization of these microbial consortia in bioleach operations remains a goal of scientific and engineering research (Rawlings 1997; Watling 2006; Norris 2007; Schippers 2007; Schippers et al. 2014).

Microorganisms involved in bioleaching have minimal requirements for reproduction and catalytic functions. Carbon for cellular synthesis is provided by CO_2 available in air. Their energy sources for growth derive from oxidation of ferrous iron and sulfur. The terminal electron acceptor for the energy yielding oxidation reactions is O_2 in the atmosphere. Soluble phosphate (PO_4^{3-}), potassium (K^+), ammonium ion (NH_4^+), and a few trace elements, also required for growth of these organisms, are available from the ore and gangue minerals in their habitat. In heap bioleach operations, phosphate can originate from a variety of gangue and ore minerals, and potassium comes from clay minerals. Ammonium ion, which is the organism's nitrogen source, is typically available in sufficient quantities as a residue of blasting (Zaitsev et al. 2008).

Bioleaching microorganisms are acidophilic (acid-loving), requiring a sulfuric acid environment of less than pH 2.5, with some of these organisms functioning actively at pH values less than 1. The acidic environment ensures that ferrous iron and certain reduced sulfur compounds, serving as energy sources for the organisms, remain in solution and available for oxidation. Metals of value, such as copper, zinc, nickel and others extracted from sulfide minerals during the bioleach process, also remain soluble in the acidic solution, making them readily recoverable by conventional metallurgical methods. Although the solution in which the bioleach organisms thrive is acidic (usually pH 2 and lower), the intracellular portion of the microorganisms remains at a neutral pH, resulting in a huge proton gradient across the microorganism's cell membrane.

Two reactions (1 and 2) embody the microorganism's task in bioleaching. The microorganisms oxidize ferrous iron as an energy source for biosynthesis and growth under acidic conditions in the presence of oxygen supplied by air. The end product, ferric iron, is an effective oxidant for sulfidic minerals:

$$2Fe^{2+} + 2H^+ + 0.5O_2 \rightarrow 2Fe^{3+} + H_2O \qquad \textbf{(EQ 1)}$$

Sulfur is also an energy source for some bioleaching microbes:

$$2S^0 + 3O_2 + 2H_2O \rightarrow 2H_2SO_4 \qquad \textbf{(EQ 2)}$$

The end product of this oxidation is sulfuric acid, which maintains acidic conditions to keep metal ions in solution and reduces the amount of acid that may need to be added to meet acid demand of the ore or concentrate. Microbial sulfur oxidation also requires oxygen derived from air.

Bioleaching microorganisms tolerate most heavy metal cations to 50 g/L or even higher concentrations (Watling 2011) via natural adaptation or intentional adaptation in the laboratory. There are, however, a few cationic metals/metalloids that can be toxic, and for these constituents to be toxic, they must be soluble. Mercury and silver, generally thought to be toxic, are usually not problematic because silver has a low solubility in acidic leach solutions and mercury adsorbs to gangue materials, mitigating its toxic effect. Arsenic in the form of arsenate (AsO_4^{3-}) anion has low toxicity; bioleaching microorganisms are resistant to 20 g/L arsenate (Van Aswegen et al. 2007). Arsenic as arsenite (AsO_3^{3-}) elicits a toxic effect; 1.5 g/L As^{3+} retards microbial activity with 5g/L As^{3+} completely inhibiting the organisms (Breed et al. 2000). Arsenite is effectively oxidized to arsenate by ferric iron (Reaction 3), which is present in the bioleach environment, thus mitigating potential As^{3+} toxicity. However, attention is necessary to ensure that the ferric iron concentration is sufficient and conditions are suitably oxidizing for this reaction to proceed.

$$H_3AsO_3 + 2Fe^{3+} + H_2O \rightarrow H_3AsO_4 + 2Fe^{2+} + 2H^+ \qquad \textbf{(EQ 3)}$$

Nitrate (NO_3^-) at concentrations of 1.25–2.5 g/L slows microbial cell division and suppresses ferrous iron oxidation by bacteria and archaea. There is some evidence that mesophilic and moderately thermophilic bacteria exhibit some adaptation to NO_3^- at these concentrations and resume ferrous iron oxidation, but iron oxidation by archaea remains suppressed (Shiers et al. 2014; Watling 2015).

Chloride (Cl^-) inhibits iron oxidation and growth of biomining microorganisms (Watling 2011) with inhibition occurring in the range of 5–7 g/L NaCl. The search for salt-tolerant biomining organisms has been ongoing for at least 20 years. Norris et al. (2010) discovered a bacterium capable of iron oxidation in 20 g/L NaCl at a temperature just below 50°C; however, growth of this organism was difficult to establish on copper sulfide ore in column tests. More recently, organisms capable of tolerating 70 g/L of sea salts have been obtained from acidic saline drains, lakes, and sediments in Western Australia (Rea et al. 2015); research to determine these salt-tolerant organisms' effectiveness for bioleaching is ongoing. Chloride toxicity is a significant problem for biomining in regions of the world where the water supply is brackish or only seawater is available for processing. In some mining areas, atacamite, an acid-soluble copper chloride mineral, is prevalent. Care must be taken to ensure that atacamite is excluded from ore destined for commercial bioleaching circuits; otherwise, chloride concentrations in the leach solution can easily exceed the tolerance level of the bioleaching organisms.

Fluoride (F^-) can be problematic in bioleaching, but not because the anion is toxic. Below pH 3.45, fluoride occurs predominantly as hydrofluoric acid (HF), and HF crosses the cell membrane of the acidophilic microorganisms as an uncharged molecule. When it is inside the microbial cell, HF dissociates because, as noted earlier, the intracellular portion of the microbes is a neutral pH. HF dissociation releases protons acidifying and killing the cell (J.A. Brierley and Kuhn 2010). F^- toxicity levels are directly related to the presence of other ions, particularly aluminum, in solution. Fluoride complexes with aluminum, significantly reducing the toxicity of F^- to the biomining microorganisms, because the complexed molecule is too large to cross the organism's cell membrane (J.A. Brierley and Kuhn 2010; Watling 2015).

Sulfate (SO_4^{2-}) concentrations in stirred-tank and heap leach solutions easily reach concentrations of 150 g/L, and although high sulfate concentrations have been implicated with the slowing of iron oxidation by the bioleaching bacteria (Watling 2013), there is little research data to support this contention.

Chemistry of Sulfide Mineral Bioleaching and Microorganism–Mineral Interactions

Bioleaching is based on oxidation–reduction reactions. Microorganisms oxidize ferrous iron and sulfur compounds, as expressed in Reactions 1 and 2, with the ferric iron and sulfuric acid providing the oxidizing agent for dissolution of sulfide minerals and acid maintaining metals in solutions, respectively. From ferrous iron oxidation, the microorganisms benefit by obtaining energy for their cell division and biochemical functions. Therefore, microorganisms serve as catalysts generating and regenerating the oxidant (Fe^{3+}) and supplying acid. In purely chemical systems, a chemical oxidant, such as hydrogen peroxide, ozone, or oxygen (usually under pressure), must be added to oxidize iron and sulfur, and the chemical oxidant is not regenerated, requiring recurrent addition of the oxidant.

Oxidation–reduction (redox) potential is controlled by the ratio of ferric iron to ferrous iron (Fe^{3+}:Fe^{2+}) in solution; the greater the Fe^{3+}:Fe^{2+}, the higher the redox potential. Different minerals go into solution at differing redox potentials, referred to as the rest potential of the minerals. The secondary copper sulfide mineral, chalcocite, is oxidized rapidly by the microbially generated ferric iron at a relatively low redox potential (~550–600 mV vs. SHE [standard hydrogen electrode]). Consequently, chalcocite is often considered to be acid soluble. Nevertheless, ferric iron is consumed in the oxidation reaction (4), and this consumption must be accounted for in the process design of commercial operations.

$$Cu_2S + 2Fe^{3+} \rightarrow Cu^{2+} + 2Fe^{2+} + CuS \qquad \textbf{(EQ 4)}$$

One of the products of the kinetically fast oxidation of chalcocite is blaubleibender covellite (CuS). The oxidation of CuS (Reaction 5) requires a much higher redox potential, that is, a higher Fe^{3+} to Fe^{2+} ratio (~850–870 mV vs. SHE), and is kinetically slow (Bustos et al. 1999).

$$CuS + 2Fe^{3+} \rightarrow Cu^{2+} + 2Fe^{2+} + S^\circ \qquad \textbf{(EQ 5)}$$

The resulting elemental sulfur is then oxidized to sulfuric acid by microorganisms according to Reaction 2.

Ferric iron oxidation of pyrite (Reaction 6) yields substantial acid and ferrous iron, with the ferrous iron serving as a source of energy for the microorganisms as it is reoxidized to ferric iron:

$$FeS_2 + 14Fe^{3+} + 8H_2O \rightarrow 15Fe^{2+} + 2SO_4^{2-} + 16H^+ \qquad \textbf{(EQ 6)}$$

Pyrite oxidation produces 0.41 kg acid/kg FeS_2 oxidized and generates significant heat (12,884 kJ/kg pyrite oxidized). In some commercial applications, the acid generated from pyrite oxidation offsets the purchase of acid; in other operations, particularly stirred-tank bioleaching/biooxidation of concentrates, pyrite oxidation may require the addition of a neutralizing agent (typically limestone, $CaCO_3$) to maintain an optimum pH for the microorganisms. Heat generation in heap leach operations is usually beneficial, because increasing the temperature improves microbial activity, increases chemical rates of reaction, and enhances primary copper sulfide mineral leaching, if these minerals are present. However, pyrite oxidation in stirred-tank reactors necessitates cooling to ensure that the temperature does not rise above the maximum temperature tolerated by the consortium of microorganisms used in the reactor system.

The microbes in bioleach/minerals biooxidation plants are both present in the leach solution (referred to as planktonic cells) and attached to mineral surfaces. The number of organisms attached to the minerals far exceeds the number in solution, and techniques for enumerating attached microbes are limited and often fraught with error. Attachment also has serious implications when using microbial counts in solution as an indicator of microbial "health" of an operating bioleach plant; the number of microbes in solution is often far lower than the actual number of microbes present in the leach system that includes the ore or concentrate. Microbes favor attachment to sulfur-rich zones and dislocation sites, such as cracks, mineral grain boundaries, and crystal defects. Initial attachment is likely because of physicochemical factors, such as electrostatic and hydrophobic forces between mineral particles and the microorganisms. After microbes are attached to the minerals, a biofilm is formed consisting of extracellular polymeric substances. The conditions and chemical reactions occurring within and at the biofilm–mineral interface remain under investigation, but the prevailing theory (Rohwerder and Sand 2007) is that high concentrations of ferric iron under the biofilm react with the sulfide minerals. Metal sulfide oxidation occurs via two pathways mediated by microbial catalysis of the ferrous-ferric cycle: the "thiosulfate pathway" for acid-insoluble metal sulfides, such as pyrite and molybdenite; and the "polysulfide pathway" for acid-soluble metal sulfides, such as sphalerite, arsenopyrite, and chalcopyrite. For the latter pathway, the metal sulfides are dissolved by the combined action of electron extraction by ferric iron and proton attack (Schippers et al. 2014).

Advances in understanding the complexity of microbial communities in bioleaching processes are facilitated by developments in modern molecular biology techniques that allow analysis of complex microbial populations in commercial bioleach plants. Methods to accomplish this have been described by Roberto (2008), Chávez et al. (2011), Watling (2011), and Demergasso et al. (2011). These protocols lead to a comprehensive definition of the diversity of the microbial populations present in mineral biooxidation and bioleaching operations. Genetic manipulation has been considered for improving the bioleach microorganisms. To date this has not succeeded in demonstrable improvements in leaching. However, there is significant value in understanding the genetics of the bioleaching microbes. In the near term, these findings are important in the indirect improvement of bioleaching through enhanced understanding of the physiology of the microbes rather than by developing and adding genetically manipulated microorganisms to commercial processes (Gericke 2011).

More detailed information on the microbiology and chemistry of bioleaching can be found in the following publications: Rossi 1990; Watling 2006, 2015; Donati and Sand 2007; Rawlings and Johnson 2007; Gericke 2011; Sobral et al. 2011; Vera et al. 2013; C.L. Brierley and Brierley 2013; Schippers et al. 2014; C.L. Brierley 2016.

BIOLEACHING/MINERALS BIOOXIDATION TESTING

Testing to evaluate bioleaching/biooxidation requires the same level of metallurgical rigor that is applied to the evaluation of alternative technologies such as pressure oxidation and chemical leaching. Bioleaching is unique only in using microbial agents as catalysts for the extraction process. All other aspects of the technology entail classic extractive metallurgical considerations. This section is a guide to evaluate bioleaching for treating sulfidic ores and concentrates.

Analyses

Implementation of a successful bioleach project is dependent on comprehensive geometallurgical characterization of the ore (McFarlane et al. 2011), the concentrate (Dew et al. 1997; Van Aswegen et al. 2007), or both. Geometallurgical characterization must be carried out on representative samples over life-of-mine, and extensive analyses are of paramount importance to preclude poor and misleading interpretation of the test results. Reliable geological, mineralogical, chemical, and metallurgical data are used for process design and economic modeling. Table 2 outlines key parameters to be determined using the best possible representative ore and concentrate samples for analyses.

Detailed mineralogical evaluation is essential. QEMSCAN (Quantitative Evaluation of Minerals by Scanning Electron Microscopy) is a fully automated microanalysis system for quantitative chemical analysis and generation of high-resolution mineral maps and images and porosity structure. The data are used to determine the size, shape, mineral associations, and liberation of mineral particles (Greenwood-Smith et al. 1985). Mineral liberation analysis (MLA) establishes mineral quantity and mineral phases (Miranda and Seal 2008). MLA and QEMSCAN are powerful analytical tools that can be utilized for samples of ores, concentrates, tailings, and leached samples. Other sophisticated analytical techniques are available for assessing mineral and chemical structure, morphology, surface and bulk impurities, and particle identification: time-of-flight laser ionization mass spectrometry (TOF-LIMS), time-of-flight secondary ion mass spectroscopy (TOF-SIMS), X-ray photoelectron spectroscopy (XPS), electron probe microanalysis with automated image analysis (EPMA-AIA), and others. With comprehensive mineralogical information of representative samples, the variables for process development, including bioleaching, become defined and manageable.

Table 2 Key analyses of ores and concentrates in preparation for bioleach testing

Chemical Analyses
• Metals: Identify possible toxic metals/metalloids—inductively coupled plasma (ICP) spectrometry; other analyses for anions of concern
• S species: Total sulfur, sulfide-sulfur, sulfate-sulfur—LECO combustion analysis
• Carbon: Total carbon, carbonate, organic carbon—LECO combustion analysis
Mineralogy
• Characterize ore and gangue constituents.
• Identify and quantify metal associations with sulfides or other constituents (e.g., organic carbon).
• Ascertain and thoroughly characterize any encapsulation of ore minerals in gangue or other sulfides—ore microscopy, petrographic analyses, X-ray diffraction (XRD), X-ray fluorescence (XRF).
• Use sophisticated analytical techniques for metals dissemination in sulfide minerals—QEMSCAN, mineral liberation analysis (MLA), scanning electron microscopy (SEM), mass spectroscopy, other spectroscopic analyses for speciation determination.
Metallurgical
• Sequential leach procedure
• Size fraction determinations
• Bulk density, etc.

Mineralogical and chemical analyses are not limited to head samples of ores and concentrates. These analyses are conducted on bioleached residues to assess the effectiveness of leaching, determine which minerals may not have been effectively leached, and identify secondary mineral formation (e.g., jarosite precipitation).

A useful tool in metallurgical amenability evaluation is the semiquantitative sequential (diagnostic) leach procedure (Parkinson and Bhappu 1995), which entails partial dissolution of copper minerals in sequential solutions of sulfuric acid, cyanide, and aqua regia. Copper oxide species have approximately 100% dissolution in acid and cyanide solutions. The secondary copper sulfide minerals have little dissolution in acid with approximately 100% dissolution in cyanide. Primary sulfides (e.g., chalcopyrite) have no dissolution in either solution. The sum of copper dissolved in the sulfuric acid and cyanide steps indicates the amount of copper that may be leachable (recoverable). A ferric sulfate solution step can be added as an indicator of enhanced copper extraction by bioleaching. However, this ferric leach step associated with the sequential (diagnostic) leach procedure will not reveal potential microbial toxicity issues that might be associated with the sample.

Preg-robbing is the adsorption of gold–cyanide complexes by carbonaceous material in the ore or concentrate; certain sulfide minerals (e.g., pyrrhotite, chalcopyrite, and pyrite) have also been implicated in the adsorption of gold from solution (Rees and van Denter 2000; Zhao et al. 2016).

It is prudent to determine whether preg-robbing is a factor with gold ores and concentrates, because this phenomenon can result in substantially lower than expected recoveries of gold in cyanide leaching. Simple and widely used tests for determining preg-robbing have been described by Hausen and Bucknam (1985) and Schmitz et al. (2001).

Bioleaching/Biooxidation Amenability Determination in Stirred-Tank Reactors

Amenability testing assesses whether a particular metal sulfide ore or concentrate is likely to undergo dissolution at a rate and to an extent that justifies consideration of bioleaching as an option for a commercial metal extraction process (Olson and Harvey 2011). Amenability testing is conducted in well-aerated, stirred batch reactors typically 5 L or less in size with a solids concentration of 15%–20% (w/v) in a nutrient solution (Lee et al. 2011). Stirred-tank batch reactor testing has advantages over simple flask reactor testing, which is sometimes used for amenability testing, because greater liquid volumes and mass of ore or concentrate provide more biooxidized material for subsequent analyses and process evaluation. Solution samples can be withdrawn during the test to monitor the progress of the bioleach process without interfering with the test. Stirred-tank reactors also provide excellent mixing and good aeration for microbial activity during the evaluation phase.

The selection of microorganisms for amenability testing is dependent on the mineralogy of the ore as well as conditions anticipated for application. For example, a heap with sulfidic refractory gold ore in which the gold is embedded in pyrite would be expected to heat. Therefore, amenability testing should include evaluation of consortia of mesophilic, moderately thermophilic, and extremely thermophilic (archaea) microorganisms. With the exception of the extremely thermophilic archaea, microbial consortia used for amenability testing are often cultured from the ore samples to be evaluated and adapted to increasing solids concentration through subculturing. For amenability tests run at temperatures between about 49° and 65°C, enrichment cultures of moderately thermophilic microorganisms can be obtained from mine sites that have stockpiles or heap leach operations close to these temperatures (Norris et al. 2012). Enrichment cultures are obtained using a growth medium containing ferrous iron, elemental sulfur, or both at pH ~1.5 and incubated at temperatures suitable for growth of the desired group of microorganisms (i.e., 30°–45°C for mesophilic microorganisms, with 55°C preferable for moderately thermophilic microorganisms). The extremely thermophilic archaea typically are not naturally occurring in base and precious metal ores and therefore cannot be cultured from ore samples. The extremely thermophilic archaea are available from type culture collections, such as the American Type Culture Collection or the Deutsche Sammlung von Mikroorganismen und Zellkulturen GmbH, which is the German type culture collection; the extremely thermophilic archaea can also be cultured from hot acid springs and solfatara. Laboratories that routinely perform amenability testing usually maintain consortia of mesophilic, moderately thermophilic, and extremely thermophilic microorganisms and adapt these consortia to the ore and/or concentrate undergoing evaluation through a subculturing process.

Amenability tests in batch reactors can be conducted in a relatively short period (several months). These tests allow for evaluating several variables (e.g., different temperature groups of microbes) and are fairly inexpensive yet provide critical information as to whether next-stage testing (continuous stirred-tank reactors for concentrates and column tests for ores) should be pursued or whether the ore/concentrate has characteristics (e.g., excessive amounts of acid-consuming gangue minerals, or excessive cyanide and/or lime consumption for refractory gold ores) that preclude bioleaching/biooxidation as a process option. Table 3 lists the purposes of amenability testing. Careful interpretation of amenability test results is imperative, because the testing is conducted on ground ore and concentrates under batch conditions that do not reflect the conditions of whole ore heap leaching or concentrate leaching in continuous stirred-tank reactors (Olson and Harvey 2011). Interpretation of acid production and consumption and rates of oxidation of ores and concentrates in amenability tests needs particular attention because of the particle size and because oxidation rates are usually slower in batch reactors.

Table 3 Laboratory bioleach batch reactor amenability testing with concentrate and pulverized ore

1. Determine amenability of feed to bioleaching.
2. Identify potential toxicity of ore/concentrate feed and site water.
3. Determine ultimate metal extraction.
4. Approximate oxidation rates of various sulfide minerals.
5. Quantify gold recovery vs. percent sulfide–sulfur oxidation for precious metal ores/concentrates.
6. Obtain order-of-magnitude reagent consumptions.
7. Ascertain performance of different suites of microorganisms:
   - Mesophilic microbes for low-temperature conditions
   - Thermophilic microbes for high-temperature conditions
8. Estimate acid consumption/production.
9. Determine mass loss.

Continuous Stirred-Tank Reactor/Mini Pilot-Plant Testing of Concentrates

The next stage of evaluating concentrates for bioleaching/biooxidation after aerated and stirred batch reactor amenability testing is continuous stirred-tank reactor (CSTR) testing. For CSTR tests involving microorganisms, typically three to four aerated and agitated reactors in series are employed. The first stage is usually larger in size or involves two reactors in parallel for retaining solutions for a longer period to allow approximately 60% or more of the sulfide-sulfur to be oxidized in this stage. CSTRs should minimally be several liters in size; small CSTRs are usually constructed of glass and are water-jacketed to allow temperature control for testing various suites of microorganisms. Small CSTRs are used to confirm amenability of the concentrate to bioleaching, adapt the microbial consortium to higher metal and total dissolved solids (TDS) concentrations, confirm metal recovery versus percent sulfide-sulfur oxidation, and quantify residence time in each reactor stage for the bioleaching/biooxidation process to achieve acceptable metal recoveries.

Mini-CSTR pilot plants are usually stainless-steel (316 L), temperature-controlled reactors equipped with agitators and sparged with air. Mini pilot plants of 120-L size can be used to design industrial plants for biooxidation of sulfidic refractory gold concentrates. These mini pilot plants are composed of one 60-L primary reactor and three 20-L secondary reactors in series. A continuous mini pilot-plant run of four to six months can require nearly 1 t (metric ton) of concentrate.

CSTR runs of this duration confirm amenability of the concentrate sample to biooxidation pretreatment; adapt the microbial culture to improve oxidation performance; obtain process data for design of the industrial plant including residence time of the concentrate in each reactor stage; ensure that waste products are environmentally acceptable; establish optimal cyanidation conditions for gold recovery from the biooxidized product; and establish optimal reagent consumptions during biooxidation and subsequent countercurrent decantation, neutralization, and cyanidation of the biooxidized product.

CSTR campaigns are run on concentrate samples representative of life-of-mine. Among the many considerations are the types and amounts of sulfide minerals, the total sulfide-sulfur content, the percentage of sulfide-sulfur that must be oxidized to achieve acceptable metal recoveries, optimal particle size, ideal solids density, and concentrate throughput. These factors, among others, influence the retention time of the solids in each stage, the amount of oxygen required, and the amount of heat generated. These in turn determine the design of the aeration and cooling systems, the reactor sizes and configuration, and ultimately the capital and operating costs of the full-scale industrial plant. The ultimate plant design is based on a combination of factors including the highest sulfide-sulfur content coupled with tonnage throughput (i.e., the maximum metric tons of sulfur loading per day), loading of ions such as arsenic, and variations of feed through life-of-mine.

Performing successful mini-CSTR pilot-plant campaigns requires extensive sampling and analysis over a period of months. Each reactor is analyzed daily for total iron, iron speciation, redox potential, pH, dissolved oxygen, and temperature; solution from the last-stage reactor is analyzed for a suite of metals. The feed to the CSTR is assayed for metals, total iron, total sulfur and sulfur species, total carbon and carbon species, silica, and specific gravity. The oxidized solids and leach liquor from the final reactor are assayed daily to complete a mass balance. A mass balance is performed on each concentrate tested. A key microbiological element of CSTR operation and design of the commercial-scale plant is determining optimum solids retention time in the various stages, particularly in the first-stage reactor, which sees the highest sulfide-sulfur concentrations. Avoiding microbial "washout" is critical to CSTR performance. Washout occurs when the solids retention time in a reactor stage is less than the time required for the microorganisms to divide (i.e., divide by binary fission to produce two daughter cells). Biooxidation/bioleaching ceases if washout occurs and is manifested almost immediately by a rapid decline in the redox potential of the reactor suffering the washout.

Laboratory Column Reactor Testing of Ores

The next step for evaluating ores for heap bioleaching/biooxidation is column testing, after batch reactor tests are completed and the ore is shown to be amenable to the process. Column tests can be carried out in sizes ranging from small columns that may be 1 m high and 15 cm in diameter to 10 m high and more than 1 m in diameter. Column diameters should be a minimum of 4×, preferably 6–10×, the diameter of the largest particle size to avoid wall effects and solution channeling in the column. The objectives of column testing are summarized in Table 4.

Table 4 Purpose of bioleach laboratory column testing

1. Determine optimum ore particle size for maximizing extraction.
2. Verify acid consumption for respective particle sizes.
3. Ascertain rates of oxidation for sulfide minerals by particle size.
4. Understand iron balance.
5. Reveal potential toxic conditions.
6. Evaluate rates and effects of bioleaching at different temperatures.
7. Assess competence of ore.

Column tests are performed on different crush sizes of ore, which have been thoroughly analyzed for metal concentrations. It is advisable to agglomerate the ore with a microbial culture, adapted to the ore being tested. It is advantageous to use the mesophilic, moderately thermophilic, and extremely thermophilic cultures that have been adapted to the ore during the batch reactor amenability tests.

Columns are aerated from the base and may be irrigated initially with a microbial nutrient solution (Lee et al. 2011). Barren raffinate from downstream solution processing is often used for irrigation of columns when copper ores are being column leached. Columns should be operated in closed-loop solution circulation, which allows constituents to build up in solution; the purpose is to evaluate microbial tolerance to increasing ion concentrations and to identify constituents that may be problematic to the microorganisms. pH control is often necessary to maintain the pH in the desired range. When testing copper sulfide ores, the effluent solution, referred to as pregnant leach solution (PLS), is processed periodically with a small solvent extraction plant to remove dissolved copper; the resulting raffinate is recycled for column irrigation. Column tests are used to compare metal recoveries and rates of leaching at different particle sizes, evaluate leaching under different degrees of acidity, and assess rest–leach cycles. Column tests are useful in assessing the performance of different microbial populations (see the "Case Study 1" section), such as thermophilic microbes (Norris et al. 2012), when heating of the proposed ore heap is predicted based on sulfide grade of the ore and rates of sulfide oxidation and for primary copper sulfide ores (e.g., chalcopyrite and enargite). Such tests are performed using temperature-controlled columns (Acar et al. 2005).

Column testing of sulfidic refractory precious metal ores requires additional consideration. The microorganisms oxidize the sulfide minerals, typically pyrite and arsenopyrite, releasing iron and arsenic into solution. The gold and most of the silver, if present, remain with the solids. It can be difficult to determine when sufficient sulfide has been oxidized to achieve maximum gold recoveries, because there are no methods of determining this effectively by assaying the effluent solution. Dissolved iron and arsenic are not good indicators because of precipitation in the column of jarosite, possibly scorodite and other ferric arsenate compounds. Consequently, replicates of the column test are set up and periodically a column is terminated and the leached ore analyzed for sulfide-sulfur and other constituents, and the oxidized ore is subjected to bottle-roll cyanidation testing to determine gold recovery. For very large column tests, provisions are made for periodic sampling, preferably through portals constructed on the side of the column at various depths.

Head samples and leached residues of column tests are fully analyzed mineralogically and chemically. Effluent solutions are analyzed systematically for redox potential, pH, metal/metalloid concentrations, total iron, iron species, and TDS; effluent solution volumes are measured. Makeup water must be added because of evaporation. A mass balance that considers the solution volumes and metals concentrations of samples removed for analyses and solvent extraction removal of copper is completed on termination of each column test (Olson and Harvey 2011).

Lee (2011) outlined testing for a sulfidic refractory whole ore biooxidation pretreatment. The ratio of sulfide oxidation to increased gold extraction along with metal extraction achievable at different crush sizes are fundamental to evaluating the potential for practical application of a whole ore biooxidation pretreatment process. Other factors include cyanide consumption for leaching gold after biooxidation pretreatment as well as reagents required for neutralization of the ore prior to gold leaching and disposal of acidic solutions generated in the process.

Iron precipitation as jarosite is often extensive in column tests and can be particularly voluminous when column leaching is at elevated temperatures with moderately thermophilic and extremely thermophilic microorganisms. Jarosite precipitation is of no concern because it is fine grained and does not impact solution percolation. It does increase acidity, as jarosite precipitation is an acid-producing reaction and it affects the iron mass balance. Care must be exercised when interpreting iron assays in the leached residues because precipitated iron will be measured.

Hydraulic testing of the ore should also be an imperative during column leaching. Hydraulic testing involves compression tests to ascertain the ore dry bulk density and hydraulic conductivity as a function of ore stacking depth. Other hydraulic tests determine moisture retention and agglomerate quality, ore porosity, pore pressure, percent saturation, and drain-down curves. These data are useful to predict full-scale heap behavior and metal recovery (Robertson et al. 2013).

The modeling and predicting of column bioleach/biooxidation results to performance of full-scale, crushed ore heap leach operations requires applying scale-up factors. Scaling crushed ore column tests to run-of-mine (ROM) stockpile bioleaching is difficult, but column tests do provide useful information on the feasibility of ROM leaching.

Case Study 1: Temperature Selects Predominant Groups of Microorganisms

Column testing using a pyritic ore has demonstrated the variability of the microbial populations with change of temperature (J.A. Brierley 2003). A stepwise increase of temperature from 22°–60°C resulted in decimation of mesophilic iron-oxidizing bacteria with an increase of moderately thermophilic microbes, followed by the thermophilic archaea iron-oxidizing microbes. Reversal of the temperature series from 60°–22°C resulted in return to dominance of mesophilic bacteria with a concurrent decrease in the populations of moderately thermophilic microorganisms and the thermophilic archaea. This column test demonstrated that temperature was the overriding factor that selected for the dominant groups of microbes.

Pilot/Demonstration-Scale Heaps

The type of ore being considered for heap bioleaching/biooxidation often determines whether pilot or demonstration heaps are operated. Secondary copper sulfide ores may undergo crib tests at the mine site. These are essentially large-scale column tests, and the same protocols applied to laboratory column tests pertain to crib tests. Often, secondary copper sulfide ores are not subjected to pilot- or demonstration-scale testing after successful column and crib tests, principally because secondary copper sulfide heap bioleaching is considered a mature technology for porphyry-type supergene ores. It is prudent, however, to pilot-test secondary copper sulfide ores that differ in ore genesis from porphyry-type supergene ores as these ores may possess very different hydraulic characteristics and vary significantly from porphyry-type supergene ores in their response to heap bioleaching. Primary copper sulfide ores, such as chalcopyrite and enargite, and sulfidic refractory gold ores should be evaluated in pilot- or demonstration-scale tests at the mine site. Primary copper sulfide ore heap and ROM stockpile bioleaching is an emerging technology; the precise conditions and the engineering of the heap/stockpile to effect primary copper ore leaching are still under development. Heap biooxidation of sulfidic refractory gold ore was commercialized (Logan et al. 2007) at one mine site after pilot- and demonstration-scale testing. Sulfidic refractory gold ores from other sources may perform differently, which necessitates pilot- or demonstration-scale testing. An important benefit of mine-site pilot or demonstration plant testing is to provide a learning opportunity for mine operators in handling microorganisms and conducting the bioleach process.

J.A. Brierley et al. (1995) described pilot- and demonstration-scale testing of a sulfidic refractory gold ore that entailed bioleaching pyrite to expose encapsulated gold. Six pilot plants varying in size from 432–25,900 t were used to test different ores and gold leach methods after biooxidation pretreatment of the ore. After the pilot-plant test, a much larger demonstration plant of 708,000 t ore was constructed and operated (Shutey-McCann et al. 1997). Inoculation of the ore was found to decrease the time required for natural microbes to establish an active population in the ore. Also, the heap temperature increased substantially because of oxidation of pyrite, which required the addition of moderately thermophilic and extremely thermophilic microbes to the ore heap, as the succession of microbes in heaps that heat is important and complex (C.L. Brierley 2001). A process for microbial inoculation of ore was developed (J.A. Brierley and Hill 1993). Other processes for inoculating heaps are described by Gericke (2011).

Laboratory amenability and column tests, pilot- and demonstration-plants, and commercial bioleach plants are always subject to colonization by naturally occurring (transient, wild-type) microbes. Therefore, trying to control conditions for specific microbes, which may be intentionally added to these systems, is difficult, if not impossible, given current understanding and technology know-how. The physicochemical environment (e.g., pH, TDS, temperature, metal concentration, and mineralogy) of bioleach systems selects the resident microflora. Inoculation of a heap by a select monoculture or a specialized laboratory consortium can initiate a bioleach process, but ultimately process conditions will select the best microbes for the operation. The only known exception is the inoculation of a heap with the extremely thermophilic archaea, when the heap temperature exceeds 60°C (Logan et al. 2007).

Solution and solids analyses performed for column tests are also carried out for pilot- and demonstration-scale plants. Probes to measure various parameters (e.g., pH, redox potential, temperature, and moisture) should be placed at various depths and locations in the pilot or demonstration heap. Periodically, the heap should be cored to obtain samples at different depths and locations for chemical, mineralogical, and microbial analyses. Samples can be examined for direct counts of microorganisms, analyzed for the numbers of iron- and sulfur-oxidizing microorganisms at different temperature regimes, and subjected to metagenomics sequence analysis to determine the distribution of different types of organisms.

Pilot and demonstration heaps offer opportunities to evaluate seasonal effects, precipitation events, rest periods, irrigation rates, aeration, evaporation rates and effects, ore competency under extended leach times and lift heights, irrigation methods, heap covers, and other factors. The extensive data obtained from these tests can be used to develop models to predict performance and economic viability (Marsden and Botz 2017).

Successful testing and the decision to commercialize a bioleach operation is a lengthy process. The time required from initiation of testing at laboratory scale to commercialization can be about 10 years. Regardless of the time required, thorough testing is mandatory to prevent possible failure.

ROM STOCKPILE LEACHING AND COMMERCIAL HEAP

Run-of-Mine Stockpile Bioleaching

Stockpile bioleaching in which ROM ore is used has been referred to as "dump" leaching. The mining industry has employed this process for about 50 years to recover copper from oxides and secondary sulfide, submarginal-grade ROM ores (Sheffer and Evans 1968). Initially, dissolved copper was recovered by the process of cementation using scrap iron (Rossi 1990) until solvent extraction technology became widely used in the 1970s. Most early operations employed surface solution ponds where bacteria oxidized acidic ferrous iron. The resulting ferric iron percolated into the ore bed. Limited oxygen reached the bacteria in the ore bed, because of the solution impoundments on the surface. Early investigation (see the "Case Study 2" section), of the bioleaching bacterial population indicated a limited distribution of the microbes within a sulfidic copper leaching operation (Bhappu et al. 1969).

ROM stockpile leaching of submarginal-grade copper ores remains an economically vital practice at many open pit copper mining operations. Some stockpiles have been under leach for decades with new ore continually being added. As mines become deeper, new ore added to stockpiles has trended increasingly toward copper sulfides, including chalcopyrite. Although little or no engineering efforts were undertaken in the early history of ROM stockpile leaching to enhance bioleaching activity, this has changed in recent years, principally to enhance the bioleaching of primary copper sulfide minerals.

Recognizing that ROM stockpiles were oxygen limited, which significantly diminishes microbial development and activity, techniques have been introduced to increase natural air convection, and some operations have instituted forced aeration. These include (Schlitt 2006)

- Finger dumps, which entail stacking ROM ore in patterns that allow maximum exposure of stockpile slopes to the atmosphere;
- Angled holes drilled into the stockpiles through which pressurized air is introduced;
- Placement of perforated pipes under newly stacked ore through which pressurized air is introduced;
- Air injection through vertical drilled holes; and
- Extended rest periods, which promote stockpile drainage, allowing more air to enter the pore spaces.

As early as the 1960s, heating of stockpiles was known (Beck 1967) with temperatures of 60°–80°C recorded. Higher temperatures in portions of the heap induced convective airflow, particularly in stockpiles under rest.

Engineered stockpiles are becoming increasingly common with primary copper sulfide ores being placed in stockpiles and the recognition that moderate and extreme thermophiles at temperatures higher than 50°C are required to leach these hypogene ores. Engineered stockpiles at Morenci, Arizona, United States (Ekenes and Caro 2012), incorporate aeration lines under newly stacked ore, and forced aeration is employed to ventilate the heaps. Microorganisms native to the ore are introduced onto newly stacked ore via raffinate irrigation to kick-start leaching. Extensive sampling and analysis of solutions and solids have been instituted on engineered stockpiles to evaluate the effectiveness of chalcopyrite bioleaching (Ekenes and Caro 2012). Engineered stockpiles have shown increased microbial pyrite oxidation, which has resulted in higher stockpile temperatures; increased acidity reducing external acid addition to maintain an acceptable operating pH; and, most importantly, improved copper extraction from primary copper sulfide minerals (Ekenes and Caro 2012).

Microbial populations in the PLS from the commercial-scale, engineered ROM stockpile leach facility at Escondida, Chile, were monitored over a period of six years using culture-dependent and independent approaches. The Escondida stockpile, which incorporates forced aeration, contains copper sulfide ores including chalcopyrite. As the height of the stockpile increased, microbial populations changed from predominantly mesophilic to thermotolerant and moderately thermophilic microorganisms—with *Leptospirillum ferriphilum*, a thermotolerant iron-oxidizing bacterium, and *Sulfobacillus thermosulfidooxidans*, a moderately thermophilic, iron- and sulfur-oxidizing bacterium—predominating the microbial population (Acosta et al. 2014).

Extensive microbial analyses are limited on commercial-scale stockpiles under leach, particularly on leached solids from these operations, so it is somewhat premature to draw conclusions about operating conditions in relation to the identified microbial populations. The numbers of microorganisms attached as biofilms to the ore within the stockpile far exceed those in the PLS, and the microbial composition may differ as well. Nevertheless, continuing to identify microorganisms and determine their distribution on the ore within the stockpile and in the PLS will eventually lead to a better understanding of the dynamic changes in microbial populations over time and to correlating changes in populations with chemical and physical changes within the stockpile.

Case Study 2: Early Investigation of Bacterial Distribution in an ROM Stockpile Under Leach

Samples of the stockpile were collected from the surface and at depth using a pneumatic drill providing mixed samples from 8-ft intervals within the stockpile. Iron-oxidizing bacteria were present at the surface in concentrations from 10^3 to 10^7 cells/g dry weight. Table 5 shows the decline of iron-oxidizing bacteria with depth. No bacteria were present below 88 ft.

The bacterial population was likely affected by lack of moisture or limited wetting of ore in some areas of the stockpile and lack of oxygen within the stockpile. PLS samples from the stockpile contained 10^6 cells/mL of iron-oxidizing bacteria suggesting that (1) portions of the stockpile were

Table 5 Iron-oxidizing bacteria detected at depth

Interval Below Surface, ft	Bacteria Concentration, cells/g dry weight
8–16	10^3
24–32	10^2
40–48	10^3
56–64	10^1
72–80	10^1
88–144	None detectable

favorable for growth of the bacteria, or (2) the bottom area of the stockpile, where large boulders segregate as a result of the way stockpiles are constructed, allowed air to enter, resulting in good bacterial activity at the base of the stockpile.

Commercial Heap Bioleaching of Base Metal Sulfides and Biooxidation of Sulfidic Refractory Gold Ores

Commercial-scale heap bioleaching, also referred to as percolation leaching, of secondary copper sulfide ores began in the mid-1990s. The technology was a merger of sulfuric acid leaching of copper oxides, developed in the 1960s; ROM stockpile leaching of copper sulfide ores; and "thin-layer" leaching with ferric iron (Bustos et al. 1993). Today commercial-scale heap bioleaching of copper sulfide ores is widely practiced in many locations throughout the world and has been developed for other sulfide minerals as well (Saari and Riekkola-Vanhanen 2011).

Biooxidation pretreatment of sulfidic refractory gold ores, developed in the 1990s, has also been practiced commercially (Logan et al. 2007). Biooxidation pretreatment is a process for enhancing gold recovery by oxidizing sulfide minerals, principally pyrite and arsenopyrite (Reactions 7 and 8), in which gold is embedded. Biooxidation pretreatment exposes the gold for subsequent extraction by conventional metallurgical technology, typically cyanide leaching. Significant improvements in gold recovery are achievable using biooxidation pretreatment.

$$FeS_2 + 14Fe^{3+} + 8H_2O \rightarrow 15Fe^{2+} + 2SO_4^{2-} + 16H^+ \quad \textbf{(EQ 7)}$$

$$FeAsS + 5Fe^{3+} \rightarrow As^{3+} + 6Fe^{2+} + S^\circ \quad \textbf{(EQ 8)}$$

Heap "reactors" are technically and economically (Acevado and Gentina 1993) suitable for pretreating refractory, lower-grade (e.g., 1.0–2.4 g Au/t) sulfidic whole ores that may not be amenable to flotation; are too low-grade to economically process by roasting, pressure oxidation, or stirred-tank biooxidation processes; or have mineralogical characteristics that preclude other processing methods. Biooxidation heap pretreatment, requires a period of 270–380 days depending on sulfide-sulfur grade, particle size, and other factors. After biooxidation pretreatment, the heap can be neutralized with lime/limestone and gold is extracted using conventional cyanide leaching practices. Alternatively, ammonium thiosulfate can be used for gold extraction after biooxidation of double-refractory sulfidic ores containing preg-robbing carbon (Wan and Brierley 1997).

One reported project used a heap approach for biooxidation pretreatment of refractory gold ore. The Newmont Biopro technology (Logan et al. 2007; Roberto 2017) pretreatment heap was operated using the same principles as a bioleach heap for base metal ores. The goal is to maximize the biooxidation of pyrite (Reaction 7) and arsenopyrite (Reaction 8) and other sulfide minerals that occlude gold. Inoculated and agglomerated ore is stacked on a high-density polyethylene-lined pad with a base layer of crushed rock placed on the liner. Within the crushed rock layer is an array of perforated pipes for forced aeration of the ore pile and a system of pipes for collecting the circulating leach solution. Low-pressure fans provide air for the ore heap. The heap is subjected to leach–rest cycles to conserve heat within the heap and to ensure drainage of the heap to allow void spaces to fill with air.

The refractory sulfidic ore is inoculated and agglomerated with an acidic solution containing mesophilic, moderately thermophilic, and extremely thermophilic microorganisms. A consortium of microbes for inoculation is required because the ore heats during the biooxidation process as a result of pyrite and arsenopyrite oxidation. Initially, the stacked ore is at ambient temperature favoring the growth of mesophilic microbes. As the temperature increases in the ore bed, those microbes active at higher temperatures dominate the population. Eventually, the extremely thermophilic archaea dominate in those areas of the heap reaching temperatures of more than 55°–60°C (J.A. Brierley 2003). Metallurgical processing of the biooxidized gold ore entails either neutralization of the heap followed by cyanide leaching of the heap or dismantling the heap, and neutralizing the biooxidized residue and leaching the product using carbon-in-leach cyanidation. The Biopro technology processed 8.8 Mt of sulfidic refractory gold ore, resulting in recovery of 120,000–180,000 oz/yr of gold between 2000 and 2005 (Roberto 2017).

The commercial success of heap bioleaching/biooxidation requires attention to physical design factors based on ore characteristics and operating methodology to ensure optimal permeability for solutions and air (John 2011). These factors are critical to favorable solution chemistry and microbial development and activity. Based on successful column testing, ores are crushed, typically to 12 mm or larger, agglomerated with acid and in some cases with an acidic microbial culture, and the ore is stacked on prepared pads. Depending on the ore permeability, forced aeration is implemented with placement of perforated air pipes beneath the ore in a gravel bed to provide oxygen and carbon dioxide to the microorganisms. Some agglomerated ore heaps are sufficiently permeable that forced aeration is not necessary, relying on natural air convection through the top and exposed slopes of the heap. In implementing forced aeration, it is important to make sure the air pipes are not in the phreatic head to avoid the air pipes filling with solution. This is usually managed by placing drain lines in the gravel bed beneath the air lines (Schnell 1997; Schlitt 2006). Air addition is based on the amount of sulfide to undergo oxidation, sulfide mineralization, oxidation rates of the sulfide minerals, heap height, elevation of the plant, and efficiency of oxygen utilization. Copper sulfide heap leach operations are drip-irrigated with raffinate, intermediate leach solution—often with additional acid added to meet demand—and solution application is controlled to ensure wetting of the heap without saturation and ponding; this facilitates airflow throughout the heap. Copper sulfide heaps of porphyry-type supergene ore do not generally heat; sulfide grades are typically low, and pyrite, although present, does not effectively oxidize because the redox potential required to oxidize pyrite is seldom achieved in the heap. Consequently, these commercial secondary copper heap bioleach operations in high desert regions often suffer from cool temperatures that slow microbial and chemical reaction rates. Thermal covers are sometimes employed to reduce evaporative losses and minimize solution cooling because of evaporation. Rest cycles are typically instituted to enhance microbial oxidation as the heaps drain of solution and to mitigate heat loss through solutions exiting the heap. Other measures that have been used to increase heap temperature are heating the raffinate and heating the ore.

Ores that contain higher sulfide grades or necessitate the microbial oxidation of pyrite or arsenopyrite to enhance gold recovery do require the use of forced aeration (Logan et al. 2007) and can benefit from implementation of other approaches to improve natural ventilation, including "kit-kat" structures, whereby heaps are separated by valleys or ditches to expose larger areas of the heap to the atmosphere (Saari and Riekkola-Vanhanen 2011).

Effective wetting of heaps undergoing bioleaching biooxidation is often a problem because of the stacking of clay materials or ores that create excessive fines during leaching, compaction of heaps by truck stacking of ore rather than automated stacking devices, compaction from multiple lift emplacement, and solution channeling because of poor agglomerate formation. Solution channeling can result in significant metal reserves left in the leach pad because of large volumes of under-leached ore. Subsurface leaching is now being employed to deliver leach solution to these areas resulting in recovery of metal inventories from under-leached ore volumes (Rucker et al. 2017). Compaction can lead to saturated conditions within the heap; solution saturation prevents effective aeration of the heap resulting in limited ferrous iron oxidation by the microorganisms and ultimately poor extraction of the metals of value.

Total iron concentrations in the PLS of many commercial heaps bioleaching supergene copper ores from porphyry-type deposits are often low; 1–2 g/L in solution is common. The pH is maintained by acid addition in the 1.2–2 range so metals, including iron, remain in solution. These operations are dependent on naturally occurring microorganisms, principally mesophilic bacteria and archaea, developing throughout the heap and maintaining the iron in the ferric form. Nutrients for the microorganisms are not added in commercial heap leach operations, as sufficient nutrients are generally available from the ore. The success of microbial development and activity in the heap is ascertained by measuring the redox potential and the ferric-to-ferrous iron ratio in the PLS. The first mole of copper from chalcocite (Cu_2S) is readily leachable at a low redox potential and occurs rapidly in the heap. The second mole of copper from the blaubleibender covellite product of chalcocite leaching requires a higher redox potential and leaches slowly. The redox potential in the heap will not increase to the millivolt level required for dissolution of blaubleibender covellite (CuS) until all available Cu_2S is oxidized to Cu^{2+} and CuS (Reaction 4).

In heap leach operations with arsenic minerals as a component of the ore, it is particularly important that the redox potential is sufficiently high and adequate ferric iron is present to ensure that As^{3+} is oxidized to As^{5+}, because the former is toxic to the microorganisms. Some arsenic minerals, such as realgar and orpiment, are acid soluble and arsenic will go into solution as arsenite (AsO_3^{3-}); ferric iron readily oxidizes this to arsenate (AsO_4^{3-}). Special attention during commercial plant start-up is necessary to accomplish this oxidation reaction.

Microbial population developments in heap leaching of supergene porphyry-type copper ores have not been extensively studied. Most microbial evaluations have looked at populations of mesophilic iron- and sulfur-oxidizing bacteria in PLS and raffinate from these operations. Microbial numbers are usually in the 10^4–10^6 per milliliter range in PLS with slightly lower numbers in raffinate. However, it is not clear how these microbial numbers and populations relate to microbial development and activity within the ore bed.

Several microbiological studies have been done on secondary copper operations that are not porphyry ores. The secondary copper ore at Zijinshan mine in China represents a quartz-alunite type epithermal deposit with high pyrite (5.8%) content. A coarse crush (–4 cm) sulfidic ore is stacked, and heaps rely on convective ventilation, not forced aeration. The oxidation of the pyrite results in an estimated internal heap temperature of 70°C, which enhances convective aeration. The PLS has a pH <1, acidity of 30 g/L, total dissolved iron concentration of 50 g/L, and Eh (redox potential) 700–740 mV versus SHE. Reportedly, the severe conditions of the leach solution have limited the microbial population of the PLS to 10^4/mL. The microbial population is dominated by bacteria of the genus *Leptospirillum* near the heap surface with increasing distribution of thermotolerant and moderately thermophilic microorganisms with depth (Liu et al. 2010). Archaea of the genus *Ferroplasma* were also detected (Ruan et al. 2011), which is tolerant to low pH values and high dissolved iron concentrations.

The Monywa (Myanmar) secondary copper sulfide ores are from a high-sulfidization system. In the heap bioleach operation, total iron concentrations in the PLS approach 25 g/L with nearly all as ferric iron with a redox potential of 750–775 mV versus SHE and high acidity (~20 g/L). Solid and leachate samples from the Monywa operation were subjected to direct cell counts and polymerase chain reaction–denaturing gradient gel electrophoresis (PCR-DGGE) analysis of the 16sRNA gene in 2006 (Hawkes et al. 2006). At the time of this analysis, portions of the heap temperature at Monywa reached 46°C and the PLS and raffinate pH ranged from 0.95–1.3. Organisms identified by Hawkes and colleagues (2006) at Monywa were closely related to *L. ferriphilum* and *A. caldus*; a new archaea species, *Ferroplasma cyprexacervatum*, was isolated. *Sulfobacillus* species were also detected.

Bioleaching using a heap reactor has been developed for recovery of nickel, zinc, cobalt, and copper from a black schist ore (Puhakka et al. 2007; Saari and Riekkola-Vanhanen 2011) at the Talvivaara Sotkamo mine located in eastern Finland. The internal heap temperatures range from 10° to 90°C; heating is attributed to rapidly oxidizing pyrrhotite in the ore. A complex consortium of bioleaching microorganisms is present in the heap. The reported microbes span the range of operating temperatures with the mesophilic and moderately thermophilic bacteria and archaea, and the extremely thermophilic archaea.

Crushed Ore Heap Bioleaching of Hypogene Copper Sulfides

A significant challenge for heap bioleaching is effective leaching of primary copper sulfide minerals, principally chalcopyrite and enargite. Effective bioleaching technology to accomplish this is under development.

Lee et al. (2011) compared mesophilic and thermophilic bioleaching of complex sulfidic ore composites containing primary enargite and covellite. The results verified the advantage of high-temperature bioleaching; 80%–95% copper extraction at 65°C compared with 8%–20% copper extraction at 20°–22°C was observed. Advances in high-temperature bioleaching will lead to eventual demonstration of the process using field-site pilot plants.

Instrumented laboratory columns (Robertson et al. 2011) and large laboratory column (6 m) testing have shown the potential for effective high-temperature bioleaching of chalcopyrite (Dew et al. 2011). At temperatures higher than 50°C and up to 80°C, copper recovery from low-grade (0.35%–0.7% copper) primary chalcopyrite ores increased from typical limiting values of about 30% to more than 60% (Dew et al. 2011). Application of chalcopyrite bioleaching in a commercial-scale, crushed ore heap is dependent on autogenous heat generation from oxidation of contained pyrite in the ore. Pyrite content and rates of pyrite oxidation will control which groups of microbes should be tested and considered for bioleaching of primary copper sulfide ores (Dew et al. 2011). Significant engineering considerations are also necessary to maximize and maintain heap temperatures to attain optimal conditions for microbial activity of thermophiles.

COMMERCIAL-SCALE STIRRED-TANK BIOLEACHING/BIOOXIDATION OF CONCENTRATES

Testing of biooxidation pretreatment used for refractory sulfidic gold concentrates with CSTR systems (Van Aswegen et al. 1991) resulted in development of commercial operating Biox plants throughout the world (Van Aswegen et al. 2007). The first commercial engineered bioreactor system for biooxidation pretreatment of a refractory sulfidic gold concentrate was implemented by Gold Fields in 1986 (Gericke et al. 2009) at the Fairview mine, South Africa; this Biox plant remains in operation today. Thirteen Biox plants have been successfully commissioned (Outotec 2017). In addition, four CSTR plants for biooxidation pretreatment of refractory gold concentrates have been built and operated by other technology developers and one CSTR plant developed by the Bureau de Recherches Géologiques et Minières (BRGM) operated for bioleaching cobalt from pyritic tailings (Morin and d'Hugues 2007).

Based on mini-CSTR pilot-plant runs, a process design package, which includes process specifications, mass and heat balances, flow diagrams, valve and instrument diagrams, equipment list, and detailed equipment specifications, is completed and the plant is constructed by an engineering company. A typical commercial plant has a layout similar to the mini pilot plant with two or more primary reactors in parallel followed by up to four secondary stages. Each reactor is equipped with impellers to disperse air and suspend concentrate in the reactors and cooling coils to maintain reactor temperatures in the 40°–45°C range. The microbial inoculum used in the mini pilot-plant runs is brought to the plant site and scaled up. Each reactor is started in batch and then converted to continuous operation during commissioning. Because of the large amount of sulfide-sulfur present in the typical 20% pulp density used in stirred-tank reactors, many microorganisms (10^8–10^{10}/mL) are present in the reactors. To maintain these organisms and to ensure rapid cell division, nutrients are supplied (Table 6). Limestone is typically added to maintain a pH of 1.1–1.5 across all reactor stages. Air is added to each reactor via a sparge ring located just under the impeller.

Table 6 Nutrients for mineral biooxidation processes

Nutrient	Amount, kg/t concentrate	Source	Reference
Nitrogen	1.7	Ammonium sulfate	van Aswegen et al. 2007
Phosphorus	0.9	Ammonium phosphate	
Potassium	0.3	Potassium salts	
NH_4^+	8.4		Dew et al. 1997
PO_4^{3-}	1.56		
K^+	1.42		

Residence time of the solids across all stages of commercial CSTRs is usually four to five days. After the final stage of biooxidation/bioleaching, the contents of the final reactor are subjected to solid–liquid separation. When bioleaching base metal concentrates, the liquor contains the metals of value and this PLS is processed by conventional metallurgical technologies to recover the metals. In biooxidizing precious metal concentrates, the gold and silver remain with the solids, which are water-washed in countercurrent decantation thickeners and then neutralized with limestone and lime; the precious metals are solubilized with cyanide.

The only commercial CSTR plant for base metals is in Uganda. This technology, developed by BRGM, was first used at the Kasese Cobalt Company in 1988 to bioleach cobalt from pyritic tailings (Morin and d'Hugues 2007).

OTHER BIOLEACH TECHNOLOGIES

Other bioleach technologies have been proposed and described by Carretero et al. (2011) as alternative commercial applications for metals leaching. Several of the process technologies have been field tested. Only one of these technologies, the Bacox process, resulted in construction of commercial plants.

Bioleaching of ore particles sized to 0.95–1.9 cm in a vat reactor has been tested at laboratory scale. No data are available as to the effectiveness of this approach on a commercial scale.

The Geocoat and Geoleach technologies are based on bioleaching of sulfidic concentrates coated on a crushed ore substrate (Harvey and Bath 2007). The Geocoat process was evaluated at pilot-plant scale for biooxidation pretreatment of refractory sulfidic gold concentrates. The concentrate was inoculated with microorganisms and coated on the ore for biooxidation in heaps. The Geoleach technology was designed to bioleach metal sulfide concentrates coated on an ore substrate and stacked in heaps. The basis of this process was to control the heap leach variables to facilitate heating via biooxidation of the sulfide minerals and to maintain elevated temperature within the heap reactor. This process could be used for bioleaching primary copper sulfide concentrates at elevated temperatures with thermophilic microorganisms. Neither technology has been used on a commercial scale.

BioCop is a bioleach technology for copper extraction from primary copper concentrates in a stirred-tank reactor process (Carretero et al. 2011). This technology is based on thermophilic bioleaching at 80°C. A 2,300-m^3 demonstration plant was constructed and operated in Chile. The extremely corrosive conditions of high temperature, acidic pH, and high ferric iron concentration necessitated use of tanks with ceramic lining. Apparently, the technology proved viable for bioleaching, but the plant was closed and the process abandoned. Although the reasons for not implementing a commercial operation were not disclosed, economic consideration is a probable cause.

BioNic/BioZinc technology is a stirred-tank reactor process for bioleaching nickel and zinc sulfide concentrates. There is no information to indicate the process has advanced to the commercial development stage (Carretero et al. 2011).

The Brisa process has been tested for bioleaching copper from primary chalcopyrite (Carretero et al. 2011). The technology is based on using a microbial ferric iron generator with a separate chemical leach reactor for contacting the chalcopyrite with the ferric iron to which chloride and silver is added to enhance copper extraction. There is no information indicating this technology has been advanced to pilot-scale testing.

BacTech Environmental Corporation (BacTech 2017) developed a biooxidation pretreatment process (Bacox) for refractory sulfidic gold concentrates. This process, similar to the Biox technology, employs a CSTR system. Bacox was advanced to commercial scale with commissioning of three plants (J.A. Brierley 2014).

REFERENCES

Acar, S., Brierley, J.A., and Wan, R.Y. 2005. Conditions for bioleaching a covellite-bearing ore. *Hydrometallurgy* 77:239–246.

Acevado, F., and Gentina, J.C. 1993. Bioleaching of minerals—A valid alternative for developing countries. *J. Biotechnol.* 31:115–123.

Acosta, M., Galleguillos, P., Ghorbani, Y., Tapia, P., Contador, Y., Velásquez, A., Espoz, C., Pinilla, C., and Demergasso, C. 2014. Variation in microbial community from predominantly mesophilic to thermotolerant and moderately thermophilic species in an industrial copper heap bioleaching operation. *Hydrometallurgy* 150:281–289.

BacTech. 2017. Home page. www.bactechgreen.com/s/Home .asp. Accessed August 2017.

Beck, J.V. 1967. The role of bacteria in copper mining operations. *Biotechnol. Bioeng.* 9:487–497.

Bhappu, R.B., Johnson, P.H., Brierley, J.A., and Reynolds, D.H. 1969. Theoretical and practical studies on dump leaching. *AIME Trans.* 244:307–320.

Breed, A.W., Dempers, C.J.N., and Hansford, G.S. 2000. Studies on the bioleaching of refractory concentrates. *J. SAIMM* 100(7):389–398.

Brierley, C.L. 1974. Molybdenite-leaching: Use of a high-temperature microbe. *J. Less-Common Met.* 36:237–247.

Brierley, C.L. 2001. Bacterial succession in bioheap leaching. *Hydrometallurgy* 59:249–255.

Brierley, C.L. 2016. Biological processing of sulfidic ores and concentrates—Integrating innovations. In *Innovative Process Development in Metallurgical Industry*. Edited by V.I. Lakshmanan, R. Roy, and V. Ramachandran. Switzerland: Springer International.

Brierley, C.L., and Brierley, J.A. 2013. Progress in bioleaching: Part B applications of microbial processes by the minerals industries. *Applied Microbiol. Biotechnol.* 97:7545–7552.

Brierley, J.A. 1966. Contribution of chemoautotrophic bacteria to the acid thermal waters of the Geyser Springs group in Yellowstone National Park. Ph.D. dissertation, Montana State University.

Brierley, J.A. 2003. Response of microbial systems to thermal stress in biooxidation-heap pretreatment of refractory gold ores. *Hydrometallurgy* 71:13–19.

Brierley, J.A. 2014. Biohydrometallurgy innovations—Discovery and advances. In *Mineral Processing and Extractive Metallurgy—100 Years of Innovation*. Edited by C. Anderson, R. Dunne, and J. Uhrie. Littleton, Colorado: SME. pp. 447–455.

Brierley, J.A., and Brierley, C.L. 1978. Microbial leaching of copper at ambient and elevated temperatures. In *Metallurgical Applications of Bacterial Leaching and Related Microbiological Phenomena*. Edited by L.E. Murr, A.E. Torma, and J.A. Brierley. New York: Academic Press. pp. 477–490.

Brierley, J.A., and Brierley, C.L. 1986. Microbial mining using thermophilic organisms. In *Thermophiles: General, Molecular, and Applied Microbiology*. Edited by T.D. Brock. New York: John Wiley and Sons. pp. 279–305.

Brierley, J.A., and Hill, D.L. 1993. Biooxidation process for recovery of gold from heaps of low grade sulfidic and carbonaceous sulfidic ore materials. U.S. Patent 5,246,486.

Brierley, J.A., and Kuhn, M.C. 2010. Fluoride toxicity in a chalcocite bioleach heap process. *Hydrometallurgy* 71:13–19.

Brierley, J.A., and LeRoux, N.W. 1977. A facultative thermophilic *Thiobacillus*-like bacterium: Oxidation of iron and pyrite. In *Conference Bacterial Leaching 1977*. Edited by W. Schwartz. Weinheim, Germany: Verlag Chemie. pp. 55–66.

Brierley, J.A., and Lockwood, S.J. 1977. The occurrence of thermophilic iron-oxidizing bacteria in a copper leaching system. *FEMS Microbiol. Lett.* 2:163–165.

Brierley, J.A., Wan, R.Y., Hill, D.L., and Logan, T.C. 1995. Biooxidation-heap pretreatment technology for processing lower grade refractory gold ores. In *Biohydrometallurgical Process*. Vol. I. Edited by T. Vargas, C.A. Jerez, J.V. Wiertz, and H. Toledo. Santiago: University of Chile. pp. 253–262.

Brock, T.D., Brock, K.M., Belly, R.T., and Weiss, R.L. 1972. Sulfolobus: A new genus of sulfur-oxidizing bacteria living at low pH and high temperature. *Archiv fur Mikrobiologie* 84(1):54–68.

Bryner, L.C., Beck, J.V., Davis, D.B., and Wilson, D.G. 1954. Microorganisms in leaching sulfide minerals. *Ind. Eng. Chem.* 46:587–592.

Bustos, S., Castor, S., and Montealegre. R. 1993. The Sociedad Minera Pudahuel bacterial thin-layer leaching process at Lo Aguirre. *FEMS Microbiol. Rev.* 11:231–236.

Bustos, S., Espejo, R., González, C., and Scheffel, R.E. 1999. Bacterial heap leaching of covellite. In *Proceedings of Copper 99-Cobre 99 International Conference. Vol. IV: Hydrometallurgy of Copper*. Edited by S.K. Young, D.B. Dreisinger, R.P. Hackl, and D.G. Dixon. Warrendale, PA: The Minerals, Metals & Materials Society. pp. 69–95.

Carretero, E., Sobral, L.G.S., de Souza, C.E.G., and de Oliveira, D.M. 2011. Bioleaching of metal sulphides—Ores and concentrates. In *Biohydrometallurgical Processes: A Practical Approach*. Edited by L.G.S. Sobral, D.M. de Oliveira, and C.E.G. de Souza. Rio de Janeiro, Brazil: Centre for Mineral Technology. pp. 125–140.

Chávez, P., Obreque, J., Vera, J., Quiroga, D., and Castro, J. 2011. Microorganisms counting techniques in hydrometallurgy. In *Biohydrometallurgical Processes: A Practical Approach*. Edited by L.G.S. Sobral, D.M. de Oliveira, and C.E.G. de Souza. Rio de Janeiro, Brazil: Centre for Mineral Technology. pp. 71–76.

Colmer, A.R., and Hinkle, M.E. 1947. The role of microorganisms in acid mine drainage: A preliminary report. *Science* 106(2751):253–256.

Colmer, A.R., Temple, K.L., and Hinkle, M.E. 1950. An iron-oxidizing bacterium from the acid drainage of some bituminous coal mines. *J. Bacteriol.* 59(3):317–328.

Demergasso, C., Soto, P., Zepeda, V., and Galleguillos, P. 2011. Scientific monitoring of industrial bioleaching process. In *Biohydrometallurgical Processes: A Practical Approach*. Edited by L.G.S. Sobral, D.M. de Oliveira, and C.E.G. de Souza. Rio de Janeiro, Brazil: Centre for Mineral Technology. pp. 89–102.

Dew, D.W., Lawson, E.N., and Broadhurst, J.L. 1997. The BIOX® process for biooxidation of gold-bearing ores or concentrates. In *Biomining: Theory, Microbes and Industrial Processes*. Edited by D.E. Rawlings. Berlin: Springer-Verlag and Landes Bioscience. pp. 45–80.

Dew, D.W., Rautenbach, G.F., van Hille, R.P., Davis-Belmar, C.S., Harvey, I.J., and Truelove, J.S. 2011. High temperature heap leaching of chalcopyrite: Method of evaluation and process model validation. In *Percolation Leaching: The Status Globally and in Southern Africa*. SAIMM Symposium Series 269. Johannesburg: Southern African Institute of Mining and Metallurgy. pp. 201–219.

Donati, E.R., and Sand, W., eds. 2007. *Microbial Processing of Metal Sulfides*. Dodrecht, The Netherlands: Springer.

Ehrlich, H.L. 2004. Beginnings of rational bioleaching and highlights in the development of biohydrometallurgy: A brief history. *Eur. J. Miner. Process. Environ. Prot.* 4(2):102–112.

Ekenes, J.M., and Caro, C.A. 2012. Improving leaching recovery of copper from low-grade chalcopyrite ores. SME Preprint No. 12-099. Littleton, CO: SME.

Gericke, M. 2011. Review on the role of microbiology in the design and operation of heap bioleaching processes. In *Percolation Leaching: The Status Globally and in Southern Africa*. SAIMM Symposium Series 269. Johannesburg: Southern African Institute of Mining and Metallurgy. pp. 165–181.

Gericke, M., Neale, J.W., and van Staden, P.G. 2009. A Mintek perspective of the past 25 years in minerals bioleaching. *J. SAIMM* 108:567–585.

Golovacheva, R.S., and Karavaiko, G.I. 1978. A new genus of thermophilic sporeforming bacteria, *Sulfobacillus*. *Mikrobiologiya* 47:815–822.

Golyshina, O.V., Pivovarova, T.A., Karavaiko, G.I., Kondrat'eva, T.F., Moore, E.R.B., Abraham, W.R., Lunsdorf, H., Timmis, K.N., Yakimov, M.M., and Golyshin, P.N. 2000. *Ferroplasma acidiphilum* gen. nov., sp. nov., an acidophilic, autotrophic, ferrous-iron-oxidizing, cell-wall-lacking, mesophilic member of the *Ferroplasmaceae* fam. nov., comprising a distinct lineage of the Archaea. *Int. J. Syst. Evol. Microbiol.* 50:997–1006.

Greenwood-Smith, R., Reid, A.R., and Jackson, B.R. 1985. Quantitative mineral image analysis—A new tool in the mineral processors armory. In *Complex Sulfides Processing of Ores, Concentrates and By-Products*. Edited by A.D. Zunkel, R.S. Boorman, A.E. Morri, and R.J. Wesely. Warrendale, PA: TMS-AIME. p. 813.

Harvey, T.J., and Bath, M. 2007. The GeoBiotics GEOCOAT® technology—Progress and challenges. In *Biomining*. Edited by D.E. Rawlings and B. Johnson. Heidelberg, Germany: Springer-Verlag. pp. 97–112.

Hausen, D.M., and Bucknam, C.H. 1985. Study of preg-robbing in the cyanidation of carbonaceous gold ores from Carlin, Nevada. In *Applied Mineralogy: Proceedings of the Second International Congress on Applied Mineralogy*. Edited by W.C. Park, D.M. Hausen, and R.D. Hagni. Warrendale, PA: TMS-AIME. pp. 833–856.

Hawkes, R.B., Franzmann, P.D., and Plumb, J.J. 2006. Moderate thermophiles including "*Ferroplasma cupricumulans*" sp. nov. dominate an industrial-scale chalcocite heap bioleaching operation. *Hydrometallurgy* 83:229–236.

John, L.W. 2011. The art of heap leaching—The fundamentals. In *Percolation Leaching: The Status Globally and in Southern Africa*. SAIMM Symposium Series 269. Johannesburg: Southern African Institute of Mining and Metallurgy. pp. 17–42.

Kelly, D.P., and Wood, A.P. 2000. Reclassification of some species of *Thiobacillus* to the newly designated genera of *Acidithiobacillus* gen. nov., *Halothiobacillus* gen. nov., and *Thermithiobacillus* gen. nov. *Int. J. Syst. Evol. Microbiol.* 50:511–516.

Lee, J. 2011. Biooxidation of gold ores. In *Biohydrometallurgical Processes: A Practical Approach*. Edited by L.G.S. Sobral, D.M. de Oliveira, and C.E.G. de Souza. Rio de Janeiro, Brazil: Centre for Mineral Technology. pp. 191–208.

Lee, J., Acar, S., Doerr, D.L., and Brierley, J.A. 2011. Comparative bioleaching and mineralogy of composited sulfide ores containing enargite, covellite and chalcocite by mesophilic and thermophilic microorganisms. *Hydrometallurgy* 105:213–221.

Liu, X., Chen, B., Wen, J., and Ruan, R. 2010. *Leptospirillum* forms a minor portion of the population in Zijinshan commercial non-aeration copper bioleaching heap identified by 16S rRNA clone libraries and real-time PCR. *Hydrometallurgy* 104:299–403.

Logan, T.C., Seal, T., and Brierley, J.A. 2007. Whole-ore heap biooxidation of sulfidic gold-bearing ores. In *Biomining*. Edited by D.E. Rawlings and B. Johnson. Heidelberg, Germany: Springer-Verlag. pp. 113–138.

Marsden, J.O., and Botz, M.M. 2017. Heap leach modeling—A review of approaches to metal production forecasting. *Miner. Metallurg. Process.* 334(2):53–64.

McFarlane, A., Koziy, L., and Barclay-Martin, J. 2011. Geometallurgical ore characterization for heap leaching. In *Percolation Leaching: The Status Globally and in Southern Africa*. SAIMM Symposium Series 269. Johannesburg: Southern African Institute of Mining and Metallurgy. pp. 81–96.

Miranda, P., and Seal, T. 2008. MLA: Current projects at the center for advanced mineral and metallurgical processing. In *Hydrometallurgy 2008: Proceedings of the Sixth International Symposium*. Edited by C.A. Young, P.R. Taylor, C.G. Anderson, and Y. Choi. Littleton, CO: SME. pp. 455–461.

Morin, D.H.R., and d'Hugues, P. 2007. Bioleaching of a cobalt-containing pyrite in stirred reactors: A case study from laboratory scale to industrial application. In *Biomining*. Edited by D.E. Rawlings and D.B. Johnson. Heidelberg, Germany: Springer-Verlag. pp. 35–55.

Norris, P.R. 2007. Acidophile diversity in mineral sulfide oxidation. In *Biomining*. Edited by D.E. Rawlings and D.B. Johnson. Heidelberg, Germany: Springer-Verlag. pp. 199–216.

Norris, P.R., Davis-Belmar, C.S., Nicolle, J., Le, C., Calvo-Bado, L.A, and Angelatou, V. 2010. Pyrite oxidation and copper sulfide ore leaching by halotolerant, thermotolerant bacteria. *Hydrometallurgy* 104:432–436.

Norris, P.R., Calvo-Bado, L.A., Brown, C.F., and Davis-Belmar, C.S. 2012. Ore column leaching with thermophiles: I. Copper sulfide ore. *Hydrometallurgy* 127–128:62–69.

Olson, G.J., and Harvey, T.J. 2011. Bio-oxidation amenability testing. In *Biohydrometallurgical Processes: A Practical Approach*. Edited by L.G.S. Sobral, D.M. de Oliveira, and C.E.G. de Souza. Rio de Janeiro, Brazil: Centre for Mineral Technology. pp. 103–124.

Outotec. 2017. Outotec Biox® process. www.outotec.com/products/leaching-and-solution-purification/biox-process/. Accessed August 2017.

Parkinson, G.A., and Bhappu, R.B. 1995. The sequential copper analysis method—Geological, mineralogical and metallurgical implications. SME Preprint No. 95-90. Littleton, CO: SME.

Puhakka, J.A., Kaksonen, A.H., and Riekkola-Vahnen, M. 2007. Heap leaching of black schist. In *Biomining*. Edited by D.E. Rawlings and D.B. Johnson. Heidelberg, Germany: Springer-Verlag. pp. 139–149.

Rawlings, D.E. 1997. Mesophilic, autotrophic bioleaching bacteria: Description, physiology and role. In *Biomining: Theory, Microbes and Industrial Processes*. Edited by D.E. Rawlings. Berlin: Springer-Verlag and Landes Bioscience. pp. 229–245.

Rawlings, D.E., and Johnson B.D., eds. 2007. *Biomining*. Berlin: Springer Verlag.

Rea, S.M., McSweeney, N.J., Degens, B.P., Morris, C., Siebert, H.M., and Kaksonen, A.H. 2015. Salt-tolerant microorganisms potentially useful for bioleaching operations where fresh water is scarce. *Miner. Eng.* 75:126–132.

Rees, K.L., and van Denter, J.S.L. 2000. Preg-robbing phenomena in the cyanidation of sulphide gold ores. *Hydrometallurgy* 58(1):61–80.

Roberto, F.F. 2008. Use of real-time PCR assays to characterize and monitor acidophilic microbe populations. In *Hydrometallurgy 2008: Proceedings of the Sixth International Symposium*. Edited by C.A. Young, P.R. Taylor, C.G. Anderson, and Y. Choi. Littleton, CO: SME. pp. 494–505.

Roberto, F.F. 2017. Commercial heap biooxidation of refractory gold ores—Revisiting Newmont's successful deployment at Carlin. *Miner. Eng.* 106:2–6.

Robertson, S.W., van Staden, P.J., and Seyedbagheri, S. 2011. Advances in high temperature heap leaching of refractory copper sulphide ores. In *Percolation Leaching: The Status Globally and in Southern Africa*. SAIMM Symposium Series 269. Johannesburg: Southern African Institute of Mining and Metallurgy. pp. 97–109.

Robertson, S., Guzmán, A., and Miller, G. 2013. Implication of hydrodynamic testing for heap leach design. Presented at Hydroprocess 2013, 5th International Seminar on Process Hydrometallurgy, July 10–12, 2013, Santiago, Chile. https://gecamin.com/hydroprocess/2013/english/proceedings. Accessed September 2015.

Rohwerder, T., and Sand, W. 2007. Mechanisms and biochemical fundamentals of bacterial metal sulfide oxidation. In *Microbial Processing of Metal Sulfides*. Edited by E.R. Donati, and W. Sand. Dordrecht, The Netherlands: Springer. pp. 35–58.

Rossi, G. 1990. *Biohydrometallurgy*. Hamburg: McGraw-Hill Book Company.

Ruan, R., Xingyu, L., Zou, G., Chen, J., Wen, J., and Wang, D. 2011. Industrial practice of a distinct bioleaching system operated at low pH, high ferric concentration, elevated temperature and low redox potential for secondary copper sulfide. *Hydrometallurgy* 108:130–135.

Rucker, D.F., Zaebst, R.J., Gillis, J., Cain IV, J.C., and Teague, B. 2017. Drawing down the remaining copper inventory in a leach pad by way of subsurface leaching. *Hydrometallurgy* 169:382–392.

Saari, P., and Riekkola-Vanhanen, M. 2011. Talvivaara bio-heap leaching process. In *Percolation Leaching: The Status Globally and in Southern Africa*. SAIMM Symposium Series 269. Johannesburg: Southern African Institute of Mining and Metallurgy. pp. 53–65.

Schippers, A. 2007. Microorganisms involved in bioleaching and nucleic acid-based molecular methods for their identification and quantification. In *Microbial Processing of Metal Sulfides*. Edited by E.R. Donati, and W. Sand. Dordrecht, The Netherlands: Springer. pp. 3–33.

Schippers, A., Hedrich, S., Vasters, J., Drobe, M., Sand, W., and Willschier, S. 2014. Biomining: Metal recovery from ores with microorganisms. In *Advances in Biochemical Engineering/Biotechnology—Geobiotechnology. I: Metal-Related Issues*. Edited by A. Schippers, F. Glombitza, and W. Sand. Berlin: Springer-Verlag. pp. 1–47.

Schlitt, W.J. 2006. The history of forced aeration copper sulfide leaching. SME Preprint No. 06-019. Littleton, CO: SME.

Schmitz, P.S., Duyvesteyn, S., Johnson, W.P., Enloe, L., and McMullen, J. 2001. Adsorption of aurocyanide complexes onto carbonaceous matter from preg-robbing Goldstrike ore. *Hydrometallurgy* 61:121–135.

Schnell, H.A. 1997. Bioleaching of copper. In *Biomining: Theory, Microbes and Industrial Processes*. Edited by D.E. Rawlings. Berlin: Springer-Verlag and Landes Bioscience. pp. 21–43.

Sheffer, H.W., and Evans, L.G. 1968. *Copper Leaching Practices in the Western United States*. Information Circular 8341. Washington, DC: U.S. Bureau of Mines.

Shiers, D.W., Ralph, D.E., and Watling, H.R. 2014. The effects of nitrate on substrate utilization by some iron(II)- and sulfur-oxidising bacteria and archaea. *Hydrometallurgy* 150:259–268.

Shutey-McCann, M.L., Sawyer, F.P., Logan, T., Schindler, A.J., and Perry, R.M. 1997. Operation of Newmont's biooxidation demonstration facility. In *Global Exploitation of Heap Leachable Gold Deposits*. Edited by D. Hausen. Warrendale, PA: The Minerals, Metals & Materials Society. pp. 75–82.

Sobral, L.G.S., Oliveira, D.M., and Souza C.E.G., eds. 2011. *Biohydrometallurgical Processes: A Practical Approach*. Rio de Janeiro, Brazil: Centro de Tecnologia Mineral.

Temple, K.L., and Colmer, A.R. 1951. The autotrophic oxidation of iron by a new bacterium *Thiobacillus ferrooxidans*. *J. Bacteriol.* 62(5):605–611.

Van Aswegen, P.C., Godfrey, M.W., Miller, D.M., and Haines, A.K. 1991. Developments and innovations in bacterial oxidation of refractory ores. SME Preprint No. 91-75. Littleton, CO: SME.

Van Aswegen, P.C., van Niekerk, J., and Olivier, W. 2007. The BIOX® process for the treatment of refractory gold concentrates. In *Biomining*. Edited by D.E. Rawlings and D.B. Johnson. Heidelberg, Germany: Springer-Verlag. pp. 1–33.

Vera, M., Schippers, A., and Sand W. 2013. Progress in bioleaching: Fundamentals and mechanisms of bacterial metal sulfide oxidation. Part A. *Appl. Microbiol. Biotechnol.* 97:7529–7541.

Wan, R.Y., and Brierley, J.A. 1997. Thiosulfate leaching following biooxidation pretreatment for gold recovery from refractory carbonaceous-sulfidic ore. *Min. Eng.* 49(8):76–80.

Watling, H.R. 2006. The bioleaching of sulphide minerals with emphasis on copper sulphides—A review. *Hydrometallurgy* 84:81–108.

Watling, H.R. 2011. Adaptability of biomining organisms in hydrometallurgical processes. In *Biohydrometallurgical Processes: A Practical Approach*. Edited by L.G.S. Sobral, D.M. de Oliveira, and C.E.G. de Souza. Rio de Janeiro, Brazil: Centre for Mineral Technology, pp. 39–70.

Watling, H.R. 2013. Chalcopyrite hydrometallurgy at atmospheric pressures: 1. Review of acid sulfate, sulfate-chloride and sulfate-nitrate process options. *Hydrometallurgy* 140:163–180.

Watling, H.R. 2015. Review of biohydrometallurgical metals extraction from polymetallic mineral resources. *Minerals* 5:1–60.

Woese, C.R. 1982. Archaebacteria and cellular origins: An overview. In *Archaebacteria*. Edited by O. Kandler. Stuttgart: Gustav Fischer Verlag. pp. 1–17.

Zaitsev, G., Mettänen, T., and Langwaldt, J. 2008. Removal of ammonium and nitrate from cold inorganic mine water by fixed-bed biofilm reactors. *Miner. Eng.* 21(1):10–15.

Zhao, C., Huang, D., Chen, J., Li, Y., Chen, Y., and Li, W. 2016. The interaction of cyanide with pyrite, marcasite and pyrrhotite. *Miner. Eng.* 95:131–137.

Zimmerley, S.R., Wilson, D.G., and Prater, J.D. 1958. Cyclic leaching process employing iron oxidizing bacteria. U.S. Patent 2,829,964.

CHAPTER 10.8

Agitated Bioleach Reactors

Paul Miller

BRIEF HISTORY OF BIOPROCESSING USING AGITATED REACTORS

This chapter describes the design and operational aspects of agitated reactors using bacteria for minerals processing and focuses on the processing of refractory gold concentrates, as there have been few commercial applications using bioreactors for treatment of base metal concentrates. Several terms are commonly used to describe bio-hydrometallurgical processes, with *biooxidation* or *bacterial oxidation* and *bioleaching* or *bacterial leaching* being favored. Biooxidation or bacterial oxidation generally refers to the use of bacteria for solubilization of gangue elements, such as arsenopyrite or pyrite, to expose inert metals, such as gold, for subsequent recovery. In biooxidation processing, the bacteria are only responsible for removing mineralization that occludes the inert metals. The term *bioleaching* or *bacterial leaching* is generally used for bioprocesses in which the metals of interest are made soluble in the bioleach liquor (e.g., copper, nickel, or zinc from chalcopyrite, pentlandite, and sphalerite, and so on). For the discussions in this chapter concerning agitated bioreactors, these definitions are used as interchangeable terms unless specified otherwise.

The role of bacteria in the oxidation of minerals and metal extraction was identified many years prior to their commercial application in the agitated bioreactors currently used in minerals processing operations (Colmer and Hinkel 1947). The use of the Bacfox process at the Buffelsfontein uranium plant in South Africa during the 1970s was perhaps the first case of engineering an environment in which a bacterial process of a few days' residence time formed part of an overall commercial process (James 1976). The commercial development of an agitated reactor process using bacteria for the treatment of high-value arsenical gold concentrates was driven by the need to introduce an economic and environmental alternative to roasting. Biooxidation using reactors could satisfy both of these criteria as a suitable method of oxidation. Not only was the process economic, but also, from an environmental perspective, the arsenic made soluble by biooxidation could be neutralized with low-cost limestone and stabilized downstream as solid ferric arsenate. In 1986, the first biooxidation plant for refractory gold concentrate treatment was commissioned at Fairview in South Africa (Van Aswegen et al. 1989). It was followed by other plants throughout the world. Currently, more than 20 refractory gold projects worldwide use biooxidation with agitated, aerated reactors. These plants are listed in Table 1 with their locations and various operational parameters. As a result of more than 30 years of commercial application, biooxidation is now well accepted as a pretreatment method for removal of refractory components from gold concentrates prior to cyanidation.

The adoption of the technology for base metal recovery and other elements such as radio-actinides has not yet found widespread commercial use. An excellent review by Watling cites examples of treatment applications for polymetallic concentrates, tailings, black shales, and mine wastes, indicating the limits of commercial bio-hydrometallurgical processes, other than gold, for which agitated reactors have been used (Watling 2015). Base metal processes have been developed to pilot or demonstration scale for the bioleaching of copper (Van Staden et al. 2000), nickel (Dew and Miller 1997), and zinc sulfide concentrates (Sandstrom and Petersson 1997) using mesophiles, moderate thermophiles, and thermophiles, and a process for the extraction of cobalt from pyrite concentrate was commercialized in 1998 (Briggs and Millard 1997). Bioleaching using agitated reactors for the removal of arsenic as a contaminant of nickel concentrates has been proposed (Fewings et al. 2015), and more recently, a project has proceeded to construction status (Neale et al. 2015), but the operational status and related information for these projects has yet to be reported.

Paul Miller, Managing Director, Sulphide Resource Processing Pty Ltd., Perth, Western Australia, Australia

Table 1 Chronology of recognized commercial bacterial oxidation plants for refractory gold processing

Plant and Location	Operating Years	Design Capacity, stpd concentrate	Total Reactor Volume, m³	Approximate Percentage S^{2-} in Feed	Technology Supplier	Reason for Closure	References
Fairview, South Africa	1986–present	10 (1986) 35 (1991) 40 (1994) 55 (1999)	1,260	18–24	Biox	N/A	Van Aswegen 1993; Van Aswegen et al. 2007
Tonkin Springs, Nevada, USA	1989–1990	1,500 (whole ore)	8,800	1.3	Own installation	Funding constraints and technical issues	Foo et al. 1990
Sao Bento, Brazil	1990–present	150 (1990) 300 (1994) 380 (1997)	1,300	19	Biox	Bacterial oxidation section shut down for energy saving	Suttill 1990; Van Aswegen et al. 2007
Harbour Lights, Western Australia	1992–1994	40	980	18	Biox	Ore depleted	Brierley 1995
Wiluna, Western Australia	1993–present	115 (1993) 158 (1995)	4,230	24	Biox	N/A	Brown et al. 1994; Stephenson and Kelson 1997; Odd et al. 1993
Sansu, Ashanti, Ghana	1994–present	720 (1994) 960 (1995)	21,600	11	Biox	N/A	Nicholson et al. 1994
Youanmi, Western Australia	1994–1998	120	3,000	28 (partial oxidation)	BacTech	Ore depleted (high mining costs)	Budden and Bunyard 1994; Miller 1997
Olympiada, Siberia	1997–present	200 (approx.)	N/A	N/A	BioNord	N/A	Sovmen et al. 2008
Proano, Tamboraque, Peru	1999–2003	60	1,572	30	Biox	Closed but reopened	Loayza and Ly 1999
Beaconsfield, Tasmania	2000–present	60–70	2,241	27–34	BacTech Mintek	N/A	Pinches et al. 2000
Laizhou, Shandong Province, China	2001–2010	100 (2001) 200 (2008)	4,050	21–25	BacTech Mintek	Change of owner (on care and maintenance)	Miller et al. 2004
Suzdal, Kazakhstan	2005–present	196 (2005) 520 (2009)	7,800	12	Biox	N/A	Van Aswegen et al. 2007
Fosterville, Australia	2005–present	211	5,400	21	Biox	N/A	Whincup and Binks 2004; Amiconi et al. 2004
Bogoso, Ghana	2006–present	825	21,000	20	Biox	N/A	Van Niekerk 2012
Jinfeng, China	2007–present	790	16,000	9.4	Biox	N/A	Van Niekerk 2012
Kokpatas, Uzbekistan	2009–present	2,138	43,200	20	Biox	N/A	Van Niekerk 2012
Agnes, South Africa	2010–present	20	396	30	Biox	N/A	Van Niekerk 2012
Runruno, Philippines	2014–present	404	10,800	17	Biox	N/A	Olivier and Jardine 2014

Adapted from Miller and Brown 2005
N/A = not applicable.

GENERIC FLOW SHEET SHOWING UNITS AND BASIC PROCESSING

Bioleaching and biooxidation processing using agitated reactors has been confined to concentrate treatment, with a single exception using whole ore as feed at Tonkin Springs (Nevada, United States) in 1989 (Foo et al. 1990). A block flow sheet of typical unit operations involved in biooxidation for treatment of refractory gold concentrate is shown in Figure 1, whereas a flow-sheet example of bioleaching for recovery of base metals (copper) and precious metals from a polymetallic concentrate is given in Figure 2.

The upstream unit operations for concentrate treatment to provide feed for biooxidation or bioleach processing are common to all projects in terms of crushing, grinding, gravity gold recovery if appropriate (Holder 2007), and concentrate production by flotation. Concentrate feed slurry is drawn continuously from the stock tanks and diluted to the working pulp density for bioprocessing. A typical biooxidation plant consists of one or more trains of six agitated, aerated reactors with three reactors operating in parallel as primary reactors and three operating in series as secondary reactors. The reaction is exothermic and reactors are cooled with water to maintain a set temperature, generally between 42°C and 50°C, depending on the bacterial culture in use (Brown et al. 1994; Budden and Bunyard 1994).

After bioprocessing, a countercurrent decantation (CCD) circuit can be used to wash the solids and separate the liquor for further treatment. Filtration and washing can be used as an alternative to a CCD circuit (Miller et al. 2004). For biooxidation, the liquor containing acidic ferric and arsenic elements is neutralized in stages by addition of lime or limestone to produce a stable ferric arsenate cake or slurry for disposal and a clean water for reuse within the plant (Nyombolo et al. 2000). The biooxidized solids are neutralized with lime and subjected to cyanidation treatment for recovery of gold and silver values. In the case of bioleaching, the liquor containing the metals of value is subjected to routines of selective precipitation and/or solvent extraction electrowinning (SX-EW) to produce salable metal products (Briggs and Millard 1997).

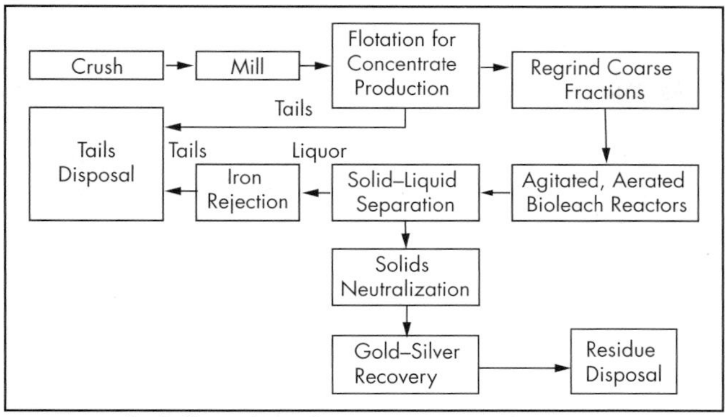

Figure 1 Block flow sheet for precious metal concentrate treatment using biooxidation in reactors

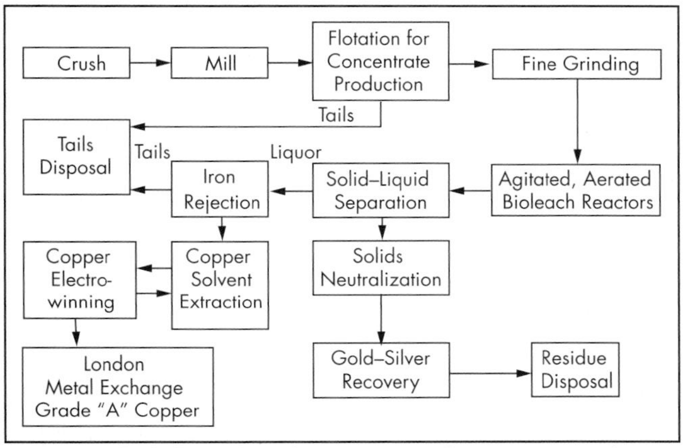

Figure 2 Block flow sheet for copper and precious metal concentrate treatment using bioleaching in reactors

If the bioleached solids contain precious metals, the solids can be subjected to cyanidation or otherwise disposed.

DESIGN CRITERIA FROM TEST WORK

The emphasis on test work is to deliver design criteria of required accuracy in a timely and cost-effective manner. Various phases of test-work investigations are normally undertaken through several sequential steps. Whether it is a requirement that every step be undertaken is project specific and typically depends on the novelty of the project and potential differences from existing commercial bioprocessing operations. An example of the sequence of test steps together with typical schedules and material requirements is given in Table 2.

Some biooxidation projects have been successfully scaled up using criteria derived from conducting only small-scale laboratory batch tests (Pinches et al. 2000). Often this is undertaken because insufficient material is available for more-detailed test studies or because sufficient confidence and knowledge is available from treating similar ore types of similar composition from other projects that use biooxidation (Miller et al. 2004). In such cases, the design criteria are based on the results from available batch testing combined with assumptions that allow the design envelope to be sufficiently flexible to cater to less well-characterized feedstocks.

Table 2 Phases of test work showing typical sample requirements and schedules for project development stages

Test Phase	Deliverable	Scope of Work	Sample	Schedule, months
1	Process amenability	Screening and testing different mixed cultures with ore or concentrate samples	2–5 kg	3–4
2	Scoping study data	Batch testing using bioreactors of a few liters' volume and first metal recovery tests	5–20 kg	4–6
3	Scoping–prefeasibility study data	Batch testing using larger bioreactors, giving greater amounts of liquor and residues for metal recovery and environmental testing	20–100 kg	6–8
4	Prefeasibility–feasibility study data	Semicontinuous–continuous testing using a train of laboratory-scale bioreactors with larger-scale batch downstream testing	100–500 kg	6–12
5	Feasibility	Fully integrated, continuous demonstration-scale pilot plant in purpose-built facility	20 t+	9–18

Residence time calculations for a commercial plant can be modeled from continuous pilot-plant campaigns using commercial feedstock by producing a series of operating curves (Van Aswegen et al. 2007). The theoretical modeling of biooxidation and bioleaching processes for design purposes can be complex, partly because of the partition of the bacterial population between solids and liquid; both phases also supply substrate for growth and division, complicating modeling exercises (Gonzalez et al. 2004; Dinkla et al. 2013).

FEEDWATER QUALITY, DILUTION, AND RESIDENCE TIME

An important criterion for bioprocessing is the need for a reasonable quality of makeup water. Although this can be true for certain metallurgical processing operations, largely because of issues of corrosion and scaling, the possibility of certain ions being detrimental because of toxicity is a further consideration in bioprocessing. Water enters the process from several sources:

- Water associated with the concentrate from upstream processing
- Dilution water added to achieve the working pulp density in the reactors
- Water used for reagent makeup, such as nutrients and limestone slurry

An early area for concern was the potential toxicity that residual flotation reagents associated with the concentrates may have on bacterial activity. Conducting toxicity testing using the intended flotation reagents is standard protocol during project development phases. Although there has been little evidence that carryover of these reagents can cause toxicity issues, many operations have facilities for a concentrate regrind or a re-upping step, and this would serve to dilute reagents from flotation. Cyanide, which is highly toxic in all forms to the bacteria, can be used as a pyrite depressant in flotation and should be avoided when producing concentrate feeds for bioprocessing. Van Aswegen et al. (2007) list the following reagents that may show toxic effects toward the culture:

- Cyanide and thiocyanate
- Grease and oil
- Detergents, solvents, and degreasing components
- Biocide descaling reagents and other water treatment reagents
- Certain flocculants, flotation reagents, and nutrients

Chloride is the most common detrimental ion that can be present at elevated levels in some waters. Historically, some biooxidation processes have operated well in arid regions where water makeup is limited and only relatively saline water can be used (Bell and Quan 1997). Van Aswegen et al. (2007) suggest that 5,000 ppm chloride is a maximum acceptable limit for chloride in the biooxidation operation using Biox technology with a total dissolved solid of 2,100 ppm and that arsenic (V) concentrations of 15–20 g/L are tolerated.

Accurate metering of dilution water to achieve the required pulp density of feed to biooxidation is critical, as the pulp density of the operation is a key parameter in dictating the rate of oxidation and operational efficiency. The parameters of feed pulp density, residence time, and pulp feed rate are the major process parameters for control. If the operating pulp density and/or pulp flow rate is too low, the efficiency of the plant suffers as oxidation is achieved well within the residence time and the mass transfer capabilities of the reactors are underutilized. Conversely, if the pulp density or/and throughput are increased, only a partial oxidation is achieved within the residence time and final metal recovery may decrease. It is unusual for the pulp density of biooxidation feed to exceed 20%, as above this value and irrespective of concentrate type, the rate of oxidation appears to decrease (Bailey and Hansford 1993).

TANK CONFIGURATIONS AND MATERIALS OF CONSTRUCTION

The process objective is to maximize the effectiveness of reactor volume, with limiting factors in primary reactors typically being related to oxygen transfer and in secondary reactors being particle size or inhibition because of increasing ionic composition of the liquor. Reactors are normally of similar dimensions but with variations in agitator, cooling, and aeration duty. A typical tank configuration consists of half of the entire reactor volume being primary reactors operating in parallel with feed being divided equally. The remaining tanks consist of reactors operating in series in which the partially oxidized pulp exiting the primary reactors is combined and fed to the remaining reactors operating in series. A configuration of this type ensures that sufficient residence time is available in the primary reactors for growth and division of the bacterial population. For many plants, a three-day residence time in the primary reactors is common. Some plants will amend the tank configurations to operate four reactors as primary reactors and only two reactors as secondary reactors, or to install seven reactors per train with four as primary reactors and three as secondary reactors (Van Aswegen et al. 2007; Miller and Brown 2005). Such variations can allow a greater tonnage throughput, resulting in an overall increase in metal production but may sacrifice recovery if residence time is too low. Operating a plant with a low residence time requires a greater fineness of control to ensure a stable operation.

The acidic nature of the pulp and often the consideration of water quality require the use of rubber-lined steel or specialized steels as materials of construction. This includes SAF 2205 and other high-grade stainless steels (Ritchie and Barter 1997). The largest tank sizes believed to be in use are 1,500 m^3 in size at the Kokpatas project in Uzbekistan commissioned in 2009. Typically, the water quality in terms of chloride concentration, in combination with the acidity, dictate the grade of stainless steel required. Specialty steels with higher tensile strength are becoming more favored. Although they are more expensive on a weight basis, they can be more cost-effective because a thinner plate thickness is required.

Acid-resistant construction materials are also required in the interface operations of solid–liquid separation, residue washing, and liquor neutralization. Thickeners used in countercurrent washing require acid proofing, and primary neutralization tanks are normally constructed from stainless steel. The bioreactors are generally contained in a single concrete bund area sized to contain the entire volume of a single reactor for emergency draining purposes. This concrete must also be acid-proofed, as must all sump areas likely to be in contact with acidic pulp. A high-quality epoxy coating can be used for concrete protection. Galvanized steels are not usually permitted in the biooxidation area because of the possibility of arsenic solutions reacting with the zinc to produce toxic arsenious gas. Walkways and areas of the superstructure located above the reactors also require corrosion protection with relevant coatings. The use of plastics in certain areas of some newer plants has become more common in recent years as a lower-cost alternative to steels. Smaller tankage, such as that in nutrient makeup or acid, is constructed from epoxy or equivalent plastics.

FEED PARTICLE SIZE AND FINE GRINDING OF CONCENTRATE FEEDS

Because coarse fractions are present in most concentrate feeds, it is normal practice to regrind coarse components as part of the feed preparation before concentrate is stored in stock tanks. Components coarser than 120 µm are generally reground to avoid the following:

- Solids settling in the bioreactors forming large fillets over time between the baffles at the base of the reactor. This gives a considerable metal "lock-up" within the biooxidation process overtime.
- Coarse components may contain higher gold values with a higher specific gravity that settle more readily in reactors.
- When a substantial amount of the reactor volume is occupied by settled solids, these solids becomes severely compacted and are difficult to resuspend. The reactor must be taken off-line and drained, and the solids must be removed manually.
- An extensive buildup of solids reduces the reactor volume, making it impossible to achieve the required throughput and meet design specification.

Regrinding even larger proportions of the concentrate to improve oxidation kinetics is a favored method for either increasing plant throughput above design specifications or ensuring the complete oxidation of more refractory minerals. Although theoretically the throughput design basis of primary reactors is fixed, the conservative warranties offered for oxygen transfer are usually exceeded in practice, making particle size the rate-limiting factor for oxidation. Without regrinding the concentrate, most of the fractions would readily meet the criteria for suspension by the specified agitation duty, but particle size can be the rate-limiting step in oxidation kinetics (Holder 2007).

AGITATION AND AERATION

Agitators are used to both suspend solids and disperse air. Air is typically provided using low-pressure blowers and is delivered to a ring main beneath the agitator for dispersion. The air exiting the blowers is at an elevated temperature and is usually precooled before delivery to the reactors. Because the oxidation process results in a decreased solids density through each stage of the process, accompanied by lower air demands for reaction, each reactor stage will have different agitation and aeration needs. Air is therefore metered to reactors according to the oxidation requirements for a reactor.

Agitation with solids suspension and high aeration in large reactors have created a unique duty for the type of agitators used in biooxidation. The hydrofoil designs developed in the 1980s have been traditionally favored for this three-phase mixing duty (Lally 1987; Fraser 1993; Kubera and Oldshue 1992). Gas-transfer coefficients are similar or superior to turbines but with lower power requirements, whereas solids suspension is homogeneous. However, large airflows are a recognized characteristic of bacterial oxidation reactors, owing to typical utilization factors of only between 30% and 40%, combined with the high demand for oxygen in the reaction. Air-to-sulfide ratios of between 25,000 and 30,000 nm^3/t (normal cubic meters per metric ton) of sulfur are common values in design (Miller and Brown 2005). The agitator must disperse the air while ensuring that flooding is avoided and solids suspension is maintained. The size of the agitator is thus a function of the air-expansion thermodynamics. A specific agitator power of 30–35 nm^3/h of air per installed kilowatt is often used for first basis agitator sizing, and it has been implied that the conventional power requirement for agitation on traditional plants is between 0.2–0.3 kW/m^3 of reactor volume (Miller and Brown 2005).

Airflows and agitator power requirements in design are sized accordingly to the requirements for each reactor stage in the process. It is normal practice for the first secondary reactor to have an agitator of comparable size to the primary reactors to allow for maintenance, such that the secondary reactor can also have the duty of a primary reactor. Typically, two-thirds of the oxidation is completed in the primary reactor stage. Traditionally, the agitators have used fixed speed, with adjustments made only by changing the drive belt and the gear wheels. The use of variable-speed drives is now gaining favor in new plants because it allows the optimization of the mixing duty of each reactor. Although it may add to the plant capital cost, the operating cost savings from optimizing power draw for each individual reactor can be considerable. The failures of agitators and associated equipment such as gearboxes in biooxidation plants has also been higher than expected because of high bending moment, torque, and thrust. More recently, the use of dual impellers has been favored and reported as being more cost-effective while achieving greater operational robustness. Such improvements to agitation/aeration systems are claimed to reduce agitator power requirements by 20% while achieving the same oxygen mass transfer requirements (Olivier and Jardine 2014). Experts suggest that this dual-impeller system also offers a reduction in bending moment, torque, and thrust that will improve the mechanical life of the agitator system.

Several considerations are necessary in the design specification and operation of the low-pressure air blowers used to deliver air to the reactors. First, design calculations for the total air blower requirement should include the site elevation of the project to account for the lower oxygen content of air at heights above sea level. Second, one must determine the number of blowers to be used to deliver the total air required. Optimization of capital costs suggests the use of a fewer number of large blowers. By contrast, optimization of operating costs suggests the use of a larger number of smaller blowers, such that the number of blowers operating can be managed according to variations in air demand because of feed variations. The turndown ratio of blowers is generally limited, representing a fixed operating cost when in use. The siting of blowers also needs consideration. Usually the blowers are separately housed for noise abatement and positioned away from areas prone to high dust levels. Cable routings may also need consideration to reduce the potential for electrical interference.

COOLING SYSTEMS AND WATER REQUIREMENT

The heat generated from the exothermic reactions is often a few megawatts of energy. Heat is removed in situ using cooling water to retain a set-point temperature conducive to bacterial activity. The set-point temperature chosen depends on the microbial population used in the process as specified by the technology vendor. The optimum set-point temperature for many plants using mesophilic cultures is generally between 40° and 42°C, but for plants using moderate thermophiles, a temperature of about 50°C is typically the optimum set point.

Bundles of cooling water tubes inside the reactor are favored for heat removal and often positioned at the walls of the tank to act as baffles. A less sophisticated method applied at one project has been to use a film of cascading water on the outside of the reactors (Budden and Bunyard 1994).

Cooling water bundles are prefabricated off-site from a similar grade of stainless steel to be used in the reactors. The cooling water bundles contribute a significant cost to the overall capital cost of the reactor. This is because of the substantial number of coils required, combined with the labor content for custom fabrication.

Some heat is removed from the reactor by evaporative loss, by the effects of aeration, as well as by heat being lost through the tank walls. The amount of heat removed in this way depends on the local climate but is generally low. Concerns are often raised that operating reactors in extremely cold climates is untenable because of high heat losses from tank walls, preventing the temperature from being maintained. Heat balance calculations, as well as data from commercial plants that operate in extremely cold climates with high wind chill factors, has shown this not to be the case and heat removal is always required. An exception may be the final reactor where oxidation is nearing completion and heat generation is comparatively low. These reactors may require external jacketing to retain more heat or, depending on plant design, hot cooling water exiting the primary reactors' cooling water circuit can be redirected to these reactors before being returning to the cooling tower.

The heat generated is low grade and is not generally useful to transfer to other areas of the plant. The heat is removed from the cooling water using conventional forced-draft towers in a closed circuit. Failure of the cooling water supply can have a severe effect, with an excursion of temperature above the set point significantly impairing the process. In general terms and by empirical observation, a temperature rise between approximately 3° and 5°C is the maximum tolerance before significant process impairment occurs. Because the cooling towers are open to atmosphere, biocide and other chemicals such as antiscalants must generally be added to mitigate fouling and maintain the heat transfer duty. The particulate load of the water also increases because of the scrubbing action of the water as it passes through the tower. In-line ultraviolet irradiation devices can reduce the need for biocides (Holder 2007), and sand filters can reduce the particulate matter collected in the water over time. For some projects operating in regions of elevated temperature and humidity, the use of cooling towers is much less effective because of the cooling requirement operating close to the ambient wet-bulb temperature, and refrigeration may be a preferable concept for consideration. Closed-circuit refrigeration can also be favored for reducing needs for water treatment associated with cooling water towers and occupies a smaller footprint.

FEED SUPPLY, REAGENT ADDITIONS, SPILLAGE HANDLING, AND PULP TRANSFER

Primary reactors receive diluted concentrate feed slurry from a header feedbox. Acid or limestone additions are made, if required, to each reactor to adjust the operating pH value to a set point. The header box located above the nest of primary reactors is fitted with a splitter arrangement such that feed can be distributed equally into the primary reactors. A timed actuator is often fitted to allow feed rates to be adjusted between reactors when required. Nutrient solution is also dosed to the header box to mix with the concentrate before entering the primary reactors. Nutrients are predissolved in a mixing tank before reporting to a nutrient holding tank ready for use. The nutrients resemble an agriculture fertilizer mix consisting of ammonium, phosphate, magnesium, and potassium salts and can be delivered to the site either as agricultural-grade chemicals in bags or as a specified premixed pellet for dissolution. The bacteria utilize carbon from carbon dioxide to form new cells, and the carbon dioxide formed by the reaction of acid with the carbonates present in the concentrate may be an important supplement to the natural supply of carbon dioxide that is present in the air supplied to the reactors for oxidation. A very low percentage of carbonates is ideally required in the feed, which is neutralized by the natural acidity generated from oxidation in the primary reactors to supply this need (Van Aswegen et al. 2007). If the concentrate feed contains excessive carbonates, then acid addition will be made to the header feedbox to neutralize excess alkalinity. Concentrated sulfuric acid is typically metered and pumped directly from an acid storage vessel to the header feedbox.

The ability to add limestone slurry to all reactors is common, with limestone being supplied through a ring main. Limestone additions are made to reactors if the final acidity level produced by oxidation becomes too high and the pH value is below the optimal value for bacterial activity as specified by the technology vendor. A typical range of pH values for primary reactor operations may be 1.5–2.5 depending on feedstock and culture type. A typical range of pH values for oxidized pulp emerging from the last secondary reactor may be 0.8–1.8, again depending on technology vendor specification for optimum microbial activity and the type of concentrate feedstock being treated.

To reduce reagent additions, and when feasible, many operating plants aim to minimize the need for pH adjustments and achieve a good operating acid balance in which the natural acidity generated from oxidation is used to neutralize any excess carbonate in the concentrate feed. Excessive acid addition to the concentrate feed will require excessive use of limestone in downstream neutralization. Excessive limestone addition to reactors also gives rise to excessive gypsum and ferric hydroxide precipitates, creating an extra mass of solid to be carried forward into downstream processing. There has been little evidence that such precipitates hinder subsequent gold extraction by cyanidation.

Formation of a stable foam on the surface of reactors tends to occur when changes are made to the process. Such changes may include a sudden variation in feed rate or type of feed, or introduction of a toxin, all of which may provoke a microbial response, resulting in a highly stable foam. Antifoams can be used for foam suppression, but the choice of antifoam is limited to those considered nontoxic. Such antifoams are likely to be more expensive, and the amount that can be used may be limited. Many plants have a high freeboard allowance on reactors to allow foam to be contained in the reactors as part of normal operational practice. Some plants have deep launders that transfer pulp between reactors, which helps to remove foam from the primary reactors where it is often more prevalent. Water sprays positioned on top of reactors or launders have also proved successful in combatting excessive foaming. If foaming is not controlled and there is extensive overflow from the top of reactors, the foam collects in the bund area and is pumped back to the reactors. The reason for the occurrence of foam must be understood and addressed, as it should only be a temporary feature of an operation while undergoing change from a steady-state condition.

Various methods have been adopted for transferring pulp between reactor stages, with airlifts or gravity flow by launders being favored. It is important to maintain a consistent level of pulp in reactors while achieving good pulp transfer, and airlifts can be problematic for this reason (Holder 2007). Launders are a simpler, effective alternative, subject to being correctly designed to prevent solids settlement during longer sections. Slurry transfer using centrifugal pumps has also been used in two plants (Spencer et al. 1997). During plant operation, occasionally significant quantities of pulp must be transferred from one reactor to another. For example, this is required when starting up a reactor after scheduled emptying and maintenance, or for cross-inoculation of reactor contents to aid recovery from an unscheduled shutdown. For this reason, it is useful to be able to connect reactors with temporary pipework and valving arrangements at the reactors' base to rapidly transfer large amounts of pulp between the reactors.

SOLID–LIQUID SEPARATION, LIQUOR NEUTRALIZATION, AND WATER REUSE

After biooxidation, the liquor is separated from the oxidized solids, which are also washed in preparation for lime guarding and cyanidation. CCD or thickening followed by wash filtration can be used. CCD circuits were favored in early plants, but filtration with washing is also considered as an established alternative. Some plants have experienced gold losses from entrained solids reporting to the thickener or CCD liquor overflow ahead of liquor neutralization, resulting in the use of a polishing filter to mitigate such losses (Holder 2007). Current generally accepted practice is to include a polishing filter or separate settlement catchment tank to capture any entrained solids that would represent a gold loss. This polishing filter or catchment tank is also important to prevent solid losses occurring in the overflow for periods when the CCD circuit is unstable.

Neutralization of bioliquor is typically conducted in four sequential reactor stages with a total nominal residence time of about six hours. Reactors are agitated and a small amount of aeration supplied to convert any residual ferrous iron present to the ferric form (Nyombolo et al. 2000). Limestone is generally the reagent of choice, but the superior neutralizing qualities of lime can be more economical if reagent transportation costs are a significant factor for the project. Typical limestone use ranges from 400 to 800 kg/t of concentrate feed, depending on the sulfide mineralogy and sulfide concentrations present in the feed. Despite the high limestone use, local limestone is usually available to reduce supply costs. Alternative sources of alkalinity have been used on some projects, including the use of flotation tails high in calcite providing a source of carbonate that reduces the total quantity of commercial limestone or lime required (Van Aswegen et al. 2007). When using commercial limestone, it is either delivered finely ground and pneumatically discharged into an on-site silo or delivered as crushed limestone and stockpiled. In the latter case, on-site grinding is conducted prior to preparation of a slurry, which is then delivered to the ring main for use in liquor neutralization or pH adjustment of bioreactors.

In general, when arsenic is present in the liquor, a minimum molar ratio of iron to arsenic of 3:1 is necessary during neutralization to produce a ferric arsenate compound with long-term stability for disposal (Ringwood 1995; Nyombolo et al. 2000). Monitoring of such ferric arsenic wastes over time has verified the stability of these wastes (Kraus and Ettel 1989; Broadhurst 1993; Van Aswegen et al. 2007). The solids produced from neutralization of biooxidation liquor contains gypsum as the major compound, with solid ferric hydroxide species in lesser amounts and ferric arsenate present as a minor compound in the total waste.

The neutralized liquor containing the precipitated solids reports to a thickener for settling followed either by filtration or discharging the thickened underflow directly to impoundment. The clarified solution from the thickener together with filtrates or return water after impoundment is transferred to process water storage for reuse. Creating a high underflow density of the precipitate together with poor filtration characteristics can cause problems in plant operations. Plate-and-frame filters are generally favored for filtration duty, and slow filtration times result in relatively large filtration areas. Although the final filter cake is quite stable, large amounts of water are contained within the chemical matrix; this usually represents the major source of permanent water loss from the circuit. The neutralization process generally results in clean water that is of good quality but that is high in calcium and has an increased tendency for scaling.

Whether water is reclaimed from the ferric precipitate impoundment area or retrieved from filtration at plant level is project specific and related to arrangements and siting for tailings disposal. The expense of construction and the permitting of tailings impoundments can lead to severe restrictions in volumes available for tailings disposal, which encourages filtration and stacking of tailings to reduce the final volume for impoundment.

In many plants, the water circuit for biooxidation is separate from the water circuit for cyanidation. This is necessary to prevent contamination of toxic cyanide species to the biooxidation circuit (Van Aswegen et al. 2007). Technologies are available for tertiary detoxification for degradation of thiocyanates. Use of such technologies is likely to increase because they involve a single water circuit of less complexity and reduce costs (Van Buuren 2014).

RESIDUE TREATMENT PRIOR TO CYANIDATION AND BASE METAL RECOVERY FROM BIOLIQUOR

As biooxidation residue often contains some transient sulfur species, lime guarding prior to cyanidation is a critical step to reduce downstream cyanide consumption (Hackl 1989; Van Aswegen et al. 2007; Miller and Brown 2005). Methods to protect residues from consuming excessive cyanide include more intensive lime guarding, or extension in biooxidation residence time (converting transient sulfur species to sulfates); and fine grinding of the feed, which may also improve the kinetics of oxidation as well as give a more benign oxidized residue for cyanidation.

For base metal bioleaching, these elements become soluble in the leach for recovery with conventional SX-EW or selective precipitation routines. An example of this was shown by the Kasese Cobalt operation with sequential removal of iron, zinc, and copper from the bioleach liquor, followed by cobalt SX-EW (Briggs and Millard 1997). These downstream metal rejection and recovery routines contrast with the biooxidation for precious metal processing, in which gold, silver, and platinum group elements are exposed by removal of sulfides but remain inert in the oxidized residue for later downstream recovery.

TAILINGS DISPOSAL AND OTHER DISCHARGE CONSIDERATIONS

Two distinct types of tailings are produced from bioprocessing. One is the final leached residue after metal extraction, which may have also included cyanidation and detoxification treatment. The second tailing is the ferric precipitate from liquor neutralization, which is likely to contain stabilized ferric arsenate if derived from refractory gold concentrate processing. Disposal methods of each of the tailings will vary according to their characteristics and the specific project requirements to meet permitting and monitoring requirements. Separate disposal facilities are generally favored. Leach residues are particulate, which is relatively easy to filter and of smaller volume. By contrast, the ferric or ferric arsenate precipitate is voluminous, contains significant amounts of water, and is more difficult to thicken to a high density and to filter. Depending on local laws, stabilized ferric arsenate meeting U.S. Environmental Protection Agency or equivalent disposal requirements may be suitable for landfill, whereas cyanidation residues are more likely to require lined dams with greater monitoring and reporting requirements.

A further issue that favors separate disposal of the two tailings is water reclamation. Even after a typical detoxification process, the cyanide residue will still contain thiocyanate species that are toxic to bioprocessing, and this water cannot be reused in bioprocessing unless tertiary treatment is used to degrade thiocyanate species. For traditional plants, reclamation of any water from a tails dam containing cyanide residues is limited for reuse in downstream processing. Conversely, water derived from the bioleach liquor after neutralization can be reclaimed and used in all process areas.

GENERIC CAPITAL AND OPERATING COSTS

Capital Costs

Published information on costs for biooxidation or bioleaching is limited. Recently, Van Niekerk reported the percentage of total capital cost that various areas of equipment contribute to the total cost for bacterial oxidation plants using Biox technology (Van Niekerk 2015). This information is portrayed in Table 3, indicating as expected that the tanks, blowers, and agitators are the major contributors to equipment cost.

Biooxidation reactors typically contribute about half of the equipment cost (Van Aswegen et al. 2007). The author's experience suggests that the capital cost of a reactor can be divided into three components with roughly equal contributions: (1) the materials of reactor construction and fabrication, (2) the cost of the internal fixtures and fabrication for cooling and air provision, and (3) the combined cost of the agitator and gearbox.

In 1996, published information on the capital costs of various biooxidation plants using different technology providers showed a range of costs according to concentrate throughput and sulfur content (Poulin and Lawrence 1996). These historic costs varied from US$2.4 million for a plant treating 40 stpd (short tons per day) concentrate rated at 7 stpd sulfur oxidation to US$25 million for a plant treating 760 stpd rated at 85 stpd sulfur oxidation. These were cited as installed costs for the total biooxidation plant. Table 4 indicates the cost values for four biooxidation facilities given in this publication and escalated to current 2017 values by factoring. Cost factoring is an inferior method for cost escalation because of the time span involved, but it still suggests a broad first estimate of possible costs for biooxidation plants. Most biooxidation plants are considered on a common cost basis in terms of the metric tons of sulfide treated per annum. Using this basis and factored data from published information, a capital cost range of US$1,100 million to US$1,700 million per annual metric ton of sulfide oxidized is shown in Table 4.

In support of these estimates, the author can offer an indicative range of costs based on values obtained by examining a portfolio of biooxidation projects or cost studies for potential projects conducted at feasibility level over a 20-year period. Projects showing extreme variations for capital costs were discounted because of distortion factors unique to certain projects. The following table shows the result from this exercise factored to 2017 values, with battery limits from receipt of thickener underflow to discharge of oxidized solids from washing in preparation for cyanidation:

Capital Cost per Annual Metric Ton of Sulfide Oxidized, US$	Capital Cost per Annual Metric Ton of Concentrate Treated at 25% S, US$
1,400–2,200	350–550

The capital cost of neutralization of bioleach liquor is also included. Assuming the treatment of a concentrate containing 25% sulfur, the cost per annual metric ton of concentrate is also indicated. Although the preceding table is only intended to provide a first-estimate indicative cost guide, it is somewhat supportive of the published costs for biooxidation plants when factored to a 2017 cost basis.

Operating Costs

Various authors have highlighted the contribution made by different areas to biooxidation operating costs. Three such studies are compared in Table 5, indicating reagents are the largest contributor to oxidation costs (Ritchie and Barter 1997; Batty and Post 1999; Van Aswegen et al. 2007). The use of reagents is project specific; limestone for neutralization typically is the largest contributing factor because of the large amounts required.

When examining the operating cost contributions, one must recognize that many of the operating costs are fixed costs; regardless of tonnage throughput, the annual cost does not vary significantly. For example, most equipment has minimal turndown capability, so that power usage is not directly related to throughput, and labor requirement and maintenance costs are often insensitive to throughput rate.

Relatively little published data are available concerning operating costs per metric ton of concentrate or metric ton of sulfide oxidized. The 1996 study referred to earlier (Poulin and Lawrence 1996) highlights two such costs from different biooxidation plants, which when converted to current values, suggest cost equivalents of US$67 and US$110 per metric ton

Table 3 Breakdown of capital cost as a function of equipment for a typical Biox project

Equipment Area	Contribution to Equipment Cost, %
Tanks	28
Blowers	23
Agitators	19
CCD thickeners	11
Stainless-steel piping	6
Waste residue tailings thickener	5
Pumps	4
Cooling towers	2
General	2

Table 4 Results from escalation of published capital costs for biooxidation plants to 2017 values

Project Number	Capacity, t/d concentrate	Sulfide, % concentrate	Oxidation, % concentrate	Sulfide Oxidation, t/a sulfide	Capital Costs		
					Original, million US$	Escalated for 2017, million US$	Escalated for 2017, t/a sulfide, million US$
1	40	20	88	2,441	2.4	4.1	1,700
2	115	24	90	8,613	5.6	9.3	1,100
3	760	11	98	29,442	25.0	40.5	1,400
4	120	28	32	3,728	3.9	6.3	1,700

Table 5 Comparison of published information showing contribution of cost components to biooxidation operating costs

Operating Cost Area	Study by		
	Batty and Post 1999, %	Van Aswegen et al. 2007, %	Ritchie and Barter 1997, %
Blowers	10	—	15
Agitators	5	—	9
Other power	6	—	6
Total power	21	34	30
Reagents	50	33	58
Labor	14	13	7
Maintenance	4	15	5
Analyses	6	—	—
Other	5	5	—
Total	**100**	**100**	**100**

of concentrate for the two different facilities. If the sulfide content and extent of oxidation are considered, these values in terms of cost per metric ton of sulfide oxidized can be recalculated to give values of US$310 and US$463 per metric ton of sulfide oxidized, respectively.

Based on experience but not verified by others, the author can offer an indicative range of operating costs obtained by examining a portfolio of biooxidation projects or cost studies for potential projects conducted at feasibility level over a 20-year period. The output from this exercise is given in the following table, suggesting a range of operating costs of between US$75 and US$175 per metric ton of concentrate, assuming a 25% sulfide content. This is equivalent to US$300–US$700 per metric ton of sulfide oxidized. Much of this variation is due to different reagent requirements and the unit power costs for the project.

Operating Cost per Ton of Sulfide Oxidized, US$	Operating Cost per Ton of Concentrate Treated at 25% S, US$
300–700	75–175

REFERENCES

Amiconi, T., Satalic, D., and Bain, C. 2004. Fosterville Gold Project—Taking the project from the feasibility study through the detailed design phase. In *Bac-Min 2004: Conference Proceedings*. Melbourne, Victoria: Australian Institute of Mining and Metallurgy. pp. 157–159.

Bailey, A.D., and Hansford, G.S. 1993. Factors affecting biooxidation of sulphide minerals at high concentration of solids: A review. *Biotechnol. Bioeng.* 42:1164–1174.

Batty, J.D., and Post, T.A. 1999. Bioleach reactor development and design. Billiton Process Research Presentation at ALTA 1999 Conference, Brisbane, Queensland.

Bell, N., and Quan, L. 1997. The application of BacTech (Australia) Ltd. technology for processing refractory gold ores at Youanmi gold mine. In *International Biohydrometallurgy Symposium, BIOMINE '97*. Melbourne, Victoria: Australian Mineral Foundation. pp. M2.3.1–M2.4.1.

Brierley, C.L. 1995. Bacterial oxidation. *Eng. Min. J.* 196:42–45.

Briggs, A.P., and Millard, M. 1997. Cobalt recovery using bacterial leaching at the Kasese Project Uganda. In *International Biohydrometallurgy Symposium, BIOMINE '97*. Melbourne, Victoria: Australian Mineral Foundation. pp. M2.4.1–M2.4.11.

Broadhurst, J.L. 1993. The nature and stability of arsenic residues from the BIOX process. In *International Conference and Workshop: Applications of Biotechnology to the Minerals Industry, BIOMINE '93*. Glenside, South Australia: Australian Mineral Foundation. pp. M2.4.1–M2.4.11.

Brown, A., Irvine, W., and Odd, P. 1994. Bioleaching—Wiluna operating experience. In *International Biohydrometallurgy Symposium, BIOMINE '94*. Melbourne, Victoria: Australian Mineral Foundation. pp. 16.1–16.8.

Budden, J.R., and Bunyard, M.J. 1994. Pilot plant test work and engineering design for the BacTech bacterial oxidation plant at the Youanmi gold mine. In *International Biohydrometallurgy Symposium, BIOMINE '94*. Melbourne, Victoria: Australian Mineral Foundation. pp. 4.1–4.8.

Colmer, A.R., and Hinkle, M.E. 1947. The role of microorganisms in acid mine drainage: A preliminary report. *Science* 106(2751):253–256.

Dew, D., and Miller, D. 1997. The BioNIC process: Bioleaching of mineral sulphides for nickel recovery. In *International Biohydrometallurgy Symposium, BIOMINE '97*. Melbourne, Victoria: Australian Mineral Foundation. pp. M.7.1.1–M.7.1.9.

Dinkla, I.J.T., Gonzalez-Contreras, P.A., Gahan, C.S., Weijma, J., Buisman, C.J.N., Henssen, M.J.C., and Sandström, A. 2013. Quantifying microorganisms during biooxidation of arsenite and bioleaching of zinc sulphide. *Miner. Eng.* (48):25–30.

Fewings, J., Seet, S., McCreedon, T., and Fitzmaurice, C. 2015. Bioheap application at Cosmic Boy Nickel Concentrator—An update. In *Proceedings of the ALTA 2015 Nickel Cobalt Copper Sessions*. Brisbane, Queensland: ALTA Metallurgical Services. pp. 356–372.

Foo, K.A., Reid, W.W., and Young, J.L. 1990. Designing very large bio oxidation plants. In *Randol Conference 1990*. Golden, CO: Randol International. pp. 3723–3729.

Fraser, G.M. 1993. Mixing and oxygen transfer in mineral bioleaching. In *International Conference and Workshop Applications of Biotechnology to the Minerals Industry, BIOMINE '93*. Glenside, South Australia: Australian Mineral Foundation. pp. 16.1–16.11.

Gonzalez, R., Gentina, J.C., and Acevedo, F. 2004. Biooxidation of a gold concentrate in a continuous stirred tank reactor: Mathematical model and optimal configuration. *Biochem. Eng. J.* 19:33–42.

Hackl, R.P. 1989. What to be aware of in cyanidation of biooxidised products. In *Randol Conference 1989*. Golden, CO: Randol International. pp. 143–144.

Holder, R. 2007. Beaconsfield Gold Mine—Ironing out the bugs. Presented at the Ninth Mill Operators Conference, Freemantle, West Australia, March 19–21.

James, H.E. 1976. Recent trends in research and development work on the processing of uranium ore in South Africa. Presented at IAEA Advisory Group Meeting on Uranium Ore Processing, Washington, DC, November 24–26.

Krause, E., and Ettel, V.A. 1989. Solubilities and stabilities of ferric arsenate compounds. *Hydrometallurgy* 22:311–337.

Kubera, P.M., and Oldshue, J.Y. 1992. Advanced impeller technologies match mixing performance process needs. Lightnin Technical Article No. 171.00. Rochester, NY: Lightnin.

Lally, K.S. 1987. A-315 axial flow impeller for gas dispersion. Lightnin Technical Article No. 144.00. Rochester, NY: Lightnin.

Loayza, C., and Ly, M.E. 1999. Biooxidation of arsenopyrite concentrate for industrial plant Tamboraque using acid mine drainage. In *International Biohydrometallurgy Symposium, BIOMINE '99*. Melbourne, Victoria: Australian Mineral Foundation. pp. 162–167.

Miller, P.C. 1997. The design and operating practice of bacterial oxidation plant using moderate thermophiles. In *Biomining: Theory, Microbes and Industrial Processes*. Edited by D.E. Rawlings. Berlin: Springer Verlag.

Miller, P., and Brown, A. 2005. Bacterial oxidation of refractory gold concentrates. In *Developments in Minerals Processing: Advances in Gold Ore Processing.* Edited by M.D. Adams. Burlington, MA: Elsevier Science. pp. 371–402.

Miller, P., Jiao, F., and Wang, J. 2004. The bacterial oxidation (BACOX) plant at Laizhou Shandong Province, China—The first three years of operation. In *Bac-Min 2004: Conference Proceedings*. Melbourne, Victoria: Australian Institute of Mining and Metallurgy.

Neale, J., Gericke, M., Pawlik, C., and Van Aswegen, P. 2015. The Mondo Minerals nickel sulphide bioleach project from test work to design. In *Proceedings of ALTA 2015 Nickel Cobalt Copper Sessions*. Brisbane, Queensland: ALTA Metallurgical Services. pp. 373–396.

Nicholson, H.M., Smith, G.R., Stewart, R.J., and Kock, F.W. 1994. Design and commissioning of Ashanti's Sansu BIOX plant. In *International Biohydrometallurgy Symposium, BIOMINE '94*. Melbourne, Victoria: Australian Mineral Foundation. pp. 2.1–2.8.

Nyombolo, B.M., Neale, J.W., and Van Staden, P.J. 2000. Neutralisation of bioleach liquors. In *Colloquium on Bacterial Oxidation for the Recovery of Metals.* Johannesburg: Southern African Institute of Mining and Metallurgy.

Odd, P.A.R., Craven, B., and Irvine, W. 1993. Bioleaching a feasible process for Wiluna Refractory gold ores. In *International Conference and Workshop Applications of Biotechnology to the Minerals Industry, BIOMINE '93*. Glenside, South Australia: Australian Mineral Foundation. pp. 16.1–16.11.

Olivier, J.W., and Jardine, J.G. 2014. Application of BIOX GIII principles during Runruno BIOX plant development. MINEX, Moscow, Russia, October 7–9.

Pinches, A., Neale, J.W., and Deeplaul, V. 2000. The Beaconsfield bacterial oxidation gold plant. In *Randol Conference 2000*. Golden, CO: Randol International.

Poulin, R., and Lawrence, R.W. 1996. Economic and environmental niches of biohydrometallurgy. *Miner. Eng.* 9(8):799–810.

Ringwood, K. 1995. *Arsenic in the Gold and Base Metal Mining Industry.* Australian Minerals and Energy Environment Foundation Occasional Paper No. 4. Melbourne, Victoria: Australian Minerals and Energy Environment Foundation.

Ritchie, I., and Barter, J. 1997. Process and engineering design aspects of biological oxidation plants. In *International Biohydrometallurgy Symposium, BIOMINE '97*. Melbourne, Victoria: Australian Mineral Foundation. pp. M14.4.1–M14.4.8.

Sandstrom, A., and Petersson, S. 1997. Bio-oxidation of a complex zinc sulphide ore: A study performed in continuous bench and pilot scale. In *International Biohydrometallurgy Symposium, BIOMINE '97*. Melbourne, Victoria: Australian Mineral Foundation. pp. M.1.1–M.1.11.

Sovmen, V., Savushkina, S., Stragis, J., Pavlyuchenko, M., and Krivomazova, G. 2008. Mineralogy of technological process at operating factory in Olympiada Mine. In *Proceedings of the XXIV International Minerals Processing Congress*, Vol. 1. Edited by Wang Dian Zuo et al. Beijing: Beijing Science Press.

Spencer, P., Satalic, D.M., Baxter, K.G., and Pinches, T. 1997. Key aspects in the design of a bacterial oxidation reactor. In *International Biohydrometallurgy Symposium, BIOMINE '97*. Melbourne, Victoria: Australian Mineral Foundation. pp. M.4.1.1–M.4.1.8.

Stephenson, D., and Kelson, R. 1997. Wiluna BIOX Plant expansion and new developments. In *International Biohydrometallurgy Symposium, BIOMINE '97*. Melbourne, Victoria: Australian Mineral Foundation. pp. M.4.1.1–M.4.1.8.

Suttill, K.R. 1990. Sao Bento plans BIOX. *Eng. Min. J.* 6:30–35.

Van Aswegen, P.C. 1993. Bio-oxidation of refractory gold ores: The Genmin Experience. In *International Conference and Workshop Applications of Biotechnology to the Minerals Industry, BIOMINE '93*. Glenside, South Australia: Australian Mineral Foundation. pp. 15.1–15.14.

Van Aswegen, P.C., Godfrey, M.W., Miller, D.M., and Haines, A.K. 1989. Design and operation of a commercial bacterial oxidation plant at Fairview. In *Perth International Gold Conference*. Golden, CO: Randol International. pp. 144–127.

Van Aswegen, P., Van Niekerk, J., and Olivier, W. 2007. The BIOX Process for the treatment of refractory gold concentrates. In *Biomining*. Edited by D.E. Rawlings and D.B. Johnson. Berlin: Springer-Verlag.

Van Buuren, C. 2014. Biomins novel integrated technologies. Presented at MINEX, Moscow, Russia, October 7–9.

Van Niekerk, J. 2012. Recent advances in the BIOX technology. Presented at ALTA Metallurgical Conference, Perth, Western Australia, May 2012.

Van Niekerk, J. 2015. Factors affecting the selection of BIOX as the preferred technology for the treatment of refractory gold concentrate. *Adv. Mater. Res.* no. 1130:191–196.

Van Staden, P.J., Rhodes, M., Pinches, A., and Martinez, T.E. 2000. Process engineering of base metal concentrate bioleaching. Randol Copper Hydrometallurgy Roundtable, Tuscon, Arizona, September 5–8.

Watling, H.R. 2015. Review of biohydrometallurgical metals extraction from polymetallic mineral resources. *Minerals* 5:1–60.

Whincup, P., and Binks, M. 2004. A case study of the development engineering and construction of a biological oxidation plant for refractory pyrite arsenopyrite gold ore. In *Bac-Min 2004: Conference Proceedings*. Melbourne, Victoria: Australian Institute of Mining and Metallurgy.

CHAPTER 10.9

Evaporation and Crystallization

Alison Lewis

This chapter covers the recovery of solids in crystalline form from hydrometallurgical solutions or wastewaters via either solvent removal (evaporation) or cooling (cooling crystallization). A novel application of crystallization (eutectic freeze crystallization [EFC]) for the recovery of both salts and water from hydrometallurgical brines is also covered.

Evaporation and crystallization are typically used in a hydrometallurgical process when the recovery of a moderately soluble compound is desired. The advantage of both of these processes is that they simultaneously separate and purify the product. Although the equipment is generally relatively simple, the crystallization process is governed by very complex, interacting variables. It is essentially a simultaneous heat- and mass-transfer process with a strong dependence on fluid and particle mechanics (Ullmann and Gerhartz 1998).

Energetically, cooling crystallization is the most favorable process and is operationally very simple (Lewis et al. 2015). Indirect cooling can be employed but must be carefully controlled to minimize scaling. For systems that are prone to severe scaling, indirect cooling cannot be used. In such cases, flash cooling or direct cooling is appropriate. For cooling crystallization, if several salts are dissolved in solution, as long as the solubilities of the salts are different, selective crystallization is possible (Hayes and Gray 1985).

Evaporative crystallization is more energy intensive than cooling crystallization and is not suitable for temperature-sensitive compounds, since evaporative crystallizers operate at higher temperatures. However, evaporation is suitable for compounds whose solubility is not a strong function of temperature.

CRYSTALLIZATION PRINCIPLES AND TECHNIQUES

This section covers the basic principles and techniques governing crystallization processes. The performance of any crystallizer, including cooling and evaporative crystallizers, is always governed by the phase equilibria, or solubility considerations. This will determine whether or not crystals will form in the process and will influence the mode of process to be selected.

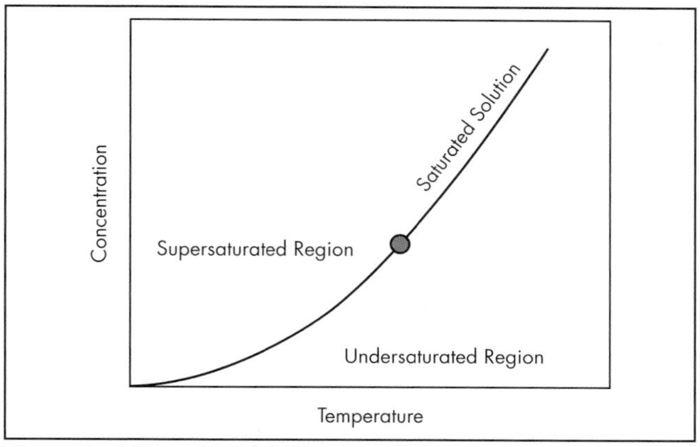

Figure 1 Schematic solubility curve

Solubility

The solubility of a substance in a solvent is the maximum concentration that can exist at equilibrium at a given set of conditions. The solubility most commonly increases with increase in temperature but, especially in hydrometallurgical systems, can decrease as the temperature increases. This is termed *retrograde solubility*.

For crystallization from solution, the phase diagram is usually represented in the form of a solubility curve, as illustrated schematically in Figure 1. On this diagram, the following thermodynamic states can be identified.

- Saturated solution: The state of the solution is represented by any point on the solubility curve.
- Undersaturated region: The solution is capable of dissolving more crystals.
- Supersaturated region: The solution contains solute in excess of that predicted by the equilibrium condition.

SELECTION OF CRYSTALLIZATION METHOD

The solubility curve for a particular solute in a solvent is a useful tool in selecting an appropriate crystallization method.

Alison Lewis, Dean, Faculty of Engineering & Built Environment, Dept. of Chemical Engineering, University of Cape Town, Cape Town, South Africa

Figure 2 illustrates how the selection of the crystallization method is partly based on solubility curve considerations.

Cooling Crystallization

Cooling crystallization is the preferred method in the case of compounds with a moderate solubility (in the range of 100 to 300 kg/m) and a temperature-solubility curve with a steep positive slope (i.e., dWeq/dT > 0.005°C^{-1}, where Weq = weight fraction in solution at equilibrium and T = temperature, in °C). Cooling crystallization is also suitable for compounds that do not have a strong tendency to scale. For scaling compounds, vacuum cooling or flash cooling can be used, as these avoid the necessity for heat transfer surfaces.

Evaporative Crystallization

For moderately to highly soluble compounds (10–90 wt %) with shallow or flat temperature-solubility curves, where cooling crystallization will result in a very low yield, evaporative crystallization is the method of choice. Theoretically, evaporative crystallization, vacuum evaporation, and multistage evaporation can be applied irrespective of the slope of the solubility curve.

The graph in Figure 3 shows the solubility data for several inorganic salts. From this graph, it is apparent that potassium nitrate (KNO_3), for example, would be a good candidate for recovery by cooling crystallization, whereas sodium chloride (NaCl) would not. Because of its extremely flat solubility curve, NaCl would be suitable for recovery by evaporative crystallization instead.

The graph in Figure 4 illustrates that sodium sulfate (Na_2SO_4) would also be a good candidate for recovery by cooling crystallization due to the steep positive slope of the solubility curve. The discontinuity in the solubility data given in the figure also indicates the presence of a phase change. In other words, the solid phase that will crystalize out from an aqueous solution of Na_2SO_4 below 32.4°C will be the decahydrate salt, whereas above 32.4°C, the solid phase will be the anhydrous salt (Mullin 2001).

Eutectic Freeze Crystallization

EFC is a suitable technique for the treatment of brines and reverse osmosis retentates. It is a combination of melt cooling and solution cooling crystallization and results in the simultaneous crystallization of ice and salt. By cooling the solution slightly below the eutectic temperature, ice is formed and simultaneously salt is crystallized because of the elimination of solvent (as ice) from the solution.

Since the density of ice crystals is lower than that of the solution, the formed ice crystals float, whereas the salt crystals, which are denser than the solution, sink to the bottom of the vessel. So an integrated process of cooling crystallization of ice and salt and gravity separation from the solution can be achieved (Lewis et al. 2015). Because thermodynamically, it is cheaper to freeze 1 kg of water (333 kJ) than to evaporate it (2,300 kJ), EFC can be a very economical process. In addition, because a pure salable salt can be produced simultaneously with the ice, the EFC process can become profitable. It can also become an attractive solution when zero discharge of waste streams is mandatory. In those cases, the often-dilute solution is first preconcentrated by reverse osmosis and can then be treated in an EFC process.

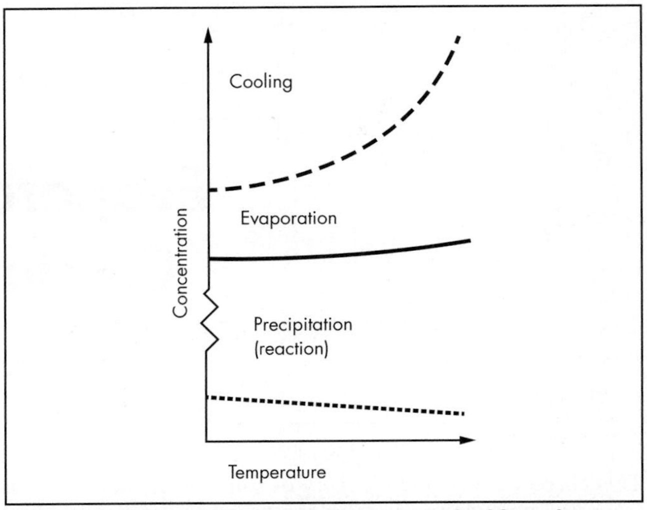

Figure 2 Selection of the crystallization method based on solubility curve considerations

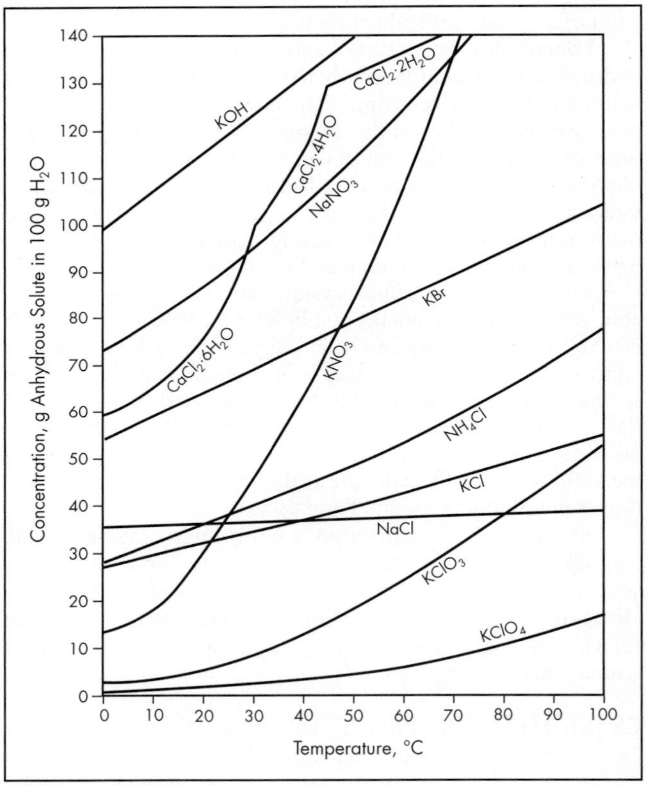

Source: Cheremisinoff 1994

Figure 3 Solubility data for several common inorganic compounds

THEORETICAL CRYSTAL YIELD

If the solubility information for a substance in a solvent is known, then the theoretical crystal yield can be calculated. This is the maximum yield of pure crystals that could be obtained through cooling or evaporation. For example, Figure 4 indicates that if a saturated solution containing Na_2SO_4 is cooled

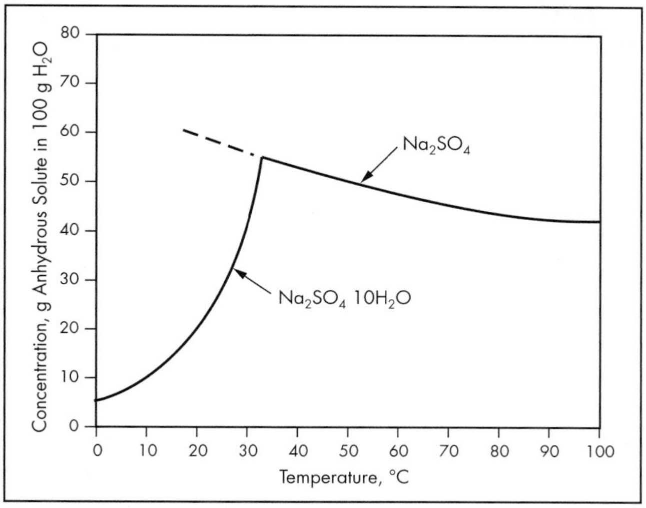

Source: Cheremisinoff 1994

Figure 4 Solubility data for Na_2SO_4

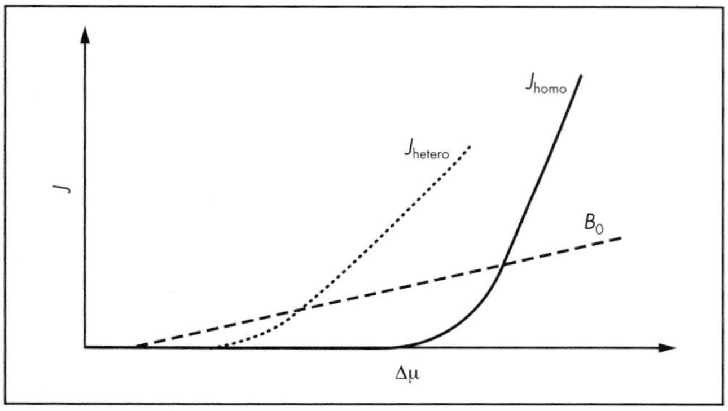

Source: Lewis et al. 2015, © Cambridge University Press 2015

Figure 5 Effect of supersaturation on the type of nucleation

from 30°C to 10°C, then the maximum possible theoretical crystal yield will be

$$\begin{aligned} c_*@30°C - c_*@10°C &= 40\text{ g } Na_2SO_4 \text{ in } 100\text{ g } H_2O \\ &\quad -10\text{ g } Na_2SO_4 \text{ in } 100\text{ g } H_2O \\ &= 30\text{ g } Na_2SO_4 \text{ in } 100\text{ g } H_2O \end{aligned} \quad \textbf{(EQ 1)}$$

where c_* is the saturation concentration of solute in solvent at that temperature.

Likewise, the solubility curve also indicates the solute that is *not* recoverable by cooling or evaporative crystallization. In the previous example, the unrecoverable Na_2SO_4 is the saturation concentration at the final temperature of 10°C, that is, 10g Na_2SO_4 in 100 g H_2O.

PRINCIPLES

Crystallization uses the creation of supersaturation of the compound in solution as the driving force for the recovery of the compound. Once this supersaturation has been created, either by temperature difference in the case of cooling crystallization or by solvent removal in the case of evaporation, the formation of the solid phase occurs via two main processes: nucleation and growth. Secondary processes include aggregation and attrition.

Nucleation

The first stage of crystallization is the nucleation stage, in which new crystal matter is formed. The nucleation rate is a function of the supersaturation and can be expressed as an empirical relationship in the form of a power law.

$$J = k_n S^n \quad \textbf{(EQ 2)}$$

where

J = nucleation rate, number/s
k_n = nucleation rate constant
S = supersaturation
n = power law exponent for nucleation, usually >2 for nucleation

The type of nucleation is a function of the level of supersaturation, with low supersaturation levels favoring secondary nucleation. Secondary nucleation is the desired form of nucleation for cooling crystallization processes, and it can take place only if crystals of the species under consideration are already present.

As the supersaturation levels increase, moderate supersaturation levels favor heterogeneous nucleation, whereas the highest supersaturation levels favor homogeneous nucleation. Homogeneous nucleation is the least desirable form of nucleation, as it is unpredictable and often leads to a product with undesirable particle characteristics.

Figure 5 shows the relationship between supersaturation and the different types of nucleation, with B_0 representing secondary nucleation, J_{hetero} representing heterogeneous nucleation, and J_{homo} representing homogeneous nucleation.

Growth

Growth is the mechanism by which nuclei are enlarged and is the mechanism ultimately responsible for the final crystal habit, or shape. The classical crystal growth model is by a two-stage mechanism:

1. Mass transport, either by diffusion or convection, of the solute molecule from the bulk solution to the crystal face, followed by
2. Deposition of the solute molecule onto the crystal surface, with accompanying integration of the growth unit into the crystal lattice.

Analogously to the nucleation rate, the growth rate can also be modeled as a simple power law, as follows:

$$G = k_g S^g \quad \textbf{(EQ 3)}$$

where

G = growth rate, ms^{-1}
k_g = growth rate constant, ms^{-1}
S = supersaturation, –
g = power law exponent for growth, usually ±2 for spiral growth, $2 > g > 1$ for birth-and-spread growth, and 1 for rough growth

Again, the type of growth that occurs is a function of the supersaturation, with lower supersaturation values leading to spiral growth, moderate supersaturation values leading to birth-and-spread growth, and the highest supersaturation values leading to rough growth. This is illustrated in Figure 6, which shows the effect of supersaturation on the type of growth. Figure 7

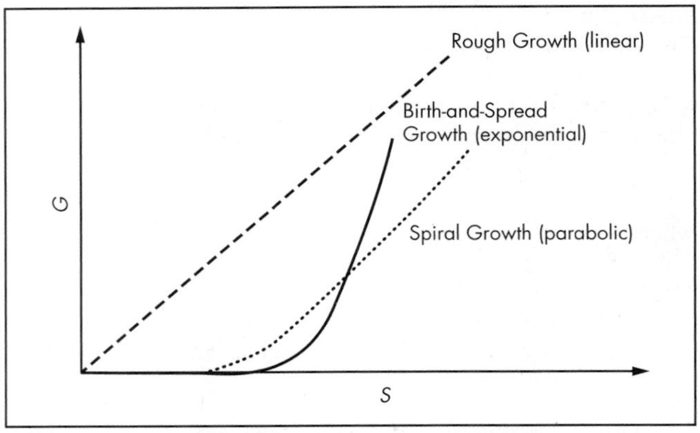

Source: Lewis et al. 2015, © Cambridge University Press 2015

Figure 6 Effect of supersaturation on the type of growth

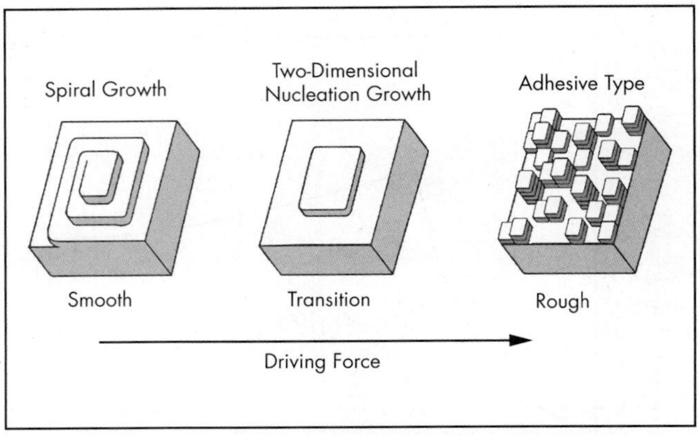

Source: Sunagawa 2005, © Cambridge University Press 2005

Figure 7 Effect of driving force on crystal morphology

demonstrates how the crystal morphology is affected with an increase in driving force.

Secondary Mechanisms

Secondary mechanisms include aggregation and attrition. Aggregation (sometimes called agglomeration) takes place when suspended crystals collide in a supersaturated solution. If the colliding crystals stay in contact for a sufficient time, then crystalline bridges between them can be formed by crystal growth. Agglomeration occurs in the submicrometer and micrometer ranges and is less important for particles larger than 50 µm (Mersmann 2001). The resulting particles are called agglomerates.

Attrition occurs during crystallization of large crystals (>100 µm) in systems that have high solubility. The supersaturation of the system needs to be low to avoid activated nucleation (Mersmann 2001). The attrition rate is a function of the size of the parent crystals, as well as their collision velocity.

Particle Size Distribution

Crystallization processes produce a product that usually exhibits a range of particle sizes. The very extensive topic of particle size distributions is beyond the scope of this chapter. Suffice to mention that one of the two most important product characteristics of an industrially crystalized product is the particle size distribution (the other is purity).

Reporting the size of a product in terms of its crystalline product distribution can be unwieldy, and therefore the distribution is usually summarized in terms of its average particle size and the spread of the distribution (Jones 2002). The average particle size can be defined in terms of

- The mode, which is the maximum density value of the mass density distribution; and
- The median size, which is the size that splits the distribution into two equal parts. For a mass-based distribution, the median mass means that half the mass of the particles is smaller than the median size and half larger. For a size-based distribution, the median size means that half the particles are smaller than the median and half larger (Lewis et al. 2015).

The spread of the distribution can be defined in terms of the coefficient of variation (CV). Because there are many different definitions of this term, it is important to be aware of which one is being used.

The usual definition of CV is as follows:

$$CV = \frac{\text{standard deviation } \sigma}{L_{4,3}} \quad \textbf{(EQ 4)}$$

where $L_{4,3}$ is the mass-based mean size.

A second definition, called the *quartile ratio*, is defined as

$$CV = \ln\frac{L_{75}}{L_{25}} \quad \textbf{(EQ 5)}$$

where

L_{75} = upper quartile
L_{25} = lower quartile

Lastly, another definition of the width uses *quintiles*, which are the division points when the data is divided into fifths. The upper quintile is equivalent to the 80th percentile, and the lower quintile is equivalent to the 20th percentile. The coefficient of variation defined using quintiles is as follows:

$$CV = \ln\frac{L_{80} - L_{20}}{2L_{50}} \quad \textbf{(EQ 6)}$$

Crystal Purity

As the growth rate increases, the purity of the formed crystal tends to decrease, as faster growth rates favor the development of macrosteps, instabilities, and the inclusion of mother liquor. All of these decrease the purity of the final crystal product. Therefore, there is a trade-off between faster production rates (and increased impurities) and slower production rates (and decreased impurities). Figure 8 illustrates how the formation of macrosteps leads to an increase in mother liquor inclusions and thus impurities.

Caking

Small crystals are more prone to cake than are large crystals because of the greater number of contact points per unit mass, but actual size is less important than size distribution and shape (Ullmann and Gerhartz 1998). A narrow crystal size distribution is linked to lower caking rates.

Faster growth rates are also linked to caking since these cause macrosteps and mother liquor inclusions. The mother

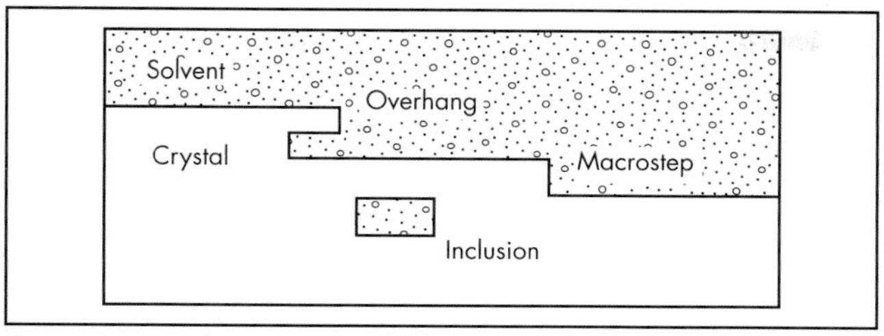

Source: Lewis et al. 2015, © Cambridge University Press 2015

Figure 8 Macrosteps, overhangs, and mother liquor inclusions formed under conditions of high supersaturation

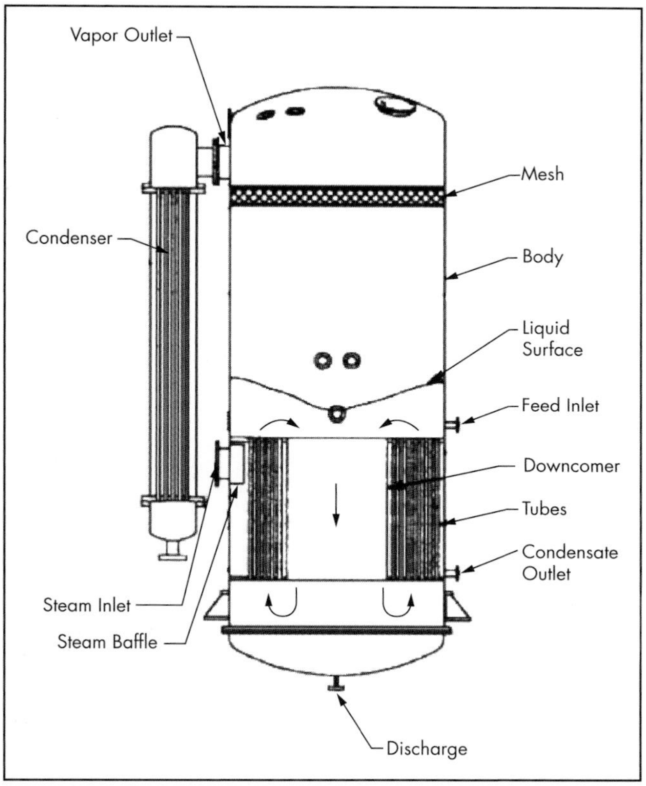

Source: Bennett 2004

Figure 9 Thermo-syphon crystallizer: Swenson calandria evaporator

liquor inclusions act as "micro-crystallizers" and cause recrystallization during storage.

TECHNIQUES

Techniques for solution and evaporative crystallization range from the very simple (solar ponds for evaporation) to more sophisticated arrangements such as cascades of draft-tube baffled (DTB) and forced circulation crystallizers. The choice of crystallizer type depends on the material to be crystallized and the choice of solvent, the method of crystallization, the production rate, the product specifications, the necessity for flexibility of the design, and whether or not a single crystallizer or cascade will be selected. Generally, as far as the production rate is concerned, a batch-wise system should be selected for a production rate of <5 kt/yr, whereas a continuous system should be selected for a production rate of >20 kt/yr.

Solar Ponds

The simplest option is the solar pond, although the lack of control and design can make the product quality extremely variable, as the size and hardness of the final crystals depends on the wind and other weather conditions (Lewis et al. 2015). Improvements on stagnant solar ponds include forced heat transfer; using a stirrer or pump to suspend the crystals; and/or exerting some control of the hydrodynamic flow pattern, for example through a draft tube. This reduces mixing and energy input and lowers the secondary nucleation rate. Another improvement would be to use subatmospheric operation in case of evaporation for energy efficiency. This would entail using a closed vessel.

Thermo-Syphon crystallizer

The next level of sophistication as far as equipment is concerned is a thermo-syphon crystallizer (Figure 9). This crystallizer has no stirrer but uses circulation through natural convection. The design includes a heat exchanger and a draft tube, and the crystallizer is operated at subatmospheric pressures. The advantages include low investment and maintenance costs, simple operation, and minimal attrition of crystals. The heat exchanger simultaneously operates as an effective fines destruction device.

Stirred Draft-Tube Baffled Crystallizer

A stirred draft-tube baffled crystallizer with an internal heat exchanger (Figure 10) is very similar to the thermo-syphon crystallizer, but with the addition of a stirrer inside the bottom of the draft tube to aid mixing and heat transfer. A steady movement of magma and feedstock up from the feedpoint to the surface of the liquor produces uniform boiling. DTB crystallizers are designed to make a large, uniform crystal product (De Leer 1981) through the maintenance of a very low degree of supercooling (<1°C), and without violent vapor flashing that can cause excessive nucleation and salt buildup on the inner walls (Mullin 2003).

Forced Circulation Crystallizer

The forced circulation crystallizer (Figure 11) is the most widely used crystallizer, mostly for vacuum evaporative crystallization of substances with a flat solubility curve, such as NaCl. It is the cheapest vacuum crystallizer, especially in cases where large amounts of solvent need to be evaporated to achieve a large production rate (Mersmann 2001). Operation is usually stable, it has a large capacity, and the design is straightforward. The disadvantages include attrition of crystals by the pump and the fact that there is no external fines destruction.

Fluidized Bed or Oslo Crystallizer

The primary advantage of the Oslo crystallizer (Figure 12) is the ability to grow crystals in a fluidized bed, which is not subject to mechanical circulation methods. This means that the since there is clear liquor circulation, the attrition of crystals is minimized. The Oslo crystallizer has the advantage of a coarse product due to the long residence time and classification, and it has the capacity for fines removal and destruction.

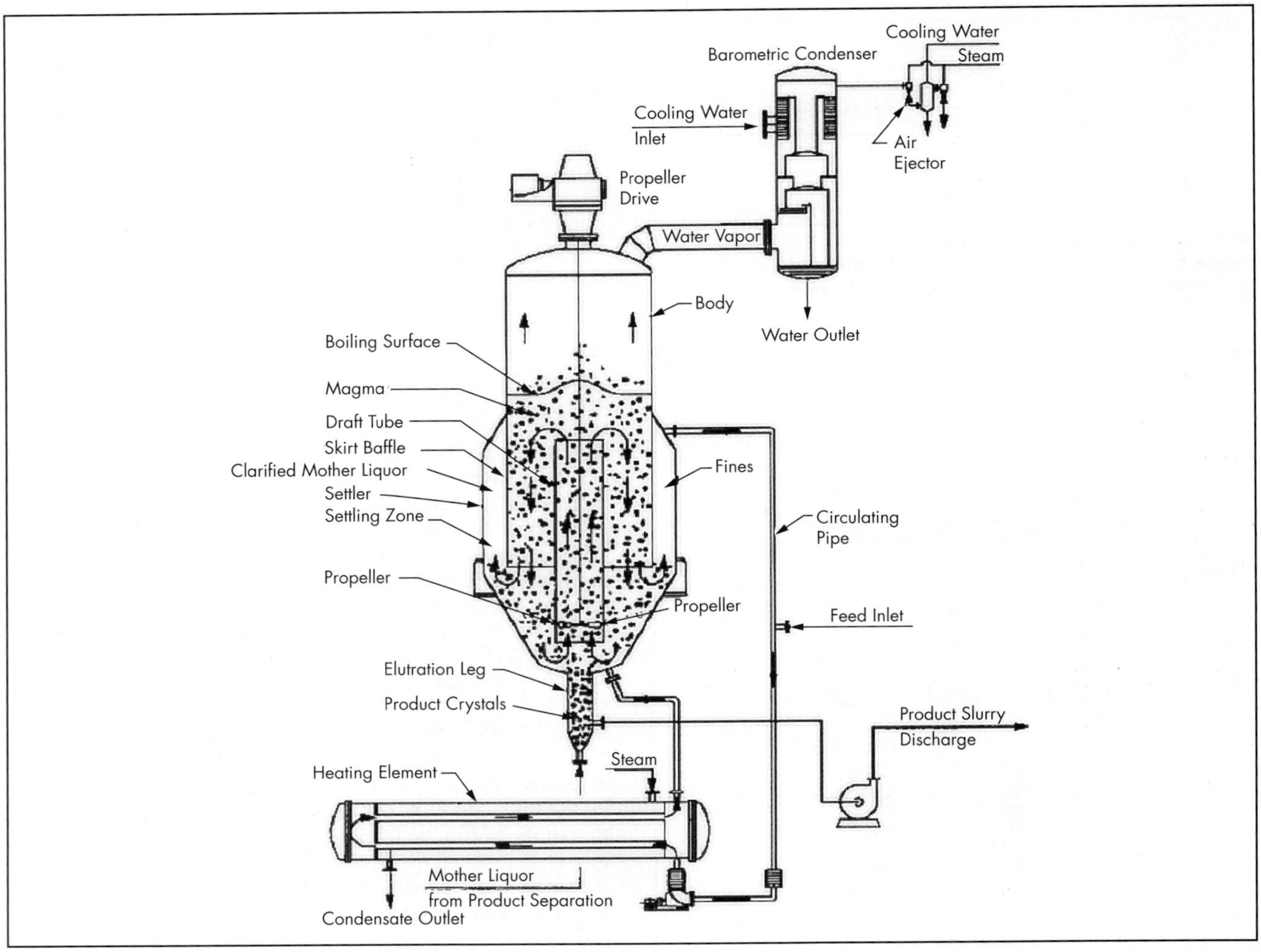

Source: Bennett 2004

Figure 10 Stirred draft-tube baffled crystallizer

The disadvantages include a low production rate of product, the risk of blockage around the downcomer, and numerous crystal–crystal collisions.

Spray Evaporative Crystallizer

A spray evaporative crystallizer is suitable for applications where waste heat at very low temperature is available and where the product will not be contaminated by atmospheric air nor contaminate the air itself (Myerson 2001). This type of crystallizer is essentially a spray dryer and is usually economical to operate, since waste heat can be used. However, the product quality can be low and there is a tendency to form agglomerates.

Direct Cooling Crystallizer

The direct cooling crystallizer uses a refrigerant, such as liquid propane or carbon dioxide, which is mixed with the suspension and then evaporates. It can be applied for very low temperatures and is advantageous since encrustation on heat exchanger tubes is avoided. Once used, the refrigerant vapor leaves the surface of the crystallizer and is compressed, condensed, and recirculated to the crystallizer (Myerson 2001). Direct contact cooling crystallizers are often used in melt crystallization applications (Mersmann 2001). A DTB or forced circulation crystallizer without the heat exchanger can also serve as a direct cooling crystallizer.

Surface Cooling Crystallizer

A surface cooling crystallizer (Figure 13) has the advantages of forced heat transfer, minimal scaling on the heat transfer surfaces, and limited stirring action (only off bottom). The disadvantages include no internal pumping action (unlike crystallizers fitted with stirrer and draft tube).

For more information about crystallizers and crystallizer design, refer to Mullin (2001), Beckmann (2013), Myerson (2001), Mersmann (1988), Lewis et al. (2015), Jones (2002), and Chianese and Kramer (2012).

The websites of the equipment manufacturers can also be extremely informative, for example, Swenson Technology (www.swensontechnology.com/) and GEA (www.geav.com/en/index.jsp).

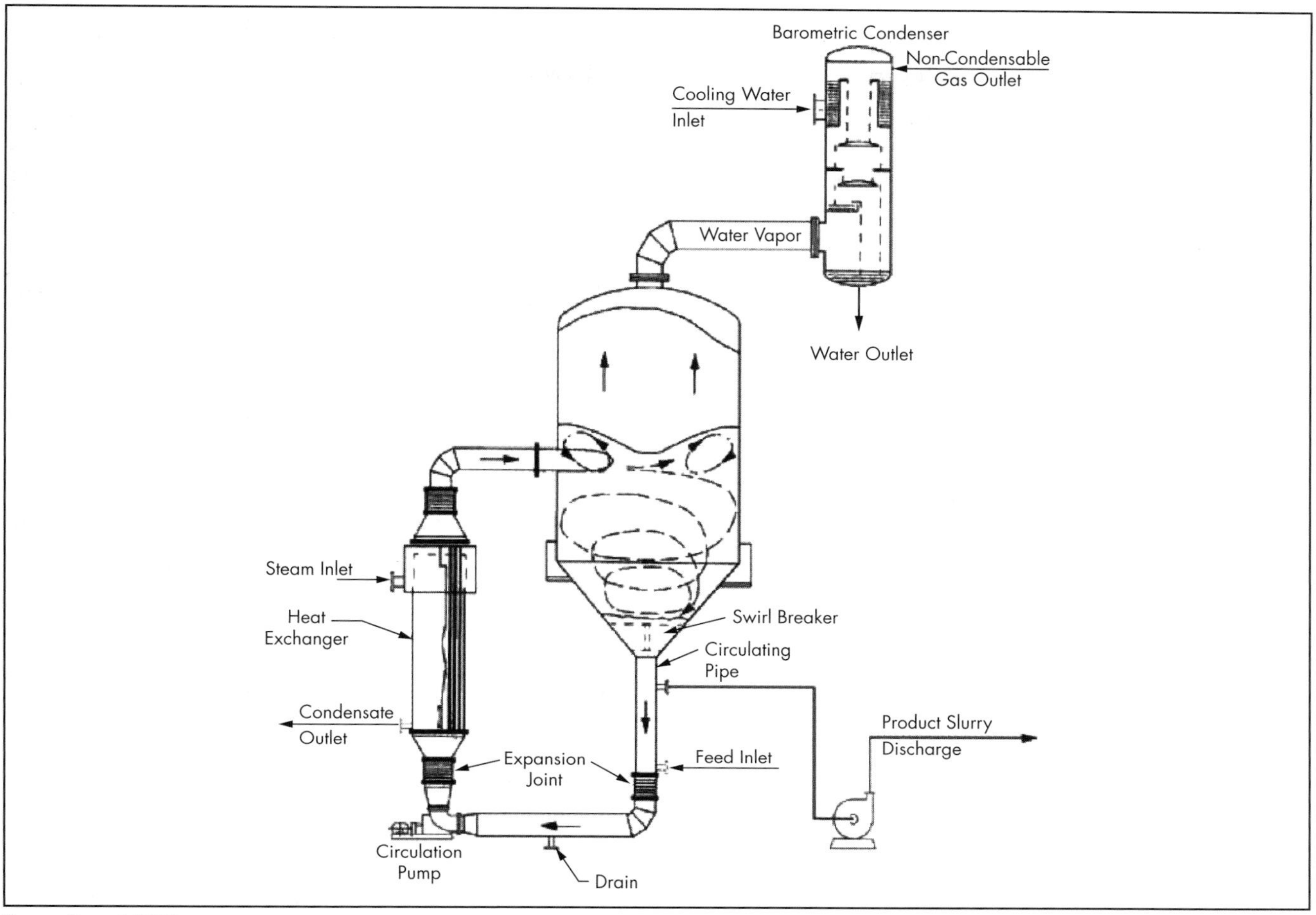

Source: Bennett 2004

Figure 11 Forced circulation crystallizer

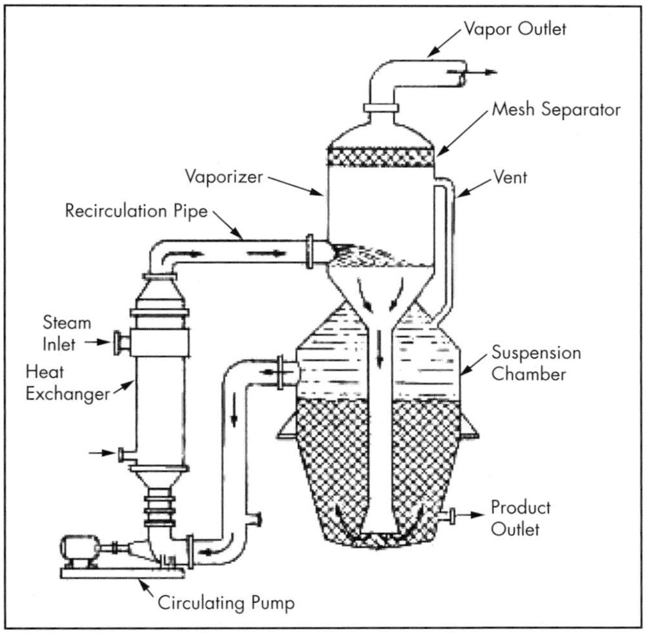

Source: Partlow 2017

Figure 12 Oslo crystallizer

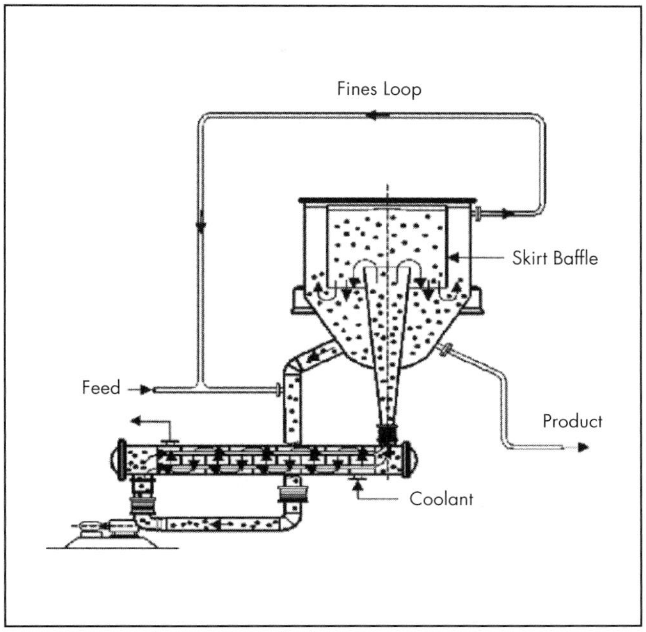

Source: Lewis et al. 2015, © Cambridge University Press 2015

Figure 13 Surface cooling crystallizer

CONCLUSIONS

This chapter has focused on the various techniques, selection of method, and types of commercially available crystallizers for evaporative and cooling crystallization. It does not attempt to be complete but rather highlights the important considerations when selecting each type of crystallizer, with the aim of highlighting the different features of various types of crystallizers. Most important, it is the products to be manufactured that determine the equipment design, and not vice versa.

REFERENCES

Beckmann, W. 2013. *Crystallization: Basic Concepts and Industrial Applications*. Weinheim, Germany: Wiley-VCH Verlag.

Bennett, R.C. 2004. Crystallization and evaporation. In *The Engineering Handbook*. Edited by R.C. Dorf. Boca Raton, FL: CRC Press.

Cheremisinoff, P.N. 1994. *Handbook of Water and Wastewater Treatment Technology*. New York: Marcel Dekker.

Chianese, A., and H.J. Kramer. 2012. *Industrial Crystallization Process Monitoring and Control*. Weinheim, Germany: Wiley-VCH Verlag.

De Leer, B.G.M. 1981. *Draft-Tube-Baffle Crystallizers: A Study of Stationary and Dynamicly-Behaving Crystal-Size Distributions*. Delft, Netherlands: Delft University of Technology.

Hayes, P.C., and Gray, P.M.J. 1985. *Process Selection in Extractive Metallurgy*. Brisbane, Queensland: Hayes Publishing Company.

Jones, A.G. 2002. *Crystallization Process Systems*. Amsterdam, Netherlands: Butterworth-Heinemann.

Lewis, A.E., Seckler, M.M., Kramer, H., and Van Rosmalen, G.M. 2015. *Industrial Crystallization: Fundamentals and Applications*. Cambridge, UK: Cambridge University Press.

Mersmann, A. 1988. Design of crystallizers. *Chem. Eng. Process. Process Intensification* 23(4):213–228.

Mersmann, A. 2001. *Crystallization Technology Handbook*. New York: Marcel Dekker.

Mullin, J.W. 2001. *Crystallization*. Amsterdam, Netherlands: Butterworth-Heinemann.

Mullin, J.W. 2003. Crystallization and precipitation. In *Ullmann's Encyclopedia of Industrial Chemistry*. Edited by F. Ullmann and W. Gerhartz. Weinheim, Germany: Wiley-VCH Verlag.

Myerson, A.S. 2001. *Handbook of Industrial Crystallization*. Amsterdam, Netherlands: Butterworth-Heinemann.

Partlow, W. 2017. *Civil Engineering Handbook: Solids Concentration*. www.civilengineeringhandbook.tk/solids-concentration/fig.html. Accessed February 2018.

Sunagawa, I. 2005. *Crystals: Growth, Morphology and Perfection*. Cambridge: Cambridge University Press.

Ullmann, F., and Gerhartz, W. 1998. *Ullmann's Encyclopedia of Industrial Chemistry*. Weinheim, Germany: Wiley-VCH Verlag.

CHAPTER 10.10

Precipitation

Alison Lewis

Precipitation is typically used in a hydrometallurgical process as the final separation and purification step. The advantage of precipitation is that, although it causes a phase change of the product (from the aqueous to the solid phase), it simultaneously effects the purification of the remaining liquor. Precipitation also has an advantage in that it can be used in cases where large volumes of dilute liquor are to be treated. Thus, precipitation is most widely used either to recover metal values (as pure metals or as their salts) or as a means of removing impurities from process streams and wastewaters.

WHAT IS PRECIPITATION?

Precipitation is also referred to as reactive crystallization, which describes how the unit operation occurs. The process is usually effected by adding two reagents together, causing a chemical reaction and forming a sparingly soluble compound, usually referred to as the precipitate. An example of a precipitation reaction is the removal or recovery of copper from a sulfate solution by the precipitation of copper sulfide.

$$CuSO_4 + H_2S \rightarrow CuS\ (\downarrow) + H_2SO_4 \quad \textbf{(EQ 1)}$$

PRECIPITATION METHODS

Further examples of the various precipitation methods and the metals to which they are applied are given in Figure 1.

Purification of Liquor

As shown in Figure 1, precipitation can be used as a method of both liquor purification and recovery of metal value. In the category of liquor purification, where the product is the liquor itself, two methods are available: electrochemical reduction (sometimes referred to as cementation) and ionic precipitation.

Electrochemical Reduction

In the case of electrochemical reduction, generally called *cementation* in the hydrometallurgy industry, the metal to be removed is the more electropositive or noble metal, and it is the one that is reduced. The metal that is used for the reduction is the more electronegative or the sacrificial metal, and it is the one that is oxidized. For example, in the purification of zinc solutions, the more electropositive cuprous ion (Cu^{2+}) is reduced to elemental copper by the addition and simultaneous oxidation of zinc, which is more electronegative.

$$Cu^{2+} + Zn \rightarrow Cu\ (\downarrow) + Zn^{2+} \quad \textbf{(EQ 2)}$$

The advantage of this technique is that the addition of zinc to remove copper from a zinc solution has the effect of avoiding the addition of any foreign ions. A table of standard electrode potentials is used to determine the hierarchy of electropositive/electronegative metals. The same purification method can also be used for the removal of Au, Ag, and Cd.

Ionic Precipitation

In the case of ionic precipitation for the purification of liquor, the metals to be removed include Fe, Cu, Co, Ni, and Zn, and they are usually removed in the form of sulfide or hydroxide salts, but also as phosphates, arsenates, carbonates, halides, oxalates, oxides, and oxometallates. The presence of arsenic in a leach solution, for example, is undesirable because it contaminates the final product. It can also cause disposal problems because of its toxicity (Habashi 1993). It can be effectively removed as a calcium or an iron arsenate as follows:

$$2AsO_4^{3-} + 3Ca^{2+} \rightarrow Ca_3(AsO_4)_2 \quad \textbf{(EQ 3)}$$

$$AsO_4^{3-} + Fe^{3+} \rightarrow FeAsO_4 \quad \textbf{(EQ 4)}$$

Hydroxide Precipitation

Hydroxide precipitation is effected by the addition of an alkaline agent to alter the pH and thus remove the metal of interest as a metal hydroxide. The precipitation reaction for a divalent (n = 2) or trivalent (n = 3) metal ion will be

$$M^{n+} + nOH- \Leftrightarrow M(OH)_n\ (\downarrow) \quad \textbf{(EQ 5)}$$

Magnesium, calcium, sodium, and ammonia salts are commonly used as a source of hydroxide ion. Basic conditions can also be obtained by the addition of sodium carbonate (Na,CO_3). Some of the most commonly used alkaline agents are quick lime (CaO) and slaked lime ($Ca(OH)_2$), which are cheap and easy to use.

Alison Lewis, Dean, Faculty of Engineering & Built Environment, Dept. of Chemical Engineering, University of Cape Town, Cape Town, South Africa

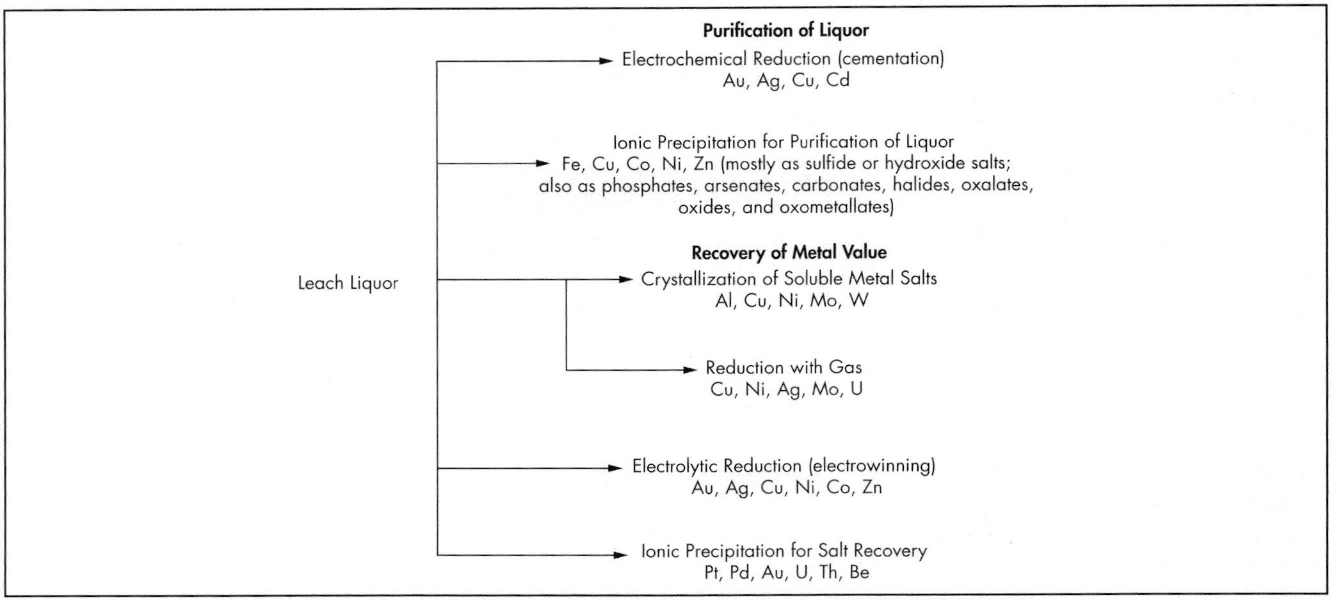

Adapted from Gupta and Mukherjee 1990

Figure 1 Summary of various precipitation methods and metals to which they are applied

Compared to sodium hydroxide (NaOH), lime has the advantages that it is cheaper, the solids settle faster, the precipitate (often referred to as sludge) has a higher solids concentration, and the sludge is easier to dewater. Most of these factors are because the lime acts as a coagulant during the precipitation/settling process. The disadvantages of lime precipitation are that it takes longer to neutralize the liquor, there is a more complicated feed system, and the volume of sludge produced is greater. In general, although metal hydroxide precipitation is cheap and easy to implement, it also has its challenges, the most important of which is the tendency of hydroxides to form gelatinous precipitates that are difficult to thicken or filter.

Figure 2 illustrates the effect of changing the pH on the solubility of a range of metal hydroxides. From the figure, clearly most metal hydroxides exhibit amphoteric behavior. For example, $Al(OH)_3$ is an amphoteric hydroxide; that is, it is soluble in both an acid (below pH = 6) and a base (above pH = 7) (Habashi 1993). However, most metal hydroxides have their points of minimum solubility under more basic condition (i.e., pH >9).

Sulfide Precipitation

Sulfide precipitation is effected by the addition of a sulfide source agent to cause the precipitation of the metal as a sulfide salt. The recovery of metals as metal sulfide salts is common in copper and nickel hydrometallurgy (Gupta and Mukherjee 1990). Sources of sulfide include hydrogen sulfide gas (H_2S), NaHS, and Na_2S. Although the precipitation reaction for a divalent metal ion is usually written as given in the following equation, this is an oversimplification.

$$M^{2+} + S^{2-} - \Leftrightarrow MS\ (\downarrow) \qquad \textbf{(EQ 6)}$$

In reality, any sulfide source will dissolve and speciate in aqueous solution. For example, H_2S is a weak acid and dissociates in two steps according to the following:

$$H_2S \Leftrightarrow H^+ + HS^- \Leftrightarrow 2H^+ + S^{2-} \qquad \textbf{(EQ 7)}$$

The dissociation behavior of H_2S as a function of pH is illustrated in Figure 3. The figure shows that between the pH values of 9 and 16, the H_2S is available in the form of HS^-. Therefore, the reaction when using H_2S as the sulfide source in this pH range will be as follows:

$$HS^- \Leftrightarrow MS^- + H^+ \qquad \textbf{(EQ 8)}$$

Therefore, metal sulfide precipitation is invariably accompanied by a decrease in pH of the solution, as the precipitation reaction itself produces protons that lower the pH.

Although hydroxide precipitation is widely used, sulfide precipitation has several advantages, including the lower solubility of metal sulfide precipitates, which allows very high extents of removal/recovery (see Figure 4).

Other advantages include the potential for selective metal removal, fast reaction rates, a lower tendency for gel formation, and potential for reuse of sulfide precipitates by smelting. The lower solubilities of sulfides as compared to metal hydroxides have the advantage of generating (at least theoretically) very low residual metal concentrations in the treated wastewater. However, the sparingly soluble nature of metal sulfides also means that the precipitates produced often have poor particle properties (Lewis 2010). The explanation is as follows.

The driving force for precipitation is supersaturation (S), which is calculated using

$$S = \frac{a_+ a_-}{K_{sp}} \qquad \textbf{(EQ 9)}$$

where

a_+ = activity of the cation in solution
a_- = activity of the anion in solution
K_{sp} = solubility product of the precipitating salt

From the equation, it follows that the less soluble the precipitating salt, the lower the K_{sp} value and thus the higher the

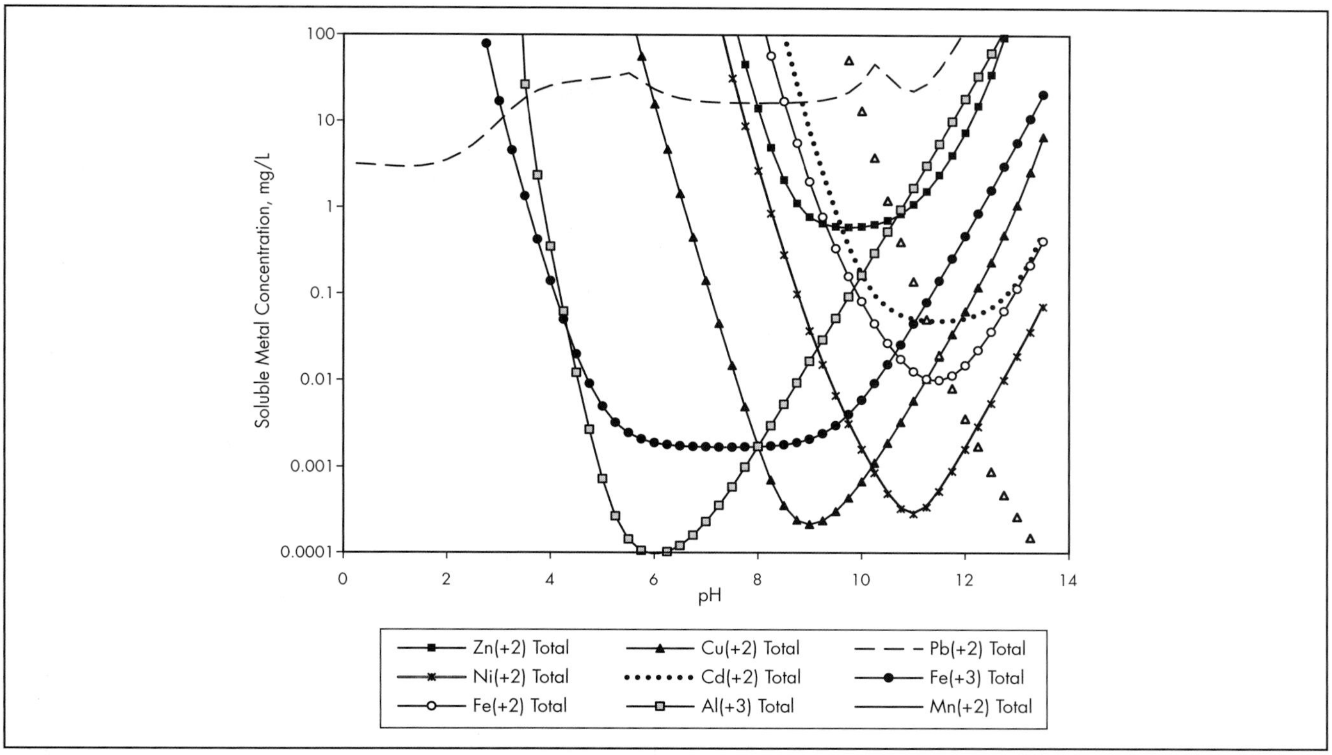

Source: Lewis 2010

Figure 2 Effect of pH on solubility of metal hydroxide ions

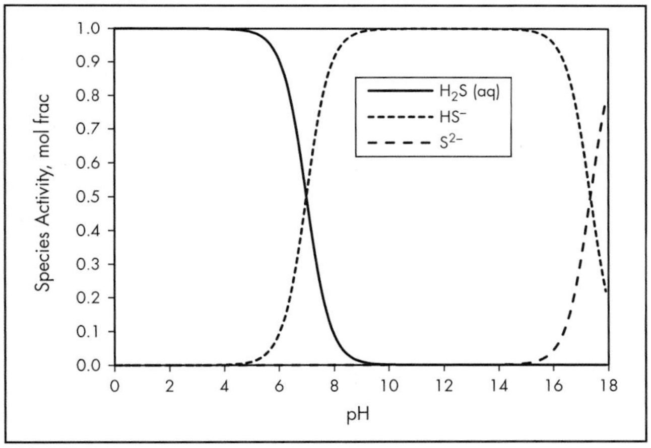

Source: Lewis 2010

Figure 3 Effect of pH on speciation of H_2S

supersaturation. A high supersaturation means that the driving force is large and that fast processes, such as nucleation, are favored over slow ones, such as growth.

Figure 5 illustrates how the supersaturation affects the dominant mechanisms in the precipitation process. At low supersaturation values, precipitation is dominated by growth, and larger crystal sizes can be achieved. At higher supersaturation values (which are the norm for precipitation processes), nucleation dominates and the crystal size decreases. This results in the precipitation process generating numerous extremely small (often submicrometer) particles that are usually colloidal in nature and thus difficult to filter and dewater, as they tend to clog filters. CuS, in particular, does not settle very well (Nduna et al. 2014). At high supersaturation values, agglomeration also begins to be an important mechanism, as it is a function (among other things) of the number and density of particles in solution. Because agglomerates retain mother liquor in the interstitial spaces between primary particles, the purity of agglomerates is usually low.

Recovery of Metal Value

As shown in Figure 1, precipitation can also be used as a method of recovery of metal value. The first method of recovering metal value is via crystallization of soluble metal salts. This is covered in Chapter 10.9, "Evaporation and Crystallization." The remaining three methods are reduction with gas, electrochemical reduction (also called electrowinning), and ionic precipitation for salt recovery.

Reduction with Gas

Reduction of a metal with a gas, such as hydrogen, can be used for recovery of metals that are more noble than hydrogen. The generalized reaction can be written as follows:

$$M^{2+} + H_2 \rightarrow M + 2H^+ \qquad \textbf{(EQ 10)}$$

where M represents metals such as Cu, Ni, and Zn. However, metals such as Ni and Co can be recovered at pH values of 5 and higher from ammoniacal solutions using hydrogen reduction. This method of metal recovery offers an alternative to electrowinning, as it uses less energy, has a higher productivity, and offers the possibility of controlling the properties of

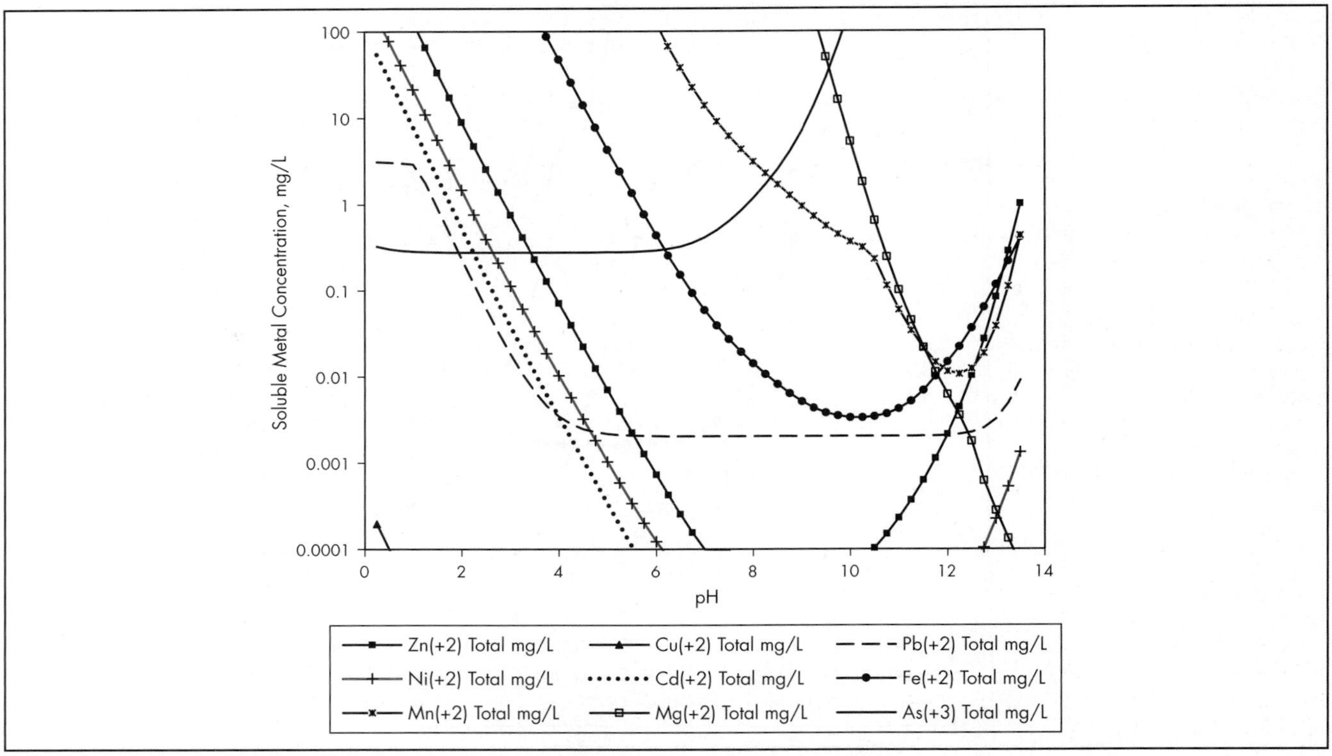

Source: Lewis 2010

Figure 4 Effect of pH on solubility of metal sulfide ions

the produced powder by using surface-active addition agents (Saarinen et al. 1998; Ntuli and Lewis 2006).

Although the reduction is thermodynamically favored, the kinetics of the reduction are extremely slow, and temperatures of >150°C and pressures of >30 bar are required for feasible operation. However, the particulate properties of the elemental metal produced using this method are usually purity and amenability to downstream processing.

Besides the recovery of metals in pure form, metals such as uranium, vanadium, and molybdenum can be recovered from solution in oxide form by using hydrogen gas (Merritt et al. 1985).

Electrolytic Reduction

Electrolytic reduction (or electrowinning) to recover ultrapure metals from solution is usually one of the last processes in a train of hydrometallurgical unit operations. Technically, electrolytic reduction fits into the broader category of electrochemical reduction, which also includes the processes of cementation and hydrogen reduction, so in this description, the term *electrowinning* is used.

The central feature of the electrowinning operation is the electrolytic cell, which consists of alternate plate anodes and cathodes. The metal to be recovered is deposited onto a "starter sheet" at the cathode. In the case of zinc electrowinning, the starter sheet is aluminum, which ensures that the cathode is perfectly flat and undistorted. This ensures a constant anode/cathode spacing and hence a uniform current density over the whole of the cathode surface (Hayes and Gray 1985).

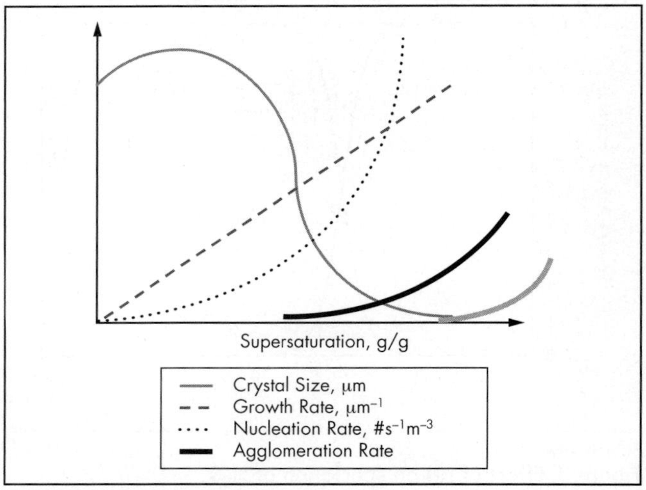

Source: O'Grady 2011

Figure 5 Effect of supersaturation on crystal size as well as the rates of nucleation, growth, and agglomeration

A version of electrowinning called *electrorefining* can also be used, for example, to recover copper. In this process, the anodes consist of large, cast slabs of impure copper. These anodes dissolve during refining to produce pure copper, which is deposited at the cathode onto titanium or stainless-steel sheets.

Although electrowinning is usually seen as distinct from precipitation, it is in fact an electrochemical precipitation reaction, with the applied current providing the driving force, or the supersaturation.

Ionic Precipitation for Salt Recovery

Ionic precipitation to recover metals as their salts is used in the production of platinum, iridium, palladium, gold, uranium, thorium, and beryllium. In the recovery of the platinum group metals (PGMs; platinum, iridium, and palladium), their nobility (their resistance to dissolution in media that dissolve almost all of the base metals) is overcome by dissolving them in chloride media and thus using their ability to form chloride complexes. All of the PGMs, except ruthenium, in their tetravalent oxidation state form hexachloro complex anions in strong chloride media and thus can all be precipitated from aqueous chloride solution as their ammonium salts, which enables their recovery from solution as follows:

$$2NH_4Cl + (MCl_6)^{2-} \Leftrightarrow (NH_4)_2(MCl_6)\ (s) + 2Cl^- \quad \textbf{(EQ 11)}$$

If the metal is present as M^{2+}, it must first be oxidized to the M^{4+} by, for example, chlorine gas. The solubilities of the hexachloro complexes of the PGMs are highly temperature dependent, and this property is usually exploited in the precipitation of the PGM salts from chloride solutions (Ullmann and Gerhartz 1998). Again, the main challenge encountered in the precipitation of the PGMs using this method is the difficulty of precipitating crystals that are sufficiently large for filtration (Butler et al. 2001). This is due to the large driving force, or supersaturation, generated by the mixing of the two reagents for the precipitation reaction.

In Figure 6, the effect on nucleation of changing either the p[Me] and/or the p[X] is illustrated, where, in this case, p[Me] is the log of the cation or metal concentration in solution and p[X] is the log of the anion concentration in solution. Zone 1 is the undersaturated zone (at high p[Me] and p[X] values, i.e., very low concentrations of both metal and anion); b is the solubility curve; Zone 2 is the zone of metastable supersaturated solution; c is the limit of the metastable zone; Zones 3 and 4 are regions of heterogeneous nucleation, Zone 4 being the region of dendritic growth; Zone 5 is the region of homogeneous nucleation and transient precursor formation; Zone 6 is the region of positively charged colloids; and Zone 7 is the region of negatively charged colloids.

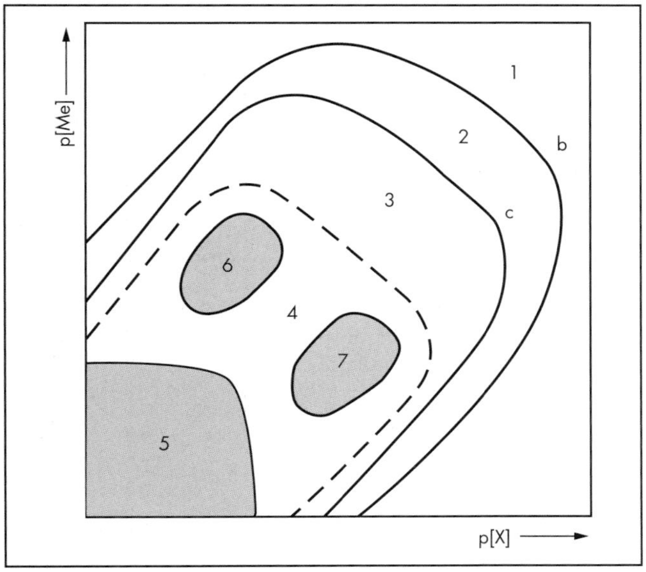

Source: Söhnel and Garside 1992
Figure 6 Precipitation diagram

In other words, as the concentration of either the metal cation or the anion increases, the potential for undesirable nucleation events and poor particle properties increases.

Operational Considerations

Ionic precipitation is a highly effective method of recovering or removing metals (or impurity elements) from solution, because the precipitates formed are usually very sparingly soluble substances (solubility in the range of 0.001–1 kg/m^3). This allows for high recovery of the product or a high degree of removal of species. However, this benefit comes at a cost: the sparingly soluble nature of the precipitates mean that the processes are operated at very high supersaturation levels. This leads to the formation of solid phases with a low product quality (small primary crystals and agglomerates) as well as other unexpected phases, such as polymorphs and amorphous materials (Lewis et al. 2015).

In addition, the combination of the high supersaturation and fast kinetics means that there are often significant concentration variations in the precipitation vessel, accompanied by locally steep supersaturation gradients near the inlet point(s).

In any precipitation process, control of supersaturation is the key to controlling product quality. If the supersaturation is too high, then the produce quality suffers in two ways:

1. The nucleation rate is high.
2. The type of growth is rough and uncontrolled.

When these phenomena are encountered together, there is a cumulatively adverse effect on the product quality.

In precipitation processes, supersaturation is often created by the feeding of a reagent in a semi-batch fashion into an agitated solution of the other reagent; consequently, there is a competition between the mixing and the reacting processes. This is because the chemical reaction is usually faster than the mixing of the reagents at the molecular level. Therefore, because the solubility of the product is low, the supersaturation at the feed point is very high and the process is carried out under mixing limited conditions (Torbacke and Rasmuson 2001). Thus, in a precipitation process, the mixing process is often rate limiting and to generate a product with desirable properties, the supersaturation in the reactor must be controlled by controlling the mixing regime.

Determining the Rate Limiting Process: Mixing or Reaction

To determine whether the precipitation process is mixing or reaction limited, the characteristic time for the various mixing processes must be compared to the characteristic time constant for the reactive precipitation process, which can be calculated for an nth order chemical reaction as follows:

$$\tau_R = \frac{1}{\left(k_n (C_{A0})^{n-1}\right)} \quad \textbf{(EQ 12)}$$

where

τ_R = characteristic time constant for the chemical reaction, s
k_n = reaction rate constant (m^{3}kg^{-1}s^{-1} for a first-order reaction)
C_{A0} = initial concentration of reagent A, kg/m^{-3}
n = reaction order

The characteristic time constants for each of the scales of mixing can then be compared with the characteristic time constant for the chemical reaction. If the characteristic time constant for the reaction is smaller than that for the mixing (i.e., the reaction is faster than any of the mixing processes), then the precipitation process is mixing limited. If the characteristic time constant for the reaction is larger than that for the mixing (i.e., the reaction is slower), then the precipitation process is reaction limited.

The time constants for the three scales of mixing are given in the following sections.

Macromixing

Macromixing occurs at a scale equivalent to the size of the reactor and can be described as bulk dispersion of the liquids throughout the reaction volume. For a well-baffled tank and fully developed turbulence, the macromixing time, τ_{macro}, is between three and five times the circulation time, τ_c (Baldyga and Bourne 1999):

$$\tau_{macro} = 4\tau_c \quad \textbf{(EQ 13)}$$

where

τ_{macro} = macromixing time, s
τ_c = circulation time, s

For a vessel using an impeller with four pitched blades, Roelands and coworkers (2002) express the circulation time as follows:

$$\tau_c = \frac{V}{q_c} \quad \textbf{(EQ 14)}$$

where

V = vessel contents, m^3
q_c = pumping capacity of the impeller, m^3/s

The pumping capacity of the impeller is given as

$$q_c = N_q N d^3 \quad \textbf{(EQ 15)}$$

where

N_q = flow number (taken as 0.73 for a pitched blade turbine [Nienow 1997])
N = impeller speed, rps
d = impeller diameter, m

Mesomixing

The term *mesomixing* describes mixing on a scale comparable to the size of the reagent feed pipe, that is, larger than micromixing (on the molecular scale) but smaller than that of macromixing (on the scale of the whole reactor) (Bałdyga and Bourne 1999). Two mesomixing mechanisms have been identified:

1. Inertial-convective disintegration of large eddies in the course of dispersion
2. Turbulent dispersion, where a feed stream spreads out transverse to its local streamline

The time constant for mesomixing due to the inertial-convective disintegration of large eddies is determined from the Torbacke and Rasmuson (2001) relationship:

$$\tau_{meso} = a\left(\frac{\Lambda^2}{\varepsilon}\right)^{1/3} \quad \textbf{(EQ 16)}$$

where

a = 1–2 (2 for fully developed turbulence in liquids [Baldyga et al. 1997])
Λ = macroscale turbulence, m
ε = local energy dissipation rate, m^2/s^3

The macroscale turbulence is estimated by Torbacke and Rasmuson (2001) to be

$$\Lambda = \sqrt{\frac{Q_f}{\pi u}} \quad \textbf{(EQ 17)}$$

where

Q_f = feed flow rate, m^3/s
u = fluid velocity, m/s

The local energy dissipation rate is given by the Torbacke and Rasmuson (2001) relationship:

$$\varepsilon = \frac{N_p N^3 d^5}{V} \quad \textbf{(EQ 18)}$$

where

N_p = power number (taken as 1.5 for a pitched blade [Uhl and Gray 1996])
N = impeller speed, rps
d = impeller diameter, m

$$N_p = \frac{P}{\rho N^3 D^5} \quad \textbf{(EQ 19)}$$

where

P = power, W
ρ = liquid density, kg/m^3
N = impeller speed, rps
D = impeller diameter, m

The time constant for mesomixing due to turbulent dispersion is determined from

$$\tau_d = \frac{Q_F}{UD_i} \quad \textbf{(EQ 20)}$$

where

τ_d = time constant for turbulent dispersion, s
Q_F = volumetric flowrate of feed solution, m^3s^{-1}
U = liquid velocity, ms^{-1}
D_i = turbulent diffusivity, m^2s^{-1}

Micromixing

Micromixing is homogenization on the molecular scale through diffusion and occurs at the Kolmogorov scale, which is the size of the smallest eddy. Since micromixing happens on the molecular scale, it directly influences the chemical reaction and thus nucleation and crystal growth, which are essentially molecular-level processes (Bałdyga and Bourne 1999). The micromixing time can be determined from the local specific energy dissipation rate, in terms of Kolmogorov's turbulence theory (Baldyga and Bourne 1989):

$$\tau_{micro} = 17.24\sqrt{\frac{\nu}{\varepsilon}} \quad \textbf{(EQ 21)}$$

where

ν = kinematic viscosity, m^2/s
ε = local specific energy dissipation rate, m^2/s^3

Micromixing takes place by

- Molecular diffusion,
- Laminar deformation of striations below the Kolmogorov scale, and
- Mutual engulfment of regions having different compositions.

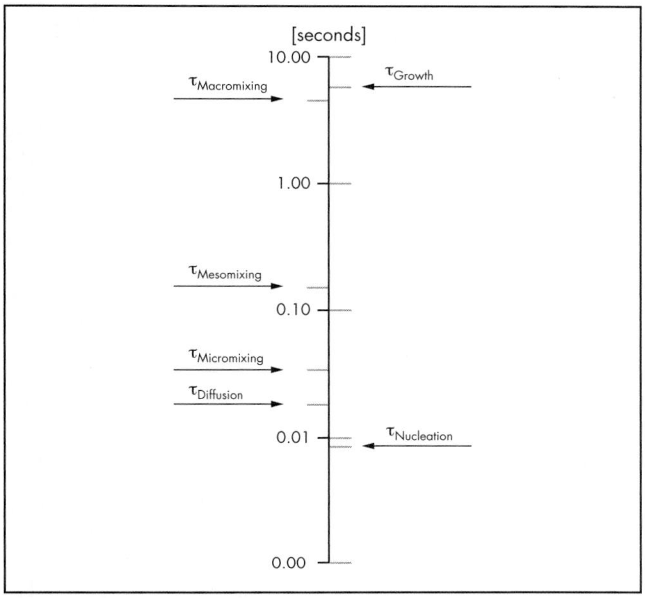

Source: Bałdyga and Bourne 1999

Figure 7 Competition between reaction and mixing: Example of identification of the rate limiting step

Engulfment is often the limiting one of these three. The time constant for engulfment is given by

$$\tau_e = \left(\frac{\nu}{\varepsilon}\right)^{0.5} \quad \textbf{(EQ 22)}$$

where

ν = kinematic viscosity, m^2/s

ε = local turbulent energy dissipation rate, m^2/s^3

After the four time constants have been calculated, the reaction time can be compared to each of the mixing time (macro, meso, and micromixing) to determine which one dominates, as illustrated in Figure 7. In this example, the nucleation (i.e., reaction) time is far shorter than any of the calculated mixing times. Therefore, this system is mixing limited.

Practical Approach to Mixing Configurations in Stirred Vessels

From a practical point of view, when precipitation is carried out in a stirred tank, not much can be done about the high local supersaturation, since this is caused by the solubility of the precipitate as well as by the introduction of the reagents at the feed points. However, to some extent, the mixing configuration can be adjusted to minimize the effect of high local levels of supersaturation in the reaction vessel. Methods of adjusting the mixing configuration include changing the feed configuration and changing the stirring speed or type of stirrer. This is illustrated in Figure 8. The figure illustrates five precipitation configurations that are carried out by mixing two streams, A and B. At the left-hand side of the upper figure, the local supersaturation is relatively small, and larger particle sizes can be produced. At the right-hand side of the upper

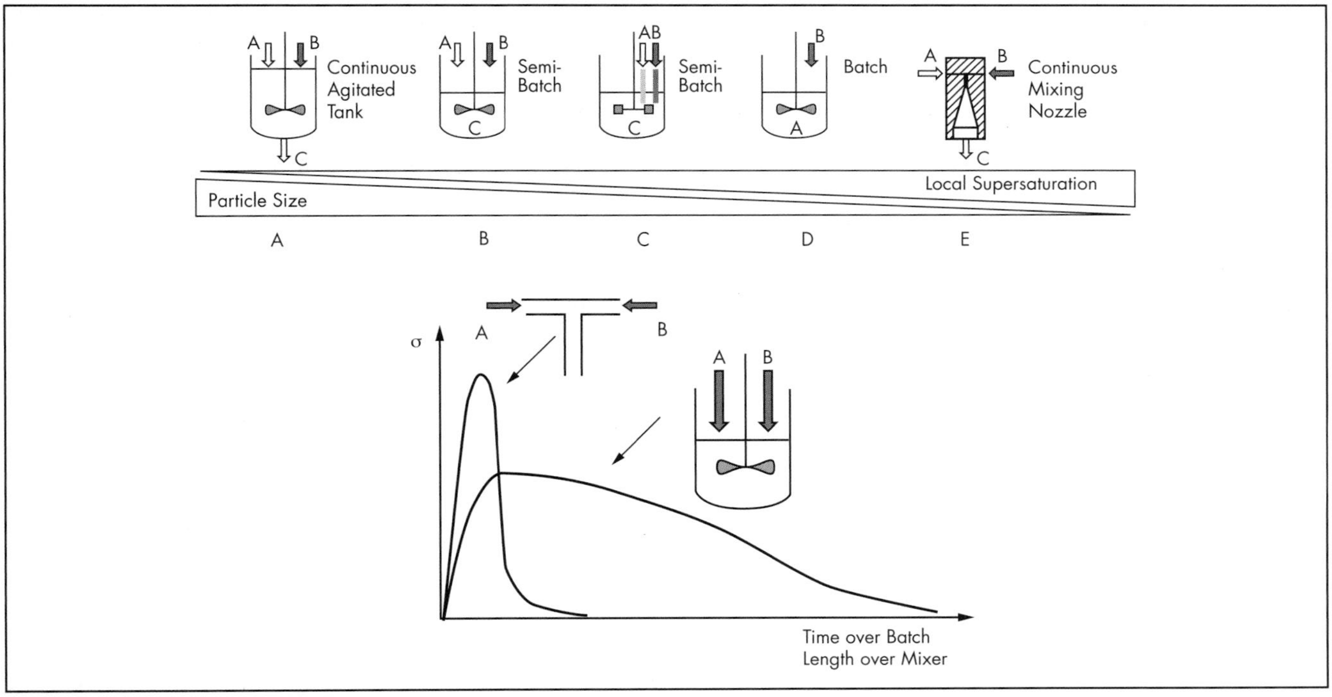

Source: Beckmann 2013

Figure 8 Feed point configuration modifications for moderately soluble (top left) and sparingly soluble (top right) substances

figure, the local supersaturation is very high, and smaller particle sizes are produced.

Geometry A shows reagents A and B being added in a continuous system to an agitated tank to produce product C. Geometry B illustrates the semi-batch version. Geometry C illustrates another semi-batch version. All three of these geometries, which involve feeding the both reagent streams A and B into the mother liquor, have the beneficial effect of reducing the local supersaturation.

In Figure 8, geometry D is the most commonly used, but not optimal, configuration, in which reagent B is added to a vessel that already contains reagent A. At the point of addition, the local supersaturation is relatively high, leading to moderately small crystals.

Geometry E shows the two reagents first being mixed in a T mixer nozzle to produce product C. Mixing the two reagent streams in a nozzle substantially increases the local supersaturations but has the beneficial effect of narrowing the particle size distribution. This is illustrated in the lower part of the figure. For the nozzle geometry (geometry E), small particles are produced, but the particle size distribution is relatively narrow. For geometry B, larger particles can be produced, but the particle size distribution is relatively wide.

CONCLUSION

In summary, precipitation is a process that has the potential to be highly effective in both metal recovery and removal from aqueous solution. However, because of the usually large driving forces involved, attention needs to be paid to process design and operational considerations.

REFERENCES

Baldyga, J., and Bourne, J.R. 1989. Simplification of micromixing calculations. I. Derivation and application of new model. *Chem. Eng.* 42(2):83–92.

Baldyga, J., and Bourne, J.R. 1999. *Turbulent Mixing and Chemical Reactions*. Chichester, UK: Wiley.

Baldyga, J., Bourne, J.R., and Hearn, S.J. 1997. Interaction between chemical reactions and mixing on various scales. *Chem. Eng. Sci.* 52(4):457–466.

Beckmann, W. 2013. *Crystallization: Basic Concepts and Industrial Applications*. Weinheim, Germany: Wiley-VCH Verlag.

Butler, B., Centurier-Harris, J., and Lewis, A.E. 2001. Improving platinum precipitation processes. *Miner. Eng.* 14(8):905–909.

Gupta, C.K., and Mukherjee, T.K. 1990. *Hydrometallurgy in Extraction Processes*. Oxfordshire, UK: Taylor and Francis.

Habashi, F. 1993. *A Textbook of Hydrometallurgy*. Quebec City, QC: Metallurgie Extractive Quebec.

Hayes, P.C., and Gray, P.M.J. 1985. *Process Selection in Extractive Metallurgy*. Cincinnati: Hayes Publishing Company.

Lewis, A.E. 2010. Review of metal sulphide precipitation. *Hydrometallurgy* 104(2):222–234.

Lewis, A.E., Seckler, M.M., Kramer, H., and van Rosmalen, G.M. 2015. *Industrial Crystallization: Fundamentals and Applications*. Cambridge, UK: Cambridge University Press.

Merritt, R.C., Seidel, D.C., Burnham, D.A., and Bush, P. 1985. Recovery from solution. In *SME Mineral Processing Handbook*. Edited by N.L. Weiss. Littleton, CO: SME-AIME. pp. 51–59.

Nduna, M.K., Lewis, A.E., and Nortier, P. 2014. A model for the zeta potential of copper sulphide. *Colloids Surf. A* 441:643–652.

Nienow, A.W. 1997. On impeller circulation and mixing effectiveness in the turbulent flow regime. *Chem. Eng. Sci.* 52(15):2557–2565.

Ntuli, F., and Lewis, A.E. 2006. The effect of a morphology modifier on the precipitation behaviour of nickel powder. *Chem. Eng. Sci.* 61(17):5827–5833.

O'Grady, D. 2011. Supersaturation: Driving force for crystal nucleation and growth [blog]. Mettler Toledo; 2011. http://blog.autochem.mt.com/2011/03/supersaturation-driving-force-for-crystal-nucleation-growth/. Accessed July 2018.

Roelands, C.P.M., Derkson, J.J., ter Horst, J.H., Kramer, H.J.M., and Jansens, P.J. 2002. Mixing time scale analysis for the measurement of nucleation rate of concomitant polymorphs. *Chem. Eng. Trans.* 1:29–34.

Saarinen, T., Lindfors, L.-E., and Fugleberg, S. 1998. A review of the precipitation of nickel from salt solutions by hydrogen reduction. *Hydrometallurgy* 47(2-3):309–324.

Söhnel, O., and Garside, J. 1992. *Precipitation: Basic Principles and Industrial Application*. Boston: Butterworth-Heinemann.

Torbacke, M., and Rasmuson, Å.C. 2001. Influence of different scales of mixing in reaction crystallization. *Chem. Eng. Sci.* 56(7):2459–2473.

Uhl, V.W., and Gray, J.B. 1996. *Mixing: Theory and Practice*. Vol. 1. New York: Academic Press.

Ullmann, F., and Gerhartz, W. 1998. *Ullmann's Encyclopedia of Industrial Chemistry*. Weinheim, Germany: Wiley-VCH.

CHAPTER 10.11

Carbon Adsorption

Matthew Jeffrey

In extractive metallurgy, the dominant use of activated carbon is in the adsorption of the gold cyanide complex from aqueous solution following the cyanide leaching of gold ores. The historical development of activated carbon adsorption in gold metallurgy has been well described by Marsden and House (2006), who highlight the very rapid take-up of the carbon-in-pulp (CIP) technology from the 1970s throughout the United States (with the Homestake mine in South Dakota being regarded as the birth of modern CIP), South Africa and Australia. By the 1990s, most new gold processing plants constructed used some form of the carbon adsorption technology, either as CIP, carbon-in-leach (CIL), or carbon-in-column (CIC). CIP and CIL both make use of carbon directly in a leach slurry, whereas CIC circuits are used to recover gold from pregnant solutions (usually from heap leach operations). Figure 1 shows a block flow diagram of an oxide gold mill with a CIL circuit, where carbon is added to each leach tank to adsorb the gold cyanide complex. The carbon is recovered from pulp by screening and then acid washed to remove inorganic contaminants (e.g., carbonates) prior to recovery of gold from the carbon by elution and electrowinning.

Although this chapter exclusively deals with the application of activated carbon for gold adsorption, readers are reminded that activated carbon is also a very powerful adsorbent for off-gas treatment applications, for example, the removal of volatile organic compounds (Shepherd 2001). In addition, several modern mineral processing plants employ sulfur-impregnated activated carbon (SIAC) for the capture of mercury in off-gas treatment systems, especially those servicing gold rooms and carbon regeneration furnaces. In Newmont-affiliated facilities, SIAC systems can be found at operations with mercury present in the ore, for example, Twin Creeks and Carlin in Nevada, United States. For further information on SIAC systems, the reader is referred to Krumins et al. (2014).

The intent of this chapter is to provide a short summary of background theory and then focus on the metallurgical testing and principles used in modern gold processing plant design. Finally, some key examples of operating plants are presented that deal with a range of different and complex ore feeds. The detailed design of the complete carbon adsorption, desorption, and gold recovery circuits used in modern plants is covered in Chapter 10.12, "Carbon-in-Leach and Carbon-in-Pulp Circuit Design."

BACKGROUND THEORY

Activated carbon is produced through the high-temperature activation of a carbonaceous source, producing an extremely high surface area, porous product. The most common activated carbon used in gold processing applications is derived from coconut shell typically having a BET surface area (internal and external surface area measured by gas adsorption according to the method of Brunauer, Emmett, and Teller) of 900–1,000 m^2/g (Marsden and House 2006). Several good publications outline the theory of carbon adsorption of gold cyanide; examples recommended for further reading include Marsden and House (2006), Staunton (2005), and Fleming (2002). Hence, only a summary is presented here.

Adsorption and Desorption of Gold

Although there has been debate in the literature, it is now widely accepted that gold cyanide adsorption onto activated carbon occurs via an ion-pair mechanism, which explains the increased affinity for gold in the presence of cations commonly present in cyanide leach solutions such as Ca^{2+} (McDougall et al. 1980). The adsorption reaction is also reversible and exothermic, with the loading capacity decreasing with increasing temperature and free cyanide concentration as well as with decreasing ionic strength. These factors are key in the design of elution circuits, whereby high temperatures and either high-cyanide or low-ionic-strength solutions are adopted to strip or desorb the gold from carbon.

Kinetic and Equilibrium Considerations

The loading of gold onto carbon, a heterogeneous reaction, occurs by several key reactions steps:

1. Mass transport of the gold cyanide complex (or ion pair) from the bulk solution to the outside carbon. This reaction is first order and dependent on the mass transfer coefficient, which is primarily governed by agitation.

Matthew Jeffrey, Process Manager, Newmont USA Limited, Englewood, Colorado, USA

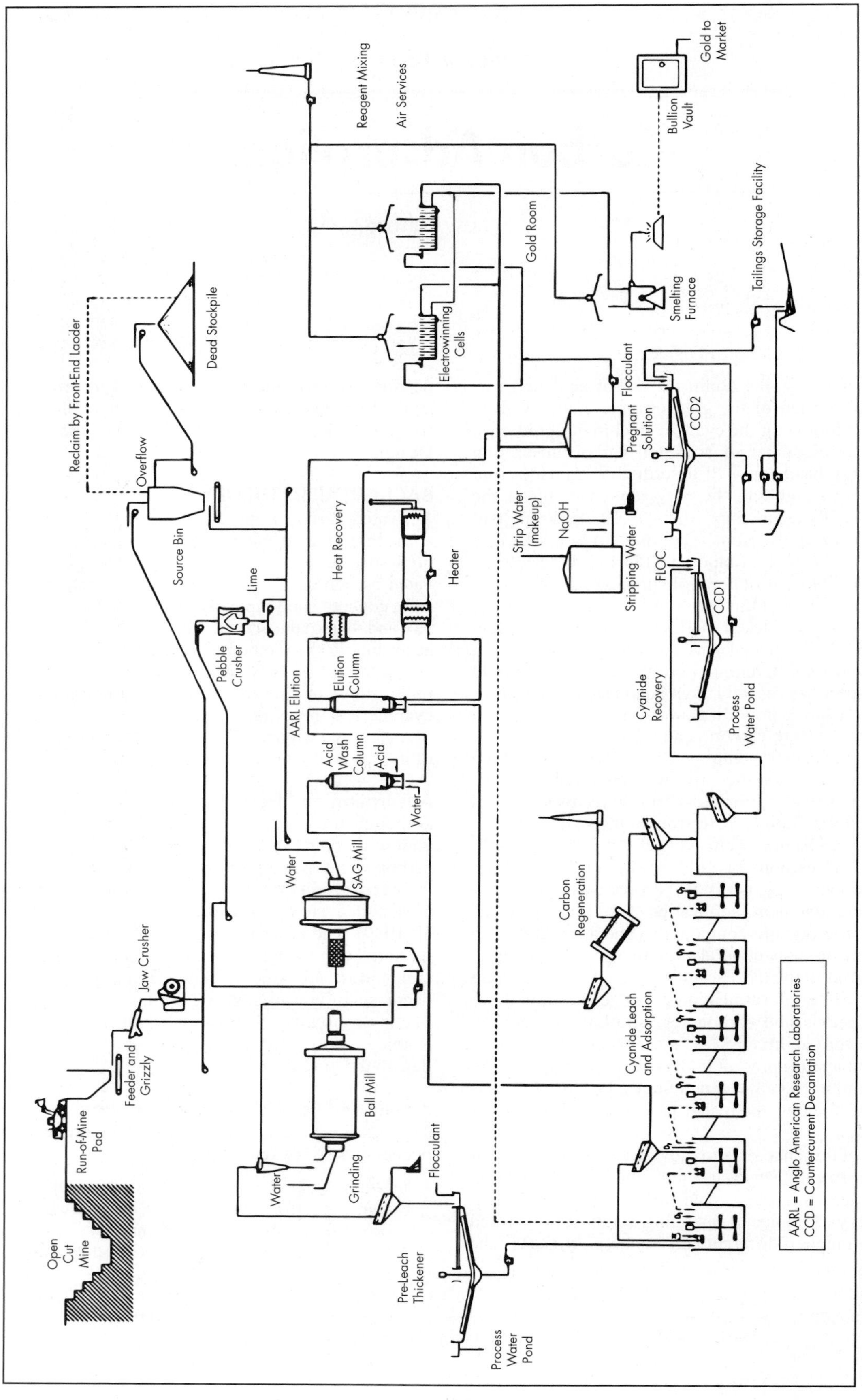

© Newmont USA Limited

Figure 1 Block flow diagram of a typical gold processing circuit using carbon adsorption

2. Adsorption of the gold complex onto the carbon surface, which is a chemical reaction for which the reaction rate is also first order with respect to gold concentration.

These two steps account for the loading of gold around the perimeter of the carbon particle. However, after the outside surface sites of the carbon are occupied, internal surface sites must be used for further adsorption, either by the solution or surface diffusion of gold down into the small-diameter pores. This step is slow in comparison to the first two steps, which account for loading of gold on the outside of the carbon. X-ray tomography images have shown that at 2,500 g/t gold loading, the adsorption is primarily isolated to the outside surface of the carbon to a depth of nominally 200 µm (Pleysier et al. 2008). Vegter and Sandenbergh (1996) have argued that the internal structure of the carbon becomes loaded via the surface diffusion mechanism, which was based on a comparison of mass transfer rates for surface diffusion versus solution diffusion down very narrow pores. X-ray tomography at high loadings (25,000/t) is consistent with this theory, as the internal structure was very evenly loaded and independent of the differences in carbon pore sizes.

The adsorption of gold on carbon ultimately reaches an equilibrium loading, for which the gold loaded on carbon ($[Au]_C$) is a function of the concentration of gold remaining in solution ($[Au]_{aq}$), and the empirical Freundlich isotherm shown in Equation 1 has been demonstrated to represent the equilibrium very well:

$$[Au]_C = K[Au]_{aq}^{\,n} \qquad \textbf{(EQ 1)}$$

where K and n are the Freundlich isotherm parameters. At 1 mg/L gold in solution, the gold on carbon is K g/t.

Modeling of Carbon Adsorption

The well-known Nicol–Fleming model (Nicol et al. 1984) has been used for modeling gold adsorption for several years. The model is based on the adsorption rate being a steady-state first-order reaction that approaches an equilibrium constraint as described by a Freundlich isotherm, as shown in the following equation:

$$d[Au]_{aq}/dt = k[C]([Au]_{aq} - [Au]_E) \qquad \textbf{(EQ 2)}$$

where

k = rate constant (mass transfer coefficient) for adsorption
$[C]$ = concentration of carbon, g/L
$[Au]_E$ = equilibrium solution gold concentration at the equilibrium gold loading on carbon

Substituting for the isotherm and applying to a well-mixed continuous stirred tank reactor (CSTR), the following equation is obtained:

$$[Au]_0 - [Au]_{aq} = k\tau\,[C]([Au]_{aq} - K[Au]_{aq}^{\,n}) \qquad \textbf{(EQ 3)}$$

where

$[Au]_0$ = gold in solution concentration available for adsorption, which includes the gold leached in the contactor
τ = retention time of the contactor

This is combined with the mass balance for gold, resulting in two equations with two unknowns ($[Au]_{aq}$ and $[Au]_C$). The model is simple to implement and has been shown to represent CIL/CIP plants very well, particularly at the lower gold loadings obtained in most processing operations. The model is also the basis of the SIMCIL model discussed by Staunton (2005) and used in the AMIRA P420 project (a collaborative industry project coordinated by AMIRA International, which commenced in 1990 and was still running at the time of publication). The SIMCIL model is a steady-state mass balance that incorporates the CSTR adsorption model into a web-based platform. To improve the model, Dai et al. (2010a) have published corrections to the equilibrium isotherms based on the impact of ionic strength of the solutions and the competitive adsorption of copper, which can have a significant impact on the gold loading. More sophisticated dynamic and population balance models have been developed that account for the periodic nature of carbon transfer in processing plants (Stange et al. 1990), but these have not found widespread use by operators.

Co-Adsorption and Fouling

Carbon adsorption operations become more complex when the feed

- Contains high silver that needs to be recovered,
- Contains high levels of cyanide-soluble copper,
- Is a flotation concentrate or tails that contain flotation reagents (organics), or
- Contains other carbonaceous matter that contributes to the phenomenon known as preg-robbing.

The operation of such facilities is described later in this chapter, and the background theory for these more complex feeds includes the following:

- The adsorption of the silver cyanide complex formed when leaching silver from the ore is significantly weaker than observed for the gold cyanide complex. This is one reason the Merrill–Crowe process is still used for many high-silver applications. If carbon adsorption is to be used, higher rates of carbon transfer are required for effective operation, resulting is larger elution circuits. This is a second reason for considering Merrill–Crowe over carbon alternatives. Modeling of gold–silver circuits has met with mixed success because of the dependence of silver loading on gold loading, as gold can displace silver on carbon at higher loadings.
- The different complexes of copper (the di-, tri-, and tetra-cyanocopper species) have an affinity for carbon that decreases as the coordination number increases. For this reason, carbon may be used to adsorb the $Cu(CN)_2^-$ complex if desired, but in gold operations the cyanide strength is usually maintained such that the $Cu(CN)_3^{2-}$ and $Cu(CN)_4^{3-}$ complexes are dominant to minimize the adsorption of copper. Modeling of these circuits is even more complex than gold–silver circuits, as demonstrated by Dai et al. (2010b) in the development of a mechanistic copper cyanide loading model.
- Organics, such as flotation frothers and collectors, readily adsorb onto activated carbon, which results in "fouling" of the carbon (La Brooy et al. 1984). Therefore, higher rates of carbon transfer are usually targeted when feeding a CIP or CIL circuit with flotation concentrates or tails, resulting in lower gold loadings. Under such exposure to organics, maintaining the correct carbon reactivation conditions becomes paramount to ensure organic degradation.

- Preg-robbing ores are very common and the key consideration when processing is to ensure that the gold is adsorbed onto carbon as quickly as possible after leaching by maintaining a high carbon concentration in the tanks and typically using low carbon loadings. Dunne et al. (2013) have presented a detailed summary of a range of operations treating preg-robbing ores.

METALLURGICAL TESTING AND MODELING

The following sections deal with the key metallurgical working, including laboratory testing and modeling, which are recommended for the design of new carbon adsorption circuits or for characterization/optimization of existing circuits.

Laboratory Testing for New Circuits

In designing a carbon circuit, three key parameters need to be determined:

1. The gold on loaded carbon, which sets the carbon transfer rate and size of the elution circuit.
2. The gold in barren solution, which is termed "solution loss" and contributes to the economics of the project.
3. The carbon concentration in the tanks, which determines the gold hold-up on carbon in circuit and the required carbon inventory.

Unfortunately, it is very difficult to determine all of the preceding design criteria from laboratory testing alone, for the following reasons:

- The carbon and slurry (or solution) in a plant move countercurrently, with the retention time of carbon usually 10–15 times that of the slurry.
- The mass transfer coefficient for adsorption is important in determining the kinetics of gold adsorption on carbon, and at bench scale it is usually significantly higher than that observed in a full-scale plant.
- Standard CIL/CIP bottle roll tests conducted with typical 10-g/L carbon produce loadings that are only a fraction of that observed in a plant. For example, a 1-g/t gold ore with 90% of the gold being cyanide extractable will produce loaded carbon of around 48 g/t at 40% solids. This is closer to the conditions normally seen in the back end of a CIL/CIP plant.
- Virgin carbon, which is almost always more active than plant carbon, is usually used in laboratory tests.

To address some of these issues, tests will often be conducted with various carbon concentrations, which will produce different points on an isotherm of gold on carbon versus gold in solution. This method is often used to identify whether fouling of the carbon is an issue, as evidenced by a poor isotherm plot. The downside to this method is that if any preg-robbing material is present in the ore, the data will be unduly impacted, because the tests are carried out at low carbon concentrations. If this is the case, the staged contact approach is often adopted, whereby several batches of slurry are leached consecutively with the same batch of carbon to slowly load the carbon at the expected or design carbon concentration (as occurs from the back to the front of a full-scale carbon circuit). Obviously, this approach entails a significant quantity of work and a higher demand of sample for each sample tested, and it is usually used only if the simpler testing shows issues with loading.

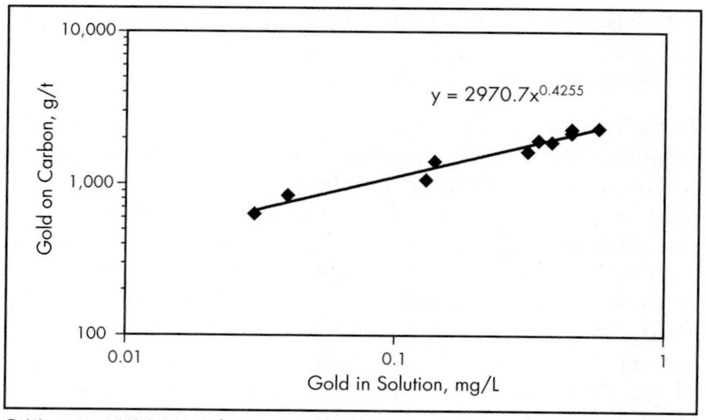

© Newmont USA Limited

Figure 2 Laboratory-determined gold loading isotherm for a greenfield project

One word of caution: Although all the testing methods give good information on adsorption isotherms, they do not represent the carbon adsorption kinetics well because of mass transfer differences between laboratory equipment and a full-scale plant. Thus, modeling and relevant benchmarking are highly recommended for supplementing the design of a carbon circuit, with information on isotherms obtained from testing providing an input to the model. In addition, the determination of laboratory isotherms using solutions alone should only be used in CIC applications; that is, testing for CIL/CIP must use slurry, as this does impact the adsorption isotherm. In addition, because carbon adsorption is dependent on the solution constituents and ionic strength, testing solutions should represent as much as possible the likely process solution.

An example where carbon adsorption testing was conducted on a greenfield project is shown in the data in Figure 2, the adsorption isotherm for two of the key ore types overlaid. In this case, the isotherm is much flatter than usually observed, possibly because of the saprolite ore containing fine clays. Even without detailed modeling, this data can be used to determine a starting point for the loaded gold on carbon. At the design percentage solids and retention time, the pregnant solution feeding the first carbon contactor contains 0.6 mg/L gold. A good rule of thumb is that the first contactor should load 30%–40% of the gold. For this example, if we want to load 40% of the gold in the first CIL tank, the gold solution grade in the first contactor would be 0.36 mg/L, with an equilibrium gold on carbon from Figure 2 of 1,923 g/t. The actual gold on carbon would be less than this, as equilibrium is not reached within the contactor because of the short retention time of 2–3 hours (in CIL), as compared to the 72-hour retention time used to obtain the equilibrium isotherm. Based on the rate equation for adsorption in a well-mixed CSTR, the loaded gold on carbon can be estimated, as shown in Equation 3.

For a simple CIL circuit with good carbon activity, a mass transfer coefficient $k = 0.1$ L/g/h has been found to model the adsorption kinetics well from an extensive database of plant surveys from CIL circuits. At a 2-hour retention time per contactor, and 10 g/L carbon, the value of $kt[\mathrm{C}] = 2$, that is, $[\mathrm{Au}]_{\mathrm{C}} = 1{,}728$ g/t. However, in this example the gold isotherm indicates poorer adsorption performance, and the actual laboratory kinetic testing gave $k = 0.1$ L/g/h, indicating that the plant k will be lower, because in a full-scale CIL plant the mass transfer coefficient is always less than that obtained in a very well-agitated laboratory reactor. Therefore, in this example, choosing $k = 0.05$ L/g/h, the gold on loaded carbon is calculated to be 1,200 g/t from Equation 3, which is a good starting point for the design of the circuit. For this project, after extensive modeling and optimization, a loaded carbon grade of 1,150 g/t was planned. An estimate of gold on loaded carbon enables the preliminary sizing of the carbon transfer and hence elution/regeneration circuit.

Laboratory Testing for Plant Performance

Testing of plant carbon properties is important to understand the performance of the plant carbon circuit. It is important to consider the impact that an operating environment has on these same properties when using test data for plant design and optimization. The types of tests adopted are well covered by Staunton (2005) and involve measuring the rate of adsorption (k) of gold from a solution using a reproducible laboratory experimental setup. Similar to the earlier discussions in this chapter concerning the mass transfer coefficient, the rate constant obtained is somewhat meaningless, which is why the percentage carbon activity is reported as the rate constant for the carbon being tested divided by the rate constant for fresh carbon tested under exactly the same experimental conditions. Staunton (2005) also outlines other carbon characterization tests that have been used in the past, such as abrasion resistance and iodine number. These, however, are not routinely used in industry but are more important to carbon suppliers.

Carbon Adsorption Circuits

The conventions used in distinguishing CIP from CIL in this chapter are as follows:

- **CIP:** A circuit for which leaching is essentially complete prior to contact with carbon. Usually, the retention time of solids in each carbon contactor is low, resulting in smaller tanks than used for leaching. For example, the Kemix pump cell contactors may have as low as 15 minutes per stage retention time (Dippenaar and Proudfoot 2005). To compensate for short retention time of solids, a higher carbon concentration in each contactor is used (typically >20 g/L, and up to 50–60 g/L). The product of retention time and carbon concentration ($C\tau$) is typically as low as 15–20 hours g/L.
- **CIL:** A circuit for which leaching is occurring during contact with carbon. Usually, the retention time of solids in each contactor is high to allow complete leaching, resulting in large carbon contactors (often the same size as leach tanks). Typically, lower carbon concentrations in each contactor are used (<20 g/L). A very common CIL configuration is two leach-only tanks and six CIL tanks. The mass transfer coefficient in CIL is usually lower than CIP, with $C\tau$ needing to be higher (e.g., 30–40 hours g/L) to compensate.

During early-stage studies for gold plants, often the starting point for scoping the carbon adsorption circuit is estimating the rate at which carbon is taken out of the circuit, stripped, reactivated, and placed back into the circuit, which is determined from the gold on loaded carbon. It is common to use a benchmarked "upgrade ratio" in choosing a suitable gold on loaded carbon, where the upgrade ratio is concentration of gold on carbon divided by the gold in solution feeding the first carbon contactor. This method is only recommended as a starting point, with detailed modeling required to support the detailed design (discussed in Chapter 10.12, "Carbon-in-Leach and Carbon-in-Pulp Circuit Design"). For CIL applications, the upgrade ratio should also consider the gold that leaches in that stage. It is also common to report upgrade ratios as gold on carbon divided by gold in the feed ore, but this approach depends on the percentage of solids in the contactors and the gold extraction from ore to solution. Hence in benchmarking, it is better to convert to an equivalent pregnant solution grade to calculate the upgrade ratio. A typical CIL plant is usually designed with an upgrade ratio around 1,500–3,000. In the greenfield site discussed earlier, the upgrade ratio was 1,920. Obviously, if the modeling produced an upgrade ratio outside of the range of most commercial plants, more work would be warranted.

Mass transfer is a very important aspect of the design of a commercial plant, and in practice, the power input into agitation to maintain solids in suspension (given the slurry density and viscosity) effectively fixes the mass transfer coefficient, unless a more vigorous agitation system is used, such as the Kemix pump cells often used in CIP plants (see, e.g., Rogans and McArthur [2002]). Thus, the key driver for increasing the adsorption rate if the number of stages is fixed is to increase the carbon concentration in the contactor, as shown in Equation 3. When designing a plant, the objective is to optimize the carbon concentration so that the adsorption kinetics are fast enough to produce a low gold in barren solution. This is often a trade-off between the number of contact stages (capital cost) and the carbon concentration, with lower carbon (and hence gold) inventory achieved with more stages.

The physical aspects of CIL/CIP design, such as tank agitation, interstage carbon transfer, and interstage screening, are described by Altman et al. (2002) and Altman and McTavish (2002) and discussed in Chapter 10.12, "Carbon-in-Leach and Carbon-in-Pulp Circuit Design."

CIC Circuits

Kappes (2002) reported that in a survey of 28 CIC operations, the average pregnant leach solution (PLS) grades were 0.7 mg/L gold, and average loaded gold on carbon was 3,900 g/t. This corresponds to an upgrade ratio on average of 5,570, which is significantly higher than that obtained for CIL applications. Ultimately, the upgrade ratio used will depend on the desired barren solution grade that is recycled back to the leach pad, the number of CIC contactors in series, and the activity of the carbon (i.e., fouling and regeneration frequency). Similar to CIL and CIP, detailed modeling will be used for the design. Most commonly, four to six contactors are used. For example, at the Newmont Carlin operations, six CIC contactors are used in series, with two to three parallel trains, depending on the PLS solution flow.

Modern CIC plants are usually composed of standard off-the-shelf columns to accommodate the correct solution flow rate. This is an important factor, as a well-chosen column will produce some fluidization and bed expansion of carbon in the column. Bed expansion/fluidization should not too be so vigorous as to cause issues with plugging of overflow screens if present, or carryover of carbon if not. A typical maximum value of 30% bed expansion can be used to provide conditions with good mass transfer, with Ritson (2013) showing correlations for bed expansion with upward flow velocity as a function of carbon particle size, and Kappes (2002) showing the bed expansion as a function of temperature. The packed bed density is typically around 0.5 t carbon/m^3, which can be used to determine the carbon inventory in each contactor based on the column height and allowing for bed expansion. This information will usually be provided by the CIC manufacturer or engineering company and enables the carbon retention time per contactor to be determined, with two to three days per contactor often being regarded as a reasonable maximum to maintain carbon activity and very high upgrade ratios. For example, at one of the historical Newmont operations, an upgrade ratio of around 11,000 was obtained from a feed of nominally 0.6 mg/L gold, with six columns containing 1.8 t carbon each. Carbon was transferred at 1 t/d, giving around 1.8 days per contactor carbon retention time.

CIL/CIP Circuits

Details of designing CIL/CIP circuits are provided in Chapter 10.12, "Carbon-in-Leach and Carbon-in-Pulp Circuit Design"; however, an example of the importance of modeling and carbon adsorption parameters for predicting performance and optimizing CIL/CIP circuits should not be overlooked. To demonstrate the power of modeling, the following example has been developed.

A simple oxide mill/leach/CIL circuit with 50,000 t/d and 1 g/t gold has been designed. The feed to the circuit is 50% solids, and no preg-robbing was observed during the testing campaign, so a simple circuit with three leach tanks and seven adsorption tanks has been selected. A 20-t elution column has been selected, giving a loaded carbon grade of approximately 2,000 g/t. The barren carbon is expected to contain 40 g/t gold, which is typical for a well-operated elution circuit. All the tanks are 5,000 m^3 each to provide the required retention time required to leach all the cyanide-extractable gold. In this example, standard adsorption parameters (for the isotherm and adsorption rate based on the database available of operating simple oxide circuits) have been used for demonstration purposes, and the calculated barren solution grade is 0.012 mg/L at a carbon concentration of 13.7 g/L for each tank. Benchmarking this example to other similar operations and ongoing adsorption testing and circuit profile sampling is required to validate the CIL modeling, especially because ore-type changes can have a significant impact on performance, as discussed in the earlier example where clays impacted the adsorption of gold.

The impact of operating parameters on the predicted solution tail are shown in Table 1. In the first three cases, the number of CIL stages was decreased, but the total carbon inventory remained the same (with higher carbon concentrations for the cases with fewer contactors). The total leach retention time was kept constant by increasing the number of leach tanks when decreasing the number of CIL stages. The model shows essentially no difference in barren solution grade with number of stages of contact, which is consistent with industry practice that has shown that six operational CIL stages are adequate to achieve low barren gold grades. However, the example with six CIL contactors has slightly higher gold inventory because of more carbon in the first contactor (highest loaded carbon grades). In the next two cases with seven CIL stages (the base design), the model clearly shows that the barren solution grades are a function of carbon concentration (inventory) because of the faster adsorption at higher carbon concentrations, but this improvement comes at a cost in terms of the gold in circuit inventory. The last case shows one of the benefits of having seven adsorption tanks when treating an ore at twice the average feed grade. Because the elution frequency usually cannot be varied significantly for an operating plant, higher loaded carbon grades are required, resulting in higher solution losses. In this case, with seven CIL stages and 20 g/L carbon in each stage, the solution loss can be maintained at 0.012 mg/L. A plant with six CIL stages would struggle under these conditions. Therefore, in designing a CIL/CIP circuit, consideration of the variability in the mill feed over the life of mine is always required.

Table 1 Modeled variation of gold in barren solution for a typical CIL plant

CIL Variable	Barren Solution Grade, mg/L
12 g/L carbon, 8 CIL contactors	0.012
13.7 g/L carbon, 7 CIL contactors	0.012
16 g/L carbon, 6 CIL contactors	0.013
20 g/L carbon, 7 CIL contactors	0.008
10 g/L carbon, 7 CIL contactors	0.019
20 g/L carbon, 7 CIL contactors, 2 g/t feed grade, same elution frequency	0.012

Elution/Regeneration Circuits

The design of elution and reactivation circuits is also very important because the activity and gold concentration on the barren carbon both have a significant impact on the solution tails grade. Further details on the design of elution and reactivation facilities can be found in Von Becknamm and Semple (2002), Hosford and Wells (2002), and Chapter 10.12, "Carbon-in-Leach and Carbon-in-Pulp Circuit Design."

PLANT PRACTICE

The following sections present operation data from carbon adsorption circuits with varying levels of complexity.

CIC Circuits

Much of the data available on CIC circuit performance is based on heap leach solutions with high gold grades and very good upgrade ratios. The two examples that follow are somewhat different. The first is a heap leach where carbon is not regenerated/reactivated, and hence lower carbon activities result. The second example represents the growing trend with heap leach in that lower-grade ore is placed and leached, with the higher-grade ore feeding a mill, and thus the pregnant solution grades seen in the feed to CIC are very low.

In the first example, Figure 3 shows the upgrade ratio obtained for gold and silver (over two separate months of operation; the left graph is one month and the right graph is a separate month) for the CIC circuit where carbon is not regenerated/reactivated. An additional complication for this circuit is that the pregnant solution also contains variable copper, and thus the upgrade ratio is low because of high carbon movement to compensate for both the low activity and the copper on loaded carbon. The data shown in Figure 4 is the gold and silver recovery from solution over the same period as the data shown on the right side of Figure 3; the gold recovery is typically around 90%, with the barren gold solution grades ranging from 0.07 to 0.1 mg/L. The silver recovery is lower, as its adsorption is weaker than gold. This example illustrates very

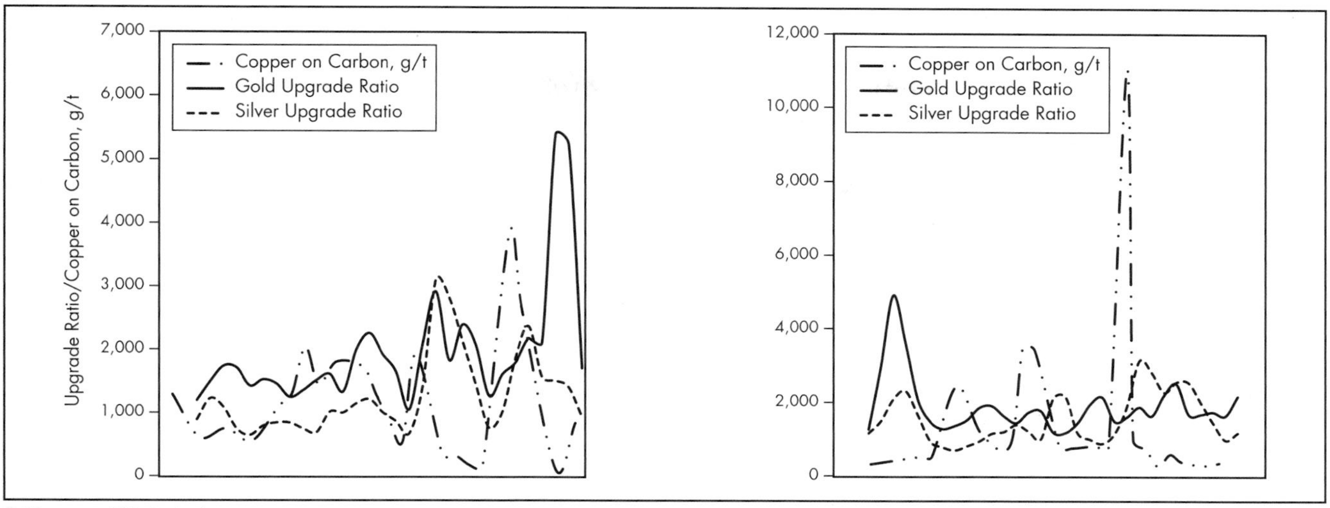

© Newmont USA Limited

Figure 3 CIC circuit gold and silver upgrade ratios for a circuit without carbon regeneration/reactivation and variable copper in the pregnant solution

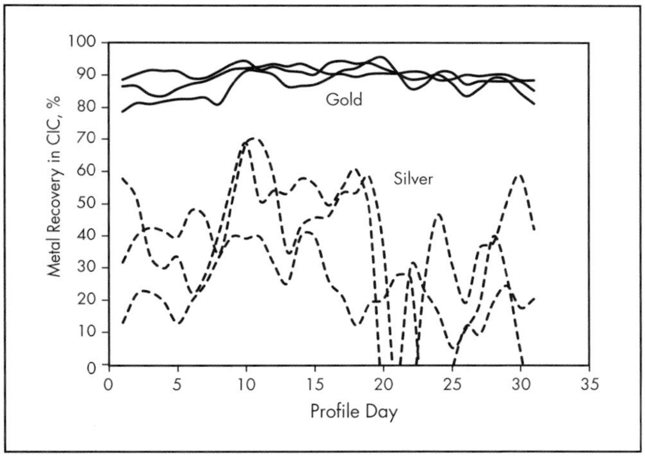

© Newmont USA Limited

Figure 4 Gold and silver CIC solution recoveries (three separate CIC trains) for a circuit without carbon regeneration/reactivation

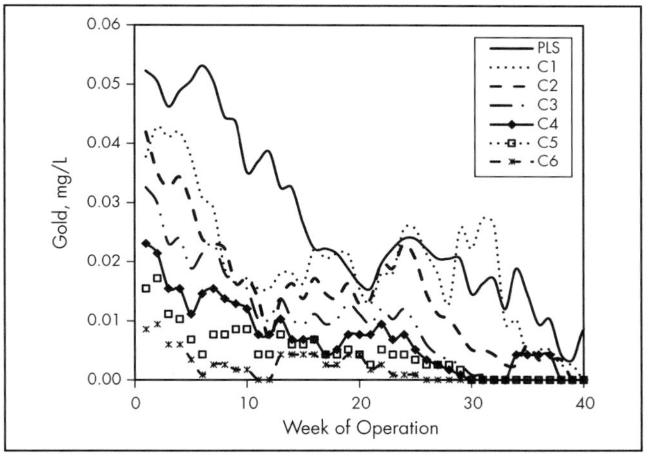

© Newmont USA Limited

Figure 5 Low-grade heap leach CIC solution profiles

well the impact of low carbon activity, with lower than typical CIC recoveries at lower upgrade ratios. However, other factors preclude carbon regeneration for this site, and the copper in the feed makes this circuit even more complicated to operate. Despite this fact, the circuit still recovers gold and silver, with the metal values in the barren solution being recycled to the heap leach, and hence remaining in inventory. This example highlights the robust nature of the CIC technology.

The second example is an oxide heap leach, and Figure 5 shows the CIC solution profile samples for nearly one year of operation (C1 is the gold solution concentration exiting CIC contactor 1). For this operation, the carbon activity is high because of regeneration/reactivation in the mill facility. Clearly, CIC is an exceptional technology; even as the PLS grade decreases to 0.01 mg/L gold, the adsorption efficiency remains very high. This is also demonstrated in the upgrade ratio (Figure 6), which remains high despite the decrease in PLS grade. Only the first 15 weeks are shown, as assay accuracies at <0.02 mg/L gold distort the data.

Simple CIL/CIP Circuits

Simple CIL/CIP circuits are relatively straightforward to design and operate, with significant benchmarking and historical knowledge in the industry. For example, the profiles for gold-in-solution and gold on carbon are shown in Figures 7 and 8, respectively, for an oxide mill processing around 2.8 g/t gold ore at 48% solids. The loaded carbon for the three separate surveys averages 4,170 g/t, and the final gold in solution averages 0.001–0.002 mg/L with nine stages of CIL. Note that the gold-in-solution graph shows 10 stages, because the grinding and preleach thickener contains cyanide, and hence stage 1 represents some leaching prior to the CIL circuit. This is a well-designed and operated circuit, producing high upgrade ratios and very low gold loss to tails.

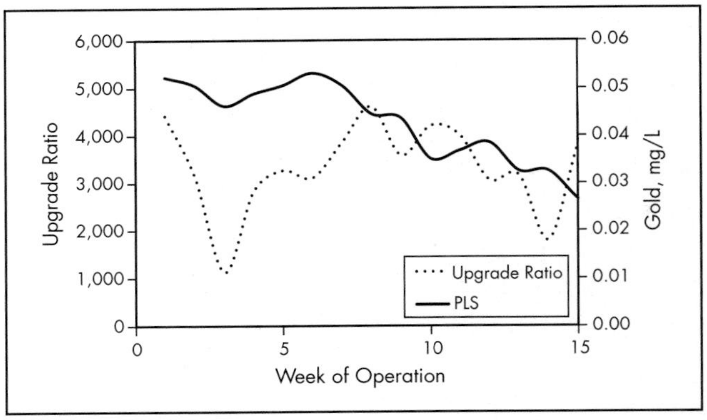

© Newmont USA Limited

Figure 6 Low-grade heap leach CIC gold upgrade ratio

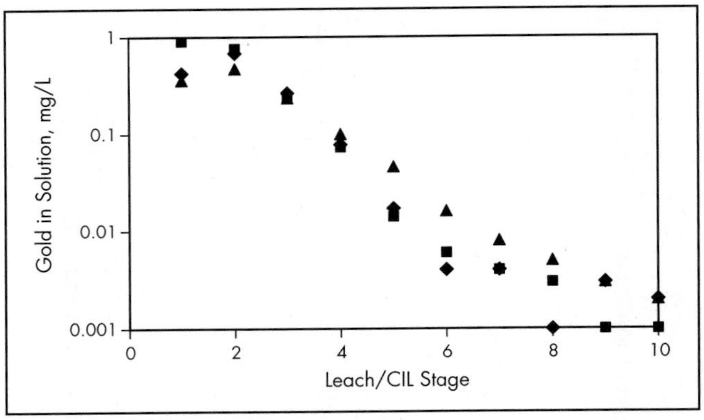

© Newmont USA Limited

Figure 7 Oxide mill CIL application gold in solution profiles (nine stages of CIL; different symbols represent circuit profiles taken at different times)

The graph in Figure 9 shows the gold on carbon versus gold in solution, along with the equilibrium isotherm and the modeled data for the plant (using a spreadsheet model). This demonstrates how powerful the modeling of the CIL circuits can be, and gold metallurgists should possess this capability when dealing with CIL/CIP plants. Circuit survey data can be used to continually update the model parameters and can be vital in understanding the potential impact of plant changes. For example, in one operation, CIL modeling has been used to justify the capital for an additional stage of CIL, and in another, the implementation of oxygen in the first leach tank has been used to increase the leach rate and hence gold on loaded carbon in the first carbon contactor.

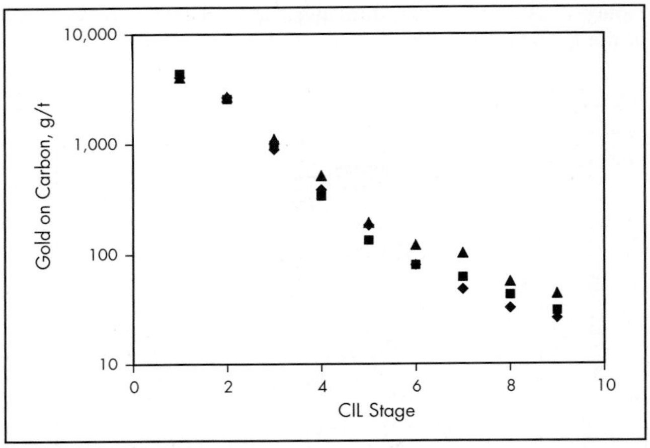

© Newmont USA Limited

Figure 8 Oxide mill CIL application gold on carbon profiles (nine stages of CIL)

High-Silver CIL/CIP

High-silver ores are not normally processed by CIL/CIP circuits as high-carbon transfers are required and adsorption is not as effective as it is with gold. Silver ore processing is more typically undertaken using countercurrent decantation or filtration for solid–liquid separation, followed by zinc cementation of the precious metals from the pregnant liquor (Merrill–Crowe). Such flow sheets are the preferred technologies when the Ag:Au ratio exceeds 10.

Profile data follows for a CIL operation during a two-month period that processes a gold ore containing higher silver. The first four tanks are leach only, and Figures 10 and 11 show the solution profiles for gold and silver, respectively. Figures 12 and 13 show the carbon profiles for gold and silver, respectively. The average gold grade feeding the first adsorption tank is 1.15 mg/L, with a maximum of 1.5 mg/L. For silver, the average and maximum grades were 8.6 mg/L and 16.7 mg/L, respectively, and hence this plant is operating close to the recommended maximum of Ag:Au = 10 for carbon adsorption. Despite this, the plant still operates well, with gold in solution tails around 0.01 mg/L and silver in tails in the range of 0.3 to 0.8 mg/L, that is, ~95% silver recovery to carbon. Obviously, maintaining good carbon activity is key to the successful operation, as the loaded metal is extremely high. Taking into account the difference in atomic mass of silver to that of gold, the maximum loadings seen in Figures 12 and 13 are equivalent to carbon loaded with 14,700 g/t gold. Another important consideration is the additional stripping and electrowinning capacity required to handle the very high total precious metal movements.

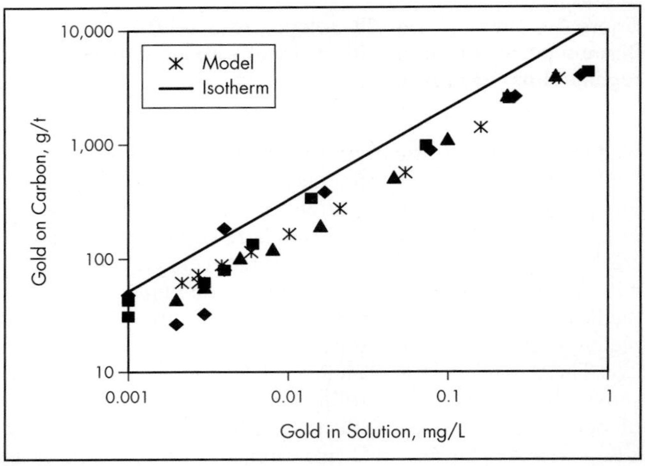

© Newmont USA Limited

Figure 9 Oxide mill CIL application gold loading equilibrium isotherm, plant data, and CIL model data (different symbols represent different sampling campaigns)

Because the carbon from this plant is dominantly loaded with silver, an isotherm for silver loading can be constructed from the profile data (which also include operational data from a separate window of processing high-silver ore). Figure 14 shows that the Freundlich-type model for silver loading is $2{,}066[Ag]_{aq}^{0.727}$. The actual equilibrium loadings may be higher. However, these data provide an estimation of the maximum loadings obtainable in a CIL/CIP circuit for high-silver applications. In this operation, very little copper or other species affect the adsorption of metals, so when designing for CIL/CIL applications with this complexity, testing is critical to define the expected metal loadings.

High-Copper CIL/CIP

For facilities processing high-copper ores, the key to successful operation is to minimize the copper on loaded carbon. The best way to accomplish this is to operate the circuit with high free cyanide, which can be more difficult than anticipated, especially when the copper feed grades in the ore vary and the cyanide solubility of the copper varies as a function of variable mineralogy. A common complication is that the cyanide addition is limited by cyanide code compliance (e.g., 50 ppm weak acid dissociable [WAD] cyanide at the spigot discharge), detox capacity, or even cyanide supply constraints. This often leads to periods of operation with low free cyanide, which results in very good loading of copper.

An example of a high-copper feed is a flotation plant that leaches the cleaner scavenger tail, a higher copper/gold stream with variable copper grade and mineralogy, which is dependent on the flotation performance. Figure 15 shows circuit profile information including the copper in solution in the first carbon contactor, the copper on loaded carbon, and the free cyanide concentration. The free cyanide had been

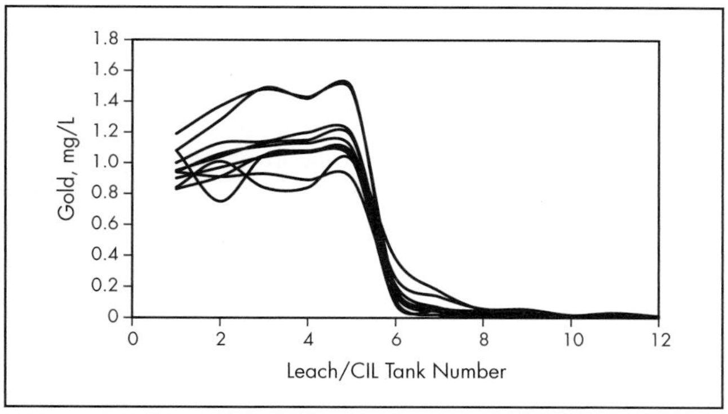

Figure 10 High-silver mill CIL application gold in solution profiles (seven stages of CIL)

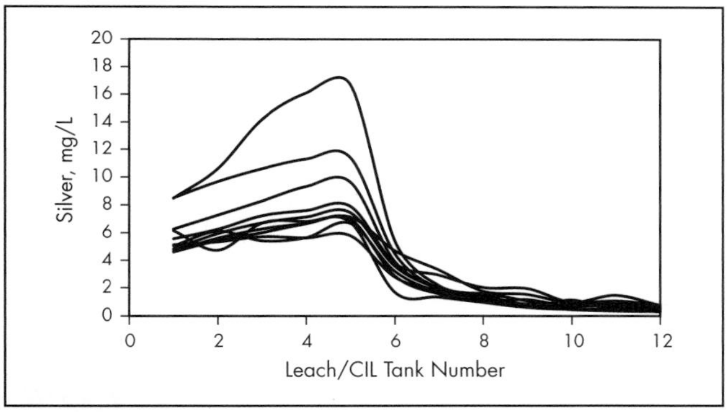

Figure 11 High-silver mill CIL application silver in solution profiles (seven stages of CIL)

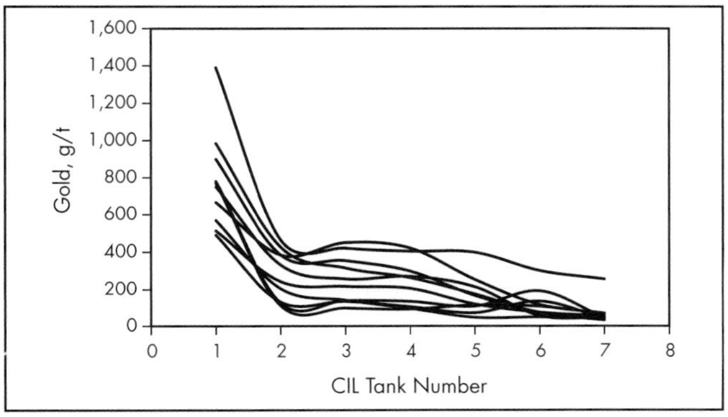

Figure 12 High-silver mill CIL application gold on carbon profiles (seven stages of CIL)

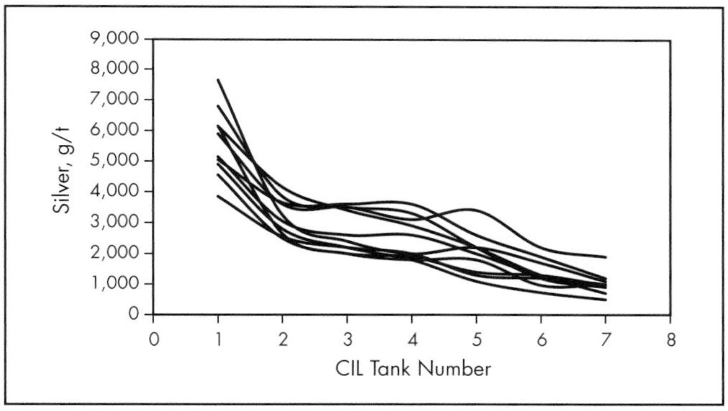

Figure 13 High-silver mill CIL application silver on carbon profiles (seven stages of CIL)

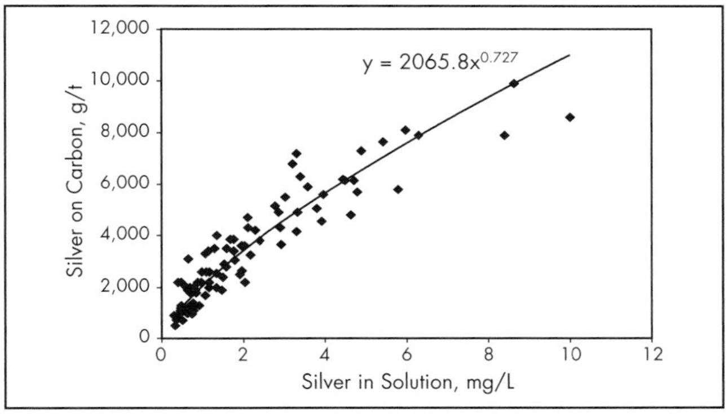

Figure 14 High-silver mill CIL application silver loading isotherm from plant data (i.e., not at equilibrium)

measured using titration, and also calculated from the copper and a WAD cyanide measurement (using a flow injection–amperometric detection analyzer). This plant operates very well with copper loadings usually less than 1,000 g/t for copper in solution averaging around 400–500 mg/L. However, when the free cyanide drops, there is a large spike in copper loading on carbon.

Figure 16 shows the gold and silver loadings on the activated carbon during the same period. Although this data is somewhat scattered because of the variation in feed grades, the impact on gold and silver loading appears minimal because of the low overall metal loadings on carbon. The low loadings are necessary because of the presence of copper (similar to the CIC application discussed earlier) and the presence of organic flotation reagents that lead to fouling. Figure 17 shows the impact of higher copper loadings obtained during one week of operation with higher copper feed grades and limitations on cyanide addition. Clearly, significantly reduced gold and silver loadings on the carbon are obtained at the very high copper loadings.

Handling of carbon highly loaded with copper can be challenging, and the industry standard practice is to undertake a cold cyanide strip. By exposing the loaded carbon to an environment of high free cyanide, a proportion of the copper is selectively eluted over gold at the lower temperatures. The strip solution can then be treated with a SART (sulfidization–acidification–recycle–thickening) process to precipitate copper as copper sulfide and recycle the cyanide. Often, though, the strip solution is bled back into the leach circuit to make use of the high cyanide, which then relies on good CIL/CIP operation to ensure that this copper ends up going to detoxification. Any copper that goes into electrowinning can also

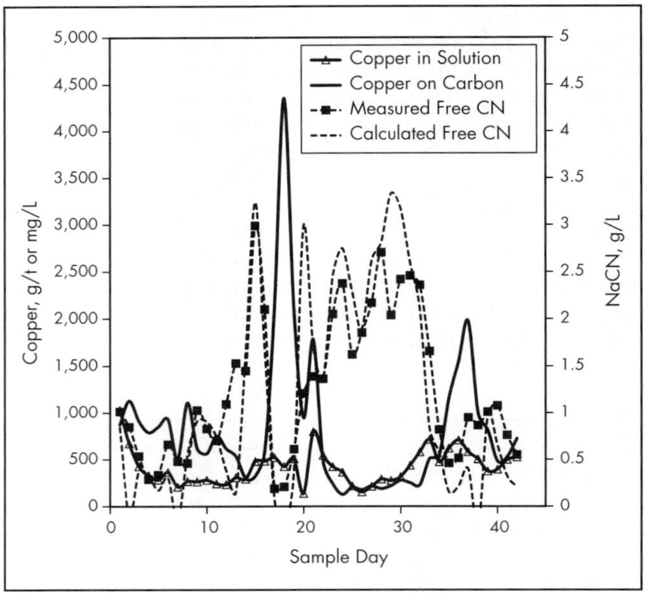

Figure 15 High-copper CIL first-stage copper in solution and on carbon over 45 days of operation. Also shown is the measured and calculated free cyanide in the CIL stage.

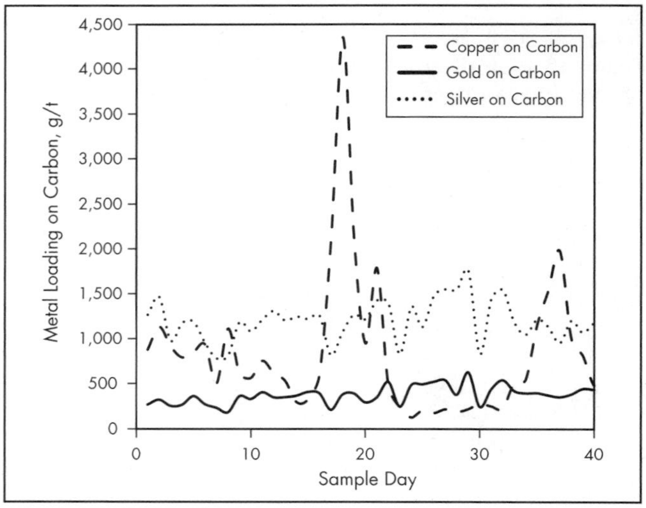

Figure 16 High-copper CIL first-stage copper, gold, and silver on carbon over the same period of operation as shown in Figure 15

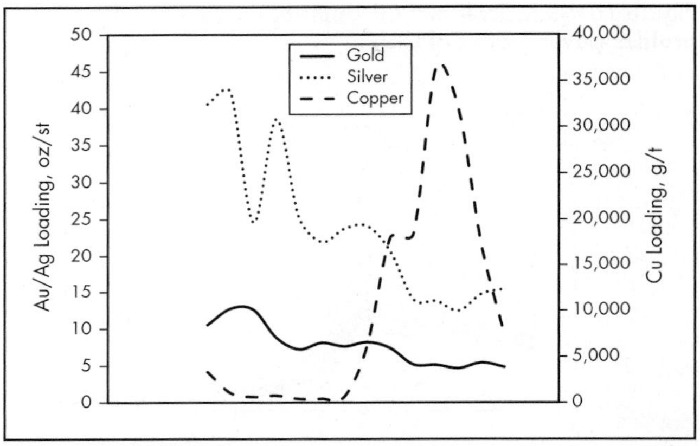

Figure 17 High-copper CIL first-stage copper, gold, and silver on carbon over a separate period of operation with higher copper in feed

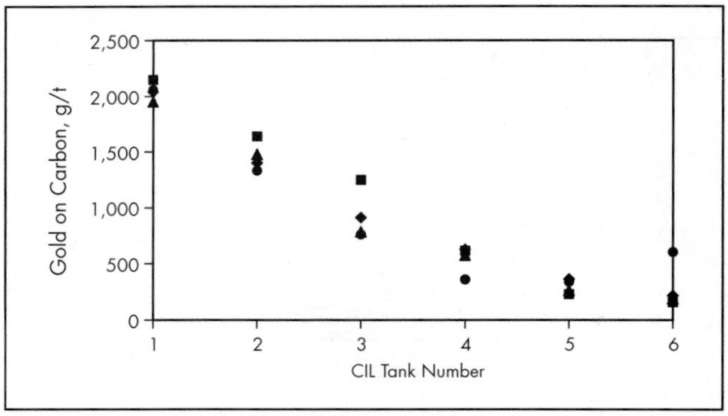

Figure 18 Gold on carbon for a CIL plant processing preg-robbing ore (different symbols represent different sampling campaigns)

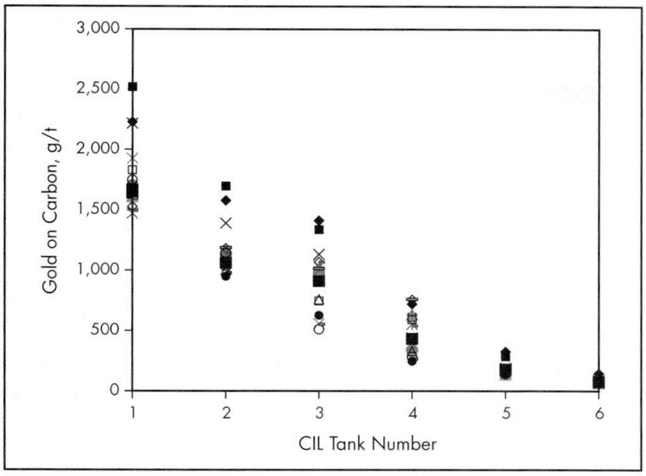

© Newmont USA Limited

Figure 19 Gold on carbon for a CIL plant processing a flotation concentrate (different symbols represent different sampling campaigns)

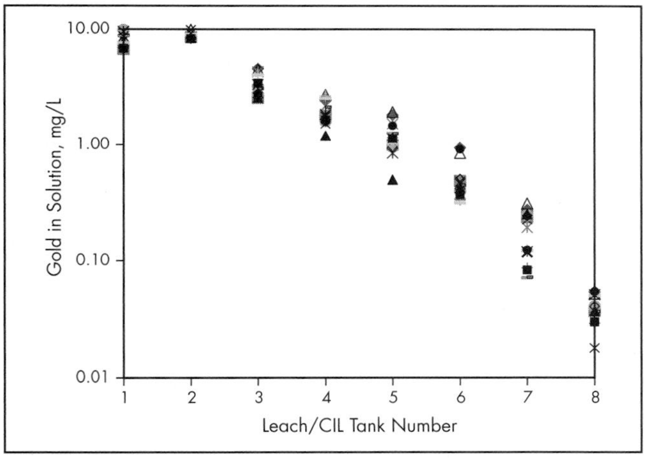

© Newmont USA Limited

Figure 20 Gold in solution for a CIL plant processing a flotation concentrate (different symbols represent different sampling campaigns)

be problematic, and similar to CIL/CIP, high cyanide concentrations can be adopted to minimize the plating of copper. However, some copper does report to cathode, and hence the electrowinning design needs to take this extra plated metal into account.

Preg-Robbing and High-Fouling CIL/CIP Circuits

When processing ores containing carbonaceous material that preg-robs gold from solution (adsorbs the gold cyanide complex by the same mechanism as activated carbon), an important operational strategy is to minimize the gold in solution and maximize the rate of adsorption. This is accomplished by having carbon in all the leach tanks, by reducing the loading of gold on carbon (remembering that according to the isotherm, gold on carbon is a function of gold in solution), and by operating at higher carbon populations. Consequently, the upgrade ratios will be lower than for conventional circuits. Figure 18 shows the gold-on-carbon profiles for an operation treating mildly preg-robbing ore for which the average gold feed grade was 4.6 g/t gold and the average loaded carbon was 2,080 g/t.

The operation of CIL/CIP circuits treating either flotation tails or concentrates, which usually result in high fouling of the carbon, is very similar to preg-robbing circuits in that carbon must be transferred faster, with lower loadings of gold. This is because both the equilibrium isotherm and the carbon adsorption rate (activity) are impacted by the organic reagents (collectors and frothers). The circuit profiles for a flotation concentrate leach circuit for a four-month period are shown in Figure 19 (gold on carbon) and Figure 20 (gold in solution). The circuit consists of two leach and six adsorption tanks, and the average gold in solution feeding the first adsorption tank was 10.3 mg/L, producing a loading of 1,757 g/t gold on average, that is, an average upgrade ratio of only 171 from solution. The tails solution grades achieved average 0.04 mg/L, representing 99.6% gold recovery from solution.

CONCLUSIONS

Carbon adsorption is a very robust technology for recovering gold from cyanide leach solutions, whether it be via CIP, CIL, or CIC processes. This has led to carbon adsorption–based processes being the dominant technology in gold hydrometallurgy, with the traditional Merrill–Crowe process usually adopted only for circuits with very high silver (Ag:Au > 10). The implementation of CIP/CIC/CIL circuits for simple gold ores is relatively straightforward, with significant operating data and modeling available to inform on this application. For complex ores, often lower gold loadings need to be targeted to account for either fouling or co-loading of copper and/or silver. For preg-robbing ores, lower gold loadings result from the requirement to minimize the dissolved gold in solution. The examples provided for treating complex ores provide a basis for understanding the potential impact of these more complex feeds, which are the norm for modern gold processing facilities. These complex ores require significantly more metallurgical work to define the carbon adsorption behavior, and detailed modeling to design the CIL/CIP circuit. Despite the challenging applications, carbon adsorption is an exceptional technology.

REFERENCES

Altman, K.A., and McTavish, S. 2002. CIP/CIL/CIC adsorption circuit equipment selection and design. In *Mineral Processing Plant Design, Practice, and Control*. Vol. 2. Edited by A.L. Mular, D.N. Halbe, and D.J. Barratt. Littleton, CO: SME.

Altman, K.A., Schaffner, M., and McTavish, S. 2002. Agitated tank leaching selection and design. In *Mineral Processing Plant Design, Practice, and Control*. Vol. 2. Edited by A.L. Mular, D.N. Halbe, and D.J. Barratt. Littleton, CO: SME.

Dai, X., Breuer, P.L., and Jeffrey, M.I. 2010a. Modeling the equilibrium loading of gold onto activated carbon from complex cyanide solutions. *Min. Metall. Process.* 27(4):190.

Dai, X., Jeffrey, M.I., and Breuer, P.L. 2010b. A mechanistic model of the equilibrium adsorption of copper cyanide species onto activated carbon. *Hydrometallurgy* 101(3):99.

Dippenaar, A., and Proudfoot, M.R. 2005. The performance of the AAC pumpcell circuits at the Gold Fields Limited Driefontein and Kloof Operations. In *Randol Perth Forum 2005*. Golden, CO: Randol International.

Dunne, R., Staunton, W.P., and Afewu, K. 2013. A historical review of the treatment of preg-robbing gold ores—What has worked and changed. In *Proceedings World Gold 2013*. Melbourne, Victoria: Australasian Institute of Mining and Metallurgy.

Fleming, C.A. 2002. CIP/CIL/CIC adsorption circuit process selection. In *Mineral Processing Plant Design, Practice, and Control*. Vol. 2. Edited by A.L. Mular, D.N. Halbe, and D.J. Barratt. Littleton, CO: SME.

Hosford, P., and Wells, J. 2002. Selection and sizing of elution and electrowinning circuits. In *Mineral Processing Plant Design, Practice, and Control*. Vol. 2. Edited by A.L. Mular, D.N. Halbe, and D.J. Barratt. Littleton, CO: SME.

Kappes, D.W. 2002. Precious metal heap leach design and practice. In *Mineral Processing Plant Design, Practice, and Control*. Vol. 2. Edited by A.L. Mular, D.N. Halbe, and D.J. Barratt. Littleton, CO: SME.

Krumins, T., Zunti, L., and Frischmuth, R. 2014. Mercury removal from pressure oxidation vent gas. In *Seventh International Symposium on Hydrometallurgy*. Edited by E. Asselin, D.G. Dixon, F.M. Doyle, D.B. Dreisinger, M.I. Jeffrey, and M.S. Moats. Westmount, QC: Canadian Institute of Mining, Metallurgy and Petroleum.

La Brooy, S.R., Hosking, J.W., Muir, D.M., Ruane, M., Smith, I.M., and Hinchliffe, W.D. 1984. Studies on the fouling and regeneration of CIP carbon. In *Proceedings of AusIMM Regional Conference: Gold—Mining, Metallurgy and Geology*. Melbourne, Victoria: Australasian Institute of Mining and Metallurgy.

Marsden, J.O., and House, C.I. 2006. *The Chemistry of Gold Extraction*. Littleton, CO: SME.

McDougall, G.J., Hancock, R.D., Nicol, M.J., Wellington, O.L., and Copperthwaite, R.G. 1980. The mechanism of the adsorption of gold cyanide on activated carbon. *J. S. Afr. Inst. Min. Metall.* 80(9):344–356.

Nicol, M.J., Fleming, C.A., and Cromberge, G. 1984. The absorption of gold cyanide onto activated carbon. II: Application of the kinetic model to multistage absorption circuits. *J. S. Afr. Inst. Min. Metall.* 84(3):70–78.

Pleysier, R., Dai, X., Wingate, C.J., and Jeffrey, M.I. 2008. Microtomography based identification of gold adsorption mechanisms, the measurement of activated carbon activity, and the effect of frothers on gold adsorption. *Miner. Eng.* 21(6):453–462.

Ritson, G. 2013. Optimum metal recovery from solutions: Design considerations. In *Proceedings of the Heap Leach Solutions Conference 2013*. Vancouver, BC: InfoMine.

Rogans, J., and McArthur, D. 2002. The evaluation of the AAC pump-cell circuits at Anglogold's West Wits operations. *J. S. Afr. Inst. Min. Metall.* 102(4):181.

Shepherd, A. 2001. Activated carbon adsorption for treatment of VOC emissions. Presented at the 13th Annual EnviroExpo, Boston, MA, May.

Stange, W., King, R.P., and Woollacott, L. 1990. Towards more effective simulation of CIP and CIL processes. 2. A population-balance-based simulation approach. *J. S. Afr. Inst. Min. Metall.* 90(11):307.

Staunton, W.P. 2005. Carbon-in-pulp. *Dev. Miner. Proc.* 15:562.

Vegter, N.M., and Sandenbergh, R.F. 1996. Rate-determining mechanisms for the adsorption of gold di-cyanide onto activated carbon. *J. S. Afr. Inst. Min. Metall.* 96(3):109.

Von Becknamm, J., and Semple, P.G. 2002. Selection and design of carbon reactivation circuits. In *Mineral Processing Plant Design, Practice, and Control*. Vol. 2. Edited by A.L. Mular, D.N. Halbe, and D.J. Barratt. Littleton, CO: SME.

CHAPTER 10.12

Carbon-in-Leach and Carbon-in-Pulp Circuit Design

Stuart Smith

Several factors influence the design of a carbon adsorption circuit outside the realm of the pure process engineer's task of providing a functional design. These include capital expenditure (CAPEX), operating expenditure (OPEX), and risk. The design has to provide the desired financial outcome, not the optimum process solution.

The design of a facility is often compromised by the need to minimize entry capital. The discipline of process engineering is similarly compromised. A successful design needs to address the engineered solution as a holistic blend of process engineering and fiscal engineering.

To effect a design, a set of process design criteria (PDC) is required. Much of the PDC comes from metallurgical test work. Other components, such as design throughput, come from analysis undertaken during the various phases of study and project definition.

PROCESS DESCRIPTION—CIL AND CIP

An example of a generic carbon-in-pulp (CIP) process flowsheet diagram (PFD) is presented in Figure 1. The only difference between the CIP and carbon-in-leach (CIL) flow sheet is that CIL would not have the three leach tanks prior to the carbon tanks.

Ground or scrubbed ores in slurry form are sent to the trash screen. Typically, the pH of the slurry has previously been adjusted to provide protective alkalinity prior to the addition of cyanide. The pulp density of the slurry usually lies in the range of 48% to 50% solids w/w if subjected to pre-leach thickening. If the slurry results directly from a cyclone overflow, it can be expected to be in the range of 40% to 42% solids w/w.

The trash screen scalps out oversize material and directs it to an oversize bunker. Trash screen undersize slurry is directed to the first of the CIL/CIP tanks via a distributor, which allows the slurry to be directed to the second tank should the first tank be off-line. The distributor often receives cyanide solution, pH modifier, and other streams such as sump pump spillage return or spent eluate return.

Tanks are typically agitated with axial flow mechanical agitators submerged in the slurry. The tanks are generally configured with a height-to-diameter aspect ratio of around unity. Agitators are centrally and vertically mounted in the open tank, with three or more baffles provided around the tank periphery to stop vortexing.

Air and/or oxygen gas is injected into the tanks to provide oxygen for leaching. Various arrangements exist, from open pipes simply submerged in the tank to down-agitator shaft installations and more elaborate diffusers. Some ores respond favorably to elevated oxygen concentrations in that leach kinetics and/or extraction is improved. Cyanide consumption may be reduced if reactive sulfide minerals are present. Some ores require such high levels of oxygen that the volumes of air that would be required to satisfy the demand would impede the agitator operation, in which case, use of oxygen gas or some other oxidant is mandatory.

The slurry passes from one tank to another usually by means of a series of launders. The launders are configured so that tanks can be bypassed. A staggered tank arrangement simplifies this bypass arrangement and is the most common layout applied. An alternative is a carousel arrangement, which is described briefly below.

In each of the carbon contactors, some form of inter-tank screen resides to retain the activated carbon. The inter-tank screen aperture is coarser than the trash screen so that all of the ore particles should be able to pass the inter-tank screen. The flow of slurry through the inter-tank screen holds the carbon against the screen surfaces; consequently, some form of "sweeping" is required to keep the screen area open.

As the precious metals leach from the ore, the slurry passes from one tank to the next. Transfer is usually by gravity flow. In some instances, flow is assisted or totally dependent on the use of pumping inter-tank screens.

In the pure CIL case, the slurry is contacted with carbon at all times as the precious metals leach. This results in lower solution tenors of precious metals, as the carbon is present at all stages of leaching. The slurry continues to flow down the tank farm, through the inter-tank screens, to eventually discharge over the carbon safety screen. The carbon safety screen's function is to capture activated carbon that may have

Stuart Smith, Technical Director, Metifex Pty Ltd., Brisbane, Queensland, Australia

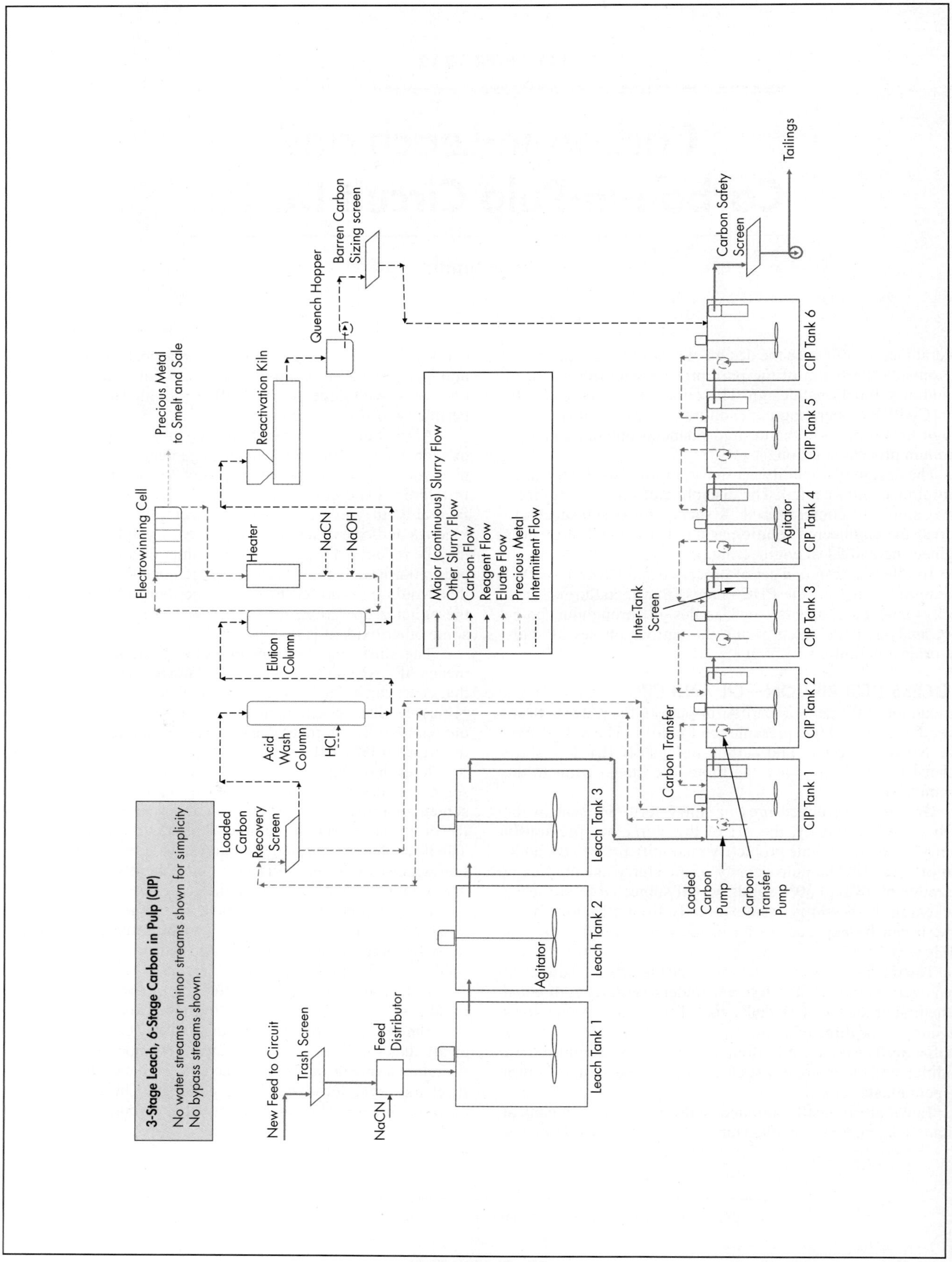

Figure 1 Generic CIP process flow-sheet diagram

escaped the tank farm because of a hole or a badly seated inter-tank screen.

In the pure CIP case, the slurry flows through the leach tanks before entering the carbon contactors. The precious metal leaching is effectively complete when the slurry contacts the carbon. The resulting solution tenors are much higher than a comparative CIL case. This results in higher carbon loadings of precious metals.

Barren carbon (ex-elution/reactivation or new) is added to the last contactor. This carbon, having a low precious metal loading, will continue to adsorb the low-solution concentrations in the last tank. As this carbon loads up, it will start to approach the equilibrium loading associated with the solution tenor of the tank in which it resides, becoming ineffective. Carbon is transferred to the next upstream tank, where the solution tenor is higher and the carbon loading is now lower than the equilibrium loading. Hence, the carbon will adsorb the precious metals until it again starts to approach equilibrium. The countercurrent transfer of carbon continues until the carbon reaches the first carbon contactor and becomes fully loaded with precious metals.

Carbon is transferred by pumping slurry and the carbon contained in it from the body of the downstream contactor up and into the upstream contactor. This carbon transfer is usually achieved by use of an airlift or a recessed impeller pump. The transfer results in slurry being transferred upstream as well. This results in dilution of the upstream tank solution tenor and this is reflected in the inefficiency of the contactor circuit. This back-mixed slurry will pass back through the inter-tank screen and return to the tank from which it originated.

Loaded carbon in the first contactor is recovered by use of an airlift or a recessed impeller pump. The slurry is pumped to the loaded carbon recovery screen. Here, the loaded carbon is separated and washed while the slurry component flows back to the tank from where it came. The washed loaded carbon gravitates to the acid-wash column/hopper. Some operations perform the acid wash after elution, others beforehand. The PFDs presented here assume that acid washing is pre-elution.

Carbon is acid washed and then transferred to the elution column as a water–carbon slurry. The elution circuit is generally an Anglo American Research Laboratories– (AARL) or Zadra–based process. These processes are described in more detail below. Both utilize elevated temperatures and high-sodium-concentration liquors to desorb the precious metals off the carbon and back into a high-grade solution.

The high-grade solution is passed through electrowinning cells where the precious metals are precipitated onto the cell cathodes. The metal is periodically recovered from the cathodes and smelted to produce doré bars of varying purity as final mine product. Refining is typically conducted by specialist refining companies off-site.

Eluted carbon is transferred as a water–carbon slurry to the reactivation kiln. The carbon is dewatered over a screen and reports to the kiln feed hopper. Heat and a reducing steam environment is applied by the kiln to increase the carbon activity. The carbon is typically quenched in good-quality water and then transferred to the last contactor as a water–carbon slurry.

The carbon is screened prior to reentering the last carbon contactor to remove fines. Some operations screen the kiln discharge dry. This is a more efficient sizing operation than the wet process.

FACILITY DESIGN

The following sections discuss the various aspects that need to be considered when selecting a design philosophy for a CIL or CIP facility. The advantages and disadvantages are noted for consideration.

The design process needs to consider the variability of the ores that will be fed to the carbon circuit. Considerations include

- Variable metal flow due to differential ore throughput for, say, oxides compared to harder primary ores;
- Preg-robbing tendencies that would indictate CIL over CIP;
- Ores that are sensitive to the actual solution tenor present (Le Chatelier's principle), suggesting that CIL may be advantageous over CIP;
- A CIP facility that requires less carbon to be eluted, as there is more precious metal adsorbed per unit mass of carbon;
- A CIP having less carbon and precious metal inventory;
- The financing strategy to be undertaken influencing the process design;
- Downstream processes such as cyanide detoxification, tailings deposition, and water balance considerations.

Pulp Density Considerations

The pulp density of the slurry to be leached and processed will influence the design of the carbon circuit. In some low-CAPEX designs, the comminution circuit hydrocylone (cyclone) classification is compromised and runs at high overflow pulp densities. The benefit is the deletion of a pre-leach thickener.

The advantages of low-pulp-density circuits are

- Reduced entry capital with regard to thickening,
- Enhanced precious metal and reagent recycle if a tailings thickener is employed,
- Reduced viscosity, and
- Reduced preg-robbing (reduced solution tenor).

Elevated pulp-density circuits have advantages of

- Higher solution tenors increasing carbon loadings,
- Reduced tank volumes,
- Reduced reagent consumption,
- Reduced solution losses, and
- Improved tailings deposition.

Trash Screening

The task of the trash screen is to remove the near-size particles that may block the inter-tank screens as well as coarse oversize that can cause tank sanding issues and elution circuit problems.

Trash screens are typically linear vibrating screens (high frequency) fitted with long-life polymer decks. Wire decks are sometimes used but are generally higher wear and harder to maintain. Linear belt screens are also used, particularly when large quantities of vegetation is present. Trash should not be directed to the carbon circuit bund, as cleanup of spills will result in the trash entering the carbon circuit.

The unit process of trash screening is one that is often compromised, both in design and in operation. Design compromises include

Figure 2 Contaminated activated carbon

- Using minimal size equipment to reduce CAPEX but also to address engineering considerations such as vibrating load; and
- Use of slotted polymer screens to reduce the total screen dimensions, as the slots provide increased open area over square apertures.

Slotted decks should be avoided when processing ores with vegetation present, significant nonelectric explosive waste (nonel), or flaky minerals such as micas. Slotted trash screen decks should not be used when the inter-tank screen apertures are square.

Sieve bends should never be used as a trash screen. The action of a sieve bend is such that to cut at, say, 600 µm, a 1,000-µm slot is required. A sieve bend will allow a minor percentage of material up to the slot size to pass, which is only exacerbated as the screen wears.

Given the recirculating nature of the carbon circuit, any minor bypassing of trash will quickly result in a buildup in the circuit, which is difficult to reduce and almost impossible to totally remove. The issue with coarse buildup in the circuit can be a major operational issue and one not to be underestimated.

Figure 2 presents the deck of a loaded carbon recovery screen. It is apparent that the activated carbon has a lot of coarse sands mixed in and the screen apertures are blinded with sands as well as fine carbon. The sand in the carbon causes issues in the acid wash, elution, and reactivation circuits as do ore slimes that are carried over because of poor screen efficiency. Silicate fouling of carbon can result, and this has an irrecoverable impact on the carbon circuit. The heart of the carbon circuit is the carbon itself, and the cleanliness of this carbon is key.

Tankage and Agitators

CIL/CIP tanks are most commonly agitated with axial flow mechanical agitators. Various impeller configurations are employed. Other less common agitation alternatives are

- Draft tubes, which involve elevated CAPEX but can have OPEX benefits over open tank agitation;
- Air sparge, which is a high-OPEX option and is impractical for large tanks; and
- A Pachuca tank (air), which encompasses high OPEX and is not suited to large tanks.

As tank height-to-diameter ratios increase above unity, the effective cost of the tankage increases per unit volume. This includes costs assocated with structural steel, electrical systems, and in supporting vibrating loads. Open agitated tanks having a height-to-diameter ratio nearing unity have a capital advantage over draft-tube or pachuca/air-agitated tanks. The latter agitation methods require tanks of greater height-to-diameter ratio.

Tank materials of construction are almost exclusively mild steel. The high pH of the process provides corrosion protection for unlined or uncoated mild steel. Under some circumstances, anaerobic biological oxidation can occur. The biomass establishes a favorable environment under an isolating film on the inside of the tank wall, which results in pitting corrosion. Elevated pH values of +10.5 are often necessary to reduce the impacts of bacterial attack (Mardhiha et al. 2014).

Coarse abrasive slurries can wear the lower strakes of mild steel tanks. Poor trash screening practices exacerbate this wear issue. Such coarse material can also settle out in the corners of the tank. This can result in entrapment of activated carbon and exacerbate corrosion activity.

Carbon tanks can be configured in a carousel arrangement such that the tanks are fed by a distributor to the first contactor. Slurry then flows to the next tank in the carousel, and so on. Instead of transferring carbon, the slurry flow path is changed; consequently, what was the second tank now becomes the first tank in the series. What was the first tank is drained or carbon transferred for elution. The tank is then reestablished with the addition of barren carbon and then the tank is inserted to the end of the carousel as the last tank. The cycle repeats and the tanks are repeatedly moved up in the sequence to maintain the carbon contact with higher and higher solution tenors.

Oxygen Source and Introduction

The leaching of gold and silver requires oxidation of the metals prior to complexation with cyanide. Oxygen is the typical oxidant and can be supplied by the ambient atmosphere, aeration into the tanks, oxygen gas introduction to the tanks, or disassociation of hydrogen peroxide introduced to the tanks.

Part of the metallurgical test-work phase of a project should include measurement of the oxygen demand of the ore (Fraser 1984). Equilibrium dissolved-oxygen concentration decreases with rising temperature and increasing altitude, and so the site conditions also need to be considered. If oxygen gas is used over air, there may be advantages such as

- Faster leach kinetics (elevated carbon loadings and reduced tankage volume),
- Improved circuit efficiency (reduced back-mixing),
- Improved final extractions,
- Reduced soluble losses (steeper solution tenor profile), and
- Reduced cyanide consumption.

Disadvantages can include

- Increased capital cost,
- The need to observe specific design considerations,
- Elevated operating cost potential, and
- Gold passivation and elevated cyanide consumption if the conditions are too aggressive.

Oxygen adsorption into the slurry from the ambient atmosphere will only provide adequate oxygen supply to the most benign of oxygen consumers. As modern plant tankage has increased in size, the lower surface area of the tank per unit volume of slurry contained has made this option impractical. Air/oxygen sparging options include

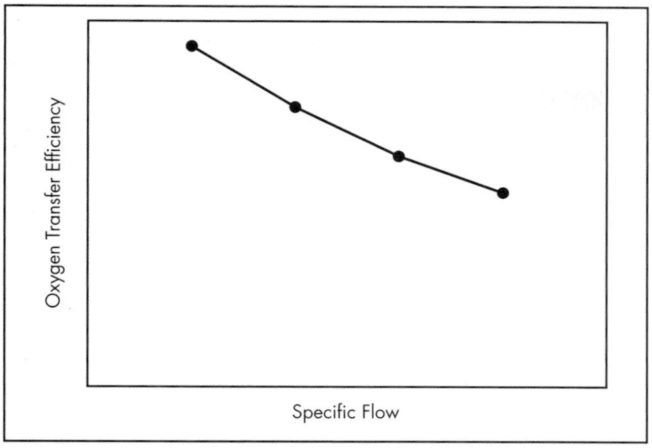

Figure 3 Aeration efficiency

- Proprietary lance-based sparging methods,
- Open-lance-based options,
- Gas introduced down the center of the tank agitator shaft,
- Sparge rings under the tank agitator, and
- Shear reactors.

Figure 3 shows test-work data for a given open-sparge system as the airflow increases. A nominal 100% increase in the airflow results in a decrease in transfer efficiency of around 25%. Doubling the air addition, therefore, only provides a 50%–60% increase in oxygen mass transfer.

Air-based systems may achieve 5%–15% utilization for oxygen transfer, although vendors of sparging equipment may claim higher values. Oxygen-gas-based systems often achieve utilizations of greater than 80%, even for moderately basic lance systems. Oxygen transfer is more efficient because the volumes introduced are lower than the air equivalent, and the equilibrium concentration of oxygen systems is some five times greater than air.

The design of the oxygen supply needs to consider the following:

- The oxygen demand, temperature, and viscosity of the slurry
- Cost of generation of air or oxygen
- Oxygen supply purity
- Altitude
- Utilization/transfer efficiency

Inter-Tank Screens and Launders

The task of the inter-tank screen is to retain the coarser activated carbon in the carbon contactor while allowing the slurry to pass. Various configurations of inter-tank screens have been developed over the years and have included options such as air-swept woven or wedge-wire screens. Cylindrical, flat panel, and V-shaped options exist. Mechanically swept screens are the most common. These are generally vertical cylinder screens of woven or wedge wire and have a set of external noncontacting, rotating wiper arms.

Some mechanically wiped screens include what is effectively a low-head/high-volume pump impeller to provide the small amount of head necessary to overcome the head loss of the inter-tank screens and launders. Capital cost and complexity of the screen are off-set by savings in the tankage and tank-top steel costs.

The inter-tank screen working area has to be estimated with consideration of the volume of slurry being processed, but also the viscosity and the back-mixed volume resulting from carbon transfer. The inter-tank screen apertures are normally square for woven wire or slotted for wedge wire. The woven-wire aperture is normally 1 × 1 mm, whereas wedge-wire slots may be finer in the short dimension, around 800–900 μm.

Inter-Tank Carbon Transfer

The two dominant methods for transfering carbon between the carbon contactors are airlifts or recessed impeller pumps. The advantages of the airlift are low CAPEX, simplicity, and relatively low maintenance. Recessed impeller pumps are higher CAPEX but are less abrasive on the carbon and are capable of higher flow rates.

Carbon transfers are often conducted in short periods of time because it is believed that this minimizes the upset conditions due to back-mixing. It is also common to paritally empty a contactor of carbon prior to transferring carbon to it from downstream to minimize short-circuiting of carbon. Both of these practices actually reduce the efficiency of the carbon circuit and are to be avoided as a basis of design or operation. Carbon transfers should be designed to be continuous but with adequate capacity to catch up in the event of circuit disturbances. This diminishes CAPEX of the transfer equipment, and inter-tank screen capacity is reduced.

Carbon Safety Screen

It is common practice to use a vibrating screen for the carbon safety screen duty. It should also be the same model and size as the trash screen to allow commonality of spares. The carbon safety screen aperture is often the same as the inter-tank screens or, preferably, slightly coarser. Slotted polyurethane panels are most commonly used.

Screen Aperture Selection

Several factors must be considered regarding screen aperture shape, materials of construction, and wear. The plant designer must allow for practical considerations and not just design for a "new" plant where reliable aperture dimensions are present.

Screen aperture shapes for the trash, inter-tank, and carbon safety screens, and even the loaded carbon recovery screen, need to be considered holistically:

- The trash screen aperture must be the finest used of all the various applications.
- Square apertures may be necessary to deal with certain trash types.
- If a square aperture inter-tank screen is to be used, then the trash screen aperture must be square to save passing of flaky material that will not pass the inter-tank screen.
- The inter-tank screen aperture should be at least 100 μm coarser than the trash screen. The trash screen will wear and pass coarse particles with time that a new inter-tank screen cannot.
- The carbon safety screen should be 100–200 μm or even coarser than the inter-tank screen. As inter-tank screens wear, they will pass trash and coarse minerals that have entered the circuit, resulting in excessive fines being caught by the carbon safety screen.
- The loaded carbon recovery screen should be the same aperture as the inter-tank screen at most, but preferably 100 μm or so finer to minimize the recirculation of

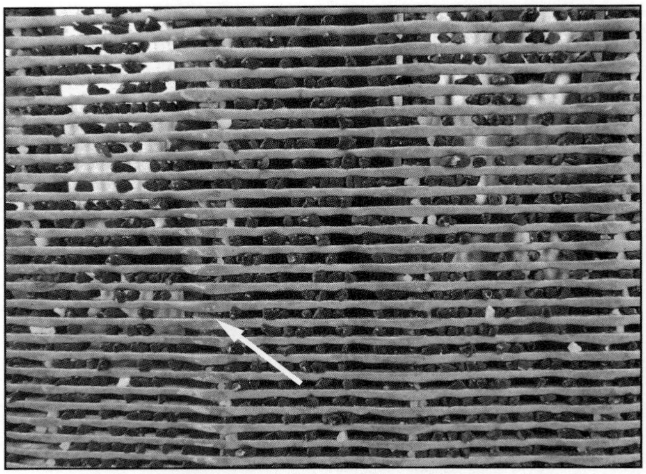

Figure 4 Blinded and worn inter-tank screen

near-size carbon too coarse to pass the inter-tank screen but too fine to be retained by the partially worn loaded carbon screen.

- The barren carbon sizing screen should be the aperture of the inter-tank screen plus 200 µm. This deals with unavoidable inter-tank screen wear.

Figure 4 presents an example of a partially blinded and badly worn wedge-wire inter-tank screen. The effective short dimension of these apertures is approximately 2 mm, double the original new-screen aperture. The white arrow points to areas retaining the original sub-1-mm aperture adjacent to one of the internal support wires.

Even if the allowances for wear in the various screening applications suggested above were applied, such high wear would still result in losses of fine carbon from the contactors and excessive amounts of carbon and trash being recovered on the carbon safety screen. The recovered carbon is too fine to use back in the circuit unless the inter-tank screens are replaced. This is an example of why practical considerations need to be made when selecting screen apertures.

LOADED CARBON RECOVERY, ACID WASH, AND ELUTION

Loaded Carbon Recovery

The first stage of the acid wash and elution process is the loaded carbon recovery step. Loaded carbon recovery is usually undertaken in 4–6 hours. The first contactor should have adequate carbon present so that partial removal takes the population of carbon down to nominal design levels to minimize the impact to contactor performance.

Recessed impeller pumps are the most common means of loaded carbon recovery. The loaded carbon recovery screen location and height usually preclude the use of airlifts. The loaded carbon recovery screen must produce a very clean carbon product. The efficiency of the acid wash, elution, and even reactivation process are all impacted by the quality of the carbon recovered. Carbon containing a large percentage of fines can

- Consume additional acid,
- Generate soluble silicates that permanently foul the carbon and cause electrowinning issues,
- Result in corrosion issues,
- Aggravate heat exchanger and solution heater blockages,
- Reduce heat transfer in the elution and reactivation kiln due to accrection, and
- Make cell cleanout and final smelting onerous.

The slurry containing the loaded carbon should be sent back to the tank it originated from to avoid short-circuiting (downstream return) or total residence time compromises (upstream return).

Acid Wash

As most CIL and CIP operations use quicklime or hydrated lime for pH control, the activated carbon can foul with calcium carbonate. The acid wash is for removing the calcium carbonate and other acid-soluble foulants, typically prior to elution to keep these ions out of the elution circuit.

The acid wash vessel is normally configured to receive the high-value loaded carbon directly from the loaded carbon screen discharge chute to avoid abrasion from transfers. Acid washing can be conducted in open-top vessels, low-aspect closed tanks, or columns that often have an aspect ratio the same or similar to the elution column. Many operations have used a rubber-lined column to undertake both acid washing and then elution. Although cost-effective with regard to equipment, there remains a risk of acid and cyanide mixing should the acid wash/elution sequencing be upset.

Acid washing is usually conducted using a nominal 3% w/w hydrochloric acid solution. Materials of construction therefore need to be resistant to low pH and also chloride attack, which negates the use of typical stainless-steel wetted parts. Fiberglass, high-density polyethylene and rubber-lined steel are common materials.

Dilute acid is recirculated, generally upflow, through the bed of carbon, typically for 2 hours at 2 BV/h (bed volumes per hour). Raw water is then used to rinse the carbon and reduce the pH. The water rinse continues until the pH of the acid wash discharge is greater than 5 or preferably higher.

It has become more common to pass a final dilute sodium hydroxide rinse through the carbon to ensure that the pH is alkaline at the end of the wash-rinse cycle. Once the acid wash is complete, the carbon can either be transferred to the elution column, or if the plant is single column, the carbon remains in situ, ready for elution.

Elution Options

Several methods are used to elute precious metals from activated carbon. The three most common methods discussed in this section are atmospheric Zadra, pressure Zadra, and AARL elution. Other methods employed include the Micron Research method (alcohol-based elution), sodium sulfide–based elution (Adams 1994), variants that use alcohols and other accelerants, and variants that use pressurised electrowinning (Fast 1989).

Atmospheric Zadra

The Zadra process was developed by J.B. Zadra and colleagues while working for the U.S. Bureau of Mines in the 1950s (Zadra et al. 1952). Atmospheric pressure Zadra stripping is considered to be the first commercially successful process developed for stripping gold from carbon. The Zadra process consists of continuous (upflow) circulation of a nominal 1% sodium hydroxide and 0.1%–0.5% sodium cyanide solution at as high a temperature as possible before the solution starts to boil or flash, which is generally around 98°C, depending on the altitude of the installation. Flow rates are in the 2–2.5 BV/h rate, and the circuit needs to operate for 2 to 3 days to achieve low barren-carbon values.

The reagents are mixed in a sump and circulated as the solution is heated. The eluate is pumped upflow through the elution column. The aspect ratio of the column should be at a ratio of 6:1 height to diameter to facilitate plug flow. As the solution circulates, the precious metals are desorbed and enter the eluate primarily due to the sodium saturation coupled with the high temperature as well as the influence of the elevated pH.

The eluate discharges the top of the elution column and is directed to electrowinning. The electrowinning cell (or cells) results in the cathodic precipitation of the precious metals from solution. The elevated pH provides protection to the anodes from acid attack due to the evolution of hydrogen. A supplementary sodium hydroxide dose is common during the elution process to replace consumed reagent due to acid consumption.

The barren eluate is pumped through a heater/heat exchanger to replace any heat lost and then recirculated back up through the column. This cycling of the eluate is continued until the pregnant solution has a gold grade on the order of 5–10 mg/L and/or the electrowinning discharge is less than 5 mg/L. Heating is stopped and the solution continues to circulate to lower the temperature of the eluate and ensure that the heating circuit has no hot spots, which can result in damage to the heater. Once cooled, the carbon can be pumped or pressure transferred to the reactivation kiln. Pressure transfer is the preferred method, as it results is less carbon attrition.

Figure 5 shows the more complex pressure Zadra flow sheet, which is described in the following section. However, this flow sheet is effectively the same as the atmospheric Zadra, except for the heat exchangers shown in gray. The advantage of the atmospheric Zadra is low-entry capital and simplicity of operation, resulting in high reliability and utilization. The disadvantages include low elution rate, sensitivity to elevation (elution temperature), low electrowinning efficiency, elevated barren carbon grades, and low carbon activity post-elution.

Pressure Zadra

The pressure Zadra process is basically the same process as the atmospheric Zadra except that the heating and column components are pressurized to allow temperatures in excess of the solution atmospheric boiling point to be utilized, typically 110°–135°C. This greatly accelerates the elution rate such that the actual elution stage can be conducted in 12–16 hours. Allowing for transfers of carbon to and from the column and heat-up time, a pressure Zadra elution can generally be completed in 18–20 hours.

The use of pressurized equipment has increased mechanical design demands, which also adds to the CAPEX and complexity. In some jurisdictions, the temperature and pressure may influence what operator and maintenance expertise is necessary to operate a pressure Zadra (or AARL) plant. It is necessary for the designer to be familiar with such regulations because they may influence design and process selection.

Although there is additional cost to a pressure Zadra, a much smaller-capacity facility can be utilized compared to an atmospheric unit, given that three times the elution frequency is achieved. Pressure Zadra has the following advantages over atmospheric Zadra:

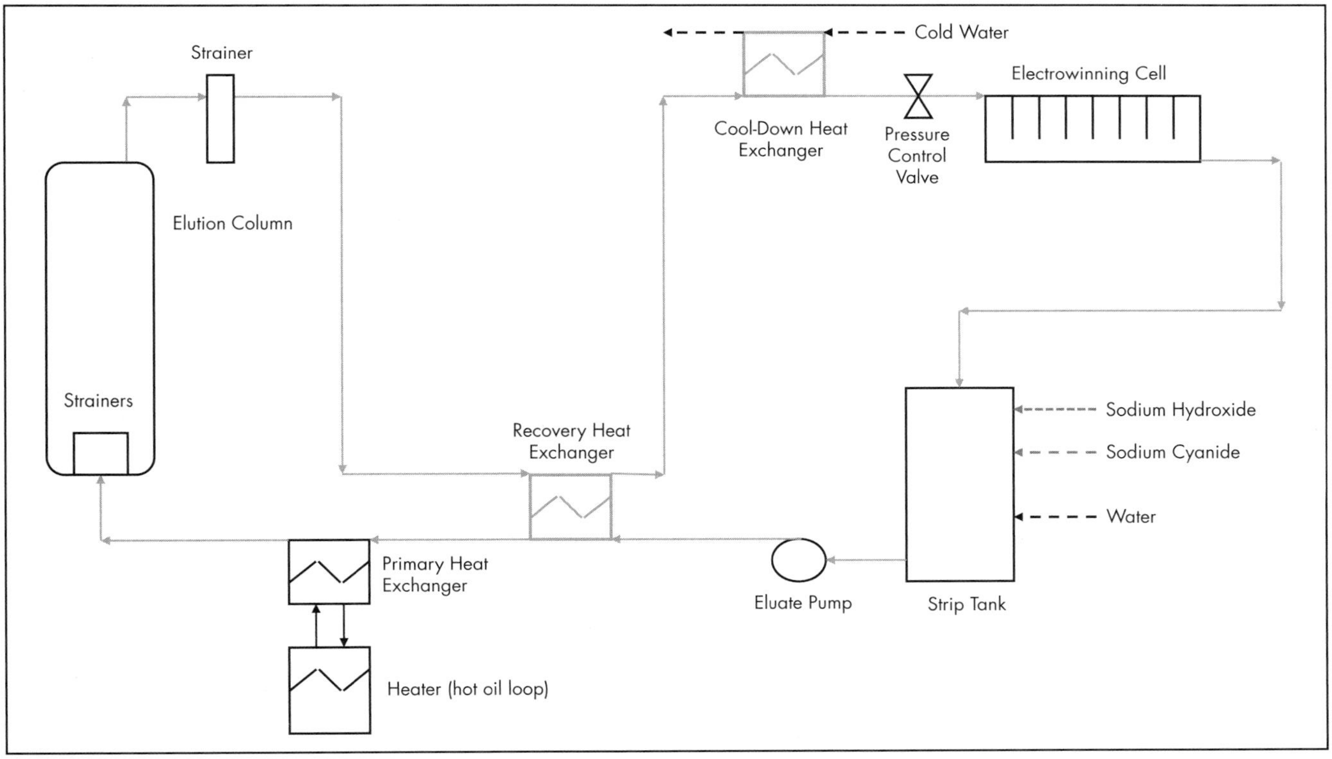

Figure 5 Pressure Zadra flow sheet

- Capital per unit mass of carbon processed per unit time is typically lower.
- Smaller batches of carbon reduce carbon inventory.
- CIL/CIP inventory management is simplified.

Consequently, most modern Zadra facilities are of the pressure variety. Figure 5 presents a typical configuration.

Given that electrowinning is conducted at atmospheric pressure, there is a need to reduce the temperature of the pressure Zadra eluate as it leaves the top of the elution column prior to electrowinning. This requires the use of additional heat exchangers and/or a flash vessel to cool the eluate.

The barren eluate from the electrowinning cell is circulated via the eluate tank to a heat exchanger and is preheated by the pregnant eluate discharging the column on its way to electrowinning. The barren eluate then passes through the heating equipment (electric, gas, or diesel-fired heater/boiler) where it is heated to the final elution temperature. Eluate is then directed to the base of the elution column and flows upward through the carbon. The pregnant eluate discharges through the column top and the cycle is repeated.

To achieve a rapid cooldown of a pressure Zadra, it is common to add a cooldown heat exchanger employing cold raw water. Once the circuit is cooled below the flash point, the carbon can be transferred to the reactivation kiln.

Anglo American Research Laboratories Elution

The AARL elution process was developed in the 1970s and gained acceptance in the 1980s. The process is somewhat different from the Zadra-based processes in that the elution and electrowinning steps are isolated from each other. The AARL process uses low-ionic-strength (deionized) water to undertake elution post heat-up and pretreatment with reagents.

An example of an AARL flow sheet is presented in Figure 6. The AARL process has three basic steps. The first is the presoak step, which involves circulating a sodium hydroxide and cyanide solution in a closed loop through the column and the solution heating circuit to achieve temperature. Typically 1–2% w/v of sodium hydroxide and similar concentrations of sodium cyanide are used in the presoak. It is common for operations to reduce the sodium cyanide concentrations to 0.5% or even lower post-commissioning. This presoak stage is very similar to the initial heat-up stage of a Zadra-based process. The flow rate of presoak solution and eluate is on the order of 2–2.5 BV/h.

Once the elution column discharge is ~95°C, the flow path of the solution is diverted and the second stage of actual elution commences. Low-ionic-strength water (eluate) is pumped through a recovery heat exchanger, which scavenges heat from the presoak solution exiting the column. The cooled presoak solution is directed to the eluate tank via a pressure control valve.

The preheated water (eluate) now passes through the primary heat exchanger (heater) where it is heated to the elution temperature, typically in the 110° to 140°C range. The circuit is pressurized to stop the solution from flashing off. The water (eluate) is pumped upflow through the column, desorbing the precious metals from the carbon. Pregnant eluate exits the top of the column and passes through the recovery heat exchanger, giving up heat to the cold, new water entering the circuit, before being directed to the pregnant eluate tank. The efficiency of the circuit is such that typically only 8 BV of

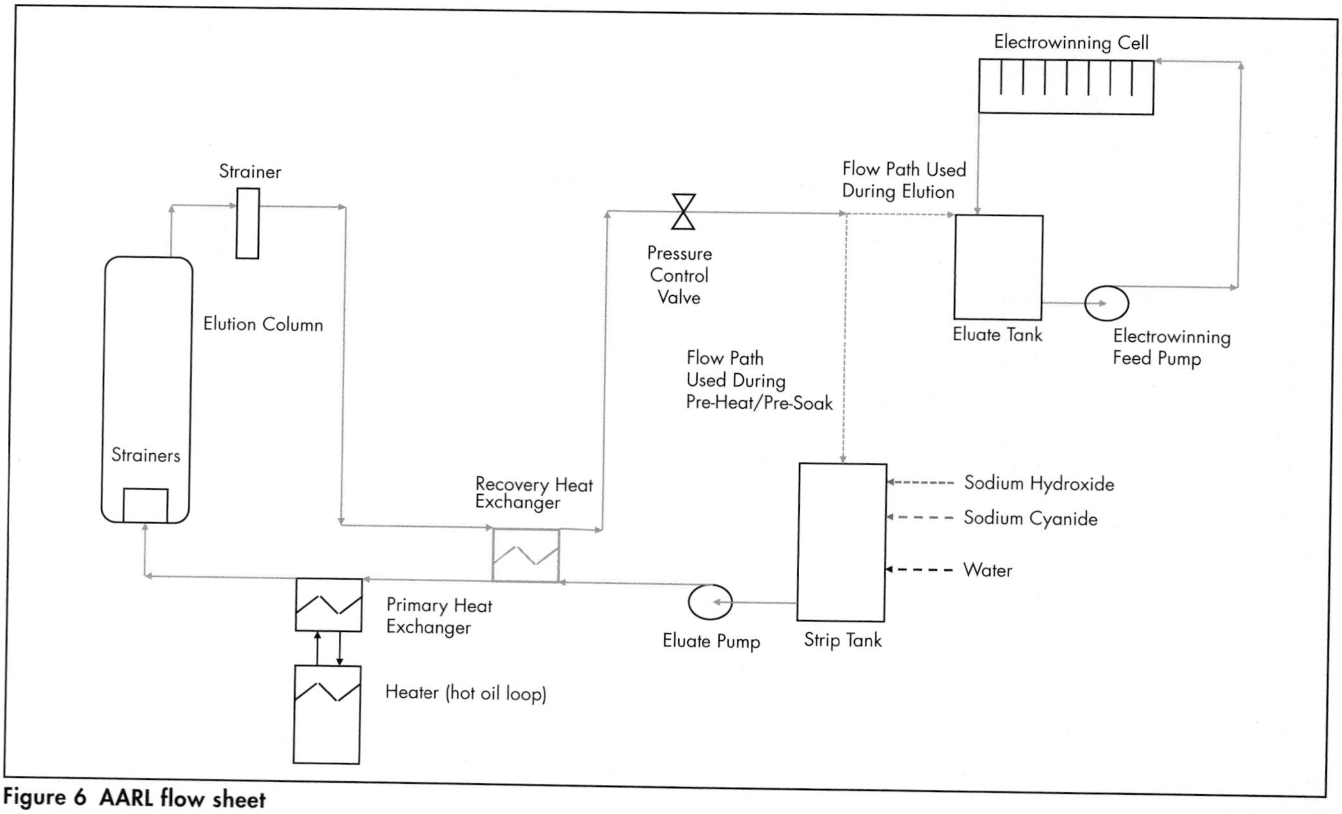

Figure 6 AARL flow sheet

eluate are required to complete an elution, taking only 4 or so hours post-presoak/preheat.

The last stage of the elution is the cooldown stage. Once cooled, the carbon can be transferred to the reactivation kiln. An AARL circuit can process up to three batches of carbon per day at 100% utilization. However, circuits are generally designed to process between 1 and 1.5 batches per day and sized accordingly. In practice, a maximum of 2.5–2.8 elution cycles per day are achievable.

The eluate is stored in the eluate tank from which it is pumped to the electrowinning circuit. The electrowinning tail recirculates back to the eluate tank, and the solution is recirculated until the required barren eluate grade is achieved. As the electrowinning cycle is independent of elution, it is possible to utilize multiple electrowinning tanks to allow multiple elution cycles to be run. One eluate tank can be filled or emptied while another tank is being used for the actual electrowin cycle itself.

The efficiency of the process is very dependent on the quality of the elution water used. If water having elevated levels of sodium, calcium, or magnesium are applied, the elution rate is significantly impacted, as is the final barren carbon grade. A reverse osmosis or nanofiltration plant is sometimes required to provide the necessary volume of quality water.

This is where Zadra-based processes have an advantage over AARL: The Zadra process recirculates around 1–2 BV of eluate, whereas the AARL process requires 8–10 BV of good-quality water for elution. This water demand can add to CAPEX and OPEX associated with the AARL option.

Split AARL

A variation of the AARL process is the split AARL process, which

- Conserves water (saves 3–4 BV),
- Generates higher-grade eluate,
- Makes for more efficient electrowinning, and
- Reduces recycle of values in barren eluate return to the adsorption circuit.

The circuit effectively operates the same as a conventional AARL except that the last few BVs of low-grade eluate are directed to a "starter" or "lean" eluate tank instead of the eluate tank.

The presoak step and preheat of the next elution cycle are conducted with new water. Once the elution steps from presoak/preheat to elution proper are completed, the low-grade "starter" eluate is then used. Once the starter eluate is expended, new water is consumed to complete the elution. The increased complexity and cost of the split AARL is offset by savings in water treatment and a minor offset in electrowinning capital.

Advantages and Disadvantages

When considering the two most common elution options, pressure Zadra or AARL variants, the following advantages and disadvantages should be considered:

- **Water quantity and quality.** The AARL process requires a larger volume of quality water.
- **Upgrade/expansion.** An AARL facility is typically easier to expand.
- **Flexibility.** Projects that have wide variation in feed grades or throughput are better serviced with the more-flexible AARL option.
- **Water balance.** The AARL option can influence the process water balance.
- **Capital cost.** An AARL circuit has a higher capital cost *for the same batch capacity* than a Zadra equivalent. If the AARL is sized for multiple elutions per day, the effective CAPEX can be less than a Zadra circuit.
- **Performance.** An AARL circuit will typically provide lower barren carbon grades than a Zadra circuit.
- **Reactivation.** The AARL barren carbon contains less-soluble salts. This generally results in elevated carbon activity prior to and after reactivation.
- **Operating cost.** The operating costs are so similar that when all facets are considered, there is often no clear advantage between AARL and Zadra. Often the AARL is suggested to have a higher OPEX, but AARL advantages can be shown to reduce operating costs in other areas.

ELECTROWINNING

It is assumed in this section that eluates are derived from the most common Zadra or AARL elution variants and not alcohol or other elution processes. Electrowinning is usually conducted in square or slightly rectangular cross-sectional cells. The anodes and cathodes are placed at right angles to the eluate flow. The eluate is introduced at one end of the cell and passes through the anodes and cathodes, exiting the cell partially or near fully depleted of precious metals. A direct current potential of around 3–5 V is applied between the anodes and cathodes. Current demands vary depending on the size of the cells and may range from a few hundred amperes up to 6,000 amperes or beyond.

Circular (radial flow), tubular (axial flow), and other cell variants have been tested over the years with varying success. The rectangular cell remains the most common, mainly due to the ease of cleaning, the layout affording close packing of cathodes and anodes, and simple control of the eluate flow path.

Anodes are generally stainless-steel mesh or punched plate. Cathodes are either mild steel wool or knitted stainless-steel wire. The mild-steel-wool options provide a high surface area. This affords a low current density and a competent precious metal deposit with good adherence to the wool. Stainless wire cathodes present a higher current density due to the reduced area. This results in a poorly adhering precious metal deposit that can be relatively easily removed from the cathode by brushing or hosing.

Mild steel cathodes are subjected to oxidation or acid treatment prior to smelting. Direct smelting is also practiced. As the mild steel wool is consumed, it must be replaced after each cycle. Stainless-steel cathodes are washed to dislodge the precious metal. The precious metal sludge is collected, filtered, and dried, and then the sludge is smelted to produce doré. The washed cathodes are returned to the electrowinning cell for reuse.

Design Considerations

Elution Process

The type of elution being undertaken needs to be considered during the electrowinning cell design process. A Zadra variant requires the barren electrowinning eluate to be as low a grade as reasonably practical to increase the efficiency of the elution itself. Therefore, the electrowinning cell should have a pass efficiency of 90% or higher.

AARL processes are not as dependent on the pass efficiency, as the eluate is just recycled back to the eluate tank. As long as the electrowinning circuit is designed to recirculate an adequate flow and electrowin the precious metal in the allotted time, the number of cathodes is not generally a constraint.

Cell Flow Rate

The cell flow rate should be in the range of 250 to 500 L/m^2 of cross-sectional area per minute (Filmer 1982). Too high a flow rate can result in abrasion of the precious metal precipitate such that it dislodges and forms a slime in the bottom of the cell, settles in the eluate tank itself, or is recirculated to the elution.

Current Demands

The electrowinning process results in oxygen evolution at the anode according to the following equation:

$$4OH^- \rightarrow O_2 + 2H_2O + 4e^-$$

At the cathode, evolution of hydrogen occurs according to this equation:

$$2e^- + 2H_2O \rightarrow H_2 + 2OH^-$$

Gold (or other metal) deposition occurs at the cathode according to the following equation:

$$e^- + Au(CN)^-_2 \rightarrow Au + 2CN^-$$

There is an imbalance in the amount of hydroxide consumed and generated, such that supplementary sodium hydroxide is dosed part way thought the electrowinning process or elution to maintain alkalinity.

Other metal precipitation at the cathode, including silver, copper, and even mercury, results in allowances being made in the current supply to the cell. Many designers assume a fixed current efficiency for gold and silver when calculating the rectifier capacity. Values of 10% for gold and 15% for silver are common values applied independent of eluate grade. However, the current efficiency relevant to the metals is a function of the solution grade and is not a constant. Figure 7 presents field data plotted on a linear axis for a rectangular electrowinning cell fitted with stainless-steel knitted cathodes. The constituents of the eluate, relative concentrations, and influence they may have on current efficiency should be considered prior to deciding how current efficiency is to be applied to sizing a rectifier.

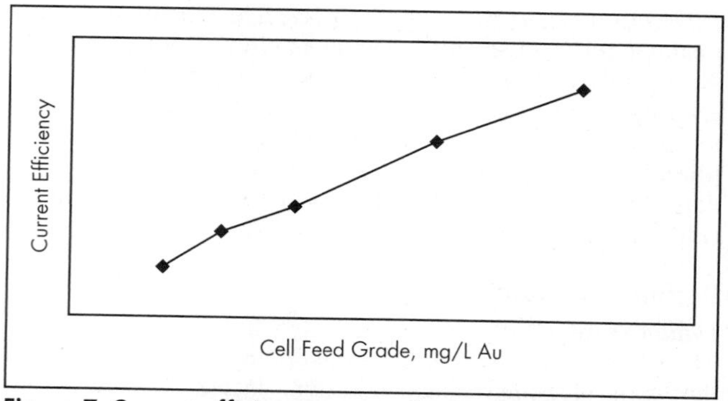

Figure 7 Current efficiency versus electrowinning cell feed grade

Other Design Features

Electrowinning cells should have anodes and cathodes spaced as tightly as possible without risk of short-circuiting of fluid flow or electrical current. Cell bodies need to be able to be easily cleaned to remove metal sludge and mineral fines, and cells should be located at a height to allow free draining. In addition, mercury will elute from the carbon and present to the electrowinning cell. Capture and control of emissions of mercury is required if present (retort, sulfur-impregnated carbon adsorption, wet scrubbing).

Eluate tanks necessitate a means of easy cleaning to remove precious metal and mineral sludge with consideration of a potential toxic atmosphere. Eluate tanks also require consideration of mercury presence.

REACTIVATION AND REGENERATION OF CARBON

Reactivation and regeneration are two different processes that are often referred to erroneously as one and the same (Claflin et al. 2015). *Regeneration* is the displacement of adsorbed species present on the carbon that does provide an increase in carbon activity. Regeneration may not provide the necessary return to activity of the fouled carbon, particularly if pyrolysis products remain. *Reactivation* is considered the process where the conditions are such that the water–gas reaction is initiated, typically at temperatures greater than 700°C (up to 750°C) in the presence of H_2O (steam):

$$C + H_2O \rightarrow H_2 + CO$$

The conditions need to be retained for a period of time until the pyrolysis products are removed. Additional time is potentially disadvantageous should the carbon hardness be impacted or the micropores be allowed to grow to mesopores (Claflin et al. 2015). It is therefore important that the carbon post-elution is subjected to the correct conditions to achieve reactivation and not just regeneration. The designer needs to specify the reactivation kiln such that the *temperature of the carbon is achieved and maintained for the appropriate time at this temperature.* Too often, it is the nominal kiln residence time and temperature of the kiln shell that are accepted as proxies for the required carbon conditions. That is, the conditions needed to be experienced by the carbon in the kiln are not the same as the kiln operating conditions.

Reactivation Circuit Design

The type of feed to the CIL/CIP, water quality, and other factors such as potential oil contamination from the mine will determine the need and frequency of reactivation. Circuit feed emanating from flotation processes will contain elevated levels of reagents that can easily foul the carbon. Water high in calcium or other species can similarly foul the carbon chemically or physically (carbonate deposition). Whereas acid washing may remove such carbonate buildup, other contaminants such as humic acids and tannins will remain.

Unless the ore is particularly clean, the water quality is excellent or the demand for reduced capital is critical, designing for less than 100% reactivation of the elution capacity is to be avoided. On exiting the reactivation kiln, the carbon needs to be cooled in an inert atmosphere or, more often, it is quenched in water. It is important that the quench water is of good quality. Poor-quality water can itself result in reduced activity.

Carbon Sizing

The reactivated carbon has to be screened to remove any fines prior to introduction back into the CIL/CIP. Fines introduced to the CIL/CIP can partially load up with precious metals and then discharge to tails, resulting in losses.

Quenched carbon is usually pumped as a slurry from the quench tank to the carbon sizing screen located on top of the CIL/CIP. The screen oversize drops into the last CIL/CIP tank. Wet carbon tends to ball up once dewatered on a vibrating screen. Clusters of wet carbon migrate along the screen and the resulting sizing efficiency is low. The use of water sprays and, in some instances, wash troughs provide some benefit.

The most effective sizing method is to screen the dry carbon post-reactivation. Minor fine ash and carbon will adhere to the oversize, but after the carbon is slurried and pumped to the CIL/CIP, a simple dewatering screen will remove most of these misreporting fines.

DESIGN OF A CARBON CIRCUIT—CASE STUDY

The following sections describe *one method* for selecting the basis of a basic carbon circuit design. Where appropriate, comment is made as to alternative methods that can be applied. The design assumes free milling ore with low silver grade, and simple metallurgy with no complicating issues such as poor water quality, high viscosity, or high cyanide-soluble copper levels.

Data Required

The design of a carbon circuit requires a selection of process design criteria, much of which is the result of metallurgical test work undertaken on representative ore samples. Process design criteria required include:

- A proposed production schedule reporting head grade and ore types to be processed as a function of time.
- Viscosity data at various pulp densities and shear rates and at the appropriate pH using the appropriate pH modifier for the various ore types.
- Leach data at the appropriate conditions of pulp density, pH, oxygenation/aeration, and reagent dose.
- Carbon characterization data. The methodology used to design the circuit will dictate the data required. Typically, carbon kinetic data are required as a minimum (Fleming k and n values). Kinetic data are required for all metals of interest.
- Carbon equilibrium data should also be determined.

Carbon characterizations must be determined at the actual size distribution of the carbon to be applied to the design and after the carbon has been attritioned and equilibrated with a sodium cyanide solution. Use of certain carbon adsorption procedures that specify pulverizing the carbon are not applicable to CIL/CIP (or heap leach) design. It is preferable to use a slurry and not a solution to determine the carbon characteristics, as a slurry will typically provide lower values with regard to both kinetics and equilibrium loadings.

Example Design Case Basis

The design cases evaluated are based on the following:

- Seven cases evaluating a throughput of 4 Mt/yr (million metric tons per year) over 8,000 operating hours per year on an ore having a head grade of 3 g/t Au, insignificant silver, and a moderate gold leach rate.
- Seven additional cases evaluating a throughput of 4 Mt/yr over 8,000 operating hours per year on an ore having a head grade of 4 g/t Au, insignificant silver, and a slower gold leach rate.
- Ore specific gravity of 2.7.
- CIL/CIP feed at 42% solids w/w.
- Split AARL elution.
- Return eluate having a grade of 3 mg/L Au.
- Process water having a grade of 0.01 mg/L Au.
- 24-hour residence time.
- No leaching has occurred prior to the tank farm.
- Kinetic leach data available and leach model fitted.
- Equilibrium Freundlich isotherm data available.
- Decay parameters applied to carbon kinetic parameters based on benchmarked data.
- Tails solution grade of 0.010 mg/L targeted.

Model Structure

Several methods are applied to model CIL and CIP performance. The most basic models consider only the leaching and carbon adsorption processes. Constant carbon kinetics are considered and a final loaded carbon grade is assumed based on benchmarking or the use of "factors" or generic equations. Other models, including the model applied herein, consider carbon equilibria, carbon kinetic decay, and other influences such as back-mixing from carbon-transfer operations.

A leach equation is fitted to the test-work data to provide an estimate of the gold leaching from the solids phase and into solution as a function of time.

Carbon performance decays as the carbon progresses up the tank train. The loadings in the plant will not be as high as determined in the laboratory using fresh carbon. The age of the carbon and the other species that may be present (silver and copper as well as water quality) influence carbon performance. The Fleming k and n values are discounted from the test work using factors based on operating experience and decayed as a function of carbon residence time.

While the model used herein considers a constant rate of decay in carbon characteristics, it is often found that the first carbon contactor is the one in which the carbon suffers the most due to the exposure of contaminants in the new circuit feed.

The carbon loading in each contactor is compared to the loadings predicted by the application of a Freundlich isotherm for the solution concentration calculated for each contactor. In the example herein, it has been assumed that the carbon loading should not exceed 70% of the equilibrium loading determined by test work. The use of such a discount is based on plant observations and experience.

Such adjustments provide some allowance for real-world plant operational issues compared to the laboratory environment, such as uneven carbon populations from one tank to another.

The CIL/CIP model applied herein considers the influence of back-mixing due to carbon transfer. In those cases where carbon populations are low, the amount of slurry that has to be back-mixed (transferred) to the preceding tank will be higher than those cases where populations are higher. This back-mixing influences the residence time distribution of the ore and will flatten the concentration profile of the gold in solution and therefore the corresponding carbon loadings.

Lower final loaded carbon grades require higher elution rates to recover a given mass of precious metal and therefore require elevated carbon transfer rates. Therefore, higher final carbon loadings will provide a benefit with regard to elution, electrowinning, and the reactivation plant by reducing the required capacity.

The model also considers the influence of precious metal in the process water entering the facility due to dilution (water sprays) as well as in the actual feed stream from the comminution circuit itself. Recycled eluate is also considered. The model assumes a split AARL circuit with a return barren eluate grade of 3 g/m^3. The amount of recycle is a function of the elution type and the elution rate required, and consequently a function of the loaded carbon grade achieved for each case.

Design Exercise for 3-g/t Head Grade

Cases Evaluated

The model was applied to assess seven plant arrangements, or cases. All cases have a total residence time of 24 hours. The cases are summarized as follows:

- **Case 1.** Pure CIL comprising six tanks. All tanks have carbon present.
- **Case 2.** Hybrid CIL comprising six tanks. The first tank does not contain carbon but all subsequent tanks do. Tank sizes and configuration are the same as for case 1.
- **Case 3.** Pure CIL comprising seven tanks. All tanks have carbon present. The tanks are smaller than those modeled for cases 1 and 2.
- **Case 4.** Hybrid CIL comprising seven tanks. The first tank does not contain carbon but all subsequent tanks do.
- **Case 5.** Pure CIL comprising eight tanks. All tanks have carbon present. The tanks are smaller than those modeled for cases 3 and 4.
- **Case 6.** Hybrid CIL comprising eight tanks. The first tank does not contain carbon but all subsequent tanks do.
- **Case 7.** CIP comprising four leach tanks (residence time of 18 hours) and six adsorption tanks (residence time of 1 hour each for a total of 6 hours), having a total residence time of 24 hours. This circuit is not a "pure" CIP, because if it were, the leach residence would be 24 hours alone. This configuration modeled can be considered a "practical" CIP design, where the minor leaching in the latter stages of the processing are accepted to reduce plant capital cost.

All cases assume the same carbon population in each tank as relevant to the case evaluated and constant carbon transfer. Design is based on 100% utilization, as is necessary to correctly size a facility.

Performance

Three plots of the solids grade, solution grade, and the carbon grade by tank are presented as Figure 8, Figure 9, and Figure 10 for select cases.

Figure 8 shows typical profiles for the various phases of solids, solution, and carbon. A carbon loading of just under 2,400 g/t has been achieved. When compared to Figure 9, one leach and five contactors, it can be seen that case 2 has achieved slightly higher carbon loading of just under 2,500 g/t. The contact solution grade (the solution grade equilibrating with the first carbon contact) for case 1 is just under 0.48 g/m^3, whereas for case 2 it is 0.54 g/m^3, allowing for a higher carbon loading.

As a result, case 2 has a slightly lower elution demand, which is advantageous in both elution capacity required and the amount of back-mixing that will be experienced. Consequently, case 2 may be considered preferable over case 1. However, the practical consideration is that the carbon profile is very steep for case 2 compared to case 1. If carbon populations are not properly managed, or if a tank were to be taken off-line for maintenance as will happen in normal operations, the impact this would have on the case-2 arrangement will be much more significant than it would be to case 1. Therefore, the consideration as to the most appropriate arrangement needs to consider the practical aspects and not just the theoretical and steady-state model output.

Figure 10 presents the CIP case. As there is no carbon present in the first four tanks, the solution grade can be seen to climb in excess of 1.9 g/m^3. The resulting solution grade in contact with the carbon in the first contactor is around 0.8 g/m^3 compared to the values of around 0.5 g/m^3 seen in case 1 and case 2. This results in higher carbon loadings, reduced elution demands, and reduced back-mixing associated with carbon transfers.

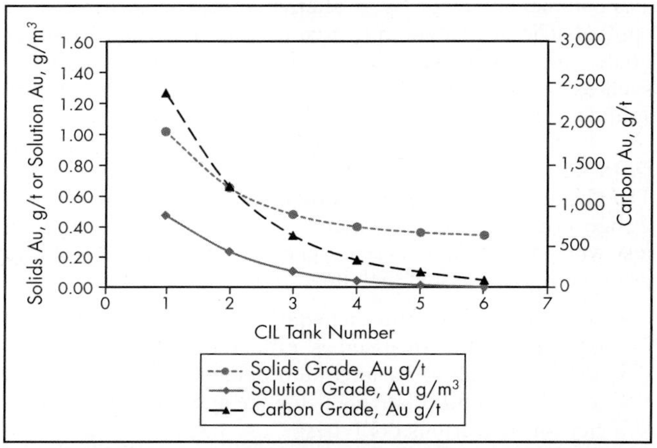

Figure 8 Case 1—Pure CIL, 3-g/t head grade, and six tanks

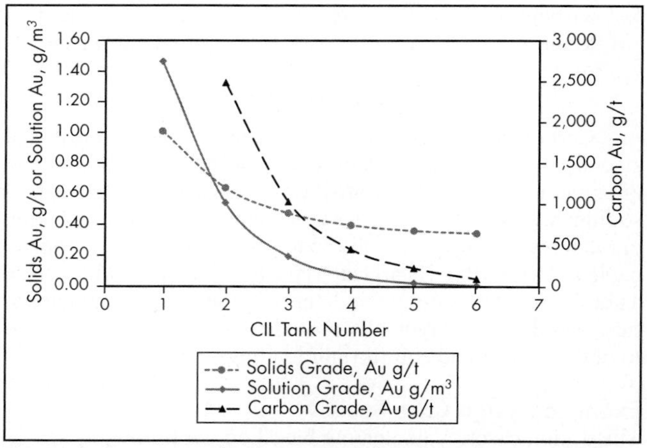

Figure 9 Case 2—Hybrid CIL, 3-g/t head grade, one leach tank, and five carbon contactors

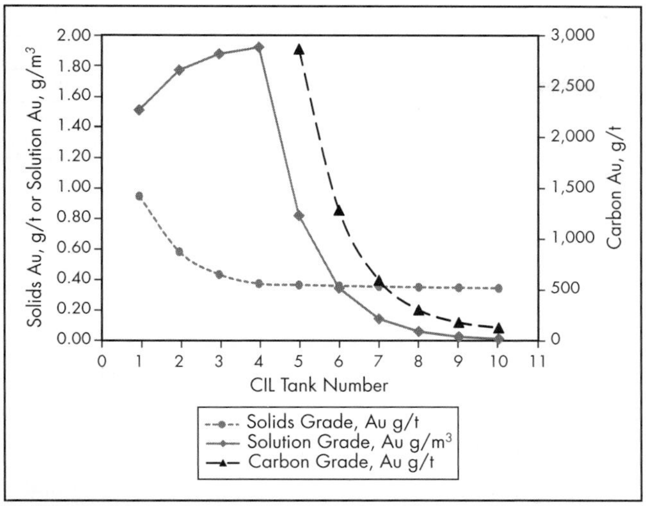

Figure 10 Case 7—CIP, 3-g/t head grade, four leach tanks, and six carbon contactors

The carbon profile is steep, and the loss of a carbon contactor due to maintenance will impact the profile as a consequence. However, in a CIP operation, it is easier to recover from such upsets by adding the small amount of displaced carbon to the other tanks or, in fact, adding new carbon, because the carbon inventory is so much lower than that of an equivalent CIL that the inventory is more easily managed.

Table 1 presents several inputs and outputs that allow the performance of the various cases to be evaluated. The tonnage rate, head grade, residence time, and pulp density have all been held constant in each of the cases modeled, as has the barren carbon grade, which is equivalent to the introduced carbon grade. Notably, the lower the introduced carbon grade, the lower the elution rate required for a fixed flow of precious metal, given that the loaded carbon grade is "fixed" by the contact solution grade. A more effective elution that provides lower barren carbons also means a smaller elution demand, and this is an important consideration when evaluating the type of elution circuit to be applied.

The model has been run to achieve the same soluble loss of 0.010 g/m^3 for each case. Given that the tailings pulp density is the same for each case, then the soluble loss for each case is also the same. In effect, the carbon circuit options are comparable based on the solution losses.

Referring to Table 1:

- The contact solution grade increases as the number of tanks is increased. Having a leach tank present escalates the contact solution grade and provides for an elevated carbon loading.
- The elution rate diminishes as the number of tanks increases/contact solution increases/loaded carbon grade increases.
- The carbon population decreases for the CIL cases as the number of tanks increases, as does the carbon residence time. However, the back-mixing due to carbon transfers increases as the number of CIL tanks increases.
- The back mixing results in a loss of extraction, which increases with the number of CIL tanks.
- A single leach tank in the CIL cases reduces the back-mixing and improves the extraction.
- Both the gold and carbon inventory decrease as the number of CIL tanks increases. The use of a single leach tank in the CIL cases escalates the gold inventory for comparable cases 1 and 2, but decreases the gold inventory for the comparable case 3 and 4 as well as comparable cases 5 and 6.

Table 1 Case performance summary for 3-g/t head grade

	Case 1	Case 2	Case 3	Case 4	Case 5	Case 6	Case 7
	CIL	Hybrid	CIL	Hybrid	CIL	Hybrid	CIP
Parameter	6 Adsorption	1 Leach 5 Adsorption	7 Adsorption	1 Leach 6 Adsorption	8 Adsorption	1 Leach 7 Adsorption	4 Leach 6 Adsorption
Tonnage rate, t/h	500	500	500	500	500	500	500
Head grade, Au g/t	3.00	3.00	3.00	3.00	3.00	3.00	3.00
Residence time, h	24	24	24	24	24	24	24
Tails pulp density, % solids w/w	41.8	41.8	41.8	41.8	41.8	41.8	41.8
Contact solution grade, mg/L	0.48	0.54	0.51	0.60	0.54	0.64	0.82
Solution loss (liquor grade) mg/L	0.0100	0.0100	0.0100	0.0100	0.0100	0.0100	0.0100
Carbon final loading, Au g/t	2,382	2,484	2,434	2,570	2,474	2,635	2,863
Barren carbon loading, Au g/t	100	100	100	100	100	100	100
Elution rate, kg/h	590	559	577	538	568	523	480
Elution rate, t/d	14.2	13.4	13.8	12.9	13.6	12.6	11.5
Carbon population, g/L slurry	9.0	10.7	7.5	8.4	6.5	7.0	16.5
Carbon residence time, days	13	14	11	12	10	10	8
Back-mixing (carbon transfer), m^3/h	65.8	52.1	77.2	64.2	86.8	74.6	29.1
Extraction from solids, no back-mix, %	88.58	88.58	88.58	88.58	88.58	88.58	88.58
Extraction with back-mixing, %	88.44	88.49	88.41	88.47	88.38	88.44	88.57
Loss due to back-mixing, %	0.14	0.09	0.17	0.11	0.20	0.14	0.01
Gold inventory, oz	5,040	5,260	4,093	4,086	3,509	3,403	2,495
Carbon inventory, t	189	189	158	152	138	130	87

Table 2 Financial considerations for 3-g/t head grade

Parameter	Case 1 CIL 6 Adsorption	Case 2 Hybrid 1 Leach 5 Adsorption	Case 3 CIL 7 Adsorption	Case 4 Hybrid 1 Leach 6 Adsorption	Case 5 CIL 8 Adsorption	Case 6 Hybrid 1 Leach 7 Adsorption	Case 7 CIP 4 Leach 6 Adsorption
Extraction loss (8,000 h/yr), oz/yr	552	339	665	444	766	541	31
Loss in revenue, US thousand $/yr	662	407	798	532	920	649	37
Gold inventory, oz	5,040	5,260	4,093	4,086	3,509	3,403	2,495
Value of gold inventory, US million $	6.05	6.31	4.91	4.90	4.21	4.08	2.99
Incremental value over CIP case 7, US million $	3.05	3.32	1.92	1.91	1.22	1.09	0
Carbon inventory (including elution and reactivation), t	232	229	199	190	179	167	122
Value of carbon inventory, US thousand $	139	137	120	114	107	100	73

- The CIP case presents with reduced elution rates, lowest gold and carbon inventory, and, importantly, the lowest back-mixing and highest extraction.

The capital cost of the tank farm will grow as the number of tanks increases for a given residence time. The hybrid CIL options (cases 2, 4, and 6) would be a marginally lower cost than the corresponding pure CIL options (cases 1, 3, and 5) because there is minor savings in inter-tank screen costs and transfer pumping.

The CIP case would have the higher capital associated with the tank farm. There will be savings in some areas compared to the CIL options (screen access and transfer pumping), but the increased tankage costs will generally negate these benefits.

The capital cost of the elution, electrowinning, and reactivation kiln facilities will cost less as the number of tanks increases. This provides some capital cost offset when considering CIL/CIP tank configurations for a new project.

Although the pure capital cost is one of the most important considerations when evaluating plant design options, other factors will significantly influence the cash flow of the project, particularly in the all-important start-up and finance payback period. These factors can be considered as working capital and can be expected to influence the terms of any finance agreements thay may be required to fund a project; the establishment of the inventory will delay payments to the financiers but will require operating costs to be sunk.

Table 2 summarizes some of these factors and quantifies them based on a gold price of US$1,200/oz and a carbon cost of US$600/t. The "extraction loss," based on an 8,000-hour operating year, is the difference in the theoretical leach extraction without back-mixing compared to what is actually achieved when back-mixing is considered. The CIP case has a minor loss in revenue because of this inefficiency, whereas the CIL options, particularly the higher number of tanks options, suggest elevated losses.

The value of the gold inventory is a significant cost. As the tank numbers increase, the inventory is reduced. This is a "one-off" cost to the project, because once the inventory is established, it remains until the end of the mine life. However, this can be considered as cash flow that is sunk in the first month or two of the project life. The CIP presents the best case with regard to inventory lockup, as it presents the lowest value.

The lower-capital-cost CIL cases (cases 1 and 2) in reality have an additional inventory-based cash-flow burden compared to the more elaborate CIL cases or the CIP case. So although cases 1 and 2 may have the lowest capital and the least loss due to back-mixing of the CIL cases, cases 1 and 2 also have the highest inventory to contend with. It is likely that the cost of inventory for cases 1 and 2 would exceed the incremental capital associated with cases 3 and 4.

The hybrid CIL described in case 2 may still appear appealing as the lowest entry capital cost option compared to the most expensive CIP option, case 7. Case 2 will have a loss of nominally US$400,000/yr in extraction and require an additional US$3.3 million to be sunk as gold inventory compared to CIP case 7. This is nominally US$4.5 million over a three-year non-discounted payback period, which is a significant disadvantage to the CIL case. In addition, the CIP will be more robust to operate and will require lower cost elution, electrowinning, and reactivation facilities with associated reduced operating costs.

It is therefore necessary to consider not just the basic metallurgical demands and the entry capital cost when designing a CIL or CIP facility, but also the operability, inventory, and losses due to the operating mechanisms of the design cases to reach the most appropriate financial case.

A further consideration in circuit selection is the variability of the ore blends that the plant will experience. Particularly in the payback period of the operation, it is common for operations to experience soft oxides at start-up, potentially of lower grade than deeper sulfide ores. Then as the open pit develops, the sulfides present and require different operating conditions or a different CIL/CIP design. They may require more comminution but may also have different leaching, adsorption, and carbon management demands. It is therefore paramount for the designer to fully understand the production schedule implications with regard to what the design must contend with as a function of time.

The influence of carbon inventory itself (i.e., cost of carbon) are insignificant compared to the capital cost and gold inventory considerations for cases 1 to 7.

Design Exercise for 4-g/t Head Grade

To present the influence that variable parameters can have on the CIL/CIP design, a second series of cases was run. The same plant configurations were used as had been designed for cases 1–7 presented earlier. However, the head grade of the ore was increased to 4 g/t and the leach rate moderated, as could be expected if the various plant configurations were to experience a change in ore type.

The cases here are designated with an "H" to infer "high-grade." The 3-g/t head grade case 1 is directly comparable with the 4-g/t case 1H.

In summary:

- **Case 1H.** Pure CIL comprising six tanks.
- **Case 2H.** Hybrid CIL comprising six tanks.
- **Case 3H.** Pure CIL comprising seven tanks.
- **Case 4H.** Hybrid CIL comprising seven tanks.
- **Case 5H.** Pure CIL comprising eight tanks.
- **Case 6H.** Hybrid CIL comprising eight tanks.
- **Case 7H.** CIP comprising four leach tanks (residence time of 18 hours) and six adsorption tanks (residence time of 1 hour each for a total of 6 hours), having a total residence time of 24 hours.

Table 3 presents the inputs and outputs so the cases can be compared. The same general trends as were found for the 3-g/t modeling are presented here. Some notable differences resulted from slower leaching and higher quantity of precious metal to be processed, the most significant of which are listed here:

- The carbon residence times for the CIL options have all increased from nominally 10–13 days (cases 1 to 6) up to 20–28 days for cases 1H to 6H. These carbon residence times are considered too high and, as such, these CIL options would need to have lower carbon loadings applied and much higher elution rates enforced in reality.
- The slower leach rate and higher grade is pushing more gold to the tail end of the circuit. This is advantageous to the CIP case. The contact solution grades for the CIL cases are lower here than what they were for the 3-g/t cases even though the head grade is higher, resulting in lower carbon loadings but still having to process around 30% more precious metal.
- The CIL carbon populations have increased significantly by around a factor of 3, whereas the CIP population has increased by a factor of 30%, in line with the incremental precious metal flow.
- Gold inventory for the CIL options has increased dramatically.

Cases 1H to 6H show that if a circuit had been selected based on the case 1–6 models, then the circuit would have been compromised by the higher-grade but slower-leaching ore. The likely practical outcome would have been to maximize the elution rate, maximize carbon population to the point of maximum carbon residence time, and suffer the soluble losses that would occur if the ore change had not been allowed in the design. This is where an AARL elution circuit would prove advantageous over a Zadra facility, as increased elution rates would be more easily achieved.

Table 4 summarizes the same values as had been presented in Table 2. The loss in revenue per year due to back-mixing has jumped for all seven cases, but is most significant for the CIL cases. The gold inventory has increased slightly for CIP case 7H, but for the CIL cases, the inventory has increased by a factor of 3. There is now a very high proportion of working capital to be considered if the lower-capital-cost CIL options are to be considered.

The CIP (case 7H) in this example has handled the change in characteristics of the ore with little impact. The higher-grade and slower-leaching ore is easily managed. However, the CIL options, although probably competitive for the 3-g/t scenarios, would in reality not be applicable to the 4-g/t cases without significant modification.

If this 4-g/t ore were to be harder than the 3-g/t ore, and it was to be processed at a lower throughput, then the CIL options would be expected to perform more effectively.

Table 3 Case performance summary for 4-g/t head grade

Parameter	Case 1H CIL 6 Adsorption	Case 2H Hybrid 1 Leach 5 Adsorption	Case 3H CIL 7 Adsorption	Case 4H Hybrid 1 Leach 6 Adsorption	Case 5H CIL 8 Adsorption	Case 6H Hybrid 1 Leach 7 Adsorption	Case 7H CIP 4 Leach 6 Adsorption
Tonnage rate, t/h	500	500	500	500	500	500	500
Head grade, Au g/t	4.00	4.00	4.00	4.00	4.00	4.00	4.00
Residence time, h	24	24	24	24	24	24	24
Tails pulp density, % solids w/w	41.7	41.7	41.7	41.7	41.7	41.7	41.8
Contact solution grade, mg/L	0.35	0.47	0.35	0.49	0.35	0.50	0.94
Solution loss (liquor grade), mg/L	0.0100	0.0100	0.0100	0.0100	0.0100	0.0100	0.0100
Carbon final loading, Au g/t	2,142	2,370	2,140	2,401	2,139	2,419	3,000
Barren carbon loading, Au g/t	100	100	100	100	100	100	100
Elution rate, kg/h	899	801	903	791	906	786	619
Elution rate, t/d	21.6	19.2	21.7	19.0	21.7	18.9	14.9
Carbon population, g/L slurry	28.1	30.5	23.6	25.2	20.6	21.9	20.7
Carbon residence time, days	28	28	23	24	20	21	7
Back-mixing (carbon transfer), m^3/h	32.0	26.3	38.2	31.4	44.1	36.0	29.9
Extraction from solids, no back-mix, %	89.70	89.70	89.70	89.70	89.70	89.70	89.70
Extraction with back-mixing, %	89.45	89.54	89.40	89.49	89.35	89.46	89.67
Loss due to back-mixing, %	0.25	0.16	0.30	0.21	0.35	0.24	0.03
Gold inventory, oz	15,630	14,885	12,590	12,018	10,602	10,271	3,110
Carbon inventory, t	593	537	499	456	435	404	110

Table 4 Financial considertations for 4-g/t head grade

Parameter	Case 1H CIL 6 Adsorption	Case 2H Hybrid 1 Leach 5 Adsorption	Case 3H CIL 7 Adsorption	Case 4H Hybrid 1 Leach 6 Adsorption	Case 5H CIL 8 Adsorption	Case 6H Hybrid 1 Leach 7 Adsorption	Case 7H CIP 4 Leach 6 Adsorption
Extraction loss (8,000 h/yr), oz/yr	1,272	838	1,549	1,058	1,791	1,249	168
Loss in revenue, US million $/yr	1.53	1.01	1.86	1.27	2.15	1.50	202
Gold inventory, oz	15,630	14,885	12,590	12,018	10,602	10,271	3,110
Value of gold inventory, US million $	18.76	17.86	15.11	14.42	12.72	12.33	3.73
Incremental value over CIP case 7H, US million $	15.02	14.13	11.38	10.69	8.99	8.59	—
Carbon inventory (incuding elution and reactivation), t	658	594	564	513	500	460	154
Value of carbon inventory, US thousand $	395	357	339	308	300	276	92

If leach kinetics enhancement could be undertaken, for instance, by adding oxygen to the leach, then again the CIL performance would improve. More of the gold would leach earlier and this would result in much lower carbon inventory and higher loaded carbon grades. This point also highlights why increased leach kinetics may be beneficial even if the final extraction is the same—the adsorption circuit is more efficient, and both elution demands and back-mixing are reduced.

If the period of time that the 4-g/t ore were to be experienced was short or the 4-g/t material could be blended, then the CIL options again may be positive. These points all highlight the need for the designer to understand the ores to be processed and how the ore variability may influence the design.

Although the scenarios run for the 4-g/t head grade support CIP, it is often the case that for faster leaching and perhaps lower-grade ores there will be a bias back to CIL preference over CIP. If ores show any preg-robbing tendencies, then full CIL will be the preferred flow sheet and the designer will need to establish how a CIL facility can be designed/configured to deal with issues such as elevated grades and slow-leaching components. Use of different carbon populations in the different tanks and higher elution rates at lower carbon loadings are options that could be explored.

In summary, the designer should evaluate the plant configuration options with due consideration of the ore characteristics, how these characteristics vary, and the production schedule. Assuming a configuration and simply applying it to a life-of-mine composite or some other "general" ore can be expected to result in a suboptimal outcome. This is particularly true if the pure entry capital of the tank farm is used in isolation as a decision-making tool.

The designer also has to be aware of the financial environment. Cost of finance, finance type and terms, influence of working capital (inventory), and capital cost all playing a part in the flow-sheet decision process. Consideration as to operating cost influences (reduced stripping rate due to higher carbon loadings) also play a part in the financial process considerations. The process is not being designed to be the most metallurgically efficient or to have the lowest capital or operating cost, but to make the best or most robust return under the prevailing financial constraints.

Silver, Copper, and Other Species

The cases presented previously assumed no complications from other species in solution. In some instances, such as if there are high values of silver and/or copper in solution, there will be increased complexity in the modeling and design process, as these metals may adsorb and retard gold adsorption, or even desorb and free up sites for the gold under certain conditions. This then infers additional steps that would need to be considered in the modeling of the gold adsorption itself.

If low levels of silver are present and they have been included in the carbon characterization test work for gold, then the design process remains relatively simple. The circuit can be designed for the gold component, and the silver balance (loaded carbon grade of silver) can be back-calculated based on the carbon flow rate required to manage the gold. Only if the equilibrium loadings of silver are exceeded is there a need to increase the carbon flow rate over that required for the gold component alone such that the silver is managed. In such cases, the silver loadings will drive the elution circuit capacity and not the gold.

Higher silver-grade ores can become problematic. The carbon flow rates may become so great as to make the circuit impractical. Examples exist in industry where silver desorption is witnessed as the carbon is progressed or transferred upstream. Gold adsorption still occurs by utilizing the sites vacated by the silver, but the interim desorption of the silver results in the gold having an apparent reduction in kinetics. The silver is displaced downstream where it loads onto carbon again, and thus an internal recycle of silver presents as it is again moved forward with the carbon. There will be a high soluble loss of silver under such scenarios, as this behavior occurs when the carbon circuit cannot deal with the excessive quantities of silver present.

In such cases, consideration should be given to countercurrent-decantation/Merrill–Crowe-type flow sheets or hybrid flow sheets instead of carbon-only-based flow sheets. Alternatively, additional carbon characterization test work is required and a comprehensive understanding of the silver variability in the plant feed is necessary to retain carbon-based process routes.

Copper presents high adsorption kinetics but can also be controlled to a degree with free cyanide concentration. Elevated cyanide concentration results in the $Cu(CN)_4^{3-}$ complex being dominant over the $Cu(CN)_3^{2-}$ complex. The $Cu(CN)_4^{3-}$ complex has a low affinity for carbon and, if adsorbed, will give up sites for gold and silver adsorption. There is a cost associated with the cyanide demand, and if copper levels are too high, the copper will continue to dominate the carbon and adversely impact the gold and silver loadings. High cyanide-soluble copper ores require detailed evaluation before effective circuit modeling and confidence in design can be achieved. Cyanide consumption for both copper dissolution and carbon management often make such high copper ores uneconomic.

The site water itself can influence the design of a carbon-based circuit. The cations Ca^{2+}, Na^{+}, and Mg^{2+} are all known to influence carbon adsorption characteristics as will the CN^{-} radical itself. pH also impacts carbon adsorption characteristics. For these reasons, it is key that the designer has considered the influences with adequate evaluation at the laboratory stage of development to ensure that carbon performance is understood.

Model Alternatives

The CIL/CIP model described in this chapter is just one of many methods used to estimate circuit performance or develop a design. Simpler models are used and are applied, as are more complex models. The simplest models have their place but are generally limited to study-level evaluations or are applied to ores for which the confidence in predicted performance is high. Examples include a project expansion on the same ore type, a project restart where operating history from the past is available, or processing an ore from a region of reliable characterization, such as many of the oxide ores from the Western Australian Goldfields.

Models often do not consider the equilibrium loadings. The design needs to consider the kinetics to understand how much carbon is required and how quickly it needs to be progressed through the plant. The design also needs to ensure that the loadings do not exceed what is chemically possible with regard to loading—and this implies that equilibrium data and assessment are key in optimizing a design.

Simple models include the following:

- Carbon loading factored based on the head grade of the ore. This type of model does not allow for pulp density influences, tank configuration, or competing species. As presented according to the previous modeling, the grade of the solution in contact with the carbon in the first contactor influences the final loading; consequently, models that do not estimate this solution value result in escalated risk when estimating the achievable final carbon loading.
- Carbon loading factored on solution grade. The precious metal leached is estimated to provide a solution grade that is fed to the first carbon contactor. A factor is then applied to give the anticipated carbon loadings. The issue with this method is that it is the solution tenor in the first contactor in contact with the carbon that dictates the carbon loading, not the grade of the introduced solution. Carbon activity, population, and how much additional leaching occurs in the first contactor will influence the actual solution tenor in contact with the carbon.
- Carbon loading based on a "generic" algorithm that is based on a head-grade or solution-grade assumption. Some models use an equation based on a large number of historical observations to estimate carbon loadings, for both gold and silver. Such models assume the ore is therefore generic. If the ore being assessed is an outlier, then deviation from the generic algorithm can be expected. This may be positive or negative in outcome with regard to loading achieved, but even a positive/conservative outcome means an over-designed plant and is therefore suboptimal.
- Stage efficiency. Most circuits are observed to operate with a solution-stage efficiency of around 50%. That is, tank $n + 1$ will have a solution tenor of 50% of tank n. The various cases presented (cases 1 to 7H) had a range of stage efficiency of 40%–65%, suggesting that a fixed-stage efficiency is invalid. The leach rate, number of tanks, and configuration all give different efficiencies to achieve the same soluble loss of 0.01 g/m^3.

Such methods have their place, but the designer needs to appreciate that the method applied to design a CIL or CIP circuit may have great bearing on the confidence and reliability of the outcome. Reliable design requires reliable inputs. Representative test work remains the key to reliably estimating plant performance, particularly for the more complex and atypical ore types.

REFERENCES

Adams, M.D., 1994. The elution of gold from activated carbon at room temperature using sulphide solutions. *J. S. Afr. Inst. Min. Metall.* (August):187–198.

Claflin, J., La Brooy, S., and Preedy, D. 2015. Lessons learnt and performance—Installing and commissioning an Ausenco carbon reactivation kiln in Africa. In *MetPlant 2015.* Melbourne, Victoria: Australasian Institute of Mining and Metallurgy.

Fast, J.L. 1989. Carbon stripping: The practical alternatives. Presented at the International Gold Expo, Reno, NV, September 7.

Filmer, A. 1982. The electrowinning of gold from carbon in pulp eluates. In *Carbon-in-Pulp Technology for the Extraction of Gold: Proceedings*. Melbourne, Victoria: Australasian Institute of Mining and Metallurgy.

Fraser, G. 1984. Aeration design for gold ore leaching. In *Gold Mining, Metallurgy and Geology: Proceedings*. Melbourne, Victoria: Australasian Institute of Mining and Metallurgy.

Mardhiha, I., Norhazilan, N., Nordin, Y., Arman, A., Rosilawati, R., and Ahmad, S.A.R. 2014. Effect of pH and temperature on corrosion of steel subject to sulphate-reducing bacteria. *J. Environ. Sci. Technol.* 7(4)209–217.

Zadra, J.B., Engel, A.L., and Heinen, H.J. 1952. *Process for Recovering Gold and Silver from Activated Carbon by Leaching and Electrolysis*. USBM Report of Investigation 4843. Washington, DC: U.S. Bureau of Mines.

CHAPTER 10.13

Ion Exchange Technologies

Brandon Kern

Ion exchange processes can be used for extractive metallurgy in situations where ionic concentrations are relatively low, generally with a practical upper bound in the thousands of parts per million (tenths of a percent) of metal concentration in solution. Some overlap of application space exists between solvent extraction and ion exchange for metals extraction; at lower concentrations where solvent extraction encounters diminishing return, ion exchange processes find increased removal efficiency. As such, in many applications, ion exchange processes are considered for final polishing applications; however, some key areas have found good success in utilizing ion exchange for primary recovery. These applications are discussed in greater detail herein.

The techniques that use ion exchange technology in mining are multiple and diverse. Technology for mining applications is usually decided according to the recovering techniques being used, including in situ leach (ISL), alkaline or acid leaching, resin-in-pulp (RIP), and heap leaching. Each technique requires specific characteristics for the ion exchange resin (IER) technology.

The bulk of ion exchange processes for primary recovery generally focus in one of three different areas: base metals mining (copper, cobalt, nickel), uranium mining, and gold recovery. However, the application of ion exchange also extends to the recovery of high-value, low-volume products, such as rare earth metals or platinum group metals, and more recently in the processing of lithium-containing brines. Depending on the desired separation, different IER types and chemistries are deployed. Generally, the major design parameters are resin morphology (gel versus macroporous), resin bead size and size distribution, and resin functional group.

The copolymer backbone of IERs (usually styrenic, but occasionally phenolic or acrylic polymers) dictate many of their physical properties. An exhaustive summary of the different types of ion exchangers and their properties is outside the scope of this handbook but is extensively reported by other sources (Dorfner 1991; Helfferich 1962; Kunin 1972). Important to mining applications, however, is the distinction between gel copolymer matrices and macroporous polymers. Gel polymers consist of a continuous polymer phase that can be swollen with water or process fluid to occupy roughly 50% of the resin's volume. This water of hydration is responsible for shuttling ionic species into and out of the resin bead to facilitate contact with the charged ion exchange functional groups contained therein. Macroporous IERs are produced of similar polymer that has microchannels and voids present throughout; these channels provide for increased access for contact between ions in solution and the resin's functional groups. In practice, gel copolymer IERs tend to offer higher total exchange capacity than their macroporous counterparts, whereas macroporous resins tend to have a higher overall mechanical stability and fouling resistance. For these reasons, gel resins are often the best choice for treatment of clarified solutions where their higher capacity can be utilized. Macroporous resins are best suited to applications where physical stability and fouling resistance are important, for example, RIP applications. Figure 1 illustrates some of the physical structure differences between macroporous and gel copolymers.

A wide variety of different ion exchange functional groups can be utilized for mining processes, each with their own specific advantages and applications. The most typical standard anion and cation exchange functional groups are shown in Table 1. A separate family of IERs use chelating agents that are bound to the copolymer matrix; these chelating resins can utilize a wide variety of different chemistries to selectively bind certain classes of metals and can have specific utility in mining applications.

Most hydrometallurgical applications of ion exchange involve the use of strong base anion exchangers because many of the metals of interest form stable anionic species in the leachate stream. Some metals separations, however, may require the use of selective IERs (generally referred to as *chelating* resins) that incorporate different chemical functionality tailored to selectively remove one particular species, frequently in the presence of high background concentrations of undesired species. Many of the developments in this area involve the creation of new resins with different functionality, higher capacity, and better selectivity, which may involve the use of different chelating chemistry or adsorbent technology.

Brandon Kern, Technical Service Specialist, Ion Exchange, The Dow Chemical Company, Midland, Michigan, USA

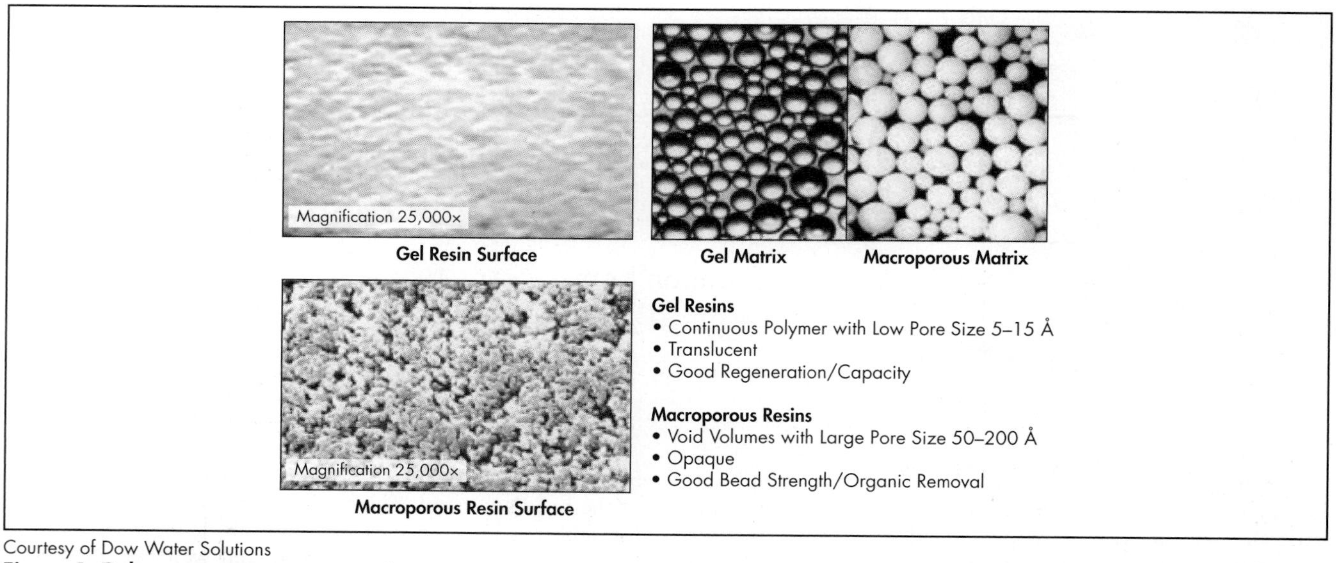

Courtesy of Dow Water Solutions

Figure 1 Gel versus macroporous resins

Table 1 Typical functional group families of ion exchange resins

Ion Exchange Resin Class	Typical Structure
Strong acid cation exchanger	~~~~ SO_3H
Weak acid cation exchanger	~~~~ COOH
Weak base anion exchanger	~~~~ NR_2
Strong base anion exchanger (chloride form)	~~~~ $NR_3{}^+Cl^-$
Copolymer backbone (styrenic, acrylic, or phenolic)	~~~~

Table 2 Characteristics of resin families used in metallurgy

Chelating Resins	Strong or Weak Base Anion Resins	Strong Acid Cation Resins
Highly selective	Used when total anion removal needed	Used for total cation removal
Work in concentrated feeds	Effective for separation of metallochloride complexes in chloride matrices	Macroporous, Gaussian particle size distribution generally favored (fouling resistance)
Selectivity is *tunable* with pH	Frequently used for $AuCN^-$ and uranium applications	Medium to high levels of sulfonation (high capacity)

Some general descriptions of the different resin families used in metallurgical processes are described in Table 2.

URANIUM EXTRACTION USING ION EXCHANGE

Strong and weak base anion exchangers have been employed in uranium recovery from ISL systems since the 1950s (Benes et al. 2001). Early divergence in ISL philosophy led to the development of mainly carbonate leach systems in North America, whereas sulfuric acid leaching techniques have generally been preferred in other geographies. The role of the IER is to selectively fix uranium from the leach solution and then recover it by eluting from the resin for further processing. Strong base anion resins are now mainly used to recover uranium, both from acid leach and from carbonate leach circuits,whereas use of a cation exchange resin would permit loading of the $UO_2{}^{2+}$ directly, and loading competition would exist between the uranium and other cations (iron, calcium, magnesium, etc.), which are likely at much higher concentrations. Recovery of anionic uranyl sulfates and uranyl carbonates allows for a higher-purity product as it mitigates interference from these other contaminants. The specific resin functional group and the polymer matrix can be varied according to the solution composition to further limit the influence of interfering anions.

Carbonate Leaching

Most common in the United States are ISL systems involving dissolution of uranium into sodium carbonate solutions, as opposed to sulfuric acid solutions that are practiced in other geographies. The following reactions are relevant to uranium ion exchange in carbonate leach systems:

$$UO_2{}^{2+} + O_2 \Leftrightarrow UO_3 \qquad \textbf{(EQ 1)}$$

$$UO_3 + 2Na_2CO_3 + H_2O \Leftrightarrow Na_4UO_2(CO_3)_3 + 2NaOH \qquad \textbf{(EQ 2)}$$

$$2(R^+)_2CO_3{}^{2-} + UO_2(CO_3)^{4-} \Leftrightarrow (R^+)_4UO_2(CO_3)_3{}^{4-} + 2CO_3{}^{2-} \qquad \textbf{(EQ 3)}$$

Alkaline conditions result in the formation of a quadrivalent anionic carbonate complex. This approach has dual advantages of both yielding a complex with very strong selectivity for both strong and weak base IERs as well as the avoidance of loading many metal impurities that do not form anionic complexes. Leaching of loaded resin is typically performed using an eluent composed of a mixture of sodium carbonate and sodium bicarbonate salts (Zaganiaris 2009).

Acid Leaching

Acid leach is practiced using sulfuric acid (H_2SO_4). Uranium in the leach solution forms uranyl sulfate complexes according to the following reactions:

$$UO_2{}^{2+} + SO_4{}^{2-} \Leftrightarrow UO_2(SO_4) \qquad \text{Log } K_1 = 2.98 \qquad \textbf{(EQ 4)}$$

$$UO_2{}^{2+} + 2SO_4{}^{2-} \Leftrightarrow UO_2(SO_4)_2{}^{2-} \qquad \text{Log } K_2 = 3.83 \qquad \textbf{(EQ 5)}$$

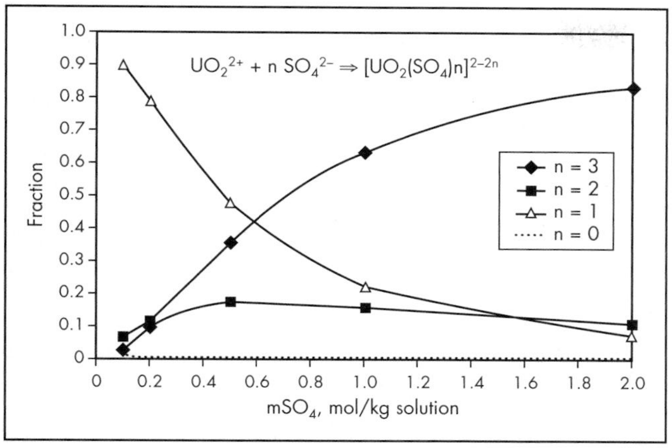

Courtesy of Dow Water Solutions

Figure 2 Uranyl complex ionic form as a function of sulfate concentration

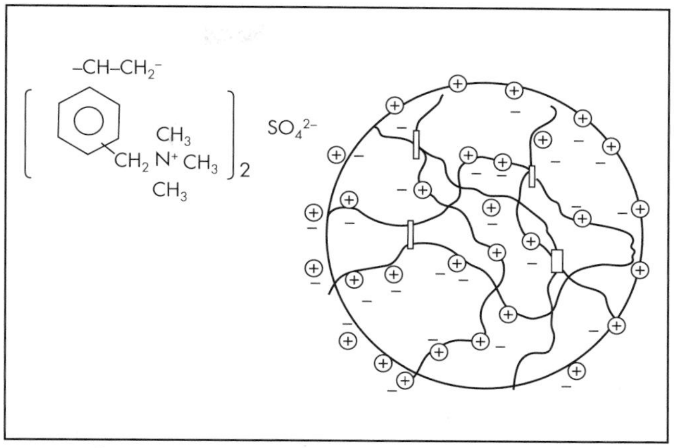

Courtesy of Dow Water Solutions

Figure 3 Donnan potential

$$UO_2^{2+} + 3SO_4^{2-} \Leftrightarrow UO_2(SO_4)_3^{4-} \quad \text{Log } K_3 = 3.45 \quad \textbf{(EQ 6)}$$

And in addition,

$$HSO_4^- \Leftrightarrow H^+ + SO_4^{2-} \quad K_d = 0.012 \quad \textbf{(EQ 7)}$$

Both divalent and quadrivalent anionic sulfate complexes are formed with uranium in acidic leach systems. Optimum pH for recovery of uranium is around 2 with a concentration in sulfate of approximately 0.25 mol/kg in solution; under these conditions, formation of the quadrivalent complex is preferred. Loading of the uranyl sulfate complexes onto anion exchange resins takes place according to the following reaction:

$$2(R^+)_2\ SO_4^{2-} + UO_2(SO_4)_3^{4-} \Leftrightarrow (R^+)_4UO_2(SO_4)_3^{4-} + 2SO_4^{2-} \quad \textbf{(EQ 8)}$$

ISL circuits in sulfate service are most commonly eluted using sulfuric acid. From Equations 1 to 3, the relative fractions of the different species can be calculated for a given concentration of uranium and SO_4^{2-} in solution. Figure 2 shows the ionic form of the uranyl as a function of the concentration of sulfate. This illustrates that the uranium enters the resin as a neutral species. At higher concentration in sulfate, the loading of uranium in the resin decreases, which corresponds to the point where uranium is in its negatively charged form.

The explanation for this phenomenon resides in the concept of the Donnan potential, described graphically in Figure 3. The Donnan potential has one immediate consequence for electrolyte sorption: It repels co-ions from the exchange resin and thus prevents the internal co-ion concentration from rising beyond an equilibrium value, which is usually much smaller than the concentration in the external solution. Thus, only the neutral species can easily enter the resin matrix while charged molecules or ions are repelled. This concept is the basis for the concept known as *ion exclusion*.

Chloride Effect

Uranyl sulfate enters in the matrix and forms a stable complex with the sulfate attached to the functional group. So the mechanism of uranium fixation can also be seen as a chelating mechanism. Chloride impairs the loading of the resin despite a lower affinity in comparison with the sulfate, which can be explained as the chloride causing an interruption in the formation of uranyl sulfate complexation with nearby sulfate groups. As such, modest increases in chloride concentration can result in significant impairment to loading capacity.

Silica Fouling

In many mining applications, silica fouling becomes critical. If silica fouling is not overcome, specific issues, such as irreversible fouling, impaired kinetics, resin coalescence and pressure drop, and so forth, ultimately may lead to shutdowns of production. Silica fouling is considered a priority in RIP technology because the ore is in contact with the lixiviant solution and the resin, the ore contains silica, and the concentration of silica is more important.

The mechanism of silica fouling is not completely understood, but many hypotheses exist. Silica is not very soluble at the low pH commonly used in the mining process. As pH of a system increases above about 8, silica solubility increases dramatically. Perhaps even more important is the morphology of the silica with varying pH; silica has many somewhat soluble gel phases at different pH levels.

Consideration of these factors, along with the Donnan effect described earlier, leads to an interesting discussion of the various portions of the polymer bead. If one dissects the main layers of the bead into the Nernst, or unstirred, layer and maintains the remainder of the bead as internal to the Nernst layer, a plausible hypothesis emerges relating to the depth of silica fouling on the resin. Silica generally accumulates in IERs in a layer very near the surface of the bead, consistent with the location of the Nernst film, as shown graphically in Figure 4.

Contacting Systems

These ion exchange systems can utilize different forms of resin contactors, depending on the application. Most common for uranium processing are fixed-bed systems configured in a lead–lag arrangement. In these systems, clarified leach liquor is passed through a lead column, followed by a polishing column, while a third column is being regenerated. As the lead column becomes exhausted, it is removed from service and eluted; consequently, the lag column takes the lead position and the third, freshly regenerated column is placed in the lag position. This arrangement ensures lower leakage (loss of uranium in the barren solution) and also results in improved loading of uranium onto the lead bed.

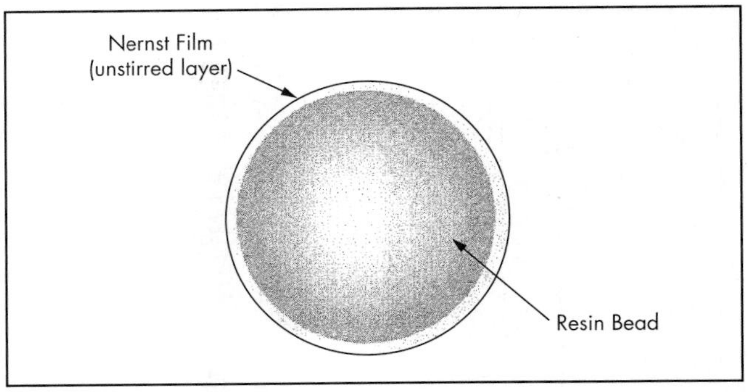

Adapted from Dorfner 1991

Figure 4 Layers of the ion exchange bead

In some instances, uranium ore bodies are of sufficiently high grade as to permit use of traditional grinding and leaching processes for uranium extraction. For such operations, uranium-containing ore is ground and leached with sulfuric acid, which leaches the uranium in the form of uranyl sulfate complexes. The resultant slurry may then be clarified prior to ion exchange by means or resin-in-leach (RIL) contacting, or the slurry (pulp) may be contacted directly with the ion exchange media via an RIP process. RIL applications may make use of many different fluidized-bed continuous ion exchange contacting systems, for example, the Higgins Loop (Higgins 1957) or the Porter system (Porter 1975). RIP processes generally use a series of stirred-vessel contactors through which the resin and pulp are exchanged in a countercurrent fashion. Uranium-rich pulp is fed into subsequent contactors from left to right, decreasing in grade through ion exchange in each of the contacting vessels. IER is screened and transferred from right to left, countercurrent to the pulp, such that as the resin loads, it is transferred into sequentially higher-concentration pulps, thus enhancing its loading efficiency. With RIP contactors, the physical stability and robustness of the selected IER is especially important, as physical contact with the pulp can crush and abrade resins with time (Yahorava et al. 2009). Additionally, larger-bead resin products are preferred as they are easier to separate from the pulp via screening.

RECOVERY OF BASE METALS

Base metal recovery can be attained using a few different ion exchange approaches. Many of the transition metals react in the presence of high halogen concentrations to form anionic metal–halogen complexes. This feature can be exploited in the presence of hydrochloric acid, where metal–chloride complexes form based on their affinity for chloride complex formation and the concentration of hydrochloric acid in the solution. This topic is covered in greater detail elsewhere (Kraus 1956). Separation of chloride-complexed metals can be effectively achieved through the use of anion exchangers; some control of the metals that are removed versus those that stay behind in solution can be influenced by changing the concentration of hydrochloric acid (HCl) being used in the process.

Other options for base metals recovery include the use of IERs that possess selective chelating ligands (Mendes and Martins 2004). Some chelating ligands, such as aminomethylphosphonic acid or iminodiacetic acid can be characterized as having improved selectivity for divalent and multivalent cations over monovalent species such as sodium and potassium. As a result, these resins can be used for selective metals recovery in the presence of brines. Other chelating ligands have high specificity for select classes of metals. For example, bis-picolylamine (BPA) functionalized chelant resins have demonstrated exceptional selectivity for copper and nickel (Grinstead 1984); and hydroxypropyl picolylamine (HPPA) functionality offers similar selectivity but improved regeneration characteristics (Grinstead 1979). The chemical structures of these functional groups are illustrated in Figure 5.

Lateritic Nickel Recovery by Ion Exchange

The hydrometallurgical processing of laterites to melting-grade nickel is complicated because of the high concentration of iron present in the pregnant leach solution (PLS). The use of chelating IERs for the selective removal of nickel from this high-iron PLS offers several advantages over more traditional techniques; for example, it allows for the selective removal of nickel from lateritic PLS with most iron reporting to the barren. Contaminating iron is then efficiently scrubbed from the resin using reducing agents and polished from the eluate with iron-selective resins. The resulting high-purity nickel eluate is ready for downstream electrowinning or precipitation without the need for further concentrating. Figure 6 details the individual development steps for a typical nickel laterite project and where iron removal is required. This approach to lateritic nickel recovery is facilitated by a BPA-functionalized ion exchanger that has particularly high selectivity for nickel and

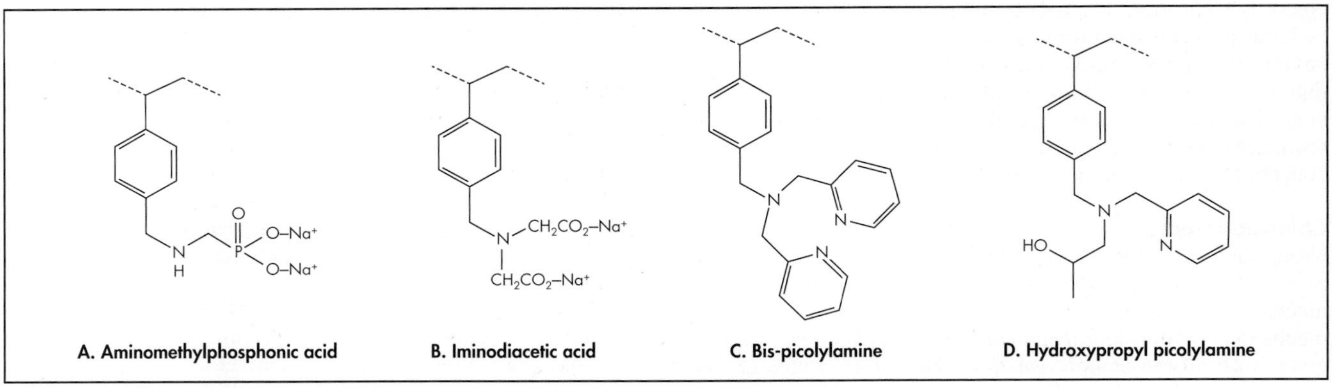

Source: Kern and Hwang 2014

Figure 5 Chemical structures of chelating functional groups

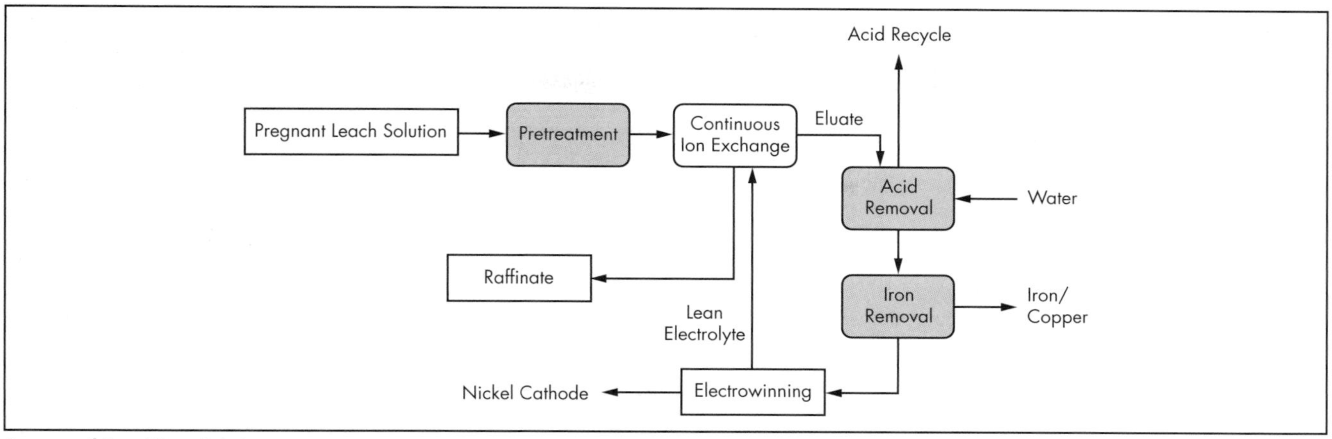

Courtesy of Dow Water Solutions

Figure 6 Laterite process utilizing ion exchange

copper present in acidic feeds. The picolylamine functional group binds copper especially strongly, such that any copper loaded from the nickel laterite ore does not interfere with the nickel elution step, which can utilize up to 20% sulfuric acid as an eluent. Stripping of copper from the resin requires use of an alkaline ammonium sulfate solution. The ammonia competitively binds the copper ions into a coordination complex that interrupts the interaction of the BPA functional group with the copper ions, facilitating its removal from the resin.

Continued development of resins for this application has introduced the HPPA functionality, which functions in a similar capacity. This resin possesses a somewhat lower affinity for copper than the BPA functionality, which in turn permits stripping of copper from the resin using 20% sulfuric acid. This feature makes the sequential elution of copper, nickel, and cobalt from the resin more challenging, as the difference in selectivity between the different metals is less pronounced. However, the ability to acid-strip copper makes this resin significantly more convenient to use in conventional acidic copper-processing applications.

Ion Exchange for Processing Gold

In the operation of many cyanide-leached gold mining facilities, activated carbon is used as an economical means for separation of gold using carbon-in-pulp contacting approaches. However, IERs offer some advantages over carbon for gold processing. Specifically, current gold-selective IERs have a higher selectivity of gold cyanides over base metal cyanides, thereby allowing for higher-purity gold products as a result of processing. Additionally, the higher capacity and faster kinetics of ion exchange lend themselves to improved processing (Lewis and Bouwer 2000). These benefits are especially pronounced in the processing of poor-grade or carbonaceous (preg-robbing) ores.

In cyanide leach systems, gold exists as an aurocyanide complex $[Au(CN)_2]^-$. Being an anion, these species can be readily removed from solution via capture onto an anion exchange resin. Protonated tertiary amine weak base anion resins can be used for this application, and the sorbed aurocyanide species can be eluted from the resin using a solution of sodium hydroxide. Following elution of the weak base moiety using base, the resin needs to be reprotonated through treatment with dilute acid. Alternatively, a quaternary amine strong base anion resin can be employed. Although these resins lack the requirement of post-elution acid treatment, they can be more difficult to regenerate. Regeneration requires an acidified thiourea solution to break the ionic interaction between the aurocyanide complex and the functional group.

One potential drawback to standard anion exchange resins (especially weak base resins) is that other anionic metallocyanide complexes compete for loading sites with the aurocyanide complex. As such, specially functionalized anion exchange resins have been developed that optimize preferential selectivity for loading of aurocyanide species over other metallocyanide anions (Johns and Marsh 1993). These resins allow for a significant improvement in the quality of gold extraction from comparably poor quality leachate.

ION EXCHANGE SYSTEM DESIGN

Although the applications of ion exchange for liquid separations are numerous and varied, there are some general guidelines that hold true across multiple applications. Because the ion exchange process is fundamentally an equilibrium interaction between the resin's functional sites and the solution's ionic composition, flow characteristics that impact residence time and contact efficiency between the resin and the solution are important parameters to understand. Although there are, of course, exceptions to every rule, the following are some general guidelines that broadly apply to a variety of ion exchange applications.

Column Design and Flow Considerations

Ion exchange depends on efficient contact between solution and resin. The kinetics of the ion exchange process itself are generally quite rapid, so the rate-limiting step is often related to the efficiency with which the resin is contacted by solution. Generally, these flow characteristics consist of a combination of bulk flow rate (expressed in bed volumes per hour [BV/h] or empty-bed contact time) and linear flow velocity. Most IER systems operate in a broad range between 5 and 25 BV/h, depending on the characteristics of the solution being treated. For example, a system designed to treat a flow rate of 50 m^3/h with a bulk flow rate of 10 BV/h would require 5 m^3 of installed resin.

Another component of flow rate is the liner velocity, which refers to how quickly the solution is passing a given point in the resin bed. Flow velocities generally cover a broad range between 5 and 60 m/h. In a fixed ion exchange column, there is a stagnant film or unstirred layer of liquid around the individual resin beads, even as the bed is under flow. The size of this stagnant film layer depends on the solution velocity across the resin bed, with higher velocities generally resulting in smaller film layers. It is diffusion of ionic species into and out of this stagnant film layer that accounts for a sizable portion of the kinetic rate for ion exchange. Because smaller stagnant films yield shorter diffusion lengths to reach the ion exchange functionality of the resin bead, higher flow velocities can yield more efficient ion exchange kinetics. However, this effect must be balanced with other factors such as IER bead size, overall kinetic hindrance, and the size of ionic species being removed. Generally, most systems operate within a broad linear flow range of 5 to 60 m/h. Slow flow rates run the risk of inefficiency because of solution channeling, while operation on the fast end of the range risk excessive pressure drop across the resin bed. For a defined bulk flow rate, the linear flow rate can be modified by changing contact equipment dimensions. For example, a long, narrow IER bed has a faster linear flow velocity than a wide, short bed that contains the same amount of resin.

Most IERs suggest a minimum resin bed depth ranging from 0.75 to 1 m. Maximum resin depth is usually dictated by operational limitations on pressure drop across the vessel and functional limitations on equipment size. Generally, commercial ion exchange column dimensions are optimal at a height-to-width ratio (H:W) between 2:1 and 4:1. Pilot systems and other applications with lower flow rates generally have a H:W ratio exaggerated beyond 4:1 to maintain the necessary minimum resin depth as specified.

Other Treatment Factors

Besides flow rate and treatment volume, other system design considerations include composition of both the feed solution and the desired regenerant, as these dictate any materials compatibility or corrosion concerns when specifying equipment. In general, most IERs are sensitive to oxidative environments and undergo de-crosslinking or other oxidative damage in their presence. As such, potential oxidizing agents (hypochlorite, permanganate, etc.) should be avoided for contact with ion exchange. Nitric acid in particular can be damaging to IERs and under certain circumstances can react to form unstable compounds. Therefore, the use of ion exchange for treating solutions containing nitric acid is not advised because of safety concerns.

With the exception of RIP systems and fluidized beds, some type of pretreatment is generally required for the removal of suspended solids. Suspended solids can quickly become entrapped in ion exchange beds, increasing pressure drop and compromising contact between the solution and the resin. Sand filters or some other backwashable media filtration are generally preferred for many hydrometallurgy applications, though the use of bag filters or replaceable cartridge filtration may also be feasible depending on the application and filtration needs. Higher solids applications may require the upstream use of a technique capable of handling higher solids content, such as a hydrocyclone or dissolved air flotation.

Cost Considerations

Ion exchange systems for mineral processing applications can vary greatly in cost and scope, from simple systems treating small flow rates to huge, multi-vessel automated systems treating hundreds of cubic meters of solution per hour. As such, system capital expenditure for treatment in the mining space can range from thousands or tens of thousands of dollars to tens of millions of dollars, simply based on the size of the system and its complexity. Extent of process automation, equipment design (fixed bed versus moving bed, RIP, etc.), flow rate for treatment, materials compatibility and corrosion control, pretreatment requirements, and process redundancy are all factors that have a significant impact on the capital requirements for ion exchange system implementation. Furthermore, regeneration frequency and efficiency, chemical handling, and waste and wastewater management factors need to be considered in determining operational costs for an ion exchange facility. Discussions with original equipment manufacturers regarding their ion exchange systems early in the planning phase can help manage cost expectations through evaluation of these relevant factors.

SUMMARY

IERs can be used for a variety of different separations in extractive metallurgy, particularly in cases where low-grade ores are being processed or otherwise low concentrations of metal ions are being recovered. The three major application areas (uranium, gold, base metals) have been covered in greater detail in this chapter. Other application areas in mining use IERs, although in many of these cases, the ion exchange processes tend to be used for secondary recovery of side streams, as opposed to the main technology for primary recovery. However, in some of these applications, ion exchange has found a comfortable niche, including the processing of platinum group metals, fractionation of rare earth metals, and lithium extraction from brines. As lower-grade ore-processing gains momentum as a result of depletion of higher-grade stocks, ion exchange technologies will likely find continued adoption for their ability to effectively process the lower-concentration leachates that result.

REFERENCES

Benes, V., Boitsov, A.V., Fuzlullin, M., Hunter, J., Mays, W., Novak, J., Slezak, J., Stover, D.E., Tweeton, D., and Underhill, D.H. 2001. *Manual of Acid In Situ Leach Uranium Mining Technology.* IAEA TecDoc 1239. Vienna, Austria: International Atomic Energy Agency.

Dorfner, K., ed. 1991. *Ion Exchangers*. Berlin: Walter de Gruyter.

Grinstead, R.R. 1979. Copper-selective ion-exchange resin with improved iron rejection. *J. Met.* 31(3):13–16.

Grinstead, R.R. 1984. Selective absorption of copper, nickel, cobalt and other transition metal ions from sulfuric acid solutions with the chelating ion exchange resin XFS 4195. *Hydrometallurgy* 12(3):387–400.

Helfferich, F.G. 1962. *Ion Exchange*. Mineola, NY: Courier Corporation.

Higgins, I.R. 1957. Counter-current liquid-solid mass transfer method and apparatus. U.S. Patent 2,815,322A.

Johns, M.W., and Marsh, D. 1993. A technical and economic comparison between the carbon-in-pulp process and the MINRIP resin-in-pulp process. In *Proceedings of the Randol Gold Forum*. Golden, CO: Randol International.

Kern, B., and Hwang, J. 2014. Overview of selected ion exchange uses in mining applications. Presented at Central Regional Meeting of the American Chemical Society. Dow Water Solutions.

Kraus, K.A. 1956. Anion exchange studies of the fission products. In *Proceedings of the International Conference on Peaceful Use of Atomic Energy*. Vol. 7. New York: United Nations. pp. 113–125.

Kunin, R. 1972. Ion exchange resins. In *Ion Exchange Resins*. Malabar, FL: Krieger.

Lewis, G.O., and Bouwer, W. 2000. Resin-in-leach: An effective option for gold recovery from carbonaceous ores. In *Proceedings of the Randol Gold and Silver Forum*. Golden, CO: Randol International. pp. 25–28.

Mendes, F.D., and Martins, A.H. 2004. Selective sorption of nickel and cobalt from sulphate solutions using chelating resins. *Int. J. Miner. Process.* 74(1-4):359–371.

Porter, R.R. 1975. Continuous ion exchange process and apparatus. U.S. Patent 3,879,287A.

Yahorava, V., Kotze, M. H., and Scheepers, J. 2009. Evaluation of various durability tests to assess resins for in-pulp applications. Presented at the 5th Southern African Institute of Mining and Metallurgy Base Metals Conference.

Zaganiaris, E.J. 2009. *Ion Exchange Resins in Uranium Hydrometallurgy*. Paris: Books on Demand.

CHAPTER 10.14

Solvent Extraction

Kenneth N. Han and Matthew Soderstrom

Solvent extraction (SX), often referred to as *liquid–liquid extraction*, is a separation process in which a chosen metal is transferred from one immiscible liquid phase to another. Metals dissolved in the aqueous phase are extracted into an organic phase containing a suitable extractant.

In the late 1800s, Berthelot and Jungfleisch (1872) first identified the principle by which a metal ion can be extracted from the aqueous phase to the organic phase, eventually forming an equilibrium constant of the metal distributed between two immiscible liquids, which has led to the concept of the distribution ratio. During 1940–1950, the U.S. Manhattan Project spurred in-depth studies on SX and ion exchange for recovery and upgrading of uranium and other metals, mainly actinide group metals. In recent years, this technology has become a major force in the separation and/or production of various metals from leach liquors, including Cu, Ni, Co, U, Th, Zn, Au, Pt group metals; rare earth metals; and many more.

In SX, a chosen extractant is dissolved in an organic carrier, usually a hydrocarbon-based solvent diluent. The extractant and carrier solvent are contacted with an immiscible aqueous phase, in which metal ions are dissolved. Because of the affinity between the extractant and metal ion in the aqueous phase, the metal ion transfers from the aqueous to the organic phase.

In this chapter, definitions and nomenclatures, various kinds of extractants and diluents, and reaction mechanisms involved in SX will be presented and discussed. The chapter will also include SX thermodynamics to help expand the knowledge in development of better reagents and processes in recovering and purifying various industrial metals.

DEFINITIONS AND NOMENCLATURE

As shown in Equations 1 and 2, a metal ion M^{n+} or a metal salt MS in the aqueous phase reacts with an extractant HR or R in the organic phase to form an organic soluble metal complex. Here n stands for the valence of the metal M and is an integer.

$$M^{n+}(a) + nHR(o) = MR_n(o) + nH^+(a) \quad \textbf{(EQ 1)}$$

$$MS(a) + nR(o) = (MS)R_n(o) \quad \textbf{(EQ 2)}$$

The symbols (a) and (o) represent the aqueous phase and the organic phase, respectively. It should be noted that extractants are placed in the organic phase and, therefore, the chemical interaction between a metal ion and an extractant can take place when these two chemical moieties are in direct contact either in the aqueous phase, in the organic phase, or at the interface. There are various ways in which the chemical reaction can take place, dependent on the extractant utilized. Examples given in the preceding equations represent two different mechanisms. The reaction shown in Equation 1 takes place when the metal ion replaces the hydrogen ion associated with the extractant, while the reaction shown in Equation 2 represents a neutralized metal salt interacting with the extractant.

Figure 1 shows a simple flowchart depicting an industrial SX process. As seen in the figure, leach liquor containing various metals is mixed with the organic phase containing the extractant. In extraction, the metal of interest transfers into the organic phase, leaving a metal-depleted aqueous phase often referred to as *raffinate*. This raffinate may be subjected to further purification and/or metal recovery processes, or returned to leach. The organic phase loaded with metals may then be subjected to a scrubbing step to eliminate impurities from the organic followed by a stripping step to transfer the metal of interest from the organic phase to the aqueous. The organic phase depleted of metal may be recycled back to the extraction step.

Efficiency of metal removal is often described in terms of percent recovery as shown in Equation 3. An additional frequently used term is the distribution ratio, D, shown in Equation 4. These are defined as follows:

$$\%\ \text{recovery} = (c_f - c_r)/c_f \times 100 \quad \textbf{(EQ 3)}$$

where c_f and c_r are the concentration of metal in the feed and raffinate streams.

$$D = \frac{C_o}{C_a} = \frac{M_o/V_o}{M_a/V_a} = \frac{V_a}{V_o} \times \frac{M_t - M_a}{M_a} \quad \textbf{(EQ 4)}$$

Kenneth N. Han, Professor Emeritus, South Dakota School of Mines & Technology, Rapid City, South Dakota, USA
Matthew Soderstrom, Business Director Metal Extraction Products, Solvay S.A., Tempe, Arizona, USA

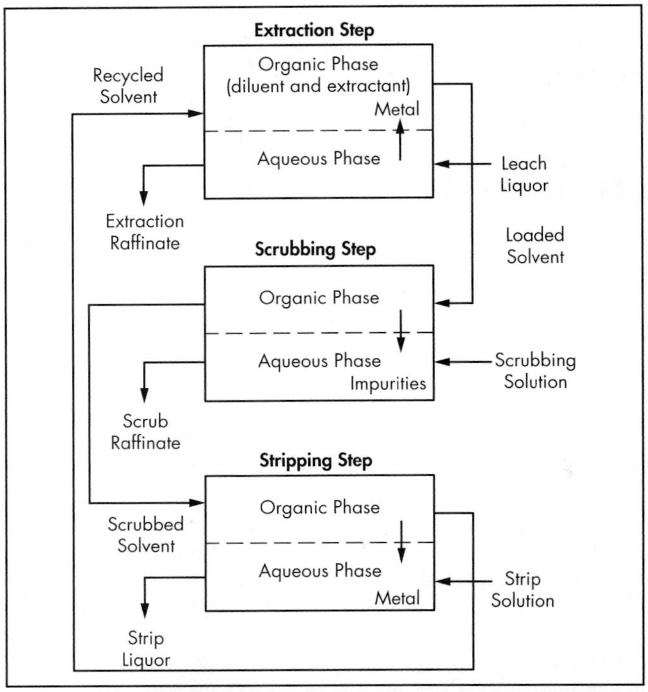

Figure 1 Flowchart of a simple solvent extraction operation

where C_o and C_a are the concentrations of the metal in the organic phase and the aqueous phase, respectively; M_o and M_a are the mass of metal in the organic phase and aqueous phase, respectively; V_o and V_a are the volumes of the organic phase and the aqueous phase, respectively; and M_t is the mass of the total metal ($M_a + M_o$). The distribution ratio is often referred to as *distribution coefficient* and *partition coefficient*.

Another commonly used term is the *separation factor* β as defined here:

$$\beta = \frac{D_A}{D_B} \quad \textbf{(EQ 5)}$$

where D_A and D_B are distribution ratios of the species A and B, respectively.

The use of these equations and mass balance concepts significantly aid in the evaluation of the various SX extractants and in the design of circuits to treat the various feed streams. For example, D values may be used to estimate the number of stages required to achieve a given recovery. Multiple authors have covered these mathematical relationships and their interpretation (see Han 2002; Han and Fuerstenau 2003).

VARIOUS TYPES OF SOLVENT EXTRACTANTS

There are four common types of SX mechanisms: cationic, solvation, anionic, and chelation. These and one other less common method are discussed next.

Cationic

The cationic type, which can also include chelating extractants that release protons in exchange for cations, involves cations interacting with solvents as shown in the following equation (Mason et al. 1978; Zheng et al. 1991).

$$M^{n+}(a) + nHA(o) = MA_n(o) + nH^+(a) \quad \textbf{(EQ 6)}$$

In the preceding equation, M represents an n valence metal ion, and HA represents a protonated extractant. The following equation shows the equilibrium constant between the reactants and products:

$$K = \frac{[MA_n](o)\ [H^+]^n(a)}{[M^{n+}](a)[HA]^n(o)} \quad \textbf{(EQ 7)}$$

Using the D value and rearranging the terms, one can show

$$D = \frac{C_M(o)}{C_{M^{n+}}(a)} = K\frac{[HA]^n(o)}{[H^+]^n(a)} = \frac{\gamma_{M(o)}}{\gamma_{M^{n+}(a)}} \quad \textbf{(EQ 8)}$$

$$\log D = \log K + n \log[HA](o) + n\,\text{pH} \quad \textbf{(EQ 9)}$$

Here, K refers to the equilibrium constant; [ ] refers to the thermodynamic activity of each component; D is the distribution ratio; $C_M(o)$ and $C_{M^{n+}}(a)$ represent the molar concentration of the metal complexes in the organic phase and aqueous phase, respectively; and γ is the activity coefficient. Equation 9 assumes that the activity coefficients of the metal compounds in the organic phase and the aqueous phase cancel each other out, which may not be true in practice. This aspect will be discussed further in a later section.

$pH_{1/2}$ values are also commonly used to assist in the design or evaluation of SX flow sheets. $pH_{1/2}$ is defined as the pH when $D = 1$ or the pH when the metal extraction to the organic phase is 50%. It can easily be shown that $\log K = -n\,pH_{1/2} - n \log [HA](o)$. Because $\log D = \log K + n \log[HA](o) + n\,pH$,

$$\log D = n(pH - pH_{1/2}) \quad \textbf{(EQ 10)}$$

As indicated by Equation 10, the lower the $pH_{1/2}$, the larger D will be and, as a result, the greater the metal extraction. The following are examples in this group:

- Carboxylic acids (e.g., Versatic 10, Versatic 911, and naphthenic acids)
- Phosphorous acids
  - Phosphoric acids, such as bis(2-ethylhexyl)phosphoric acid (DEHPA)
  - Phosphonic acids (e.g., EHEHPA, P507, PC-88A, and Ionquest 801)
  - Phosphinic acids (e.g., P229; Cyanex 272, 301, 302; and PIA-8)
- Aryl sulfonic acids (e.g., Synex 1051)
- Some oximes, such as ACORGA and LIX reagents

Ionquest 801 represents 2-ethylhexyl phosphonic acid; P229, di(2-ethylhexyl)phosphinic acid; Cyanex 272, bis(2,4,4-trimethyl pentyl)phosphinic acid; Cyanex 302, bis(2,4,4-trimethyl pentyl)thiophosphinic acid; PIA-8, bis(2-ethylhexyl) phosphinic acid; ACORGA, reagents based on C_9 aldoxime and C_9 ketoxime; and LIX, reagents based on alphatic and aromatic oximes. See Table 1 for others.

Solvation

The solvation exchange mechanism occurs in SX when the neutral metal salt complexes with the organic solvent (Peppard et al. 1957; Reddy et al. 1999; El-Nadi 2012). Some examples using tri-n-butyl phosphate (TBP) are shown in the following equations:

$$M^{3+} + 3NO_3^- + 3TBP(o) = M(NO_3)_3(TBP)_3(o) \quad \textbf{(EQ 11)}$$

Table 1 Some extractants used in solvent extraction

Extractant Type	Name	Structural Formula	Trade Name	Applications	References
Acidic Carboxylate	Versatic acids $R_1 + R_2 = C7$: 10 $R_1 + R_2 = C6–C8$: 911	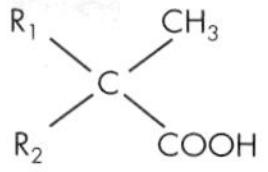	Versatic 10 Versatic 911	Cu, Zn, Cd, Ni/Co, REE	Habashi 1993, Flett 2005
	Naphthenic acids $R_1–R_4$: Varying alkyl groups	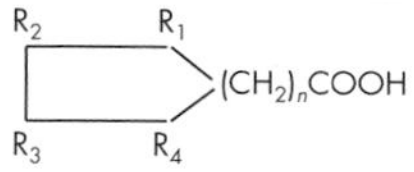		Cu, Zn, Co, Ni, REE	Habashi 1993, Flett 2005
Phosphorus	Phosphoric acids: Di(2-ethylhexyl) phosphoric acid	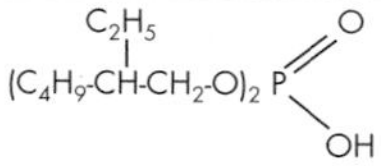	D2EHPA P 204	U, REE, Co/Ni, Zn	Habashi 1993, Flett 2005
	Phosphonic acids: 2-Ethylhexyl phosphonic acid mono-2-ethylhexyl ester	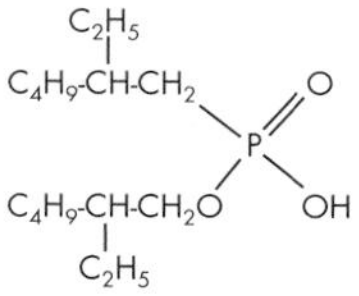	HEH/EHP PC-88A P 507	Co/Ni, Zn, Th, U, REE	Habashi 1993, Flett 2005
	Phosphinic acids: Organophosphinic acid	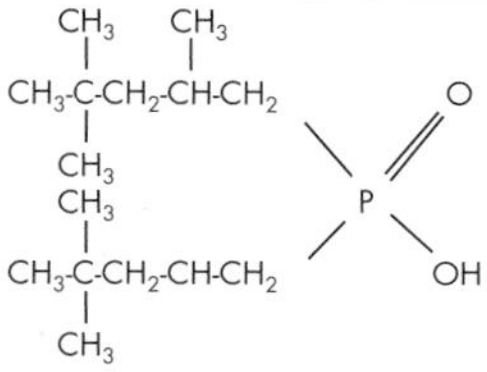	Cyanex 272 PIA-8	Co/Ni, Zn, Mn, Ca REE	Cytec Industries 2010
	Organodithiophosphinic acid	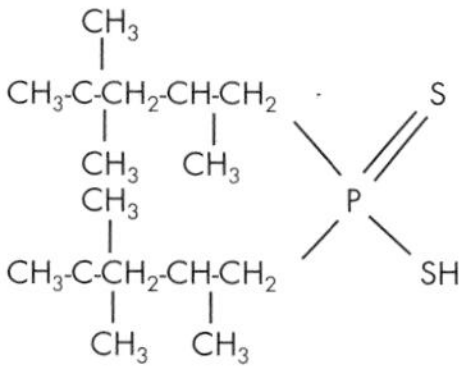	Cyanex 301	Ni, Co, Mn	Cytec Industries 2010, Cox 2010
	Aryl sulfonic acids R: Aryl	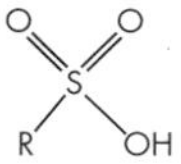	Syanex 1051	Mg	Cox 2010
Solvation Phosphorous ester	Tributyl phosphate	$(C_4H_9O)_3\text{-}P = O$	TBP	Th, Ce^{4+}, U, REE	Cox 2010
	Dibutyl phosphate	$C_{12}H_{27}O_3P$	DBBP	U, Fe, Zr/Hf, Au, REE	Cox 2010
Phosphinic oxides	Trioctylphosphinic oxide	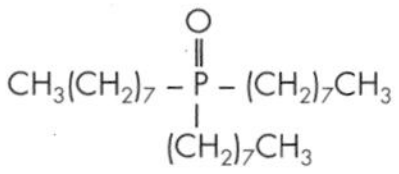	TOPO Cyanex 921	U, Zr/Hf, REE, Au	Cytec Industries 2010, Cox 2010
Hydroxyoxime derivatives	α-Alkarylhydroximes	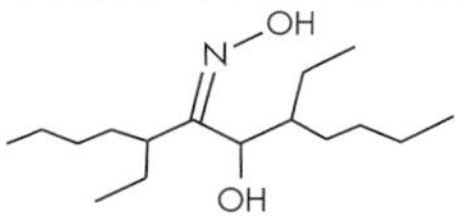	LIX 63	Cu, Ni	Cox 2010
	β-Alkylarylhydroxyoximes	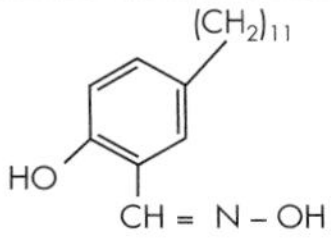	LIX 860	Cu, Ni	Cox 2010
Hydroxyoxine derivatives	8-Hydroxy-7-(4-ethyl-1-methyloctyl)-8-hydroxyquinonline	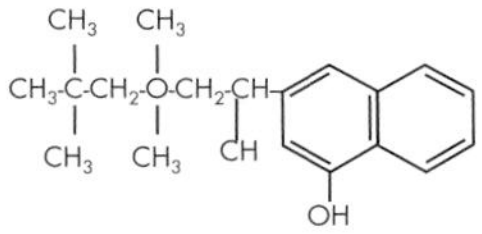	Kelex 100	Cu, Zn	Habashi 1993, Flett 2005

(continues)

Table 1 Some extractants used in solvent extraction (continued)

Extractant Type	Name	Structural Formula	Trade Name	Applications	References
Various alcohols	Methyl isobutyl ketone	$CH_3-\overset{O}{\overset{\|}{C}}-CH_2-\underset{CH_3}{\overset{CH_3}{CH}}$	MIBK	Nb/Ta, Hf/Zr, Au, Li	Habashi 1993, Flett 2005
Anion					
	Primary amines	$R-\underset{R}{\overset{R}{C}}-NH_2$	Primene JMT	Fe(III), REE, Mo, Th	Habashi 1993, Cox 2010
	Secondary amines: Di-tridecyamine	$R_1, R_2 = CH_3(CH_2)_{12}NH$	Adegen 283	Th, U, Cr, Cu, Cd	Habashi 1993, Cox 2010
	Tertiary amines: Tri-octylamine	$CH_3(CH_2)_6CH_2-N\overset{CH_2(CH_6)_6CH_3}{\underset{CH_3(CH_2)_6CH_3}{}}$	TOA/Alamine	U, V, W, Mo, Cu, Co, REE	Habashi 1993, Cox 2010
	Tri-isooctylamine		TiOA/Adogen 381		
	Quaternary amines: Tri-octyl methyl ammonium chloride	$R-\underset{R}{\overset{R}{N}}\overset{R}{\underset{CH_3}{}}$ Cl	Aliquat 336	U, Mo, V, W, REE	Habashi 1993, Cox 2010

$$M^{4+} + SO_4^{2-} + 2HSO_4^- + 2B(o)$$
$$= M(SO_4)(HSO_4)_2 2B(o) \qquad \textbf{(EQ 12)}$$

Other examples may include the interaction of Ce(IV) and Th(IV) with Cyanex 923 (Liao et al. 2001). Examples belonging to this group are

- Phosphorous ester, such as TBP and dibutyl-butyl phosphonate (DBBP);
- Phosphinic oxides, such as Cyanex 921 (TOPO) and Cyanex 923;
- Esters and thio analogs, such as Cyanex 471X; and
- Various ketones, such as methyl isobutyl ketone (MIBK) and di-isobutyl ketone (DIBK).

Cyanex 923 represents a mixture of four tryalkyl phosphine oxides; Cyanex 921, tri-n-octyl phosphine oxide; and Cyanex 471X, tri-isobutylphosphine sulfide.

Anionic

Anion-type exchangers (Rice and Stone 1962; El-Yamani and Shabana 1985; Cerna et al. 1992; Xie et al. 2014) extract metal ions as anionic complexes and therefore are only effective in the presence of strong anionic ligands. An example for an amine extractant is shown in the following equations:

$$2RNH_2(o) + H_2SO_4 = (RNH_3)_2SO_4(o) \qquad \textbf{(EQ 13)}$$

$$2La(SO_4)_3^{3-} + 3(RNH_3)_2SO_4(o)$$
$$= 2(RNH_3)_3La(SO_4)_3^{3-}(o) + 3SO_4^{2-} \qquad \textbf{(EQ 14)}$$

Examples in this group are

- Primary amines, such as Primene JMT and N1923;
- Tertiary amines, such as Alamine 336; and
- Quaternary amines, such as Aliquat 336 and Adogen 464.

N1923 represents primary amine; Alamine 336, trialkyl amine; Aliquat 336, trialkylmethylammonium; Adogen 464, a high molecular weight quaternary ammonium salt. See Table 1 for others.

Chelation

A metal ion forms a chelation compound with an organic anion, which is the chelation agent, for example, Cu^{2+} and ethylenediamine:

$$\begin{array}{ccc} H_2C & - & CH_2 \\ / & & \backslash \\ H_2N & & NH_2 \\ \backslash & & / \\ & Cu^{2+} & \end{array}$$

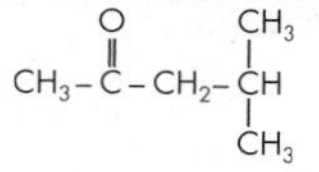

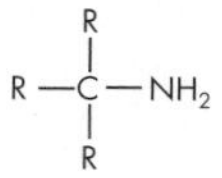

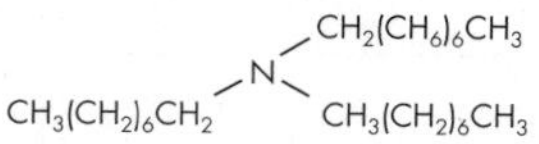

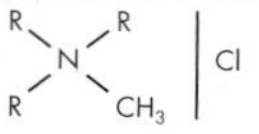

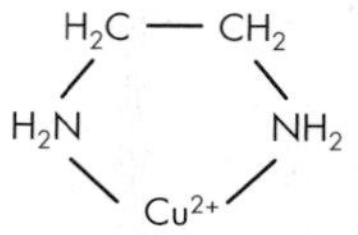

Ionic Liquid

Ionic liquids are a special kind of extractant, which is in the early stages of investigation. Ionic liquids are organic salts that consist entirely of ions typically having a melting point below 100°C (Seddon 1997; Earle et al. 2006). These types of special compounds have recently gained interest in SX applications because of their unique properties, such as low flammability (Smiglak et al. 2006), low vapor pressure (Earle et al. 2006), high thermal stability (Fox et al. 2003), and potential electrochemical applicability. Many investigators have attempted to use such liquids as viable extractants in SX applications for metal ions (Bartsch et al. 2002; Cieszyńska et al. 2007; Campos et al. 2008; Egorov et al. 2010; Okuno et al. 2006; Regel-Rosocka 2009; Srncik et al. 2009; Sun et al. 2010).

The most frequently applied in SX applications include 1-alkyl-3-methylimidazolium hexafluorophosphate $[C_n mim]$ $[PF_6]$, tetrafluoroborate $[C_n mim][BF_4]$, and bis(trifluoromethylsulphonyl) imide $[C_n mim][N(SO_2CF_3)_2]$ (an anion also abbreviated as $[Tf_2N]$) (Regel-Rosocka and Wisniewski 2011). A series of bifunctional ionic liquid extractants composed of A336 cation and phosphonic acid or carboxylic acid anions have also been tried for rare earth element extraction (Sun et al. 2010; Lachwa et al. 2006; Fukaya et al. 2007).

In addition to being used as potential extractants, significant interest has been shown by the industry in the use of ionic liquids as alternative carrier solvents.

Table 1 lists some extractants frequently used in SX practices and/or laboratory tests.

STRIPPING

In its simplest form, stripping is often (but not always) the reverse of the extraction reaction in SX. Using Equation 1 as an example,

$$M^{n+}(a) + nHR(o) \rightarrow MR_n(o) + nH^+(a) \qquad \textbf{(EQ 1)}$$

The reverse of Equation 1, as shown in Equation 15, represents stripping.

Table 2 Various stripping efficiencies for ferric iron under different stripping conditions

Reagent	Concentration, M	Ferric Iron Stripping, %
Nitric acid	6	1.8
Sulfuric acid	6	32.5
Hydrochloric acid	6	48.2
Oxalic acid	0.95	72.4
Sodium oxalate	0.4	12.2
Ammonium oxalate	0.8	65
Sodium + oxalate	0.4	50
Oxalic acid + ammonium oxalate	0.95	75

Adapted from Akhlaghi et al. 2010

$$nH^+(a) + MR_n(o) \rightarrow M^{n+}(a) + nHR(o) \qquad \textbf{(EQ 15)}$$

Stripping can be significantly more complicated and in some cases require an entirely different reaction.

Example

Stripping can also be very complicated depending on the type of metal complexes formed. As an illustration of this point, consider the stripping of ferric iron, Fe(III) DEHPA (Akhlaghi et al. 2010). In this extraction, Fe(III) may form a complex with DEHPA, such as $Fe(OH)R_2(HR)_2$ and/or $Fe(HR_2)_3$, resulting in a stable metal/extractant complex that is difficult to strip. Table 2 shows the different stripping efficiencies with different media.

Regardless of the stripping methodology used, it is critical that the SX reagent is sufficiently regenerated to allow it to be returned to the extraction section of the plant for further metal extraction.

SOLVENT EXTRACTION THERMODYNAMICS

There are many factors to consider when developing or designing an SX process flow sheet. This section discusses some of the relevant considerations and reviews some of the calculations that may be used to better understand the chemistry and assess a conceptual flow sheet.

In the case of uranyl ion (UO_2^{2+}) extraction with TBP, TBP is first dissolved in an organic phase with metallurgical-grade kerosene as the diluent. The reaction between UO_2^{2+} and TBP is supposed to take place as shown in the following equation:

$$UO_2^{2+} + 2NO_3^- + 2TBP(o) = UO_2(NO_3)_2(TBP)_2(o) \qquad \textbf{(EQ 16)}$$

A possible schematic presentation of this reaction is given in Figure 2. Case 1 assumes TBP is partially soluble in water. In fact, the solubility of TBP in water is about 6 mL/L of water, and therefore the reaction mechanism suggested in Figure 2 is quite possible. However, if the solubility of the extractant in water is zero, then the following mechanism could be proposed as shown in Figure 3.

Cations in water, such as UO_2^{2+}, are surrounded by water molecules to form a hydrated species, such as hexahydrate uranyl ion ($UO_2(H_2O)_6^{2+}$), and therefore, they are very hydrophilic. As a result, the case shown in Figure 2 may be more likely than the case shown in Figure 3.

Because of the chemical affinity between UO_2^{2+} and TBP, TBP in the aqueous phase would react with UO_2^{2+} in water. Once they are combined, UO_2^{2+} would become hydrophobic due to the attachment of TBP, which makes the passage to the organic phase natural. Another important aspect on the interaction between TBP and UO_2^{2+} is that UO_2^{2+} should combine with NO_3^- before interaction with TBP. Therefore, this type of interaction is referred to as *solvation*. In solvation-type exchange mechanisms, the speciation of metal ions in the presence of various anions in the solution plays an important role in the interaction reaction. This aspect will be detailed further in the following section.

Like the TBP example, it is important to know in each system where or how the reaction is taking place. This can have a direct impact on how the system should be designed or provide important information for improving the extraction or stripping reactions.

The equilibrium constant K for the reaction given by the following equation would be

$$K = \frac{[UO_2(NO_3)_2(TBP)_2](o)}{[UO_2^{2+}](a)[NO_3^-]^2(a)[TBP]^2(o)} = D\frac{1}{[NO_3^-]^2(a)[TBP]^2(o)} \qquad \textbf{(EQ 17)}$$

It should be noted that the preceding treatment assumes that activity coefficients of the various species involved are assumed to be 1 (which can be misleading).

On this basis, $D = K\ [NO_3^-]^2[TBP]^2(o)$ and

$$\log D = \log K + 2\log[NO_3^-] + 2\log[TBP] \qquad \textbf{(EQ 18)}$$

From the preceding equation, one can see that the factors affecting the value D (the partition coefficient) are the concentration of NO_3^- and also the concentration of TBP.

Note, as shown in Equation 18, that the assumed stoichiometric coefficient of 2 for TBP needs justification. In fact, this value is often determined experimentally. In reality, this value can be any value—1, 2, 3—or some fraction, such as 1.5, and so on. The variation of n is primarily caused by the valence of the metal ion concerned and also by polymerization of the extractant during the course of SX.

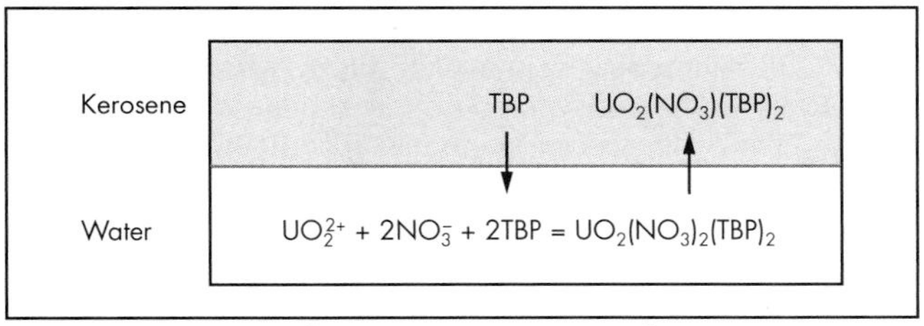

Figure 2 Case 1: Schematic presentation of a reaction between UO_2^{2+} and TBP

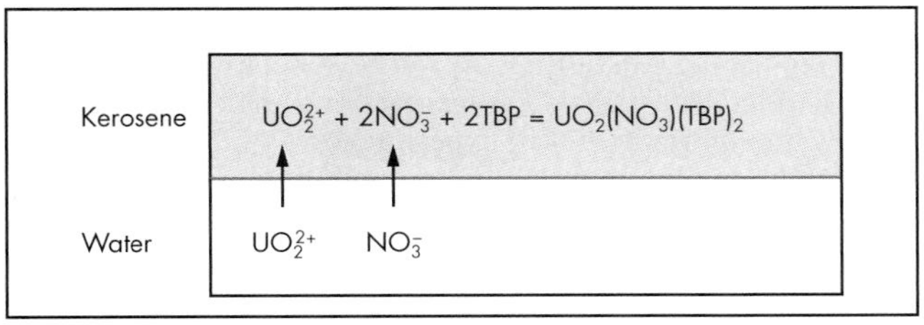

Figure 3 Case 2: Schematic presentation of a reaction between UO_2^{2+} and TBP

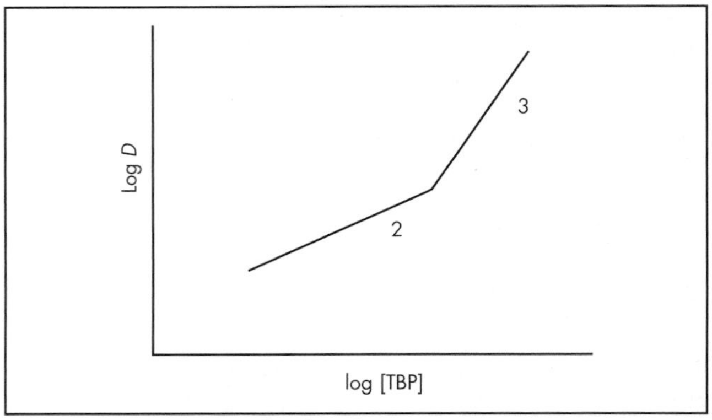

Figure 4 Log *D* versus log [TBP]

The stoichiometric value may be estimated by running a series of experiments, in which the value *D* is calculated as a function of the concentration of TBP in the organic phase. In such tests, the concentration of NO_3^- is held constant. The plot of log *D* versus log [TBP] as seen in Equation 17 produces a curve with a slope corresponding to the moles of extractant involved in the reaction. One often finds that the slope changes with the concentration of extractant. For example, the slope can be 2 up to a certain concentration, and it becomes 3 as the concentration increases beyond a certain point (see Figure 4).

Similar methodology as used for TBP (a solvation-type extractant) can also be used for a cation-type extractant such as di-(2-ethylhexyl) phosphoric acid (DEHPA). Abbreviating DEHPA as HA, the acid type of extractants will react with the metal ions in solution releasing H^+ as shown in Equation 19:

$$M^{3+}(a) + n\text{HA}(o) = \text{MA}_n(o) + nH^+(a) \tag{EQ 19}$$

$$K = \frac{[\text{MA}_n](o)\cdot[H^+]^n(a)}{[M^{n+}](a)\cdot[\text{HA}]^n(o)} \tag{EQ 20}$$

$$D = \frac{C_M(o)}{C_{M^{n+}}(a)} = K\frac{[\text{HA}]^n(o)}{[H^+]^n(a)}\frac{\gamma_{M(o)}}{\gamma_{M^{n+}(a)}} \tag{EQ 21}$$

Because DEHPA is a cation type extractant that exchanges protons, the numerical values of *K* and *D* will be very much affected by the pH of the solution as shown in Equations 20 and 21. The value *n* can be found experimentally by making a log-log plot as shown in Figure 4. However, in such experiments, the pH of the solution should be kept constant, which is not easy because the reaction will generate H^+. As a result, some researchers carry out experiments in a pH buffer solution.

It can be seen clearly that the *D* value for the TBP system is a function of NO_3^- and TBP, while that for DEHPA is a function of pH and DEHPA concentration.

In addition to using *D* values to better understand the stoichiometry of the reaction(s), the Gibbs standard free energy at room temperature $\Delta G^\circ_{R,25}$ for the reaction can be calculated by the K value through a set of experiments.

$$\Delta G^\circ_{R,25} = -RT \ln K_{25} \tag{EQ 22}$$

However, there are some careful considerations required in conducting such experiments and interpreting the results. The equilibrium constant K can be easily calculated, provided the activities of the species involved in Equation 19 are known. The activity of H^+ can be measured relatively easily with the help of pH electrodes when the pH of the solution is relatively high, such as 2 or above. However, the accuracy of the activity of H^+ suffers when the pH of the solution is low (below 1) because the pH meter cannot serve to provide this information in a strong acidity. In such cases, the concentration H^+ should be obtained by titration, and then the corresponding activity should be found after obtaining the activity coefficient.

There are numerous equations for determining activity coefficients, which is outside the scope of this work.

Temperature Effects

When the SX reaction is studied at different temperatures, one can obtain the equilibrium constant K as a function of temperature. Such information is valuable because the establishment of K as a function of temperature (*T*) will allow one to predict the extraction reaction at any temperature using the Van't Hoff equation:

$$\ln\frac{K_2}{K_1} = -\frac{H^\circ_R}{R}\left(\frac{1}{T_2} - \frac{1}{T_1}\right) \tag{EQ 23}$$

When ΔH°_R and ΔG°_R are known, ΔS°_R can be determined by the following equation:

$$\Delta G^\circ_R = \Delta H^\circ_R - T\,\Delta S^\circ_R \tag{EQ 24}$$

As a result, any K at any temperature can be calculated.

Synergistic Effect

Many SX formulations have been blended and found to have better performance in combination than would have been expected based on the individual components. For example, when TBP and DEHPA are mixed (for rare earth applications), the resulting SX is much better than TBP or DEHPA alone. Such a synergistic effect (SE) is defined by the following equation:

$$\text{SE} = \frac{D_{\text{overall}}}{\sum D_i} \tag{EQ 25a}$$

For two extractant systems, D_1 and D_2,

$$\text{SE} = \frac{D_{\text{overall}}}{D_1 + D_2} \tag{EQ 25b}$$

where

SE > 1, synergism is present
SE = 1, no synergism
SE < 1, antagonistic effect is present

CHOICE OF DILUENTS

There are many diluents that can be used in SX test work. Some of the major properties considered in choosing the diluent may include water immiscibility, density, boiling and melting points, flash points, and also the dielectric constant. In addition, viscosity and price should be included in this decision-making process. Table 3 lists some of the important and frequently used diluents used in laboratory assessments.

In commercial and/or industrial applications, the most frequently used diluents are metallurgical-grade kerosene-based solvents. These tend to be a mixture of many organic compounds with branched- and straight-chain alkanes and naphthenes and aromatic hydrocarbons. The diluents listed in Table 3 are generally used in laboratory experiments,

Table 3 Physical properties of some diluents

Common Diluents	Boiling Point, °C	Melting Point, °C	Density, g/mL	Solubility, g/100 g water	Dielectric Constant	Flash Point, °C
Benzene	80.1	5.5	0.8765	0.18	2.28	–11
1-Butanol	117.7	–88.6	0.8095	6.3	17.8	37
Carbon tetrachloride	76.8	–22.6	1.594	0.08	2.24	N/A*
Chlorobenzene	131.7	–45.3	1.1058	0.05	5.69	28
Chloroform	80.7	–63.4	1.4788	0.795	4.81	N/A
Cyclohexane	80.7	6.6	0.7739	0.1	2.02	–20
1,2-Dichloroethane	83.5	–35.7	1.245	0.861	10.42	13
Dodecane	214	–10	0.78	—	2.0	N/A
Diethylene glycol	246	–10	1.1197	10	31.8	124
Diethyl ether	34.5	–116.2	0.713	7.5	4.267	–45
Di-butyl ether	142.4	–97.9	0.769	0.03	—	25
Ethyl acetate	77	–83.6	0.895	8.7	6	–4
Heptane	98	–90.6	0.684	0.01	1.92	–4
Hexane	69	–95	0.659	0.014	1.89	–22
Methylene chloride	39.8	–96.7	1.326	1.32	9.08	1.6
Nitro methane	101.2	–29	1.382	9.50	35.9	35
Pentane	36.1	–129.7	0.626	0.04	1.84	-49
Propyl alcohol	97	–126	0.803	Miscible	21.8	22
Tetrachloroethane	121.1	–19	1.622	0.15	2.5	Not flammable
Toluene	110.6	–93	0.867	0.05	2.38	4
Tri-ethyl amine	88.9	–114.7	0.728	0.02	2.4	–11
o-Xylene	144	–25.2	0.897	Insoluble	2.57	32
p-Xylene	138.4	13.3	0.861	Insoluble	2.27	27
m-Xylene	139.1	–47.8	0.868	Insoluble	2.37	27

Adapted from Reichardt 2003
*N/A = not applicable.

Table 4 Pertinent characteristics of ShellSol 2046 AR

Property	Unit	Range
Distillation, final boiling point	°C	230–275
Aromatics	% Volume	8
Flash point, Pensky-Martens closed cup	°C	80
Viscosity at 40°C	mm^2/s	1.60–2.00

Source: Shell Chemicals 2016

especially when the SX chemistry is studied. However, kerosene and its derivatives are currently used in industry. Some of the industrial diluents that have been well defined are ShellSol, Orform SX, and Escaid-based solvents.

ShellSol 2046 AR is a special kerosene with a mixture of paraffins, naphthenes, and aromatics, which can be characterized as exhibiting a high flash point, low evaporation rate, and low vapor pressure. Some typical characteristics of this diluent are given in Table 4.

Orform SX, an industrially used SX diluent, is a petroleum distillate and applied in extraction of metals such as copper, cobalt, nickel, platinum, palladium, uranium, cadmium, beryllium, and lanthanide series metals. There are typically three different branches: Orform SX-11, Orform SX-12, and Orform SX-80. Some salient chemical properties are given in Table 5.

Table 5 Pertinent characteristics of Orform SX series

Property	SX-11	SX-12	SX-80
Gravity at 15.6°C, g/mL	0.795	0.82	0.82
Flash point, °C	114	76.7	76.7
Composition, wt %			
Alkyl aromatics	0.2	4–23	4–23
Cycloparaffins	0.01	42	42
Paraffins	97	34–53	34–53
Viscosity, mm^2/s			
25°C	6.5	2.4	2.4
10°C	15.2	3.4	3.4
0°C	—	4.3	4.3

Source: Chevron Phillips Chemical Company 2013

LABORATORY TESTING METHODS AND ANALYSIS

In this section, various laboratory testing methods on SX will be briefly introduced and discussed. Laboratory tests are essential for the design or optimization of a successful SX operation. Such tests allow evaluation of both the physical and metallurgical properties of the system and their dependence on process variables such as the ratio of organic-to-aqueous phases, contact time, number of stages, and throughput as well as SX chemistry. Benchtop laboratory tests are typically completed to better understand the equilibrium conditions (extractant loading, stripping characteristics, and stoichiometry), selectivity, kinetics, solubility, and phase separation properties of the system before further testing of the system under dynamic conditions.

Once the chemistry (stoichiometric requirements, conditions under which the reaction will proceed, and desired metal separation) has been defined, laboratory test work is completed to validate the reagent capability to achieve the desired recovery or separation.

S Curves or Loading Curves

Loading curves are often generated to determine what metals within the feed solution are likely to load and under which conditions. The metal loading is often evaluated as a function of pH (for chelating or cation exchange–type reactants) or as a function of some other anion in the case of solvating reagents. The curves can be generated with the actual leach liquor (containing all the metals of interest) or can be generated with single metal solutions. These loading curves are used to determine the relative affinity the extractant may have for the various ions in solution. Dependent on what is present and what separation is desired, the aqueous feed may require a pretreatment to remove undesired impurities prior to the SX step or may require adjustment during the extraction or stripping process.

For example, consider the separation of Co from acidic leach liquors using phosphinic acid–based extractants. In this case, the reaction proceeds following Equations 26 and 27 as shown here:

$$M^{n+}(a) + nHA(o) = MA_n(o) + nH^+(a) \quad \textbf{(EQ 26)}$$

$$Co^{2+} + 2HA \leftrightarrow CoA_2 + 2H^+ \quad \textbf{(EQ 27)}$$

The extractant chelates Co at a relatively high pH (pH >4) and exchanges the metal cation for protons. In addition to loading Co, the extractant would be expected to load many elements within the pH range under which Co loads (in this case Ca, Mg, Zn, Cu would also be expected to load based on the S curves). Some of these elements would not be stable at the pH at which cobalt extraction takes place. For these reasons, the leach liquor is often treated prior to the Co SX step to precipitate out Fe, Al, and Mn. Following impurity removal, the reaction can proceed; however, the metal loading would be limited by the generation of acid during extraction. For the extractant to continue loading metals, the acid that is generated because of the proton exchange would require neutralization as shown in the following equation:

$$Co^{+2} + 2HA + 2NaOH + SO_4^{=} \rightarrow CoA_2 + 2H_2O + Na_2SO_4 \quad \textbf{(EQ 28)}$$

Isotherm Generation

Simple and quick shake tests are often completed using beakers with overhead stirrers, jars with magnetic stirrers, separating funnels, or other agitation devices to contact the organic and aqueous phases together at various organic-to-aqueous ratios to determine the metal distribution between the phases as a function of the organic-to-aqueous volumetric ratio. SX is usually mass transfer controlled, and therefore, it normally takes less than 30 minutes to reach equilibrium.

Dependent on the type of extractant being used, these distribution curves may be generated directly by mixing the two phases (the aqueous feed and the extractant diluted to the correct stoichiometric reagent concentration) at the various volumetric ratios. In other cases, the distribution curves may need to be generated under more controlled conditions (e.g., if the pH needs to be held constant via the addition of base).

Once generated, the equilibrium curves can then be used to help design the circuit using McCabe–Thiele techniques (discussed later).

Kinetics Tests

Typical SX test work also includes an evaluation of the reaction kinetics: The organic and aqueous phases are mixed together at a set organic-to-aqueous ratio, the dispersion is sampled, and phases are separated and analyzed to assess metal transfer as a function of time. For example, organic and aqueous are mixed together at a 1:1 ratio using a baffled beaker with an overhead stirrer at 1,000 rpm. Samples are taken at 30, 60, 120, 180, 300, and 900 seconds. The phases are allowed to separate, and the aqueous and/or organic are then subjected to chemical analysis using such devices as atomic absorption spectrometers, induction-coupled plasma devices, and visible and/or ultraviolet photometers. The test gives an indication of the mixing time required for the solutions to reach equilibrium and can be used to help design the mixers and understand the mixer retention time requirements.

Phase Separation Tests

Another typical test involves mixing the aqueous and organic together at a set mixing condition and for a set period of time (e.g., 3 minutes), then stopping the agitation and allowing the phases to separate, while recording the separation time. This test provides information on the settling time (at meters per second) and can be used to size the dynamic equipment. Phase separation tests are usually evaluated under both aqueous and organic continuity (because continuity frequently impacts phase separation characteristics). The longer it takes for the phases to separate, the larger the settler requirement (assuming mixer-settler equipment will be used). In addition to determining the time for the phases to separate, the phase separation tests can give an indication of the quality of the separation. For SX to work well, the phases should separate completely, leaving a clean interface. Presence of a third phase, a stable emulsion, or a crud layer (solid stabilized emulsion) would raise questions on the operability of the process and would require further study. Phase separation tests can also be used to obtain a first indication of the likelihood of entrainment of one phase or the other being an issue. Signs of a secondary haze or small droplets of organic in the aqueous phase may suggest the system would have high organic losses, or samples of the organic phase may be analyzed for entrained aqueous.

Each SX process is different, but there are common challenges that should be considered when developing a flow sheet for a potential SX process. These challenges can and should be identified through the initial benchtop testing. The common issues to investigate include the following:

- **Solubility.** Ensure that the phases are sufficiently immiscible so one phase does not leave with the other as a dissolved loss. Ensure that the ligand and/or metal ligand complex solubility is not exceeded, leading to preferential loss of the reagent. If soluble losses are likely to occur, then understanding the conditions to recover the dissolved species should be considered.

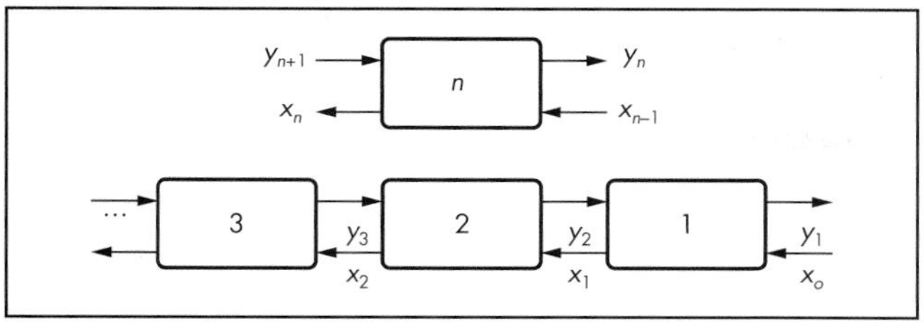

Source: Han 2002; Han and Fuerstenau 2003

Figure 5 Schematic showing countercurrent extraction process

- **Selectivity.** Ensure that the metal of interest is extracted preferentially or can be separated from other impurities through proper adjustment of the process conditions.
- **Stripping.** Ensure that the metals that load can be removed efficiently from the organic and/or ensure that separate stripping regimes are implemented to remove impurities that have a high affinity for the extractant. Making sure there is no *poisoning* of the ligand caused by minor metal loading is an important component of testing, and organics should be taken through multiple extract and strip cycles to fully understand the impact of any circulating load.
- **Stability.** Ensure that there are no elements that load and cannot be removed or there are no conditions that result in the ligand permanently losing its extraction capabilities.
- **Speed/kinetics.** Ensure that the targeted metal separation can be achieved with reasonable and practical mixer retention time.
- **Separation.** Similar to solubility, insufficient phase separation can lead to significant issues in the design and operation of the plant. To achieve the metal separation targets, the phases need to separate. Poor separation can also lead to excessive crud formation, which can also significantly impact the operability of the process. SX designs should include process equipment to allow easy removal of crud from the circuit as well as methodology or equipment to limit the influx of solids into the plant.

Of the issues highlighted, many can be overcome or sufficiently addressed using engineering controls if they are recognized and understood early within the process.

More extensive tests may include the use of mixer-settlers, columns, and centrifugal contactors to better assess the solution properties under dynamic conditions. Piloting tests are ideally used to confirm the metallurgical and physical behavior expected from the shake tests. There are a variety of sizes and shapes in each of these devices. The amounts of liquid samples used in these tests are also multiple because of the type of the equipment and purpose in carrying out such tests.

When the SX chemistry process is studied in a laboratory testing situation, cleaning the extractants and/or diluents (organic liquid) prior to the testing is often critical. The impurities in these chemicals can greatly affect SX chemistry. In some cases, a pretreatment of the extractant or diluent is advised.

SOLVENT EXTRACTION HARDWARE

Mixer-Settlers

The mixer-settler device is the most commonly used piece of SX equipment and consists of a mixing chamber followed by a settling compartment where the phases separate because of the gravity differences. This device can be used as a single unit as well as multiple units in series. Commercially available devices are available in the market or can be easily built in-house.

Columns

Generally, there are two types of columns used in SX: packed columns and pulse types with plates or trays for mixing. These devices are easy to use, but because columns do not have discrete stages, it is not convenient to study pertinent variables acting in the SX process. These devices are commercially available in the market or can be individually manufactured.

Centrifugal Contactors

Centrifugal contactors are discrete-stage units much like mixer-settlers, but the primary difference is the separation of the two-liquid phase mixture. Centrifugal contactors utilize a centrifugal force via a spinning rotor to enable fast and effective separation. Such devices are available commercially.

CONTINUOUS PROCESS

SX operations are often run continuously in a countercurrent mode as shown in Figure 5 (Han 2002; Han and Fuerstenau 2003).

Typically, the goal is to purify and concentrate a certain metal ion. In its simplest form, the process consists of two reaction sections:

1. **Extraction cascade.** The metal is selectively transferred from the aqueous phase to the organic phase in anywhere from one to multiple extraction stages.
2. **Stripping cascade.** The metal is transferred from the loaded solvent phase back to an aqueous phase (usually while concentrating the metal of interest), utilizing anywhere from one to multiple strip stages.

SX plants generally consist of mixer-settler units arranged in series with countercurrent flow. Both the aqueous and organic phases enter the mixer, where one phase is dispersed in the other to allow mass transfer. The dispersion then overflows the mixer into a settler where the phases are allowed to separate. The dispersion is discharged into the settler where it flows through a distribution fence and then separates by gravity as it passes down the length of the settler. The organic phase flows over the upper weir and into a collection launder while the aqueous phase passes over a lower weir, into a similar launder. Although mixer-settlers are the most common, other types of equipment—for example, pulsed columns, sieve-plate columns, centrifugal contactors, and in-line mixers—are also used or have been contemplated. Plants are typically designed and the operating conditions are optimized through the utilization of experimentally generated isotherms and McCabe–Thiele techniques.

Equilibrium and Distribution Isotherms

Isotherms define the extractant's metal loading capacity in both the extraction and stripping sections of the plant.

The *extract isotherm* defines the maximum amount of metal that may be removed from the feed liquor for each organic-to-aqueous (O/A) volumetric ratio. Organic (extractant plus diluent) is mixed with the feed liquor at various O/A ratios until equilibrium is obtained. The organic and aqueous are separated, and the metal concentration in each phase is analyzed. The data are graphed, with organic metal concentration on the y-axis and aqueous metal concentration on the x-axis.

The *strip isotherm* defines the maximum amount of metal that may be removed from the organic for each O/A ratio. Organic (loaded with metal) is mixed with the strip liquor at various O/A ratios until equilibrium is reached. The organic and aqueous are separated and the metal concentration in each phase is analyzed. The data are graphed, with aqueous metal on the y-axis and organic metal on the x-axis.

Once the extract and strip isotherms have been generated, it is possible to use McCabe–Thiele techniques to determine the number of stages required to achieve a given recovery or to produce a specific rich liquor. It is an iterative graphical procedure used to ensure that mass balance conditions are met between extraction and stripping cascades. Full explanation of McCabe–Thiele techniques can be found elsewhere (McCabe and Thiele 1925; Perry and Green 1984).

Equation 29 represents the mass balance for a metal ion being extracted from the aqueous phase to the organic phase through this process:

$$Ax_o + Oy_{n+1} = Ax_n + Oy_1 \ldots \quad \textbf{(EQ 29)}$$

where A and O represent flow rates of aqueous phase and organic phase, respectively; x_o indicates the fractional composition of ion in aqueous phase entering and x_n represents the fractional composition of the ion in the aqueous phase exiting the countercurrent extraction operation; and y_{n+1} indicates the fractional composition of the ion in organic phase entering with y_1 representing the fractional composition of the ion in the organic phase exiting the countercurrent extraction process.

The y-axis of Figure 6 represents the concentration of the ion in the organic phase, and the x-axis represents the concentration of the same ion in the aqueous phase. The entering aqueous solution contains the ion to be extracted at x_0 denoted by port a in the graph. As this solution enters the first tank—stage 1—the ion will be subjected to transfer into the oil phase, and the final concentration of this ion in the stage will approach x_1 given a sufficient time, which is denoted by port b in the diagram. This represents the concentration of the ion in the new feed solution entering stage 2. This process will continue until a satisfactory composition of this metal has been reached in the nth stage. Such a plot is often referred to as a McCabe–Thiele diagram.

The loaded organic exiting the extraction cascade (y_1) is then fed to a stripping cascade to recover the metal of interest from the organic phase and to regenerate the organic solvent for further extraction.

Equation 29 can be rearranged to yield the following:

$$y_1 = \frac{A}{O}(x_o - x_n) + y_{n+1} \ldots \quad \textbf{(EQ 30)}$$

Figure 7 shows the distribution isotherm line for an ion between the aqueous phase and organic phase together with the operating line given by Equation 30. It also shows the stripping isotherm for an ion between the aqueous phase and organic phase together with the operating line.

The y-axis of Figure 7 represents the concentration of the ion in the aqueous phase, and the x-axis represents the concentration of the same ion in the organic phase. The entering organic solution contains the ion to be stripped at y^*_1 denoted by port a in the graph. As this solution enters the

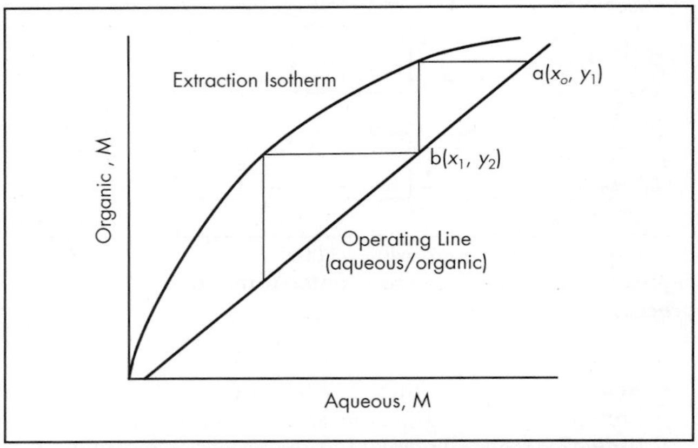

Figure 6 McCabe–Thiele diagram showing relationship between distribution isotherm and operating line

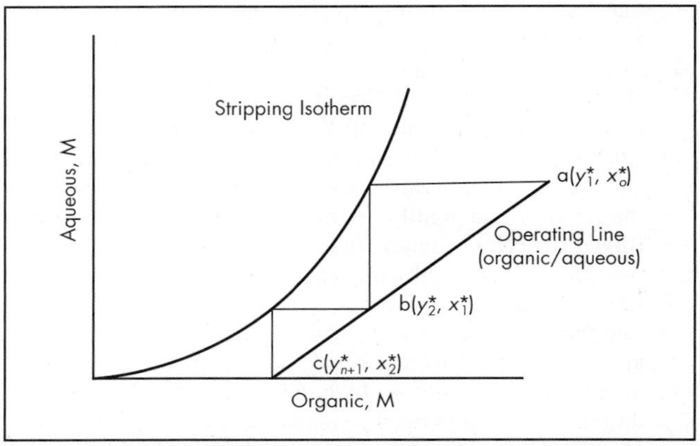

Figure 7 McCabe–Thiele diagram showing relationship between distribution isotherm and operating line for a two-stage stripping cascade

first stage—stage 1—the ion will be subjected to transfer into the aqueous phase, and the final concentration of this ion in the tank will approach y^*_2 given a sufficient time, which is denoted by port b in the diagram. This represents the concentration of the ion in the organic solution entering stage 2. This process will continue until a satisfactory composition of this metal has been reached in the nth stage.

Equation 31 represents the mass balance for a metal ion being transferred from the organic phase to the aqueous phase in the process.

$$Ax^*_{n+1} + Oy^*_n = Ax^*_n + Oy^*_{n+1} \cdots \quad \textbf{(EQ 31)}$$

Equation 31 can be rearranged to yield Equation 32:

$$x_1 = \frac{A}{O}(y_n - y_{n=1}) + x_{n \neq 1} \ldots \quad \textbf{(EQ 32)}$$

The McCabe–Thiele technique will have achieved a satisfactory result when the composition in y^*_1 is equivalent to y_1 and y^*_{n+1} is equivalent to y_{n+1}.

In industrial practice, many different SX configurations are used, depending on the needs of the operation. Although countercurrent flow is the standard operation, it is not uncommon for the organic, feed liquor, and strip liquor streams to be operated in series or parallel flow. Operators also often

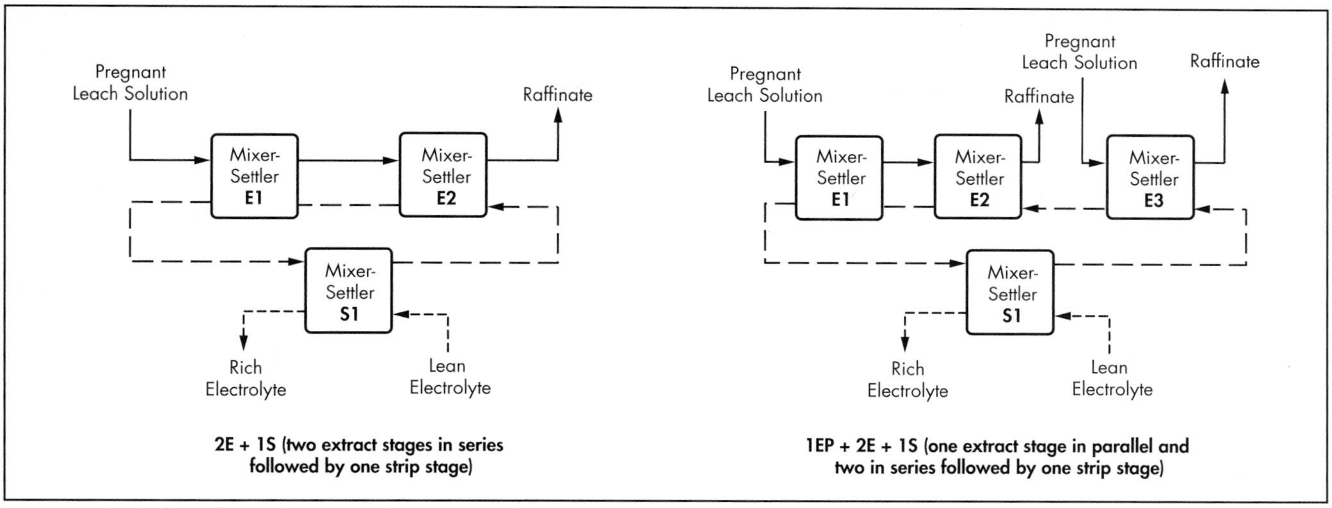

Figure 8 Typical configurations used in copper solvent extraction

Table 6 Typical design parameters for a mixer-settler unit

Mix box organic-to-aqueous ratio	1:1
Mix box retention time	2–3 minutes
Mixer continuity	Variable*
Specific settler flow	3-8 $m^3/(m^2/h)$
Organic phase velocity	3 cm/s

*Choose continuity to minimize entrainment.

consider options such as splitting the streams to redistribute flow or adding streams to enhance recovery or production. Depending on the extractant properties, additional stages may also be required for pre-neutralization or regeneration of the extractant.

One or more wash or scrub stages may be used if the feed solution contains high concentrations of impurities, which may be entrained into the strip liquor or if the extractant is not sufficiently selective for the ion of interest and the impurities need to be chemically removed.

SX configurations vary depending on the separation required and the extractant properties. In base metal separations, it is common to use one to five extract stages, one to three scrub and wash stages, and one to three strip stages. In rare earth separations, an SX plant can have hundreds of stages between extract scrub and strip with significant metal recycle from the rich liquor back to the scrub and extract cascades.

Typical flowcharts for copper SX with some flow information are shown in Figure 8. Extractants, extractant concentrations, and O/A ratios are chosen to achieve the desired metal separations and to concentrate and purify the metal of interest. Physical considerations are also extremely important in a commercial operation. The two immiscible fluids need to be mixed together to achieve the mass transfer, but the phases must also sufficiently separate to minimize entrainment of one phase into the other and to prevent the formation of solid stabilized emulsions. Typical design parameters for a mixer-settler unit are given in Table 6.

REFERENCES

Akhlaghi, M., Rashchi, R., and Vahidi, E. 2010. Stripping of Fe(III) from D2EHPA using different reagents. In *XXV International Mineral Processing Congress (IMPC) 2010: Smarter Processing for the Future.* Melbourne, Victoria: Australasian Institute of Mining and Metallurgy. pp. 255–262.

Bartsch, R.A., Chun, S., and Dzyuba, S.V. 2002. Ionic liquids as novel diluents for solvent extraction of metal salts by crown ethers. In *Ionic Liquids: Industrial Applications for Green Chemistry.* Edited by R.D. Rogers and K.R. Seddon. Washington, DC: American Chemical Society. pp. 58–68.

Berthelot, M., and Jungfleisch, J. 1872. Sur les lois qui president ou partage d'un corps entre deue dissolvants. *Ann. Chim. Phys. [París].* 26:396.

Campos, K., Vincent, T., Bunio, P., Trochimczuk, A., and Guibal, E. 2008. Gold recovery from HCl solutions using Cyphos IL101 (a quaternary phosphonium ionic liquid) immobilized in biopolymer capsules. *Solvent Extr. Ion Exch.* 26(5):570–601.

Cerna, M., Volaufova, E., and Rod, V. 1992. Extraction of light rare earth elements by amines at high inorganic nitrate concentration. *Hydrometallurgy* 28(3):339–352.

Chevron Phillips Chemical Company. 2013. *Orform SX Solvent Extraction Diluents.* The Woodlands, TX: Chevron Phillips Chemical Company.

Cieszyńska, A., Regel-Rosocka, M., and Wiśniewski, M. 2007. Extraction of palladium(II) ions from chloride solutions with phosphonium ionic liquid Cyphos IL101. *Pol. J. Chem. Technol.* 9(2):99–101.

Cox, M. 2010. Liquid-liquid extraction and liquid membranes in the perspective of the twenty-first century. In *Solvent Extraction and Liquid Membranes: Fundamentals and Applications in New Materials.* Edited by M. Aguilar and J.L. Cortina. Boca Raton, FL: CRC Press.

Cytec Industries. 2010. *Mining Chemicals Handbook*, version 2. Woodland Park, NJ: Cytec Industries.

Earle, M.J., Esperanca, J.M.S.S., Gilea, M.A., Canongia Lopes, J.N., Rebelo, L.P.N., Magee, J.W., Seddon, K.R., and Widegren, J.A. 2006. The distillation and volatility of ionic liquids. *Nature* 439:831–834.

Egorov, V.M., Djigailo, D.I., Momotenko, D.S., Chernyshov, D.V., Torocheshnikova, I.I., Smirnova, S.V., and Pletnev, I.V. 2010. Task-specific ionic liquid trioctylmethylammonium salicylate as extraction solvent for transition metal ions. *Talanta* 80(3):1177–1182.

El-Nadi, Y.A. 2012. Lanthanum and neodymium from Egyptian monazite: Synergistic extractive separation using organophosphorous reagents. *Hydrometallurgy* 119-120:23–29.

El-Yamani, I.S., and Shabana, D.L. 1985. Solvent extraction of lanthanum(III) from sulfuric acid solutions by Primene JMT. *J. Less-Common Met.* 105(2):255–261.

Flett, D.S. 2005. Solvent extraction in hydrometallurgy: The role of organophosphorus extractants. *J. Organomet. Chem.* 690(10):2426–2438.

Fox, D.M., Awad, W.H., Gilman, J.W., Maupin, P.H., de Long, H.C., and Trulove, P.C. 2003. Flammability, thermal stability, and phase change characteristics of several trialkylimidazolium salts. *Green Chem.* 5:724–727.

Fukaya, Y., Sekikawa, K., Murata, K., Nakamura, N., and Ohno, H. 2007. Miscibility and phase behavior of water–dicarboxylic acid type ionic liquid mixed systems. *Chem. Commun.* 29:3089–3091.

Habashi, F. 1993. *A Textbook of Hydrometallurgy*. Sainte-Foy, QC: Metallurgie Extractive Quebec.

Han, K.N. 2002. *Fundamentals of Aqueous Metallurgy*. Littleton, CO: SME.

Han, K.N., and Fuerstenau, M.C. 2003. Hydrometallurgy and solution kinetics. In *Principles of Mineral Processing*. Edited by M.C. Fuerstenau and K.N. Han. Littleton, CO: SME.

Lachwa, J., Szydlowski, J., Makowska, A., Seddon, K.R., Esperanca, J.M.S.S., Guedes, H.J.R., and Rebelo, L.P.N. 2006. Changing from an unusual high-temperature demixing to a UCST-type in mixtures of 1-alkyl-3-methylimidazolium bis[(trifluoromethyl)sulfonyl]amide and arenes. *Green Chem.* 8(3):262–267.

Liao, W., Yu, G., and Li, D. 2001. Solvent extraction of Ce(IV) and F from sulphuric acid leaching of bastnasite by Cyanex 923. *Solvent Extr. Ion Exch.* 19(2):243–259.

Mason, G.W., Bilobron, I., and Peppard, D.F. 1978. Extraction of U(VI), Th(IV), Am(III) and Eu(III) by bis para-octyl phosphoric acid in benzene diluents. *J. Inorg. Nucl. Chem.* 40(10):1807–1810.

McCabe, W.L., and Thiele, E.W. 1925. Graphical design of fractionating columns. *Ind. Eng. Chem.* 17(6):605–611.

Okuno, M., Hamaguchi, H., and Hayashi, S. 2006. Magnetic manipulation of materials in a magnetic ionic liquid. *Appl. Phys. Lett.* 89:132506.

Peppard, D.F., Driscoll, W.J., Siromen, S.J., and Mason, G.W. 1957. Fractional extraction of the lanthanides as their dialkyl orthophosphates. *J. Inorg. Nucl. Chem.* 4:334–343.

Perry, R.H., and Green, D.W. 1984. *Perry's Chemical Engineers' Handbook*, 6th ed. New York: McGraw-Hill.

Reddy, M.L.P., Bosco Bharathi, J.R., Peter, S., and Ramamohan, T.R. 1999. Synergistic extraction of rare earths with bis(2,4,4-trimethyl pentyl) dithiophosphinic acid and trialkyl phosphine oxide. *Talanta* 50(1):79–85.

Regel-Rosocka, M. 2009. Extractive removal of zinc(II) from chloride liquors with phosphonium ionic liquids/toluene mixtures as novel extractants. *Sep. Purif. Technol.* 66(1):19–24.

Regel-Rosocka, M., and Wisniewski, M. 2011. Ionic liquids in separation of metal ions from aqueous solutions. In *Applications of Ionic Liquids in Science and Technology*. Edited by S. Handy. Rijeka, Croatia: InTech.

Reichardt, C. 2003. *Solvents and Solvent Effects in Organic Chemistry*, 3rd ed. Weinheim, Germany: Wiley-VCH Verlag.

Rice, C., and Stone, C.A. 1962. *Amines in Liquid-Liquid Extraction of Rare Earth Elements*. Report of Investigations 5923. Washington, DC: U.S. Bureau of Mines.

Seddon, K.R. 1997. Ionic liquids for clean technology. *J. Chem. Technol. Biotechnol.* 68:351–356.

Shell Chemicals. 2016. ShellSol 2046 AR. Technical Datasheet 448, March. p. 3.

Smiglak, M., Reichert, W.M., Holbrey, J.D., Wilkes, J.S., Sun, L., Thrasher, J.S., Kirichenko, K., Singh, S., Katritzky, A.R., and Rogers, R.D. 2006. Combustible ionic liquids by design: Is laboratory safety another ionic liquid myth? *Chem. Commun.* 24:2554–2556.

Srncik, M., Kogelnig, D., Stojanovic, A., Körner, W., Krachler, R., and Wallner, G. 2009. Uranium extraction from aqueous solutions by ionic liquids. *Appl. Radiat. Isot.* 67(12):2146–2149.

Sun, X., Ji, Y., Zhang, L., Chen, J., and Li, D. 2010. Separation of cobalt and nickel using inner synergistic extraction from bifunctional ionic liquid extractant (Bif-ILE). *J. Hazard. Mater.* 182(1–3):447–452.

Xie, P., Zhang, T.A., Dreisinger, D., and Doyle, F. 2014. A critical review on solvent extraction of rare earths from aqueous solutions. *Miner. Eng.* 56:10–28.

Zheng, D., Gray, N.B., and Stevens, D.W. 1991. Comparison of naphthenic acid, versatic acid, and D2EHPA for the separation of rare earths. *Solvent Extr. Ion Exch.* 9(1):85–102.

Cementation

J. Brent Hiskey

Cementation is a general term used in the hydrometallurgical industry for the recovery of any dissolved metal from aqueous solution by contact reduction. More precisely, cementation or metal displacement is described as the electrochemical precipitation of a metal from solution by another more electropositive metal. The precipitation of copper on iron is a classic example of metal extraction by cementation:

$$Cu^{2+} + Fe^{\circ} \rightarrow Cu^{\circ} + Fe^{2+} \quad \textbf{(EQ 1)}$$

Cementation has been practiced since time immemorial. A process employing this reaction was successfully carried out by the ancient Chinese (Pu 1982; Lung 1986), and in the 17th century a license was granted by the Spanish Crown for the use of copper cementation on iron on a commercial scale at Rio Tinto, Spain (Nash 1912). Another classic example of the use of cementation in metal extraction is the recovery of gold from cyanide solution by zinc, which was part of the original MacArthur–Forrest process (MacArthur et al. 1887, 1889a, 1889b). Gold cementation onto zinc from cyanide solution was subsequently improved and is commonly known as the Merrill–Crowe process for gold recovery. For the most part, the early application of these processes was performed with limited understanding of the fundamental aspects of the reactions and of the role of important side reactions.

There are principally three areas of hydrometallurgy where cementation is used:

1. Primary metal recovery
2. Electrolyte purification
3. Treatment of dilute aqueous effluents

Prior to the advancements and adoption of modern techniques in copper solvent extraction and electrowinning, cementation accounted for the majority of the copper recovered from the leaching of low-grade ores in dumps and heaps throughout the world. One of the world's largest copper cementation plants was operated by Kennecott Utah Copper at the Bingham Canyon mine near Salt Lake City, Utah (United States). This facility was reviewed by Schlitt et al. (1979) and had two modules, each containing 13 large cone-type precipitators designed for producing 180 m/d of cement copper from low-grade copper leach solutions. This plant could handle a solution flow rate of 190 m^3/min.

Besides metal recovery, cementation is widely used for the purification of electrolytes. One important example is the removal of impurities in the electrolytic zinc process (EZP). This process accounts for approximately 90% of the world's zinc production. Contaminant ions such as cadmium, copper, nickel, and cobalt cause problems during zinc electrolysis and are removed from the electrolyte using zinc dust cementation.

Cementation is an extremely attractive method for removing hazardous metallic ions from process streams and effluents. Some examples include the precipitation of silver from photographic processing discharges, the cementation of cadmium by zinc in wastewaters, and the removal of mercury(II) from chloroalkali effluents using Zn and Fe.

The stoichiometry for a typical cementation reaction, as illustrated by Equation 1, is relatively simple, and the reaction kinetics frequently follow straightforward diffusion of the depositing ion to the surface of the substrate. In actuality, cementation is an extremely complex process, involving intricate interactions between electrochemical parameters and metal crystallization steps. Furthermore, the deposit structure and morphology can exhibit a significant effect on the overall efficiency of the process, and surface passivation often inhibits the reaction.

THERMODYNAMICS AND REACTION EQUILIBRIA

Cementation or contact reduction occurs between aqueous metal ions and a metal substrate according to the following general overall equation:

$$M_1^{z_1+} + \frac{z_1}{z_2}M_2^o = M_1^o + \frac{z_1}{z_2}M_2^{z_2+} \quad \textbf{(EQ 2)}$$

where $M_1^{z_1+}$ is the precipitating ion and M_2^o is the more electropositive metal surface. A rudimentary thermodynamic evaluation of the cementation reaction can be performed by a simple examination of the electromotive force series of the elements to determine which metal is more electropositive. A little more detail is provided by superimposing the respective oxidation potential–pH (Pourbaix) diagrams. This provides

J. Brent Hiskey, Professor Emeritus, Mining & Geological Engineering, University of Arizona, Tucson, Arizona, USA

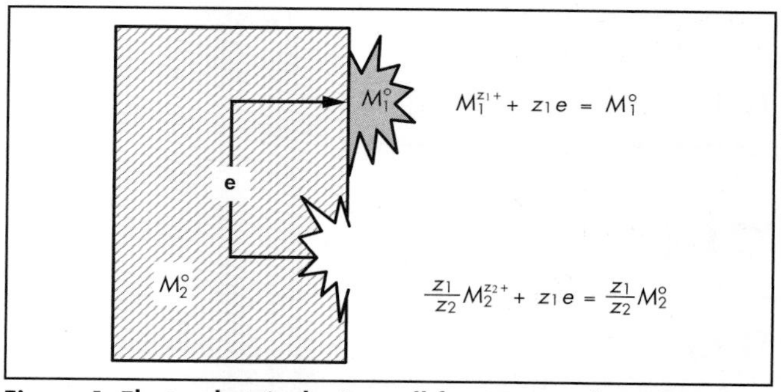

Figure 1 Electrochemical microcell for a general cementation reaction

a direct measure of the thermodynamic driving force for the reaction and the stable phases expected under mixed potential conditions. Ritchie (2003) explained this carefully for the Cu^{2+}/Zn system. Unfortunately, this approach does not predict the equilibrium concentrations of the individual metals.

The process is shown schematically in Figure 1 with separate half-cell reactions expressed as reduction steps. This is clearly an electrochemical system composed of two separate electron-transfer steps divided spatially from each other. These half-cell reactions when short-circuited represent a net cathodic process for the deposition of M_1^o and a net anodic process for the dissolution of M_2^o with electrons flowing from anode to cathode. The potential $E°$ for the short-circuited system at equilibrium is the same for both the cathode and the anode. Therefore, the respective Nernst reduction potentials become equal $E_{o1} = E_{o2}$, as expressed by the following equality:

$$E_{o1}^{\circ} + \frac{2.303\ RT}{z_1 F}\log a_{M_1^{z_1+}} = E_{o2}^{\circ} + \frac{2.303\ RT}{z_1 F}\log a_{M_2^{z_2+}}^{\frac{z_1}{z_2}} \qquad \text{(EQ 3)}$$

where R is the molar gas constant, T is the absolute temperature, and F is the Faraday constant. From here it is possible to calculate the ratio of the activities of the metal ions under equilibrium conditions in the absence of complexing ligands. The corresponding ratio of $M_1^{z_1+}$ to $M_2^{z_2+}$ at 298 K is given by

$$\frac{(a_{M_1^{z_1+}})^{z_2}}{(a_{M_2^{z_2+}})^{z_1}} = \frac{[M_1^{z_1+}]^{z_2}}{[M_2^{z_2+}]^{z_1}} = 10^{\frac{-z_1 z_2 (E_{o1}^{\circ} - E_{o2}^{\circ})}{0.059}} = \psi^{z_1 z_2} \qquad \text{(EQ 4)}$$

where $[M_1^{z_1+}]$ and $[M_2^{z_2+}]$ represent the equilibrium concentrations of the metal ions for the corresponding cementation reaction, and Ψ is a function of ΔE_o°, the thermodynamic driving force and equals $10^{-\Delta E_o^{\circ}/0.059}$. From stoichiometry it follows that

$$[M_2^{z_2+}] = \frac{z_2}{z_1}([M_1^{z_1+}]_o - [M_1^{z_1+}]) \qquad \text{(EQ 5)}$$

For many cementation systems (e.g., Cu^{2+}/Fe, Cu^{2+}/Zn), the equilibrium concentration $[M_1^{z_1+}]$ is much less than the initial concentration $[M_1^{z_1+}]_o$, therefore Equation 4 simplifies to the following relationship:

$$[M_1^{z_1+}] = \left(\frac{z_2}{z_1}[M_1^{z_1+}]_o\right)^{\frac{z_1}{z_2}} \psi^{z_1} \qquad \text{(EQ 6)}$$

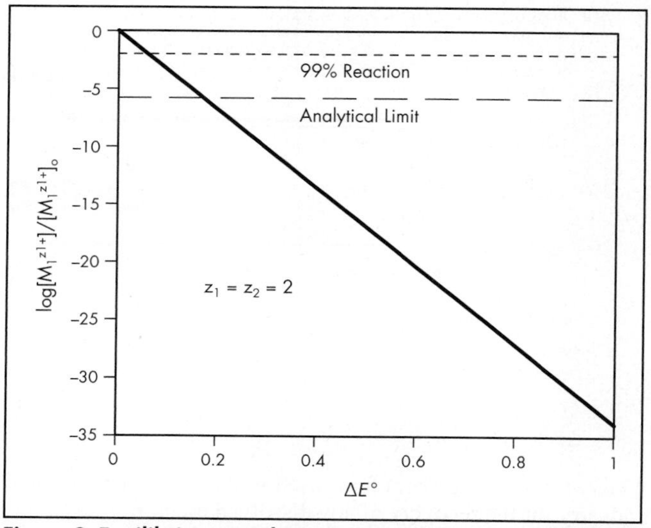

Figure 2 Equilibrium metal concentrations as a function of ΔE_o°

It is possible to calculate the ratio of $[M_1^{z_1+}]/[M_1^{z_1+}]_o$ for a select system at a given value of $[M_1^{z_1+}]_o$. A plot of $\log[M_1^{z_1+}]/[M_1^{z_1+}]_o$ as a function of ΔE_o° is shown in Figure 2. This figure was constructed for a hypothetical system where $z_1 = z_2 = 2$ and the value of $[M_1^{z_1+}]_o$ equals 0.001 M. As depicted in Figure 2, when the value of ΔE_o° increases, the equilibrium concentration of $M_1^{z_1+}$ moves to extremely low levels. In this figure, the top dashed line represents 99% reduction (for practical purposes, complete reduction) and even a small value of ΔE_o° equal to only 0.059 V corresponds to 99% reduction. Pang and Ritchie (1982) studied the mercury(I)–silver displacement reaction, which has an extremely low $\Delta E_o^{\circ} = -0.003$ V. Measurable reaction was observed at 298 K and $[Hg_2^{2+}] = 1.32 \times 10^{-2}$ M. When the value of ΔE_o° is greater than 0.2 V, the system comes to equilibrium below the analytical limit for dissolved metals. In actual systems, metal displacement falls short of the ideal theoretical equilibrium. For example, copper cementation on iron ($\Delta E_o^{\circ} = 0.777$ V) would have a predicted $\log[M_1^{z_1+}]/[M_1^{z_1+}]$ value equal to −26.34 at equilibrium. Experience has proven that the $\log[M_1^{z_1+}]/[M_1^{z_1+}]_o$ is seldom less than −3.0, even for extended cementation times. Several factors influence the final reactant concentration including surface passivation, caused by the formation of a stable second phase, and surface blockage by the depositing metal. In some cases, the formation of metal oxides and hydroxides can cause passivation (Ho and Ritchie 2002). Furthermore, when solubility limits are exceeded within the pore structure of the deposit, certain salts such as $FeSO_4$ (ferrous sulfate) can retard the reaction (Schalch et al. 1976).

A similar theoretical treatment can be performed for cementation systems containing metal complexes. For cementation processes involving metal complexes formed by type (L^-) ligands such as CN^- and Cl^-, the general overall stoichiometry follows this equation:

$$M_1(L)_{\alpha}^{(\alpha - z_1)-} + \frac{z_1}{z_2}M_2 = M_1 + M_2(L)_{\beta}^{(\beta - z_2)-} + \left(\alpha - \frac{z_1}{z_2}\beta\right)L^- \qquad \text{(EQ 7)}$$

Table 1 Equilibrium concentration relationships for selected metal complex cementation systems

System	Equilibrium Concentration Relationship	ψ	K_α	K_β
$Au(CN)_2^- / Zn$	$[Au(CN)_2^-] = \sqrt{[Au(CN)_2^-]_o}\frac{(K_\beta C_L^4+1)\psi}{\sqrt{2(K_\alpha C_L^2+1)}}$	1.72×10^{-42}	1.32×10^{39}	2.51×10^{31}
$Ag(CN)_3^{2-} / Zn$	$[Ag(CN)_3^{2-}] = \sqrt{[Ag(CN)_3^{2-}]_o}\frac{(K_\beta C_L^4+1)\psi}{\sqrt{2(K_\alpha C_L^3+1)}}$	3.10×10^{-27}	2.08×10^{22}	2.51×10^{31}
$Cu(CN)_4^{3-} / Zn$	$[Cu(CN)_4^{3-}] = \sqrt{[Cu(CN)_4^{3-}]_o}\frac{(K_\beta C_L^4+1)\psi}{\sqrt{2(K_\alpha C_L^4+1)}}$	1.87×10^{-22}	3.49×10^{30}	2.51×10^{31}

Note: The ψ, K_α, K_β values were calculated from HSC Chemistry software (Outotec 2009).

Several examples of practical importance that illustrate the role of complexation in cementation reactions are as follows:

$$Au(CN)_2^- + ½Zn = Au + ½Zn(CN)_4^{2-} \quad \textbf{(EQ 8)}$$

$$Ag(CN)_3^{2-} + ½Zn = Ag + ½Zn(CN)_4^{2-} + CN \quad \textbf{(EQ 9)}$$

$$Cu(CN)_4^{3-} = ½Zn = Cu + ½Zn(CN)_4^{2-} + 2CN \quad \textbf{(EQ 10)}$$

Sedzimir (2001) developed expressions analogous to Equation 6 for cementation equilibria comprising selected cyanide complexes. Table 1 presents relationships for the equilibrium concentrations of the depositing metal as a function of ligand concentration in stoichiometric excess.

The ratio of $[M_1(L)_\alpha^{(\alpha-z_1)-}]/[M_1(L)_\alpha^{(\alpha-z_1)-}]_o$ was calculated according to the relationships and data shown in Table 1 for the cementation of gold, silver, and copper from the respective cyanide complexes. These results are plotted in Figure 3 for $\log[M_1(L)_\alpha^{(\alpha-z_1)-}]/[M_1(L)_\alpha^{(\alpha-z_1)-}]_o$ as a function of $[CN^-]$.

These results exhibit very little sensitivity to cyanide concentration in the range shown. In general, the equilibrium concentration for metal complexes is greater than that for the free ions. It is not possible to compare this for the complexes of Au^+(aurous) and Cu^+(cuprous) ions since they are not thermodynamically stable in aqueous solution. They have a strong tendency of undergoing disproportionation. On the other hand, the equilibrium concentration of free Ag^+(argentous) ion cementation on zinc can be determined. The log $[Ag^+]/[Ag^+]_o$ would be about –24.7. Miller et al. (1990) examined the effect of [NaCN] on gold cementation from cyanide solution by particulate zinc. At 10^{-2} and 10^{-1} *M* NaCN the equilibrium gold concentration could have conceivably reached the analytical limit. At 10^{-3} *M* NaCN (sodium cyanide) the cementation reaction was retarded and log $[Au(CN)_2^-]/[Au(CN)_2^-]_o$ approached only –1.3. As observed by Barin et al. (1980), low cyanide ion concentrations inhibit the overall cementation reaction by the formation of $Zn(OH)_{2(solid)}$.

KINETICS

There is overwhelming empirical evidence that cementation reactions obey a first-order rate law. For an irreversible cementation reaction, the rate equation for surface area (*A*) and solution volume (*V*) is

$$\frac{d[M_1^{z_1+}]}{dt} = -\frac{k_o A}{V}[M_1^{z_1+}] \quad \textbf{(EQ 11)}$$

If the specific rate constant (k_o) is independent of concentration, the rate equation may be integrated to give the usual first-order form:

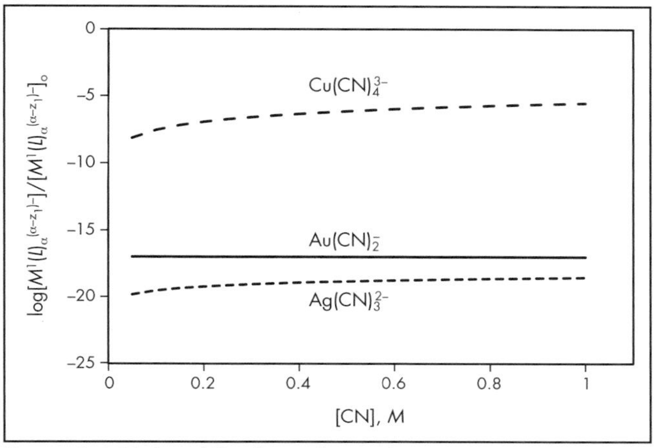

Figure 3 Initial and final metal complex concentrations as a function of cyanide concentration

$$\log\frac{[M_1^{z_1+}]}{[M_1^{z_1+}]_o} = -\frac{k_o A}{2.303\,V}t \quad \textbf{(EQ 12)}$$

where $[M_1^{z_1+}]_o$ and $[M_1^{z_1+}]$ equal the initial concentration and the concentration at time (*t*) of the noble metal ion. A plot of $\log[M_1^{z_1+}]/[M_1^{z_1+}]_o$ versus time would yield a straight line with a slope equal to $-(k_oA/2.303\ V)$ from which the value of (k_o) can be easily determined.

It can conveniently be shown that under conditions of boundary layer diffusion control that Equation 12 becomes

$$\log\frac{[M_1^{z_1+}]}{[M_1^{z_1+}]_o} = -\frac{A\,D}{2.303\,V\delta}t \quad \textbf{(EQ 13)}$$

where *D* is the coefficient of diffusion (cm^2/s) and δ is the boundary layer thickness (cm). It is apparent that k_o equals D/δ, which is identified as the apparent mass transfer coefficient. For cementation under strong agitation, the value of k_o would be of the order of 10^{-2} cm/s.

Evans Diagram

The mechanism of cementation reactions can be conveniently discussed in terms of Evans diagrams. Evans diagrams were originally applied as a graphical means for interpreting the rate of corrosion reactions (Bockris and Reddy 1973) and have been successfully adapted to the interpretation of cementation reactions. Power and Ritchie (1976) provided an excellent treatment of the use of Evans diagrams in the analysis of metal displacement (cementation) reactions. These diagrams are basically the superimposition of the two independent current–potential (polarization) curves of the cementation constituent half reactions. Miller (1979) presented a thorough theoretical development and construction of a representative diagram. It is instructive to begin with the typical polarization curve for a general metal as shown in the log *i/E* plot in Figure 4. Three distinct regions are observed: at very low current densities, the potential is equal to the reversible electrode potential (E_o); at intermediate current densities, the potential varies linearly with log *i* (Tafel region); for sufficiently negative potentials, the current density assumes a maximum value (limiting current density) where diffusion to the electrode surface is controlling.

A characteristic Evans diagram for the cementation reaction between Cu^{2+} ions and Fe metal is depicted in Figure 5. This idealized diagram was constructed by Power and Ritchie (1976) from electrochemical data of Hurlen (1960, 1961). The anodic curve for iron dissolution is approximately a mirror image for that of copper deposition except diffusion is not typically observed until fairly high current densities are obtained.

The point of intersection of the two curves establishes the mixed potential for the cementation system. For this example, it indicates that the reaction is under mass transfer control. As shown, the mixed potential (E_m) for the conditions used in this plot is below the reversible potential for the copper electrode and the reaction should be under diffusion control. By contrast, the Evans diagram for the Fe^{2+}/Zn system (ΔE_o = 0.30 V) for 10^{-1} *M* Fe^{2+} reveals a point of intersection in the Tafel region and the rate would be expected to follow chemical control.

An idealized Evans diagram developed by Power (1975) and discussed by Miller (1979) clearly illustrates the point at which the cementation rate switches from mass transfer control to chemical control. The polarization diagram showing the shift from diffusion to chemical control is depicted in Figure 6.

As reasoned by Power (1975), cementation will likely be controlled by mass transfer when the equilibrium electrode potentials ($\Delta E_o = E_{o,c} - E_{o,a}$) of the component half-cell reactions differ by more than 0.36 V. It was also noted that if ΔE_o is less than 0.06 V, the system could be considered controlled by an electrochemical surface reaction. It is reasonable to accept that between these limits cementation will follow mixed kinetics. A simplifying assumption involves the use of the difference between the standard electrode potentials (ΔE_o°) for the component half-cell reactions as an approximation for ΔE_o.

Miller (1979) tabulated kinetic parameters for several cementation systems. Of the 15 systems listed, 13 had ΔE_o° values greater than 0.36 V, all of which exhibited rate constants and activation energies characteristic of diffusion control. The two systems for which $\Delta E_o^\circ < 0.36$ V, Ni^{2+}/Fe and Pb^{2+}/Fe, revealed activation energies typical of chemical control. Furthermore, the Ni^{2+} reduction by Fe exhibited a comparatively small rate constant of approximately 10^{-4} cm/s. Similar results were observed by Strickland and Lawson (1971, 1973).

It is important to recognize that, for various reasons, certain examples fail to conform to Evans diagram criterion. Annamalai and Hiskey (1978) examined the kinetics of copper cementation on aluminum. The Cu^{2+}/Al system has an extremely large ΔE_o° value equal to 2.04, which would certainly suggest diffusion control according to the Evans diagram. The naturally existing aluminum oxide film has a profound influence on the cementation reaction. The presence of chloride ion will attack the resistive oxide layer and allow copper cementation to occur. At temperatures <33°C, the experimental activation energy E_a = 20.4 kcal/mol which indicates reaction control for the system. This value is close to that obtained by MacKinnon and Ingraham (1970) for the same system (E_a = 14.7 kcal/mol at T < 40°C). Another example is the Cu^{2+}/Ni system where ΔE_o° = 0.57 V.

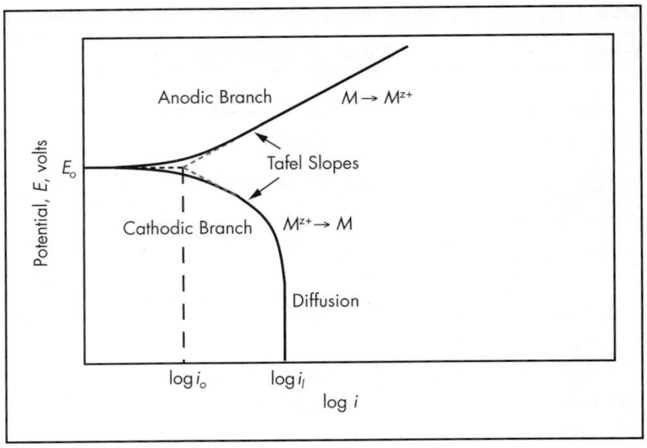

Adapted from Miller 1979

Figure 4 Typical polarization curves for a metal

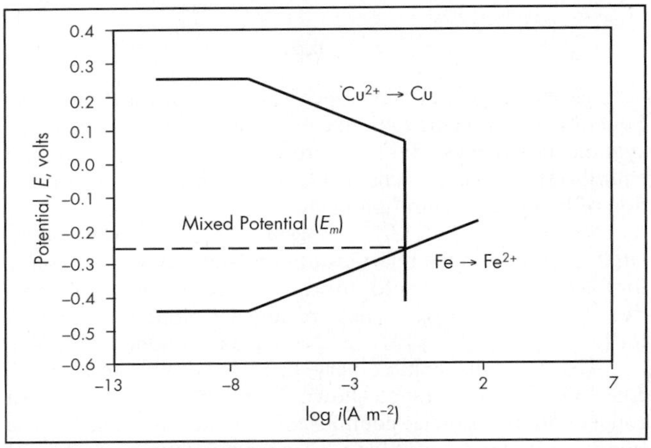

Source: Power and Ritchie 1976

Figure 5 Evans diagram for the Cu^{2+}/Fe system showing diffusion control. Polarization curves are for 10^{-3} *M* Cu^{2+} and Fe metal.

DEPOSIT STRUCTURE AND MORPHOLOGY

The crystal structure and morphological features of metal cementation deposits vary widely. Surface deposits have been observed to occur as powders, spongy clusters, fine and coarse dendrites, botryoidal groups, and uniform foil-like films. Metal deposits produced by cementation are similar to those formed during electrolytic deposition. This is certainly reasonable since both processes are electrochemical in nature. The physical characteristics of cementation surface deposits are influenced by such factors as hydrodynamic conditions, temperature, metal ion concentration, complexing ligands, and pH. Electrolytic deposits associated with electrorefining and electroplating are strongly dependent on current density

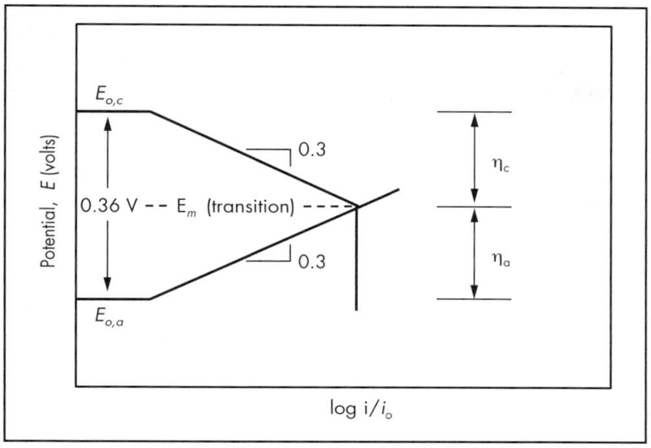

Source: Power and Ritchie 1976

Figure 6 Idealized Evans diagram showing the transition between diffusion and chemical control. Tafel slopes have been assumed to be 0.03 V per decade of log i/i_o and the exchange current densities are equal ($i_{o,c} = i_{o,a}$).

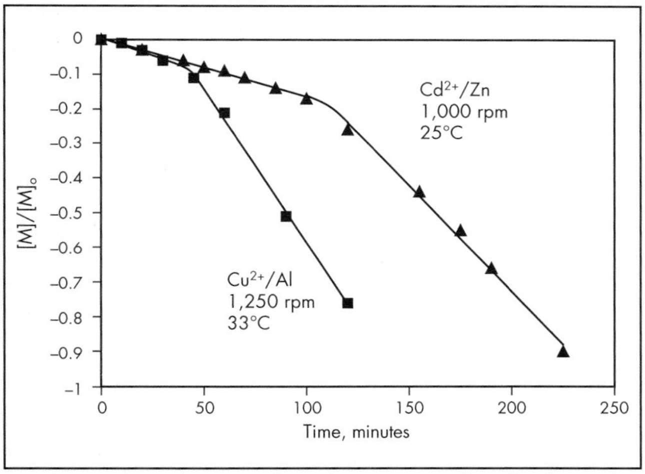

Figure 7 First-order rate plots showing initial and enhanced regions of cementation

and additives (i.e., smoothing agents). Current density is not an independent experimental or operational parameter for cementation reactions.

Deposit structure can have a pronounced effect on cementation kinetics. Some deposits can result in an enhancement of reaction rate while others can totally block the surface and severely retard the reaction. In many systems, two distinct kinetic regions are identifiable in the first-order rate plots. An initial slow period is followed by a second stage where the rate is noticeably enhanced. The initial stage is associated with deposition on a smooth surface and is relatively slow. As the deposit increases in thickness, surface roughness is developed. Surface features protrude into the boundary layer and cause localized turbulence that increases the mass transfer rate. Strickland and Lawson (1971) have discussed this in detail. They suggest that the enhanced rates are a result of surface roughness and/or an increase in the cathodic surface area.

First-order rate plots showing the initial and enhanced regions for cementation on rotating disks are provided in

Table 2 Values of the specific rate constant k_o

	Mass Transfer Coefficient, cm/s	
Cementation System	**Initial Stage**	**Enhanced Stage**
Cu^{2+}/Al	7.8×10^{-3}	3.5×10^{-2}
Cd^{2+}/Zn	6.4×10^{-3}	2.3×10^{-3}

Figure 7 for the Cu^{2+}/Al and Cd^{2+}/Zn systems. Values of the specific rate constant k_o are presented in Table 2 for the initial and enhanced stages for the data shown in Figure 7.

Kinetic enhancements corresponding to 4.5 and 3.6 times for Cu^{2+}/Al and Cd^{2+}/Zn, respectively, were found for these systems. Strickland and Lawson (1971) have reported various degrees of enhancement ranging from 2 to 7 times for Cu^{2+}/Zn, Cu^{2+}/Fe, Pb^{2+}/Zn, Ag^{+}/Zn, Ag^{+}/Cu, and Ag^{+}/Cd in sulfate solutions.

Some cementation systems exhibit surface blockage by the deposit. This results in passivation of the surface and restriction of the anodic half-cell reaction. This can occur when the surface is covered by a massively heavy type of deposit that impedes diffusion within the pores of the deposit. Alternately, deposits that are dense, compact, and adherent can essentially block the reaction. There are many examples where surface blockage has affected the rate of the cementation reaction. Under certain conditions, cementation from cyanide solutions has demonstrated a propensity for blockage-type deposits. Ritchie (2003) has listed some general factors favoring surface blockage:

- Deposition from a complex ion
- When the reactant concentration exceeds some critical value
- Low reaction temperature
- Compatible structures for depositing and displacing metals

Von Hahn and Ingraham (1966) investigated the cementation of silver on copper in perchloric acid and alkaline cyanide solutions. For rotating disk geometry, they found kinetic enhancement in the perchloric acid system and a deposit structure that was composed of loosely adhering powder. On the other hand, in the alkaline cyanide system, silver cementation was inhibited by the deposit, which was smooth, dense, and strongly adherent.

ALLOY FORMATION

An interesting phenomenon associated with certain cementation reactions is the formation of metal alloys. This is especially interesting since cementation is carried out at relatively low temperatures (approximately room temperature). The theoretical and practical considerations involving the electrodeposition of alloys from aqueous solution have been examined for many years. Two classic examples are the electrolytic deposition of brass and bronze alloys. Brenner (1963) presented a considerable amount of information regarding the electrodeposition of alloys and reviewed an extensive list of examples. A limited number of alloys are formed during the electroless plating. The formation of alloys during cementation is even rarer.

One of the first investigations to observe alloy formation during cementation was that by Straumanis and Fang (1951). They studied deposits associated with the displacement of Cu, Au, Ni, and Ag on Zn. Copper deposits contained 9–13 at. %

Zn and were confirmed by the X-ray patterns to be α-brass. Annealing the deposit at 900°C resulted in sharp X-ray lines for α-brass and an increase in Zn content. Von Hahn and Ingraham (1966) examined the Pd^{2+}/Cu system and by X-ray-diffraction data clearly established the formation of Cu-Pd alloys and not pure Pd. The individual alloy crystallites of Cu-Pd ranged from 20 to 45 at. % Pd. Lee and Hiskey (2003) and Hiskey and Lee (2003) detected Au-Cu alloys during the cementation of gold on copper from ammoniacal thiosulfate solutions. The alloy composition was dependent on the initial Cu–Au ratio in solution. Intermetallic alloys were predominantly Cu_3Au in composition. Underpotential deposition theory is a possible mechanism for explaining the formation of these alloys.

LABORATORY EXPERIMENTS

Conceivably there are many experimental methods that might be used to investigate cementation reactions. Most investigations are aimed at obtaining kinetic and mechanistic data, and information about the efficiency of the reaction. This information is helpful in developing a chemical model for the system and in process scale-up. All studies basically depend on the type of reactor and the nature of the substrate. The two most common methods are performed in batch reactors using either suspended particles or a controlled geometry metallic substrate. Both systems can independently study such parameters as temperature, noble metal ion concentration, pH, ionic strength, complexing ligands, additives, and hydrodynamics.

Particulate System

A stirred batch reactor is typically used to suspend particles of different size in the solution. Particles are normally screened into narrow size fractions, and the geometric mean size for each particle size interval is used to calculate the total surface area. The particulate sample can be treated to remove oxide films and other debris from the surface prior to being introduced into the reactor. Solutions containing the noble metal ion are adjusted to the proper experimental conditions, and the experiment commences when the particles are added to the reactor. Agitation can be briefly stopped at set intervals to allow particles to settle and solution aliquots to be removed to analyze for the dissolved metal. In a suspension of spherical particles, it is possible to correlate the mass transfer coefficient to the agitation intensity according to the Harriott (1962) relationship.

Controlled Geometry

Most often, fundamental kinetic and mechanistic studies of cementation employ controlled geometry experiments. In this system, once again, batch cementation tests are carried out using metal strips under static conditions or a rotating disc or rotating cylinder apparatus. Rotating disc and cylinder systems create more agitation and thus reduce the diffusion boundary layer thickness. A disc geometry is very popular, not only because of simple fabrication, but importantly, the hydrodynamics of the system can be analyzed and modeled mathematically (Levich 1962). A rotating disc can be conveniently used to analyze the effect of surface roughness on the cementation reaction. Furthermore, electrochemical measurements using a rotating disc electrode assembly can be correlated directly to the rotating disc kinetic data. Another important feature of the rotating disc is the direct transfer of the disc for surface product examination by optical and scanning electron microscopy techniques. Prior to use, metal discs are polished using a successively finer grit size of silicon carbide paper. After thoroughly washing and rinsing, the disc is immersed into the test solution to start the reaction. Solution aliquots are periodically moved to analyze for the dissolved metal.

INDUSTRIAL APPLICATIONS OF METALLURGICAL CEMENTATION

As noted earlier, there are three main areas of hydrometallurgy where cementation is of industrial importance: primary metal recovery, electrolyte purification, and the treatment of dilute aqueous effluents. It should be emphasized that the general thermodynamic, kinetic, and electrochemical principles apply irrespective of application and method.

Metal Recovery

Copper

The cementation of copper on iron has been practiced for millennia with very little change to the basic operating principles. At Rio Tinto, Spain, copper cementation was outlined in the patent or license granted to Don Alvaro Alonso de Garfias in 1661 by the Spanish Crown; however, the installation of an engineered system was not achieved until 1752 (Nash 1904). The process employed channels (launders) and tanks loaded with iron sheets or scrap. During these early periods, relatively high-quality copper was produced at Rio Tinto: the poorest grade, *cobre negro*, contained about 90% Cu; and the finest grade, *cobre fino*, about 98% Cu. These precipitates were equal in quality to that obtained by smelting. In 1850, a smelter was constructed for the specific purpose of refining *cobre negro* precipitates.

Gravity-flow launders charged with shredded scrap iron remained the industrial standard for copper cementation plants associated with heap and dump leaching operations even into the 1980s. Spedden et al. (1966) indicate that processing 3,785 L/min of copper-bearing leach solution would require a launder section measuring 152.4 m long, 1.2 m wide, and 1.2 m deep. A large leaching operation would require several sections. A typical plant would also need the necessary pumping system, holding tanks, gantry or crane, trammel screens, thickeners, and drying pans. One variation used perforated plastic pipes positioned in the launder floor to evenly distribute pregnant leach solution (PLS) on the iron substrate. The normal operating cycle would consist of the following steps (Jacky 1967):

1. Bedding the section with iron
2. Introducing and controlling the PLS
3. Draining, washing, and excavating the copper precipitates
4. Washing the section and punching orifices

This system would effectively yield 90% copper recovery. These plants were relatively simple to construct and operate. Unfortunately, they suffered from poor hydrodynamics, high iron factors (2 to 4 times theoretical), a lot of manual labor, and impure copper precipitates that required additional refining. In an attempt to overcome some of the drawbacks of conventional launders, Kennecott designed a cone-type precipitator that was installed at their operations in the late 1960s. Figure 8 shows an illustration of a cone precipitator. A variation of this cementation reactor was developed by Kennecott that utilized sponge iron or particulate iron as the precipitant instead of shredded iron scrap (Back 1967).

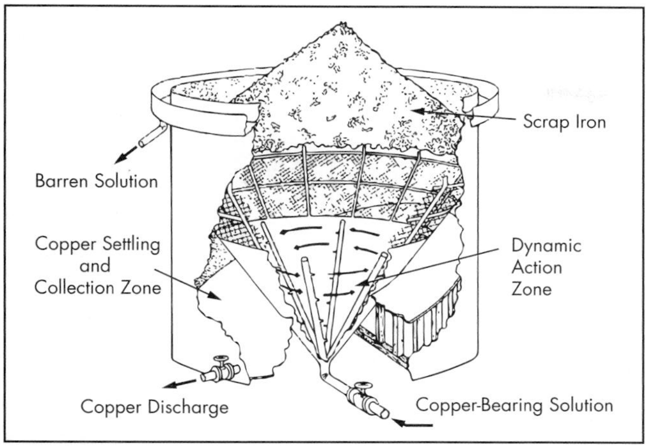

Adapted from Back 1967

Figure 8 Pictorial representation of Kennecott cone precipitator

Copper cementation was eventually phased out at most leaching operations with the successful development and implementation of copper solvent extraction (Hiskey 2014).

Gold

The cementation of gold by zinc was an integral part of the cyanide process as developed by MacArthur et al. (1887, 1889a, 1889b). Process improvements to the cementation unit operation are generally known today as the Merrill–Crowe process for gold recovery. The overall reaction for this process is shown in the following equation:

$$Au(CN)_2^- + \frac{1}{2}Zn = Au + \frac{1}{2}Zn(CN)_4^{2-} \quad \textbf{(EQ 14)}$$

The Merrill–Crowe cementation process is mainly employed for gold/silver recovery from dilute sodium cyanide solutions and is the preferred recovery method for gold cyanide solutions containing high silver concentrations. In addition, Merrill–Crowe is a process alternative to electrowinning for gold recovery from carbon-in-column and carbon-in-pulp eluates. A simplified process flow sheet is depicted in Figure 9.

For efficient precipitation, it is critical that the PLS be perfectly free of all suspended solids. This is normally accomplished by clarifying filters. Following clarification, the solution undergoes deaeration under vacuum to remove dissolved oxygen. Without this step there would be excessive zinc consumption during the cementation reaction. A properly operated deaeration tower can produce solution with <1 ppm O_2. Zinc dust is added along with barren cyanide solution to a small mixing cone. The zinc dust slurry is then combined with the clarified, deaerated PLS and the cementation reaction is initiated. This is followed by a plate-and-frame filter press where the cementation reaction is allowed to continue and the precipitated metals collected. A continuous cementation process will have several filter presses operating in parallel. Chi et al. (1997) reported that a typical cycle requires five to seven days for completion. At the end of each cycle, the filter cake is blown dry with air and then discharged for subsequent furnace refining. Depending on gold and sodium cyanide concentration, controlled additions of lead nitrate can serve as an activator or accelerator for the cementation reaction.

Electrolyte Purification—Electrolytic Zinc Process

The removal of trace levels of impurities from zinc sulfate electrolyte by cementation is a critical step in zinc electrowinning. Metals more electropositive than zinc (e.g., copper, nickel, cobalt, and cadmium) can exhibit detrimental effects during electrowinning, including co-deposition and redissolution of the cathode. These effects decrease current efficiency and result in poor product quality. Cementation of impurities by zinc dust has been an electrolyte purification strategy for many years. Important zinc does not add any impurities ions into the electrolyte. Metal cations are precipitated according to the following general reaction:

$$M^{2+} + Zn^o \rightarrow M^o + Fe^{2+} \quad \textbf{(EQ 15)}$$

Raghavan et al. (1999) reviewed several zinc dust purification technologies for various EZP producers in the world. The cold–hot purification method is a common approach in many plants. This treatment option is shown schematically in Figure 10.

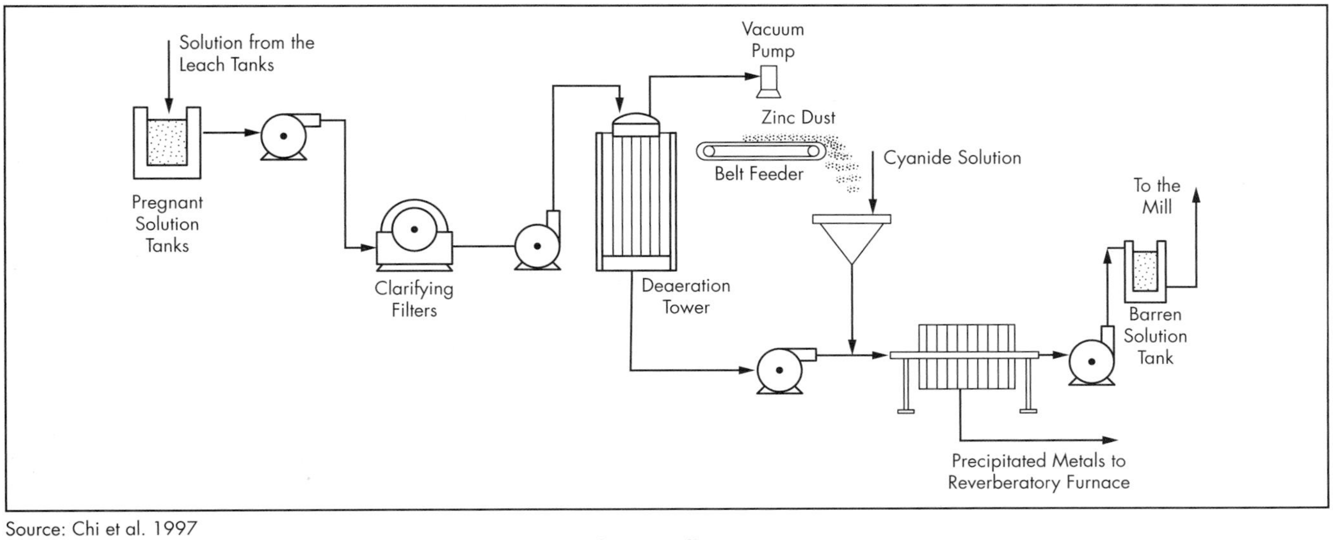

Source: Chi et al. 1997

Figure 9 Generalized flow sheet for gold/silver recovery by Merrill–Crowe cementation

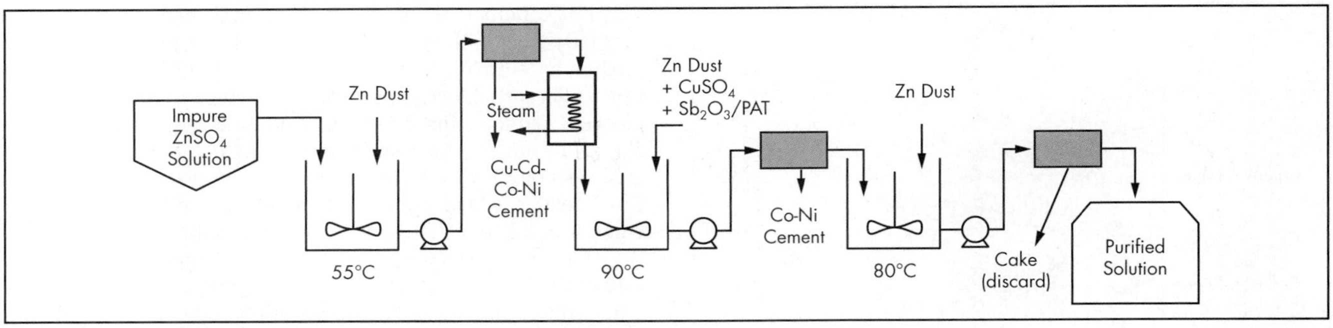

Adapted from Raghavan et al. 1999

Figure 10 Cold–hot EZP electrolyte purification method

More recently, Krause (2014) provided a comprehensive study aimed at optimizing the zinc cementation purification process. It was noted that a typical cold–hot purification scheme involves initial removal of copper at low temperature (60°–65°C), followed by high-temperature (70°–100°C) removal of cobalt and nickel using activators, and a final low-temperature (60°–65°C) cadmium cementation step. Because of slow cementation kinetics with cobalt and nickel, activators such as As_2O_3 (arsenic(III) oxide), Sb_2O_3 (antimony(III) oxide), antimony metal, or potassium antimony tartrate (PAT) are required (Singh 1996).

Treatment of Dilute Aqueous Effluents

Cementation is one of the most effective and economic techniques for removing toxic and dilute metal values from industrial waste solutions. It has been demonstrated to be a feasible technique because of its relative simplicity, ease of control, low energy consumption, and the ability to recover valuable metals and to stabilize toxic materials. Most of the initial interest was directed at waste streams containing dilute concentrations of copper. This was sensible given the extremely long history and operating experience using this technique for primary metal recovery. More recently, there as has been considerable interest in the treatment of effluents containing toxic heavy metals such as cadmium. The cementation of cadmium on zinc has been extensively studied in the laboratory (Younesi et al. 2006; Nosier 2003; Taha and Ghani 2005; Aurousseau et al. 2004). As early as 1975, Richard and Brookman (1975) proposed the use of cementation for the removal of mercury from wastewaters. Mercury removal from aqueous solution by zinc cementation has been an ongoing area of research (Ku et al. 2002).

REFERENCES

Annamalai, V., and Hiskey, J.B. 1978. A kinetic study of copper cementation on pure aluminum. *Min. Eng.* (June):650.

Aurousseau, M., Pham, N.T., and Ozil, P. 2004. Effects of ultrasound on the electrochemical cementation of cadmium by zinc powder. *Ultrason Sonochem.* 11(1):23.

Back, A.E. 1967. Precipitation of copper from dilute solutions using particulate iron. *J. Met.* (May):27.

Barin, I., Barth, H., and Yaman, A. 1980. Electrochemical investigation of the kinetics of gold cementation by zinc from cyanide solutions. *Erzmetallurgy* 33:399.

Bockris, J.O'M., and Reddy, A.K.N. 1973. *Modern Electrochemistry*. Vol. 2. New York: Plenum Press.

Brenner, A. 1963. *Electrodeposition of Alloys: Principles and Practice*. New York: Academic Press.

Chi, G., Fuerstenau, M.C., and Anderson, P.A. 1997. Study of Merrill–Crowe processing. Part II: Regression analysis of plant operating data. *Int. J. Miner. Process.* 49:185.

Harriott, P. 1962. Mass transfer to particles: Part I. Suspended in agitated tanks. *AIChE J.* 8(1):93–102.

Hiskey, J.B. 2014. Innovative strategies for copper hydrometallurgy. In *Mineral Processing and Extractive Metallurgy: 100 Years of Innovation.* Edited by C.G. Anderson, R.C. Dunne, and J.L. Uhrie. Englewood, CO: SME.

Hiskey, J.B., and Lee, J. 2003. Kinetics of gold cementation on copper in ammoniacal thiosulfate solutions. *Hydrometallurgy* 69:45.

Ho, E.M., and Ritchie, I.M. 2002. Cementation of copper ions by nickel metal. Paper presented at the Fifth International Symposium on Electrochemistry in Mineral and Metal Processing. The Electrochemical Society.

Hurlen, T. 1960. Electrochemical behavior of iron. *Acta Chem. Scand.* 14:1533.

Hurlen, T. 1961. On the kinetics of the $Cu/Cu^{2+}{}_{aq}$ electrode. *Acta Chem. Scand.* 15:630.

Jacky, H.W. 1967. Copper precipitation methods at Weed Heights. *J. Met.* (April):22.

Krause, B.J. 2014. Optimisation of the purification process of a zinc sulfate leach solution for zinc electrowinning. M.E. thesis, Department of Materials Science and Metallurgical Engineering, University of Pretoria, South Africa.

Ku, Y., Wu, M.H., and Shen, Y.S. 2002. Mercury removal from aqueous solutions by zinc cementation. *Waste Manage.* 22:721–726.

Lee, J., and Hiskey, J.B. 2003. Alloy formation during the cementation of gold on copper from ammoniacal thiosulfate solutions. In *Hydrometallurgy 2003—Fifth International Conference in Honor of Professor Ian Ritchie –Volume 2: Electrometallurgy and Environmental Hydrometallurgy*. Edited by C.A. Young. Warrendale, PA: The Minerals, Metals & Materials Society.

Levich, V.G. 1962. *Physicochemical Hydrodynamics*. Englewood Cliffs, NJ: Prentice-Hall.

Lung, T.N. 1986. The history of copper cementation on iron—The world's first hydrometallurgical process from medieval China. *Hydrometallurgy* 17:113.

MacArthur, J.S., Forrest, R.W., and Forrest, W. 1887. Process of obtaining gold and silver from ores. British Patent 14,174.

MacArthur, J.S., Forrest, R.W., and Forrest, W. 1889a. Process of obtaining gold and silver from ores. U.S. Patent 403,202.

MacArthur, J.S., Forrest, R.W., and Forrest, W. 1889b. Process of separating gold and silver from ores. U.S. Patent 418,137.

MacKinnon, D.J., and Ingraham, T.R. 1970, Kinetics of copper (II) cementation on pure aluminum disc in acidic sulfate solutions. *Can. Metall. Q.* 9:443.

Miller, J.D. 1979. Cementation. In *Rate Processes of Extractive Metallurgy.* Edited by H.Y. Sohn and M.E. Wadsworth. New York: Plenum Press.

Miller, J.D., Wan, R.Y., and Parga, J.R. 1990. Characterization and electrochemical analysis of gold cementation from alkaline cyanide solution by suspended zinc particles. *Hydrometallurgy* 24:373.

Nash, W.G. 1904. *The Rio Tinto mine: Its history and romance.* London: Simpkin, Marshall, Hamilton, Kent.

Nash, W.G. 1912. Precipitation of copper from mine waters. *Min. Sci. Press* 104(5):212.

Nosier, S.A. 2003. Removal of cadmium ions from industrial wastewater by cementation. *Chem. Biochem. Eng. Q.* 17(3):219.

Outotec. 2009. HSC Chemistry software, version 7.0. Outotec.

Pang, J.T.T., and Ritchie, I.M. 1982. The reaction between mercury ions and silver: Dissolution and displacements. *Electrochim. Acta* 27(6):683.

Power, G.P. 1975. Cementation reactions. Ph.D. thesis, University of Western Australia, Perth, Western Australia.

Power, G.P., and Ritchie, I.M. 1976. A contribution to the theory of cementation (metal displacement) reactions. *Aust. J. Chem.* 29:699.

Pu, Y.D. 1982. The history and present status of practice and research work on solution mining in China. In *Interfacing Technologies in Solution Mining*. Edited by J.W. Schlitt and J.B. Hiskey. Littleton, CO: SME-AIME.

Raghavan, R., Mohanan, P.K., and Verma, S.K. 1999. Modified zinc sulphate solution purification technique to obtain low levels of cobalt for the zinc electrowinning process. *Hydrometallurgy* 51:187.

Richard, M.D., and Brookman, G. 1975. The removal of mercury from industrial waste waters by metal reduction. *Eng. Bull. Purdue Univ.* II, 713.

Ritchie, I.M. 2003. Some aspects of cementation reactions. In *Hydrometallurgy 2003—Fifth International Conference in Honor of Professor Ian Ritchie—Volume 2: Electrometallurgy and Environmental Hydrometallurgy*. Edited by C.A. Young. Warrendale, PA: The Minerals, Metals & Materials Society.

Schalch, E., Nicol, M.J., Balestra, P.E.L., and Stapleton, W.M. 1976. *An Electrochemical Investigation of Copper Cementation on Iron.* NIM Report No. 1799. Johannesburg: National Institute of Metallurgy.

Schlitt, W.J., Ream, B.P., Haug, L.J., and Southard, W.D. 1979. Precipitating and drying cement copper at Kennecott's Bingham Canyon facility. *Min. Eng.* (June):671.

Sedzimir, J. 2001. Precipitation of metals by metals (cementation) equilibria in absence or in presence of complexes in the solution. *Arch. Metall.* 46(1):3.

Singh, V. 1996. Technological innovation in the zinc electrolyte purification process of a hydrometallurgical zinc plant through reduction in zinc dust consumption. *Hydrometallurgy* 40:247.

Spedden, H.R., Malouf, E.E., and Prater, J.D. 1966. Cone-type precipitators for improved copper recovery. *J. Met.* 18(10):1137.

Straumanis, M.E., and Fang, C.C. 1951. Structure of metal deposits obtained by electrochemical displacement upon zinc. *J. Electrochem. Soc.* 98(1):9.

Strickland, P.H., and Lawson, F. 1971. The cementation of metals from dilute aqueous solution. *Proc. Aust. Inst. Min. Met.* 237:71.

Strickland, P.H., and Lawson, F. 1973. The measurement and interpretation of cementation rate data. In *International Symposium on Hydrometallurgy*. Edited by D.J.I. Evans and R.S. Shoemaker. New York: AIME.

Taha, A.A., and Ghani, S.A. 2005. Effect of surfactants on the cementation of cadmium. *J. Colloid Interface Sci.* 280(1):9.

Von Hahn, E.A., and Ingraham, T.R. 1966. Kinetics of PdII cementation on sheet copper in perchlorate solutions. *Trans. Metall. Soc. AIME* 236:1098.

Younesi, S.R., Alimadadi, H., Alamdari, E.K., and Marashi, S.P.H. 2006. Kinetic mechanisms of cementation of cadmium ions by zinc powder from sulphate solutions. *Hydrometallurgy* 84:155.

CHAPTER 10.16

Electrowinning

Michael S. Moats and Michael L. Free

Electrowinning is the process of electrolytically *winning* or recovering metal from a solution containing dissolved metal. The process of electrowinning is used to recover aluminum, copper, gold, magnesium, manganese, nickel, silver, zinc, and other metals from liquids on a commercial scale. For aluminum and magnesium, the process of electrowinning involves high temperatures and molten salt environments. For copper, gold, magnesium, manganese, nickel, silver, and zinc, electrowinning is performed in water-based, aqueous liquids. Electrowinning and related principles are also important to other applications such as integrated circuit fabrication and circuit board printing, electroforming of precision parts, and battery recharging. This chapter is designed to give a brief overview of the fundamentals of electrowinning of base metals in aqueous media. A more complete overview of related fundamentals is available elsewhere (Free 2013).

For electrowinning to occur, the desired metal must be dissolved in an ionic form. Common base metal ions are Cu^{2+}, Ni^{2+}, and Zn^{2+}. Each of these metal ions has a 2+ charge because of the loss of two electrons from the metal. Metal ions are usually created by breaking atomic bonds in minerals during leaching to release the metals as ions. The metal ions are often concentrated in aqueous solution and mixed with acid, most commonly sulfuric acid, to form the electrolyte. The electrolyte is the solution in which electrowinning is performed.

Water plays a key role in electrowinning of base metals. Water can be electrolytically broken into oxygen and hydrogen gas ($2H_2O \leftrightarrow 2H_2 + O_2$). The process of electrolytic water splitting or water electrolysis requires considerable energy. Water electrolysis has oxidation ($2H_2O \leftrightarrow 4H^+ + 4e^- + O_2$) and reduction ($4H^+ + 4e^- \leftrightarrow 2H_2$) components that can balance. Each of these reactions is known as a *half-cell reaction* because an opposite half-cell reaction is needed to form a complete electrochemical cell. The water oxidation reaction is the primary anode reaction for electrowinning metals from sulfate-based electrolytes. Metal reduction on the cathode is the primary cathodic reaction, an example of which is $Cu^{2+} + 2e^- \leftrightarrow Cu$. The reduction of hydrogen ions to form hydrogen gas can occur on cathodes that are at low potentials. Hydrogen ion reduction consumes electrons that would otherwise go to metal reduction in lower-potential metals such as nickel and zinc.

The polar and open structure of water molecules readily accommodates ion dissolution and mobility, making the combination of water, acid, and other dissolved ions a very useful electrolyte. The dissolved ions in the electrolyte are necessary for the conduction of current through ion movement. Although dissolved ions are necessary to conduct electrical current between the anode(s) and cathode(s), many dissolved ions do not participate in electrochemical reactions.

BACKGROUND FUNDAMENTALS

Chemical and electrochemical reaction equilibria are generally determined using thermodynamics. Chemical and electrochemical reactions can be expressed as

$$ne^- = \sum_j v_j X_j^{z_j} \quad \textbf{(EQ 1)}$$

where *e* represents an electron with a superscripted minus symbol to indicate its charge; *n* is the number of electrons ($n = 0$ for nonelectrochemical reactions) in half-cell reactions; v_j is the stoichiometric coefficient of the species (positive values represent products and negative values represent reactants, which are generally shown on the left-hand side of the equation); *X* is the chemical formula of species "*j*"; and z_j is the charge of species *j*. Half-cell reactions are written as reduction reactions to be consistent with most traditional electrochemistry. An example of the application of the general chemical/electrochemical formula equation is

$$Cu^{2+} + 2e^- = Cu \quad \textbf{(EQ 2)}$$

in which $v_{Cu2+} = -1$, $z_{Cu2+} = 2$, $v_{Cu} = 1$, $z_{Cu} = 0$, and $n = 2$.

The free energy of a reaction, which identifies reaction direction (negative is forward) and quantifies available energy, is expressed as

$$\Delta G_r = \Delta G_r^\circ + RT\ln\prod_j a_j^{v_j} \quad \textbf{(EQ 3)}$$

Michael S. Moats, Associate Professor, Missouri University of Science & Technology, Rolla, Missouri, USA
Michael L. Free, Professor, University of Utah, Salt Lake City, Utah, USA

in which R is the gas constant (8.314 J/mol·K), T is the absolute temperature (K), and a_j^{vj} is the activity of species j. The standard reaction free energy, ΔG_r°, can be written as

$$\Delta G_r^\circ = \sum_j v_j \Delta G_{f_j}^\circ \quad \textbf{(EQ 4)}$$

in which the subscript f denotes the formation of the compound from the elements under standard conditions.

The standard free energy values for chemical compounds are available in data tables that are commonly based on the reaction of pure elements under standard conditions and unit activities. Pure solid substances and liquids have unit activities. Dissolved ions have unit activities under ideal conditions at a concentration of 1 molal, although in practice, a non-unity activity coefficient is used to correct for deviations from ideality. Pure gasses have unit activities at 100 kPa unless they are associated with substances that have liquid standard states.

At equilibrium, the free energy of the reaction equals zero and the equilibrium constant may be used to represent the equilibrium activity relationship between the species:

$$\Delta G_r^\circ = -RT \ln \prod_j a_{jeq}^{v_j} \quad \textbf{(EQ 5)}$$

in which each species j activity is the equilibrium activity as denoted by the *eq* subscript.

The activity of an ion in aqueous medium can be expressed as

$$a_j = \gamma_j m_j \quad \textbf{(EQ 6)}$$

in which γ_j is the activity coefficient of species j, and m_j is the molality (moles per 1,000 g of water) of species j. The activity coefficient is determined using a variety of correlations and equations and is generally significantly less than one in most common solutions. The activity coefficient is often determined using equations such as the Davies equation, which is

$$-\log \gamma_j = A z_j^2 \left(\frac{\sqrt{I_{str}}}{1 + \sqrt{I_{str}}} - 0.2 I_{str} \right) \quad \textbf{(EQ 7)}$$

in which $A = 0.2409 + 9.01 \times 10^{-4}T$, (valid from 273 to 333 K) (equal to 0.5094 at 298 K), and ionic strength, I_{str}, is determined using the following relationship:

$$I_{str} = \frac{1}{2} \sum_j^n \frac{m_j z_j^2}{m^\circ} \quad \textbf{(EQ 8)}$$

where m_j is the molality of species j, z_i is the charge of species j, and m° is the reference molality (1 molal) for aqueous systems. Other equations are needed for solutions with I values exceeding 0.3.

Electrochemical Principles

The equilibrium electrochemical half-cell potential is determined based on free energy using the Nernst equation:

$$E_{eq} = E^\circ - \frac{RT}{nF} \ln \prod_j a_j^{v_j} \quad \textbf{(EQ 9)}$$

in which F is the Faraday constant (96,485 C/mol [coulombs per mole]) and the standard potential, E°, can be calculated from standard free energy values using the following equation:

$$E^\circ = -\frac{\Delta G_r^\circ}{nF} \quad \textbf{(EQ 10)}$$

Application of the Nernst equation to the water oxidation equation (written as a reduction reaction)

$$4H^+ + O_2 + 4e^- \leftrightarrow 2H_2O \quad \textbf{(EQ 11)}$$

is

$$E_{eq} = E^\circ - \frac{RT}{nF} \ln \prod_j a_j^{v_j}$$
$$= \frac{-\Delta G_r^\circ}{nF} - \frac{RT}{nF} \ln \frac{a_{H_2O}^2}{a_{H^+}^4 a_{O_2}} \quad \textbf{(EQ 12)}$$

The standard reaction free energy is (2(–237,141) – [4(0) + (0)]) – 474,282 J/mole. After conversion to a base 10 logarithm (log) (2.303 log = ln), combined with the assumption of unity for the activity of water and oxygen, results in

$$E_{eq} = \frac{-(-474{,}282)}{4(96{,}485)} - \frac{2.303(8.314)(298)}{4(96{,}485)} \log \frac{1^2}{a_{H^+}^4 (1)}$$
$$= 1.229 - 0.0148 \log \frac{1}{a_{H^+}^4} \quad \textbf{(EQ 13)}$$

Substitution of the relationship for pH [pH = $-\log(a_{H^+})$] and rearrangement leads to

$$E_{eq(H_2O/O_2)} = 1.229 - 0.0591\,\text{pH} \quad \textbf{(EQ 14)}$$

When the potential of an aqueous process, such as occurs on typical anodes, is higher than this equilibrium potential for water at the pH of the environment, the resulting oxidizing environment will cause water electrolysis or electrochemical splitting into hydrogen ions and oxygen gas. Consequently, this relationship between potential and pH is a practical upper limit for water stability. It is also a minimum potential necessary for typical anodic oxidation of water. This upper potential limit for aqueous systems indicates water will begin to decompose above it.

Water also decomposes into hydrogen gas and hydroxide ions as previously indicated. The corresponding equilibrium potential equation for this reaction ($4e^- + 4H_2O \leftrightarrow 2H_2 + 4OH^-$ or $4H^+ + 4e^- \leftrightarrow 2H_2$) as a function of pH is

$$E_{eq(H_2O/H_2)} = 0 - 0.0591\,\text{pH} \quad \textbf{(EQ 15)}$$

These relationships can be used with relationships for other species, such as metal ions, metallic metals, metal sulfides, and metal oxides to produce phase stability or Pourbaix diagrams. These diagrams are commonly used to identify potential and pH conditions needed for metal extraction and electrowinning.

An example of a stability diagram for nickel is shown in Figure 1. The phase stability diagram for nickel shows that a high potential is needed to stabilize dissolved nickel ions as Ni^{2+} ions in an electrolyte. It also shows that the electrolyte has reasonable stability in the acidic range. Furthermore, if the pH approaches the neutral range near pH 7, the nickel will precipitate out of the electrolyte. The diagram also indicates that the potential must be below –0.25 V when the pH is 0 to allow the metal to be electrowon from solution.

Note that the production of nickel occurs below the water stability line associated with hydrogen gas production when the pH is below about 4. Correspondingly, some hydrogen gas is produced at the cathode during nickel production.

It is possible to significantly reduce the electrochemical potential of an electrochemical reaction through complexation. A good example of this is gold complexation with thiosulfate. Under standard conditions, the electrodeposition of gold from its single ion valence ($Au^{3+} + 3e^- \leftrightarrow Au$) requires a potential of 1.498 V. However, if the gold is complexed with chloride ($AuCl_4^- + 3e^- \leftrightarrow Au + 4\ Cl^-$) under standard conditions, the standard potential for the reaction drops to 1.002 V.

A list of standard electrochemical potentials is given in Table 1. The actual potentials in applications can be calculated using the Nernst equation, which uses the standard potential and adjusts the potential based on actual species activities, instead of assuming all activities are one. The data in Table 1 provide a relative electromotive force (*emf* or *E*) or potential scale for standard atmospheric conditions to compare metal and other elements and compounds tendencies to acquire electrons. The higher the potential, the easier it is to cause the reaction to proceed as a reduction reaction. Thus, metals with high standard potentials, such as gold, are easily electrowon. Metals such as zinc are more difficult to electrowin than metals such as iron. Consequently, in zinc electrowinning plants, iron and other metals with higher potentials than zinc must be removed from the electrolyte prior to zinc electrowinning to prevent the undesired metals from codepositing with the zinc.

Table 1 also lists reactions such as the ferric reduction reaction ($Fe^{3+} + e^- \leftrightarrow Fe^{2+}$), which can occur as an undesired electron consuming reduction reaction on cathodes as well as in the reverse direction as an undesired oxidation reaction on anodes that resupplies the ferric ions for continued reduction on the cathode.

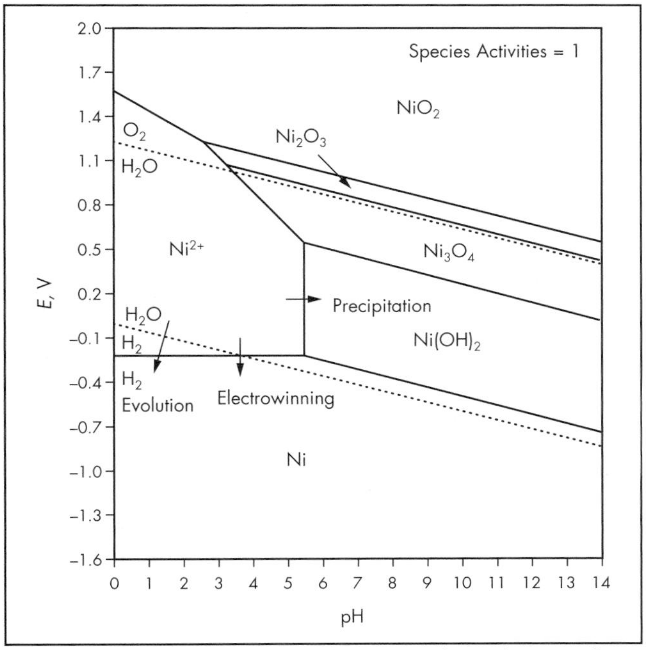

Figure 1 Simplified phase stability or Pourbaix diagram for nickel in water

Table 1 Selected standard half-cell reactions and standard potentials (E°)*

Reaction	E°, V
$Au^+ + e^- \leftrightarrow Au$	+1.692
$Au^{3+} + 3e^- \leftrightarrow Au$	+1.498
$Cl_2 + 2e^- \leftrightarrow 2Cl^-$	+1.358
$Pt^{3+} + 3e^- \leftrightarrow Pt$	+1.358
$Ag^+ + e^- \leftrightarrow Ag$	+0.799
$Fe^{3+} + e^- \leftrightarrow Fe^{2+}$	+0.770
$O_2 + 2H^+ + 2e^- \leftrightarrow H_2O_2$	+0.680
$Cu^{2+} + 2e^- \leftrightarrow Cu$	+0.337
$2H^+ + 2e^- \leftrightarrow H_2$	0.000
$Pb^{2+} + 2e^- \leftrightarrow Pb$	–0.126
$Sn^{2+} + 2e^- \leftrightarrow Sn$	–0.136
$Ni^{2+} + 2e^- \leftrightarrow Ni$	–0.250
$Co^{2+} + 2e^- \leftrightarrow Co$	–0.277
$Cd^{2+} + 2e^- \leftrightarrow Cd$	–0.403
$Fe^{2+} + 2e^- \leftrightarrow Fe$	–0.410
$Zn^{2+} + 2e^- \leftrightarrow Zn$	–0.763
$Mn^{2+} + 2e^- \leftrightarrow Mn$	–1.180
$Al^{3+} + 3e^- \leftrightarrow Al$	–1.662
$Mg^{2+} + 2e^- \leftrightarrow Mg$	–2.363
$Na^+ + e^- \leftrightarrow Na$	–2.712
$Li^+ + e^- \leftrightarrow Li$	–3.050

*Assumes all reactions involving hydrogen are at pH = 0.

The data in Table 1 show that metals such as magnesium have standard potentials that are far below that of hydrogen ion reduction. The potentials for aluminum, magnesium, sodium, and lithium are so far below that of hydrogen that hydrogen ion reduction to hydrogen gas is generally more rapid than that of the metal, which creates a potentially dangerous environment in addition to making the metal reduction very inefficient. Thus, commercial production of very low-potential metals, including aluminum and magnesium, is performed in molten salts rather than water.

Overall cell voltage between the cathode and anode can be measured using a simple multimeter. The electrochemical potential for half-cell reactions can be measured and characterized along with associated reaction rate kinetics using a potentiostat and an electrochemical cell, such as depicted in Figure 2 with three electrodes. The electrode that is evaluated is the working electrode. The working electrode is often rotated to control mass transport conditions. The counter or auxiliary electrode is used to produce or receive electrons for the working electrode. The auxiliary electrode is often made of inert materials, and the main auxiliary reaction is commonly water electrolysis. The current supplied to the auxiliary electrode is measured by the potentiostat.

A reference electrode is used as the third electrode. The reference electrode provides a steady reference potential from a half-cell reaction system that is inside the reference electrode but in ion contact with the solution through a porous medium. A common reference electrode half-cell system consists of silver and silver chloride solids in a solution that is saturated with potassium chloride (KCl). The associated half-cell reaction ($AgCl + e^- \leftrightarrow Ag + Cl^-$) remains at a constant equilibrium potential of 0.199 V because of the unit activities of the solid silver chloride (AgCl) and Ag and the constant activity

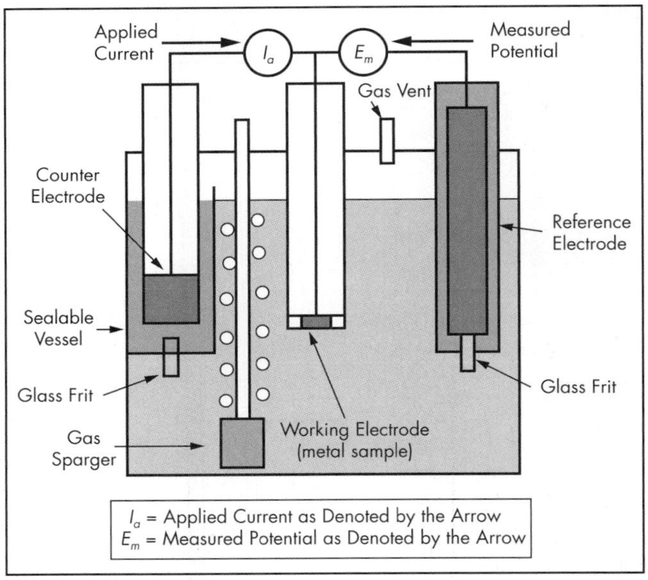

Adapted from Free 2013

Figure 2 Schematic diagram of a typical three-electrode electrochemical testing cell

of Cl^- in a KCl-saturated solution. Calculation of the working electrode (WE) potential relative to the thermodynamic electrochemical potential or standard hydrogen electrode (SHE) is accomplished based on the measured (meas) and reference (ref) potentials using the following equation:

$$E_{\mathrm{WE(vs.\,SHE)}} = E_{\mathrm{ref}} + E_{\mathrm{meas}} \quad \textbf{(EQ 16)}$$

Solution resistance in the electrolyte is calculated as

$$R = \frac{1}{k}\frac{d}{A} \quad \textbf{(EQ 17)}$$

where R is the resistance, k is the specific conductivity of the electrolyte, d is the separation distance between electrodes, and A is the cross-sectional area of the electrodes. Solution resistance results in a voltage drop proportional to the current. Energy consumption is directly proportional to potential (or voltage), current, and time (energy = EIt).

Electrochemical Kinetics

The rate or kinetics of electrochemical reactions determines the commercial utility of electrowinning. Electrochemical kinetics is a function of species availability, potential, and electron exchange kinetics. The rate of electrochemical half-cell reactions is well-described using the Butler–Volmer equation (Bockris and Reddy 1973):

$$i = i_o[\exp\left(\frac{\alpha_a F\eta}{RT}\right) - \exp\left(\frac{-\alpha_c F\eta}{RT}\right) \quad \textbf{(EQ 18)}$$

in which i is the current density, i_o is the equilibrium exchange current density, α_a is the anodic charge transfer coefficient, η is the surface overpotential or the difference between the surface potential (E_{surf}) and the equilibrium reaction potential (E_{eq}), and α_c is the cathodic charge transfer coefficient. This form of the Butler–Volmer equation is based on bulk concentrations of reacting species, which are generally different than the concentrations at the surface where electron exchange occurs. A more useful form of the Butler–Volmer equation that is explicit in terms of surface and bulk solution concentrations that contribute to the exchange current density can be expressed as (Newman and Thomas-Alyea 2004)

$$i = k'[C_{ba}\left(\frac{C_{sa}}{C_{ba}}\right)^{\lambda_a}\exp\left(\frac{\alpha_a F\eta}{RT}\right) - C_{bc}\left(\frac{C_{sc}}{C_{bc}}\right)^{\lambda_c}\exp\left(\frac{-\alpha_c F\eta}{RT}\right) \quad \textbf{(EQ 19)}$$

where k' is a constant that is directly related to the equilibrium exchange current density, C_b is the bulk concentration, C_s is the surface concentration, and λ is a factor that depends on reaction mechanisms and related factors and is usually between 0.25 and 1. (The additional subscript a denotes anodic, and c denotes cathodic.) Only the first term is generally needed for reactions occurring anodically and the second term for reactions occurring cathodically provided that the reactions are not near (generally within approximately 0.15 V) their equilibrium potentials.

The system of Butler–Volmer equations that applies to a given reaction scenario is subject to the constraint that

$$\sum I = 0 \quad \textbf{(EQ 20)}$$

When all currents are associated with the same homogeneous surface,

$$\sum i = 0 \quad \textbf{(EQ 21)}$$

The mass (m) changes in processes such as electrodeposition or corrosion are based on Faraday's law, which can be mathematically expressed as

$$m = \frac{ItA_w}{nF} = \frac{iAtA_w}{nF} \quad \textbf{(EQ 22)}$$

in which I is the current, t is time, A_w is the molecular or atomic weight of the species that is depositing or dissolving, A is the electrode area, and the other terms as defined previously.

Ion Mass Transport

All electrochemical reactions require mass transport of associated ions to supply the electrode reactions as well as to conduct current. The conduction of current is caused by the migration of ions in response to an electrochemical potential gradient between the positive and negative electrodes. Because some ions are more mobile than others, they are capable of carrying more current. As an example, hydrogen ions are approximately four times more mobile than chloride ions. Consequently, a solution of hydrochloric acid will have approximately four times more current carried by hydrogen ion migration than by chloride ion migration.

Ions that migrate to an electrode must be eventually consumed by the reaction or be transported away from the electrode by a combination of diffusion and/or convection. Ions that migrate from an electrode need to be eventually replenished by diffusion and/or convection.

The combination of the ion transport processes on current flow can be expressed by the following equation (Newman and Thomas-Alyea 2004):

$$
\begin{aligned}
i &= F\sum_j z_j N_j \\
&= -F^2 \nabla E \sum_j z_j^2 u_j c_j \\
&\quad + FV\sum_j z_j c_j \\
&\quad - F\sum_j z_j D_j \nabla c_j
\end{aligned} \quad \textbf{(EQ 23)}
$$

where N_j is the flux density of species j, c_j is the concentration of species j, u_j is the ion mobility of species j, V is velocity, D is diffusivity, and all other variables as previously defined. The ion transport current density of a reacting species at an electrode surface must balance the electrochemical reaction current density for the same species under steady-state conditions.

In most electrowinning scenarios, the background electrolyte ions, such as hydrogen ions and sulfate ions, carry most of the electrolyte current flow by ion migration. Correspondingly, the first term of the ion transport current density equation can usually be neglected for the depositing metal. The second term of the mass transport current density equation can also be neglected if the convective flux of ions is small. These are oversimplifications for accurate modeling, but they, along with a simplified concentration gradient, result in a simplified equation that is very useful in electrowinning:

$$i = -nFD\frac{(C_b - C_s)}{\delta} \quad \textbf{(EQ 24)}$$

in which δ is the boundary layer thickness, which is related to the fluid flow and geometry of the electrodes and the cell. If the electrochemical reaction is fast compared to diffusion, the surface concentration of reacting species will be small, and the current density will approach its limiting current density. The limiting current density, i_l, can be estimated as

$$i_l = \frac{-nFDC_b}{\delta} \quad \textbf{(EQ 25)}$$

The limiting current density of the metal that is being electrodeposited is important to product quality. Generally, if the metal is deposited at a rate that is higher than about 40% of the limiting current density, it is difficult to maintain smooth surfaces—even with smoothing additives. This is because the mass transport is significantly restricted by the limitations of diffusion, which forces undesired surface protrusions, such as nodules, to acquire significantly more depositing metal than flat areas, because of better access to incoming diffusing ions, thereby accelerating the growth of protrusions and increasing roughness.

Fluid Flow

The effects of fluid flow, which influence ion transport to electrodeposition surfaces, can be characterized by the Navier–Stokes equation (Potter and Foss 1982):

$$\rho\left[\frac{\partial V}{\partial t} + (V \cdot \nabla)V\right] = -\nabla p - \rho g \nabla h + \mu \nabla^2 \quad \textbf{(EQ 26)}$$

in which ρ is the density of the fluid, V is the velocity, p is the pressure, g is the gravitational acceleration, h is height, and μ is viscosity.

Conservation of mass leads to the equation of continuity (Potter and Foss 1982):

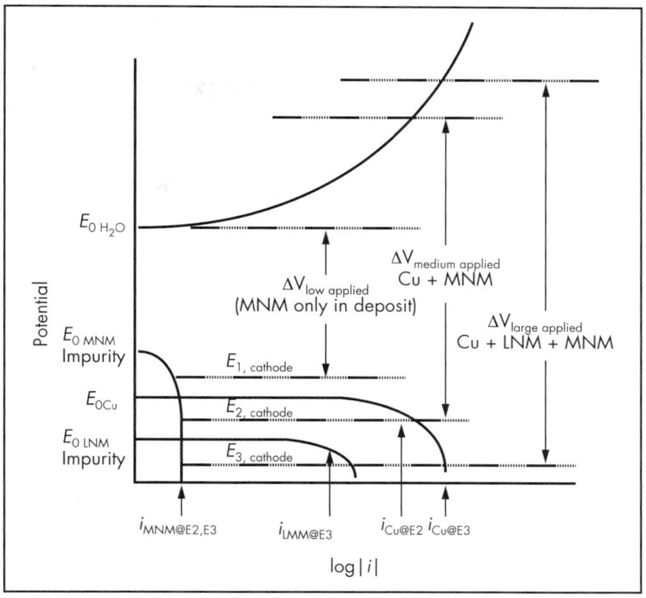

Figure 3 Comparison of potential and the logarithm of the absolute value of current density for a scenario involving copper, less noble metal (LNM), and more noble metal (MNM)

$$\frac{\partial \rho}{\partial t} + \nabla \cdot (\rho V) = 0 \quad \textbf{(EQ 27)}$$

The Navier–Stokes equation and the equation of continuity are generally simplified to facilitate practical fluid flow calculations. They can be used to determine local velocities that control convective ion transport and influence the diffusion boundary layer thickness.

Applied Fundamentals

Real-world applications such as electrowinning involve a combination of the fundamental equations that have been described. The rate of electrodeposition is related to the applied voltage, which is the driving force for the resulting current. The flow of electrolyte influences the voltage requirement as well as the quality of the resulting deposit. A comparison of voltage and rate for electrochemical reactions for a typical electrowinning scenario is shown in Figure 3.

Electrowinning operations are generally controlled on a current basis, but it is the voltage of the half-cell reactions that drives the current. The applied voltage is closely related to the current that is applied. If the applied voltage results in a cathode potential that is below the electrochemical potential of any metal, that metal will deposit at a rate given by the Butler–Volmer equation.

The rates of deposition for each of these metals depend on the dissolved metal ion concentration and associated half-cell potential for metal deposition. At $E_{1,\text{cathode}}$, the cathode potential is below that of the more noble metal (MNM) and above copper. Correspondingly, only MNM will deposit. The applied potential at $E_{1,\text{cathode}}$ is relatively low but significant, primarily caused by the difference in the potentials of the water oxidation reaction at the anode and the reduction of MNM on the cathode. At $E_{1,\text{cathode}}$, no copper will deposit because the cathode potential is above that the copper deposition half-cell reaction. Note that the rate of MNM deposition is very low because of the low dissolved concentration of MNM impurity. Generally, electrolytes are purified to eliminate nearly all MNM impurities to reduce codeposition of the MNM on the cathode.

At $E_{2,\text{cathode}}$, the rate of deposition of MNM is relatively small compared to that of copper because it is in low concentration in solution. An example of this is silver deposition with copper, which generally occurs at a very low rate, which is equal to its diffusion-based limiting current density, because of its very low concentration in copper electrowinning tankhouses. Thus, the limiting current density is important in impurity control.

At $E_{3,\text{cathode}}$, MNM, copper, and the less noble metal deposit. An example of this could be silver, copper, and nickel. However, if the nickel concentration is not high and the cathode potential is not unusually low, its potential is far enough below that of copper to prevent it from depositing in most industrial electrowinning scenarios.

The effect of MNM impurity on cathode purity can be calculated based on the MNM limiting current density and the overall current density on the cathode, provided that some basic information is available. The following equation is for lead in zinc, which can also be applied to other metals (Free 2013):

$$C_{\text{MNM, in dep.metal}} = \frac{i_{l,\text{MNM}}\, n_{\text{dep.metal}} A_{w,\text{MNM}}}{i_{\text{total}} \frac{\beta}{100} n_{\text{MNM}} A_{w,\text{dep.metal}}} \quad \textbf{(EQ 28)}$$

where A_w is the atomic weight, β is the current efficiency in percent, and dep. is deposited.

Correspondingly, the boundary layer thickness, which can be used to calculate the limiting current density of the desired metal, can be calculated as

$$\delta = \frac{n_{\text{dep. metal}} F D_{\text{MNM}} C_{b,\text{MNM}} A_{w,\text{MNM}}}{C_{\text{MNM, in dep. metal}}\, i_{\text{total}} \frac{\beta}{100} A_{w,\text{dep. metal}}} \quad \textbf{(EQ 29)}$$

As an example, if the concentration of lead in a zinc cathode is 10 ppm in a cell operating at 395 A/m^2 and 89% efficiency, the associated diffusivity of Pb^{2+} ions is 1×10^{-5} cm^2/s, and the concentration of Pb^{2+} ions in solution is 0.133 mg/L or 6.4×10^{-7} mol/L, then the boundary layer thickness is (Free 2013) as follows:

$$\delta = \frac{(2)\frac{9{,}6485\ \text{coul}}{\text{mol}}\left(\frac{1\times10^{-5}\text{cm}^2}{\text{s}}\right)\frac{6.4\times10^{-7}\text{mol}}{1{,}000\ \text{cm}^3}\left(\frac{207.2\ \text{g}}{\text{mol}}\right)}{(10\times10^{-6})395\frac{A}{\text{m}^2}\frac{\frac{\text{coul}}{\text{s}}}{A}\frac{1\ \text{m}^2}{10{,}000\ \text{cm}^2}\frac{89}{100}\left(\frac{65.4\ \text{g}}{\text{mol}}\right)} = 0.011\ \text{cm}$$

Figure 3 shows the applied voltage. The applied voltage is a key variable that determines the energy consumption. The applied potential (V_{applied}) can be expressed mathematically as

$$V_{\text{applied}} = E_{\text{anodic}} - E_{\text{cathodic}} + \eta_{\text{anodic}} + \eta_{\text{cathodic}} + IR_{\text{solution}} + IR_{\text{other}} \quad \textbf{(EQ 30)}$$

A simplified version of the Butler–Volmer equation that can be applied to find the charge transfer–based portion of the overvoltage when there is more than 150 mV from the half-cell potential for a given reaction is known as the Tafel equation, and it is expressed as

$$\eta = \beta \log \frac{i_M}{i_{0M}} \quad \textbf{(EQ 31)}$$

in which η is the charge transfer overpotential; β is the Tafel slope, which is usually between 0.04 and 0.2 V per decade; i_M is the current density of the metal reduction for metal *M*; and i_{0M} is the equilibrium exchange current density for metal *M*. The anodic Tafel slope can be calculated from the following equation:

$$\beta = \frac{2.303RT}{\alpha F} \quad \textbf{(EQ 32)}$$

in which α is the charge transfer coefficient (usually between 0.25 and 0.75). The cathodic Tafel slope is calculated using the same equation with a negative sign. In some cases, a concentration overvoltage is needed, but that is beyond the scope of this chapter, so the reader is invited to use other sources for more details.

The energy consumption (in kilowatt-hour per metric ton [kW·h/t]) needed to electrowin the metal is expressed as

$$\text{Energy}(\text{kW·h/t}) = \frac{E(n)26{,}800}{A_w\left(\frac{\beta(\%)}{100}\right)} \quad \textbf{(EQ 33)}$$

As an example, the energy consumption per metric ton of zinc in an electrowinning operation with 92% efficiency, $n = 2$, and a voltage drop of 3.5 V across the cell can be calculated as follows:

$$E(\text{kW·h/t}) = \frac{E(n)26{,}800}{A_w\left(\frac{\beta}{100}\right)} = \frac{3.5(2)26{,}800}{(65.38)\left(\frac{92}{100}\right)} = 3{,}120\ \text{kW·h/kg}$$

The current efficiency is calculated based on a comparison of the metal weight deposited divided by the weight that should have theoretically deposited based on Faraday's law, as shown in the following equation:

$$\beta = \frac{m_{\text{dep}} nF}{ItA_w} \quad \textbf{(EQ 34)}$$

As an example, the current efficiency in an operation that deposits 500 t of nickel per day using a current of 21,000,000 A (current supply by rectifier multiplied by the number of cells) is calculated as shown:

$$\beta = \frac{(100\%)(500\ \text{t})(2)(96{,}485\ \text{C/mole})(1{,}000\ \text{g/kg})(1{,}000\ \text{kg/t})}{(21{,}000{,}000\ \text{C/s})(1\ \text{day})(24\ \text{h/d})(3{,}600\ \text{s/h})(58.7\ \text{g/mole})} = 90.6$$

Another factor that is worth considering is the solution resistance. The significance of resistance is often in relation to the associated voltage drop, which requires combining the equation with Ohm's law ($E = IR$). As an example, the solution resistance (in ohms) and voltage drop associated with 1-m^2 electrodes, spaced 3 cm apart that operate at 300 A/m^2 in a solution that has a specific conductivity of 1 Ω^{-1}cm^{-1} are calculated as follows:

$$R = \frac{d}{kA} = \frac{(3\ \text{cm})}{(1\ \Omega^{-1}\text{cm}^{-1})(1\ \text{m}^2)(10{,}000\ \text{cm}^2/\text{m}^2)} = 0.0003\ \Omega$$

$$E = IR = (300\ \text{A/m}^2)(1\ \text{m}^2)(0.0003\ \Omega)(\text{V/A}\ \Omega) = 0.09\ \text{V}$$

Although that voltage drop is relatively small, it has a significant impact on the overall power consumption.

COMMERCIAL ELECTROWINNING

The purpose of commercial electrowinning is to produce high-quality metal cathodes with utmost safety and low cost. Modern electrowinning facilities are designed to operate with high efficiency in terms of energy consumption, labor productivity, and online availability.

To produce metal, electrolyte, which is a high-conductivity solution rich in metal ions, enters from prior processing steps. Direct electrical current is applied, and metal is won from solution. Figure 4 shows a typical flow diagram where electrowinning is interrogated into a hydrometallurgical flow sheet. Electrowinning requires electrolyte of suitable purity to produce high-quality cathode. This purity is commonly achieved through the application of solvent extraction (SX) (typical for copper) and/or precipitation and cementation (typical for zinc). Producing high-quality cathode with low costs from low-quality electrolyte is difficult. Therefore, adequate solution purification prior to electrowinning is absolutely necessary to achieve the goals of high purity and low cost.

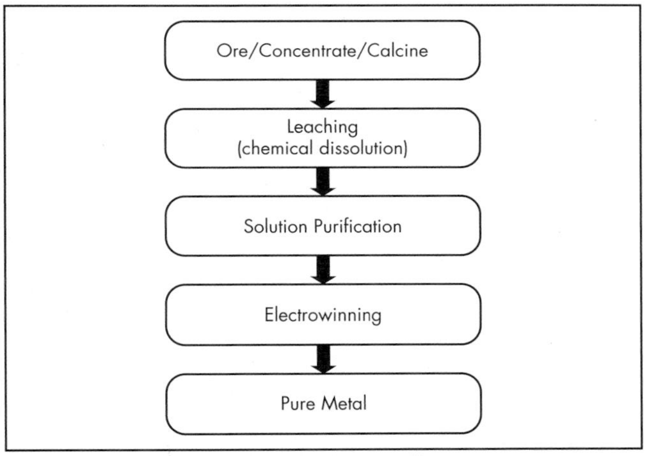

Figure 4 Flow diagram of a hydrometallurgy process to recover pure metal by electrowinning

CELLHOUSE DESIGN

Cellhouse design involves the trade-off between producing large quantities of commercial quality cathode with minimal energy input, high productivity, and safe working conditions and the need to minimize capital expenditures (Anderson et al. 2009; Honey and Watson 1985).

Cells

The largest area of any electrowinning facility is used for the electrowinning cells. Cells are where the metal is electrodeposited from electrolyte that flows through them onto cathodes that hang in them. A tankhouse (copper) or cellhouse (zinc) contains many cells. Modern cells are made from monolithic cast polymer concrete, as shown in Figure 5 (Robinson et al. 2012).

Most modern electrowinning facilities use the Walker system; that is, cells are connected electrically in series (Figure 6) while the electrolyte passes through the cells in parallel (Figure 7). Each cell has multiple anodes and cathodes. Incoming and outgoing electrolyte pipes enter the tankhouse near the electrical null point to minimize stray currents. Images of a commercial copper tankhouse and zinc cellhouse are shown in Figures 8 and 9, respectively.

The number of cells needed can be calculated using Faraday's law, a reasonable current efficiency, the projected rectifier output, and the desired design capacity of the facility. For example, a copper electrowinning facility with a design capacity of 50,000 t/a (metric tons per year) using a 40,000-A rectifier would need 138 cells. This is calculated first by using Faraday's law where I = 40,000 A; t = 1 year converted to seconds; A_w = 63.546 g/mol for copper; n = 2, because copper in a sulfate solution is divalent; F = 96485 C/mol e; and grams is converted into metric tons.

$$m = \frac{ItA_w}{nF} = \frac{40{,}000 \times 365 \times 24 \times 3{,}600 \times 63.546}{2 \times 96{,}485 \times 1 \times 10^{-6}}$$

$$= 415.4 \frac{t}{\text{cell} \times y}$$

Faraday's law calculates the theoretical maximum. To calculate a more practical output, current efficiency and time efficiency need to be incorporated. In this calculation, a current efficiency of 92% and time efficiency of 95% are used for demonstration purposes. Thus, each cell is expected to produce approximately 363 t/a. The number of cells needed is then calculated by dividing the design capacity by the production rate of each cell. The final number of cells is the next nearest even number greater than the calculated value because electrowinning facilities are almost always designed with cell lines of equal length. Hence, the calculated number of cells needed is 138 for this example.

$$415.4 \frac{t}{\text{cell} \times y} \times 0.92 \times 0.95 = 363.1 \frac{t}{\text{cell} \times y}$$

$$50{,}000 \frac{t}{\text{cell} \times y} / 363.1 \frac{t}{\text{cell} \times y} = 137.7\ \text{cells}$$

Most of the detailed information for designing electrowinning tankhouses is held by companies that perform the design and construction of these facilities, and it is not available in the open literature. However, most commercial electrowinning cells are clearly designed to conform to general industry practices as described in each of the commercial practice sections of this chapter. In the case of copper electrowinning, the following are some general design criteria that are discussed in the open literature (Beukes and Badenhorst 2009; Wells and Snelgrove 1990):

- Cathode-to-cathode center spacing is 0.1 m.
- Cell width is 1.3 m.
- Cell height is 1.5 m.
- Cell length is 5.5 m.
- Cathode face solution flow is 0.05–0.1 $m^3/h/m^2$.
- Individual cell spacing is 0.045 m.
- Distance between cell banks is 2–3 m.
- Cell walkway construction materials are nonconductive.
- Cell construction material is polymer concrete.

Source: Robinson et al. 2012

Figure 5 Large polymer concrete cell

Similar general practices are often used for zinc and nickel electrowinning. One notable difference from cells used for copper is the increased height of the cells used for zinc electrowinning. For nickel, the infrastructure is designed to accommodate cathode and anode bags, which are not present in cells for copper.

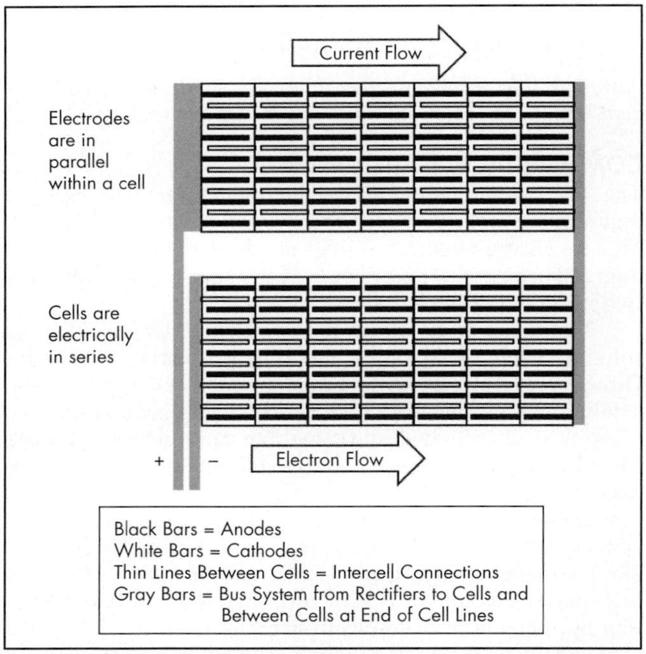

Figure 6 Schematic diagram illustrating current flow through cells in an electrowinning facility

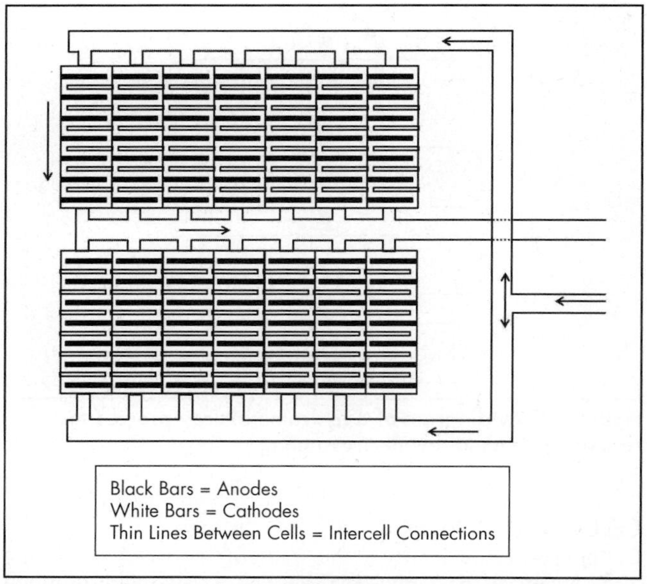

Figure 7 Schematic diagram illustrating electrolyte flow through cells in an electrowinning facility

Electrical

Electrowinning facilities use large oil-filled or dry-type rectifier transformers or air- and/or water-cooled thyristor-controlled rectifiers. These rectifiers are housed in prefabricated buildings with associated equipment such as switchgear, HVAC equipment, cooling water equipment, DC disconnect switches, heat exchangers, control equipment, alarms, and so forth (Brown 1990). The transformers convert high-voltage alternating current into lower-voltage direct current, and electrical power is provided to the electrowinning cells by a rectifier or rectifiers. Many older facilities use two rectifiers because of concerns about rectifier downtime (Brown 1990; Robinson et al. 2013a). Modern rectifiers are now reliable enough to allow designs with only one rectifier (Wiechmann et al. 2000).

Direct current is fed from the rectifier(s) to the cells using copper bus systems (Brown 1990). The bus system is fabricated out of rectangular bars of copper. The copper is typically 1.27-cm-thick copper bar with a height that is dependent on the current carrying capacity needed (15–35.6 cm). For copper electrowinning, six 1.27-cm by 35.6-cm copper bars with 1.27-mm spacing between each bar for ventilation and heat dissipation are used in an open-air configuration. The bars are plated with approximately 40 μm of silver at all copper contacting surfaces to reduce electrical resistance and oxidation.

This design gives a cross-sectional area of 271.3 cm^2 (e.g., $1.27 \times 35.6 \times 6$). Using a current density design value of between 116.3 and 131.8 A/cm^2 of cross-sectional area gives a total current capacity of 31,500–35,700 A DC. The lower current density figure is used for initial design to allow for a maximum heat increase limit of 40°C over ambient conditions and to allow for future increases of the current carrying capacity of the system without major DC bus system design changes. To carry additional current (e.g., 40,000 A used in previous cell number calculations), either more bars or bars with a greater cross-sectional area would be needed. A final decision will be based on economics and after review by an electrical engineer.

Current passes from the bus system into the cells through the use of copper contact bars. These are bars of various designs, depending on the current being used in the system. In zinc electrowinning, the copper bars are often water cooled to reduce the weight of copper needed for the large currents employed. Other metal systems use lower currents

Figure 8 Commercial copper tankhouse with acid mist capture hoods

Source: Horsehead Holding 2012
Figure 9 Commercial zinc cellhouse

and thus use solid copper contact bars. Intercell contact bars were reviewed by Boon et al. (2013). A close-up image of an Outotec double-contact system is shown in Figure 10.

Cells contain many anodes and cathodes. The anodes in the cell are all at one electrical potential (hence, in parallel electrically). The cathodes are all at another (lower) potential. There is always one more anode than cathode in an electrowinning cell. Because current flows through the path of least resistance, even spacing of electrodes is essential.

Uniform electrode spacing is achieved through the use of cell top furniture. Cell furniture is typically made of fiber-reinforced plastic. Notches are molded so electrodes fit securely in the furniture. Spacing is also accomplished through the use of in-cell spacers typically attached to each anode. These spacers are made of various plastics and produced in many styles, including hairpins, buttons, A-shape, and delta blades. More sophisticated in-cell spacing systems are available.

Piping

The electrolyte used in electrowinning is, by design, highly conductive. To reduce the likelihood of stray currents and to

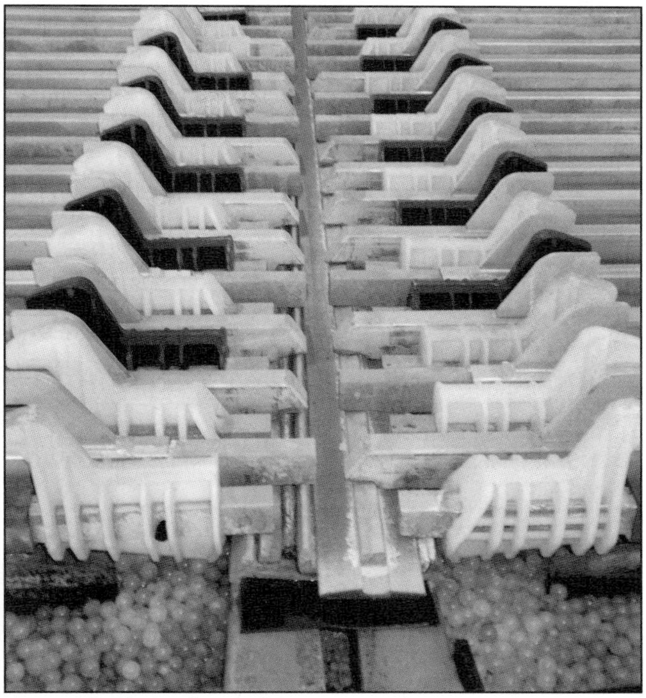

Figure 10 Electrodes resting on a contact system in a copper electrowinning facility

protect personnel, nonconductive piping is used throughout and around the electrowinning facility. These pipes are typically high-density polyethylene. Metallic valves and flanges should be avoided because of corrosion and electrical safety issues.

Electrodes

In aqueous electrowinning, metal is deposited at the cathode while an oxidation reaction occurs at the anode. Two types of cathode systems are used: permanent blanks and starter sheets. There are two types of anodes used: Pb-alloy and mixed metal oxide (MMO)–coated titanium. Regardless of the type, each electrode consists of a sheet or mesh of material, where the electrochemical reactions occur, attached to a header bar. The header bar makes electrical contact with the surrounding system and provides mechanical support to the sheet or mesh.

In permanent cathode blank systems, the metal being produced is electrodeposited onto a substrate of a different material, such as aluminum, stainless steel, or titanium. Plastic edge strips are used to ensure that the metal deposit does not grow around the cathode blank, and the strips allow for deposit removal. The metal is typically deposited for 2–10 days and then harvested. Harvesting occurs by removing the deposited metal along with the cathode blank from the cell. The deposit metal is then removed from the blank either manually or by machine (Lenz and Ducharme 2000; Sabau and Bech 2007). The blank is then returned to the cell for reuse. Harvesting is done by a manual or computer-controlled overhead crane. The crane lifts a rack of cathodes, which can be a third, half, or all of the cathodes in the cell.

The life of a cathode blank is 1.5 to more than 10 years, depending on the electrowinning system and material of construction. Permanent blanks are used in the electrowinning of

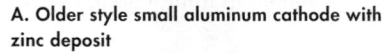

A. Older style small aluminum cathode with zinc deposit B. Stainless-steel blank with copper cathode C. Copper starter sheet

Figure 11 Commercial electrowinning cathodes

copper, nickel, zinc, and other metals. Images of aluminum and stainless-steel blanks used in zinc and copper electrowinning are presented in Figures 11A and 11B, respectively. To facilitate lifting, some electrodes have hooks (Figure 11A) and some have *windows* under the header bar (Figure 11B).

Starter sheets are produced by plating (for one or two days) onto a permanent *mother* blank of stainless steel or titanium. Polymer edge and bottom strips are fitted on the blanks to permit easy stripping of the starter sheets. The stripping is done manually or by automated machinery. Trimming and support loop cutting and attaching are also done manually or with automated machinery. Starter sheets are then placed into *commercial* cells to continue plating until full thickness is achieved (Figure 11C).

Materials Handling

In the design of electrowinning facilities, materials handling and specifically crane movement must be considered. Harvesting of cathodes and replacement of anodes are time-consuming events. Therefore, significant planning occurs on the ability to move electrodes around the plant.

In the design of plants (Beukes and Badenhorst 2009), travel distance between the stripping machine and the end of the cell line is considered. This can lead to the decision to have multiple cranes and centralized stripping machines. Obviously the decision between one crane and multiple cranes is based on economics.

Cranes are also designed to carry a third to half of the electrodes of one type to the cells. To achieve this, strongbacks or other carrying devices are used. Because the weight of lead anodes and the strongback can be significant, the weight capacity of the crane should be considered during the design phase.

Cranes can be fully or partially automated (Kuusisto et al. 2005). The automation of cranes can improve the efficiency of crane movement and remove personnel from the electrowinning environment.

Capital Expenses

Very limited cost information is specific to electrowinning. The general estimate of the capital cost for a small electrowinning tankhouse facility (100 t/d) is \$550–\$900 t/a capacity (Wells and Snelgrove 1990). If this is adjusted for inflation to 2015 using the Chemical Engineering Plant Cost Index (CEPCI), the range is \$890–\$1,460 t/a capacity. Roughly 30% of the capital cost is for the cells, 15% for electrical infrastructure, 15% for building infrastructure, and the rest for remaining expenses (Wells and Snelgrove 1990). Based on information from overall operations for some of the newer combined copper mining/leaching/SX/electrowinning facilities for which overall capital cost information is available, such as those for the Safford mine (\$550 million in capital cost for 120,000 t/a capacity; International Mining 2006), adjusted to \$637 million for inflation using CEPCI data for 2015 (Cheresources 2017), the overall capital cost is anticipated to be about \$5,310 t/a capacity. Thus, using inflation-adjusted values, the fraction of an overall project's capital cost that can be attributed to electrowinning is likely to range from about 17% to 28% for a midsize plant.

Operating expenses for electrowinning for a 36,500-t/a facility were estimated to be \$0.130 per kg of copper, which when adjusted for inflation is \$0.211 in 2015 (Cheresources 2017; Wells and Snelgrove 1990). Because the power consumption for copper production in electrowinning operations is a large portion of the operating expenses and because of variations in power costs, there are notable differences in individual plant-site expenditures. Thus, in areas with high power costs, the operating costs will be substantially higher.

PROCESS CONTROL

The ability of an electrowinning facility to produce low-cost, high-quality cathode is driven by electrolyte control, operational details, current density, acid mist abatement, and impurity treatment.

Electrolyte Control

Modern electrowinning can produce higher than 99% purity for many metals. One of the keys to producing this high quality is electrolyte control. The concentration of metal to electrowin is usually maintained at a high level and only a small amount is removed during a single pass of the electrolyte through the cell. For copper and zinc, concentrations of

35–45 g/L are typical with a plating solution depletion (inlet concentration minus outlet concertation) of 5 g/L. The electrolyte conductivity is also maintained at high levels to reduce electrical consumption by the inclusion of a supporting electrolyte, such as sulfuric acid or sodium sulfate. Temperature is carefully controlled as it can affect metal deposit structure (e.g., higher temperature in nickel electrowinning produces less stressed deposits), side reactions (e.g., zinc electrowinning maintains lower electrolyte temperature to minimize hydrogen evolution), and the interaction with the electrolyte with upstream processing (e.g., electrolyte contacting the organic phase in SX).

Contamination of inorganic and organic impurities is minimized. Each metal electrowinning system has different impurity requirements. Metals with more negative standard reduction potentials have more stringent requirements. The cause for this is the increased chance for hydrogen evolution at the cathode and lower current efficiency. Because of the differences in requirements, some operations analyze electrolyte composition hourly while other may analyze once or twice per shift.

In most electrowinning systems, inorganic and/or organic additives are used to facilitate the production of dense, smooth deposits that are low in impurities. Organic smoothing agents, such as glue, gelatin, guar, or starch, are used. Inorganic ions, such as cobalt and manganese, are used to minimize lead anode corrosion in copper and zinc, respectively. Boric acid is used as a buffer in nickel electrowinning.

Operational Details

As with any large operation, attention to details is critical in metal electrowinning. Operations should strive for uniform current distribution within a cell to produce deposits of consistent thickness. This will decrease the likelihood of uneven growth and short circuiting (e.g., physical contact between anodes and cathodes). Short circuits result in current bypassing the electrochemical reactions, which decreases the productivity of the cell and increases the energy cost to produce metal.

Because cathodes or anodes are electrically in parallel within a cell to minimize the variation in current distribution, anodes and cathodes need to be straight, evenly spaced, and well aligned. This is achieved thorough the maintenance of cell top furniture, cathodes, and anodes and consistent electrode placement in the cells. Removal and correction of bent electrodes is necessary to operate low-cost electrowinning facilities.

Electrowinning is best performed under steady-state conditions. Thus, disruptions to electrical power and/or electrolyte flow, upsets in electrolyte quality, and changes to harvesting schedules will cause operational difficulties. These will likely cause increased costs and a lower-quality product.

Current Density

The final operational control for electrowinning is current density or applied current. In most facilities, the applied current may be adjusted based on external factors, such as metal availability from the mine or availability and/or cost of electrical power. Each metal electrowinning system operates best within a certain range of current densities. If the current density is too high, deposit quality usually suffers and the likelihood of short circuiting increases. If current density is too low, then capital utilization is reduced or current efficiency may suffer, as in the case with zinc electrowinning. Effort should be made to maintain current density within the normal range found in similar operations.

Acid Mist Mitigation

The anode in aqueous electrowinning operations creates oxygen bubbles that rise and collapse at the top of the highly acidic electrolyte. McGinnity and Nicol (2014) recently reviewed the fundamentals of acid mist generation and practical methods to control it in industrial practice. The bubble collapse causes the formation of tiny droplets of the acidic electrolyte that produces an acid mist in electrowinning tankhouses. Most countries regulate the acid mist that is allowable in the air in tankhouses to 1 mg/m^3 or less (OSHA 1985; Otero et al. 2003). Consequently, electrowinning operations use one or more methods of reducing acid mist in addition to using personal protection equipment to limit worker exposure to acid mist.

Many industrial sites use ventilated hoods to capture acid mist above the cells (Mella et al. 2003) or forced-air ventilation for the entire facility (Davis and De Visser 2000).

One method of reducing acid mist is the use of plastic balls or beads (Robinson et al. 2013b). The layer of balls or beads that is a few centimeters thick floats on the electrolyte. The balls or beads provide a surface on which acid droplets collect and drain back into the electrolyte. Some sites use barriers such as brushes or wipers attached to anodes, which generally require some periodic spraying of water to keep them clean, or polymeric pads or sheets to cover the cells.

Another method involves the use of chemicals to reduce surface tension and/or create a foam to suppress the formation of acid mist. In the copper industry, it is common to use a suppressant such as a fluoropolymer (e.g., 3M's FC-1100) at a dosage level of 10–20 ppm, which does not interfere with SX phase disengagement. In the zinc industry, it is common to use compounds such as licorice to create a foam to mitigate acid misting (Cheng et al. 2004).

Impurity Treatment

Electrowinning follows processing that is used to remove impurities, so key impurities that cause problems in electrowinning are removed prior to electrowinning as part of solution purification (SX, precipitation, cementation, or ion exchange). Consequently, most related impurity removal technologies are described in other chapters. If a problematic impurity does enter the electrolyte, it is usually controlled by electrolyte bleeding. Bleeding is commonly employed in copper electrowinning and is used to control the concentration of iron, manganese, and/or chloride in the electrolyte. A small, continuous flow of electrolyte is typically sent back to solution purification (often the first stage of extraction in SX) to allow for copper to return to the tankhouse while reducing the transfer of the offending element.

Only the removal of lead and/or manganese are exceptions. Most electrowinning uses lead anodes that corrode during the process and generate lead corrosion products that fall to the bottoms of cells and can become suspended. To control lead contamination, cells must be periodically cleaned to remove lead-bearing sludge. The sludge can be manually removed or eliminated through the use of diaphragm pumps. The lead-bearing sludges are decanted, drummed, and returned for recycling of the lead content. Many electrowinning facilities also filter electrolyte to reduce the concentration of suspended lead-bearing particles. In zinc electrowinning, strontium carbonate is added to the electrolyte entering each cell to precipitate soluble lead as an insoluble salt to allow for the production of special high-grade zinc.

In zinc electrowinning, manganese dioxide will also form in significant quantity on the anode, cell walls, and within pipes. This requires periodic cleaning of cells, pipes, and anodes. The resulting sludge is used as an oxidant in solution purification.

SPECIFIC METALS

The preceding sections provided general descriptions of industrial metal electrowinning for aqueous solutions; however, many specific operating details are dependent on the metal being recovered. The specific details are caused by the electrochemistry of the system, the integration of electrowinning with upstream unit operations, and/or the market for the metal. In the following sections, more details are presented regarding the electrowinning of zinc, copper, nickel, and gold. Other metals, such as manganese, cobalt, cadmium, chromium, antimony, and tellurium, are also electrowon commercially (Habashi 1998).

Zinc

Zinc is the largest tonnage metal electrowon from water-based solutions. In 2014, approximately 12 Mt (million metric tons) of electrolytic refined zinc were produced (International Lead and Zinc Study Group 2015; International Zinc Association, n.d.). The electrowinning of zinc occurs after leaching with sulfuric acid and solution purification by neutralization, and zinc dust cementation produces a high-purity electrolyte. Traditionally, sphalerite concentrates are roasted to produce a zinc oxide calcine and sulfur dioxide (which will be converted in sulfuric acid or a sulfate-bearing fertilizer). The calcine is then leached using sulfuric acid. The resulting solution is purified through a multistep process by pH control to precipitate an iron-bearing residue and zinc dust cementation to remove inorganic contaminants. Finally, the neutral leach solution is mixed with return electrolyte from the cellhouse to produce commercial electrolyte for electrowinning. This process is called roast-leach-electrowin (Habashi 1998). In the past few decades, several zinc facilities have added or replaced roasting with direct leaching of concentrates using high temperature and pressure conditions. Pressure leach produces an elemental sulfur product, which reduces the need to find local markets for sulfuric acid or fertilizer (Sutherland 1988; Masters et al. 1989).

Electrowinning operating data from a survey of zinc smelters (Moats et al. 2010) is presented in Table 2. Each operation uses hundreds of cells. The cells are typically made of polymer concrete (Moats et al. 2008). Each cell contains 30–50 cathodes. The distance between cathodes is 64–90 mm. The size of cells and electrodes are different and reflect a drive toward higher productivity. Modern cellhouses use 3.0–3.4-m^2 plating area cathodes while older cells use 2.0- and 2.6-m^2 plating area cathodes (Martin et al. 2000).

Moats et al. (2008) surveyed 10 zinc cellhouses in more detail. From their data, cathodes are commercially pure (grade 1050 or 1070) strain-hardened aluminum alloys welded to a 6000 series (Al-Mg-Si) header bar. To facilitate zinc stripping, 60% of the cellhouses use separation aids. Cathode life varied from 10 to 33 months, and a correlation of cathode life and fluoride concentration in the electrolyte feed was found and is shown in Figure 12. Fluoride concentrate is also known to lead to pitting of the aluminum surface, which leads to stickers. *Stickers* are zinc deposits that stick to the aluminum mother blank. Stickers are minimized by routine brushing of the aluminum blanks.

Most anodes employed in zinc electrowinning are binary Pb-Ag alloys with Ag content ranging from 0.45% to 1%. Anodes are rolled or cast. Most operations precondition their anodes prior to their initial use employing either physical (sandblasting) or chemical/electrochemical means. Anode service life is impacted by anode composition, fabrication method (rolled or cast), current modulation (operation at too low or too high a current density), cell monitoring (i.e., how quickly short circuits are corrected), anode pretreatment method, anode cleaning cycles, and electrolyte composition. Anode lifetimes range from 24 to 72 months. Manganese concentration in the electrolyte is maintained at 2.5–5 g/L to facilitate the deposition of manganese dioxide (MnO_2) on the lead anodes and reduce their corrosion (Zhang and Cheng 2007).

Zinc cellhouses typically operate at cathode current densities around 500 A/m^2. To lower electricity costs, some operations will lower their current during times of high-cost, peak-power demand and increase them during the night or off-peak. Current densities can range from 240 to 690 A/m^2. Current efficiency is usually 88%–91%. Cell voltages and electrical power consumption are typically 3.2–3.5 V and 2,800–3,400 kW·h/t of zinc, respectively (Moats et al. 2008).

The electrolyte specifications for zinc production are quite stringent because of zinc's low standard reduction potential (–0.77 V vs. SHE). The concentrations of zinc and sulfuric acid are closely monitored with most plants operating with 45–65 g/L Zn and 160–200 g/L sulfuric acid (H_2SO_4). Electrolyte temperature is also controlled between 35° and 45°C with preference given to 37°–40°C. Electrolyte is heated by electrical resistance within each cell. To maintain electrolyte temperature, all cellhouses cool their electrolytes. Porter (1991) estimates that 3.5–4.0 GJ/t of zinc cathode is wasted through electrolyte heating.

The presence of too many impurities can lead to product contamination and/or low current efficiency caused by hydrogen evolution or re-solution of the zinc deposit. The effects of common impurities from Robinson and O'Keefe (1976) are listed in Table 3. An example of an electrolyte specification (Houlachi 2008) is presented in Table 4 along with reasons for the requirement. Impurities can

- Cause cathode purity issues (Pb, Cd, Cu),
- Lower current efficiency because of hydrogen evolution and re-solution of zinc (Ge, Sb, Co, Ni),
- Disrupt operations by enhancing corrosion (Cl),
- Cause zinc to stick to the aluminum mother blanks (F), and
- Lead to scaling of pipes and equipment (Ca).

Plating times are maintained between 24 and 72 hours to avoid re-solution caused by the deposition of impurities with lower hydrogen overvoltage such as nickel and cobalt.

Inorganic and organic additives are used to control and facilitate the production of special high-grade zinc. Organic compounds such as glue/gelatin, gum Arabic, and licorice are used. Glue/gelatin/gum Arabic are used to control the structure of the zinc deposit. Licorice is used to create a foam layer on top of the electrolyte and reduce the presence of

Table 2 Zinc electrowinning operating data from selected operations

Location	Hiroshima, Japan	Iijima, Japan	Kidd, Canada	Trail, Canada	Budel, Netherlands	Kokkola, Finland	Plovdiv, Bulgaria	Baiyin, China	San Luis Potosi, Mexico
Number of cells	360	680	660	548	432	840	468	416	384
Cell dimensions, m									
Depth	1.6	1.526	1.50	2.18	2.07	3.6	1.55	1.7	1.4
Width	0.9	0.89	0.99	1.22	1.47	1.97	0.94	0.95	1.2
Length	2.9	3,250/3,998	3.84	4.93	4.75	0.845	2.5	4.1	3.5
Cathodes per cell	33	48/58	40	50	44	44	27	44	44
Cathode–cathode distance, mm	75	64	75	89	90	75	75	75	73
Size of cathodes, mm									
Height	1,220	1,150	1,156	1,650	1,600	1,275	1,100	1,190	1,165
Width	800	800	660	1,025	902	698	666	800	765
Thickness	6	6	7	6.35	7	6	4	6	5
Cathode life, months	20	27	22	17	12	18	10	15	18
Number of stickers/d	126	N/D*	79	13	2	250	40	N/D	N/D
Plating area, m^2	1.67	1.67	1.35	3.07	2.6	1.55	1.28	1.16	1.8
Deposit time, h	24–48	24/48	30, max.	72	32	38	N/D	24	24
Current density, A/m^2	70–600	50–690	544.4	440	491	564	520	450	474
Cell voltage, V	2.8–3.5	2.6–3.5	3.5	3.41	3.26	3.5	3.6	3.2–3.4	3.3
Current efficiency, %	89.5	91.5	89–90	90–92	92.5	91.6	93	88–90	90.5
Power, kW·h/t	3,205	3,171	3,309	3,200	3,247	3,205	3,200	3,100–3,200	2,990
Stripping method	Automatic	N/D	Automatic	Automatic	Automatic	Automatic	Manual	Manual	Manual
Anodes per cell	34	49/59	41	51	45	45	28	45	45
Size of anodes, mm									
Height	1,164	1,070	1,053	1,570	1,605	1,245	1,055	1,170	1,122
Width	730	745	590	965	828	630	620	728	715
Thickness	10	6.5	8	10.75	14	8	8	7.3	8
Anode life, months	48	72	36	36	48	36	24	24	24
Anode material									
% Ag	0.75	0.8	0.75	0.75	0.5	0.7	1	1	0.75
Electrolyte									
Temperature, °C	40	44	37	37	45	39	34–36	35–42	42
Zn, g/L	50	53	65	55	50	53.8	50–60	45~55	65
H_2SO_4, g/L	190	160–190	203	165	180	170	160–175	160–180	185
Mg, g/L	10	15	1.6	3	10	15.7	3.3	2.5	4.5
Mn, g/L	15	5.5	3	2.4	5	5.5	2	5~10	7
Cl, mg/L	500	N/D	84	225	600	210	160	⊠100	45
F, mg/L	10	N/D	5	30	5	31	25	⊠20	9
Additives used									
Glue, kg/d	N/D	N/D	210–1,300	25.2	N/D	2.87	8	80	75
Gelatin, kg/d	N/D	66	22.5	N/D	30	N/D	0	N/D	N/D
Licorice, kg/d	N/D	N/D	0.19	63	7	N/D	0	N/D	N/D
Strontium carbonate, kg/d	100	531	5.6	800	350	580	150	330	6.5
K-Sb-tartrate, g/d	N/D	18	67.5	N/D	35	91	N/D	N/D	N/D

Source: Moats et al. 2010
*N/D = no data provided.

acid mist within the cellhouse. Most operations add strontium carbonate to precipitate soluble lead from the electrolyte to reduce its contamination in the cathode. Some operations add soluble antimony to improve stripping and/or affect zinc electrodeposition.

Zinc cathode produced by electrowinning is rarely sold to market. Zinc refineries melt and cast their cathodes to produce ingots of various shapes, sizes, and composition. The ingots are produced to meet customer specifications.

Copper

Copper is the second largest tonnage metal electrowon from aqueous solutions. In 2016, nearly 3.9 Mt of electrowon copper were produced (International Copper Study Group 2017).

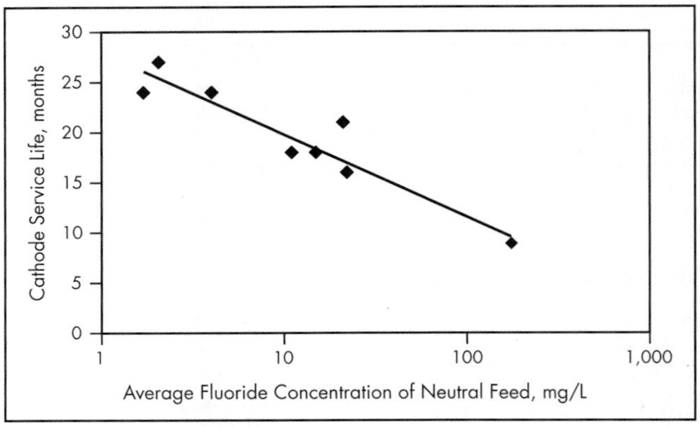

Source: Moats et al. 2008

Figure 12 Cathode service life versus fluoride concentration of the neutral feed

Copper electrowinning typically occurs after SX has concentrated copper while excluding most impurities from a leach solution. Detailed operating data from a selected number of newer copper tankhouses are presented in Table 5 (Robinson et al. 2013a). Newer tankhouses utilize polymer concrete cells, and older facilities can still use concrete lined with polyvinyl chloride (PVC) or lead. Newer cells contain 60–80 cathodes with older cells using less. The distance between cathodes is 95–105 mm. Most tankhouses use ~2.0-m^2 plating area cathodes.

Robinson et al. (2013a) surveyed 34 tankhouses. Most facilities use stainless-steel blanks while a few older plants are still using starter sheets. 316L is the most common grade of stainless steel used. 316L plates are inserted in copper hanger bars and typically welded into place. Blank life is a function of chloride concentration and less than 25 mg/L is recommended for 316L. Some facilities are using duplex stainless, which has low nickel levels, partially because of lower costs when the nickel price is high. Numerous older facilities have converted from starter sheets to stainless-steel blanks (Addison et al. 1999). As stainless-steel cathodes represent a multimillion-dollar capital investment, repairs are needed and can be done on-site (Valentine 2002).

Most copper electrowinning anodes use a rolled Pb-Ca-Sn plate soldered into a copper hanger bar that is covered in Pb. The typical Sn and Ca contents are 1.35% and 0.07%, respectively. Lead anode lifetimes are typically greater than 5 years, and some report lifetimes up to 10 years. Lead anode lifetime is affected by temperature, current density, power consistency, cobalt concentration, chloride concentration, and short circuiting (Rasmussen 1995).

Three tankhouses have converted to MMO-coated titanium anodes. MMO anodes use an iridium dioxide-tantalum oxide (IrO_2-Ta_2O_5) coating on commercial pure titanium mesh. Pieces of coated mesh are welded to titanium-clad copper conductor bars that are press fit into copper hanger bars. MMO anodes operate at lower potential than lead anodes (reported to be 13%–15%) and do not need cobalt in the electrolyte. MMO anodes, however, initially cost more and can suffer damage from short circuiting. Conversion to MMO anodes from Pb-alloy anodes will significantly depend on the electrical power costs of a facility (Moats 2008).

Table 3 Impurity effects on current efficiency and zinc purity

Elements	Effect	Comment
Al, Mg, Ca, Na	Little effect	Standard reduction potential more negative than zinc
Cd, Pb	Decrease cathode purity	Marginally more positive potentials than zinc; high hydrogen overpotential
Pt, Ag, Au, Fe, Co, Ni, Cu	Decrease current efficiency	More positive reduction potentials than zinc; low hydrogen overpotential
Sb, As, Ge, Se, Te	Complex, but usually a decrease in cathode efficiency	Can lead to localized corrosion cells and zinc redissolution

Source: Robinson and O'Keefe 1976

Table 4 Example of a zinc electrowinning electrolyte specification

Element	Value	Reason
Mg, g/L	<6	Higher solution viscosity and cell voltage
Mn, g/L	2.5–5.0	Protect anode coating, lower Cl_2 gas evolution
Fe, mg/L	<5	Indicator of purification problems
SiO_2, mg/L	<100	Settling and filtration problems
Ge, mg/L	<0.01	Poor current efficiency
Sb, mg/L	<0.02	Poor current efficiency
As, mg/L	<0.02	Poor current efficiency and potential arsine generation
Cu, mg/L	<0.1	Poor current efficiency and product contamination
Cd, mg/L	<0.3	Product contamination
Co, mg/L	<0.2	Poor current efficiency; re-solution after an induction period
Ni, mg/L	<0.2	Poor current efficiency; re-solution after an induction period
Tl, mg/L	<0.3	Product contamination, stripping
Ca, mg/L	<450	Gypsum buildup
F, mg/L	<30	Pitting aluminum, stickers, frequent brushing
Cl, mg/L	<250	Enhanced anode and cathode corrosion

Adapted from Houlachi 2008

Copper tankhouses typically operate at cathode current densities of 250–350 A/m^2, but some facilities can operate up to 400–450 A/m^2. Current efficiencies are usually between 88% and 92%. Cell voltages are typically 1.7–1.8 V for cells with MMO anodes and 1.9–2.1 V for cells with lead anodes, resulting in energy consumptions of 1.7–1.8 or 1.8–2.1 MW·h/t of copper, respectively (Robinson et al. 2013a).

The electrolyte specifications for copper production are less stringent than most other metals because copper's standard reduction potential (0.34 V vs. SHE) is more positive than most other metal impurities potentials as well as that of hydrogen. The concentrations of copper and sulfuric acid are monitored usually once or twice a shift with most plants operating with 35–45 g/L Cu and 160–200 g/L H_2SO_4. Electrolyte temperature will depend on whether the tankhouse is downstream from SX. If it is, then the temperature will be 35°–50°C to avoid overheating of the organic phase. If it is not, then higher temperatures are permissible with temperatures approaching 60°C at some plants. Because the current density employed at copper tankhouses is not as high as zinc electrowinning, electrolyte heating is less of an issue and no active cooling is necessary.

Common impurities that cause production issues in copper electrowinning electrolytes are iron, manganese, chloride, and SX organic. The reduction of ferric to ferrous ($E° = 0.77$ V vs. SHE) is thermodynamically preferred to copper electrodeposition. The presence of iron in the electrolyte causes a decrease in current efficiency, but not cathode contamination. The best methods to control iron concentration in the electrolyte is proper operation of the SX circuit or bleeding electrolyte back to the extraction circuit to recover copper and exclude iron.

Manganese in the electrolyte is oxidized by the anode. Oxidized manganese can cause (1) MnO_2 formation on MMO anodes, which could shorten anode lifetime if a blinding deposit occurs; (2) MnO_2 formation on lead anodes, which can disrupt the lead dioxide (PbO_2) scale, leading to cathode contamination and/or enhanced anode corrosion; or (3) permanganate formation, which can oxidize and damage the organic in SX (Cheng et al. 2000; Zhang and Cheng 2007). The best methods to mitigate manganese issues is to control manganese concentration by proper SX operation or electrolyte bleed and to maintain an Fe/Mn ratio ≥10 to avoid permanganate formation (Miller 2011).

Chloride in copper electrowinning electrolyte can be a problem for stainless-steel cathodes as it can cause pitting corrosion. Most cathode suppliers recommend chloride concentrations less than 25 mg/L. As with iron or manganese, chloride concentration is controlled by proper SX operation or electrolyte bleed. In operations where the chloride concentration of the pregnant leach solution is very high, SX requires a wash stage or two to mitigate chloride carryover to the tankhouse.

The last common impurity is organic from the SX circuit. Most operations have sand filters, dual-media filters, and/or Jameson cells to capture organic that leaves the SX circuit with the electrolyte. Filters should be routinely backwashed to minimize organic carryover. Organic in the cell can cause *organic burn* of the cathode (a blackish oxide near the solution line) and is a fire hazard.

Additives are used to control and facilitate the production of grade A copper cathode. Organic compounds such as guar, modified starch, or polyacrylamide are used as smoothing agents to produce dense copper cathode. A polyacrylamide has also been shown to inhibit MnO_2 formation on MMO anodes (Sandoval et al. 2013). Cobalt sulfate is added to maintain the cobalt concentration in the electrolyte near 150 mg/L to minimize lead anode corrosion and reduce cell voltage.

Acid mist generation is a concern in many copper tankhouses because a foaming agent cannot be used. A foaming agent would cause a significant phase disengagement problem in SX. Therefore, copper tankhouses use mist suppression balls, hoods, crossflow ventilation, and FC-1100 to reduce the generation of acid mist.

Electrowon copper cathode is often sold to market. Therefore, cathodes typically need to meet the London Metal Exchange or Commodity Exchange grade A specifications. These include composition and appearance criteria.

Nickel

Unlike zinc and copper, nickel electrowinning is always performed in separated cells (Crundwell et al. 2011). A diaphragm is needed to ensure that the pH near the cathode surface remains relatively high (pH >3) to avoid significant evolution of hydrogen gas. A diaphragm or bag is placed around a frame that contains either the cathode or anode. Anode bags are always used in chloride electrolyte. Cathode bags are typically used in sulfate electrolyte. To minimize migration of high-acidity electrolyte produced at the anode, higher pH catholyte is added to the top of each cathode. The solution flows through the diaphragm to the anode and is then removed as anolyte. Boric acid is added to the electrolyte as a buffer to help control the pH. Sodium lauryl sulfate is added at some locations to help release H_2 bubbles from the cathode surface, thus improving the quality of the deposit.

Nickel electrowinning is performed from chloride- or sulfate-based electrolytes. Electrowinning of nickel from chloride electrolyte is done with 0.8-m-wide, 1-m-deep anodes and cathodes interleaved in an electrolyte-filled rectangular cell. Nickel is electrowon on starter sheets or as *crowns* on masked stainless blanks. Nikkelverk's electrowinning plant produces nickel crowns as well as full cathode plates (Stensholt et al. 1988) that are used by the nickel-plating industry. They are produced by masking titanium mother blanks with polymer, except for unmasked circles where nickel electrodeposition can occur.

In nickel chloride electrowinning, coated titanium anodes are used. Lead alloy anodes are not used because they corrode quickly. Modern anodes are titanium coated with ruthenium and/or iridium oxide and other non–precious metal oxides. They are conductive but inert. Each anode is placed in a permeable polyester cloth bag topped with a solid polypropylene gas-collection *lid*. Chlorine and chlorine-saturated electrolyte (anolyte) are continuously drawn from each anode compartment by vacuum through an individual polymer tube.

Cell currents are typically 20–30 kA (~240 A/m^2 of cathode, both sides). Plating typically occurs for seven days and then the nickel cathodes are harvested. Current efficiency is 98%–99%. Each cell operates at ~3 V.

Industrial nickel from sulfate electrowinning is similar to industrial nickel from chloride electrowinning. The major difference is that chlorine is not generated at the anode. This means that lead alloy anodes can be used and anode bags do not have to be used. Anglo American Platinum commissioned a new nickel sulfate electrowinning tankhouse in 2011 using state-of-the-art equipment and technology at Rustenburg, South Africa (Hagemann et al. 2016). Permanent cathodes, mechanical stripping, anode skirts, and cell hoods are used to reduce worker exposure to aerosols and solutions that contain nickel sulfate and acid. The development and testing of the technology was performed in-house. Operational details from Rustenburg are presented in Table 6.

Gold

Gold electrowinning is usually performed in relatively small vessels or cells that are about 1 m long and contain less than 1 m^3 of solution. A schematic drawing and picture of gold electrowinning cells are provided in Figure 13. These cells often contain several cathodes and anodes, although there is one extra anode. The cathodes commonly consist of steel wool (30 kg/m^3) supported by a stainless-steel frame, and anodes are generally perforated stainless-steel plates or woven stainless-steel mesh (Wells et al. 1992). The cells generally operate between 2 and 5 V at a current density between 30 and 100 A/m^2 (Marsden and House 2006). The cells are designed to effectively recover most of the gold from solutions containing around 0.2 kg of dissolved gold per cubic meter of

Table 5 Copper electrowinning operating data from selected operations

Operation	Collahuasi	Gaby	Minera El Abra	Safford	Chino	Kansanshi
Location	Chile	Calama, Chile	Calama, Chile	Safford, AZ, USA	Santa Rita, NM, USA	Solwezi, Zambia
Date of survey	2013	2010	2013	2013	2013	2012
Date of commissioning	1997	2008	1996	2008	1986	2004
Nameplate cathode capacity, t/a	60,000	150,000	225,000	105,545	68,040	139,000
Actual cathode production, t	36,808	150,000	153,424	87,653	21,047	104,677
ANODES						
Type	Pb alloy	—*	Transitioning from Pb to Ti mixed metal oxide	Pb/Ca/Sn	Ti mixed metal oxide	Pb/Al/Ca
Lead anodes						Supplier: Castle Lead Works
Composition	98.55	98.6–98.4	98.585	Balance	—	Remainder
Pb, %	—	—	—	—	—	—
Sn, %	1.375	1.25–1.50	1.35	1.30–1.45	—	1.25–1.75
Ca, %	0.075	0.06–0.10	0.065	0.065–0.085	—	0.05–0.10
Sb, %	—	—	—	<0.0005	—	—
Other	—	0.005–0.02 Al	—	0.015 min Al	—	0.0005–0.02 Al
Rolled or cast	Rolled	Rolled	Rolled	Rolled	—	Cross rolled
Life before replacement, years	6	5	6	7–10	—	5
Blade length × width × t, mm	990 × 940 × 7.5	1,080 × 938 × 6	1,074 × 940 × 6	1,319 × 968 × 6.35	—	1193 × 922 × 6
Hanger bar material	Cu	Pb-covered Cu	Pb-covered Cu	Cu	—	Cu
Hanger bar protection	—	Soldered	—	Pb	—	6 mm Pb/6% Sb cast over bar and extending 54 mm below the bar
Bar/blade connection	Welded	—	Soldered	Welded	—	Full penetration weld
Anodes hung symmetrically or asymmetrically	Asymmetric	Symmetrically	Symmetrically	Symmetrically	—	Symmetrically
Anode/cathode spacers	—	Delta separator	Abra style	Hairpin insulator	—	3 PVC buttons per side
Titanium anodes						
Coating description	—	—	Mixed metal oxide	—	Mixed metal oxide	—
Mesh length × width × t, mm	—	—	935 × 940	—	952.5 × 889 × 3.175	—
Number of conductor bars	—	—	8	—	6	—
Header/mesh connection	—	—	Press fit	—	Press fit	—
Anodes hung symmetrically or asymmetrically	—	—	Symmetrically	—	Asymmetrically	—
Anode/cathode spacers	—	—	Abra style	—	3 PVC hairpins	—

(continues)

Table 5 Copper electrowinning operating data from selected operations (continued)

Operation	Collahuasi	Gaby	Minera El Abra	Safford	Chino	Kansanshi
CATHODES						
Plating time, days	6	6.5	5	7	7	6–7
Stainless steel or starter sheets	Stainless steel	Stainless steel	Stainless steel	316L stainless steel	Electrorefined starter sheets	Stainless steel
Cu starter sheets						
Length × width × thickness, mm	—	N/A	—	—	952 × 946 × 1.6	—
Mass after plating, kg	—	N/A	—	—	85–100	—
Hanging method	—	—	—	—	2 hanger strap loops	—
Stainless-steel blanks						
Technology provider	Kidd process	IsaKidd process	Kidd process	ISA process	—	Cathodex
Length × width × t, mm	1,220 × 1,040 × 3	1,132 × 1,018 × 3.25	1,018 × 1,232 × 3	1,300 × 1,000 × 3.25	—	1,193 × 922 × 3
Type of stainless steel	316L	316L 2B	Duplex	316L	—	316 L
Side edging	Strip	—	Xstrata-ICL	Side edge strip	—	Yes
Bottom edging	None	—	None	None	—	Yes
Hanger bar						
Type	—	Rectangular	Rectangular	Rectangular	—	Rectangular
Material	Cu	Cu-covered stainless steel	Cu	Cu-plated 316L SS	—	Cu
Height × width × length, mm	—	—	38 × 25 × 1,430	44 × 31 × 1,294	—	50 × 50 × 6,645
Joint between bar and blade	—	—	Welded	Intermittent weld	—	None
Mass each side, kg	35–40	—	32.5–37.5	50–60	—	40
Stripping method						
Manufacturer	Kidd	—	EPCM, AISCO, and MESCO	Outokumpu	—	MESCO
Water temperature, °C	80	70	65–85	70	—	70
Blanks/hour	300	—	300	320	—	50–70
Features	—	—	No.1—No. 2—No. 3	—	—	—
Washing	Yes	Yes	Yes—Yes—Yes	Yes	—	Yes
Stripping	Yes	Yes	Yes—Yes—Yes	Yes	—	Yes
Stacking	Yes	Yes	Yes—Yes—Yes	Yes	—	Yes
Sampling	Yes	Yes	No—No—No	No	—	Yes
Weighing	Yes	Yes	Yes—Yes—No	Yes	—	Yes
Strapping	No	Yes	—	Yes	—	Yes
Blank buffing	No	—	—	—	—	Yes
Cathode Cu						
Pb, ppm	<1		<3	<1	<0.3	1–4
S, ppm	6		<8	4	<8	2–8

(continues)

Table 5 Copper electrowinning operating data from selected operations (continued)

Operation	Collahuasi	Gaby	Minera El Abra	Safford	Chino	Kansanshi
CELLS						
Number	188	504	3 banks in operation, actually 486 cells	182	210 (80 in current operation)	320
Length × width × depth, m	6.3 × 1.27 × 1.4	6.42 × 1.30 × 1.386	6.57 × 1.30 × 1.40	7.63 × 1.18 × 1.63	6.8 × 1.2 × 1.3	7.06 × 1.144 × 1.57
Height above cell floor, m	0.49	—	1–2	0.052	0.76–0.91	1.5
Construction material	Polymer concrete	Polymer concrete	Polymer concrete	Polymer concrete	160 paralined concrete, 50 polymer concrete	Precast vinyl-ester resin concrete
Lining material	19-mm polypropylene	—	None	—	—	Vinyl-ester resin concrete
Capping board material	—	Polymer concrete and fiber-reinforced polymer	Polymer concrete	Protruded vinyl-ester fiberglass	Polyester, polyvinyl	Concrete
Intercell busbar	—	Triangular	Double-double Abra style	Double-double contact	Double-double contact	Floating null point, non-grounded system
Cross-sectional area, mm^2	—	—	855	2,904	1,140	—
Current density, A/mm^2	—	276.5/333.3	—	—	—	280–320
Anodes/cathodes per cell	61/60	60/61	67/66	145	—	70/69
Cathode–cathode centerline distance, mm	100	100	95	101.6	—	95
Cell inspection system	Gaussmeter	Yes	Infrared camera	—	Infrared camera	Manual
Inspections	Once per day	—	Once per shift	3 per day	2 per day	Daily
Mist suppression system						
Hoods	No	—	—	No	—	—
Balls	Yes	Yes	Yes	Yes	Removed	Yes
Ventilation	Yes	Outokumpu	Yes	No	Yes	—
Mist suppression reagents						
FC-1100	Yes	No	Yes	Yes	Yes	Yes
Mistop	No	—	—	—	—	No
Other	No	—	—	—	—	No
Concentration, mg/L	3	—	—	—	—	10
Consumption, kg/t cathode	0.13	—	—	—	—	0.15
Cell cleaned after (days)	130	120	—	90–100	No cell cleaning	180
ELECTRICAL						
Rectifiers, kA/V	38/207	4 at 40/491–40/539	4 of 45/400	0.135	4 at 18, 2 at 20, 2 at 8	21–55/160–175
Current per cell, kA	20	33	32–40	47	40 or 44 maximum	—
Cathode current density, A/m^2	170	276.5/333.3	255–315	253	250–300, 360 maximum	280
Cathode current efficiency, %	93	90	92	89.1	93	85
Anode–cathode voltage, V	—	1.95/2.14	1.89	1.877	1.8	2
Total AC energy conserved (MW·h/t)	1.9	—	1.74	1.8	1.80	2.63

(continues)

Table 5 Copper electrowinning operating data from selected operations (continued)

Operation	Collahuasi	Gaby	Minera El Abra	Safford	Chino	Kansanshi
ELECTROLYTE						
Total volume in plant, m^3	3,000	—	1,1700	5,800	2,410	5,000
Flow per cell, m^3/min	0.3	0.29	0.32	0.6	~0.18	0.25
Flow control system	—	Own design	One flowmeter per 340 cells	—	In-line ball valves on each section of cells	Total flow control with individual cell valves
Electrolyte distribution system and hole diameters	Manifold, 3 mm	—	No	Cell manifold, 6.35–9.525 mm	50-mm PVC inlet pipe, no manifold	Ring main, 4 mm
Electrolyte entering cells						
Cu, g/L	39	50	>36	36	38	50–60
H_2SO_4, g/L	190	157	170–200	185	170–180	160–190
Temperature, °C	47	44	40–44	47	32	50–55
Temperature control method	—	—	Thermometer	Heat exchangers	Heat exchangers	Heat exchanger
Circulation pumps	—	—	Horizontal centrifugal	—	Durco centrifugal	Yes
Electrolyte leaving cells						
Cu, g/L	37	35	>34	33.8	35	32–38
H_2SO_4, g/L	195	180	195	188.6	180–190	180–200
Temperature, °C	48	45	40–43	47	33	40
Fe^{2+}, g/L	0.2	—	0.5	0.55	1.0–1.5	1–2
Fe^{3+}, g/L	0.8	—	1.2	2.09	2.0–2.5	3–4
Co^{2+}, mg/L	150	150	180	160	15	100–130
Mn, mg/L	60	—	<40	66	300–400	350
Cl^-, mg/L	25	<30	<25	16	25	20
Smoothing reagent	DXG-F7	—	HydroStar	HydroStar	Cyquest N-900	Guar/starch
Addition rate, g/t cathode	175	250	>300	347–716	650	700
Cobalt addition, g/t cathode	1,300	365	Maintain 180 ppm	113	None	400
Electrolyte bleed, m^3/d	144	—	384	98	136	5
Bled to control	Mn	—	Fe^{3+}, Mn, Cl	Mn and Cl	Fe, Mn	Fe
Destination	E1	—	Solvent extraction wash and others	Solvent extraction E1, E2, or E3	Solvent extraction E1	Leach

Adapted from Robinson et al. 2013a
*A dash indicates no data provided.

solution. Most of the applied current is used by side reactions such as hydrogen production from water. The solution from the electrowinning cell is recirculated back to the stripping process to replenish the dissolved gold content. The steel wool cathodes that contain the deposited gold are harvested for subsequent refining after the weight of deposited gold exceeds the weight of the initial cathode.

LABORATORY AND PILOT TESTING

Sizing

Laboratory-scale or pilot testing of electrowinning can be performed using electrodes of 1 cm^2 to full commercial-size electrodes in cells of widely various sizes. As electrochemical reactions are surface area dependent, they typically adjust to changes in scale (electrode size). Thus, results in small experiments can predict large-scale operations well. Other features such as hydrodynamics, fluid flow, gas bubble movement, and/or particle settling do not necessarily scale well. Therefore, the selection of electrode and cell size is dependent on the phenomenon being studied.

Electrochemical Measurements

Basic electrochemical measurements use a potentiostat/galvanostat connected to a three-electrode cell as described earlier in the "Applied Fundamentals" section. A potentiostat/galvanostat is a high-impedance power supply that is often computer controlled. Depending on the experiment, potential is controlled and current is measured or vice versa. There are numerous electrochemical experiments that can be performed to study electrowinning. For more details, readers are directed to consult a text specific to electrochemical methods such as Bard and Faulkner (1980).

Pilot Testing

Electrowinning pilot testing is typically performed if (1) longer term phenomena are being studied, such as anode corrosion or deposit stress, or (2) the integration of electrowinning with upstream processing is being evaluated. Pilot electrowinning cells can also vary in size and shape. Sizing is typically dependent on the purpose of the test and the cost of operating the pilot cell. Usually, the smallest possible cells that accomplish the purpose of the test are used to minimize costs.

For pilot testing, a cell with electrodes, an electrolyte reservoir, a heat source, pump, power supply, and data recording system are the minimum equipment needed. Control of the electrolyte needs to be evaluated prior to the experiment. A decision concerning single-pass electrolyte use versus

Table 6 Operational parameters for a modern nickel electrowinning plant from sulfate electrolyte

Location: Rustenburg, South Africa, 2009	
Cathode production, t/a	20,000
Sulfate Electrolyte	
Ni, g/L	80 in, 50 out
Addition agents	Boric acid, guar gum
Temperature, °C	60–65
pH	3.5
Anodes	
Material	Lead alloy (0.6% Sn, 0.05% Sr)
Cathodes	
Material	Nickel starter sheets
Center-to-center distance apart, m	0.16
Number per cell	41 anodes, 40 cathodes
Plating time	6 + 2 days making starter sheets on Ti mother blanks
Cathode Compartment (construction materials)	Yes
Frame	Oregon pine
Membrane	Woven terylene
Cells	
Materials	Precast concrete
Inside length × width × height, m	6.6 × 1.2 × 1.2
Electrical	
Cell potential, V	3.6–3.9
Cathode current density, A/m^2	220
Current efficiency, %	97

Source: Crundwell et al. 2011

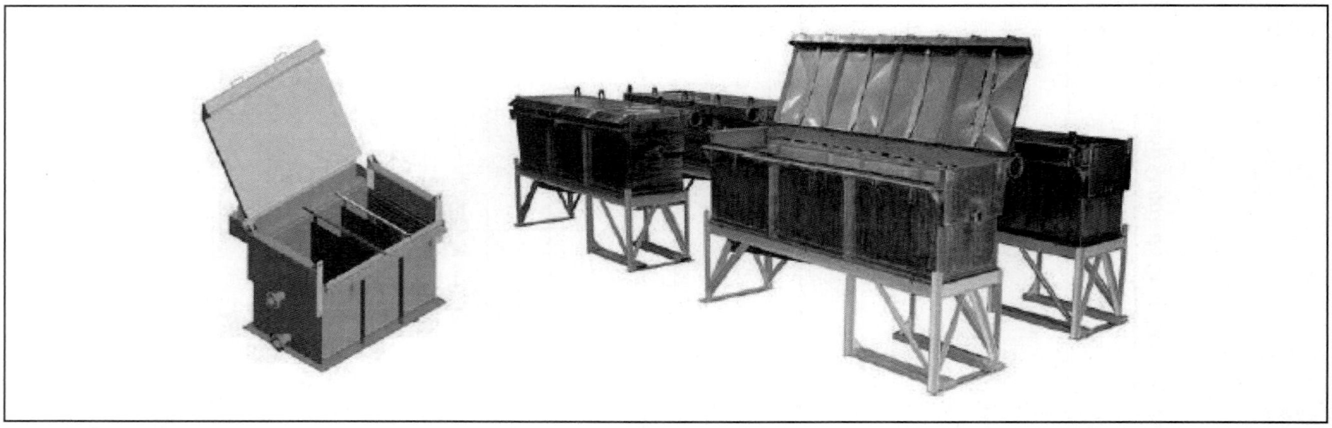

Courtesy of FLSmidth

Figure 13 Schematic drawing and photograph of gold electrowinning cells

electrolyte recirculation with ion replenishment and acid neutralization needs to be made. The continuous addition of smoothing or polarizing compounds is highly advised if metal deposition is being studied.

In an electrowinning cell, anodes are connected to the positive terminal of the power supply. Cathodes are attached to the negative. Cells and electrodes should be electrically operated similar to industry; that is anodes and cathodes are in parallel within a cell. Cells are connected in series.

Electrolyte flow rate to each cell should be controlled. Flow rate is usually selected to be similar to commercial cells in terms of average residence time, specific flow rate (electrolyte flow rate divided by surface area of cathodes or anodes), or decrease in metal concentration being plated.

The power supply should be sized to provide enough current and voltage to operate the pilot system. As most electrowinning systems operate with constant current, the power supply should be set to maximum voltage and controlled using the current setting.

Data recording can include cell voltage, oxidation-reduction potential, pH, and temperature. Measurement frequency will depend on the phenomena being studied and the duration of the pilot testing. Measurements can occur manually or automatically with a computer-controlled system.

Safety

As with any experiment, care must be given from a safety standpoint as electricity and highly conductive electrolytes and electrodes are used in electrowinning research. Additionally, most electrowinning solutions are highly corrosive and may contain hazardous compounds. Care must be given to safe and proper handling and disposal. Depending on the size and location of a pilot system, proper ventilation or mist capture may also be needed.

REFERENCES

Addison, J.R., Savage, B.J., Robertson, J.M., Kramer, E.P., and Stauffer, J.C. 1999. Implementing technology: Conversion of Phelps Dodge Morenci, Inc. central EW tankhouse from copper starter sheets to stainless steel technology. In *Proceedings of the Copper 99–Cobre 99 International Conference*. Warrendale, PA: The Minerals, Metals & Materials Society. pp. 609–618.

Anderson, C.G., Giralico, M.A., Post, T.A., Robinson, T.G., and Tinkler, O.S. 2009. An update: Selection, equipment sizing and flowsheet applications in copper solvent extraction and electrowinning. *Recent Advances in Mineral Processing Plant Design*. Edited by D. Malhotra, P.R. Taylor, E. Spiller, and M. LeVier. Littleton, CO: SME. pp. 287–319.

Bard, A., and Faulkner, L. 1980. *Electrochemical Methods: Fundamentals and Applications*. New York: Wiley.

Beukes, N.T., and Badenhorst, J. 2009. Copper electrowinning: Theoretical and practical design. In *SAIMM Southern African Hydrometallurgy Conference 2009*. Johannesburg: Southern African Institute of Mining and Metallurgy.

Bockris, J., and Reddy, A.K.N. 1973. *Modern Electrochemistry*. New York: Plenum Press.

Boon, C., Fraser, R., Johnston, T., and Robinson, D. 2013. Comparison of intercell contact bars for electrowinning plants. In *Ni-Co 2013*. Edited by T. Battle, M. Moats, V. Cocalia, H. Oosterhof, S. Alam, A. Allanore, R. Jones, N. Stubina, C. Anderson, and S. Wang. Hoboken, NJ: Wiley.

Brown, R.R. 1990. Rectifier and DC bus system design for the copper electrowinning industry. *IEEE Trans. Ind. Appl.* 26(6):1116–1119.

Cheng, C.Y., Hughes, C.A., Barnard, K.R., and Larcombe, K. 2000. Manganese in copper solvent extraction and electrowinning. *Hydrometallurgy* 58(2):135–150.

Cheng, C.Y., Urbani, M.D., Miovski, P., and San Martin, R.M. 2004. Evaluation of saponins as acid mist suppressants in zinc electrowinning. *Hydrometallurgy* 73(1-2):133–145.

Cheresources. 2017. Chemical engineering plant cost index (CEPCI) (discussion forum and user blog). www.cheresources.com/invision/topic/21446-chemical-engineering-plant-cost-index-cepci/page-2.

Crundwell, F., Moats, M., Ramachandran, V., Robinson, T., and Davenport, W.G. 2011. *Extractive Metallurgy of Nickel, Cobalt and Platinum Group Metals*. Oxford: Elsevier.

Davis, J.A., and De Visser, J. 2000. Cellhouse ventilation. *Lead-Zinc 2000*. Edited by J.E. Dutrizac, J.A. Gonzalez, D.M. Henke, S.E. James, and A.H.-J. Siegmund. Warrendale, PA: The Minerals, Metals & Materials Society. pp. 579–588.

Free, M.L. 2013. *Hydrometallurgy Fundamentals and Applications*. New York: Wiley.

Habashi, F. 1998. *Principles of Extractive Metallurgy*. Vol. 4, Amalgam and Electrometallurgy. Sainte-Foy, QC: Metallurgie Extractive Quebec.

Hagemann, J., Ndlovu, J., Nelson, L., and Gilmore, M. 2016. Full deposit nickel electrowinning at Rustenburg base metal refiners. In *Proceedings of the XXVIII International Mineral Processing Congress (IMPC 2016)*. Westmount, QC: Canadian Institute of Mining, Metallurgy and Petroleum.

Honey, R.N., and Watson, R.H. 1985. Cominco's new zinc electrolytic and melting plant at Trail, BC. *J. Miner. Met. Mater. Soc.* 37(8):47–50.

Horsehead Holding. 2012. Investor Presentation. August. Horsehead Holding Corporation.

Houlachi G. 2008. Short course: Zinc and lead metallurgy [notes]. Conference of Metallurgists (COM), Winnipeg, MB, Canada, August 24–27.

International Copper Study Group. 2017. *World Copper Factbook*. Lisbon: International Copper Study Group.

International Lead and Zinc Study Group. 2015. Lead and zinc statistics. www.ilzsg.org/static/statistics.aspx?from=1.

International Mining. 2006. All American copper giant. https://im-mining.com/2006/11/20/all-american-copper-giant.

International Zinc Association. n.d. Zinc basics. www.zinc.org/basics/.

Kuusisto, R., Pekkala, P., and Karcas, G. 2005. Outokumpu SX EW technology package. Presented at the Third Southern African Base Metals Conference: Southern Africa's Response to Changing Global Base Metals Market Dynamics, Kitwe, Zambia, June 26–29.

Lenz, J., and Ducharme, D. 2000. Zinc autostripping at Falconbridge Limited Kidd Metallurgical Division. *Lead-Zinc 2000*. Edited by J.E. Dutrizac, J.A. Gonzalez, D.M. Henke, S.E. James, and A.H.-J. Siegmund. Warrendale, PA: The Minerals, Metals & Materials Society. pp. 563–578.

Marsden J.O, and House, C.I. 2006. *The Chemistry of Gold Extraction*, 2nd ed. Littleton, CO: SME.

Martin, F.S., Tamargo, F., and Lefevre, Y. 2000. Asturiana de zinc expansion at the San Juan de Nieva plant for a zinc production of 440,000 tonnes per year. *Lead-Zinc 2000*. Edited by J.E. Dutrizac, J.A. Gonzalez, D.M. Henke, S.E. James, and A.H.-J. Siegmund. Warrendale, PA: The Minerals, Metals & Materials Society. pp. 555–562.

Masters, I., Doyle, B., and Weir, D. 1989. Direct pressure leaching of Australian zinc concentrates. In *Non-Ferrous Smelting Symposium: 100 Years of Lead Smelting and Refining in Port Pirie*. Melbourne, Victoria: Australasian Institute of Mining and Metallurgy.

McGinnity, J.J., and Nicol, M.J. 2014. Sulfuric acid mist: Generation, suppression, health aspects, and analysis. *Miner. Process. Extr. Metall. Rev*. 35(3):149–192.

Mella, S., Villarroel, R., and Lillo, A. 2003. Copper electrowinning in the absence of acid mist. In *Electrochemistry in Mineral and Metal Processing VI: Proceedings of the International Symposium*. Edited by F.M. Doyle, G.H. Kelsall, and R. Woods. Pennington, NJ: Electrochemical Society. pp. 379–390.

Miller, G. 2011. Methods of managing manganese effects on copper solvent extraction plant operations. *Solvent Extr. Ion Exch*. 29(5-6):837–853.

Moats, M.S. 2008. Will lead-based anodes ever be replaced in aqueous electrowinning? *JOM* 60(10):46–49.

Moats, M., Guerra, E., and Gonzalez, J.A. 2008. Zinc electrowinning—operating data. In *Zinc and Lead Metallurgy*. Edited by L. Centomo, M. Collins, J. Harlamovs, and J. Liu. Montreal, QC: Metallurgy and Materials Society of the Canadian Institute of Mining, Metallurgy and Petroleum.

Moats, M., Guerra, E., Siegmund A., and Manthey, J. 2010. Primary zinc smelter operating data survey. In *Lead-Zinc 2010*. Edited by A.H-J. Siegmund, L. Centomo, C. Geenen, N. Piret, G. Richards, and R.I. Stephens. Hoboken, NJ: Wiley.

Newman, J., and Thomas-Alyea, K.E. 2004. *Electrochemical Systems*, 3rd ed. Hoboken, NJ: Wiley.

OSHA (Occupational Safety and Health Administration). 1985. OSHA Method ID-165SG. *Acid Mist in Workplace Atmospheres*. www.osha.gov/dts/sltc/methods/inorganic/id165sg/id165sg.html.

Otero, J.F., San Martin, R.M., and Cruz, A. 2003. Successful industrial use of quillaja saponins (quillaja saponaria molina) for acid mist suppression in copper electrowinning process. In *Electrometallurgy and Environmental Hydrometallurgy*, vol. 2. Edited by C. Young, A. Alfantazi, C. Anderson, A. James, D. Dreisinger, and B. Harris. Warrendale, PA: The Minerals, Metals & Materials Society.

Porter, F. 1991. *Zinc Handbook: Properties, Processing, and Use in Design*. Boca Raton, FL: CRC Press.

Potter, M.C., and Foss, J.F. 1982. *Fluid Mechanics*. Okemos, MI: Great Lake Press.

Rasmussen, S.A. 1995. Operating characteristics and anode reactions in Morenci's electrowinning tankhouse. SME Preprint No. 95-234. Littleton, CO: SME.

Robinson, D.J., and O'Keefe, T.J. 1976. On the effects of antimony and glue on zinc electrocrystallization behaviour. *J. Appl. Electrochem*. 6(1):1–7.

Robinson, T.G., Sole, K.C., Moats, M.S., Crundwell, F.K., Moritmitsu, M., and Palmu, L. 2012. Developments in base metal electrowinning cellhouse design. In *Electrometallurgy 2012*. Edited by M. Free, M. Moats, G. Houlachi, E. Asselin, A. Allanore, J. Yurko, and S. Wang. Hoboken, NJ: Wiley.

Robinson, T., Sole, K., Sandoval, S., Siegmund, A., Davenport, W., and Moats, M., 2013a. Copper electrowinning: 2013 World tankhouse operating data. In *Copper 2013*. Vol. 5, Electrowinning/Electrorefining. Edited by R. Abel and C. Delgado. Santiago: Instituto de Ingenieros de Minas Chile.

Robinson, T., White, D., and Grassi, R. 2013b. Acid mist abatement in base metal electrowinning. In *Ni-Co 2013*. Edited by T. Battle, M. Moats, V. Cocalia, H. Oosterhof, S. Alam, A. Allanore, R. Jones, N. Stubina, C. Anderson, and S. Wang. New York: Springer International. pp. 143–153.

Sabau, M., and Bech, K., 2007. Status and improvement plans in Inco's electrowinning tankhouse. In *Copper 07–Cobre 07 Proceedings of the Sixth International Conference*. Vol. 5, Electrorefining and Electrowinning. Edited by G.E. Houlachi, J.D. Edwards, and T.G. Robinson. Montreal, QC: Canadian Institute of Mining, Metallurgy and Petroleum. pp. 439–450.

Sandoval S., Clayton C., Gebrehiwot E., Morgan J. 2013. Tankhouse parameters for transition from lead to alternative anodes. In *Hydroprocess 2013*. Edited by F. Valenzula and C. Young. Santiago, Chile: Gecamin. pp. 97–98.

Stensholt, E., Zachariasen, H., Lund, J., and Thornhill, P. 1988. Recent improvements in the Falconbridge nickel refinery. In *Extractive Metallurgy of Nickel and Cobalt*. Edited by G. Tyroler and C. Landoolt. Warrendale, PA: TMS-AIME. pp. 403–412.

Sutherland, C. 1988. Modernization of Cominco's zinc plant and lead smelter at Trail, British Columbia. *Can. Min. Metall. Bull*. 81(912):85–89.

Valentine, T. 2002. Layout and design of electrowinning cathode blank repair shops at Morenci. SME Preprint No. 02-148. Littleton, CO: SME.

Wells, J.A., and Snelgrove, W.R. 1990. The design and engineering of copper electrowinning tankhouses. In *Proceedings of the International Symposium on Electrometallurgical Plant Practice*. Edited by P.L. Claessens. Amsterdam: Elsevier.

Wells, J., Hopkins, W., and Stein, R. 1992. Chemical and electrolytic processing. In *SME Mining Engineering Handbook*. Edited by H. Hartman. Littleton, CO: SME.

Wiechmann, E.P., Burgos, R.P., and Holtz, J. 2000. Sequential connection and phase control of a high-current rectifier optimized for copper electrowinning applications. *IEEE Trans. Ind. Electron*. 47(4):734–743.

Zhang, W., and Cheng, C.Y. 2007. Manganese metallurgy review. Part III: Manganese control in zinc and copper electrolytes. *Hydrometallurgy* 89(3):178–188.

CHAPTER 10.17

Electrorefining

Michael S. Moats and Michael L. Free

Electrorefining is the purification of metal by the simultaneous electrolytic dissolution of impure metal and the electrodeposition of the same metal at a higher purity level. Electrorefining is often critical to meeting final purity specifications for metals such as copper, gold, lead, and silver that are smelted or made into a moderately pure precursor material. Electrorefining is also important to many industries, such as the electronics industry, that rely on high-purity metals.

The general background information that is discussed in Chapter 10.16, "Electrowinning," also applies to electrorefining. However, there are several key differences between electrowinning and electrorefining. The most important difference is that in electrorefining the anode consists of the metal that is to be refined. Correspondingly, the reaction at the anode is primarily metal dissolution in electrorefining instead of water oxidation, which generates bubbles and acid mist, at relatively inert anodes in electrowinning. Another key difference is the much lower energy requirement and higher current efficiency for electrorefining than for electrowinning. The use of anodes with inclusions that can become suspended particles is another important aspect of electrorefining that is not present in electrowinning.

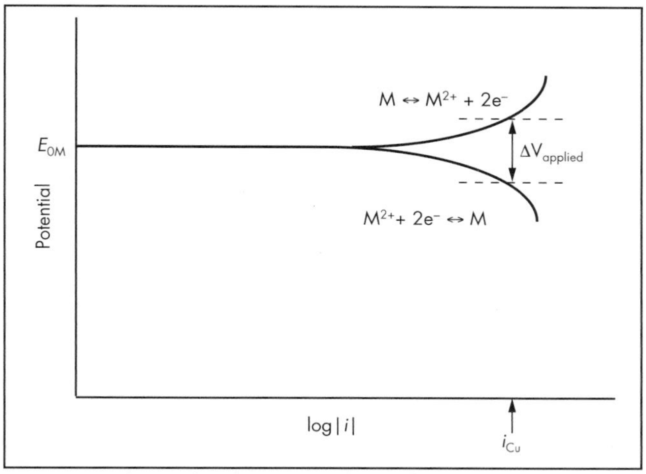

Figure 1 Relationship between potential and current density for an electrorefining cell

Figure 1 shows the relationship between electrochemical potential and the logarithm of the absolute value of the current density for an electrorefining scenario. The applied potential or voltage for electrorefining is as much as one order of magnitude lower than that for electrowinning. Thus, the corresponding energy requirement for electrorefining is much lower than for electrowinning. The main difference is the absence of a cell voltage difference between the anode and cathode half-cell reactions. The other difference is the absence of a relatively large overvoltage needed to drive the water oxidation reaction that is often utilized for electrowinning but not electrorefining.

The anodes used in electrorefining contain impurities that are distributed between the atoms of the main metal as interstitial impurity atoms based on their solubility in the metal when the metal solidifies. Impurity atoms that are present at levels higher than their solubility limit in the main metal are also found in separate crystals of a wide variety of secondary or tertiary phase compounds. Consequently, examination of anode cross-sections using scanning electron microscopy reveals a variety of inclusions that are made up of the impurity-bearing compounds as illustrated in Figures 2 and 3.

Anode inclusions are typically only a few micrometers in diameter. As the surrounding matrix of the main metal is electrochemically forced to dissolve during the electrorefining process, these inclusions are dissolved in the electrolyte, released into solution, or form a layer of particles that often adheres to the anode. The released particles are referred to as *slimes* because they form a fine sludge of fine particles that is often "slimy." The layer of adhering slimes is often referred to as a *slimes layer*. Because the slimes consist of microscopic particles, a significant portion of slimes remaining in suspension become available to be transported to and incorporated in the cathode as impurity particles unless they are retained inside of anode bags, remain in the slimes layer, or settle to

Michael S. Moats, Associate Professor, Missouri University of Science & Technology, Rolla, Missouri, USA
Michael L. Free, Professor, University of Utah, Salt Lake City, Utah, USA

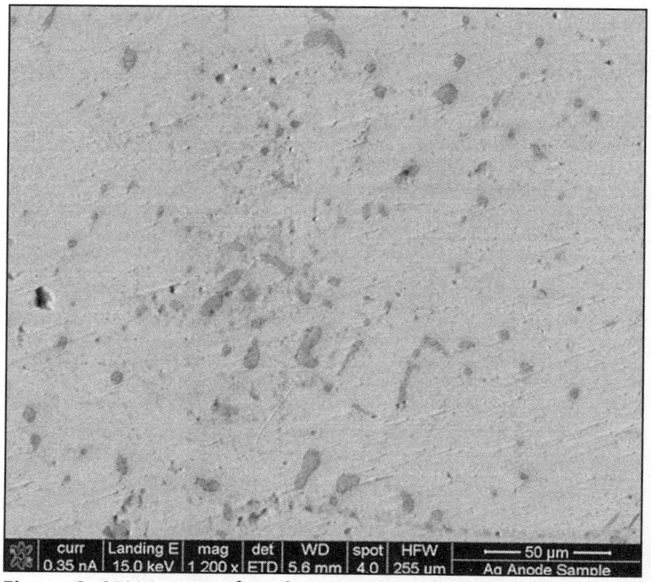

Figure 2 SEM image of a silver anode with gold- and copper-based inclusions

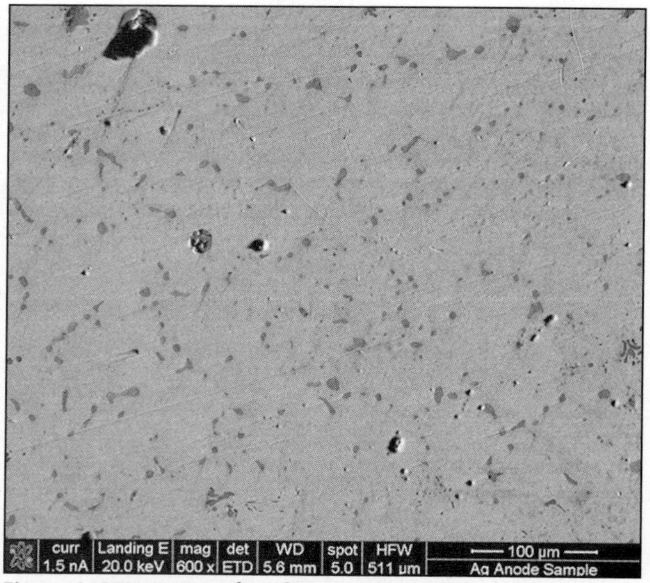

Figure 3 SEM image of a silver anode with gold-, copper-, and selenium-based inclusions

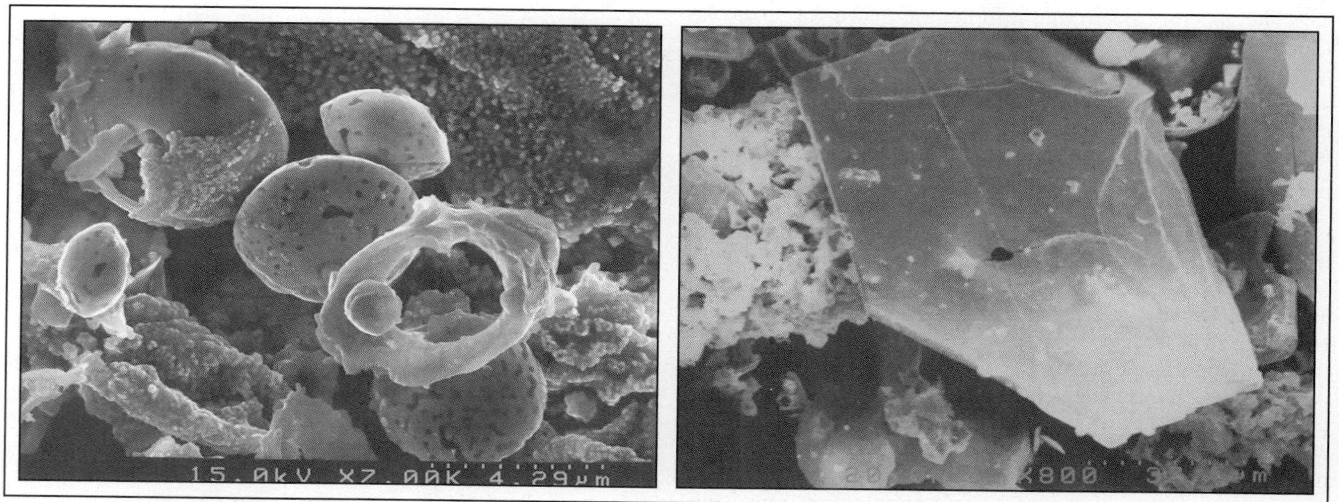

Figure 4 SEM images of copper anode slimes

the bottom of the cell. SEM images of slimes generated during the dissolution of commercial copper anode are shown in Figure 4. These images reveal the presence of cuprous selenide (Cu_2Se) rings and spheres, a kupferglimmer platelet, lead sulfate ($PbSO_4$) crystals, and silver crystallites.

In copper and lead electrorefining, anode bags are not normally used. In silver and nickel refining, anode bags are commonly used. In most cases, a significant portion of the slimes particles that are generated either fall to the bottom of the cell (or anode bag) where they are periodically collected or remain in adherent slimes layers attached to the anode. However, residual suspended slimes particles are often a major source of cathode contamination.

COMMERCIAL PRACTICE

The electrorefining of high-purity metals has been practiced commercially for cobalt, copper, gold, indium, lead, nickel, silver, and tin. Copper is the largest tonnage metal that is electrorefined. As such, industrial aspects of copper electrorefining will be discussed in detail. Lead, nickel, silver, and gold will also be discussed in this chapter.

Copper

Most of the world's refined copper is produced by electrorefining. The source material for electrorefineries is both primary ore and recycled scrap. Following pyrometallurgical processing, fire-refined blister copper is cast into anodes, which are fed into the electrorefinery. These anodes are electrolytically dissolved into a solution of primarily copper sulfate and sulfuric acid. Simultaneously with anode dissolution, copper is electrodeposited at the cathode. Cathodes are harvested and either sent to market or to a downstream processing facility.

Copper electrorefining operating data has been collected periodically and reported every three to four years at international copper conferences since 1987. Detailed operating data from seven refineries collected in 2013 are presented in Table 1 (Moats et al. 2013).

Most anodes are cast in molds on rotating wheels or discs at the smelter. Improved control of anode weight and dimensions has occurred over the years through automation. Anodes once received at the refinery are almost always prepared by a machine that presses the anodes flat, turns the ears slightly to improve their ability to hang straight in the cell, and mills the contact surface where the ears will touch the intercell contact bar.

Anodes are usually placed in an empty cell following draining, removal of slimes, and cleaning. Some newer facilities place anodes using automated cranes, while older plants use manual cranes. Most tankhouses now employ polymer concrete for the cell material while older facilities use lined concrete. After the anodes are placed, cathodes are inserted between the anodes, although in some facilities, cathodes and anodes are inserted together. Finally, the electrolyte is returned to the cell. Once the cell (or section of cells) reaches a specific temperature (varies by operation), the cell is electrified and electrorefining commences.

An anode cycle is typically 21–28 days depending on the thickness of the anode and the current density employed. Each anode cycle produces two or three crops of cathodes. Some residual anode material is removed from the cell at the end of the electrorefining cycle to avoid parts of the anode breaking free and falling to the bottom of the cell. In 2013, the average amount of the residual anode material that was removed as scrap after the process was completed was 13.7% (Moats et al. 2013). The changing of cathodes and anodes, as well as the cleaning of the cells, contributes to downtime and less than 100% time efficiency.

Anode composition can affect the performance of the electrorefinery. The average anode composition from the 2013 survey is presented in Table 2. During dissolution, more noble elements (Au, Ag), compounds that dissolve slowly or are insoluble (Cu_2Se, nickel oxide [NiO], cassiterite [SnO_2]), and compounds that form by precipitation (antimony arsenate [$SbAsO_4$], $PbSO_4$) form a slimes layer on the anode. These slimes are valuable because of their gold and silver content, but they are also a source of possible contamination of the cathode. They are collected when the anodes are removed from the cell and sent to a separate facility for processing.

Soluble elements that are less noble than copper (Ni, As, Sb, Bi) and compounds that dissolve easily (cuprous oxide [Cu_2O]) report to the electrolyte. Dissolution of other elements and Cu_2O increases copper and impurity levels that are not balanced by the cathode deposition. This requires a continual bleed of electrolyte to remove excess copper from solution and other impurities such as Ni, As, Sb, and Bi. Copper is liberated or removed from the bleed stream by electrowinning in a process referred to as *liberation*. Liberation is typically conducted in stages to decrease the copper content of the electrolyte bleed stream. In the last stage when copper concentrations are low, As, Sb, and Bi are co-deposited. The contaminated product from the last liberation stage is returned to a smelter for reprocessing. During the last stage of liberation, the production of arsine gas is possible, which requires monitoring and the use of either hooded cells or cells placed outside. Ni is recovered at several plants by evaporation or precipitation after liberation. Black acid (solution after liberation) or acid recovered by ion exchange is returned to the electrorefinery as makeup acid.

Arsenic, antimony, and bismuth are considered problem elements from a hygiene (As) or contamination (Sb, Bi) perspective. The problems related to antimony and bismuth are mitigated if the anodes contain an As/(Sb + Bi) molar ratio ≥ 2 (Krusmark et al. 1995; Noguchi et al. 1995; Wesstrom 2014). Maintaining the proper ratio is believed to mitigate the problems associated with Sb and Bi by promoting the precipitation of $SbAsO_4$ and bismuth arsenate ($BiAsO_4$) in the slimes layer. Improper anode chemistry can lead to floating slimes, pipe scaling, and anode passivation (Wesstrom 2014). If antimony concentration becomes too high in the electrolyte, it will lead to the formation of colloidal particles that can float on the electrolyte surface. These *floating slimes* often become incorporated in the cathode, leading to nodulation and contamination near the top of the cathode. These nodules become a source of short circuits, which lower current efficiency. High antimony levels can also lead to scale formation on cell walls and pipes, which can cause flow issues within a plant.

Anode passivation is another potential problem. It is caused by trying to dissolve the anode too fast relative the chemistry of the anode, electrolyte composition, and temperature. Anode passivation leads to uneven anode dissolution and/or poor cathode deposits. Anode passivation can be alleviated by operating at lower current density, higher temperature, higher chloride concentration, or higher arsenic concentrations in the anode (Moats and Hiskey 2010).

Cathodes are deposited for 7–14 days. Newer facilities use stainless-steel sheets that are often referred to as *mother blanks*, whereas older plants use very thin *starter* sheets of copper as the cathodes on which the electrorefined copper is deposited. Starter sheet cathode tankhouses usually plate at lower current density (on average at 278 A/m^2) for longer times (10–14 days) at lower current efficiency (95.5%) than stainless-steel cathode facilities (312 A/m^2, 5–7 days, 96.6%; Moats et al. 2013). For these economic reasons, many electrorefineries have upgraded to stainless-steel cathodes over the past two decades.

The high-purity (>99.99%) copper that deposits on the cathode sheets is usually smooth and dense in plate or sheet form with enough ductility to be harvested by automated mechanical stripping. To achieve this, electrolysis parameters are well controlled. These parameters include current density, temperature, electrolyte composition, electrolyte flow, and the use of additives. Average current density at copper electrorefineries in the 2013 survey was 310 A/m^2 (Moats et al. 2013). This value has been increasing over the past two decades. Some electrorefineries operate at 400 A/m^2 with the use of parallel electrolyte flow (Mettop-BRX technology). Electrolyte temperature and composition are monitored and controlled through the use of heat-retention blankets or cell covers, heat exchangers, and electrolyte bleeding. Copper electrolyte refineries typically operate near 65°C with an average electrolyte composition of 49 g/L Cu, 173 g/L H_2SO_4 (sulfuric acid), 5.5 g/L As, 0.3 g/L Sb, 0.2 g/L Bi, and 13.0 g/L Ni. Glue, thiourea, chloride, and occasionally Avitone A (a sodium alkyl sulfonate) are used as additives to control the surface morphology and crystal structure.

In electrorefining, the flow of electrolyte is maintained to ensure even flows to each cell, to provide additives at the correct concentrations, and to ensure that there is no turbulent flow, which would otherwise increase oxygen dissolution from the air and reduce current efficiency because of the reduction of oxygen at the anodes.

Table 1 Operating data from seven copper electrorefineries

	Refinery						
	Brixlegg, Austria	Las Ventanas, Chile	Hamburg, Germany	Naoshima, Japan	Onsan Refinery 1, Korea	Magna, UT, USA	El Paso, TX, USA
Nameplate cathode capacity, t/a	120,000	400,000	416,000	234,000	360,000	281,000	455,000
2012 production, t	113,600	398,000	365,706	213,996	352,192	190,000	195,080
Anodes							
Analysis	Typical	(4 brands)					
Cu, %	>99.2	99.5–99.7	98.5–99.6	99.2	99.5	99	98.9–99.7
Ag, ppm	400	190–450	200–2,000	856	772	462	160–1,200
Au, ppm	6	1–20	10–150	40.5	50.2	52	1 -15
S, ppm	8	10–40	10–70	30	10	38	7–30
Se, ppm	<10	110–220	200–600	480	404	784	200–1,000
Te, ppm	<10	1–10	20–200	180	146	118	10–100
As, ppm	100	500–900	400–1,500	1,330	1,456	1,113	30–200
Sb, ppm	200	20–300	100–600	220	229	46	30–400
Bi, ppm	30	5–20	20–250	380	250	679	5–100
Pb, ppm	2,000	10–300	200–2,000	2,340	195	2,121	10–600
Fe, ppm	25	10–40		10	Not analyzed	13	5–100
Ni, ppm	3,000	80–400	1,000–5,000	1,940	758	271	50–600
O, ppm	1,000	1,400–2,000	1,000–2,000	1,160	1,223	1,709	1,050–2,400
Conventionally or continuously cast	Conventional	Conventional	Mold on wheel	Conventional	Mold on wheel	Conventional	Mold on wheel
Automatic weight control	Yes	Yes	Yes	Yes	Yes	Yes	Yes
Anode preparation machine	Yes	Yes	Yes	Yes	Yes	Yes	Yes
Weighing	Yes	Yes	Yes	No	No	Yes	Yes
Straightening	Yes	Yes	Yes	Yes	Pressed	Yes	Yes
Lug milling	Yes	Yes	Yes	No	Yes	Yes	Yes
L × W × T, mm	990 × 945 × 38/45	915 × 895 × 39	950 × 905 × 55	980 × 960 × 44	1,183 × 1,170 × 50	1,038 × 938 × 39	940 × 921 × 51
Mass, kg	292/350	276–280	397	373	415	335	388.2
Life, days	14	16	21	23	21	20	28
Scrap, %	13.5	18–19	10–12	14	11	18	13
Anode Slimes							
kg/t of anode	5	1.9	5–8	6.4	2.6	9.7	2–4
Removed after no. days	14	16	21	23	21	20	28
Anode slimes analysis, %							Autoclaved assays
Cu	<1 (after leach)	25–27	12–19	12.3	18.9	30	1
Ag	4–8	140–180	10–20	8.1	13.3	5	20
Au	0.05–0.1	3–4	0.4–1.0	0.58	0.79	0.5	0.2
S		5–6				0	12
Se	0.1–0.2	8–9	4–8	4.24	13.6	5	20
Te		0.5–2	1–2	1.83	2.2	1	0.4
As	0.5–1.0	5–6	2–3	2.83	2.9	5	2
Sb	1.0–5.0	4–5	3–5	1.31	0.6	1	4

(continues)

Table 1 Operating data from seven copper electrorefineries (continued)

	Refinery						
	Brixlegg, Austria	Las Ventanas, Chile	Hamburg, Germany	Naoshima, Japan	Onsan Refinery 1, Korea	Magna, UT, USA	El Paso, TX, USA
Bi		0.2–0.3	1–2	2.51	0.9	3	0.7
Pb	25–40	5–6	11–18	21.23	14.5	30	5
Fe		0.1–0.2				0.25	0.04
Ni	1.0–4.0	0.5–1	1–4	1.23		0.05	0.05
Other					13.1% Ba		
Electrolytic Cells							
Commercial cells	312	1,880	1,080	808	1,140	1,400	2,504
Stripper cells		126		88	0	0	216
Liberator cells	15	32	3–8	36	276	24	5 plus 16 at nickel plant
Cell construction	KVK polymer concrete	Polymer concrete	Polymer concrete	Reinforced concrete	Polymer concrete	Polymer concrete	CTI polymer concrete
Lining material		No		Semiflexible PVC	None	None	N/A
L × W × D, mm	5,900 × 1,180 × (1,370–1,560)	4,101 × 1,080 × 1,280	6,350 × 1,171 × (1,330–1,500)	5,400 × 1,200 × (1,250–1,400)	4,970 × 1,066 × (1,460–1,317)	4,812 × 1,194 × 1,524	4,456 × 1,099 × 1,194
Anodes/cathodes per cell	57/56	40/39	63/64	54/53	50/49	47/46	38/39
Anode spacing, center to center, mm	100	100	95	97	98	98	102
Electrode spacing, anode face to cathode face, mm							
Crop 1	28/26	28	19	26	20	28	25.5
Crop 2	35/31		27	35	30	40	14.75
Crop 3	–/35	—	35		40		N/A
Busbar area, A/mm^2	1.0	—		1.18		0.55	0.72
Cell Inspection							
Times/day	14 h/d	2	15	1	3~9	1	1
Infrared camera, handheld, or crane	Yes, handheld	Handheld	Handheld camera	Handheld	Handheld	Handheld	Handheld
Infrared gun	No	Check	No	No	No	No	Training
Cell voltage monitoring	Yes	Section voltage	Yes	No	Yes	No	Every 4 h on section
Gauss meter	Yes	Yes	Yes	No	Yes	Yes	Yes
Electrical							
Amperage per cell	38,000/44,800	21,600–22,000		27,500	30,900	25,500	17,500–18,000
Cathode current density, A/m^2	340/400	310	335	260	316 (Design 325)	285	255–271
Cell voltage, mV	300–400	300–450	330	330	340	280	
AC/DC converting equipment	Thyristor	Siemens rectifier	3 Si rectifiers	2 Thyristor rectifiers	4 Rectifiers	4 Thyristor rectifiers	SCR rectifier
Converter capacity, kA and V	38 kA,140 V/ 50 kA, 100 V		40 kA, 160–180 V	33 kA/183 V	35 kA (200 V, 150 V, 100 V, 30 V)	27,600 kW; 269 & 360 V	6,000 kW, 325 V DC
Cathode current efficiency, %	95–99	93–94	93–97	97.5	98.3	94.7	97
Rectifier output energy, kW·h/t of cathode		360		280	329	340	290
Periodic reverse current?	Yes	No	No	No	No	No	No

(continues)

Table 1 Operating data from seven copper electrorefineries (continued)

	Refinery						
	Brixlegg, Austria	**Las Ventanas, Chile**	**Hamburg, Germany**	**Naoshima, Japan**	**Onsan Refinery 1, Korea**	**Magna, UT, USA**	**El Paso, TX, USA**
Section switches and type	Hundt & Weber	Manually operated	Hundt & Weber	Pneumatic	Pneumatic	Yes, air brake	No
Time efficiency, %	98.5	95–96	96.8	98	98.2	91	96
Electrolyte							
Cu, g/L	45–50	46–52	45–49	51.4	45~47	45	40–50
Free H_2SO_4, g/L	160–180	165–185	170–200	182	170~175	170	170–200
As, g/L	0.2–0.3	8–10		7	6.1	12.8	9 -10
Sb, g/L	0.4–0.5	0.50–0.6		0.3	0.28	0.06	0.3–0.55
Bi, g/L	0.01	0.01–0.03		0.3	0.36	0.04	0.05–0.08
Ni, g/L	18–22	2.5–3.3		20.7	14.7	4.3	13–16
Fe, g/L	0.1–0.2	0.6–1.0			0.42	0.4	1.1
Cl, g/L	0.04	0.04–0.05		68	0.04	0.04	0.02–0.06
Cell inlet temperature, °C	65		63	67	65	65	65.5
Cell outlet temperature, °C	65	64–66	63	66	63	62	63–65.5
Circulation, L/min/cell	25–75	62–64		35	25	20	20–26.5
Distribution system	Conventional, Mettop-BRX	18–20	No	Manifold	Manifold (bottom-in, top-out)		Head tank manifold
Filtered	Yes	10%–20%	No	Ultrafilter	Scheibler	Yes, 15% of total volume	Slimes return only
Circulating pumps	Centrifugal	Centrifugal	Horizontal centrifugal	Vertical	Vertical	Horizontal centrifugal	Horizontal centrifugal
Purification method	Liberator cells, nickel sulfate plant	Bleed to hydrometallurgical process	Cu liberation, evaporation, electrowinning; Ni, Fe, As removal	Liberators & $NiSO_4$ by concentration	Activated carbon, decopper	Cu liberation, bleed to hydrometallurgy plant for impurity control	Liberator cells, acid purification unit, nickel carbonate, sodium sulfate
Addition Agents							
Glue, g/t of cathode	45	48		120	85	73	58
Thiourea, g/t of cathode	20	62		120	111	115	18
Avitone A, g/t of cathode	10	33		4		0	15
HCl, NaCl, g/t of cathode	HCl to 0.04 g/L	55 (NaCl)		170 (HCl)	145 (HCl)	73 as NaCl	to 0.035 gpL
Automatic sensing of glue or thiourea?	No	No	Collamat (glue)	Yes	No	Collamat	No
Cathode Description	Stainless steel	Cu starting sheet	Stainless steel	Cu starting sheet	Stainless steel	Stainless steel	Cu starting sheet
Copper Starting Sheets							
L × W × T, mm		960 × 965 × 1		1,000 × 1,000 × 1.3			953 × 953 × 0.75
Mass, kg		5.2		14			6.8
Starting blanks		Titanium		Stainless steel			Ti
Cathode pressing		No		No			Yes
After no. of hours							72–76
Permanent Stainless Steel	ISA process	—	ISA process		Kidd process	Kidd process	
Stainless steel		—			316L	316L	
L × W × T, mm	1,132 × 1,046 × 3.25	—	965 × 975 × 3.25		1,070 × 932 × 3.25	990 × 990 × 3	
Mass, kg	39	—	37		40	37	
Hanger bar	ISA, Outotec	—	Hollow rectangular				

(continues)

Table 1 Operating data from seven copper electrorefineries (continued)

	Refinery						
	Brixlegg, Austria	**Las Ventanas, Chile**	**Hamburg, Germany**	**Naoshima, Japan**	**Onsan Refinery 1, Korea**	**Magna, UT, USA**	**El Paso, TX, USA**
Hanger bar material	Stainless steel + copper	—	Stainless steel, copper	Copper with steel inside	Cu-solid bar with stainless-steel sheath	Cu	
Hanger bar, L × W × T	1,342 × 30 × 43, 1,342 × 25 × 50	—	43 × 30 × 1,311	1,330 × 30 × 30	1,170 × 38 × 25	36.5 × 25.4 × 25.4	
Blade to hanger bar joint	Weld	—	Weld	Welded	Weld	Weld	
Blade side edge strip	ABS, PE*	—	ABS	Mitsubishi Snapjaws	MPPO†	PP‡	
Cathode Analysis, ppm	Typical						
Ag	7	08–09	10–17	12	7.6	10	6–15
S	6	3.0–4.0	<5	3	8.8	<4	3
Se	0.2	0.5–0.6	<0.5	<1	0.13	0.3	<1.0
Te	0.1	<0.5	<0.5	<1	0.11	0.1	<0.1–1.4
As	0.8	0.5–0.6	<1	<1	0.39	0.5	<0.1–1.9
Sb	0.5	0.5–0.7	<1	<1	0.1	0.2	<1.0
Bi	0.6	0.1–0.2	<0.5	<1	0.1	0.3	0.1–0.9
Pb	1.2	0.5–0.6	<1	<1	0.11	1.5	<0.5
Fe	0.1	0.5–1.0		<1	0.36	0.6	1–2
Ni	0.6	0.5–0.6	1	<1	0.1	<1	<1.0
Cathode Operations							
Plating time, days	7/4-5-5	8	6–8	11 or 12	7	10	13.5
Final Cu mass, kg	60–65/42–45, 53–56	120	40–76 per plate	170	120	65 per side	175
Productivity, tankhouse labor hours per ton of cathode			0.36	0.35	1.2	0.42	0.7
Washing method	Hot water	Tanks	Spray	Hot water spray	Hot water (75°C) spray	Aisco washing machine	Spray chamber
Stripping machine	Outokumpu Wenmec	No	Outokumpu Wenmec	Mitsubishi (starting sheet)	Yes	Aisco	
Machine features							
Mother blanks/hour	~600	—	500	650	432	500 maximum	
Washing	Yes	—	Yes	Yes	Yes	Yes	
Stripping	Yes	—	Yes	Yes	Yes	Yes	
Stacking	Yes	—	Yes	Yes	Yes	Yes	
Sampling	No	—	No	No	Yes	Yes	
Weighing	Yes	—	Yes	No	Yes	Yes	
Strapping	Yes	—	No	No	Yes	Yes	
Other		—			Vision system		

Source: Moats et al. 2013
* ABS = acrylonitrile butadiene styrene; PE = polyethylene.
† MPPO = modified polyphenylene oxide.
‡ PP = polypropylene.

Table 2 Average copper anode composition

Element	Concentration
Cu	99.3 %
Ag	750 ppm
Au	30 ppm
S	30 ppm
Se	400 ppm
Te	110 ppm
As	890 ppm
Sb	350 ppm
Bi	180 ppm
Pb	1,170 ppm
Fe	60 ppm
Ni	2,040 ppm
O	1,460 ppm

Source: Moats et al. 2013

Lead

The electrorefining of lead by the Betts' process is one of the first large-scale usages of electrometallurgy with the first refinery installed in 1902 at Trail, British Columbia, Canada. It involves refining 96%–99% Pb anodes produced by pyrometallurgical methods to produce >99.99% lead. The main advantages of the electrolytic route are (1) removal of bismuth, (2) production of high-purity lead, and (3) production with minimal lead dusts and fumes. The main disadvantages are (1) impurity removal/disposal from the residues and solution and (2) higher costs than pyrometallurgical refining (Thornton et al. 2001). A thorough description of the process is given by Gonzalez-Dominguez et al. (1991).

The Betts' process involves the electrolytic dissolution of lead from cast 200–300-kg anodes into the electrolyte. The electrolyte is typically based on fluorosilicic acid (H_2SiF_6) although a few plants use fluoroboric (HBF_4) or sulfamic (H_3NSO_3) acid. Both fluorosilicic and fluoroboric electrolyte plants require adequate ventilation because of the potential generation of toxic hydrogen fluoride fumes above the electrolyte. Sulfamic electrolyte can decompose into ammonia and sulfate, which can cause environmental issues. For these reasons, the use of a less toxic electrolyte, such as methanesulfonic acid, has been investigated (Jin and Dreisinger 2016).

Pb^{+2} is transported across the cell through the electrolyte. Pb plates onto Pb starter sheets produced by a highly automated process of rolled and embossed refined lead (15 ppm of Sb is added to improve stiffness). Cathode morphology is controlled by the proper levels of additives, temperature, and current density.

Cast anodes are made of lead bullion. Anode microstructure is critical. A *honeycomb* structure with uniform size grains surrounded by impurities is desired. As Pb electrolytically dissolves, a skeleton of slimes resembling a honeycomb remains. Nonuniformity of the lead grains leads to nonadherences of the slimes and uneven dissolution, which causes production difficulties. Anode microstructure is strongly influenced by As, Sb, Bi, and Ag. Slimes adhesion increases when Bi is greater than 0.23%. Slimes adhesion increases when eutectic-forming elements (As and Ag) are present. Water quenching significantly improves slimes adhesion. Control methods are used for cooling rates and casting techniques to produce homogeneous properties (Gonzalez-Dominguez et al. 1991).

The use of fluorosilicic electrolyte is most widely practiced. Electrolyte composition is 0.2–0.5 M $PbSiF_6$ (lead fluorosilicate) and 0.5–0.8 M H_2SiF_6 with milligrams-per-liter concentrations of Bi, Cu, As, Sb, and Sn. The electrolyte temperature is usually 40°C. Although noble elements are more electrochemically noble, they can still electrochemically dissolve to a small extent. Au, Ag, Cu, Sb, Bi, and As report to the anode slimes if the anode overpotential is controlled. The slimes are treated to recover high-value elements and dispose of low-value elements. Similar to copper electrorefining, lead anodes contain oxide inclusions. Because the oxide inclusions dissolve chemically, the apparent anode current efficiency, which is based on the electrolytic and chemical dissolution rates, is greater than 100%. This results in an increase of dissolved lead concentration in the electrolyte with time. Lead concentration is controlled by electrowinning or precipitation. The allowed concentrations of impurities depend on product specifications. Plants operate under conditions to avoid dissolution of noble impurities. Impurities are removed from the electrolyte by cementation on lead granules in a purification column or electrodeposition in a separate electrolysis circuit.

Lead electrorefining is performed with current densities of 120–230 A/m^2, which produce cell voltages of 0.3–0.6 V. Current efficiency is 90%–98%. The electrical energy consumption is 120–195 kW·h/t. Lead is deposited for 4–7 days in cells with 16–43 cathodes.

A flat smooth lead deposit is achieved by adding reagents to the electrolyte. These additives increase the cathodic overpotential, which is necessary to avoid rough and porous deposits. Glue, lignin sulfonate, and aloes are used. Glue is added at rates of ~600 g/t. Lignin sulfonate is introduced at 200–1,000 g/t. Aloes are added at the Trail operation at 170 g/t. The Trail controls their operation by measuring overpotential (Kerby and Jankola 1990). Betts' lead electrorefining operational data is reported by Gonzalez-Dominguez et al. (1991).

Nickel

High-purity nickel is produced by electrorefining at a few locations. Electrorefining is an older technology and has been mostly supplanted by hydrogen reduction or electrowinning as the preferred choices to produce commercial pure nickel. Electrorefining of nickel starts with either matte or impure nickel anodes. Impure metal anodes are used at Norilsk, Russia; Pechenga, Russia; and Jilin, China (Crundwell et al. 2011). An overview of the processes is given by Vignes (2013). Historic details of both types of electrorefining are provided by Boldt and Queneau (1967).

Typically, matte anodes are ~75% Ni, ~20% S with the rest as impurities. Anodes are loaded into woven polypropylene bags and placed in a cell. Nickel starter sheets are placed in a compartment between two adjoining anodes. The compartment is also covered with a woven cloth. A current density of ~250 A/m^2 is applied to electrolytically corrode nickel from the anodes and plate pure nickel metal on the starting sheet cathodes. After approximately 10 days, the enlarged cathodes are pulled and replaced with new starting sheets. Anodes are replaced after 16–17 days. Anode scrap and corrosion products are processed to recover high-value elements, such as Ni, Co, Cu, Ag, Au, and platinum group metals (PGMs). The solution inside the anode bag (i.e., anolyte) flows out of the cell and is purified and replenished in an electrolyte purification plant before being returned to the cathode bags as pure catholyte. The catholyte is returned to the cathode compartments in such a way that flow is always away from the cathodes into the anolyte, preventing catholyte contamination. Physically, this is indicated by a *head* of catholyte in the cathode compartments above the anolyte. The reader is referred to Boldt and Queneau's (1967) seminal work, *The Winning of Nickel*, for more details regarding the electrorefining of matte anodes.

Crude nickel anodes are made by casting after pyrometallurgical reduction of nickel oxide. Electrorefining occurs similarly as with nickel matte anodes. Typically, cobalt, copper, iron, lead, zinc, and arsenic chemically and/or electrochemically dissolve and report to the anolyte. These are removed during anolyte treatment prior to the return of the solution to the catholyte compartment. Silver, gold, and PGMs remain as solids and collect in the slimes, which are processed to recover these high-value elements.

Silver

The purpose of silver electrorefining is to (1) produce >99.9% silver from an impure silver material and (2) collect all of the gold and PGMs in the anode slimes for further processing and eventual recovery. Pyrometallurgical methods typically produce doré with 98%–99.5% silver. The doré is cast into small anodes, placed into anode bags, and then electrolyzed in either a Moebius cell or Balbach-Thum cell. The silver dissolves into a silver nitrate–sodium nitrate–nitric acid electrolyte. The electrolyte will contain ~150 g/L Ag with a pH of 1.0–1.5 at a temperature of ~35°C. The bags facilitate the collection of the high-value slimes. Copper, lead, and palladium will accumulate in the electrolyte. The electrolyte is treated once the concentrations of these elements reach 10 g/L Cu, 10 g/L Pb, or 1 g/L Pd. The electrolyte is processed to recover the silver and then disposed (Habashi 1998).

Because of its high exchange current density, electrodeposition of silver produces needles. These needles are periodically broken either manually or automatically to avoid short-circuiting the cell. The cells are designed to allow the breaking of the needles. Moebius cells are more commonly used now because they utilize less floor space and are easier to mechanize. Schematic diagrams for Moebius and Balbach-Thum cells can be found elsewhere (Habashi 1998; Pletcher and Walsh 1990). After removing the needles from the cell, they are washed and melted to produce silver bullion bars.

Gold

Most gold bullion is pyrometallurgically refined using the Miller chlorination process to produce 99.6% purity. For high-purity gold (99.99%) needed for coinage, electrolytic refining is required. Electrorefining of gold using the Wohlwill process removes trace amounts of silver, copper, zinc, and PGMs. Partially refined gold bullion anodes are electrolytically corroded into 60°C, 80–100 g/L Au, 80–100 g/L HCl electrolyte by the application of a current density of ~800 A/m^2. Impurities accumulate in the electrolyte or as anode sludge. The electrolyte is bled and treated to recover the gold. The sludge is recycled back into pyrometallurgical refining. Gold deposits at the cathode. The cathodes are washed with a hot sodium thiosulfate solution to remove entrained impurities (Marsden and House 2006).

REFERENCES

Boldt Jr., J.R., and Queneau, P. 1967. *The Winning of Nickel*. Princeton, NJ: Van Nostrand.

Crundwell, F., Moats, M., Ramachandran, V., Robinson, T., and Davenport, W.G. 2011. *Extractive Metallurgy of Nickel, Cobalt and Platinum-Group Metals*. Oxford: Elsevier.

Gonzalez-Dominguez, J.A., Peters, E., and Dreisinger, D.B. 1991. The refining of lead by the Betts process. *J. Appl. Electrochem*. 21(3):189–202.

Habashi, F. 1998. *Principles of Extractive Metallurgy*. Vol. 4, Amalgam and Electrometallurgy. Sainte-Foy, QC: Metallurgie Extractive Quebec.

Jin, B., and Dreisinger, D.B. 2016. A green electrorefining process for production of pure lead from methanesulfonic acid medium. *Sep. Purif. Technol*. 170:199–207.

Kerby, R.C., and Jankola, W.A. 1990. Monitoring of organic additives in electrolyte at Cominco's lead/zinc operations. In *Proceedings of the International Symposium on Electrometallurigical Plant Practice*. Edited by P.L. Claessens and G.B. Harris. Montreal, QC: Canadian Institute of Mining, Metallurgy and Petroleum.

Krusmark, T., Young, S., and Faro, J. 1995. Impact of anode chemistry on high current density operation at Magma copper's electrolytic refinery. In *Copper 1995—Cobre 95*. Vol. 3, Electrorefining and Hydrometallurgy of Copper. Edited by W.C. Cooper, D.B. Dreisinger, J.E. Dutrizac, H. Hein, and G. Ugarte. Montreal, QC: Canadian Institute of Mining, Metallurgy and Petroleum.

Marsden, J.O., and House, C.I. 2006. *The Chemistry of Gold Extraction*. Littleton, CO: SME.

Moats, M., and Hiskey, J.B. 2010. How anodes passivate in copper electrorefining. In *Copper 2010 Proceedings*. Vol. 4, Electrowinning and -refining. Claustahl-Zellerfeld, Germany: GDMB Society for Mining, Metallurgy, Resource and Environmental Technology.

Moats, M., Robinson, T., Wang, S., Filzwieser, A., Siegmund, A., and Davenport, W. 2013. Global survey of copper electrorefining operations and practices. In *Copper 2013*. Vol. 5, Electrowinning/Electrorefining. Edited by R. Abel and C. Delgado. Santiago, Chile: Instituto de Ingenieros de Minas de Chile.

Noguchi, F., Itoh, H., and Nakamura, T. 1995. Effect of impurities on the quality of electrorefined cathode copper, behavior of antimony in the anode. In *Copper 1995—Cobre 95*. Vol. 3, Electrorefining and Hydrometallurgy of Copper. Edited by W.C. Cooper, D.B. Dreisinger, J.E. Dutrizac, H. Hein, and G. Ugarte. Montreal, QC: Canadian Institute of Mining, Metallurgy and Petroleum.

Pletcher, D., and Walsh, F.C. 1990. *Industrial Electrochemistry*. New York: Chapman and Hall.

Thornton, I., Rautiu, R., and Brush, S. 2001. *Lead—The Facts*. London: IC Consultants.

Vignes, A. 2013. *Extractive Metallurgy 2: Metallurgical Reaction Processes*. Hoboken, NJ: Wiley.

Wesstrom, B. 2014. The effects of high antimony in electrolyte. In *Proceedings of Bill Davenport Honorary Symposium of [Conference of Metallurgists] COM 2014*. Westmount, QC: Metallurgy and Materials Society of the Canadian Institute of Mining, Metallurgy and Petroleum.

CHAPTER 10.18

Gaseous Reduction in Aqueous Solutions

Jari Aromaa, Mari Lundström, and Olof Forsén

Gaseous reduction is one means of recovering a metal from aqueous solution. The method is also sometimes called *hydrothermal* or *solvothermal* reduction. A reducing gas such as hydrogen, sulfur dioxide, or carbon monoxide is used to reduce dissolved metal ions. The first application was used for nickel production. Leaching and recovery processes were developed by Chemical Construction Corporation and Sherritt Gordon Mines in the 1950s (Schaufelberger 1956; Mackiw et al. 1957).

Nickel and other metals are industrially produced by the reduction of ammoniacal solution by hydrogen under pressure at elevated temperatures in mechanically agitated autoclaves. The processes are commercially confidential and site specific. Gaseous reduction of metals from salt solutions is an alternative to electrowinning that uses electric energy for metal deposition. Gaseous reduction is also useful in production of specialty products such as metal powders.

The most important applications are reduction of nickel and cobalt by hydrogen (Crundwell et al. 2011). Sulfur dioxide has found limited use. Carbon monoxide has been studied in recovery of silver, copper, nickel, and cobalt, but no processes have been developed (Burkin 2001). Hydrogen is relatively inexpensive to produce on a large scale. The reaction products in an aqueous solution are either hydrogen ions or hydroxyl ions. Neither creates by-products that would cause operating or disposal problems. Sulfur dioxide is toxic and causes sulfuric acid formation in the solution and production contamination. Carbon monoxide has also shortcomings, as it is toxic, explosive, difficult to synthesize, and expensive, and it changes the acid–base balance. Sulfur dioxide and carbon monoxide can be used for precipitation of metal powders from alkaline and ammonia (NH_3) solutions (Naboychenko 2009).

THERMODYNAMICS OF REDUCTION

The gaseous reduction is based on the electrochemical potential difference between dissolved metal and the reducing gas. The standard electrode potentials (E°) show thermodynamic limits for gaseous reduction. The gaseous reduction is possible only when the equilibrium potential of the metal to be reduced is higher than that of the reducing gas that in turn becomes oxidized. Some relevant standard electrode potentials are shown in Table 1. By comparing these standard electrode potentials, it is seen that theoretically, only copper can be reduced by carbon monoxide or hydrogen; and copper, copper–ammonia complexes, ferric iron, lead, and tin by sulfur dioxide.

The operating conditions are never in the standard state described by the standard electrode potentials shown in Table 1. The equilibrium potential of the metal at any temperature and reacting species concentration is given by the Nernst equation (Equation 1). For a simple metal ion reduction reaction $Me^{z+} + ze^- = Me$, the equilibrium potential is given by the following equation:

$$E = E^\circ - RT/zF \cdot \ln(1/[Me^{z+}]) \qquad \textbf{(EQ 1)}$$

where E is the equilibrium potential (V), E° is standard electrode potential (V), R is general gas constant (8.3143 J mol^{-1} K^{-1}), T is absolute temperature (K), z is the number of electrons in the reaction, F is the Faraday constant (96,485 C mol^{-1}), and $[Me^{z+}]$ is the concentration of metal ions in the solution (mol dm^{-3}).

The equilibrium potential of the metal should be as high as possible to provide stronger driving force for the reaction. The equilibrium potential calculated by Equation 1 increases with increasing metal ion concentration. When calculating the equilibrium potential, it must be noted that the activity coefficient of metal in solution usually decreases with increasing concentration. The activity coefficient changes in the typical concentration, ranging from 0.1 to 1 M, are not large (Meddings and Mackiw 1964; Burkin 2001), but the decrease in activity coefficient loses some of the beneficial equilibrium potential increase. As the activity coefficients are less than 0.1 (Meddings and Mackiw 1964; Burkin 2001), the metal equilibrium potentials usually decrease slightly with increasing temperature.

The equilibrium potential of the reducing gas is also calculated from the Nernst equation. To maintain a high driving

Jari Aromaa, Senior Lecturer, Aalto University, Espoo, Finland
Mari Lundström, Professor, Aalto University, Espoo, Finland
Olof Forsén, Professor Emeritus, Aalto University, Espoo, Finland

Table 1 Standard electrode potentials of reactions of interest in metal gaseous reduction

Reaction	Standard Electrode Potential, E° (V)	Reaction Type
$Ag^+ + e^- \leftrightarrow Ag$	0.799	Metal ion reduction
$[Ag(NH_3)_2]^+ + e^- \leftrightarrow Ag + 2NH_3$	0.375	Metal complex reduction
$Cu^{2+} + 2e^- \leftrightarrow Cu$	0.337	Metal ion reduction
$CO + H_2O \leftrightarrow CO_2 + 2H^+ + 2e^-$	0.104	Reducing gas oxidation
$H_2 \leftrightarrow 2H^+ + 2e^-$	0.000	Reducing gas oxidation
$Fe^{3+} + 3e^- \leftrightarrow Fe$	–0.037	Metal ion reduction
$[Cu(NH_3)_4]^{2+} + 2e^- \leftrightarrow Cu + 4NH_3$	–0.023	Metal complex reduction
$[Cu(NH_3)_2]^+ + e^- \leftrightarrow Cu + 2NH_3$	–0.124	Metal complex reduction
$Pb^{2+} + 2e^- \leftrightarrow Pb$	–0.126	Metal ion reduction
$Sn^{2+} + 2e^- \leftrightarrow Sn$	–0.136	Metal ion reduction
$SO_2 + 2H_2O \leftrightarrow SO_4^{2-} + 4H^+ + 2e^-$	–0.157	Reducing gas oxidation
$Ni^{2+} + 2e^- \leftrightarrow Ni$	–0.250	Metal ion reduction
$Co^{2+} + 2e^- \leftrightarrow Co$	–0.277	Metal ion reduction
$Cd^{2+} + 2e^- \leftrightarrow Cd$	–0.403	Metal ion reduction
$[Ni(NH_3)_6]^{2+} + 2e^- \leftrightarrow Ni + 6NH_3$	–0.497	Metal complex reduction

Source: Outotec 2015

force, the equilibrium potential of the gas oxidation reaction should be as low as possible. Some of the theoretical oxidation reactions of reducing gases are given by the following equations:

$$H_2 = 2H^+ + 2e^- \quad \textbf{(EQ 2)}$$

$$SO_2 + H_2O = SO_4^{2-} + 4H^+ + 2e^- \quad \textbf{(EQ 3)}$$

$$CO + H_2O = CO_2 + 2H^+ + 2e^- \quad \textbf{(EQ 4)}$$

The Nernst equation to calculate equilibrium potential for hydrogen reaction (Equation 2) is shown in the following equation:

$$E = E° - (RT/2F)\cdot\ln\{p(H_2)/a(H^+)^2\} \quad \textbf{(EQ 5)}$$

where $p(H_2)$ is the partial pressure of H_2 in atmospheres and $a(H^+)$ is the activity of a hydrogen ion. The equilibrium potentials of sulfur dioxide oxidation (Equation 3) and carbon monoxide oxidation (Equation 4) are given in Equations 6 and 7, respectively.

$$E = E° - (RT/2F)\cdot\ln\{p(SO_2)/(a(SO_4^{2-})\cdot a(H^+)^4)\} \quad \textbf{(EQ 6)}$$

$$E = E° - (RT/2F)\cdot\ln\{p(CO)/(p(CO_2)\cdot a(H^+)^2)\} \quad \textbf{(EQ 7)}$$

The equilibrium potentials of the gas oxidation reactions (shown in Equations 2 through 4) decrease with increasing pH and increasing gas pressure. The equilibrium potential of the sulfur dioxide oxidation reaction (Equation 3) decreases with decreasing sulfate ion concentration and that of the carbon monoxide oxidation reaction (Equation 4) with decreasing carbon dioxide partial pressure.

During reduction, the metal concentration decreases from about 1 M solution to 0.1 M (Burkin 2001). This means that the thermodynamic activity of the metal deposition reaction will decrease. At the same time, the reducing gas pressure will decrease as it is consumed, and also the pH of the solution declines as hydrogen ions are produced. These changes will decrease the driving force. Figure 1 shows equilibrium potential values for nickel, copper, and hydrogen at T = 100°C. The metal equilibrium potentials have been calculated using Equation 1 and the activity coefficients from Meddings and Mackiw (1964) and Burkin (2001), and the hydrogen potentials using Equation 5.

The thermodynamic conditions necessary for metal reduction can also be estimated by using potential–pH or E–pH diagrams. These diagrams show the stability areas of selected compounds. Figure 2 shows a potential–pH diagram for copper and nickel. The reduction conditions must be selected so that the metal is the stable compound. In Figure 2, the lower dashed line is the hydrogen reaction equilibrium. As shown in Figures 1 and 2, the potential of hydrogen reaction must be lower than the reaction between solid and dissolved metal so that the metal reduction is thermodynamically possible. Reduction of copper is possible at the whole pH range shown, but reduction of nickel is possible only when pH is higher than approximately 5.5.

During the reduction process, the pH of the solution decreases because hydrogen ions are formed. The driving

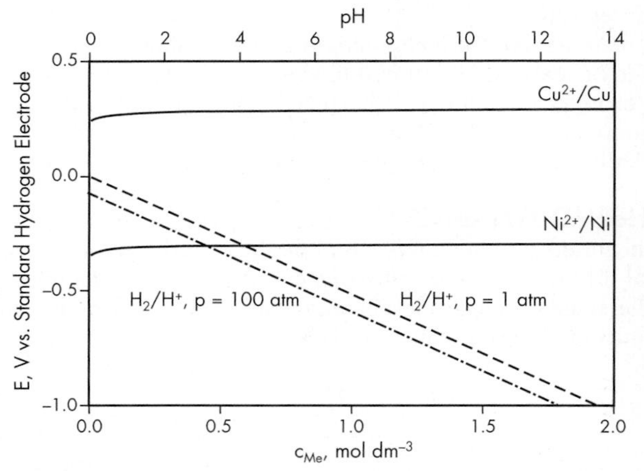

Figure 1 Equilibrium potential values of nickel and copper reduction and hydrogen oxidation

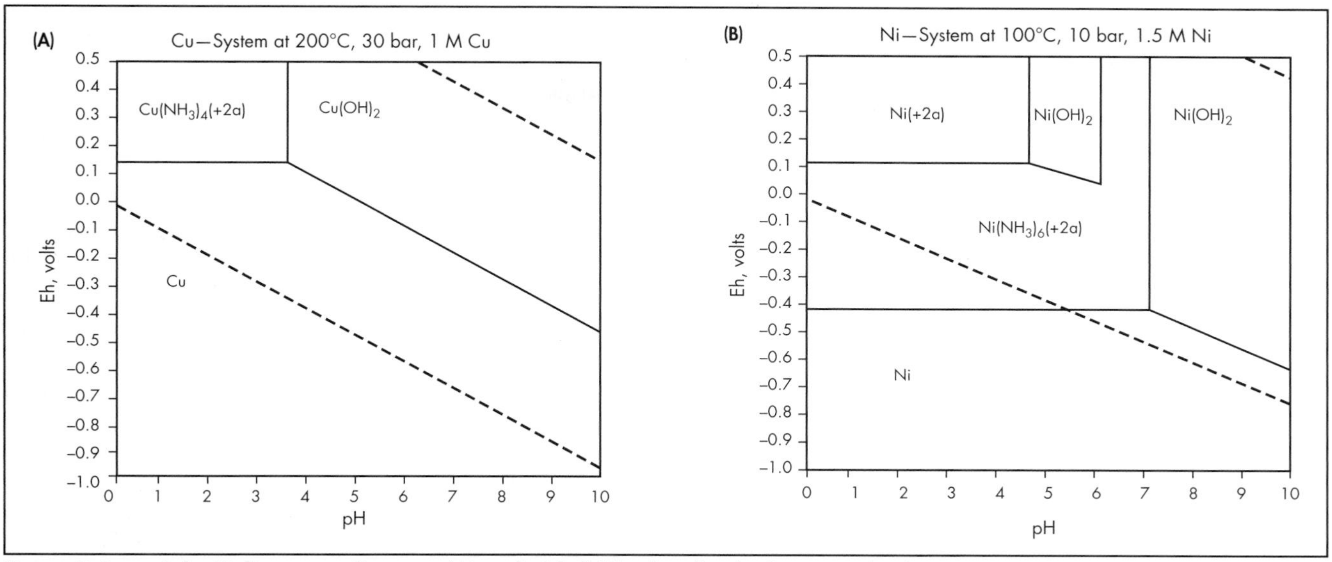

Figure 2 Potential–pH diagrams of copper (A) and nickel (B) related to hydrogen reduction

force of metal reduction decreases and finally metal deposition ceases. One approach to maintain the thermodynamic conditions is to neutralize the acid. This has been done by using anhydrous ammonia (Burkin 2001), but hydroxides and carbonates can also be used. The solubility of many metal salts first increases but then decreases with increasing temperature. Above a certain pH value, the hydrolysis of the metal becomes significant, causing formation of metal hydroxides that will not reduce to metals. To keep metals in solution, it is necessary to complex the cations. The equilibrium potential equations and information shown in Figure 1 must be modified to include the effect of the complexing ligands. In a complex system, the equilibrium (Equation 8) prevails, and its equilibrium constant β is given by Equation 9. In the following equations, n is the maximum coordination number of the metal ion with ligand L (Burkin 2001).

$$Me^{z+} + nL = MeL_n^{z+} \quad \textbf{(EQ 8)}$$

$$\beta = a(MeL_n^{z+})/\{a(Me^{z+})\cdot a(L)^n\} \quad \textbf{(EQ 9)}$$

Using (Nernst) Equation 1 and equilibrium constants of the complexes, it is possible to calculate the equilibrium potential of the metal–metal complex as follows:

$$E = E - RT/zF\cdot\Sigma(\beta_i\cdot a_i) \quad \textbf{(EQ 10)}$$

For many metals the most common method to complex metal cations and raise pH is to use ammonia to produce metal amines $[Me(NH_3)_x]^{z+}$. As the ammonia concentration increases, the ratio of higher-to-lower complexes increases. The addition of ammonia will decrease the equilibrium potentials of metal–metal complexes and the hydrogen reaction but not in the same ratio. Thus, there will be a metal-to-ammonia ratio where the driving force of metal reduction has its maximum value. For nickel reduction, this will be in the ammonia-to-nickel ratio of 2.0–2.5 (Meddings and Mackiw 1964; Burkin 2001); for cobalt reduction, the relevant ratio is 2.5–3.0 (Burkin 2001); and for copper reduction, 2.2–2.5 (Naboychenko et al. 2009b).

KINETICS OF REDUCTION

The thermodynamic conditions can be estimated by using equilibrium potentials, taking into account the concentrations of reacting species, temperature, complex formation and changes in the metal concentration, reducing gas pressure and fugacity, and pH decrease. The reaction steps of metal reduction by reducing gases are nucleation, particle growth, and agglomeration. At first, very small stable particles of metal nucleate; this stage may need seeding. Clusters containing a few atoms are thermodynamically unstable and become stable after growing to particle size 1–5 nm (Naboychenko 2009). The nuclei grow by deposition of metal on particle surfaces. Particles can agglomerate and be permanently covered by overgrowing metal deposit. In general, the precipitation of metal by a reducing gas includes the following four stages (Burkin 2001; Naboychenko 2009):

1. Gas dissolution and saturation of the solution by the gas
2. Adsorption of the gas on the nuclei or seed surface and gas activation with the formation of an intermediate compound
3. Electrochemical metal ion reduction
4. Reduced metal remaining in solution and providing new adsorption site for reducing gas

The solubility of gas increases with increasing gas pressure and decreases with increasing concentration of dissolved species. The absorption of the reducing gas into solution takes place in two sequential steps: Dissolution of gas into the solution at the gas–liquid interface followed by diffusion of dissolved gas through the solution boundary layer accompanied by a chemical reaction. The gas adsorption rate increases with increasing stirring speed. The gas adsorption flux increases up to a certain stirring speed after which it remains constant. The main purpose of liquid stirring is to supply a constant flow of the gaseous reactant to metal surfaces.

The adsorption of gas and its activation result in active intermediate compounds. Molecular hydrogen is chemically inert and for metal reduction reactions, activation of the hydrogen molecule is necessary. Activation of the hydrogen results in intermediate metal hydrides. The activation can follow a homolytic activation path, producing two hydrogen atoms, or a heterolytic path, producing a proton H^+ and a hydride H^- (Osseo-Asare 2003; Naboychenko 2009). The homolytic path requires two metal cations, as shown in Equation 11; but the heterolytic path requires only a single metal cation, as shown in the next equations (Halpern 1959; Osseo-Asare 2003):

$$2Ag^+ + H_2 \rightarrow 2AgH^+ \qquad \textbf{(EQ 11)}$$

$$Ag^+ + H_2 \rightarrow AgH^\circ + H^+ \qquad \textbf{(EQ 12)}$$

$$Cu^{2+} + H_2 \rightarrow CuH^+ + H^+ \qquad \textbf{(EQ 13)}$$

Some of the metal hydride intermediates are unstable and they will react in reverse reactions to regenerate hydrogen gas or in reactions involving reduction of the metal ion or another aqueous species. For example, copper hydride (CuH^+) can react with cupric ions to produce cuprous ions after Equation 14, and the cuprous ions can then react to metallic copper and cupric ions in disproportionation reaction (Equation 15) (Von Hahn and Peters 1965). The reaction takes place at sufficiently high cuprous ion concentrations.

$$CuH^+ + Cu^{2+} \rightarrow 2Cu^+ + H^{+a} \qquad \textbf{(EQ 14)}$$

$$2Cu^+ \rightarrow Cu + Cu^{2+} \qquad \textbf{(EQ 15)}$$

As the cupric ions can react directly with molecular hydrogen to produce metallic copper (Equations 13–15), the copper reduction process is homogeneous. Deposition of Ni^{2+} or Co^{2+} ions, however, requires the presence of a solid catalyst, and their reduction process is heterogeneous (Osseo-Asare 2003). Deposition by heterogeneous mechanism is electrochemical with separate anodic and cathodic reactions. The electrochemical reduction of metal ions in solution follows the principles of a corrosion cell (Figure 3). Oxidation of hydrogen gas releases electrons, which are used in the reduction of dissolved metal ions to solid metal. As the reactions in gaseous reduction of metals are electrochemical, the kinetic factors can be studied using Evans diagrams. The Evans diagram examples in Figure 4 show that to enhance the kinetics of the reactions, the electrochemical potential difference between the anodic gas oxidation and cathodic metal reduction should high, the exchange current densities of the reactions should be high, the polarization of the reactions should be low, and the concentration of the reacting species should be high to avoid mass transfer limitations.

Solutions for gaseous reduction are usually alkaline with complexing and buffering agents to maintain high metal concentration and to prevent increase in H^+ concentration that would decrease the driving force for reduction (Figure 1). The deposition processes proceed on solid material surfaces. To produce initial particles, catalysts are often used. The catalysts are chemical nucleating agents or metallic seed particles. Chemical nucleating agents act as homogeneous catalysts to nucleate metal particles, for example, ferrous sulfate in nickel reduction. Addition of nucleating agent results in formation

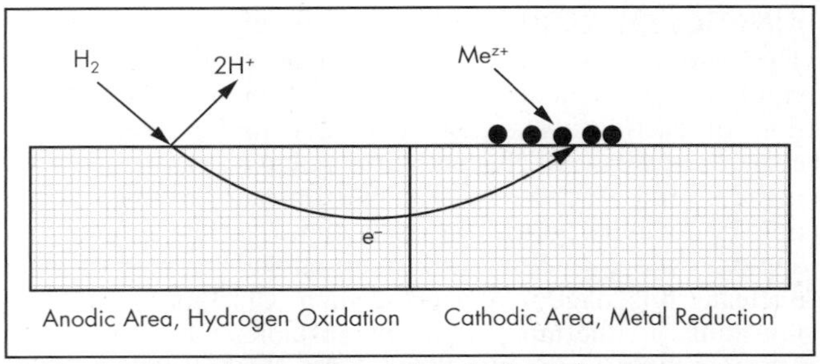

Figure 3 Spontaneous electrochemical cell for heterogeneous metal deposition

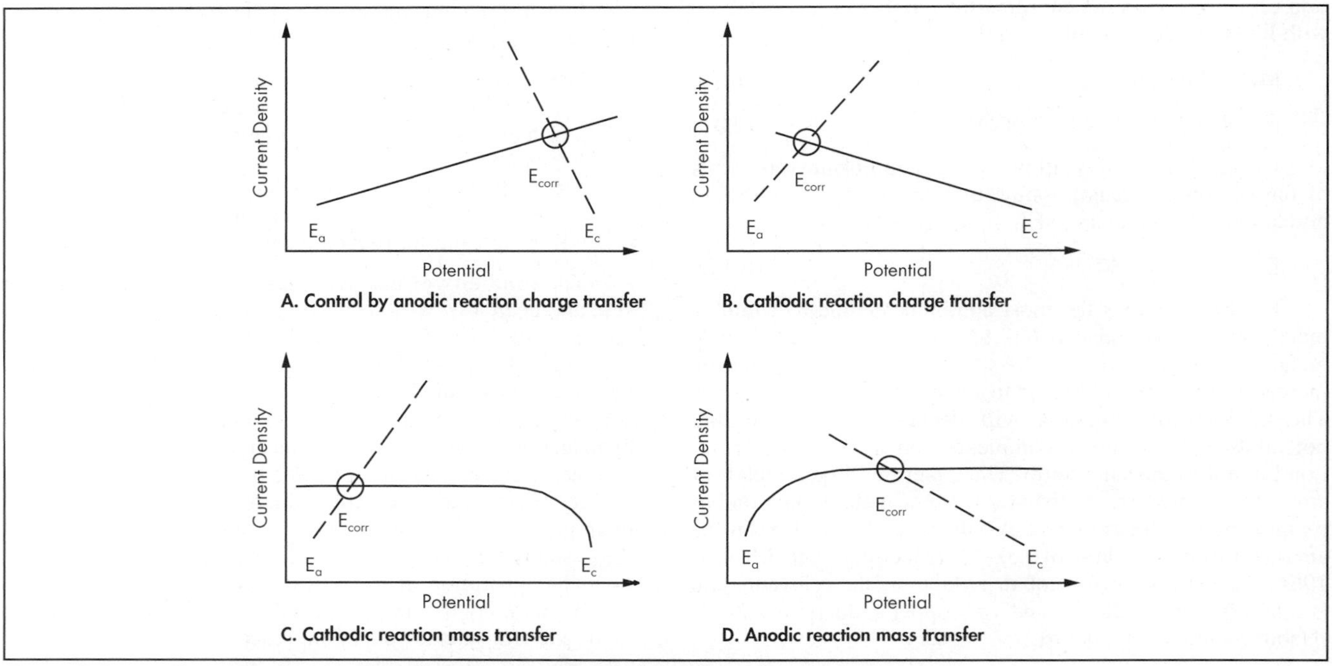

Figure 4 Evans diagrams showing the effect of polarization of anodic and cathodic reactions

of finely divided compounds on which metal nucleates. These particles act as seeds for further reduction.

The reaction kinetics strongly vary depending on metal, complexing agent, base solution, and their respective concentrations. For some systems, such as copper in ammonia solution, the reaction rate does not depend on metal concentration. Rate laws for the hydrogen reduction of metal ions are typically of the form given in the following equation:

$$r = -d[Me]/dt = k \cdot [Me]^x \cdot P_{H2}{}^y \cdot S \cdot \exp(-E_a/RT) \quad \textbf{(EQ 16)}$$

where r is reaction rate; t is time; k is an apparent rate constant; [Me] and P_{H2} represent dissolved metal concentration and hydrogen pressure, respectively; and x can range from 0 to 1 and y from 0.5 to 2. The factor S describes the catalyst surface area, and the effect of temperature is taken into account by the activation energy of the reaction, E_a (Osseo-Asare 2003). Depending on the presence of surface-active agents and inhibiting agents, the whole surface area of the catalyst does not necessarily support reduction.

The particle growth includes nucleation and metal reduction on the nucleus followed by metal deposition as shells on the growing particles. Growth involves diffusion of metals and reducing gas to the surface, adsorption, reduction reactions, particle deposition and incorporation in metal lattice, and desorption of reaction products. In the beginning of the deposition process, aggregation of small particles can be important to produce larger particles that can then grow to the desired size (Taty Costodes et al. 2006). When a critical particle size has been reached, aggregation ceases and particles grow by continuing deposition of shells. Particles can also break to smaller ones during the process because of particle collisions, surface attrition, and solution turbulence. In practice, the deposition process is finished when particles become too large to remain in suspension and the reactive area of the slurry becomes too small to maintain a meaningful reduction rate.

PROCESS DEVELOPMENT

The basic chemistries of gaseous reduction systems are well known. The product suitability for various applications depends on particle size distribution and particle morphology. For process development and optimization, several aspects can be studied, including solution composition, the effect of additives on nucleation, reduction, agglomeration and powder surface properties, the changes during densification steps, and particle size distribution and particle morphology.

Research and scale-up of deposition processes is often difficult because of complex particle forming including nucleation, growth, aggregation, and breakage processes. Also design of stirring and mixing to maintain gas distribution and slurry suspension is challenging when scaling from laboratory-scale to process equipment. Laboratory autoclaves usually measure less than 1 dm^3, pilot plant autoclaves are tens of cubic decimeters, and large-scale autoclaves in nickel production measure more than 22 m^3 (Crundwell et al. 2011).

For studying gaseous reduction, an autoclave is needed. The autoclave must be capable of operating at 150°–250°C and at pressures of several megapascals. Construction materials are stainless steel or titanium. Electric heating or steam coils are used for temperature control. Depending on the system, high-pressure gas lines for reducing gas and nitrogen might be needed. Change of gas flow between reducing gas and nitrogen must be simple and secure. As the reducing gases H_2, sulfur dioxide (SO_2), and carbon monoxide (CO) are potentially hazardous, the exit gas lines and treatment of exit gases must be designed according to relevant safety rules. The autoclave must have good stirring to dissolve gas and to maintain metal particles in suspension. Removable polytetrafluoroethylene reaction vessels can be used to prevent metal deposition on autoclave walls.

A typical test procedure starts with preparation and preheating of the solution, charging and sealing the autoclave, and heating the autoclave with stirring and possibly with nitrogen blanketing. On reaching the test temperature, flow of reduction gas is started. To maintain correct operating pressure and ensure the flow of reducing gas, the gas outlet valve must be partially open when operating with constant gas flow, or reducing gas is added when pressure in the autoclave decreases indicating consumption of the gas. Autoclave pressure monitoring cannot be used when CO is the reducing gas, as it produces carbon dioxide (CO_2) and no pressure change is noticed. A system for sampling of solution or slurry through a dip pipe from a pressurized autoclave to a sample vessel can be constructed to get samples for kinetic analysis. However, if the autoclave volume is small, then sampling can change the conditions inside. After the test, the autoclave is cooled and flushed with nitrogen before opening the vessel. The metal slurry is then collected, washed, and dried.

Tests for gaseous reduction include solution preparation, nucleation, reduction, and control of reduced metal powder properties. Solution preparation includes dissolving desired initial metal and complex forming compounds, adjusting the complex-to-metal molar ratio, pH adjustment and buffering, and addition of nucleation agents or metal seed and surface-active agents to control deposit morphology. In addition to the solution characteristics, reducing gas pressure, temperature, and stirring can affect reactions. All these system parameters can affect either nucleation or reduction of metal:

- Increasing metal concentration will increase the equilibrium potential of metal reduction reactions and thus the thermodynamic driving force; it will also increase the reduction rate.
- The complex-to-metal ratio will affect the thermodynamic driving force that will decrease with both very low and very high values. Too low a complex-to-metal ratio will cause too rapid a pH decrease and too high a ratio will decrease the concentration of active metal. Both will eventually stop the reduction process.
- An increase in acid concentration will decrease the driving force of the reduction reaction and reaction rate. As the oxidation reactions of gases produce hydrogen ions, buffering compounds are needed.
- Nucleation agent or metal seed is needed to control the nucleation stage. Nucleation and first densification steps will affect particle size and shape and plating of metal on vessel walls.
- Surface-active agents control deposition rate and aggregation and affect the shape of particles.
- Increasing the reducing gas pressure will provide more thermodynamic driving force, allow operation at lower pH, and enhance reduction kinetics.
- Increasing temperature will increase reaction rate and increase gas solubility under autoclave conditions.

- Intense stirring is needed to maintain good gas–liquid transfer and to keep the solids suspended for maximum reduction rate.

Studies on nucleation and first solids precipitation are essential. Too little precipitate does not provide enough surface for further nucleation, and the precipitate will deposit on autoclave internals. Too much precipitate, however, will result in too strong aggregation and decrease available surface area for subsequent densification steps (Saarinen et al. 1998). Reproducing the nucleation step without seeding is difficult. Therefore, tests to determine deposit growth rate, including aggregation and breakage, have to be done with controlled seeding.

Reaction rate measurements are usually based on solution analyses. As the metal is precipitated, the solution concentration decreases with time. Pressure decrease has also been used to record progress of metal reduction with H_2. Analysis of particle formation rates can be made using particles collected after repeated tests (Ntuli and Lewis 2007) or using representative slurry samples (Ntuli and Lewis 2006). Reaction rate coefficients and reaction order can be calculated from the solution analysis –d[Me]/dt data. The same data can be used to describe the factors that affect reaction kinetics, such as solution composition, reducing gas pressure, catalyst, and temperature as shown in Equation 16. The fraction of precipitated metal can be described by the following general equation:

$$X = 1 - \exp(-kt^n) \qquad \textbf{(EQ 17)}$$

where X is conversion or the fraction of reduced metal at time t, and k and n are specific rate constants for nucleation and growth. Analysis of X versus t data can be undertaken using a double logarithmic plot (Equation 18) or surface reaction model akin to the shrinking core model shown in the next equation.

$$\ln(-\ln(1 - X)) = \ln(k) - n \cdot \ln(t) \qquad \textbf{(EQ 18)}$$

$$1 - (1 - X)^{1/3} = k \cdot t \qquad \textbf{(EQ 19)}$$

Plotting the $\ln(-\ln(1 - X))$ versus $\ln(t)$ or $1 - (1 - X)^{1/3}$ versus t gives constant values that can be used to estimate the effects of test parameters on reduction. The double logarithmic plot (Equation 18) can mask small changes in the reaction kinetics, and therefore it must be used cautiously.

The metal powder properties are determined by the relative rates and mechanisms of nucleation, growth, particle aggregation, and breakage. Open particles are more desirable than closed ones because they have larger surface area giving faster rates, they do not accumulate solution as impurity, and they are easier to turn to briquettes. The characterization of deposited metal includes determination of particle properties and bulk powder properties. The individual particle properties include size and shape. The bulk powder properties include chemical composition, particle size distribution, specific surface area, apparent density, and flow rate. Some of the factors that can influence the morphology of deposited particles include temperature, pH, reducing gas pressure, and the presence and concentration of impurities or contamination in solution or in reducing gas. The surface-active additives can change particle morphology by activation or passivation of the growing particle surface.

EQUIPMENT AND OPERATION

The gaseous reduction processes are typically batch autoclave processes. They operate at temperatures 120°–270°C, total pressure of 3–4 MPa, and reduction time of one cycle of 10–60 minutes. The metal concentration in feed solution is 40–100 g dm^{-3} and in the final solution, a few grams per cubic decimeter. For large-capacity hydrogen precipitation of metals, horizontal autoclaves with mechanical stirring and coils for heating or cooling are used. The autoclave body is usually made of steel and lined inside with a corrosion-resistant material, such as high-alloyed stainless steel, nickel alloy, or titanium. Schematics of an autoclave are shown in Figure 5. The autoclaves typically operate in batch mode and are constructed without baffles. Continuous production has been proposed since the 1960s; in the batch mode, the autoclave is under gas pressure and depositing metal less than half of the operating time. Continuous mode is more complicated than batch mode because of the risk of hydrogen leakage and the complexity of solution supply and especially of product discharge. The control of particle size and density is also more difficult (Naboychenko 2009). In reduction autoclaves, the prepared metal-bearing solution and wash solution are fed through a dip pipe. The dip pipe is also used for solution and gas removal under pressure to flash tanks. The process gases enter through a common header. Line blinds and valves are used to isolate the process gases not in use. Finished metal powder slurry is discharged through flash tanks or drained from the bottom at ambient pressure.

A schematic autoclave process is shown in Figure 6. The autoclave feed is prepared using metal-containing pregnant solution and compounds to control solution pH and complex formation, such as ammonia. The solution can be based on ammonia, sulfuric acid, hydrochloric acid, or organic media. An emerging trend is to use pulps containing metal hydroxides

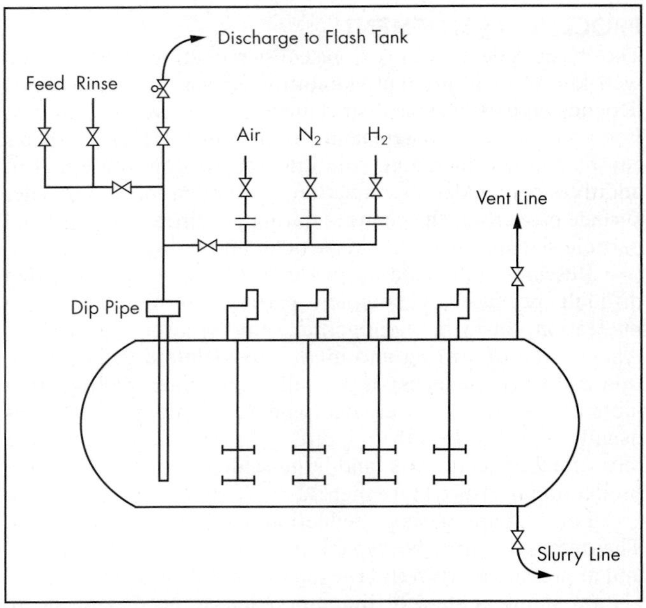

Adapted from Benson and Colvin 1964; Naboychenko 2009; Crundwell et al. 2011

Figure 5 Autoclave

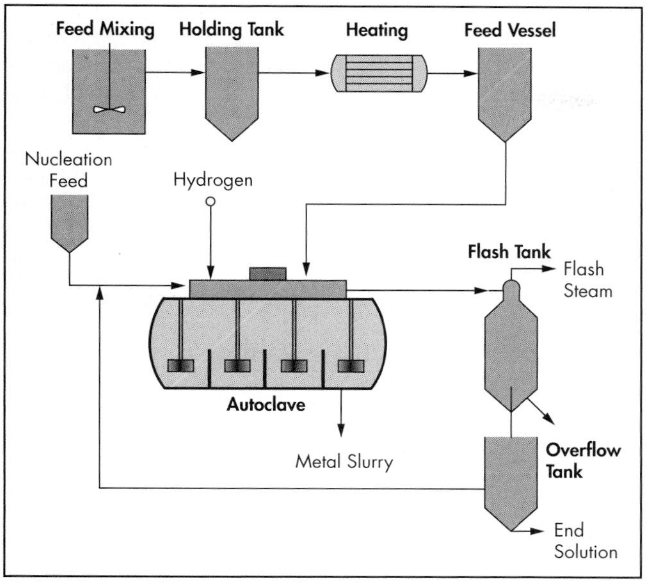

Figure 6 Autoclave process flow sheet

or oxides. Preparation of the feed solution requires control of metal concentration, impurities, and additives such as seeding and activating compounds. Deposition rate and metal recovery usually increase with increasing metal concentration in the feed as long as there is a sufficient amount of reducing gas available. Depending on the raw material composition and leaching method, the pregnant solutions contain various impurities. In preparation of autoclave feed, the suspended particles must be filtered away to prevent their action as seed. To prevent coprecipitation of impurity metals and product contamination, solution purification is required. Dissolved impurity metals are removed by chemical precipitation, ion exchange, or solvent extraction.

Most of the metals reduced by gases from solution require seeding, such as deposition of nickel and cobalt. Copper is an exception and no seeding is necessary as the reaction is self-nucleating through the copper disproportionation reaction. For copper, the metal surface has no significant effect. The deposition of nickel and cobalt are heterogeneous reactions, and they are affected by the catalytic properties of the metal surface. Seeding by additives or solid particles provides active sites for reducing gas, and the induction period is eliminated. Seeding does not only accelerate the nucleation, but helps to control particle size. Surfactants prevent metal deposition on the autoclave surfaces, and they affect deposited powder shape, texture, and dispersion. The surfactants are organic substances such as polyacrylic acid, polyacrylates, thioacetamide, and anthraquinone and its derivatives. Some of the surfactants are strong reducing compounds, promoting nucleation and deposition (Naboychenko 2009).

The feed is heated and pumped into the autoclave. Additives or small seed particles to promote nucleation, possible surface-active additives, and reducing gas are fed to start the nuclei deposition process. After the nuclei forming step, the depleted solution is discharged, leaving the reduced metal slurry in the autoclave. The reduced metal of the first batch has small particle size and high surface area, and it is used as seed for the reduction in the actual deposition batches. The fresh process solution is fed into the autoclave, and metal is reduced by gas growing as a shell on the nuclei. When solution becomes depleted and reduction has ended, the mixers are stopped, the metal powder is allowed to settle to the bottom of the autoclave, and the spent solution is removed from on top of the settled particles. With repeated batch reductions, the metal particle size increases and particles become more compact. This results in a powder with higher apparent density, and thus each batch reduction is termed *densification*. The particle size increase often follows a logarithmic law. To maintain the densification process, it is necessary to increase gas pressure, temperature, and time as the number of cycles increases (Saarinen et al. 1998).

The rate of gaseous precipitation of metals in the autoclave is determined by hydrodynamics, pressure of reducing gas, temperature, and solution composition as shown in rate law Equation 16. Saturation of the liquid by the reducing gas and uniform composition of the slurry are achieved by sufficient stirring. Stirring intensity is limited by increasing energy consumption, stirrer seal design, and solution foaming. When the gas pressure is increased, the concentration of dissolved gas in the solution increases proportionally. This increases the reduction rate. Maximum pressure is limited by design of the autoclave and gas lines and plant safety. The use of higher temperature increases reaction rates, but it leads to higher energy consumption, more expensive construction because of higher pressure, and sometimes higher risk of corrosion. Also, the composition of the solution may change with temperature because of hydrolysis, decreasing solubility of the dissolved species, and sintering of metal particles. For example, some sulfates are known to crystallize at high temperatures and dissolve back when the temperature is lowered.

When the size of the metal particles in the autoclave becomes too large to maintain a suspension and to serve as an effective catalyst, the gas is flashed from the autoclave, and both the powder and depleted solution are discharged. In the flash tank, the temperature of the gas decreases rapidly. For an efficient flash system, several stages with different pressure decrease can be built as a flash train. The slurry with deposited metal powder is taken to further processing, and evaporated gas goes to a gas scrubbing system. After this, a new cycle is prepared. Any metal deposited to the internal surfaces of the autoclave is dissolved away between batches. Control of seeding for nucleation is essential to minimize plating on autoclave internals and downtime between deposition batches. In nickel production, the leaching time can be 6–7 hours (Crundwell et al. 2011). To maintain continuous operation, parallel autoclave lines are needed to stagger the operating stages. This way, continuous feed from solution preparation can match the deposition and cleaning stages of autoclaves (Benson and Colvin 1964; Saarinen et al. 1998). Typically, the number of parallel autoclaves is six to seven in nickel production and two to three in cobalt production.

APPLICATIONS

Gaseous reduction is a special production method. The most important hydrometallurgical applications are production of nickel and cobalt using ammoniacal solutions. Gaseous reduction is also used in the production of special metal powder products, and it has been suggested as a method to remove excess metals in purification of hydrometallurgical process solutions.

Nickel Production

Nickel powders are used as an alloying element, in powder metallurgy, for catalysts, in rechargeable batteries, as conductive fillers for electrical applications, and in thermal spray coatings. Nickel powders are produced by carbonyl process, gaseous reduction, atomization, solid phase reduction, reduction with organic compounds, and electrolysis. Some of these methods can also produce nickel-clad powders for special applications. The nickel deposition is mainly accomplished using nickel ammoniacal sulfate solutions, but basic nickel carbonate, nickel ammonium acetate, and reduction of nickel oxide and nickel hydroxide have been proposed as alternatives. Nickel powder produced by hydrogen reduction is less pure than electrowon nickel, but suitable for most applications. Hydrogen reduction requires less plant space, energy, and labor force than electrowinning. The operating costs depend on the cost of chemicals, mainly on the source and production of hydrogen (Crundwell et al. 2011).

Nickel cannot nucleate homogeneously from a solution, and therefore solid particles are needed on which nickel can deposit. A catalyst is used to start the nucleation and precipitation in the first reduction batch. Known nickel-reduction catalysts include ferrous sulfate, ferrous sulfate modified with aluminum sulfate, platinum chloride, palladium chloride, and silver sulfate (Saarinen et al. 1998). As the reduction proceeds, the particles become denser and larger. The rate of reduction and the yield are functions of temperature, hydrogen pressure, seed addition, mass transfer, amount of alkali used, and the identity of the catalyst and its concentration. The pH of the solution determines the reduction rate and the morphology of the particles to a great degree (Saarinen et al. 1998). The overall reaction is shown in the following equation. Free ammonia is necessary to neutralize the hydrogen ions arising during reduction.

$$Ni(NH_3)_nSO_4 + H_2 \rightarrow Ni^O + (NH_4)_2SO_4 + (n-2)NH_3 \qquad \textbf{(EQ 20)}$$

The principal products of nickel reduction are powder with a particle size of 0.1–0.2 mm and ammonium sulfate, $(NH_4)_2SO_4$, which can be purified and sold as fertilizer. The nickel powder is washed and dried, and most of the nickel is briquetted with a polyacrylic acid binder and sintered. Deposited nickel is removed from autoclave internals by leaching with ammonia–ammonium sulfate solution and compressed air (Crundwell et al. 2011).

Nickel hydrogen reduction was originally developed to recover metal from air–ammonia leaching of sulfide concentrates. Nowadays, the plants process sulfide mattes and intermediate sulfide precipitates. Industrially, hydrogen reduction of nickel is practiced at BHP Billiton in Kwinana, Australia; Minara Resources in Murrin Murrin, Australia; Sherritt in Fort Saskatchewan, Canada; Ambatovy Joint Venture in Toamasina, Madagascar; Norilsk Nickel in Harjavalta, Finland; and Impala Platinum in Springs, South Africa (Crundwell et al. 2011). All these plants use the process developed by Sherritt Gordon during the 1950s and use nickel–ammonium sulfate solutions. The world nickel production by hydrogen reduction is about 240,000 t/yr (metric tons per year) (Crundwell et al. 2011). Of these plants, the BHP Billiton, Sherritt, Norilsk Nickel, and Impala Platinum use sulfide ore mattes and precipitates as raw material, whereas Minara Resources, Ambatovy, and partly Sherritt use lateritic ores that are converted to sulfides and re-leached before metals reduction. Leaching of sulfides is done either with air and ammonia or oxygen and sulfuric acid. The acid liquors are ammoniated for metals recovery.

The Sherritt Fort Saskatchewan plant uses purified solution from precipitate leaching containing 70–75 g dm^{-3} Ni, 35–40 g dm^{-3} NH_3, and 320–340 g dm^{-3} $(NH_4)_2SO_4$. The pregnant leach liquor is purified by separating cobalt by precipitation as cobalt hexamine salt, removing copper by deposition as sulfide, and converting all sulfur compounds to sulfates by oxidation and hydrolysis in autoclave (Naboychenko et al. 2009c). The operating temperature is 185°C and pressure 3.0–3.5 MPa. The NH_3/Ni ratio is approximately 1.9. The seed is 10-mm nickel particles. To produce particles of 0.1–0.2-mm size, 60–80 densification steps are needed, and a single densification step lasts up to an hour. The Fort Saskatchewan plant operates seven 7–22-m^3 autoclaves to produce 32,000 t/yr of nickel (Crundwell et al. 2011).

Minara Resources' Murrin Murrin plant uses laterite ores as raw material. The Ni-Co ore is first leached in autoclaves. From the pregnant solution, a mixed nickel and cobalt sulfide is precipitated by hydrogen sulfide gas. The mixed sulfide is then leached in an autoclave with oxygen to produce a sulfate solution. Impurities, such as Zn, Fe, and Cu, are removed, and cobalt and nickel are separated using solvent extraction. The nickel is precipitated in six parallel autoclaves. The nickel powder particle size is 250–300 mm at maximum. The powder is dried before forming briquettes that are sintered in a furnace. The product is more than 99.9% Ni with some Co, Fe, and Si as impurities. The Murrin Murrin plant produces 40,000 t/yr of nickel.

The Impala Platinum Springs plant uses purified solution from leaching of nickel matte containing 55–65 g dm^{-3} Ni and 330–350 g dm^{-3} $(NH_4)_2SO_4$. The operating temperature is 180°C and pressure 2.8 MPa. The NH_3/Ni ratio is 1.8–1.95. The time for one densification step is 10 minutes. The Springs plant produces 13,000 t/yr of nickel (Crundwell et al. 2011).

Cobalt Production

Cobalt is usually present in nickel and copper ores and is recovered during production of those metals. The world production of cobalt was 86,000 t in 2013 as metal, oxides, hydroxides, and other compounds. About half of the cobalt is produced as solid metal or powders (Fisher 2011). Cobalt and cobalt-alloy powders are used in rechargeable batteries, in the powder metallurgy industry in production of permanent magnets, and as high-temperature superalloys and wear-resistant alloys. Cobalt powder is used also in tool and die steels and as a binder in cemented carbides. Other applications of cobalt include catalysts, drying agents, and magnetic recording media. Cobalt metal can be produced by gaseous reduction, solid phase reduction, atomization, carbonyl process, and electrolysis. The gas reduction process is well known, but electrowinning is normally preferred because of lower capital and operating costs, higher metal recovery, and easier and less stringent operability and maintenance (Fisher 2011). Gaseous reduction is usually practiced along the Sherritt nickel production process. Cobalt metal is produced by Sherritt in Fort Saskatchewan, Minara Resources in Murrin Murrin, Impala Platinum in Springs, and Ambatovy Joint Ventures in Toamasina. Cobalt is produced from solution coming from cobalt–nickel separation in nickel production. The cobalt must be removed before gaseous reduction of nickel to ensure nickel purity. The cobalt was previously deposited as complex cobalt hexamine sulfate and redissolved, but currently, solvent extraction is used for cobalt–nickel separation. The cobalt-containing stream is then processed in its own gaseous reduction process.

The cobalt reduction mechanism is heterolytic and seed is needed. Cobalt seed is made by precipitation from cobaltous diamine solution using sodium sulfide particles and sodium cyanide and catalyst (Crundwell et al. 2011; Naboychenko et al. 2009a). The cobalt reduction reaction is shown in the following equation:

$$[Co(NH_3)_2]SO_4 + H_2 = Co + (NH_4)_2SO_4 \qquad \textbf{(EQ 21)}$$

Cobalt reduction is slightly faster than nickel reduction, and the resulting cobalt particles are slightly larger and more irregular because of particle agglomeration. Cobalt plated on the autoclave is dissolved by cobaltic hexamine solution (Crundwell et al. 2011).

The Sherritt plant in Fort Saskatchewan uses purified cobalt-rich solution from nickel production. The production is 4,000 t/yr (Crundwell et al. 2011). The feed liquor contains 75–80 g dm^{-3} Co, 550 g dm^{-3} $(NH_4)_2SO_4$, and 50 g dm^{-3} NH_3. The plant operates two 7-m^3 stainless-steel autoclaves. The operating pressure is 3.5 MPa and temperature 175°C. The seed is 0.01 mm, and 60–80 densifications, each taking approximately 45 minutes, are used to produce 0.2-mm particles (Crundwell et al. 2011).

Copper Production

Copper powders are used in powder metallurgy, when high electrical and thermal conductivities, ductility, and corrosion resistance are needed for the parts. Copper powders are also used in paints, coatings, inks, metal-polymer composites, and brazing pastes. Copper powder is produced mainly by reduction of copper oxide, atomization of molten metal, chemical precipitation, and electrolytic deposition. Copper powders can be precipitated from ammoniacal, sulfurous, and organic media by reduction with hydrogen, carbon monoxide, and sulfur dioxide, as respectively shown by the following three equations (Naboychenko et al. 2009b):

$$Cu(NH_3)_2^{2+} + H_2 \rightarrow Cu + 2NH_4^+ \qquad \textbf{(EQ 22)}$$

$$Cu(NH_3)_2^{2+} + CO + H_2O \rightarrow Cu + CO_2 + 2NH_4^+ \qquad \textbf{(EQ 23)}$$

$$Cu(NH_3)_2^{2+} + SO_2 + 2H_2O \rightarrow Cu + SO_4^{2-} + 2NH_4^+ + 2H^+ \qquad \textbf{(EQ 24)}$$

Hydrogen reduction in an autoclave from the leach liquors of ores and concentrates was reported in 1957 and reduction of copper from the bleed stream of copper electrorefining and various leach solutions since 1983 (Agrawal et al. 2006). Contrary to nickel and cobalt production, there is no copper metal production using hydrogen reduction. Hydrogen reduction is used for production of copper powders, and it has been tested for purification of copper electrolysis bleed solutions (Agrawal et al. 2006) and complex solutions containing several metals (Park et al. 2015). The copper production process is similar to the hydrometallurgical methods of nickel and cobalt production. The main differences are that reduction of copper does not need seeding as copper precipitation proceeds by the disproportionation reaction of Cu^+ ions (Equation 15) and the number of densification cycles is small. The reduction of copper with hydrogen can be achieved under pressure from both acidic and ammoniacal solutions (Agrawal et al. 2006).

Noble Metal Production

The noble metals include silver, gold, and platinum group metals (PGMs). Production methods of noble metal powders include chemical precipitation, thermal decomposition of metal salts, electrolysis, hydrogen reduction, and melt atomization (Naboychenko et al. 2009d). For noble metal reduction, hydrogen and carbon monoxide can be used. The reduction of noble metals is less critical on solution conditions than the reduction of base metals such as nickel, cobalt, and copper.

In gaseous reduction, the dissolved metals are typically chloride complexes. The general reaction of PGM reduction is as follows:

$$H_yMeCl_x + y/2H_2 \rightarrow Me + xCl^- + 2yH^+ \qquad \textbf{(EQ 25)}$$

As the PGMs are noble metals, their reduction can be accomplished using relatively low temperatures and hydrogen gas pressures. The metal ion concentration is typically grams per liter, but it can be as low as a few tens of milligrams per liter (Naboychenko et al. 2009d). To maintain complex stability, the chloride content is high, for example, 1 M hydrochloric acid (HCl). Use of an autoclave is not always necessary and reduction can happen at atmospheric pressure. Selective precipitation can be carried out by controlling temperature and hydrochloric acid concentration.

SUMMARY

Metal production by gaseous reduction started in the 1950s when Chemical Construction Corporation and Sherritt Gordon Mines developed a method to produce nickel. Much of the theoretical background was developed in the 1950s and 1960s. Gaseous metal reduction from aqueous solutions is mainly used in the production of nickel and cobalt, and approximately 30% of nickel metal is produced by reduction with hydrogen. The main method for nickel and cobalt production is electrowinning. Another application is production of metal powders for various applications. In this field, gaseous reduction is just one of many alternative methods. The reduction techniques are site specific. The general principles are the same, but details in solution preparation, practice of reduction stage operation, use of additives, and so forth, vary.

REFERENCES

Agrawal, A., Kumari, S., Bagchi, D., Kumar, V., and Pandey, P.D. 2006. Hydrogen reduction of copper bleed solution from an Indian copper smelter for producing high purity copper powders. *Hydrometallurgy* 84(3-4):218.

Benson, B., and Colvin, N. 1964. Plant practice in the production of nickel by hydrogen reduction. In *Unit Processes in Hydrometallurgy*. Edited by M.E. Wadsworth and F.T. Davis. New York: Gordon and Breach.

Burkin, A.R. 2001. *Chemical Hydrometallurgy*. London: Imperial College Press.

Crundwell, F.K., Moats, M.S., Ramachandran, V., Robinson, T.G., and Davenport, W.G. 2011. *Extractive Metallurgy of Nickel, Cobalt and Platinum-Group Metals*. Amsterdam: Elsevier.

Fisher, K.G. 2011. Cobalt processing developments. In *6th Southern African Base Metals Conference*. Johannesburg: Southern African Institute of Mining and Metallurgy.

Halpern, J. 1959. Homogeneous catalytic activation of molecular hydrogen by metal ions and complexes. *J. Phys. Chem.* 63(3):398.

Mackiw, V.N., Lin, W.C., and Kunda, W. 1957. Reduction of nickel by hydrogen from ammoniacal nickel sulfate solutions. *J. Met.* 9(6):786.

Meddings, B., and Mackiw, V.N. 1964. The gaseous reduction of metals from aqueous solutions. In *Unit Processes in Hydrometallurgy*. Edited by M.E. Wadsworth and F.T. Davis. New York: Gordon and Breach.

Naboychenko, S.S. 2009. Reduction methods of powder production. In *Handbook of Non-Ferrous Metal Powders—Technologies and Applications.* Edited by O.D. Neikov, S.S. Naboychenko, I.V. Murashova, V.G. Gopienko, I.V. Frishberg, and D.V Lotsko. Amsterdam: Elsevier.

Naboychenko, S.S., Murashova, I.B., Neikov, O.D. 2009a. Production of cobalt and cobalt-alloy powders. In *Handbook of Non-Ferrous Metal Powders—Technologies and Applications.* Edited by O.D. Neikov, S.S. Naboychenko, I.V. Murashova, V.G. Gopienko, I.V. Frishberg, and D.V. Lotsko. Amsterdam: Elsevier.

Naboychenko, S.S., Murashova, I.B., Neikov, O.D. 2009b. Production of copper and copper-alloy powders. In *Handbook of Non-Ferrous Metal Powders—Technologies and Applications.* Edited by O.D. Neikov, S.S. Naboychenko, I.V. Murashova, V.G. Gopienko, I.V. Frishberg, and D.V. Lotsko. Amsterdam: Elsevier.

Naboychenko, S.S., Murashova, I.B., Neikov, O.D. 2009c. Production of nickel and nickel-alloy powders. In *Handbook of Non-Ferrous Metal Powders—Technologies and Applications.* Edited by O.D. Neikov, S.S. Naboychenko, I.V. Murashova, V.G. Gopienko, I.V. Frishberg, and D.V. Lotsko. Amsterdam: Elsevier.

Naboychenko, S.S., Murashova, I.B., Neikov, O.D. 2009d. Production of noble metal powders. In *Handbook of Non-Ferrous Metal Powders—Technologies and Applications.* Edited by O.D. Neikov, S.S. Naboychenko, I.V. Murashova, V.G. Gopienko, I.V. Frishberg, and D.V. Lotsko. Amsterdam: Elsevier.

Ntuli, F., and Lewis, A.E. 2006. The effect of a morphology modifier on the precipitation behavior of nickel powder. *Chem. Eng. Sci.* 61(17):5827.

Ntuli, F., and Lewis, A.E. 2007. The influence of iron on the precipitation behaviour of nickel powder. *Chem. Eng. Sci.* 62(14):3756.

Osseo-Asare, K. 2003. Hydrogen reduction of metal ions: An electrochemical model. In *Hydrometallurgy 2003*, Vol. 2. Edited by C.A. Young, A.M. Alfantazi, C.G. Anderson, D.B. Dreisinger, B. Harris, and A. James. Warrendale, PA: The Minerals, Metals & Materials Society.

Outotec. 2015. HSC Chemistry software database. www.outotec.com/products/digital-solutions/hsc-chemistry/. Accessed July 2018.

Park, K.H., Mishra, D., Nama, C.W., Sahub, K.K., and Agrawal, A. 2015. Selective reduction of Cu from aqueous solutions through hydrothermal route for production of Cu powders. *Hydrometallurgy* 157(1):280.

Saarinen, T., Lindfors, L.-E., and Fugleberg, S. 1998. A review of the precipitation of nickel from salt solutions by hydrogen reduction. *Hydrometallurgy* 47(2-3):309.

Schaufelberger, F.A. 1956. Precipitation of metal from salt solution by reduction with hydrogen. *J. Met.* 8(5):695.

Taty Costodes, V.C., Mausse, C.F., Molala, K., and Lewis, A.E. 2006. A simple approach for determining particle size enlargement mechanisms in nickel reduction. *Int. J. Miner. Process.* 78(2):93.

Von Hahn, E.A., and Peters, E. 1965. The role of copper(I) in the kinetics of hydrogen reduction of aqueous cupric sulfate solutions. *J. Phys. Chem.* 69(2):547.

CHAPTER 10.19

Aqueous-Phase Redox Precipitation

Courtney A. Young and Hsin-Hsiung Huang

Precipitation is a commonplace, hydrometallurgical process used to synthesize metal-bearing solids from solution. It is accomplished by the various techniques shown in Figure 1, which is similar to the original illustration presented by Habashi (1999). As indicated, these techniques are classified as either physical or chemical precipitation and are described elsewhere in this handbook. In general, the techniques on the left tend to be heterogeneous reactions that are used to produce compounds, and those on the right tend to be homogeneous reactions that are usually used to make metal. They are briefly reviewed here to set the stage for discussing aqueous-phase reduction–oxidation (redox) precipitation.

In physical precipitation, also known as crystallization, cooling and/or evaporation are used to make compounds normally in the absence of chemical addition. By comparison, chemical precipitation is accomplished by adding reagents, thereby inducing one of four processes: hydrolytic, ionic, substitution, and redox precipitation. These four processes are also used to yield compounds; however, redox precipitation is more commonly used to produce metal. For example, hydrolytic precipitation is primarily used to make compounds as hydroxides (OH^-) and oxides (O^{2-}) by simply adjusting the solution pH. Although hydrolytic precipitation is a form of ionic precipitation, the latter generally refers to compounds precipitated upon addition of sulfides (S^{2-}), halides (e.g., F^- and Cl^-), and oxyanions (e.g., SO_4^{2-}, CO_3^{2-}, and PO_4^{3-}) among many other anions. Likewise, substitution precipitation occurs when ligands, including waters of hydration, are coordinated with a transition metal and are displaced by other ligands, causing the transition metal to precipitate. Such reactions are dependent on coordination chemistry and ligand size. Substitution precipitation often occurs when one or more cations are precipitated by one or more anions, yielding either a compound in stoichiometric proportion such as a double salt (as is often the case for cyanide, CN^-) or in nonstoichiometric composition such as a solid solution (as is often the case for silicates, SiO_4^{2-}). Finally, in redox precipitation, a change in oxidation state occurs, requiring electrons to be transferred either electrolytically or electrochemically in a process typically used to make metal.

With electrolytic precipitation, an electrical current is supplied externally via batteries, generators, and power plants, depending on the scale needed. Electrowinning and electrorefining are used almost exclusively for metal reduction. Electron transfer can also be done internally via electrochemical reactions involving solid, gas, and aqueous phases. Solid-phase redox precipitation is predominantly used as a reductive process more commonly referred to as cementation. Gas-phase redox precipitation is also predominantly reductive and is exemplified with the use of hydrogen. With aqueous-phase redox precipitation, ionic and nonionic species are dissolved in solution and used as either oxidants or reductants. Because solid- and gas-phase reactions involve solid and gas phases as reactants, they are heterogeneous. By comparison, when the reactants are all dissolved species, the electrochemical reactions are homogeneous.

In the literature, emphasis has predominantly been on "ionic" reductive precipitation; however, this terminology does not apply to similar reactions involving nonionic species as well as oxidative precipitation. Furthermore, this terminology is easily confused with "ionic" chemical precipitation where redox reactions do not occur. Consequently, in keeping with the names of solid- and gas-phase redox precipitation, it is more appropriately referred to herein as aqueous-phase redox precipitation. Hence, this chapter examines several ionic and nonionic redox precipitation processes, including organic reactants. Example thermodynamic calculations using StabCal software (Huang 2018) are illustrated with E_H–pH (Pourbaix) diagrams for selected case studies. Thermodynamic data were predominantly taken from the National Bureau of Standards (NBS; Wagman 1982); however, data taken from other sources were made consistent to the NBS data so calculation errors would be prevented. The sources are identified within as needed. Additional case studies are also presented to demonstrate aqueous-phase redox precipitation for other metals, but associated E_H–pH diagrams are not presented, to minimize duplication.

Courtney A. Young, Department Head & Professor, Metallurgical & Materials Engineering, Montana Tech, Butte, Montana, USA
Hsin-Hsiung Huang, Professor, Department of Metallurgical & Materials Engineering, Montana Tech, Butte, Montana, USA

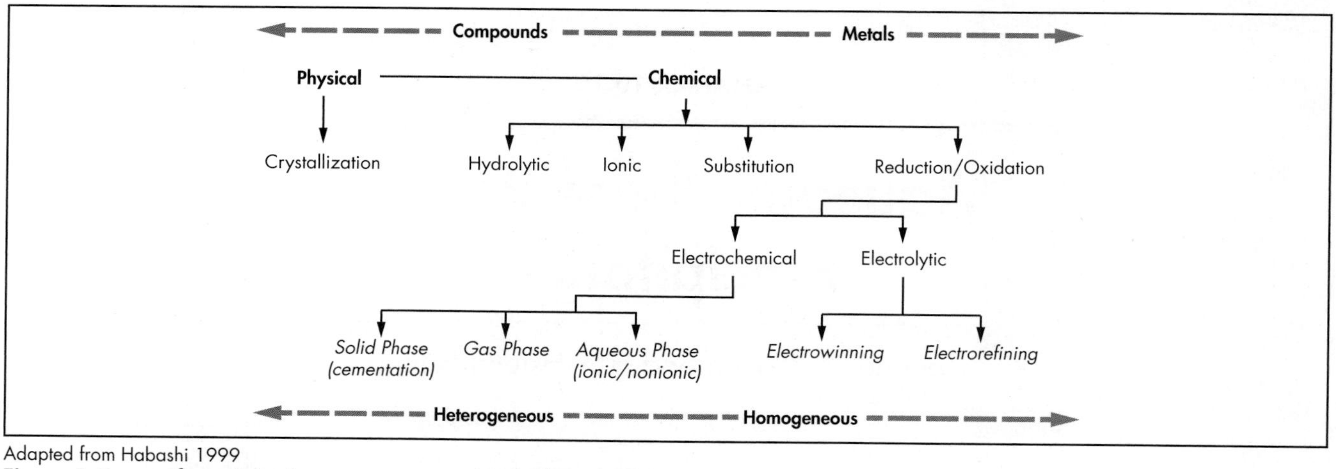

Adapted from Habashi 1999

Figure 1 Types of precipitation processes used in hydrometallurgy

INDUSTRIAL AND LABORATORY PRACTICES

Aqueous-phase redox precipitation is essentially no different than traditional precipitation. Reactions are therefore conducted in the same vessels conducted in either batch or continuous-flow mode. Reactors would therefore include but not be limited to stirred tanks/mixers/agitators, thickeners/clarifiers, tube/pipe reactors, airlifts/bubble columns, crystallizers, fluidized beds, and filter chambers. Two issues may need to be addressed to improve reaction efficiencies. First, potential control may be necessary such that, for example, solutions can be deaerated to remove dissolved oxygen, particularly in the case when aqueous-phase reduction is being done. Reactors may therefore need special features to allow this or be preceded with a conditioning tank where it is done. Second, as is the case for all precipitation vessels, special consideration may be needed for solid–liquid separation to remove the precipitate from the solution. This is done simultaneously with filter chambers and thickeners/clarifiers. It can also be aided by seeding the precipitate so that large particles are obtained and/or by using vessels with appropriate designs, as exemplified with inward or outward flows on stirred-tank reactors.

GENERAL THEORY

Redox Reactions

Because electrochemical reactions involve changes in oxidation state, they can be represented as half-cell reactions to illustrate the electron transfer that occurs. In this regard, a principal goal of hydrometallurgical processing is to precipitate the metal (M°) of interest from its cation (M^{2+}) by reduction such that the reducing agent (R^+) simultaneously undergoes oxidation:

$$\text{Cathodic reduction:} \quad M^{2+} + 2e^- = M^\circ \qquad \text{(EQ 1)}$$

$$\text{Anodic oxidation:} \quad R^+ = R^{2+} + e^- \qquad \text{(EQ 2)}$$

Obviously, the two half-cell reactions are balanced regarding the number of electrons being transferred and then summed to yield the overall electrochemical process:

$$\text{Redox reaction:} \quad M^{2+} + 2R^+ = M^\circ + 2R^{2+} \qquad \text{(EQ 3)}$$

Thus, in this case, two dissolved species electrochemically react to precipitate the more noble metal. It is preferred that the other product (R^{2+}) does not precipitate and therefore remains in solution or is evolved off as a gas.

Butler–Volmer Equations

Because electrode kinetics can be used to model these reactions, the Butler–Volmer equations for the cathodic and anodic reactions apply. To illustrate, the theoretical redox couple, R^+/R^{2+}, used in Equations 2 and 3 is exemplified:

Cathodic reduction ($R^{2+} + e^- = R^+$):

$$i_c = i_o\exp[(-\alpha)nF\eta/RT] \qquad \text{(EQ 4)}$$

Anodic oxidation ($R^+ = R^{2+} + e^-$):

$$i_a = i_o\exp[(1-\alpha)nF\eta/RT] \qquad \text{(EQ 5)}$$

where α is the transfer coefficient that is used as a measure of symmetry between the half-cell reactions of the redox couple; n is the number of electrons involved (mol); F is Faraday's constant (23,060 cal/V); R is the gas constant (1.986 cal/mol/K); T is temperature (K); i_c is the cathodic current (amp); i_a is the anodic current (amp); i_o is the exchange current (amp) at $\eta = 0$ and where $i_a = -i_c$; and η is the potential (V) difference ($E - E_{eq}$), such that E_{eq} is the equilibrium potential (V) determined from thermodynamics according to the Nernst equation and E is the measured or applied potential (V). Thus, $+\eta$ is referred to as the overpotential and $-\eta$, the underpotential. By summing the cathodic and anodic currents, the total current results in the following:

$$\text{Total faradic current: } i = i_c + i_a \qquad \text{(EQ 6)}$$

However, because each half-cell reaction involves dissolved species, they are dependent on concentration and will eventually become mass-transfer (diffusion) controlled. This occurs at undervoltages less than $-\eta$ for the cathodic reaction and at overvoltages greater than $+\eta$ for the anodic reaction. At these potentials, the limiting currents of i_{Lc} and i_{La} are reached for both half-cell reactions, respectively. All of these parameters are illustrated by the current–potential plot in Figure 2. By taking the log of the absolute values of i, i_c, and i_a, the Tafel plot shown in the inset at the bottom right of Figure 2 results.

Applying this concept to the cathodic reaction of the more noble metal (Equation 1) and the anodic reaction of

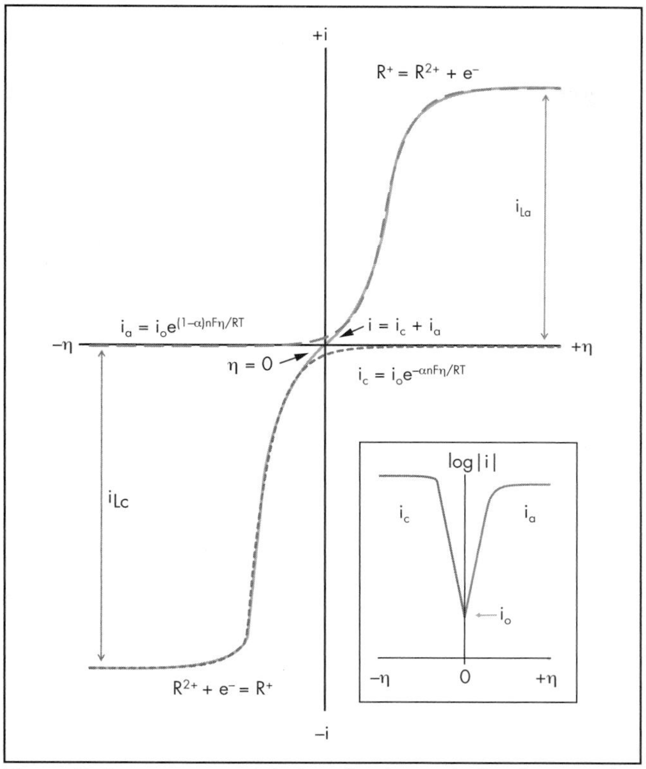

Figure 2 Current–potential plot of a theoretical redox couple R^{2+}/R^+ (inset shows the resulting Tafel plot)

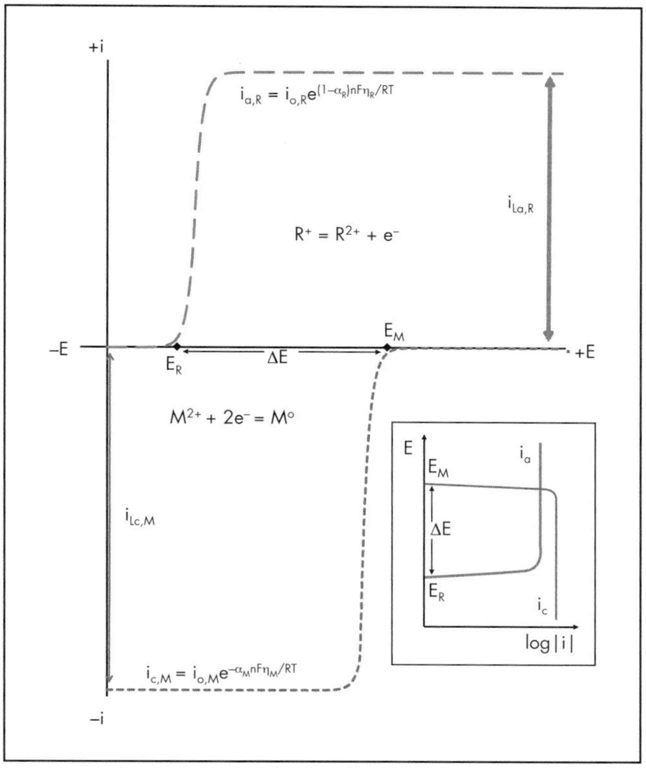

Figure 3 Current–potential plot of a theoretical redox precipitation reaction involving aqueous phases (inset shows the resulting Evans diagram)

the more reactive reductant (Equation 2) yields the current–potential plot in Figure 3. This causes the cathodic reaction in Figure 2 to shift to higher potentials, resulting in an overvoltage ($\eta = \Delta E = E_M - E_R$) that drives the redox reaction (Equation 3). As with cementation reactions, if this overvoltage exceeds approximately 0.36 V, the reaction is expected to become mass-transfer controlled, as determined by the half-cell reaction with the lower limiting current. The rate of the other half-cell reaction should then be chemically controlled. This is illustrated by the Evans diagram inset at the bottom right of Figure 3, which shows, in this example, mass transfer being controlled by the anodic oxidation reaction and chemical control by the cathodic reduction reaction.

While the overvoltage is critical, many other parameters play significant roles in the kinetics and include the faradic and exchange currents (i and i_o), temperature (T), reactant concentrations (R^+ and M^{2+}), product concentration (R^{2+}), and the charge of the dissolved species. The latter is not often considered in the literature, but, as exemplified in Reaction 3, if cations are reacting, repulsion forces between them will slow the reaction; however, if the reacting ions are oppositely charged, attractive forces should increase the reaction. Furthermore, if one reactant is neutral, it should not affect the reaction at all. Similar analogies can also be made regarding the dissolved products that form.

E_H–pH Diagrams

The beauty of redox reactions is that they can be illustrated with Pourbaix diagrams. Because these diagrams plot potential (E_H) as a function of pH, they are also referred to as E_H–pH diagrams. Like other speciation diagrams, they are determined from thermodynamic calculations to show which species are predominantly stable under particular conditions referred to as predominant stability regions. Furthermore, a line that separates two regions represents the conditions needed for a reaction to occur at equilibrium between the predominant species in the regions. Hence, E_H–pH diagrams show the predominant species and reactions with neighboring species under the influence of E_H and pH. However, the effects of temperature as well as ligand type and concentration can also be illustrated. To calculate a line, for example, the redox reaction and free energy of the reaction (dG_{rxn}) between species A and B undergoing redox change in the presence of two complexing ligands C and D are as follows:

$$\text{Redox reaction: } aA \leftrightarrow bB + cC + dD + hH^+ + ne^- \quad \textbf{(EQ 7)}$$

$$\text{Free energy: } dG_{rxn} = \Sigma(\nu_i \times dG_i^\circ) \quad \textbf{(EQ 8)}$$

where lowercase letters and ν_i represent the stoichiometric coefficients for each species from the reaction and, of course, H^+ is the hydrogen cation. Species shown on the left-hand side of the equation will have negative coefficients and vice versa. The letter n represents the number of electrons (e^-) involved, and dG_{rxn} relates to the reaction equilibrium constant and can be calculated from free energy of formation dG° of species with their stoichiometric coefficients. According to the Nernst equation for the redox reaction in Equation 7, the equilibrium between A and B is linear with $E_H = a + b \times pH$ and can be expressed as

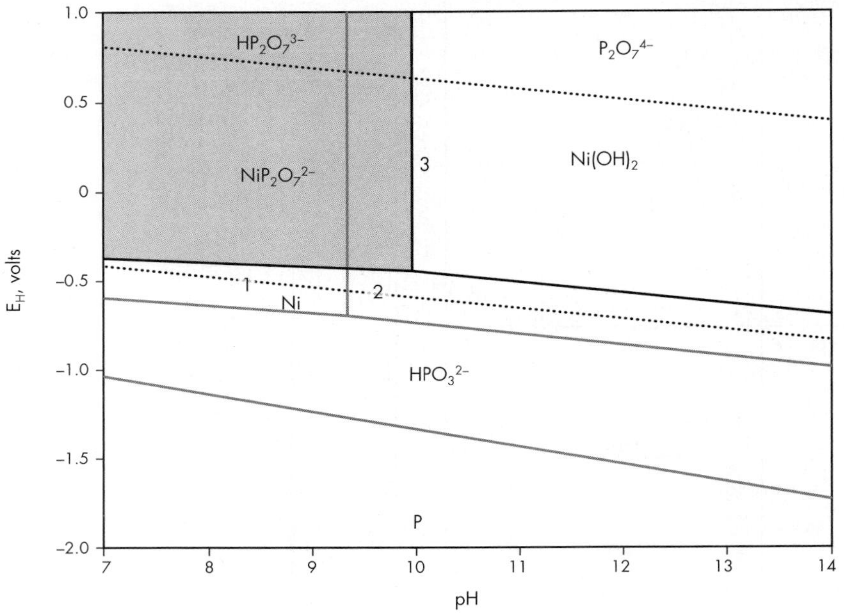

Note: Thermodynamic data are from Wagman 1982.

Figure 4 E_H–pH diagram of nickel illustrating the boundary lines between $NiP_2O_7^{2-}$ (in shaded area) and surrounding species (Ni and $Ni(OH)_2$) as influenced by P ligands of $HP_2O_7^{3-}$ and $P_2O_7^{4-}$ (labeled in gray)

Table 1 E_H–pH equations determined from thermodynamics for equilibrium reactions of $NiP_2O_7^{2-}$ with Ni and $Ni(OH)_2$

Line	Chemical Reaction	dG_{rxn}, kcal	E_H–pH Equation
1	$Ni(s) + HP_2O_7^{3-} \leftrightarrow 2e^- + NiP_2O_7^{2-} + H^+$	−7.696	E_H = −0.17240–0.02958 pH
2	$Ni(s) + P_2O_7^{4-} \leftrightarrow 2e^- + NiP_2O_7^{2-}$	−20.411	E_H = −0.44809
3	$2H^+ + Ni(OH)_2 + P_2O_7^{4-} \leftrightarrow NiP_2O_7^{2-} + 2H_2O$	−26.878	pH = 9.94439

$$E_H = E^\circ_H + \frac{\ln(10)RT}{(n\times F)}\left[\log\left(\frac{\{B\}^b\{C\}^c\{D\}^d}{\{A\}^a}\right)\right] - \frac{\ln(10)RT}{(n\times F)}\times hpH \qquad \textbf{(EQ 9)}$$

where

$$E^\circ_H = \frac{dG_{rxn}}{(n\times F)}$$

and, for non-redox reactions between A and B where n = 0, the equation would be a vertical line at x = pH expressed as

$$pH = \frac{1}{h}\times\left\{-\log K + \left[\log\left(\frac{\{B\}^b\{C\}^c\{D\}^d}{\{A\}^a}\right)\right]\right\} \qquad \textbf{(EQ 10)}$$

where

$$-\log K = \frac{dG_{rxn}}{\ln(10)RT}$$

such that {species} denotes the species activity and other terms are as defined previously. It is understood that solid species have unit activity and aqueous species have activities determined by the product of concentration in molarity (M) and an activity coefficient γ. It is understood that horizontal lines can also be determined for redox reactions in cases where h = 0.

To demonstrate, the E_H–pH diagram of nickel complexed with phosphorus at 25°C is shown in Figure 4 using the same conditions noted later in case study 1 (i.e., [P] = 0.40, [Ni] = 0.13 M, T = 298 K). The predominant area of $NiP_2O_7^{2-}$ (shaded) is complexed by $HP_2O_7^{3-}$ at lower pH and $P_2O_7^{4-}$ at higher pH. $NiP_2O_7^{2-}$ is also surrounded by two solids: Ni and $Ni(OH)_2$. The chemical reactions, free energies (dG_{rxn}), and resulting equations for $NiP_2O_7^{2-}$ in equilibrium with Ni and $Ni(OH)_2$ as affected by the different ligands are listed in Table 1. The equations match those highlighted in the figure.

ILLUSTRATED CASE STUDIES

Case Study 1—Ionic Reduction of Nickel

Electroless plating is one of the best examples of ionic reductive precipitation. The technology has been used on a variety of base and precious metals, including but not limited to Ni, Cu, Co, Au, Pt, Ag, and their alloys (Mallory and Hadju 1990). It was first developed by Brenner and Riddell (1946) for nickel plating on steel and has since been well studied and documented (Sudagar et al. 2013). In one example, Mallory (1990) examined and compared four reductants, namely, nonionic hydrazine (N_2H_4) and dimethylamine borane [$(CH_3)_2NHBH_3$] as well as ionic borohydrite (BH_4^-) and hypophosphite

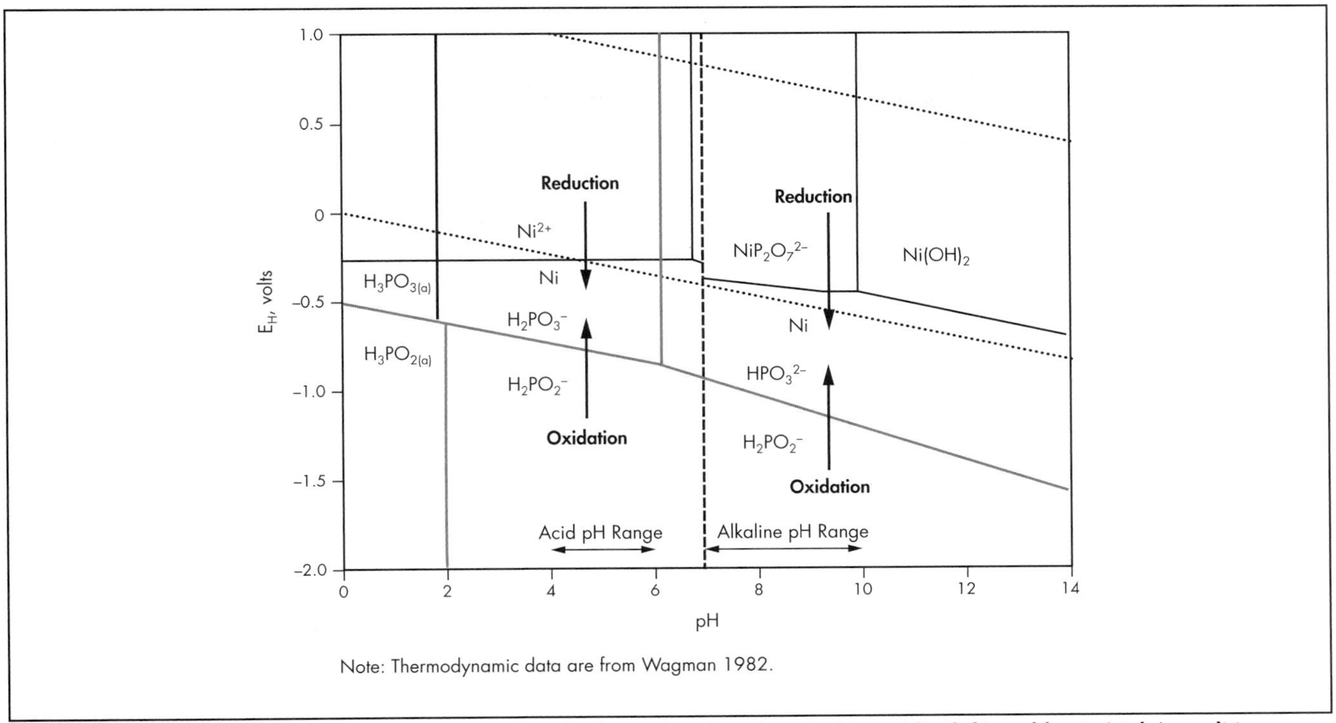

Figure 5 E_H–pH diagram for nickel electroless plating by hypophosphite ($H_2PO_2^-$) in acidic (left) and basic (right) conditions

($H_2PO_2^-$) with the following "primary" redox reactions occurring, respectively:

Anodic reduction: $Ni^{2+} + 2e^- = Ni°$ **(EQ 11)**

Cathodic oxidation: $N_2H_4 = N_2 + 4H^+ + 4e^-$ **(EQ 12)**

$(CH_3)_2NHBH_3 + 3H_2O = (CH_3)_2NH_2^+ + H_3BO_3 + 5H^+ + 6e^-$ **(EQ 13)**

$BH_4^- + 4H_2O = B(OH)_4^- + 8H^+ + 8e^-$ **(EQ 14)**

$H_2PO_2^- + H_2O = H_2PO_3^- + 2H^+ + 2e^-$ **(EQ 15a)**

such that dimethylammonium [$(CH_3)_2NH_2^+$], boric acid (H_3BO_3), tetrahydroxyborate [$B(OH)_4^-$], and byphosphite ($H_2PO_3^-$) are also produced. However, the reductants also reduce water to hydrogen gas that, in turn, can react with the reductants themselves, causing, for example, elemental B and P to also precipitate out. These side reactions are ignored here so that the primary reactions can be easily illustrated with thermodynamics using StabCal (Huang 2018). In this case, hypophosphite is examined under acidic and basic conditions.

Acidic Hypophosphite

The resulting E_H–pH diagram at 298 K in Figure 5 (at the left) depicts the ionic reduction of 7.5 g/L Ni^{2+} by 25 g/L $H_2PO_2^-$, which are typical concentrations identified by Mallory (1990). The diagram shows that the best conditions for electroless plating are between pH 5 and 6 with the Ni° forming near –0.3 V and $H_2PO_3^-$ near –0.8 V. Because the reaction produces acid (Reactions 15a, 15b), it will likely be best at the higher pH with pH control; furthermore, the higher pH will also yield the greater potential difference ($\Delta E = 0.6$ V) to help drive the reaction even more.

Alkaline Hypophosphite

At the right in Figure 5, the E_H–pH diagram also shows that nickel could be electrolessly plated under alkaline conditions. In this case, Ni^{2+} would exist as a diphosphate complex ($NiP_2O_7^{2-}$) and would undergo reduction to produce both metal and metaphosphite (HPO_3^-):

Anodic reduction:
$NiP_2O_7^{2-} + 4H^+ + 6e^- = Ni° + 2HPO_3^{2-} + H_2O$ **(EQ 16)**

Cathodic oxidation:
$H_2PO_2^- + H_2O = HPO_3^{2-} + 3H^+ + 2e^-$ **(EQ 15b)**

The half-cell reactions would occur near –0.4 V and –1.1 V, respectively. By comparison, the ΔE of 0.7 V is slightly higher and the reaction occurs in a wider pH range (7 to 10), making pH control easier. Side reactions would also occur, causing, for example, elemental P to form. In addition, hypophosphite ($H_2PO_2^-$) consumption would be three times greater.

Case Study 2—Inorganic Reduction of Gold

Reductive precipitation has been used historically for detecting the presence of metals, particularly gold (Au) and other precious metals. With this technology, a sample is often treated with aqua regia (1:3 ratio of HNO_3:HCl) at elevated temperature to solubilize any gold that may be present as gold chloride ($AuCl_4^-$). The resulting solution is then treated with inorganic chemicals like stannous chloride ($SnCl_2$) and ferrous sulfate ($FeSO_4$), resulting in a precipitate if gold is present. Today, this technology has been adapted to many reductants, with ionic inorganic compounds being emphasized in this case study and organic compounds in case study 3.

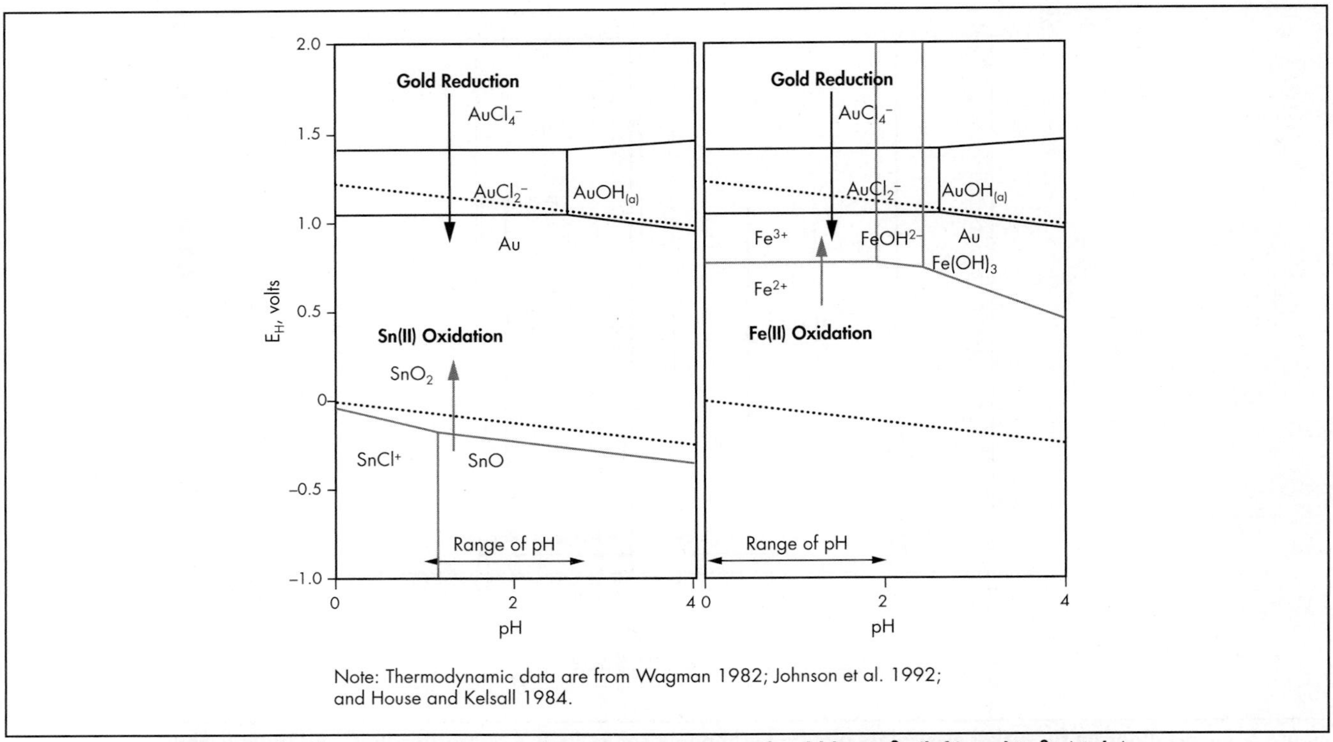

Figure 6 E_H–pH diagrams for the aqueous-phase reductive precipitation of gold by Sn^{2+} (left) and Fe^{2+} (right)

Purple of Cassius

When $SnCl_2$ is used, a purple colloid or precipitate forms, depending on the amount of gold present. Although knowledge and use of the precipitate dates to Egyptian records during the Roman-Greco times nearly 2,000 years ago, it is named purple of Cassius after Andreas Cassius, who used it as a pigment for making ruby-colored glass and porcelain during the late 1600s (Hunt 1976; Habashi 2016). The electrochemistry has since become well known:

Anodic reduction: $Au^{3+} + 3e^- = Au^o$ **(EQ 17)**

Cathodic oxidation: $Sn^{2+} = Sn^{4+} + 2e^-$ **(EQ 18)**

Resulting gold is nano-sized and adsorbs on colloidal tin dioxide (SnO_2) which forms when the stannic (Sn^{4+}) hydrolizes. The two precipitates therefore coagulate and effectively yield a nanocomposite with an apparent stoichiometry of $Au_2(SnO_2)_3$.

The process is depicted by the E_H–pH diagram in Figure 6 at the left with the following concentrations: 5 mM Au^{3+}, 50 mM Sn^{2+}, and 120 mM Cl^-, as studied by Tarozaite et al. (2006). Clearly, the best conditions require a pH of < 1 so that Sn^{2+} is dissolved as $SnCl^+$ and oxidizes to Sn^{4+} (Reaction 18) and then precipitates SnO_2 at a potential near –0.2 V. However, in order for Au^o to form from $AuCl_4^-$ solutions, reduction to $AuCl_2^-$ must first occur at 1.4 V. In this regard, Au^o does not form until the potential reaches approximately 1.05 V; however, this still creates a significant overpotential ($\Delta E = 1.25$ V) to drive the overall redox reaction.

Green Copperas

Ferrous sulfate ($FeSO_4$) is a teal-colored compound that has been referred to since ancient times as both green vitriol and green copperas (Gee 1920). When it is dissolved in solution and added to $AuCl_4^-$, a dark brown precipitate of Au^o forms (Hoke 1982). This gold precipitation saw significant use for recovery in the Plattner chlorination/leach process in the mid-1800s prior to cyanidation taking over the gold mining industry (Marsden and House 2006). Today, because its redox chemistry is well understood, the technology is used in gold recycling, refining, and electroless plating as well as the making of nanoparticles, nanowires, microcircuits, and mirror/optical coatings:

Anodic reduction: (See Equation 17)

Cathodic oxidation: $Fe^{2+} = Fe^{3+} + e^-$ **(EQ 19)**

These half-cell reactions are depicted by the E_H–pH diagram on the right side of Figure 6 at 0.01 M Au^{3+}, 0.01 M Fe^{2+}, and 0.5 M Cl^-. These concentrations encompass the range used in the literature (Parinayok et al. 2011; Wojnicki et al. 2015). In this case, the pH is kept below 2 and the potential difference of the Fe^{2+}/Fe^{3+} redox couple at ~0.75 V and the $AuCl_2^-/Au^o$ redox couple at ~1.05 V is low ($\Delta E = 0.3$ V), suggesting that the redox reaction will likely be chemically controlled.

Case Study 3—Organic Reduction of Gold

Gold and other precious metals can also be reduced with several organic reductants (Newman and Blanchard 2006). Perhaps the most commonly used are oxalic acid ($C_2H_2O_4$) and citric acid ($C_6H_8O_7$), which are the focus for this case study; however, there are a host of other organics that can also be used: formate, ascorbate, formaldehyde, ethylenediaminetetraacetic acid (EDTA), glutarate, acetone, alcohols, amines, and sugars, to name a few. It is understood that some organics are considered ionic because they speciate depending on

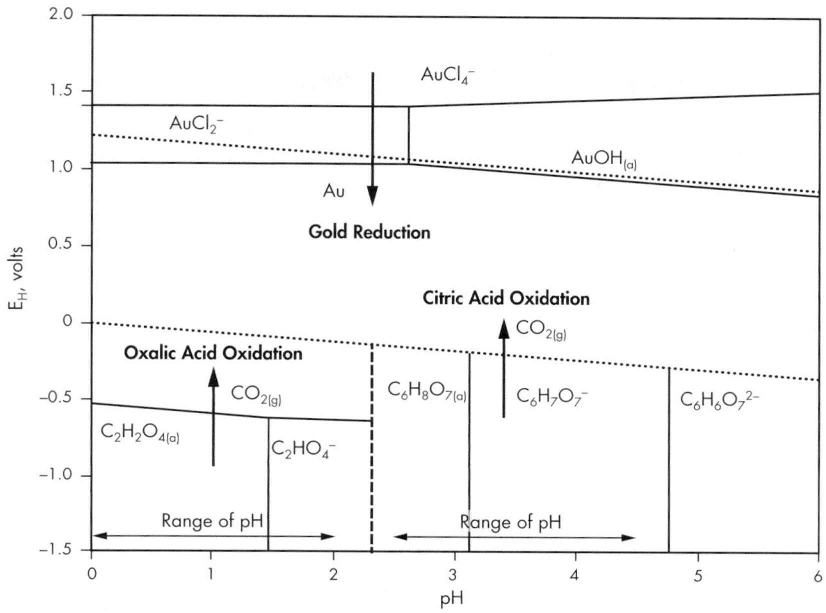

Figure 7 E_H–pH diagram for the aqueous-phase reductive precipitation of gold by oxalic acid and hydrogen oxalate (left) and citric acid, dihydrogen citrate, and monohydrogen citrate (right)

the pH, even if one of the species is uncharged. For example, oxalic acid and citric acid are neutral but, when fully unprotonated, become ionized and are referred to as oxalate and citrate, respectively. Other organics are nonionic because they are neither an acid nor a base (e.g., a polyol such as glucose sugar) and therefore do not normally protonate or hydroxylate.

Oxalic Acid

Oxalate, ferrous, and stannous ions are likely the most common reductants used for the ionic reductive precipitation of gold. Oxalate is predominantly used in refining and recycling, often as a polishing step, to produce high-purity gold due to its selectivity (Hoke 1982; Feather et al. 1997). It has also seen some use in making various nanostructures (Navaladian et al. 2007; Al-Thabaiti et al. 2015). It is favored because its redox chemistry is simple, yielding carbon dioxide (CO_2) gas as the only additional product:

Anodic reduction: (See Equation 17)

Cathodic oxidation: $C_2O_4^{2-} = 2CO_2 + 2e^-$ **(EQ 20)**

Because the CO_2 gas evolves off and gold solid precipitates out, the reaction can feasibly go to completion with high efficiency. However, it is also important to note that oxalate is not a true organic because it lacks C-H bonds, but it does behave is if it were.

On the left side of Figure 7, the E_H–pH diagram for the oxalate system is illustrated assuming Au = 0.01 M, Cl = 0.5 M, $C_2H_2O_4$ = 0.1 M, and P_{CO2} = 0.01 atm. As indicated, oxalic acid ($C_2H_2O_4$) and hydrogen oxalate ($C_2HO_4^-$) oxidize under acidic conditions less than pH 2.5 near −0.55 V at pH 0 and −0.65 V at pH 2.5. With the Au° anodic reaction occurring at ~1.05 V, a huge overpotential (ΔE) between 1.6 and 1.7 V results, depending on the pH. This large overpotential is a great example of the value of using organic reductants.

Citric Acid

Citric acid ($C_6H_8O_7$) deprotonates to citrate ($C_6H_5O_7^{3-}$) through two species: dihydrogen citrate ($C_6H_7O_7^-$) and monohydrogen citrate ($C_6H_6O_7^{2-}$). It is currently being used to prepare nanosized gold particles and wires, which require special chemicals, called stabilizers, to help control their size, shape, and growth direction. Example stabilizers include but are certainly not limited to gelatin, polyvinyl pyrrolidone, polyvinyl alcohol and acetate, polyethylene glycol, polyethyleneimine, polyamidoamine, polydithiafulvene, and chitosan (Dzimitrowicz et al. 2015; Ayoub et al. 2016). Turkevich et al. (1951) pioneered the fundamental study about the nucleation and growth of nano gold using citrate to produce nearmonosized particles. Since then, a plethora of research (Wuithschick et al. 2015) confirmed the redox chemistry:

Anodic reduction: (See Equation 17)

Cathodic oxidation:

$C_6H_5O_7^{3-} + 5H_2O = 6CO_2 + 15H^+ + 18e^-$ **(EQ 21)**

However, research efforts have shown that the reactions follow a multistep process and Reaction 21 may not go to completion, depending on how much citrate is added relative to the initial gold concentration (i.e., at, above, or below stoichiometry):

$$C_6H_5O_7^{3-} + 4H_2O = 4.5CO_2 + 10H^+ + 13e^- + CH_2O + \tfrac{1}{2}HCOOH$$ **(EQ 22)**

$$C_6H_5O_7^{3-} = 2CO_2 + H^+ + 2e^- + C_5H_4O_5^{2-}$$ **(EQ 23)**

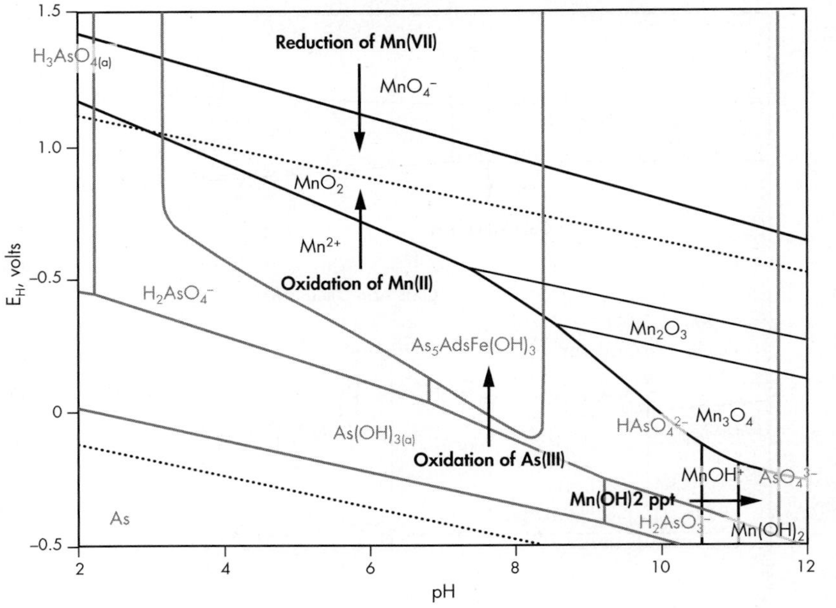

Note: Thermodynamic data are from Wagman 1982.

Figure 8 E_H–pH diagram for the aqueous-phase oxidative precipitation of Mn^{2+} (manganous) and As^{3+} (arsenite as $As(OH)_3$) with MnO_4^- (permanganate)

where, in Reaction 22, CH_2O is acetone and HCOOH is formic acid (Leng et al. 2015); and, in Reaction 23, $C_5H_4O_5^{2-}$ is 3-ketoglutarate (Verma et al. 2014). Acetone, formic acid, and glutarate are also reductants but cannot react further if all $AuCl_4^-$ was already consumed.

Figure 7 also reveals, at the right, the E_H–pH diagram for the citrate system, assuming Au = 0.005 M, Cl = 0.12 M, $C_6H_8O_7$ = 0.1 M, and P_{CO2} = 0.01 atm. As indicated, citric acid ($C_6H_8O_7$), dihydrogen citrate ($C_6H_7O_7^-$), and monohydrogen citrate ($C_6H_6O_7^{2-}$) oxidize under acidic conditions less than pH 6 and essentially at the same potentials as hydrogen evolution for water. Under these conditions, before converting to Au°, $AuCl_4^-$ will first reduce to aqueous AuOH (pH 2.5–6), primarily at the same potentials as oxygen evolution for water. Consequently, the potential difference will not be as large (ΔE = 1.2 V) as for oxalate; however, H_2 gas is not evolved, so side reactions with citrate in this regard are avoided (e.g., compare to the case study on nickel electroless plating). The best pH range for producing nanosized gold by citric acid reductive precipitation is near pH 3 and for maximizing precipitate is near pH 5.

Case Study 4—Oxidation of ARD Species

Acid rock drainage (ARD) refers to water that reacts with oxygen and sulfide-bearing rock, resulting in an acidic discharge due to the formation of sulfuric acid (H_2SO_4). Because of the acidity as well as the types of sulfide minerals involved, many metals are also solubilized including but not limited to those that are primary to, for example, the Berkeley pit lake in Butte, Montana (United States): Al, As, Cd, Cu, Fe, K, Mg, Mn, Na, and Zn (Davis and Ashenberg 1989). ARD treatment by lime (CaO) addition raises the pH and causes many of the metals to undergo hydrolytic precipitation (Zick et al. 2004); however, two stages are often needed to prevent metals from resolubilizing (Huang et al. 2005). Furthermore, to be effectively remediated, other metals must be oxidized (Twidwell and McCloskey 2011) and/or treated at high pH levels (Freitas et al. 2013), as is the situation for As^{3+} and Mn^{2+}, which are the subject of this case study. In this regard, their oxidation by permanganate (MnO_4^-) is presented and discussed:

Anodic reduction:

$$MnO_4^- + 4H^+ + 3e^- = MnO_2 + 2H_2O \quad \textbf{(EQ 24)}$$

Cathodic oxidation:

$$As(OH)_3 + H_2O = HAsO_4^{2-} + 4H^+ + 2e^- \quad \textbf{(EQ 25)}$$

$$Mn^{2+} + 2H_2O = MnO_2 + 4H^+ + 2e^- \quad \textbf{(EQ 26)}$$

$$Fe^{2+} + 3H_2O = Fe(OH)_3 + 3H^+ + e^- \quad \textbf{(EQ 27)}$$

According to Reaction 24, permanganate (MnO_4^-) will reduce to manganic oxide (MnO_2) and thereby oxidize arsenite (as $As(OH)_3$) to arsenate (as $HAsO_4^{2-}$) per Reaction 25, and manganous ion (Mn^{2+}) to manganic oxide (MnO_2) per Reaction 26. Ferrous (Fe^{2+}) oxidation to ferrihydrite ($Fe(OH)_3$) is included as Reaction 27 because resulting arsenate must adsorb at its surface to be remediated. These reactions are depicted by the E_H–pH diagram in Figure 8, which was created using approximate total concentrations for each metal found at 100-m depth in the Berkeley pit lake ARD: As = 1.0 ppm, Mn = 200 ppm, and Fe = 1,000 ppm (Davis and Ashenberg 1989).

Arsenic Remediation

Adsorbed As^{5+} (as arsenate) is labeled "$As_5AdsFe(OH)_3$" in Figure 8 and is shown to occur between pH 3 and 9 where the As^{5+} concentration in solution is less than 0.1 ppm and where the Fe:As mole ratio in solution is at least 10:1. The Drinking Water Standard for arsenic of 0.01 ppm (EPA 2012) is met between pH 4 and 7 and, according to this thermodynamic model, is maximum between pH 5 and 6. Using the average at pH 5.5, Reaction 24 for permanganate will occur at a potential near 1.15 V. Because Reaction 25 for arsenic occurs at a potential near 0.15 V, a strong potential difference ($\Delta E = 1.0$ V) between the two half-cell reactions is observed, suggesting that the overall redox reaction will be mass-transfer/diffusion controlled. Although not depicted for clarity purposes, ferrihydrite does not form at this pH level until approximately 0.2 V is exceeded, and while arsenate adsorption will occur at this point, its concentration will not go below 0.1 ppm until a potential near 0.35 V is reached.

Manganese Remediation

The Mn^{2+} stability region in Figure 8 denotes the E_H–pH conditions where its concentration falls below the manganese National Secondary Drinking Water Regulation of 0.05 ppm (EPA 2012). Its ionic precipitation as $Mn(OH)_2$ is also shown to occur at pH 11. While attainable, it is not cost-effective due to the amount of lime that would be needed; furthermore, the resulting water would have to be treated even more to meet the pH-8.5 requirements for discharge. Hence, its remediation is best accomplished through aqueous-phase oxidative precipitation. Considering pH 5.5 per the arsenic discussions, Reaction 26 for manganous ion oxidation occurs near 0.75 V, and with Reaction 24 occurring at 1.15 V, a moderate potential difference ($\Delta E = 0.4$ V) will occur. Thus, the overall reaction is also expected to be mass-transfer/diffusion controlled, albeit on the threshold of being chemical reaction control. Although the process is known to work, the only issue is that remediation is dependent on precision: too much or not enough permanganate will still leave manganese in solution. In this regard, the preferred oxidant is nonionic hydrogen peroxide (H_2O_2).

OTHER HYDROMETALLURGICAL PROCESSES

In this section, other examples of aqueous-phase precipitation processes are given but are not illustrated with E_H–pH diagrams. Although there is an abundance of other examples, only five current industrial and research examples are emphasized, with minor discussions given as needed. While the same concepts apply, it is only necessary to show applications with other types of redox reagents and/or metals. This is made simpler by showing only the overall redox reactions instead of half-cell reactions.

Tollens' Process

Tollens' process is an electroless plating technique commonly applied to coating glass for optical purposes. It is used for making silver mirrors and thus referred to as silvering (Antonello et al. 2012). Because the reaction is similar to nonionic organic reduction, it is also being used to make nanoparticles, nanowires, and microcircuits (Purdy and Muscat 2016). The reaction is frequently used as a demonstration in chemistry labs around the world to coat the inside of flasks or the surface of glass slides:

$$2Ag(NH_3)_2^+ + 3OH^- + RCHO = 2Ag^\circ + RCOO^- + 4NH_3 + 2H_2O \quad \textbf{(EQ 28)}$$

where RCHO and $RCOO^-$ are usually aldehydes and carboxylates such as glocose and gluconic acid, respectively.

Se/Te Removal Process

Because sulfate is considered the most stable form of sulfur, sulfur-containing compounds with lower oxidation states than sulfate tend to be excellent reductants. In this regard, sulfurous acid (H_2SO_3) is used to remove selenite (SeO_3^{2-}) and tellurite (TeO_3^{2-}) from copper-bearing leach solutions prior to copper electrowinning, particularly in platinum group metal refineries (Lottering et al. 2012):

$$2Cu^{2+} + SeO_3^{2-} + 4H_2SO_3 + H_2O = Cu_2Se^\circ + 4H_2SO_4 + 2H^+ \quad \textbf{(EQ 29)}$$

where the Cu_2Se° is an alloy that is filtered and sold. TeO_3^{2-} reacts the same way and consequently is sold with the Cu_2Se°. High temperatures may be needed for both reactions, particularly if both Se and Te are present in higher oxidation states.

Sherritt Gordon Process

Following ammoniacal leaching of Ni-Co sulfide concentrates, a typical leach solution at the Fort Saskatchewan operations in Alberta, Canada, would contain ~100 g/L free ammonia (NH_3), 5–10 g/L sulfur mostly as thiosulfate ($S_2O_3^{2-}$), 40–50 g/L Ni^{2+}, 0.7–1 g/L Co^{2+}, and 5–10 g/L Cu^{2+} (Wadsworth 1987). As soon as the free ammonia was thermally removed by steam injection, the thiosulfate would disproportionate into sulfide and sulfate, with the former causing the cuprous (Cu^{2+}) to precipitate:

$$Cu^{2+} + S_2O_3^{2-} + H_2O = CuS + SO_4^{2-} + 2H^+ \quad \textbf{(EQ 30)}$$

In this case, it is the change in state of the sulfur that leads to the metal being ionically precipitated. Importantly, dithionite ($S_2O_4^{2-}$) and other sulfoxy compounds are known to cause reductive precipitation of cupric to metallic copper but is not observed here, presumably because the chemical environments are distinctly different.

Cyanide Destruction Processes

Similarly, the oxidation of cyanocomplexes can also lead to metal precipitation (Young et al. 2001). This includes reactions with several dissolved oxidants including but not limited to hydrogen peroxide (H_2O_2), Caro's acid (H_2O_5), bleach/hypochlorite (OCl^-), bisulfite (HSO_3^-), and metabisulfate ($S_2O_7^{2-}$). Because the reactions are similar, only that involving hypochlorite is depicted:

$$M(CN)_4^{-2} + 4OCl^- + 2OH^- = 4OCN^- + 4Cl^- + M(OH)_2 \quad \textbf{(EQ 31)}$$

where, in this case, M represents a common divalent weak-acid-dissociable cyanide such as Zn^{2+} and Cd^{2+}, for example. It is understood that additional hypochlorite will form carbonate (CO_3^{2-}) and either nitrite (NO_2^-) or nitrate (NO_3^-). This particular process also works for free cyanide and thiocyanate but not strong-acid-dissociable cyanide. Furthermore, it is not normally used on a large scale but is commonly used for cyanide destruction in most chemical and research labs.

ARD Schwertmannite Process

Schwertmannite is a solid solution found as a precipitate in ARD systems. It has an approximate stoichiometry of ($Fe_8O_8(OH)_{4.8}(SO_4)_{1.6}$) and can be synthesized by contacting ARD with hydrogen peroxide (H_2O_2):

$$8Fe^{2+} + 1.6SO_4^{2-} + 2H_2O_2 + 8.8H_2O = Fe_8O_8(OH)_{4.8}(SO_4)_{1.6} + 12.8H^+ \quad \textbf{(EQ 32)}$$

As with permanganate, hydrogen peroxide oxidizes ferrous (Fe^{2+}) to ferric (Fe^{3+}) to induce the precipitation. Furthermore, arsenite (AsO_3^{3-}) will also oxidize to arsenate (AsO_4^{3-}), which can then substitute for sulfate in the compound (Liu et al. 2015).

REFERENCES

Al-Thabaiti, S.A., Obaid, A.Y., and Khan, Z. 2015. Gold nanocomposites: Kinetic, mechanistic and structural effects of reducing agents on the morphology, *Can. Chem. Trans.* 3(1):12–28.

Antonello, A., Jia, B., He, Z., Buso, D., Perotto, G., Brigo, L., Brusatin, G., Guglielmi, M., Gu, M., and Martucci, A. 2012. Optimized electroless silver coating for optical and plasmonic applications. *Plasmonics* 7.

Ayoub, A.H., Safadi, N., Inibtawi, M., and Shadafny, S. 2016. Formulation of small size gold nano-particles stabilized with novel polyethylene glycol (PEG)-n-acetyl cysteine (NAC) conjugate, *J. Chem.* 5(4):16–21.

Brenner, A., and Riddell, G.E. 1946. Nickel coating on steel by chemical reduction. *J. Res. Nat. Bur. Std.* 37(1):31–34.

Davis, A., and Ashenberg, D. 1989. Aqueous geochemistry of the Berkeley Pit, Butte, Montana, U.S.A. *Appl. Geochem.* 4:23–36.

Dzimitrowicz, A., Jamroz, P., Greda, K., Nowak, P., Nyk, M., and Pohl, P. 2015. The influence of stabilizers on the production of gold nanoparticles by direct current atmospheric pressure glow microdischarge generated in contact with liquid flowing cathode. *J. Nanopart. Res.* 17(4):185–192.

EPA (U.S. Environmental Protection Agency). 2012. *2012 Edition of the Drinking Water Standards and Health Advisories.* EPA 822-S-12-001. Washington, DC: EPA.

Feather, A., Sole, K.C., and Bryson, L.J. 1997. Gold refining by solvent extraction—The Minataur™ process. *J. SAIMM* 105(July/August):169–173.

Freitas, R.M., Perilli, T.A.G., and Ladeira, A.C.Q. 2013. Oxidative precipitation of manganese from acid mine drainage by potassium permanganate. *J. Chem.* Vol. 2013.

Gee, G.E. 1920. *Recovering Precious Metals from Waste Liquid Residues: A Complete Workshop Treatise Containing Practical Working Directions for the Recovery of Gold, Silver and Platinum from Every Description of Waste Liquids in the Jewellery, Photographic, Process Workers, and Electroplating Trades.* New York: Spon and Chamberlain.

Habashi, F. 1999. *A Textbook of Hydrometallurgy*. Saint Foy, QC: Métallurgie Extractive Québec.

Habashi, F. 2016. Purple of Cassius: Nano gold or colloidal gold? *Eur. Chem. Bull.* 5(10):416–419.

Hoke, C.M. 1982. *Refining Precious Metal Wastes: Gold, Silver, Platinum Metals; A Handbook for the Jeweler, Dentist, and Small Refiner*. New York: Metallurgical Publishing.

House, C.I., and Kelsall, G.H. 1984. Potential–pH diagrams for the Sn/H_2O-Cl system. *Electrochim. Acta* 29(10):1495–1464.

Huang, H-H. 2018. *StabCal: Stability Calculation for Aqueous System, Programs and User Manual.* Butte, MT: University of Montana, Metallurgical and Materials Engineering.

Huang, H.H., Twidwell, L.G., Anderson, C.G., and Young, C.A. 2005. Chemical titration simulation—An equilibrium calculation approach. In *Proceedings of the International Symposium on Computational Analysis in Hydrometallurgy*. Edited by D.G. Dixon and M.J. Dry. Montreal, QC: Canadian Institute of Mining, Metallurgy and Petroleum.

Hunt, L.B. 1976. The true story of purple of Cassius: The birth of gold-based glass and enamel colours. *Gold Bull.* 9(4):134–139.

Johnson, J.W., Oelkers, E.H., and Helgeson, H.C. 1992. SUPCRT92: A software package for calculating the standard molal thermodynamic properties of minerals, gases, aqueous species, and reactions from 1 to 5000 bar and 0 to 1000°C. *Comp. Geosci.* 16(7):899–947.

Leng, W., Pati, P., and Vikesland, P.J. 2015. Room temperature seed mediated growth of gold nanoparticles: Mechanistic investigations and life cycle assessment. *Environ. Sci. Nano* 2(5):440–453.

Liu, F., Zhou, J., Zhang, S., Liu, L., Zhou, L., and Fan, W. 2015. Schwertmannite synthesis through ferrous ion chemical oxidation under different H_2O_2 supply rates and its removal efficiency for arsenic from contaminated groundwater. *PLOS One* 10(9):e0138891.

Lottering, C., Eksteen, J.J., and Steenekamp, N. 2012. Precipitation of rhodium from a copper sulphate leach solution in the selenium/tellurium removal section of a base metal refinery. *J. SAIMM* 112(April):287–94.

Mallory, G.O. 1990. The fundamental aspects of electroless nickel plating. In *Electroless Plating: Fundamentals and Applications.* Edited by G.O. Mallory and B.J. Hajdu. New York: American Electroplaters and Surface Finishing Society. pp. 1–56.

Mallory, G.O., and Hajdu, B.J., eds. 1990. *Electroless Plating: Fundamentals and Applications.* New York: American Electroplaters and Surface Finishing Society.

Marsden, J.O., and House, C.I. 2006. *The Chemistry of Gold Extraction.* Littleton, CO: SME.

Miller, S.L., and Smith-Magowan, D. 1990. The thermodynamics of the Krebs cycle and related compounds, *J. Phys. Ref. Data* 19(4):1049–1073.

Navaladian, S., Janet, C.., Viswanathan, B., Varadarajan, T.K., and Viswanath, R.P. 2007. A facile room-temperature synthesis of gold nanowires by oxalate reduction method. *J. Phys. Chem.* 111(38):14150–14156.

Newman, J.D., and Blanchard, G.J. 2006. Formation of gold nanoparticles using amine reducing reagents. *Langmuir* 22(13):5882–5887.

Parinayok, P., Yamashita, M., Yonezu, K., Ohashi, H., Watanabe, K., Okaue, Y., and Yokoyama, T. 2011. Interaction of Au(III) and Pt(IV) complex ions with Fe(II) ions as a scavenging and a reducing agent: A basic study on the recovery of Au and Pt by a chemical method. *J. Colloid Interface Sci.* 364(1):272–275.

Purdy, S.C., and Muscat, A.J. 2016. Coating nonfunctionalized silica spheres with a high density of discrete silver nanoparticles. *J. Nanoparticle Res.* 18(3):70–92.

Sudagar, J., Lian, J., and Sha, W. 2013. Electroless nickel, alloy, composite and nano coatings—A critical review. *J. Alloys Compd.* 571:183–204.

Tarozaite, R., Juskenas, R., Kurtinaitiene, M., Jagminiene, A., and Vaskelis, A. 2006. Gold colloids obtained by Au(III) reduction with Sn(II): Preparation and characterization. *Chemija* 17(2-3):1–6.

Turkevich, J., Stevenson, P.C., and Hillier, J. 1951. A Study of the Nucleation and Growth Processes in the Synthesis of Colloidal Gold. *Discuss. Faraday Soc.* 11:55–75.

Twidwell, L.G., and McCloskey, J.W. 2011. Removing arsenic from aqueous solution long-term product storage. *JOM* 63(8):94–110.

Verma, H.N., Singh, P., and Chavan, R.M. 2014. Gold nanoparticle: Synthesis and characterization. *Vet. World* 7(2):72–77.

Wadsworth, M.E. 1987. Leaching—Metals applications. In *Handbook of Separation Process Technology*. Edited by R.W. Rousseau. Norwich, NY: Knovel.

Wagman, D.D. 1982. *The NBS Tables of Chemical Thermodynamic Properties: Selected Values for Inorganic and C_1 and C_2 Organic Substances in SI Units.* New York; Washington, DC: American Chemical Society and American Institute of Physics for the National Bureau of Standards.

Wojnicki, M., Zabieglinska, K., and Luty-Blocho, M. 2015. Influence on experimental condition on silver nanoparticle synthesis process in aqueous solutions. *Ores Nonferrous Met.* 60(3):103–109.

Wuithschick, M., Birnbaum, A., Witte, S., Sztucki, M., Vainio, U., Pinna, N., Rademann, K., Emmerling, F., Kraehnert, R., and Polte, J. 2015. Turkevish in new robes: Key questions answered for the most common gold nanoparticle synthesis. *ACS Nano* 9(7):7052–7071.

Young, C.A., Anderson, C.G., and Twidwell, L.G., eds. 2001. *Cyanide: Social, Industrial and Economic Aspects.* Warrendale, PA: The Minerals, Metals & Materials Society.

Zick, R.L., Velegol, D.A. Jr., Hess, M.W., and Foote, M. 2004. Butte mine flooding water treatment facility: Implementation of major component of selected remedy for historic contamination at Berkeley Pit site. In *Proceedings of the Joint Conference of the 21st Annual Meeting of the American Society of Mining and Reclamation and 25th West Virginia Surface Mine Drainage Task Force Symposium.* Edited by R.I. Barnhisel. Lexington, KY: American Society of Mining and Reclamation. pp. 2079–2104.

PART 11

Pyrometallurgy

CHAPTER 11.1

Pretreatment for Pyrometallurgy

Timothy C. Eisele and S. Komar Kawatra

This chapter covers the various pretreatments used to prepare mineral concentrates and direct-shipping ores so that they are suitable feed for smelting operations. These pretreatments are necessary to ensure that the smelter feed can be handled easily, has the proper chemical composition, and has the correct physical form to behave well in the smelting operation. These properties will vary depending on the smelting operation. Specifically, blast furnaces require feed particles that are large enough not to be entrained in the airstream and blown out of the furnace, while providing enough bed permeability that the reducing gases can flow through easily (Poveromo 1999). At the other extreme, flash smelting furnaces require a dry, free-flowing material with particles small enough to react quickly with the furnace gases.

Mineral concentrates are typically moist powders, and consist of specific minerals that may contain chemically bound volatiles such as carbon dioxide (CO_2) or water. The basic operations that are used to prepare the concentrates for smelting are those that remove volatiles and adjust the particle size and particle chemistry. These processes are

- Drying and calcination,
- Precombination with reductants and fluxes, and
- Agglomeration and other processes that alter physical properties.

CALCINATION

Calcination processes consist of heating minerals to decompose them into a gas phase and a solid phase. The most common applications of calcination are heating of hydrates to drive off water, and heating of carbonate minerals to drive off carbon dioxide.

Chemically Bound Water

A wide range of minerals contain water of hydration, with water included in their crystal structure. Removing this water requires higher temperatures than simple removal of surface water, which evaporates fully at only 100°C. Examples of common minerals that contain chemically bound water are gypsum ($CaSO_4 \cdot 2H_2O$) and goethite (FeO-OH, which is a composition equivalent to $Fe_2O_3 \cdot H_2O$), along with many others. Removal of this water requires heating and results in changes in the crystalline structure of the mineral. The products from removal of chemically bound water tend to be either powders or highly porous solids.

Carbon Dioxide

Carbonate minerals contain the CO_3^{-2} anion as part of their crystalline structure. Prominent examples are $CaCO_3$, $CaMg(CO_3)_2$, and $FeCO_3$, among many others. Upon heating, these carbonates decompose to their corresponding oxides, releasing CO_2 gas. The decomposition temperature is highly variable, with some carbonates decomposing at a few hundred degrees whereas others (such as $CaCO_3$) require temperatures above 900°C.

Other Volatile Components

Ore concentrates may also contain volatile organics, either natural or added in the course of processing. Many of these can be removed by simple vaporization, although in an oxidizing atmosphere some will partially combust, leaving a carbonaceous residue that will need to be burned away to remove it.

BLENDING AND FLUXING

If the concentrate is to be sent to a smelting operation, it is often beneficial to pre-blend it with components that will be necessary reactants in the smelter. By pre-blending, it is possible to achieve better contact of the ore with its reactants, resulting in more complete reactions and shorter reaction times.

Fluxes

Fluxes are materials that will improve melting of gangue minerals and removal of impurities during smelting. The most common of these are CaO and MgO, which are readily available from the minerals limestone ($CaCO_3$) and dolomite ($CaMg(CO_3)_2$). Because limestone and dolomite are insoluble, they can be ground and combined with the mineral concentrate slurry before the slurry is filtered, which provides

Timothy C. Eisele, Assistant Professor, Department of Chemical Engineering, Michigan Technological University, Houghton, Michigan, USA
S. Komar Kawatra, Professor & Chair, Department of Chemical Engineering, Michigan Technological University, Houghton, Michigan,USA

uniform blending with the ore (Malysheva and Chemyshev 1974; Roorda et al. 1981; Abouzeid et al. 1985; Bleifuss and Goetzman 1991). When the ore concentrate is then either heated to fire pellets, or smelted, the carbonate minerals decompose to their oxides, which then react with the silica and alumina gangue to produce slags with sufficiently low melting points to be practical. One issue to be careful of is that these minerals can reduce filtration rates by changing the surface chemistry of the mineral particles, so a mineral concentration plant that is filtration limited may have much reduced capacity when it adds flux to its concentrate slurry before filtration.

Integral Reductants

When the mineral concentrate is an oxidized form that needs to be reduced to metal, it can be helpful to combine the ore with a solid reductant (Goksel et al. 1987). In the case of iron oxides, carbonaceous reductants combined directly with the ore concentrate reduce the time taken for reduction to only 10–15 minutes, compared to several hours using gaseous reductants such as CO. The reason for the more rapid reduction with the combined reductant is that the carbonaceous solids are already where they are needed, and they do not need to diffuse in from the gas phase. If a mixture of ore concentrate and reductant is formed into pellets and heated, the pellets will self-reduce to produce metal and gases, even in an oxidizing atmosphere, because the reducing gases produced inside the pellet flow outward, preventing oxygen from entering the pellet.

Most work using integral reductants uses either coal or coke breeze, which are nearly pure carbon, low-cost, and stable in contact with moisture. Many other organics that are not essentially pure carbon will also perform well, including starches, flour, agricultural by-products like sugarbeet pulp and bagasse, molasses, wood pulp, plastics, and a broad spectrum of similar substances. These other materials will begin to decompose at lower temperatures than coke breeze, which can lead to more rapid reduction of the metal oxide in the pellet. Many of them are also capable of acting as binders, which makes them very useful additives if the ore concentrate is to be pelletized or briquetted. They can also generate volatile organic acids and soot, which must be considered in the construction of smelters using such materials as reductants.

Although there are advantages of including integral reductants in pellets produced for smelter feed, it can sometimes be a problem to accomplish this. Many pelletization processes depend on heating the pellets to harden them, and during the heating any integral reductant will burn away before it ever has a chance to contribute to reduction in the smelter. Use of reductants to make self-reducing pellets therefore depends on the use of a cold-bonding process to harden the pellets, rather than high-temperature hardening.

AGGLOMERATION

Agglomeration processes form fine particulates into larger shapes, and then bind them together to give the agglomerates sufficient mechanical strength that they can be handled and processed. These shapes can be spherical balls, cylinders, briquettes, or random-shaped masses, depending on the processes used to produce them. The bonding can be due to addition of specific binders to hold the fine particles together inside the agglomerates, or due to thermal treatments that cause the particles to sinter and fuse together (Poveromo 1999).

Sinter and Nodule Plants

Sinter plants produce agglomerates by heating the fine particles sufficiently that they fuse together, and no supplemental binder is used. These are primarily used for iron oxide concentrates that consist of particles of about a few millimeters in diameter. The particles to be sintered are combined with a solid fuel (typically coal or coke breeze) and placed on a heat-resistant traveling grate while drawing air down through the particle bed. The bed of particles is then ignited on the surface, causing it to burn at a temperature of approximately 1,300–1,480°C. As the bed of particles is carried along by the traveling grate, the flame front travels down through the bed, progressively sintering it throughout its entire depth. After the fuel burns out, the resulting fused, porous material is cooled and screened into appropriate sizes for use as furnace feed, typically 25 mm × 6 mm. Agglomerates larger than this can be crushed to size, and fines can be recirculated back through the process.

The sinter product is subject to abrasion and breakage during handling, so sinter plants are typically located very close to the furnaces that they feed. In addition to agglomerating ores, they are also used for recycling in-house fines such as furnace dusts and mill scale. It is difficult to run a sinter plant without leakage of combustion gases, so they are not favored in areas where emissions regulations are stringent.

Nodulizing is similar to sintering in that it does not require the addition of binders, but it differs in that the concentrate is not directly combined with a fuel. In a nodulizing plant, fine concentrate is fed to a rotary kiln that is operated at a sufficiently high temperature that portions of the fines are beginning to melt.

As the charge is tumbled in the kiln, it rolls into roughly spherical nodules that are bonded together by the molten portion of the fines. After the nodules exit the kiln, they are very durable and high strength. The process has the advantage of being largely insensitive to feed variables such as moisture content and particle size. However, it also has the disadvantages of high fuel consumption, difficulty in preventing the formation of "rings" of sintered material adhering inside the kiln, nonuniform nodule size, and low nodule porosity leading to poor reducibility. As a result, nodulization has fallen out of general use.

Pelletization

Pelletization is a two-step process, where the fine particles are first agglomerated by rolling them into a "green ball" that is held together by moisture and a binder material. The green balls are then hardened in a second step, usually by drying and heating sufficiently for the individual particles in the ball to sinter together. The resulting pellets are uniform in size, durable, and sufficiently porous to be easily smelted (Sastry and Fuerstenau 1972; Abouzeid et al. 1979).

The particles making up pellets are held together by capillary forces, adhesional and cohesional forces, electrostatic and van der Waals forces, and mechanical interlocking of grains. The relative strengths of these forces are shown in Figure 1. Note that the bonding strength increases as the particle size decreases, due to the increased particle surface available over which bonding can occur (Stone and Cahn 1968).

Different forces dominate at different points in the pelletization process. Freshly made wet pellets are primarily bonded together by capillary forces, with small contributions by van der Waals and electrostatic forces. After the pellets are

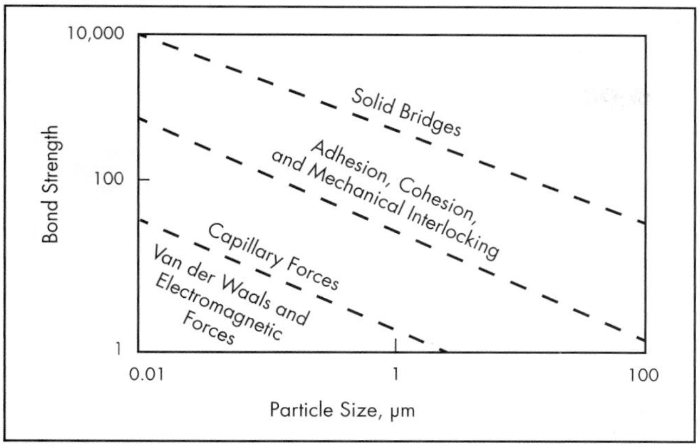

Adapted from Sastry 1996a

Figure 1 Relative strengths of different bonding mechanisms as a function of particle size

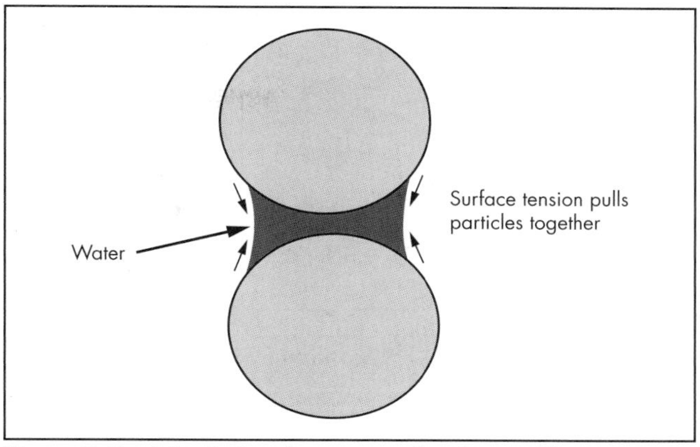

Figure 2 Particles being drawn together by capillary action

dried, the loss of moisture eliminates the water responsible for the capillary forces, and if the pellets are to hold together, the water is replaced by adhesional and cohesional forces from the binder. In the case of certain cold-bonding binders, the particles can also be held together by covalent bonds that result from chemical reactions. Firing of pellets forms solid bridges between mineral grains, giving the maximum strength. Mechanical interlocking is usually not significant unless the pellets are composed of ductile particles, and the pellets are compressed with sufficient force for the particles to deform around one another and lock together.

For pelletization to be successful, the feed ore must (1) have a sufficiently fine particle size distribution; (2) be moist enough that the individual mineral grains can stick together into pellets, but not be so moist that the ore becomes "muddy"; and (3) contain a binder to hold the pellet together during the period between drying and final hardening.

Binders

The availability of a suitable binder is key to making the pelletization process work. It is possible to form green balls using only water to bind the balls together, with the surface tension of the water drawing the mineral grains together as shown in Figure 2. This occurs when there is enough water present to bridge between particles, but not so much that the pore space is entirely filled with water, which, depending on the size distribution of the ore, typically occurs at approximately 9%–12% moisture by weight. However, the pellets must be dried before they are fed to a pyrometallurgical process, and it is necessary to have a non-evaporating binder to keep the balls from immediately falling apart again after they are dried. Binders also help to make the moist ore more plastic than it would be otherwise, resulting in more uniform and controllable pellet growth. Many different binders are possible, depending on the nature of the material being pelletized and the desired final pellet properties.

Binders can be classified into five groups, based on their mechanism of action (Holley 1982):

1. **Inactive film.** The binder is an adhesive material that adheres to the mineral grain surfaces and binds them together through capillary forces, adhesional forces, and cohesional forces. Green pellets bonded with inactive film binders can be broken up and re-agglomerated because the binding action is reversible.
2. **Chemical film.** The binder film undergoes an irreversible chemical reaction on the particle surfaces, causing the pellet to harden.
3. **Inactive matrix.** The binder is added at a high enough dosage that it forms a continuous matrix that the particles are embedded in. The binder is a material such as tar, pitch, or wax that is either made into a fluid emulsion in water, or heated enough to make it fluid. It then hardens upon drying or cooling. The high viscosity of the binder may require high compaction forces to form the pellets.
4. **Chemical matrix.** Similar to the inactive matrix, except that the hardening of the binder is an irreversible chemical reaction, such as the hardening of portland cement.
5. **Chemical reaction.** The binder and mineral particles undergo a chemical reaction to bond very strongly together. This could include dissolution and recrystallization of the mineral particles.

In most cases, ore pelletization uses film binders because they are used at low dosages. They may be inactive films such as clays or organic binders, or active films such as cements, as shown in Figure 3.

Bentonite. Bentonite clay is the most widely used binder for ore concentrate pelletization, due to its ease of handling, effectiveness at low dosages, low cost, and ability to continue acting as a binder all the way up to sintering temperatures. It is normally used as an inactive film type of binder. Bentonites are primarily composed of the clay mineral montmorillonite, which consists of aluminosilicate layers with a net negative charge, held together by exchangeable cations between the layers. There are two primary types of bentonite: sodium bentonites, where the exchangeable interlayer cations are mainly Na^+ ions; and calcium bentonites, where the interlayer cations are Ca^{2+} ions (Grim 1968). Clays other than bentonite can be used, although they are somewhat less effective due to their lesser swelling ability in water (Ehrlinger et al. 1966; Haas et al. 1987).

Sodium bentonites have a very high water-absorption capacity, because the monovalent sodium ions are readily solvated by water molecules. This allows water to be drawn between the layers, with the clay absorbing as much as 900% of its weight in water. Calcium bentonites have much lower water absorption capacity, as illustrated in Figure 4. The

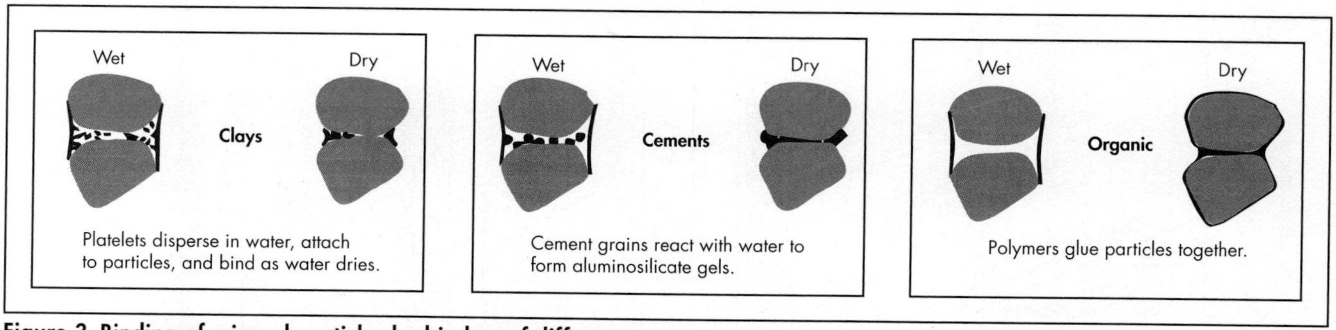

Figure 3 Binding of mineral particles by binders of different types

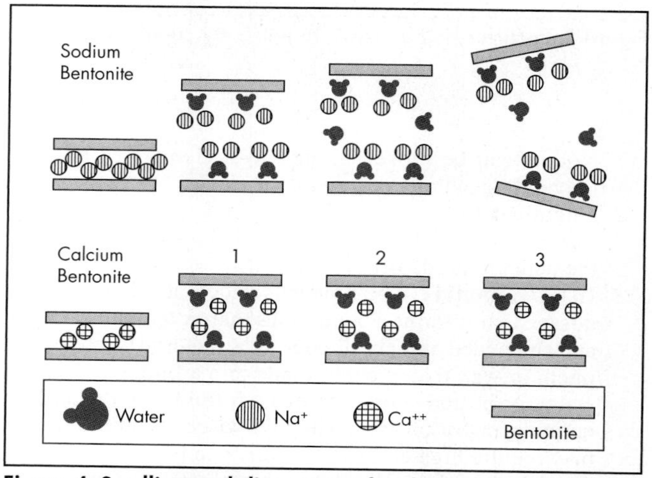

Figure 4 Swelling and dispersion of sodium bentonite versus calcium bentonite

divalent calcium ions bond more strongly than the monovalent sodium ions; therefore, the calcium bentonite does not hydrate as readily and has much less water absorption capacity.

The expansion of the bentonite allows the clay platelets to be spread over the surface of the particles being bonded. The platelets then bridge between the particles. The effectiveness of the binder can be enhanced by altering the mixing conditions. For example, Figure 5 shows glass beads bonded by bentonite. In Figure 5A, the beads and bentonite were combined using a conventional mixer, whereas the material in Figure 5B was mixed by passing it between a set of compression rolls 20 times. The compressive shear mixing used to produce sample B clearly spread the bentonite much more thoroughly over the surfaces of the particles being bonded. This improvement in mixing approximately doubles the green pellet strength produced by the bentonite at any given dosage (Kawatra and Ripke 2001, 2002).

Bentonite has several applications other than as a binder for mineral agglomerates, so many different tests for measuring the quality were developed for each of these applications. These tests that may be reported by bentonite suppliers include the following:

- Batch balling
- Enslin water absorption
- Alumina plate water absorption
- Grit content
- Moisture content
- Size distribution
- Marsh funnel
- Gel strength
- Colloid content
- Chemical analysis
- Methylene blue uptake
- Free swelling
- Exchangeable cations by atomic absorption spectroscopy
- Glycolated layer expansion

Not all of these tests are relevant to mineral agglomeration, as many of them relate to the use of bentonite in moisture absorption, gelling, and water purification applications. Further details on these tests are provided elsewhere (Eisele and Kawatra 2003).

The most useful test for evaluating the value of bentonite as a binder is the batch balling test, where the bentonite is actually used to make pellets using the exact mineral that is to be agglomerated. The procedure for batch balling has never been fully standardized, so each ore producer typically uses its own ore concentrate, apparatus, and specific procedures, which makes it difficult to compare bentonite performance between different laboratories. The most nearly standardized procedure is that developed by the Bentonite Users Committee (1982a, 1982b), which is as follows.

Equipment

- A 600-6 airplane tire (40.6 cm diameter × 15.24 cm wide), mounted to rotate at 52 rpm, with access to one side for adding feed mix and removing pellets
- Model no. 1, Cincinnati Muller, 30.48 cm diameter
- Screens: 4.75 mm, 3.35 mm, 13.2 mm, 12.5 mm, and either 2.0 mm or 1.7 mm
- Spray bottle filled with distilled water
- Balance accurate to 0.1 g, with 3-kg capacity
- Scoop for removing balls from tire
- Airtight, moisture-resistant containers for seeds and balls

Materials

- At least 2,500 g of mineral to be agglomerated (dry weight). This should be at the appropriate moisture content for pelletization, which varies from 8.5% moisture to 12% moisture depending on particle size distribution and mineralogy.
- Dry, powdered bentonite, in sufficient quantity to reach the desired dosage in the material to be agglomerated. For iron ores, a typical dosage is near 0.5%–0.66% bentonite by weight.

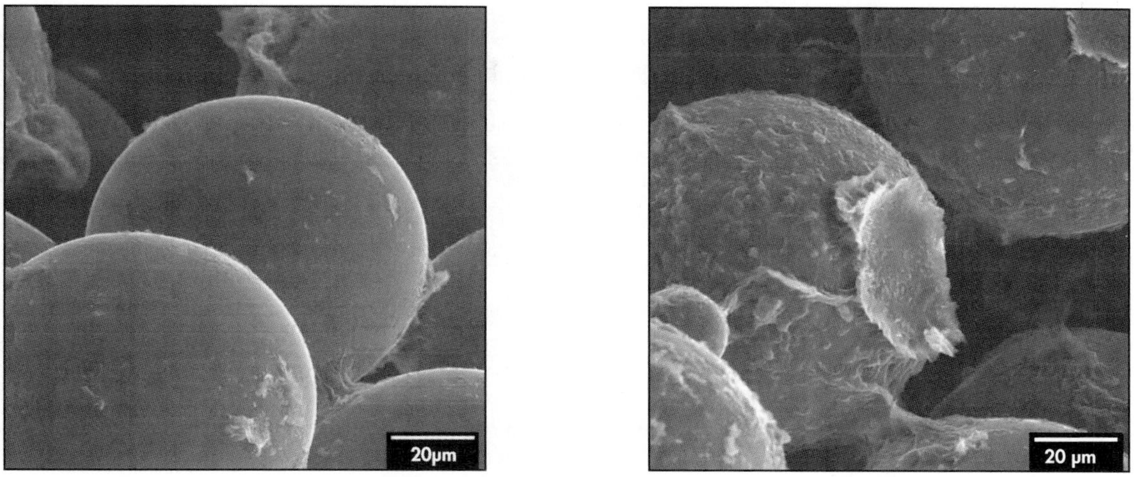

A. 0 Passes B. 20 Passes

Source: Kawatra and Ripke 2001

Figure 5 Effect of compressive shear mixing using compression rolls on the distribution of bentonite over glass bead surfaces

Procedure

- Weigh out the appropriate amount of bentonite for the quantity of ore being agglomerated.
- Spread the concentrate uniformly in the muller, and distribute the bentonite uniformly over the top of the concentrate.
- Mix for 3 minutes. If the mix is clearly too dry, additional water can be slowly added ahead of the muller wheel after 1 minute of mixing.
- Remove material from the muller, and force through either 2.0-mm or 1.7-mm screen to delump the material.
- Start the balling tire rotating, and take approximately 700 g of the ore/bentonite/water feed mixture to use to produce seeds. Add a small amount of the feed mixture, spraying lightly with water to start seed formation.
- When the top size of the seeds approaches 4.75 mm, use the scoop to remove the seeds from the tire, and screen them at 4.75 mm and 3.35 mm. Discard the +4.75-mm material, save the 4.75 × 3.35-mm seeds in a sealed container, and return the –3.35-mm material to the balling tire. Resume adding feed mixture and spraying with water, and repeat the screening procedure until approximately 34 g of seeds have been produced.
- Return the 34 g of 4.75 × 3.35-mm seeds to the balling tire, and slowly add more feed mixture over a 6-minute period, spraying water as needed to cause the seeds to enlarge into pellets. When the pellets approach the target size, remove the pellets and screen them at 13.2 mm and 12.5 mm. Discard the +13.2-mm material, retain the 13.2 × 12.5-mm balls for testing, and if additional balls are needed, return the –12.5-mm balls to add additional feed mixture and enlarge them to the target size.

Pellets produced using this procedure can then be tested as green pellets, or dried and fired for additional tests as appropriate.

Bentonite is an extremely popular binder due to its acceptable cost, availability, and effectiveness. The primary limitation to its use is the silica content, as a typical bentonite is approximately 35% SiO_2. After processing the ore concentrates to remove silica, it is somewhat counterproductive to then re-add silica as a component of the bentonite. Much of the interest in replacing bentonite as a binder is therefore to use a binder that is at least lower in silica, or even entirely free of it.

Other inorganic binders. Inorganics other than bentonite clay that can potentially be used as binders include sodium hydroxide, potassium hydroxide, soluble salts, portland cement, lime, calcium carbonate, magnesia, and sodium silicate. Only a few of these have been used industrially for ore concentrate agglomeration, due to issues with cost, contamination, and low strength. Most do not have performance as good as bentonite clay. The inorganic binders that have been used are primarily portland cement, lime, and calcium carbonate.

Portland cement. Portland cement is primarily used for cold-bonding applications, where it is intended to form pellets that reach high strength at room temperature (Dutta et al. 1992, 1997; Go and Tate 1982; Iyengar et al. 1968). It is most effective as an "active matrix" binder, where it forms a continuous matrix bonding the mineral grains together. This requires a dosage of approximately 10% portland cement by weight, with sufficient moisture for the cement to chemically react with the moisture and harden. Cement has been used in the Grancold and COBO processes for forming iron ore pellets. The Grancold process formed wet pellets that were bedded in loose iron ore concentrate powder for 3–6 days to prevent them from sticking together while hardening. The pellets were then stockpiled for an additional 30 days to reach a high enough strength for processing (Svensson 1969). Although the pellets produced were reported to be adequately fluxed by the lime in the cement and to have good reduction properties, the long curing time made it uncompetitive (Linder and Thulin 1973). The COBO process used steam curing to accelerate the hardening, but the added cost of the steam made this process uncompetitive as well (Doughty 1975).

Lime. Lime can be used alone as a binder, as it will hydrate in water and bond mineral particles together (De Souza 1976). This action is enhanced if silica is present, as lime and silica will react in the presence of water to form calcium hydrosilicates. This can make use of the silica gangue already present in the ore concentrate, and therefore does not necessarily add more silica. If the natural silica is not sufficiently reactive, it is possible to add pozzolanic materials such as coal fly ash to react with the lime, producing a material very similar to portland cement (Eisele et al. 1998a, 1998b).

The process can be considerably accelerated by operating under hydrothermal conditions. The basic process is to combine ore concentrate, CaO, sufficient water for pelletizing, and any other additives desired in the pellet. After the green pellets are formed, they are autoclaved for 1–2 hours in 2,070 kPa steam at approximately 200°C to develop the amorphous calcium hydrosilicates (Goksel et al. 1987) The disadvantage of this approach is the need to convey pellets into and out of a pressurized autoclave, which adds considerable process complexity and cost.

Calcium carbonate binder. Calcium hydroxide reacts with carbon dioxide to form calcium carbonate, which is stable up to approximately 900°C. If ore is combined with approximately 7.5% calcium hydroxide, along with sufficient water to make the material adhesive, it can be pelletized using standard equipment. The pellets are then treated with CO_2 gas, so that the carbonation reaction can proceed:

$$Ca(OH)_2 + CO_2 * CaCO_3 + H_2O$$

Although this produces pellets with compressive strengths in the range of 200–500 lb/pellet, it has not been widely pursued industrially because of the cost of the CO_2 treatment and the need for producing large quantities of calcium hydroxide.

Organic binders. A broad range of organic materials exist that are sufficiently "sticky" that they have been at least considered as ore binders. Those that were reported as having been tested for this application are listed in Table 1.

Although using low-cost organics such as dairy wastes, molasses, or paper by-products as binders is very attractive, in general their usefulness is limited both by difficulty in transporting and storing them without rotting, and by their generally poor performance compared to purpose-manufactured binders. As a result, ore pelletization with organic binders has been limited to only a few products. The organic binders that have been commercialized are the Alcotac series (Ciba Specialty Chemicals), which are acrylic copolymers, and the Peridur series (AkzoNobel), which are cellulose derivatives (Kortman et al. 1987).

The primary advantage of organic binders is that they do not contribute silica to the finished pellet, making it easier to meet the product silica specification and increasing plant productivity. They are also reported to produce more uniform pellets that are less likely to spall during preheating. Some of these binders can also be added at considerably lower dosage than bentonite while still achieving comparable strength. For example, Peridur at a dosage of 0.1 wt % gives the same pellet dry compressive strength as bentonite at 0.5 wt %. This is largely offset by the higher cost of the organic binders compared to bentonite.

A serious problem with organic binders is the fact that, when the green pellets are heated to harden them, organic binder ignites and burns away starting at approximately 250°C, whereas ores such as magnetite do not begin to oxidize and recrystallize to form interparticle bonds until over 600°C. Other ores, such as hematite, require even higher temperatures to bond together, over 1,100°C. As a result, organic binders are not able to hold pellets together over most of the temperature range during preheating. This causes pellets to be excessively fragile, leading to high levels of dust production and pellet breakage.

Table 1 Organic materials that have been tested as binders for ore pelletization

Paper by-products • Hemicellulose • Caustic leonardite • Naphthalene sulfonate • Lignin sulfonates	Dairy wastes • Lactose • Whey permeate • Whey	Starches • Alkalized • Nongelled • Gelled
Cellulose derivatives • Carboxymethyl cellulose • Hydroxyethyl cellulose • Carboxyhydroxyethylcellulose	Starch/acrylic • Guar gums	Natural gums • Guar gums • Xanthan gums
Tars	Molasses	Humic acids
Bitumens	Ethylene oxide	Sawdust/wood flour
Polyacrylamides	Polyvinyl alcohol	Peat moss

There are options for reducing the problems with binders burning away during preheating. One is to increase the ignition temperature of the binder. Unfortunately, carbon compounds are not sufficiently stable in oxidizing environments, and will need significant advances in technology to find a "backbone" molecule that will outperform the existing molecular structures. Another option is to use "straight-grate" firing kilns that handle pellets more gently than the more common "grate-kiln" systems. The downside of this is that straight-grate firing has more issues with mechanical reliability and variation in pellets quality than the grate-kiln systems do.

Binder blends. The most promising approach to reducing the thermal stability issues with organic binders is to blend them with inorganic additives. These additives may be binders such as bentonite or lime, or may be compounds that are not binders by themselves but that melt or react with the ore particles to bond them together while the organic portion burns away.

AkzoNobel experimented with sodium tetraborate as a "pellet strengthener" additive for Peridur binder. When added at a rate of 4 parts sodium tetraborate/1 part Peridur, the pellet quality during preheating increased markedly, due to the ability of the sodium tetraborate to melt at a moderately low temperature (743°C). More recently, calcined colemanite ($Ca_2B_6O_{11}$) in combination with carboxymethylcellulose binder has given pellet strengths at 1,000°C that are comparable to the strengths using bentonite, and roughly twice the strength achieved using organic binders alone (Sivrikaya et al. 2013).

Funa, an organic binder produced by alkali extraction of humic acids from coal (Qiu et al. 2003), is considered a blend of organic and inorganic materials, since it contains approximately 40%–50% humic acid and 20%–25% inorganics such as silica, alumina, calcium salts, and iron salts. These inorganic components remain after the humic acid portion burns away, and as a result the Funa binder has excellent high-temperature performance. Because it is more hydrophobic than bentonite, the pelletization performance with Funa binder is reported to be very different, with slower pellet growth rate than when bentonite or other organic binders are used.

Bentonite can also be combined with organic binders such as starch, where the starch reduces the quantity of bentonite needed to produce satisfactory pellet strengths while the bentonite provides strength to the pellet during heating. This combination has occasionally been used on an industrial scale. In most cases the performance improvement has not been considered sufficient to warrant the added complexity.

In some cases, binders must not be blended because their modes of action are fundamentally incompatible. For example, lime cannot be combined with bentonite because the lime is a cementitious binder that works by reacting calcium hydroxide with silicates (Ripke and Kawatra 2000). The effectiveness of bentonite as a binder depends on its ability to absorb moisture and expand, and this ability depends on it having sodium ions present as its exchangeable cations. The calcium from lime displaces the sodium from the bentonite, converting it into the much less effective calcium bentonite. As a result of this phenomenon, experiments combining bentonite with a lime-based binder showed that the performance of the blended binder was dramatically worse than either the bentonite or the lime-based binder alone.

Balling Devices and Operations

The two most common types of pelletization devices are balling disks and balling drums. A balling disk consists of a shallow pan, tilted at an angle from the vertical, and rotating around its axis, as shown in Figure 6.

Moistened feed and binder are added to the angled, rotating pan. As the pan rotates, the forming pellets roll over the fresh feed, layering material on their surfaces and growing. Material that is not picked up by a rolling pellet is scraped off of the pan surface, producing more "seed" pellets. The forming pellets sort themselves by size, with the largest pellets rising to the top, so the pellets of the finished size discharge over the edge of the disk. The pellet size is adjusted by varying the feed rate.

A balling drum consists of a sloping cylindrical drum rotating around its long axis, with feed and recirculated pellets entering the uphill end and pellets discharging onto a screen at the downhill end, as shown schematically in Figure 7. Unlike the disk pelletizer, the screen in the drum pelletizer circuit provides positive control of pellet size. These are normally operated with a significant circulating load, with pellets passing through the drum two to three times before reaching the finished pellet size.

The formation of pellets from fine particles proceeds through three mechanisms, as shown in Figure 8. During nucleation, individual particles adhere together to form "seed" pellets a few millimeters in diameter. These seeds are large enough to begin rolling in the pelletizer. As they roll, they can coalesce with other, similarly sized pellets merging into a single pellet, which leads to a rapid increase in size while reducing the number of pellets present. Or, they can accumulate layers of unagglomerated fines onto their surfaces, which leads to slow, steady growth in size while keeping the number of pellets constant. During continuous operation of a pelletizer, all three of these mechanisms proceed simultaneously (Sastry and Fuerstenau 1972; Sastry 1996a, 1996b).

The forces compacting the individual grains into the pellets depend on the pellet mass and the size of the contact point between the pellet and the material being picked up. For example, consider the growing pellet shown in Figure 9. If we assume the mass of the pellet is 1 g, and the ore grain being picked up is 25 μm in diameter, then the entire weight of the pellet is being supported by the cross section of the 25-μm grain, which is equal to $\pi(0.0025/2)^2 = 4.91 \times 10^{-6}$ cm^2. The force exerted by the weight of the 1 g pellet is equal to its mass multiplied by gravitational acceleration, which is 0.001(9.81) = 0.00981 N. The pressure exerted by the pellet on the ore grain is $0.00981/(4.91 \times 10^{-6}) = 1{,}998$ N/cm^2, which is equal to 19,980 kPa. Although the actual pressures developed are likely to be lower than this value, clearly the pressure can nevertheless be quite high. The high point pressures developed lead to a firm, well-compacted final pellet.

Green Ball Quality Determination

The quality of pellets for firing depends on their ability to survive the handling and drying process while they are transported from the pelletizing device to the kiln. The three primary quality tests are shown in Table 2 (ASTM 1997). The virtue of these tests is that they can be performed very quickly and are therefore useful for quality control.

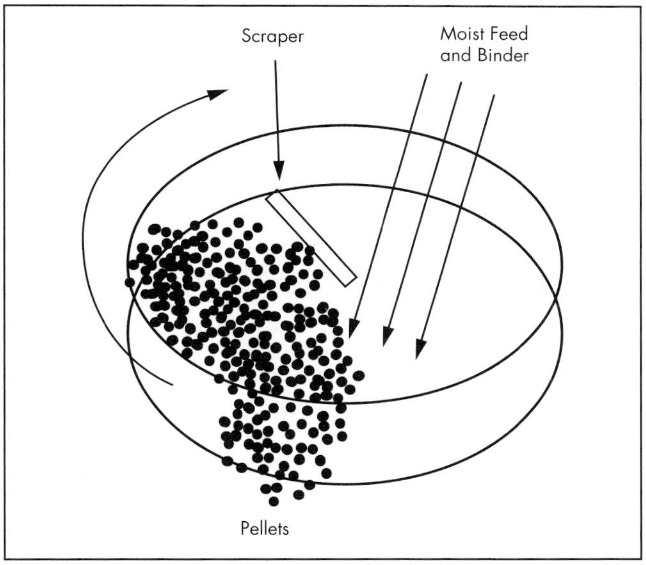

Figure 6 Disk pelletizer

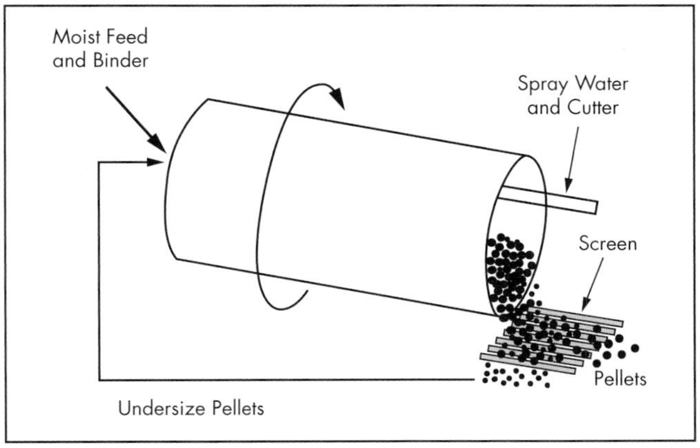

Figure 7 Drum pelletizer

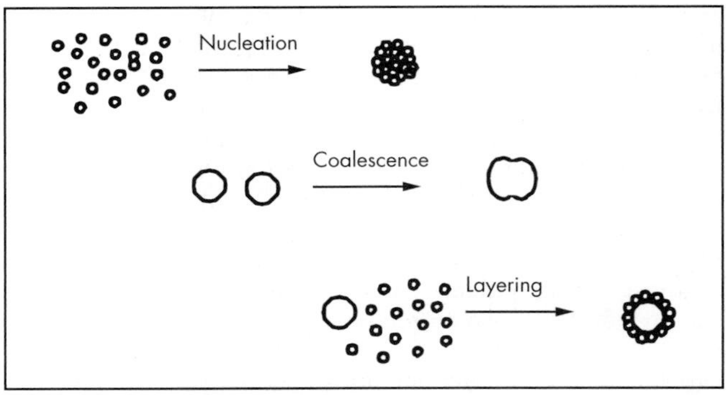

Figure 8 Pellet growth mechanisms

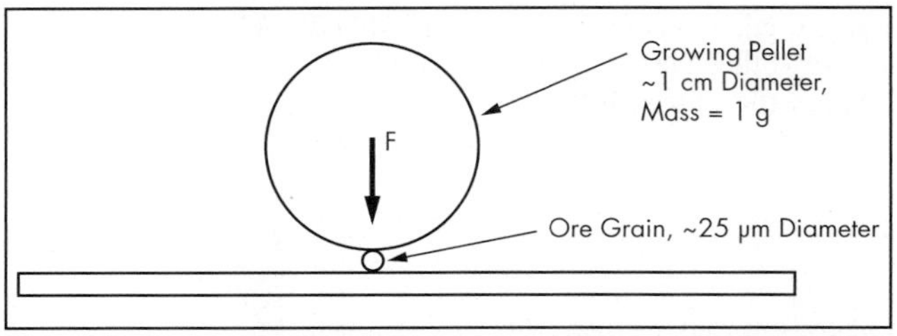

Figure 9 Estimation of compaction pressure during formation of a rolled pellet

Pellet Firing

The moist green pellets are primarily held together by capillary forces and typically have crushing strengths of about 10–15 N/pellet. This is not strong enough for handling and processing, which typically require strengths of about 1,800 N/pellet or higher. To achieve these strengths, the pellets are normally heated sufficiently that the mineral grains recrystallize and sinter together. This heating process must be carried out using appropriate heating rates to fully develop the final pellet strength.

Drying. The drying stage removes the moisture from the pellet. This must be done slowly so that the water vapor can escape without developing high internal steam pressure that can cause the pellets to spall their surfaces or explode. The drying stage is carried out on a moving grate and takes approximately 4 minutes.

Preheating. The preheating stage brings the pellet temperature up to the point where sintering can begin to occur. If there are carbonate minerals present, such as siderite, ankerite, dolomite, or calcite, the preheating stage will also calcine them to drive off their chemically bound CO_2. During the preheating stage, the pellet strength is entirely due to the binder, since water has already been evaporated and sintering has not yet begun. The preheating stage typically takes approximately 3 minutes.

Sintering. Pellets begin to sinter at approximately 930°C and are heated to 1,260°C for full sintering. If the ore is a mineral that can be oxidized, such as magnetite (Fe_3O_4), it will fully oxidize during the sintering step.

Sintering requires approximately 3 minutes to allow the pellet to fully harden. If the pellets are heated too quickly they may not sinter properly, leading to a shell of material surrounding a loose core, which results in a very weak pellet.

Table 2 Test procedures for "green" pellets, before firing

Test	Procedure	Use of data
Wet-knock	Repeatedly drop undried pellets 46 cm onto a steel plate, and count the number of drops needed for visible fracturing to occur. Repeat for at least 20 pellets, and calculate an average value.	Measures the ability of the wet pellet to remain intact during handling. Satisfactory values are between 5 and 10 drops.
Wet-crush	Undried pellets are crushed using a compression testing apparatus, with a compression speed of 40 mm/min. The peak load before fracture is recorded. Repeat for at least 20 pellets, and calculate an average value.	Measures the ability of the pellets to be stacked on each other without fracturing. Satisfactory values are greater than 11 N/pellet.
Dry-crush	Pellets are dried at 105°C for at least 1 hour (weigh before and after drying to determine moisture content). Crush as for the wet-crush test. Repeat for at least 20 pellets and calculate an average value	Measures the ability of the pellets to maintain their strength during the early stages of drying. Satisfactory values are greater than 22 N/pellet.

Cooling. Pellets are cooled with air. Allowing 5 minutes for cooldown will minimize the thermal shock to the pellets, and also allow the recovery of heat to be recycled to the process. The hot air exiting the coolers is then used as the air to feed the sintering, preheating, and drying stages, with combustion in the sintering stage used to provide heat for the whole process. The airflows must be designed to ensure that the air entering the drying stage is not saturated with moisture from fuel combustion, as this would interfere with drying.

Determining final pellet quality. The final pellets must have sufficient mechanical strength to survive intact during handling and be sufficiently free of fine particles that they will perform well in the pyrometallurgical operation. The three most important parameters are the fired crushing strength, the tumble test results, and the Q index (Sportel et al. 1997).

Fired crushing strength is determined by placing a single fired pellet on a compression testing machine and applying an increasing load until the pellet fractures. This is repeated for at least 20 pellets and the results averaged. For blast furnace feeds, this should be greater than 1,780 N/pellet.

Tumble tests are performed using ASTM Test Procedure E279. It consists of placing 11.33 kg of fired pellets into a steel drum of the dimensions specified in the standard. It is then rotated at 24 rpm for 200 revolutions. The pellets are then removed from the drum and screened to determine the percent coarser than 6.3 mm (the Tumble Index), and the percent finer than 600 μm (the Abrasion Index). The tumble test results are also used to calculate the Q index by multiplying the wt % coarser than 12.7 mm by the Tumble Index.

Briquetting

Like pelletization, briquetting typically requires a binder to hold the particles together in the agglomerate (Messman 1977). The key difference is that briquetting applies considerable mechanical force to the material to compress it together, rather than depending on only the weight of a growing pellet to provide compression (Koizumi et al. 1988). Briquetting is useful for materials that do not have an appropriate particle size distribution for pelletization. The limitation of briquetting is that the devices are considerably lower capacity than

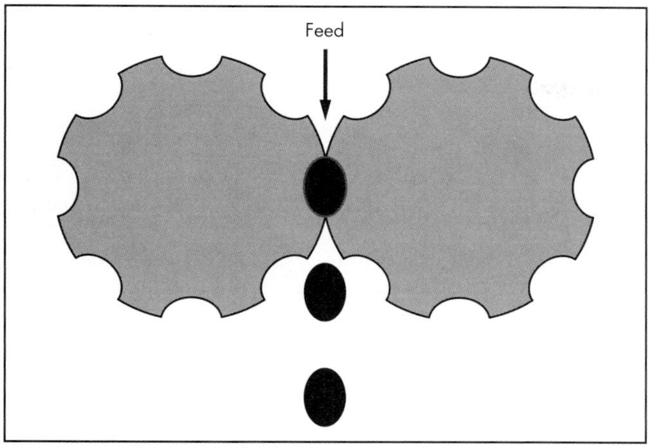

Figure 10 Schematic of a briquetting roll compressing powder into briquettes

pelletization drums and disks, and costs are higher due both to the energy needed for the compression and to the higher mechanical wear. The compression forces in briquetting are most commonly applied using briquetting rolls, which consist of rolls with pocketed surfaces that compress the initial particulates, as shown schematically in Figure 10. The size of the briquettes depends on the shape of the pockets in the briqueting rolls.

REFERENCES

Abouzeid, A.-Z. M., Seddik, A.A., and El-Sinbawy, H.A. 1979. Pelletization kinetics of an earthy iron ore and the physical properties of the pellets produced. *Powder Technol.* 24(2):229–236.

Abouzeid, A.-Z. M., Negm, A.A., and Kotb, I.M. 1985. Iron ore fluxed pellets and their physical properties. *Powder Technol.* 42(3):225–230.

ASTM E 382. 1997. *Standard Test Method for Determination of Crushing Strength of Iron Ore Pellets.* West Conshohocken, PA: ASTM International.

Bentonite Users Committee. 1982a. Batch Balling Results: Round Robin. February 5.

Bentonite Users Committee. 1982b. Revised Batch Balling Procedure for Reproducibility Tests. April 22.

Bleifuss, R.L., and Goetzman, H.E. 1991. Replacement of limestone and dolomite with lime–dolomite hydrate for the production of fluxed pellets. In *Proceedings of the Annual Meeting—Minnesota Section AIME, 64th Annual Meeting of the Society for Mining, Metallurgy & Exploration*, January 16–17. Littleton, CO: SME. pp. 195–206.

De Souza, R.P. 1976. Production of pellets in CVRD using hydrated lime as binder, is growing up fast. In *Ironmaking Proceedings of The Metallurgical Society of AIME Iron and Steel Division, 35th Annual Ironmaking Conference*, St. Louis, MO, March 29–April 1. pp. 182–196.

Doughty, F.T.C. 1975. COBO: A low-cost cold bond process. In *Proceedings of the 14th Biennial Conference of the IBA*, Hyannis, MA, August. pp. 173–182.

Dutta, D.K., Bordoloi, D., Gupta, S., Borthakur, P.C., Srinivasan, T.M., and Patil, J.B. 1992. Investigation on cold bonded pelletization of iron ore fines using Indian slag-cement. *Int. J. Miner. Process* 34(1-2):149–159.

Dutta, D.K., Bordoloi, D., and Borthakur, P.C. 1997. Investigation on reduction of cement binder in cold bonded pelletization of iron ore fines. *Int. J. Miner. Process.* 49(1-2):97–105.

Ehrlinger, H.P. III, Mirza, M.B., Camp, L.R., and Jackman, H.W. 1966. *Illinois Clays as Binders for Iron Ore Pellets: A Further Study.* Illinois State Geological Survey, Industrial Minerals Notes No. 28, November 1966.

Eisele, T.C., and Kawatra, S.K. 2003. A review of binders in iron ore palletization. *Miner. Process. Extr. Metall. Rev.* 24(1):1–90.

Eisele, T.C., Kawatra, S.K., and Banerjee, D.D. 1998a. High-carbon fly-ash binders for iron ore pellets. In *ICSTI-Ironmaking Conference Proceedings*. pp. 1175–1181.

Eisele, T.C., Kawatra, S.K., and Banerjee, D.D. 1998b. Binding iron ore pellets with fluidized-bed combustor fly ash. *Miner. Metall. Process.* 15(2):20–23.

Go, H., and Tate, M. 1982. Softening and melting behavior of cold bond pellets of iron ores with the use of portland cement as binder. *J. Iron Steel Inst. Jpn.* 69(14):1889–1895.

Goksel, A.M., Scott, T.A., Weiss, F., and Coburn, J. 1987. PTC cold bond agglomeration process and its various applications in the iron and steel industry. In *Proceedings of the 20th Biennial Conference of the IBA*, Orlando, FL. pp. 191–213.

Grim, R.E. 1968. *Clay Mineralogy*, 2nd ed. New York: McGraw-Hill.

Haas, L.A., Aldinger, J.A., Blake, R.L., and Swan, S.A. 1987. *Sampling, Characterization, and Evaluation of Midwest Clays for Iron Ore Pellet Bonding.* RI 9116. Washington, DC: U.S. Bureau of Mines.

Holley, C.A. 1982. Binders and binder systems for agglomeration. Presented at the *Institute for Briquetting and Agglomeration Biennial Conference.*

Iyengar, M.S., Dutta, S.N., Jana, B.C., Awatramoni, I.K., and Saikia, P.C. 1968. A new process for making iron ore pellets from hematite ore fines and/or from blue dust with cement as binder. Indian Patent No. 119163.

Kawatra, S.K., and Ripke, S.J. 2001. Developing and understanding the bentonite fiber bonding mechanism. *Miner. Eng.* 14(6):647–659.

Kawatra, S.K., and Ripke, S.J. 2002. Effects of bentonite fiber formation in iron ore pelletization. *Int. J. Miner. Process.* 65:141–149.

Koizumi, H., Yamaguchi, A., Doi, T., and Noma, F. 1988. Fundamental development of iron ore briquetting technology. *J. Iron Steel Inst. Jpn.* 74(6):962–969.

Kortmann, H.A., Bock, W., Van den Boogaard, V., and Kater, T. 1987. Peridur: A way to improve acid and fluxed taconite pellets. *Skillings Min. Rev.* (January 3):4–8.

Linder, R., and Thulin, D. 1973. Grancold-Pelletierung Von Eisenerzkonzentrat (Pelletizing of iron ores by the Grangcold process). *Aufbereitungs-Technik* 14(12): 799–802.

Malysheva, T.Y., and Chernyshev, A.M. 1974. Comparative study of the mechanism of the formation of fluxed and unfluxed pellets. *Steel USSR* 4(5):349–350.

Messman, H.C. 1977. Binders for briquetting and agglomeration. In *Proceedings of the 15th Biennial Conference of the IBA*, Montreal, QC. pp. 173–178.

Poveromo, J.J. 1999. Chapter 8: Iron ores. In *The Making, Shaping, and Treating of Steel*, 11th ed. Edited by D.H. Wakelin. Pittsburgh: Association for Iron and Steel Technology. pp. 569–664.

Qiu, G., Jiang, T., Fa, K., Zhu, D., and Wang, D. 2003. Interfacial characterizations of iron ore concentrates affected by binders. *Powder Technol.* 39(1):1–6.

Ripke, S.J., and Kawatra, S.K. 2000. Iron ore pellet binding mechanism, pozzolanic vs. physical binders. In *Proceedings of the XXI International Mineral Processing Congress*, Vol. A. Edited by P. Massaci. Amsterdam, Netherlands: Elsevier. pp. A4 120–127.

Roorda, H.J., Burghardt, D., Cortmann, M.A., and Kater, T. 1981. Production of fluxed iron-ore pellets for direct reduction using cellulose binders. *Preprints, 3rd International Symposium on Agglomeration*, May 6–8, 1981, Nurnberg, Germany. pp. B152–B164.

Rumpf, H. 1962. The strength of granules and agglomerates. In *Agglomeration*. Edited by H. Rumpf and W.A. Kneppe. New York: Interscience Publishers.

Sastry, K.V.S. 1996a. Role of bentonite in the balling of iron ore concentrates. Presented at the Bentonite Task Force Meeting of the Empire Pellet Plant, January 17.

Sastry, K.V.S. 1996b. Process engineering principles of iron ore balling circuits. Presented to Cliffs Mining Services Co., April 3.

Sastry, K.V.S., and Fuerstenau, D.W. 1972. Ballability index to quantify agglomerate growth by green pelletization. *J. Petrol. Technol.* 252(3):254–258.

Sivrikaya, O., Arol, A.I., Eisele, T.C., and Kawatra, S.K. 2013. The effect of calcined colemanite addition on the mechanical strength of magnetite pellets produced with organic binders. *Miner. Process. Extr. Metall. Rev.* 34(4): 210–222.

Sportel, H., Beentjes, P., Rengersen, J., and Droog, J. 1997. Quality of green iron ore pellets. *Ironmaking Steelmaking* 24(2):129–132.

Stone, R.L., and Cahn, D.S. 1968. Ultrafine-particle concentration and the strength of unfired iron ore pellets. *Trans. Inst. Min. Metall.* 241(December):533–537.

Svensson, J. 1969. The Grancold process. *Steel Times* (May):362.

CHAPTER 11.2

Molten Salt Electrolysis

Halvor Kvande and Judith C. Vidal

BACKGROUND

Electrochemistry Basics

Electrochemistry is a branch of chemistry that deals with the chemical changes accompanying the passage of an electric current. Perhaps the simplest definition may be to bring about a chemical reaction by using the potential energy gradient (electrode voltages), with the amount of reaction being dependent on the amount of electric charge transferred (current multiplied by time) to sustain that reaction.

Electrochemistry studies chemical reactions that take place in a solution at the interface of an electron conductor (a metal or a semiconductor) and an ionic conductor (the electrolyte), and involve electron transfer between the electrodes and the electrolyte or species in solution. Electrochemical processes necessitate energy addition, when the introducing energy is not able to provide sufficient energy for the reaction to occur. Energy can be added electrochemically, provided the media used (electrolyte) is more stable. The overall process is split into two parts: one half reaction at anode potential and the other half reaction at cathode potential.

For some metallic elements, electrolysis is the only practical way of extracting the metal. This is done by enabling heterogeneous chemical reactions to occur that are energetically unfavorable. Consequently, energy has to be transferred into the materials, and simultaneously, recombination of the products must be prevented. Energy can only be transferred as *heat* (through a temperature gradient) or by doing *work* (mechanically or electrically).

Electrolytic Cells

Electrolytic cells are the focus of this chapter. Electrolytic cells do not have a favorable energy gradient to naturally proceed, and they require a voltage gradient to be imposed to enable it, plus they must have a favorable electrochemical reaction mechanism.

Based on this concept and in order to build an electrochemical cell, the following components need to be incorporated in the process:

- A reactor
- Electrodes—anode and cathode
- An electrolyte
- A completed electric circuit
- A process for melting and dissolution of the raw material
- Separation systems for the products in the reactor
- The impact of design and operations

The Reactor

For the context of this chapter, the reactor is a medium-to-high-temperature furnace that has a refractory lining. For the reactor a container is needed for the molten electrolyte and the liquid metal that is produced. Thus, the electrolyte and the metal are contained in a preformed carbon lining, which also has refractory and thermally insulating materials inside a steel shell. The container must be electronically conducting to enable electric energy to be brought to the reacting interfaces. The container should not be corroded by the metal or the electrolyte. The requirements for a cell lining material are that it lasts a long time and minimizes any metal contamination. Carbon comes closest to meet these requirements. Many types of cells use a carbonaceous-based lining for containing the metal and introducing the cathodic potential.

Preventing failure and obtaining a reasonable lifetime of the cathode lining are extremely important—some cells only last for several months before they begin operating inefficiently and consequently produce an unacceptable quality of metal. In the most serious case, they can also disintegrate and leak electrolyte and metal from the cell.

For aluminum electrolysis cells, the challenge is building a leak-proof container that holds the liquid metal and electrolyte with a large, flat, horizontal area and has a reasonable lifetime. The materials used are prebaked carbon cathode blocks, silicon carbide (SiC) sidewall bricks, and carbonaceous ramming paste. The containing carbonaceous lining material has embedded steel rods that are electronically conductive to introduce the current and thereby enable the energy to be brought to the reacting interface. The cathode blocks and sidewall materials are shaped and baked by manufacturers, while the ramming paste is formed directly in the cathode by ramming and is baked when the cathode is preheated and the cell is put into its early operation. For magnetic reasons, the

Halvor Kvande, Professor Emeritus, Norwegian University of Science & Technology, Trondheim, Norway
Judith C. Vidal, Building Energy Science Group Manager, National Renewable Energy Laboratory, Golden, Colorado, USA

container should not conduct current through the sidewalls. Frozen electrolyte (ledge) is used to seal and protect the sidewalls and conserve heat.

Electrodes—Anode and Cathode

Electrolysis involves at least three material phases—one must be a connecting ionic conductor (electrolyte), and the two others must be physically separated electronically conducting interfaces (electrodes). The electrodes can be inert and thereby do not contaminate the metal produced. The electronic conducting electrodes can also participate in the respective oxidation or reduction reaction (as the carbon anode does in aluminum electrolysis), or they can be passive (as is the aluminum metal pad cathode).

Aluminum production utilizes a consumable carbon anode that is horizontally orientated, and the carbon surface must be continuously replenished by gradually lowering the anodes. Anode change can cause a large operating disturbance, and it is a challenge to perform the anode replacement or replenishment without disturbing the continuous cell operation. The requirement for carbon anode design is that it will enable continuous electrolysis and minimize the disturbance that the anode replenishment introduces.

With consumable carbon anodes, the dominant electrochemical reaction is to form carbon dioxide by utilizing the oxide ions of the dissolved alumina. In industrial cells, these anodes have to be changed typically after about 25 days.

The Electrolyte

A solvent electrolyte (ionic) is used to dissolve the consumable (i.e., a metal oxide) adequately. This solvent must be electrochemically more stable than the metal oxide solute. The electrolyte can be a reactant or simply a solvent for the reactant.

Maintaining a density difference between the electrolyte and the metal is necessary to keep the two liquids apart. Direct common contact of the metal with the anode gas must be avoided. Different components of the electrolyte (such as those that are added to modify the melting point) could cause the density of the electrolyte to increase or decrease and thereby exceed or become less than that of the metal.

The main functions of the electrolyte are

- To pass electricity from the anode to the cathode;
- To be a solvent for the raw material to enable its electrolytic decomposition, forming metal at the cathode and a gas at the anode;
- To provide a physical separation between the cathodically produced metal and the anodically evolved gas to prevent back-reaction between them; and
- To act as a heat-generating resistor that allows the cell to be self-heating.

Optimizing cell performance and operations are important because all reactions have optimum conditions. To maintain the electrolyte at the optimum temperature and composition a control system and philosophy are needed to ensure that the electrolyte temperature and composition are within the control band.

Completed Electric Circuit

An electrolysis cell requires an adjustable power supply connecting the two electronic conductors. An electric energy supply is needed to enable an energy gradient to be established and then cause the electrochemical reactions. The direct current (DC) energy supply is at high voltage, because there is often a large number of industrial electrolysis cells (several hundred) connected in series. Electrochemical reactions are slow, so both high electric current and large electrode areas are needed.

Electrolysis is a continuous process, and a cell cannot be stopped and restarted easily. The requirement is a good and stable electricity supply. If the production is interrupted by a power outage for more than a few hours, the molten electrolyte in the cells will start to solidify. This results in a significantly higher electrical resistance and higher power line voltage, which then makes restarting cells almost impossible when the electrical power is restored. In aluminum electrolysis, the consumable carbon anodes lower the energy required and "complete the circuit."

Melting Process

With molten salt electrolysis there is a need for an energy control system to ensure that both reactions and temperature are maintained. Within material supply or removal limits, the rate of the electrochemical reaction is controlled by the imposed electric energy gradient at each of the reaction interfaces. An electric energy input is required to enable the reactions and must be at a rate that will control the reactions and simultaneously maintain the reacting temperature. Heat is generated by the cell ohmic resistance of the electrolyte and electrodes, and it is also generated by the electrode polarization. Heat is only supplied by the voltage between the anode assembly and the cathode. However, for small cells, external heating can be used to keep the cells hot.

Good cell operation also requires a method of replenishing the added raw material. It is a challenge to add the raw material in a manner that ensures that its dissolved concentration stays within the values that will enable the desired electrode reactions to occur without introducing contaminants to the metal product, and while not introducing cell operational disturbances through excessive amounts of undissolved material in the bottom of the cell.

The cell temperature is usually measured manually once a day using a thermocouple. Continuous temperature measurement is not easy because of the corrosive nature of most of the molten salt electrolytes. The easiest way to change the temperature is to vary the cell voltage, which implies an increase or decrease of the anode–cathode distance in the cell.

Separation Systems

Many molten metals are easily reoxidized, and it is therefore crucial to avoid having the anode and cathode products come in physical contact with each other in the cell. This is easiest when the metal has a higher density than the electrolyte, because then the metal moves downward and the gas upward. However, for example in magnesium electrolysis, the metal is lighter than the electrolyte and therefore moves upward and floats on top of the surface of the electrolyte. The anode gas, chlorine, also moves upward. This requires that the liquid magnesium is kept in a compartment in the cell where a chlorine atmosphere is not present.

The metal produced has to be removed from the cell periodically. Extraction of the molten metal is called tapping. This is a labor-intensive manual routine operation. The spout of a vacuum ladle or crucible is dipped into the metal pad in the cell, and the liquid metal is then siphoned into a ladle by suction from an air ejector system. Then the metal is analyzed, weighed, and transported to a furnace in the cast house. It is then an operational challenge to ensure that the metal has the required purity.

Impact of Design and Operations

Electrolysis is a continuous process with spatial batch operations. This results in conditions in the cell that are not uniform. The problems are aggravated by the limited number of measurements that can be made and the accuracy of the measurement because of the spatial and temporal variations. Development in the industry is clearly toward cells with higher amperage and at the same time attempting to reduce the electric energy consumption. The larger the cell and the lower the energy input, the worse it gets.

The chemical composition of solvent electrolytes would enable desired oxidation reactions to occur without introducing contaminants to the product. A major weakness of electrochemical reactors is that the potential gradient is not selective to the reaction. If it is too high, undesirable parallel reactions can occur (e.g., anode effect in aluminum electrolysis). This can seriously reduce the cell productivity.

Some manually interfaced work practices are unavoidable. A method is needed for replenishing the alumina that is continuously depleted. The alumina replenishment needs to be done in a measured manner so that the alumina concentration stays within a target band. The replenishment of the alumina should cover the full cross-sectional area of the cell to maintain uniform concentration distribution. Multiple volumetric feeders and "breakers" are required to ensure that the alumina goes into the electrolyte. Mixing and dissolving the alumina and distributing it uniformly are very important. Undissolved alumina getting beneath the metal pad is causing instability through changed current flow. Dense or clumped alumina can form sludge.

In aluminum electrolysis cells, a method is needed for introducing the consumable carbon anodes and replenishing them to a new target position. Multiple electrodes are used to minimize disturbance. Positioning of the anodes must be adjustable to maintain the cell voltage and anodic current distribution. Because the carbon anodes are hot, a technique is needed for protecting them from air oxidation and heat loss. Therefore the newly set anodes need to be re-covered and integrity maintained. The anode cover material must not introduce contamination; hence, a mixture of alumina and recycled solid electrolyte is used.

MOLTEN SALTS

Molten salts and their mixtures form a unique type of liquids, which are different from both aqueous solutions and organic liquids. In molten salts there is an array of electrostatically charged particles, negative anions and positive cations, which are in close contact. Because of the strong attractive forces of charges with an opposite sign and the strong repulsive forces between particles of the same sign, a cation will preferentially have anions as its nearest neighbors, and vice versa for anions. The structural units are either simple ions or associated groups such as paired ions and complex ions.

In the solid state, these ions are fixed in space, but when the material is molten, it is accompanied by an increase in volume, which is typically 20% for alkali halides. This expansion allows for ions to migrate without much restriction and it may be attributed to the presence of unoccupied positions in the molten salt.

Because of the ionic nature of the molten salts, they have been modeled after theories of gases and solid-state materials. In the molten salt, the vacancy model assumes that Schottky defects arise in the ionic crystal when the ions move out from their lattice positions. In the hole model, empty space is generated by thermal expansion of the cell when localized density changes. The cell model assumes that the ions in the molten salt will move within a defined cell space, which is considerably smaller than the volume it would have in the gas phase, and it defines the short-range order of the liquid. The liquid free-volume model states that the liquid has by itself a free volume and it is statistically distributed throughout the short-range cells (Bloom and Bockris 1964).

Using X-ray neutron diffraction, the interaction of ions has been modeled in what is known as radial distribution functions. This method determines a probability of finding two ions or atoms separated by a given distance. When an alkali halide material is molten, unlike ions are brought closer together while similar ions are separated from each other. At the same time, the long-range order is disrupted, while some short-range order remains. Because of the expansion of the volume, vacancies appear in the structure (Bloom 1967).

When the raw material for electrolysis, usually a metal oxide or a halide, goes into solution, it can form either ionic or molecular species. Ionic forms of the metal to be extracted are required for high extraction efficiency. If molecular (neutral) species of the dissolved raw material are formed, then the electrochemical reduction at the cathode is not possible. The electrolyte conductivity decreases because of the presence of neutral complex molecules. To avoid this, halide molten salts are usually used as electrolytes, because in general they are strong ionic solvents and tend to form single ionic species with the raw material. They have high electrical/ionic conductivity, high ionic mobility, and low viscosity and thus provide faster extraction kinetics. Some disadvantages of halide molten salts are their high vapor pressure and corrosiveness, and inert atmospheres are required in many cases.

During the electrolysis of molten salts, the metal produced has a tendency to be slightly soluble in the halide. In fact, all metals are soluble in their molten salts. In the case of alkali metals, the solubility increases with the atomic number, thus $Li < Na < K$. For alkaline earth metals, the tendency is the same: $Mg < Ca < Sr$. The solubility in either case is increased if a displacement reaction occurs, where the metal reduced reacts to form a different halide salt. Typically, in a mixture of salts, the salt of a less noble metal will reduce the solubility of the other metal compared to its pure salt. Also, the interfacial tension between the metal being deposited and the molten salt will dictate its solubility. Higher surface tensions lead to lower solubility (Corbett 1964; Bredig 1964).

The main characteristics that the electrolyte molten salt must have are low melting point; high thermal stability; low viscosity; low vapor pressure; good ionic conductivity; high solvent power of the raw material; high decomposition voltage; low corrosiveness with cathode, anode, and vessel; easy purification; proper density; and low cost (Minh 1985). Low melting points will allow the system to operate at lower temperatures, and thus the operating costs can decrease.

ELECTROLYSIS

The three types of electrolysis are based on the electrolyte temperature range stability: (1) aqueous solutions at low temperatures (<100°C), (2) ionic liquids at mid-temperatures (100°–300°C), and (3) molten salts at high temperatures (>300°C). These fluids, once in the liquid form, must form ions in single and complex forms. The more ionic strength, the more efficient the electrolysis is, because it is a measure of the ionic conductivity of the electrolyte and thus the mobility of the species dissolved.

Aqueous solutions electrolysis is less expensive, because of the lower temperatures, but many metals cannot be extracted in this medium because the required amount of electrical energy will decompose the electrolyte instead. When higher temperature electrolytes are required, the operating costs will increase because of the temperature requirement, so high efficiency must be met to have a cost-effective technology. The same fundamental concepts used for low-temperature electrolysis, to reduce metals from aqueous electrolytes, apply in molten salt electrolytic processes.

Electrochemistry Principles

The thermodynamic behavior of electrochemical reactions will be dependent on the activities of each component in the electrolyte and the temperature. This can be described by the use of the Nernst equation:

$$E(T) = E°(T) - (RT/nF) \ln K \quad \textbf{(EQ 1)}$$

where E(T) is the reversible potential of the reaction at the specified activity and temperature; E°(T) is the standard electrode potential for unit activity of the species; R is the universal gas constant; T is absolute temperature; n is the number of electrons transferred; F is Faraday's constant; and K is the equilibrium constant for the electrochemical reaction.

Electrodes

The material and the stability of the electrodes are key aspects in any electrochemical process. Typically, they are categorized as working electrodes and reference electrodes. The materials of choice for the cathode and anode are of paramount importance. It must be relatively inert under the harsh environment of the molten salt. Platinum, nickel, molybdenum, and tungsten, among others, are common for fabricating the electrodes. In the case of fluoride systems, Inconel and nickel are typically used.

To obtain information about the electrochemical reactions (half-cell) occurring at the anode and cathode, it is necessary to determine the value of each single-electrode potential (E°). The use of reference electrodes is the only way to obtain the electrode potential based on a reference state. Then, equilibrium thermodynamic data can be obtained because

$$\Delta G° = -n F E° \quad \textbf{(EQ 2)}$$

Here, ΔG° is the standard Gibbs energy for the electrochemical reaction. Varying the electrode potential (potentiodynamic scans) at a selected temperature and composition would allow for the determination of the electrode kinetics of the cathodic and anodic reactions.

In the case of aqueous solutions at 25°C, the standard hydrogen electrode (SHE) is employed as the conventional reference electrode in performing electrode potential measurements. Thus, the assigned potential for this (standard) reference electrode is zero. When using other reference electrodes, the measurement of a selected electrode potential yields a value that is offset from the tabulated values, because the potential of the other reference electrode is not the same as the SHE. The minimum cell voltage required in the electrolytic process (ε_{cell}) is the difference between the equilibrium electrode potentials of each half-cell reaction (anode and cathode) and is given by the following equation:

$$\varepsilon_{cell} = E_{cathode} - E_{anode} \quad \textbf{(EQ 3)}$$

This value corresponds to the situation when no current is flowing in the circuit. The evaluation of the electrode kinetics is performed by the application of an overpotential to the electrolytic cell, thereby inducing an external current flow in the system.

A reference electrode must not influence the current and the potential changes occurring at the half-cell reaction. It must have basic requirements like stability, reversibility, good reproducibility, and simplicity of construction. Measurement of the electrode potential not only requires a suitable reference electrode but also a high impedance voltmeter (electrometer). In molten salts, the use of reference electrodes is very limited and rather troublesome. No universal electrode works in these electrolytes because of chemical interactions of different electrodes with different salts (i.e., corrosion of metal, reactivity of gases, etc.). The use of metals as reference electrodes in molten salts is well known. The use of silver as an electrode is very common (Ag/Ag^+) for chlorides, carbonates, nitrates, and sulfates. For fluorides, the use of Ni/Ni^{2+} is commonly used. Solid oxide electrodes are also employed as reference electrodes. In this case, the insoluble electrode forms a cell: $M/M_xO_y/O^{2-}$. Some of these solid oxide electrodes are chemically stable under high temperatures and highly corrosive environments. Gases, such as hydrogen, chlorine, and even oxygen, can be used as reference electrodes. For gas reference electrodes to work, they must have an inert support material like porous carbon.

Common anodes are made of graphite because of its high electrical conductivity, and it is chemically stable with the molten salt. The drawback of this material is that the anodic products, such as oxygen and chlorine, react with carbon and form carbon dioxide (CO_2) and toxic Cl-O compounds. To avoid these reactions, inert anodes are being considered and utilized. General guidance on the development of inert anodes, in particular metallic-based for molten chlorides, is provided elsewhere (Wang and Xiao 2013). Thermodynamic considerations and experimental evaluations in novel molten-salt electrolysis processes are reported.

The quality of the electrodeposited metal strongly depends on the temperature and cathodic current density (the amount of current per cathodic surface area unit). The larger the number of electrons available, the faster the deposition kinetics. If the metal atoms need to be reduced and deposited as a solid phase over the cathode, then its morphology (dense film, faceted phases, columnar morphology, and dendrites) will depend on the cathodic current density.

Electric Double Layer

When surfaces are charged electrically, excess charges will tend to accumulate on each side of the interface, and the electric double layer refers to the two parallel layers of charge surrounding the object. In the case of a cathode immersed in an ionic solution, the metal side of the interface will have

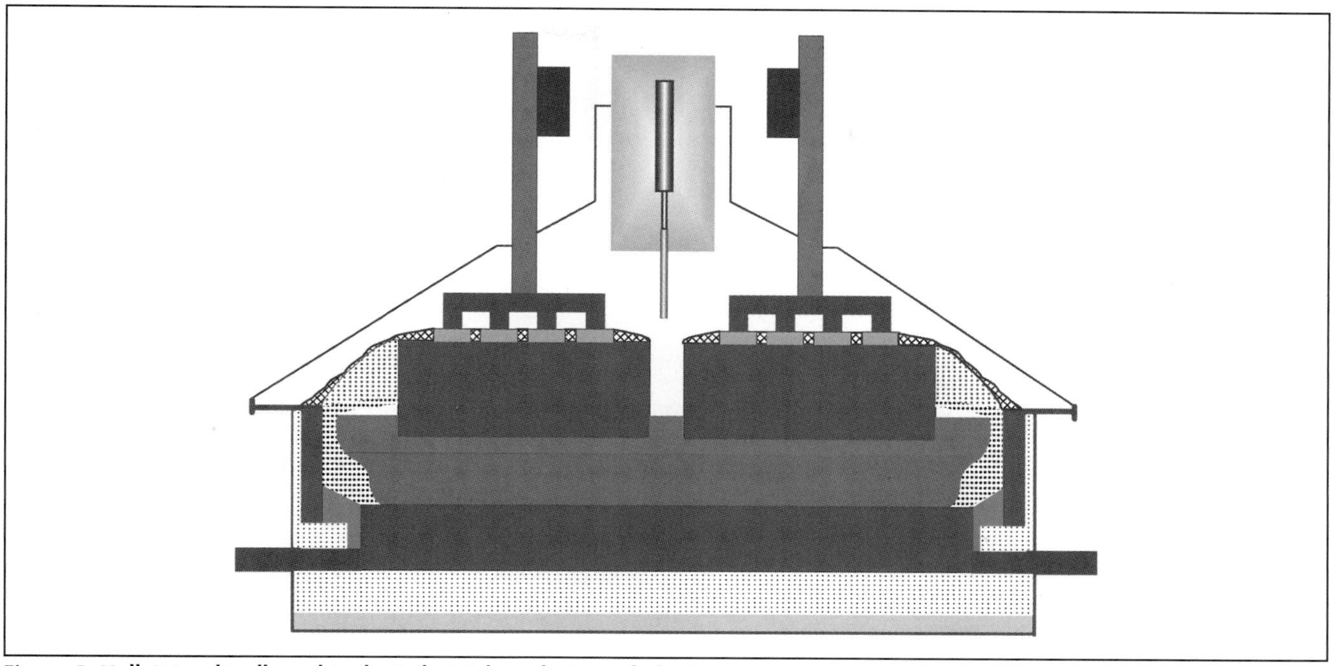

Figure 1 Hall–Héroult cell used in the industrial production of aluminum

electrons that are facing cations, which will accumulate on the electrolyte side. This effect was first suggested by Hermann von Helmholtz in 1879, after whom the "Helmholtz double layer" terminology was introduced, which refers to when a capacitance term is created at the interface. This first approximation of a double layer is very simplistic and does not take into account the effect of thermal motion of the liquid ions. In reality, the ions distribute themselves in a diffusive layer, and thus the potential drop across it decays exponentially.

Faraday's Laws

The mass of a material produced at an electrode during electrolysis is directly proportional to the quantity of electrical charge (the number of coulombs) transferred at that electrode. Also, the mass of the material formed at an electrode is directly proportional to the element's equivalent mass. The equivalent mass of a substance is the amount of an element that reacts, or is involved in reaction, with one mole of electrons. So, for each Faraday of electric charge (96,485 coulombs) one gram-equivalent of material will be produced or destroyed.

Faraday's laws assume a perfect system that does not include problems typically found in molten salt systems. One example of such deviations includes discharging of more than one type of ion simultaneously, for which the charge used will be consumed or given by more than one species. The formation of intermediate species near the cathode, which will then alter the effective charge used, also causes deviations. Electronic charges are also lost in some instances through conducting material in the electrolyte, or even by semiconductor types of transfers. Finally, the loss of deposited material at the cathode is the most common problem. Some materials deposited could redissolve partially in the electrolyte, or they could evaporate or simply be removed mechanically by fluid flow near the cathode. These deviations then reduce the effectiveness of the electrolysis process, thus reducing the amount of electrical current effectively used. The current efficiency of a cell is thus determined by obtaining the ratio of quantity of the substance effectively obtained to that predicted theoretically by Faraday's laws. One can say that the current efficiency in industrial processes is the percentage of the electricity that is used to produce the metal, while the rest of the current produces heat given off to the surroundings.

COMMERCIAL MOLTEN SALT ELECTROLYSIS

Several metals are currently being commercially extracted using molten salt electrolysis: aluminum (Hall–Héroult process), magnesium, sodium, calcium, and lithium.

Aluminum

Aluminum is second only to steel in annual production volume, and more aluminum is produced than all of the other nonferrous metals combined. The total world production of primary aluminum was 63 Mt (million metric tons) in 2017; in addition, around 20 Mt of aluminum were recycled (IAI 2018).

Aluminum is remarkable for its low density and ability to resist corrosion through the phenomenon of passivation. Aluminum and its alloys are vital to the aerospace industry and important in transportation and structures, such as building facades and window frames.

The process for electrowinning of aluminum revolves around the independent discoveries in 1886 and the subsequent patents of Hall and Héroult. The Hall–Héroult process is the electrolytic reduction of aluminum oxide (alumina) dissolved in a molten electrolyte of mainly cryolite (Na_3AlF_6) at a temperature of ~960°C (Figure 1).

Cryolite, a chemical compound composed of sodium, aluminum, and fluorine, constitutes about 80% of the electrolyte (the current-conducting medium) in the aluminum smelting operation. Naturally occurring cryolite was once mined in Greenland, but the compound is now produced synthetically for use in the production of aluminum. Aluminum fluoride is

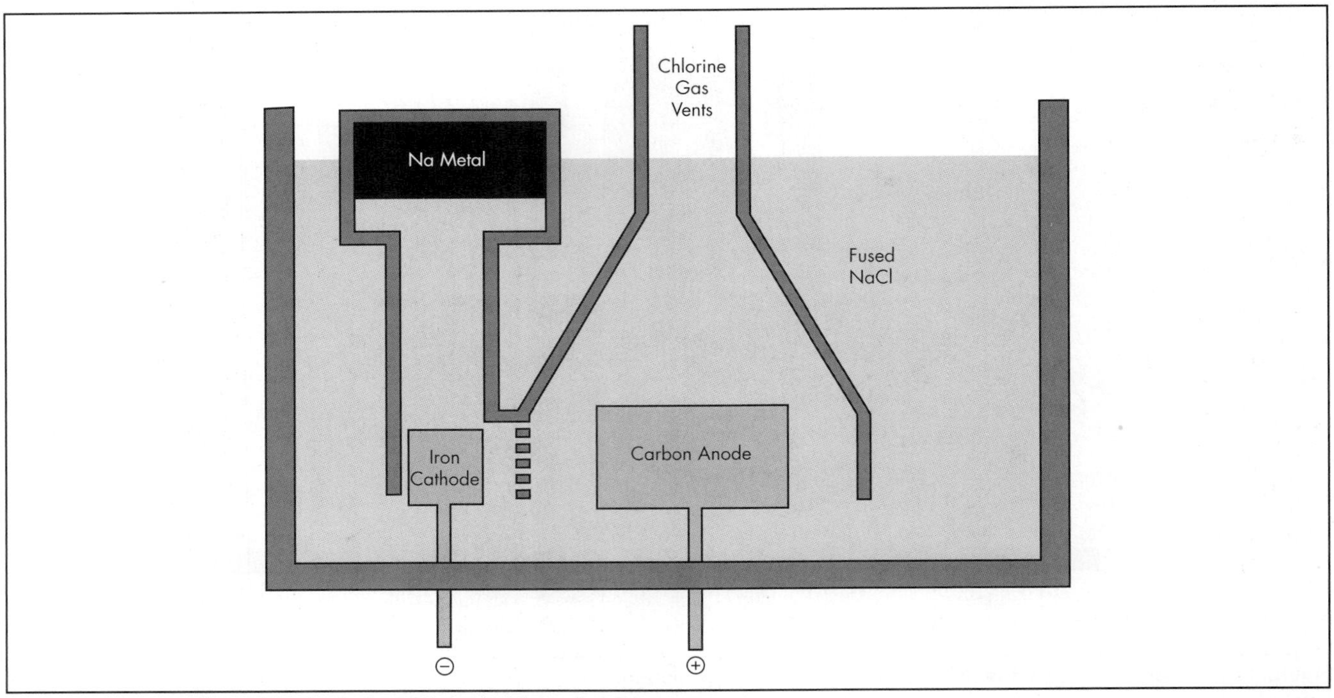

Figure 2 Downs cell used in the industrial production of sodium

added to lower the melting point of the electrolyte, but the main reason for addition of aluminum fluoride is to keep the electrolyte composition constant by neutralizing the sodium oxide impurity in the alumina feed. Modern cells typically have a target of 10 wt % aluminum fluoride in the electrolyte.

The anodes are made of carbon, prebaked or Soderberg, from a mixture of calcined petroleum coke and coal-tar pitch. The cathode is the molten Al (99%). The overall reaction is

$$Al_2O_3(\text{dissolved}) + 3/2C(s) \rightarrow 2Al(l) + 3/2CO_2(g) \qquad \textbf{(EQ 4)}$$

Thus, the other major ingredient used in the smelting operation is carbon. Carbon electrodes transmit the electric current through the electrolyte. During the smelting operation, some of the carbon is consumed as it combines with oxygen to form carbon dioxide. In fact, about 0.4 kg C is used for every kilogram of aluminum produced. Most of the carbon used as anodes in aluminum smelting is a petroleum coke, a by-product of oil refining, and additional carbon is obtained from coal-tar pitch.

Because alumina is a very stable compound and aluminum smelting involves passing an electric current through a molten electrolyte, it requires large amounts of electrical energy. As a global average, production of 1 kg Al requires 13.4 kW·h of electric energy. The cost of electricity represents about one-third of the cost of smelting aluminum.

Magnesium

Magnesium is the third most used metal in construction (after iron and aluminum). Nearly 70% of the world production of magnesium is used to make alloys, which have a very low density, comparatively high strength and excellent machinability. Other important uses are in the manufacture of titanium and high grade steel for construction.

The annual world production is now about 900,000 t. The industrial process involves two stages:

1. Production of pure magnesium chloride from sea water or brine
2. Electrolysis of molten magnesium chloride

The electrolyte composition in the cells varies considerably among the different magnesium producers. The so-called NaCl-KCl (sodium chloride–potassium chloride) type of electrolyte consists of 50 wt % NaCl, 35 wt % KCl, and 15 wt % $MgCl_2$ (magnesium chloride), which is kept at about 700°C. During electrolysis, magnesium and chlorine are produced. The molten metal is removed from the cell and cast into ingots. The high specific energy consumption (12–16 kW·h/kg) has been a motivation to improve the process with respect to energy and current efficiency. The cathodic process in dilute solutions of Mg(II) species, typically 0.1–10 wt % $MgCl_2$, is controlled by diffusion of the reactant (Børresen et al. 1997; Corrosion Doctors 2018).

Sodium

Sodium is the most important of the alkali metals, with annual world production of more than 200,000 t. It is used for manufacturing tetraethyl lead as an additive to gasoline ($[Pb(C_2H_5)_4]$, and also for metal reduction, particularly during titanium and tantalum production. Sodium is industrially produced by the Downs' process, in which molten NaCl is electrolyzed in a special apparatus called the Downs cell that uses a carbon anode and an iron cathode (Figure 2). Calcium chloride, $CaCl_2$, is added to lower the melting point of the electrolyte to below 600°C. The products (liquid sodium and chlorine) are kept from coming in contact to avoid NaCl reformation. Oxidation of elemental sodium is dangerous because it is explosive. To avoid oxidation, elemental sodium must therefore be prevented from contact with oxygen. Downs cells

Table 1 Cathode and anode materials and products, plus some cell operational data for production of aluminum and magnesium, and some alkali and alkaline earth metals

Cathode Product	Anode Product	Anode Material	Cathode Material	Amperage, kA	Voltage, V	Temperature, °C	Electrolyte Composition
Aluminum	CO_2	Carbon	Aluminum (carbon)	100–600	4.0–4.5	955–965	80% cryolite 10% AlF_3 5%–7% CaF_2 2%–4% Al_2O_3
Magnesium	Cl_2	Graphite	Steel	100–400	5	670–700	50% NaCl 30% KCl 20% $MgCl_2$
Sodium	Cl_2	Multi-electrode	Multi-electrode	40–80	6.5–7.0	590–600	46%–53% $BaCl_2$ 23%–26% $CaCl_2$ 24%–28% NaCl
Lithium	Cl_2	Graphite	Mild steel	12–40	6	400–460	40%–60% LiCl 60%–40% KCl
Calcium	Cl_2	Graphite	Liquid Ca–Cu alloy	20	7–10	650–715	80%–85% $CaCl_2$ 15%–20% KCl
Strontium	Cl_2	Graphite	"Contact-cathode"	—	8–12	680–800	84% $SrCl_2$ 16% KCl

operate at 25–40 kA and at potentials of 6.5–7.0 V. With a multi-electrode arrangement, modern Downs cells can have amperages up to 80 kA.

Lithium
All of the world's primary lithium is produced by molten salt electrolysis, with an annual world production of about 43,000 t. Lithium is the lightest of all metals and is used in a variety of applications, including the production of organolithium compounds, as an alloying addition to aluminum and magnesium, and as the anode in rechargeable lithium–ion batteries. The electrolysis cell has a central cathode of mild steel on which molten lithium is produced. Opposing graphite plates serve as the anodes, where chlorine is formed. A bell-shaped structure positioned above the cathode collects the rising liquid metal and prevents it from reacting with the rising chlorine gas. Anhydrous lithium chloride is the cell feed, and potassium chloride is the solvent. At 400°C the electrolyte composition is about 35–45 mole % KCl. The electric energy consumption is high, about 35 kW·h/kg Li (Kipouros and Sadoway 1998).

Operating Data for Industrial Electrolytic Processes
Table 1 summarizes data from industrial production of the so-called light metals aluminum and magnesium, and also for some alkali and alkaline earth metals. It is seen that graphite anodes and chlorine gas evolution is common for several of these metals, and chloride melts are then used as electrolytes at temperatures in the range from 450°C to 800°C. Amperage and voltage vary considerably between the various processes. Notably, aluminum production is unique, because it uses consumable carbon anodes and a fluoride-containing electrolyte.

PROPOSED PRODUCTION OF METALS IN MOLTEN SALTS

Boron
Many metals have been proposed to be produced by electrolysis in molten salts. Boron is one of the semi-metals and is primarily used in chemical compounds. About half of all boron produced globally is used as an additive in glass fibers of boron-containing fiberglass for insulation and structural materials. The next leading use is in polymers and ceramics in high-strength, lightweight structural and refractory materials. Elemental boron has been produced in molten MgF_2-NaF-LiF mixtures with B_2O_3 as the source of boron. The morphology of the product was agglomerated boron spheres with 0.5–5 μm of average particle size depending on the cell potential applied. The highest grade or purity of 88.4 wt % B was obtained at 700°C and 1.5 V, while the highest recovery of 91.7 wt % B was obtained for the experiment at 700°C, 2.0 V, and electrolyte premelted at 1,330°C (Zhang et al. 2007).

Ruthenium
Electrolysis of ruthenium (Ru) from LiCl–KCl molten salts using $RuCl_3$ as the raw material has been investigated. Effects of process parameters such as current density, time, and agitation of the electrolyte on the thickness and morphology of white/gray deposit of ruthenium layers were studied. The highest cathodic current efficiency (η) of 99.68% was achieved at a current density of 3 mA/cm^2 after 2 hours of electrodeposition time with rotating cathode speed of 50 rpm (Yan and Pang 2011).

Alloys of Lithium with Aluminum and Magnesium
Aluminum–lithium alloys are used in the aerospace industry because of the weight advantage they provide. The lithium content in the commercial alloys is low, only about 2.5%. The alloys can be produced by electrolysis of molten lithium chloride with a graphite anode and aluminum (or some aluminum alloys) as the consumable cathode.

Magnesium–lithium (Mg–Li) alloys with low lithium content (~25 wt % Li) for aerospace and aircraft applications were prepared at laboratory scale via electrolysis at low temperature in a molten salt electrolyte of LiCl:KCl = 1:1 (wt %). The optimum conditions, in which maximum current efficiency was obtained, were 480°C and cathode current density of 1.13 A/cm^2 (Sireli 2014).

Rare Earth Metals
Nowadays, rare-earth metals (REMs) are produced in the most convenient and economic way by molten salt electrolysis. Using fluoride–oxide molten salts is advantageous because

they provide higher current efficiencies than the chloride melts (Zhu 2014). It is more difficult to study the REM electrode processes in fluoride–oxide systems than the aluminum ones. This is because REM electrolysis is carried out at higher temperatures and REM fluorides are more aggressive. Some researchers have reported the electrode reactions and anodic gases composition of neodymium electrolysis in LiF-NdF_3-Nd_2O_3 melts (Wang et al. 2007).

DIRECT OXIDE ELECTROCHEMICAL REDUCTION

Direct oxide electrochemical reduction (DOER) is the electro-deoxidation of a solid metal oxide in molten salts. Chen et al. (2000) first achieved the direct reduction of titanium dioxide (TiO_2) into titanium in a molten chloride salt composed of $CaCl_2$ from 800°C to 900°C with a cell potential of 3.0–3.1 V. This innovative method is now often referred as the FFC (Fray–Farthing–Chen) Cambridge process. A TiO_2 reduction pathway has been proposed (Schwandt and Fray 2005; Suzuki 2005; Nagesh and Ramachandran 2007) and several titanium sub-oxides (Ti_4O_7, Ti_3O_5, Ti_2O_3, and $CaTi_2O_4$) have been characterized

The overall reaction is the oxygen removal from a solid M_xO_y oxide at the cathode and the formation of $CO_2(g)$ at the carbon anode, according to the following reactions:

$$\text{Cathode: } M_xO_y(s) + 2y\ e- = x\ M(s) + y\ O^{2-} \quad \textbf{(EQ 5)}$$

$$\text{Anode: } 2O^{2-} + C(s) = CO_2(g) + 4\ e- \quad \textbf{(EQ 6)}$$

The FFC process in molten LiCl- or $CaCl_2$-containing salts has been intensively studied worldwide for reduction feasibility studies of

- High purity (solar grade) silicon (Yasuda et al. 2007),
- Reduction of rare earth oxides (Hirota et al. 1999; Song et al. 2015),
- Spent nuclear fuel (UO_2) (Sakamura et al. 2006; Choi et al. 2011),
- Triuranium octoxide (U_3O_8) (Jeong et al. 2006),
- Uranium dioxide-plutonium dioxide (UO_2-PuO_2) mixed oxides (Iizuka et al. 2007),
- Spent fuel (Herrmann and Li 2010), and
- Pure metals or alloys (reduction of Nb_2O_5, Fe_2O_3, NiO–TiO_2, and TiO_2-MnO_2) (Jeong et al. 2007; Wang et al. 2008; Zhu et al. 2006, 2015).

However, no process has yet reached the industrial scale. In some cases, this may be because of the purity of the product, which is usually contaminated with carbides produced as a by-product during the electrolysis.

The use of inert anodes has been proposed to avoid the carbide formation. Catalytic materials, such as gold, are required to allow the oxygen evolution electrochemical reaction:

$$2O^{2-} = O_2(g) + 4\ e^- \quad \textbf{(EQ 7)}$$

However, this reaction is difficult to control in molten chloride salts because of the close potentials of $Cl_2(g)$ and $O_2(g)$ evolution. Molten fluorides have been proposed with the use of inert gold anodes to avoid the formation of anodic halide gases. Several eutectic compositions have been proposed, such as LiF–NaF (+2 wt % Li_2O) at 750°C for SnO_2 and Fe_3O_4, and LiF–CaF_2 (+2 wt % Li_2O) at 850°C for TiO_2, TiO (Gibilaro et al. 2011a), and UO_2 (Gibilaro et al. 2011b).

The process has been demonstrated for the electro-deoxidation of oxides of

- Titanium (Chen et al. 2000; Mohandas et al. 2011b),
- Tantalum (Jeong et al. 2007),
- Niobium (Vishnu et al. 2013; Song et al. 2012),
- Niobium carbide (Song et al. 2015),
- Zirconium (Mohandas and Fray 2009),
- Chromium (Chen et al. 2004),
- Chromium carbide (Dai et al. 2015),
- Iron (Li et al. 2009),
- Tungsten (Erdogan and Karakaya 2010),
- Aluminum (Yan and Fray 2009),
- Nickel (Descallar-Arriesgado et al. 2011),
- Silicon (Yasuda et al. 2007),
- Uranium (Gourishankar et al. 2002; Mohandas et al. 2011a; Sri Maha Vishnu et al. 2012), and so forth.

THE ANODE EFFECT

The anode effect (AE) can occur with carbon or inert anodes and in both molten salts and aqueous electrolytes. This is a characteristic and important phenomenon in molten salt electrolysis, and in particular in the primary production of aluminum by the Hall–Héroult process. There may be two causes of AEs: (1) the initiation of an AE is due to the formation of some intermediate compounds on the surface of the anode, or (2) the initiation of an anode effect is simply due to accumulation of gas bubbles on the anode. It is known that during the AE, other anodic reactions take place and the anodic gas composition changes because of the simultaneous discharge of oxygen and fluoride ions. It has been proposed that some events prior to the AE occur, such as the following (Zhuxian 1992; Qiu and Zhang 1987):

- A change in wettability of the electrolyte because of the electrolyte composition change. In this case, the anodic gas produced can accumulate at the anodic surface because the molten salt is not wetting the anode.
- A change in the anodic process because of the depletion of oxygen ions (e.g., low alumina concentration) so other ions such as fluorides can be oxidized at the anode and fluoride-containing gases are formed.
- An increase in the anodic overpotential because of the depletion in oxygen anions. The surface tension of the electrolytes thus increases and allows the formation of large bubbles at the surface, decreasing the electrical contact between the electrolyte and the anode.
- Electrostatic attraction of gases by the anode at high anodic potential.

During an AE in aluminum electrolysis cells, a gas film under the anodes is formed, which consists of perfluorocarbon gases (PFCs), mainly CF_4 and smaller amounts of C_2F_6 (Nissen and Sadoway 1997). These compounds are strong greenhouse gases with high global warming potentials, and they cause a much more serious effect on the environment than carbon dioxide. The U.S. Environmental Protection Agency and the primary aluminum producers in the United States have established the Voluntary Aluminum Industrial Partnership to substantially reduce the PFC emissions (Dolin 1999). There is a certain current density, called critical current density (CCD), in which the AE can be induced. The CCD depends on many factors such as type of electrode materials, anodic size and shape, electrolyte composition (in particular, the alumina concentration), and temperature, among others. Zhu and Sadoway (1999, 2000, 2001) suggested that PFC generation can be avoided completely in small laboratory cells by simply stepping down the cell current in small increments, because its formation kinetics is controlled by an interfacial step rather than by mass transfer.

Thus, an AE occurs when the electrolyte in the cell becomes depleted of alumina below a critical level, causing the cell voltage to fluctuate and increase several folds due to the presence of a gas film under the working surface of the anode. Previously it was believed that PFC gases were evolved only during AEs when the cell voltage was high (above 8 V). However, recently it has been discovered that large modern aluminum electrolysis cells can have PFC emissions also at much lower cell voltages. At high anodic current densities and for modern cell control strategies in high amperage cells, PFC emissions can start well below the normal cell voltage of 4.0–4.5 V. This may be due to poor dissolution and mixing of alumina in the electrolyte in these cells.

For several years now, there has been research and development work going on to reduce both the frequency and duration of AEs in industrial aluminum electrolysis cells. This reduces the total greenhouse gas emissions and also improves the cell operational stability. With modern cell control, AEs do not really have any useful purpose, and perhaps the future aluminum electrolysis cells can be operated without AEs. An excellent presentation of understanding AEs has recently been given by Dorreen et al. (2017).

REFERENCES

Bloom, H. 1967. *The Chemistry of Molten Salts.* New York: W.A. Benjamin.

Bloom, H., and Bockris, J.O'M. 1964. Structural aspects of ionic liquids. In *Fused Salts.* Edited by B.R. Sundheim. New York: McGraw-Hill. pp. 1–62.

Børresen, B., Haarberg, G.M., and Tunold, R. 1997. Electrodeposition of magnesium from halide melts—Charge transfer and diffusion kinetics. *Electrochim. Acta* 42:1613–1622.

Bredig, M.A. 1964. Mixture of metals with molten salts. In *Molten Salt Chemistry*. Edited by M. Blander. New York: Interscience Publishers. pp. 367–425.

Chen, G.Z., Fray, D.J., and Farthing, T.W. 2000. Direct electrochemical reduction of titanium dioxide to titanium in molten calcium chloride. *Nature* 407:361–364.

Chen, G.Z., Gordo, E., and Fray, D.J. 2004. Direct electrolytic preparation of chromium powder. *Metall. Mater. Trans. B* 35:223–233.

Choi, E.Y., Hur, J.M., Choi, I.K., Kwom, S.G., Kang, D.S., Hong, S.S., Shin, H.S., Yoo, M.A., and Jeong, S.M. 2011. Electrochemical reduction of porous 17 kg uranium oxide pellets by selection of an optimal cathode/anode surface area ratio. *J. Nucl. Mater.* 418:87–92.

Corbett, J.D. 1964. Fused salt chemistry. In *Fused Salts.* Edited by B.R. Sundheim. New York: McGraw-Hill. pp. 91–154.

Corrosion Doctors. 2018. Sodium production by electrowinning: Downs' process. http://corrosion-doctors.org/Electrowinning/Sodium.htm. Accessed October 2015.

Dai, L., Lu, Y., Wang, X., Zhu, J., Li, Y., and Wang, L. 2015. Production of nano-sized chromium carbide powders from Cr_2O_3/C precursors by direct electrochemical reduction in molten calcium chloride. *Int. J. Refract. Met. Hard Mater.* 51:153–159.

Descallar-Arriesgado, R.F., Kobayashi, N., Kikuchi, T., and Suzuki, R.O. 2011. Calciothermic reduction of NiO by molten salt electrolysis of CaO in $CaCl_2$ melt. *Electrochim. Acta* 56:8422–8429.

Dolin, E.J. 1999. PFC emissions reductions: The domestic and international perspective. *Light Met.* pp. 56–67.

Dorreen, M.M.R., Hyland, M.M., Haverkamp, R.G., Metson, J.B., Jassim, A., Welch, B.J., and Tabereaux, A.T. 2017. Co-evolution of carbon oxides and fluorides during the electrowinning of aluminium with molten NaF-AIF_3-CaF_2-$A1_2O_3$ electrolytes. *Light Met.* pp. 533–539.

Erdogan, M., and Karakaya, I. 2010. Electrochemical reduction of tungsten compounds to produce tungsten powder. *Metall. Mater. Trans. B* 41:798–804.

Gibilaro, M., Pivato, J., Cassayre, L., Massot, L., Chamelot, P., and Taxil, P. 2011a. Direct electroreduction of oxides in molten fluoride salts. *Electrochim. Acta* 56:5410–5415.

Gibilaro, M., Cassayre, L., Lemoine, O., Massot, L., Dugne, O., Malmbeck, R., and Chamelot, P. 2011b. Direct electrochemical reduction of solid uranium oxide in molten fluoride salts. *J. Nucl. Mater.* 414:169–173.

Gomez, J.C. 2009. An evaluation of elemental boron production using molten salt electrolysis. PhD thesis, Colorado School of Mines, Golden, CO.

Gourishankar, K., Redey, L., and Williamson, M. 2002. Electrochemical reduction of metal oxides in molten salts, *Light Met.* p. 1075.

Herrmann, S.D., and Li, S.X. 2010. Separation and recovery of uranium metal from spent light water reactor fuel via electrolytic reduction and electrorefining. *Nucl. Technol.* 171(3):247–265.

Hirota, K., Okabe, T.H., Saito, F., Waseda, Y., and Jacob, K.T. 1999. Electrochemical deoxidation of RE–O (RE=Gd, Tb, Dy, Er) solid solutions. *J. Alloys Compd.* 282:101–108.

IAI (International Aluminium Institute). 2018. Statistics: Primary aluminium smelting energy intensity. www.world-aluminium.org/statistics/primary-aluminium-smelting-energy-intensity/#data. Accessed August 2018.

Iizuka, M., Inoue, T., Ougier, M., and Glatz, J.-P. 2007. Electrochemical reduction of (U, Pu)O_2 in molten LiCl and $CaCl_2$ electrolytes. *J. Nucl. Sci. Technol.* 44(5):801–813.

Jeong, S.M., Park, S.B., Hong, S.S., Seo, C.S., and Park, S.W. 2006. Electrolytic production of metallic uranium from U_3O_8 in a 20 kg batch scale reactor. *J. Radioanal. Nucl. Chem.* 268(2):349–356.

Jeong, S.M., Jung, J.Y., Seo, C.S., and Park, S.W. 2007. Characteristics of an electrochemical reduction of Ta_2O_5 for the preparation of metallic tantalum in a LiCl–Li_2O molten salt. *J. Alloys Compd.* 440:210–215.

Kipouros, G.J., and Sadoway, R.D. 1998. Toward new technologies for the production of lithium. *JOM* 50(5):24–26.

Li, G., Wang, D., and Chen, Z. 2009. Direct reduction of solid Fe_2O_3 in molten $CaCl_2$ by potentially green process. *J. Mater. Sci. Technol.* 25(6):767–771.

Minh, N.Q. 1985. Extraction of metals by molten salt electrolysis: Chemical fundamentals and design factors. *J. Met.* 37(1):28–33.

Mohandas, K.S., and Fray, D.J. 2009. Electrochemical deoxidation of solid zirconium dioxide in molten calcium chloride. *Metall. Mater. Trans. B* 40:685–699.

Mohandas, K.S., Sanil, N., Shakila, L., Sri, D., Vishnu, M., and Nagarajan, K. 2011a. Studies on the electrochemical deoxidation of uranium dioxide in molten calcium chloride. In *Metals and Materials Processing in a Clean Environment. Vol. 3: Molten Salts and Ionic Liquids.* Edited by F. Kongoli. Mont-Royal, QC: Flogen. p. 239–251.

Mohandas, K.S., Shakila, L., Sanil, N., Sri Maha Vishnu, D., and Nagarajan, K. 2011b. Galvanostatic studies on the electro-deoxidation of solid titanium dioxide in molten calcium chloride. In *Metals and Materials Processing in a Clean Environment. Vol. 3: Molten Salts and Ionic Liquids.* Edited by F. Kongoli. Mont-Royal, QC: Flogen. p. 253–267.

Nagesh, C., and Ramachandran, C.S. 2007. Electrochemical process of titanium extraction. *Trans. Nonferrous Met. Soc. China* 17:429–433.

Nissen, S.S., and Sadoway, D.R. 1997. Perfluorocarbon (PFC) generation in laboratory-scale aluminum reduction cells. *Light Met.* p. 159–164.

Qiu, Z., and Zhang, M. 1987. Studies on anode effect in molten salts electrolysis. *Electrochim. Acta* 32:607–613.

Sakamura, Y., Kurata, M., and Inoue, T. 2006. Electrochemical Reduction of UO_2 in Molten $CaCl_2$ or LiCl. *J. Electrochem. Soc.* 153(3):D31–D39.

Schwandt, C., and Fray, D.J. 2005. Determination of the kinetic pathway in the electrochemical reduction of titanium dioxide in molten calcium chloride. Electrochim. Acta 51:66–76.

Sireli, G.K. 2014. An investigation of ruthenium coating from LiCl–KCl eutectic melt. *Appl. Surf. Sci.* 317:294–301.

Song, Q.S., Xu, Q., Tao, R., and Kang, X. 2012. Cathodic phase transformations during direct electrolytic reduction of Nb_2O_5 in a $CaCl_2$–NaCl–CaO melt. *Int. J. Electrochem. Sci.* 7:272.

Song, Q., Xu, Q., Meng, J., Lou, T., Ning, Z., Qi, T., and Yu, K. 2015. Preparation of niobium carbide powder by electrochemical reduction in molten salt. *J. Alloys Compd.* 647:245–251.

Sri Maha Vishnu, D., Sanil, N., Murugesan, N., Shakila, L., Ramesh, C., Mohandas, K.S., and Nagarajan, K. 2012. Determination of the extent of reduction of dense UO_2 cathodes from direct electrochemical reduction studies in molten chloride medium. *J. Nucl. Mater.* 427:200–208.

Suzuki, R.O. 2005. Calciothermic reduction of TiO_2 and in situ electrolysis of CaO in the molten $CaCl_2$. *J. Phys. Chem. Solids* 66:461–465.

Vishnu, D.S., Sanil, N., Shakila, L., Panneerselvam, G., Sudha, R., Mohandas, K.S., and Nagarajan, K. 2013. A study of the reaction pathways during electrochemical reduction of dense Nb_2O_5 pellets in molten $CaCl_2$ medium. *Electrochim. Acta* 100:51–62.

Wang, D., and Xiao, W. 2013. Inert anode development for high-temperature molten salts. In *Molten Salts Chemistry: From Lab to Applications.* Edited by F. Lantelme and H. Groult. Amsterdam: Elsevier. pp. 171–186.

Wang, G., Wang, X., and Zhu, H. 2007. Electroanalytical study of electrode processes on carbon anode in lithium fluoride and neodymium fluoride melt. In *Proceedings of Rare Metal Technology.* Warrendale, PA: The Minerals, Metals & Materials Society. p. 533.

Wang, S.I., Haarberg, G.M., and Kvalheim, E. 2008. Electrochemical behavior of dissolved Fe_2O_3 in molten $CaCl_2$-KF. *J. Iron Steel Res.* 15:48–51.

Yan, S., and Pang, S. 2011. Development on molten salt electrolytic methods and technology for preparing rare earth metals and alloys in China. *Chin. J. Rare Earths* p. 442.

Yan, X.Y., and Fray, D.J. 2009. Direct electrolytic reduction of solid alumina using molten calcium chloride–alkali chloride electrolytes. *J. Appl. Electrochem.* 39:1349–1360.

Yasuda, K., Nohira, T., Hagiwara, R., and Ogata, Y.H. 2007. Direct electrolytic reduction of solid SiO_2 in molten $CaCl_2$ for the production of solar grade silicon. *Electrochim. Acta* 53:106–110.

Zhang, M.L., Yan, Y.D., Hou, Z.Y., Fan, L.A., Chen, Z., and Tang, D.X. 2007. An electrochemical method for the preparation of Mg–Li alloys at low temperature molten salt system. *J. Alloys Compd.* 440:362–366.

Zhu, H. 2014. Rare earth metal production by molten salt electrolysis. In *Encyclopedia of Applied Electrochemistry.* New York: Springer Science. p. 1765–1772.

Zhu, H., and Sadoway, D.R. 1999. The electrode kinetics of perfluorocarbon (PFC) generation, *Light Met.* p. 241–246.

Zhu, H., and Sadoway, D.R. 2000. An electroanalytical study of electrode reactions on carbon anodes during electrolytic production of aluminum. *Light Met.* p. 257–263.

Zhu, H., and Sadoway, D.R. 2001. Towards elimination of anode effect and PFC emissions via current shunting, *Light Met.* p. 303–307.

Zhu, J., Dai, L., Yu, Y., Cao, J., and Wang, L. 2015. A direct electrochemical route from oxides to $TiMn_2$ hydrogen storage alloy. *Chin. J. Chem. Eng.* 23(11)1865–1870.

Zhu, Y., Ma, M., Wang, D., Jiang, K., Hu, X., Jin, X., and Chen, G.Z. 2006. Electrolytic reduction of mixed solid oxides in molten salts for energy efficient production of the TiNi alloy. *Chin. Sci. Bull.* 51(20):2535–2540.

Zhuxian, Q. 1992. A new contribution to the theory of anode effect in aluminum electrolysis. *Chin. J. Met. Sci. Technol.* 8:15–19.

CHAPTER 11.3

Calcination

Jannette L. Chorney

Calcination involves the application of thermal energy to dissociate and remove water from hydrates, carbon dioxide from carbonates, and/or other chemically bound gases from ores or green coke. Commercial calcination operations are generally conducted at higher temperatures than drying; because the reactions are more endothermic, more heat must be supplied to sustain the reaction at relatively high temperatures.

Calcination is frequently performed in directly fired rotary kilns. However, shaft kilns, fluidized-bed reactors, flash calcining systems, and other equipment are also suitable for calcination operations. When a controlled atmosphere is required, calcination can also be performed using indirectly heated furnaces.

Applications for calcination products are wide ranging, and the impact of those applications extends beyond the products themselves. Magnesia (MgO) uses include environmental remediation applications and refractories for steel production. Calcined petroleum coke is used in aluminum and steel production. Other calcined materials fulfill many other chemical and industrial needs. *Calcine* comes from the Latin word for lime, *calx*, and lime (CaO) is a widely used calcination product.

Few mineral products have as many or as varied applications as lime. Most lime is produced specifically to meet customer specifications. The largest consumer of lime is the steel industry, where lime is used as a flux for impurity removal in steel production (basic oxygen furnaces, electric arc furnaces, and secondary refining). Other major uses of lime include environmental remediation applications, such as water treatment and flue gas desulfurization (FGD); stabilization in construction; pH control in mining and mineral processing; sugar production; and for the production of precipitated calcium carbonate (PCC), which is used for paper manufacturing, medicinal applications, and so forth. The United States has 75 commercial lime plants; some of these are lime kilns producing lime for internal (captive) use, such as in the sugar industry (USGS 2009; National Lime Association, n.d.).

Limestone is a sedimentary rock that is the third most abundant sediment in the earth's crust (behind shale and sandstone). It occurs in varying degrees of purity, and most deposits are not suited to the production of high-grade lime. Mineralogically, four calcium or magnesium carbonate minerals are important as constituents of limestone. Two calcium carbonates ($CaCO_3$) are of interest: calcite, which has a rhombohedral structure; and aragonite, which has an orthorhombic structure. The magnesium carbonates of interest are dolomite ($CaCO_3 \cdot MgCO_3$) and magnesite ($MgCO_3$); both dolomite and magnesite have a rhombohedral structure (USGS 2009; Downey 2013).

Limestone is classified according to the amount of magnesium carbonate that it contains. High-calcium limestone contains less than 3% magnesium carbonate. High-calcium limestone is used to produce chemical grade lime for applications such as for PCC and sugar production. Magnesian limestone contains 5%–35% magnesium carbonate and is used to produce magnesian quicklime. Dolomitic limestone contains 35%–46% magnesium carbonate and is used to produce dolomitic quicklime. Glass manufacture is one application for dolomitic lime. The most significant impurities in limestone include silica (quartz, chert, clay, shale, feldspar), alumina, hematite, sulfur (often in pyrite), and sulfate (gypsum and anhydrite). Magnesium carbonate ($MgCO_3$) occurs in practically all limestone deposits (Downey 2013).

Most limestone mining operations in the United States are quarries, but there are a few underground mining operations. After quarrying, the ore is crushed and sized before calcining. Crushing for rotary kiln calcination is optimized to reduce the formation of fines, which are not normally a salable product. Off-gases from calcination are treated to remove particulate matter, and air-quality monitoring is conducted.

HYDRATED LIME

Hydrated lime, an important lime product, is often used in neutralization, construction and pharmaceutical applications, water treatment, FGD, and for soil stabilization. Lime hydration is a highly exothermic reaction commonly known as *slaking*. The heat of reaction for the following reaction was calculated from the appropriate heats of formation (Kubaschewski et al. 1967). Quicklime readily hydrates to form calcium hydroxide, $Ca(OH)_2$, when it is slurried in water:

Jannette L. Chorney, Adjunct Professor, Montana Tech, Butte, Montana, USA

$$CaO + H_2O = Ca(OH)_2 \quad \text{(EQ 1)}$$
$$\Delta H^\circ_{R\ at\ 298.15\ K} = -15.400\ \text{kcal}$$

Hydrated lime may be marketed as a dry solid, which is sold as both bulk material and bagged, or as milk of lime. The $Ca(OH)_2$ slurry is called *milk of lime*. Wet hydration is conducted in a single stage. Dry hydration can be performed in one, two, or three stages, but the three-stage process described here is becoming more prevalent.

A more rapid hydration tends to produce finer particles, because there is less time for agglomeration to occur. It is important to balance the initial lime particle size with reaction considerations, because a smaller particle, while producing fine hydrated lime caused by the rapid reaction, may react so quickly as to expose the particles to excessive temperatures, resulting in a delay in the hydration. Important parameters for optimal size $Ca(OH)_2$ production therefore include proper feed sizing, adequate mixing, sufficient hold times, good water-spray distribution, and adequate monitoring and controls. To produce hydrated lime with the highest surface area (>40 m^2/g), which is used in FGD, additives are employed (Collarini 2012).

The atmospheric hydration reaction occurs in three phases: In the first few seconds after the lime is contacted by the water, a strong rapid reaction proceeds, then there is a brief period where the hydrated lime formed on the surface of the lime particle impedes contact between the lime particle underneath and the water; finally, the $Ca(OH)_2$ crystals separate from the particle and the remaining lime reacts with the water. The hydrator is therefore constructed with three chambers. The first chamber is a reaction compartment, where the lime and the water are vigorously agitated; the lime/hydrate then proceeds to a further mixing compartment and finally advances to a seasoning compartment. The resulting product of this process is a dry powder, which is the predominant commercial form (Collarini 2012; Cimprogetti 2017).

CALCINATION REACTIONS AND THERMODYNAMICS

Thermodynamics allow the determination of operating parameters where a calcination reaction will be spontaneous. In the van't Hoff reaction isotherm shown in Equation 2, ΔG_R is the Gibbs free energy of the reaction, and ΔG_R° is the Gibbs free energy in standard state of the reaction. From Equation 3, R is the ideal gas constant, T is the temperature of interest, Q is the activity quotient, and K is the equilibrium constant. The activity quotient is equal to the product of the activities of the reaction products divided by the product of the activities of the reaction reactants. The activity of a pure condensed substance in its standard state is equal to 1, and the activity of a gas is essentially equal to the partial pressure of the gas.

$$\Delta G_R = \Delta G_R^\circ + RT\ln Q \quad \text{(EQ 2)}$$

$$\Delta G_R^\circ = -RT\ln K \quad \text{(EQ 3)}$$

The van't Hoff reaction isotherm equation can be manipulated to the form shown in the following equation:

$$\Delta G_R = RT\ln(Q/K) \quad \text{(EQ 4)}$$

The van't Hoff reaction isotherm in this form allows several determinations to be made about the reaction. If Q = K, then the system is at equilibrium, and $\Delta G_R = 0$. If $Q/K > 1$, ΔG_R will be positive, the reaction will not proceed as written, and the reverse reaction is favored. If $Q/K < 1$, ΔG_R will be negative,

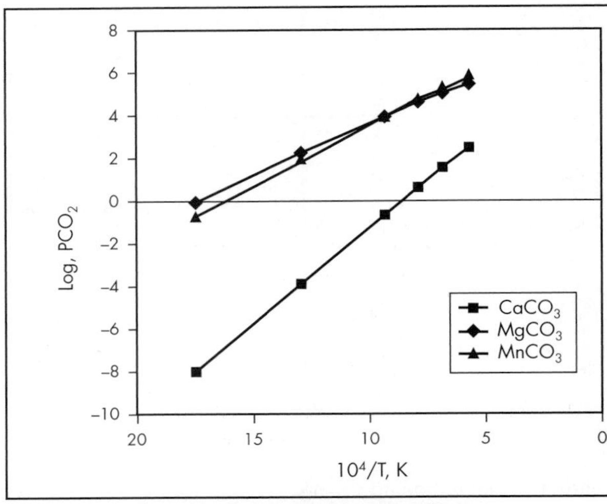

Figure 1 Carbonate decomposition equilibria

and the reaction will therefore be spontaneous. It is possible, theoretically, to then manipulate the operating parameters, so that the value of Q is less than K. These operating conditions would allow the desired calcination reaction to proceed. The general carbonate calcination reaction is shown in the following equation, and the van't Hoff reaction isotherm for this reaction is shown in Equation 6.

$$MeCO_{3(s)} \rightarrow MeO_{(s)} + CO_{2(g)} \quad \text{(EQ 5)}$$

$$\Delta G_R = \Delta G^\circ_R + RT\ln P_{CO2} \quad \text{(EQ 6)}$$

Manipulating the van't Hoff equation yields the following equation. Further, for the carbonate reaction, it yields Equation 8. It is then useful to plot $\ln P_{CO2}$ (or $\log P_{CO2}$) as a function of 1/T.

$$d\ln K/d(1/T) = -\Delta H^\circ_R/R \quad \text{(EQ 7)}$$

$$d\ln P_{CO2}/d(1/T) = -\Delta H^\circ_R/R \quad \text{(EQ 8)}$$

A van't Hoff diagram, shown in Figure 1, can be used to determine the temperature at which the decomposition reaction is spontaneous. Temperature increases from left to right in the diagram; as a point of reference, 10 on the abscissa is equivalent to 726.85°C. The temperature necessary for the decomposition pressure to reach 1 atm varies considerably from one substance to the other. The van't Hoff diagram can also be used to select temperatures for selective decomposition.

Calcination proceeds according to the shrinking core model. As a piece of carbonate decomposes, increasing temperatures are required to calcine further into the particle, through the layer of oxide that has formed, as shown in Figure 2, because CaO, for example, has a lower thermal conductivity than $CaCO_3$. As a carbonate particle is calcined, it is enveloped in carbon dioxide (CO_2) gas film. If the ambient pressure (P_{TOTAL}) is 1 atm, then the CO_2 partial pressure around the particle will also be 1 atm. For rapid carbonate decomposition to occur, the equilibrium vapor pressure of the CO_2 must exceed the CO_2 partial pressure in the surrounding atmosphere (Downey 2013).

The purpose of limestone calcination is to convert the calcium carbonate mineral to calcium oxide, CaO. Calcium oxide is a very important industrial commodity that is known

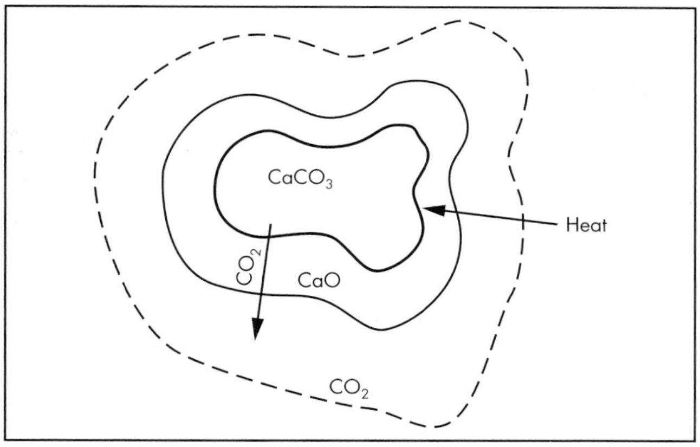

Figure 2 Calcination of calcium carbonate

Table 1 Common carbonate reactions and reaction heats

Material	Reaction	Heat of Calcination Reaction*
Calcium carbonate	$CaCO_3 \rightarrow CaO + CO_{2(g)}$	$\Delta H°_{R\ at\ 298\ K}$ = 42.750 kcal
Magnesium carbonate	$MgCO_3 \rightarrow MgO + CO_{2(g)}$	$\Delta H°_{R\ at\ 298\ K}$ = 24.250 kcal
Manganese carbonate	$MnCO_3 \rightarrow MnO + CO_{2(g)}$	$\Delta H°_{R\ at\ 298\ K}$ = 27.850 kcal
Dolomite	$CaMg(CO_3)_2 \rightarrow CaO{\cdot}MgO + 2CO_{2(g)}$	$\Delta H°_{R\ at\ 298\ K}$ = 67.000 kcal

*Data from Kubaschewski et al. 1967

as *burned lime*, *quicklime*, or just *lime*. Where the heat of reaction for each of the following reactions is calculated from the appropriate heats of formation (Kubaschewski et al. 1967), the calcium carbonate calcination reaction is

$$CaCO_3 \rightarrow CaO + CO_{2(g)} \qquad \textbf{(EQ 9)}$$
$$\Delta H°_{R\ at\ 298.15\ K} = 42.750\ \text{kcal/mole}$$

The calcination product of dolomite, $CaCO_3{\cdot}MgCO_3$, is called *dolomitic lime* or *dolime*. For dolomite, the reaction is as follows:

$$CaMg(CO_3)_2 \rightarrow CaO{\cdot}MgO + 2CO_{2(g)} \qquad \textbf{(EQ 10)}$$
$$\Delta H°_{R\ at\ 298.15\ K} = 67.000\ \text{kcal}$$

Table 1 shows the heats of reaction for some of the most common carbonate calcination reactions. The theoretical limestone dissociation temperature is 848°C. At 900°C, the CO_2 partial pressure reaches atmospheric pressure (760 mm Hg) in a pure CO_2 atmosphere. The theoretical dissociation temperature of dolomite is 725°C (Downey 2013). In all cases, the actual dissociation temperature is influenced by the impurities present. Practical calcining temperatures are considerably higher than their theoretical values because the reaction and heat transfer rates are not rapid enough for economical operation. The effects of the operating parameters on the product are very similar for the different carbonates.

For limestone calcination, high-reactivity lime is produced when the surface temperature of the material does not exceed 1,180°C as it passes through the kiln's calcination zone. The corresponding gas temperatures are typically 90–150 degrees higher than the surface temperature of the charge materials. Crystal size, pore volume, specific surface area, and reactivity are important and closely related lime properties. Other carbonates have similar temperature parameters to produce a product with high reactivity and other desired properties. Optimum conditions should be determined for each type of stone; an optimum residence time exists for each feed size distribution and temperature.

At constant residence time, increasing temperature generally reduces reactivity, causes the grains to coarsen, and reduces surface area and porosity. Temperature is more important than residence time for these qualities. Inorganic impurities (silica [SiO_2] and alumina [Al_2O_3]) have a similar effect on crystal size, pore volume, specific surface area, and reactivity.

The analyses of the feed and the product are related in a certain ratio, which depends on the amount of CO_2 removed during calcination. Loss-on-ignition (LOI) tests are performed on representative samples of the burned product as a quality control measure. Residual CO_2 content in the product is part of the LOI, which can also include relatively small amounts of moisture and organic matter in the product.

Important properties include grade, purity, and reactivity. Using lime as an example, *grade* refers to the percent CaO in the calcined product, whereas purity refers to the available CaO determined by ASTM C25-17. Reactivity is the capacity of the CaO to react with water during hydration and is tested with the ANSI EN 459-2 reactivity test (commonly referred to as the T60 test), which measures the temperature rise as a function of time when reacted with water.

The main quality control issue in limestone calcination is the balance between unreacted core versus overburned lime on the calcined particles' surfaces. The terms *soft burned*, *medium burned*, and *hard burned* relate to the lime reactivity; just as *caustic* (or *light burned*), *hard burned*, and *dead burned* refer to magnesia reactivity (Martin Marietta 2018). *Soft lime* is produced when the surface temperature of the material does not exceed 1,180°C as it passes through the kiln's calcination zone. Reactivity primarily depends on the calcination temperature, and to a lesser extent, on the calcination time and impurities (which can flux with CaO).

Compared with hard-burned lime, soft-burned lime has smaller crystal grain size, larger specific surface area (>1.0 vs. <0.3 m^2/g determined via BET [Brunauer–Emmett–Teller] analysis), smaller pores, but larger total pore volume (46%–55% vs. <34%), lower apparent density (1.5–1.8 vs. >2.2), and a higher chemical reactivity (>20°C/min vs. <2°C/min) (Downey 2013).

The time available (in a particular calcining operation) to generate the heat transfer determines the maximum particle size that can be effectively calcined in the system. Additionally, the feed particle size distribution will determine the uniformity of the product quality; wider size distributions naturally lead to greater quality variations. A narrow feed size distribution is required to maintain uniform product quality.

Magnesite is calcined to produce caustic (light-burned) magnesia, hard-burned magnesia, or dead-burned magnesia. Following calcination at temperatures ranging between 700° and 900°C, the MgO is further heated to ~1,750°C to shrink the particles. This action increases the particle density and renders the MgO less reactive. Magnesia reactivity is measured as a function of time with the citric acid test. Dead-burned MgO produced by magnesite calcination has little reactivity and is often destined for use in refractory applications (i.e., basic refractory brick for magnesite, mag-chrome, chrome-mag, etc.) or in ceramic manufacturing (Downey 2013). Hard-burned magnesia is calcined at temperatures ranging from 1,000° to 1,500°C for applications requiring low reactivity, such as phosphate cement (Martin Marietta 2018). Caustic magnesia is produced by calcining magnesite at temperatures ranging from 700° to 1,000°C for applications including environmental remediation.

Other minerals are calcined as well, such as magnesium hydroxide, $Mg(OH)_2$, and magnesium sulfate, $MgSO_4$, to produce MgO. Magnesium sulfate decomposition proceeds according to the following equations (Kubaschewski et al. 1967):

$$MgSO_4 \rightarrow MgO + SO_3 \quad \text{(EQ 11)}$$
$$\Delta H^\circ_{R\text{ at }298.15\text{ K}} = -57.200\text{ kcal}$$

$$SO_3 \rightarrow SO_2 + \tfrac{1}{2}O_2 \quad \text{(EQ 12)}$$
$$\Delta H^\circ_{R\text{ at }298.15\text{ K}} = -33.650\text{ kcal}$$

Petroleum coke (green coke) is calcined to remove volatile matter, moisture, and impurities. Calcining also increases the density and improves the electrical conductivity of the coke within the coke feed quality limitations. Coke calcining is performed on-site (e.g., at the anode plant) or at the petroleum refinery. Applications for the calcined petroleum coke include anodes for aluminum production, for the production of titanium, and in the steel industry for carbon and graphite electrodes and as a recarburizing agent (Aminco 2013; Edwards 2014).

Calcining of petroleum coke includes drying, devolatilization, and densification stages. Most coke calcination is performed in rotary kilns at 1,200°–1,350°C under an oxygen-deficient atmosphere. Liquid, gas, or solid fuel is utilized for process heat, which can be supplemented by the controlled combustion of a small amount of the volatiles and coke fines using secondary and tertiary combustion air. Important operating parameters include temperature and residence time. After calcination, the coke is cooled in a rotary cooler, normally by water quenching. The kiln off-gas proceeds to an afterburner, where the volatiles are combusted with injected air, and the heat is recovered for steam production (Edwards 2014).

Shaft kilns and rotary hearth kilns are also used to a lesser extent for coke calcination. Shaft kilns are predominately utilized in China. Feed for a shaft kiln can contain a higher percentage of volatile matter, because the green coke can be easily blended with calcined coke to adjust the amount of volatile matter to the approximately 11%–12% required in the feed. Green coke is fed from the top and proceeds through the kiln by gravity; the residence time of the coke is controlled by a discharge valve or gate at the bottom of the kiln, below a cooling jacket. The volatile matter that is released from the coke flows from near the top of the shaft to the surrounding flue, where it is combusted and indirectly heats the calcine. The heating rate of the calcine is very slow, ~1°C/min. Residence time in a shaft calciner is 28–36 hours, whereas a rotary calciner has a residence time of ~1 hour. The shaft calciner produces a higher yield of coke with a higher density and larger particle size (Edwards 2011).

CALCINATION EQUIPMENT

The original carbonate calcining systems were simple pot kilns. The U.S. lime industry currently favors rotary kilns, whereas shaft furnaces prevail in Europe, but some of the most recent kiln installations in the United States have been parallel shaft regenerative kilns (USGS 2009). Equipment choice is dependent on many factors: the product required by the customer (e.g., reactivity), sales volume, resource utilization efficiency (using fines as well as larger stone), capital cost, maintenance costs, and environmental impact among others. Both of these alternatives are fully automated, high-capacity systems, but the parallel flow regenerative shaft kilns are significantly more energy efficient. A rotary kiln with a preheater is approximately 55% energy efficient, where a parallel shaft regenerative kiln may be more than 80% efficient. Process control technologies have become the norm in commercial lime production. Energy costs, product uniformity, and decrepitation are important considerations in the choice of kiln. Table 2 summarizes the capacity, fuel consumption, and applicable feed size range for the main commercial lime kilns.

Coke calcining is typically conducted in long rotary kilns, as shown in Figure 3, but shaft kilns, which are indirectly heating by combusting the volatile matter released as the coke is calcined, and rotary hearth furnaces are utilized as well.

Rotary Kilns

In general, rotary kilns produce uniform, high-quality products, have high availability and production capacity, require relatively low labor costs, but have relatively high fuel consumption. Rotary kilns can be operated utilizing most fossil fuels, including coal, natural gas, oil, petroleum coke, and some waste fuels. If a low sulfur lime is required, it may therefore be produced in a rotary kiln. Rotary kilns without a preheater (long rotary kiln) are used for coke calcination and can also be useful for lime that decrepitates (Rosenqvist 1974). A conventional long rotary kiln can have a capacity up to 907–1,134 t/d (metric tons per day), utilizing feed size of ~3–55 mm.

Rotary kilns are essentially heat exchangers and conveyors—heat is transferred from the combustion flame and gases to the bed as the material moves from the feed to the discharge end of the kiln. Radiation from the flame and the refractory accounts for most of the heat transfer to the charge, so it is important for the material to be continually agitated to expose fresh surface area to facilitate the heat transfer. Conduction from the refractory bricks accounts for a substantially smaller percentage of the heat transferred to the charge, and convection accounts for a minor percentage of heat transfer, as the hot gas stream contacts the charge (Downey 2013).

A rotary kiln is a cylindrical reactor that rotates about its longitudinal axis. The tube is usually sloped down roughly 1 cm per 0.24 m from the feed end to the firing end (from the horizontal). Commercial reactors are typically 2.4–5.2 m in diameter, and the length-to-diameter ratio generally ranges from 4 to >10. The charge in the kiln is normally less than 15% of the cross-sectional area. In some applications, sections with expanded diameter are included to increase burden depth

Table 2 Kiln capacity, fuel consumption, and feed size by type

Kiln Type	Capacity, t/d	Fuel Consumption, GJ/t limestone	Feed, mm
Annular shaft	200–600	3.64–3.94	15–200
Rotary kiln with preheater	300–1,200	5.1–6.3	3–55
Parallel shaft regenerative (rectangular)	100–400	3.64–3.93	30–120
Parallel shaft regenerative (circular)	300–800	3.64–3.93	30–160
Parallel shaft regenerative (fine lime)	200–400	3.31–3.56	15–40

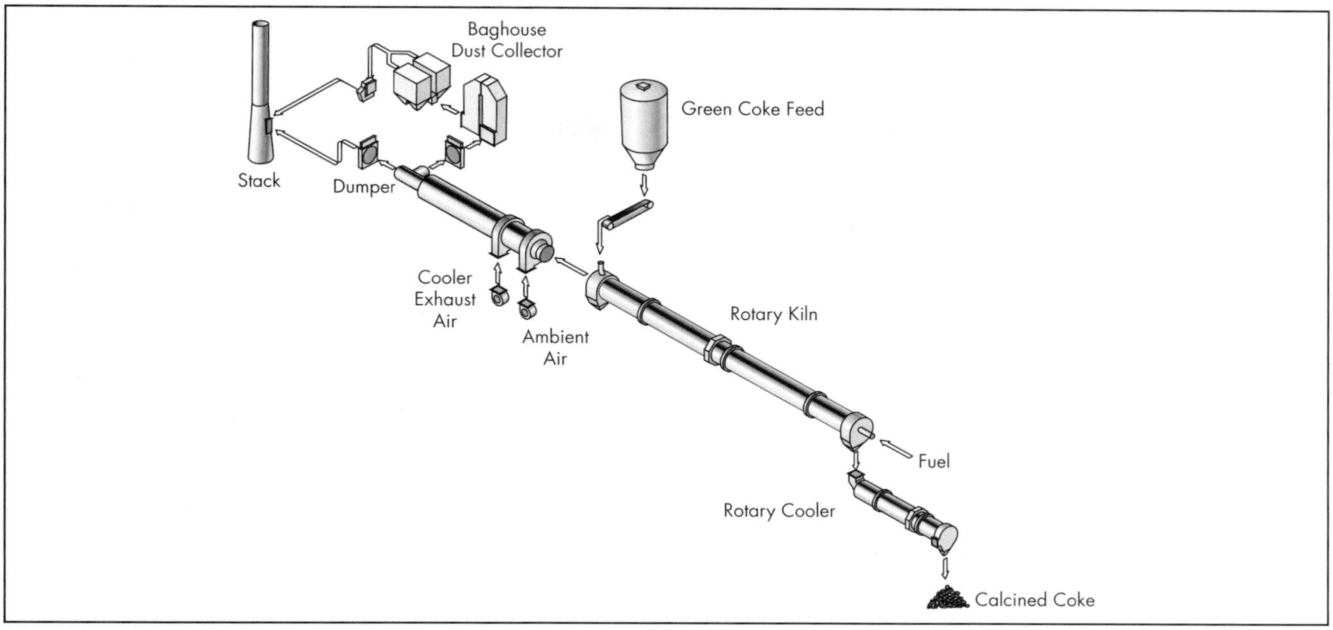

Adapted from Metso 2017
Figure 3 Rotary coke calciner

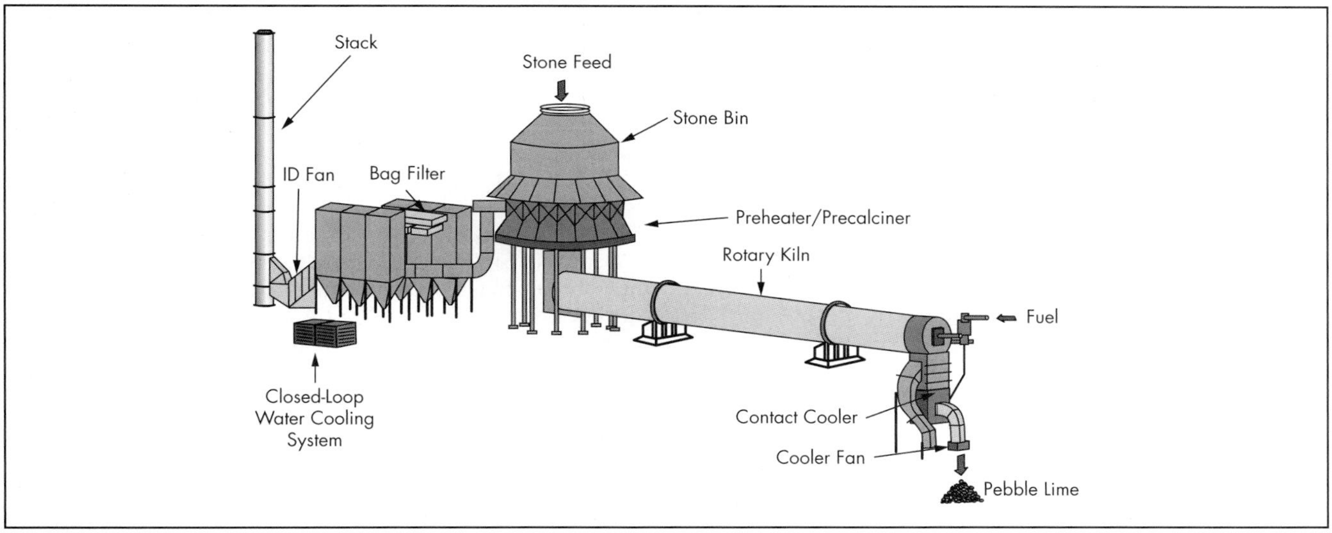

Courtesy of Metso
Figure 4 Preheater rotary lime kiln with baghouse

and residence time. Dams may also be installed to increase residence time. The steel shell is refractory lined to reduce heat loss and to protect the steel shell.

The tube ends are sealed, typically by leaf seals, to prevent gas leakage. Maintaining a slight negative pressure (0.3–2.5 mm of water) in the kiln helps control emissions. Draft is provided by induced-draft (ID) fans that are located beyond the feed end of the kiln (typically after the baghouse). Feed enters the higher end of the cylinder and works its way through the cylinder because of the feed rate; rotational speed (measured in revolutions per minute), which is controlled by a variable speed drive; and slope. These factors dictate the residence time of the burden in the tube, but only the rotational speed is controllable by the operator. Finished product discharges at the lower end of the tube to a cooler. Countercurrent gas-to-solids flow also improves heat transfer efficiency (Downey 2013). Gas analyzers are usually installed at the feed end to monitor oxygen and carbon monoxide (CO) in the off-gas.

Because of the fuel consumption issue, rotary kilns are usually equipped with preheaters and other fuel-saving devices. Major savings can be achieved by external preheaters, particularly if some of the carbonate decomposition takes place in the preheater. A rotary kiln with a preheater is shown in Figure 4. Coke is usually calcined in a rotary kiln with drying and preheating occurring in the first section of the kiln.

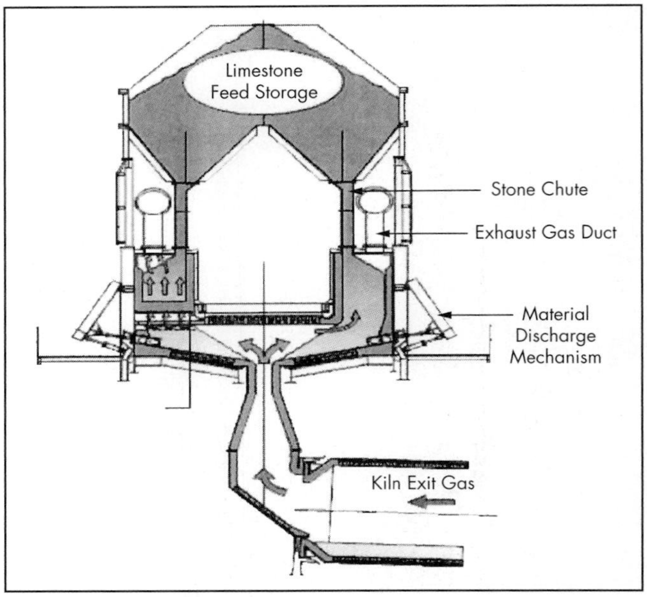

Courtesy of Metso
Figure 5 KVS/Metso preheater

Common types of preheaters used with a rotary kiln include the FLSmidth, which is similar to the Metso but has an internal stone bin, and the Kennedy Van Saun (KVS) now distributed by Metso, shown in Figure 5, which takes advantage of the ~1,000°–1,100°C kiln off-gas to partially calcine the charge, releasing 20%–25% of the CO_2 before the stone enters the kiln. In this type of preheater, stone is fed from an integral storage bin through vertical chutes, where it is heated, moisture is evaporated, and the decomposition begins. Discharge to a transfer chute to the kiln feed end is achieved by hydraulic rams on the preheater perimeter (Metso, n.d.). Preheater off-gas averages ~230°–260°C, so filter media choice must take this into consideration if a baghouse is utilized for dust collection.

Gas suspension preheaters, consisting of multiple cyclones, are commonly used in the cement industry. Dry feed is injected into the side of the uppermost riser pipe and cyclone where it contacts the rising exhaust gas; the feed proceeds down through the other cyclones, countercurrent to the rising exhaust gas; and the preheated feed is ultimately fed into the kiln.

Shaft Kilns

Vertical Shaft Kilns

Vertical shaft kilns are frequently utilized to calcine green petroleum coke and are also used for lime calcination for captive production. The vertical shaft kilns for coke calcination are normally indirectly heated by the combustion in the flue that surrounds the shaft of the volatiles released from the coke feed (Metso 2017). The vertical shaft kilns can process between 15 and 400 t/d of lime using coke or coal as fuel. For limestone calcination, the coal or coke is blended with the feed and is conveyed to the kiln. Lime is extracted with vibrating extractors at the bottom of the kiln, and CO_2 is recovered from the top of the kiln to be used in other processes.

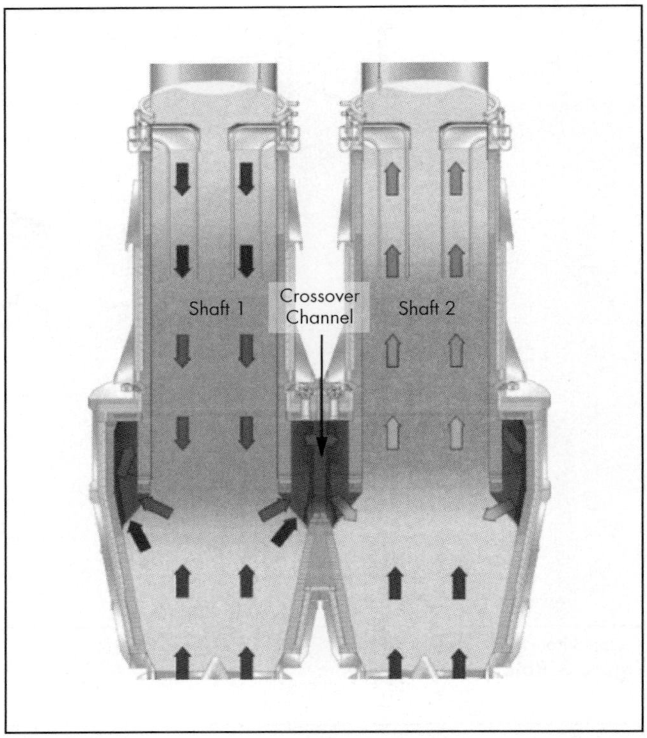

Courtesy of Maerz
Figure 6 Gas flow in a PFR lime kiln

Annular Shaft Kilns

Annular shaft kilns have some commercial utilization for both limestone and dolomite. With capacities of 200–600 t/d of 15–200-mm stone, fuel consumption averages ~870–940 kcal/kg for limestone and 820–900 kcal/kg for calcination of dolomite. In gigajoules per metric ton, that is 3.64–3.94 GJ/t of limestone and 3.43–3.76 GJ/t of dolomite. These external combustion kilns have two combustion levels with four to five combustion chambers per level and normally utilize gas, fuel oil, or solid fuels (ThyssenKrupp 2016).

Parallel-Flow Regenerative Shaft Kilns

PFR shaft kilns have two vertical shafts connected by a crossover channel. The parallel flow of limestone and hot gases preheat the combustion air in one shaft (the burning shaft), and concurrently, a mixture of combustion gases and air from the lime cooler preheats the limestone in the second shaft (the nonburning shaft), as shown in Figure 6. At intervals of ~10–15 minutes, each shaft cycles through the burning and nonburning mode (Maerz, n.d.).

The parallel flow principle is ideal for producing high-reactivity quicklime at the highest thermal efficiency of all modern lime kilns. PFR kilns can be fired with gaseous, liquid, and pulverized coke, or pulverized coal with a low swell index (<1%). Accurate control of key operating parameters, such as temperature and pressure within the kiln, fuel flow rate, combustion and cooling airflow rates, stone level, and discharge rate of product are crucial to achieve the high thermal efficiency in the PFR kilns (Maerz, n.d.).

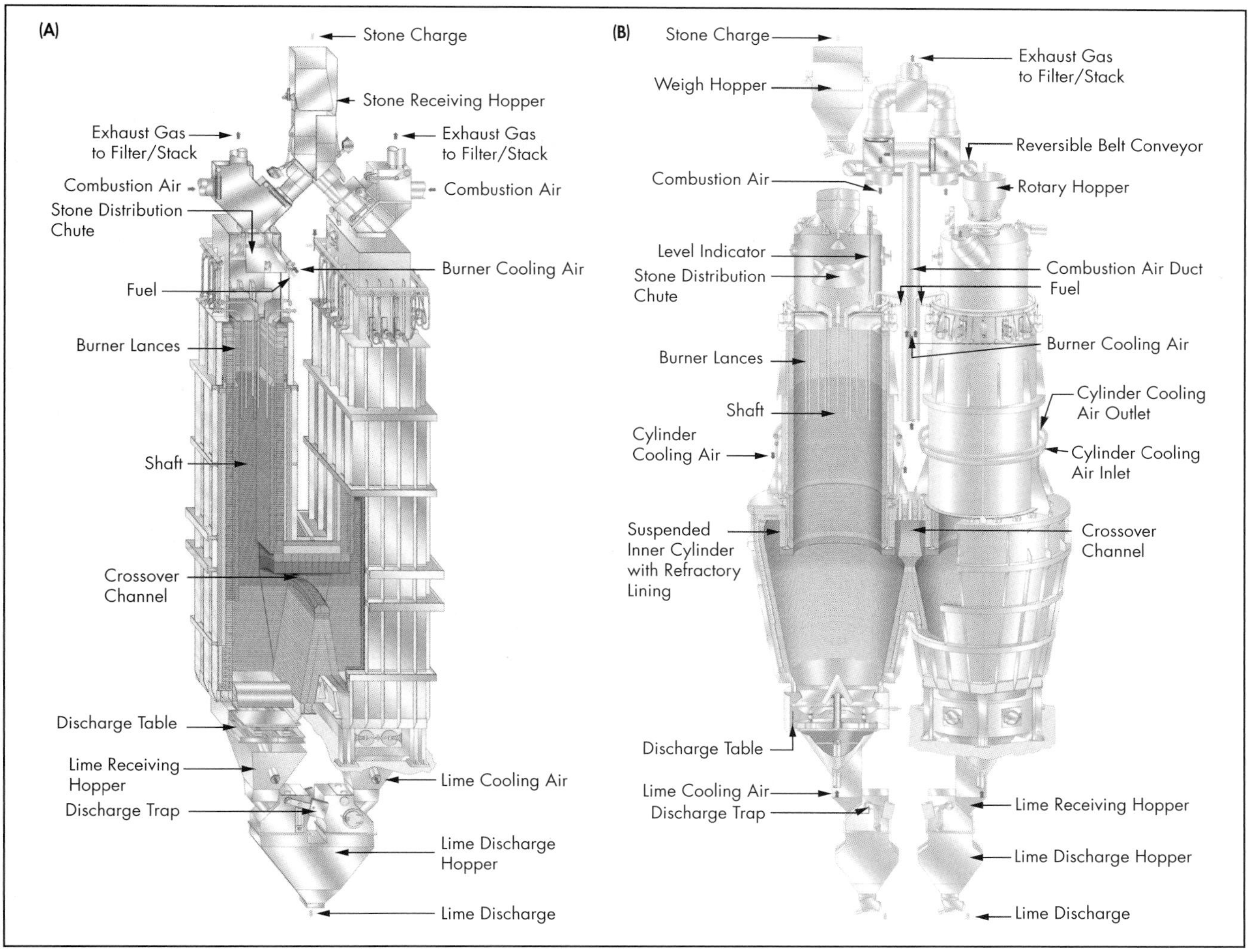

Courtesy of Maerz

Figure 7 (A) Rectangular PFR and (B) circular PFR kilns

Cimprogetti and Maerz are among the major manufacturers of PFR (or twin-shaft regenerative) kilns. Three types of PFR kilns are typically used:

1. The Maerz rectangular PFR kiln, shown in Figure 7A, is capable of producing 100–400 t/d of high-reactivity lime, handles 30–120-mm stone, with 810–870 kcal/kg or 3.64–3.93 GJ/t heat input.
2. The circular PFR kiln, shown in Figure 7B, is capable of producing 300–800 t/d of high-reactivity lime, handles 30–160-mm stone, with 810–870 kcal/kg or 3.64–3.93 GJ/t heat input.
3. The Maerz Finelime PFR kiln is capable of producing 200–400 t/d of high-reactivity lime, handles 15–40-mm stone, with 790–850 kcal/kg or 3.31–3.56 GJ/t heat input (Maerz, n.d.).

In North America, there are two additional types of lime kilns that are in very limited use, mostly in specialty applications: a rotary hearth kiln (calcimatic kiln) and a fluidized-bed kiln. They are applicable for lime that decrepitates but have limited fuel options.

Flash Calciners

Flash calciners are utilized for cement, gypsum, and alumina calcining among other applications. Gas suspension calciners (GSCs), or flash calciners, such as those from FLSmidth or Calix, utilize a series of high-efficiency cyclones that are low maintenance, have low nitrogen oxide (NO_x) emissions, and can be fueled by natural gas, fuel oil, coal, coke, or alternate fuels. Capacity for the FLSmidth GSC is 1–800 t/h of fine ore (FLSmidth 2013).

Rotary Hearth Calciners

Rotary hearth calciners are sometimes used for coke calcination because they minimize decrepitation and produce consistent coke quality with low energy consumption. Feed enters from the top of the sloped rotating hearth and is moved toward the discharge edge by rabbles.

Calcimatic Kilns

Calcimatic kilns utilize a circular refractory hearth divided into heating zones. This type of kiln is often operated at 25–85 rpm, although these large-diameter hearths can be

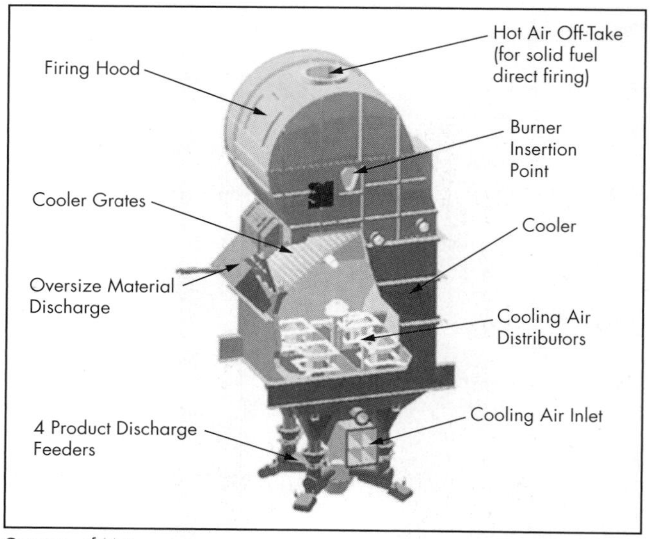

Courtesy of Metso

Figure 8 Metso Niems cooler

operated at up to 200 rpm. Fuel choices are limited, as coal cannot be used as a fuel for calcimatic kilns (Boynton 1966).

Fluidized-Bed Calciners

A fluidized-bed reactor is a vertical, usually cylindrical carbon steel vessel that is lined with insulating and refractory firebrick. It has four main sections. The wind box is a plenum that delivers fluidizing gas to the bed through a constriction (or distribution) plate. The constriction plate acts as a distributor for the fluidizing gas, supporting the bed and preventing backflow of solids into the wind box. The third and fourth main sections are the fluidized bed itself, in which the bed may consist of the calcine product; and the freeboard, where larger solids disengage from the upwardly moving gas stream (Downey 2013). A fluidized-bed kiln produces highly reactive soft-burned lime. Fine feed is required for this type of kiln, and it is often used for lime that decrepitates. Fuels that can be utilized in the fluidized-bed kiln are gas, fuel oil, finely ground coal, or fine coke. The Fluo-Solids kiln is an adaptation of a conventional fluidized-bed reactor that has multiple beds (Cimprogetti 2017; Boynton 1966).

Coolers

A rotary cooler with a water quench is typically used for coke calcined in a rotary kiln. Most carbonate kilns utilize a contact cooler such as a Niems. A Niems cooler, shown in Figure 8, has four quadrants. Ambient air is blown uniformly through stainless-steel louvers into the bed of hot lime. Vibratory feeders control the bed level in the cooler by controlling the lime product discharge. Heat recovery efficiencies approaching 90%, with an airflow equivalent to the secondary combustion requirements for calcination are achieved by the Niems cooler (Metso 2018).

Some plants utilize a satellite cooler or a grate cooler. A satellite cooler is the least efficient of the three types of lime coolers. The satellite cooler consists of tubes attached to the kiln in a parallel configuration that rotate with it. Normally, the kiln ID fan provides the air for the cooler. Satellite coolers discharge gas temperatures are low, ~80°C, so they are not usually used for secondary air. Discharge gas for grate coolers is approximately 540°C and is utilized for secondary air, but it produces more than is required for the process, and the high off-gas temperature is problematic for the baghouse or electrostatic filtration systems that are most common (Metso 2018).

REFERENCES

Aminco Resources. 2013. Calcined petroleum coke. www.amincoresources.com/calcined-petroleum-coke/. Accessed August 2018.

ANSI EN 459-2. *Building Lime—Part 2: Test Methods*. Washington, DC: American National Standards Institute.

ASTM C25-17. *Standard Test Methods for Chemical Analysis of Limestone, Quicklime, and Hydrated Lime*. West Conshohocken, PA: ASTM International.

Boynton, R. 1966. *Chemistry and Technology of Lime and Limestone*. New York: Interscience. p. 238.

Cimprogetti. 2017. *Lime Hydrating Process*. Dalmine, Italy. www.cimprogetti.com/2017_02_depliant_hydration_EN_EMAIL.pdf.

Collarini, O. 2012. Lime hydration process development. Presented at LimeCon, December 2012. www.cimprogetti.com/H_2012_EN_004.pdf. Accessed July 2018.

Downey, J. 2013. Processing of elevated temperature systems. Lecture notes, Montana Tech, Butte, MT.

Edwards, L. 2011. Quality and process performance of rotary kilns and shaft calciners. In *Light Metals 2011*. Edited by S.J. Lindsay. Hoboken, NJ: Wiley.

Edwards, L. 2014. Pelletization and calcination of green coke using an organic binder. U.S. Patent 8,864,854 B2.

FLSmidth. 2013. Flash calciner system for minerals: Gas suspension calcination (GSC). www.flsmidth.com/~/media/PDF%20Files/Pyroprocessing/FLSmidth_FlashCalciner_Brochure_email.pdf. Accessed August 2018.

Kubaschewski, O., Evans, E.L., and Alcock, C.B. 1967. *Metallurgical Thermochemistry*, 4th ed. Oxford: Pergamon Press.

Maerz. n.d. PFR [parallel flow regenerative] kilns for soft-burnt lime. www.maerz.com/portfolio/pfr-kilns-for-soft-burnt-lime/. Accessed August 2018.

Martin Marietta Magnesia Specialties. 2018. Magnesia grades. http://magnesiaspecialties.com/magnesia-grades/. Accessed August 2018.

Metso. 2017. Petroleum coke calcining. www.metso.com/products/petroleum-coke-calcining. Accessed August 2018.

Metso. 2018. Pebble lime cooler. www.metso.com/products/pyro-process/pebble-lime-cooler/. Accessed July 2018.

Metso. n.d. Preheater-kiln lime calcining systems. Brochure No. 1330-03-02MPR. Danville, PA: Metso Mineral Industries.

National Lime Association. n.d. *Lime: The Essential Chemical*. Arlington, VA: National Lime Association.

Rosenqvist, T. 1974. *Principles of Extractive Metallurgy*. New York: McGraw-Hill.

ThyssenKrupp. 2016. Annular shaft kiln: The art of modern calcining. Product sheet. Beckum, Germany: ThyssenKrupp Industrial Solutions.

USGS (U.S. Geographical Survey). 2009. *Mineral Industry Surveys*. Reston, VA: USGS.

CHAPTER 11.4

Vapor-Phase Extraction

Jerome P. Downey

Vapor-phase extraction technologies exploit differences in the vapor pressures of various substances to separate the more volatile constituents from other condensed matter. Volatile species of interest enter the vapor phase and are transported from the furnace vessel, condensed downstream, and recovered as value-added products or process intermediates or, in other cases, rejected as impurities. Key process parameters such as temperature, pressure, furnace atmosphere (gas composition), and vessel residence time require precise control to maximize separation selectivity.

In the chemical process industries, vapor-phase extraction is employed as a means of separating organic species in fractional distillation operations; petroleum refining and spirits alcohol production are familiar examples. Vapor-transport processes are also used in the materials science and engineering field as a means of effecting materials synthesis, materials separation and purification, and coating substrate materials. In extractive metallurgy, vaporization techniques are employed in many primary and secondary metal-extraction processes as well as in certain refining processes. Historical examples of metal extraction include mercury and zinc from ores, concentrates, or scrap while examples of the elimination of deleterious constituents in metal refining operations include expulsion of undesirable volatile contaminants such as arsenic and antimony from refractory gold ores and base metal concentrates (Alcock 1976; Rosenqvist 1974). Other industrially significant vapor-transport applications include coal gasification, flame and plasma spray coating, and chemical vapor deposition (Hastie and Hager 1990).

FUNDAMENTALS

For a given vapor metallurgy application, the most important process parameters are temperature, pressure, the compositions of the applicable condensed and gas phases, and residence time of the condensed phase at temperature. Selection of the appropriate combination of process parameters depends on the physical and chemical characteristics of the target constituents as well as those of the other matrix components. Vapor-phase chemistry can be extraordinarily complex, replete with uncommon valence states and the evolution of multicomponent and/or polymeric gas species. Multicomponent condensed-phase solutions present unique challenges because they exhibit nonideal thermodynamic behavior and are possibly subject to undesired reactions or interactions between components.

The most obvious and certain way to vaporize a pure substance is simply to boil it, which requires processing at an elevated temperature such that the vapor pressure of the target substance equates to one atmosphere (or one bar, depending on the convention) at the substance's normal boiling point. However, it is not always practical or economical to boil a substance to effect vapor-phase transport. Fortunately, many substances are sufficiently volatile that vaporization can be effected at temperatures well below their normal boiling points. Water, for example, readily evaporates at temperatures well below its boiling point in arid, low-humidity regions.

The equilibrium vapor pressure of a pure element "i" in its standard state is established by the free-energy relationship in Equation 1, where $\Delta G°$ represents the standard Gibbs free energy of the vaporization reaction, R is the ideal gas constant, T is the temperature of interest (kelvin), and P_i represents the equilibrium vapor pressure (atm) of pure substance i. P° represents the standard pressure, which is typically 1 atm (or 1 bar); when the standard pressure value is unity, it is frequently omitted from thermodynamic calculations. Two of the more frequently used values of the gas constant are 8.314 J/mole-K and 1.987 cal/mole-K.

$$\Delta G° = RT\ln P_i/P° \qquad \textbf{(EQ 1)}$$

The Van't Hoff reaction isotherm, shown in Equation 2, expresses the relationship between the free energy of reaction and the equilibrium constant. This equation is especially useful for identifying the theoretical process conditions or "windows of operating parameters" necessary to vaporize a given substance.

$$\Delta G_R = \Delta G_R° + RT\ln Q \qquad \textbf{(EQ 2)}$$

In this equation, ΔG_R represents the Gibbs free energy of the vaporization reaction, and Q represents the reaction activity quotient, which is expressed as the ratio of the partial pressure of substance (P_{Me}) as a vapor to the activity of the substance

Jerome P. Downey, Professor of Extractive Metallurgy, Montana Tech, Butte, Montana, USA in a condensed phase (a_{Me}), as seen in Equation 3. The activity of a pure condensed substance in its standard state is unity.

$$Q = P_{Me}/a_{Me} \quad \text{(EQ 3)}$$

The numerical value of the equilibrium constant, K, equates to the reaction quotient under equilibrium conditions. The Van't Hoff reaction isotherm equation is frequently presented in the form shown in the following equation:

$$\Delta G_R = RT\ln(Q/K) \quad \text{(EQ 4)}$$

Equation 4 ranks among the most useful relationships in chemical thermodynamics because it enables one to establish whether a reaction is possible under a given set of process conditions or, alternatively, to precisely define the operating conditions that must be imposed on the system to achieve a process goal. When the values of Q and K are identical, the ΔG_R value is 0 and the system is at equilibrium. If the Q:K ratio is less than 1, then ΔG_R has a negative value, indicating that the reaction is spontaneous. Finally, if the Q:K ratio is greater than 1, ΔG_R has a positive value, indicating that the reaction is not possible and that the reverse reaction is favored.

To create the thermodynamic driving force required to vaporize a given substance, it is necessary to define and impose process conditions under which the value of the reaction activity quotient becomes and remains less than the value of the corresponding equilibrium constant or, in other words, create conditions wherein *the equilibrium vapor pressure of the specie of interest exceeds its partial pressure above the condensed phase*. Thus, at temperatures below the normal boiling point of a substance, it is sometimes possible to create and sustain the necessary thermodynamic driving force by minimizing the partial pressures of the vapor species in the furnace atmosphere. The partial pressure of the nascent gas species can be minimized in several ways: (1) by reducing the total system pressure and thereby reducing the partial pressure of the vapor species; (2) by introducing an inert diluent gas to lower the concentration of the nascent gas species; (3) by continuously purging the system to remove the vapors as they evolve; or (4) by some combination of methods 1, 2, and 3.

Vaporization reactions are exclusively endothermic, so increasing temperature will always increase the vapor pressure of a substance. However, practical limitations often dictate the maximum operating temperature of a system. Although exceptions exist, vapor-phase extraction is generally practiced on substances with boiling points less than 1,000°C (Moore 1981). Downstream, the vaporized substances are condensed by controlled cooling of the process gas stream. Care must be taken to avoid undesired back reactions with other gas-phase constituents and/or entrained particulate matter, particularly if the purity of the condensed substance is of concern. Particulate matter is frequently omitted by mechanical separation (e.g., cyclones, baghouses, and electrostatic precipitators) prior to cooling to the point of incipient condensation (i.e., the dew point). In certain applications, it may be necessary to modify and control the chemical composition of the gas phase to condense a pure element or a particular compound.

In many applications, the target element or compound does not exist discretely in the matrix. Rather, it is a component in a solid or liquid solution. In such cases, the activity of the target substance cannot be assumed to be unity. In an ideal solution, the activity of a solute is equal to its mole fraction at all concentrations. However, few of the non-gas-phase solutions encountered in metallurgical applications exhibit ideal behavior; very dilute solutions are the exception. In non-ideal or "real" solutions, solute activity (a_i) generally relates to its concentration and, as shown in Equation 5, is the product of the mole fraction of the solute (X_i) and a system-specific proportionality constant or activity coefficient that is designated by the symbol "γ_i":

$$a_{\underline{i}} = \gamma_i * X_i \quad \text{(EQ 5)}$$

Activity coefficient values typically change as the solute concentration varies, except in the dilute solution region. The dilute solution region is regarded as the solute concentration range over which the tangent to the activity curve coincides with the activity curve, as shown for a generic system in Figure 1 (not drawn to scale). The tangent deviates from the activity curve as the solute concentration increases. When the tangent extends below the activity curve, as shown in Figure 1, the solute is said to exhibit positive deviation from Henry's law and negative deviation from Raoult's law, meaning that the attraction between solvent and solute atoms exceeds that between like atoms. Conversely, positive deviation from Raoult's law means that the solute atoms are repulsed by the solvent atoms so that their "escape tendency" increases, a favorable situation when vapor-phase extraction of the solute is desired. The Henrian standard states are convenient for thermodynamic computations that involve dilute solutions and are conveniently defined as either 1 at. % (atomic percent) or 1 wt % (weight percent) of the solute dissolved in solvent. The activity constant in the dilute solution region is termed the *Henry's law constant*, which is designated by the symbol "f_i" in the following equation:

$$h_i = f_i * C_i \quad \text{(EQ 6)}$$

In Equation 6, h_i represents the Henrian activity and C_i represents concentration, which may be conveniently expressed in units of weight percent or atomic percent. If the standard state is selected within the extremely dilute solution region, then the Henrian activity coefficient for the solute specie (f_i) is one and the solute activity equals its concentration.

The solubility of certain impurity elements in metallic solutions is quite limited, and this statement holds particularly true for dissolved gases. Gases rapidly dissolve in high-temperature liquid metal melts, and these melts quickly approach equilibrium. For this reason, thermodynamic calculations are often highly effective for evaluating dilute solution behavior. When diatomic gases are involved, the concentration of dissolved gas varies according to the square root of the partial pressure (P_{P2}) of the gas above the melt. This relationship, commonly known as Sieverts' law, is expressed by the following equation:

$$h_{\underline{G}} = K * P_{G2}^{0.5} \quad \text{(EQ 7)}$$

where G_2 is a diatomic gas, and $\underline{G}$ represents the concentration of G in the melt. For example, consider the dissolution of diatomic phosphorous gas in liquid silicon according to the reaction in Equation 8. The free-energy expression permits calculation of the Gibbs free energy of reaction ($\Delta G°_R$, J/mol) for the dissolution of diatomic phosphorous gas (Takahiro et al. 1996).

$$\tfrac{1}{2}P_{2(g)} = \underline{P}_{\text{ in silicon}} \quad \text{(EQ 8)}$$

$$\Delta G°_R = -139{,}000\ (\pm 2{,}000) + 43.4\ (\pm 10.1)\ T\ \text{(J/mol)}$$

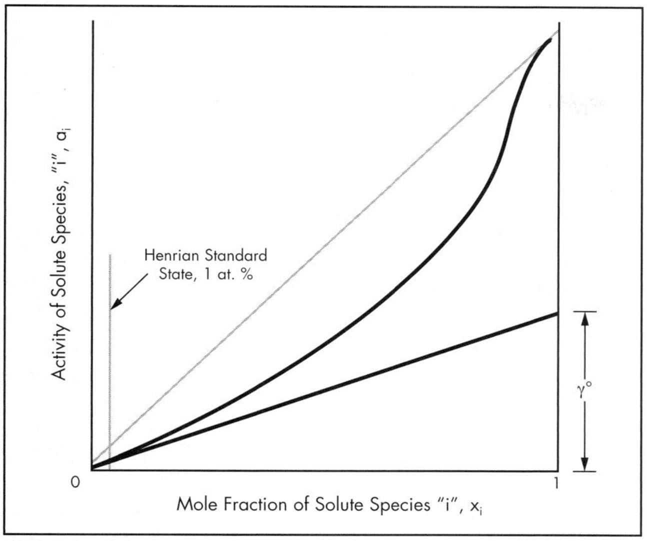

Figure 1 Activity of solute species "i" as a function of mole fraction

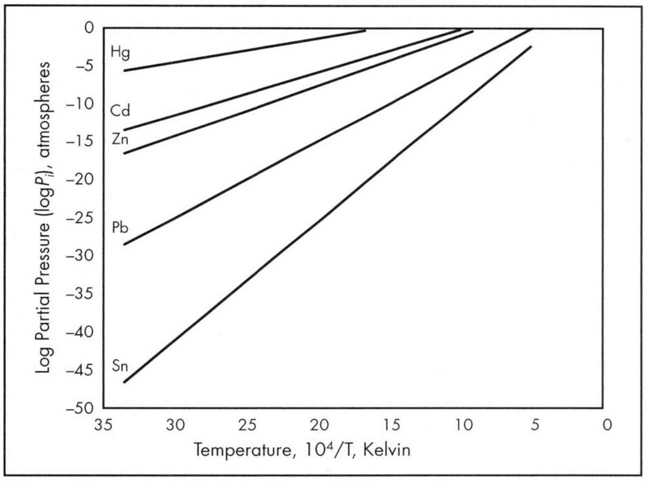

Figure 2 Van't Hoff diagram to illustrate the vapor pressures of five metals as a function of temperature (1 atm total pressure)

Experimental evidence that confirms Henry's law is valid for dissolved phosphorous concentrations up to 0.1%. The dissolution reaction in the following equation prevails for phosphorous concentrations less than 0.005% (Takahiro et al. 1996):

$$P_{(g)} = \underline{P}_{\text{in silicon}} \quad \textbf{(EQ 9)}$$

$$\Delta G°_R = -387{,}000\ (\pm 2{,}000) + 103\ (\pm 10)\ T\ (\text{J/mol})$$

In either case, a numerical value for the equilibrium constant can be obtained from the appropriate free-energy equation. The equilibrium constant can then be written in terms of the dissolved phosphorous activity and the partial pressure of the phosphorous gas (P or P_2) above the melt. For the reaction involving diatomic phosphorous gas, the equilibrium constant will take the following form:

$$K = a_{\underline{P}}/P_2^{0.5} \quad \textbf{(EQ 10)}$$

Presented in various formats, phase stability diagrams are used to define optimum process conditions. A Van't Hoff plot, such as the one shown in Figure 2, provides a simple example. These types of diagrams are constructed to illustrate the equilibrium vapor pressure of a specie of interest (expressed as the natural logarithm or base 10 logarithm) as a function of temperature (expressed as the reciprocal of temperature, K). The diagram can be used to identify temperatures at which complete vaporization is ensured (i.e., the temperature at which P_i exceeds atmospheric pressure). A second type of phase stability diagram is constructed to illustrate the stability regions of the condensed phases for the system in question. The axes of these two-dimensional isothermal diagrams typically represent two gaseous species that define the relevant potential of the atmosphere (e.g., carbon monoxide partial pressure versus carbon dioxide partial pressure can be used to define the oxygen potential of a system). Isobars that reflect the equilibrium vapor pressures of the relevant gas species can be superimposed atop the condensed phase fields. The diagrams are used to identify an "operating point" or a range of operating conditions that favor formation of the gas species. Kellogg presented a treatise on the construction, interpretation, and practical application of these diagrams (Kellogg 1966).

Another popular method for illustrating stable equilibrium phases are diagrams generated as a result of so-called free-energy minimization software, such as HSC Chemistry 9 (Outotec Technologies 2016). This software enables the user to define an initial composition and the temperature or pressure range (the other variable is held constant). Based on this input, the algorithm iteratively calculates the minimum free-energy curves and produces product plots of equilibrium composition versus temperature (or pressure).

Kinetic, mass transport, and heat transfer considerations can significantly influence behavior in vapor-phase extraction processes. Turkdogan et al. (1963) proposed a means to enhance diffusion-limited metal vaporization rates in metal/oxide systems. Diffusion-limited metal vaporization is complex and involves several system-specific and species-specific factors, the discussion of which falls outside the scope of this chapter. Subjectively, vapor pressures in excess of 0.1 atm typically result in rapid vaporization, 0.01 atm corresponds to slow vaporization, and 0.001 atm corresponds to a very slow vaporization rate (Kellogg 1966).

VAPORIZATION OF METALS

In general, the group 1 alkali metals and the group 2B metals, which include cadmium, mercury, and zinc, are quite volatile at relatively modest temperatures and are, therefore, excellent candidates for vapor-phase extraction. Conversely, the group 6A metals, which include chromium, molybdenum, and tungsten, exhibit characteristically low vapor pressures and are extremely poor candidates (Rosenqvist 1974). The vapor pressure of pure metals increases with increased temperature and follows an Arrhenius relationship, presented in general form in Equation 10, where A and B represent substance-specific constants and T is temperature (K).

$$\text{Log } P_i = A/T + B \quad \textbf{(EQ 11)}$$

The simplest vapor-phase extraction operations involve direct vaporization of a single element. Although it is not exclusively the case, most metals evaporate as monatomic vapors. Mercury is a case in point:

$$Hg_{(c)} \rightarrow Hg_{(g)} \quad \textbf{(EQ 12)}$$

Because mercury is a noble metal that is relatively easy to produce, some of the earliest vapor-phase metal extraction applications involved mercury extraction and refining. Although mercury extraction can and has been done in a variety of furnace types and configurations, including rotary tube and multiple-hearth furnaces, mercury extraction is now commonly performed under vacuum in indirectly heated vessels called retorts. Retorts typically operate in the batch mode when processing scrap or, in environmental applications, a mercury-contaminated matrix such as sludge or soil. The mercury-bearing feed materials are sized and placed in trays before admission to the retort. The normal boiling point of mercury is 357°C, so it is advantageous to heat the charge above this relatively low temperature to maximize extraction. Mercury vaporizes, exits the retort, and is subsequently condensed and recovered downstream in chilled condensers and carbon filters. Process parameters may be controlled to dictate whether the mercury is recovered as metal for recycle or condensed as a specific compound in preparation for disposal. The U.S. Environmental Protection Agency (EPA 1990) lists retorting as the best demonstrated available technology under their land disposal restrictions.

While most metals tend to vaporize as monomers, many metalloids and compounds vaporize as polymeric species. For example, arsenic is a troublesome impurity that is often vaporized as part of a metal refining process such as copper and nickel converting and fire refining. Elemental arsenic vaporizes as a combination of As, As_2, and As_4 molecules.

VAPORIZATION OF COMPOUNDS

Vaporization is applied to a wide range of compounds, including carbides, nitrides, tellurides, halides, oxides, and sulfides of many common and exotic metals (Kellogg 1966). As with elemental mercury, most common mercury compounds are quite volatile and, therefore, amenable to vapor-phase extraction. In small-scale operations, mercury retorts process mercury-bearing materials in batch mode, whereas continuous rotary kilns or multiple-hearth furnaces are preferred for larger industrial operations. As high as 95% recovery as commercial-grade (99.9%) mercury is achieved through careful selection, application, and control of process conditions. Full or partial dissociation of a compound frequently leads to the formation of the metal vapor. Accordingly, mercury oxide decomposes when it is heated above 450°C according to the following reaction:

$$2HgO = 2Hg_{(g)} + O_{2(g)} \quad \textbf{(EQ 13)}$$

This mode of behavior, dissociation leading to two separate gaseous species, is consistent with the behavior of several other metals compounds. In a second example, zinc sulfide (ZnS) dissociates to zinc and sulfur gases, as shown in Equation 14. Upon cooling, the reverse reaction occurs to condense zinc sulfide.

$$ZnS_{(s)} = Zn_{(g)} + 0.5S_{2(g)} \quad \textbf{(EQ 14)}$$

Evaporation may occur due to partial dissociation, as is the gas with the dissociation of silica to form silicon monoxide gas and oxygen, as shown in the following equation:

$$SiO_{2(g)} = SiO_{(g)} + 0.5O_{2(g)} \quad \textbf{(EQ 15)}$$

Some volatile compounds directly evaporate without dissociation. Arsenic trioxide (As_2O_3) evaporates according to the following reaction:

$$2As_2O_{3(s)} \rightarrow As_4O_{6(g)} \quad \textbf{(EQ 16)}$$

A metal that occurs in a compound may be vaporized by means of chemical reaction with a gas in a process generally known as "roasting." Relatively nonvolatile metals can be extracted at temperatures far below their boiling points if gaseous compounds with suitable properties can be formed. For example, elemental mercury can be recovered from cinnabar (HgS), the most abundant mercury mineral, by roasting in an oxidizing atmosphere, as shown in the following equation:

$$HgS_{(s)} + O_{2(g)} = Hg_{(g)} + SO_{2(g)} \quad \textbf{(EQ 17)}$$

Vapor metallurgical refining of nickel provides an interesting example of volatile compound formation in a refining application. Gaseous nickel tetracarbonyl ($Ni(CO)_4$) forms when nickel and carbon monoxide react at relatively low temperatures (between 40° and 90°C) according to the reaction in Equation 18. The reverse reaction (nickel tetracarbonyl decomposition) occurs when the gas is heated into the 150°–300°C range (Boldt 1967). Thus, nickel can be selectively extracted by exposing the feed matrix (typically impure granules that contain pre-reduced nickel metal) to carbon monoxide to form the volatile nickel tetracarbonyl, which exits in the vessel discharge gases. The gas is filtered to remove entrained particulate, and then condensed downstream for recovery. In one industrial configuration, the nickel condensed on a circulating stream of nickel pellets and a bleed stream of the product pellets advanced to market.

$$Ni_{(s)} + 4CO_{(g)} = Ni(CO)_{4(g)} \quad \textbf{(EQ 18)}$$

Polymeric compound families include oxides, sulfides, and chlorides, to name just a few.

TECHNOLOGIES

As with most other aspects of pyrometallurgy, no universally applicable vapor-phase extraction or refining technologies exist. Most industrial processes are highly commodity specific, because they have been developed to deal with the unique physical and chemical properties associated with a particular metal, its matrix (ore, concentrate, scrap), and its naturally occurring impurities. Vapor-phase extraction is performed using heated rotary calciners, rotary kilns, multiple-hearth furnaces, static-bed and fluidized-bed reactors, reverberatory furnaces, blast furnaces, electric arc furnaces, among other furnace types. Technological distinctions include the choice of indirect versus direct heating, batch versus semi-batch versus continuous processing, furnace atmosphere control and limitations, and cocurrent versus countercurrent gas-to-solids flow. Process gas handling system requirements must be decided on a case-specific basis. Some common equipment applied for capture and recovery of vapors include absorption towers/scrubbers, baghouses, and electrostatic precipitators. In this section, some of the more popular vapor-phase extraction vessels are summarily described.

Rotary tube furnaces are applied to a very wide variety of thermal processing applications that include drying, calcination, pyrolysis, roasting, thermal desorption (selective removal of volatile contaminants), thermal oxidation, and vitrification. Rotary tube furnaces are amenable to many vapor-phase extraction applications. A rotary tube furnace is a cylindrical reactor that rotates about its longitudinal axis. The tube is usually inclined to the horizontal. Commercial reactors are typically 0.3–3.6 m diameter and the length-to-diameter ratio generally ranges from 4 to greater than 10. In some applications, sections with expanded diameter are included to increase burden depth and residence time. The tube ends are sealed by various means to prevent gas leakage, and maintaining slight negative pressure (0.02–0.20 mm of mercury) in the furnace also helps control emissions. Draft is provided by exhaust blowers and gas fans that are typically located downstream of the furnace. Ideal feed materials are granular, free-flowing solids that do not soften, agglomerate, or melt at the process temperature. Feed enters the higher end of the cylinder and works its way through the cylinder because of head (feed rate), rotational speed (rpm), and slope. These factors dictate the residence time of the burden in the tube. Finished product discharges at the lower end of the tube.

Rotary tube furnaces can operate in batch, semicontinuous, or continuous mode. Process gases may flow in the same direction (cocurrent) or in the direction opposite (countercurrent) to the solids through the rotary tube. The gas flow rate and direction affects solids residence time in the tube. In applications that involve chemical reactions between constituents in the gas and condensed phases, kinetic implications are associated with each configuration, with countercurrent gas versus solids flow providing advantages with respect to reactant concentration at both ends of the tube. Countercurrent gas-to-solids flow also improves heat transfer efficiency. Cocurrent flow is used to dry heat-sensitive materials at higher inlet gas temperature because of rapid gas cooling during the initial evaporation of surface moisture.

Rotary tube furnaces may also be classified as directly or indirectly heated, although hybrid units exist. *Directly heated* or "direct heat" means that the solid charge is heated by combustion or preheated gases that pass directly over the surfaces of the charge solids. Directly heated rotary kilns are among the most important industrial furnaces used for continuously conducting relatively high-temperature operations. In a directly heated rotary kiln, the interior surface of the metal tube (shell) is lined with insulating block and/or refractory brick. Lengths in excess of 50 m are common. Diameters may be constant or vary along the length of the kiln; some kilns have an expanded diameter to increase residence time in a particular heating zone.

Directly heated units are usually more economical to operate but are generally unsuited for applications that require close furnace atmosphere control. In some situations, diluting the process gas with combustion gas may be desirable. Gas volumes and velocities are relatively high (~30 m/min) in direct-fired units because the heat is transferred from combustion gas to the solid charge. The high gas velocity causes fine solid particles to be entrained in the cylinder discharge gas, which requires downstream dust capture-and-control measures.

Indirectly heated means that the heating medium does not directly contact the solid charge. Indirectly heated rotary calciners are metal cylinders surrounded by stationary, refractory-lined gas-fired or electrically heated furnaces or "fireboxes." Indirectly heated calciners require a minimum purge gas flow through the cylinder and are well suited for gas-sealed operation under controlled oxidizing, reducing, or inert atmospheres with minimal dusting. The temperature that these units can sustain mainly depends on the cylinder material: carbon steel is satisfactory for maximum temperatures between 375° and 425°C; stainless steel is satisfactory for 540°–760°C; Inconel and other alloys can sustain higher temperatures; and graphite is reserved for extremely high temperatures (>2,000°C). Lifters may be longitudinally positioned on the inner tube surface to prevent charge material from sliding. Thermal efficiency can range from 40% to 70%.

Multiple-hearth (sometimes termed "multi-hearth") furnaces are vertical steel towers that contain a series of circular refractory hearths. The feed material is introduced in the top hearth and progresses downward through a series of hearths. The center shaft slowly rotates rabble arms on each hearth to turn over the charge and push it across the hearth to drop holes, which are located near the outside of one hearth and near the center shaft on the next hearth. Gases typically flow upward and countercurrent to the charge for improved mass and heat transfer.

Suspension roasting is essentially a modification of multiple-hearth roasting wherein feed is dried and preheated on the upper hearths and then falls through the center section of the furnace, countercurrent to the upwardly flowing oxidizing furnace gas. Flash or autogeneous roasting is a variation of this theme; preheated ore is injected along with preheated air. The process is effective with sulfide minerals that oxidize exothermally and require little or no supplemental fuel. However, the benefits of countercurrent operation are sacrificed.

Fluidized-bed roasting is attractive because very large mass and heat transfer coefficients exist between solid and gas. Gases pass upward through a solid (0.02 to 2 mm in diameter) particle bed. At low gas flow rates, the gas permeates the bed without moving the particles, and the pressure drop across the bed is proportional to gas flow rate. As gas velocity approaches a critical value, the bed expands as the effective particle masses are balanced by the drag forces imposed by the gas stream. Over a narrow range of velocities, the particles become individually suspended with a downward velocity relative to the gas stream approximately equal to its terminal velocity. Bed expansion occurs as gas velocity continues to rise, so the distance between particles increases and the pressure drop across the bed diminishes. At even higher gas velocities, bed expansion ceases and the pressure drop is independent of gas velocity. The bed assumes fluid characteristics and its appearance resembles that of a boiling liquid with bubbles rising through the bed and bursting at the surface. Within this state of aggregative fluidization, loose particle clusters (dense phase) are dispersed in a gas atmosphere that contains individual particles (lean phase). Gas-to-solid contact is exceptionally high because of the turbulent bed conditions while the relative velocity between gas and solid particles is within the laminar flow range. Heat transfer and mass transfer rates are very high in fluidized beds. Effective thermal conductivities of fluidized beds have been measured to exceed about 100 times that of silver. Convective heat transfer ensures uniform bed temperatures.

APPLICATIONS AND UNIT OPERATIONS

Numerous examples of historical and current industrial implementation of vapor-phase extraction technologies are available. The following discussion is not intended to be comprehensive. It offers specific examples that have commercial precedents and, in most instances, reflect prevailing industrial practices.

Drying is unquestionably the most frequently practiced industrial vapor-phase extraction process. In many applications, a substantial portion of the bulk water content is removed by sedimentation, filtration, centrifugation, or other mechanical means. However, the processed material invariably retains a considerable moisture content. Drying is typically the preferred alternative when circumstances demand a low to negligible free-moisture content. In general, drying is an evaporative process that is conducted at relatively low temperatures (typically 90°–200°C). On the industrial scale, drying is performed in many equipment types, and equipment selection depends on factors such as the physical and chemical characteristics of the material to be dried, the physical and chemical characteristics of the liquid phase, and the initial and final moisture contents. Because drying operations generally do not require tight furnace atmosphere control, directly heated fossil-fuel-fired rotary tube furnaces are frequently the technology of choice.

Selective vaporization and distillation exploits differences in vapor pressure to separate components in a solution. When a significant vapor pressure difference exists between two or more components in a solution, the components can be separated by selective volatilization of the component with the greater vapor pressure. As the matrix is heated to an appropriate temperature, the vapor phase becomes enriched with the more volatile component as the liquid phase is depleted of the same. Classically, selective volatilization and distillation is performed in batch and continuous (or fractionating) reflux columns with preference for continuous systems in high-throughput operations. In such systems, the liquid is introduced near the center of a tall vertical column. Within the column, a series of trays is positioned to allow liquid to cascade from one tray to the next while gas rises from upward through series. Temperature in the lower portion of the column is maintained below the boiling point of the less volatile constituent and above the boiling point of the more volatile constituent. The temperature in the upper portion of the column must also be low enough to permit the less volatile element to condense. Consequently, the less volatile element accumulates and is removed from the bottom of the column (i.e., "bottoms") while the more volatile element exits the column as vapor and is condensed and recovered downstream.

Cadmium and zinc have normal boiling points of 765° and 905°C, respectively. Thus, cadmium can be selectively separated from zinc by virtue of its greater volatility. Magnesium–lead alloy is an example of a metallurgical azeotropic system. Simple distillation is incapable of separating the two components in a binary azeotropic mixture, wherein an intermediate compound vaporizes instead of an individual component. In this example, liquid Mg_2Pb vaporizes without change in composition; this behavior is analogous to a congruent melting point in a condensed phase system. Azeotropic behavior can sometimes be managed by altering the system pressure or by adding another component that has a strong affinity for one of the original components.

On occasion, it is desired to remove a metal or compound that does not exhibit a sufficiently substantial vapor pressure at elevated temperature. Under such circumstances, it may be expedient to instigate a chemical reaction to form a volatile compound that is better suited for vapor-phase extraction. In extractive metallurgy, the term *roasting* applies to a wide variety of thermal processes that are typically performed as a pretreatment process to render some component(s) in the feed material amenable to subsequent processing in a separate unit operation. Roasting is performed for the specific purpose of causing a chemical change in one or more selected feed components. It is usually performed at elevated temperatures that do not cause an appreciable degree of charge fusion or sintering, as these phenomena are usually detrimental to the process objective. Ideally, the roaster product, termed *calcine*, is a free-flowing, loosely agglomerated powder. Roasting processes are classified in many ways, often according to objectives such as oxidizing, reducing, chloridizing, sulfatizing, or magnetizing. In some instances, the sole purpose of the roasting step may be to eliminate an undesired volatile species such as arsenic. Once in the gas phase, these materials exit the reactor in the discharge gas stream and condense downstream as the gas cools.

Carbothermic reduction and vaporization was once the primary method for zinc extraction and refining. Zinc vapors were produced by processing zinc concentrates in horizontal and vertical retorts. The condensate was refined in similar manner to achieve the specified grade. This technology is considered obsolete, having been supplanted by the roast-leach-electrowinning processes. However, vapor metallurgy processes are currently used in the secondary industries to extract zinc from primary smelting slags (slag fuming) and from electric arc furnace (EAF) dust.

Slags produced in primary nonferrous smelting operations often contain appreciable concentrations of volatile and potentially valuable constituents such as lead, zinc, and/or tin. Slag fuming operations are generally performed in a separate, dedicated batch furnace where the zinc-bearing feed material and reductant (e.g., metallurgical coke or fine pulverized coal) are injected into molten slag. The slag is blown with air to create a strongly reducing environment conducive to carbothermal reduction of zinc via contact with the solid reductant or, more probably, the carbon monoxide that forms via the Boudouard reaction, which is shown in Equation 19. Lead, zinc, or tin are reduced to their respective metals and then vaporize and exit with the furnace discharge gas. The metals reoxidize and form fine particulate matter upon cooling and are recovered for further processing to produce metal or marketable compounds. Slag fuming is characterized by high process temperatures (~1,300°C), low oxygen potential, and good bath mixing.

$$C_{(s)} + CO_{2(g)} = 2CO_{(g)} \quad \textbf{(EQ 19)}$$

EAF dust contains relatively high concentrations of zinc that come from galvanized scrap in the EAF charge. The Waelz kiln process has emerged as the predominant method for recycling zinc contained in EAF dust. The contained zinc is present as franklinite (ZnO) and zinc ferrite ($ZnFe_2O_4$). To effect the zinc reduction and vaporization, a pelletized mixture of EAF dust, a carbonaceous reducing agent (coke), and slag formers (sand and lime) are charged to the rotary kiln. Typical kiln dimensions are 50 m long by 3.6 m in diameter. The kiln operates with countercurrent gas-to-solids flow and provides 4–6 hours of reaction time. The zinc is removed by roasting the dust in the presence of the carbonaceous reductant according to the reactions shown as Equations 20 and 21. Iron and the other nonvolatile EAF dust constituents form an iron-rich slag phase.

$$ZnO_{(s)} + C_{(s)} = Zn_{(g)} + CO_{(g)} \quad \textbf{(EQ 20)}$$

$$ZnFe_2O_{4(s)} + 4C_{(s)} = Zn_{(g)} + 2Fe_{(s)} + 4CO_{(g)} \quad \textbf{(EQ 21)}$$

Once the zinc vapor enters the kiln atmosphere, it is oxidized by the oxygen in the combustion gas, as indicated by the following reaction:

$$Zn_{(g)} + \frac{1}{2}O_{2(g)} = ZnO_{(s)} \qquad \textbf{(EQ 22)}$$

Other volatile constituents, such as lead and cadmium, enter the vapor phase during carbothermal reduction and are reoxidized in the furnace atmosphere. The crude Waelz oxide assays 60%–65% zinc and contains impurities that include cadmium oxide and lead oxide. The oxides may be subject to further thermal or hydrometallurgical processing to produce marketable zinc oxide or zinc metal (Mager and Meurer 2000).

In extractive metallurgy, volatile gaseous halides figure prominently in transporting material from one solid to another. Most metal halides are volatile or have relatively low decomposition temperatures, which makes volatilization appear to be an expedient way to separate certain refractory metals, provided that it is possible to exercise adequate control of the halide potential. Halide stability decreases in the following order: iodide, bromide, chloride, fluoride.

A commonly effective treatment scheme involves performing a chloridizing roast to convert certain target metals to their volatile chlorides and a subsequent or simultaneous volatilizing roast. The pressure range of interest for establishing the behavior of volatile metal chlorides ranges from 10^{-10} to 1 atm. Volatile metal compounds, particularly chlorides, are extracted by vapor-transport processes, purified by fractional distillation, and reduced to metal. Titanium production is an example of chlorination roasting applied to abet vapor-phase transport, but the process is also applied in the production of hafnium, niobium, uranium, and zirconium (among other metals).

Titanium oxide (TiO_2) in the form of natural or synthetic rutile is converted to titanium tetrachloride ($TiCl_4$) in static- or fluidized-bed chlorinators. A static-bed chlorinator accepts feed in the form of briquettes made of rutile and a carbonaceous reductant such as metallurgical coke, while a fluidized-bed chlorinator is fed with rutile and ground coke. In both cases, chlorine (Cl_2) gas is injected at the bottom of the chlorinator and flows upward through the bed that is heated to between 800° and 1,000°C by the enthalpy of the exothermic reaction supplemented by electrical resistance (carbon electrodes) heating. Titanium tetrachloride gas forms according to the following carbochlorination reaction:

$$2TiO_2 + 3C + 4Cl_{2(g)} = 2TiCl_{4(g)} + 2CO_{(g)} + CO_{2(g)} \qquad \textbf{(EQ 23)}$$

The reaction requires excess chlorine because chlorine efficiency is between 75% and 85%. The carbon tetrachloride exits with the chlorinator discharge gas stream that also includes carbon dioxide, carbon monoxide, unreacted chlorine, and entrained particulate matter. Following coarse particulate removal in a cyclone, the chlorinator discharge gas enters a direct-contact cooling tower. A circulating stream of liquid titanium tetrachloride is sprayed into the freeboard to reduce gas temperature and condense titanium tetrachloride liquid, which collects at the bottom of the tower. Following separation from sludge and other insoluble matter, the crude titanium tetrachloride contains iron, vanadium, silicon, tin, and other metallic impurities that were also converted to chlorides. Most of these chlorides are insoluble in titanium chloride and are readily separated by fractional distillation. Vanadium oxychloride ($VOCl_3$) is an exception that requires chemical precipitation for removal. The resulting titanium tetrachloride is approximately 99.8% pure and is stored under inert gas, pending further processing to produce titanium oxide pigment or titanium sponge via the Kroll process or the Hunter process (Gill 1980).

Magnesium production via the Pidgeon process involves a silicothermic reduction process that is performed in nickel-chromium steel retorts. The retort charge consists of a 5:1 ratio of calcined dolomite and ferrosilicon (FeSi) that has been ground, mixed, briquetted, and preheated to approximately 700°C (Gill 1980). Inside the retort, silicothermic magnesium reduction proceeds according to the following reaction:

$$MgO{\cdot}CaO_{(s)} + FeSi_{(s)} = 2CaO{\cdot}FeSiO_{2(l)} + 2Mg_{(g)} \qquad \textbf{(EQ 24)}$$

The cylindrical retorts are horizontally oriented in a fossil-fuel-fired furnace such that one end of the retort extends outside of the furnace. While the heated zone of the retort is maintained at approximately 1,150°C, a vacuum connection draws the nascent magnesium vapor into the water-cooled extension, where magnesium crystals condense on a removable steel sleeve. At the end of the retorting cycle, the resulting 99.98% magnesium ring or "crown" is withdrawn, melted, and cast. Calcined dolomite (dolime) is the preferred feed material (versus magnesium oxide) because the calcium oxide in the dolime improves magnesium yield by displacing magnesium oxide taken up in the dicalcium silicate slag (Gill 1980).

Vapor-phase transport is extensively employed in modern steelmaking operations to expel dissolved impurities such as hydrogen and nitrogen, which exist in their atomic forms in liquid steel. Unless the concentrations are reduced, these impurities generally cause deleterious effects on the steel product characteristics. Multiple variations on the theme have been practiced, but, in essence, a ladle that contains a charge of molten metal to be refined is placed within a chamber where the total pressure can be lowered to minimize the partial pressure of the target gaseous impurity above the melt. Commercial vacuum systems operate at pressures as low as 0.1 torr.

Vapor-phase removal of hydrogen and oxygen take place in accordance with Sieverts' law, which states that the dissolved gas concentration is proportional to the square root of the partial pressure of the gas above the melt. The relevant reactions are expressed as Equations 25 and 26. Modern degassing operations are also effective in decarburizing molten steel because the vacuum promotes reaction between carbon and oxygen to produce carbon monoxide gas, as indicated by the reaction in Equation 27.

$$\underline{H} = \frac{1}{2}H_{2(g)} \qquad \textbf{(EQ 25)}$$

$$\underline{N} = \frac{1}{2}N_{2(g)} \qquad \textbf{(EQ 26)}$$

$$\underline{C} + \underline{O} = CO_{(g)} \qquad \textbf{(EQ 27)}$$

Vacuum degassing of steel is often performed by situating the ladle bearing the molten steel near a vacuum chamber such that snorkels can extend from the chamber into the melt. Steel is degassed at nominal temperature and total pressure of 1,600°C and 1 torr, respectively (Ghost 2001).

Vacuum refining is also applied in nonferrous metal refining operations. Refining lead bullion produced via primary or secondary smelting operations entails a series of batch operations that are typically conducted in large steel kettles. In the usual sequence, these operations include drossing (copper removal), softening (antimony and arsenic removal), desilverizing, dezincing, and debismuthizing. Dezincing is generally accomplished by vacuum refining. A kettle fitted with a bell-shaped vacuum lid is charged with desilverized lead. Vacuum is applied as the molten lead is heated to, and maintained at, approximately 600°C. Over several hours, dissolved zinc is drawn into the vapor phase and subsequently condenses on the inner surface of the bell. The zinc returns to the desilverizing process (Parkes process), where it is used to remove precious metals, mostly silver, from the lead.

REFERENCES

Alcock, C.B. 1976. *Principles of Pyrometallurgy*. London: Academic Press.

Boldt, J.R., Jr. 1967. *The Winning of Nickel*. Princeton, NJ: Van Nostrand.

EPA (U.S. Environmental Protection Agency). 1990. *Final Best Demonstrated Available Technology (BDAT) Background Document for Mercury-Containing Wastes D009, K106, P065, P092, and U151*. Report PB90-234170. Washington, DC: EPA.

Ghost, A. 2001. *Secondary Steelmaking Principles and Applications*. Boca Raton, FL: CRC Press.

Gill, C.B. 1980. *Nonferrous Extractive Metallurgy*. New York: John Wiley and Sons.

Hastie, J.W., and Hager, J.P. 1990. Vapor transport in materials and process chemistry. In *1990 Elliott Symposium Proceedings*. Warrendale, PA: Iron and Steel Society. pp. 301–321.

Kellogg, H.H. 1966. Vaporization chemistry in extractive metallurgy. *Trans. Metall. Soc. AIME* 236:602–615.

Mager, K., and Meurer, U. 2000. Recovery of zinc oxide from secondary raw materials: New developments of the Waelz process. In *Recycling of Metals and Engineered Materials*. Edited by D.L. Steward and J.C. Daley. Warrendale, PA: The Minerals, Metals & Materials Society. pp. 329–340.

Moore, J.J. 1981. *Chemical Metallurgy*. London: Butterworths.

Outotec Technologies. 2016. HSC Chemistry 9, version 9.1.1. Outotec Technologies.

Rosenqvist, T. 1974. *Principles of Extractive Metallurgy*. New York: McGraw-Hill.

Takahiro, M., Morita, K., and Sano, N. 1996. Thermodynamics of phosphorus in molten silicon. *Metall. Mater. Trans. B* 27(6):937–941.

Turkdogan, E.T., Grieveson, P., and Darken, L.S. 1963. Enhancement of diffusion-limited rates of vaporization of metals. *J. Phys. Chem.* 67(8):1647–1654.

CHAPTER 11.5

Smelting

Phillip J. Mackey and David George-Kennedy

The word *smelt* means to heat, fuse, and melt or react an ore at temperature to chemically extract a metal—in effect, to "obtain a metal by this process" (Oxford English Dictionary 2017). Smelting is an ancient technique of extracting a metal from ore or concentrate by the use of high temperature, involving both melting and chemical reactions. Today, the technique has evolved into a highly developed technology worldwide to produce metals used in everyday life. Metals remain indispensable to the modern world; hence, the importance of smelting and extractive metallurgy to society cannot be overstated.

Sometimes a particular metal can be produced as a crude metal directly in one step—typically, a *crude* or *impure metal* here implies a form of metal containing several trace or impure elements and that requires further refining to obtain a final product for the market. Examples include the production of iron metal called pig iron in the iron blast furnace (which requires refining to make usable steel), the smelting of calcined nickel laterite ore in an electric furnace to produce ferronickel (which requires further processing to make stainless steel), or direct lead smelting of lead sulfide concentrate to produce lead metal (which requires refining to remove unwanted elements and also to extract contained silver that is normally present). In other cases, the primary smelting step can produce an intermediate metal or metallic product (called *matte* in the case of copper or nickel-copper smelting), requiring further treatment. With a focus on the more practical aspects of the art of smelting, this chapter briefly describes the historical evolution, design, and operation of the main smelting technologies in use today, with less emphasis on the theoretical and physicochemical aspects since they are covered elsewhere in this handbook.

The main minerals of the metal to be extracted and present in the feed to a given smelting operation are typically either as the sulfide or as the oxide; such minerals are invariably associated with several other minerals, including the gangue or waste rock minerals. The product of smelting is typically a metal or intermediate phase and a slag consisting of the gangue oxides.

The furnace temperature for the actual smelting step is typically determined by the range of melting points of the gangue and/or the metal, or its sulfide or oxide. The common minerals of the key metals extracted by smelting discussed in this chapter are included in Table 1. Therefore, smelting reactions when treating sulfides will typically involve a high-temperature oxidation reaction using air or oxygen-enriched air to produce the metal (or intermediate product) by the following simplified reaction (where M is the metal and air (21% O_2) is assumed, although commonly, oxygen-enriched air is now used):

$$MS + O_2 + 79/21N_2 = M + SO_2 + 79/21N_2 \qquad \textbf{(EQ 1)}$$

On the other hand, treatment of essentially an oxide mineral to recover the metal involves a high-temperature reduction step—typically with carbon in the form of coal or coke (or sometimes as natural gas), or by CO produced by the reaction of carbon with oxygen—expressed in simple terms as follows:

$$2MO + C = 2M + CO_2 \qquad \textbf{(EQ 2)}$$

$$MO + CO = M + CO_2 \qquad \textbf{(EQ 3)}$$

More than 30 key elements, as illustrated in Figure 1, can be identified as being extracted today by at least one smelting step (followed, in most cases, by a subsequent pyrometallurgical refining step, an electrorefining step, or a hydrometallurgical refining step). The melt temperature in smelting is usually set by the slag melting point in question. Temperatures can range from around 1,200°C in copper or nickel smelting to more than 1,700°C in ferrochrome smelting.

HISTORICAL BACKGROUND

Partly because of ready occurrence and lower melting points, the first metals smelted were likely lead, tin, and then later copper, with evidence of early smelting activities dating back more than 8,000 years; the smelting of iron followed much later. Thus, there is the Stone Age (from the very early millennia to about 6,000 BC), then the Bronze Age (about 3,000 BC to 600 BC)—with an earlier pre-Bronze Age sometimes referred to as the Chalcolithic or Copper Age around the fifth and the

Phillip J. Mackey, President, P.J. Mackey Technology Inc., Kirkland, Quebec, Canada
David George-Kennedy, Chief Advisor, Smelting & Refining, Rio Tinto Copper & Diamonds, Salt Lake City, Utah, USA

Table 1 Key minerals of interest in smelting—sulfides and oxides

Mineral	Theoretical Composition	Metal Content, wt %
	Sulfides	
Copper (% Cu indicated)		
Chalcopyrite	$CuFeS_2$	34.6
Chalcocite	Cu_2S	79.9
Bornite	Cu_5FeS_4	63.3
Covellite	CuS	66.4
Pyrrhotite	FeS	63.5 (% Fe)
Pyrite	FeS_2	46.6 (% Fe)
Nickel (% Ni indicated)		
Pentlandite	$Ni_{4.5}Fe_{4.5}S_8$	34.2
Millerite	NiS	64.7
Heazlewoodite	Ni_3S_2	73.3
Siegenite	$(Co, Ni)_3S_4$	28.9
Lead (% Pb indicated)		
Galena	PbS	86.6
Zinc (% Zn indicated)		
Sphalerite	ZnS	67.1
Tin (% Sn indicated)		
Stannite	Cu_2FeSnS_4	27.6
Iron (% Fe indicated)		
Pyrite	FeS_2	46.5
Pyrrhotite	Fe_7S_8	60.4
	Oxides and Carbonates	
Copper (% Cu indicated)		
Malachite	$CuCO_3{\cdot}Cu(OH)_2$	34.6
Azurite	$2CuCO_3{\cdot}Cu(OH)_2$	79.9
Chrysocolla	$CuSiO_2{\cdot}2H_2O$	63.3
Antlerite	$Cu_3SO_4(OH)_4$	66.4
Nickel (% Ni indicated)		
Garnierite	$(Ni, Mg)_6Si_4O_{10}(OH)_8$	Pure mineral by formula ~38.9% Ni; typically found in diluted form by gangue, with actual ores much lower ~1%–3% Ni.
Nickeliferous limonite	$(Fe, Ni)O(OH){\cdot}5H_2O$	Ni content in pure mineral 24.7%, typically found in diluted form by gangue, with actual ores much lower ~1%–1.7% Ni.
Lead (% Pb indicated)		
Cerussite	$PbCO_3$	77.5
Zinc (% Zn indicated)		
Smithsonite	$ZnCO_3$	52.1
Tin (% Sn indicated)		
Cassiterite	SnO_2	78.8
Iron (% Fe indicated)		
Hematite	Fe_2O_3	77.7
Magnetite	Fe_3O_4	72.4
Limonite	$(Fe)O(OH){\cdot}nH_2O$	62.9% with $n = 0$. Note that n can range from 0 to ~3 (at 3, this corresponds to ~37.8% H_2O). *Limonite* is a synonym for *goethite*, also a form of limonite.
Siderite	$FeCO_3$	48.2
Chromium (% Cr indicated)		
Chromite	$FeCr_2O_4$	46..5
Niobium (% Nb indicated)		
Pyrochlore	$(Na,Ca)_2{\cdot}Nb_2O_6{\cdot}(OH, F){\cdot}2H_2O$	38.6
Titanium (% Ti indicated)		
Rutile	TiO_2	59.9
Ilmenite	$FeTiO_3$	31.6
Aluminum (% Al indicated)		
Alumina	Al_2O_3	52.9
Bauxite	Consists of several minerals including gibbsite ($Al(OH)_3$) with other oxides or hydroxides of aluminum plus iron oxides, typically goethite and hematite.	15%–25% Al (typically)

Note: In the past, copper oxide ores were smelted with coal or coke in a blast furnace to produce metallic copper, often called "black copper" because it contained some residual iron. Since about the 1880s, copper oxide ores have been leached with sulfuric acid and the copper recovered from the resulting solution originally by cementation—essentially precipitation by iron. Since 1968, however, the copper has been recovered from solution by solvent extraction–electrowinning (Schlesinger et al. 2011). The copper oxide minerals are included here for completeness.

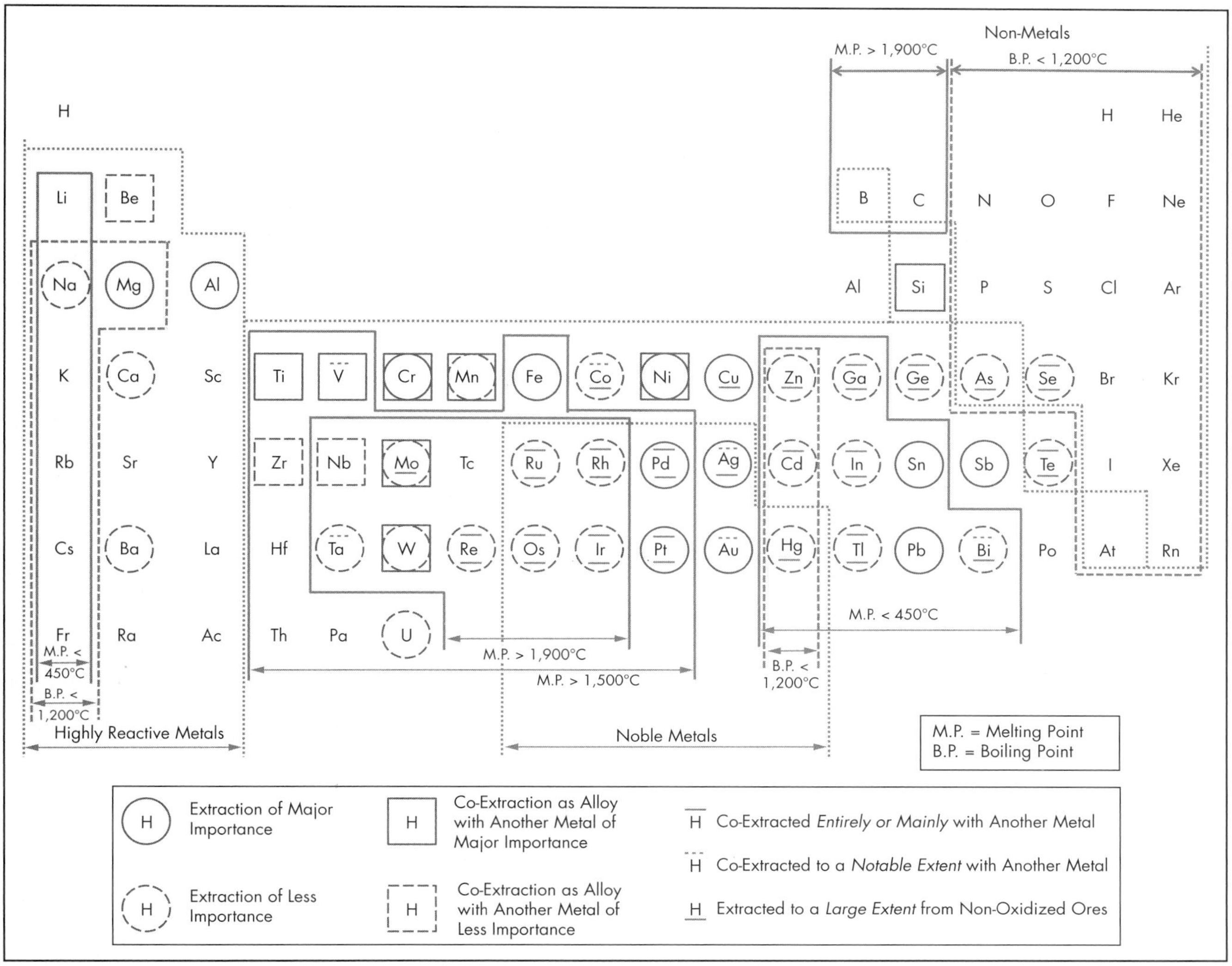

Adapted from Dannatt and Ellingham 1948

Figure 1 Abridged periodic table of the elements highlighting those metals subjected to smelting and common co-extracted element associations

fourth millennia BC—and following the Bronze Age, the Iron Age (1200 BC to 1 BC). Some writers consider up to the present to be the "Iron Age," for example, Smil (2016).

Numerous examples abound of early smelting activities discovered around the world—this includes sites of small bowl-shaped furnaces in the Timna Valley in present-day Southern Israel (Rothenberg 1978) and the Rudna Glava site in eastern Serbia, which is one of the oldest ancient copper smelting sites in Europe. The best-known site in China is the large Tonglushan site near Huangshi in Hubei Province. Military demands often drove the early need for metals (Tylecote 1976). Mining and metallurgical techniques of the early European and Renaissance Period have been famously documented by skilled observers such as Vannoccio Biringuccio (1540) and Georgius Agricola (1556), among others. The scale of operation during this era can be seen in the inset for Figure 2. By raising the height of the small furnace using refractory materials and blowing air at the bottom, the blast furnace was subsequently developed for smelting. One good example of an early iron blast furnace (dating from 1730 to 1883) is the beautifully restored furnace at the Forges Du Saint-Maurice near Cap-de-la-Madeleine in Quebec, Canada (Wikipedia 2018).

The blast furnace for copper smelting was particularly well developed in Germany by 1700–1800. However, over the same period in the United Kingdom, the reverberatory furnace was favored over the blast furnace for copper smelting because of the ready availability of coal, in particular, at Swansea, Wales. Swansea emerged as the major copper smelting center of the world by the early 1800s. A single reverberatory furnace measuring about 2.7 m wide by 4 m long handling 30 t (metric tons) of copper per day consumed about 175 kg/h of coal (Percy 1861). Overall, several separate furnace steps were required to produce finished copper from ore, corresponding to an overall fuel consumption of about 50,000 MJ/t Cu product.

The lead blast furnace remains in use at some plants today, for example, at the Belledune lead smelter in New Brunswick, Canada (Longval et al. 2008; Glencore Canada 2018). The blast furnace is no longer used for copper or nickel smelting.

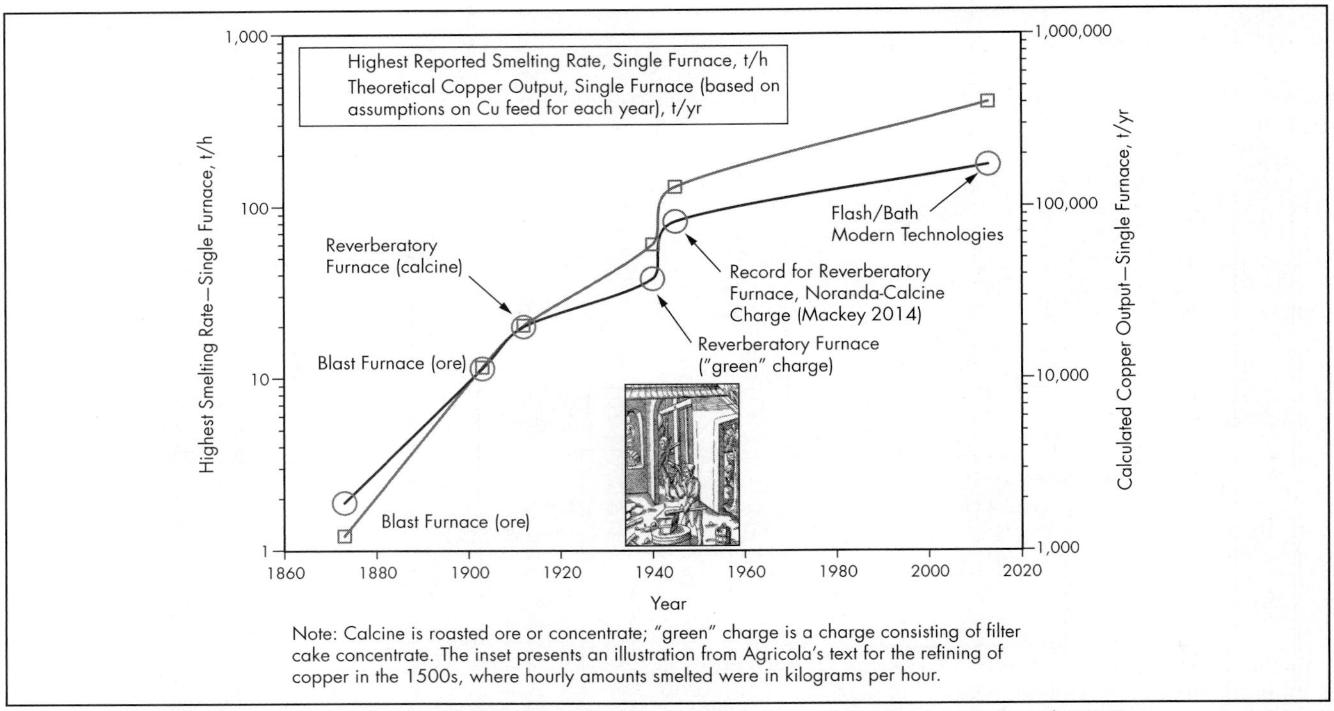

Source: Mackey 2014. Copyright 2014 by The Minerals, Metals & Materials Society. Used with permission. (Inset source: Agricola 1556)

Figure 2 Highest reported smelting rate of copper feed in a single furnace (of varying types) by year over time from 1870 to 2013

Table 2 World metal production by smelting in 2017

Metal	World Production, Mt/yr	Comments
Steel	1,691.2	Pig iron (a major component of steel production) totaled 1,170 Mt in 2017, mostly by blast furnace smelting. Scrap and direct reduced iron mainly accounted for the balance. Worldwide, about 74% of steel was produced via oxygen-blown converting and about 25% by the electric furnace. China accounted for 49% of world steel in 2017.
Aluminum	63.4	All primary aluminum was produced by the Hall–Héroult electrolytic process. China accounted for 41% of world aluminum in 2017.
Copper	18.9	This is copper produced by smelting (includes recycled scrap) and represents ~80% of total copper produced in 2017 (total copper: ~23.1 Mt). The balance is copper produced by leaching and electrowinning.
Ferrochrome	12.9	Ferrochrome is a key ingredient in stainless steel (along with nickel). World stainless-steel production approached 48 Mt in 2017.
Nickel	1.8	This is nickel produced by smelting based on an estimated 87% of world nickel production by pyrometallurgy (Mackey 2009). World nickel production was 2.1 Mt in 2017.

The increase in the size of reverberatory furnaces from about 1800 to the middle of the 20th century is described by Mackey (2014). The trend for the increase in instantaneous smelting rate for a single furnace (of various types) is illustrated in Figure 2. One factor influencing this increase was the availability of a concentrate produced by milling and flotation since about the early 1900s. The large reverberatory furnace (typically, 33 m long by 10 m wide [dimensions inside hearth] and firing 3,600 L of heavy fuel oil per hour [~4,000 MJ/t of charge]) gradually gave way to newer oxygen smelting technologies.

WORLD SMELTING INDUSTRY

Smelting, which produced 1,800 Mt (million metric tons) of metal in 2017, is the largest technological segment of the world metal production industry. Steel is, by far, the major metal produced worldwide, representing 1,692 Mt or 94% of all metals in 2017. The world production of a number of metals produced by smelting is summarized in Table 2 (for 2017). Of interest, the global growth rate for production of these metals has been quite robust, averaging 4.5% per year in the decade 2002–2011 (See et al. 2014) and at a rate that continues approximately to the present time. Much of the recent growth in metals has occurred in China.

FERROUS METALLURGY—THE IRON BLAST FURNACE AND STEEL CONVERTER

The ironmaking and steelmaking chapters in this handbook specifically deal with iron- and steelmaking; however, a brief introduction is included here. The blast furnace is a tall, vertical furnace, generally circular in section, used for smelting. In the case of ironmaking, feed materials consisting of iron ore (approximately ~62% Fe) typically compacted as sinter or pellets, coke, and limestone flux are introduced to the top, while air is introduced under pressure countercurrently at the bottom (hence the term *blast*). The temperature at the lower section is such that liquid iron—referred to as pig iron, containing 4%–5% C and ~1% Si—and molten slag form and are tapped out. Iron blast furnace slag is typically 35% SiO_2, 40% CaO, 15% Al_2O_3, and 7% MgO. The main chemical reactions occurring involve reduction Reactions 2 and 3 given previously.

Over time, the height and diameter of the iron blast furnace has increased (Mackey 2014). By about the 1950s, a typical iron blast furnace was sized about 5–6 m in diameter and 20–25 m high (representing an internal volume of 600–1,000 m^3). Such a plant produced pig iron of 1,000–1,500 t/d (Michard et al. 1961). Today, the largest blast furnaces in Japan have a hearth diameter of about 25 m with an internal volume of 5,500 m^3 and produce ~10,000 t/d of pig iron (Mackey 2014). Additional details are provided in Peacey and Davenport (1979) and Biswas (1981).

Pig iron is generally converted to steel using an oxygen converter. This vessel uses tonnage oxygen introduced by a top-mounted lance to oxidize unwanted elements such as carbon, silicon, and phosphorus and is essentially a modern version of the Bessemer converter that used bottom-blown air (Smil 2016). Today, a 400-t charge of pig iron is converted to steel in about 45 minutes.

UNIT OPERATIONS FOR NONFERROUS METALLURGY

Significant progress was made in the mid-20th century in improving smelting productivity through the adoption of integrated roasting and smelting technologies, oxygen enrichment, and larger vessels. For copper and nickel, the use of reverberatory and blast furnaces were steadily phased out in favor of alternative technologies with much higher energy and associated cost efficiencies; lead followed a similar, but delayed, path. In the case of ferroalloy smelting, significant efficiencies were introduced by the use of a larger electric furnace and improved feed-preparation techniques.

Suspension smelting technologies for copper and nickel, such as the Outotec and Inco flash furnaces, paved the way globally for a steady industry transition away from early practice. Submerged-tuyere smelting concepts such as the Noranda/El Teniente furnaces were later developed based on the well-established Peirce–Smith converter. These technologies were widely adopted in South America as a means to upgrade older facilities. Top lance concepts were first applied at a significant scale for copper through the development of the Mitsubishi process (commercialized in 1974), and a decade or so after this, top submerged lance (TSL) technologies were developed and commercialized for both lead and copper. More recently, SKS/BBS technology derived from submerged tuyere practice has enjoyed widespread success in China for lead and copper. SKS refers to a smelting process developed in China and piloted at the Shuikoushan (SKS) plant in China, and BBS refers to bottom-blowing smelting (China). Advanced electric furnace smelting operations now dominate ferroalloy production. The smelting industry is now offered a wide range of technologies to consider for both captive (smelter at the mine site) and custom (feed shipped to a distant smelter) applications.

Other types of technologies are utilized for treatment of scrap metal and the smelting of iron and aluminum. It has been estimated that 87% of world nickel production in 2009 was by means of a pyrometallurgical step (Mackey 2009), whereas a pyrometallurgical step is involved in the production of more than 80% of world copper (ICSG 2017). All iron, steel, aluminum, titanium, and the major ferroalloys (Fe-Cr, Fe-Mn, Fe-Ni, Fe-Nb, Fe-Si, and Fe-V) are produced via smelting.

Suspension/Flash Smelting Furnace

Suspension or flash smelting (Cu and Cu-Ni) involves the combustion of fine sulfide minerals above a molten bath where the combustion heat supports furnace operation at the temperatures required. The first patent describing this approach was granted in the late 1800s (Bridgman 1897) and was conceived based on reverberatory furnace design at the time:

> The material to be treated, pulverized, say, to the fineness of forty to sixty mesh, is fed as a spray of dust into the highly-heated and strongly-oxidizing atmosphere of a furnace chamber. In its finely-divided condition it is rapidly and thoroughly oxidized in passing through this heated atmosphere on the hearth; and the heat also melts the more or less oxidized ore as it falls, giving the well-known reaction of the sulfid on the oxid.

Several subsequent patents reviewed by Bryk et al. (1958) describe other efforts advancing this concept (Klepinger et al. 1915; Freeman 1932 and Zeisberg 1937).

In 1931, Anaconda Copper Mining Company performed trials of a vertical reaction shaft mounted over a reverberatory furnace (Laist and Cooper 1934). Similar trials were conducted in the Soviet Union in 1935 and Yugoslavia in 1937 (Bryk et al. 1958). Driven by a shortage of post-war electricity supply in Finland, in 1947 Outokumpu (Särkikoski 1999) successfully piloted a shaft–reverberatory combination very similar in concept to the earlier Anaconda trials. This was quickly followed in 1949 by a fully commercialized furnace using preheated air at Harjavalta, subsequently known as the Outokumpu flash furnace. In Canada, Inco Ltd. developed the Inco flash furnace, with a commercial operation commencing in 1952 (Norman 1936; Gordon et al. 1954; Queneau and Marcuson 1996).

Figure 3 illustrates the increase in the adoption of the Outokumpu flash furnace over the period from 1930 to 1990; other technologies are also shown. This trend also reflects the impact of environmental regulations and energy costs on the adoption of suspension/flash smelting furnace technology. Because of its very high oxygen enrichment and consequent lower off-gas volume/higher SO_2 strength, Inco technology was initially viewed as superior to Outokumpu's. As the latter adopted higher enrichment levels, these economic differences were reduced. Korhonen and Valikangas (2014) suggest that the reason for a marked industry preference for Outokumpu technology (now owned by Outotec) is due to differences in licensing models utilized by the two companies—Outokumpu strongly marketed the technology; Inco did not. A much later type of flash smelting process in Russia led to the development of the Kivcet process for lead-zinc and copper (Ashman et al. 2000).

In an updated assessment of the comprehensive industry survey reported by Kapusta in 2004, flash smelting represented 49% of global copper capacity; it remains the dominant technology for sulfide concentrate smelting.

Outotec Flash Furnace

The general adoption of oxygen enrichment for Outotec technology did not start until successful implementation at both

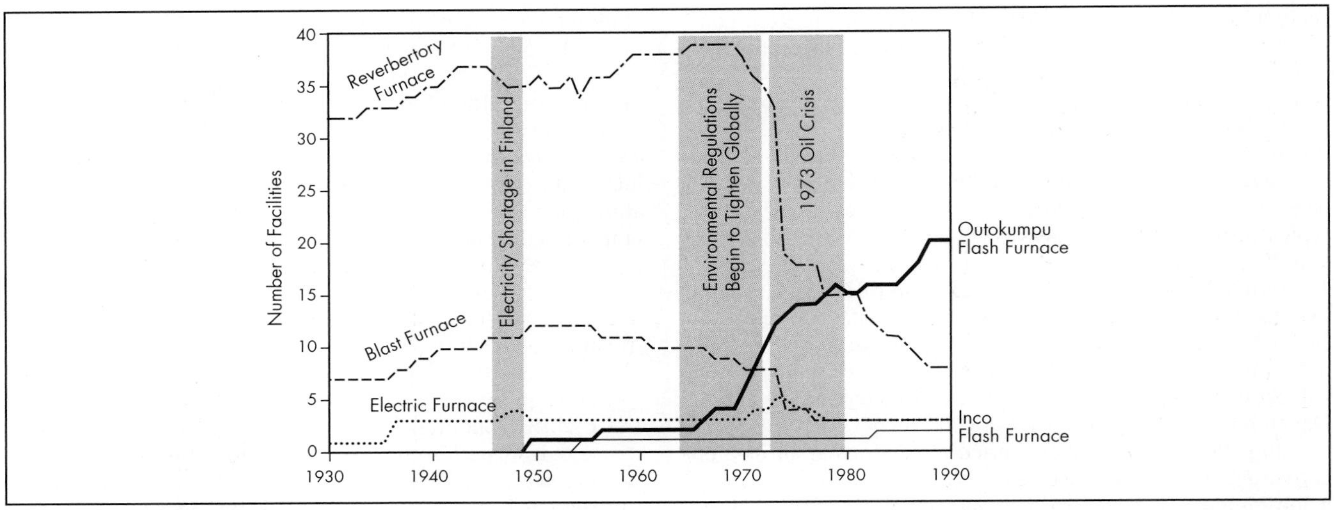

Source: Korhonen and Valikangas 2014

Figure 3 Smelter facility technology use

Harjavalta and Hitachi in 1972 (Harkki and Juusela 1974; Yasuda 1974; Särkikoski 1999). The use of air preheating gradually declined in preference to oxygen enrichment and improved burner design (Suzuki et al. 1998; Colins 1987).

In 1960, Harjavalta modified their second flash furnace for nickel smelting (Paakkonen and Mattelmaki 1996). The Kalgoorlie (Australia) and Selebi-Phikwe (Botswana, Africa) nickel smelters were commissioned in 1972 and 1973, respectively; the Norilsk, Russia, smelter in 1981; and a fifth in 1992 at the Jinchuan smelter in China. The Kalgoorlie and Jinchuan smelters uniquely included an electrically heated slag-cleaning appendage.

For copper matte-making furnaces, target matte grades have progressively increased from 45% Cu to a typical 65% Cu, and as high as 70% Cu, constrained by subsequent conversion operations and higher oxidic losses in the slag. Nickel matte-making furnaces typically target 40% Ni+Cu+Co, constrained by nickel and cobalt oxidic losses in slag. In the early 1990s, Outokumpu developed the DON (direct Outokumpu nickel) process, motivated by a desire to eliminate nickel converting and the need to treat high-magnesia feeds. The modified Harjavalta furnace and ancillary operations were commissioned in 1995, taking nickel matte iron tenors down from 18% to 4% (Makinen and Taskinen 1998). The DON process was subsequently deployed at the Forteleza smelter, Brazil, in 1998, which closed in 2013.

Motivated by a challenging (and unique) chalcocite/bornite/carbonate concentrate, and following extensive test work at Outotec's Pori pilot facility in Finland, the direct-to-blister process was successfully commercialized at Glogow, Poland, in 1978 (Czernecki et al. 1998). The direct-to-blister process and improved furnace designs have subsequently been deployed at Olympic Dam, Australia, in 1988 (and expanded in 1999) and Konkola, Africa, in 2008, both utilizing silicate slag.

Flow sheet and process description. The Salt Lake City (Utah, United States) Kennecott smelter includes both flash smelting and flash converting for the treatment of copper concentrate. As illustrated by the flow sheet in Figure 4, the facility receives concentrate from the nearby Bingham Canyon mine and concentrator for stockpiling and transfer to feed bins. Reclaimed concentrate is delivered to wet feed bins and continuously blended with granulated converter slag and crushed silica flux as feed to a rotary kiln dryer. Nitrogen (with some air) is used to temper the dryer combustion gas inlet to mitigate concentrate sulfide pyrolysis and oxidation, a baghouse cluster returns entrained dry product, and caustic scrubbing of the off-gas ensures that sulfur dioxide emissions are low. In more recent or upgraded facilities, steam coil dryers are used.

Dry feed and dust are fed into the smelting furnace reaction shaft burner using independent loss in weight feeders and distribution systems plus oxygen-enriched air. Copper matte is tapped and delivered to two granulators; this material is screened and conveyed to a matte storage dome for subsequent converting to blister. Granulated nickel matte produced in the DON process is treated hydrometallurgically.

Slag is tapped and slow cooled for metal recovery by milling and flotation. In contrast, nickel slag is cleaned in electric furnaces with coke/coal reductant for metal recovery, such as in the DON process. This is because nickel oxide in slag is non-recoverable by slag milling.

For direct-to-blister copper processes, the flash smelting furnace is configured similarly to a flash converting furnace whereby the blister is tapped and laundered to anode furnaces for subsequent refining (see Chapter 11.6, "Conversion and Refining"). The high oxidic copper primary slag is reduced using an electric furnace (from ~20% to 4% Cu) with further copper recovery achieved in a crush/grind/float facility.

Off-gas from the furnace at ~1,350°C is directed to the waste heat boiler consisting of radiative and convective sections and cooled to ~350°C before treatment in the acid plant.

Furnace configuration. A typical, modern Outotec flash furnace for copper smelting is well represented in Figure 5. Extensive use of high-quality copper cooling elements above and below the bath has been the general trend in the last few decades. Details of furnace construction are provided by Janney et al. (2007). Table 3 presents key dimensional parameters for exemplar Outotec smelters.

Ancillary equipment. Concentrate burner design, performance, and operation are critical to the overall success of an Outotec flash smelting furnace. Figure 6 shows a typical

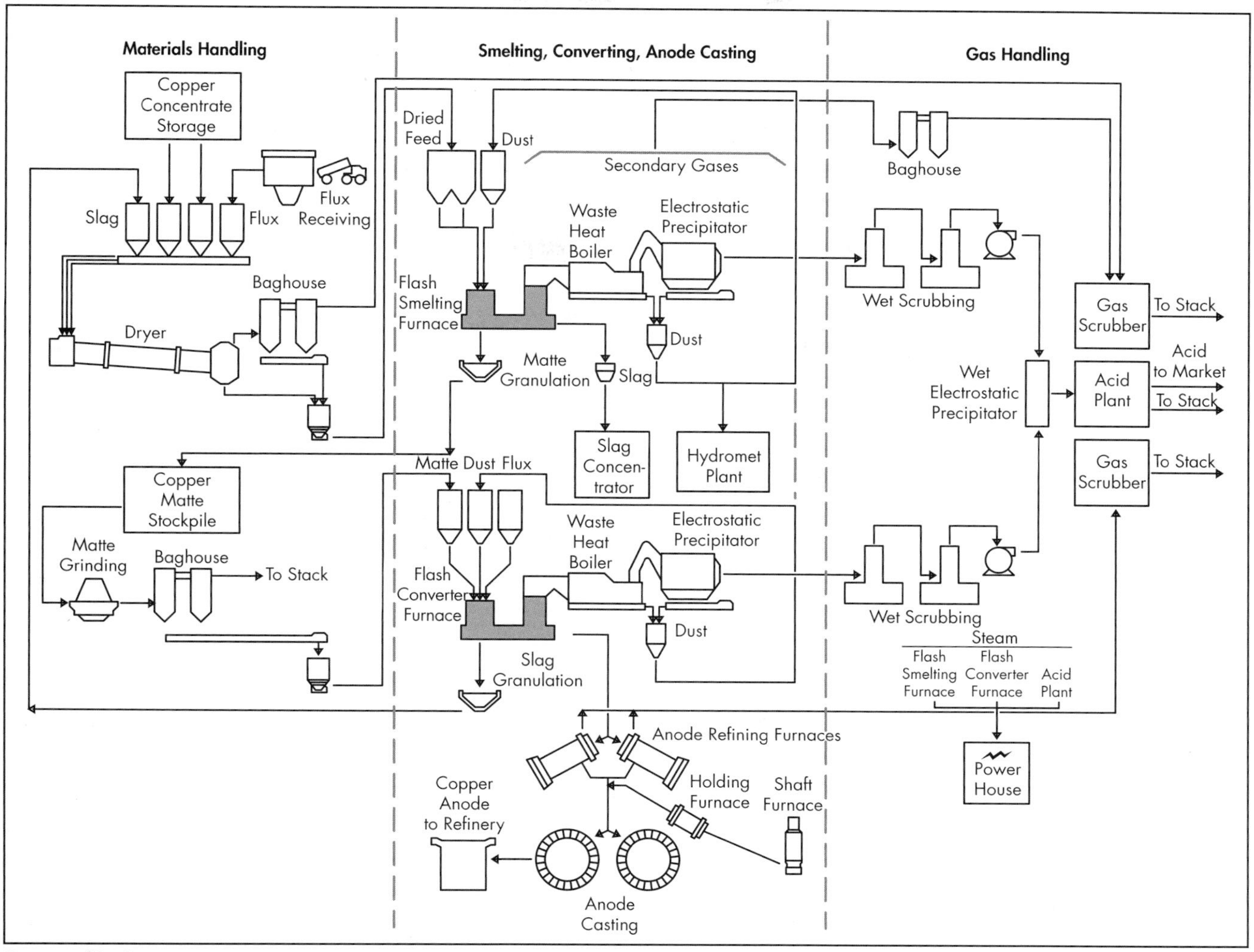

Courtesy of Rio Tinto

Figure 4 Integrated flash smelting and converting process flow sheet at the Salt Lake City, Utah, smelter

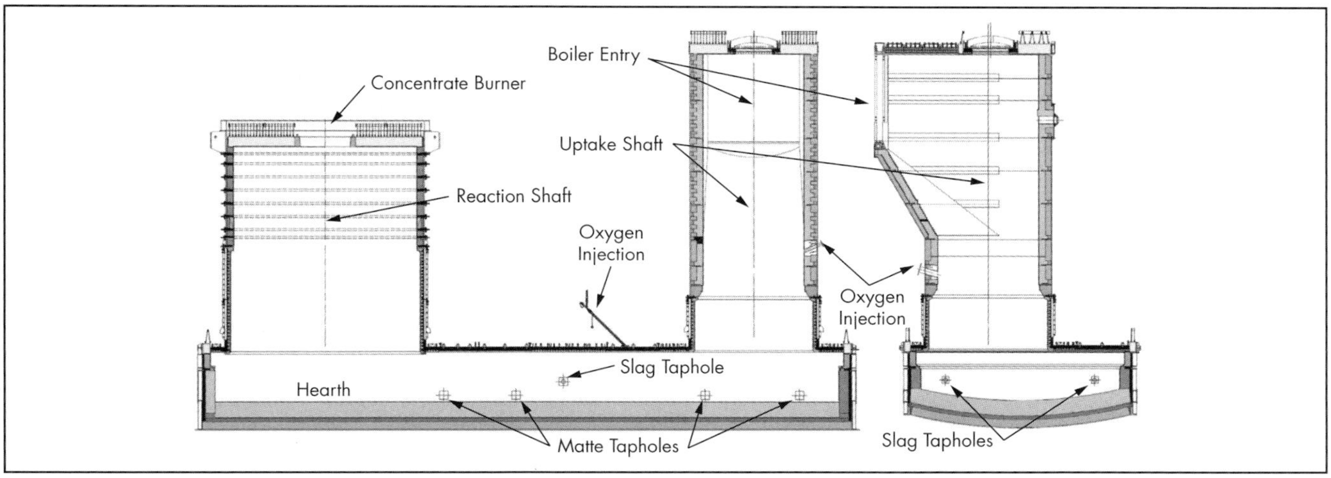

Courtesy of RHI, with permission from Outotec and Rio Tinto

Figure 5 Flash smelting furnace elevation longitudinal and lateral sections from the Salt Lake City smelter

Table 3 Physical parameters of exemplar Outotec flash smelting furnaces

Item	Harjavalta Original	Harjavalta DON	Konkola Direct-to-Blister	Tongling
Year started	1949	1995	2008	2013
Shaft height (inside dimensions), m	7.9	7.6	6	7.9
Shaft diameter (inside dimensions), m	3.7	4.6	6	7
Settler length (outside dimensions), m	19.5	19.5	24	25.9
Settler width (outside dimensions), m	7	7	7.4	9.2
Matte/blister production, t/d	160	116	750	2000
Off-gas volume, Nm^3/h	10,000	16,000	46,000	85,000

Source: Bryk et al. 1958; Zhou and Sun 2013; Outotec (personal communication); Warner et al. 2007

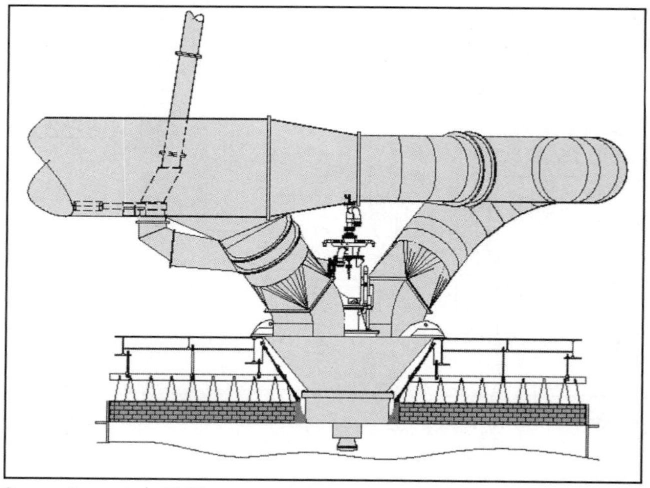

Source: Legg et al. 2009

Figure 6 Typical concentrate burner arrangement

arrangement of a modern burner. The oxygen–air mixture is delivered to a conical plenum that directs the gas to a vertical annulus at the reaction shaft entry. The annulus area is throttled by mechanical means to maintain a narrow range of velocity at a wide range of gas volume. The vertical velocity is typically 100 m/s. The concentrate/flux/coolant solid blend is delivered to the burner, typically on opposing sides, with internal baffling to ensure symmetrical annular discharge of the material inside the gas annulus.

Figure 7 contrasts the well-established burner concept shown in Figure 7A with some alternatives. The concept shown in Figure 7B involves imparting a rotation to the gas, thereby achieving both vertical and horizontal momentum at the shaft entry. The concept depicted in Figure 7C involves imparting a rotation to the gas and locating the symmetrical annular discharge of solid on the outside of the gas annulus.

Operational aspects. Servo and regulatory process control of a flash furnace involves maintaining matte grade, slag composition, and settler temperature through manipulation of oxygen, flux, and auxiliary fuel, respectively. In addition to fuel, manipulation of enrichment and dust charged can be used for temperature control. All of these parameters are manipulated at the furnace burner entry except flux, which is usually located upstream at the dryer entry. The proportions of concentrate and recycled "burden" materials fed to the dryer must result in a blend that can be smelted at the desired rate and provide sufficient flexibility for parameter manipulation at the furnace.

Energy flows. An example mass and heat balance of the Kennecott flash smelting furnace is shown in Table 4 and is typical of the internal control calculations utilized in real time. The table also shows the waste heat boiler mass and heat balance for the same example. The change in formation enthalpy, Hf, is due to dust sulfation (in unsulfated, out sulfated); the sum of this and the change in sensible enthalpy, Hs, equals the steam sensible enthalpy. Also note the contribution of dust sulfation to the amount of steam produced in the boiler at ~19% (4,469 MJ/h). Dust sulfation is the final oxidation of sulfidic dust to a sulfate.

Inco Flash Furnace

Means to economically separate oxygen from nitrogen in air on a sufficiently large scale for industrial use began to develop early in the 20th century. By the 1920s, the production of gaseous oxygen by cryogenic separation techniques at a reasonable cost for metallurgical applications was becoming closer to reality (Davis 1923).

A 500-t/d copper concentrate Inco flash furnace, based on the use of commercial-grade cryogenic oxygen, commenced operations at the Inco Copper Cliff smelter (near Sudbury, Ontario, Canada) in 1952. This followed many years of testing and piloting using a 20-t/d unit. The furnace design resembled a rectangular reverberatory furnace but with a central off-gas uptake in the roof and feed fired with commercial-grade oxygen entering from each end wall (Boldt and Queneau 1967). The original concept of using commercial-grade oxygen for smelting had been described earlier by Norman (1936) in an undergraduate thesis at the University of Toronto (Ontario, Canada).

The oxygen for the new process was supplied by a 300-t/d Air Liquide "Oxyton" plant, the third largest tonnage oxygen plant in the world at the time. The high SO_2 off-gas was initially converted to liquefied SO_2 for the market; later, a sulfuric acid plant was installed. Within two years, Inco commissioned a larger 1,000-t/d flash furnace (subsequently enlarged) that replaced the earlier 500-t/d unit. The plant went through several changes, and two new furnaces were built for bulk nickel-copper concentrate in 1993; however, by the end of 2017, Vale (the former Inco) was operating a single nickel flash furnace; a separate copper concentrate is now shipped to third-party smelters for treatment.

Despite these interesting developments in Canada in the 1950s, the use of tonnage oxygen elsewhere for nonferrous smelting in general, and copper smelting in particular, was slow to catch on. This was in part due to the close link of the cost of oxygen with the cost of electricity, the main component of the cost of tonnage oxygen. In some parts of the world, the cost of electricity was relatively high. In Eastern Canada in the

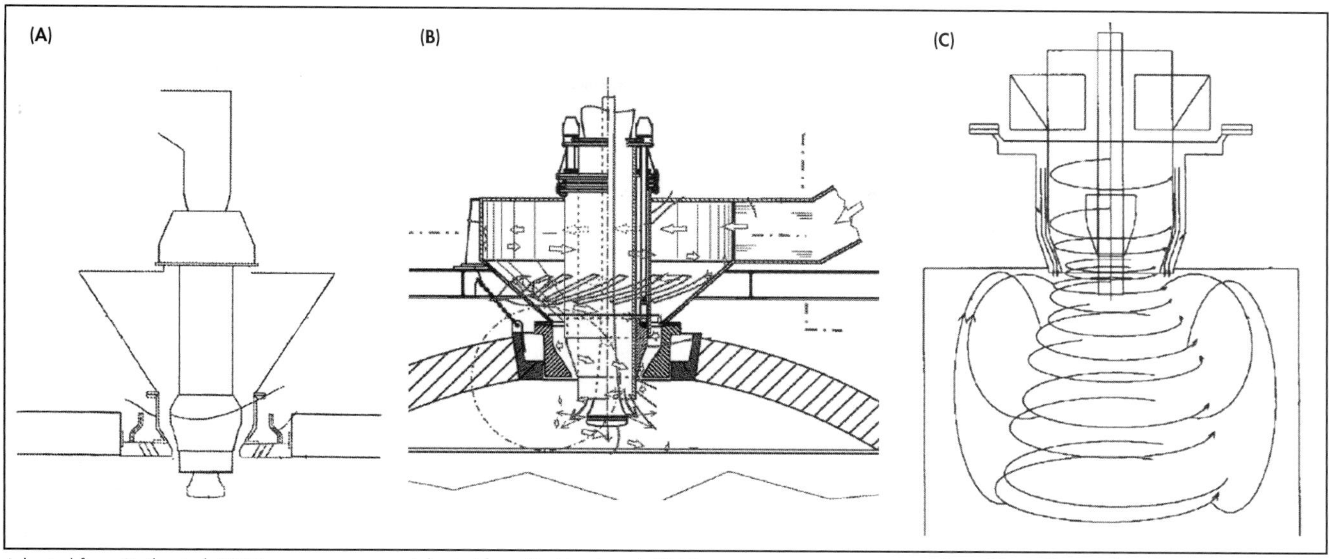

Adapted from Sipila et al. 2012 Adapted from Bjorklund et al. 2013 Adapted from Zhou and Liu 2014

Figure 7 Recently patented burner concepts

1950s, energy costs were low and fairly stable; heavy fuel oil was about 1 cent/L and electricity was around 0.5 cents/kW·h. Another factor is believed to be the generally conservative nature of the mining industry in embracing change.

The significant impact of the decrease in fossil fuel requirement in smelting as the $\%O_2$ enrichment is increased was illustrated by Tarassoff and Mackey (1983) for typical Eastern-Canadian concentrates of the era (~65% $CuFeS_2$, 12% FeS_2, and 8% FeS); see Figure 8. The positioning of several smelting technologies is also indicated. An Inco flash furnace smelting dry concentrate at ~95% oxygen enrichment is autogenous at 55% Cu matte. Regarding levels of oxygen enrichment for group 2 processes shown in Figure 8, much higher enrichments are currently used, given the higher proportion of "burden" in the feed.

At one time, there were five Inco flash furnaces in operation worldwide (one at the Chino smelter in New Mexico and one at the Hayden smelter in Arizona, both in the United States; two units at Copper Cliff in Canada; and a version at the Almalyk smelter in Uzbekistan). As of mid-2018, with Chino closed since 2002, one unit at Copper Cliff closed since mid-2017, and the unit at Almalyk replaced by a Vanyukov furnace in 2017, there are currently only two Inco furnaces in operation: the one at Hayden and one at Copper Cliff. The data in Table 5 for the Copper Cliff furnace are, however, representative of the former arrangement with two Inco flash furnaces treating a nickel-copper concentrate. Some data are also taken from Crundwell et al. (2011).

Kivcet Furnace

This process was developed in Russia for copper and lead smelting. The acronym KIVCET stands for "oxygen vortex, cyclone, and electrothermic" in Russian. The idea to smelt sulfide concentrates in a high-temperature cyclone reactor for fast reactions was first tested in Russia in the 1950s on copper and zinc-lead concentrates. The furnace arrangement included the cyclone reactor mounted above a reverberatory-type furnace plus a connected electric furnace arrangement for slag cleaning; a 350-t/d plant based on this approach was built in Kazakhstan.

A Kivcet lead plant was partly built 1984 at Karachipampa, Bolivia, but not completed. In 1987, Snamprogetti commenced operation of a Kivcet lead smelter at Porto Vesme in Italy. In 1997, Cominco (now Teck Resources Ltd.) started up a Kivcet lead smelter as a plant retrofit at its Trail (British Columbia, Canada) smelter, where operations had originally begun in 1900. The new plant has operated very well since that time and is presently treating ~450,000 t/yr of feed. Of interest, the Kivcet facility replaced an earlier retrofit based on the QSL process, which, however, did not live up to expectations and was closed about a year after its 1990 start-up (Ashman et al. 2000).

At the Trail operation, the feed to the Kivcet furnace consists of new concentrate and zinc plant residue for a total feed rate of about 55 t/h (Figure 9). Lead bullion is tapped, while slag is tapped from a taphole located after the electric furnace arrangement used to clean the slag of metal values (the slag section is separated from the reaction zone by a partition wall). Off-gas leaves the vessel and, after cleaning, is directed to the acid plant. Teck has introduced several innovations, including allowing the operation to proceed with a coke "checker" layer on the slag to facilitate efficient slag reduction and cleaning, and also operating in tandem with a continuous lead drossing plant.

Submerged Tuyere Furnaces

The use of forced air blown into a melt contained in a furnace to effect high-temperature smelting or refining of metals reaches back to antiquity. Air bellows were used to produce a stream of air for refining silver, as described in the Old Testament (Jeremiah 1985), while the Agricola woodcut included as an inset earlier in Figure 2 shows the use of bellows to force air to refine copper. In 1878, Holloway in the United Kingdom successfully employed a modified Bessemer converter to test the autogenous smelting of a copper-iron sulfide ore along with added flux (Mackey and Campos 2001; Wraith and Mackey 2010). Holloway's work was the first successful, large-scale smelting test of this nature and is now referred to as "bath smelting" (Themelis and Kellogg 1983).

Table 4 Example of flash smelting furnace mass and heat balance

Material	In or Out	Flow	Temperature, °C	Hf, MJ/h*	Hs, MJ/h*
Flash Smelting Shaft					
Bedded concentrate	In	164.541 t/h	25	–445,135.7	0.0
Coolant blend	In	11.430 t/h	25	–57,454.0	0.0
Silica flux	In	14.537 t/h	25	–209,284.1	0.0
Dust charged	In	17.146 t/h	25	–80,632.0	0.0
Shaft air-oxygen	In	34,962 Nm^3/h	25	0.0	0.0
Disperser air-oxygen	In	3,002 Nm^3/h	25	0.0	0.0
Shaft product	Out	171.731 t/h	1,559	–659,782.3	215,338.7
Shaft flux	Out	14.537 t/h	1,559	–209,284.1	23,746.8
Shaft off-gas	Out	32,168 Nm^3/h	1,559	–275,127.4	100,887.3
Losses	Out	—	—	—	11,715.2
Sum of (out – in)	—	—		–351,687.9	351,687.9
Shaft natural gas	In	0 Nm^3/h	25	0.0	← Net H
Shaft natural gas off-gas	Out	0 Nm^3/h	1,559	—	0.0
Flash Smelting Settler					
Shaft product	In	171.731 t/h	1,559	–659,782.3	215,338.7
Shaft flux	In	14.537 t/h	1,559	–209,284.1	23,746.8
Shaft off-gas	In	32,168 Nm^3/h	1,559	–275,127.4	100,887.3
Slag	Out	101.618 t/h	1,300	–756,383.3	149,290.8
Matte	Out	69.910 t/h	1,275	–42,999.2	58,041.0
Dust	Out	11.518 t/h	1,335	–38,438.4	11,497.4
Off-gas	Out	33,254 Nm^3/h	1,335	–289,607.7	88,263.9
Losses	Out	—	—	—	19,037.2
Sum of (out – in)	—	—	—	16,765.1	–13,842.4
Shaft natural gas off-gas	In	0 Nm^3/h	1,559	—	0.0
	Out	—	1,335	—	0.0
Settler injection gas	In	170 Nm^3/h	25	0.0	0.0
	Out	—	1,335	0.0	333.6
Air ingress	In	850 Nm^3/h	25	0.0	0.0
	Out	—	1,335	0.0	1,602.8
Settler natural gas	In	340 Nm^3/h	25	4,859.2	← Net H
Settler natural gas off-gas	Out	3,701 Nm^3/h	1,335	—	5,150.4
Waste Heat Boiler					
Furnace off-gas	In	37,974 Nm^3/h	1,335	–72,503.2	23,705.5
Dust	In	11.518 t/h	1,335	–9,187.0	2,748.0
Sulfation air	In	0 Nm^3/h	25	0.0	0.0
Air ingress	In	2,549 Nm^3/h	25	0.0	0.0
Off-gas to electrostatic precipitator	Out	36,518 Nm^3/h	400	–68,299.7	6,071.5
Dust to electrostatic precipitator	Out	11.173 t/h	400	–12,197.5	637.1
Dust from waste heat boiler	Out	5.088 t/h	250	–5,662.4	169.5
Losses	Out	—	—	—	500.0
Sum of (out – in)	—	—	—	–4,469.3	–19,075.5
Steam production	—	44.847 t/h	—	—	–23,544.8

*Hf = estimated standard formation; Hs = sensible enthalpies.

With the later development of the pneumatic copper converter by Pierre Manhès in France (Pelletier et al. 2009) in 1880, it was soon found that during converting, added silica flux smelted well and was able to form good-quality slag when introduced to the converter while blowing. The larger Peirce–Smith converter was developed in the United States in 1909 (Southwick 2009). Several attempts were made in the 1950s and 1960s to smelt sulfide feed material as well as flux in the Peirce–Smith converter—for example, the pioneering Hitachi converter process (Tsurumoto 1961) and later work at Kennecott (Messner and Kinneberg 1969). These technologies did not catch on, given that the copper converter is a batch process while, ideally, the smelting of concentrate should be continuous. Consequently, trying to do both in one—copper matte converting plus concentrate smelting—was seen as not offering any great advantage.

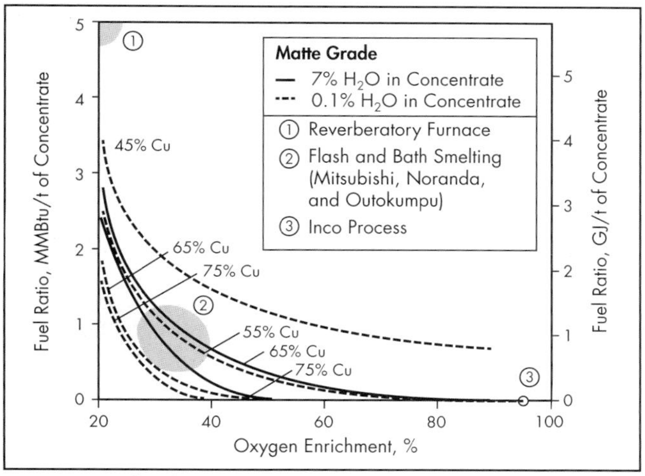

Source: Tarassoff and Mackey 1983. Copyright 1983 by The Minerals, Metals & Materials Society. Used with permission.

Figure 8 Effect of oxygen enrichment and matte grade on the fuel requirement in smelting

Noranda Process

The first major commercial breakthrough in continuous bath smelting occurred in 1965 with the successful small-scale piloting of a new continuous copper smelting and converting process, later called the Noranda process, by N.J. Themelis and co-workers in Canada (Figure 10). The initial small-scale piloting (1.8 t of concentrate per day) was carried out at the Noranda Research Centre in Pointe-Claire (Montreal, Quebec, Canada) and later on a 90-t/d concentrate pilot plant built and successfully operated at the Horne smelter of Noranda Mines Limited (Themelis et al. 1972; Mills et al. 1976; Diaz et al. 2011).

The Noranda process breakthrough essentially involved two new, innovative features:

1. The use of an extended converter vessel with submerged tuyeres, which was known to provide for a well-mixed bath for efficient smelting
2. The addition of tapholes—one for the metal product (copper or high-grade matte) and one for slag (to permit continuous operation of the process)

Table 5 Technical data for the Inco flash furnace plants (typical as of 2016)

Company/Plant (and reference)	Location	Year Commenced	Vessel Size (length × width × height), m	Blowing Rate, Nm^3/h	Capacity, t/yr concentrate
Vale Inco (Donald and Scholey 2005; Crundwell et al. 2011)	Copper Cliff, Ontario, Canada	First plant: 1952; current furnaces: 1993 (the Copper Cliff smelter began in the 1880s)	30 × 9 × 8	14,000 (96% O_2)	500,000 (per furnace)
Asarco Hayden (Schlesinger et al. 2011; Marczeski and Aldrich 1985)	Hayden, Arizona, United States	2001 (the smelter began in 1912)	22 × 5.5 × 5	~11,000 (95% O_2)	629,000

Note: The data for the Vale Inco operation represent the arrangement with two Inco flash furnaces at Copper Cliff smelting a nickel-copper bulk concentrate. In mid-2017, the company switched to producing separate nickel and copper concentrates and to a one-furnace operation at the smelter treating the nickel concentrate; the copper concentrate is now shipped to third-party smelters.

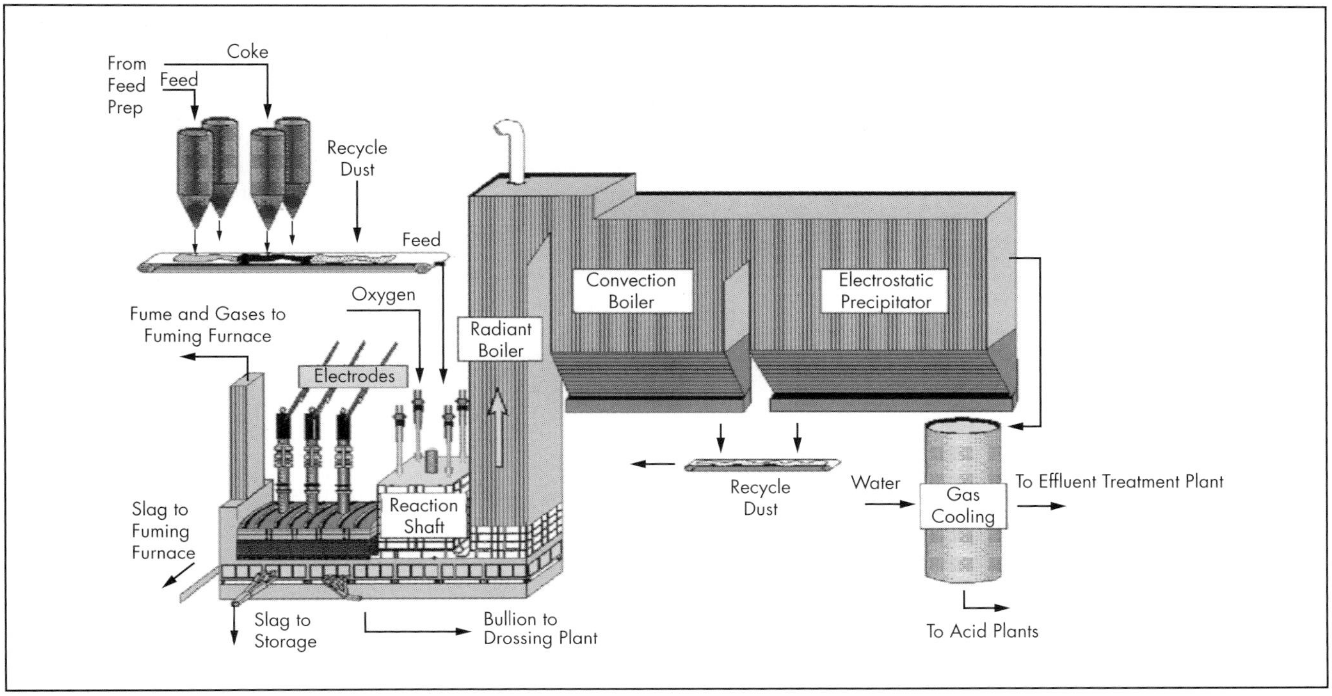

Source: Rioux et al. 2014

Figure 9 Kivcet plant at Trail, British Columbia

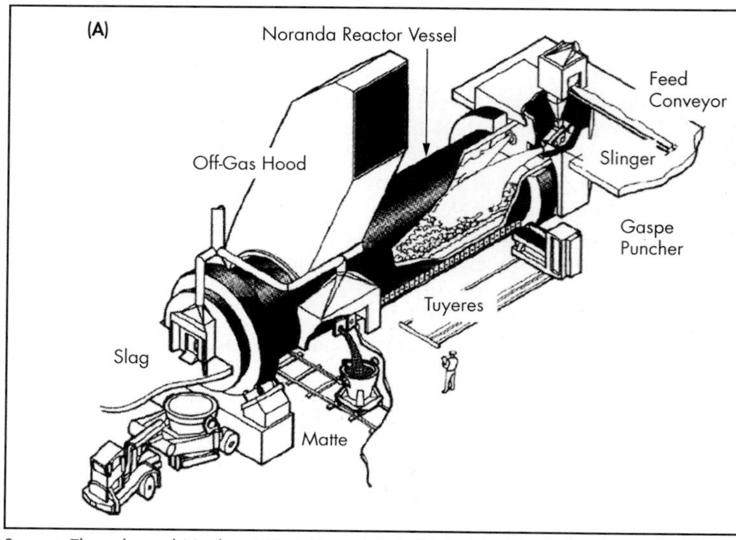

Source: Themelis and Mackey 1992. Copyright 1992 by The Minerals, Metals & Materials Society. Used with permission.

Source: Noranda Mines Limited 1974

Figure 10 (A) Cutaway view of the Noranda process and (B) testing the Noranda process technology in the early 1970s at the Noranda Research Centre in Canada

In this manner, the process became a truly continuous operation (continuous feeding and blowing with periodic tapping).

Following the successful pilot campaign, the first commercial Noranda process plant began at the company's Horne smelter in March 1973 (Figure 10). At 5.1 m in diameter and 21.3 m long, this was the world's largest bath smelting vessel at the time (Themelis et al. 1972; Mills et al. 1976). Designed for a capacity of 726 t/d copper concentrate, plant capacity was soon increased in stages by introducing improvements, the chief one being the use of oxygen-enriched air (Prevost et al. 2013).

Capacity is currently 840,000 t of feed material per year, or approximately 2,550 t/d, an increase of more than three times the original capacity (Themelis et al. 1972; Mills et al. 1976; Prevost et al. 2013). The plant remains in operation today and now handles a considerable quantity of recycled material such as e-scrap and similar materials (Prevost et al. 2013).

The Noranda process (Table 6) operates as follows:

- Either filter cake concentrate (as at the Horne plant in Canada) or bone dry concentrate (as at the Altonorte plant in Chile) are introduced to the smelting vessel. Oxygen-enriched air is blown through a series of submerged tuyeres, reacting with the added feed in the well-mixed bath. High-grade matte and a slag are tapped periodically.
- The matte is transferred for converting to blister copper, while the slag is allowed to slow cool in the same ladles used for tapping. When cool, the slag is removed from the ladles, crushed, and milled to recover the contained copper by flotation.
- The off-gas containing sulfur dioxide from the smelting reactions is withdrawn via the off-gas mouth and, after gas cooling and cleaning of particulates, is directed to the acid plant.

In 2018, two Noranda process plants were in operation—one at the original Horne smelter (now Glencore) in Canada and one at the Altonorte plant in Chile; refer to Table 6 (Mackey and Campos 2001; Prevost et al. 2013). Over the years, three Noranda process vessels operated from 1977 to 1995: one at the Salt Lake City, smelter; one Noranda process reactor operated at the Port Kembla smelter (later, Southern Copper) in New South Wales, Australia, from 1988 to 2003 (closed because of small plant capacity); and one vessel at the Daye smelter in Hubei Province, China (the latter was closed to make way for a large new replacement plant). Interestingly, Noranda process vessels were delivered in the 1990s to a plant in Canada (Flin Flon, Manitoba) and to a plant in China (Shenyang, Liaoning Province), but neither of these facilities were completed because of poor economic situations at the respective plants.

El Teniente Process

The El Teniente process had its origin at the Caletones smelter in Chile in the early 1970s. At the time, the basic units at the smelter—which was then owned by Kennecott of the United States and which typically had the capacity to handle up to ~600,000 t/yr of concentrate—consisted of three rotary dryers, two oil-fired reverberatory furnaces (typically consuming 600–700 L of oil per ton of copper), seven Peirce–Smith converters, and associated copper furnaces. To meet higher-capacity requirements, the plant had implemented batch smelting of concentrates in the Peirce–Smith converters based on experience at Kennecott and also in Japan (Munoz et al. 1982; Mackey and Campos 2001). However, achieving the required tonnages with this approach was problematic.

At about the end of 1970, the copper companies were nationalized by the new leftist government in Chile, a situation further pressuring the difficult situation at Caletones. Plant personnel, however, continued investigations and developed the notion that concentrate smelting in a converter would be workable if it could be made continuous by adding a taphole on an end plate. Following pilot-plant testing of this alternative approach during the period 1972–1974, the so-called El Teniente converter for continuous smelting in a vessel 4 m

Table 6 Technical data on the Noranda process plants

Company/Plant (reference)	Location	Year Commenced	Vessel Size (diameter × length), m	Blowing Rate, Nm^3/h	Tonnage/yr concentrate
Glencore Horne (Mills et al. 1976; Prevost et al. 2013)	Rouyn-Noranda, Quebec, Canada	1973 (the smelter began in 1927)	5.1 × 21.3	55,000 (42% O_2)	840,000
Glencore Altonorte (Mackey and Campos 2001; Schlesinger et al. 2011)	La Negra, Region II, Chile	2001 (the copper smelter began in 1993)	5.3 × 26.4	76,000 (39% O_2)	1,160,000

Table 7 Technical data on the El Teniente process plants in Chile

Company/Plant (reference)	Location	Year Commenced	Vessel Size (diameter × length), m	Blowing Rate, Nm^3/h	Tonnage/yr concentrate
Codelco, Chuquicamata (Diaz et al. 1995)	Calama	1991	5 × 22	62,000 (40.5% O_2)	550,000
Codelco, Potrerillos (Diaz et al. 1995)	Potrerillos	1994	5 × 22	59,000 (37% O_2)	630,000
Enami Hernan Videla, Paipote (Diaz et al. 1995)	Copiapo	1993	3.8 × 14	32,000 (39% O_2)	360,000
Codelco, Las Ventanas (Diaz et al. 1995; Medina et al. 1993)	Puchuncavi	1984	4 × 15	36,000 (37% O_2)	360,000 (after improvements)
Codelco, Caletones (Diaz et al. 1995)	Caletones, El Teniente	1977	5 × 22	67,000 (36.5% O_2)	1,140,000 (two units)

Note: Data are also from Schlesinger et al. 2011, Sanhueza et al. 2003, and industry information.

in diameter by 17.5 m long started operation at Caletones in 1977. The process as developed consisted of smelting filter cake concentrates in the vessel that had been originally charged with reverberatory furnace matte called seed matte to assist in the heat balance (Munoz et al. 1982; Diaz et al. 1995; Kellogg and Diaz 1992). A second unit was installed in 1978.

Many design, operational, and maintenance improvements were introduced, in particular, relocation of the Garr Gun feeding system from the barrel top to the end plate, and the development of a removable panel to effect rapid, hot refractory repairs.

Resembling the Noranda process vessel in many aspects, it was an independent process development. After some discussions, Codelco-Chile and Noranda Mines Limited struck an agreement to collaborate on technology topics in 1983 (Mackey et al. 1985).

In 1984, two new, larger bath smelting vessels (5 m in diameter by 22 m long) were installed along with dry concentrate addition by injection through four injection tuyeres, replacing the older, smaller units (Diaz et al. 1995). Each vessel was able to smelt 1,600 t/d of concentrate.

Beginning in 1989, the Caletones smelter was then operated with the two El Teniente vessels, initially treating together a total of about 70% of the smelter capacity (the balance by a single reverberatory furnace). The reverberatory furnace was later closed with the introduction of the El Teniente slag cleaning process (Diaz et al. 1995).

The success achieved at the Caletones smelter over this period motivated other Chilean smelters to follow the Caletones example, and currently five of the seven smelters in Chile now successfully operate the El Teniente process, having a total concentrate capacity of about 3.5 Mt/yr, or about 4%–5% of world smelter capacity, a considerable amount for a single process (Table 7). In addition, the technology was licensed in the period of the 1980s to the early 2000s at plants in Peru, Mexico, Zambia, and Thailand. However, due to a range of different factors, these facilities have since closed.

Vanyukov Process

The Vanyukov furnace and process, a stationary submerged tuyere smelting process, so named after the inventor Andrey V. Vanyukov (1917–1986), was developed through laboratory and pilot tests in the period from the 1950s to the 1970s in the former USSR. The Vanyukov process employs a stationary, rectangular furnace with a series of tuyeres located near the hearth along the side walls blowing into the slag phase. The tuyeres can be closed rapidly, as required, by special plugs. The furnace can operate at high levels of oxygen enrichment (Mechev 1991; Engineering Dobersek 2017). The lower section of the furnace is refractory-lined while the upper wall sections are water-cooled, similar to that in a slag fuming furnace (Ashman et al. 2000). The Vanyukov process operates with filter cake concentrate and flux charged to the melt through ports in the furnace roof. The produced matte and slag are tapped periodically. Process gas high in SO_2 content leaves via the gas uptake and, after cooling and cleaning, is directed to the acid plant (Kellogg and Diaz 1992).

The first commercial Vanyukov plant for the treatment of nickel-copper concentrates was built in Norilsk in present-day Russia in 1977. Subsequently, three additional furnaces were built at Norilsk, with two Vanyukov furnaces built at the Balkhash smelter in present-day Kazakhstan. In 2017, a Vanyukov furnace was built at the Almalyk smelter in Uzbekistan to treat copper concentrates, replacing the former oxygen flash furnace (essentially a version of the Inco flash furnace). This unit is 2.5 m wide at the tuyere line and 9.6 m long and fitted with 14 tuyeres along each side (Engineering Dobersek 2017).

There are presently seven Vanyukov furnaces in operation: four at Norilsk, two at Balkhash, and one at Almalyk. Technical details on some of these units are given in Biswas and Davenport 1994 and Kellogg and Diaz 1992. A version of the Vanyukov furnace has been developed in China where several such plants now operate (Jiang et al. 2012; Yao and Jiang 2014).

Table 8 Technical data on two of the QSL lead smelting plants in operation

Company/Plant (reference)	Location	Year Commenced	Vessel Size (diameter × length)	Capacity, t/yr Pb
Berzelius Stolberg (Fischer 1983; Meurer and Ambroz 2010)	Duisburg, Germany	1990; subsequently, capacity was expanded with same vessel	3.5 m oxidation zone, 3 m reduction zone, 33 m long	150,000
Korea Zinc (Deininger et al. 1994)	Onsan, South Korea	1991	4.5 m oxidation zone, 4 m reduction zone, total 41 m long	60,000

Table 9 Smelters in China based on the SKS/BBS technology (2016)

No.	Plant	Company	Location (city, province)	Capacity, t/yr Cu	Type of Converter
1	Fangyuan I	Dongying Shandong	Dongying, Shandong	100,000	Peirce–Smith
2	Fangyuan II	Dongying Shandong	Dongying, Shandong	200,000	SKS converter
3	Yantai Humon	Shandong Humon Smelting Company	Yantai, Shandong	60,000	SKS converter
4	Huading	Huading Copper Development Company	Baotou, Inner Mongolia	100,000	Peirce–Smith
5	Hongtoushan	Liaoning Hongtoushan	Fushun, Liaoning	100,000	Peirce–Smith
6	Yuanqu	Zhongtiaoshan	Yuanqu, Shanxi	130,000	Peirce–Smith
7	Yuguang	Yuguang Gold and Lead Company	Jiyuan, Henan	100,000	SKS converter
8	Yimen	Yunnan Copper Company	Yimen, Yunnan	50,000	Peirce–Smith
9	Zhongyan Huangjin	China Gold Company	Sanmenxia, Henan	300,000	Flash converter
10	Qinghai	China Western Mining	Xining, Qinghai	100,000	Peirce–Smith
11	Lingbao	Lingbao Gold Company Ltd.	Lingbao, Henan	100,000	Peirce–Smith
12	Shuikoushan	Shuikoushan Nonferrous Metals Group Ltd.	Hengyang, Henan	120,000	SKS converter
13	Zhongtiaoshan	Zhongtiaoshan	Houma, Shanxi	200,000	SKS converter
Total capacity				**1,560,000**	

Adapted from Liang 2016

QSL Process

The QSL process is a continuous smelting process that was originally developed in 1973 by two distinguished professors in the United States (P. Queneau and R. Schuhmann), responding to concerns about lack of preparedness in the U.S. metallurgical industry of the challenges that they saw ahead: energy usage, environmental issues, and industry competitiveness (Queneau and Schuhmann 1974). Originally called the "QS" process, but, following a technology agreement concluded with Lurgi in Germany and Air Liquide in Canada and France, the letter L was added to the acronym, QSL.

Although developed generally for nonferrous metals, the focus has been mainly for lead smelting. The use of the patented Savard–Lee shrouded tuyere, which provides for refractory protection, was key to the QSL process development (Mackey and Brimacombe 1992). Following laboratory and pilot-plant test work in the late 1970s, a 10-t/h (lead concentrate) demonstration plant was built in 1981 at the Berzelius Stolberg lead smelter in Duisburg, Germany (Fischer 1983).

The QSL process features a long, horizontal cylindrical furnace with the capability to rotate through about 90 degrees; the pilot vessel at Duisburg was 2.5 m in diameter by 22 m long. The molten bath generally consists of lead bullion on the furnace hearth together with a layer of slag. Feed enters the vessel by means of a port in the upper barrel located at one end where feed is oxidized; lead bullion is tapped from this end. The QSL reactor has a partition wall to separate the slag into two zones such that the high lead slag generated at the feed (oxidizing) end flows toward the reduction zone where the lead content of the slag is reduced by reductant addition prior to slag tapping at the end opposite the feed/bullion tapping end (Fischer 1983).

Following successful demonstration of plant performance, a commercial-scale QSL plant with a design capacity of 100,000 t/yr of lead bullion and sized at 3.5 m diameter by 33 m long was built at Duisburg, commencing operations in 1990 (Berzelius Metall 2014). Since then, the capacity has been increased to 150,000 t/yr of lead bullion. In 1990, a new QSL plant with a capacity of 60,000 t/yr lead bullion (~170,000 t/yr feed or ~21 t/h feed) commenced operations at Korea Zinc in South Korea (Table 8.) A smaller QSL lead plant is apparently operating in Gansu, China.

In 1986, Cominco Ltd. in Trail, after evaluation of the Kivcet and QSL technologies, selected the latter to replace the aging Trail blast furnaces for environment improvements as required by law. This new plant, designed to treat about 1,000 t/d of lead concentrate and zinc plant residues (~42 t/h or four times the demonstration plant capacity) featured a 42-m-long reactor vessel, 4.5-m diameter in the oxidation zone and 4-m diameter in the reduction zone (Walker et al. 1990). The plant started up in 1990; however, serious difficulties soon surfaced regarding the smelting ability to handle the generally inert zinc plant residues in the feed and poor slag reduction behavior. Less than 12 months later, the plant closed, which, incidentally, also occurred during a period of poor metal prices. Cominco was forced to restart the blast furnaces amid a new review of suitable technology options whereupon the Kivcet technology was subsequently chosen.

SKS Process

The SKS/BBS process is a new method developed in China for lead and copper smelting with initial test work and piloting carried out from 1990 to 2000. The first commercial SKS plant for lead smelting was built in the late 1990s, followed by

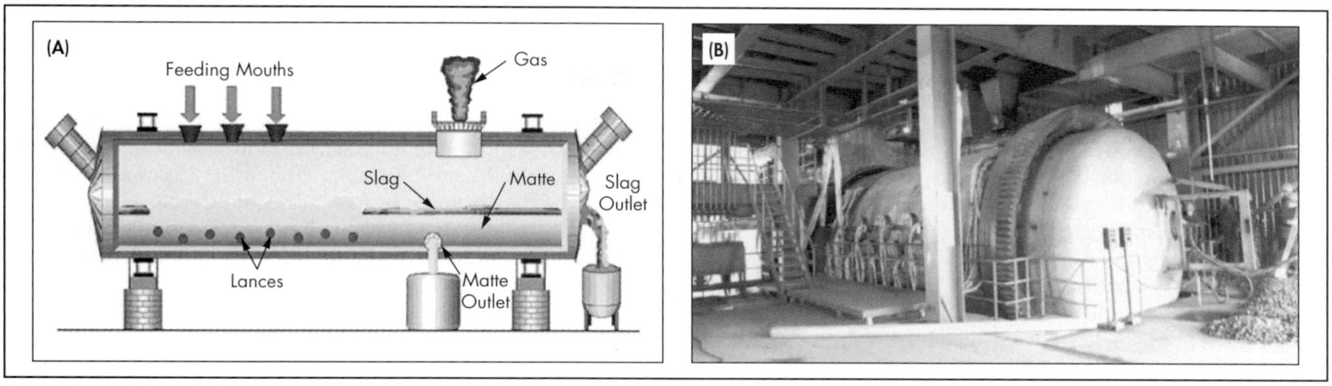

Source: Yao and Jiang 2014
Courtesy of Dongying Fangyuan

Figure 11 (A) SKS smelting vessel and (B) installation of the SKS vessel at the Dongying I smelter in China

a significant number of new lead smelters based on the technology. The first commercial plant for copper smelting with a capacity of 12,400 t Cu/yr was installed at Shenquan, Vietnam, in 2001 (Liet al. 1999; Jiang et al. 2012; Mackey 2013).

The process was conceived at the the Beijing Central Engineering Institute for Nonferrous Metallurgical Industries (now the China ENFI Engineering Corporation, or China ENFI) in Beijing, China, as an improved bath smelting technology and uses a horizontal cylindrical furnace for both lead and copper smelting. The vessel operates with high-pressure shrouded tuyeres akin to the Savard–Lee tuyeres developed in Canada in the 1960s (Mackey and Brimacombe 1992; Kapusta 2017) instead of the more conventional lower-pressure single-pipe tuyeres. The technology was then jointly developed through piloting by Shuikoushan Mining Corporation (Hunan Province, China) and ENFI, hence the acronym SKS. The SKS pilot vessel was sized 2.23 m in diameter by 7.98 m long, nominally rated at 50 t/d concentrate and tested over the period 1990 to 1993; the pilot total blowing rate was ~850 Nm^3/h of 50% O_2 air injected via four shrouded tuyeres.

About 2.3 Mt Pb was produced in China based on the new SKS technology in 2016, representing about half of China's total lead output, which corresponds to about 20% of world lead production (based on 2016 data). The description of the SKS reactor given in the paragraphs that follow for copper smelting is somewhat similar to that used for lead smelting. Additional information on the SKS technology for lead smelting is available in the paper by Gao et al. (2012) and the compilation by Siegmund et al. (2010).

The second commercial plant for copper with a capacity of 120,000 t/yr Cu commenced in 2008 at the new Dongying Fangyuan smelter in Dongying City, Shandong Province, China. A significant number of new copper smelting plants based on this technology have subsequently been built in China, as listed in Table 9. This indicates a total copper output capacity of 1.56 Mt/yr Cu using this process, corresponding to about 8% of world smelted copper (based on 2016 data). The Dongying Fangyuan smelter has continued development work independently and, as discussed in the following paragraphs, now refers to the smelting vessel as the submerged lance smelting (SLS) process.

In concept, the SKS/BBS copper smelting process is "similar to the Noranda furnace," as noted by Beijing General Research Institute of Mining and Metallurgy authors Jiang et al. (2012). The SKS/BBS process thus ingeniously incorporates different features of existing technologies: the Noranda process (Canada), the El Teniente process (Chile), the Savard–Lee concentric tuyere (Canada), and the QSL process (United States and Germany) (Mackey 2013).

The SKS/BBS vessel is a cylindrical, refractory-lined furnace, as illustrated in Figure 11. Three ports are located on the upper part of the shell toward one end to introduce feed to the furnace. There is also an opening on the upper part of the shell toward the other end for exhausting process off-gas. Provision is made for matte tapping from the barrel (as in the Noranda process) and slag tapping from an end plate at the same end as the off-gas mouth (also as on the Noranda process reactor and on the El Teniente converter). Provision is made for the use of high-pressure shrouded tuyeres located near the bottom of the vessel. Coursol et al. (2015) provide several illustrations of these tuyeres.

In operation, filter cake quality concentrate, slag concentrate, flux, and reverts are introduced to the vessel via the feed ports, while oxygen-enriched air is introduced at high pressure (at about 550 kPa) via the shrouded tuyeres. These tuyeres are arranged in two rows, a lower row aligned at 7 degrees from the bottom and an upper row aligned at 22 degrees from the bottom. Reactions in the bath between the injected oxygen and added concentrate, flux, and other feeds produce a high-grade matte, a fayalite slag, and process off-gas.

The off-gas is exhausted through the mouth and is cooled in a waste heat boiler, followed by cleaning in an electrostatic precipitator, and is then treated in an acid plant for sulfur recovery. The high-grade matte is tapped into a ladle and typically transferred to a Peirce–Smith converter for converting to blister copper, followed by anode refining and then electrorefining of the produced anodes. The tapped slag is slow-cooled in the same ladles used for slag tapping, then milled and subjected to flotation to recover the copper as a slag concentrate that is recycled to the smelting vessel.

In a recent Chinese development, a new continuous converter based on the use of the same shrouded tuyeres as the smelting vessel has commenced commercial operation in China. The unit is somewhat akin to the Noranda continuous converter in operation in Canada, but using the same type of oxygen concentric tuyeres as for the smelting vessel. The converting vessel is discussed in Chapter 11.6, "Conversion and Refining."

Process data for the large SLS vessel at Dongying II are given in Table 10. The nominal capacity of the SLS vessel at this plant is 1,500,000 t/yr of concentrate. This vessel is the smelting unit of the Fangyuan Dongying two-step process (described in Chapter 11.6).

Top Submerged Lance Smelting Furnace

Top submerged lance (TSL) smelting involves the continuous smelting of a wide range of nonferrous primary and secondary materials in a stationary, vertical, refractory-lined vessel with a lance inserted into the molten slag bath. The first technology patent was granted in 1981 after extensive research and development work performed at the Commonwealth Scientific and Industrial Research Organisation in Australia in the 1970s (Floyd 1981). Ausmelt (now owned by Outotec, one of the major suppliers of the technology today) has its early history linked to the original inventor. IsaSmelt (owned by Glencore, the other major supplier) has its history linked to the early industry support of the research and commercialization at Mount Isa, Australia, for lead and copper.

Between the early 1980s through 2015, Ausmelt has reported more than 60 commercial applications of their TSL furnaces for tin, lead, and copper smelting and zinc slag fuming (Outotec 2015). Their first major copper smelting application was at the Houma, China, smelter in 1999 (which also included a converting application); their largest to date is at the Daye, China, smelter, commissioned in 2010 (Table 11). Their largest tin and lead smelting furnaces have been supplied to the Gejiu, China, smelter in 2002 and 2010, respectively.

Table 10 Recent operating data for the SLS vessel of the two-step copper smelting process SLS furnace sized 5.5 m diameter × 28.8 m equipped with 23 tuyeres

Item	2017 Data
Concentrate rate, t/h	170
Total feed rate, t/h	210 (maximum reached ~240 t/h)
Coal rate, t/h	0
Furnace temperature, °C	1,180
Oxygen blowing rate, Nm^3/h	28,890
Air blowing rate, Nm^3/h	14,940
Total blowing rate, Nm^3/h	43,830
Oxygen enrichment of blowing air, % O_2	73
Matte grade, % Cu	73–78
Matte output, t/h	50–60
Dust produced, % of concentrate	<2
Iron-silica ratio in slag	1.6–1.8
Copper content of slag, % Cu	1.9–3.1
Sulfur capture, %	>99.5

Courtesy of Dongying Fangyuan

IsaSmelt, when it was owned by Mount Isa Mines, conducted pilot-scale work at their Mount Isa lead and copper smelter before commercializing the technology for lead smelting in 1991 (Alvear 2016). A copper smelting demonstration plant at the same facility in the late 1980s supported commercialization for copper smelting as well as sale of the technology to the Miami (Arizona, United States) copper smelter in 1992. The technology has subsequently been adopted by several smelters for lead and copper smelting (IsaSmelt 2015). Their largest copper smelting furnaces were supplied to the Tuticorin (India), Ilo (Peru), and Kansanshi (Zambia) smelters in 2005, 2007, and 2015, respectively. Their largest lead smelting furnace was supplied to the Ust-Kamenogorsk (Kazakhstan) copper and lead smelter.

Flow Sheet and Process Description

The Kansanshi, Zambia, smelter is shown in Figure 12 as a typical recent TSL smelting flow sheet (de Vries et al. 2016) and is discussed along with the Daye, China, smelter (Wan et al. 2013) and other facility features (Alvear 2016). Concentrates received from the nearby Kansanshi mine are filtered and bed blended with concentrates from the more distant Sentinel mine, also in Zambia. The blended concentrate is reclaimed and mixed with silica and lime fluxes and recycled materials in a paddle mixer with water added to adjust total feed moisture to about 10%; this is common practice for more recent TSL smelters; pelletizers performed this step in earlier smelters. De Vries et al. (2016) state that because of the relatively high carbonaceous content of the Kansanshi smelter feed, a large amount of crushed slag combined with revert material must be recycled, substantially more proportionally than other large TSL smelters. The feed is conveyed to a feed port at the furnace roof, which then drops into the bath where it is smelted in a bath agitated by the submerged lance.

Off-gas from the furnace is cooled by boiler tube panels in the roof and passed into the radiative waste heat boiler, then cooled using evaporative sprays prior to entry into the electrostatic precipitator. Some facilities use an additional convective waste heat boiler (Köster and Martinez 2016), and some older facilities use a "flux flow" boiler. Air is added via the feed port and downstream in the radiative boiler to facilitate post-combustion and ensure satisfactory sulfation of furnace dust. Total dust collected in the boiler and precipitator is typically less than 1% of the feed and is recycled back to the feed paddle mixer.

The furnace is tapped on a batch basis via launders into an electric furnace where the matte and slag separate. Other facilities such as the Mount Isa, Tuticorin, and Ilo smelters use rotary holding furnaces for this duty (Alvear 2016). An underflow weir is used at the Daye smelter that allows continuous

Table 11 Physical parameters of TSL smelting furnaces

Item	Ausmelt (Gejiu, China)	IsaSmelt (Ust-Kamenogorsk, Kazakhstan)	Ausmelt (Daye, China)	IsaSmelt (Kansanshi, Zambia)
Year started	2010	2012	2010	2015
Height (inside dimensions), m	17	Not available	16.5	15.7
Diameter (inside dimensions), m	5	3.6	5	5.5
Lance diameter (inside dimensions), m	0.35	0.2	0.5	0.3 and 0.45
Nominal concentrate throughput	520 t/d Pb concentrate	1,200 t/d Pb concentrate and other materials	4,100 t/d Cu concentrate	3,300 t/d Cu concentrate

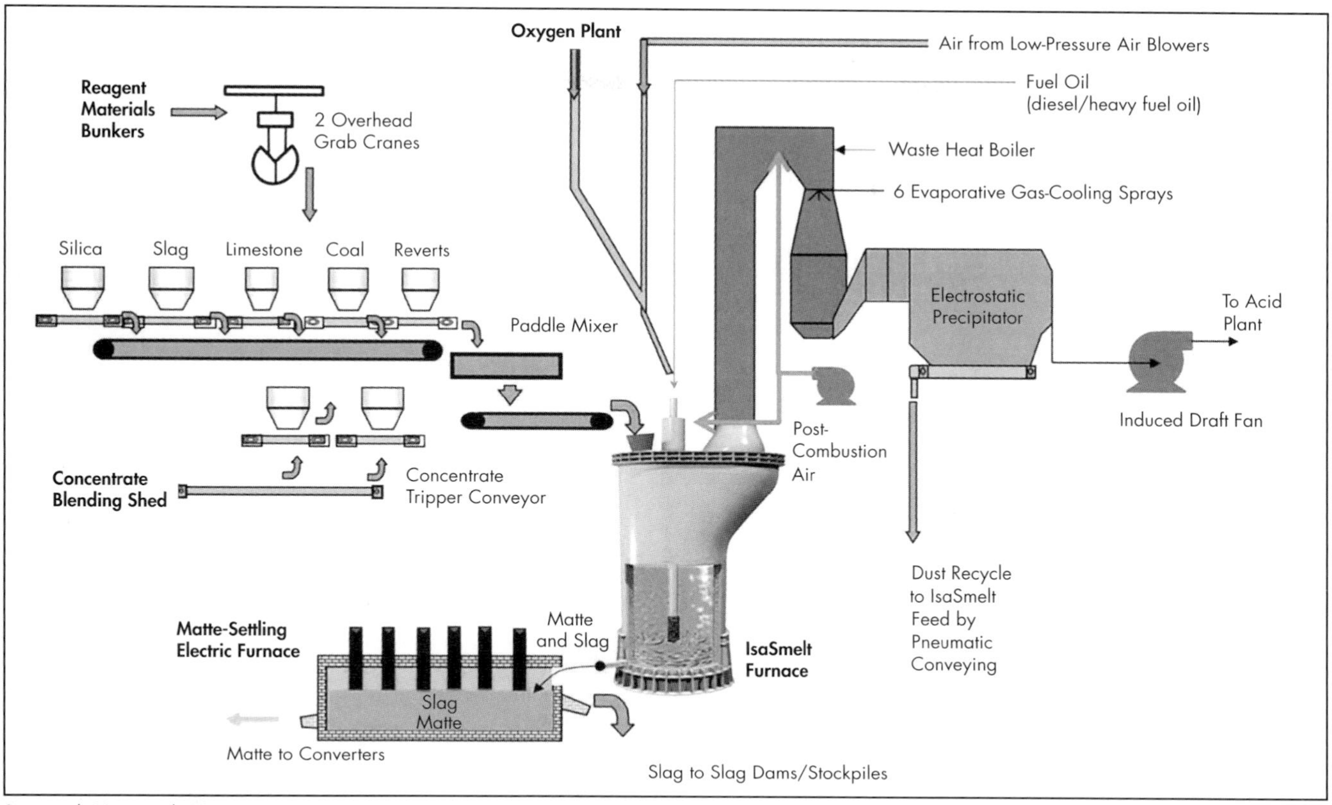

Source: de Vries et al. 2016

Figure 12 IsaSmelt plant at Kansanshi, Zambia

flow of furnace matte and slag to the electric furnace (Wan et al. 2013); it is claimed that this has resulted in more-stable furnace operation. Additional data are given in Table 11.

TSL furnaces can be designed with a traditional refractory-lined system, with or without exterior water cooling of the steel furnace shell or, alternatively, may incorporate water-cooled copper cooling elements behind the working refractory lining (such as integrated at the Miami smelter). Alvear 2016 reports that the Mount Isa and Tuticorin smelters have achieved four-year campaigns, with the Ilo smelter planning to achieve three-year campaigns. All of these smelters, including the Kansanshi smelter, use a traditional refractory-lined system. Time between repair or replacement of the lance can be as long as 30 days, but some smelters have reported times as short as 2 days.

De Vries et al. (2016) describe the use of accelerometers mounted at locations on the TSL furnace to detect abnormal turbulent conditions or to detect when an accretion has fallen into the bath. They also describe the use of these devices mounted on the lance carriage to monitor lance–bath interaction.

Top Lance Furnace (Mitsubishi Process)

In Japan in the 1950s–1960s, the post-war demand for metals such as copper and nickel was increasing. With a new awareness of the need for pollution control, Mitsubishi Metals Corporation, who operated the Naoshima smelter, realized that the reverberatory furnaces then in use not only consumed vast quantities of expensive fuel oil, but the older technology could not cope with new requirements for environmental control. The company subsequently embarked on a quest for a significantly improved copper smelting process.

The first commercial plant, called the Mitsubishi process, was designed for a capacity of 600 t/d concentrate, or 48,000 t/yr Cu, and began in 1974 at the Naoshima smelter in Japan (Nagano and Suzuki 1976). This followed several years of development work, including piloting on a 230-t/d concentrate semicommercial plant at the Onahama smelter (MMC 2012; Nagano and Suzuki 1976; Schlesinger et al. 2011).

Schlesinger et al. (2011) provide a flow sheet of the Mitsubishi process. Pre-dried concentrate stored in feed bins is pneumatically injected through several submerged lances in the circular smelting furnace. Matte at between 65% and 70% Cu content flows continuously by launder to the circular converting furnace and is converted by oxygen-enriched air injected by a number of submerged lances, along with the production of a calcium ferrite slag using limestone as the flux (see Chapter 11.6, "Conversion and Refining").

Smelting slag is cleaned of its copper content in the electric slag cleaning furnace, while solidified converter slag is recycled to the smelting furnace. Furnace off-gases are first cooled in separate waste heat boilers and, after dust removal, the gases are directed to the acid plant. The arrangement of the stationary furnaces provides for good fugitive gas collection, while the high SO_2-strength process gas allows for a high level of sulfur dioxide fixation as sulfuric acid.

In 1981, a second commercial Mitsubishi process plant commenced operations at Kidd Creek, Ontario, Canada

Table 12 Technical data on operating Mitsubishi process facilities

Plant (Country)	Company	Year of Start-up	Capacity, t/yr Cu	Features of Smelting Furnace
Smelters Based on Mitsubishi Process				
Naoshima (Japan)	Mitsubishi Materials	1974 (original plant; now closed) 1991 (expanded plant)	48,000 270,000	7 m × 10.5 m (oval) 10 m (circular) Lance rate: 33,000 Nm^3/h, 56.1% O_2
Onsan (South Korea)	LS-Nikko	1998	270,000	10 m (circular) Lance rate: 38,000 Nm^3/h, 55% O_2
Gresik (Indonesia)	PT Smelting	1998	300,000	10 m (circular) Lance rate: 37,030 Nm^3/h, 53.2% O_2
Dahej (India)	Birla	2005	250,000	10 m (circular) Lance rate: 38,600 Nm^3/h, 56% O_2
Smelter Using the Mitsubishi Smelting Furnace				
Onahama (Japan)	Mitsubishi Materials	2007	200,000	10 m (circular) Lance rate: 31,300 Nm^3/h, 56% O_2
Total capacity			**1,290,000**	

Source: Schlesinger et al. 2011; MMC 2004, 2012
Note: The total copper capacity by the current Mitsubishi process plants worldwide (1.29 Mt/yr Cu) represents about 7% of world smelter capacity in 2016.

(closed in 2010). Over the years, apart from the Naoshima plant, four additional Mitsubishi process plants were built. In addition, a stand-alone Mitsubishi smelting furnace was installed at Onahama to replace a reverberatory furnace for concentrate smelting, with the latter furnace then acting as a slag cleaning vessel/holding furnace (MMC 2012; Prayoga and Sundana 2016; Taniguchi et al. 2006). Thus, currently in operation, there are four Mitsubishi process smelters and one plant using a single stand-alone Mitsubishi smelting furnace as the main smelting unit.

Table 12 provides technical data on the current Mitsubishi process plants. Information by Newman et al. (1987) provide operating data and furnace heat and mass balance data. The following references include an extensive reference list covering many aspects of the Mitsubishi process and its operations: Newman et al. 1987; Shibasaki et al. 1993; MMC 2004; Taniguchi et al. 2006; and Mizuta et al. 2012.

The Mitsubishi process, given its well-stirred bath achieved by multi-lance gas injection, has excellent abilities to treat electronic scrap (e-scrap) and other scrap materials. The furnaces at Naoshima have been reconfigured to handle a range of scrap materials including e-scrap, spent anodes, pressed copper scrap, and other precious metal scrap materials (Mizuta et al. 2012). A new rotary kiln sized 5 m in diameter and 14 m long plus associated gas and liquid effluent handling facilities was also installed at Naoshima to pretreat combustible scrap materials, including e-scrap. By 2016, the Mitsubishi process at Naoshima was treating about 80,000 t/yr of e-scrap and other materials. Of interest, the plant at Onahama was handling ~60,000 t/yr of e-scrap materials (Mizuta et al. 2012).

The Mitsubishi process was a significant breakthrough in copper smelting technology when the commercial plant was installed at Naoshima in 1974. It remains a major technology today.

Smelting in the Electric Furnace

The principle of heating a substance by direct current (DC) electricity evolved soon after the discovery of the electric battery reported by Alessandro Volta in 1800. The heating and smelting of mineral ores on a significant scale required more power than batteries could realistically provide; hence, it would wait many decades—until the development of the dynamo in 1867—before electric smelting of ore, initially iron ore on a small scale, was found to be feasible (Borchers 1897; Stansfield 1914). Notable too were the tests by Hall (United States) and Héroult (France) independently conducted in 1886, leading to the use of electricity for producing aluminum by electrolysis of a fused salt. The Hall–Héroult process remains the main method of producing primary aluminum today. Other applications of fused salt electrolysis soon followed. Héroult was also one of the first pioneers who applied the electric furnace for smelting. This work led to tests on the use of the electric furnace for the smelting of a range of materials, including sulfide concentrates, iron ore (an oxidic ore), and a range of other oxidic ores under reducing smelting conditions, leading to the production of a range of ferroalloys (such as ferrochromium, ferromanganese, ferronickel, and ferrosilicon, etc.). Gradually, the alternating current (AC) configuration evolved for electric furnace smelting.

The evolution of the electric furnace for the smelting of Cu-Ni sulfide concentrates is briefly described to illustrate these developments. Following test work in 1912 on a modified Héroult furnace with sizes ranging from 200 kW to 900 kW, the Sulitjelma Copper Company of Norway subsequently built a 3,000-kW rectangular furnace for smelting copper and nickel concentrates (Witherell and Skougor 1922). A 9,000-kVA electric furnace for copper smelting was built by Outokumpu at Imatra in southern Finland in 1928 to treat concentrates from the Outokumpu mine in northern Finland. This was the largest electric furnace based on cheap hydropower for smelting at the time. Later in the 1930s, two 12,000-kVA furnaces were built by Inco of Canada at Petsamo in northern Finland for the smelting of nickel concentrates. Inco later built three 18,000-kVA furnaces to smelt nickel concentrates at Thompson, Manitoba, Canada, in 1960. Barth (1960) established the principles for modern electric furnace smelting based on AC configuration for smelting of sulfide concentrates or oxide ores. Large DC furnaces are still used today (Geldenhuys 2013).

During the 1970s, three large furnaces were commissioned for copper smelting: Mufulira in Zambia at 30,000 kVA; and Anaconda at 30,000 kVA and Inspiration at 50,000 kVA, both in the United States. Interestingly, the Mufulira and Inspiration furnaces were later replaced by oxygen smelting, while the

Table 13 Main categories of AC electric furnace operating modes for the smelting of a range of materials

No.	Category	Examples	Typical Temperatures	Reference
1	Slag resistance (Figure 13A)	Ni-Cu concentrate smelting	Matte: 1,250°C; slag: 1,350°C	McKague and Norman 1984
		PGM (platinum group metal) concentrate smelting (normal PGM concentrate)	Matte: 1,200°C; slag: 1,360°C	Mostert and Roberts 1973
		PGM concentrate smelting—high $Cr_2O_3{\cdot}MgO$	Matte: 1,500°C; slag: 1,650°C	Hundermark et al. 2006
2	Submerged or shielded arc (Figure 13B)	Ferronickel laterite smelting	Metal: 1,450°C; slag: 1,550°C	Voermann et al. 2004, Warner et al. 2006
		Ferrochrome chrome ore smelting	Metal: 1,700°C; slag: 1,800°C	Geldenhuys 2013
3	Open arc	Iron- and steelmaking	Metal: 1,500°C; slag: 1,550°C	Fruehan et al. 1998

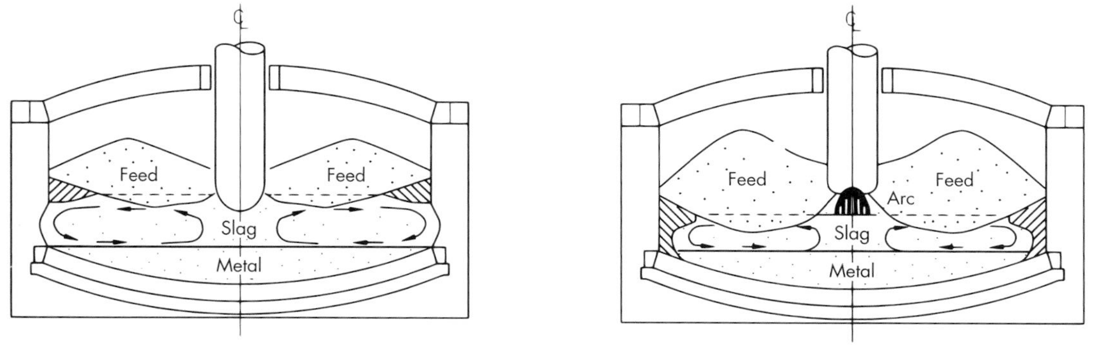

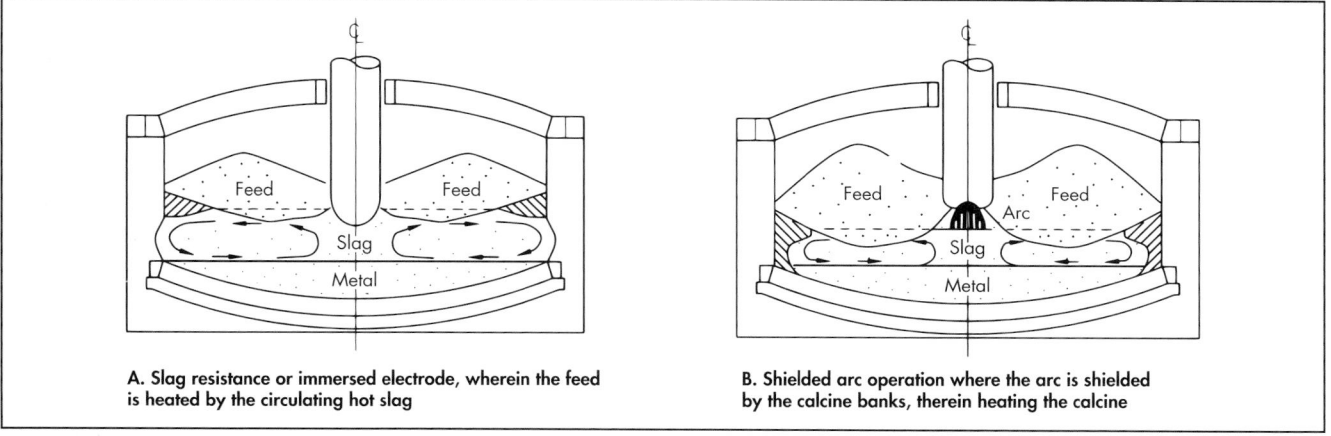

A. Slag resistance or immersed electrode, wherein the feed is heated by the circulating hot slag

B. Shielded arc operation where the arc is shielded by the calcine banks, therein heating the calcine

Adapted from Voermann et al. 2004

Figure 13 Two modes of electric furnace operation

Anaconda plant closed down. Electric furnace smelting of nickel-copper concentrates was developed by the former Falconbridge (now Glencore) in the 1970s in Canada and also in Rustenburg, South Africa; these plants remain in operation today (Senne 1977 [Anaconda]; Herneryd et al. 1954 [Boliden]; Biswas and Davenport 1976 [Inspiration and others]; Mostert and Roberts 1973 [Rustenburg]; Blair et al. 1976 [Mufulira]; McKague et al. 1984 [Falconbridge]; Crundwell et al. 2011 [several plants]).

Electric furnace smelting of sulfide concentrates largely involves heating and melting with separation of the sulfide minerals to form a matte phase and the gangue minerals forming a slag. On the other hand, the reduction smelting of oxidic ores (after first calcining or pelletizing/sintering) involves both smelting plus significant chemical reduction of the feed by added coal, which, along with iron reduction, forms a ferroalloy.

The first large-scale electric furnace for the smelting of calcined nickel laterites began with the construction of four 12,500-kVA electric furnaces in New Caledonia in the 1950s (Thurneyssen et al. 1960). The production of a range of ferroalloys by electric smelting followed a similar pattern. Today, such furnaces operate at nearly ten times this capacity, to more than 100 MVA.

The smelting characteristics of the modern electric furnace treating a range of materials from sulfidic concentrates, oxidic ores fed as calcined or sintered feed, or metallic scrap, such as iron scrap, can be considered in three main categories, listed in Table 13 and illustrated in Figure 13. Plasma furnace technology, which utilizes highly charged gases in the electric arc to boost temperature, is briefly covered later in the "Other Types of Furnaces" section.

Normally, an electric furnace operating in the slag resistance mode illustrated in Figure 13A operates continuously. The concentrate feed as either a bone-dry or calcined material is introduced continuously; matte as the typical metallic product and slag are tapped periodically. The small stream of furnace off-gas is continuously withdrawn and cleaned in gas handling facilities. For the treatment of Ni-Cu or platinum group metal (PGM) concentrates, the power requirements are typically of the order of 450–680 kW·h/t of charge, the actual value in part depending on the particular operating temperature, which in turn is generally governed by the slag composition—the higher $Cr_2O_3{\cdot}MgO$ PGM materials operate at a higher temperature and require more power, or in some cases, temperature may be set by the grade of matte or metal alloy.

The submerged or shielded arc furnace illustrated in Figure 13B is typically used for the production of a range of ferroalloys by reduction smelting of pre-reduced and/or calcined ore (Habashi 1997). The process is continuous with the charge material normally added hot as a result of the pre-reduction or calcining stage. The average power consumption for the production of ferronickel is approximately 450–500 kW·h/t of calcine. Lateritic calcine is fairly low grade at about 2%–2.5% Ni; expressed on a per-metric-ton basis of ferronickel metal product, the unit power consumption would accordingly be much higher (Warner et al. 2006). The six main ferroalloys produced in this type of electric furnace include ferrochrome (45%–75% Cr), ferromanganese (75%–90% Mn), ferronickel (20%–40% Ni), ferroniobium (55%–65% Nb), ferrosilicon (30%–60% Si), and ferrovanadium (35%–75% V). Fe-Nb is also produced by alumino-thermic smelting.

Ferrochrome furnaces operate at higher temperatures, and the unit power consumption, taking into account the higher grade of feed (35%–40% Cr_2O_3)—relative to that for ferronickel production—is therefore a higher value. Typically, the power consumption for chrome production is approximately 1,500 kW·h/t of moderate-grade chrome feed, or roughly 3,500–4,000 kW·h/t of ferrochrome product (55%–60% Cr).

The open arc operation used for smelting scrap iron and scrap ferroalloys—such as stainless steel, for example—is typically operated batch-wise (Fruehan et al. 1998). In the case of iron- and steelmaking, the capacity of a modern electric arc furnace can range from 100 t to more than 400 t per charge (or heat). A typical cycle involves charging, melting, refining/slag removal, and tapping the final charge.

The time per charge can range from 45 to 60 minutes, depending on the type of charge. The power requirement for scrap steel smelting typically ranges from 440 kW·h/t of charge to 540 kW·h/t of charge. About one quarter of the world's steel is now produced by electric arc smelting.

Electric Furnace Smelting of Ilmenite

As indicated in Table 1, titanium mainly occurs as the minerals rutile and ilmenite. Several pyrometallurgical steps are involved in the processing of these minerals to produce titanium metal or pure TiO_2; however, a smelting operation is only carried out on ilmenite ores, as briefly described in the following paragraph.

Beneficiated ilmenite ore, such as mined at the Allard Lake operation in Quebec, Canada, and containing some 30% Ti and 36% Fe, is smelted in an electric furnace at temperatures higher than 1,650°C to produce a titanium-rich slag (40%–50% Ti or 70%–80% TiO_2) and high-quality pig iron. After tapping and cooling, the slag is processed by leaching techniques to produce high-grade TiO_2, which is used as a pigment. The pig iron is used to produce iron products.

Rutile can be processed to produce either (1) metal by the Kroll process involving high-temperature reduction in the presence of chlorine to generate $TiCl_4$, which is then reduced at elevated temperatures to Ti metal by magnesium; or (2) pigment-grade TiO_2 by leaching operations.

Electrolytic Reduction

The Hall–Héroult cell consists of a refractory-lined, rectangular steel cell filled with molten cryolite (Na_3AlF_6 at ~950°–960°C) and fitted with a carbon liner on the bottom acting as a cathode, with several carbon blocks suspended at the top of the melt acting as the anode (Figure 14). Aluminum oxide (alumina) produced from refining bauxite (Table 1) is periodically added to the cell and dissolves in the molten cryolite. By the action of the current, the bath is electrolytically reduced to aluminum metal that settles on the carbon liner and is periodically tapped out. The tapped metal is transferred to a holding furnace for casting.

A typical large, modern Hall–Héroult cell, about 3 m by 10 m, operates with a current ranging from 100 kA to 200 kA at 4–4.5 V. A cell operating at 200 kA would produce about 65 kg/h. A 100,000-t/yr plant would therefore require about 200 such cells having a typical electrical consumption of 13 kW·h/kg Al. The Hall–Héroult cell is used for the production of most of the aluminum produced worldwide (USGS 2017). A typical large plant may consist of 1,000 cells.

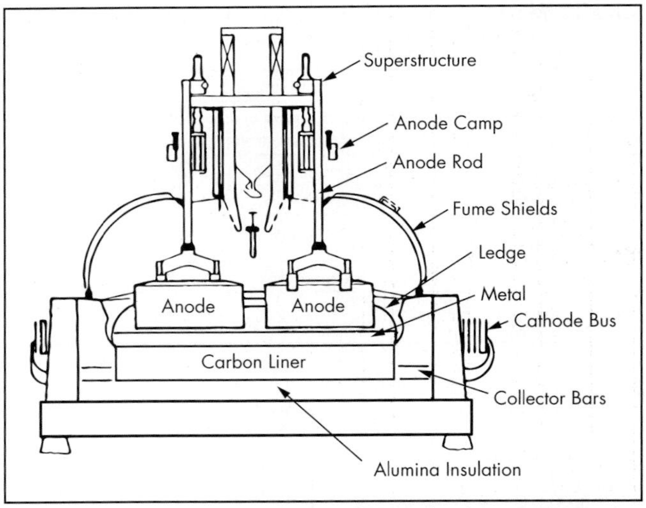

Source: Banerjee and Evans 1990. Copyright 1990 by The Minerals, Metals & Materials Society. Used with permission.

Figure 14 Hall–Héroult cell

OTHER TYPES OF FURNACES

A safe and effective furnace for smelting requires the following key features: an arrangement for proper feeding of the charge material to the hot zone of the furnace, an ability to provide good heat transfer to effect smelting, and, finally, allowance for discharge of the products.

A well-developed process control system is also required, and the furnace design and furnace life need to be adequate for the duty required. A wide range of other furnace types and configurations, in addition to those discussed in earlier sections of this chapter, have been developed by the industry to provide these conditions for a range of diverse feed materials for small or large throughputs and for either batch or continuous operation. This section briefly lists and summarizes these other furnaces used for smelting.

After their useful lives, metals have always been recycled as scrap. In the 1800s, the main use of copper was as a protective sheathing material on sailing ships, and worn or used sheathing was recycled by melting in small reverberatory furnaces. Today, the recycling rate of metals such as aluminum, copper, nickel, chromium (the latter two mainly as stainless steel), cobalt, lead, iron, and steel can range from 30% to 50%. Copper, aluminum, and lead are typically recycled using specialized reverberatory or rotary furnaces. The various types of furnace used for smelting and scrap recycling are also included here in abbreviated form.

Short Rotary Furnace

Short rotary furnaces are very flexible and widely used. Applications cover a wide range of smelting duties, such as metallic scrap materials, intermediates, and residues. These furnaces are commonly used for handling lead residues and similar materials (Forrest and Wilson 1990; Schlesinger et al. 2011).

Major features of a short rotary furnace include the following:

- A cylindrical drum furnace typically fitted with mechanism and rollers to rotate slowly (Boeckenhauer and Kaczmarczyk 2018).
- A version of the rotary furnace that does not rotate but can tilt is also common for smelting scrap aluminum, copper, and other materials.
- Includes fuel/air/oxygen-enriched arc burner for heating or smelting; this is normally a batch operation.
- The length/diameter ratio typically ranges from ~1.2 to 1.5.
- A 3-m-diameter furnace may handle a charge up to 12 t; a 4-m-diameter furnace can handle 20–35 t.

A new version of the rotary furnace has been developed in China that employs submerged high-pressure tuyeres to aid smelting and refining.

Plasma Furnace

A plasma furnace is a type of high-temperature electric furnace for smelting a wide range of complex materials. The DC arc is generated to provide high temperatures to maximize destruction of hazardous components in scrap and waste materials. Capacities are generally lower relative to the large AC or DC arc furnaces for ore smelting or iron and steel production. Applications include the treatment of e-scrap, precious metal materials, spent pot line materials (aluminum), and other waste materials (Tetronics International 2018).

Ladle Furnace

A ladle furnace is a refractory-lined, vertical furnace resembling a ladle operating in dual function as both a ladle transfer vessel and a refining or smelting unit, the latter carried out at a ladle station. The ladle station includes a set of furnace electrodes that can be lowered for providing heating (for refining and/or smelting). Ladle units range from 30 t (~3 m in diameter, 6 MW) used in ferroalloy production to more than 200 t used in the steel industry. Ladle furnaces are used for a range of metallurgical refining operations (examples are provided in Zamalloa et al. 2009).

Aluminothermic Furnace and Process

An aluminothermic furnace is a type of furnace arrangement that provides smelting heat by a unique energetic chemical reaction involving aluminum metal, iron oxide, and the oxidic ore or mineral of the metal to be produced as a ferroalloy. A mixture of the oxide of the metal to be produced (this includes iron oxide plus, for example, an ore such as one containing chromic oxide, manganese oxide, or niobium oxide) along with aluminum pellets are placed in a suitably sized refractory-lined ladle, a cover is fitted, and the charge is ignited by an electrical spark or magnesium wire.

As shown in Figure 15, the smelting reaction is proceeding in the refractory-lined ladle positioned inside the hooded area. An energetic chemical reaction occurs, with the temperature temporarily rising to higher than 2,000°C, thus reducing the metal oxide and the iron to form a ferroalloy that sinks to the bottom of the crucible and a molten slag formed from gangue components in the ore. The contents of the crucible are allowed to cool and are then removed.

The following equations describe the main reactions (M is the metal to be produced and added as an oxide ore):

$$Fe_2O_3 + 2Al = Al_2O_3 + 2Fe$$

Courtesy of Niobec

Figure 15 Aluminothermic furnace and arrangement for smelting to ferroalloy

$$3MO + 2Al = Al_2O_3 + 3M$$

Formerly used for the production of ferroalloys such as ferrochrome, ferromanganese, and tungsten, the aluminothermic furnace has now been replaced by electric furnace smelting, but it is still used to produce ferroniobium (De Fuccio et al. 1992).

Reverberatory Furnace for Primary and Scrap Metal Smelting

Reverberatory furnaces are used as a flexible furnace for batch processing a range of scrap metals and for melting alloys such as brass, bronzes, and so forth. Sizes range from 50 t to more than 300 t. A reverberatory furnace uses an oxy-fuel burner to melt the charge and maintain temperature during refining and casting. A typical size for a 200–250-t charge is 5 m wide by 8.5 m long. However, the length/width ratio can range from 1.5 to 2.5 (Figure 16).

Typical fuel consumption (assuming natural gas) for the preceding size is approximately 0.06 Nm^3 of natural gas per kilogram of metal (assuming copper and employing a bath temperature of ~1,200°C; lower temperatures and less fuel are used for handling aluminum scrap, ~700°C) or lead scrap smelting (as used to treat lead battery scrap); however, temperatures may be ~1,200°C to allow for a liquid slag. Off-gases are typically used to generate steam.

OTHER ISSUES RELATING TO SMELTING

The technologies described in this chapter require careful consideration during design and operation to ensure workplace safety and hygiene, and to achieve compliant environmental impact. Sustainable deployment of these technologies is essentially underpinned by success in these areas. Sustainability also depends on how significant the technology costs are to the businesses utilizing them. Energy efficiency can influence environmental impact as well as costs. All of these issues are expanded on in this section.

Safety

In addition to general heavy-industry workforce safety risks, one must consider manual handling ergonomics, thermal exposure, and acute exposure to gases such as sulfur dioxide and carbon monoxide. The National Institute for Occupational

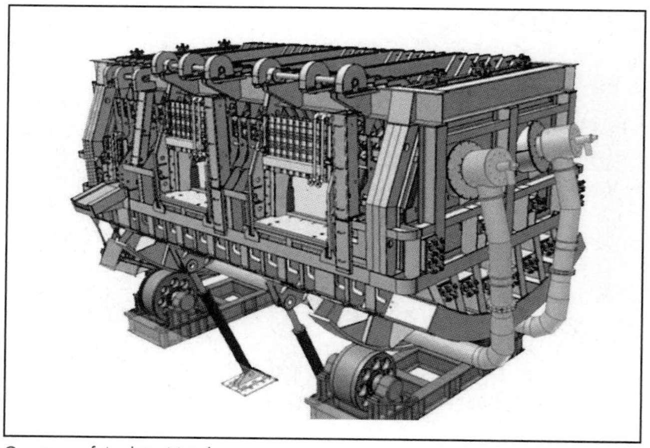

Courtesy of Andritz Metals

Figure 16 Modern reverberatory furnace (furnace can be tilted, as shown, or fixed)

Safety and Health (NIOSH) has established exposure concentrations that it considers to be "immediately dangerous to life and health," or IDLH. NIOSH assigns IDLH values of 100 ppm and 1,200 ppm to sulfur dioxide and carbon monoxide, respectively (NIOSH 2007).

Materials handling of molten products of smelting generally involve tapping, laundering, and associated maintenance activities. Workers can be exposed to substantial ergonomic risks when removing launder buildup using unsophisticated hand tools such as crowbars and jackhammers. This risk is often mitigated through considered selection of launder materials that allow easy release of buildup and avoidance of drop and splash points. Launder covers combined with heating (burners or, less commonly, electric heating) are also used for some molten materials or long launder lengths. Other worker activities may involve periodic cleaning of portals for feeding materials, sampling, or accommodating burners. Depending on material characteristics, frequency, and general layout, remotely controlled equipment is often used to mitigate worker manual-handling activities.

Acute exposure to high-temperature gases and also hazardous gases is possible during worker activities such as tapping and portal cleaning. Generally these risks are substantially mitigated through ergonomic design, furnace draft control, and administrative procedures. For tasks where the possibility of exposure remains significant, additional personal protective equipment is used (such as aluminized coats and self-contained breathing apparatuses) to significantly mitigate or eliminate consequences.

The general trend in the smelting industry today is to continually try to eliminate tasks that place a worker in an area with these potential acute exposures. Improvements in ancillary equipment design, process control, and the use of remote/automated equipment are contributing to progress in this area.

Hygiene

The Occupational Safety and Health Administration (OSHA) sets permissible exposure limits (PELs) for regulated substances. Time-weighted average (TWA) concentrations for OSHA PELs should not be exceeded over a worker's 8-hour shift. Tapping, laundering, and other molten materials handling activities can result in nominally elevated workplace levels of gases such as sulfur dioxide (PEL TWA: 5 ppm), and fumes such as cadmium (PEL TWA: 0.005 mg/m^3), arsenic (PEL TWA: 0.010 mg/m^3), lead (PEL TWA: 0.050 mg/m^3), and copper (PEL TWA: 1 mg/m^3) (NIOSH 2007).

Capture of point-source secondary emissions from tapholes, launders, and stationary ladles is generally achieved through the use of hoods and covers ventilated via ductwork to a central gas cleaning system before discharge to the atmosphere. Typical volumes range from 5,000 to 15,000 Nm^3/h per taphole, depending on hood efficiency. Open-area emissions are minimized to keep gas volumes manageable and still achieve reasonable hood velocities (which dictate hood capture efficiency). Some technologies require large areas of capture, such as furnace roofs—volumes range from 15,000 to 50,000 Nm^3/h. Even when excluding fugitive emissions capture requirements, secondary emission gas-ventilation volumes can be more than 500,000 Nm^3/h.

For workplace areas where exposure is nominally above permissible levels, additional personal protective equipment is used to ensure that worker biological exposure remains within limits. A large range of cartridge-based respirators offer protection from gases and fumes. For some contaminants, government regulators require engineering plans to ensure that the best available technology is properly deployed to maintain workplace levels below exposure limits.

Environment

Smelting facilities can impact the environment through air and water emissions. Development and adoption of the best available technology has been motivated by both economic and regulatory forces to minimize loss of value and environmental impacts.

Furnace off-gases are cooled, de-dusted, and sent to sulfuric acid recovery prior to discharge. A material fraction of sulfur dioxide, 1%–3% depending on technology, is recovered as weak sulfuric acid that must be neutralized. Additional capture is required for some smelting technologies to manage significant fugitive emissions, and this gas can be combined with either secondary emission hygiene gas for de-dusting and scrubbing or the primary off-gas prior to sulfuric acid recovery.

Emissions of sulfur dioxide and heavy metals from point sources such as main stacks and individual baghouses/scrubbers are typically subject to robust and well-defined local regulatory limits. Additionally, semiquantitative visible opacity observations can be required by regulators to report transient diffuse emissions above facility buildings.

The largest effluent stream associated with smelting technologies is usually the off-gas weak sulfuric acid stream. This stream is often amenable to by-product recovery such as indium and rhenium. Neutralization to pH 2 produces relatively clean gypsum, which can be sold in some locations. A nonhazardous effluent is achieved by raising the pH to 9, which can then be recycled for internal use (such as scrubber makeup water or evaporative cooling applications) or further treated if necessary along with other effluent streams to make compliant with disposal limits. The recovered heavy metal sludge is recycled, sold, or impounded.

Energy

Smelting facilities consume energy in the form of electric power and direct-fired fossil fuel. Power is utilized to drive process equipment, with a large fraction typically used to generate tonnage oxygen. Fuel is used to supplement the pyrometallurgical processes, launder heating, and steam generation/superheating. Tables 14 and 15 show the distribution between these major forms of energy consumed by an average global copper smelter and the Salt Lake City smelter.

The energy consumption numbers calculated by Coursol et al. (2010) correspond to a carbon footprint of 0.63 t CO_{2eq} per metric ton of copper in anode (assuming natural gas as the fossil fuel). Grimes et al. (2008) discuss copper primary smelting energy consumption in the context of comparison with secondary production. They report a carbon footprint of 1.02 t CO_{2eq} per metric ton of copper in anode (higher presumably because they assume a weighted mix of natural gas, oil, and coal for power generation). Their figure increases after electrorefining to 1.25/t of copper cathode.

Nilsson et al. (2017) report total carbon footprint assessments with a range from 2 to 4 t CO_{2eq} per metric ton of mined copper, which provides an indication of the copper smelting contribution to overall primary copper carbon footprint. Grimes et al. (2008) report the following carbon footprints for smelting to product from raw material (in metric tons of CO_{2eq} per metric ton of product): 1.97 for steel, 1.63 for lead, and 8.5 for nickel.

Energy efficiency in smelting is primarily motivated by the local cost of power and fuel. Carbon footprint reduction is influenced by promulgation of government policy or market-priced carbon credits. Both are complementary but constrained by process complexity and cost of capital. Most smelting technologies have achieved higher energy efficiency compared to that in the past through the use of oxygen generated cryogenically or with pressure swing absorption. This has significantly increased capital efficiency and reduced the fuel requirement (see Figure 8 and its associated discussion). The cost of power required to produce the oxygen is more than offset by the reduction in unit consumption of fuel. High-pressure saturated steam (6–7 MPa) via waste heat boilers is typically produced as part of furnace off-gas cooling, and this is usually used to generate power by means of a back-pressure turbine generator (5%–6% thermal efficiency) followed by a condensing turbine generator (15%–17% thermal efficiency) or more efficiently applied in condensing thermal applications such as steam coil dryers, facility (or local domestic) heating, and so on. For facilities with sulfuric acid production, some low-pressure steam (1–2 MPa) can be usefully generated using economizers ahead of absorption or indirectly through exchange with the converter beds. Typically, most of the energy liberated during acid absorption is transferred to cooling towers as waste heat. The copper smelters at Sanmenxia, China, and Salt Lake City both operate heat recovery systems to recover absorption energy to low-pressure steam that is passed to a condensing turbine generator (Puricelli et al. 1998). The Salt Lake City smelter uses furnace off-gas high-pressure steam (after superheating with fuel) to drive compression of process gas at the acid plant.

The turbine compressor steam discharge adds to low-pressure steam flow to the condensing turbine generator. Table 15 provides a comparison of this smelter's energy consumption compared to the global copper smelter shown in Table 14. The fossil fuel energy at this facility is much larger due to auxiliary high-pressure steam production and superheating (5,300 MJ/t) and concentrate drying kiln (2,400 MJ/t), neither of which is considered in the Coursol et al. (2010) analysis (which assumed the use of steam coil dryers). Some of the former is offset by integrated power generation, and electrical energy base assumptions may be slightly different, but overall, the high use of fossil fuel results in higher net energy consumption of 16,930 MJ/t compared to the data from Coursol et al. (2010), 11,409 MJ/t.

Table 14 Calculated energy data for an average global copper smelter*

Electrical Energy		Fossil Fuel,	Total,
kW·h/t anode Cu	MJ/t anode Cu	MJ/t anode Cu	MJ/t anode Cu
918†	3,305	2,712	6,016
‡	8,697	2,712	11,409

Source: Coursol et al. 2010
*Taken from the average of four processes examined.
†Integrated power generation has been included in this imported power.
‡Electrical energy is converted directly to megajoules. The average thermal efficiency of imported (or purchased) power generation is assumed to be 38%.

Table 15 Energy consumption in copper smelting for the Salt Lake City smelter

Electrical Energy		Fossil Fuel,	Total,
kW·h/t anode Cu	MJ/t anode Cu	MJ/t anode Cu	MJ/t anode Cu
1,300	4,700	10,900	15,600
630*	2,290	10,900	13,190
†	6,030	10,900	16,930

Courtesy of Rio Tinto
*Integrated power generation reduces imported power.
†Electrical energy is converted directly to megajoules. The average thermal efficiency of imported power generation is assumed to be 38%.

Costs

The operating cost structure for smelting facilities has a "variable" component, representing the incremental cost of production, and a "fixed" component, representing costs independent of production. Both are a function of the technology utilized, the local unit costs, and business structure. The sustaining cost structure for a smelting facility consists of replacement and overhaul project expenditures that are primarily a function of the technology employed. Schlesinger et al. (2011) provide a summary of capital and operating cost data for a range of copper operations.

Variable costs largely consist of power (production of oxygen, processing of smelting slag), major consumables (smelting and converting flux, lime/limestone, and caustic soda for gas scrubbing and effluent treatment), refractory (molten material launders and ladles), and other smaller but numerous items. Some variable costs can be optimized over time with improvement and modernization programs. Fixed costs largely consist of labor (operating, maintenance, and site management), services (operating and maintenance), corporate overheads, and other smaller but numerous items. Usually, some fixed costs can be viewed as flexible over the longer term in response to production and market expectations. For copper smelters, the global average fixed costs represent about

51% of total direct costs (labor, maintenance, and on-site services; Wood Mackenzie 2016). Regional labor rates influence this proportion significantly, with fixed cost as low as 25% and as high as 75%.

Sustaining costs largely consist of regular overhauls required by smelting and converting furnaces and acid plants. The frequency and cost of furnace overhauls tend to vary between technologies. Smaller, more intense furnaces need less material but tend to achieve shorter campaigns between overhauls. Acid plants treating metallurgical gases tend to require more frequent overhauls compared to plants producing acid in a sulfur-burning duty.

CONCLUDING REMARKS

The world has seen a steady increase in metal consumption over a very long period of time up to the present day as living standards have improved. Ore or concentrate processing by a pyrometallurgical smelting step produces more than three-quarters of all metals used by society, thus highlighting the importance of smelting technologies discussed in this chapter. The technology used today is vastly superior in terms of safety, productivity, and environmental footprint compared to those in the past. It is believed that this trend will continue into the future. Present-day smelting technologies for a wide range of metals and materials are described, including the treatment of recycled scrap, an increasingly important segment of the metals industry.

It is thought that widespread adoption of a sufficiently flexible hydrometallurgical process to handle a range of different concentrates, as presently handled by custom smelters (who, in total, process more than 60% of the world's copper concentrate), would not occur until quite some time into the future. Perhaps development of a suitable hydrometallurgical process for implementation at the mine site may also occur at some time in the future. Such an installation would, however, increase the capital cost and complexity of a mine-mill operation, which is often located in a remote area and already stressed by the treatment of increasingly low ore grades.

ACKNOWLEDGMENTS

The authors acknowledge the assistance of Bob Drew of RHI, Gary Walters of Hatch, Markku Lahtinen of Outotec, Gerardo Alvear of Aurubis (formerly of Glencore), and Katsuhiko Abe of Mitsubishi Materials.

REFERENCES

Agricola, G. 1556. *De Re Metallica.* Translated by H.C. Hoover and L.H. Hoover. 1950 edition. New York: Dover Publications. pp. 530–536.

Alvear, G.R.G. 2016. IsaSmelt: The leading bath smelting technology in the non-ferrous smelting. In *Extractive Metallurgy of Non-Ferrous Metals*. Edited by R. Raghavan. India: Vijay Nicole Imprints.

Ashman, D.W., Goosen, D.W., Reynolds, D.G., and Webb, D.J. 2000. Cominco's new lead smelter at Trail Operations. In *Lead-Zinc 2000.* Edited by J.E. Dutrizac, J.A. Gonzalez, D.M. Henke, S.E. James, and A.H. Siegmund. Warrendale, PA: The Minerals, Metals & Materials Society. pp. 171–185.

Banerjee, S.K., and Evans, J.W. 1990. Measurements of magnetic fields and electromagnetically driven melt flow in a physical model of a Hall–Héroult cell. *Met. Trans. B* 21B:59–69.

Barth, O. 1960. Electric smelting of sulfide ores. In *Extractive Metallurgy of Copper, Nickel and Cobalt.* Edited by P. Queneau. New York: Interscience Publishers. pp. 241–262.

Berzelius Metall. 2014. *Sustainable Battery-Recycling: We Close the Loop.* Accessed October 2017: www.ecobatgroup.com/ecobatgroup-n/information/brochures.php?navid=13.

Biringuccio, V. 1540. *Pirotechnia.* Translated by C.S. Smith and M.T. Gnudi. 1959 edition. New York: Basic Books.

Biswas, A.K. 1981. *Principles of Blast Furnace Ironmaking: Theory and Practice.* Brisbane, Queensland: Cootha Publishing.

Biswas, A.K., and Davenport, W.G. 1976. *Extractive Metallurgy of Copper*. New York: Pergamom Press.

Biswas, A.K., and Davenport, W.G. 1994. *Extractive Metallurgy of Copper.* New York: Pergamon Press.

Bjorklund, P., Peltoniemi, K., Jafs, M., Ahokainen, T., Pienimaki., K., and Pesonen, L.P. 2013. Suspension smelting furnace and a concentrate burner. U.S. Patent 2013-0099431 A1.

Blair, I.S., Humphriss, J., Joyce, B.P., and Warnock, J.L. 1976. The introduction of electric furnace smelting technology to Mufulira, Zambia. In *Extractive Metallurgy of Copper: Vol. I, Pyrometallurgy and Electrolytic Refining.* Edited by J.C. Yannopoulos and J.C. Agarwal. New York: TMS-AIME. pp. 167–196.

Boeckenhauer, K., and Kaczmarczyk, T. 2018. Modern furnaces for aluminum scrap recycling. https://www.secowarwick.com/wp-content/uploads/2017/03/MODERN-FURNACES-FOR-ALUMINUM-SCRAP-RECYCLING-AP.pdf.

Boldt, J.R., and Queneau, P. 1967. *The Winning of Nickel: Its Geology, Mining and Extractive Metallurgy.* Princeton, NJ: Van Nostrand.

Borchers, W. 1897. *Electric Smelting and Refining.* London: Charles Griffin.

Bridgman, H.L. 1897. Process of reducing ores. U.S. Patent 578,912.

Bryk, P., Ryselin, J., Honkasalo, J., and Malmström, R. 1958. Flash Smelting Copper Concentrates. *J. Met.* 10(6):395–400.

Colins, D. 1987. Coal utilisation at the Kalgoorlie nickel smelter. In *Coal Power '87.* Melbourne, Victoria: Australasian Institute of Mining and Metallurgy. pp. 25–30.

Coursol, P., Mackey, P.J., and Diaz, C. 2010. Energy consumption in copper sulphide smelting. In *Proceedings of Copper 2010.* Vol. 2, Pyrometallurgy I. Clausthal-Zellerfeld: GDMB. pp. 649–668.

Coursol, P., Mackey, P.J., Kapusta, J.P.T., and Cardona Valencia, N. 2015. Energy consumption in copper smelting: A new Asian horse in the race. *JOM* 67(5):1066–1075.

Crundwell, F.K., Moats, M.S., Ramachandran, V., Robinson, T.G., and Davenport, W.G. 2011. *Extractive Metallurgy of Nickel, Cobalt and Platinum-Group Metals.* New York: Elsevier.

Czernecki, J., Smieszek, Z., Gizicki, S., Dobrzanski, J., and Warmuz, M. 1998. Problems with elimination of the main impurities in the KGHM Polska Miedz S.A. copper concentrates from the copper production cycle (shaft furnace process, direct blister smelting in a flash furnace). In *Sulfide Smelting '98: Current and Future Practices.* Warrendale, PA: The Minerals, Metals & Materials Society. pp 315–343.

Dannatt, C.W., and Ellingham, H.J.T. 1948. Roasting and reduction processes—A general survey. *Discuss. Faraday Soc.* 4:126–139.

Davis, F.W. 1923. *The Use of Oxygen or Oxygenated Air in Metallurgical and Allied Processes.* Report of Investigation 2502. Washington, DC: U.S. Bureau of Mines.

De Fuccio, R., Jr., De Fuccio, E.A., Betz, E.W., and De F Sousa, C.A. 1992. A semi-continuous autothermic reduction process for the production of ferroniobium. In *Proceedings of Infacon 1992*. Johannesburg: Southern African Institute of Mining and Metallurgy. pp. 279–283.

De Vries, D., Hunt, S., Dyussekenov, N., Milovanov, M., and Chikashi, H.M. 2016. Design, commissioning and operation of an IsaSmelt furnace at the Kansanshi copper smelter. Presented at the Copper 2016 Conference, Kobe, Japan, November 13–16, 2016.

Deininger, L., Choi, K.C., and Siegmund, A. 1994. Operating experience with the QSL plants in Germany and Korea. In *EPD Congress 1994*. Edited by G. Warren. Warrendale, PA: The Minerals, Metals & Materials Society. pp. 477–501.

Diaz, C., Schwarze, H.., and Taylor, J.C. 1995. The changing landscape of copper smelting in the Americas. In *Proceedings of Copper 95-Cobre 95 International Conference: Vol. IV, Pyrometallurgy of Copper.* Edited by W.J. Chen, C. Diaz, A. Luraschi, and P.J. Mackey. Montreal, QC: Canadian Institute of Mining, Metallurgy and Petroleum. pp. 3–28.

Díaz, C.M., Levac, C., Mackey, P.J., Marcuson, S.W., Richards, G., Schonewille, R., and Themelis, N.J. 2011. Innovation in nonferrous metals pyrometallurgy 1961–2011. In *The Canadian Metallurgical Landscape 1960 to 2011—Golden Anniversary of the Conference of Metallurgists.* Edited by J Kapusta, P.J Mackey, and N. Stubina. Westmount, QC: Canadian Institute of Mining, Metallurgy and Petroleum. pp. 333–360.

Donald, J.R., and Scholey, K. 2005. An overview of Inco's Copper Cliff operations. In *Nickel and Cobalt 2005: Challenges in Extraction and Production.* Edited by J.R. Donald and R. Schonewille. Montreal, QC: Canadian Institute of Mining, Metallurgy and Petroleum.

Engineering Dobersek. 2017. *Completed Plants 2016.* Moenchengladbach, Germany: https://www.dobersek.com/fileadmin/media/Download/ED_Fertige_Anlagen_2016.pdf.

Fischer, P. 1983. The QSL process: A new lead smelting route ready for commercialization. In *Advances in Sulfide Smelting*. Vol. 2. Edited by H.Y. Sohn, D.B. George, and A.D. Zunkel. Warrendale, PA: TMS-AIME. pp. 513–527.

Floyd, J.M. 1981. Submerged injection of gas into liquid-pyrometallurgical bath. U.S. Patent 4,251,271.

Forrest, H., and Wilson, J.D. 1990. Lead recycling utilizing short rotary furnaces. In *Lead-Zinc '90.* Edited by T.S. Mackey and R.D. Prengaman. Warrendale, PA: The Minerals, Metals & Materials Society. pp. 971–978.

Freeman, H. 1932. Process of smelting finely divided sulphide ores. U.S. Patent 1,888,164.

Fruehan, R.J., United States Steel Company, and American Society of Metals. 1998. *The Making, Shaping and Treating of Steel, Vol. 2: Steelmaking and Refining Volume*, 11th ed. Pittsburgh, PA: American Iron and Steel Institute.

Gao, W., Wang, C., Yin, F., Chen, Y., and Yang, W. 2012. Situation and technology progress of lead smelting in China. *Adv. Mater. Res.* 581-582:904–911.

Geldenhuys, I.J. 2013. Aspects of DC chromite smelting at Mintek—An overview. In *Proceedings of Infacon XIII: The 13th International Ferroalloys Congress.* Karaganda, Kazakhstan: P. Dipner.

Glencore Canada. 2018. Who we are: Zinc—Brunswick lead smelter. www.glencore.ca/en/who-we-are/Pages/zinc.aspx. Accessed September 2018.

Gordon, J.R., Norman, G.H.C., Queneau, P.E., Sproule, W.K., and Young, C.E. 1954. Autogenous smelting of sulfides. U.S. Patent 2,668,107.

Grimes, S., Donaldson, J., and Gomez, G.C. 2008. *Report on the Environmental Benefits of Recycling*. London: Centre for Sustainable Production and Resource Efficiency, Imperial College.

Habashi, F. 1997. *Handbook of Extractive Metallurgy: Vol. 4, Ferroalloy Metals and Alkali Metals.* New York: Wiley-VCH.

Harkki, S.U., and Juusela, J.T. 1974. New developments in Outokumpu flash smelting method. Paper No. A74–16. New York: TMS-AIME.

Herneryd, O., Sundstrom, O.A., and Norro, A. 1954. Copper smelting in Boliden's Ronnskar works described. *J. Met.* 6(3):330–337.

Hundermark, R., de Villiers, B., and Ndlovu, J. 2006. Process description and short history of Polokwane smelter. In *Proceedings of the Southern African Pyrometallurgy 2006 International Conference*. Edited by R.T. Jones. Johannesburg: Southern African Institute of Mining and Metallurgy.

ICSG (International Copper Study Group). 2017. *World Copper Fact Book.* Lisbon, Portugal: ICSG.

IsaSmelt 2015. Welcome to IsaSmelt. www.isasmelt.com/en/Pages/home.aspx. Accessed October 2018.

Janney, D., George-Kennedy, D., and Burton, R. 2007. Kennecott Utah copper smelter rebuild over 11 million tonnes of concentrate smelted. In *Cu2007: Sixth International Copper/Cobre Conference.* Vol. III, Book 2. Montreal, QC: Canadian Institute of Mining, Metallurgy and Petroleum. pp. 431–443.

Jeremiah. 1985. Book 6, Verse 29. In *The NIV Study Bible* (New International Version). Grand Rapids, MI: Zondervan. p. 1132.

Jiang, K., Li, L., Feng, Y., Wang, H., and Wei, B. 2012. The development of China's copper smelting technologies. In *The T.T. Chen Honorary Symposium on Hydrometallurgy, Electrometallurgy and Materials Characterization.* Edited by S. Wang, J.E. Dutrizac, M. Free, J.Y. Hwang, and D. Kim. Warrendale, PA: The Minerals, Metals & Materials Society. pp. 167–176.

Kapusta, J.P.T. 2004. JOM world nonferrous smelters survey, Part I: Copper. *JOM* 56(7):21–27.

Kapusta, J.P.T. 2017. Submerged gas jet penetration: A study of bubbling versus jetting and side versus bottom blowing in copper bath smelting. *JOM* 69(6):970–979.

Kellogg, H.H., and Diaz, C. 1992. Bath smelting processes in non-ferrous metallurgy—An overview. In *Proceedings of the Savard/Lee International Symposium on Bath Smelting*. Edited by J.K. Brimacombe, P.J. Mackey, G.J.W. Kor, C. Bickert, and M.G. Ranade. Warrendale, PA: The Minerals, Metals & Materials Society. pp. 39–65.

Klepinger, J.H., Krejci, M.W., and Kuzell, C.R. 1915. Process of smelting ores. U.S. Patent 1,164,653.

Korhonen, J.M., and Valikangas, L. 2014. Constraints and ingenuity: The case of Outokumpu and the development of flash smelting in the copper industry. In *Handbook of Research on Organizational Ingenuity and Creative Institutional Entrepreneurship*. Edited by B. Honig, J. Lampel, and I. Drori. Cheltenham: Edward Elgar.

Köster, S., and Martínez, A. 2016. Optimization of the efficiency of waste heat recovery in the copper-industry. Presented at the Copper 2016 Conference, Kobe, Japan, November 13–16.

Laist, F., and Cooper, J.P. 1934. An experimental combination of shaft roasting and reverberatory smelting. *Trans. AIME* 106:104–110.

Legg, A.C., Ntsipe, L., Bogopa, M., and Dzinomwa, G. 2009. Modernization of the BCL smelter. In *Base Metals Conference 2009*. Johannesburg: Southern African Institute of Mining and Metallurgy. pp. 53–67.

Li, C., Wang, J., Wang, Z., Jiang, J., and Huang, Q. 1999. The SKS copper smelting process. In *Proceedings of the Copper 99-Cobre 99 International Conference: Vol. VI, Smelting Technology Development, Process Modelling and Fundamentals*. Edited by C. Diaz, C. Landolt, and T. Utigard. Warrendale, PA: The Minerals, Metals & Materials Society. pp. 67–82.

Liang, S. 2016. A review of oxygen bottom blowing process for copper smelting and converting. Paper PY16-1. Presented at Copper 2016, Kobe, Japan, November.

Longval, A., Shannon, T., Jiao, Q., and Davis, B. 2008. Operational improvements at Brunswick's lead blast furnace. In *Zinc and Lead Metallurgy*. Edited by L. Centomo, M. Collins, J. Harlamovs, and J. Liu. Montreal, QC: Canadian Institute of Mining, Metallurgy and Petroleum. pp. 29–40.

Mackey, P.J. 2009. Introduction to nickel. Nickel Pyrometallurgy Short Course. COM Conference, Sudbury, August, Montreal, QC, Canadian Institute of Mining, Metallurgy and Petroleum.

Mackey, P.J. 2013. Copper smelting technologies in 2013 and beyond. Paper presented at Copper 2013, Santiago, Chile.

Mackey, P.J. 2014. Evolution of the large copper smelter—1800s to 2013. In *Celebrating the Megascale: Proceedings of the Extraction and Processing Division Symposium on Pyrometallurgy in Honor of David G.C. Robertson*. Edited by P.J. Mackey, E.J. Grimsey, R.T. Jones, and G.A. Brooks. Warrendale, PA: The Minerals, Metals & Materials Society. pp. 17–37.

Mackey, P.J., and Brimacombe, J.K. 1992. Savard and Lee—Transforming the metallurgical landscape. *Proceedings of the Savard/Lee International Symposium on Bath Smelting*. Edited by J.K. Brimacombe, P.J. Mackey, G.J.W. Kor, C. Bickert, and M.G. Ranade. Warrendale, PA: The Minerals, Metals & Materials Society. pp. 3–28.

Mackey, P.J., and Campos R. 2001. Modern continuous smelting and converting by bath smelting technology. *Can. Metall. Q.* 40(3):355–376.

Mackey, P.J., Hallett, G.D., Porcile, F., and Themelis, N.J. 1985. Use of the Noranda and Codelco processes for expansion and modernization of copper smelters. In *Extraction Metallurgy '85*. London: Institution of Mining and Metallurgy. pp. 1101–1124.

Makinen, T., and Taskinen, P. 1998. Physical Chemistry of Direct Nickel Matte Smelting. In *Sulfide Smelting '98: Current and Future Practices*. Warrendale, PA: The Minerals, Metals & Materials Society. pp. 59–68.

Marczeski, W.D., and Aldrich, T.L. 1985. Retrofitting Hayden plant to flash smelting. Paper presented to the AIME South-West Branch, Tucson, Arizona.

McKague, A.L., and Norman, G.E. 1984. Operation of Falconbridge's new smelting process. *CIM Bull.* (June):1–7.

Mechev, V.V. 1991. On status and trends of copper industry development in the U.S.S.R. In *Proceedings of Copper 91-Cobre 91. Vol. IV, Pyrometallurgy of Copper.* Edited by C. Diaz, C. Landolt, A. Luraschi, and C.J. Newman. Montreal, QC: Canadian Institute of Mining, Metallurgy and Petroleum. pp. 91–106.

Medina, A., Richter, G., Toro, C., and Sanchez, 1993. Teniente technology with acid plant operation and impurities distribution at Las Ventenas smelter. In *The Paul E. Queneau International Symposium—Extractive Metallurgy of Copper, Nickel and Cobalt: Vol. II, Copper and Nickel Smelter Operations*. Edited by C.A. Landolt. Warrendale, PA: The Minerals, Metals & Materials Society. pp. 1429–1439.

Messner, M.E., and Kinneberg, D.A. 1969. Direct converter smelting at Utah using oxygen. *J. Met.* 21(7):23–29.

Meurer, U., and Ambroz, H. 2010. Primary lead production with the QSL process—An ecological and economical advanced technology. In *Lead-Zinc 2010*. Edited by A. Siegmund, L. Centomo, C. Geenen. N. Pirit, G. Richards, and R. Stephens. Warrendale, PA: The Minerals, Metals & Materials Society. pp. 429–437.

Michard, J., Dancoisne, P., and Chanty, G. 1961. Blast furnace practice with 100 Pct low grade self-fluxing sinter. In *Proceedings of the Blast Furnace, Coke Oven and Raw Materials Conference*. Vol. 20. New York: Interscience Publishers. pp. 329–350.

Mills, L.A., Hallett, G.D., and Newman, C.J. 1976. Design and operation of the Noranda continuous smelter. In *Extractive Metallurgy of Copper: Vol. I, Pyrometallurgy and Electrolytic Refining*. Edited by J.C. Yannopoulos and J.C. Agarwal. New York: TMS-AIME. pp. 458–487.

Mizuta, Y., Oguma, N., Iwahori, S., and Sato, H. 2012. Promotion of recycling business by combination of a pretreatment plant and the Mitsubishi Process at Naoshima smelter and refinery. In *International Smelting Technology Symposium: Incorporating the 6th Advances in Sulphide Smelting Symposium*. Edited by J.P. Downey, T.P. Battle, and J.F. White. Warrendale, PA: The Minerals, Metals & Materials Society. pp. 125–132.

MMC (Mitsubishi Materials Corporation). 2004. *The Mitsubishi Process: Copper Smelting for the 21st Century.* Tokyo: Mitsubishi Materials Corporation. www.mmc.co.jp/sren/MI_Brochure.pdf

MMC (Mitsubishi Materials Corporation). 2012. History: The Mitsubishi process. www.mmc.co.jp/sren/History.htm. Accessed September 2018.

Mostert, J.C., and Roberts, P.N. 1973. Electric smelting at Rustenburg Platinum Mines Limited of nickel-copper concentrates containing platinum group metals. *J. S. Afr. Inst. Min. Metall.* (April):290–299.

Munoz, G., Vera, G., and Salazar, H. 1982. *Codelco-Chile: A Realistic Way to Increase Copper Smelting Capacity, Copper Smelting—An Update.* Edited by D.B. George and J.C. Taylor. Warrendale, PA: TMS-AIME. pp. 143–163.

Nagano, T., and Suzuki, T. 1976. Commercial operation of the Mitsubishi continuous copper smelting and converting process. In *Extractive Metallurgy of Copper: Vol. I, Pyrometallurgy and Electrolytic Refining.* Edited by J.C. Yannopoulos and J.C. Agarwal. New York: TMS-AIME. pp. 439–457.

Newman, C.J., Macfarlane, G., and Molnar, K. 1987. Oxygen usage in the Kidd Creek smelter. In *Proceedings of the International Symposium on the Impact of Oxygen on the Productivity of Non-Ferrous Metallurgical Processes.* Edited by G. Kachaniwsky and G.J. Newman. Toronto, ON: Pergamon Press. pp. 259–268.

Nilsson, A.E., Aragonés, M.M., Torralvo, F.A., Dunon, V., Angel, H., Komnitsas, K., and Willquist, K. 2017. A review of the carbon footprint of Cu and Zn production from primary and secondary sources. *Minerals* 7(9):168.

NIOSH (National Institute for Occupational Safety and Health). 2007. *NIOSH Pocket Guide to Chemical Hazards.* Publication No. 2005-149. Cincinnati, OH: Department of Health and Human Services.

Noranda Mines Limited. 1974. *Noranda Reactor* (brochure). Toronto, ON: Noranda Mines Limited.

Norman, T.E. 1936. Autogenous smelting of copper concentrates with oxygen-enriched air. *Eng. Min. J.* 137(10):499–502; 137(11):562–567.

Outotec 2015. Outotec Ausmelt TSL furnace. https://www.outotec.com/products/smelting-and-converting/ausmelt-tsl-furnace/. Accessed October 2018.

Oxford English Dictionary. 2017. Smelt. Oxford: Oxford University Press.

Paakkonen, E., and Mattelmaki, M. 1996. The direct Outokumpu nickel smelting (DON) process and the Harjavalta expansion. In *Nickel '96.* Melbourne, Victoria: Australasian Institute of Mining and Metallurgy. pp. 235–241.

Peacey, J.G., and Davenport, W.G. 1979. *The Iron Blast Furnace Theory and Practice*, 1st ed. Oxford: Pergamon Press.

Pelletier, A., Mackey, P.J., Southwick, L.M., and Wraith, A.E. 2009. Before Peirce and Smith—The Manhés converter: Its development and some reflections for today. In *The International Peirce–Smith Converting Centennial Symposium.* Edited by J.P.T. Kapusta and A.E.M. Warner. Warrendale, PA: The Minerals, Metals & Materials Society. pp. 29–49.

Percy, J., 1861. *Metallurgy—The Art of Extracting Metals from their Ores and Adapting Them to Various Uses of Manufacture: Fuel, Fireclays, Copper, Zinc, Brass, etc.* London: John Murray.

Prayoga, A., and Sundana, D. 2016. Recent improvements at Gresik smelter. Presented at the Copper 2016 Conference, Kobe, Japan, November 13–16.

Prevost, Y., Bedard, M., and Levac, C. 2013. The Noranda Process. In *Proceedings of Copper 2013, International Copper Conference: Vol. III, The Nickolas Themelis Symposium on Pyrometallurgy and Process Engineering.* Edited by R. Bassa, R. Parra, A. Luraschi, and S. Demetrio. Santiago, Chile: Chilean Institute of Mining Engineers. pp. 265–277.

Puricelli, S.M., Grendel, R.W., and Fries, R.M. 1998. Pollution to power: A case study of the Kennecott sulfuric acid plant. In *Sulfide Smelting '98: Current and Future Practice.* Warrendale, PA: The Minerals, Metals & Materials Society.

Queneau, P.E., and Marcuson, S.W. 1996. Oxygen pyrometallurgy at Copper Cliff—A Half century of progress. *JOM* 48(1):14–21.

Queneau, P.E., and Schuhmann, R. 1974. The Q–S oxygen process. *J. Met.* 26(8):14–16.

Rioux, D.J., Moore, T.A., McTeer, G.J., and Verhelst, D.L. 2014. Teck's Kivcet lead tapping experience. In *Furnace Tapping Conference 2014.* Johannesburg: Southern African Institute of Mining and Metallurgy.

Rothenberg, B. 1978. Excavations at Timna Site 39: A chalcolithic copper smelting site and furnace and its metallurgy. *J. Archaeo-Metall.* 1:1–15.

Sanhueza, J.A., Rojas, O.C., and Rojas, P.A. 2003. Recent developments at ENAMI's Hernan Videla Lira Smelter. In *Proceedings of the Copper 2003—Cobre 2003 5th International Conference, Vol. 4, Book 1: Pyrometallurgy of Copper.* Edited by C. Diaz, J. Kapusta, and C. Newman. Montreal, QC: Canadian Institute of Mining, Metallurgy and Petroleum. pp. 307–323.

Särkikoski, T. 1999. *A Flash of Knowledge. How an Outokumpu Innovation Became a Culture.* Espoo, Finland: Outokumpu Oyj.

Schlesinger, M.E., King, M.J., Sole, K.C., and Davenport, W.G. 2011. *Extractive Metallurgy of Copper*, 5th ed. Amsterdam: Elsevier.

See, J., Robertson, D., and Mackey, P.J. 2014. Current and suggested focus on sustainability in pyrometallurgy. In *Celebrating the Megascale: Proceedings of the Extraction and Processing Division Symposium on Pyrometallurgy in Honor of David G.C. Robertson.* Edited by P.J. Mackey, E.J. Grimsey, R.T. Jones, and G.A Brooks. Warrendale, PA: The Minerals, Metals & Materials Society. pp. 429–446.

Senne, M.A. 1977. Start-up of the electric furnace at Anaconda. Preprint No. A77-16. New York: AIME.

Shibasaki, T., Hayashi, M., and Nishiyama, Y. 1993. Recent operation at Naoshima with a larger Mitsubishi furnace line. In *The Paul E. Queneau International Symposium—Extractive Metallurgy of Copper, Nickel and Cobalt: Vol. II, Copper and Nickel Smelter Operations.* Edited by C.A. Landolt. Warrendale, PA: The Minerals, Metals & Materials Society pp. 1413–1428.

Siegmund, A., Centomo, L., Geenen, C., Piret, N., Richards, G., and Stephens, R. (eds.). 2010. *Pb-Zn 2010.* Warrendale, PA: The Minerals, Metals & Materials Society.

Sipila, J., Peltoniemi, K., Bjorklund, P., and Xia, J. 2012. Concentrate burner. U.S. Patent 8,206,643 B2.

Smil, V. 2016. *Still the Iron Age—Iron and Steel in the Modern World.* New York: Elsevier.

Southwick, L. 2009. William Peirce and E.A. Cappelen Smith, and their amazing converting machine. In *International Peirce–Smith Converting Centennial.* Edited by J. Kapusta and T. Warner. Warrendale, PA: The Minerals, Metals & Materials Society. pp. 3–27.

Stansfield, A. 1914. *The Electric Furnace—Its Construction and Use.* New York: McGraw-Hill.

Suzuki, Y., Suenaga, C., Ogasawara, M., and Yasuda, Y. 1998. Productivity increase in flash smelting furnace operation at Saganoseki smelter and refinery. In *Sulfide Smelting '98: Current and Future Practices.* Warrendale, PA: The Minerals, Metals & Materials Society. pp. 587–595.

Taniguchi, T., Matsutani, T., and Sato, H. 2006. Technological innovations in the Mitsubishi process to achieve four years campaign. In *Sohn International Symposium, Advanced Processing of Metals and Materials: Vol. 8, International Symposium on Sulphide Smelting.* Edited by F. Kongoli and R.G. Reddy. Warrendale, PA: The Minerals, Metals & Materials Society. pp. 261–273.

Tarassoff, P., and Mackey, P.J. 1983. New and emerging technologies in sulfide smelting. In *Advances in Sulfide Smelting.* Vol. 2. Edited by H.Y. Sohn, D.B. George, and A.D. Zunkel. Warrendale, PA: TMS-AIME. pp. 399–426.

Tetronics International. 2018. *About Tetronics.* http://tetronics.com/assets/Tetronics-Corporate-Brochure1.pdf.

Themelis, N.J., and Kellogg, H.H. 1983. Principles of sulphide smelting. In *Advances in Sulphide Smelting: Vol. I, Basic Principles.* Edited by H.Y. Sohn, D.B. George, and A.D. Zunkel. Warrendale, PA: TMS-AIME. pp. 1–29.

Themelis, N.J., and Mackey, P.J. 1992. Solid–liquid and gas–liquid interactions in a bath smelting reactor. In *Proceedings of the Savard/Lee International Symposium on Bath Smelting.* Edited by J.K. Brimacombe, P.J. Mackey, G.J.W. Kor, C. Bickert, and M.G. Ranade. Warrendale, PA: The Minerals, Metals & Materials Society. pp. 353–375.

Themelis, N.J., McKerrow, G.C., Tarassoff, P., and Hallett, G.D. 1972. The Noranda process. *J. Met.* 24(4):25–32.

Thurneyssen, C.G., Szczeniowski, J., and Michel, J. 1960. Ferronickel smelting in New Caledonia. In *Extractive Metallurgy of Copper, Nickel and Cobalt.* Edited by P. Queneau. New York: Interscience Publishers. pp. 287–300.

Tsurumoto, T. 1961. Copper smelting in the converter. *J. Met.* 13(11):820–824.

Tylecote, R.F. 1976. *A History of Metallurgy.* London: Metals Society.

USGS (U.S. Geological Survey). 2017. Aluminum. In *Mineral Commodity Summaries.* Washington, DC: USGS.

Voermann, N., Gerritsen, T., Candy, I., Stober, F., and Matyas, A. 2004. Developments in furnace technology for ferronickel production. In *Infacon X: Proceedings of the 10th International Ferroalloys Congress.* Johannesburg: Southern African Institute of Mining and Metallurgy. pp. 455–465.

Walker, M.J., Reynolds, D.R., and Lee, G.W. 1990. The new Cominco lead smelter at Trail. In *Lead and Zinc '90.* Edited by T.S. Mackey and R.D. Prengaman. Warrendale, PA: The Minerals, Metals & Materials Society. pp. 919–932.

Wan, J., Luo, Y., Chen, B., Swayn, G.P., Wood, J.T., and Creedy, S.J. 2013. Design and commissioning of an Outotec Ausmelt copper smelter for Daye Non-Ferrous Metals Company Ltd. In *Copper-Cobre 2013 International Conference.* Vol. 3. Westmount, QC: Canadian Institute of Mining, Metallurgy and Petroleum. pp. 379–365.

Warner, A.E.M., Díaz, C.M., Dalvi, A.D., Mackey, P.J., and Tarasov, A.V. 2006. Worldwide nickel smelting survey, part III: Nickel: Laterite. *JOM* 58(4):11–20.

Warner, A.E.M., Díaz, C.M., Dalvi, A.D., Mackey, P.J., Tarasov, A.V., and Jones R.T. 2007. JOM world non-ferrous smelter survey, part IV: Nickel: Sulfide. *JOM* 59(4):58–72.

Wikipedia. 2018. Forges du Saint-Maurice. https://en.wikipedia.org/wiki/Forges_du_Saint-Maurice. Accessed August 2018.

Witherell, C.S, and Skougor, H.E. 1922. The Westly electric furnace for smelting copper concentrates. *Eng. Min. J.* 113(9):356–362.

Wood Mackenzie. 2016. *Global Copper Smelter and Refinery Supply Summary.* Wood Mackenzie

Wraith, A.E., and Mackey, P.J. 2010. Sulphide bath smelting: 19th century concept and Holloway's legacy. In *Proceedings of Copper 2010: Vol. 3, Pyrometallurgy II.* Clausthal-Zellerfeld: GDMB. pp. 945–960.

Yao, S., and Jiang, C. 2014. Modern copper smelting and converting technology applications, development, and improvement in China copper industry. In *COM 2014—Conference of Metallurgists Proceedings.* Westmount, QC: Canadian Institute of Mining, Metallurgy and Petroleum.

Yasuda, M. 1974. Recent developments of copper smelting at Hitachi smelter. Preprint No. A74-8. New York: TMS-AIME.

Zamalloa, M., Reyes, A., Fañas, J.J., and Frias, J.R. 2009. New developments in ferronickel refining at Xstrata Nickel's Falcondo smelter. In *Pyrometallurgy of Nickel and Cobalt 2009.* Edited by J. Liu, J. Peacey, M. Barati, S. Kashani-Nejad, and B. Davis. Montreal, QC: Canadian Institute of Mining, Metallurgy and Petroleum. pp. 195–208.

Zeisberg, F.C. 1937. Ore roasting. U.S. Patent 2,086,201.

Zhou, S., and Liu, W. 2014. Floating entrainment metallurgical process and reactor. U.S. Patent 8,663,360 B2.

Zhou, J., and Sun, L. 2013. Flash smelting and flash converting process and start-up at Jinguan Copper. In *Proceedings of Copper 2013.* Santiago, Chile: Chilean Institute of Mining Engineers. pp. 377–384.

CHAPTER 11.6

Conversion and Refining

David George-Kennedy and Phillip J. Mackey

This chapter introduces the pyrometallurgical processes of conversion (also called "converting") and refining, and discusses how they are broadly applied, with a focus on current primary metallurgical practice. To convey the relative importance and specific goals of conversion and refining, these processes are placed in the context of the periodic table of elements and with smelting processes covered in Chapter 11.5, "Smelting." The main sections covering unit operations include a brief description of key developments up to the present day and typical modes of operation of furnace technologies used at modern smelting plants for the major metals.

The chapter is organized based on some specific definitions. *Conversion* is defined as the pyrometallurgical removal of elements from a melt originally produced by a smelting process to allow subsequent refining of the crude metal so produced. *Refining* is defined as the removal of generally low levels of elements deemed deleterious to the refined metal properties or subsequent steps required to refine the metal to commodity or market specifications.

METALS REQUIRING CONVERSION AND REFINING

Conversion processes cover the nonferrous practice of removing iron and sulfur from copper-nickel mattes produced in any of the relevant smelting processes described in Chapter 11.5. In practice, conversion is based on reacting oxygen or oxygen-enriched air, either with comminuted solid matte using suspension or flash technology or introduced to a molten matte bath using submerged tuyere, top submerged lance, or top lance furnace technologies to produce crude metal/alloy or enriched sulfide. Conversion generally lowers the levels of major deleterious elements, but final control of such elements must usually be dealt with in subsequent refining steps (Figure 1).

Refining processes generally implemented at primary smelting facilities involve the removal of deleterious elements from crude metal/alloy resulting from the conversion of matte or are produced directly from concentrate. Elements are considered deleterious if they affect final metal quality marketability or if they negatively impact further metal refining steps, such as in copper electrorefining to cathode, or final metal/alloy quality. The key processes involved in refining are oxidation (slagging and gasifying of oxidized impurities), liquation (solidification of higher melting point impurities), drossing, vacuum (vaporizing impurities), and stirring/injection (slagging of impurities and inclusions).

HISTORICAL BACKGROUND

This section provides a short historical sketch on conversion and refining, with particular reference to copper and nickel through the late 1800s. Up until the introduction of the bottom-blowing Bessemer converter in 1856 (Weeks 1896; Bessemer 1905; Wilder 1949), conversion and refining operations in iron and steel and nonferrous metallurgy were carried out in relatively small furnaces, such as the reverberatory-type furnaces (Bloxman 1871), along with manual manipulation of the melt by metal bars and/or paddles. These early operations evolved over centuries from small bowl-shaped furnaces, some originally operated with hand-operated bellows.

Following the success of the Bessemer process for steelmaking, similar approaches were attempted for copper and nickel sulfide conversion. Among other challenges, the high thermal conductivity of copper caused freezing on the vessel bottom. Working at the Éguilles copper plant in France in 1880 with a small Bessemer converter, metallurgists Pierre Manhès and Paul David modified the placement of air-injection tuyeres from the bottom of the vessel to the side and above the copper melt, making conversion of the melt possible without freezing (Pelletier et al. 2009). They subsequently developed a horizontal barrel converter. Toward the end of the 1800s, both horizontal and upright configurations were being used with acidic (silica-based) refractory linings, which also served as the source of flux (Figure 2). The potential benefits of using basic refractory linings were long appreciated, but commercial success was proving elusive (Mathewson 1913). The breakthrough occurred with development of the Peirce–Smith converter in 1909, still in use in modernized form today (Southwick 2009). Early nickel conversion furnaces were similar to those utilized for copper and were deployed in

David George-Kennedy, Chief Advisor, Smelting & Refining, Rio Tinto Copper & Diamonds, Salt Lake City, Utah, USA
Phillip J. Mackey, President, P.J. Mackey Technology Inc., Kirkland, Quebec, Canada

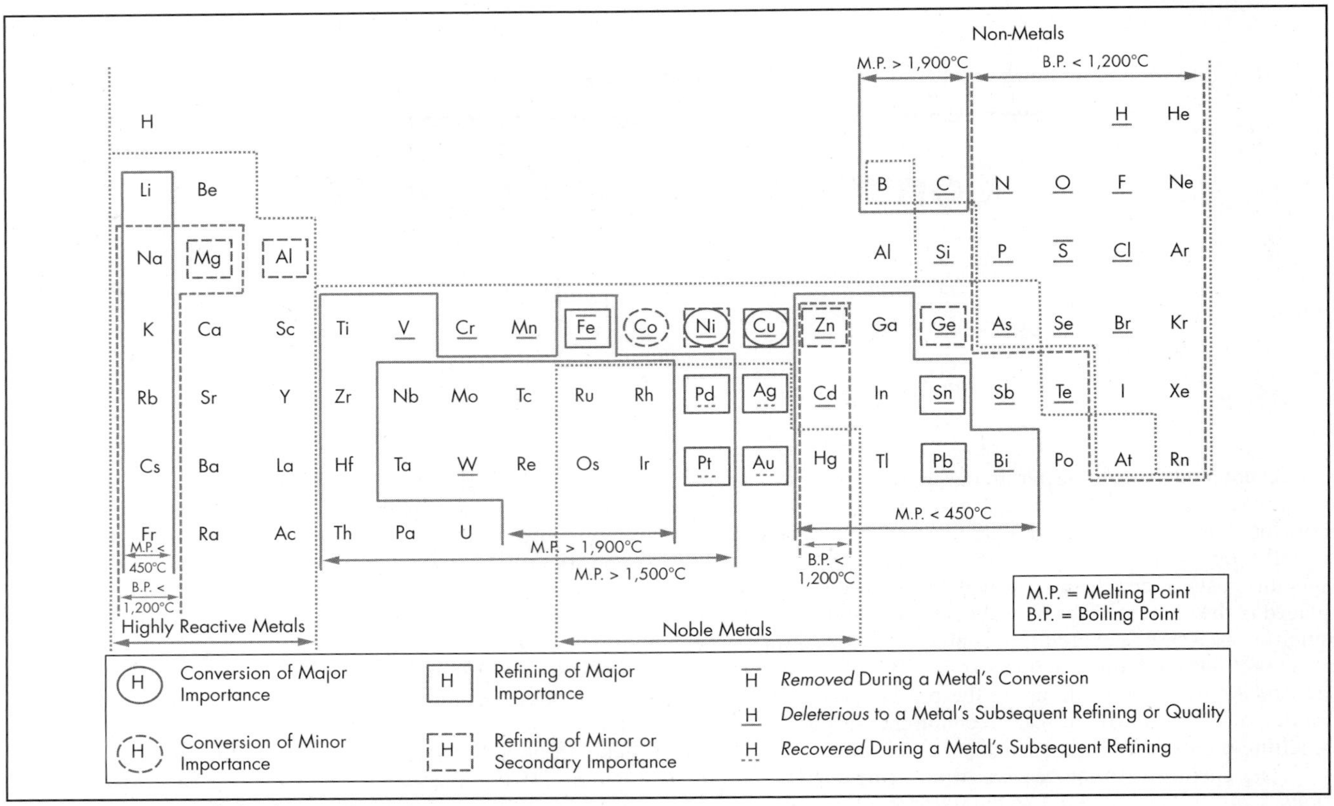

Adapted from Dannatt and Ellingham 1948

Figure 1 Abridged periodic table of the elements highlighting those metals subjected to conversion and refining and common deleterious element associations

Norway and Canada in the late 1800s (Royal Ontario Nickel Commission 1917).

Lead metallurgical evidence can be found as far back as the seventh millennium BC. Early pit/bowl furnaces were used to smelt oxidic and sulfidic lead ores with charcoal. Given that the crude lead nearly always contained a significant concentration of silver (0.01% to 0.5%), "cupellation" was gradually carried out to separate lead and other impurities by oxidation, to produce silver and a litharge (lead oxide) slag. By 1850 the "Parkes process" was developed in the United Kingdom, which uses zinc additions to remove the silver in a more concentrated form. This process remains in use today.

THEORETICAL ASPECTS OF MODERN CONVERSION AND REFINING

Conversion Theory

This section provides a short overview of the common theoretical concepts of conversion for copper and nickel.

Copper Conversion

For suspension or flash conversion of copper matte (George 2002), feed is ground to a size sufficient to ensure early ignition and combustion, typically P_{80} <75 µm. Sufficient oxygen is introduced to ensure near complete overall sulfur elimination after reactions in the bath below the flash shaft. The main reactions are presented here, together with the corresponding heat of reaction:

$$Cu_2S(s) + 2O_2(g) \rightarrow 2CuO(\text{l solution}) + SO_2(g) \quad -2.6\ \text{MJ/kg Cu} \qquad \textbf{(EQ 1)}$$

$$2FeS(s) + 7/2\ O_2(g) \rightarrow Fe_2O_3(\text{s and l solution}) + 2SO_2(g) \quad -0.8\ \text{MJ/kg Cu (at Fe/Cu 0.1)} \qquad \textbf{(EQ 2)}$$

$$Cu_2S(s) + O_2(g) \rightarrow 2Cu(l) + SO_2(g) \quad -0.4\ \text{MJ/kg Cu} \qquad \textbf{(EQ 3)}$$

$$FeS(s) + 3/2O_2(g) \rightarrow FeO(\text{s and l solution}) + SO_2(g) \quad -0.6\ \text{MJ/kg Cu (at Fe/Cu 0.1)} \qquad \textbf{(EQ 4)}$$

A further reaction of FeO with silica and/or lime flux also occurs to make a converter slag. In addition, several subreactions also occur, as discussed in more detail by Davenport et al. (2001).

Reactions 1–4 can also be used in general to describe batch or continuous top lance conversion and batch submerged tuyere conversion (known as the Peirce–Smith converter). In top lance conversion, the feed copper matte is generally molten and the oxygen is transferred to the slag/matte emulsion via a top-blowing lance (at ~1,230°C) with an efficiency of >90%. Lime/limestone as flux (or silica flux) is continuously injected to allow for formation of the conversion slag. Because of the agitated melt, the sulfur dioxide partial pressure is lower compared to suspension or flash conversion. Consequently, sulfur

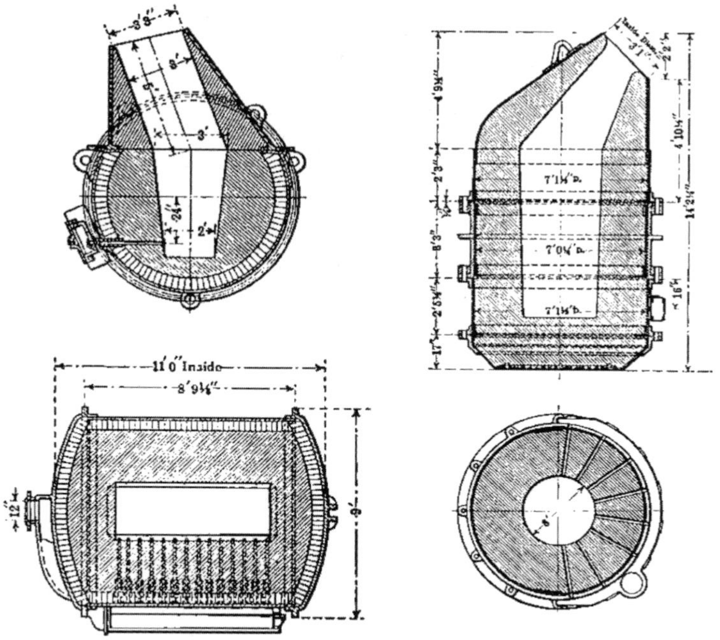

Note: Units of measurement are in feet and inches.

Source: Hofman 1914

Figure 2 Historic horizontal and upright conversion furnaces with acidic refractory linings

in the blister produced from these operations tends to be lower than that in flash conversion for the same copper level in slag.

Nickel Conversion

Submerged tuyere conversion for nickel matte is very similar to the copper matte blow (at ~1,300°C) but typically with a higher initial Fe/(Ni + Cu + Co) ratio. Generally, the end point in nickel conversion is a low-iron nickel (plus copper) matte, which is subsequently treated hydrometallurgically to recover the contained metals.

Refractory and Containment

For all conversion processes, high-quality chrome-magnesia refractories are invariably utilized. Common to all applications is the wear associated with contacting the refractory with slag containing oxidic copper (calcia and silica based). For the continuous processes, substantial external cooling of the shell and refractory is required to control steady-state liner geometry.

Refining Theory

This section provides some brief comments on the theoretical concepts of refining for ferrous and nonferrous applications.

Iron Refining

Iron produced via the blast furnace must be refined for the production of steel. Liquid iron from the blast furnace typically contains carbon, silicon, manganese, phosphorus, and sulfur levels that must be reduced. The level of these elements will depend on blast furnace iron ore and coke quality (Turkdogen and Fruehan 2012).

Ferroalloy Refining

Ferronickel, as produced in electric furnace smelting of calcined laterite ore, is refined typically in 25-to-30-t (metric ton) batches using a ladle refining system adapted from that used in steel refining. Several refining agents are added to the ferronickel melt along with oxygen lancing to remove elements such as phosphorus, silicon, carbon, and sulfur that may be present. Phosphorous removal is achieved by adding lime with oxygen injection, producing a calcium phosphate slag; carbon and silicon are also removed by oxygen lancing. Sulfur is removed by the addition of calcium carbide, producing a calcium sulfide slag (Thurneyssen et al. 1960, Zamalloa et al. 2009).

Copper Refining

Crude or "blister" copper produced either directly from smelting or, more typically, via a conversion process must have the sulfur and oxygen levels reduced to permit the casting of flat anode shapes suitable for subsequent electrorefining. The oxidation blow for final sulfur elimination in this copper raises the melt oxygen to a level that must subsequently be reduced to allow casting of quality anode shapes. This "reduction step" is executed through the same tuyeres using either natural gas, ammonia, or, more efficiently, superheated steam/natural gas reduction.

Lead Refining

Lead produced directly in a blast furnace or other smelting furnace is subjected to many fire-refining processes to achieve required final metal quality. Typically, these are carried out sequentially in a series of heated ladle-type furnaces referred to as "kettles." In this way, elements such as copper (also nickel and cobalt), arsenic, tin, zinc, antimony, and bismuth are removed by chemical flux additions and skimming off the resulting dross or slag that is formed (Tan and Vix 2006; Dawson 1989; Ashman et al. 2000; Leroy et al. 1970; Blanderer 1984; Kapoulitsas et al. 2000).

Tin Refining

Crude tin produced in either reverberatory or submerged lance furnaces is cooled to approximately 400°C, with concurrent air agitation for refining. Elements such as copper, lead, and zinc are also removed by chemical flux addition and removal of the resulting dross (Smith 1996).

Gold, Silver, Platinum, and Palladium Refining

Precious metals such as gold, silver, platinum, and palladium tend to deport to copper and nickel mattes in pyrometallurgical smelting, conversion, and refining processes (Avarmaa et al. 2015). Consequently, many of the processes associated with copper and nickel primary production are also utilized for the recovery of these precious metals. In these processes, matte and alloy target compositions are often dictated by the refining processes used to recover the precious metals (Barrett and Knight 1989).

UNIT OPERATIONS FOR NONFERROUS CONVERSION

Significant progress was made in the early 20th century in improving conversion productivity through the successful adoption of larger vessels, high-quality basic refractories, improved process control, and better primary and secondary hooding. Although the batch Peirce–Smith converter dominated, alternative designs modeled after the submerged tuyere concept were also introduced. The Hoboken converter (Belgium) was developed to give improved off-gas capture with some measure of success (Johnson et al. 1979) but has not been widely adopted. The Noranda converter (Canada) and the Shuikoushan/bottom-blowing converter (SKS/BBC) and submerged lance conversion and refining (SLCR) converters (China), which are also based on the submerged tuyere concept, were recently developed for continuous conversion, thus offering an improved conversion approach compared to conventional batch conversion. Nevertheless, the classic Peirce–Smith converter remains the predominant submerged tuyere configuration, representing almost three-quarters of the current global reported copper and nickel conversion capacity.

The last half of the 20th century also saw the steady development and commercialization of top lance and suspension or flash conversion, such as the Mitsubishi Materials Corporation (MMC) and Outotec–Kennecott processes applied to copper (Kapusta 2004). More recently, batch and continuous top submerged lance conversion have been deployed, such as the IsaSmelt and Ausmelt processes on copper and high-platinum-group-metal (PGM) nickel mattes (Warner et al. 2007).

An inspection of Table 1 shows that in the last 25 years, 36% of the newly installed or modernized 11.0 Mt (million metric tons) Cu/yr global conversion capacity was based on alternatives to Peirce–Smith converters. The last decade saw a similar level of alternative converter technology adoption, amounting to 36% of the installed capacity of 6.8 Mt Cu/yr. The pace of new technology adoption in the last five years has increased to represent 54% of the 4.0 Mt Cu/yr of newly installed, modernized, or under-construction capacity.

Submerged Tuyere Conversion Furnace

The submerged tuyere conversion furnace refers to furnace technology whereby the process gas is injected into the melt via submerged tuyeres to effect the required metallurgical reaction. The vessel is fitted with a turning mechanism so it can be turned about 50° to bring the tuyeres out of the melt for operations such as charging matte or skimming (by rotating the vessel further).

Included in this category of furnaces are the copper and nickel-copper converters, commonly known as Peirce–Smith converters, the Noranda converter, the SKS/BBC (Yie 2014), and the SLCR (Cui et al. 2016) in China. The Peirce–Smith converters typically operate in batch mode, as will be described here; however, several versions on new submerged tuyere converter furnaces operating continuously have been recently developed. These include the Noranda converter developed in Canada and the SKS/BBC and SLCR developed in China. Today, there are more than 200 Peirce–Smith converters worldwide at more than 60 smelters (considering those smelters that produce over 50,000 t Cu/yr).

Peirce–Smith Converter Furnace

Since development of the Manhès converter, vessels were gradually increased in size (Wraith et al. 1994; Mackey 2014). A major breakthrough in converter design and operation came in 1909 with development in the United States of the Peirce–Smith horizontal converter, so named after the two developers. This vessel, which was a huge success and widely adopted, included magnesite brick lining, silica flux feeding, means for tuyere punching, and a robust turning mechanism.

This unit became the main converter "workhorse" of the copper industry from that time to the present day, and it is expected to operate well into the future. Up until the 1990s, all major copper smelters used a Peirce–Smith converter for copper conversion. These vessels are also used in nickel-copper conversion. The units increased in size to match the ever-increasing demand for copper through the 20th century (Wraith et al. 1994; Mackey 2014). By about the 1940s, the industry generally adopted a "standard" vessel size of 4.0 m diameter by 9.1 m—first introduced in about 1910—and vessels this size still operate today at some smaller plants.

Most large smelters now use bigger vessels—a large converter today is typically 4.5 m diameter by 13.8 m long, having an inside volume 100 times that of the first units developed in France in the 1880s. As an example, such large vessels were installed in 2000 at the Ronnskar smelter, Sweden; typical blowing rates are ~46,000 Nm^3/h. The modern converter today has similar overall features as the original barrel units, such as submerged air injection, flux addition, tuyere punching, and so forth, but has developed into a robust, high-efficiency unit with advanced process controls and very efficient fugitive gas capture (Figure 3).

Key ancillary features of the Peirce–Smith converter include

- Air delivery (single-stage compressor, 110–130 kPa),
- Drive mechanism and emergency turnout (generally electric drive),
- Hood (see Figure 3),
- Automatic tuyere punching machine,
- Off-gas handling, and
- Advanced process control instruments and controls.

Table 1 Conversion technology copper plus nickel/PGM new or modernized capacity

Company	Country	Plant (Name) Location	Annual Production, t/yr Cu	Converter Furnace Type	Start and Last Expansion Year
Mitsubishi Materials Corporation	Japan	Naoshima	270,000	**Top lance—Mitsubishi**	(1974) 1991
Mount Isa Mines	Australia	Mount Isa	232,000	Peirce–Smith	1992
Freeport-McMoRan	USA	Miami, Arizona	175,000	Hoboken	1992
Glencore	Chile	(Altonorte) La Negra	375,000	Peirce–Smith	1992
Rio Tinto Kennecott Copper	USA	(Kennecott) Salt Lake City, Utah	272,200	**Suspension/flash—Outotec**	1995
Anglo American	Chile	Chagres	180,000	Peirce–Smith	1995
Pan Pacific Copper	Japan	Saganoseki	472,650	Peirce–Smith	(1973) 1996
Aurubis	Bulgaria	Pirdop	210,000	Peirce–Smith	1996
Glencore	Canada	(Horne) Rouyn-Noranda, Quebec	188,145	**Noranda converter**	1997
Jinlong Copper Company (JV Sumitomo)	China	Tongling, Anhui	300,000	Peirce–Smith	1997
Hindalco Birla Copper	India	Dahej, Gujarat	250,000	Peirce–Smith	1998
PT Smelting (Mitsubishi Materials Corporation JV)	Indonesia	Gresik	271,000	**Top lance—Mitsubishi**	1998
LG-Nikko Copper Inc.	South Korea	Onsan II	262,000	**Top lance—Mitsubishi**	1998
Zhongtiashan Non-Ferrous Metals Company	China	Houma, Linfen, Shanxi	75,000	**Top submerged lance—Ausmelt**	1999
Boliden Mineral	Sweden	Rönnskär	230,000	Peirce–Smith	2000
Yunnan Copper Company	China	Kunming, Yunnan	220,000	Peirce–Smith	2002
Anhui Tongdu Copper	China	Tongling, Anhui	187,500	Peirce–Smith	2003
National Iranian Copper	Iran	Khatoon Abad	80,000	Peirce–Smith	2004
Hindalco Birla Copper	India	Dahej, Gujarat	250,000	**Top lance—Mitsubishi**	2005
Sterlite Industries	India	Tuticorin	325,000	Peirce–Smith	2005
Mopani Copper Mines	Zambia	Mufulira	162,500	Peirce–Smith	2006
Russian Copper Company	Russia	(Karabash) Chelyabinsk	137,500	Peirce–Smith	2006
Anglo American	South Africa	(Waterval) Rustenburg (Ni/PGM)	22,000	**Top submerged lance—Ausmelt**	(2003) 2006
Jiangxi Copper Company	China	Guixi, Yingtan, Jiangxi	700,000	Peirce–Smith	(1985) 2007
Mitsubishi Materials Corporation	Japan	Onahama	180,000	Peirce–Smith	(1965) 2007
Chifeng Jinfeng Copper Industry Company	China	Chifeng, Inner Mongolia	100,000	Peirce–Smith	2007
China Nonferrous Metal Mining Company	Vietnam	Bát Xát, Lào Cai	10,000	Peirce–Smith	2007
Grupo Mexico	Peru	(Southern Copper) Ilo	316,500	Peirce–Smith	2007
Jinchuan Non-Ferrous Metals Corporation	China	Jinchuan, Jinchang, Gansu (Ni/Cu)	125,000	Peirce–Smith	2008
Chifeng Jinjian Copper Group	China	Chifeng, Inner Mongolia	100,000	Peirce–Smith	2008
Dongying Fangyuan Nonferrous Metals Company	China	Dongying, Shandong	125,000	Peirce–Smith	2008
Yunnan Copper Company	Zambia	Chambishi	170,000	Peirce–Smith	2009
Yunnan Copper Company	China	Chuxiong, Yunnan	100,000	Peirce–Smith	2009
Daye Nonferrous Metals Company	China	Daye, Huangshi, Hubei	300,000	Peirce–Smith	2010
Shandong Humon Smelting Company	China	Laishan, Yantai, Shandong	75,000	Peirce–Smith	2010
Yanggu Xiangguang Copper Company	China	Yanggu, Liaocheng, Shandong	400,000	**Suspension/flash—Outotec**	(2007) 2011
Yunnan CopperCompany	China	Liangshan, Sichuan	100,000	Peirce–Smith	2011
Zijin Copper Company Ltd	China	Shanghang, Longyan, Fujian	200,000	Peirce–Smith	2011
Chifeng Fubang Copper Industry Company	China	Linxi, Chifeng, Inner Mongolia	60,000	Peirce–Smith	2011
Xinjiang Xinxin Mining Industry Company	China	Xinxin, Xinjiang	90,000	Peirce–Smith	2011
Kazzinc Ltd	Kazakhstan	Ust-Kamenogorsk	70,000	Peirce–Smith	2011
Baotou Huading Copper Development Company	China	Baotou, Inner Mongolia	100,000	Peirce–Smith	2012
Yunnan Tin Corp	China	Gejiu, Honghe, Yunnan	100,000	**Top submerged lance**	2012
Tongling Non-Ferrous Metals Group	China	Tongling, Anhui	400,000	**Suspension/flash - Outotec**	2012
Wuxin Copper Company Ltd.	China	Ürümqi, Xinjiang	100,000	Peirce–Smith	2013
Jinchuan Non-Ferrous Metals Company	China	Fangchenggang, Guangxi	400,000	**Suspension/flash—Outotec**	2013
Yuguang Gold and Lead Company	China	Jiyuan, Henan	100,000	**Submerged tuyere—SKS/BBC ENFI**	2013
Zhongtiashan Group	China	Yuanqu, Shanxi	125,000	Peirce–Smith	2014
Huludao Copper Group	China	Huludao, Liaoning	100,000	Peirce–Smith	2014
First Quantum Metals	Zambia	Kansanshi	300,000	Peirce–Smith	2014
Rudarsko Topioicarski Basen	Serbia	Bor	100,000	Peirce–Smith	(1961) 2015
Henan Zhongyuan Gold Smelter Company	China	Sanmenxia, Henan	375,000	**Suspension/flash—ENFI**	2015
Dongying Fangyuan Nonferrous Metals Company	China	Dongying, Shandong	375,000	**Submerged tuyere—SKS-SLCR**	Commission
China Minmetal Corporation	China	Hengyang, Hunan	100,000	**Peirce–Smith**	Construction

Note: Alternatives to Peirce–Smith conversion are identified in bold; operations that have been shut down are not included in the table.

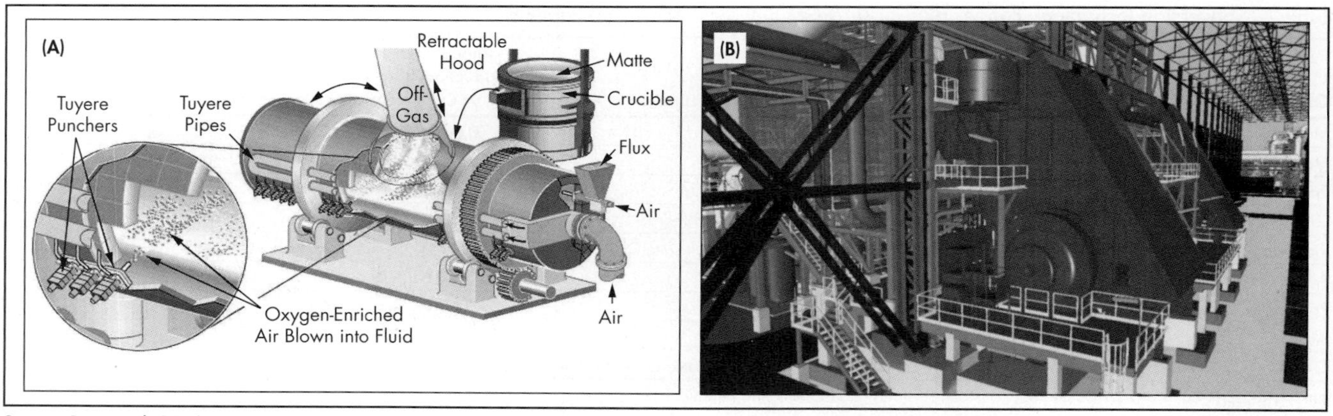

Source: Butts et al. 1999 Source: Dundee Precious Metals 2014

Figure 3 (A) Cutaway view showing the main operating features of the Peirce–Smith converter; and (B) modern, fully enclosed hooding devices to capture fugitive emissions

Flow sheet and process description. This section describes the typical process sequence in the operation of the copper Peirce–Smith converter. With the vessel "out of blowing"—turned so the tuyeres are out of the melt—the charge of molten matte produced in the smelting furnace is added by ladle through the mouth of the vessel (Figure 3). Typical copper matte today contains about 60% Cu, 15% Fe, 23% S, and 1%–2% O; there may also be small quantities of trace elements present, such as lead, zinc, and so on. The vessel is then turned to blowing position with the air turned on along with the introduction of silica flux, and the conversion operation commences. Blowing continues in stages until the level of iron is reduced by forming an iron-silicate slag; this blowing cycle is referred to as the "slag blow." The slag is skimmed off and additional ladles of matte added. This blowing–skimming sequence continues two or three times until the iron level is reduced to a few percent, the vessel is considered full, and all the slag has been skimmed. The resulting matte is then essentially present as copper sulfide (also referred to as "white metal"). The "copper blow" then commences, whereby the blowing air oxidizes all the sulfur to form metallic copper with the final melt referred to as blister copper. The bath temperature increases over this cycle because of oxidation of Cu_2S, and cold material such as copper scrap material can be added.

Great skill is required to detect the "end point," and even though new instrumentation is now available to measure the end point accurately, many plants still rely on the long-ago method of sampling the copper bath near the end of the cycle (by an iron bar thrust through a tuyere) and observing the surface of the sample upon cooling. The Peirce–Smith converter effectively handles the elimination of minor elements present in matte; published data are available on this topic (Schlesinger et al. 2011).

The converter off-gas is cooled and cleaned of dust in an electrostatic precipitator and directed to the sulfuric acid plant. Plants in Europe and Japan have high sulfur capture, in particular, and typically capture 99.3%–99.8% of sulfur input to the plant. The converter slag is cooled and copper typically recovered by a grinding-flotation process (Nesset 2011). The resulting high-copper slag concentrate is recycled to the smelting furnace. At some smelters, converter slag is cleaned in an electric furnace.

A typical heat balance for the copper blow of a Peirce–Smith converter is presented in Table 2. This is given in terms of both 1,000 kg of white metal and for a typical 4 m-by-9 m vessel, which until recently was the standard vessel size used in industry. The data in the table can be scaled to any size of vessel (and correcting for heat losses). Generally today, software tools such as HSC Chemistry (Outotec) or FactSage (Thermfact and GTT-Technologies) are used for converter computations. As the converter process is autogenous, fossil fuel energy requirements are low—the main energy requirement is the electrical power to drive the air blowers and fans to control off-gases.

A typical flow sheet for the Peirce–Smith converter within a modern smelter complex is illustrated in Figure 4. It can be seen that, in addition to treating matte from the flash furnace, the Peirce–Smith converters here also treat matte from the electric furnace and the Kaldo/TBRC (top-blown rotary converter) plant. A smelter normally has two or more converters; typically, one (or more) is blowing and one is charging (one will also be down for brick re-lining). The usual arrangement of converter vessels is in the area of the plant known as the "converter aisle."

Furnace size and productivity. Selected physical parameters of a large, modern submerged tuyere converter (Peirce–Smith converter) are presented in Table 3. The total cycle time for a vessel such as the one shown in the table (from empty converter ready to take matte to empty converter after skimming blister copper) is approximately 7–8 hours. There may be two or more such blows per day, equivalent to over 600 t Cu/d. Careful scheduling is required to maximize crane movement for the highest plant efficiency. Specialized software developed for the type of discrete event scheduling encountered here is helpful in optimizing plant operations (Coursol et al. 2009).

Furnace configuration. The Peirce–Smith converter is lined with chrome-magnesite refractory bricks. This lining slowly wears over the course of a vessel campaign; the 4.5 m by 13.4 m vessel presented in Table 3 can have a campaign life of about 3 months, representing more than 130 charges, after which a new refractory lining is installed.

Process control. The modern converter vessel is equipped with advanced process controls for the highest efficiency. All input and output materials are recorded, and all process and

Table 2 Heat balance in the copper blow of a Peirce–Smith converter

Heat Input			Heat Output*		
	Heat Amount, million kcal			Heat Amount, million kcal	
Item	1,000 kg White Metal (Cu_2S)	Charge in Industrial Converter (2-hour blow)	Item	1,000 kg White Metal (Cu_2S)	Charge in Industrial Converter (2-hour blow)
Sensible heat in white metal†	0.209	21.6	Sensible heat in blister copper	0.148	15.3
Heat of reaction	0.326	33.8	Sensible heat of off-gases	0.329	34.0
			Converter heat losses	0.024	2.5
			Heat absorbed by cold copper	0.034	3.5
Totals	**0.535**	**55.3**	**Totals**	**0.535**	**55.3**

*Heat capacity data is from Schuhmann 1952.
†White metal is assumed here at 1,300°C and the blast air is assumed to be at 25°C (heat input).

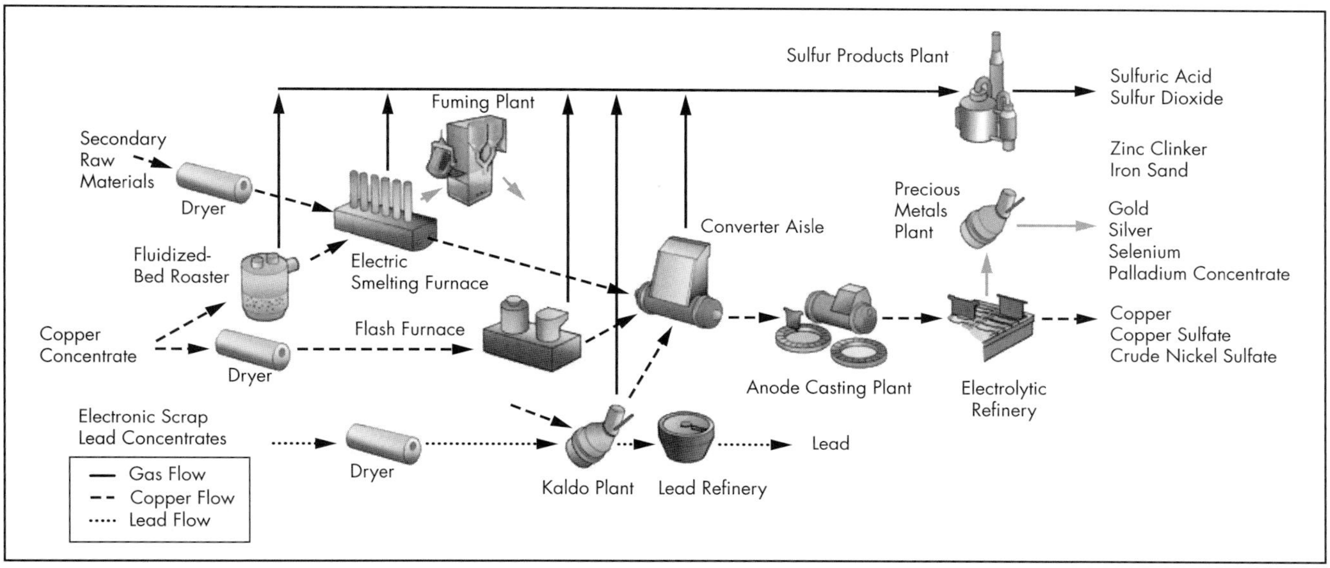

Adapted from Sundqvist 2010

Figure 4 Peirce–Smith converters in a typical smelter flow sheet

off-gas flows as well as temperatures are also monitored and recorded. Modern instruments are now available to monitor off-gases by optical analysis techniques, typified by the Semtech instrument (Ek and Olsson 2005), to monitor and control the converter end point. The vessels are fully hooded for process and fugitive gas collection.

Advanced process modeling tools such as FactSage or HSC Chemistry are now used to accurately model and track the process during the blow. The paper by Tan (2009) illustrates the good correlation between the observed vessel temperature and that predicted by the model during a slag blow at the Mount Isa smelter in Queensland, Australia. The model can generally predict temperature changes due to stoppages to add matte and/or skim slag over the converter cycle.

Continuous Conversion—Submerged Tuyere Processes

A major advancement in conversion by submerged tuyere technology is the development of a continuous conversion process whereby air or oxygen-enriched air blowing is continuous, with matte additions typically added on an intermittent but regular basis. Blister copper is produced and removed continuously or tapped intermittently. Three commercial processes based on submerged tuyere technology are the Noranda continuous converter, the SKS/BBC, and the SLCR processes.

Noranda continuous converter. The Noranda converter commenced on a commercial scale at the Glencore Horne smelter in Canada in 1998 following several years of testing and piloting (Boisvert et al. 1998; Prevost et al. 1999). The Noranda converter vessel is 4.5 m in diameter and 19.8 m long and is fitted with up to 42 submerged tuyeres for introducing air or oxygen-enriched air, an end-wall port for feeding solid charge, a matte feed mouth, an off-gas mouth, two copper tapholes in the vessel barrel, and a single slag taphole (Figure 5). The operation of the vessel is continuous with air or oxygen-enriched blown continuously, copper and slag tapped periodically while blowing, and high-grade matte from the adjacent Noranda process reactor is transferred by crane and added periodically. The copper is transferred for finishing, then handled conventionally in the anode furnace. After tapping, the slag is treated for copper recovery by milling and flotation. The off-gas, rich in sulfur dioxide, is treated in the smelter acid plant.

Typical metallurgical data for the Noranda converter are presented in Tables 4 and 5. Table 4 shows a typical mass balance, and Table 5 presents a typical heat balance for slightly different operating conditions. In Table 5, the molten matte charge was 1,166.4 t/d (48.6 t/h), the solid charge was 165.6 t/d (6.9 t/h), and the solid carbon addition was 24 t/d (1 t/h). The copper tapped was 864 t/d (36 t/h), and the slag tapped was 182.4 t/d (7.6 t/h). The tuyere blowing rate was 34,750 Nm^3/h, and there was 5,000 Nm^3/h of air at the feed port. The tuyere oxygen enrichment at 23.5% was lower than shown in Table 4; hence, relatively less solid charge could be charged.

Oxygen bottom-blowing continuous converter. The SKS/BBC and the SLCR converter were developed in China over the past few years to provide for continuous conversion by submerged tuyere injection using a horizontal cylindrical furnace akin to a Noranda reactor vessel. The SKS/BBC was first commercialized at the Yuguang copper smelter in China in 2013 (Figure 6; Yie 2014). The smelter incorporates an SKS smelting vessel and the new SKS/BBC. The continuous converter is 4.1 m in diameter and 18 m long with a design capacity equivalent to 100,000 t/yr of cathode copper. The vessel is fitted with the same type of concentric tuyeres as on the SKS smelting vessel. The single continuous conversion vessel is positioned at a slightly lower elevation compared to the SKS smelting furnace so that molten matte at about 74% Cu content flows directly by launder to the conversion vessel, entering via a port in the end wall (Figure 6). Matte addition by launder is generally continuous but can be intermittent for a short period. Several similar continuous conversion vessels have since been commissioned in China.

A continuous bottom-blowing converter process (called SLCR) with a similar concentric tuyere design has recently been implemented in a different mode of operation and on a larger scale at the Dongying Phase II smelter in China. Here, combining conversion and refining in a single vessel, two such vessels are installed that operate in tandem to handle matte from one SKS type smelting vessel, now referred to as the submerged lance smelting (SLS) process. The new converter vessels, 4.8 m diameter by 23 m with 17 tuyeres, referred to as the SLCR process, carry out conventional conversion and then anode refining in the same vessel (Figure 7) so that the final product is anode copper (Luraschi and Schachter 2015; Cui et al. 2016). The Dongying Phase II plant commenced operations in September 2015.

Table 3 Selected physical parameters of a modern, large Peirce–Smith converter treating matte of about 65% Cu

Item	Generic Furnace
Charge capacity, total matte/charge, t	350
Diameter (inside steel dimension), m	4.5
Length (inside steel dimension), m	13.4
Number of tuyeres	60
Tuyere type	Single pipe
Refractory weight, t	210
Blister production,* t/charge	310
Nominal blowing volume, Nm^3/h	46,200
Nominal off-gas volume, Nm^3/h	110,000

Source: Lehner and Wiklund 2001; Ek and Olsson 2005
*Includes the recycle of some scrap copper.

Table 4 Typical mass balance operating data for the Noranda converter

Item	Data
Molten matte from Noranda reactor, t/d	887
Solid coolant addition (solid matte, scrap, etc.), t/d	551
Silica flux, t/d	78
Solid carbon addition, t/d	20
Tuyere blowing rate, Nm^3/h	18,600
Oxygen enrichment, %	41.6
Number of blowing tuyeres in operation	13
Operating temperature, °C	1,218
Copper production rate, t/d	887
Slag production rate, t/d	252

Source: Prevost et al. 2013

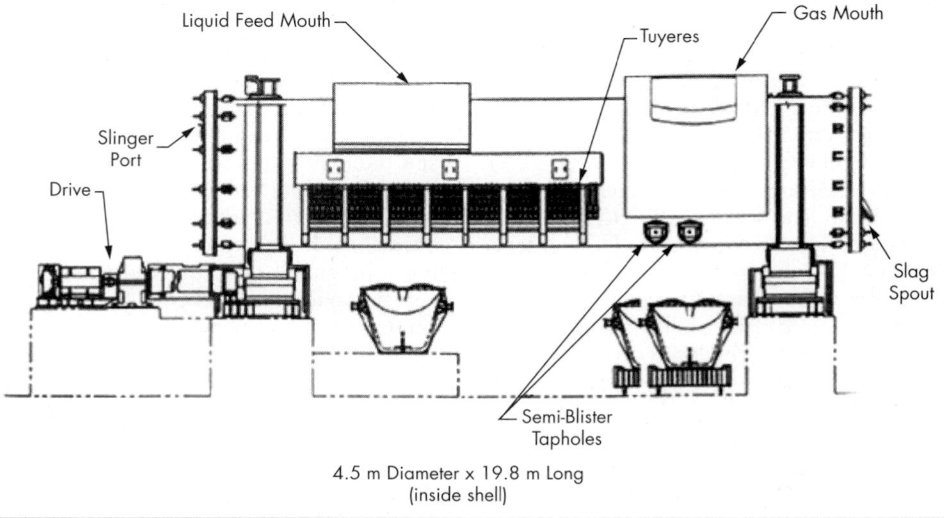

Source: Boisvert et al. 1998. Copyright 1998 by The Minerals, Metals & Materials Society. Used with permission.
Figure 5 Longitudinal section of Noranda continuous converter

Table 5 Typical heat balance operating data for the Noranda converter

Heat Input		Heat Output	
Item	**Million kcal/h**	**Item**	**Million kcal/h**
Sensible heat in molten matte and injected air	8.91	Sensible heat in tapped copper	6.27
Heat of reactions (including coal combustion)	21.22	Sensible heat of off-gases	18.50
		Sensible heat of slag	2.13
		Vessel heat losses	3.28
Total	**30.13**	**Total**	**30.13**

Source: Mackey et al. 1995

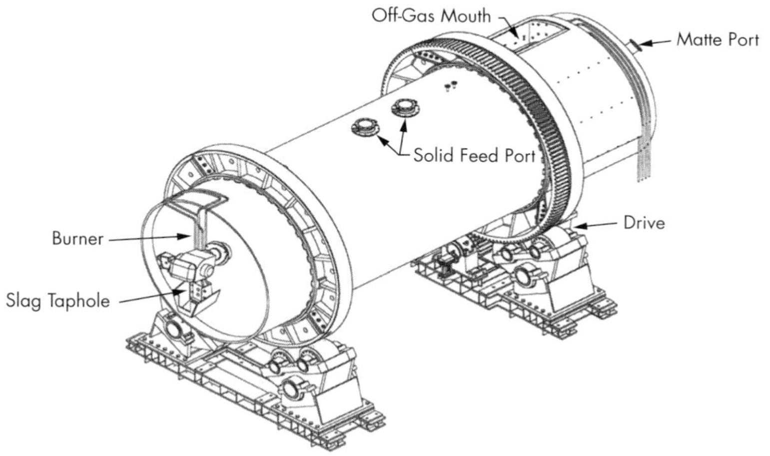

Note: Oxygen injectors and copper taphole are behind vessel.

Adapted from Yie 2014
Figure 6 SKS/BBC continuous converter at the Yuguang smelter, China

At Dongying Phase II, matte from the SLS (also known as SKS/BBS) furnace is laundered to one of two SLCR units. The operation of these vessels is as follows: One of the two units starts to be filled with matte to a level that allows concurrent conversion to commence by blowing oxygen-enriched air. After ~400 t of contained copper is received, the matte is redirected to the other unit, converting slag is removed, and refining commences. The sulfur level is lowered, then natural gas and air are used to transform the blister copper to anode copper. Following casting, the unit is readied to take a new charge of matte when the other unit has completed its conversion cycle. Typical operating data for the SLCR process are given in Table 6.

Suspension/Flash Conversion Furnace

Suspension or flash conversion (FC) involves continuous combustion of comminuted solid matte above a molten bath where the combustion heat supports furnace operation at the temperatures required. The first patent to directly describe the general principles of this approach was granted to Haglund in 1940; however, it would be about 55 years later before a commercial plant treating copper matte would be commissioned.

In the late 1940s, Inco investigated applying the principles of their existing suspension smelting technology to conversion (Victorovich 1993). Chalcocite concentrate containing about 2%–4% Ni derived from the selective flotation of copper-nickel matte leaves a mushy nickel oxide residue when converted in the submerged tuyere Peirce–Smith furnace. The objective was to flash-convert the chalcocite concentrate at higher temperature to reduce residue formation. Further work was to wait until the late 1970s, with deployment in 1985 at the Copper Cliff (Ontario, Canada) smelter. The technology used a surplus Peirce–Smith converter modified to convert 3% nickel chalcocite concentrate by a type of FC arrangement and was operated for eight years before being replaced by a larger unit (Queneau and Marcuson 1996).

In the late 1970s, Kennecott started development of the so-called solid matte oxygen conversion process (Richards et al. 1983). Kennecott subsequently entered into a development and marketing agreement with Outotec in the early 1980s (George 2002). In early 1991, a large-scale granulation and subsequent pilot conversion program was initiated to mature the process and furnace technology. Outotec's success with the commercial application of direct-to-blister furnace technology at Glogow (Poland) and Olympic Dam (Australia) increased confidence in scale-up and operation.

In 1995, Rio Tinto Kennecott commenced operation of the first commercial application of suspension conversion using FC technology at their Salt Lake City (Utah, United States) smelter. Outotec FC technology was commissioned at the Yanggu smelter in 2007 (expanded in 2011), the Tongling smelter in 2012, and the Fangchenggang smelter in 2013 (all in China). See Table 7 showing key physical parameters for exemplar FC furnaces. The Sanmenxia (China) smelter commissioned a modified FC technology in 2015 that utilizes a cyclonic burner design developed at the Yanggu smelter.

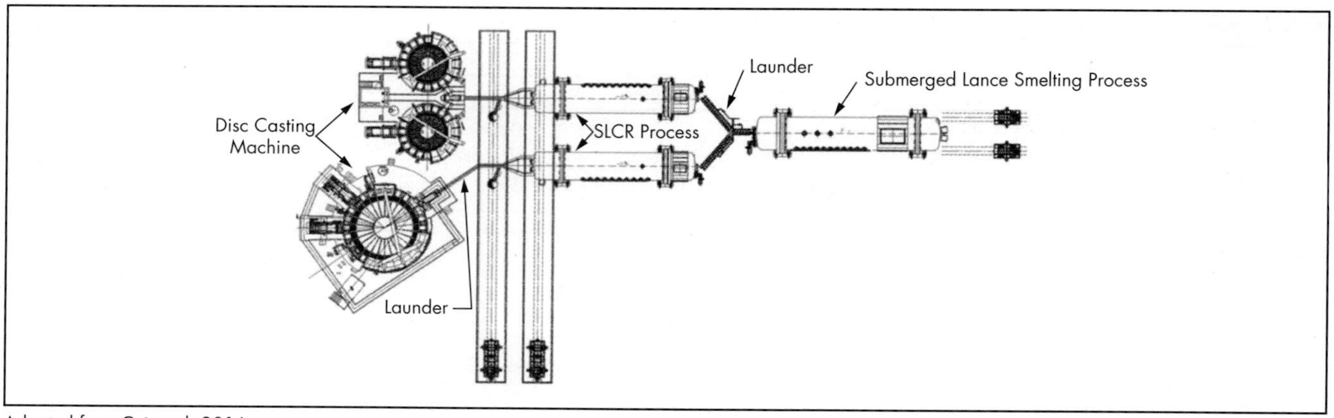

Adapted from Cui et al. 2016

Figure 7 SLCR conversion furnaces at the Dongying Phase II smelter

Table 6 Operating data for the SLCR process at the Dongying Phase II smelter, China

Item	Data
Matte charging rate by launder from SLS (73%–78% Cu), t/h	50–60
Solid charge addition	Variable
% O_2 in conversion air	21–30
Tuyere pressure, MPa	0.65–0.75
Copper content of SCLR slag (iron-silicate type), %	8–10
Typical conversion/reduction cycle time, hours	16
Casting rate, t/h	100
Casting time, hours	4
Copper temperature, °C	1,250

Source: Cui et al. 2016, with permission from Dongying Fangyuan

Table 7 Physical parameters of exemplar Outotec flash conversion furnaces

Item	Salt Lake City (Utah, United States)	Tongling (China)
Year started	1995	2012
Shaft height (inside dimensions), m	6.5	7
Shaft diameter (inside dimensions), m	4.25	5
Settler length (inside dimensions), m	18.75	21.8
Settler width (inside dimensions), m	6.5	6.7
Blister produced, t/d	1,100	1,240
Off-gas volume, Nm^3/h	16,000	38,000

Source: George et al. 1995; Zhou and Sun 2013

Flow Sheet and Process Description

At Kennecott, screened, granulated matte from the smelting process is conveyed to a matte storage dome where it is radially stacked and reclaimed in a continuous operation. Stockpiled matte can be introduced and stacked with new matte or physically added to the stacked pile using mobile equipment. Capacity can range from 7,000 to 10,000 t. Reclaimed matte is delivered to a combined dryer and grinding circuit consisting of an air heater, vertical roller mill with integrated classifier (similar to those used to pulverize cement and coal), and baghouse cluster to collect product. The dried and ground matte is pneumatically conveyed to a feed storage bin located above the FC furnace. A second storage bin receives pneumatically conveyed burnt lime flux, and a third obtains recycled dust collected in the furnace off-gas system.

Matte, flux, and dust are fed into the furnace reaction shaft burner using independent loss-in-weight feeders and distribution systems. Oxygen is supplied to the burner (can be greater than 96% in the plenum) to satisfy process requirements, and distribution air is provided at the shaft entry to control plume geometry (effectively reducing process enrichment to ~85% O_2). Blister copper is tapped from the settler into one of two anode furnaces via heated and ventilated refractory-lined and covered launders and catch basins. Slag is tapped and delivered to a granulator via water-cooled copper launders with steel covers to ensure adequate ventilation. Granulated FC slag is screened and conveyed to a storage bin for recycle to the smelting furnace. Provision is made for stockpiling and introduction of this slag as process conditions dictate.

Off-gas from the furnace is directed to the waste heat boiler consisting of radiative and convective sections. The cooled, partially de-dusted off-gas passes by means of a vertical duct to the dry electrostatic precipitator where the majority of remaining entrained dust is removed. The partially sulfated dust that settles in the boiler sections is collected using drag conveyors, commingled with the more fully sulfated dust from the precipitators, and pneumatically conveyed to the furnace dust bin. Depending on process requirements, some of this dust may be diverted to the smelting part of the process.

The cooled and de-dusted gas is ducted to wet gas scrubbers to humidify and cool the gas for subsequent ducting to wet electrostatic precipitators before treatment in the acid plant. System draft is controlled independently using a fan either ahead of or after the wet gas scrubbers.

Furnace Configuration

Since start-up of the first flash converter at Salt Lake City, furnace containment design has seen many changes to the original arrangement. The first major permanent changes were performed primarily in response to major melt containment design deficiencies. The changes made are described in more detail by Kennecott authors and are reflected in Figure 8 (Newman and Weaver 2002). Blister taphole design is a critical feature of these furnaces, and Figure 9 contrasts the end wall

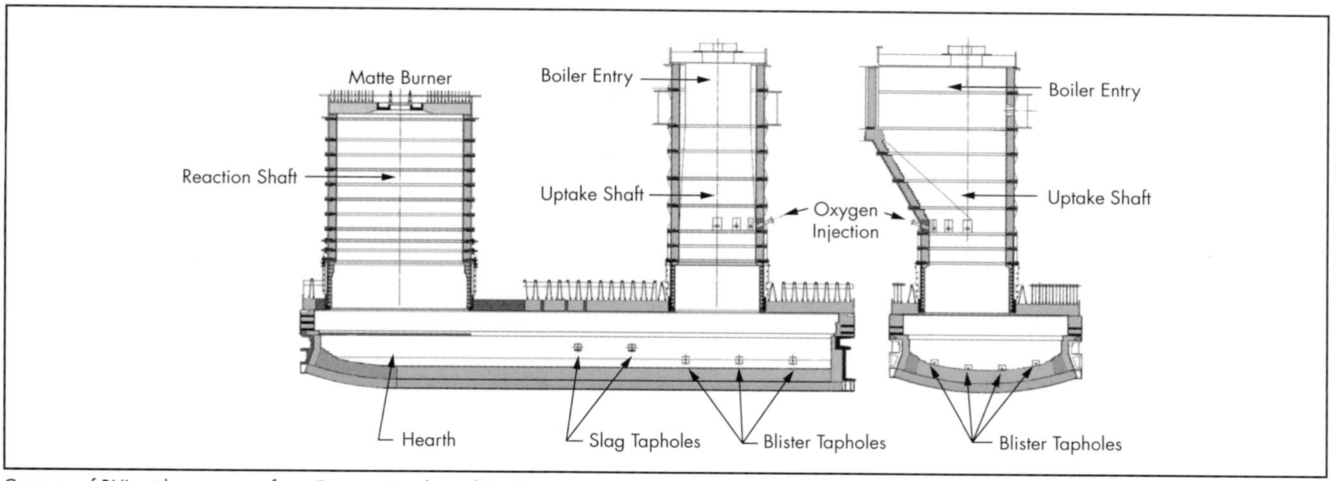

Courtesy of RHI, with permission from Outotec, Hatch, and Rio Tinto

Figure 8 Flash conversion furnace elevation longitudinal and lateral sections from the Salt Lake City smelter

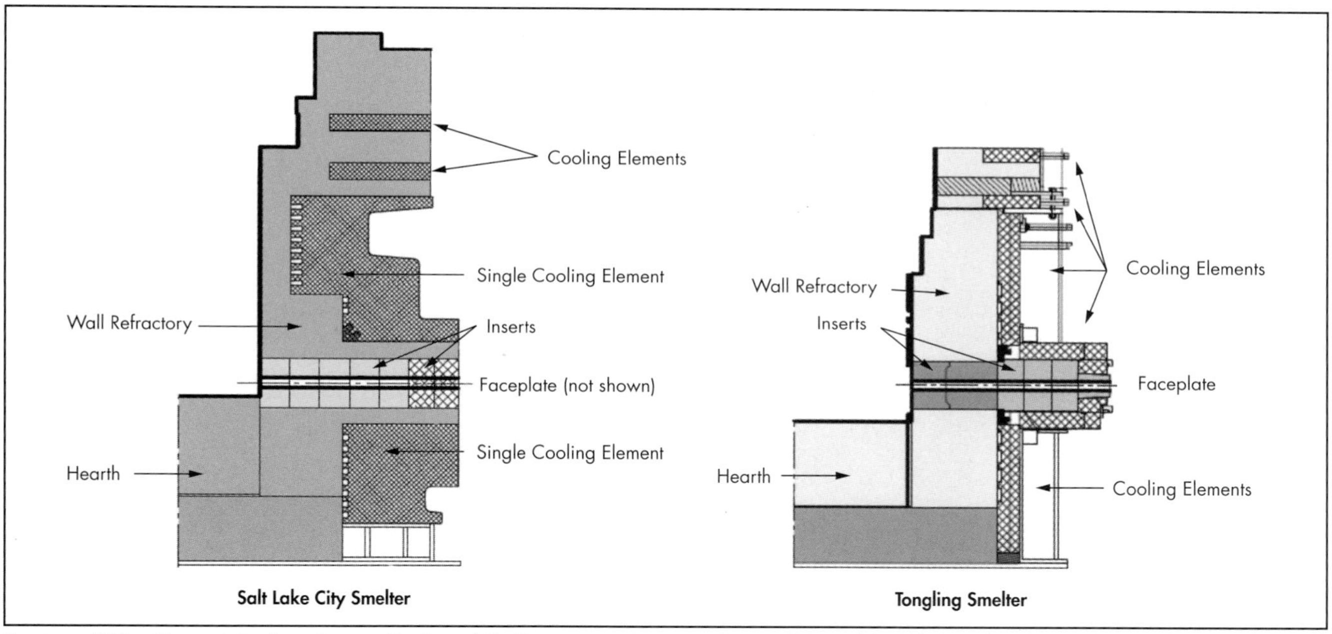

Courtesy of RHI, with permission from Outotec, Hatch, and Rio Tinto   Source: Zhou and Sun 2013, with permission from RHI

Figure 9 Flash conversion furnace blister taphole elevation longitudinal sections from two smelters

blister taphole designs for the Salt Lake City and Tongling converters. Both designs are considered reasonably robust.

Ancillary Equipment

Granulation of batch tapped matte and converter slag are also critical support unit operations for flash conversion. Granulated product is typically sized at P_{80} 1 mm. Details of the particular matte and converter slag granulator systems are available (George et al. 2006; Lengyel and Loveless 2015).

Operational Elements

Uniform distribution of matte is important in flash conversion, and the system at Kennecott has evolved to meet this requirement. Blister and slag are typically batch tapped with timing coordinated to ensure that slag layer thickness is maintained as the blister level raises and lowers. Under normal operation, no hearth buildup exists; however, during start-up after an extended outage, the blister layer is cooler and diligence is required in achieving a clear first tap, particularly if sidewall buildup is significant.

Servo and regulatory process control of an FC furnace involves maintaining blister sulfur (set indirectly by control of slag copper content), slag composition, and settler temperature through manipulation of oxygen, flux, and auxiliary fuel, respectively. In addition to fuel, manipulation of enrichment and dust charged can be used for temperature control. These parameters are manipulated at the furnace burner entry with close to constant auxiliary fuel on the settler. To accommodate

Table 8 Example flash conversion furnace mass and heat balance

Material	In or Out	Flow	Temperature, °C	Hf, MJ/h*	Hs, MJ/h*
Flash Converter Shaft					
Dome matte	In	70.000 t/h	25	–47,130.5	0.0
Lime flux	In	1.261 t/h	25	–13,791.6	0.0
Dust charged	In	7.480 t/h	25	–32,287.4	0.0
Shaft air-oxygen	In	10,907 Nm³/h	25	0.0	0.0
Disperser air	In	1,711 Nm³/h	25	0.0	0.0
Shaft product	Out	67.259 t/h	1,648	–80,303.4	74,249.4
Shaft flux	Out	1.263 t/h	1,648	–13,800.0	1,924.7
Shaft off-gas	Out	10,762 Nm³/h	1,648	–119,134.9	39,000.8
Losses	Out	—	—	—	12,552.0
Sum of (out – in)	—	—	—	–120,028.8	127,726.9
Shaft natural gas	In	293 Nm³/h	25	7,698.1	← Net H
Shaft natural gas off-gas	Out	938 Nm³/h	1,648	—	4,220.8
Flash Converter Settler					
Shaft product	In	67.259 t/h	1,648	–80,303.4	74,249.4
Shaft flux	In	1.263 t/h	1,648	–13,800.0	1,924.7
Shaft off-gas	In	10,762 Nm³/h	1,648	–119,134.9	39,000.8
Slag	Out	10.047 t/h	1,250	–42,890.9	11,274.2
Blister	Out	47.330 t/h	1,240	–1,568.3	36,591.6
Dust	Out	4.550 t/h	1,300	–9,874.8	3,921.9
Off-gas	Out	13,011 Nm³/h	1,300	–148,918.7	36,650.3
Losses	Out	—	—	—	16,736.0
Sum of (out – in)	—	—	—	9,985.7	–10,000.9
Shaft natural gas off-gas	In	938 Nm³/h	1,648	—	4,220.8
	Out	—	1,300	—	3,465.2
Settler injection gas	In	255 Nm³/h	25	0.0	0.0
	Out	—	1,300	0.0	485.9
Air ingress	In	850 Nm³/h	25	0.0	0.0
	Out	—	1,300	0.0	1,556.3
Settler natural gas	In	85 Nm³/h	25	1,271.5	← Net H
Settler natural gas off-gas	Out	925 Nm³/h	1,300	—	2,189.2
Waste Heat Boiler					
Furnace off-gas	In	15,978 Nm³/h	1,300	–164,190.3	44,346.9
Dust	In	4.550 t/h	1,300	–9,874.8	3,921.9
Sulfation air	In	0 Nm³/h	25	0.0	0.0
Air ingress	In	2,549 Nm³/h	25	0.0	0.0
Off-gas to electrostatic precipitator	Out	15,558 Nm³/h	350	–155,753.7	10,816.5
Dust to electrostatic precipitator	Out	4.581 t/h	350	–18,301.8	1,065.7
Dust from waste heat boiler	Out	2.244 t/h	250	–8,979.7	349.3
Losses	Out	—	—	—	1,255.2
Sum of (out – in)	—	—	—	–8,970.1	–34,782.1
Steam produced	—	19.918 t/h	—	—	–43,752.3

*Hf = estimated standard formation; Hs = sensible enthalpies.

these requirements, control systems and procedures similar to those described in Chapter 11.5, "Smelting," for flash smelting furnaces are utilized. Monitoring for process and furnace/equipment integrity is typically similar to that adopted for flash smelting furnaces.

Energy Flows

An example of a mass and heat balance of the Kennecott FC furnace and boiler is shown in Table 8 and is typical of the internal control calculations utilized in real time, discussed earlier. In these examples, separate balances for the reaction shaft and settler sections of the furnace are shown, assuming a nominal shaft sulfur elimination of 80% and shaft oxygen fixed according to Reaction 1. Under these conditions, the calculated shaft exit temperature is quite high compared to the settler. In practice, all control variables are manipulated at the reaction shaft except settler freeboard heating, which is normally targeted to be held constant, resulting in a variable actual reaction shaft exit temperature. This effectively reduces the required balance to a combined, simplified one for control purposes.

Top Lance Conversion Furnace

Top lance conversion involves either batch processing of a charge of solid or molten matte or continuous conversion of a molten matte stream. Top-blown rotary converter (TBRC) furnaces derived from the single-lance Kaldo steel process are used for the batch processes, Outotec and Andritz Maerz being the major technology suppliers. Top lance continuous conversion uses multiple-lance technology with a stationary furnace developed and supplied by MMC.

In 1958, Inco successfully trialed the use of a small Kaldo steel furnace at the Domnaverts Steel Works in Sweden to convert nickel matte to crude nickel (Queneau and Marcuson 1996), leading to subsequent piloting at their Port Colborne (Canada) facility. In 1971, the Copper Cliff smelter in Canada commissioned two TBRCs, producing nickel for subsequent refining.

In 1974, the Tennant Creek smelter in Australia operated two 25-t TBRCs in the belief that good bismuth elimination could be achieved given the efficient mixing, gas contact, and higher operating temperature. While some measure of success was achieved with the conversion operation, the smelter closed shortly after start-up because of operating difficulties coupled with low metal prices. A single 73-t TBRC was installed at the Afton smelter in British Columbia, Canada, for combined smelting and conversion when the smelter commenced in 1978. The plant was generally deemed to have been successful but closed in 1983 because of low copper prices (Nickel et al. 1980).

In 1990, the Stillwater PGM smelter in Montana (United States) commenced operation with two 1.5-t TBRC vessels for converting solid matte produced by an electric furnace (Roset et al. 1992). The plant has expanded several times since then, and larger TBRCs are now used for matte conversion.

TBRCs are now widely employed to treat precious metal–containing alloys from lead production, anode slime from copper electrorefining, and copper scrap/e-scrap. A 10-t TBRC is used at Glencore's Montreal (Quebec, Canada) refinery to process de-coppered anode slimes to produce doré metal (Lessard 1989).

The copper scrap processing facility of Metallo-Chimique NV (now Metallo Belgium) in Beerse, Belgium, treats a range of copper and precious metal scrap materials. Smelting and conversion is successfully carried out in one of the five TBRC units at the plant, including three 150-t units and two smaller ones (Goris 2013).

MMC began development of continuous top lance smelting and conversion processes in 1959 using a small slag fuming furnace at their Naoshima (Japan) facility. A subsequent pilot to demonstration scale-up program from 1968 through the early 1970s at their Onahama (Japan) smelter helped to resolve lance service life and intermediate product transfer issues (MMC 2012).

The Naoshima smelter commissioned the first commercial continuous top lance conversion furnace as part of the integrated Mitsubishi process in 1974 (and expanded in 1991). This was followed by a second captive smelter at Kidd Creek (Canada) in 1981 (shut down in 2010). In 1998, two Mitsubishi-based smelting and conversion facilities were commissioned: one at the Gresik (Indonesia) smelter and one at the Onsan II (South Korea) smelter. The Port Kembla (Australia) smelter replaced their Peirce–Smith converters with a single top lance conversion furnace in 2000 (shut down in 2003). The Dahej (India) smelter adopted the integrated Mitsubishi technology in 2005. See Table 9 showing key physical parameters for exemplar Mitsubishi converters.

Table 9 Physical parameters of Mitsubishi conversion furnaces

Item	Naoshima (Japan)	Gresik (Indonesia)
Year started (expanded)	1974 (1991)	1998
Height (inside dimensions), m	3.6	3.7
Diameter (inside dimensions), m	8.0	9.0
Lance diameter (inside dimensions), m	0.05	0.05
Number of lances	10	10
Blister produced, t/d	950	875
Off-gas volume, Nm^3/h	28,800	27,000

Source: Schlesinger et al. 2011

Mitsubishi Converter

Flow sheet and process description. In a conversion furnace, matte is continuously siphoned from the electric furnace and laundered to the conversion furnace. In the case of a stand-alone converter, the matte is supplied by means of a holding furnace. Oxygen-enriched air (30%–35%) along with limestone flux and granulated slag is, via charge tanks, blown through up to ten rotating process lances to produce blister copper and a calcia-based slag. In addition to using granulated converter slag as a coolant, bundled and shredded scrap is charged to the furnace. The slag continuously overflows from the syphon and is laundered to a granulator for recycling to both the smelting and conversion furnaces. Blister flows through the siphon and is laundered directly to either anode furnaces or a holding furnace.

Furnace size and productivity. Mitsubishi employs advanced furnace design and operation to minimize refractory wear. This includes the installation of specially designed water-cooled elements in critical areas for the furnace (MacRae et al. 1998).

Operation and control. The continuous flow of molten matte to the converter is inferred from measurements made of the launder flowing stream. Total oxygen delivered to the lances is adjusted to achieve target sulfur in blister produced of 0.7%–0.9% indirectly through control of the copper content in slag (14%–15% Cu). Lower sulfur levels are avoided because of the corresponding higher slag aggressiveness to refractory. Expert systems are also used (Goto et al. 1998). Typically, an overall melt depth of 1–1.1 m with 0.12–0.15 m of slag is maintained. A relatively thin layer of slag is targeted to maximize oxygen utilization for a normal lance position of 0.3 m above the bath (Newman et al. 1991).

Top-Blown Rotary Converter

Configuration. The TBRC provides a high degree of mixing of the molten phases and the injected gas with considerable process flexibility. It consists of a rotating vessel (up to 40 rpm) that is tilted to improve mixing and accommodate an inclined oxygen lance (Figure 10).The typical furnace vessel working volume for matte and anode slime applications ranges between 0.8 and 2 m^3. (*Working volume* is defined as the melt volume and is shown in the Figure 10 inset.) Copper and electronic scrap (and also lead concentrate smelting) furnace working volumes are significantly larger and range between 4 and 16 m^3.

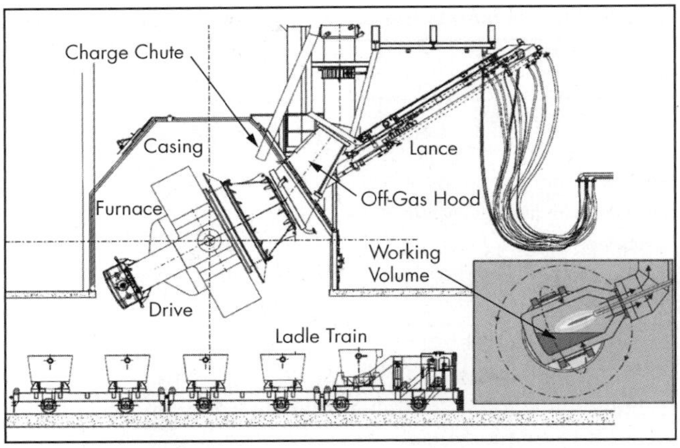

Courtesy of Outotec

Figure 10 Elevation of TBRC general arrangement with furnace schematic (inset)

The general TBRC assembly consists of a furnace that is rotated and tilted using hydraulic power transmission, furnace enclosure or casing and charging equipment ventilated to gas scrubbing, water-cooled conversion lance delivering air/oxygen, water cooled off-gas hood, and ladle train to receive process products. Some facilities use a water-cooled utility lance for flame ignition and process monitoring.

Operation. Because the entire operating vessel is encapsulated inside the ventilated casing, the working environment of a TBRC plant is generally very good. Process off-gases are usually wet gas cleaned. If dust is produced as a filter cake material, such material cannot be introduced to an operating unit (normally added to empty vessel and time must be allowed to dry). Smelting and conversion of anode slime starts with charging of free-flowing material via a chute from charging bins with the furnace vessel in the upright position. This type of unit has the flexibility to achieve both conversion and reduction steps as required. After final refining, the doré metal bath is ready, with levels of antimony, bismuth, selenium and tellurium typically less than 0.01% of each (Lessard 1989).

Roset et al. 1992 describe the Stillwater smelter early experiences with using silica-based slags for nickel-copper-PGM matte conversion. Due to operational issues associated with the frequency and magnitude of excessive melt foaming, practice evolved toward using lime as a flux. This resulted in better melt stability throughout the conversion stages. This plant also now handles considerable quantities of electronic scrap (Roset et al. 2014).

Top Submerged Lance Conversion Furnace

Top submerged lance (TSL) conversion involves either batch processing of a charge of molten matte or continuous melting and conversion of comminuted solid matte. The process has evolved as a logical progression from the analogous smelting applications, and in fact was described in the first patent for the so-called SiroSmelt submerged lance (Floyd 1981). The lance and furnace technology is currently supplied by IsaSmelt (owned by Glencore) and Ausmelt (owned by Outotec).

In 1995, the Bindura (Zimbabwe) nickel-copper smelter commissioned an Ausmelt-supplied TSL conversion furnace (shut down in 2000) to process high copper-nickel leach residue in a three-stage residue smelt, matte convert, and slag reduction operation. The first large-scale Ausmelt conversion furnace treating primary sourced material was commissioned in 1999 at the Houma smelter in China (Mounsey et al. 1999). This facility initially operated on a batch basis, with a separate slag blow and a subsequent copper blow.

In 2002, the Rustenberg (South Africa) PGM smelter commissioned a large TSL converter distinguished by matte fed down the lance and described below (Viviers and Hines 2005). Ausmelt converters were commissioned at the Birla (India) smelter in 2003 (shut down shortly after because of design and operational problems) and the Gejiu (China) smelter in 2012. Matusewicz et al. (2007) describe Ausmelt's "continuous copper conversion (C3)" process and discuss the advantages and disadvantages of silica, calcia/silica, and calcia-based slags. Development work conducted at the Houma smelter led to the selection of ferrous calcia silicate slag for the Gejiu converter. Table 10 shows key physical parameters for exemplar Ausmelt-supplied converters.

In 1997, the Hoboken (Belgium) secondary smelter commissioned an IsaSmelt-supplied TSL converter to process complex blast furnace solid matte (Edwards and Alvear 2007). The operation is performed on a batch basis using a high-lead, silica-based slag in the slag blow with a copper blow or conversion stage to produce blister copper, which is granulated and subjected to sulfate leach/electrowinning to recover the copper and precious metals. IsaSmelt also supplied a copper alloy–producing furnace to the Lünen (Germany) secondary smelter, which was commissioned in 2002. This operation processes copper residues and copper alloy scrap in a smelting step to produce a "black" copper (high tin and lead content) and high-zinc slag. IsaSmelt has described concepts (its so-called IsaConvert process) for large-scale continuous TSL conversion using calcia-based slags for copper (Edwards and Alvear 2007) and nickel-PGM smelters (Bakker et al. 2011).

Flow sheet and process description. The Rustenburg PGM flow sheet (Figure 11) involves the granulation of nickel matte and drying/conveying in a spray dryer via heated air to a bin ahead of the conversion furnaces (one duty, one cold standby). Additionally, matte received via road from other PGM smelters is pneumatically conveyed to dedicated bins. The blended matte is pneumatically conveyed into the converter via the lance. The product matte is slow cooled in molds for subsequent processing to refined precious metals and nickel. Converter slag is granulated and dried for subsequent cleaning to recover contained metal.

The furnace off-gas leaves the converter by means of a Foster Wheeler high-pressure water membrane freeboard and uptake and is then de-dusted and cooled in an evaporative cooler and sent to an acid plant. In the integrated flow sheet at the Gejiu smelter (Figure 12), matte produced from the smelter is separated from the slag in a settling furnace and transferred to the TSL converter either directly via launder (as at Houma) or granulated and subsequently fed through a bin.

Containment principles. The Houma and Gejiu conversion furnaces of Ausmelt design use a chrome magnesite refractory and water-cooled furnace shell. Refractory life of 18 months has been reported with this arrangement. Viviers and Hines (2005) describe the Rustenberg smelter rationale for adopting high-intensity Hatch copper-cooler technology and a water-cooled freeboard for its Ausmelt technology-based converter (Figure 13). Service life has been very good with the changes adopted.

Operational elements. The Houma smelter is reported to operate their converter in batch mode with two stages. The first stage involves laundering molten matte from the settling furnace with concurrent continuous conversion to white metal via lance/bath action. Upon maximum operating level being achieved, feed is interrupted and the slag tapped. The second stage involves conversion of the white metal to blister copper as a batch. It is reported that to accommodate this batch operation, the lance turndown ratio is designed to be 3:1 (Mounsey et al. 1999). Prior to 2007, the slag used was a silica-based, or fayalite, slag. After joint development with Ausmelt, a ferrous calcia silicate slag was adopted, the C3 process (Abbott 2007).

The Gejiu smelter based their converter design and initial operation on the Houma experience. Following commissioning it was decided to change from the two-stage batch process to an almost fully continuous blister-making process, whereby the granulated matte feed is interrupted during slag tapping to allow settling of any entrained copper (Outotec, personal communication).

The Rustenberg PGM smelter reports that its ability to satisfy a very tight tolerance of 3.0 ± 0.5% Fe in final nickel matte was significantly enhanced through a continuous conversion (Viviers and Hines 2005). Operation of the converter was initially intended to be in batch mode with two stages. The first stage involved concurrent feeding and conversion to 13% Fe, the intention being to minimize oxidic metal losses in the initial slag. The second stage involved interruption of feed and subsequent batch conversion to the final nickel matte target composition. This mode of operation proved operationally difficult because of melt stratification, making it difficult to sample while in operation and also resulting in frequent, and violent, bath foaming. Additionally, variable starting conditions for the second stage made it difficult to achieve final target composition. Continuous operation of the converter was adopted because of these problems and involves concurrent feeding and conversion to the final target of 3% Fe with nickel in slag used to control air/oxygen flow rates via the lance. Matte is tapped before the slag, with matte removed until slag is detected to ensure the matte level does not approach the slag taphole. This mode of operation has proven to be a success because of easier lance level control, elimination of matte carryover–induced granulator explosions, product quality that is not impacted by feed rates changes, achievement of constant gas flow to the acid plant, and reduced cycle time through feeding without interruption. However, oxidic losses to the slag are significantly higher, requiring a change in operation of the cleaning furnace.

Table 10 Physical parameters of Ausmelt top submerged lance conversion furnaces

Item	Houma, China	Rustenberg, South Africa	Gejiu, China
Year started	1999	2002	2012
Height (inside dimensions), m	12.6	14.8	~16
Diameter (inside dimensions), m	4.4	4.4	4.4
Lance diameter (inside dimensions), m	0.35	0.4	0.4
Metal/matte produced	~230 t/d metal	~650 t/d matte	~300 t/d metal

UNIT OPERATIONS FOR FERROUS AND NONFERROUS REFINING

A massive increase in primary steel production occurred during the 20th century. Pig iron production in 2014 was 1,190 Mt (USGS 2015), up from about 30 Mt in 1900, a 40-times increase. Most of the pig iron produced in 2014 was refined to various grades of steel using a combination of top lancing and submerged injection of oxygen and inert gases such as a basic oxygen furnace and derivative technologies. Ferronickel

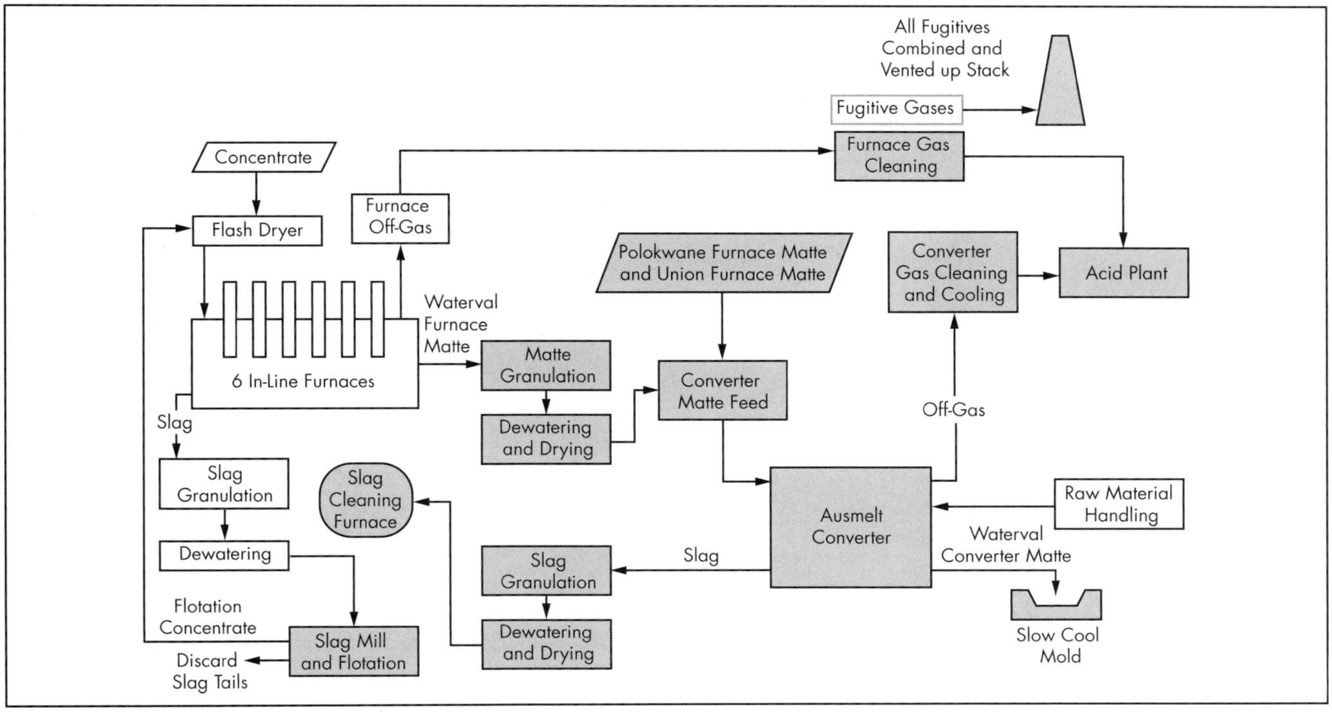

Adapted from Viviers and Hines 2005

Figure 11 Flow sheet for the Rustenberg, South Africa, PGM smelter

was not a production commodity in 1900. The first large-scale electric furnace smelting of calcined laterite ore to produce ferronickel commenced in the 1950s (Thurneyssen et al. 1960). Steel-grade consistency and range also advanced with the adoption of secondary steel refining ladle and degassing technologies. The reader is referred to Barker et al. (2012), Kor and Glaws (2012), and Miller et al. (2012) for a comprehensive treatise on ferrous unit operations.

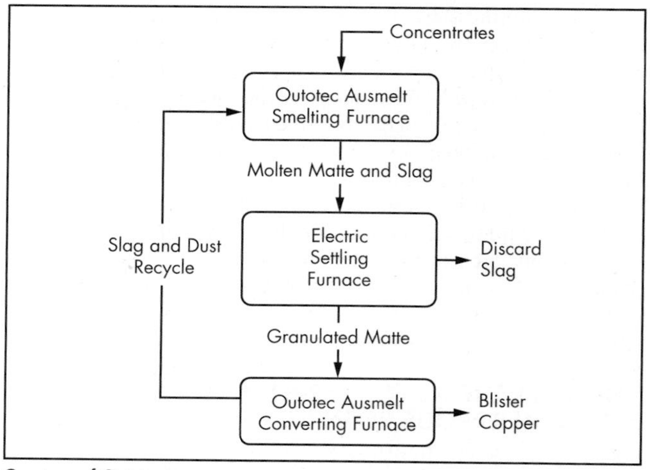

Courtesy of Outotec

Figure 12 General integrated flow sheet used at the Gejiu, China, smelter

Primary copper production in 2014 was 18.7 Mt, up from about 0.6 Mt in 1900, about a 30-times increase. More than 80% of copper produced in 2014 was electrolytically refined from anode copper. Current submerged tuyere anode refining technology was adopted around the middle of the 20th century and has remained dominant given its relative simplicity along with improvements—some in part as a result of steel industry technology influences such as porous plugs and oxygen/fuel burners as now used in copper operations.

Primary lead and tin in 2014 was 5.5 Mt/yr and 0.3 Mt/yr, respectively, mostly derived from batch drossing processes using large kettle technologies. High-productivity continuous drossing technologies have also been employed for lead refining. The reader is referred to Fern and Shaw (1983), Knight and Reader (1996), and Moor (2000) for descriptions of lead unit operations.

Copper Anode Furnace

At the start of the 20th century, demand for high-quality electrolytically refined copper for electrical applications was rapidly expanding, placing greater demand on the throughput and quality of anode copper production. Although rotary furnaces had been in use for some time, one of the first large-size rotary anode furnaces was the 3.9 m diameter by 9.1 m long unit installed in 1949 at the New Cornelia smelter at Ajo, Arizona, United States (Byrkit 1953), but installation and operation of the reverberatory furnace remained common up to the 1950s–1960s. Of interest is the conventional 130-t reverberatory furnace installed at the Gaspe smelter in Canada as late as 1955 when the smelter started operations; however, it was

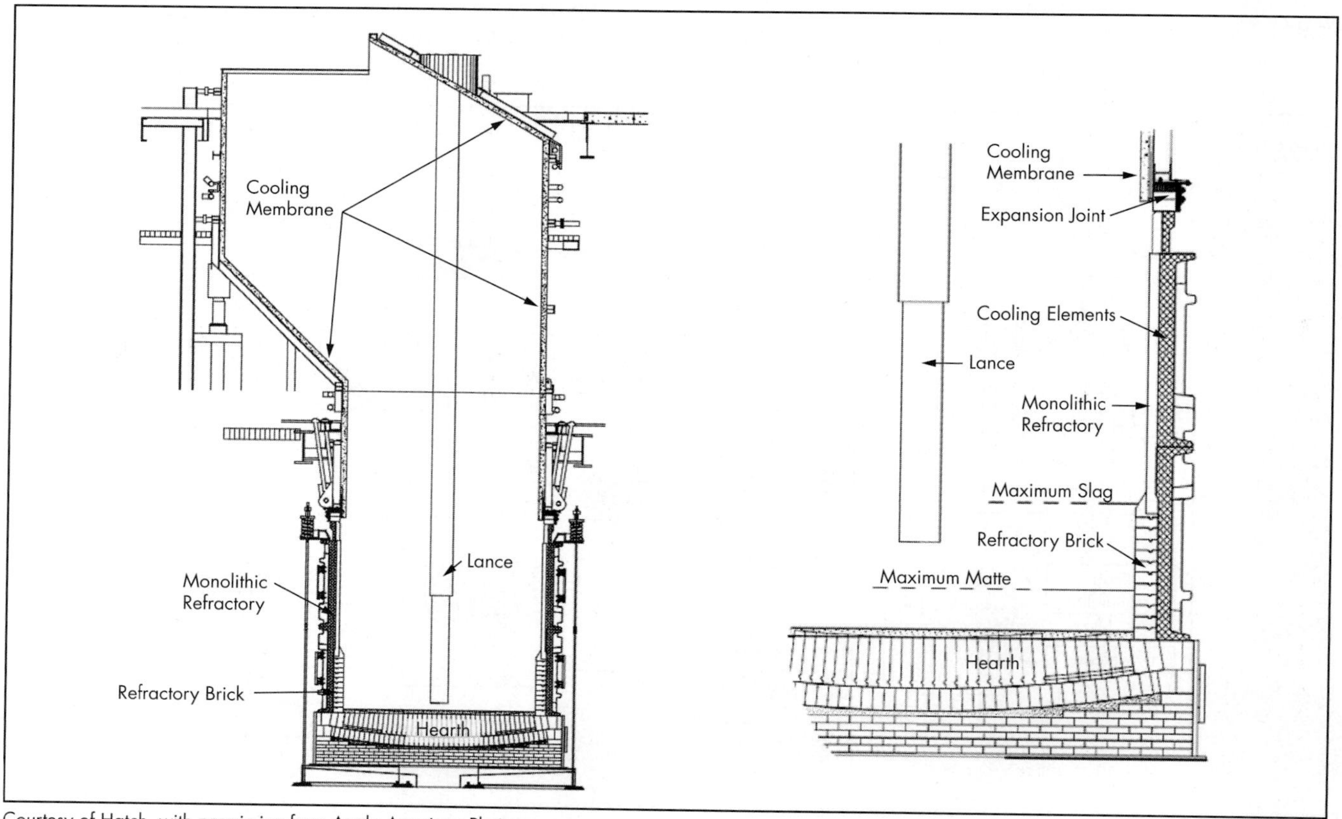

Courtesy of Hatch, with permission from Anglo American Platinum

Figure 13 Anglo conversion process furnace elevation section general arrangement from the Rustenberg smelter

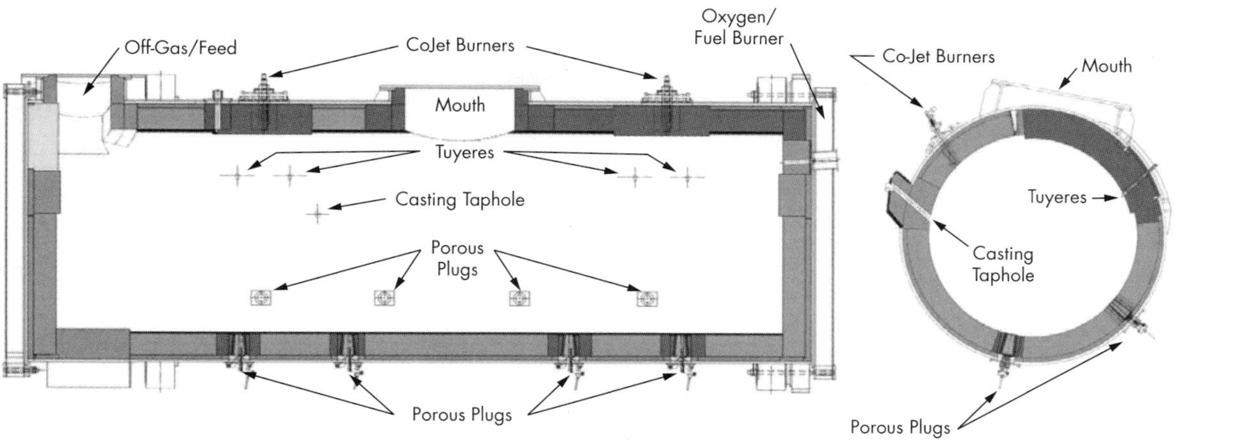

Courtesy of RHI, with permission from Rio Tinto

Figure 14 Anode refining furnace elevation longitudinal and lateral sections from the Salt Lake City smelter

later replaced with a rotary anode furnace (the Gaspe smelter closed in 2002).

The initial operating practices for rotary anode furnaces reflected the then-established practice of using green birchwood poles thrust into the melt for deoxidation (referred to as "poling"). The first use of gaseous reductant occurred in 1961 at the Douglas smelter in Arizona (McKerrow and Pannell 1972). Today, rotary anode furnace technology is basically the only choice available for refining blister copper, regardless of the source.

Configuration and Layout

The electrolytic copper refining process requires a flat, relatively smooth anode surface to be cast for best results. To achieve this, casting operations require copper having low sulfur (<0.003% S) and low oxygen (<0.12% O) contents.

The first step in blister copper refining is blowing with air using a submerged tuyere to remove the last traces of sulfur, followed by a deoxidation or reduction step involving the injection of natural gas or fuel oil, sometimes along with superheated steam, to remove the oxygen picked up by the oxidation step. The finished melt is cast as anodes typically on a circular casting wheel. Anodes today may be up to 950 mm long by 900 m wide by 50–55 mm thick and can weigh ~400 kg.

The submerged-tuyere anode furnace consists of a refractory-lined rotary furnace equipped with a fuel (oil or gas) burner and port, furnace mouth for charging blister copper from Peirce–Smith converters via crane as well as to act as an off-gas port, a copper tapping spout, one to four submerged tuyeres and ancillaries, porous plugs for melt agitation, and the furnace driving mechanism. For furnaces receiving laundered copper from converter alternatives to the batchwise Peirce–Smith converter, a separate off-gas port, also acting as a feed port, is usually used (Figure 14). Slag removal from the mouth is usually more frequent, and porous plug agitation more critical.

Generally, scrap charging through the mouth must also be practiced in these facilities. Some furnaces, such as those deployed at the Salt Lake City smelter, employ Praxair CoJet burners to assist with efficient scrap melting and temperature control.

Tuyere operation involves high-pressure air injection (typically 400–500 kPa) for oxidation and introduction of reductant, typically high-pressure natural gas or fuel oil, for de-oxidation. The tuyere consists of a 25-to-30-mm-diameter stainless-steel pipe (often concentrically set within a larger outer pipe).

Submerged-tuyere anode furnaces in use today are commonly sized about 4.3 m diameter and about 10 m long and hold a 270-to-300-t copper charge. The airflow for oxidation is typically 360–400 Nm^3/h, and the oxidation cycle can range from 0.5 to 0.75 hours. When natural gas is used for deoxidation, the typical flow rate is about 600 Nm^3/h per tuyere, with the deoxidation cycle lasting 3 hours (with two tuyeres operational).

As smelters have increased in size, the size of the anode furnace has also increased. Recent large FC plants have installed anode furnaces up to 4.9 m diameter by 14.2 m long, with a copper charge capacity of more than 550 t.

CONCLUDING REMARKS

In this chapter, the authors have attempted to balance description of the current major technologies employed for conversion and refining for the ferrous and nonferrous industries in the context of a historical narrative. In doing so, they hope to have demonstrated that changes in practice have been, and in all likelihood will continue to be, directed by end-use market demands, raw materials and energy availability and efficiency, production cost, and environmental and workplace hygiene imperatives. Historically, there are many examples where these influences have been so strong that change has been quite rapid (e.g., post–Second World War steel markets worldwide and the adoption of basic-oxygen-furnace technology). However, it is far more frequent in conversion and refining that these forces result in a steady direction away from less-efficient established practice to better alternatives—for example, the vastly greater productivity of horizontal rotary vessels (Peirce–Smith converters and anode refining furnaces) over reverberatory furnaces, development of better capture of fugitive gases from Peirce–Smith converters, adoption of porous plug agitation and CoJet steelmaking technologies by the copper-nickel industry in anode refining, large-scale continuous copper conversion in an FC furnace, and so on.

As practices improve, the ability of the broader conversion and refining industry to adopt these improvements depends on the technical risk, complexity, and the incremental capital cost involved balanced against gains in the cost of production. These factors evolve much more rapidly in an industry that is actively expanding. This is why Europe was the center of change in the 1800s, both North America and Europe were centers of activity in the 20th century, and China has been, and will probably continue to be, the most active in the 21st century.

It is instructive to look at the development of SKS/BBC and SLCR technologies as a 21st-century example of what can result from the "churn" that comes from rapid industry expansion. These technologies are an integration of largely existing ferrous and nonferrous western technologies (e.g., the Savard–Lee shrouded tuyere; Kapusta et al. 2015) specifically adapted in China to meet a strategic need for an alternative for retrofitting the country's numerous modestly sized copper smelters. The apparently successful commercial application in smelting (SKS/bottom-blowing smelting) clearly reduced the perceived technical risk involved in adopting the same approach for copper conversion (BBC) in several new, and large, facilities. And along the same trajectory, this has recently been extended to integrating conversion and refining in the same vessel, known as the SLCR process developed in China (Cui et al. 2016).

History reveals that there is significant value in maintaining cross-industry awareness of developments. The scale of pig iron–derived steel refining is 80 times larger than that of copper conversion and refining. Operational practices and vendor development technologies of the larger industry can probably continue to benefit an astute operator in the smaller industry. The final matte produced in the smelting/conversion of nickel ores/concentrates is now treated hydrometallurgically, and zinc is now almost exclusively won and refined hydrometallurgically after roasting the sulfide concentrates. Can the same trend be expected for copper concentrate treatment?

At the current scale of copper concentrate smelting, with its substantial by-production of sulfuric acid, relatively benign slag, and excellent environmental performance, it seems unlikely that the large tonnage processes of smelting and conversion will be replaced in the foreseeable future. However, given that the majority of concentrates are now shipped long distances to third-party custom smelters, could these facilities be motivated to adopt alternative hydrometallurgical treatment routes for blister copper using processes that lower energy requirements and reduce working inventory costs?

ACKNOWLEDGMENTS

The authors wish to acknowledge the assistance of Rob Knight (retired), Bob Drew of RHI, Gary Walters of Hatch, Markku Lahtinen of Outotec, Gerardo Alvear recently of Glencore (now Aurubis), and Katsuhiko Abe of Mitsubishi Materials Corporation.

REFERENCES

Abbott, P. 2007. Ausmelt signs agreement for continuous copper conversion. March 28th. http://member.afraccess.com/media?id=CMN&filename=20070328/00706614.pdf.

Ashman, D.W., Goosen, D.W., Reynolds, D.G., and Webb, D.J. 2000. Cominco's new lead smelter at Trail Operations. In *Lead-Zinc 2000*. Edited by J.E. Dutrizac, J.A. Gonzalez, D.M. Henke, S.E. James, and A.H. Siegmund. Warrendale, PA: The Minerals, Metals & Materials Society. pp. 171–185.

Avarmaa, K., O'Brien, H., Johto, H., and Taskinen, P. 2015. Equilibrium distribution of precious metals between slag and copper matte at 1250–1350°C. *J. Sustain. Metall.* 1:216–228.

Bakker, M.L., Nikolic, S., Burrows, A.S., and Alvear, G.R.F. 2011. ISACONVERT™—continuous conversion of nickel/PGM mattes. *JOM* 63(5):60–65.

Barker, K.J., Paules, J.R., Rymarchyk, N., and Jancosko, R.M. 2012. Oxygen steelmaking furnace mechanical description and maintenance considerations. *The Making, Shaping and Treating of Steel*, 11th ed., Vol. 2. Warrendale, PA: Association for Iron and Steel Technology.

Barrett, K.R., and Knight, R.P. 1989. Operation of the bottom blown oxygen cupel at Britannia Refined Metals Ltd. In *Precious Metals '89*. Warrendale, PA: The Minerals, Metals & Materials Society. pp. 455-473.

Bessemer, H. 1905. *Sir Henry Bessemer, F.R.S.: An Autobiography*. London: Offices of Engineering.

Blanderer, J. 1984. The refining of lead by oxygen. *J. Met.* 36(12):53–54.

Bloxam, C.L. 1871. *Metals: Their Properties and Treatment*, 2nd ed. London: Longmans, Green and Company.

Boisvert, M., Janneteau, G., Landry, J-P., Levac, C.A., Perron, D., McGlynn, F., Zamalloa, M., and Poretta, F. 1998. Design and construction of the Noranda converter at the Horne smelter. In *Sulphide Smelting '98: Current and Future Practices*. Warrendale, PA: The Minerals, Metals & Materials Society. pp. 569–583.

Butts, A., Schlechten, A.W., and Taylor, J.C. 1999. Copper processing. In *Encyclopaedia Britannica.* www.britannica.com/technology/copper-processing/images-videos/A-side-blown-copper-nickel-matte-converter/113914.

Byrkit, J.W. 1953. Operations at New Cornelia copper smelter of Phelps Dodge Corporation. *J. Met.* 5(5):633–642.

Coursol, P., Mackey, P.J., Morissette, S., and Simard, J.M. 2009. Optimization of the Xstrata Copper–Horne Smelter Operation using discrete event simulation. *CIM Bull.* 102(1114):1–10.

Cui, Z., Wang, Z., Wang, H., and Wei, C. 2016. Two-step copper smelting process at Dongying Fangyuan. Paper No. PY16-4. Presented at the Copper 2016 Conference, Kobe, Japan, November 13–16.

Dannatt, C.W., and Ellingham, H.J.T. 1948. Roasting and reduction processes—A general survey. *Discuss. Faraday Soc.* 4:126–139.

Davenport, W.G., Jones, D.M., King, M.J., and Partelpoeg, E.H. 2001. *Flash Smelting: Analysis, Control and Optimization.* Warrendale, PA: The Minerals, Metals & Materials Society.

Dawson, R.D. 1989. Technical developments from 100 years of lead smelting at Port Pirie. In *Non-Ferrous Smelting Symposium: 100 Years of Lead Smelting and Refining in Port Pirie*. Melbourne, Victoria: Australasian Institute of Mining and Metallurgy.

Dundee Precious Metals. 2014. Understanding the Tsumeb Smelter [Presentation]. March 20, 2014. Toronto, ON: Dundee Precious Metals.

Edwards, J.S., and Alvear, G.R.G. 2007. Conversion using IsaSmelt™ technology. In *Proceedings of Copper 2007*. Vol. 3. Montreal, QC: Canadian Institute of Mining, Metallurgy and Petroleum. pp. 17–28.

Ek, M., and Olsson, P. 2005. Recent developments on the Peirce–Smith conversion process at the Ronnskar smelter. In *Conversion and Fire Refining*. Warrendale, PA: The Minerals, Metals & Materials Society. pp. 19–26.

Fern, J.H., and Shaw, R.W. 1983. Copper separation and recovery at the B.H.A.S. lead smelter, Port Pirie, South Australia. In *MMIJ/AusIMM Joint Symposium*. pp. 71–82.

Floyd, J.M. 1981. Submerged injection of gas into liquid-pyrometallurgical bath. U.S. Patent 4,251,271.

George, D.B. 2002. Continuous copper conversion—A perspective and view of the future. In *Sulfide Smelting 2002*. Warrendale, PA: The Minerals, Metals & Materials Society.

George, D.B., Gottling, R.J., and Newman, C.J. 1995. Modernization of Kennecott Utah Copper smelter. In *Proceedings of Copper 1995*. Vol. 4. Montreal, QC: Canadian Institute of Mining, Metallurgy and Petroleum. pp. 41–52.

George, D.B., Nexhip, C., George-Kennedy, D., Foster, R., and Walton, R. 2006. Copper matte granulation at the Kennecott Utah Copper smelter. In *EPD Congress 2006*. Warrendale, PA: The Minerals, Metals & Materials Society. pp. 577–589.

Goris, D. 2013. Do You Know Metallo-Chemique NV. Recycling Metals from Industrial Waste. Short course and workshop, Colorado School of Mines, Golden, CO, June 25–27.

Goto, M., Oshima, E., and Hayashi, M. 1998. Control aspects of the Mitsubishi continuous process. *JOM* 50(4):60–64.

Haglund, T.R. 1940. Roasting process. U.S. Patent No. 2,209,331.

Hofman, H.O. 1914. *Metallurgy of Copper*. New York: McGraw-Hill.

Johnson, R.E., Themelis, N.J., and Eltringham, G.A. 1979. A world-wide survey of copper conversion practice. *J. Met.* 31(6):28–36.

Kapoulitsas, P., Giunti, M., Hampson, R., Cranley, A., Gray, S., Kretschmer, B., Knight, R., and Clark, I. 2000 Commissioning and optimisation of the new lead and silver refinery at the Pasminco Port Pirie smelter. In *Lead-Zinc 2000*. Warrendale, PA: The Minerals, Metals & Materials Society. pp. 187–201.

Kapusta, J.P.T. 2004. JOM world nonferrous smelters survey, Part I: Copper. *JOM* 56(7):21–27.

Kapusta, J.P.T., Larouche, F., and Palumbo, E. 2015. Adoption of high oxygen bottom blowing in copper matte smelting: Why is it taking so long? Presented at the 2015 Conference of Metallurgists, Metallurgical Society of CIM, Toronto, ON, August 2015.

Knight, R.P., and Reader, R.J. 1996. Refining of lead bullion and precious metals recovery at Britannia Refined Metals Ltd. In *Lead into the Future Conference*. London: Institution of Mining and Metallurgy. pp. 53–73.

Kor, G.J.W., and Glaws, P.C. 2012. Ladle refining and vacuum degassing. In *The Making, Shaping and Treating of Steel*, 11th ed., Vol. 2. Warrendale, PA: Association of Iron and Steel Technology.

Lehner, T., and Wiklund, J. 2001. The new Ronnskar smelter. In *Proceedings of EMC 2001*. Clausthal-Zellerfeld: GDMB. pp. 49–59.

Lengyel, A.J., and Loveless, M. 2015. Improved maintenance from copper matte granulation nozzle design at the Kennecott Utah Copper smelter. Database No. 621040. www.researchdisclosure.com/.

Leroy, J.L., Lenoir, P.J., and Escoyez, L.E. 1970. Lead smelter operation at N.V. Metallurgie Hoboken S.A. In *Extractive Metallurgy of Lead and Zinc*. New York: AIME. pp. 824–852.

Lessard, B. 1989. Anode slimes treatment in a top blown rotary converter at the CCR division of Noranda Minerals Inc. In *Precious Metals 1989*. Warrendale, PA: The Minerals, Metals & Materials Society. pp. 427–440.

Luraschi, A., and Schachter, H. 2015. Challenges in competitiveness for the Chilean smelters. Presentation at Smelter Seminar, Instituto de Ingenieros de Minas de Chile, October.

Mackey, P.J. 2014. Evolution of the large copper smelter—1800s to 2013. In *Celebrating the Megascale: Proceedings of the Extraction and Processing Division Symposium on Pyrometallurgy in Honor of David G.C. Robertson*. Edited by P.J. Mackey, E.J. Grimsey, R.T. Jones, and G.A. Brooks. Warrendale, PA: The Minerals, Metals & Materials Society. pp. 17–37.

Mackey, P.J., Harris, C., and Levac, C. 1995. Continuous conversion of matte in the Noranda Converter, Part I: Overview and Metallurgical Background. In *Proceedings of Copper 95-Cobre 95 International Conference: Vol. IV, Pyrometallurgy of Copper*. Edited by W.J. Chen, C. Diaz, A. Luraschi, and P.J. Mackey. Montreal, QC: Canadian Institute of Mining, Metallurgy and Petroleum. pp. 337–349.

MacRae, A., Wallgren, M., Wasmund, B., Lenz, J., Majumdar, A., Zuliani, P., and Elvestad, P. 1998. Conversion furnace upgrades at the Kidd Metallurgical Division copper smelter. In *Sulfide Smelting '98: Current and Future Practices*. Warrendale, PA: The Minerals, Metals & Materials Society. pp. 387–397.

Mathewson, E.P. 1913. Development of the basic-lined converter for copper mattes. In *Trans. AIME*. pp. 1033–1037.

Matusewicz, R., Hughes, S., and Hoang, J. 2007. The Ausmelt continuous copper conversion (C3) process. *Proceedings of Copper 2007*. Vol. 3. Montreal, QC: Canadian Institute of Mining, Metallurgy and Petroleum. pp. 29–47.

McKerrow, G.C., and Pannell, D.G. 1972. Gaseous deoxidation of anode copper at the Noranda smelter. *Can. Metall. Q.* 11:629–633.

Miller, T.W., Jimenez, J., Sharan, A., and Goldstein, D.A. 2012. Oxygen steelmaking processes. In *The Making, Shaping and Treating of Steel*, 11th ed., Vol. 2. Warrendale, PA: Association of Iron and Steel Technology.

MMC (Mitsubishi Materials Corporation). 2012. History: The Mitsubishi process. www.mmc.co.jp/sren/History.htm. Accessed September 2018.

Moor, P.J. 2000. Recent developments in the lead refining operations at Britannia Refined Metals Ltd. In *Lead-Zinc 2000*. Warrendale, PA: The Minerals, Metals & Materials Society. pp. 345–359.

Mounsey, E.N., Li, H., and Floyd, J.W. 1999. The design of the Ausmelt Technology smelter at Zhong Tiao Shan's Houma smelter, People's Republic of China. In *Proceedings of Copper 99-Cobre 99 International Conference*. Vol. 5. Warrendale, PA: The Minerals, Metals & Materials Society. pp. 357–370.

Nesset, J.E. 2011. Significant Canadian Developments in Mineral Processing Technology—1961–2011. In *The Canadian Metallurgical and Materials Landscape 1960 to 2011*. Westmount, QC: Canadian Institute of Mining, Metallurgy and Petroleum. pp. 241–293.

Newman, C.J. and Weaver, M.M. 2002. Kennecott flash conversion furnace design improvements—2001. *Sulfide Smelting 2002*. Warrendale, PA: The Minerals, Metals & Materials Society. pp. 317–328.

Newman, C.J., MacFarlane, G., Molnar, K., and Storey, A.G. 1991. The Kidd Creek smelter—An update on plant performance. In *Proceedings of the Copper 91–Cobre 91 International Symposium*. Vol. 4. New York: Permamon Press. pp. 65–80.

Nickel, W.P.T., Siewert, F., and Thornton, G.W. 1980. Commissioning the Afton smelter. Paper No. A80-55. Presented at the 109th AIME Annual Meeting, Las Vegas, Nevada, February.

Pelletier, A., Mackey, P.J., Southwick, L., and Wraith, A.E. 2009. Before Peirce and Smith—The Manhès converter: Its development and some reflections for today. In *International Peirce–Smith Conversion Centennial*. Warrendale, PA: The Minerals, Metals & Materials Society. pp. 29–49.

Prevost, Y., Lapointe, R., Levac., C.A., and Beaudoin, D. 1999. First year of operation of the Noranda converter. In *Proceedings of Copper 99-Cobre 99 International Conference*. Vol. 5. Warrendale, PA: The Minerals, Metals & Materials Society. pp. 269–282.

Prevost, Y., Bedard, M., and Levac, C. 2013. First 15 years of Operation of the Noranda Converter. In *Ralph Lloyd Harris Memorial Symposium*. Westmount, QC: Canadian Institute of Mining, Metallurgy and Petroleum. pp. 115–124.

Queneau, P.E., and Marcuson, S.W. 1996. Oxygen pyrometallurgy at Copper Cliff—A half century of progress. *JOM* 48(1):14–21.

Richards, K.J., George, D.B., and Bailey, L.K. 1983. A new continuous copper conversion process. In *Advances in Sulfide Smelting*. Vol. 2. Warrendale, PA: TMS-AIME. pp. 489–498.

Roset, G.K., Matousek, J.W., and Marcantonio, P.J. 1992. Conversion practices at the Stillwater precious metals smelter. *JOM* 44(4):39–42.

Roset, G.K, Flynn, D., and Schumacher, K. 2014. Modifications to a smelter to accommodate recycled materials. In *Celebrating the Megascale: Proceedings of the Extraction and Processing Division Symposium on Pyrometallurgy in Honor of David G.C. Robertson*. Edited by P.J. Mackey, E.J. Grimsey, R.T. Jones, and G.A Brooks. Warrendale, PA: The Minerals, Metals & Materials Society. pp. 237–241.

Royal Ontario Nickel Commission. 1917. *Report of the Royal Ontario Nickel Commission*. Toronto, ON: A.T. Wilgress.

Schlesinger, M.E., King, M.J., Sole, K.C., and Davenport, W.G. 2011. *Extractive Metallurgy of Copper*, 5th ed. Amsterdam: Elsevier.

Schuhmann, R., Jr. 1952. *Metallurgical Engineering: Volume I, Engineering Principles*. Cambridge, MA: Addison-Wesley Press. pp. 59–60.

Smith, R. 1996. An analysis of the processes for smelting tin. In *Mining History: The Bulletin of the Peak District Mines Historical Society*. Vol. 13. pp. 91–99.

Southwick, L. 2009. William Peirce and E.A. Cappelen Smith, and their amazing conversion machine. In *International Peirce–Smith Conversion Centennial*. Edited by J. Kapusta and T. Warner. Warrendale, PA: The Minerals, Metals & Materials Society. pp. 3–27.

Sundqvist, R. 2010. *Welcome to Boliden Capital Markets Day*. Presentation at Capital Markets Day, November 1–2.

Tan, P. 2009. Application of thermo-chemical and thermo-physical modelling in copper converter industries. In *International Peirce–Smith Conversion Centennial*. Edited by J. Kapusta and T. Warner. Warrendale, PA: The Minerals, Metals & Materials Society. pp. 273–295.

Tan, P., and Vix, P. 2006. Thermodynamic modeling of copper drossing process in lead refining. In *EPD Congress 2006*. Warrendale, PA: The Minerals, Metals & Materials Society. pp. 509–515.

Thurneyssen, C.G., Szczeniowski, J., and Michel, J. 1960. Ferronickel smelting in New Caledonia. In *Extractive Metallurgy of Copper, Nickel and Cobalt*. Edited by P. Queneau. New York: Interscience Publishers. pp. 287–300.

Turkdogen, E.T., and Fruehan, R.J. 2012. Fundamentals of iron and steelmaking. In *The Making, Shaping and Treating of Steel*, 11th ed., Vol. 1. Warrendale, PA: Association of Iron and Steel Technology.

USGS (U.S. Geological Survey). 2015. *Mineral Commodity Summaries*. Washington, DC: USGS.

Victorovich, G.S. 1993. Oxygen flash conversion for production of copper. In *The Paul E. Queneau International Symposium—Extractive Metallurgy of Copper, Nickel and Cobalt*. Vol. 1. Warrendale, PA: The Minerals, Metals & Materials Society. pp. 501–529.

Viviers, P., and Hines, K. 2005. The New Anglo Platinum conversion project. In *First Extractive Metallurgy Operators' Conference*. Melbourne, Victoria: Australasian Institute of Mining and Metallurgy. pp. 101–108.

Warner, A.E.M., Díaz, C.M., Dalvi, A.D., Mackey, P.J., Tarasov, A.V., and Jones R.T. 2007. JOM world nonferrous smelter survey, part IV: Nickel: Sulfide. *JOM* 59(4):58–72.

Weeks, J.D. 1896. The invention of the Bessemer process. In *American Manufacturer and Iron World*. Feb. 21. pp. 259–261.

Wilder, A.B. 1949. The Bessemer converter process. *J. Met.* (November):22–27.

Wraith, A.E., Harris, C.L., and Mackey, P.J. 1994. On factors affecting tuyere flow and splash in converters and bath smelting reactors. In *From Agricola to the Present—A Meeting Between East and East: Responsibility for the Future—Metallurgy I, EMC '94*. Clausthal-Zellerfeld: GDMB. pp. 50–78.

Yie, Y. 2014. Development of the new oxygen-enriched bottom blowing continuous copper conversion process. In *The Davenport Honorary Symposium*. Westmount, QC: Canadian Institute of Mining, Metallurgy and Petroleum.

Zamalloa, M., Reyes, A., Fañas, J.J., and Frias, J.R. 2009. New developments in ferronickel refining at Xstrata Nickel's Falcondo smelter. In *Pyrometallurgy of Nickel and Cobalt 2009*. Edited by J. Liu, J. Peacey, M. Barati, S. Kashani-Nejad, and B. Davis. Montreal, QC: Canadian Institute of Mining, Metallurgy and Petroleum. pp. 195–208.

Zhou, J., and Sun, L. 2013. Flash smelting and flash conversion process and start-up at Jinguan Copper. In *Proceedings of Copper 2013*. Santiago, Chile: Chilean Institute of Mining Engineers. pp. 377–384.

PART 12

Processing of Selected Metals, Minerals, and Materials

CHAPTER 12.1

Alumina

Benny E. Raahauge, Ken Evans, Manfred Bach, Daniel Thomas, and Peter-Hans ter Weer

HISTORICAL CONTEXT

In 1988, the Bayer process celebrated its 100-year anniversary since its invention by the Austrian chemist Karl Josef Bayer. In the intervening years, significant events have shaped the present alumina industry (Gerard and Stroup 1962; Cundiff 1985; Hudson 1987; Donaldson and Raahauge 2013).

More energy-efficient digestion designs have evolved, with a shift toward indirect slurry heating technologies in preference to either indirect liquor heating or direct steam injection. The introduction and use of synthetic flocculants, combined with improved thickener designs, has reduced the size of settler and washer tanks, decreased losses of soda and alumina, and increased the solids concentration of the bauxite residue. Improved bauxite residue management guidelines on dry mud stacking emerged from 1986 onward. A tragic accident with a breached dam in Hungary in 2010 highlighted the importance of adhering to these criteria. Many refineries converted their precipitation process from floury to sandy alumina during the 1980s and 1990s. Sandy alumina has a coarser particle size distribution and higher specific surface area (SSA) for fluoride capture at the smelters. Increasing oil prices made rotary kilns obsolete, and they were replaced by stationary calciners, with an almost 30% reduced fuel consumption.

Economy of scale is the major driver for justifying new refineries outside China. China, however, has emerged as the major alumina-producing country in the world, increasing production from approximately 4 Mt (million metric tons) in 2000 to more than 70 Mt in 2017. Growth was predominantly achieved through replication of standardized process trains, with 0.3- to 0.5-Mt capacity each only achieving 65%–70% capacity utilization. In the same period, production in the rest of the world increased from approximately 48 to about 56 Mt, mainly through installation of new process trains of 0.7- to 2.0-Mt capacity each. Closures of poorly performing assets in this period were partially offset by debottlenecking existing refineries.

Total world bauxite demand in 2013 was estimated at 325 Mt, with China accounting for about 135 Mt. Decreasing the ratio of alumina to silica (Al_2O_3/SiO_2) in Chinese bauxite increased the imported bauxite price to about US$50–60/t CIF (cost, insurance, and freight) in 2015. This led to investigation of new bauxite beneficiation technologies, with flotation introduced commercially in China. The alumina price decoupled from the traditional percent linkage to the 3-month London Metal Exchange price of primary aluminum and was replaced with its own Alumina Price Index (API). The API reflects alumina cost price fundamentals.

Despite the increasing deficit of alumina on an estimated world demand of 126 Mt/yr alumina in 2017, price volatility remains an issue. For example, a high of US$600/t fob (free on board) in Australia in 2005/2006 to a low of US$200/t in early 2016 has been seen. Such price volatility makes investment in new alumina refineries very risky in the near term.

Alumina Uses and Properties

Annual production of aluminum hydroxide globally in 2017 was 189 Mt. Of this, about 178 Mt was converted to aluminum oxide for conversion to aluminum metal (termed *smelter-grade alumina* [SGA]); the remaining approximately 11 Mt of aluminum hydroxide is termed *nonmetallurgical-grade alumina* or *chemical-grade alumina* (CGA). This CGA market is comprised almost equally between uses of the aluminum hydroxide and specialty aluminum oxides. Aluminum hydroxide is used to make water-treatment chemicals, such as aluminum sulfate, polyaluminum chloride, and sodium aluminate. It is also used as fire retardant and to make pseudoboehmites for catalysts, for titanium dioxide coatings, for zeolite production, or as feedstock for speciality aluminum oxides and aluminum hydroxides. The specialty aluminum oxides include calcined alumina, activated alumina, fused alumina, and tabular alumina. Specialty calcined alumina is used in refractories, engineering ceramics, electronics, electrical insulators, spark

Benny E. Raahauge, General Manager, Alumina & Pyro Technology, FLSmidth, Copenhagen, Denmark
Ken Evans, Formerly with Rio Tinto Alcan (Retired), Evans Technical Consulting Limited, Chalfont St. Peter, United Kingdom
Manfred Bach, Senior Sales Manager Alumina, FLSmidth Wiesbaden GmbH, Walluf, Hessen, Germany
Daniel Thomas, Principal Consultant, Alumina, Advisian, WorleyParsons Group, Brisbane, Queensland, Australia
Peter-Hans ter Weer, Director, TWS Services and Advice Bauxite & Alumina Consultancy, Huizen, The Netherlands

plugs, wear parts, whitewares, and polishing; many of these applications require a high alpha-alumina content and are normally still made using rotary kiln calcination. The specialty applications are driven by the characteristics of aluminum oxide, which impart good temperature resistance, hardness and abrasive properties, chemical resistance, thermal shock resistance, electrical resistance, and transparency to radiofrequency radiation. As well as the alpha-alumina content, the other critical requirements for specialty aluminum oxides are controlled soda content, particle size, particle size distribution, and morphology.

Activated alumina may be obtained either by low-temperature (LT) calcination of aluminum hydroxide at 200°–400°C or flash activation at approximately 1,100°C. The material has a high SSA—up to 350 m^2/g—and the pore size distribution can be carefully controlled. Major applications include acting as a desiccant, a Claus catalyst, catalyst removal agent in polyethylene production, a fluoride and arsenic removal agent for water, and a catalyst bed support.

Fused alumina is manufactured by electrofusing calcined aluminum oxide to more than 2,000°C in a furnace and then allowing the fused melt to slowly cool. After cooling, it is crushed and graded to different grain sizes and primarily used for the manufacture of abrasives; minor amounts are used in refractories.

Tabular alumina is manufactured from calcined alumina by initially pelletizing finely milled calcined alumina into spheres ~15–30 mm in size. The spheres are then sintered in a shaft kiln at about 1,900°C. The spheres are sometimes used as produced, or more frequently, they are crushed and separated into various size fractions. The main use is in high-alumina refractories for many industries with secondary uses in the filtration of molten aluminum metal used for can stock, as a support medium in desiccant beds, heat exchangers (HEs), or catalysts. The name *tabular* derives from the large tablet-shaped alumina crystals that typically measure 50–200 µm with a density of at least 3.45 g/cm^3 (Sleppy et al. 1991).

Table 1 Chemical composition of typical raw bauxites

Bauxite-Grade Applications	Major Oxides, %			
	Al_2O_3	SiO_2	Fe_2O_3	TiO_2
Metallurgical	31–52	1.5–15	5–30	1–6
Cement	45–55	Max. 6	20–30	3
Abrasive	Min. 55	Max. 5	Max. 6	Min. 2.5
Chemical	Min. 55–58	Max. 5–12	Max. 2	0–6
Refractory	Min. 59–61	Max. 0.5–5.5	Max. 2	Max. 2.5

Source: Mahadevan 2015

Bauxite Uses and Composition

Aluminum represents about 8% of the earth's crust by mass but is never found isolated as the metal. Bauxite of various chemical compositions is used in different industries (Table 1) but not in quantities comparable to bauxite used for production of metallurgical-grade alumina or SGA. The bauxite application as a ceramic proppant in the hydraulic fracturing industry, however, is quickly growing (estimated to be about 12% per year), reaching 65 Mt/yr by 2017 (O'Driscoll 2013).

Bauxite is the ore with the highest content of alumina of the raw material considered for making alumina (Table 2), which is subsequently reduced to primary aluminum. The major mineral is gibbsite, $Al(OH)_3$, also named alumina trihydrate (ATH) $Al_2O_33H_2O$. The other common minerals are boehmite and diaspore, AlO(OH), also named alumina monohydrate (AMH), $Al_2O_3H_2O$.

BAUXITE AND SMELTER-GRADE ALUMINA

The Bayer process using bauxite as raw material is the most economical process currently known to be applicable to ores containing hydrated alumina minerals for the production of SGA and thus the focus of the remainder of this alumina section.

Bauxite Origin, Mineralogy, and Composition

Bauxite is named after the deposit found in Les Baux, France, in 1821, which was discovered by geologist Pierre Berthier. Figure 1 shows bauxite reserves in selected countries.

There are two basic types of bauxite—lateritic and karst—referring to their geological formation over thousands of years (Komolossy 2010). Lateritic bauxites are mostly found in or close to the tropics and account for approximately 90% of total bauxite deposits of about 67 Mt (Best 2013). Lateritic bauxite is formed as a residue after weathering/acid leaching away of silicates from the original alumina-silicate formation, leaving behind aluminum hydroxides. Many elements and minerals can be found in tropical bauxite (Authier-Matin et al. 2001; Table 3).

Karst-type bauxite is residual sediment accumulated on karstified carbonate rocks. This type of bauxite accounts for about 10% of total bauxite deposits and can be found in

Table 2 Mineralogy of aluminous ores

Ore/Rock	Mineral	Formula	Al_2O_3 in Mineral, %	Al_2O_3 in Ore, %
Bauxite	Gibbsite	$Al(OH)_3$	65	
	Boehmite	AlO(OH)	85	35–67
	Diaspore	AlO(OH)	85	
Clay	Kaolin	$Al_2O_32SiO_22H_2O$	40	20–35
Anorthosite	Albite	$Na_2O\ Al_2O_36SiO_2$	20	
	Anorthite	$CaO\ Al_2O_32SiO_2$	37	22–30
Nepheline syenite	Nepheline	$(Na,K)\cdot(Al,Si)_2O_4$	36	
	Feldspar	$KAlSi_3O_8$	18.4	20–25
Alunite	Alunite	$K_2SO_4Al_2(SO_4)_34Al(OH)_3$	37	6–15
Shale	Clay + illite	$(OH)_4K_2(Si_6Al_2)(Mg, Fe)O_{20}$	38.5	15–30
Aluminous phosphate	Wavellite	$Al_3(PO_4)_2(OH)_35H_2O$	37	
	Millisite	$(NaK)CaAl_6(PO_4)_4(OH)_9$	37	5–15

Source: Milne 1982

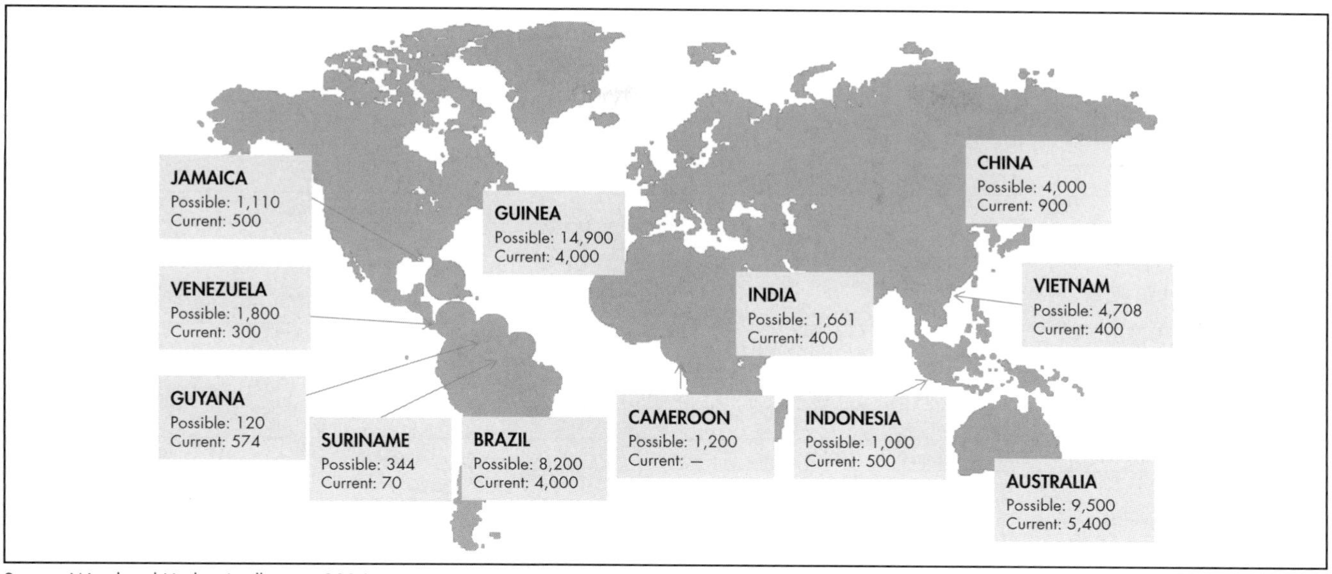

Source: Wood and Harbor Intelligence 2014

Figure 1 Bauxite reserves per million metric tons

Table 3 Mineralogical composition of tropical bauxite

Element	Mineral	Chemical Composition
Major		
Aluminum	Gibbsite	$Al(OH)_3$ or $Al_2O_3{\cdot}3H_2O$
Silicon	Boehmite	$AlO(OH)$ or $Al_2O_3{\cdot}H_2O$
Iron	Quartz	SiO_2
Titanium	Kaolinite/halloysite	$Al_2Si_2O_5(OH)_4$ or $Al_2O_32SiO_22H_2O$
	Hematite	Fe_2O_3
	Aluminian goethite	$(Fe,Al)O(OH)$ or $(Fe,Al)_2O_3H_2O$
	Anatase	TiO_2
	Rutile	TiO_2
Minor		
Carbon	Organic carbon	Humic materials
Phosphorus	Wavellite	$Al_3(PO_4)_2(OH)_35H_2O$
Calcium	Crandallite-H	$CaAl_3(PO_4)(PO_3OH)(OH)_6$
Potassium	Calcite	$CaCO_3$
Manganese	Crandallite-H	$CaAl_3(PO_4)(PO_3OH)(OH)_6$
Magnesium	Illite	$KAl_2(Si_3AlO_{10})(OH)_2$
Sodium	Lithophorite	$(Al, Li)MnO_2(OH)_2$
Strontium	Magnesite	$MgCO_3$
Sulfur	Dolomite	$CaMg(CO_3)_2$
Zinc*	Dawsonite	$NaAlCO_3(OH)_2$
Chromium*	Celestite	$SrSO_4$
Vanadium*	Woodhouseite	$CaAl_3(PO_4)(SO_4)(OH)_6$
Zirconium	Pyrite	FeS_2
	Gahnite	$ZnAl_2O_4$
	Chromite	$FeCr_2O_8$
	Schubnelite	$Fe_2(V_2O_5)2H_2O$
	Zircon	$ZrSiO_4$

Source: Authier-Matin et al. 2001. Copyright 2001 by The Minerals, Metals & Materials Society. Used with permission.

* At least some of the zinc, chromium, and vanadium have been established to be present in alumina goethian lattice.

Jamaica, Russia, China, and Mediterranean countries, where diaspore is the main Al_2O_3 mineral (Komolossy 2010).

Smelter-Grade Alumina Requirements

Alumina is a commodity traded on world markets with changing specifications over time (Table 4). Many smelters may buy on long-term contracts or buy from the spot market. To minimize transportation cost, refineries are finding it economical to trade alumina with each other. As the smelters have continually improved in energy efficiency, environmental performance, and increased pot size and life, the quality requirements of the alumina have changed. The types of aluminum alloys now produced and the use of recycled aluminum has added to lower levels of impurities being required in the product. SGA is needed that has the required transport characteristics of flowability and attrition resistance (S.J. Lindsay 2014; Archer 1983); that is, it is not dusty and is able to absorb hydrogen fluorides from the off-gases passing through the dry scrubbers (high surface area) without breaking down (attrition resistance). Further, a particle sizing is required such that the alumina will disperse freely in the pots and enable complete dissolution of the alumina (Homsi 2001; Welch and Kuschel 2007). The chemical requirements of SGA address the application as raw material for primary aluminum production.

Chemical Requirements

The chemical requirements of SGA are shown in Table 5 (middle column) compared with bauxite composition (right-hand column). A high level of *soda* is tolerated as the electrolyte is cryolite. Too high a soda content, however, means excess bath (electrolyte) is produced, so there is an upper limit that decreases as pot life is increased (J.B. Lindsay 2012). *Lime* is added to the bath and thus calcium is not such a critical component in SGA. Neither the soda nor the calcium end up in the metal. The Bayer process can generally meet the required impurity-level specifications, provided the pregnant liquor is filtered to remove residual particulates prior to precipitation. Historically, *silica* was an issue with especially low-reactive silica bauxites and for certain refineries because of their design and processing conditions. It was often necessary to add reactive silica (with consequently higher caustic consumption) or redigest some of the precipitated gibbsite to meet product specifications. With the introduction of predesilication, not only can silica in liquor (and consequently product alumina) be controlled to the required level, but it is

Table 4 Smelter-grade alumina properties over time

SGA Property	Unit	Floury Alumina	Sandy Alumina	Typical 2015
Purity	wt %	>99.3	>99.3	99.3–99.7
Loss on ignition, 300°–1,000°C	wt %	0.1–0.3	0.4–0.8	<1.0
Gibbsite	wt %	No data	No data	<0.5
Bulk density, loose	g/L	800	950	950–1,000
Angle of repose	degree	45	30–32	30–32
Particle size distribution				
>100 mesh/150 μm	wt %	No data	No data	5–10
<325 mesh/45 μm	wt %	45–60	5–10	Maximum 10
Superfines <20 μm	wt %	No data	1.3	1–2
Specific surface area, BET (Brunauer–Emmett–Teller)	m^2/g	5	50–80	60–80
Alpha content	%	85–96	10–30	<5–10
Attrition index*	wt %	No data	13–23	<20
Flow funnel test*	second	No data	No data	85–95
Dust index*	kg/t	No data	1.03–1.37	0.5–1.3

*Not in all certificates of analyses; see text for references and additional comments.

Table 5 Typical chemical compositions of smelter-grade alumina and tropical bauxite*

Constituent	In SGA, wt % oxide	In Bauxite, wt %
Al_2O_3* as gibbsite	<0.5%	30–60
Al_2O_3 as boehmite	No data	<0.2–20
Al_2O_3	99.3–99.7	35–67
Fe_2O_3	0.005–0.020	1–30
SiO_2 clay	No data	0.3–6
SiO_2 quartz	0.005–0.025	0.2–10
Na_2O	0.30–0.50	No data
CaO	<0.005–0.040	0.1–2
TiO_2	0.001–0.008	<0.5–10
ZnO	<0.001–0.010	0.002–0.10
P_2O_5	<0.0001–0.0015	0.02–1.0
V_2O_5	<0.001–0.003	0.01–0.10
Ga_2O_3	0.005–0.0015	0.004–0.013
Cr_2O_3	No data	0.003–0.30
BeO	<1 ppm	No data
MnO	<0.001	No data
K_2O	<0.002	No data
SO_3/S	<0.05–0.20	0.02–0.10
F	No data	0.01–0.10
Hg (ppm)	No data	50–1.000
Organic carbon (as C)	No data	0.02–0.40

Source: Authier-Matin et al. 2001; S.J. Lindsay 2005b; B.J. Welch (personal communication)

*Typically, alpha-Al_2O_3 content ranges from 2% to 30%.

also possible to process lower-reactive silica bauxites while still maintaining the required quality. *Iron*, present as iron oxide in the bauxite, is sparingly soluble in Bayer liquor. Any iron that is dissolved in the liquor precipitates and ends up in the alumina. Iron can be an issue for high-temperature (HT) digestion refineries and those refineries processing low-iron bauxite. In some instances, it has required low-iron bauxite to be blended with high-iron bauxite to control the iron in alumina. There are other ways that iron can be controlled, and generally, the smelter requirements can be met.

Titanium also behaves like iron, but it can be removed from Bayer liquor prior to precipitation using lime. Lime is also used for phosphate control, but that is often as much for processing issues within the refinery rather than product quality. Gallium cannot be controlled, as it reacts similarly to aluminum, and thus refineries processing high-gallium bauxite produce high-gallium alumina. The gallium in the liquor can be, and at some refineries is, extracted for gallium production. The large amount of gallium in Bayer liquor is far greater than the world's current requirement, and generally, it is not profitable to extract it. Beryllium is a worker exposure concern at smelters in the United States and Quebec and is currently a watch list item for alumina producers (S.J. Lindsay 2005a).

Physical Requirements

The physical characteristics required for SGA are shown in Table 6 together with the reasons for their importance. The modern Bayer refinery can readily meet the particle size requirements that are predominantly related to the handling properties (flowability) of the alumina and its impact on anode effect and current efficiency in the smelters (Lindsay 2014). Content of fines (<325 mesh) and superfines (<20 μm) is of special importance in this respect, but also because of its perceived impact on dustiness (Perra 1984; Hsieh 1987; Authier-Matin 1989). Fines and superfines are produced in precipitation (Audet et al. 2012) and from the breakdown of the alumina particle in stationary calciners (Klett et al. 2010; Raahauge and Devarajan 2015), as well as in certain types of dry scrubbers at the smelters (Taylor 2005; S.J. Lindsay 2011). Transportation of SGA has also added to the need for attrition-resistant particles (Wind et al. 2012; Saatci et al. 2004) that do not easily break down. Laser sizers, the flow funnel (Hsieh 1987), and attrition tests (Forsyth and Hertwig 1949) are major enablers to characterize these properties. A marked improvement in alumina properties by reducing the fines content of the alumina and making a more robust particle (Sang 1987; Bhasin and Schenk 1992) has reduced many of the issues at the smelters and enabled the cells to be run at ever higher efficiencies.

The Alumina Quality Workshops held in Australia since 1993 (S.J. Lindsay 2005a) bring smelting and alumina refiners

Table 6 Smelter-grade alumina properties and their importance

SGA Property	Role/Importance of Property in Smelter
Purity	Na and Ca for AlF_3 consumption, other metallic impurities
Loss on ignition	0°–300°C adsorbed readily; releases volatiles (can form *volcanos* for point feeders) 300°–1,000°C predominantly Al-O-H compounds that are major contributors to hydrogen fluoride (HF) evolution
Bulk density	Accuracy of correlation between volumetric feed and assumed mass
Angle of repose	Covering quality if using only Al_2O_3, flow properties, ability to fill storage vessels, actual volumetric feed transfer
Particle size distribution	Consistency of properties, tendency to release dust (fines), cell performance perception of impact on solubility (coarse)
Specific surface area, BET (Brunauer–Emmett–Teller)	Potential capacity for HF gas absorption
Alpha content	Measure of conversion of the calcined $Al(OH)_3$ (*trihydrate*) to the most stable alumina phase, a secondary measure of surface area for a given calcination method; crusting tendency
Attrition index	Tendency for the polygranular alumina to disintegrate to a finer particle size range
Flow funnel test	The flowability of alumina
Dust index	Qualitative assessment of ability to become suspended in the cell environment and be transported away

Source: Welch and Kuschel 2007. Copyright 2007 by The Minerals, Metals & Materials Society. Used with permission.

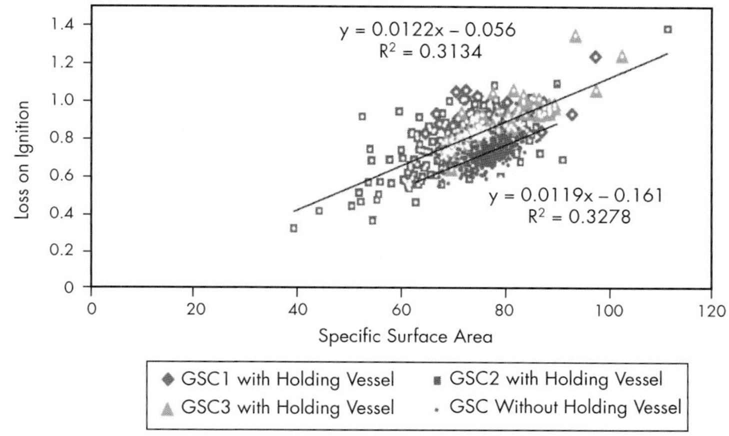

Source: Raahauge and Devarajan 2015

Figure 2 Properties of smelter-grade alumina from gas suspension calciner (GSC)

together and facilitate the understanding of refinery issues and smelter requirements. The major issue now is that small amounts of superfines (<20 µm) are sufficiently deleterious, as they can segregate in bins and then have a significant effect on handling and dusting. Some of those superfines come from the refinery, and some are produced at the smelter.

The use of dry scrubbers at the smelters to clean the off-gases has necessitated producing a high SSA SGA product for hydrogen fluoride (HF) capture (S.J. Lindsay 2009). To produce the SSA required, calciners at the refineries are operated at an overall lower temperature. This results in a higher loss on ignition (LOI) as the two properties are interlinked, which is illustrated in Figure 2 showing LOI versus SSA from a gas suspension calciner (GSC) without and with a fluidized holding vessel. With current calcination technology, it is difficult to produce higher SSA without increasing the LOI content.

Measurements of Smelter-Grade Alumina Properties

There are agreed-on methods for measuring the various alumina properties, for example, Standards Australia. Chemical analysis uses either inductively coupled plasma spectrometry or X-ray fluorescence (XRF) (Figure 3). Particle size analysis uses sieve shaker screens (down to and including 325-mesh/45-µm sieve) although most sizing for control utilizes lasers. There are agreed-on standard tests for flowability and attrition resistance but, to date, none for dusting (Kimmerle 1999).

BAUXITE, THE BAYER PROCESS FLOW SHEET, AND UNIT OPERATIONS

Bauxite is widely traded, which will continue, but currently, there is no single international approach to determine the price.

Commercial Bauxite Assessment

The key Bayer plant processing parameters are extractable alumina, mud factor, chemical caustic consumption, lime consumption, and goethite-to-hematite ratio, as well as total organic content.

Over the years, several assessment methods have been employed of varying accuracy, performance time, and cost (Hill and Robson 1981). Prior to 1990, mostly wet chemical analytical methods were used, while the tendency today is toward using advanced analytical techniques such as X-ray diffraction, XRF, and thermogravimetric analysis in

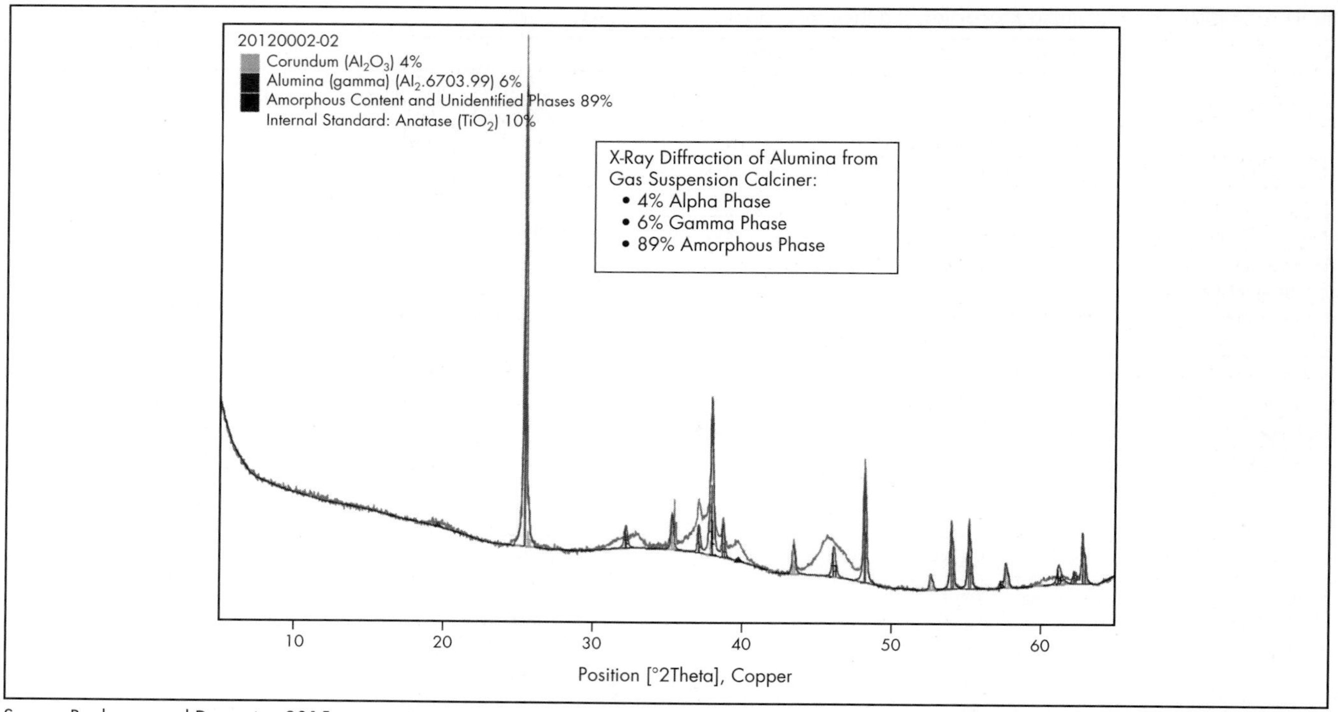

Source: Raahauge and Devarajan 2015

Figure 3 Chemical analysis of SGA from GSC by X-ray diffraction

combination with mathematical models (Ferret 2014). New bauxite deposits are analyzed based on representative composite samples obtained from detailed drilling programs and feasibility studies carried out to decide if a deposit represents a valuable asset with respect to investing in a new alumina refinery (Wehrli and Jiang 2010).

Other Raw Materials and Fuels

The main reagent and consumable is *caustic soda* reacting with silica minerals in the bauxite (primarily kaolin) and precipitate as desilication products (DSPs). The high-strength caustic can be used to remove alumina scales from tanks and piping, before adding that liquor to the process as caustic makeup.

Refinery boilers are fueled with coal, heavy fuel oil, or natural gas, which produce steam and power. Calciners are fueled by heavy fuel oil or natural gas to maintain the purity of the alumina. In China, sinter kilns are fueled by coal, and most calciners are fueled by coal gas derived from coal gasification.

Mining and Beneficiation

Nowadays the utilized bauxite deposits are nearly all-surface or near-surface deposits, and open cut mining is used, making mining very easy. The average thickness of bauxite deposits ranges from 2 to 20 m with a production average of 5 m (Wagner 2013). There is rarely any need to use blasting unless a hard cap has formed in the upper layer. The overburden thickness, which has to be removed before excavating the bauxite, can vary from 0.4 to 12 m with an average of 2 m (Wagner 2013).

Some deposits can be quite deep (e.g., in Brazil), requiring high stripping ratios that significantly increase the mining cost. For surface bauxites, the mining costs per metric ton of alumina are low and not a major contributor to the cost of the alumina; however, those costs increase sharply for the deeper deposits. Several of these deposits—about 50%—contain a high level of clay minerals that need to be washed from the bauxite to reduce reactive silica, increase alumina content, and improve the handling characteristics. The clay has a severe impact on both the mining and transport of the material to the washing stage because of its sticky nature.

For those bauxites requiring washing, the washing operation is relatively straightforward, which, at most, requires some crushers to generally break up the material. The washer/scrubber stage is followed by further size classification with material <1 mm in size usually being discarded to tailings ponds with >80% clay. Grade control is relatively clearcut for recoverable alumina and reactive silica, although in some areas, the organics content in the bauxite also needs to be considered. No milling is required, and consequently there is not a large power requirement for the mining/washing operations readily done at the mine site. For some operations, the tailings are filtered, and the filtered mud is deposited to the tailings area with such areas rehabilitated. Given that many of these deposits are in equatorial and tropical areas, there are water containment issues.

China has very rapidly developed to become the largest alumina producer in the world, consuming approximately 130 Mt/yr bauxite, of which about 50% is imported (Jiang et al. 2016). The reason for the relatively high import of bauxite is that most Chinese bauxite deposits are diasporic with a decreasing alumina-to-silica (A/S) ratio over time, resulting in an increasing bauxite-to-alumina ratio and increased soda consumption. For high-silica bauxites with A/S = 3 to 5, froth flotation techniques have been developed and introduced into the flotation Bayer process. In China, flotation technology is also being developed for desulfurization of bauxite, and reverse flotation has been tested in Brazil at Companhia Brasileria de Aluminio's plant at Mirai, Minas Gerais (Chaves 2010). Because many of the high-quality, easily accessible bauxite deposits have been exploited for years, more sophisticated bauxite beneficiation technologies are emerging. Subject to grain size and iron content, jigs for coarse or spiral classifiers for fines classification in combination with drum separators or wet high-intensity magnetic separators may enter future bauxite beneficiation flow sheets (Bautenbach 2008). Thermal activation of boehmitic bauxite for removing organics (Hollit et al. 2002) and roasting high-silica bauxites (Smith and Xu 2010) have been tested but have still not reached industrial-scale application.

Transport of the bauxite to the refinery can be a significant cost, as many refineries are situated some distance from the mines. Each ton of bauxite is transported 54 km on average from the mine to the local refinery stockpile or shipping point (Wagner 2013). Some examples are the Worsley refinery in Western Australia receiving bauxite from its Mount Saddleback mining hub on a 51-km-long conveyor line (Cochrane et al. 2011). In some cases, a slurry pipeline is the most economical solution, for example, the 245-km-long, 13.5-Mt/yr Mineração Bauxita Paragominas pipeline in Brazil (Grandhi et al. 2008). The average energy consumption ranges from 40 to 470 mJ/t dry bauxite mined with an average of 153 mJ/t, encouraging mine operators to adopt energy-efficient strategies and reduce emissions (Wagner 2013). Bauxite bought commercially is invariably significantly higher in cost than for a refinery-owned mine and, together with transport costs, can be a major part of alumina cost.

Introduction to the Bayer Process and Terminology

A typical schematic block flow sheet of the alumina refinery Bayer process for LT and HT digestion of bauxite is shown in Figure 4, naming the various unit operations in the process. The *Bayer liquor loop* or *Bayer wheel* comprises a circulating caustic liquor flow with dissolved alumina (Al_2O_3) measured in grams per liter (g/L). The following liquor definitions apply:

- **Spent liquor.** This occurs after precipitation and before digestion. Spent liquor is the most common name for this part of the Bayer process liquor loop.
- **Test tank liquor.** This is the spent liquor just before digestion. Liquor analysis is used to calculate the bauxite quantity to be charged to the process. *Digester blow-off* is cooled by flashing before settling.

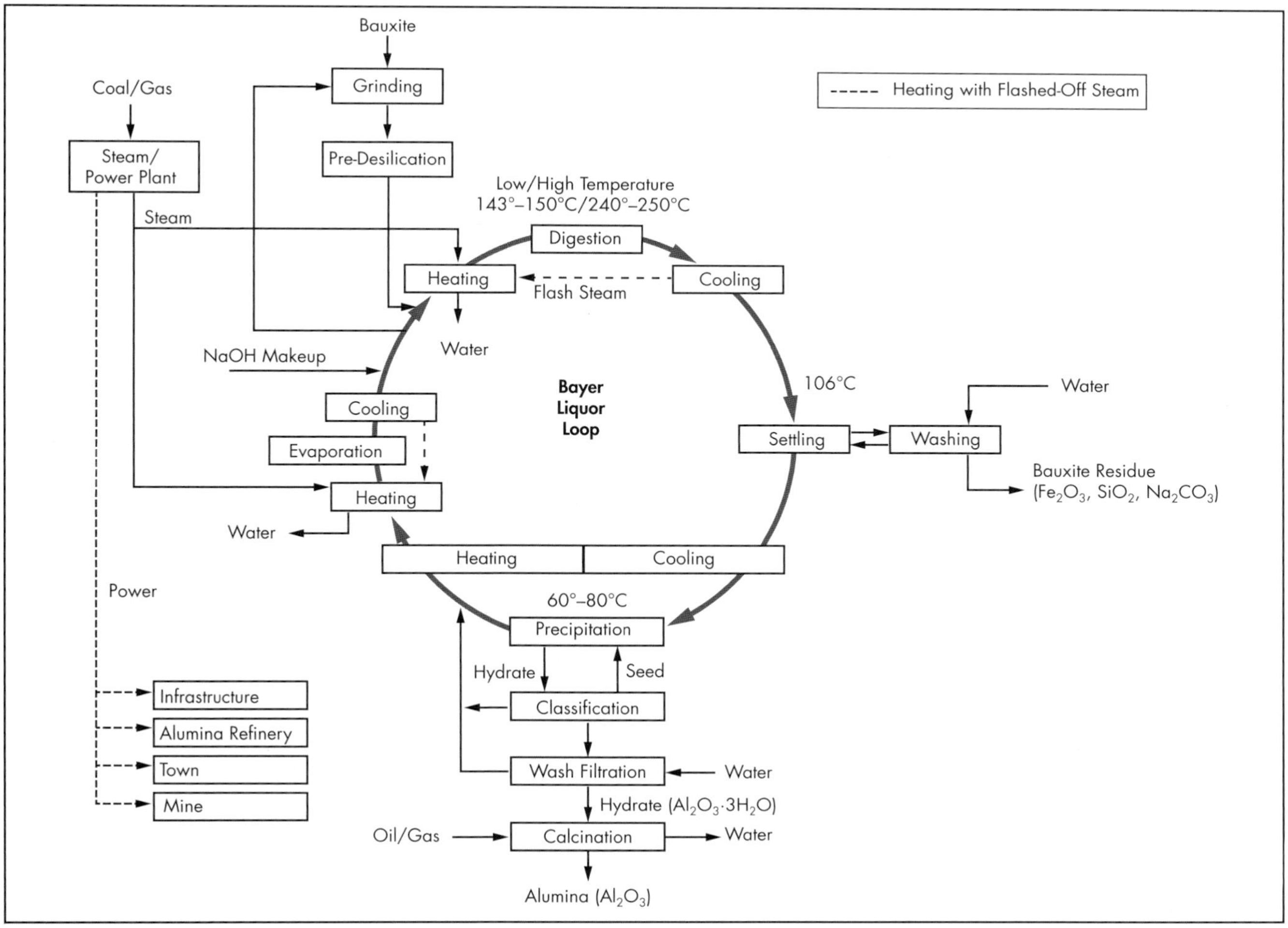

Source: P.J.C. ter Weer 2014

Figure 4 Bayer process block flow sheet

- **Pregnant liquor.** This occurs after digestion/settling and before precipitation; also named *green liquor*. Pregnant liquor is polished by filtration to remove solids before precipitation of $Al(OH)_3$.
- **Caustic concentration.** The caustic concentration of the circulating liquor is determined by thermotitration and is expressed in various ways, originating in the old North American and European practices of Bayer process design and operation. The following conversions apply:

$$Na_2CO_3\ (g/L) = (106/80) \times NaOH\ (g/L) = (106/62) \times Na_2O\ (g/L) \quad \textbf{(EQ 1)}$$

Total caustic (TC). TC or C includes sodium present as $[OH^-] + [Al(OH)_4^-]$ and is expressed as g/L sodium carbonate (Na_2CO_3).

Total alkali (TA). TA includes sodium present as $[OH^-] + [Al(OH)_4^-] + [CO^{3-}]$ and is expressed as g/L Na_2CO_3.

Total soda. This is measured by atomic absorption, including carbonate ions (CO_3^{2-}) and sodium with other anionic impurities ($C_2O_4^{2-}$, Cl^-, SO_4^{2-}, etc.) of which sodium oxalate ($Na_2C_2O_4$) plays an important role.

The alumina-to-caustic (A/C) ratio is also expressed in various ways:

$$A/C = Al_2O_3\ (g/L)/Na_2CO_3,\ \text{caustic}\ (g/L) \quad \textbf{(EQ 2)}$$

$$RP = Al_2O_3\ (g/L)/Na_2O,\ \text{caustic}\ (g/L) = (106/62) \times A/C \quad \textbf{(EQ 3)}$$

$$MR = Na_2O,\ \text{caustic}\ (\text{mole}/L)/Al_2O_3\ (\text{mole}/L) = (102/106)/(A/C) = (102/62)/RP \quad \textbf{(EQ 4)}$$

where
RP = ratio ponderel
MR = mol ratio

The circulating liquor density ranges from 1.25 g/mL up to as high as 1.35 g/mL, with an average of 1.30 g/mL or 1,300 g/L. The liquor density, D, can be estimated from the following formula (as corrected by Raahauge; Ikkatai and Okada 1963):

$$D\ (g/cm^3) = 1.027 + 0.8985\cdot10^{-3}\cdot[Na_2O] - 0.25\cdot10^{-6}\cdot[Na_2O]^2 - 0.395\cdot10^{-3}\cdot T - 0.2\cdot10^{-5}\cdot T^2 + 0.49\cdot10^{-3}\cdot[Al_2O_3] \quad \textbf{(EQ 5)}$$

The "red side" flow-sheet variations depend on the type of bauxite used, whereas the "white side" flow-sheet variations with precipitation are almost standard, regardless of the type of bauxite. The overall refinery production and economics are very dependent on the liquor productivity or yield of alumina expressed as g/L Al_2O_3 and calculated as follows:

$$\text{yield} = C_{in}\cdot [(A/C)_{in} - (A/C)_{out}] \quad \textbf{(EQ 6)}$$

The green liquor $(A/C)_{in}$ ratio from digestion is typically in the range of 0.72 to 0.77. The yield of alumina ranges from 50 to 90 g/L Al_2O_3 with an average of 70 g/L Al_2O_3.

Using the preceding average numbers, the circulating liquor equals 18.6 g liquor/g Al_2O_3 produced or 18.6 t liquor/t Al_2O_3 or 14.3 m^3/t Al_2O_3 produced.

The alumina production is given by the following equation:

$$\text{alumina production} = \text{liquor flow (green/pregnant)} \times \text{yield (g/L alumina)} \times \text{availability (\%)} \quad \textbf{(EQ 7)}$$

Because of the closed-loop liquor circulation, impurities build up in the liquor. The effect of the impurities is to increase the end ratio $(A/C)_{out}$ and thus decrease the yield and productivity (Lectard and Nicolas 1983). Keeping the level of impurities in the liquor circuit under control is a key enabler to maximize the yield/productivity of a modern Bayer refinery.

All refineries have carbonate removal systems. Some remove sulfate and others remove oxalate and other organics from the liquor. The equipment used is bauxite specific. Some refineries use *salting out* for carbonate and oxalate removal, utilizing evaporators (falling film). Others remove or control carbonate by side stream *causticizing*. Lime is added to wash liquor, reacting with the liquor in a series of agitated tanks, returning the causticized liquor to the washing operation. Some refineries precipitate sodium oxalate from the spent liquor and filter the slurry, and then treat the filter cake to recover the caustic by various methods and equipment. Organics removal is required at some refineries, subject to organic content in the bauxite and years of refinery operation in order to maintain yield/productivity. As a rule of thumb, more than 10 g/L total organic carbon is critical. Total organic removal can be achieved by *liquor burning*. Using a rotary kiln or GSC with associated equipment is one option, where stack gases are generated that must be thoroughly controlled and cleaned (Pulpeiro et al. 1998, 2000). *Wet oxidation* is another option, which is less efficient with respect to oxalate removal, but more cost-effective (Costine et al. 2010; Clegg 2005).

Bayer Process Unit Operations

The weathered nature of the tropical bauxite means that there are no physiochemical accessibility issues related to digesting the gibbsite.

Comminution

The bauxite is typically ground from F80 = 10–25 mm to P80 = 200–400 µm (Filidore and Hadawy 2010), with Bond work indices ranging from 1 to 34 kW·h/t. Some bauxites (notably Jamaican bauxites) almost disintegrate down to slime-sized material that results in flowability issues. Normally the percent solids in the mill can be of the order of 50% without any issues, but for Jamaican bauxite, this reduces to 30% because of the high viscosity of the slurry. The particles are generally a finely interdispersed mixture of gibbsite, iron oxides, and clay. Grinding is done to reduce the size such that the slurry can be readily pumped through the digesters and quartz separated as sand particles afterward. Only in a few instances, such as with diasporic bauxite, is there a need to specifically grind the bauxite to a very fine size for extraction. All bauxite grinding is done in spent liquor at around 70°C.

Rod, ball, rod-ball mills, and some semiautogenous mills are used with systems being open or closed circuits. Where closed-circuit milling is practiced, this is often associated with hard components in the bauxite, and then classification by DSM (sieve bend) screens is used.

Pre- and Post-Desilication Chemistry

The ground bauxite is fed to slurry storage tanks that are also used for pre-desilication prior to digestion. Most refineries use pre-desilication, but at least one refinery has used post-desilication (Harato et al. 1996) to reduce the soda consumption almost 50% compared to a conventional pre-desilication process. Post-desilication is not further discussed (Banovolgyi and Siklosi 1998; Songqing 2008).

The ultimate objective of pre-desilication is to reduce the silica present in the liquor by dissolving and precipitating reactive silica as DSPs. The dynamic desilication process is illustrated in Figure 5. The process involves kaolinite ($Al_2O_3{\cdot}2SiO_2{\cdot}2H_2O$), which is reactive silica that reacts readily with sodium hydroxide (NaOH) and dissolves in the liquor at approximately 100°C (Roach and White 1988). Silica as quartz needs temperatures higher than 200°C to be significantly attacked by caustic (Oku and Yamada 1971).

Following the dissolution of kaolinite (Equation 8), more complex reactions take place by seeded precipitation forming various compositions of DSP (Equation 9) (Authier-Matin et al. 2001), where *X* may be a mix of $\frac{1}{2}CO_3^{2-}$, $\frac{1}{2}SO_4^{2-}$, $2AlO_2^-$, Cl^-, and so forth (Milne 1982).

$$Al_2O_3{\cdot}2SiO_2{\cdot}2H_2O(s) + 6NaOH(aq) \rightarrow 2NaAlO_2(aq) + 2Na_2SiO_3(aq) + 5H_2O \quad \textbf{(EQ 8)}$$

$$6NaAlO_2(aq) + 6Na_2SiO_3(aq) + Na_2X(aq) + 12H_2O \rightarrow 3(Na_2O{\cdot}Al_2O_3{\cdot}2SiO_2{\cdot}nH_2O)(s) + Na_2X + 12NaOH \quad \textbf{(EQ 9)}$$

Pre-Desilication Unit Operations

On its way to the storage tanks, the slurry is heated, normally close to 100°C, to dissolve all the clay and form the DSP. These large tanks are agitated by rakes, and there are often three or four in series to reduce short-circuiting and give the residence time required (normally 6–24 hours) (Thomas and Pei 2007; Zirnsak et al. 1999).

The solids density of slurry is normally of the order of 50%, resulting in a plug-flow-type passage of the tanks with limited back-mixing (Oku and Yamada 1971). Exactly how much desilication is required is bauxite specific and also depends on digestion conditions, but modern Bayer refineries have little trouble controlling the silica to meet customer specifications.

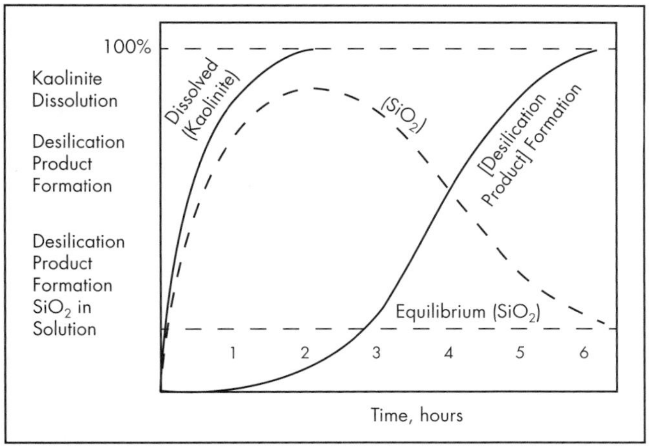

Source: Kotte 1981. Copyright 1981 by The Minerals, Metals & Materials Society. Used with permission.

Figure 5 Typical desilication process

Digestion Chemistry

The digestion process is to dissolve all the hydrated alumina in the bauxite and react all the clay minerals that then precipitate as DSP and thus desilicate the liquor. The digestion process is conducted at different temperatures, pressures, and caustic concentrations depending on the mineralogy of the bauxite.

Bauxite containing mainly trihydrated alumina (ATH), with traces of monohydrated alumina (AMH), as boehmite, can be digested at atmospheric pressure and moderate temperatures (100°–107°C) with caustic concentration between 150 and 250 g/L (Grubbs 1987; Bitsch and Lamerant 1995), compared to standard Bayer digestion of gibbsite (140°–145°C) at less than 10 atm. The chemical reaction dissolving the gibbsite is shown in the following equation:

$$2Al(OH)_3(s) + 2NaOH(l) \rightarrow 2NaAl(OH)_4(l) \quad \textbf{(EQ 10)}$$

The boehmite (230°–245°C) and diaspore (230°–280°C) with the addition of calcium oxide (CaO) need higher digestion temperatures and pressures up to 80 atm (Figure 6), with caustic concentrations of 180–200 g/L Na_2CO_3 and 230–350 g/L Na_2CO_3, respectively, to dissolve together with other impurities such as quartz, goethite, and anatase:

$$AlOOH(s) + NaOH(l) + H_2O \rightarrow NaAl(OH)_4(l) \quad \textbf{(EQ 11)}$$

$$SiO_2(s) + 2NaOH(l) \rightarrow Na_2SiO_3(l) + H_2O \quad \textbf{(EQ 12)}$$

$$2FeOOH(s) \rightarrow Fe_2O_3(s) + H_2O \quad \textbf{(EQ 13)}$$

$$TiO_2(s) + 2NaOH(l) \rightarrow Na_2TiO_3(l) + H_2O \quad \textbf{(EQ 14)}$$

During HT digestion with optimized lime addition, many complex DSPs are formed and precipitated (Jiang et al. 2016).

Digestion Flow Sheet and Equipment

The bauxite slurry from pre-desilication and spent liquor is heated and sent to the digesters. This is either done as two separate streams or in some refineries by mixing the bauxite slurry and spent liquor streams before pumping the mixed streams through the slurry heaters. The slurry or liquor heaters can be shell and multitube HEs or jacketed pipe heaters (JPHs) with the slurry or liquor passing inside the pipes in either case. In dual- or two-stream plants, the spent liquor is normally heated by flash and live steam to a higher temperature than the digester temperature and then mixed with the slurry either in the digester or earlier in a contact live steam heater (LSH) (Figure 7A), where steam is added directly to achieve digestion temperature.

As live steam adds dilution, the trend is to minimize direct live steam uses and heat the slurry mixed with spent liquor as a single stream. Since 2002, JPHs have gained a foothold in HT digestion (Songqing and Zhonglin 2004; Kelly et al. 2006). The latest refineries built outside China are using a single stream with JPH (Haneman and Rae 2015; Figure 8). Several stages are required to maximize energy efficiency, matching the flash vessel stages so that export of excess live steam is minimized or avoided (Kelly et al. 2006). Some refineries use internal heating coils in the digesters after the heaters. A very attractive and energy-efficient system for diasporic bauxite is to use molten salt as the final heating stage to avoid final steam heating. This was used in the VAW (Vereinigte Aluminium-Werke) tube digesters (Bielfeldt and Winkhaus 1981) in Germany. Subsequently, this technology has been vastly improved in Chinese refineries where it has been a

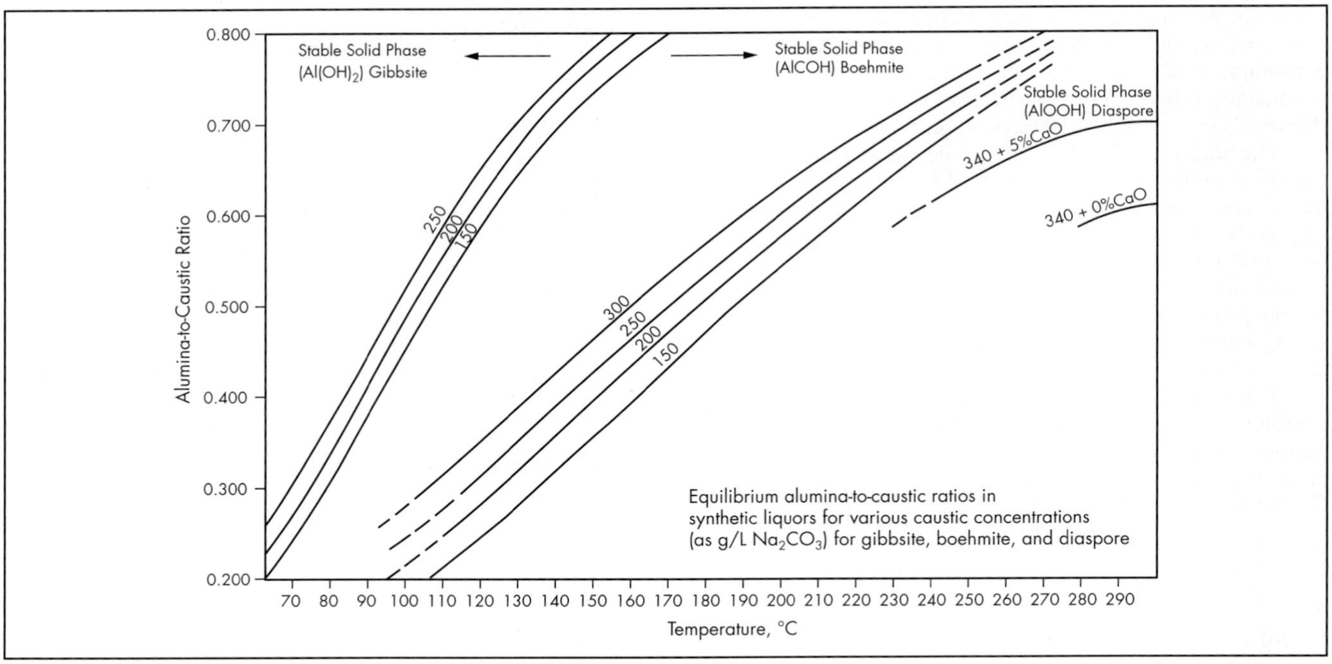

Source: Kotte 1981. Copyright 1981 by The Minerals, Metals & Materials Society. Used with permission.

Figure 6 Equilibrium A/C ratio in synthetic caustic liquors

major enabler in terms of energy reduction in processing such bauxites (Songqing and Zhonglin 2004).

The digesters are a series of tanks: self-agitated (without internals) or with mechanical agitators. Vertical autoclaves or horizontal tube digesters are used today. Autoclaves are economically used for HT digestion of bauxites that require more than 5–10 minutes of retention time. Normally, there are three to five autoclaves arranged in series to minimize short-circuiting and achieve the residence time required. For HT plants needing long digestion times, there can be up to 10 autoclave units. The slurry feed may be as high as 20% solids, but that becomes less than 10% at the final digester. For pumping, positive displacement pumps are used because of the higher pressure in HT refineries processing boehmitic or diasporic bauxite.

The digested slurry is let down to atmospheric pressure through a series of flash tank/HE stages, about 3–4 for LT digestion and 10–12 for HT digestion. Letting down the slurry in incremental stages is required to (1) provide steam at varying pressures and temperatures to heat the incoming bauxite/slurry/spent liquor to maximize heat transfer efficiency and (2) reduce wear and carryover through the flash vessels. In HT plants, with evaporation economy of about 7 t water/t steam compared to only about 3–4 t water/t steam (Donaldson 2011), enough flashing steam is recovered to almost eliminate evaporation at the expense of installing more flashing stages. The final flash vessel is essentially a vented tank with little steam exiting, as any steam exiting is a direct energy waste. The last flash stage(s) may be connected to a water-cooled condenser instead of a liquor heater. Heat exchange between the spent liquor heating side of the process and the digested slurry cooling side is critical to ensure minimum energy usage. Multistages are used to maximize heat recovery. Because of the evaporation requirement to remove dilution from the main liquor circuit, most heat interchange is through flashing of the hot liquor/slurry. This includes some vacuum stages in the lower temperature end to produce steam, which is used to heat the incoming cool liquor. At some of the lowest temperatures, plate heat exchangers (PHEs) are used, but there can be scaling and other issues associated with such equipment. Much of the inefficiency in the heat recovery system comes from scaling of heater tubes by DSP. This may become much less of an issue by using a chemical additive to control the growth of such DSP, which has shown great potential (Spitzer et al. 2005). Another major inefficiency is that the *shell side* of the HEs can scale up from carryover in the flash tanks. The shell side is very difficult to clean. Heat exchange is still the major energy inefficiency in the Bayer process and will continue to improve with development of chemicals and design to reduce the impact of scaling. Double digestion in combination with a pressure decanter is an alternative to tube digestion for bauxite containing ≥5% boehmite (Kumar et al. 2008).

Solid–Liquid Separation and Clarification

Cooled slurry from the digestion area is fed to the settler/washer thickener train for separation of the solids from the pregnant liquor, washing the solids almost caustic free in a countercurrent thickening process (Laros and Baczek 2009). In the initial thickener (typically called a *clarifier*, *decanter*, or *settler*), it is important to achieve an overflow with high clarity (typically <0.1 g/L solids content) and an underflow slurry with maximum solids concentration (typically >30%). Both requirements are achieved by carefully selecting the most suitable flocculants or combination of flocculants, in combination with a proper feed-well design. The overflow liquor from the settler is sent to a security filtration area, whereas the mud from the underflow is fed to the washer train.

The washer train is a series of thickeners (commonly termed *washer*) arranged in a countercurrent decantation scheme for removing caustic liquor from the mud and

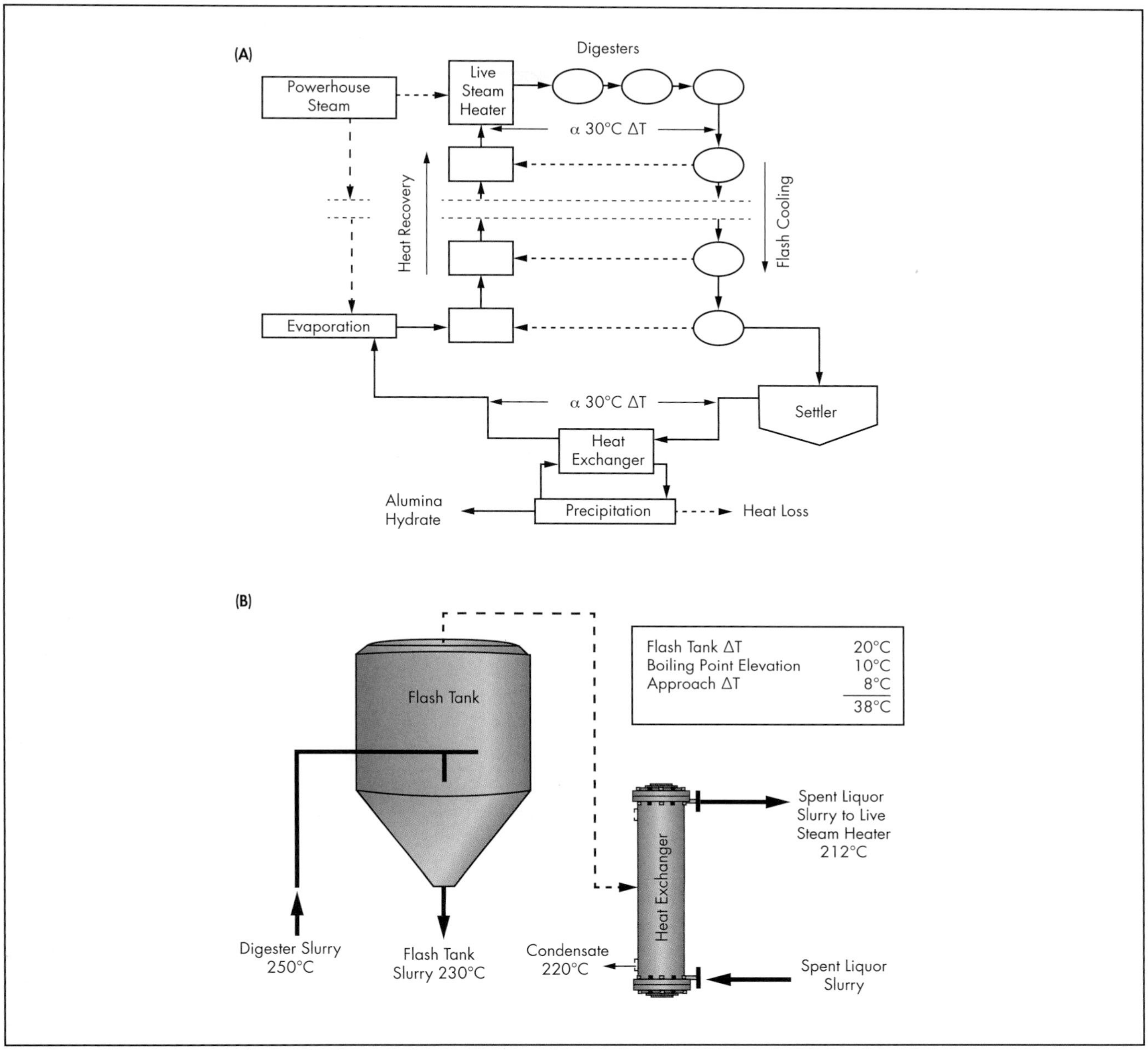

Source: Donaldson 2011. Copyright 2011 by The Minerals, Metals & Materials Society. Used with permission.
Figure 7 Temperature difference across digestion and precipitation

recovering aluminate liquor. This is achieved by the addition of caustic-free hot condensate to the feed of the final washer stage together with the underflow mud from the upstream washer. The same process is repeated several times (typically four to five times), depending on the underflow solids concentration, which can reach up to 65% and higher. When the final washing stage is followed by a filtration stage, the number of washers can be further reduced. For washer services, underflow solids concentration becomes the most important criteria because the increase in underflow solids concentration is key to reducing the number of washing stages. The mud discharged with the underflow of the last washer is either directly pumped to the mud disposal area or fed to an additional filtration stage using red mud drum filters or filter presses.

For several decades, the predominant continuous thickener design used in alumina refineries was the flat bottom/large diameter/outward raking concept, using starch as an additive to promote solids settling and overflow clarification. Today, with the development of modern synthetic flocculants and new feed-well designs, this approach has changed significantly. The use of synthetic flocculants allows achievement of higher settling rates in combination with higher underflow solids concentrations, resulting in thickener designs with greatly reduced tank diameter. By selecting suitable tank parameters, including tank height, thickener rake, and rake drive design, such thickeners produce mud at the high viscosity suitable for paste disposal (common supplier terms are Hi-Rate Thickener or Deep Cone Thickener). Mainly for washer services,

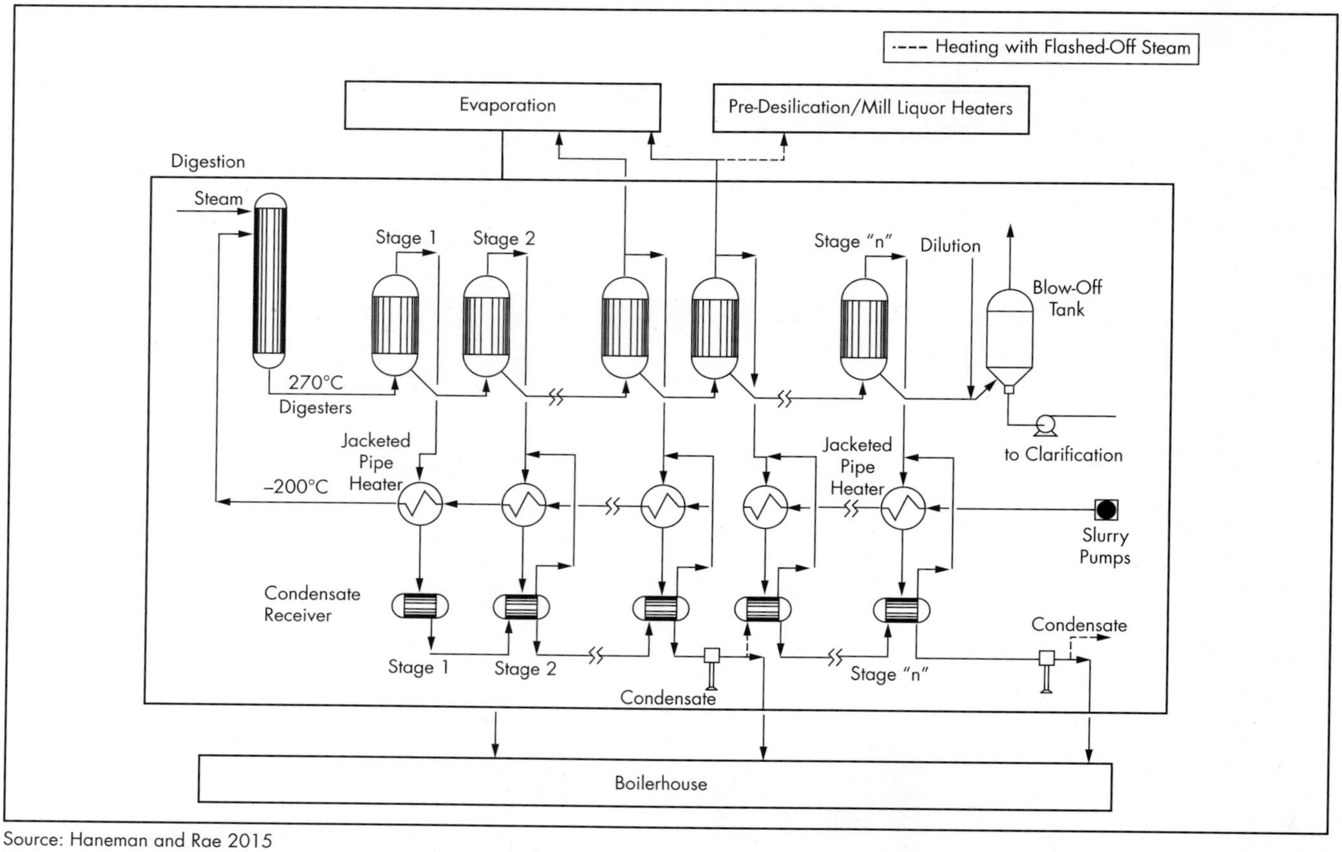

Source: Haneman and Rae 2015

Figure 8 Ma'aden single-stream hybrid digestion flow sheet

autodilution feed-well designs are provided, suitable for adjusting the feed solids concentration at a level of minimum flocculent consumption and optimum thickener performance. Modern rake designs are designed for high raking torques and to promote underflow density. Rake lifting devices are provided for improving scale removal during tank cleaning.

De-scaling is an important issue to be considered in thickener tank designs, because depending on the tank designation, the tank has to be taken out of service more or less frequently for removal of scales growing on surfaces in contact with the liquor. Scale removal is carried out by chemical cleaning using concentrated caustic liquor or mechanical cleaning based on hydroblasting technology with blasting pressures in the range of 1,500 bar and higher. Tank design is playing a major role in accelerating scale removal during mechanical cleaning.

Many settler/washer thickener designs installed in alumina refineries require the installation of an upfront sand removal system to avoid raking problems. Removal of sand (>150 μm particles) is typically provided by a combination of hydrocyclones followed by a series of rake or screw classifiers in a countercurrent arrangement to wash the sand free of caustic. Because of the higher mud underflow solids concentrations that are achieved in modern thickener designs, those thickeners are tolerating higher sand loads and typically no longer warrant sand removal systems.

A key feature of any settler/washer is a properly designed mud-level transmitter measuring the vertical solids profile inside the thickener tank. The measurement is used to control the mud bed height by controlling the underflow pump rotational speed. The mud-level instrument can also be used to control the flocculent dosage based on the solids inventory in the supernatant.

The clarified aluminate liquor discharged with the overflow of the settler is fed to the security filtration area for further reduction of the solids concentration (<20 mg/L) to avoid quality losses of the hydrated alumina produced in the precipitation area. For decades, horizontal leaf pressure filters (e.g., Kelly filters) have been the standard type of technology for security filtration and are still operated in large numbers in alumina refineries around the globe. These filters have to be drained and opened after each filtration cycle for removal of the accumulated solids and cloth wash. A limited number of refineries used sand filters for security filtration to overcome some of the issues of horizontal leaf filters. The sand filter concept, however, was limited to only a few plants and is no longer considered in modern flow sheets. The predominant contemporary technology is the vertical leaf pressure filter technology, operating with liquor backflush for filter cake removal. This concept allows operating the filter for a large number of cycles with intermittent caustic and sometimes also condensate cleaning to remove nonreacted flocculants, oxalate, and precipitated scale from the filter media. The backflush vertical leaf–type pressure filter typically needs to be opened for changing the filter bags only. Both pressure filter concepts, however, require the addition of tricalcium aluminate (made by reacting milk of lime with aluminate liquor) as a filter aid to support the removal of the fine particles (1–2 μm) that are carried over with the settler overflow. Several vertical pressure

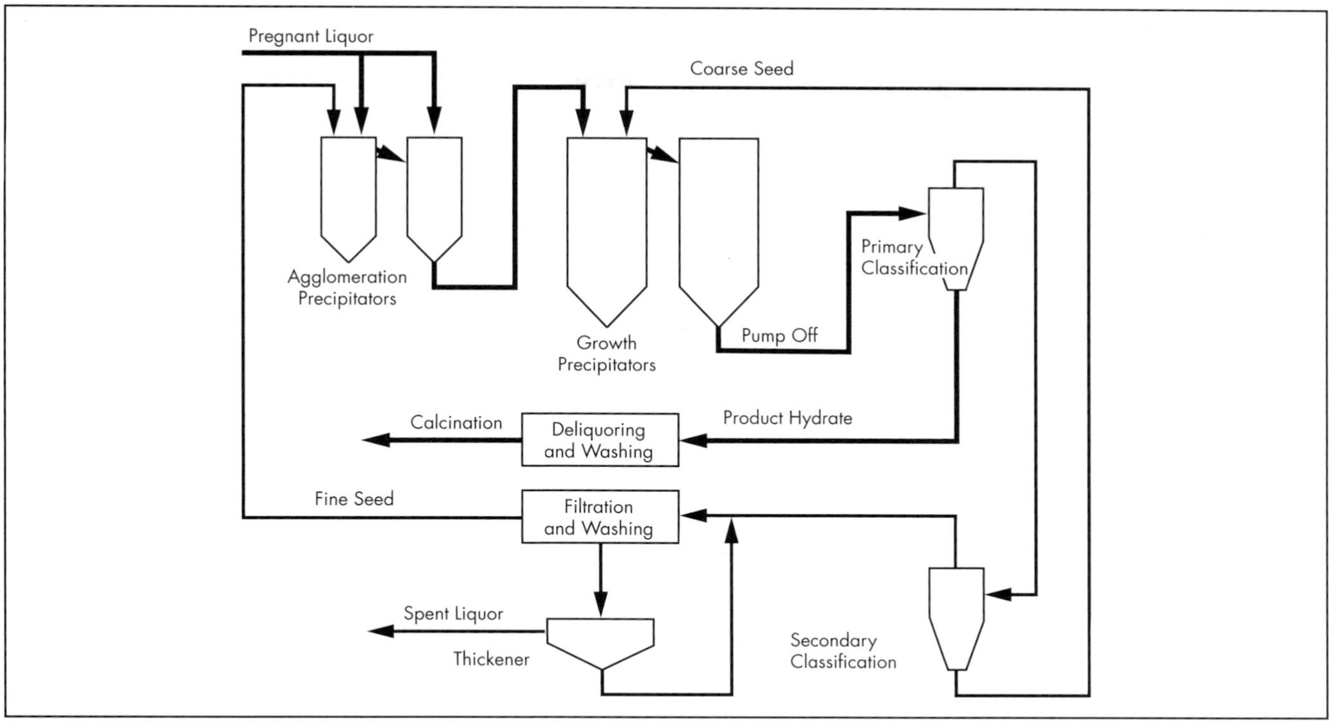

Source: Howard 1988. Copyright 1988 by The Minerals, Metals & Materials Society. Used with permission.

Figure 9 Swiss aluminum precipitation flow sheet

filters are operated in parallel to achieve the desired plant filtration capacity. Standby filters need to be provided for cases when filters are taken out of service for changing filter bags.

Hydrate Precipitation Chemistry

The polished liquor is sent to hydrate precipitation after being cooled through further heat exchange. The chemistry is the reverse of the ATH digestion:

$$2NaAl(OH)_4(l) \rightarrow 2Al(OH)_3(s) + 2NaOH(l) \quad \textbf{(EQ 15)}$$

Although the chemistry appears simple, the design, operation, and control of the population balance of the precipitation process is rather complicated:

- Agglomeration of fine seed particles reduce the number of particles in suspension.
- Formation of new nuclei/crystals and breakage of weak agglomerates increase the number of particles in suspension.
- Agglomerated particles grow to final particle size distribution and process yield.

The driving force and linear growth rate of $Al(OH)_3$ precipitation from the caustic solution depends on seed surface area, temperature, the degree of supersaturation ($A - A_{eq}$), and the caustic concentration C(g/L) of the liquor phase:

$$G \propto Sa\ Exp(-E/RT)\ (A/C - A_{eq}/C)^g/C^{1/g} \quad \textbf{(EQ 16)}$$

where

G = linear growth rate of $Al(OH)_3$, μm/h
Sa = seed area of solid $Al(OH)_3$ in liquor, m^2/L
E = activation energy, kJ/mole
R = universal gas constant, kJ/mole/Kelvin
T = absolute temperature, Kelvin
A = dissolved alumina concentration, g/L
A_{eq} = equilibrium concentration, g/L

The equilibrium concentration of dissolved $Al(OH)_3$ can be predicted (Roesenberg and Healy 1996; McCoy and Dewey 1982). The rate of agglomeration and nucleation is also proportional to the seed area (Sa) of the particle population in the liquor. Because both the rate of precipitation and equilibrium concentration of alumina increase with liquor temperature, there will be an optimum temperature profile across the precipitation flow sheet to maximize the yield of alumina (Chaubal 1990). Therefore inter-tank slurry coolers are installed between the growth tanks (Audet et al. 2012) to control the supersaturation profile.

Precipitation Flow Sheet and Unit Operations

The agglomeration process is critical to understand. Favorable conditions are low to medium charge of fine seed and high initial ratio. Fine seed hydrate from the second classification stage is added to facilitate agglomeration of fine seed particles into coarser particles (Anjier and Roberson 1985) (Figure 9). The agglomeration process is completed after 4–6 hours of retention time (Tschamper 1981).

The slurry with the agglomerated hydrate particles from the agglomeration tank is mixed with coarse hydrate seed particles from the first classification stage and added to the first precipitation growth tank. Published growth tank conditions are shown in Tables 7 and 8. The residence time required to maximize the yield is of the order of 30–40 hours. Nowadays, the precipitators are mechanically agitated.

The optimum precipitation conditions with respect to achieving high yield or liquor productivity and a strong product have been suggested (Sang 1989) as shown in Figure 10. The *floury* precipitation process employed prior to the 1980s achieved a high yield but did not produce particles suitable for modern smelters. At that time, the objective of the successful conversion from floury to sandy alumina (Tschamper 1981; Minai et al. 1978) was to maintain the high yield of the European floury alumina process while producing a coarser and stronger hydrate particle suitable for calcination in stationary calciners. As an example, those objectives were met by the Gove alumina refinery in Australia (Bhasin and Schenk 1992):

- Liquor productivity: 70–80 g/L Al_2O_3
- Hydrate: 6–7 wt % <45 µm or 325 mesh
- Alumina attrition index: 13%–15% on a 45-µm sieve

Impurities, however, also impact the yield or liquor productivity and in a declining direction. Both mineral (sodium chloride [NaCl], sodium sulfate [Na_2SO_4], SiO_2) and organic compounds increase the end point A/C ratio, and thus decrease the yield (Lectard and Nicolas 1983).

Precipitation is one of the most critical areas in the Bayer process, as the amount of alumina removed from the liquor directly relates to the productivity of the circuit and the quality of the final calcined product with respect to purity, particle size, and toughness or attrition resistance. Design and operation of the hydrate precipitation area is driven by economics, including the resulting liquor productivity or yield (P-H. ter Weer 2014; Thomas and Iroside 2015).

The tankage requirement and the large recycle requirement means that the precipitation area in the Bayer process is a highly capital-intensive area.

Table 7 Agglomeration conditions

Agglomeration Parameters	Initial A/C Ratio	Temperature, °C	Seed Charge, g/L
Alcoa Chemie	0.65–0.70	65–80	15–70
Kaiser Alumina	0.575–0.700	75–85	70–140
Sumitomo	0.534*	65–80	30–150
Swiss Aluminium	No data	46–77	7–25*

Source: Anjier and Roberson 1985. Copyright 1985 by The Minerals, Metals & Materials Society. Used with permission.
* Al_2O_3 supersaturation (g/L/m²) seed surface.

Table 8 Growth conditions

Growth Parameters	Initial A/C Ratio	Temperature, °C	Seed Charge, g/L
Alcoa Chemie	0.490	70	300–900*
Pechiney	0.590–0.620	50–75	700–2,000
Kaiser Aluminum	0.575–0.700	75–85	120–300
Norsk Hydro	0.600–0.650	60–75	500–600*
Sumitomo	0.400–0.480	45–65	400–1,500*
Swiss Aluminium	No data	46–77	130–400

Source: Anjier and Roberson 1985. Copyright 1985 by The Minerals, Metals & Materials Society. Used with permission.
*Solids retention precipitators.

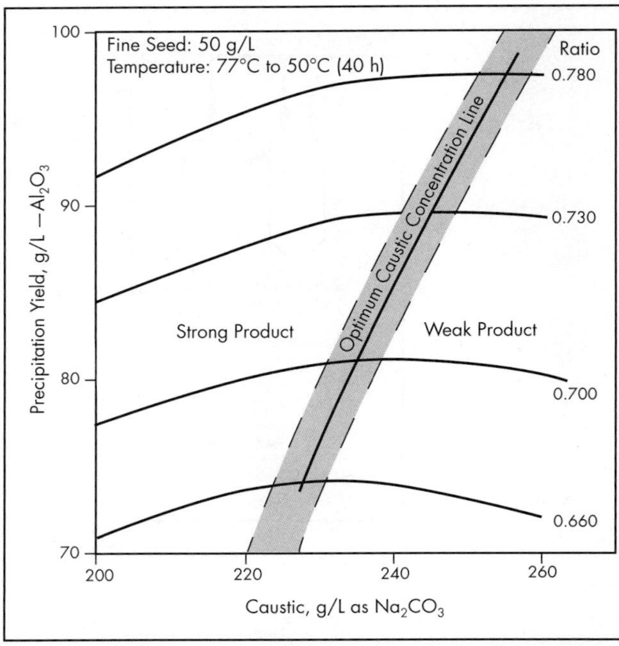

Source: Sang 1989. Copyright 1989 by The Minerals, Metals & Materials Society. Used with permission.

Figure 10 Yield versus caustic concentration and initial ratio

Hydrate Classification

At the end of precipitation, the slurry is classified in hydrocyclones. The amount of seed area has to be carefully controlled; otherwise, too many fines will be produced, and a weak agglomerated product will be achieved rather than the desired tough product. The primary hydrocyclone is used to remove product size material with an average particle size of approximately 100 µm. The overflow, returned as coarse and fine seed, is split to improve the strength of the product by classification in a secondary hydrocyclone. The fine seed from the overflow of the secondary hydrocyclone is concentrated in a fine seed thickener. The fine and coarse seed slurry is accompanied by a significant amount of liquor, hence the seed is often filtered to prevent dilution of the incoming liquor. There are also precipitation circuits with a single stage of hydrocyclone classification, which produce a coarser fine seed for the agglomeration process (Audet et al. 2012). This results in a weaker alumina particle with lower bound soda in the alumina (Raahauge and Devarajan 2015).

Scaling Chemistry

Bayer liquor is supersaturated with respect to many components of the liquor, and one thing it does is form scale. No part of the refinery circuit is immune to scaling, and it requires a large amount of resources to ensure that such scaling does not result in significant production losses. In particular, valves can readily scale and thus become an issue when used for isolation, which extends maintenance shutdowns. When shutdowns for scale removal are frequent, spare equipment is used. Scale removal is accomplished via chemical and mechanical cleaning.

The scale that forms is generally categorized as either *growth* scales or *settled* scales. The former are invariably hard and grow at predictable rates; the latter, which primarily occur in slurry lines and tanks, result in slurry particles being cemented together.

Settled/cemented scales can grow very rapidly and unpredictably, both in rate and location. Throughout the long history of the Bayer process, various ways and means to minimize the negative impact of scale on productivity, energy efficiency, and safety have been a major focus, and that focus continues. There are two major types of growth scale.

Gibbsite scales form after the pregnant liquor has been clarified. Such scales can be removed in most locations using caustic wash systems utilizing fresh caustic with the resulting wash liquor added to the process liquor stream.

DSP scales form in the spent liquor heaters. These are normally removed using an acid cleaning system. A recent breakthrough was the production of a scale inhibitor reagent for such DSP scales. This can almost eliminate the growth of DSP in two-stream digester flow sheets (Spitzer et al. 2005). The solid in the slurry in single-stream digester flow sheets will absorb the reagent and render it ineffective.

Heat Interchange and Evaporation

Solids streams exiting the Bayer process are washed for economic, environmental, and product-quality reasons. These washing operations result in water addition to the Bayer circuit, and water is also added via rainfall, instrument purge water, pump gland water, bauxite moisture content, steam injection heating, and/or other chemical additives.

To remain in balance, all of the added water must be removed from the Bayer circuit. Part of this removal requirement is met by flash evaporation in digestion. A small amount leaves with the washed bauxite residue and washed gibbsite feeding calcination. The remainder of the water balance requirement is met by spent liquor evaporators. The heating ΔT_h (30°C in LSH in Figure 7A) is dictated by the heat recovery efficiency from the digester slurry, bauxite type, and boiling point elevation (Dewey 1981) and approach as shown in Figure 7B (Donaldson 2011). The cooling ΔT_c (HE) is dictated by the precipitation temperature profile controlling the supersaturation as discussed earlier.

Pregnant liquor typically leaves the red area at around 105°C and returns as spent liquor at around 50°–60°C. Heat interchange between these two streams helps reduce overall energy consumption. Two different types of heat exchange systems are typically used for this purpose: vacuum flash cooling or PHEs. Vacuum flash systems typically consist of three to five flash stages feeding vapor to shell-and-tube spent liquor heaters. Condensate collected from the heaters is generally of good quality and may be used for product washing, instrument/gland water, or boiler feed water. PHE systems exchange heat between pregnant and spent liquor in one or more plate packs. The final stage of either system may be water-cooled rather than exchanging the heat with spent liquor, in order to provide better temperature control and/or to achieve lower pregnant liquor exit temperature. Both vacuum and plate systems accumulate gibbsite scale and require periodic chemical cleaning with hot caustic solution and occasional acid and/or mechanical cleaning if other scales or deposits build up. Vacuum flash systems can generally operate for longer periods before cleaning is required. The vacuum flash alternative generally requires higher capital cost and has a slightly unfavorable impact on alumina supersaturation for precipitation. However, its positive impact on water balance may reduce evaporation plant capital and operating costs.

Two main technologies are employed for evaporation: multi-effect and multi-flash; some configurations may employ a combination thereof.

Water Balance and Energy Consumption

Energy consumption varies widely throughout the alumina industry, ranging from 7 to 32 GJ/t alumina, but both refineries with LT (gibbsitic bauxite) and HT (boehmitic bauxite) digestion plants can be <10 GJ/t in total energy consumption (Hendricks 2010). Energy consumption for the Bayer circuit (excluding calcination) is predominantly used for digestion and evaporation.

Calcination accounts for about 2.79–3.0 GJ/t alumina (Monteiro et al. 2011). An efficient refinery water balance will lead to reduced evaporation requirements and hence reduced energy consumption. Drivers for this improvement include

- Refinery layouts to minimize rainfall capture to the Bayer circuit,
- Design and operation to reduce need for hose-up operations,
- Reduction of moisture in bauxite and other feedstocks,
- Selection of digestion and/or heat interchange options that maximize water removal,
- Employing high-efficiency mud-washing processes such as pressure filtration/washing,
- Selection of instruments with low water addition rates, and
- Selection of pump seal technologies with low water addition rates.

Optimization of a new or existing refinery involves trade-off considerations between capital cost, operating philosophy, and operating cost.

Cogeneration and Power Consumption

The Bayer circuit, particularly if processing gibbsitic bauxites, is a large consumer of medium-pressure steam. This makes it well suited to cogeneration of electricity using steam turbines. Steam is produced at a higher temperature and pressure than that required by the Bayer process, then passed through a steam turbine to generate electricity and reduce pressure to the required level used by the Bayer process. A refinery processing gibbsitic bauxite may be able to generate close to 100% of its internal electrical requirements by this means. The quantity of electricity generated by the steam turbines is typically around 80% of the incremental fuel consumption (relative to operation of the refinery without a steam turbine). This is nearly double the efficiency of conventional (non-cogeneration) electricity generation and so is generally an attractive option. If natural gas is available and attractively priced, a gas turbine cogeneration configuration may be employed. Electricity is generated in the gas turbine, and the waste heat from the gas turbine feeds a heat recovery steam generator (HRSG), which supplies steam to the Bayer circuit. In this configuration, the electricity generation can be significantly higher than the refinery's electrical requirements. A steam turbine can be used in combination with the HRSG to further boost the electricity generation. Selection of the appropriate cogeneration configuration for a refinery must take into account a range of economic, logistic, and commercial considerations.

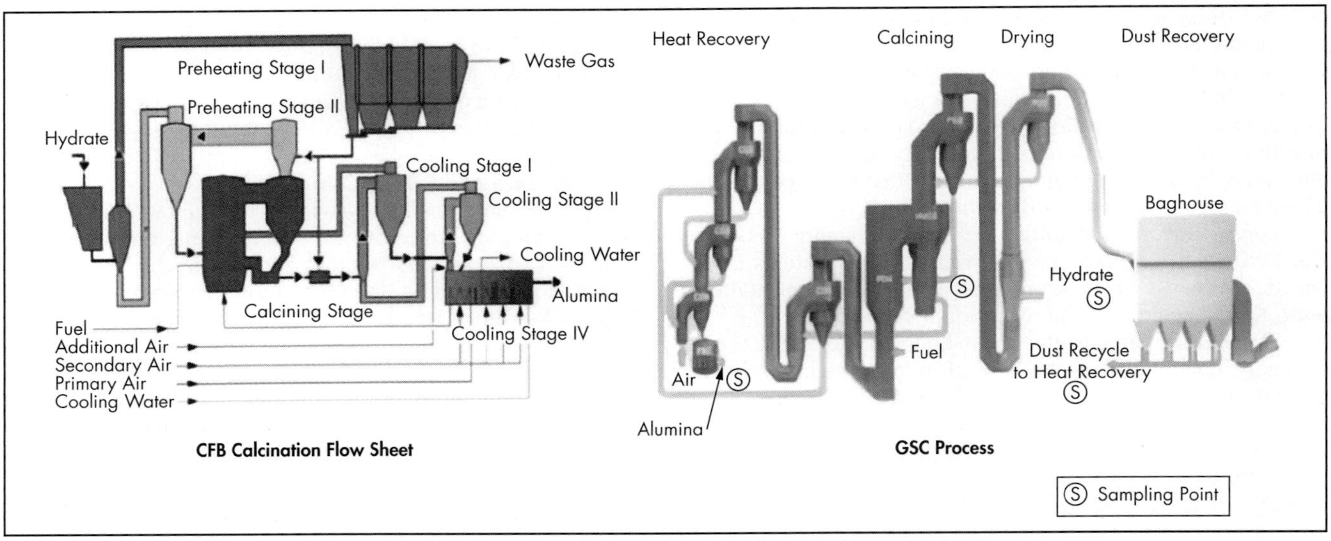

Source: Klett et al. 2010. Copyright 2010 by The Minerals, Metals & Materials Society. Used with permission.

Source: Raahauge and Devarajan 2015

Figure 11 Circulating fluid bed and gas suspension calciner technologies

Materials of Construction

Most plant equipment is exposed to caustic soda solutions at moderate temperatures and pressures. Experience has shown that appropriately selected carbon steel is adequate for these services. Areas exposed to a combination of high temperatures, high velocities, high caustic concentration, vibration, and/or frequent chemical cleaning may give unsatisfactory service life in carbon steel and may warrant application of duplex alloys or nickel lining. Equipment for preparing and circulating acid cleaning typically employs chemical-resistant linings or uses more exotic alloy materials. Pumps and control valves exposed to slurry service are vulnerable to erosion and may employ high chrome-iron liners and/or have their design customized to reduce local velocities. In the calcination area, refractory steels are employed for anchoring the refractory materials used for thermal insulation and protection of vessels and ducts against high-process stream temperatures.

Process Control and Simulation

Modern refineries employ a high level of automatic control. Control hierarchy typically consists of large numbers of individual control loops, some of which may be integrated and optimized by supervisory controls. The supervisory controls may range from simple cascade controllers to sophisticated fuzzy-logic and/or model-based controllers. A key prerequisite for good process control is reliable instrumentation. This remains a significant challenge in the Bayer refinery because of the high tendency for scale growth. Solid strategies to address this challenge include customized instrument design, flushing systems, duty/standby configurations, and preventive instrument maintenance. A heat and mass balance simulation for the Bayer circuit is an important tool for design, operation, and optimization. There are several commercial simulation platforms that contain chemical and process database information to support modeling of the Bayer circuit. Alternatively, some operating companies develop in-house simulations using their own software platform. A dynamic simulation of part or all of the refinery can be used as the back end for a training simulator to train control room operators. The simulation can also be used to develop and test process control strategies, or as part of a model-based control scheme.

Calcination

The product hydrate from precipitation is filtered and washed, normally on pan filters (Petersen et al. 2009), with the final wash step aided by steam and sometimes dewatering aids. Being quite coarse (50–150 µm), the hydrate is readily filtered to low moisture contents (6%) and then fed to the calciner to dehydrate and convert into alumina at temperatures around 1,000°C. Initially, such calcination was done in rotary kilns. Since the 1950s, fluid-flash and circulating fluid bed (CFB) stationary calciners with several heat recovery stages were developed (Fish 1974; Williams and Schmidt 2012) and designed to be significantly more energy efficient than rotary kilns. Since then, there has been significant further development of this concept originating from GSC technology used in the cement industry.

There are many flow-sheet commonalities between the CFB and GSC technologies for alumina (Figure 11), such as

- Feeding moist hydrate into venturi dryers,
- Direct drying and preheating/calcination of feed hydrate by hot combustion products,
- Calcination to alumina in direct-fired HT furnace/reactors,
- Preheating air for combustion by direct heat exchange with hot alumina in staged cyclone/fluid bed–type HEs,
- Indirect and final cooling of alumina with air and/or water in fluid bed coolers, and
- Dedusting of the exhaust gas in electrostatic precipitators or fabric filter/baghouses.

The major differences between the CFB and GSC furnace technologies are summarized in Table 9 and shown in Figure 12. The CFB furnace is a multipass calcination process operated at a pressure exceeding atmospheric pressure. The GSC furnace is a low-velocity (5–7 m/s) single-pass calcination process operated below atmospheric pressure. The operation of the GSC furnace is inherently safe with respect to avoiding emission of hot alumina from holes/openings generated by internal wear of the refractory-lined calciner vessels and/or ducts over time.

Table 9 Typical stationary calciner furnace operating parameters

Circulating Fluid Bed Furnace		Gas Suspension Calciner Furnace		
Holding Vessel	**Not Applicable**	**Holding Vessel**	**No**	**Yes**
Nominal capacity, t/d	3,500	Nominal capacity, t/d	4,500	3,500
Temperature, °C	950	Temperature, °C	1,050	<960
Pressure, kPa (absolute)	>101.3	Pressure, kPa (absolute)	<101.3	<101.3
Retention time, minutes	3–5	Retention time, seconds	10–12	≥200

Source: Raahauge and Devarajan 2015

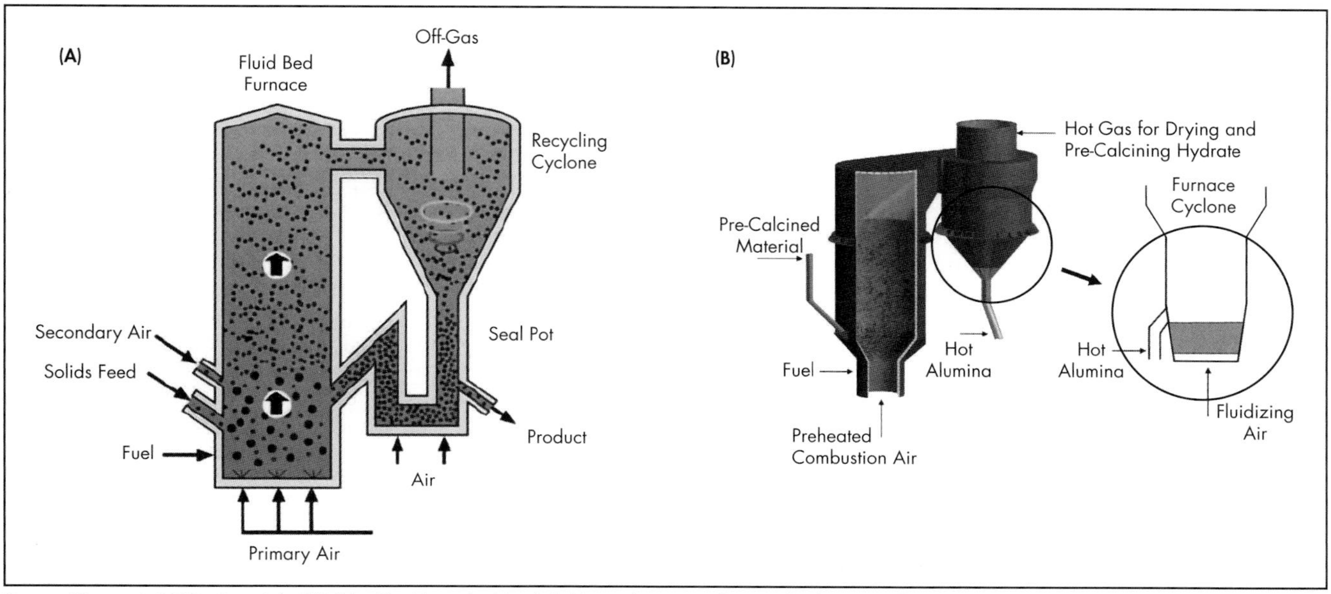

Source: Klett et al. 2010. Copyright 2010 by The Minerals, Metals & Materials Society. Used with permission.

Source: Raahauge and Devarajan 2015

Figure 12 (A) Circulating fluid bed furnace and (B) gas suspension calciner furnace with and without holding vessel

A major calciner design requirement has been the need for stationary calciners to minimize the breaking of particles (Klett et al. 2010; Raahauge and Devarajan 2015) during calcination to meet the smelter requirement of maximum 10% <45 µm or 325 mesh and minimize generation of superfine (<20 µm) particles.

Smelters use the alumina for dry-scrubbing HF in the exhaust gases (S.J. Lindsay 2009). The calcination temperature is therefore regulated to meet the SSA (m^2/g) requirement of SGA. There is much current research on the pore size distribution of SGA (Agbenyegah et al. 2014) and its impact on HF absorption efficiency.

Modern stationary calciners are fuel efficient, running on either gas or heavy fuel oil, with the latter causing acid dew-point issues to be considered in the design (Wind and Raahauge 2009; Missalla et al. 2011).

The thermal energy requirement in calcination is about 30% of the total energy requirement of a modern refinery, corresponding to less than 2,750 MJ lower heat value (LHV)/t alumina in GSC units. Energy is supplied from the combustion of fossil fuel releasing significant amounts of CO_2.

The rapid expansion of the Chinese alumina capacity in recent years is mostly based on coal as an energy source. The GSC technology in China uses coal gas as fuel, making this technology the most frequently applied calcination technology worldwide.

Environmental Control in Stationary Calciners

The electrostatic precipitators and/or baghouses collect fine particles from the stack gases. Compared to electrostatic precipitators, dust emission from baghouses do not increase in cases of a power failure because they offer an absolute barrier against the dust emission. Modern stationary calciners operate with relatively low exit stack temperatures in the range of 150° to 180°C. This allows dedusting the exhaust gases using baghouses rather than electrostatic precipitators as introduced to the 4,500-t/d GSCs replacing the old rotary kilns at Queensland Alumina (Fenger et al. 2004).

The collected dust particles can be added back to the calcination process, or to the alumina product, often with significant impact on the flow properties and dusting of the resulting SGA. In addition, gibbsite in the dust increases HF generation in the smelters. Limited emissions of nitrogen oxides, CO, and sulfur oxides within local national standards are normally not a problem.

Composition, Management, and Uses of Bauxite Residue

Composition. The residual material remaining at the end of the Bayer process is termed *bauxite residue* or *red mud*. The mineralogical and chemical composition of this residue can be varied and is dependent on the nature of the bauxite used and the extraction conditions. It is mainly composed of metallic oxides that have not been attacked by the caustic soda; the main components typically lie within the following ranges:

- Fe_2O_3, 5%–60%
- Al_2O_3, 5%–30%
- TiO_2, 0.3%–15%
- CaO, 2%–14%
- SiO_2, 3%–50%
- Na_2O, 1%–10%

Small amounts of a wide range of other components may also be present, normally as metallic oxides, for example, arsenic, beryllium, cadmium, chromium, copper, gallium, lead, manganese, mercury, nickel, potassium, scandium, thorium, uranium, vanadium, zinc, zirconium, and rare earth elements (REEs). Nonmetallic elements that may occur in the bauxite residue are phosphorus, carbon, and sulfur. Sodium may be present in a sparingly soluble form as DSP or in a soluble form predominantly as a mixture of sodium aluminate and sodium carbonate.

The DSP ($3Na_2O{\cdot}Al_2O_2{\cdot}2SiO_2{\cdot}0\text{-}2H_2O{\cdot}2NaX$ where X could be $CO_3{}^{2-}$, Cl^-, OH^-, $SO_4{}^{2-}$, or $Al(OH)_4{}^-$) arises from the reaction between sodium aluminate and soluble sodium silicates. Small quantities of the soluble sodium compounds resulting from the sodium hydroxide used in the extraction process will remain depending on the dewatering and washing systems used.

All Bayer alumina refineries try to maximize the recovery of the valuable caustic soda from the residues to reuse it during the extraction process. A variety of organic compounds can also be present, arising from vegetable and organic matter in the bauxite/overburden or the use of crystal growth modifiers or flocculants. Compounds present include carbohydrates, alcohols, phenols, and the sodium salts of polybasic and hydroxyacids such as humic, fulvic, succinic, and acetic or oxalic acids.

Quantities. The global average for the production of bauxite residue per metric ton of alumina is between 1 and 1.5 t. Based on the annual production of SGA and CGA in 2017 of 126 Mt, it is estimated that some 160 Mt of bauxite residue is produced annually. This is generated at some 80 active Bayer plants of which approximately 30 are in China; in addition, there are at least another 50 closed legacy sites. Thus the combined stockpile of bauxite residue at active and legacy sites is estimated at 3,000 Mt.

Disposal and storage. The early history of bauxite residue disposal and storage involved using estuaries or land impoundment areas adjacent to the factory. Disposal into rivers, estuaries, or the sea was common for many years, but this has now ceased. Storage of bauxite residue as a dilute slurry (typically 20% solids) in old mines, impounded areas, or dammed valleys was widely practiced until the mid-1980s, but since then, there has been a growing trend to higher solids storage methods. Almost all modern plants use dry stacking technology to produce higher solids and more compact residue. More recently, to produce an even higher solids residue, filter presses have become increasingly common.

Uses. In one of Bayer's early patents in the 1890s, he described the benefits of recovering the iron values of the bauxite residue. Since that period, there has been an enormous amount of work within the alumina producers, research institutes, universities, and by entrepreneurs to extract value from the bauxite residue. Despite all this work, the estimated total usage is between 3 and 4 Mt/yr. The following is a list of the most successful large-scale uses:

- Cement (500,000–1,500,000 t)
- Raw material/additive in iron and steel production (200,000–1,500,000 t)
- Roads/landfill capping/soil amelioration (200,000–500,000 t)
- Construction materials such as bricks, tiles, or ceramics (100,000–300,000 t)
- Other, for example, refractory, adsorbent, or acid mine drainage treatment catalysts

Currently, there is particular interest in recovering REEs, which are essential constituents of modern permanent magnets; nickel–metal hydride batteries and lamp phosphors, with dramatic growth expectations driven by increasing demand for electric and hybrid cars; wind turbines; and compact fluorescent lights. Another area of focus is the recovery of scandium, which has a growing demand in high-strength aluminum alloys. The tragic accident at the Ajka alumina plant in Hungary in 2010 in which caustic red mud slurry inundated a village following a breached dam, the growing cost of remediating closed residue disposal sites, and the availability of EU and other funding has encouraged a resurgence of work at various universities and institutions as well as collaborative development activities between industry and universities.

ALUMINA PLANT COSTS AND ECONOMY OF SCALE

Economic evaluations of an alumina refinery project require generating (discounted) annual cash flows over its lifetime and applying several criteria to ascertain if it meets threshold levels (net present value [NPV], internal rate of return [IRR], etc.).

Key Inputs to Project Economics

Key inputs to annual cash flows are (P.J.C. ter Weer 2006)

- On the revenue side, the amount of product generated and its sales price; and
- On the cost side, capital cost, operating cost, and tax payable.

The quality of the bauxite deposit and the design of the refinery also significantly affect an alumina refinery project's capital and operating costs.

Bauxite Deposit Quality Criteria

The main bauxite deposit quality criteria affecting alumina project economics are (P.J.C. ter Weer 2014)

- Country infrastructure capital cost;
- Distance from the bauxite resource to the export port (bauxite, alumina);
- Disturbed acreage per metric ton of alumina produced;
- Material handled per metric ton of alumina, consisting of
  - Material mined and
  - Residue to disposal;
- Alumina in boehmite;
- Total caustic consumption per metric ton of alumina;
- Ratio of the extractable organic carbon to available Al_2O_3; and
- Resource-contained alumina.

Alumina Refinery Design Criteria

The main plant design criteria affecting economics are as follows (P.J.C. ter Weer 2015):

- Liquor productivity/yield
- Digestion temperature
- Digestion technology
- Bauxite residue settling and washing technology
- Heat interchange technology
- Precipitation technology
- Power- and steam-generation technology
- Calcination technology
- Bauxite residue disposal technology
- Overall plant
  - Design and layout
  - Production capacity
  - Equipment and additives
  - Control equipment

Economy of Scale

Increasing plant production capacity is achieved by debottlenecking the plant area that limits plant capacity (this is often, but not always, the digestion area). This may be accomplished by adding another *train/unit* with commoning up in some areas, but also through higher flows of the liquor circuit, additional tank volume, pumps, piping, and so forth, and through increased liquor productivity/yield (approximately twofold over the past 40 years).

The design of alumina refinery production capacities of greenfield projects outside China have evolved from approximately 0.5–1.0 Mt/yr alumina 25–30 years ago to 1.4–3.3 Mt/yr alumina for more recently constructed and future planned projects. A refinery project represents a major investment with an expected operating life of 30–50 years.

Actual production capacities of existing projects have significantly increased as a result of brownfield expansions, debottlenecking, and improved process efficiencies and operations performance. Rationale for this trend is the economy of scale: An increased production capacity is required to improve the economics of greenfield projects to meet corporate economic criteria.

Economy of Scale and Operating Cost

Economy of scale affects operating costs (P.J.C. ter Weer 2006), which include variable and fixed costs.

Variable Costs

Variable costs include bauxite, caustic soda, coal, and so forth. Overall plant online time of an alumina refinery with more than one production unit/train (e.g., a digestion unit) is generally slightly higher than a plant with only one unit (indicatively 0.2%–0.5% absolute) because of increased flexibility in equipment operation and maintenance. As a result of an increase in production units, the plant operates with fewer interruptions, and operating efficiencies improve (e.g., consumption of bauxite, caustic soda, and energy; indicatively by 0.5%–3%).

Fixed Costs

Fixed costs include labor, maintenance materials, contract services, administration, and so forth. Economy of scale may have a significant impact because of the dilution of fixed annual expenses by a larger production volume (drop in cost per metric ton of alumina produced). If the increase in production capacity includes an increase in the number of production units, the dilution effect is reduced. In addition, the requirements of complex and large alumina refineries may result in disproportionate increases of overhead costs.

Table 10 illustrates this concept for a greenfield project evaluated at different production capacities. The larger refinery capacity shown in Example 1 is mainly the result of more production units in several areas (e.g., digestion), causing a limited improvement of fixed costs per metric ton of alumina. The decrease of variable costs is mainly caused by improvements in efficiencies from operating two production units instead of one.

In another case (Table 10, Example 2), a greenfield project increased its design capacity from 2.8 to 3.3 Mt/yr maintaining the number of production units. This change represented an increase in equipment size at similar process conditions. This had two major consequences: virtually unchanged variable operating costs and significantly diluted fixed costs. Example 2 shows a drop in fixed costs ($3/tA [metric ton of alumina] at a production increase of 0.5 Mt/yr), which in relative terms is much larger than the drop in fixed costs of Example 1 ($5/tA at a production increase of 1.8 Mt/yr).

In summary, economy of scale has a significant effect on fixed operating costs, especially for plant-capacity increases resulting from an increase in equipment size rather than equipment number. The effect on variable operating costs is generally small, unless the increased production capacity results in an increase in production trains, particularly going from one to two trains, because of an improved online time.

Table 10 Economy of scale effect on operating expenditures (OPEX)

Example 1: Increased Number of Operating Units		
Plant Capacity, Mt/yr	**1.4**	**3.2**
Variable Costs		
Bauxite	14	14
Energy	90	88
Consumables*	32	31
Other variable costs†	19	18
Total Variable Costs, $/tA‡	**155**	**151**
Fixed Costs		
Maintenance (materials and contract services)	18	18
Labor	12	8
Other fixed costs	8	7
Total Fixed Costs, $/tA	**38**	**33**
Total OPEX, $/tA	**193**	**184**
Example 2: Increased Equipment Size		
Plant Capacity, Mt/yr	**2.8**	**3.3**
Total Variable Costs, $/tA	**152**	**151**
Fixed Costs		
Maintenance (materials and contract services)	20	18
Labor	9	8
Other fixed costs	7	7
Total Fixed Costs, $/tA	**36**	**33**
Total OPEX, $/tA	**188**	**184**

Source: P.J.C. ter Weer 2013
* Caustic soda, lime, etc.
† Including transport costs.
‡ tA = metric tons of alumina.

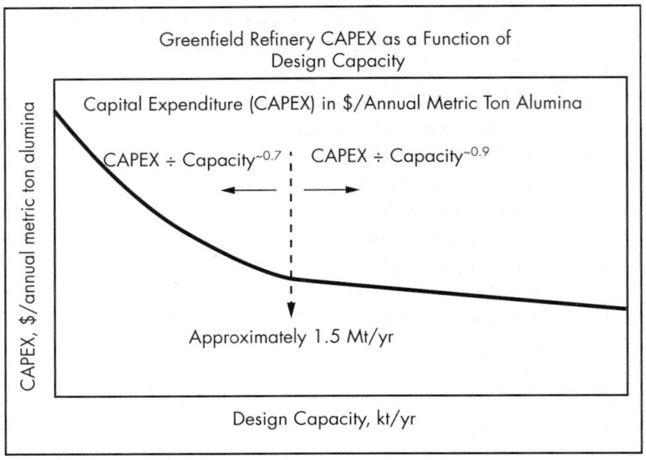

Source: P.J.C. ter Weer 2013

Figure 13 Capital expenditure versus design capacity

Economy of Scale and Capital Cost

The overall impact of the economy of scale is a drop in capital cost per metric ton of alumina produced at higher production capacities, although its impact on equipment and plant infrastructure differ (P.J.C. ter Weer 2007, 2013). In most cases, increases in plant/project capacity are a combination of increases in equipment size and equipment number (e.g., increased number of production trains).

In addition, an increased project scope adds (sometimes disproportionally) to project complexity. As a result, actual capital cost per metric ton of alumina produced as a function of plant capacity will not show a smooth curve. Canbäck et al. (2006) refer to a study of 20 industries showing that at the plant level, beyond a minimum optimum scale, few additional economies of scale can be exploited.

Available information for the alumina industry suggests that with respect to the relationship between the refinery capital expenditure (CAPEX) and design capacity, a differentiation may be made in two ranges as illustrated in Figure 13:

1. Up to ~1.5 Mt/yr with a power factor of ~0.7 ± 0.05
2. Above ~1.5 Mt/yr with a power factor of ~0.9 ± 0.1

The production capacity of ~1.5 Mt/yr is consistent with the current maximum production capacity of a refinery with one digestion unit/train. This appears reasonable because this area is generally taken as a plant design bottleneck because of its high unit CAPEX and its requirement for constant flow to achieve optimum performance levels.

In summary, the impact of economy of scale on alumina refinery CAPEX appears most pronounced for design production capacities up to ~1.5 Mt/yr, with less potential to improve capital cost per metric ton of alumina at larger capacities. Note that 1.5 Mt/yr is not a fixed number, but indicative only (range ~1.4–1.7 Mt/yr) and may increase over time as equipment sizes increase.

Infrastructure Costs and Overall Economics

Several new and future projects have been designed with production capacities well above 1.5 Mt/yr. The reason for this is that greenfield projects have infrastructure requirements that may include access roads, bridges, railway lines, port facilities, and so forth. When requirements are extensive, the related CAPEX is significant and has a disproportional impact on the economics of greenfield projects with relatively small production capacities.

An example illustrates this point for two greenfield project options at the same location: 1.5 and 3 Mt/yr alumina design capacity. The following infrastructure requirements have been assumed: 100-km railway line, railway wagons and locomotives, port jetty and wharf, port ship loading/unloading, and storage and handling facilities. Table 11 provides indicative numbers for capital, operating, and sustaining capital costs for these two options, as well as their indicative economics.

Table 11 shows that despite the refinery CAPEX per annual metric ton of alumina for the two options essentially following the trend illustrated in Figure 13, overall project economics flip from a significant positive NPV (8%), with an IRR of 8.6%, payback period of 10 years, and a value over investment ratio (VIR) of 5.7% for the 3-Mt/yr case to a significant negative NPV (8%), with an IRR of 6.7%, payback period of 11.5 years, and a VIR of –10.5% for the 1.5-Mt/yr case. The major contributor to this is the disproportional increase in cost per metric ton of alumina of the infrastructure CAPEX for the smaller project.

The reasoning could also be turned around: A significant increase in project scale is required to achieve acceptable overall project economics.

Summarizing, economy of scale and infrastructure requirements are two key design criteria to consider when deciding on the production capacity of a greenfield (bauxite and) alumina project.

Alternative Project Approaches

Economy of scale and infrastructure requirements have resulted in ever-increasing production capacities for recently built and future planned greenfield refinery projects. As a consequence, the complexity of greenfield projects has significantly increased, and capital cost in many cases has grown to several billion dollars, with vital consequences:

Table 11 Economy of scale effect on capital expenses and overall economics (indicative)

Plant Capacity, Mt/yr	1.5	3
Capital Cost		
Mine	115	200
Refinery	1.800	3,200
Infrastructure	430	500
Total Capital Cost, million $	**2,345**	**3,900**
$/Annual tA*	1,563	1,300
Operating Cost, $/tA†	**193**	**186**
Sustaining Capital, $/tA	**8**	**8**
Economics (indicative)‡		
Net present value (8%), million $	–248	221
Internal rate of return, %	6.7	8.6
Payback period, yr§	11.5	10
Value over investment ratio, %**	–10.5	5.7

Source: P.J.C. ter Weer 2013

* tA = metric tons of alumina.

† Including infrastructure operating expenditures.

‡ Alumina price at $360/tA; 30 years' operation.

§ After start of operations.

** Capital efficiency ratio net present value (8%) as a percentage of capital expenditure.

- Project owners aim at risk reduction through project financing and formation of joint ventures, further complicating project implementation (Kjar 2015).
- Globally, only a limited number of (large) companies have the human and financial resources to develop greenfield bauxite and alumina projects.
- Only a limited number of engineering firms have the skills and experience required to successfully implement these megaprojects.
- Only (very) large bauxite deposits get developed.

The uncompetitive capital cost of recent Western-developed greenfield alumina projects is a result of (among other reasons) large project size and increased project complexity (Kjar 2010). To quickly and more cost-effectively build a large plant/project, increasing capacity in small increments and replicating modern plant design is recommended (Kjar 2010).

A new approach from a slightly different angle aims at designing an alumina refinery for a dedicated production capacity without provisions for future expansions (P.J.C. ter Weer 2011). This enables optimizing plant layout for the targeted production capacity, for example, with respect to positioning similar equipment close to each other, use of common spare equipment, and so forth. Possible future (brownfield) capacity expansions should justify themselves on their own economic merits.

Both of these alternative approaches result in improved project economics.

OTHER BAUXITE/ALUMINA PROCESSES

In China, there has been an enormous growth in alumina production capacity from about 2–4 Mt/yr in 2000 to approximately 80 Mt/yr in 2017, with an improved utilization rate from 66% in 2009 to 76% in 2013 (Clark 2014). China, however, is facing the problem that more than 95% of its domestic bauxite reserve is diasporic bauxite with a decreasing A/S ratio (Songqing 2008). Such bauxites require very aggressive digestion conditions of high caustic concentration, temperatures >260°C, and fine grinding of the abrasive bauxite. To achieve such digestion temperatures, direct steam injection using very high-pressure steam is required, which becomes economically inefficient. Consequently, many alternative processes using diasporic bauxite have been developed in China (Songqing 2008).

Processing Diasporic Bauxite

The lime-soda sinter process (Figure 14) has been used for many years to produce alumina from diasporic bauxites. Bauxite with soda ash and lime is sintered at approximately 900°C, and the solid discharge is leached to recover the sodium aluminate. The alumina is then recovered from the sodium aluminate solution, not by seeding, but by carbonation of the liquor that results in very high supersaturations and in virtually all the alumina being precipitated. The precipitation process is rapid, as it is essentially a nucleation rather than a growth process. The spent liquor, now primarily a sodium carbonate solution, is then recycled to the kiln. The addition of high amounts of lime is needed to recover the alumina that can be lost to silica minerals—the CO_2 from lime production being the source of CO_2 for the precipitation. Consequently,

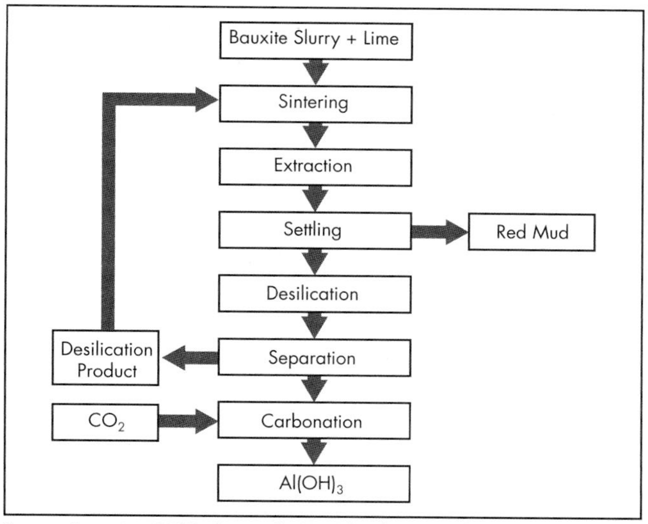

Source: Songqing 2008. Copyright 2008 by The Minerals, Metals & Materials Society. Used with permission.

Figure 14 Sintering process

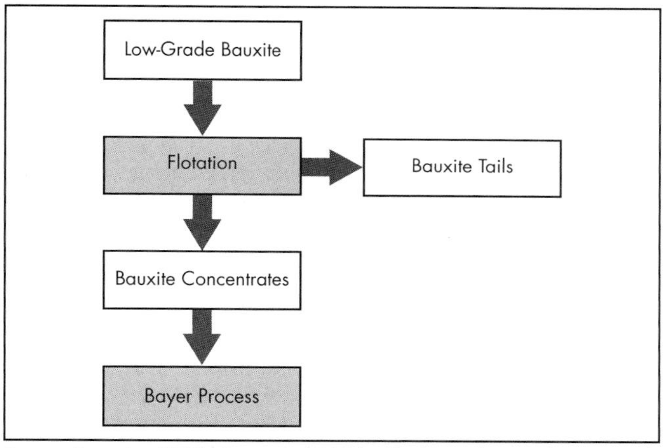

Source: Songqing 2008. Copyright 2008 by The Minerals, Metals & Materials Society. Used with permission.

Figure 15 Flotation Bayer process

there is a high energy usage of the order of 35 GJ/tA (Wanxing 2011), or more than three times the current usage of the modern Bayer process (<10 GJ/tA).

The 800,000-t/yr flotation Bayer process (Figure 15), was introduced after successful removal of silica by flotation as described earlier, reducing the energy consumption to about 15 GJ/tA (Wanxing 2011). However, there is more investigation to be done with respect to optimizing the flotation process to further increase the yield of alumina and lower the energy consumption. The trend has been to add Bayer digesters to sinter plants (using either internal coil heaters in the digesters or tube digesters) to increase capacity. Although the sinter process does a good job of recovering alumina and soda from the residue, especially with the decrease in the bauxite A/S ratio, the overall energy consumption will always be greater with a combined process than with a straight Bayer process.

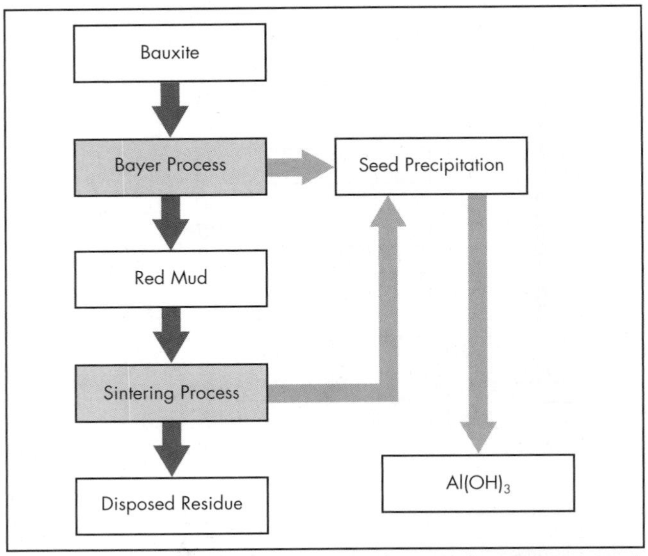

Source: Songqing 2008. Copyright 2008 by The Minerals, Metals & Materials Society. Used with permission.

Figure 16 Series Bayer-sinter process

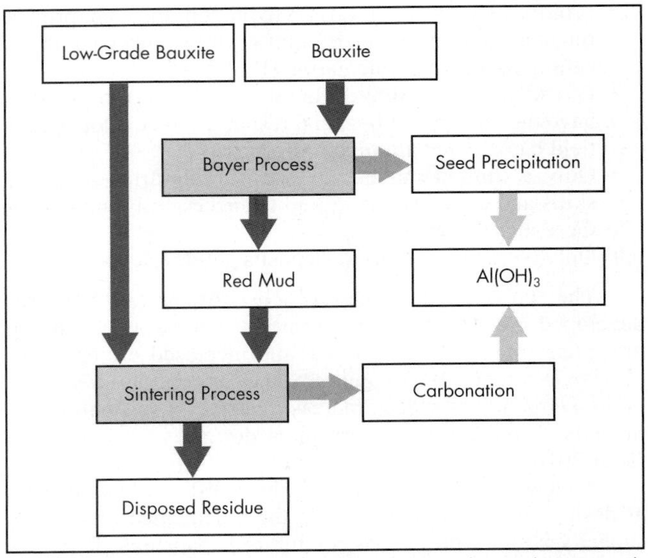

Source: Songqing 2008. Copyright 2008 by The Minerals, Metals & Materials Society. Used with permission.

Figure 17 Combined Bayer-sinter process

The *series Bayer-sinter process* (Figure 16) was developed for bauxite that is first treated using digesters to recover the easy-to-dissolve alumina. The residue (together with some additional bauxite and lime), which still contains significant amounts of alumina (as well as a lot of soda tied up as DSP), is used to feed the sinter kilns. The alumina is recovered as well as the soda via reaction of the DSP with the lime. This serial combination process reduces the overall energy requirement to about 25 GJ/tA.

The *combined Bayer-sinter process* (Figure 17) has a higher alumina output than the serial Bayer process, and thus slightly higher energy consumption as well. One benefit of this combination process is that the kiln destroys both organic and inorganic carbon, and the resulting sodium aluminate liquor is very clean; that is, it is an effective impurity removal process (although costly). Consequently, *white hydrate* can be produced from such liquor that is required in many non-SGA markets, and those markets are less concerned about the sizing and toughness of the product.

Processing Other Alumina-Silicate Raw Materials

Because of the strategic nature of aluminum and because many countries do not have viable bauxite resources, there has been a large amount of work investigating production of SGA from other aluminum-rich resources. The most notable is recovery of alumina from various aluminosilicates, such as nepheline (Smirnov 1969; Skaarup et al. 2013), clays, feldspars, and fly ash, of which only nepheline and fly ash (Shen 2012) have reached proven commercial-scale operation. Production of alumina from fly ash would certainly be good, as there are enormous deposits of fly ash available from coal-fired power plants (Shen 2012).

Many other processes (Nunn and Chuberka 1979; Primeau and Gilbert 2012) have been studied—some acid processes, some chlorination-type processes—but to date, none have come to commercial fruition in terms of SGA production.

Most such processes do not regenerate their reagent as does the Bayer process, and construction materials would be significantly more costly. Such processes, although producing a chemically refined alumina, do not produce SGA with the physical requirements of the modern smelter, namely flow characteristics, particle size, and toughness. Such processes can produce low-soda alumina, something that is impossible to achieve directly from the Bayer process.

ACKNOWLEDGMENT

The authors thank Gerald I.D. Roach, formerly from Alcoa, for providing the first draft of this manuscript.

REFERENCES

Agbenyegah, G.E.K., McIntosh, G.J., Hyland, M.M., and Metson, J.B. 2014. Changes of pore size distribution of smelter grade alumina during HF scrubbing. Presented at the 11th Australasian Aluminium Smelting Technology Conference, Dubai, United Arab Emirates.

Anjier, J.L., and Roberson, M.L. 1985. Precipitation technology. In *Light Metals 1985*. Edited by H.O. Bohner. Warrendale, PA: TMS-AIME. pp. 367–375.

Archer, A.M. 1983. Considerations in the selection of alumina for smelter operation. *J. Met.* 35(9):43–46.

Audet, D.R., Little, R., and Hofstee, N.J. 2012. Stabilisation of product quality at Yarwun alumina refinery. In *Proceedings of the 9th International Alumina Quality Workshop (AQW)*. Perth, Western Australia: AQW. pp. 304–307.

Authier-Matin, M. 1989. Alumina handling dustiness. In *Light Metals 1989*. Edited by P.G. Campbell. Warrendale, PA: The Minerals, Metals & Materials Society. pp. 103–112.

Authier-Matin, M., Forte, G., Ostap, S., and See, J. 2001.The mineralogy of bauxite for producing smelter grade alumina. *JOM* 53(12):36–40.

Banovolgyi, G., and Siklosi, P. 1998. The improved low temperature digestion process (ILTD): An economic and environmentally sustainable way of processing gibbsitic bauxite. In *Light Metals 1998*. Edited by B. Welch. Warrendale, PA: The Minerals, Metals & Materials Society.

Bautenbach, S. 2008. Mineral processing technologies in the bauxite and alumina industry. Presented at the 8th International Alumina Quality Workshop, Darwin, Northern Territory.

Best, J. 2013. Introduction to laterite bauxite: From exploration through exploitation. Presented at the International Committee for the Study of Bauxite, Alumina and Aluminum (ICSOBA) Congress, Krasnoyarsk, Russia, September 3–6.

Bhasin, A.B., and Schenk, H.J. 1992. Productivity and efficiency improvements at the Gove alumina plant. In *Light Metals 1992*. Edited by E. Cutshall. Warrendale, PA: The Minerals, Metals & Materials Society. pp. 203–207.

Bielfeldt, K., and Winkhaus, G. 1981. Challenge to alumina production technology in the 80s. Presented at the International Committee for the Study of Bauxite, Alumina and Aluminum (ICSOBA) 9th Symposium: Alumina production until 2000, Tihany-Balaton, Hungary.

Bitsch, R., and Lamerant, J-M. 1995. Optimization of gibbsitic bauxite treatment. In *Light Metals 1995*. Edited by J.W. Evans. Warrendale, PA: The Minerals, Metals & Materials Society

Canbäck, S., Samouel, P., and Price, D. 2006. Do diseconomies of scale impact firm size and performance? A theoretical and empirical overview. *ICFAI J. Manage. Econ.* 4(1):27–70.

Chaubal, M.V. 1990. Physical chemistry considerations in aluminum hydroxide precipitation. In *Light Metals 1990*. Edited by C. Bickert. Warrendale, PA: The Minerals, Metals & Materials Society.

Chaves, A.P. 2010. Bauxite upgrading practices in Brazil. *Recl. Trav. Chim. Pays-Bas*. 35(39):110–116.

Clark, A. 2014. China's appetite for bauxite and its implication for the world: The idea of "value-in-use" bauxite pricing. Presented at the 20th Bauxite and Alumina Conference, Miami, FL, February 24–26.

Clegg, R. 2005. Development of liquor purification at Alcan Gove. Presented at the 5th International Alumina Quality Workshop, Perth, Western Australia.

Cochrane, D., Mibus, T., and Oakeley, P. 2011. Worsley Alumina Pty Ltd, BHP Billiton. In *Australasian Mining and Metallurgical Operating Practices*. Monograph 28. Edited by W.J. Rankin. Melbourne, Victoria: Australasian Institute of Mining and Metallurgy.

Costine, A.D., Loh, J.S.C., McDonald, R.G., and Power, G. 2010. Unique high-temperature facility for studying organic reactions in the Bayer process. In *Light Metals 2010*. Edited by J.A. Johnson. Warrendale, PA: The Minerals, Metals & Materials Society.

Cundiff, W.H. 1985. Section 19: Alumina. In *SME Mineral Processing Handbook*. Edited by N.L. Weiss. Littleton, CO: SME-AIME.

Dewey, J.L. 1981. Boiling point raise of Bayer plant liquors. In *Light Metals 1981*. Edited by G.M. Bell. Warrendale, PA: TMS-AIME.

Donaldson, D. 2011. A perspective on Bayer process energy. In *Light Metals 2011*. Edited by S.J. Lindsay. Hoboken, NJ: Wiley–The Minerals, Metals & Materials Society.

Donaldson, D., and Raahauge, B.E., eds. 2013. *Essential Readings in Light Metals*. Vol. 1, Alumina and Bauxite. Hoboken, NJ: Wiley–The Minerals, Metals & Materials Society.

Fenger, J., Raahauge, B.E., and Wind, C.B. 2004. Experience with 3 × 4500 TPD gas suspension calciners (GSC) for alumina. In *Light Metals 2004*. Edited by A.T. Tabereaux. Warrendale, PA: The Minerals, Metals & Materials Society.

Ferret, F.R. 2014. Challenges in estimating the commercial value of bauxite in the expanding market. Presented at the 20th Bauxite and Alumina Conference, Miami, FL, February 24–26.

Filidore, A., and Hadawy, J. 2010. Bauxite grinding practices and options. In *Light Metals 2010*. Edited by J.A. Johnson. Warrendale, PA: The Minerals, Metals & Materials Society.

Fish, W.M. 1974. Alumina calcination in the fluid-flash calciner. In *Light Metals 1974*. Edited by H. Forberg. New York: TMS-AIME. pp. 673–682.

Forsyth, W.L., and Hertwig, W.R. 1949. Attrition characteristics of fluid cracking catalysts. *Ind. Eng. Chem.* 41(6):1200–1206.

Gerard, G., and Stroup, P.T., eds. 1962. *International Symposium on the Extractive Metallurgy of Aluminum*. New York: AIME.

Grandhi, R., Weston, M., Talavera, M., Pereira Brittes, G., and Barbosa, E. 2008. Design and operation of the world's first long distance bauxite slurry pipeline. In *Light Metals 2008*. Vol. 1, Aluminum and Bauxite. Edited by D.H. DeYoung. Hoboken, NJ: Wiley–The Minerals, Metals & Materials Society.

Grubbs, D.K. 1987. Reduction of fixed soda losses in the Bayer process by low temperature processing of high silica bauxites. In *Light Metals 1987*. Edited by R. Zabreznik. Warrendale, PA: TMS-AIME.

Haneman, B., and Rae, B.L. 2015. Design, start-up and operation of the new digestion facility at the Ma'aden alumina refinery. Presented at the International Committee for the Study of Bauxite, Alumina and Aluminum (ICSOBA) Symposium, Dubai, United Arab Emirates.

Harato, H., Ishida, T., Kato, H., and Ishibashi, K. 1996. The development of a new Bayer process that reduces the desilication loss of soda by 50% compared to the conventional process. Presented at the 4th International Alumina Quality Workshop, Darwin, Northern Territory.

Hendricks, L. 2010. The need for energy efficiency in Bayer refining. In *Light Metals 2010*. Edited by J.A. Johnson. Warrendale, PA: The Minerals, Metals & Materials Society.

Hill, V.G., and Robson, R.J. 1981. The classification of bauxite from the Bayer plant standpoint. In *Light Metals 1981*. Edited by G.M. Bell. Warrendale, PA: TMS-AIME. p. 15.

Hollit, M., Kisler, J., and Raahauge, B. 2002. The Comalco bauxite activation process. In *Proceedings of the 6th International Alumina Quality Workshop*. Brisbane, Queensland: The Workshop.

Homsi, P. 2001. Alumina requirements for smelting. In *Seventh Australasian Aluminium Smelting Technology Conference and Workshops*. Edited by M. Skyllas-Kazacos and B.J. Welch. Sydney, New South Wales: Centre for Electrochemical and Minerals Processing. pp. 426–455.

Howard, S.G. 1988. Operation of the Alusuisse precipitation process at Gove, N.T. Australia. In *Light Metals 1988*. Edited by B. Welch. Warrendale, PA: TMS-AIME. pp. 125–128.

Hsieh, H.P. 1987. Measurement of flowability and dustiness of alumina. In *Light Metals 1987*. Edited by R. Zabreznik. Warrendale, PA: TMS-AIME. pp. 139–150.

Hudson, L.K. 1987. Alumina production. In *Production of Aluminium and Alumina*. Edited by A.R. Burkin. Hoboken, NJ: Wiley.

Ikkatai, T., and Okada, N. 1963. Viscosity, specific gravity, and equilibrium concentration of sodium aluminate solutions. In *Extractive Metallurgy of Aluminum*. Vol. 1, Alumina. New York: Interscience.

Jiang, T., Pan, X., Yu, H., Hou, X., Tu, G., Zhang, R., and Lu, Y. 2016. Effect of lime addition during digestion on stability of digested liquor of diasporic bauxite. In *Light Metals 2016*. Edited by E. Williams. Cham, Switzerland: Springer International.

Kelly, R., Edwards, M., Deboer, D., and McIntosh, P. 2006. New technology for bauxite digestion. In *Light Metals 2006*. Vol. 1, Alumina and Bauxite. Edited by T.J. Galloway. Hoboken, NJ: Wiley. The Minerals, Metals & Materials Society. pp. 59–64.

Kimmerle, F. 1999. Globalisation of analysis of smelter grade alumina. Presented at the 5th International Alumina Quality Workshop, Bunbury, Western Australia.

Kjar, A. 2010. A case for replication of alumina plants. In *Light Metals 2010*. Edited by J.A. Johnson. Warrendale, PA: The Minerals, Metals & Materials Society. pp. 183–190.

Kjar, A. 2015. Joint ventures—Old risks and new challenges. Presented at the International Committee for the Study of Bauxite, Alumina and Aluminum (ICSOBA) 2015, Dubai, United Arab Emirates.

Klett, C., Missalla, M., and Bligh, R. 2010. Improvement of product quality in circulating fluidized bed calcination. In *Light Metals 2010*. Edited by J.A. Johnson. Warrendale, PA: The Minerals, Metals & Materials Society. pp. 33–38.

Komolossy, G. 2010. Review on global bauxites: Resources, origin and types. *Recl. Trav. Chim. Pays-Bas*. 35(39):96–109.

Kotte, J.J. 1981. Bayer digestion and predigestion desilication reactor design. In *Light Metals 1981*. Edited by G.M. Bell. Warrendale, PA: TMS-AIME.

Kumar, S., Hogan, B., Leredde, P-J., Laurier, J., and Forte, G. 2008. Double digestion process: An energy efficient option for treating boehmitic bauxite. In *Proceedings of the 8th International Alumina Quality Workshop*. Perth, Western Australia: AQW.

Laros, T.J., and Baczek, F.A. 2009. Selection of sedimentation equipment for the Bayer process—an overview of past and present technology. In *Light Metals 2009*. Edited by G. Bearne. Warrendale, PA: The Minerals, Metals & Materials Society. pp. 107–110.

Lectard, A., and Nicolas, F. 1983. Influence of mineral and organic impurities on the alumina tri-hydrate yield in the Bayer process. In *Light Metals 1983*. Edited by R.M. Adkins. Warrendale, PA: TMS-AIME.

Lindsay, J.B. 2012. Customer impacts of Na_20 and CaO in smelter grade alumina. In *Light Metals 2012*. Edited by C.E. Suarez. Cham, Switzerland: Springer International. pp. 163–167.

Lindsay, S.J. 2005a. SGA properties and value stream requirements. In *Proceedings of the 7th International Alumina Quality Workshop*. Perth, Western Australia: AQW.

Lindsay, S.J. 2005b. SGA requirements in coming years. In *Light Metals 2005*. Edited by H. Kvande. Warrendale, PA: The Minerals, Metals & Materials Society. pp. 117–122.

Lindsay, S.J. 2009. Dry scrubbing for modern pre-bake cells. In *Light Metals 2009*. Edited by G. Bearne. Warrendale, PA: The Minerals, Metals & Materials Society. pp. 275–280.

Lindsay, S.J. 2011. Attrition of alumina in smelter handling and scrubbing systems. In *Light Metals 2011*. Edited by S.J. Lindsay. Hoboken, NJ: Wiley–The Minerals, Metals & Materials Society. pp. 163–168.

Lindsay, S.J. 2014. Key physical properties of smelter grade alumina. In *Light Metals 2014*. Edited by J. Grandfield. Cham, Switzerland: Springer. pp. 597–601.

Mahadevan, H. 2015. Bayer process. Presented at the International Bauxite, Alumina and Aluminium Society (IBAAS); China Aluminum International Engineering Corporation Limited (CHALIECO); and Suzhou Research Institute for Nonferrous Metals (SINR): The Development and Future of Aluminum in China: Reality and Dream, Suzhou, China, November 25–27.

McCoy, B.N., and Dewey, J.L. 1982. Equilibrium composition of sodium aluminate liquors. In *Light Metals 1982*. Edited by J.E. Andersen. Warrendale, PA: TMS-AIME.

Milne, D.J. 1982. The influence of mineralogy on alumina processing. *Chem. Eng. Aust.* ChE7(2):38–42.

Minai, S., Kainuma, A., Fujiike, M., Yoshida, O., and Kanehara, M. 1978. Sandy alumina conversion. In *Light Metals 1978*. Edited by J.J. Miller. New York: TMS-AIME.

Missalla, M., Schmidt, H., Ribeiro, J., and Wischnewski, R. 2011. Significant improvements of energy efficiency at Alunorte's calcination facility. In *Light Metals 2011*. Edited by S.J. Lindsay. Hoboken, NJ: Wiley–The Minerals, Metals & Materials Society. pp 157–162.

Monteiro, A., Wischnewski, R., Azevedo, C., and Moraes, E. 2011. Alunorte global energy efficiency. In *Light Metals 2011*. Edited by S.J. Lindsay. Hoboken, NJ: Wiley–The Minerals, Metals & Materials Society. pp. 179–184.

Nunn, R.F., and Chuberka, P. 1979. The comparative economics of producing alumina from S.S. non-bauxitic ores. In *Light Metals 1979*. Edited by W.S. Peterson. Warrendale, PA: TMS-AIME.

O'Driscoll, M. 2013. Proppant prospects for bauxite. Presented at the 19th Bauxite and Alumina Conference, Miami, FL, March 13–15.

Oku, T., and Yamada, K. 1971. The dissolution rate of quartz and the rate of desilication in the Bayer liquor. In *Light Metals 1971*. Edited by T.G. Edgeworth. New York: TMS-AIME.

Perra, S. 1984. Measurement of sandy alumina dustiness. In *Light Metals 1984*. Edited by J.P. McGeer. Warrendale, PA: TMS-AIME. pp. 269–286.

Petersen, B., Bach, M., and Arpe, R. 2009. The world's largest hydrate pan filter: Engineering improvements and experiences. In *Light Metals 2009*. Edited by G. Bearne. Warrendale, PA: The Minerals, Metals & Materials Society.

Primeau, D., and Gilbert, M-M. 2012. The orbite process: An acid-based technology for extraction alumina from clay and alternative feedstock. Presented at the 19th International Committee for the Study of Bauxite, Alumina and Aluminum (ICSOBA) Symposium, Belém, Brazil, October 29–November 2.

Pulpeiro, J.G., Fleming, L., Hiscox, B., Fenger, J., and Raahauge, B.E. 1998. Sizing an organic control system for the Bayer process. In *Light Metals 1998*. Edited by B. Welch. Warrendale, PA: The Minerals, Metals & Materials Society.

Pulpeiro, J.G., Fleming, L., Hiscox, B., Fenger, J., and Raahauge, B.E. 2000. A year of operation of the solid-liquid calcination (SLC) process. In *Light Metals 2000*. Edited by R.D. Peterson. Warrendale, PA: The Minerals, Metals & Materials Society.

Raahauge, B.E., and Devarajan, N. 2015. Experience with particle breakdown in gas suspension calciners. Presented at the International Committee for the Study of Bauxite, Alumina and Aluminum (ICSOBA) Conference, Dubai, United Arab Emirates.

Roach, G.I.D., and White, A.J. 1988. Dissolution kinetics of kaolin in caustic liquors. In *Light Metals 1988*. Edited by B. Welch. Warrendale, PA: TMS-AIME.

Roesenberg, S.P., and Healy, S.J. 1996. A thermodynamic model for gibbsite solubility in Bayer liquors. In *Proceedings of the 4th International Alumina Quality Workshop*. Southbank, Victoria: Allied Colloids.

Saatci, A., Schmidt, H.-W., Stockhausen, W., Stroeder, M., and Sturm, P. 2004. Attrition behavior of laboratory calcined alumina from various hydrates and its influence on SG alumina quality and calcination design. In *Light Metals 2004*. Edited by A.T. Tabereaux. Warrendale, PA: The Minerals, Metals & Materials Society. pp. 81–86.

Sang, J.V. 1987. Factors affecting the attrition strength of alumina products. In *Light Metals 1987*. Edited by R. Zabreznik. Warrendale, PA: TMS-AIME. pp. 121–127.

Sang, J.V. 1989. Fines digestion and agglomeration in at high ratio in Bayer precipitation. In *Light Metals 1989*. Edited by P.G. Campbell. Warrendale, PA: The Minerals, Metals & Materials Society.

Shen, Y. 2012. China's alumina market and development trend. Presented at the 18th Bauxite and Alumina Seminar, Miami, FL, March 6–8.

Skaarup, S.B., Gordeev, Y.A., Volkov, V.V., and Sizyakov, V.M. 2013. Dry sintering of nepheline—A new more energy efficient technology. Presented at the International Committee for the Study of Bauxite, Alumina and Aluminum (ICSOBA) Congress, Krasnoyarsk, Russia, September 3–6.

Sleppy, W.C., Pearson, A., Misra, C., and MacZura, G. 1991. Non-metallurgical use of alumina and bauxite. In *Light Metals 1991*. Edited by E. Rooy. Warrendale, PA: The Minerals, Metals & Materials Society. pp. 117–124.

Smirnov, M.N. 1969. Physical-chemical fundamentals of alumina production from nepheline. In *Bauxite-Alumina-Aluminum: Proceedings of the Second International Symposium of the International Committee for the Study of Bauxites, Oxides and Hydroxides of Aluminum (ICSOBA)*. Budapest, Hungary: Research Institute for Non-Ferrous Metals.

Smith, P., and Xu, B. 2010. Options for high silica bauxite—the roast leach process-mechanism. Presented at the International Committee for the Study of Bauxite, Alumina and Aluminum (ICSOBA) Symposium, Zhengzhou, China.

Songqing, G. 2008. Chinese bauxite and its influence on alumina production in China. In *Light Metals 2008*. Vol. 1, Aluminum and Bauxite. Edited by D.H. DeYoung. Hoboken, NJ: Wiley–The Minerals, Metals & Materials Society.

Songqing, G., and Zhonglin, Y. 2004. Preheaters and digesters in the Bayer digestion process. In *Light Metals 2004*. Edited by A.T. Tabereaux. Warrendale, PA: The Minerals, Metals & Materials Society.

Spitzer, D.P., Rothenberg, A.S., Heitner, H.I., Kula, F., Lewellyn, M.E., Chamberlain, O., Dai, Q., and Franz, C. 2005. A real solution to sodalite scaling problems. In *Proceedings of the 7th International Alumina Quality Workshop (AQW)*. Perth, Western Australia: AQW.

Taylor, A. 2005. Impacts of the refinery process on the quality of smelter grade alumina. In *Proceedings of the 7th International Alumina Quality Workshop (AQW)*. Perth, Western Australia: AQW. pp. 103–107.

ter Weer, P-H. 2014. Relationship between liquor yield, plant capacity increases and energy savings in alumina refinery. *JOM.* 66(9):1939–1943.

ter Weer, P.J.C. 2006. Operating cost—issues and opportunities. In *Light Metals 2006*. Edited by T.J. Galloway. Warrendale, PA: The Minerals, Metals & Materials Society. pp. 109–114.

ter Weer, P.J.C. 2007. Capital cost: To be or not to be. In *Light Metals 2007*. Edited by M. Søorlie. Warrendale, PA: The Minerals, Metals & Materials Society. pp. 43–48.

ter Weer, P.J.C. 2011. New development model for bauxite deposits. In *Light Metals 2011*. Edited by S.J. Lindsay. Hoboken, NJ: Wiley–The Minerals, Metals & Materials Society. pp. 5–11.

ter Weer, P.J.C. 2013. Economies of scale and alumina refining. *Aluminium Int. Today.* (March/April):51–54.

ter Weer, P.J.C. 2014. Sustainability and bauxite deposits. In *Light Metals 2014*. Edited by J. Grandfield. Cham, Switzerland: Springer. pp. 149–154.

ter Weer, P.J.C. 2015. Sustainability and alumina refinery design. In *Light Metals 2015*. Edited by M. Hyland. Cham, Switzerland: Springer. pp. 137–142.

Thomas, D., and Iroside, L. 2015. To yield, or not to yield. Presented at the International Committee for the Study of Bauxite, Alumina and Aluminum (ICSOBA), Dubai, United Arab Emirates.

Thomas, D., and Pei, B. 2007. Chemical reaction engineering in the Bayer process. In *Light Metals 2007*. Edited by M. Søorlie. Warrendale, PA: The Minerals, Metals & Materials Society.

Tschamper, O. 1981. Improvements by the Alusuisse process for producing coarse aluminum hydrate in the Bayer process. In *Light Metals 1981*. Edited by G.M. Bell. Warrendale, PA: TMS-AIME. pp. 103–115.

Wagner, C. 2013. Sustainable bauxite mining—a global perspective. Presented at the International Committee for the Study of Bauxite, Alumina and Aluminum (ICSOBA) Congress, Krasnoyarsk, Russia, September 3–6.

Wanxing, L. 2011. Feeding China's appetite for bauxite and its implications for the global producers. Presented at the 17th Bauxite and Alumina Conference, Miami, FL, March 7–9.

Wehrli, J.T., and Jiang, Y. 2010. Aurukun bauxite—A high reactive silica, monohydrate bauxite and advantages and disadvantages for process design and operation. Presented at the International Committee for the Study of Bauxite, Alumina and Aluminum (ICSOBA) Symposium, Zhengzhou, China.

Welch, B.J., and Kuschel, G.I. 2007. Crust and alumina powder dissolution in aluminum smelting. *JOM.* 59(5):50–54.

Williams, F., and Schmidt, H.W. 2012. Flash- and CFB calciners, history and difficulties of development of two calcination technologies. In *Light Metals 2012*. Edited by C.E. Suarez. Cham, Switzerland: Springer International.

Wind, S., and Raahauge, B.E. 2009. Energy efficiency in gas suspension calciners. In *Light Metals 2009*. Edited by G. Bearne. Warrendale, PA: The Minerals, Metals & Materials Society.

Wind, S., Jensen-Holm, C., and Raahauge, B.E. 2012. Development of particle breakdown and alumna strength during calcination. In *Proceedings of the 9th International Alumina Quality Workshop*. Perth, Western Australia: AQW.

Wood, A., and Harbor Intelligence. 2014. Keeping the alumina and bauxite industry profitable: Challenges and opportunities. Presented at the 20th Bauxite and Alumina Conference, Miami, FL, February 24–26.

Zirnsak, M., Stegink, D., and Teodosi, A. 1999. Agitation improvements in predesilication tanks at the EurAllumina refinery. Presented at the 5th International Alumina Quality Workshop, Bunbury, Western Australia.

CHAPTER 12.2

Aluminum Extraction

Barry Welch and Halvor Kvande

More aluminum is produced today than all of the other nonferrous metals combined. The total world production of primary aluminum was close to 63 million t (metric tons) in 2017, and, in addition, around 20 million t were recycled. The country with the largest smelter production of aluminum is China, with about 56% of the global production, while the five largest aluminum producers in 2016 were Hongqiao (China), UC Rusal (Russia), Rio Tinto Alcan (Canada), Shandong Xinfa (China), and Chalco (China). There has been a significant growth in demand for aluminum in recent years, and the more conservative predictions are a growth at or above 5% per annum over the next decade. As a curiosity, of the 1 billion t of aluminum ever produced, it is estimated that about 75% is still in productive use.

Aluminum is second only to steel in annual production volume. With a density of 2.7 kg/dm^3 at room temperature, aluminum has only about one-third of the density of iron and steel and belongs to the so-called light metals. Besides the light weight, the properties that have enhanced the growth in its use include the ease of forming into simple or complex shapes by rolling, extrusion, or casting of a diverse range of alloys. Thus, the main cast house products are foundry alloys, extrusion ingots, sheet ingots, remelt ingots, and wire rod. The ability to form decorative and corrosion-protective surfaces gives a long low-maintenance life to the finished forms. Furthermore, aluminum has high electrical and thermal conductivity and excellent strength-to-weight ratio. These benefits are capped by the recyclability of almost all products formed from aluminum and its alloys. These properties have made it the world's most widely used nonferrous metal and have impacted and improved the quality of life for everyone. Usage examples range from the expansion in aircrafts and safe air travel that has been enabled by aluminum, the use of aluminum cables for bringing power to new productive and sustainable communities with lower energy losses, lightweight and easily manageable recreational marine and pleasure craft, and complex extruded shapes that have enabled architectural advances in energy-saving green building construction. Protective aluminum packaging enables preservation of precious nutritional and pharmaceutical resources, while the recyclability of aluminum beverage cans is well known. Although production of the metal is energy intensive, the net greenhouse gas emissions are lowered when used in the transport industry because of the reductions in weight and fuel consumption.

PRODUCTS SHIPPED FROM ALUMINUM SMELTERS

Pure aluminum is soft and ductile with limited strength, but by alloying with small amounts of other metallic elements, the properties of aluminum products can be modified to suit a wide diversity of applications. However, it is important for smelters to produce the aluminum to a high purity, so that the desired properties of the finished product after adding the selected alloy agents are not compromised. Typical impurities that can be co-deposited during the smelting process include the alkali metals present as components of the electrolyte, Fe and Si from either impurities in the smelter-grade alumina (SGA) or through corrosion of the cell components, and the metals of other oxide impurities that are precipitated with the alumina in the Bayer process. Although the alkali metals can be readily removed from the metal tapped from the cells—usually by a displacement reaction with aluminum fluoride—it is important to ensure that the SGA and work practices do not introduce the other reducible impurities. This is emphasized in Table 1, which summarizes the designation of the products shipped from the aluminum smelters, the purity requirements, and primary uses.

The two main groups of alloys are cast aluminum and wrought aluminum alloys. The main alloying elements used to impart specific properties on the finished product are copper, iron, manganese, magnesium, silicon, and zinc. These alloying elements impart different properties, making the product more suitable with respect to strength, weldability, casting, machinability, and corrosion resistance. For wrought aluminum alloys, the internationally accepted designation series are given in Table 2.

In Table 2, the first number in the series is simply to identify the key alloying element. The number 5 in the 5XXX

Barry Welch, Professor Emeritus, University of Auckland, Auckland, New Zealand
Halvor Kvande, Professor Emeritus, Norwegian University of Science & Technology, Trondheim, Norway

Table 1 Designation and purity specifications of aluminum shipped from smelters

As-Cast Products*	Purity	Primary Use	Shape/Size	Comments
P1020A	≥99.70%	Remelt for a wide range of products and alloys	Most common are T-shaped ingots of 20 kg to 750 kg in mass. Occasionally as square or round billet.	Commercial-grade purity; standard on the London Metals Exchange
P0610A	≥99.84%			Purity grade with little or no premium
P0507A	≥99.88%			High purity with premiums of tens of dollars per metric ton
P0506A	≥99.89%			
P0406A	≥99.90%			
P0404A	≥99.92%	Aerospace applications		Very high purity with premiums more than $100 per metric ton
P0303A	≥99.94%			
P0202A	≥99.96%			Super purity with premiums of hundreds of dollars per metric ton
99.97%–99.98%	≥99.97%–99.98%	Electronics, capacitor foil		Highest super-purity grade that can be produced in standard reduction cells
P1535A	≥99.5%	Common foil		Off-grade metal that is discounted by tens of dollars per metric ton
Letter grade	<99.5%	Common foil additive		Low-grade metal with discounts of more than $100 per metric ton for metal with >1% iron
Foundry alloy	Varies, typically >7% Si	Wheels for trucks and automobiles		Premium product that is high in silicon content
Conductor rod	Typically <0.03% Si	Conductor wire, transmission cable	Coiled rod	Premium product that is low in trace metals that detract from electrical conductivity

Data courtesy of the Aluminium Association

*P0202A through P1535A and others not shown in this table are commonly designated grades by the Aluminium Association. The first two numbers identify the maximum Si content. The last two numbers identify the maximum Fe content of the metal. Thus, P1020A has a maximum of 0.10% Si and 0.20% Fe. Other impurities including Zn, Ga, V, and "all others" may also have limits, especially on the higher-purity metal grades.

series simply means that the main alloying element is magnesium. In the 1XXX series of pure aluminum, the third and fourth numbers indicate the purity percentage of the aluminum above 99%. A1030 aluminum will then be 99.3% pure. In the higher-numbered series, the last two numbers are an identifier for a specific alloy.

As with other metal products, treatment changes the basic properties of the alloy. The alloys that can be heat treated are the 2XXX, 6XXX, and 7XXX series.

MODERN ALUMINUM ELECTROWINNING CELLS

The established process for electrowinning of aluminum revolves around the independent discoveries in 1886 and the subsequent patents of C.M. Hall (1886) and P.L.T. Héroult (1886). Fundamentally, the core features of the process have remained unchanged since, other than design, cell size, and refinements to operating conditions to enhance the energy efficiency and eliminate potentially harmful environmental impact.

The successful development of the aluminum industry has been achieved by merging two different industrial processes, namely, the extraction of a high-purity alumina from bauxite by the Bayer process, and the Hall–Héroult process for directly electrowinning a high-purity metal capable of satisfying the requirements for the range of applications. Resources of bauxites, the raw material for aluminum, are only located in seven areas: Western and Central Africa (mostly Guinea), South America (Brazil, Venezuela, and Suriname), the Caribbean (Jamaica), Oceania and Southern Asia (Australia and India), China, the Mediterranean (Greece and Turkey), and the Urals (Russia). The Bayer process is described in a separate chapter in this handbook but is integral to the success of the industry.

Aluminum does not occur in nature but rather in extremely stable combined forms, particularly oxides of which the weathered bauxite ore bodies present the easiest path for

Table 2 Widely used alloy designation system for wrought aluminum alloys

Series	Major Alloying Elements	Changed Properties
1XXX	Al of minimum 99% purity	
2XXX	Al–Cu	Increased strength and improved machinability
3XXX	Al–Mn	Increased strength and hardness
4XXX	Al–Si	Improved ductility
5XXX	Al–Mg	Improved strength and corrosion resistance
6XXX	Al–Mg–Si	Improved stability to cast and corrosion resistance
7XXX	Al–Zn–(Mg)–(Cu)	Reduced ability to be cast and improved strength
8XXX	Al–Other elements	
9XXX	Unused series	

Data courtesy of the Aluminium Association

extraction in a pure form. Hence, the extraction of Al is reliant on the following general path:

$$Al_2O_3 + \text{reductant/energy} \Rightarrow 2Al + [O_2 \text{ or oxidized reductant}] \quad \textbf{(EQ 1)}$$

This process should be performed in a manner that prevents by-products interacting with the aluminum product, a constraint best satisfied using an electrochemical path since it provides a simple mechanism for separating the aluminum from the other products. Chemical conversion to intermediates such as aluminum chloride and aluminum oxycarbides to enable alternative paths for the final extraction step have been researched at length, but until now, both energetically and environmentally, the conditions are best satisfied by using

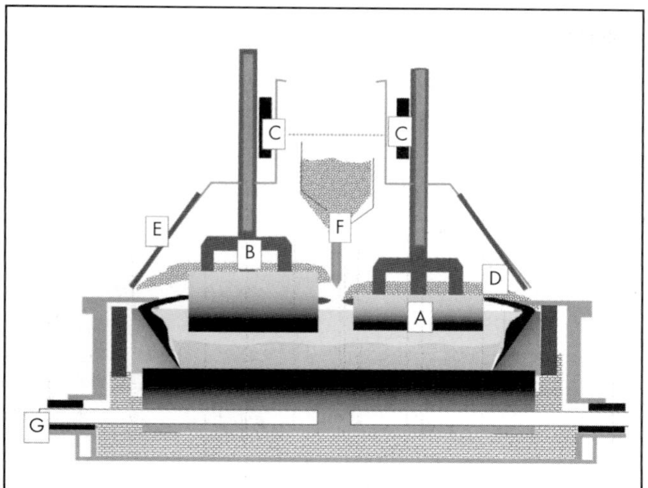

Figure 1 Common features of a modern aluminum electrolysis cell

carbon as a co-reductant. Operating the electrochemical cell at as low temperature as possible, but above the melting point of aluminum, is important because this enables the product to subsequently be delivered in an appropriate form.

Some of the very early electrowinning cells were externally heated and cylindrical in shape and, consequently, they were referred to as "pots," an expression that continues today. The process revolves around the electrochemical decomposition of alumina dissolved in a molten fluoride electrolyte that is dominated by sodium fluoride and aluminum fluoride proportioned to give stoichiometric compositions on the aluminum fluoride–rich side of cryolite, $3NaF{\cdot}AlF_3$. Consumable carbon anodes are used to lower the electrical energy requirements, and hence the operating electrochemical potentials. Because of the operating temperature that is necessary, it involves two liquid layers, the solvent electrolyte and the liquid aluminum cathode. Consequently, one of the challenges is to ensure that the density of the electrolyte is sufficiently lower than that of the metal to avoid inversions through magnetic or other forces that can be generated. A basic cross section showing the common features applicable to all modern aluminum cell designs is illustrated in Figure 1.

The objective is to produce the metal according to the following two reactions:

$$Al_2O_3\text{ (dissolved)} + 3C(s) = 2Al(l) + 3CO(g) \qquad \textbf{(EQ 2)}$$

$$Al_2O_3\text{ (dissolved)} + 1.5C(s) = 2Al(l) + 1.5CO_2(g) \qquad \textbf{(EQ 3)}$$

The process conditions are selected so that a maximum amount of metal is produced according to Equation 3, with the dominant electrochemical oxidation occurring on the horizontally oriented undersurface. The consumption of the anodes necessitates an engineering design to both steadily lower the anode assembly and also to periodically replace the spent anode "butt" (A in Figure 1) with a new rodded anode assembly. The assembly comprises a combination of a new carbon block, and the supporting aluminum rod stem plus a steel yoke assembly (B in Figure 1) that is cast into recesses in the top of the block for good electrical contact. Typically, only 70%–80% of the anode carbon introduced to the cell is consumed with the residual butt being reprocessed for recycling. Limiting the minimum size of the anode butt prevents electrolyte overflowing the upper surface and subsequently corroding the assembly contaminating the metal. The newly introduced cold anode gives a thermal disturbance and does not carry its normal share of current until the freeze formed during introduction has melted and the interface of the carbon assumes the correct anode–cathode distance. Consequently, the anodes are changed at regular time intervals and those within a cell are of varying ages so that there is a minimum disturbance to the cell at any one time. However, within a cell there will therefore be some spatial and temporal variations in properties.

As the metal pad height increases, the estimated production during the time interval is periodically siphoned out of the cell and simultaneously the anodes are lowered by the vertically movable anode support beam (C in Figure 1) to enable maintenance of the interelectrode distance at or near the target value. Typically, several metric tons of metal are siphoned from each cell at intervals between one and three days, the amount being dependent on the design height variation limits.

Prevention of air burn of the upper surfaces of the hot anodes has been a challenge, because there is the requirement to produce a high-purity metal, and potential contaminants must therefore be avoided. Today, the industry tends to use a powdered anode cover mixture (D in Figure 1) comprising a blend of alumina and electrolyte components. The spillage into the electrolyte does not introduce major disruptions or quality issues. The cover material blend also serves as a good thermal insulation enabling minimization of cell heat loss, and the interaction between vapors from the electrolyte and the electrolyte components in the cover material tends to sinter and minimize the potential for air access.

The SGA supplied by modern refineries contains approximately 5 mol % alumina monohydrate, boehmite, or, more correctly, the oxyhydroxide $Al_2O_2(OH)_2$. Much of the remaining alumina is of the gamma structure, which tends to absorb moisture. Consequently, there is some direct and indirect hydrolysis of the electrolyte-liberating hydrogen fluoride with the cell gases following alumina additions. It has been unequivocally established (Patterson 2002) that the hydroxyl ion is stable within the electrolyte, but during electrolysis, hydrogen fluoride is co-evolved at the anode according to the following equation:

$$1.5Al_2O_3{\cdot}H_2O + Na_3AlF_6(l) + 3C(s) = 4Al(l) + 3HF(g) + 3CO_2(g) + 3NaF(l) \qquad \textbf{(EQ 4)}$$

To prevent the release of these gases to the atmosphere, modern cells are hooded (E in Figure 1) and the cells operate at a slightly reduced pressure to take the gases to an emissions scrubbing unit (a dry scrubber).

Maintenance of the dissolved concentration of alumina in the electrolyte is achieved by adding small amounts regularly according to a predictive control strategy aimed at achieving complete dissolution of the oxide in the electrolyte before it can sink beneath the metal pad. Consequently, all modern cells are fitted with point feeders (F in Figure 1), which add a fixed volume of powder alumina equivalent to a specified weight between 1 and 2 kg per shot. The point feeders are located at regular intervals along the center channel of the cell with the number of feeders installed being influenced by the cell amperage.

The computer-generated model of a modern high-amperage cell technology given in Figure 2 demonstrates that the modern cell is vastly different to the early pots, with the lower section more closely resembling a "matchbox" with an overlying superstructure supporting the anodes, the alumina reservoir containers (bins), and the fume capture covers. That the cells have high length-to-width ratio is because the width is constrained by being limited to two rows of anodes.

Also shown in Figure 2 is the cell-to-cell interconnection busbars. Current is introduced to the cell by the four anode risers, and after passing through the electrolyte, it exits the cell via the cathode current collector bars (G in Figure 1), which are connected to the flat aluminum bars that surround the lower part of the cell. This cathode busbar provides interconnection to the anode risers of the next cell, as seen in the extreme right-hand side.

The voltage drop through the external busbar represents wasted energy, and consequently—subject to design and cost constraints—the cross-sectional areas are maximized and lengths minimized.

Each cell is self-heating and operated within the energy input that will maintain the electrolyte and electrode interfaces within a narrow band of the target temperature. However, it must also provide the thermal energy to preheat all materials introduced to the cell and the conversion energies for the various reactions that occur—not only for Equations 1 to 3 but also for secondary chemical reactions due to operation and control limitations and reactions of impurities introduced with the materials. Therefore, the following relationships constrain the operation and productivity of the cells:

$$Q_{cell}\,(W) = V_{cell\ to\ cell} * I = V_{at\ cell} * I + I^2 * R_{external\ busbar} \qquad \textbf{(EQ 5)}$$

where

$V_{cell\ to\ cell}$ = repetitive voltage between cells connected in series (typically between 4.0 and 4.6 V)
$V_{at\ cell}$ = voltage gradient between the anode beam and the cathode busbar
$R_{external\ busbar}$ = ohmic resistance of the cell-to-cell interconnection in busbars
I = current passing through each cell (while all are maintained constant in potline operation)

The energy introduced to the cell environment ($V_{at\ cell} * I$) needs to have an average energy input rate to satisfy the following general relationship because the materials are all introduced at room temperature:

$$V_{at\ cell} * I = Q_{preheat\ materials\ added} + Q_{all\ reactions} + Q_{heat\text{-}loss\ of\ cell} \qquad \textbf{(EQ 6)}$$

The horizontal orientation of both anode and cathode means that the productivity per cell is low, with the largest cells (~600 kA) producing less than 4.6 t of aluminum per day, yet each of these cells requires an operating footprint of 100 m^2. Therefore, it is not surprising that the cell heat loss, $Q_{heat\text{-}loss\ of\ cell}$, accounts for approximately half the energy input.

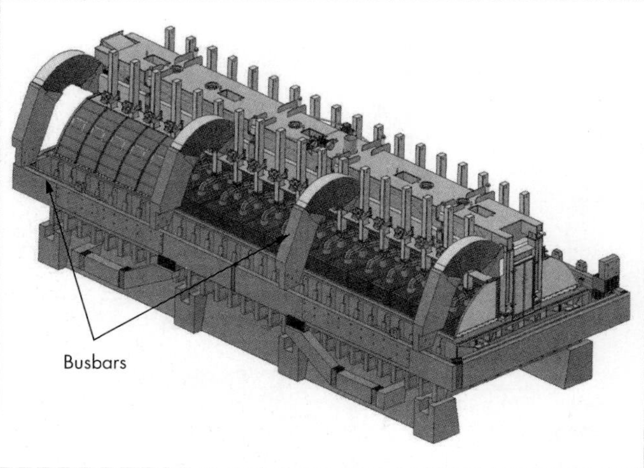

Courtesy of Emirates Global Aluminium

Figure 2 Computer-generated pre-construction design model of the DX+ cell technology of Emirates Global Aluminium, displaying the cell-to-cell interconnection with anode and cathode busbars

INDUSTRIAL POTLINES

Aluminum is a globally traded metal, so smelters are usually located close to a good transport system—often shipping—to enable the smelter to both receive the materials resources and deliver the product. However, as the user of an immense amount of electrical energy, the majority of operating smelters are also located a modest distance away from the primary source of electrical energy (such as hydroelectric dams). Consequently, the most efficient way of transmitting the energy to the smelter site is as AC at a high voltage. Commonly, it is also interconnected to a national grid for security of supply.

The AC power is converted to high-voltage DC power by rectifying transformers with efficiencies ranging from 97% to 98.5%. Because of the heat generated within the rectifier units, their current capacity is limited and usually several are interfaced in parallel to safely provide the necessary line current.

As illustrated in Figure 3, the cathode of one cell is electrically connected to the anode of the next cell to form a cell line (in industry jargon, it is called a *potline*). Series connection allows the use of high-voltage rectifier transformers with the rectified voltage, varying from 400 to 2,000 V DC. The number of cells interconnected in series is set to enable the potline to operate at a line voltage about 50 V below the supply design limit. This voltage constraint is to provide a reserve if any cell requires extra voltage due to process excursions, for starting newly reconstructed cells, or for other emergencies such as providing the extra heat when the line current has to be reduced through failures of rectiformers. Process excursions occur in an unpredicted manner when the alumina concentration of an individual cell is depleted to the extent that the electrochemical reactions at the anode change to include formation of perfluorocarbons, which form a resistive intermediate on the surface and the ohmic resistance can lift the cell voltage in excess of 20 V. Today, smelters interfaced to the national electricity distribution grid are expected to reduce their consumption to make energy available to the community, and this is done by reducing the line current, but, as evident from Equation 6, to maintain the cells at temperature, an increase in the voltage of each cell then becomes necessary.

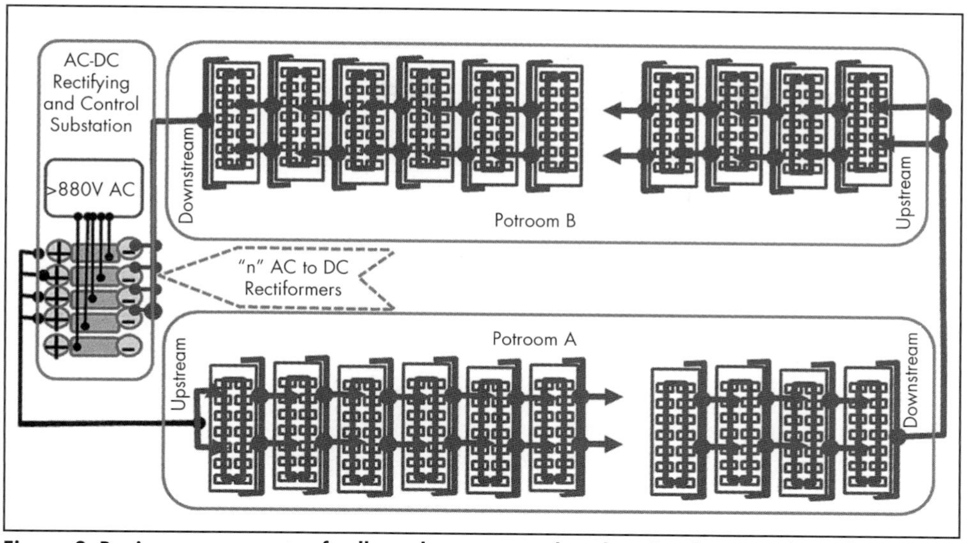

Figure 3 Basic arrangements of cells and power supply substation for a modern aluminum smelter

For many years, aluminum smelter potlines typically had about 200 cells in series (as a consequence of having 880-V supply available), split into two rooms to circumvent the need of a return busbar. More recently, potlines with up to 440 cells have been constructed. In the latter case, the line voltage can reach as high as 2,000 V DC.

Subject to satisfactory workspace and ventilation, the side-by-side cells are positioned as close as possible, with typical modern cell-to-cell spacing between 6 and 6.5 m, depending on design and technology.

The potline interconnection of cells in series is not without risk, as failure of one cell can interrupt the whole potline. Consequently, the cell design features include ways of rapidly bypassing an individual unit. Figure 3 also highlights two other common terms used in operations—the more positive side or end of a cell is referred to as being *upstream* while the more negative side or end is referred to as being *downstream*.

The DC power plant needs to meet requirements for supply reliability, because an interruption in power supply for more than a few hours will result in cooling of the electrolyte to the extent that it can totally freeze, losing its electrical conductivity, and thereafter it is not possible to pass current to restart the cell. In the intermediate time zone, the forces generated by partial freezing can also damage the carbonaceous and refractory materials of cell construction. All will result in serious problems to the aluminum production process. Power plant designs therefore include spare capacity in the form of rectifying transformer units, so the installed power plant is oversized for the smelter. This is called the n-1 principle and is illustrated in Figure 3, where it is noted that one of the rectifying transformers is not connected to the cell busbar network, although it can be readily cut into circuit if needed. It also has the advantage that, at any necessary time, a single unit can be cut out of service for routine maintenance and servicing. However, with the growing demand for the metal but variable supply of electrical energy from National Grids (which can include wind or solar energy), the spare rectifying transformer can also find intermittent use by increasing the current through the potline for a short term, thus utilizing energies that would otherwise be wasted for production of metal.

Courtesy of Norsk Hydro

Figure 4 Modern side-by-side 300+ kA cell line in Sunndalsøra, Norway

Because of the similarities in designs and materials of construction of each cell in the potline, they all tend to operate with voltages within ±50 mV of the average value, with most of the fluctuations being randomly distributed because of the work practices and control strategy. Accordingly, the total line voltage has little variability, and hence, the potline current tends to be constant. Today, power plants have stabilizing units to ensure that a constant potline current is maintained unless there is a failure of the equipment or an operational need to make variation. The modern side-by-side cell potline room shown in Figure 4 also highlights several other features that need to be incorporated in the design and construction of aluminum smelters.

In Figure 4, on the left-hand side of the building, a duct can be seen that is used to convey the emissions and cell gases to a central collection and fume treatment plant so that there are minimal process emissions. Perhaps not surprisingly, this end of the cell is referred to as the duct end. For safety reasons, this duct is electrically insulated from the cells, but outside the building, they all go to a common duct. Collection fans are generating the reduced pressure necessary. The wide aisle on the right-hand side of the cell (referred to as the tap end) is for service vehicles such as replacement anode delivery, and the metal "tapping crucible" is used to vacuum-siphon the metal production from the cell. Overhead cranes are used for assisting these work practices and also for transporting "ore bins" filled with SGA for replenishing the supply in the cells storage hopper. However, modern smelter designs use enclosed air-assisted transport systems that maintain a reliable supply of the alumina at the cell and eliminate the spillage and dust generation inherent in the earlier alumina replenishment systems. The piping of such a modern alumina transport system is evident just to the right of the fumes duct.

In the right foreground of Figure 4, another service unit can be seen, this being the anode butt cooling station. When spent anodes are removed from the cell, they are at temperatures in excess of 900°C and coated with liquid and solid electrolyte, which can evolve fluoride vapors, and gaseous and particulate hydrolysis products from interaction with the moisture of the air. This cooling station enables used anode cooling and simultaneously captures the undesirable emissions.

NECESSARY MATERIALS AND SUPPORT SERVICES

Even though the potroom becomes the nerve center of aluminum production, aluminum smelters also require a continuing supply of DC electrical energy, and alumina. Support services are also required for the supply and removal of carbon electrodes and for transforming the metal siphoned from the cell to a deliverable product for customers. The metal treatment center is referred to as the cast house. Prior to 1960, almost all aluminum smelters extracted their own alumina from bauxite within their smelting complex, but most smelters now purchase the refined SGA from specialized refineries. Consequently, a modern aluminum smelter is usually comprised of the following sections:

- Storage facilities and transport systems for distributing the alumina to the cells
- Two or more potrooms with the cells interfaced in series—these potroom pairs are connected at one end to a stabilized DC power source
- A high-voltage AC power supply or nearby generation from natural gas, coal, or hydroelectric dam
- An anode production facility comprising a paste- and shape-forming unit (green carbon), baking facility, and "anode rodding room" for preparing the anodes for application in the cell
- A "tapped metal treatment station" for metal quality upgrade
- A cast house for product shape and chemical composition adjustment to produce the metal grades appropriate to the application
- A gas treatment center for recovery of volatiles and particulate emissions from the cell

Historically smelters have had a bad reputation for their impact on the environment, but today almost all the gaseous hydrogen fluoride emissions are captured in a "dry scrubber" (or fume treatment plant or gas treatment center) and recycled back to the cell. This is done by adsorption of the gaseous hydrogen fluoride on the SGA. Nevertheless, to achieve efficiency, the Brunauer–Emmett–Teller (BET) surface area and the phase structure of the alumina are compromised. This sequence of operating units, excluding the anode production and rodding, is integrated according to the flow sheet in Figure 5 of a modern aluminum smelter showing the basic range of products prepared for shipment to customers.

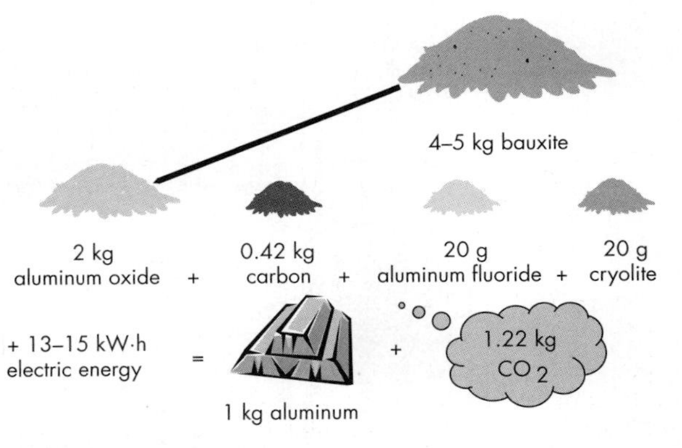

Note: To produce 1 kg of aluminum, the above amounts of raw materials are needed.

Figure 5 Flow sheet of a modern aluminum smelter

ENERGY RESOURCES

Although most of the earlier smelters were located within reasonable transmission range for electrical energy generated from the potential energy of hydroelectric dams, with a few exceptions, the global availability of such resources has been exhausted. Geopolitical or infrastructure resources mitigate against using the few remaining ones that are of sufficient size and located in regions where there is little demand from the human consumer sector. Consequently, there has been a gradual shift to using thermal energy conversion systems for the electrical energy supply.

Data for the mix of power sources for aluminum production in 2016, reported to the International Aluminium Institute (IAI 2017), show the percentages given in Table 3, including Chinese aluminum production.

Table 3 contrasts with a situation in 2000 when the industry was dominated by hydroelectricity generation, whereas coal is now the dominant resource for electrical energy used in aluminum production. This change has an important global environmental consequence because of the high carbon dioxide (CO_2) emissions from coal-fired power plants, many of which have a carbon footprint several times as high as the smelting process itself. However, the modern shift has been to use natural gas resources, which have a lower CO_2 footprint and higher thermal efficiencies. This is particularly so in the Middle East region where the lower-grade energy is also used successfully for producing potable water via desalination.

SMELTER-GRADE ALUMINA

In the electrowinning process, alumina is consumed according to the stoichiometric ratios given in Equations 2 or 3, leading to a theoretical consumption of 1.89 kg of Al_2O_3 per kg Al produced. However, in practice, the industrial value quoted is

Table 3 Four main sources of electric power for 2016 global aluminum production

Primary Energy Source	Percentage
Coal	61
Hydroelectric	27
Natural gas	10
Nuclear	1.5

Data from IAI 2017
Note: In Iceland, some thermal energy for aluminum production is extracted from geothermal resources while in China some 7% of the electrical energy used for smelters comes from wind power or solar panels.

usually in the range of 1.92 to 1.93 kg of Al_2O_3 per kg Al, the difference due to the impurities in the alumina, a feature often ignored in mass balance calculations.

Although the primary purpose of the alumina is to make aluminum in the smelting process, its properties must

- Be compatible with the alumina handling systems integrated into the smelter (such as the transport systems and volumetric point feeders used to dose alumina into the electrolyte);
- Enable the cell to make metal to the required purity and not introduce harmful alloying metals;
- Have a minimal amount of alkali and alkali earth elements, as these also accumulate in the electrolyte, harming the required physicochemical properties;
- Have adequate, assessable surface area to chemisorb the gaseous hydrogen fluoride released during the smelting process so that it can be returned to the process (as aluminum fluoride). This recovery process is referred to as dry scrubbing; and
- Also be used to both insulate the top of the anodes to conserve the heat and simultaneously prevent access of air to avoid unwanted oxidation of the hot carbon. Accordingly, there are a range of physicochemical properties that are sought by smelters, and these are usually encompassed in quality specifications, as shown in Figure 6.

The extensive range of properties listed in Figure 6 fall into the following five categories:

1. **Those pertinent to the dry scrubbing process and generation of emissions.** They include specific surface area or BET surface area, alpha content (the thermodynamically stable phase), and moisture content including monohydrate, $Al_2O_2(OH)_2$. During the dehydration and transition from the alumina trihydrate precipitated in the Bayer process to the finished product, various transition phases are formed, and these generate much of the surface area available for the adsorption of the gaseous hydrogen fluoride in the fume treatment process. However, this surface area is at the expense of having the residual $Al_2O_2(OH)_2$, which is typically present in the final product at levels of 4–5 mol %, and it has been demonstrated that the dissolved species generate much of the hydrogen fluoride that it is necessary to adsorb (Patterson 2002). Hence, there is often debate on the optimum specifications for this purpose.
2. **Handling properties, such as particle size and distribution, attrition index, angle of repose, and flowability.** Having high proportion of fine particles (less than ~45 μm) adversely affects the consistency of flow and transport properties, leading to irregularities in delivery via the transport system and inconsistencies in the volumetric mass discharged at each point feeder in the cell.

Certificate of Analysis—Alumina

Ship: ____________ Shipment Date: ____________

Description of Goods: Alumina in Bulk

Quantity Loaded: _____ t (metric tons)

1. Chemical Analysis

1.1	SiO_2	0.011	%	1.7	Ga_2O_3	0.009	%
1.2	Fe_2O_3	0.008	%	1.8	V_2O_5	<0.002	%
1.3	TiO_2	0.003	%	1.9	ZnO	<0.001	%
1.4	Na_2O	0.36	%	1.10	MnO	<0.001	%
1.5	CaO	0.014	%	1.11	K_2O	<0.002	%
1.6	P_2O_5	<0.0005	%				

2. Physical Analysis

2.1	Size		
	+100 mesh	12.2	%
	+200 mesh	71.9	%
	+325 mesh	92.8	%
	Superfines (–20 μm)	0.7	%
2.2	Flow time, 2 mm	3.7	minutes
2.3	Flow time, 4 mm	67.1	seconds
2.4	Surface area (BET)	81.4	m^2/g
2.5	Loss on ignition: 300°–1,000°C	0.92	%
2.6	Attrition index	19.1	%
2.7	Loose bulk density	1.000	g/cm^3
2.8	Angle of repose	29.7	°
2.9	Alpha alumina	2.9	%

Figure 6 A set of typical alumina specifications as a certificate of analysis

The superfines (arbitrarily defined as being less than 20 μm) can become airborne during the feeding process because of the high velocity of cell gases exiting the feeder hole. This leads to further inconsistencies. Hence, the greatest attention is usually devoted to the fines content when assessing the quality of the shipment.

3. **Chemical demand of the electrolyte.** This group includes impurities that are stable in the electrolyte and not reduced by the aluminum metal. The alkali and alkaline earth metal oxide groupings fall into this category. They tend to accumulate in the electrolyte by reacting with aluminum fluoride forming the respective metal fluorides.
   - An alumina with 5,000 ppm sodium oxide will require the addition of more than 20 kg of aluminum fluoride to the electrolyte for every ton of metal produced, with the resulting reaction product being a surplus of the cryolite electrolyte according to the following reaction:

   $$3Na_2O + 4AlF_3 = 2Na_3AlF_6 + Al_2O_3 \quad \textbf{(EQ 7)}$$

   - Similarly, calcium oxide produces calcium fluoride that will be produced by the following reaction:

   $$3CaO + 2AlF_3 = 3CaF_2 + Al_2O_3 \quad \textbf{(EQ 8)}$$

   - If the amount of CaO exceeds about 12% of the Na_2O, it will result in both an increase in CaF_2 concentration and an increase in the density of the electrolyte, sometimes even to the point where metal and electrolyte inversions could occur.
4. **Impurities that impact the metal quality.** This encompasses most other metal oxides and includes any oxide that is completely reduced by aluminum and alloys with it. Silicon and iron oxides represent the greatest amounts in most SGA.
5. **Impurities that impact cell performance.** Usually these impurities are only being partially reduced by aluminum and therefore able to undergo cycling oxidation and reduction processes within the cell, lowering the faradic efficiency. Phosphorous oxides are the ones of greatest concern.

In the majority of smelters in the world, prior to delivering the alumina to the smelting cells, it passes through a fume treatment plant to adsorb the hydrogen fluoride and capture the fine particulate matter emitted from the cell, and typically this increases the feeding requirements at the cell to approximately 1.98 kg of (impure) alumina per kilogram of aluminum produced. Subject to the primary alumina having a low attrition index, the dry scrubbing process does not modify the handling properties significantly.

ANODE CARBON

Calcined Petroleum Coke

On the basis of Equations 2 and 3, carbon is considered a raw material in aluminum production, because carbon is consumed by the anode reaction.

Typically, between 390 and 430 kg of carbon are used per ton of metal produced by a combination of the reactions according to Equations 2 and 3, as well as excesses through faradic efficiency losses and secondary reactions and processes. This consumption figure is referred to as the net carbon consumption. However, because of metal purity requirements, only between 70% and 80% of each anode introduced to the cell is consumed during that cycle. This leads to the gross anode consumption figure, which ranges between 500 and 600 kg/t, but the residual amount is subsequently recycled.

The fundamental requirement for any carbon to be used as an anode is that it must be of adequate purity and be able to conduct electrical energy, and thus be potentially graphitizable. Although there is no naturally occurring material that satisfies these requirements, the coke residue from petroleum refining satisfies these requirements because the original starting material is usually complex ring structured. However, it usually carries with it chemically bound sulfur and trace elements, including nickel and vanadium. The green coke residue remaining after crude oil is initially refined by heat treating in the distillation units at about 450°C, and 4 to 5 bar pressure has to be heat-treated further by calcining to temperatures in excess of 1,100°C, and preferably >1,300°C. This evolves residual volatile constituents and converts the structure to ordered and better-conducting graphitic crystallites.

The specifications presented in Table 4, show that some of the impurities carry through to the calcined product. Of these, sulfur, vanadium, and nickel are considered to be undesirable. The vanadium and nickel contents impact the metal quality, while the sulfur introduces another gaseous pollutant that requires treatment. Sodium is a proven catalyst for oxidation, leading to higher carbon consumption through secondary reactions, and, consequently, it is highly undesirable.

The broader specifications listed in Table 4 have emerged as a consequence of a growing shortage of low-sulfur coke and the unfortunate link between vanadium and sulfur contents.

The linkage between the sulfur and vanadium in calcined petroleum coke (CPC) is illustrated in Figure 7, which correlates a wide range of petroleum coke sources. This figure demonstrates a growing trend to higher vanadium content within the coke sources available.

Coal Tar Pitch

For application within the smelting cell, the CPC, together with the clean, recycled residual material from the previously used anode (butt), have to be bound into a solid, low-

Table 4 Typical specifications for impurities carry through to the calcined coke

Property	Maximum or Minimum	Typical Specification	Broader Specification
Moisture, %	Maximum	0.3	0.5
Ash, %	Maximum	0.3	0.5
Sulfur, %	Maximum	3.0	3.5
Vanadium, ppm	Maximum	350	500
Nickel, ppm	Maximum	250	300
Iron, ppm	Maximum	300	400
Calcium, ppm	Maximum	200	250
Silicon, ppm	Maximum	250	300
Sodium, ppm	Maximum	150	200
Real density, g/cc	Minimum	2.05	2.03
Vibrated bulk density (28 × 48 mesh = 0.3–0.6 mm)	Minimum	0.85	0.82
Particle size (+4 mesh = 4.75 mm)	Minimum	25	25
De-dust oil, %	Maximum	0.3	0.5

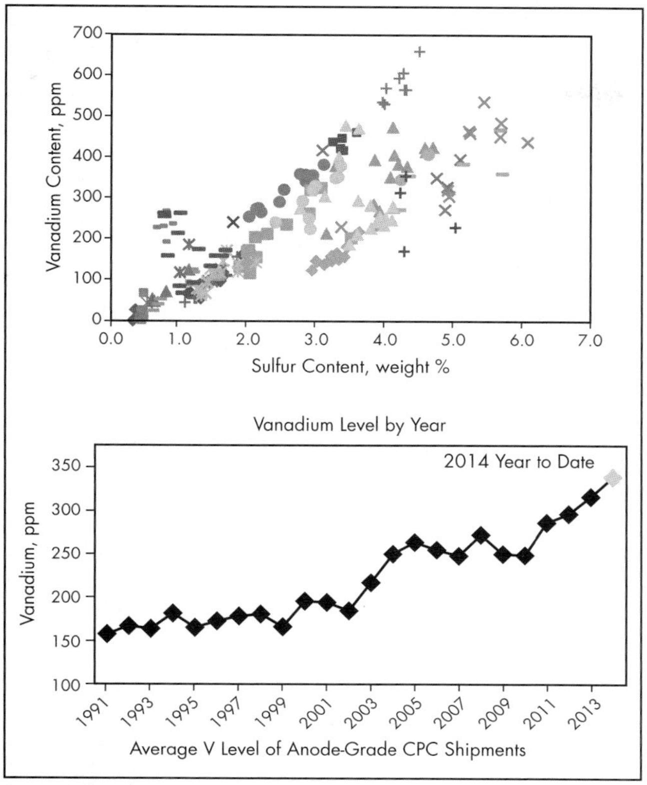

Source: Edwards 2015

Figure 7 Vanadium and sulfur contents in calcined petroleum coke shipments

porosity, compacted mass, and this introduces the requirement of a binder material.

The most satisfactory binder material for electrode production is coal tar pitch (CTP), the less-volatile components of the materials distilled from bituminous coking coals when producing coke for the iron and steel industry. Typically, carbon electrode production requires about one part by weight of pitch per six parts of CPC.

Softening point, viscosity versus temperature relationship, and coking value are all important properties sought in determining satisfactory CTP. The residual coke formed from the thermal decomposition of CTP should also assume a similar graphitic structure to CPC, and once heat-treated, it should have a similar reactivity.

THE ELECTROLYTE AND ITS OPTIMIZATION

The molten electrolyte in the cells has several important functions:

- To be the solvent for alumina to enable its electrolytic decomposition to produce aluminum at the cathode and carbon oxides at the anodes.
- To be a good ionic conductor for passing electricity from the anodes to the cathode in the cell.
- To provide a physical separation between the produced aluminum metal and the evolved carbon oxide gases to avoid the so-called back reaction.
- To provide a heat-generating resistor that enables the cell to be self-heating.
- To have an electrolyte with low enough density to cover and protect the metal from oxidation by the air.

In addition to these functions, it should be thermochemically more stable than the dissolved oxides in the environment (especially the reducing potential of aluminum and dissolved metal, and oxidizing potential of CO_2 gas) and electrochemical potential range applicable to the process at the electrode–electrolyte interfaces. This combination of requirements can only be satisfied by a selected group of molten fluorides.

The congruently melting cryolitic salts, K_3AlF_6, Na_3AlF_6, and Li_3AlF_6, all exhibit adequate solvency properties for alumina, and the resulting solutions all have melting points significantly above that of liquid aluminum. Of these, K_3AlF_6 dissolves the highest amount of alumina. Adequate but decreasing solubility of alumina is also exhibited if the alkali-fluoride-to-aluminum-fluoride molecular ratio is varied from the 3:1 proportions. Each of the cryolitic salts has a minor disadvantage linked with the fact that alkali metal has a small solubility in aluminum and co-deposits during the electrolysis process. The co-deposited alkali metal can be removed during subsequent metal treatment by a displacement reaction with pure aluminum fluoride. K_3AlF_6 has not found commercial applicability because the co-deposited potassium undergoes damaging secondary reactions with the electrode contact media, the graphitized carbon block that also serves as part of the container construction. This interaction causes a swelling and rapid deterioration of the carbon, limiting the cell life.

All alkali metals exhibit a significant solubility in their pure molten fluoride (Dworkin et al. 1962), with the solubility of potassium in molten KF exceeding 5 mol % at 850°C, and for Na in NaF it exceeds 2.5 mol % at 1,000°C. The co-deposited alkali metals also exhibit significant solubility in molten alkali fluorides, forming dissolved species that are readily oxidized if they come in contact with cell gases. This provides a mechanism for lowering the faradic efficiency for metal production. Again, of these, K_3AlF_6 is the worst, and the industry has standardized on using Na_3AlF_6 as the primary solvent component.

Electrolyte Maintenance

As seen from Equations 2 and 3, the electrolyte itself does not take a direct part in the main reactions for aluminum production. Fortunately, the solvent fluorides used to dissolve the alumina and other oxides present in the raw material are fluorides that are electrochemically more stable in the presence of a carbon anode under modest electrode potentials. There are a wide range of oxides that exhibit moderate solubility in the fluorides of their own salts, but salt melts containing the cryolitic anion AlF_6^{3-} formed from melts with the stoichiometric proportions of 3 alkali F·AlF_3 have been found to exhibit the highest solubilities of alumina. After considering other properties, such as electrical conductivity of the resulting melt and the electrochemical tendency to co-deposit the alkali metal with the aluminum, cryolite (Na_3AlF_6) has become the solvent of choice, although occasionally blends of other alkali fluorides have been used.

The main losses of electrolyte occur through it penetrating into the porous materials of construction of the cell and gaseous or particulate fluorides not captured and recycled to the cell. Stringent environmental regulations exist that typically limit the fluoride loss to less than 1 kg/t of metal produced. Leading-edge technologies have losses as low as 0.20 kg of fluoride per ton of metal.

Using Aluminum Fluoride as a Solvent Electrolyte Modifier

Tingle et al. (1981) demonstrated that the sodium solubility in cryolite could be reduced dramatically by increasing the excess aluminum fluoride content. The sodium content of aluminum in equilibrium with pure cryolite at 1,000°C is 175 ppm, whereas increasing the excess aluminum fluoride concentration to 11 wt % reduces it to less than 50 ppm. Given the following two interfacial equilibrium processes established in the NaF-AlF_3-Al, the sodium, and hence the dissolved metal solubility, are substantially reduced by this electrolyte modification:

$$3NaF\text{ (melt)} + Al(l) \rightleftharpoons 3Na\text{ (in Al)} + AlF_3\text{ (melt)} \quad \textbf{(EQ 9)}$$

$$Na\text{ (Al)} \rightleftharpoons Na\text{ (dissolved in electrolyte)} \quad \textbf{(EQ 10)}$$

There has been speculation and model predictions in the literature that aluminum also dissolves as a subvalent species, but the reliable, detailed measurements of metal solubility that have been conducted have demonstrated that sodium is the prominent dissolved species for all electrolyte compositions, and, therefore, this equilibrium is sufficient for expressing the dissolved content.

Lithium Fluoride as an Electrolyte Modifier

Since lithium fluoride (LiF) is thermodynamically stable and gives rise to a small mobile cation and it is also stable within the cell environment, it has been introduced as an additive to the molten electrolyte to increase the electrical conductivity, thus enabling operating at a lower voltage for the same interelectrode distance. When introduced to the electrolyte (in the form of lithium carbonate), it forms lithium cryolite by reaction with excess aluminum fluoride. When LiF/Li_3AlF_6 is present in the molten salt system, it also lowers the melting point, with the beneficial effect of lowering the vapor pressure. These property changes were considered beneficial in some technology and electrolytes containing up to 3.5 wt % LiF were applied. However, the co-deposition of lithium, which can form the metastable Al_3Li phase (δ'), introduced problems for some of the applications of aluminum even though such precipitates strengthen the metal by impeding dislocation motion during deformation. With the advent of dry scrubbers for emissions control and the development of better cell alumina feed control, the disadvantages generally started outweighing the advantages, and lithium fluoride is now rarely used as a deliberate additive. However, some bauxite deposits have lithium oxide associated with them and this can carry through to the SGA. In such situations, a steady-state lithium fluoride concentration becomes established in the electrolyte—typically less than 1.5 wt % LiF equivalent. In the rare situations where it exceeds 2 wt %, it becomes necessary to treat the liquid metal to decompose the metastable Al_3Li phase.

Calcium Fluoride as an Electrolyte Modifier

In early development of the smelting process, many additives were tried to improve the electrolyte, especially to lower the operating temperature. These included magnesium and calcium fluorides, and calcium fluoride (CaF_2) was found to be beneficial for that purpose and became a common additive. However, there has never been any demonstration of other beneficial properties from calcium fluoride additions. The one harmful result is when the concentration exceeds approximately 7 wt %, the density of the electrolyte can increase sufficiently to risk metal/electrolyte inversions. It also reduces the alumina solubility. When this reaches a critical limit, its addition can be at the sacrifice of using aluminum fluoride, which has proven beneficial effects.

A commonly accepted compromise for these two additives is that the total weight percent of the two should not exceed 17. Therefore, many smelters have chosen 5 wt % CaF_2 as an ideal target. Depending on the quality of the Bayer processing and the bauxite mineral origin, oxides of sodium, calcium, lithium, and magnesium are sometimes present in the SGA, and consequently, these end up as fluorides within the melts, following reactions with regularly added aluminum fluoride, as illustrated in Equation 7. Simultaneously, cryolite is made by Equation 7 and in order for the concentration not to change significantly, the weight ratio of sodium oxide to calcium oxide in the alumina should be between 9 and 12. If the calcium oxide results in a lower ratio, extra electrolyte needs to be added to maintain the concentration.

Reducible Impurities

Both the anode carbon and alumina introduce trace amounts of other elements that are also transferred to the electrolyte via similar reactions to Equations 7 and 8. These include Ti, V, Be, Mg, Li, K, Ni, Si, Fe, and P. Initially, they are transferred to the electrolyte phase as a fluoride. If present in the anode, this could be electrochemically generated, for example:

$$4F^- + Si\text{ (anode)} \Rightarrow SiF_4\text{ (electrolyte, at electrolyte–gas interface)} \quad \textbf{(EQ 11)}$$

This generates a concentration in the electrolyte where two loss processes occur:

$$3SiF_4\text{ (in electrolyte)} + 4Al(l) \rightleftharpoons 4AlF_3\text{ (in electrolyte)} + 3Si\text{ (in Al)} \quad \textbf{(EQ 12)}$$

and vaporization, at a rate proportional to its concentration:

$$SiF_4\text{ (in electrolyte)} \rightleftharpoons SiF_4(g) \quad \textbf{(EQ 13)}$$

Most of the impurities listed either co-deposit or are virtually quantitatively reduced by the liquid aluminum degrading the metal quality. A few others are "partitioned" according to this exemplifying illustration. The less reducible impurities, such as the alkali and alkali earth impurities, accumulate as fluorides in the electrolyte.

Solvent Composition and Saturated Alumina Solubility

It is evident that the cryolite anion, AlF_6^{3-}, is essential for dissolution of the alumina, with it dissolving by a chemical reaction that can proceed to equilibrium. Many reactions have been proposed, including the following:

$$Na_3AlF_6 + Al_2O_3 \rightleftharpoons 3NaAlOF_2 \quad \textbf{(EQ 14)}$$

$$2Na_3AlF_6 + 2Al_2O_3 \rightleftharpoons 3Na_2Al_2O_2F_4 \quad \textbf{(EQ 15)}$$

$$4Na_3AlF_6 + Al_2O_3 \rightleftharpoons 3Na_2Al_2OF_6 + 6NaF \quad \textbf{(EQ 16)}$$

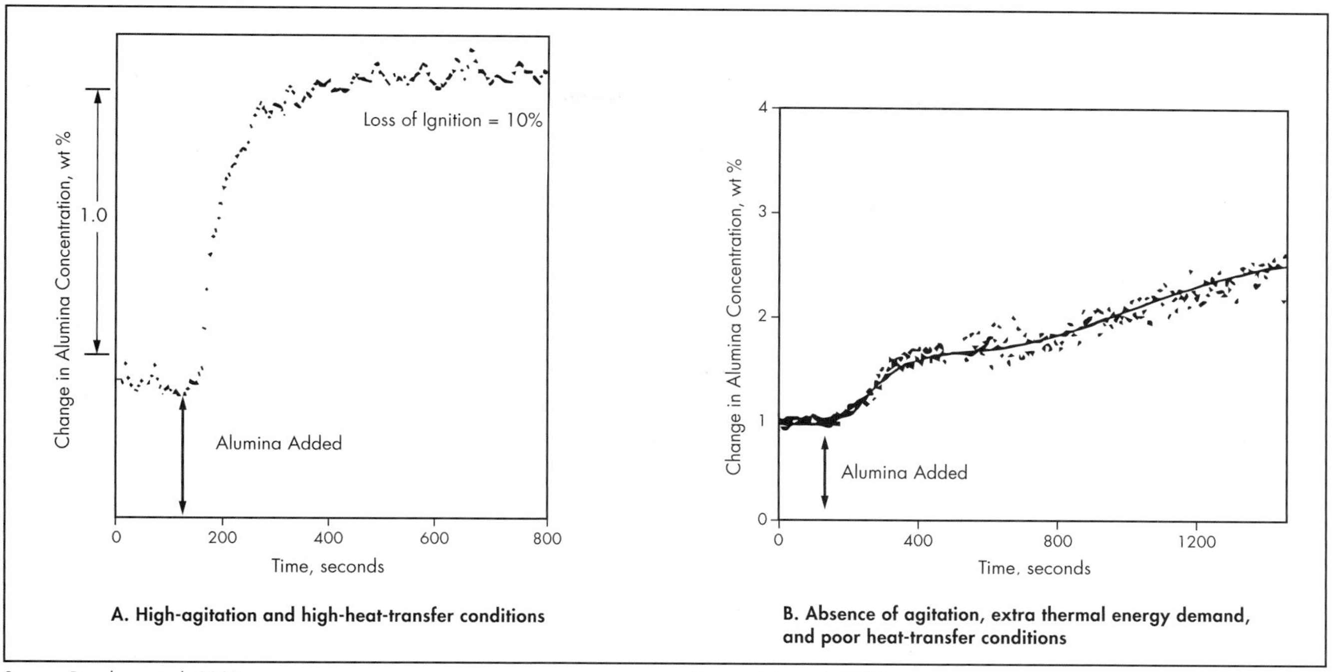

Source: Bagshaw et al. 1985 Source: Jain et al. 1983

Figure 8 Differences in alumina dissolution and mixing in an electrolyte

Key features that arise from various studies include the fact that dissolution is very endothermic, with a heat of dissolution exceeding 108 kJ/mol, and the cryolite anion AlF_6^{3-} concentration in the melt limits the alumina solubility.

Alumina Dissolution Kinetics and Mixing

Specifying conditions for adequate dissolution kinetics is like writing a dissertation on the optimum way of sweetening one's coffee by adding sugar. We know qualitatively how it is achieved: not adding too much, adding it in a powder form, coupled with rapid agitation to disburse, and having a hot solvent. We also know that if it is not done correctly, it does not all dissolve and forms a residue in the bottom of the cup, despite not having exceeded the saturation solubility.

The endothermic dissolution of alumina has many parallels, but it has two added complications. The first is the powder added is cold (near room temperature), and basically, this doubles the energy requirements for dissolution. The second is that because of the cell design and materials of construction, the operating temperature of the electrolyte should be controlled near its liquidus point (<15°C), so we potentially have the situation analogous to dipping the ice cream in hot chocolate—a protective coating is formed on addition. Practically, therefore, heat transfer becomes an important consideration. The significance of this cannot be underestimated, because essentially, each kilogram of primary alumina added requires more than 0.54 kW·h/kg to achieve preheat and dissolution. If this were achieved by extracting from 10 kg of electrolyte, it would need to lower the superheat by 10°C and simultaneously freeze one-third of the electrolyte mass.

The aluminum smelting industry has been concerned about dissolution kinetics because of the need for maintaining a minimum concentration to sustain the electrochemical processes, and also to prevent the formation of a resistive layer of undissolved alumina underneath the metal pad and thus disrupting uniform current flow. Although the results and interpretation of many of the investigations in laboratories for the dissolution kinetics of alumina are often a consequence of the experimental design, several general trends have been confirmed from detailed analysis of published results. These include the following:

- At low superheat of the electrolyte, the process is controlled by heat transfer.
- The dissolution speed is retarded for electrolytes using high excess aluminum fluoride concentration.
- SGA that releases volatiles on addition tends to disperse better in the electrolyte and have faster dissolution kinetics.
- Agitation has a similar impact.
- The dissolution rate is reduced as the electrolyte approaches saturation.

The two dissolution curves shown in Figure 8 illustrate the extremes with the dissolution times for an addition that should increase the concentration by ~1.0 wt %. It ranges from less than 20 seconds to more than 10 minutes when the superheat is low and there is an absence of agitation (Bagshaw et al. 1985; Jain et al. 1983).

All the observations and measurements of the dissolution kinetics are consistent with standard interfacial heat and mass transfer controlled processes, with the speed decreasing as the concentration of the cryolite anion is reduced (hence, the aluminum fluoride concentration impact) or as the interface achieves its equilibrium saturation value. And it is increasing as the interfacial contact area (through dispersion or agitation) is increased. Therefore, the following generalized mass transfer rate equation is useful for describing the process, regardless of whether it is mass or heat transfer controlled:

$$\text{rate } Al_2O_3 \text{ dissolving} = k_{m/h} \cdot (\text{area in contact}) [Na_3AlF_6 \text{ (liquid)}]^n \cdot [C_{Al_2O_3, \text{ saturated}} - C_{Al_2O_3, \text{ bulk electrolyte}}] \quad \textbf{(EQ 17)}$$

where the applicable heat or mass transfer rate constant, $k_{m/h}$, is used, and while the order of the dependence of concentration of the cryolite anion, "n", represents the inadequately determined order for the cryolite concentration influence. Moving to electrolytes using ~11 wt % excess aluminum fluoride became dependent on developing point feeders so that alumina concentration, dissolution kinetics, and mixing could be optimized.

Prior to the development of point feeders, massive additions of alumina were made to the cell—amounts ranging from 50 to 250 kg per addition; therefore, having a high solubility of alumina was crucial, even though undissolved alumina would invariably be found within the cell. Subsequently, the shift to small addition as typified by point feeders, and the development of low alumina concentration sensing feeding logic, enabled operation of the cells at lower concentrations under normal conditions while ensuring good alumina solubility. This facilitated a shift away from the cryolite stoichiometry on the aluminum fluoride–rich side of the phase diagram to reduce the metal solubility in the electrolyte and thus raise the faradic efficiency of the cells. Consequently, between 1960 and 1990, the trend in the industry was to gradually increase the AlF_3 concentration in the electrolyte from about 3 wt % up to the level of 10–12 wt % that is used now.

Operating Temperature and Electrolyte Maintenance

The liquidus temperature of the modern point-fed electrolyte is typically in the band 942° to 955°C. For operational and cell life reasons, it is usually desirable that the operating temperature be no more than a few degrees above the liquidus temperature to ensure retention of a frozen cryolite layer around the sidewalls. This layer acts as a thermostat (through freezing or melting if there is transient imbalance in the energy content of the electrolyte) and also forms a leak protection barrier while preventing undesirable horizontal current flow between the end of the metal pad and the otherwise electrically conducting sidewall components.

Consequently, the operating temperature should be between 955° and 965°C. This temperature and composition band also ensure that the density of the electrolyte remains within the band of approximately 2.08 to 2.12 kg/dm^3 (compared to 2.30 kg/dm^3 for the molten metal), ensuring that liquid inversion is avoided in the normal stability range of the liquid aluminum metal pad.

The total mass of electrolyte used in the cell varies with design, which particularly influences the free volume between the electrodes and the sidewalls of the cell. During the last 50 years, the average cell size of the leading-edge technologies has increased dramatically, with the operating line current typically being three times that of the first generation of mechanically fed cells. Through extensive mathematical modeling of the impact of the magnetic field on the metal pad, design variations have also reduced the distortion of the metal surface as well as lowering the metal pad velocity, thus enabling operation at reduced interelectrode distance—the zone occupied much of the electrolyte. Simultaneously, to maximize productivity, the size of the anodes has been increased within the cell by reducing the surrounding free volume presented by the gaps between each anode block and the gap at both the center channel and between the outer edge and the cell sidewall. Consequently, the mass of electrolyte per cell kiloamps has been reduced quite dramatically.

The steady reduction in electrolyte mass per kiloamp tightens the requirements for alumina dissolution and mixing, and accordingly, the control of the chemical composition of the electrolyte is also of greater importance.

ELECTROLYTE COMPOSITION CONTROL

The composition of the electrolyte is selected for optimum overall cell performance; therefore, smelters generally take routine samples of the electrolyte for analysis so that corrective actions can be taken to adjust the composition to the target control band. While there have been a range of sensors developed to assist solvent composition control, the impact of the operating dynamics plus limited accuracies of the sensors generally result in limited sampling for analysis of individual cells by a laboratory with resulting slow feedback. Although the solvent electrolyte is fairly inert to the primary electrochemical processes, losses of electrolyte components occur through secondary processes within the cell, as summarized later in Table 6. For example, cells typically emit about 20 kg of hydrogen fluoride for every ton of aluminum produced, although the amount of this is dependent on the climate and quality of alumina used and can vary by 50%. Most of the hydrolyzing material originates from moisture or structural water introduced by the alumina. Component losses also occur through vaporization, with the main volatile component being sodium tetrafluoroaluminate ($NaAlF_4$). More than 95% of these emissions are captured and recycled back to smelting cells, with each cell receiving the average amount lost rather than the amount lost through its individual process conditions prevailing. Consequently, limited variation for individual cells can occur through these mechanisms.

The components of the electrolyte derived from aluminum fluoride also become the dominant variable, normally requiring replenishment. This is because of its reaction with the alkali and alkaline earth metal oxides present in the SGA to form stable fluorides, as illustrated by Equations 7 and 8. As seen in Table 5, the amount of these oxides can vary considerably. The amount of aluminum fluoride necessary is greater than given by Equations 7 and 8 because an additional amount is needed to readjust the newly formed cryolite to the target proportions. A simple material balance correlation for the amount of aluminum fluoride consumed by the electrochemically stable SGA impurities is given by the following equation:

$$\text{kg } AlF_3/\text{ton Al} = 34.3 \cdot [\%Na_2O] + 19 \cdot [\%CaO] + 0.49 \cdot [\%Na_2O] \cdot [\%AlF_3] \quad \textbf{(EQ 18)}$$

where the percentages of the oxides are based on weight, as is the percentage of excess aluminum fluoride above the stoichiometric cryolite composition.

The AlF_3 maintenance needs for "new cells" (i.e., less than a year old) are invariably less than predicted by Equation 18, and it can even be nonexistent in the first 60 days of operation. This is a consequence of the carbonaceous material used for the cathode lining that takes up a large amount of sodium, forming sodium–carbon intercalation compounds until localized saturation at the interfaces is reached. The source of the sodium is from either direct electrodeposition or via co-deposition with the aluminum. However, Equation 18 serves

Table 5 Thermodynamic analysis of the electrochemical reactions established in smelting cells (based on thermochemical database)

Important Electrochemical Reactions Occurring in the Cell	Equation No.	$\Delta H^\circ_{reaction}$, kJ	$\Delta G^\circ_{reaction}$, kJ	$E^\circ_{enabling}$, V	$E^\circ_{revised\ estimate}$, V ±0.04	$E^\circ_{complete}$, V
Normal Metal Production						
$Al_2O_3 + 3C = 2Al + 3CO(g)$	2	1,351	622	1.074	1.10	2.333
$Al_2O_3 + 1.5C = 2Al + 1.5CO_2(g)$	3	1,099	690	1.191	1.22	1.898
$Al_2O_3 + 3C + 3S = 2Al + 3COS(g)$	19	1,203	568	0.980	1.03	2.103
Generation of Gaseous Emissions						
$1.5Al_2O_3{\cdot}H_2O + Na_3AlF_6(l) + 3C = 4Al + 3HF(g) + 3CO_2(g) + 3NaF(l)$	4	2,617	1,239	1.041	1.07	2.644
Perfluorocarbon Formation Initiation Reaction						
$Al_2O_3 + 3C + 2Na_3AlF_6(l) = 4Al + 3COF_2(g) + 6NaF(l)$	20	2,877	2,157	1.863		2.485
$Al_2O_3(s) + 4.5C + 2Na_3AlF_6(l) = 4Al + 3CO(g) + 1.5CF_4(g) + 6NaF(l)$	21	3,061	2,085	(1.811)	1.85	2.644
$4Na_3AlF_6 + 3C = 4Al(l) + 12NaF + 3CF_4(g)$	22	3,496	2,986	2.578		3.020
$2Na_3AlF_6 + 2C = 2Al(l) + 6NaF + C_2F_6(g)$	23	1,813	1,618	2.796		3.132

as an excellent predictor of the aluminum fluoride demand for mature cell operation.

Since the sodium oxide content of the SGA from almost all alumina refineries normally exceeds 0.3 wt %, the demand justifies installing an aluminum fluoride feeder that adds aliquots in controllable proportions to help maintain the target concentration band. Thereafter, limited variations are made based on cell condition variations and sample analysis. A range of predictive algorithms exist for estimating the aluminum fluoride additions required. The predictive challenge arises because of the lack of knowledge of how much of the frozen side ledge—which is more than 97% pure cryolite—has transferred into or out of the electrolyte through small temperature fluctuations, or what the actual liquid electrolyte mass or volume actually is.

Electrolyte Mass Control

Varying and unpredictable electrolyte losses can also occur as a consequence of the anode change and metal tapping work practices, and these losses occur into crack openings in the cell lining and create leakages into the insulating material below. In the opposite direction, if the sodium oxide content exceeds 0.3 wt %, usually the cells become electrolyte producers because of the amount of aluminum fluoride being added. A typical common feature linked to the mass of the electrolyte is the total electrolyte depth in the cell, and this becomes the control measure for the mass of electrolyte in the cell. The depth is determined by dipping a steel rod into the electrolyte and metal pad and, after a short time (typically 5–10 seconds), removing it and measuring the length of frozen electrolyte on the rod. Because of differences in heat and mass transfer in the metal pad, there is an identifiable step in the thickness of the frozen layer at and below the metal pad surface. Typically, operating technologies target a depth between 15 and 20 cm, but because of the limited accuracy of measurements, the electrolyte depth control usually has a 2-cm control band. The height variation only affects a limited mass because about 80% of the electrolyte is in the planar surface underneath the anodes.

The source of the electrolyte material, when needed, is usually from another cell because of the tendency to overproduce electrolyte. Depth maintenance is important for keeping the true current density constant, and this is important for controlling electrode reactions.

Electrochemical and Chemical Reactions Occurring in the Cell

During normal operation of aluminum smelting cells, the cathode product is a sodium–aluminum alloy with the sodium content varying between approximately 50 and 200 ppm, depending on process conditions. The alloy is also contaminated with impurities that are present in the anode or in the SGA and whose oxides can be reduced by the reaction with aluminum at the metal pad interface. While acknowledging that impurities could be electrodeposited, their amounts, as well as the amount of sodium co-deposited, is very small and insufficient to influence the potential of the cathodic electrode. Accordingly, it is common to assume there is only one cathode reaction, namely, the reduction of the aluminum ions from the electrolyte solution to form aluminum. Consequently, it is common in the literature to set the potential of the cathode reaction at 0 V and look at the combined electrode potentials as representative of the anode reaction potential. This is the equivalent of the normal hydrogen electrode in aqueous electrochemical systems. Therefore, in the subsequent discussion, the calculated potentials are generally assigned to the anode process.

A more diverse range of gaseous products have been evolved and identified at the anode with the proportions of different gases varying with operating conditions. The identified anode gases that have evolved include hydrogen fluoride (HF), carbon monoxide (CO), CO_2, carbonyl fluoride (COF_2), carbonyl sulfide (COS), carbon tetrafluoride (CF_4), and dicarbon hexafluoride (C_2F_6). Although all of them have some undesirable property environmentally, carbon dioxide is the most desirable one because it leads to a minimum amount of greenhouse gas emissions and forms an important part of the life-cycle chain. Consequently, the challenge in operating the smelting cell is to minimize the formation of the less desirable products by control of electrode potentials or other conditions.

While there is a range of potential constituents for giving rise to these reactions and the specifications for the raw materials, the series of unique chemical and electrochemical reactions that can form these products is presented in Table 5. This table also includes the predicted change in standard enthalpy and Gibbs energies of each reaction at 960°C. In Table 5 the standard state selected for alumina is the saturated solution, so that data for pure solid alpha alumina can be used. The reacting fluoride components $Na_3AlF_6(l)$ and NaF(l) have been selected because they have reliable thermochemical databases. Although all the reactants will be at a reduced thermodynamic activity, the selected reactants would be at the higher concentrations based on the Temkin model for molten salts. Accordingly, the standard-state values should be a close representation of the applicable ones in normal smelting electrolytes. However, adjustments are made using published activity models to minimize errors in enabling electrode potentials. In this electrowinning process, since products are not normally present and therefore reversibility cannot be established, the potential calculated from given energy is referred to as an enabling one.

For electrochemical reactions to proceed in a cell operating essentially at 1 bar pressure, the following three conditions must be satisfied:

1. The magnitude of the interfacial potential gradient at the electrode *must exceed* that predicted from the Gibbs energy for the reaction to achieve reversible conditions according to the following relationship:

$$E_{reversible} = -\Delta G_{reaction}/zF \qquad \textbf{(EQ 24)}$$

where

z = number of electrons transferred per g·mole of reacting species

F = Faraday's proportionality constant

2. A favorable reaction mechanism must exist, combined with an increase in the electrode potential gradient, to maintain the kinetics (current density) of the reactions to be displaced from reversibility. This increase in electrode potential is usually referred to as overpotential or polarization.
3. The total energy transferring across the reacting interface must satisfy the enthalpy of the following reaction:

$$\Delta H_{reaction} = \Delta G_{reaction} + T\Delta S_{reaction} \qquad \textbf{(EQ 25)}$$

The extra entropic energy $T\Delta S_{reaction}$ is often overlooked in thermodynamic predictions of energy requirements for this and alternative processes. The entropic energy can be provided by the electrode overpotential, but if this is insufficient, it must be supplied by heat transfer. In the aluminum electrowinning process, ultimately, the energy comes from the voltage. For this reason, the Gibbs and enthalpy energies are converted to equivalent voltages in the last two columns of Table 5. Comparing the last two columns of voltages highlights how the cell anode reactions can transgress to co-evolution of perfluorocarbons, since the enabling potential gradients (calculated from the Gibbs energy) can be exceeded before sufficient energy is provided by the total electrode potential necessary to complete the production of carbon oxides evolutions without extra heat transfer.

THE CATHODIC METAL DEPOSITION REACTIONS

Of the three cations present in the electrolyte, the deposition of aluminum has the lowest potential, and calcium requires the largest (more negative) potential gradient for deposition. Although the sodium cation is more stable, it will co-deposit in the parts-per-million region, typically forming solutions with between 50 and 200 ppm sodium. The reference equilibrium value (no electrolysis) for an electrolyte at 960°C and the equivalent excess aluminum fluoride concentration of 10.5 wt % is 50 ppm Na in Al, whereas a pure cryolite electrolyte with no excess aluminum fluoride at 1,000°C has 170 ppm Na in Al.

Expressing the respective aluminum and sodium deposition reactions in terms of the species present in the highest concentration according to Table 6 produces two respective deposition reactions:

$$AlF_6^{3-} + 3e^- \Rightarrow Al + 6F^- \qquad \textbf{(EQ 26)}$$

$$Na^+ + e^- \Rightarrow Na \text{ (in Al)} \qquad \textbf{(EQ 27)}$$

The respective deposition potentials at zero current will be

$$E_{(AlF_6^{3-}/Al)} = E^\circ_{(AlF_6^{3-}/Al)} - [RT/3F] \cdot \ln a_{Al} + [RT/3F] \cdot \ln (a_{AlF6}^{3-}/(a_{F^-})^6) \qquad \textbf{(EQ 28)}$$

and

$$E_{(Na^+/Na \text{ in } Al)} = E^\circ_{(Na^+/Na \text{ in } Al)} - [RT/F] \cdot \ln X_{(Na^+/Na \text{ in } Al)} + [RT/F] \cdot \ln (a_{Na}{}^+/(\Upsilon_{(Na^+/Na \text{ in } Al)}) \qquad \textbf{(EQ 29)}$$

where

a_i = respective thermodynamic activities of the species, expressed as molecular or ionic fractions

X = mol fractions

Υ_{Na+} = activity coefficient of the sodium in aluminum

However, the two metal deposition potentials at zero current must be equal, and they also have been arbitrarily set at zero:

$$E_{(AlF_6^{3-}/Al)} = E_{(Na^+/Na \text{ in } Al)} = 0 \qquad \textbf{(EQ 30)}$$

For a cell under electrolysis conditions, the cathode will be subject to polarization, and the potentials of each of the reactions will shift by the same amount. Since the sodium dissolved in aluminum is in a very dilute solution, Henry's law will apply and the activity coefficient of the sodium dissolved in the aluminum will be constant. Hence, during electrolysis:

$$\eta_{cathode \text{ at } i = i} = E_{(Na^+/Na \text{ in } Al)}\ i = i - E_{(Na^+/Na \text{ in } Al)}\ i = 0 \qquad \textbf{(EQ 31)}$$

and therefore applying Equation 28 to these two scenarios produces the following equation:

$$\eta_{cathode \text{ at } i = i} \cong [RT/F] \cdot \ln (X_{(Na^+/Na \text{ in } Al)}\ i = i/X_{(Na^+/Na \text{ in } Al)}\ i = 0)) \qquad \textbf{(EQ 32)}$$

where $\eta_{cathode}$ is the cathode overpotential and because of the reference potential equal to the cathode potential as a consequence of the operating current density and cell conditions.

The sodium cation is present in the highest concentration and the ionic fractions of the species will not change substantially in the operating cell because of the controlled electrolyte composition. Therefore, the cathode polarization is readily given by the change in the sodium content of the aluminum produced. Based on the measured sodium analysis in the aluminum for various operating conditions, it is evident that the cathode polarization will seldom, if ever, exceed –150 mV. Consequently, a value of –100 mV has been allowed for the cathode potential in all thermochemical calculations.

Table 6 Energy required to sustain the chemical and electrochemical processes occurring in an aluminum smelting cell

	Pure Materials; Stoichiometric CO_2 as the Only Electrochemical Product		Typical Materials and Performance	
	kW·h/ kg Al	kJ/mol Al_2O_3	kW·h/ kg Al	kJ/mol Al_2O_3
Preheat materials	0.67	130	0.79	154
Conversion reactions	5.67	1,097	5.815	1,130
Performance band	0	0	±0.075	±15
Total process	6.34	1,227	6.605	1,284
Equivalent voltage needed, V	2.128	±0.050	2.216	±0.050

The sodium cation is also the smallest ion present in the solution, and there is no evidence of any tendency for it to form complexes. Therefore, it is not surprising that research has shown that the sodium cations are the dominating species for transporting electric charge through the electrolyte, and this results in a buildup of sodium ion concentration on the electrolyte side of the electrode potential gradient for metal deposition. In some operating technologies, this has been so great that it changes the liquidus temperature to the point that a photographic recording has been made of a thin layer of pure cryolite crystallizing on the metal pad surface. This is expected to occur before the sodium concentration in the metal becomes 180 ppm Na in Al based on the equilibrium data of Tingle et al. (1981).

GENERAL ISSUES OF ANODIC REACTIONS

There are two major differences at the anode:

1. The anode is consumed, and therefore all constituents within the anodic electrode undergo a process change.
2. All electrode products that leave the electrode surface in the gaseous state require significant entropic energy for the phase change for the final transformation. The transformation energy is supplied by the interfacial electrode potential and, if in deficit, it is based on the overall enthalpy of reaction supplied by heat transfer from the electrolyte.

A major consequence of the second point, as is evident by the equivalent potentials listed in Table 5, is that other reactions become enabled before the energy requirements can be satisfied by the electrode potential gradient alone; therefore, competitive mechanisms of parallel reactions can occur. The conductive heat loss through the anode provides competition for supplying the deficit in energy. For an average age anode, the required conductive heat loss across the interface approaches 10 kW/m^2.

Because of the interaction with the anode material, all anode reactions are heterogeneous, and therefore, other than the charge transfer step, the reaction rate control can be controlled by normal heterogeneous kinetic factors. These include surface coverage and interfacial concentrations as well as being subject to the energy supply to provide the total enthalpy.

Electrochemical Formation of Carbon Oxides

As seen in Table 6, there are two possible cell reactions, both of which involve the same anion source. Based on the thermodynamic data CO(g), formation is enabled at a lower interfacial anode potential than $CO_2(g)$. But there is the unique dichotomy that, once the current density raises the anode potential above that enabling $CO_2(g)$ formation, the total energy required for the product is much less.

During normal electrolysis of an aluminum smelting cell, which is usually at a current density between 0.7 and 0.9 $A{\cdot}cm^{-2}$, the carbon dioxide content of the anode gas exceeds that of carbon monoxide by a factor of between 3 and 5.5. However, some of the carbon monoxide can arise through secondary reactions, such as oxidation of the metal dissolved in the electrolyte, leading to current efficiency losses, or direct chemical reaction between carbon dioxide and the anode carbon according to the following equation:

$$C(s) + CO_2(g) \rightleftharpoons 2CO(g) \quad \textbf{(EQ 33)}$$

which, while being endothermic (ΔH°_{960C} = 168 kJ/mol CO_2), has a favorable standard Gibbs energy change (ΔG°_{960C} = –45 kJ/mol CO_2) that would result in the reverse gas composition ratios to those observed.

The interaction between the electrochemically evolved $CO_2(g)$ with the anode carbon in the pores and emergent surface of the anode according to Equation 33 changes the net reaction for more of the metal; effectively it is being produced by carbon monoxide rather than carbon dioxide. This reaction can be minimized during the anode production process by maximizing the density (hence, minimizing porosity) and increasing the heat treatment or baking temperature to reduce the reactivity of the carbon.

Simplifications in teaching have led to the general interpretation that $CO_2(g)$ is the "primary product" with carbon monoxide formation being kinetically inhibited. Based on this understanding, the literature quotes the theoretical carbon consumption as being 0.333 kg C per kg Al produced.

Such a concept is extremely questionable, however, since both gases originate from the same ionic species in the electrolyte, and both electrochemical reactions would involve similar surface intermediate products. It is also contrary to numerous research findings, such as the fact that the proportion of CO(g) evolved during electrolysis increases as the current density is reduced under design conditions that minimize carbon monoxide formation by a chemical reaction (Hume et al. 1992).

The mechanistic interpretation of reactions has been based on fitting conceptual mechanisms to Tafel curves obtained in laboratory measurements of anode interfacial potential changes as a function of current density (i.e., overpotential). Three sets of published data are presented in Figure 9 and they show an inflection indicative of two electrode processes. Although not shown here, Richards and Welch (1964) found that the curves showed a hysteresis, with the inflection between the two processes being minimized when the polarization curve is obtained by reducing current from a high value, and the electrode has been subjected to electrolysis. A fourth set of data that is not reproduced here but is the one on which most of the interpretation against electrochemical formation of CO(g) is based, is consistent with the illustrated data below 0.03 $A{\cdot}cm^{-2}$ and also above 0.4 $A{\cdot}cm^{-2}$ but is totally devoid of data at current densities between these points.

Each of these sets of data is consistent with two electrochemical processes as proposed by Dewing and van der Kouwe (1975), with each reaction displaying fast charge transfer but with the tendency to develop some form of

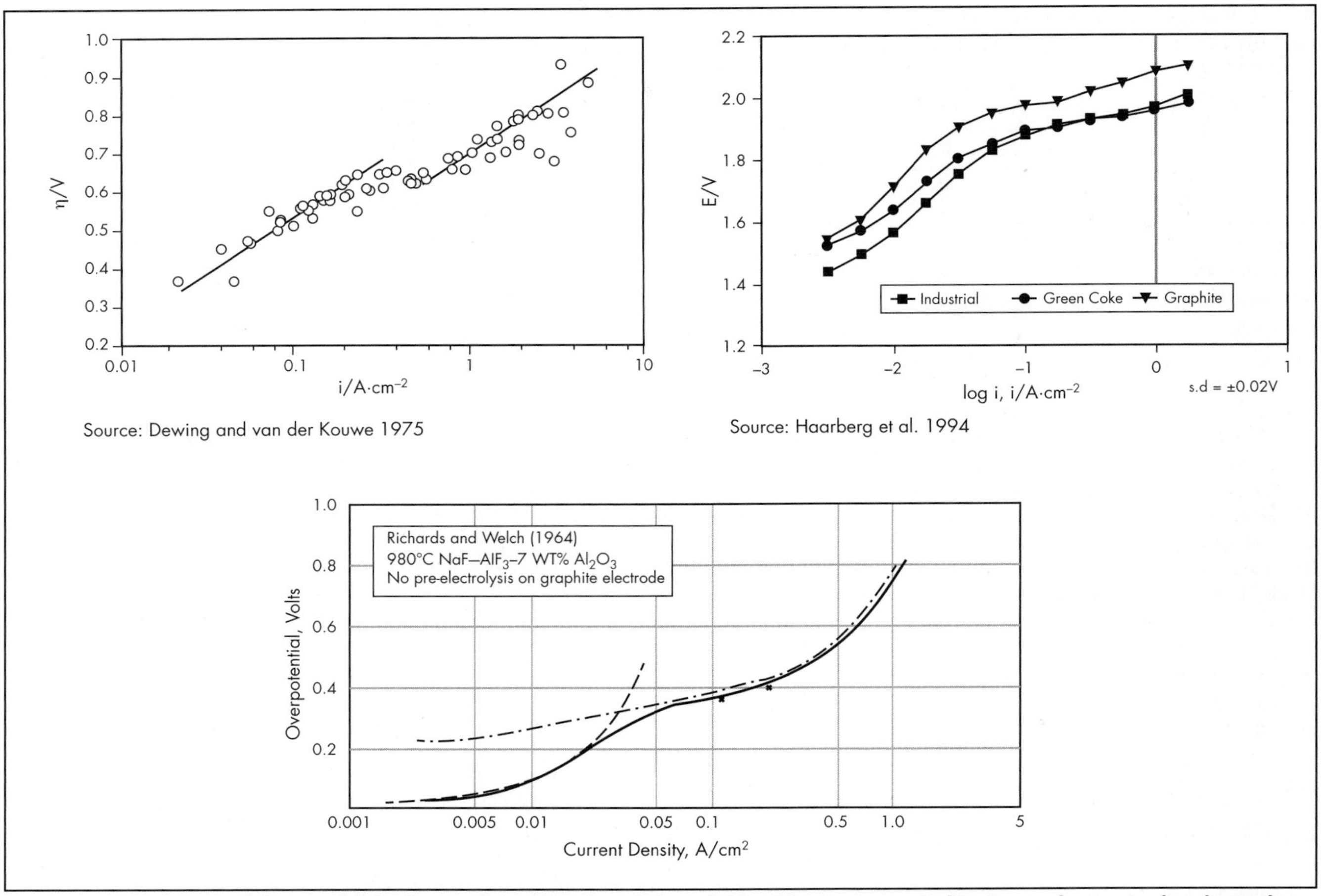

Figure 9 Comparison of three published sets of anode polarization which all have curves indicating a change in the electrode process while in the carbon oxide formation electrode potential region

concentration polarization. In the case of the first process, the formation of CO(g), the polarization would arise from surface coverage of the carbon anode, reducing the number of available sites for further discharge. It is extremely unlikely that it would be due to depletion of the oxide ions since they are involved in CO_2(g) formation as well. Such an interpretation would theoretically conform to the two superimposed (dashed) curves that have been annotated on the data graph of Richards and Welch (1964). However, for the first electrode process, since the same oxide anion is also involved in the second process, the concentration polarization is more likely due to buildup and surface coverage of a reaction intermediate on the carbon electrode. Furthermore, the Tafel slope and the shape of the polarization curve change once the electrode potential has increased enough to enable CO_2(g) formation. This is mechanistically consistent with the electrode polarization at high current densities being due to depletion of the oxide concentration, which is well known. However, one cannot eliminate a parallel contribution to the magnitude of the polarization being from surface coverage/concentration buildup of the reaction intermediate.

The existence of stable surface intermediates is supported by the dangerous back electromotive force that exists in operating cells that have been disconnected from their normal electrolysis power supply and are in the open-circuit mode. There have been observations of "residual gas evolution" continuing for a short time afterward as well. All research that has encompassed a range of different carbon structures for the anode (by heat treatment or graphitization) has found that the polarization—particularly of the first electrode process—is less with the more disordered carbon structure, and the proportion of carbon monoxide then increases.

Transitions from CO(g) formation at low current densities in the range 0.02 to 0.05 A·cm^{-2} to co-evolution of CO_2(g) at higher current densities is not only consistent with practical observations and carbon consumption data but also mechanistically explicable.

Electrochemical Formation of CO(g)

For modern operating current densities and slotted anode designs, between 4% and 8% of the current is being carried by electrochemical formation of carbon monoxide.

The most consistent mechanistic sequence of reactions would be as follows:

1. Local interfacial equilibrium dissociation:

$$AlOF_2^- + 2F^- \rightleftharpoons AlF_4^- + \tfrac{1}{2}O^{2-} \quad \textbf{(EQ 34)}$$

2. Charge transfer with bonding of the oxide onto carbon surface:

$$O^{2-} + x\,C \rightleftharpoons O = C_x \quad \textbf{(EQ 35)}$$

3. Additional entropic energy transfer and/or associated surface structural rearrangement to form carbon monoxide gas:

$$O{=}C_x \Rightarrow CO(g) + (x-1)\ C \quad \textbf{(EQ 36)}$$

$$rate = k_1\ [CO{=}C_x] \quad \textbf{(EQ 37)}$$

The rate constant k_1 for carbon monoxide formation may be controlled by interfacial energy transfer, but this will also depend on the crystalline structure of the carbon with the more disordered carbon structures having a higher rate constant, and lower surface coverage and associated polarization as a given current density.

Electrochemical Formation of Carbon Dioxide

The low Tafel slope resulting when the two processes are enabled supports a second oxide anion being discharged on either the surface intermediate (formed by step 2) or onto the adjacent carbon atom. This would be followed by additional entropic energy transfer and/or an associated surface structural rearrangement:

$$2O{=}C_x \Rightarrow CO_2(g) + (2x-1)\ C \quad \textbf{(EQ 38)}$$

$$rate = k_2\ [CO{=}C_x]^n \quad \textbf{(EQ 39)}$$

The residual entropic energy required for completion of the reaction for forming carbon dioxide is less than half the amount required for forming carbon monoxide at current densities above 0.7 $A{\cdot}cm^{-2}$, while the strained intermediate structure that could result on the carbon crystal lattice would also contribute to the rate constant for carbon dioxide formation being significantly larger.

The combined considerations of the electrode polarization and analysis of anode gases and reaction mechanisms clearly indicate that for normal quality anodes used in the aluminum smelting industry, approximately 5% of the metal is formed by the carbon monoxide mechanism. Therefore, the minimum theoretical carbon consumption is closer to 0.35 kg C per kg Al rather than the 0.333 commonly quoted. Generally, the better baked the carbon anode (higher baking and heat treatment temperature), the less will be the proportion of carbon monoxide formed.

Other Reactions Within the Cell

Table 5 also lists reactions associated with the major impurities in the alumina and carbon raw materials. The dissolved water or hydroxyl arising from the incomplete calcination of the alumina precipitated in the Bayer process is the major source of hydrogen fluoride emitted during continuous operation of the cell. Its electrochemical formation (Equation 4) is enabled at all practical operating current densities, but completion of the reaction requires a greater amount of energy and formation of carbon dioxide. The primary problem of this reaction is the environmental harm that gaseous fluoride emissions cause on plant and animal life. However, the hydrogen fluoride is captured with high efficiency by modern cell designs and nearly qualitatively removed in the dry scrubbing process, so the main impact of this reaction is on the capital cost of the smelter.

The sulfur bound in the anodes is electrochemically oxidized and released as COS(g), but no accurate mass balances have been presented. Other sulfur-containing gases have also been detected while SO_2(g) becomes the dominant sulfurous gas after combustion with air.

Operating Implications of the Cell Reactions

Five major consequences may arise that impact cell operating conditions:

1. The anode potential needs to be constrained below ~1.8 V to prevent the co-evolution of perfluorocarbons.
2. The constraint on the anode potential also introduces constraints on the operating current density and minimum alumina concentration of the cell because of their impact on the electrode polarization and, hence, anode potential.
3. Within the electrode potential constraint operating region, an energy deficit remains for completion of the cell reactions, and consequently, the heat transfer rate from the electrolyte to the anode can be a limiting feature.
4. It is important to minimize spatial variations in cell conditions and properties, as temperature or alumina concentration gradients can change the current density and potential of individual anodes enabling localized perfluorocarbon evolution with minimal change in cell voltage.
5. The quality of raw materials also has an impact on the total energy consumption, because they can enhance the proportion of the electrochemically enabled but thermodynamically unfavorable secondary reactions.

VOLTAGE "AT CELL" AND CELL CONTROL

Operating cells are self-heating, and, therefore, the voltage between the anode beam and cathode current collector bars must be high enough to provide the energy required, when obviously multiplied by cell current, for

1. The energy to preheat all materials—alumina and impurities, gross anode carbon and its impurities, make-up chemicals, and anode cover material—introduced to the cell;
2. The energy required for the state changes of materials that are removed from the cell—such as the hot residual anode butt, and reacted hot anode cover material;
3. The energy for all the conversion reactions; and
4. The heat loss of the cell, which is dependent on cell size but decreases per unit metal production when operating current is increased.

The summation of the first three is the requirements for the conversion process of the raw materials to their final products. As demonstrated in the discussion of the multiplicity of reactions, there are several unavoidable processes compared with the simple value usually assumed for aluminum extraction according to the reaction forming carbon dioxide, which is usually quoted as the base reference situation. As seen in Table 5, these increase the process energy by more than 4%, while the value can also increase with minor variations for different quality raw materials and faradic efficiencies. In the industrial environment, it is more common to discuss energy consumption in terms of kilowatt-hours used per unit production. Through Faraday's law, there is also a correlation between energy and voltage. Consequently, in constructing, three forms of units are presented. Expressing in the terms of voltage is usable for understanding the cell design and operating constraints.

The equivalent voltage needed for the process forms the base requirement, and the other energy input requirement comes from the cell heat loss. Two important components of the cell voltage that need to be allowed for are the

1. Anode potential—which depends on current density and dissolved alumina concentration, but ideally it should not exceed 1.75 V; and
2. Cathode potential—which is current density and electrolyte composition dependent, but it seldom exceeds 0.1 V.

Beyond these components, the rest of the energy can be provided from heat generated through ohmic resistance of the electrode materials and electrolyte used. As seen in Figure 10, there are three major contributions to the ohmic resistance of a cell, but there are several components contributing to each one. The three contributors are

1. The ohmic voltage drop as a consequence of the resistive components of the anode and its assembly;
2. The ohmic voltage drop as a consequence of the resistance of the cathode lining, its jointing to the steel current collector bar and its resistance; and
3. The ohmic resistance of the electrolyte filling the gap between the two electrode surfaces.

In Figure 10, the electrochemical voltages are shown as voltage-dependent resistance because although current flows across the interface, its magnitude is not proportional to Ohm's law. Also, the electrode has an inductance to maintain the potential gradient, hence the capacitor in parallel.

The final voltage at the cell once the line current is set is determined to give the cell heat balance for the metal productivity to be achieved. Once the various resistive components are set by design (according to constraints discussed below), the adjustable variable that enables the required voltage to be achieved is the resistance of the electrolyte, which is varied by changing the interelectrode separation.

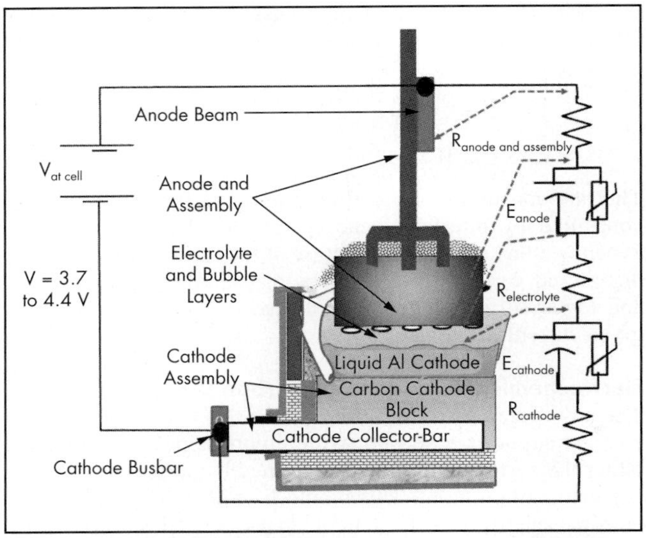

Figure 10 Breakdown of the ohmic resistance components contributing to overall cell voltage

Anode Resistance

Total anode resistance depends on anode size, but also the cross-sectional design of the current-carrying anode assembly, and the interfacial contact resistance generated by the casting between the anode carbon and the terminal anode "stubs" of the current-carrying assembly. The resistance of the anode carbon itself is a variable because of the consumption of the carbon, which brings about a substantial change in the height or depth (typically more than 70% reduction during the life cycle) and lesser changes in the cross-sectional area facing the cathodic surface of the metal pad. Modern anodes tend to have length-by-width geometric areas between 1.0 and 1.4 m^2, and for a 1-m^2 anode base, the total assembled resistance would drop from about 44 to 25 $\mu\Omega$ (micro-ohms) during its life.

Cathode Assembly Resistance

Direct electrical contact with the molten metal cathode is through a series of preformed carbonaceous blocks running across the width of the cell and jointed with the carbonaceous paste that solidifies during the preheating and start-up. All structural metals that withstand the temperature required have too high solubility in aluminum to enable formation of marketable quality aluminum or to give any reasonable service life. Of the electronically conducting refractories, only the diborides of titanium and zirconium have adequate resistance to attack by the metal or electrolyte. But both the costs and the challenges presented by using a refractory material have until now prevented their introduction to operating cells despite many innovative attempts.

Unfortunately, because of the open porosity of the carbon blocks (typically >15%) and penetration in the pores with electrolyte, a series of chemical and electrochemical reactions occur, particularly near the metal interface, causing the carbon block to slowly wear. Consequently, the cross-sectional area of each of these carbon blocks is typically 45–50 cm deep by 45–50 cm wide, but they have low-resistance steel "cathode current collector bars" embedded in them, starting approximately 25 cm below the surface, that are in contact with the liquid aluminum. To minimize the electrical resistance of the total assembly, some of the embedded steel bars also have inserts of copper. The carbonaceous materials used for forming the blocks vary from pitch-bonded calcined anthracite to high-temperature, fully graphitized material, so the resistance of the material varies widely. Very approximately, the cross-sectional area of the preformed carbon blocks matches that of the anode surface, and with design variations the resistance of the completed cathode assembly ranges from 18 to 55 $\mu\Omega/m^2$.

Electrolyte Resistance

Today, the industry is standardizing on electrolyte compositions of 4.5–6 wt % CaF_2, 9.5 to 12 wt % excess aluminum fluoride, and approximately 2.5 wt % dissolved alumina operating at a temperature between 955° and 965°C. For the average value, the electrical resistivity is approximately 0.45 Ω-cm, but inevitably, the cells have chemistry and temperature swings around this band, and after allowing for that, the range could vary between 0.427 and 0.466 Ω-cm. However, because of the horizontally oriented electrodes, which carry the dominant part of the current, the resistance is split into two parts—a partially bubble-filled layer immediately under the anode surface and a conventional layer for the remainder of the gap between the anode and cathode. Although many texts display the bubble as being dispersed randomly in the electrolyte, the hydrodynamic pressures generated by the higher density of the electrolyte tend to keep the bubbles extremely close to the electrode surface. Simulation and modeling studies have shown that these bubbles coalesce and release by rolling along but adjacent to the anode surface in surges. Measurements in laboratory cells have demonstrated that, typically, the average surface coverage of the horizontally oriented electrodes when operated at 0.75 A·cm^{-2} is approximately 35%–40%, and the average depth of the bubble layer is approximately 6 mm (Aaberge et al. 1997). However, the coverage reduces substantially with a small inclination of the electrode surface away from horizontal or a reduction in the distance of the travel path. Slotted anodes have been introduced to achieve these benefits and thus reduce the resistance. Ultimately, the anode-to-cathode spacing and cell heat-balance requirements set the working electrolyte resistance.

CONCLUDING REMARKS

The inherent problems of the present technology for aluminum smelting can be directly linked with limitations of materials available for reactor design resulting in a limited cell life (typically less than six years), design constraints associated with both the need to replenish the consumable anodes, and having two liquid layers that are separated by a small difference in density, causing stability problems of the metal layer in the intense magnetic field, plus the slow reaction rate necessitating a large reaction surface area and, consequently, a large number of cells. These lead to the high capital cost and low energy efficiency because of the large heat loss area. These various constraints have also presented barriers to developing alternative processes despite numerous attempts and extensive research and development.

The economic situation for many aluminum producers has been difficult in recent years. But there are positive signs too. The world will continue to need aluminum, and demand will eventually catch up with production. The current weakness and uncertainty in the world economy are weighing on the aluminum price, but there is continued solid future growth in aluminum demand, and about 5% or 6% growth is expected in the next several years. Global demand for primary aluminum is estimated to reach 70 million t/yr by 2020, which would mean *doubling* since 2010.

ACKNOWLEDGMENT

The authors acknowledge the valuable assistance provided by Stephen Lindsay, technical and operating specialist for Alcoa, who reviewed the manuscript draft and assisted with obtaining accurate, up-to-date data.

REFERENCES

Aaberge, R.J., Ranum, V., Williamson, K., and Welch, B.J. 1997. The gas under anodes in aluminium smelting cells part II: Gas volume and bubble layer characteristics. In *Light Metals 1997*. Warrendale, PA: The Minerals, Metals & Materials Society. pp. 341–346.

Bagshaw, A.N., Kuschel, G.I., Taylor, M.P., Tricklebank, S.B., and Welch, B.J. 1985. Effect of operating conditions on the dissolution of primary and secondary (reacted) alumina powders in electrolytes. In *Light Metals 1985*. Warrendale, PA: TMS-AIME. pp. 549–559.

Dewing, E., and van der Kouwe, Th. 1975. Anodic phenomena in cryolite-alumina melts: I. Overpotentials at graphite and baked carbon electrodes, electrochemical science and technology—Technical papers. *J. Electrochem. Soc.* 122(3):358–363.

Dworkin, A.S., Bronstein, H.R., and Bredig, M.A. 1962. Miscibility of metals with salts. VI. Lithium–lithium halide systems. *J. Phys. Chem.* 66:572–573.

Edwards, L. 2015. The history and future challenges of calcined petroleum coke production and use in aluminum smelting. *JOM* 67:308–321.

Haarberg, G.M., Solli, L.N., and Sterten, Å. 1994. Electrochemical studies of the anode reaction on carbon in NaF-AlF_3-Al_2O_3 melts. In *Light Metals 1994*. Warrendale, PA: The Minerals, Metals & Materials Society. pp. 227–231.

Hall, C.M. 1886. U.S. Patent No. 400,766. Serial No. 207,601.

Héroult, P.L.T. 1886. French patent No. 175,711.

Hume, S.M., Utley, M.R., Welch, B.J., and Perruchoud, R.C. 1992. The influence of low current densities on anode performance. In *Light Metals 1992*. Warrendale, PA: The Minerals, Metals & Materials Society. pp. 687–692.

IAI (International Aluminium Institute). 2017. Primary Aluminium Smelting Energy Intensity, Date of Issue: June 30, 2017. World Aluminium website. www.world-aluminium.org/statistics/primary-aluminium-smelting-energy-intensity/#data. Accessed February 2018.

Jain, R.K., Tricklebank, S.B., Welch, B.J., and Williams, D.J. 1983. Interaction of aluminas with aluminium smelting electrolytes. In *Light Metals 1983*. Warrendale, PA: TMS-AIME. pp. 609–621.

Patterson, E. 2002. Hydrogen fluoride emissions from aluminium electrolysis cells. PhD thesis, University of Auckland, New Zealand.

Richards, N.E., and Welch, B.J. 1964. Electrochemistry. In *Proceedings of the 1st Australasian Conference*. Oxford: Pergamon Press.

Tingle, W.H., Petit, J., and Frank, W.B. 1981. Sodium content of aluminium in equilibrium with NaF-AlF_3 melts. *Aluminium* 57:286–288.

CHAPTER 12.3

Antimony Production and Commodities

Corby G. Anderson

Globally, the primary production of antimony is now isolated to a few countries and continues to be dominated by China. As such, antimony is currently deemed a critical and strategic material for modern society.

BACKGROUND

Antimony is a silvery, white, brittle, crystalline solid classified as a metalloid that exhibits poor conductivity of electricity and heat. Alchemists were fascinated by a property of antimony to form a crystalline star (i.e., the star Regulus) under certain conditions. For alchemists, of course, that symbolized the quintessence of matter (Van der Krogt 2010). In the Western world, antimony was first isolated by Vannoccio Biringuccio and he first described it in 1540. In 1604, Basilius Valentinus (1893) wrote a monograph on antimony translated as *The Triumphal Chariot of Antimony*, regarded as the first monograph devoted to the chemistry of a single metal. Other authors have followed suit (Wang 1909, 1912; Zhao 1988; Anderson 2012). The formal chemical symbol for antimony is Sb, credited to Jöns Jakob Berzelius and taken from the Greek word *stibium*. Its alchemy elemental symbol is a circle with a cross above it, often depicted as ♁. It has an atomic number of 51, an atomic weight of 122 g/g-mol, and a density of 6.697 g/cm^3 at 26°C (Miller 1973). Antimony metal, also known as regulus, melts at 630°C and boils at 1,380°C. Molten antimony upon solidification expands. It does not readily oxidize and keeps its luster even in moist air and at elevated temperatures in the range of 100° to 250°C. At temperatures above its melting point, powdered antimony ignites and burns with a white-green flame. Antimony is resistant to attack by dilute hydrochloric acid and concentrated hydrofluoric acid. Antimony and the natural sulfide of antimony were known as early as 4000 BC. It was used as a coating for copper between 2500 BC and 2200 BC. The sulfide was used as eyebrow paint in early biblical times, and a vase found at Tello, Chaldea, was reported to be a cast of antimony.

Antimony is seldom found in nature as a native metal because of its strong affinity for sulfur and metals such as copper, lead, and silver. In fact, the word *antimony* (from the Greek *anti* [opposed] plus *monos* [solitude]) means "a metal not found alone." Metallic antimony is too brittle to be used alone and, in most cases, must be incorporated in an alloy or compound. For this reason, its development was slow until military applications created new markets and the Russo–Japanese War of 1905 triggered demand. Antimony was widely used at that time and during World War I because it was found to be the best alloying material for lead to produce munitions that were capable of armor plate penetration. A variety of compounds containing antimony as the major constituents were also used for other ammunition types such as detonators, tracer bullets, and armory. In an analysis of strategic mineral supplies prior to World War II, it was stated that "Antimony has more uses of direct military character than other members of the strategic group and probably more important uses than any of the others except mercury" (Roush 1939). The start of mass production of automobiles gave a further boost to antimony, as it is a major constituent of lead–acid batteries. The major use for antimony is now as a trioxide for flame retardants.

OCCURRENCE GEOLOGY AND MINERALOGY

The abundance of antimony in the earth's crust is approximately 0.2 mg/kg. Antimony is a chalcophile, occurring with sulfur and the heavy metals of copper, lead, and silver as a sulfosalt. It occurs as either a distinct antimony-bearing sulfide, such as stibnite; sulfosalt minerals, such as tetrahedrite; rare antimonides, such as aurostibite; or as minor or trace constituent in sulfides such as arsenopyrite and pyrite (Butterman and Carlin 2004). More than 100 minerals of antimony are found in nature. Some of the more common ones are listed in Table 1 (Anderson 2012). Industrially, stibnite is the predominant ore of interest and importance. Stibnite deposits are usually found in quartz veins. The most significant antimony mineral deposits occur in geologic environments with a thick sequence of siliciclastic sedimentary rocks in areas with significant fault and fracture systems. China has the bulk of the world's identified resources. Bolivia, Canada, Japan, Mexico, Russia, South Africa, Tajikistan, and Turkey also have demonstrated antimony resources. Antimony deposits frequently contain minor amounts of gold, silver, and mercury sulfides. Figure 1

Corby G. Anderson, Harrison Western Professor, Dept. of Metallurgical & Materials Engineering, Colorado School of Mines, Golden, Colorado, USA shows global antimony deposits and production locations (Seal et al. 2017). Most antimony deposits occur in orogenic gold provinces in collisional belts (Bliss and Orris 1992; Dail 2014; Seal et al. 1995). They can have linear zones parallel to sutures but are usually inboard of porphyry Cu and Sn-W. They are often found in distinct Sb-Au and Sb-Au-W belts such as the Murchison Range Antimony Line and the Bolivian Sb-Au belt. In several cases, there are also Carlin-like gold-bearing belts. The ultimate metal sources are assumed to be from subducted metal-enriched basinal shales assimilated into ascending reduced magmas under thick continental crustal contamination conditions. But some belts exhibit sedimentary exhalative (SedEx) deposits, distal volcanogenic massive sulfide (VMS), and high-temperature Mississippi Valley–type (MVT) basinal brine system characteristics or hybrids.

The host rocks are often platform or shelf carbonates, carbonaceous clastics, or silty calcareous units. There are few Archean deposits, but numerous Neoproterozoic and Paleozoic deposits. Antimony mobility is often controlled by alkaline hydrothermal solutions and antimony deposition by acidification and/or a drop in temperatures. Most common antimony deposits are Sb-Au-enriched low-temperature quartz veins but, based on contained metals, are insignificant. These vein deposits are strongly structurally controlled, narrow, and occur in discrete belts, with little wall rock alteration and are usually, but not always, small. This is attributed to orogenic processes inboard of subduction zones. Larger districts and deposits typically have higher-temperature assemblages, rock alteration, often with tungsten and likely magmatic input. Several larger deposits have an Sb-rich basinal brine context, including many Chinese and some Variscan and Andean deposits and could represent variants of MVT, Irish-style, or capital regional district.

By far, China dominates in antimony resources and reserves. The Southeast China orogenic belt is the world's dominant resource for antimony. It hosts more than 500 known deposits. It hosts the supergiant Xikuangshan antimony deposit in Xiangzhong Basin in Devonian carbonates. It also hosts numerous other large deposits in the Xuefeng Range with distinctive W-Au-Sb association. There is a country rock antimony endowment in region 20-40 X Clarke and at Xikuangshan on the order of 100,000 X Clarke concentration. These metals—likely sourced from older Proterozoic "basement" and genesis deposits uncertain in the Xiangzhong Basin but possibly variants of high-temperature MVT Xuefeng Range deposits—appear to be sourced from reduced W-Au-Sb–bearing granitic magmas and/or are orogenic Sb-Au (W) vein systems from remobilization.

The largest known deposit, Xikuangshan in Lengshuijiang, China, is a "Supergiant" antimony deposit (i.e., the world's largest past producer) with 801,000 t (metric tons) of antimony produced from 1897 to 1990. As of 2002, it hosted reserves of 570,000 t of antimony. These are now mostly depleted, and the mine has been shut down. It had four major deposits hosted in a series of antiforms in a 2 km wide × 9 km long belt. There are no igneous rocks for 25 km^2 around this ore field, except for a small ultramafic dike. The deposit is mostly stratiform, carbonate-hosted, and, to a lesser extent, black shale-hosted. It has collapsed breccias in limestones at shale contacts, important for lithologic control with silicification as the most common alteration. This has resulted in several prominent formations from early high temperature (200°–300°C) progressing to lower temperature (100°–200°C). These are classified as (1) quartz-stibnite, (2) calcite-stibnite, (3) barite-quartz-stibnite, and (4) fluorite-quartz-stibnite. They have many characteristics like some MVT or Irish-type Pb-Zn systems (but the metal is Sb vs. Pb-Zn).

Another major Chinese deposit is Woxi in Xiangxi District. It is a significant past producer of antimony since 1875 with estimated reserves of 94,600 t Sb in 1998. It is a series of deposits hosted in antiforms along the Woxi thrust. These are hosted in arkosic clastics, red and green shales, and slates—unconformity controls on fluid flow. It is mostly stratiform but also includes cross-cutting vein systems and

Table 1 Common primary antimony minerals

Mineral Name	Compound
Andorite	$AgPbSb_3S_6$
Annivite	$Cu_{12}(Sb,Bi,As)_4S_{13}$
Aurostibite	$AuSb_2$
Berthierite	$FeSb_2S_4$
Boulangerite	$Pb_5Sb_4S_{11}$
Bournonite	$PbCuSbS_3$
Bournonite	$PbCuSbS_3$
Breithauptite	NiSb
Breithauptite	NiSb
Cervantite	Sb_2O_4
Cylindrite	$Pb_3Sn_4Sb_2S_{14}$
Dyscrasite	Ag_3Sb
Falkmanite	$Pb_3Sb_2S_6$
Famatinite	Cu_3SbS_4
Franckeite	$Pb_5Sn_3Sb_2S_{14}$
Freibergite	$(Cu,Ag)_{12}Sb_4S_{13}$
Gabrielite	$Tl_6Ag_3Cu_6(As,Sb)_9S_{21}$
Geocronite	$Pb_5(As,Sb)_{12}S_8$
Gerstleyite	$Na_2(Sb,As)_8S_{13}\cdot 2H_2O$
Gudmundite	FeSbS
Horsfordite	Cu_6Sb
Jamesonite	$Pb_4FeSb_6S_{14}$
Kermesite	Sb_2S_2O
Livingstonite	$HgSb_4S_7$
Meneghinite	$Pb_4Sb_2S_7$
Ramdohrite	$Ag_2Pb_3Sb_3S_9$
Romeite	$5CaO_3\cdot Sb_2O_5$
Senarmontite	Sb_2O_3
Stenhuggarite	$CaFeSbAs_2O_7$
Stephanite	Ag_5SbS_4
Stibiconite	$Sb_2O_4\cdot H_2O$
Stibiobismuthine	$(Bi,Sb)_4S_7$
Stibiocolumbite	$Sb(Nb,Ta)O_4$
Stibiodomeykite	$Cu_3(As,Sb)$
Stibioenargite	Cu_3 $(As,Sb)_4$
Stibioluzonite	$Cu_3(Sb,As)S_4$
Stibio-tellurobismutite	$(Bi,Sb)_2Te_3$
Stibnite	Sb_2S_3
Tetrahedrite	$Cu_{12}Sb_4S_{13}$
Ullmannite	NiSbS
Valentinite	Sb_2O_3
Zinckenite	$PbSb_2S_4$

Source: Anderson 2012

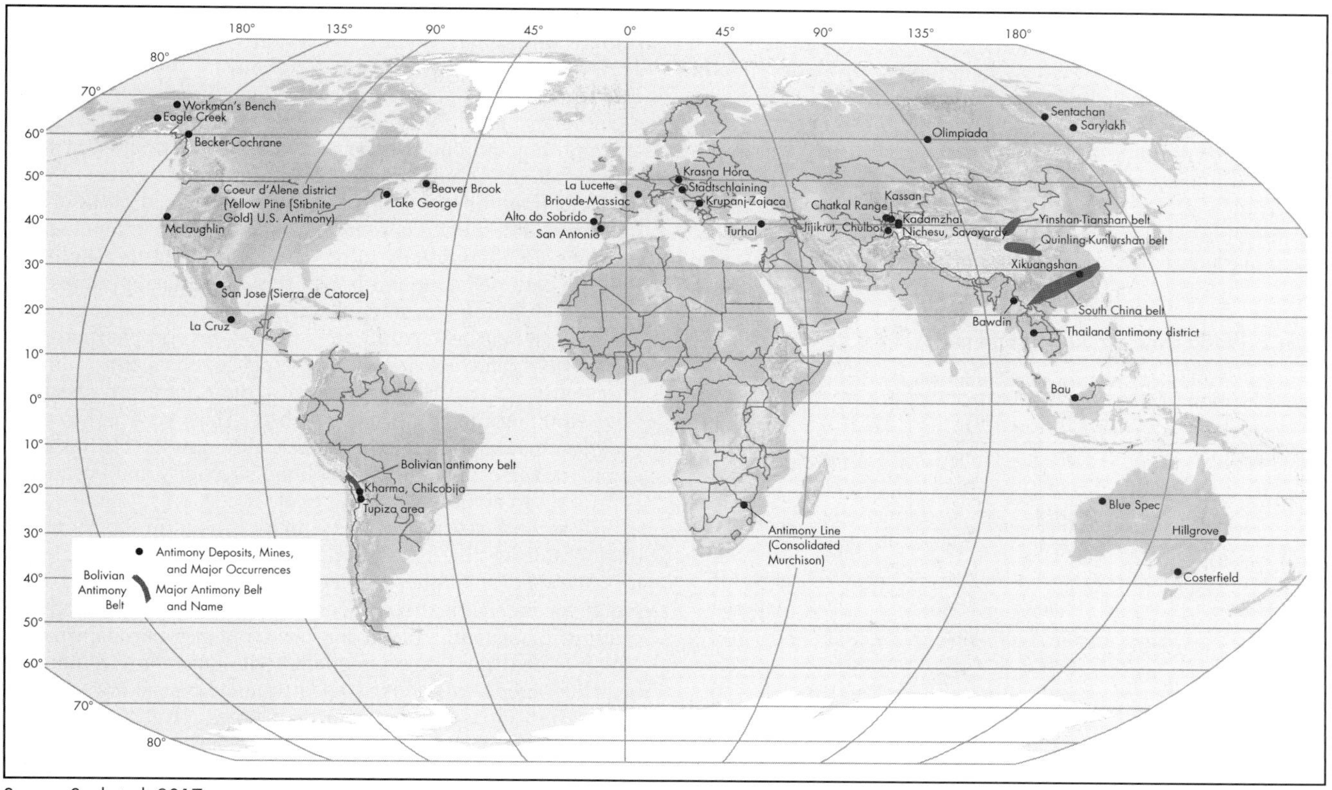

Source: Seal et al. 2017

Figure 1 Locations of selected antimony deposits, mines, and major mineral occurrences

replacement-style mineralization. There are no igneous rocks in the district. Likely fluids with deep crustal origin sourced the deposit in underlying basement, but with cryptic magmatic input. These are early high temperature (350°C) progressing to lower temperature (250°C). Typical deposit styles include (1) planar conformable quartz-stibnite veins, (2) sheeted quartz-scheelite veins, and (3) irregular quartz-scheelite-stibnite stockworks.

Other known Chinese deposits include Xujiashan, Songxi, and Qinling-Dabie. The Xujiashan deposit is in Hubei Province, South China. It has a genesis perhaps as a variant of an MVT. It is a carbonate-hosted stratiform located on the flank of a "basement" knob. It is hosted in dolomitized carbonates near lacustrine-dolomite transition with capping shales. Its isotopic signature is as a classic MVT and is very similar to the Dzhizhikrut deposit in Tajikistan and others in the Balkans and the Iberian Peninsula.

The Songxi deposit in Eastern Guangdong is an Ag-Sb association stratiform in black shale with high total organic carbon. There are volcanics, or sills, below the ore section. It is isotopic, and chemistry data suggest a hybrid nature. Its genesis could be as SedEx or possibly a distal VMS.

The Qinling-Dabie orogenic belt has more than 50 significant deposits. These are orogenic Au-Sb vein-hosted lodes in brittle-ductile shear zones. They have greenschist facies comprised of Devonian clastic rocks. It is a developing belt, but most antimony will be sacrificed to recovery of gold.

Antimony has been produced from ores in more than 15 countries. As shown in Table 2, world reserves of antimony are estimated to be about 1.5 Mt (million metric tons), and in 2017, global antimony production was estimated to be about 150,000 t (Guberman 2018). China, the leading global producer, accounts for over 70% of the world's mined production and most of the reserve base. (Reserves consist of demonstrated resources that are currently economic.) The reserve base consists of reserves plus marginally economic reserves and resources that are subeconomic. Moreover, China together with four other minor producing countries—Bolivia, the Republic of South Africa, Tajikistan, and Russia—currently account for more than 95% of total world antimony production. Table 3 lists domestic U.S. statistics including pricing for a recent seven-year period. As seen, the United States has a very high import reliance on antimony of more than 80% (Guberman 2017).

USES AND APPLICATIONS

Antimony compounds have been known since ancient times and were powdered for use as medicine and cosmetics, often known by the Arabic name, *kohl*. Antimony trioxide, Sb_2O_3, is now the most important antimony compound produced. Today, antimony trioxide is produced by controlled volatilization of antimony metal in an oxidizing furnace. It is used in halogen-compound flame-retarding formulations for adhesives, sealants, plastics, paints, textiles, and rubber. Major markets for flame retardants include electronics, plastics, and fabrics used in making children's clothing; aircraft and automotive seat covers; and bedding. During World War II, the fireproofing compound antimony trichloride ($SbCl_3$) saved the lives of many American troops when it was applied to tents and vehicle covers. In a fire, antimony and chlorine recombine to form unstable compounds that remove oxygen from the air, thereby smothering the flames (Gibson 1998; Eyi 2012).

Most commercial grades of antimony trioxide contain between 99.2% and 99.5% Sb_2O_3 with varying amounts of impurities such as arsenic, iron, and lead. Commercial suppliers offer various grades of antimony trioxide based on the relative tinting strength of their products, which is a function of particle size. Antimony trioxide also finds growing use as a catalyst for PET (polyethylene terephthalate) production. Antimony pentoxide and sodium antimonate, $NaSb(OH)_6$, are also used as flame retardants. Antimony compounds, primarily sodium antimonate, are also used in decolorizing and refining agents for optical glass and CRT (cathode ray tube) glass. In the electronics industry, antimony use grows for diodes. Antimony metal is also used for production of antimonial lead, which is an important product of the secondary lead smelter. A blast furnace charge containing used or discarded battery plates, type metal, and bearing metal is reduced to lead bullion. The bullion is then refined in reverberatory furnaces to meet specifications. Lead-antimony alloys are used in starting-lighting-ignition batteries, ammunition, corrosion-resistant pumps and pipes, tank linings, roofing sheets, solder, cable sheaths, and antifriction bearings. Antimony also has several pharmaceutical uses. Organic antimony compounds, such as antiprotozoal drugs, are used in the treatment of certain parasitic diseases. Potassium antimony tartrate ($K_2Sb_2(C_4H_2O_6)_2$) was formerly used as a nauseant and expectorant, although its use has been abandoned because of the compound's toxicity and because it is difficult to administer (Abdi et al. 2003).

Very high-purity antimony metal is used in the semiconductor industry in silicon wafers for making infrared detectors, diodes, and other devices. Antifriction bearings, type metal for mechanical type setting, and solder and decorative castings of Britannia metal and pewter also contain antimony. Graphite bearings are impregnated with antimony to increase heat tolerance. Antimony is used in nuclear reactors together with beryllium as start-up neutron sources. "Antimony black" is finely ground metallic antimony used in bronzing for metals and plaster casts.

Antimony sulfide will combust in the presence of oxygen. It is a key combustion-supporting ingredient in the manufacture of ammunition primers, detonators, smoke-screen generators, visual range-finding shells, tracer bullets, and the striking surface of safety matches. It is also used to provide the "glitter" effect in fireworks. The rubber industry uses antimony as a vulcanizing agent. Antimony is also used in ceramics and

Table 2 World mine production and reserves for antimony

	Mine Production, t			Estimated Mine
Country	**2015**	**2016**	**2017**	**Reserves, t**
United States	NA*	NA	NA	60,000
Australia	3,700	5,000	5,000	140,000
Bolivia	4,200	2,670	2,600	310,000
Burma	3,000	3,000	2,000	NA
China	110,000	108,000	110,000	480,000
Guatemala	NA	25	25	NA
Iran	NA	200	200	NA
Kazakhstan	NA	573	570	NA
Laos	NA	242	240	NA
Mexico	NA	196	200	18,000
Pakistan	NA	114	100	NA
Russia (recoverable)	9,000	8,000	8,000	350,000
South Africa	—	1,200	1,200	27,000
Tajikistan	8,000	14,000	14,000	50,000
Turkey	2,500	4,000	3,500	100,000
Vietnam	1,000	643	600	NA
World total (rounded)	**142,000**	**148,000**	**150,000**	**1,500,000**

Source: Guberman 2017, 2018
*NA = not available

Table 3 Annual salient antimony statistics for the United States

	2011	2012	2013	2014	2015	2016	2017 (estimated)
Production							
Mine	—*	—	—	—	—	—	—
Smelter							
Primary, t	W†	W	W	W	W	W	W
Secondary, t	2,860	3,050	4,410	4,230	3,850	3,780	3,700
Imports for consumption							
Ores and concentrates, t	NA‡	380	342	378	308	119	100
Oxide, and metal, t	23,500	22,300	24,300	23,800	22,500	23,300	24,000
Exports of metal, alloys, oxide, waste, and scrap, t	4,170	4,710	3,980	3,240	3,200	1,950	2,200
Apparent consumption, t	22,300	20,600	24,700	24,900	23,200	25,300	25,000
Price, U.S. cents/lb	650	565	463	425	327	335	401
Stocks, year-end, t	1,430	1,430	1,470	1,400	1,290	1,090	1,100
Plant employees, no.	24	24	24	27	27	27	27
Net import reliance, %	87	85	82	83	83	85	85

Source: Guberman 2017, 2018
*No domestic production.
†Information withheld.
‡NA = not available.

Table 4 Estimated plant capacities of historic leading producers of refined antimony

Company	Location	Total Capacity, t/yr as Sb	Products
Hsikwangshan Mining Administration	China	30,000	Metal, trioxide, pentoxide, sodium antimonate
Kadamjaisk Antimony Combine	Kyrgystan	20,000	Metal, trioxide
Amspec Chemical Corporation	United States	15,000	Trioxide
Laurel Industries Inc.	United States	12,500	Trioxide
Société Industrielle et Chimique de L'Aisne	France	12,000	Metal, trioxide
Campine	Belgium	10,000	Trioxide
Dachang Mining Administration	China	10,000	Metal
Mines de la Lucette	France	9,500	Metal
Enal	Bolivia	9,300	Trioxide
Great Lakes Chemical (Anzon)	United States	6,000	Trioxide
Union Minière	Belgium	6,000	Sodium antimonate
Guzhou Dushan Dongfeng	China	4,000	Metal, trioxide
Hubei Chongyang	China	4,000	Metal, trioxide
United States Antimony Corporation	Mexico and United States	1,500	Trioxide, sodium antimonate, metal
Sunshine Mining and Refining	United States	1,500	Metal, sodium antimonate
Total		**140,500 (estimated)**	

glassmaking. Sodium antimonate is used as a flame retardant as well as for removing bubbles from glass.

Currently, secondary production accounts for about 20% of total antimony production and takes place mainly through the recycling of lead–acid car batteries and, to a much smaller extent, through the valorization of antimony-containing residues from copper, lead, and gold production. These secondary sources of antimony are expected to become increasingly important because copper, lead, and gold processors must deal with lower-quality ores and tighter environmental regulations, making it more difficult simply to discard or stockpile antimony-containing residues (Dupont and Binnemans 2017). Accordingly, the major conservation practice within the antimony industry is the recycling of the metal in used lead–acid storage batteries, type metal, and Babbitt metal. Also, antimonial lead and antimony metal are recovered from intermediate smelter products such as slags, drosses, flue dusts, and residues generated at copper and lead smelters. The supply of secondary antimony substantially exceeds that from primary sources for antimonial lead applications. In other minor uses, antimony oxides are used as white pigments in paints, whereas antimony trisulfide and pentasulfide yield black, vermillion, yellow, and orange pigments. Camouflage paints contain antimony trisulfide, which reflects infrared radiation. Antimony trisulfide is also used in the liners of automobile brakes as well as in safety-match compositions. In the production of red rubber, antimony pentasulfide is used as a vulcanizing agent. Antimony compounds are also used in catalysts, pesticides, ammunition, and medicines. As well, a variety of compounds containing antimony as the major constituent are also used for other ammunition such as detonators, tracer bullets, and armory.

Flame retardants continue to drive growth demand for antimony trioxide (Wintzer and Guberman 2015). The historic capacity of the world's major established antimony product producers is listed in Table 4. Table 5 lists a recent compilation of production capacities of global mining operations (Roskill Consulting Group Limited 2011). New production facilities are alleged to be under construction in Russia, Tajikistan, and Oman. According to their website, "Tri-Star [Resources] is a partner in the Oman Antimony Roaster Project, an antimony and gold production facility with capacity of up to 50,000 dry tonnes per annum of antimony concentrate, producing up to 20,000 tonnes of antimony and 60,000 ounces of gold" (Tri-Star Resources 2018). Overall, it is estimated that the distribution of antimony uses and consumption worldwide is flame retardants 72%, transportation including batteries 10%,

Table 5 Estimated historic world mine capacity by main producing companies

Country	Company	Capacity, t/yr as Sb
Australia	Mandalay Resources	2,750
Bolivia	Various	5,460
Canada	Beaver Brook Antimony Mine	6,000
China	Hsikwangshan Twinkling Star	55,000
	Hunan Cheznu Mining	20,000
	China Tin Group	20,000
	Shenyang Huachang Antimony	15,000
Kazakhstan	Kazzinc	1,000
Kyrgyzstan	Kadamshai	500
Laos	SRS	500
Mexico	United States Antimony Mexico	500
Myanmar	Various	6,000
Russia	GeoProMining	6,500
South Africa	Consolidated Murchison	6,000
Tajikistan	Anzob	5,500
Thailand	Various	600
Turkey	Cengiz and Ozdemir Antimuan Madenleri	2,400
Total		**153,710 (estimated)**

Source: Roskill Consulting Group Limited 2011. Adapted courtesy of Roskill Information Services Limited.

Table 6 Typical Chinese stibnite ore mineral processing results

Operation	Ore Grade, % Sb	Concentrate Grade, % Sb	Tailings Grade, % Sb	Sb Recovery, %
Hand sorting	2.25	7.80	0.12	95.95
Heavy media separation	1.58	2.65	0.18	95.11
Flotation	3.19	47.58	0.21	93.97
Average	2.68	19.44	0.18	94.11

Source: Xikuangshan Administration of Mines 1979

chemicals 10%, ceramics and glass 4%, and other 4%. The United States, Japan, and Western Europe, which together account for around 70% of world demand, dominate world consumption of antimony.

MINERAL PROCESSING

Originally, antimony was mined and hand sorted to effect concentration. Since the bulk of primary production is in China, where labor is plentiful and cheap, surprisingly, hand sorting still finds a large application. However, in recent years, many other unit operations are used in the mineral processing of antimony. These include primarily conventional crushing and grinding, followed by gravity concentration and flotation. Given that stibnite is the predominant mineral and China is the predominant producer, the industrial mineral processing of a typical Chinese stibnite ore will be elucidated.

A typical stibnite deposit may contain antimony sulfide along with pyrite and a gangue consisting of quartz, calcite, barite, kaolin, and gypsum. Ore grading about 2.7% Sb is treated by a combination of hand sorting, heavy medium separation, and flotation (Xikuangshan Administration of Mines 1964, 1979; Chen 1964; Huang et al. 1965). First, the ore is hand sorted on the feed belt to crushing for a finished concentrate of antimony lumps. Crushing takes place in two-stage closed circuit consisting of a jaw and cone crusher. After crushing to –150 mm, a two-stage hand-sorting treatment is carried out to give antimony lumps and tailings that are –150+35 mm. The –35+10-mm fraction is heavy medium separated to discard a portion of gangue. Then the hand-selected low-grade concentrate and the heavy product from the heavy media separation are ground to –10 mm. This is carried out in three separate sections. The first and second sections are combined in a closed circuit consisting of ball mills and spiral classifiers. The third section uses a ball mill in conjunction with a spiral classifier and provides the final ore feed to a fineness of 60% –200 mesh. Ore ranging from –35 to +10 mm is fed to a drum heavy media separator. Separation density is regulated at 2.62–2.56 g/cm^3 using ferrosilicon. The light product is used as backfill for the mine. The heavy product is sent to the grinding section where it is prepared for flotation. Bulk flotation is performed with natural pH on the ore ground to 60% –200 mesh. Flotation reagents used are butyl xanthate as a collector (100 g/t), sodium diethyldithiocarbamate as a collector (75 g/t), shale oil as a collector (300 g/t), lead nitrate as an activator (150 g/t), pine oil as a frother (100 g/t), and kerosene (65 g/t) as a collector. Scavenger tailings are combined and delivered to a separate bank of cells for final, tertiary scavenging. Rougher concentrate is thrice cleaned to make a pure stibnite concentrate. Overall, 33% of the ore is treated by hand sorting, 7% by heavy media separation, and 60% by flotation. The typical separation results are given in Table 6 and a typical flow sheet is shown in Figure 2 (Xikuangshan Administration of Mines 1979).

PYROMETALLURGY

For primary production, the antimony content of the ore has traditionally determined the pyrometallurgical method of recovery. In general, the lowest grades of sulfide ores containing 5%–25% Sb are volatilized to antimony trioxide, 25%–40% Sb ores are smelted in a blast furnace, and 45%–60% Sb ores are treated by liquation or iron precipitation. A description of each of these, as well as oxide reduction, follows (Motoo 1974; Vladislav 1981; Xiao and Zhou 1981).

Oxide Volatilization

Removal of antimony as the volatilized trioxide is the only pyrometallurgical method suitable for low-grade ores. Combustion of the sulfide components of the ore supplies some of the heat; hence, fuel requirements are minor. There are many variations of the volatilization process, the principles employed being the same but the equipment differing. As shown in Equation 1, the sulfur is burned away at about 1,000°C and removed as a waste gas, whereas the volatile antimony trioxide is recovered in flues, condensing pipes, a baghouse, a Cottrell precipitator, or a combination of the above. Roasting and volatilization are affected almost simultaneously by heating the ore, mixed with coke or charcoal, under controlled conditions in equipment such as a shaft furnace, rotary kiln, converter, or roaster. If the volatilization conditions are too oxidizing, the nonvolatile antimony tetroxide, Sb_2O_4, may form and the recovery of antimony, as antimony trioxide, is diminished. However, special attention to the choice of charge, volatilization conditions, and selection of product results in a high-grade oxide that is suitable for use in ceramics and other applications. Now, this is primarily accomplished by oxidation of antimony metal, which is illustrated in Equation 2.

$$2Sb_2S_3 + 9O_2 \rightarrow 2Sb_2O_3 + 6SO_2 \quad \textbf{(EQ 1)}$$

$$2Sb + 1.5O_2 \rightarrow Sb_2O_3 \quad \textbf{(EQ 2)}$$

Liquation

Antimony sulfide is readily but inefficiently separated from the gangue of comparatively rich sulfide ore by heating to 550°–600°C in perforated pots placed in a brick furnace. This is illustrated in Equation 3. The molten sulfide is collected in lower containers. A more efficient method uses a reverberatory furnace and continuous liquation; however, a reducing atmosphere must be provided to prevent oxidation of antimony sulfide and loss by volatilization. The oxide volatilization process to recover additional antimony usually treats the residue containing 12%–30% Sb. The liquated product, called crude, liquated, or needle antimony, is sold as such for applications requiring antimony sulfide or is converted to metallic antimony by iron precipitation or careful roasting to the oxide followed by reduction in a reverberatory furnace.

$$Sb_2S_{3(s)} \rightarrow Sb_2S_{3(l)} \quad \textbf{(EQ 3)}$$

Oxide Reduction

The oxides of antimony are reduced to metal with charcoal in reverberatory furnaces at about 1,200°C. This is illustrated in Equations 4 and 5. An alkaline flux consisting of soda, potash, and sodium sulfate is commonly used to minimize

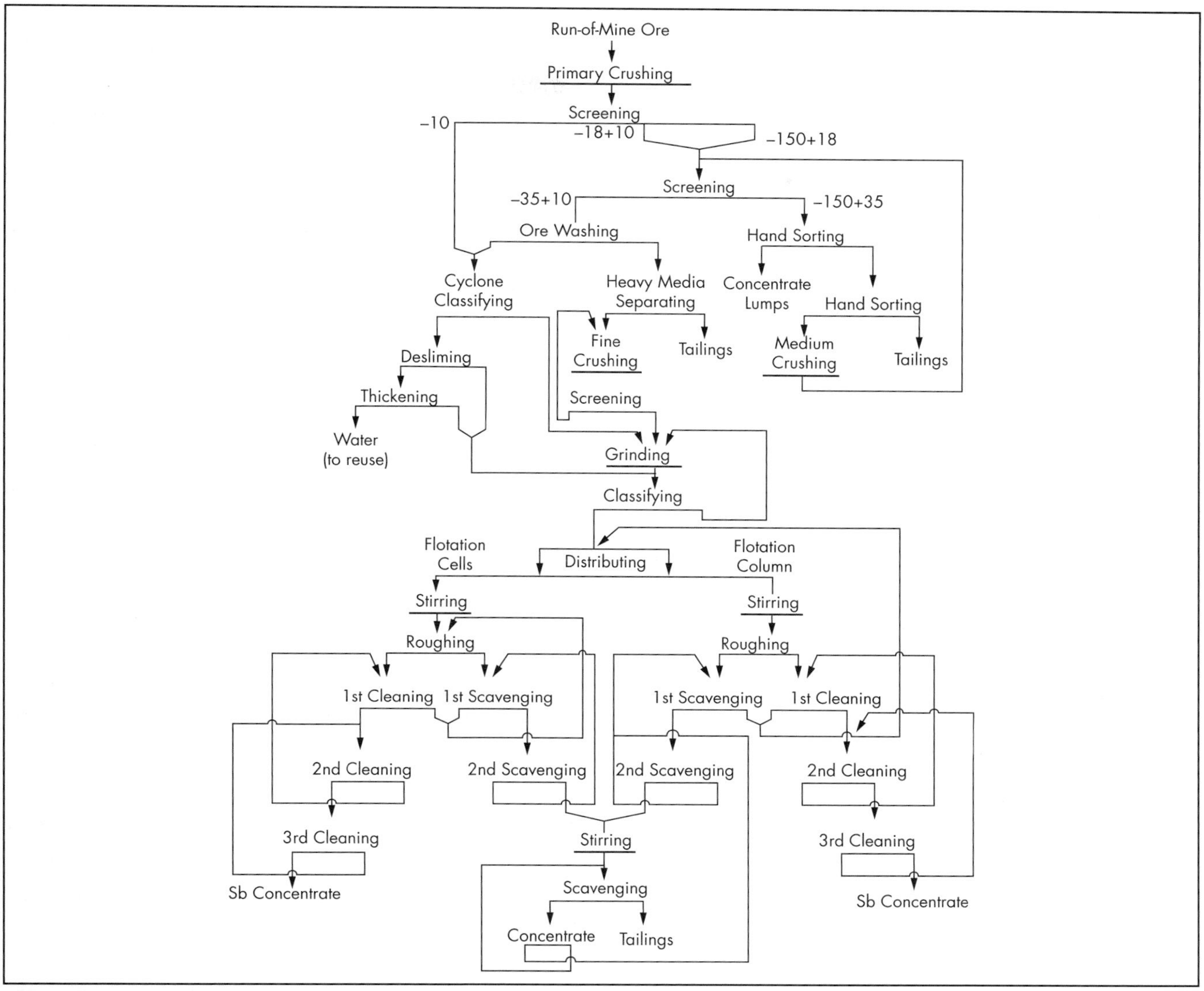

Source: Xikuangshan Administration of Mines 1979

Figure 2 Typical Chinese antimony concentrator flow sheet

volatilization and dissolve residual sulfides and gangue. Part of the slag is frequently reused. Loss of antimony from the charge by volatilization is high (12%–20% or more), even with the use of ample slag and careful control. This necessitates the use of effective Cottrell precipitators or baghouses, and considerable recycling of oxide.

$$Sb_2O_3 + 3CO \rightarrow 2Sb + 3CO_2 \quad \textbf{(EQ 4)}$$

$$CO_2 + 3C \rightarrow 6CO \quad \textbf{(EQ 5)}$$

Iron Precipitation

Rich sulfide ore or liquated antimony sulfide (crude antimony) is reduced to metal by iron precipitation. This is illustrated in the following equation:

$$Sb_2S_3 + 3Fe \rightarrow 2Sb + 3FeS \quad \textbf{(EQ 6)}$$

This process consisting of heating molten antimony sulfide in crucibles with slightly more than the theoretical amount of fine iron scrap depends on the ability of iron to displace antimony from molten antimony sulfide. Sodium sulfate and carbon are added to produce sodium sulfide, or slag is added to form a light, fusible matte with iron sulfide and to facilitate separation of the metal. Because the metal so formed contains considerable iron and some sulfur, a second fusion with some liquated antimony sulfide and salt follows for purification.

Blast Furnace Smelting

Intermediate grades of oxide or sulfide or mixed ores, liquation residues, mattes, rich slags, and briquetted fines or flue dusts are processed in water-jacketed blast furnaces at 1,300°–1,400°C. Theoretically, this process is illustrated in Equations 7 and 8. In general, the same blast-furnace practice used for lead is followed, employing a high-smelting column, comparatively low air pressure, and a separation of slag and metal in a forehearth. It is the favored mode of smelting for all materials containing 25%–40% Sb that can be mixed with fluxes to give a charge sufficiently poor in metal to hold down

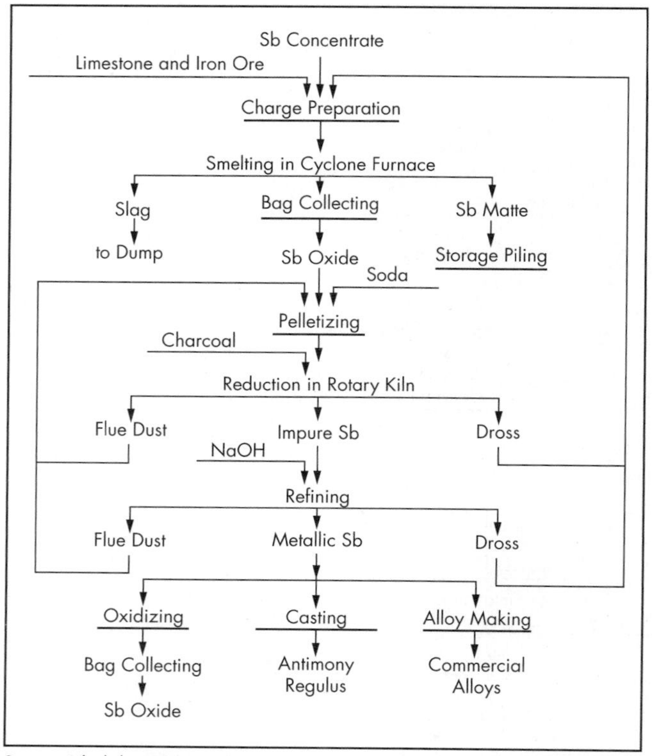

Source: Vladislav 1981

Figure 3 Flow sheet of the Bolivian Vinto antimony smelter

volatilization. As well, slag, usually running under 1% Sb, is desired because it tends to reduce volatilization losses.

$$Sb_2S_3 + 9O_2 \rightarrow 2Sb_2O_3 + 6SO_2 \quad \textbf{(EQ 7)}$$

$$2Sb_2O_3 + Sb_2S_3 \rightarrow 6Sb + 3SO_2 \quad \textbf{(EQ 8)}$$

A flow sheet for a primary pyrometallurgical production facility in Bolivia is shown in Figure 3 (Vladislav 1981). This facility was nationalized from Glencore by the Bolivian government.

HYDROMETALLURGY

Hydrometallurgical methods can be employed for simple antimony materials as well as complex ones containing any number of metals. Normal industrial antimony hydrometallurgical practices call for a two-step process of leaching followed by electrodeposition. There are only two solvent systems used in antimony hydrometallurgy. These are the alkaline sulfide system and the acidic chloride system. The alkaline sulfide system predominates.

Alkaline Sulfide System

The alkaline sulfide system has been employed industrially in the former Soviet Union, China, Australia, and the United States (Anderson et al. 1992, 1994; Anderson and Krys 1993; Nordwick and Anderson 1993). Essentially, the lixiviant is a mixture of sodium sulfide and sodium hydroxide, and when it is applied to stibnite, a solution of sodium thioantimonite is formed through a complexation reaction. This can be illustrated as follows:

$$Na_2S + Sb_2S_3 \rightarrow 2NaSbS_2 \quad \textbf{(EQ 9)}$$

$$NaSbS_2 + Na_2S \rightarrow Na_3SbS_3 \quad \textbf{(EQ 10)}$$

However, dissolution of elemental sulfur in sodium hydroxide is also used as a lixiviant for alkaline sulfide leaching of antimony. The combination of sodium hydroxide and elemental sulfur results in the formation of species other than just sulfide (S^{-2}). Both sodium polysulfide (Na_2S_X) and sodium thiosulfate ($Na_2S_2O_3$) are created along with sulfide. This is illustrated simplistically in the following scenario (all species in Equations 11–13 are aqueous species, except for elemental sulfur):

$$4S^\circ + 6NaOH \rightarrow 2Na_2S + Na_2S_2O_3 + 3H_2O \quad \textbf{(EQ 11)}$$

$$(X - 1)S^\circ + Na_2S \rightarrow Na_2S_X \text{ (where X = 2 to 5)} \quad \textbf{(EQ 12)}$$

Because of the oxidizing power of polysulfide on sodium thioantimonite, the major species in solution is normally sodium thioantimonate (Na_3SbS_4). This can be viewed as follows:

$$Na_2S_X + (X - 1)Na_3SbS_3 \rightarrow (X - 1)Na_3SbS_4 + Na_2S \quad \textbf{(EQ 13)}$$

The electrodeposition of the antimony from the alkaline sulfide solution to cathode metal is normally carried out via electrowinning in either diaphragm or non-diaphragm cells.

The primary anode reactions are as follows:

$$4OH^- \rightarrow 4e^- + 2H_2O + O_2 \quad \textbf{(EQ 14)}$$

$$S^{-2} \rightarrow 2e^- + S^\circ \quad \textbf{(EQ 15)}$$

The primary cathode reaction is as follows:

$$SbS_3^{-3} + 3e^- \rightarrow Sb^\circ + 3S^{-2} \quad \textbf{(EQ 16)}$$

The cathodic metal antimony product may attain a grade of more than 99.5% purity after washing. A typical industrial alkaline sulfide hydrometallurgical leach plant flow sheet is shown in Figure 4 (Ackerman et al. 1993). This industrial alkaline sulfide hydrometallurgical plant ran successfully for almost 60 years.

Acidic Chloride System

The other methodology for antimony hydrometallurgy is the acidic chloride system. While the alkaline sulfide system predominates, much research and pilot-scale work has been undertaken to utilize chloride-based technology (Kim et al. 1975; Su 1981; Tang et al. 1988; Li and Xu 1984; Thibault et al. 1997). In the acidic chloride hydrometallurgical antimony system, hydrochloric acid, HCl, often in conjunction with ferric chloride, $FeCl_3$, is commonly used as the solvent for sulfide mineral such as stibnite. This is illustrated as follows:

$$6FeCl_3 + Sb_2S_3 \rightarrow 2SbCl_3 + 6FeCl_2 + 3S^\circ \quad \textbf{(EQ 17)}$$

$$Sb_2S_3 + 6HCl \rightarrow 2SbCl_3 + 3H_2S \quad \textbf{(EQ 18)}$$

In the aqueous solution, $FeCl_3$ is both an oxidizer and a chloridizing agent to convert the antimony of the sulfide mineral into a chloride complex while producing elemental sulfur. In cases where the antimony is already oxidized, it may be leached directly with HCl without the need for $FeCl_3$. This is illustrated as

$$Sb_2O_3 + 6HCl \rightarrow 2SbCl_3 + 3H_2O \quad \textbf{(EQ 19)}$$

As with the alkaline sulfide system, the solubilized antimony chloride can be produced by electrowinning from

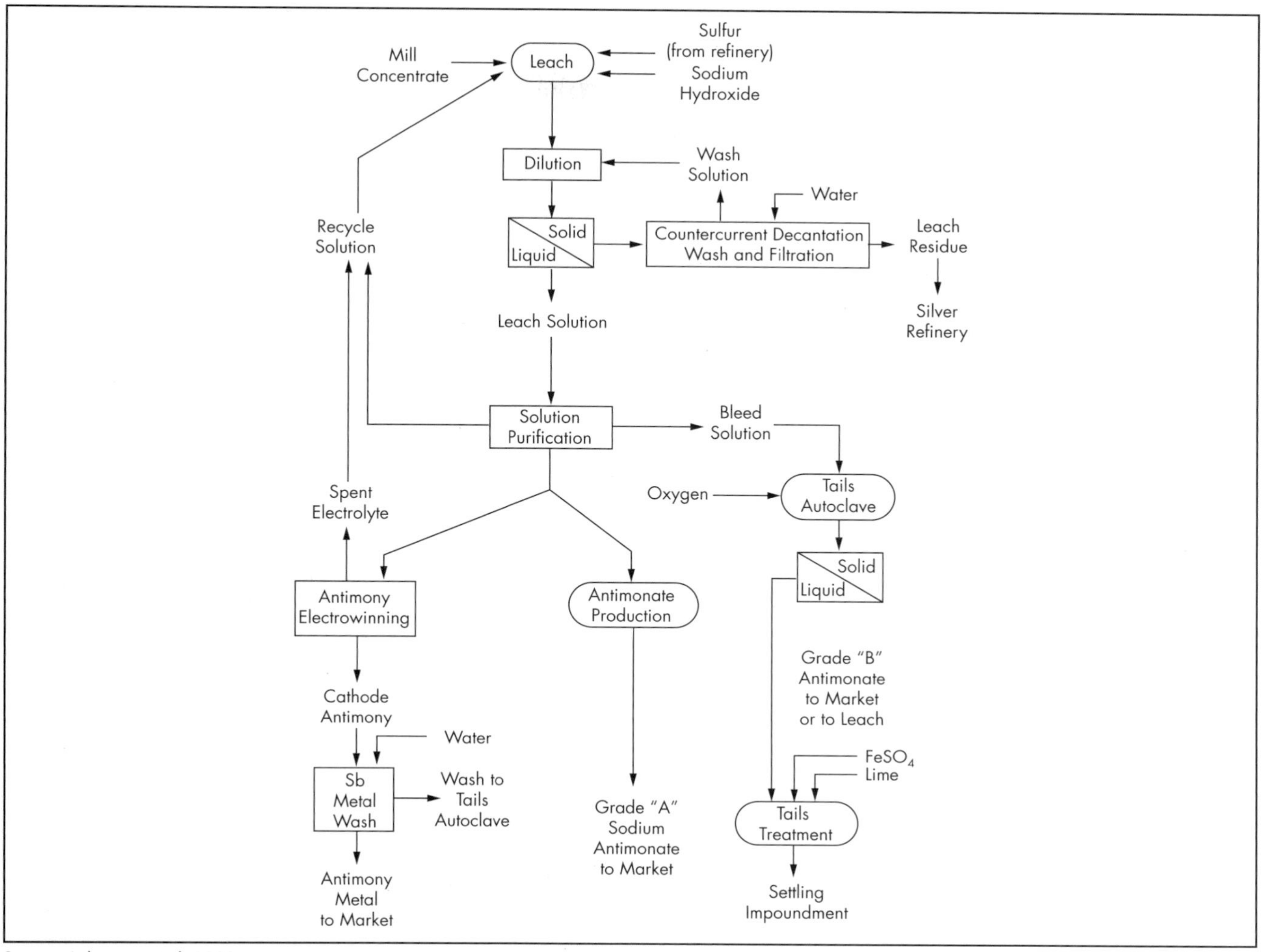

Source: Ackerman et al. 1993

Figure 4 Sunshine (Idaho) industrial hydrometallurgical alkaline sulfide antimony process

solution in diaphragm cells. This produces cathode antimony metal. The primary cathode reaction is

$$Sb^{+3} + 3e^- \rightarrow Sb \qquad \textbf{(EQ 20)}$$

The primary anode reactions are as follows:

$$Fe^{+2} \rightarrow e^- + Fe^{+3} \qquad \textbf{(EQ 21)}$$

$$3Cl^- + Fe^{+3} \rightarrow FeCl_3 \qquad \textbf{(EQ 22)}$$

Alternatively, the antimony chloride solution can be treated by hydrolysis precipitation of the antimony from solution as a solid oxychloride. Then the precipitated solid is treated with ammonia to produce a pure antimony oxide. This is illustrated in the following reactions:

$$SbCl_3 + H_2O \rightarrow SbOCl + 2HCl \qquad \textbf{(EQ 23)}$$

$$SbOCl + H_2O \rightarrow Sb_4O_5Cl_2 + 2HCl \qquad \textbf{(EQ 24)}$$

$$2SbOCl + 2NH_4OH \rightarrow Sb_2O_3 + NH_4Cl + H_2O \qquad \textbf{(EQ 25)}$$

$$Sb_4O_5Cl_2 + 2NH_4OH \rightarrow Sb_2O_3 + NH_4Cl + H_2O \qquad \textbf{(EQ 26)}$$

In summary, there are two predominant hydrometallurgical systems for antimony. Alkaline sulfide technology is by far the most utilized because of its inherent antimony selectivity and its ease of full-scale application due to minimization of the corrosion issues that are associated with the chloride system.

PRODUCT QUALITY AND MARKETING

In ancient times, in extracting antimony from sulfide ore by the addition of iron, a starlike, fern-shaped crystalline pattern appeared during solidification. This was dubbed the "Philosopher's Signet Star" in 1610 and, until recently, the shape of these stars had commonly served to indicate the grade of the metallic antimony. Indeed, when antimony metal contains impurities like sulfur, arsenic, lead, or iron to any appreciable extent, its surface shows the presence of these foreign elements by specks, a leaden appearance, or a poorly defined appearance of the crystalline pattern. This starring phenomenon is also produced by cooling antimony metal, or so-called regulus under cover of a layer of a properly prepared starring mixture, or coverture (Wang 1918), which is a slag that has a melting point lower than that of antimony metal. While starring was the original indicator of antimony quality,

Table 7 Analysis of typical impure antimony production products

Element	Impure Sb from Reduction Smelting, %	Impure Sb from Precipitation Smelting, %	Hydrometallurgical Cathode Sb Metal, %
Sb	96–97	80–90	98–98.5
As	0.2–0.3	0.2–0.3	0.02–0.1
Pb	0.1–0.25	0.1–5	—
C	0.001–0.01	0.03–0.2	0.004–0.005
Fe	0.01–0.5	3–15	0.005–0.01
Na	0.02–0.1	0.02–0.1	0.1–1.0
Sn	0.01–0.1	0.01–0.1	0.01–0.03
S	0.1–1.0	0.2–0.3	0.15–0.40

Source: Xiao and Zhou 1981

Table 8 Chinese national standards for antimony metal purity

Grade Classification	Sb%	As%	Fe%	S%	Cu%	Total Impurities, %
I	99.85	<0.05	<0.02	<0.04	<0.01	<0.15
II	99.65	<0.10	<0.03	<0.06	<0.05	<0.35
III	99.50	<0.15	<0.05	<0.08	<0.08	<0.50
IV	99.00	<0.25	<0.25	<0.20	<0.20	<1.00

Source: Xiao and Zhou 1981

Table 9 Typical quality of antimony trioxide

Assay	Percentage
Sb_2O_3 content	99.5
As	<0.03
Pb	<0.08
F	<0.01
–325 Mesh	99.99

Source: Xiao and Zhou 1981

with the enhanced analytical techniques now available, specifications are now better defined. Tables 7, 8, and 9 (Xiao and Zhou 1981) indicate the typical quality and specifications for various antimony products.

The marketing and sales of antimony metal and its compounds is subliminal. Entities such as the Chinese Fanya Metal Exchange were supposed to offer some insights, but the timeliness and reliability of the data have been suspect and called into question. According to Penfold World Trade, Metal Bulletin does publish a twice-weekly quotation for two metal grades, but there is no quotation for trioxide (B. Goldstein, personal communication). In China, several sources publish Chinese prices in local currency that may be indicative of pricing trends rather than of actual transactions. Unlike commodities such as copper or gold, there are no global exchanges for pricing and specifications. As such, the antimony market and its few entities are not transparent, and its trade aspects are generally closely held knowledge.

GLOBAL OUTLOOK

The European Union has studied and deemed antimony as a material of growing criticality, as illustrated in Figure 5 (European Commission 2018). In the figure, antimony is circled and it is near the top of the supply risk. This risk has grown in recent years.

Roskill (2018) provides a current projection of the imminent future for antimony:

> Antimony is mostly consumed in flame retardants and lead-acid batteries. Together these end uses accounted for 84% of antimony demand in 2017. Consumption trends in these two critical applications for antimony thus shape overall demand dynamics.
>
> In both cases, a similar situation prevails: while overall demand (for flame retardants and lead-acid batteries) has been steadily increasingly, the antimony loading within these applications has been reducing. In flame retardants, this is mainly because of high antimony prices prompting substitution of antimony, and legislative and requirements forcing changes to flame retardant formulas. In batteries, lead-calcium-tin alloys are increasingly being used instead of antimonial lead in battery grids for sealed-for-life maintenance-free automotive batteries, also called valve-regulated lead-acid (VRLA) batteries.
>
> Owing in part to these trends, demand for antimony in both non-metallurgical and metallurgical applications has stagnated in recent years. Consumption peaked in 2010 at around 201kt but has since declined year-on-year. Roskill expects that long-term demand for antimony in lead-acid batteries will continue to decline, with tin increasingly used instead of antimony, and lead-acid batteries themselves likely to fall victim to the EV boom over the longer term. While the outlook for construction and plastics is broadly positive, which suggests demand for flame retardants will increase, the future of antimony demand in flame retardants will remain highly sensitive to prices and the regulatory environment. Considering these factors, Roskill expects overall demand to grow at less than 1%py over the period to 2027.
>
> Importantly, Roskill believes the market may soon experience a fundamental shift. Antimony enters the supply chain in two ways: primary mine production, and secondary recovery of antimonial lead. In the mid-2020s, secondary supply from antimonial lead will be sufficient to meet metallurgical demand. As such, little or no primary supply will be required for metallurgical applications, making the metallurgical side of the market effectively "self-sufficient".
>
> With a modest growth in demand for antimony in non-metallurgical end uses expected, Roskill envisages that demand could start to outstrip current supply levels over the longer term. This would create a situation in which additional feedstock would be required, for supply to meet demand. This would need to be met from an increase in mine output from existing producers or new supply from potential producers. Alternatively, processors might look to utilize the growing surplus of secondary antimony available from antimonial lead although such recovery is not yet typical on a commercial scale.
>
> Global mine supply has been declining in recent years, mostly because of falling demand. However, there are considerable uncertainties over the future of mine production in China, with reports of dwindling reserves, falling grades and mine closures

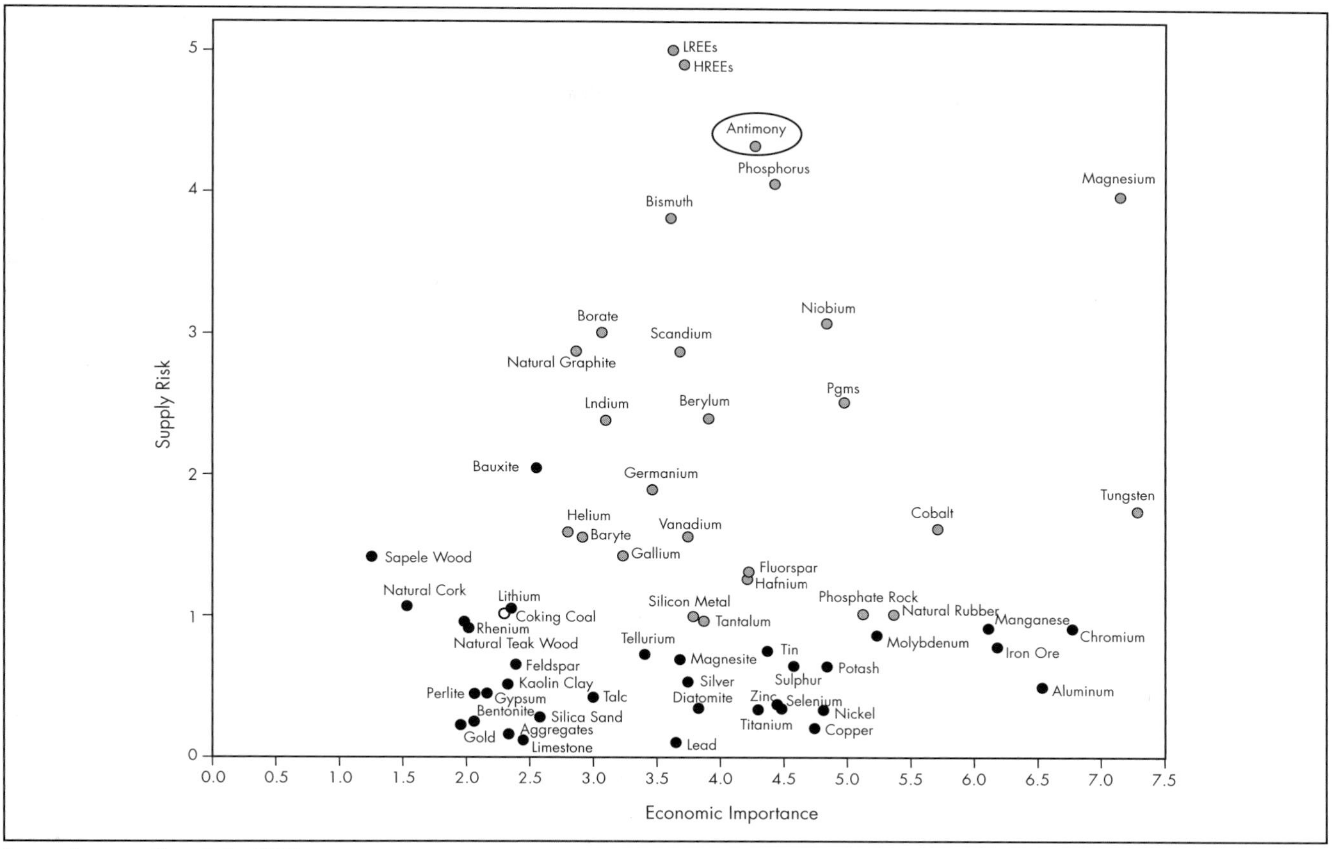

Source: European Commission 2018

Figure 5 European commission critical raw materials matrix as of 2017

related to environmental inspections. Mine production has fallen from 115kt in 2010 to 80kt in 2017 according to Roskill estimates. Falling Chinese feedstock supply has been partially offset by increases in Tajikistan, Russia, and Australia. Environmental inspections and challenging market conditions have also led to a considerable reduction in Chinese smelter capacity. This could present an opportunity for new entrants into the market, although China will remain the dominant power in the antimony market for the foreseeable future.

SUMMARY

Globally, the primary production of antimony is now isolated to a few countries and is still dominated by China. Global resources and reserves are being depleted while demand is growing. Hence, antimony is currently deemed a critical and strategic material.

ACKNOWLEDGMENTS

Portions of this chapter are from "The Metallurgy of Antimony" by Corby G. Anderson in *Chemie der Erde—Geochemistry*, 72(Suppl. 4):3–8, Copyright © 2012 Elsevier, and are used with permission.

REFERENCES

Abdi, Y.A., Gustafsson, L.L., Ericsson, O., and Hellgren, U. 2003. *Handbook of Drugs for Topical Parasitic Infections*, 2nd ed. Bristol, PA: Taylor and Francis.

Ackerman, J.B, Anderson, C.G., Nordwick, S.M., and Krys, L.E. 1993. Hydrometallurgy at the Sunshine mine metallurgical complex. In *Proceedings of the Fourth International Symposium on Hydrometallurgy*. Edited by J.B. Hiskey and G.W. Warren. Littleton, CO: SME.

Anderson, C.G. 2012. The metallurgy of antimony. *Chem. Erde* 72(Suppl. 4):3–8.

Anderson, C.G., and Krys, L.E. 1993. Leaching of Antimony from a Refractory Precious Metals Concentrate. In *Proceedings of the Fourth International Symposium on Hydrometallurgy*. Edited by J.B. Hiskey and G.W. Warren. Littleton, CO: SME.

Anderson, C.G., Nordwick, S.M., and Krys, L.E. 1992. Processing of antimony at the Sunshine mine. In *Residues and Effluents: Processing and Environmental Considerations*. Edited by R.G. Reddy, W.P. Imrie, and P.B. Queneau. Warrendale, PA: The Minerals, Metals & Materials Society.

Anderson, C.G., Nordwick, S.M., and Krys, L.E. 1994. Antimony separation process. U.S. Patent No. 5,290,338.

Bliss, J., and Oriss, G. 1992. Descriptive model of simple Sb deposits—Model 27d. In *Mineral Deposit Models*. Edited by D.P. Cox and D.A. Singer. USGS Bulletin 1693. Denver, CO: U.S. Geological Survey.

Butterman, W.C., and Carlin Jr., J.F. 2004. *Mineral Commodity Profiles: Antimony*. Open-File Report 03-019. Reston, VA: U.S. Geological Survey.

Chen, J. 1964. *Floatability of Xikuangshan Stibnite Ore.* Test Report.

Dail, C. 2014. Antimony (Sb): Deposit types, distribution and frontiers for new resources. Presented at SME Critical Minerals Symposium, Denver, CO.

Dupont, D., and Binnemans, K. 2017. Preventing antimony from becoming the next rare earth. *Recycl. Technol.* October. rt2017_research_18_20_preventing_antimony_pdf_19455.pdf.

European Commission. 2018. Critical raw materials. http://ec.europa.eu/growth/sectors/raw-materials/specific-interest/critical_en. Accessed May 2018.

Eyi, E. 2012. *Antimony Uses, Production, Prices*. London: Tristar Resources.

Gibson, R.I. 1998. Type-cast Antimony, the metallic sidekick, sets the world on fire and puts it out. *Geotimes* February. p. 58.

Guberman, D.E. 2017. Antimony. In *Mineral Commodity Summaries 2017*. Reston, VA: U.S. Geological Surveys.

Guberman, D.E. 2018. Antimony. In *Mineral Commodity Summaries 2018*. Reston, VA: U.S. Geological Surveys.

Huang, M., et al. 1965. *The Continuous Heavy Medium Separation Process on Antimony Sulfide Ores from Xikuangshan.* Test Report. China.

Kim, S.S., et al. 1975. Leaching of antimony with ferric chloride. *Taehan Kwangsan Hakoe Chi [J. Korean Inst. Miner. Min. Eng.]* 12(4):35–39.

Li, W., and Xu, B. 1984. A trial process for direct hydrometallurgical production of antimony white. *Hunana Metall.* 4:20–23.

Miller, M.H. 1973. Antimony. In *United States Mineral Resources*. Edited by D.A. Brost and W.P. Pratt. USGS Professional Paper 820. Washington, DC: U.S. Government Printing Office. pp. 45–50.

Motoo, W. 1974. Equilibrium in reduction of antimony oxide with carbon monoxide. *Jpn. Phys. Chem. Inst. Compilation* 8(12).

Nordwick, S.M., and Anderson, C.G. 1993. Advances in antimony electrowinning at the Sunshine mine. In *Proceedings of the Fourth International Symposium on Hydrometallurgy*. Edited by J.B. Hiskey and G.W. Warren. Littleton, CO: SME.

Roskill Consulting Group Limited. 2011. *ANCOA Ltd.: Study of the Antimony Market*. July. London: Roskill Information Services Limited.

Roskill. 2018. *Antimony: Global Industry, Markets and Outlook 2018*. https://roskill.com/market-report/antimony/. London: Roskill Information Services Limited.

Roush, G.A. 1939. *Strategic Mineral Supplies.* New York: McGraw-Hill.

Seal, R.R., II, Bliss, J.D., and Campbell, D.L. 1995. Stibnite-quartz deposits. In *Preliminary Compilation of Descriptive Geoenvironmental Mineral Deposit Models.* Edited by E.A. du Bray. Open-File Report 95-831. Denver, CO: U.S. Geological Survey.

Seal, R.R., II, Schulz, K.J., and DeYoung Jr., J.H. 2017. Antimony. In *Critical Mineral Resources of the United States—Economic and Environmental Geology and Prospects for Future Supply.* Edited by K.J. Schulz, J.H. DeYoung Jr., R.R. Seal II, and D.C. Bradley. USGS Professional Paper 1802-C. Reston, VA: U.S. Geological Survey. pp. C1–C17.

Su, G. 1981. *New Hydrometallurgical Process for Antimony.* Research Report. China.

Tang, M. et al. 1988. New techniques for treating the Dachang Jamesonite concentrate. *J. Cent. South Inst. Min. Metall.* 198(4):18–27.

Thibault, J., et al. 1997. Process for producing antimony trioxide. International Patent Application No. PCT/CA97/00659.

Tri-Star Resources. 2018. Home page. London: Tri-Star Resources. http://tri-starresources.com/. Accessed May 2018.

Valentinus, B. 1893. *Triumph-Wagen des Antimonij* [*The Triumphal Chariot of Antimony*]. London: James Elliott.

Van der Krogt, P. 2010. Stibium antimony. In *Elementymology and Elements Multidict.* www.vanderkrogt.net/elements/element.php?sym=Sb. Accessed May 2018.

Vladislav, H. 1981. Apparatus and methods for thermal treatment of various raw materials for nonferrous metallurgy in a cyclone furnace. Ger. Offen. 2906584.

Wang, C.Y. 1909. *Antimony: Its History, Chemistry, Mineralogy, Geology, Metallurgy, Uses.* London: Griffin's Scientific.

Wang, C.Y. 1912. The antimony industry of China. *Eng. Min. J.* 94(17):777–778.

Wang, C.Y. 1918. Starring mixture for antimony smelting. U.S. Patent No. 1284164.

Wintzer, N.E., and Guberman, D.E. 2015. *Antimony—A Flame Fighter.* Fact Sheet 2015-3021. Reston, VA: U.S. Geological Survey.

Xiao, Y.F., and Zhou, W.T. 1981. *Improved Smelting and Refining Processes for Production of Antimony and its Oxides in Xikuangshan, China.* Beijing: Xikuangshan Bureau of Mines.

Xikuangshan Administration of Mines. 1964. *Volatilization Smelting of Antimony Sulfide Flotation Concentrate in Blast Furnace.* Summary Test Report. China.

Xikuangshan Administration of Mines. 1979. *A Brief Account of Xikuangshan Ore Processing.* China.

Zhao, T.-C. 1988. *The Metallurgy of Antimony.* Changsha: Central South University of Technology Press.

CHAPTER 12.4

Arsenic Production, Commodities, and Fixation

Larry G. Twidwell

Mineral processing and extractive metallurgical operations have created and are creating appreciable arsenic-bearing wastewater and waste solid products that have to be handled, treated for recycle, or treated for environmentally safe disposal. Intense research and operational activities are being conducted to provide the best viable processing procedures to ensure that the mineral processing and extractive metallurgical industries are profitable and environmentally secure. The focus of this chapter is on the element arsenic, even though many other deleterious elements may also be present in ores and concentrates. Numerous base metal resources contain arsenic-bearing minerals, especially resources containing mineral sulfides. Information on presently treated metal-bearing resources and potential new resources is voluminous, especially for those containing arsenic mineralization. The influence of elevated arsenic concentrations in the treatment of copper–arsenic sulfides and to a lesser extent the treatment of copper–gold–arsenic sulfides are considered. Not all treatment processes are discussed; however, examples are provided to illustrate arsenic problems and industrial solutions. The major emphasis of this chapter is the present state of the art for arsenic immobilization/fixation and long-term storage considerations. A more detailed review is available to the reader in Twidwell (2018).

BACKGROUND

Arsenic is the 20th most abundant element in the earth's crust. It rarely is found in its elementary form but is often associated with oxygen and sulfur. It is widely found in commonly processed base metal ores such as copper, lead, and zinc and is often associated with gold- and silver-bearing ores. Arsenic minerals may be found as arsenides, sulfides, sulfosalts, oxides, and arsenates. Arsenic release has occurred originating from human activities, including metal smelting, chemical production, coal combustion, waste disposal practices, and widespread application of pesticides and fungicides (SME 2015):

> Atmospheric arsenic emissions from smelting represent the largest contribution of arsenic from the mining and metals industry by far and have been the focus of pollution control technologies and increasingly stringent regulations. Like other industries, the mining, mineral processing and extractive metallurgical industries are strictly regulated and monitored by multiple government departments, agencies, and bureaus at the local, state and federal levels. Regulations, including EPA's Resource Conservation and Recovery Act (RCRA) and Land Disposal Restrictions (LDR), have both narrative and numerical criteria and standards for protection of human health, aquatic life, air quality, endangered and threatened species, disposal of solid wastes and the environment.

ARSENIC CONSUMPTION AND PRODUCTION IN THE UNITED STATES

Statistical information for arsenic production and commodities is provided yearly by the U.S. Geological Survey in its publication, *Mineral Commodity Summaries* (Brininstool 2017):

> Arsenic trioxide and primary arsenic metal have not been produced in the United States since 1985. The principal use for arsenic trioxide was for the production of arsenic acid used in the formulation of chromated copper arsenate (CCA) preservatives for the pressure treating of lumber used primarily in nonresidential applications. Only three companies produced CCA preservatives in the United States in 2016. Ammunition used by the U.S. military was hardened by the addition of less than 1% arsenic metal, and the grids in lead-acid storage batteries were strengthened by the addition of arsenic metal. Arsenic metal was also used as an antifriction additive for bearings, to harden lead shot, and in clip-on wheel weights. Arsenic compounds were used

Larry G. Twidwell, (Retired) Emeritus Professor of Metallurgical Engineering, Metallurgy/Materials Eng. Dept., Montana Tech, Butte, Montana, USA

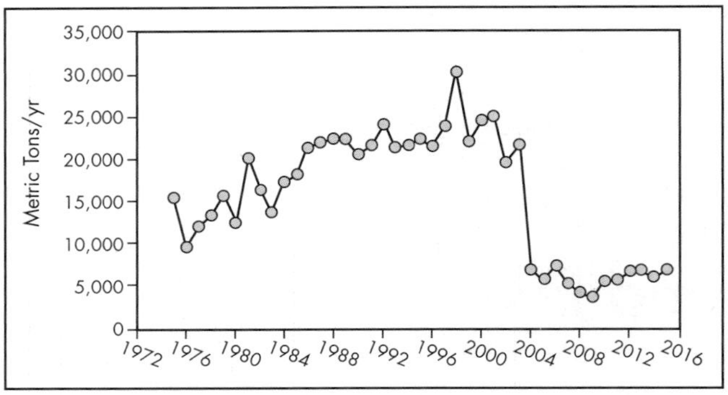

Data from Brininstool 2017

Figure 1 Apparent U.S. consumption of arsenic as a function of years

in herbicides and insecticides. High-purity arsenic (99.9999%) was used by the electronics industry for GaAs semiconductors that are used for solar cells, space research, and telecommunications. Arsenic also was used for germanium-arsenide-selenide specialty optical materials. Indium-gallium-arsenide was used for short-wave infrared technology.

The trend for total U.S. consumption (primarily arsenic and arsenic trioxide) is illustrated in Figure 1 covering the period 1975–2016. Most of the U.S. consumption has been based on imported arsenic trioxide. There has been a major industrial shift away from recovery of arsenic and arsenic compounds that can be marketed to controlling their removal/immobilization/and environmentally safe long-term storage.

World Resources

The world production and reserves of arsenic trioxide are summarized in Table 1. In addition, according to Brininstool (2017):

> Arsenic may be obtained from copper, gold, and lead smelter flue dust as well as from roasting arsenopyrite, the most abundant ore mineral of arsenic. Arsenic has been recovered from realgar and orpiment in China, Peru, and the Philippines; has been recovered from copper-gold ores in Chile; and was associated with gold occurrences in Canada. Orpiment and realgar from gold mines in Sichuan Province, China, were stockpiled for later recovery of arsenic. Arsenic also may be recovered from enargite, a copper mineral.

ENVIRONMENTAL REGULATIONS

A companion chapter in this handbook (Chapter 9.3, "Treatment of Effluent Waste") focuses on treatment technologies and environmental regulations that are important to mineral and extractive metallurgical processing, including the Hazardous and Solid Waste Amendments to the Resource Conservation and Recovery Act (RCRA). These regulations directed the U.S. Environmental Protection Agency (EPA) to aggressively manage some hazardous waste, and the act restricted (banned) some hazardous waste from disposal without further stabilization; that is, the Land Ban Restrictions (LBR) were initiated.

Table 1 World production and reserves (arsenic trioxide)

	Production* (arsenic trioxide)		
	2015	2016†	Reserves
Belgium	1,000	1,000	World reserves data are unavailable but are thought to be more than 20 times world production.
Bolivia	50	50	
China	25,000	25,000	
Japan	45	45	
Morocco	6,900	7,000	
Namibia	1,960	1,900	
Russia	1,500	1,500	
United States	—	—	
World total (rounded)	**36,500**	**36,500**	

Source: Brininstool 2017

*Chile, Mexico, and Peru were believed to be significant producers of commercial-grade arsenic trioxide but have reported no production in recent years.

†Estimated.

Other environmental regulations that have significant consequences to the mineral processing and extractive metallurgical industries have been formulated through the Clean Water Act and its amendments; the Comprehensive Environmental Response, Compensation, and Liability Act (Superfund); and the Clean Air Act and its amendments.

TREATMENT OF ARSENIC-CONTAINING RESOURCES

The growing demand for copper, depletion of high-grade ores, and a more robust regulatory environment have required the metallurgical industry to treat more complex ores. In general, the complex ores that must be treated in the copper and gold industries contain arsenic-bearing minerals. The presence of these minerals in mined copper sulfide deposits has, for decades, been processed by roasting and smelting processes. The arsenic presence was not considered a detriment in the past because arsenic trioxide was recovered as a marketable product which, of course, is not true today. Today the presence of arsenic in concentrates is considered a detriment, and copper smelters require that their concentrates contain <0.5% As.

Complex ores containing higher levels of arsenic minerals, such as enargite, luzonite, tennantite, arsenical pyrite, and arsenopyrite, must be considered for processing. Safarzadeh and Miller (2016) have summarized the need and problem facing the mineral processing and extractive metallurgical copper and gold industries: "the downstream processing of high-arsenic copper concentrates represents a significant metallurgical challenge in terms of both arsenic separation and also its stabilization in an environmentally benign form that fulfills the current and future environmental policies." Therefore, a flurry of current publications address and discuss current processing capabilities and the areas where additional emphases should be placed. Recent literature reviews are especially important to evaluate the state of the art; define future processing treatment technologies; and ensure that the handling, immobilization, and disposal practices are appropriate and are conducted in an environmentally safe way. The following review publications are recommended: Filippou et al. 2007; Long et al. 2012; Safarzadeh et al. 2014a; Lane et al. 2016; and Nazari et al. 2017.

Table 2 Common arsenic-containing minerals

Type	Mineral	Formula
Sulfides and sulfosalts	Aresenopyrite	$FeAsS$
	Aresenical pyrite	$Fe(As,S)_2$
	Cobalite	$CoAsS$
	Enargite	Cu_3AsS_4
	Gerdorffite	$NiAsS$
	Orpiment	As_2S_3
	Proustite	Ag_3AsS_3
	Realger	As_4S_4
	Tennantite	$(Cu,Fe)_{12}As_4S_{13}$
Arsenides	Domeykite	Cu_3As
	Löllingite	$FeAs_2$
	Nickeline	$NiAs$
	Rammelsbergite	$NiAs_2$
	Safflorite	$CoAs_2$
	Sperrylite	$PtAs_2$
As(III) oxides	Arsenolite	As_2O_3
	Claudite	As_2O_3
	Gebhardite	$Pb_8(As_2O_5)_2OCl_6$
	Leiteite	$ZnAs_2O_4$
	Reinerite	$Zn_3(AsO_3)_2$
	Trippkeite	$CuAs_2O_4$
As(V) oxides	Austinite	$CaZnAsO_4OH$
	Conichalcite	$CaCuAsO_4OH$
	Erythrite	$CO_3(AsO_4)_2{\cdot}8H_2O$
	Hörnesite	$Mg_3(AsO_4)_2{\cdot}8H_2O$
	Johnbaumite	$Ca_5(AsO_4)_3OH$
	Mansfieldite	$AlAsO_4{\cdot}2H_2O$
	Oliverite	$Cu_2(AsO_4)OH$
	Sarmientite	$Fe_2AsO_4SO_4OH{\cdot}5H_2O$
	Scorodite	$FeAsO_4{\cdot}2H_2O$

Source: Nazari et al. 2017

Arsenic Mineral Sulfide Associations

Arsenic sulfide minerals are often associated with other sulfide minerals, especially in copper, gold, and silver deposits. A list of arsenic-bearing sulfide minerals is presented in Table 2 (Nazari et al. 2017). The breakdown of known arsenic compounds is: 60% arsenates, 20% sulfides/sulfosalts, and 10% oxides; the remainder are arsenite, arsenides, and native arsenic.

Occurrence of Arsenic-Bearing Minerals

Enargite, tennantite (copper–arsenic–sulfur), and arsenopyrite (iron–arsenic–sulfur) are common copper–arsenic and iron–arsenic minerals present in many of the world's base metal sulfide ores, especially copper- and lead-bearing ores. Gold–silver ores are often associated with pyrite, auriferous arsenopyrite, tetrahedrite, tennantite, and in some cases enargite and tennantite.

Safarzadeh et al. (2014b) have listed several enargite and/or tennantite containing properties; for example, enargite was first identified as an abundant mineral at Butte, Montana (United States). The Anaconda Smelter in Anaconda, Montana (closed in 1980), processed concentrates from the Berkeley Pit ores in Butte, Montana, containing arsenic contents of more than 10% for many years because the company produced arsenic trioxide as a saleable product. More than 400,000 t (metric tons) of stockpiled arsenic-bearing waste were stabilized by cement/lime encapsulation with subsequent storage in a membrane-lined repository. Other smelters have treated high arsenic-bearing concentrates back when there was a market for arsenic trioxide; for example, the El Indio plant in Chile (now closed) treated concentrates containing up to 10% As until it closed in 2002. U.S. arsenic-bearing deposits are not uncommon; for example, such deposits exist at Goldfield and Pyramid, Nevada, and Summitville, Colorado. Other important world deposits include Cheolpec, Bulgaria; Recsk, Hungary; Sardinia, Italy; Lepanto, Philippines; Frieda River, Papua New Guinea; La Coipa, Chile; and Bor, Serbia (Safarzadeh et al. 2014b).

Pyrometallurgical/Hydrometallurgical Processing

The control of arsenic during the pyrometallurgical and hydrometallurgical processing unit operations are briefly discussed in this section.

Copper and Gold–Silver Production

Present-day conventional copper production technologies are described in Chapters 12.8 and 12.9, "Copper Hydrometallurgy" and "Copper Pyrometallurgy," respectively. For the reader who is interested in the details of conventional copper production and gold processing technologies, two texts are recommended: the *Extractive Metallurgy of Copper* (Schlesinger et al. 2011) and the *Gold Ore Processing Handbook* (Adams 2016).

Most conventional treatment to recover copper from sulfide concentrates is via pyrometallurgical processing. Approximately 75%–80% of primary copper is or has been produced by this method for decades. The other 20%–25% of the world's copper is recovered by hydrometallurgical processing of copper oxide ores.

Present-day pyrometallurgical processing of copper sulfides include the use of various ambient and elevated temperature unit operations, for example, flotation concentration of sulfide minerals, roasting, smelting, converting, sulfuric acid production, casting, and electrorefining. Arsenic (and other undesirable elements) and its compounds may be present in the concentrates, dusts, unreacted minerals, and condensed fumes from baghouses, dry and wet electrostatic precipitators, wet gas scrubbers, and sulfuric acid production. The majority of the arsenic is volatized during the elevated temperature processing as arsenic trioxide (As_2O_3) and/or compounds of arsenic sulfide (As_2S_3) in the flue dust and gas cleaning products.

The arsenic sulfide minerals distribute to concentrates along with other non-arsenic-sulfide-bearing minerals, and if the arsenic presence is greater than 0.2% (Tayebi-Khorami et al. 2017), there are smelter penalty charges; if >0.5%, the concentrate is likely to be rejected from treatment via smelting. Collection of arsenic from the various unit operations, and the associated environmental regulations and health issues resulting from the handling, transportation, and impoundment/immobilization requirements, have caused most currently operating smelters to shy away from treating resources with appreciable arsenic content.

Treatment of Low-Arsenic-Bearing Copper Concentrates

For treatment of concentrates with acceptable arsenic concentration, the collected arsenic and other impurity-bearing dusts and smelting residues are normally recycled and blended with the incoming feed concentrate. Therefore, the arsenic bleed from the smelting system is primarily dissolved arsenic in the slag phase and there is no need to form disposable arsenic-bearing waste products. An example of the distribution of arsenic to the slag phase for four technologies treating low arsenic-copper concentrates is presented in Table 3. Normally, the arsenic content of smelting slags varies, but, in general,

Table 3 Arsenic removal distribution for four copper smelting technologies

Technology	Reference	As in Feed, %As	Oxygen Enrichment, $\%O_2$	Matte Grade, %Cu	*D*Matte, %	*D*Slag, %	*D*Gas, %
Teniente converter	Font et al. 2005	0.30	36	70–75	12–20	13–38	75–42
Mitsubishi process	Surpunt and Hasegawa 2003	0.30	48	68	19.0	27.2	53.8
Flash furnace	Font et al. 2005	0.30	60	60–65	35	32	33
IsaSmelt	Present work	0.26	60	60	5.5	4.2	90.3

Source: Alvear et al. 2006

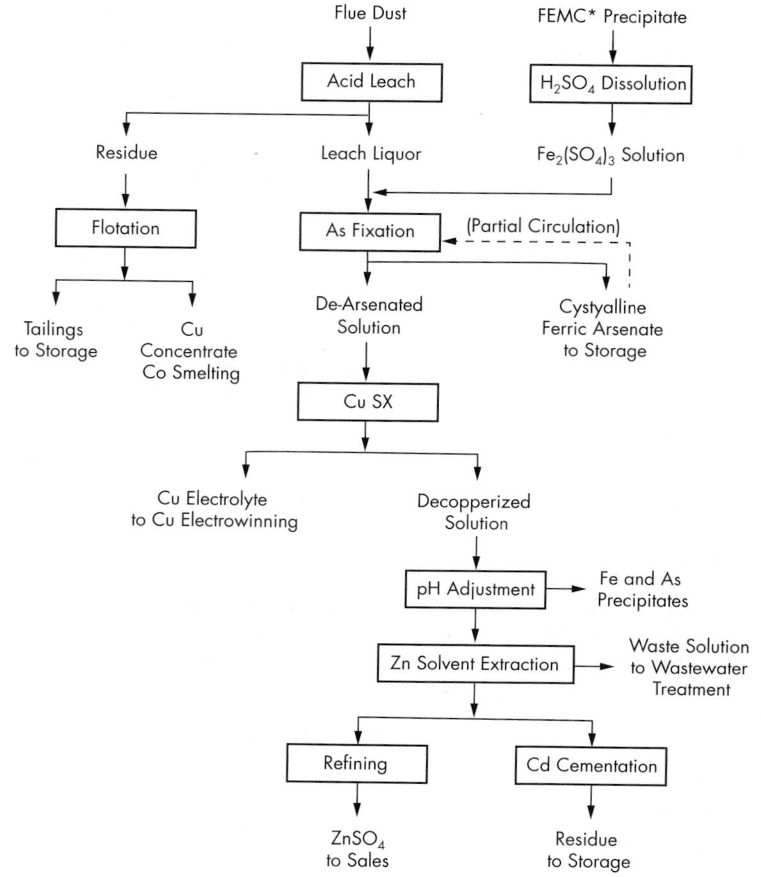

*FEMC is a source of iron to produce the ferric sulfate leach reagent.

Source: Ichimura et al. 2007

Figure 2 Process flow sheet for treatment of Teniente furnace dust

it is less than 0.1%, and therefore the slags can be safely disposed of in the smelter storage site.

Example Treatment of Elevated Arsenic Concentrates

Ichimura et al. (2007) describe the development of a process to treat smelter dust from a Teniente furnace containing ~11% As. The goal of the project was to recover copper for recycle to a smelting operation and to stabilize arsenic as scorodite ($FeAsO_4 \cdot 2H_2O$). The project demonstrated that copper could be recovered and treated in an electrorefining circuit and arsenic could be immobilized by the formation of scorodite. The authors' proposed dust treatment flow sheet is presented in Figure 2. The following conditions were reported for successfully removing arsenic from the final sulfuric acid leach solution: ambient pressure, temperature 95°C, initial pH 1.25, time 12 hours, Fe/As molar ratio 1.3. These conditions favor the formation of scorodite rather than amorphous ferrihydrite (described later). The scorodite, as a function of Fe/As molar ratio (1.2–3.6), was leached using the EPA's toxicity characterization leach procedure (TCLP) regulatory test (EPA 1992).

The results showed an arsenic extraction of 0.6–12.6 mg/L; most results were less than 5 mg/L. The TCLP test results, for the waste to be considered nonhazardous, must be less than 5 mg/L As.

According to the *International Mining* newsletter (iMining 2017), there are only four copper smelters in the world that can presently treat very high-arsenic concentrates: "Tsumeb, Namibia; Altonorte, Chile; Guixi, China; and Horne, Canada. Tsumeb is the only option for treating large volumes of very high-arsenic concentrates with >1% As." Also stated in the newsletter is that the arsenic content in mines that are producing complex concentrates include Marcapunta, Peru, 8% As; Chelopech, Bulgaria, 6% As; and Chuquicamata, Chile, 1.2% As.

Treatment of High-Arsenic Complex Concentrates

The mineral processing and hydrometallurgical treatment of arsenic containing complex concentrates are discussed in the following paragraphs.

Mineral processing for arsenic removal from copper sulfide concentrates. In most cases it is desirable to remove or partially remove arsenic-bearing minerals from a sulfide concentrate before the concentrate enters the pyrometallurgical processing sequence. Filippou et al. (2007) describe possible approaches for separating enargite and tennantite from non-arsenic-containing copper sulfides. A few generalities are presented here: "The separation of copper/arsenic minerals from noncopper gangue minerals (pyrite, pyrrhotite, sphalerite and galena) is easy because enargite does not exhibit natural floatability. However, what is more difficult is the separation of enargite and tennantite from other copper sulfides, such as chalcocite, covellite, and chalcopyrite." The arsenic minerals end up in the processed concentrate, which then enters the smelting unit operations.

Numerous study results have demonstrated that successful flotation separation of enargite and tennantite (and other arsenic-sulfide minerals) from copper sulfides. What is apparent from the literature is that the arsenic mineral separations are dependent on several factors, including mineralogical makeup of the concentrate, particle size distribution, preoxidation of the flotation pulp (or not), type of collector, selection and use of depressants, gas sparging, controlled dissolved oxygen content, and pH. However, the flotation separations for some arsenic mineral makeups are especially dependent on proper E_H (solution potential) to control the surface oxidation characteristics of the arsenic mineral sulfide or associated metal sulfides and metal oxides (Filippou et al. 2007; Chimonyo et al. 2017). These authors present brief descriptions of the separations. Details of the separations are presented in the individual study publications. Recommended publications include

- Enargite, tennantite separation from covellite, chalcocite, and chalcopyrite (Fornasiero et al. 2001);
- Enargite, tennantite from chalcocite (Huch 1994);
- Enargite separation from chalcopyrite; and
- Tennantite from chalcopyrite and bornite (Smith and Bruckard 2007).

Filippou et al. (2007) present Figure 3 as an example to illustrate the possible separation of enargite from chalcopyrite to show the effect of E_H and pH (pH 9 and 10). The authors note that enargite can be floated from chalcopyrite at an E_H potential of 0 mV (pH 9) or the reverse can be achieved at a potential of +350 mV.

Two additional example studies are reported here to illustrate the importance of pulp potential control. Senior et al. (2006) investigated single mineral flotation using pH and E_H control to form a basis for the possible beneficiation of Tampakan deposits in the Philippines. Their investigation was conducted using single mineral pulps (enargite, chalcocite, cuprite, and chalcopyrite) over the E_H range –500 to + 500 mV and pH values of 8 and 11. The authors' conclusions were that pH and E_H conditions could be set for effective separations of two-component copper sulfides. This flotation work was

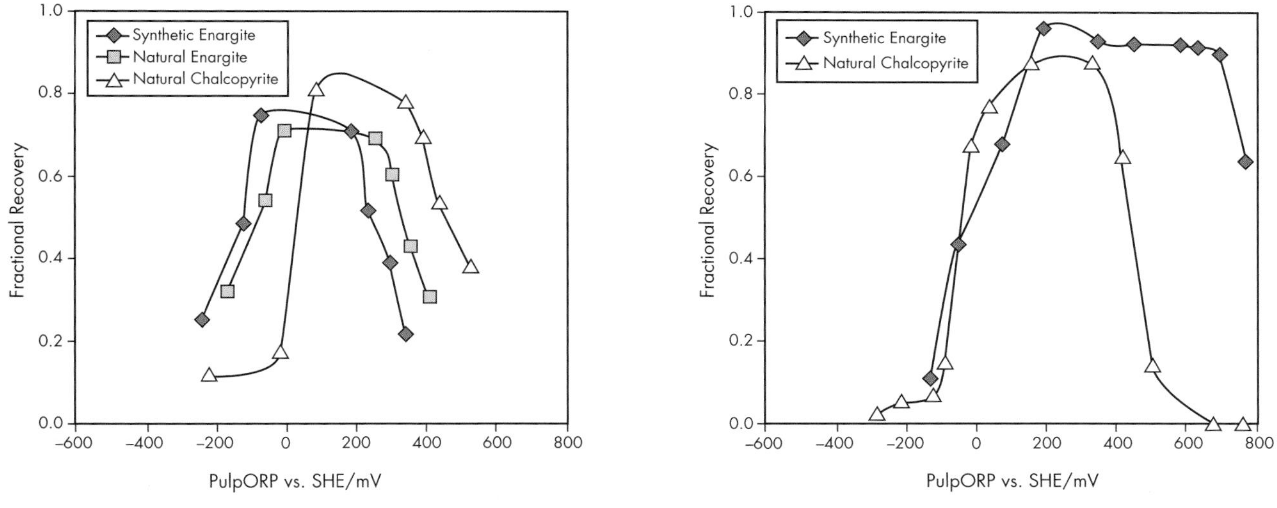

A. pH 9.0, 20 mg/L potassium amyl xanthate, Na_2S and $KMnO_4$ used for ORP adjustment

B. pH 10.0, 14 mg/L potassium amyl xanthate, Na_2S and NaOCl used for ORP adjustments

Source: Filippou et al. 2007

Figure 3 Enargite–chalcopyrite flotation recoveries against pulp oxidation–reduction potential (ORP)

followed by studies on an actual Tampakan porphyry copper–gold ore (not on single minerals) primarily containing enargite, chalcopyrite, and bornite (Tayebi-Khorami et al. 2017).

Studies for separation of enargite or tennantite by flotation from other common sulfide constituents is a currently active research area. Successful flotation separations may produce a "clean" concentrate that can be recycled to the conventional smelting circuit and a "dirty" concentrate. The dirty concentrate may be bled with incoming concentrate or it may be treated to recover additional copper and to produce an arsenic product that can be disposed of in an environmentally acceptable manner.

Mineral processing flotation for refractory gold ores. The mineral flotation concentration of pyrite and arsenical sulfide minerals is well established and widely practiced. Some gold ores show refractoriness when subjected to conventional cyanide leaching; for example, they may have less than 80% gold recovery. The refractoriness is because the gold is encapsulated in sulfide minerals such as pyrite (FeS_2), pyrrhotite (FeS), and arsenopyrite and, in some cases, contain carbonaceous material. Treatment is necessary because arsenopyrite is not susceptible to cyanide leaching (Marsden and House 2006). Dunne (2005) presents an excellent overview for flotation of refractory gold ores, including detailed discussions of collectors, frothers, activators, depressants, and flotation practices for refractory gold ores; arsenopyrite, pyrrhotite, and pyrite ores; and copper–gold ores. Dunne states that there are three Brazilian refractory gold flotation plants, four in North America, six in the Australasian region, and three on the African continent. The gold–sulfide concentrates are treated in a variety of ways: bacterial leaching, pressure leaching, roasting, or supplied to a smelter. Deportment of arsenic to final disposal is discussed in the upcoming sections, especially in the "Fixation of Arsenic" section.

Pretreatment of concentrates. Selective separation of a high arsenic concentrate is not always possible by flotation treatment and other approaches may be necessary. Pretreatment options in copper and gold processing include roasting, atmospheric pressure leaching, and autoclave leaching.

Roasting. Enargite is considered a refractory copper mineral, and its concentrates are not amenable to conventional extraction technologies; therefore, ores and concentrates containing enargite often need preoxidation before further treatment (Safarzadeh and Miller 2016). Roasting is a viable option for the removal of arsenic from enargite concentrates.

Historically, enargite- and tennantite-bearing concentrates have been treated by roasting in a neutral or low-oxygen environment. The classic examples are the El Indio treatment process in Chile and the Anaconda smelter in Anaconda, Montana. Both facilities used multi-hearth roasters (500°–700°C) to treat copper concentrates containing ~8%–10% As. The roaster product calcine arsenic content was less than 0.3% and was, therefore, appropriate for use in the conventional copper smelting process. Arsenic oxide was condensed and collected from the roaster and smelting unit operations. The final arsenic trioxide product contained >90–97.5 As_2O_3. The Anaconda smelter was closed in 1985 and the El Indio was shut down in 2002 because the markets for arsenic trioxide disappeared.

An example of a present-day concentrate pretreatment roasting process is Outotec's partial roaster using a two-stage fluidized bed reactor. The arsenic is volitized in the first bed under a controlled low-oxygen condition (partial roasting) at 400°–575°C to produce arsenic oxide, and then the bed is subsequently treated in the second stage at a higher temperature (dead roasting >600°C) to convert the the iron to hematite and to remove sulfur. The arsenic oxide is then stripped from the roaster gas through a series of cleaning operations and is further treated for arsenic immobilization as scorodite (ambient pressure, elevated temperature process described later) and long-term storage in a permitted site. The Outotec process is applicable to copper–arsenic ores and concentrates and also to copper–iron–gold ores and concentrates. Codelco uses the Outotec partial roaster at its Ministro Hales mine in the Calama region, Chile, to produce a high-arsenic copper concentrate and a low-arsenic calcine. Process water is treated in an integrated effluent plant to produce calcium arsenite. The low-arsenic calcine can be utilized in a conventional smelting circuit. The high-arsenic concentrate may be blended with third-party low-arsenic concentrates at the Ocean Partners' concentrate blending facility in Taiwan (iMining 2017).

The two-stage roaster treatment is appropriate for treating arsenopyrite–pyrite–gold concentrates. Examples of where conventional roasting operations are used include Fairview (South Africa), La Belliere (France), Getchell (Nevada), Mount Morgan (Australia), and Campbell Red Lake and Giant Yellowknife (Canada) (Marsden and Sass 2014).

Hydrometallurgical Processing of Arsenic-Bearing Sulfides

Hydrometallurgical processes account for about 20% of the world primary production of copper; Chile produces about five times more than the United States (Hiskey 2014). This illustrates the importance of a hydrometallurgical approach. Safarzadeh et al. (2014a) suggest that hydrometallurgical processes can be grouped into two categories as shown in Figure 4: selective arsenic leaching or collective leaching of both copper and arsenic. The listed treatments are labeled according to whether they are commercial or laboratory/pilot processes; for example, commercial processes that dissolve

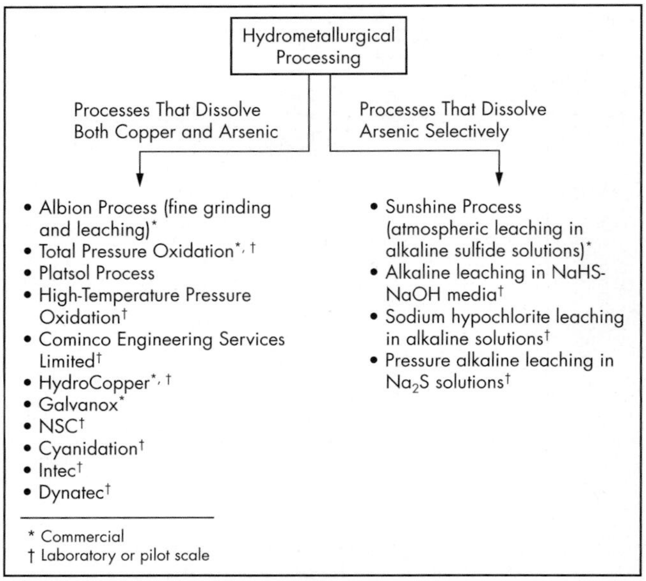

Source: Safarzadeh et al. 2014a

Figure 4 Summary of the processes that have been demonstrated for the treatment of enargite concentrates

Table 4 Summary of operating parameters used in ASL technologies

	Sunshine Process	Equity Process	Melt Process
References, as cited in Lane et al. 2016	Ackerman et al. (1993) Anderson et al. (1991)	Dayton (1982) Edwards (1985, 1991)	Baláž (2000b, 2003) Baláž and Achimovičová (2006) Baláž and Dutková (2009)
Commercial (C)/ pilot plant (PP)	C (1940s–2001)	C (1981–1994)	PP (1990s)
Mode of operation	Batch	Batch	Continuous
Production	20 t/d^{-1}	90 t/d^{-1}	0.5 t/d^{-1}
Retention/ residence time	8–12 hours	8 hours	1 hours
Temperature	104°C	107°C	88°–105°C
Pressure	Atmospheric	Atmospheric	Atmospheric
Minerals in feed	Tetrahedrite, pyrite, galena, bornite	Tetrahedrite, tennanite, chalcopyrite, pyrite, sphalerite, galena, arsenopyrite	Tetrahedrite, pyrite, chalcopyrite, siderite, quartz
Slurry density	200 g L^{-1}	Up to ~600 g L^{-1}	300 g L^{-1}
Initial particle size	70% <200 mesh	No data	70% <200 mesh
Reagent additions	Na_2S: 100 g L^{-1} (equivalent)* NaOH: 15 g L^{-1} Na_2CO_3: 25 g L^{-1}	NaHS† NaOH: not determined	Na_2S: 300 g L^{-1} NaOH: 50 g L^{-1}

Source: Lane et al. 2016
*Elemental sulfur is dissolved in NaOH to produce Na_2S.
†Addition of NaHS is adjusted according to the concentrations of Sb and As in the feed. The molar ratio of S^{2-}/(Sb + As) was set to ~2.

both copper and arsenic include the Albion, total pressure oxidation, HydroCopper, and Galvanox technologies. Only the Sunshine process is listed as commercial (but is not operating today). Each of the listed treatments is discussed in the review publications. Two of the hydrometallurgical treatments have been selected for further comments in this chapter because they show "copper extraction and arsenic fixation." The selections include the alkaline sulfide leach (ASL) to illustrate a selective leach; and the high-temperature pressure oxidation for processes dissolving copper and arsenic.

Alkaline Sulfide Leach

ASL of enargite and other arsenic-bearing minerals results in selectively dissolving arsenic and leaving a copper residue, and, under some treatment conditions, a gold-bearing residue. The reagents, sodium sulfide (Na_2S) or sodium bisulfide (NaHS), dissolve arsenic from enargite, tennantite, realgar (As_2S_2), orpiment (As_2S_3), and arsenic trioxide but not from arsenopyrite (Anderson 2016).

The general process conditions for the ASL treatment are illustrated in Table 4 (Lane et al. 2016). The treatment is conducted at ambient pressure, elevated temperature near boiling, Na_2S, NaOH reagents, high slurry density, and in most cases long contact times. Most studies show that copper is almost completely retained in the leach residue. Leaching enargite, tennantite, and tetrahedrite produces predominately covellite (CuS) and chalcocite (Cu_2S). Safarzadeh et al. (2014a) suggest the process flow sheet presented in Figure 5. Note that the stabilization of As(V) is via ambient pressure scorodite precipitation (discussed later). The flow sheet is based on the results from the previously operated Sunshine Mining and Refining Company's antimony plant in Idaho, United States (Anderson et al. 1991; Anderson and Nordwick 1995); the Equity Silver Plant in British Columbia, Canada (Lane et al. 2016); the MELT pilot process in Krompachy, Slovakia (Balaz and Dutkova 2009); and multiple research papers.

Recent pilot-scale demonstrations for the application of an ASL process (designated the Toowong Process) to flotation concentrates (Tampakan copper–gold project in the Philippines) has been reported by Rohner et al. (2016). The process has been demonstrated to be applicable to the dissolution of enargite, tennantite, and other arsenic sulfide minerals. The authors suggest that the Toowong process has several advantages over other ASL applications; for example, more rapid leaching kinetics, selective arsenic dissolution, and lower reagent additions are required. The authors state that the leach does not dissolve gold or arsenopyrite. The other ASL processes leach a portion of the gold, but the Toowong Process does not leach the gold. Example results from recent Toowong pilot demonstration studies show arsenic extracted values of 90%–92%. The dissolved arsenic resulting from the leach can be precipitated/stabilized by ferrihydrite precipitation or by the ambient pressure, elevated temperature precipitation of scorodite. To date (early 2018) the Toowong Process has not been commercialized.

Although most of the ASL processes listed note that gold is only partially solubilized (or not at all), two investigations report that gold can be solubilized in an alkaline sulfide solution by changing the conditions so that polysulfides are formed (Anderson 2016; Wassink et al. 2005). Polysulfides are formed by the addition of elemental sulfur. The polysulfides perform as an oxidizer to the gold with subsequent formation of aqueous gold sulfide complexes. If arsenic is present, it dissolves in the alkaline solution as sodium arsenite, which is oxidized by the polysulfides to sodium arsenate. Anderson et al. (2016) state that the recovery of arsenic can be accomplished by ferrihydrite adsorption or formation of scorodite.

Alkaline Hypochlorite Leach

Many studies show that arsenic can be effectively removed from copper–arsenic sulfide concentrates by sodium hypochlorite leaching. The general conditions include ambient to 60°C temperatures at alkaline pH values of 12–12.5. Lane et

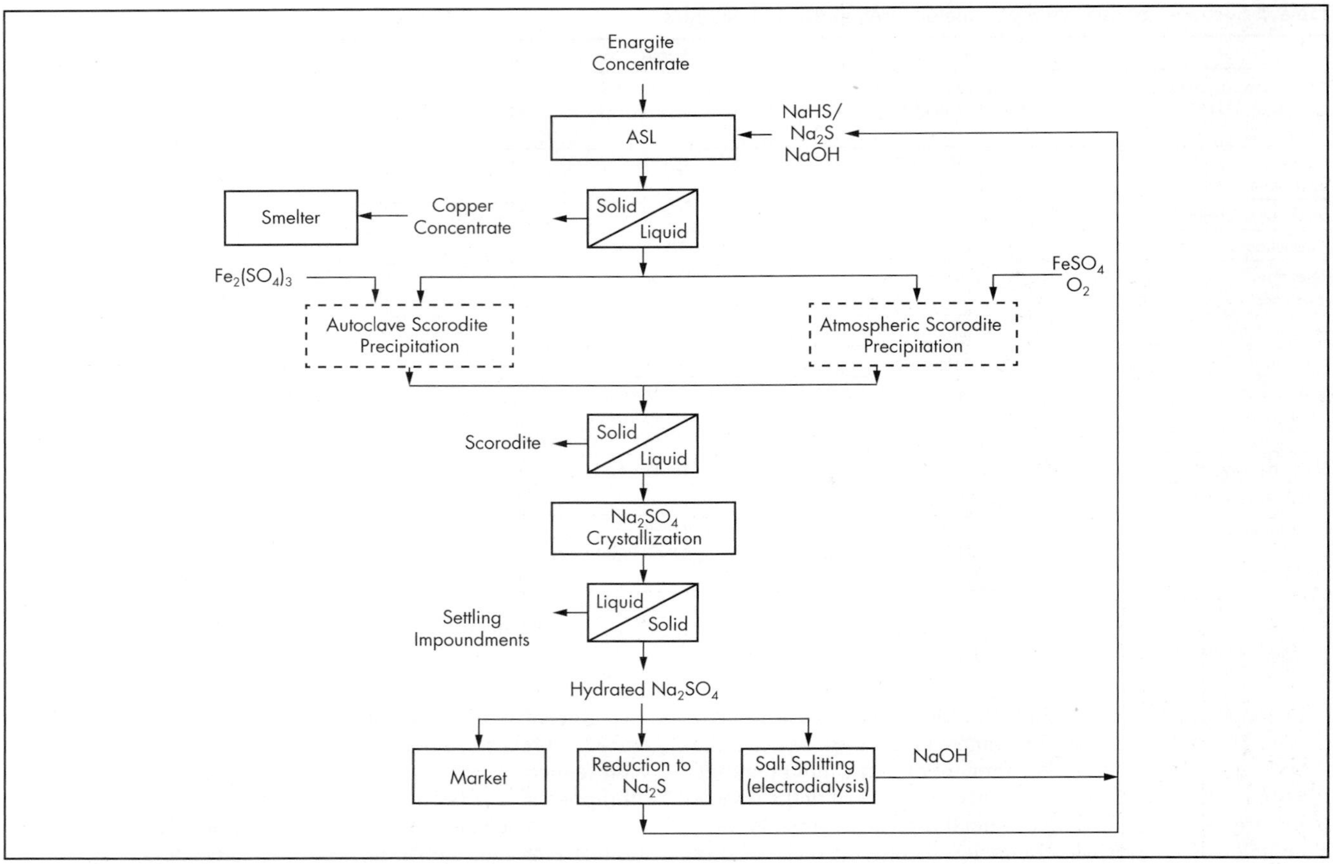

Source: Safarzadeh et al. 2014a

Figure 5 ASL flow sheet proposed for enargite concentrates

al. (2016) note that "hypochlorite leaches has been largely limited to laboratory-scale studies and unlike ASL has not been employed on an industrial scale." The major reasons for this are (1) hypochlorite is not very selective for non-arsenic copper sulfides, and (2) exceedingly high additions of hypochlorite are required for effective removal of arsenic; for example, 17.5 moles of sodium hypochlorite are required per mole of enargite. Most of the hypochlorite leach studies do not discuss the final deportment of arsenic.

Pressure Oxidation

Outotec provides autoclave technology for treating copper–arsenic and gold–arsenic–pyrite concentrates (Ruonala et al. 2011). The process is based on the atmospheric oxidation/precipitation of amorphous ferric arsenate. The precipitation conditions include oxidation of As(III) and Fe(II) at atmospheric pressure, an Fe/As mole ratio of 1–3.5, and a pH of 1.5–4.5. The amorphous ferric arsenate is then converted to crystalline scorodite by autoclave treatment using the operating conditions: 160–200°C, pH 1.5–4.5, and an Fe/As mole ratio 1–1.5. The stated advantage of this treatment is that only the ferric arsenate residue need be treated in an autoclave and not the entire solution.

Cominco Engineering Services Limited (CESL) has patented and demonstrated its pressure oxidation (POX) process to treat copper–gold–arsenic sulfide concentrates (Salomon-de-Friedberg et al. 2017). The CESL process is an intermediate temperature, pilot-scale autoclave treatment that has been applied to more than 18 high-arsenic copper–gold concentrates (up to 18% As). Autoclave conditions include ~150°C, ~14 bar pressure, 60- to 90-minute reaction time to oxidize copper sulfides and As(III) to As(V). The oxidized arsenic reacts with Fe(III) and precipitates as scorodite. The product residue contains scorodite and appreciable copper. The residue is subjected to atmospheric acidic leaching to recover copper. The copper-free arsenic-bearing leach residue is washed through multiple contact stages and disposed of as scorodite. The TCLP test was applied to many residue samples during the pilot studies, and the results were quoted by the authors to always be less than 0.15 mg/L As(V). Stability of the scorodite has been evaluated by contacting the residue with 20 times its weight in water using the EPA's synthetic precipitation leaching procedure (SPLP; Method 1312) and observing the dissolution of arsenic as a function of time. The results have shown leachability of less than the British Columbia limit of <2.5 mg/L when aged over a three-year period. The process has not, at this time, progressed to industrial use.

Hydrometallurgical Processing of Arsenic-Bearing Gold Concentrates

Similar hydrometallurgical processes for pretreatment of arsenic-copper concentrates exist for arsenic-bearing gold concentrates and include high-pressure acid oxidation, low-pressure acid oxidation, alkaline oxidation processes, and

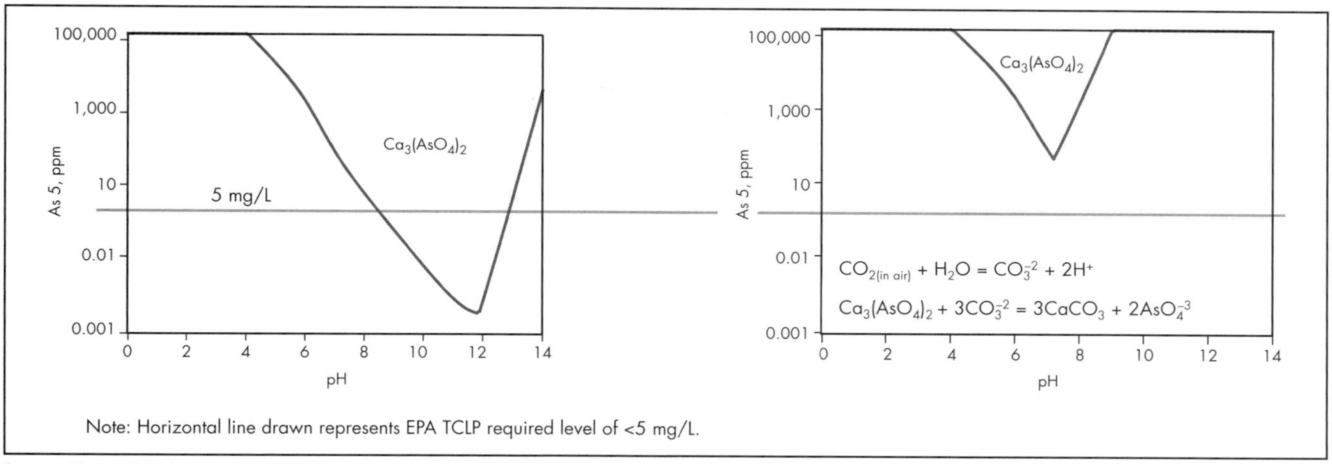

Source: Huang 2016

Figure 6 Conversion of calcium arsenate to calcium carbonate by carbon dioxide in air

bacterial leaching processes (Marsden and House 2006; Nazari et al. 2017; Strauss et al. 2017).

High-Pressure Acid Oxidation

Many high-pressure acid oxidation facilities are in operation. In general, the reaction conditions include oxygen pressures of 1,800–2,000 kPa and temperatures of 180–210°C. The product residue after pressure oxidation is a mixture of scorodite, hematite, basic ferric sulfates, jarosites, and arsenical ferrihydrite, as well as sulfur and gangue residues (Robins and Jayaweera 1992). Robins and Jayaweera have examined products from several high-pressure autoclaving operations and state that very fine amorphous ferrihydrite particles (100–500 nm) exist that were formed along with calcium products. The autoclave residues are composed of relatively stable mineral forms that are considered safe for disposal (scorodite, ferric arsenate); however, because of the presence of ferrihydrite and the presence of calcium arsenates and arsenites formed during the neutralization stage, Robins and Jayaweera (1992) have concerns in regard to long-term stability in storage: "The long term stability of these ferric materials is poor, but could lead to the acceptance of a slow release option rather than complete containment of residues."

In the mid-1980s, several autoclave reactors to treat gold ores or concentrates were commissioned and operated. "This established pressure oxidation as a viable method for treating a range of refractory ores and concentrates, with high gold recovery" (Marsden and Sass 2014). Plants were installed "between 1988 and 2000 at Goldstrike, Getchell, Lone Tree, and Twin Creeks (all in Nevada), Campbell Red Lake and Con (Canada), Lihir and Porgera (Papua New Guinea), and Macraes (New Zealand)" (Marsden and Sass 2014). An advantage of autoclave oxidation is that the arsenic is fixed as scorodite.

Bacterial Oxidation

"Bacterial oxidation is now considered to be a proven commercial technique for the treatment of refractory sulfide gold concentrates. ... testimony for the continuing emergence and acceptance of bacterial oxidation is provided by evidence that the number of bacterial-oxidation plants for gold concentrate treatment now rivals that for pressure leaching" (Miller and Brown 2005).

BIOX is an industrial example of bacteria biohydrometallurgy applied to oxidize refractory gold ores containing pyrites, arsenopyrite, and copper sulfides. The process is currently being used at 12 sites. The bacteria are mesophilic (40–45°C) and acidophile (pH 1.2–1.8) microorganisms that can function in arsenic concentrations up to 20 g/L of As(V) and 6 g/L As(III) (Gonzalez-Contreras 2012). Scorodite is formed as the arsenic-bearing disposable product. Additional examples of bacterial biohydrometallurgy are presented in the following section.

FIXATION OF ARSENIC

Additional information on this subject is presented by Twidwell (2018).

Past Practice

For decades, the major practice for the disposal of arsenic-bearing solutions was lime addition to form calcium arsenate/calcium arsenite with placement in containment ponds or tailings impoundments. Robins was the first to alert industry that calcium arsenate is unstable when exposed to carbon dioxide in air (Figure 6) and is therefore not suitable for storage of calcium As(III) or As(V) compounds (Robins and Tozawa 1982). Note that in Figure 6 there is a range of pH values where calcium arsenate is stable in the absence of carbon dioxide (at a TCLP value of 5 mg As/L) and that in the presence of carbon dioxide calcium arsenate is not thermodynamically stable at 5 mg/L. Riveros et al. (2001) have demonstrated that calcium arsenate sludge can leach up to 4,400 mg As/L using the TCLP leach test. Nishimura and Umetsu (1985) have shown that crystalline calcium arsenate can be formed by calcination and that its solubility is greatly decreased. Nevertheless, dissolution of arsenic will occur with time. However, Nazari et al. (2017) state that two smelters in Chile (Codelco's Chuquicamata and Noranda's Altonorte plant) still employ lime neutralization to form calcium arsenite/gypsum and calcium arsenate/gypsum. The resulting residues are stored and monitored in permitted hazardous waste landfills.

Brief Summary of Current Industrial Practice

Three ferric/arsenic precipitation removal technologies are presently practiced by industry throughout the world: (1) ambient temperature arsenic adsorption/co-precipitation to form arsenical ferrihydrite (Fh); (2) autoclave precipitation of scorodite ($FeAsO_4 \cdot 2H_2O$); and (3) more recently an ambient pressure, elevated temperature precipitation of scorodite.

The ambient temperature Fh technology is relatively simple, and the presence of commonly associated metals (aluminum, copper, lead, zinc) and gypsum have a stabilizing effect on the long-term stability of the outdoor storage of the product. The disadvantages of the adsorption technology are that a relatively large amount of waste material is created (Fe/As mole ratio varies but is usually ~3–4 and can be as high as 10); the product is difficult to filter; the arsenic must be present in the fully oxidized state (arsenate); the presence of competitive associated anionic species may negatively influence the adsorption of arsenate; and the long-term stability of the product in the presence of reducing substances in anoxic and/or bacterial environments is in question.

The second technology practiced at several copper smelting facilities is arsenic removal by precipitation of scorodite. The advantages of the scorodite process over the Fh technology are less waste (Fe/As molar ratio of 1), greater density (better filterability), and better thermodynamic stability (under some conditions). The disadvantages of autoclave scorodite precipitation are that the treatment process is more capital and energy intensive, the compound may dissolve incongruently to form arsenical Fh if the pH is greater than 3–4, and its long-term storage may not be stable under reducing and/or anaerobic bacterial conditions. Today the Fh/arsenate process is more often applied throughout the world. However, the third technology is the elevated temperature, ambient pressure scorodite process, which is likely to be widely adopted in the future (Demopoulos 2008).

Ferrihydrite/Arsenic Treatment

The EPA promulgated regulations for the best demonstrated available technology (BDAT) to be used for removal of arsenic (Rosengrant and Fargo 1990). The specified BDAT for treatment of effluent solutions is adsorption on Fh. This technology has also been selected by the EPA as one of the best available technologies for removing arsenic from drinking waters and its application is widespread.

What Is Ferrihydrite?

Fh is a ferric oxyhydroxide. The accepted formula is $5Fe_2O_3 \cdot 9H_2O$ (Paktunc et al. 2008). It is a large surface area solid phase often referred to as an amorphous material, but it is actually a metastable nanocrystalline material. Important reviews detailing conditions for formation and the stability of Fh are presented by Jambor and Dutrizac (1998) and Cornell and Schwertmann (2003). These reviews are excellent sources of information on Fh occurrence, structure, chemical composition, adsorptive capacity for cations and anions, transformation rate, and a summary of the factors that influence its transformation to hematite or goethite.

Fh is characterized by X-ray diffraction as having a two-line structure, which relates to the number of broad peaks present. Two-line Fh is formed by rapid hydrolysis to pH 4–7 at ambient temperature and is the form usually precipitated in industrial treatment systems. Crystallite sizes have been reported to be 2–4 nm. The surface area of freshly precipitated two-line Fh is 150–340 m^2/g (Paktunc et al. 2008). Hohn (2005) has demonstrated arsenic-loaded 7% (As(V)) Fh prepared at pH 4 and 7 self-flocculate to a mean agglomerate size of 5–10 μm.

Ferrihydrite Transformation

Fh is considered a metastable phase that transforms to hematite or goethite with time. The rate of transformation has been investigated in great detail and is a function of time, temperature, pH, and the presence of adsorbed anions and cations. For example, conversion of two-line Fh to hematite at 25°C is half complete in 280 days at pH 4 but is completely converted at 100°C in 4 hours. Transformation results in a relatively large change in surface area. For example, freshly prepared two-line Fh showed a surface area of ~150 m^2/g that, when converted to goethite at 25°C, was reduced to 92 m^2/g; when converted to goethite at 90°C, the particulate surface area was reduced to 9 m^2/g (Cornell 2003). The fact that conversion occurs reasonably rapidly and that it results in a significant decrease in surface area may hold important negative consequences for long-term outdoor storage stability for adsorbed arsenic. However, in real industrial systems, the Fh conversion rate may be mitigated (changed from days, years, or decades) by the presence of other species and solution conditions during precipitation and subsequent storage. General factors that decrease the rate of conversion to more crystalline forms include lower pH and temperature and presence of adsorbed arsenate, silicate, aluminum, manganese, and heavy metals.

Ferrihydrite/Arsenic (Arsenical Fh)

The structural relationships for Fh adsorption of arsenate are via the formation of inner-sphere complexes rather than simple surface adsorption. The exact nature of the adsorption is controversial, but the use of extended X-ray absorption fine structure spectroscopy has shown that the adsorption is by bidentate corner-sharing surface complexes without the formation of monodentate corner sharing (Sherman and Randall 2003).

The terms *scorodite*, *ferric arsenate*, and *arsenical Fh* are often used throughout the arsenic literature, sometimes incorrectly. Paktunc and Bruggerman (2010) experimentally investigated the structure of scorodite, ferric arsenate, and arsenical Fh and have clarified the distinctions between the three forms. For example, scorodite is a fully crystallized phase containing an Fe/As molar ratio of 1: ($FeAsO_4 \cdot 2H_2O$); ferric arsenate is an amorphous product ($FeAsO_4 \cdot 4\text{-}7H_2O$); and arsenical Fh is arsenate absorbed within the Fh ($5Fe_2O_3 \cdot 9H_2O$) structure. Ferric arsenate forms at low pH and is transformed rapidly to scorodite at pH levels below ~1.7. Above that, pH ferric arsenate and arsenical ferrihydrite form up to a pH of ~4.5 for Fe/As ratios from 1 to 10.

Removal of Arsenic from Aqueous Solutions

A relatively wide range of arsenic removal results, by Fh precipitation/adsorption, are reported in the literature; this is to be expected because several experimental factors influence the removal. The factors include the method of Fh/arsenic contact, the iron/arsenic molar ratio, pH, time, initial concentration of arsenic, arsenic valence, and the presence of other anionic species.

Method of contact. Two approaches are currently used in industry: Co-precipitation or adsorption. *Co-precipitation* occurs when iron and arsenic are present as dissolved species in the solution phase. When the solution pH is raised, Fh precipitates in situ throughout the solution with intimate contact between the solid and the arsenic in the solution phase. The second approach is termed *adsorption* and occurs when a ferric containing solution is added to an arsenic solution at the desired pH; upon addition of the dissolved ferric, Fh forms and supplies surfaces for arsenic adsorption. The loading densities achieved by the two approaches are very different. For example, co-precipitation results in a greater loading density (mole As/mole Fe is 0.7–1.0); adsorption results in a loading density of 0.3–0.5.

Fe/As molar ratio, pH, initial arsenic concentration. The influence of Fe/As molar ratio as a function of pH is illustrated in Figure 7.

Valence state. The presence of arsenite, As(III), needs to be considered when selecting an appropriate Fh technology. The normal approach when considering Fh removal of arsenic is to consider ways to oxidize the As(III) to As(V). It is often stated that As(V) is much more effectively removed by Fh than As(III). However, the relative removal of As(V) and As(III) depends on the Fe/As ratio, pH, and whether the arsenic species are present individually or as mixtures. As(III) is often found in appreciable concentration in ambient temperature metallurgical operation flue-dust leaching solutions, acid blowdown solutions, wastewater, groundwater, and surface waters. The influence of Fe/As molar ratio, arsenic valence state, and pH are illustrated in Figure 8.

Oxidation of As(III). The oxidation of As(III) has been the focus of many studies. Nazari et al. (2017) present detailed discussions concerning what oxidants have been studied and a summary of the application conditions. With respect to industrial applications, Nazari et al. state that: "As(III) bearing streams obtained from hydrometallurgical and pyrometallurgical processing frequently consist of a high concentration of As(III). Hydrogen peroxide, permanganate, ozone and SO_2/O_2 gas mixture have been typically employed in industrial scale to oxidize As(III) to As(V)." The oxidation of As(III) is an important consideration because successful removal of arsenic by Fh adsorption or scorodite precipitation requires that the arsenic be present as arsenate. This is especially true when forming scorodite.

Presence of associated ions. The presence of associated ions such as phosphate, sulfate, carbonate, and dissolved organic species can *greatly* influence the removal of arsenic and the long-term stability of Fh. A review of the effect of associated ions is beyond the scope of this chapter, and the

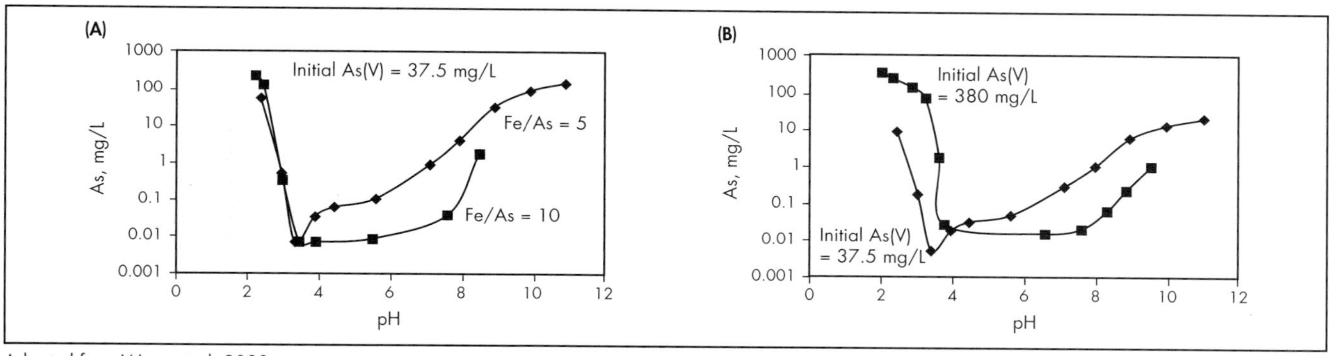

Adapted from Wang et al. 2000

Figure 7 (A) Influence of Fe/As molar ratio and pH on the removal of arsenic by Fh co-precipitation and (B) the influence of initial concentration

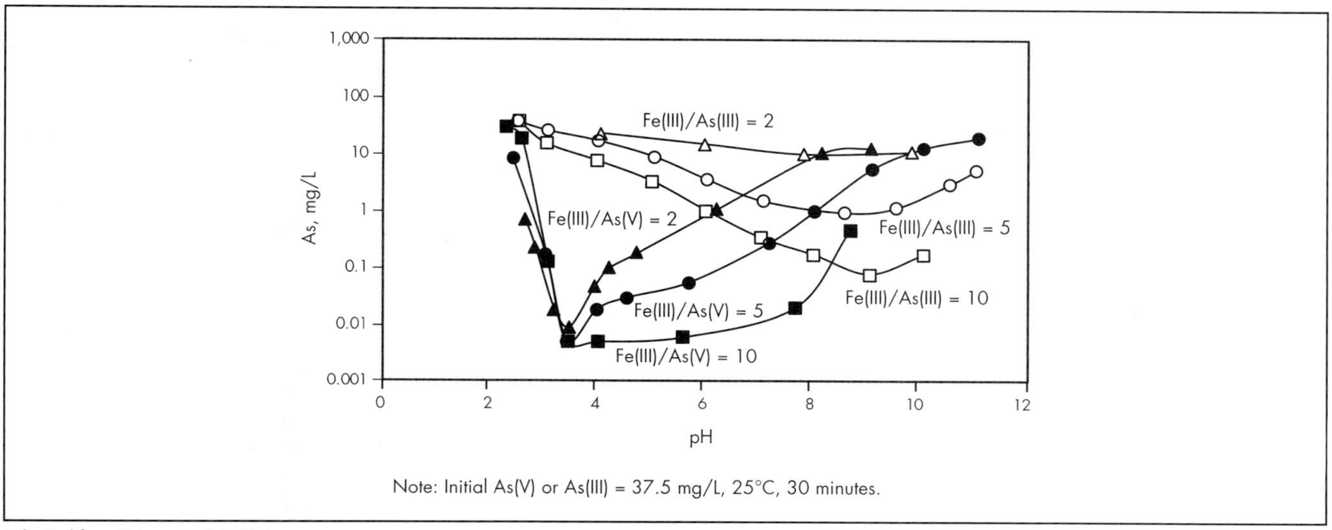

Adapted from Wang et al. 2000

Figure 8 Influence of Fe/As molar ratio, arsenic valence, and pH on the removal of arsenic by Fh co-precipitation

reader is referred to Jain and Loeppert (2000) and Viloante et al. (2003). The presence of cations, such as Al(III), enhances the removal arsenic and increases the long-term stability of Fh (De Klerk et al. 2012; Twidwell and McCloskey 2011). Krause and Ettel (1989) have shown that the presence of divalent cations, Zn, Cd, Pb, Ca, and Mg, extend the Fh formation and stability range from pH values of 4–7 to ~4–10.

Stability of Ferrihydrite and Example Industrial Applications

Many investigations have focused on the formation and stability of Fh (Ford 2002; Harris 2000, 2001; Hohn 2005; Twidwell et al. 2007). A few examples are presented here.

Ford (2002) investigated the storage stability of two-line Fh and arsenate-loaded Fh for periods up to 112 days at 40°C and pH 6. He found that the stability with respect to rate of conversion to hematite (Fe_2O_3) was a function of arsenate loading and that for some lower-level arsenate loadings, the arsenic release from the solid phase did not occur even though appreciable Fh was converted to Fe_2O_3 as a function of time. Ford also demonstrated, for arsenate loadings less than the maximum loading capacity of Fh, that a significant fraction of the retained arsenate was lost from the originally co-precipitated Fh but was not released to the solution phase. That is, the arsenate transferred to the crystalline Fe_2O_3 phase that was created during the aging process.

Twidwell et al. (2007) investigated the stability of Fh and aluminum-modified Fh (AMF, Al/Fe molar ratio of 1) under a variety of conditions, including the [Al+Fe]/As molar ratio, temperature (25° and 70°C), initial As(V) concentration (0.1 to 10 mg/L), and aging time (30 minutes to 500 days). The partitioning of arsenate after 500 days at 70°C to the AMF formed hematite was significantly greater than for the Fh solid, whereas the extent of conversion for the As-loaded AMF solids was much lower (17%) than for the As-loaded Fh (73%) solids, showing that greater As loading to the hematite conversion product occurred. Partitioning of arsenate from the amorphous solids to hematite or to the solution phase at ambient temperature did not occur over 500 days for either AMF or Fh.

Removal of arsenic from solution by ferric precipitation has been or is practiced at numerous extractive metallurgical facilities, for example, Xstrata's Horne smelter in Canada; the Giant mine, the Con mine, and the Teck-Corona mine in Canada; the Kennecott Utah smelter in Utah, United States; Placer Dome Lonetree (Arizona) and Getchell mines (Nevada) in the United States; Barrick's gold mining operations in Nevada, United States; and the Saganoseki and Kosaka copper smelters in Japan (Valenzuela 2000). Harris (2000) has tabulated worldwide industrial operating practice (as of 2000/2001) for removal and stabilization of arsenic by the Fh, autoclave, and lime neutralization processes.

Nazari et al. (2017) noted several industrial facilities that control their arsenic removal by arsenical Fh adsorption and/or ferric arsenate formation. Examples presented included the Xstrata Horne smelter, a uranium mill in northern Saskatchewan, Canada (Demopoulos 2014), and the Boliden Harjavalta plant in western Finland (Salokannel et al. 2013). The arsenic disposal practices are summarized: Xstrata's product includes ferric arsenate, arsenical Fh, and ferric arsenite. The products are mixed with smelter slag and are placed in their smelter slag tailings impoundment. Arsenic release is less than 1 mg/L as required by the province of Quebec. The uranium mill product is ferric arsenate, which is placed in its tailings facility. Arsenic release is below Saskatchewan regulations of 2 mg/L. The Boliden products are ferric arsenate and metal hydroxides; placement was not specified.

Many investigations have considered Fh stability under oxic conditions. However, Doerfelt et al. (2016) have recently considered the stability of Fh and AMF in the presence of sulfide as a reducing agent at anoxic conditions at relatively high pH values. Their investigation focused on molar ratios of Fe(III)/As(V) = 4 and Fe(III)/Al(II)/As(V) = 2/2/1 co-precipitated products. The products were subjected to "excess" sulfide (molar ratio of sulfide/Fe(III) = 1) at "extremely reducing" conditions in a nitrogen protected environment. The results of the study were very encouraging and demonstrated that Fh and aluminum Fh were stable under these severe treatments. The investigators stated: "It was found that the ferric-arsenate co-precipitates could retain up to 99% of its arsenic content despite the high pH (10.5) and extremely reducing (E_H <–200 mV) environment. There was no significant reduction of arsenate and only 45% of ferric iron was reduced by 50% (down to 15 mg/L) hence mixed Fe(III)/Al(III)-arsenate co-precipitates may offer better resistance to reductive destabilization over the long term than all iron co-precipitates." Additional data are presented in their publication. Refer to Chapter 9.3, "Treatment of Effluent Waste," for further information concerning the long-term stability of arsenical Fh presented by Paktunc et al. (2008); Riveros et al. (2001); and Drahota and Filippi (2009).

Ferrihydrite Stability Concerns

In general, arsenical Fh passes the EPA TCLP toxicity test, and the waste products do not have to be subjected to further stabilization as required by the LBR. However, an important unknown at this time is whether the product from Fh adsorption of arsenic will be stable if storage conditions are anaerobic or may become anaerobic or contain microbial agents. For example, Chatain and Sanchez (2005) investigated the effect of controlling the solution redox potential (E_H) and pH using sodium ascorbate (–7 to 345 mV) and sodium borohydride (–500 to 140 mV) to treat an arsenic-bearing gold mining soil (2.8% As, 1.8% on Fh). The release of arsenic from the soil under oxidizing conditions (410 mV) showed the normal Fh release of arsenic (V) (i.e., ~0.3 mg/L); whereas the treatment with 0.046 mol/L sodium ascorbate at an E_H = –7 mV (pH ~6) released ~80 mg/L As(III).

The effect of bacterial reduction of Fh and arsenate can be extensive. Kocar et al. (2010) found that the effect of sulfate-reducing bacteria (that produces dissolved sulfide species) was to reduce Fh to other iron solids along with the reduction of arsenate to arsenite. Chatain and Bayard (2005) investigated the influence of anaerobic conditions (at pH ~7) with indigenous bacterial activity on the release of arsenic (and other metals) from a contaminated mining soil (3% As, 0.3% on Fh). The results showed <4 mg/L arsenic release from baseline soil/water leaches (80 days) and ~100 mg/L As(III) for nutrient-fed indigenous bacteria.

Scorodite Formation

The second technology practiced at several copper and gold facilities is arsenic removal by precipitation of scorodite.

What Is Scorodite?

Scordite is a naturally occurring iron–arsenic mineral, $FeAsO_4{\cdot}2H_2O$. It has a low solubility in the water environment and has 1 mole of arsenic/mole of iron; that is, it contains 25%–30% As, whereas the maximum that Fh can contain is 0.5%–7%, depending on the required Fe/As ratio (usually 3–10). Natural scorodite is often associated with arsenopyrite and enargite and is found in copper- and gold-bearing deposits. It is relatively stable at pH values of 2.8–5.3 (Riveros et al. 2001) and passes the TCLP solubility test.

Several technologies can be used to form scorodite:

- Autoclave hydrothermal precipitation of scorodite from acidic solutions (pH ~1, ~150°C) containing Fe(III) and As(V) (Gomez et al. 2011a, 2011b; and many others)
- Elevated temperature, ambient pressure precipitation from acidic solutions (pH ~1, 90-95°C) containing Fe(III) and As(V) or As(III) (Demopoulos 2008, 2005; and many others)
- Intermediate temperature, ambient pressure precipitation by in situ oxidation of Fe(II) in the presence of As(V) from acidic solutions (pH ~1, ~70°C, 95°C) (Fujita et al. 2010, 2012)
- Intermediate temperature, ambient pressure precipitation by biogenic in situ oxidation of Fe(II) in the presence of As(V) from acidic solutions (pH ~1, ~70°C) (Okibe et al. 2013; Gonzalez-Contreras 2014)
- Intermediate temperature, ambient pressure precipitation by biogenic in situ oxidation of Fe(II) and As(III) from acidic solutions (pH ~1, ~70°C) (Okibe et al. 2013, 2014)

Autoclave Applications

There have been many autoclave studies demonstrating the successful formation of scorodite and scorodite-like phases. The results of an extensive study of the formation and characteristics of three phases formed during the hydrothermal precipitation of Fe(III) and As(V) are presented by Gomez et al. (2011a). The phases formed included sulfate substituted scorodite $[(FeAsO_4)_{1\text{-}0.67x}(SO_4)_x{\cdot}2H_2O]$ $x \leq 0.16$; FASH (also referred to by Swash and Monhemius [1994] as Type 1 $[(FeAsO_4)_{0.998}(SO_4)_{0.01}{\cdot}0.72H_2O]$); and BFAS (also referred to by Swash and Monhemius [1994] as Type II $[(FeAsO_4)_{1-x}(OH)_x{\cdot}wH_2O$ $x = 0.3$ to $0.7]$). The conditions for temperature and initial molar ratio of Fe(III)/As(V)) to form the various products are summarized by the authors in their publication. The authors also report that TCLP stability tests were conducted on the products during short- and long-term aging. Short-term (multiple TCLP solution contacts for 24 hours at pH ~5) results were as follows: FASH was slightly more soluble than scorodite and BFAS; all gave <1 mg/L after seven contacts. Long-term (>8 months at pH 3, 5, and 7.5) test results were as follows: BFAS and scorodite about equivalent; FASH had a higher release. At pH 3, all were <1 mg/L; at pH 5, FASH was 2.5 mg/L, BFAS was ~0.1 mg/L, sulfate substituted scorodite was 0.6 mg/L. At pH 7, all were relatively high, >1 mg/L. The authors recommend that the BFAS may be the best form for storage at pH <7.

The most recent autoclave scorodite formation (205°C, 5 bar O_2, 50 minutes) test work has been reported by Strauss et al. (2017). The study investigated the stability of the BFAS product before and after cyanidation. "BFAS precipitates, as well as their cyanidation residues were found to pass the respective environmental tests in terms of arsenic release." The stability tests were short-term TCLP (EPA Method 1311) and SPLP (EPA Method 1312, a water leach test at a pH of ~4.8) exposures. The results were: before cyanidation 0.55 mg/L (SPLP) and 0.56 mg/L (TCLP); after cyanidation 0.16 mg/L (SPLP) and 0.16 mg/L (TCLP). Long-term test work was not reported.

Atmospheric Formation of Scorodite

Atmospheric scorodite formation has been investigated for more than 20 years and has now advanced to industrial application. Filippou and Demopoulos (1997) and Demopoulos (2009) have described the process: ferric ions are fed to a reactor at ~80–95°C containing arsenate at a pH of ~0.9 to form amorphous ferric arsenate; crystallization is accomplished by slow addition stepwise neutralization over a pH range of 0.9 to 4 in the presence of scorodite seed crystals with the result being that the ferric arsenate formed crystalline scorodite. Fujita et al. (2010) have shown that scorodite can be formed at lower temperatures by in situ oxidation of ferrous ions in arsenate solutions.

It is often stated in the literature that scorodite can only be formed from elevated arsenic containing solutions (>10 g/L); however, Caetano et al. (2009) have demonstrated that scorodite can be formed from dilute arsenic-bearing solutions, for example, 0.1 to 1.1 g/L under ambient pressure, elevated temperature treatment. The advantage of nonautoclave processing is that it is less costly and more energy efficient than the autoclave process and produces less waste material to be disposed of when compared to the Fh process.

Stability of Scorodite

Scorodite formed by all the aforementioned processes pass the short-term EPA TCLP test. Example longer-term test results are summarized in Table 5. The EPA TCLP test is not considered a reasonable measure of stability by many investigators. The test is an acetate pH buffered environment (pH ~5) designed to simulate co-disposal conditions in a municipal landfill. It is a test used to determine if a solid waste should be considered to be hazardous (for arsenic the measured concentration must be less than 5 mg/L). It is a test conducted under oxidizing conditions at only one pH, one solid/liquid ratio, one temperature, and one reaction time. However, an industrial waste may be stored under reducing conditions, microbiological conditions, changing pH conditions, changing oxidation–reduction potentials, and temperature. The TCLP results are biased by not considering reaction kinetics, particulate size, time and susceptibility to reagent complexation, and valence state.

Example Industrial Applications That Produce Scorodite

Blanchard et al. (2017) investigated arsenic speciation in the JEB Tailings Management Facility (TMF) at McClean Lake, Saskatchewan, Canada, to verify that atmospheric precipitated scorodite is stored in its repository. The facility uses an oxidizing sulfuric acid leach of uranium ore that contains appreciable arsenic (as nickel arsenides), for example, ~300 to 50,000 µg/g. The arsenic is leached as As(V) and As(III) in concentrations of ~100 mg/L to ~10,000 mg/L. The arsenic removal process is performed on its process raffinate solution: ferric sulfate is added to provide a Fe(III)/As molar ratio of at least 3; conditions are E_H +680 mV, pH ~1; pH is adjusted with lime to pH 4 then to 7.5. The precipitated product is thickened along with other residues and pumped to the tailings disposal site. Laboratory studies showed that poorly

Table 5 Examples of long-term stability test work for scorodite

Reference	Study Period	Results, mg As/L	Comments
Demopoulos 2005; Bluteau 2004 (thesis); Bluteau and Demopoulos 2007	20 weeks at 75°C; 66 weeks at 22°C	For the 22°C 66-week water solubility tests, the results were: pH 5, 0.35 mg/L; pH 6, 0.97 mg/L; pH 7, 5.89 mg/L. The 66-week tests showed that scorodite "undergo slow incongruent dissolution yielding a highly metastable nano-sized two-line ferrihydrite phase" (2004, 2005). "The growth and re-crystallization of ferrihydrite was apparently retarded by arsenate adsorption" (2007).	At pH levels above ~4, incongruent dissolution of scorodite will slowly form nanosize Fh. After 20 weeks at 75°C or 66 weeks at 22°C, the authors saw no signs of growth or transformation of the amorphous arsenical Fh.
Gomez et al. 2011a	>8 months at pH 3, 5, and 7	Short-term (multiple TCLP solution contacts for 24 hours at pH ~5): FASH slightly more soluble than scorodite (SR) and BFAS all gave <1 mg/L after 7 contacts. Long-term (>8 months): • pH 3: all were <0.1 mg/L • pH 5: SR 0.6 mg/L; FASH 2.5 mg/L; BFAS ~0.1 mg/L • pH 7: all high >1	Tested the short- and long-term stability of scorodite (SR), FASH (Type I); BFAS (Type II), autoclave products in TCLP.
Lagno et al. 2010 (See also Katsarock 2011)	10 days, 6 weeks	Results: Oxic—the coating was protective (reduced As release from 1.5 mg/L to ~0.15 mg/L at pH 4; 45 mg/L to 1.5 mg/L at pH 8). Anoxic—pH 7, 100 mV, not as protective, Fe(III) and As(V) partially reduced to Fe(II) and As(III). Uncoated scorodite showed ~200 mg/L As at 6 weeks (pH 8); coated showed <10 mg/L (values not given by authors, data from their Figure 13).	Scorodite was encapsulated with aluminum phosphate and then subjected to oxic and anoxic water aging. Oxic aging was conducted for 10 days and anoxic aging for 6 weeks. Four materials were evaluated: scorodite; one, two, and three layers of aluminum phosphate on scorodite.
Bluteau et al. 2009	Up to 57 weeks	The gypsum-saturated equilibrium arsenic concentration was 3.6 mg/L at pH 7; without gypsum the value (as reported by Bluteau and Demopoulos 2007) was 5.9 mg/L.	Scorodite dissolution tests were conducted in deion water saturated with gypsum (4–8 g/200 mL) at 22°C, pH 5, 7, 9 for up to 57 weeks.
Gonzalez-Contreras 2012; Gonzalez-Contreras et al. 2012	Up to 1 year	Aging time after formation was 10 or 22 days. The TCLP exposure was for 1 year: the 10-day aged product leached 1.5–2 mg/L; the 22-day aged product leached 1 mg/L to 0.16 µg/L. The most stable crystals were formed at pH 1.2, aged 22 days and when exposed to synthetic landfill conditions for 1 year leached 16 µg/L.	Bioscorodite stability tests were conducted in EPA TCLP Method 1311 test conditions.
Salomon-de-Friedberg et al. 2017 (CESL)	3 years	Water leachability was less than the British Columbia limit of <2.5 mg/L for 3-year aging. TCLP was applied to "many residue samples during the pilot studies"; results were always <0.15 mg/L.	Stability of scorodite was evaluated by contacting the residue with 20 times its weight in water and observing the dissolution of arsenic as a function of time.
Leetmaa et al. 2016	169 days 900 days	A sulfate gel/scorodite system (at a low Al(III)/As(V) molar ratio of 0.1 and 0.2) was aged in a water environment at pH ~7 for up to 900 days. The reported results were that arsenic dissolution was <2 mg/L for unwashed gel/scorodite and <0.5 mg/L for washed (TCLP solution). Also, the authors state "There was only 0.2 mg/L As released from the sulfate gel sample/scorodite system (Al:As = 1 equilibrated at pH 7.3 for 169 days; that is 50 times lower than the solubility of the control scorodite."	Aluminum hydroxide gel encapsulated scorodite was evaluated by long-term aging in a water leach environment.

crystalline scorodite formed up to a pH of 3.2. Excess iron then precipitates as amorphous arsenical Fh during the near-neutral pH adjustment. Prior to this study, scorodite had not been identified in the TMF. However, using X-ray absorption near-edge structure spectroscopy showed the following: the tailings samples consisted of "scorodite and poorly crystalline iron-containing arsenates." The iron-containing arsenates were assumed to be Fh, "arsenate adsorbed on ferrihydrite is likely present given the abundance of ferrihydrite in the TMF."

CESL demonstrated its POX process to treat copper–gold–arsenic sulfide concentrates (Salomon-de-Friedberg et al. 2017).

Outotec provides POX technology for treating copper–arsenic and gold–arsenic–pyrite concentrates (Ruonala et al. 2010).

EcoMetales's copper flue dust treatment plant (known as PTPA) and the Arsenic and Antimony Abatement Process (known as PAAA) near Calama, Chile, treat smelter flue dust, refinery effluent, and other solid hazardous waste from Codelco's Chuquicamata, Potrerillos, and Ventanas smelter and refinery complexes (iMining 2017). Dusts are acid leached, and leach residues are recycled to the smelters. The As(III) specie in the leach solution is oxidized and a ferric reagent solution (Fe/As mole ratio ~1) is added to precipitate scorodite under conditions of pH values in the range of 1 to 1.2, at an ambient pressure, and a temperature of 80°–85°C to form scorodite (Demopolous 2014). The product is disposed of in an authorized/permitted nearby site.

The Plaques (Huisman et al. 2011) ASENOTEQ process forms scorodite using a biological oxidation process. As(III) solutions are oxidized to As(V) by H_2O_2 prior to being fed into a reactor at ~70°C, pH 1.2, and an Fe/As ratio of 1.5. Ferrous ions are oxidized biologically to ferric ions in situ and scorodite forms (Gonzalez-Contreras 2012). The preferred application is to solutions containing arsenic concentrations >1,000 mg/L.

BIOX is a well-commercialized example of bacteria biohydrometallurgy applied to oxidize refractory gold ores containing pyrites, arsenopyrite and arsenical pyrites, and copper sulfides.

Biogenic scorodite or bioscorodite formation has also been described by Gonzalez-Contreras (2012) in her doctoral thesis and also in related publications (Gonzalez-Contreras et al. 2012; Gonzales-Contreras 2014). The process is based on the biological oxidation of ferrous ions in the presence of arsenate, and the formation of scorodite was demonstrated in a continuous stirred two-tank reactor system. The conditions were: 72°C, 2.8g/L As(V), and 2.4 g/L ferrous were fed to the reactor system at a pH 1.2. A 10% by volume biomass (thermoacidophilic microorganisms) oxidized the ferrous to ferric in situ with the As(V). The reported rate of As(V) removal was 1 g/L/d and the formation of scorodite was 3.2 g scorodite/L/d. The study conclusion was that 99% of the incoming arsenic was removed and that bioscorodite formed. Testing in TCLP solutions showed good stability; for example, only 0.4 mg/L As(V) was leached in 100 days of exposure. When jarosite was present, the stability test resulted in a solution As(V) concentration of 0.8 mg/L. Gonzalez-Contreras et al. (2012) note that the relative stability of the bioscorodite is dependent on several factors, including the rate of precipitation and aging time. They present aging stability data; for example, a leaching test (conditions not given) showed 0.1 mg/L of dissolved arsenic after 40 days. The authors suggest that the process could be utilized for treating acid plant effluents, copper electrorefining electrolyte bleed streams, and leach solutions from treatment of arsenic oxide dusts and contaminated soils.

Conceptual Flow Sheets for Formation of Arsenical Fh and Scorodite

Conceptual flow sheets for forming arsenical Fh and scorodite are presented in Figures 9 and 10.

Other Fixation Possibilities

Encapsulated Scorodite

Scorodite is unstable in a variety of conditions, including alkaline solutions. Several studies have been designed to increase scorodite stability at high pH levels, for example, the use of surface coatings to encapsulate and provide a protective barrier. Encapsulation of scorodite particles with hydroxyapatite (HAP, $Ca_{10}(PO_4)_6(OH)_2$) and hydroxyfluoapatite (FAP) showed arsenic leachability of <1 mg/L for HAP and ~8 mg/L for FAP when exposed to an anoxic environment at pH 9 for 40 days (Katsarock 2011). Uncoated scorodite released 22 mg/L in the same environment. However, chemical and mechanical stability was not sufficient to protect the scorodite (Demopoulos 2014; Lagno et al. 2010).

Leetmaa et al. (2014) have investigated the use of an aluminum hydroxyl gel made from aluminum sulfate salts that have protective properties. The gel procedure was applied to ambient pressure precipitated scorodite. Several variables were investigated, including type of gel, Al(III)/As(V) molar ratio, liquid/solid ratio, and prewashing technique. The sulfate gel/scorodite system (at low Al(III)/As(V) molar ratios of 0.1 and 0.2) was aged in a water environment (initially at a pH of 8 and allowed to drift to 7) for up to 900 days. The reported results were that arsenic dissolution was <2 mg/L for unwashed gel/scorodite and <0.5 mg/L for washed (TCLP solution). Additional data are presented and discussed in the investigator's publication.

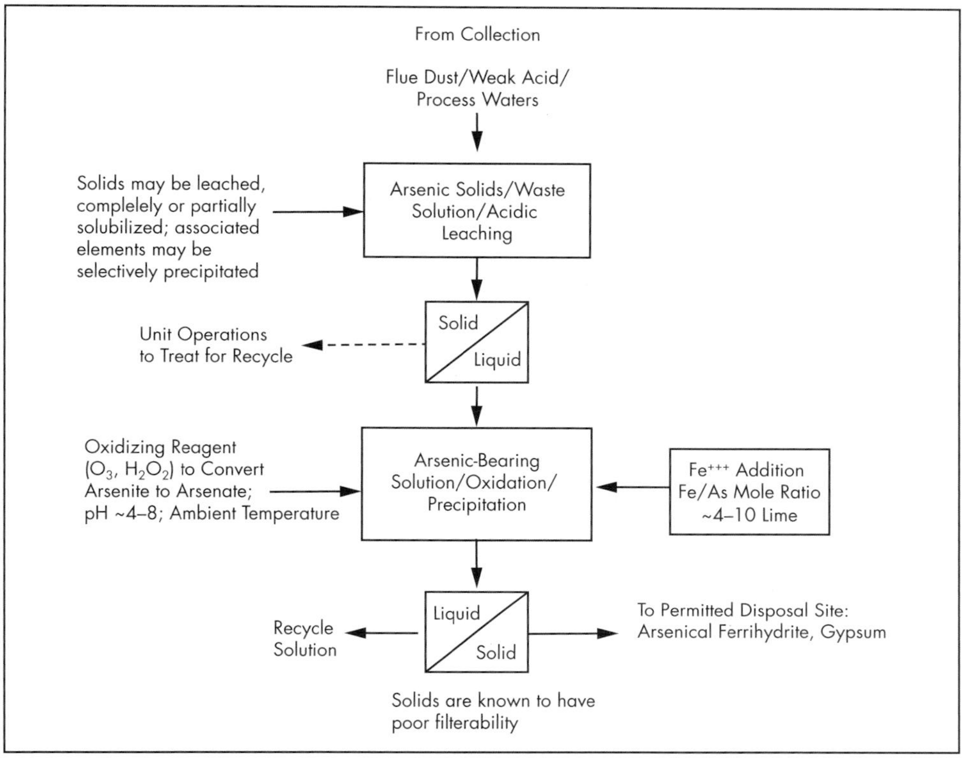

Figure 9 Conceptual flow sheet for forming arsenical Fh

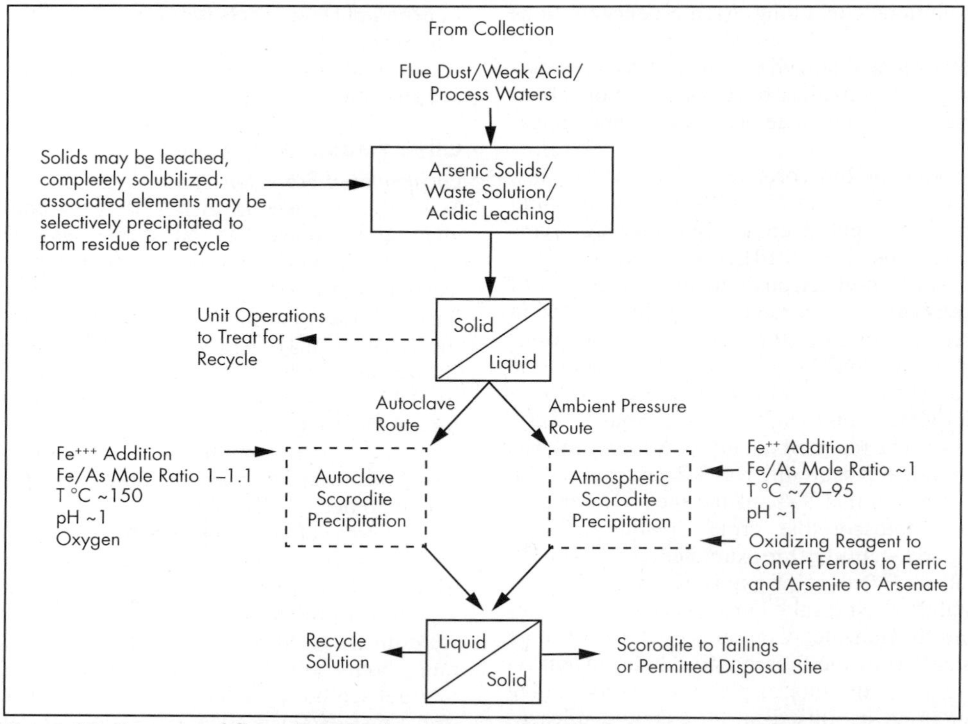

Figure 10 Conceptual flow sheet for forming scorodite by autoclave or ambient pressure, elevated temperature treatment

Arsenic Trisulfide

Arsenic(III) trisulfide, As_2S_3, has a very limited solubility in water, for example, <1 mg/L at pH <4, and is relatively stable under anoxic and reducing conditions. It contains a very high arsenic content, ~60%, which makes it a desirable storage product. However, it is susceptible to atmospheric and bacterial oxidation. Therefore, it cannot be stored in landfill disposal sites. An example of the production of arsenic trisulfide is sulfidization of As(III) solutions practiced at the Saganoseki copper smelter; the arsenic-bearing solution is treated by lime neutralization and NaHS to form arsenic trisulfide. The arsenic trisulfide is then stabilized by treating it in a non-oxidative autoclave at 200°C, 20 atm, to polymerize and densify the product. The product is compact, contains ~60% As and 1% H_2O, and is stored in concrete (Nazari et al. 2017; Valenzuela et al. 2006). Another example for a treatment that produces arsenic trisulfide is described by Gabb and Davies (1999) at the Kennecott Utah smelter. Leach solutions are treated to control the distribution of copper, arsenic, and cadmium. Copper is first selectively precipitated by pH and E_H control (at 40°–60°C) into a residue solid that is recycled to the smelter. Arsenic and cadmium are then precipitated as sulfides by addition of either H_2S gas or NaHS. The solids are recovered and are routed to their EPA-permitted hazardous waste disposal facility.

Plaques (Huisman et al. 2011) has developed and patented the THIOTEQ process for forming arsenic trisulfide. The process has two stages, one chemical and the other biological. Bisulfide (HS^-) is produced in a biological reactor external to the chemical reactor. Elemental sulfur and ethanol are fed to the biological reactor to form bisulfide. The HS^- solution is fed to the chemical reactor at pH 1–2, 60°–90°C, to precipitate As(III) as orpiment (As_2S_3). Very acidic solutions in the biological reactor are desirable; for example, pH <3 is required to attain a final arsenic concentration of <0.2 mg/L. Orpiment must be stored in suitable permitted repositories. The authors state that the THIOTEQ technology is used at more than 10 industrial plants for the reduction of sulfur compounds. The preferred application is to solutions containing high arsenic concentrations but <5 g/L.

Arsenate Phosphate Hydroxyapatites

Formation and long-term stability of arsenate phosphate hydroxyapatites [APHAP; $Ca10(As_xP_yO_4)_6(OH)_2$] have been investigated by McCloskey et al. (2008). These compounds were investigated because apatite compounds are very stable in the presence of carbon dioxide in air. The studies demonstrated excellent stability for arsenate–apatite containing a PO_4/AsO_4 molar ratio greater than 7; for example, the APHAP compound showed an arsenic release using the water-based EPA Method 1312. The aging solubility results were <1.4 µg/L and <0.8 µg/L (at pH 12.3) after 4 years and 8 years, respectively.

Symplesite

McCloskey and Twidwell (2008) designed and conducted a full-scale operation using a two-stage treatment system at an industrial site in Emeryville, California, United States, to lower As(V) concentrations from 100 mg/L to <25 µg/L. The plant was operated for a four-year period to clean contaminated water to less than the drinking water standard. The first stage of treatment was precipitation of symplesite

[$Fe_3(AsO_4)_2$:$8H_2O$] by the addition of lime and ferrous sulfate (at pH at 7). The arsenic was lowered to <6 mg/L. The resulting water was then treated by Fh precipitation using hydrogen peroxide as the oxidant for Fe(II) conversion to Fe(III) at an Fe/As molar ratio of 4. The residues from the two stages were mixed, dried, and sent to a permitted disposal facility. Symplesite does not pass the TCLP test requirement and must therefore be subjected to stabilization as required by the RCRA–LDR regulations.

Yukonite

Bluteau et al. (2009) and Gomez et al. (2010) have proposed that the mineral yukonite [$Ca_2Fe_3(AsO_4)_4(OH)\cdot 12H_2O$] likely forms from poorly crystalline scorodite with aging time, especially at pH levels of 7 and higher. Yukonite has been identified in mine waste and cyanidation tailings, and near neutral pH in natural environments (Drahota and Filippi 2009). Stability studies have shown positive results, such as an arsenic solubility of <5 mg/L at pH 5–10 in an oxidizing, gypsum-containing environment for 66 days. Values were pH 5, 3.1 mg/L; pH 7, 8, 10 were 0.4, 0.45, 1.12 mg/L, respectively (Bohan et al. 2014). Jia and Demopoulos (2008) demonstrated that a poorly crystalline ferric arsenate (Fe(III)/As(V) molar ratio of 2), formed by lime neutralization at 22°C, pH 8, does, indeed, transform to yukonite [$Ca_2Fe_3\ (AsO_4)_4(OH)\cdot 12H_2O$] when aged at 75°C for 7 weeks.

Tooeleite

Tooeleite ($Fe_6(AsO_3)_4SO_4(OH)_4\cdot 4H_2O$) is a ferric arsenite oxyhydroxy-sulfate hydrate nano-crystalline compound found in waste at the U.S. Mine Gold Hill in Tooele County, Utah (Opio 2013). Tooeleite has been found in acid mine drainage area and is thought to be formed via bacterial oxidation of Fe(II) (Morin et al. 2007; Egal et al. 2009). Tooeleite is similar to scorodite and is known as the trivalent arsenic form of scorodite (Nazari et al. 2017). It has been proposed to be an As(III) storage compound (Nishimura and Robins 2008). However, Raghav et al. (2013) showed appreciable leachability in the TCLP test (which is conducted at a nominal pH ~5): "tooeleite and silica amended tooeleite often was as least an order of magnitude higher than the TC (toxicity concentration)." Opio (2013) noted that preparation of tooeleite can be formed at ambient temperature at pH values in the range 2–3.5 but is rapidly converted to poorly crystalline ferric arsenite at pH >4. Calcination at 600°C produced a ferric arsenate calcine with a TCLP solubility of <5 mg/L. Synthesis at 95°C showed no improvement for arsenic extraction in TCLP solutions. Their conclusion was "the resultant precipitation of tooeleite from an As(III)-bearing weak acid and calcination of the resultant precipitate may offer a new process for As(III) fixation from copper smelter weak acid effluents." Long-term stability testing was not reported. Choi and Olvera (2017) suggests that Yukonite forms from Fh using conditions pH 1.8–4.5, initial As(III) concentration >0.75 g/L, Fe/As mole ratio of 0.8–2, and ambient temperature. The authors report that they tested the short-term stability of tooeleite in TCLP solutions as a function of pH (1.8–9). They present a figure but do not provide the numerical values for the leach results except in the graph. The arsenic concentration values for pH values 1.8–4.5 appear to be greater than 5 mg/L and at pH 5 and greater show hundreds of milligrams per liter of arsenic.

CONCLUDING REMARKS

Mineral processing and extractive metallurgical operations have created and are creating appreciable arsenic-bearing wastewater and waste solid products that have to be handled, treated for recycle, or treated for environmentally safe disposal. This is particularly true as the base metals and precious metals processing industries are faced with the necessity of treating more complex arsenic-bearing mineral ore bodies. As a result of the need to treat complex ores, and the need to meet present and future environmental regulations, research and operational emphases are presently being placed on the ways to produce stable arsenic-bearing residual products that can be safely stored in appropriate impoundments. These efforts must be continued.

REFERENCES

Adams, M.D. 2016. *Gold Ore Processing: Project Development and Operations*, 2nd ed. New York: Elsevier.

Alvear, G.R., Hunt, S.P., and Zhang, B. 2006. Copper IsaSmelt: Dealing with impurities. In *Proceedings of the Sohn International Symposium Advanced Processing of Metals and Materials*. Vol. 8. Edited by F. Kongoli and R.G. Reddy. Warrendale, PA: The Minerals, Metals & Materials Society.

Anderson, C.G. 2016. Alkaline sulfide gold leaching kinetics. *Miner. Eng.* 92:248–256.

Anderson, C.G., and Nordwick, S.M. 1995. Pretreatment using alkaline sulfide leaching and nitrogen species catalyzed pressure oxidation on a refractory gold concentrate. In *EPD Congress 1996*. Edited by G.W. Warren. Warrendale, PA: The Minerals, Metals & Materials Society. pp. 323–341.

Anderson, C.G., Nordwick, S.M., and Krys, L.E. 1991. Processing of antimony at the Sunshine mine. In *Residues and Effluents: Processing and Environmental Considerations*. Edited by R.G. Reddy, W.P. Imrie, and P.B. Queneau. Warrendale, PA: The Minerals, Metals & Materials Society. pp. 349–366.

Balaz, P., and Dutkova, E. 2009. Fine milling in applied mechanochemistriy. *Miner. Eng.* 22(7):681–694.

Blanchard, P.E.R., Van Loon, L.L., Reid, J.W., Cutler, J.N., Rowson, J., Hughes, K.A., Brown, C.B., Mahoney, J.J., Xu, L., Bohan, M., and Demopoulos, G.P. 2017. Investigating arsenic speciation in the JEB Tailings Management facility at McClean Lake, Saskatchewan using X-ray absorption spectroscopy. *Chem. Geol.* 466:617–625.

Bluteau, M.-C. 2004. The incongruent dissolution of scorodite-solubility, kinetics and mechanisms. M.Eng. thesis, McGill University, Quebec.

Bluteau, M.-C., and Demopoulos, G.P. 2007. The incongruent dissolution of scorodite-Solubility, kinetics and mechanism. *Hydrometallurgy* 87:163–177.

Bluteau, M., Becze, L., and Demopoulos, G.P. 2009. The dissolution of scorodite in gypsum-saturated waters: Evidence of Ca-Fe-AsO_4 mineral formation and its impact on arsenic retention. *Hydrometallurgy* 97:221–227.

Bohan, M., Mahoney, J., and Demopoulos, G.P. 2014. *The Synthesis, Characterization, and Stability of Yukonite: Implications in Arsenic Mobility*. Internal report. Montreal, QC: McGill University.

Brininstool, M. 2017. Arsenic. In *Mineral Commodity Summaries 2017*. Reston, VA: U.S. Geological Survey.

Caetano, M.L., Ciminelli, V.S.T., Rocha, S.D.F. Spitale, M.C., and Caldeira, C.L. 2009. Batch and continuous precipitation of scorodite from dilute industrial solutions. *Hydrometallurgy* 95(1-2):44–52.

Chatain, V., and Bayard, R. 2005. Effect of indigenous bacterial activity on arsenic mobilization under anaerobic conditions. *Environ. Int.* 31(2):221–226.

Chatain, V., and Sanchez, F. 2005. Effect of experimentally induced reducing conditions on the mobility of arsenic from a mining soil. *J. Hazard. Mater*. B122:119–128.

Chimonyo, W., Corin, K.C., Wise, J.G., and O'Connor, C.T. 2017. Redox potential control during flotation of a sulphide mineral ore. *Miner. Eng.* 110:57–64.

Choi, Y., and Olvera, O. 2017. Enhanced leaching of arsenopyrite concentrates as a pretreatment for gold recovery from refractory ores. Presented at the ALTA Gold-PM Conference, May 25–26.

Cornell, R.M., and Schwertmann, U. 2003. *The Iron Oxides, Structure, Properties, Reactions, Occurrence and Uses*, 2nd ed. New York: VCH.

De Klerk, R.J., Jia, Y., Daenzer, R., Gomez, M.A., and Demopoulos, G.P. 2012. Continuous circuit co-precipitation of arsenic (V) with ferric iron by lime neutralization: Process parameter effects on arsenic removal and precipitate quality. *Hydrometallurgy* 111:65–72.

Demopoulos, G.P. 2005. On the preparation and stability of scorodite. In *Proceedings of Arsenic Metallurgy*. Edited by R.G. Reddy and V. Ramachandran. Warrendale, PA: The Minerals, Metals & Materials Society. pp. 25–50.

Demopoulos, G.P. 2008. Advanced hydrometallurgical technology for the immobilization of arsenic. Presented at the Arsenic Stabilization Workshop, Santiago, Chile, May.

Demopoulos, G.P. 2009. Aqueous precipitation and crystallization for the production of particulate solids with desired properties. *Hydrometallurgy* 96(3):199–214.

Demopoulos, G.P. 2014. Arsenic immobilization research advances: Past, present and future. Presented at the 53rd Annual Conference of Metallurgists, Vancouver, BC, September 28–October 1.

Doerfelt, C., Feldmann, T., Roy, R., and Demopoulos, G.P. 2016. Stability of arsenate-bearing Fe(III)/Al(III) coprecipitates in the presence of sulfide as reducing agent under anoxic conditions. *Chemosphere* 151:318–323.

Drahota, P., and Filippi, M. 2009. Secondary arsenic minerals in the environment: A review. *Environ. Int.* 35:1243–1255.

Dunne, R. 2005. Flotation of gold and gold-bearing ores. In *Proceedings of the Developments in Mineral Processing—Advances in Gold Ore Processing*. Edited by M.D. Adams. New York: Elsevier pp. 309–344.

Egal, M.C., Casiot, G., Morin, M., Parmentier, O., Gruneel, S., Lebrun, F., and Elbaz-Poulichet 2009. Kinetic control on the formation of tooeleite, schweretmannite and jarosite by acidithioibacillus ferrooxidans strains in an As(III)-rich acid mine water. *Chem. Geol.* 265:432–441.

EPA (U.S. Environmental Protection Agency). 1992. Toxicity characteristic leaching procedure (TCLP). www.epa.gov/osw/hazard/testmethods/sw846/pdfs/1311.pdf. Accessed December 15, 2017.

Filippou, D., and Demopoulos, G.P. 1997. Arsenic immobilization by controlled scorodite precipitation. *JOM* 49(12):52–55.

Filippou, D., St-Germain, P., and Grammatikopoulos, T. 2007. Recovery of metal values from copper-arsenic minerals and other related resources. *Miner. Process. Extr. Metall. Rev*. 28(4):247–298.

Font, G., Alvear, G., Moyano, A., and Caballero, C. 2005. Fractional distribution of arsenic in the Teniente continuous converting process. In *Proceedings of Arsenic Metallurgy*. Edited by R.G. Reddy and V. Ramachadran. Warrendale, PA: The Minerals, Metals & Materials Society. pp. 195–205.

Ford, R.G. 2002. Rates of hydrous ferric oxide crystallization and influence of co-precipitated arsenate. *Environ. Sci. Technol.* 36:2459–2463.

Fornasiero, D., Fullston, D., Li, C., and Ralston, J. 2001. Separation of enargite and tennantite from non-arsenic copper sulfide minerals by selective oxidation or dissolution. *Int. J. Miner. Process.* 61:109–119.

Fujita, T., Shibata, E., and Nakamura, T. 2010. Solubility of scorodite synthetized by oxidation of ferrous ions. In *Proceedings of Copper 2010, Hamburg, Germany*. Vol. 7. Claustal-Zellerfeld, Germany: GDMB. pp. 2923–2934.

Fujita, T., Fujeda, S., Shinoda, K., and Suzuki, S. 2012. Environmental leaching characteristics of scorodite synthesized with Fe(II) iron. *Hydrometallurgy* 111-112:87–102.

Gabb, P.J., and Davies, A.I. 1999. The industrial separation of copper and arsenic as sulfides. *JOM* (September):8–9.

Gomez, M.A., Becze, L., Blyth, R.I.R., Cutler, J.N., and Demopoulos, G.P. 2010. Molecular and structural investigation of yukonite (synthetic and natural) and its relation to Arseniosiderite. *Geochim. Comsmochim. Acta* 74(2010):5835–5851.

Gomez, M.A., Becze, L., Cutler, J.N., and Demopoulos, G.P. 2011a. Hydrothermal reaction chemistry and characterization of ferric arsenate phases precipitated from Fe2(SO4)3–As2O5–H2SO4 solutions. *Hydrometallurgy* 107:74–90.

Gomez, M.A., Becze, L., Celikin, M., and Demopoulos, G.P. 2011b. The effect of copper on the precipitation of scorodite (FeAsO4·2H2O) under hydrothermal conditions: Evidence for a hydrated copper containing ferric arsenate sulfate-short lived intermediate. *J. Colloid Interface Sci.* 360:508–518.

Gonzalez-Contreras, P.A. 2012. Bioscorodite: Biological crystallization of scorodite for arsenic removal. Doctoral thesis, Wageningen University, Wageningen, Netherlands.

Gonzalez-Contreras, P.A. 2014. The biological method to immobilize arsenic bearing waste as scorodite. In *Proceedings of the Conference of Metallurgists*. Westmount, QC: Canadian Institute of Mining, Metallurgy and Petroleum.

Gonzalez-Contreras, P.A., Weijma, J., and Buisman, J.N. 2012. Continuous bioscorodite crystallization in CSTRs for arsenic removal and disposal. *Water Res.* 46:5883–5892.

Harris, B. 2000. The removal and stabilization of arsenic from aqueous process solutions: Past, present, future. In *Minor Elements 2000*. Edited by C. Young. Littleton, CO: SME. pp. 3–21.

Harris, B. 2001. The removal and arsenic from process solutions: Theory and industrial practice. In *Proceedings of the Fifth International Symposium on Hydrometallurgy*. Edited by C. Young, A. Alfantazi, C. Anderson, A. James, D. Dreisinger, and B. Harris. Warrendale, PA: The Minerals, Metals & Materials Society. pp. 1889–1902.

Hiskey, J.B. 2014. Innovative strategies for copper hydrometallurgy. In *Mineral Processing and Extractive Metallurgy: 100 Years of Innovation*. Edited by C.G. Anderson, R.C. Dunne, and J.L. Uhrie. Englewood, CO: SME. pp. 347–360.

Hohn, J.W. 2005. Modified ferrihydrite for enhanced removal of arsenic from mine wastewater. M.Sc. thesis, Department of Metallurgical/Materials Engineering, Montana Tech of the University of Montana, Butte, MT.

Huang, H.H. 2016. Stability Calculations Software. Montana Tech, Department of Metallurgy and Materials, Butte, MT.

Huch, R.O. 1994. Method for achieving enhanced copper-containing mineral concentrate grade by oxidation and flotation. U.S. Patent 5,295,585.

Huisman, J., Weghuis, M.O., and Gonzalez, P. 2011. Biotechnology-based processes for arsenic removal. Presented at Clean Mining 2011, Ninth International Conference on Clean Technologies for the Mining Industry, April 10–12, Santiago, Chile.

Ichimura, R., Tateiwa, H., Almendares, C., and Sanchez, G. 2007. Arsenic immobilization of Teniente furnace dust. In *Proceedings of Cu 2007*. Vol. 6, Sustainable Development, HS&E, and Recycling. Edited by D. Rodier and W. Adams. Montreal, QC: Canadian Institute of Mining, Metallurgy and Petroleum. pp. 275–288.

iMining. 2017. High-arsenic copper concentrates. *International Mining Newsletter*. https://imining.com/2016/02/23/high-arsenic-copper-concentrates. Accessed December 7, 2017.

Jain, A., and Loeppert, R.H. 2000. Effect of competing anions on the adsorption of arsenate and arsenite by ferrihydrite. *J. Environ. Qual.* 29(5):1422–1430.

Jambor, J., and Dutrizac, J.E. 1998. Occurrence and constitution of natural and synthetic ferrihydrite, a widespread iron oxyhydroxide. *Chem. Rev.* 98:2549–2585.

Jia, Y., and Demopoulos, G.P. 2008. Co-precipitation of arsenate with iron(III) in aqueous sulfate media: Effect of time, lime as base and co-ions on arsenic retention. *Water Res.* 42(93):661–668.

Katsarock, L. 2011. Synthesis and application of hydroapatite and fluorapatite to scorodite encapsulation. M.Sc. thesis, McGill University, Montreal, QC.

Kocar, B., Borch, T., and Fendorf, S. 2010. Arsenic repartitioning during biogenic sulfidization and transformation of ferrihydrite. *Geochim. Cosmochim. Acta* 74:980–994.

Krause, E., and Ettel, V.A. 1989. Solubility and stabilities of ferric arsenate compounds. *Hydrometallurgy* 89(22):311–337.

Lagno, F., Rochab, S.D.F., Chryssoulisc, S., and Demopoulos, G.P. 2010. Scorodite encapsulation by controlled deposition of aluminum phosphate coatings. *J. Hazard. Mater*. 181:526–534.

Lane, D.J., Cook, N.J., Grano, S.R., and Ehrig, K. 2016. Selective leaching of penalty elements from copper concentrates: A review. *Miner. Eng*. 98:110–121.

Leetmaa, K., Guo, F., Becze, L., Gomez, M.A., and Demopoulos, G.P. 2016. Stabilization of iron arsenate solids by encapsulation with aluminum hydroxyl gels. *J. Chem. Technol. Biotechnol*. 91:408–415.

Long, G., Peng, Y., and Bradshaw, D. 2012. A review of copper-arsenic mineral removal from copper concentrates. *Miner. Eng*. 36-38:127–138.

Marsden, J.O., and House, C.I. 2006. *The Chemistry of Gold Extraction*, 2nd ed. Littleton, CO: SME. pp. 258–260.

Marsden, J.O., and Sass, S.A. 2014. Innovations in gold and silver extraction and recovery. *Mineral Processing and Extractive Metallurgy: 100 Years of Innovation*. Edited by C.G. Anderson, R.C. Dunne, and J.L. Uhrie. Englewood, CO: SME. pp. 361–388.

McCloskey, J., and Twidwell, L.G. 2008. *Removal of Arsenic from Sherwin-Williams Ground Water in Emeryville, CA*. Internal Report. Butte, MT: MSE.

McCloskey, J., Twidwell, L.G., and Lee, M. 2008. *Arsenic Removal from Mine and Process Waters by Lime Phosphate Precipitation: Pilot Scale Demonstration*. Mine Waste Technology Program. Report MWTP 305. Butte, MT: MSE.

Miller, P., and Brown, A.R.G. 2005. Bacterial oxidation of refractory gold concentrates. In: *Developments in Mineral Processing: Advances in Gold Ore Processing*. Edited by M.D. Adams. New York: Elsevier. pp. 371–402.

Morin, G., Rousse, G., and Elkaim, E. 2007. Crystal structure of tooeleite, $Fe_6(AsO_3)_4SO_4(OH)_4 \cdot 4H_2O$, a new iron arsenite oxyhydroxysulfate mineral relevant to acid mine drainage. *Am. Mineral.* 92:193–197.

Nazari, A.M., Radzinske, R., and Gahreman, A. 2017. Review of arsenic metallurgy: Treatment of arsenical minerals and the immobilization of arsenic. *Hydrometallurgy* 174:258–281.

Nishimura, T., and Robins, R.G. 2008. Confirmation that tooeleite is a ferric arsenite sulfate hydrate, and is relevant to arsenic stabilization. *Miner. Eng*. 21(4):246–251.

Nishimura, T., and Umetsu, Y. 1985. Removal of arsenic from wastewater by addition of calcium hydroxide and stabilization of arsenic-bearing precipitate by calcination. *CIM J.* 78(878):75.

Okibe, N., Koga, M., Sasaki, K., Hirajima, T., Heguri, S., and Asano, S. 2013. Simultaneous oxidation and immobilization of arsenite from refinery waste water by thermoacidophillic iron-oxidizing archaeon, Acidianus brierleyi. *Miner. Eng*. 48:26–134.

Okibe, N., Koga, M., Morishita, S., Tanaka, M., Heguri, S., and Asano, S. 2014. Microbial formation of crystalline scorodite for treatment of As(III)-bearing copper refinery process solution using acidianus brierleyi. *Hydrometallurgy* 143:34–41.

Opio, F.K. 2013. Investigation of Fe(III)-As(III) phases and their potential for arsenic disposal. Doctoral thesis, Queen's University, Kingston, ON.

Paktunc, D., and Bruggeman, K. 2010. Solubility of nanocrystalline scorodite and amorphous ferric arsenate: Implications for stabilization of arsenic in mine wastes. *Appl. Geochem*. 25:674–683.

Paktunc, D., Dutrizac, J.E., and Gertsmann, V. 2008. Synthetic and phase transformations involving scorodite, ferric arsenate and arsenical ferrihydrite: Implications for arsenic mobility. *Geochim. Cosmochim. Acta* 72:249–272.

Raghav, M., Shan, J., Saez, A.E., and Ela, W.P. 2013. Scoping candidate minerals for stabilization of arsenic-bearing solid residues. *J. Hazard. Mater*. 263:525–532.

Riveros, P.A., Dutrizac, J.E., and Spenser, P. 2001. Arsenic disposal practices in the metallurgical industry. *Can. Metall. Q*. 40(4):395–420.

Robins, R.G., and Jayaweera, L.D. 1992. Arsenic in gold processing. *Miner. Process. Extr. Metall. Rev.* 9(1-4):255–271.

Robins, R.G., and Tozawa, K. 1982. Arsenic removal from gold processing waste waters: The potential ineffectiveness of lime. *CIM Bull.* 75(4):171–174.

Rohner, P., Turner, D., Walker, D., and MacDonald, L. 2016. The Toowong Process: Developing a viable solution for copper's dirty problem. In *Proceedings of ALTA 2016 Nickel-Cobalt-Copper*. Blackburn North, Victoria: ALTA Metallurgical Services.

Rosengrant, L., and Fargo, L. 1990. Final best demonstrated available technology (BDAT) background document for K031, K084, K101, K102, As wastes (D004), Se wastes (D010), and P and U wastes containing As and Se listing constituents. EPA/530/SW-90/059A, Washington, DC: U.S. Environmental Protection Agency.

Ruonala, M., Leppinen, J., and Miettinen, V. 2011. Method for removing arsenic as scorodite. Outotec Patent EP 2398739A1.

Safarzadeh, M.S., and Miller, J.D. 2016. The pyrometallurgy of enargite: A literature guide. *Int. J. Miner. Process.* 157:103–110.

Safarzadeh, M.S., Moats, M.S., and Miller, J.D. 2014a. Recent trends in the processing of enargite concentrates. *Miner. Process. Extr. Metall. Rev.* 35:283–367.

Safarzadeh, M.S., Moats, M.S., and Miller, J.D. 2014b. An update: Recent trends in the processing of enargite concentrates. *Miner. Process. Extr. Metall. Rev.* 35(6):390–422.

Salokannel, A., Nevatalo, L.M., Martikainen, M., Meer, T., and McCulloch, J. 2013. Technology selection for removal of contaminants from an industrial wastewater. Presented at Metallurgical Plant Design and Operating Strategies, Perth, Western Australia.

Salomon-de-Friedberg, H., Mayhew, K., Lossin, A., and Omaynikova, V. 2017. Hydrometallurgical considerations in processing arsenic-rich copper concentrates. Presented at ALTA Conference, Perth, Western Australia, May 23–25.

Schlesinger, M.E., King, M.J., Sole, K.C., and Davenport, W.G. 2011. *Extractive Metallurgy of Copper*, 5th ed. New York: Elsevier.

Senior, G.D., Guy, P.J., and Bruckard, W.J. 2006. The selective flotation of enargite from other copper minerals: A single mineral study in relation to beneficiation of the Tampakan deposit in the Philippines. *Int. J. Miner. Process.* 81:15–26.

Sherman, D.M., and Randall, S.R. 2003. Surface complexation of arsenic(V) to iron(III) (hydro)oxides: Structural mechanism from ab initio molecular geometries and EXAFS spectroscopy. *Geochim. Cosmochim. Acta* 67(22):4223–4230.

SME (Society for Mining, Metallurgy & Exploration). 2015. The Role of Arsenic in the Mining Industry. Englewood, CO: SME. www.smenet.org/SME/media/Government/The-Role-of-Arsenic-in-the-Mining-Industry.pdf. Accessed April 2015.

Smith, L.K., and Bruckard, W.J. 2007. The separation of arsenic from copper in a NorthParkes copper-gold ore using controlled-potential flotation. *Int. J. Miner. Process.* 84:15–24.

Strauss, J.A., Yahorava, V., and Gomez, M.A. 2017. Pressure oxidation in gold circuits: Basic ferric arsenate sulphate and basic ferric sulphate behavior in downstream processing. In *Proceedings of the World Gold and Nickel Cobalt, Conference of Metallurgists*. Westmount, QC: Canadian Institute of Mining, Metallurgy and Petroleum.

Surpunt, S., and Hasegawa, N. 2003. Distribution behavior of arsenic, antimony and bismuth in the smelting stage of the Mitsubishi process. *Proceedings of the Yazawa International Symposium*. Vol. I. Edited by F. Kongoli, K. Itagaki, C. Yamauchi, and H. Sohn. Warrendale, PA: The Minerals, Metals & Materials Society. pp 375–381.

Swash, P.M., and Monhemius, A.J. 1994. Hydrothermal precipitation from aqueous solutions containing iron(III), arsenate and sulphate. *Hydrometallurgy* 1994:177–190.

Tayebi-Khorami, M., Manlapig, E., Forbes, E., Bradshaw, D., and Edraki, M. 2017. Selective flotation of enargite from copper sulphides in Tampakan deposit. *Miner. Eng.* 112:1–10.

Twidwell, L.G. 2018. Treatment of arsenic-bearing minerals and fixation of recovered arsenic products: A review (presentation). School of Mines, Metallurgy/Materials Engineering Department, Montana Tech of the University of Montana, Butte, MT.

Twidwell, L.G., and McCloskey, J.W. 2011. Removing arsenic from aqueous solution long-term product storage. *JOM* 63(8):94–110.

Twidwell, L.G., Hohn, J.A., and Robins, R.G. 2007. *Modified Ferrihydrite for Enhanced Removal of Arsenic from Mine Wastewater*. Mine Waste Technology Program, Activity IV, Project 39, MSE-TA, Butte, MT.

Valenzuela, A. 2000. Arsenic management in the metallurgical industry. Doctoral thesis. National Library of Canada, Acquisitions and Bibliographic Services.

Valenzuela, A., Balladares, E., Cordero, D., and Sanchez, M. 2006. Arsenic management in the metallurgical industry: The Chilean experience. In *Proceedings of the Sohn International Symposium: Advanced Processing of Metals and Materials*. Vol. 9: Legal, Management and Environmental Issues. Warrendale, PA: The Minerals, Metals & Materials Society. pp. 407–422.

Violante, A.M., Ricciardella, M., and Pigna, M. 2003. Adsorption of heavy metals on mixed Fe-Al oxides in the absence or presence of organic ligands. *Water Air Soil Pollut.* 45(1):289–306.

Wang, Q., Nishimura, T., and Umetsu, Y. 2000. Oxidative precipitation of arsenic removal in effluent treatment. In *Minor Elements 2000*. Edited by C. Young. Littleton, CO: SME. pp. 39–52.

Wassink, B., Dreisinger, D., West-Sells, P., and Fisher, N. 2005. Leaching of a gold ore using the hydrogen sulfide-bisulfide-sulfur system. In *Proceedings of the Treatment of Gold Ores. First International Symposium.* Edited by G. Deschenes, D. Hodouin, and L. Lorenzen. Montreal, QC: Canadian Institute of Mining, Metallurgy and Petroleum.

CHAPTER 12.5

Beryllium

Edgar E. Vidal

Beryllium is one of 12 "strategic and critical materials" as defined by section 12(1) of the Stock Piling Act of the National Defense Stockpile program that is recommended to be stocked in 2015 at all times. A strategic and critical material is defined as: "(i) would be needed to supply the military, industrial, and essential civilian needs of the United States during a national emergency and (ii) are not found or produced in the United States in sufficient quantities to meet such need" (Under Secretary of Defense for Acquisition, Technology and Logistics 2015).

Beryllium metal and alloys are used in many military and aerospace applications. Beryllium metal has the highest specific stiffness of any structural metal and incredibly high thermal conductivity and heat capacity. The United States continues to be the largest extractor, producer, fabricator, and consumer of beryllium in the world. There is only one producer in the United States that is fully integrated. Materion Corporation has the largest reserves of beryllium ore in the form bertrandite, controlling roughly 65% of the reserves and covering 85% of the world's needs. A yearly extraction of 200–300 t (metric tons) of beryllium ore has been enough to cover the world's demand after the end of the Cold War (Jaskula 2016). Other producers in China, Russia, and Kazakhstan contribute to the world production of beryllium, mostly supplying the Asian markets.

Beryllium metal and its composites are mostly used in defense, space, and scientific applications. Beryllium alloys in the form of strip and bulk products (such as copper-beryllium [Cu-Be]) and beryllium oxide ceramics, however, are typically used in industrial and commercial applications. Figure 1 shows the distribution of beryllium use based on market application.

Beryllium compounds and alloys have been known to have detrimental health effects on some people. For example, beryllium metal, beryllium oxide, and beryllium-copper in the form of very small dust particles may sensitize an individual if inhaled and may lead to a condition known as chronic beryllium disease. Other compounds, such as solutions containing beryllium fluoride if in contact with the skin, may produce a condition known as acute beryllium disease. It is highly recommended that safety data sheets for beryllium-containing materials be reviewed carefully when handling.

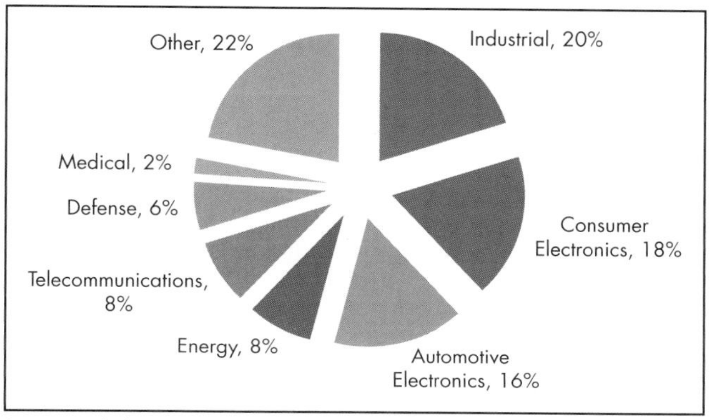

Source: Jaskula 2016

Figure 1 Beryllium metal, alloys, and composites distribution based on sales revenues

BERYLLIUM MINERALS AND ORES

The common sources for beryllium are bertrandite and beryl minerals. Beryl is a beryllium-aluminum-silicate containing roughly 67 wt % silica, 19 wt % alumina, and 14 wt % beryllium oxide. Examples of beryllium-rich beryl include emeralds and aquamarine. Economically minable grades of beryl ore typically contain 6–12 wt %. BeO. Beryl ore mining is not a primary operation in most cases, and rather, it is a secondary operation from the extraction of mica and lithium minerals. The concentration of beryl ores is typically done by hand sorting, as the amounts involved are not significant enough to justify large investment of capital equipment. Sources of beryl ores include Brazil, South Africa, and Mozambique (Jaskula 2016).

The other source of beryllium of high importance is bertrandite, which is a beryllium-silicate-hydroxide mineral with composition $Be_4Si_2O_7(OH)_2$. In the United States, the development of the large Spor Mountain bertrandite reserve in Utah decreased the amount of beryl required from international sources. The tuffs of rhyolitic nature in the Utah source

Edgar E. Vidal, Director of Business Development for Defense, Nuclear & Science, Materion Beryllium & Composites, Mayfield Heights, Ohio, USA contain 0.1–1.0 wt % of BeO. Depending on the market value of beryllium, the economically minable amounts can vary, beginning at 0.20–0.70 wt %. BeO. Despite the bertrandite ore grade being lower than that of beryl ore, the recovery of beryllium from these ores is less energy intensive (Montoya et al. 1962).

BENEFICIATION OF BERYLLIUM-CONTAINING ORES

Hand picking is currently the method of choice for concentrating beryl and bertrandite ores. The investment of capital equipment cannot be justified to concentrate these ores because of the relatively low demand for beryllium. Portable *beryl pickers* or *berylometers* have been made to aid in the location and selection of high beryllium-containing ores. The suspected beryllium-containing ore body is irradiated with a gamma source. The beryllium contained in the minerals has the natural isotope Be^9, which emits a neutron and forms the Be^8 isotope under gamma radiation. Within the same berylometer, there is a Geiger counter that quantifies the amount of neutrons emitted and correlates it with a precalibrated concentration of beryllium (Darwin and Buddery 1960).

The relatively low density of beryl and bertrandite makes them difficult to concentrate using density means versus quartz and other gangue minerals. The magnetic properties are also nondistinguishable from gangue minerals. The most promising beneficiation process is flotation. In the 1960s, the following process was defined (Havens 1963):

1. Phenacite, bertrandite, and beryl are separated from associated minerals, such as quartz, calcite, manganocalcite, fluorite, sericite, apatite, and micas, using fatty acid flotation.
2. Fluoride ions and polyphosphate ions depressants are added to prevent the entry of undesired minerals into the froth. Crushed ore in the form of an aqueous pulp is treated with a water-soluble fluoride, such as sodium fluoride, and a water-soluble polyphosphate, such as sodium hexametaphosphate with and without tetrasodium pyrophosphate.
3. A flocculating agent, such as alum or aluminum sulfate, is useful in flocculating the concentrate or froth for fast removal from the flotation cell.
4. After the conditioning time, a collector oil, such as oleic acid, tall oil fatty acid, or fish oil fatty acids, is added to the pulp. After an additional conditioning period, air is passed into the pulp, and the froth concentrate is recovered.
5. The residual pulp is reconditioned with sodium hexametaphosphate and fatty acid collector and froth floated again, recovering a second concentrate that may be treated separately.
6. The process is carried out at room temperature and atmospheric pressure conditions.

More recently, Bulatovic (1988) patented a process for the beneficiation of bertrandite ores. Figure 2 shows a general schematic of the following process:

1. The ore is ground to a P_{80} of –400 mesh (37 μm).
2. Sulfuric acid is added as a preconditioner, adjusting the pH between 5 and 5.5. After the pretreatment, the slurry is thickened to 65 wt % solids.

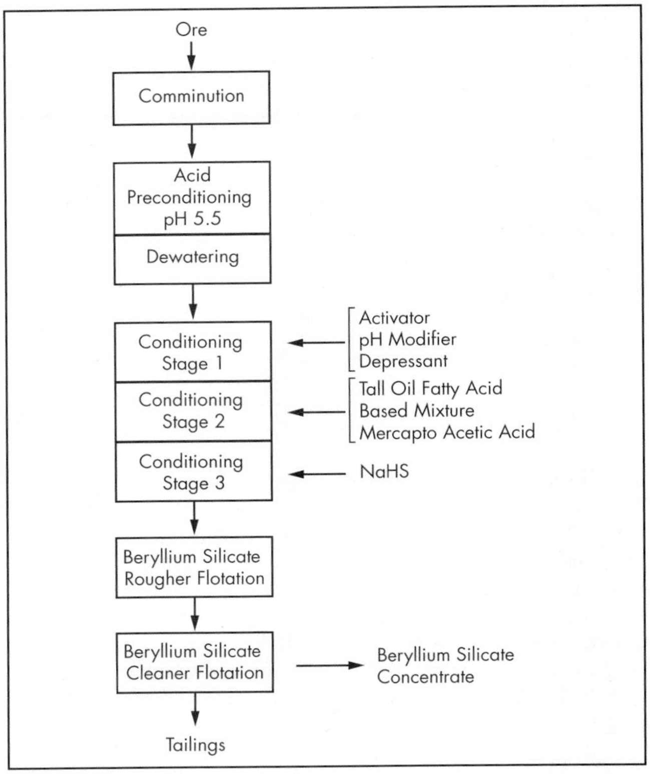

Source: Bulatovic 1988

Figure 2 Flotation process for the concentration of phenacite and bertrandite ores

3. The pretreated slurry is treated with sodium carbonate to adjust the pH to 9.5. Sodium fluoride is typically used as an activator, but sodium silico-hexafluoride can also be used.
4. Sodium hexametaphosphate and carboxymethyl cellulose are used as depressants for albite, mica, carbonates, fluorite, and siliceous gangues. The conditioning stage is followed by the addition of the collector mixture.
5. The collector mixture consists of cresylic acid, an aliphatic alcohol, and kerosene.
6. The conditioning is followed by a conventional rougher and cleaner flotation stages.

There has been a renewed interest in developing processes for the beneficiation of bertrandite, particularly in China. Some conditions are very similar, with variations in the chemicals used (Wei et al. 2009).

EXTRACTION OF BERYLLIUM

Many processes have been suggested for the extraction of elemental beryllium from ore concentrates (Walsh 2009). In this chapter, only the process currently in production and most economically viable is reviewed. Other important extraction methods include the electrolysis of beryllium-containing molten salts.

BERYLLIUM HYDROXIDE PRODUCTION

The current preferred method of extraction is the sulfate process for decomposing the ore and recovering the beryllium

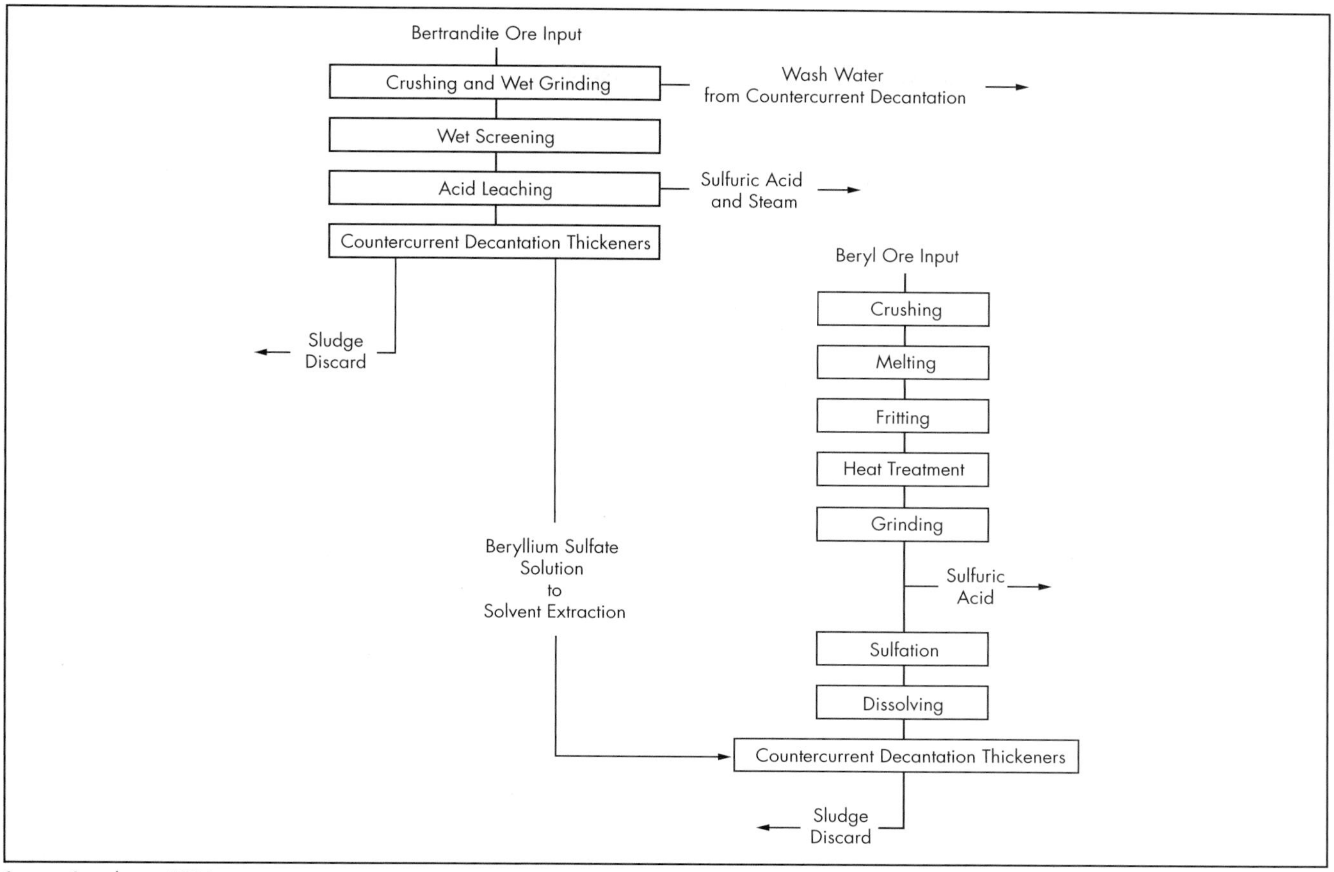

Source: Stonehouse 1986

Figure 3 Beryllium sulfate solution production from bertrandite and beryl concentrates

oxide component (Sawyer and Kjellgren 1935; Kjellgren 1946). In the sulfate process for beryllia production, water-soluble beryllium and aluminum sulfates are formed by reacting beryl or bertrandite with sulfuric acid at 400°C. The process starts by activating the beryl ore by melting and rapid quenching into a frit material using cold water. The energy consumption imposes an economic trade-off, forcing the use of high-grade beryl ore containing 6–10 wt % beryllium oxide. Bertrandite, conversely, does not require this frit-forming step because it reacts readily with the sulfuric acid solution. Silica is precipitated out as an insoluble compound. The sulfate reactions can be described as follows:

$$3BeO \cdot Al_2O_3 \cdot 6SiO_2 + 6H_2SO_4 = 3BeSO_4 + Al_2(SO_4)_3 + 6SiO_2 + 6H_2O \quad \textbf{(EQ 1)}$$

$$Be_4Si_2O_7(OH)_2 + 4H_2SO_4 = 4BeSO_4 + 2SiO_2 + 5H_2O \quad \textbf{(EQ 2)}$$

Figure 3 shows a flowchart of the process that uses both beryl and bertrandite ore concentrates down to the beryllium sulfate solution (Stonehouse 1986).

The slurry is passed through a series of countercurrent decantation vats for washing and recovering the soluble beryllium sulfate from the siliceous solids. The typical overflow leach solution contains roughly 12 g/L beryllium in approximately 0.5 M free sulfuric acid. The leach solution typically contains more than 90% of the beryllium content of the input concentrate (Stonehouse 1986).

The next steps consist of a series of purification processes to reject the impurities before beryllium hydroxide recovery. Several methods have been suggested for separating iron, aluminum, and other heavy metals from the beryllium sulfate solution, but the reaction with ammonium or potassium sulfate to form an alum is the process used nowadays.

The free sulfuric acid leach liquor is neutralized with aqueous ammonia to pH = 1, creating enough ammonium sulfate to form ammonium alum. By cooling below 15°C, the crystallization of ammonium alum, $Al_2(SO_4)_3 \cdot (NH_4)_2SO_4 \cdot 24H_2O$ is promoted. The crystals produced are removed by centrifugation. Organic chelating agents, such as the sodium salts of ethylenediaminetetraacetic acid (EDTA) and triethanolamine, are added to the centrifugate. The sequential additions of sodium hydroxide then yield the intermediate reactions:

$$BeSO_4 + 2NaOH = Na_2SO_4 + \alpha - Be(OH)_2 \quad \textbf{(EQ 3)}$$

$$\alpha - Be(OH)_2 + 2NaOH = Na_2BeO_2 + 2H_2O \quad \textbf{(EQ 4)}$$

The sodium beryllate solution is passed through a heat exchanger that hydrolyzes it and precipitates it into β-beryllium hydroxide:

$$Na_2BeO_2 + 2H_2O = 2NaOH + \beta - Be(OH)_2 \quad \textbf{(EQ 5)}$$

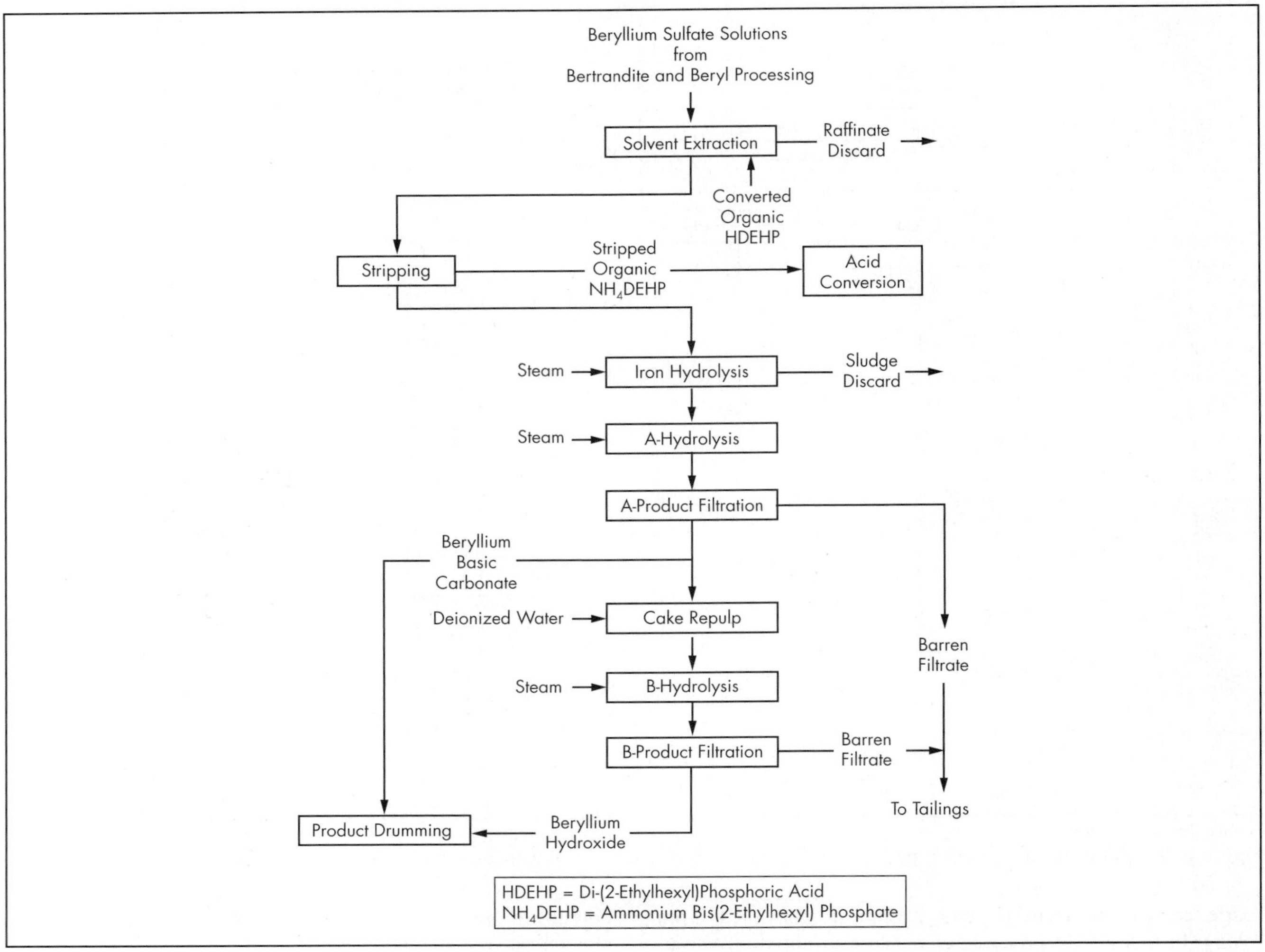

Source: Stonehouse 1986

Figure 4 Purification and precipitation of beryllium hydroxide from beryllium sulfate

The granular hydroxide produced is separated in continuous centrifuges, where it is washed and dewatered to less than 10 wt % free water.

Figure 4 shows the purification steps and precipitation of the beryllium hydroxide just described (Stonehouse 1986).

The original Sawyer–Kjellgren process just described was subsequently modified using solvent extraction technology with the sulfate leach solution. Foos and Maddox (1966) mention the recovery of beryllium from the organic extraction solvent by separation with aqueous ammonium carbonate. This ammonium beryllium carbonate solution is purified and boiled to precipitate beryllium basic carbonate. The beryllium hydroxide is then formed by a pressure hydrolysis step at 165°C. As before, the granular beryllium hydroxide is recovered by filtration.

Once the beryllium hydroxide has been produced, different processes exist to produce beryllium metal, copper-beryllium alloys, and beryllium oxide ceramics.

BERYLLIUM OXIDE

In a straightforward approach, beryllium oxide is produced from beryllium hydroxide directly by calcining. One drawback, however, is that the BeO powders contain elevated levels of fluorine or other impurities, which are detrimental to some of the high-density beryllium oxide ceramics. One alternative process proposed by Gadeau (1933) describes beryllium hydroxide preparation from beryl ore that is reacted with sodium fluorosilicate. The reaction forms a solution of sodium fluoroberyllate. The sodium fluoroberyllate is dried and mixed with sodium carbonate and calcined above 525°C. This second calcination product is leached in water to solubilize the sodium fluoride. A wet paste of beryllium oxide is obtained after solid–liquid separation and several water washings of the solid. Unfortunately, this wet paste of beryllium oxide retains elevated levels of fluoride and sodium, which are unsuitable for beryllium oxide ceramics.

Another commercial process for producing beryllium oxide involves a two-step process where the conversion of beryllium hydroxide to beryllium sulfate tetrahydrate is followed by the calcination of the hydrated sulfate to the oxide at temperatures higher than 1,100°C. This process produces SO_2, which requires secondary recovery processing to convert the acidic off-gas to a nonhazardous usable form. This process can be modified by using urea, ammonium oxalate, ammonium acetate, or ammonium sulfate. These chemicals are selected because of their ability to be completely removed during calcination. The addition of these substances is described by Schwenzfeier and Pomelee (1965).

The production of beryllium oxide makes possible the conversion of this material into usable ceramic materials, incorporation of beryllium into copper alloys, and preparation of metal matrix composites with beryllium metal.

BERYLLIUM METAL

In the past, the two commercial methods that were practiced for producing beryllium were based on converting the beryllium hydroxide into a chloride or a fluoride. Today, the preferred production process for beryllium metal extraction is the reduction of beryllium fluoride with magnesium (Stonehouse 1986). The production of beryllium fluoride from hydroxide takes place in two steps. First, the beryllium hydroxide is reacted with ammonium bifluoride, producing ammonium fluoroberyllate; then, the fluoroberyllate is thermally decomposed at elevated temperatures (>125°C) to produce beryllium fluoride and volatile ammonium fluoride:

$$Be(OH)_2 + 2NH_4 \cdot HF_2 = (NH_4)_2BeF_4 + 2H_2O \quad \textbf{(EQ 6)}$$

$$(NH_4)_2BeF_4 = BeF_2 + 2NH_4F \quad \textbf{(EQ 7)}$$

The beryllium fluoride produced is then mixed with magnesium metal, which is heated to at least the melting point of beryllium metal (1,286°C). The magnesiothermic reduction kinetics increase considerably above 850°C, after both the magnesium and the beryllium fluoride are molten.

$$BeF_2 + Mg = MgF_2 + Be \quad \textbf{(EQ 8)}$$

To enhance the yield of beryllium produced, an excess amount of beryllium fluoride helps in fragmenting the slag and releasing the entrapped beryllium globules. Once the temperature goes above the melting point of beryllium, fairly large pieces of beryllium coalesce.

After the beryllium metal has been cleaned and slag removed, it typically goes to a vacuum melting process to produce beryllium billets, which are then turned into chips and subsequently powder, which feeds the powder metallurgy process for making basic shapes or near-net shapes.

COPPER-BERYLLIUM ALLOYS

Copper-beryllium alloys are still one of the principle uses for beryllium, especially in a structural material. Copper-beryllium alloys with 0.008–5 wt % Be have been produced both by vacuum casting plus hot rolling and by vacuum casting and extrusion (London 1979). Copper and nickel alloys benefit from the addition of a few percentage points of beryllium, which significantly improves their mechanical properties.

Early in the 1900s, master alloys of copper-beryllium were produced by electrolysis methods. In 1939, Sawyer and Kjellgren replaced this process by the now more common carbothermic reduction of BeO in the presence of molten copper. An arc furnace is charged with copper and BeO, which reacts with carbon to reduce the BeO to beryllium metal, which then incorporates into the molten copper. The process operates in the temperature range from 1,800° to 2,100°C (Cribb and Ali 1994).

REFERENCES

Bulatovic, S. 1988. Beryllium flotation process. U.S. Patent 4,735,710.

Cribb, W.R., and Ali, S. 1994. Production of Cu-Be master from beryllium oxide. *Miner. Process. Extr. Metall. Rev.* 13:157–164.

Darwin, G.E., and Buddery, J.H. 1960. *Beryllium*. New York: Academic Press.

Foos, R.A., and Maddox, R.L. 1966. Process for producing basic beryllium material of high purity. U.S. Patent 3,259,456.

Gadeau, R.A. 1933. Process for the preparation of oxide of beryllium. U.S. Patent 1,925,920.

Havens, R. 1963. Flotation process for concentration of phenacite and bertrandite. U.S. Patent 3,078,997.

Jaskula, B.W. 2016. Beryllium. In *Mineral Commodity Summaries, 2016*. Reston, VA: U.S. Geological Survey.

Kjellgren, B.R.F. 1946. The production of beryllium oxide and beryllium copper. *J. Electrochem. Soc.* 89(1):247–261.

London, G.J. 1979. *Alloys and Composites, Beryllium Science and Technology*. Vol. 2. New York: Plenum Press. pp. 297–318.

Montoya, J.W., Havens, R., and Bridges, D.W. 1962. *Beryllium-Bearing Tuff from Spor Mountain, Utah: Its Chemical, Mineralogical, and Physical Properties*. Report of Investigations 6084. Washington, DC: U.S. Bureau of Mines.

Sawyer, C.B., and Kjellgren, B.R.F. 1935. Beryllium and aluminum salt production. Canadian Patent CA 354,080.

Sawyer, C.B., and Kjellgren, B.R.F. 1939. Production of alloys containing beryllium. U.S. Patent 2,176,906.

Schwenzfeier, C.W., and Pomelee, C.S. 1965. Production of high-purity sinterable beryllium oxide. U.S. Patent 3,172,728.

Stonehouse, A.J. 1986. The physics and chemistry of beryllium. *J. Vac. Sci. Technol. A* 4(3):1163–1170.

Under Secretary of Defense for Acquisition, Technology and Logistics. 2015. *Strategic and Critical Materials 2015 Report on Stockpile Requirements*. Ref ID 9-05A8E24. Washington, DC: U.S. Department of Defense.

Walsh, K.A. 2009. *Beryllium Chemistry and Processing*. Edited by D.L. Olson, E.E. Vidal, E. Dalder, A. Goldberg, and B. Mishra. Materials Park, OH: ASM International.

Wei, S., Zhiqiang, R., Yuehua, H., and Hongjun, H. 2009. Floating and collecting agent of bertrandite beryllium ores and application thereof. Chinese Patent CN101716559.

CHAPTER 12.6

Coal Preparation

Gerald H. Luttrell, Aaron Noble, and Rick Q. Honaker

Coal preparation involves the upgrading of run-of-mine (ROM) coals using low-cost solid–solid and solid–liquid separation processes. The processes remove undesirable impurities such as waste rock and water from carbonaceous material. For coal-fired power stations, these inorganic impurities reduce coal heating value, leave behind an undesirable ash residue, contribute to particulate and gaseous emissions, and increase transportation costs (Doherty 2006). The presence of impurities can also influence the suitability of coal for high-end uses such as the manufacture of metallurgical coke or the generation of petrochemicals and synthetic fuels. Unwanted surface moisture, which increases transportation costs and decreases heating value, can also create handling and freezing issues for industrial consumers. Consequently, all coal supply agreements impose strict limitations on the purity levels of purchased coal.

The methods and processes used to upgrade coal have changed dramatically during the past century (Carris 2007). In the early 1900s, manual sorting methods such as handpicking were used by the industry for quality improvement. This labor-intensive approach was soon made obsolete by mechanized processes that reduced misplacement and increased productivity. Many of these early facilities cleaned only the coarser particles and either recombined untreated fines back into the coal product or discarded the fines as waste. These historic periods were followed by many decades of technology development that allowed operators to partially or completely upgrade the entire size range of mined coals (Osborne 2012). Today's modern coal processing facilities, some of which handle thousands of short tons per hour (stph) of ROM coal, are as functionally complex as their counterparts in either the mineral beneficiation or chemical processing industries. This level of design sophistication is dictated by increasingly competitive markets that require plants to operate at very high levels of processing efficiency to remain profitable.

PROCESS FLOW SHEET

The push for enhanced efficiency has forced plant designers to develop flow sheets that incorporate several parallel processing circuits, often three or more, which are specifically configured to optimize the upgrading of each size fraction. The multiple circuits are needed to maintain efficiency since the processes employed in coal processing each have a limited range of applicability in terms of particle size. For example, Figure 1 provides a simplified flow sheet for a modern plant that includes four independent circuits for treating coarse (>10 mm), intermediate (10–1 mm), fine (1–0.15 mm) and ultrafine (<0.15 mm) coal. In the United States, the coarse and intermediate size fractions are typically treated using dense medium vessels and dense medium cyclones (DMCs), respectively. In some cases, the dense medium vessel can be eliminated by reducing the top size of the feed coal. The fine coal fraction, which cannot be handled by dense medium processes because of practical constraints associated with recovery of the circulating medium, are typically treated by water-based gravity separation processes that include water-only cyclones (WOCs), spirals, teeter-bed separators, or a combination of these unit operations. The ultrafine size fraction, which is too small for gravity separation processes, is usually upgraded by froth flotation. The entire ultrafine coal slurry may be treated by flotation, deslimed to remove a substantial portion of <45-µm solids prior to flotation, or discarded as waste without treatment (Bethell and Luttrell 2005). Dewatering of the coal products is typically performed using centrifugal basket-type dryers for the coarser factions and screen-bowl centrifuges for the finer fractions. Vacuum filters are used in a few plants but are difficult to justify unless a thermal dryer is available to ensure that moisture specifications can be met. Figure 2 illustrates the range of particle sizes that are typically treated by commercially available processes for the sizing, cleaning, and dewatering of coal feedstocks. Unfortunately, site-specific variations in the liberation characteristics and mineralogical makeup of different ROM coals preclude the adoption of a standard flow sheet that is appropriate for all commercial sites. Differences in circuit layouts are typically justified by designers based on both technical and financial considerations that attempt to balance the necessity to accommodate a

Gerald H. Luttrell, E. Morgan Massey Professor, Mining & Minerals Engineering, Virginia Tech, Blacksburg, Virginia, USA
Aaron Noble, Associate Professor, Mining & Minerals Engineering, Virginia Tech, Blacksburg, Virginia, USA
Rick Q. Honaker, Professor, Mining Engineering, University of Kentucky, Lexington, Kentucky, USA

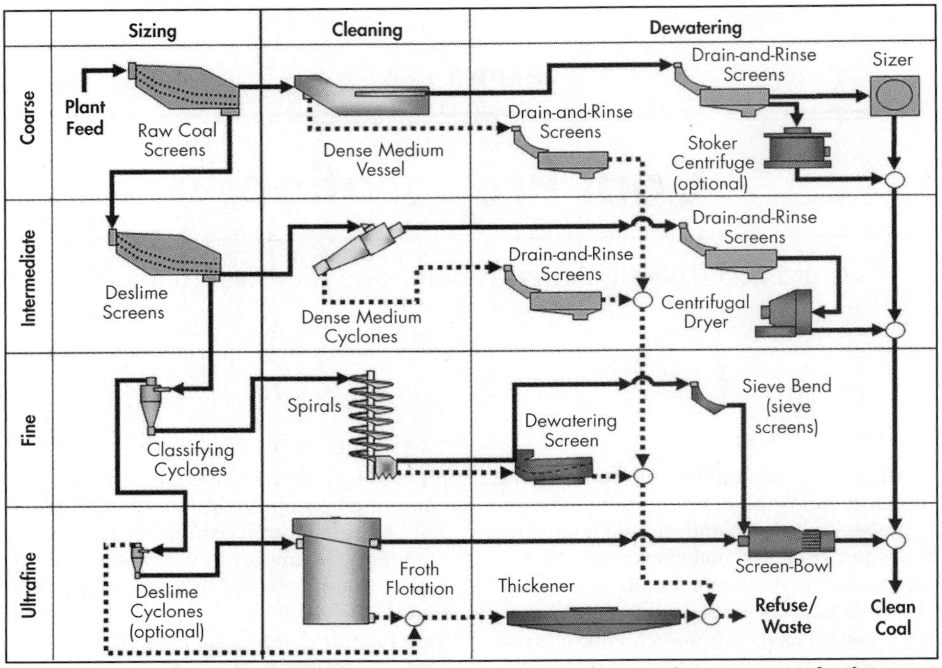

Figure 1 Example of a simplified flow sheet used by modern coal preparation facilities

specific raw coal size distribution or washability against any undesirable increase in capital costs or operating and maintenance costs. Operator preferences and vendor biases also contribute to the large variations in how coal is sized, cleaned, and dewatered. Additional descriptions and details related to modern coal preparation practices and trends can be found in two informative books: *Designing the Coal Preparation Plant of the Future* (Arnold et al. 2007) and *Challenges in Fine Coal Processing, Dewatering, and Disposal* (Klima et al. 2012).

Particle Size Separations

Coarse Particle Screening

The first step in coal preparation involves crushing and sizing ROM coals into acceptable size classes. The selected top size should balance improvements in liberation with reductions in performance created by the need to treat a higher percentage of the feed tonnage in less-efficient fine coal circuits. The vast majority of equipment used in coal-sizing operations is mature technology and includes well-known unit operations such as screens, sieves, and classifying cyclones. Figure 2A shows the typical sizes of particles that can be produced by these common types of industrial sizing operations. As indicated by this chart, large particles are typically sized using one or more types of screening systems. Screens are simply mechanical sizing devices that use a mesh or perforated plate to sort particles into fine and coarse fractions; that is, particles either pass through the screen openings or are retained on the screen surface. Vibrating screens use an off-balance rotating mechanism to create oscillations/vibrations that segregate particles and move material along the screen surface. Traditional designs of inclined and horizontal vibrating screens have been displaced in recent years by multi-slope (banana) screens, largely because of the high unit capacity and improved sizing efficiency offered by these machines (Figure 3). The steep slope at the feed end creates a thin bed of particles that can more rapidly pass through the screen openings, while the shallow slope at the discharge end provides a lower velocity and repetitive attempts for high-efficiency sorting of near-sized particles.

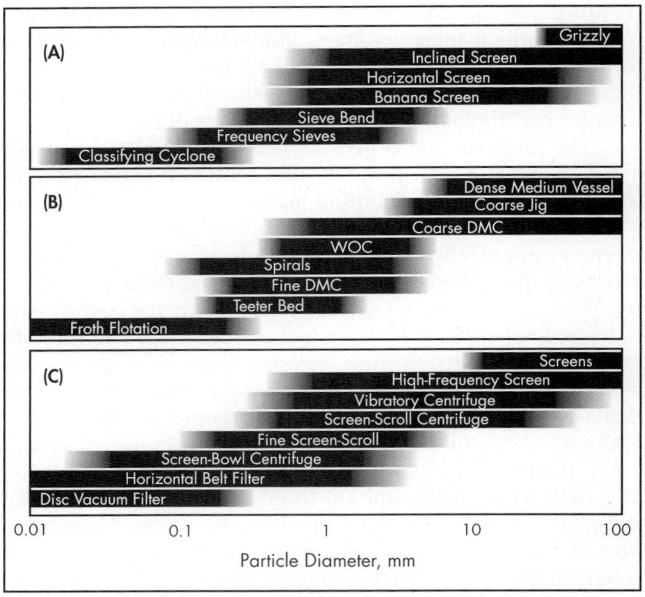

Source: Arnold et al. 2007

Figure 2 Particle size ranges for (A) sizing, (B) cleaning, and (C) dewatering equipment used in coal preparation

Vibrating screens are used in the coal industry for feed scalping, raw coal sizing, and drain-and-rinse (D&R) applications. Typical operating conditions are given in Table 1. When necessary, heavy-duty scalping screens are used remove large unbroken rock from ROM feeds. After scalping, raw coal screens and deslime screens are used to size feeds

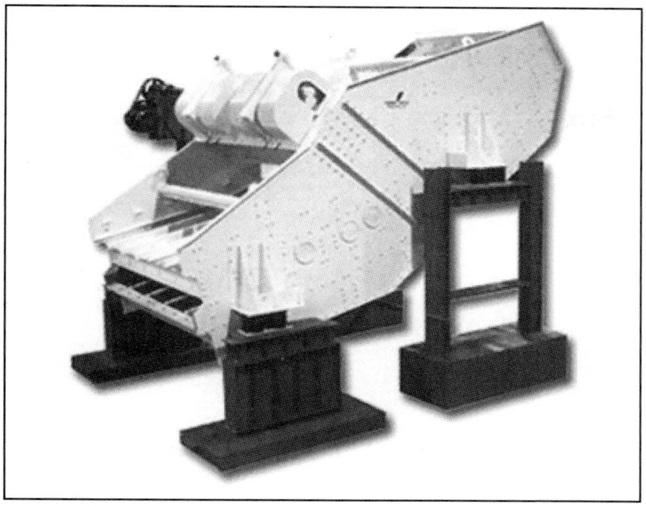

Courtesy of Conn-Weld Industries Inc.
Figure 3 Multi-slope banana screen

Table 1 Typical operating conditions for vibrating screens*

Screen Type	Speed, rpm	Throw, in.	Motion	Water
Scalping	700–800	3/8–1/2	Circular	No
Raw coal	700–900	5/16–7/16	Circular	Yes
Deslime	900–950	3/8–1/2	Straight	Yes
Drain-and-rinse	900–950	3/8–1/2	Straight	Yes

*These are approximate guidelines; actual demands may vary depending on site conditions.

Table 2 General recommendations for water addition to screens*

Screen Type	Recommendation
Raw coal	5–7 gpm of spray water per stph of dry feed
Deslime	4–7 gpm of dilution water per stph of dry feed with 100 gpm of spray water per foot of screen width
Drain-and-rinse	3–5 gpm of spray water per stph of dry feed (or at least 30–50 gpm per foot of screen width)
Fine sieves	Add dilution water to ensure <25% feed solids

*These are approximate guidelines; actual demands may vary depending on site conditions.

for dense medium vessel and DMC circuits, respectively. A double-deck screen with two different deck openings is often used for both sizing steps. These types of screening systems require water sprays/boxes to ensure efficient removal of fines. Recommended water addition rates are listed in Table 2. D&R screens are used to recover and rinse magnetite medium from products generated by dense medium separators. These screens must be sufficiently long to ensure time is available to drain medium, rinse medium, and dewater the solid products. D&R applications often require a combination of flume/banana screens or sieve bend/horizontal screens for efficient medium removal.

Fine Particle Sieving

Screening of very fine particles is typically achieved using static sieve screens (sieve bends), which use a curved panel equipped with slotted bars perpendicular to the flow to slice material from the flowing stream. Sieve screens are often

Courtesy of Conn-Weld Industries Inc.
Figure 4 Fine-wire sieve bend

necessary because of the low capacity and increased likelihood of plugging openings in vibrating screens. Sieves use profile wires turned perpendicular to the flow to serve as cutting edges to slice water/fines from the primary flow (Figure 4). Since the sizing is done by water partitioning, it is important to maintain the feed solids content to less than 25%–30% by weight. Recently, operators have begun to install sieves in two sequential stages to further reduce fines misplacement to the oversize product (Bethell and Luttrell 2005). Sieve cut size can be estimated from a rule of thumb that states that the required slot size (in millimeters) can be found by dividing the target cut size (in mesh) into 21. Based on this equation, a target 28 mesh cut size would dictate the use of a 0.75-mm slot opening. Sieve capacity, which is normally reported as flow capacity per unit of width, can be estimated from the values provided in Table 3. A well-designed feed distribution box and properly balanced feed gate are essential to ensure that the full width of the sieve receives feed slurry.

High-frequency screens are also used in the coal industry, but only for the dewatering of fine waste solids prior to disposal. While this type of screen is inefficient for particle sizing, the thick bed of solids created by the high vibrational frequency is ideally suited for solids retention and water rejection. Also, when highly efficient fine sizing is required, coal operators are forced to consider advanced sieving technologies such as the Derrick StackSizer, which can efficiently segregate particles at sizes as fine as 0.075 mm via the use of integrated washing/repulping channels placed at strategic points down the sieve surface (Hollis 2006; Brodzik 2007; Mohanty et al. 2002, 2009).

Table 3 Approximate sieve bend capacities*

Slot, mm	gpm/ft for Specified Wire Designation in inches				
	0.03	0.045	1/16	3/32	1/8
¼	100	60	30	—	—
⅓	115	75	40	—	—
½	135	125	90	70	55
¾	—	—	175	135	110
1	—	—	215	170	135
1¼	—	—	235	190	165
1½	—	—	260	215	190
1¾	—	—	280	230	210
2	—	—	300	250	225

Courtesy of Conn-Weld Industries Inc.
*Arc angle = 60°, radius = 40 in. Actual capacity varies depending on feed solids, type and amount of liquid, and desired efficiency.

Fine Particle Classification

Conventional screening and sieving of very fine particles is impractical because of capacity limitations (Firth and O'Brian 2003). Therefore, plant designers are forced to employ classifying cyclones (Figure 5) for many fine sizing applications including raw coal (0.15–0.25 mm) sizing and deslime (0.04–0.05 mm) sizing. Classifying cyclones are also used in the coal industry as clean coal cyclones for thickening and desliming ahead of dewatering units and as effluent cyclones for recovering solids from streams for the purpose of clarification. Despite the perceived advantages, classifying cyclones are typically not used in two stages in the coal industry because particle density effects unavoidably concentrate ash-bearing minerals in the underflow product (Honaker et al. 2007). Table 4 lists typical capacity ranges and cut sizes for coal applications. As expected, cyclone diameter is the variable that primarily determines both the particle cut size and capacity. Most commonly, cyclone diameters of 20, 15, and 6 in. are used to produce nominal cut sizes of 0.25, 0.15, and 0.045 mm for raw coal sizing and desliming applications. Feed solids are generally held below 12%–15% solids by weight for 15- and 20-in. raw coal cyclones and below 6%–8% solids by weight for 6-in. deslime cyclones. Inlet pressures are held in the 15–25 psi range for most applications, with the higher-pressure region more common for smaller cyclones. Because of the lack of moving parts, cyclones require little maintenance in coal plants. When problems do occur, potential causes for losses of coarse particles in the overflow include plugged apex, low inlet pressure, misaligned liner, worn vortex finder, and plugged vent pipe. Similarly, primary causes for excess fines and/or water in the underflow product include worn apex and high feed solids.

Solid–Solid Separations

Dense Medium Separation

The most important function of a coal processing facility is the separation of carbonaceous material from waste rock. This separation is typically accomplished using low-cost processes that segregate coal and rock based on differences in physical properties such a size, density, and wettability. As shown in Figure 2B, the effectiveness of different solid–solid separation processes is generally limited to a relatively narrow size range. For coarse particles, the most commonly used property difference for separation is density because

Reprinted with the permission of FLSmidth A/S
Figure 5 Classifying cyclone bank

Table 4 Approximate classifying cyclone capacities and cut-size values*

Diameter, in.	Cut Size, mesh	Solids, stph	Pulp, gpm
6	200–270	2–5	80–200
10	150–200	5–10	200–400
15	100–150	11–21	420–800
15†	65–150	16–29	600–1,000
20	65–100	18–34	700–1,300
20†	48–65	29–47	1,100–1,800
26	48–65	39–78	1,500–3,000

Reprinted with the permission of FLSmidth A/S
*These are approximate guidelines; actual demands may vary depending on site conditions.
†Extended body cyclone design.

pure organic coal has a density in the range of 1.26–1.32 SG (specific gravity), while waste rock has a much higher density of 2.2–2.7 SG. Therefore, one of the most popular and efficient methods for treating ROM coals coarser than ¼-in. (6.3–12.7 mm) is the dense medium vessel. This high-capacity separator (Figure 6) consists of a large open tank through which a dense suspension of finely micronized magnetite in water is circulated. Low-density coal particles introduced into the suspension float to the surface of the dense medium where they are transported by the overflow medium onto a collection screen, while high-density waste rock sinks to the bottom of the vessel where it is collected and discarded by a series of mechanical scrapers called *flights*. The washed coal and rock products pass over D&R screens to wash the magnetite medium from the surfaces of the products and to dewater the particles (Figure 7). As shown in Table 5, the clean coal capacity for a dense medium vessel can be estimated from the overflow weir length and medium density. The refuse capacity can be estimated from the flight height and width as shown in Table 6. The medium is circulated through the vessel at a rate of approximately 250 gpm/ft of weir length.

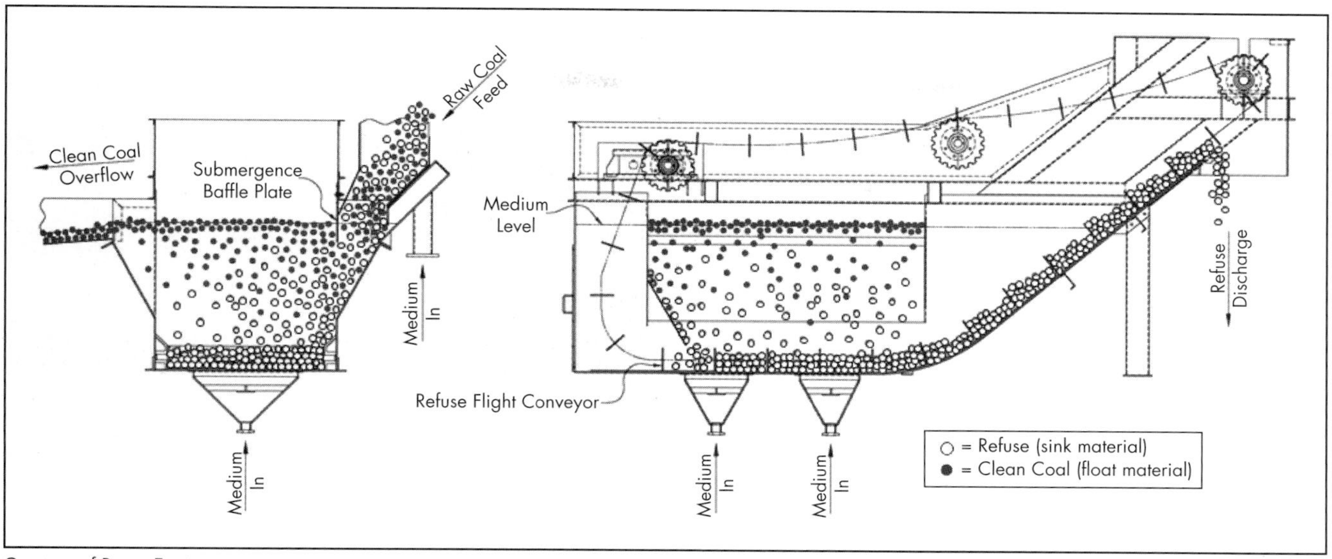

Courtesy of Peters Equipment

Figure 6 Dense medium vessel

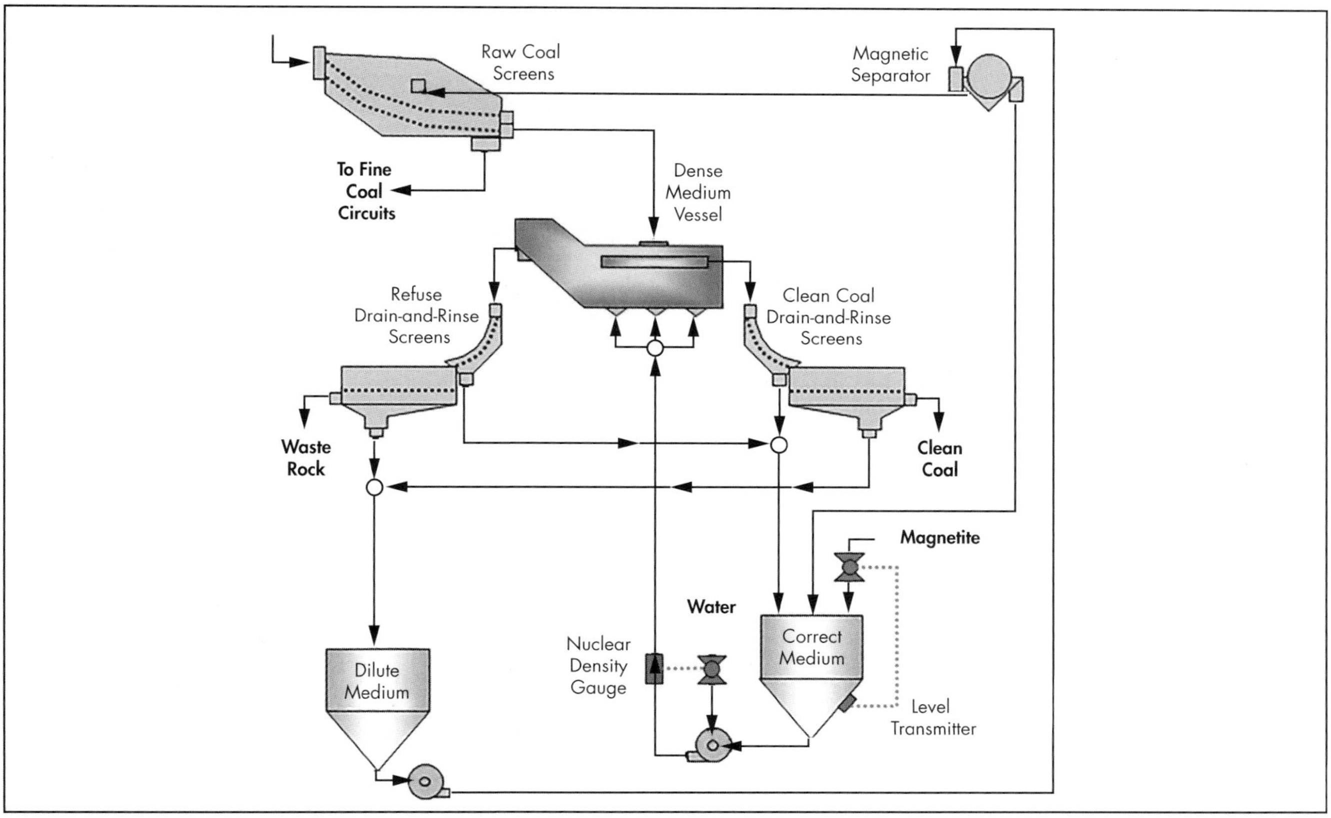

Figure 7 Dense medium vessel circuit

Dilute medium rinsed from the D&R screens is recovered and recycled back through the circuit using a drum-style magnetic separator (Figure 8). Modern magnetic separators incorporate stronger magnets (950 gauss interpole) and a wider magnetic arc (up to 140°) to increase capacity and to reduce losses of magnetic solids down to <0.5 g/gal of treated slurry (Bethell and Barbee 2007). Typical capacities for the standard 36- and 48-in.-diameter drums are 70–120 and 100–170 gpm of dilute medium and 5.5 and 8.0 stph of magnetite solids per foot of drum width, respectively.

Particles too small to be efficiently treated by a static dense medium vessel are upgraded using DMCs (Figure 9).

Table 5 Approximate clean coal capacities for a dense medium vessel*

Medium, SG	Specific Capacity, stph/ft	Clean Coal Capacity, stph, for Specified Weir Length			
		4 ft	8 ft	12 ft	16 ft
1.4	1.44	100	200	300	400
1.5	1.21	120	240	360	480
1.6	1.17	140	280	420	560
1.7	1.15	160	320	480	640

Courtesy of Peters Equipment
*These are approximate guidelines; actual demands may vary depending on site conditions.

Table 6 Approximate reject capacities for a dense medium vessel*

Flight Height, in.	Specific Capacity, stph/ft	Reject Capacity, stph, for Specified Flight Width			
		4 ft	4½ ft	5 ft	5½ ft
8	25	261	294	326	359
9	30	316	355	394	433
10	35	370	416	462	508
11	40	424	477	530	583

Courtesy of Peters Equipment
*These are approximate guidelines; actual demands may vary depending on site conditions.

Courtesy of Eriez Magnetics
Figure 8 Drum-style magnetic separators

These high-capacity units make use of the same basic principle as dense medium vessels, that is, an artificial magnetite–water medium is used to separate low-density coal from high-density rock. In this case, however, the rate of separation is increased by the centrifugal effect created by passing medium and coal through one or more cyclones. Modern DMCs are capable of treating particle sizes as large as 3–3.5 in. (75–90 mm), which can in many cases eliminate the need for a static vessel (Ziaja and Yannoulis 2007). Nominal capacities and operating limits for DMCs of different diameters are listed in Tables 7 and 8, respectively. Typically, apex and vortex diameters are selected to provide, a two-thirds split of medium to overflow and a medium-to-coal ratio in the feed stream of about 4:1 by volume. The diameters of the inlet, vortex finder, and apex are typically a function of cyclone diameter (Dc) and fall in the ranges of 0.2–0.3 × Dc, 0.43–0.50 × Dc, and 0.3–0.4 × Dc, respectively (De Korte and Engelbrecht 2007).

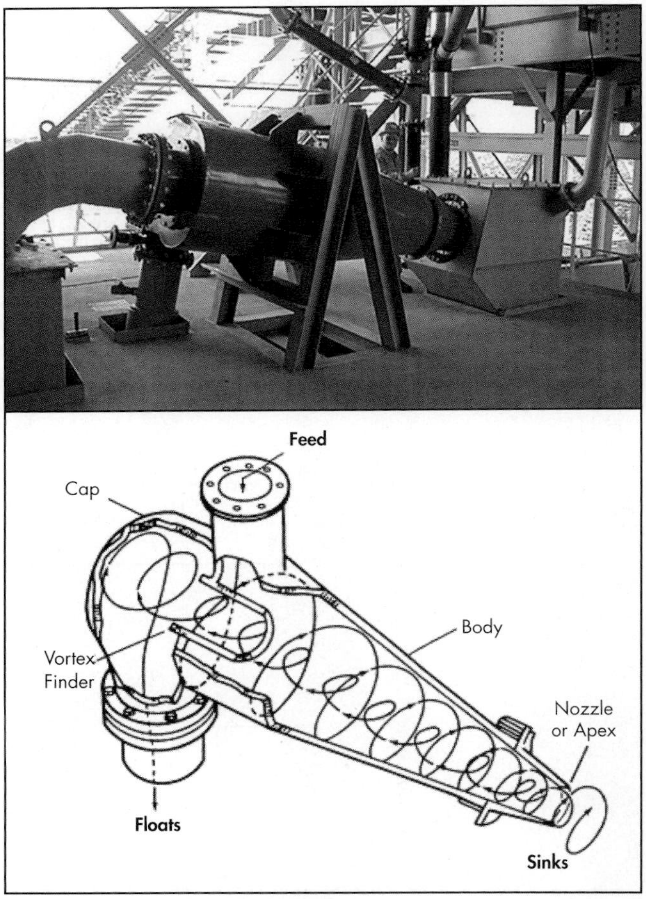

Reprinted with the permission of FLSmidth A/S
Figure 9 Dense medium cyclone

Gravity Separation

Coal particles that are smaller than 1 mm are too small to be effectively handled by the D&R screens required by dense medium processes. Therefore, particles in this size range are commonly processed using water-based gravity separators. As shown in Figure 2B, there are several alternatives for cleaning this size of material, including WOCs, spirals, and teeter-bed separators. WOCs are similar to classifying cyclones but are configured with a truncated wide-angled conical bottom to emphasize density partitioning and suppress classification effects (Figure 10). WOCs of larger diameter are typically required to process coarser solids, as shown in Table 9. The self-generated medium of solids and water within the WOC allows a separation to be made between low-density coal and high-density rock, although the separation is usually not efficient because coarser particles tend to separate at a much lower density than finer particles (Majumder 2011). As a result, WOCs have been largely displaced in the U.S. coal industry by spiral separators. Spiral technology consists of a corkscrew-shaped helical trough supported along a central pole. As slurry passes down the trough, the resultant flowing

Table 7 Approximate capacity ratings for dense medium cyclones*

Diameter, in.	Top Size, in.	Solids, stph	Slurry, gpm
20	¾	85	1,050
26	1½	160	2,045
30	1½	225	2,830
33	2	290	3,650
40	2½	460	5,760
44	3	580	7,280
48	3	725	9,100
55	4	985	12,300

Reprinted with the permission of FLSmidth A/S
*These are approximate guidelines; actual demands may vary depending on site conditions.

Table 8 Operating conditions for dense medium cyclones*

	Approximate Inlet Pressure (psi) at Specified Medium Specific Gravity		
Diameter, in.	1.5 SG	1.6 SG	1.7 SG
20	10–12	11–13	12–14
26	13–16	14–17	15–18
30	15–18	16–20	17–21
33	17–20	18–21	19–23
40	20–24	21–26	23–28
44	22–27	23–28	25–30
48	24–29	25–31	27–33
55	27–33	29–35	31–38

Reprinted with the permission of FLSmidth A/S
*These are approximate guidelines; actual demands may vary depending on site conditions.

film segregates particles according to density, although variations in particle size and shape also influence the separation. Because of the low unit capacity (<2–4 stph per spiral), spirals are usually arranged in groups that are fed by an overhead distributor (Figure 11). Spirals have the advantage of separating coal at a more consistent density across the size range between 0.2 and 1.0 mm but are generally unable to efficiently separate at densities lower than about 1.65 SG. To maintain acceptable cut points and separation efficiencies, spirals typically have to be run at less than 2–3 stph per spiral and at volumetric slurry rates of about 35–40 gpm per spiral. Also, protection sieves are often used ahead of the spiral circuit to keep oversize solids (>1.5 mm) out of the feed.

Both WOCs and spirals are often employed in two stages or in combination with each other to improve separation efficiencies and reduce misplacement (Luttrell et al. 1998, 2007). In recognition of this limitation, spiral manufacturers now offer compound spirals that incorporate a two-stage rougher–cleaner arrangement along the spiral run. The efficiencies afforded by this configuration rival those of dense medium separation for this size fraction (Zhang et al. 2008). Several plants have also begun to incorporate spiral technology as desulfurization units to remove fine pyrite from fine feed slurries prior to flotation treatment (Chafin et al. 2012). This pretreatment minimizes the recovery of pyritic sulfur, and particularly pyrite containing inclusions of carbon, that often floats in flotation processes that rely on differences in particle wettability for selectivity.

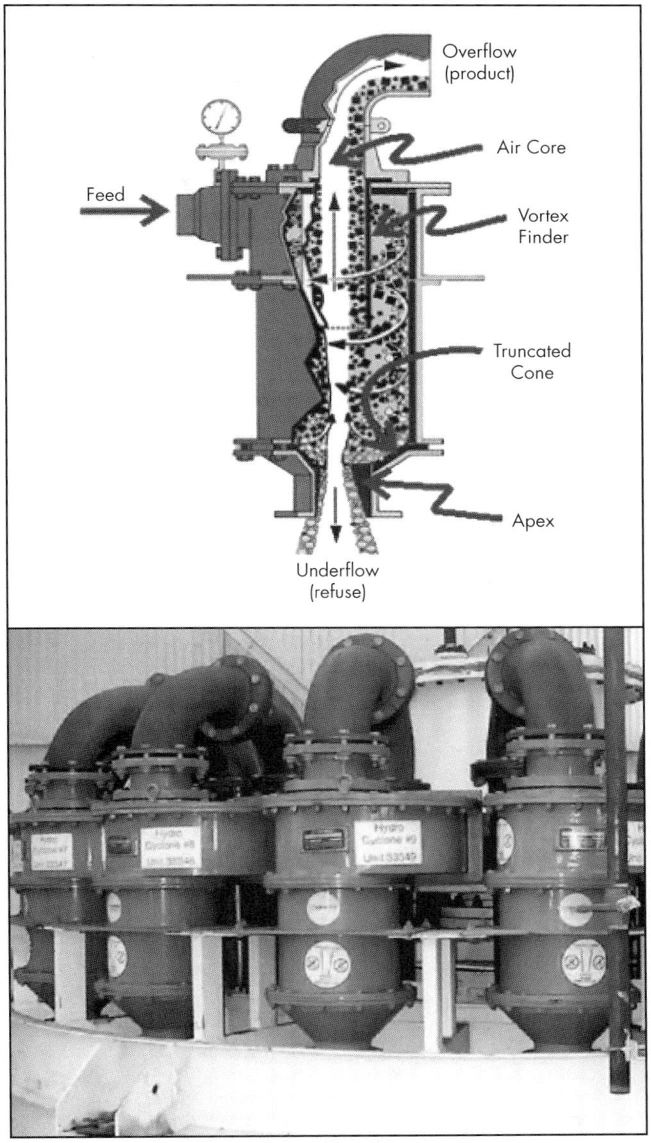

Reprinted with the permission of FLSmidth A/S
Figure 10 Water-only cyclones

In recent years, hindered-bed separators have begun to reemerge in the coal industry. New technologies, such as the CrossFlow separator (Figure 12) and the REFLUX™ Classifier, incorporate innovative design features that improve the overall separation efficiency of this historic coal upgrading method (Kohmuench et al. 2002; Galvin 2012; Ghosh et al. 2012). Hindered-bed separators inject fluidization water across the base of the separator through a distribution network. Solids settle against the upward flow of fluidization water and form a dense bed. Consequently, light/small particles report to overflow, while dense/large particles report to underflow. Bed density is automatically controlled using an actuated underflow valve that adjusts in response to signals from pressure transmitters located in the teeter bed. Sizing of the feed solids is critical because fines report to overflow regardless of density. Compared to spirals, this technology has the advantage of being able to operate at substantially lower density cut points (Luttrell et al. 2007). Full-scale industrial units are capable of feed rates >100 stph for feeds as fine as 1 × 0.25 mm (Galvin et al. 2002). Feed solids contents in the range of 30%–40% by weight are typically preferred to maintain separation efficiency. Fluidization water rates vary depending on the application but are typically in the range of 4–9 gpm/ft^2 of cross-sectional area. Also, as with any gravity separator, the presence of excessive oversize solids in the feed should be avoided.

Froth Flotation Separation

Froth flotation is the most common method used to upgrade ultrafine coal particles <0.15 mm. This application of flotation, which has been recently described by Laskowski et al. (2007), exploits inherent differences between hydrophobic carbonaceous matter and hydrophilic mineral matter. Most coal feedstocks can be floated using only a frothing agent, although hydrocarbon collectors, such as diesel fuel or fuel oil, are often added to improve flotation kinetics (Laskowski 2001). The addition rates are very small and typically about

Table 9 Operating conditions for water-only cyclones*

	Diameter, in.			
Condition	**10**	**15**	**20**	**26**
Rate, stph	7–9	20–30	50–60	95–120
Flow, gpm	325–375	775–825	1,000–1,250	1,800–1,250
Top size, mm	½	1	6.4	19
Feed solids, %	8–10	10–15	15–18	18–22
Pressure, psi	8–12	8–15	15–18	18–22

Reprinted with the permission of FLSmidth A/S

*These are approximate guidelines, actual demands may vary depending on site conditions.

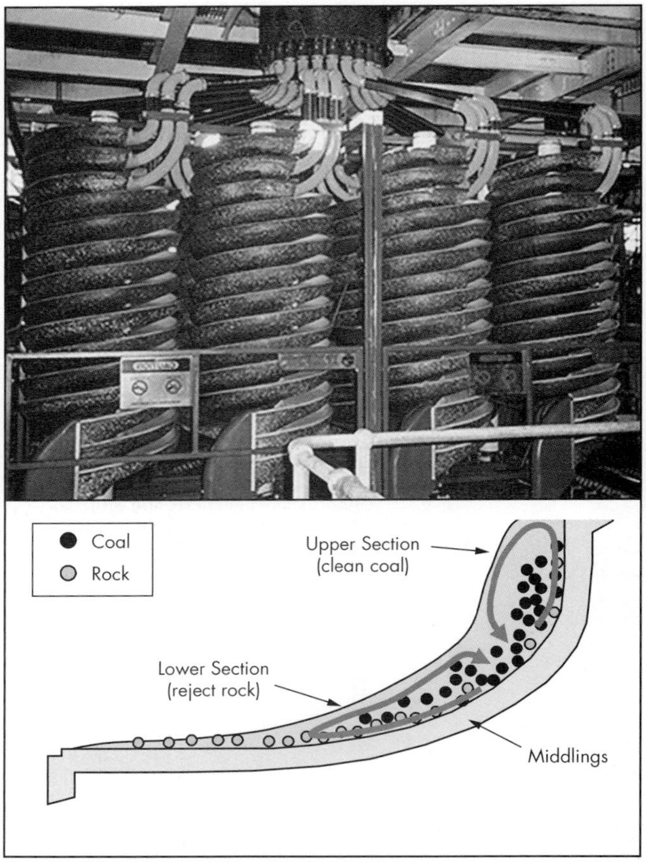

Photo courtesy of Eriez Flotation Division

Figure 11 Spiral separator

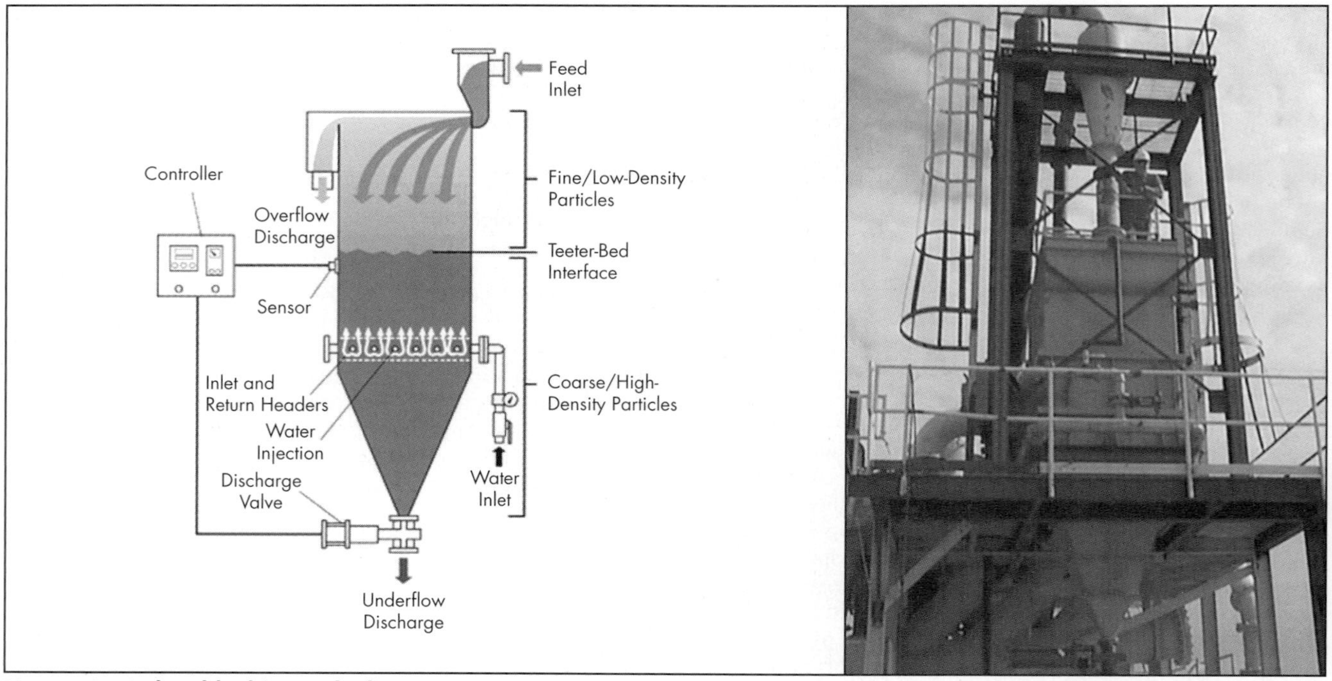

Figure 12 Hindered-bed (teeter-bed) separator

Courtesy of Eriez Flotation Division

Figure 13 Installation of column flotation cells

0.1–0.5 lb of reagent per ton of coal feed. A variety of environmentally friendly collectors, such as blends of canola, vegetable, and soybean oils, have become more common in the coal industry for applications involving underground injection of fine wastes (Skiles 2003). Near-neutral pH levels (6.5–8) are generally acceptable for the flotation of most coals.

In addition to reagent type and dosage, a wide variety of other parameters can impact the flotation of coal. For example, while particles as large as 0.5 mm can be treated in coal flotation plants, the industry typically uses this process only for particles with top sizes smaller than about 0.1–0.2 mm to attain maximum separation efficiencies. Deslime flotation circuits, in which <45-µm particles are largely removed and discarded prior to flotation, have become popular because of the low value, high moisture, low capacity, and high processing cost of ultrafine slimes (Bethell 2012). Both mechanical stirred-tank machines and flotation columns are used in the U.S. coal industry. Stirred (conventional) machines are commonly used in banks of four to five cells in series. Pulp residence times of 4–5 minutes are often sufficient to achieve good recoveries of combustible material for these types of circuits. Since the late 1990s, column-type flotation cells (Figure 13) have become favored over conventional machines largely because of the ability of columns to remove high-ash clays from the froth product via the addition of wash water to a relatively deep froth (Davis et al. 1995). The main downside is that columns typically require longer pulp residence times, with 10–15 minutes or more being used in industrial installations. These installations require water flow rates of 3–4 gpm/ft^2 and gas flow rates of 3–5 scfm/ft^2. The majority of column flotation installations incorporate either static spargers such as SlamJet gas injectors or dynamic spargers such as CavTube (Kohmuench et al. 2012) or Microcel (Yoon et al. 1992) bubble generators. More recently, a compact column known as the StackCell technology has made inroads in the coal industry by offering column-like performance in a low-profile design that eliminates the high foundation loads and larger gas compressors typically associated with column cells (Kohmuench et al. 2010).

Solid–Liquid Separations

The final step in coal preparation is the removal of water from the plant products. Solid–liquid separations include dewatering units for the removal of unwanted moisture from plant products and thickening/clarification units for the removal of solids from process water. For clean coal products, unwanted surface moisture must be removed to avoid lower heating values, handling/freezing problems, and higher transportation costs. Excess moisture must also be removed from coal wastes to avoid downstream handling and disposal issues associated with coarse and fine refuse streams. Figure 2C provides an overview of the range of particle sizes that are commonly treated by the various solid–liquid separation processes used in the coal preparation industry. As should be expected, several different types of mechanical dewatering methods are required to attain optimal performance for each size class (Arnold 1999). The removal of water from coarser (>5 mm) coal is typically carried out using simple screens, while finer particles typically require the use of more intense solid–liquid separation processes such as centrifuges, filters, and thermal dryers (Meenan 2005).

Centrifugal Dewatering

Popular centrifuge-style dewatering machines used in the coal industry include centrifugal dryers and screen-bowl centrifuges. Two designs of centrifugal dryers are the vibratory centrifuge (Figure 14) and screen-scroll centrifuge (Figure 15). The vibratory centrifuge uses a reciprocating motion to induce the flow of solids across the surface of the dewatering basket. The screen-scroll centrifuge uses a screw-shaped scroll that rotates at a slightly different speed than the basket to positively transport the solids along the basket surface. The vibratory centrifuge, which is typically is the least costly to operate, is commonly used to dewater coarser particles in the size range of 3 × ¼ in. The largest unit (56-in. diameter) can reach a capacity of more than 300 stph for a single machine. Both the screen-scroll and the vibratory centrifuges are appropriate for intermediate sizes, although the screen-scroll is more commonly used for ½ in. × 28 mesh solids. Screen-scroll machines also typically provide lower product moistures than vibratory machines for the same size of feed. The largest screen-scroll centrifuges (48-in. diameter) can provide capacities of up to 160 stph.

For finer particles <1 mm, the screen-scroll centrifuge has been displaced in modern facilities by the screen-bowl centrifuge (Figure 16). A screen bowl is a horizontal centrifuge that consists of a solid bowl section and a screen section. Slurry is introduced into the bowl section where most of the solids settle out under the influence of the centrifugal field. A rotating scroll transports the settled solids up a beach and across the screen section where additional dewatering takes place prior to product discharge. Solids that pass through the screen section are circulated back to the machine feed. This unit is capable

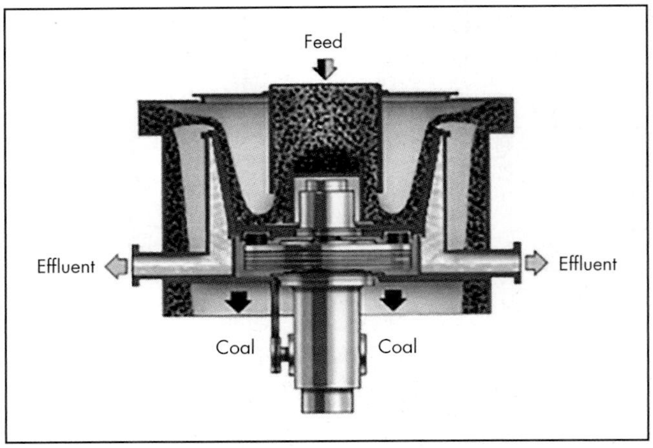

Courtesy of Elgin Separation Solutions
Figure 14 Vibratory centrifuge

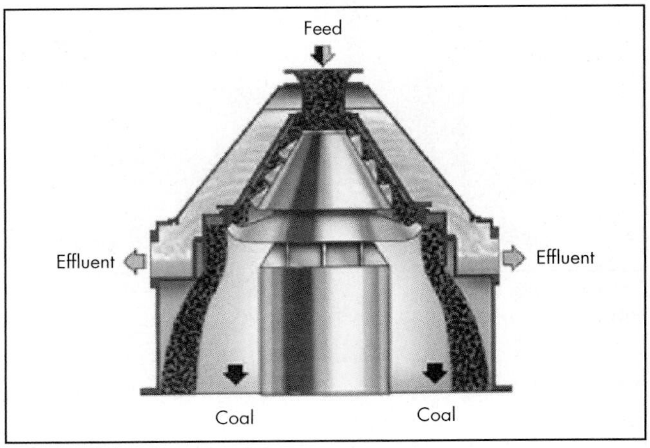

Courtesy of Elgin Separation Solutions
Figure 15 Screen-scroll centrifuge

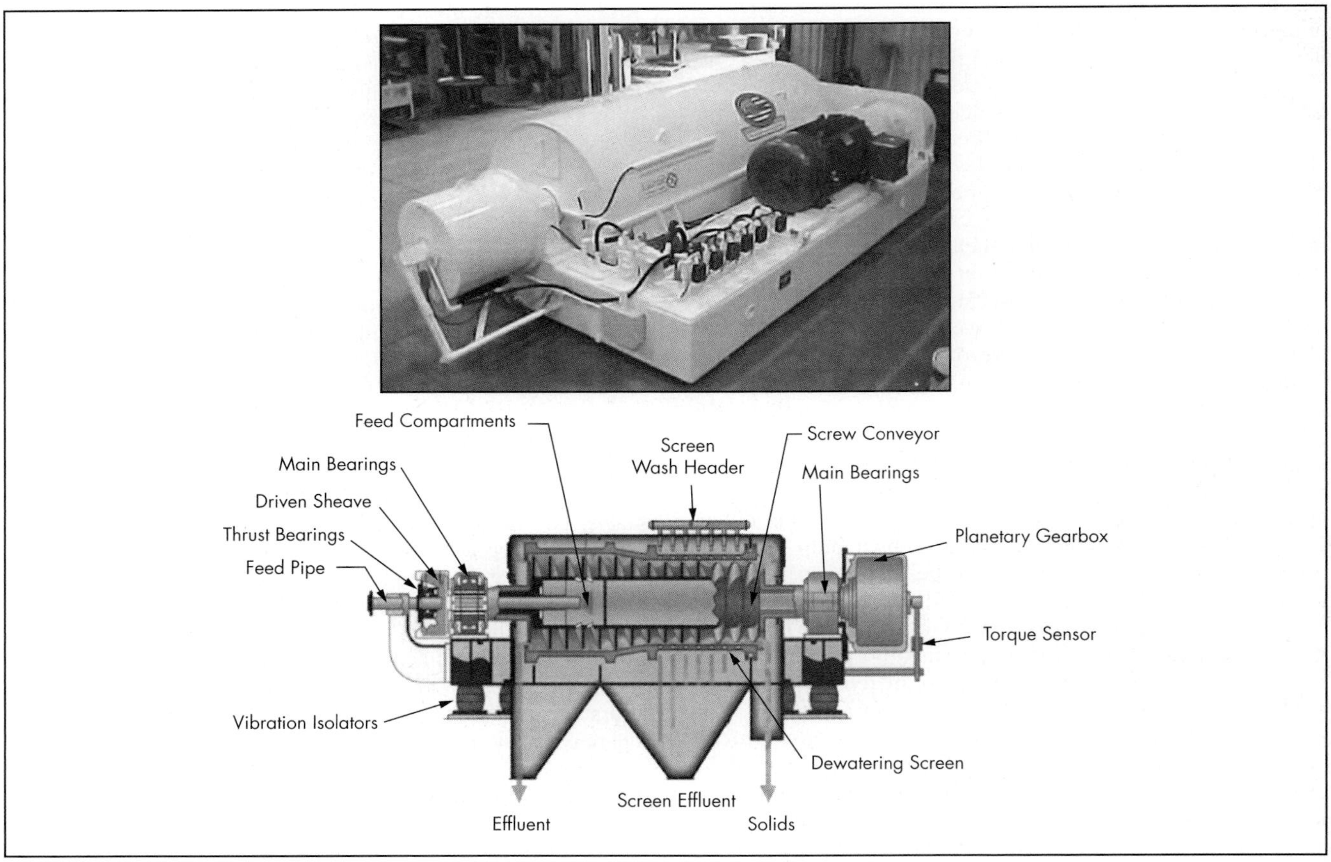

Reprinted with the permission of FLSmidth A/S
Figure 16 Screen-bowl centrifuge

Table 10 Capacity ratings for screen-bowl centrifuges*

Machine Type	Throughput Capacity Coarse, stph†	Fine, stph‡	Slurry, gpm
36 × 72	35–40	20–25	400
36 × 72H§	40–50	25–30	400
36 × 96	35–50	25–28	400
40 × 72	45–65	25–30	500
44 × 132	75–80	50–60	800
44 × 132H§	80–100	60–65	800

Reprinted with the permission of FLSmidth A/S
*These are approximate guidelines; actual demands may vary depending on site conditions.
†Coarse = 1 × 0.1 mm and 35%–50% feed solids
‡Fine = Minus 0.1 mm and 20%–25% feed solids
§H = High capacity

of providing low moistures, although some ultrafine solids are lost with the main effluent discarded from the solid bowl section (Mohanty et al. 2009). An industry rule of thumb is that screen-bowl centrifuges recover +98% of the +0.1 mm solids and about 50% of particles finer than 45 µm (Schultz et al. 2008). An estimate of product moisture from a screen-bowl centrifuge can be obtained from the following equation:

$$\text{moisture, \%} = 9.2 + 0.01(S)^2 \quad \text{(EQ 1)}$$

where S is the percentage of –45 µm solids in the feed. For maximum coal dryness, screen-bowl centrifuges should be run using a effluent weir setting that provides a low pool depth of 50 mm (2 in.) or less. For maximum recovery of solids, screen-bowl centrifuges should be run at a weir setting that provides a pool depth of 76–114 mm (3–4.5 in).

Approximate production capacities for screen-bowl centrifuges are listed in Table 10. The load placed on a screen-bowl centrifuge is monitored online using torque sensors and amp meters. The torque sensor is used to monitor the solids loading on the scrolling mechanism. An increase in torque indicates that more solid material is being fed to the unit. An amp meter installed on the drive is used to monitor the overall load generated by the rotating assembly. Higher amps typically indicate that more feed is going into the machine, although blinding of one of the discharge hoppers can also increase the amp reading.

Filtration Dewatering

If high coal recovery is desirable, higher-cost filtration processes may be used to dewater fine coal (Meenan 2005). Historically, flotation concentrates were dewatered in the coal industry using disc vacuum filters. Disc filters consist of a series of circular filtration discs that are constructed from independent wedge-shaped segments (Figure 17). At the beginning of a cycle, a segment rotates down into the filter tub filled with fine coal slurry. Vacuum is applied to the segment so that water is drawn through the filter media and a particle cake is formed. The resultant cake continues to dry as the segment rotates out of the slurry and over the top of the machine. The drying cycle ends when the vacuum is removed and the cake is discharged using scrapers and a reverse flow of compressed air. Filters typically produce products that are higher in moisture (25%–35%) than screen-bowl centrifuges but also achieve higher solids recoveries (95%–99%) than screen-bowl centrifuges (80%–90%). Cake capacities for disc

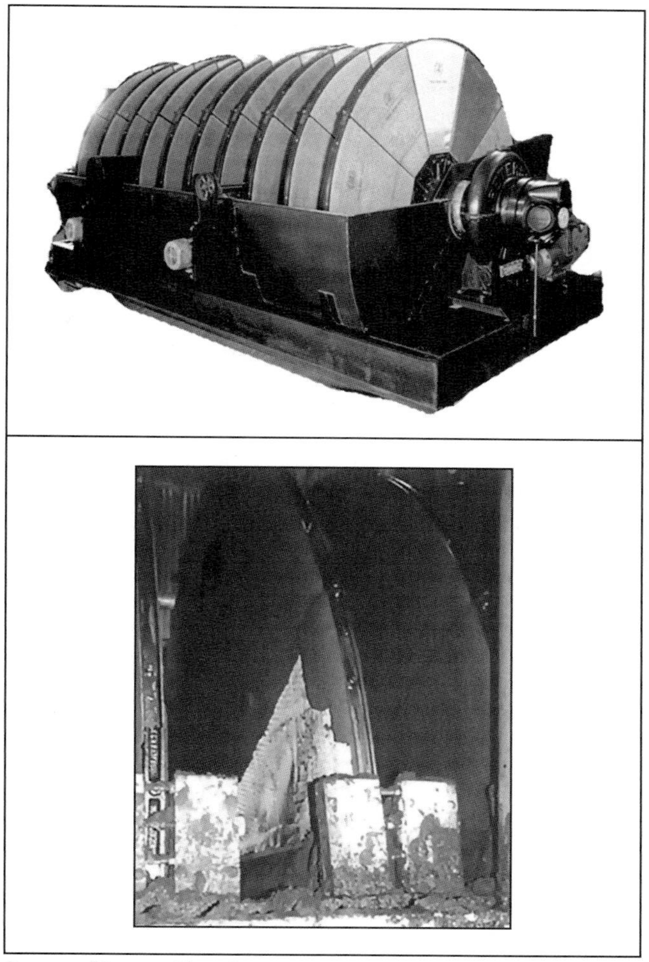

Courtesy of Peterson Filters Corp.
Figure 17 Coal disc filter

filters typically fall in the range of 40–50 lb/h of dry solids per square foot of filter disc area for clean coal. These units, which typically operate at vacuums of 13–15 in. Hg, require vacuum pumps that deliver 5–7 cfm of gas flow per square foot of disc area. Belt filters are commonly used in countries such as Australia but are not typically employed in the mountainous U.S. coalfields because of footprint constraints. In extreme applications, plate-and-frame pressure filters have been placed into service in the coal industry for dewatering ultrafine particles. These batch-type units are effective, but their use is difficult to justify across the coal industry because of high capital/operating costs and high-pressure gas safety concerns.

Thickening/Clarification

Solid–liquid separation is a key issue for water conservation and waste disposal and typically involves clarification technologies such as thickeners, belt filter presses, and plate-and-frame pressure filters (National Research Council 2002). Thickening is performed in all coal plants to clarify dilute slurry rejected from fine coal circuits. A thickener is a large settling tank that produces a clarified overflow that can be reused in the plant as process water and a thickened underflow containing 20%–35% solids by weight that is typically discarded into a waste impoundment (Figure 18). Chemical additives are introduced

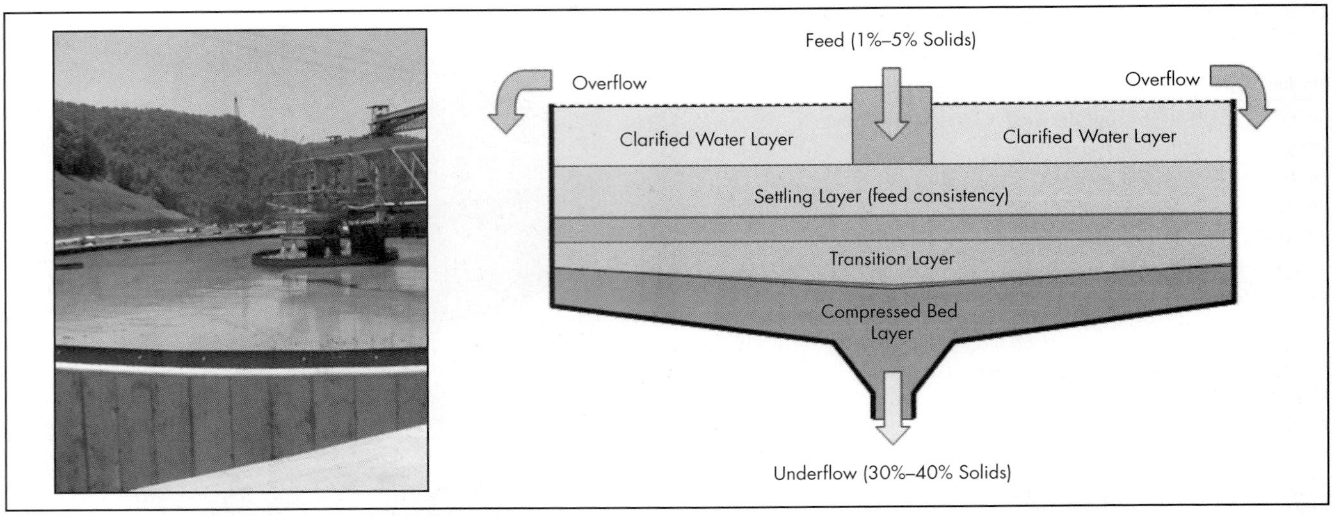

Figure 18 Conventional coal waste thickener

to the thickener feed to promote aggregation of ultrafine particles. These additives include coagulants, which neutralize particle charges so that self-aggregation occurs, and flocculants, which consist of long-chain high-molecular-weight molecules that bridge particles together. Proper chemical dosing is needed to ensure that good water clarity and proper settling rates are maintained. Control of slurry pH is also important, with an acceptable range being 6.5–8 for most sites. An estimate of 1 ft^2 of surface area per gallon per minute of feed slurry is often used for first-pass designs of coal thickeners. Coal thickeners require constant monitoring to ensure that overflow clarity and underflow density are maintained and the rake mechanism is not overloaded. In some cases, the underflow may be further dewatered using belt filter presses or plate-and-frame pressure filters to meet strict disposal requirements imposed by some waste disposal permits.

Table 11 Example of particle size data for a ROM coal

[1] Passing, mm	[2] Retained, mm	[3] Mass %	[4] Cumulative % Retained
102	50.8	6.76	6.76
50.8	25.4	15.38	22.14
25.4	12.7	18.58	40.72
12.7	9.53	8.44	49.16
9.53	1.5	34.58	83.74
1.5	1.0	1.90	85.64
1.0	0.25	8.96	94.60
0.25	0.15	2.29	96.89
0.15	0.075	1.60	98.49
0.075	0.044	0.65	99.14
0.044	0.00	0.86	100.00

CHARACTERIZATION

Particle Size Analysis

The most important characterization procedure used in coal preparation is particle size analysis. Sizing is typically performed using standard square-grid screens/sieves with shaking/rapping times of <10 minutes to avoid degradation of friable coals. The mass of dry solids retained on each screen is then cumulated as shown in Table 11. Columns 1 and 2 are the passing and retained screen sizes, respectively. Column 3 is the individual weight percentage retained on each screen or sieve. The cumulative values for the weight retained for each screen size is tabulated in column 4. For coal, the cumulative mass retained can often be plotted as a linear relationship using Rosin–Rammler graph paper based on the following equation:

$$Y = 100\exp(-D/K)^m \quad \textbf{(EQ 2)}$$

where Y is the cumulative weight percent retained on the screen or sieve, D is the screen or sieve opening (particle size), K is the size constant, and m is the distribution constant. The Rosin–Rammler plot is obtained by plotting the cumulative percent retained (column 4) versus the retained opening size (column 2). An example is provided in Figure 19. The solid line corresponding to K = 14.6 and m = 0.75 represents the

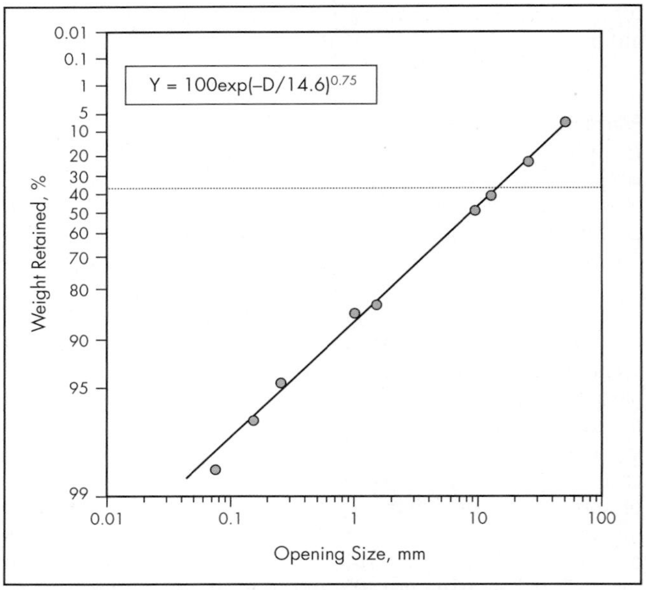

Figure 19 Rosin–Rammler graph for a ROM coal

Table 12 Washability data for three size fractions of ROM coal

Specific Gravity		Individual			Cumulative Float			Cumulative Sink		
[1]	[2]	[3]	[4]	[5]	[6]	[7]	[8]	[9]	[10]	[11]
Sink	Float	Weight %	Ash %	Sulfur %	Weight %	Ash %	Sulfur %	Weight %	Ash %	Sulfur %
A. 50 × 10 mm (56.82% of feed)										
1.26	1.30	10.16	5.67	0.56	10.16	5.67	0.56	100.00	62.01	0.60
1.30	1.40	16.24	11.37	0.71	26.40	9.18	0.65	89.84	68.38	0.61
1.40	1.50	3.20	21.68	1.27	29.60	10.53	0.72	73.60	80.96	0.58
1.50	1.60	2.56	31.53	1.49	32.16	12.20	0.78	70.40	83.66	0.55
1.60	1.80	2.83	40.74	2.39	34.99	14.51	0.91	67.84	85.63	0.52
1.80	2.00	2.60	55.90	2.04	37.59	17.37	0.99	65.01	87.58	0.44
2.00	2.40	62.41	88.90	0.37	100.00	62.01	0.60	62.41	88.90	0.37
B. 10 × 1 mm (24.21% of feed)										
1.26	1.30	39.20	5.53	0.57	39.20	5.53	0.57	100.00	30.71	0.83
1.30	1.40	23.06	11.33	0.72	62.26	7.68	0.63	60.80	46.94	1.00
1.40	1.50	5.47	19.42	1.15	67.73	8.63	0.67	37.74	68.70	1.17
1.50	1.60	2.46	28.93	1.65	70.19	9.34	0.70	32.27	77.06	1.17
1.60	1.80	1.79	39.07	2.28	71.98	10.08	0.74	29.81	81.03	1.13
1.80	2.00	2.40	51.22	2.43	74.38	11.41	0.80	28.02	83.71	1.06
2.00	2.40	25.62	86.75	0.93	100.00	30.71	0.83	25.62	86.75	0.93
C. 1 × 0.1 mm (18.97% of feed)										
1.26	1.30	43.05	3.51	0.64	43.05	3.51	0.64	100.00	22.38	0.78
1.30	1.40	26.92	9.95	0.74	69.97	5.99	0.68	56.95	36.65	0.88
1.40	1.50	6.06	19.14	1.03	76.03	7.04	0.71	30.03	60.59	1.01
1.50	1.60	2.45	28.14	1.52	78.48	7.69	0.73	23.97	71.06	1.00
1.60	1.80	2.50	34.76	1.78	80.98	8.53	0.76	21.52	75.95	0.94
1.80	2.00	1.55	52.46	2.52	82.53	9.36	0.80	19.02	81.37	0.83
2.00	2.40	17.47	83.93	0.68	100.00	22.38	0.78	17.47	83.93	0.68

best fit through the coarse end of the size distribution. More than one line segment may be necessary to represent sizing data from upstream material handling operations that degrade the feed coal from its natural size distribution or from feed streams that contain particles from multiple sources with different inherent breakage properties.

Washability Analysis

The ability of coal preparation to improve coal quality varies widely from site to site because of variations in the liberation characteristics of ROM coals. The degree of liberation, and ultimately the coal cleanability, can be quantified in the laboratory using washability (float–sink) analysis. Washability tests are performed by sequentially passing a sized coal sample through flasks/tanks of dense liquids of increasingly higher densities, as described by ASTM D4371-06 (ASTM International 1994). Table 12 shows size-by-size washability data collected for a ROM coal feed crushed to a 2-in. (50-mm) top size. Columns 3–5 are the experimental mass, ash, and sulfur values for each density class listed in columns 1–2. These individual class values are cumulated in columns 6–8 and 9–11 for the cumulative float and cumulative sink products, respectively. Based on this data, the coarsest 50 × 10 mm fraction floated at a density of 1.6 SG would produce a clean coal product containing 12.20% ash and 0.78% sulfur at a clean coal yield of 32.16%. The reject product, which consists of 1.6 SG sink, would consist of the remaining 67.84% reject yield and would contain 85.63% ash and 0.52% sulfur.

The size-by-size washability analyses are typically cumulated across all size fractions to obtain total yield and quality data expected for a given feed coal. For example, Table 13 shows the cumulated data for the size-by-size washabilities listed in Table 12 for a plant processing 1,200 stph of dry ROM feed. Because of the near-linear relationship between mean specific gravity and ash, the maximum total clean coal yield is usually obtained when all size classes are separated at the same density (Abbott 1982). Therefore, as shown in Table 13, section A, separating all three size fractions at identical 1.60 SG produces 549 stph of clean coal at a 9.76% ash content. Separation at different SG values in an attempt to independently produce a 9.76% ash for all three size fractions results in a suboptimal operating point that provides only 540 stph of clean product at 9.76% ash, as indicated in Table 13, section B. The net difference of 9 stph represents a potential revenue gain of more than $2.5 million annually (i.e., 9 stph × $50/st × 5,700 h/yr = $2.57 million/yr). The selection of optimal SG values for cases involving quality specifications other than ash can be handled using an optimization method known as micropricing, which is discussed elsewhere in the literature (Luttrell et al. 2014).

Flotation Release/Kinetics Analysis

Froth flotation separates particles based on differences in surface wettability and not on differences in density. Therefore, the recommended method for determining the washability of fine coal is a technique known as release analysis. Although several versions of this laboratory method exist, the short-form

Table 13 Combined yields and qualities for three size factions of ROM coal separated at (A) constant density and (B) constant cumulative clean coal ash

[1] Size, mm	[2] Size Split %	[3] Feed, stph	[4] Cut Point, SG	[5] Clean Yield %	[6] Clean, stph	[7] Clean Ash %	[8] Clean Sulfur %
A. Operation at identical densities of 1.60 SG for all size classes							
50 × 10	52.09	625.1	1.600	31.90	199.4	11.99	0.79
10 × 1	22.15	265.8	1.600	69.93	185.9	9.22	0.70
1 × 0.15	17.40	208.8	1.600	78.48	163.9	7.66	0.73
Total	**91.64**	**1,200.0**	—	**45.76**	**549.1**	**9.76**	**0.74**
B. Select SG to provide the same cumulative ash content from all size classes							
50 × 10	52.09	625.1	1.439	28.13	175.8	9.76	0.67
10 × 1	22.15	265.8	1.744	71.62	190.4	9.76	0.73
1 × 0.15	17.40	208.8	2.106	83.15	173.6	9.76	0.81
Total	**91.64**	**1,200.0**	—	**44.99**	**539.8**	**9.76**	**0.74**

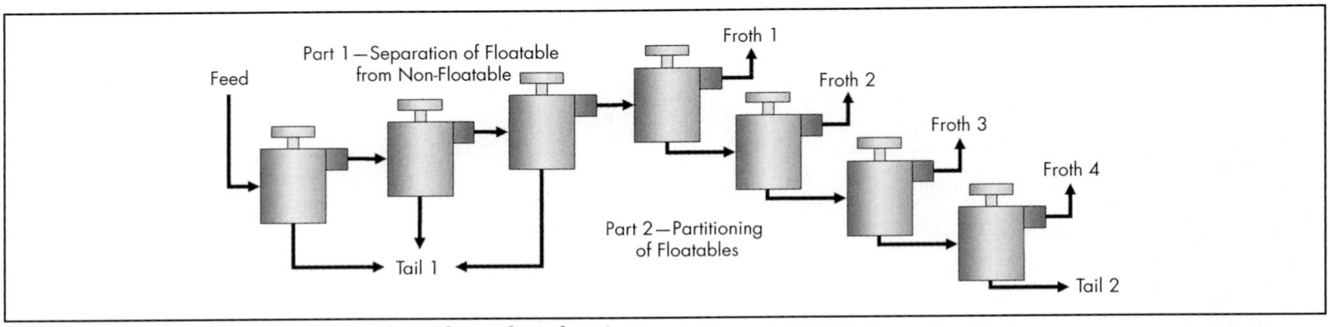

Figure 20 Flotation release analysis procedure (short form)

method is the most popular in the United States. As shown in Figure 20, the short-form method involves two parts. In part 1, feed coal is placed in a laboratory froth flotation cell and floated to exhaustion. The froth concentrate is diluted with fresh water and refloated. This procedure is repeated several times (usually three or four) until essentially no non-floatable solids remain in the flotation cell. All of the tailings from part 1 are combined as the primary tailings (Tail 1). In part 2, the cumulative froth concentrate is floated again using low air rates and impeller speeds. Froth products are taken at intervals that provide roughly the same amount of froth solids in each froth product. This generates several samples of clean coal (Froth 1–4) and a secondary tailing (Tail 2). The data from the release tests is then cumulated as shown in Table 14 under Release Analysis Test. An evaluation of the traditional procedures used for flotation release analysis has been discussed extensively in the literature (Forrest et al. 1994; Mohanty et al. 1998).

In many cases, flotation release tests are performed in conjunction with a timed kinetics test in which froth products are taken at timed intervals, usually ¼, ½, 1, 2, 4, and 8 minutes (see Table 14 under Kinetics Flotation Test). A comparison of flotation kinetics and release data allows designers to ascertain whether column cells equipped with a froth washing system can be justified and to estimate residence time requirements for a given flotation feed slurry.

PERFORMANCE ASSESSMENT AND PREDICTION

Partition Analysis for Density Separations

Data obtained from washability analyses represent ideal separations that cannot be duplicated in industrial practice because of the unavoidable misplacement of particles in real-world separation processes. In the coal industry, this misplacement is commonly quantified by means of a partition curve analysis (Wills 1997). For density separators, partition analyses are often carried out by first collecting representative samples of the feed, clean, and refuse streams around a dense medium or gravity separator. After they are collected, the samples are dried and representative splits are subjected to ash analysis. From the ash values, the ratio (Z) of clean tonnage to reject tonnage is calculated using this equation:

$$Z = (A_r - A_f)/(A_f - A_c) \quad \textbf{(EQ 3)}$$

where A_r, A_f, and A_c, and are the refuse ash, feed, and clean contents, respectively. Next, float–sink tests are performed on representative samples of the clean coal and refuse products over an appropriate range of densities. Columns 1–3 in Table 15 provide an example of the types of information obtained from such an exercise. From these data, the partition factors (P) for the separation can be calculated from the following equation:

$$P = 1/[1 + Z(c/r)] \quad \textbf{(EQ 4)}$$

where c and r are the respective weight percentages of clean coal and reject in each density class. The partition factor represents the probability that a feed particle in a given density class will report to the refuse stream. As shown in Figure 21, the partition curve can now be constructed by plotting the partition factors (column 4) as a function of the mean SG of each density class (column 1). The *cut point* (SG_{50}) is defined as the separating SG that corresponds to $P = 0.5$. The *steepness* of

Table 14 Comparison of flotation release analysis and timed-flotation kinetic data

Release Analysis Test							
	Individual		Cumulative Froth		Cumulative Tail		Combined
Float Product	Weight %	Ash %	Weight %	Ash %	Weight %	Ash %	Recovery
Froth 1	18.23	3.09	18.23	3.09	81.77	51.99	31.04
Froth 2	18.23	3.21	36.46	3.15	63.54	65.99	62.04
Froth 3	9.32	5.22	45.78	3.57	54.22	76.44	77.56
Froth 4	7.23	9.07	53.01	4.32	46.99	86.80	89.11
Tail 2	4.23	31.23	57.24	6.31	42.76	92.30	94.22
Tail 1	42.76	92.30	100.00	43.08	0.00	—	100.00
Kinetics Flotation Test							
	Individual		Cumulative Froth		Cumulative Tail		Combined
Time, minutes	Weight %	Ash %	Weight %	Ash %	Weight %	Ash %	Recovery
0.50	31.30	6.20	31.30	6.20	68.70	58.16	50.53
1.00	15.40	9.23	46.70	7.20	53.30	72.30	74.59
2.00	7.70	11.23	54.40	7.77	45.60	82.61	86.35
4.00	3.80	17.32	58.20	8.39	41.80	88.55	91.76
8.00	1.90	22.30	60.10	8.83	39.90	91.70	94.30
Tail	39.90	91.70	100.00	41.90	0.00	—	100.00

Table 15 Example of partition calculations for a dense medium separator*

[1]	[2]	[3]	[4]
Specific Gravity	Clean, wt %	Refuse, wt %	Partition Factor
1.28	13.13	0.00	0.00
1.35	61.48	0.00	0.00
1.43	16.16	0.00	0.00
1.48	5.58	0.01	0.00
1.53	2.44	0.08	0.05
1.58	1.17	0.59	0.43
1.65	0.04	1.17	0.98
1.75	0.00	1.99	1.00
1.90	0.00	3.96	1.00
2.10	0.00	9.03	1.00
2.20	0.00	83.17	1.00
—	100.00	100.00	—

*Given that A_f = 51.69%, A_c = 10.62%, and A_r = 78.89.
Z = (78.89 – 51.69/(51.69 – 10.62) = 0.66

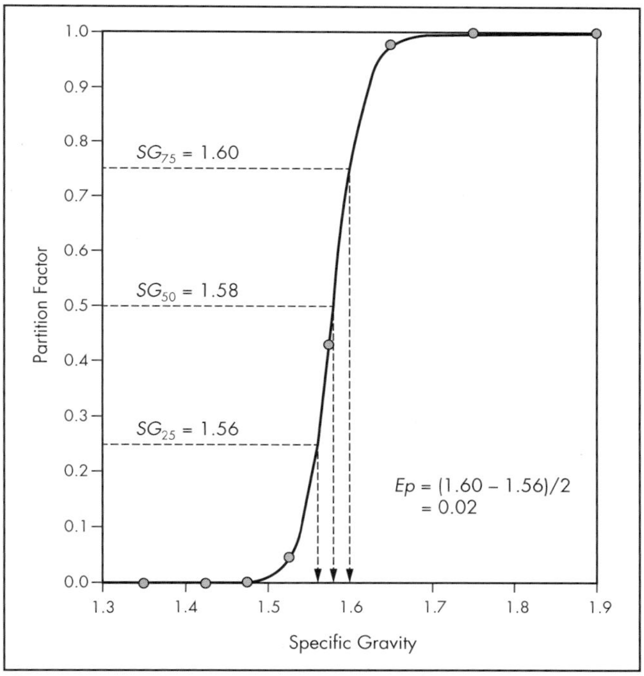

Figure 21 Density separation partition curve

the partition curve, which reflects the sharpness of the separation, is typically defined by an *Ep* value calculated from the following equation:

$$Ep = (SG_{75} - SG_{25})/2 \qquad \textbf{(EQ 5)}$$

where SG_{75} and SG_{25} are the densities corresponding to *P* values of 0.75 and 0.25, respectively. A lower *Ep* is desirable since it means that fewer particles are misplaced during the separation. For the example shown, SG_{50} = 1.58 and Ep = 0.02. The *Ep* value can be compared with historic values to determine whether the level of separation performance is acceptable. The overall shape of a partition curve also provides operators with a useful diagnostic tool. Ideally, the head and tail of the partition curve should close out at *P* values of 0 and 1, respectively. Values that do not close out at these upper and lower limits represent gross misplacement of particles that adversely deteriorate the performance of the separation process. This undesirable bypass can often be eliminated through better operating practices, improved maintenance, or upgraded equipment and circuit designs. The extent of coal misplacement is also commonly reported in the coal industry as an *organic efficiency*, which is defined as the actual yield of clean coal divided by the theoretical maximum yield of clean coal that could be achieved at the same ash content according to a washability analysis. Organic efficiencies may be in the high-90 percentiles for well-designed and well-run operations, although lower values are not uncommon for problematic plants.

Table 16 Example of partition analysis for a classifying cyclone bank

[1] Pass, mm	[2] Retain, mm	[3] Mean, mm	[4] Feed, wt %	[5] Oversize, wt %	[6] Undersize, wt %	[7] (u – o)	[8] (u – f)	[9] Partition Factor
8	4	6	14.6	26.5	0.0	–26.5	–14.6	1.00
4	2	3	17.9	32.1	0.4	–31.7	–17.5	0.99
2	1	1.5	10.7	17.2	2.8	–14.4	–7.9	0.88
1	0.5	0.75	16.7	14.4	20.5	6.1	3.8	0.47
0.5	0.25	0.375	14.1	5.1	25.1	20.0	11.0	0.20
0.25	0.15	0.2	8.5	1.8	15.9	14.1	7.4	0.12
0.15	0.045	0.0975	10.4	1.7	21.2	19.5	10.8	0.09
0.045	0	0.0225	7.1	1.2	14.1	12.9	7.0	0.09
–	–	**Totals**	**100.0**	**100.0**	**100.0**	–	–	–

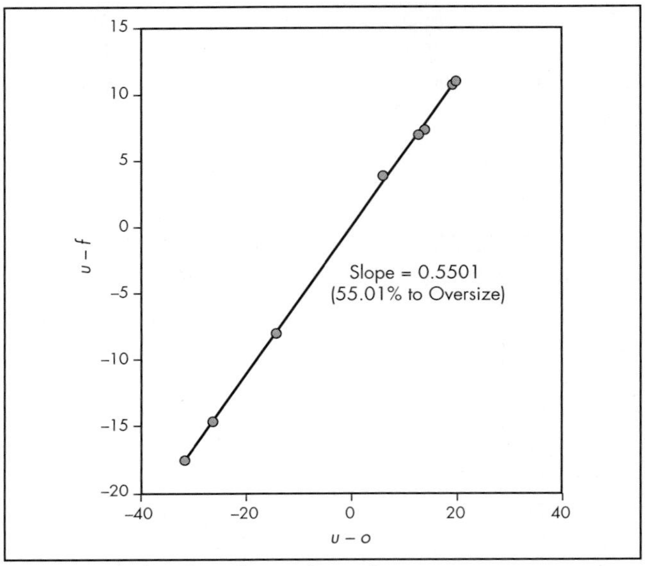

Figure 22 Estimation of oversize split

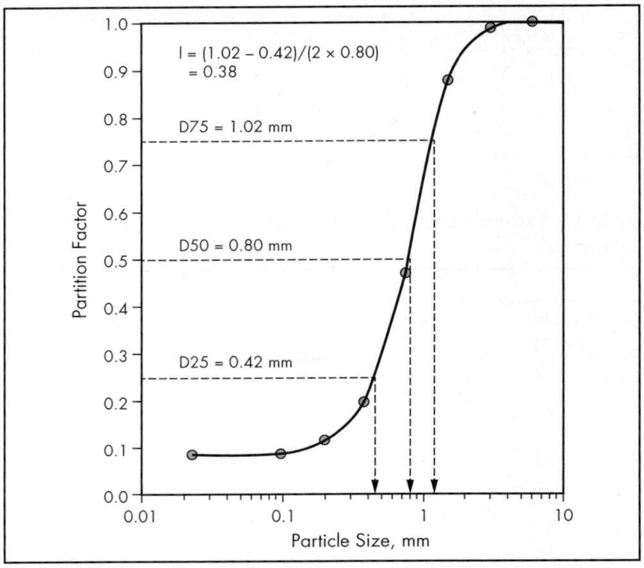

Figure 23 Particle sizing partition curve

Partition Analysis for Size Separations

A different procedure is required for constructing partition curves for screening and classification units since ash values cannot be relied upon for the determination of the *Z* ratio. For this type of separation, the partition analysis is performed by collecting samples of the feed, oversize, and undersize products from sizing devices such as screens, sieves, or classifying cyclones. Each sample is then sized in the laboratory to determine the weight percentage in each size class. An example of such a data set is provided in Table 16. Columns 7 and 8 are calculated by plotting the differences in undersize minus oversize ($u - o$) versus undersize minus feed ($u - f$). As shown in Figure 22, the slope of the resultant line is the mass yield of product reporting to the oversize, which is 55.01% for the current example. Data for size fractions that produce points that do not fall along the line generally represent data that are not reliable and, as outliers, they should be eliminated from the slope calculation. After the oversize yield has been determined, the oversize partition factor (P) for each size class can be calculated using the following equation:

$$P = Y(o/f) \quad \textbf{(EQ 6)}$$

where Y is the yield to oversize, and o and f are the respective weight percentages in the oversize and feed. Using this expression, the partition factors for each size class are determined as listed in column 9 of Table 16. Plotting of the mean size (column 3) versus the partition factor (column 9) provides a partition curve for the sizing device (Figure 23). This plot indicates a cut size (D_{50}) of 0.80 mm for this example. The sharpness of the sizing separation is also often determined from the partition curve using the *imperfection* (I) defined as

$$I = (D_{75} - D_{25})/(2 \times D_{50}) \quad \textbf{(EQ 7)}$$

where D_{75} and D_{25} are the particle sizes corresponding to P values of 0.75 and 0.25, respectively. A lower imperfection is desirable. Typically, imperfection values for classifying cyclones range from 0.20 to 0.80 (Heiskanen 1993), with values of 0.35–0.45 common for raw coal classifying cyclone applications. It is also normal for classifying cyclone partition curves to display a low-tail bypass (B_{lo}) because of short-circuiting of finer particles with water to the oversize product. Low-tail bypass values of 5%–15% are not uncommon for raw coal classifying cyclones and may exceed 35%–40% for coal deslime cyclone applications. For the plot provided in

Figure 23, $I = 0.38$ and $B_{lo} = 9\%$. After this is determined, databases of partition curves such as these can be used to estimate size distributions for oversize and undersize products for different feed coals.

Misplacement in Froth Flotation

Because standard float–sink tests cannot be performed on flotation products, coal operators are forced to use the release analysis procedure to determine the misplacement in froth flotation circuits. While this laboratory technique is normally used to characterize the ultimate separation of feed coals, the procedure can also be used to quantify misplacement in clean coal and refuse samples. For example, Table 17 shows the results of a release analysis test performed on a sample of clean coal from a bank of conventional flotation machines. Although the clean coal contained 9.44% ash, 93.58% of the solids in the froth product contained less than 3.89% ash. The remaining 6.42% of the mass had a very high ash content of 90.37%. The high-ash fraction is typically associated with the hydraulic entrainment of high-ash clay slimes with the water reporting to the froth concentrate. These data suggest that multistage recleaning or column cells equipped with froth washing systems would be a better technology choice for this flotation feed. Typically, well-run column cells will allow less than 1% of the water present in the feed to report to the clean coal product. Release analysis can also be used as a yardstick against which the separation efficiency for flotation can be estimated. The *flotation efficiency* is defined as the actual yield of clean coal in the froth product divided by the theoretical maximum yield of clean coal that could be achieved at the same ash content according to release analysis. Flotation efficiencies are typically lower (80%–90%) than the organic efficiencies observed for well-run dense medium separators.

Table 17 Release analysis on a clean froth product

	Individual, %		Cumulative, %		
Product	**Mass**	**Ash**	**Mass**	**Ash**	**Recovery**
Clean 1	26.89	2.01	26.89	2.01	29.10
Clean 2	25.25	2.30	52.14	2.15	56.34
Clean 3	26.73	3.60	78.87	2.64	84.79
Clean 4	13.58	8.15	92.45	3.45	98.56
Tail 2	1.13	39.72	93.58	3.89	99.32
Tail 1	6.42	90.37	100.00	9.44	100.00
Feed	100.00	9.44	—	—	—

Process Simulation

Historic data compiled from partition analyses can also be used to simulate the performance of coal processing circuits (King 1999). This approach is commonly used to evaluate the cleanability of new feed coals in existing plants or to design new

Table 18 Simulation of circuit performance using partition modeling

Input Data

[1]	[2]	[3]	[4]	[5]	[6]	[7]	[8]	[9]	[10]	[11]	[12]	[13]
Specific Gravity Class	**Feed, % of Class**			**Ash, % of Class**			**Partition to Oversize**			**Partition to Refuse**		
	50 × 10	**10 × 1**	**1 × 0.1**	**50 × 10**	**10 × 1**	**1 × 0.1**	**50 × 10**	**10 × 1**	**1 × 0.1**	**50 × 10**	**10 × 1**	**1 × 0.1**
1.26 × 1.30	10.16	39.20	43.05	5.67	5.53	3.51	0.99	0.92	0.07	0.00	0.00	0.00
1.30 × 1.40	16.24	23.06	26.92	11.37	11.33	9.95	0.99	0.92	0.07	0.00	0.00	0.00
1.40 × 1.50	3.20	5.47	6.06	21.68	19.42	19.14	0.99	0.92	0.07	0.05	0.05	0.05
1.50 × 1.60	2.56	2.46	2.45	31.53	28.93	28.14	0.99	0.92	0.07	0.50	0.50	0.50
1.60 × 1.80	2.83	1.79	2.50	40.74	39.07	34.76	0.99	0.92	0.07	0.95	1.00	1.00
1.80 × 2.00	2.60	2.40	1.55	55.90	51.22	52.46	0.99	0.92	0.07	1.00	1.00	1.00
2.00 × 2.40	62.41	25.62	17.47	88.90	86.75	83.93	0.99	0.92	0.07	1.00	1.00	1.00
Total	**100.00**	**100.00**	**100.00**	**62.01**	**30.71**	**22.38**	—	—	—	—	—	—

Output Data

[14]	[15]	[16]	[17]	[18]	[19]	[20]	[21]	[22]	[23]	[24]	[25]	[26]
Specific Gravity Class	**Feed, % of Total**			**Oversize, % of Total**			**Reject Product**			**Clean Product**		
	50 × 10	**10 × 1**	**1 × 0.1**	**50 × 10**	**10 × 1**	**1 × 0.1**	**50 × 10**	**10 × 1**	**1 × 0.1**	**50 × 10**	**10 × 1**	**1 × 0.1**
1.26 × 1.30	5.77	9.49	8.17	5.71	8.73	0.57	0.00	0.00	0.00	5.71	8.73	0.00
1.30 × 1.40	9.23	5.58	5.11	9.13	5.14	0.36	0.00	0.00	0.00	9.13	5.14	0.00
1.40 × 1.50	1.82	1.32	1.15	1.80	1.22	0.08	0.09	0.06	0.00	1.71	1.16	0.00
1.50 × 1.60	1.45	0.60	0.46	1.44	0.55	0.03	0.72	0.27	0.02	0.72	0.27	0.02
1.60 × 1.80	1.61	0.43	0.47	1.59	0.40	0.03	1.51	0.40	0.03	0.08	0.00	0.03
1.80 × 2.00	1.48	0.58	0.29	1.46	0.53	0.02	1.46	0.53	0.02	0.00	0.00	0.02
2.00 × 2.40	35.46	6.20	3.31	35.10	5.71	0.23	35.10	5.71	0.23	0.00	0.00	0.23
Total	**56.82**	**24.21**	**18.97**	**56.25**	**22.27**	**1.33**	**38.89**	**6.97**	**0.31**	**17.36**	**15.30**	**0.31**
		Weight, %	100.00		Weight, %	79.85		Weight, %	46.17		Weight, %	32.96
		Ash, %	46.91		Ash, %	52.62		Ash, %	83.56		Ash, %	10.87

Calculation Legend:

Col[15] = 56.82 × Col[2]
Col[16] = 24.21 × Col[3]
Col[17] = 18.97 × Col[4]
Col[18] = Col[15] × Col[8]
Col[19] = Col[16] × Col[9]
Col[20] = Col[17] × Col[10]
Col[21] = Col[18] × Col[11]
Col[22] = Col[19] × Col[12]
Col[23] = Col[20] × Col[13]
Col[24] = Col[18] – Col[21]
Col[25] = Col[19] – Col[22]
Col[26] = Col[20] – Col[23]

processing facilities based on compiled coal characterization data. The simulation calculations involve applying projected partition factors for particle sizing and density separation to the size-by-size washability for a feed coal sample. The resultant values for the individual size and density classes can then be cumulated to obtain the overall performance in terms of mass yield and product quality. An example of such a series of calculations is provided in Table 18. These calculations assume that the partition curve is independent of the feed coal washability. This assumption is often not valid for water-based density separators in which particle–particle interactions influence the partitioning response (Tavares and King 1995). For these unit operations, the overall shape of the partition curve can also be greatly impacted by water splits and other operational parameters (Rao et al. 2003). The approach can also introduce significant discretization errors if the upper and lower SG ranges for any density class are too large (Mohanta and Mishra 2009). Therefore, considerable care should be exercised when applying partitioning data from one operation to another application. Several software programs are commercially available that can be used for this type of simulation. Commercial programs include Limn, MetSim, and ModSim, as well as custom programs created by a wide variety of individuals and companies (Rong and Lyman 1985; Rong 1992). In most cases, the simulation routines use empirical partition functions that belong to a mathematical family of transition functions (Scott and Napier-Munn 1992; King 2001).

SUMMARY

Coal preparation has played, and will continue to play, an essential role in supplying high-quality feedstocks for coal-fired power stations, metallurgical coking operations, and industrial boiler systems. Modern coal preparation plants incorporate a complex array of solid–solid and solid–liquid separation processes to remove unwanted impurities such as ash, sulfur, and moisture from ROM coals. Separation processes used by these facilities include screening, classification, dense medium separation, gravity concentration, froth flotation, centrifugation, filtration, and thickening. These operations require constant monitoring to ensure that optimal separations are maintained. Monitoring typically involves the routine collection of representative samples from the various plant circuits to ensure that product specifications have been met and that the misplacement of coal and/or rock is maintained within strict limits. Troubleshooting of performance problems often requires additional laboratory analyses that may include particle sizing, float–sink testing, and flotation release analysis. These analytical procedures often make it possible for plant operators to identify inefficiencies that can be corrected through better operating procedures, improved maintenance programs, or redesign of plant equipment or circuitry. In many cases, these improvements generate significant revenue from the recovery of usable coal from waste streams, which provides a strong financial incentive for management to develop and support efficiency monitoring programs.

REFERENCES

Abbott, J. 1982. The optimisation of process parameters to maximise the profitability from a three-component blend. In *Optimisation of Coal Recovery: Proceedings of the First Australian Coal Preparation Conference*. Newcastle, New South Wales: Coal Preparation Societies of New South Wales and Queensland. pp. 87–105.

Arnold, B.J. 1999. Simuation of dewatering devices for predicting the moisture content of coals. *Coal Prep.* 20(1-2):35–54.

Arnold, B.J., Klima, M.S., and Bethell, P.J. 2007. *Designing the Coal Preparation Plant of the Future*. Littleton, CO: SME.

ASTM International. 1994. Standard test method for determining the washability characteristics of coal. In *Annual Book of ASTM Standards*, Vol. 5.05, Standard No. D4371. West Conshohocken, PA: ASTM International.

Bethell, P.J. 2012. Dealing with the challenges facing global fine coal processing. In *Challenges in Fine Coal Processing, Dewatering, and Disposal*. Edited by M.S. Klima, B.J. Arnold, and P.J. Bethell. Englewood, CO: SME. pp. 33–45.

Bethell, P.J., and Barbee, C.J. 2007. Today's coal preparation plant—A global perspective. In *Designing the Coal Preparation Plant of the Future*. Edited by B.J. Arnold, M.S. Klima, and P.J. Bethell. Littleton, CO: SME. pp. 9–20.

Bethell, P.J., and Luttrell, G.H. 2005. Effects of ultrafine desliming on coal flotation circuits. Paper No. 38. In *Proceedings of the Centenary of Flotation Symposium*. Melbourne, Victoria: Australasian Institute of Mining and Metallurgy. pp. 719–728.

Brodzik, P. 2007. Application of Derrick Corporation's StackSizer technology in clean coal spiral product circuits. In *Designing the Coal Preparation Plant of the Future*. Edited by B.J. Arnold, M.S. Klima, and P.J. Bethell. Littleton, CO: SME. pp. 89–96.

Carris, D.M. 2007. A historic perspective. In *Designing the Coal Preparation Plant of the Future*. Edited by B.J. Arnold, M.S. Klima, and P.J. Bethell. Littleton, CO: SME. pp. 3–8.

Chafin, C., Bethell, P.J. DeHart, G., and Corder, R. 2012. Maximizing fine pyrite rejection at the Arch Coal Leer Plant. In *Challenges in Fine Coal Processing, Dewatering, and Disposal*. Edited by M.S. Klima, B.J. Arnold, and P.J. Bethell. Englewood, CO: SME. pp. 67–78.

Davis, V.L., Jr., Bethell, P.J., Stanley, F.L., and Luttrell, G.H. 1995. Plant practices in fine coal column flotation. In *High-Efficiency Coal Preparation*. Edited by S.K. Kawatra. Littleton, CO: SME. pp. 237–246.

De Korte, G.J., and Engelbrecht, J. 2007. Dense medium cyclones. In *Designing the Coal Preparation Plant of the Future*. Edited by B.J. Arnold, M.S. Klima, and P.J. Bethell. Littleton, CO: SME. pp. 61–71.

Doherty, M. 2006. Coal quality impact on unit availability and emissions. Presented at the American Electric Power Site Visit, Asia Pacific Partnership on Clean Development and Climate, Columbus, OH, October 30–November 4.

Firth, B., and O'Brian, M. 2003. Hydrocyclone circuits. *Coal Prep.* 23:167–183.

Forrest, W.R., Adel, G.T., and Yoon, R.-H. 1994. Characterizing coal flotation performance using release analysis. *Int. J. Coal Prep. Util.* 14(1-2):13–27.

Galvin, K.P. 2012. Development of the REFLUX™ Classifier. In *Challenges in Fine Coal Processing, Dewatering, and Disposal*. Edited by M.S. Klima, B.J. Arnold, and P.J. Bethell. Englewood, CO: SME. pp. 159–185.

Galvin, K.P., Doroodchi, E., Callen, A.M., Lambert, N., and Pratten, S.J. 2002. Pilot plant trial of the REFLUX Classifier. *Miner. Eng.* 15(1-2):19–25.

Ghosh, T., Patil, D., Honaker, R.Q., Damous, M., Boaten, F., Davis, V.L., and Stanley, F.S. 2012. Performance evaluation and optimization of a full-scale REFLUX Classifier. In *Proceedings of Coal Prep 2012*. New York: Penton Publishing. pp. 67–82.

Heiskanen, K. 1993. *Particle Classification*. Amsterdam, The Netherlands: Springer.

Hollis, R. 2006. Blue Diamond Coal Company's Leatherwood preparation plant undergoes major circuit modification. In *Proceedings of the 24th International Coal Preparation Conference*. New York: Penton Publishing. pp. 95–112.

Honaker, R.Q., Boaten, F., and Luttrell, G.H. 2007. Ultrafine coal classification using 150 mm gMax cyclone circuits. *Miner. Eng.* 20(13):1218–1226.

King, R.P. 1999. Practical optimization strategies for coal-washing plants. *Coal Prep.* 20:13–34.

King, R.P. 2001. *Modeling and Simulation of Mineral Processing Systems*. Woburn, MA: Butterworth-Heinemann.

Klima, M.S., Arnold, B.J., and Bethell, P.J. 2012. *Challenges in Fine Coal Processing, Dewatering, and Disposal*. Englewood, CO: SME.

Kohmuench, J.N., Mankosa, M.J., Luttrell, G.H., and Adel, G.T. 2002. Process engineering evaluation of the CrossFlow separator. *Miner. Metall. Process.* 19(1):43–49.

Kohmuench, J.N., Mankosa, M.J., and Yan, E.S. 2010. Evaluation of the StackCell technology for coal applications. *Int. J. Coal Prep. Util.* 30(2-5):189–203.

Kohmuench, J.N., Yan, E.S., and Christodoulou, L. 2012. Column and nonconventional flotation for coal recovery—Circuitry methods and considerations. In *Challenges in Fine Coal Processing, Dewatering, and Disposal*. Edited by B.J., Arnold, M.S. Klima, and P.J. Bethell. Englewood, CO: SME. pp. 187–209.

Laskowski, J.S. 2001. Coal flotation and fine coal utilization. In *Developments in Mineral Processing*. Amsterdam, The Netherlands: Elsevier Science Publishing.

Laskowski, J.S., Luttrell, G.H., and Arnold, B.J. 2007. Coal flotation. In *Froth Flotation: A Century of Innovation*. Edited by M.C. Furstenau, G. Jameson, and R.-H. Yoon. Littleton, CO: SME.

Luttrell, G.H., Kohmuench, J.N., Stanley, F.L., and Trump, G.D. 1998. Improving spiral performance using circuit analysis. SME Preprint No. 98-208. Littleton, CO: SME.

Luttrell, G.H., Honaker, R.Q., Bethell, P.J., and Stanley, F.L. 2007. Design of high-efficiency spiral circuits for coal preparation plants. In *Designing the Coal Preparation Plant of the Future*. Edited by B.J. Arnold, M.S. Klima, and P.J. Bethell. Littleton, CO: SME. pp. 73–87.

Luttrell, G.H., Noble, A., and Stanley, F. 2014. Micro-price optimization of coal processing operations. SME Preprint No. 14-127. Englewood, CO: SME.

Majumder, A.K. 2011. Processing of coal fines in a water-only cyclone. *Fuel* 90(2):834–837.

Meenan, G. 2005. Implications of new dewatering technologies for the coal industry. Invited lecture at the Center for Advanced Separation Technologies (CAST) Workshop, Virginia Tech, Blacksburg, VA, July 26–28.

Mohanta, S., and Mishra, B.K. 2009. On the adequacy of distribution curves used in coal cleaning: A statistical analysis. *Fuel* 88:2262–2268.

Mohanty, M.K., Honaker, R.Q., Patwardhan, A., and Ho, K. 1998. Coal flotation washability: An evaluation of traditional procedures. *Int. J. Coal Prep. Util.* 19(1-2):24–32.

Mohanty, M.K., Palit, A. and Dube, B. 2002. A comparative evaluation of new fine particle size separation technologies. *Miner. Eng.* 15:727–736.

Mohanty, M.K., Zhang, B., and Geilhausen, R. 2009. *On-Site Evaluation of the Stack Sizer in Conjunction with the Falcon Concentrator.* Final technical report to Department of Commerce and Economic Opportunities/Illinois Clean Coal Institute, Project No. 07-1/9.1B-2, April.

National Research Council. 2002. *Coal Waste Impoundments: Risks, Responses, and Alternatives*. Washington, DC: National Academies Press.

Osborne, D. 2012. Milestones in fine coal cleaning development. In *Challenges in Fine Coal Processing, Dewatering, and Disposal*. Edited by M.S. Klima, B.J. Arnold, and P.J. Bethell. Englewood, CO: SME. pp. 3–32.

Rao, B.V., Kapur, P.C., and Rahul, K. 2003. Modeling the size-density partition surface of dense-medium separators. *Int. J. Miner. Process.* 72(1-4):443–453.

Rong, R.X. 1992. Optimization of complex coal preparation flowsheets. *Int. J. Miner. Process.* 34(1-2):53–69.

Rong, R.X., and Lyman, G.J. 1985. Computational techniques for coal washery optimization—Parallel gravity and flotation separation. *Coal Prep.* 2:51–67.

Schultz, W., Jahnig, R., Bratton, R.C., and Luttrell, G.H. 2008. Operating and maintenance guidelines for screenbowl centrifuges. *Proceedings of the 25th Annual International Coal Preparation Exhibition and Conference*. New York: Penton Publishing. pp. 169–177.

Scott, I.A., and Napier-Munn, T.J. 1992. Dense-medium cyclone model based on the pivot phenomenon. *Trans. Inst. Min. Metall.* 101(January-April):C61–C76.

Skiles, K.D. 2003. Search for the next generation of coal flotation collectors. *Proceedings of Coal Prep 2003, 20th Annual International Coal Preparation Exhibition and Conference*. New York: Penton Publishing. pp. 187–204.

Tavares, L.M., and King, R.P. 1995. A useful model for the calculation of the performance of batch and continuous jigs. *Coal Prep.* 15:99–128.

Wills, B.A. 1997. *Mineral Processing Technology*. 6th ed. Oxford: Butterworth-Heinemann.

Yoon, R.-H., Luttrell, G.H., Adel, G.T., and Mankosa, M.J. 1992. The application of Microcel column flotation to fine coal cleaning. *Coal Prep.* 10:177–188.

Zhang, B., Yang F., Sahoo, P.K., Mohanty M.K., and Zhang, X. 2008. Optimization of a compound spiral for cleaning high sulfur fine and ultrafine coal. *Proceedings of the XXIV International Mineral Processing Congress*. Vol. 2. Beijing: Science Press. pp. 653–662.

Ziaja, D., and Yannoulis, G.F. 2007. Is there anything new in coarse of intermediate coal cleaning? In *Designing the Coal Preparation Plant of the Future*. Edited by B.J. Arnold, M.S. Klima, and P.J. Bethell. Littleton, CO: SME. pp. 43–59.

CHAPTER 12.7

Copper Mineral Processing

Adam Johnston, David G. Meadows, and Frank Cappuccitti

Copper is a metal that exhibits the highest electrical conductivity besides silver. It is regarded as a relatively noble metal. Because of its electrical conductivity, copper is used extensively in the electrical industry for wiring, motors, and cabling. It is also used in plumbing, industrial machinery, and construction materials because of its durability, machinability, corrosion resistance, and ability to be cast with high precision and tolerances.

MINERAL PROCESSING IN COPPER PRODUCTION

This chapter focuses on the use of mineral processing extraction techniques to recover copper minerals from ores. Details of copper extraction by hydrometallurgy and pyrometallurgy are provided elsewhere in the handbook. Overwhelmingly, the mineral processing of copper ores refers to the use of crushing, grinding, and froth flotation to recover copper sulfide minerals from ores to concentrates. A few plants use copper oxide ore flotation, but these are not discussed in this chapter. A typical copper sulfide concentrator flow sheet is shown in Figure 1.

Since large semiautogenous grinding (SAG) mills were popularized in copper concentrators in the 1980s and 1990s, and with the treatment of lower-grade deposits, SAG mills have prevailed and are included in many of the concentrator flow sheets with fewer larger pieces of equipment being employed. In addition, larger gyratory crushers, cone crushers, ball mills, and circular tank flotation machines have also been included as a feature of these flow sheets, allowing copper concentrators to employ fewer pieces of equipment and at lower overall operating costs.

Several of concentrators built before the 1990s often had different approaches for the comminution and flotation circuits. Since then, the complexity and variation in flow-sheet design and configuration has been significantly reduced. Except where unique mineralogical or metallurgical considerations exist, most copper flotation concentrators use a standard approach, like that shown in Figure 2.

Emerging comminution technologies such as high-pressure grinding rolls (HPGRs) and new-generation flotation cells are challenging this norm; however, most deviations from this arrangement are usually due to complications in metallurgy related to ore mineralogy, including

- Oxidation and alterations of copper or gangue minerals;
- Recovery of credit elements such as gold, zinc, or molybdenum;
- Management of penalty species such as arsenic, bismuth, zinc, or talc; and
- Significantly harder ores requiring different comminution approaches (e.g., Bond ball mill work index >20 kW·h/t and low JK drop weight test A×b parameters)

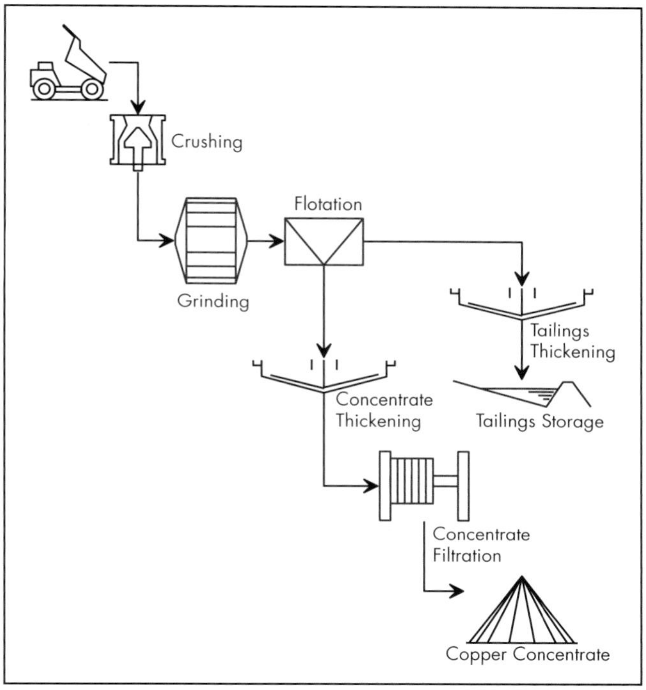

Figure 1 Typical copper sulfide concentrator arrangement

Adam Johnston, Chief Metallurgist, Transmin Metallurgical Consultants, Lima, Peru
David G. Meadows, Manager of Global Process Technology, Bechtel Mining and Metals, Phoenix, Arizona, USA
Frank Cappuccitti, President, Flottec LLC, Boonton, New Jersey, USA

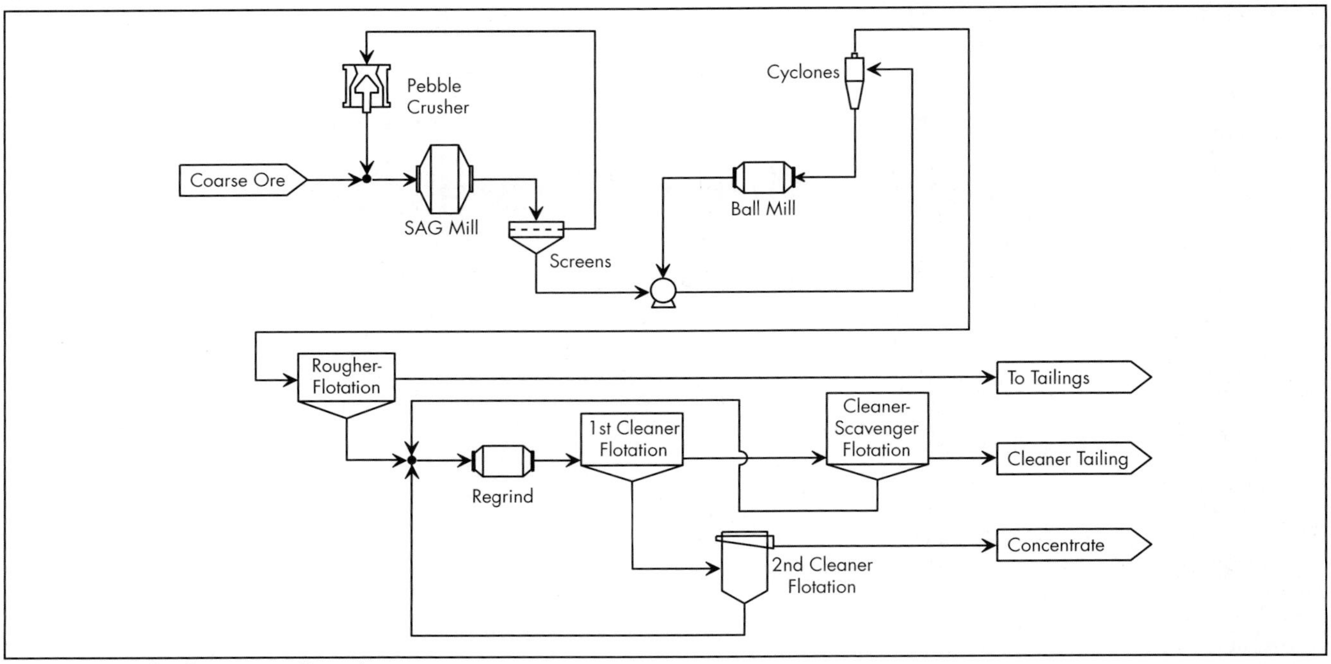

Figure 2 Standard recovery flow sheet

SUSTAINABILITY OF COPPER PRODUCTION

In 2016, 70% of the copper produced in Chile was by flotation to concentrates. The Comisión Chilena del Cobre, or Cochilco, projects that percentage will increase to more than 90% by 2028 (Cochilco 2017). Oxide ores tend to be found near the surface in arid areas and have bright blue and green mineral colors. Consequently, they are easier to find than the deeper sulfide ores. Over the last 30 years, they have been cheaper to exploit via the low-cost heap leaching–solvent extraction and electrowinning process.

Very few new large sulfide deposits have been found since the 1990s. Most of the new copper ore reserves have come from brownfields exploration of existing deposits. Therefore, the deposits with higher grades, amenable infrastructure, social and environmental factors, and high-quality concentrates are already being mined and processed.

Since copper demand is projected to continue to increase, the existing operations have been compensating by increasing throughput, but there has been scant technological innovation to reduce specific energy consumption and water consumption or to improve the metallurgical response of refractory ores.

The Copper Alliance, whose members accounted for about half of the copper in ore mined in 2015, reports that while the mass of copper ore treated has been increasing, the carbon dioxide (CO_2) emissions have also increased from 3.4 to 4.1 (t CO_2/t Cu), a 20% increase, between 2011 and 2015. While worker safety increased, other sustainability measures fell. Water consumption increased, and the total number of workers employed decreased (ICA 2018). Longer-term issues facing the mining industry that could impact copper production in the future include

- Fewer large resource discoveries,
- Lower copper grades,
- Harder ores with complex mineralogy,
- Increasing levels of hazardous elements in copper ores,
- More-stringent environmental regulations,
- Increasing anti-mining sentiment,
- More-stringent tailings disposal regulations,
- Less access to water for mining activities,
- More-stringent quality requirements for copper concentrates, and
- Shortage of experienced skill works in the mining industry.

MINERALOGY AND GEOLOGY OF COPPER ORE DEPOSITS

Copper ores constitute a large and geologically diverse group of ores, including some of the world's largest known ore bodies. Many are dominated by copper as the primary metal of production, others have co-product or by-product gold or molybdenum, while others are part of polymetallic copper-zinc-lead ores.

Mineralogy of Copper Ores

The mineralogy of copper ores is summarized in Table 1. Primary (hypogene) copper ores usually contain only chalcopyrite, and sometimes bornite, as the main copper minerals. These are accompanied by variable amounts of iron sulfide, iron oxide, or copper-iron-arsenic sulfide minerals. Native gold is present in some ores, but gold is often found in solution in the sulfides. Primary ores are often overlain by oxide and supergene zones, which may be of economic importance because of higher enriched copper grades. The mineralogy of these oxide and supergene zones can be complex, both in mineralogical assemblage and in copper mineral textures. Common copper ore minerals are listed in Table 2.

Classification of Ore Types

Given that the first ores to be sent to a concentrator from the mine are near the surface, the early financial success during start-up of a new concentrator often depends on the correct classification of oxide, mixed, and supergene ore zones. These zones have often undergone phyllic and argillic alterations, which can further complicate flotation performance due to viscosity and surface coatings caused by clays. Different techniques can be used to classify drill core and samples into these categories, including the following:

1. **Visual inspection.** Gossan rocks have brown colors due to clays and iron oxides such as goethite and jarosite, which may indicate oxide zones. Care must be taken when logging copper mineralogy visually because the bright blue and green copper oxide minerals on rock surfaces may appear more abundant than what is actually inside the ore. Sulfide minerals may be inside the rock fragments and may be harder to see with the naked eye.
2. **Sequential copper assay.** This is discussed in more detail in the "Sequential Leach Analysis" section that follows.
3. **Geochemical indicators:**
   - Cobalt tends to report to the pyrite in copper sulfide deposits. Where the ore and pyrite has been oxidized, the cobalt will be leached away. Reviewing the cobalt grade down a drill hole can indicate the oxidation profile and grade of pyrite.
   - S/(Fe+Cu) ratio. Copper, iron, and sulfur grades cannot be used alone because copper and gangue mineralogy can overlap, so these grades alone do not indicate copper mineralogy or extent of oxidation. The S/(Fe+Cu) ratio, however, can show contrast between sulfide areas with high values and oxide areas where the sulfur has been depleted by oxidation. Where pyrite is abundant, as compared to copper, a simplified S/Fe ratio can work just as well.

Table 1 Mineralogy of copper deposits

Ore Zone	Mineralogy
Oxide	Cuprite, native copper, native gold, malachite, azurite, goethite, hematite
Supergene	Chalcocite, digenite, covellite, native gold
Hypogene	Pyrite, pyrrhotite, chalcopyrite, bornite, enargite, tetrahedrite, molybdenite, magnetite, hematite, native gold

Table 2 Common copper ore minerals

Name	Specific Gravity, g/cm³	Molecular Formula	% Cu	Cu/S Mass Ratio	Hardness
Chalcocite	5.6–5.8	Cu_2S	79.9	3.96	2.5–3.0
Covellite	4.6	CuS	66.5	1.98	1.5–2.0
Chalcopyrite	4.1–4.3	$CuFeS_2$	34.6	0.99	3.5–4.0
Bornite	4.9–5.1	Cu_5FeS_4	63.3	2.48	3.0–3.3
Digenite	5.6	Cu_9S_5	78.1	3.57	2.5–3.0
Tennantite	4.6–4.7	$Cu_{12}As_4S_{13}$	51.6	1.83	3.5–4.5
Enargite	4.4	Cu_3AsS_4	48.4	1.49	3.0
Tetrahedrite	4.6–5.2	$Cu_{12}Sb_4S_{13}$	45.8	1.83	3.5–4.0
Stannoidite	4.3	$Cu_8Fe_2ZnSn_2S_{12}$	38.3	1.32	4.0
Mawsonite	4.7	$Cu_6Fe_2SnS_8$	43.9	1.49	3.5–4.0
Wittichenite	6.3–6.7	Cu_3BiS_3	38.4	1.98	2.0–3.0
Djurleite	4.2	$Cu_{31}S_{16}$	79.6	3.84	2.3–3.0
Cubanite	4.7	$Cu_2S{\cdot}Fe_4S_5$	23.4	0.66	3.5–4.0
Luzonite	4.4	Cu_3AsS_4	58.8	1.98	3.5

Sequential Leach Analysis

Sequential leach analysis is a geochemical procedure where a small mass of pulverized sample is leached in a series of different, increasingly aggressive lixiviants. After each leach, the solution is assayed for copper. The final residue is digested with strong acids to determine the remainder, or residual, copper. The results can indicate the proportion of different copper minerals in the sample.

A common regime is sulfuric acid–soluble copper (CuAS), sodium cyanide–soluble copper (CuCN), and residual copper (CuR). Other regimes can include weaker acids such as citric or acetic acid to improve contrast between oxide copper minerals and enriched copper minerals, which are partially soluble in sulfuric acid. Because of the different procedures and varying reagents, leach times, temperatures, redox modifiers, and sample preparation, care should be taken when comparing results from different sources.

The proportion of the sequential copper assays can be plotted on a ternary diagram (Figure 3). The exact proportion

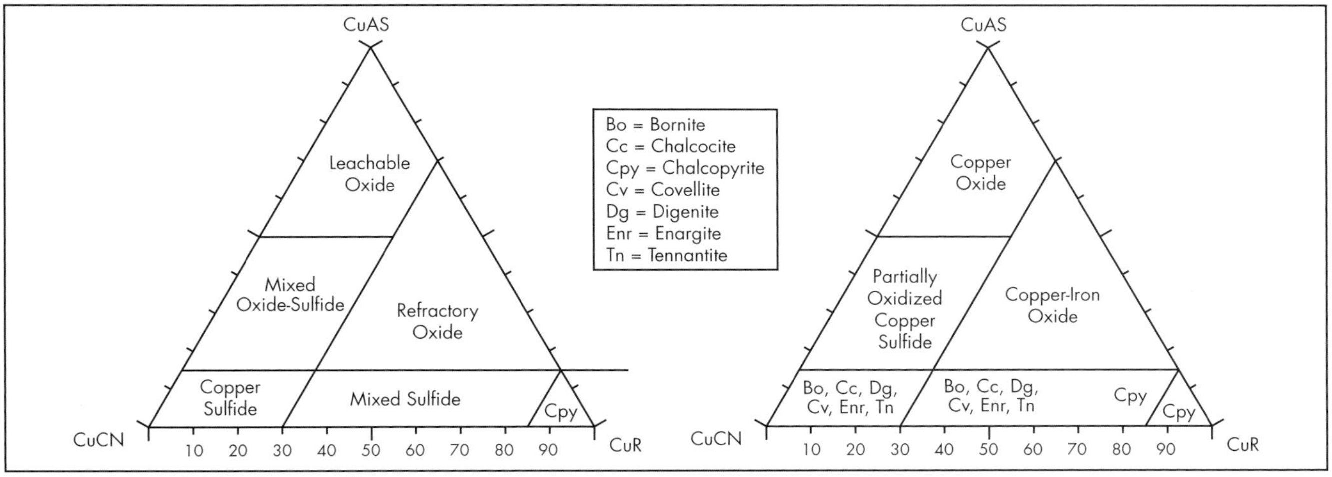

Figure 3 Sequential leach assay diagrams

of each component in each ore type varies by deposit and by sequential leach procedure.

Given the many pitfalls of using sequential leach procedures, the user should pay special attention to the following:

- **Fresh samples.** Old or oxidized samples will show erroneously high acid-soluble values. For fine samples, a low-drying oven temperature should be used to avoid copper sulfide oxidation in the samples.
- **Standard procedure.** The same method should be used for exploration assays and metallurgical sample assays. This is important in the geometallurgical analysis that aligns metallurgical test-work results with the resource block model.
- **Caution with metallurgical products.** Tailing samples will often be ground and dried in a hot oven before assay, partially roasting the sample. Consequently, a sequential assay will show erroneously high acid-soluble values.
- **Oxidative leach solutions.** Enriched copper sulfide minerals such as chalcocite will partially leach in weak sulfuric acid solutions that also contain oxidants such as ferric or chloride ions. Samples that show less than 30% of the copper present as acid soluble could be sulfide samples with no oxides present.
- **Normalize with total copper assays.** The sequential copper assays are less accurate than primary metal assays. Therefore, it is common to normalize the sequential assay by the ratio of the components rather than using the absolute sequential assay grade.

Therefore, the solubilities of single copper minerals in acid and cyanide (Tables 3 and 4) are indicative but not accurate for mine samples.

Chalcopyrite

Chalcopyrite (cpy) is a relatively unreactive copper iron sulfide that is usually associated with pyrite (py). Porphyry hypogene copper mineralogy is almost exclusively chalcopyrite. Chalcopyrite flotation is sensitive to oxidation potential, as it needs to oxidize slightly to be collected by xanthate collectors. High-pyrite deposits (>5% py) can have high oxygen consumption, requiring the use of other collectors, such as dithiophosphate or thionocarbamate (Figure 4).

Bornite

Bornite is a high-grade copper iron sulfide mineral that can be floated with xanthate collectors at lower pH, and with dithiophosphates or thionocarbamates, to form very-high-grade copper concentrates. Small amounts of sodium sulfide (100–200 g/t) can be used to promote bornite flotation, but an excess can lower dissolved oxygen in the slurry, depressing flotation of the other copper sulfides.

Chalcocite, Covellite, Digenite, Enargite, and Tennantite

The copper sulfide minerals—chalcocite, covellite, digenite, enargite, and tennantite—are readily floated using xanthates, dithiophosphates, thionocarbamates, and mercaptobenzothiazole collectors. These minerals can produce high-grade copper concentrates but often have lower copper recoveries due to partial oxidation, clays and weathered gangue minerals, and poor liberation due to complex mineral textures.

Table 3 Solubility of copper minerals in cyanide

		Percent Total Copper Dissolved	
Mineral	**Molecular Formula**	**23°C**	**45°C**
Azurite	$2CuCO_3 \cdot Cu(OH)_2$	94.5	100.0
Malachite	$CuCO_3 \cdot Cu(OH)_3$	90.2	100.0
Chalcocite	Cu_2S	90.2	100.0
Copper metal	Cu	90.0	100.0
Cuprite	Cu_2O	85.5	100.0
Bornite	Cu_5FeS_4	70.0	100.0
Enargite	Cu_3AsS_4	65.8	75.1
Tetrahedrite	$4Cu_3S \cdot Sb_3S_3$	21.9	43.7
Chrysocolla	$CuSiO_3$	11.8	15.7
Chalcopyrite	$CuFeS_2$	5.6	8.2

Source: Hedley and Tabachnick 1958

Table 4 Solubility of copper minerals in sulfuric acid and cyanide

Mineral Species	Approximate Composition	Approximate Dissolution in Sulfuric Acid Solution	Approximate Dissolution in Sodium Cyanide Solution
Oxides			
Atacamite	$Cu_2Cl(OH)_3$	100	100
Azurite	$2CuCO_3Cu(OH)_2$	100	100
Chrysocolla	$CuSiO_32(H_2O)$	100	45
Cuprite	Cu_2O	70	100
Malachite	$CuCO_3Cu(OH)_2$	100	100
Native copper	Cu	5	100
Tenorite	CuO	100	100
Secondary Sulfides			
Chalcocite	Cu_2S	3	100
Covellite	CuS	5	100
Primary Sulfides			
Bornite	Cu_5FeS_4	2	100
Chalcopyrite	$CuFeS_2$	2	7

Source: Hedley and Tabachnick 1958
Note: Samples are generally finely ground (–150 mesh), and reaction time is generally one hour or less.

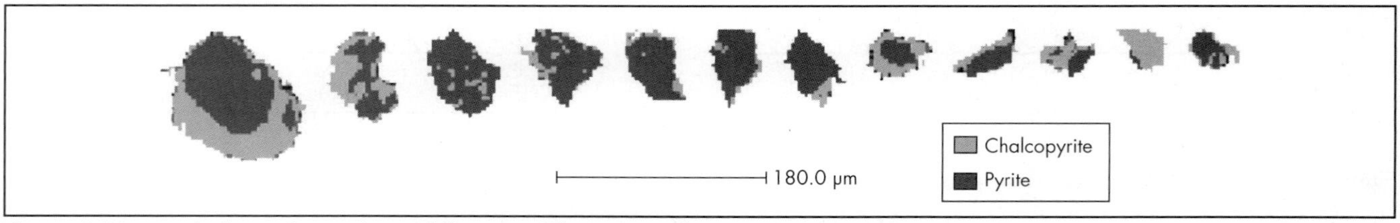

Figure 4 Binary chalcopyrite-pyrite middling particle textures from a copper porphyry ore

Mixed Copper Mineralogy

Supergene copper mineralogy can be complex but generally presents in three textures:

1. Copper sulfides replacing pyrite, either completely or as rims (Figure 5).
2. Copper sulfides replacing chalcopyrite, either completely or as rims (Figure 6).
3. Copper sulfides replacing other copper sulfides.

FLOW SHEETS

Process Selection

Most copper concentrators use the same arrangement for the process:

1. Comminution. A combination of crushing, autogenous grinding mills, SAG mills, and/or HPGRs reduce the ore particle size to P_{80} 1–4 mm, suitable to be fed to ball mills. The ball mills reduce the particle size further to P_{80} 106–250 µm.
2. Classification. Hydrocyclones are, by far, the most popular equipment for classifying the mill discharge. Fine screen decks have been used in smaller copper concentrators in recent years.
3. Rougher flotation. A high recovery froth flotation stage of 15–40 minutes of residence time produces a rougher concentrate in 5%–15% mass pull from the feed. The rougher tailings are usually sent to final tailings storage.
4. Regrind. The regrind circuit can be fed combinations of rougher concentrate, rougher-scavenger concentrate, or cleaner-scavenger concentrate.
5. Cleaner flotation. Two or three stages of cleaning, 5–15 minutes of residence time each, are used to separate the copper minerals from non-copper sulfides and gangue in the rougher concentrate to produce commercial concentrate quality.
6. Cleaner scavenger flotation of 10–20 minutes of residence time is used to capture slow-floating particles from the first cleaner tailings. The cleaner-scavenger tailings usually report to the final tailings.
7. The copper concentrate is thickened and filtered to produce a cake with 8%–10% moisture for shipment to smelters.
8. The combined final tailings are usually thickened to recover water and reagents before being sent to the tailings storage. In some cases, the water is recovered from the tailings storage facility without using a thickener.

This arrangement is shown in the simplified flow sheet in Figure 7.

Despite efforts to develop a standardized design for copper concentrators, new projects have continued to embrace variations of this conceptual arrangement due to factors that include the following:

- Oxidation and alteration of the sulfides or gangue minerals.
- By-products. Economic benefits are realized from recovering secondary products such as molybdenum, zinc, lead, gold, silver, magnetite, uranium, and nickel.
- Penalty elements. Recovery of deleterious components such as arsenic, bismuth, cadmium, uranium, fluorine, talc, antimony, lead, zinc, and iron can result in severe economic penalties for concentrate sales.
- Geography. Plants in difficult terrain may separate the main processes geographically, changing the mass and water balances and the resulting arrangement.
- Tailings storage and water consumption limits. Water recovery from tailings, tailings volume, and geotechnical stability are related. As concerns about tailings stability and scrutiny regarding freshwater consumption increase, tailings deposition strategies are trending toward paste and dry-stacked technologies.
- Preconcentration. Ore sorting, dense media, gravity concentration, and coarse particle flotation have had little industrial success, but as head grades decrease, there is growing interest in preconcentrating the copper sulfide minerals at coarser particle sizes before grinding to the customary rougher flotation particle size.

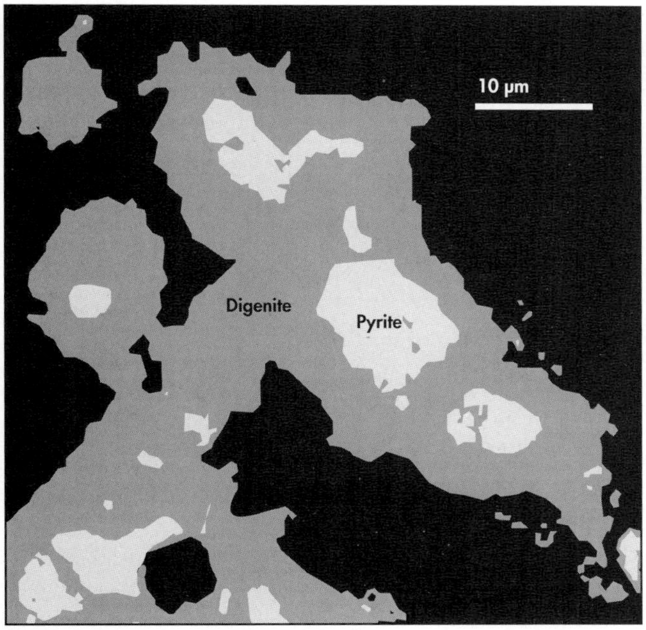

Figure 5 Digenite rimming pyrite from a manto deposit

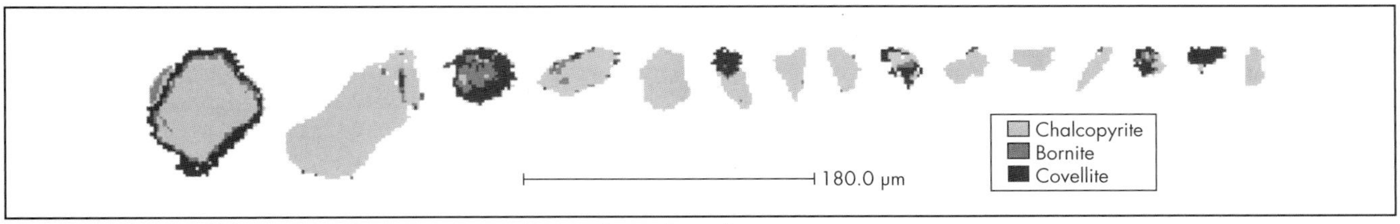

Figure 6 Mixed sulfide textures from a supergene porphyry deposit

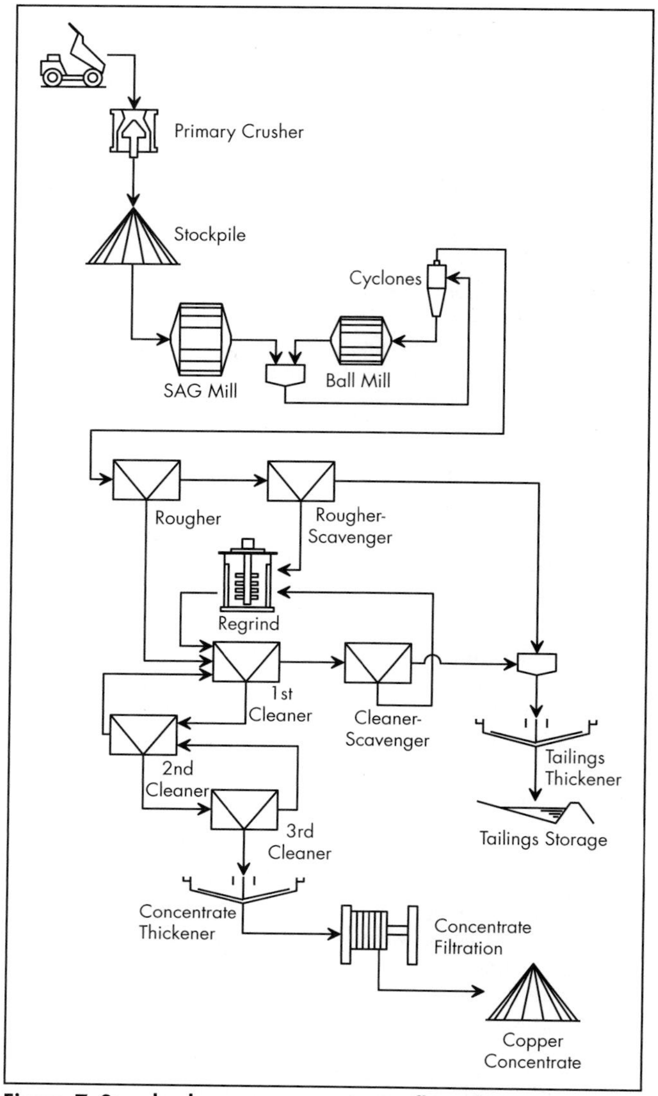

Figure 7 Standard copper concentrator flow sheet

Particle Size Selection

The conventional method of selecting particle size for plant design has been to run batch grinding tests in a laboratory mill, followed by kinetic batch rougher flotation tests. A cumulative grade–recovery curve is developed from the results. A pragmatic economic decision is then made by comparing the marginal copper recovery projected at each grind size versus the marginal capital and operating costs of grinding. This method does not consider the effect that a continuous grinding circuit with cyclone classification will have on the liberation of copper sulfides.

Copper and iron sulfides have higher specific gravity than most nonsulfide gangues. Consequently, a coarser industrial rougher flotation feed size can be considered versus the laboratory tests where cyclones are used for classification. This effect is lost where screens are used as classifiers or where the gangue has a high specific gravity.

Coarse particle recovery processes such as centrifugal concentrators have been used to recover gold and native copper from cyclone underflow streams. Klimpel (1993) reported that recovery optimization efforts for coarse grinds should focus on the coarse particles, whereas fine grinds should focus recovery optimization of the fine particles. This can be seen when reviewing the efforts at the Grasberg mine in Indonesia (P_{80} 210 µm primary, 38 µm regrind) and Prominent Hill mine in South Australia (P_{80} 100 µm, 15 µm regrind). In both cases, mechanical agitation, froth hydrodynamics, and reagent regimes were tested. Different opportunities were identified in each case for copper recovery improvement routes for coarse and fine particles.

Comminution

Copper ore particles are reduced in size to liberate the copper minerals from the gangue so that they can be floated selectively. Comminution is expensive because it consumes a lot of energy; consequently, concentrator designs are configured to preconcentrate the copper at a coarse grind size. The low-grade concentrate is then typically reground to a finer size to achieve liberation from the gangue. Coarse tailings are preferred because they can also reduce tailings storage costs, underground backfill costs, water consumption, and the capital and operating costs associated with crushing and grinding.

In the 1980s, semiautogenous milling replaced three stages of crushing as the preferred approach to comminution for large grinding circuits (Figure 8). This was due to SAG mills having higher capacities, simpler flow sheets, lower operating costs, higher availabilities, reduced maintenance complexities, and the ability to treat wet and sticky ores.

Although most new concentrators continue to favor the SAG mill approach, some existing concentrators are retrofitting the circuit with secondary crushing to increase throughput through the SAG mills. A significant portion of new concentrators are using HPGRs to reduce energy consumption, particularly where the ore is competent or where the energy costs are high. The largest HPGR units have lower capacity than the largest SAG mills; thus, high tonnage is achieved by installing multiple units in parallel (Figure 9).

Flotation

Copper is upgraded from 0.3%–2% Cu in the ore to 22%–45% in a concentrate using the froth flotation process. The flotation circuit design is different in every plant. There is no standard for the flow-sheet arrangement or circuit parameters. There are, however, a few common approaches to pragmatic flow-sheet development.

Reagents

Since all ores are different, most concentrators also have different reagent regimes for copper flotation. Flotation reagents vary depending on factors that include

- Copper mineral assemblage and textures;
- Precious metal (gold, silver) and other valuable base metals (lead, zinc, molybdenum, nickel);
- Iron sulfide (pyrite, marcasite, pyrrhotite) abundance, texture, and associations;
- Nonsulfide gangue interactions (clays, chlorite, talc, carbonates);
- Penalty element mineralogy (arsenic, bismuth, antimony, cadmium); and
- Ore variability.

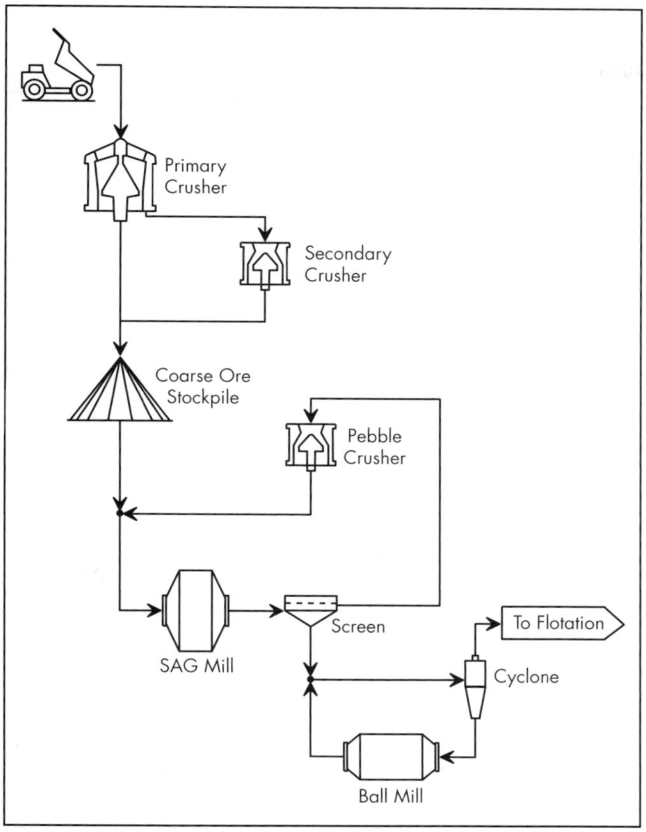

Figure 8 Common semiautogenous milling arrangement

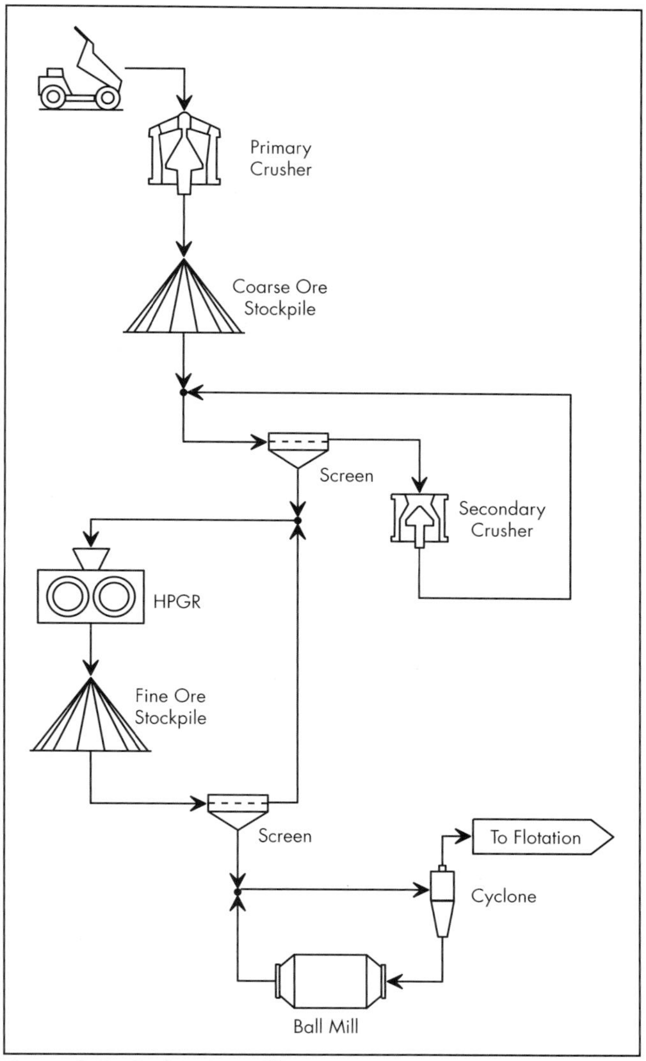

Figure 9 Typical HPGR milling arrangement

The objective is to choose a reagent regime that can consistently recover the valuable minerals into a clean concentrate, at a low cost, within the variations of daily mining feed blends to the concentrator. Table 5 provides a list of reagents commonly used in copper sulfide mineral flotation.

There has been little change in the molecules available for copper flotation in the last 30 years. Reagent manufacturers offer blends of molecules directed at certain ore types, marketing them as number codes rather than by their chemistry. Since there are several reagent suppliers, each with several candidate collectors, a metallurgist may be faced with dozens of different potential reagents to test. Many of the competing reagents have similar chemistry; therefore, testing them is redundant.

It is unusual for a reagent selection to be made on the cost of the reagent alone. Usually the metallurgical performance (selective floatability of the valuable metals) is the key reagent selection criteria. However, reagent schemes can be changed because of other factors, including occupational and health issues.

The optimum reagents for rougher flotation, including collectors, modifiers, and frothers, can be different. Rougher flotation requires strong floatability and low selectivity to achieve high recovery for large, middling particles. Cleaner flotation requires weaker, more-selective floatability to achieve a high-quality concentrate from fine particles. It is costly to change the water or adsorbed reagents between stages, so this practice is limited to separations on final concentrates, such as copper–molybdenum separation, chalcopyrite–enargite separation, and copper–lead separation.

The usual practice is to balance the reagent selection for rougher and cleaner performance. Since rougher flotation is higher tonnage, using lower pH and fewer modifier additions is preferred. It is cheaper to add depressants, pH modifiers, and collectors in the cleaner stage.

Milk of lime is used to increase pH. The lime is prepared in slaking mills and dosed to the grinding circuit and regrind circuits to achieve pH 8–10 in the rougher flotation circuit and pH 9–11 in the cleaners. Lime consumption depends on the ore acidity and the target pH but is generally 0.4–1.5 kg/t lime per metric ton of ore treated. Lime consumption is higher in plants that use seawater, so collectors that allow selectivity at a lower pH are used.

Sulfidization agents such as sodium metabisulfite and sodium sulfite (Na_2S) can be used to sulfidize unstable minerals such as bornite, semi-oxidized minerals, or oxide minerals such as malachite, forming a copper-sulfide coating that can be collected for flotation. Carboxymethylcellulose (CMC)

Table 5 Reagents used in copper sulfide mineral flotation

Reagent Type	Reagent Name	Typical Addition Rate, g/t	Reagent Function
Collector	Xanthate	5–350	Provide the appropriate level of collector for the mineral concerned (e.g., low levels of ethyl xanthate for selective flotation of chalcopyrite, but higher additions of propyl xanthate for bulk flotation)
	Dithiophosphate	10–250	
	Thionocarbamate	10–15	
	Xanthogen formate	2–25	
	Xanthic ester	2–25	
	Mercaptobenzothiazole	25–250	
	Thiocarbanilide	25–75	
Frother	Synthetic alcohols (e.g., methyl isobutyl carbinol, or MIBC)	25–50	Provide the appropriate type of froth (e.g., wet or dry; brittle or persistent; with or without some collecting power)
	Polyglycols (e.g., polypropylene glycol)	5–100	
	Polyglycol ethers	5–25	
	Alkoxy compounds (e.g., triethoxybutane, or TEB)	5–25	
	Pine oil	15–100	
	Eucalyptus oil	25–100	
	Cresylic acid	25–100	
pH Modifier	Quick lime (CaO) and slaked lime ($Ca(OH)_2$)	200–2,400	Provide high pH (>11) for depression of iron sulfides
	Sodium carbonate	500–2,500	Provide pH levels about 8–9
	Sulfuric acid	250–2,500	Provide acid solutions (pH < 6)
	Sulfurous acid and sodium metabisulfite	NA	Provide pH about 6, and depression of galena and sphalerite
Activator	Copper sulfate	100–2,500	Activates depressed sphalerite
	Sodium sulfide and sodium hydrosulfide	250–500	Activates tarnished copper sulfides and some oxide copper minerals
Depressant	Sodium sulfite and SO_2	250–2,000	Depresses sphalerite and galena
	Sodium sulfide and sodium hydrosulfide	200–500	Depresses chalcopyrite in Mo–Cu separation
	Sodium cyanide and Aero cyanide	5–250	Depresses most sulfide minerals, but especially iron sulfides
	Zinc sulfate	50–1,500	Depresses sphalerite
	Sodium dichromate	100–2,500	Galena depressant in Cu–Pb separation
	Sodium silicate	50–1,000	Gangue dispersant
	Nokes reagent (P_2S_5 + NaOH)	5,000	Chalcopyrite depressant in chalcopyrite–molybdenite separation
	Carboxy methyl cellulose (CMC)	5–500	General gangue depressant
	Organic colloids (e.g., glue, gelatin, starch, tannin, and quebracho)	5–500	Gangue depressant for minerals such as talc, mica, and carbonaceous matter
	Dextrin	29–120	Carbonaceous gangue depressant
Other	Fuel oil	250–2,000	Froth modifier
	Nitrogen	NA	Enhances copper and nickel flotation
	Air, oxygen	NA	Oxidizes reductants in the pulp and sulfide mineral surfaces
	Carbon dioxide	NA	Inert carrier gas and pH control

Source: Woodcock et al. 2007
NA = Not available.

and guar gum are used to depress naturally hydrophobic magnesium silicate minerals, such as talc and pyroxene, whereas dextrin and starch are used to depress carbonaceous minerals.

Flow-Sheet Development Test Work

Mineralogy and batch kinetic flotation tests are used to determine the relationship between laboratory rougher grind size. Typically, P_{80} 75, 106, 150, and 212 μm are used for the first trial, and narrower ranges can be tested after that. An economic evaluation is required to determine the optimum grind size for an industrial plant, since the capital and operating costs for achieving the grind must be compared to the value of the metal recovered at each grind size.

Once the grind size has been established, grade–recovery curves for different reagents should be used to select a reagent suite. Finally, collector dose requirement can be determined by stage, adding small amounts of collector in stages, like a titration. The grade and recovery of the first kinetic concentrates will aid in assessing whether the first rougher concentrates can be sent straight to cleaning or if regrinding will be required first. Regrinding and cleaning tests are performed in a similar manner.

Where possible, the flow-sheet and reagent regime should be simple, be similar for all ore types, and have few low middling and recirculating streams. Once a flow-sheet and reagent regime has been established, a locked-cycle flow sheet should be run to check the performance when middling streams reach steady state. Figure 10 shows an example of a selectivity curve comparison where, as more collector is added, the flotation becomes less selective for copper over iron. Figures 11 to 13

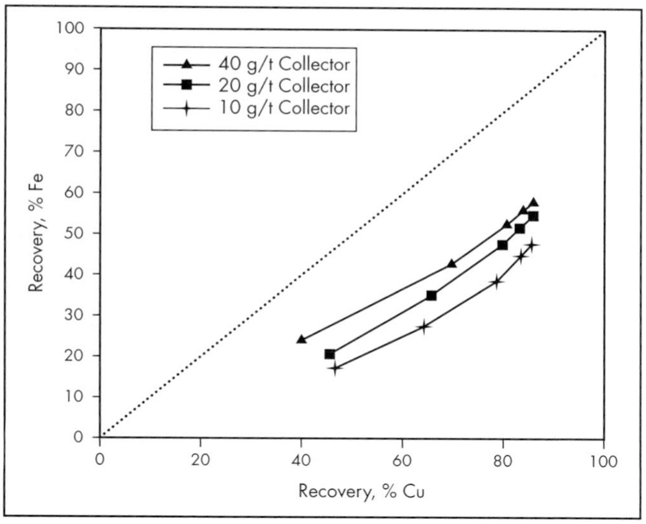

Figure 10 Iron–copper selectivity versus collector dose

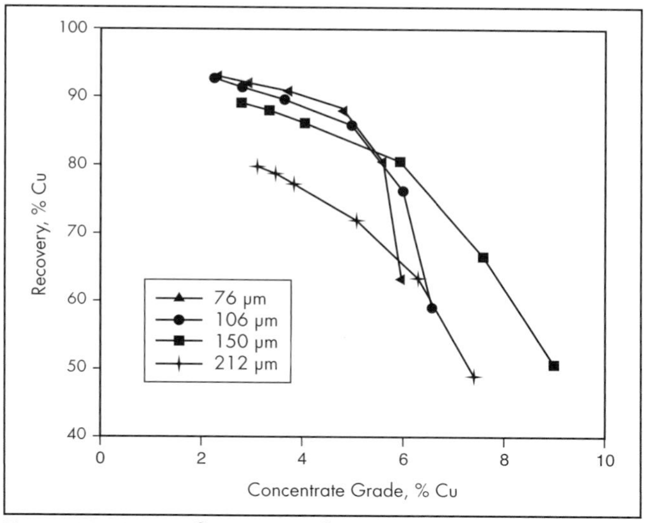

Figure 12 Copper flotation performance versus grind size

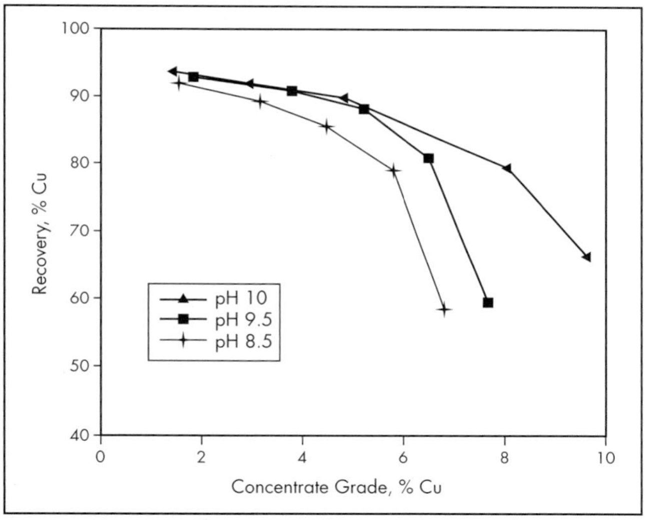

Figure 11 Copper flotation performance versus pH

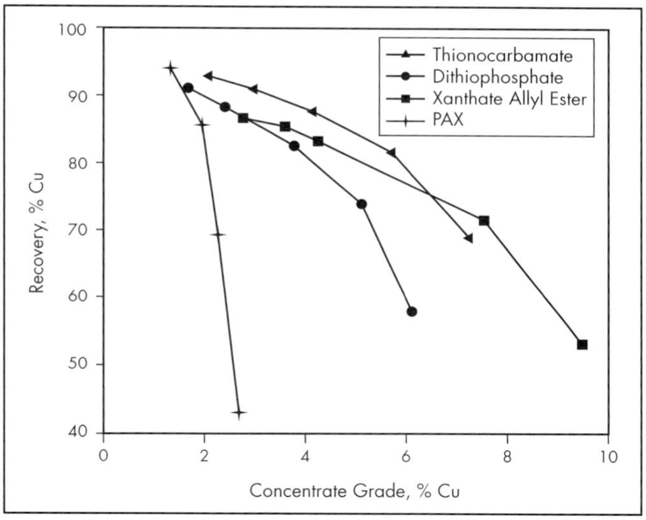

Figure 13 Copper flotation performance versus collector type

show examples of grade–recovery curve comparisons of flotation development test-work results to evaluate pH, grind size, and collector type for copper ores, respectively. The curves that are closer to the top right of each graph indicate better copper floatability and selectivity over gangue.

Grinding Media Effects on Flotation

The forged steel balls in a grinding mill with chalcopyrite and pyrite can create conditions that have a strong impact on copper flotation performance (Greet et al. 2004). Iron from the balls enters the solution because of corrosion. While ball mill ball consumption for low-sulfide, alkaline ores is primarily due to abrasion, in high-pyrite ore grinding, more than 50% of the ball consumption can be attributed to corrosion. All the resulting iron in solution can form iron oxide coatings over the chalcopyrite surfaces, reducing its floatability. The corrosion of the grinding media also has an anodic effect on copper minerals, leaching copper into solution. This copper can partially activate the pyrite, reducing the effect of collectors, crowding the froth, and diluting the concentrate. Operators will typically act to maintain the copper concentrate quality, reducing the copper recovery.

Several approaches can be used to reduce the impact of the grinding media effect on copper flotation performance:

- High pH. Increasing the pH in the grinding mill can passivate steel surfaces and precipitate cations before they affect surface chemistry.
- Autogenous grinding.
- Inert media. High-chrome balls can be used in SAG or ball milling, and high-chrome, sand, or ceramic media can be used in regrind milling.

Conditioning

Large copper concentrators have generally removed conditioning tanks from the circuit. Collectors and lime are added to the SAG mill discharge, leaving the residence time of the ball mill and cyclones for conditioning. Frothers do not require conditioning and are therefore added to the flotation cell feed. Conditioning tanks can act as a buffer to reduce the impact of abrupt changes in flow or grade on flotation performance.

Equipment Selection

Self-aerating and forced-air mechanical machines are used successfully for rougher and rougher-scavenger stages. Either mechanical cells or column flotation cells can be used for cleaner and recleaner stages. Column cells are often favored in ores with high viscosity and clay content, as froth washing is required to reduce gangue entrainment to the final concentrate. Cleaner-scavenger cells are typically mechanical cells to ensure high recovery of fine and coarse middling particles. Typical rougher residence times are between 20 and 30 minutes.

The growing interest in using new-generation flotation machines facilitates high-shear bubble attachment and froth washing in a single unit. These units include the Jameson cell, the staged flotation reactor, and the StackCell, for example. Currently, these units have not replaced rougher cells; however, they have been used in cleaning circuits in place of mechanical and column cleaner cells.

Outotec advises that the range of froth carrying rate should be

- 0.8–0.15 t/m^2/h for rougher cells,
- 0.3–0.8 t/m^2/h for scavenger cells, and
- 1–2 t/m^2/h for cleaner cells.

The froth carrying rate is the dry mass of concentrate that is recovered per hour per square meter of cell froth area (t/h/m^2). If the carrying rate is low, the froth will tend to collapse, returning the concentrate to the flotation cell before it is pushed into the launders, reducing recovery. This can be resolved using froth crowders. If the carrying rate is too high, then there will likely not be enough bubble capacity to recover all of the concentrate required, reducing recovery. Also, the froth will not have enough time to drain adequately, reducing the quality of the concentrate. This can be resolved by adding more flotation cells to get more froth area or by reducing the concentrate load though reagent or flow-sheet modification.

As flotation cells and plant throughputs get larger, the issue of high carrying-rate loads in the cleaner circuits are more common. This is because the volume of flotation cells grows to the cube power compared to the diameter, whereas the froth area only grows to the square power.

Outotec recommends that the lip loading not exceed 1.5 t/h/m. The lip loading is the dry mass of concentrate that is recovered per hour per linear meter of froth overflow to a launder (t/h/m). If the lip loading is too high, then concentric or radial launders can be added to the cells, although that will reduce the froth area for the carrying rate calculation.

Copper Flotation Flow-Sheet Strategies

Basic Copper Flotation

A simple flow sheet has become commonplace in plants that treat low-grade, low-complexity copper ores (Figure 14). Such plants are used to concentrate a high recovery of valuable metals (copper, molybdenum, gold) into a rougher concentrate at a coarse grind size. The primary grind is often between P_{80} 150 and 250 µm. The rougher concentrate is reground to produce a feed to the cleaner circuit of P_{80} 20–56 µm.

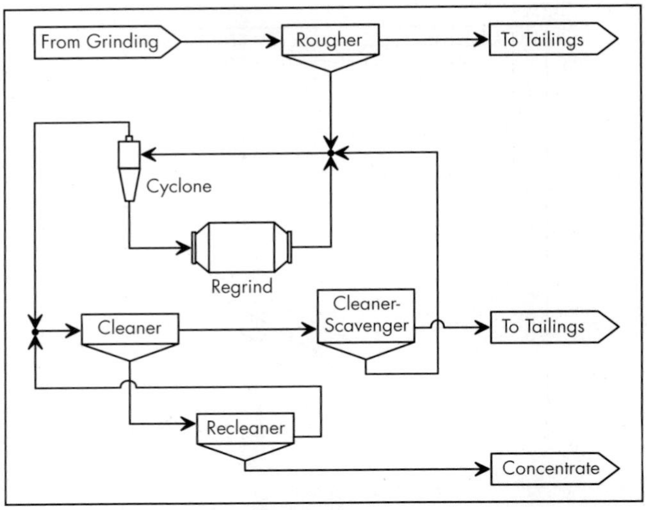

Figure 14 Basic copper flotation arrangement

The first cleaner tailings are scavenged in a high-recovery cleaner-scavenger stage. The tailings from the cleaner-scavenger stage are sent to final tailings rather than returning to the rougher feed. This is to avoid recirculating loads of fine middling and weakly activated pyrite particles between the roughers and cleaners. This recirculating load would destabilize the rougher stage, reducing the residence time. Also, the fine reground particles would degrade the hydrodynamics of the rougher flotation bubbles and froth. Most older plants that used to return the cleaner tailings to the rougher feed have been modified to send the cleaner tailings directly to final tailings.

The first cleaner concentrate can be cleaned further in one or two additional cleaning stages, arranged in countercurrent, where concentrates advance to the next stage while the tailings report to the feed of the previous stage.

Concentrators that use this arrangement include Chuquicamata, Disputada, Escondida, Esperanza, Los Bronces, Los Pelambres, and Ministro Hales (all in Chile); Antapaccay, Constancia, and Las Bambas (all in Peru); Alumbrera (Argentina); Bagdad, Metcalf, and Ray (all in the United States); Boddington (Australia); Sossego (Brazil); Grasberg (Indonesia); Lumwana (Zambia); and Oyu Tolgoi (Mongolia).

Split Rougher Flotation Flow Sheet

The enriched copper minerals, including chalcocite, bornite, and covellite, can cause problems when they are reground: They are brittle and can slime, resulting in recovery losses to the slow-floating –15 µm fraction; and they can inadvertently activate iron sulfides, decreasing the selectivity of the cleaners and reducing the final concentrate grade. Consequently, plants that treat these minerals can often direct the first rougher concentrate to the second cleaner feed, bypassing the regrind and first cleaner stages. Entrained gangue and coarse middling particles will then have to be classified back to the cleaner-scavenger for reject to tailings or to the cleaner-scavenger concentrate for regrinding. The liberated copper particles will report directly to the final concentrate without having been reground. The slower-floating rougher-scavenger concentrate is reground, as per the basic copper flow sheet.

Concentrator design and layout will influence the operator's ability to implement or change a split rougher flow sheet in an existing plant (Figure 15). Bypassing regrind from the first rougher cells requires that these cells be operated at high-grade concentrates. Achieving these high grades requires understanding of the mass pull–grade relationship and how to control it. This can be achieved by hydrodynamics or reagents.

Plants that operate split rougher flow sheets include Batu Hijau (Indonesia) and Salobo (Brazil).

High- and Low-Grade Regrind Circuits

The two Cerro Verde concentrators in Peru use a variation of the split rougher flow sheet. The rougher and rougher-scavenger concentrates are reground separately in high-grade and low-grade regrind mills. This flow sheet has more unit operations than the basic flow sheet, but capital intensity is still low since the two concentrators treat a combined 360,000 t/d (Figure 16). Somewhat similar approaches are used at Antapaccay and Las Bambas (Peru), and Bingham Canyon (United States).

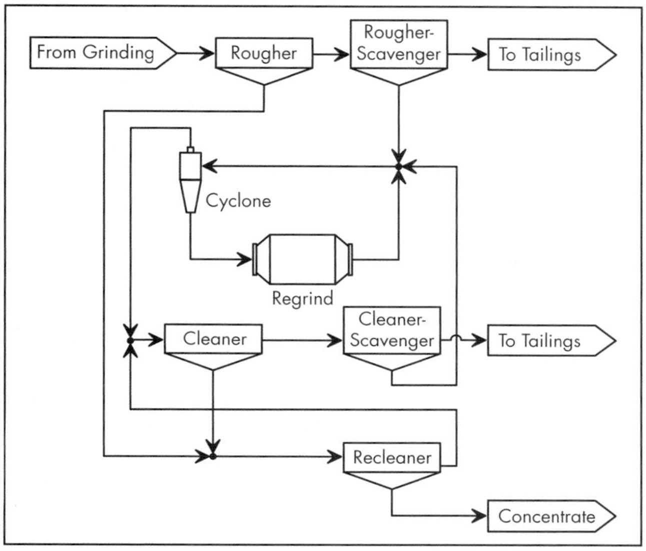

Figure 15 Split rougher copper flotation arrangement

High-Grade Copper Treatment Flow Sheet

Plants that treat high-grade copper ores may find that the cleaner circuit cells' carrying capacity or concentrate lip loading may be high, lowering cleaner circuit recovery. To reduce the burden of the fast floating, liberated copper particles from the cleaner circuit, a cleaner-scalping stage can be used. A high-capacity, high-shear flotation machine with froth washing capabilities is ideal for this stage (Figure 17). The cleaner-scalper concentrate reports to the final concentrate, while the tailings report to the normal cleaning circuit, but with reduced copper tonnage.

Plants that use this arrangement include Carmen de Andacollo (Chile), Phu Kham (Laos), Prominent Hill, and New Afton (Canada). Carmen de Andacollo scalps final concentrate directly from the rougher concentrate using mechanical flotation cells before regrind. Phu Kham and Prominent Hill scalp final concentrate from the regrind discharge using Jameson cells. New Afton scalps final concentrate from the regrind discharge using a bank of staged flotation reactors.

Sand-Slime Circuits

Split sand-slime circuits can be used where clays in the slimes are found to hinder the collection and recovery of the coarse particles. This option is expensive because it requires two rougher flotation circuits to be built and run in parallel but has the advantage of being able to focus the reagent and froth recovery arrangements for coarse and fine particles separately (Figure 18).

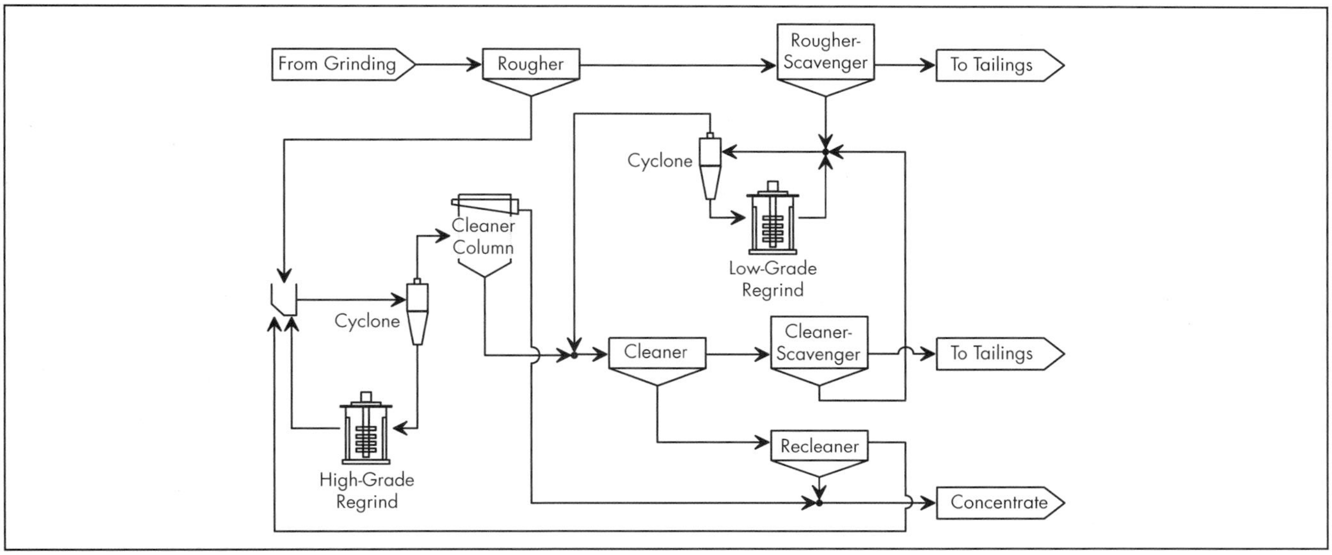

Figure 16 High- and low-grade regrind copper flotation arrangement

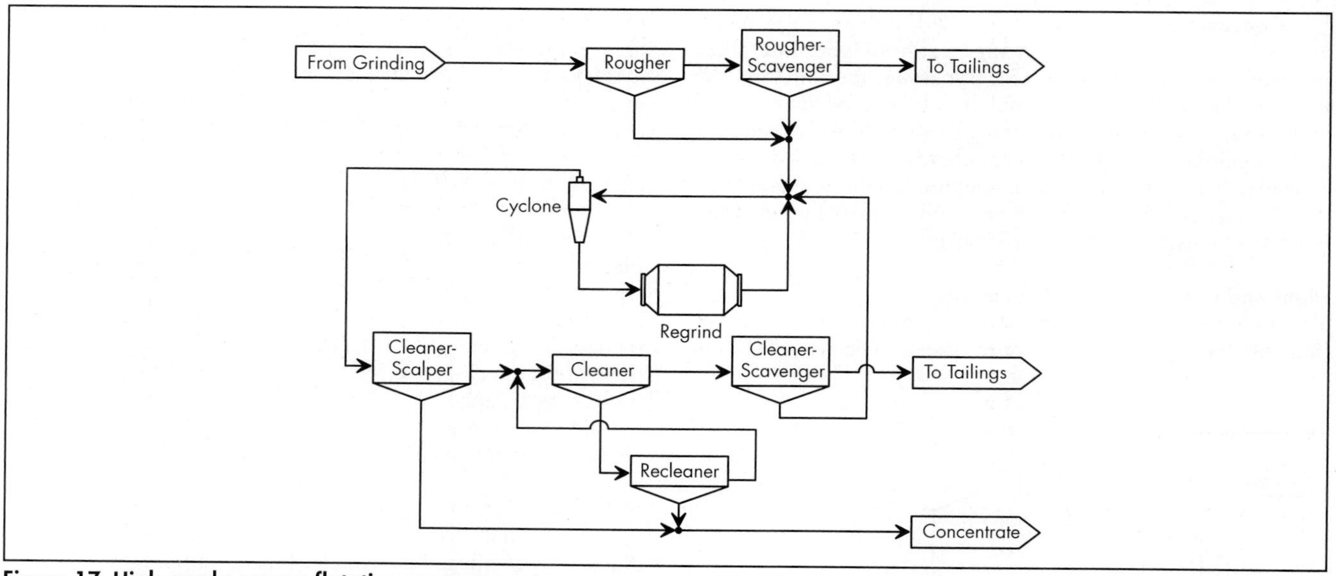

Figure 17 High-grade copper flotation arrangement

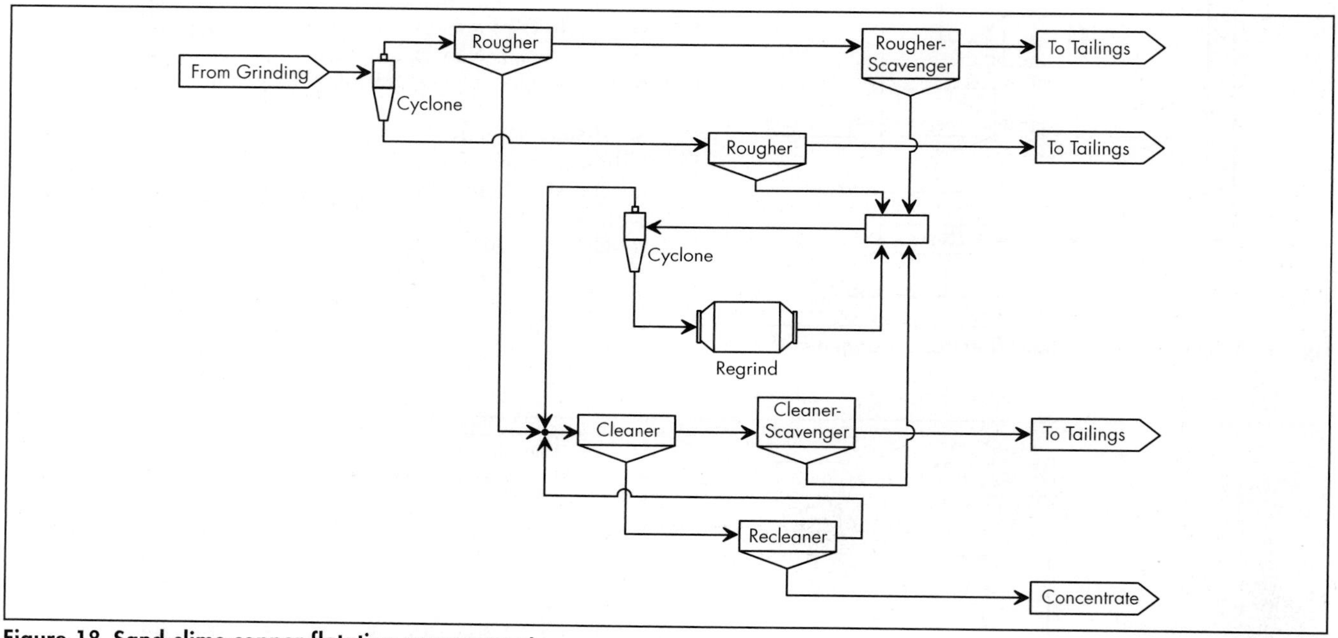

Figure 18 Sand-slime copper flotation arrangement

Cuajone in Peru uses this arrangement. Collahuasi in Chile has a similar approach, where the rougher tailings are cycloned to recover coarse particles to the underflow, which are scavenged by flotation, then reground for cleaning.

Removing Metallic Copper in the Milling Circuit

Where present, it can be favorable to remove metallic copper or gold particles from the circulating loads in a grinding circuit. The density of copper is 8.95, and gold is 19.3; therefore, particles of these metals will be ground to a very fine size before reporting to the cyclone overflow, since the cyclone classifies particles by density and by size. This approach is useful for other heavy minerals, such as galena, for example, as it has a density of 7.2–7.6 and is fragile, so will slime in a closed grinding circuit.

Smaller gold operations often use gravity bowl concentrators, but these devices are less efficient for sulfide ores, have limited capacity, and add water dilution to the circuit. Several larger concentrators treating copper-gold ores use a combination of flash flotation and gravity separation to enhance gold recovery (Figure 19). A portion of the primary cyclone underflow stream is passed through the flash flotation cell to recover liberated, fast-floating gold particles. This prevents them from returning to the ball mill, where they will be overground with the potential of not floating them in the downstream rougher circuit.

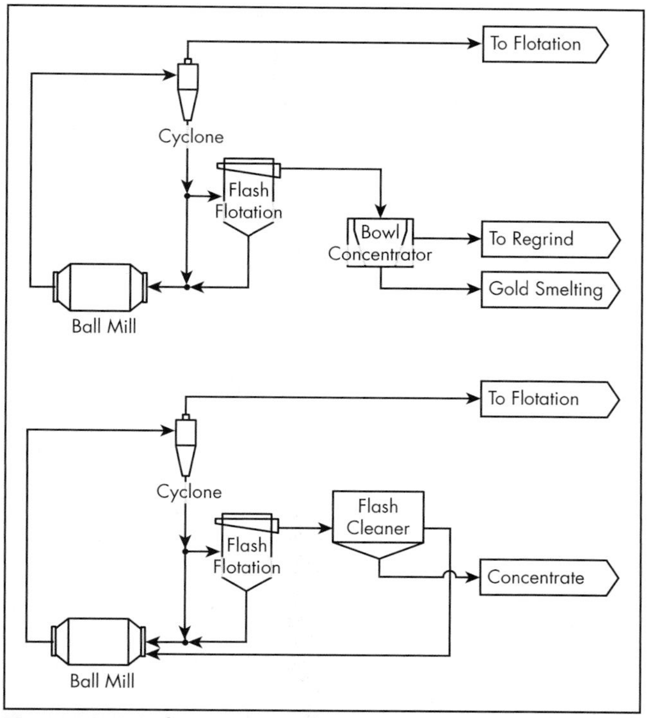

Figure 19 Grinding circuit scalping alternative arrangements

A disadvantage of removing metals in the grinding circuit is that the primary cyclone overflow no longer represents the plant feed for metallurgical accounting purposes. Care needs to be taken in frother selection using a flash cell in the grinding circuit. The flash circuit in grinding normally requires strong frothers to work properly because of the coarse grind. Successful implementation requires that the strong frother for flash flotation be compatible with the subsequent rougher circuit.

Tintaya (Peru) used gravity concentrators in place of flash flotation. Boddington; Cadia and Telfer (both in Australia); and New Afton use flash flotation, but the flash flotation concentrates are handled in different ways, depending on the proportion of copper sulfides, gold, and pyrite, and their associations. Cadia does not have a cyanidation circuit, so the gold metal is recovered from the flash flotation concentrate via gravity concentration and is then upgraded and smelted to doré. Telfer, on the other hand, has both gravity concentration for gold and flash flotation for copper and copper sulfides in the grinding circuit.

Copper Concentrate Treatment to Recover By-Products

Copper can often be present in ores with other metals that are co-collected to a bulk concentrate for subsequent separation. In Chile, this arrangement is known as collective flotation (bulk copper-molybdenum) and selective flotation (copper–molybdenum separation). Other metals can include lead, bismuth, molybdenum, nickel, and arsenic. A generalized flow sheet for this arrangement is shown in Figure 20. Depressants for copper separation flotation are listed in Table 6.

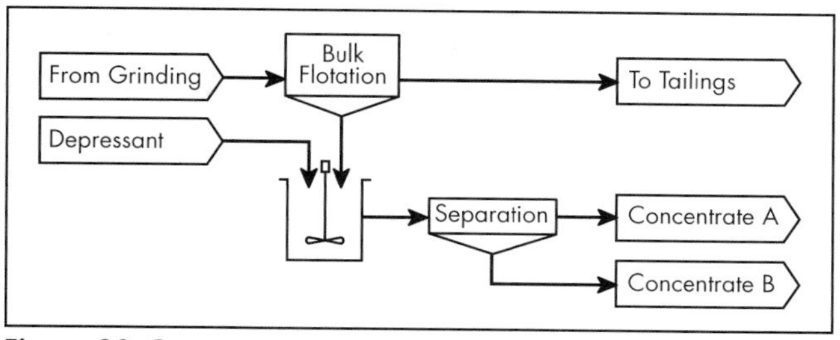

Figure 20 Copper concentrate separation arrangement

Table 6 Depressants for copper separation flotation

Bulk Concentrate	Concentrate A	Concentrate B	Depressant
Cu-Mo	Copper	Molybdenum	NaHS
Cu-Bi	Copper	Bismuth	Cyanide
Cu-Ni	Nickel	Copper	Oxidant
Cu-As	Copper	Arsenic	Reductant
Cu-Pb	Lead	Copper	Dichromate or sodium metabisulfite
Cu-Pb	Copper	Lead	Cyanide

Ore-Specific Flow-Sheet Strategies

Copper-Gold Ores

Copper and gold are often found together in the same deposits, though in different proportions and mineral assemblages. Gold that reports to a commercial copper concentrate is readily recovered by the smelter, and so has favorable payment terms. Cyanide-soluble copper that reports to a gold cyanidation plant will increase cyanide consumption and augment the cost of gold recovery from the leach solution.

Gold that is associated with pyrite can be high or low grade. In South America, the pyrite associated with porphyry copper ores has low gold grades. The copper-gold deposits in Southeast Asia and Australia tend to have a high gold grade in the pyrite, which can be recovered as a separate concentrate for downstream treatment to recover the gold.

If the pyrite gold grade is low, then it cannot be co-floated with the copper, as the decrease in copper concentrate grade will incur more additional transport and treatment charges than the gold will attract in credits. If the gold-to-copper ratio is high, then a bulk gold-pyrite-copper concentrate may attract favorable commercial terms at copper smelters.

Gold that in a low-grade pyrite concentrate or in copper flotation tailings may be recovered by cyanidation, if amenable. If the gold is associated with pyrite or arsenopyrite and is refractory to cyanidation, then oxidation would be required before it can be leached. Where cyanide-soluble copper is present in the feed to a gold cyanidation plant, copper flotation may be primarily used to reduce cyanide consumption. Liberated native gold and electrum can be recovered from the grinding circuit using flash flotation or gravity concentrators.

Copper-Lead-Zinc Ores

Polymetallic base metal mines often contain copper, lead, zinc, and other minor metals such as silver, gold, arsenic, and antimony. These ores are typically heterogeneous, and so a

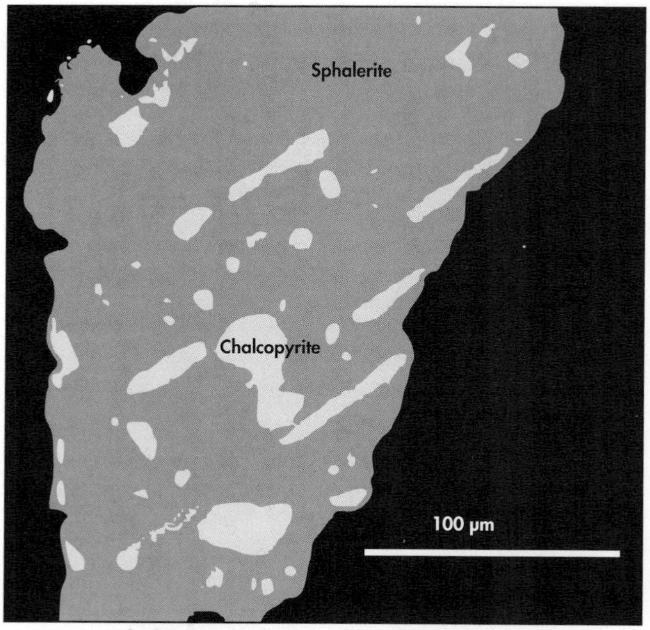

Figure 21 Chalcopyrite disease in sphalerite from a skarn deposit

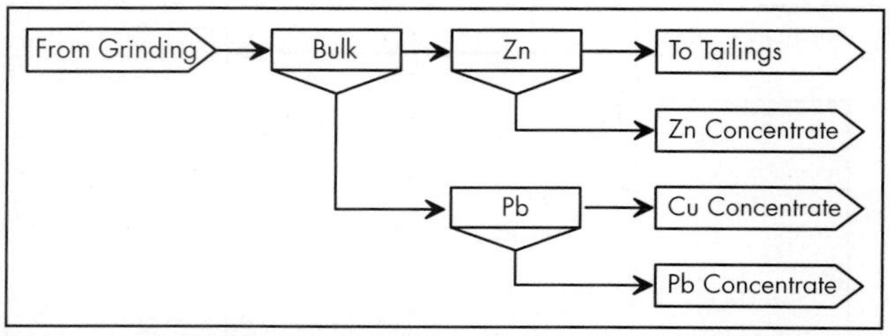

Figure 22 Bulk flotation approach to base metal flotation

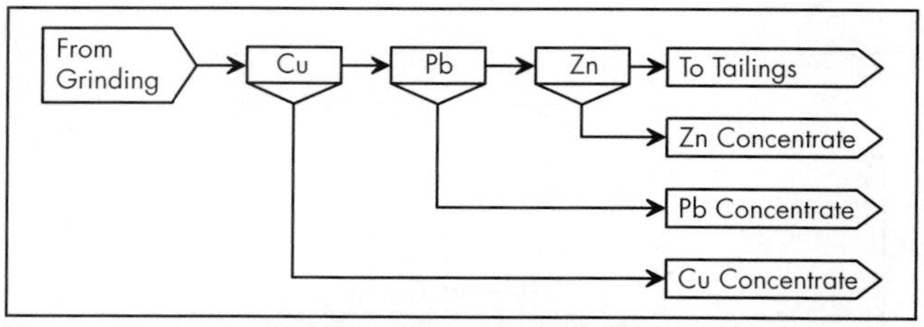

Figure 23 Sequential flotation approach to base metal flotation

treatment plant design can follow several approaches, including campaign ore and mining blends:

- **Campaign ore.** Antamina (Peru) is one of the world's largest base metal mines and runs one- to three-week campaigns of eight ore types based on relative copper, zinc, bornite, and bismuth grades. This is done to facilitate high recoveries and concentrate grades for copper, lead, zinc, and molybdenum concentrates. Antamina can achieve this through detailed geometallurgical test work and careful open pit mine planning.
- **Mining blends.** Most mixed base metal concentrators receive the ore from underground mines. Since underground mining is expensive compared to open pit mining and concentrator operating costs, the underground production rate is the bottleneck and there is little opportunity for optimizing the concentrator feed for metallurgical response. Therefore, the concentrator flow sheets and reagent regimes are designed to be flexible rather than optimized. Multiple fine ore bins can be used to blend ore types, but inevitably the concentrator feed varies, affecting the recovery and product quality.

Copper and zinc are commonly found together in volcanogenic massive sulfide, skarn, stratabound, and massive sulfide deposits. It is usual to find chalcopyrite rather than enriched copper sulfide minerals associated with sphalerite.

Chalcopyrite disease in sphalerite is a particularly difficult texture to treat by flotation (Figure 21). The chalcopyrite inclusions in the sphalerite are too fine to liberate by grinding, and since a portion of the particle surface is chalcopyrite, the particle will be impossible to selectively float from chalcopyrite.

When working with deposits that have copper sulfide minerals and sphalerite in different domains, combining the two in grinding may result in activation of the sphalerite by the copper sulfides. For example, bornite from a porphyry domain mixed with sphalerite from a neighboring skarn domain could cause an increase in zinc displacement to the copper concentrate, from 5% to 90%, that cannot be controlled by depressants. This will reduce zinc recovery and reduce the quality of the copper concentrate.

The two basic approaches to copper-zinc and copper-lead-zinc flow-sheet design are bulk and sequential. The bulk flotation process is, by far, the most popular, because it decouples the difficult copper–lead separation flotation stage from the simple zinc flotation stage (Figure 22). This reduces reagent costs and operating complexity.

One of the issues with the bulk flotation process is that once copper in depressed, it is difficult to activate again. This means that any impurities in the bulk concentrate can be concentrated in the copper concentrate, reducing its quality. Therefore, for some deposits it is advantageous to selectively float a copper concentrate first, especially where gold or silver recovery to the copper concentrate can be improved. This is done using the sequential flotation arrangement (Figure 23). Copper is collected by using reagents that have a high selectivity of copper over galena, sphalerite, and iron sulfides. Lead is next collected using zinc cyanide complex to depress the zinc, and weak xanthate, thionocarbamate, or Aerophine to float the galena. Finally, the zinc is recovered by activating the sphalerite with copper sulfate and collecting with a thionocarbamate at pH 9–10 or a xanthate collector at pH 10–12.

As long as the copper mineralogy is chalcopyrite and no cyanide-soluble copper minerals are present, then cyanide can be used to depress the copper, allowing for the relatively simple separation of copper and lead by floating the galena. If cyanide-soluble copper minerals and/or precious metals are present, or if the ore is not otherwise amenable to copper depression by cyanide, then lead depression can be used. In the past, lead depression was achieved by adding dichromate reagent. Unfortunately, this effective reagent poses risks to health and the environment; consequently, it is often no longer approved. In such cases, a sulfoxy species such as

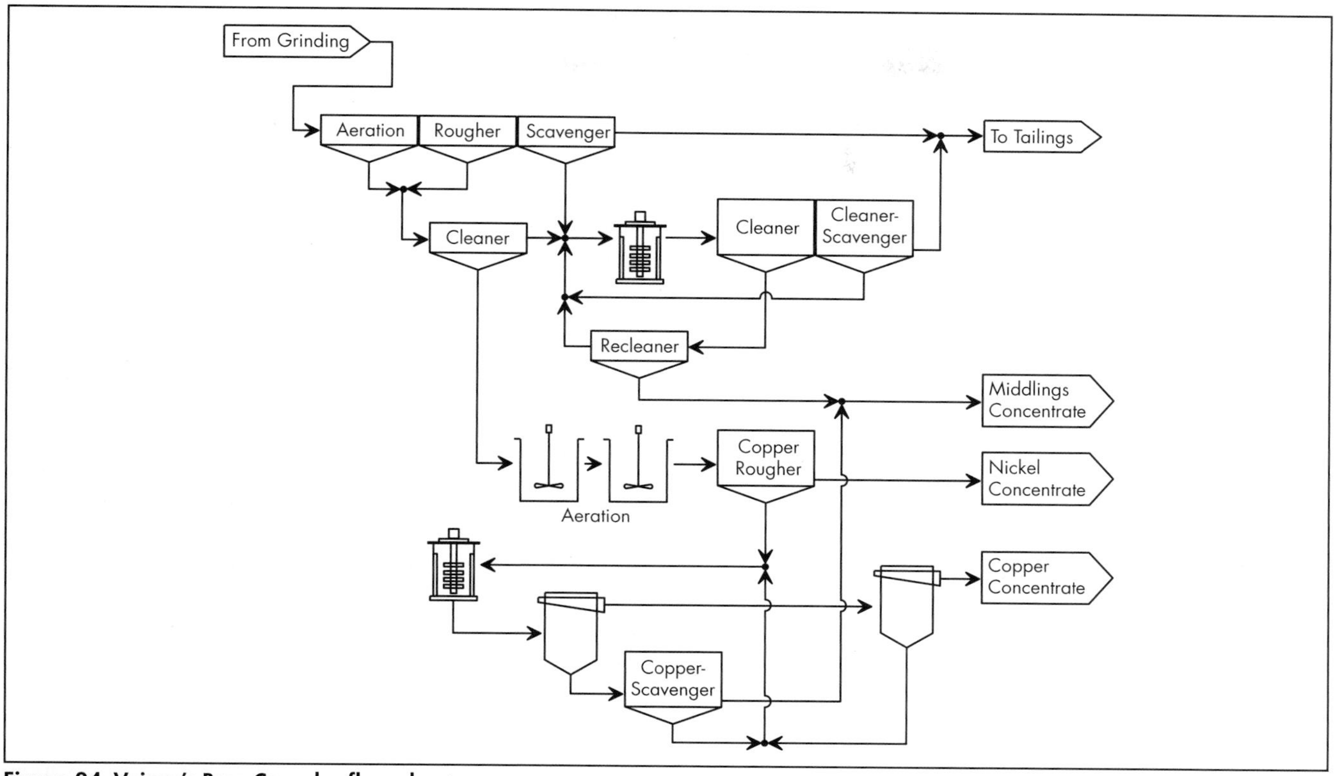

Figure 24 Voisey's Bay, Canada, flow sheet

sulfur dioxide (SO_2) gas or sodium metabisulfite can be used. Variations on these processes can include the following:

- **Carbon pre-float.** Organic carbon can rob reagents from flotation, flatten froth beds, and dilute concentrates. A pre-flotation stage to remove the carbon can be used, either without reagent addition or with a small amount of fuel oil added to grinding.
- **Talc pre-float.** Zinc ores can contain naturally hydrophobic talc that can slime in the regrind mill and dilute the concentrate grade. Pre-float of talc without reagents can help at high grades, but at lower grades, CMC addition can aid in depressing the talc.
- **Scalping lead from grinding.** Because of its high specific gravity, galena tends to recirculate in the cyclone underflow, eventually sliming, reducing lead recovery, and possibly inadvertently activating the sphalerite to float in the bulk concentrate. The use of flash flotation in the grinding circuit can be used to recover the galena when it is still coarse.

Clearly, several of the preceding processes are incompatible. The variety of flow sheets and reagent suites in copper-lead-zinc concentrators is much greater than for other kinds of copper concentrators.

Copper-Nickel Ores

Copper-nickel ores typically contain copper as chalcopyrite and nickel as pentlandite. These minerals are usually accompanied by pyrrhotite and talc, which impede the production of high-quality concentrates. The focus of the flow sheet and reagent additions will depend on the ratio of copper to nickel.

In Australia, the copper head grades are low, and the separation is made as a cleaning process for a nickel concentrate, whereas in Canadian operations, the separation results in two commercial concentrates (Figure 24). Chalcopyrite, pentlandite, and some pyrrhotite are recovered to a bulk concentrate using a strong xanthate collector. The bulk concentrate is typically reground to liberate the chalcopyrite from the pentlandite. The copper and nickel in the bulk concentrate can be separated by two regimes:

1. The bulk concentrate is oxidized by aeration in the presence of lime for 30 minutes to depress the pentlandite.
2. The pentlandite is depressed by raising pH higher than 10 using lime and by adding cyanide, dextrin, or starch.

The chalcopyrite and pyrrhotite are selectively floated, producing a nickel concentrate in the flotation tailings. The chalcopyrite-pyrrhotite concentrate is then upgraded using SO_2, Na_2S, or triethylenetetramine to depress the pyrrhotite. Talc can be dispersed in the rougher flotation using soda ash or depressed using CMC. Although modern reagent regimes are more efficient at rejecting pyrrhotite, some older operations continue to remove it from final concentrates using magnetic separation.

Copper Ores High in Arsenic

Arsenic mineralogy can be difficult to identify and map in a copper deposit. The recovery of arsenic can be affected by the arsenic mineralogy, mineral textures, and reagent regime. Enargite and tennantite are copper sulfide minerals and, consequently, are extremely difficult to separate from other copper minerals by flotation. Limited success has been reported by

Chuquicamata using an oxidation process and by Northparkes (Australia) using a reductive process. In any case, separating the enargite/tennantite from the other copper minerals also results in a corresponding separation of copper. The Ministro Hales concentrator at Chuquicamata treats high-enargite ore through a conventional flotation concentrator, and the resulting high-arsenic copper concentrate is treated by roasting.

Orpiment and realgar are uncommon in copper sulfide ores but are readily separable by flotation. Arsenopyrite can be separated from copper sulfides by using the same strategies as for selective flotation against iron sulfides, such as using a selective collector, high pH, or cyanide. Tetrahedrite can also contain arsenic, copper, and precious minerals. The amount of arsenic in the tetrahedrite is usually small compared to the precious metal credits and therefore is not of concern. Tetrahedrite has strong floatability under normal copper flotation conditions.

Currently 0.5% As is the commercial limit for copper concentrates, and many copper head grades are 0.5% Cu or less. Assuming that copper and arsenic recoveries are similar, and with a copper upgrade ratio of 60:1, an arsenic head grade of more than 0.008% or 80 g/t As may result in a high-arsenic copper concentrate. If the arsenic can be depressed, then the acceptable head grade of arsenic can be higher. Therefore, mine planners should calculate the copper concentrate arsenic grade using the copper and arsenic recoveries, and copper and arsenic head grades.

Geometallurgical Planning

Plant operators should be well informed of the expected metallurgical performance from the geometallurgical model. If an ore type enters production that fouls the concentrate grade, the operators may act to maintain the concentrate grade by cutting recovery, unnecessarily. Conversely, if a lower recovery is expected, the operators should know to avoid over-adding reagents that may result in a lower-than-necessary concentrate grade. Geometallurgical planning can be used to avoid unexpected excursions in rock hardness, pulp viscosity, and penalty element grades that could affect the throughput, metallurgy, and profitability.

The performance of copper flotation can be predicted by bench-scale flotation tests. Kinetics flotation tests are used to determine the rougher recovery, whereas kinetics and locked-cycle tests are used to predict the cleaning circuit losses and concentrate quality for each ore type.

The metallurgical performance of mining blends can often be calculated from the sum of the parts. That is, the fine metal recovery and concentrate grade are calculated separately for each ore type in the blend, then the concentrates are combined to total the fine metal recovery and a weighted average concentrate grade for the blend. This does not apply where there are toxic effects between ore types. Examples of such toxic effects could include

- Clays that reduce floatability for the entire blend,
- Pyrite or sphalerite in one ore type that is activated by enriched copper sulfides in another, and
- High-pyrite ore types that may reduce the ability to recover slow-floating components such as gold or molybdenum from another ore type.

On the other hand, if one ore type contains high penalty element grades, such as arsenic or bismuth, then blending it with other cleaner ore types can result in a cleaner final concentrate that does not attract any penalty.

Thickening and Filtration

Concentrate is thickened and filtered before being fed to a smelter. Conventional or high rate concentrate thickeners are used to thicken flotation concentrate to 60%–68% solids prior to filtration. Entrained air can cause froth to form on the thickener surface. Vertical and horizontal plate pressure filters are most commonly used for filtration, achieving 8%–10% moisture in the filter cake.

Usually the smelters are distant from the concentrator, so the concentrate is transported there by any combination of rail, truck, pipeline, or ship. Where pipelines are used, some unit processes can be relocated from the concentrator site to the pipeline discharge site. For example, Los Bronces, Collahuasi, and Esperanza have the copper-molybdenum separation plants, concentrate thickening, and concentrate filtration at sea level where the concentrate pipelines discharge at the coast.

Antamina, Escondida, Grasberg, Los Pelambres, Ridgeway (Australia), and Bingham Canyon transport their concentrates via pipeline, before thickening and filtration. Thermal drying of copper concentrates was common in older plants that used disk vacuum filters but is uncommon in newer plants that use pressure filters.

Seawater Washing

When seawater or saline water has been used for flotation, the final concentrate needs to be washed with clean water to avoid smelter penalties for chloride levels. Batu Hijau uses three stages of countercurrent decantation to wash the concentrate. Fresh water captured from rain is used. Esperanza uses a washing stage in the horizontal plate pressure filtration of final copper concentrates. Reverse osmosis water consumption per metric ton of final concentrate is 0.25 m^3/t.

Emerging Processes

The energy and grinding steel used in comminution is a major contributor to the overall operating costs (20%–30% of operating expenses) and carbon footprints of the mining operation (15–30 kg C/t ROM [run-of-mine]). Furthermore, storage of finer tailings typically consumes more water than that of coarser tailings particles. Therefore, companies are developing new technologies to recover copper at coarser sizes.

- **Ore sorting.** Advanced sensors characterize individual particles and classify them.
- **Crossbelt analysis.** Crossbelt analysis by neutron activation determines if the ore is above the economic cutoff grade. Low-grade ore is rejected before entering the mill.
- **Coarse particle flotation.** Where liberation is achieved at coarse particle sizes, a low-quality concentrate can be recovered, decreasing the energy required to reduce all of the ore to the same size. The coarse concentrate is then reground and treated in a traditional flotation circuit. Cadia has announced that it plans on using the Eriez HydroFloat equipment for this process.
- **Rejection of pebbles.** Where pebbles are low grade, they can be rejected directly or rejected by ore sorters on the recycling stream.

Another tactic is to spend less specific energy to achieve the same final product size. Strategies include the following:

		Flotation Kinetics	Retention Time	Bubble Size (D_b)	Gas Holdup (E_g)	Recovery	Grade	Favored Particle Size	Water Recovery
J_g	↑	↑	↓	↑	↑	↑	↓	Coarse	↑
Froth Depth (forced air)	↑	↓	↓	—	—	↓	↑	Fines	↓
Froth Depth (self-aeration)	↑	?	↓	↑	↑	?	?	?	?
Frother	↑	↑	↓	↓	↑	↑	↓	Fines	↑
Collector	↑	↑	—	—	—	↑	↓	All Sizes	↓
Water	↑	↓	↓	—	↓	↓	↑	Fines	↓

Note: Kinetics will only increase until air recovery begins to drop. J_g is the superficial gas rate (volumetric air flow/cell area).

Courtesy of Flottec

Figure 25 Flotation operating strategy summary

- Intensive blasting. Mine operations use more intensive drilling and blasting of the ROM, creating increased fracturing, reducing the mill's feed particle size and grinding work index.
- Secondary crushing
- HPGRs
- Removal of tramp grinding ball steel using magnets on the mill discharge to reduce the amount of steel in the recirculating load.

PLANT OPERATING STRATEGIES

Although it is the honor-bound duty of every metallurgist to maximize recovery, investors in a mining operation are more interested in the return on capital. Most copper mines operate with the primary comminution equipment as the throughput bottleneck, and so these units are force fed, regardless of plant design criteria or optimum metallurgical recovery. Invariably a coarser primary grind results, giving a slightly lower copper recovery, but the higher tonnage disproportionately increases revenue. For example, an expansion of 10% tonnage may increase the primary grind P_{80} by 25 μm and decrease the flotation rougher residence time by 10%. Combined, they may decrease the rougher recovery by 2% and the final recovery by 1.5%, and may increase the cash flow of the operation by 8.5%.

In the short term, the other areas have to deal with the grade supplied by the mine and the tonnage and grind determined by the grinding area. The regrind and cleaner areas are often operated on a minimum concentrate specification basis. The cleaner flotation pH, reagent addition, and mass pull are configured to achieve high recoveries, only as long as the concentrate quality is met. In the medium and long term, plant modifications can be engineered to improve circuit performance, but budget priority is often given to increasing throughput before metallurgical recovery. As a result, it is important that plant metallurgists ensure that the flotation flow sheet, reagents, and hydrodynamics are optimized for selectivity and kinetics (Figure 25).

Instrumentation

Central control rooms fitted with distributed control system or supervisory control and data acquisition system interfaces are where the plant operators can monitor plant performance. Common instrumentation and control systems can include the following:

- Status, speed, or power load of mechanical equipment such as mills, pumps, and feeders.
- Monitor froth height, froth velocity, and air addition rates for flotation cells.
- Monitoring of major streams for flow rates, tonnages, and slurry densities.
- Monitoring pH and redox potential to control reagent addition.
- In-stream X-ray fluorescence (XRF) analysis of most major streams to chemical analyses (copper, zinc, lead, iron).
- Some plants have particle size indicators on the flotation feed stream.
- Closed-circuit television cameras on conveyor transfer points, sumps, crusher chambers, and other areas.
- Tailings and concentrate pipeline systems, including leak or blockage detection.
- Expert systems are commonplace for optimizing throughput of stable comminution circuits. They are also useful in maintaining froth heights by controlling the cascading surges in flow rates between cells by manipulating the intercell valves. Expert systems have not yet taken holistic operating control of flotation concentrators.

Concentrate Panning

Operators of smaller and older concentrators still practice the art of panning the concentrates to assess the separation efficiency and contaminant content to make process control decisions. Operators at larger and modern plants rely on in-stream XRF analysis of the concentrate streams and do not pan the concentrates.

Froth Cameras

Froth cameras are used extensively to monitor and control the froth removal rate (linear speed of overflowing froth) from the flotation cells. Control strategies vary, but most use a combination of airflow rate and froth height control to maintain a target froth removal velocity. Although the cameras can also monitor bubble size, bubble collapse rate, froth color, and texture, there has yet to be a general implementation of a control loop to use these parameters industrially.

Rougher Mass Pull

Capacities for the regrind and cleaner circuits are usually designed based on locked-cycle test results and scale-up from laboratory test work. Changes in tonnage, ore types, grind size, and liberation profiles mean that the duty of the industrial circuit can be quite different to design. In some cases, extra regrind capacity is required, but more often, regrind requirements are reduced, possibly because of a preferential overgrinding of the dense sulfide minerals due to overclassification in the cyclones. Generally, as long as the regrind and cleaning circuit retention times can be maintained, the rougher mass pull should be maximized to capitalize on metal recovery.

Control of froth height, frother addition, and air addition should be maintained to ensure constant removal of froth from all cells. Self-induced aeration cells can form a wave on the cell surface that can lead to unmineralized slurry overflowing into the concentrate, which should be avoided. Since the flotation feed grades of copper and iron sulfides vary, operating the rougher flotation to a target concentrate grade will lead to a loss in copper recovery.

Cleaner-Scavenger Recirculating Load

Since the cleaner flotation stages are typically run in closed circuit between high-recovery scavenger cells and high-grade concentrate recleaner cells, a recirculating load will form that must be managed. It is critical that the floatability copper minerals in the scavenger concentrate be increased to allow it to be recovered in the cleaner stages. The floatability of these particles can be increased by either

- Regrinding coarse middlings to improve the liberation of the copper minerals, or
- Adding additional collectors to increase the hydrophobicity of the particles.

Regrinding will not improve the floatability of the fine copper losses, which are often a substantial portion of the copper losses. Adding strong collectors that are less selective between copper and gangue minerals, such as pyrite, will increase the floatability of the copper and gangue, ultimately reducing the grade of the final concentrate.

The use of column cells for the final stages of concentrate cleaning are popular because a deep froth bed and froth wash can reduce entrained gangue minerals from the froth, decreasing the insoluble gangue grade of the final concentrate. However, traditional column cells can have trouble recovering fine particles because of the low shear mixing of the bubbles and slurry. Also, heavy particles such as native gold and copper, and particles of coarse copper sulfides, can have trouble maintaining bubble attachment up the column and through the froth.

CONCENTRATE MARKETING

Some of the older copper producers have integrated processes that produce copper sulfide concentrates from their own mines and process it into copper using their own pyrometallurgical (smelting or roasting) or hydrometallurgical processing (leaching) to produce London Metal Exchange–quality cathodes for sale to copper users. Most new copper concentrators focus on producing commercial-grade copper concentrates, which are then sold to custom smelters. These smelters pay for the concentrate based on the agreed copper price, copper content, copper grade, penalty element grades, credit element content, treatment, and refining fees. The net smelter return (NSR) is the amount paid by to the concentrate producer and is often used by mine planners in determining the value of a block of ore, expressed as NSR $/t.

Most commercial copper concentrates have between 22% and 32% Cu, but the penalty and credit element contents of the concentrates can alter the value significantly. The commercial terms are usually expressed in U.S. dollars per dry metric ton (US$/dmt).

Smelters and concentrate producers typically enter into medium- to long-term take-off contracts, where all or a portion of concentrate produced will be purchased by the smelter using previously agreed terms. A concentrate specification will be agreed to and deviation from that specification can result in penalties, renegotiation, or refusal to purchase the concentrate.

Downstream costs for concentrate producers can include the following:

- Transport to port
- Port fees
- Shipping fees
- Insurance
- Marketing fees

The transport and port fees are imposed on the wet concentrate mass, or wet metric tons (wmt). Insurance and marketing fees are often estimated as a percentage of the concentrate value.

The commercial terms with the custom smelter usually include the following:

- Payable metal
- Treatment charge ($/dmt)
- Refining charge ($/lb Cu)
- Precious metal credits (gold, silver)
- Penalty element deductions

The payable copper content is often the lower amount between 1% deduction in grade or 96.5% of the metal contained. For example, 1 t at 30% Cu:

$$1 \text{ t} \times (30 - 1)\% = 0.290 \text{ t Cu}$$

$$1 \text{ t} \times 96.5\% \times 30\% = 0.2895 \text{ t Cu}$$

Therefore, the payable copper is 0.2895 t Cu, because it is the lower amount.

Table 7 Concentrate terms

Species	Penalties, in US$
As	$2–6/dmt concentrate for each 0.1% above 0.1% $10–20/dmt concentrate for each 0.1% above 0.5%
Bi	$2–8/dmt for each 0.01% above 0.02% Bi
Cd	$2–6/dmt for each 0.01% above 0.02% Cd
Cl	$0.5–1.0/dmt concentrate for each 0.01% above 0.05% Cl
F	$1/dmt concentrate for each 0.01% above 0.03% F
Hg	$2–4/dmt concentrate for each 1.0 g/t above 10 g/t Hg
Pb + Zn	$2.5–5.0/dmt for each 1.0% above 6% (Zn + Pb)
Sb	$2–6/dmt for each 0.1% above 0.1% Sb

The payable gold content is often the lower amount between 1 g/dmt deduction in grade or 96% of the metal contained. In this example, if a concentrate contains <1 g/t Au, then no gold credits are paid. The payable silver content is often the lower amount between a 30-g/dmt deduction in grade or 90% of the metal contained. In this example, if a concentrate contains <30 g/t Ag, then no silver credits are paid.

Treatment charges are designed to pay for the smelting process and are typically between $60 and $120/dmt of concentrate. Refining charges are designed to pay for the electrorefining of the copper metal and are often dependent on the prevailing electrical energy cost, which is typically between $0.06 and $0.12/lb of payable copper. Refining charges can also apply to gold and silver credits too, typically between $5 and $10/oz for gold and $0.5 and $1.0/oz for silver.

Penalties are charged by smelters for undesired concentrate components that cause operational or environmental issues for the smelters. These penalties are usually expressed as a charge per unit grade above a threshold. There are upper limits for most of the species, beyond which sale of the concentrate can be difficult or prohibitively expensive (Table 7).

China has many of the world's custom smelters and has mandated that copper concentrates with more than 0.5% As cannot be imported into China. Few smelters will purchase copper concentrates with over 0.5% As. Producers have the following options:

- Sell the concentrate at a low price to a trader who will mix it with low arsenic concentrate stocks.
- Treat the concentrate using roasting or hydrometallurgical processes, then deal with the disposal issues of the arsenic by-products.

As the quality of concentrates continues to decline, while environmental standards increase, it is possible that it will become harder for concentrate producers to market their concentrates.

Moisture

Copper concentrate marketing can be affected by the moisture content of the concentrate to be shipped. If the moisture content is too low, potential dusting can lead to fire, safety, and environmental risks. If the moisture content is too high, it will increase transport charges and it may pass the transport moisture limit (TML). Above the TML, the concentrate may fluidize in a moving truck or ship, causing the vessel to become unstable and topple. Concentrate over this limit can be subject to costly drying before it can be shipped, slowing the sale of the concentrate.

Other Provisions

Other cost provisions include marketing (the personnel, legal, commercial, and logistical costs of concentrate sales), insurance, royalties, taxes, and losses (theft, spillage, dust).

WATER

Process water management for copper flotation concentrators can be complicated by reagents added, acid mine drainage, raw water quality, and soluble salts in the ore itself. Contaminated water must be either treated or excluded from the recycled process water regime. The use of seawater over potable water is becoming favored in arid areas to avoid conflict with competing water uses (Schumann et al. 2003).

Raw water sources include the following:

- **Fresh water from wells or water courses.** Where fresh water is plentiful and does not have competing uses for agriculture or population centers, fresh water is preferred for copper flotation because it is low cost, does not interfere with reagents, and does not pose issues for safety or maintenance. When fresh water is only available seasonally, storage in reservoirs can be required to ensure year-long production.
- **Seawater.** Chile has established legislation requiring new copper concentrators to only use water from the ocean. In other areas, seawater can also be chosen where fresh water is not available or socially acceptable. Seawater has its own set of hydrodynamic character. For example, seawater or water with high ionic strength generates small bubbles, even without frother. Seawater can buffer lime so that pH <10 is difficult to achieve. Seawater can collapse the double layer, causing agglomeration problems. There are two approaches to the use of seawater:
  1. **Desalinated seawater via reverse osmosis (RO).** This approach achieves the previously mentioned benefits of fresh water, but with the added cost of the RO process, power transmission to the RO plant, and transport of the fresh water from the coast to the concentrator. In Chile, this can account for more than 20% of project capital cost for mines at more than 3,000 m above sea level. Desalinated water is used at Cerro Lindo in Peru, and Candelaria and Escondida, both in Chile.
  2. **Use of seawater directly in flotation.** Since lime is used to raise the pH in flotation, the salinity of process water is significantly lower than seawater. Several major operations have successfully used seawater for the flotation of copper and other base metals. Seawater is used at Texada (Canada), Batu Hijau, and Esperanza. The disadvantages include increased lime consumption, increased maintenance costs due to corrosion, increased complexity of molybdenum flotation, and the environmental risks associated with transporting the water and storing saline tailings in perpetuity. The higher specific gravity of seawater can increase transport costs to high altitudes, justifying the cost of desalination, and lower thickening rates. The concentrates also require washing to reduce the halide concentration before smelting.
- **Sewerage water.** Sewerage water can have a low social cost but can present operating difficulties and reduce metallurgical performance, given the presence of salinity,

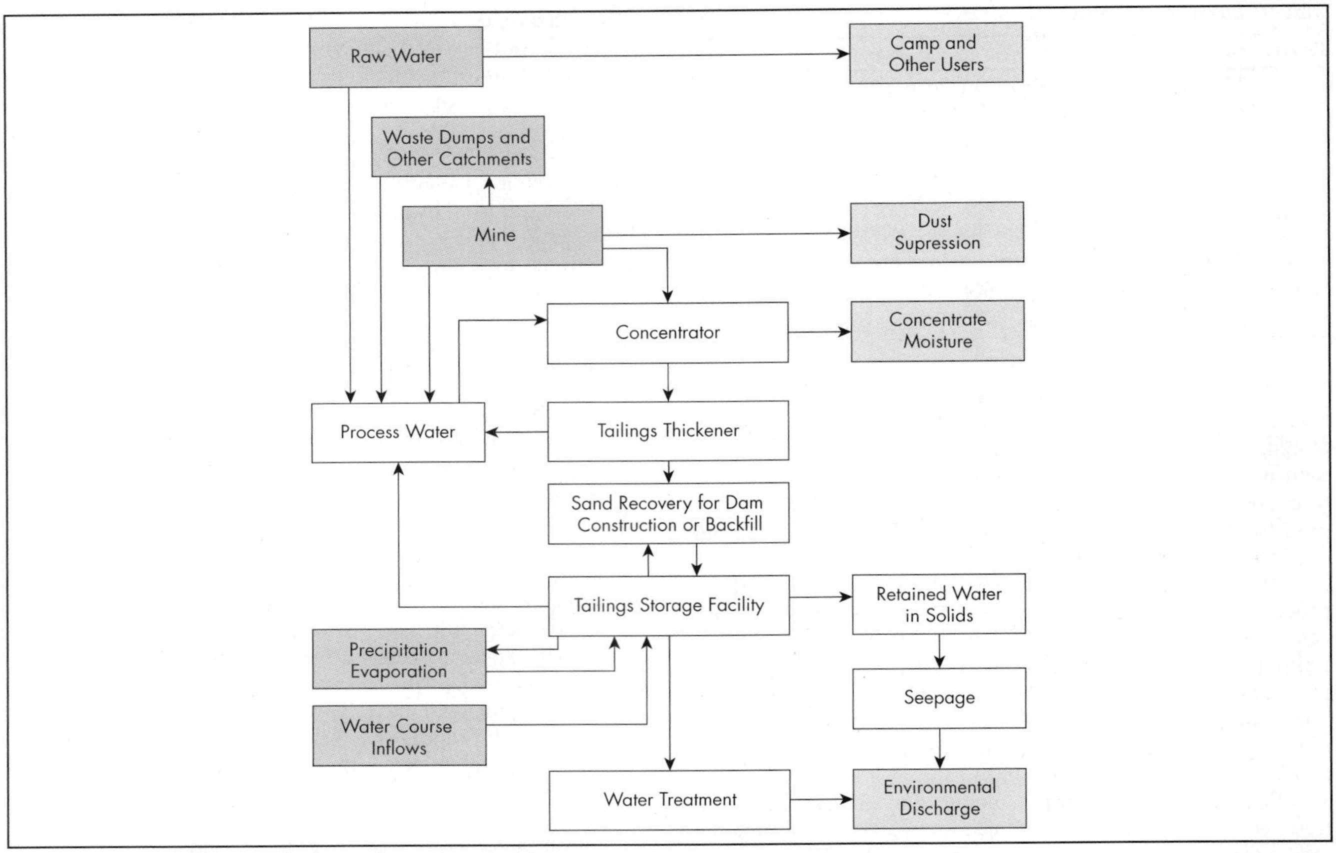

Figure 26 Site-wide water balance for a copper mine

detergents, chemical and biological oxygen demand, and colloids. Sewerage water is used at Cerro Verde and Cadia.

Further consideration should be given to other factors that can affect process water quality that may be missed during project development:

- **Acid generation from the mine, waste rock stockpiles, or tailings facility.** Iron sulfides can oxidize when exposed to air, adding ferric, sulfate, and other ions to water, potentially increasing lime consumption and affecting flotation recovery and selectivity.
- **Accumulation of reagents.** Water-soluble frothers, collectors, depressants, or their decomposition products may inhibit the easy recycling of water between flotation processes. Water from copper-molybdenum separation plants and copper-lead separation plants, for example, can be problematic in copper flotation.
- **Increased concentration of dissolved salts due to evaporation.** Evaporation in the tailings facility can concentrate the dissolved salts in the process water.
- **Ores in arid climates.** Ores in arid climates may have soluble components, such as copper, iron, and acid salts that may impact the metallurgical response.
- **Harsh waters.** Harsh waters can also affect reagent solution preparation as well. Many flocculants and organic depressants will not dissolve well in harsh waters, so fresh water is needed for preparation.
- **Buildup of ions.** The buildup of specific ions in the water can affect flotation. This is sometimes evident between winter and summer when the temperature differences can affect the solubility of various materials.

The water balance within the concentrator is usually of little consequence to the overall water consumption of the operation. Water consumption is typically dictated by the tailings disposal and storage arrangement (Figure 26). The balance between evaporation and precipitation in the mine, waste rock storage, and plant and tailings catchment areas will determine if the operation will have a positive or negative water balance. A negative balance will require raw water to be supplied. A positive water balance will require water to be treated to meet regulations and subsequently be discharged into the environment.

Where water is scarce, a reduction in the raw water requirement can be achieved by reducing water lost from tailings to evaporation and to encapsulation in the deposited tailings. The approximate net water consumptions per metric ton of ore treated are as follows:

- Conventionally subarial deposited tailings, >0.75 m^3/t
- Thickened tailings, 0.5–0.75 m^3/t
- Paste-thickened tailings, 0.25–0.5 m^3/t
- Filtered tailings, 0.15–0.25 m^3/t

LARGE-SCALE COPPER CONCENTRATORS

Antamina Copper-Zinc Mine, Peru

Antamina is a copper-zinc mine located in the Department of Ancash in the high Andes of Peru. The original concentrator was designed for 70,000–100,000 t/d, depending on ore types. A simple flow sheet for the Antamina mine is shown in Figure 27, and the major process equipment are listed in Table 8.

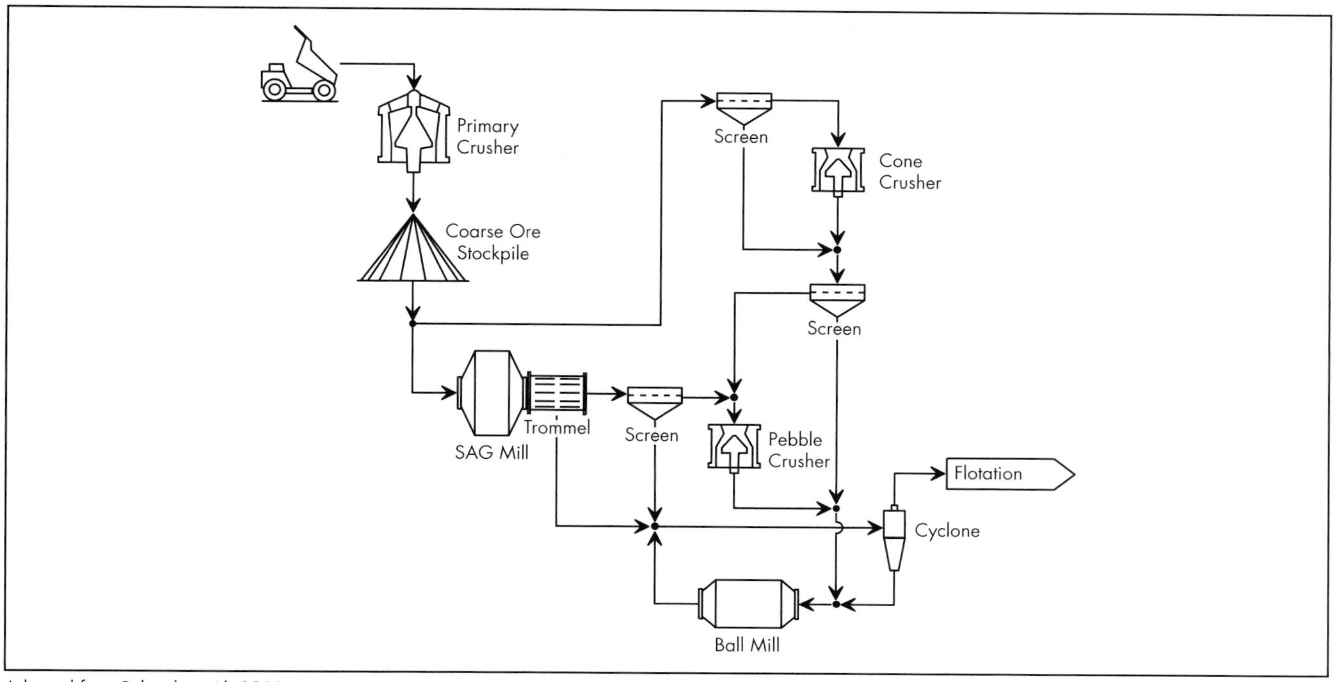

Adapted from Rybinski et al. 2011

Figure 27 Simplified Antamina comminution flow sheet

Table 8 Major process equipment, Antamina

Equipment	Quantity	Equipment Description
Primary crusher	1	1.52 m × 2.26 m gyratory crusher, 750 kW
SAG mill	2	38 ft diameter × 19 ft effective grinding length, 20.1 MW
Ball mill	4	24 ft diameter × 36 ft, 11.2 MW
Pebble crushing	2	Cone crushers, 800 hp
Pebble crushing screen	2	10 ft × 24 ft double-deck screens
Cyclone	4 clusters	660-mm cyclones, 14 per cluster
Flotation cell, rougher, copper	32	Four rows of 8 cells of 130-m^3 forced-air machines
Flotation cell, cleaner-scavenger	5	One row of 5 cells of 130-m^3 forced-air machines
First cleaner cell, copper	6	4.3 m diameter × 14 m high column cells
Second cleaner cell, copper	4	4.3 m diameter × 14 m high column cells
Regrind mill, copper	2	VTM-1000, 1,000 hp
Flotation cell rougher, zinc	32	Four rows of 8 cells of 130-m^3 forced-air machines
Flotation cell, cleaner-scavenger	5	One row of 5 cells of 130-m^3 forced-air machines
First cleaner cell, zinc	7	4.3 m diameter × 14 m high column cells
Second cleaner cell, zinc	3	4.3 m diameter × 14 m high column cells
Regrind mill, zinc	2	VTM-1000, 1,000 hp
Concentrate thickener, copper	1	35 m diameter
Concentrate thickener, zinc	1	35 m diameter
Concentrate thickener, bulk	1	25 m high rate
Concentrate thickener, molybdenum	2	6 m diameter
Concentrate filter	4	Pressure filters PF144

Batu Hijau Copper Mine, Indonesia

Batu Hijua is a copper-gold mine located on the Indonesian island of Sumbawa in Southeast Asia. The original concentrator was designed for 80,000 t/d and uses seawater. A simple flow sheet for the Batu Hijau mine is shown in Figure 28, and the major process equipment are listed in Table 9.

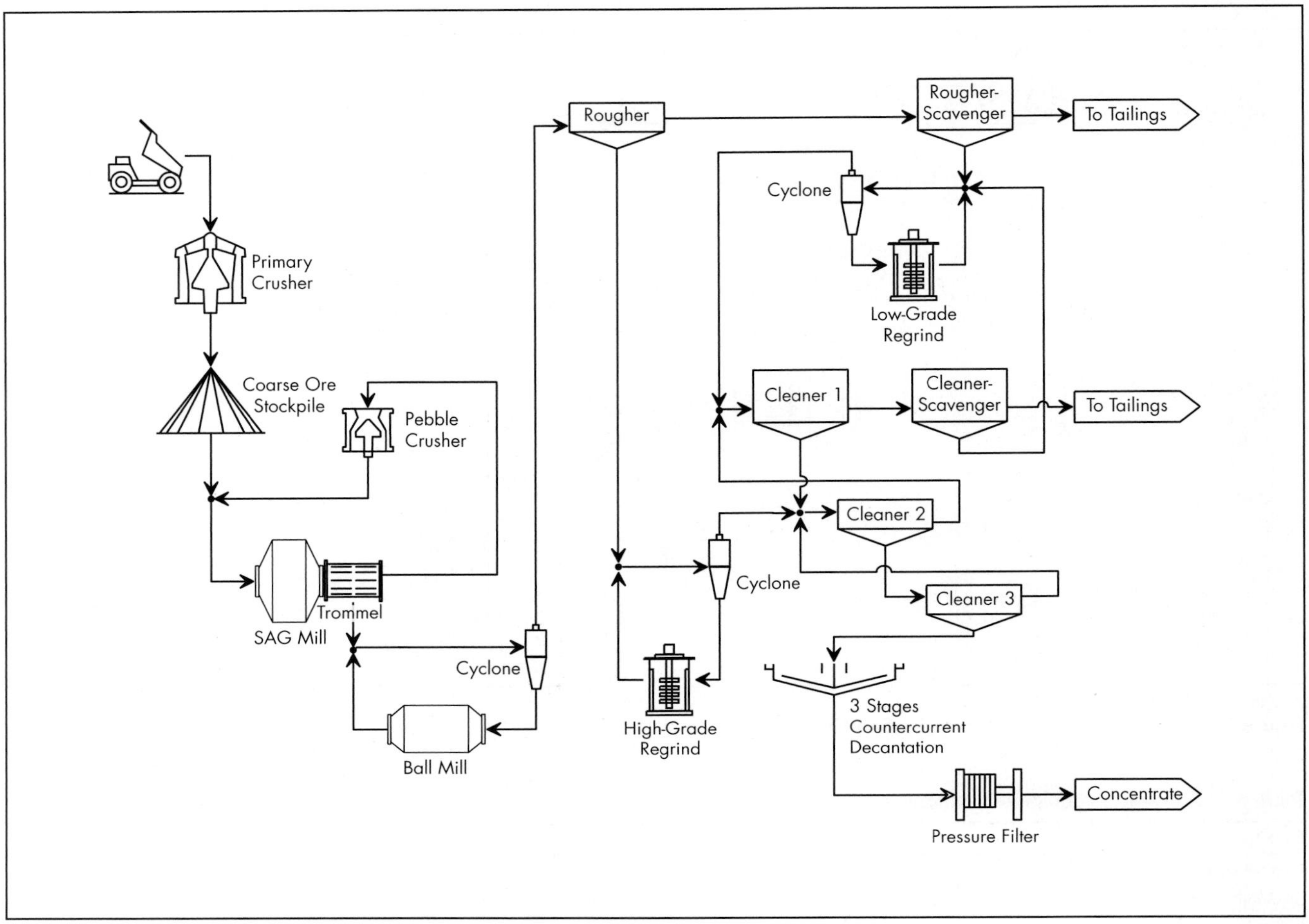

Adapted from DeMull et al. 2001

Figure 28 Simplified Batu Hijau flow sheet

Table 9 Major process equipment, Batu Hijau

Equipment	Quantity	Equipment Description
Primary crusher	2	60 in. x 89 in. gyratory, 750 kW
Pebble crusher	2	1,000-hp cone crushers
SAG mill	2	36 ft × 19 ft, 13,423 kW
Ball mill	4	20 ft × 33.5 ft, 6,711 kW
Cyclone	4 clusters	33-in.-diameter cyclones
Flotation cell	50	127-m^3 self-aspirating cells
First cleaner	4	42.5 m^3
Second cleaner	10	14.2 m^3
Third cleaner	4	14.2 m^3
Cleaner-scavenger cell	4	42.5 m^3
Regrind	2	VTM-1250
Polishing mill	1	500 hp
Thickener	3	25 m diameter
Filter press	2	Pressure filters

Boddington Copper Gold Mine, Western Australia

Boddington is a copper-gold mine northwest of Boddington in Western Australia. The concentrator was designed for approximately 100,000 t/d. In 2017, the plant produced 787,000 oz Au and 80 million lb Cu. The comminution circuit incorporates three-stage crushing with HPGRs and ball mills. A simple flow sheet for the Boddington mine is shown in Figure 29, and the major process equipment are listed in Table 10.

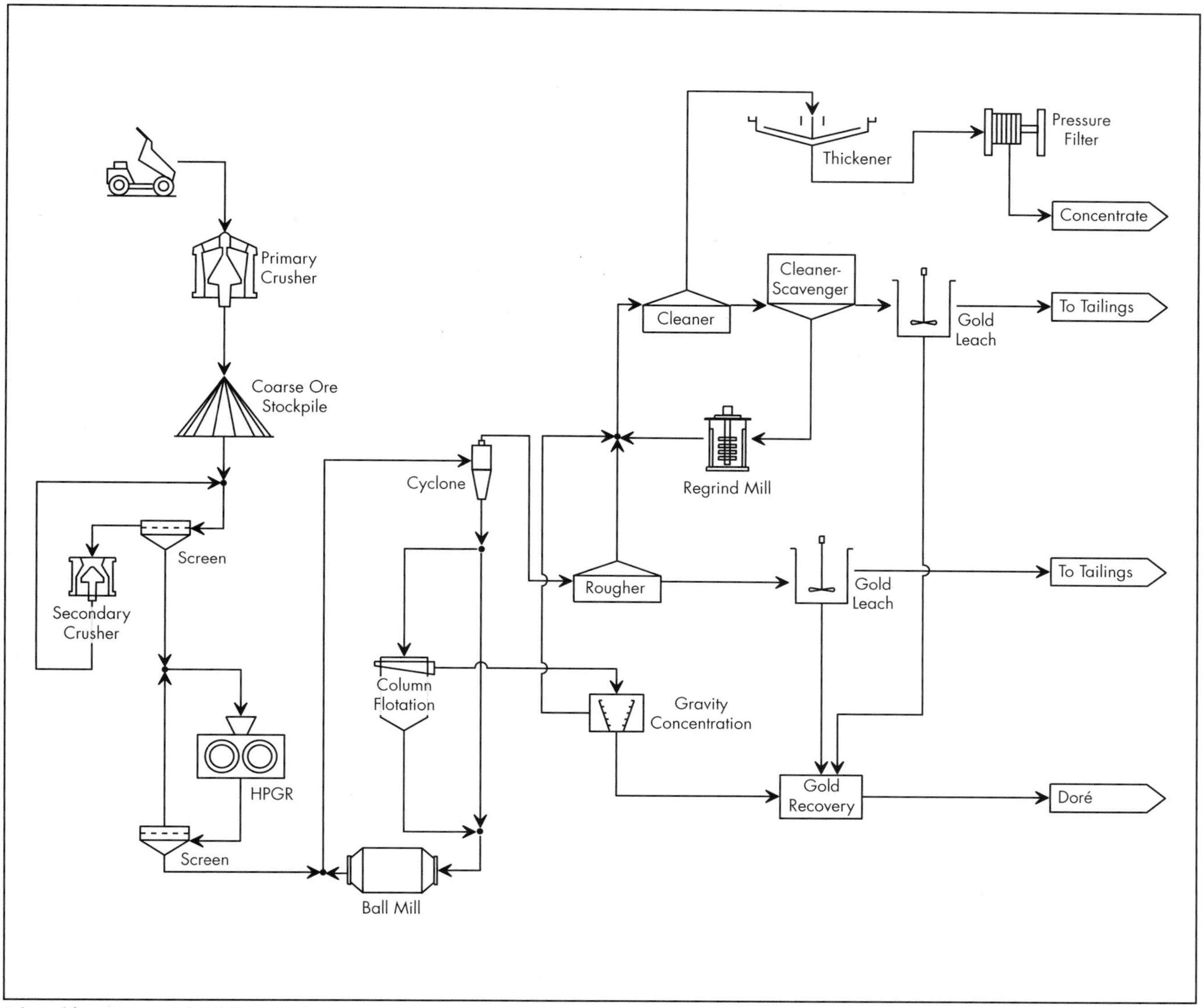

Adapted from Dunne et al. 2007

Figure 29 Simplified Boddington flow sheet

Table 10 Major process equipment, Boddington

Equipment	Quantity	Equipment Description
Primary crusher	2	60 in. × 113 in. gyratory
Coarse ore stockpile	1	40,000 t live
Banana screen	4	3.6 m × 7.3 m single-deck
Secondary crusher	6	1,000 hp cone crusher, 746 kW
HPGR	4	2.4 m diameter × 1.65 m 2 × 2.8 MW drives
Wet screen	8	3.66 m × 7.93 m (aperture of 9–10 mm)
Ball mill	4	7.9 m × 13.4 m, 16 MW
Primary cyclone	26	12 cyclones per cluster

Cadia Copper-Gold Mine, Australia

Cadia is a copper-gold mine located close to Orange, New South Wales, in Australia. The original concentrator was designed for ~47,000 t/d. The flash flotation cells in the comminution circuit treat ~50% of the circulating load, and the gold in the flotation concentrate is recovered by a centrifugal gravity separator. A flow sheet for the Cadia mine is shown in Figure 30, and the major process equipment are listed in Table 11. Cadia had the world's first 12.2-m SAG mill, which has been the basis for important SAG milling models.

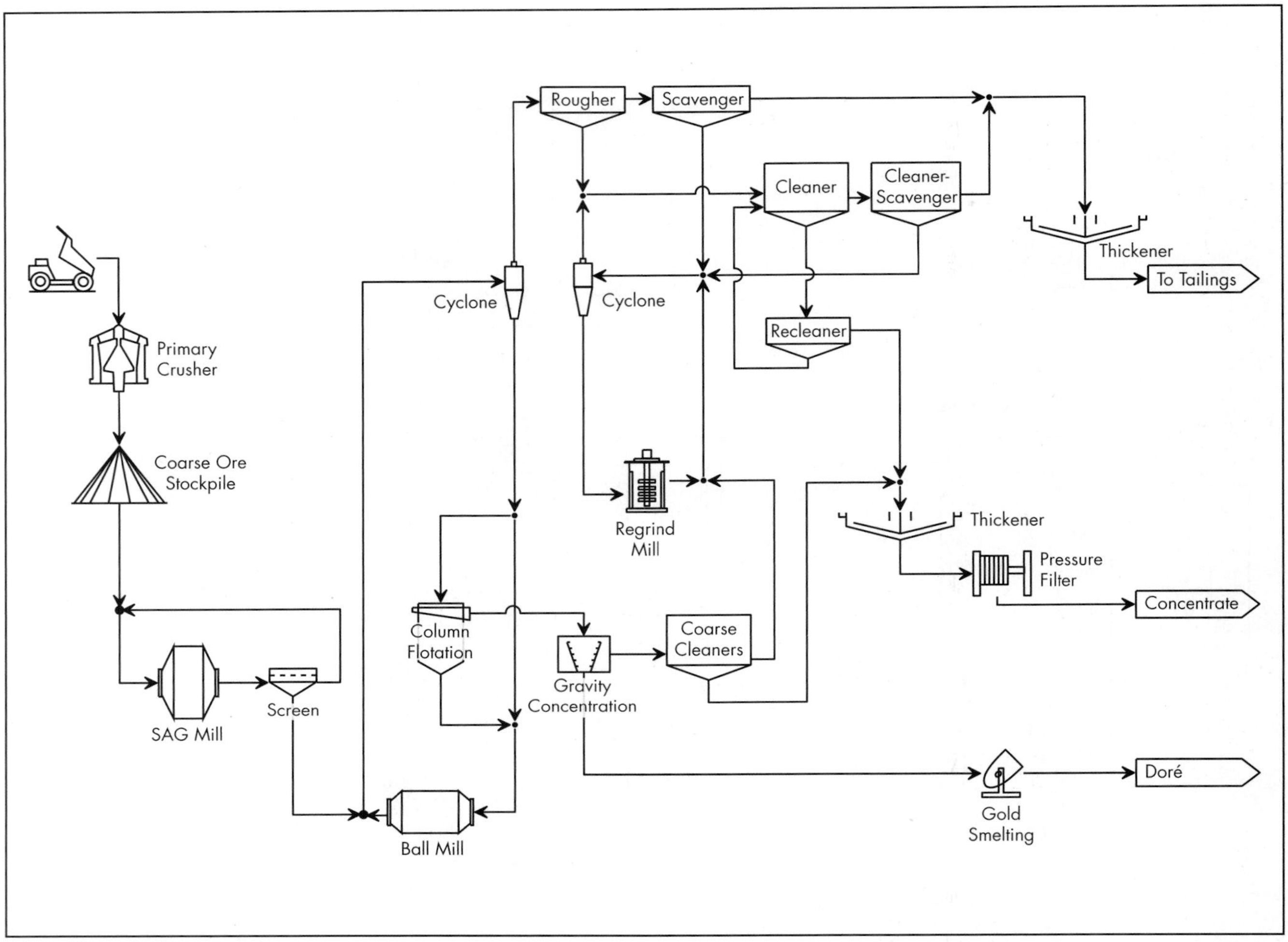

Adapted from Cesnik 2009

Figure 30 Simplified Cadia flow sheet

Table 11 Major process equipment, Cadia

Equipment	Quantity	Equipment Description
Primary crusher	1	60 in. × 110 in. gyratory
SAG mill	1	12.2 m × 6.1 m effective grinding length, 20 MW
Ball mill	2	6.7 m × 10.97 m; 8,760 kW
Rougher flotation	14	Two rows of seven 150-m^3 forced-air machines
Regrind mill	1	Vertimill, VTM400
Flash flotation cell	1	SK1200 flash float
Cleaner flotation	6	30-m^3 forced-air machines
Scavenger-cleaner flotation	4	30-m^3 forced-air machines
Recleaner flotation	4	8.5-m^3 forced-air machines
Tailings thickener	1	53 m diameter
Concentrate thickener	1	12 m diameter
Concentrate filter	1	Plate-and-frame, VPA1540/24 press

Cerro Verde Copper Mine, Peru

Cerro Verde is a copper-molybdenum mine located close to Arequipa southwest of Peru. The original concentrator was designed for 108,000 t/d, and a second plant was built to treat an additional 240,000 t/d. The comminution circuit consists of three-stage crushing with HPGRs and shell-mounted ball mills. A simple flow sheet for the Cerro Verde mine is shown in Figure 31, and the major process equipment are listed in Table 12.

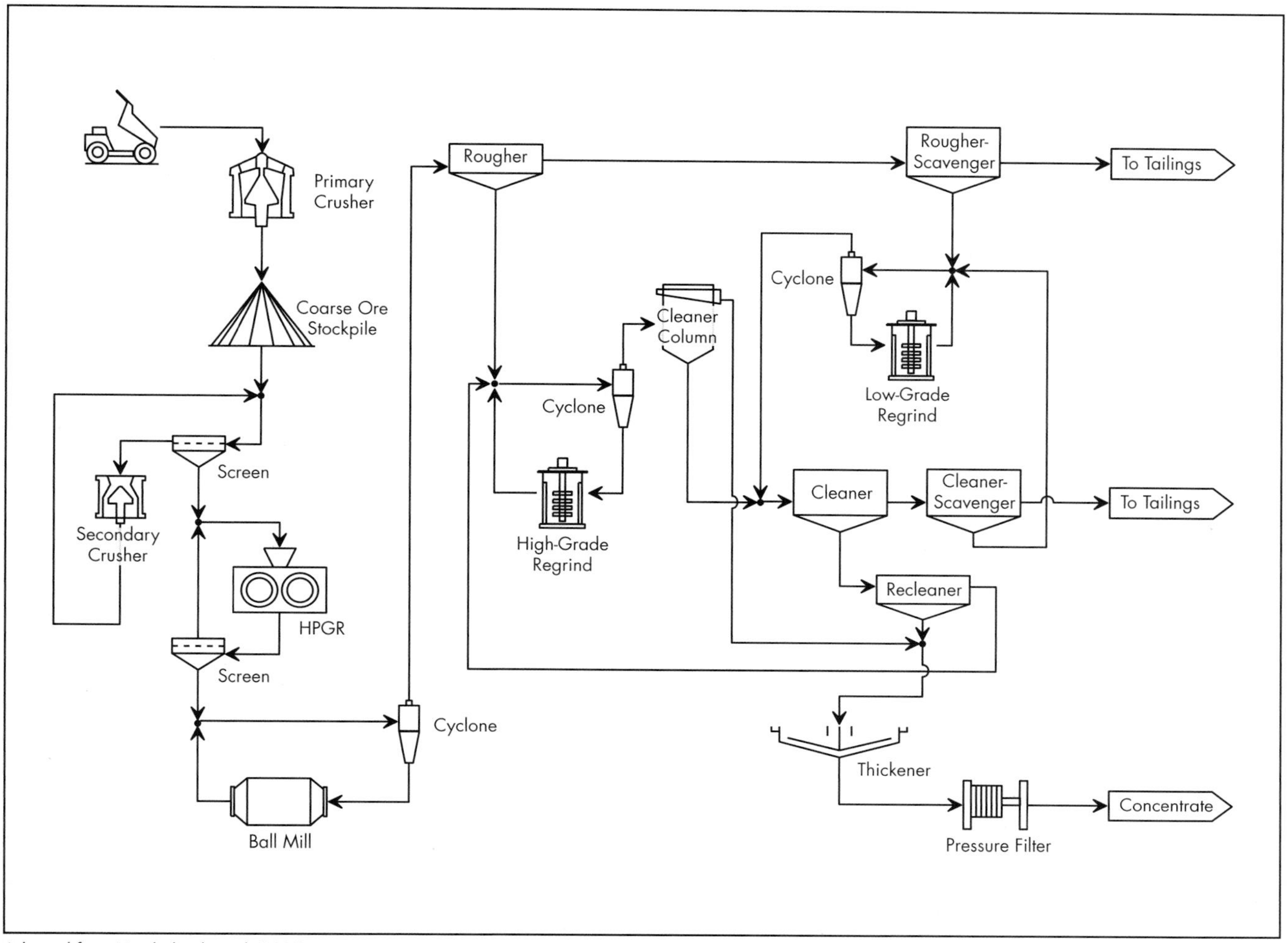

Adapted from Vanderbeek et al. 2017

Figure 31 Simplified Cerro Verde Concentrator II flow sheet

Table 12 Major process equipment, Cerro Verde Concentrator II

Equipment	Quantity	Equipment Description
Primary crusher	2	1.5 m × 2.9 m gyratory crushers, 746 kW
Secondary screen	8	3.6 × 7.9 m double-deck banana screens
Secondary crusher	8	Cone crushers, 1,250 hp each
Tertiary crushing, HPGR	8	2.4 m diameter × 1.65 m wide, 2.5-MW twin drives
HPGR screen	12	3.66 × 8.5 m double-deck wet screens
Ball mill	6	8.2 m × 14.6 m, shell-supported 22-MW gearless drives
Primary cyclone	6	Sixteen 33-in.-diameter cyclones
Rougher flotation	6 rows	Nine 257-m^3 self-aspirating mechanical cells per row
Cleaner-scavenger cell	12	Two rows of six 257-m^3 cells
Final column cleaner	6	4.88 m diameter × 12 m high
Concentrate thickener	1/1/1	35 m diameter for copper and 45 m diameter for bulk concentrate, plus 35 m diameter for clarifier
Tailings thickener	4	80 m diameter

Prominent Hill Copper Mine, South Australia

Prominent Hill is a copper-gold mine located in Australia. It processes approximately 8 million t/yr of copper ore and has Jameson flotation cells in the rougher circuit. A simple flow sheet for the Prominent Hill mine is shown in Figure 32, and the major process equipment are listed in Table 13.

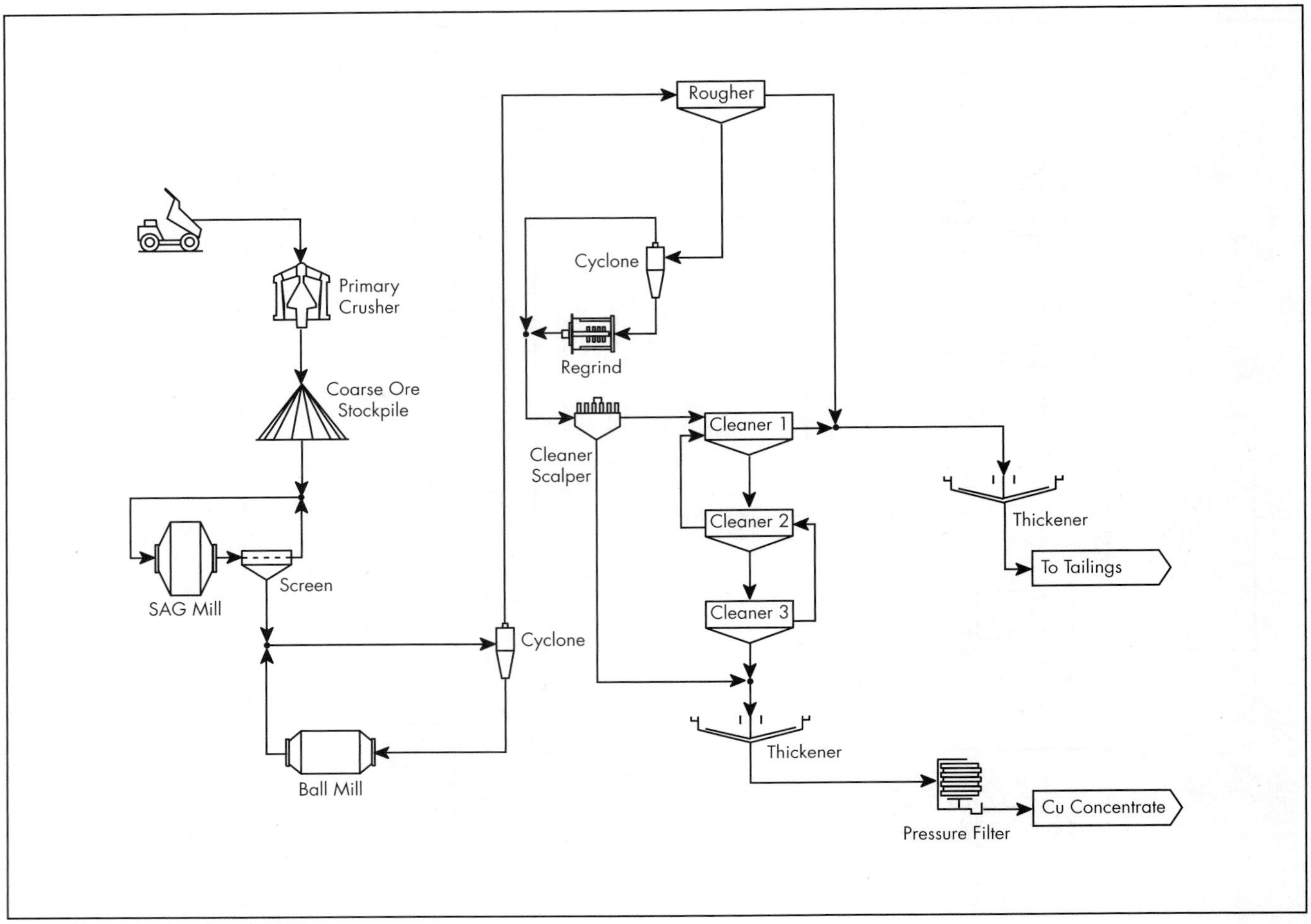

Adapted from Woodward et al. 2013

Figure 32 Simplified Prominent Hill flow sheet

Table 13 Major process equipment, Prominent Hill

Equipment	Quantity	Equipment Description
Primary crusher	1	60 in. × 89 in. gyratory, 600 kW
Coarse ore stockpile	1	30,000 t live, 130,000 t total
SAG mill	1	10.36 m diameter × 5.18 m effective grinding length, 12 MW
SAG discharge screen	12	3.66 m × 8.5 m double deck wet
Ball mill	1	7.3 m diameter × 10.5 m, 12 MW
Primary cyclone	1 cluster	Cluster of 16 × 840 mm cyclones
Rougher flotation	6	150-m^3 forced-air tank cells
Rougher concentrate regrind mill	1	M10,000 IsaMill
Concentrate scalper	1	J5400/18 Jameson cells
First cleaner flotation	8	50-m^3 forced-air tank cells
Second cleaner flotation	6	20-m^3 forced-air tank cells
Third cleaner flotation	4	20-m^3 forced-air tank cells
Tailings thickener	1	45 m high rate
Concentrate thickener	1	23 m high rate
Concentrate thickener	1	Horizontal plate pressure filter, 156 m^2

Other Concentrators

Other large copper concentrators include the following (including sources where additional information can be obtained):

- Alumbrera copper mine, Argentina (Morrell and Munro 2010)
- Candelaria copper mine, Chile (Marsden and Ogonowski 1999)
- Constancia copper mine, Peru (Klohn et al. 2016)
- Escondida copper mine, Chile (Bergholz and Schreder 2004)
- Grasberg copper-gold mine, Indonesia (Veloo and Coleman 1996)
- Lumwana copper mine, Zambia (Araya et al. 2014)
- Phu Kham copper mine, Laos (Hoyle et al. 2013)
- Salobo copper mine, Brazil (Godoy et al. 2010)
- Sossego copper mine, Brazil (Mendonça et al. 2015)

REFERENCES

Araya, R., Cordingley, G., Mwanza, A., and Huynh, L. 2014. Necessity driving change and improvement to the cleaner circuit at Lumwana copper concentrator. Presented at the 46th Annual Canadian Mineral Processors Operators' Conference, Ottawa, ON, January 21–23.

Bergholz, W., and Schreder, M. 2004. Escondida—Phase IV Grinding Circuit. In *Proceedings of the 36th Annual Meeting of the Canadian Mineral Processors*. Montreal, QC: Canadian Institute of Mining, Metallurgy and Petroleum.

Cesnik, F. 2009. Improvements in flotation cell operation and maintenance at Newcrest Cadia Valley Operations. In *Proceedings of the 10th Mill Operators' Conference*. Melbourne, Victoria: Australasian Institute of Mining and Metallurgy.

Cochilco (Comisión Chilena del Cobre). 2017. *Proyección de la producción de cobre en Chile 2017–2028 [Projection of Copper Production in Chile 2017–2028]*. Santiago, Chile: Cochilco.

DeMull, T., Spenceley, J., and Hickey, P. 2001. Planning and teamwork lead to successful start-up at Batu Hijau. SME Preprint No. 01-159. Littleton, CO: SME.

Dunne, R., Hart, S., Parker, B., and Veillette, G. 2007. Boddington gold mine—An example of sustaining gold production for 30 years. In *Proceedings of World Gold 2007*. Melbourne, Victoria: Australasian Institute of Mining and Metallurgy.

Godoy, M., Bergerman, M., Godoy, P., and Rosa, M.A. 2010. Development of the Salobo project. In *Proceedings of the CIM Conference and Exhibition 2010*. Montreal, QC: Canadian Institute of Mining, Metallurgy and Petroleum.

Greet, C., Small, C., Steinier, P., and Grano, S. 2004. The Magotteaux Mill: Investigating the effect of grinding media on pulp chemistry and flotation performance. *Miner. Eng.* 17:891–896.

Hedley, N., and Tabachnick, H. 1958. Mineral dressing notes 23. In *Chemistry of Cyanidation*. New York: American Cyanamid Company.

Hoyle, A., Bennett, D., and Walker, P. 2013. The increased recovery project at the Phu Kham copper-gold operation, Laos PDR. In *Proceedings of MetPlant 2015*. pp. 246–265. Melbourne, Victoria: Australasian Institute of Mining and Metallurgy.

ICA (International Copper Association). 2018. Measuring our contribution to sustainable development. https://sustainablecopper.org/ica-indicators/. Accessed September 2018.

Klimpel, R.R. 1993. The interaction of grind size, collector dosage, and frother type in industrial chalcopyrite rougher flotation. SME Preprint No. 93-80. Littleton, CO: SME.

Klohn, S., Stephenson, D., and Granados, H. 2016. Constancia project process plant design and start up. In *Proceedings of the 48th Annual Meeting of the Canadian Mineral Processors Operators' Conference*. Westmount, QC: Canadian Institute of Mining, Metallurgy and Petroleum.

Marsden, J., and House, I. 1992. *The Chemistry of Gold Extraction*. New York: Ellis Horwood.

Marsden, J.O., and Ogonowski, D.L. 1999. The Candelaria concentrator expansion project. SME Preprint No. 99-139. Littleton, CO: SME.

Mendonça, A., Luiz, D., Souza, M., Oliveira, G., Delboni, H., and Bergerman, M. 2015. Sossego SAG mill—10 years of operation and optimizations. In *Proceedings of SAG 2015*. Vancouver, BC: University of British Columbia.

Morrell, S., and Munro, P.D. 2010. Increased profits through mine and mill integration. In *After 2000—The Future of Mining*. Melbourne, Victoria: Australasian Institute of Mining and Metallurgy.

Rybinski, E., Ghersi, J., Davila, F., Linares, J., Valery, W., Jankovic, A., Valle, R., and Dikmen, S. 2011. Optimisation and continuous improvement of Antamina comminution circuit. In *Proceedings of SAG 2011*. Vancouver, BC: University of British Columbia.

Schumann, R., Levay, G., Dunne, R., and Hart, S. 2003. Managing process water quality in base metal sulfide flotation. In *Proceedings of Water in Mining*. Melbourne, Victoria: Australasian Institute of Mining and Metallurgy. pp. 251–260.

Vanderbeek, J., Gelfi, P., and Enriquez, J. 2017. The Cerro Verde 2 Concentrator: Design, start up and operation. SME Preprint No. 17-115. Englewood, CO: SME.

Veloo, C., and Coleman, R. 1996. P.T. Freeport Indonesia's SAG plant: Flowsheet, start-up and training issues. Presented at Canadian Mineral Processors meeting. www.onemine.org/document/document.cfm?docid=231854.

Woodcock, J.T., Sparrow, G.J., Bruckard, W.J., Johnson, N.W., and Dunne, R. 2007. Plant practice: Sulfide minerals and precious metals. In *Froth Flotation: A Century of Innovation*. Edited by M.C. Fuerstenau, G. Jameson, and R.-H. Yoon. Littleton, CO: SME.

Woodward, P., Muhamad, N., and Ly, T. 2013. Optimisation of the Prominent Hill flotation circuit. Presented at Metallurgical Plant Design and Operating Strategies, Perth, Western Australia, July 15–17.

CHAPTER 12.8

Copper Hydrometallurgy

Martin C. Kuhn and Russell D. Alley

ORIGINS OF HYDROMETALLURGY

Habashi (2014) provides an excellent introduction for the origins of hydrometallurgy:

> The roots of hydrometallurgy may be traced back to the period of alchemists when the transmutation of base metals into gold was their most important occupation. Some of these operations involved wet methods. For example, when an alchemist dipped a piece of iron into a solution of copper vitriol, i.e. copper sulphate, the iron was immediately covered by a layer of metallic copper. The major question, however, that remained unanswered was: how can the transmutation of iron or copper into gold be effected? Gold, the most noble of all metals was insoluble in all acids or alkalis known at that time.
>
> The discovery of *aqua regia* by Jabir Ibn Hayyan (720–813 AD), the Arab alchemist, may be considered as the first milestone marking the beginning of hydrometallurgy. ... In the Middle Ages, certain soils containing putrefied organic matter were leached to extract saltpeter (salt of stone, potassium nitrate), a necessary ingredient for the manufacture of gunpowder. The process was fully described by Vannoccio Biringuccio (1480–1539) in his *Pirotechnia* published in 1540.
>
> In the 16th Century, the extraction of copper by wet methods received some attention. Heap leaching was practiced in the Harz Mountains area in Germany and in Rio Tinto mines in Spain. In this operation, pyrite containing some copper sulfphide minerals was piled in the open air and left for months to the action of rain and air, whereby oxidation and dissolution of copper took place. A solution containing copper sulphate was drained from the heap and collected in a basin. Metallic copper was then precipitated from this solution by scrap iron, a process that became known as "cementation process." This is the same process that was already known to the alchemists and was in operation to an appreciable extent until recently when it was replaced by solvent extraction-electrowinning.

ELECTROMETALLURGY

The first application of Alessandro Volta's voltaic pile in metallurgy was in 1807 when Humphry Davy at the Royal Institution in London built a large battery and was able to decompose caustic potash and caustic soda to identify potassium and sodium as metals for the first time. This was followed in the next year, using the same technique, by the identification of barium, calcium, magnesium, and strontium.

Michael Faraday who was Davy's assistant, continued this type of research further and in the early 1830s found the relationships between current used and quantity of metal deposited. He was the first to have clear ideas concerning the quantity of metal deposited and intensity of electricity (i.e., the quantities now measured with regard to amperes and volts). Faraday introduced the terms *ion*, *cation*, *anion*, *electrode*, *electrolyte*, and so on, which are commonly used today.

REFINING METALS

In 1876, copper was refined electrolytically in Germany by the Norddeutsche Affinerie in Hamburg. This was followed by gold refining in 1878 using the Wohlwill process, named after German chemist Emil Wohlwill. The first copper refinery built in America was at Newark (New Jersey) in 1883 by the Balbach Smelting and Refining Company. In 1884, German metallurgist Bernhard Möbius (1852–1898) invented a process for silver refining.

HYDROMETALLURGY IN COPPER PRODUCTION

In 1910, Albert C. Burrage, a Boston (Massachusetts, United States) capitalist, obtained options on the entire mining territory at Chuquicamata (Chile) and offered his options to the firm of M. Guggenheim's Sons in New York (United States). In 1912, the Guggenheims organized the Chile Exploration Company in New Jersey to take over the options and incorporated the Chile Copper Company in Maine (United States).

Martin C. Kuhn, Chief Executive Officer, Minerals Advisory Group LLC, Tucson, Arizona, USA
Russell D. Alley, President, Minerals Advisory Group LLC, Tucson, Arizona, USA

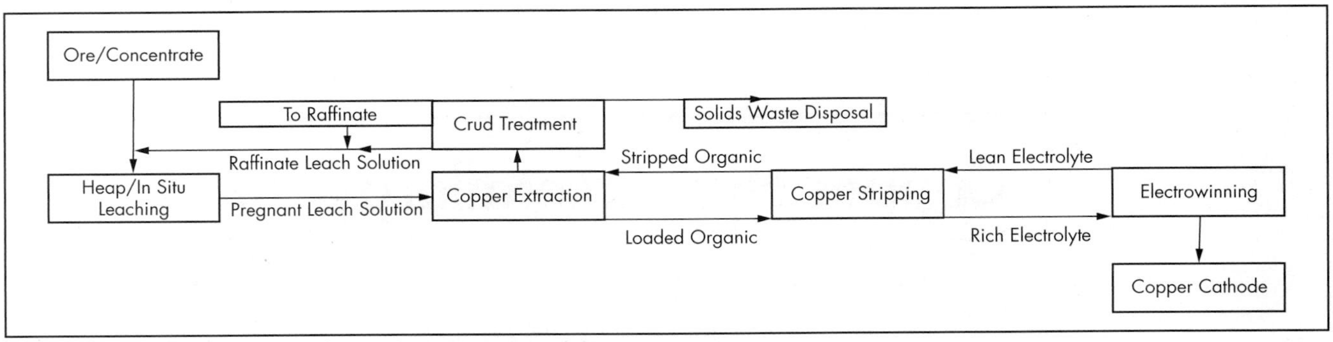

Figure 1 Block flow diagram of the solvent extraction/electrowinning process

The low-grade ore could not be smelted in the ordinary way because of its chlorine content, which meant that an entirely new process had to be devised. This was accomplished by E.A. Cappelen Smith, a consulting metallurgist for the Guggenheims, and a new chapter in the history of copper extraction was recorded. Instead of treating the ore by concentration and smelting, as was the usual procedure, it was subjected to leaching of the copper in huge vats and precipitating copper by electricity. The Chuquicamata ore was the first of the porphyry copper ores to be leached using electricity to precipitate copper metal (Marcosson 1957).

Large-scale leaching of oxide and secondary copper sulfide ores and pioneering electrochemical technology was developed and commercialized in North and South America between 1912 and 1919. General John C. Greenway, general manager of Calumet and Arizona Mining Company (C&A) and New Cornelia Copper Company developed a highly oxidized zone of copper mineralization in the Little Ajo Mountains in Arizona (United States). Utilizing the acid from the C&A copper smelter in Douglas, Arizona, General Greenway and L.D. Ricketts reported to the C&A board that they had developed a process for leaching the New Cornelia Copper Company carbonate ores with sulfuric acid. Construction of a large leaching plant and related facilities was begun in March 1916.

The leaching plant had 11 lead-lined leaching tanks. After the copper had been leached from the ore and the resulting tailings washed and drained, the tails were sent to a tailings dump at the mine site and the solutions (pregnant leach solution, or PLS) were treated in cementation launders to precipitate copper metal as a powder. The cement copper was washed from the launder and reabsorbed into an acid solution and returned to the electrolytic plant. The electrolytic plant consisted of 102 electrolytic tanks, with lead/antimony anodes and copper cathode starting sheets (Rickard 1996).

The next great leap of hydrometallurgical processing in the heap/dump/vat leaching followed by electrolytic copper was developed by General Mills Chemicals laboratories in the late 1950s. The first work was based on the theory and application developed by the U.S. Atomic Energy Commission at Oak Ridge, Tennessee (United States), for the recovery of uranium. General Mills successfully developed and marketed trialkyl tertiary amine for use in the uranium industry. Thus was born from this technology copper solvent extraction (or, as originally called, liquid ion exchange, or LIX).

Using the basic principles of LIX, General Mills began researching the recovery of copper from dilute solutions containing copper. The result of this work was the development of an organic reagent first introduced in 1963 under the trademark name LIX 63. In 1964, LIX 64 was introduced and a new and revolutionary technology was presented to the industry. The major copper producers continued to manufacture copper by conventional means, and the risks inherent with new technology were procured by a small entrepreneurial company, Ranchers Exploration and Development, at its Bluebird mine in Arizona in 1968. Ranchers Exploration and Development successfully installed and operated the new process and led the industry into a huge expansion of hydrometallurgical operations worldwide. Between 1968 and 1994, more than 42 mining operations adopted the solvent extraction and electrowinning (SX/EW) process. From 1968 to 2017, approximately 4.8 Mt (million metric tons) of copper were produced by SX/EW (Kuhn 1993).

Demonstrated But Not Commercialized Sulfuric Acid–Based Processes

"Solvent extraction and electrowinning (SX/EW) is a two-stage hydrometallurgical process that first extracts and upgrades copper ions from low-grade leach solutions into a solvent containing a chemical that selectively reacts with and binds the copper in the solvent. The copper is extracted from the solvent with strong aqueous acid which then deposits pure copper onto cathodes using an electrolytic procedure (electrowinning)" (Wikipedia 2016).

An example of the SX/EW process for copper is given in Figure 1. In addition to the basic diagram, a unit operation of crud removal and treatment is shown. In general, *crud* comprises ore fines, dissolved silica, and precipitates that mix with an aqueous phase, as well as organic materials that present operating problems and the loss of expensive organic reagents. Crud is removed by vacuuming the mass and centrifuging to separate the liquid and organic from the solids. A portion of the organic can be recovered and regenerated for return to the system.

WORLD COPPER MINE PRODUCTION AND COPPER MINING CAPACITY

World copper mine production from 1900–2017 is shown in Figure 2, which demonstrates and relative proportion of tonnage produced as either smelted copper from copper concentrates or copper recovered by SX/EW from heap, run-of-mine (ROM) ore, or other sources. Almost all copper is treated electrolytically during its production from ore. It is electrorefined from impure copper anodes or electrowon from leach–solvent extraction solutions. Considerable copper scrap is also electrorefined (ICSG 2017b).

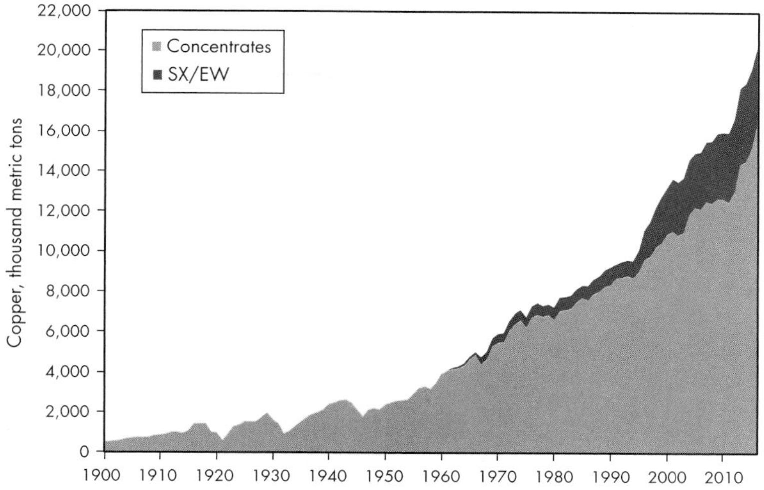

Note: Since 1900, when world production was less than 500,000 t of copper, world copper mine production has grown by 3.2% per year to 20.2 Mt (million metric tons) in 2016. SX/EW production, virtually nonexistent before the 1960s, increased to 3.9 Mt in 2016.

Adapted from ICSG 2017b

Figure 2 World copper mine production, 1900–2016

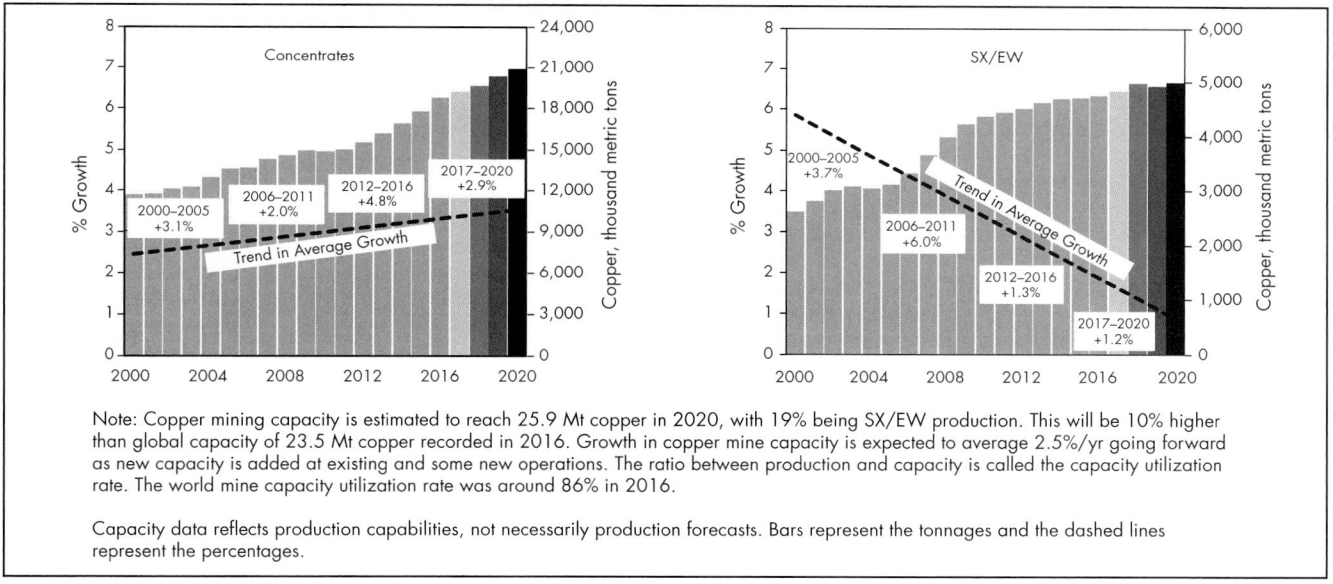

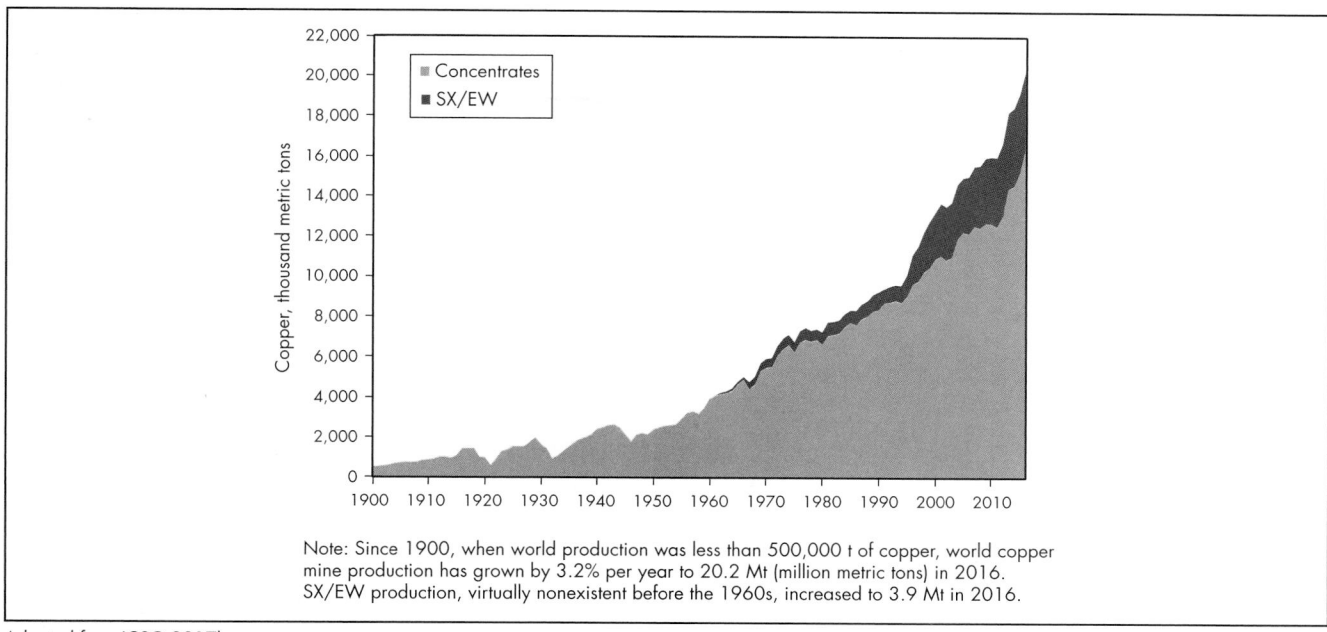

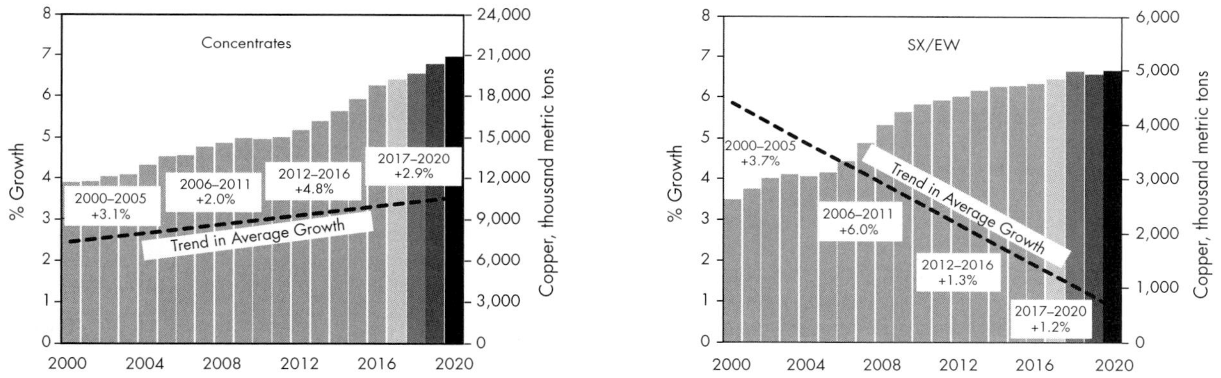

Note: Copper mining capacity is estimated to reach 25.9 Mt copper in 2020, with 19% being SX/EW production. This will be 10% higher than global capacity of 23.5 Mt copper recorded in 2016. Growth in copper mine capacity is expected to average 2.5%/yr going forward as new capacity is added at existing and some new operations. The ratio between production and capacity is called the capacity utilization rate. The world mine capacity utilization rate was around 86% in 2016.

Capacity data reflects production capabilities, not necessarily production forecasts. Bars represent the tonnages and the dashed lines represent the percentages.

Adapted from ICSG 2017a

Figure 3 Trends in copper mining capacity, 2000–2020

Figure 2 does not demonstrate all copper recovery by hydrometallurgical processes. Prior to 1968, heap and dump leaching operations recovered significant tonnages of copper as cement copper (a hydrometallurgical process); however, cement copper characteristically was shipped to a smelter for conversion into anode copper for electrorefining using copper starter-sheet cathodes or stainless-steel cathodes. Smelted concentrates and cement copper are both converted to copper metal and electrorefined to cathode copper. Electrorefining is a hydrometallurgical process. It could be assumed that all of the copper shown in Figure 2 required some hydrometallurgical processing.

Figure 3 demonstrates a trend of increasing production from concentrates and a decrease in production from SX/EW of copper. This trend suggests decreasing availability or development of oxide and secondary sulfide deposits and increasing production from primary copper sulfide sources.

The copper industry has actively worked since the late 1960s to develop hydrometallurgical processes to compete with smelters and reduce off-gas emissions from smelting operations. The industry response to the U.S. Clean Air Act prompted significant improvements in off-gas handling of smelter emissions, and the improvements resulted in more efficient smelting operations. Numerous hydrometallurgical

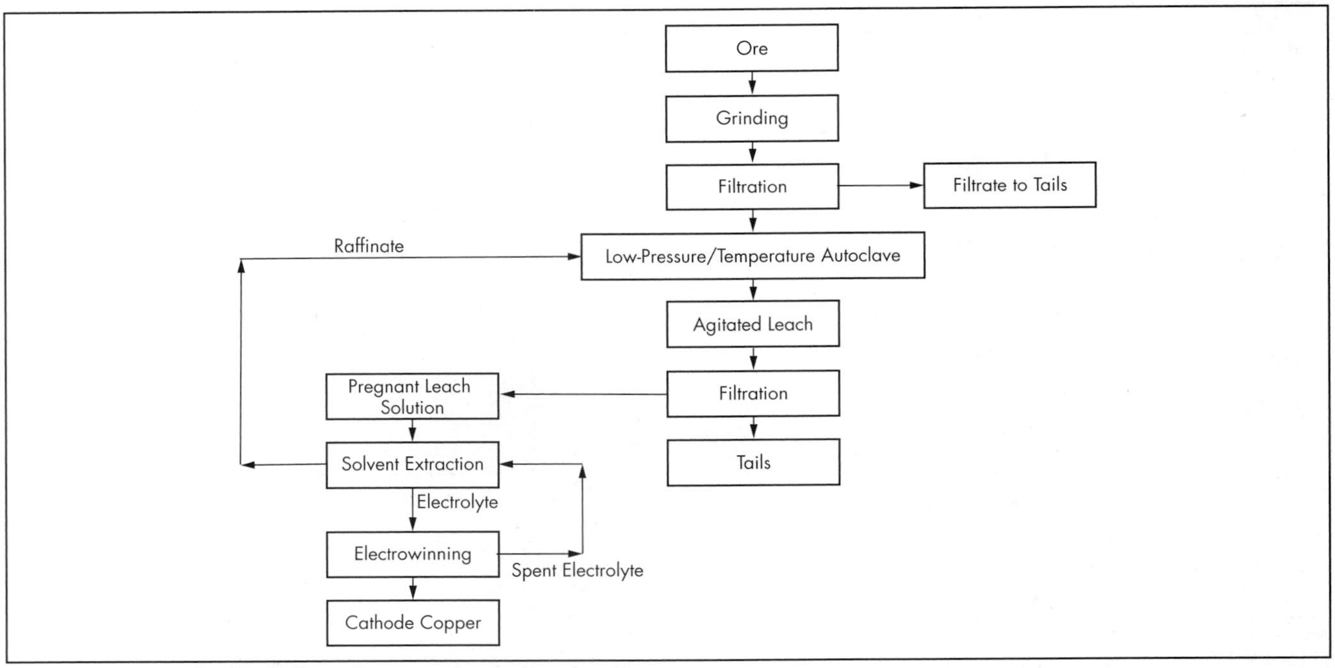

Figure 4 Gunpowder block flow diagram of autoclave leach process

processes to treat copper concentrates have been developed, tested, and in some cases commercialized in the past 60 years. The decision to smelt concentrates or use leach/SX/EW depends upon concentrate mineralogy and the need for sulfuric acid in heap/dump leaching on oxide/secondary sulfide ores. Various hydrometallurgical processes in this discussion are given as examples:

- Commercial hydrometallurgical processes for the recovery of copper
- Other commercialized hydrometallurgical processes for copper production from ores and concentrates
- Tested but not commercialized high-pressure and high-temperature autoclave processes
- Demonstrated but not commercialized sulfuric acid–based processes
- Demonstrated but not commercialized chloride processes

Commercial Hydrometallurgical Processes for the Recovery of Copper

Most hydrometallurgical processes today use similar technologies but are generally developed based on copper mineralogy, copper grade, and concentration processes. Copper from ROM ore can be recovered using sulfuric acid as a leachate on as-mined feed at as-mined material sizes for dump leaching, or crushed/agglomerated ore for heap or on–off pad leaching.

The following six examples of commercial operations using hydrometallurgical processes are discussed in this chapter:

1. **Mount Gordon mine (Australia):** Autoclave leaching of high-grade copper ores
2. **Sepon copper mine (Laos):** High-temperature and high-pressure autoclave, total pressure oxidation (TPO) leaching of copper concentrates
3. **Albion Process:** Low-temperature and atmospheric leaching of ultrafine ground copper concentrates
4. **El Abra mine (Chile):** Typical heap/pad leach of crushed whole ore by sulfuric acid leachate and copper production by SX/EW
5. **Cerro Verde mine (Peru):** Heap leaching of secondary copper mineralization using bacterially assisted oxidation of copper mineralization
6. **Kansanshi mine (Africa):** High-temperature and high-pressure autoclave leaching TPO of primary copper mineralization

Mount Gordon Mine

Operations at the Mount Gordon (formerly Gunpowder) mine in Queensland, Australia, provide an excellent example of whole ore leaching of high-grade secondary copper mineralization using low temperature and low-pressure leach conditions in an acidic environment.

Copper was discovered at the Mammoth ore body in 1927. According to Hespe (2001), "Small operators produced a total of 7,000 tonnes copper in high-grade ore from shallow workings." In 1969–1977, Consolidated Goldfields Ltd. and Mitsubishi Ltd. produced flotation concentrates for smelting ranging from 4,000 to 11,000 t/yr (metric tons per year). Between 1978 and 1982, "Renison Goldfields Consolidated Ltd. experimented with heap leaching and in situ leaching of ... ore to produce copper cement and copper cathode by SX/EW" (Hespe 2001). In 1988, the plant closed and operations were suspended.

Between 1996 and 2000, a low-pressure and low-temperature autoclave technology was developed and successfully implemented for the recovery of copper as cathode copper. Mined high-grade copper ore containing principally carbonate copper and chalcocite (Cu_2S) was ground, filtered, and fed to a low-pressure and low-temperature autoclave followed by agitation leaching to recover copper from chalcocite/carbonate-type mineralization. The leachate was sulfuric acid

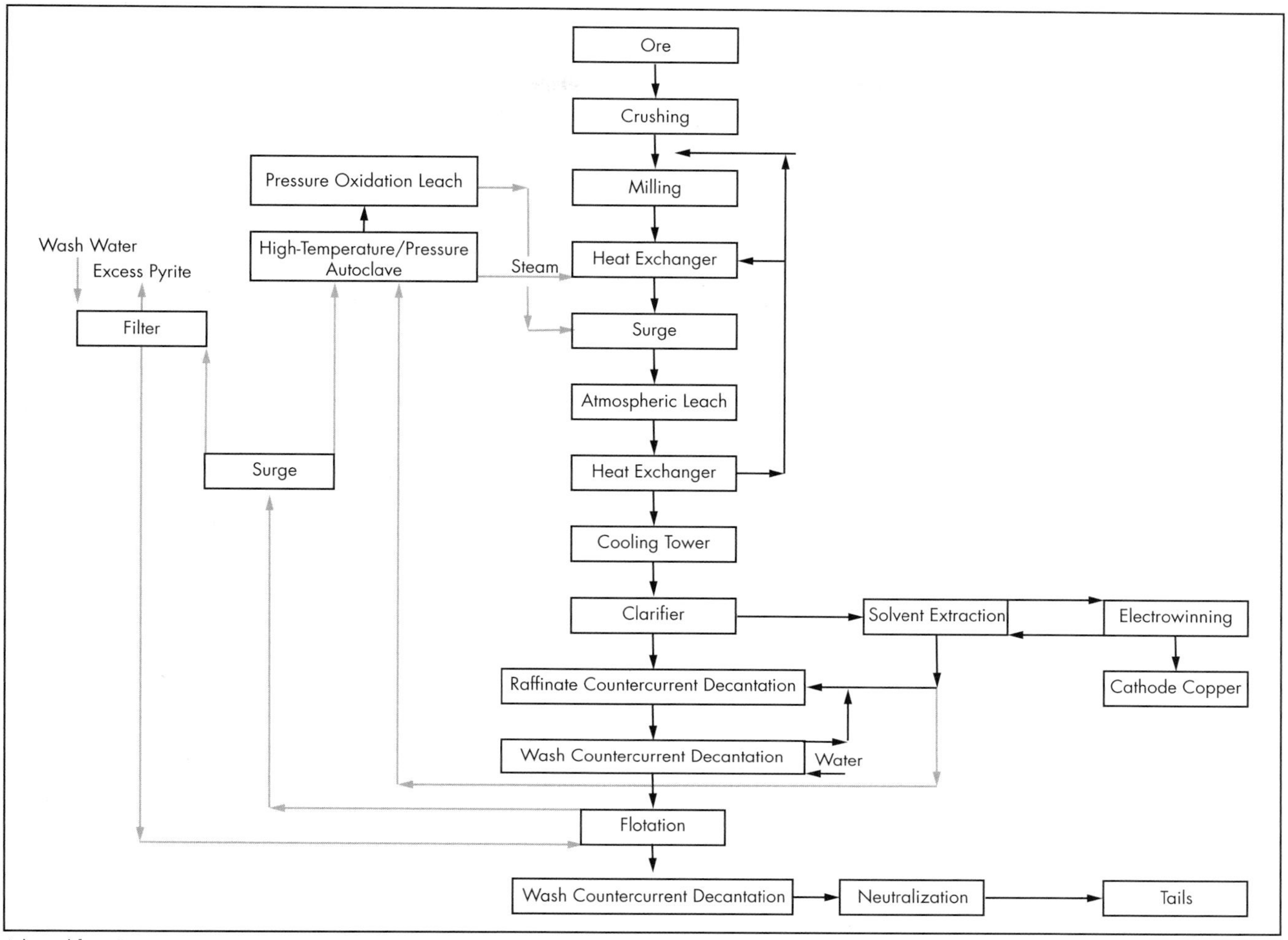

Adapted from Dreisinger 2003

Figure 5 Block diagram of the Sepon process

based and the copper liberated as aqueous copper sulfate for recovery in an SX/EW electrolytic plant. The plant "was completed in late 1998," and "in 1999, as the plant was brought on line, 32,133 tonnes of cathode copper were produced at a cash cost of US43.9 cents/lb" (Hespe 2001). Figure 4 displays a simplified flow diagram of the process steps used at the Gunpowder mine. The operation was closed in 2003.

Sepon Copper Mine

The Sepon open pit copper mine is 40 km north of the town of Sepon in the Savannakhet Province of southern Laos (Figure 5). The mine uses high-temperature and high-pressure autoclave TPO leaching of copper concentrates. The mine is operated by Lane Xang Minerals—90% owned by MMG and the remaining 10% by the Lao government. Production began in 2005 (Thomaz 2007; MMG Limited 2017).

Albion Process

The Albion Process was developed in 1998 by MIM Holdings (Brisbane, Queensland, Australia). The process uses low-temperature and atmospheric leaching of ultrafine ground copper concentrates. The technology was developed in 1994 and is patented worldwide. When applied to the processing of copper minerals, a low-grade rougher or bulk flotation concentrate is finely ground to a particle size in the range 80% passing 8–18 µm using a stirred media mill, such as the IsaMill. The IsaMill is a horizontally inclined stirred mill and is the most advanced ultrafine grinding technology available today, with 36 mills installed worldwide and a total grinding capacity in excess of 60 MW.

The finely ground concentrate is then leached in a sulfuric acid solution to recover copper to the leach solution. The oxidative leaching stage is carried out in agitated tanks operating at atmospheric pressure, with a residence time of 24–30 hours. A copper recovery in excess of 97% would be expected across the leach. The leaching tanks are conventional baffled open reactors operating with centrally mounted impellers, similar in design to a gold leach vessel. Oxygen is introduced to the leach slurry for oxidation and is added as a gas.

The Albion Process is applied in gold, copper, and zinc concentrate processing facilities. A typical flow sheet for a circuit treating copper concentrate is shown in Figure 6.

El Abra Mine

This section describes a typical heap/pad leach of crushed whole ore by sulfuric acid leachate and copper production

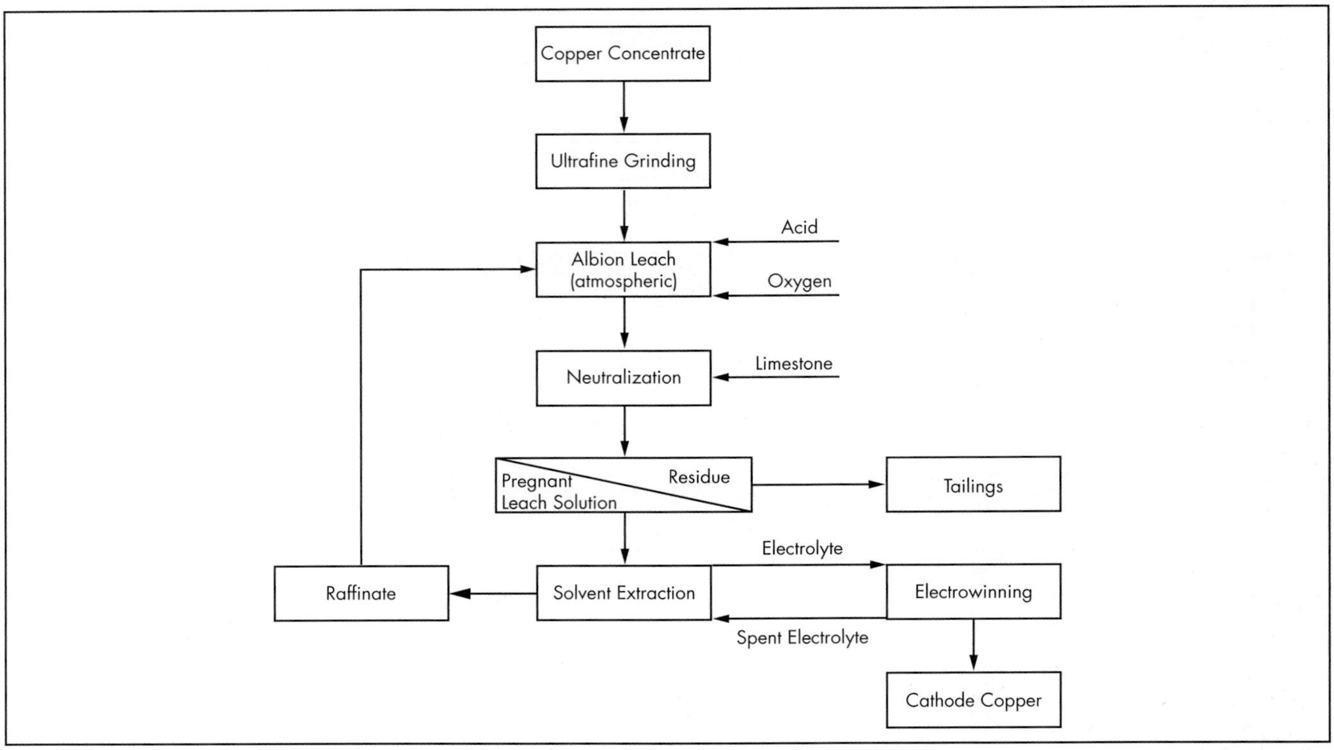

Source: MIM Holdings 1998

Figure 6 XStrata Technology's Albion Process for copper concentrates

by SX/EW. The El Abra mine is owned by the Sociedad Contractual Minera El Abra, which is a joint venture between U.S.-based company Freeport-McMoRan (51%) and Chilean state company Codelco (49%). The mine is located in El Loa Province, 55 km north of Calama in northern Chile. The ore body is located at an elevation of ~3,900 m (Freeport-McMoRan, n.d.).

El Abra is a porphyry copper-type deposit that occurs in a narrow (30–50 km wide), north-south-trending mineral belt that also contains the Chuquicamata and Radomiro Tomic porphyry copper deposits. The oxide copper minerals are mainly chrysocolla and pseudomalachite. The sulfide mineralization is predominantly bornite (60%), with chalcopyrite (25%) and chalcocite (15%) (Freeport-McMoRan, n.d.).

El Abra processed oxide copper ores in its hydrometallurgical plant from 1996 to 2010. The plant has a crushing circuit, agglomeration drums, a dynamic heap (on–off), and SX/EW facilities. The average grade considered is 0.55% of total copper, with a throughput of 92–100 kt/d. A ROM stockpile leaching was integrated to the process in 2001 to treat low-grade oxides. The SX/EW design capacity is 225 kt/yr of copper in cathode form. Production as originally designed was achieved in 2003 (Figure 7) (Freeport-McMoRan, n.d.).

In 2011, as the oxide ore was diminishing, El Abra's Sulfolix project started to treat sulfide ores. The sulfides are currently processed in a permanent heap. The leaching operation is assisted by aeration and bacteria augmentation. Plant throughput increased to 125 kt/d and production decreased to an average of 140 kt/y of copper in cathodes.

Cerro Verde Mine

This section describes heap leaching of secondary copper mineralization using bacterially assisted oxidation of copper mineralization at the Cerro Verde mine in Peru. According to Freeport-McMoRan (2018), in the 1800s, "Spaniards mined high-grade oxide copper ore from Cerro Verde. ... Anaconda Copper owned the property from 1916 until 1970, when the mine was acquired by the Peruvian government. The government mined Cerro Verde's oxide ores and built one of the world's first SX/EW facilities in 1972. Cyprus Amax purchased the operation in 1994" and upgraded it. In late 1999, Cyprus-Amax was acquired by Phelps Dodge, which in turn was acquired by Freeport-McMoRan in 2007. Sociedad Minera Cerro Verde S.A.A., a Peruvian mining company, now operates the open pit copper, molybdenum, and silver mining complex.

Cerro Verde is an open pit copper and molybdenum mining complex located 32 km southwest of Arequipa, Peru, at an elevation of 2,700 m. The ore body is part of the Southern Peru belt of copper porphyry deposits. These porphyry deposits stretch from the Chilean border through the Toquepala and Cuajone deposits to some distance west of Arequipa. Small-scale mining activity is believed to have taken place at Cerro Verde in Inca times and subsequently during Spanish colonization. Between 1868 and 1879, Chilean contractors worked the concessions and reportedly mined thousands of metric tons of oxide ore with grades as high as 15% until the Pacific War of 1879. Thereafter, various small operators worked the site. Between 1880 and 1915, minor activity occurred at the property with <20,000 t of material mined.

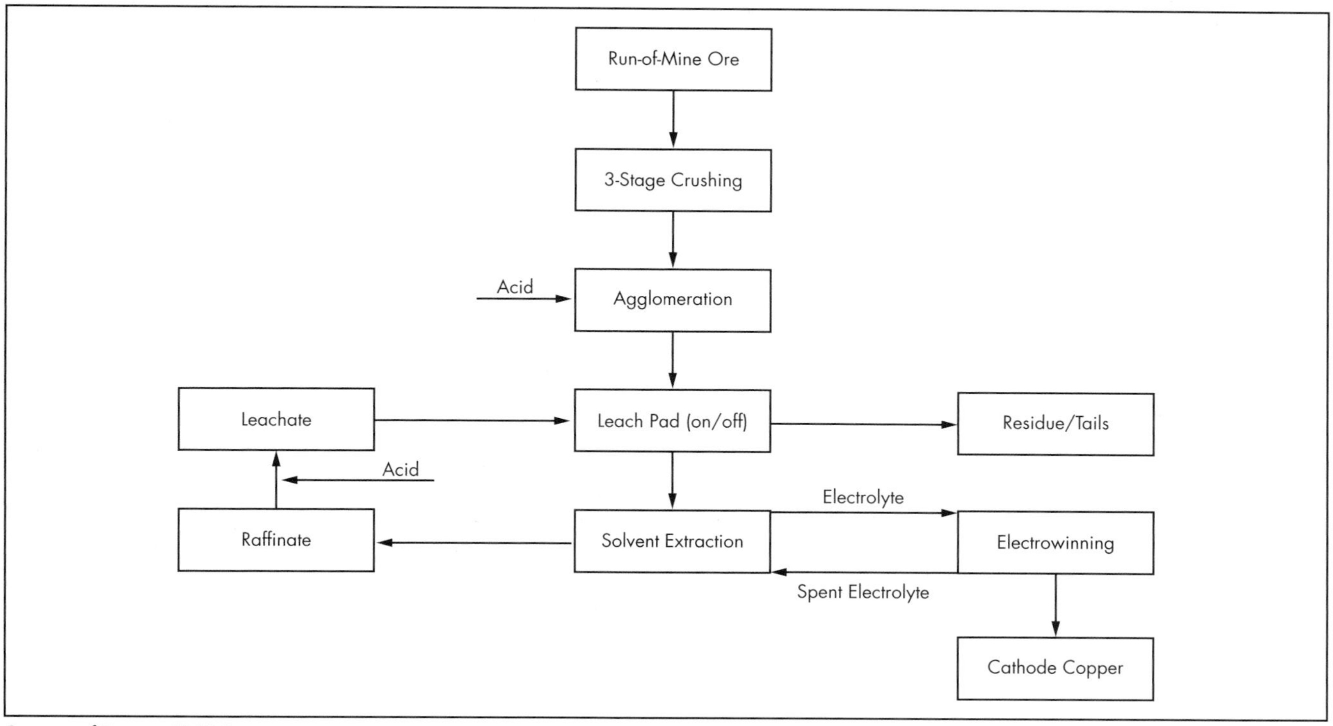

Courtesy of Freeport-McMoRan

Figure 7 El Abra mine flow sheet

During 1916 to 1970, Andes Exploration Company (Anaconda) explored Cerro Verde, drilling for mineral and water extending to the adjacent Santa Rosa deposit. In 1970, the concessions reverted to the Peruvian State, allocated to Minero Perú, formed to promote development of under-exploited resources. In 1977, industrial operation of the leach/SX/EW plant started, one of the world's first SX/EW facilities. In 1994, Cerro Verde was acquired by Cyprus Amax in a privatization process. During 1999, Phelps Dodge acquired Cyprus Amax and assumed ownership of the mine. In 2007, Freeport-McMoRan Copper and Gold acquired Phelps Dodge, becoming the largest publicly traded copper company in the world. Today, the mine's stockholders are Freeport-McMoRan (53.56%), Sumitomo Metal Mining (21%), Compañía Minera Buenaventura S.A.A. (19.58%), and Peruvian shareholders (5.86%).

The Cerro Verde deposit is comprised of porphyritic quartz monzonite stocks and breccias emplaced in granodiorite. There are three open pits with oxide and secondary sulfide deposits: Cerro Verde, Santa Rosa, and Cerro Negro. Cerro Verde and Santa Rosa are single-mineralized ore body deposits separated by waste at the surface with a common root at depth. The Cerro Negro mine contains oxide copper for leaching process.

Cerro Verde's hydromettalurgical plant is formed by three stages of crushing plant, with a capacity of 39 kt/d, and agglomeration drums. Then the ore is delivered to a high-grade leaching pad, using one overland belt of 4 km and piled in a leach pad using the radial stacker.

The irrigation leaching is in cascade, the raffinate solution irrigates ROM leach pads, and the intermediate leach solution (ILS) contains ferric iron and copper. This ILS mixes with the raffinate solution and irrigates high-grade leach pads, and those pads produce the PLS to the SX plant. The ferric iron in the ILS allows additional ferric to the leach pads, thereby allowing better and more rapid leaching of sulfide minerals in the heaps. The SX plant has five trains and the EW plant has 230 electrolytic cells. Copper is produced in cathodes of 100% high grade at a rate of ~140 million lb/yr (Figure 8).

In 2014, several lenders provided a 5-year unsecured loan for Cerro Verde to undergo a major expansion. The total projected capital cost was US$4.4 billion. Expansion of the mine and mill was completed in 2015 and increased the concentrator plant capacity from 120,000 t/d of ore to 360,000 t/d, making it the world's largest concentrating facility by the end of 2015. The project was expected to increase annual production to ~272,155 t of copper and 6,804 t of molybdenum. At the end of 2015, Cerro Verde produced 47,673 t of copper in cathodes, 199,309 t of copper in concentrates, and 3,298 t of molybdenum (Business News Americas 2018).

Kansanshi Mine

High-temperature and high-pressure autoclave leaching TPO of primary copper mineralization is used at the largest mine in Africa, the Kansanshi mine. The mine is located ~10 km north of Solwezi and 180 km northwest of Chingola. It is owned and operated by Kansanshi Mining PLC, which is 80% owned by a First Quantum subsidiary (Figure 9).

To produce copper in concentrate, sulfide ore is treated by crushing, milling, and flotation. As oxide ore was depleted and sulfide ore grades were beginning to fall, a decision was made to expand the sulfide milling circuit, which was commissioned at the end of 2008. While the expansion circuit is dedicated to the treatment of sulfide ore, mixed ore is treated through the original sulfide circuit. Oxide ore is treated by crushing, milling, flotation, leaching, and the SX/EW process, which produces a sulfidic and gold-bearing flotation

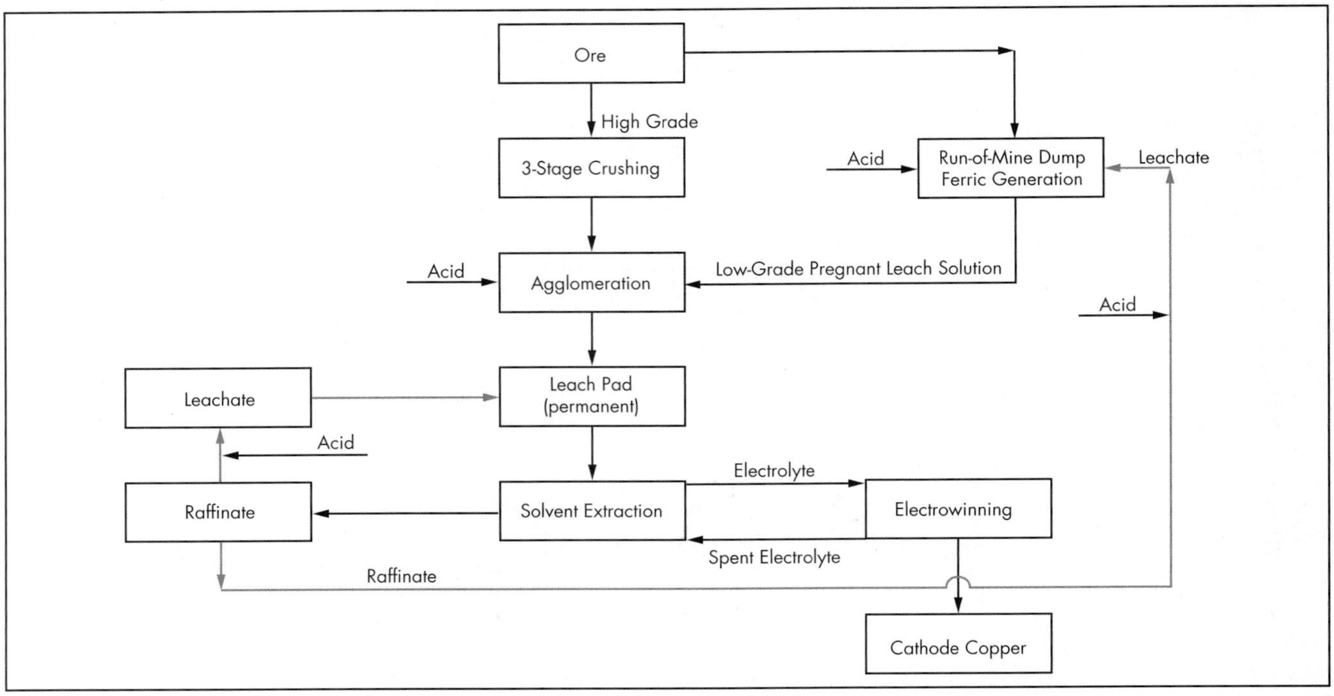

Courtesy of Freeport-McMoRan

Figure 8 Hydrometallurgical flow process

concentrate as well as electrowon cathode copper. In 2012, the oxide treatment capacity was increased to 14.5 Mt/yr by installing equipment from the Bwana Mkuba copper SX/EW plant. Additional leach, SX, electrowinning, and countercurrent decantation thickeners were commissioned in 2013 and 2014 (First Quantum Minerals 2017).

High-pressure and high-temperature autoclave leaching of flotation concentrates allowed the oxidation and copper leaching from primary copper mineralization. Leached concentrates were advanced to atmospheric leaching and liquid solids separation. PLS liquid fed solvent extraction and copper electrowinning as copper cathodes.

Gold recovery by gravity was expanded with four new gravity concentrators in April 2010. This provided two concentrators per milling train and increased gold recovery from all ore types. Gemini tables were installed to treat the gravity concentrates and produce a high-grade concentrate for direct smelting to gold bullion. The Kansanshi smelter was commissioned in the third quarter of 2014, and first anodes, from melted cathodes, were poured by December of that year. The smelter began commercial production in July of the following year. Primary processing steps include smelting, converting, fire refining, and casting. The smelter has a nominal capacity of 1.2 Mt/yr of concentrate to produce more than 300,000 t/yr of copper metal and more than 1 Mt/yr of sulfuric acid as a by-product. Acid production from the smelter permits the Kansanshi process plant to process mixed cleaner tails through the leaching circuit and produce additional cathode (First Quantum Minerals 2017).

Other Commercialized Hydrometallurgical Processes

In Situ Leaching of Copper Mineralization

The Chinese performed in situ leaching (ISL) of copper by 977 AD, and perhaps as early as 177 BC. Copper is usually leached using sulfuric acid or hydrochloric acid and then recovered from solution by SX-EW or by chemical precipitation (Wikipedia 2018).

According to Hiskey (1994), "The recovery of mineral and metal values by solution mining has been practiced for centuries. What appears at first glance to be a simple, empirical technique, is in actuality a very complicated process involving a large number of critical parameters that encompass several scientific and engineering disciplines. Hydrometallurgy, hydrology, geology and geochemistry, rock mechanics, chemistry, and environmental engineering and management are a few of the specialties utilized by modern operators."

Approximately 25% of the primary copper produced in the United States is derived by leaching. Most of this is a result of huge dump and heap leaching activities from large open pit mines. A small fraction of copper is produced from modest ISL operations treating low-grade rock left from earlier mining activities. Because of its inherent advantages over conventional mining and milling, ISL technology is being considered as a sustainable choice for recovering copper from both oxide and sulfide deposits. Both physical and chemical parameters must be integrated to successfully predict ISL performance (Hiskey et al. 1992).

"Ores most amenable to leaching include the copper carbonates malachite and azurite, the oxide tenorite, and the silicate chrysocolla" says Gorain (2016). "Other copper minerals, such as the oxide cuprite and the sulfide chalcocite may require addition of oxidizing agents such as ferric sulfate and oxygen to the leachate before the minerals are dissolved. The ores with the highest sulfide contents, such as bornite and chalcopyrite will require more oxidants and will dissolve more slowly. Sometimes oxidation is speeded by the bacteria *Thiobacillus ferrooxidans*, which feeds on sulfide compounds."

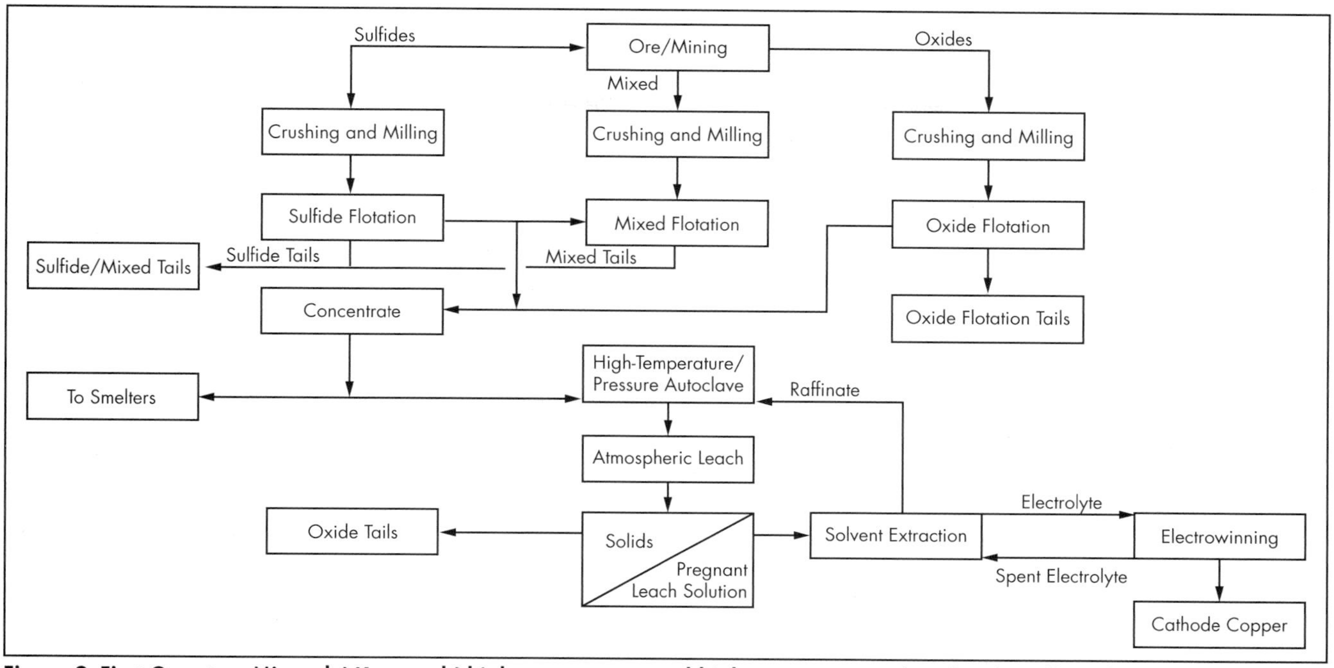

Figure 9 First Quantum Minerals' Kansanshi high-temperature and high-pressure autoclave leaching flow sheet

Copper ISL is often carried out by stope leaching, where broken low-grade ore is leached in a conventional underground mine. The leaching may take place in backfilled stopes or caved areas. In 1994, stope leaching of copper was reported at 16 mines in the United States. At the San Manuel mine (Arizona), ISL was originally used by collecting the resultant solution underground. Then in 1995, it was converted to a well-to-well recovery method and has been proposed for other copper deposits in Arizona (Wikipedia 2018). Two in situ projects for the extraction of copper are currently in the development phase in Arizona.

Commercial Ammoniacal Leaching of Copper Concentrates

According to Arbiter and McNulty (1999):

> The first ammonia leaching plants, applied to copper carbonate and native copper tailings in 1915, were followed more recently by research and development of flowsheets for ammonia leaching of sulfide concentrates. These were applied to two commercial plants. Anaconda's Arbiter Plant started up in 1974 with a design capacity of 36,000 tons/year of cathode copper, to be produced by ammonia leaching with oxygen, followed by solvent extraction and electrowinning. The plant shut down in late 1977 as a result of high maintenance and operating costs, partly due to harsh winters; to complication associated with sulfate disposal; and to changes in mineralogy. BHP's Coloso plant in Chile was designed to produce 80,000 tons/year of cathodes by leaching part of Escondida's concentrate production. Using a similar flowsheet but with air and low temperatures to avoid sulfate production, it started up in late 1994 and shut down in mid-1998, after failing to reach cathode design capacity and experiencing problems with its technology.

Complete ammoniacal leaching of copper from chalcocite and covellite concentrates was demonstrated by Anaconda and BHP. Two areas of processing affected both operations. The recovery and recycling of ammonia and the transfer of ammonia in SX to the electrolyte caused operational problems.

Other Commercial High-Pressure/High-Temperature Autoclave Leach Operations

Freeport (FMI) operated a total pressure oxidation plant in Bagdad, Arizona, and is operating a very large total pressure oxidation plant at its Morenci mining operation in Arizona (Marsden et al. 2003). Both plants successfully produced cathode copper and sulfuric acid for heap leaching. The ability to produce and use sulfuric acid for heap leaching provides a significant financial incentive for the combined hydrometallurgical process. The current Morenci plant utilizes TPO technology. FMI initially built a medium-temperature autoclave plant but recently converted to high temperature.

Commercial Low-Pressure/Low-Temperature Leach Processes

Cobre Las Cruces is owned by First Quantum Minerals, a Canadian company that produces copper, nickel, gold, zinc, and platinum group metals. CLC operates an open pit mine and hydrometallurgical process plant. The project began exploration in 1992 and production started in mid-2009. The site is located within the towns of Gerena, Guillena and Salteras in the province of Seville, in southern Spain, and the pit and plant occupy 946 ha. Anticipated annual production averages 72,000 t Cu, which equates to 25% of Spanish internal demand. A total of 1 Mt Cu is estimated for the 15-year operating period. Spain is the fourth largest consumer of copper in the European Union and ranks 14th in the world (CLC, n.d.).

Table 1 Hydrometallurgical process parameters

	Metallurgical Process							
Unit Operation	**Albion**	**BioCOP**	**CESL**	**HTPOX**	**HydroCopper**	**Galvanox**	**Intec**	**Nikko**
Grinding	Ultrafine	Fine	Fine	Cleaner	Fine	Cleaner	Fine	Fine
Particle size P_{80}, μm	5–20	42	37	53	37	53	40	40
Leachate	H_2SO_4	H_2SO_4	H_2SO_4/Cl^{-1}	H_2SO_4	Cl^{-1}	H_2SO_4	Cl^{-1}	Cl^{-1}/Br^{-1}
Leach temperature, °C	85	65–85	150	220	90	80–220	85	85
Oxidant	O_2	Thermophiles	O_2	O_2	O_2/Cl_2	O_2	Br_2Cl/air	Br_2Cl/air
Iron stabilization	FeOOH	Fe_2O_3	Fe_2O_3	$Fe_2O_3/2FeOHSO_4$	Fe_2O_3/FOOH	Fe_2O_3	Fe_2O_3	Fe_2O_3/FOOH
Precious metals	Cyanide	Cyanide	Cyanide	Cyanide	Cu	Cyanide	Hg/Cu	Carbon
Residence time, hours	48	50	48	48	30	48	20	26
Copper recovery	SX-EW	SX-EW	SX-EW	SX-EW	H_2 to sponge	SX-EW	EW dendrites	SX-EW
Arsenic fixation	$FeAsO_4{\cdot}2H_2O$	$FeAsO_4{\cdot}2H_2O$	$FeAsO_4{\cdot}2H_2O$	$FeAsO_4{\cdot}2H_2O$	$FeAsO_4{\cdot}2H_2O$	$FeAsO_4{\cdot}2H_2O$	$FeAsO_4{\cdot}2H_2O$	$FeAsO_4{\cdot}2H_2O$
S^0 oxidation, %	10	100	5–30	100	2	2	2	2
FeS_2 oxidation, %	40	100	65	100	<5	2	<5	<5

Other Processes

These high-pressure and high-temperature autoclave processes have been tested but not commercialized:

- Cominco Engineering Services Ltd. copper process from Teck Resources Limited
- Galvanox, a galvanically assisted atmospheric leaching technology for copper concentrates (Dixon et al. 2008)

The following sulfuric acid–based processes have been demonstrated but not commercialized:

- Activox process
- Sherritt-Cominco copper process
- Sherritt Gordon low-temperature leach process
- Total pressure oxidation process/placer dome process
- Anaconda super concentrate process
- BacTech biological leach process
- Hecla (Cyprus) roast/leach process
- CQG process

These chloride processes have been demonstrated but not commercialized:

- CLEAR process (40,000 t/yr copper plant operated from 1976 to 1978 and was closed because of materials issues)
- Outotec HydroCopper process
- Canada Centre for Mineral and Energy Technology ferric chloride leach process
- Cuprex metal extraction process
- Cymet process
- Dextec GmbH process
- GCM/HBMS process
- Intec copper process
- Minemet Recherche process for copper
- University of British Columbia–Cominco process
- U.S. Bureau of Mines process for chalcopyrite

Table 1 generally summarizes probable operating conditions for eight potentially viable processes for the recovery of copper from concentrates. The extracted data shown has been summarized by the Minerals Advisory Group from existing published literature (Kuhn 2018).

REFERENCES

Arbiter, N., and McNulty, T. 1999. Ammonia leaching of copper sulfide concentrates. In *Proceedings of the Fourth International Conference, Copper 99-Cobre 99*. Vol. IV. Warrendale, PA: The Minerals, Metals & Materials Society. pp. 197–213.

Business News Americas. 2018. Sociedad Minera Cerro Verde S.A.A. https://www.bnamericas.com/company-profile/en/sociedad-minera-cerro-verde-saa-cerro-verde. Accessed July 2018.

CLC (Cobre Las Cruces). n.d. Who we are. www.cobrelascruces.com/index.php/quienes-somos/?lang=en. Accessed July 2018.

Dixon, D.G., Mayne, D.D., and Baxter, K. 2008. Galvanox™—A novel galvanically-assisted atmospheric leaching technology for copper concentrates. *Can. Metall. Q.* 47(3):327–336.

Dreisinger, D. n.d. New developments in the atmospheric and pressure leaching of copper ores and concentrates [presentation]. Vancouver, BC: University of British Columbia. pp. 22–27, 50. https://www.convencionminera.com/perumin31/encuentros/tecnologia/miercoles18/1500-David-Dreisinger.pdf. Accessed July 2018.

First Quantum Minerals. 2017. Kansanshi: Mining and processing. https://www.first-quantum.com/Our-Business/operating-mines/Kansanshi/default.aspx. Accessed June 2018.

Freeport-McMoRan. n.d. Global locations: South America—El Abra. http://fmjobs.com/global-locations/south-america.aspx#tab-chile. Accessed July 2018.

Freeport-McMoRan. 2018. South America: Cerro Verde. https://www.fcx.com/operations/south-america.

Gorain, B.K. 2016. Physical processing: Innovations in mineral processing. In *Innovative Process Development in Metallurgical Industry: Concept to Commission*. Edited by V.I. Lakshmanan, R. Roy, and V. Ramachandran. Cham, Switzerland: Springer International.

Habashi, F. 2014. History of innovations in extractive metallurgy. In *Mineral Processing and Extractive Metallurgy: 100 Years of Innovation*. Edited by C.G. Anderson, R.C. Dunne, and J.L. Uhrie. Englewood, CO: SME. pp. 29–48.

Hespe, A.M. 2001. Recent developments in geology, exploration and production at the Mt. Gordon copper mine. In *AIG Journal.* Paper 2001-02. West Perth, Western Australia: Australian Institute of Geoscientists.

Hiskey, J.B. 1994. In-situ leaching recovery of copper: What's next? In *Hydrometallurgy '94*. Dordrecht: Springer. pp. 43–67.

Hiskey, J.B., Oner, G., and Collins, D.W. 1992. Recent trends in copper in situ leaching. *Miner. Process. Extr. Metall. Rev.* 8(14):1–16.

ICSG (International Copper Study Group). 2017a. *Directory of Copper Mines and Plants*. Lisbon, Portugal: ICSG.

ICSG (International Copper Study Group). 2017b. *World Copper Factbook 2017.* Lisbon, Portugal: ICSG.

Kuhn, M.C. 1993. The revitalization of U.S. copper mining [presentation]. Edison Electric Institute, The Electrotechnology Marketing Group. pp. 1–15.

Kuhn, M.C. 2018. Hydrometallurgical Process Parameters [unpublished]. Minerals Advisory Group.

Marcosson, I.F. 1957. Anaconda in Chile. In *Anaconda*. New York: Dodd, Mead. pp. 194–219.

Marsden, J.C., Brewer, R.E., and Hazen, N. 2003. Bagdad concentrate leach plant; Freeport total pressure oxidation at Bagdad, Arizona. In *Hydrometallurgy* 2003. Warrendale, PA: The Minerals, Metals & Materials Society.

MIM Holdings. 1998. The Albion Process. Copyrighted and trademarked in Brisbane, Queensland, Australia, May 1998.

MMG Limited. 2017. Our operations: Mining operations—Sepon. www.mmg.com/en/Our-Operations/Mining-operations/Sepon.aspx. Accessed July 2018.

Rickard, F.R. 1996. Chapters VI–VII. In *The Development of Ajo Arizona*. Arizona: Rickard. pp. 29–32.

Thomaz, C. 2007. Sepon mine, Laos. In *Mining Weekly*. November 30. www.miningweekly.com/article/sepon-mine-2007-11-30/rep_id:3650. Accessed July 2018.

Wikipedia. 2016. Solvent extraction and electrowinning. https://en.wikipedia.org/wiki/Solvent_extraction_and_electrowinning. Accessed July 2018.

Wikipedia. 2018. In situ leach: Copper. https://en.wikipedia.org/wiki/In_situ_leach#Copper. Accessed July 2018.

CHAPTER 12.9

Copper Pyrometallurgy

Mark E. Schlesinger

About three-quarters of the world's mined copper ore is processed using a combination of beneficiation (using minerals processing techniques described elsewhere in this volume) and high-temperature technology that ultimately produces purified copper for sale. Figure 1 illustrates the generalized flow sheet for pyrometallurgical-based copper production, which includes four high-temperature unit processes: smelting, converting, fire refining and anode casting, and remelting. This flow sheet is the descendent of technology first developed a century ago, and although some of the unit processes have seen considerable advances in recent decades, others are ready for changes that might come about soon. This chapter presents a snapshot of current technology; it is a summary of material published in more extensive form elsewhere (Schlesinger et al. 2011).

CONCENTRATES

There are several inputs to the copper smelting and refining process. The most important is the copper concentrate produced by the beneficiation process described elsewhere. Other inputs include oxygen-enriched air, carbonaceous fuel, flux, converting slag and reverts, and purchased copper scrap.

Copper concentrates primarily consist of copper and iron sulfide minerals, along with gangue minerals from the original ore. Their grade (wt % copper) ranges between 22 and 35, depending on mineral processing strategy and practice. In addition to copper, concentrates contain varying levels of impurities, usually in solid solution in the sulfide minerals; these include valuable impurities such as gold, silver, selenium, and tellurium, and undesirable impurities, in particular antimony, arsenic, and bismuth. Concentrates also contain 8%–10% moisture, in part to reduce the risk of fire. The percentages of these impurities vary considerably from ore to ore, and thus from concentrate to concentrate. Most smelters blend concentrates obtained from numerous mines; some custom smelters purchase all of their concentrates from independent mining and milling operations.

The value of copper concentrates is determined by a smelter schedule, an example of which is shown as Table 1 (Hustrulid et al. 2013). The percentages refer to the daily market price of each metal (copper, gold, or silver). The only penalties charged by this smelter are for mercury and fluorine content; most copper smelters also charge penalties for antimony, arsenic, and bismuth; for excess moisture content; and for excessive gangue levels. The prices quoted in this schedule vary substantially from smelter to smelter, and are the result of negotiation between the smelting facility and the mine supplying the concentrate.

In addition to concentrates, many smelting operations make extensive use of purchased scrap as a source of copper. The scrap is added to converting furnaces as cold dope to absorb excess heat from the converting reaction. Some smelting operations purchase large amounts of electronic waste as well. The e-waste contains a high percentage of copper; it also contains gold, silver, and platinum-group metal content that can be recovered during the refining process. Copper is also obtained from reverts (skulls from ladles and trimmings from casting operations) and from the converter slag that is often added to smelting furnaces.

Smelting reactions use the oxygen in air to oxidize the sulfur, iron, and other impurities in copper concentrate and transfer them to a separate phase. Over recent decades, it has become common to enrich the air to 24%–28% O_2 by adding pure oxygen. This increases reaction temperatures and thus process kinetics, reduces nitrogen flow-through and the heat losses that result from it, and reduces off-gas volumes and the size of the equipment required to handle it. Copper smelting operations are usually big enough to afford oxygen production facilities of their own; smaller smelters instead rely on purchased oxygen supplied by larger-scale producers. The other gas purchased by copper smelting operations is natural gas, used for providing energy to smelting and refining furnaces. This is typically supplied by pipeline. On rare occasions fuel oil is used instead.

Copper smelting and converting furnaces require a flux to react with the iron oxide generated during oxidation of concentrates and produce a low-melting-point slag. This is usually silica, produced from nearby quarries. The silica contains a few percent of other mineral oxides (Al_2O_3, CaO, Fe_2O_3, MgO) as impurities. Furnaces that use a calcium ferrite slag

Mark E. Schlesinger, Professor of Metallurgical Engineering, Missouri University of Science & Technology, Rolla, Missouri, USA

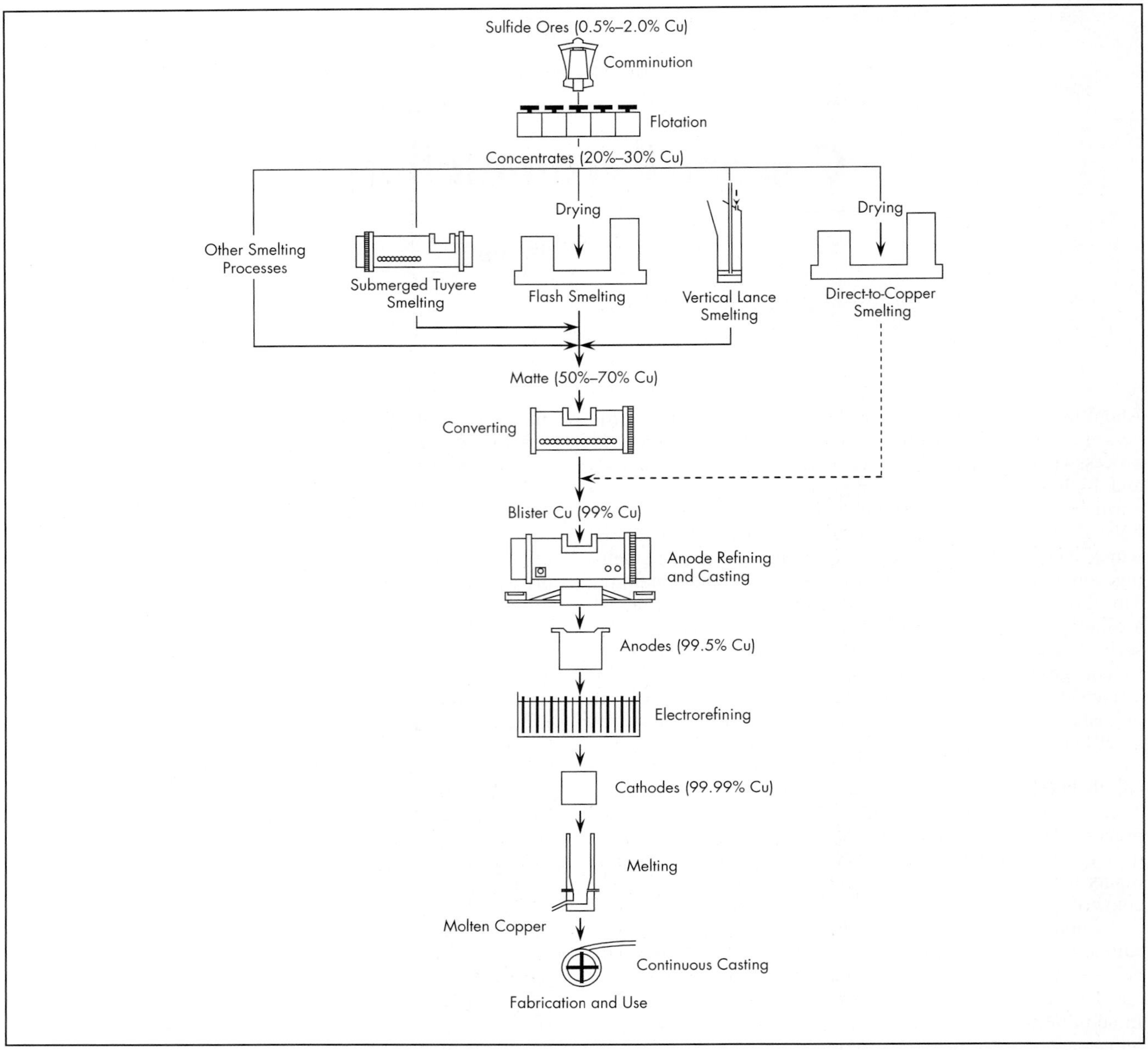

Source: Schlesinger et al. 2011

Figure 1 Generalized flow sheet for pyrometallurgical processing of copper concentrates

in place of the typical fayalite slag use burnt lime (CaO) as a flux instead.

PREPROCESSING

Copper concentrates are sometimes fed directly to smelting furnaces. However, for flash smelting, thermal preprocessing is performed. The common unit operation is drying, which reduces the moisture content to 0.5 wt % or less. This is done at smelters using flash furnaces. Drying keeps clogs from forming in concentrate burners and improves ignition of concentrates. Drying is normally performed using indirect heating of concentrates. A common type of dryer is the steam dryer shown as Figure 2. The steam (generated using the hot off-gas from smelting) is fed at 16–20 bar gauge pressure through tubing bundles in the sides of the vessel; the vessel rotates to ensure good contact and faster drying. The length of these dryers is 9.0–12.5 m, and the diameter is 3.5–4.5 m. Processing capacities range from about 100 to 240 t/h of concentrate. Steam dryers like these are replacing the rotary and flash dryers that were previously used. The use of indirect heating by the steam flowing through the tubing bundles reduces the amount of off-gas produced by the dryer, and eliminates CO_2 and SO_2 emissions that result from the combustion of fossil fuel and reaction of off-gases with the sulfide concentrate. Dust emissions are reduced as well.

The high concentration of impurities (particularly arsenic) in some new sources of copper concentrate has revived interest in partial roasting as a pretreatment process. The

Table 1 Model smelter schedule for copper concentrates*

Payments	Copper	Pay for 95% to 98% of the copper content at market value. Minimum deduction of unit (1% of a ton) per dry metric ton for copper concentrates grading below 30%. Unit deductions and treatment charges may be higher for concentrates about 40% Cu.
	Gold	Deduct 0.03 to 0.05 oz/dmt (troy ounces per dry metric ton) and pay for 90% to 95% of the remaining gold content at market value.
	Silver	Deduct 1.0 oz/dmt and pay for 95% of the remaining silver content at market value.
Deductions	Treatment/refining charges	Treatment changes in annual contracts for 2011 varied between \$56.00 and \$90.00/t (per metric ton) concentrate with refining charges of \$0.056 to 0.090/lb Cu.
	Price participation	Prior to mid-2006, contracts often included price participation clauses. Most contracts settled after late 2006 either eliminated price participation clauses or have capped the price participation to \$0.04–\$0.10/lb Cu at a basis of \$1.20/lb Cu or above. In 2011 none of the contracts reported a price participation clause.
	Complex concentrates	Treatment charges for "complex" concentrates are typically \$15 to \$25/t higher than charges for clean concentrates. In addition, copper concentrates grading over 40% may be charged up to \$10/t more for treatment. This extra charge varies depending on the tightness of the concentrate market. Currently, no smelters are imposing this charge on high-grade concentrates.
	Refining charges	Range from \$6.00 to \$8.00/oz of payable gold and \$0.50 to \$0.75/oz of payable silver.
Penalties	Deleterious element assessments	Copper concentrates containing excessive amounts of the following elements may be penalized or rejected: lead, zinc, arsenic, antimony, bismuth, nickel, alumina, fluorine, chlorine, magnesium oxide, mercury. Lead, zinc, and arsenic levels about 2% each often result in rejection. For fluxing ores, iron must be less than 3% and the alumina at low levels so that the available silica fluxing content remains high.
	Moisture	High moisture content may also be penalized due to material handling difficulties.

Source: Infomine USA 2012, as cited in Hustrulid et al. 2013

*Spot treatment charges since 2008 have fluctuated widely from treatment charges between \$5–\$120/t and \$0.005–\$1.20/lb refining. Mid-year 2011 spot contracts were reported lower at \$42–\$44/t.

Adapted from LS-Nikko 2007

Figure 2 Kumera stream dryer for copper concentrates

primary purpose of this roasting process is to react arsenide minerals with oxygen to generate As_2S_3 vapor:

$$2Cu_3AsS_4(s) + heat = Cu_2S(s) + 4CuS(s) + As_2S_3(g)$$

The arsenic sulfide vapor is then oxidized to produce arsenic trioxide (As_2O_3), which is then condensed and recovered for further processing. Some oxidation of sulfide minerals also occurs:

$$3FeS_2(s) + 8O_2 = Fe_3O_4(s) + 6SO_2(g) + heat$$

$$4CuFeS_2(s) + 7O_2 = 2Cu_2S(s) + FeS(s) + Fe_3O_4(s) + 5SO_2(g) + heat$$

The heat produced by these combustion reactions is used to decompose the arsenides in the concentrate.

Figure 3 illustrates the partial roasting of high-arsenic copper concentrates (Holmström et al. 2014). The fluid bed roaster at left is a replacement for the rotary kilns and multiple-hearth roasters used in previous decades. Typical temperatures are 680°–720°C, and oxygen input is controlled at less than 5% of the input sulfur rate to minimize oxidation of the sulfur in the concentrate. The roasting process also volatilizes mercury from the concentrate, allowing its capture; this minimizes subsequent mercury emissions during smelting and converting. Roasting of this sort is currently uncommon but may become more widespread as dirtier concentrates are processed.

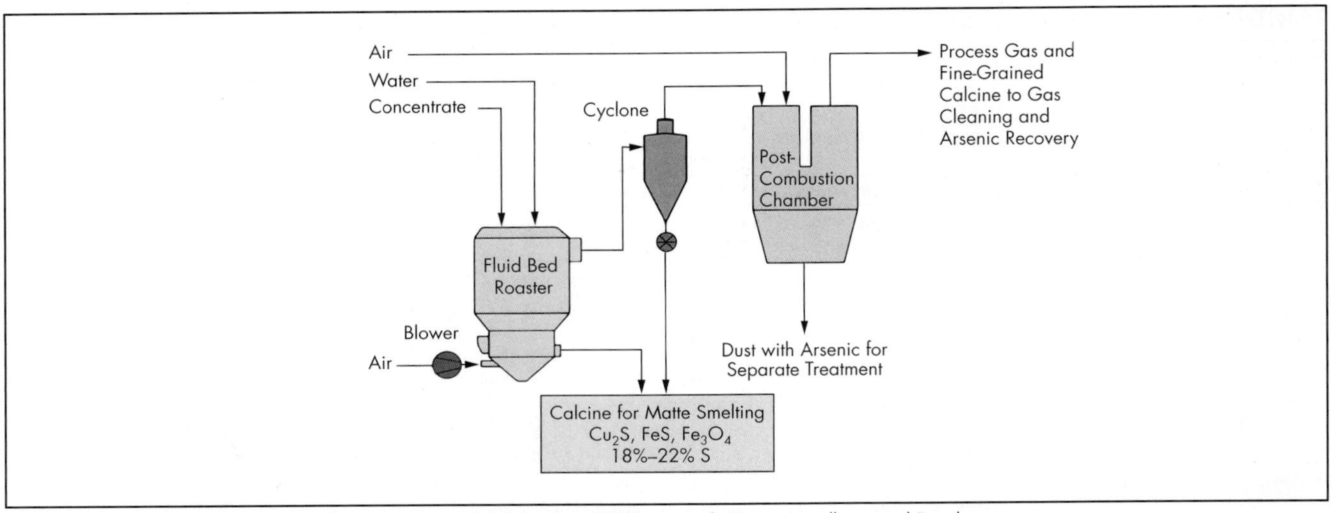

Source: Holmström et al. 2013, reprinted with permission from the Canadian Institute of Mining, Metallurgy and Petroleum

Figure 3 Partial roasting process for high-arsenic concentrates

SMELTING

The primary goal of copper production is to separate the copper in minerals such as chalcopyrite ($CuFeS_2$) and bornite (Cu_5FeS_4) from the iron and sulfur, achieving a recovered copper stream with (1) high purity, (2) high overall recovery of the input copper, (3) minimized cost, and (4) minimized emissions of pollutants. Theoretically, it is possible to directly react copper minerals with oxygen to produce metallic copper, iron oxide, and sulfur dioxide (SO_2) gas:

$$4CuFeS_2(s) + 11O_2(g) = 4Cu(l) + 2Fe_2O_3(s) + 8SO_2(g)$$

The overall reaction is exothermic and should generate enough heat to produce a molten copper phase that can be separated from the hematite (iron oxide) produced by the oxidation of copper. However, in practice this reaction does not work. The primary reason is that much of the copper metal reacts with oxygen to generate cuprite (Cu_2O), which is lost to the process. A means is needed to remove as much of the iron as possible without oxidizing the copper, to minimize losses when the final oxidation step is performed.

Figure 4 shows the Cu_2S-FeS-SiO_2 phase diagram, which is the basis of matte smelting. The diagram shows a large miscibility gap between a molten oxide–based iron silicate slag and a sulfide-based melt known as a matte. When SiO_2 flux is added to concentrate along with air, the resulting reaction oxidizes some of the iron to form a molten slag, leaving nearly all of the copper in a molten Cu_2S-FeS matte. This removes most of the iron in the concentrate, as well as some of the sulfur, without oxidizing the copper at the same time. Nearly all sulfide concentrates are smelted in this manner, using a variety of technologies that are discussed later in this chapter.

Matte smelting is a balancing act. It would be nice to eliminate as much of the iron in the concentrate as possible, producing a matte with a high grade (% Cu). However, as the fraction of copper sulfide in the matte grows, an exchange reaction with iron oxide in the slag is pushed to the right:

$$Cu_2S\ (matte) + FeO\ (slag) = Cu_2O\ (slag) + FeS\ (matte)$$

This results in higher levels of copper dissolved in the slag, and thus greater losses of copper from the concentrate. Figure 5 shows the percent of copper in slag in equilibrium with matte of varying grades. As can be seen, producing matte with more than 65% Cu has drawbacks.

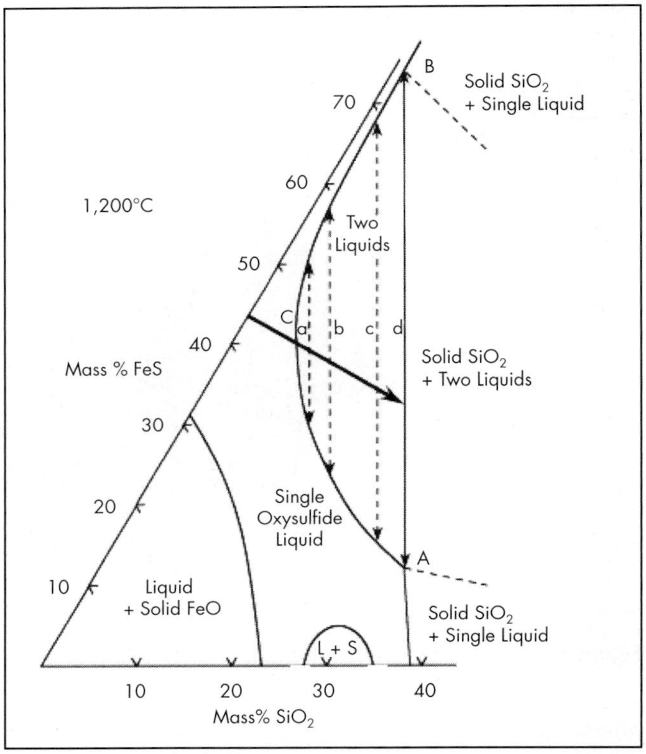

Source: Yazawa and Kameda 1953

Figure 4 Cu_2S-FeS-SiO_2 phase diagram

Temperature control of matte smelting is also a balancing act. Typical matte temperatures are 1,200°–1,250°C. Low temperatures increase refractory life. However, at low temperatures, wüstite (FeO) dissolved in the slag will oxidize to form magnetite, which is less soluble and precipitates as a solid:

$$3FeO\ (slag) + ½O_2 = Fe_3O_4(s)$$

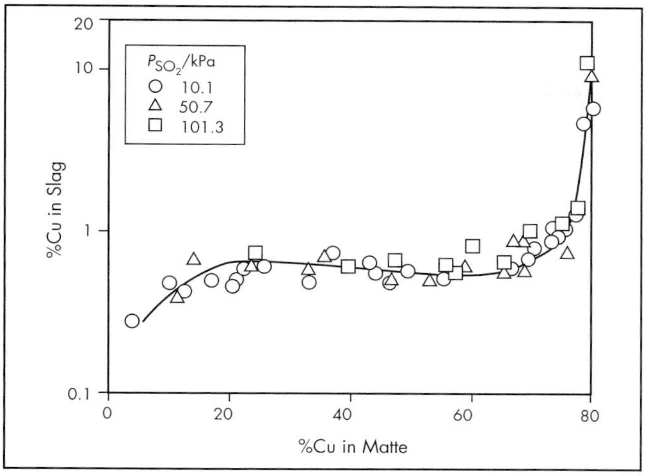

Source: Roghani et al. 2000

Figure 5 Copper content in smelting slag as a function of matte grade

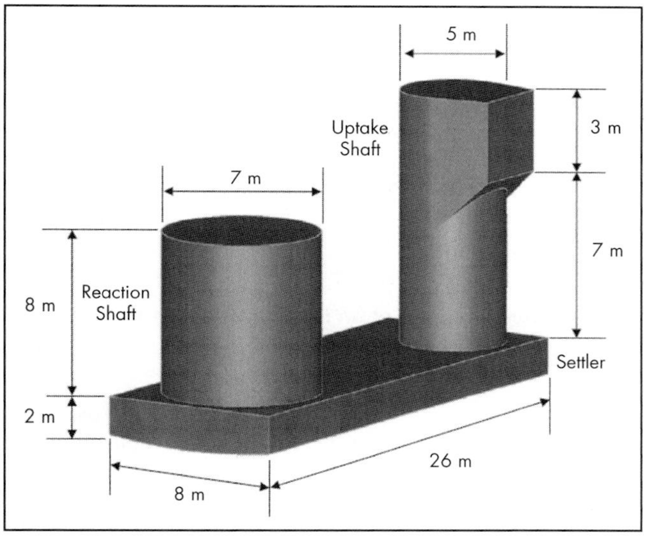

Source: Schlesinger et al. 2011

Figure 6 Outotec (Outokumpu) flash furnace

This can block air flow into the melt through the furnace tuyeres and is difficult to remove from the furnace. As a result, a higher temperature is often used to reduce most of the magnetite and keep it dissolved in the slag.

The heat generated by oxidation of sulfur and iron during smelting should be sufficient to make the process autothermal—sufficient to produce molten matte, slag, and off-gas from room-temperature concentrate, flux, and air. However, in practice this is difficult because reverts such as converter slag and e-waste are often added to the charge. These generate little or no energy when smelted and require heat from the rest of the melt. In addition, heat losses from the system impact the energy balance. As a result, the energy balance needs to be improved. This can be done by burning fossil fuels such as natural gas or coal, or by enriching the air with added oxygen. The reduced amount of nitrogen in enriched air reduces heat losses from the system and raises the temperature in the smelting furnace. Improving the reaction kinetics also helps the smelter heat balance, in addition to improving furnace productivity. As a result, current smelter design is centered around technology that improves smelting kinetics. Several designs of smelting furnaces are in use around the world; however, most are one of two types: flash furnaces or bath smelters.

Flash Smelting

Figure 6 illustrates the basic design of the Outokumpu (now Outotec) flash furnace, developed in the late 1940s. The furnace features (1) a concentrate burner mounted in the roof of the reaction shaft at left; (2) an uptake shaft where hot off-gases from the smelting reaction are removed, allowing dust to settle back into the reactor; and (3) a settling area where molten matte and slag are allowed to separate into two immiscible layers. Matte is removed through several tapholes located near the bottom of the settling unit, and slag is removed through a separate taphole located closer to the top of the settling area. The interior of the furnace is lined with direct-bonded magnesia-chrome refractories, backed by water-cooled copper cooling jackets in the walls to reduce wear. Flash furnaces have a range of sizes and capacities but can process up to 4,200 t/d of concentrate.

Concentrates are fed as a dry powder to the concentrate burner, to avoid clumping. They are usually mixed with dusts recovered from the smelting operation, sludges from wastewater treatment, and reverts (skulls and oxide crusts from ladles and other molten-material containers). Flux consists primarily of crushed and ground quartzite rock added to the concentrate before being fed to the burner. The air mixed with the concentrate is enriched to 45%–80% O_2, depending on concentrate grade and smelting strategy; this adjusts the energy balance so that fossil fuels are rarely needed. Controlling the oxygen percentage in the enriched air and the total O_2-to-concentrate ratio allows the smelter to set the output temperature of the matte and slag, as well as the matte grade. (The more iron and sulfur in the concentrate oxidized, the higher the matte grade and temperature.)

The smelting reactions that occur when the concentrate is combusted generate droplets of molten matte and slag, which fall through the reaction shaft to the settling area. As this molten mixture flows toward the other end of the reactor, these separate into two molten layers, the less dense slag on top. Higher temperatures encourage quicker settling; lower temperatures increase the immiscibility of the two phases, reducing the percentage of copper in the slag. Lower temperatures also encourage the deposition of solid magnetite on the bottom and sides of the reactor; although too much is undesirable, a layer of magnetite protects the refractory from the molten matte and slag, lengthening its life. Matte and slag are typically tapped at 1,250° and 1,220°C, respectively.

The molten matte tapped from the furnace has a grade of 60%–70% Cu (75%–87% Cu_2S). The remainder is mostly FeS, but other minor elements in the concentrate also wind up in the matte. These include gold, silver, lead, nickel, and some of the antimony, arsenic, bismuth, selenium, and tellurium. Matte is tapped into ladles and taken away for charging to the converter.

Molten smelter slag is based on fayalite (Fe_2SiO_4) but contains a small amount of sulfur and up to 2% copper. (Less than 1% Cu is more common.) If the percentage is high, the slag may be subjected to mineral processing after it solidifies to try and recover some of the copper; if not, the slag will be dumped, used as roadfill, or mixed into cement.

The off-gas from the process consists primarily of SO_2 produced by the smelting reaction and N_2 from the combustion air. Since smaller droplets of matte and slag settle slowly, a tall off-gas chamber at the right is provided to minimize the dust carried out of the furnace. Even so, some dust remains, and a series of gas-treatment steps are applied to recover it. This dust will be charged back to the furnace. The off-gas, which has a strength (% SO_2) of up to 70%, will be treated to produce sulfuric acid. The higher strength makes this process easier, and was one of the significant advantages of the flash furnaces over the reverberatory smelting furnaces originally used for processing copper concentrates.

Flash smelting is still the most common method for producing matte from copper concentrates, especially in larger smelting operations. Other technologies, particularly bath smelting, are competing to be the long-term future for this unit process.

Bath Smelting

This technology was first developed in Australia in the 1970s, and several types are now in use. Figure 7 illustrates the best known, the top submerged lance (TSL) furnace. This furnace is sold commercially as IsaSmelt or Ausmelt technology (the two types are similar) and is the fastest-growing smelting device. The furnace shown is 3.5–5 m diameter and 12–16 m high, and is lined with the same type of mag-chrome refractory used in flash furnaces. TSL furnaces have somewhat higher capacities than flash furnaces (up to 5,000 t/d of concentrate) but a much smaller footprint per unit capacity. The furnace also features a lance consisting of a stainless-steel outer pipe (up to 0.5 m diameter) and a carbon-steel inner pipe. Air and oxygen are mixed with hydrocarbon fuels and blown through this lance into the molten bath inside this furnace; the air is typically enriched to 50%–65% O_2. A separate feed port is provided for solids, including concentrates, converter slag, reverts, and powdered coal for additional combustion energy. Efforts are being made to include shredded copper scrap in the feed as well.

Pelletized concentrates are fed moist to TSL furnaces, to avoid dust generation; as a result, no drying is needed. The feed falls into and dissolves in a molten bath consisting of an emulsion of matte and slag, kept agitated by the gas blown in through the lance. The reaction of the air with the concentrate dissolved in the bath produces high smelting kinetics; the height of the shaft allows dust or droplets of molten emulsion to settle back into the bath. Periodically, some of the emulsion is tapped through the bottom either to a ladle or directly into a separate settling furnace where matte and slag are allowed to separate. The high bath temperature (1,250°C) and agitation wear the lance away, and periodically a new lance is attached at the top to make up the loss. Off-gases are lower strength than those produced by a flash furnace, because of the additional air needed to burn the fuel and coke. The agitation also results in greater refractory wear, which is made up for by installing cooling panels or using thicker refractory.

Smelting in a TSL furnace is currently a semicontinuous process, as in a flash furnace. However, efforts are being made to make the process continuous by linking furnace output directly to a settling furnace and a converter. The ability to produce impure molten copper directly from concentrate has long been a goal of copper producers. The need to use a reducing environment for smelting, followed by an oxidizing environment for converting, is what has made this goal difficult to accomplish.

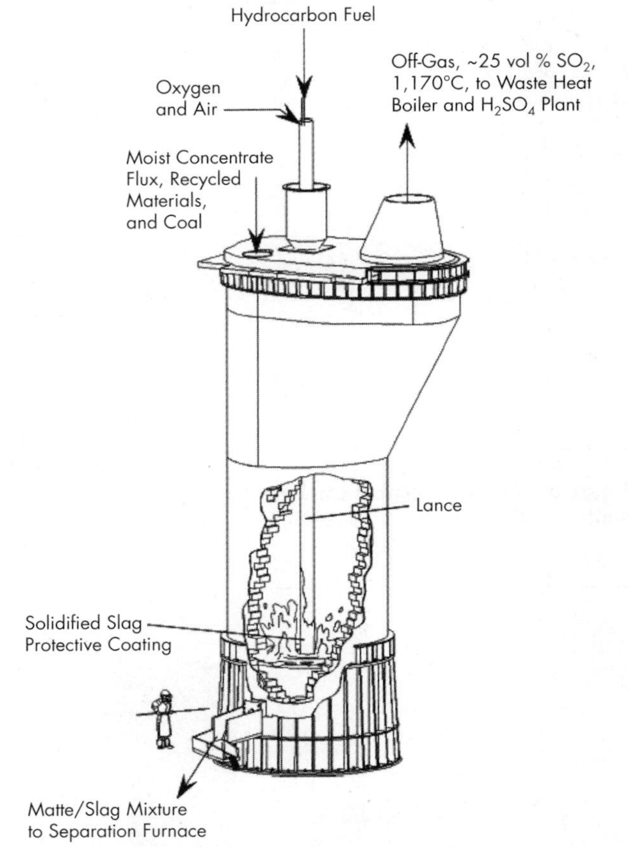

Source: Schlesinger et al. 2011

Figure 7 Cutaway view of an IsaSmelt TSL bath smelting furnace

Figure 8 illustrates an early process designed to accomplish direct-to-copper smelting, the Mitsubishi process. This process uses three linked furnaces. The first, the S-furnace, is where smelting takes place. Lances are used to inject both concentrate and enriched air onto the surface of a molten bath; because the lance is not inserted into the bath, it tends to wear at a lower rate than in a TSL furnace. Smelting reactions produce matte and slag as before; these are continuously tapped and flow to the second cleaning furnace, or CL-furnace. Matte and slag are allowed to settle in the CL-furnace and tapped separately; the slag is handled as described previously, whereas the matte flows to the third C-furnace, where conversion to molten impure copper is completed. This process eliminates the need for ladles and tapping, and should produce matte and slag of consistent quality and temperature. In practice, it is difficult to keep the three furnaces operating in tandem, and as a result, the number of Mitsubishi-type furnaces worldwide is small.

A third type of bath smelter is more recent. Figure 9 depicts the bottom-blown oxygen smelting furnace that has been installed over the past 10 years by several Chinese copper producers (Cui et al. 2013). This design is a long, cylindrical, refractory-lined device (4.4 m diameter × 16.5 m length)

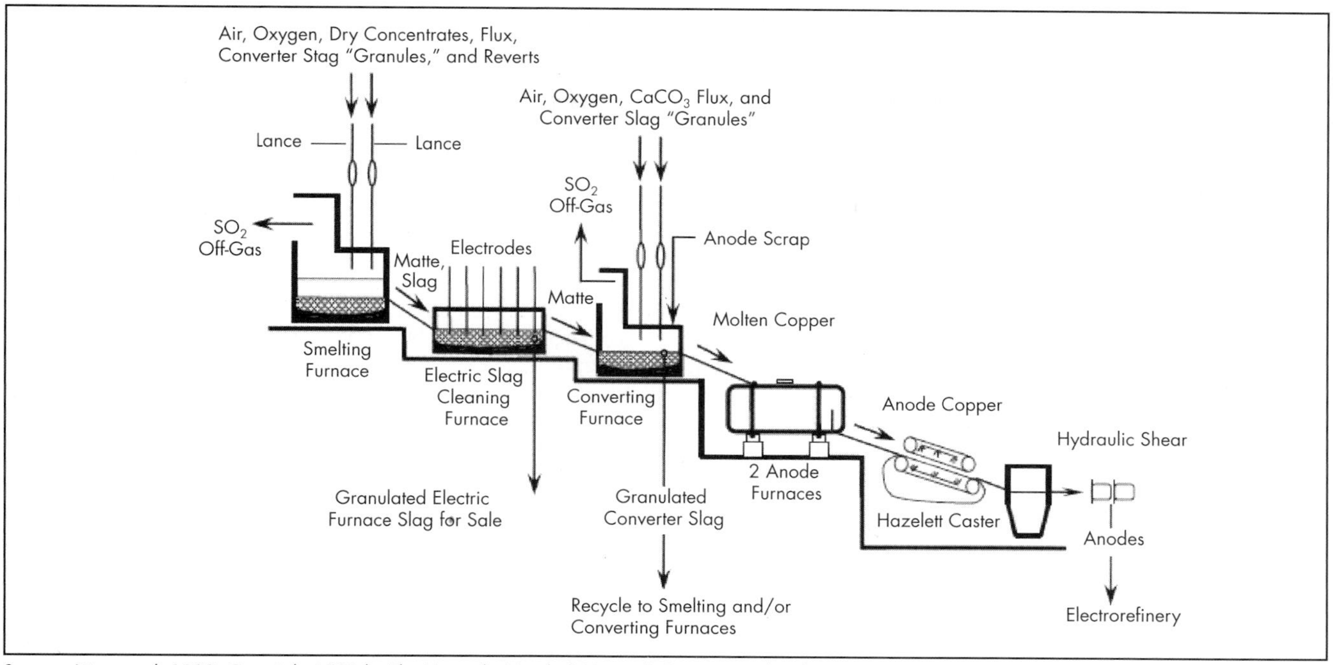

Source: Ajima et al. 1999. Copyright 1999 by The Minerals, Metals & Materials Society. Used with permission.

Figure 8 Process flow sheet for the Mitsubishi process

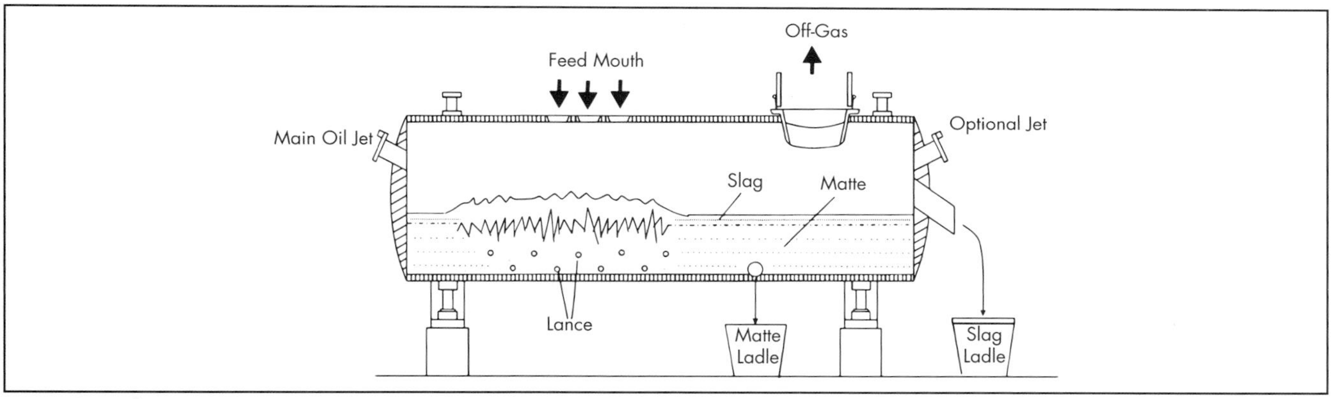

Source: Cui et al. 2013

Figure 9 Chinese bottom-blown smelting furnace

with openings at the top for solid feed input and off-gas removal. The feed is moist pelletized concentrate, along with plant reverts, slag, and the concentrate produced by beneficiation of smelter and converter slags. Enriched air (70% O_2) is blown into the molten matte/slag emulsion in the furnace through lances inserted through the refractory at the bottom; the lances are similar in construction to those used in TSL furnaces. Some fuel oil is burned at the top to keep the slag molten. The lances are located toward the left side of the vessel; this allows matte and slag to settle as they flow toward the right. Slag is tapped from the right side of the vessel; matte is tapped from the bottom. The off-gas is high–strength (68% SO_2), reducing the volume of gas to be handled. The smelter processes slightly less than 2,000 t/d of concentrate.

Operating practice with this device produces a high-grade (70% Cu) matte. This means that the slag has higher levels of copper (2%–3%) than other smelting devices. The bottom-blown smelter was designed to handle lower-grade concentrates (~25% Cu), which have higher Fe/Cu ratios than other concentrates. This means that much more FeO is generated during smelting. The smelter produces a slag with higher Fe/SiO_2 ratios than other smelting furnaces, and so magnetite formation is a concern. The claim is made that enriched air injection through the matte layer limits oxidation of FeO in the slag, minimizing magnetite. This smelter also operates at lower temperatures (1,150°–1,170°C), which reduces refractory wear but makes the tapping of slag more difficult. Future practice may include using dried feed instead of moist, which would improve the energy balance and further improve off-gas strength (Jie 2013). This device appears to be more suited for smaller smelting operations.

Submerged-Tuyere Smelting

This type of smelting technology is a descendant of the Peirce–Smith converter and operates on similar principles. The three main types are the older Noranda and Teniente smelters, and the newer Vanyukov smelting furnace. They are much less common than flash furnaces and bath smelters.

Figure 10 illustrates a typical Teniente smelting furnace. The length of these furnaces ranges from 14 to 23 m, and the inside diameter is 4–5 m. As with other smelting furnaces, mag-chrome brick is used as the lining. The right-hand side of this view shows 35–50 tuyeres, 5–6 cm diameter; the left-hand side is a "quiet zone" where matte and slag settle in separate layers for tapping. The tuyeres are mounted into panels for easier replacement as they wear. The device is mounted on trunnions so that it can be rotated to expose the tuyeres for inspection and repair. Tapping of slag is performed at the left end, and tapping of matte at the right. Tapping temperatures are typical: 1,240°C for slag, 1,250°C for matte. A variety of materials can be fed to the Noranda smelter, including moist or dry concentrates, flux, dust, reverts, and in some cases considerable solid scrap. Oxygen-enriched air (40% O_2) is blown in through the tuyeres, and ordinary air is used to blow in dry concentrates. Because the degree of enrichment is lower than in other smelting furnaces, more nitrogen reports to the off-gas stream. This dilutes the strength of the off-gas to 20% SO_2. Teniente furnaces typically process 2,000–2,500 t/d of concentrate.

Teniente smelters were designed to produce very-high-grade matte (72%–74% Cu). This minimizes the amount of time required to convert it and reduces the amount of converter slag. However, it also means the smelter produces very-high-copper slag (~8% Cu). This slag must be subsequently processed to recover the copper, increasing costs elsewhere in the process. No water cooling of the refractory is needed, because of the reduced agitation of the bath compared with flash furnaces or bath smelters.

The design of Noranda smelters is similar to that of Teniente furnaces, but there are some differences. Matte is tapped from the same side of the furnace as slag. Dry concentrate is fed through the tuyeres rather than from the top, but other feed (reverts, flux, slag concentrate) is input through a slinger belt that throws the material into the furnace, ensuring better distribution. About 10% of the feed consists of scrap, including shredded electronic scrap. Recent installations have a higher smelting capacity (3,000 t/d) than most Teniente furnaces.

Figure 11 shows the cross section of a Vanyukov smelting furnace, invented in the former Soviet Union and installed in a

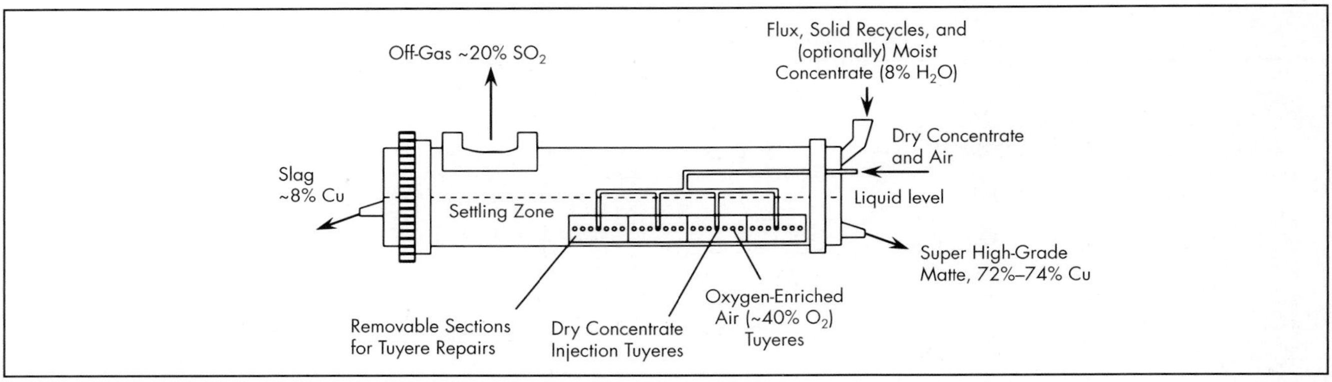

Source: Schlesinger et al. 2011

Figure 10 Schematic of Teniente smelting furnace

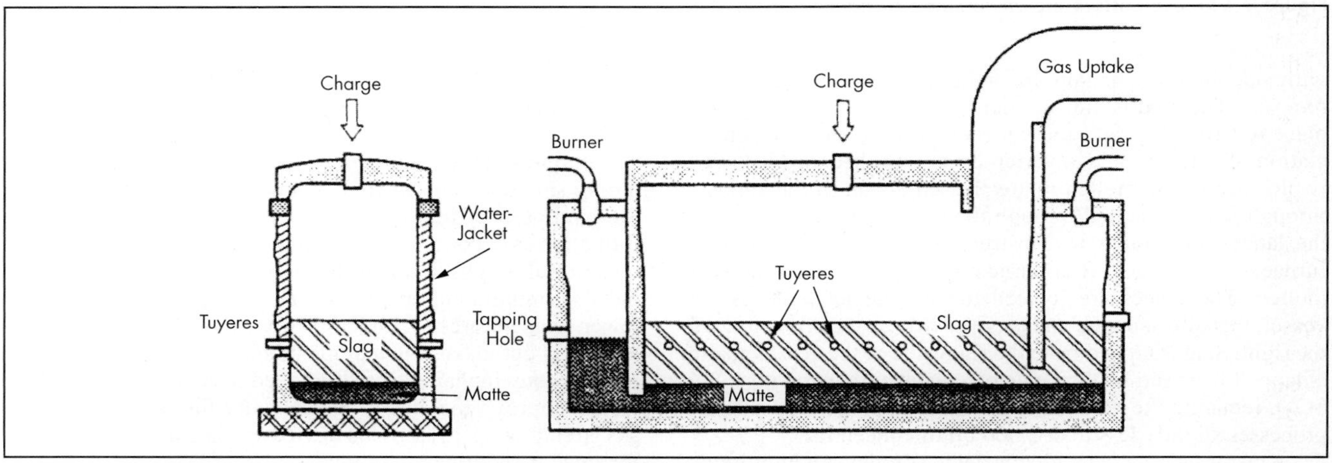

Source: Kellogg and Diaz 1992. Copyright 1992 by The Minerals, Metals & Materials Society. Used with permission.

Figure 11 Sketch of Vanyukov smelting furnace

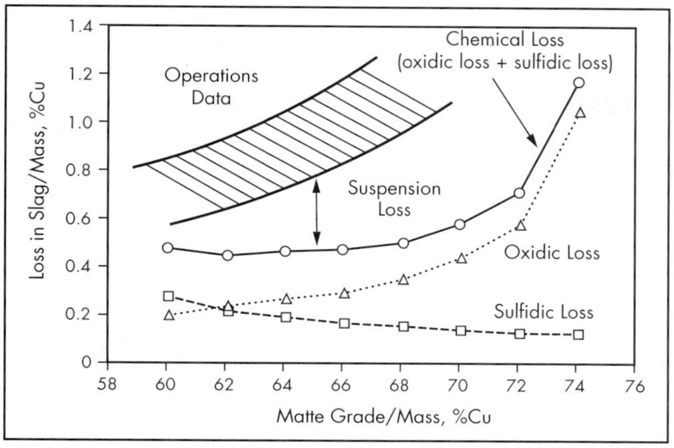

Source: Furuta et al. 2006. Copyright 2006 by The Minerals, Metals & Materials Society. Used with permission.

Figure 12 Copper loss in slag as a function of matte grade

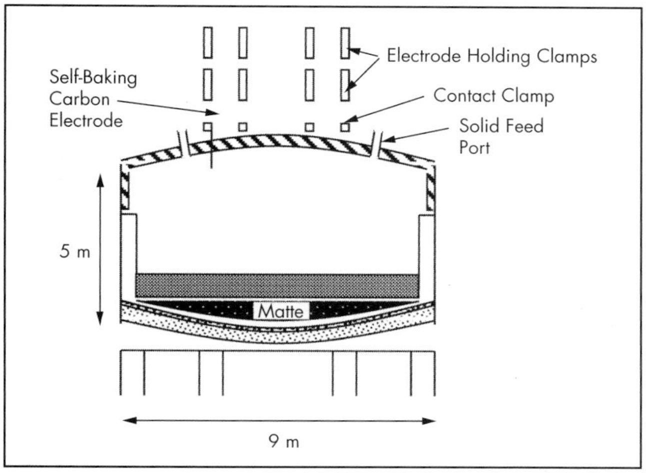

Source: Schlesinger et al. 2011

Figure 13 Cross section of a slag cleaning furnace

limited number of locations. Unlike the Noranda and Teniente smelters, it does not rotate to clear the tuyeres for cleaning or replacement. Feed is input through a hole in the top, and enriched air is blown in through the tuyeres. An advantage of the Vanyukov furnace is the use of weirs to separate the matte and slag tapping zones from the rest of the furnace. This allows better settling and reduces the copper of the slag to 2% or less. A variety of matte grades are produced in this furnace, depending on concentrate grade and Cu/Fe ratio.

SLAG PROCESSING

Slag is produced by both smelting and converting furnaces. The amount of copper in these slags depends on the type of furnace, the grade of matte or molten copper being produced, and the oxidation potential of the furnace environment. Figure 12 illustrates this point (Furuta et al. 2006). The graph shows the percentage of copper in smelting slags as a function of matte grade. Copper occurs in slag in three forms: dissolved copper sulfide, dissolved copper oxide, or entrained particles of sulfide matte that have not had sufficient time to settle out of the slag. The concentration of dissolved sulfidic copper decreases slightly as the matte grade increases. This is the result of increasing immiscibility between the matte and slag phases, which decreases overall sulfur solubility in the slag. The concentration of oxidic copper increases rapidly, because of the shifting equilibrium of the reaction described earlier:

$$Cu_2S \text{ (matte)} + FeO \text{ (slag)} = Cu_2O \text{ (slag)} + FeS \text{ (matte)}$$

As the increasing grade of the matte increases the activity of Cu_2S, the activity (and thus the concentration) of Cu_2O goes up as well. The fraction of suspended matte in the slag is the difference between the overall percentage of copper in the slag (the shaded area in Figure 12) and the percentage of dissolved copper. Factors that increase the level of suspended matte in the slag include high slag viscosity (which increases required settling time) and inadequate settling time because of poor furnace design. The lengthy settling area of flash furnaces minimizes the amount of suspended matte in slags produced by flash furnaces; the low operating temperature of Chinese bottom-blown smelters increases slag viscosity and thus dissolved suspended matte in these slags.

Smelter slags can contain 0.5%–8% or more copper; converter slags analyze 4%–12% copper. This percentage, multiplied by the mass of slag produced in these furnaces, means that a considerable fraction of the copper in the input raw materials (concentrates, reverts, dusts, scrap) would be lost to the slag if it were simply disposed of. As a result, most smelters treat their process slags to recover as much of this as is feasible, using one or more of several techniques.

If the slag is still molten, simply giving it more time to settle may remove most of the suspended matte. Figure 13 shows a typical electric slag-cleaning furnace, popular since the 1960s. These furnaces are commonly used to treat both high-copper smelting slags from submerged-tuyere smelters like the Noranda and Teniente furnaces and the matte-slag emulsions tapped from TSL furnaces (Ausmelt, IsaSmelt, Mitsubishi). This is an AC furnace with three carbon graphite electrodes, usually the self-baking Søderberg type. Heat is generated by passing current through the slag layer at the top, which minimizes turbulence and improves settling. Power use is 15–70 kW·h/t of slag treated, depending primarily on the furnace input and the required residence time for complete settling. The furnace sidewalls are typically water cooled to reduce refractory wear and increase furnace life.

Besides molten slag or emulsion, inputs to the slag-cleaning furnace may include converter slag or reverts, along with a reducing agent such as coal, coke, natural gas, or ferrosilicon. The reducing agent minimizes dissolved copper oxide by reducing it to metal, which is slightly soluble in matte:

$$C + Cu_2O \text{ (slag)} = CO + 2Cu \text{ (matte)}$$

Adding reducing agent also reduces magnetite (Fe_3O_4) in the slag to wüstite (FeO). This decreases slag viscosity and increases settling rates. Fuel-fired slag-cleaning furnaces are sometimes used instead of electric, depending on costs.

Most furnaces are tapped on a semicontinuous basis, although the slag-cleaning furnaces for TSL smelters are more likely to operate on a continuous basis. The products of slag-cleaning furnaces are low-copper slag (0.6%–1.3%), and normal-grade matte (65%–70% Cu).

When high-copper converter slags cool, the dissolved copper oxide is partially reduced to metallic copper:

$$Cu_2O + 3FeO = 2Cu(l) + Fe_3O_4$$

This copper exsolves as small metallic particles as the solid solidifies. Crushing and grinding the converter slag by typical comminution technology and using froth flotation to recover the metallic copper eliminates the need for a slag-cleaning furnace. This is especially true in plants using a flash furnace, which normally produce a smelter slag with sufficiently low copper content for disposal without further processing.

CONVERTING

The molten matte produced by copper smelting and matte settling consists mostly of Cu_2S and FeS, along with minor amounts of several impurity elements. Metallic copper can be produced by reacting the matte with oxygen, a process known as converting:

$$Cu_2S\text{ (matte)} + O_2 = 2Cu(l) + SO_2$$

The FeS in the matte oxidizes to form FeO, which dissolves in the slag. Some of the copper also oxidizes to form dissolved Cu_2O in the slag:

$$Cu_2S\text{ (matte)} + \tfrac{3}{2}O_2 = Cu_2O\text{ (slag)} + SO_2$$

This produces a slag with high dissolved copper content, 4%–8%. However, because most of the iron in the concentrate was oxidized during smelting, the amount of slag produced during converting is relatively small, so the amount of copper lost in it is relatively small as well. The reprocessing of the converter slag described in the previous section will recover most of this copper, minimizing overall losses to the disposal slag. Converting of copper matte has been practiced for over a century, using technology that is only now undergoing significant changes.

Peirce–Smith Converting

Figure 14 illustrates a Peirce–Smith converter, used to produce most of the world's copper. This is a cylindrical vessel mounted on trunnions so it can be rotated back and forth for charging, blowing, and tapping. Dimensions vary, but a typical size is 4.5 m diameter × 12 m length. Like smelting furnaces, the vessel is lined with mag-chrome brick. The vessel treats 600–1,000 t/d of matte and produces 500–900 t of molten impure blister copper in the process. It features numerous tuyeres throughout its length, through which enriched air (24%–29% O_2) is blown into molten matte. A large opening in the top is where molten matte is charged, and where molten matte and copper are drained from the furnace after converting; it is also the exit from which off-gas (a mixture of N_2 and SO_2) is emitted during the process. A hood is placed over the opening to collect the off-gas during converting; this hood is removed during tapping and emptying, which is one of the disadvantages of the Peirce–Smith furnace. Converting is

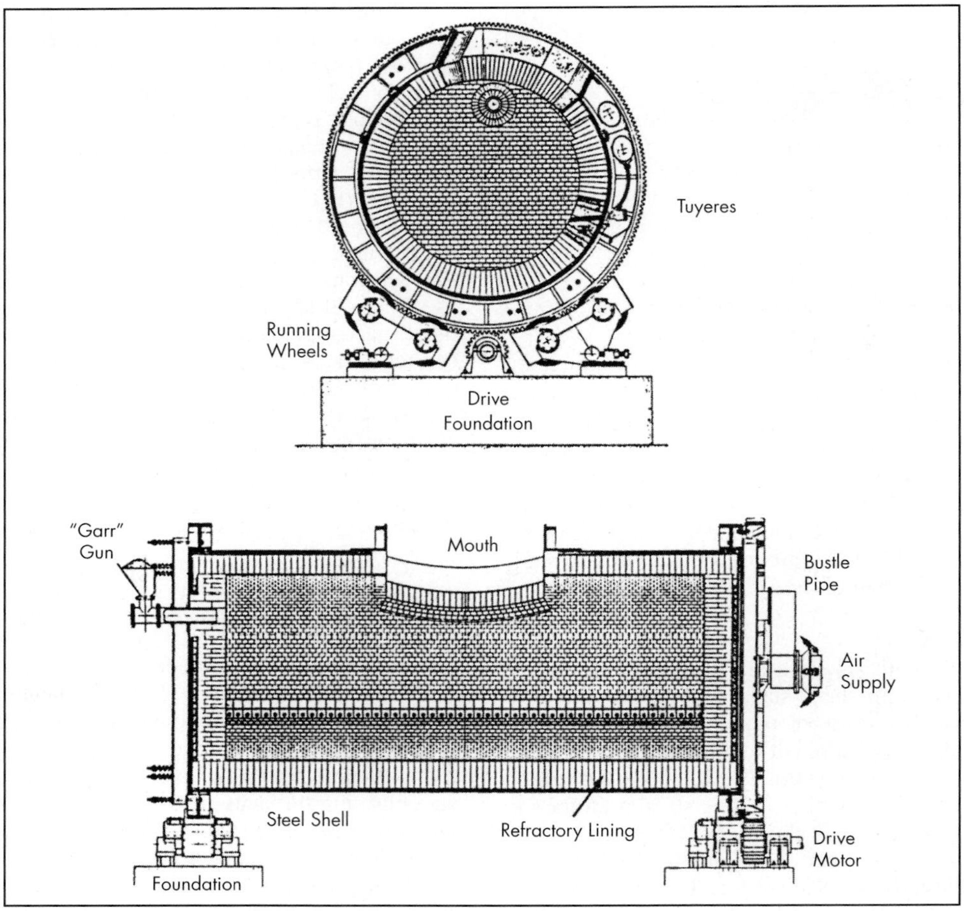

Courtesy of Harbison–Walter Refractories, Pittsburgh, Pennsylvania

Figure 14 Peirce–Smith converter

operated as a batch process, so the vessel is completely emptied after a 6- to 12-hour cycle.

The converting process involves several separate steps. The first is charging. During this step, the vessel is rotated about 15 degrees from vertical to allow molten matte to be poured from the ladle by which it was transported from the smelter. Most of the charge consists of matte, but other material is also added: flux (either silica or calcium carbonate, depending on the type of slag being produced), purchased scrap, and in some cases plant reverts. The reactions that take place during converting are exothermic and generate more heat than is needed to produce blister copper and slag, so the added scrap and reverts known as cold dope absorb this excess heat.

When the vessel has been filled, it is rotated back to the vertical position and the hood is replaced. At this point, blowing through the tuyeres begins. The oxidation of iron sulfide is kinetically favored, so this occurs first, producing converter slag and almost-pure molten Cu_2S, better known as white metal. When this is complete, the vessel is rotated by 80 degrees and the converter slag is skimmed off the top into a ladle for copper recovery. Typical slag temperature is 1,220°C.

When the vessel is rotated back to vertical, the final blowing step begins. This oxidizes the white metal and produces molten copper with 0.01%–0.05% dissolved sulfur and 0.1%–0.8% dissolved oxygen. Some impurities in the matte also wind up in the copper; gold, silver, selenium, and tellurium are desirable, whereas antimony, arsenic, bismuth, and lead are not. When the final step is complete, the vessel is rotated to a horizontal position, and the copper (~1,200°C) is emptied into a ladle for fire refining.

The strength (% SO_2) of the off-gas varies throughout the process but averages about 10%. The two main reasons are (1) the low level of enrichment of the air blown through the tuyeres, which means high nitrogen throughput, and (2) air leakage into the hood during converting. The hood is operated under low pressure to prevent fugitive vapors from leaking out into the plant environment, and this causes outside air to be sucked into the hood, further diluting the off-gas. One reason for preferring higher matte grades in the smelter is that it reduces the length of the converting process, thus reducing the amount of this low-strength off-gas being produced.

Traditionally, converter slag has been produced by adding silica and generating the same type of fayalite ($FeO\text{-}SiO_2$) slag produced in smelting furnaces. In recent years, some converting furnaces have begun to replace the silica with limestone ($CaCO_3$) or burnt lime (CaO), which results in a calcium ferrite ($CaO\text{-}Fe_2O_3$) slag. This slag has lower solubility for Cu_2O, which reduces copper losses in the slag. It also absorbs more antimony, arsenic, and bismuth, removing more of these undesirable impurities from the molten copper. However, it is more aggressive toward refractories and requires more careful control to avoid magnetite precipitation.

Peirce–Smith technology has been used to convert copper mattes for more than a century. During that time, several shortcomings of the process have become apparent. These include slow processing time, low off-gas strength, fugitive vapors, vessel maintenance, frequent refractory replacement, and the inefficiency of a batch process. As a result, numerous modifications have been tried over the years, including the following:

- The Hoboken converter, which uses a syphon attached to one side of the unit as a semipermanent seal to eliminate the need for sucked-in air.
- The Mitsubishi converter, which uses lances to replace the tuyeres. This eliminates the need for the vessel to rock, allowing a permanent seal with the off-gas ductwork.
- The Noranda continuous converter, which is similar in construction to the Teniente/Noranda smelter shown earlier. This converter is actually semicontinuous, with periodic charging and tapping. The device is never entirely drained of converter slag and therefore never has to be turned for skimming. This eliminates the need for rotation, again allowing a better seal with the off-gas system.

Over the long term, none of these devices has been successful enough to take the place of the Peirce–Smith furnace. However, in recent decades, other converting furnaces have been introduced that may be the future of this process. The first of these is a flash converter, based on the flash furnace smelter described earlier. It uses solidified and crushed matte from the smelter, mixed with lime flux and sprayed through a burner in the top to react with highly enriched air (82%–88% O_2). As the matte "burns," slag and molten copper form; these are given a chance to settle in the same sort of arrangement as a flash smelting furnace. A second tower gives dust a chance to settle and allows evacuation of the off-gas. The off-gas has higher strength (30%–40%) and a steadier composition. Dusts are also recycled to the converter. Molten slag and blister copper are periodically tapped from the side and bottom, similar to a smelter. Because the flash converter uses solidified matte, it is decoupled from the smelter and can be operated on stored matte when the smelter is down for maintenance or repair. The flash converter was designed to use a calcium ferrite slag, and therefore lime is charged with the matte to the burner.

Perhaps the most promising new converting technology is based on the TSL bath smelting furnace described earlier. Both Ausmelt and IsaSmelt have developed converters based on this furnace, with the Ausmelt C3 converter more widely used at this time. Bath converting is performed in a vessel similar to that shown in Figure 7. The vessel is charged through the feed port with molten matte, along with flux and small amounts of cold dope. Enriched air (30%–40% O_2) is blown in through the lance, and an emulsion of slag and molten copper is tapped from the bottom and allowed to settle. Because no separate slagging step is used, copper losses in the slag can be high; therefore, high matte grades are encouraged to reduce the overall amount of converter slag produced. Bath converting has many potential advantages. A sealed system eliminates fugitive vapor and produces stronger off-gas; semicontinuous operation means that the vessel is operating more often than the batch-process Peirce–Smith converter; consistent operating conditions mean consistent off-gas strength and furnace temperatures. Efforts are being made to link TSL converting to TSL smelting and settling to produce a truly continuous copper extraction process.

FIRE REFINING

As mentioned previously, the impure copper produced by smelting and converting contains 0.01%–0.05% dissolved sulfur and 0.1%–0.8% dissolved oxygen. If this copper is cast directly, the dissolved sulfur and oxygen react as the metal cools:

$$\underline{S} + 2\underline{O} = SO_2\ (\text{gas})$$

The SO_2 bubbles formed by this reaction leave behind small bubbles called blisters as they rise to the surface of the solidifying copper, hence the name *blister copper*. Blister copper also contains numerous additional impurities that are in violation of product-quality standards. As a result, the copper must be refined before it can be sold. For copper produced by smelting and converting, there are two refining steps: fire refining (covered here) and electrorefining (see Chapter 10.17, "Electrorefining"). The primary purpose of fire refining is to remove the dissolved sulfur and oxygen; electrorefining primarily removes/recovers the other impurities.

Figure 15 illustrates a typical anode furnace used for fire refining of blister copper. The vessel design is roughly similar to a Peirce–Smith converter. The biggest difference is the reduced number of tuyeres; there are two instead of the 30–50 used for a converter. The reactions taking place in an anode furnace are not exothermic enough to keep the contents warm, so a fossil fuel—typically natural gas, occasionally propane or naphtha—is burned. Both the fuel combustion air and the air blown in through the tuyeres are unenriched. The vessel tilts in the same way as a Peirce–Smith converter to accept input through the hole at the top, and tilts over the same way to let purified molten copper flow out after refining is complete. A hood is provided for the removal of product gases, although the strength (% SO_2) is negligible.

The refining cycle in an anode furnace begins when enough blister copper (250–300 t) has been charged. This is equal to several ladles of molten metal. At this point, air is blown in to oxidize most of the remaining sulfur in the metal:

$$\underline{S} + O_2 = SO_2$$

This happens until the dissolved sulfur is reduced to ~30 ppm (0.03 wt %). During this one-hour process, the concentration of dissolved oxygen in the molten blister increases, so a second refining step is needed to remove the oxygen. This is accomplished by blowing in hydrocarbons through the tuyeres. The hydrocarbons react with dissolved oxygen to produce water vapor and carbon monoxide, which leave the furnace. This effectively removes the $\underline{O}$, without dissolving any new impurities in the metal. This second step often requires about two hours to complete. When the process is complete, the vessel is rolled over and the molten anode copper is poured into a trough that leads to the casting wheel.

Figure 16 shows the final step in the pyrometallurgical processing of copper, the casting of molten copper into molds that will become anodes for electrorefining upon solidification. The anodes (16–32, depending on plant capacity) are placed in a continuously rotating casting wheel. Each is roughly 1 m × 1 m, and thick enough to produce an anode 360–410 kg in mass. They are sprayed with a barite slurry before being rotated under the spoon from which molten metal is poured into them, to make removal of the solidified anode easier. When the wheel has rotated about 270 degrees after being filled, an automatic pin and raising mechanism picks the anode up and removes it from the mold. The now-empty mold is sprayed with more barite and rotated back under the spoon to be filled again. About 50–100 t/h copper is cast this way. The process is designed to produce very flat anodes of uniform thickness and mass to ensure good results from electrorefining. Some copper producers use a continuous anode casting line that produces an anode-thick slab of solidified copper that is then sheared into anodes. This eliminates the need to maintain the molds, but maintaining uniform results is difficult. As a result, its use is not widespread.

Courtesy of Palabora Mining Company

Figure 15 Anode furnace in operation

Courtesy of Miguel Palacios, Atlantic Copper, Huelva, Spain

Figure 16 Anode casting wheel

REFERENCES

Ajima, S., Koichi, K., Kanamori, K., Igarashi, T., Muto, T., and Hayashi, S. 1999. Copper smelting and refining in Indonesia. In *Copper 99—Cobre 99, Vol. 5: Smelting Operations and Advances*. Edited by D.B. George, W.J. Chen, P.J. Mackey, and A.J. Weddick. Warrendale, PA: The Minerals, Metals & Materials Society. pp. 57–69.

Cui, Z., Wang, Z., and Zhao, B. 2013. Features of the bottom blown oxygen copper smelting technology. In *Copper 2013, Vol. 3: Nickolas Themelis Symposium on Pyrometallurgy and Process Engineering*. Edited by R. Bassa, R. Parra, A. Luraschi, and S. Demetrio. Westmount, QC: Canadian Institute of Mining, Metallurgy and Petroleum. pp. 923–933.

Furuta, S., Tanaka, S., Hamamoto, M., and Inada, H. 2006. Analysis of copper loss in slag in Tamano type flash smelting furnace. In *Sohn International Symposium, Vol. 8: International Symposium on Sulfide Smelting*. Edited by F. Kongoli and R.G. Reddy. Warrendale, PA: The Minerals, Metals & Materials Society. pp. 123–133.

Holmström, A., Hedström, L., Lundholm, K., Andersson, M., and Nevatalo, L. 2014. Partial roasting of copper concentrate with stabilization of arsenic and mercury. In *46th Annual Canadian Mineral Processors Operators Conference*. Edited by T. Crowie and J. Zinck. Westmount, QC: Canadian Institute of Mining, Metallurgy and Petroleum. pp. 251–264.

Hustrulid, W.A., Kuchta, M., and Martin, R.K. 2013. *Open Pit Mine Planning and Design*, 3rd ed. Vol. 1. Boca Raton, FL: CRC Press. p. 98.

Jie, Y. 2013. Latest development of oxygen bottom-blowing copper smelting technology. In *Copper 2013, Vol. 3: Nickolas Themelis Symposium on Pyrometallurgy and Process Engineering*. Edited by R. Bassa, R. Parra, A. Luraschi, and S. Demetrio. Westmount, QC: Canadian Institute of Mining, Metallurgy and Petroleum. pp. 873–887.

Kellogg, H.H., and Diaz, C. 1992. Bath smelting processes in non-ferrous pyrometallurgy: An overview. In *Proceedings of the Savard/Lee International Symposium on Bath Smelting*. Edited by J.K. Brimacombe, P.J. Mackey, G.J.W. Kor, C. Bickert, and M.G. Ranade. Warrendale, PA: The Minerals, Metals & Materials Society. pp. 39–65.

LS-Nikko. 2007. LS-Nikko Copper CDM Project Overview [As Dry Steam (Steam Dryer)]. www.lsnikko.com/download/cdm.pdf. Accessed February 2007.

Roghani, G., Takeda, Y., and Itagaki, K. 2000. Phase equilibrium and minor element distribution between FeO_x-SiO_2-MgO–based slag and Cu_2S-FeS matte at 1573 K under high partial pressure of SO_2. *Metall. Mater. Trans. B* 31B:705–712.

Schlesinger, M.E., King, M.J., Sole, K.D., and Davenport, W.G. 2011. *Extractive Metallurgy of Copper*, 5th ed. New York: Elsevier.

Yazawa, A., and Kameda, A. 1953. Copper smelting. I. Partial liquidus diagram for FeS-FeO-SiO_2 system. *Technol. Rep. Tohoku Univ.* 16:40–58.

CHAPTER 12.10

Diamonds

Graham Popplewell

AGE, ORIGIN, AND EMPLACEMENT OF DIAMONDS

Diamonds originate in kimberlites (and lamproites, a subtly different but related rock type), which are deeply derived and volcanically emplaced structures. Diamonds also occur in alluvial deposits, which are formed when such primary deposits weather over millennia, releasing diamonds (and other heavy minerals) to be subsequently concentrated again through river and marine currents.

Until the 1980s, there was little comprehensive understanding of the age, origin, and manner of emplacement of diamonds into minable environments. No plausible mechanisms had been proposed that could explain how diamonds were created within the kimberlite or lamproite deposits. Subsequent advances in geochemical analytical methods, however, led to studies of the geochemistry of inclusions found in diamonds. What emerged were conclusions that are now widely accepted (Kirkley et al. 1991):

- Diamonds are derived from either of two rock types, eclogite and peridotite, typical of carbon sources associated with subducted oceanic crust (basalt) and upper mantle zones, respectively.
- Diamonds are almost always much older (1,000–3,300 million years) than the kimberlite or lamproite in which they are found (>50 million years but varies widely).
- The temperature and pressure at which the diamond form of carbon is preserved intersect at favorable locations approximately 150 km below stable continents (the "diamond stability field"). Oceanic areas have a geothermal gradient that is too high, pushing the zone of diamond stability farther down.
- Kimberlite pipes originate from areas deeper than the diamond stability field (except in oceanic areas), and the kimberlite rock serves simply as the transport vehicle for diamonds. It collects diamonds and conveys them upward, expanding explosively as it approaches the surface. The journey from the mantle to the surface must be at a velocity that is fast enough to prevent the diamonds from converting to graphite or carbon dioxide or dissolving in the magma itself. Russell et al. (2012) suggest that extremely high magma velocities are achieved by the interaction of alkaline carbonates, native to the kimberlite, with acidic silicate rocks entrained during the ascent. The acid–base chemical interaction releases carbon dioxide, which then acts in a manner similar to a rocket propellant.

NATURAL DIAMOND PRODUCTION

Reported data (Table 1) show that in the period 2014–2016, just over half of the global rough diamond production by value was from just two countries, Russia and Botswana. If Canadian, South African, and Angolan production are added to Russia and Botswana, five countries accounted for more than 80% of global natural diamond production in 2016.

Natural Gem Diamonds

The main factor driving the diamond mining industry is the market for diamond jewelry. Concerted marketing efforts over many decades have focused on creating an emotionally driven attitude to the acquisition of diamond jewelry, underpinned by celebration of personal life events. Attempts to commoditize diamonds as tradable investments have been generally unsuccessful for many reasons, including the essentially "non-fungible" nature of diamonds. No two are the same, diamond grading systems are complex, and the difference in value can vary significantly between adjacent grades.

Other Uses of Diamonds

The application of diamonds to a wide variety of uses is underpinned by their physical properties, some of which are also exploited in the diamond recovery process. These physical properties include the following:

- **Hardness:** Diamond is the hardest known natural material, leading to its use as an abrasive and enabling it to survive for millions of years through geological weathering processes. The hardness of diamond relative to other ore components is particularly employed in the liberation process.

Graham Popplewell, Technical Director & Senior Fellow, Diamond Processing, Fluor Corporation, Perth, Western Australia, Australia

Table 1 Rough diamond production by country

	2014			2015			2016		
	Volume, carats	Value, US$/carat	Value, US$	Volume, carats	Value, US$/carat	Value, US$	Volume, carats	Value, US$/carat	Value, US$
Russia	38,303,500	97.47	3,733,262,920	41,912,390	101.15	4,239,585,340	40,322,030	88.75	3,578,732,550
Botswana	24,668,091	147.84	3,646,952,179	20,778,642	143.73	2,986,469,130	20,501,000	138.82	2,845,948,820
Canada	12,011,619	166.78	2,003,267,161	11,677,472	143.52	1,675,936,000	13,036,449	107.18	1,397,308,512
South Africa	7,430,956	164.76	1,224,311,494	7,218,463	192.57	1,390,033,447	8,311,674	150.26	1,248,912,618
Angola	8,791,340	149.86	1,317,456,072	9,016,343	131.11	1,182,128,882	9,021,467	119.65	1,079,411,359
Namibia	1,917,690	602.57	1,155,536,792	2,053,095	591.08	1,213,539,148	1,717,658	532.60	914,827,141
Democratic Republic of the Congo	15,652,015	8.72	136,505,486	16,016,332	8.28	132,539,972	23,207,443	10.63	246,700,973
Australia	9,288,232	32.76	304,319,165	13,563,935	22.73	308,356,848	13,957,722	15.50	216,337,288
Sierra Leone	620,181	357.50	221,713,243	500,000	308.51	154,253,129	549,086	289.34	158,872,778
Zimbabwe	4,771,637	50.00	238,581,841	3,490,881	50.00	174,544,058	2,102,873	50.00	105,143,675
Other	1,323,213	388.37	513,900,646	1,171,810	362.04	424,240,130	1,343,284	453.04	608,560,880
Total	**124,778,474**	**116.17**	**14,495,806,999**	**127,399,363**	**108.96**	**13,881,626,084**	**134,070,686**	**92.49**	**12,400,756,594**

Source: KPCS n.d.

- **Density:** Diamond has a specific gravity (sg) of 3.53. Graphite has a specific gravity of 2.26. Because the host rocks within which diamonds are found typically have a density less than 2.65 g/cm^3, the relatively high density of diamond is the basis for its concentration using gravity separation techniques.
- **Dispersion:** Because of their high refractive index (2.42, versus glass, 1.52), diamonds separate the colors of light to a greater extent, providing the "fire" in a cut gemstone.
- **Transparency:** Diamond is transparent from the ultraviolet (225 nm) to the far infrared range, which it retains at very high temperatures and radiation intensities. This makes it an ideal material for use as a protective "window" for optical devices in extreme conditions (e.g., lasers and sensing devices in missile nose-cone assemblies).
- **Electrical conductivity:** Although a few diamonds are electrical semiconductors, the vast majority are electrically nonconducting.
- **Thermal conductivity:** The thermal conductivity of diamonds is four times that of copper at room temperature, leading to their use as heat sinks in powerful electronic devices.
- **Fluorescence:** Approximately 98.5% of all diamonds will fluoresce when exposed to X-rays. This is used as a basis for final diamond recovery. Type II diamonds are unique in that they do not fluoresce reliably under X-ray stimulation but are recoverable using X-ray transmission (XRT).

Grade of Diamond-Bearing Deposits

A good diamond deposit may contain less than 0.2 ppm diamond or 1 carat/t (per metric ton) ore mined. Diamond ore concentrations are so low that they are often quoted in carats per 100 t. This is far below concentration levels typical of base metal mines, and hence very high efficiencies are required to render the operation economic. Diamonds are not fungible; the value increases exponentially with size and varies with color and clarity. Therefore, the fundamental processing task is to remove 999999.8 ppm of non-diamond without losing or damaging the 0.2 ppm that represents the value.

MINERAL PROCESSING FOR DIAMOND RECOVERY

Mineral processing in the diamond industry follows four basic steps:

1. Diamond liberation (through comminution)
2. Concentration (gravity separation)
3. Recovery (X-ray, grease)
4. Cleaning/removal of non-diamond material (acid, caustic)

Given both the correlation between diamond size and value, and the very high recovery efficiencies required for successful operation, the principles of "diamond value management" are woven throughout the diamond recovery process design and operation. This involves a suite of analytical tools serving to identify the "total content curve" and hence value by incremental size range of the deposit. The tools are populated and calibrated to a deposit by ore sampling and the application of statistical methods. The tools are then verified and improved by monitoring the actual value recovered into the final diamond production. Target recoveries of at least 99% of the contained value are typical, and investigation and mitigating action will follow if unmet, to determine where in the process chain the value is being lost and why (Rider and Roodt 2003).

A typical flow sheet for a modern kimberlite treatment plant for the recovery of diamonds in the most common size range of 1–25 mm is shown in Figure 1. Although diamonds outside this size range are recovered in some installations, the core size range represents the overwhelming proportion of value for most operations (Popplewell 2007).

A typical flow sheet for the recovery of diamonds from an alluvial diamond deposit is shown in Figure 2. This is a simpler process, not requiring any comminution circuit to liberate diamonds. The following sections describe processes pertaining to this common size range, unless otherwise noted.

Not all diamond recovery operations utilize all processing steps shown. Combining primary and secondary scrubbing into a single operation and omitting reconcentration and single-particle X-ray sorters are both common variations. Although the use of dense medium separation (DMS) as the principal separation process is almost ubiquitous outside Russia, the use of bulk X-ray sorters (which are, in contrast, extensively used

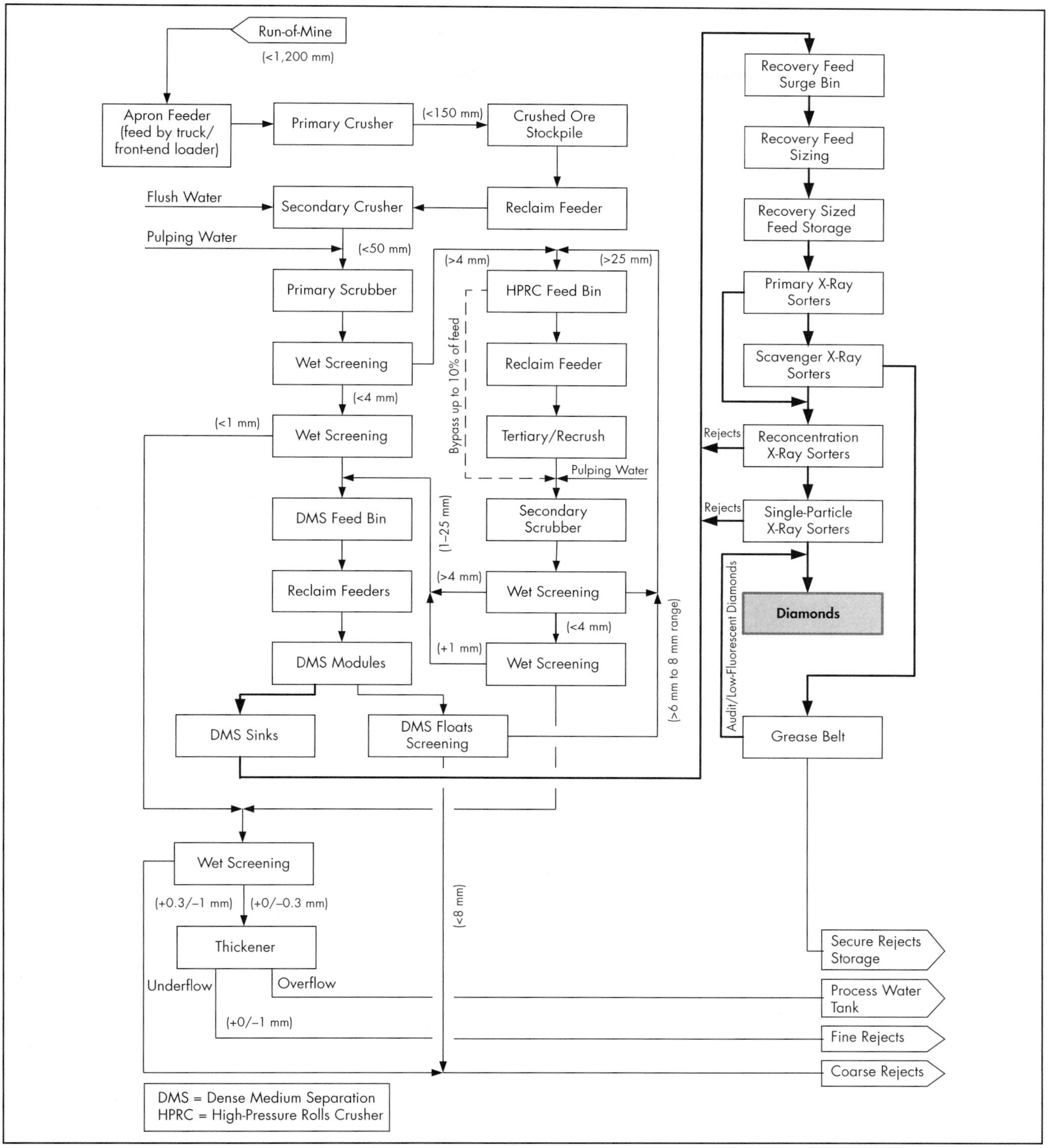

Figure 1 Kimberlite diamond recovery flow sheet

in Russian diamond recovery plants) is emerging as an alternative, where the deposit is statistically determined to contain economically recoverable large diamonds (generally accepted as >25 mm). DMS processing is then restricted to the size fraction below approximately 8 mm, for which X-ray sorting is not currently considered cost-effective. Increasingly sophisticated geological modeling has supported the introduction of such "large" diamond recovery processes in some recently developed mines. These are exclusively based on XRT sorting. Diamonds that would previously have been broken during size reduction before DMS, through a lack of knowledge of their potential presence, are now being recovered (Van Niekerk et al. 2016).

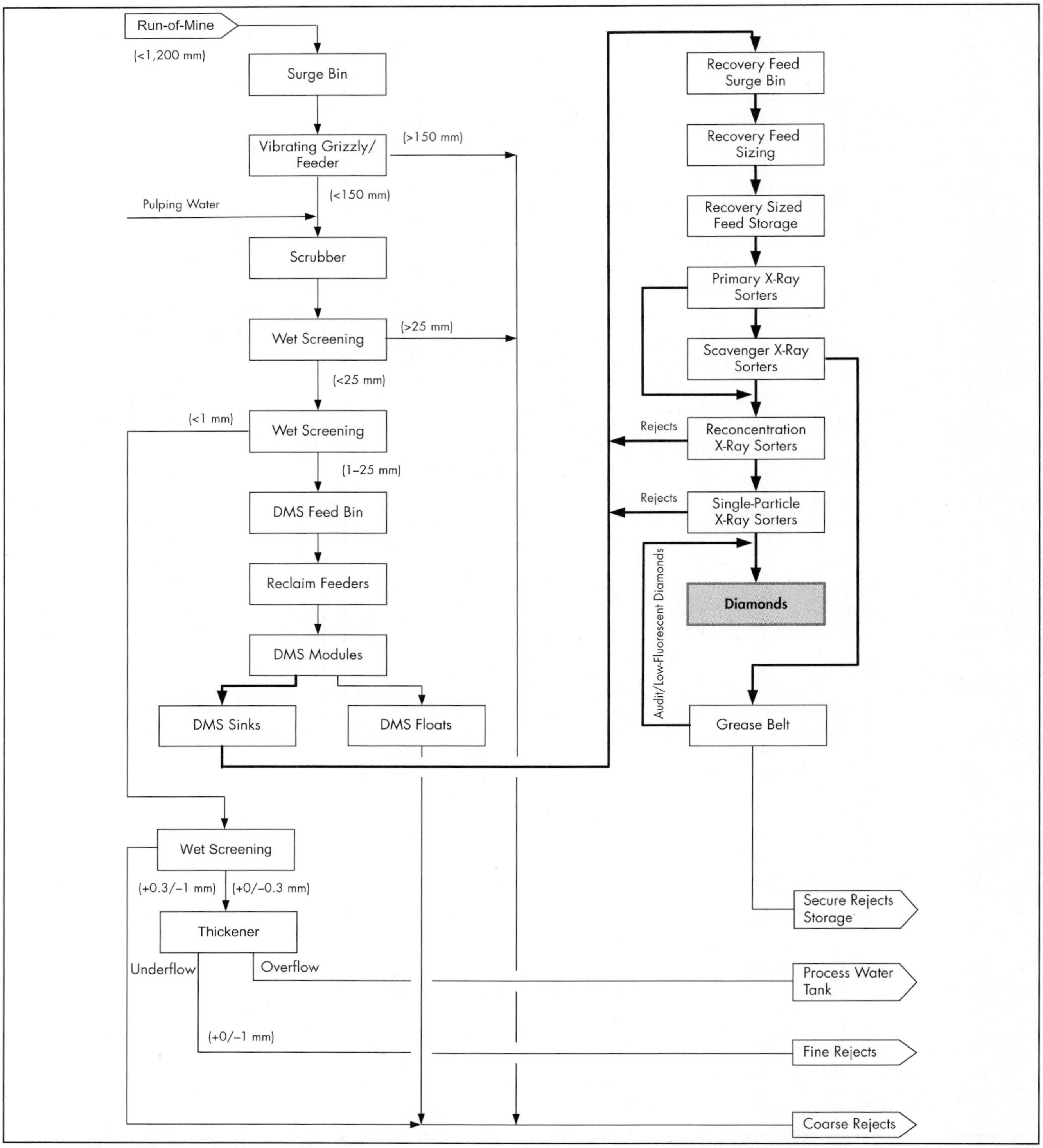

Figure 2 Alluvial diamond recovery flow sheet

Diamond Liberation

Only "free" diamonds are cut and sold. Free diamonds exist naturally, in alluvial deposits, but most have to be liberated from the kimberlite rock host. Beyond this basic requirement, the purpose of the comminution section of a diamond recovery plant is to prepare the ore for a separation process that extracts contained diamonds, either directly via X-ray luminescence or XRT sorting, or via a gravity separation process where the difference in density between the diamonds and the other host rock minerals is exploited. (The diamonds contained in the concentrate are typically also recovered using X-ray sorters.) Currently, gravity separation is by far the most common separation process outside Russia, and DMS is the dominant technology.

Following comminution, the ore feed contains particles, including diamonds, from the target diamond top size down to "dust." However, the smallest rough gem diamonds that are considered marketable are approximately 0.8 mm (DTC [Diamond Trading Company] no. 1 sieve size). Therefore, although DMS applications for other minerals range to below 0.5 mm, for diamonds the bottom size limit is almost always in the range of 0.8–1.5 mm, the actual value being deposit specific or the result of variation in the market for the diamond product.

Although some ore deposits either have been well-washed by nature (a few alluvial diamond deposits) or consist of competent rock (some primary diamond deposits, e.g., Argyle in Western Australia or Letlhakane in Botswana), most contain clay-forming minerals, which complicate the process of establishing a reasonably sharp cutoff at the desired bottom size. Without disagglomerating such clay-bound ore prior to screening, clumps of small particles stick together and report to the dense medium cyclone, where they are likely to break up under the high shear forces present. The result is contamination of the circulating medium with low-density fine material, and this leads to higher medium viscosity, which compromises the separation.

Primary Crushing

A variety of comminution devices are available. Run-of-mine primary crushing unit selection is largely driven by throughput capacity and feed hardness, with high-capacity gyratory crushers giving way to lower-capacity jaw crushers, and roll "mineral" sizers or even "roadheaders," capable of a wide range of capacities on relatively soft, friable ore. The purpose of primary crushing is to produce a conveyable ore stream. The key output from primary crushing is a maximum lump size for the downstream process. This is usually around 250 mm, with a P80 between 100 and 150 mm.

Following primary crushing, further size reduction is intended to limit the material to a desired top size. This is the size of the largest diamond that statistical interpretation of exploration results suggests is present in economically recoverable quantity. The top size is therefore deposit specific and can range up to 60–75 mm, although the vast majority of installations limit the top size for diamond recovery to 20–32 mm. Material characteristics become more important, particularly clay and moisture content, and the overall competence of the ore, as described by A*b and Ta parameters from drop weight testing (www.jktech.com.au). With ores such as iron ore, manganese, or andalusite, "grade" varies from lump to lump and hence is a notionally continuous concept that can be traded off against "recovery" and is typically represented as such on a grade–recovery diagram. Higher-grade product generally requires sacrifice of mass recovery and vice versa. The product specification for these other ore type examples is a bulk specification. With diamonds, it is different. Either a particle is a diamond or it is not. There is no bulk specification, and a "grade–recovery" curve does not apply. Simply put, a 2-carat diamond is desirable, whereas a 2-kg lump of gray rock containing two carats of tiny diamonds is not.

Part of the process of reducing the ore top size is therefore about liberating as many of the diamonds as possible so they can follow a density-specific pathway through the dense medium cyclone. However, broken diamonds are almost always worth less than unbroken diamonds, so care must be taken while reducing the top size to the desired value. In addition, a high size–reduction ratio is desirable, as particles reduced to below the bottom size for diamond recovery following comminution are screened out to tailings and therefore reduce the load on the DMS section of the plant.

Secondary Crushing

Secondary crushers, usually cone crushers, are employed to reduce primary crushed product to a P80 in the range of 25–32 mm. Secondary crusher product is screened to scalp off material greater than the diamond top size, with oversize returned to the crusher and undersize reporting to the DMS. The need to avoid damage to larger diamonds passing through the crusher means that the crusher gap setting must be greater than the diamond top size, so that a recirculating load around the crusher is inevitable.

Tertiary Crushing

The feed to the tertiary crushing (often referred to as recrushing) section is derived from the DMS floats product, where the coarser fraction is returned for further size reduction to liberate additional diamonds not released in the secondary crushers. The target crushed product top size is usually in the range of 6–8 mm, defined as the DMS floats size below which the remaining "locked" diamond value is not significant. Prior to the widespread acceptance of high-pressure rolls crusher (HPRC) for tertiary crushing duties, cone crushers were employed for this role. For the same reason of avoiding diamond damage, crusher operating gaps were relaxed, and relatively high circulating loads (200%–300%) were common, with consequent adverse capital and operating cost impacts.

The use of HPRC (Figure 3) in a mineral processing application (as opposed to the cement industry) was pioneered in the diamond industry in South Africa, and HPRCs are used for tertiary crushing at nearly all larger kimberlite operations across Southern Africa and Canada. Early smooth and profiled roll surfaces have largely been replaced by rolls with studded surfaces (with the first installation in 1998 at Ekati diamond mine in Canada). Ore accumulates between the studs, forming an autogenous surface, which protects the studs from wear

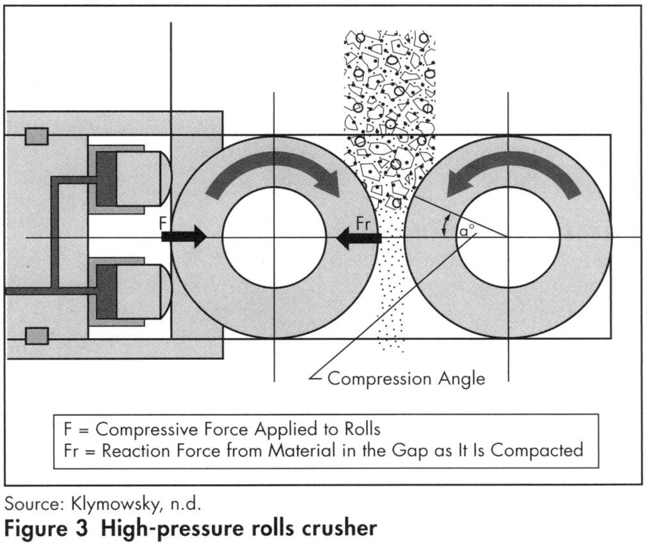

Source: Klymowsky, n.d.

Figure 3 High-pressure rolls crusher

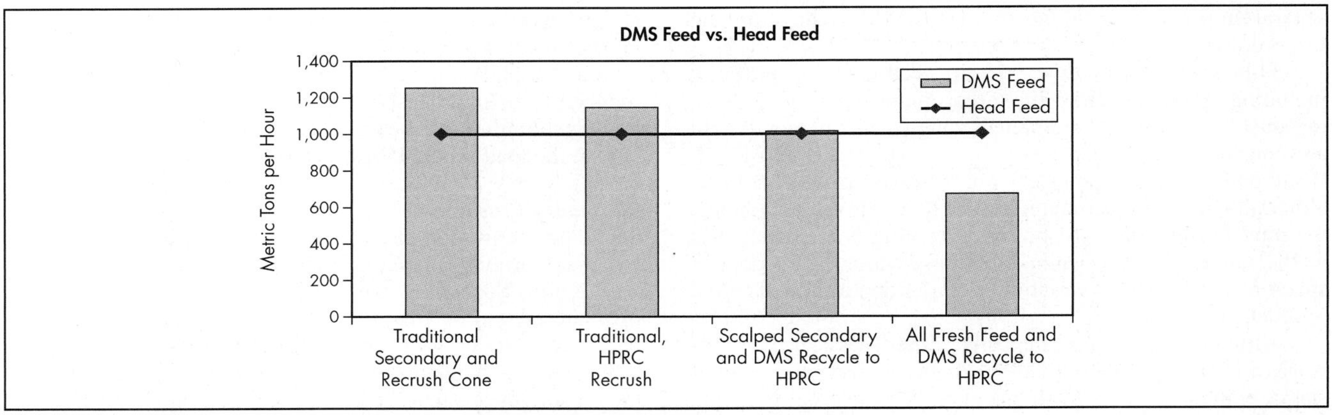

Figure 4 Benefit of maximizing comminution by HPRC on DMS capacity requirements

while increasing friction. The greater the friction angle, the greater the ability of the rolls to draw in material, resulting in a wider gap between the rolls in operation and hence higher capacity. The wider the operating gap, the less the risk of damage to any diamonds present, which obviously have to be smaller than the operating gap to avoid being crushed themselves. Since diamonds present will be the hardest component of the ore stream to the HPRC, non-diamond material is crushed around the diamonds, liberating them preferentially. The typical product from an HPRC operating in a tertiary crushing application is finer than that from a cone crusher, reducing the volume of feed to the DMS section, and drastically reducing crusher circulating load relative to cone crushers (Figure 4; Fyfe 2005).

Vertical shaft impact (VSI) crushers have been used for many years in operations where the ore is sticky and for marine operations where large amounts of shell are encountered. Although the relatively low capital cost is attractive, concerns about diamond damage greatly limited their acceptance for many years. However, recent trials conducted by one of the major diamond producers suggest that, properly controlled, VSI crushers can be a very effective comminution device in a tertiary crushing application where any liberated diamonds have already been removed. The operating principle (Figure 5) involves separation of the inflowing ore into two streams, with one stream forming a bed into which the second stream is accelerated horizontally via a vertically oriented rotor. This avoids impact of the ore stream on the crusher components, promoting instead interparticle comminution between the two streams. By adjusting the relative proportions of the two streams and the rotor speed, the crushing performance can be optimized for the particular ore being treated.

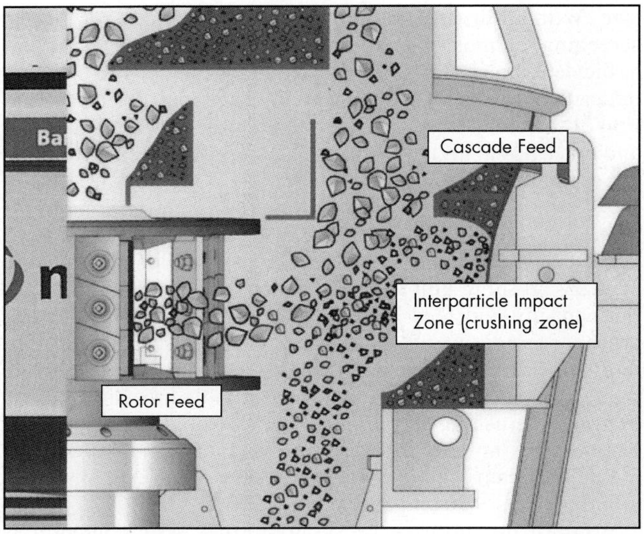

Adapted from Metso Minerals 2009

Figure 5 Interparticle crushing in a vertical shaft impact crusher

DISAGGLOMERATION

Although some ores have so little clay content that disagglomeration can be achieved by simple wet screening with washing sprays, this is relatively rare. Typically, a degree of energy input is required to loosen clay bonds (Figure 6). If jet pumps are used for material transport, the intense shear forces in the mixing chamber can result in a substantial "scrubbing" effect, and in some cases, no further disagglomeration effort is required, although, more commonly, jet pumping would complement other disagglomeration equipment. Log washers can also be used where clay bonding is weak to moderate. However, most commonly, disagglomeration is carried out in wet rotary vessels. The disagglomeration process is referred to as "scrubbing," which describes the essential nature of the process. Tumbling the ore with water in a drum disperses clay minerals, which are screened out and discarded along with any material smaller than the bottom diamond size (Figure 7). If not removed, such fine material contaminates the dense medium, increasing viscosity and adversely affecting separation cyclone performance.

Wet rotary scrubbers used in production installations vary in size from a minimum of perhaps 1.2 m up to 5 m diameter. Scrubber sizing should be based on pilot-scale test work, to determine energy input, design, and operating parameters. Shell diameter is driven by overall throughput capacity considerations, with the length selected to provide the retention time required for effective disagglomeration. Scrubber working volume is dictated by the diameter of the discharge dam ring. Scrubber absorbed power is dictated by the working volume, liner configuration, scrubber drum rotational speed, and feed slurry density. For effective scrubbing, the feed must contain sufficient competent larger ore particles (100–150 mm) to

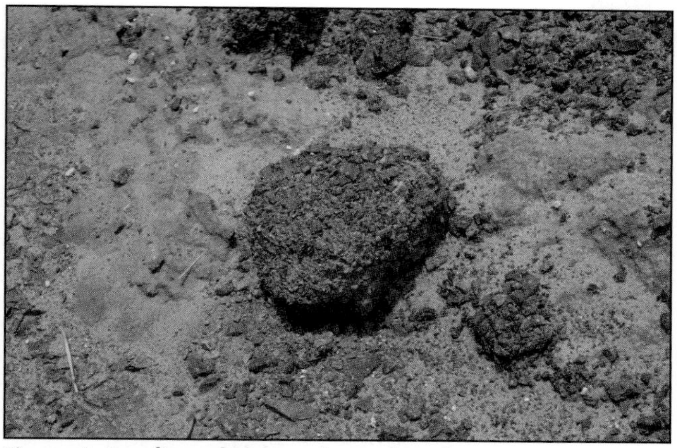

Figure 6 Pre-disagglomerated weathered ore

Figure 7 Disagglomerated and washed ore

transfer energy to the charge. If not present in sufficient quantity, a separate charge is needed, which requires a discharge grate to be fitted to retain the coarse particles while passing the rest of the charge.

Autogenous Milling

Autogenous milling is widely practiced in the Russian diamond mining industry and performs the dual roles of disagglomeration and true size reduction, enhancing diamond liberation. Elsewhere, autogenous milling as a pre-DMS preparation stage at diamond mines has, until recently, been restricted to a relatively narrow range of softer ores and is more accurately described as "super scrubbing." Where amenable, however, the use of autogenous milling presents an opportunity to simplify the overall comminution flow sheet. The potential for a "single stage of comminution," post-primary crushing, resulting in a product consisting mainly of barren kimberlitic fines along with some grits, pebbles, indicator minerals as well as the prized undamaged diamonds, is of obvious value. Recent success in introducing autogenous mills to several Western diamond mines highlights the need to consider both scrubbing and milling as options for diamond liberation and pre-DMS mass reduction (Daniel et al. 2016). This requires a thorough understanding of the range of ore characteristics, scrubber/mill sizing, and operating parameters through pilot-scale test work.

DEGRITTING AND DESLIMING

Washed ore from the scrubbing section reports to a screening section where unwanted fines (smaller than the bottom size for diamond recovery) are removed. Sometimes, this is preceded by a scalping screen intended to generate a midsize split, to optimize the size ranges for separate DMS circuits dedicated to coarser and finer particle stream, respectively. The separate DMS circuits would typically employ different sized cyclones with different operating parameters. This scalping duty is relatively straightforward, and since the size split between the coarse and fines streams is typically driven (within a limited range) by overall mass balance considerations, the efficiency of the sizing split is not particularly important.

Although horizontal vibrating screens are still used widely, particularly for higher-capacity installations, multislope vibrating screens have become the norm for the degritting/desliming duty. The main reason is that, at the relevant apertures (to achieve a size cutoff in the range 0.8–1.5 mm), the screening duty is driven largely by drainage. A rule of thumb that has been used for many years is that the solids content of the degritting screen underflow should not exceed 7% of the total underflow stream on a volume basis. This requires that the screen passes a large quantity of water to achieve acceptable removal of undersized material from the screen feed. The design of a multislope screen is such that the high velocity and changing direction of the slurry promotes dewatering. There is a parallel between this conventional screen comparison and the greater drainage capacity performance of a circular sievebend deck relative to an inclined flat drainage panel of similar area.

Material of Construction

With a screening size usually in the range of 0.8–1.5 mm, there are few exceptions to the selection of polyurethane-based modular screening panels for feed preparation duties. A sizing split to provide coarse and fine DMS feed streams, in a more varied range from perhaps 4–10 mm, allows the alternative of rubber to also be considered. Rubber, although generally more expensive than polyurethane, has a longer wear life and greater mobility, reducing the effects of aperture blinding on efficiency.

GRAVITY CONCENTRATION

Diamonds, with a specific gravity of 3.53, occur in primary (kimberlite/lamproite) and secondary (alluvial) deposits where the density of accompanying rocks and minerals is usually much lower. Most primary deposits have average densities <2.6 sg, with usually very little (perhaps 0.2%–0.8%) >3.0 sg (meaning, in the context of a heavy medium separation, that this will report to the same product as the diamonds). Alluvial deposits are formed when primary deposits weather over millennia, releasing diamonds and other heavy minerals to be subsequently concentrated again through river and marine currents. These deposits may be reworked geologically in various ways such that the volume of accompanying higher-density material varies considerably. Even so, most alluvial deposits contain a heavy mineral content <2% by volume, with occasional "spikes" up to perhaps 10% by volume.

Concentration of diamond-bearing deposits through gravity separation is therefore an obvious technique, and gravity separation technology based on jigging and rotary washing pans has been practiced for more than a century and is still in use today in smaller operations. However, both methods rely on relatively weak enhancement of the competing effects of gravity and hindered settling, leading to relatively inefficient separations and hence the need to "cut-low" and accept contamination of the diamond-bearing product to avoid diamond loss. Both also required continuous operator attention, particularly rotary washing pans. With the evolution of heavy medium separation pioneered by the Dutch State Mines in the 1950s using magnetite (5.3 sg) for the washing of coal, interest turned to parallel application for the recovery of diamonds. Coal washing utilizes separation densities of typically 1.4–1.8 sg, which would clearly not be applicable to diamond recovery where a nominal target separation density is usually 3.0–3.1 sg. Such densities are not achievable using magnetite alone, as the magnetite content of the medium slurry would need to be so high that the resulting slurry viscosity would render the separation impractical. Ferrosilicon (FeSi) powders were therefore developed, initially as a by-product of the steelmaking industry, but they evolved over time into a specific product of 6.8 sg, based on a 14%–16% w/w silicon content. Although heavy medium separation using FeSi has been applied to many commodities (particularly iron ore, manganese, chrome, and industrial minerals such as andalusite), the production of FeSi with reliable specifications with respect to density, size analysis, magnetic susceptibility, and corrosion resistance was driven by the rapid adoption of heavy medium separation in the diamond mining industry, where consistent and predictable performance is essential. Although jigs and rotary washing pans are still employed by some small and marginal diamond mining operators, heavy media separation using FeSi medium has come to dominate the more significant production operations across the Western world.

Type of Separators

Cyclones dominate in the diamond industry. Although the steady development of high-capacity X-ray sorting technology (both X-ray luminescence and XRT) has essentially removed the need for a "coarse" heavy medium separation technology for diamonds, being essentially a "single-particle" discrimination technology requiring the feed to be presented in a monolayer, X-ray processing becomes progressively more expensive as the bottom size for treatment reduces and the required monolayer becomes ever thinner (with the unit capacity decreasing accordingly). Currently, the economic bottom size limit for bulk X-ray sorting is in the range of 6–8 mm, with heavy medium cyclones the overwhelming choice below this limit. New diamond recovery installations are therefore likely to feature combinations of X-ray separators and heavy medium cyclones for primary ore processing. The most common heavy media cyclone sizes in use are made from high-chrome cast iron, with 20-degree cone angle the overwhelming standard. The cyclone configuration is based on the original Dutch State Mines equipment specifications, with the component dimensions as shown in Table 2.

Specialist cyclone manufacturers offer units with minor variations to the standard dimensions. The density of the circulating medium required to generate the desired "cut density" of 3.0–3.2 sg varies with size of cyclone, type of medium, and the cyclone feed pressure and is usually in the range of 2.4–2.7 sg, giving a "density differential" of 0.3–0.8 sg. A narrower density differential is usually considered to represent superior separation efficiency, as the separation is spread over a larger area of the cyclone. In reality, since primary diamond separations generate very little concentrate (i.e., sinks), narrower differentials are of more interest in reconcentration duties, where much higher concentrate volumes are present and where the majority of the particles are high specific gravity. The attraction of a high-density differential lies in the fact that it allows operation at a lower cyclone feed density and this represents a lower volume of FeSi in circuit.

Table 2 Standard DMS cyclone configuration

Parameter	Relative to Diameter	Production Example
Cyclone diameter	1	356 mm (nominal 350 mm)
Inlet diameter	0.2	70 mm
Vortex finder diameter	0.41	146 mm
Cone angle	Not applicable	20 degrees
Spigot diameter	0.21–0.25	76 mm, 83 mm, 89 mm

Table 3 Gravity- versus pump-fed cyclone features

Parameter/Feature	Gravity Feed	Pump Feed
Cyclone feed pressure	Fixed and reliable	Subject to fluctuation with impellor vane blockages, etc. Pressure monitoring is essential.
	Cannot be varied to suit changes in feed characteristics.	Can be varied to suit either feed changes or general impellor wear through use of variable-speed drive.
Power failure	Most medium drains back to the circulating medium tank. No gravel solids in medium are dumped to the floor while purging the circulating medium pump.	A greater volume of medium, including feed gravel, must be dumped to the floor to avoid cyclone feed pump blockage.
Building size	High	Compact

Cyclones with 40-degree cone angles have been utilized in the past for feed containing a large number of flat particles (e.g., on marine gravels with high shell content). However, such high cone angles generate high-density differentials, leading to the need to operate the cyclone with unusually low feed densities (as low as 1.8 sg) to maintain a cut density of <3.2 sg. The high differential in turn promotes retention of coarse dense particles that do not immediately leave the cyclone but linger between the cyclone spigot zone and the vortex finder zone. This situation can potentially lead to the ejection and loss of a large diamond to the float stream.

Gravity Versus Pump-Fed Cyclones

Both gravity- and pump-fed cyclone feed arrangements are in common use. There is a tendency for small- to medium-capacity operations to pump-feed the dense medium cyclones, with medium- to high-capacity installations using gravity feed. Each has advantages and disadvantages that influence the selection in each case. Although debate is influenced by various perceptions of potential benefits and drawbacks, some of which are exaggerated, the principal differentiating features are listed in Table 3.

The most common cyclones in use and their associated capacities at different geometric feed head (multiples of the cyclone diameter) and typical feed size ranges for primary diamond recovery are shown in Table 4. Note that, when

Table 4 Diamond recovery cyclones

Nominal Cyclone Diameter, mm	Capacity at 9D Head*		Capacity at 14D Head*		Capacity at 18D Head*		Typical Feed Size Range, mm‡	M/O Ratio	FeSi Grade
	m^3/h	t/h†	m^3/h	t/h†	m^3/h	t/h†			
250	31	10	39	13	44	15	0.5–10	7:1	270D
360	61	23	77	29	87	33	0.8–20	6:1	270D
420	94	36	118	45	133	50	1–25	6:1	150D
510	134	59	168	74	190	84	2–30	5:1	150D
610	187	83	235	104	N/A§	N/A	6–32	5:1	100D

*Capacities for standard Dutch State Mines cyclone configuration. Capacity can be increased approximately 10% if an extended barrel section is fitted.
†Solids capacity at M/O ratio and average solids of 2.65 sg.
‡Cyclone size is selected based on the minimum size of feed.
§610-mm cyclones are very rarely used above 14D head.

Table 5 Ferrosilicon specification for diamond applications

Property	100D	150D	270D
Type	Milled	Milled	Milled
% passing 20 μm	25–35	40–50	52–62
% passing 45 μm	61–69	73–81	85–93
% passing 75 μm	90–95	94–98	97–100
Silicon, %	14–16	14–16	14–16
Carbon, %	1.3 maximum	1.3 maximum	1.3 maximum
Iron, %	80 minimum	80 minimum	80 minimum
Sulfur, %	0.05 maximum	0.05 maximum	0.05 maximum
Phosphorus, %	0.15 maximum	0.15 maximum	0.15 maximum
Rust index, %	1.2 maximum	1.2 maximum	1.2 maximum
% Nonmagnetics (Davis tube)	0.75 maximum	0.75 maximum	0.75 maximum
Density, sg	6.7–7.1	6.7–7.1	6.7–7.1

Data from DMS Powders (Pty) Ltd.

employed in a reconcentration duty, medium/ore (M/O) ratios are typically two to three times higher relative to a primary separation. The FeSi grade listed is described further in Table 5. The capacity of large installations (higher than 200 t/h feed capacity) is provided by parallel streams.

Media Type and Specification

FeSi medium for diamond recovery is usually milled with a 14%–15% silicon content. Atomized FeSi can be used, but it is more expensive than milled and the separation densities required for diamond recovery do not warrant the lower slurry viscosity of the atomized product. Atomized FeSi is used for higher-density separations (>3.6 sg) for iron ore manganese and other high-specific-gravity minerals. The FeSi designation or grade is based on the particle size distribution. Generally, coarser grades are slightly cheaper and also result in lower medium losses, as finer particles are less efficiently recovered in screen washing and are less efficiently recovered by magnetic separators.

The three grades represented in Table 5, which refer to DMS Powders products manufactured in South Africa, cover at least 90% of FeSi used in diamond heavy medium processes.

Media Recovery Systems and Losses

Diamond recovery heavy medium cyclone floats and sinks products are processed in the same way as other heavy medium separation commodities, with the exception that the sinks product represents a very small fraction of the feed (usually <1% by mass) but contains at least 99.5% of the value of any diamonds present in the feed. Physical barriers protect the sinks product from unauthorized access, and strict security procedures and remote monitoring protect personnel from unwarranted suspicion or accusations.

In a well-run installation, the total of process plus corrosion losses should be in the range of 150–300 g FeSi per metric ton of feed solids. Smaller installations, where compromises in equipment sizing have been made to reduce capital cost, may incur losses as high as 300–500 g/t of feed solids.

Performance Monitoring

The density of separation is the key control variable to maintain plant performance. The circulating medium density should be monitored continuously by density gauges. Other common online performance measurements are the cyclone feed pressure and, naturally, the rate of feed of fresh gravel to the cyclone(s).

Regular testing of the separation efficiency is undertaken using colored beads (tracers) of closely controlled density and different sizes representing the gravel feed size range. In diamond recovery plants, the principal purpose is to confirm that diamonds are not being lost. Introducing tracers of different sizes but the same density as diamond (3.53 sg), tests can be performed integral with the gravel feed, as the volume of sinks is usually very small (<1% of cyclone feed), and the tracers hence are recovered relatively easily by placing a mesh basket over the end of the sinks screen. The expectation is that 100% of all such tracers larger than 2 mm will be recovered, with at least 95% of 2-mm tracers recovered.

Attempts have been made over many years to develop an automated system of detecting and recovering tracers, but these have not found widespread use to date. Early testing of irradiated diamonds (irradiation turns them green) has given way to potentially promising technologies such as the use of tracers containing radio-frequency identification (RFID) tags. However, the smallest size at which RFID tags are currently practical and detectable (perhaps 10 mm) is too large for measuring cyclone DMS efficiency. The presence of water on the products and rinsing screens is also a significant barrier to this technology, as water masks the radio signal.

FINAL DIAMOND RECOVERY

The final diamond recovery process usually involves the following key steps:

1. Bulk mass reduction using wet or dry high-intensity magnetic separation

2. Primary (bulk) X-ray recovery
3. Single-particle X-ray recovery
4. Grease scavenging/auditing
5. Final cleaning

Magnetic Separation

With few exceptions, diamonds are nonmagnetic. The heavy minerals that accompany the diamonds from the previous gravity concentration process (usually DMS) may have paramagnetic properties that provide a mechanism to greatly increase the concentration of diamonds in a low-cost dry process.

Although not essential in the coarser size fractions (>8 mm), the feed is usually dried because this eliminates drag and surface tension effects from entrained water. The feed is also separated into narrow size ranges (maximum 3:1 ratio) to allow the separator operating parameters to be optimized.

Either rare-earth rotating magnetic drum separators are used or a very thin continuous Kevlar belt is used by passing it over a head pulley constructed of an array of high-intensity magnets (e.g., MagRoll, Permroll). In either case, diamonds are unaffected by the magnetic field and are projected into a forward trajectory by the speed of the drum or belt. The magnetic minerals are attracted to the magnet array and follow a downward trajectory when out of the influence of the magnet. A simple splitter arrangement is used to direct the diamonds and heavy minerals into different bins. However, as the high-strength magnets are also high-gradient, this means that the magnetic field falls away very fast with distance from the magnet surface. The farther away the center of mass of a "magnetic" particle is from the surface of the magnet, the less attractive force is experienced, and only a weak change of trajectory is achieved on leaving the device. A mass rejection of 70%–90% is not uncommon in the <4-mm size fraction, falling off as the particle size increases. Design and control of typical units is described by Gehauf (2004).

X-Ray Sorting

It is in the final recovery section of a modern diamond processing plant that technology developments associated with sensor-based sorting have had a major impact. The fact that almost all diamonds (except most Type II) fluoresce under X-ray stimulation has been exploited since the 1970s, with steady advances in sorter sensitivity and discrimination. Much faster signal processing and intelligent algorithms, along with developments such as dual-wavelength sorters, have greatly improved the ability to recover smaller diamonds without unacceptable contamination from other minerals through poor discrimination or electronic noise. X-ray fluorescence (XRF) sorters have predominantly been confined to operation on DMS concentrates. However, many Russian plants use sorters in preference to DMS and up to very coarse sizes. DMS is a bulk separation process where particles follow a stream through the separation vessel determined by their specific gravity, whereas XRF is a single-particle discrimination process where each diamond must be identified and captured individually. For the sorter optical sensors to "see" the diamond, the stream of feed flowing through the sorter must also be a single-particle layer, so that diamonds are not obscured by nonfluorescing particles. As the particle size becomes smaller and the bed depth thinner, the capacity of the sorter diminishes rapidly and additional sorters are required to handle the throughput required. It is therefore unlikely that XRF will completely replace DMS in new plants, except for very small tonnage operations. Current trends suggest that DMS will remain the choice for sizes smaller than 6 mm, with XRF becoming established for sizes greater than 10 mm. Ore- and mine-specific factors will dictate whether XRF or DMS is the best selection for material in the 6–10 mm range.

A more recent development is the introduction of XRT sorting, which does not rely on detection of emitted light from the diamonds. The diamonds are detected using a dual-energy discrimination method targeting the atomic mass of the constituent carbon. This means that all diamonds are potentially recoverable. Since there is no need for the sensor to directly "see" the diamond, higher capacities are possible, as a monolayer is not essential, although bed-depths must still be controlled. An extension of this is the potential ability to detect diamonds that are unliberated (Riedel and Dehler 2010). Currently, however, signal processing limits the bottom size of diamonds recoverable by XRT to about 6 mm. Although this appears to be a retrograde step compared to XRF sorters that routinely process material down to 1 mm, the Type II diamonds that XRT sorting can also recover represent more typically the largest and most valuable diamonds in the deposit, which will hence be more reliably recovered (along with essentially all other diamonds in the same size range). Also of interest is the fact that XRT can detect diamonds that are still partially or fully encapsulated in host kimberlite particles, with the ability to detect a 10-mm diamond within a 30-mm kimberlite particle shown to be feasible.

Although sensor-based sorters using CCD color cameras (based on reflection and transparency) and photometric sorters (based on refection/absorption) have been used in diamond recovery, they have not found widespread use.

A final step in the application of sensor-based sorting is the use of laser-based sorters using the raman-shift phenomenon produced by the diamond crystal lattice to identify diamonds. This is a particularly diamond-specific technology, and it can be used as a final "confirmation" to provide certainty of final product integrity. However, the signal strength produced by the raman shift from a small diamond is very weak and requires long signal acquisition times, limiting throughput. As a defalsifying application, laser-raman sorters are also true single-particle sorters, again resulting in very low throughput rates (of the order of 250 g/h for 1–2 mm size diamonds)

Grease Scavenging and Auditing

The majority of diamonds exhibit natural hydrophobic behavior (not wettable) in a flowing stream of water), which has been exploited to recover them for more than 100 years. The diamond surface would rather adhere to grease than stay wetted in water. Most of the other minerals in diamond concentrates are wettable. This means that grease capture is a very highly specific method for recovering free diamonds because very little waste material is captured along with the diamonds.

Washed and sized gravel containing liberated diamonds is flushed in a controlled manner over a screen, continuous moving belt (rubber or plastic), or a rotating drum. The separation surface is coated with a thin film of petroleum grease. Although the vast majority of the material is flushed over the grease and discarded, most diamonds stick to the grease. The loaded grease is removed either on a batch basis by hand or, in more modern equipment, automatically. The grease is melted, and the trapped diamonds are released for cleaning in a solvent or detergent. The grease is reapplied to the separating device surface for reuse. Grease is specifically formulated for this application and is commercially available for diamond recovery use from major producers such as Shell and Mobil. It is a blend containing sufficient wax to provide a surface that is stiff enough to avoid penetration by wetted particles, but "sticky" enough to avoid adhered diamonds from being flushed away in the flowing stream of water and waste particles. Since these properties are sensitive to temperature, the temperature of the water flowing over the grease needs to be maintained within limits, usually around 25°C.

Diamond recovery using grease is, in its simplest form, a very low-cost method. However, there are drawbacks to relying purely on grease as the final recovery method of choice:

- Not all diamonds will stick to grease. The reason is usually that they have become coated over time with a mineralized layer. This can sometimes be removed using attritioners, with or without conditioning chemicals.
- Separating very small diamonds from the grease can be a lengthy process because of the high viscosity of the molten grease.
- From security principles, it is considered undesirable in most major installations to allow direct contact between personnel and diamonds, which implies it is necessary to automate the whole grease process.
- Automated grease belts and drums are relatively complex to control and maintain to a high level of performance that ensures continuous and reliable recovery of degreased diamonds, and hence are no longer as cheap and simple as the concept implies. Capital costs can be the same as for an XRF sorter treating the same magnitude feed stream.
- Failures of the system, apart from resulting in spillage of diamond-bearing material, usually result in contamination of peripheral equipment. Surrounding structures can become coated with grease over time if grease is overheated during the melting process.

However, grease is still considered to have a legitimate role as a scavenging, or auditing, process following removal of essentially all the diamonds using XRF or XRT sorters.

Diamond Cleaning

An important part of ensuring the integrity of the diamond value management chain is in ensuring that what is called a diamond product is 100% diamond and nothing else. This means that diamonds must be cleaned, preferably at the earliest opportunity, and certainly before they are certified to be "diamond." Along with the removal of any remaining discrete particles of non-diamond origin, it is also necessary to ensure that no coatings remain on the diamonds. Several processes are employed, most commonly caustic fusion followed by a "deep-boil" in a combination of strong acids. Although time-consuming, the latter has the advantage of potentially removing a very thin outer skin from the diamonds, which represents the "interface" between the diamond and the environment over many millions of years. Removing this "damaged" layer can result in a net higher value being attributed to the diamond, despite the small loss in mass.

REFERENCES

Daniel, M.J., Bellingan, P., and Rauscher, M. 2016. The modelling of scrubbers and AG mills in the diamond industry and when to use them. Presented at the SAIMM Diamonds Still Sparkling Conference, Gaborone, Botswana.

Fyfe, G. 2005. The High-Pressure Roll Crusher: Unlocking the potential of the modern diamond plant. Presented at the Diamonds Source to Use Conference, SAIMM, Johannesburg, South Africa.

Gehauf, R. 2004. A practical guide on selecting and optimizing rare earth magnetic separators. SME Preprint No. 04-47. Littleton, CO: SME.

Kirkley, M.B., Gurney, J.J., and Levinson, A.A. 1991. Age, origin and emplacement of diamonds: A review of scientific advances in the last decade. *CIM Bull.* 85(956):48–57.

Klymowsky, R.I.B. n.d. *High Pressure Grinding Rolls for Minerals*. Beckum, Germany: Polysius AG, a company of ThyssenKrupp Technologies. p. 3.

KPCS (Kimberley Process Certification Scheme). n.d. Public statistics area. https://Kimberleyprocessstatistics.org/public_statistics. Accessed March 2018.

Metso Minerals. 2009. *Barmac VSI Crushers: Barmac B-Series VSI* (brochure). www.deltagroup.com.eg/360download.php?file_name=pdf/metso/crushing/barmac.pdf. Accessed June 2018.

Popplewell, G. 2007. Small diamonds, engineering at the margin. Presented at the Diamonds Source to Use Conference, SAIMM, Johannesburg, South Africa.

Rider, P.J., and Roodt, A. 2003. Diamond value management: Knowledge management and the measurement of value addition. *J. S. Afr. Inst. Min.* Metall. 103(9):551.

Riedel F., and Dehler, M. 2010. Recovery of unliberated diamonds by X-ray transmission sorting. Presented at the Diamonds Source to Use Conference, SAIMM, Gaborone, Botswana.

Russell, J.K., Porritt L.A., Lavallee Y., and Dingwell, D.B. 2012. Kimberlite ascent by assimilation-fueled buoyancy. *Nature* 481(7381):352.

Van Niekerk, L.M., Olivier, A., Armstrong, J., and Sikwa, N.A. 2016. Pioneering large diamond recovery at Karowe diamond mine. Presented at the SAIMM Diamonds Still Sparkling Conference, Gaborone, Botswana.

CHAPTER 12.11

Fluorspar

Peter Huxtable

Fluorspar is the name for the naturally occurring mineral fluorite (calcium fluoride, CaF_2). It is the predominant commercial source of fluorine, a nonmetallic element and the lightest of the halogens, and provides 93% of the global demand for fluorine.

Another significant but much smaller (7%) source of fluorine is from phosphate rock (fluorapatite, $Ca_5(PO_4)F$) as fluorosilicic acid (FSA). This is primarily used to produce some (15%) of the alumimun fluoride (AlF_3) used as the flux for aluminum smelting, and in water fluoridation and toothpaste manufacture.

Cryolite (Na_3AlF_6) is a minor source of fluorine found in Greenland, Europe, but minable reserves were depleted by 1962. Other fluorine-bearing minerals are of relatively little significance and include sellaite (MgF_2), topaz ($Al_2SiO_4(FOH)_2$), villiaumite (NaF), and a rare earth bastnaesite ((Ce,La)(CO_3)F).

Georgius Agricola described the use of fluorspar as a metallurgical flux in 1556. Fluorspar mining started in England around 1775 and in the United States between 1820 and 1840, and industrial demand increased for fluorspar as a fluxing agent in the late 1880s as the steel industry expanded. In 1886, the development of the electrolytic process for aluminum production led to a further increase in demand.

In the 1930s, fluorocarbons were developed using hydrofluoric acid (HF) and were used as refrigerants in household and commercial cooling equipment. This application was greatly expanded in the 1970s and 1980s, and fluorocarbons' use as aerosol propellants was also developed. Concerns about the apparent depletion of the stratospheric ozone layer pointed to the chlorine in chlorofluorocarbons as a primary suspect and led to a phaseout of their use through the international Montreal Protocol treaty of 1987 and subsequent amendments (UNEP 2017). More ozone-friendly replacements—hydrochlorofluorocarbons and hydrofluorocarbons (HFCs)—were developed that contained less chlorine and required more fluorspar to manufacture, although not-in-kind replacements (mainly the more flammable hydrocarbons) replaced all, except for some specialty medical, aerosol applications.

The United Nations Framework Convention on Climate Change in Kyoto (UN 1998) imposed further restrictions on HFCs, perfluorocarbons, and sulphur hexafluoride in 1997 because of their global warming potential (GWP) impact—along with carbon dioxide, methane, and nitrous oxides. These restrictions have become increasingly stringent; the European Union (EU) 2014 F-Gas regulations (EU 2014) became effective in January 2015, and the Significant New Alternatives Policy (SNAP) program from the U.S. Environmental Protection Agency (EPA) became effective in 2016 (EPA 2015). This has led to further developments of almost zero GWP hydrofluoroolefins and low-GWP HFC blends.

The trend in annual world production output since 1913 (Figure 1) shows an average production of 6 Mt from 2009 through 2016.

GEOLOGY, MINERALOGY, AND MINING

Fluorine has an estimated crystal abundance of 0.065%, and among the elements it ranks 13th in order of abundance.

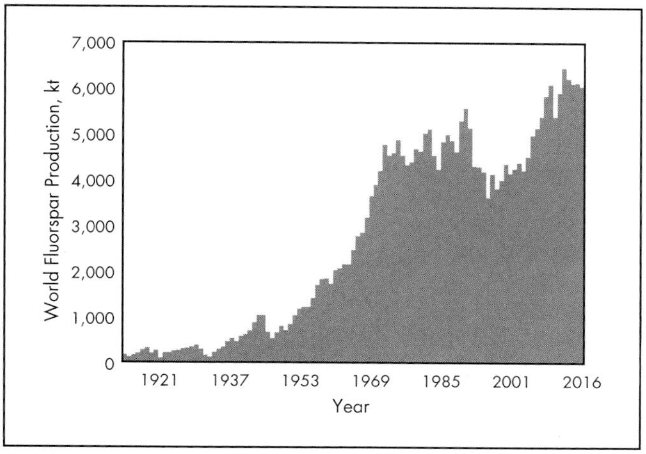

Data from USGS 2016; Roskill 2013, 2015–2016, 2017

Figure 1 World fluorspar production output 1913–2016

Peter Huxtable, Principal, Huxtable Associates, Calver, Derbyshire, United Kingdom

In its pure form, fluorspar contains 51.1% calcium and 48.9% fluorine, and has a specific gravity of 3.18 and a hardness of 4 on the Mohs scale. Pure fluorite melts at 1,378°C and is relatively insoluble in water (0.146 g/L at 25°C). Crystalline fluorspar has a very low index of refraction (1.433–1.435) and low dispersion. It is isotropic and has the unusual ability to transmit ultraviolet light. This makes fluorspar useful in optical systems and high-quality, special-purpose lenses.

Fluorite belongs to the cubic system mineralogically and tends to occur in well-formed isometric crystals, forming cubes and octahedrons. It also occurs in both massive and earthy forms, and as crusts or globular aggregates with radial fibrous texture. Crystalline fluorspar exhibits a great range of colors from colorless to yellow, blue, purple, green, rose, red, bluish and purplish black, and brown.

The natural mineral is commonly associated with other minerals, such as quartz, calcite, barites, galena, siderite, celestite, sphalerite, chalcopyrite, sulfide, or phosphate minerals.

The U.S. Geological Survey (USGS) estimates world reserves in 2017 at 260 Mt (100% CaF_2 basis) with 41 Mt in South Africa, 32 Mt in Mexico, 40 Mt in China, and 22 Mt in Mongolia (USGS 2017). Identified world resources are approximately 500 Mt. In addition, large quantities of fluorine are present in phosphate rock worldwide, estimated at the equivalent of 4.9 billion tons 100% fluorspar equivalent (USGS 2017).

Fluorite occurs in a wide variety of geological environments, with deposition under an extended range of physical and chemical conditions. The deposits occur worldwide, and the average ore grade of sources mined ranges from a low of 8% CaF_2 (Vietnam) to 85% CaF_2 (Mexico). China contributed 65% of world output in 2016 with another 23% from Mexico, Mongolia, South Africa, and Vietnam. The most important modes of fluorite occurrence from an economic view are as follows:

1. Fissure veins commonly occur along faults or shear zones and are found in igneous, metamorphic, and sedimentary rocks. Mineralization occurs as lenses or ore shoots separated by barren zones. Ore shoot widths of 0.5–10 m and lengths of 60–300 m are common, and CaF_2 content typically ranges from 25% to 80%. Vein systems may be several kilometers long and ore presents to depths of 300 m or more below the surface. Important deposits include China, northeast Spain (Osor), Morocco (El Hammam), United States (Rosiclare, Illinois), United Kingdom (Derbyshire), and Italy (Sardinia).
2. Stratiform, manto, or bedded deposits occur in carbonate rocks. Certain beds are replaced along or adjacent to structural breaks such as joints or faults, frequently with sandstone, shale, or clay capping. Well-developed deposits are found in Mexico (northern Coahuila) and South Africa (Zeerust). CaF_2 content ranges from 11% upward.
3. Replacement deposits in carbonate rocks are well developed along contacts with intrusive igneous rocks and include some of the largest and highest grade fluorspar deposits, including those in Mexico (Rio Verde, San Luis Potosi).
4. Stockworks and fillings in shear and breccia zones are generally lower grade, such as those in South Africa (Buffalo); often they are not economic.
5. Fluorspar is a common mineral in *alkaline rock* although not often sufficiently abundant to be economic. Examples can be found in Namibia (Okorusu) and India (Amba Dongar).
6. Concentrations of fluorspar resulting from the weathering of primary vein and replacement deposits are generally sources of metallurgical-grade material. These *residual deposits* can extend to depths of 30 m or more. Examples can be found in Spain (Asturias), Thailand, South Africa (Marico), the United States, and the United Kingdom.
7. Fluorspar occurs in lead–zinc veins in many parts of the world. Economically viable *gangue mineral* deposits average 10%–20% CaF_2.

Table 1 World consumption of fluorspar (kt) by application and grade, 2013–2016

Industry	Acidspar	Metspar	Total
Chemicals	2,300	–	2,300
Steel	–	2,100	2,100
Aluminum	1,400	–	1,400
Other	–	400	400
Total	**3,700**	**2,500**	**6,200**

Data from Roskill 2013, 2015–2016, 2017; Corathers and Machamer 2006

Fluorspar is mined by both open-pit surface and underground methods depending on the geology and location of the deposit. Surface or near-surface deposits are mined using standard open-pit methods ranging from small-scale operations at some of the smaller Chinese mines, to draglines and scrapers, to power shovels. Open-pit methods are employed in China, Kenya, South Africa, Thailand, and Vietnam.

Deeper deposits are extracted using underground techniques by deep-shaft or adit access using rail, haul trucks, or conveyors. Techniques are typically room-and-pillar for bedded deposits, and shrinkage or open stoping for deeper vein deposits.

USES, MARKETING, AND GEOGRAPHIC SOURCES

There are two grades of commercial fluorspar. One is a predominantly metallurgical-grade gravel material (*metspar*) that is typically 85%–92% CaF_2, 5–20 mm in size, and produced by simple crushing and gravity separation. This material is primarily used as a flux in steelmaking, and in smaller quantities in cement clinker and the glass industry.

A higher (>97%) CaF_2 acid-grade product (*acidspar*) is produced by further processing, fine grinding, and froth flotation. This material is used in the chemical industry where the starting point is the manufacture of HF. It is also used as a flux in the production of aluminum after being used to produce AlF_3.

These two grades conform to the international harmonized tariff codes 252922 (acidspar) and 252921 (metspar).

Typical consumption of fluorspar worldwide from 2013–2016 by grade and application is shown in Table 1. The chemical industry is the largest consumer, using 37% of the total, followed by the steel industry with 34% and the aluminum industry with 23%.

Refrigeration and air-conditioning account for around 50% of fluorocarbon production, followed by foam-blowing agents that improve the insulation properties of plastics. These are non-feedstock applications and are subject to environmental controls under the Montreal Protocol (UNEP 2017) and GWP international treaties (UN 1998; EU 2014; EPA 2015). The development of fluorine-based replacements has enabled

Table 2 World production of fluorspar (kt) by country and grade, 2016

Country	Acidspar	Metspar	Total	% Total
China	2,200	1,500	3,700	65.0
Mexico	430	230	660	10.5
Mongolia	10	270	280	5.0
South Africa	220	10	230	4.0
Vietnam	200	—	200	3.5
Spain	110	10	120	2.0
Other	290	270	560	10.0
Totals	**3,460**	**2,290**	**5,750**	**100.0**

Data from Roskill 2013, 2015–2016, 2017

this market to continue to be supplied without reductions in acidspar requirements. Polymer precursors are feedstock fluorocarbons that are not affected by these environmental considerations, and they are the starting blocks for a range of fluoroelastomers and fluoropolymers with outstanding chemical and thermal stability, and are electrically inert and nonflammable. This is a fast-growing area in which polytetrafluoroethylene (PTFE) accounts for 60% of the market. (PTFE includes the nonstick surface coatings on cookware.) Polyvinylidene fluoride is used to manufacture lithium-ion batteries for mobile phones, laptops, and hybrid and electric cars. Diverse industrial applications include electronics, crystal glass, uranium manufacture, and herbicides and detergent manufacture.

The use of fluorspar to produce AlF_3 as a flux for aluminum manufacture is also rapidly increasing because of the high growth rate of aluminum production worldwide. Although about 15% of AlF_3 is produced from by-product FSA, this is not a serious contender for further market erosion of the use of natural fluorspar.

The principal production of fluorspar by leading countries and grade is shown in Table 2. China, North America, and Russia/Commonwealth of Independent States are self-sufficient for their fluorspar needs. Europe has a net import requirement of 0.5 t per year.

Fluorspar prices are not internationally quoted because little material is freely traded and most is either used in-house or sold on long-term supply arrangements and contracts. *Industrial Minerals* magazine quotes typical price ranges of both acidspar and metspar by principal exporter and by main end-use importer (*Industrial Minerals* 2017). Price data are shown in Figure 2 for acidspar wet filtercake exports free on board (fob) China, which is the predominant benchmark for price indications, and for metspar fob Mexico as the main world exporter of the product.

The EU has recognized fluorspar as a critical raw material as a result of its 2008 Raw Materials Initiative (European Commission 2008) and as shown in its latest 2017 update (European Commission 2017). It has also been identified on the British Geological Survey Risk List 2015 (BGS 2015).

PROCESSING

All processing is by physical beneficiation whereby unwanted waste rock (gangue) is separated from the host fluorite by a series of standard techniques tailored to the character of the mined ore feed and the specifications for end use. For acidspar, fine grinding and froth flotation are required to produce the >97% CaF_2 specification material, which adds considerably to the overall processing costs because of high electric power costs and the use of flotation reagents.

Metspar processing in less-developed countries can be performed by hand-sorting of high-grade lump ore followed by crushing and screening to remove most of the fines. For lower-grade ores or where there is relatively coarse interlocking of minerals, gravity separation using jigs, tables, or heavy media separation is used.

Heavy media cone and drum separators are particularly effective in the size range of 0.5–3.8 cm either for metspar production or to preconcentrate the crude ore for flotation feed. For finer sizes, a heavy media cyclone or a Dynawhirlpool (DWP) is frequently used and the sinks can be sold as metspar, the sands for flotation feed, and the floats for road gravel or concrete aggregate. Ores as low as 8%–10% CaF_2 can be processed this way to produce a flotation feed of 40% CaF_2 or more. Ferrosilicon and magnetite are used as the heavy media solution and cleaned off the surfaces of products with high-pressure water, and recycled through magnetic separators.

Washing plants are also commonly used ahead of flotation to remove any fine clays. Water supply to the plant is often recycled from the process tailings with fresh water input only to the flotation section.

Froth flotation takes advantage of the varying surface properties of different materials enabling separation of valuable product from gangue tailings in flotation cells. Fine grinding in rod or ball mills enables liberation of the fluorite from other minerals present and the host silica or limestone. The circuit is chosen to minimize the generation of extremely fine sizes that are not amenable to efficient flotation recovery.

The flotation feed at suitable pulp density is fed to conditioning tanks where suitable reagents are added and the pH modified with soda ash for optimal recovery. In some operations, the pulp is heated to greater than ambient temperature to improve recovery rates, although the additional cost is not often justified by the potential extra recovery. The pulp is agitated in the presence of injected air and a frother reagent, which creates minute air bubbles that attach preferentially to the desired mineral that has been coated with a collector, a suitable flotation reagent, and is then floated off as a product in a froth. Any metallic sulfides (lead or zinc) in the host ore are removed first using a xanthate or dithiophosphate collector in a circuit before the fluorspar extraction. After rougher

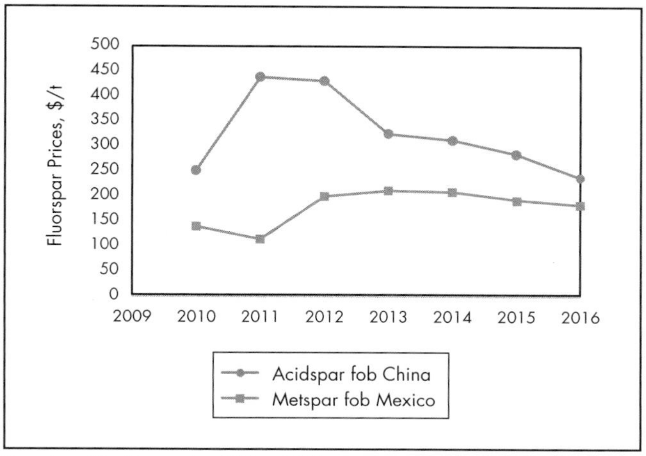

Data from UN Comtrade 2017a, 2017b

Figure 2 Fluorspar world prices, $/t fob source port

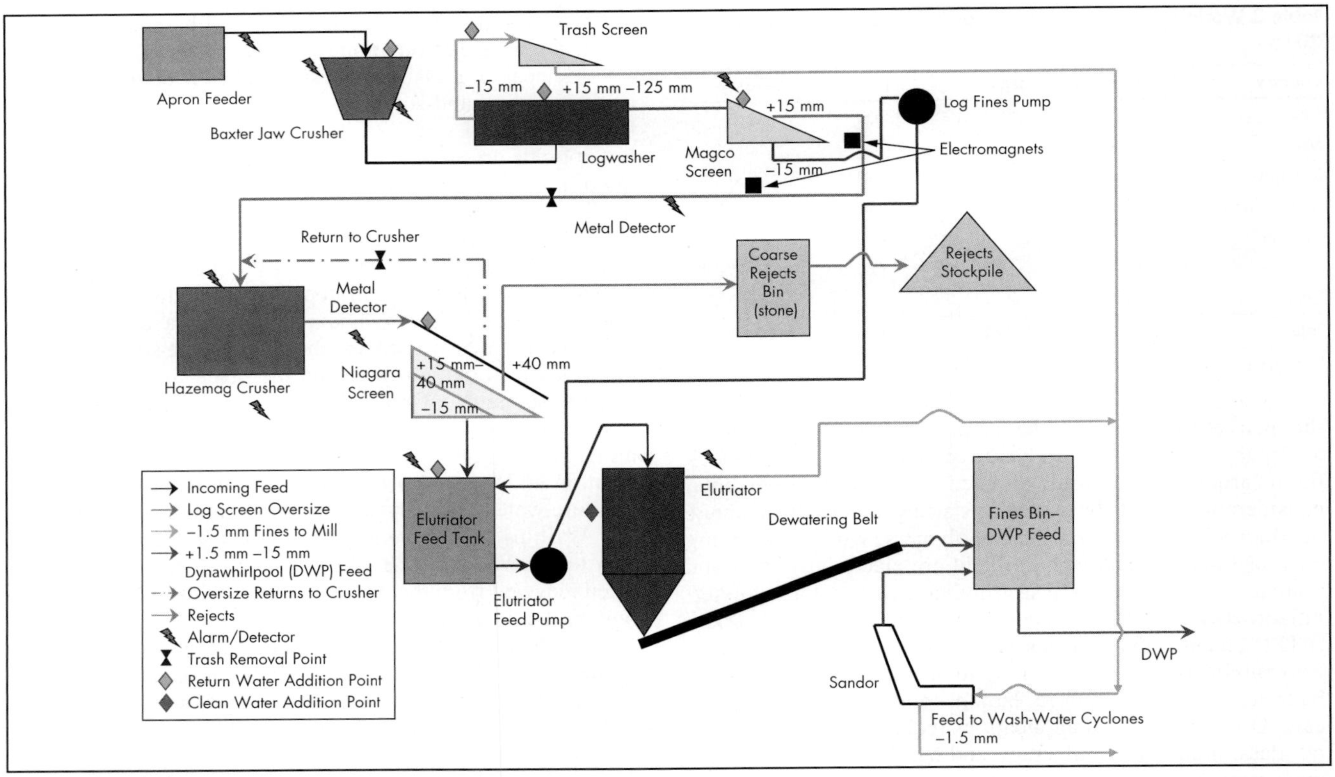

Courtesy of British Fluorspar

Figure 3 Cavendish Mill crushing and washing circuit

extraction to maximize recovery, the product goes to cleaner cells to maximize the finished grade of concentrate.

Fluorspar is then extracted in rougher cells using suitable collectors, normally fatty acids (such as oleic), along with quebracho or tannin to depress calcite or dolomite; sodium silicate, starch, or dextrine to depress silicates or any iron; and dichromates to depress any barites in the ore. The rougher concentrate then proceeds to a cleaner circuit, which may include up to six circuits to ensure >97% CaF_2 product and often includes a small regrind mill to liberate interlocking grains of fluorspar and gangue and to maximize recovery—and minimize the levels of calcium carbonate and silica where product specifications typically require less than 1.5% and 1%, respectively. The final concentrate slurry with a pulp density of around 25% solids goes to a thickener where the density is raised to about 55% prior to filtration to produce a filter cake with a moisture below 10%. This material is either shipped directly to the customer or dried first prior to shipment in bulk powder tankers or in 25-kg or 1-t big bags for use by smaller-volume end users.

Tailings disposal from the flotation process is either (1) to a tailings lagoon, direct or via a thickener, or (2) using a filter press or belt filter. Residual waters will recirculate for use in the washing circuits.

The processing of the Cavendish Mill, Glebe Mines, Derbyshire operations (currently owned by British Fluorspar, Ltd., part of the Fluorsid Group) were described in some detail along with process flow sheets at a symposium (Huxtable 1989). The current circuit has evolved, and those changes are shown in Figures 3 through 7.

REFERENCES

Agricola, G. 1556. *De Re Metallica*. Translated by Herbert C. Hoover and Lou Henry Hoover. 1950 edition. New York: Dover Publications.

BGS (British Geological Survey). 2015. Risk List 2015: Current supply risk for chemical elements or element groups which are of economic value. Keyworth, UK: National Environment Research Council. www.bgs.ac.uk/downloads/start.cfm?id=3075.

Corathers, L.A., and Machamer, J.F. 2006. Fluorspar. In *Industrial Minerals and Rocks*, 7th ed., Vol. 2. Edited by J.E. Kogel, N.C. Trivedi, J.M. Barker, and S.T. Krukowski. Littleton, CO: SME.

EPA (U.S. Environmental Protection Agency). 2015. Significant New Alternatives Policy (SNAP): SNAP Regulations. https://www.epa.gov/snap/snap-regulations. Accessed August 2017.

European Commission. 2008. The Raw Materials Initiative—Meeting our critical needs for growth and jobs in Europe. http://ec.europa.eu/growth/sectors/raw-materials/policy-strategy/index_en.htm. Accessed August 2017.

EU (European Union). 2014. Regulation (EU) No. 517/2014 of the European Parliament and of the Council of 16 April 2014 on fluorinated greenhouse gases and repealing Regulation (EC) No. 842/2006 Text with EEA relevance. http://eur-lex.europa.eu/legal-content/EN/ALL/?uri=uriserv:OJ.L_.2014.150.01.0195.01.ENG. Accessed August 2017.

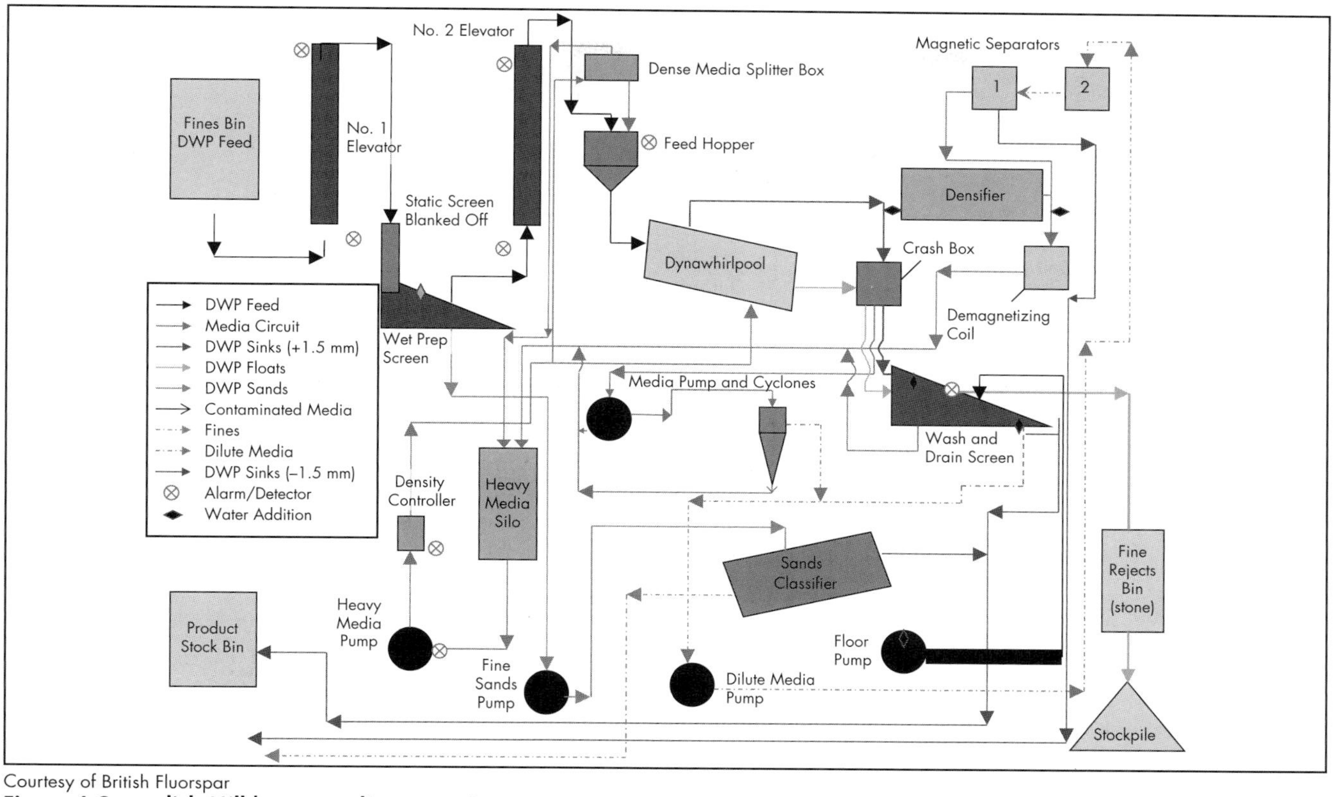

Courtesy of British Fluorspar

Figure 4 Cavendish Mill heavy media separation

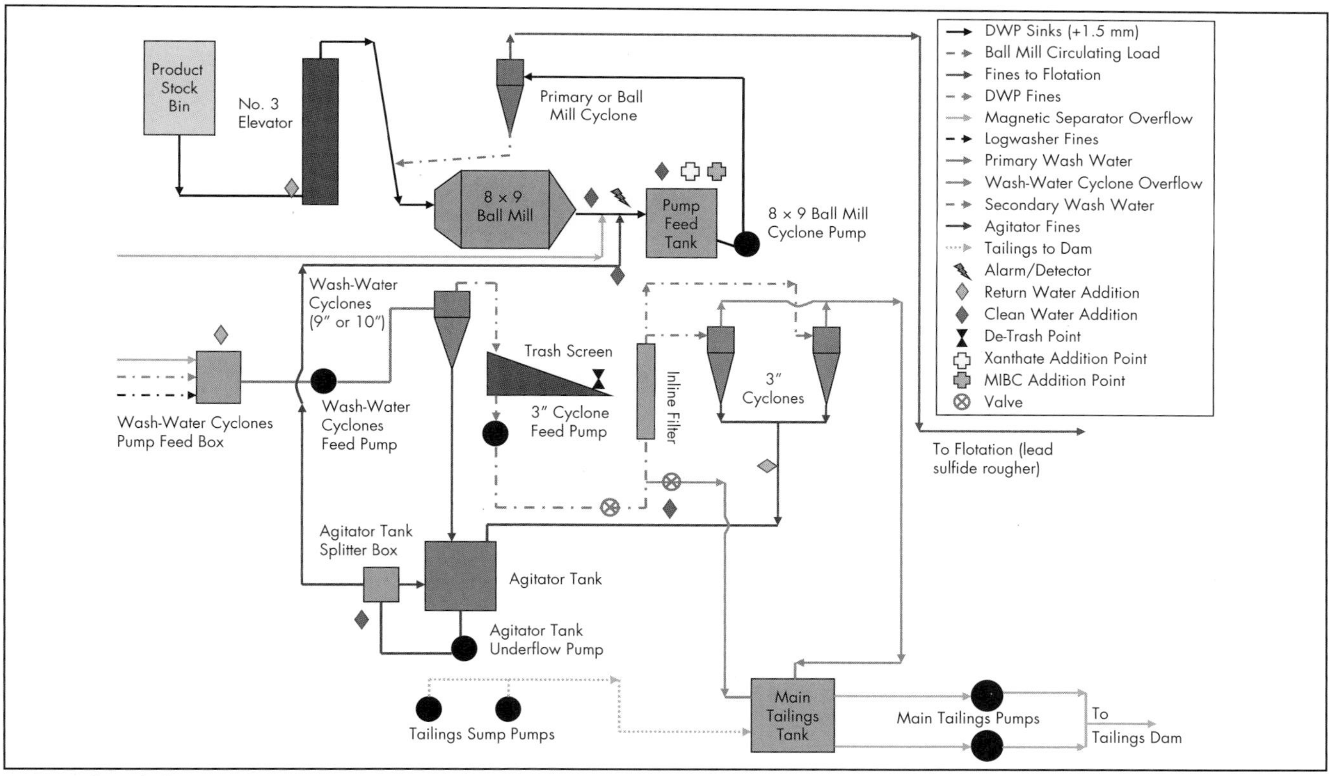

Courtesy of British Fluorspar

Figure 5 Cavendish Mill grinding circuit

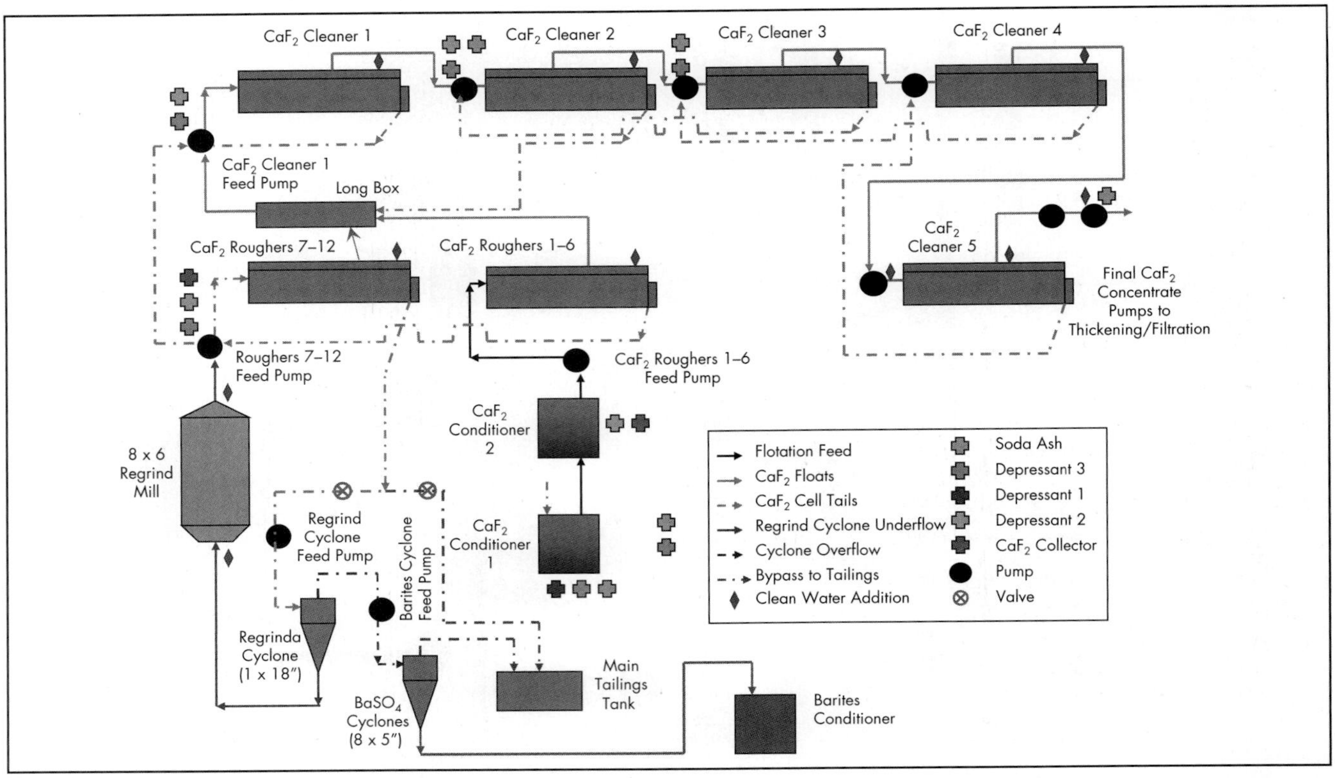

Courtesy of British Fluorspar

Figure 6 Cavendish Mill fluorspar flotation°

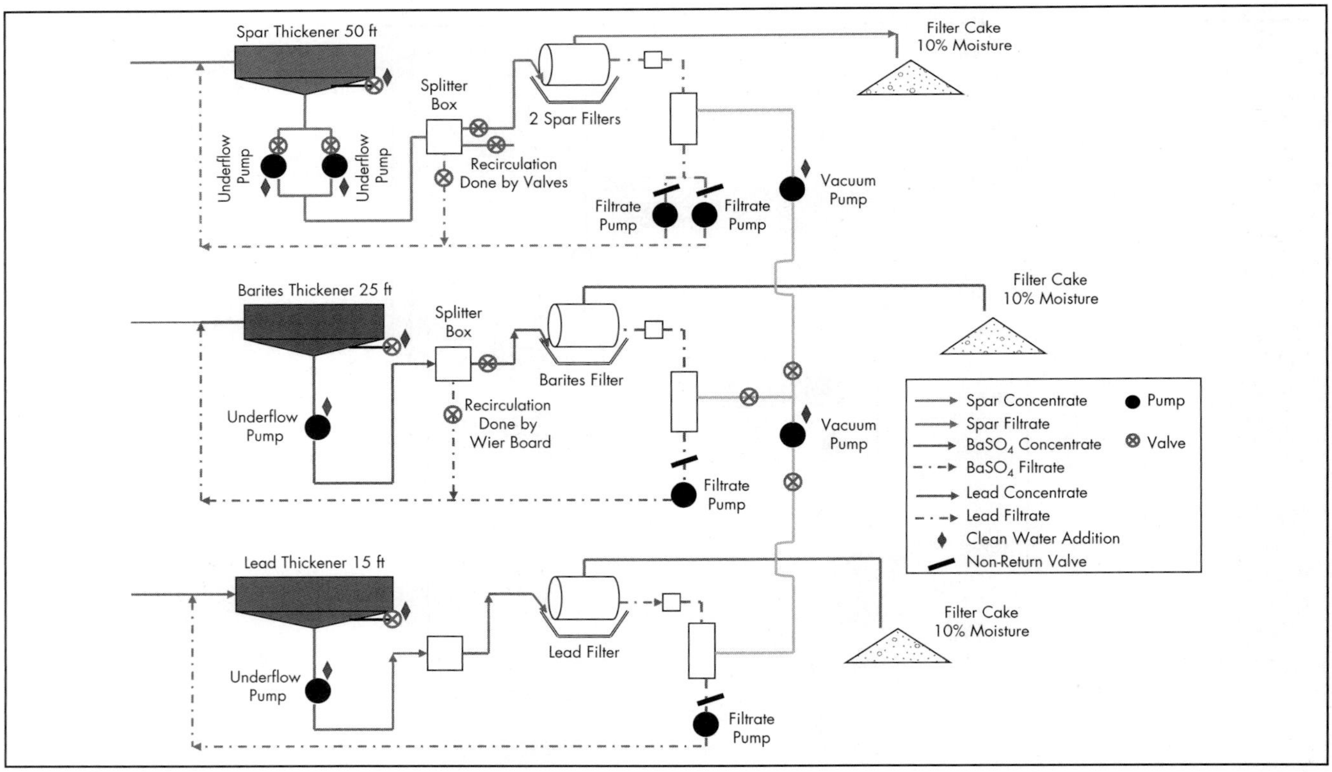

Courtesy of British Fluorspar

Figure 7 Cavendish Mill thickening and filtration

European Commission. 2017. Communication from the Commission to the European Parliament, the Council, the European Economic and Social Committee and the Committee of the Regions on the 2017 list of Critical Raw Materials for the EU. http://ec.europa.eu/transparency/regdoc/rep/1/2017/EN/COM-2017-490-F1-EN-MAIN-PART-1.PDF. Accessed October 2017.

Huxtable, P.L. 1989. The one hundred percent solution. In *Proceedings of Mineral Processing in the UK*. London: Institution of Mining and Metallurgy.

Industrial Minerals. 2017. Fluorspar online price information (subscriber only). www.indmin.com/Pricing.html. Accessed August 2017.

Roskill. 2013. *Fluorspar: Global Industry Markets and Outlook*, 11th ed. London: Roskill Information Services Limited.

Roskill. 2015–2016. *Fluorspar: Industry Briefing*. London: Roskill Information Services Limited.

Roskill. 2017. *Fluorspar: Global Industry Markets and Outlook to 2021*, 12th ed. London: Roskill Information Services Limited.

UN (United Nations). 1998. Kyoto Protocol to the United Nations Framework Convention on Climate Change. http://unfccc.int/resource/docs/convkp/kpeng.pdf. Accessed August 2017.

UN Comtrade (United Nations Commodity Trade Statistics Database). 2017a. Acid-grade fluorspar. https://comtrade.un.org/db/mr/daCommoditiesResults.aspx?px=H3&cc=252922. Accessed August 2017.

UN Comtrade (United Nations Commodity Trade Statistics Database). 2017b. Metspar. https://comtrade.un.org/db/mr/daCommoditiesResults.aspx?px=H3&cc=252921. Accessed August 2017.

UNEP (United Nations Environment Programme). 2017. *The Montreal Protocol on Substances That Deplete the Ozone Layer,* 10th ed. http://ozone.unep.org/en/treaties-and-decisions/montreal-protocol-substances-deplete-ozone-layer. Accessed August 2017.

USGS (U.S. Geological Survey). 2016. Fluorspar. In *Historical Statistics for Mineral and Material Commodities*. Washington, DC: USGS.

USGS (U.S. Geological Survey). 2017. Fluorspar. In *Mineral Commodity Summaries*. Washington, DC: USGS.

CHAPTER 12.12

Gold and Silver

John O. Marsden

Many of the methods currently used for gold and silver extraction are based on techniques that have been known, or established, for centuries. Gravity concentration, amalgamation, cyanide leaching, roasting, zinc precipitation, and carbon/charcoal adsorption are all processes that have been used for at least 100 years, and combinations of these remain as the basis for most gold and silver recovery flow sheets in the modern era. The ingenuity of the early pioneers in developing technology with such longevity is admirable. However, the past four decades (1973–2017) have witnessed a period of exceptional innovation following the end of the official fixed gold price in the United States. These developments have helped to make gold and silver extraction safer, more efficient, and with reduced impacts and risks to human health and the environment. This chapter discusses the major processes that are used for gold and silver extraction worldwide. Selected case studies are provided to illustrate the major industrial applications.

All data are in metric units unless otherwise stated, and all references to ounces (oz) refer to troy ounces (31.1035 g).

SUPPLY, DEMAND, AND PRICE

Gold Production, Consumption, and Price

Gold, silver, and copper were the first metals used by humans because of their native occurrence, and their inherent malleable and ductile properties, which meant they could be easily worked with primitive tools. Gold is an exceptional conductor of heat and electricity and is resistant to corrosion over a wide range of conditions. These properties, combined with its relative scarcity on earth, have made it a metal of great value. Also, given the high density of gold (19.3 g/cm^3), its intrinsic value in the eyes of humankind, and its resistance to oxidation/degradation under ambient conditions, it has proven to be an extremely effective method of storing wealth, be it as jewelry, coins, bars, or in other forms.

The earliest uses of gold were in the Middle East during the Neolithic period, where gold was collected from streambeds either manually or by crude gravity concentration methods. In Egypt, during the reign of Menes in 3050 BC, gold was used as a means of monetary payment in the form of grains and small bars. Since that time, gold has been used as a medium of exchange worldwide, providing a basis for international trade. However, the major uses for gold are in jewelry, decoration, and as a store of wealth.

A summary of gold demand by application for 2014 is provided in Table 1. Approximately 53% of the physical demand for gold was in jewelry, with China and India consuming much of this. Physical bars and coins as investment vehicles accounted for 26%. A further 11% was acquired by central banks and other institutions, through official purchases. Uses of gold for electronics, dentistry, and other industrial applications consumed the remaining 10%.

Gold production grew slowly through the ages, reaching around 400 t per year by 1900. In 1934, the U.S. government set an official gold price of $35/oz. While production continued to increase, the industry suffered in the 1950s and 1960s as costs increased worldwide with inflation, and consequently profitability decreased. Between 1971 and 1973, the U.S. dollar was devalued, effectively increasing the gold price. In 1973, the official dollar-gold price conversion was terminated, leading to a period of dramatically increased prices, reaching a peak of $850/oz at one point in 1980. These economic conditions improved gold mine profitability and increased exploration and production efforts worldwide. The gold price

Table 1 Gold demand/consumption by application for 2014

Application	Gold, t	Proportion, %
Jewelry	2,213	53.2
Technology		
• Electronics	279	6.7
• Other industrial	87	2.1
• Dentistry	34	0.8
Retail Investment		
• Physical bars	829	19.9
• Coins	251	6.0
Central banks and other institutions	466	11.2
Total physical demand	**4,159**	**100.0**

Data from GFMS, Thomson Reuters 2015

John O. Marsden, President & Consultant, Metallurgium, Phoenix, Arizona, USA

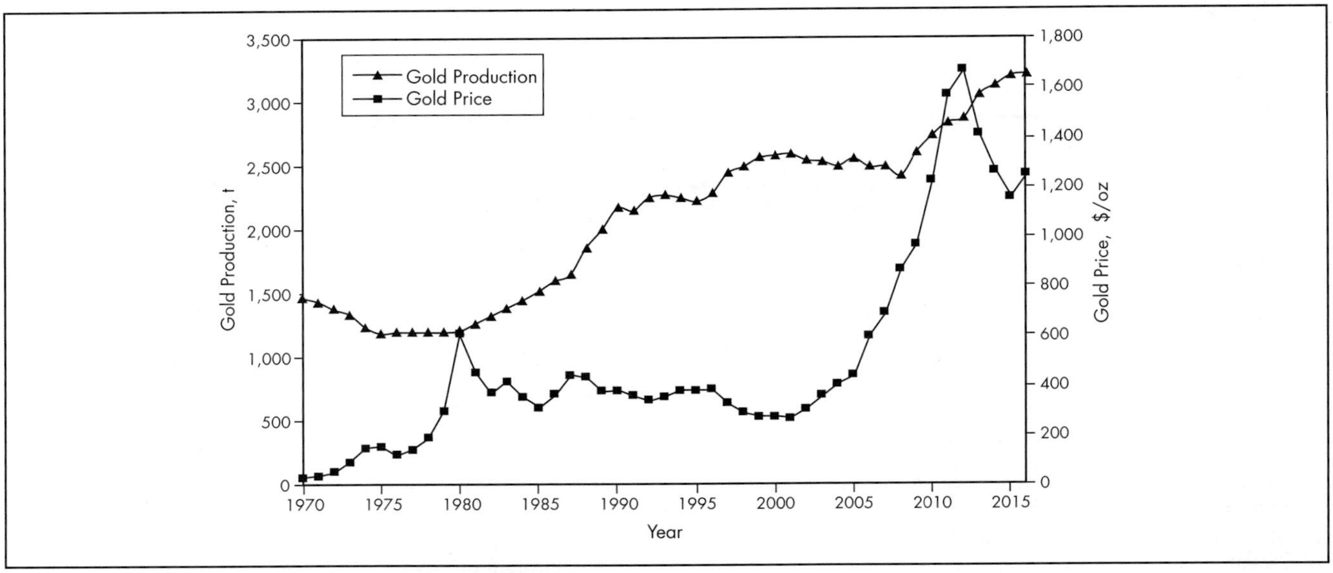

Data from USGS 1970–2016; GFMS, Thomson Reuters 1984–2015
Figure 1 Gold production and price between 1970 and 2015

stagnated between 1984 and 2001 before a strong run-up in price reaching $1,791/oz in 2012.

It is estimated that all of the gold produced by humankind up to 2014 is approximately 184,000 t. Most of this is still in circulation today in one form or another. This amount of gold, if collected in one place, would fill a cube with a side length of just over 21 m, highlighting its relative scarcity. In addition to mine supply (3,133 t in 2014), a significant amount of gold is recycled from scrap (1,125 t in 2014). Mine production and scrap recycling are differentiated for the purposes of assessing supply and demand.

Gold production and price data are shown in Figure 1. Notably, gold production grew at an average annualized rate of 1.7% between 1900 and 1973, while price increased by 2.3% per year. Between 1973 and 2014, gold production grew at 2.1% per year, while price increased by a staggering 6.4%, proving its long-term worth as a hedge against inflation and economic uncertainty.

Gold production by country and region is shown in Table 2. The dominance of South Africa as the world's largest gold producer in the 1970s, 1980s, and 1990s has declined to the point where, in 2014, South Africa was ranked sixth behind China, Australia, Russia, United States, and Peru.

Gold ore grades declined significantly between 2005 and 2014, from approximately 1.85 g/t to 1.30 g/t—a 30% reduction. This is due not only to declining average ore grades of reserves globally but also to the sharp gold price increase between 2004 and 2012, which significantly reduced revenue-based cutoff grades at existing operations and resulted in processing of peripheral or marginal material. Gold ores to be mined in the future comprise increasingly complex mineralogy and lower metal grades. This presents some real challenges for maintaining production growth rates that will be required to satisfy demand.

The costs of gold production were relatively constant between 1994 and 2004, averaging $247/oz cash cost and $303/oz total production cost (GFMS, Thomson Reuters 1984–2015). However, operating costs increased dramatically in the following years, reaching $750/oz (cash cost) and $983/oz (total production cost), respectively. The average "all-in sustaining cost" statistic, which was introduced in 2007 to try to provide more consistency and transparency in reporting across the gold industry, reached $1,314/oz in 2014, with a corresponding average gold price of $1,266/oz. Operating and capital costs for gold and silver production are discussed later in the "Economic Considerations" section.

A listing of the major gold-producing companies globally in 2014, 2004, and 1994 and their reported production is provided in Table 3.

Silver Production, Consumption, and Price

Silver has been treasured since early times and was used by the Romans as the basis of their monetary system. It was widely mined during the era of the Roman Empire and later in Spain, Germany, and Eastern Europe. Discovery of the "New World" in 1492 brought an enormous increase in silver production through development of the rich deposits in Bolivia, Peru, Mexico, and the western United States.

Silver, like gold, has attractive properties of malleability and ductility, high heat and electrical conductivity, and corrosion resistance. It also forms silver salts (e.g., silver chloride, iodide, and bromide) that are sensitive to light. Film and paper coating with a silver-halide emulsion have the capacity for recording and reproducing images that have been used by the photographic industry since the late 1800s, although this use is now in decline. Although not as good a conductor of electrons as gold, silver is widely used in electronics and electrical applications because of its lower cost. Silver is also used in jewelry and silverware (plated and solid Sterling silver), and as physical bars, coins, and medals. A summary of silver demand by application in 2014 is provided in Table 4.

Silver production and price data between 1975 and 2014 are summarized in Figure 2. Mine production of silver grew at an annual rate of 2.8%, reaching 27,290 t in 2014, while the price increased by an average of 3.7%. Notably, the daily silver price peaked in 2011 at $48.70/oz. Similar to the gold

Table 2 Summary of gold production by country and region between 1974 and 2014*

Country/Region	2014	2004	1994	1984	1974
Europe					
• Russia	262	182	165	269	NA†
• Turkey	32	5	NA	NA	NA
• Other	24	19	25	16	12
Total Europe	319	206	190	285	12
North America					
• United States	205	260	331	66	35
• Canada	154	129	146	86	52
• Mexico	118	22	15	8	4
Total North America	477	411	492	160	91
South America					
• Peru	173	181	38	11	3
• Brazil	81	43	75	62	14
• Argentina	60	29	1	NA	NA
• Chile	44	40	43	22	4
• Colombia	43	24	26	21	8
• Dominican Republic	36	NA	NA	11	NA
• Other	114	83	54	21	5
Total South America	550	399	238	147	33
Asia					
• China	462	217	130	59	NA
• Indonesia	116	114	55	4	NA
• Uzbekistan	80	84	64	NA	NA
• Kazakhstan	49	15	NA	NA	NA
• Mongolia	31	19	4	NA	NA
• Philippines	43	32	31	34	17
• Other	75	79	62	11	10
Total Asia	856	560	347	108	28
Africa					
• South Africa	164	363	584	683	759
• Ghana	108	58	45	12	19
• Mali	47	40	6	NA	NA
• Tanzania	46	48	7	NA	NA
• Burkino Faso	39	2	3	NA	NA
• Democratic Republic of Congo	40	5	8	10	4
• Other	144	63	47	30	12
Total Africa	588	578	699	735	794
Oceania and other					
• Australia	273	258	256	39	16
• Papua New Guinea	58	76	61	19	21
• Other	13	16	14	2	2
Total Oceania and other	344	350	331	60	39
World total	**3,133**	**2,504**	**2,296**	**1,495**	**996**

Data from GFMS, Thomson Reuters 1984–2015
*For 1994 and 1984 data, Russia, China, and other former Soviet Union countries were reported as non-Western gold production. For 1974, only noncommunist production was reported.
†NA = not available

market, a significant amount of silver is recycled from scrap—5,240 t in 2014.

Silver is produced from primary sources (31% in 2014), as a by-product of gold production (13%), as a by-product of copper production (20%), as a by-product of lead and zinc production (35%), and with 1% coming from other sources. As such, the silver supply–demand balance and, to a significant extent, price are driven by factors related to lead, zinc, copper, and gold production.

Silver production by country and region is summarized in Table 5. The top five producing countries are Mexico, Peru, China, Australia, and Chile. A list of the major silver-producing companies and their production in 2014 and 2004 is provided in Table 6.

OCCURRENCES, ORE TYPES, AND MINERALOGY

The characteristics of a gold/silver deposit and its mineral assemblages determine the mining method(s), extraction process requirements, and, in particular, the performance of all unit operations involved in an extraction flow sheet. Consequently, a good understanding of the ore types and mineralogy is important for effective design or operation of gold and silver extraction processes.

Gold Properties and Gold Minerals

The gold mineralogy within each ore deposit is unique, due to the variations in the following:

- Mineralogical mode of occurrence of gold
- Gold grain size distribution
- Host and gangue mineral type
- Host and gangue mineral grain size distribution
- Mineral associations
- Mineral alterations
- Variations of the preceding characteristics within a deposit or with time

Because gold is inert at ambient temperatures and pressures, there are very few naturally occurring compounds of the metal. The average concentration of gold in the earth's crust is approximately 0.005 g/t, which is much lower than most other metals, for example, silver (0.07 g/t) and copper (50 g/t). The predominant occurrence of gold is as native metal, often alloyed with up to 15% Ag. Other gold minerals include alloys with tellurium, selenium, bismuth, mercury, copper, iron, platinum, rhodium, other platinum group elements, and alloys containing >15% silver. There are no common naturally occurring gold oxides, silicates, carbonates, sulfates, or sulfides. Therefore, gold generally occurs in a mineral form different to most other elements, which, inter alia, often allows selective extraction from other mineral mixtures.

Native gold grains have been known to contain up to 99.8% Au, but most vary between 85% and 95% Au content, with silver as the main impurity. Pure gold has a density of 19.3 g/cm^3, though native gold typically has a density of 15–18 g/cm^3 due to alloying with other elements, for example, silver (10.5 g/cm^3) and copper (8.9 g/cm^3). Hence, if liberated from gangue minerals, it can be readily recovered at particle sizes above approximately 20 μm by gravity concentration—a major method of recovery of gold employed throughout history. Gravity concentration can be very selective, as the most common gangue minerals (e.g., quartz and other silicates) have densities in the range of 2.7 to 3.5 g/cm^3. The freezing and boiling points of gold are 1,064°C and 2,808°C, respectively. Gold has a metallic luster, and the color is a distinctive deep yellow (golden) but may be light yellow or orange-yellow with high silver and copper contents, respectively. Its distinctive high reflectivity and low hardness can be

Table 3 Major gold-producing companies in 2014, 2004, and 1994

Rank	Company	2014 Gold Production, t	Company	2004 Gold Production, t	Company	1994 Gold Production, t
1	Barrick Gold	194	Newmont Gold	212	Anglo American	252
2	Newmont Gold	151	AngloGold Ashsanti	188	Gold Fields SA	124
3	AngloGold Ashanti	138	Barrick Gold	154	Barrick Gold	72
4	Goldcorp	89	Gold Fields	129	Gencor	65
5	Kinross Gold	82	Placer Dome	114	Placer Dome	54
6	Newcrest Mining	72	Harmony	102	JCI	52
7	Navoi MMC	73	Navoi MMC	58	Newmont Gold	52
8	Gold Fields	64	Buenaventura	51	Homestake	40
9	Polyus Gold	53	Kinross Gold	50	Rio Tinto Zinc	40
10	Sibanye Gold	49	Rio Tinto	48	Anglovaal	39

Data from GFMS, Thomson Reuters 1984–2015

used in its identification by microscopic examination of polished sections. Pure gold is an excellent electrical and thermal conductor.

Gold is very soft, ductile, and malleable (e.g., 1 oz of gold can be beaten into an area of 30 m^2) with Vickers and Mohs hardness numbers of 40–95 kg/mm^2 and 2.5–3.0, respectively. These unusual physical properties are a result of the face-centered cubic crystal structure. Native gold rarely occurs in its cubic crystal form, and the famous rounded masses, known as nuggets, are now only found occasionally. Gold does not display cleavage on breakage. There are many terms to describe the various, and often distinctive, forms of native gold, including sponge gold, flakey gold, grain gold, foil gold, moss gold, and tree gold.

Gold usually occurs alloyed with some silver. When the silver content is between 25% and 55%, the mineral is called electrum. Electrum has a pale yellow color, due to the silver content and a lower density than gold (i.e., 13–16 g/cm^3). A term commonly used to express the purity of gold is *fineness*, which is the gold content in parts per thousand (e.g., 99% gold content is 990 fineness).

Gold forms many alloys with tellurium. The more common gold-bearing tellurides are sylvanite ($(Au,Ag)_2Te_4$), calaverite ($AuTe_2$), and petzite (Ag_3AuTe_2), with krennerite ($AuTe_2$), montbrayite (Au_2Te_3), and kostovite ($CuAuTe_4$) less common. Gold-telluride occurrence is widespread and is often associated with some free gold and sulfide minerals. The densities of gold tellurides (8.0–10.0 g/cm^3) are lower than native gold, and the colors are less distinctive shades of white, gray, and black. The silver mineral, hessite (Ag_2Te), is commonly encountered in gold-telluride ores.

Gold occasionally occurs with bismuth in the mineral maldonite (Au_2Bi). Maldonite has a density of 15.5 g/cm^3 and a lower Mohs hardness (1.5 to 2.0) than gold. It has very low solubility in cyanide solutions. Gold and copper (Cu) form the extremely rare intermetallic compounds auricupride ($AuCu_3$) and tetra-auricupride (AuCu). The copper content imparts a deep red color. Gold tellurides and gold-bismuth minerals present special problems for processing due to their low/slow solubility in alkaline cyanide solutions.

Gold can occur as ultrafine ("invisible") solid–solution inclusions within sulfide mineral grain structures (Chryssoulis and Cabri 1990; Petruk 2000). For example, the gold content within sulfide mineral structures has been measured as follows:

Table 4 Silver demand/consumption by application for 2014

Application	Silver, t	Proportion, %
Jewelry	6,693	20.2
Coins and bars (investment)	6,096	18.4
Silverware	1,888	5.7
Industrial fabrication		
• Electrical and electronics	8,207	24.7
• Brazing alloys and solders	2,056	6.2
• Photography	1,418	4.3
• Photovoltaic	1,863	5.6
• Other Industrial uses	4,957	14.9
Total demand/uses	**33,178**	**100.0**

Data from Thomson Reuters and The Silver Institute 1975–2015

- Arsenopyrite: <0.2 to 15,000 g/t
- Pyrite: <0.2 to 132 g/t
- Tetrahedrite: <0.2 to 72 g/t
- Chalcopyrite: <0.2 to 7.7 g/t

Such occurrences are important because, for example, in an ore containing 1% by weight of arsenopyrite and a gold ore grade of 10 g/t, all of the gold could be present in solid–solution (i.e., invisible) within the arsenopyrite lattice. In this case, the arsenopyrite would only need to have an average gold content of 1,000 g/t, much less than the upper end of the range indicated, and concentrations in this range (i.e., 250 to 1,500 g/t) are commonly encountered in practice. Hence, the occurrence of gold in solid–solution in sulfide minerals (especially arsenopyrite, pyrite, and pyrrhotite) is of major importance in many refractory gold systems.

Classification of Ore Types

Primary and secondary gold- and silver-bearing materials can be classified into 15 mineral processing–based categories, which are related to their mineralogical and historical characteristics (McQuiston and Shoemaker 1975; Marsden and House 2006):

- Primary ores
  - Placers
  - Free-milling ores
  - Oxidized ores
  - Silver-rich ores
  - Iron sulfides

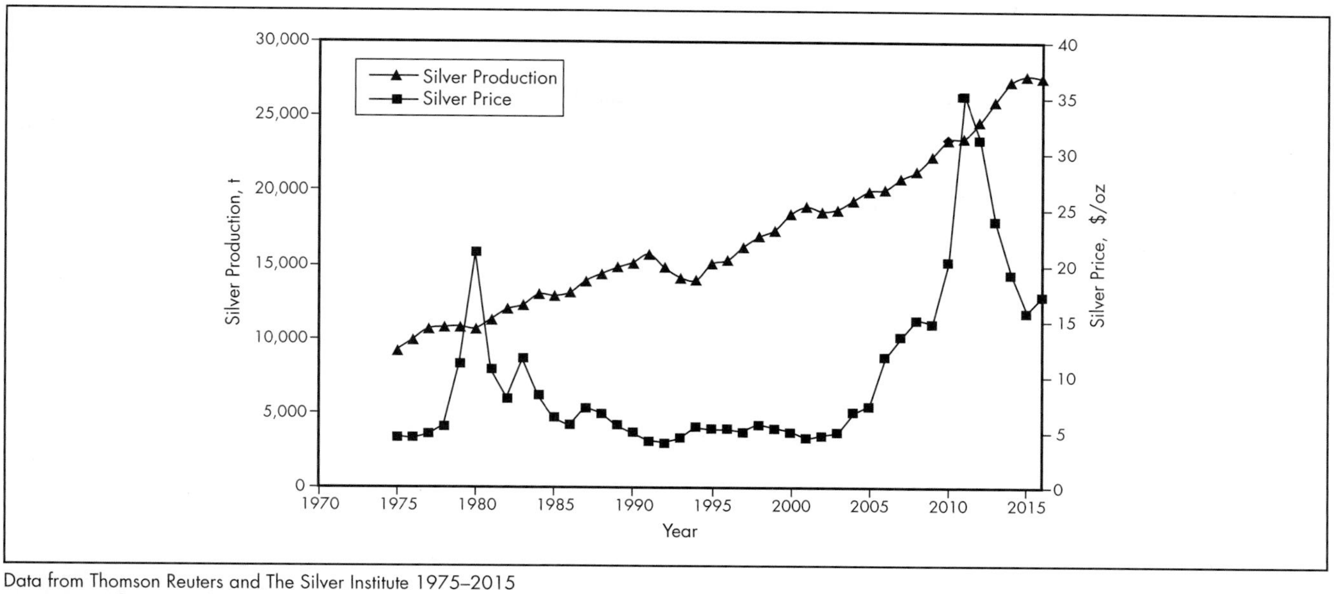

Data from Thomson Reuters and The Silver Institute 1975–2015

Figure 2 Silver production and price between 1975 and 2015

- Arsenic sulfides
- Copper sulfides
- Antimony sulfides
- Tellurides
- Carbonaceous ores
- Secondary materials
  - Gravity concentrates
  - Flotation concentrates
  - Tailings
  - Refinery materials
  - Recycled gold

Each of these classes of gold- and silver-bearing materials has special mineralogical characteristics, which may affect their processing (Chryssoulis and McMullen 2005; Chryssoulis and Cabri 1990; Baum 1988). More detailed information for each ore type as well as information on the characteristics of secondary materials is provided in the literature (Marsden and House 2006).

PROCESS DEVELOPMENT AND SELECTION

Process selection is the systematic development of the optimum metal extraction route for a particular feed material using the most appropriate technology. In the case of gold and silver, this procedure has two main objectives, as follows:

1. Optimization of project economics, principally a function of gold and silver recovery, throughput rate, and processing costs (capital and operating)
2. Development of a process that satisfies all of the project requirements, including, for example, environmental, financial, and sociopolitical considerations

Initial testing is required to determine the nature of gold mineralization within a deposit. This work starts in the assay laboratory. Individual drill core, reverse circulation or surface samples are analyzed for gold, silver, and other elements, based on the geological assessment of the deposit type and characteristics. In addition, all or a portion of the individual samples are subjected to cyanide solubility tests using 10–100 g splits from pulverized assay pulp samples, typically at <106 μm (150 mesh) size. The tests can either be performed hot (e.g., 85°–95°C, 2 hours) or cold (ambient, 24 hours) at pH 11 with 5–10 g/L NaCN (sodium cyanide) at 30% solids. The solubility of various minerals and elements that can interfere with gold (and silver) extraction varies depending on the method applied. As such, this testing approach does not necessarily provide an accurate indication of metallurgical response for a potential commercial process, but it does give clues to amenability by cyanide leaching, potential for refractory behavior, and indications of gangue interferences. This can be used to provide a comprehensive database of gold and silver cyanide solubility within a deposit, and it gives an indication of variability of response within and across ore types (McClelland and McPartland 2002).

Low gold extractions indicate that the ore contains refractory components that do not respond well to direct cyanide leaching. This could be due to any or a combination of the following reasons:

- Gold locked in sulfides
- Gold locked in silicates (finer grinding required)
- Gold associated with "preg-robbing" or "preg-borrowing" carbonaceous matter
- High consumption of cyanide due to reactive sulfides, copper, or zinc present
- Coarse gold that is incompletely leached within 24 hours under the conditions applied

Depending on the nature of the deposit, the grades of metal values, and the ore type(s) identified, the next phase of more definitive metallurgical testing may include bottle roll cyanide leaching, column cyanide leaching, gravity concentration, and flotation. Such testing is usually performed on representative drill-hole samples (i.e., split core or splits of reverse circulation cuttings) from various locations within the deposit. This testing may include the following:

- Standardized bottle roll cyanide leaching tests, which are typically run using 1 kg of sample, with leaching conducted at pH 10.5–11 (using lime) and 0.5–1.0 g/L NaCN in an open-to-atmosphere rolling bottle for up to 96 hours. Sampling is conducted at intervals through the test. A series of tests may be run initially to determine the particle size to be used in the standardized bottle roll test. Such testing provides information on gold and silver extractions and extraction rate, cyanide and lime consumption. Gold and silver extractions and extraction rate can be applied directly to commercial operations and facilities design, although allowances for process inefficiencies and losses should be considered.

- Column cyanide leaching tests (to determine heap leach performance) are conducted on samples at the proposed crush size in columns ranging in size from 100 mm diameter × 1 m high (15-kg sample) to 3 m diameter × 10 m high (125-t sample). The column diameter to largest particle size should be a minimum of 4:1, and preferably 6:1. Typical crush sizes used for column leach testing include 80%–100% <8–9 mm (quaternary crush), <12–13 mm (tertiary crush), <25–50 mm (secondary crush), <125–150 mm (primary crush), and run-of-mine (ROM) or "pseudo" ROM (truncated top size). The need for agglomeration is also considered during this testing, based on the amount of fines present and the load-permeability characteristics of the ore. Column leach testing provides information on gold and silver extractions. However, the ultimate gold and silver extraction must be adjusted using a scale-up factor to allow for losses and inefficiencies in a commercial heap leaching operation (Marsden and Botz 2017).
- Diagnostic mineralogical assessment may be required to determine the reasons for low gold (and silver) extractions by cyanide leaching. If the ore provides a good response to standard cyanide leaching, process development can proceed through grind size optimization, cyanide concentration optimization, carbon-in-pulp or carbon-in-leach (CIP/CIL) testing, and thorough variability testing on multiple samples throughout the deposit. Classification of the deposit into different ore types may be required, with appropriate and representative variability testing within and across all ore types.
- If the ore contains coarse gold, or free gold that is expected to be recoverable by gravity concentration, then gravity concentration testing is required. Standardized gravity recoverable gold (GRG) testing should be performed using a laboratory-scale centrifugal concentrator. The scale-up of results of GRG testing requires effective interpretation and analysis (Grewal et al. 2009).
- If the ore has sulfide refractory components, then batch-scale rougher flotation testing is required. This will determine the amenability to flotation concentration. If acceptable recoveries can be achieved into a flotation concentrate (i.e., 90%–95% into 10%–20% by weight of feed), then flotation may assist in providing a smaller mass of material for more intensive treatment. More intensive treatment may include pretreatment steps ahead of cyanide leaching of the concentrate, for example, regrinding (or ultrafine grinding), preaeration, roasting, pressure oxidation, or biological oxidation. Alternatively, consideration may be given to selling the concentrate to a third party for treatment by smelting. If neither direct cyanide leaching (with or without gravity) nor flotation are effective, then alternative whole ore treatment approaches must be considered (i.e., roasting and pressure oxidation).
- More specialized testing is required for effective development of oxidative pretreatment techniques (e.g., roasting, pressure oxidation, and biological oxidation). Batch, semicontinuous, and continuous pilot-plant testing methods are available. Effective testing programs must be developed on a case-by-case basis.
- In all cases, the presence of elements and minerals that can interfere with processing should be identified early in the metallurgical test-work program. Such elements and minerals include, but are not limited to, the following: S (total, sulfide, and elemental), Cu, Zn, Ni, Hg, As, Sb, Te, Se, Bi, U, Th, Cl, F, carbonate, total organic carbon (TOC), and clay minerals.

Table 5 Summary of silver production by country and region for 2014 and 2004

Country/Region	2014, t	2004, t
Europe		
• Russia	1,334	1,166
• Poland	1,263	1,362
• Sweden	395	292
• Other	367	94
Total Europe	3,359	2,914
North America		
• Mexico	6,000	2,569
• United States	1,169	1,250
• Canada	482	1,294
Total North America	7,651	5,113
South America		
• Peru	3,779	3,061
• Chile	1,574	1,359
• Bolivia	1,344	435
• Argentina	905	162
• Guatemala	858	—
• Other	268	62
Total South America	8,728	5,079
Asia		
• China	3,568	1,966
• Kazakhstan	544	703
• India	261	65
• Other	759	815
Total Asia	5,132	3,549
Africa		
• Morocco	277	230
• Other	202	153
Total Africa	479	383
Oceania and other		
• Australia	1,848	2,221
• Other	93	34
Total Oceania and other	1,941	2,255
World total	**27,290**	**19,293**

Data from Thomson Reuters and The Silver Institute 1975–2015

The need for more rigorous metallurgical testing, for example, including bulk sampling and larger scale pilot-plant

Table 6 Major silver-producing companies in 2014 and 2004

Rank	Company	2014 Silver Production, t	Company	2004 Silver Production, t
1	KGHM Polska Miedž	1,257	BHP Billiton	1,546
2	Fresnillo	1,257	Industrias Peñoles	1,384
3	Goldcorp	1,145	KGHM Polska Miedž	1,344
4	Glencore	1,085	Grupo Mexico	603
5	BHP Billiton	1,058	Kazakhmys	551
6	Polymetal Int.	893	Barrick Gold	538
7	Pan American Silver	812	Polymetal Int.	538
8	Volcan Cia. Minera	700	Rio Tinto	460
9	Tahoe Resources	631	Buenaventura	442
10	Codelco	619	Coeur d'Alene	439

Data from Thomson Reuters and The Silver Institute 1975–2015

testing (e.g., 1–100 kg/h), depends on the characteristics of the deposit, mineralogical complexity, and metallurgical response to processing. Also, the risk tolerance of the developer, the method of financing the project, and the development time line may have an impact on metallurgical testing requirements to support a feasibility study. Simple process flow sheets using agitated cyanide leaching with or without gravity concentration often do not require pilot-plant testing. More complex flow sheets (e.g., incorporating flotation and oxidative pretreatment) are likely to require more extensive process development work, including a continuous pilot plant on representative bulk samples.

In general, the risk associated with the development of gold projects can be minimized through well-planned metallurgical test-work programs and effective geometallurgical characterization of the deposit, coupled with careful consideration of all the specific project requirements, including available capital, target profitability, acceptable levels of risk, and specific environmental/regulatory factors.

EXTRACTION AND RECOVERY METHODS

Historical Developments

For information on historical aspects of gold extraction technology development, the reader is referred to the original texts (Agricola 1556; Eissler 1891; Rose 1896, 1935; Julian and Smart 1904; Bain 1910; Clennell 1915) and other historical reviews (Marsden and House 2006; Shoemaker 1984; Habashi 1987; McNulty 1989). Several excellent texts on the history of gold and silver metallurgical practice in South Africa during the 20th century are available in the literature (King 1949; Dorr and Bosqui 1950; Adamson 1972; Stanley 1987).

Process Flow Sheets

There are many and varied flow sheets in commercial use for gold and silver extraction. The flow sheet developed is dictated by the ore mineralogy and characteristics, but it may also be impacted by other factors, including location, elevation, topography, climate, environmental concerns, regulations and permitting, safety and health, and other sociopolitical factors. By far, the major method for gold extraction is the cyanide leaching process, applied successfully for more than 125 years.

The approximate proportion of gold production by extraction and recovery method is provided in Table 7. Gold extraction technology is dominated by milling, agitated cyanide leaching, CIP or CIL, and electrowinning. Gravity concentration is commonly incorporated into the operations, typically applied within the grinding circuit. However, the application of heap leaching, to both crushed and ROM ores, has grown significantly since the 1980s and represents a low-cost option for ores that are amenable to this approach. Flotation, treatment of flotation concentrates, intensive leaching of gravity concentrates, and oxidative pretreatment processes (i.e., roasting, pressure oxidation, and biological oxidation) have all evolved significantly and form important components of many flow sheets. The following sections provide information on how unit operations are combined into effective flow sheets for gold and silver recovery.

Comminution

Comminution of gold ores and concentrates is primarily required to liberate gold, gold-bearing minerals, and other metals of economic value to make them amenable to subsequent gold extraction steps. However, this preparation may also be necessary to facilitate materials handling between stages. The degree of comminution required depends on many factors, including the liberation size of gold, the size and nature of the host minerals, and the method(s) to be applied for gold recovery. The optimum particle size is dictated by economics: a balance between gold recovery, processing costs (i.e., reaction kinetics and reagent consumptions), and comminution costs for a given processing method. Other factors, such as particle fluidization requirements (agitated leaching, flotation, CIP), permeability (e.g., the effect of fines on heap leaching), and solid–liquid separation efficiency, may also play an important role in particle size optimization. The type of comminution equipment selected can have an important impact on subsequent processing steps, for example, impact crushing, semiautogenous grinding (SAG), ball milling, and high-pressure roll crushing or high-pressure grinding rolls.

The major uses of comminution in gold extraction flow sheets are for the following:

- Gold liberation before leaching (i.e., by crushing prior to heap leaching and crushing and grinding before agitated leaching)
- Sulfide mineral liberation before flotation
- Gold liberation before flotation and/or gravity concentration
- Optimization of sulfide mineral particle size prior to oxidative pretreatment

Table 7 Estimated distribution of gold production by extraction and recovery method applied in 2004

Extraction Method	Recovery Method	Approximate Proportion of Global Production, %
Heap leaching—cyanide	Carbon-in-columns	5
Heap leaching—cyanide	Zinc precipitation	5
Agitated cyanide leaching (with/without gravity)	Carbon-in-pulp/carbon-in-leach	28
Agitated cyanide leaching (with/without gravity)	Zinc precipitation	4
Agitated cyanide leaching (with/without gravity)	Resin-in-pulp/resin-in-leach	4
Gravity concentration	Smelting or intensive leaching	9
Flotation concentration	Smelting of concentrates	4
Flotation and gravity concentration	Smelting of concentrates	6
Flotation and agitated cyanide leaching	Carbon-in-pulp/carbon-in-leach	3
Whole ore roasting, agitated cyanide leaching	Carbon-in-pulp/carbon-in-leach	2
Whole ore pressure oxidation, agitated cyanide leaching	Carbon-in-pulp/carbon-in-leach	5
Flotation, roasting of concentrate, agitated cyanide leaching	Carbon-in-pulp/carbon-in-leach	2
Flotation, pressure oxidation of concentrate, agitated cyanide leaching	Carbon-in-pulp/carbon-in-leach	<1
Flotation, biological oxidation of concentrate, agitated cyanide leaching	Carbon-in-pulp/carbon-in-leach	1
Unaccounted for		22

Adapted from Marsden and House 2006

- Regrinding of flotation and gravity concentrates or tailings for gold liberation and surface preparation
- Regrinding of roaster calcine for gold liberation and surface preparation
- Ultrafine grinding of gold-bearing sulfide concentrate prior to leaching

Physical aspects of comminution (i.e., procedures and equipment) for treatment of gold ores are considered in Part 3 of this handbook. However, comminution processes, and grinding in particular, have several important implications for gold and silver extraction, as follows:

- Because of their high density, gold particles tend to stay in grinding circuits longer than gangue minerals. As a result, and because gold is soft and malleable, gold particles become flattened, and hard gangue mineral particles (e.g., quartz) may become embedded in gold surfaces. This reduces particle density and hydrophobicity, hindering the response to gravity concentration and flotation, respectively. The problem is usually limited to ores with coarse (>50 µm) gold and may be overcome by the early removal of gold by gravity concentration and/or flash flotation within the grinding circuit.
- Grinding mills, slurry pumps, pipelines, and classification equipment can provide good mixing and valuable residence time for leaching reactions, and the addition of cyanide to grinding circuits can achieve >80% gold dissolution for some ores prior to the dedicated leaching stage. This also reduces lockup of coarse gold in grinding equipment.

Cyanidation

The use of cyanide leaching to dissolve gold and silver from ores and concentrates has been applied commercially on a large scale for more than 100 years. The overall reactions for gold and silver dissolution are expressed as follows:

$$4Au + 8CN^- + O_2 + 2H_2O = 4Au(CN)_2^- + 4OH^- \quad \textbf{(EQ 1)}$$

$$4Ag + 8CN^- + O_2 + 2H_2O = 4Ag(CN)_2^- + 4OH^- \quad \textbf{(EQ 2)}$$

Cyanide leaching can be applied using agitated tank systems (open or closed), heap or stockpile systems, or in vats. In situ leaching is generally unsuitable for the application of the cyanide reagent scheme because of its toxicity and environmental requirements.

Agitated Leaching

Commonly applied to a wide range of ore types, agitated cyanide leaching is typically performed in steel tanks, and the solids are kept in suspension either by air or mechanical agitation. Air agitation in conical-bottomed leach tanks (Browns or Pachuca tanks) was widely practiced in the early years of cyanidation (up until the 1980s) but has largely been superseded by more efficient mechanical agitation using flat-bottomed cylindrical tanks with improved mixing efficiency. Typical energy input for mechanically agitated tanks is 0.04–0.06 kW·h/m^3. Well-designed baffled systems with downcomers can approach perfectly mixed flow conditions in a single reactor, which help to optimize reaction kinetics and make the most of available leaching equipment. A generalized flow sheet for the application of agitated cyanide leaching with CIP for gold and silver recovery is provided in Figure 3.

Particle size. The material to be leached is ground to a size that optimizes gold recovery and comminution costs, typically between 80% <150 µm and 80% <45 µm. In a few cases, whole ore may need to be ground as fine as 80% <20 to 25 µm for optimal processing, either by oxidative pretreatment and/or leaching. Agitation leaching is rarely applied to material at sizes coarser than approximately 150–180 µm because it becomes increasingly difficult to keep coarse solids in suspension, and abrasion rates increase. Increasingly, agitation leaching is being considered to treat very finely ground materials and, with the advances in stirred milling equipment, concentrates have been ground to 80% <7 to 10 µm to liberate gold contained in refractory and semi-refractory sulfide mineral matrices prior to processing by agitation leaching and/or oxidative pretreatment.

Slurry density. Leaching is usually performed at slurry densities of between 35% and 50% solids, depending on the

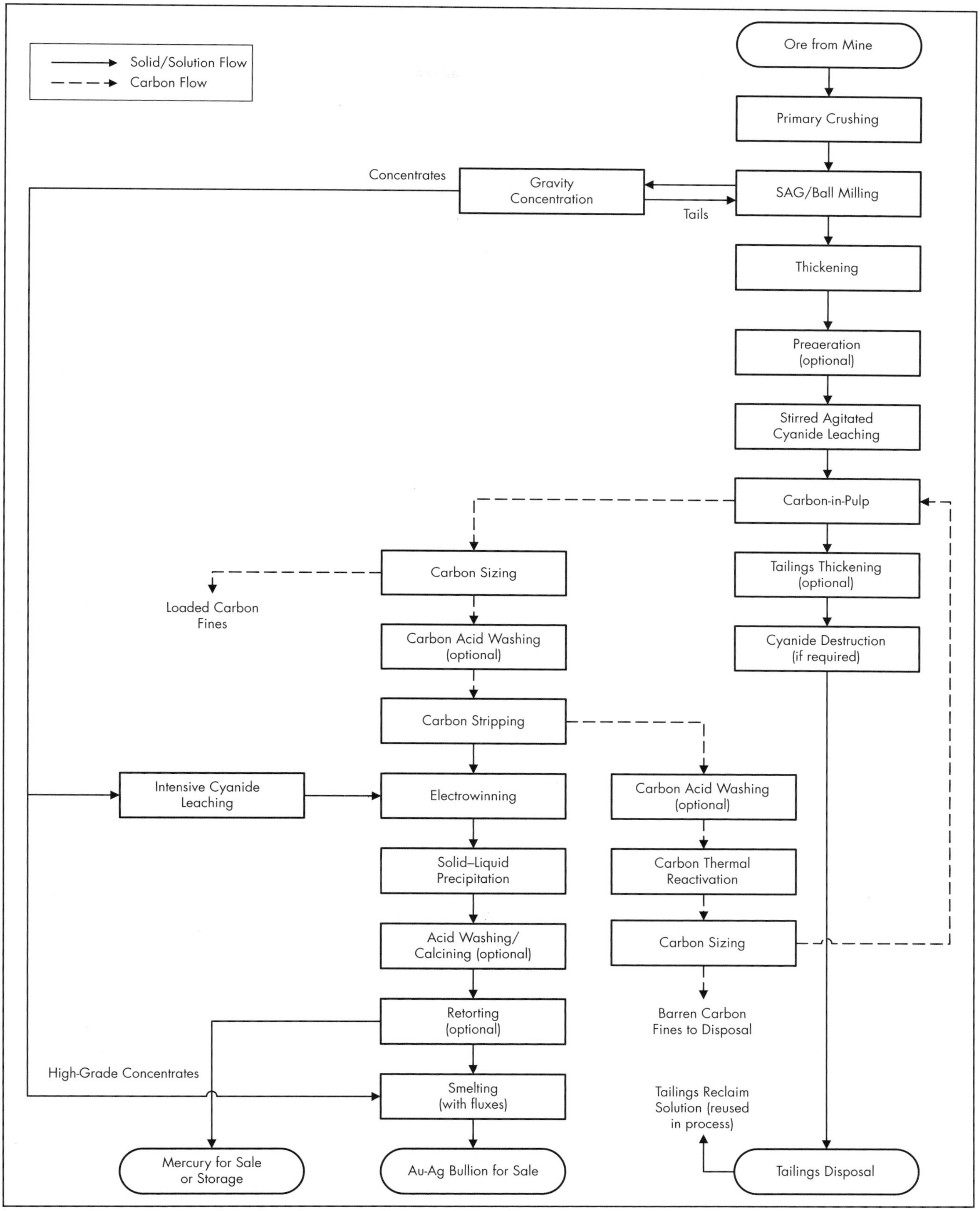

Figure 3 Schematic process flow sheet for a typical agitated cyanide leaching with CIP circuit, incorporating gravity concentration and intensive cyanide leaching

solids' specific gravity, particle size, and the presence of minerals that affect slurry viscosity (e.g., clays and clay-forming minerals). Mass transport phenomena are maximized at low slurry densities. However, solids retention time in a fixed volume of leaching equipment increases as the density increases. In addition, reagent consumptions are typically minimized by maximizing slurry density, since optimal concentrations can be achieved at lower dosages.

pH modification. Alkali, required for slurry pH modification and control, must always be added before cyanide addition to provide protective alkalinity to prevent the loss of cyanide by hydrolysis. Most leaching systems operate between pH 10 and 11, and preferably 10.2–10.5. Staged addition of alkali may be required throughout the leaching circuit to maintain the desired operating pH, particularly when treating ores containing alkali-consuming materials. Calcium hydroxide (slaked lime, $Ca(OH)_2$) is generally used for pH modification. Unslaked lime (CaO) is sometimes used because it is less costly than slaked lime, but it is less effective for pH modification. Unslaked lime must generally be added in solid form into the mill feed. For nonacidic- or non-alkali-consuming ores, calcium hydroxide concentrations of 0.15 to 0.25 g/L are typically required to achieve the desired pH range for leaching (i.e., pH 10–11). This represents typical lime consumptions of 0.15 to 0.5 kg/t for nonacidic ores. Sodium hydroxide is rarely used as a pH modifier because of its higher cost and potential adverse effects on downstream processing. When lime is used as the pH modifier, anti-scaling reagents are generally required to be added into the process solution to manage calcium carbonate scaling.

Cyanide. Cyanide may be added to agitated leaching systems either prior to the leaching circuit (during grinding) or, more commonly, in the first stage of leaching. Subsequent reagent additions can be made into later stages of leaching to maintain or boost cyanide concentrations to maximize gold dissolution. In the absence of cyanide-consuming minerals in the ore or concentrate to be leached, cyanide concentrations used in practice range from 0.05 to 0.5 g/L NaCN and are typically between 0.15 and 0.30 g/L NaCN. Typical cyanide consumptions observed in agitated leaching systems for free-milling ores vary from about 0.25–0.75 kg/t. In cases where the feed material contains significant amounts of cyanide consumers and/or high silver content (i.e., >20 g/t), higher cyanide concentrations may be applied (e.g., 2–10 g/L NaCN. In such cases, cyanide consumptions may vary from 1 to 2 kg/t and in some cases much higher, depending on the nature and amount of cyanide-consuming minerals. Cyanide concentrations are usually monitored by manual titration techniques or less commonly by online cyanide analyzers, based on titrimetric, colorimetric, potentiometric, and ion-specific electrode techniques.

Oxygen. Oxygen is typically introduced into leaching systems as compressed air, either sparged into tanks as the primary method of agitation or supplied purely for aeration. In either case, crude sparging systems are usually sufficient to provide satisfactory bubble dispersion and to ensure that adequate dissolved oxygen concentrations are maintained. Typically, dissolved oxygen concentrations can be maintained at, or close to, calculated saturation levels with effective air sparging (i.e., 8.2 mg/L O_2 at sea level at 25°C). In some cases, particularly when treating ores that contain oxygen-consuming minerals, pure oxygen or hydrogen peroxide have been added to increase dissolved oxygen concentrations above those attainable with simple air sparging systems.

Residence time. Residence time requirements vary depending on the leaching characteristics of the material treated and must be determined by test work. Leaching times applied in practice vary from a few hours to several days, but typically 12–36 hours. Leaching is usually performed in three to six stages, with the individual stage volume and number of stages dependent on the slurry flow rate, required residence time, the efficiency of mixing equipment used, and engineering design preferences.

Heap Leaching

Heap leaching is a low-cost method that is most suitable for treatment of low-grade materials that do not justify the higher costs of grinding and agitation leaching. Ores can be treated either at a ROM size or as crushed material, with the optimum size determined as a trade-off between gold recovery and crushing costs. A schematic flow sheet for ROM/crushed ore heap leaching is shown in Figure 4.

The first large-scale cyanide heap leach for gold and silver was commissioned at Newmont Gold's Carlin mine (Nevada, United States) in 1971 (Pizarro et al. 1974), processing material with grades below the established mill cutoff grade. This was followed by applications at Cortez and Smoky Valley/Round Mountain (all in Nevada) in the late 1970s (McQuiston and Shoemaker 1980; McClelland and Eisele 1982). Thereafter, the use of heap leaching boomed throughout the western United States.

Heap leaching is usually performed on lined leach pads, which not only contain the gold-bearing leach solutions produced but also protect the ground and groundwater beneath and around the leach pad. A variety of effective leach pad and solution pond liner systems have been developed and further information is available in the literature (Strachan and Van Zyl 1988). Either Merrill–Crowe zinc precipitation or carbon adsorption processes can be used for recovery of gold and silver from cyanide solution.

Materials handling, ore placement, and agglomeration. Heap leach feed material is placed on the lined leach pad by one of several possible methods, including conveyor stacking, truck dumping, loader-assisted stacking, or combinations of these methods. The method selected depends on the ore properties and specific project conditions, such as the processing rate, location, climate, and mining method. Material is typically stacked in lift heights ranging from 5 m to >15 m, depending on ore permeability, production schedule requirements, and liner area availability. Lift heights of 5–10 m are generally applied for crushed ores and 10–15 m for ROM ores. Multiple lifts may be used to achieve the ultimate pad height, commonly up to about 100 m. In some cases, ultimate heap heights have approached 200 m (e.g., Zortman, Montana, United States). The maximum pad height depends principally on the ore properties, such as permeability and the structural stability of the stacked material, but also on other local factors, for example, topography, climate, and regulatory requirements (Marsden et al. 1995). Ores with high proportions of fines, particularly clays and fine silts, may require special stacking methods, such as conveyor stacking or loader-assisted stacking, to ensure that any detrimental segregation or compaction of fine particles is avoided. In cases where significant fines and/or clay are present, the ore may need to be agglomerated prior to stacking on the leach pad (Arnold et al. 1990; Heinen et al. 1978, 1979; McClelland and McPartland 2002). Agglomeration is achieved by the addition of water and lime

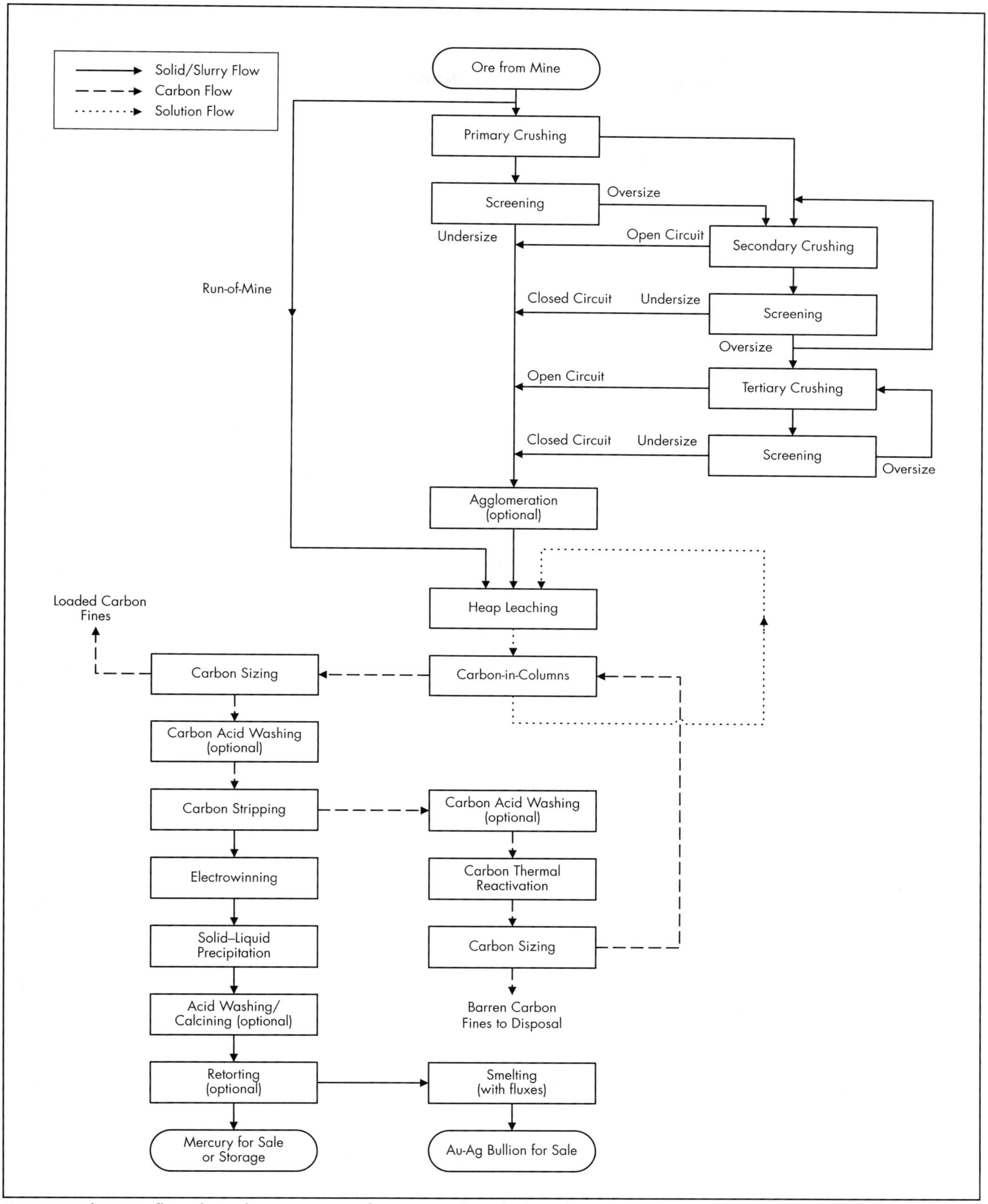

Figure 4 Schematic flow sheet showing options for ROM and crushed ore heap leaching

and/or cement to the ore followed by physical agglomeration, using either a series of short conveyor belts and drop points or rotating drum agglomeration equipment. Typical cement additions are 5–15 kg/t. During the agglomeration procedure, the fine particles adhere and bind to larger particles as agglomerates, which reduces the potential for migration in the heap. The porosity, long-term stability, and mechanical strength of the agglomerates should be optimized by test work.

Solution application. Alkaline cyanide solution is applied to the top of the heap by a suitable distribution system, such as agricultural sprinkling or drip irrigation. Application rates vary depending on ore properties and the gold dissolution rate but typically range from 0.1 to 0.3 L/m^2/min. Solution must be applied at a slower rate than the permeability of the leach pad to prevent buildup of solution on the surface of the pad. This is important to minimize any threat to wildlife (due to ponding on the heap surface) and to maintain the stability of stacked ore. Solution sprays or sprinkler systems and drip irrigation systems have all been used with good success at heap leaching operations worldwide.

pH modification. Alkali, which is required for pH modification and control, may either be added to the leach solution and/or introduced to the ore prior to leaching. The latter practice is essential when treating ores that have high alkali consumption, and the reagent may be added directly to haulage trucks or onto conveyors during crushing or agglomeration. Calcium hydroxide (slaked lime) is typically the most economic alkali to use for this purpose. When pH modification and control is required in leach solutions, calcium hydroxide is generally used. Unslaked lime is sometimes used because it is less costly than slaked lime, but it is less effective for pH modification. For nonacid- or non-alkali-consuming ores, calcium hydroxide concentrations of 0.15 to 0.25 g/L are generally required to achieve the desired pH range for leaching (i.e., pH 10–11).

Dissolution rate. Gold leaching rates in heap leaching environments are highly dependent on the mass transport of cyanide and oxygen to the exposed gold surfaces. The concentration of cyanide is depleted as the leach solution percolates through the heap and cyanide reacts with precious metals and other ore constituents, and as other cyanide-consuming reactions take place. However, in the absence of significant amounts of oxygen-consuming minerals, there is little evidence to suggest that dissolved oxygen concentrations are depleted significantly, and it is apparent that in most oxide heaps, sufficient oxygen is drawn into the heap by the flow of solution to keep the solution at, or close to, oxygen saturation. Consequently, gold dissolution rates can often be maximized by maintaining the cyanide concentration in leach pad runoff solutions above approximately 0.05–0.1 g/L NaCN.

Leaching efficiency. Gold extractions obtained by heap leaching are generally in the range of 50% to 80% and depend on the following:

- Degree of gold liberation achieved at the heap feed particle size
- Efficiency of solution contact with ore, which is a function of the uniformity of solution application and the homogeneity of material on the heap
- Chemistry of the solution applied
- Relationship between dissolution rate and the time allowed for leaching

For large-scale heap leaching operations, many of these factors are difficult to control precisely, and gold extractions achieved are often quite variable, even within a single ore type. Effective modeling of the heap is required to project gold and silver dissolution and recovery rates as accurately as possible (Marsden and Botz 2017). Careful monitoring of process solutions is critical to successful heap leach operation, particularly early in the leaching cycle when reactions involving the most reactive ore constituents occur. In addition to gold (and silver) concentrations, monitoring of pH, cyanide concentration, dissolved oxygen concentration, base metal ion concentrations, and temperature may need to be considered. Cyanide consumptions for crushed ore heap and ROM ore stockpile leaching operations are generally lower than those experienced in agitated leaching systems and, for non-refractory ores, typically range from 0.1 to 0.5 kg/t. Lime consumptions are variable, depending on the ore type and properties, but typically range from 0.15 to 0.75 kg/t. Some heap leach operations experience lime consumption in excess of 5 kg/t due to reactive gangue mineral components.

Vat Leaching

Vat leaching is usually performed in large wooden or concrete structures (vats) or steel tanks, or may be accomplished by heap leaching in a valley-fill configuration, where the heap can be flooded. In either case, the ore is completely immersed in the leach solution either throughout the leaching cycle or for portions of it. This has the advantage of efficiently wetting all surfaces of the ore to be leached and, to a degree, aiding mass transport. The proposed advantages are that channeling of solutions and the development of dead zones, which may be experienced during heap leaching, are avoided. The process is now little used, because of the relatively high capital cost associated with vat construction and the high operating cost, with only minor (if any) recovery benefits over standard heap leaching practice. However, the method was employed successfully to treat sands at Homestake Lead (South Dakota, United States) for about 100 years until the operation closed in 2000.

Carbon Adsorption

The application of activated carbon for the recovery of gold and silver from cyanidation plant slurries gained worldwide commercial acceptance in the 1980s. In 1973, the Homestake mine (South Dakota) successfully replaced the conventional slime treatment and zinc precipitation circuit with CIP becoming the largest scale application of carbon adsorption at the time. In 1980, large plants were installed in South Africa at President Brand (to treat calcine), Randfontein Estates, Western Areas, RM3 (Rand Mines), and Beisa (Boydell 1984; Laxen 1982). Widespread application followed rapidly throughout South Africa, the United States, Australia, and the rest of the world. By 1989, approximately 50% of world gold production was produced by carbon (Marsden and House 1992).

The typical operating conditions that applied for CIP processes are as follows:

- 30%–50% solids, depending on slurry viscosity
- 6 × 12 mesh activated carbon granules typically used, occasionally 8 × 16 mesh
- 5–40 g/L carbon concentration
- Interstage screening at 0.7–0.9 mm
- pH 10–11
- Cyanide concentration allowed to decay through the CIP circuit, depending on the specific leach response of the ore

- Carbon transfer rates designed to optimize stripping and regeneration cycle
- Gold on loaded carbon concentrations of 1,500–5,000 g/t, and as high as 10,000 g/t in some cases
- Gold on stripped carbon concentrations of 30–100 g/t, sometimes <30 g/t
- Retention time of 1–2 hours per stage, six to eight stages

Interstage screening is used to separate ground slurry from activated carbon granules. Initial CIP plants relied on external screens that required pumping of the entire slurry flow from tank to tank. The development of screens located within agitated slurry tanks provided many advantages. The key issue was blinding of the screen over time with particulate material and carbonate scale. Self-cleaning, air-swept, and mechanically swept devices and both in-tank and external screens were employed with varying results. NKM (North Kalgoorlie Mines), modified-NKM, and Derrick screens are now largely preferred over the earlier Kambalda, equal-pressure air-sparged ("EPAC"), and other screen configurations.

A modification of the CIP process is CIL, whereby the separate agitated tank leaching and CIP steps are combined. The advantage of this process is the introduction of carbon to the slurry at the same time cyanide is added for leaching, which counteracts the "preg-borrowing" or "preg-robbing" potential of carbonaceous ore constituents. However, the disadvantages of CIL include increased carbon inventory and gold-in-circuit inventory, lower gold loadings, and potentially higher carbon losses (with the contained gold).

The sharing of facilities for treatment of CIP/CIL and carbon-in-columns (CIC)-generated carbon resulted in substantial capital and operating savings (Mansanti et al. 1987). For small operators, stripping and reactivation of loaded carbon was often available through third parties in nearby locations. Large operators with multiple operations, such as Newmont Gold and Barrick Gold in Nevada, established centralized carbon handling facilities to further reduce costs and improve security.

Excellent reviews of the development of CIP/CIL technology are available in the literature (Anonymous 1982; Stanley 1987; SAIMM 1991; Staunton 2005).

Merrill–Crowe Precipitation

The use of zinc precipitation to recover gold and silver from clarified process solutions has been applied commercially on a large scale since 1890. The overall reaction is described by the following equation:

$$2Au(CN)_2^- + Zn = 2Au + Zn(CN)_4^{2-} \quad \textbf{(EQ 3)}$$

The rate of this reaction is first order with respect to cyanide concentration, indicating that sufficient free cyanide must be present for the reaction to proceed rapidly. The process is referred to as Merrill–Crowe precipitation after the pioneers who developed and implemented the technology. It is used to treat solutions at temperatures close to ambient, ranging from just above freezing to 40°C, depending on climate and processing methods, and which vary in gold concentration between approximately 0.5 and 10 g/t. Overall gold recoveries from solution are typically >98% and sometimes as high as 99.5%. A generic flow sheet for a typical cyanide leaching circuit with Merrill–Crowe recovery of gold and silver is provided in Figure 5. Zinc precipitation is also used occasionally to treat hot, concentrated carbon strip/eluate solutions.

Solution preparation. Precipitation from cold, dilute gold cyanide solutions can only be achieved effectively from clarified feed solution containing <10 mg/L suspended solids, and preferably <5 mg/L. Consequently, leached slurries must be filtered or thickened and then clarified before precipitation. The gold-bearing solutions produced directly by other processes, such as heap and vat leaching, usually require clarification to produce an acceptable feed for precipitation. Clarification may be achieved in a variety of equipment, depending on the quantity and quality of solution. Such equipment includes plate-and-frame, leaf, candle and disc filters, sand-bed clarifiers, pinned bed, settling tanks, and "double-vee" (or "hopper") clarifiers.

Deaeration. The solution must be deaerated to a dissolved oxygen concentration <1.0 mg/L, and preferably <0.5 mg/L, to avoid undesirable side reactions that reduce precipitation efficiency and consume excess zinc. This is usually achieved mechanically by applying a vacuum to a deaeration vessel (e.g., Crowe tower) through which the solution is allowed to cascade. The vacuum required to sufficiently deaerate the solution varies, depending on the temperature and the altitude of the operation. The vacuum requirement decreases with increasing temperature and with increasing altitude above sea level. For example, at 20°C at sea level, a vacuum of 27.4 in. Hg is required to reduce the dissolved oxygen concentration to 0.5 mg/L, compared with only 22.0 in. Hg at 1,640 m above sea level.

Precipitation. Zinc is added at a rate between 5 and 30 times the stoichiometric precious metals (Au+Ag) requirement, depending on the solution composition and process operating efficiency. For example, a solution containing 5 g/t gold would require a zinc addition rate of 8 g/t solution, using 10 times the stoichiometric requirement. The zinc dust is either added directly to the gold-bearing (pregnant) solution, sometimes called "emulsifying," or may be premixed with cyanide solution, to prepare the zinc surfaces for precipitation, and added as a slurry. Typical zinc consumptions vary from about 1 g/g precious metal precipitated to about 10 g/g, depending on solution conditions and operating efficiency (Marsden and Fuerstenau 1993). A soluble lead salt (e.g., lead nitrate) may be added to the solution after clarification to aid precipitation. Lead is usually added at a rate of approximately 1/20 to 1/10 that of zinc by mass. The retention time allowed for precipitation varies greatly at different operations. The reaction kinetics are such that, in the absence of any interfering species (i.e., oxygen, sulfide ions), precipitation is complete within a few minutes of zinc addition. Reaction vessels with 2–5 minutes of retention time are occasionally provided within circuits, but usually the reaction is achieved in a length of piping between the zinc addition point and the filtration system used for precipitate–solution separation. Further reaction may occur within the filter bed itself.

Filtration. Several different filtration systems have been used to remove the precipitate from the zinc-solution mixture, including various types of plate-and-frame, leaf, disc, and candle filters operating over a range of pressures and flow rates. Diatomaceous earth is often used to aid filtration, either added to the solution during precipitation or used to precoat the filter to create a suitable filter medium, or a combination of the two. Filters may also be pre-coated with zinc.

Process efficiency. The process is capable of producing a "barren" tailings solution containing <0.03 g/t Au, with some operations achieving <0.01 g/t Au. Consequently, depending

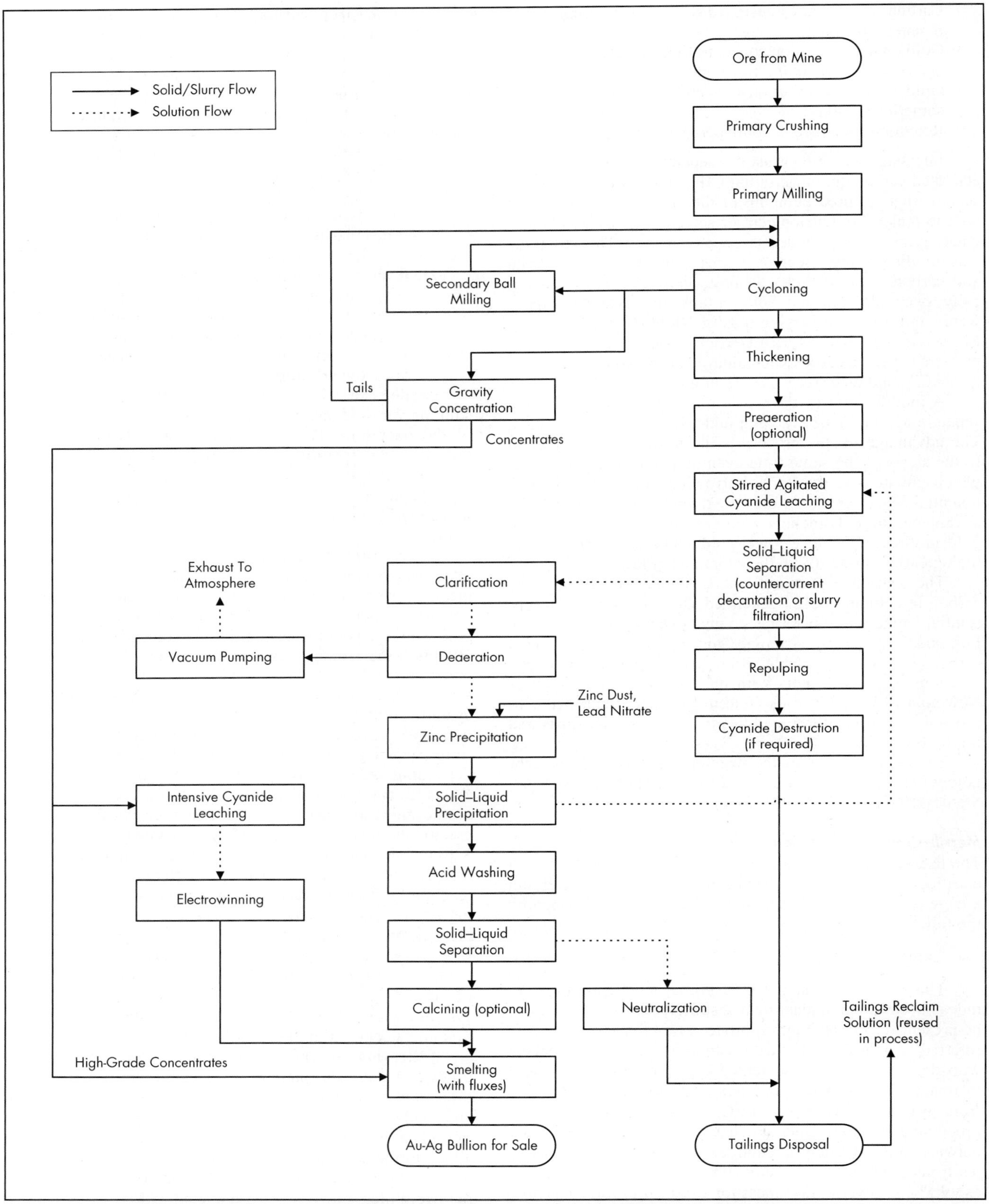

Figure 5 Schematic flow sheet showing a typical configuration for agitated cyanide leaching with zinc precipitation (Merrill–Crowe process) for recovery of gold and silver

on solution feed grade, precipitation efficiency is usually consistently >98%. The barren tailing solution produced by filtration is either recycled to the process or bled to the final tailings disposal system. The solid products of precipitation vary greatly, depending on the original solution composition, the efficiency of precipitation, and the method of solids removal from solution. The most important components and their typical approximate ranges are gold and silver (25%–75%), zinc (3%–20%), diatomaceous earth (0%–25%), silica/silicates (1%–10%), and other metals such as lead, mercury, and copper (1%–10%).

Resin Recovery

Metallurgists in the former U.S.S.R. implemented resin-in-pulp technology in the late 1960s and 1970s, whereby ion exchange resin beads were used to perform the same role as activated charcoal. Several plants were reportedly built in Russia, Uzbekistan, Kazakhstan, and Kyrgyzstan. Complex elution and regeneration systems were required for the removal of base metals, the recovery of gold and silver, and for the removal of other contaminants. Several of these plants were retrofitted with carbon adsorption systems. Subsequently, significant work has been conducted on the development of effective resin systems with considerable success (Fleming and Cromberge 1984). As a result, there have been commercial installations of resin systems at Golden Jubilee (South Africa), Penjom (Malaysia), and Gedabek (Azerbaijan). Barrick Gold commissioned a large ion exchange resin adsorption system for gold recovery from thiosulfate leach slurry at Goldstrike (Nevada) in 2015. This latter application is considered further in the case study later in this chapter.

Effluent Treatment

Many alternatives exist for detoxification of cyanide-containing solutions. Where applicable, the preferred method is to allow the cyanide concentration to decay naturally through the leaching and CIP/CIL circuit to the point at which it reaches levels acceptable for discharge to the tailings containment facility. Many operations are able to meet strict discharge limits to tailings facilities without the need for any form of cyanide destruction other than natural degradation. However, these operations carefully manage cyanide concentrations within the leaching and CIP/CIL circuit, as well as control wash ratios in thickeners using recirculated, reclaimed, and/or fresh water in the circuit. The cyanide degrades further over time in the tailings facility, and the understanding of such degradation mechanisms has improved significantly over the past 25–30 years (Whitlock and Mudder 1986).

Where the above methods are not sufficient to meet the desired target (e.g., the Cyanide Code guideline of ≤50 mg/L WAD [weak-acid-dissociable] cyanide in tailings storage facilities [ICMI 2002]) and/or other imposed regulatory environmental requirements, other methods of detoxification must be used. In the early 1980s, Inco (Sudbury, Ontario, Canada) developed and patented a method using sulfur dioxide and air, with copper as a catalyst, which was applied successfully at several Canadian operations. The primary reaction is as follows:

$$CN^- + SO_2 + O_2 + H_2O = CNO^- + H_2SO_4 \qquad \textbf{(EQ 4)}$$

To achieve an acceptable reaction rate, addition of 30–90 g/t of copper (usually copper sulfate) as a catalyst is generally required, unless the solution contains copper or a longer reaction time can be tolerated in the circuit. Either sulfur dioxide gas or sodium/ammonium metabisulfite can be used as the sulfur dioxide source. Air is injected into the slurry to achieve the required SO_2/O_2 ratio. In practice, air flow rates of 1–2 L/min per liter of solution are required with 3–4 kg SO_2 required per kg CN, and sometimes higher than this. The oxidation is performed between pH 7.5 and 9.5, with pH 9 close to optimal. Lime must be added for pH control due to the formation of sulfuric acid.

After almost 30 years of application at operations throughout the world, the use of sulfur dioxide–air has become the generally preferred and most cost-effective method of cyanide destruction where natural degradation is not adequate.

Many other methods have been tested and used commercially with varying degrees of success. Caro's acid (a mixture of hydrogen peroxide and sulfuric acid) and hydrogen peroxide systems have both been used successfully for cyanide destruction at a variety of operations. The advantages are that free and WAD cyanides are converted to cyanate very rapidly and completely because of the strongly oxidizing conditions. This can be useful in situations where rapid detoxification is required or where there is limited retention time available in existing equipment to achieve detoxification. The disadvantage is the relatively high cost of reagents, which is typically much more costly than the sulfur dioxide–air process to achieve similar residual-free and WAD cyanide concentrations. Caro's acid is used at several Newmont operations in North America.

Biological oxidation for cyanide destruction in low-cyanide-concentration applications was used very successfully at the Homestake mine (South Dakota) since about 1984 and was subsequently applied at several other Homestake operations (Whitlock and Mudder 1986).

Cyanide Recovery and Recycling

Some degree of cyanide recovery and recycling has been practiced for many years in the form of reclaiming solutions for reuse in the leaching process (e.g., from tailings facilities and from tailings thickeners). However, technology for the separation of free cyanide from solution to allow recycling in a concentrated form has been known for more than a century. The method of acidifying cyanide-bearing solutions to generate hydrogen cyanide gas, the removal of such hydrogen cyanide from solution by volatilization with air, followed by adsorption and recovery of hydrogen cyanide into a caustic scrubber system is referred to as the AVR (acidification, volatilization, recovery) process. This approach was applied in the early 1990s at Golden Cross (New Zealand) and DeLamar silver mines (Idaho, United States). Despite the viability and cost-effectiveness of the technology, it has not been applied widely throughout the industry. This is largely a result of the effectiveness of the simpler cyanide recycling procedures described previously (e.g., tailings thickening, reclaim solution collection, and recycling) and the effective management and control of cyanide concentrations within agitated leaching, CIP/CIL, and heap leaching circuits.

Special Requirements for Cyanide-Soluble Copper

The need to be able to effectively treat ores containing significant amounts of cyanide-soluble copper has been recognized for many years, even back to the earliest days of cyanidation in the early 1900s. In the modern era (1972–2012), the Telfer (Australia), Ok Tedi (Papua New Guinea), and El Indio (Chile) operations were all learning how to deal with the presence of significant copper in the feed to their cyanide leaching circuits in the late 1970s and early 1980s. These plants processed ores with variable cyanide-soluble copper content, with up to 500–700 g/t cyanide-soluble copper in feed. The ability to prevent copper loading onto activated carbon by maintaining a sufficiently high free-cyanide concentration by keeping the majority of the copper as the $Cu(CN)_4^{3-}$ complex was recognized in the 1980s (Marsden 2006). Where this was not sufficient to avoid significant copper loading on carbon, other techniques were required. One approach applied at El Indio was the use of a cold cyanide elution step to remove copper loaded onto carbon prior to hot elution, avoiding the co-elution of copper with gold and silver in the conventional (Zadra/Anglo American Research Laboratories, or AARL) elevated-temperature elution step. This allowed the copper eluate solution to be treated and disposed of separately, and avoided the bulk of the copper from reporting to the doré bullion. This approach is used at many operations worldwide.

An important development for the treatment of copper-bearing ores was the development of the sulfidization, acidification, recycling, and thickening (SART) process (MacPhail et al. 1998; Barter et al. 2001). The process involves precipitation of copper sulfide (theoretically Cu_2S) by the addition of sulfuric acid and sodium hydrosulfide (NaSH), which releases the complexed cyanide as hydrogen cyanide (HCN). The solids are separated from the solution phase by thickening and filtration to produce a copper sulfide solid for sale or for further treatment. The thickener overflow solution is re-neutralized with lime and the contained hydrogen cyanide is converted back to calcium cyanide, suitable for recycling and reuse in the cyanide leaching circuit. The first commercial application was at the Telfer operation in 2005 to treat barren solution from leaching of a pyritic flotation concentrate. SART plants have subsequently been installed at Yanacocha (Peru), Maricunga (Chile), Lluvia de Oro (Mexico), and Gedabek (Azerbaijan) to treat pregnant gold-bearing solutions produced by heap leaching.

Alternatives to Cyanide

Since the inception of the cyanidation process in the late 1800s, cyanide has been used almost universally because of its relatively low cost, great effectiveness for gold and silver dissolution, selectivity for gold and silver over other metals, and effectiveness for loading onto (and stripping from) activated carbon. Also, despite some concerns over the toxicity of cyanide, it can be applied with little risk to human health and the environment. The oxidant most commonly used in cyanide leaching is oxygen, usually supplied from air, which contributes to the attractiveness of the process.

Since the mid-1970s, alternative leaching reagent schemes to cyanide have been investigated for any, or combinations of, the following reasons:

- Environmental pressures, and in some cases restrictions or limitations, may make the application of cyanide difficult in certain locations.
- Some alternative reagent schemes provide faster gold (and/or silver) leaching kinetics.
- Several can be applied in acidic media, which may be more suitable for refractory ore treatment.
- Some are more selective than cyanide for gold and silver over other metals (e.g., copper).

Some of the more important reagent systems that have been investigated (or reinvestigated) include chlorine-chloride, thiosulfate, thiocyanate, thiourea, ammonia, ammonia-cyanide, alkaline sulfide, bromine-bromide, iodine-iodide, and other halide combinations. All of the alternative reagent schemes have some significant disadvantages, and, at this time, none appear to be widely applicable without further advances in the technologies. However, thiosulfate has emerged as the front-runner of the alternative schemes, with the specific currently recognized potential application to treat carbonaceous preg-robbing ore types. The reason for the particular applicability of the thiosulfate system is that the thiosulfate gold and silver complexes are not readily adsorbed onto carbonaceous constituents of the ore (Breuer and Jeffrey 2003). This property in itself presents a problem for metal recovery from thiosulfate leach solution in that carbon adsorption is not suitable. However, other options have been investigated with considerable success, including resin adsorption/desorption and direct precipitation from solution (e.g., using copper powder). The application of thiosulfate leaching for gold recovery is discussed further later in this chapter.

Gravity Concentration

Gravity concentration is widely used for the recovery of free gold and gold associated with heavier minerals (e.g., many sulfide and titanium minerals). A variety of equipment is available for this, and recent developments have enabled the recovery of free gold down to about 20 μm in size, and occasionally down to 10 μm. The resulting concentrates may be treated by direct cyanidation, smelting, amalgamation, intensive cyanide leaching, or occasionally, by flotation depending on their mineralogy.

Up until the late 1970s, gravity concentration used jigs, strakes, rockers, sluices, corduroy blankets, spirals, tables, and other devices. More efficient centrifugal concentrators, such as the Knelson and Falcon units, were developed and introduced in the late 1980s and 1990s (Laplante et al. 1990; Knelson and Edwards 1990; Marsden et al. 1993). The centrifugal concentrator was designed to accept minus-6-mm feed and was capable of recovering gold particles down to about 25 μm. In some applications, gold particles of 10 μm diameter and below were recovered (albeit with lower efficiency). Successful installations have included Rosebery, Sunrise Dam, and Kundana (all in Australia), Ashanti-Obuasi (Ghana), Morila (Mali, West Africa), Freeport-Grasberg (Indonesia), Porgera (Papua New Guinea), Dome and Campbell (Canada), Serra Grande (Brazil), Rio Narcea (Spain), Alumbrera (Argentina), Telfer and Cadia (Australia). The Knelson and Falcon centrifugal concentrators have provided the operator with excellent options for gravity gold (and silver) recovery.

Centrifugal concentrators can be installed in several possible configurations in gold extraction flow sheets, including treatment of the following process streams and/or intermediate products:

- All or a portion of cyclone underflow slurry within the primary grinding circuit
- All or a portion of a grinding mill discharge
- All or a portion of the cyclone underflow stream within a regrinding circuit (i.e., after rougher flotation)
- Flotation concentrates or intermediate products

- Fine portion of an ore feed to a mill (e.g., following coarse and fines separation using screening and/or cycloning)
- All or a portion of the coarse cyclone underflow of a reclaimed tailings stream

The use of centrifugal gravity concentrators in conjunction with flotation where the ore contains a significant proportion of the gold in gravity recoverable gold (GRG) form (i.e., approximately >40% to 50% GRG at the target grind size) often increases the overall gold recovery by 2%–5%, depending on the ore mineralogy and the gold particle size. Gravity concentration is particularly useful when there is significant coarse gold present (>150 μm diameter particles) that is more difficult to recover effectively by flotation. The combination of gravity concentration and flotation is particularly effective for ores containing a wide gold size distribution. The application of gravity concentration prior to a chemical treatment process (e.g., cyanide leaching) can often be beneficial to overall gold recovery, because the coarse gold particles are recovered prior to leaching and can be treated separately (i.e., by shaking tables, intensive cyanidation, etc.) for maximum recovery. Coarse gold particles take the longest to leach during atmospheric cyanide leaching, and their removal can reduce the leaching retention time and/or increase the overall gold recovery.

Leaching of Gravity Concentrates

An important development that contributed to the success of the centrifugal gravity concentrator technology was the parallel development of intensive cyanide leaching processes to replace the conventional, and out of favor, amalgamation process. Intensive leaching operates on the principle of achieving fast leaching rates by achieving elevated dissolved oxygen concentrations using a pressurized system, the addition of a suitable oxidant, elevated temperature, application of relatively high cyanide concentration, and combinations of these conditions. During the 1990s and early 2000s, two commercial processes emerged: the Consep Acacia reactor and the Gekko inline leach reactor (Watson and Steward 2002; Longley et al. 2003). As an example, the Acacia process uses cyanide concentrations of up to 15–25 g/L, 3–4 g/L caustic, and 2–10 g/L of a suitable oxidant. Leaching is carried out at 50°–65°C.

These intensive leaching systems have several important advantages over (mercury) amalgamation, including

- The elimination of mercury (and the associated safety, health, and environmental risks),
- The ability to achieve high gold (and silver) recoveries within relatively short leaching times (i.e., 18–24 hours, and sometimes less),
- Compatibility of the resulting gold- and silver-bearing solution with electrowinning systems to recover precious metals from high-grade carbon strip solutions, and
- The ability to supply modular intensive cyanide leaching systems that can be preassembled and delivered to remote locations quickly and efficiently.

The development of centrifugal concentrators and effective, intensive cyanide leaching systems, coupled with Laplante's approach to the evaluation of ores for their amenability to gravity concentration techniques (Laplante 2000), have dramatically changed the approach to gravity gold and silver recovery over the past 25 years.

Amalgamation

In the amalgamation process, relatively pure metal particles (such as gold or silver) are incorporated into liquid mercury and separated from gangue minerals. Metallic gold is readily amalgamated, and throughout history this has been used as an effective means of separating gold from host rock. In the mid-1800s amalgamation accounted for almost 50% of South African gold production. For human health and environmental reasons, amalgamation of ROM ore has gradually diminished. With the exception of its use by informal miners in Indonesia, Russia, China, Brazil, central and West Africa, and some other Latin American countries, the process is now virtually obsolete. Amalgamation is still used for the treatment of gravity concentrates and has the advantages of being a simple and cost-effective process in applications where health and environmental aspects can be effectively controlled. The use of amalgamation is likely to continue to be used for informal placer production and at small-scale operations in some lesser developed countries. However, the application of amalgamation for the treatment of gravity concentrates is rapidly being superseded by intensive cyanide leaching processes and other technologies.

Flotation

Flotation has been applied as an effective method of preconcentration prior to cyanide leaching and/or oxidative pretreatment processes since the earliest days of flotation in the 1910s and 1920s. The large-scale application of flotation at Paracatu (Brazil) is a good example of this, where approximately 130,000 t/d (50 Mt/yr [million metric tons per year]) of ore with a head grade of about 0.4 g/t are processed using a combination of gravity concentration, flotation, and cyanide leaching (Anonymous 2012). Flotation concentrates are leached in dedicated hydrometallurgical (CIL) facilities while the flotation tailings are discarded without leaching.

With the exception of some evolutionary developments in reagents and flotation conditions and progressively larger flotation cells, the basic flotation process has remained largely unchanged from the early configurations. Xanthates are still used as the primary collector for gold- and silver-bearing sulfides. Several special collectors have been developed for gold extraction, and these have been applied with success in many applications (e.g., dithiophosphates). An excellent review of flotation for gold and silver recovery is available in the literature (Kappes et al. 2011).

The application of inert gas-assisted flotation (using nitrogen) by Santa Fe Pacific Gold at the Lone Tree operation (Nevada) reportedly provided a 5%–15% boost in gold recovery to the concentrate and increased the rate of recovery (Kappes 2007).

A generic flow sheet illustrating the use of flotation within a gold extraction flow sheet for a non-refractory or mildly refractory sulfide ore is provided in Figure 6.

Treatment of Flotation Concentrates

The ability to economically grind ores and concentrates to fine sizes (e.g., 80% passing 10–20 μm) provides the opportunity to liberate precious metals from refractory sulfide ores without the need for oxidative treatment. In 2001, Kalgoorlie Consolidated Gold Mines (Western Australia) installed an ultrafine grinding circuit to treat refractory sulfide gold ore to supplement the roaster capacity. The ground product was cyanide leached, achieving a gold recovery of more than 90% (Ellis 2003). Ultrafine

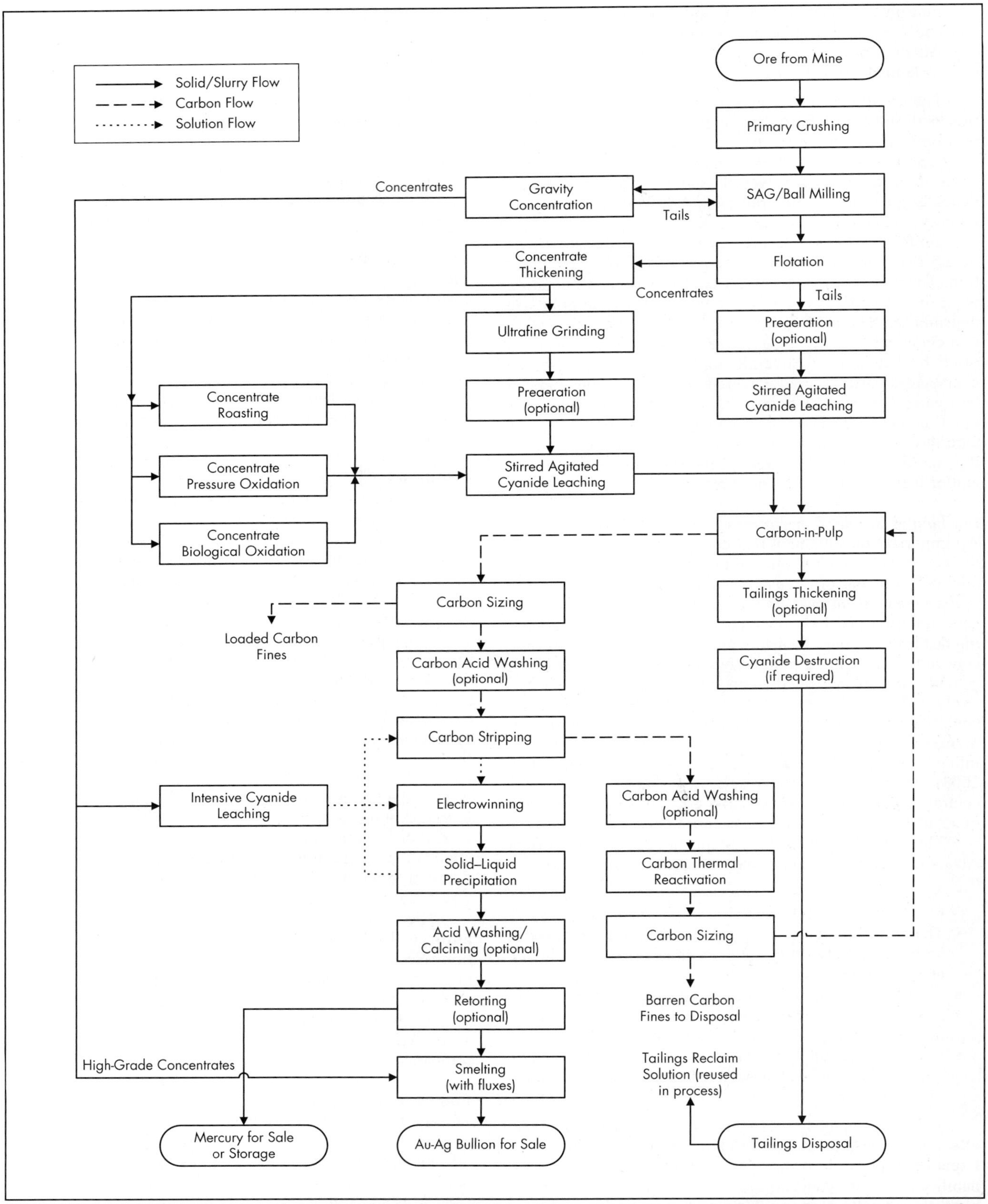

Figure 6 Schematic flow sheet showing options for treatment by flotation and oxidation or ultrafine grinding prior to agitated leaching of the concentrates and optional leaching of tailings

grinding was subsequently installed at Kumtor (Kyrgyzstan) in 2005, Pogo (Alaska, United States) in 2006, and Lake Cowal (Australia) in 2006. The product grind sizes are 80% <12 µm (95% <20 µm), 80% < 10 µm (95% <17 µm), and 80% < 15 µm, respectively, for these operations. Specific energy requirements to grind pyrite concentrate from approximately 80% <75 µm to 80% <10–15 µm vary between about 30 and 80 kW·h/t, depending on the concentrate mineralogy and the specific product size. Typically, Colorado River sand or equivalent, or suitable spherical ceramic product, is used as the grinding media.

Oxidative Pretreatment

Four options are available to pretreat sulfidic and carbonaceous refractory ores (i.e., materials that are not effectively treated by simple cyanidation, flotation, or gravity concentration techniques): roasting, atmospheric oxidation, pressure oxidation, and biological oxidation. These are discussed in the following sections. A schematic flow sheet illustrating the use of flotation and oxidative pretreatment of the concentrates, followed by agitated cyanide leaching, is provided in Figure 6.

Roasting

In the 1980s, significant improvements were made in roasting technology, which brought low-cost roasting back into favor for treatment of refractory gold and silver ores and concentrates. The implementation of fluidized bed (FB) and circulating fluidized bed (CFB) reactors with effective gas scrubbing and cleaning systems resulted in plants with significantly improved temperature control, better control of gas composition, and much lower particulate and sulfur dioxide emissions than conventional roasting technology, for example, open hearth roasting (Marsden and Sass 2014). Between 1988 and 1990, whole-ore FB roasters were commissioned at Big Springs, Cortez, and Jerritt Canyon (all in Nevada), and a CFB roaster to treat concentrates was installed at Gidji/Kalgoorlie Consolidated Gold Mines (Western Australia). A large whole-ore CFB roaster using dry grinding was installed at Newmont's Carlin operation (Nevada) in 1994, and other CFB roaster installations followed at Syama (Mali, West Africa), Minahasa (Indonesia), and Goldstrike (Nevada).

The main advantages of roasting over pressure oxidation include

- Destruction of carbonaceous (preg-robbing) components of the feed,
- Near-complete oxidation of sulfide minerals in very short retention times, and
- The ability of the CFB technology to avoid over-roasting of fine particles due to the nature of the cyclone hot-gas separation technology employed.

However, the disadvantages of the new, improved roasting technology include

- The energy and cost required to eliminate the moisture from the ore or concentrate (i.e., by dry grinding, drying, or slurry feed to the roaster),
- The need to fix sulfur dioxide in the bed and/or gas scrubbing requirements,
- The need to either fix arsenic in the roaster bed or remove it from the gas phase, and
- Cost of gas handling and scrubbing systems to remove undesirable constituents.

The roasting operations of Barrick Gold and Newmont Gold in Nevada are described in more detail later in this chapter, with more detailed flow sheets provided.

Atmospheric Oxidation

Dissolved oxygen in solution under ambient conditions is capable of oxidizing some sulfide minerals. This can be applied as a simple, low-cost, preaeration step before cyanide leaching to oxidize and/or passivate the surfaces of some of the more reactive reagent-consuming sulfides such as pyrrhotite and marcasite. This treatment is often only capable of partial (surface) oxidation of sulfides and is usually unsuitable for the treatment of ores where gold is intimately associated with sulfides.

Air or oxygen is sparged into agitated tanks, and sufficient retention time, typically 4 to 24 hours, is provided to allow adequate oxidation and/or passivation of cyanide-consuming minerals. Air is a considerably cheaper source of dissolved oxygen than pure oxygen. However, in some cases, pure oxygen can substantially increase oxidation rates and improve the degree of sulfide oxidation obtained with a corresponding further decrease in cyanide consumption. Soluble lead salts and pH modifiers, such as lime and cement, may be added to the first stage of preaeration or, more typically, in the grinding circuit. Low-pressure (atmospheric) oxidative pretreatment has been applied at many operations for the treatment of ores (e.g., Homestake Lead [United States], Pamour Porcupine [Canada], and East Driefontein [South Africa]) and for concentrates (e.g., Agnico Eagle [Canada]).

Pressure Oxidation

The use of pressure oxidation to treat gold- and silver-bearing sulfide ores was investigated in the 1970s and 1980s. This work built upon the established use of pressure oxidation technology that had been applied successfully for many years in the zinc and nickel industries. In 1985, an acidic pressure oxidation plant using horizontal multi-compartment autoclaves was installed at the Homestake McLaughlin mine (California, United States) to treat a pyritic ore. This was followed in 1986 by a pressure oxidation circuit to treat gold-bearing sulfide flotation concentrates at Sao Bento (Brazil). These plants established pressure oxidation as a viable method for treating a range of refractory ores and concentrates, capable of oxidizing >95% of the contained sulfide sulfur and achieving high gold recoveries by subsequent cyanide leaching (i.e., >90%). As a result, several pressure oxidation plants and expansions were commissioned between 1988 and 2000, including Goldstrike, Getchell, Lone Tree, and Twin Creeks (all in Nevada); Campbell Red Lake and Con (Canada); Lihir and Porgera (Papua New Guinea); and Macraes (New Zealand). The most recent pressure oxidation project is the large, 24,000-t/d operation at Pueblo Viejo (Dominican Republic), which achieved commercial production in 2013.

The advantages of pressure oxidation compared to roasting include the following:

- Off-gas handling requirements (vent gas handling rather than roaster off-gas handling and treatment) are greatly reduced.
- Encapsulation of gold values in the calcine is avoided.
- Arsenic is fixed as stable ferric arsenate (assuming sufficient iron present), rather than being volatilized in the roasting process.

- Typically (but not always), higher gold recoveries are achieved.
- The ability to accommodate slurry feed and avoid the need for dry grinding is realized in some cases.

However, the disadvantages of pressure oxidation are the loss of most of the silver to jarosite formation (which is essentially insoluble in cyanide solution), and the higher capital and operating cost.

In the late 1980s, a hot lime boil process was developed to address the issue of silver jarosite formation (Berezowsky and Weir 1989). This consisted of agitating the pressure oxidation discharge slurry for 1–2 hours at 80°–90°C with lime to maintain a pH of about 11. Silver extractions were significantly increased, typically from 5%–20% to over 80%. In some cases, this procedure also increased gold extraction due to the release of gold occluded in jarosite and iron precipitates. However, there was a cost associated with this process (energy for heat and agitation, lime consumption), although the heat could be obtained from the pressure oxidation step if the feed contained sufficient sulfur. Several enhancements to this process were made subsequently, including a modified hot lime process that was adopted at Pueblo Viejo following a hot cure step, which reduced lime consumption. Other options to enhance silver extraction following pressure oxidation have been proposed (Simmons and Gathje 2003).

Case studies for the application of pressure oxidation in gold extraction flow sheets are provided later in this chapter.

Biological Oxidation

Throughout the 1970s and 1980s, researchers at Cardiff University (Wales), University of British Columbia (Canada), University of New Mexico (United States), and Gencor (South Africa), among many others, investigated bacterial oxidation to treat sulfide refractory gold and silver ores. In 1986, a 10-t/d plant was commissioned by Gencor at Fairview (South Africa) to process flotation concentrates, and this facility operated successfully for many years. A biological oxidation circuit was installed at Sao Bento (Brazil) in 1991 to partially oxidize sulfide concentrate ahead of the pressure oxidation circuit to increase throughput. A larger, 720-t/d agitated biological oxidation system was installed to treat sulfide flotation concentrates at Ashanti-Obuasi (Ghana) in 1994. This application was successful and was subsequently expanded to 960 t/d, firmly establishing biological oxidation as a viable option for refractory gold and silver sulfide concentrates. Additional installations followed at Harbour Lights, Wiluna, Youanmi, Fosterville and Beaconsfield (all in Australia); Coricancha (Peru); Bogoso (Ghana); Suzdal (Kazakhstan); Kokpatas (Uzbekistan); Olimpiada (Russia); and Laizhou and Jinfeng (China) (Brierly 2014).

An important attribute of materials that are most suitable for treatment by biological oxidation is the ability to liberate a substantial portion of the contained gold (and silver) without needing to achieve complete, or essentially complete, oxidation of the refractory sulfide minerals. Ores and concentrates containing gold and silver mineralization, either hosted within more reactive sulfide minerals or located along the surfaces, within fractures and grain boundaries of sulfide minerals, are most amenable to biological oxidation. Unlike pressure oxidation and roasting, which achieve complete or near-complete oxidation within short retention times (minutes to hours), biological oxidation in agitated tank systems require several days (typically three to five) to achieve acceptable levels of sulfide oxidation and gold-silver liberation. For biological heap leaching systems, retention times of at least 90 days are required.

The major advantages of biological oxidation over pressure oxidation and roasting include the following:

- Operation at atmospheric pressure and moderate temperatures
- Relatively simple flow sheet and less complex equipment
- Lower capital cost for unit of throughput (although this is strongly dependent on the retention time requirements)
- Potentially lower operating costs

However, the disadvantages are the typically lower gold extraction achieved and the less robust operating control capability due to the longer residence time and slower process response to changes.

Other Methods

A significant number of alternative methods and processes have been proposed, and in some cases applied, for processing of refractory gold ores. These include the use of chlorination, as applied at Jerritt Canyon and Carlin (both in Nevada) in the 1980s, and nitric acid/NO_x oxidation. At the current time, these processes have limited commercial potential. More information is available in the literature (Marsden and House 2006).

Smelting and Refining

Refining processes are used to upgrade the intermediate products of earlier recovery processes, principally, loaded cathodes, electrowinning sludge, zinc precipitate and mercury–gold amalgam, all of which generally contain >10% Au. In addition, many by-products of gold extraction and refining processes are treated further for gold recovery (e.g., loaded carbon fines, refinery slags, high-grade dusts, old crucibles and furnace liners, and refinery floor sweepings). The methods applied depend on the nature of the material, and the type and amount of impurities present.

Refining is usually performed in two stages, as follows:

1. Treatment at the point of production (e.g., at the mine site) to produce a crude bullion (typically 90%–99.5% total precious metals), sometimes containing lead, copper, zinc and iron
2. Refining of crude bullion from the first stage to produce high-purity gold and silver for sale

The first stage can be applied at relatively low cost at the mine site, even for small quantities of precious metal production. This process involves smelting with modest amounts of fluxes (borax, silica, sodium carbonate, sodium nitrate, etc.) yielding a low-volume product that is easy to transport and can be sampled representatively, either in the molten state or as homogeneous solid bullion for accurate accounting purposes. In some cases, pretreatment may be required prior to smelting, for example, acid washing of gold-silver loaded mild steel wool or zinc precipitates and/or retorting of precious metal products containing mercury. Alternatively, lower-grade products of recovery processes (e.g., zinc precipitate, cathodes, electrowinning sludge) may be shipped directly to an independent third-party refinery, although this is rarely done.

The second stage is usually performed by dedicated refineries, often located close to major gold/silver-producing regions, which can process the doré bullion more economically on a bigger scale; for example, the Rand Refinery Limited (Germiston, South Africa), Johnson Matthey (Salt Lake City, Utah, United States), Handy and Harman Electronic Materials (North Attleboro, Massachusetts, United States), and the Perth Mint (Perth, Western Australia). Several general methods are available to upgrade crude bullion to these high-grade products, namely, pyrorefining, hydrorefining, and electrorefining. Pyrorefining techniques are used for the direct production of 99.6% pure gold. This product is then further treated by electrorefining to produce ≥99.99% pure gold. Hydrorefining methods are used by some refineries to produce 99.6%–99.9% gold. The purity of the final product depends on the ultimate use: typically 99.6% for jewelry and monetary bullion, and ≥99.99% for production of coins.

Further information on bullion smelting and refining processes is available in the literature (Marsden and House 2006)

Tailings Disposal

Gold extraction operations generate a variety of waste products, which must be disposed of responsibly, in compliance with environmental regulations, and as economically as possible. The control and treatment of gold extraction by-products and effluents must be considered as an integral part of all gold extraction processes. Tailings from cyanide processes are generally stored in lined containment facilities with effective measures maintained for the control and management of such storage facilities. The specific liner requirements depend on regulatory requirements (e.g., local, regional, national) as well as established industry guidelines and must be considered on a case-by-case basis by environmental specialists.

In some cases, there may be metallurgical incentives for treating effluents prior to disposal, such as economic benefits from recycling of reagents and/or the recovery of metal values, or the need to remove components that have an adverse effect on the primary process when a portion of the effluent stream is recycled. In such cases, effluents can be (1) treated for reagent and/or metal recovery prior to disposal in the containment facility, (2) treated in situ, or (3) a portion of the effluent can be returned for further treatment (e.g., tailings impoundment decant solution pumped back to the primary process).

The need for detoxification depends on the threat to wildlife, the possibility (or requirement) of a release to the environment, the risk of contaminating surface water or groundwater, or a combination of these factors. The threat to wildlife is a recognized problem in many locations. The greatest threat is presented by toxic cyanide species, that is, free cyanide and WAD cyanide species. However, extremes of pH and high concentrations of other metals are also potentially harmful. The threat to wildlife can generally be significantly reduced by the following procedures:

- Minimizing the exposed surface area of solution, which reduces the attractiveness of the water body to birds and other wildlife. This applies to solution ponds, tailings ponds, and heap leach pads (i.e., prevention of solution ponding on pads).
- Preventing wildlife access to toxic solutions, for example, by enclosing effluent containments (fencing, tanks versus ponds) and covering solution ponds with netting or other suitable material (e.g., polyethylene), plastic balls, or overlapping plastic panels.
- Minimizing the concentration of potentially harmful species in solution by good process design and control of reagents (cyanide, lime, etc.) within the process.

Where the application of such techniques is impractical, detoxification may be required to reduce toxic constituents to safe levels. The possibility of contaminating surface water or groundwater depends on the location, the type and effectiveness of containment device used, the operating philosophy and efficiency, the level of toxicity of the material, and the various (local, regional, national) environmental regulations under which the operation falls, as well as any application industry guidelines. These factors vary greatly, and any need for detoxification must be considered on a case-by-case basis.

The need to discharge effluent into the environment depends on the overall water balance of the operation, which is determined by many factors, including the processing method and water requirements, climate (evaporation and precipitation), the amount of water generated during mining operations, and the method of effluent disposal. Operations that have a positive water balance, where the combined mining and processing operations generate more water than they consume, must discharge solution into the environment. Operations with a large negative water balance may need to discharge effluent under certain circumstances, for example, following a period of high rainfall or as a result of process equipment failure. Effluents that are discharged may have the potential to contaminate surface water or groundwater and may need to be treated to meet environmental regulations prior to discharge.

ECONOMIC CONSIDERATIONS

Operating Cost

Operating costs for gold and silver extraction and recovery processes vary significantly depending on the specific ore mineralogy, the metallurgical response, and the flow sheet selected. Other factors, such as location, climate, environmental and permitting requirements, financing availability, and sociopolitical factors, may affect the operating cost. Table 8 provides a generic indication of operating costs for 13 flow sheets applied around the world for gold and silver recovery. These generic costs have been developed using the following major assumptions:

- Basis: 2015, U.S. dollars
- Throughput rate = 10,000 t/d (10,750 t/h at 93% availability)
- Throughput rate = 3.65 Mt/yr
- Ore gold grade = 3 g/t (used to calculate gold production for cost estimation purposes only)
- Ore sulfur grade (oxidation processes) = 4.0%, unless otherwise stated
- SAG mill specific energy consumption = 7 kW·h/t
- Bond ball mill work index = 14 kW·h/t
- Abrasion index = moderate
- Wear steel price = $1,200/t
- Sodium cyanide price = $2.80/kg
- Lime price = $0.10/kg
- Power unit price = $0.065/kW·h

Labor rates have been estimated using a range-based approach. The upper end of the range considers western U.S. market rates with 35% burden overhead. The lower end of the range reflects labor rates in regions of the world where market rates are significantly lower (e.g., parts of Africa and Asia).

Table 8 Generic operating cost estimates for selected gold and silver extraction processes, 2015 basis, 10,000 t/d processing rate, western United States location*

Item	Run-of-Mine Heap Leaching, CIC	Crushing Heap Leaching, CIC	Fine Crushing Agglomeration, Heap Leaching, CIC	Milling, Gravity Concentration, Flotation, Third-Party Smelting	Milling, Gravity Concentration, Flotation, Concentrate Leaching, CIP	Milling, Gravity Concentration, Flotation, Concentrates and Tailings, CIP	Milling, Gravity Concentration, Leaching, CIP
Throughput rate, t/d	10,000	10,000	10,000	10,000	10,000	10,000	10,000
Throughput rate, Mt/yr	3,650,000	3,650,000	3,650,000	3,650,000	3,650,000	3,650,000	3,650,000
Feed gold grade, g/t	1.00	1.00	1.00	2.00	2.00	2.00	2.00
Feed sulfur grade, %	<1	<1	<1	<1	<1	<1	<1
Feed carbonate grade, %	NA†	NA	NA	NA	NA	NA	NA
Flotation recovery assumption, %	0	0	0	90	90	90	0
Leach extraction assumption, %	50	75	82	0	95	95	92
Overall recovery, %	50	75	82	86	86	90	92
Gold production, oz/yr	58,676	88,014	96,229	200,672	200,672	210,060	215,928
Operating Cost, $/t							
Labor	0.35–1.06	0.55–1.67	0.71–2.15	0.68–2.05	0.77–2.34	0.83–2.50	0.83–2.50
Power	0.12	0.31	0.41	1.04	1.43	1.69	1.56
Wear steel	0.00	0.24	0.36	1.50	1.59	1.59	1.50
Reagents							
Oxygen (excluding power)	0.00	0.00	0.00	0.00	0.00	0.00	0.00
Cyanide	0.42	0.56	0.56	0.00	0.28	1.12	1.12
Lime	0.10	0.10	0.10	0.05	0.08	0.10	0.10
Flotation	0.00	0.00	0.00	0.80	0.80	0.80	0.00
Sulfuric acid	0.00	0.00	0.00	0.00	0.00	0.00	0.00
Flocculant	0.00	0.00	0.00	0.25	0.25	0.38	0.38
Sodium metabisulfite/ ammonium metabisulfite	0.00	0.00	0.00	0.00	0.05	0.13	0.13
Other	0.20	0.20	1.20	0.20	0.30	0.30	0.30
Subtotal reagents	0.72	0.86	1.86	1.30	1.76	2.82	2.02
Maintenance supplies	0.12	0.25	0.33	0.33	0.45	0.48	0.45
Operating supplies	0.09	0.19	0.25	0.24	0.34	0.36	0.34
General and administrative	0.25	0.50	0.50	0.50	0.50	0.50	0.50
Off-site (refining, transport)	0.19	0.29	0.32	10.66	0.68	0.73	0.71
Total process operating cost, excluding labor	1.49	2.65	4.02	15.57	6.74	8.18	7.08
Total process operating cost, low labor	1.84	3.20	4.73	16.25	7.51	9.01	7.91
Total process operating cost, high labor	2.55	4.32	6.17	17.62	9.08	10.68	9.58

(continues)

Labor costs are strongly site specific and it may be difficult in such cases to provide all of the required expertise, capacity, and experience necessary to operate more complex facilities (e.g., those using oxidative pretreatment) with such a labor structure. Each operation/project should evaluate labor requirements and costs on a rigorous basis based on the process requirements and the local/regional labor market and rates.

The operating cost estimates include processing only. Costs for mining; mine development; geological, geotechnical, and hydrological support; tailings disposal facility operation; tailings/water supply pumping requirements over an extended distance; environmental monitoring and controls; and site-wide general and administrative costs are excluded from the operating cost estimates provided.

Process operating costs may vary outside the ranges specified in Table 8 due to variation in key assumptions (i.e., SAG mill grindability, Bond ball mill work index, ore abrasion index, cyanide and lime consumptions) and commodity unit prices (e.g., steel, cyanide, lime, and power). Also, the specific site location may have a significant effect on operating cost due to excess freight, logistics, taxes, fees, and duties incurred. As such, the operating costs should be considered as indicative only and they are not suitable for application to any particular project without further assessment, evaluation, and engineering work.

Capital Cost

Generic capital cost estimates for 13 gold and silver extraction and recovery process flow sheets are provided in Table 9. These were developed based on an assessment of more than 20 gold and silver projects developed between 2010 and 2015. The cost estimates are reported on a 2015 U.S. dollar

Table 8 Generic operating cost estimates for selected gold and silver extraction processes, 2015 basis, 10,000 t/d processing rate, western United States location* (continued)

Item	Milling, Gravity Concentration, Flotation, Roasting, Leaching, CIP	Milling, Gravity Concentration, Flotation, Pressure Oxidation, Leaching, CIP	Milling, Gravity Concentration, Flotation, Biooxidation, Leaching, CIP	Milling, Gravity Concentration, Roasting, Leaching, CIP (4% S)	Milling, Gravity Concentration, Roasting, Leaching, CIP (2% S)	Milling, Gravity Concentration, Pressure Oxidation, Leaching, CIP (4% S)
Throughput rate, t/d	10,000	10,000	10,000	10,000	10,000	10,000
Throughput rate, Mt/yr	3,650,000	3,650,000	3,650,000	3,650,000	3,650,000	3,650,000
Feed gold grade, g/t	3.00	3.00	3.00	3.00	3.00	3.00
Feed sulfur grade, %	4.0	4.0	4.0	4.0	2.0	4.0
Feed carbonate grade, %	4.0	4.0	4.0	4.0	2.0	4.0
Flotation recovery assumption, %	90	90	90	0	0	0
Leach extraction assumption, %	88	90	88	88	88	90
Overall recovery, %	79	81	79	88	88	90
Gold production, oz/yr	278,828	285,165	278,828	309,809	309,809	316,850
Operating Cost, $/t						
Labor	1.25–3.80	1.25–3.80	1.15–3.49	1.46–4.41	1.46–4.41	1.46–4.41
Power	5.89	5.89	5.58	6.32	4.83	6.32
Wear steel	1.59	1.59	1.59	1.50	1.50	1.50
Reagents						
Oxygen (excluding power)	0.65	0.65	0.54	0.72	0.40	0.72
Cyanide	0.28	0.28	0.28	1.40	1.40	1.40
Lime	1.20	2.50	1.50	1.50	1.50	3.00
Flotation	0.80	0.80	0.80	0.00	0.00	0.00
Sulfuric acid	0.00	0.40	0.00	0.00	0.00	2.50
Flocculant	0.63	0.63	1.13	1.13	0.63	1.25
Sodium metabisulfite/ ammonium metabisulfite	0.13	0.13	0.25	0.25	0.25	0.25
Other	0.75	0.75	0.75	0.75	0.75	0.75
Subtotal reagents	3.80	6.13	5.25	5.75	4.93	9.87
Maintenance supplies	2.64	2.88	2.19	3.01	2.79	3.17
Operating supplies	1.88	2.05	1.46	2.15	1.99	2.27
General and administrative	1.00	1.00	1.00	1.00	1.00	1.00
Off-site (refining, transport)	0.63	0.65	0.65	0.69	0.69	0.72
Total process operating cost, excluding labor	17.43	20.19	17.70	20.42	17.74	24.85
Total process operating cost, low labor	18.68	21.44	18.85	21.88	19.20	26.31
Total process operating cost, high labor	21.23	23.99	21.19	24.83	22.15	29.26

*Units are in US$/t processed, unless otherwise stated.
†NA = not applicable

basis. The estimates are presented as a range. The direct costs consider mechanical equipment price and installation cost (materials and direct labor hours). The indirect portion considers a factor of 38% of direct costs to cover engineering, procurement, and construction management (EPCM), any applicable EPCM fee, mobile cranes, construction camp and facilities, freight, taxes, duties and fees, vendor support, initial reagent/material fills, spare parts, commissioning, and start-up support. Owners' costs are excluded from the estimates. The estimates exclude mining equipment, mine development, mine truck shop and support facilities, access road to site, main power line to site, tailings disposal facility, and any long-distance conveying and/or piping requirements due to specific site location.

The indirect portion of capital cost can vary significantly depending on the approach used to execute a particular project. The factor applied to develop the estimates in Table 9 considers a mid-tier to major EPCM contractor executing the project on behalf of a client. Lower cost approaches are potentially available to clients using internal project management resources and hybrid project management teams (client and third-party resources combined). Alternative approaches have proven to be effective for some projects.

Table 9 Generic capital cost ranges for selected gold and silver extraction processes, 2015 basis, 10,000 t/d processing rate (US$, millions)*

Area (all values in US$ millions), 2015 Basis	Run-of-Mine Heap Leaching, CIC	Coarse Crushing Heap Leaching, CIC	Fine Crushing Agglomeration, Heap Leaching, CIC	Milling, Gravity Concentration, Flotation, Third-Party Smelting	Milling, Gravity Concentration, Flotation, Concentrate Leaching, CIP	Milling, Gravity Concentration, Flotation, Concentrates and Tailings, CIP	Milling, Gravity Concentration, Leaching, CIP
Site earthworks	7–11	9–14	11–17	11–17	16–24	16–24	16–24
Primary crush, stockpile, reclaim	0	15–23	15–23	15–23	15–23	15–23	15–23
Secondary crush	0	13–20	13–20	0	0	0	0
Tertiary crush, agglomeration	0	0	17–26	0	0	0	0
Grinding	0	0	0	28–42	28–42	28–42	28–42
Gravity concentration	0	0	0	5–8	5–8	5–8	5–8
Flotation	0	0	0	6–9	6–9	6–9	0
Regrinding	0	0	0	3–5	3–5	3–5	0
Thickening/filtration	0	0	0	4–6	2–3	4–6	4–6
Oxidative pretreatment	0	0	0	0	0	0	0
Neutralization and thickening	0	0	0	0	0	0	0
Agitated leaching/heap leaching	10–15	10–15	10–15	0	3–5	6–9	5–8
CIP/CIL/CIC	5–8	5–8	5–8	0	3–5	8–12	8–12
Carbon stripping and regeneration	6–9	8–12	8–12	0	13–20	13–20	13–20
Electrowinning and smelting	1–2	2–4	2–4	0	3–5	3–5	3–5
Cyanide destruction	0	0	0	0	2–3	2–3	2–3
Tailings thickening	0	0	0	2–3	2–3	2–3	2–3
Reagents—oxygen plant	0	0	0	0	0	0	0
Reagents—limestone, lime, other	1–2	2–3	2–3	2–3	4–6	5–8	5–8
Ancillaries and support facilities	5–10	8–12	10–15	8–12	11–17	11–17	11–17
Services—power, air, water	3–5	5–8	6–9	5–8	6–15	6–15	6–15
Laboratory, shop	3–6	3–6	3–6	4–8	4–8	4–8	4–8
Subtotal directs	41–68	80–125	102–158	93–144	126–201	137–217	127–202
Midpoint direct capital	55	103	130	119	164	177	165
Indirect costs (at 38% of directs)	21	39	49	45	62	67	63
Total directs and indirects	62–89	136–186	151–207	138–189	188–263	204–284	190–265

(continues)

Capital costs are strongly dependent on specific characteristics and properties of the ore, site location characteristics, as well as the approach to process engineering design, equipment selection and sizing, and the safety/capacity factors applied during the design stage. As such, the generic estimates provided in Table 9 are intended as high-level indicative guidance only and are not suitable for the planning or design of any specific project or operation. It is specifically noted that there are operations and projects that fall outside of the estimate ranges. Each individual project should be advanced through a logical progression of conceptual, preliminary, and definitive feasibility studies, with capital cost estimates developed by qualified engineering professionals using the available metallurgical testing data and site-specific information. Capital cost estimates are typically developed to the following levels of accuracy:

- Conceptual or scoping stage: ±30%–40%
- Preliminary feasibility study: ±20%–25%
- Feasibility study: ±15% (sometimes +15%, –10%)

ENVIRONMENTAL, HEALTH, AND SAFETY ASPECTS

Environmental Regulations

In parallel with the resurgence of the gold mining industry between 1972 and 1990, there was a period of increasing environmental awareness and a major movement toward control of factors affecting the environment, particularly in developed countries but also worldwide. A major effort with respect to the gold and silver extraction industry was the publication of *International Cyanide Management Code for the Manufacture, Transport, and Use of Cyanide in the Production of Gold* (Cyanide Code) in 2002, to which a large number of gold and silver producers that use cyanide have committed to follow (ICMI 2002). The Cyanide Code was developed by the International Cyanide Management Institute (ICMI), a nonprofit organization set up under the United Nations Environment Programme and the International Council on Metals and the Environment. All of this activity represented significantly increased emphasis on the control and treatment

Table 9 Generic capital cost ranges for selected gold and silver extraction processes, 2015 basis, 10,000 t/d processing rate (US$, millions)* (continued)

Area (all values in US$ millions), 2015 Basis	Milling, Gravity Concentration, Flotation, Roasting, Leaching, CIP	Milling, Gravity Concentration, Flotation, Pressure Oxidation, Leaching, CIP	Milling, Gravity Concentration, Flotation, Biooxidation, Leaching, CIP	Milling, Gravity Concentration, Roasting, Leaching, CIP (4% S)	Milling, Gravity Concentration, Roasting, Leaching, CIP (2% S)	Milling, Gravity Concentration, Pressure Oxidation, Leaching, CIP
Site earthworks	20–30	20–30	20–30	20–30	20–30	20–30
Primary crush, stockpile, reclaim	15–23	15–23	15–23	15–23	15–23	15–23
Secondary crush	0	0	0	0	0	0
Tertiary crush, agglomeration	0	0	0	0	0	0
Grinding	28–42	28–42	28–42	28–42	28–42	28–42
Gravity concentration	5–8	5–8	5–8	5–8	5–8	5–8
Flotation	6–9	6–9	6–9	0	0	0
Regrinding	3–5	3–5	3–5	0	0	0
Thickening/filtration	2–3	2–3	2–3	6–9	6–9	6–9
Oxidative pretreatment	54–80	74–110	46–69	70–105	60–90	83–127
Neutralization and thickening	5–8	5–8	5–8	11–17	11–17	11–17
Agitated leaching/heap leaching	3–5	3–5	3–5	6–9	6–9	6–9
CIP/CIL/CIC	3–5	3–5	3–5	9–13	9–13	9–13
Carbon stripping and regeneration	13–20	13–20	13–20	13–20	13–20	13–20
Electrowinning and smelting	3–5	3–5	3–5	3–5	3–5	3–5
Cyanide destruction	2–3	2–3	2–3	2–3	2–3	2–3
Tailings thickening	2–3	2–3	2–3	3–5	3–5	3–5
Reagents—oxygen plant	6–9	6–9	6–9	6–9	4–6	6–9
Reagents—limestone, lime, other	18–27	18–27	18–27	20–30	14–21	20–30
Ancillaries and support facilities	13–20	13–20	13–20	13–20	13–20	13–20
Services—power, air, water	8–24	8–24	8–24	9–27	9–27	9–27
Laboratory, shop	4–8	4–8	4–8	4–9	4–9	4–9
Subtotal directs	213–337	233–367	205–326	243–384	225–357	256–406
Midpoint direct capital	275	300	266	314	291	331
Indirect costs (at 38% of directs)	105	114	101	119	111	126
Total directs and indirects	318–442	347–481	306–427	362–503	336–468	382–532

*Excludes costs for mining equipment, mine development, mine truck shop and support facilities, access road to site, main power line to site, tailings disposal facility, and any long distance conveying/piping requirements due to specific site location. Also excludes any specific contingency requirements.

of gold extraction by-products and effluents, which should be considered as an integral part of gold extraction processes.

Cyanide Safety and the Cyanide Code

The mining industry has used cyanide for gold and silver extraction and recovery for more than 125 years. Cyanide is generally the preferred reagent to use for the extraction and recovery of gold and silver because it is highly selective for dissolution of these metals over many other elements that are present in the ores, it is available at competitive cost, and it naturally decomposes into significantly less toxic and ultimately nontoxic species over time. Although cyanide is an acutely toxic chemical in certain concentrations, with the proper management and control of its use, it can be applied safely and without harm to humans, wildlife, aquatic life, and the environment in general.

Care and diligence is required for the safe and effective use of cyanide. Methods, procedures, and management controls are available to gold and silver producers that use cyanide in the Cyanide Code publication (ICMI 2002).

The Cyanide Code provides a set of principles and standards of practice to be followed by signatories. The principles are as follows:

1. **Production.** Encourage responsible cyanide manufacturing by purchasing from manufacturers who operate in a safe and environmentally protective manner.
2. **Transportation.** Protect communities and the environment during cyanide transport.
3. **Handling and storage.** Protect workers and the environment during cyanide handling and storage.
4. **Operations.** Manage cyanide process solutions and waste streams to protect human health and the environment.

Table 10 Key process statistics for the Carlin gold quarry (2014 basis, approximate data)

Description	Mill 5 Flotation, Tails Leaching, and CIP	Mill 6 Roasting and CIL	Heap Leaching and CIC	Carlin Operations Total*
Ore throughput rate, t/d	13,700	10,000	Variable	—
Ore throughput rate, Mt/yr	4.7	3.2	Variable	—
Ore grade, g/t	—	—	—	1.7 open pit/ 8.0 underground
Plant availability, %	94	88	100	—
Gold recovery, %	62	87 (82–90)	63 (55–70)	—
Gold production, oz/yr	—	—	—	907,000
Gold production, kg/yr	—	—	—	28,210

*Ore grade data are reserve grades as reported by Newmont Gold (2014).

5. **Decommissioning.** Protect communities and the environment from cyanide through development and implementation of decommissioning plans for cyanide facilities.
6. **Worker safety.** Protect workers' health and safety from exposure to cyanide.
7. **Emergency response.** Protect communities and the environment through the development of emergency response strategies and capabilities.
8. **Training.** Train workers and emergency response personnel to manage cyanide in a safe and environmentally protective manner.
9. **Dialogue.** Engage in public consultation and disclosure.

Gold mining companies and the producers and transporters of cyanide used in gold and silver mining operations can become signatories to the Cyanide Code. By becoming a signatory, a company commits to follow the Cyanide Code's principles and implement its standards of practice, or in the case of producers and transporters, the principles and practices identified in their respective verification protocols. Cyanide Code signatories' operations are required to be audited by an independent third-party auditor to verify their compliance with the Cyanide Code. The Cyanide Code is administered by the ICMI. A list of signatories to the Cyanide Code is maintained by the ICMI (www.cyanidecode.org).

Mercury

The process of amalgamation, whereby mercury is used to recover free gold and silver from ores and concentrates, has been applied historically. Mercury presents a hazard to human health and to the environment; therefore, the control and management of mercury in gold and silver operations is an important factor (Van Zyl and Eurick 2000). Although the use of amalgamation is declining, the process is still applied in some parts of the world. Also, mercury occurs naturally in many gold and silver ores and is often liberated, partially extracted, and recovered in many of the processes applied for gold and silver recovery. In fact, a significant portion of mercury metal production that is supplied to meet industrial and chemical needs globally comes as a by-product from gold and silver production.

Mercury is a liquid at ambient temperatures with an appreciable vapor pressure (1.3×10^{-3} mm at 20°C) and significant solubility in air-free water (6.4×10^{-7} g/L). Because of the high vapor pressure, mercury in ores and concentrates can volatilize and accumulate in certain areas of processing facilities. In such cases, special design procedures must be followed, including the provision of contained air space, with air extraction and scrubbing facilities. Also, steps to prevent accumulation of condensed mercury beads within and around processing equipment must be taken, especially in gold rooms and concentrates handling facilities. This is accomplished by coating walls, ceilings, and other structural components with a nonporous epoxy, resin, or other suitable product. Furthermore, mercury traps should be provided for capture of mercury under water, with easy and efficient systems for removal and containment of accumulated mercury.

Because of its high vapor pressure, mercury is well suited to recovery by retorting. In this case, mercury-containing products (typically electrowinning cathodes and sludges, zinc precipitates, or products of amalgamation processes) are heated at 425°–550°C in a specially designed chamber (retort) to drive off mercury as a vapor. The exhaust gas is collected, cooled, and the mercury is recovered by simple distillation. This process can be extremely efficient with respect to the removal of mercury from the source material as well as collection of mercury from the vapor (Gale and Weldon 2002).

Locations that use mercury (i.e., for amalgamation) or where mercury occurs naturally within the material to be treated should implement monitoring systems at locations where mercury may be volatilized or may accumulate. Routine checking of workers in areas where they may be exposed to mercury is also required. Other more intensive procedures and preventive measures may be required in certain situations and for specific feed materials.

Industrial systems in use for the removal of mercury are also discussed in the case studies that follow.

PLANT PRACTICE

Case Study 1: Carlin Gold Quarry

The Carlin gold quarry is owned and operated by Newmont Gold (Denver, Colorado, United States). The quarry is located 10 km north of Carlin, Nevada. Newmont has been producing gold in the Carlin District since 1965, and the area has a rich history of gold production and technology. Ore is treated through two main processing facilities: Mill 5 and Mill 6 using pressure oxidation and roasting, respectively (see Table 10). Mill 5 processes sulfide refractory ore and oxide ore. Mill 6 processes material with both sulfide and carbonaceous refractory components, sometimes referred to as double-refractory ores. In the Carlin District refractory ores, gold occurs primarily associated with pyrite, arsenian pyrite, and silicates, with minor arsenopyrite. Carlin ores generally contain modest amounts of silver (0.1–0.3:1 Ag/Au).

Lower-grade oxide ore is also heap leached at ROM size at various facilities within the Carlin District. These facilities use strategically located CIC systems for gold and silver recovery from solution, with loaded carbon transferred by truck to centralized stripping and reactivation facilities.

Mill 5 Plant (Flotation, Tails Leaching, and CIP)

A simplified process flow sheet for the Mill 5 plant is included in Figure 7, showing the major equipment sizing information. ROM refractory ore containing 1.3%–2.0% sulfide sulfur and 1.5–2.5 g/t gold is fed to the SAG mill circuit. The SAG screen undersize is sent to the ball mill circuit to generate a product at 80% passing 150 μm. Rougher and scavenger flotation is performed at pH 5.5–6. The scavenger concentrate is cleaned (four stages) and the cleaner concentrate is combined with the rougher concentrate. Cleaner flotation is performed at pH 5.5–7. The final concentrate is thickened and filtered. The main reagents used are Cytec frother F-507 (5 g/t), PAX, or potassium amyl xanthate (120 g/t), Cytec MX-6205 (50 g/t), and lead nitrate (100 g/t). Between 700 and 1,200 t/d of concentrate are generated (representing 5%–9% mass pull), with 40%–50% gold recovery and 80% sulfide sulfur recovery into the final concentrate. The gold and sulfide sulfur grades of the final concentrate are 8–16 g/t and 14%–18%, respectively. The concentrates are used to supplement feed to the roaster (Mill 6) and Twin Creeks pressure oxidation circuits, as required. This is important, as the concentrate contains significant sulfide sulfur fuel value and iron for arsenic fixation at both of these locations.

The scavenger flotation tailings are thickened to 48%–50% solids and then sent to leaching and CIP with a total leach retention time of 20–22 hours. Cyanide is added into the leach tanks to maintain a concentration of 0.15 g/L NaCN. Milk of lime is added into the slurry to achieve an operating pH of 10.2–10.5. The cyanide concentration is then allowed to decay through the CIP circuit, reaching approximately 0.05 g/L NaCN in the final tank in each train. Air is injected into the two leach tanks and oxygen is sparged into the first two CIP tanks. Dissolved oxygen concentration is maintained around 15 mg/L in the first two CIP tanks, declining to 4–5 mg/L in the last tank in each train. Gold recovery in the leaching and CIP circuit is approximately 35%–40%, equivalent to 15%–20% of the gold in the mill feed. Cyanide and lime consumptions in the leaching circuit are 0.13 kg/t and 7–8 kg/t, respectively. Overall gold recovery to the flotation concentrate and from leaching-CIP treatment of the flotation tailings averages 62%. Caro's acid is added as required into the tailings booster pump sump to control WAD cyanide below 15 mg/L. The discharged tailings typically contain 1–5 mg/L WAD cyanide.

Mill 6 Plant (Gold Quarry Roaster)

The Mill 6 processing plant flow sheet is provided in Figure 8, including major equipment sizing information. The plant treats materials containing both sulfide and carbonaceous refractory components. The typical ore feed composition is 6–7 g/t Au, 1.8% sulfide sulfur, 0.4%–0.65% TOC (0.65% maximum), 3.0% carbonate (CO_3, 3.6% maximum), and 4%–6% moisture. In addition, some of the ore and concentrate feed materials contain arsenic. The maximum amount of arsenic that can be accommodated in the roaster feed is 0.3%.

Ore is crushed in two stages (primary and secondary) to 80% passing 22 mm. Crushed ore is then dry-ground in a double-rotator mill, which consists of a drying chamber, coarse ore grinding chamber, and fine ore grinding chamber. Each chamber in the double-rotator mill is 6.1 m diameter and 25.6 m long. The mill is fitted with an 11.2-MW wraparound drive. Prior to entry into the grinding chambers, the ore is dried to <0.5% moisture using air heated by a natural gas burner. The primary (coarse) grinding chamber uses 90-mm-diameter steel balls, and the secondary (fine) chamber uses 64-mm-diameter steel balls. The final product generated by the dynamic separators is collected in a cluster of four baghouses and then discharged to a fine ore bin. The final product particle size distribution is 80% passing 75 μm. The fine ore storage bin has a capacity of 6,000 t, sufficient to feed the roasters for a period of about 8 hours when the mill is down. Trona (soda ash, or anhydrous sodium carbonate) is occasionally added to the roaster feed to fix sulfur dioxide formed within the dry grinding process.

Ground ore is removed from the fine ore bins and delivered by air slide to one of two parallel roasting circuits. Each roasting circuit can treat approximately 200–225 t/h. The ore is preheated to 175°C in a circulating fluid bed preheater with a retention time of 2–5 minutes. Preheated ore is transferred to the circulating fluid bed roaster, which is operated as a single-stage roast. Each roaster train consists of a 4.0-m internal diameter roaster, cyclones, two seal pots, fluidizing air blowers, oxygen preheater, induction burner, two calcine coolers, and two quench tanks. The roaster operates at 510°C with a retention time of 6–8 minutes. Fluidizing air is heated by natural gas and delivered to the roaster at a flow rate of about 30,000 Nm^3/h. Gaseous oxygen is injected into the fluidizing gas stream to maintain an atmosphere of 32% oxygen. The oxygen concentration in the gas phase is controlled for optimum roaster performance. More than 95% of the sulfide sulfur and 90%–95% of the organic carbon (TOC) are oxidized in the fluid bed and subsequent process equipment. The ore leaves the roaster and enters the calcine cooler, which provides an additional 6–8 minutes of retention time at or close to the roasting temperature, 515°C. Oxygen is injected into the calcine cooler to assist with further oxidation. It is noteworthy that between a quarter and a half of the organic carbon is oxidized in the roaster and most of the remainder is oxidized or passivated in the calcine coolers. Oxygen consumption is approximately 35 kg/t in the roaster and 22 kg/t in the calcine cooler. The total consumption in the roasters is 570 t/d. The calcine cooler product is quenched at 15%–20% solids, producing a slurry at approximately 65°–70°C. The roaster off-gas is cooled from 515°C to 375°C in a waste heat boiler, followed by particulate removal in an electrostatic precipitator to generate a de-dusted gas phase.

The de-dusted off-gas streams from the two roaster circuits are combined for treatment in a single gas cleaning plant. Fluorine is removed in a dedicated absorption tower using sacrificial silica saddles to prevent deterioration of the acid plant catalyst, followed by another stage of wet electrostatic precipitation. Mercury is removed by scrubbing the gas with mercuric chloride solution to generate calomel (mercurous chloride)—a Boliden Norzink process. Excess calomel is treated by electowinning to produce metallic mercury as a product. Some calomel is converted back to mercuric chloride by chlorination, which is returned to the process.

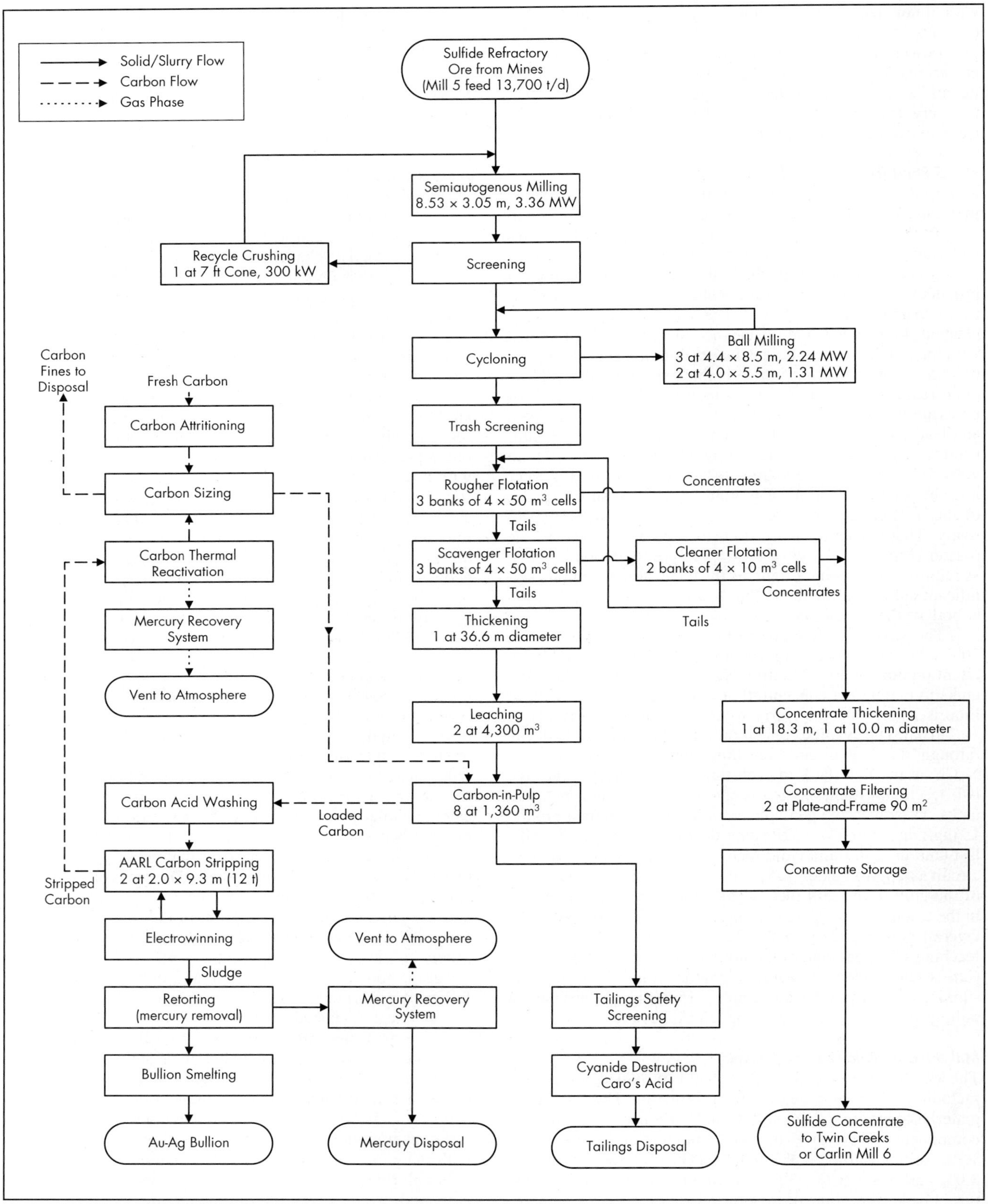

Data courtesy of Newmont Mining Corporation

Figure 7 Flow sheet for Carlin Mill 5 processing operations (Newmont Gold)

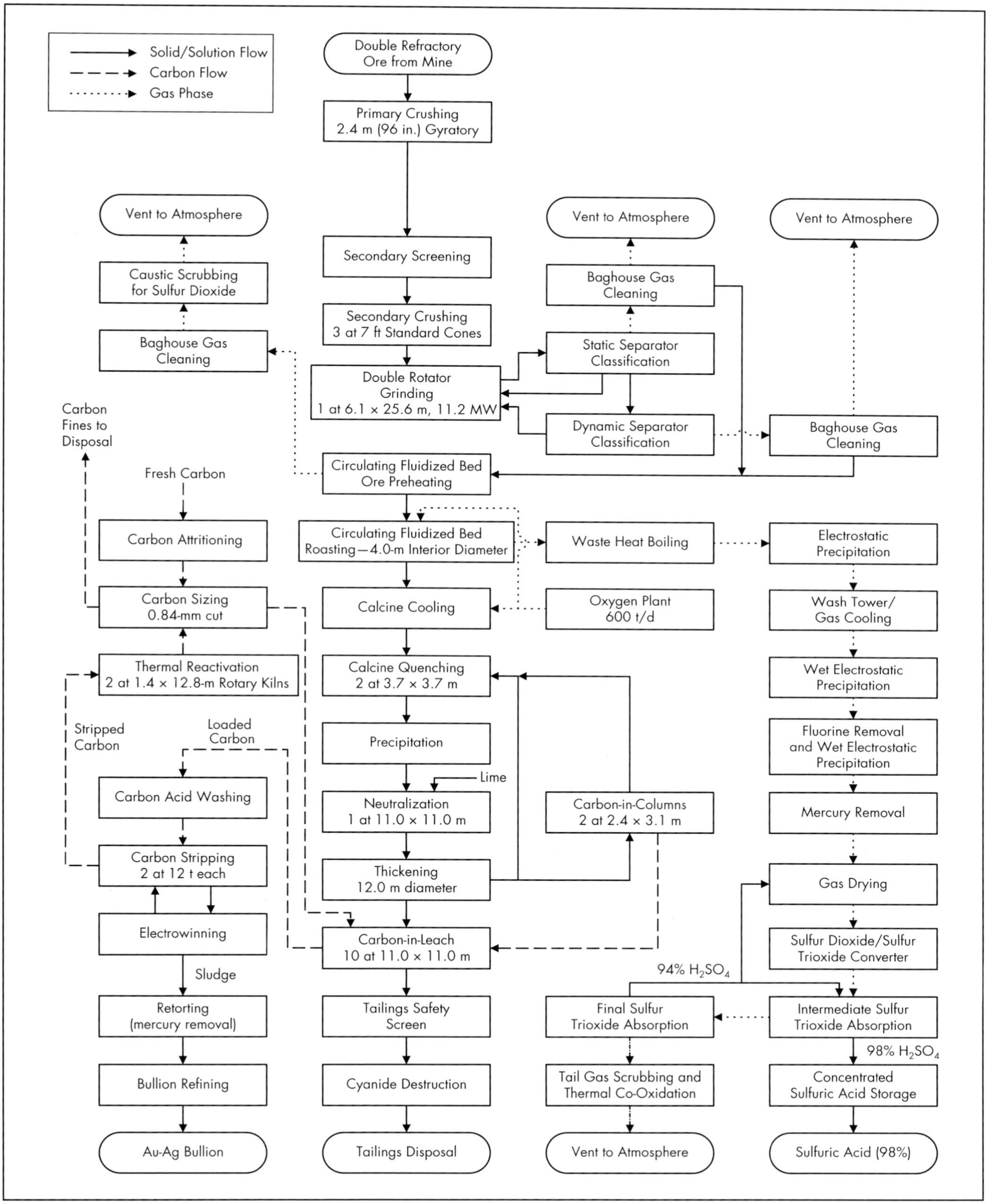

Data courtesy of Newmont Mining Corporation

Figure 8 Flow sheet for Carlin Mill 6 processing operations (Newmont Gold)

Finally, the gas is dried (using 94% sulfuric acid in a drying tower) and then passed through a catalytic converter using a vanadium pentoxide catalyst to convert sulfur dioxide to sulfur trioxide. The sulfur trioxide is then absorbed in a two-stage system to generate a >98% sulfuric acid product. Lower concentration sulfuric acid from the final absorption stage is recirculated back to the first ("intermediate") absorption stage, from which concentrated sulfuric acid is produced. The sulfuric acid is used primarily at the nearby Newmont Phoenix mine copper dump leach and the Newmont Twin Creeks pressure oxidation plant (which requires 3 kg/t sulfuric acid consumption), but is also available for sale to third parties. The final tail gas is subjected to thermal treatment at 815°C using a natural gas burner to reheat the gas stream, to convert any remaining carbon monoxide to carbon dioxide. Finally, the gas stream is scrubbed prior to venting to the atmosphere. The gas handling systems accompanying the roasting circuits are significant in scale, complex, and materials/maintenance intensive.

The slurry exiting the calcine cooler is neutralized with milk of lime (~4 kg/t CaO) and then fed to a thickener. The thickener underflow (46%–48% solids) is fed to the CIL circuit for gold extraction and recovery. The first two tanks of the CIL circuit receive slurry at approximately 40°C. Milk of lime is added to adjust and maintain the slurry pH around 10.5. Cyanide is added into the first two CIL tanks to achieve a concentration of 0.10–0.12 g/L NaCN. Cyanide concentration is allowed to decay naturally to 0.03–0.04 g/L NaCN in the final (No. 10) CIL tank. Cyanide consumption is approximately 0.45 kg/t and total lime consumption is 5 kg/t. The slurry density decreases from 44% solids in the first CIL tank to about 42% in the final CIL tank. The retention time in CIL is 13–14 hours. Oxygen is sparged into the first four CIL tanks at a rate of 50 Nm^3/h, achieving a dissolved oxygen concentration of 16 mg/L. Oxygen consumption in CIL is 5 kg/t. Compressed air is sparged into the remaining CIL tanks (No. 5–10), and the dissolved oxygen concentration is maintained around 4 mg/L. The CIL circuit uses 6 × 12 mesh (1.68 × 3.36 mm) activated carbon. Carbon concentration of 18–20 g/L is maintained throughout the CIL circuit. Both Kemix and NKM screens are used for interstage slurry transfer in the CIL circuit, one per tank. The screens are fitted with 0.84-mm (20-mesh) aperture stainless-steel cloths.

Approximately 36 t of carbon is transferred to the stripping circuit each day in 12-t batches, with typical gold on carbon loading of 2,200 g/t. Acid wash and strip take place in the same vessel. The strip vessel is rubber lined to allow hot acid wash followed by neutralization and stripping. The hot pressurized acid wash is completed in 1 hour with 2.5% hydrochloric acid solution. The acid-washed carbon (and the resulting solution) is neutralized with sodium hydroxide. The loaded carbon is stripped using a solution containing 0.2% sodium cyanide and 0.5% sodium hydroxide at 110°C and a system pressure of 69 kPa. An eluate rise velocity through the carbon bed of 20 m/h is maintained to complete a 6 bed-volume strip cycle within 5–6 hours (excluding loading and unloading of the vessel). The carbon is washed with 2 bed volumes of water after stripping is complete. Final gold loading on carbon after stripping is approximately 50–60 g/t.

The stripped carbon is reactivated in one of two horizontal rotary kilns, maintaining 600°C for approximately 30 minutes. The reactivated carbon is water quenched and screened at 0.65 mm (28 mesh) before being returned to the CIL circuit. Overall carbon consumption averages 40 g/t.

Gold, silver, and mercury are recovered from the strip eluate solution using 16 electrowinning cells. Mild steel wool cathodes are used. The precious metal–bearing cathodes are removed at the end of each cycle and are acid washed using 94% sulfuric acid. The acid wash removes much of the iron (mild steel wool), and the wash solution is directed back to the Mill 6 process for neutralization. The acid-washed sludge is retorted for 18 hours at 650°C to remove mercury. Approximately 0.2 t of mercury is recovered per month. Special precautions are taken in the gold room to prevent exposure of mercury to the air and to scrub gas streams prior to exhaust to atmosphere (e.g., electrowinning cells are covered and vented to scrubber, final retort gas phase is scrubbed after condensation). The retorted sludge is smelted with fluxes in one of two 0.06-m^3 induction furnaces at 1,370°C to produce gold-silver bullion. Approximately 125 kg Au and 10 kg Ag are produced during each smelt.

Slurry leaving the final CIL tank passes through safety screens fitted with 0.65-mm (28-mesh) apertures. The final tailings are treated for cyanide destruction by the addition of Caro's acid into the CIL tailings surge tank. The tailings are then pumped to a lined subaerial tailings disposal facility. Typically, the WAD cyanide concentration at the tailings disposal spigot discharge point is 1–10 mg/L.

Heap Leaching Operations

ROM ore has been processed at various locations in the north and south Carlin areas since late 1979. Ore is typically placed on double-lined heap leach pads in 8–10-m-high lifts up to maximum heap heights of 90 m. Ore is leached by applying barren solution containing ~0.1 g/L NaCN at pH 10.5. Typical cyanide consumption is 0.15 kg/t. Solution pH control is achieved by adding lime directly to ROM ore in trucks in the range of 1–2 kg/t. A milk of lime system was added to treat 1,800–2,700 m^3/h of low-pH solutions in the Carlin District, raising the pH from 6–8 to 10.5. This system used thickening and reseeding of the thickener feed with underflow sludge to achieve effective settling rates for the gypsum produced.

Heap leaching operations in the Carlin North area typically process material containing 0.30–0.35 g/t gold. The amount of material placed on leach pads varies depending upon availability of ore in the mine operations. Gold recoveries in the range of 50%–60% are achieved from ROM material. The Emigrant heap leach operation, located 20 km south of Carlin, treats 11–13 Mt/yr with a gold grade of 0.5 g/t. Ultimate gold recovery is projected to be 63% at the completion of leaching. Newmont operates four sets of CICs in the Carlin District; three of these can process up to 900 m^3/h each, and the fourth treats 570 m^3/h. Loaded carbon is transferred by truck in 6-t batches to centrally located carbon stripping (and reactivation/regeneration) facilities in Carlin. Typical gold concentration on loaded carbon is 1,200–1,500 g/t.

Case Study 2: Goldstrike Property

The Goldstrike property is owned and operated by Barrick Gold Corporation (Toronto, Ontario, Canada). It is located 25 km north of Carlin, Nevada. The Goldstrike property, discovered in the 1960s, was acquired by Barrick in 1987. The property was initially developed in 1988 with an oxide ore milling and cyanide leaching operation. This was converted to a refractory ore treatment operation in 1989 using pressure oxidation technology. In 2014 and 2015, the pressure oxidation

Table 11 Key process statistics for the Goldstrike property

Description	Roasting, Cyanide Leaching, and CIL (2014 basis)	Pressure Oxidation and Thiosulfate Leaching (plant design basis)
Ore throughput rate, t/d	17,400	11,800
Ore throughput rate, Mt/yr	6.4	3.7
Ore gold grade, g/t	10.0	5.0
Cutoff grade, g/t	Revenue based	Revenue based
Plant availability, %	91	86
Gold recovery, %	88	80
Gold production, oz/yr	1,804,000	475,000
Gold production, kg/yr	56,100	14,800
All-in sustaining cost, $/oz	854	Not available

Data courtesy of Barrick Gold Corporation
Note: Gold production by roasting includes Cortez production (902,000 oz in 2014).

and cyanide leaching circuit was modified into a facility using thiosulfate leaching (see Table 11).

At the Goldstrike property, there are four main mining operations: the Betze-Post open pit (~40,000 t/d total tons mined) and the Miekle, Rodeo, and Betze underground operations, mining approximately 10,000 t/d at each location. The open pit ore grade is approximately 5–7 g/t and the underground ore grades are in the range of 8 to 12 g/t. All of the ores mined contain either sulfide refractory or carbonaceous plus sulfide refractory mineralization, which have a significant impact on processing and ore routing. A minor amount of oxide ore is sent to the Newmont oxide mill (Mill 5 at Gold Quarry) for toll processing.

Gold mineralization is present primarily in colloidal form within pyrite, marcasite, and arsenical pyrite. The parameters measured for ore control purposes include Au, Ag, S^{2-}, CO_3^{2-}, As, TOC, preg-robbing capacity, and bleach-leach response. The preg-robbing capacity is measured using two testing procedures, the first a direct cyanide leaching test and the second consisting of a cyanide leaching test with solution spiked with gold. The preg-robbing capacity is the ratio of the two gold extraction results. The bleach-leach test is comprised of a hypochlorite leaching step followed by a standard cyanide leaching test. Both of these tests provide an indication of preg-robbing capacity by carbonaceous constituents. For preg-robbing values below about 30% (gold adsorbed), direct cyanide leaching and activated carbon recovery can compete favorably with thiosulfate leaching and resin recovery.

Roasting Circuit

A simplified flow sheet for the Goldstrike roasting circuit is shown in Figure 9, including sizing information for the major processing equipment. Roaster feed is primary crushed and delivered to a covered coarse ore stockpile. Significant amounts of ore are maintained in stockpiles ahead of the primary crusher to provide blending options for the roaster feed to ensure efficient and consistent operation. Ore is reclaimed by vibrating feeders and delivered to the dry grinding circuit. Ore is dried and ground by two Krupp Polysius 5.8-m-diameter by 21-m effective grinding length double-rotator grinding mills fitted with 7.5-MW-drive motors. The Bond ball mill work index is approximately 14.5–16 kW·h/t, with excursions up to 22.5 kW·h/t.

Ore ground to 80% passing 90–95 µm is fed to two circulating fluidized bed roasters. The average sulfide sulfur content of the feed is maintained at 1.6% ±0.2% and the total carbonaceous material (TCM) content is maintained at 0.9% ±0.1% (1% maximum allowable in roaster feed). The TCM analysis is equivalent to the TOC analysis employed commonly within the gold industry. Elemental sulfur is added as required to maintain the required sulfur content for suboptimal ore feed composition. Oxygen is fed into the roaster, supplied by a dedicated 900-t/d oxygen plant at the site. An atmosphere of 65%–75% oxygen is maintained in the gas phase with a superficial gas velocity of 0.25–0.30 m/s.

The first stage of the roaster is controlled to 540°C ±25°C, and the second stage at 450–600°C ±25°C, depending on the feed composition. The retention time is 22 minutes in the first stage and 18 minutes in the second stage. The gas phase from the two roasters is combined into a single gas stream and is cooled to 75°C prior to solid particle removal. A wet electrostatic precipitator stage removes remaining fine solids particles and a significant portion of the mercury in the gas phase. Mercury is removed in a Boliden Norzink adsorption tower, forming mercury chloride.

A significant portion of the sulfur dioxide is fixed in the ore during roasting. The proportion of sulfur dioxide that is fixed depends on the carbonate content of the feed. For low carbonate feeds (e.g., Betze-Post, <1% CO_3), approximately 30%–40% is fixed. For higher carbonate feeds (e.g., Rodeo, North and West Betze, 5–15% CO_3), 70%–80% is fixed, and up to 90% in some cases. Residual sulfur dioxide in the gas phase is scrubbed using a dual-alkali system, with soda ash and milk of lime added to the scrubber system. The use of soda ash at pH 10 produces sodium sulfite that reacts with lime to form gypsum.

Roasters convert a portion of the organic carbonaceous/carbonate constituents in the feed to carbon monoxide, rather than carbon dioxide. As a result, the solids and sulfur-depleted gas phase is directed to a post-combustion system to convert carbon monoxide to carbon dioxide. To achieve this, the gas phase is thermally oxidized at 1,450°C using natural gas. Ore feed materials with high arsenic can cause complications for the roasting process. The intermediate and final ferric arsenate products can occlude gold, resulting in elevated losses to the residue. Feed material containing more than about 2,000 g/t As (0.2%) were historically processed by pressure oxidation followed by agitated cyanide leaching.

The solids are discharged from the roaster into a quench tank and then delivered to the neutralization circuit. The slurry pH is adjusted to 10.5 using lime addition to the neutralization tank and the resulting slurry is fed to a thickener. The neutralization thickener underflow is delivered to the CIL and gold recovery circuit at a density of approximately 46%–48% solids. The CIL process consists of six stages (2,600 m^3 per stage), operating at a slurry density of 32%–34% solids and with a total retention time of 20 hours. A cyanide concentration of 0.1 g/L NaCN is maintained in the first two CIL tanks and air is sparged into all of the CIL tanks. NKM interstage screens are used between stages. Approximately 20–24 t loaded carbon containing approximately 7 kg/t are transferred each day out of the lead CIL tank and delivered to the carbon elution and regeneration circuit in the pressure oxidation plant. Loaded carbon is acid washed at pH 2 for 0.5 hours and then gold is stripped using a modified pressure Zadra procedure

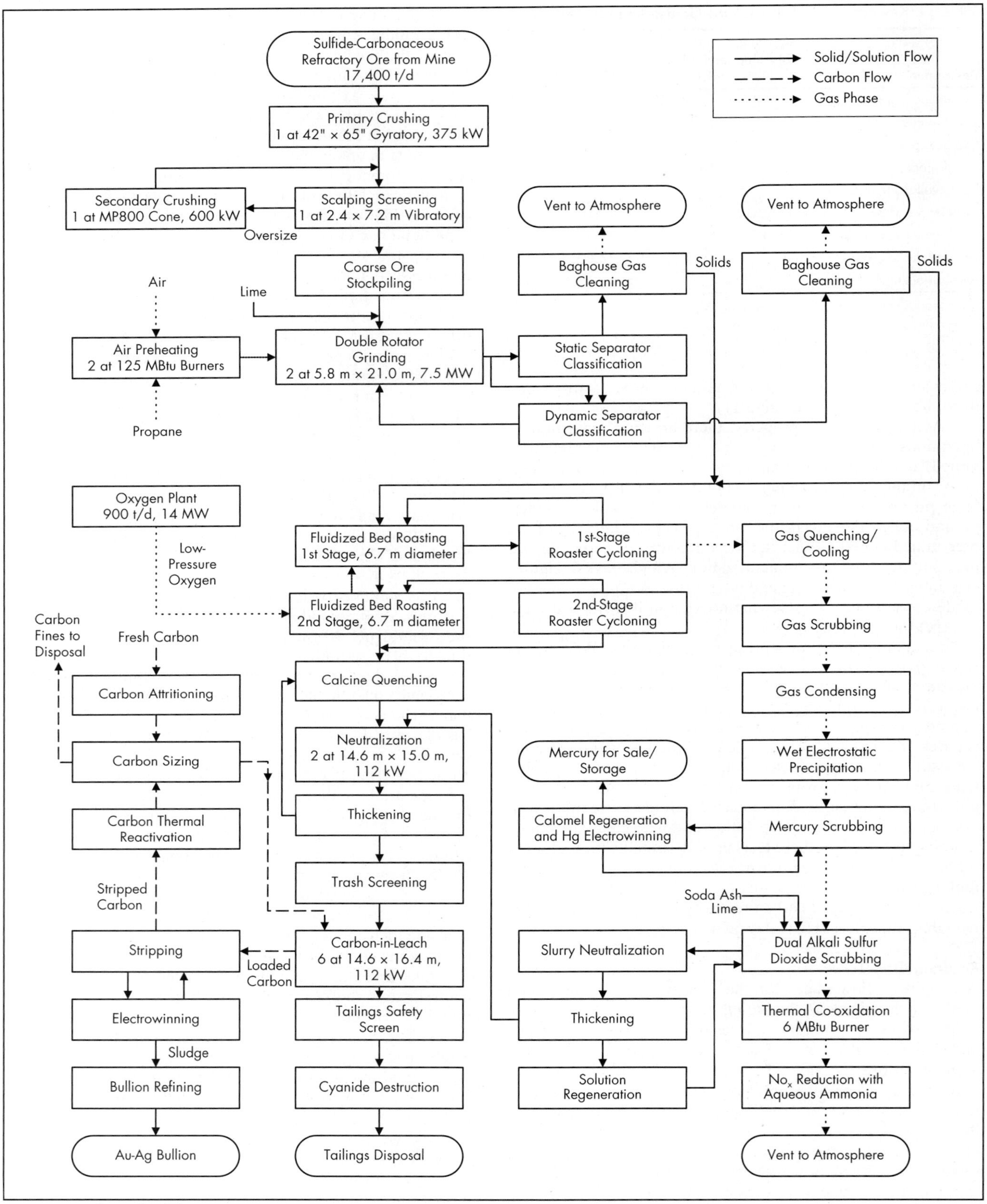

Data courtesy of Barrick Gold Corporation

Figure 9 Flow sheet for Goldstrike roaster plant (Barrick Gold)

with freshwater makeup. The stripping cycle takes between 8 and 12 hours to complete.

The CIL tailings are treated for cyanide destruction using a sulfur dioxide–air system, with roaster quench solution containing sulfur dioxide/sulfur trioxide being used as the main reagent. Automatic samplers are used on the CIL feed and tailings, and dual mass flow measurement (nuclear density and slurry flowmeters) are used to estimate CIL feed weight from the roaster plant.

Pressure Oxidation Circuit

A simplified flow sheet for the pressure oxidation circuit is shown in Figure 10, including sizing information for the major processing equipment. The original milling circuit, constructed in 1988 and expanded in 1991–1992, was designed to feed three autoclaves. This is shown on the left side of Figure 10. The original (smaller) autoclave was subsequently shut down, leaving two autoclaves in operation. The second and larger milling circuit, constructed in 1992–1993, was designed to feed three additional autoclaves (Pekrul 2012). This circuit is shown on the right side of Figure 10.

The combined processing capacity of the two grinding circuits and six autoclaves was approximately 16,000 t/d when operated in conjunction with the two cyanide leaching and CIL plants, that is, 1994–2008. Concurrently with the construction and commissioning of the new thiosulfate leaching and resin gold recovery circuit, the throughput target was modified to 11,800 t/d due to the need to provide a finer product to the downstream process, 80% minus 74 μm (and 100% minus 150 μm for efficient resin screening in resin-in-leach, or RIL).

The pressure oxidation circuits can either be operated in acidic or "alkaline" (neutral pH) configurations, depending on the carbonate content and the carbonate-to-sulfide-sulfur ratio. Processing of high-limestone ore (greater than approximately 7%–8% carbonate content) requires operation in the alkaline configuration, based on an acid price of $180/t. In mid-2015, two of the autoclaves were operating in acidic configuration and three were operating in alkaline mode, processing a combined total of approximately 8,000 t/d as part of an ongoing ramp-up of production.

For the acidic configuration, the ore is acidulated prior to pressure oxidation. Acidulation is performed to reduce the carbonate-to-sulfide-sulfur ratio to 1:1. The ore feed to the plant averages 2% sulfide sulfur, and therefore carbonate content must be reduced to 2%. No slurry or solution is recycled back from the pressure oxidation discharge, and all acid requirements for acidulation are met using concentrated acid. For the alkaline (nonacidic) configuration, the ore is not pre-acidulated. A consequence of running the autoclaves in alkaline mode is that anhydrite tends to form, resulting in substantial scaling within the autoclaves, lower sulfide sulfur oxidation achieved, and lower gold extractions. The autoclaves are controlled to 225°C. Oxygen is supplied to the autoclaves from a dedicated 1,080-t/d oxygen plant at the site. Sulfide sulfur oxidation varies from >90% for the acidic pressure oxidation process to between 50% and 70% for the alkaline (nonacidic) process. As a result, gold extractions are typically 8%–10% lower for the alkaline process compared to the acidic process. The Goldstrike ore contains mercury, and scrubbing of mercury from the gas phase in the autoclave circuit is an important component of the operations.

Silver forms jarosite in the Goldstrike autoclaves under acidic or alkaline conditions and is not recovered to any appreciable extent. The ratio of silver to gold in the ore feed is approximately 1:1 and is not sufficient to justify the use of a hot lime boil step to increase the extraction of silver in the subsequent leaching step. This is compared to Pueblo Viejo feed ore that contains ~60 g/t Ag and for which a hot lime boil step is justified.

Thiosulfate Leaching Plant

Barrick has developed and commercially implemented thiosulfate leaching technology to treat highly carbonaceous and sulfide refractory (so-called double refractory) ores to supplement the roasting capacity at the Goldstrike site. One of the key drivers was the decline of sulfide refractory ore reserves with lower organic carbon content (i.e., weaker preg-robbing characteristics) that were suitable for processing by pressure oxidation and cyanide leaching, resulting in available capacity in the pressure oxidation circuit, which was scheduled to shut down in 2008. The use of thiosulfate leaching technology presented an opportunity as a replacement for the cyanide leaching circuit, allowing carbonaceous-sulfidic refractory ore to be processed by the circuit. The advantage of the thiosulfate leaching regime in this case was that the gold–thiosulfate complex is not adsorbed onto the carbonaceous ore constituents to any significant extent. The pressure oxidation–thiosulfate circuit treats similar ore feed materials to the roaster–cyanide leaching circuit; however, the highest gold-grade material typically is fed to the roaster that achieves higher gold recovery. The chemistry of the thiosulfate gold leaching system has been covered in detail in the literature (Marsden and House 2006) and is not discussed further here.

The process flow sheet is shown in Figure 10. Slurry from the pressure oxidation circuit (at 65%–70% sulfide sulfur oxidation) is neutralized using lime to adjust the pH to 8–8.5. The slurry is then fed to the trash screens fitted with 0.3 × 7 mm apertures to remove any oversize material. This step is extremely important to avoid plugging the interstage screens within the RIL circuit and to ensure that expected throughput rates can be maintained. This is an important difference to CIP/CIL systems, which typically use 6 × 12 mesh carbon granules, significantly coarser than the resin.

The calcium thiosulfate leaching circuit consists of seven stages of resin-in-leaching (four 2,600-m^3 tanks followed by three 950-m^3 tanks). Calcium thiosulfate was selected in preference to ammonium and sodium thiosulfate because of lower cost and similar effectiveness for leaching. Calcium thiosulfate and copper sulfate are added into the leach tanks. The RIL circuit is operated at a slurry density of 34%–36% solids. Purolite A-500 resin sized at 0.8 × 1.0 mm is used in the circuit for gold adsorption. Resin makeup requirements are 50,000 L per month, or 0.16 L/t ore processed. Resin concentrations are maintained at approximately 20 mL/L in the first RIL tanks, decreasing to 10–15 mL/L in the downstream tanks. Each tank is equipped with three NKM interstage screens fitted with stainless-steel wedge wire screens with 0.5-mm apertures. The resin is loaded up to 1,400–2,000 g/t. Batches of 30 m^3 of resin (resin density = 1.08 t/m^3) are transferred to the elution and regeneration circuit. Up to eight resin elution cycles are required daily to meet the design gold production requirements. The leach circuit tailings pass over a tailings safety screen fitted with 0.5 × 7 mm apertures to recover resin and other trash. The tailings are thickened to 52% solids and then pumped to a lined tailings storage facility.

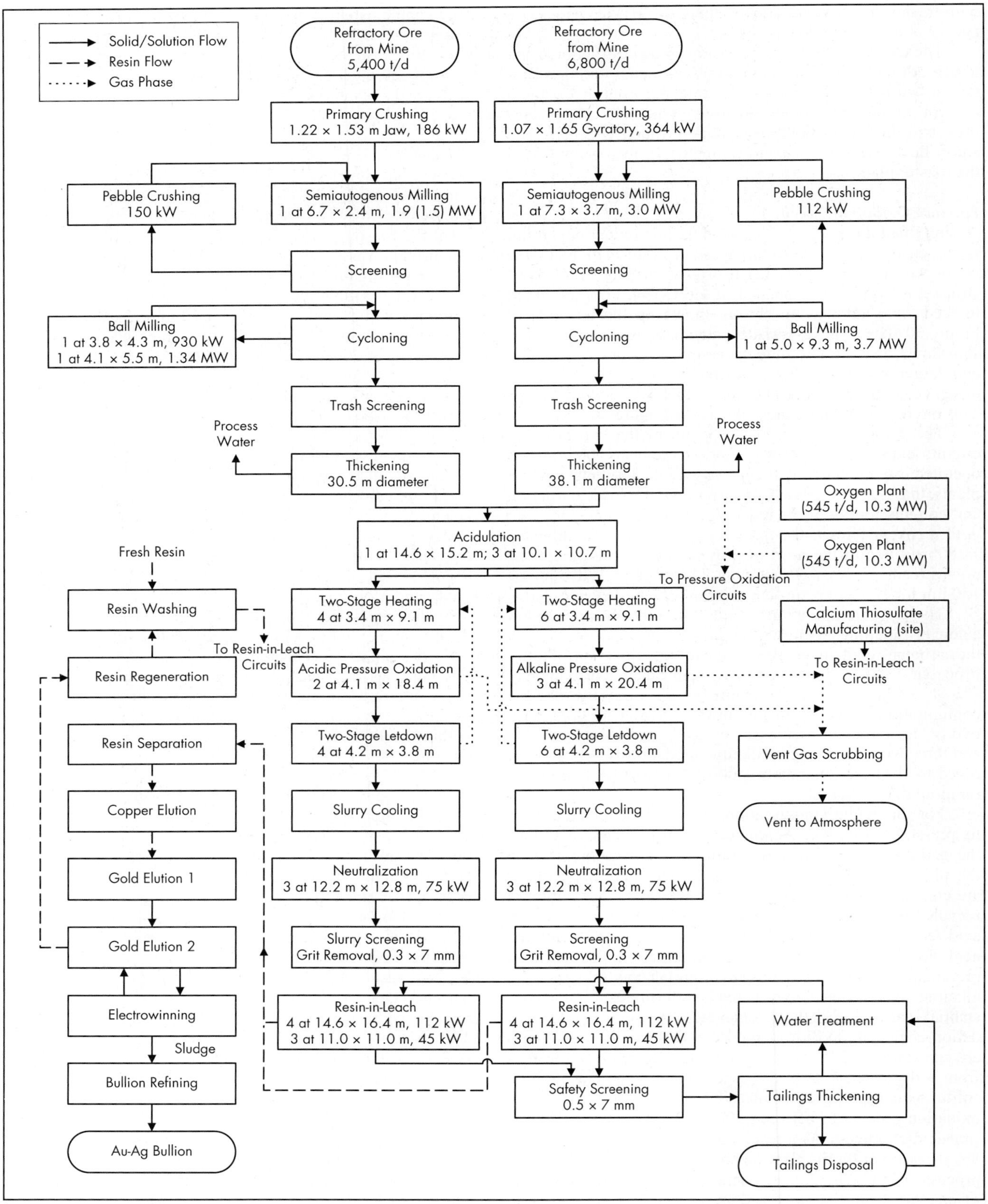

Data courtesy of Barrick Gold Corporation

Figure 10 Flow sheet for Goldstrike pressure oxidation and thiosulfate leaching plant (Barrick Gold)

Table 12 Key process statistics for the Twin Creeks operation (2015 basis, approximate data)

Description	Oxide Milling and CIL	Sulfide Milling and CIL	ROM Heap Leaching and CIC	Total
Ore throughput rate, t/d	3,150	9,700	10,000–30,000	–
Ore grade, g/t	2.0	3.4	0.75	–
Cutoff grade, g/t	0.6	1.70	0.25	–
Plant availability, %	95.0	91.5	100	–
Gold recovery, %	75	84 (77–90)	70 (68–72)	–
Silver recovery, %	40	40	Not available	–
Gold production, oz/yr	52,000	297,000	90,000	439,000
Gold production, kg/yr	1,617	9,250	2,800	13,667

Data courtesy of Newmont Mining Corporation

It is important to recognize that the thiosulfate leaching plant (or TCM plant) processes material containing 2%–3% TOC, significantly higher than the average roaster feed. The maximum organic carbon content that can be processed by the roaster plant is 1% TOC.

Unlike carbon adsorption for gold recovery in the cyanide leaching system, the resin used in the thiosulfate leaching system has fixed capacity. This places a constraint on the gold on resin transfer capability. This factor is considered in the process plant design. Desorption of gold from the resin and resin regeneration is completed in a complex series of steps, summarized as follows:

1. Removal of copper
2. Two stages of gold elution using trithionate, with the first-stage pregnant solution going to electrowinning and the second stage recycled
3. Washing of the resin
4. Resin regeneration with sodium sulfide
5. Washing of the resin prior to recycling to RIL
6. Regeneration of trithionate with sodium thiosulfate
7. Several stages of water treatment to allow recycling of process solutions

The desorption/elution and resin regeneration process takes approximately 12 hours to complete. Trithionate concentration must be actively managed within the system as it loads onto the resin, reducing gold loading capacity and adversely affecting adsorption/desorption kinetics. The resin desorption and regeneration scheme is proprietary Barrick Gold technology.

A key feature of the design of the RIL circuit is the need to use 316 stainless steel (or equivalent) for all of the tanks, agitators, and associated screens, piping, and pumping systems because of the highly corrosive nature of the leach solution.

Case Study 3: Twin Creeks

The Twin Creeks facility is owned and operated by Newmont Gold. It is located 24 km north of Golconda, Nevada. The operation treats ore mined both at the site and from other locations (including Deep Post and Deep Star at Carlin, the nearby Pinson and Getchell mines, and from further afield in Mexico, Peru, and Greece). Ore is treated through several processing facilities, summarized as follows:

- Juniper processing plant—SAG and ball milling, agitated cyanide leaching and CIP (originally the Chimney Creek plant, commissioned in 1987)
- Sage processing plant—SAG and two-stage ball milling, acidic pressure oxidation, neutralization, agitated cyanide leaching and CIL (commissioned in 1997)
- ROM heap leaching of lower-grade oxide ore followed by CIC for recovery of gold and silver (started in 1987)

Most of the ore mined is refractory in nature, with both sulfide and carbonaceous (preg-robbing) components present. The mining and processing operations at Twin Creeks require effective ore characterization and the application of rigorous material routing specifications. The key characterization parameters include gold, silver, sulfide sulfur, carbonate, TOC content, and, when necessary, arsenic and antimony concentrations. There is a strong relationship between gold grade, gold recovery, and TOC. Ore materials containing >1% TOC are sent to the Newmont Carlin operations for processing in the roaster circuit (see Table 12).

Juniper Processing Plant

A simplified process flow sheet for the Juniper plant is included in Figure 11, showing the major equipment sizing information. Oxide ore feed through a 500-mm grizzly and ground by SAG and ball milling (one stage) to generate a product at 80% passing 75 μm. The slurry is then transferred to the Sage plant neutralization circuit, where it is blended with the Sage pressure oxidation (autoclave) discharge. The carbonate contained in the Juniper ore is used to neutralize acid generated by sulfide oxidation in the pressure oxidation circuit. The carbonate content of the oxide ore feeding the Juniper mill varies up to 20% (as CO_3). From this point forward in the circuit, the oxide ore is co-processed together with the oxidized refractory ore in the Sage circuit, described in the following section.

Sage Processing Plant

The Sage processing plant flow sheet is provided in Figure 11, including major equipment sizing information. The plant treats sulfide refractory ore by semiautogenous milling and two stages of ball milling in closed circuit with cyclones to generate a product at 80% passing 27 μm. The fine grind is required to adequately liberate the gold-bearing sulfide minerals prior to pressure oxidation. It is noteworthy that relaxation of the grind size to 80% passing 45 μm would incur a gold recovery loss of approximately 2.5% as well as a small decrease in throughput. The plant processes approximately 9,700 t/d at an average gold grade of 3.4 g/t. The cutoff grade to the process is approximately 1.5 g/t. Gold recovery from refractory ores processed in the Sage plant vary from 77%

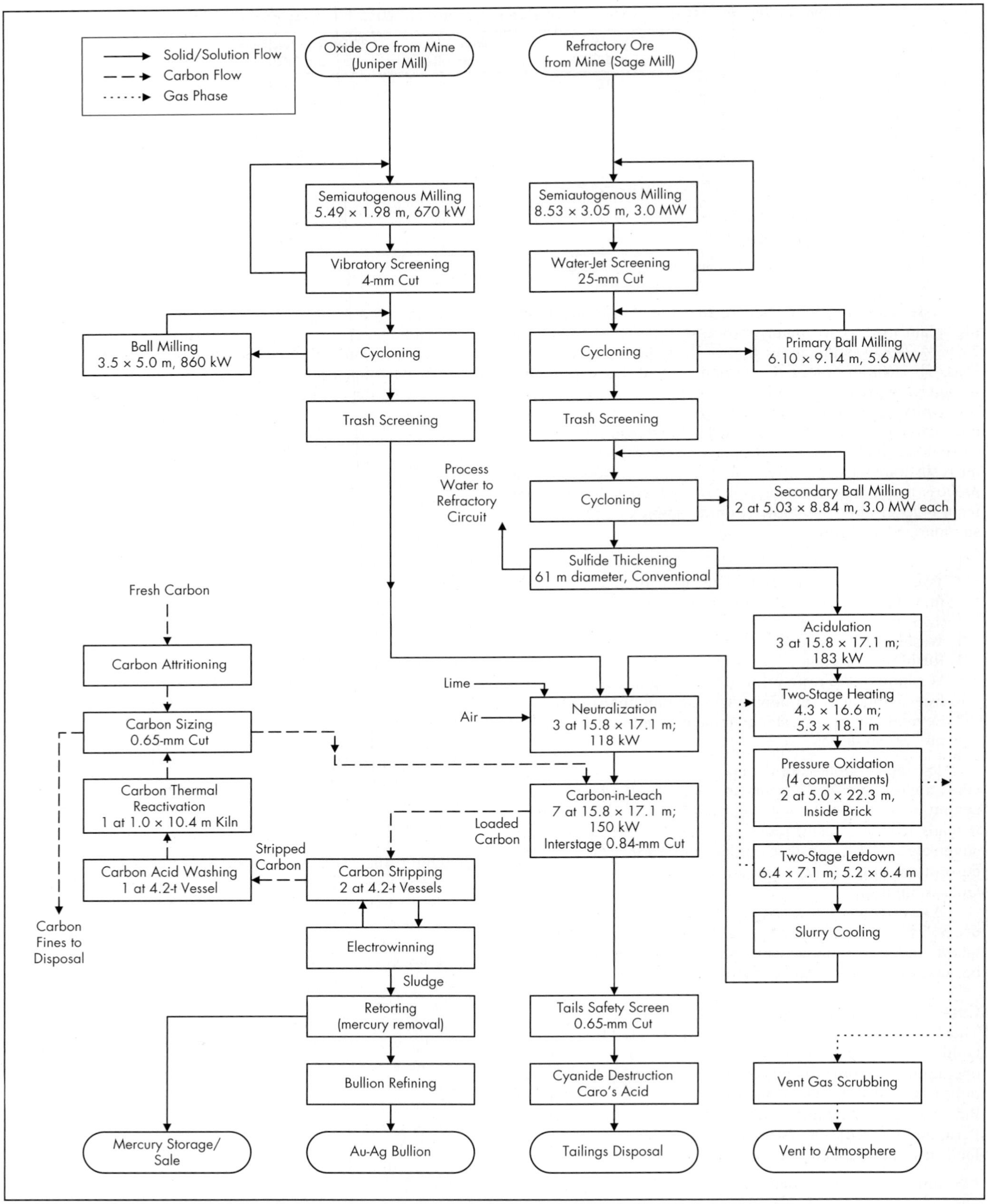

Data courtesy of Newmont Mining Corporation

Figure 11 Flow sheet for the Twin Creeks processing operations (Newmont Gold)

(for higher carbonaceous content materials, 0.6%–1.0% TOC) to more than 90% (for lower carbonaceous content materials, 0.1%–0.3% TOC). Silver recovery is approximately 40%.

The target sulfide sulfur content is 3.8%–4.2%. In practice, feed ore is blended to between 3.2% and 4.5% sulfide sulfur. The pressure oxidation circuit can accept ore with a sulfide sulfur/carbonate ratio of 0.9:1.0, or higher. The maximum carbonate that can be accommodated by the autoclave venting system is 6% (due to carbon dioxide generation); however, the carbonate concentration is typically maintained at about 4.5%–4.6%. The process feed ore composition is adjusted using sulfide concentrates from external sources. The ore also contains arsenic, present in the form of arsenopyrite, realgar, and orpiment.

After grinding, the ore is screened for trash removal and then thickened to a slurry density of 46%–48% solids. The ore is pre-acidulated in agitated tanks (three stages, average 15 hours of retention time) with sulfuric acid added to neutralize a portion of the carbonate. The acidulation step is controlled to maintain a free-acid concentration of 20–35 g/L in the autoclave discharge. The acidulated slurry is then fed via Geho diaphragm pumps (three per autoclave) into two autoclaves (5.0 m diameter × 22.3 m long, inside brick), each with four compartments. The retention time in the autoclaves varies between 48 and 67 minutes. The autoclaves are operated at 225°C and 3,172 kPa. The temperature varies from 218°C in the first compartment to 227°C in the third compartment. Up to 1,360 t/d of oxygen (typically 1,200–1,270 t/d is used) is added into the autoclaves for sulfide sulfur oxidation, with 690 kPa oxygen overpressure maintained. Typically, 92%–94% sulfide sulfur oxidation is achieved.

Slurry exits the autoclaves and pressure is released in two stages of flash letdown and passes into a surge tank. The flashed slurry temperature is 90°C. The flash and autoclave vent gas is scrubbed prior to release to atmosphere to ensure that environmental requirements are met. The slurry is cooled to approximately 40°C in shell and tube heat exchangers. At this point, the solution phase contains about 5 g/L Fe_{Tot}, 3 g/L Fe^{3+}, and 2 g/L Fe^{2+}. The slurry is neutralized with lime and preaerated ahead of CIL to convert a significant portion of the Fe^{2+} to Fe^{3+}, thereby reducing cyanide consumption in CIL. Three tanks (~3,000 m^3 each) in series are used for this purpose. Approximately 1,700 m^3/h of air is sparged into the three tanks. Lime and cyanide consumptions are 20–30 and 0.5 kg/t, respectively. No lead salts are added during preaeration.

The CIL circuit comprises seven stages, providing an average residence time of 16.5 hours. Slurry is fed to the CIL circuit at a density of 42% solids. Interstage slurry screening is provided by eight 2.4-m-diameter Kambalda screens (6–7 operating) in each tank, providing a maximum screen area of 27 m^2 per tank. The interstage screens are fitted with 0.84-mm aperture wire cloths. The first CIL tank uses a Derrick vibrating screen for carbon transfer.

The CIL circuit uses 6 × 12 mesh (1.68 × 3.36 mm) activated carbon. A carbon concentration of 50–60 g/L is maintained in the first tank, decreasing to 30 g/L in the final CIL stage. Cyanide concentration of 1.4–1.6 g/L NaCN is maintained in the first CIL tank, decreasing to 1.2–1.4 g/L in the second CIL tank. Cyanide concentration is allowed to decay naturally to 0.03–0.04 g/L NaCN in the final (No. 7) CIL tank. Air is sparged into all CIL tanks, and the dissolved oxygen concentration ranges from 2–4 mg/L in the first tank to 8 mg/L in the final tank. The introduction of a preaeration stage ahead of CIL more than doubled these dissolved oxygen concentrations.

Approximately 30–35 t of carbon is transferred to the stripping circuit each day in 4.2-t batches, with typical gold on carbon loading of 1,200–1,500 g/t. Carbon is acid washed with 2.5% nitric acid solution after stripping in dedicated vessels, removing approximately 90% of carbonate scale. The acid-washed carbon (and the resulting solution) is neutralized with sodium hydroxide and then transferred to one of two stripping vessels. The loaded carbon is stripped using a solution containing 0.2% sodium cyanide and 0.5% sodium hydroxide at 150°C and 460 kPa. An eluate rise velocity through the carbon bed of 20 m/h is maintained to complete a 12-bed-volume strip cycle within 5–6 hours (including loading and unloading of the vessel). The carbon is washed with 2 bed volumes of water after stripping is complete. Final gold loading on carbon after stripping is approximately 50–60 g/t.

The stripped carbon is reactivated in a horizontal rotary kiln, maintaining 600°C for approximately 30 minutes. The reactivated carbon is water quenched and screened at 0.65 mm before being returned to the CIL circuit. Overall carbon consumption averages 40 g/t.

Gold, silver, and mercury are recovered from the strip eluate solution using six electrowinning cells. Five of the cells are configured in parallel followed by a single scavenger cell. Stainless-steel wool cathodes are used. The precious metal–bearing sludge is washed off the cathodes at the end of each cycle. The sludge is retorted for 18 hours at 650°C to remove mercury. Approximately 1 t of mercury is recovered per month. Special precautions are taken in the gold room to prevent exposure of mercury to the air and to scrub gas streams prior to exhaust to atmosphere (electrowinning cells are covered and vented to scrubber, final retort gas phase is scrubbed after condensation, etc.). The steel-wool cathodes are acid washed every 90 days. The retorted sludge is smelted with fluxes in a 0.06-m^3 induction furnace at 1,370°C to produce gold-silver bullion. Approximately 125 kg Au and 70 kg Ag are produced during each smelt.

Slurry leaving the final CIL tank passes through safety screens fitted with 0.65-mm (28-mesh) apertures. The final tailings are treated for cyanide destruction by the addition of Caro's acid into the CIL tailings surge tank. The tailings are then pumped to a lined subaerial tailings disposal facility. Typically, the WAD cyanide concentration at the tailings disposal spigot discharge point is 10–20 mg/L. Tailings return water is treated using ammonium metabisulfite to remove minor amounts of residual-free and WAD cyanide, mercury, copper, and zinc before reuse in the processing facilities. A copper concentration of 20–30 mg/L is required in the solution for effective cyanide destruction with ammonium metabisulfite. The WAD cyanide concentration in the treated return water is <1 mg/L and averages 0.6 mg/L.

Heap Leaching Operations

ROM ore is leached on four lined leach pads. Ore is placed on the leach pads at a variable rate (4–12 Mt/yr), depending on the availability of suitable oxide ore from the pit. The average ore grade is approximately 0.75 g/t, and the cutoff grade is 0.25 g/t. The ore is placed in 12-m-high lifts with a maximum of five lifts on each heap (i.e., 6 m maximum heap height, although the liner is designed and permitted for 10 m height).

Typical gold recovery from ROM heap leaching is 68%–72%. Gold production varies from 5 to 18 kg/d, averaging about 2,800 kg/yr.

Lime is added to trucks at a rate ranging from 0.75 kg/t for sandy limestone/sandstone ore types to 1.25 kg/t for higher clay-bearing ore. The ore is leached with barren solution containing 0.15 g/L NaCN. The solution application rate varies depending on the ore type and its permeability: 2.5 L/h/m^2 for higher clay materials; 10 L/h/m^2 for sandy limestone/sandstone materials. Drip emitters are used for solution application with either 1-, 2-, or 4-L/h emitters arranged on 0.8 × 0.9-m centers. Drip emitter lines are flushed on a regular basis to remove solids that have accumulated in the lines. Sprinklers are used for solution application on side slopes. The net cyanide consumption ranges from 50 to 65 g/t, depending on ore type. For the sandy limestone/sandstone ore types, the leach cycle is 180–25 days. For the higher clay-bearing ores, the leach cycle is 36 days.

Gold and silver are recovered from pregnant solution by seven CIC systems of conventional design. Three of these systems treat 570 m^3/h each (2.6 m diameter × 2.9 m high, five stages, 4–5 t carbon per stage), another three treat up to 225 m^3/h each (1.8 m diameter × 4.3 m high, five stages, 2.5 t carbon per stage), and one small system treats an additional 140 m^3/h, as required. Up to 2,500 m^3/h of pregnant solution can be processed, and 6 × 12 mesh carbon is used with 110% bed expansion achieved. A dedicated carbon stripping circuit at the Pinion mill processes approximately 10 t/d loaded carbon from the combined CIC systems. Strip eluate solutions are processed using zinc precipitation and filtration, and the precipitate is then refined at the main gold room refining facilities at the Juniper/Sage milling facility.

SILVER

Silver Extraction and Recovery Methods

The majority of silver production is as a by-product from lead-zinc sulfide operations. Processing typically involves crushing and grinding, and flotation to produce separate lead and zinc sulfide concentrates for treatment by lead and zinc smelters. Silver is recovered following smelting in the electro-refinery anode slimes and is further processed by standard silver-refining techniques. In some cases, lead-zinc flotation tailings are processed by cyanide leaching to recover silver (and gold, if present).

The Cannington Ag-Pb mine (Queensland, Australia) was the largest primary silver deposit in 2014, producing 24.7 million oz. The major ore minerals are galena and sphalerite, with silver present as freibergite (a complex sulfide, $Ag_6Cu_4(Fe,Zn)_2Sb_4S_{13}$). Other major primary silver deposits include Escobal (Guatemala), Fresnillo and Saucito (Mexico), Dukat (Russia), and Uchucchacua (Peru). These six mines accounted for 13% (112 million oz) of global silver production in 2014.

Silver also occurs in many copper sulfide ores, although generally the value of contained silver is significantly less than that of gold. For such ores, a significant portion of the silver may be recovered into the flotation concentrate and is recovered downstream in the smelting and refining process as anode slimes. The anode slimes are further processed by standard silver refining techniques (Marsden and House 2006). Two hundred million oz of silver were recovered as a by-product of copper production in 2014.

Table 13 Summary of treatment methods available for various silver ores and mineral occurrences

Mineral Occurrence	Treatment Method(s)
Native silver, electrum	Gravity concentration, intensive leaching (or amalgamation), and direct cyanide leaching
Silver sulfides	Direct cyanide leaching with long retention time and high cyanide concentration
Silver chloride	Direct cyanide leaching
Silver with manganese	1. Caron process (reduction of MnO_2), cyanide leaching 2. Segregation for Mn rejection; cyanide leaching
Silver with Pb-Zn sulfides	Flotation followed by recovery during smelting and refining
Silver with oxidized Pb-Zn ores	Cyanide leaching of flotation tailings
Silver with Cu sulfides	Flotation followed by recovery during smelting and refining
Silver with silica	1. Salt roasting, regrinding, cyanide leaching 2. Fine grinding, cyanide leaching

As discussed in this chapter, silver is recovered as a by-product or co-product of gold operations. In some cases (e.g., Coeur Rochester in Nevada), silver is the primary product with gold as a by-product. All of these operations use cyanide leaching for silver and gold co-recovery. A summary of treatment methods for a variety of silver ore types is provided in Table 13.

Processing Techniques for Enhanced Silver Recovery from Gold Ores

A few of the more important innovations related to silver extraction and recovery as a by-product or co-product of gold using the methods described in this chapter are summarized here:

- Use of elevated cyanide concentrations to treat ores and concentrates with high silver content (e.g., 1–5 g/L NaCN, and occasionally higher)
- Use of hot lime boil following pressure oxidation to break down silver jarosite and render a significant portion of the silver to be leachable
- Favored use of Merrill–Crowe zinc precipitation to treat solutions containing >10:1 Ag/Au ratio, rather than CIP/CIL or CIC processes
- Size reduction targeted at silver minerals and their effective liberation
- Use of mineralogical techniques to identify and characterize associations of silver minerals in feed and residues from processing operations

Further information is available in the literature (Marsden and House 2006).

ACKNOWLEDGMENTS

The author thanks Barrick Gold and Newmont Gold for their permission to publish operating data, and also Andy Bolland, Andy Cole, John Cole, Mike Eiselein, Jacqueline Graham, Bill Janhunen, Debbie Laney, Michael McGlynn, John Pekrul, Mike Schaffner, Jim Sigurdsen, Greg Theis, and Don Wilhite for providing key operating information and reviewing relevant sections of the chapter. The author thanks GFMS, Thomson Reuters, and The Silver Institute for permission to use their data.

REFERENCES

Adamson, R.J. 1972. *Gold Metallurgy in South Africa.* Johannesburg: Chamber of Mines of South Africa.

Agricola, G. 1556. *De Re Metallica.* Translated by Herbert C. Hoover and Lou Henry Hoover. 1912 edition. London: The Mining Magazine.

Anonymous. 1982. *Carbon-in-Pulp Technology for the Extraction of Gold.* Melbourne, Victoria: Australasian Institute of Mining and Metallurgy.

Anonymous. 2012. Paracatu mine, Brazil. In *Crusher News*, July 21, 2012.

Arnold, J.R., Keane, J.M., Lofftus, V.G., and Ahlness, J.K. 1990. Gold heap leach information exchange. In *Gold '90: Proceedings of the Gold '90 Symposium.* Edited by D.M. Hausen et al. Littleton, CO: SME. pp. 283–311.

Bain, H.F. 1910. *More Recent Cyanide Practice.* San Francisco: Mining and Scientific Press.

Barter, J., Lane, G., Mitchell, D., Kelson, R., Dunne, R., Trang, C., and Dreisinger, D. 2001. Cyanide management by SART. In *Cyanide: Social, Industrial and Economic Aspects.* Edited by C.A. Young. Warrendale, PA: The Minerals, Metals & Materials Society. pp. 549–562.

Baum, W. 1988. Mineralogy related processing problems of epithermal gold ores. In *Proceedings of Process Mineralogy VIII.* Edited by D.J.T. Carson et al. Warrendale, PA: TMS-AIME. pp. 3–20.

Berezowsky, R.M.G.S., and Weir, D.R. 1989. Refractory gold: The role of pressure oxidation. In *Proceedings of the World Gold 1989 Symposium.* Edited by R.B. Bhappu and R.J. Harden. Littleton, CO: SME-AIME. pp. 295–304.

Boydell, D.W. 1984. Developments in plant practice for the recovery of gold in South Africa. In *Gold Mining, Metallurgy and Geology: Perth and Kalgoorlie Branches Regional Conference.* Melbourne, Victoria: Australasian Institute of Mining and Metallurgy.

Breuer, P.L., and Jeffrey, M.I. 2003. A review of the chemistry, electrochemistry, and kinetics of the gold thiosulfate leaching process. In *Hydrometallurgy 2003: Proceedings of the 5th International Symposium Honoring Professor Ian M. Ritchie.* Vol. 1. Warrendale, PA: The Minerals, Metals & Materials Society. pp. 139–154.

Brierly, J.A. 2014. Biohydrometallurgy innovations—Discovery and advances. In *Mineral Processing and Extractive Metallurgy: 100 Years of Innovation.* Edited by C.G. Anderson, R.C. Dunne, and J.L. Uhrie. Englewood, CO: SME. pp. 447–456.

Chryssoulis, S.L., and Cabri, L.J. 1990. Significance of gold mineralogical balances in mineral processing. *Trans. Inst. Min. Met.* 99:C1–C10.

Chryssoulis, S.L., and McMullen, J. 2005. Mineralogical investigation of gold ores. In *Developments in Mineral Processing.* Vol. 15. Edited by M.J. Adams. Amsterdam: Elsevier B.V.

Clennell, J.E. 1915. *The Cyanide Handbook*, 2nd ed. New York: McGraw-Hill Book Company.

Dorr, J.V.N., and Bosqui, F.L. 1950. *Cyanidation and Concentration of Gold and Silver Ores*, 2nd ed. New York: McGraw-Hill.

Eissler, M. 1891. *The Metallurgy of Gold*, 3rd ed. London: Crosby, Lockwood and Son.

Ellis, S. 2003. Ultra fine grinding—A practical alternative to oxidative treatment of refractory gold ores. In *Proceedings of the Eighth Mill Operators' Conference.* Melbourne, Victoria: Australasian Institute of Mining and Metallurgy. pp. 11–17.

Fleming, C.A., and Cromberge, G. 1984. The extraction of gold from cyanide solutions by strong and weak base anion exchange resins. *J. S. Afr. Inst. Min. Metall.* 84(5):125–137.

Gale, C.O., and Weldon, T.A. 2002. Bullion production and refining. In *Mineral Processing Plant Design, Practice, and Control.* Edited by A.L. Mular, D.N. Halbe, and D.J. Barratt. Littleton, CO: SME. pp. 1747–1759.

GFMS, Thomson Reuters. 1984–2017. *GFMS Gold Survey [various years].* London: Thomson Reuters.

Grewal, I., Van Kleek, M., and McAlister, S. 2009. Gravity recovery of gold from within grinding circuits. In *Recent Advances in Mineral Processing Plant Design.* Edited by D. Malhotra, P.R. Taylor, E. Spiller, and M. LeVier. Littleton, CO: SME. pp. 499–506.

Habashi, F. 1987. One hundred years of cyanidation. *CIM Bull.* 80(905):108–114.

Heinen, H.J., Peterson, D.G, and Lindstrom, R.E. 1978. *Processing Gold Ores Using Heap Leach–Carbon Adsorption Methods.* Washington, DC: U.S. Bureau of Mines.

Heinen, H.J., McClelland, G.E., and Lindstrom, R.E. 1979. *Enhancing Percolation Rates in Heap Leaching of Gold-Silver Ores.* Report of Investigation 8388. U.S. Bureau of Mines.

ICMI (International Cyanide Management Institute). 2002. *International Cyanide Management Code for the Manufacture, Transport, and Use of Cyanide in the Production of Gold.* Washington, DC: ICMI.

Julian, H.F., and Smart, E. 1904. *Cyaniding Gold and Silver Ores*, 1st ed. London: Charles Griffin.

Kappes, R. 2007. N2TEC flotation technology—A review and update. Presented at the Precious Metals Processing—Advances in Primary and Secondary Operations Symposium. Littleton, CO: SME.

Kappes, R., Fortin, C., and Dunne, R. 2011. The current status of the chemistry of gold flotation in industry. In *World Gold 2011: Proceedings of the 50th Conference of Metallurgists.* Westmount, QC: Canadian Institute of Mining, Metallurgy and Petroleum. pp. 385–396.

King, A. 1949. *Gold Metallurgy on the Witwatersrand.* Johannesburg: Transvaal Chamber of Mines.

Knelson, B., and Edwards, R. 1990. Development and economic application of Knelson concentrators in low grade alluvial gold deposits. In *Proceedings of the AusIMM Annual Conference—The Mineral Industry in New Zealand.* Melbourne, Victoria: Australasian Institute of Mining and Metallurgy. pp. 123–128.

Laplante, A.R. 2000. Testing requirements and insight for gravity gold circuit design. In *Randol Gold and Silver Forum 2000.* Golden, CO: Randol International. pp. 73–83.

Laplante, A.R., Liu, L., and Cauchon, A. 1990. Gold Recovery at the mill of Les Mines Camchib Inc. In *Proceedings of the 22nd Annual Meeting of the Canadian Mineral Processors.* Ottawa, ON: Canadian Mineral Processors. pp. 397–414.

Laxen, P.A. 1982. Carbon-in-pulp in South Africa. In *Symposium on Carbon-in-Pulp Technology.* Melbourne, Victoria: Australasian Institute of Mining and Metallurgy.

Longley, R.J., McCallum, A., and Katsikaros, N. 2003. Intensive cyanidation: Onsite application of the InLine Leach Reactor to gravity gold concentrates. *Miner. Eng.* 16:411–419.

MacPhail, P.K., Fleming, C.A., and Sarbutt, K.W. 1998. Cyanide recovery by the SART process for the Lobo-Marte Project, Chile. In *Randol Gold and Silver Forum '98.* Golden, CO: Randol International. pp. 319–323.

Mansanti, J.G., Arnold, J.R., Gourdie, J.H., and Marsden, J.O. 1987. Double-thickener circuit at Gold Fields' Chimney Creek mine. *Miner. Metall. Process. J.* November.

Marsden, J.O. 2006. Overview of gold processing techniques around the world. *Miner. Metall. Process. J.* 23:121–125.

Marsden, J.O., and Botz, M.M. 2017. Heap leach modeling—A review of approaches to metal production forecasting. *Miner. Metall. Process. J.* 34(2):53–64.

Marsden, J.O., and Fuerstenau, M.C. 1993. Comparison of Merrill–Crowe precipitation and carbon adsorption for precious metals recovery. In *Proceedings of the XVIII International Mineral Processing Congress.* Melbourne, Victoria: Australasian Institute of Mining and Metallurgy.

Marsden, J.O., and House, C.I. 1992. *The Chemistry of Gold Extraction*, 1st ed. Chichester: Ellis Horwood.

Marsden, J.O., and House, C.I. 2006. *The Chemistry of Gold Extraction*, 2nd ed. Littleton, CO: SME.

Marsden, J.O., and Sass, S.A. 2014. Innovations in gold and silver extraction and recovery. In *Mineral Processing and Extractive Metallurgy: 100 Years of Innovation.* Edited by C.G. Anderson, R.C. Dunne, and J.L. Uhrie. Englewood, CO: SME.

Marsden, J.O., Mansanti, J.G., and Sass, S.A. 1993. Innovative methods for precious metals recovery in North America. *Min. Eng.* 9:1144–1151.

Marsden, J.O., Todd, L.C., and Moritz, R. 1995. Effect of lift height, overall heap height and climate on heap leaching efficiency. In *Proceedings of the Randol Gold Forum Perth '95.* Golden, CO: Randol International. pp. 272–283.

McClelland, G.E., and Eisele, J.A. 1982. *Improvements in Heap Leaching to Recover Silver and Gold from Low-Grade Resources.* Report of Investigation 8612. Washington, DC: U.S. Bureau of Mines.

McClelland, G.E., and McPartland, J.S. 2002. Bench-scale and pilot plant tests for cyanide leach circuit design. In *Mineral Processing Plant Design, Practice, and Control.* Edited by A.L. Mular, D.N. Halbe, and D.J. Barratt. Littleton, CO: SME. pp. 251–263.

McNulty, T. 1989. A metallurgical history of gold. Paper presented at American Mining Congress, San Francisco, CA, September 20.

McQuiston, F.W., and Shoemaker, R.S. 1975. *Gold and Silver Cyanidation Plant Practice.* Vol. I. AIME Monograph. New York: AIME.

McQuiston, F.W., and Shoemaker, R.S. 1980. *Gold and Silver Cyanidation Plant Practice.* Vol. II. AIME Monograph. New York: AIME.

Newmont Gold. 2014. *Annual Report.* Greenwood Village, CO: Newmont.

Pekrul, J. 2012. Refractory gold pressure oxidation—A comparison in operating practices. Presented at Pressure Hydrometallurgy, Conference of Metallurgists, Niagara, Canada. Canadian Institute of Mining, Metallurgy and Petroleum.

Petruk, W. 2000. *Applied Mineralogy in the Mining Industry.* Amsterdam: Elsevier Science B.V.

Pizzaro, R., McBeth, J.D., and Potter, G.M. 1974. Heap leaching practice at the Carlin Gold Mining Co., Carlin, Nevada. Paper presented at Annual AIME meeting, Dallas, TX, February 23–28.

Rose, T.K. 1896. *The Metallurgy of Gold*, 2nd ed. London: Charles Griffin.

Rose, T.K. 1935. *The Metallurgy of Gold*, 6th ed. London: Charles Griffin.

SAIMM (Southern African Institute of Mining and Metallurgy). 1991. Recent developments in (carbon)-in-pulp technology [by various authors]. In *Proceedings of Colloquium held at Mintek, Randburg, South Africa, on October 7–8.* Johannesburg, South Africa: Southern African Institute of Mining and Metallurgy.

Shoemaker, R.S. 1984. Gold: Quid non mort alia pectora cogis, auri sacra fames. In *Proceedings of Precious Metals: Mining, Extraction and Processing.* Edited by V. Kudryk, D.A. Corrigan, and W.W. Liang. Warrendale, PA: TMS-AIME.

Simmons, G.L., and Gathje, J.C. 2003. High temperature pressure oxidation of ores and ore concentrates containing silver using controlled precipitation of sulfate species. U.S. Patent No. 6,641,642.

Stanley, G.G. 1987. *The Extractive Metallurgy of Gold in South Africa.* Monograph M7. Johannesburg: Southern African Institute of Mining and Metallurgy.

Staunton, W.P. 2005. Carbon-in-pulp. In *Developments in Mineral Processing.* Vol. 15. Edited by M.D. Adams. Amsterdam: Elsevier B.V.

Strachan, C., and Van Zyl, D.J. 1988. Leach pads and liners. In *Introduction to the Evaluation, Design, and Operation of Precious Metal Heap Leaching Projects.* Edited by D.J. Van Zyl, I.P.G. Hutchison, and J.E. Kiel. Littleton, CO: SME. pp. 176–202.

Thomson Reuters and The Silver Institute. 1975–2014. *World Silver Survey [various years].* Washington, DC: The Silver Institute.

USGS (U.S. Geological Survey). 1970–2016. *Mineral Commodity Summaries.* Reston, VA: USGS.

Van Zyl, D.J.A., and Eurick, G.M. 2000. The management of mercury in the modern gold mining industry. www.unites.uqam.ca/gmf/globalmercuryforum.com. Accessed Feb. 2, 2016.

Watson, B., and Steward, G. 2002. Gravity leaching—The ACACIA reactor. In *Proceedings of the Symposium on Metallurgical Plant Design and Operating Strategies.* Melbourne, Victoria: Australasian Institute of Mining and Metallurgy. pp. 383–390.

Whitlock, J.L., and Mudder, T.I. 1986. The Homestake wastewater treatment process: Biological removal of toxic parameters from cyanidation wastewaters and bioassay effluent evaluation. In *Fundamental and Applied Biohydrometallurgy.* Edited by R.W. Lawrence, R.M.R. Branion, and H.G. Ebner. Amsterdam; New York: Elsevier.

CHAPTER 12.13

Graphene

Gary R. Wilson and Courtney A. Young

STRUCTURE AND PROPERTIES

Graphene is a synthetic nano-allotrope of carbon, related in structure to carbon nanotubes and other fullerenes. By its strictest definition, graphene is a single-atom-thick lattice of covalently bonded carbon atoms in a hexagonal, honeycomb-like arrangement. In reality, many products marketed as graphene have multiple layers, and these additional layers typically detract from the idealized properties of single-layer *pristine* graphene.

A single-atom-thick, pristine graphene qualifies as a two-dimensional crystal and was the first of this type of crystal to be isolated and studied. Figure 1 illustrates the close relationship between the crystal structure of graphene and that of the naturally occurring carbon allotrope graphite. Graphite is made up of many layers of graphene stacked one on the other. The individual layers are connected not by chemical bonding but by much weaker van der Waals forces. This allows the planes of graphene within graphite to slide across one another with relative ease, and even to be pried apart. The latter method was used in the first isolation of graphene in 2004 (Novoselov et al. 2004).

Each carbon in the lattice of graphene shares three sp^2-hybridized covalent bonds with adjacent carbons. The remaining p_z orbital of each carbon contributes to pi-bonding in an aromatic network that extends throughout the whole of the crystal. This system allows unparalleled electrical conductivity through the material, with pristine graphene demonstrating a charge mobility of 10,000 $cm^2V^{-1}s^{-1}$ (Novoselov et al. 2004) and a high thermal conductivity of 5,000 $W{\cdot}m^{-1}K^{-1}$ (Balandin et al. 2008).

Pristine graphene is also physically very strong, with a Young's modulus of 1 TPa, and a tensile strength of 130 GPa (Lee et al. 2008). It is highly optically transparent (97.7%) (Nair et al. 2008), possesses a high theoretical specific surface area (2,630 m^2 g^{-1}) (Stoller et al. 2008), and is impermeable to gases as small as helium (Bunch et al. 2008).

Graphene's allure as a material is perhaps best summed up by Novoselov (2011), one of the two scientists who originally isolated it: "In graphene, we have a unique combination of properties which are not seen together anywhere else: conductivity and transparency, mechanical strength and elasticity." Although the physical properties of graphene are indeed characterized by impressively large numbers, graphene is really a fundamentally new and different kind of material, and little can be said with certainty about its ultimate potential.

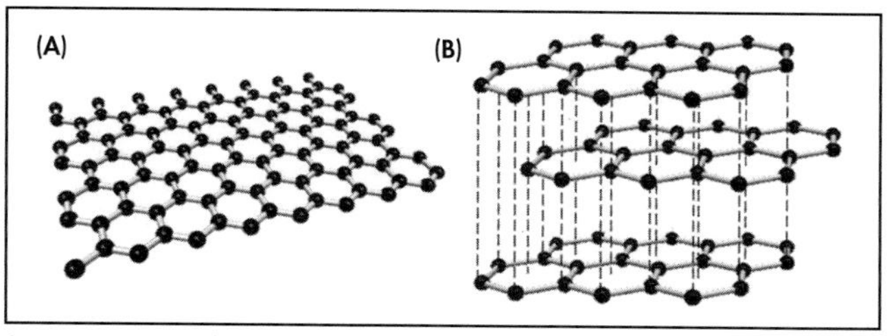

Source: Kumar and Lee 2013

Figure 1 Structural relationship between (A) graphene and (B) graphite

Characterization

Producing defect-free, single-atom-thick graphene in macroscopic quantities is still very difficult. Depending on the method of production used, the resulting graphene may have extra layers, structural deformities, or chemical impurities. Graphene cannot be called pristine if it contains defects like these. The presence of any such defects alters the properties of the material, and typically these changes will be for the worse. The grain size of the individual crystals of graphene also affects the material's properties because the grain boundaries will not be as strong or conductive as the pristine portions of the material. Methods have been developed to characterize graphene to demonstrate the quality of material produced using different techniques.

Graphene quality can be demonstrated in terms of the material's electrical properties. The ratio of charge mobility to electrical resistance is inversely proportional to the number of layers present in the graphene (Nezich et al. 2012). Reporting the sheet resistance by itself is common, as it gives

Gary R. Wilson, Graduate Student, Department of Metallurgical & Materials Engineering, Montana Tech, Butte, Montana, USA
Courtney A. Young, Department Head & Professor, Metallurgical & Materials Engineering, Montana Tech, Butte, Montana, USA

a qualitative indication of graphene quality. Lower resistance implies that the material in question more closely resembles pristine graphene (Bae et al. 2010).

Ultraviolet-visible (UV-vis) spectroscopy can be used to determine the number of graphene layers present in the material by reading the transmittance of a certain frequency. Pristine graphene has an optical transmittance of 97.1% at 550-nm wavelength. At this same wavelength, two-layer graphene allows 94.3% transmittance and six-layer allows only 83% (Sun et al. 2010). UV-vis can also be used to distinguish between graphene and the related material graphene oxide. The first absorbs at 262 nm and the second at 232 nm. Graphene oxide is discussed in greater detail later in this chapter.

The step height between individual layers of graphene is 0.34 nm, well within the detection limits of atomic force microscopy (Allen et al. 2010). Complications in reading the height profile can arise because of differential attraction between the substrate and graphene which itself is semimetallic. Difficulties can also be attributed to the fact that water can adsorb onto the graphene in a thin layer. These factors result in increased readings between 0.6 and 1.0 nm for single-layer graphene (Novoselov et al. 2004). Therefore, even though they are themselves undesirable features, folds in the graphene structure are often relied upon to calculate the number of layers in the graphene because the step height of 0.34 nm remains consistent for each layer (Allen et al. 2010).

Raman spectroscopy can be used to characterize graphene as well. Defects in the network of sp^2 bonds result in a band (the D band) around 1,350 cm^{-1} that does not present itself in pristine graphene (Allen et al. 2010). Increasing the number of layers of graphene decreases the intensity of a band at ~1,580 cm^{-1} (the G band) and decreases the intensity of a band at ~2,700 cm^{-1} (the 2D band, also called the G' band), thereby allowing the quantification of the number of layers present in the graphene (Allen et al. 2010; Wang et al. 2008).

MARKETS

In 2009, worldwide production of graphene was about 15 tons per year (Segal 2009). By 2015, annual graphene *nanoplatelet* production capacity alone increased to more than 900 tons, although it is not known how much of this capacity is actually utilized (Peplow 2015). About 27% of this production is in the Americas, 68% in Asia, and the rest in Europe, Africa, and the Middle East. A study from the same year noted that there were more than 200 companies producing graphene or graphene applications worldwide (Future Markets 2015).

China is the leading producer of graphene today. The majority of graphene produced is in the form of either nanoplatelets or thin films. The former is applied in nanocomposite materials, conductive inks and coatings, and energy storage; the latter is mainly geared toward transparent electrodes. Millipore Sigma sells graphene nanoplatelets with a variety of particle sizes and surface areas for as low at $115 per 250 mg (Millipore Sigma 2018a). They package their nanoplatelets in glass bottles as a dry powder or as a dispersion in either N,N-dimethylformamide (DMF) or N-methyl-2-pyrrolidone (NMP). For comparison, Millipore Sigma also sells monolayer thin films of graphene at $375 for a sheet about 10 cm in diameter (Millipore Sigma 2018b). Note that these films are not monocrystals; rather, they have a grain size near 10 μm. The thin film is embedded on a layer of copper foil and can be purchased coated with poly-methyl-methacrylate (PMMA) to prevent contamination.

Graphene currently sees use in many commercial products. Several companies that produce sports equipment have incorporated graphene nanocomposites into their equipment. These include Vittoria's line of graphene-enhanced bike wheels (Vittoria 2018), Head's Graphene-XT tennis rackets (Lammer 2013), and Victor's Meteor X JJS badminton racket (Victor 2018). Applications of graphene in electronics are plentiful as well. Vorbeck uses graphene in their conductive inks and coatings (Crain et al. 2012). Zap&Go produces graphene supercapacitors for quick charging of consumer electronics (Voller et al. 2017). Graphene Lighting PLC produces energy-efficient graphene-enabled lightbulbs (Lai 2017).

SOURCES

Graphene only exists naturally in the form of graphite, from which it must be isolated to take advantage of its properties. The original isolation of graphene used highly oriented pyrolytic graphite, an expensive, manufactured form of graphite used for calibrating certain instrumentation. Despite this costly parent material, the isolation was simple and effective, involving repeated peeling apart of the graphite crystals with adhesive tape (Novoselov et al. 2004), a process commonly referred to as the Scotch-tape method. Although this is an effective way to acquire high-quality graphene for laboratory testing, this method is time consuming and lacks scalability, so other methods of producing graphene have since been found.

Graphene production methods generally fall into one of two categories: top-down or bottom-up. Top-down processes utilize either naturally occurring or synthetic graphite and involve physical or chemical separation to isolate sheets of single- to multilayer graphene. The strength of top-down processes is that they more easily produce high-quality graphene, can be done at relatively low temperatures, and require less energy. The main weakness of industrial-scale top-down methods is that they do not produce very large sheets of graphene.

Bottom-up methods of graphene production necessarily involve a starting material other than elemental carbon, such as methane or ethanol. Chemical and/or physical techniques are then used to *build* graphene from these ingredients. Bottom-up methods vary in their effectiveness, but can produce arbitrarily large sheets of graphene. These sheets are not necessarily monocrystalline, and in fact, bottom-up techniques are more likely than top-down to produce defects in the resulting graphene. In this regard, however, the gap between top-down and bottom-up may be closing.

TOP-DOWN PRODUCTION TECHNIQUES

Mechanical exfoliation methods involve applying physical forces to graphite to free individual layers of graphene. These layers are connected not by chemical bonds, but by comparatively weak van der Waals forces. As shown in Figure 2, the forces applied to separate graphene layers can be either normal to the lattice plane (as when prying the layers apart with adhesive tape) or within the lattice plane as a shear force (Yi and Shen 2015). Many top-down methods utilize both forces.

A very common way to apply the normal force is by liquid-phase intercalation. In this case, van der Waals forces are overcome chemically using either an organic solvent (Hernandez et al. 2010) or water combined with surfactants (Lotya et al. 2009). For organic solvents, NMP is the most common. Solvents and water/surfactant systems used must have a similar surface tension to that of graphene to be energetically favorable (Hernandez et al. 2008), allowing liquid to

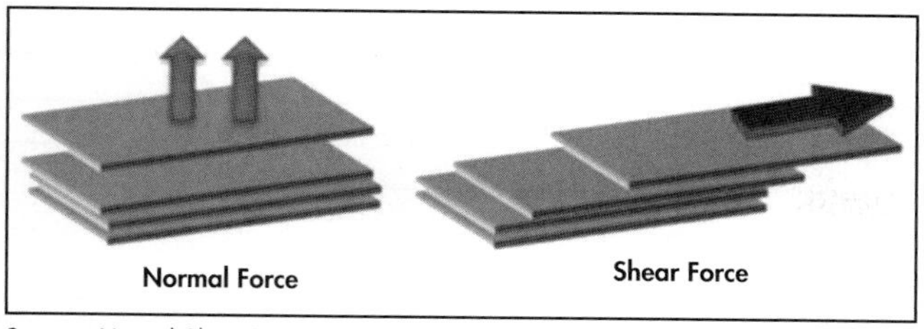

Source: Yi and Shen 2015
Figure 2 Mechanical exfoliation

enter between individual layers and force them apart. Van der Waals forces approach zero and become negligible at a separation distance of ~5 angstroms (Spanu et al. 2009).

It is common to combine this method with sonication to enhance the rate and extent of exfoliation. Sonication has, however, been shown to cause defects in the graphene produced (Bracamonte et al. 2014). Another way of enhancing exfoliation is by applying shear forces via ball milling, which can be done dry or wet and in the absence or presence of sonication (Zhao et al. 2010; Knieke et al. 2010). Dry milling is typically assisted by the presence of water-soluble salts, and the liquid used in wet milling is the intercalating solvent or water/surfactant mixture (Lv et al. 2014; Varrla et al. 2014).

Collisions during milling may as easily impart a normal force as a shear force. Normal force collisions result in some exfoliation along the graphite planes, but also fracture graphite into smaller particles, resulting in smaller diameters of graphene platelets. A way of imparting shear force that seeks to avoid fracturing does so by blending the graphene-liquid mixture to separate the layers (Varrla et al. 2014). Interestingly enough, a patent has been acquired for a graphene-producing process that quite literally uses a kitchen blender, although a precise balance of surfactant concentration and blending speed are required (Coleman and Paton 2014).

Another top-down method of graphene production involves creating the graphite oxide intermediate. This method is highly related to other forms of mechanical exfoliation, but the normal force used to push the layers of graphene apart is supplied primarily by a chemical method. Graphite is submerged in a strong acid that reacts with the graphite on the surface and in between the layers, adding carboxyl groups, hydroxyl groups, and even epoxides to the graphene. A maximum of about 40% of the carbon atoms, previously sp^2 hybridized, become sp^3 hybridized (Mkhoyan et al. 2009). The change in hybridization increases the spacing between each layer from 0.34 nm to about 0.65–0.75 nm in a period of about 96 hours (Schniepp et al. 2006).

The chemically transformed graphite, now called *graphite oxide*, is made of up of layers of graphene oxide. The graphene oxide layers are easily separated by intercalation with a polar solvent and can then be reduced to graphene by the introduction of hydrazine. The resulting graphene is of fairly low quality and is referred to as *reduced graphene oxide* rather than just graphene. This process is arguably the least-expensive scalable method of producing something that can be called graphene. Although too poor quality for electronic applications, reduced graphene oxide is suitable for use in strength-based composite materials.

BOTTOM-UP PRODUCTION TECHNIQUES

Chemical vapor deposition (CVD) begins with a carbon precursor as feed, typically a short-chain hydrocarbon such as methane or ethane (Seah et al. 2014). Unsaturated short-chain hydrocarbons, such as acetylene and ethylene, have been shown to be more effective at graphene production, allowing for lower production temperatures (Nandamuri et al. 2010). These highly flammable gases are more difficult to handle safely though, and CVD is a high-temperature process (typically 800°–1,000°C) (Seah et al. 2014). Industrially, short-chain aliphatic alcohols may be preferred, as they are much less flammable and are still capable of producing high-quality graphene (Guermoune et al. 2011).

The carbon precursor is fed, typically in a gaseous state, to the chamber with the transition metal substrate. Transition metals were originally favored because the vacancies in their *d* shells can act as an electron acceptor to facilitate chemisorption of the carbon precursor onto the surface. Copper, however, is widely used despite having a filled 3*d* shell, and has a slightly different mechanism of graphene growth from that of another common transition metal substrate, nickel (Figure 3) (Seah et al. 2014).

On a nickel substrate, the carbon precursor (1) adsorbs onto the surface, then (2) dissociates by hydrogenation to generate hydrogen gas and carbon adatoms (adsorbed atoms). The carbon adatoms (3) diffuse into the nickel foil at high temperature, and then the temperature is lowered to decrease the solubility of the carbon adatoms in the foil. In a step called *segregation*, the carbon adatoms are (4) expelled to the surface, where they (5) self-organize into graphene.

Graphene formation on a copper foil acts by a similar process, but the solubility of carbon in copper is very low, so the segregation step (2) is thought to immediately follow dehydrogenation (1) in a completely surface-mediated process. An important consequence of the difference between these two processes is that, on nickel, the segregation step will continue until the carbon dissolved in the bulk material reaches equilibrium with the carbon that is expelled to the surface, even if the feed of carbon precursor has been shut off (Seah et al. 2014). This makes it potentially more difficult to control the

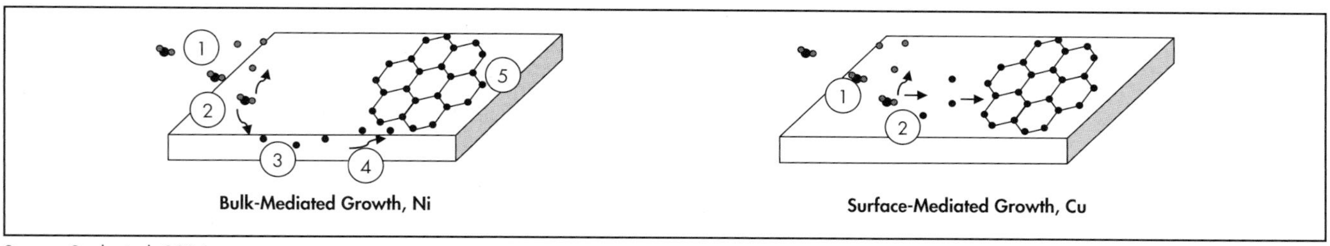

Source: Seah et al. 2014
Figure 3 Growth by chemical vapor deposition on nickel and copper

Table 1 Qualitative comparison of graphene produced by different methods

Method	Maximum Grain Size	Minimum Thickness	Crystal Quality	Cost	Applicability
Scotch tape	Large	Monolayer	Pristine	Not applicable	Laboratory
Ball milling	Small	Multilayer	High	Low	Conductive inks, composites
Blending	Small	Few-layer	High	Low	Conductive inks, composites
Reduced graphene oxide	Small	Multilayer	Low	Very low	Composites
Chemical vapor deposition (CVD)	Large	Monolayer	Pristine	High	Electronics, energy storage
Plasma-enhanced CVD	Large	Monolayer	Pristine or less	Medium high	Electronics, energy storage

number of layers of graphene produced as compared to the purely surface-mediated growth on copper, where segregation stops as soon as the feed of carbon precursor stops.

Temperature is a major parameter influencing the size and quality of graphene produced by CVD. Lower temperatures are desired both to save energy and to allow for applications in electronics, which would be damaged by higher temperatures, but a certain amount of heat is necessary to decompose the carbon precursor. However, the low solubility of C in Cu films requires *higher* temperatures (Memon et al. 2013), and these higher temperatures increase the quality and quantity of monolayer graphene produced. At temperatures above 1,050°C, however, the low melting point of Cu begins to interfere with graphene formation (Fan et al. 2011). Ni foils also require high temperatures (though not as high as Cu) because graphene formation competes with nickel carbide (Ni_2C) formation at low temperatures (Lahiri et al. 2011).

One variation of CVD that has succeeded in lowering processing temperatures is plasma-enhanced chemical vapor deposition (PECVD). A growth temperature as low at 240°C has been reported (Kalita et al. 2012). In PECVD, a plasma is produced through radio frequencies (Qi et al. 2011), magnetically enhanced arc discharges (Levchenko et al. 2010), or microwaves (Kalita et al. 2012). The plasma couples with the substrate, causing localized rapid heating and potentially removing the metal's native oxide layer, allowing for more uniform nucleation of graphene (Kumar et al. 2012). Continuous plasma bombardment can produce defects as well, and apparently, the quality of graphene produced by PECVD is still better when performed at higher temperatures (Woo et al. 2013).

Large sheets of graphene, on the order of meters, are currently produced by CVD but are ridden with defects. As of 2015, the largest monolayer single-crystal grains of graphene grown were about 1 cm in diameter (Hao et al. 2013). An important downside to CVD is the substrate. Graphene attached to a transition metal foil is usually not a useful form of the material. The substrate must be dissolved (typically in an acid) and the graphene layer transferred to the more useful substrate (Reina et al. 2008). The transfer process likely produces defects in the graphene as well.

The various features of graphene produced by each of the production methods discussed in this chapter are summarized in Table 1.

REFERENCES

Allen, M.J., Tung, V.C., and Kaner, R.B. 2010. Honeycomb carbon: A review of graphene. *Chem. Rev.* 110(1):132–145.

Bae, S., Kim, H., Lee, Y., Xu, X., Park, J.-S., and Zheng, Y., et al. 2010. Roll-to-roll production of 30-inch graphene films for transparent electrodes. *Nat. Nanotechnol.* 5(8):574–578.

Balandin, A.A., Ghosh, S., Bao, W., Calizo, I., Teweldebrhan, D., Miao, F., and Lau, C.N. 2008. Superior thermal conductivity of single-layer graphene. *Nano Lett.* 8(3):902–907.

Bracamonte, M.V., Lacconi, G.I., Urreta, S.E., and Foa Torres, L.E.F. 2014. On the nature of defects in liquid-phase exfoliated graphene. *J. Phys. Chem. C* 118(28):15455–15459.

Bunch, J.S., Verbridge, S.S., Alden, J.S., van der Zande, A.M., Parpia, J.M., Craighead, H.G., and McEuen, P.L. 2008. Impermeable atomic membranes from graphene sheets. *Nano Lett.* 8(8):2458–2462.

Coleman, J., and Paton, K. 2014. A scalable process for producing exfoliated defect-free, non-oxidised 2-dimensional materials in large quantities. U.S. Patent Application WO2014140324A1.

Crain, J.M., Lettow, J.S., Aksay, I.A., Korkut, S.A., Chiang, K.S., Chen, C.-H., and Prud'homme, R.K. 2012. Printed electronics. U.S. Patent 8278757 B2.

Fan, L., Li, Z., Li, X., Wang, K., Zhong, M., Wei, J., Wu, D., and Zhu, H. 2011. Controllable growth of shaped graphene domains by atmospheric pressure chemical vapour deposition. *Nanoscale* 3(12):4946.

Future Markets. 2015. *The Global Market for Graphene to 2025: Market Size, Production Volumes, Applications, Products, Prospects, Research and Companies.* www.rnrmarketresearch.com/the-global-market-for-graphene-to-2020-market-report.html. Accessed March 2018.

Guermoune, A., Chari, T., Popescu, F., Sabri, S.S., Guillemette, J., Skulason, H.S., Szkopek, T., and Siaj, M. 2011. Chemical vapor deposition synthesis of graphene on copper with methanol, ethanol, and propanol precursors. *Carbon* 49(13):4204–4210.

Hao, Y., Bharathi, M.S., Wang, L., Liu, Y., Chen, H., Nie, S., Wang, X., et al. 2013. The role of surface oxygen in the growth of large single-crystal graphene on copper. *Science* 342(6159):720–723.

Hernandez, Y., Nicolosi, V., Lotya, M., Blighe, F.M., Sun, Z., De, S., McGovern, I.T., et al. 2008. High-yield production of graphene by liquid-phase exfoliation of graphite. *Nat. Nanotechnol.* (9)3:563–568.

Hernandez, Y., Lotya, M., Rickard, D., Bergin, S.D., and Coleman, J.N. 2010. Measurement of multicomponent solubility parameters for graphene facilitates solvent discovery. *Langmuir* 26(5):3208–3213.

Kalita, G., Wakita, K., and Umeno, M. 2012. Low temperature growth of graphene film by microwave assisted surface wave plasma CVD for transparent electrode application. *RSC Adv.* 2(7):2815–2820.

Knieke, C., Berger, A., Voigt, M., Taylor, R.N.K., Röhrl, J., and Peukert, W. 2010. Scalable production of graphene sheets by mechanical delamination. *Carbon* 48(11):3196–3204.

Kumar, A., and Lee, C.H. 2013. Synthesis and biomedical applications of graphene: Present and future trends. In *Advances in Graphene Science*. Edited by A. Mahmood and A. London, UK: IntechOpen.

Kumar, A., Voevodin, A.A., Zemlyanov, D., Zakharov, D.N., and Fisher, T.S. 2012. Rapid synthesis of few-layer graphene over Cu foil. *Carbon* 50(4):1546–1553.

Lahiri, J., Miller, T., Adamska, L., Oleynik, I.I., and Batzill, M. 2011. Graphene growth on Ni(111) by transformation of a surface carbide. *Nano Lett.* 11(2):518–522.

Lai, C.P. 2017. LED light bulb simultaneously using as night-light. U.S. Patent 20160356461A1.

Lammer, H. 2013. Sporting goods with graphene material. U.S. Patent 8342989 B2.

Lee, C., Wei, X., Kysar, J.W., and Hone, J. 2008. Measurement of the elastic properties and intrinsic strength of monolayer graphene. *Science* 321(5887):385–388.

Levchenko, I., Volotskova, O., Shashurin, A., Raitses, Y., Ostrikov, K., and Keidar, M. 2010. The large-scale production of graphene flakes using magnetically-enhanced arc discharge between carbon electrodes. *Carbon* 48(15):4570–4574.

Lotya, M., Hernandez, Y., King, P.J., Smith, R.J., Nicolosi, V., et al. 2009. Liquid phase production of graphene by exfoliation of graphite in surfactant/water solutions. *J. Am. Chem. Soc.* 131(10):3611–3620.

Lv, Y., Yu, L., Jiang, C., Chen, S., and Nie, Z. 2014. Synthesis of graphene nanosheet powder with layer number control via a soluble salt-assisted route. *RSC Adv.* 4(26):13350–13354.

Memon, N.K., Tse, S.D., Chhowalla, M., and Kear, B.H. 2013. Role of substrate, temperature, and hydrogen on the flame synthesis of graphene films. *Proc. Combust. Inst.* 34(2):2163–2170.

Millipore Sigma. 2018a. Graphene nanoplatelets 900409. www.sigmaaldrich.com/catalog/product/aldrich/900409. Accessed March 2018.

Millipore Sigma. 2018b. Monolayer graphene film 900415. www.sigmaaldrich.com/catalog/product/aldrich/900415. Accessed March 2018.

Mkhoyan, K.A., Contryman, A.W., Silcox, J., Stewart, D.A., Eda, G., Mattevi, C., Miller, S., and Chhowalla, M. 2009. Atomic and electronic structure of graphene-oxide. *Nano Lett.* 9(3):1058–1063.

Nair, R.R., Blake, P., Grigorenko, A.N., Novoselov, K.S., Booth, T.J., Stauber, T., Peres, N.M.R., and Geim, A.K. 2008. Fine structure constant defines visual transparency of graphene. *Science* 320(5881):1308–1308.

Nandamuri, G., Roumimov, S., and Solanki, R. 2010. Chemical vapor deposition of graphene films. *Nanotechnol.* 21(14):145604.

Nezich, D., Reina, A., and Kong, J. 2012. Electrical characterization of graphene synthesized by chemical vapor deposition using Ni substrate. *Nanotechnol.* 23(1):015701.

Novoselov, K.S. 2011. Nobel lecture: Graphene: Materials in the flatland. *Rev. Mod. Phys.* 83(3):837–849.

Novoselov, K.S., Geim, A.K., Morozov, S.V., Jiang, D., Zhang, Y., Dubonos, S.V., Grigorieva, I.V., and Firsov, A.A. 2004. Electric field effect in atomically thin carbon films. *Science* 306(5696):666–669.

Peplow, M., 2015. Graphene booms in factories but lacks a killer app. *Nature* 522(7556):268–269.

Qi, J.L., Zheng, W.T., Zheng, X.H., Wang, X., and Tian, H.W. 2011. Relatively low temperature synthesis of graphene by radio frequency plasma enhanced chemical vapor deposition. *Appl. Surf. Sci.* 257(15):6531–6534.

Reina, A., Son, H., Jiao, L., Fan, B., Dresselhaus, M.S., Liu, Z., and Kong, J. 2008. Transferring and identification of single- and few-layer graphene on arbitrary substrates. *J. Phys. Chem. C* 112(46):17741–17744.

Schniepp, H.C., Li, J.-L., McAllister, M.J., Sai, H., Herrera-Alonso, M., Adamson, et al. 2006. Functionalized single graphene sheets derived from splitting graphite oxide. *J. Phys. Chem. B* 110(17):8535–8539.

Seah, C.-M., Chai, S.-P., and Mohamed, A.R. 2014. Mechanisms of graphene growth by chemical vapour deposition on transition metals. *Carbon* 70 (April): 1–21.

Segal, M. 2009. Selling graphene by the ton. *Nat. Nanotechnol.* 4(10):612.

Spanu, L., Sorella, S., and Galli, G. 2009. Nature and strength of interlayer binding in graphite. *Phys. Rev. Lett.* 103(19):196401.

Stoller, M.D., Park, S., Zhu, Y., An, J., and Ruoff, R.S. 2008. Graphene-based ultracapacitors. *Nano Lett.* 8(10):3498–3502.

Sun, Z., Yan, Z., Yao, J., Beitler, E., Zhu, Y., and Tour, J.M. 2010. Growth of graphene from solid carbon sources. *Nature* 468(7323):549–552.

Varrla, E., Paton, K.R., Backes, C., Harvey, A., Smith, R.J., McCauley, J., and Coleman, J.N. 2014. Turbulence-assisted shear exfoliation of graphene using household detergent and a kitchen blender. *Nanoscale* 6:11810–11819.

Victor. 2018. Meteor X [racket]. http://us.victorsport.com//product/10017. Accessed March 2018.

Vittoria. 2018. Vittoria graphene technology. www.vittoria.com/us/graphene-technology. Accessed July 2018.

Voller, S., Tuck, J.R., Rajendran, M., Trindade, A., Dutta, M., Rashidi, N., Walder, T., and Lamarie, Q. 2017. Battery charger. U.S. Patent 9774201B2.

Wang, Y.Y., Ni, Z.H., Yu, T., Shen, Z.X., Wang, H.M., Wu, Y.H., Chen, W., and Shen Wee, A.T. 2008. Raman studies of monolayer graphene: The substrate effect. *J. Phys. Chem. C* 112:10637–10640.

Woo, Y.S., Seo, D.H., Yeon, D.-H., Heo, J., Chung, H.-J., et al. 2013. Low temperature growth of complete monolayer graphene films on Ni-doped copper and gold catalysts by a self-limiting surface reaction. *Carbon* 64:315–323.

Yi, M., and Shen, Z. 2015. A review on mechanical exfoliation for the scalable production of graphene. *J. Mater. Chem. A* 3:11700–11715.

Zhao, W., Fang, M., Wu, F., Wu, H., Wang, L., and Chen, G. 2010. Preparation of graphene by exfoliation of graphite using wet ball milling. *J. Mater. Chem.* 20:5817.

CHAPTER 12.14

Graphite

Andrew Scogings and Daryl Evans

PROPERTIES

Graphite is an allotrope of carbon, characterized by a hexagonal structure, with weak bonds between the carbon layers that facilitate easy cleavage, which makes it one of the softest substances known. Graphite is gray to black, opaque, very soft, and has a low density and a metallic luster. It is flexible and exhibits both metallic and nonmetallic properties, making it suitable for diverse industrial applications. Physical properties include a specific gravity of 2.2 and Mohs hardness of 1–2. Metallic properties include thermal and electrical conductivity, whereas nonmetallic properties include chemical inertness, high thermal resistance, and lubricity.

Natural graphite occurs in several forms, described as *amorphous*, *flake*, and *vein*. Graphite may also be manufactured synthetically from carbon-bearing raw materials such as petroleum coke and tar pitch. Natural graphite products generally contain associated mineral impurities, which are referred to as *ash*. These impurities may include silicate and sulfide minerals such as quartz or pyrite in the case of flake graphite. Amorphous graphite may contain sedimentary rock impurities such as shale, sandstone, quartzite, or limestone.

USES

The major uses of graphite are in refractories, batteries, expandable graphite, brake linings, and steelmaking-foundry operations. The largest end market for graphite is in refractories, foundries, and crucibles for protective linings, shapes, or vessels used in high-temperature environments such as steel production. These markets are believed to account for around 35% of total graphite consumption, predominately consuming flake and vein graphite. Steelmaking-foundry operations use graphite as a carbon raiser, as foundry facings to facilitate the removal of an object from a mold, and for lubricating extrusion dies.

The refractories market requires a range of flake size distribution and purity, which vary according to application. For example, alumina magnesia carbon bricks may demand fine flake graphite <100 mesh, whereas magnesia carbon bricks typically use 90%–95% C graphite with a broad flake distribution –100 to +50 mesh.

Metallurgy is the second largest market for natural graphite, which is used in metal production, particularly as a recarburizer in steel and consuming mainly amorphous graphite or fine flake. This accounts for approximately 25% of total graphite output. Batteries are the third largest graphite market. Although only consuming around 13% of worldwide graphite production, it is potentially the fastest growing market. Chinese producers use –100 mesh (94% C) small flake for making spherical graphite (for battery anode applications). Solid lubricants, based mainly on amorphous graphite, consume approximately 10% of production, having been a classic use for centuries.

Both flake and amorphous graphite are used in parts and components. This wide range of products varies from motor vehicle brake pads to carbon brushes for electric motors to pencils. As a group, it is believed to consume roughly 10% of total output. Brake linings use an organic friction material impregnated with graphite, especially for heavy vehicles such as trucks. Expandable graphite is another market that is anticipated to grow, and is used in applications such as fire retardation to replace halogenated retardants, insulation, and heat transmission. These markets require large flake products generally >80 mesh. Graphite is typically specified at a minimum by particle size and carbon content (purity). There are no set industry specifications, although in countries such as China, the government has established national standards.

PRICES

The U.S. Geological Survey (USGS) has published prices for graphite imported into the United States (Olson 2017):

- Flake graphite cost $1,240/t (metric ton) in 2015, compared to $1,370 in 2012.
- Amorphous (Mexican) graphite cost $370/t in 2014, compared to $339 in 2012.
- Lump and chip (Sri Lankan) graphite cost $1,890/t in 2015, which has remained relatively constant since 2011.

Andrew Scogings, Principal Consultant, CSA Global Pty Ltd., Perth, Western Australia, Australia
Daryl Evans, Director of Metallurgy, IMO Pty Ltd., Perth, Western Australia, Australia

Flake graphite prices remained relatively steady for many years until approximately 2005, after which they climbed gradually to 2008 in tandem with increased Chinese steel production. Following the global financial crisis, prices declined in 2009 before resuming an upward trend and spiking dramatically during 2011–2012. Prices have since declined to 2008 levels because of excess production versus market demand. Amorphous graphite has shown similar trends. For example, flake graphite (>80 mesh, 94%–97% C) sold for about $750/t up until 2004. The price increased to a peak of around $2,500/t in 2011–2012 before falling back to $1,000/t in 2014. Currently, the price ranges from $750 to 850/t (Industrial Minerals 2018).

One graphite price not generally quoted is for *jumbo flake*, a term that describes flakes +35 mesh size or larger. Trade is relatively limited and probably only a few thousand metric tons per year, but likely prices may have been around $2,000/t in 2015. Spherical graphite is another product for which prices are not widely publicized. However, according to Chinese customs data, prices for uncoated spherical product range around $3,000/t (Scogings et al. 2015b).

GLOBAL RESOURCES

Table 1 shows the global resource of graphite and includes the top 10 countries where most of the resources are identified. The reserve numbers that are included in Table 1 are separately shown in Table 2.

PRODUCTION

Global natural graphite production has risen from about 100,000 t/yr in the early 1900s to approximately 1.2 Mt (million metric tons) in 2014, at a compounded annual growth of about 3% since 1950.

According to USGS data, China was the world's leading producer of natural graphite and supplied about two-thirds of the market in 2014 (~800,000 t/yr). Brazil (~80,000 t/yr), India (~170,000 t/yr), Canada, and North Korea were estimated to have collectively contributed an additional 27% of global production (Olson 2017). China's production has rapidly grown since 1998, having overtaken the rest of the world in 2000. Production quoted in India has been under discussion for some time, and it is believed that the tonnages reported are for run-of-mine (ROM), not for concentrate produced; hence India's production is probably closer to around 25,000 t/yr.

Leading non-Chinese flake graphite miners and producers include

- Graphit Kropfmühl (Germany, Mozambique [under construction], and Zimbabwe),
- Skaland Graphite (Norway),
- Imerys Graphite and Carbon (Namibia [under construction] and Canada),
- Nacional de Grafite and Grafite do Brasil (Brazil),
- Zavalivskiy Graphite (Ukraine),
- Tirupati Carbons and Chemical (India), and
- Bass Metals (Madagascar).

Nacional de Grafite is reportedly the biggest mine outside of China, producing about 70,000 t/yr of concentrate.

Heilongjiang is the leading flake graphite–producing province in China and is believed to host the largest flake graphite deposits in Asia. The two main graphite mining areas are Luobei and Jixi. Producers in Heilongjiang include Aoyu Graphite Group in the Luobei area. In the Jixi area, companies such as BTR New Energy Materials, Huanyu New Energy Technology, PuChen Graphite, and the Liumao graphite mine are operational (Scogings et al. 2015b).

Shandong is also a notable producing area and includes companies such as Qingdao Black Dragon Graphite Group and Qingdao Hensen Graphite, of which the latter produces flake graphite, micronized graphite, expandable graphite, and spherical graphite. Hunan Province is the world's leading source of amorphous graphite and is believed to produce about 200,000 t/yr (Scogings et al. 2015b).

Sri Lanka is a relatively small producer, although well known for its high-quality vein graphite. There are two significant vein graphite producers in Sri Lanka: Kahatagaha Graphite Lanka (government owned and operated) and Bogala Graphite Lanka, which is owned by Graphit Kropfmühl (Scogings et al. 2015b).

Table 1 Graphite resources

	Identified Resources, thousand Mt		
Country	**Amorphous**	**Flake**	**Total**
China	Not applicable	360,000	360,000
Russia with Ukraine	560,000	100,000	660,000
Madagascar	—	180,000	180,000
North Korea	31,000	2,100	33,100
Mexico	13,000	2,100	15,100
India	—	14,000	14,000
Czech Republic	—	13,000	13,000
Sri Lanka	450 (crystalline vein)	8,800	9,250
United States	5,900	280	6,180
Canada	—	5,700	5,700
Other	613,650	690,000	1,303,650
Total	**1,224,000 (includes crystalline vein)**	**1,375,980**	**2,599,980**

Adapted from Robinson et al. 2017

Table 2 Graphite reserves

	Reserves, thousand Mt		
Country	**Amorphous**	**Flake**	**Total**
China	55,000	6,000	61,000
Russia with Ukraine	1,000	6,400	7,400
Mexico	3,100	106	3,206
Sri Lanka	50 (crystalline vein)	1,800	1,800
North Korea	1,000	700	1,700
Canada	—	1,500	1,500
Madagascar	—	940	940
Czech Republic	—	900	900
India	—	735	735
Zimbabwe	—	600	600
Other	320	1,639	1,639
Total	**60,470 (includes crystalline vein)**	**21,320**	**81,420**

Adapted from Robinson et al. 2017

GEOLOGIC SETTINGS AND RESOURCES

Economic graphite deposits occur in three main geologic types: flake graphite disseminated in metamorphosed sedimentary rocks; amorphous graphite formed by metamorphism of coal or carbon-rich sediments; and veins or lump graphite filling fractures, fissures, and cavities in country rock. Although no reliable information is available about total world reserves of graphite ore, USGS estimates 230 Mt of world reserves, accounted for by Brazil (72 Mt), China (55 Mt), India (8 Mt), Madagascar (0.9 Mt), Mexico (3.1 Mt), and Turkey (90 Mt) (Olson 2017).

EXPLORATION

Graphite is explored for using methods such as field mapping, trenching, geophysics, drilling, assaying of the graphite content, and mineralogical and metallurgical testing. Data generated in this way, if successful, could lead to the estimation of a designated *mineral resource*. Reporting mineral resources conforming to international codes requires that they are classified according to confidence in geological and grade continuity, in addition to accounting for product specifications.

Several methods are used for drilling for graphite such as reverse circulation or diamond core drilling (DD), whereas auger drilling may occasionally be used to explore highly weathered, clayey mineralization. DD is by far the preferred method of exploration drilling for graphite, as the graphite flakes and host rock are relatively undisturbed when retrieved as core and can be used for extractive metallurgical tests.

Petrographic examination of polished thin sections is a relatively affordable and quick option to estimate the in situ graphite flake size distribution and likely liberation characteristics, as graphite may be associated with mineral impurities such as iron sulfides, quartz, feldspar, and mica. Polarized-light microscopy may be complemented by methods such as scanning electron microscopy (SEM), Quantitative Evaluation of Minerals by Scanning Electron Microscopy (QEMSCAN), and mineral liberation analysis (or automated SEM).

Carbon may be present in rocks in several different forms including organic carbon, carbonate minerals, or graphitic carbon. Organic carbon and carbon in carbonate minerals, such as calcite, should be removed before assaying for total graphitic carbon (TGC). These two forms of nongraphitic carbon are typically removed from the sample by acid leaching and calcination.

TGC assays quantify the amount of graphite contained within a deposit but do not indicate the amount of graphite that may be recoverable or the purity or flake size of liberated graphite and the process required to liberate and produce a graphite concentrate. Therefore, it is necessary to evaluate representative samples of mineralization from a deposit to confirm appropriate metallurgical processes and likely product mix.

The past few years have seen intensive exploration for graphite by both private and publicly listed companies, with most exploration aimed at flake graphite deposits. Australia, Canada, Mozambique, and Tanzania were the main hot spots, with the biggest resources discovered in northern Mozambique and southern Tanzania. Other target countries have included Brazil, Madagascar, Malawi, Sweden, and the United States.

This intensive exploration has added more than 4.5 billion t to the global flake graphite mineral resource base, at grades of between approximately 2% and 25% graphite. Total contained graphite is estimated to be in excess of 400 Mt and, assuming a mining and process yield of 70%, this equates to approximately 280 Mt of product of all flake sizes.

Given anticipated market growth in battery and expandable graphite applications, there is a window of opportunity for new flake graphite supply to enter the existing market. However, given the extent of new discoveries and proposed production rates, this could yield additional production of more than 1 Mt/yr of flake graphite over the next few years. This appears to be well in excess of growth opportunities, thus many projects will likely fail to reach production.

Comparing Flake Graphite Exploration and Development Projects

Graphite exploration projects may be ranked according to factors such as deposit size, contained graphite, location (country risk) and logistics, flake size distribution, product purity, offtake agreements, and expected time frame to production. Drivers for success include having a sufficiently high-grade flake graphite deposit with low stripping ratio that is economic to mine and process to yield highest-quality marketable products. In addition, jurisdiction, logistics, time frame to production, and offtake agreements must be taken into consideration.

MINING METHODS

Flake graphite deposits are typically hosted in metamorphic rocks, such as gneiss and schist, generally have tabular or lenslike geometry, and may be mined opencut, although some high-grade flake graphite deposits are mined underground in Germany, Norway, and Zimbabwe. Most flake graphite deposits being mined opencut contain between 5% and 15% graphite, whereas underground mines have grades of around 30% graphite.

Flake graphite deposits can be between 5 and 100 m thick, extend along a strike for hundreds or thousands of meters, and dip anywhere between horizontal and vertical. Deeply weathered flake graphite deposits may offer mining and processing benefits over unweathered deposits because of ease of mining and processing the soft decomposed ore. Deeply weathered deposits have been mined for many years in Madagascar and Brazil and have also been discovered in Malawi.

Vein deposits have complex geometry, are generally narrow (<1 m), and are selectively mined underground in Sri Lanka. Amorphous graphite is mined underground and usually extracted using selective room-and-pillar mining methods, similar to coal mining.

GRAPHITE BENEFICIATION

Graphite is naturally hydrophobic, and this property is conveniently leveraged via the use of flotation, targeting selective separation of hydrophobic graphite from hydrophilic gangue.

Comminution

Process flow-sheet development is predicated on specific deposit location, grade, and mineralogy. Generic flow sheets for processing hard rock and soft rock hosted ores are shown in Figures 1 and 2, respectively.

Hard Rock Hosted Graphite

Selection of the style of comminution circuit is predicated on ore competency and liberation size; selection of a dry or wet crushing circuit is heavily predicated on annual rainfall,

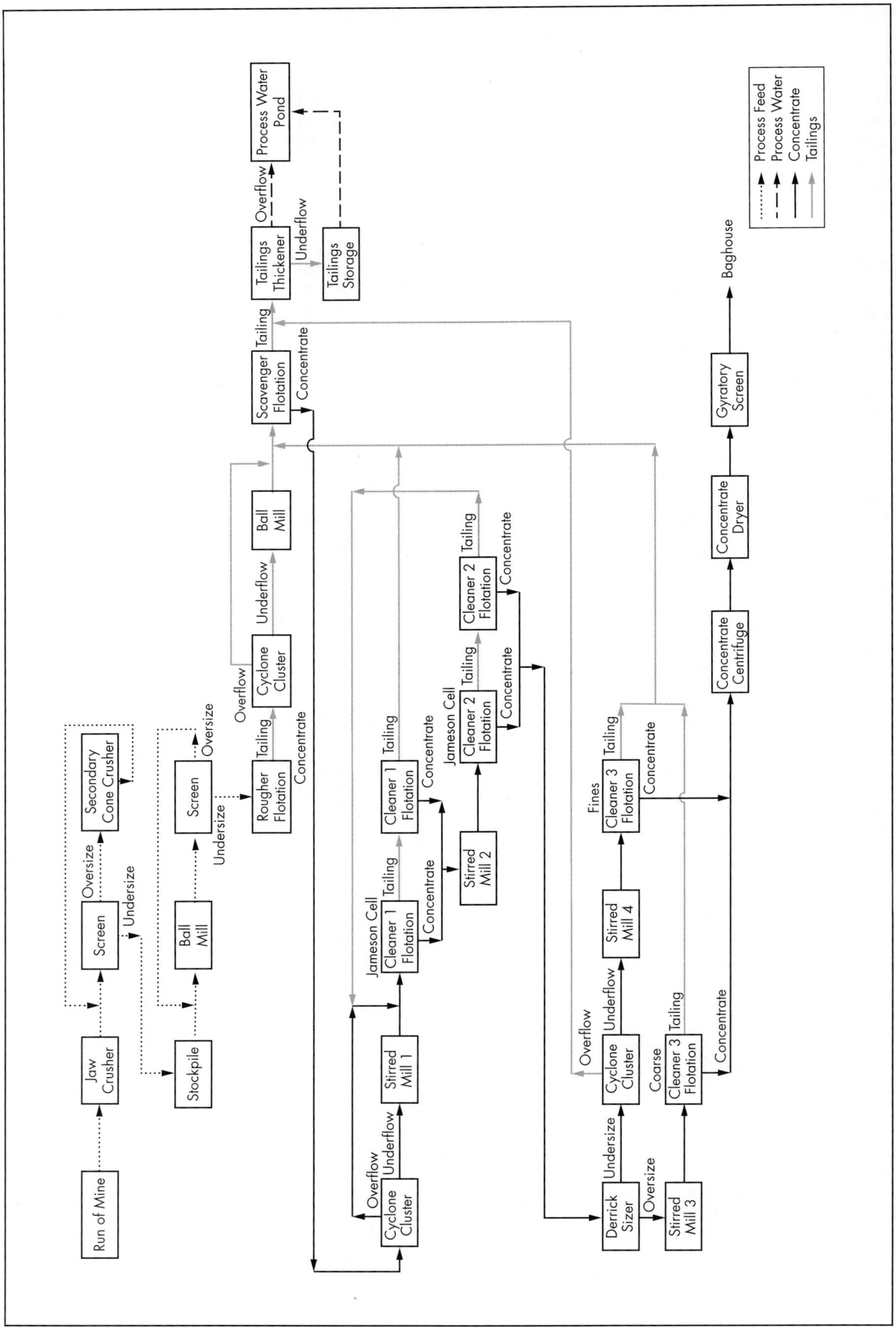

Figure 1 Competent host rock flow sheet

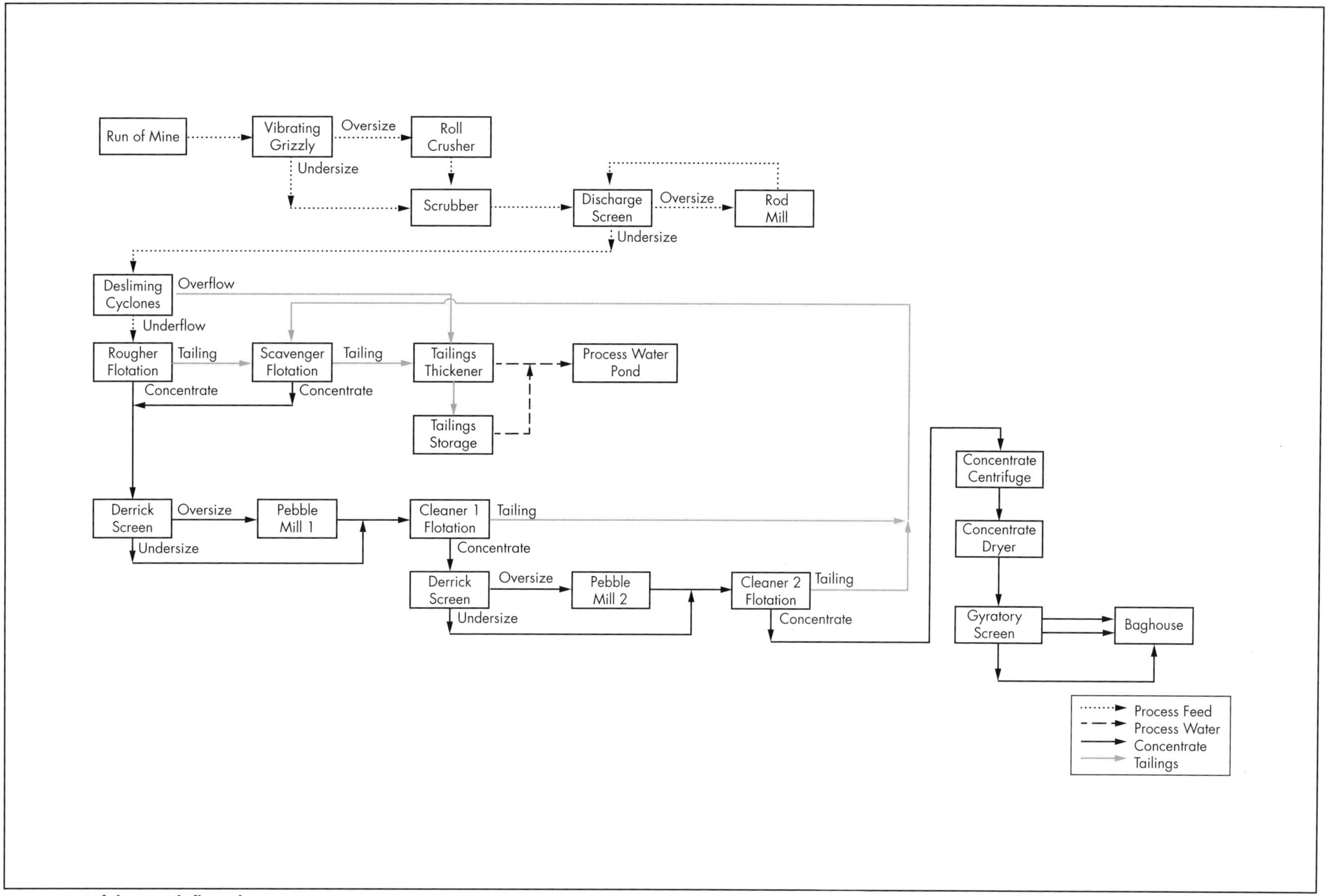

Figure 2 Soft host rock flow sheet

clay content, and location within the ore zone, either above or below the water table.

The hard rock example presented in Figure 1 incorporates the following unit processes:

- Secondary crushing based on primary jaw and closed-circuit secondary cone crushing
- Closed-circuit ball milling, targeting a moderately coarse grind (P_{80}) size, ranging from 250 to 500 µm

Tertiary crushing followed by closed-circuit rod mill grinding is a suitable alternative in cases where the graphite liberation size is coarser (nominally >700 µm). The inherent screening action characteristic of this style of milling provides a truncated mill discharge size distribution, minimizing generation of excess fines. The selection of rod milling, however, needs to be economically justifiable relative to semiautogenous grinding or coarse primary ball milling, the required throughput rate, and established ore competency.

Soft Rock Hosted Graphite

Graphite processing typical of that undertaken in western Madagascar in wet, soft, clay-rich saprolitic host is depicted in Figure 2 and includes the following:

- Coarse ROM screening using a vibrating grizzly. Grizzly oversize generally represents more competent fraction, and size reduction can usually be performed, maintaining control over top size via a coarse toothed roll crusher.
- Wet scrubbing utilizing a drum scrubber, accepting grizzly undersize and roll crushed oversize. Scrubber power requirements vary, depending on the power input, to ensure complete repulping of clay fraction. In selected cases, installation of a wave liner and the addition of a limited ball charge (2%–5%) can assist in breaking up coarse agglomerates.
- Scrubber discharge screened via a combination of coarse discharge trommel and vibrating screen. Oversize, nominally +1 mm reports for closed-circuit rod milling.
- Scrubber screen undersize, nominally –1 mm, is pumped to a hydrocyclone cluster operating at an approximate cut size of 45 µm for desliming and rejection of barren clay fraction and subgrade graphite.

Flotation

Graphite flotation is typically conducted based on initial roughing, targeting high graphite recovery. The adoption of a scavenger stage may be necessary, depending on the specific mineralogy, extent of liberation, and flotation kinetics. Rougher and scavenger concentrates are then subjected to a series of regrinding and selective cleaner flotation stages targeting high-grade concentrate >98% total carbon.

Reagents and Conditioning

Liberated graphite is naturally hydrophobic and can be floated readily using methyl isobutyl carbinol (MIBC) frother, diesel, or pine oil. A range of collectors are available that report improved gangue selectivity over more traditional reagents, also offering reduced reagent consumption based on blending with lower-cost diesel collector. Examples include proprietary reagents: IBM/D7, a hydrocarbon/turpen blend; and the Ekofol collector series, specifically Ekofol 452 G, Ekofol 452 GK, or Ekofol 452 GK 100.

Dispersants, such as sodium silicate or lactic acid, have been applied depending on gangue mineralogy and in specific cases where calcite entrainment is an issue. Lignin sulfonate can be used for dispersion.

Flotation can be conducted in the pH range 7–9 using soda ash as a pH modifier. Collector dosage depends on type; however, dosages ranging from 200 to 300 g/t flotation feed are typical. Lower primary collector dosages may be possible based on blending with lower-cost diesel. Graphite collectors tend to have some frothing properties; MIBC or proprietary reagent Senfroth are examples of typical graphite frothers.

Acceptable reagent conditioning can usually be achieved via moderate shear agitation ahead of rougher flotation based on 10-minute retention time; staged collector dosing can be applied to scavenger and selected cleaner stages as required.

Flotation Circuit Configuration

Flotation circuit configuration is ore specific; rougher flotation can be based on conventional, column, or flash flotation (depending on kinetics). Selected ores may also require the inclusion of a scavenger stage; however, in most cases, subsequent cleaning is performed based on a series of regrinding and cleaner stages.

The hard rock hosted flow-sheet example presented in Figure 1 incorporates the following:

- Conventional rougher flotation is used in which rougher tailings report for hydrocyclone classification. Cyclone underflow (coarse fraction) is deferred for ball mill regrinding and combined regrind mill discharge. Cyclone overflow is subjected to final (conventional) scavenger stage flotation.
- Preferential stirred mill grinding of coarse fraction is maintained by reclassifying combined rougher and scavenger concentrate, deferring cyclone overflow (fines) through to stirred mill discharge.
- Two stages of whole concentrate cleaning are affected based on utilization of a combination of Jameson and conventional flotation cells to promote high selectivity in a cleaner and cleaner-scavenger configuration. Column cells can be used as an alternative to Jameson-style flotation cells.
- Improved grade recovery response over whole concentrate cleaning is leveraged by screening the first-stage cleaner concentrate using a stack sizer and undertaking split size cleaner flotation, again based on separate stirred mill regrinding and staged cleaning to produce final graphite concentrate.

Cleaner Stage Regrinding

Graphite concentrate regrinding, in most cases, differs relative to other commodities, given the layered structure of graphite, requirements to maintain flake size, as well as promote liberation from gangue. The grinding (or polishing) action required is better defined as shearing rather than pure impact grinding. This is illustrated in Figure 3. Graphite ores characterized by intercalated gangue exemplify the requirement to liberate the gangue based on shearing rather than impact breakage of graphite flake.

Dominant shearing breakage is affected via use of tower or stirred mill regrind stages utilizing ceramic media. Examples of pebble mill regrinding exist in the case of

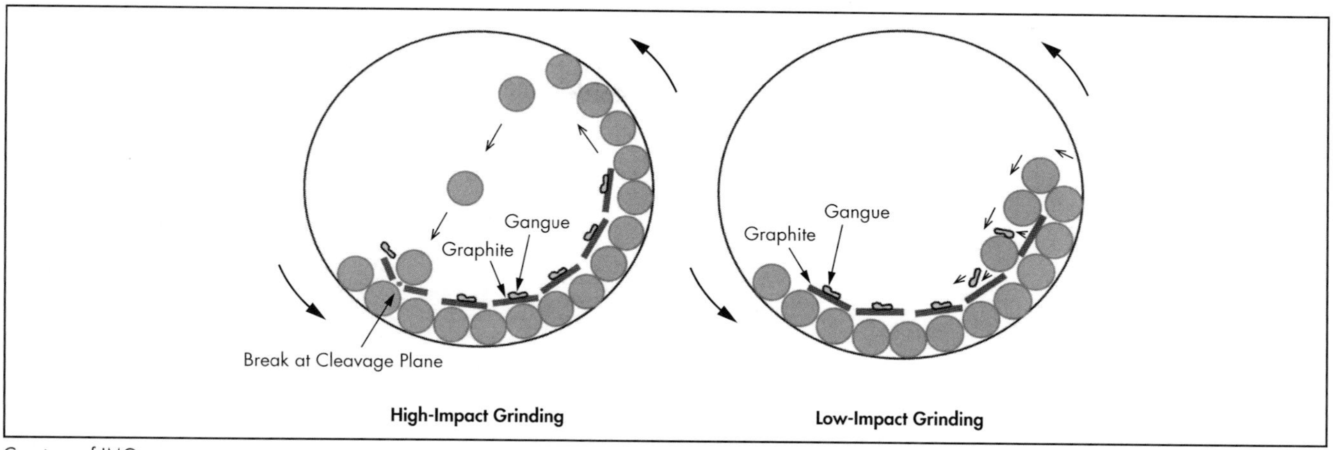

Courtesy of IMO

Figure 3 Grinding action

soft rock applications, as depicted in Figure 2. In this case, selected producers operate pebble mills based on the following conditions:

- Variable speed, grate discharge, nominally operating as low as 35%–55% critical
- Pebble charge maintained at discharge trunnion liner, nominally representing 30%–35% volume
- Mill sized to promote graphite feed flow largely through pebble charge interstices
- Discharge grate comprising outer radial slots and upper radial blank liner plates to promote graphite interstitial flow with pulp lifters used as needed to increase volumetric capacity

Concentrate Recovery

Graphite flotation concentrates tend to retain relatively high moisture content and are required to be dried prior to final screening into separately sized flake products. Concentrate flake recovery is typically performed based on the following treatment:

- Thickening and transfer of thickened underflow to concentrate holding tankage
- Dewatering based on centrifuging or pressure filtration targeting 10%–15% cake moisture
- Concentrate drying, typically 400°–500°C
- Pneumatic or screw feeder transport to multideck concentrate sizing screens
- Graphite screening to form multiple (sized) products, final bagging, and off-site transport

Graphite drying can be performed using flash, twin screw, and rotary dryers. Fluid-bed dryers may be applicable; however, the selection of drying methodology should be supported by test work to ensure that agglomerates do not form ahead of final product screening. Electric and gas- or diesel-fired dryers are available, and selection of heat source is predicated on project location and surrounding infrastructure.

Dried product moisture may vary, depending on the achievable drier feed moisture and selected drying method. Product transport through to screening and bagging is typically pneumatic with best performance at ~0.1% moisture. Screw feeding can be utilized for higher moisture content.

Graphite product screening is performed using multideck screens. Gyratory screens can be used for this purpose to prepare separate flake products for bagging. Selection of screen cut size is predicted on the markets for the graphite products, as shown in Table 3.

Table 3 Graphite particle sizes for marketing

Sizing	Market Terminology
>300 µm (+48 mesh)	Extra-large or jumbo flake
>180 µm (–48 to +80 mesh)	Large flake
>150 µm (–80 to +100 mesh)	Medium flake
>75 µm (–100 to +200 mesh)	Small flake
<75 µm (–200 mesh)	Fine flake

Source: Triton Minerals 2017

Complementary Beneficiation Methods

Although flotation is the main process employed in graphite beneficiation flow sheets, gangue mineralization is specific to deposit location, geology, and location within the oxidation profile. Other selected unit processes are employed either to improve project economics via early gangue rejection or improve the flotation circuit grade recovery response. Examples include the following:

- **Screening.** In cases where barren or below-cutoff fraction exists at coarse size, simple dry screening can be used to affect early gangue rejection.
- **Magnetic beneficiation.** Selected graphite ores may be associated with iron-bearing gangue such as magnetite and pyrrhotite. Low- or medium-intensity drum magnetic separators have been used for rejection of liberated gangue.
- **Reverse flotation.** When graphite is associated with sulfide minerals, a separate stage of flotation with copper sulfate activator and xanthate collector(s) can be used to selectively float and reject sulfides, leaving upgraded graphite reporting to the reverse flotation tailings stream.

MARKET AND PRODUCT APPLICATIONS

Graphite production has undergone a resurgence in recent times as a function of growth and demand for high-grade graphite concentrate input to lithium ion battery and fuel cell manufacture.

Product applications for graphite concentrates are wide ranging. Volume distribution across traditional and growth markets is summarized in Figure 4 (the proportions are estimated), and typical grade and size specifications are presented in Table 4.

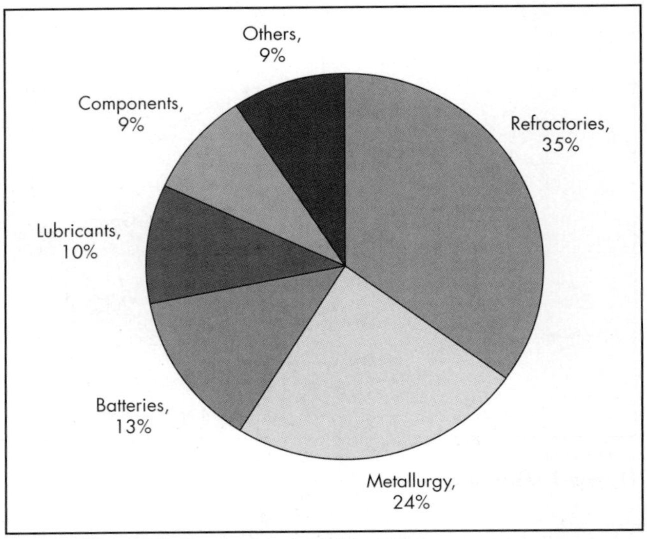

Source: Olson 2017; Scogings et al. 2015a, 2015b

Figure 4 Estimated graphite market and product applications

Marketing and Downstream Processing

The Harmonized Tariff Codes for natural graphite are 250410 (natural, in powder or in flakes) and 250490 (natural, in other forms, excluding powder or flakes) and 380110 (artificial graphite). Natural and synthetic graphite is shipped internationally in either paper bags or 1-t bulk bags and locally may also be shipped in bulk road tankers.

Downstream production is carried out by specialist processors such as Asbury Carbons, Superior Graphite, and GrafTech International in the United States. For example, Asbury processes, upgrades, and trades many types of carbons, including natural and synthetic graphite, coke, and coal, and sells to a wide range of end markets including lubricants, refractories, and foundry industries.

Superior Graphite specializes in electrothermal purification technologies, engineered graphite electrodes, advanced ceramic shapes and powders, precision particle processing, and carbon coatings. GrafTech International focuses on synthetic graphite products and markets, particularly electrodes, but the company also manufactures products based on natural graphite for thermal applications in the electronics industry and for sealants. The company buys flake graphite from sources around the world to manufacture into expandable graphite.

Table 4 Typical graphite concentrate market specifications

	Total Graphitic Carbon, %		Size, μm	
Application	**Lower**	**Upper**	**Lower**	**Upper**
Steel and Refractory Additives				
Refractories	85.0	90.00	150	710
Crucibles	85.0	90.00	75	150
Expanded graphite	90.0	98.00	200	1,700
Foundry additives	40.0	70.00	53	75
Core and mold washes	70.0	90.00	1	75
Recarburizing steel	98.0	99.00	1	5
Lubricants Automotive and Electrical				
Lubricants and releasing agents	95.0	99.00	25	300
Brake and clutch linings	95.0	98.00	1	75
Bearings	90.0	93.00	150	500
Gaskets	80.0	99.90	40	300
Carbon brushes	95.0	99.00	1	53
Electrical	93.0	95.00	150	500
Batteries and Fuel Cells				
Dry cell batteries	88.0	90.00	1	106
Alkaline batteries	90.0	99.90	5	75
Lead acid batteries	90.0	95.00	5	58
Lithium ion batteries	99.0	99.90	25	48
Spherical graphite	99.9	99.99	1	48
Fuel cells	99.9	99.99	40	150
Specialty Products				
Powder metallurgy	95.0	99.00	10	50
Polymer additives	50.0	85.00	5	15
Conductive polymers and plastics	99.0	99.99	3	38
Graphite powders	85.0	99.99	3	38

REFERENCES

Industrial Minerals. 2018. Graphite pricing news. www.indmin.com/Graphite-PricingNews.html. Accessed August 2018.

Olson, D.W. 2017. Graphite (natural). In *Mineral Commodity Summaries 2017*. Reston, VA: U.S. Geological Survey.

Robinson, G.R., Hammarstrom, J.M., and Olson, D.W. 2017. Critical mineral resources of the United States—Economic and environmental geology and prospects for future supply. Professional Paper No. 1802-J. Reston, VA: U.S. Geological Survey. p. 36.

Scogings, A., Chesters, J., and Shaw, W. 2015a. Rank and file: Assessing graphite projects on credentials. *Ind. Miner.* (August):50–55.

Scogings, A.J., Hughes, E., Shruti, S., and Li, A. 2015b. *Natural Graphite Report: Strategic Outlook to 2020: Data, Analysis, Forecasts*. www.indmin.com/Reports. Accessed July 2018.

Triton Minerals. 2017. *Additional Mineral Resource Upgrade at Ancuabe Graphite Project with Increase in Confidence Level and Shows Further Upside to Soon to Be Released DFS*. ASX announcement, December 13. West Perth, Western Australia: Triton Minerals.

CHAPTER 12.15

Gypsum

Paul J. Henkels and Roger D. Sharpe

Gypsum generally refers to the dihydrate form of calcium sulfate ($CaSO_4 \cdot 2H_2O$). Anhydrite ($CaSO_4$) is the other principal calcium sulfate mineral. This chapter includes discussion on both calcium sulfate minerals because they are strongly related. Gypsum and anhydrite are often found in close association with one another and are formed primarily from the concentration of dissolved mineral constituents in saline water. Although a few notable underground mining operations exist, gypsum is most commonly extracted from near-surface deposits by quarrying methods.

Gypsum is also co-produced in significant quantities in many industrial processes and is chemically the same as mined gypsum. These commercial processes include acid neutralization from the production of some organic acids, fertilizer manufacture, and notably flue gas desulfurization (FGD) processes.

The most significant difference between gypsum and anhydrite is that gypsum is often calcined to the hemihydrate form of calcium sulfate ($CaSO_4 \cdot ½H_2O$) through the application of a modest amount of heat (i.e., calcination). Common names for calcium sulfate hemihydrate include *stucco*, *plaster of Paris*, *molding plaster*, and *plaster*. Stucco is an intermediate product and is used in more than 90% (based on value) of all calcium sulfate products. Anhydrite, conversely, is inert at the relatively low amounts of heat applied during calcination, and its uses are limited to those few cases where calcium sulfate in the anhydrous form has an advantage.

Gypsum has a high place value. A large part of a deposit's value results from its geographic situation relative to the market area, inexpensive modes of transportation, utilities, and services. Conversely, gypsum has a low unit value. The widespread occurrence of raw gypsum gives it a low value per ton. In addition, almost 100% of extracted material is usable for manufacturing more than 400 products. Its value can be greatly increased by additional processing required for certain products.

Gypsum and anhydrite are used in many different industries. The predominant use of gypsum and anhydrite is related to construction products, including portland cement and wallboard manufacture. However, significant quantities of the minerals are also used in various agricultural applications and industrial products including glass manufacture, food and pharmaceuticals, pottery, sanitary ware, and dental plasters.

GEOLOGY AND MINERALOGY

Natural Gypsum

Gypsum and anhydrite are formed from diverse origins. Gypsum ($CaSO_4 \cdot 2H_2O$) is primarily formed as a chemically precipitated sedimentary rock in basin or sebkha environments, but it may also form as the result of solutional, karst, hydrothermal, and volcanogenic processes. Anhydrite ($CaSO_4$) is the anhydrous form of the calcium sulfate family of minerals. Anhydrite may form as a primary mineral in a sebkha depositional environment or deep-basin environments. The terms *gypsum* and *anhydrite* are commonly used interchangeably for the minerals and rocks that are composed primarily of these minerals.

Gypsum and anhydrite may be deposited simultaneously in a sedimentary environment and are easily converted from one to another under varying conditions of heat, pressure, and the presence of water. The occurrence and relationship of anhydrite and gypsum in a deposit is complex. Mineralogical, textural, crosscutting relationships, and microscopic examination may be necessary to determine the ultimate origin of anhydrite in a deposit. Some deposits exhibit multiple episodes of conversion from gypsum to anhydrite and vice versa (Sharpe and Cork 2006).

Gypsum forms monoclinic crystals with a perfect {010} cleavage and distinct cleavages along {100} and {101}. It is distinguishable from anhydrite by its lower Mohs hardness (2.0 vs. 3.5) and specific gravity (2.24 vs. 2.97 g/cm^3). Pure gypsum is colorless, but may be tinted yellow, red, and brown because of the presence of impurities. Twinning is common along {100}, forming "swallowtail twins." Gypsum is relatively soluble in fresh water (~0.2 g/100 g H_2O) and is easily dissolved or eroded in conditions of high humidity or rainfall.

Anhydrite forms orthorhombic crystals with perfect cleavages along {100} and {010} and a good cleavage along

Paul J. Henkels, Senior Technical Manager, United States Gypsum Company, Chicago, Illinois, USA
Roger D. Sharpe, Director, GeoTechnical & Mining Services, United States Gypsum Company, Chicago, Illinois, USA

Table 1 Gypsum lithology

Crystal Type	Lithology
Alabaster	A compact, fine-crystalline, translucent variety of primary gypsum that has been used by artists to produce statuary for thousands of years. Typically found in a variety of colors depending on the type and amount of impurities present, alabaster generally occurs as zones within larger gypsum deposits.
Anhydrite	Massive or granular texture. Scattered gypsum crystals commonly occur in the anhydrite groundmass and increase near the gypsum–anhydrite contact. Anhydrite is similar in appearance to rock gypsum, but usually slightly darker, with denser, finer crystals.
Gypsite	An earthy, pulverulent variety of gypsum that forms a surficial deposit in shallow saline lakes, playas, and salt pans in arid environments. The calcium and sulfur required to form gypsum are derived from the erosion and weathering of rocks and transported in surface and groundwater to closed basins. Gypsum precipitates at the surface by capillary movement and evaporation of groundwater. Gypsite is often contaminated by windblown sand or silt and clay from periodic flooding of the playas. Often called *gypsum sands*.
Satin spar	A fibrous variety of needle-shaped gypsum crystals filling fractures or along bedding planes. The needle-shaped gypsum crystals form with the C-axis oriented at a steep angle or perpendicular to the vein walls in fractured rocks undergoing deformation. It is important to recognize that satin spar is not an asbestiform mineral.
Selenite	Characterized by large, clear, euhedral crystals of gypsum. Bladed selenite crystals commonly form in fluid-filled cavities. Selenite may also form along fault zones. Cleavage fragments may be mistaken for muscovite mica. Poikilitic masses of selenite crystals, commonly known as *gypsum roses*, are formed by the crystallization of gypsum from interstitial pore fluid in unconsolidated sand.
Rock gypsum	Rock gypsum commonly consists of aggregates of gypsum crystals interbedded or mixed with mudstone, shale, siltstone, limestone, or dolomite. Gypsum rock and anhydrite may be nodular, massive, laminated, or bedded. Most rock gypsum has a medium to coarse crystalline texture, and this is the most common crystal variety of the mineral.

Source: Sharpe and Cork 2006

{001}. Anhydrite has a Mohs hardness of 3.5 and a specific gravity of 2.97 g/cm^3. Pure anhydrite is colorless, but the color is variable from colorless to dark gray (Sharpe and Cork 2006).

Commercial deposits of gypsum may be almost 100% pure or contain variable amounts of syndepositional impurities such as limestone, dolomite, clay, anhydrite, and soluble salts of potassium, sodium, and magnesium. Primary gypsum deposits consist of rock gypsum and alabaster. Selenite, satin spar, and gypsite are secondary varieties of gypsum. Anhydrite may occur as either primary or secondary minerals in a deposit, depending on its geological history. Table 1 provides the lithology of common varieties of gypsum.

Synthetic Gypsum

Synthetic gypsum originates from several industrial processes. Mainly, it is a product resulting from industrial acid neutralization processes or FGD of fossil fuel combustion in power plants and can be as pure as 98% $CaSO_4 \cdot 2H_2O$. By far, the largest amount of synthetic gypsum is obtained from the production of wet phosphoric acid from phosphate rock (phosphogypsum). However, high levels of impurities and often low levels of radon render most phosphogypsum useless (Henkels 2006).

Over the last 25–30 years, synthetic gypsum gained in popularity and use worldwide. It often replaces mined natural gypsum. In 2014, ~50% of the gypsum consumed by U.S. wallboard manufacture was from synthetic sources, primarily FGD (Crangle 2014). FGD gypsum use in the United States is expected to stabilize or decline in coming years as power companies convert electrical generation plants to natural gas to take advantage of inexpensive and plentiful shale gas. Also, as environmental concerns about coal combustion by-products are raised, leading to increased regulation, the cost of coal-fired power plants may increase, prompting reduced operating hours or closure.

Synthetic gypsum is usually available in a "wet cake" granular form (30–60 µm) with free moisture content varying from 6% to 15%. The product is normally dried by the consumer to less than 1% moisture content prior to calcining, and then handled in the same manner as natural gypsum throughout the manufacturing process (Henkels 2006). There are no significant advantages or disadvantages to using synthetic gypsum versus natural gypsum other than cost. The decision is one of economics; the cost of transportation is usually the determining factor. The increased prevalence of synthetic gypsum during the past 30 years in the United States has helped to keep the price of gypsum low (<$10/t [metric ton]).

Anhydrite

Anhydrite is considered an impurity where gypsum is the sought-after mineral. It is a common impurity in deposits of gypsum and may occur as a primary depositional mineral or the product of dehydration of deeply buried gypsum. Anhydrite can be found as a massive rock or as a mixture of gypsum and anhydrite in partially hydrated deposits. Soluble salts, such as sodium, magnesium and potassium chlorides, and sulfates are commonly concentrated in a halo in gypsum adjacent to the anhydrite contact. The soluble salts are leached from the anhydrite during hydration and precipitated in tensional features, such as fractures, many of which form during the volumetric expansion during hydration. As a contaminant, the anhydrite is harder and denser than gypsum and contains higher levels of soluble salts. Anhydrite increases the weight of finished wallboard products and is abrasive to grinding and processing equipment (Sharpe and Cork 2006).

EXTRACTION METHODS

The majority of gypsum rock is produced by surface mining operations. In North America, surface mining is by far the most common method. Only five active underground gypsum mines were in operation in 2014. Two are located in Indiana (United States), and one each in Iowa, Michigan (both in the United States), and Ontario, Canada (Sharpe and Cork 2006). In Europe, underground mining is the more commonly used extraction method.

Quarrying

Gypsum is extracted from near-surface deposits by quarrying methods. Overburden, consisting of glacial materials (till, sand, clay), shale, mudstone, siltstone, sandstone, sand and gravel, or limestone, is removed from the gypsum by various stripping methods. Stripping is performed by pan scraper, truck and excavator, front-end loader or hydraulic excavator and truck, dragline, and bulldozer. The maximum economic stripping limit (thickness) is approximately 30 m and depends on the method of stripping and the thickness of the underlying recoverable gypsum.

Final cleanup of the stripped gypsum surface is important and is determined by the final products to be manufactured. If the quarried rock is to be used as agricultural products, portland cement rock, or wallboard, then much of the impurities can be removed in the finer fractions during crushing and screening. Conversely, gypsum used for high-quality, high-value-added products requires more stringent cleaning. For example, articulated hydraulic excavators with multiple, interchangeable bucket widths scrape clay from the gypsum surface and fractures that extend deep into the gypsum.

Drilling and blasting is the primary method of quarrying gypsum. Quarry benches are generally about 8 m in height. Hydraulic rotary drilling and auger drilling are commonly used. Gypsum is soft, and penetration rates up to 7 m/min are possible. Blastholes are generally 50–100 mm in diameter and spaced relatively close together to distribute the explosive forces throughout the rock mass. The blasting components used include ammonium nitrate and fuel oil blasting agent, cast boosters, and nonelectric blast initiation systems. Bagged emulsion is used in wet holes. Approximately 1 kg of blasting agents per ton of broken gypsum is an average tonnage factor. The elastic nature of gypsum, the presence of solution-enlarged fractures, and the possible presence of water contribute to poor and inefficient rock fragmentation. An incorrectly designed blasting pattern may result in irregular fragmentation, including excessive oversized rock requiring secondary breakage, excessive fines, or irregular floor conditions that are detrimental to the efficiency and maintenance of mobile equipment. Quarry haulage trucks or over-the-road dump trucks transport the quarry-run broken gypsum from the quarry site to the primary crusher.

Another method of extracting gypsum from quarries that is gaining acceptance is the use of a surface miner. This is an adaptation of highway-resurfacing technology in which a horizontally rotating mandrel with cutting teeth chips away at the asphalt and either discharges the broken material in a windrow or directly into a haulage truck. Specialized machinery using this technology is being developed to quarry coal, gypsum, and limestone. The size, spacing, and arrangement of the cutting teeth on the mandrel are important factors in the efficient production of gypsum rock. The following are advantages over standard quarrying techniques (Sharpe and Cork 2006):

- Elimination of drilling and blasting
- Elimination of primary crushing
- Direct removal of interbedded thin waste beds or low-purity zones
- Increase in recovered gypsum purity by removal of off-specification material from windrows
- Maximization of the overall recovery of the gypsum near structures or utilities

There will be more and more use of surface miners in the future as these machines have the ability to easily avoid stripping undesired low-purity, contaminated areas and better control quality and purity compared to conventional mining methods (drill and blast).

Underground Mining

Underground mining of gypsum is far less common than quarrying. There are currently only five active underground gypsum mines in North America. The mines are located at a depth of <30 m (Hagersville, Ontario) to ~200 m (Iowa and Indiana). Access to the mine workings for workers, supplies, ventilation and escape, and production is by either vertical shafts or inclined adits (tunnels). The mined interval varies from 1.1 m (Hagersville) to 3.7 m (Iowa, Michigan, and Indiana).

Gypsum is extracted in most cases using the room-and-pillar method, in which pillars are left in place to support the roof strata, and the gypsum is removed in a checkerboard pattern. The extraction ratio, which is the proportion of material mined to material left in the supporting pillars, varies from 65% to 80%.

Drilling and blasting of gypsum in underground mines is similar to the methods used in gypsum quarries. The blasthole pattern, however, is drilled horizontally into the advancing face of a mine heading. The number of drill holes, orientation, depth, and sequence of blasting are designed to maximize breakage of the gypsum and to maintain the integrity of the adjacent pillars and immediate mine-roof strata. The roof strata are reinforced by roof-control fixtures, generally epoxy resin-grouted bolts, up to 1.5 m long. The resin-grouted bolts bind the roof strata together to form an integral beam that is stronger than the individual strata.

Continuous miner technology, popular in coal-mining operations, is also currently being deployed in a few specific instances. Although more capital investment intensive, continuous miners eliminate the need for drill and blasting and so forth and lower extraction costs.

Underground gypsum mines are generally very stable. For example, long-term measurements of roof and pillar convergence (vertical closure) at the USG Corporation mine at Sperry, Iowa, indicate that the structure should be stable for hundreds of years. An underground gypsum mine near Grand Rapids, Michigan, is currently being used as a computer network security center (Sharpe and Cork 2006). Table 2 shows a generalized list of mines by state groupings.

Rock Processing

Figure 1 shows the processing steps for gypsum rock. After mining, gypsum rock is moved through a primary crushing operation. The primary crushing is accomplished by gyratory, jaw, roll, or impact crushers, depending on the size of the mine run rock, the desired throughput, and the type of subsequent processing. Secondary crushing is done with a variety of standard units, but hammer mills, roll-type, and cone-type crushers are most commonly used. Fine grinding of uncalcined gypsum is generally accomplished by air-swept roller mills fitted with integral air separators for better particle size control, although high-energy impact mills with air classifiers also have been used.

Both primary and secondary crushing steps are usually conducted with vibrating screens in the circuit, in part to maximize crushing efficiency and to reduce the production of ultrafines, but also to recover portland cement rock, the first marketable product of gypsum processing. The particle size of portland cement rock will vary with the requirements of each individual cement plant, but in the United States, it most often falls within a range of 38–51 mm top size by 6–13 mm bottom size.

Table 2 Location of gypsum mines in the United States*

State	2014			2015		
	Active Mines	Quantity, thousand t	Value, thousand $	Active Mines	Quantity, thousand t	Value, thousand $
Arizona, Colorado, and New Mexico	7	1,210	10,900	6	1,040	9,300
Nevada and Utah	6†	W‡	W	9	2,110	25,200
Arkansas and Louisiana	2	W	W	2	W	W
California	5	689	6,030	4	690	5,930
Iowa and Indiana	5	1,410	12,500	4	995	8,550
Michigan	3	233	2,050	2	W	W
South Dakota and Wyoming	3	W	W	3	W	W
Kansas, Oklahoma, and Texas	20	8,300†	73,300†	20	8,780	71,500
Total	**51†**	**14,900†**	**132,000†**	**50**	**15,200**	**135,000**

Source: Crangle 2016
* Data are rounded to no more than three significant digits; may not add to totals shown.
† Revised.
‡ Withheld to avoid disclosing company proprietary data; included in "total."

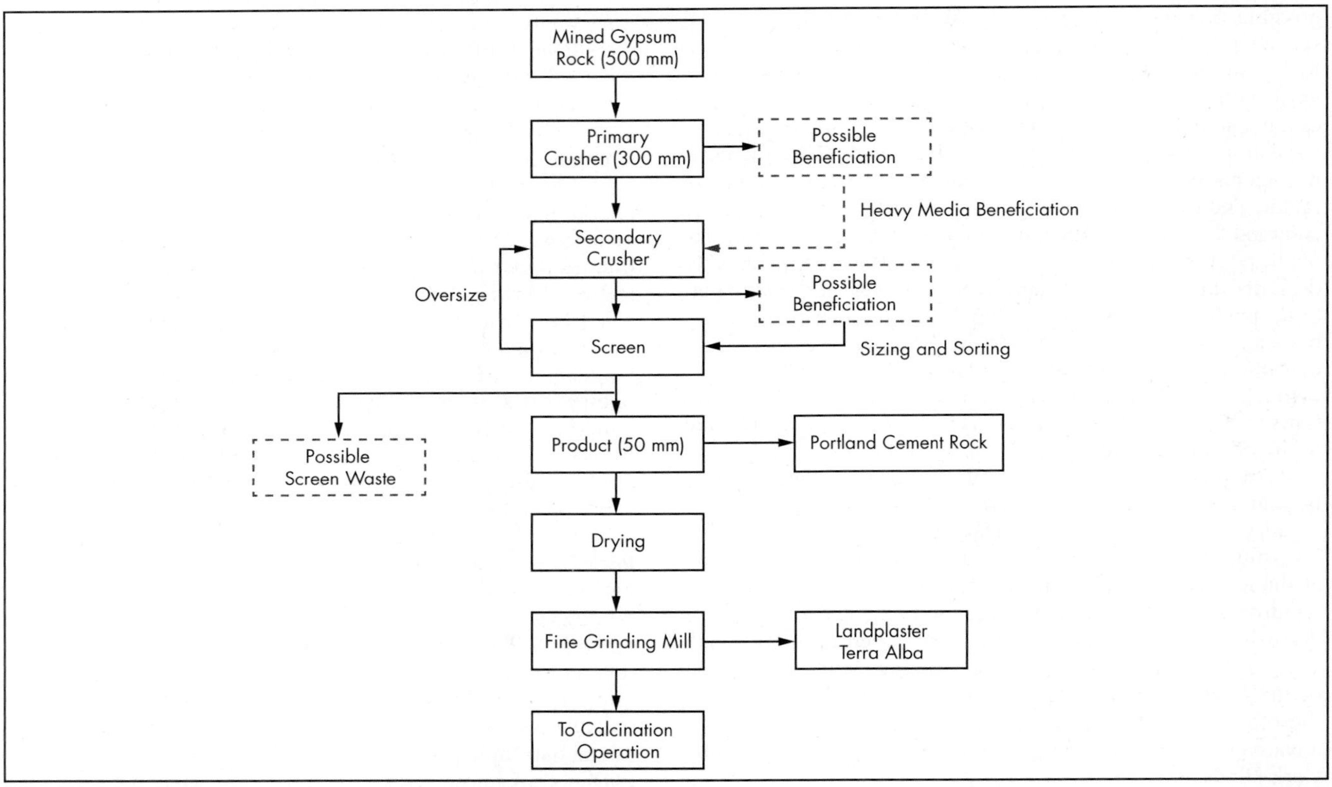

Adapted from Henkels 2006
Figure 1 Gypsum rock block flow diagram

In many cases, the purity or quality of gypsum as mined is sufficiently high that it can be used without upgrading; however, beneficiation techniques are sometimes used to meet the required specifications. The most common form of beneficiation is simple classification by particle size with dry screening, air separation, or other means wherein a size fraction containing a significant portion of impurity is discarded. In general, this method is used to reduce clay or sand (soil) impurities. In some cases, impurities that are harder than gypsum, and hence tend to concentrate in the coarser size fractions after crushing, can also be reduced by screening.

Drying (the removal of free moisture) may occur either before or after the secondary crushing stages, depending on the amount of free moisture in the mine run rock. Crushed gypsum is difficult to handle. Material measuring 5 mm is not free-flowing, particularly if wet, so drying is often required to assure the free flow of material in subsequent steps. This most often is accomplished in rotary dryers and must be carefully controlled so that the temperature of the rock does not exceed 49°C, the point at which dissociation of combined water begins to take place.

Beneficiation. Washing and/or wet screening is used in a few cases, particularly where the need for whiteness exists. Heavy media sink–float separation for the removal of impurities is currently employed at one operation in Canada. In many cases, gypsum is amenable to other gravity methods of beneficiation including froth flotation. In general, the relatively high cost of these methods and the availability of high-quality deposits and synthetic gypsum usually eliminate the need for beneficiation. Nevertheless, lower-cost beneficiation techniques continue to be investigated based on narrow rock sizing and adapting color-sorting techniques popular in waste recycling centers.

GYPSUM CHEMICAL AND PHYSICAL CHARACTERISTICS

Chemical Properties

Gypsum and anhydrite are able to undergo repeated dehydration and hydration, depending on the depth of burial and availability of water. Gypsum will begin to thermally decompose into the metastable species, calcium sulfate hemihydrate ($CaSO_4 \cdot \frac{1}{2}H_2O$), in heated air higher than 70°C and at 1 atm of pressure. Differential thermal analysis curves show two endothermic peaks between 100° and 200°C (Deer et al. 2013). The first peak represents the loss of 1½ molecules of water during the formation of calcium sulfate hemihydrate. Further heating totally dehydrates gypsum, forming dead-burned (anhydrous) anhydrite ($CaSO_4$), which is used as a filler in plastics and as a moisture absorbent. The second decomposition peak occurs when the remaining one-half molecule of water is removed. Calcining processes for the manufacture of gypsum wallboard and plasters involve heating gypsum to ~155°C for 1–3 hours, during which time calcium sulfate hemihydrate forms.

Heating gypsum liberates water vapor, which helps hinder the spread of fire. Specialized types of wallboard are formulated and manufactured for use as a firestop between multifamily housing units, in the lining of elevator shafts, and in the walls and ceilings between living spaces and garages.

Gypsum's solubility in fresh water is approximately 150 times greater than that of limestone. This chemical characteristic provides a source of calcium and sulfur when gypsum is used in agricultural applications. The ionic exchange capability of calcium for sodium prevents the buildup of alkali in soils.

Although gypsum is a very soft mineral, specialized methods of calcining have been developed to manufacture denser, harder crystals. For most uses, the gypsum is ground to a powder and heated in a continuous kettle or batch kettle at 1 atm of pressure. This method of calcining produces anhedral, rough, splintered crystals known as *beta calcium sulfate hemihydrate*. This product is used in the manufacture of wallboard and industrial and construction plasters. The calcination of 12–250-mm sized, high-purity gypsum in an autoclave at elevated pressure produces dense, euhedral crystals of *alpha calcium sulfate hemihydrate*. This product is used in specialized plasters and cements for art and statuary, architectural applications, industrial prototype modeling, and road and floor repair (Sharpe and Cork 2006).

Physical Properties

The softness of gypsum is its most prominent physical feature. It is distinguished by its softness (Mohs hardness of 2) and three distinct cleavage planes. Anhydrite is distinguishable from gypsum anhydrite and has a hardness of 3.5 on the Mohs scale. Anhydrite rock is usually darker in color than gypsum rock (Sharpe and Cork 2006).

MINERAL USAGE

The earliest known use of gypsum plaster dates back some 8,000 years with the discovery of its use in Anatolia. Around 3,700 BC, Egyptians observed that gypsum rock, when exposed to fire, broke down into a powder that, when mixed with water, would form a putty that could be plastered on a rough mud brick or stone wall to make a smooth finish. Iranians, Babylonians, Greeks, and Romans were familiar with the art of gypsum plasters as well. Examples of its use include the walls of Jericho, the pyramid of Cheops, the palace of Knossos, and the decorated interior walls of Pompeii.

Gypsum was used in a limited way for ornamental purposes down through the centuries. Gypsum plaster did not achieve wide acceptance because of its quick (25–30 minutes) setting or hardening time that made it difficult to use. The first real understanding of gypsum chemistry was developed in France about 1760, and the gradual growth of its present utilization dates from that time. However, a craft so firmly steeped in tradition was slow to accept scientific inference, so that only in the last relatively few decades has gypsum manufacture developed into a modern industry. Today, a worldwide industry has been built on the mining and processing of this versatile mineral.

The primary use of gypsum worldwide is in the manufacture of construction plasters and portland cement. However, the principal use of gypsum in North America is in various types of wallboard panels. For example, ~90% of the total gypsum produced in the United States is used in wallboard and construction plasters (Sharpe and Cork 2006).

Gypsum is a very versatile mineral that can be used in the manufacture of several hundred products. Processing methods vary from simply crushing and sizing of quarry-run gypsum to specialized methods of calcination in closed pressure vessels.

The following sections describe in more detail the process steps and various mineral uses of gypsum in general order of added-value processing.

Uncalcined Gypsum

Gypsum that has been processed only by grinding and sizing is known as *land plaster*, *portland cement retarder*, and *terra alba*. Land plaster is used for agricultural gypsum and a raw feedstock for manufacturing wallboard and plasters. Portland cement retarder is used in the manufacture of portland cement. Terra alba is used in food and pharmaceutical applications.

Uncalcined gypsum is principally used as a retarder for portland cement, soil conditioner, mineral filler, and in other minor industrial applications. Approximately 25% of the gypsum mined in the United States is used in these markets. In countries where building practices differ from those in the United States and Canada (poured concrete, block, or brick), however, the relative usage of gypsum varies widely.

Although calcium sulfate deposits are the world's largest sulfur resources, only minor quantities of gypsum and anhydrite have been used to produce sulfur or sulfur compounds. This use is accompanied by a unique, site-specific set of economics, because sulfur is generally available from nongypsum sources at lower cost.

Portland Cement Rock

Quarry-run anhydrite or a blend of gypsum and anhydrite may be used as a portland cement rock product. Portland cement manufacturers use a variety of products from gypsum, anhydrite, or an anhydrite–gypsum blend, depending on the manufacturing process. The primary function of gypsum or anhydrite in portland cement is to control the setting time of the finished product. The SO_3 content of the portland cement rock is the most important quality parameter.

Adding calcium sulfate also controls the early strength characteristics of cement and product shrinkage during drying and curing. Approximately 3–5 wt % of calcium sulfate compounds are ground with clinker to form portland cement.

Gypsum also aids in the grinding of clinker by reducing the tendency of fine particles to agglomerate and adhere to the walls of the mill and grinding media (Hansen et al. 1988). Gypsum is much softer than clinker, is easier to grind, and has a much greater fineness (surface area) than clinker. Portland cement rock is typically ground to a particle size of 6–65 mm. Calcium silicates and aluminates that constitute clinker have negatively charged oxygen ions on the crystal surfaces. The hydrogen ions in the water molecules of the gypsum particles bind to the negatively charged clinker particles. The neutralization of the electrical charges by the attraction of the gypsum to clinker particles reduces the tendency for agglomeration.

Road Rock

Where alternate economical sources are limited, mixtures of anhydrite and gypsum rock (–19 mm and –51 mm are the typical grades sold) are used (e.g., in northwest Oklahoma) for temporary roads needed for energy exploration. Temporary roads are needed to access drilling sites for natural resources (oil and natural gas) and in the development of wind farms.

Glass Batch

Relatively pure uncalcined gypsum, depending on glass-batch chemistry, can also substitute for salt cake (sodium sulfate) in glass manufacturing. Thermal decomposition of gypsum in the glass melt produces sulfur dioxide as well as oxygen. The oxygen reacts with any free or reduced sulfur to form additional sulfur dioxide. Increasing the amount of gypsum in the glass batch decreases the sulfur content in the melt and results in a lighter-colored glass. Iron pyrites and carbon or blast-furnace slag are commonly added to the melt to manufacture amber-colored glass. Green-colored glass has a natural amber hue. Adding gypsum provides a source of sulfate in soda–lime glass. Decomposition of gypsum produces sulfur dioxide, which, in low concentrations, removes seeds and forms a clear glass. Surface scum can form on molten glass because of improper flow conditions in the furnace, especially near the bridge wall. The presence of coarse sand and stratification of the raw materials in the batch may result in selective melting and the formation of a surface scum. Gypsum added to the glass melt reacts with sodium carbonate to form sodium sulfate. The sodium sulfate melts and reacts with free silica to form sodium silicate, which in turn separates from the molten glass and floats on the surface. The gypsum sold for this application is normally sized at about 1.6 mm.

Agricultural Gypsum

Gypsum provides several benefits in agriculture. The specifications of agricultural gypsum are primarily related to the degree of fineness (particle size and surface area). As gypsum dissolves, it is a source of elemental calcium (25% by weight) and sulfur (20% by weight). Gypsum has a neutral pH (7.0) and is 150 times more soluble than ground limestone. Finely ground agricultural gypsum permits rapid dissolution and absorption by plants. Long-term availability of these elements during a growing season can be accomplished by applying agricultural gypsum of multiple particle sizes. Finely ground gypsum (100% passing through a 425-mesh screen) can also be dissolved in irrigation water for easy application. Generally called land plaster, this material can be produced from either gypsum or anhydrite and is usually ground in air-swept roller mills. The principal purposes for applying gypsum on agricultural land are to

1. Improve the structure or physical condition of soil by breaking down compacted clays, which, in turn, increases porosity and aids drainage;
2. Supply neutral, soluble calcium for peanuts, tomatoes, and other crops;
3. Supply sulfur quickly from available sulfate;
4. Neutralize alkaline soils;
5. Reduce high sodium salt content in irrigation waters;
6. Improve availability and utilization of nitrogen;
7. Act as a flocculating agent to clear up muddy farm ponds; and
8. Stimulate growth of soil microorganisms.

In other agricultural uses, land plaster, either in the dihydrate or anhydrous form, is often added as an ingredient in formulating feeds and feed premixes for beef cattle, dairy cows, and sheep. As a feed supplement, it supplies total sulfur requirements in safe, easy-to-mix form; increases the efficiency of nonprotein nitrogen in urea feeds; is an ideal supplement for ensilage enhancers; and is an effective regulator for self-feeding stock on the range.

Terra Alba

Terra alba is white, high-purity, uncalcined gypsum that has numerous uses in the food and pharmaceutical industries. It is made by fine grinding and air separation of gypsum with a purity of greater than 97%. Terra alba has a minimum calcium content of 23% (by weight). It is used to buffer the pH and reduce the hardness of water in brewing beer; as a calcium supplement and enrichment in baked goods including bread; as a diluent and inert extender in pharmaceutical products such as aspirin tablets; and in canned vegetables, cheeses, and artificially sweetened jellies and preserves. Terra alba gypsum is also used as a carrier for insecticides and micronutrients; granulated and pelletized gypsum is produced in limited quantities for these purposes.

Food- and Pharmaceutical-Grade Products

Gypsum is an approved additive on the Food and Drug Administration "Generally Recognized as Safe" listing of food additives. The use of gypsum in specific food products is described in Title 21, Part 184 of the Code of Federal Regulations (21 CFR 184.1230). The permitted amount of gypsum allowable in different types of foods is defined in the Food Chemical Codex in the United States and the National Formulary in the United Kingdom (Sharpe and Cork 2006).

Calcined Gypsum

The greatest utilization of calcined gypsum is in the manufacture of wallboard and the formulation of plasters for building markets. Beta hemihydrate gypsum or stucco is by far more common than alpha hemihydrate. Beta stucco is calcined at atmospheric pressure conditions and is suited for the construction applications where high early strength development is necessary. Alpha hemihydrate is normally calcined under pressure, and it makes a denser, higher strength cast; it is preferred for industrial uses where these characteristics are important. Other factors considered in the utilization of these two forms are cost (beta is considerably less expensive to make) and water demand (alpha requires less excess water, over and above that necessary for rehydration, to make a slurry of equivalent consistency or viscosity) (Henkels 2006).

Gypsum that is chemically transformed by heat or pressure to remove three-fourths of the water of crystallization is known as calcium sulfate hemihydrate, stucco, and plaster of Paris. The chemical reaction is the same for both alpha and beta stucco and is reversible at atmospheric temperatures and pressures:

$$CaSO_4 \cdot 2H_2O + \text{heat} = CaSO_4 \cdot \tfrac{1}{2}H_2O + 1\tfrac{1}{2}H_2O$$

Many different products may be manufactured from alpha or beta hemihydrate or a mixture of both. The products are further mixed with portland cement, fiberglass, plastic resins, and other materials to produce products with high strength and density, fire and water resistance, and other specialized characteristics.

Alpha Hemihydrate

Alpha hemihydrate represents less than 1% of all gypsum hemihydrate produced. It is made by one of two methods and produces dense, euhedral crystals of calcium sulfate hemihydrate:

1. Relatively large-sized rock, anywhere from 12 to 250 mm, is loaded and calcined batchwise in an autoclave at elevated pressure (generally between 1 and 6 bar).
2. Finely ground land plaster or synthetic gypsum is mixed with water, and the resulting slurry is fed batchwise or continuously into a pressurized reactor. Reactor pressure is maintained between 1 and 4 bar. The calcined gypsum is then de-watered and dried with a filter belt or centrifuge and heated dryer.

Alpha hemihydrate is used in highly specialized plasters and cements for art and statuary, architectural applications, industrial prototype modeling, road repair, and flooring underlayment plasters.

Beta Hemihydrate

See Figure 2 for a flow diagram of a gypsum mill/beta calcination operation. Beta hemihydrate is by far more common than alpha hemihydrate and is produced by a few generalized methods, including the following:

- Finely grind gypsum (95% –100 mesh) in a roller-style mill and then calcine at temperatures between 140° and 160°C in vertically oriented, cylindrical steel kettles at atmospheric pressure. The calcination can be a continuous process or a batch process. Continuous calcining of land plaster is predominantly used for producing stucco for wallboard. Batch calcining is usually used for manufacturing construction and industrial plasters. The beta hemihydrate calcining process produces rough, fractured, fragmented particles.
- Calcine sized rock (–50 mm) in a long horizontally oriented rotary style vessel at ~150°C exit temperature and then fine grind (98% –100 mesh) the calcined rock. This is a continuous method and the stucco is generally used for construction plasters.
- The preceding calcining methods involve two separate unit operations. Another general approach involves one unit operation where the fine grinding and calcining occur simultaneously. Sized rock (typically 50 mm) is fed to an air-swept heated hammer- or roller-style grinding mill. Calcination takes place as the material is being ground and conveyed to a receiving dust collector. Control temperatures in the heated air stream after grinding are

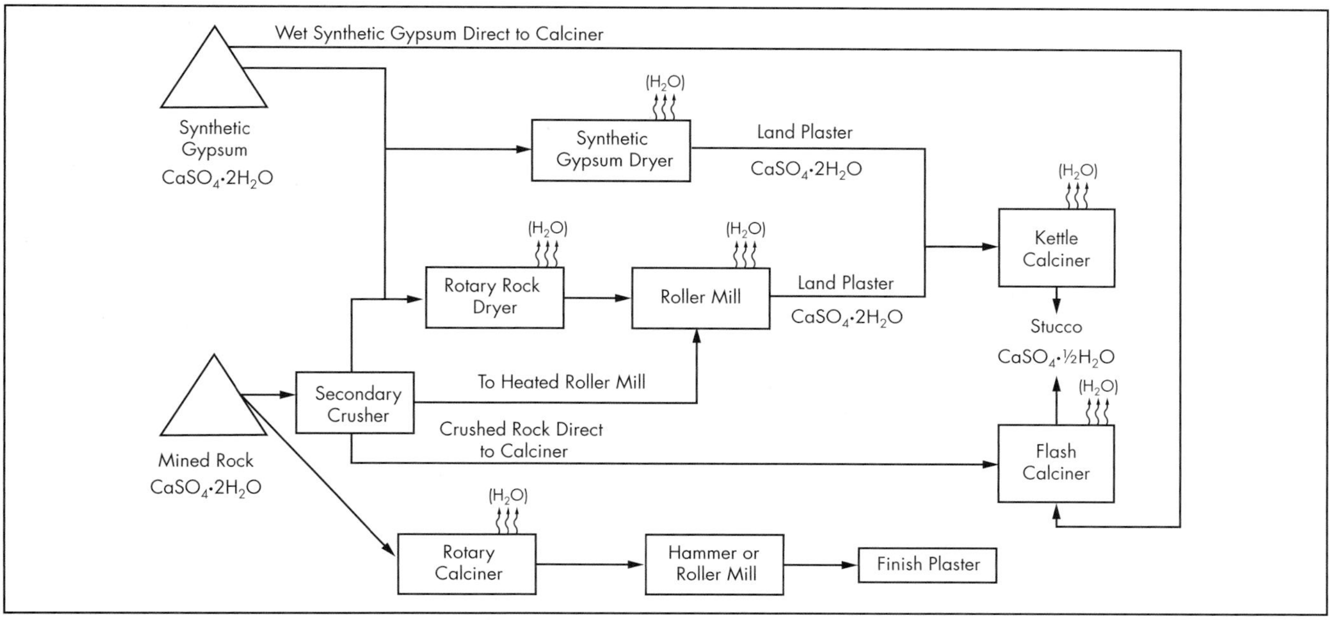

Figure 2 Gypsum mill/beta calcination operation

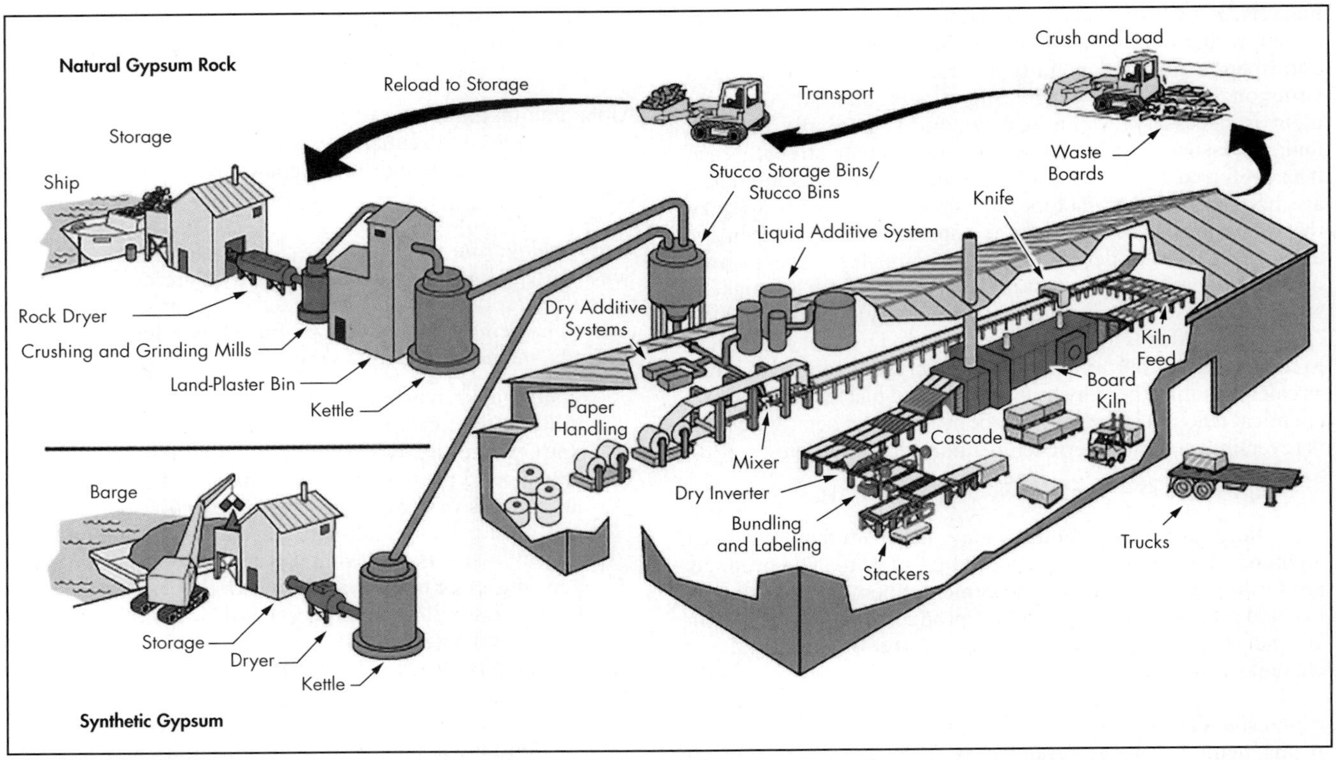

Source: Henkels 2006

Figure 3 Conceptualized drawing of a wallboard operation

typically in the 150°–165°C range. This is a continuous process and referred to as *flash calcination*.

In all beta calcination cases, as the molecular water is removed during calcination, it is released in the form of steam vapor.

Plasters. Plaster may be alpha hemihydrate, beta hemihydrate, or a mixture of both crystal phases. White art plasters used in schools or arts and crafts classes are made from beta hemihydrate.

The process of manufacturing pottery and ceramic products involves several steps from preparation of an initial model to the final product. An original block mold is manufactured from the finished model of the product to be manufactured. Then a case mold is made from the block mold. The case mold becomes a die for fabricating multiple working or production molds. The block and case molds are typically manufactured from alpha hemihydrate. The use of alpha hemihydrate allows for the production of dense, hard, strong, and durable molds that can be intricately detailed. Working molds, which are used for mass production, are manufactured from a blend of alpha and beta hemihydrate or from alpha hemihydrate. Although most construction plasters are made from beta hemihydrate, in certain cases, blends of beta and alpha are used, especially where higher compressive strengths are required.

Gypsum wallboard. Wallboard is also commonly known as *plasterboard* in markets outside of North America. Figure 3 is a conceptualized drawing of wallboard production. Three possible sources of raw material are shown: (1) natural gypsum rock delivered from an on-site quarry or an underground mine, (2) synthetic gypsum delivered from a nearby power plant or by barge, and (3) off-specification wallboard recycled into the manufacturing stream. The dry components are mixed with water and other liquid additives in a high-energy mixer designed for a continuous flow-through of material. The resulting slurry is dropped on a moving strip of paper, the edges of which are turned up to form a wide, shallow trough. This mass then moves under a second strip of paper, and the roughly formed board moves between edge guides and under a forming plate, which precisely controls the specified board thickness. The edges of the bottom paper are turned over to form square or rounded corners, and an adhesive is added to bond the top paper to the lower paper along its edges. The thickness of the board can be varied by raising the forming plate. The now fully formed board moves along on a flat conveyor belt up to a few hundred meters long. The length of the board line is designed to provide sufficient time (2–5 minutes) for the stucco to set. At that point, the moving strip is cut at specific intervals by a rotating knife to produce individual boards for drying in a kiln to remove excess free moisture. The speed of the board line normally varies between 0.5 and 3.0 m/s depending on machine design and the wallboard type being made.

Anhydrous Gypsum

In addition to the hemihydrate industrial plasters described earlier, several products are made from calcium sulfate with no water of crystallization. In this process, gypsum is calcined at higher temperatures past the hemihydrate stage, and all the chemical water is removed. The chemical reaction can occur in one step:

$$CaSO_4{\cdot}2H_2O + \text{heat} = CaSO_4 + 2H_2O$$

Or the reaction can happen in two steps in which first the gypsum is calcined to hemihydrate:

$$CaSO_4{\cdot}2H_2O + heat = CaSO_4{\cdot}\tfrac{1}{2}H_2O + 1\tfrac{1}{2}H_2O$$

Then more heat is added to convert the hemihydrate to anhydrite:

$$CaSO_4{\cdot}\tfrac{1}{2}H_2O + heat = CaSO_4 + \tfrac{1}{2}H_2O$$

As in beta hemihydrate calcination, the molecular water removed during calcination is steam vapor.

Soluble anhydrite. Soluble anhydrite is made when the calcining temperature has been increased to ~200°C. This material has a high affinity for water, making it an efficient drying agent. It is sold in gradational particle sizes for use as a desiccant in laboratory and commercial applications.

Insoluble anhydrite. When the calcination process is carried to a temperature of ~600°C, a dead-burned or insoluble anhydrite product is formed, which is used for industrial filler. Dead-burned gypsum is preferred for some filler uses over uncalcined gypsum, particularly where the temperature of the product in which the filler is used exceeds 50°C, which is the point at which natural gypsum begins to release water of crystallization. Dead-burning usually produces a whiter product and is preferred where color is important.

Dead-burned gypsum fillers are used as a source of calcium in food products, and for yeast and beer processing. They also serve as diluents or extenders in such compositions as pharmaceutical pills, rubber, artificial wood, paper, and pigments. Dead-burned gypsum is used as filler in thermoplastics such as polyvinyl chloride (PVC) products: vinyl siding, window frames, moldings, conduit, and pipe. The filler imparts acid resistance and low electrical conductivity. It is also used in food packaging and as an additive in specialty portland cement compounds.

Keene's cement. A generic name for dead-burned construction gypsum plasters, additives are used to set and harden Keene's cement after mixing with water. The usual set specifications fall in the range of 4–12 hours. Its major use is in a wall plaster where extra density, strength, and hardness are desired. It is made in only a few locations and only in small quantities.

QUALITY ASSURANCE AND QUALITY CONTROL ISSUES

The manufacturing processes for wallboard can be adjusted for a wide range of gypsum purity. Wallboard can be made from relatively low-purity (low 80s) or very high-purity (high 90s) gypsum. The key is that the purity of the gypsum supplied from the mine or quarry, as well as the calcined gypsum stucco, should be consistent. Formulating calcined stucco with air-entraining agents can decrease the weight of the finished wallboard, enhancing its purity.

Most gypsum contains 10%–15% impurities, although some deposits may be exceptionally pure (i.e., +95%) or somewhat impure (i.e., 80%). In general, the amount of impurity that can be tolerated depends on (1) the type of impurity, (2) the product being manufactured, and (3) the competitive situation. Impurities are usually separated into three categories, based on their effect on the manufacturing process and finished products:

1. Insoluble or relatively insoluble minerals such as limestone, dolomite, anhydrite, anhydrous clay, silica minerals
2. Soluble evaporite minerals, including chlorides (halite, sylvite, etc.) and sulfates (mirabilite, epsomite, etc.)
3. Hydrous but insoluble minerals (e.g., the montmorillonite group of clays)

PRODUCT GRADES AND SPECIFICATIONS

Gypsum is used in highly diversified industries and products. Therefore, grades and specifications vary according to the industry and end-use requirements of customers. Gypsum used in construction products, food, and pharmaceutical products must meet stringent regulations.

Construction Products

ASTM International defines the specifications for testing uncalcined gypsum and construction products manufactured from gypsum. ASTM C471M-17 defines the testing methods for the chemical analysis of gypsum. ASTM C472-99 defines the standard testing methods for the physical properties of gypsum, gypsum plasters, and gypsum-based concrete. In 1999, ASTM began to phase in a new international standard for interior and exterior gypsum wallboard products and veneer plasters that combined nine separate earlier standards. ASTM C1396 eliminated inadvertent inconsistencies in the separate standards.

Packaging

Gypsum is used in more than 400 products, and therefore many different forms of packaging are used for distribution to customers. Raw ground gypsum (land plaster) for agricultural uses and portland cement rock is distributed primarily by bulk truck. Wallboard panels are loaded onto trucks or railroad cars for distribution. Construction and industrial plasters are packaged predominately in paper valve bags in weights varying from 1 to 45 kg, but are also packaged in intermediate bulk containers, flexible intermediate bulk containers, and in certain instances, bulk trucks.

ECONOMIC IMPORTANCE AND OUTLOOK

Gypsum is normally mined in open pit operations from deposits widely distributed throughout the world and tends to be consumed within the many countries where it is produced. Less than 20% of the world's crude gypsum production is estimated to enter international trade. Only a few countries, such as Canada, Mexico, Spain, and Thailand, are major crude gypsum exporters. Of these, Canada and Mexico are significant exporters because of their large deposits in proximity to gypsum-consuming facilities in the United States.

Pricing

By far, China produces the most gypsum in the world. China produced more than 129 Mt (million metric tons) in 2013, and an estimated 132 Mt in 2014. The United States is the world's second-ranked producer and consumer of gypsum. Production of mined crude gypsum in the United States during 2014 was estimated to be 17.1 Mt, an increase of 5% compared with 2013 production. The average price of mined crude gypsum was $9/t. Synthetic gypsum production in 2014, most of which was generated as an FGD product from coal-fired electric power plants, was estimated to be 13.2 Mt and priced

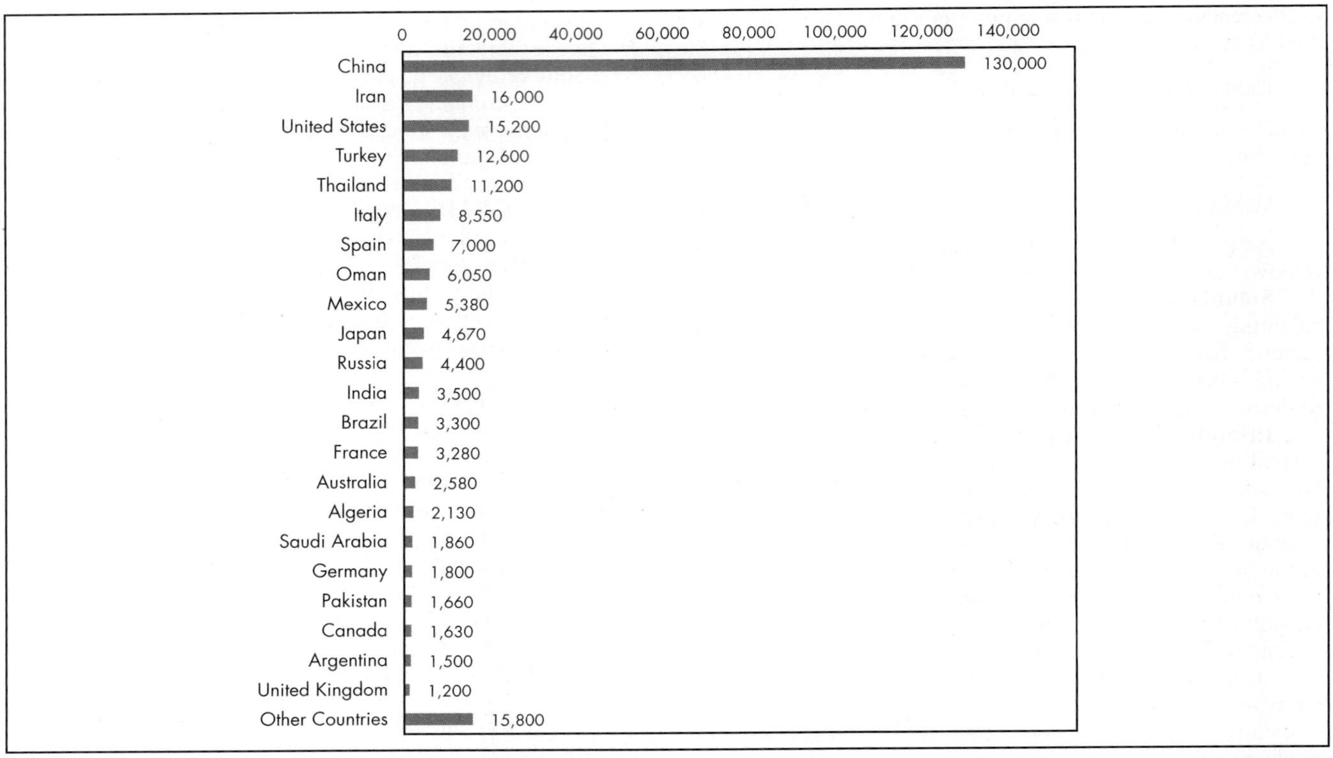

Source: Crangle 2016

Figure 4 Crude gypsum production by country in 2015, in thousand metric tons

at approximately $1.50/t. In 2014, 47 companies mined gypsum in the United States at 118 mines with plants in 17 states. U.S. gypsum exports totaled 70 kt (kiloton), and imports were 3.5 Mt (Crangle 2014). Figure 4 lists the top 22 gypsum producers by country.

Gypsum is a low unit-value, high place-value industrial mineral, and its ultimate value is based on value-added processing. The lowest prices are for ground gypsum used for portland cement and agricultural gypsum. Calcining gypsum for use in manufacturing wallboard or construction and industrial plasters increases the price significantly. The most valuable products include specialty dental, orthopedic, and industrial plasters; food and pharmaceutical-grade gypsum (terra alba); and industrial gypsum cement. Table 3 lists the average crude gypsum prices in the United States between 1994 and 2014.

There is no recognized commodity value of a ton of raw gypsum. Although mining costs are relatively low, the energy-intensive post-mining processing and ultimate end use add significant value to the products made from the gypsum. Freight costs are a significant factor in marketability of a gypsum deposit.

Gypsum deposits are distributed throughout much of the world, and gypsum is produced in more than 90 countries (Crangle 2014). The value of a deposit of gypsum depends on geological, mining, engineering, and other factors.

The utility and value of a gypsum deposit can change over time as technological innovations in mining and manufacturing occur and new products are developed. Factors that must be considered when evaluating a gypsum resource for potential development, mining, manufacturing, and marketing include the following:

- Proximity to market area—because gypsum has a low unit value, an ideal deposit of gypsum would be located within the heart of a growing metropolitan area.
- Transportation—this includes the shipment of raw and finished materials by truck, train, ship, or barge; the import of paper used in wallboard manufacturing; transportation of finished products to jobsites, distributors, and retail outlets; and the potential for backhauls of raw materials or finished products.
- Fuel and utilities including water—wallboard manufacturing is highly energy intensive and requires an infrastructure of electrical service and a source of fuel for calcination and drying of wallboard panels.

There is no foreseeable shortage of either gypsum or anhydrite resources in North America or the world. However, there are instances where it may be difficult to find gypsum that can be considered economic at a given time and location, a problem that has its roots in place value.

Because of its widespread occurrence and huge potential reserves, and also because its uses are such that it is not basic to survival in a national emergency, gypsum is not considered a strategic mineral. This has permitted natural economic factors to prevail in the development of the mineral worldwide, which overall is a healthy situation that should continue to prevail.

Gypsum mining does not result in some of the environmental issues that are commonly associated with mining, such as acid-mine drainage, heavy metal contamination of surface water or groundwater, or the use of cyanide in heap leaching. Gypsum mines and quarries are regulated by numerous local, state, and federal agencies for compliance with environmental and safety regulations (Sharpe and Cork 2006).

Table 3 Average gypsum price in the United States

Year	Average Crude Rock fob* Mine Price, $US/t
1994	6.70
1995	7.29
1996	7.10
1997	7.11
1998	6.92
1999	6.99
2000	8.44
2001	8.44
2002	7.31
2003	6.90
2004	7.21
2005	7.48
2006	8.83
2007	7.50
2008	8.70
2009	7.40
2010	6.90
2011	8.20
2012	7.70
2013	7.50
2014	8.00
2015	7.80
2016	8.00
2017 (estimate)	8.20

Data from USGS 1995–2018
*fob = free on board.

The use of synthetic gypsum is expected to continue to climb worldwide and negatively impact the demand for mined rock. In the United States, however, the availability of inexpensive natural gas is reducing the number of operating hours for coal-fired power plants and synthetic gypsum production. In addition, increased environmental regulations on coal combustion by-products may impact the availability of synthetic gypsum. It is expected that synthetic gypsum usage has reached a plateau or will decline modestly in the coming few years in the United States. These two factors should increase the need for mined rock in the United States in the future.

The use of recycled wallboard will continue to grow. Off-specification wallboard is commonly recycled back into the manufacturing stream. Environmental regulations often require the plants to be zero discharge. Therefore, all off-specification wallboard is recycled. Also, it is expected to become more commonplace for wallboard plants to accept customer waste from new construction. This will also have some impact on reducing demand for mined gypsum.

A North American trend toward reducing wallboard weight will continue and is expected to expand into construction markets worldwide. This will reduce the gypsum used per sheet of wallboard. However, as emerging markets continue adopting Western construction practices and increase gypsum wallboard use, it will more than compensate for lighter wallboard products.

REFERENCES

ASTM C471M-17. 2017. *Standard Test Methods for Chemical Analysis of Gypsum and Gypsum Products*. West Conshohocken, PA: ASTM International.

ASTM C472-99. 2014. *Standard Test Methods for Physical Testing of Gypsum, Gypsum Plasters and Gypsum Concrete*. West Conshohocken, PA: ASTM International.

ASTM C1396/C1396M-17. 2017. *Standard Specification for Gypsum Board*. West Conshohocken, PA: ASTM International.

Crangle Jr., R.D. 2014. Gypsum. In *Minerals Yearbook 2014*. Reston, VA: U.S. Geological Survey.

Crangle Jr., R.D. 2016. Gypsum. In *Mineral Commodity Summaries 2016*. Reston, VA: U.S. Geological Survey.

Deer, W.A., Howie, R.A., and Zussman, J. 2013. Gypsum. In *An Introduction to the Rock-Forming Minerals*, 3rd ed. Hertfordshire, UK: Berforts Information Press. pp. 445–447.

Hansen, W.C., Offutt, J.S., Roy, D.M., Grutzeck, M.W., and Schatzlein, K.J., eds. 1988. *Gypsum and Anhydrite in Portland Cement*, 3rd ed. Chicago: U.S. Gypsum Company.

Henkels, P.J. 2006. Construction uses: Gypsum plasters and wallboards. In *Industrial Minerals and Rocks: Commodities, Markets, and Uses*, 7th ed. Edited by J.E. Kogel, N.C. Trivedi, J.M. Barker, and S.T. Krukowski. Littleton, CO: SME.

Sharpe, R., and Cork, G. 2006. Gypsum and anhydrite. In *Industrial Minerals and Rocks: Commodities, Markets, and Uses*, 7th ed. Edited by J.E. Kogel, N.C. Trivedi, J.M. Barker, and S.T. Krukowski. Littleton, CO: SME.

USGS (U.S. Geological Survey). 1995–2018. Gypsum. In *Mineral Commodity Summaries*. Reston, VA: USGS.

CHAPTER 12.16

Iron Ore Beneficiation

S. Jayson Ripke, Joe Poveromo, Thomas P. Battle,
Henry Walqui, Howard Haselhuhn, and Mike Larson

Iron ore is a mineral substance, which when heated in the presence of a reductant, will yield metallic iron (Fe). Iron ore is the source of primary iron for the world's iron and steel industries. It is, therefore, essential for the production of steel, which in turn is essential for maintaining a strong industrial base. Almost all (98%) of the iron ore mined is used in steelmaking. Iron ore is mined in approximately 50 countries. The seven largest of these producing countries account for about three-quarters of total world production. Aside from production, when it comes to exportation, Australia and Brazil together dominate, each having about one-third of total exports (USGS, n.d.).

Iron ore is used in the form of pelletized concentrates, sintered fines, or lump ore as feed to the primary method of steelmaking, using blast furnaces (BFs) to basic oxygen furnaces in an integrated steel mill. Additional BF feeds include reductants (usually coke) and often fluxes (such as dolomite and/or limestone) to make the liquid hot metal (pig iron), which is the main ingredient in steel production. Pelletized concentrates and lump ore are also used in direct reduction (DR; usually with reformed natural gas) by the shaft furnace to make direct reduced iron and hot briquetted iron for use in electric arc furnaces (EAFs) where steel scrap is recycled in mini-mills, which is the secondary method of steelmaking. To reduce EAF slag production, iron ore pellets or lump reduced by DR for use in an EAF (called *EAF-* or *DR-grade*) usually contain fewer impurities (gangue) than BF-grade pellets or lump.

Iron was first produced circa 2000 BCE, but steel did not become a major commodity until the 1850s (Bell 2017). Iron ore reserves consist mostly of mixtures of gangue and iron oxides in the form of Fe_2O_3 (hematite) and Fe_3O_4 (magnetite, which only differs from hematite by FeO and can be written as $Fe_2O_3 \cdot FeO$), FeO(OH) (goethite), and to a lesser extent ironsands (such as $FeTiO_3$ [ilmenite] and titanomagnetites), $FeO(OH)n(H_2O)$ (limonite), or $FeCO_3$ (siderite). In addition to the iron ore minerals, terms such as *taconite*, *banded iron formation* (BIF), *itabirite*, and *laterite* are also used to describe the geologic occurrence of iron ores. Aside from iron oxides, iron sulfides such as FeS_2 (pyrite) are abundant, but because sulfur is a deleterious element to steelmaking, iron sulfides are not considered an iron ore reserve. In the United States (and other industrialized nations), the equivalent of 150 kg (330 lb) of iron ore must be provided annually per person to maintain the current modern way of life including infrastructure (beams, rebar) and transportation (cars, trucks, trains, planes) (MEC 2016).

According to stoichiometry, pure magnetite contains 72.4% Fe, pure hematite 69.9% Fe, pure goethite 62.9% Fe, and pure siderite 48.2% Fe. Typical gangue constituents are SiO_2 (silica) and Al_2O_3 (alumina). In addition to sulfur, other deleterious elements that are usually avoided or removed from iron ores include phosphorus, copper, manganese, chromium, and alkalis (such as sodium and potassium). In some cases, calcium and magnesium (usually in the form of limestone or dolomite) are added to the iron ore concentrate before balling and induration to create a *basic* or *fluxed* pellet (vs. an *acid* or *unfluxed* pellet). This is because Ca and Mg act as a basic flux for acidic gangue (silica and alumina) and therefore produce a sufficiently low-viscosity slag in the subsequent pyrometallurgical processing steps used for producing iron and steel.

Some ores, called direct shipping ores (DSOs), require little or no beneficiation and simply need to be crushed and screened to produce saleable products; other ores require significant beneficiation flow sheets that can vary widely depending on the type(s) and texture(s) of iron ore minerals associated with the ore. Goldring (2003) presented a relatively simple system that divides iron ores into eight categories, including (1–4) four types of BIF, (5) detrital, (6) fluviatile (channel iron), and (7–8) two types of hypogene magmatic (Kiruna or Andean/Magnitnaya). More than 90% of all iron

S. Jayson Ripke, Manager–R&D, Midrex Technologies Inc., Pineville, North Carolina, USA
Joe Poveromo, Raw Materials & Ironmaking Global Consulting, Self-Employed, Bethlehem, Pennsylvania, USA
Thomas P. Battle, Extractive Metallurgy Consultant, Self-Employed, Charlotte, North Carolina, USA
Henry Walqui, Director, Field Pilot Plant Programs, CiDRA Minerals Processing, Windsor, Connecticut, USA
Howard Haselhuhn, Applications Engineer–Industrial Minerals, Solvay Technology Solutions, Stamford, Connecticut, USA
Mike Larson, Applications Engineering Manager, Moly-Cop USA, Ewen, Michigan, USA

produced comes from BIFs, which from a mining perspective can often be divided into three zones:

1. DSO-associated low-grade merchant ores, which occur around the high-grade ores that can be mined concurrently, and which require minor upgrading by washing or gravity separation techniques to increase their iron content
2. The underlying iron formations, or taconite, from which most of these deposits have been derived
3. A hard, dense, low-grade material that requires extensive crushing, grinding, and concentration to produce an acceptable concentrate

COSMIC ABUNDANCE

Iron is one of the more intriguing elements in the periodic table. Because of its nuclear structure, it has a pivotal role in the generation of the heavier elements in the centers of stars. Because of its multiple valence states, and relative ease in reduction and oxidation, it has played a key role in the formation and geological history of the earth, from the creation of a molten Fe-Ni core, resulting in a magnetic field important to life, to its ready oxidation and precipitation from early oceans when oxygen first appeared in the atmosphere, leading to the formation of BIFs. Not only is iron relatively abundant in the earth's crust, but through these BIFs, it has been concentrated in huge ore deposits throughout the world. Of course, none of this would matter economically if the products of iron manufacture—iron itself, and steel—were not of fundamental importance to modern civilization.

The accepted theory of elemental formation in our universe indicates that after the big bang, only hydrogen and helium were present, along with minute amounts of slightly heavier elements. Clouds of gas began to collapse under their own gravity and eventually reached temperatures and pressures, such that four hydrogen atoms could fuse to form a helium atom. Because the mass of the products is slightly less than the mass of the reactants, energy is released, enough to stabilize the structure for millions of years, as a star. Eventually, the helium-rich core will contract further and heat up until the triple alpha process (three helium nuclei interacting) creates further energy by fusing to carbon. As core temperatures continue to increase, fusion of the heavier nuclei with alpha particles leads to yet heavier elements. If the star is large enough, this process will eventually lead to the formation of iron and nickel, but here it stops. Iron and nickel have the most tightly bound atomic nuclei of all the elements, so further fusion absorbs, not releases, energy. The star will undergo a catastrophic collapse, and then explode as a supernova. Just before and during the supernova explosion, all the elements heavier than iron are manufactured, and most of the contents of the star are blown into the interstellar medium, where they are available to the next generation of star formation (Chown 2001; Gribbin 2000).

Given its high *metal* content (to an astronomer, anything heavier than helium is a metal), the sun is felt to be at least a third-generation star. It contains roughly 35 atoms of iron for every million atoms of hydrogen and is exceeded in abundance only by H, He, O, C, Ne, N, and Mg, and equaled by Si (Choi 2017). Therefore, it played an important role in the development of the planets, especially with much of the hydrogen, helium, neon, and nitrogen remaining gaseous at temperatures where iron had already solidified, either as metallic iron (or Fe-Ni alloys) or as oxides. As the earth began to form, it also heated up, enough that the metallic iron melted and sank to the center of the planet, forming a liquid core and generating a magnetic field around the earth. There was still enough iron left after this to be a major component of the earth's crust, exceeded only by oxygen, silicon, and aluminum. But it is different from the other metals on the list because it is highly soluble in seawater in its reduced state. That did not have a significant effect on global mineralogy until a family of early organisms started releasing oxygen as a waste product; this turned the atmosphere into a more oxidizing medium, worth memorializing as the *great oxidation event*. But that meant that oceanic iron would oxidize from ferrous to ferric, and ferric is much less soluble in water than ferrous. Iron compounds crashed out of solution and settled on sea floors in bands of differing levels of iron compounds, typically colored black and red. These bands formed all over the world over hundreds of millions of years. Uplifted, twisted around, and weathered, they form the BIFs recognized from Western Australia to Minnesota (United States), from Quebec (Canada) to Brazil to India. More than 90% of the world supply of iron is mined from BIFs (Hazen 2012).

STATISTICS AND PRICING

The U.S. Geological Survey maintains up-to-date iron ore statistics in their annual *Mineral Commodity Summaries*, *Minerals Yearbook*, and monthly *Mineral Industry Surveys* that focus on the United States, but also cover the entire world (USGS, n.d.). Historical data can be found in their *Iron Ore Statistical Compendium* (Kuck 2013). Historical pricing for iron ore fines and pellets can be found at Infomine .com (Investment Mine 2016); for current index and spot pricing, see the Metal Bulletin. Iron ore is priced according to U.S. dollars (US$) or cents (US¢) per dry metric ton units. The Metal Bulletin publishes 13 individual iron ore indices based on detailed specifications that are explained in the Metal Bulletin iron ore indices (Metal Bulletin 2016).

In the United States, iron ore (i.e., pellets) is sold by the long ton (998 kg [2,200 lb]) and its end-product steel is sold by the short ton (907 kg [2,000 lb]).

TRANSPORTATION

Most iron ore (as fines, lump, pellets, or concentrate) is transported by bulk carrier (merchant ships of various sizes) or railroad. Iron ore concentrate can be transported by pipeline as a slurry, such as done by Minas-Rio (Anglo American 2015).

QUALITY

The quality requirements depend on how the iron ore is consumed. Concentrates and sinter feed must meet particle size distribution and chemistry requirements to make sufficient quality pellets or sinter. The chemistry must meet a minimum total iron content measured by titration and maximum *acid* gangue content, in particular SiO_2 and Al_2O_3 (and TiO_2). The basicity is important for downstream processing to produce a sufficiently low-viscosity slag. Basicity is a ratio of basic-to-acid components and can be increased by the addition of fluxes, such as dolomite and/or limestone, to increase the CaO and MgO. In addition to acid components, other deleterious elements must be below their maximum allowance.

- **Iron.** Total iron content should be as high as possible.
- **Acid gangue.** Acid gangue, namely $SiO_2 + Al_2O_3$, should be as low as possible, preferably below 5% for BF and below 2% for DR (to minimize slag volume in the EAF during steelmaking).
- **Alkali metals, Na, K, and Li.** Alkalis should be as low as possible because they promote pellet swelling and degradation, fume off during reduction, and can precipitate and form scaffolds within the BF.
- **Basic oxides.** Basic oxides, namely CaO + MgO, reduce the Fe content, but a limited amount of basic oxides (<3%) may displace purchased flux in ironmaking or steelmaking.
- **Phosphorus.** P should be as low as possible, preferably below 0.030%, as P must be removed in steelmaking.
- **Sulfur.** S should be below 0.020% for BF and below 0.008% for DR.
- **Manganese and titania.** Mn and TiO_2 should be as low as possible. For DR, TiO_2 should be below 0.15% to minimize slag volume in the EAF during steelmaking.
- **Zinc.** Zn is not normally present in ores.
- **Other minor elements.** Cr, Pb, Cu, Sn, Ni, Mo, As, Sb, V, and Li should all be as low as possible. For DR, the total of Cu, Ni, Cr, Mo, and Sn should be low as possible, as these are the key residuals being monitored in the scrap supply.
- **Free moisture and loss on ignition (LOI).** Free moisture and LOI as CO_2 and H_2O are undesirable in the feed material because of the extra heat load and increased volume of gas to be handled.

Pellets, sinter, and lump ore must meet certain physical and metallurgical properties, such as particle size distribution, strength (cold compression strength, drop, tumble), chemistry, and reducibility, to remain competent during handling and processing during reduction in a BF (or shaft furnace for DR of pellets or lump). Poveromo (1999) lists the chemistry and sizing for most iron ores with various grades (sinter, pellets, fines, lump) in several separate tables and describes pellet properties.

BENEFICIATION

The term *beneficiation* regarding iron ores encompasses all the methods used to process ore to improve its chemical, physical, or metallurgical characteristics in ways that will make it a more desirable feed for the ironmaking furnace (Poveromo 1999). Such methods include crushing, screening, blending, concentrating, and agglomerating.

The iron ores that fall within the preceding three categories (pellets, sinter, and lump ore) have quite different processing requirements, and discussion of the appropriate beneficiation steps have been grouped accordingly. The following brief and generalized descriptions are intended only to describe how the basic types of ore are beneficiated and are not to be interpreted as suitable for all ores.

Direct Shipping Ores

DSOs typically require the least amount of processing, such as simply crushing, screening, and blending.

Crushing and Screening

Iron ore of merchantable grade must be properly sized prior to charging to the BF. BF technology commonly requires crushing and screening of direct charge lump ore to the size range from 6 to 30 mm (¼ to 1¼ in.). The specific size selected is based on the characteristics of the ore and is specified to maintain high stack permeability and also allow sufficient time for reduction of coarser material. Consequently, crushing and screening are an integral part of ore-producing facilities.

Crushing commonly involves a primary jaw crusher with secondary gyratory or cone crushers operating in closed circuit with vibrating screens. Roll crushers, hammer mills, or vertical impact crushers are used. Equipment selection is determined largely by the friability of the ore. Most of the screening operations on high-grade ores are dry except when the fines fraction can be effectively upgraded by desliming.

The –6 mm (–¼ in.) fines produced by crushing and screening are most commonly agglomerated by sintering or are sometimes reground and pelletized. The sinter produced is also crushed and screened to meet size specifications compatible with the other charge components.

Blending

The mining program at individual mines is developed to produce a uniform product. Although there are multiple handling steps involved in most loading and shipping systems, they often do not provide enough mixing to meet quality assurance standards now required by industry, especially if both size distribution and chemistry standards are specified. Sophisticated combined blending and load-out facilities are now almost universal in the industry.

Stacking and reclaim systems are most commonly used. Stacking results in layering of the ores (Figure 1). Each successive layer represents an ore that may differ in size distribution or chemical composition from adjacent layers. The elongated pile is built up to a height limited by the stacking capability of the machine. The ore may then be reclaimed for use by bucket-wheel excavators, front-end loaders, or a scraper-cross conveyor. Removal of ore from the face of the pile results in a stream of material that is a uniform blend of ore from all the layers. Variations of this method are used extensively for the stocking and reclaiming of iron ore pellets, as well as other bulk commodities used by the steel industry, such as coal and limestone.

Low-Grade Merchant Ores

The earliest iron ore beneficiation techniques were developed to upgrade the lower-grade ores associated with DSOs. The natural ore-forming process produced layers of relatively pure iron oxides interbedded with partially decomposed silica-rich layers. Ores in which the silica layers have been completely decomposed can be easily upgraded by simple washing techniques where the fine-grained friable silica particles can be separated from the heavier, denser, and coarser iron ore particles by hydraulic classification. Ores in which the silica-rich layers have not been weathered as intensely and are still relatively cohesive have to be broken by crushing and are upgraded by gravity concentration techniques, such as jigging and heavy media separation, with the finer fractions upgraded in spiral concentrators. The technology for beneficiating low-grade ores was largely developed on the Mesabi Range in Minnesota in the 1950s and 1960s and applied worldwide to similar deposits.

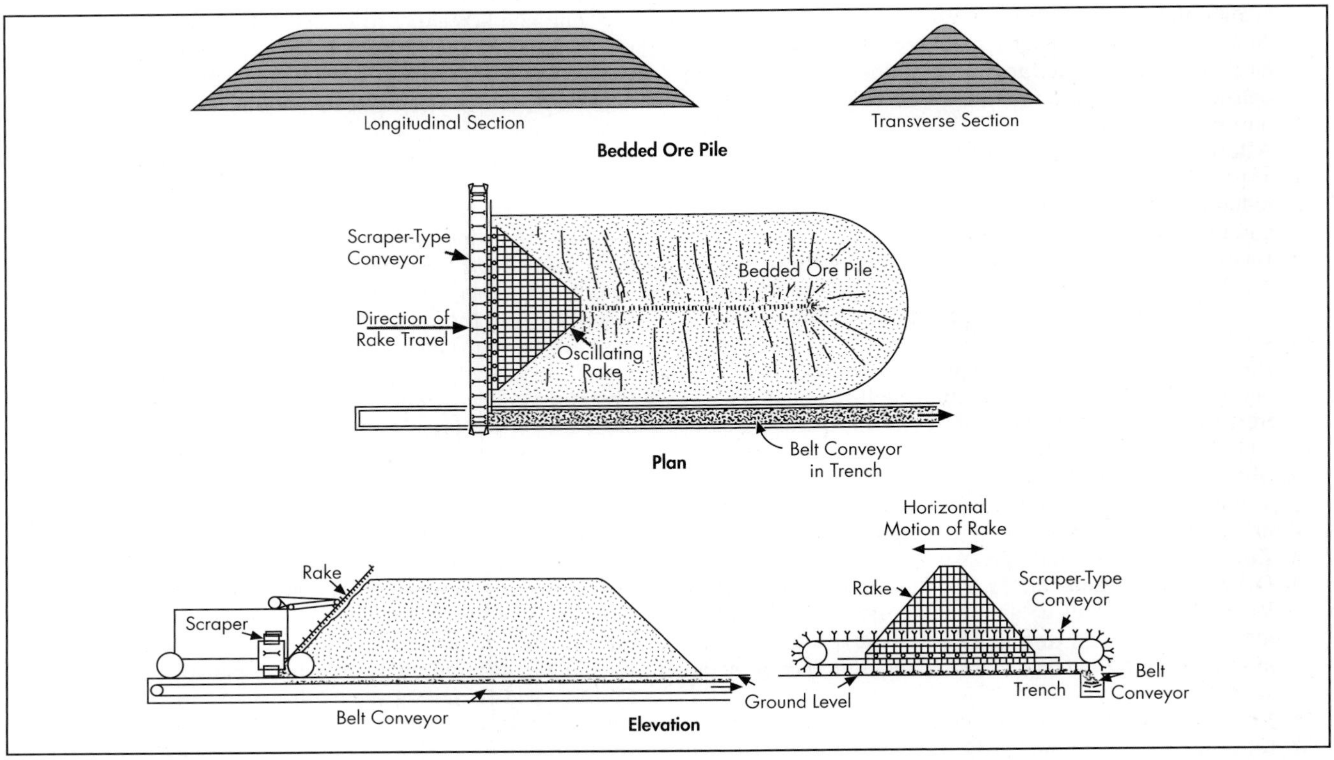

Source: Poveromo 1999

Figure 1 Principle of operation or a stacking-and-reclaiming system

Washing

Washing is the simplest iron ore concentration process and takes advantage of the high specific gravity and comparatively coarse size of the iron-bearing minerals to separate them from the finer, lighter, siliceous gangue, which is predominantly quartz and clay minerals. The ore is prepared for washing by crushing finer than 50 mm (2 in.) in one or two stages. The crushed ore is fed to log washers that were specifically developed for the wash ores of the Western Mesabi Iron Range. The intense agitation of the ore by the paddles (similar to a modern pug mill) combines with the counterflowing water to efficiently break up the ore and remove the fine silica to leave a coarse residual iron-rich product. The log washer overflow was often re-treated in rake or spiral classifiers to recover additional fine iron. Some washing plants employed spiral classifiers in one or two stages without a log washer on ores containing a minimum amount of sticky clay gangue.

Hindered settling classifiers of various types, Hydrosizers, pocket sizers, and Wilfley shaking tables were also used to recover fine iron before the development of the Humphrey spiral.

Jigging

Jigging is a more complex form of beneficiation than simple washing and is used on the more refractory ores that require crushing to break up the silica-rich layers. Jigs used for iron ore beneficiation are basically horizontal screens that carry a bed of ore 15–25 cm (6–10 in.) deep. The ore is fed at one end and is stratified by the pulsing action of water, either caused by an oscillating pump or by physical up and down movement of the jig screen itself. As the ore moves down the deck, the pulsing allows the lighter silica-rich particles to work their way to the top of the bed, while the higher-density iron-rich particles segregate along the base. The two products are separated at the end of the jig, the silica-rich particles over the top of the discharge weir and the iron ore concentrate under the bottom. Iron ore jigs work best on particles ranging in size from 25 mm to 1 mm (1 in. to 1⁄16 in.).

Heavy Media Separation

Heavy media separation devices were developed as a more effective alternative to jigging for the upgrading of the more refractory ores in the 1950s. Heavy media separation processes operate on the sink-and-float principle. A suspension of fine, –200 mesh, ferrosilicon in water is used to create a fluid medium with a specific gravity of approximately 3.0. Silica-rich particles with a specific gravity of about 2.6 will float on the surface of such a medium, while the denser and heavier iron ore particles with a specific gravity higher than 4.0 will settle to the bottom. The conventional medium for concentrating coarse ore is ferrosilicon containing 15% silicon and 85% iron. Water suspensions containing 64%–85% by weight of finely ground ferrosilicon have specific gravities ranging from 2.20 to 3.60.

The separation vessels for coarse ore (+9 mm [+3⁄8 in.]) are commonly spiral classifiers, rake classifiers, or rotating drums. Ore finer than 9 mm (3⁄8 in.) and coarser than 3 mm (1⁄8 in.) can be separated in heavy media cyclones where the high gravitational forces accelerate the settling of the heavy iron ore particles. Finely ground magnetite is used to make up the heavy media for the cyclone separators rather than ferrosilicon. The dynamics of the cyclone create the density and media fluidity required despite the lower specific gravity of the magnetite. Furthermore, the cost of magnetite is much less than ground ferrosilicon.

The fluid medium, ferrosilicon and magnetite, is washed from the sink-and-float products on fine screens equipped with wash troughs and water sprays and is recovered from the wash water with magnetic separators and recycled.

Spirals

The Humphrey spiral, first developed for the treatment of beach sands, is used in iron ore concentration to treat –6 to +100 mesh ore (efficiency below 100 mesh decreases rapidly, and spirals are ineffectual on finer materials). Spirals are widely used for the supplementary recovery of fine iron from merchant ore types and are the primary concentration device for the specular hematite ores of the Labrador trough and similar ores that can be liberated by grinding no finer than 20 mesh.

The spiral concentrator is a curved-bottom trough, wound around a vertical axis in the form of a helix (Figure 2). When fed at the top with a slurry of iron ore and gangue, the less-dense gangue, being more readily suspended by the water, attains greater tangential velocity than the iron minerals and migrates toward the outer rim of the spiral trough. Wash water added along the inside rim helps wash away the lighter gangue. After a few turns, a band of iron mineral forms along the inner rim, and the gangue forms bands toward the outer rim. Ports are spaced along the inner rim to collect and remove the iron minerals. The gangue remains in the spiral and discharges at the bottom.

Reichert Cone

The principal advantages of the Reichert cone are capacity and the ability to recover fine heavy minerals efficiently down to approximately 325 mesh, which is finer than is attainable in spirals. A single Reichert cone has a capacity of up to 100 t/h (metric tons per hour; 98.4 ltph [long tons per hour]) and can be effectively used to recover specular hematite fines as well as merchant ore fines. Desliming of the feed is desirable—and essential for merchant-type ore fines.

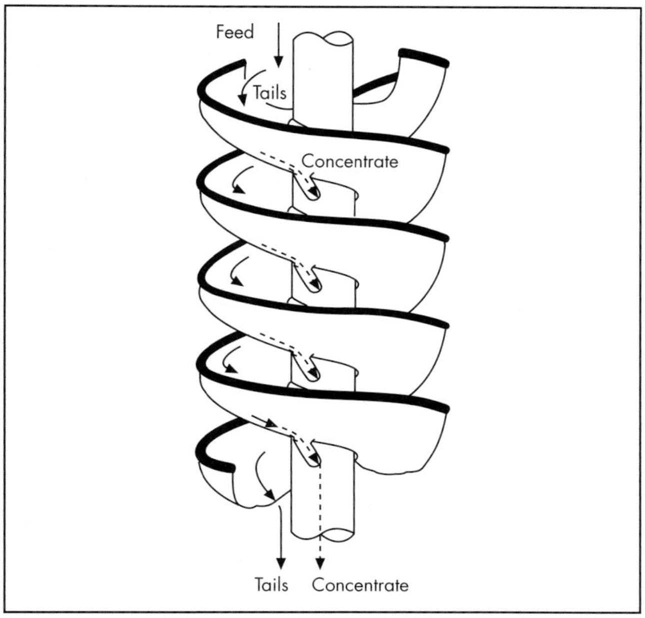

Source: Poveromo 1999

Figure 2 Spiral classifier

The Reichert cone (Figure 3) is a flowing-film concentrator. The denser particles concentrate at the bottom of a flowing film of slurry having a solids content of about 60% by weight. The separation mechanism is a combination of hindered settling of the dense particles and interstitial trickling of the fine particles. The separation element in the Reichert unit is an inward sloping 1.9-m (6.25-ft)-diameter cone. Feed pulp is evenly distributed around the periphery of the cone. As the pulp flows by gravity toward the center, the fine and the heavy particles concentrate on the bottom and are removed through an annular slot near the apex of the cone. The tailings flow over the slot and are collected at the apex or center of the cone. Because the efficiency of this separation process is relatively low, it is repeated several times within a single stacked arrangement of cones to increase the recovery. Generally, the highest-grade concentrate is produced in the primary separation cone.

Wet High-Intensity Magnetic Separation

Wet high-intensity magnetic separation (WHIMS) machines were developed to recover nonmagnetic iron units. They can be effectively applied across a wide particle size range from 10 to 500 mesh depending on the matrix used. WHIMS applications include recovery of iron from natural ore fines, upgrading of spiral concentrates for DR feed, and recovery of hematite from tailings.

In WHIMS, electromagnets produce a very high-strength magnetic field that is applied to a matrix consisting of steel

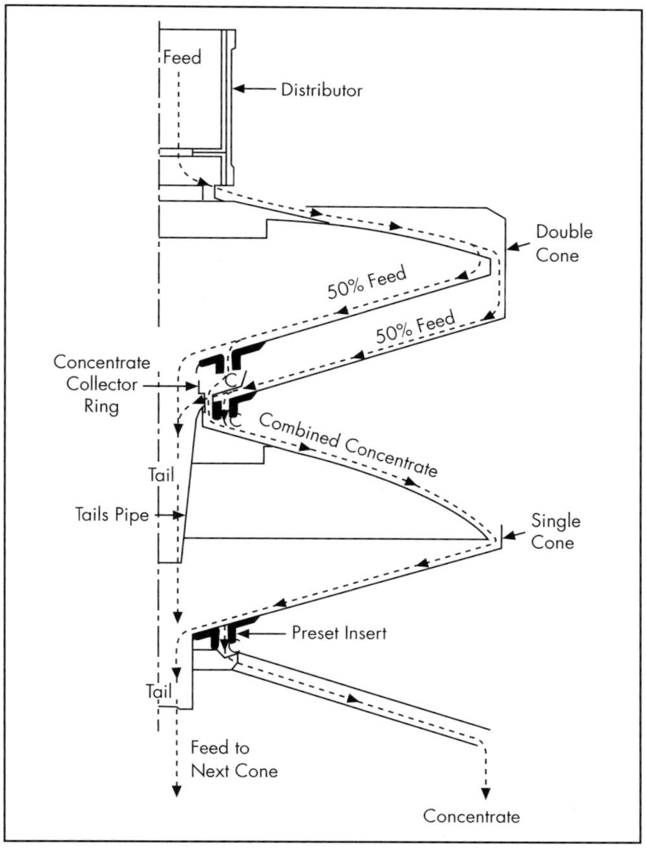

Source: Poveromo 1999

Figure 3 Reichert cone concentrator section

balls, spaced grooved plates, steel wool, or pieces of expanded metal. The matrix is contained in an annular ring that is rotated between the high-intensity magnets. The iron ore slurry is introduced at a point where the matrix is in the field. The high magnetic gradients developed around the matrix hold the iron oxide minerals while the siliceous gangue is washed through. The iron oxide concentrate is released and discharged as the matrix moves out of the magnetic field.

Primary Ores

Primary iron ores contain only 25%–35% recoverable iron that is present as either magnetite or hematite. Typically, the ore is very hard and has a high work index; in addition, it is finely disseminated so very fine grinding is required for liberation. Processing is complex, and mining, crushing, grinding, and concentration costs are very high. Mining companies have to continuously develop new processing options to reduce costs and maintain product quality. The high quality of the pellets and sinter produced from the concentrates, however, has revolutionized BF ironmaking. Concentrates containing 4.0% silica are now the norm, and the unfluxed (a.k.a. acid) or fluxed (basic) indurated pellets now dominate ironmaking in North America.

To produce a final product that is uniform in chemical and physical properties, blending of the crude ore from various shovel locations mining different grades of ore is accomplished by using sophisticated computerized mine plans.

Currently, the greatest production in North America is from the magnetite taconites of Minnesota and Michigan. This iron ore beneficiation and palletization technology was developed by E.W. Davis (1964). There is also substantial production from the specular hematite iron formations of the Labrador trough. The discussion of beneficiation techniques will use these two iron ore types as models; however, similar iron formations are present on all of the continents and are being exploited, or will be in the future.

Comminution

Comminution means to break into smaller parts. The following subsections discuss the different ways this is done.

Drilling and blasting. Comminution begins in the mine. The impact of fragmentation achieved in blasting on downstream comminution, crushing, and grinding is just beginning to be quantified. More companies now regard this as the first step in comminution and are willing to allocate more money to drilling and blasting to achieve finer fragmentation and reduce downstream costs.

Primary, secondary, and tertiary crushing. Because of the hardness of the ore, three and sometimes four stages of crushing are used in magnetite-taconite plants to reduce the ore to rod mill feed size (Figure 4). Primary and secondary crushing is almost universally done by gyratory or cone crushers. (Jaw crushers seldom have the capacity or the durability to serve as a primary crushing unit unless the ore is exceptionally soft.) Tertiary crushing to a top size of 25 mm (1 in.) is almost universally completed by shorthead crushers operating in closed circuit with screens. Relatively recently, and after having been successfully used for fine comminution in the cement industry, high-pressure grinding rolls (HPGRs) have become a common alternative to tertiary crushing and rod milling. HPGRs can reduce 60-mm (2.5-in.) crude ore to rod mill discharge size when closed with a screen. Power savings are significant, and although the capital costs are high, the metallurgical benefits are also high; therefore, the HPGR has replaced both tertiary crushing and rod mills in some taconite plants.

Autogenous milling. Autogenous grinding is an alternative to multistage crushing. Autogenous mills use the ore as the grinding media. The crude ore is crushed to a top size of 305–457 mm (12–18 in.) in one stage of primary crushing. The tumbling mills have a high diameter-to-length ratio, and the larger ore fragments are the grinding media. The benefit of the autogenous milling circuit is that it completely eliminates the secondary and tertiary crushing required by conventional rod mill and ball mill plants. Metal consumption is lower because no steel grinding media is used, but the power input per ton of product produced is higher because of the low density of the natural ore grinding media.

The mill discharge is usually sized on a relatively coarse trommel screen at 3 mesh, and the oversize pebbles are recirculated. This recirculating pebble load tends to build up in some circuits and will eventually curtail mill production. The most recent process improvement applied to autogenous milling circuits has been the addition of magnetic cobbing and pebble crushing to reduce the circulating load and significantly increase throughput. Three alternatives for pebble crushing have been or are being tested on a commercial scale. These are water flush crushing, vertical impact crushers, and HPGRs.

The final product size from autogenous milling is controlled by cyclone classification of the trommel undersize. In some plants, the autogenous mill grinds to final liberation size, –200 mesh. Other plants close the autogenous mills at a coarser size, –6 mm (–¼ in.), and achieve the final grind in conventional ball mills or pebble mills.

Semiautogenous milling. Autogenous milling requires competent ore that will provide the coarse grinding media required to achieve effective breakage. Where such competent ore is limited, or the ore tends to break to a smaller size and the smaller fragments cannot provide the energy required for breakage, steel balls are added to the mill. This is termed *semiautogenous grinding*, or *SAG*. The other components of the circuit are essentially as described for autogenous milling.

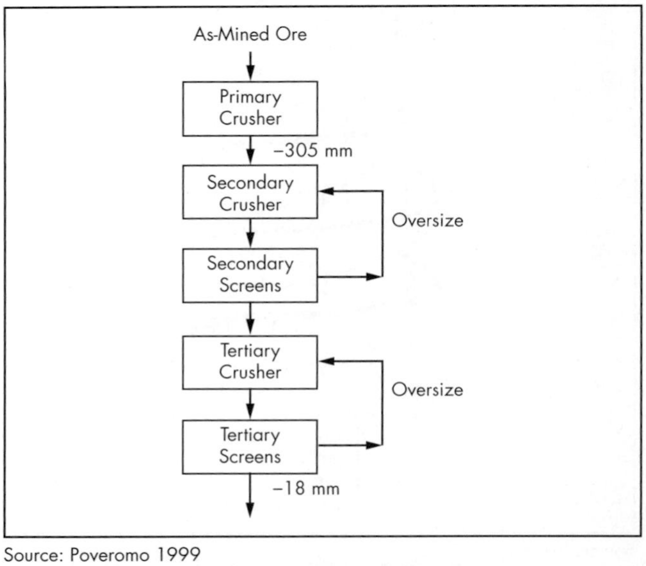

Source: Poveromo 1999

Figure 4 Flow sheet of an iron ore crushing circuit

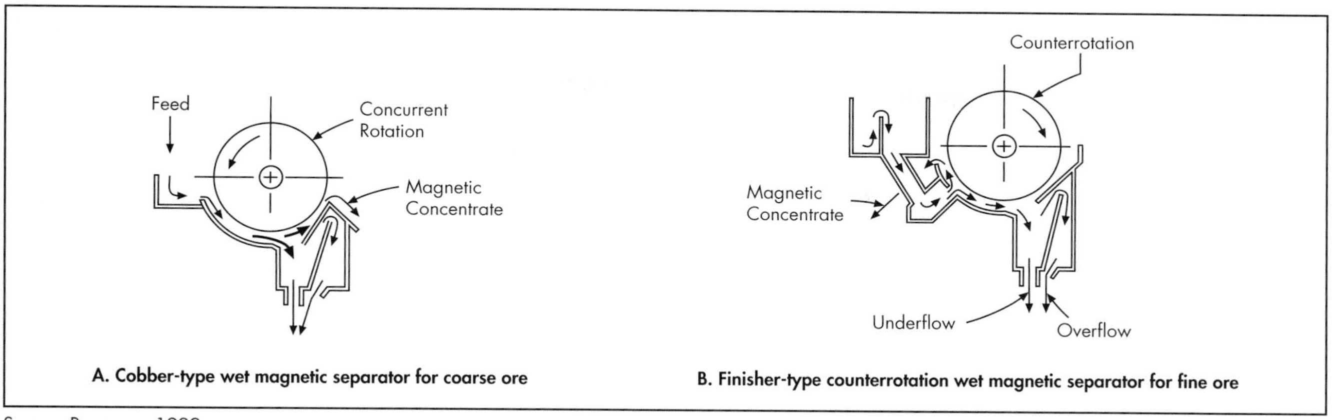

Source: Poveromo 1999

Figure 5 Concurrent separators

Rod milling. Rod mills were designed to grind nominal –35 mm (–1¼ in.) feed, usually prepared by tertiary cone crushers, to –3 mesh. They are tumbling mills with a length-to-diameter ratio of 1.5:2.0. The grinding media are steel rods up to 100 mm (4 in.) in diameter. The ore is ground wet at approximately 68%–80% solids. Rod mills are most commonly operated in open circuit. Product size is controlled by combinations of feed rate, rod charge characteristics, mill speed, slurry density, and viscosity.

Ball milling. Ball mills are the principal means of fine grinding in the iron ore industry. They are tumbling mills like the rod mill, but usually with a smaller length-to-diameter ratio, ranging up to 1.5. The grinding media are steel balls rather than rods. These mills are almost always operated in closed circuit with classification devices, such as cyclones, spiral classifiers, or density classifiers, commonly operated in combination with fine screens. Grinding efficiency is greatly affected by the classification efficiency and mill operating parameters such as mill speed, media charge, slurry density, and viscosity.

Pebble milling. Pebble mills are also used for fine grinding. They are similar to ball mills except that they are charged with 25–100 mm (1–4 in.) pebbles rather than steel balls. Their principal advantage is that steel grinding media is not required, and mill liner wear is reduced. The offset is the generally high power consumption per ton of product produced. The pebbles are most often obtained from an associated autogenous mill circuit where their diversion helps solve the primary mill recirculating load problem. However, they can also be obtained from a conventional crushing plant circuit.

Magnetic Separation

Magnetic separation is universally used for concentration of magnetite-bearing iron ores and magnetite-taconite. Because of the layered nature of the taconite—quartz-rich layers interbedded with magnetite-rich layers—partial rejection of silica can be achieved at a relatively coarse size. A taconite plant may have as many as five magnetic separation stages at successively finer sizes before a final concentrate is produced.

Coarse dry magnetic cobbing. Coarse magnetic cobbing of tertiary crusher discharge at –19 to –25 mm (–¾ to –1 in.) rod mill feed has been practiced by several plants. Such separation is usually accomplished dry, using a magnetic head pulley. Head pulleys with fixed permanent magnets in radial or axial configurations have been used as well as fixed magnets inside a rotating drum, similar to the wet magnetic cobbing drums described later. Rejection of 10%–15% of the ore while recovering more than 98% of the magnetite is possible depending on the grade of the feed. Magnetic separation at –75 mm (–3 in.) was also used at one mine to upgrade lean ore stockpiles.

Magnetic cobbing of rod mill discharge. Magnetic separation of magnetic taconite at –3 mesh is a critical step in the process. At this size, 25%–45% of the ore can be discarded as tailings, which reduces downstream grinding and concentration costs. The magnetic cobbers (Figure 5) generally consist of a fixed permanent magnet assembly inside a rotating stainless steel drum. The magnetic field strength at the drum surface is generally low, less than 0.2 T (2,000 gauss). The magnetite is pulled against the drum and carried to the discharge lip as the drum rotates. Tailings drop to the bottom of the tank and are rejected through spigots. Drums of 1.2-m (4-ft) diameter have a capacity of more than 32.7 t/h per meter of length (10 ltph per foot of length).

Finisher and cleaner magnetic separation. Finisher and cleaner magnetic separators are used after ball mill or pebble milling to treat the circulating load, or between grinding stages, or to treat the final classifier product. Because of the finer grain size, the tank configuration of finisher and cleaner magnet separators and the feed and tailings discharge systems are different than those of coarse cobbers. The feed and tailings can be held in suspension, so bottom spigot discharge is not required. Several different options for presenting the feed and discharging the tailings have been devised. The magnetic field strength of the finisher and cleaner magnetic separators is generally lower than that of the cobbers to allow rejection of fine middlings and minimize free silica entrapment. Field strengths as low as 0.08 T (800 gauss) have been specified for some plants. Capacity of 1.2-m (4-ft)-diameter separators treating fine material is also somewhat lower, generally less than 16.4 t/h per meter of width (5 ltph per foot of width).

Hydraulic Separation

Hydraulic separation is an important step in the magnetic separation plant. It is applied most commonly to the final classification product, usually ahead of the finisher or cleaner magnetic separation steps. The feed is passed through a magnetic coil that causes the individual magnetite particles to coagulate by magnetic flocculation. The magnetic floccules settle very rapidly in a hydraulic separation tank, and the silica slimes and some fine middlings will be rejected as a final tailings in the overflow. Dewatering in this stage is as important as the silica rejection, and the densified classifier discharge is then fed to the final magnetic separators.

Classification and Fine Screening

The efficiency of fine grinding circuits is greatly affected by the efficiency of the classification system used to close the circuit. Numerous variations of rake, bowl, and spiral hydraulic classifiers were used before the introduction of hydrocyclones in the 1950s. Now cyclones are universally employed because of their low operating cost and minimal space requirement. Because of the high gravitational forces in the cyclone, both particle size and particle specific gravity affects the size separation. This tends to increase the percentage of middling particles in the cyclone overflow and affects downstream processing. The addition of fine screens to treat the cyclone overflow can reduce this problem by allowing the cyclone to operate at a coarser size split and use the fine screens to control the final concentrate size. Either stationary slotted screens or vibrating sandwich deck screens can be used to make separations at approximately 325 mesh. The combination of cyclones and fine screens is almost universal in magnetite-taconite plants, and some plants are operating circuits in which the cyclone is essentially eliminated, and fine screens are used alone. A recent development in the taconite industry is the testing of density classifiers as an alternative to cyclones because they can make a sharper, cleaner size split.

Flotation

Froth flotation is effective for the concentration of fine (–100 mesh) iron ores. It can be used for the supplementary upgrading of magnetite concentrates, as currently practiced at four taconite operations in Minnesota, or as the primary recovery method for hematite ores, as currently practiced at the Tilden mine in Michigan (United States).

Flotation processes depend on the fact that certain reagents added to water suspensions of finely ground iron ore selectively cause either iron oxide minerals or gangue particles to exhibit an affinity for air. The minerals having this affinity attach to air bubbles passing through the suspension and are removed from the suspension as a froth product.

The reagents added to induce the preferential affinity for air are commonly called *collectors* or *promoters*; substances added to cause stable bubble or froth formation are known as *frothers*; other substances added for control purposes such as pH adjustment or to cause better dispersion or flocculation are known as *modifiers*, *dispersants*, and *depressants*.

Flotation collectors are of two general types: anionic and cationic. Anionic collectors ionize in solution such that the active species (that which attaches to the positively charged mineral surface) is negatively charged. Conversely, the active ionic species in cationic flotation collectors is positively charged.

The main application of anionic flotation is to float iron-bearing minerals away from gangue material. The most common collectors used are fatty acids or petroleum sulfonates. Fuel oil is often added along with the collectors to promote recovery of iron oxide particles finer than approximately 10 mm (3⁄8 in.).

Conversely, cationic flotation is used to float siliceous gangue away from finely ground crude ore and to remove small amounts of gangue material from some magnetite concentrates. Cationic collectors are primary aliphatic amines or diamines, beta-amine, or ether amines, generally in acetate form. See additional information from Silva et al. (2011).

Magnetite-taconite concentrates. Supplementary upgrading of magnetite concentrates is practiced at several taconite plants to lower the silica content for BF pellets. It will also be an essential part of the flow sheet for the production of low-silica concentrates for DR in the future. Cationic silica flotation is most commonly used for this purpose. The silica in the final concentrate may be reduced by as much as 2%–4% depending on the efficiency of the magnetic separation plant and the characteristics of the crude ore.

Conventional mechanical flotation cells, 14–43-m^3 (500–1,500-ft^3) capacity, are used in Minnesota taconite plants. Because of the carryover of fine (–500 mesh) free magnetite with the silica-rich froth, secondary froth treatment is essential. This may be complex and involves dewatering, cyclone classification and densification, regrinding, and magnetic separation. One plant has pioneered the use of column flotation for froth re-treatment because column flotation units recover –500 mesh magnetite more efficiently than the conventional mechanical cells.

Hematite ores. The basic selective flocculation, cationic silica flotation flow sheet was developed specifically to concentrate the ores of the Tilden mine, described later.

Concentration Plant Flow Sheets

The type of beneficiation process depends on the ore mineralogy (composition and distribution) and the ultimate destination (or end-use application) of the iron concentrate. Mineralogy of both the iron and gangue phases (in particular silica, phosphorus, and alumina) play a fundamental role in determining proper unit operations and flow sheet design (or selection) and was well-explained by Silva; this applies broadly to most hematite BIFs around the world (Brazil, South Africa, India, and Australia) although he focused specifically on Brazilian ores (Silva et al. 2011; Araujo et al. 2003). Examples include gravity separation (jigs and spirals) magnetic separation (rare earth wet drums, ferrous wheel, and Jones-type magnetic separators), and flotation. When it comes to end use, for example, DR feed requires a much lower silica content, necessitating a more elaborate beneficiation flow sheet than does BF feed.

MODERN FLOW SHEETS, DESIGN, AND OPTIMIZATION

The following iron ore beneficiation flow sheets represent the most modern designs in use around the world today.

United States

The 2015 status of the U.S. iron operations are shown in Table 1 when some of the plants listed with "iron ore pellets" as their primary product were idled and/or operating at reduced capacity because of a slow economy, but by 2018 they were all actively targeting full production capacity.

Magnetation

The Magnetation assets were acquired by, and are now called, ERP Iron Ore. Essar Minnesota assets were acquired by Chippewa Partners and are now called Mesabi Metallics. The

Table 1 Iron operations in the United States, 2015*

Operation and Location	Operator	Primary Product	Status	Capacity, Mt (million lt)†	Production, Mt (million lt)†	Reserves, Mt (million lt)‡
Voestalpine Texas LLC, Corpus Christi, TX	Voestalpine	Hot briquetted iron	Active	2.0 (1.9)	2.0 (1.9)	§, **
Iron Dynamics Inc., DeKalb County, IN	Steel Dynamics Inc.	Hot metal	Active	0.3 (0.29)	0.3 (0.29)	§
Reynolds Pellet Plant, White County, IN	Magnetation LLC	Iron ore pellets	Active	3.3 (3.2)	Not available	§
Nucor Steel Louisiana LLC, St. James County, LA	Nucor Corp.	Direct reduced iron	Active	2.5 (2.4)	1.1 (1.0)	§
Empire, Marquette County, MI	CCI (formerly CNR)††	Iron ore pellets	Permanent closure 2016	5.6 (5.5)	3.1 (3.0)	8.7 (8.5)
Tilden, Marquette County, MI	CCI	Iron ore pellets	Active	8.1 (7.9)	7.7 (7.6)	395 (388)
Hibbing Taconite, St. Louis County, MN	CCI	Iron ore pellets	Active	8.1 (7.9)	8.2	267 (262)
Keewatin Taconite, Itasca County, MN	United States Steel Corp.	Iron ore pellets	Active	5.4 (5.3)	1.7 (1.6)	349 (343)
Mesabi Chief Plant 1, Itasca County, MN	Magnetation LLC	Iron ore concentrates	Idled indefinitely (February 2015)	0.4 (0.39)	0.3 (0.29)	3.6 (3.5)
Mesabi Chief Plant 2, Itasca County, MN	Magnetation LLC	Iron ore concentrates	Idled indefinitely (January 2016)	1.0 (0.9)	0.9 (0.88)	18 (17.7)
Mesabi Chief Plant 4, Itasca County, MN	Magnetation LLC	Iron ore concentrates	Idled indefinitely October 2016; planned restart in 2018	2.0 (1.9)	1.4 (1.3)	‡‡
Mesabi Nugget Delaware LLC, St. Louis County, MN	Steel Dynamics Inc.	Iron nuggets	Idled temporarily (May 2015 onward)	0.4 (0.39)	Not available	§
Mining Resources LLC, St. Louis County, MN	Steel Dynamics Inc.	Iron ore concentrates	Idled temporarily (May 2015 onward)	1.0 (0.9)	0	Not available
Minntac, St. Louis County, MN	United States Steel Corp.	Iron ore pellets	Active	15 (14.7)	12 (11.8)	464 (456)
Minorca, St. Louis County, MN	ArcelorMittal S.A.	Iron ore pellets	Active	2.7 (2.6)	2.7 (2.6)	126 (124)
Northshore, St. Louis/Lake counties, MN	CCI	Iron ore pellets	Active	6.1 (6.0)	4.4 (4.3)	830 (816)
United Taconite, St. Louis County, MN	CCI	Iron ore pellets	Active	5.4 (5.3)	3.2 (3.1)	474 (466)

Adapted from Tuck 2018
* Listed operations do not include blast furnaces.
† As reported or calculated from data in company annual reports, oral communications, published online data, or U.S. Securities and Exchange Commission filings.
‡ Proven and probable reserves or equivalent, including those on owned and leased property, as reported by the company on the last publicly available date.
§ Facility does not operate an independent mine and has no reserves.
** Facility buys seaborne DR-grade pellets.
†† In 2017, Cliffs Natural Resources (CNR) returned to its previous name of Cleveland-Cliffs Inc. (CCI).
‡‡ Magnetation LLC owned mineral rights for 1,400 Mt (million metric tons; 1,378 million long tons) of unspecified iron ore equivalent resources or reserves as of April 2014.

Magnetation pellet plant was built in record time, with construction starting in June 2013 and the first pellets produced at the end of September 2014. The plant has a design capacity of 3,000,000 t (2,952,619 lt) of fluxed pellets, with the feed being mostly hematite concentrate. The plant incorporated the latest technology in environmental pollution control with the addition of low nitrogen oxide (NOx) burners and a gas suspension absorber (GSA).

Iron ore concentrate is received from the processing plants in Minnesota via rail and unloaded in Reynolds, Indiana, using a rotary car dumper, one car at a time. Unit trains with 120 cars are unloaded over a 12–18-hour time frame. The concentrate goes through a double-roll crusher (1.5 m wide and 7.3 m in diameter [6 ft wide and 24 ft in diameter]) during the winter to break frozen chunks that are present on the concentrate cars.

The unloaded concentrate is transported via a conveyor to the ore barn where a 116-m (380-ft) long shuttle conveyor spreads the concentrate the length of the ore barn (245 m [804 ft] long by 61 m [200 ft] wide) with a design capacity of 160,000 t (157,473 lt). Wheel loaders are then used to reclaim the concentrate and feed it to the grinding circuit through two reclaim hoppers into a conveyor. The feed rate is determined by the grinding demand. The reclaimed concentrate discharges into a repulper sump where it is mixed with water to target 55%–60% solids by weight. Water addition is controlled to maintain the level of the repulper sump and hit the target percent solids. The slurry then feeds two large storage tanks.

Slurry from the concentrate grinding storage tanks feed a cyclone sump, which is then pumped to a cyclone cluster with a combination of 51-cm (20-in.) and 66-cm (26-in.) cyclones.

The cyclone underflow is gravity fed to a 3,729-kW (5,000 hp) ball mill (9 m [31 ft] long and 5 m [16 ft] in diameter) running at 14.6 rpm. The mill is controlled between 3,579 and 3,654 kW (4,800 and 4,900 hp), while the target grind is controlled to 80% passing 325 mesh. The discharge from the mill goes to the cyclone sump to close the circuit. The product of the grinding circuit (cyclone overflow) discharges by gravity to a 23-m (75-ft) thickener. The filter feed thickener overflow goes to a process water sump where the water is reclaimed to be used throughout the process.

The thickened concentrate is sent to two filter feed storage tanks at 50% solids. From the storage tanks, it gets pumped to a pressurized distributor that feeds five rotary disk filters (with 4-m [12-ft] diameters). The moisture of the concentrate is controlled to 11.5%–12.0% to meet balling needs. Filter concentrate is conveyed to a 2,000-t (1,968-lt) concentrate storage bin.

Additives are also received and ground on-site. Limestone and dolomite come by truck from nearby quarries and are stored in a covered barn. They are then individually fed into raw storage bins, which feed a vertical ring roller dry grinding mill with a 75-t/h (73.8-ltph) capacity. The ratio of dolomite to limestone is controlled to meet the pellet chemistry needs. The ground and dried mix of limestone and dolomite is pneumatically transported to a storage bin. Coke breeze is used as an internal fuel to compensate for the composition of the iron ore concentrate that is mainly hematite. The raw coke arrives in trucks that are unloaded and conveyed to a raw storage bin that feeds a separate vertical ring roller dry grinding mill with a 10-t/h (9.8-ltph) capacity. The coke breeze is ground and dried, then collected in a storage bin.

The ground concentrate, limestone–dolomite blend, coke breeze, bentonite, and recycle process dust taken from stream dust collectors are then metered in with weight belt feeders in parallel lines at the required ratios. Automation in place, together with the equipment installed, ensures a very precise control of the additives, which leads to tightly controlled pellet chemistry. The mix then feeds two parallel high-intensity paddle mixers (298 kW [400 hp], 5 m^3 [176 ft^3] working capacity); here water is added to improve the performance of the balling circuit. The discharge from both mixers is combined and fed via conveyors to the balling feed bins (100 t [98.4 lt] each).

At the balling building, six balling disks (7.4 m [24 ft] in diameter, 186 kW [250 hp]) are used to produce green balls. Each balling disk has a roll screen (584 cm [230 in.] length, 231 cm [91 in.] width, 145 cm [57 in.] rolls) to remove undersize (<9.5 mm [⅜ in.]) and oversize material (>12.7 mm [½ in.]). Screen undersize and oversize are collected and recirculated to mix with the fresh feed. The on-size green balls from each of the disks are collected on a single conveyor that is then spread on a 4-m- (13-ft)-wide belt via an oscillating conveyor. Prior to entering the indurating furnace, the green balls go to a final screening stage on a roll screen feeder (4.2 m [14 ft] wide).

The green balls feed a straight-grate furnace, which measures 96 m (315 ft) long by 4 m (13 ft) wide. Hearth and side layer pellets are added prior to the green balls landing on the grate to protect the pallet cars and support strong pellet quality. The furnace is divided into six different zones: updraft drying, downdraft drying, preheat zone, firing zone, first cooling, and second cooling. The Magnetation pellet plant was the first facility to install low-NOx burners. A total of 14 are installed plus 2 smaller preheat burners. In addition, a GSA scrubber is used to treat the off gases and minimize sulfur oxides (SOx) emission.

The grate discharges to a single collecting conveyor that takes the fired pellets to a hearth layer separation bin, where a portion of the pellets are returned to the furnace to be used as the hearth layer. The final product is conveyed to two pellet load-out bins to be directly loaded into pellet trains.

Essar Steel Minnesota

Currently, the Essar Steel Minnesota project is on hold; however, it is an example of the latest concentrator design. The concentrator flow sheet and its development was presented at the 2009 Society for Mining, Metallurgy & Exploration conference in Minnesota (Murr et al. 2009). Both the concentrator and the pellet plant represent the latest design. The furnace is a recently designed straight-grate system. The low-NOx burner design for this project was pioneered by Essar (incidentally, Magnetation's furnace is the first commercial application of this low-NOx burner). The off-gas treatment system, which combines a GSA with activated carbon injection, is the first of its kind in the world on an iron ore induration furnace. It is the only plant designed to remove and capture chlorides, fluorides, SOx, and mercury and subsequently remove them from the property. Together, these advancements will make the plant the cleanest pellet plant in the world. Also noteworthy, it is a zero-discharge facility (J. Swanson, personal communication).

United States Steel Corporation

U.S. Steel operates the Minntac and Keetac mines on the Mesabi iron range in northern Minnesota.

Minntac. Minntac is located in Mountain Iron, Minnesota. In addition to the original plant (step 1), it has had two expansions (steps 2 and 3). Minntac processes magnetite ore concentrate pellets in grate-kiln-cooler induration (Figures 6 and 7). Step 1 had three lines, but lines 1 and 2 were decommissioned. Line 3 remains operational. Step 2 has lines 4 and 5, and step 3 has lines 6 and 7. They use Ovation process control software, which was upgraded from Westinghouse software (T. Colerich, personal communication).

Keetac. In addition to Essar Steel Minnesota, another example of a modern flow sheet that is also successfully proven to operate is Keetac (formerly National Steel Pellet Company [NSPC]) (Kawatra 1997). The NSPC facility, located in Keewatin, Minnesota, has produced iron ore pellets since 1966. The flow sheet has been optimized through pilot-scale testing and computer simulation studies. Energy consumption was reduced and grinding capacity increased by replacing the original hydrocyclone classification with fine screens. The combination of cyclones and fine screens is almost universal in magnetite-taconite plants, and some plants are operating circuits in which the cyclone is essentially eliminated and fine screens are used alone (Poveromo 1999).

The density effect in classification is important to understand, in particular during concentrator design. There are examples where a fine screen manufacturer is not involved in projects until after start-up when it is realized that the Fe/SiO_2 grade in the plant with cyclones does not match the Fe/SiO_2 grade achieved in a laboratory with screen classification at the same P_{80}. Cyclone classification generally requires a finer P_{80} to achieve the same Fe/SiO_2 grade as screen classification (J. Wheeler, personal communication). Wennen, Murr, and others describe this as a shift in the "grind–grade" relationship

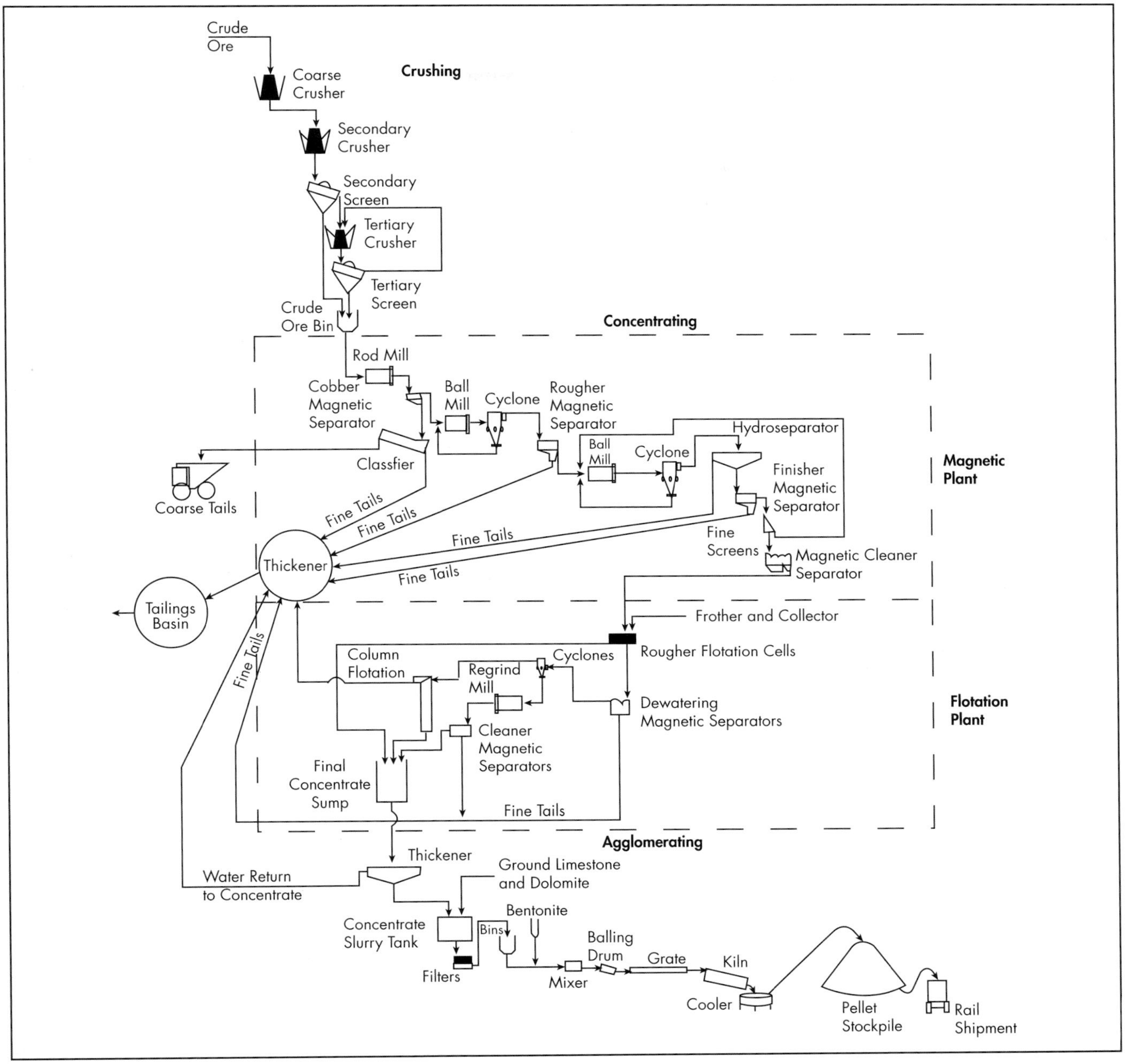

Source: Poveromo 1999

Figure 6 Minntac flow sheet

(J.E. Wennen, personal communication; Murr et al. 2009). Bleifuss (1968) has described the cause in greater detail.

Cleveland-Cliffs

In 2017, Cliffs Natural Resources reverted to its former name of Cleveland-Cliffs Inc. (CCI). CCI is the largest iron ore producer in North America with its U.S. production capacity of 19.6 t/yr (19.3 ltpy) of BF-grade iron ore pellets produced by five mines: Northshore Mining (NSM), United taconite (UTAC), and Hibbing taconite (Hibbtac) in Minnesota and Empire and Tilden in Michigan's Upper Peninsula. In addition to their BF-grade pellets, CCI has developed a DR-grade (also known as EAF-grade) pellet at its NSM operation. Aside from North America, CCI operates the *Koolyanobbing complex*, a collective term for the operating deposits at Koolyanobbing, Windarling, and Mount Jackson, located in Western Australia.

Northshore Mining. NSM produces 5.5 Mt (million metric tons; 5.4 million lt) of iron ore pellets per year. The magnetite ore is mined in Babbitt, Minnesota. Coarse crushed ore is shipped 76 km (47 mi) by rail to the fine crusher, concentrator, and pellet plant located along the shore of Lake Superior in Silver Bay, Minnesota, and was named the E.W. Davis Works in honor of this pioneer of taconite.

The concentrator plant operates 11 sections (lines) to produce the magnetite concentrate for iron ore pellet production (Figure 8). Each section had four Stearns 1 × 3-m (3 × 10-ft)

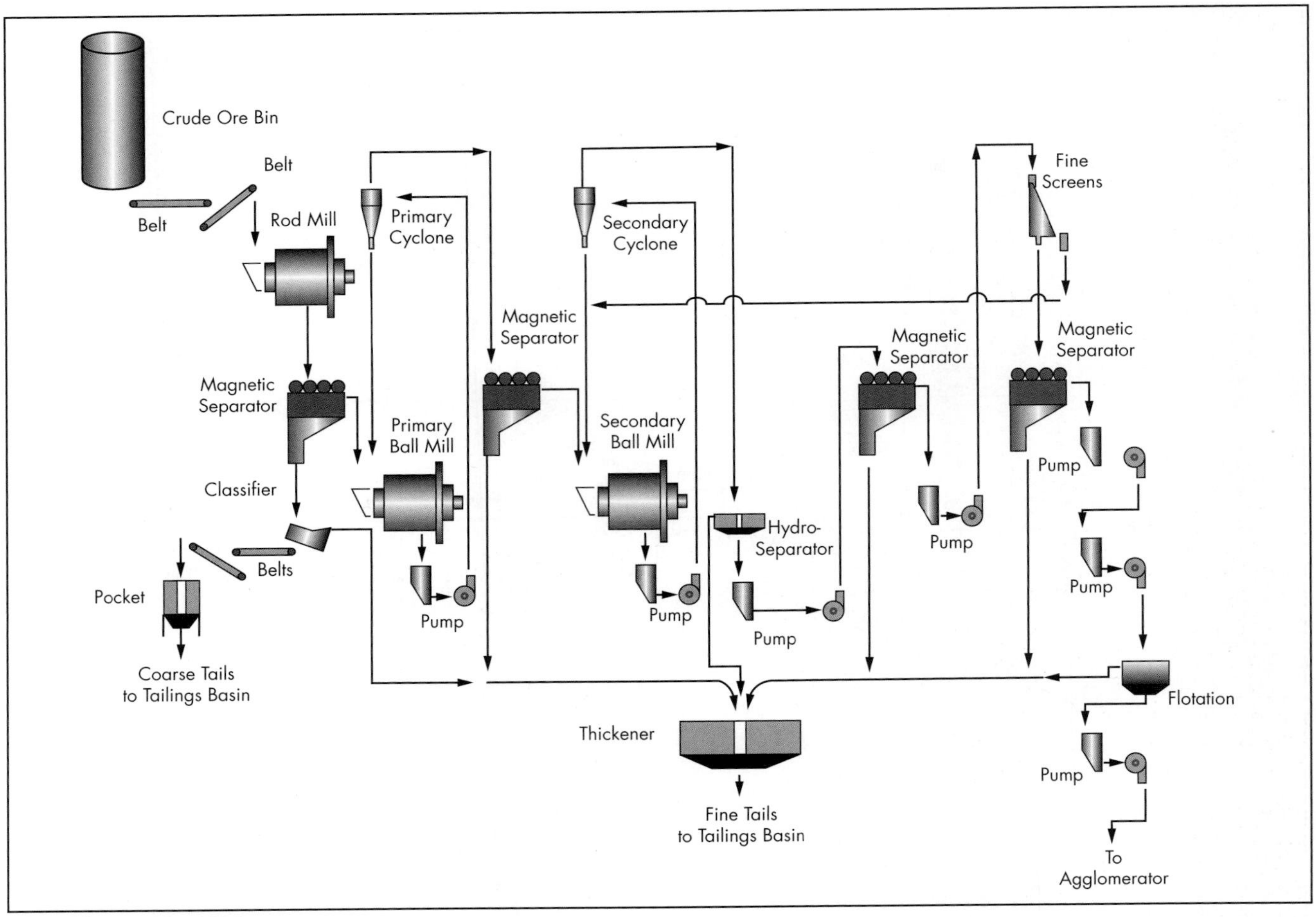

Adapted from Frosaker 2005

Figure 7 Latest Minntac flow sheet

concurrent flow rougher magnetic separators that were installed in 1979. Their typical magnetic iron recovery averaged 99.4%. Because of outdated rougher magnetic separators, 75% of unrecovered magnetite was lost in the tails. Past evaluations of upgrading the rougher magnetic separators were uneconomical, but in 2003, an economic opportunity arose when used machines from Cliffs Erie became available, which maintained concentrate grade and improved throughput and recovery (Ripke and Hoff 2005).

Tilden. The Tilden mine is owned by CCI. The mine, concentrator, and pellet plant are located in Ishpeming, Michigan, adjacent to the Empire mine, pellet plant, and concentrator. In 2014, Tilden produced 7.7 Mt (7.6 million lt) of iron ore pellets per year with its installed capacity of 8.0 t/yr (7.9 ltpy). Although hematite is the primary ore processed at Tilden, it also has the installed capability to process magnetite, although it has not done so since 2011 or 2012.

Magnetite was typically processed between December and March or April. This has an advantage because the lower temperature of the water system during winter negatively impacts hematite recovery; this is not much of an issue with magnetite. In addition, indurating furnace repairs can also be done during magnetite production because the plant becomes heavily concentrator constrained.

It takes about two days to convert the flow sheet from magnetite to hematite and less when going to magnetite because fewer grinding lines are used and the process is simpler. The screen decks on the primary mills are also swapped out: 2 mm (5⁄64 in.) for hematite; smaller, 1.6 or 1.4 mm (1⁄16 in.) for magnetite (anything bigger can plug the tailings line). When magnetite was being processed, Eriez wet magnetic separators were used for both cobbers and finishers.

So that the Cliff Drives III pit close to Tilden could be employed, series desliming was performed prior to the finishers and flotation. Using diamine and frother after cold water interferes with dispersants in desliming. And although the water system could be closed up and recycled again, it would affect water chemistry. However, significant amounts of magnetite would be stockpiled as concentrate to blend with hematite later.

Tilden hematite. The Tilden mine is a unique mine in that it processes a low-grade ore (30%–34% Fe) containing fine, disseminated hematite. Run-of-mine (ROM) ore is crushed to 225 mm (9 in.) and then ground in primary fully autogenous mills along with caustic soda to increase pH throughout the beneficiation plant (McIvor and Weldum 2004). The primary mill discharge is screened to remove –2 mm (–5⁄64 in.) (–8 mesh) particles, which are reground to 80% –74 μm (–200 mesh) in fully autogenous pebble mills. Part of the

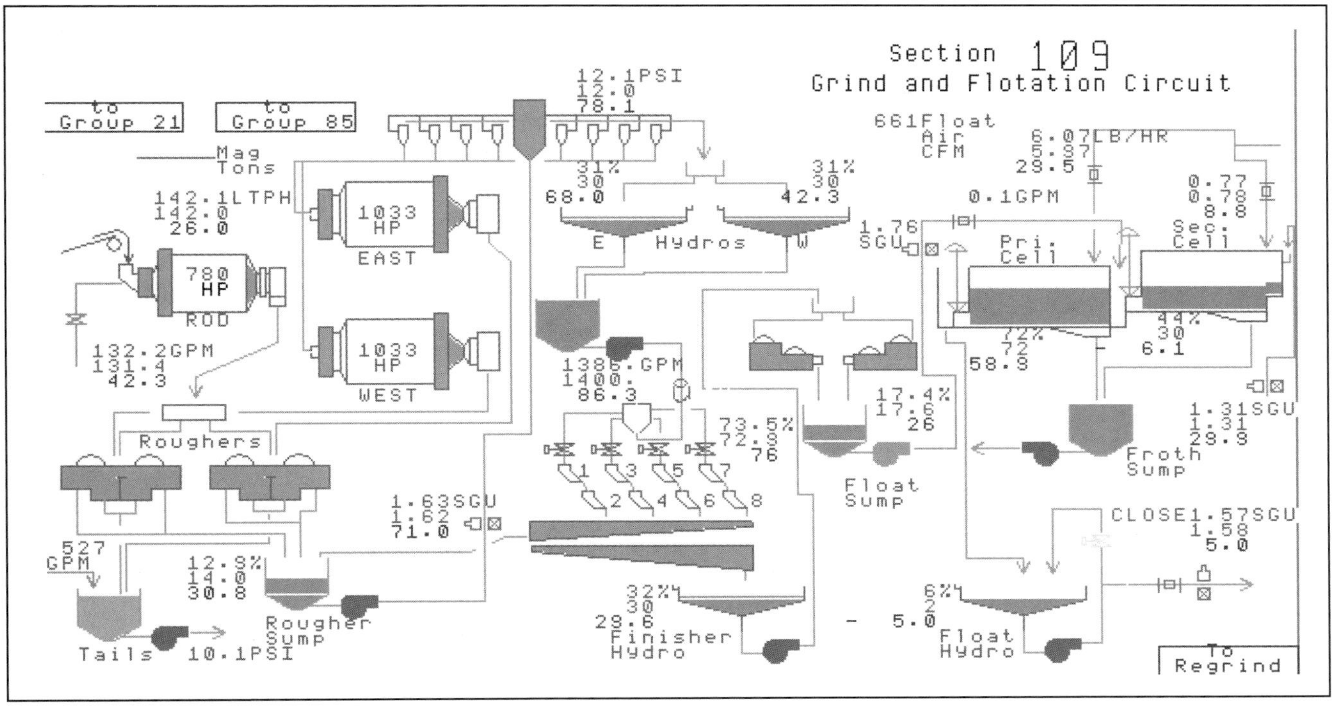

Source: Ripke and Hoff 2005

Figure 8 Display of Northshore mining concentrator flow sheet

primary mill screen oversize is used as grinding media in the pebble mills, and the remainder returns to the primary mill. The 75-mm (3 in.) by 32-mm (1¼ in.) fraction of the screen oversize is crushed using secondary cone crushers before it is returned to the grinding mill to avoid accumulation of this critical size material in the primary mill circuit (Poveromo 1999). The pebble mills are closed with cyclones with an overflow size fraction at 80%–90% passing at 25 µm. This fine-size fraction is required to achieve the product silica set point and poses many challenges in the beneficiation process. A dispersant is added to the pebble mill feed to aid in mill efficiency and in the downstream mineral separation process (Keranen 1986).

A cooked modified cornstarch is added to the cyclone overflow immediately prior to entering the deslime thickeners (Keranen 1986). The cornstarch is added to selectively flocculate the liberated hematite. After almost five decades of successful use, the mechanism of selective adsorption of starch is still debated among researchers in the field (Pradip 1994). The deslime thickeners allow selectively flocculated iron-bearing minerals as well as coarse particulates to settle into the underflow while dispersed siliceous gangue minerals are rejected through the overflow. This process defines the uniqueness of this operation and is required to economically produce high-grade iron oxide pellets from this ore. The deslime thickeners upgrade the ore from ROM to 45%–50% Fe prior to flotation.

The deslime underflow is further concentrated by flotation to at least 65% Fe using a rougher-scavenger circuit (Siirak and Hancock 1988). Starch is again added to this process to depress iron oxide surfaces while ether amine is added as a flotation collector for the reverse flotation of siliceous material not removed by the deslime thickeners (Keranen 1986). The final concentrate is then thickened, mixed with limestone flux, and filtered using disk filters. Filter cake is pelletized using a bentonite binder, balling drums, and a grate-kiln sintering furnace. A process flow diagram is included in Figure 9.

Because of the surface-chemistry-intensive nature of this process, the water chemistry is as important as the reagents used to facilitate beneficiation. The pH must be held between 10.5 and 11 throughout the process, partly because of the dispersive nature of alkaline environments with oxidized ores and partly because of the increased starch selectivity in this pH range. Of particular importance is the water hardness within the process, as free water hardness–contributing ions, particularly calcium and magnesium, within the process water act to nonselectively flocculate iron oxides and the associated gangue minerals during desliming. Water hardness contributing ions also reduce the selectivity of starch to hematite surfaces. Levels of water hardness in excess of 20 ppm can cause a substantial loss in deslime thickener underflow grade associated with an increase in deslime iron recovery (Haselhuhn and Kawatra 2015). This loss in deslime grade must be accounted for during the flotation process, which can handle excursions but results in a significant loss in flotation iron recovery. This is a significant balancing act understood by the Tilden process engineers as an optimal pH, and water hardness must be maintained to maximize overall plant iron recovery.

Hibbing taconite. CCI operates Hibbtac, which is located in Hibbing, Minnesota (Pforr 1974). The plant has a capacity of 8.2 Mt/yr (8.1 million ltpy). The concentrator flow sheet uses autogenous grinding and is shown in Figure 10 (Bymark 1985).

Essar Steel Minnesota. This project went idle in 2015 and then bankrupt. As of 2018, a new owner is vowing to be mining and processing ore by early 2020. The beneficiation flow sheet is shown in Figure 11.

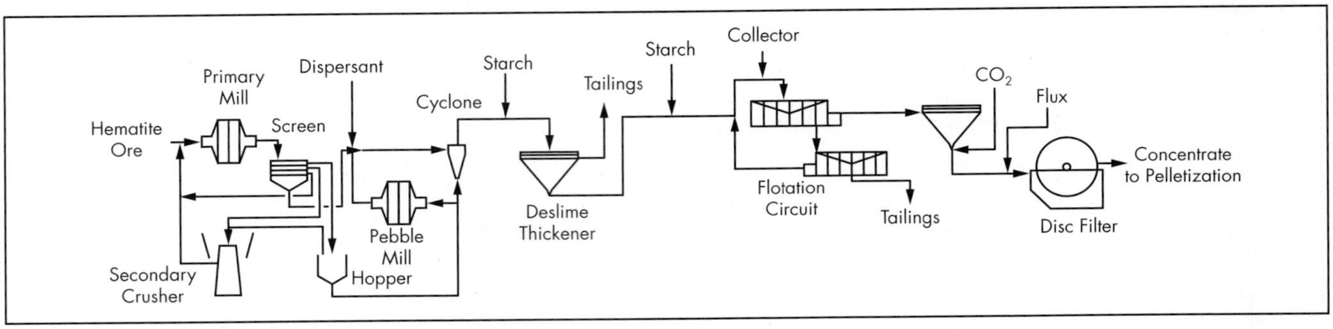

Adapted from Keranen 1986
Figure 9 Tilden flow sheet

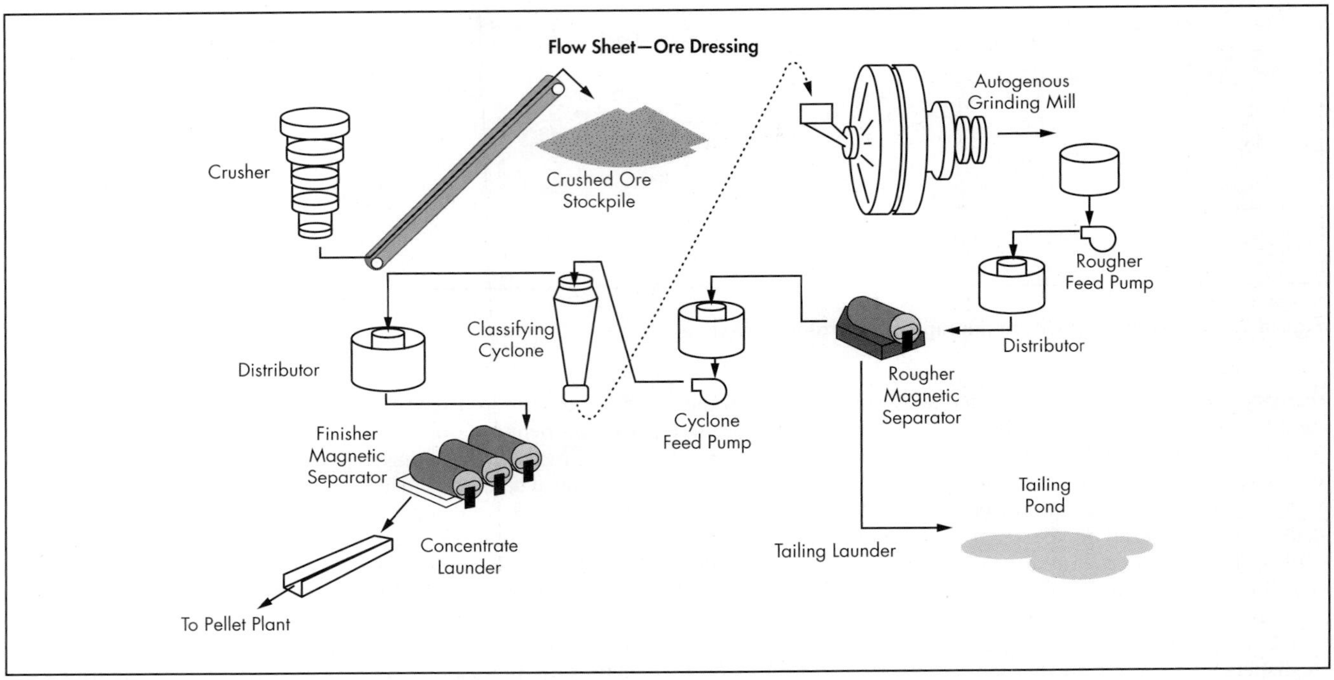

Source: Bymark 1985
Figure 10 Hibbing taconite flow sheet

Canada

In 2018, there were two commercially operating iron ore pellet operations located in Canada (ArcelorMittal Mining Canada and Iron Ore Company of Canada) with one idle (Wabush) and some additional projects under consideration.

ArcelorMittal Mining Canada

ArcelorMittal Mining Canada (AMMC), formerly Quebec Cartier Mining, is located in Mt. Wright, Quebec, Canada. AMMC is an open pit iron ore mine that produces concentrate and pellets. The former and most recent flow sheets are shown in Figures 12 and 13, respectively.

Iron Ore Company of Canada

Iron Ore Company of Canada, located in Labrador City, Newfoundland, conducted a concentrator audit in 1998, which exclusively used thousands of wash water spirals for recovery of their hematite ores. They determined that controlling this quantity of spirals was almost impossible. They sought a more simplified flow sheet and investigated nine variations before selecting and installing their choice (Hearn 2002), which was to use hindered settling separators as the primary separator before the spirals to significantly reduce the number of spirals.

Wabush Mines

Although the Wabush mines closed operation in 2014, the Wabush mines concentrator produced a relatively coarse (–1 mm) specular hematite concentrate using a unique revised flow sheet (Figure 14). The *Mining Engineering* (1970) article, "Wabush: A $300-Million Iron Operation," shows the original flow sheet with autogenous grinding, primary and secondary Humphreys spirals, drum filters, fluo-solids dryers, and high-tension electrostatic separation. In 2000, the concentrator flow sheet was modified. The primary rougher spirals were replaced with Reichert spirals, and the secondary cleaner spirals were replaced with Hydrosizers (Ripke 2007).

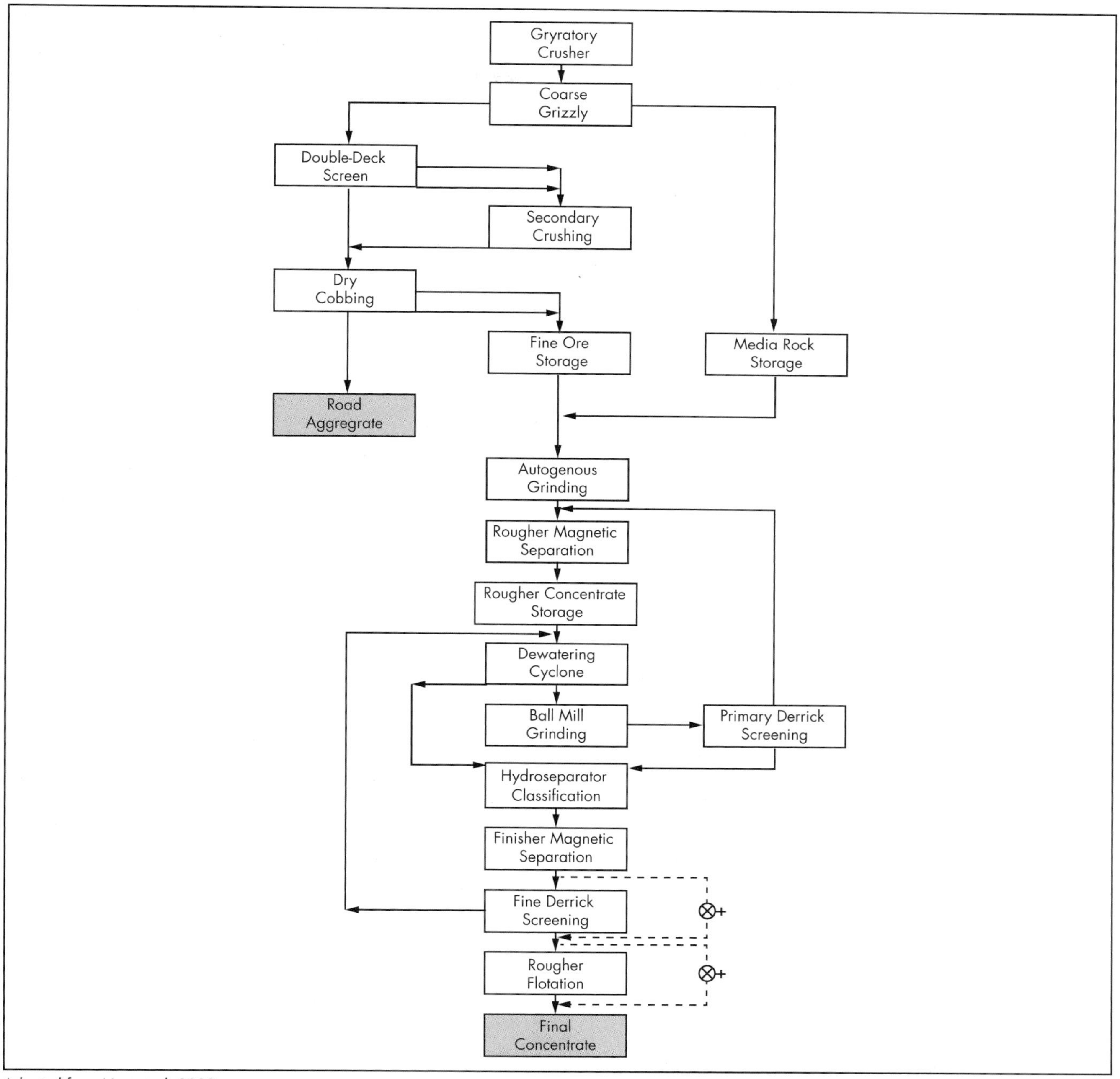

Adapted from Murr et al. 2009

Figure 11 Essar Steel Minnesota concentrator flow sheet

Mexico

Lump ore, fines, and concentrate are produced from a variety of operations in Mexico by ArcelorMittal including at Lazaro Cardenas Volcan Mines, Pena Colorada, and Las Truchas.

Brazil

Brazil is a major producer of iron ore and pellets by Vale and Samarco (a joint venture owned by Vale and BHP Billiton). Samarco is currently idled after having a tailings disaster in 2015.

Vale

Vale uses USIM PAC mineral-processing modeling software. the company's research and design strategy is to conduct in-house pilot plant tests. Vale uses ASEA Brown Boveri (ABB), Siemens, and others for concentrator process control. Vale's various pellet feeds and sinter fines are described in Table 2; each have a 10% filter cake moisture.

Current flotation practices in Brazil are explained by Silva et al. (2011) as follows: In the past, iron ore producers would only use either (1) mechanical (i.e., tank) or (2) column

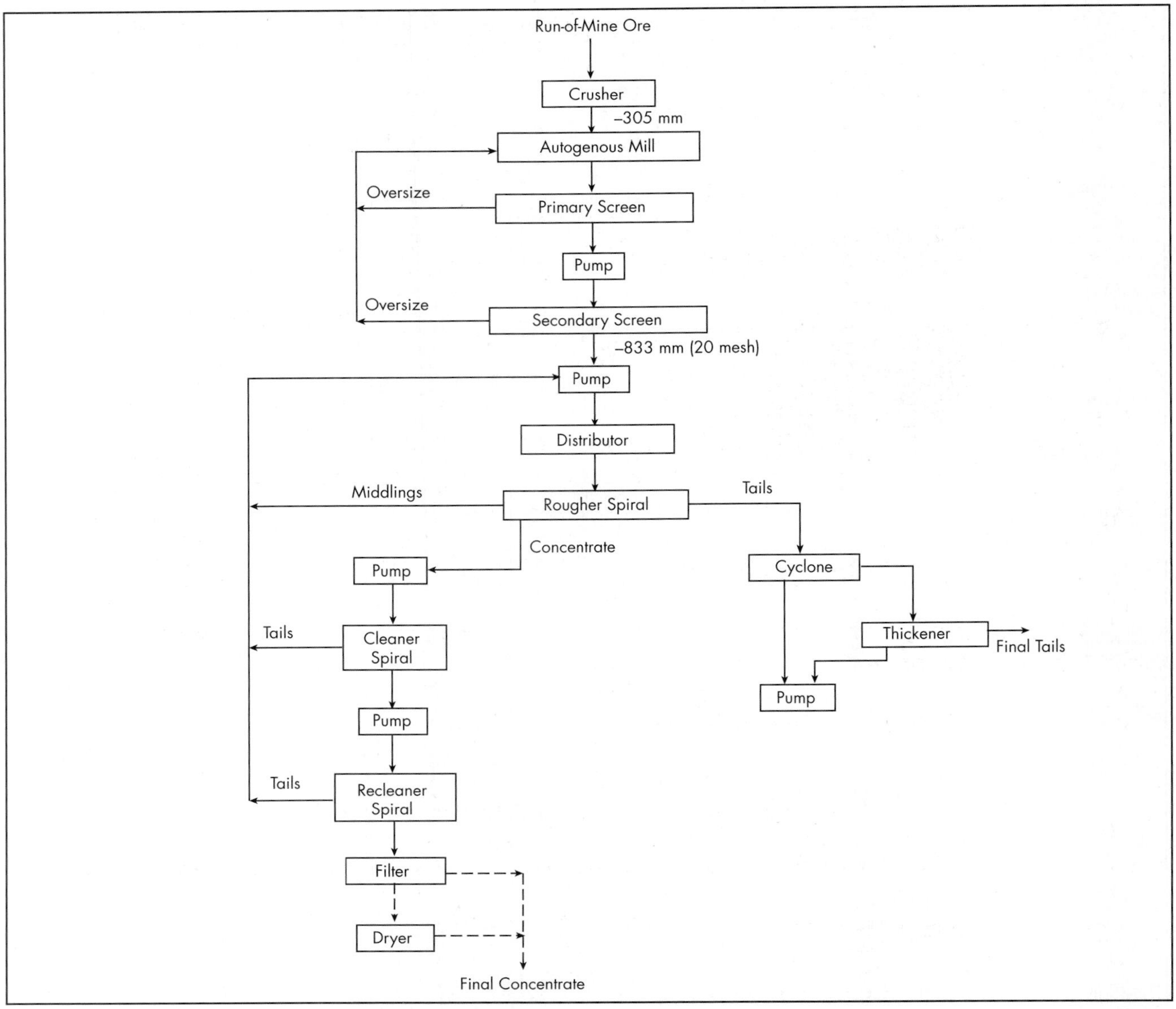

Source: Poveromo 1999

Figure 12 Past AMMC flow sheet

cells, but not combine them. The latest trend has been to combine tank and column cells. Some producers place mechanical cells first in the configuration (which is preferred), and others place their cells column cells first (and have not reconfigured them, but probably should) (Araujo et al. 2003; Silva et al. 2011). A potential alternative to tank cells or columns may be the staged flotation reactor (Swedburg et al. 2016).

Minas-Rio

Minas-Rio produces 26.5 Mt/yr (26.1 million ltpy) of pellet feed, by the flow sheet shown in Figure 15, for the BF or DR shaft furnace. It beneficiates two ore types: an itabirite and a friable ore that generates a high percentage Fe product with low contaminants (Anglo American 2015). The concentrate takes about four days to travel by pipeline as a 68% solids slurry over 529 km (329 mi). Additional information is provided by Mazzinghy et al. (2015).

Samarco Mineração

Samarco Mineração designed a specific SAG mill for the 15 different itabirites found in its projects (Rosa and Rocha 2015):

> Ore characterization is a major factor for a proper concept of grinding circuits and for equipment sizing, but there is still not a consensus if the traditional laboratory tests applied to characterize grinding behavior of ores, such as [drop weight test] DWT, SMC [Test], and Bond [work index], are the best applicable methodology for understanding the grinding performance of all ores. In this sense, the characterization of friable itabirites, a metamorphic banded iron formation with iron oxides layered with quartz, is still a challenge once their comminution behavior is incorrectly predicted by these traditional

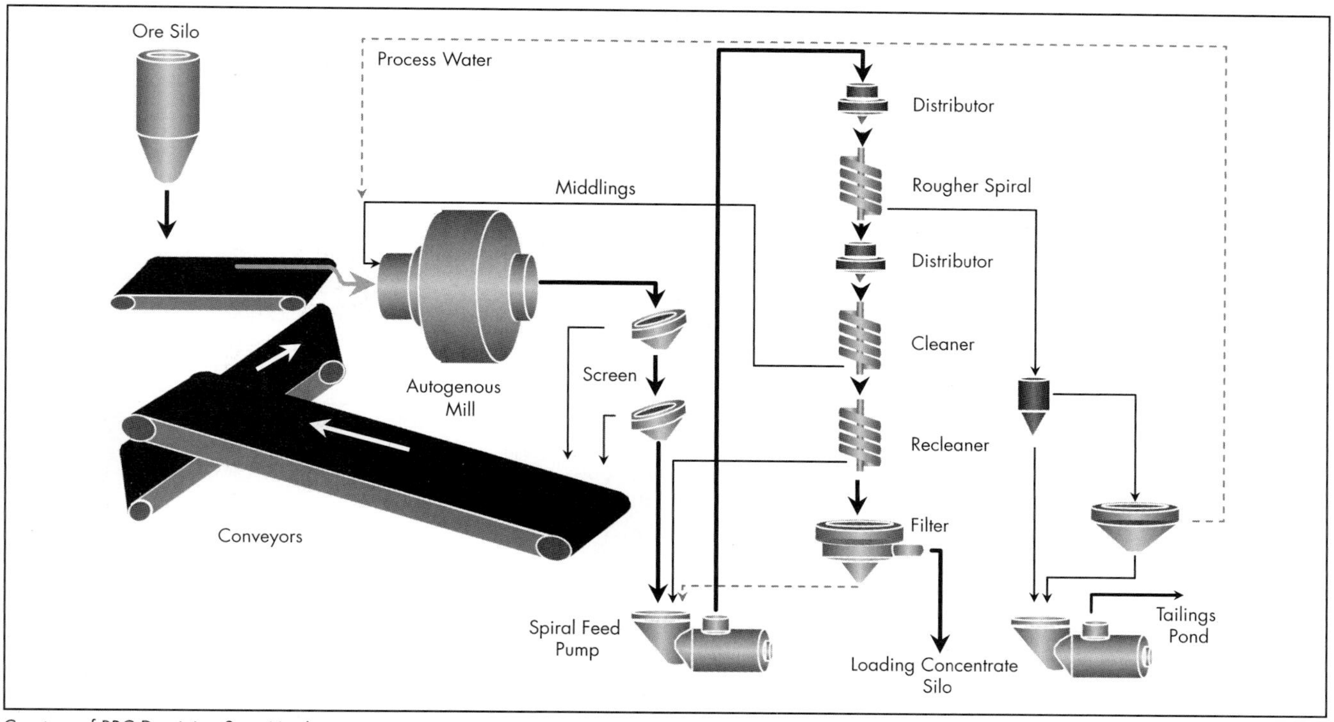

Courtesy of RBC Dominion Securities Inc.

Figure 13 Recent AMMC flow sheet

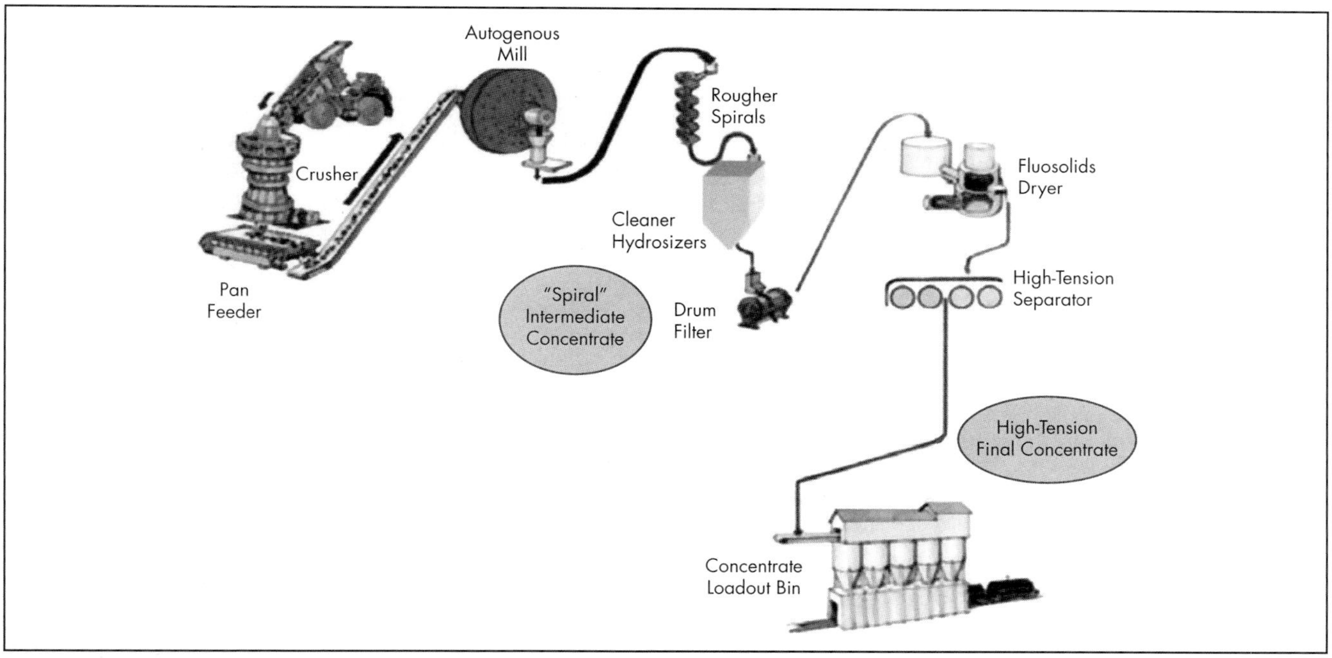

Source: Ripke 2007

Figure 14 Wabush Mines concentrator flow sheet

methodologies. Samarco Mineração S.A conducted detailed studies on the mineralogical characterization of the itabirites reserves to support the geometallurgical models of its deposits. These studies show that for friable itabirites the ore size distribution and ore hardness are not necessarily correlated. More hydrated itabirites showed lower grinding kinetics, regardless of particle size. …

Samarco [semiautogenous] milling studies raised doubts about the grinding classification of

Table 2 Vale iron ore process*

Process	Natural Moisture Beneficiation Process	Utilization of Flotation Process for Fines of Low-Grade Itabirite	Beneficiation Process of Hard and Low-Fe-Grade Itabirites	Application of High-Intensity Magnetic Separators as Scavengers on the Reverse Flotation of Quartz Aiming Reducing %Fe on the Tailings, <8% Fe	Application of High-Intensity Magnetic Separators for the Concentration of Coarse Particles, –1 +0, 15 mm (10/16 in.)
Years of operation	From 2008	From 1974	From 1974	From 2014	From 1974
Product type†	Natural fines, –16 mm (10/16 in.)	Pellet feed	Pellet feed	Pellet feed	Sinter feed
Ore type	Hematite	Low-grade itabirites	Low-grade itabirites	Low-grade itabirites	Low-grade itabirites
Tons, Mt/yr (million ltpy)	90 (88.6)	100 (98.4)	50 (49.2)	5 (4.9)	10 (9.8)
Name	Carajás mine	Iron Quadrangle	Iron Quadrangle	Iron Quadrangle	Iron Quadrangle
Crushing and screening steps	1° jaw crusher, 2° and 3° conic crushers; 1°, 2°, and 3° screening	NA‡	1° jaw or gyratory crusher, 2°, 3°, and 4° conic crushers; 1°, 2°, and 3° screening	NA	NA
Grinding	NA	NA	Ball mill in closed circuit with cyclones	NA	NA
Beneficiation steps	NA	Mechanical and column flotation cells on the reverse flotation of quartz with amine and cornstarch	Mechanical flotation rougher, cleaner, scavenger, recleaner, reverse flotation of silica with amine; vacuum disk filtration	Mechanical flotation rougher, cleaner, scavenger, recleaner, reverse flotation of silica with amine; vacuum disk filtration	One step of magnetic separation
Weight recovery, %	100	50–70	50–60	30–40	55–60
Fe recovery, %	100	70–80	75–85	70–80	60–65
Lines per plant	2	10	4	1	3
Total grade, %Fe	65	>65	>65	>65	>62
Contaminants, %	2.5 SiO_2, 1.0 Al_2O_3, 0.100 P	1.5 SiO_2, 0.5 Al_2O_3, 0.060 P	1.5 SiO_2, 0.5 Al_2O_3, 0.060 P	1.5 SiO_2, 0.5 Al_2O_3, 0.060 P	4.5 SiO_2, 0.5 Al_2O_3, 0.060 P

Courtesy of Vale
* Information obtained by the chapter authors' survey.
† The pellet feeds and sinter fines each have a 10% filter cake moisture.
‡ NA = not applicable.

ore hardness that is commonly used for itabirites. Based on our observations from pilot plant tests, it was clear that ore hardness is not always correlated to particle size distribution.

These observations led us to further investigations that correlated different ore typologies to grinding kinetics. For Samarco's ores with similar iron content, the more hydrated the ore, the higher the energy consumption in the grinding stage. Additionally, it was possible to include a grinding parameter in our geometallurgical model, which made it possible to map the ore hardness of our reserves.

This grinding parameter is also used for process control in our current concentrator, for ball mill circuits. Based on our observations from pilot plants we are confident that our internal methodology can be adjusted for [semiautogenous] milling as well.

Australia

The three major iron ore producers in Australia are Rio Tinto, BHP Billiton, and Fortescue Metals Group. However, because they produce DSO, they use little or no beneficiation, so the following additional producers are highlighted.

Arrium Mining

In 2017, Arrium was acquired by GFG Alliance and renamed SIMEC Mining. The concentrator, outside Whyalla, South Australia, installed and commissioned a 3-MW (4,023-hp) M10000 IsaMill tertiary stirred mill in late 2013. Prior to the project proceeding, extensive test work was conducted at ALS Metallurgy (formerly AMMTEC) in Perth, Australia, to compare the IsaMill option with the existing ball mill and determine the correct IsaMill layout within the existing plant. Both continuous testing and signature plot semibatch testing were completed under the supervision of Arrium personnel. Multiple signature plots were planned on different ores from around the property to investigate grinding energy variability. On start-up of the IsaMill, two months of on-site optimization took place to bring the mill to full tonnage while reducing the internal component wear rates. During this time, process guarantee sampling took place validating the previous work done at ALS Metallurgy. Once the mill configuration was set, further sampling took place as a final confirmation of the 1:1 M4 laboratory scale-up and choice of grinding media size. After liberation data showed that a coarser grind from the IsaMill than the design still resulted in an acceptable concentrate grade, work was undertaken to increase the throughput of

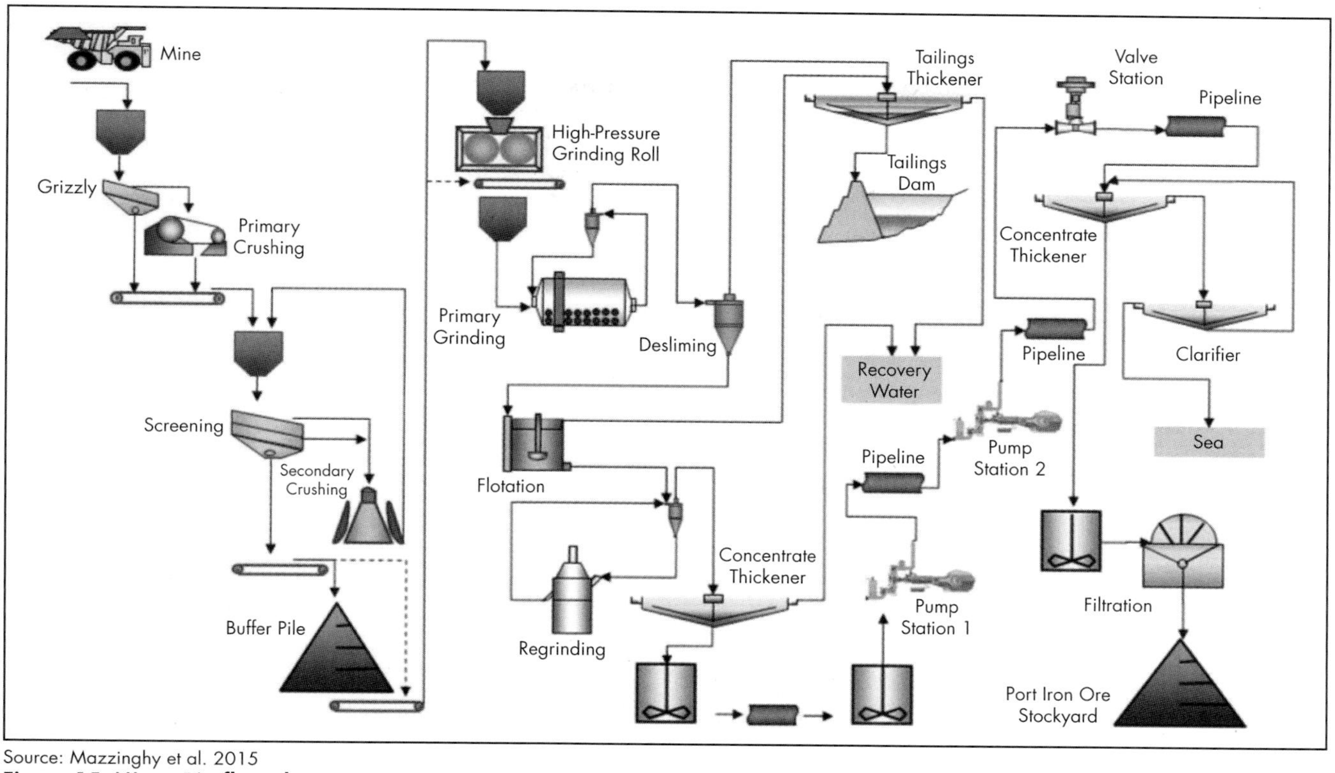

Source: Mazzinghy et al. 2015
Figure 15 Minas-Rio flow sheet

the mill past the original maximum design. The initial Arrium hematite flow sheet is shown in Figure 16.

This project has seen the Arrium magnetite operation undergo an overhaul, allowing it to produce magnetite concentrate using a lower-grade feed material at an increase in production capacity of more than 400,000 t/yr (393,682 ltpy) (Arrium Mining and Materials 2016). This throughput has continued to be expanded as the maximum capacity of the IsaMill has been pushed well beyond the original design (Figure 17). After a challenging start, handover of the plant was completed on budget and ahead of schedule. Through methodical implementation procedures, step-by-step improvements in throughput and mill wear have been realized, allowing the plant to process tonnage well beyond anything for the original design (Larson et al. 2015).

Fortescue Metals Group

Fortescue Metals Group (FMG) has become the world's fourth largest seaborne iron ore producer since being founded in 2003 and shipping its first ore to China in 2008. It currently has a capacity to produce 165 Mt/yr (162 million ltpy) of iron ore Rocket fines (59% Fe, for sinter feed) and lump from the Pilbara region of Western Australia for BF feed. Their Cloudbreak and Christmas Creek mines feed their Chichester hub with 90 Mt/yr (88 million ltpy). Their Firetail and Kings mines feed their Solomon hub with 70 Mt/yr (69 million ltpy). The >58% Fe ore is mined (in some cases with autonomous production trucks, as noted in *Mining Engineering* [2013]), screened, crushed, and desanded. "Low-phosphorous Kings ore is a stand-alone product, while higher-grade Firetail ore is blended with low-impurity Chichester ore, allowing Fortescue to maximize the benefits of both ore bodies and reduce cutoff grades" (FMG 2015). Iron ore is shipped from the mines to the hubs to the Herb Elliott port in Port Hedland by railway that consists of 620 km (385 mi) of the fastest and heaviest haul line in the world.

FMG ore processing. Clout and Rowley (2010) provide an excellent description of FMG's ore processing facilities:

> The Cloudbreak ore processing facility consists of … screening, crushing, and a beneficiation facility referred to as the "desand" … producing either fines-only or lump and fines. The ore processing facility has a nominal nameplate capacity of 45 Mt/yr [44 million ltpy] of wet product.

FMG product. Clout and Rowley (2010) provide the following description of FMG's product:

> The main Fortescue product exported to date is Rocket fines, which has 64.6 percent calcined Fe content and low contaminant levels, including low phosphorous and alumina and high loss on ignition. The Rocket fines calcined Fe is mid-way between the high-grade fines products and the premium Brockman fines products from the Pilbara, with less than about 1 percent calcined Fe separating them. Shipments of 24.6 WMt [24.2 million wlt] of Rocket fines for the first 12 months of operation were very close to the target product grade.

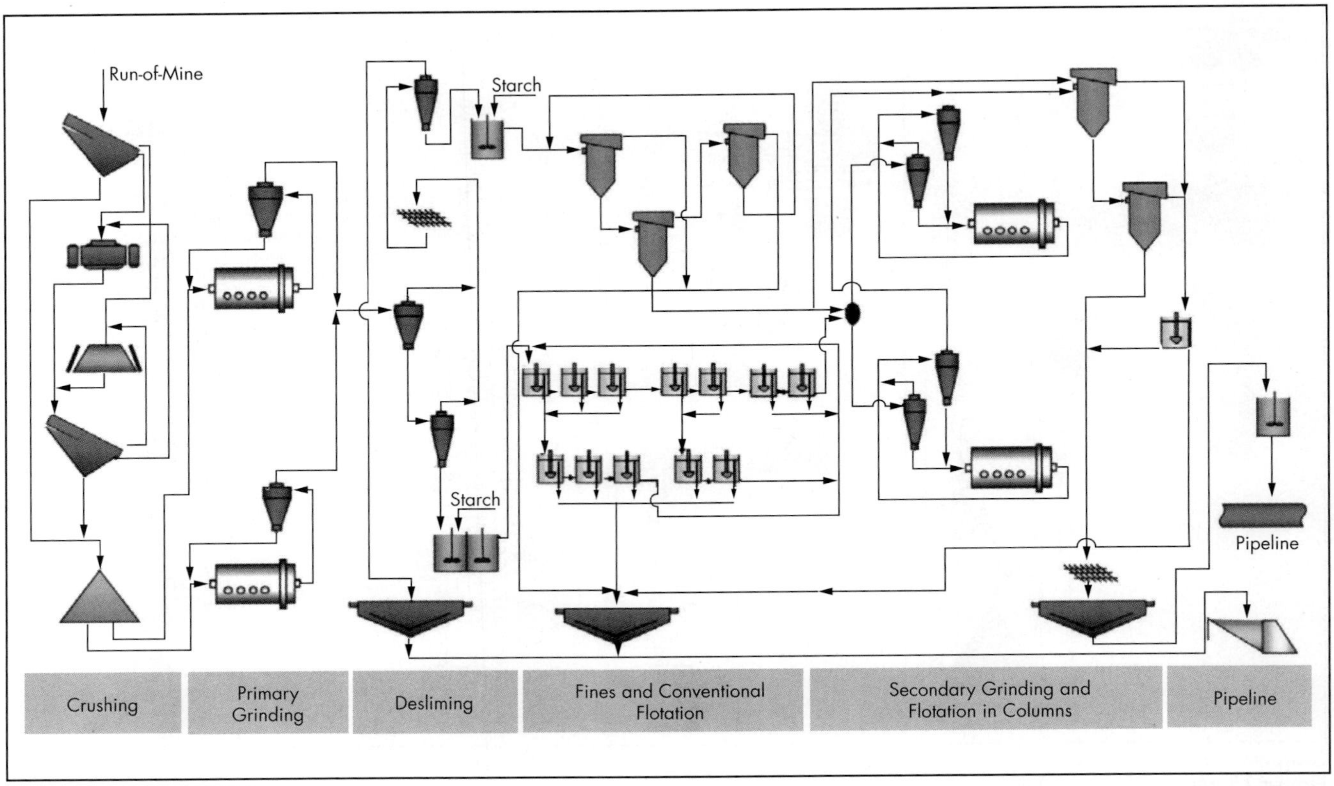

Source: Rosa and Rocha 2015
Figure 16 Arrium flow sheet

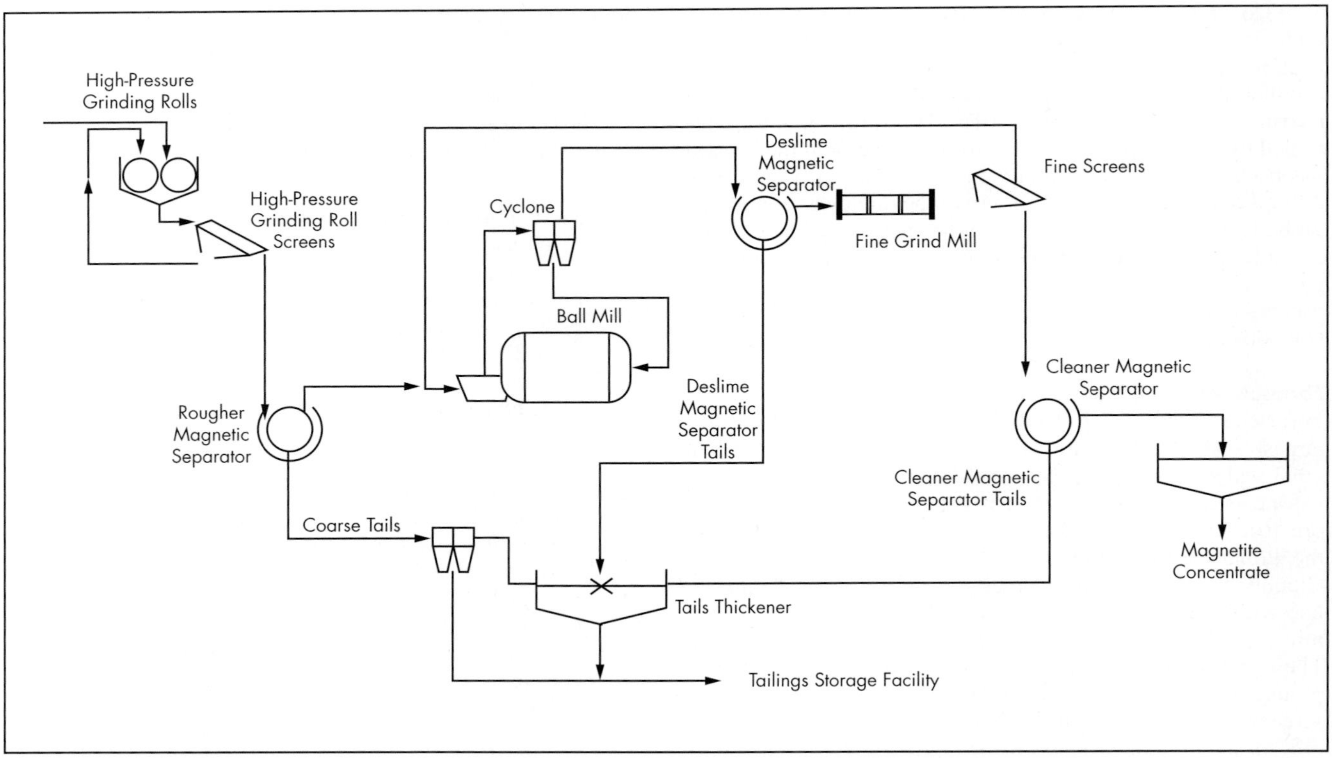

Source: Larson et al. 2015
Figure 17 Arrium modified flow sheet

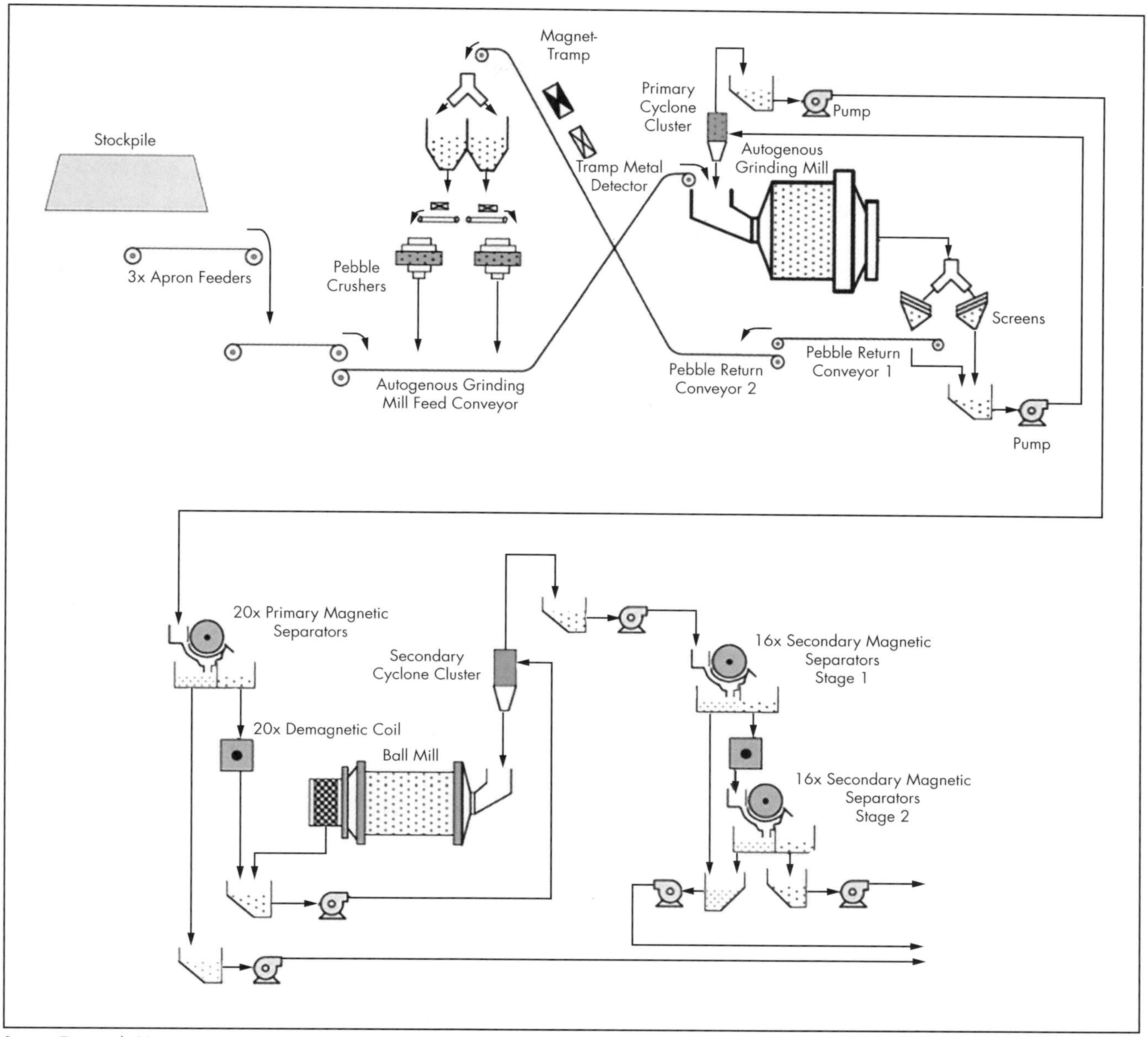

Source: Tian et al. 2015

Figure 18 Sino Iron magnetite flow sheet

Sino Iron

According to Tian et al. (2015) and J. Tian (personal communication; see Figure 18):

> Since 2015, CITIC Pacific Mining has been operating six lines of milling circuits in its Sino Iron plant, each including a 40' × 33' @ 28 MW autogenous grinding mill, the largest in operation in the world. …
>
> [Autogenous] milling is proven a more suitable than [semiautogenous] milling for magnetite processing for the Sino Iron project due to its ability to produce fine product without using grind media. This allows the plant to optimize the grind power balance between the [autogenous] milling and the following regrinding circuit in order to achieve the maximum system milling capacity, by setting the optimal grind size for the [autogenous] milling circuit.

New Zealand

New Zealand has abundant sources of iron ore in its ironsands.

New Zealand Steel

New Zealand Steel operates ironsand mining and beneficiating at its Waikato North Head deposit in Glenbrook and its Taharoa deposit in New Zealand.

Taharoa. The native ironsand dunes are mined by a dredge located within a constructed pond. The dredge houses a concentrator plant with gravity separation by spirals and cones followed by magnetic separation (Buist et al. 1973). The 57% Fe titanomagnetite concentrate, which contains about 7.8% TiO_2, is stored in a surge pile until shipping, using a specially designed ship for loading and transporting iron ore concentrate slurry. Starting in 1972, the concentrate has been reslurried and pumped to the ship through an offshore pipeline to a single mooring buoy. The system was expanded in 1978 (Cooper and Gadd 1980). In 2001, the mine's 250-t (246-lt) dredge, 500-t (492-lt) surge bin, 1,000-t (984-lt) concentrator plant, and ancillary equipment was relocated by road on trailers from the Southern Mining Region across a specially designed haul road to the Central Mining Region (Van Deventer and Rowe 2001).

Waikato North Head. Ironsand with 34% magnetics is processed at 600–2,000 t/h (590–1,968 ltph) by double-drum magnetic separation, scrubbing (to liberate clay slimes), dewatering by cyclones, and vibratory screening to reject coarse unliberated material before gravity separation by cones and spirals (Stevens and Jokanovic 2002) into a 58.8% Fe concentrate.

This concentrate is then reduced with coal by multi-hearth furnaces followed by rotary kilns and then smelted in a submerged arc furnace to produce pig iron for steelmaking. Vanadium is also recovered during the pyrometallurgical process (New Zealand Steel 2016).

Sweden

Luossavaara-Kiirunavaara AB (LKAB) has been mining in Sweden for more than 120 years (Figure 19). Most of LKAB's ore deposits must be extracted from underground mines, which are several hundreds of meters below the surface, by sublevel caving.

Luossavaara-Kiirunavaara AB

LKAB operates the world's largest and most modern underground iron ore (magnetite) mines at Kiruna, Malmberget, and Svappavaara (LKAB 2013):

> **Kiruna.** The ore body in Kiruna is about 4 km [2.5 mi] long and has a depth of 2 km [1.2 mi]. To date, more than 1 billion metric tons have been mined. A new main haulage level at 1,365 m [4,478 ft] came into operation in 2013 and will secure mining operations for another 20–30 years.
>
> **Malmberget.** The Malmberget mine consists of about 20 ore bodies. The deposits are pure magnetite ore. In the Malmberget mine, a new main haulage level at 1,250 m [4,101 ft] was inaugurated in 2012, which will considerably extend the lifetime of the mine.
>
> **Gruvberget.** The newly opened Gruvberget mine in the Svappavaara ore field is operated as an open pit mine. Production from this mine, together with the planned Leväniemi and Mertainen mines close to Svappavaara, will produce approximately 25% of the company's total output of iron and ore in 2015.

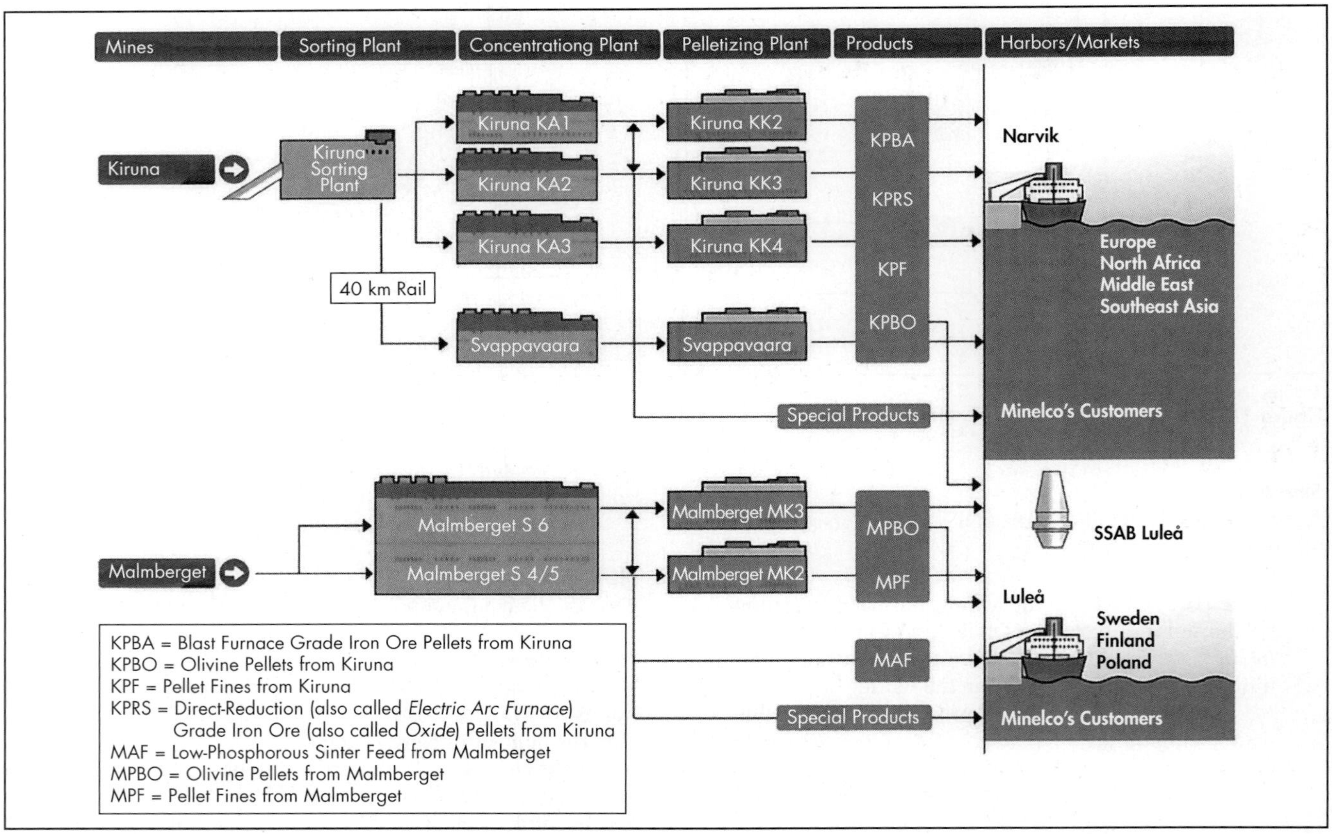

Source: Lindroos et al. 2011

Figure 19 LKAB production facilities

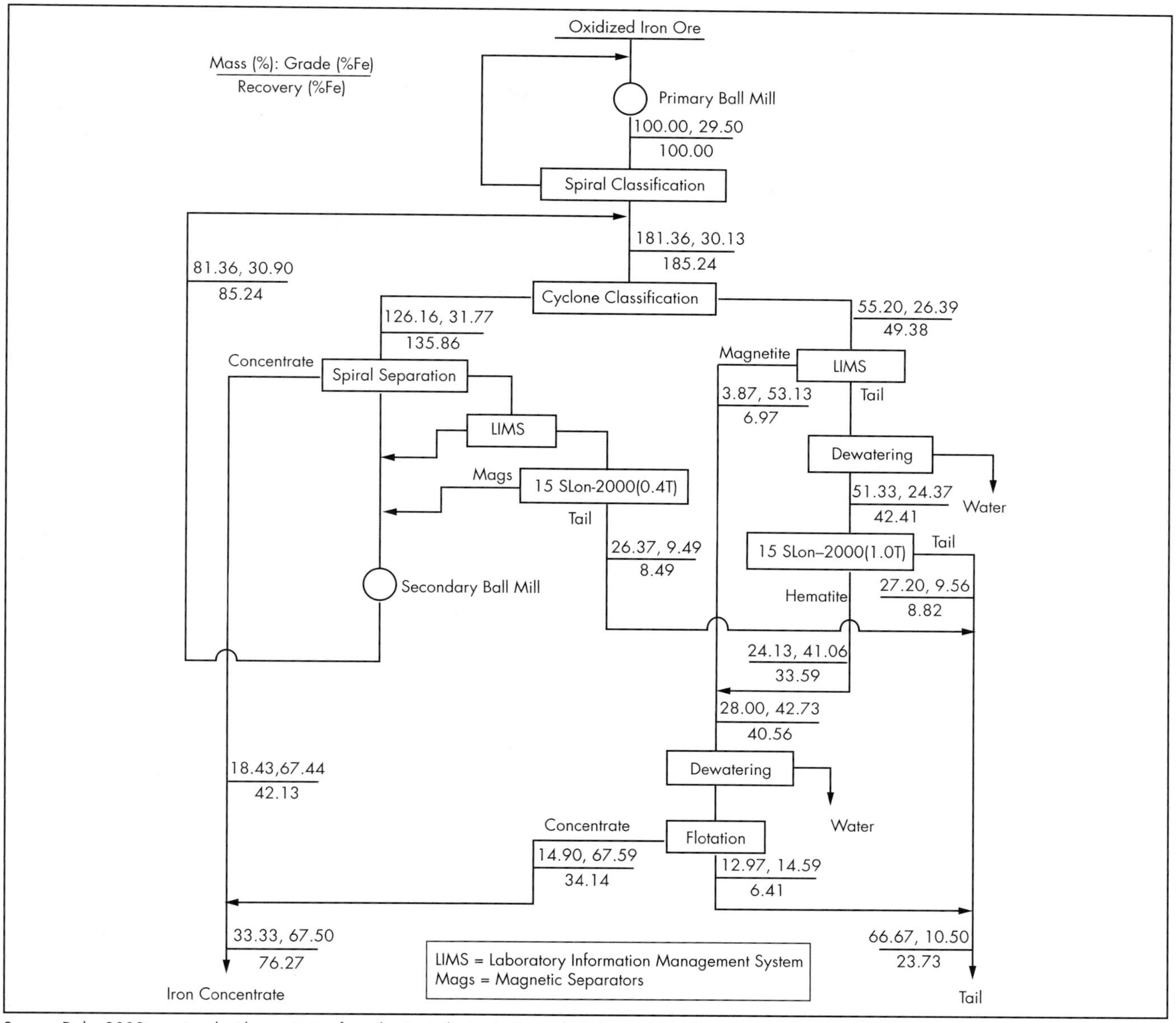

Source: Dahe 2009, reprinted with permission from the Australasian Institute of Mining and Metallurgy

Figure 20 Latest flow sheet for the Diao Jun Tai iron ore processing plant

Beneficiation includes either fully autogenous or semiautogenous grinding, wet magnetic separation, and froth flotation to reject phosphorus. The concentrates are then pelletized by balling and induration in a grate-kiln to produce BF- and EAF-grade products.

Iran

Chadormalu and Gol Gohar iron ore are the two largest iron ore mines (accounting for more than 80% of iron ore production in Iran) (Turquoise Partners 2016). Sangan iron ore mines produce 2.6 Mt/yr (2.5 million ltpy) iron ore concentrate as pellet feed from magnetite ore with 45%–55% Fe by crushing, grinding, low-intensity wet magnetic drum separation (1.3 T [1,300 gauss]), and sulfur froth reverse flotation. Grinding is done in three stages: primary autogenous (up to 700 t/h [689 ltph]), secondary ball milling to 80% –70 µm, and tertiary tower milling (5 mills) to the product size of 80% –38 µm press filtered to 8.5%–9.5% moisture (Middle East Steel 2016).

China

Hematite is the main source of iron ore in China and is difficult to beneficiate (Zou 2007). Low-grade hematite ores containing 18%–26% Fe are typically processed by gravity concentration, low-intensity magnetic separation, high-intensity magnetic separation, and reverse froth flotation (Wei et al. 2011; Liu et al. 2014). Zou (2007) and Dahe (2009) provide details for two Chinese iron ore mineral processing plants and credit two factors for their improved beneficiation: the SLon high-intensity magnetic separator (Outotec 2013) and anionic collectors for reverse silica froth flotation (Figure 20).

ACKNOWLEDGMENTS

The authors thank Jobe Wheeler, Steve Valine, Jack Swanson, Yousef Motamedhashemi, Komar Kawatra, Tony Colerich, Randy Stroop, Dean Connor, Stephen Qian, Christopher Tuck, and Ronney Silva for their contributions.

REFERENCES

Anglo American. 2015. *Minas-Rio International Media Visit—2015*. London, UK: Mining Company. www.angloamerican.com/~/media/Files/A/Anglo-American-PLC-V2/documents/minas-rio-press-conference-presentation-final.pdf.

Araujo, A.C, Amarante, S.C., Souza, C.C., and Silva, R. 2003. Ore mineralogy and its relevance for selection of concentration methods in processing of Brazilian iron ores. *Trans. Inst. Min. Metall., Sect. C.* 112(1):54-64.

Arrium Mining and Materials. 2016. GFG alliance has completed the acquisition of the Arrium mining and steel businesses. www.arrium.com/about%20us/latest%20news/fy14/major%20project%20complete%20%2024%20feb%202014.

Bell, T. 2017. A short history of steel: From the iron era to the Bessemer process and modern steelmaking. The Balance. www.thebalance.com/a-short-history-of-steel-part-ii-2340103. Accessed June 2018.

Bleifuss, R.L. 1968. The mineralogy of taconite products as related to the augmentation of magnetite middlings. In *29th Annual Mining Symposium*. Minneapolis, MN: Mines Experiment Station, University of Minnesota. pp. 131–138.

Buist, D.R., Cooper, R.H., and Terrill, I.J. 1973. Recovery of magnetite at New Zealand steel mining operations Taharoa, New Zealand. Preprint No. 73-13-364. New York: SME-AIME.

Bymark, J.V. 1985. Autogenous grinding at Hibbing Taconite Company. SME Preprint No. 85-373. Littleton, CO: SME.

Choi, C.Q. 2017. Earth's sun: Facts about the sun's age, size, and history. New York: Space.com. www.space.com/58-the-sun-formation-facts-and-characteristics.html.

Chown, M. 2001. *The Magic Furnace: The Search for the Origins of Atoms*. New York: Oxford University Press.

Clout, J.M.F., and Rowley, W.G. 2010. The Fortescue story—from exploration to the third largest iron ore producer in Australia. Section B, Applied Earth Science. *Trans. Inst. Min. Metall.*, Sect. B 119(3 Pt 2):122–131.

Cooper, R.H., and Gadd, D.E. 1980. Slurry loading of ironsand concentrate at Taharoa. In *Annual Conference, New Zealand*. Melbourne, Victoria: Australasian Institute of Mining and Metallurgy. pp. 297–306.

Dahe, X. 2009. Application of SLon magnetic separators in modernising the An Shan oxidised iron ore processing industry. In *Iron Ore 2009 Proceedings*. Melbourne, Victoria: Australasian Institute of Mining and Metallurgy. pp. 223–229.

Davis, E.W. 1964. *Pioneering with Taconite*. St. Paul, MN: Minnesota Historical Society Press.

FMG (Fortescue Metals Group). 2015. *Working Together. Delivering Results*. Annual report. East Perth, Western Australia: FMG.

Frosaker, D.M. 2005. *Minntac Pre-Classification: Hydrocyclone Classification of Flotation Fee*. Prepared for the Minnesota Department of Natural Resources Taconite Grant Technical Committee, USS—Minnesota Ore Operations, September 30. http://files.dnr.state.mn.us/lands_minerals/iocr_714.pdf. Accessed June 2018.

Goldring, D.C. 2003. Iron ore categorisation for the iron and steel industry. Section B, Applied Earth Science. *Trans. Inst. Min. Metall.* 112(1):B5–B17.

Gribbin, J. 2000. *Stardust*. New Haven, CT: Yale University Press.

Haselhuhn, H.J., and Kawatra, S.K. 2015. Role of water chemistry in the selective flocculation and dispersion of iron ore. *Miner. Metall. Process.* 32(2):69–77.

Hazen, R.M. 2012. *The Story of Earth: The First 4.5 Billion Years, from Stardust to Living Planet*. New York: Viking Penguin.

Hearn, S. 2002. The use of hindered settlers to improve iron ore gravity concentration circuits. In *Mineral Processing Plant Design, Practice, and Control*. Edited by A.L. Mular, D.N. Halbe, and D.J. Barratt. Littleton, CO: SME. pp. 939–943.

Investment Mine. 2016. Commodity and metal prices: Important industrial commodities. Vancouver, BC: Infomine. www.infomine.com/investment/metal-prices/.

Kawatra, S.K., ed. 1997. *Comminution Practices*. Littleton, CO: SME.

Keranen, C.U. 1986. Reagent preparation, distribution, and feeding systems at the Tilden mine. In *Design and Installation of Concentration and Dewatering Circuits*. Edited by A.L. Mular and M.A. Anderson. Littleton, CO: SME. pp. 308–319.

Kuck, P. 2013. *Iron Ore Statistical Compendium*. Reston, VA: U.S. Geological Survey.

Larson, M.C., Anderson, G.S., Mativenga, M.M., and Stanton, C.R. 2015. The Arrium mining isamill from inception through continuing optimization. In *Proceedings of the Sixth International Conference on Semi-Autogenous and High Pressure Grinding Technology*. Vancouver, BC: SAG Conference Foundation. pp. 1–15.

Lindroos, F., Hallberg, D., and Dahlstedt, A. 2011. LKAB Green Pellets. Westmount, QC: Canadian Institute of Mining, Metallurgy and Petroleum.

Liu, S., Zhao, Y., Wang, W., and Wen, S. 2014. Beneficiation of a low-grade, hematite-magnetite ore in China. *Miner. Metall. Process.* 31(2):136–229.

LKAB (Luossavaara-Kiirunavaara AB). 2013. Customer mines. www.lkab.com/en/Customer/Mines. Accessed June 2016.

Mazzinghy, D.B., Russo, J.F.C., Lichter, J., Schneider, C.L., Sepúlveda, J., and Videla, A. 2015. The grinding efficiency of the currently largest vertimill installation in the world. In *Proceedings of the Sixth International Conference on Semi-Autogenous and High Pressure Grinding Technology*. Vancouver, BC: SAG Conference Foundation. pp. 1–15.

McIvor, R., and Weldum, T.P. 2004. Fully autogenous grinding from primary crushing to 20 microns. In *Improving and Optimizing Operations: Things That Actually Work! Plant Operators Forum 2004*. Edited by E.C. Dowling and J.O. Marsden. Littleton, CO: SME. pp. 147–151.

MEC (Minerals Education Coalition). 2016. Mining and mineral statistics: 2015 per capita use of minerals. https://mineralseducationcoalition.org/. Accessed June 2016.

Metal Bulletin. 2016. Iron ore index: The Metal Bulletin iron ore indices. www.mbironoreindex.com/. Accessed June 2018.

Middle East Steel (MEsteel.com). 2016. *Sangan Iron Ore Mines (SIOM)*. Abstract. www.mesteel.com/countries/iran/Sangan_Iron_Ore_Mine.pdf.

Mining Engineering. 1970. Wabush: A $300-million iron operation. *Min. Eng.* 33–38.

Mining Engineering. 2013. Fortescue Metals Group moving forward with autonomous mining plans. *Min. Eng.* 65(11):43.

Murr, D.L., Wennen, J.E., and Nordstrom, W.J. 2009. Essar Steel Minnesota, concentrator flowsheet development. Presented at the 70th Annual University of Minnesota Mining Symposium, Duluth, MN.

New Zealand Steel. 2016. Waikato North Head mine site. www.nzsteel.co.nz/new-zealand-steel/the-story-of-steel/the-mining-operations/waikato-north-head-mine-site/. Accessed October 2016.

Outotec. 2013. *SLon Vertically Pulsating High-Gradient Magnetic Separator*. Espoo, Finland: Outotec.

Pforr, B. 1974. Fine screen oversize grinding at Hibbing taconite. Presented at the SME Annual Meeting, Denver, CO.

Poveromo, J. 1999. Iron ores. In *The Making, Shaping, and Treating of Steel: Ironmaking Volume,* 11th ed. Edited by R.J. Fruehan and D.H. Wakelin. Pittsburgh, PA: AISE Steel Foundation.

Pradip. 1994. Reagents design and molecular recognition at mineral surfaces. In *Reagents for Better Metallurgy*. Edited by P.S. Mulukutla, D. Malhotra, and B.A. Hancock. Littleton, CO: SME. pp. 245–252.

Ripke, S.J. 2007. Mystery of the missing grade and recovery. SME Preprint No. 07-018. Littleton, CO: SME.

Ripke, S.J., and Hoff, S. 2005. Opportunities at Northshore Mining's concentrator and pelletizer. SME Preprint No. 05-41. Littleton, CO: SME.

Rosa, A.C., and Rocha, J.M.P. 2015. SAG mill design for itabirites. In *Proceedings of the Sixth International Conference on Semi-Autogenous and High Pressure Grinding Technology*. Vancouver, BC: SAG Conference Foundation. pp. 1–17.

Siirak, J., and Hancock, B.A. 1988. Progress in developing a flotation phosphorus reduction process at the Tilden iron ore mine. In *XVI International Mineral Processing Congress*. Edited by E. Forssberg. Amsterdam: Elsevier. pp. 1393–1404.

Silva, R., Weber, A., and Foreman, D. 2011. Iron ore—A review of flotation circuits and reagents. In *Proceedings of the Iron Ore and Manganese Ore Metallurgy Conference*. Johannesburg: Southern African Institute of Mining and Metallurgy. pp. 1–14.

Stevens, F., and Jokanovic, S. 2002. An overview of the Waikato North Head mining operation. In *AusIMM New Zealand Branch Annual Conference: 150 Years of Mining*. Melbourne, Victoria: Australasian Institute of Mining and Metallurgy. p. 7.

Swedburg, K., Bennett, C., Samuels, M., and Wells, P.F. 2016. Application of the Woodgrove staged flotation reactor (SFR) technology at the New Afton concentrator. Presented at the Canadian Mineral Processors Conference, Ottawa, ON, January 19–21.

Tian, J., Zhang, C., and Wang, C. 2015. Operation and process optimization of the Sino iron ore's autogenous milling circuits: The largest in the world. In *Proceedings of the Sixth International Conference on Semi-Autogenous and High Pressure Grinding Technology*. Vancouver, BC: SAG Conference Foundation. pp. 1–12.

Tuck, C.A. 2018. Iron ore. In *Minerals Yearbook 2015*. Vol. I. Reston, VA: U.S. Geological Survey.

Turquoise Partners. 2016. Mining and base metals. *Iran Investment Monthly*. 6(62).www.turquoisepartners.com/media/1147/iim-nov11.pdf.

USGS (U.S. Geological Survey). n.d. Iron ore statistics and information. https://minerals.usgs.gov/minerals/pubs/commodity/iron_ore/. Accessed March 13, 2018.

Van Deventer, B., and Rowe, P. 2001. Taharoa minesite plant relocation. *Proceedings of the 34th Annual Conference of the New Zealand Branch of the AusIMM*. Melbourne, Victoria: Australasian Institute of Mining and Metallurgy. pp. 21–30.

Wei, D., Guan, Z., Gao, S., Liu, W., Han, C., and Cui, B. 2011. Beneficiation of low-grade haematite ores. In *Iron Ore 2011: Meeting Growing Demand*. Melbourne, Victoria: Australasian Institute of Mining and Metallurgy. pp. 545–556.

Zou, J. 2007. Advances of iron ore beneficiation in China. In *Iron Ore 2007: Proceedings*. Melbourne, Victoria: Australasian Institute of Mining and Metallurgy. pp. 31–33.

CHAPTER 12.17

Ironmaking

Basak Anameric, David Rohaus, and Tiago Ramos Riebeiro

Iron ore concentrate is reduced, smelted, and refined to produce steel by the processing routes shown in Figure 1 (Anameric and Kawatra 2008):

- Blast furnace (BF) and basic oxygen furnace (BOF)
- Direct reduction (DR) and electric arc furnace (EAF)
- Direct smelting (DS) and either BOF or EAF
- Direct melting of scrap in the EAF

Scrap can be the sole feed for an EAF, or it can used in conjunction with direct reduced iron (DRI) in an EAF. Scrap is a supplementary feed in the BOF, as well. BF and BOF processes are considered the traditional route. DR, DS, and scrap melting and EAF are considered alternative routes. EAF steel production is more widely used in the United States, but on a worldwide scale, BOF is more extensively used for steel production. Worldwide, approximately 65% of iron and steel is produced using traditional routes, and approximately 35% is produced using alternative routes.

The majority of the iron ore mined and beneficiated is used for pig iron production and subsequent crude steel production. On average, 2 billion t (metric tons) of iron ore, 1 billion t of metallurgical coking coal, and 520 million t of scrap are used to produce 1.6 billion t of crude steel (Worldsteel 2015).

This chapter gives a brief description of the ironmaking routes. The discussions are divided into the traditional ironmaking route (BF process) and alternative ironmaking routes (DR and DS processes).

BLAST FURNACE IRONMAKING

The BF is a countercurrent shaft furnace, which employs carbon, mainly in the form of coke for reduction and smelting of iron oxides to produce hot metal (pig iron) and slag. The hot metal usually contains 4%–5% carbon and 0.3%–1% silicon, and is suitable for subsequent refining into steel.

The raw materials for the BF include the following:

- Solid raw materials—iron oxides, flux, and coke
- Gas raw materials—hot airblast, oxygen, and hydrocarbon hot blast injectants

The BF operation consists of periodic charging of solid raw materials from the top of the furnace, periodic tapping of the hot metal and slag produced from the bottom, continuous injection of the hot blast with hydrocarbon injectants and oxygen enrichment through the tuyeres, and continuous removal of the top gas and dust.

A BF material balance is shown in Figure 2 (Peacey and Davenport 1979; Poveromo 2010; Lankford et al. 1985b). The iron oxide raw materials used are typically hematite (Fe_2O_3) and occasionally magnetite (Fe_3O_4). They are added as pellets, sinter, or lump ore. Depending on the particular BF, the charge may contain one, two, or more of these oxide forms.

The main gangue contained in iron oxide raw materials are silicon dioxide (SiO_2) and aluminum oxide (Al_2O_3). Flux material containing calcium oxide (CaO), calcium carbonate ($CaCO_3$, limestone), magnesium oxide (MgO), or magnesium-calcium carbonate ($MgCO_3 \cdot CaCO_3$, dolomite) can be added. Pellets are produced by upgrading low-grade mineral deposits, followed by consecutive agglomeration and induration. Pellets produced without the addition of flux are referred to as *standard pellets*, and those produced with flux addition are referred to as *fluxed pellets*. In addition to the preceding list of flux materials, magnesium-iron silicates, such as olivine $(Mg,Fe)SiO_4$, are occasionally used as flux. Sinter is produced by burning fuel with various iron oxide–containing materials to produce clinker-like agglomerates. These iron oxide–containing materials may include iron ore fines, in-plant reverts (steelmaking slag, mill scale, flue dust), and fluxes (Ricketts 2017).

Flux is added to the furnace to manipulate slag composition, enabling the formation of a fusible slag and removal of the gangue minerals. Its addition rates are maintained by use of fluxed pellets and/or sinter or its individual addition. The iron oxide raw materials and fluxes are referred to as the furnace *burden*. The furnace-loading scheme is described as the burden/coke or ore/coke ratio.

Coke is added to the furnace to supply most of the reducing gas and heat needed for reduction and smelting. It is produced by heating mixtures of powdered caking coals in the

Basak Anameric, Chief Scientist, Basak Anameric Consulting, Grand Rapids, Minnesota, USA
David Rohaus (Retired), United States Steel Corporation, Pittsburgh, Pennsylvania, USA
Tiago Ramos Riebeiro, Researcher, Laboratory of Metallurgical Processes, IPT (Institute for Technological Research), São Paulo, Brazil

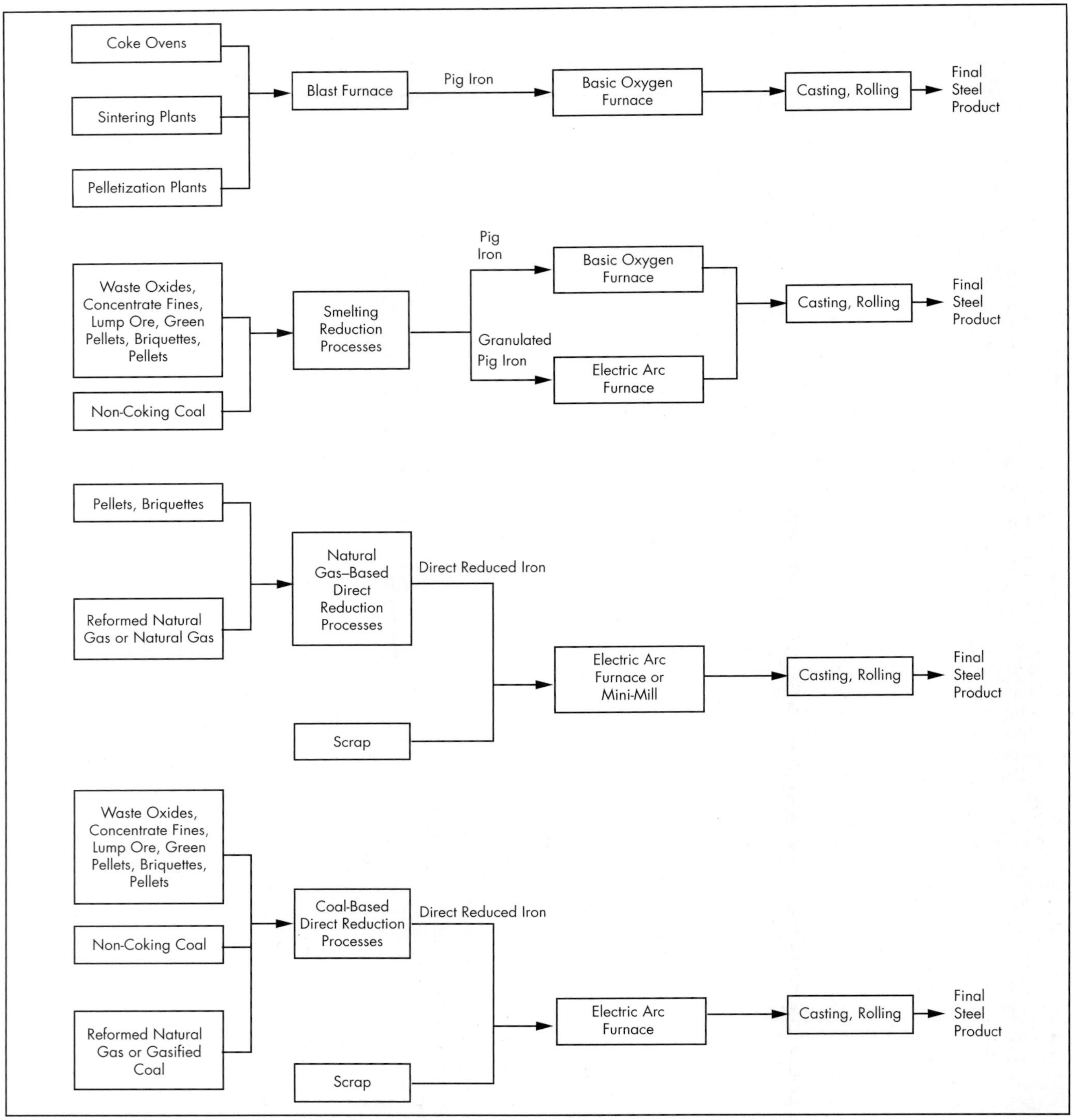

Source: Anameric and Kawatra 2008

Figure 1 Iron and steelmaking routes

absence of air (coking process). Coke is reactive at high temperatures and is strong enough to withstand being crushed even near the bottom of the BF. Coke is necessary to permit uniform gas flow through the burden as it softens and melts in the lower regions of the furnace. To date, no alternative materials to fully replace coke have been identified.

Hot airblast is used to burn coke in front of the tuyeres to provide heat for reduction, heating, and melting of the charge and products. It is preheated to 930°–1,330°C and in some cases enriched with oxygen. Gas, liquid, or solid hydrocarbon injections provide additional reducing gas (CO and H_2) for the reduction reactions. Hydrocarbon injectants include fuel oil, tar, natural gas, and pulverized coal (AIST 2018).

The main product of the BF is hot metal (pig iron). It is tapped from the furnace at regular intervals through one or several holes near the bottom of the hearth. The composition

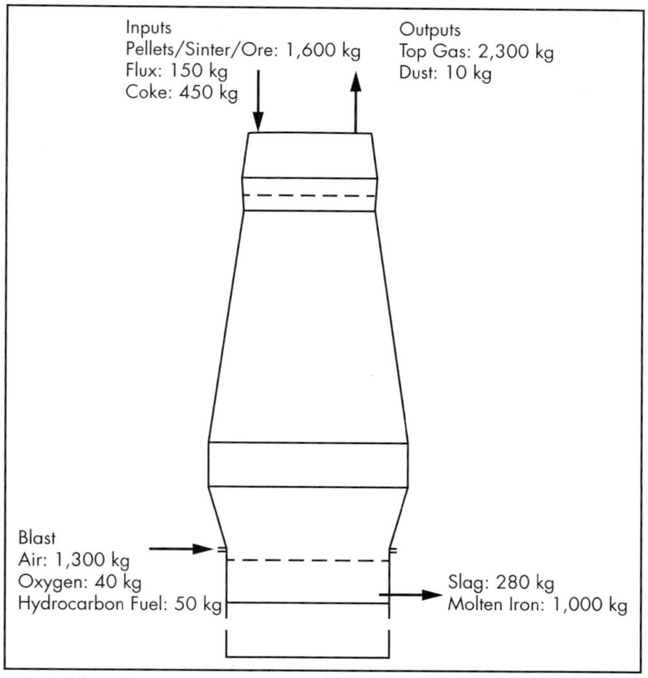

Adapted from Peacey and Davenport 1979; Poveromo 2010; Lankford et al. 1985b

Figure 2 Blast furnace materials balance

of pig iron is controlled by adjusting the slag composition and furnace temperature, particularly in the lower half of the furnace. The pig iron is transported in the molten state to an adjacent steelmaking plant, where it is refined to steel.

The by-products of the BF are slag and top gas. The slag (with an approximate composition of 30%–40% SiO_2, 5%–15% Al_2O_3, 35%–15% CaO, 5%–15% MgO, 0–1% Na_2O+K_2O, and 1%–2.5% S) contains very little iron oxide. The composition of the slag is adjusted to do the following:

- Remove SiO_2 and Al_2O_3 gangue minerals inherent to the iron oxides.
- Control the slag viscosity.
- Control the silicon content of the metal.
- Absorb potassium oxide (K_2O) and sodium oxide (Na_2O). Potassium oxide and sodium oxide enter the furnace as inherent to both coke and iron oxides. They are partially reduced to potassium and sodium vapor near the bottom of the furnace. This vapor rises to the cooler parts of the furnace, where it is re-oxidized and becomes entrapped in solid form in the descending burden. If the vapor is not absorbed by the slag, this process becomes cyclic and leads to an accumulation and buildup in the furnace. Buildup restricts the gas flow and leads to erratic descent of the charge (Peacey and Davenport 1979).
- Absorb sulfur. If the sulfur is absorbed in the slag, it does not yield in the metal. Metal with high sulfur concentrations needs to be desulfurized prior to use.

The BF top gas is collected, cleaned, and burned in auxiliary stoves to heat the hot airblast.

The main chemical events that take place in the furnace include the following steps (Lankford et al. 1985b; Peacey and Davenport 1979):

1. Burning (oxidation) of carbon by hot airblast in front of the tuyeres to generate heat and carbon dioxide (CO_2). This reaction takes place as shown:

$$C + O_2 \leftrightarrow CO_2$$

2. Endothermic reaction of carbon and carbon dioxide to produce reducing gas. This reaction takes place as shown:

$$C + CO_2 \rightarrow 2CO$$

$$CO + H_2O \rightarrow H_2 + CO_2$$

3. Reduction of iron oxides to metallic iron. These reactions take place as shown:

$$3Fe_2O_3 + CO \rightarrow 2Fe_3O_4 + CO_2$$

$$Fe_3O_4 + CO \rightarrow 3FeO + CO_2$$

$$FeO + CO \rightarrow Fe + CO_2$$

$$Fe_2O_3 + 3CO \rightarrow 2Fe + 3CO_2$$

$$FeO + C \rightarrow Fe + CO$$

4. Formation and melting of slag. The gangue minerals inherent to the iron oxides and ash inherent to coke react with the flux materials and form a fusible slag.
5. Separation of molten iron from slag. Because the density of the slag is much lower than that of the iron, the slag floats on top of the iron.

The rate at which reducing gas is generated, coke is burned, and iron oxides are reduced depends on the rate at which hot airblast is blown into the furnace. High hot airblast rates lead to rapid combustion of coke in front of the tuyeres, rapid reducing gas generation, and consequently a high rate of iron reduction.

The furnace operation is limited by the maximum allowed rate at which hot airblast can be blown into the furnace. Above this rate, the furnace gases tend to ascend through open channels in the solid charge rather than in an evenly distributed flow pattern. This causes nonuniform and/or incomplete reduction and inefficient use of carbon. Excessive ascent velocities also prevent newly melted iron and slag from descending evenly through the bosh (flooding), which may lead to an uneven descent of solid charge and erratic furnace operation (Peacey and Davenport 1979).

Blast Furnace Plant Layout

The BF plant layout shown in Figure 3 (Peacey and Davenport 1979; Carmichael 2012) integrates various types of auxiliary plants and systems required to handle the raw materials needed to make iron and its by-products. These auxiliary plants and systems include

- Raw materials receiving and storage,
- Stockhouse,
- Charging system,
- Furnace proper,
- Casthouse,
- Slag handling,
- Hot metal handling,
- Stoves and hot blast system,
- Gas plant, and
- Utilities.

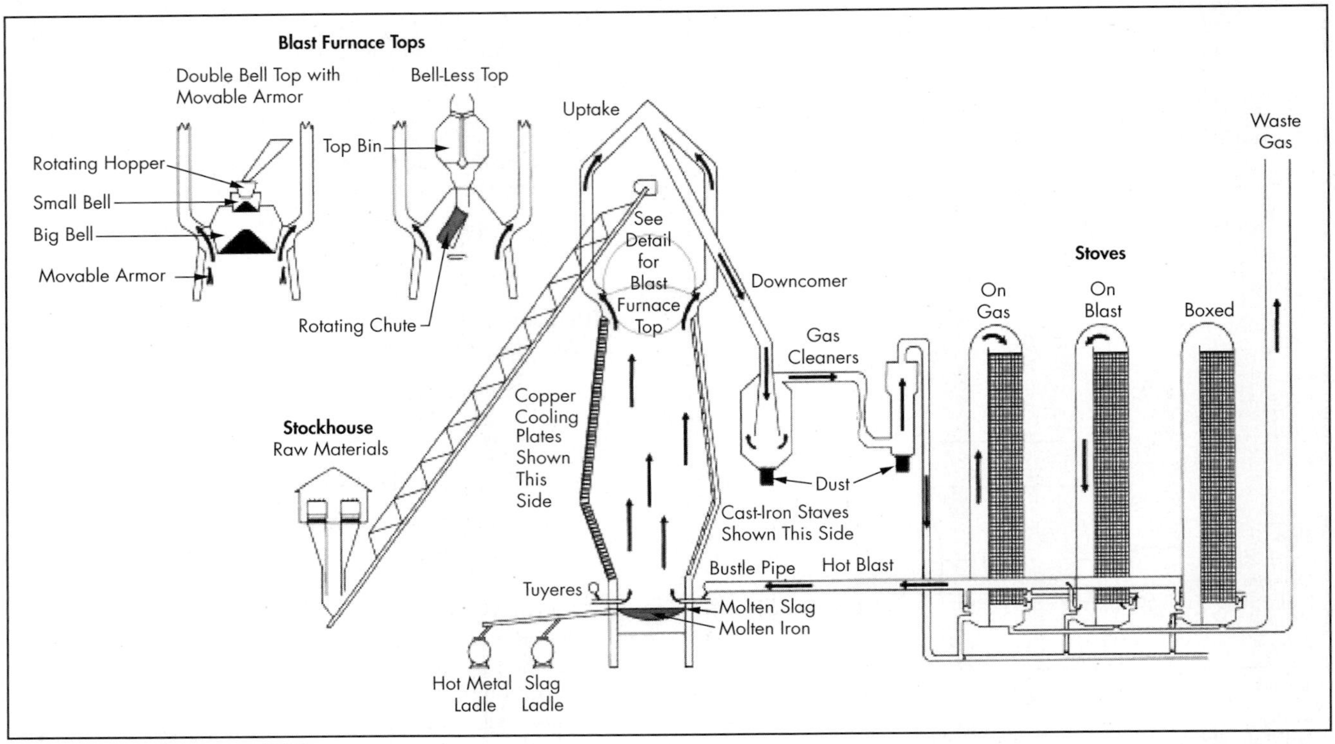

Source: Peacey and Davenport 1979

Figure 3 Blast furnace layout

The integrated functioning of these systems and plants evolved over centuries of ironmaking and enables the successful operation of a BF. Modifications to any one of these systems or plants may improve efficiencies at a large scale. Conversely, malfunction or upsets regarding any one of them may lead to long-term interruptions and a loss of productivity.

Blast Furnace Proper

The BF proper is the main vessel for ironmaking. It is designed to support reduction and smelting reactions. BFs may be either mantle-supported or free-standing. Mantle-supported furnaces have a ring girder (mantle) located at the bottom of the lower stack of the furnace. Columns resting on the main furnace foundation support the mantle. The hearth and tuyere breast are supported by the foundation.

Stockhouse

Raw materials, such as iron ore, pellets, fluxes, and coal, are delivered by bulk carriers, ship, barge, or rail car. Coke can be delivered in the same fashion or produced in the plant. Sinter can be delivered to the plant or produced in the plant using supplied ore and waste oxides (such as mill scale, pellet fines, coke breeze, sludge, etc.) generated in the plant.

Raw materials are screened, weighed, and sequenced in the stockhouse for charging (the burden) into the furnace. The raw materials are stored in individual bins and moved to the top of the furnace by skips or conveyor belts. They are charged into the furnace in separate layers. Raw materials tend to degrade and generate fines with extended outside storage and repeated handling (stockpiling, reclaiming, dumping, transfer chutes, etc.). The presence of fines decreases permeability and impairs effective gas utilization, reduces fuel efficiency, and eventually disrupts the process.

The final screening of the raw material is conducted in the stockhouse, using vibrating screens installed after the storage bins and before the weigh hoppers.

Charging System

The BF is operated with a positive top pressure. When the burden is loaded into the furnace, all the materials must pass through the furnace top without loss of either top pressure or furnace gas. BF charging systems have evolved significantly to account for larger furnaces, higher production rates, and lower fuel consumption. A bell-less top and double-bell top with movable armor charging systems are shown in Figure 3.

Hot Blast System

The primary components of the hot blast system are hot blast stoves, hot blast main, bustle pipe, tuyere stocks, tuyeres, back-draft stack, and auxiliary fuel-injection system. Most hot blast systems include three hot blast stoves; some plants use four (Helenbrook 2012).

Hot blast stoves are a regenerative heat exchange system used to preheat the cold blast to the BF. The top gas from the BF serves as the energy source (with a heating value approximately 0.1 that of natural gas) (Peacey and Davenport 1979; Lanyi et al. 2012). Top gas must be clean and dry to ensure stable operation and longevity of the stoves. It is burned to recover heat and to reduce carbon dioxide emissions. It is often enriched by other fuels, such as natural gas or coke oven gas, to attain the flame temperature required.

The heat exchange system includes sequential heating and cooling and regenerative mass of refractory. These steps (shown in Figure 3) include the following:

1. During the *on gas* stage, combustion of the gases and passage of the waste gas through the refractory take place. Air and gas are mixed in the combustion chamber and then burned. The combustion gases flow into and down through the checker refractory, heating them. The waste gases exit through the chimney to the exhaust stack.
2. During the *on blast* stage, cooling of the checker refractory and heating of the blast air take place. Cold blast air enters the bottom of the checker chamber and flows up through the checkers, in the process of which it is heated. If the temperature of the blast is higher than required, it is enriched and mixed with other gases (such as natural gas or coke oven gas).
3. During the intermediate stage, the gas system and blast system are in isolation where the stove is bottled or boxed. As the name suggests, all the valves are closed and the heat is retained in the stove.

Combustion air, hot blast enriched with oxygen, natural gas, pulverized coal, and/or other additives supplied to the furnace through the tuyeres are referred as the *wind*. Wind composition and velocity affect the kinetics of BF ironmaking. If the wind velocity is too high, downward descent of the charge is inhibited. The wind velocity can be manipulated by the furnace top pressure: Increasing top pressure increases the pressure of incoming wind, which increases the gas density and lowers the wind velocity (Carmichael 2012; Helenbrook 2012).

Tuyere Injection System

The most common BF injectants are coal, oil, tar, and natural gas. The main incentive for their use is to replace coke. This is caused by increasing coke prices, decreasing availability, and environmental constraints associated with coke production. In addition, some of these injectants help with cooling in the raceway and aid in introducing more hydrogen in the reducing gases. The tuyere injections are accompanied by an increase in blast temperature and oxygen enrichment, and a decrease in blast humidity. It is these changes that contribute most to increased coke savings and productivity (Helenbrook 2012).

Cooling System

The primary technological advances in the BF in the second half of the 20th century were related to burden distribution, high top pressure, fuel injection, and increased hot blast temperature. The resulting improved process efficiency, combined with building of larger furnaces, led to increased heat flux and consequent need for cooling to ensure a stable furnace shell (Carmichael 2012). The cooling and lining systems were developed to protect the shell and increase the campaign life.

The two main types of cooling are plate cooling and stave cooling. A typical cooling plate is manufactured from cast high-conductivity copper. Single- or multiple-pass coolers may be used, depending on their position in the furnace. In the bosh, belly, and lower stock, multiple-pass coolers are installed. Staves can be applied to all walls in all areas, including the throat. Staves offer direct protection of the BF shell by providing a homogeneous cooling shield that acts as a surface on which accretions can form.

Casthouse

The casthouse is the area where the BF is tapped and hot metal and slag are removed. BFs are typically tapped every 2–4 hours. Typical hot metal and slag tapping rates are in the range of 4–6 and 3–5 t/min. In the casthouse, hot metal and slag are directed to the appropriate handling equipment or facilities.

Various site-specific materials, equipment, and infrastructure are used for tapping and closing the furnace. These include the following:

- A tap-hole drill is used to bore a hole through the tap-hole clay into the hearth of the furnace.
- A mud gun is used to close the tap hole after casting is complete. A measured quantity of clay is pushed by the mud gun to fill the hole and to maintain an amount of clay (the *mushroom*) within the hearth. Tap-hole clays are high-temperature refractory compounds. They are very dense and require high pressure to be forced through the mud gun nozzle into the tap hole. They essentially seal of the furnace between taps.
- Trough and runner systems are used to separate iron and slag and to convey them away from the furnace. The iron trough is a refractory-lined tundish located in the casthouse floor and designed to collect iron and slag from the tap hole.

The iron flows down the trough, under a skimmer, and over a dam into an iron runner system. Trough iron level is determined by the iron dam height. Slag is lighter than iron, so it floats down the trough on top of the iron pool. The slag pools on top of the iron until there is a sufficient amount to overflow to the slag dam. It then floats down the slag runner (7% slope) into the various pieces of slag handling equipment. At the end of the cast, the slag dam is lowered so that most of the slag is drained off. The iron below the dam level remains in the trough. This reduces damage to the trough's refractory lining from oxidation and thermal shock.

The slag handling equipment includes the following:

- Pots for railway or mobile equipment are used for hauling to a dump.
- Pits are located adjacent to the furnace for air cooling and water quenching prior to excavation by mobile equipment. Pit slag is used for fill or can be crushed for use as aggregate.
- Pelletizing or granulation facilities are also located adjacent to the furnace for conversion of the slag to material suitable for backfill, aggregate, or cement replacement.

Iron is directed to hot metal transfer ladles (torpedo cars) using iron runners (3% slope) for movement to the steel shop, iron foundry, pig caster, or iron granulation unit.

Operational Zones

Numerous heterogeneous chemical reactions take place inside a BF, each serving a particular purpose. All of these reactions contribute to the goal of producing hot metal (pig iron). Physical configuration inside the furnace, gas flow pattern, and establishing a countercurrent operation are the most critical aspects for these reactions to proceed.

The BF can be conceived as five reactors (or zones), stacked one on top of another, in a single shell. These abstract reactors are referred as *regions* or *zones*. These zones exhibit large temperature gradients and use raw materials and products in a coupled fashion. Even in the same zone, temperature gradients occur, gas flow rates differ, and chemical reactions proceed differently from center to periphery.

The five zones are as follows (Geerdes et al. 2015; Lu 2012):

1. Lumpy zone, where no liquid or partial molten material exists
2. Cohesive zone (also called *softening* and *melting* zone), where alternating layers of partially reduced iron oxides and coke exists
3. Active coke zone, where an irrigated bed of coke, with hot reducing gas moving up and liquid iron and slag moving down exists
4. Raceway, where partial combustion of coke and injected fuels takes place with preheated air to produce reducing gas that contains carbon monoxide and hydrogen
5. Deadman zone where liquid products are collected, with the coke bed being pushed down into liquid pool against the buoyant force

Heat transfer, reduction, and melting are the chemical events that take place in individual zones under countercurrent conditions. These zones are different from the separation of the furnace into two regions (upper and lower furnace) based on the coke activity isotherm, for heat and mass balance calculations. For temperatures below this isotherm (950°–1,000°C), the reaction rate of carbon gasification by carbon dioxide is insignificant, so that coke may be treated as chemically inert. This region is called the *coke-inert region*. For temperatures above the isotherm, the region is called the *coke-active region* (Lu 2012; Carmichael 2012; Poveromo 2010).

Iron oxides and coke are charged in alternate layers. They retain this layered arrangement as they move down the lumpy and cohesive zones, countercurrent to the flow of the reducing gas, which preheats and pre-reduces these materials. The cohesive zone, the second zone from the top, separates the lumpy zone for gas–solid chemical reactions and the zones for processing liquids. Softening and melting of the iron oxides start toward the bottom boundary of the cohesive zone. The physical structure of the cohesive zone is important for the whole process because it provides permeability and distribution for gas flow. The physical structure, position, and direction of the cohesive zone are influenced by the raw material quality, charging practice, overall oxide/coke ratio (coke rate), and how the other reactions proceed in the furnace (Poveromo 2010; Lu 2012; Lankford et al. 1985b; Sano et al. 1997).

The reducing gas is generated in the raceway by partial combustion of coke and injectants with oxygen from the hot blast. Reducing gas ascends in the active coke zone, cohesive zone, and lumpy zone through coke slits. The momentum of the gases from the tuyeres influences the size of the raceway. Slag levels and coke quality influence the direction of the reducing gas leaving the raceway (Poveromo 2010; Wallace et al.1999).

As the molten iron and slag are produced, they drip through the active coke zone and raceway and collect in the hearth. The hearth is filled with coke. The ease of tapping hot metal and slag depends on the coke size and quality (Poveromo 2010).

Iron Oxide Processing

The majority of the iron oxides are introduced to the furnace in the form of hematite (Fe_2O_3), at ambient temperature. As they descend, they are heated and experience a series of physical and chemical changes, including reduction and metallization, crystal structure change, and softening and melting.

As iron oxides reach the lower boundary of the lumpy zone softening starts. The softening properties of the partially reduced iron oxides depend on the iron ore quality and operating conditions. The softening temperatures vary between 1,200° and 1,400°C. Softening of the iron oxides take place in the upper boundary of the cohesive zone, and melting of the iron oxides take place in the lower part of the cohesive zone (Wallace et al. 1999).

Mainly, carbon monoxide is used for reduction of iron oxides. Data from the dissection of BFs have shown that by the time iron oxides reach the cohesive zone, the degree of reduction is about 70% (Hashimoto et al. 1977). Pre-reduced iron particles stick together under the pressure of the bed above. Reducibility and size distribution of the iron oxides have a direct influence on the reduction rate and physical properties of the cohesive zone. From a point of view of reducibility, the smaller particles with large surface areas would be preferable. For gas permeability and distribution through the furnace, however, pellets and sinter with adequate size distribution are used. The iron oxide agglomerates (pellets and sinter) should provide a porous surface for reducing gases to diffuse in easily and for product gas carbon dioxide to diffuse out easily (Lu 2012; Dahlstedt et al. 2000).

The cohesive zone consists of alternating layers of partially reduced iron oxide and coke slits. The shape of the cohesive zone emulates the main path and distribution of the reducing gas. The distribution of gas flow has important consequence for efficient use of gases and protection of the lining in the lumpy zone. Because the cohesive zone separates the lumpy zone above from the liquid-processing high-temperature zone below, the size, shape, and location of the cohesive zone also play a role in the effectiveness of those zones. The location of the cohesive zone in relation to the tuyere level and the thickness of the zone depend largely on the iron oxide reducibility properties and furnace conditions (Lu 2012; Tovarovskii et al. 2009).

The gas flow in the cohesive zone can be centralized (V-shaped distribution) or wall directed (M-shaped distribution). The following factors are manipulated to obtain a centralized (V-shaped) distribution (Omori and Shimomura 1982):

- Charging pattern and difference in the angle of repose, bulk density, and size distribution for iron oxides and coke
  - The angle of repose for coke, sinter, and pellets are 35, 33, and 26 degrees, respectively.
  - The bulk density for coke, sinter, and pellets are 525 kg/m^3, 1,660 kg/m^3, and 2,150 kg/m^3, respectively.
- Segregation during charging

When centralized gas flow is observed, the iron oxides at the center of the furnace encounter large amounts of gas, thus they are heated and reduced faster than those charged at the periphery. This results in earlier formation of the cohesive layer at the center than the periphery, creating the conical shape of the cohesive zone.

Softening and melting of the iron oxides start in the cohesive zone. They are melted exclusively on the inner and lower surface of the cohesive layer. The softening temperature depends on the following (Omori and Shimomura 1982):

- Amount of slag components
- Fusion temperature of the slag components
- Amount of residual wüstite present (influenced by how fast the burden is heated and reduced)

The slag components include gangue minerals from the iron oxides, ash from coke, flux, and residual wüstite.

The shape of the cohesive zone and the softening characteristics vary when using fluxed (pellets and sinter) and standard (pellets) iron oxides. When using fluxed iron oxides, the overall fusible slag composition is basic, and fusion temperature is higher when compared with standard iron oxides. Fluxed pellets are usually more porous, which leads to faster reduction and metallization. Thus, as these pellets start to soften, there is no wüstite left in the reduced pellets to react with the slag components. When compared with fluxed iron oxides, the overall fusible slag composition for standard iron oxides is acidic and fusion temperatures are lower. This leads to early softening and formation of cohesive layers higher in the furnace, resulting in a smaller lumpy zone for prereduction (Lu 2012; Dahlstedt et al. 2000).

If the burden consists of two kinds of iron oxides with high and low melting points of the slag components, the two kinds of iron oxides melt at a different temperature. In such cases, the cohesive layer is wider (Hashimoto et al. 1977).

If the slag formed at the lower surface of the cohesive zone has a high viscosity, the void fraction in the active coke zone becomes lower. This impairs dripping of the liquid products and the descent of the bed. Thus, the melting characteristics of the iron oxides are also very important for efficient BF operation (Hashimoto et al. 1977).

Hence, it is desirable to have a cohesive zone made of cohesive layers with high softening temperatures and permeable coke slits located low in the furnace. This promotes earlier reduction for chemical efficiency, optimum shape, and minimum height, offering higher production and protection of the lining.

Coke Processing

The ascending reducing gas (bosh gas) at a temperature of 2,100°–2,300°C is generated in the raceway by partial combustion of carbon from coke and injectants with preheated air. It contains three gases: nitrogen, carbon monoxide, and hydrogen. Depending on the extent of hot blast oxygen enrichment, it contains about 50%–60% nitrogen by volume. The nitrogen in the bosh gas does not chemically react in the furnace, but it plays an important role as a heat carrier from the raceway to the lumpy zone (Lu 2012). The hydrogen content of the gas varies depending on the extent of the natural gas injection (Wallace et al. 1999).

This hot bosh gas moves countercurrent to the burden to the active coke zone, then to the cohesive zone, and finally to the lumpy zone. In the cohesive zone, bosh gas is capable of melting the iron and slag, as well as reducing the remaining wüstite. In the lumpy zone, it provides the reducing gas required for the reduction of iron oxides.

Carbon in coke may react with the carbon dioxide (or water vapor) from the reduction reactions at temperatures above the coke-active region isotherm of 950°–1,000°C. The following endothermic reaction is called the *solution loss reaction* (Peacey and Davenport 1979; Elliott and Gleiser 1960):

$$CO_2 + C \rightarrow 2CO$$

$$CO + H_2O \rightarrow H_2 + CO_2$$

As the solution loss reaction takes place, the mechanical strength of the coke is lowered (Poveromo 2010).

In the upper part of the lumpy zone (where the temperatures are below the coke activity isotherm), carbon in coke is chemically inert (solution loss reaction does not take place) and carbon dioxide is accumulated as a result of the ongoing reduction reactions. This weakens the reducing power of the rising gas. As the gas reaches the top of the furnace, it contains approximately equal amounts of carbon monoxide and carbon dioxide (Lu 2012).

As the iron oxides reach the coke activity isotherm, prereduction reactions are completed to an extent, and metallization starts. Along with the other reactions, the solution loss reaction proceeds. The carbon does not melt; it remains solid inside the furnace. It reacts with gases, dissolves in liquid iron, and reacts with some oxides in slag. In the deadman zone, it is consumed for liquid metal and slag refining reactions.

Metal and Slag Refining

For the duration between tappings or longer, liquid metal and slag are retained in the deadman zone, where they are in contact with each other for a long period. During this stay, various chemical events take place. Owing to these chemical reactions, the composition of the slag and metal vary from top to bottom. These chemical events are influenced by the slag basicity, softening temperature, and rate of iron oxide reduction (Lu 2012).

These chemical events include

- Continuing reduction of iron oxides in the slag, which results in considerable variation of the slag composition from the top to bottom of the slag pool;
- Continuing carburization of the metal;
- Sinking and coalescence of the metal droplets;
- Transfer of manganese from slag to metal;
- Transfer of sulfur from metal to slag; and
- Decrease of the silicon content of the metal.

Physical Features Required for Efficient Operation

BF operation efficiency is measured in terms of production, fuel consumption, and furnace lining life. All of these factors are limited by the countercurrent operation of the furnace and have a direct effect on the economy of the process (Hashimoto et al. 1977).

Countercurrent operation of the furnace can be achieved only by maintaining proper physical conditions and permeability in the furnace. To date, coke remains a necessary ingredient for maintaining mechanical stability and permeability and is required for good performance.

BF operations are influenced by the volume of wind and utilization rate of the reducing gases. If the wind volume is increased beyond a certain limit, the descent of the solids becomes erratic, deteriorating the countercurrent operation of the furnace. The fuel rate depends on the utilization of the gas in the furnace. The lining life depends on the thermal load to which the lining is exposed (Wallace et al. 1999; Lankford et al. 1985b; Lu 2012).

A smooth and stable BF operation and low thermal load on the lining can only be attained under stable countercurrent conditions in the furnace.

The movement of the solid bed packed with iron oxides and coke is induced by the void space created at the bottom because of melting, burning of the carbon, and tapping of the liquids. The gas flow in the opposite direction is driven by the pressure drop. The amount and rate of gas flow through the furnace dictate the rate of production. The distribution of the gas through the furnace and efficient solid–gas contact relate to efficient use of heat and reducing gases in the furnace.

Pressure drop and void fraction of the charge in high-temperature zones have the greatest effect on countercurrent condition stability. To promote this, improvements have been made to use raw materials with proper size distribution and adequate mechanical strength (Lu and Ranade 1991). The presence of fines in the bed narrows the gas path, increases the resistance to gas flow, and lowers the hot metal production rate. Raw materials lacking adequate mechanical strength generate fines in the furnace, impairing gas flow. When the conditions for adequate gas distribution in the furnace can be established, the furnace can be operated at higher wind velocities and consequently at higher top pressures. This allows for more efficient use of the reducing gas (Hashimoto et al. 1977; Peacey and Davenport 1979; Lu and Ranade 1991).

If adequate gas flow through the furnace cannot be attained and a strong localized gas flow forms in the limited cross section of the furnace, a considerable amount of gas goes to the top of the furnace without reducing the iron oxides, lowering the overall utilization of the gas. In this case, the iron oxides in other parts of the cross section have less chance to be heated and reduced by the gas. Consequently, the unreduced iron oxides proceed deeper into the furnace without being sufficiently reduced to form the cohesive layers, and the pattern of the cohesive zone changes as a result of the partial channeling of the gas in the lumpy zone.

In the cohesive layer, a dense mass of pre-reduced iron oxides impairs the gas flow. The layered charging of the iron oxides and coke allows generation of coke slits and redistribution of the gas flow in this zone. Variation in the thickness of each layer of charge over the cross section of the furnace aids in promoting the proper contact pattern for chemical reactions between solids and gas and between gas flow and the refractory lining. A sufficiently large coke slit area is necessary for stable countercurrent operation of the furnace (Hashimoto et al. 1977). A sufficiently high void fraction is necessary both for the smooth countercurrent operation in the active coke zone and for the smooth flow of iron and slag in the hearth.

ALTERNATIVE IRON UNITS AND IRONMAKING TECHNOLOGIES

Alternative ironmaking technologies include DRI, DS hot metal, and iron oxide reduction and smelting.

Direct Reduced Iron

DRI is a highly metalized solid, alternative iron unit. It still contains gangue minerals inherent to the raw materials used for its production. It is produced by solid-state reduction at temperatures lower than the melting temperature of iron, using hydrocarbon gases and/or carbon-bearing materials as reducing and carburizing agents.

During its production, only oxygen inherent to the iron oxide feedstocks is removed. Thus, DRI has a similar but more porous physical form than the raw materials (pellets, lump ore, fine ore) used for its production. Owing to this physical appearance, it is often referred as *sponge iron*. Depending on how the DRI will be used, its porous structure, physical degradation, and/or chemical stability features may be undesirable. In these situations DRI is often densified to hot briquetted iron or cold briquetted iron.

When hydrocarbon gases, such as hydrogen, carbon monoxide, or methane (natural gas), are used as fuel and reducing agents, the process is called *natural gas–based DR process*. When carbon-bearing materials, such as coal, gasified coal, or coke breeze are used as fuel and reducing agents, the process is called *coal-based DR process*. Some of the important features of natural gas–based and coal-based DR processes are described by Anameric and Kawatra (2007).

DRI can be used in EAF along with scrap for steel production; ferrous foundry operations as a new iron unit charge; EAF, foundry cupolas, and BOF as coolant; and BF to increase throughput (Anameric and Kawatra 2007).

DR processes and DRI utilization have reached commercial competency and are becoming increasingly important. This is because of

- Increasing worldwide production and demand for DRI in EAF steelmaking,
- Need for manufacturing routes involving smaller and more flexible equipment, and
- Developments in the natural gas industry.

Direct Reduction Processes

Shaft furnace, retort (batch) furnace, and fluidized bed reactors used for natural gas–based processes include (Zervas et al. 1996b; Markotic et al. 2002) the following:

- Shaft furnace processes
  - Midrex
  - Energiron III and ZR
  - Armco
  - Arex
  - Purofer
- Retort (batch) furnace processes—HyL I and II
- Fluidized bed reactor processes
  - Fluid iron ore reduction (FIOR)
  - Finmet
  - Circored
  - H-iron
  - High iron briquette (HIB)
  - Iron carbide

Rotary kiln, rotary hearth furnace, shaft furnace, multiple hearth furnace, and fluidized bed reactors used for coal-based processes include (Zervas et al. 1996b; Markotic et al. 2002) the following:

- Rotary kiln reactor processes
  - Stelco-Lurgi/Republic Steel National (SL/RN)
  - Krupp-CODIR (coal-ore-direct-iron-reduction) and Krupp-Renn
  - Direct Reduction Corporation (DRC)
  - ACCAR/OSIL (Allis-Chalmers controlled atmospheric reduction/Orissa Sponge Iron and Steel Limited)
  - Tata Iron and Steel Company (TISCO) direct reduction (TDR)
- Fluidized bed reactor process—Cirofer
- Shaft furnace processes—Kinglor-Metor and Höganäs
- Multiple-hearth furnace process—Primus

- Rotary-hearth furnace processes
  - International Metals Reclamation Company (Inmetco)
  - FastMet
  - SidComet and Comet
  - Iron Dynamics Inc. (IDI)
  - Iron Technology Mark 3 (ITmk3)
  - DRyIron

Shaft furnace. A shaft furnace is a moving bed reactor where iron oxide pellets, lump ore, or briquettes descend under gravity in a countercurrent flow to the reduction gas. The temperature and composition of the reducing gas in the furnace are controlled to maintain optimum bed temperature, degree of metallization and carburization, as well as efficient utilization of the reducing gas. Its operation efficiency depends on even distribution of the gas flow across the cross section of the shaft. This is best achieved by use of a charge of uniformly sized particles free from fines. Pellet metallurgical properties, such as low-temperature disintegration index and sticking properties, are important for shaft furnace operation. Fines generation and bridging in the furnace alter the bed permeability and impair the reduction efficiency and productivity (Zervas et al. 1996a, 1996b).

Retort (batch) furnaces. Retort (batch) furnaces operate on a similar principle to shaft furnaces. Iron oxides are charged in fixed beds in series, and the reducing gas is transported from one bed to another. Prior to injection of the reducing gas into each reactor, it is preheated and conditioned with steam. Fresh reducing gas is injected into the last retort furnace, which is the cooling furnace. Final reduction and carburization of the DRI takes place in the cooling furnace. When the reducing gas first comes in contact with the iron oxides, reduction takes place very rapidly. As the reducing gas descends in the furnace, the reaction speed slows, and reduction potential is considerably lower. This leads to uneven reduction of the charge (Zervas et al. 1996b; Anameric and Kawatra 2007).

Fluidized bed reactors. Fluidized bed reactors often consist of series of superimposed hearths in which fluidized solids flow from one to the next below in the opposite direction to the gas flow. This allows for intimate mixing of the solids with the gases. Their efficiency depends on the effective mixing of the charge, complete fluidization of the bed, mass flow, and reducing gas velocity. Fine ore, without preagglomeration and induration, is used as the iron oxide source. Fine ore size distribution is closely controlled to ensure complete fluidization and minimal carryover to the waste stream. Because the surface area of the fine ore is higher than that of the pellets, lump ore, or briquettes, the rate of reduction is faster. Fluidized bed reactors are operated at elevated pressures to ensure optimum gas velocity, that is, optimum fluidization. The sticking tendency of the highly metallized materials may cause operational problems because larger particles formed by sticking of the finer particles may interfere with fluidization of the bed (Anameric and Kawatra 2007; Chatterjee 1994; Zervas et al. 1996b).

Rotary kiln. A rotary kiln is composed of a slightly inclined rotating cylindrical kiln with an internal refractory lining. The residence time in the kiln can be adjusted with the inclination and rotation speed of the kiln. Coal and iron oxide feedstocks in the form of pellets, lump ore, or greenballs are charged together. Generation of the reducing gas and reduction of iron oxides take place simultaneously in the kiln. Controlled combustion of the coal is achieved by injecting air both at the discharge end of the kiln and through the blowers mounted along the kiln length. The burner located at the discharge end of the kiln provides the energy required for the reduction reactions. The kiln is operated with a slightly positive pressure to prevent the intrusion of unregulated air. The coal feed is adjusted to slightly excess, so that unreacted carbon can be present in the product. This prevents the product from reoxidation (Venkateswaran and Brimacombe 1977). The kiln product is cooled in a rotary cooler. Air flow in the kiln is controlled to minimize temperature fluctuations. Localized temperature increases can result in the formation of accretions on the kiln lining and ball-shaped clusters within the bed (Anameric and Kawatra 2007).

Rotary hearth furnace. Rotary hearth furnaces can be operated in multiple zones: initial reduction zone, final reduction zone, and cooling zone. Temperature, gas flow, and gas composition are controlled to provide the required conditions in each zone. The opposite rotation of the furnace hearth and gases enables countercurrent interaction of the solids and gases. The energy required for the reactions is supplied by burners located at the furnace walls. These burners are fired at the stoichiometric deficiency of air to maintain a reducing atmosphere in the furnace. The layering thickness of the solids on the hearth dictates the reduction efficiency. At low thicknesses, higher degrees of reduction can be attained and the product has more uniform properties. The production capacity of the process is directly proportional to the furnace hearth area; to achieve high capacities, proportionally large (when compared with other furnaces with the same production capacity) furnaces are needed (Zervas et al. 1996b; Anameric and Kawatra 2007).

Direct Smelting Hot Metal (Pig Iron)

DS processes aim to produce hot metal that has similar properties as the BF pig iron. They differ from DR processes in that the iron oxides are reduced and smelted. The gangue minerals inherent to the raw materials are separated from the metal via formation of a fusible slag.

DS hot metal can be used in BOF and EAF for steel production and ferrous foundry operations as a new iron unit charge.

Direct Smelting Processes

DS processes were developed as an alternative to the BF process to overcome built-in disadvantages of that process. The incentives behind development of DS processes can be summarized as follows (Anameric and Kawatra 2008):

- Beneficiation and use of widely available non-coking coals as reducing and carburizing agents, instead of coke
- Beneficiation and use of low-grade iron oxide reserves or waste oxides, which cannot be beneficiated in the BF
- Lessening environmental emissions; elimination of coke ovens, sinter plants, and induration kilns
- Lower capital and operational costs because of minimal ancillary plant and material handling requirements
- Direct use of iron ore fines or concentrate (instead of pellets, sinters, and lump ore)
- Economic operation at modest throughput rates
- Compatibility and ability to complement the existing iron- and steelworks.
- Flexible operation that can be shut down and easily restarted

- Lower energy losses and utilization of secondary energies associated with the process
- Overall lower energy consumption
- Advantages associated with the use of upgraded charge (known chemical composition, slag-free structure) for steel production

The DS process can be subdivided into two distinct procedures, based on the methods used for generation of reducing gases (Zervas et al. 1996a; Meijer et al. 2013):

1. Processes where the reducing gases are generated in the same reactor along with reduction and smelting. These processes can be subdivided as follows:
   - Complete reduction and smelting in a single-stage reactor, such as a rotary hearth furnace or linear hearth furnace, for example, ITmk3 and nodular reduced iron (NRI)
   - Complete reduction and smelting in a molten metal bath, such as direct iron ore smelting (DIOS) and high-intensity smelting (HIsmelt)
   - Complete reduction and smelting in a molten slag bath, such as Romelt and Ausiron
2. Processes where reducing gases are generated externally in two-step processing, for example, Corex, Finex, FastMet/FastMelt, RedSmelt, IDI, Inmetco, Elred, Plasmasmelt, and Inred (Smith and Corbett 1987). Corex and Finex processes are currently operating commercially.

Melter-gasifier furnace. In the melter-gasifier furnace, coal is charged from the top, where it is dried and devolatilized using the gases produced from smelting (occurs at the bottom of the furnace) and transformed into char. This char creates a bed and supports the pre-reduced iron oxides as they are smelted. The furnace operates in a countercurrent manner. The temperature at the top of the furnace is about 1,000°–1,200°C, and the temperature at the bottom of the furnace is about 1,500°C. The engineering and operational principles of this furnace resemble a BF bosh area. The commercial examples of processing schemes that use this furnace include Corex and Finex.

Electric melter-gasifier furnace. In the electric melter-gasifier furnace, reducing gas generation and final reduction and smelting of the iron oxides take place. These furnaces were designed drawing on EAF design. An early example of this process is the strategic-Udy process, in which prereduction was carried out in a rotary kiln followed by smelting in an electric furnace (Udy and Udy 1958; Udy 1959). For these furnaces, the off-gases from reduction and the electric furnace are combined and used for electricity generation. In most cases, the energy generated using the off-gases is sufficient for the process needs. The examples of processing schemes that use this furnace include Inred and Elred.

Metal bath or slag bath melter-gasifier furnace. Bath melter-gasifier furnaces include an extended bath, similar to that used in open-hearth steelmaking. This bath can be metal based or slag based. Fine ores and coal are injected into the bath where reduction, smelting, and coal gasification take place. The examples of processing schemes that use metal slag melter-gasifier furnaces include Romelt and Ausiron. The examples of processing schemes that use metal bath melter-gasifier furnaces include DIOS and HIsmelt.

Iron Oxide Reduction and Smelting Fundamentals

Ore reducibility depends on mineral properties, reaction mechanisms, and rate-controlling steps.

Iron Oxide Reduction

The kinetic aspects of iron oxide reduction include consideration of rates at which chemical reactions involved in the production of iron from its ores will take place. The reducibility of an ore, defined as the ease with which the ore can be reduced, embodies the principles of reduction kinetics. Reducibility of an ore depends on other properties such as density, porosity, particle size distribution, surface area, grain size and structure, and type and dispersion of gangue minerals. The success of the DR process depends on the reducibility of the ore and satisfying the kinetic aspects of reduction. Further discussion regarding the reduction fundamentals and kinetics is provided by Lu and Ross (Lu 1999; Ross 1972; Ross 1973).

Iron oxide reduction kinetics are important for DR processes, as they control the rate at which metallic iron is produced. This influences the production rate, product quality, economic feasibility, and competitiveness of one process with another. The rate of chemical reactions increases as the temperature increases. The reduction rate is less important in the BF process than in the DR process, because the BF operates at much higher temperatures. Rather than reaction kinetics, the BF process is limited by the availability of heat and reducing agents (i.e., concurrent gas flow and pressure drop). For DR processes, reduction takes place in the solid state, thus the maximum temperatures that can be achieved and the corresponding reduction rate are limited (Lu 1999; Ross et al. 1980; Ross 1972). Porosity and pore size distribution are the most important factors in controlling reducibility (Joseph 1936). The soft earthy hydrated ores have the best reducibility, followed by the soft hematites, the hard hematites, and finally the hard dense magnetites (Lu 1999; Joseph 1936). If a particle of a hard dense partially reduced ore is sectioned, a topochemical type of reduction can be observed. It would include a core of hematite surrounded by three concentric layers: an inner layer of magnetite, a layer of wüstite, and an outer layer of metallic iron. For porous ores, homogeneous-diffusion types of reduction with no district interfaces but a gradual transformation from iron on the outside to hematite at the center can be observed. However, at very high magnifications, the microstructure of the grains making up the porous particle will also indicate a topochemical type of reduction (Ross 1973; Bitsianes and Joseph 1955).

For the reduction reactions to take place, the reducing agents must contact the surface of the iron oxide phases. This requires the reacting gas to diffuse inward and the product gas to diffuse outward at least through the outer iron layer to react with the wüstite layer. Several different mechanisms to facilitate this have been proposed. The most widely accepted mechanisms include the following:

- For dense ore particles where the reducing gas is unable to contact the magnetite layer or hematite core, it is predicted that solid-state diffusion of ferrous ions as proposed by Edstrom takes place (Edstrom 1953; Edstrom and Bitsianes 1955). The solid-state diffusion of ferrous ions through the wüstite crystal lattice is facilitated by the presence of lattice vacancies. Some of the ferrous ions and electrons migrate to nucleation sites where they precipitate as metallic iron; other ferrous ions and electrons diffuse across the wüstite and magnetite layers where they react with magnetite and hematite to produce wüstite and magnetite. The solid-state diffusion rate of ferrous ions through wüstite is much greater than the gaseous diffusion rate of either hydrogen or carbon monoxide through the pores of the ore particles at the temperatures used for DR processes. Thus, solid-state diffusion is not usually a rate-determining step (Lu 1999; Ross 1972).

- For very porous ores, reducing gases can penetrate faster than they can react at any one interface.

Reaction Mechanisms

For reduction of hematite by a reducing gas, reduction mechanisms include (Lu 1999; Lien et al. 1971) the following:

- Diffusion of the reducing agent from the moving gas stream to the surface of the ore particle and then to the interior of the particle
- Reaction of the reducing agent with one oxygen atom from iron oxide to form carbon dioxide or water at one of the solid–solid interfaces (hematite–magnetite, magnetite–wüstite, wüstite–iron)
- Diffusion of the reduction product gases (carbon dioxide and water) to the exterior of the particle and through the main gas stream to be carried away

These steps are sequential, and the slowest step will be the rate-controlling step for complete reduction of the iron oxide.

Hydrogen reduction is endothermic; adequate heat must be supplied to the reaction site for completion of these reactions. Thus, heat transfer to the reaction site, rather than diffusion, may be the overall rate-controlling step.

The rate of endothermic reactions increases faster with an increase in temperature rather than effective gas diffusivity. Wüstite reduction by hydrogen tends to be controlled by interfacial reaction at a lower temperature, shorter reaction time, and smaller particles. At higher temperatures, longer reaction time, and larger particles, however, gas diffusion is a rate-controlling step.

Heterogeneous reactions take place only in the interfacial reaction zone. This region is located at the phase boundary with a finite thickness including the absorbed gas molecules on the solid surface in a gas–solid system. Effective diffusion through the interfacial reaction zone depends on the pore structure, which changes with time because of phase changes and sintering at high temperatures.

The changes in the pore structure can be explained by changes in crystal structure that take place during stepwise reduction of hematite through magnetite and wüstite to metallic iron (Weiss et al. 2011). In hematite, the oxygen atoms are arranged in the hexagonal close-packed structure. As the hematite reduces to magnetite and wüstite, oxygen atoms undergo a severe readjustment, to form a face-centered cubic structure. This results in approximately a 25% increase in volume (Edstrom 1953; Edstrom and Bitsianes 1955). This tends to open up the structure and facilitate further reduction reactions. In the transformation of magnetite to wüstite, the oxygen lattice remains unchanged while iron atoms diffuse in to fill the vacant sites in the iron lattice. Wüstite has a variable composition: There is a small increase in volume as its composition changes from that in equilibrium with magnetite to that in equilibrium with metallic iron (approximately 7%–13% volume increase). The nucleation and growth of iron crystals result in shrinkage, leading to an increase in porosity and greatly enhancing diffusion and reduction (Edstrom 1953; Elliott and Gleiser 1960).

For completion of iron oxide reduction reactions, kinetically the most important reaction is the reduction of wüstite to metallic iron. At the beginning of wüstite reduction, the gas–solid interface recedes as oxygen atoms are removed by the reducing gas and iron ions and electrons diffuse into the interior. Wüstite composition gradually changes across its stability field from the boundary value with magnetite toward equilibrium value with metallic iron. This interplay between interfacial reaction and solid-state diffusion in wüstite determines the mechanism of nucleation and growth of metallic iron. The mechanisms of nucleation and growth of metallic iron include the following (Lu 1999):

- When the rate of creation of excess iron atoms is much faster than solid-state diffusion to the interior of the wüstite grain, the Fe/O ratio increases rapidly, and the nucleation zones become supersaturated with iron and wüstite. This leads to iron nucleation at numerous sites and the growth of iron nuclei to overlap and cover the whole surface of the grain.
- When the accumulation of excess iron atoms near reaction sites is minimal, the Fe/O ratio is uniform and increases as the interfacial reaction proceeds. As the reaction proceeds, wüstite grain becomes supersaturated, and the nucleus of iron appears on the surface of wüstite. Then the excess iron atoms move to the nucleus, leading to its growth.

The mechanism of growth of the iron nucleus with fresh iron involves pushing up the previously nucleated iron. In the case of an area where the base is constant, the shape of iron growing out may look like whiskers (Lu 1999; Bahgat et al. 2009). In the case of an area where the base is enlarged laterally, the shape of iron growing out may look like pyramids.

Rate-Controlling Steps

Among researchers, there is a considerable diversity of opinion about the mechanism of iron oxide reduction and rate-controlling steps. Numerous viewpoints have emerged from the studies conducted, including the following (Seth and Ross 1965):

- Rate is controlled by the diffusion of gas through the boundary layer of stagnant gas. This theory is true mainly for packed beds where the flow of gas through the bed is crucial. For a single particle, the boundary layer may be prevented from being the rate-controlling step if the flow rate of the reducing gas is above the critical flow rate (Chatterjee 1994).
- Rate is proportional to the area of metal–oxide (Fe–FeO) interface. Several researchers have reported a linear advance of the Fe–FeO interface, which support the theory for reduction being controlled by surface area (porosity) (Turkdogan and Vinters 1972). However, several other researchers have reported that this theory is not always applicable and when it is applicable, it does not

hold in the final stages of reduction when the Fe–FeO interface is supersaturated.

- Rate is controlled by the transportation of reducing gas from the main stream to the metal–oxide interface and the product gas out from the metal–oxide interface to the main stream. In addition to the researchers who supported this theory, others have indirectly proved that sintering and recrystallization of iron cause a decrease in reduction rate by inhibiting the transportation of gas through the sintered layer (Udy and Lorig 1942). Several researchers have rejected this theory on the basis that they could not obtain a relationship between the particle diameter and time required to reach a certain degree of reduction.
- Rate is controlled by multiple steps with their relative contributions varying during the course of reduction.
- Several researchers have reported that in the beginning of the process, the rate-controlling step was the surface reaction, but after a layer of iron was formed, the diffusion of gas became the rate-controlling step. Toward the end, rate was controlled by porosity (Seth and Ross 1965; Turkdogan and Vinters 1971, 1972; Turkdogan et al. 1971). This is the most widely accepted theory for determination of the rate-controlling step and process control.

Reducing Gas Generation and Gas Reforming

Reducing gas generation is achieved by the gasification of a carbonaceous solid fuel or the reforming of natural gas. Natural gas is approximately 80% methane (CH_4) and composed of a mixture of hydrocarbons and in some cases contains low amounts of nitrogen. The use of natural gas directly in ironmaking faces the issue of gas decomposition and consequent carbon deposition on the iron oxide. The deposited carbon may act as a barrier to contact between the reducing gas and iron oxide, hindering the reduction reaction. Therefore, natural gas needs to be reformed.

Carbonaceous solid fuels need to meet specific requirement for use in BFs and in DR processes such as SL/RN. Coking coal with low amounts of ash and sulfur is needed for use in cokemaking and further in BFs. Reactive coals with high volatile matter content and high ash-softening temperature are the ones used in DR. If those requirements are not met, the gasification of a low-quality coal may be an alternative for ironmaking.

In both reforming and gasification, the reducing gas will be mainly composed of a mixture of CO and H_2 in different amounts. This gas is injected into a shaft furnace where reduction of iron ore is performed to produce DRI.

Reforming of Natural Gas

Following are the main reactions related to reforming of natural gas:

$$\text{Dry reforming: } CH_4 + CO_2 \leftrightarrow 2H_2 + 2CO$$

$$\text{Steam reforming: } CH_4 + H_2O \leftrightarrow 3H_2 + CO$$

$$\text{Water–gas shift reaction: } H_2O + CO \leftrightarrow CO_2 + H_2$$

Methane conversions for dry reforming are higher than for steam reforming, demonstrating a higher thermodynamic efficiency for the use of CO_2 as the reforming agent. Some dry reforming happens in the steam reforming reactor and vice versa. This is because of the water–gas shift reaction that causes the interchange between CO_2 and H_2O. Moreover, the final proportion between CO and H_2 in the product gas is also dependent on the water–gas shift reaction.

Parallel reactions include carbon deposition, which takes place according to the following reactions:

$$2CO \leftrightarrow C + CO_2$$

$$CH_4 \leftrightarrow 2H_2 + C$$

$$H_2 + CO \leftrightarrow C + H_2O$$

Inside the reduction reactor, carbon deposition and formation of iron carbide are desirable. However, in the reforming reactor, carbon deposited on the surface of the catalyst will reduce its activity, causing a drop in efficiency. In extreme cases, a blockage of the reactor may occur. Additionally, the metallic material used in the reactor walls can fail because of pit corrosion, termed *metal dusting*. This failure is caused by the carburization of the metal with initial formation of carbides that then decompose into fine particles of metal and solid carbon. This type of corrosion causes holes in the reformer tubes.

Therefore, reforming reactors are operated in conditions in which carbon deposition is avoided. In practice, steam reforming reactors operate with steam-to-carbon ratios (H_2O/C or H_2O/CH_4) greater than 1, as pressures are usually higher than 10 atm. Dry reforming is operated at temperatures as high as 1,000°C and a CO_2/CH_4 ratio of 1.

Industrial application of reforming of natural gas for DR of iron is mainly applied in the Energiron III, Energiron ZR, and Midrex processes. The only industrial application of dry reforming of natural gas is in the Midrex process. The reformer is a reactor composed of several tubes containing a bed of a catalyst of nickel supported on alumina. The Midrex reformer is a stoichiometric reformer, and hence, the amount of natural gas is controlled to react with all the CO_2 and some of the residual H_2O present in the reforming gas. In the Midrex process, the reformed gas does not need to be cooled prior to injection into the shaft furnace, which saves some energy.

The Energiron III process uses a steam reformer operated at temperatures above 700°C and pressures above 5 atm. Therefore, an excess of steam is necessary to obtain high conversions of methane. The steam-to-carbon ratio used in the reforming process is 2 or higher. Consequently, the reformed gas needs to be cooled to condense this excess steam. The cooled reformed gas is then mixed with the top gas coming from the shaft furnace. Finally, the reformed gas mixed with the top gas is directed to the gas heater to reach the necessary temperature for reduction and then directed to the shaft furnace.

The Energiron ZR process brings a different concept without a reformer. In this case, reforming reactions occur inside the shaft furnace and the DRI serves as the catalyst. The top gas from the shaft furnace is treated in the same way as in the Energiron III for removal of water and CO_2. This gas is then mixed with natural gas and humidified with controlled amounts of water. The humidified gas passes through the gas heater and is then directed to the shaft furnace.

The amount of CO in the reducing gas is higher for the Midrex process because it is produced by dry reforming. Reducing gas for the Energiron process is richer in H_2, and for the ZR concept, the amount of CH_4 is particularly high.

In addition to reforming, oxygen and natural gas may be mixed with the reformed gas before injection into the shaft furnace. Oxygen is used to partly combust the reducing gases (CO, H_2, and CH_4) for temperature increase. This will enhance the rate of reduction inside the shaft furnace. Natural gas is injected either to burn or to carburize the reduced iron.

Both the Midrex and Energiron processes have adapted versions to work with coke oven gas. Reforming is also necessary because of the high amounts of hydrocarbons in this gas. Midrex has patented a technology—the Thermal Reactor System—in which a partial oxidation of the coke oven gas is performed to generate the reducing gas rich in H_2 and CO.

ENVIRONMENTAL CONSIDERATIONS

Ironmaking processes produce solid, gaseous, and liquid emissions, which are of environmental concern.

Blast Furnace Technology

The BF ironmaking process produces solid and gaseous emissions. There may also be liquid emissions, depending on how the cooling system is designed and how the gaseous and solid by-products are treated.

The BF generates a large quantity of off-gas, approximately 1,600 m^3/t of hot metal (Lankford et al. 1985b). On a dry basis, the gas is composed of roughly 23% CO_2, 21% CO, 3% H_2, and the balance N_2. There may also be small amounts of sulfur and volatile organic compounds, depending on the charge materials. The actual BF gas volume and composition depend on the thermodynamic equilibrium in the furnace and a function of the top-charged materials as well as the fuels and oxygen that may be injected into the lower portion of the furnace through the tuyeres. The gas has a heating value of approximately 3.54 KJ/m^3, which is a function of the gas composition (Lankford et al. 1985b).

During BF operation, lighter-weight, smaller solids are entrained in the off-gas. A dry separator that forces the gases to make an abrupt direction change causes the heavier solids to fall out of the gas by centrifugal force. The solids collected by this process are dry. A wet venturi scrubber may be used to remove the smaller particulate. Venturi scrubbers use water droplets and the energy from a high pressure to force water droplets to combine with the particulate and other water droplets to form slurry drops large enough to fall out of the gas stream. The particulate concentration in the BF gas exiting the venturi scrubbers is significantly reduced, and the gas is ready to be used as an energy source or flared. The slurry collected by the venturi scrubber may be treated by a filter press or other means to separate the particulates from the water. The water may be treated and recycled to the scrubber. The removed solids are in sludge form and may contain enough carbon to be used as a solid fuel. However, its use may be limited by the presence of iron oxide or other elements originating from steel scrap or other materials charged into the furnace. The sludge may be landfilled if all regulations that apply for the sludge composition are met. Recycling the sludge in an iron ore agglomeration process or separating the carbon and iron from undesirable elements, such as zinc, has been considered, but usually the processing costs outweigh the benefits of recovering the fuel or raw materials.

Roughly up to 65% of the BF gas is used to heat the stoves. The remainder is used elsewhere in the integrated plant or flared. The solids removed from the BF gas are either dry particles or a sludge-type material. The solids may be recycled within the integrated process, in either ore agglomeration or coke making, or landfilled, depending on the composition and regulations.

The major solid by-product of the BF ironmaking process is slag. It can be used as an ingredient to make cement. It is an attractive substitute for limestone as it contains a significant amount of calcium that does not need to be calcined. Initially, slag may contain a significant amount of iron or iron oxide. The iron-bearing portion is magnetically removed after the solidified slag is crushed into manageable portions.

It should be noted that sulfur is typically part of the BF ironmaking process. A majority of the sulfur originates from the coke but some may also be introduced via the iron oxide feed materials. Nearly all of the sulfur exits the BF as part of the liquid iron or slag. In some cases, the slag chemistry may be altered to maximize sulfur pickup and minimize the amount of sulfur in the iron. The iron desulfurization capability downstream of the BF is a factor in determining the slag chemistry.

Regardless, the BF slag will likely contain sulfur in some form. Consideration should be given to the possible emission of gases containing sulfur as the slag is cooled in the slag pits. In the case of slag granulation, the water may pick up sulfur compounds and need to be appropriately treated.

The liquid iron or hot metal produced may be desulfurized before the steelmaking process. The sulfur is typically removed by injecting an alkali reagent into the liquid metal. This process generates a slag type of material that is skimmed from the liquid metal. The iron-bearing materials in the slag may be removed from the slag and recycled into the process. The sulfur content in both the iron-bearing and non-iron-bearing materials influences how those materials may be recycled, landfilled, or used outside of the process. These are additional solids emitted from the process.

BF operation requires significant cooling, which is done with water. The water-cooling system can be an open- or closed-loop system. Open systems are noncontact systems in which the cooling water is not part of the process and is confined to being discharged, usually into the original body of water. Monitoring of the amount of water consumed and the discharge temperature may be required by regulations. In the case of closed cooling systems, any blowdown discharge required to meet optimal water chemistry may also be regulated.

Any water discharged may be regulated. The dewatering process discharge may contain suspended and dissolved solids that need to be removed before being recycled into the process or discharged to maintain proper water chemistry.

Alternative Technologies

In general, the solid, gaseous, and liquid emissions from the alternative ironmaking processes are similar to the BF process. The DRI exhaust gases emerge from the top of the reactor at a temperature in the range of 250°–350°C and usually at a higher pressure, up to 6.8 atm, than BF gas (Lankford et al. 1985a; J.O.L. Morales, personal communication). The gas consists of carbon dioxide, water vapor, nitrogen, carbon monoxide, and hydrogen as a result of the iron reduction thermodynamic equilibrium. Similar to the BF process, the DRI exhaust gas has chemical energy, but it is not used as a fuel; rather, it is treated so that the reducing gases may be recycled through the process.

The gases from the top of the reactor contain dust particles that are dragged out of the reactor because of the direct contact between the gas and the solid. The exhaust gas is cleaned and quenched. Typically, most of the carbon dioxide is removed from the exhaust gas to recycle the reducing gases, hydrogen, and carbon monoxide into the process. The removed carbon dioxide may be released to the atmosphere or liquefied and sold as appropriate.

In the solid-based DRI process, the gases exiting the reactor, usually a rotary hearth furnace, are fully oxidized, hot, and near atmospheric pressure. The thermal energy may be recovered via heat exchanger. The exhaust gases will likely contain particulates, nitrogen oxides (NOx), and sulfur dioxide (SO_2). The fuel and reductant in the solid-based DRI process is usually coal that contains some sulfur. Depending on the volume and concentrations, the exhaust gas will likely be passed through a system to remove the particulates and may be treated to remove some of the SO_2 and NOx.

In both the gas- and solid-based DRI processes, it is necessary to supply thermal energy to support the chemical reactions. Depending on how the system is designed, the thermal energy may be provided by indirectly preheating the process gases before they enter the reactor. The fuel for the indirect heating is typically natural gas, and the exhaust gases from the indirect heating process will essentially be free of particulates and SO_2. However, it may be necessary to treat the exhaust gases for NOx.

Similar to the BF process, the solids removed from the exhaust gas of either the gas- or solid-based DRI processes are either dry particles or a sludge-type material. They may be recycled into the iron ore pellet process, sold, or landfilled, depending on the composition and regulations.

Fines are generated during the handling of both the iron ore pellets fed to the reactor and the DRI produced. Material handling systems equipped with highly efficient dust collecting units are employed to capture the airborne fines. These solids may be briquetted and fed into the reactor with the iron ore pellets.

Both the gas- and solid-based DRI processes require a significant amount of cooling, which is done with water. The water-cooling systems are similar to those used for the BF process, and the environmental considerations are similar.

Greenhouse Gas Emissions

Currently, there is some concern about the emission of carbon dioxide (CO_2) and other gases that may contribute to climate change. Although CO_2 emissions are not regulated by the U.S. federal government, several U.S. states and several countries around the world have established cap-and-trade or other systems in an attempt to reduce CO_2 emissions.

As with nearly all energy-intensive processes, such as the manufacture of electricity, cement, glass, paper, and chemicals, the ironmaking process emits a relatively large amount of CO_2. On an intensity basis, the BF process to make iron emits roughly 1.3 t of CO_2 per ton of liquid, carbon-saturated iron (IEA 2000). Alternatively, roughly 0.8 t of CO_2 is emitted in making a ton of liquid iron (containing some carbon) from melting DRI in an EAF. The amount of energy required to reduce and melt the iron from iron ore is the same in both processes. The lower CO_2 emissions from the DRI-EAF process can be primarily attributed to the use of natural gas as both a fuel and reductant. Although a significant amount of natural gas may be used in the BF process, the process requires some minimum amount of coke (from coal) to maintain bed porosity within the furnace.

Note that the calculation of CO_2 emissions can be complex and should be well thought out, particularly if the calculated emission intensity is used to compare processes. To understand the meaning of the value, it is important to note the basis of the CO_2 emission calculation; whether it is scope 1, scope 2, scope 3 (EPA 2018), or some other type of calculation. Briefly, the scope 1 emission only accounts for carbon that crosses the boundary in liquid or solid form. The scope 2 calculation adds the emissions associated with energy, typically electricity, which crosses the boundary either into or out of the process. Scope 3 calculations account for emissions associated with preparing the materials that cross the boundary. For example, the DRI iron ore pellets usually have a higher iron content than BF pellets, and thus, more CO_2 may have been emitted to prepare the higher-quality pellets. It may be somewhat obvious that a thorough scope 3 or other type of in-depth calculation may require at least a few assumptions and the development of standard CO_2 emission factors for various materials.

ACKNOWLEDGMENTS

The authors express their appreciation to Jane M. Grochowski for her time and efforts in reviewing this chapter and her valuable edits. Appreciation is also extended to Jack E. Grochowski for his contributions in preparing some of the schematic drawings, his time and efforts in reviewing the chapter sections, and his commendable edits.

Special thanks go to Joseph J. Poveromo, John A. Ricketts, Baojun Zhao, Jianling Zhang, Yuexin Han, Chu Mansheng, Joel Morales, Naryan Govidaswami, Ronnie Dabideen, and Mahdi Farahani for their technical input and guidance.

REFERENCES

AIST (Association for Iron and Steel Technology). 2018. *AIST Industry Roundups: North American Blast Furnace Roundup*. Warrendale, PA: AIST.

Anameric, B., and Kawatra, S.K. 2007. Properties and features of direct reduced iron. *Miner. Process. Extr. Metall. Rev.* 28(1):59–116.

Anameric, B., and Kawatra, S.K. 2008. Direct iron smelting reduction processes. *Miner. Process. Extr. Metall. Rev.* 30(1):1–51.

Bahgat, M., Abdel Halim, K.S., El-Kelesh, H.A., and Nasr, M.I. 2009. Metallic iron whisker formation and growth during iron oxide reduction: K_2O effect. *Ironmaking Steelmaking* 5(36):379–387.

Bitsianes, G., and Joseph, T.L. 1955. Topochemical aspects of iron oxide reduction. *Trans. AIME* 7(5):639–645.

Carmichael, I.F. 2012. Blast furnace design I (Lecture no. 6). Presented by D. Berdusco at the 22nd McMaster University Blast Furnace Ironmaking Course, Hamilton, ON, May 13–18.

Chatterjee, A. 1994. *Beyond the Blast Furnace*. Boca Raton, FL: CRC Press.

Dahlstedt, A., Halin, M., and Wikstrom, J-O. 2000. Effect of raw material on blast furnace performance. In *The Use of an Experimental Blast Furnace: 4th European Coke and Ironmaking Congress Proceedings*. Vol. 1. pp. 138–145.

Edstrom, J.O. 1953. The mechanism of reduction of iron oxides. *J. Iron Steel Inst.* 175(11):289.

Edstrom, J.O., and Bitsianes, G. 1955. Solid state diffusion in the reduction of magnetite. *J. Met.* 7(6):760–765.

Elliott, J.F., and Gleiser, M. 1960. *Thermochemistry for Steelmaking*. Boston: Addison-Wesley.

EPA (U.S. Environmental Protection Agency). 2018. Greenhouse gases at EPA. www.epa.gov/greeningepa/greenhouse-gases-epa. Accessed July 2018.

Geerdes, M., Kurunov, I., Lingiardi, O., and Ricketts, J. 2015. *Modern Blast Furnace Ironmaking: An Introduction*, 3rd ed. Amsterdam, Netherlands: IOS Delft University Press.

Hashimoto, S., Suzuki, A., Yoshimoto, H. 1977. Burden and gas distribution in the blast furnace. In *Ironmaking Proceedings*, vol. 36. New York: AIME.

Helenbrook, R.G. 2012. Blast furnace design II. Presented at the 22nd McMaster University Blast Furnace Ironmaking Course, Hamilton, ON, May 13–18.

IEA (International Energy Agency). 2000. *Greenhouse Gas Emissions from Major Industrial Sources. III: Iron and Steel Production*. Report No. PH3/30. Paris, France: IEA. www.ieaghg.org/docs/General_Docs/Reports/PH3-30%20iron-steel.pdf.

Joseph, T.L. 1936. Porosity, reducibility, and size preparation of iron ores. *Trans. AIME* 120:72–98.

Lankford, W.T., Samways, N.L., Craven, R.F., and McGannon, H.E. 1985a. Direct reduction and smelting processes. In *The Making, Shaping and Treating of Steel*, 10th ed. Edited by W.T. Lankford, N.L. Samways, R.F. Craven, and H.E. McGannon. Pittsburgh, PA: Association of Iron and Steel Engineers.

Lankford, W.T., Samways, N.L., Craven, R.F. and McGannon, H.E. 1985b. The manufacture of pig iron in the blast furnace. In *The Making, Shaping and Treating of Steel*, 10th ed. Edited by W.T. Lankford, N.L. Samways, R.F. Craven, and H.E. McGannon. Pittsburgh, PA: Association of Iron and Steel Engineers.

Lanyi, M.D., Cao, J., and Terrible, J.A. 2012. A sensible route to energy efficiency improvement and CO_2 management in the steel industry. Presented at the 2012 Iron and Steel Technology Conference and Exposition (AISTech).

Lien, H.O., El-Mehairy, A.E., and Ross, H.U. 1971. A two-zone theory of iron-oxide reduction. *J. Iron Steel Inst.* 209:541–545.

Lu, W-K. 1999. Kinetics and mechanism in direct reduced iron. In *Direct Reduced Iron: Technology and Economics of Production and Use*. Edited by J. Feinman and D.R. MacRae. Warrendale, PA: Iron and Steel Society.

Lu, W.K. 2012. Blast furnace reactions (Lecture no. 2). Presented at the 22nd McMaster University Blast Furnace Ironmaking Course, Hamilton, ON, May 13–18.

Lu, W.K., and Ranade, M.G. 1991. Recent advances in blast furnace ironmaking in North America. *ISIJ Int.* 31(5):395–402.

Markotic, A., Dolic, N., and Trujic, V. 2002. State of the direct reduction and reduction smelting processes. *J. Min. Metall. Sect. B* 38(3-4).

Meijer, K., Zeilstra, C., Teerhuis, C., Ouwehand, M., and van der Stel, J. 2013. Developments in alternative ironmaking. *Trans. Indian Inst. Met.* 66(5-6):475–481.

Omori, Y., and Shimomura, Y. 1982. Flow of gas, liquid and solid. In *Blast Furnace Phenomena and Modelling*. Edited by Committee on Reaction within Blast Furnace Joint Society of Iron and Steel Basic Research, The Iron and Steel Institute of Japan. London: Elsevier Applied Science.

Peacey, J.G., and Davenport, W.G. 1979. *The Iron Blast Furnace, Theory and Practice*. Oxford: Pergamon Press.

Poveromo, J.J. 2010. Coke in the blast furnace (Lecture no. 2). Cokemaking Course. McMaster University, Hamilton, ON.

Ricketts, J.A. 2017. How a blast furnace works. http://foundrygate.com/upload/artigos/How%20a%20Blast%20Furnace%20Works.pdf. Accessed July 2018.

Ross, H.U. 1972. The kinetics of iron ore reduction. www.aimehq.org/doclibrary-assets/books/Ironmaking%20Proceedings%201972/Ironmaking%20Proceedings%201972%20-%20044.pdf. Accessed July 2018.

Ross, H.U. 1973. The fundamental aspects of iron ore reduction. In *Symposium on Science and Technology of Sponge Iron and Its Conversion to Steel*. Jamshedpur, India: National Metallurgical Laboratory.

Ross, H.U., McAdams, D., and Marshall, T. 1980. Kinetics and mechanisms in direct reduced iron. In *Direct Reduced Iron: Technology and Economics of Production and Use*. Edited by J. Feinmen and D.R. MacRae. Warrendale, PA: Iron and Steel Society.

Sano, N., Lu, W.K., and Riboud, P.V. 1997. *Advanced Physical Chemistry for Process Metallurgy*. New York: Academic Press.

Seth, B.B.L., and Ross, H.U. 1965. The mechanism of iron oxide reduction. *Trans. AIME* 233:180.

Smith, R., and Corbett, M. 1987. Coal-based ironmaking. *Ironmaking Steelmaking* 14:49–75.

Tovarovskii, I.G., Bol'shakov, V.I., Togobitskaya, D.N., and Khamkhot'ko, A.F. 2009. Influence of the softening and melting zone on blast-furnace smelting. *Steel Transl.* 39(1):34–44.

Turkdogan, E.T., and Vinters, J.V. 1971. Gaseous reduction of iron oxides: Part I. Reduction of hematite in hydrogen. *Metall. Trans. B* 2(11):3175–3188.

Turkdogan, E.T., and Vinters, J.V. 1972. Gaseous reduction of iron oxides: Part III. Reduction-oxidation of porous and dense iron oxides and iron. *Metall. Trans. B* 3(6):1561–1574.

Turkdogan, E.T., Olsoon, R.G., and Vinters, J.V. 1971. Gaseous reduction of iron oxides: Part II. Pore characteristics of iron reduced from hematite in hydrogen. *Metall. Trans. B* 2(11):3189–3196.

Udy, M.C. 1959. Recent developments in strategic-Udy smelting process. In *Electric Furnace Conference Proceedings*. New York: AIME.

Udy, M.C., and Lorig, C.H. 1942. Low temperature gaseous reduction of a magnetite. Presented at the Metals Technology Meeting, AIME, Cleveland, OH.

Udy, M.C., and Udy, M.J. 1958. Production of iron and steel by the strategic-Udy process. In *Electric Furnace Proceedings*. Warrendale, PA: Association for Iron and Steel Technology.

Venkateswaran, V., and Brimacombe, J.K. 1977. Mathematical model of the SL/RN direct reduction process. *Metall. Trans. B* 8(2):387–398.

Wallace, J.P., Dzermejko, A.J., Hyle, F.W., Goodman, N.J., Lee, H.M., and Zeigler, R.W. 1999. The blast furnace facility and equipment. In *The Making, Shaping and Treating of Steel*, 11th ed. Ironmaking Volume. Edited by D.H. Wakelin. Warrendale, PA: Association for Iron and Steel Technology.

Weiss, B., Sturn, J., Voglsam, S., Strobl, S., Mali, H., and Schenk, J. 2011. Structural and morphological changes during reduction of hematite to magnetite and wüstite in hydrogen rich reduction gases under fluidized bed conditions. *Ironmaking Steelmaking* 38(1):65–73.

World Steel Association. 2015. Fact sheet: Steel and raw materials. www.worldsteel.org. Accessed July 2018.

Zervas, T., McMullan, J.T., and Williams, B.C. 1996a. Direct smelting and alternative processes for the production of iron and steel. *Int. J. Energy Res.* 20(12):1103–1125.

Zervas, T., McMullan, J.T., and Williams, B.C. 1996b. Gas-based direct reduction processes for iron and steel production. *Int. J. Energy Res.* 20(2):157–185.

CHAPTER 12.18

Kaolin

Robert J. Pruett and Mike Garska

Kaolin is clay composed of kaolin minerals that is naturally white or nearly white, or that can be beneficiated to white or nearly white. Commercial kaolin deposits are classified as primary or sedimentary. Primary deposits include residual weathering horizons, or hydrothermal alteration zones along fracture networks, hosted in feldpathic igneous rocks such as granite or feldpathic metamorphic rocks such as gneiss. Sedimentary deposits include concentrated kaolinitic clays deposited in fluvial, deltaic, or transitional marine environments, and include kaolinized sediments that were originally feldspar grains, mica grains, or rock fragments that contained labile minerals such as feldspar. Sedimentary kaolin deposits have a continuum of mineral content and textures ranging from bauxite to kaolinitic sandstone to plastic clay (Figure 1).

GOALS OF KAOLIN PROCESSING

Commercial kaolin deposits are mined and processed for wide range of uses (Table 1). Kaolin mined in the United States during 2013 was used as white pigments for paper and paint (55%); as ceramic raw materials (11%); as refractories (10%); as fillers for rubber, plastic, caulk, and sealants (7%); as fiberglass raw materials (5%); as catalysts for oil and gas refining (3%); and for other applications (9%) such as inks, cosmetics, animal feed, pesticides, pharmaceuticals, fertilizer, and roofing materials (Virta 2015). Kaolin processing is designed to obtain physical and chemical properties needed by customers within each market segment (Table 1) by providing products with consistent quality from specific ore types. Specified ranges of major oxides are needed for kaolin products supplied to the ceramics, refractories, fiberglass, and catalyst markets. Coating pigments are typically specified for their degree of whiteness (brightness and shade of color), particle size (wt % <2 μm), and rheological properties of mineral-water slurry and low abrasivity.

ECONOMIC CONCERNS

According to the U.S. Geological Survey (USGS), kaolin is produced in 60 countries (Virta 2015). The largest kaolin-producing countries in 2012 were the United States (23%), Germany (17%), China (12%), Brazil (7%), Iran (6%), Turkey (5%), United Kingdom (4%), Korea (3%), Vietnam, Czech Republic, and Mexico (each with 2%) (Anon. 2015). The largest kaolin-producing companies are Imerys, KaMin, Thiele, BASF, AKW, and Sibelco (Patel 2014). These companies mine large-volume kaolin ore deposits with uniform physical and chemical properties that are used for large-volume applications such as paper pigments. Deposits in Georgia (United States), northern Brazil, and Cornwall (United Kingdom) have the quality and large volume of ores that can be successfully developed for pigments in the context of extremely competitive market conditions, technological complexity of kaolin processing, demanding product specifications, high capital investment needed for facilities, and access to global transportation infrastructure.

The value of kaolin products relates to the beneficiation method, market, grade, and product form. The USGS provides data on total value and volume of Georgia (United States) kaolin produced by different beneficiation methods (Figure 2). The production tracked by the USGS is for kaolin beneficiated by airfloat, water-washed, delaminated, and calcined process methods, to be discussed in more detail later in this chapter. The USGS numbers show gradual increases in airfloat and hydrous (water-washed and delaminated) kaolin value per short ton over the past 20 years. Calcined kaolin values show high volatility because the product mix they represent is very broad in terms of markets served (Table 1), degree of wet and dry processing, and range of packaging and bulk shipment options.

Euromoney Institutional Investor, the publisher of *Industrial Minerals*, tracks prices for some grades of kaolin. In August 2015, its website showed that prices of no. 1 coating kaolin fob (free on board) Georgia were between $149/t and $209/t, prices of Brazilian coating kaolin slurry were between $242/t and $295/t CIF (cost, insurance, and freight) Europe, and prices of bagged <75 μm kaolin for sanitaryware

Robert J. Pruett, Formerly with Imersys Oilfield Solutions, Sandersville, Georgia, USA
Mike Garska, Senior Process Engineer, Hudson Ranch Energy Service, Calipatria, California, USA

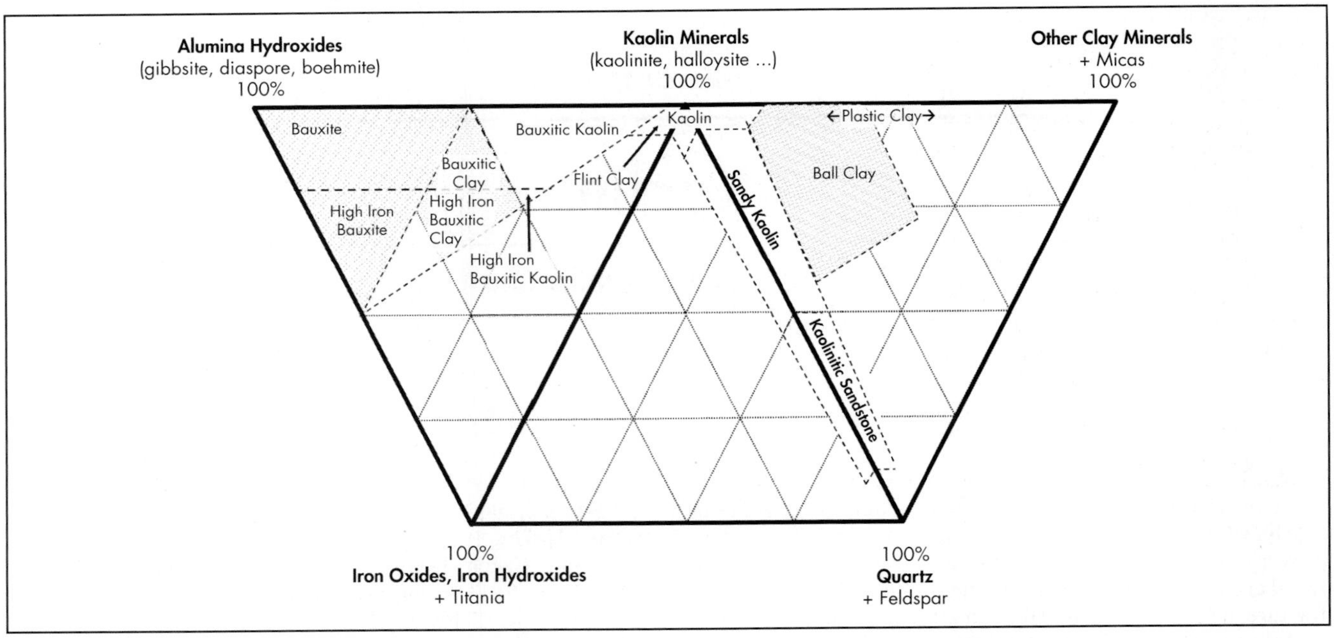

Source: Pruett 2016

Figure 1 Classification of kaolin and kaolin mineral-bearing rocks based on their mineral content

Table 1 Uses of kaolin and properties important for specified use

Properties	Refractory	Fiberglass	Ceramic	Catalyst	Rubber	Plastic	Pigment, coating	Pigment, filling	Specialty
Chemistry, major oxide	x	x	x	x					
White or near-white			x				x	x	
Refractive index							x	x	
Particle size, fineness, and range			x		x	x	x	x	
Particle shape, platy or delaminated						x	x	x	
Surface area (5–25 m^2/g)					x		x	x	
Abrasiveness, soft (2.5 Mohs)							x	x	
Hydrophilic, readily make-down			x				x	x	
Insoluble, wide pH range							x	x	See Murray 2007.
Surface charge, low (meq/g)			x		x		x	x	
Low viscosity, clay-water slurry							x	x	
Plasticity			x						
Conductivity, low (heat, electrical)	x					x			
Refractory, high melting temperature	x		x	x					
Stability, form (heat, chemical)			x				x	x	
Calcined/sintered mineral content	x		x	x	x	x	x	x	
Cost	x	x	x	x	x	x	x	x	

fob Vietnam were between \$198/t and \$209/t (*Industrial Minerals* 2015).

EXAMPLES OF KAOLIN PROCESSING OPERATIONS

Kaolin operations are categorized by dry and wet process. Dry processes include airfloat and calcined operations. Wet processes include water-washed, delaminated, and calcined operations. The distinction between dry and wet process operations is whether the kaolin is made-down into slurry with water at any point prior to shipment of final product. The general advantage with the dry process is in the lower energy and fewer chemicals needed to produce finished product. Typically, the dry process has one energy-intensive step to dry product from a moist crude or plastic state containing less than about 20% moisture. The energy intensity of wet processing kaolin is described in Pruett (2011), and it often involves dewatering and drying slurries containing between 30% and 80% moisture unless the product is shipped as slurry. Dry processed kaolins can be chemical free. Wet processed kaolin need chemicals for dispersion, beneficiation, preservation, and tailings treatment. The advantages of wet processing kaolin are the degree of purity that can be reached by removing impurities, the degree of particle engineering that can be achieved by size separation and grinding, and the overall consistency of the product blend.

Airfloat Operations

Airfloat processing was developed in the 1920s to simply dry and remove oversized grit (>45 μm) particles from kaolin used for paper filling and ceramic applications (Henry and Vaughan 1937; Pruett 2000). Dry processing for pigment and ceramic grades prior to the airfloat and wet processes often involved hand sorting the ore to remove iron-stained gangue, then drying the sorted ore on racks prior to bulk shipment (Smith 1929). The airfloat operation comprises the unit operations of mining and crude ore selection, hauling, storage, blending, feeding and crushing, drying, milling and classification, and bulk loading or packaging (Figure 3; Buie and Schrader 1982).

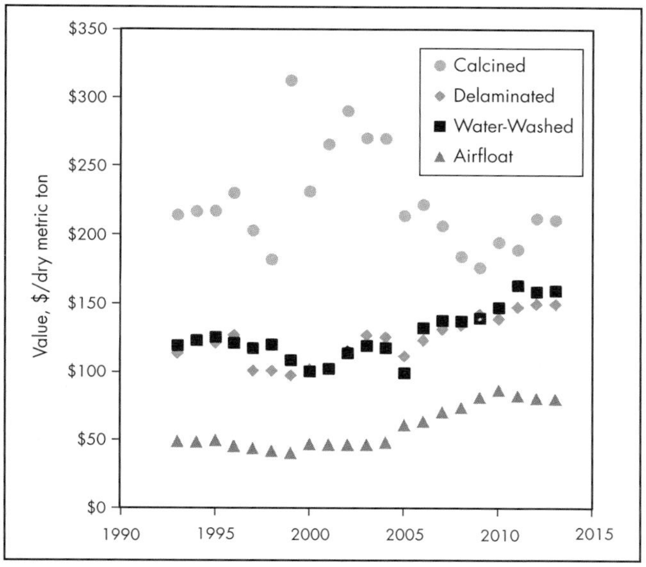

Data from USGS *Minerals Yearbook*

Figure 2 Average Georgia and U.S. kaolin values by process

Water-Washed Operations

The water-washed operation can comprise multiple unit operations to obtain high-purity pigments (Figure 4), or the operation can comprise a few unit operations such as blunging kaolin ore and degritting the kaolin slurry prior to loading the slurry into a railcar for shipment to some markets. Water-washing kaolin has been practiced for centuries, initially to separate clay-sized kaolin mineral particles from sand-sized quartz and mica particles from primary kaolin ore (Chen et al. 1997; Thurlow 2001). The first major sedimentary kaolin water-washed operation began production in 1908 at Dry Branch, Georgia, United States. Today the water-washed kaolin operation is the principal manufacturing platform for high-purity kaolins having tightly controlled particle size distributions and high brightness.

The principal reasons for water-washed kaolin processing are the benefit of complete particle dispersion in the water medium, the option to add chemicals that improve the clay's performance in terms of brightness and rheology, and the better mixing of particles from different ore sources to improve product consistency. The initial steps of mining, transport, storage, and feeding ore into the operation are similar in dry and wet processes (Figures 3 and 4). The blunging or mixing of the kaolin ore with water to form slurry with the aid of mechanical shear, pH adjustment, and chemical dispersants begins the wet part of the process. Degritting the slurry to remove coarse particles >45 μm typically follows the blunging unit operation. When the kaolin is in slurry form, it can be stored in tanks and then transported by pipeline from the mine product tanks to the beneficiation unit operations. Beneficiation can take several forms and form multiple sequences to optimally improve product purity using high-intensity wet magnetic separation, flotation, selective flocculation, ozone bleaching, and acid-reductive leaching, and change the particle size distribution to improve functional performance by blending and centrifugal classification (Murray 2007; Thurlow 2001). The

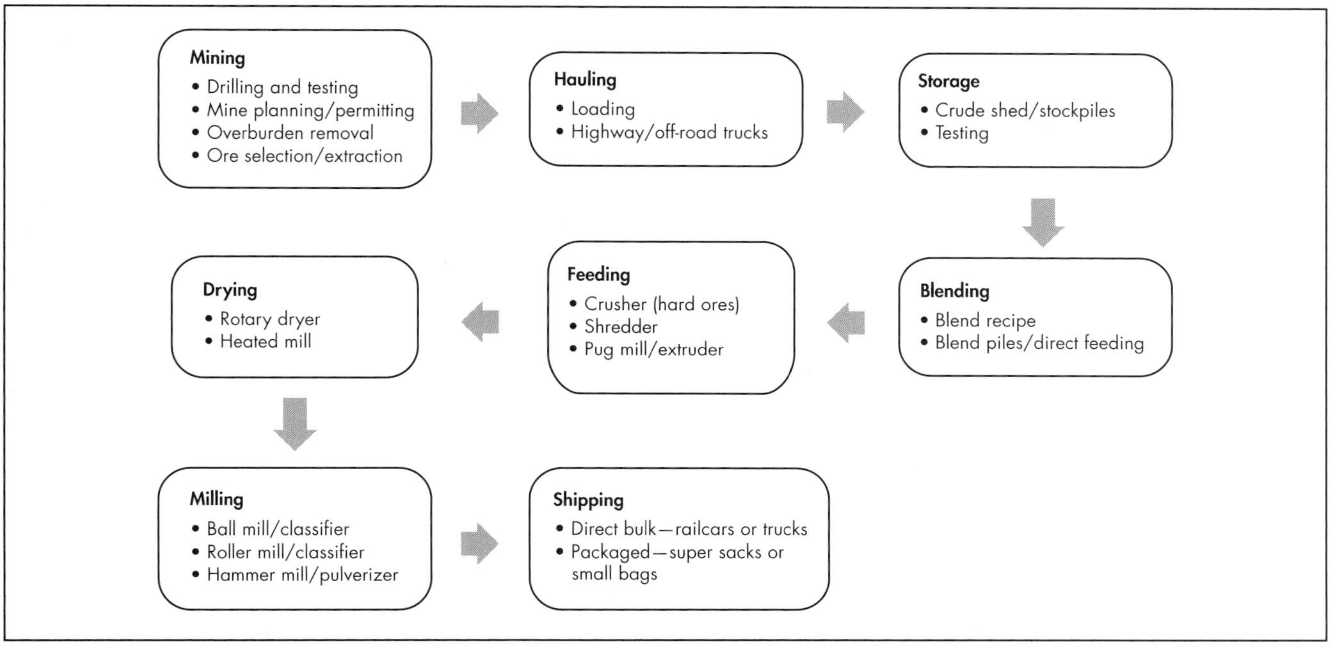

Figure 3 General flow diagram for an airfloat kaolin operation showing typical sequence of unit processes

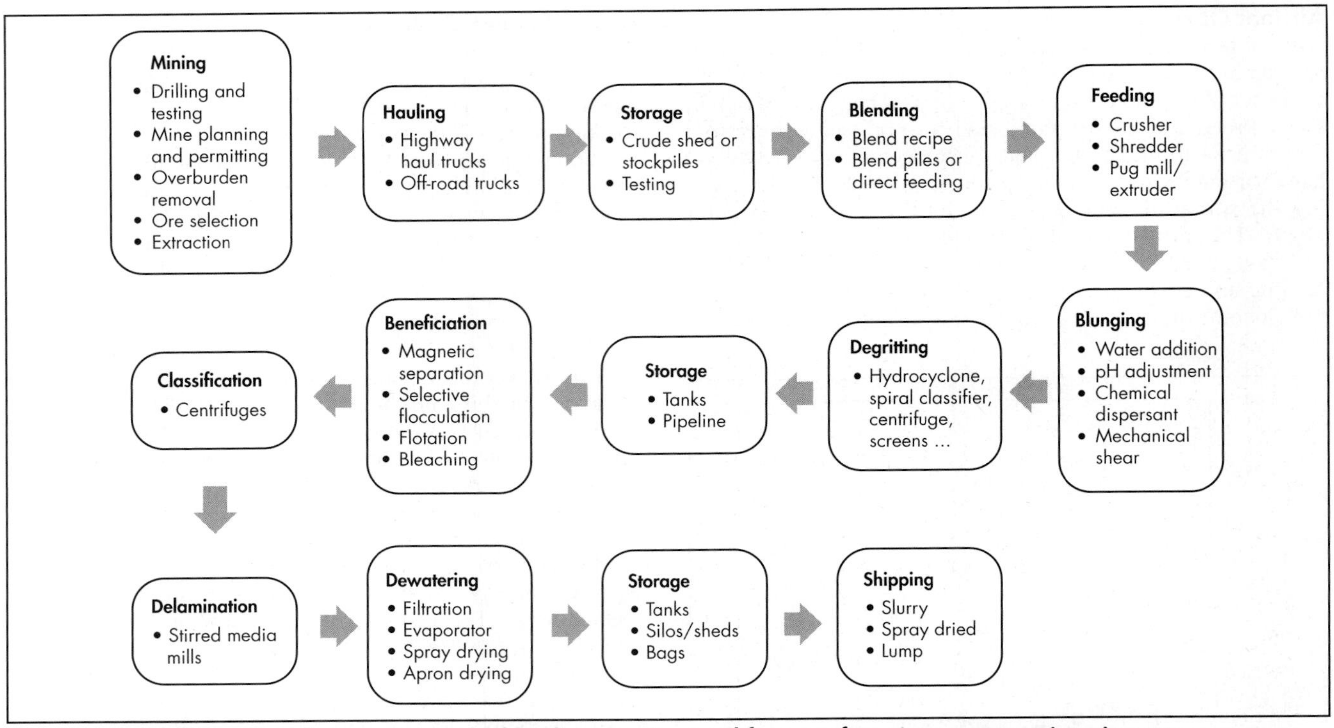

Figure 4 General flow sheet for a water-washed kaolin operation used for manufacturing paper coating pigments

beneficiation and classification processes operate more efficiently at low slurry solids, typically 15%–40%. Dewatering using filters and evaporators is needed to increase slurry solids to 60% or higher for high solids slurry shipment or to more efficiently feed drying equipment at the downstream portion of the wet process.

Delaminated Kaolin

Operations that delaminate kaolin are water-washed processes (Figure 4). Delamination was developed to address lower recoveries caused by higher market demand for fine coating kaolin beginning in the late 1930s. The lower recoveries resulted from unused coarse fractions from centrifugation filling tailings impoundments of Georgia coating kaolin producers. The coarse fractions rejected by centrifuge are mostly kaolinite stacks and kaolin agglomerates. Two developments in the middle 1950s enabled these coarse fractions to be used. First was the discovery that coarse platy particles are a good coating pigment for paper (Bundy 1993). Second was the development of delamination. Delamination is the parting or cleaving of kaolinite stacks into individual platelets (Figure 5) by extrusion or by attrition-grinding of clay slurries using stirred media mills (Feld and Clemmons 1963). The grinding media used can be quartz sand or ceramic, which is highly spherical and stronger and more wear resistant than quartz. Media sizes used for delamination are typically about 1 mm.

Calcined Kaolin

Calcined kaolin and metakaolin are thermally processed crude or beneficiated kaolin. The phase changes that occur by heating kaolin to high temperatures include (1) dehydroxilization of kaolinite about 500°C (932°F) to form metakaolin, (2) forming spinel or primary mullite at about 1,000°C (1,832°F) from metakaolin, and (3) recrystallizing the calcined kaolin mass to form secondary mullite above 1,200°C (2,192°F). Metakaolin is used as a pozzolan for cement, to make alum for water treatment, as a pigment for paper, and for a variety of specialty applications. Kaolins that are fully calcined near 1,000°C (1,832°F) are used as pigments and for mild abrasives because of their low concentrations of submicrometer mullite. Low abrasivity and high brightness are important quality parameters for calcined kaolin used for pigments. Kaolins sintered above 1,200°C (2,192°F) are used for ceramics and refractories because of their high mullite content, high strength, chemical resistance, and heat resistance.

Calcined kaolin operations are the extension of either the dry or wet process. Dry process feeds to the kiln can be crushed crude, extruded-milled product, or granulated product as in the case of proppants. Kilns used in the kaolin industry include rotary and Herreshoff. Wet process feeds to a kiln are typically dried with some additional milling or sizing before calcination to either structure the kaolin particles into porous aggregates, as in the case of pigments, or to structure the finished particle into a rounded sphere, as in the case of ceramic proppants (Figure 6). Calcined operations for pigments have milling unit operations after the calciner to produce a fine pulverized form, and some of the milled calcined kaolin can then be reslurried for shipment to pigment customers. Calcined operations for structured particles such as ceramic proppants will have screening unit operations to size the product to meet customer specifications. Most calcined kaolin operations can ship packaged products in super sacks or small bags, or in bulk form as appropriate for the market served.

UNIT OPERATIONS

The unit operations found in kaolin operations are selected and designed based on ore characteristics, product specifications,

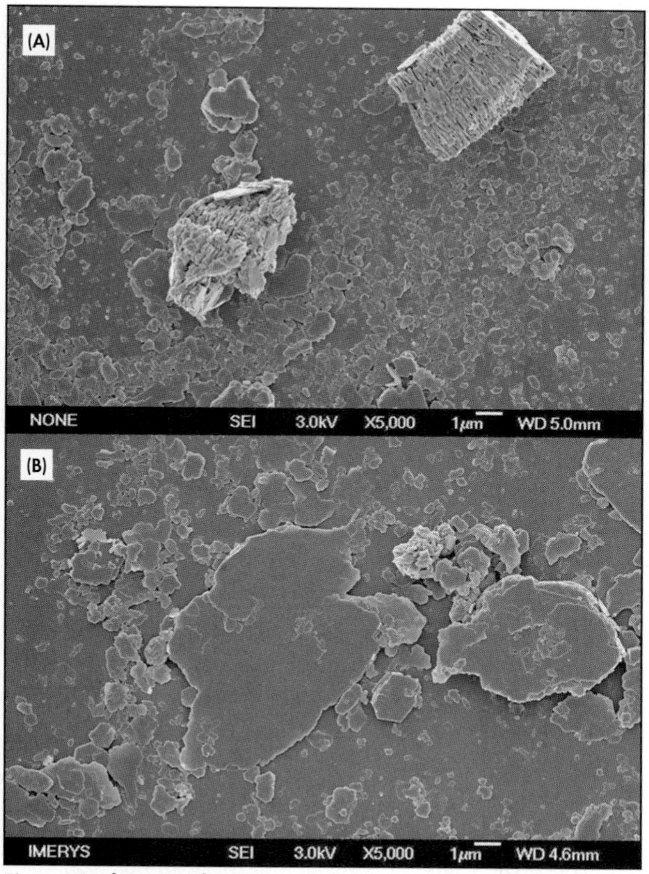

Figure 5 Electron photomicrographs showing (A) kaolinite particles contained in a slurry feeding an attrition grinder that comprise stacks and plates, and (B) kaolinite particles contained in a slurry product from an attrition grinder that comprise plates delaminated from stacks

and cost. Described in the following sections are unit operations typically used for kaolin processing.

Mining and Crude Ore Selection

Kaolin bodies are spatially delineated by extensive drilling and testing of core samples. The drill-hole spacing for exploratory work is typically about 120 m (400 ft), and drilling at 60 m (200 ft) and 30 m (100 ft) spacing typically occurs before stripping to prove reserves, prepare a mine permit, and prepare for mining. After the overburden has been removed, additional drilling at 15 m (50 ft) spacing occurs for mine quality control. The cores retrieved from drilling are described and then split into 0.3–1 m (1–3 ft) interval segments for initial quality testing for key parameters such as grit (>325 ASTM mesh) content, chemistry, brightness, and particle. The initial core intervals having similar quality are then composited into larger contiguous segments that represent one to several meters of a minable kaolin bed. The initial quality parameters are retested on the composites, and additional information is collected on quality parameters important for the intended market, such as brightness and rheology for ores intended for paper pigments, and processability parameters such as brightness response to ozone, magnetic separation, flotation, and leaching. Further details on kaolin testing are provided by Pruett and Pickering (2006).

Mining method depends on the geology and properties of the ore. The majority of kaolin mined in Georgia (United States) and northern Brazil are from sedimentary deposits. The overburden for these deposits is generally unconsolidated and can be removed by excavators and off-road trucks. Occasional hard layers can be ripped, hammered, or blasted prior to removal. The sedimentary kaolin ores can be mined using a backhoe and hauled to a stockpile or directly to a blunger. Primary kaolin hosted in altered granites such as in Cornwall and Devon (United Kingdom) may also be mined directly from the face by an excavator. Hydraulic mining was commonly practiced in the UK primary kaolins because of the high amount of unaltered matrix; the high amount of quartz, feldspar, and mica particles associated with the kaolinite; and the availability of water. Hydraulic mining involves the use of high-pressure water-jet monitors to wash the mine face to liberate kaolin clay and wash the kaolin into sumps where the kaolin slurry can be pumped to refining operations. Both dry and hydraulic mining require knowledge of the kaolin properties in the ore so that ore quality can be controlled and stockpiles with known qualities for blending can be built.

Crude Transportation and Storage

After excavation, the kaolin is transported from mine to processing plant by two basic methods: as dispersed slurry by pipeline or as solid lump clay by highway haul trucks (Pruett and Pickering 2006).

Dry

Crude kaolin ore is transported from stockpiles at the mine or directly to stockpiles near the plant by highway truck or off-road truck depending on distance and road network. Stockpiles near the plant or near a blunger site can be built on pads open to weather, or they can be built under crude sheds to prevent the clay from becoming too wet and sticky to handle during periods of wet weather. Stockpiles are typically sampled while they are built, after they are built, or both. Stockpile samples are tested similar to crude ore to confirm their quality and provide information to the quality lab for blending.

Slurry

Slurry from hydraulic mining or from blunging operations located near the mine is stored in agitated tanks to prevent sedimentation of coarse particles. Most tanks located between unit operations in water-washed kaolin operations are agitated, including final product tanks. Tanks that do not require agitation are those that contain fine particle size (>95 wt % <2 μm) dispersed kaolins or contain flocculated kaolin for solid–liquid separation or reduced-acid leaching. Transportation between unit operations in a wet plant is by pipeline. Many pipelines in Georgia exceeding 16 km (10 mi) long, and in the Rio Capim area of northern Brazil exceeding 161 km (100 mi) long, connect blunging and degritting operations located near the mine with the main beneficiation operations located near the railroad or port. The quality lab routinely tests the slurry contents of each tank for properties such as solids, pH, brightness, and particle size.

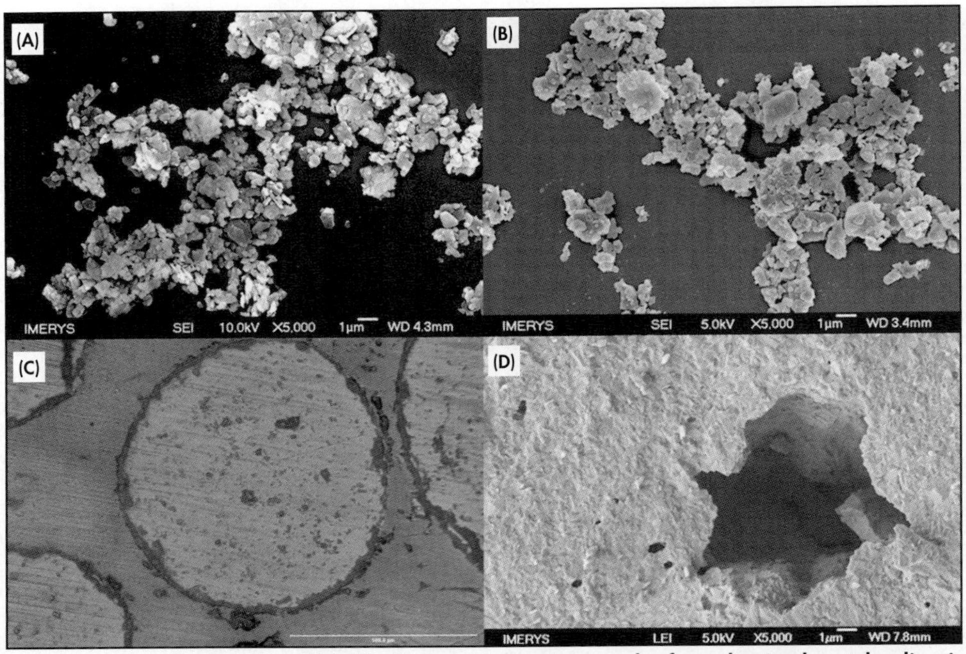

Figure 6 Examples of calcined and sintered kaolin: (A) electron photomicrograph of a pulverized metakaolin pigment, (B) electron photomicrograph of a pulverized fully calcined kaolin pigment, (C) visible-light micrograph showing a cross section of a sintered ceramic proppant, and (D) electron photomicrograph showing porcelain ceramic made by sintering kaolin to mullite

Feed Preparation

Crude kaolin ore from stockpiles must be sized for conveyor handling and to feed blungers. Soft kaolin ores may break down easily when run through a roll crusher. Hard kaolin ores may require a jaw crusher to break down boulder-sized ore into a size range to feed the process. Crushed kaolin can be fed to a shredder to further reduce top size of the ore to smaller than 5 cm (2 in.).

Dry Process

Beyond sizing the kaolin ore for handling in the operation, there is little additional feed preparation required for the dry process other than drying (Murray 1961). A blend pile can be built to help homogenize the feed to the process based on a recipe from the quality lab. Kaolin is typically dried before milling in the airfloat process using a rotary dryer. Some dry process operations can mix the wet crude ore with dry fine materials collected from the downstream process to dry the feed. Dry process calcine operations producing metakaolin for cement or chemicals will combine the crude drying and calcining using a rotary kiln.

Blunging and Slurry Preparation

Feed preparation for the wet process needs to liberate and disperse individual kaolin particles from the ore. Crushed kaolin ore is fed into a blunger (a tank or vat with a mixing shaft) to make clay slurry with water and dispersant chemicals under mechanical shear. The crushed kaolin ore is conveyed to a point where it drops into the blunger. Water is added to achieve target solids of between 30% and 70% (Murray 1980). The mixing shaft has a bottom blade or impeller for breaking up particles under high shear. The slurry must pass through a screen to prevent large lumps of clay from exiting. Downstream in the kaolin water-washed process are two other unit operations having a similar purpose. The reblunger (or repulper) is located after the filters for reslurrying filter cake with dispersant chemicals, but without additional water. The slurry (or reslurry) is located at the load-out to prepare hydrous and calcined kaolin slurries for shipment. Chemicals to disperse, stabilize, and preserve the slurries are added. Hydrous kaolin slurry solids can be raised by adding dry clay. Water is added to make calcined kaolin slurry.

Chemicals are important in the wet process beginning at the blunging unit operation. Kaolinite has a very low surface charge, unlike other clay minerals such as micas and smectite. The charge on kaolinite edges and faces are pH dependent (Yin et al. 2012). At pH <5, kaolinite particles will flocculate because of positive edge charge and negative face charge. Making a high (>50%) solids fluid slurry at low pH and without dispersant is impossible; the kaolin will have a plastic or cake-like texture. At pH >6, kaolinite particles have negative charge on their edge and faces. However, particles will not be fully dispersed until pH is >9, because some ancillary minerals such as anatase have a higher pH zero point charge. A dispersant chemical is used to amplify the negative charge and to prevent particle collision and coagulation. Chemicals used to adjust kaolin slurry pH are inexpensive and have a monovalent cation. Kaolin particles coagulate in water having divalent cations such calcium and magnesium, divalent anions such as sulfate, trivalent ions such as aluminum, and a high ionic strength that depresses the double layer surrounding particles. Typical chemicals used to raise pH during blunging and slurry make-down are sodium carbonate, sodium hydroxide, and ammonium hydroxide. Dispersant chemicals that amplify the negative charge on kaolin particle surfaces include a wide variety of phosphates such as sodium hexametaphosphate and tetrasodium pyrophosphate, sodium silicate, and sodium polyacrylate. The inorganic dispersants are short-lived at low-dose amounts, whereas sodium polyacrylate is a relatively long-lived dispersant and is used to stabilize the dispersion of slurries that need to be transported long distances by rail or ship.

Beneficiation, Wet Process

Impurity removal is one major advantage for wet processing kaolin. The dry process has few opportunities to upgrade ore quality other than by selective mining; by ore sorting; and by separating out large, hard particles. Kaolin, being a clay, has many silt-sized and clay-sized impurities disseminated through the ore and can have iron oxide (hematite and goethite) and iron sulfide (pyrite) aggregates that are fragile and prone to disaggregation and dispersion through the ore when disturbed during processing. Kaolin beneficiation unit operations take advantage of size, magnetic susceptibility, surface chemistry, chemical attack, and thermal attack.

Degritting

Degritting (or desanding) is the first beneficiation step in the wet process following blunging. The objective of this unit operation is to remove most particles that are greater than 44 µm (325 ASTM mesh) that are called grit. Removing grit particles prevents erosive wear caused by quartz particles and prevents sedimentation during periods of reduced flow in pipelines and tanks. The techniques used for degritting depend on the size and amount of unblunged kaolin mineral and nonkaolin impurities greater than 325 ASTM mesh size. Kaolin extracted from primary deposits with a monitor will have desanding performed on the slurry removed from pit by gravel pumps (Thurlow 2001). Kaolin ores that are blunged will have degritting performed adjacent to the blunger and close to the mine for easy disposal of sandy tailings. Some degritting operations have attrition scrubbers to help disperse unblunged clay particles that are rejected in the degritting process by autogenously grinding them with quartz sand in the grit.

Degritting is performed by gravity settling techniques, gravity-assisted classification techniques, and screening. Gravitational processes are most effective for removal of spherical, coarse, and dense mineral grains and rock fragments. Screens are effective for removal of platy-shaped mineral grains such as micas and for removing low specific gravity materials such as vegetable matter. The processes used for refining are a function of gangue material particle size. Process equipment that uses gravitational setline include the bucket wheel desander, spiral (or screw) classifier, dragbox (or sandbox), and hydroseparator. These gravitational processes work best on slurries having low (<30%) solids. Gravity-assisted processes are hydrocyclones and degritting centrifuges. Their advantages are a relatively small footprint and higher separation efficiency for higher (30%–70%) solids and for finer (10–100 µm) particles. Hydrocyclones are typically set up in banks having larger-diameter (≥25.4 cm [10 in.]) cones upstream of 15.2-cm and 10.2-cm (6-in. and 4-in.) banks of cones that make finer cuts. In the case of kaolin deposits with sand and silt content below 15%, a rubber-lined, continuous solid bowl decanter degritting centrifuge can achieve grit levels as low as 0.1 wt % (Adler 1999). Screens having ASTM mesh size ≥325 are used after de-sanding to remove remaining grit. Screens are located throughout the kaolin beneficiation process to remove coarse residues such as scales, paint chips, insect parts, spent grinding media, and so on, that may contaminate the slurry during its processing and storage, prior to loading for shipment. Screens used for degritting are typically rectangular or inclined; those used in process are typically rectangular or round.

Magnetic Separation

The high-intensity magnetic separator (HIMS) was developed in the mid-1960s (Iannicelli et al. 1969) for kaolin beneficiation and is now used in Georgia (United States), Brazil, Cornwall (United Kingdom), and Germany for the removal of iron-bearing discoloring impurities. Magnetic separation of kaolins requires a high magnetic field, high magnetic field gradient, multiple collecting surfaces, long retention time, well-dispersed slurry and flow, and a wet separator (Iannicelli 1976; Mills 1977). The conventional HIMS has an electromagnet surrounding a central canister that is 2.1 m (84 in.) or 3.1 m (120 in.) in diameter with an effective depth of about 51 cm (20 in.). The canister is filled with a matrix of stainless-steel wool with wire diameters between 30 and 180 µm. The field strength of a conventional HIMS is 1–2 Tesla (Stadtmuller et al. 1997). In the 1980s, superconducting magnet technology was deployed on kaolin process applications to lower energy consumption, reduce footprint, and achieve higher field strength up to 10 T. Superconducting HIMS are static canister magnets that are similar to the conventional HIMS and reciprocating canister magnets that oscillate the canister between a fixed magnetic field for processing and out of the magnetic field for flushing. The impurities removed by conventional and superconducting magnets are iron minerals such as hematite, goethite, and ilmentite; and iron-bearing titania minerals such as anatase and rutile.

Flotation

Reverse froth flotation was introduced to Georgia (United States) kaolin processing in the early 1960s to remove iron-bearing anatase using a ground limestone carrier (Greene and Duke 1962; Grounds 1964). Cundy (1976) eliminated the carrier by fully dispersing the kaolin slurry to pH 9, adding a fatty acid (oleic acid) to coat the anatase particle surfaces making them hydrophobic, and adding divalent ions (calcium or barium) to coagulate anatase particles in the conditioning process prior to reverse froth flotation. Yoon and Hilderbrand (1986) developed a noncarrier flotation process using hydroxamate collectors instead of fatty acids. Other impurities that can be removed using reverse froth flotation are graphite and organic compounds (Yildirim et al. 2013).

Selective Flocculation

Selective flocculation is another common method to remove iron-bearing anatase from kaolin. Pruett (2012) reviewed two approaches to selective flocculation used by kaolin producers in Georgia (United States). One approach is to flocculate the anatase by coagulation at moderate pH and separating with the assistance of a high-molecular-weight polymer in a thickener. The second approach is to fully disperse the kaolin slurry at high pH (>9) and add a high-molecular-weight polymer that flocculates out the kaolinite particles in the thickener to separate out dispersed colloidal impurities such as anatase, iron oxides (goethite), and phosphates.

Oxidation

Oxidation will increase brightness and whiteness of some kaolin by chemical or thermal methods. Oxidation acts on organic matter contained in gray-colored kaolin, which may drop brightness 15 (or more) units lower than an equivalent cream-colored kaolin (Schroeder et al. 2004). The chemical method typically uses ozone gas, which is generated by passing oxygen gas or dry air through a high electrical charge that ionizes the gases. The ozone is then mixed with clay slurry in a contact tower to allow the gas to react with the organic matter associated with the kaolin. Oxidizing chemicals such as hydrogen peroxide can also be used to oxidize organic matter in kaolin slurry. The second method to oxidize kaolin to increase brightness is calcination. Calcination to >400°C (752°F) will burn out most organic matter to increase brightness.

Reduced-Acid Leaching

Leaching to remove soluble iron oxides such as hematite and goethite is typically performed at ≤3 pH downstream in the wet process prior to filtration. Sulfuric acid is typically used to lower pH. Sodium hydrosulfite (sodium dithionite) is added to form the chemical-reducing environment that promotes the dissolution of iron oxides and iron hydroxides and reduction of ferric to ferrous iron associated with the kaolinite. Reduced-acid leaching improves brightness and the blue shade of kaolin products derived from oxidized kaolin ores that are cream- or pink-colored. Reduced-acid leaching is performed before dewatering with filters to enable the removal of soluble iron species. Because filtration is performed at low pH to promote kaolin flocculation, other chemicals such as alum and other filter aids are added to the low solids slurry near the addition points of sulfuric acid and sodium hydrosulfite.

Fractionation and Particle Sizing

Particle-size classification is an essential part of the water-washed kaolin process because it was integral to making coating pigments for paper coating (Table 2). The continuous bowl-type decanter centrifuge was a major advance in kaolin processing in the 1930s and enabled the production of 80% <2 µm and finer clays (Murray 1980). The disc nozzle centrifuge is used for classification to the finer cut sizes needed for ultrafine kaolin products. Cut size is the minimum particle size in the sediment underflow or maximum size escaping with the unsettled slurry product effluent (Leung 1998). Separation efficiency is a function of slurry dispersion, viscosity, and solids. Other methods to change the particle size of kaolin products are by blending finer particle size kaolin into the feed and by using stirred media mills to make finer particles.

Table 2 Particles size (equivalent spherical diameter) of paper coating pigment grades

Grade	Wt % <2 µm
Ultrafine	98
Fine No. 1	95
No. 1	90
No. 2	85

Dewatering and Drying

Water removal is required for most water-washed processes that involve beneficiation and particle sizing. The selection of dewatering and dry unit operations depends on the final product form(s) needed by the markets served (Figure 7). Kaolin slurry solids in the process start at the blunger between 35% and 70%, and the solids generally decrease through the process to between 15% and 35% at the leaching stage. Kaolin slurry from hydraulic mining is low (~5%) solids concentration.

Solid–Liquid Separation

Most approaches to separating water from clay use a slurry containing flocculated kaolin to provide an open porous structure and pathways for water to exit. Flocculating the clay to a pH <5, adding alum to coagulate particles, and possibly using a high-molecular-weight polymer to open the floc structure enable good dewatering rates for solid–liquid separation by thickeners, rotary vacuum filters, plate-and-frame filters, tube presses, and centrifuges. Thickeners can increase slurry solids up to 15% (Gwilliam 1971) or higher. Rotary vacuum filters were first used in 1950 to dewater kaolin to cake solids ranging from 50% to 65% solids (Adler 1999; Murray 1980). The plate-and-frame filter has a hydraulic ram to apply between 15 bar (220 psi) and 68 bar (1,000 psi) to achieve maximum cake solids up to 75%–80% (Thurlow 2001). Tube presses apply pressures up to 100 bar (1,450 psi) to achieve cake solids up to 82% (Gwilliam 1971).

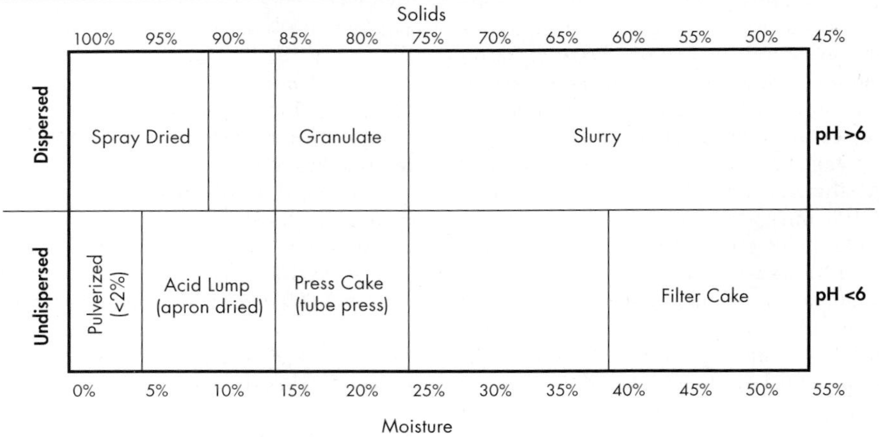

Note: Kaolin products with pH >6 and containing dispersant are slurry, granulated, and spray-dried forms. Kaolin products with pH <6 are undispersed and include filter press cake, lump, and pulverized acid lump.

Figure 7 Product forms and moisture content of kaolin products

Dewatering technologies designed for dispersed slips include centrifuges and membrane filters. Dewatering centrifuges can act on fully dispersed slurry, coagulated slurries, or fully flocculated slurries. The disc-nozzle design can increase solids into the range of 30% to 40% (Asdell 1967). Leung (1998) describes a 3,000–4,000 G decanter centrifuge that can take solids into the range of 60% to 65%. Membrane filters can take dispersed kaolin slurry solids from <20% solids to >60% solids (Adler 1999).

An important advantage of solid–liquid separation is the removal of soluble salts from the kaolin. Soluble salts are deleterious to clay-water rheological properties and can be deleterious to brightness. These salts may be inherited from the crude, from process chemicals added during blunging or postblunging, or from ferrous iron released during the leaching process.

Evaporation

High-solids slurries with about 65%–70% solids can be made using a combination of solid–liquid separation and evaporation unit operations such as evaporators or by back-mixing dried kaolin. Evaporators used to dewater kaolin can have direct flame acting on a falling curtain of kaolin slurry, or they have indirect heat-exchange with steam to heat the clay slurry prior to it passing through a low-pressure chamber to reduce boiling temperature (Willis 1987).

Kaolin slurries are dried using spray dryers by Georgia and Brazilian kaolin producers. The product formed from a spray dryer is a bead approximately <1 mm in diameter, showing face-to-face particle aggregation and having moisture content between 1 and 8 wt %.

The rotary, tray, fluid bed, and apron dryers are fed acid-flocculated filter cake or extruded noodles. The product formed from these processes are typically clay lumps or granules. Moisture contents can range up to about 15%.

Milling and Dry Classification

Milling can be performed on dried crude from the dry process, on dried water-washed or delaminated beneficiated kaolin, and on calcined kaolin. The roller mill is used in the airfloat process to reduce particle size to below 45 µm (325 ASTM mesh) and to remove some coarse-particle size impurities such as quartz (Williams 1985). The roller mill is composed of a vertical shaft suspending rollers rotating against the mill wall (bull ring). Air currents lift fine particles upward in the grinding chamber to the whizzer separator. The whizzer allows fines to pass upward, and it throws coarse particles against the wall to drop back into the grinding chamber. Other mills, such as ball mills, cage mills, and hammer mills, work on the same principle of impact grinding to degrade particles and lumps into finer sizes and using air classification to return coarser particles to the grinding chamber. Some mills have separate heat sources to dry the feed to the mill if it has not been fully dried of moisture to aid particle separation and to prevent caking on mill surfaces. The Hegman fineness of grind test is used to determine the top size of particles and effectiveness of the milling.

Finished Product

Each market application has its own specific set of properties measured and reported on the kaolin product certificate of analysis (COA). Pruett and Pickering (2006) discuss the major chemical and physical tests generally included on the COA.

The finished product from all dry and wet operations is generally stored and tested by the quality lab prior to final packaging and shipment. High-solids slurries are stored in agitated tanks. Spray-dried products and lump products are stored in silos, flat stores, or bays. Bulk dry products are loaded directly into trucks, railcars, or ships. Slurries generally need a final chemical treatment of dispersant to maintain viscosity through the logistic chain; biocide to prevent microbiological growth from causing product degradation and contamination to the customer; and thickeners to prevent sedimentation of coarse particles in un-agitated tankers, railcars, and ship holds during transportation.

Kaolin for small-volume and specialty applications or for shipment overseas can be packaged in totes or bags. Totes hold small volumes of finished kaolin slurry. Spray-dried, apron-dried, and pulverized low-moisture products can be packaged in super sacks or small bags. Bagging operations can be manual or completely automated using robots. After the bags are filled, they are typically checked, weighed, pressed to densify and remove excess air, placed on pallets, and then shrink-wrapped.

Environmental and Regulatory Considerations

Addressing liquid, solid, and gaseous effluents is required where kaolin is processed. Water-washed kaolin operations generally need large impounds or thickeners to capture tailing to clarify water reused in the process or released back to the environment. The water needs to be acidified to flocculate out kaolinite particles. The clarified water meeting local turbidity requirements needs to be neutralized to local standards prior to its release into surface waters. Some kaolin ores contain iron sulfides (pyrite) that if heated above 400°C (752°F) release SOx(g) that may require operations to scrub air released from dryers or calciners with lime. Fine dry kaolin is dusty. Capture of dust particles around dry handling areas of the plant and from emission points in the process is an important part of the process design.

Some markets have additional regulatory considerations. The Registration, Evaluation, Authorisation and Restriction of Chemicals regulations require identification of chemical components in products shipped or used in European markets. Some markets require U.S. Food and Drug Administration (FDA) or equivalent compliance. For example, chemicals used for wet processing paper pigments that are retained in the product must comply with FDA food contact rules to be accepted by some paper customers.

ACKNOWLEDGMENTS

The authors thank Imerys for permission to publish this chapter. Micrographs used in this chapter were provided by Jondahl Davis and Berenice C. Everett from the Imerys Analytical Laboratory in Sandersville, Georgia.

REFERENCES

Adler, P.E. 1999. The final frontier: Kaolin processing liquidity. *Ind. Miner.* 379(4):59–63.

Anon. 2015. Lessons from 100 years of mineral data. *Ind. Miner.* 2015(1):37–40.

Asdell, B.K. 1967. Wet processing of kaolin. *Trans. SME* 235:467–474.

Buie, B.F., and Schrader, E.L. 1982. South Carolina kaolin. In *Geological Investigations Related to the Stratigraphy in the Kaolin Mining District, Aiken County, South Carolina.* Carolina Geological Society Field Trip Guidebook. Edited by P.G. Nystrom Jr. and R.H. Willoughby. Columbia, SC: South Carolina Geological Survey.

Bundy, W.M. 1993. The diverse industrial application of kaolin. In *Kaolin Genesis and Utilization, Special Publication No. 1.* Edited by H. Murray, W. Bundy, and C. Harvey. Boulder, CO: Clay Minerals Society.

Chen, P., Lin, M., and Zheng, Z. 1997. On the origin of the name kaolin and the kaolin deposits of the Kauling and Dazhou areas, Kiangsi, China. *Appl. Clay Sci.* 12:1–25.

Cundy, E.K. 1976. Flotation of fine-grained materials. U.S. Patent 3,979,282.

Feld, I.L., and Clemmons, B.H. 1963. Process for wet grinding solids to extreme fineness. U.S. Patent 3,075,710.

Greene, E.W., and Duke, J.B. 1962. Selective froth flotation of ultrafine materials or slimes. *Min. Eng.* 14:51–55.

Grounds, A. 1964. Fine-particle treatment by ultraflotation. *Mine Quarry Eng.* (March):128–133.

Gwilliam, R.D. 1971. The E.C.C. tube filter press. *Filtr. Sep.* (March/April):1–9.

Henry, A.V., and Vaughan, W.H. 1937. *Geologic and Technologic Aspects of the Sedimentary Kaolins of Georgia.* AIME Technical Publication No. 774. New York: AIME. pp. 1–11.

Iannicelli, J. 1976. High extraction magnetic filtration of kaolin clay. *Clays Clay Miner.* 24:64–68.

Iannicelli, J., Millman, N., and Stone, W.J.D. 1969. Process for improving the brightness of clays. U.S. Patent 3,471,011.

Industrial Minerals. 2015. Kaolin. www.indmin.com/Pricing .html.

Leung, W.W.-F. 1998. *Industrial Centrifugation Technology*. New York: McGraw-Hill.

Mills, C. 1977. High gradient magnets and the kaolin industry. *Ind. Miner.* (August):41–45.

Murray, H.H. 1961. Industrial applications of kaolin. *Clays Clay Miner.* 10:291–297.

Murray, H.H. 1980. Major kaolin processing developments. *Int. J. Miner. Process.* 7:263–274.

Murray, H.H. 2007. Applied clay mineralogy: Occurrences, processing and application of kaolins, bentonites, palygorskite–sepiolite, and common clays. In *Developments in Clay Science 2*. Amsterdam: Elsevier.

Patel, K. 2014. Ceramics: An urban landscape. *Ind. Miner.* (January).

Pruett, R.J. 2000. Georgia kaolin: Development of a leading industrial mineral. *Min. Eng.* 52(10):21–27.

Pruett, R.J. 2011. The carbon footprint and lifecycle analysis of kaolin and calcium carbonate pigments in paper. *Miner. Metall. Process.* 27:17–23.

Pruett, R.J. 2012. Advances in selective flocculation processes for the beneficiation of kaolin. *Miner. Metall. Process.* 29(1):27–37.

Pruett, R.J. 2016. Kaolin deposits and their uses: Northern Brazil and Georgia, USA. *J. Applied Clay Sci.* 131:3–13.

Pruett, R.J., and Pickering Jr., S.M. 2006. Kaolin. In *Industrial Minerals and Rocks*, 7th ed. Edited by J.E. Kogel, N.C. Trivedi, J.M. Barker, and S.T. Krukowski. Littleton, CO: SME.

Schroeder, P.A., Pruett, R.J., and Melear, N.D. 2004. Crystal-chemical changes in an oxidative weathering front in a Georgia kaolin deposit. *Clays Clay Miner.* 52(2):211–220.

Smith, R.W. 1929. *Sedimentary Kaolins of the Coastal Plain of Georgia.* Bulletin No. 44. Atlanta: Georgia Geological Survey.

Stadtmuller, A., Newcombe, P., Fooks, J., Richards, D., Knoll, F., and Allen, R. 1997. Superconducting magnetic separation. *Ind. Miner.* (October):79–85.

Thurlow, C. 2001. *China Clay from Cornwall and Devon: An Illustrated Account of the Modern China Clay Industry*, 3rd ed. Exeter, UK: Cornish Hillside Publications.

Virta, R.L. 2015. Clay and shale. In *Minerals Yearbook 2013*. Reston, VA: U.S. Geological Survey. pp. 18.1–18.22.

Williams, R.M. 1985. Roller mills. In *SME Mineral Processing Handbook*. Vol 1. Edited by N.L. Weiss. Littleton, CO: SME-AIME.

Willis, M.S. 1987. Method of concentrating slurried kaolin. U.S. Patent 4,687,546.

Yildirim, I., Smith, M.D., and Pruett, R.J. 2013. Methods for purifying kaolin clays using reverse flotation, high brightness kaolin products, and uses thereof. U.S. Patent 8,501,030.

Yin, X., Gupta, V., Du, H., Wang, X., and Miller, J.D. 2012. Surface charge and wetting characteristics of layered silicate minerals. *Adv. Colloid and Interface Sci.* 179-182:43–50.

Yoon, R.-H., and Hilderbrand, T.M. 1986. Purification of kaolin clay by froth flotation using hydroxamate collectors. U.S. Patent 4,629,556.

CHAPTER 12.19

Lead and Bismuth

Andreas Siegmund

LEAD

Lead is a soft, ductile, bluish-white, dense metallic element and ranks fourth in world nonferrous metal consumption after aluminum, copper, and zinc. Lead possesses several unique properties that distinguish it from other common metals, and its ease of fabrication makes it useful in a wide range of applications. It has a comparatively high-density, low melting point; low electrical conductivity; and little mechanical strength. In addition, pure lead is soft and malleable and has outstanding corrosion resistance. The first use of lead dates far back in history to 4000 BC when a figurine was found in Egypt and is recognized as the oldest lead article. Later, because of its malleability and corrosion resistance, lead was used expansively by the Romans for water pipes, aqueducts, tank linings, and cooking pots, and then by scientists in early cosmetics, paints and pigments, and lead-rich glazes (ILA 2018). From its origins and early development in Europe, and later in North America, the lead industry has expanded enormously in the modern era. Lead also readily forms alloys and compounds and is used in many commercially important alloys. In this century, lead remains a foundation of society, but in a very different way. Although it is still used for its malleability and corrosion resistance, its chemical properties now make it a thoroughly modern metal (ILA 2018).

Notwithstanding this significantly changing pattern in the market segments, the volume of world production and consumption of refined lead has steadily grown more than tenfold in most countries over the last 100 years and particularly in the last ~45 years, according to published statistics by the International Lead and Zinc Study Group (Figure 1). In 2017, the production and consumption was 11.451 Mt (million metric tons) and refined lead was 11.594 Mt, respectively, despite the intense pressure from environmental and health legislation that has resulted in increasingly restrictive disposal and emission regulations. As one direct consequence, the usage of many lead-based products has been greatly diminished and some have been completely eliminated during the past three decades. These products include pigments based on lead (litharge and red lead, both oxides of lead, as well as white lead, a carbonate, were the most important ones), organo-lead-based antiknock compounds (tetraethyl lead) as a gasoline additive for automobiles, lead glazes for pottery, solder for various applications such as sealing the seams of steel cans, automobile body solder, soldering pipe connections for potable water systems, sealing copper radiators, lead pipes, lead-based metal bearings, lead sealants such as for drain pipes or for wine bottles, chemical tank linings, ammunition, and power and electrical cable sheathing. In most cases, the application was eliminated, or the products were substituted because of environmental pressures associated with the dispersive and toxic nature of lead in order to minimize the potential exposure of lead. The antiknock additives in gasoline were banned because lead terminated the functionality of the catalytic connectors on automobiles, which were employed to improve air quality. Most lead-bearing solder materials for steel cans and potable water systems were substituted because of a potential leaching of lead from these products.

Alloying lead with copper, steel, and aluminum improves its machinability and the alloys are used in bearing metals and castings even though all these applications are declining. Mainly to reduce vehicle weight, copper radiators have been replaced by aluminum. Roller bearings made of steel have replaced lead bearings, plastic pipes have replaced tile drain lines, and lead tank linings have been replaced by stainless steels. Although the consumption has gradually declined in recent years, owing to the development of competitive plastic and aluminum sheathing materials, lead cable-sheathing alloys are still used to protect power and communication cables mainly for underwater applications. Lead alloys are used in the manufacture of shot and other ammunition; however, steel shot is now used, and leaded steels are more and more being replaced with bismuth-containing alloys.

The successful substitution of lead-based products is mainly driven by use of materials with lower toxic properties and advanced mechanical properties such as less weight or increased strength. In addition, these substitutions restrict the release of lead to the environment. However, lead-based products will continue to exist because they are cost-effective, readily recyclable, not readily substituted, and packaged to limit exposure of lead. In addition, such products result in a

Andreas Siegmund, Principal, LanMetCon LLC, Lantana, Texas, USA

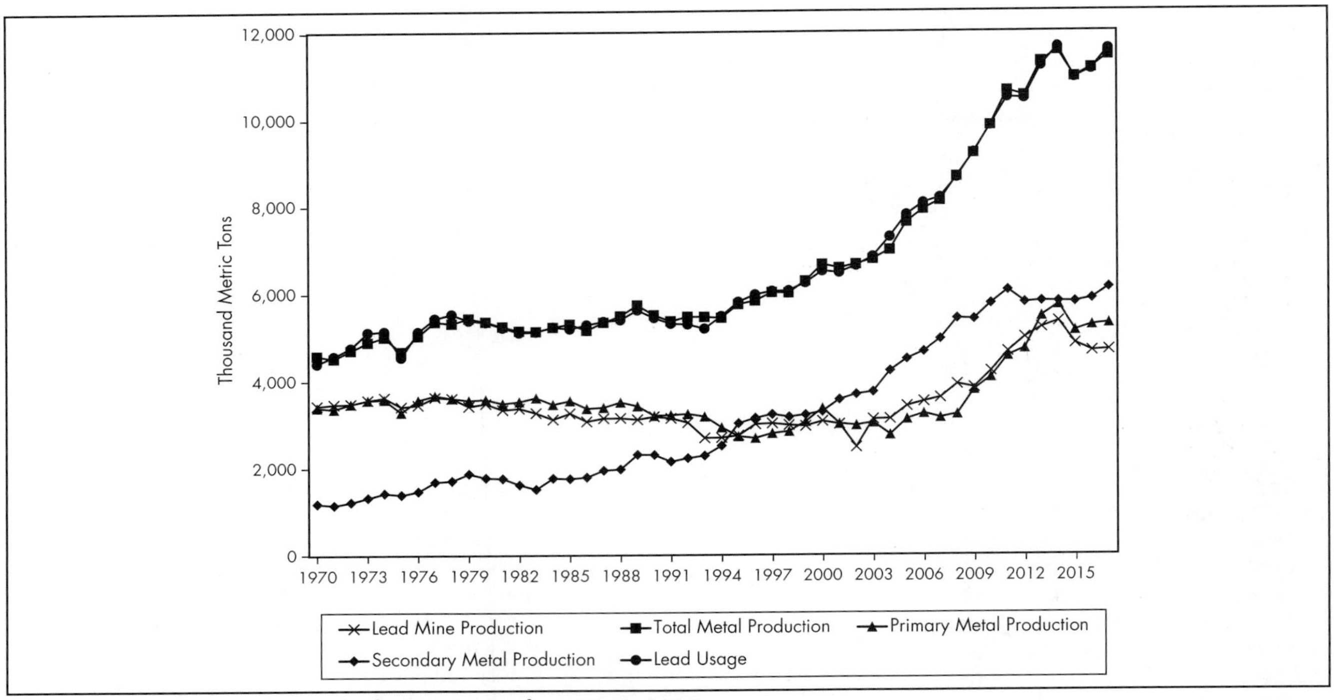

Figure 1 World lead production and consumption from 1970–2017

longer service life, use lower amounts of lead, or offer significant advantages over competitive materials.

Modern lead products have special, unique properties that enable them to compete with other materials. These special characteristics are predominantly protective and electrochemical. One of the unique properties of lead is the ability to collimate and absorb X-rays and other damaging radiation. Therefore, it will continue to find significant use in protecting the population from radiation. Applications are and will continue to exist in containers for radioactive isotopes and shielding in nuclear-powered and medical treatment facilities. Lead oxide glasses will be employed in cathode ray tubes, TV picture tubes, computer screens, and transparent shielding to protect users of these products from ionizing radiation. The degradation process of polyvinylchloride (PVC) products due to ultraviolet radiation or the exposure to other natural environmental processes will be virtually prevented by means of lead stabilizers. Lead sheet will continue to be used as waterproofing where long lifetimes are required. Therefore, semi-manufactured forms of lead products for radiation shielding, sheet for roofing, collapsible tubes, compounds in the glass and plastics industries, and insoluble anodes for metal electrowinning have survived and grown. Lead is also used as an annealing medium for certain steel alloys, a constituent of zinc galvanizing baths, and as weights and ballast. Among the newer uses that have growth potential are sound insulation, dispersion-strengthened lead, and earthquake-resistant shock absorbers in the construction industry (Prengaman 2000).

Lead is able to exist in three different states (metal, ion with a +2 charge, and ion with a +4 charge). Because of this unique property, it is used not only as electrochemical anodes to electroplate other metals from sulfuric acid solution but also can serve as an anode, cathode, or active material in lead–acid storage batteries.

The principal market for lead, however, with the most dominating and fastest growing share, is for lead–acid storage batteries for the expanding population of motorized vehicles. The battery market is characterized by relatively short product life and high recycling rates. Lead–acid storage batteries are used as starter batteries in vehicles, which supplies starting, lighting, and ignition, but also as industrial batteries in emergency systems, computer installations, fork-lift trucks, golf carts, other non-highway vehicles, and marine equipment. Moreover, they are used to a growing extent for telecommunication systems, as sealed power packages for potable electrical appliances, and for uninterruptive power sources, remote access power systems, and hybrid and electric vehicles (Siegmund 2001). The change in average end-use pattern for lead over the last five and one-half decades is illustrated in Figure 2.

More than 90% of all products using lead metals are recyclable, making lead one of the most recycled materials used today. In addition, there are a significant amount of lead-containing secondary materials available from industrial processes and environmental cleanups representing a potential raw material source. One direct consequence is the continuously rising recycling rate of lead-bearing scrap. In 2017, refined lead recovered from secondary materials totaled 6.034 Mt (Figure 1), equivalent to 50.3%—more than half of total production worldwide. In the Western World, as much as 70.6% comes from secondary production. The majority of this secondary lead is produced from spent lead–acid batteries, with most of the remainder coming from other sources such as lead pipe and sheet.

Secondary lead smelting involves the recovery of metallic lead from lead-bearing scrap material generated by industrial and consumer sources. While some secondary smelters are associated with battery manufacturers, as part of an integrated

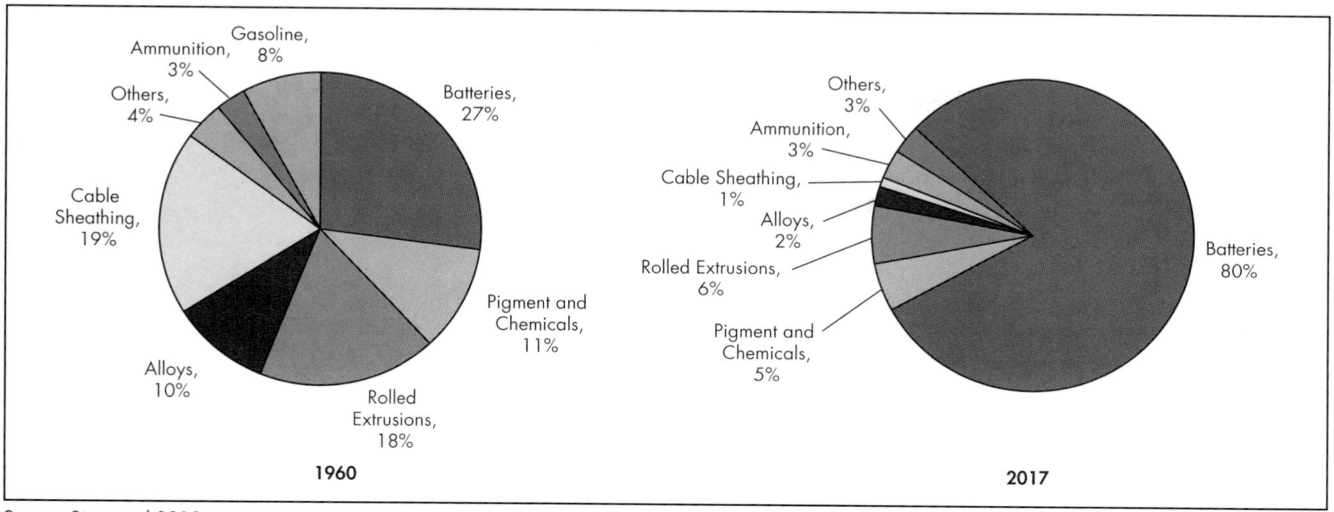

Source: Siegmund 2010

Source: ILZSG 2017

Figure 2 Lead consumption in 1960 and 2017

and closed production circuit, others are independently operated. These stand-alone smelters are purchasing scrap and selling metal or converting metal for customers on a toll basis. More and more primary smelters also tend to treat secondary materials. In conjunction with environmental concerns, the decreasing net lead mine output during the last two decades of the last century (which resulted in difficulties in sourcing adequate supplies of concentrates and led to temporarily high treatment charges) has resulted in intensive research into utilization of secondary materials.

Mineralogy and Reserves

Occurrence

Lead is among the less common metals in the earth's crust and much scarcer than other nonferrous metals. It has an average crustal abundance of between 10 and 15 ppm, which may rise up to 20 ppm in granites and shales. Lead is found in more than 200 different minerals but is an extremely chalcophile element like copper and zinc. Therefore, the majority of lead ores occurs in unoxidized primary sulfides and is of hydrothermal origin. In this process, metals are often leached simultaneously, giving rise to deposits containing combinations of metals. Two types of ore deposit—the stratabound, nonvolcanic deposits and the volcanic magmatic sedimentary of exhalative origin—form the basis of about 80% of current world lead reserves. Important examples of the nonvolcanic stratabound deposit are the Australian Cannington supergiant mine or deposits found in Missouri, United States (Doe Run Company). The major stratiform sedimentary deposits include Mount Isa, Century, and Broken Hill, all in Australia; the Russian Gores mine; and the Sullivan mine in Canada. The remaining 20% consists of multi-metal ore bodies such as Red Dog in Alaska, United States, which consists of a metamorphosed combination featuring both the stratabound and the stratiform ore bodies, or some structurally different stratabound deposits found in Cartagena, Spain, and Cerro de Pasco, Peru. Furthermore, some lead is mined from contact pyrometasomatic deposits. Examples are various mines in Mexico with Peñasquito as the largest, and some deposits in Sweden belonging to Boliden.

The most common and most important primary lead mineral is galena [PbS] with a theoretical lead content of 86.6 wt %. Other more common lead minerals are cerussite [$PbCO_3$] with 77.5% Pb, and anglesite [$PbSO_4$] with 68.3% Pb. In a few deposits, jamesonite [$Pb_4(FeSb)_7S_{14}$] with about 40% Pb and boulangerite [$Pb_5Sb_4S_{11}$] with about 58% Pb are important constituents of the ore, generally in addition to galena. Other lead minerals consist of, but are not limited to, pyromorphite [$Pb_5(PO_4)_3Cl$] with 76% Pb, vanadinite [$Pb_5(VO_4)_3Cl$], wulfenite [$Pb(MoO_4)$] with 56% Pb, and krokoite [$Pb(CrO_4)$].

Lead ores are commonly accompanied by zinc ores, with sphalerite [ZnS] predominating, but they are all gradations in relative proportions. Commonly associated sulfides in lead–zinc deposits are pyrite [FeS_2], marcasite [FeS_2], and chalcopyrite [$CuFeS_2$]. Less common are arsenopyrite [$FeAsS$], tetrahedrite [$(Cu,Fe,Ag)_{12}Sb_4S_{13}$], tennantite [$(Cu,Fe,Ag)_{12}As_4S_{13}$], enargite [$Cu_3AsS_4$], bornite [$Cu_5FeS_4$], and pyrrhotite [$FeS$]. Stibnite [$Sb_2S_3$] is an occasional constituent. Copper is often a minor constituent in lead–zinc ores.

Silver is an almost universal constituent of galena ore and may occur as argentite [Ag_2S], matildite [$AgBiS_2$], or as native silver in solid solution in the galena lattice. It may also occur in other minerals such as tetrahedrite or tennantite. The silver content of lead ores varies from <100 ppm to >4%, with an average typically between 500 ppm and 5,000 ppm. Galena with higher silver content usually has elevated concentrations of antimony and bismuth. In some instances, lead ores contain antimony, bismuth, arsenic, and selenium in recoverable concentrations.

The common gangue minerals include several carbonates, such as calcite [$CaCO_3$], dolomite [$CaMg(CO_3)$], ankerite [$Ca(Fe,Mg,Mn)(CO_3)_2$], and siderite [$FeCO_3$]. Quartz and jasperoid [SiO_2], fluorite [CaF_2], and barite [$BaSO_4$] are also common. A different assemblage of gangue minerals is found in contact metamorphic and deep vein deposits that formed at somewhat higher temperatures than most lead–zinc ores. Such deposits tend to be primarily pyrite-bearing copper deposits in which chalcopyrite and bornite are the major ore minerals, accompanied by varying but commonly substantial quantities of sphalerite, subordinate galena, and commonly

Table 1 Top world mining operations in 2014

Mine and Location	Reserves, Mt	Lead Grade, %	In Situ Pb, Mt	2014 Output, kt
Cannington (BHP Billiton), Australia	20.3	10.8	2.2	202
Doe Run, United States	34.8	5.5	1.9	188
Mount Isa (Glencore), Australia	108.8	3.0	3.3	168
Gorevsk (JSC Gorevsky), Russia	109.1	6.6	7.2	155
Red Dog (Teck Resources), United States	51.3	4.7	2.4	110
Huize Qilinchang, China	15.0	10.1	1.5	100
Magellan (Ivernia), Australia	14.7	5.6	0.8	90
Peñasquito (Goldcorp), Mexico	547.0	0.3	1.7	70
Century (MMG), Australia	58.3	1.2	0.7	64
Fankou (Zhongjin Lingnan), China	35.8	4.8	1.7	58
McArthur River (Glencore), Australia	151.6	4.1	6.3	49

a little molybdenite [MoS_2]. Characteristic gangue minerals include magnetite [Fe_3O_4], specular hematite [Fe_2O_3], and such lime-, magnesium-, and iron-bearing silicates as epidote, diopside-hedenbergite, tremolite-actinolite, garnet (andradite and grossularite), chlorite, biotite, wollastonite, and vesuvianite (Bushell et al. 1985).

Many other minerals, both sulfides and gangues, are of local occurrence in lead–zinc deposits but may be prominent in certain ore bodies. The mineral proportions and associations depend on the type of ore deposit that is formed under the varying geologic conditions imposed by the nature of the country rock, its structure, its temperature and pressure at the time of mineral deposition, and the source and composition of the ore-depositing solutions. Certain mineral associations recur regularly enough to be recognizable as distinct classes of ore deposits, such as the massive sulfide deposits in schists or metavolcanic rocks wherein sphalerite is associated with chalcopyrite and pyrite or pyrrhotite, with little or no galena or earthy gangues.

Lead ores have been found in every continent and are at currently mined in more than 40 countries. The range of ore grades exploited is normally between 2% and 10% Pb, but it also can fall to well under 1% where lead is a minor by-product. The majority of these mines are underground mines. Lead mines are typically classified into four groups of lead-producing ores, depending on the relative quantities of metal. These are lead ores, lead–zinc ores, zinc ores, and any other lead-containing ores. Whether lead is considered as a main product or by-product is dependent both on geological and economic criteria. Since lead is exploited commonly together with zinc, it is striking that the ratio between both mined metals has altered significantly. In 1960, the ratio of lead to zinc in global mine output was close to 0.7:1. This ratio decreased to 0.3:1 by 2003 and fluctuated only in a narrow range until 2010. Since then it was modestly increasing to close to 0.4:1 by the end of 2014. In the near term, the lead/zinc ratio is expected to increase further with the closure of the very large Century mine in 2016.

Global mine production of lead was about 5.364 Mt in 2014 (Figure 1), with production increases in Australia, China, Russia, and India. Taken together, the ten top mining operations, as illustrated in Table 1, accounted for almost 25% of the total lead in ore production in 2014. As shown in the table, reserves of lead are concentrated in Australia, China, Russia, Peru, Mexico, and the United States. Together these countries account for more than 85% of total reserves. This percentage may be overstated with the dynamic in determining reserve data and given the tendency in many developing countries to calculate reserves only at the time of mining development, which is done to meet business needs. Identified world lead resources total more than 2 billion metric tons. In recent years, significant lead resources have been demonstrated in association with zinc and/or silver or copper deposits in Australia, China, Ireland, Mexico, Peru, Portugal, Russia, and the United States (Alaska). Since most lead-bearing ores are complex and contain a range of metals, selective flotation is needed to permit the separation and recovery of the valuable elements.

Mineral Processing Methods

Although there have been many recent developments aimed at improving the efficiency of production and tackling ore-body problems, basic lead mining methods have been essentially unchanged. Where the deposit allows, open pit mining is carried out. However, very few lead deposits are exploited that way, and the percentage of current world lead open pit mine production is less than 10%, while a number of these mines have parallel underground mining activities. Underground mining is, by far, the dominant mining method for most of world's lead deposits. All ores require some sort of treatment. This process involves concentrating the ore by eliminating the gangue and waste material and, where required, the separation of valuable components in the mineral ore. The ore concentration makes use of the differences in physical properties between the various parts of the ore and typically consists of crushing, a pre-concentration at coarse size if possible, grinding and classification, and concentration.

Crushing. The ore, which may include some custom material, is usually crushed and physically broken down into small particles in stages. Some crushing, particularly the first stage, is often done at the mine, the remainder in the concentrator. The sizing of run-of-mine ore depends on ore type and mining procedures, and crushing practice is correspondingly affected.

Pre-concentration at coarse sizes. Most lead ore concentration is accomplished by flotation after fine grinding to achieve liberation. However, in some cases, an applied sink–float process rejects a discardable light fraction on crushed but unground ore, resulting in savings in subsequent processing.

Ferrosilicon medium is used in the majority of plants, with galena medium in some. Ferrosilicon is generally considered

Table 2 Reagents commonly used in lead and copper–lead flotation

Collectors	pH Modifiers	Zinc and Iron Depressants	Frothers	Others
Xanthates (ethyl, propyl, isopropyl, butyl amyl), Aerofloat promoters (alkali dialkyl dithio phosphates sometimes with frother), Minerec A	Lime, soda ash, sodium hydroxide, sulfuric acid, crude dry lake products like tequesquite	Sodium or calcium cyanide, zinc sulfate, sulfur dioxide or sodium bisulfite, zinc hydrosulfite	Dowfroth, cresylic acid, pine oil, MIBC (methyl isobutyl carbinol), coal tar, TEB (triethyloxybutane)	Sodium silicate

Source: Bushell and Testut 1985

to be a superior medium, but galena offers a cost advantage when indigenous. Fines must be separated from the ore by screening and washing before sink–float. The normal range of treated ores for sink–float feed is of the general order of 50 to 6 mm but in one application goes down to 0.5 mm. Jigs are sometimes used on fines to complement heavy media separation. Sink–float is not widely used. Even when a low-grade float can be produced, some of the values may be recoverable by fine grinding and flotation. The economics vary with each situation.

Grinding and classification. Crushed ore, after pre-concentration at coarse sizes if practiced, becomes the feed for grinding. If the ore is easy flowing, it may be crushed with cone crushers down to about 3 mm, which will be the feed to ball milling. Otherwise, the crushed ore is further reduced in rod mills or rolls (after pre-concentration, if practiced). Flotation feed is ground in ball mills often in one stage, sometimes in two or three stages, and sometimes with more complicated flow sheets for separate treatment of different ores or lead flotation ahead of final grinding. For lead ores, a reduction to under 0.15 mm is required. In most of the recent mills, cyclones, occasionally in two stages, are used as classifiers in grinding circuits, and in some older mills, rake and/or spiral classifiers have been replaced by cyclones. In most modern plants, the ore is ground to the coarsest size that will permit suitably low rougher tailings losses. Regrind circuits for intermediate products such as the rougher concentrate, cleaner tailings, or a scavenger product must then be employed. This procedure is followed both to minimize grinding costs and to prevent sliming, particularly of galena. Otherwise, middlings are returned to primary grinding. The overflows stay in their respective circuits.

Concentration. Most lead ores are fine grained (<0.15 mm) and are concentrated mainly by froth flotation. Flotation results are affected by (1) degree of oxidation, (2) abundance in nature of iron sulfides, (3) presence of copper minerals, and (4) nature of a non-sulfide gangue. If necessary, separate streams may be provided for different ore types. Makeup water is normally fresh water but seawater can be used.

Sulfide concentration. The lead mineral is galena, PbS, with a specific gravity of 7.5. Coarse-grained ores are sometimes treated by gravity concentration, but a flotation section is usually justified for the treatment of slimes and reground middlings. Straight flotation is the most common practice. The use of jigs in combination with flotation is employed in some plants. Reagents used in lead and copper–lead flotation are shown in Table 2. In addition, various other reagents are sometimes used as a result of local customs or unusual features of the ore. The optimum reagent balance must be found by experimenting on each ore.

With lead–zinc ores, lead (and copper, if present) are floated before zinc. Lead and copper are usually floated and cleaned together, but sometimes a copper concentrate is floated first. Occasionally a pre-float removes "natural floaters" such as talc or graphite that would otherwise contaminate the lead concentrate.

The amounts of pyrite, marcasite, and pyrrhotite greatly affect flotation and make the differentiation of lead and zinc more difficult. When galena and sphalerite-marrnatite are the principal sulfides, the zinc minerals are substantially inactivated and lead can be floated without depressant or only a little cyanide and with no or little alkali. When iron sulfides are present, lime must be used; cyanide, often with zinc sulfate, is used to depress zinc; and sulfites are occasionally added to depress iron. Lead flotation is generally at a pH in the range of 7.5 to 10, with the pH in the cleaners sometimes higher than in the roughers. The addition of lime to the cleaners without further addition of xanthate increases the xanthate concentration in solution. The consequent desorption of collector depresses marginally floatable particles to the tailings. If the ore contains reactive pyrrhotite, the pH should not exceed 11.2, as more alkaline solutions will liberate sulfide ions to solution and affect flotation adversely. The number of cleaning stages required varies; one or two are often sufficient, but complex ores such as Brunswick (Canada), Sullivan, and others employ more. Sometimes lead rougher tailings are re-floated to produce a scavenger concentrate that is generally recycled to the roughers, often after a regrind. Lead cleaner tailings might also be reground before recycling to the roughers. Flotation times in lead flotation are noted in some individual concentrators; 10–20 minutes is typical in roughing.

The bulk lead–copper concentrate is cleaned to the optimum grade and then is split into separate lead and copper concentrates. Since these are both usually end products, the need to achieve optimum grade in the bulk concentrate is obvious (sometimes the lead product is further concentrated). Two orders of lead–copper separation are used. In one, the lead is floated while the copper is depressed, and in the other, the copper is floated while the lead is depressed. When lead is floated, the copper is depressed with cyanide. Cyanide loses its effectiveness readily, particularly in the presence of cyanide-consuming constituents in the ore, so the reagent is generally stage added. Cyanide depression of copper minerals is pH sensitive, so pH should be controlled closely. The optimum range varies with individual concentrators but usually is between 10.0 and 11.5. When copper is floated, lead is depressed through use of sodium dichromate, sulfur dioxide, or starch while additional collector such as Z-200 and frother are added as required. Examples for alternative (but not common) lead concentration is the depression of copper with an ammoniated zinc cyanide complex and the addition of activated carbon to adsorb excess bulk flotation reagents. Another concentration method for a copper–lead concentrate is conditioning with a guar-dextrine depressant for separation

Table 3 Reagents commonly used in oxidized lead flotation

Collectors	pH Modifiers	Zinc Depressants	Sulfidizing Agents	Frothers	Others
Xanthates	Sodium carbonate	Sodium cyanide	Sodium sulfide	Pine oil	Sodium silicate
Amylxanthates	Sodium hydroxide	Zinc sulfate	Sodium sulfhydrate	Cresylic and xylenic acids	Oil
Thiocarbamates		Sodium sulfide		Dowfroth	Creosotes
Mercaptobenzothiazol				Speld 1333	Coal tar
Long-chain thioalkyl acids				Aerofroth 65	Carboxymethyl cellulose
Aerofloat					Kerosene
Oleic acid					Lignosulfonates
Refined tall oil					
Reagent 404					

Source: Bushell and Testut 1985

and then conditioning with sulfur dioxide at pH 4.5. Copper concentrate is floated with Z-200 and cleaned three times. The copper separation tailing is floated with xanthates at pH 5.8–6.0 to concentrate lead and depress pyrite and sphalerite. The lead concentrate is cleaned three times. Alternatively, the copper–lead separation can be accomplished with sulfur dioxide and sodium dichromate at pH 5.2 to depress galena.

Sometimes appreciable zinc reports to the lead concentrate from the final stage of normal cleaning and needs to be removed. This zinc apparently floats with the lead because of activation by lead or copper ions. "Dezincing" is analogous to lead–copper separations in that the lead concentrate is separated into two end products—a final lead concentrate and a zinc concentrate, which is added to the zinc concentrate obtained by normal zinc flotation. This practice is followed at the Sullivan mine where the material transferred contains about 15% Pb and 44% Zn. The economic desirability of this transfer follows directly from the economic considerations of separating within reasonable limits; lead in zinc concentrates going to the Teck Metals smelter is preferable to zinc in lead concentrates. In this process, the lead recleaner concentrate is heated to 38°C and conditioned at pH 11.6 with added copper sulfate. In theory, this process removes the xanthate from the lead, allowing it to be absorbed by the activated zinc. Desorption of xanthate depresses lead to the tailings, which become the final lead concentrate, and the concentrate (zinc) is added to the final zinc concentrate.

At other lead–zinc mines, the first-stage flotation concentrate, containing both lead and zinc, is refloated with activated carbon, caustic soda, maize starch, and sodium dichromate to yield tailings that become part of the final lead concentrate and a float that goes to the low-grade zinc circuit. If the feed is partly oxidized, a bulk float is first made and then separated into lead, zinc, and iron concentrates.

Oxidized lead ores. The lead minerals considered first are cerussite ($PbCO_3$) and anglesite ($PbSO_4$). Both are soft minerals that usually slime badly during comminution.

Cerussite ores are floated with a powerful sulfide collector after the cerussite has been sulfidized. Cerussite usually sulfidizes readily. Sodium sulfide is often used. However, if a large amount of sulfidizing agent is required, a mixture of sodium sulfide and sodium hydrosulfide may be preferred in order not to increase the pH too much. A succession of flotation steps is sometimes used, each step comprising a quick sulfidizing reaction, followed by flotation. With inefficient conditioning, sodium sulfide consumption is sometimes reported to be very high. Slime gangue dispersion by sodium silicate is usual. The sodium silicate is often added with the sodium sulfide or a mixture of sodium sulfide and sodium sulfate. After conditioning, the pH has to be generally close to 8.5 for good flotation. The conditioning time is critical. When the temperature rises, conditioning time must be shortened. Alternatively, sodium carbonate may be added to slow down and control the sulfidizing action in the conditioner. Aeration in the conditioner is also helpful. The conditioning time then is not so critical and the floatability of the sulfidized cerussite lasts longer after conditioning. A table of widely used reagents are listed in Table 3. Potassium amylxanthate and Reagent 404 are often used as collectors. On some ores, Reagent 404 (either alone or mixed with a xanthate or an Aerofloat collector) has given good results without sulfidizing. An insoluble oil (a creosote, coal tar, or oil–xylenol mixture) is often used to reinforce the soluble collector action. It can be added with advantage as an emulsion with the frother. The combination speeds up flotation in both roughing and cleaning stages and the soluble collector consumption is decreased. Two or three cleaning stages are usual. Conditioning time is in the range of 2 to 3 minutes and flotation time is 15–45 minutes in roughing.

Anglesite is more difficult to float by sulfidizing than cerussite, and the permissible pH range is narrower. The sliming tendency of oxidized lead ores is a source of difficulties with long chain collectors. However, where the gangue is acid, fatty acids are sometimes used as collecting agents. Flotation with fatty acids loses selectivity in the presence of limestone or dolomite.

Pyromorphite, $Pb_5(PO_4)_3Cl$, and mimetite, $Pb_5(AsO_4)_3Cl$, are the extremes of an isomorphous series, mimetite being less abundant than pyromorphite. They are generally finely disseminated and associated with cerussite and anglesite. They can be concentrated by flotation. Desliming before fine grinding may be necessary to reject the primary slimes.

Two flotation schemes are possible: either a flotation of cerussite and anglesite first after sulfidizing, followed by pyromorphite and mimetite flotation with fatty acids; or all minerals are floated together with fatty acids. Oleic acid and refined tall oil containing 2%–10% resin acids are the preferred collectors. They are used either as soaps or emulsified with nonionic emulsifying agents. Slime and pH control arc obtained with sodium silicate and sodium carbonate. Sulfidizing before fatty acid flotation may be beneficial.

Lead–zinc ores with both sulfides and oxides. Oxidized ore with insufficient oxidized minerals to justify separate recovery nevertheless poses special problems in flotation. With straight lead–zinc ores, much of the trouble results from activation of zinc and possibly iron sulfides by lead ions coming from oxidized lead minerals. Washing the ore before grinding can be helpful. Cyanide is useless for overcoming lead activation, but addition of soda ash or lime to the grinding circuit can be helpful.

Mixed sulfide–oxide ores with basic gangue can sometimes be treated for recovery of oxidized minerals as well as sulfides. Effective depression of sphalerite, which has been activated by lead ions, is essential. Sodium sulfite and bisulfite, zinc sulfate combined with ammonium sulfate, zinc sulfate and sodium silicate, or reagents that precipitate lead ions like sodium sulfide and zinc powder are used. However, sodium sulfide sometimes depresses sphalerite permanently.

Flotation of ores without iron sulfides may be conducted in either of two sequences, galena-sphalerite-cerussite or galena-cerussite-sphalerite. If smithsonite is present, it is floated last after desliming. With the order galena-sphalerite-cerussite, flotation is usually conducted in a neutral circuit. To depress sphalerite, zinc sulfate has to be used sparingly, otherwise the consumption of sulfidizing agent in the cerussite circuit would be increased. Sphalerite is floated with minimum copper sulfate, as an excess can harm cerussite flotation. Cerussite is floated after sulfidizing with sodium sulfide or sodium sulfide plus sodium hydrosulfide. Sodium silicate is nearly always essential. With the flotation order galena-cerussite-sphalerite, or if galena and cerussite are floated together, the difficulty is to avoid permanent depression by sodium sulfide. Another adverse effect of sodium sulfide is that it is sometimes adsorbed by gangue minerals that later float after activation by copper sulfate. Potential remedies include addition of ferrous sulfate, which can precipitate the soluble sulfide, or the pulp can be thickened and the water rejected. Desliming the pulp is more effective, but it entails losses of fine sphalerite.

When mixed oxide–sulfide ores contain iron sulfides, flotation is complicated further. The classic flotation sequence is galena-sphalerite-pyrite-cerussite-smithsonite. If smithsonite is not recovered and pyrite flotation is not justified, a simpler two-circuit scheme—galena plus cerussite-sphalerite—is preferred, if it works. Pyrite is usually depressed with cyanide and lime (Doyle 1970).

Lead Concentrates: Extraction Processes and Schedules

The three major factors determining revenue from ores and concentrates produced are the prevailing prices for refined lead metal, freight to the purchaser's location, and the schedules under which the materials are purchased.

Lead metal prices have, by far, the greatest impact on the price negotiable for the corresponding concentrates, since in all smelter schedules, payments for contained metal values are based on current metal price. In general, the operating margin and profit potential available to the smelter through the common schedule have been small, although a cost is added to treatment charge for emission control cost.

Under its schedule of purchase, the price offered by a smelter for an acceptable raw material is dependent on several factors. These include the specific smelting and subsequent refining processes, the limitations of a given plant in flexibility, state of repair and proximity to metal markets, and the normal metallurgical practice and costs that relate to production targets and the type of raw material input. Each concentrate has a unique specification, based both on its lead content and the type and relative quantities of recoverable and penalty elements that it contains. Lead concentrates vary from about 30% to 80% of contained lead. The quality of concentrates is also very different, ranging from clean, pure concentrates to highly complex and impure concentrates. The latter materials can be either more valuable by having a high precious metals content or can contain significantly more deleterious impurities that are subject to a penalty.

In addition, the combination of environmental concerns, the actual trend of increasing recycling rate, the rather stagnating net lead mine output, and an increase in smelter capacity as a result of growing lead demand means that many smelters have been forced to alter their feed mix to include a greater share of secondary materials. Today, only 80% of the lead produced by primary smelters is from lead and/or bulk concentrates, down from more than 90% two decades ago.

Primary Lead Smelter Schedules

Payments. In complex concentrates, primary lead smelters typically offer payment for most of contained lead, silver, gold, and copper, and sometimes for minor metals such as zinc, bismuth, antimony, and platinum, if recovered. Very infrequently, payment may be offered for other metal values. Only a part of the contained value will be paid for, the remainder representing the expected process loss. The price applicable will be somewhat below the market price of metal to allow for refining, shipping, and sales costs borne by the smelter. The precise formula varies from schedule to schedule, but these provisions are always contained. Very often, no payment will be offered for minor metal values that the smelter can and does recover. Both the list of recoverable metals and the recovery attainable are related to the smelting process and plant, and to the composition of the charge.

Penalties and bonuses. A penalty or bonus for a raw material constituent is a compensation for its effect in the smelter operation on recovery of metal values and incurred costs. These may be specified directly in amount and level of application but frequently are calculated as an adjustment to treatment charge. Common impurities in lead concentrates are bismuth, arsenic, antimony, cadmium, mercury, iron, and halides. These impurities cause the lead smelters the greatest problems during processing and are designated penalty elements. Excessive moisture content and additional handling costs related to shipment size or screen size are also usually compensated in this way.

Treatment charge. This is a deduction from the gross payment for values designed to cover all smelting operation costs, overhead, and depreciation, and allow for profit. It usually takes the form of a fixed deduction per dry ton of material processed, and thus its impact is greater on lower-grade materials. A minimum treatment charge per lot is specified. Some smelters determine the treatment charge with reference to the total payment for main metal values. This formula relates it to metal price as well as to grade. As smelting costs rise, with emission control a major contributor, the increase will be reflected by higher treatment charges (Rich 1994).

Primary Lead Smelting Processes

The recovery of lead from concentrates and other raw materials is carried out almost exclusively through pyrometallurgical processes due to the insolubility of many lead compounds in water and most common acidic solution. Smelting of most primary lead from sulfide materials is based on the roast reduction process. The conventional process included two main steps. The first stage consisted of roasting and sintering of lead sulfide and fluxes on an updraft sinter machine to produce lead oxide sinter, which is followed by a reduction process in a blast furnace. The latter can be a traditional shaft furnace or an Imperial Smelting type. Other technologies include electric furnaces or rotary furnaces. Lead oxide raw materials are smelted directly under a reducing atmosphere. The residual lead from the produced slag can be further reduced through a subsequent fuming step. Figure 3 depicts the main reactions taking place at the various stages of lead production (Stephens 2010).

About 50% of the installed capacity is still producing lead using Dwight-Lloyd updraft sinter machine and blast furnace operations, which was introduced at the beginning of the 1900s. This old practice experiences more and more difficulties regarding compatibility with today's requirements in respect to efficiency and environmental protection. Therefore, extensive research studies have been performed in different parts of the world since the beginning of the 1970s, resulting in the development of several new technologies for primary lead smelting. Most of these processes were based on direct and mainly continuous smelting technologies, but only some, like the TSL (top submerged lance), QSL (Queneau–Schuhmann–Lurgi), SKS (Shuikoushan), Kivcet, and Kaldo processes, have been realized on an industrial scale. The commercial implementation of these novel technologies toward the end of the 20th century successfully challenged the conventional method of lead smelting, and most of them have proven to be economically and environmentally viable. The intensification of the metallurgical reaction by applying the bath or flash smelting principle in conjunction with oxygen use resulted in cost savings and a higher flexibility with respect to raw materials and additives usage. By synchronizing individual auxiliary plant sections with the smelting process, a virtually waste-free production and low-emission mode of operation with optimum energy exploitation is achieved (Siegmund 2000a, 2000b).

Conventional technology (sinter machine–blast furnace). The conventional process consists of the two steps of sinter roasting and reduction. Both process steps are carried out in independent aggregates. A typical process block diagram of a sinter machine–blast furnace plant is shown in Figure 4.

The objective of the sinter roasting process is to oxidize virtually the entire amount of prevailing lead sulfides to lead oxide and convert the concentrate into an agglomerated feed with suitable mechanical properties (mainly size, hardness, and porosity) and chemistry requirements for the subsequent reduction step in the blast furnace. This sintering of the concentrates is accomplished by adequate addition and mixing with lime-, silica- and iron-containing materials as fluxes and return of sintering fines. The solid–gas reactions during the oxidizing roast of lead concentrates is more complex than for other nonferrous metals. The roasting reaction forms lead sulfate, which itself reacts with still available lead sulfide to form lead sulfates/oxide compounds. Therefore, the sequence of the roasting reactions can be assumed to be $PbS \rightarrow PbSO_4 \rightarrow PbSO_4 * PbO \rightarrow PbSO_4 * 2PbO \rightarrow PbSO_4 * 4PbO \rightarrow PbO$. In addition, the sinter process is hampered by the formation of low-melting-point eutectic compounds such as $PbS\text{-}Sb_2S_3$ (426°C), $PbS\text{-}Cu_2S$ (540°C), and $PbO\text{-}SiO_2$ (714°C). Another reason is the higher combustion heat availability from such concentrates, which could cause fusing of the sinter charge without added, and costly, recirculation of sinter.

SMELTING

Oxidation of Sulfides

$PbS + 1{,}5O_2(g) \rightarrow PbO + SO_2(g)$	(EQ 1)
$ZnS + 1{,}5O_2(g) \rightarrow ZnO + SO_2(g)$	(EQ 2)
$FeS + 1.5O_2(g) \rightarrow FeO + SO_2(g)$	(EQ 3)
$FeS_2 + 2.5O_2(g) \rightarrow FeO + 2SO_2(g)$	(EQ 4)

Decomposition of Sulfates and Ferrites

$PbSO_4 \rightarrow PbO + SO_2(g) + 0.5O_2(g)$	(EQ 5)
$ZnSO_4 \rightarrow ZnO + SO_2(g) + 0.5O_2(g)$	(EQ 6)
$ZnFe_2O_4 \rightarrow ZnO + Fe_2O_3$	(EQ 7)

Formation and Oxidation of Metallic Lead

$PbS + 2PbO \rightarrow 3Pb + SO_2(g)$	(EQ 8)
$Pb + 0.5O_2(g) \rightarrow PbO$	(EQ 9)

Equilibrium with Ferric Iron

$PbO + 2FeO \rightarrow Pb + Fe_2O_3$	(EQ 10)

Fuel Combustion

$C + O_2(g) \rightarrow CO_2(g)$	(EQ 11)
$C + 0.5O_2(g) \rightarrow CO(g)$	(EQ 12)

REDUCTION AND SLAG FUMING

$PbO + C \rightarrow Pb + CO(g)$	(EQ 13)
$PbO + 0.5C \rightarrow Pb + 0.5CO_2(g)$	(EQ 14)
$PbO + CO(g) \rightarrow Pb + CO_2(g)$	(EQ 15)
$ZnO + CO(g) \rightarrow Zn(g) + CO_2(g)$	(EQ 16)
$Fe_2O_3 + C \rightarrow 2FeO + CO(g)$	(EQ 17)
$Fe_2O_3 + CO(g) \rightarrow 2FeO + CO_2(g)$	(EQ 18)
$C + CO_2(g) \rightarrow 2CO(g)$	(EQ 19)

AFTERBURNING AND SULFATION

$Zn(g) + 0.5O_2(g) \rightarrow ZnO$	(EQ 20)
$Pb(g) + 0.5O_2(g) \rightarrow PbO(g)$	(EQ 21)
$PbO(g) \rightarrow PbO(l)$	(EQ 22)
$PbO(g) + SO_2(g) + 0.5O_2(g) \rightarrow PbSO_4$	(EQ 23)
$ZnO + SO_2(g) + 0.5O_2(g) \rightarrow ZnSO_4$	(EQ 24)
$CO(g) + 0.5O_2(g) \rightarrow CO_2(g)$	(EQ 25)

Note: The chemistry of all the direct lead smelting processes are similar but differ slightly in the temperatures or oxygen potentials used.

Source: Stephens 2010

Figure 3 Main chemical reactions during the recovery of lead at various production stages

Sintering of lead blast furnace feed can be performed either on updraft or downdraft machines. For the sinter roasting process, usually an updraft Dwight-Lloyd sinter machine, as depicted in Figure 5, is preferred for many reasons. Downdraft sinter plants are generally less able to handle the contained sulfur in lead concentrates in an ecologically acceptable way and are therefore more likely to reject higher-sulfur concentrates, notably pyritic concentrates. Dwight-Lloyd sinter machines consist of a continuous system of pallets with a grate bottom, which convey the feed mix through the machine. After a thin layer of feed material (40–60 mm) has been ignited, the main amount of feed material is added, forming a sinter bed, which then enters the

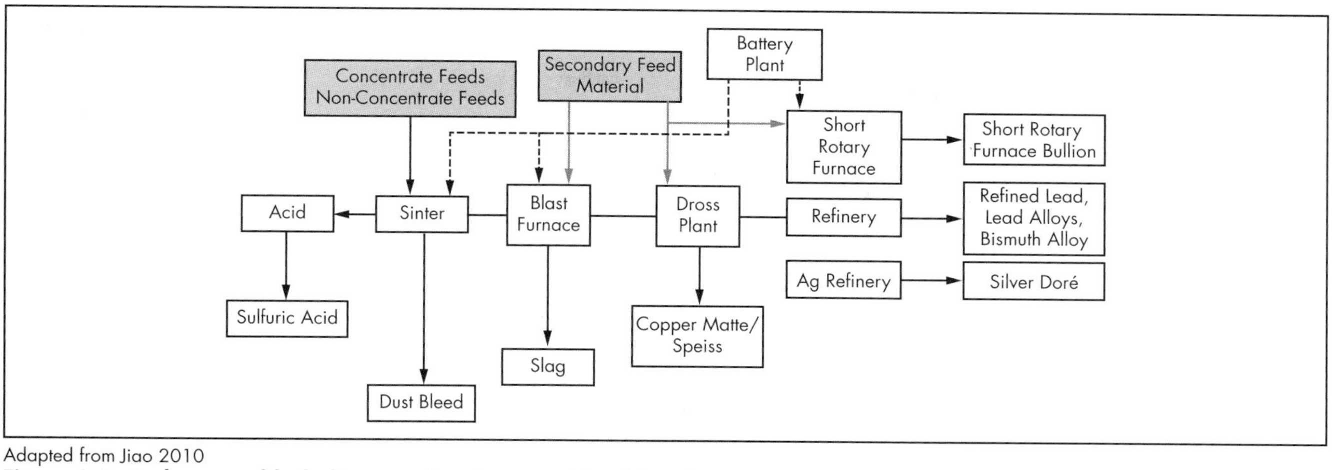

Adapted from Jiao 2010

Figure 4 Typical process block diagram of a sinter machine–blast furnace operation

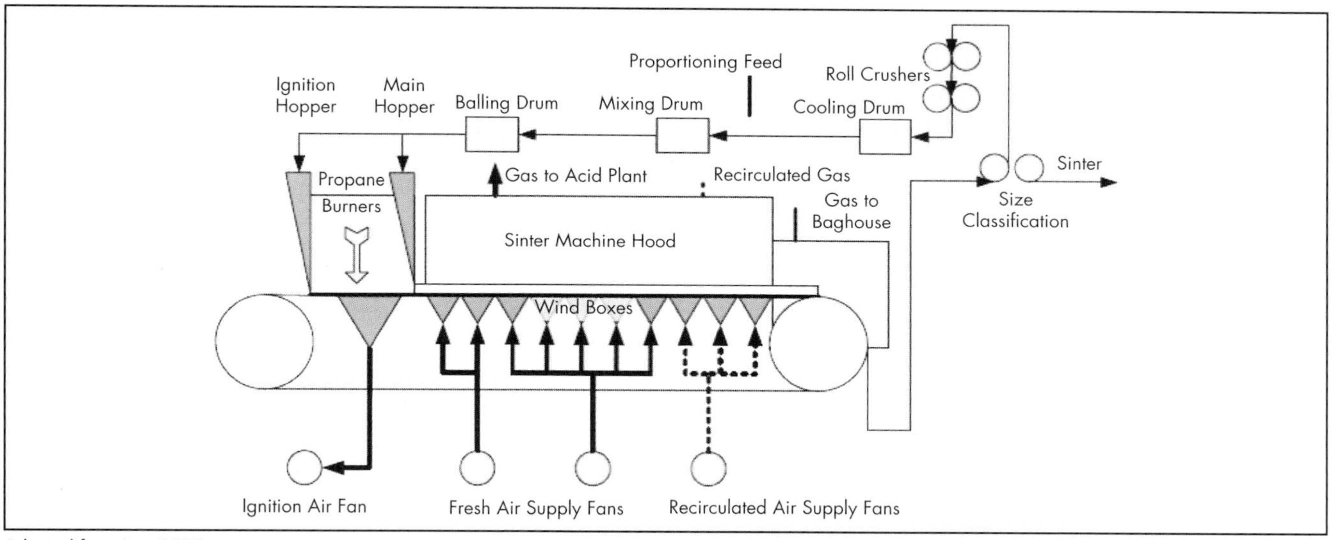

Adapted from Jiao 2010

Figure 5 Updraft sinter machine

main reaction chamber. Here, air is pushed upward through the permeable charge driving the main reaction layer from the bottom of the sinter bed to the top. At the end of the sinter process, about 10% of the lead is already present in metallic form. The high temperatures in the reaction layer aid in the agglomeration process of the feed particles. After discharge from the sinter machine, the sinter is crushed and screened. Sinter with the appropriate size is fed to the blast furnace operation while fines are returned to the sinter machine. During the oxidation process on the sinter machine, sulfur dioxide is generated that must be recovered. Most operations convert the sulfur dioxide into sulfuric acid. The advantages of this operating mode include no metallic lead lost by dripping into the wind boxes, high specific sintering capacities, the ability to efficiently separate gases and produce an SO_2-containing gas of strength compatible with sulfuric acid production, recirculation of process gases with low-sulfur-dioxide concentrations, minimum wear of the travelling grates, and reduced burden for the blast furnace operation. Typical operating data for sinter machines are shown in Table 4 (Jiao 2010).

The reduction of lead oxide is achieved through reduction smelting in either the more common vertical blast furnace or an Imperial Smelting process (ISP) zinc–lead furnace. Conventional lead blast furnaces normally have a rectangular cross section. The majority of lead blast furnace smelters process lead sulfide concentrates of fairly high grade (>60% Pb) and operate with a fairly constant raw material charge. However, most accept lower-grade lead ores in moderation. More and more plants incorporate zinc plant residues and other secondary lead-bearing materials into the blast furnace charge, operating with high zinc slags from which zinc is subsequently recovered by fuming. Commonly, zinc in slag is held below 12% and there is no subsequent recovery (Young et al. 2010). Capacity increases in existing blast furnaces have recently been accomplished through improvements in the sinter quality (hardness, porosity, size distribution), closed furnace tops, feeders with better charge distribution, preheating,

Table 4 Selected sinter plant operating data

Item	Brunswick (Canada)	Port Pirie (Australia)	Peñoles (Mexico)	Chigirishima (Japan)
Hearth area, m^2	120	83.6	120	33
Capacity, t/h	45	90	60	22
Percent of secondary	42	23	0	20
Feed moisture, %	6	11	6	4.5
Ignition fuel	Propane	Natural gas	Natural gas	Heavy oil
Percent SO_2 in exhaust gas	4.5	6	5.5	6
Sulfur burning rate, t/m^2/d	1.4	2.8	1.4	1.4
S in sinter, %	1.5	1.5	2.5	2.5

Source: Jiao 2010

and oxygen enrichment of air. Typical schematic drawings of different blast furnace types are shown in Figure 6.

The reduction of the lead oxide sinter is achieved through the addition of recycled intermediates and secondary materials, fluxes, and metallurgical coke. The charge is fed through either an open or a sealed furnace top. Air, or oxygen-enriched air to reduce coke consumption, is blown through tuyeres into the furnace at the bottom of the shaft. The coke from the feed mix is combusted in this area, forming carbon monoxide for lead oxide reduction. This reduction reaction has already started at temperatures of 160°C. Although the direct reduction of lead oxide is limited to the melting and slag zones in the middle and lower part of the furnace, a considerable recirculation rate of evaporated and condensed phases (lead compounds, zinc, and others) are observed as the charge progressively descends in the furnace. The charge transforms into the main components: lead bullion and slag. Depending on the charge composition, matte and speiss can also be formed. Matte is typically formed at sulfur contents above 2% in the charge and contains up to 35% Cu, 10%–25% Pb, up to 15% Fe, up to 8% Zn, 25%–1,200 g/t Ag, and some arsenide. The generated amount of matte may vary between 2.5% and 10% of the charge. Speiss is formed when increased amounts of arsenic, antimony, nickel, and cobalt are contained in the charge. Resulting speiss consists of 30%–40% Cu, 20%–35% Pb, 5%–10% As, up to 10% Sb, 5%–12% Ni, 3%–15% Fe, and 2%–3% S. All products accumulate at the base of the furnace and are continuously tapped into a forehearth. In general, the formation of speiss is not desired since the recovery of precious metals is complex and expensive. Furthermore, the operation of the forehearth is more labor and energy intensive due to the high melting point and density of speiss. Physical separation can be easily achieved here due to the different densities of the products (lead bullion ≈ 11 t/m^3, matte ≈ 5–6 t/m^3, speiss ≈ 7–8 t/m^3, slag <4 t/m^3). The temperature in the upper part of the hearth (slag zone) is about 1,200°C and in the lower part (lead zone) about 900°C. Typically, 95%–96% of the lead is recovered in the lead bullion (Longval et al. 2008).

The slag constituents are mainly iron silicate and lime but may contain up to 2% Pb and 10%–22% Zn. FeO in blast furnace slags ranges from 25% to 35% in worldwide practice, depending on related contents of SiO_2 (20%–25%), CaO (15%–22%), MgO (1%–4%), Al_2O_3 (3%–8%), and ZnO. High zinc slags require more iron, which is often added as a separate ironstone flux. Quite commonly, iron in the charge exceeds the requirement for the most economical slag formulation. In such a case, the additional iron demands more lime and silica flux and more thermal energy, and as a result, the smelter schedule will penalize iron in lead concentrates purchased. Pyritic iron in concentrates is generally unwelcome because of the increased ignition heat it represents and additional sulfur handling necessitated. Typical operating data for lead blast furnaces are shown in Table 5 (Jiao 2010).

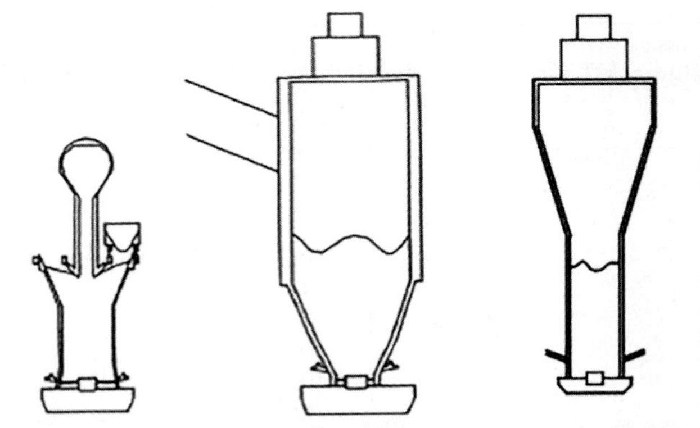

A. Lead "thimble top" blast furnace
B. Imperial Smelting blast furnace
C. Modified Imperial Smelting blast furnace for lead smelting

Source: Longval et al. 2008

Figure 6 Different types of blast furnaces

Imperial Smelting process. The Imperial Smelting process (ISP) technology is a variant of the conventional blast furnace and is designed to recover zinc and lead simultaneously. Lead content is less critical; it regularly processes lead concentrates below 50% Pb content, particularly with appreciable copper content, and mixed (bulk) lead–zinc concentrates of as low as 10% Pb, depending on the zinc content. Usually, lead concentrates are incorporated into the charge. Lead has an important hardening effect on sinter, substituting for silica, which is the practical alternative. SiO_2 flux requires compensatory lime flux addition—thus, lead replacing silica makes for lower slag production with benefits in increased fuel efficiency and lower zinc loss. Mixed lead–zinc concentrates of Pb + Zn content as low as 30% have been treated on occasion but presumably in conjunction with higher-grade materials. The process will also handle zinc ash, waelz oxides, and oxidic Pb–Zn ores as charge components. Considerable flexibility is available in charge-compounding, and the zinc blast furnace accepts properly compensated changes in feed blends without difficulty.

Four major process steps make up the ISP technology to produce lead and zinc (Figure 7). As in the blast furnace

Table 5 Operating data of several lead blast furnaces worldwide

Item	Brunswick (Canada)	Port Pirie (Australia)	Peñoles (Mexico)	Chigirishima (Japan)
Furnace type	With Imperial Smelting furnace charging	Open top, two-row tuyere	Single-row tuyere	Single-row tuyere
Hearth area, m^2	13.3	18.8	10.2	9.4
Number of tuyeres	44	92	42	56
Sinter smelted, t/h	40	58	19	19
Blowing rate, Nm^3/h	20,000	26,000	10,000	9,000
Oxygen enrichment, %	27.5	27	22.5	22
Blast preheating	No	No	No	200°C
Coke rate, %	10.5	9.5	14	—
Bullion production, t/yr	84,000	230,000	59,000	64,000
Pb in slag, %	1.2	2.3	1.5	2.8

Source: Jiao 2010

process, sintering is conducted in a Dwight-Lloyd machine. The feed mix consists of complex lead–zinc concentrates, zinc-rich residues and dusts, recirculated drosses, limestone, silica fluxes, and returned sinter fines. The produced sinter containing ~15%–22% Pb as well as 42%–45% Zn is screened and mixed with secondary zinc materials and coke preheated to 800°C. The lead and zinc is reduced through the injection of preheated air to metal. Zinc is produced as vapor, which is then condensed in a spray of molten lead. Liquid zinc separates on cooling and is either sold directly or refined. As in a blast furnace, lead collects in the hearth together with molten slag and is tapped. The solvent action of metallic lead makes it possible to recover therein the silver, gold, and bismuth content of concentrates together with a substantial part of the copper. The process cannot be confined to zinc recovery alone. Some lead in the charge is necessary for recovery of minor values as well as for reasons of economy in sintering and furnacing.

Because of the lack of raw materials and for economic reasons, many Imperial Smelting furnace plants have been permanently closed down or converted to a pure lead blast furnace. Today, only nine plants are still in operation: five in China, two in Japan, one in India, and one in Poland (Fallas 2010).

Modern Applied Technologies

Kivcet smelting technology. The technology of the Kivcet process employs the principles of simultaneous flash smelting of lead-bearing materials using technically pure oxygen. In conjunction with electrical energy, carbon reduces the lead oxide in the slag. The process was developed by Vniitsvetmet in Ust-Kamenogorsk, Kazakhstan, and transferred into industrial scale.

Today, two smelters are operating based on the Kivcet technology. The first commercial-scale smelter was constructed for Enirisorse (now Glencore) at Portovesme, Sardinia. The smelter was commissioned in 1987, and its current annual production capacity is 100,000 t/yr Pb, predominantly based on treating materials from primary sources. The second smelter was built at Teck Metals in Trail, Canada, and started operations in 1997. The annual capacity of the Trail smelter is up to 120,000 t/yr Pb. To meet the needs of Trail Operations as an integrated zinc–lead smelter, the smelter is designed to treat a high-residue charge. Metal and co-product recoveries are maximized along with reduced energy and labor costs and reduced environmental impact (Rioux et al. 2008).

A simplified flow sheet of the Teck Metals smelter is shown in Figure 8. The Kivcet process is able to treat a wide variety of lead-bearing raw material. At the Portovesme plant, the treated ratio of primary to secondary materials is about 3.5:1, while it is 1:2 at Teck Metals. The majority of the Teck Metals raw material consumption consists of internally generated zinc plant iron residues plus lead concentrates and precious metal concentrates as well as battery scrap. The flash smelting process requires drying of the feed material to <1% moisture content prior to being charged into the furnace. The exiting dry product is conveyed from the drying furnace into a ball mill, grinding it to –1 mm.

The Kivcet furnace mainly consists of three primary areas: a smelting shaft, an adjoining electric furnace that is separated by a water-cooled partition wall, and the uptake shaft (Figure 9). Concentrate burners at the top of the reaction shaft (four at Teck Metals, two in Portovesme) serve to mix the pre-dried feed material with oxygen to spontaneously initiate the roasting and smelting reactions as the charge travels down the reaction shaft. The reaction shaft is made entirely of water-cooled copper tubes cast in copper elements from the centerline of the air-cooled hearth. The reaction shaft is also equipped with a natural gas auxiliary burner (12,000 Mcal/h at Teck Metals) for heat during feed interruptions. The feed-to-oxygen weight ratio is crucial for the control of the desired reducing condition in the flame. An abundance of oxygen will increase the lead content in the slag and tend to form magnetite deposits under the reaction shaft, whereas an oxygen deficiency will result in unburned feed and formation of a matte phase and sulfide deposits in the uptake shaft. Moreover, a rather erratic and varying feed composition is detrimental to the process, making it extremely difficult to achieve stable operating conditions. Fluctuations in the fixed carbon content of the fuel coal should be minimized. The implementation of a residue-blending device and storage bins for flue dust with loss-in-weight discharge are required to avoid poor blending of different residues and variations in the amount of recirculated flue dust.

The flame temperatures are monitored frequently under each concentrate burner and are used to adjust the feed calorific value. Flame temperatures lower than 1,380°C indicate that the charge is under-fueled. Flame temperatures higher than 1,420°C indicate an excess of coal in the charge. Smelt samples taken under each burner are analyzed for sulfur. The goal for typical residual sulfur remaining in the smelt

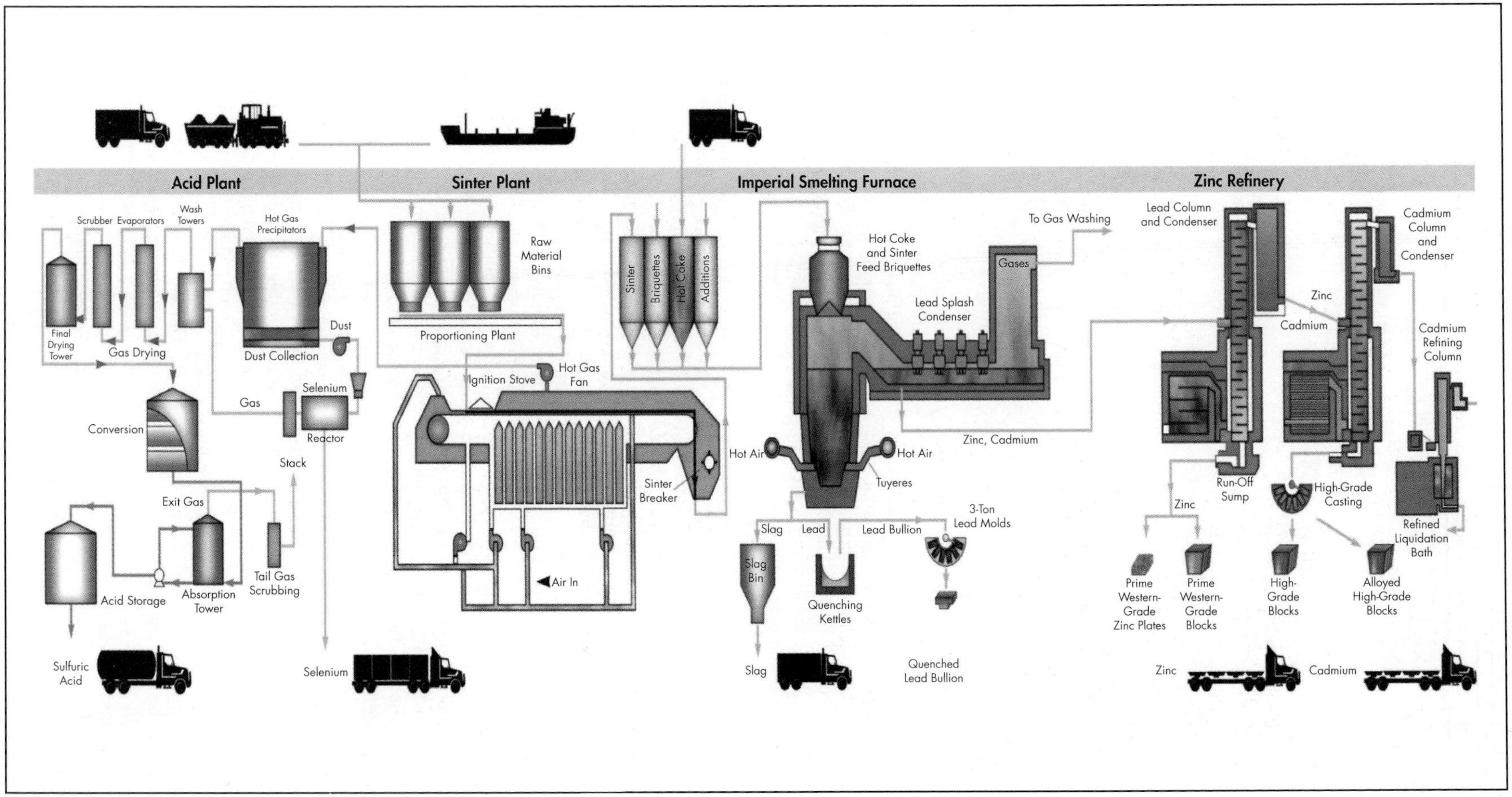

Source: Fallas 2010

Figure 7 ISP process flow sheet

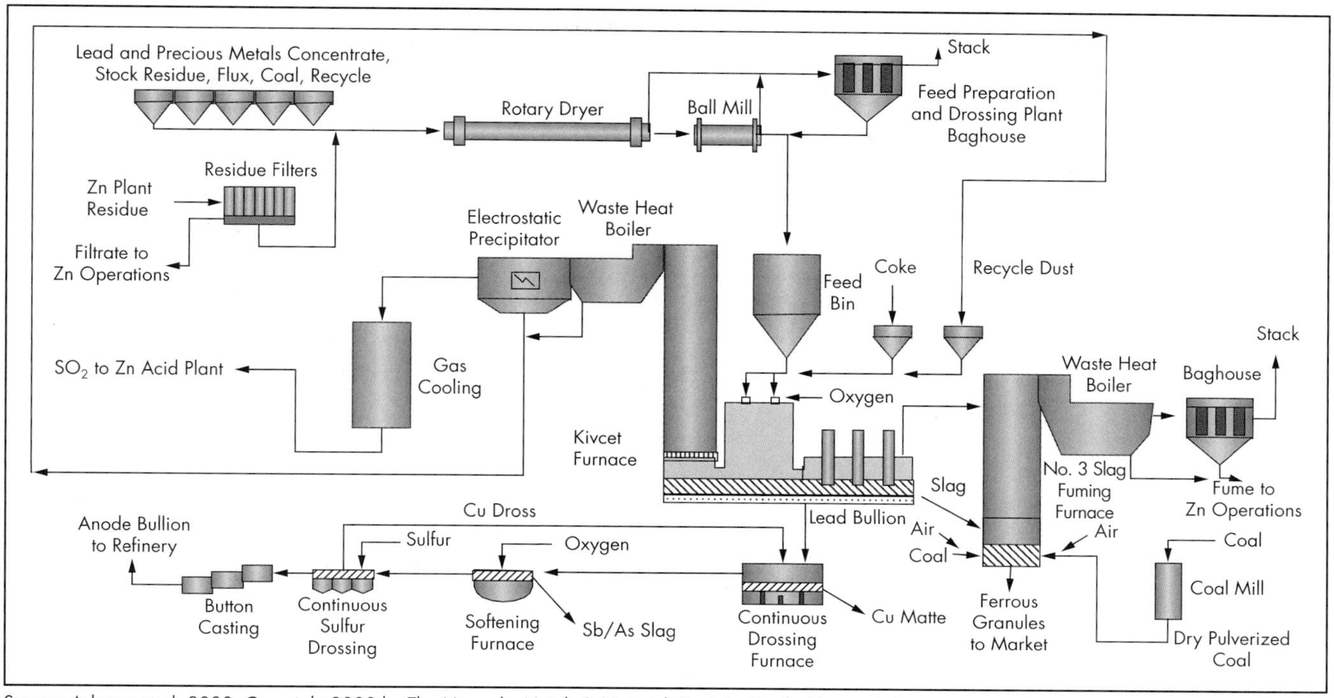

Source: Ashman et al. 2000. Copyright 2000 by The Minerals, Metals & Materials Society. Used with permission.

Figure 8 Flow sheet of the Teck Metals lead smelter

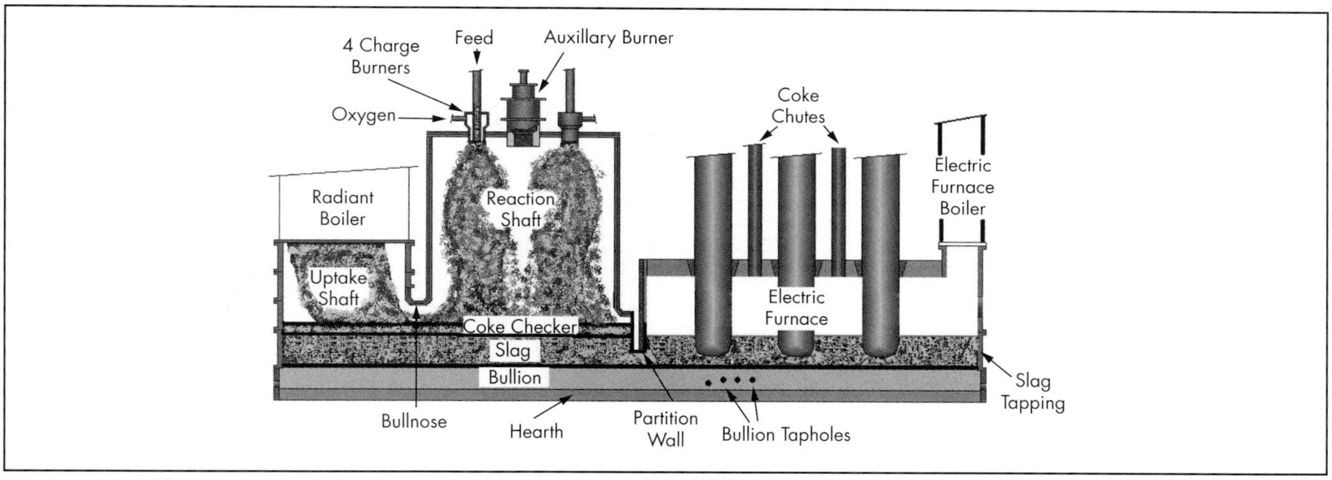

Source: Goosen and Martin 2000

Figure 9 Cross section of Kivcet furnace

are 0.5%–1.2%. Coarser coke (5–15 mm) enters the furnace together with the charge to float on the slag layer in the reaction shaft area forming a "coke checker." The added coke rate is adjusted to maintain a coke layer of ~100–150 mm thick to efficiently reduce the generated lead oxide as the smelt flows through the coke. One control parameter of the coke-layer thickness is maintaining the temperature of the coke checker between 1,100°C and 1,200°C. Coke checker temperatures higher than 1,200°C indicate a thin layer of coke in the furnace; temperatures lower than 1,100°C mean a thick layer of coke or coke piling. Poor coke checker performance may result in accretion formation because of operation with a more reducing flame to lower the lead content in the slag. These accretions mainly consist of lead sulfide. They can grow so large that they start blocking the gas passage at the bullnose (Figure 9). The bullnose accretion condition is monitored by measuring the draft differential between the reaction shaft and the boiler on a continuous basis. Accretion management and the directly dependent performance of the coke checker is one of the most critical control parameters to the entire operation.

The molten and virtually reduced slag flows under a submerged water-cooled copper partition wall, which separates the electric furnace (55 m^2 at Teck Metals) from the reaction shaft. Three in-line electrodes deliver the required power (5–7 MW at Teck Metals) to complete the slag reduction and to maintain the desired slag temperature between 1,320°C

Table 6 Main typical operating data of direct lead smelting technologies at various plants

Item	Kivcet (Portovesme, Sardinia)	Kivcet (Teck Metals, Canada)	QSL (Berzelius Stolberg, Germany)	QSL (Korea Zinc, South Korea)	TSL (Nordenham, Germany)	Kaldo (Boliden Metall, Sweden)
Lead production, t/yr	100,000	85,000	105,600	130,000	125,000	35,000
Raw material, t/yr	150,000	325,000	176,000	314,000	190,400	43,000
Primary/Secondary ratio	3.5:1	1:2	1:1	2:1	1:2	1
Types of secondaries	Battery paste, EAF-dust, zinc oxide, waelz oxide	Recycled battery, zinc plant iron residues, Pb plant drosses	Battery paste, zinc plant residues, slags, scrap	Pb-residues, Pb-Ag residues, battery paste	Battery paste, residues, scrap	
Dryer type	Rotary drum	Rotary drum	N/A*	N/A	N/A	Rotary drum
Feed rate, t/h	35	58				30
Inlet moisture, %	8–10	13				5
Outlet moisture, %	0.1	<1.0				<0.1
Off-gas volume, Nm³/h	24,000	71,000				15,000–20,000
Feed rate, t/d	600	1,220	612	1,500–1,600	670	Batchwise
Moisture, %	<1	<1	8–10	10–12	10–12	<1
Silica, kg/t bullion	92.8	170	74	20	300–600	
Lime, kg/t bullion	154.4	480	230	117	500–1,300	203
Iron, kg/t bullion	23.2	540	150	148	500–1,000	217
Coal, kg/t bullion		440	115	131		
Coke, kg/t bullion	50	110	23		1,000–3,000	48
Oxygen volume, Nm³/h	6,500	9,100	3,570	9,400–9,900	1,600	
Air volume, Nm³/h		1,500	200	1,500	7,000	
Oxygen enrichment in air			67		30–40	Enrichment
Off-gas volume, Nm³/h	12,000	22,000	23,500	33,000–38,000	30,000	25,000
SO_2–off-gas, %	22–25	~15	14.1	10–15	6–7	4–5
Steam, t/d	300–350	670	300	480	340	
Bullion, t/d	300	260	370	440	139,000	Batchwise
Matte, t/d	10	13	16.5		5,800	
Slag in t/d	120	590	274	300	46,000	Batchwise
Pb in slag, %	4	5.9	3–4	4–6	50	4.2
FeO in slag, %	26	27.9	29.2	27–29	11–12	40
SiO_2 in slag, %	22	21.6	20	20–22	5–7	21
CaO in slag, %	20	13.3	20.8	14–15	2–3	28
Zn in slag, %	9	18	8.8	15–16	6–10	4.6

Source: Adapted from Hayes et al. 2010
*N/A = Not available.

and 1,360°C. The majority of the zinc contained in the raw materials reports to the slag in the reaction shaft area. After entering the electric furnace compartment, the zinc is volatilized, depending on the adjusted reducing atmosphere, and recovered as fume. Slag is discharged from the Kivcet furnace at regular intervals and can be granulated or transferred to a subsequent slag-fuming furnace. The bullion is discharged at 850°–950°C through specially designed tapping inserts. Any matte generation in the furnace causes significant operational problems with bullion tapping. Its formation is minimized by controlling reaction shaft desulfurization and by tapping bullion at high temperatures to keep sulfides in solution. The bullion is transferred to a downstream refinery for further refining (Goosen et al. 2000).

Compared to the conventional two-step process, the off-gas volumes are much smaller (21,000 Nm³/h at Teck Metals; 12,000 Nm³/h at Portovesme) and contain SO_2 concentration of 12%–15%. This can be attributed to the application of nearly pure oxygen and the sealed design of the furnace. The off-gases enter a waste heat boiler system. The outlet gas temperature from the waste heat boiler is approximately 350°C. Typically the boiler system produces steam (23–25 t/h at Teck Metals; 12–15 t/h at Portovesme). In the Portovesme plant, the thermal energy is converted in a turbine into electricity, which is utilized in the process minimizing outside energy requirements. The collected dust in the waste heat boiler is, together with the dust precipitated in the subsequent hot-gas electrostatic precipitator, recirculated back to the process. The most recent available operating data of both Kivcet plants in operation are listed in Table 6 (Siegmund 2000a; Hayes et al. 2010).

QSL/SKS technology. The QSL (Queneau–Schuhmann–Lurgi) technology is a continuous-operating single-vessel bath smelting process. The process principles were developed and patented by two Americans, Professor Paul E. Queneau and Professor Reinhardt Schuhmann Jr. Based on a patent developed by the German companies Lurgi AG and Berzelius Metall GmbH, the QSL process for the recovery of lead advanced

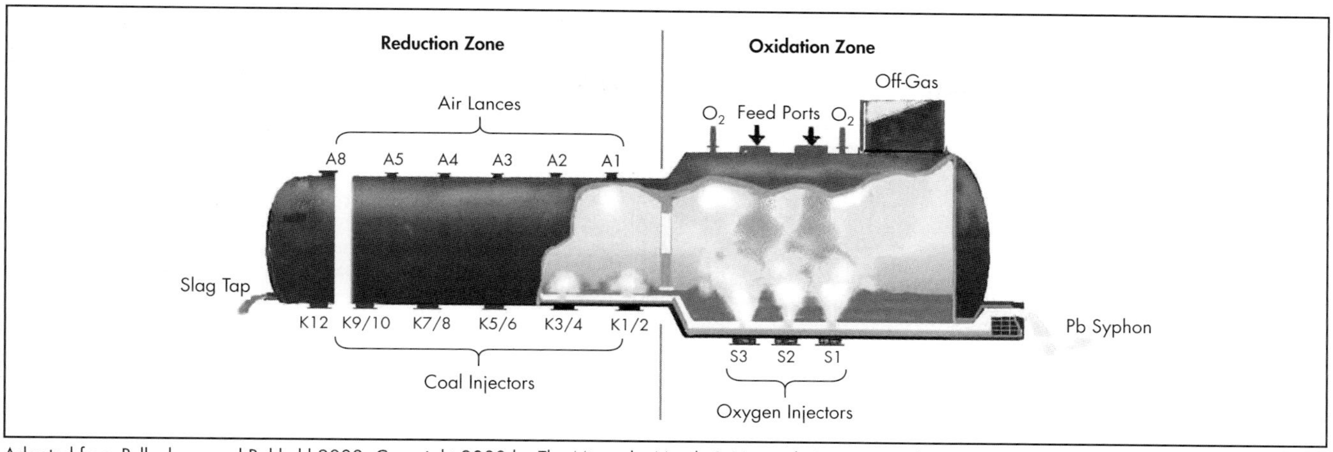

Adapted from Pullenberg and Rohkohl 2000. Copyright 2000 by The Minerals, Metals & Materials Society. Used with permission.

Figure 10 QSL converter at Berzelius Stolberg GmbH

to an industrial scale after operating a semicommercial plant for several years (Queneau and Siegmund 1996). Today, three plants based on the QSL technology are in operation: two at Korea Zinc in Ulsan, South Korea, with annual lead production of 250,000 t, and one at Berzelius Stolberg GmbH in Stolberg, Germany, with annual lead production of ~150,000 t.

The entire converting process is incorporated in one slightly inclined (0.5%), closed, kiln-like smelting unit. The converter is divided by a partition zone into smelting and slag reduction zone (Figure 10). It can be tilted by 90 degrees about its longitudinal axis when the process is interrupted. The process is designed to treat all grades of lead concentrates as well as secondary materials, applying the bath smelting principle with submerged high-pressure injection of tonnage oxygen and fossil fuels (Siegmund et al. 1998). Two converter sizes are applied. The QSL-converter at Korea Zinc has a diameter of 4–4.5 m and an overall length of 41 m, while the converter at Berzelius has a diameter of 3–3.5 m and a length of 33 m. The current ratio of primary to secondary material at the smelter in Ulsan is 2:1 and in Stolberg almost 1:1. Both operations, however, report a high flexibility of the process in terms of feed composition. They can deal with a wide variety of feed materials and quickly adapt to changes in the feed composition. Rates of secondary materials of more than 60% have been charged (Siegmund 2000b).

The feed preparation is relatively simple, consisting only of bins for material storage, weighing devices for composing the anticipated charge mixture, and a mixer to thoroughly blend and agglomerate the charge to satisfy steady-state feed requirements. Lead scrap materials can be directly charged into the converter. The moist and agglomerated feed mixture is then charged homogeneously and weight controlled through feed ports located in the roof of the smelting compartment into the converter. It falls into a dispersed molten mixture consisting of primary slag and crude lead while oxygen is injected through submerged gas/water aerosol-cooled tuyeres at the bottom of the converter. In the liquid bath, an autogenous roast reaction smelting is initiated, converting some of the lead compounds directly into low-sulfur lead bullion and forming a primary slag with a lead oxide content of 40%–50%. The exothermic reaction takes place at 1,050°–1,100°C. As fuel, low-cost available materials like bituminous coal or petroleum coke can be applied. The slag flows through a submerged opening in the partition wall into the slag reduction zone where the lead oxide is gradually reduced to metallic lead using pulverized coal as the reducing agent while the slag flows to the opposite end of the converter. The coal is injected with a carrier medium into the melt through a series of bottom-blowing gas-cooled injectors together with a controlled amount of supplementary oxygen to adjust the desired reduction potential in the melt. The independently and accurately controlled gas analysis injectors are spaced sufficiently apart to form a series of proper reaction gas-slag bubble plumes of sequentially decreasing oxygen activity and increasing temperature until slag is discharged. These chemically active mixing areas are separated by passive lead settling zones. There is virtually no interaction of adjacent bubble plumes, thus creating a continuous mixer–settler configuration (six).

At the end of the slag reduction zone, a larger settling area is provided to accomplish sufficient separation between the low-lead slag and the lead. The settled metallic lead flows countercurrently to the slag back into the smelting zones. It combines with the primarily produced lead bullion before being tapped through a siphon system for subsequent conventional refining. The final slag is tapped at the end of the slag reduction zone and granulated. The carbothermic reaction is endothermic and takes place at 1150°–1,250°C. Desired post-combustion of the process gas in the atmosphere of the slag reduction zone is accomplished with oxygen-enriched air, which is injected via lances located in the roof of the converter. The gas temperature is adjusted to 100°–150°C above the slag temperature to transfer heat back into the slag and contributes to the required temperature increase in the slag (Siegmund 2000a).

The QSL process applies nearly pure oxygen, which contributes to a comparatively small volume of highly SO_2-rich off-gas. The actual off-gas volume for the plant in Stolberg is 23,500 Nm^3/h with more than 18% SO_2, depending on the type of concentrates. For the Korea Zinc plant, the off-gas volume is approximately 35,000 Nm^3/h with SO_2 concentration between 10% and 15%. In the Stolberg converter, the off-gas from the reduction zone is combined with the generated off-gas in the smelting zone through an opening for the passage of the process gas, whereas the converter at Korea Zinc has a completely closed partition wall separating the off-gases from both zones. Depending on the adjusted gas atmosphere in

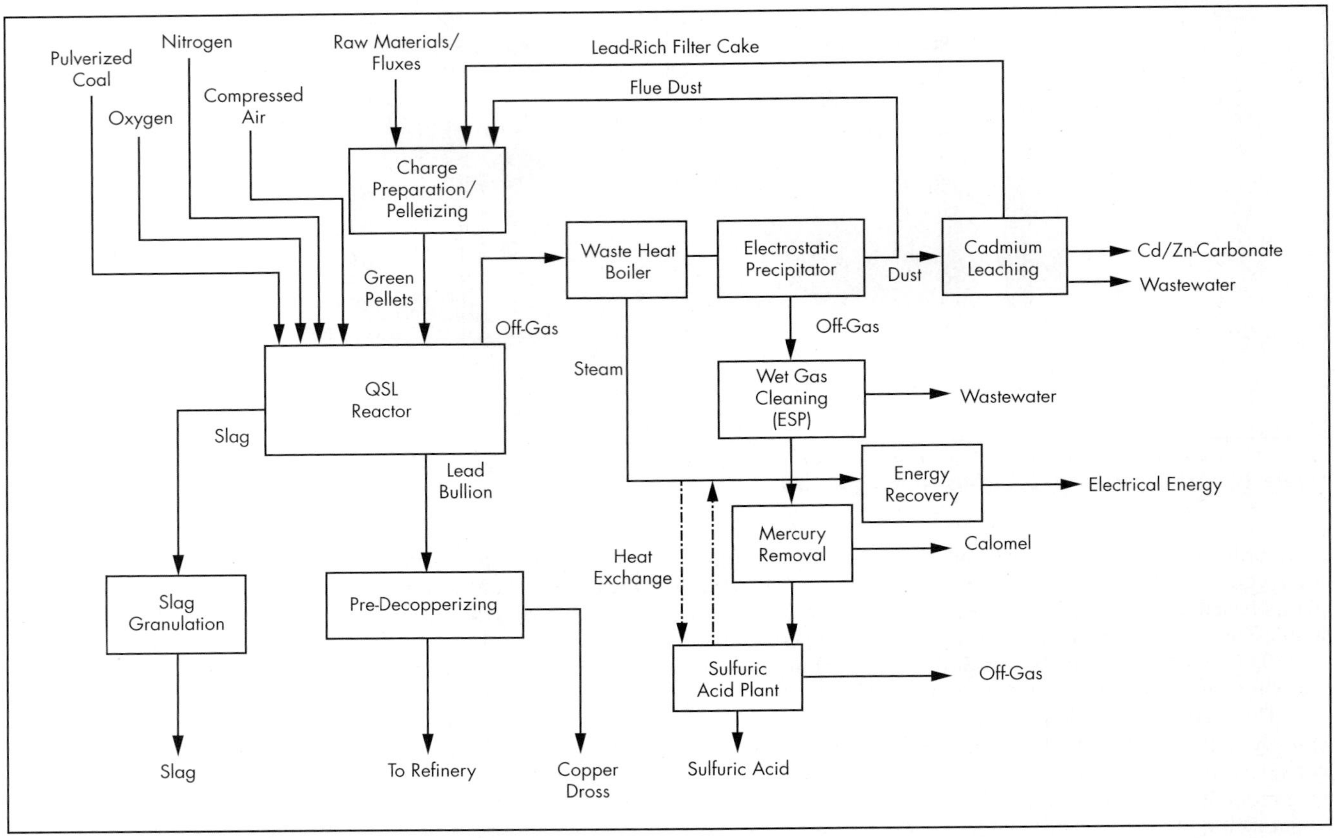

Source: Queneau and Siegmund 1996

Figure 11 Process block diagram of a QSL plant

the slag reduction zone, the resulting process gas contains zinc fumes. The gas exits the converter through a second uptake and the fumes are collected for subsequent zinc recovery. This lowers the cost of the slag-fuming operation carried out in a separate furnace.

The produced slags contain 5% Pb on average in the Korea Zinc plant and 2%–4% Pb in Stolberg. It was demonstrated already that discharge slag of well less than 2% are obtainable, but only when zinc is simultaneously fumed at strong reduction potentials. In Stolberg, the zinc is then collected in the flue dust. This is undesirable because the flue dust is being directly recirculated to the converter and it would contribute to a substantial proportion of total feed by weight. Slags containing more than 15% Zn generally become too viscous at adjusted converter temperature. Heat is recovered from the off-gas in both plants and utilized to minimize outside energy requirements. In general, the off-gases are cooled in a waste heat boiler system consisting of a vertical radiation channel and a subsequent convective pass. In the vertical uptake, the gases are cooled to below 700°C. The exiting temperature of the off-gas after the convective pass is between 350°C and 380°C. Collected dust from the primary gas system is immediately recirculated back to the process. The SO_2-rich off-gas enters a double-catalysis sulfuric acid plant to produce sulfuric acid at conversion rates of 99.5% (Meurer and Ambroz 2010).

The general process block diagram of a QSL plant is shown in Figure 11 and the current main operating data of both QSL facilities are summarized in Table 6 (Püllenberg and Rohkohl 2000).

The concept, design, and operation of the SKS process is very similar to the QSL technology. It is a semicontinuous bath smelting process applying submerged cooled injector technology. It was developed in the 1990s mainly by Shuikoushan (SKS) Mining Bureau and Beijing Central Engineering and Research Institute for Non-Ferrous Metallurgical Industries. Opposite to the QSL process, the SKS technology consists of two independently operating refractory-lined cylindrical vessels, which are connected through a launder. Today, more than 20 lead smelters in China and one in India operate on the SKS technology concept. The current process went through three development stages. Initially the technology comprised only the smelting stage as a substitution of the sinter machine, followed by the slag reduction step in a blast furnace. The second generation of the technology employed the smelting reactor from where the generated slag was discontinuously discharged into a converter-type vessel with side-blowing tuyeres. The third and current generation consists of a combination of smelting reactor and a horizontally aligned slag reduction converter with bottom-blowing injectors, which are connected through a launder. By adopting the bottom-blowing electrothermal reductive furnace concept and utilizing the heat from the molten lead oxide slag, the reductant and fuel consumption efficiency is further improved compared to the second generation (Li and Yang 2010).

The feed preparation is identical to the QSL process and the charge (lead concentrates, secondaries, and additions) is

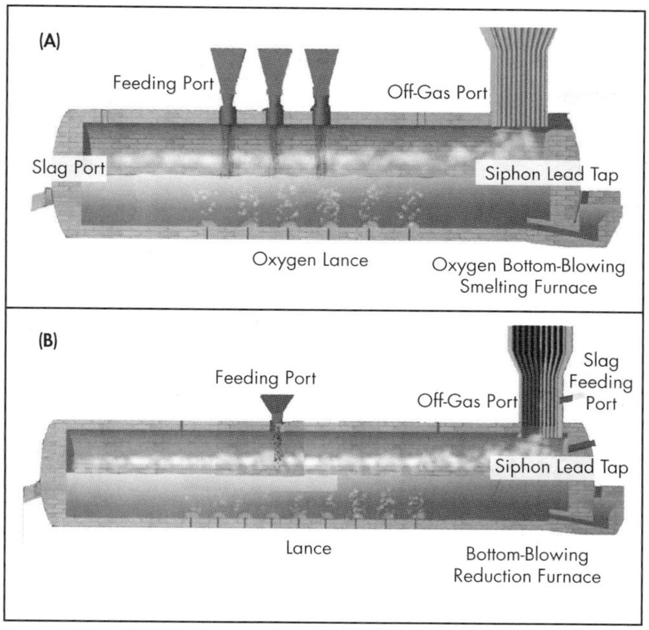

Source: China ENFI Engineering Corporation 2010

Figure 12 (A) SKS and (B) RSKS furnaces

fed from the top. Oxygen, nitrogen, and atomized water are blown into the SKS furnace through bottom-blowing supersonic injectors accomplishing effective mixing and highly efficient smelting at excellent mass and heat transfer conditions (Figure 12A). The lead-bearing materials are rapidly oxidized and the exothermic reaction from the oxidation of sulfides provides sufficient heat to facilitate autogenous smelting. The majority of the lead is oxidized and reports to the lead-rich slag with some direct primary lead production. This amount depends on the composition of the charge and the adjusted oxygen potential in the bath. The production of a low-calcium slag allows furnace operation at relatively low temperatures, which are controlled at about 1,050°C. To ensure the appropriate properties of the slag, some limestone is added in the smelting stage. Volatile dusts from the oxidation stage containing lead, zinc, arsenic, cadmium, antimony, and other elements are collected in appropriate dust collection systems and recirculated to the smelting step. The dust generation depends upon the prevailing metal compound and its associated vapor pressure under oxidizing atmosphere and temperature. To minimize the generated dust amount, the oxygen potential is maintained at an appropriate level to minimize the presence of lead sulfide, which has a significantly higher vapor pressure at prevailing temperatures than that of lead bullion and lead oxide. Furthermore, the smelting temperature is kept as low as possible to minimize the vapor pressure of the various lead-bearing materials. The injection of pure oxygen into the furnace reduces the process gas volume, resulting in less mechanical entrapment and volatile dust formation (China ENFI Engineering Corporation 2010).

When the SKS furnace reaches a bath level of about 1.6 m, lead-rich slag will be tapped into an independently operating reduction SKS (RSKS) furnace through a hot-feed port (Figure 12B). Lump coal in small amounts is fed to the furnace through a cold-feed port. Oxygen and pulverized coal are injected into the furnace through bottom-blowing injectors. The bath temperature and the desired reducing atmosphere for the reduction process is controlled by predominately adjusting the feed rate of coal (lump and pulverized coal) and oxygen. When lead in the RSKS furnace slag is reduced to about 2% or less, the reduction process is stopped and slag is tapped. The furnace temperature fluctuates within a range of 1,200°C to 1,280°C, mainly based on the concentration of ZnO in the slag. The slag temperature was initially regulated through electrodes submerged into the bath from the top, ensuring a low slag viscosity and smooth tapping. The continuous or discontinuous use of electrodes in the RSKS furnace makes it more suitable to treat high-Zn materials with high melting points (Li and Yang 2010). Meanwhile, through further optimization, it was possible to eliminate the requirement of applied electrodes introduced from the top of the furnace, simplifying its operation.

Batch feeding and batch slag tapping in the RSKS furnace has advantages compared to continuous feeding and slag tapping. Lower lead contents in the final slag are easier to achieve at an increased lead recovery rate. Moreover, reduction takes place in an independent furnace with a single-gas atmosphere. The overall process flow diagram as shown in Figure 13 is very similar to the QSL flow sheet (Figure 11), with the exception of the slag fuming furnace addition.

TSL technology. The TSL technology is derived from the Sirosmelt top submerged lance process. That technology has been developed over the past 20 years in laboratory- and pilot-plant-scale studies on the recovery of almost the complete range of nonferrous, ferrous, and precious metal materials and is successfully applied in the copper, nickel, silver, tin, and zinc industries (Creedy et al. 2010). The initial design for lead smelting consisted of a two-furnace arrangement and is marketed today by the companies Outotec under the name Ausmelt and Glencore under the name IsaSmelt. Concepts for both technologies are very similar with only a few differences. For primary lead smelting, four IsaSmelt and six Ausmelt furnaces are commercially in operation. The first TSL lead plant was commissioned in 1996 at Recylex in Germany with the intention to perform the two-stage operation for lead recovery in a single furnace. The smelter has a design production rate of 90,000 t/yr of lead based on a certain ratio of primary to secondary raw materials (Karpel 1998). However, today, only the first smelting stage is conducted in the furnace. The high-lead-containing slag is processed elsewhere (Kerney 2010). Korea Zinc contracted with Ausmelt for the second lead smelter. It was commissioned in 2000 and processes lead oxide fume, sulfate residue, and battery paste. The process concept and the principles of the furnace design are illustrated in Figure 14 (Errington et al. 2010).

Raw materials and additives are stored in bins from where a feed delivery and weigh control system collects and composes the desired charge composition and transports it to a mixing device for blending and homogenizing before being fed into the furnace. As in the QSL/SKS technology, the materials do not have to be dried or reduced to a specified material size before charging, and the principal process flow sheet is similar. The coarse, moist, and agglomerated charge is dropped through a feed port located in the inclined roof hood of the furnace and incorporated into the melt. Required air, process fuel, and fine materials are injected beneath the surface of a liquid slag bath through a top-entry submerged lance system. Oxygen-enriched air to levels up to 50% may be used with the specially designed lance, injecting the gas

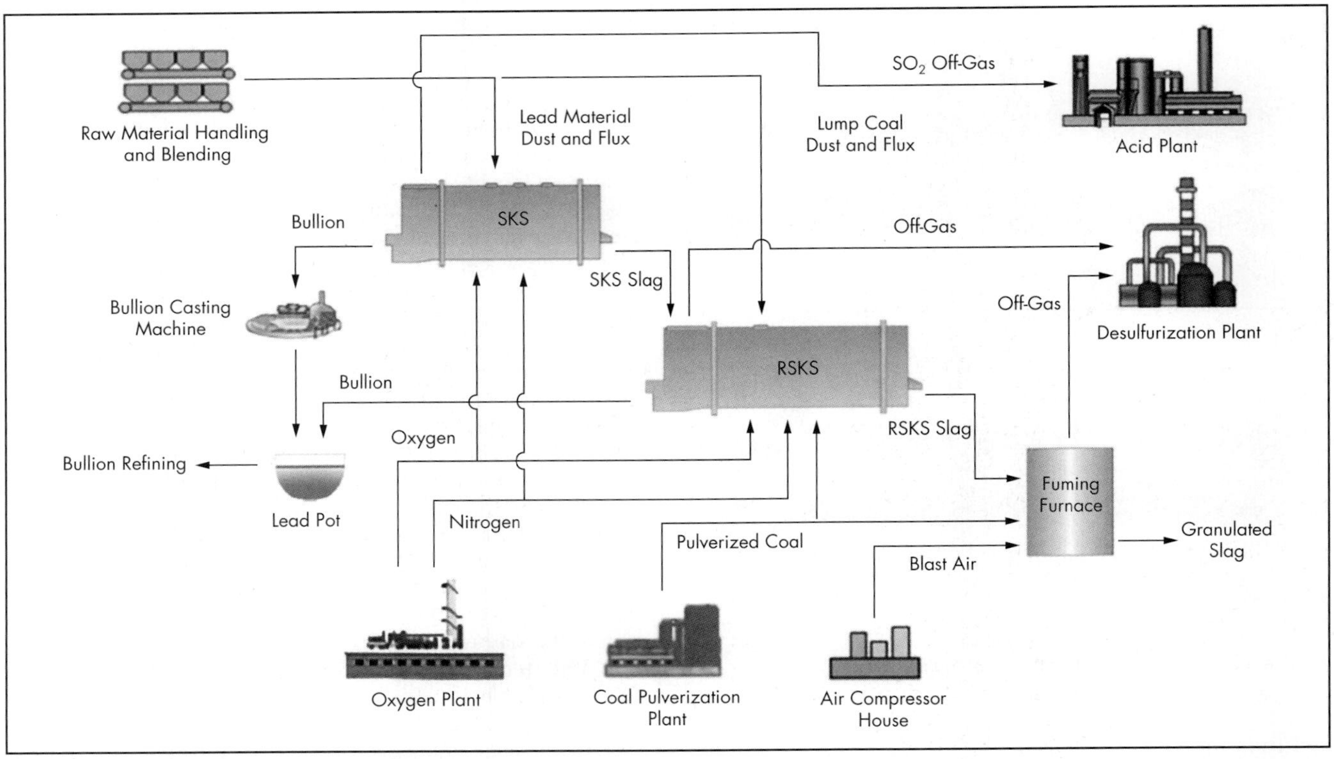

Source: China ENFI Engineering Corporation 2016

Figure 13 Typical process flow sheet SKS-RSKS technology

through the outer annulus while the fuel and the fine materials are conveyed down the lance through an inner tube. The stainless-steel lance is protected from the furnace contents by a coating of frozen slag to minimize lance wear. The coating is established by carefully lowering the lance into the furnace with an intermittent stop above the bath permitting slag to splash and freeze on the lance before submerging the tip of the lance. During operation, the latter is always below the theoretical static slag line. The media are therefore injected into the slag, creating very turbulent conditions in the furnace, promoting very rapid reactions, and resulting in high smelting capacities in small areas. The furnace is usually constructed as a refractory-lined cylinder applying a high-quality chrome-magnesite refractory lining. Where very aggressive slags, mattes, or metals are present or higher temperatures are employed, water cooling is applied behind the bricks to ensure acceptable refractory life.

A slag reduction process to produce bullion and discard slag follows the oxidation smelting process step. Precise control of the system oxidation state is a key feature of TSL technology. This, together with the addition of a reductant, such as coal, allows the reduction step to be precisely tailored to meet specific requirements. A low oxygen potential in the order of 10^{-9} to 10^{-10} atm is maintained. The liquidus temperature of lead-oxide-containing slags is raised from 1,000°C to 1,100°C to 1,150°C to 1,250°C as the lead in slag drops from more than 40% lead to about 10%. Fluxing with limestone may be needed to substitute lead oxide by calcium oxide and to prevent the precipitation of zinc ferrite or magnetite from zinc- and iron-rich slags with lead levels in the region of 15%–40% Pb in the temperature region of 1,200°C. Slag reduction is a less economic exercise than smelting of high-lead-content material directly to bullion, particularly with solid slag processed on a campaign basis. Lead oxide fume is produced in increasing quantities with increasing temperature, collected in a baghouse or electrostatic precipitator and recycled to the smelting furnace, with or without external fume treatment for impurity removal. In both stages, smelting and reduction, air is introduced into the top space of the furnace through a shroud pipe attachment on the TSL to post-combust any volatilized species. Up to 40% of the energy generated by these reactions is recovered for utilization in the bath (Kaye et al. 2008).

The overall process concept can consist of either single, batch, two-furnace continuous, two-furnace continuous and batch, or three-furnace continuous process options. The preferred approach for lead smelting is the two-stage approach where slag flows continuously from the smelting to the reduction furnace via weirs, siphons, and launders. A process block diagram of this alternative is shown in Figure 15. In general, the second furnace is smaller than the primary furnace, particularly when smelting high-grade feed. Even though the second furnace is smaller, it requires a separate gas handling system, typically incorporating an evaporative gas cooler and baghouse, and this increases plant capital cost above that of the single-furnace, two-stage operation. The final slag product from the second furnace typically contains low levels of lead but may well contain appreciable levels of zinc (8%–12% Zn). The slag can be granulated and held for long-term storage or a second slag reduction can be carried out to recover the zinc as a high-grade (>50% Zn) fume, as previously described, and produce a discard slag containing low levels of lead (<1% Pb)

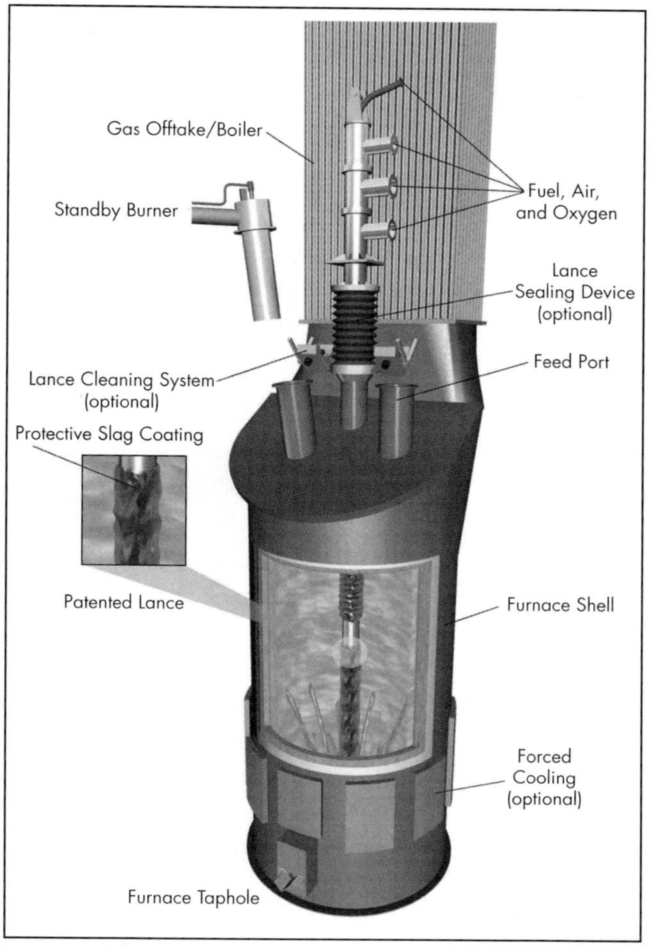

Source: Stephens 2010

Figure 14 TSL furnace design including schematic section

and zinc (<3% Zn). The main operating data of the TSL operation in Nordenham, Germany, are summarized in Table 6.

Kaldo technology. Boliden Metall AB developed the Kaldo process, and its operation in Rönnskär, Sweden, is the sole application in the lead industry. The plant is used not only for lead smelting but also for the treatment of secondary raw materials containing copper, precious metals, and plastics. The process is based on the flash smelting principle and operates on a discontinuous/batchwise basis and consists of a top-blowing rotary converter. It has special features compared to the other processes. The batchwise operating mode seems to make the process not very suitable for large-scale operation, but the furnace is a very flexible unit and can treat a wide range of secondary materials, including battery scrap, residues, and recycled dust among lead concentrates. It was exclusively using concentrates from its own Laisvall mine as lead-bearing raw material but recently has been treating other concentrates. As in other flash smelting applications, the feed material has to be dried to moisture contents less than 1% before being blown through a lance into the furnace, where it reacts in the flash with the oxygen-enriched air that is introduced. Iron and lime fluxes are also added to bind the silica in the concentrate, generating a liquid slag. The generated slag contains high concentrations of lead oxide, which is reduced to average lead contents of ~4% by coke. The lead bullion and the slag are then discharged into separate ladles. The lead is subsequently treated in a refinery, while the slag is returned to a mine for final underground disposal. The converter is completely encapsulated in a vented enclosure, generating off-gas volumes of 25,000 Nm^3/h with SO_2 concentration between 4% and 5%. The off-gases are combined with process gas from the adjacent copper smelter to produce sulfuric acid (Siegmund 2000a). Figure 16 illustrates the process flow sheet of the Kaldo process, and the main operating data are summarized in Table 6.

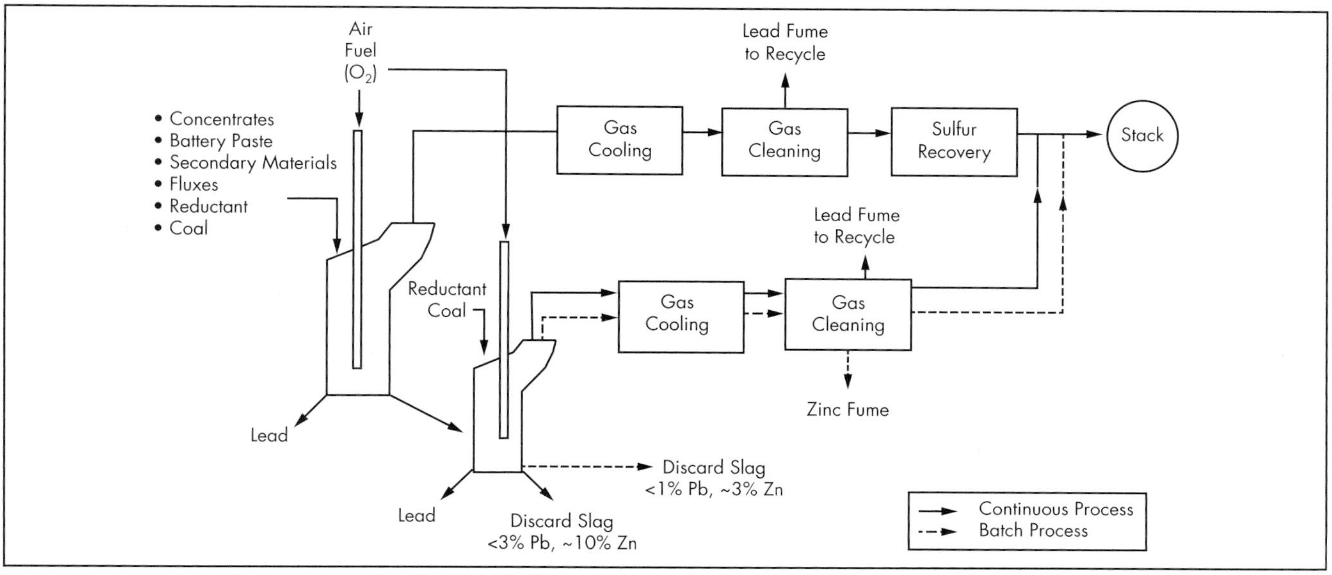

Source: Mounsey et al. 2000. Copyright 2000 by The Minerals, Metals & Materials Society. Used with permission.

Figure 15 TSL two-furnace process without zinc recovery

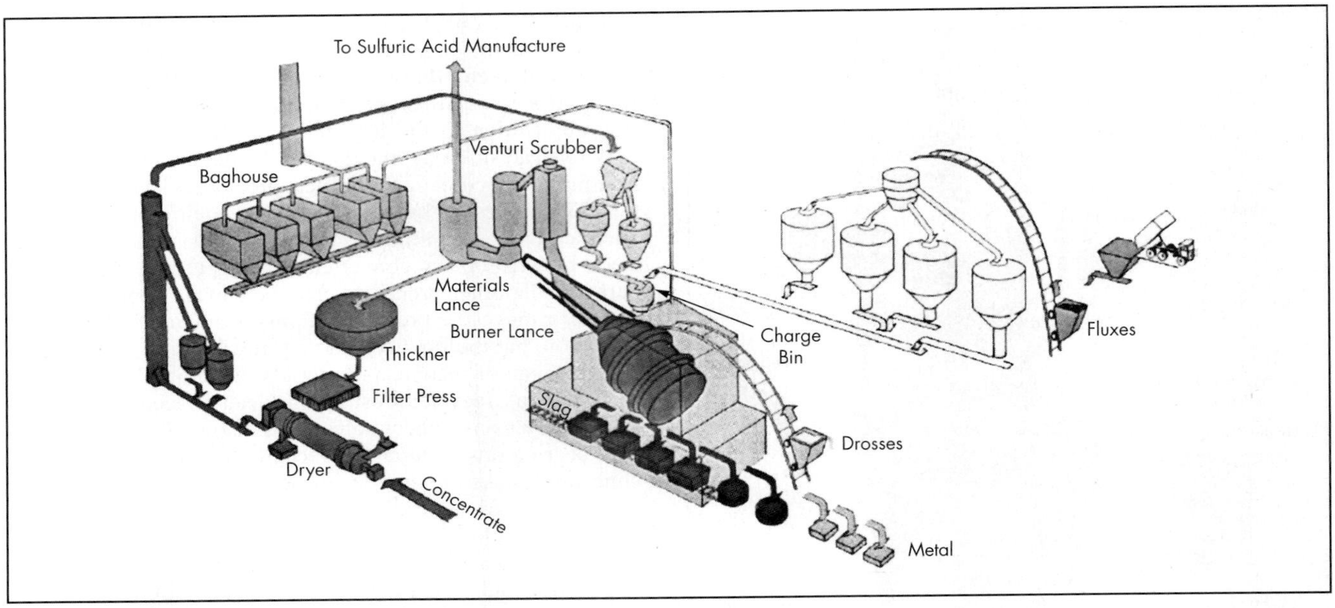

Source: Siegmund 2000a

Figure 16 Isometric process flow sheet of the Kaldo technology

Secondary Lead Smelting

Process configurations for secondary smelters vary greatly from plant to plant, region to region, country to country, and continent to continent. There are always site-specific constraints, which affect the decisions of equipment, process, and design configurations. In Europe, the configuration of a secondary lead smelter is typically a battery-breaking plant in conjunction with a discontinuously operating short rotary furnace technology and a refinery. A few smelters apply blast furnace operation treating the entire, unbroken battery, while one smelter uses a lance-type furnace. Several lance-type furnaces have also been placed into operation in Australia and the Pacific Rim. The smelters in the United States employ reverberatory furnaces coupled with either blast furnaces or electric furnaces on a continuous basis (Siegmund 2001).

In the past, secondary lead smelters were almost exclusively producing lead and lead alloys, and controlling their emissions of flue dust, lead, and SO_2. By-products were either sold or discarded. During the last two decades, lead recycling experienced many technological innovations, mainly by the addition of physical separation and/or hydrometallurgical processing steps, resulting in a remarkable improvement in the utilization of the components contained in batteries. Additional advancements were motivated by the introduction of more stringent environmental legislation and health and safety regulations, the lead industry's recognition of its environmental responsibility, as well as economic considerations. Today, one of the objectives is not to eliminate or significantly reduce the generation of waste materials and by-products but to convert them into value-added products and return them to the battery manufacturer. Necessary modifications and expansions to the plant process flow sheet have been made at smelters. Waste sulfur is recovered in a crystallizer as sodium sulfate crystals, or ammonia solutions are generated for sale. Polypropylene is washed and cleaned for extrusion purposes and subsequent utilization of the granules in battery manufacturers' injection molding operations for battery cases and tops. Further process additions are, for example, the affiliation of an incineration plant to a German secondary lead smelter incinerating the nonrecyclable components of the scrap batteries obtained from the secondary smelter together with other combustible waste materials subject to recovering energy. Another German smelter produces high-quality polypropylene compounds with specific physical and mechanical properties, which is carried out by the accurate dosing of organic as well as inorganic additives. In the United States, one smelter purifies the sulfuric acid by-product from battery breaking for reuse in new batteries by employing a solvent extraction process (Siegmund 2010). The only real waste product is still the slag from the final reduction furnace, which is generated in a nonhazardous form. The current process flow sheet of a modern U.S. operation resembles Figure 17 (Leiby 1999).

Battery breaking and component separation. Typically, the primary material processed in a secondary lead smelter is the spent automotive lead–acid battery, which accounts for ~80%–90% of the materials received from off-site. This percentage may change depending on supply. Exact specifications for components of an auto battery vary from manufacturer to manufacturer. A typical auto lead–acid battery may have the average compositions shown in Table 7 (Siegmund 2001).

Until the early 1970s, batteries were broken manually, which involved sawing off the battery tops and dumping the contents into the furnaces for processing purposes. Casing components were either processed in a separation system to recover polyethylene/propylene or discarded. During the 1980s, the industry started to replace the existing system by installing automated facilities, which consisted of breaking the batteries and subsequently physically separating each spent automobile battery into its components by efficient gravity units (Queneneau et al. 1993).

Batteries are shipped to the plant in an environmentally safe manner in dump trucks or on pallets where they are unloaded and, if necessary, stored in an acid-proof warehouse. If acid recovery is desired, then the batteries must first

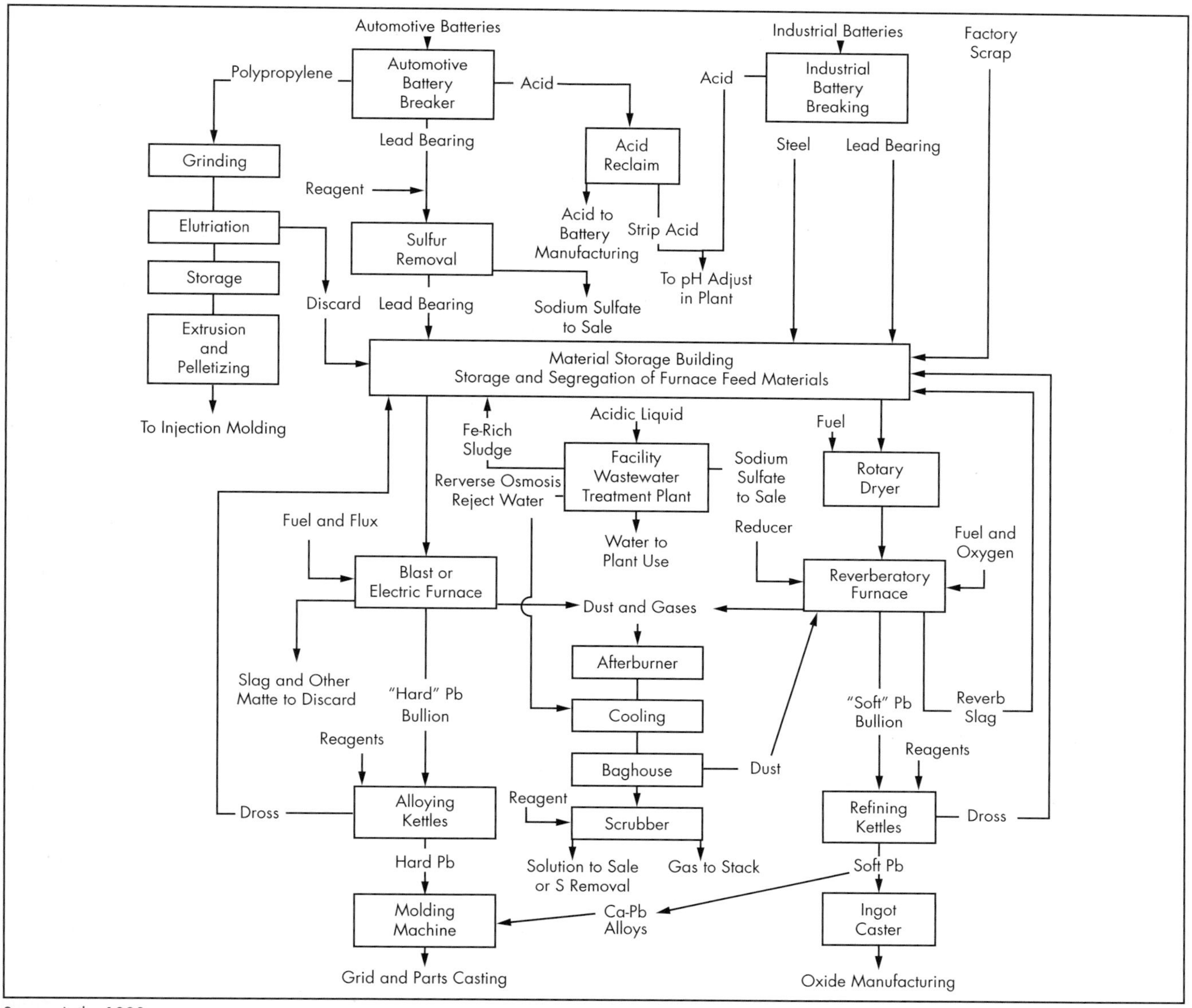

Source: Leiby 1999

Figure 17 Typical process flow sheet for a secondary lead smelter in the United States

Table 7 Typical vehicle lead–acid starter (starting, lighting, ignition) battery

Components	Compound	Percentage	Total, %
Metallic lead (grids and poles)			25
Lead paste	$PbSO_4$	50–60	38
	PbO_2	15–35	
	PbO	5–10	
	Metallic lead	2–5	
	Other	2–4	
Polypropylene cases			5
Separators, hard rubbers, etc.			10
Sulfuric acid (~15%)			22

Source: Siegmund 2001

be perforated or pre-crushed to allow the acid to be drained and collected. The batteries are reclaimed and loaded into a battery breaker, either a crusher with a tooth-studded drum or a swinging-type hammer mill. In the breaker, the batteries are shredded into small pieces for subsequent ergonomic separation into its components. Water is added to the breaker to slurry the paste. Efficient separation and concentration of the sulfur-rich paste is accomplished by screening in a rotating drum or on a vibrating screen, applying high-pressure water spray nozzles. The paste suspension is then pumped into tanks for chemically removing the sulfur or straight to a filtering device (press or belt) to recover paste. The remaining battery components are separated employing gravity-based hydroflotation separators. In these separators, the metallic lead sinks countercurrently to a water flow into a screw conveyor for removal from the system. The casing materials are carried away with the water and pass into a sink–float tank where polypropylene is separated by gravity from polyethylene, ebonite, and other plastic materials (PVC, glass fabrics, etc.). The entire process results in four intermediate products: (1) lead paste for smelting, (2) metallic lead for melting or smelting, (3) polypropylene for recycling purposes, and (4) lead-contaminated plastic fraction, which is disposed or, if permitted, can be charged to a furnace. Figure 18 illustrates a typical flow sheet of a battery-breaking plant applying the described process steps (Behrendt 2000).

Several smelters have installed a leach circuit to chemically remove sulfur from the paste prior to the smelting process to overcome three disadvantageous objectives during smelting: (1) reduction of SO_2 emission, (2) reduction of matte generation, and (3) minimization of the amount of slag. The sulfur removal can be achieved by using various reagents. The most common reagents are either caustic or soda ash producing lead hydroxide and sodium sulfate (Equation 26) or lead carbonate and sodium sulfate (Equation 27), respectively:

$$PbSO_4(paste) + 2NaOH(aq) = PbO(paste) + Na_2SO_4(aq) + H_2O \quad \textbf{(EQ 26)}$$

$$PbSO_4(paste) + Na_2CO_3(aq) = PbCO_3(paste) + Na_2SO_4(aq) \quad \textbf{(EQ 27)}$$

Both reactions decrease the paste's sulfur content from about 6%–7% to <1%. The process can be designed on a continuous or discontinuous basis. In both cases, the pH is closely monitored for accurate addition of reagent because of detrimental effects of the furnace operation. The reacted slurry is then sent to filter presses and the sodium sulfate solution to a crystallizer. Some facilities combine the liquid from their post-baghouse scrubber with the liquid from the reaction. The scrubber solution, however, requires oxidation to ensure that all the sulfites have been converted to sulfate prior to entering the crystallization process. The sodium sulfate solution is cleaned, crystallized, and dehydrated for sale.

In an alternative procedure, the battery pieces are conveyed from the breaker to a density gradient sink–float vessel with a media specific gravity of 1.4 g/cm^3. Plastic, hard rubber, and separators float and are conveyed to the plastic recovery system. Paste and grid material are moved by drag-chain conveyor and discharged onto a wedged-wire shake screen. Metallic material discharges into a screw conveyor and is transported to the raw material storage building. The paste slurry collected from the shake screen flows to a reactor tank for desulfurization, as described previously, or straight to a filtering unit. The desulfurized paste is filtered through a dewatering press with the effluent pumped to the on-site wastewater treatment plant. Desulfurized paste is discharged into the raw material storage building. The plastic recovery system consists of a hammer mill, primary and secondary sink–float washers, and pneumatic transfer to trucks for shipment. By-products (hard rubber and separators) are conveyed to the raw material storage building.

Smelting of battery scrap material. Ten years ago, materials for smelting were stored outside uncovered. Today, they are stored in an environmentally state-of-the art containment building. The materials may include slag, paste, and metallics from battery breaking, factory scrap, and some other raw materials that can be fed directly to the furnace, together with the lead-bearing material coming out of the battery breaker. Many buildings are equipped with a leak detection system and conform to all state and federal regulations. Coke products that are used as reducing agents and fluxes are also stored inside the containment building.

In the United States, the predominantly applied smelting technology consists of a multistep process consisting of selective reduction smelting of the raw materials in a reverberatory furnace coupled with slag reduction in either a blast furnace or electric furnace. This combination allows producing low-antimony "soft" lead bullion from the reverberatory furnace. The slag reduction unit treats the PbO-rich "reverb" slag as well as recycled dross from the refinery to concentrate the alloying elements (mainly antimony, tin, and arsenic) in the "hard" lead bullion (Stout 1993). This is used to generate various alloys required for different structural parts in the batteries. The raw materials are typically mixed and passed through a drying system before being charged to the reverb. A typical charge consists of 85% battery breaker material, 5% mud from the desulfurization process and wastewater plant, 5% refinery dross, and 5%–10% internal scrap. In addition to these materials, all the plastic recovery system by-products are fed in the mixture working as reducing agents and possessing some calorific value inside the furnace. One major secondary producer in the United States uses an electric furnace rather than a blast furnace, at two locations. This allows them to directly connect both furnaces, providing an overall continuous operation. Higher slag temperature in the electric furnace also permits the production of an environmentally more-stable slag and results in lower flux consumption, decreasing slag volume by about one-third. Recovery of lead, antimony, and particularly tin is increased. The discard slag from both reduction furnaces is characterized by an iron–lime matrix and has to pass an elution test to demonstrate leachability of the RCRA metals below the permitted toxicity levels defined in the Resource Conservation and Recovery Act of 1976. Typically, the slag is ground and combined with stabilization agents like cement or phosphates prior to going to the landfill.

In Europe, lead bullion is most commonly produced in short rotary furnaces. They provide a high flexibility in handling feed materials but are batch-type furnaces. The metallic lead grids, desulfurized or non-desulfurized lead paste, scrap metal, and lead-rich residues (flue dust, slimes, dross) can be either together, separate, or in each combination smelted at temperatures of about 1,000°C. Depending on the selected charge composition, "soft" or "hard" lead bullion can be produced. In the past almost exclusively, a soda-based slag was generated to capture the remaining sulfur, minimizing emissions. To produce environmentally more stable slags, many secondary lead smelters have changed the slag composition to a soda-containing slag with a silica matrix, or entirely to a fayalitic slag. Discard slag is disposed in a landfill. Recent investigations are focusing on stabilizing the slag so that it can be used as construction material in combination with other materials, avoiding high disposal costs. Alternatively, long rotary furnaces can also be applied treating secondary feed material. One long rotary furnace, for example, operates in Canada. Some plants in Europe and Asia smelt the secondary lead material in a blast furnace without breaking the batteries in a breaker system. The batteries may be pre-crushed with a front-end loader before being charged to the furnace, mainly to drain the sulfuric electrolyte contained in the battery. Only

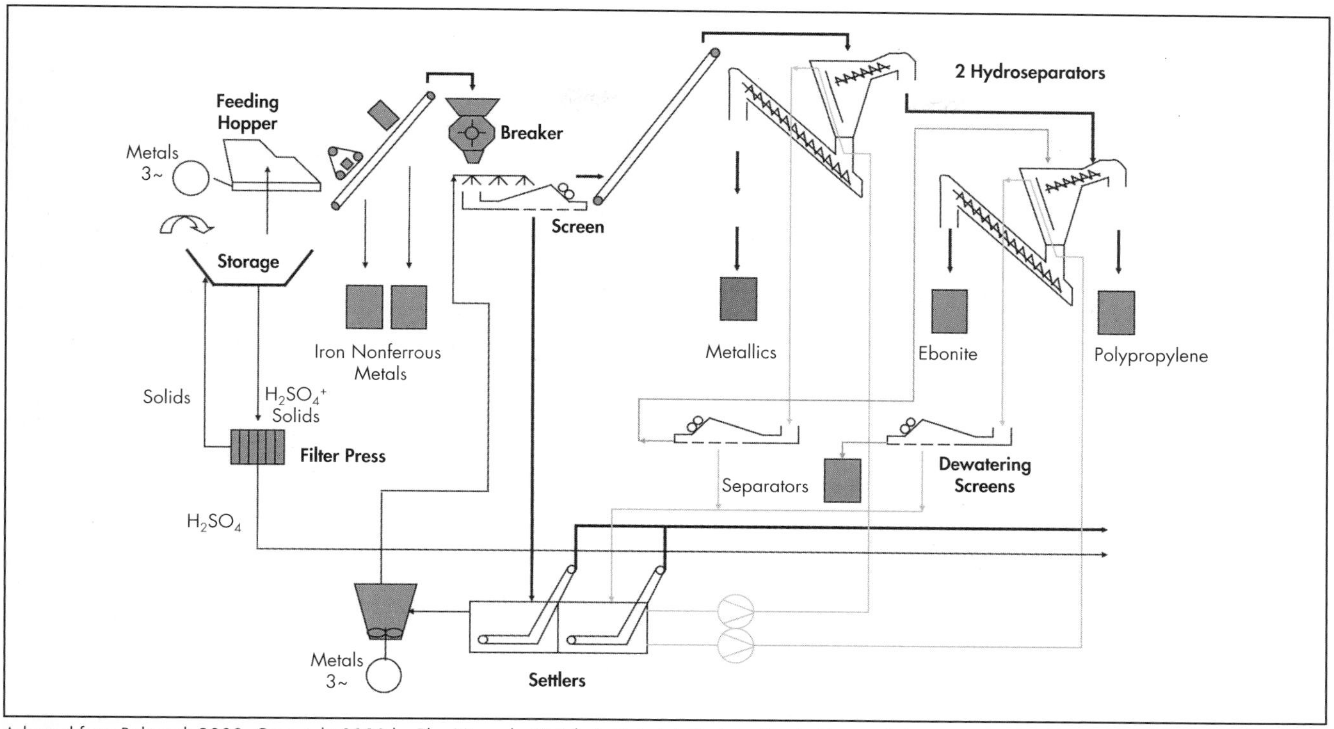

Adapted from Behrendt 2000. Copyright 2000 by The Minerals, Metals & Materials Society. Used with permission.

Figure 18 Process flow sheet for a battery-breaking plant

one type of bullion is produced, and the generated slag is primarily iron-silica-lime containing slag. Sulfur is captured as a matte. A one- or two-step TSL furnace system is also employed as a suitable smelting concept for battery scrap after battery breaking (Siegmund 2001).

Since all secondary smelters are required to control their sulfur dioxide emissions, sulfur is either removed prior to smelting in a desulfurization step of the paste or by scrubbing sulfur dioxide from the off-gas or tying it up as a matte. Some smelters are scrubbing with caustic and lime, combining the resulting solution with dilute acid from the raw material storage, making gypsum. Others are scrubbing with caustic and oxidizing the solution prior to making sodium sulfate crystals or handling it in the wastewater treatment facility. At one smelter, sulfur dioxide is scrubbed with ammonia. After several purification steps, the resulting anhydrous ammonia is sold as a raw material to the fertilizer industry. All smelters are well equipped with efficient fabric filter systems to minimize and control the dust and heavy metal emissions. In addition, many smelters have meanwhile implemented a complete enclosure of the buildings, which further minimizes emissions. Efficient filter systems, integrated dust cleaning and flue gas cleaning systems continuously remove both dust and lead to comply with the local environmental regulations on a permanent basis.

Recovery of polypropylene. As battery manufacturers became more vertically integrated, polypropylene became a valuable by-product for the smelters. They implemented compounding plants producing recycled polypropylene granules for their battery customers. The raw material for the production of polypropylene compounds is the shredded polypropylene fraction from the separation of the casings of lead–acid battery scrap in the battery-breaking area. Before being transferred to the compounding plant, the collected polypropylene chips actually undergo intensive preparation steps. In a milling unit, they are washed to remove any remaining paste and dust, shredded to a smaller and more homogeneous fraction in a knife mill, and dried to evaporate all remaining moisture. Based on the prevailing quality of the polypropylene fraction from the milling unit, it can be either processed in a single or twin extruder to a product suitable for injection molding of new battery cases or upgraded to a highly valuable product by the application of specific refining procedures. One smelter in Germany concluded, after a detailed economic and ecological analysis, that the high valuable product route was the more suitable application of the two technically viable choices for its purposes. Main equipment of the plant is a twin-screw extruder, a filter system, a granulator, and a dryer. The clean polypropylene pieces and the additives are fed continuously to the twin-screw extruder. The extruder is designed in particular for the requirements of the polypropylene material and consists of several individual heated chambers to maintain desired extruding conditions. The charge mixture is melted under carefully controlled operating parameters and subsequently mixed with additional components according to predetermined specifications to obtain a homogeneous melt.

Today, the continued improvement of the refining process of recycled polymers results in an economic and environmentally compliant solution producing high-quality polypropylene compounds, which can be applied in many diversified industrial areas. The most dominant product group, however, consists of components for the car industry (85%). Popular applications in that area are fender liner, headlight and air suction filter casings, V-belt covers, cable covers, and so forth. But Seculene PP (polypropylene) materials are also used in the electronics industry as components for washing machines, dryers, vacuum cleaners, and dishwashers (Martin and Siegmund 2000).

Alternative Technologies for Secondary Lead Smelting

During the last two decades, substantial research and pilot-scale testing has been carried out, and some technologies are still under development.

Electrowinning. At least three leach–electrowin processes have been tested on a pilot-plant scale in the past. All three processes have a different approach, but none of these technologies have yet been applied in a commercial operation (Leiby et al. 2000; Olper et al. 2000). This can be attributed to constant technological and ecological improvements in the existing smelting processes as well as for economic reasons. Today, Engitec Technologies is the only company actively pursuing to establish a commercial-sized operation of an adapted version of the electrowinning process in combination with elemental sulfur recovery in a bioprocess. After breaking the battery scrap in a conventional breaker, the paste is treated in a biosulfidization, which converts all lead-containing salts and/or oxides in the paste into PbS by means of anaerobic bacteria. The precipitated PbS and metallic lead content in the paste is then leached with the ferric fluoborate solution. Lead is dissolved and an elemental sulfur-based residue is generated. The grids-and-poles fraction is leached separately but also with ferric fluoborate solution. This promotes the distinguished production of refined lead and antimony–lead alloys. Subsequently, lead is recovered from both solutions by electrolysis in a FLUBOR diaphragm divided cell to produce lead cathodes in the cathodic compartment and to oxidize the ferrous ion to ferric ion in the anodic compartment, regenerating the leaching solution. As described in the literature, the process seems to have certain advantages but will require controls for the extremely aggressive acid that is employed. This technology was tested in the research lab of the Doe Run Company, treating primary lead sulfide concentrate with very promising results but has not been commercially implemented for economic reasons (Maccagni and Nielsen 2010).

Plasma technology. Several companies have developed technologies involving plasma arc furnaces that operate at very high temperatures. These have the potential to treat scrap material with low levels of contaminants, producing a discard nonhazardous slag. There have been two test furnaces built and operated, but the economics have not yet been successful.

Pyrometallurgical Lead Refining

Smelting processes produce lead bullion that must be further refined to remove a range of impurities derived from ores, fluxes, and reagents. The maximum amount of impurities are up to 4,000 g/t Ag, up to 12 g/t Au, 1.5% Cu, 3.0% Sb, 0.5% As, 0.3% Sn, 0.3% Zn, 0.35% Fe, 0.003% Ni, 0.1% Bi, and 0.15% S. There are two principal lead-refining techniques practiced. Either a pyrometallurgical or electrolytic refining method is employed, mainly depending on the quality of refined metal required, the necessity to remove bismuth, the relative power cost, and the efficiency and cost of by-product recovery.

The most commonly used method is the pyrometallurgical refining process, which consists of multiple steps incorporating decopperization; softening by air/oxygen blowing or alternatively using a treatment with sodium salts (Harris process) to remove antimony, arsenic, tin, and tellurium; Parkes process removal of silver and gold with subsequent dezincing; and final caustic treatment of bullion. Where bismuth is found in the charge, the Kroll–Betterton debismuthizing process, or some variant of it, will be included as the penultimate step in the molten lead refining process. The various removal operations are conducted in a certain sequence and batch methods are usually applied to avoid re- or cross contamination of previous or subsequent drosses (Richards et al. 2010). The sequence of treatments, or the number of these steps, is governed by the impurity content of the bullion. However, the developments and implementation of continuous processes for each refining step have resulted in the successful implementation of semicontinuous refining operations in a few smelters. In secondary smelters, the applied refining method is very similar, with the exception of not employing the Parkes process for precious metal removal. By-product metals recovered in salable form are copper as matte or matte speiss, antimony as antimonial lead, and silver and gold by treatment of the Parkes process zinc crusts. A typical sequence of stages of pyrometallurgical lead refining in primary and secondary lead smelters is shown in Figure 19.

Electrolytic refining by the Betts process is a preferred method for lead bullion where bismuth is present in large quantities. In this process, lead bullion anodes are electrolyzed in a lead fluosilicate-hydrofluosilicic acid electrolyte. Impurities in the lead remain as anode slimes and include bismuth, antimony, arsenic, silver, and gold. The slimes are processed for recovery of bismuth and precious metals and of antimony as antimonial lead. The drossing treatment of the bullion before electrolysis ensures the proper balance of antimony and arsenic for adherent slimes to be obtained. Usually the Sb + As content in anode lead is 1.25%–1.75% (Gonzalez 2010).

Copper Drossing

Independently of the applied process route, the lead bullion is passed through a drossing stage prior to being further refined. This first stage of refining entails the cooling of lead bullion from the furnace(s) to about 450°–500°C to lower the

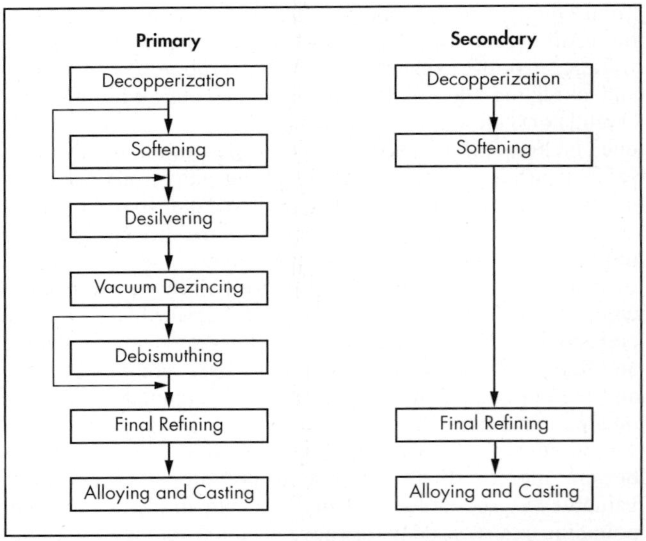

Adapted from Martin and Siegmund 2000. Copyright 2000 by The Minerals, Metals & Materials Society. Used with permission.

Figure 19 Main process steps in pyrometallurgical lead refining at primary and secondary smelters

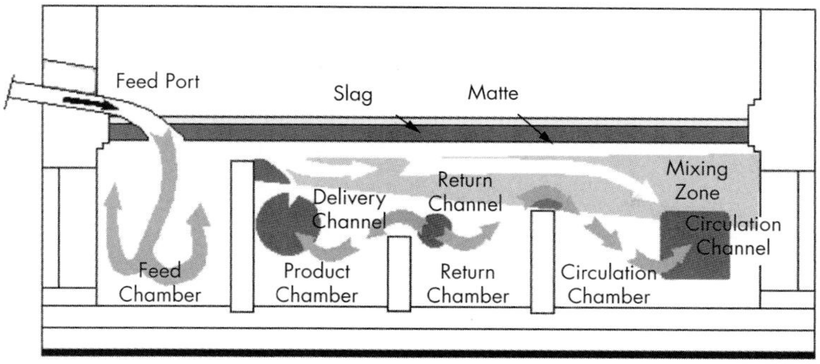

Adapted from Richards 2010

Figure 20 Continuous drossing furnace internals

solubility of elements such as copper, iron, nickel, and zinc and precipitating them out as dross. Iron and zinc are usually almost completely rejected as oxides in the form of magnetite or spinels. According to the prevailing concentration of sulfur, arsenic, antimony, and tin in the bullion, copper precipitates as sulfide, arsenide, antimonide, or stannide. Nickel, if present, tends to follow the copper. Lead sulfide also forms at lower temperatures and enters into the dross. As copper dross begins to separate at about 750°C, the kettle will already contain a certain amount of dross at a working temperature of 450°–500°C. The formation of a solid dross cover is avoided by adequate movement of the bullion. The bath agitation also promotes the separation of entangled lead and "drying" of the dross. Carbonaceous materials such as coke, bitumen, or sawdust may be added to assist in the separation process of the entangled lead and to obtain dry dross. The drosses are transported to another building and processed further to copper matte and speiss.

Lead Decopperizing

Lead bullion after drossing will contain equal or less than 0.5% Cu depending on the prevailing impurity concentration. An additional decopperizing step is required to reduce the copper content in lead bullion to levels of about 50 ppm, as required subsequently in the silver removal step.

Sulfur or pyrite/sulfur decopperizing is almost universally used for this purpose because of the strong affinity of copper to sulfur. The copper removal is accomplished by adding either lump elemental sulfur or a 4:1 pyrite sulfur mixture to the bullion at 400°–450°C in the vortex generated by a stirrer. Dross containing Cu_2S and mostly PbS crystals with entangled lead is formed. Care must be taken not to move the molten bath too long after the sulfur addition is completed or to raise the temperature. Otherwise, copper will revert back to the lead according to

$$(Cu_2S)\text{dross} + Pb \rightarrow (2Cu)\text{bullion} + (PbS)\text{dross} \quad \textbf{(EQ 28)}$$

The kettle is then mechanically skimmed by an automated dross removal unit. The collected dross is being returned to the raw material storage for treatment in the smelter.

The two-step copper removal process is usually carried out batchwise, whereby the bullion from the furnace(s) is poured in a kettle and allowed to cool for drossing. For mechanical reasons (avoiding thermal stresses and premature failure of the kettles), a sump of drossed and cool lead always remains in the kettle. In both batchwise copper removal steps, a stirrer is applied to generate the required bath movement. After the formation of dry dross, the stirrer is replaced by an automated drossing machine, which removes the dross from the lead surface. Alternatively, a continuous decopperizing process has been developed in Australia and is applied in some smelters. The technology employs a reverberatory type furnace and combines both operating steps, drossing and decopperizing of the bullion. It is heated through roof burners and cooled in the hearth to generate the required temperature gradient for copper removal. The furnace illustrated in Figure 20 is divided into three compartments by partition walls and consists of three main areas: a compartment for the inflow of the bullion, a compartment for discharging matte and slag, and a circulation launder for cooling the lead and sustaining a bullion temperature at the bottom of the furnace slightly above the melting point of lead.

Softening

Antimony, arsenic, and tin, which are elements responsible for hardening of lead, are removed from molten lead by oxidation. As those metals are less noble than lead and have a larger affinity to oxygen at elevated temperature, their removal is accomplished by preferential oxidation. Air and/or oxygen is used as the oxidation agent and blown into the liquid lead at temperatures starting around 550°C. Since the oxidation reaction is exothermic, the final temperature of the softened lead may reach 600°–650°C. Generated slag consisting of liquid lead antimony with lead antimonite and lead arsenate containing some solid lead stannate are immediately removed from the kettle by means of a suitable mechanical device due to its aggressive and corrosive nature. The collected slag is solidified and shipped to another plant location to further processing into a marketable lead alloy.

At higher prevailing tin contents, tin is sometimes removed in a separate process step by the addition of ammonium chloride or high chloride containing flue dusts and recovered as a solid dross. With high arsenic and tin levels in the bullion, many smelters are using the Harris process for removal of those metals, plus antimony. This involves treatment of lead with sodium salts (NaOH, $NaNO_3$) at temperatures between 400°C and 500°C, forming sequentially sodium arsenate (Na_3AsO_4), sodium stannate (Na_2SnO_3), and sodium antimonate (Na_3SbO_4). To lower the melting point of the molten salts, a small quantity of sodium chloride is added with the sodium hydroxide. The compounds can be recovered virtually lead free since lead is not converted by this reaction. The produced oxidation products are insoluble but finely dispersed and suspended as fine solids in the molten salt. The molten salt can contain up to 18% As, 20% Sn, and 26% Sb without an increase in viscosity. Other impurities such as selenium and tellurium are also removed from the bullion and report to the molten salts in this process. Although the Harris process has the advantages of lead-free products, no precious metals losses, selective recovery of tin and antimony products, which can be sold and up to 95.5% recovery rate of the sodium salts, it is less and less used in lead smelters because of labor- and cost-intensive treatment of the generated sodium slag, and high equipment cost with high wear rate and maintenance cost.

Parkes Process

The recovery of precious metals (silver, gold) contained in the lead bullion is regarded as one of the most important steps of the entire refining procedure, as the value of the precious metals content in the lead bullion represents a substantial part of the overall metal value. Therefore, it is imperative to minimize the silver and gold amounts bound in the process and to recover them as marketable products as efficiently as possible. Desilverizing is accomplished by the so-called Parkes process and carried out as a two-step process. Silver as well as other precious metals and most residual copper remaining after sulfur decopperizing is removed from the lead by precipitation with zinc as an intermetallic compound.

Lead bullion is pumped from the previous softening step into the desilverizing kettles which already contain crust (low-grade silver–zinc crystals) from the second desilverizing stage. The temperature of the lead bullion is adjusted to about 455°–480°C. The melt is well stirred for about 20–30 minutes so that the lead becomes saturated with zinc. While the molten lead is slightly cooling, Ag–Zn crystals are precipitated and rise to the top together with entangled lead. The formed intermetallic compounds have a higher melting point and a lower specific gravity than the lead and are virtually insoluble in lead saturated with zinc. In this first stage, a so-called "rich" crust is formed and skimmed off the kettle. After removal of the rich crust, fresh zinc as well as recovered zinc from the subsequent vacuum dezincing process and the rich rust treatment is added to the lead bath by stirring. The addition of zinc has to be accurately determined so that the cooling curve crosses the 330°C isotherm above the eutectic concentration of 0.5% Zn (see Figure 21). The charge is transferred to the second stage kettle which contains a quantity of frozen crust from the previous charge. After holding at a temperature of 455°–480°C to melt the crust, the completely dissolved zinc once again forms Ag–Zn crystals in equilibrium with the lead. The crust is then skimmed and returned to the first stage. Desilverizing is completed by cooling the bullion down to nearly freezing at 320°C. Forced cooling of the second-stage desilverizing process is employed to minimize cooling time. A "poor" low-grade silver crust is precipitated out and allowed to freeze on the surface. A slowly rotating mixer is used during cooling, which keeps open a central hole in the crust and also provides sufficient agitation of the lead to prevent freezing of crusts on the kettle walls. This poor crust remains in the kettle being recycled to the next stage desilverizing operation to increase the silver content and also to maintain the maximum throughput for the process.

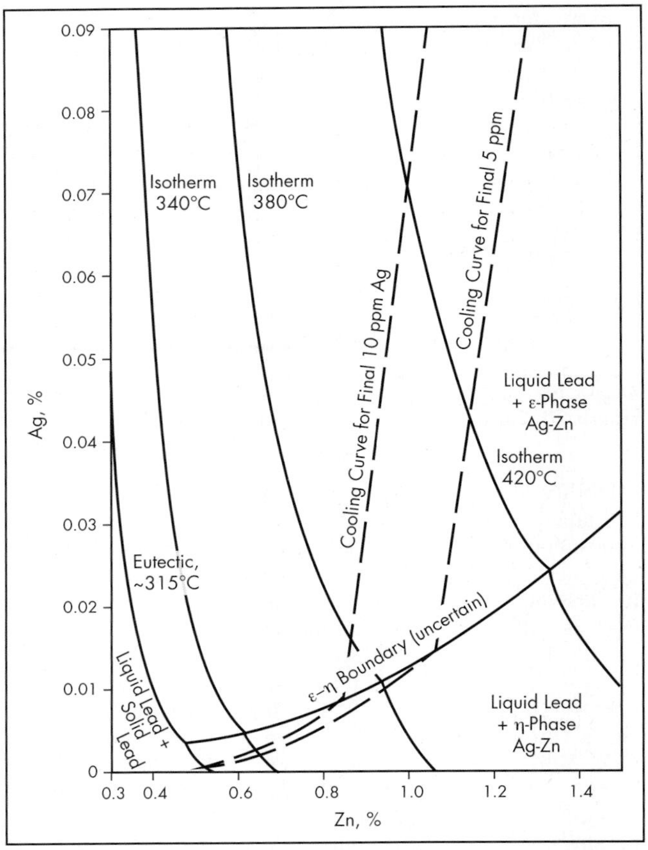

Source: Davey 1980

Figure 21 Lead corner in Pb–Ag–Zn system

The rich silver crust from the first stage of desilverizing contains a substantial amount of entangled lead. Its treatment involves three operations (liquation, dezincing, and cupellation) before either being processed to doré silver in the smelter or sold to a precious metals producer. During liquation, the entire silver-rich crust is melted, whereby it separates into two conjugate layers: the upper one zinc and silver rich, and the lower one rich in lead and comparatively poor in zinc (≈2%) and silver (1%–2%). A flux of borax may be added to prevent atmospheric oxidation, which would hinder the liquation. In the absence of oxides, the crust is melted at 650°–750°C, yielding the low-lead zinc–silver alloy and liquated lead, which is returned to the desilverizing kettles. Liquation is usually carried out in narrow but deep kettles, which are heated preferentially at the top. The rich alloy is dipped from the top while liquated lead is syphoned out from the bottom. Blocks of rich crust are added at intervals and allowed to melt.

The partially de-leaded rich zinc–silver alloy is next subjected to distillation to recover most of its zinc content for reuse. A retort or vacuum induction furnace is employed to distill zinc from the rich alloy. The induction furnace operates in the range of 900°C to 1,100°C. Therefore, it is desirable to retain considerable lead in the alloy so that it does not freeze when most of the zinc has been removed. A clay graphite crucible is contained in a refractory-lined vacuum chamber. The crucible shell is cooled by means of air passed through tubes adjacent to the crucible. The top of the furnace is provided with a water-cooled rubber seal and can be removed for charging. The furnace is connected to the refractory-lined condenser by means of a flanged pipe, which can be disconnected during the tilting of the furnace. The resulting rich silver-alloy ingots are then further treated in a silver refinery in-house for recovering of the precious metals or sold to a precious metals producer. The resulting lead bullion will normally contain less than 10 ppm of silver and is pumped to the next refining step, which is vacuum dezincing (Davey 1980).

Dezincing

Desilverized lead bullion contains about 0.55% Zn in accordance with the ternary eutectic point in the Pb–Zn–Ag system. This remaining zinc content is usually removed by kinetically controlled vacuum dezincing. A specifically designed kettle incorporating a water-cooled rim is used, which provides the sealing face for the rubber gasket between the kettle and the dezincing unit. The dezincing unit is a circular water-cooled vacuum hood equipped with a rubber gasket around the outer rim. The latter rests on the water-cooled rim. A stirrer is provided for agitating the lead during distillation. The vacuum offtake is through the top part of the kettle, which allows the vacuum pipework to remain permanently connected. The operating cycle consists of a heating stage followed by a distillation stage. The lead is heated from 320°C to the operating temperature of 590°–600°C. The dezincing unit is then positioned and the vacuum pumps started up to adjust an ultimate vacuum of <5 Pa. Up to 95% of the zinc contained in the bullion is condensed on the vacuum bell and recovered for reuse in desilverizing.

Alternative methods of dezincing through either a caustic soda treatment or selective removal with chlorides exist but are not applied anymore because of labor- and cost-intensive operations with lower zinc recovery rates.

Debismuthizing

After dezincing of the lead, the only impurity remaining is bismuth. Bismuth has a metallurgical deportment very similar to lead and follows the lead almost quantitatively through smelting and refining. The removal of bismuth to concentrations <0.01% can be accomplished through the Kroll–Betterton process. This process, which uses the same principles as de-silvering is typically a one- or two-step procedure, where bismuth is removed through the addition of a lead–calcium alloy and magnesium, a calcium–magnesium master alloy, or a mix of calcium and magnesium scrap. The reagents are added at 420°–480°C while the lead is vigorously stirred. Bismuth is converted to a bismuthide according to the following equation:

$$Ca + 2Mg + 2Bi \rightarrow CaMg_2Bi_2 \qquad \textbf{(EQ 29)}$$

The lead is subsequently cooled down to about 330°C. At prevailing bismuth concentrations of 0.05%–0.09% in the lead, a solid crust is formed at the surface with about 1% Bi. The crust is pulled from the kettle and either collected and again treated in a separate kettle with other crusts to enrich the bismuth up to 10% before being sold or treated for bismuth metal recovery. The debismuthized lead undergoes a final refining to remove residual impurities such as tin, arsenic, antimony, and zinc but also calcium and magnesium. This treatment is typically done with caustic soda before the refined lead is cast into ingots or blocks according to required specification or alloyed before casting. Process modifications over recent years have included automation of several of the refining steps, size increases of the refining kettles, improved by-product recovery, and more efficient environmental controls. Research is mainly focusing on the further development of continuous refining methods to improve efficiency and further mitigate environmental emissions.

Electrolytic Lead Refining

Many different electrorefining technologies have been proposed, but commercially, only electrolytes based on hydrofluosilicic acid, hydroborofluoric acid, or amidosulfonic acid have been employed. Today, electrolytic lead refining with soluble lead anodes is almost exclusively based on the Betts process because of economic reasons. The Betts process is the preferred refining method at bismuth contents in the raw materials of >1% and applied in several lead smelters around the world but predominantly in Asia. It is carried out utilizing a hydrofluosilicic electrolyte containing excess lead fluosilicate to electrorefine impure lead. During the electrorefining operation, the lead anodes are electrolytically corroded with the simultaneous deposition of lead with a higher purity on the cathodes. Electrochemically more noble impurities like silver, gold, copper, bismuth, antimony, arsenic, and germanium are largely retained on the anode scrap as an adherent slime. Less noble impurities like iron, nickel, and zinc will be dissolved at the anode and report to the electrolyte, but their content is usually so low that an enrichment in the electrolyte virtually does not occur. Only tin, with an electrochemical potential close to that of lead (–0.14 V versus 0.126 V) is electrochemically deposited at the cathode together with lead. These advantageous process conditions ensure that even at high impurity levels of the anodes, a cathodic lead with purities higher than 99.99% can be obtained. Furthermore, lead can be refined with a low-energy requirement of about 120 kW·h/kg Pb due to its high electrochemical potential (Gonzalez 2010).

During the electrorefining process, lead is dissolved from the centers of the grains, leaving a skeletal structure of impurities to form the metallic slime layer. Since bismuth forms a solid solution with lead at concentrations normally found in lead bullion, it is imperative for the anode slimes to adhere so that they can cement out the bismuth from the electrolyte after it has corroded with lead. If bismuth does not corrode with lead, it would enrich at the anode surface until lead corrosion is stopped. If insufficient impurities are present in the bullion, in particular antimony, this honeycomb slime layer will be incomplete with low mechanical strength to adhere. The metallic slimes tend to be soft in this case and will often slough off the anodes. Conversely, a high concentration of impurities and here, in particular, copper concentrations >0.05% will result in a very firm, passivating layer that is difficult to remove from the anodes. Therefore, a minimum content of 1.25% of combined antimony, arsenic, and bismuth concentrations in the anode is desired to form a stable and adherent slimes layer.

The electrolyte, consisting of 100–185 g/L hydrofluosilicic acid and 60–100 g/L lead as lead fluosilicate, is kept in constant circulation through the cells at flow rates of about 40 L/min. At higher rates, the anode slime can be eroded and cause contamination of the cathodic lead deposit. At much lower circulation rates (<30 L/min), the cathodic deposits become uneven because of insufficient flow, creating adequate electrolyte mixing. Employed current density is adjusted to 120–180 A/m^2. The electrolyte temperature is adjusted between 38°C and 43°C. In the absence of additives, lead deposits with a very small overpotential and tends to form rough, porous, or dendritic deposits that result in short circuits. To obtain flat, smooth deposits and minimize short-circuit occurrences, organic reagents (glue and lignin sulfonate) are added to the electrolyte. These reagents increase the cathodic overpotential and change the kinetic parameters and consequently need to be closely monitored.

The anodes are cast at a thickness between 18 and 25 mm based on the prevailing bismuth content and replaced typically in 6–8-day intervals to limit the anode corrosion period and prevent the voltage drop across the adhering slime layer from exceeding a critical value above which impurities dissolve and contaminate the cathodic deposit. Cathodes are produced from electrolytically refined lead with a thickness between 0.8 and 1 mm. They are manufactured by means of submerging a water-cooled rotating drum in a molten lead bath. One cathode cycle is typically 4–8 days.

To prevent current losses and to maintain the electric power consumption at high current efficiencies and cathode quality, it is important to closely monitor cell voltage and operate at minimum levels. The electrochemistry of the process is illustrated in Figure 22. This graph shows the operating potentials as a function of current density and time. The cell voltage tends to slightly increase during the duration of the refining cycle. Besides the voltage of the electrodes' contact and resistance, main contributors to these voltage changes are the slime resistance voltage and the electrolyte voltage. The Betts process depends on the formation of adherent, porous anode slimes. The slime layer formed undergoes structural and chemical changes during its growth, resulting in a voltage increase. In addition, the resistance of the electrolyte contributes significantly to the increase in voltage during the refining cycle because of a loss in free acid. A control value at 90 g/L or higher for the free acid concentration in the electrolysis is set and maintained. For steady operation at 160–180 A/m^2, the cathode overpotential is controlled at about 90 mV. The anode overpotential with new anodes is about 10 mV, and the plating cycle must end when the anode overpotential reaches 200 mV above the reversible potential for lead. At this point, the anode potential is also within 200 mV of the more positive bismuth potential. To avoid dissolving bismuth at the anode, the anode potential must not exceed +0.76 V (SHE, or standard hydrogen electrode) or 200 mV anode overpotential. From the diagram, it is obvious that some antimony will contaminate the cathodic lead deposit at that potential.

Overall current efficiencies of 98% are obtained in the Betts process. However, the slightly higher current efficiency at the anode compared to the cathode results in a lead accumulation in the electrolyte. Therefore, a bleed of the electrolyte is frequently sent to a lead removal tank where the lead in the electrolyte is precipitated as lead sulfate crystals by adding sulfuric acid or recovered in metallic form through an electrowinning process. The clarified de-leaded solution flows into a bismuth removal tank filled with lead shot where bismuth is removed by cementation before it returns to the electrolyte circuit.

During the electrorefining operation, fluosilicic acid is consumed by entrapment in the anode slimes and by volatilization from the surface of the electrolyte. The acid develops significant vapor pressures through the volatile decomposition

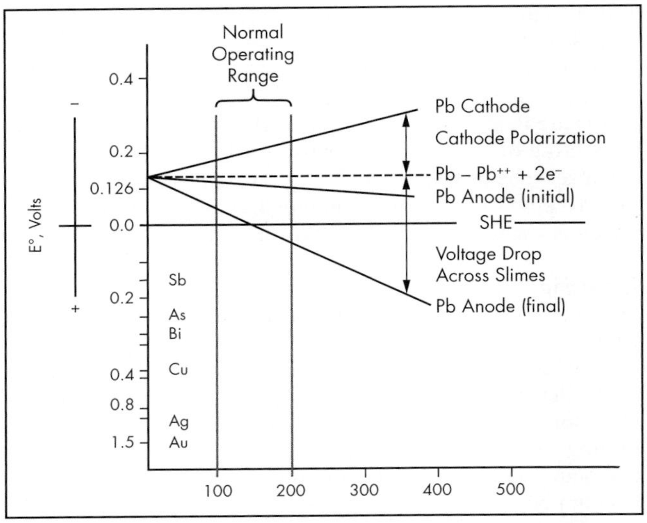

Source: Krauss 1990

Figure 22 Betts process electrochemistry

products HF and SiF_4. Both are corrosive and toxic and are removed from the cellhouse atmosphere by adequate ventilation.

The slimes from the used and corroded anodes are removed through the application of scrapers, mechanical brushes, or the impact of high-pressurized water. The cleaned scrap anodes with a remaining weight of about 30% of the original anodes are returned to the smelter while the slimes are collected and transported for further treatment. The slimes, representing only 4%–5% of the anode weight, can be treated by a variety of processes to recover silver, gold, antimony, copper, bismuth, and sometimes tin and indium. The lead cathodes are washed, melted, and cast into ingots or blocks after removing residual tin and antimony through a caustic treatment.

BISMUTH

Mineralogy and Occurrence

The bismuth contents in the earth's crust are estimated to be 0.2 ppm. Elemental bismuth may occur naturally in up to 99% Bi purity, although its sulfide (Bi_2S_3 with 81.3% Bi), oxide (Bi_2O_3), and bismutite ($nBi_2O_3 * m\ CO_2 * H_2O$) form important commercial ores.

The total number of minable ore deposits is low, and only a small amount of bismuth ores are economically processed to metal. Bismuth minerals are predominantly extracted as by-products from copper, lead, tin, tungsten, molybdenum, cobalt, nickel, silver, and gold ores. Significant amounts of bismuth are recovered from anode sludge generated during copper and lead electrorefining as well as bismuth crust from the pyrometallurgical refining of lead or flue dusts and precipitation products. The largest bismuth deposits can be found in Bolivia, China, Spain, and Peru, where significant amounts of bismuth are recovered from intermediate products during the metallurgical recovery of copper, lead, and tin.

Properties

An overview of the most important properties can be found in Table 8. Bismuth is a pentavalent post-transition metal and chemically resembles arsenic and antimony. It is a brittle metal with a silvery white color when freshly produced but

Table 8 Properties of bismuth

Property	Units
Atomic mass	208.98
Density, g/cm^3	9.78
Melting point, K	544.98
Boiling point, K	1,837
Heat of fusion, kJ/mol	11.3
Heat of vaporization, kJ/mol	151
Heat capacity (25°C), J/(mol*K)	25.52
Electrical resistivity (20°C), μΩ*m	1.29
Thermal conductivity (300 K), W/(m*K)	7.97
Thermal expansion (25°C), μm/(m*K)	13.4
Brinell hardness, MPa	94.2

Source: Kammel 1989

is often seen in air with a pink tinge, owing to surface oxidation. Bismuth is the most naturally diamagnetic element and has one of the lowest values of thermal conductivity among metals. The electrical resistance of solid bismuth is larger than that of its liquid phase. When deposited in sufficiently thin layers on a substrate, bismuth is a semiconductor, despite being a post-transition metal (Kammel 1989).

Elemental bismuth is denser in the liquid phase than the solid, a characteristic it shares with antimony, germanium, silicon, and gallium. Bismuth expands 3.32% on solidification; therefore, it was long a component of low-melting typesetting alloys, where it compensated for the contraction of the other alloying components to form almost isostatic bismuth–lead eutectic alloys. It needs to be emphasized that bismuth, unlike most other metals, solidifies with a volume increase of 3.32%. Bismuth is more noble than lead and does not really react with air, water, and nonoxidizing acids. With nitric acid, it reacts forming nitrogen oxide, while sulfuric acid dissolves it to bismuth sulfate while generating sulfur dioxide. At elevated temperatures, the metal instantly reacts with halogens and sulfur. Its compounds tend to hydrolyze in aqueous solutions at a pH > 0.5.

The recovery of bismuth from ores, concentrates, or other raw materials is predominantly carried out by means of roast reduction technology, or precipitation or segregation processes. Because of the low solubility of bismuth compounds in aqueous solutions, the application of hydrometallurgical processes is only used for the treatment of intermediate products or waste products, the enrichment of low-bismuth-bearing raw materials, or for the removal of other metals. To recover bismuth, a mixture of sulfides, oxides, metallic bismuth-containing ores, and intermediate products is typically treated because of the limited availability of one single raw material. All pyrometallurgical processes are conducted at relatively low temperatures to minimize higher evaporation losses due to the high vapor pressure of metallic bismuth and bismuth sulfide at elevated temperatures. In addition, the extremely varying compositions of the different ores and intermediate products demand that each smelter adapt the applied process technology to the special requirements of the treated raw materials.

Enrichment of Raw Materials

The pyrometallurgical recovery of bismuth is only economical with raw materials containing more than 10% Bi. The low metal content in most bismuth ores requires mechanical processing in aqueous solutions or flotation to increase the metal concentration. If these technologies are not viable methods, it is possible to increase the metal content through distillation or precipitation from aqueous solutions. This applies particularly for bismuth containing flue dusts but also to treat low-bismuth-bearing oxides. In precipitation technology, the bismuth is dissolved through leaching with concentrated hydrochloric acid. From the resulting chloride solution, bismuth is precipitated either with iron or it is precipitated by means of hydrolysis as oxychloride as follows:

$$BiCl_3 + H_2O = BiOCl + 2HCl$$

The precipitation step in this process technology can be accomplished through dilution with water or through the addition of soda or milk lime. The majority of the bismuth separation is already completed at a pH value of around 2 with <1 g/L Bi remaining in solution.

Crude Bismuth

The roast reduction process is the most common extraction process and is considered for sulfidic and sulfide-oxide ores, which are jointly molten under reducing atmosphere to crude metal after calcination at moderate temperatures.

The melting of the calcine generated from oxychlorides or other oxidic raw materials takes place in a reverberatory furnace or a crucible furnace with the addition of soda, lime, fluorspar, and recirculated slag that returns to form low-melting slag. Carbon to adjust a reducing atmosphere is added between 3% and 6% of the charge. Since the roasting of ores is incomplete at low temperatures, iron turnings are added for converting the remaining Bi_2S_3 that had not reacted during the roasting process. Simultaneously, this also converts arsenic and antimony, forming a speiss. When processing the bismuth-bearing materials in a reverberatory furnace, the new feed material is charged into the liquid slag and initially melted only at moderate heat under a reducing flame. Only after complete reduction has been accomplished is the operating temperature increased to tap and discharge the weak acidic alkaline silicate slags in a liquid form with very low viscosity. Because of the extremely low viscosity of the slag and the low melting point of the crude bismuth, the process is carried out in a refractory-lined and mobile furnace similar to the type of the English (Treib) furnace, which can be easily replaced. In smelters with small throughput capacities, crucible furnaces are employed as the preferred equipment. These furnaces are installed in a ring-type arrangement. The crucibles are constructed from refractory clay mixed with about 20% graphite. Although the operating costs of this technology are higher, the overall evaporation losses are less than in a reverberatory furnace, resulting in higher yields.

The recovered crude bismuth from the roast reduction process is further refined to bismuth metal. Simultaneously generated matte with typically 5–8% Bi is ground, roasted, and in most cases its bismuth content is further enriched through a hydrometallurgical treatment. If the slag from the refining process contains bismuth in concentrations larger than 0.1%, it is recirculated to the smelting step. Otherwise, it is disposed to an appropriate landfill.

In precipitation technology, crude bismuth is recovered from sulfidic raw materials by adding iron turnings with the following reaction taking place:

$$Bi_2S_3 + 3Fe = 2Bi + 3FeS$$

Although this technology has the advantage of not requiring a roasting step, it was almost completely replaced by the roasting process under reduced atmosphere because of the substantial amount of co-generated matte. Today, it is exclusively applied for processing Bi_2S_3 ores or ores that are rich in copper, arsenic, and antimony. The precipitation technology is mostly performed in crucibles, which produces a matte with low bismuth contents. It is required to add sodium sulfate, gypsum, soda, and coal as flux reagents and fuel to the overall process.

Bismuth-bearing by-products from lead refineries are most often initially reduction treated to form a lead–bismuth alloy. This is followed by either a selective oxygen, sulfur, or chloride treatment. Alternatively, the lead–bismuth alloy can be leached with hydrochloric acid or electrolytically concentrated in the anode sludge during the lead electrolysis. Subsequently, the products are oxidized in a converter where bismuth reports to the slag. Crude bismuth is then generated from the slag in a smelting step under reducing atmosphere.

Refining of Crude Bismuth

The bismuth contents of crude bismuth and bismuth-enriched alloys are varying and can contain up to 70% Bi. The preferred refining methods are pyrometallurgical processes. On one hand they are more flexible and allow the treatment of more impure crude bismuth or bismuth alloys, and on the other hand they also produce final bismuth metal with higher purity. The application of electrorefining demands a bismuth metal high in bismuth (>97%) and low in arsenic, antimony, and copper, which is then cast into anodes.

The process steps in the pyrometallurgical refining of bismuth are comparable to the refining of lead bullion. Copper is removed by stirring sulfur or sulfur-containing additives into the melt at temperatures between 400°C and 500°C. Subsequently, arsenic and antimony are decreased to levels of 0.1 ppm through the addition of sodium hydroxide. Precious metals are removed in accordance to the Parkes process as a foam. Compared to this refining step for lead, the required amounts of zinc are higher with about 12.5 kg Zn/t_{metal} or 2 kg Zn/kg Ag. Zinc and lead are removed through the injection of chlorine gas or stirring chlorides into the melt.

By looking at the standard reduction potential for the elements, it is obvious that only lead and silver can be effectively removed of all the prevailing impurities from bismuth ($Bi/Bi^+ = +0.2$ V) in the electrorefining process. Other impurities such as, for instance, antimony ($Sb/Sb^{3+} = +0.24$ V) are co-deposited at the cathode. Copper is collected partly in the anode sludge but, to some extent, also dissolved and reduced at the cathode. Therefore, it necessary to virtually remove all the copper, antimony, and arsenic in a pyrometallurgical refining step prior to electrorefining for the recovery of bismuth with high purities. Electrorefining of bismuth is mainly carried out in $BiCl_2$—solutions with about 100 g/L Bi and 100 g/L free hydrochloric acid. During the electrorefining process, which is performed at ambient temperature and with a current density of about 130 A/m^2, the top of the electrolyte is covered with a layer of oil or paraffin to prevent oxidation of copper and iron. The cell voltage of electrolysis is, at a maximum, 0.1 V. At elevated levels of antimony or lead in the anode, a diaphragm is employed. A silver plate is used as a permanent cathode sheet. This ensures that the brittle bismuth deposits can be easily stripped from the cathode. The generated anode sludge may contain up to 40% Bi.

Alternatively, an electrolytic refining process employing a hydrofluosilicic acid solution is used. In this process a copper blank is used as a permanent cathode sheet. The anodes are covered with anode bags made of polypropylene. The electrolyte contains 40 g/L Bi and 330–350 g/L of free hydrofluosilicic acid. In addition, 1 g of glue is added per ton of bismuth produced. The electrolysis is operated at a current density of 60 A/m^2, at which the cell voltage is about 0.2 V and current efficiencies of up to 93% can be accomplished. The anode and cathode cycle is commonly 8 days, after which the anode scrap is about 50%–55% and the anode slime rate is about 0.4%.

In both electrorefining methods, the sludge from the anode scrap will be removed by mechanical scraping. The collected high-silver-containing anode sludge is typically treated further to recover the silver, while low-silver-containing anode sludge is recirculated for anode production. The scrap anodes are further washed before being returned to the melting kettle for recycling.

After washing with water, the electrolytic bismuth is stripped of the cathodes manually or in an automated stripping machine. The cathode blanks are returned to the electrolytic cells while the refined bismuth is melted in a kettle, cleaned by a caustic soda treatment, and cast into ingots by a continuous casting machine. The refined bismuth metal typically has a purity of >99.99%. This is sufficient for the production of bismuth compounds. The production of pharmaceutical products demands bismuth that is free of arsenic, lead, and tellurium, at purity levels of >99.99%. Ultrapure bismuth of 99.9999% purity is used for the semiconductor and nuclear industry. This requires a subsequent refining step, which can be either zone melting or vacuum distillation.

Bismuth compounds account for about 60% of the production of bismuth. They are used in cosmetics, pigments, and a few pharmaceuticals, notably Pepto-Bismol, used to treat diarrhea. Bismuth's unusual propensity to expand upon freezing is responsible for some of its uses, such as in casting of printing type. Multiple-component alloys with cadmium, indium, lead, tin, and tellurium are characterized by melting points as low as 47°C. In conjunction with the unusual contraction (at bismuth contents of <47%) and expansion (at bismuth contents of >55%) properties, these alloys are also used as casting cores and forms as well as solder materials. As a metal, it is not used in many applications because of its poor mechanical properties, except for electrodes in pH measuring units, thermostat liquids for high-temperature applications, heat exchanger fluids, and catalysts. In addition, bismuth has unusually low toxicity for a heavy metal. As the toxicity of lead has become more apparent in recent years, there is an increasing usage of bismuth alloys (presently about one-third of bismuth production) as a replacement for lead. Bismuth compounds such as bismuth chloride are used as pigments, while bismuth trioxide is used as a pigment but is also used in the porcelain industry (Kammel 1989).

REFERENCES

Ashman, D.W., Goosen, D.W., Reynolds, D.G., and Webb. D.J., 2000. Cominco's new lead smelter at Trail Operations. In *Lead-Zinc 2000*. Edited by J.E. Dutrizac, D.M. Henke, S.E. James, and A. Siegmund. Warrendale, PA: The Minerals, Metals & Materials Society.

Behrendt, H.-P. 2000. Technology of processing of lead-acid batteries in Muldenhuetten Recycling und Umwelttechnik GmbH. In *Proceedings of the Fourth International Symposium on Recycling of Metals and Engineered Materials*. Edited by D.L. Stewart, R. Stephens and J.C. Daley. Warrendale, PA: The Minerals, Metals & Materials Society. pp. 79–92.

Bushell, C.H.G., and Testut, R.J. 1985. Mineral processing methods. In *SME Mineral Processing Handbook*. Edited by N.L. Weiss. Littleton, CO: SME-AIME. pp. 15-4–15-8.

Bushell, C.H.G., Testut, R.J., and Milner, E.F., eds. 1985. Lead and zinc. In *SME Mineral Processing Handbook*. Edited by N.L. Weiss. Littleton, CO: SME-AIME. pp. 15-1–15-49.

China ENFI Engineering Corporation. 2010. SKS and SKSR Lead Smelting Process [PowerPoint presentation].

China ENFI Engineering Corporation. 2016. SKS Lead Smelting Project [PowerPoint presentation].

Creedy, S., Reuter, M., Hughes, S., Swayn, G., Andrews, R., and Matusewicz, R. 2010. The versatility of Outotec's Ausmelt process for lead production. In *Lead-Zinc 2010*. Edited by A. Siegmund, L. Centomo, C. Geenen, N. Piret, G. Richards, R. Stephens. Warrendale, PA: The Minerals, Metals & Materials Society. pp. 439–450.

Davey, T.R.A. 1980. The physical chemistry of lead refining. In *Lead-Zinc-Tin '80: Proceedings of the World Symposium on Metallurgy and Environmental Control*. Warrendale, PA: TMS-AIME. pp. 477–507.

Doyle, E.N. 1970. The sink-float process in lead-zinc concentration. In *AIME World Symposium on Mining and Metallurgy of Lead and Zinc*. Edited by D.O. Rausch and B.C. Mariacher. New York: AIME. pp. 814–850.

Errington, B., Hawkins, P., and Lim, A. 2010. IsaSmelt for lead recycling. In *Lead-Zinc 2010*. Edited by A. Siegmund, L. Centomo, C. Geenen, N. Piret, G. Richards, R. Stephens. Warrendale, PA: The Minerals, Metals & Materials Society. pp. 345–414.

Fallas, D. 2010. ISP Technology, Pb-Zn short course: Lead and Zinc Processing. COM 2010. Vancouver, BC.

Gonzalez, A. 2010. Pb Electrorefining—Betts Process, Pb-Zn short course: Lead and Zinc Processing. COM 2010. Vancouver, BC.

Goosen, D.W., and Martin, M.T. 2000. Application of the Kivcet smelting technology at Cominco. Presented at the Mining Millennium 2000 Convention, Toronto, ON.

Hayes, P., Schlesinger, M., Steil, U., and Siegmund, A. 2010. Lead smelter survey. In *Lead-Zinc 2010*. Edited by A. Siegmund, L. Centomo, C. Geenen, N. Piret, G. Richards, R. Stephens. Warrendale, PA: The Minerals, Metals & Materials Society. pp. 345–414.

ILA (International Lead Association). 2018. Lead facts: History of lead. https://www.ila-lead.org/lead-facts/history-of-lead. Accessed July 2018.

ILZSG (International Lead and Zinc Study Group). 2017. End uses of lead. www.ilzsg.org/static/enduses.aspx?from=2.

Jiao, Q. 2010. Introduction to the Sinter Plant/Blast Furnace Process Technology for Primary Lead Smelting, Pb-Zn short course: Lead and Zinc Processing. COM 2010. Vancouver, BC.

Kammel, R. 1989. Nichteisen—Schwermetalle: Wismut [Nonferrous-Heavy Metals: Bismuth]. In *Ullmann's Encyclopedia of Industrial Chemistry*. Weinheim: VCH. pp. 350–354.

Karpel, S. 1998. Greater flexibility and lower costs at Nordenham. *MBM Met. Bull. Mon.* (December):40–45.

Kaye, A., Hughes, S., Matusewicz, R., and Reuter, M.A. 2008. Ausmelt technology: Developments in lead and zinc processing. In *Zinc and Lead Metallurgy*. Edited by L. Centomo, M. Collins, J. Harlamovs, and J. Liu. Montreal, QC: Canadian Institute of Mining, Metallurgy and Petroleum. pp. 63–76.

Kerney, U. 2010. Recylex Pb production in Nordenham with Bath smelting technology: An Update. In *Lead-Zinc 2010*. Edited by A. Siegmund, L. Centomo, C. Geenen, N. Piret, G. Richards, R. Stephens. Warrendale, PA: The Minerals, Metals & Materials Society. pp. 415–428.

Krauss, C.J. 1990. Electrometallurgical practice: Zinc and lead. Fundamentals and Practice of Aqueous Electrometallurgy. [short course notes], 20th Annual Hydrometallurgical Meeting, CIM, Montreal, QC, October 20–21.

Leiby, R. 1999. Secondary Lead, Recycling Metals from Industrial Waste short course. Colorado School of Mines, Golden, CO.

Leiby, R., Bricker, M., and Spitz, R.A. 2000. The role of electrochemistry at East Penn Manufacturing. In *Proceedings of the Fourth International Symposium on Recycling of Metals and Engineered Materials*. Edited by D.L. Stewart, R. Stephens, and J.C. Daley. Warrendale, PA: The Minerals, Metals & Materials Society.

Li, D., and Yang, G. 2010. Oxygen bottom-blowing smelting lead smelting technology by bottom blowing electrothermal reduction. In *Lead-Zinc 2010*. Edited by A. Siegmund, L. Centomo, C. Geenen, N. Piret, G. Richards, R. Stephens. Warrendale, PA: The Minerals, Metals & Materials Society. pp. 483–486.

Longval, A., Shannon, T., Jiao, Q., and Davis, B. 2008. Operational improvements at Brunswick's lead blast furnace. In *Zinc and Lead Metallurgy*. Edited by L. Centomo, M. Collins, J. Harlamovs, and J. Liu. Montreal, QC: Canadian Institute of Mining, Metallurgy and Petroleum. 29–40.

Maccagni, M., and Nielsen, J. 2010. The CX process: New approach for making the process even more environmental friendly. In *Lead-Zinc 2010*. Edited by A. Siegmund, L. Centomo, C. Geenen, N. Piret, G. Richards, R. Stephens. Warrendale, PA: The Minerals, Metals & Materials Society. pp. 1125–1134.

Martin, G., and Siegmund, A. 2000. Recovery of polypropylene from lead-acid battery scrap. In *Proceedings of the Fourth International Symposium on Recycling of Metals and Engineered Materials*. Edited by D.L. Stewart, R. Stephens and J.C. Daley. Warrendale, PA: The Minerals, Metals & Materials Society. pp. 93–101.

Meurer, U., and Ambroz, H. 2010. Primary lead production with the QSL process: An ecological and economical advanced technology. In *Lead-Zinc 2010*. Edited by A. Siegmund, L. Centomo, C. Geenen, N. Piret, G. Richards, R. Stephens. Warrendale, PA: The Minerals, Metals & Materials Society. pp. 429–438.

Mounsey, E.N., and Piret, N.L. 2000. A review of Ausmelt technology for lead smelting. In *Lead-Zinc 2000*. Edited by J.E. Dutrizac, D.M. Henke, S.E. James, and A. Siegmund. Warrendale, PA: The Minerals, Metals & Materials Society.

Olper, M., Maccagni, M., Buisman, C.J.N, and Schultz, C.E. 2000. Electrowinning of lead battery paste with production of lead and elemental sulfur using bioprocess technology. In *Lead-Zinc 2000*. Edited by J.E. Dutrizac, D.M. Henke, S.E. James, and A. Siegmund. Warrendale, PA: The Minerals, Metals & Materials Society. pp. 803–813.

Püllenberg, R., and Rohkohl, A. 2000. Modern lead smelting at the QSL-plant Berzelius Metall in Stolberg, Germany. In *Lead-Zinc 2000*. Edited by J.E. Dutrizac, D.M. Henke, S.E. James, and A. Siegmund. Warrendale, PA: The Minerals, Metals & Materials Society.

Queneau, P.E., and Siegmund, A. 1996. Industrial scale lead making with the QSL continuous oxygen converter. *JOM* 48(4):38–44.

Queneau, P.B., Hansen, B.J., and Spiller, D.E. 1993. Recycling lead and zinc in the United States. In *Proceedings of the Fourth International Symposium on Hydrometallurgy*. Edited by M.E. Wadsworth. New York: AIME.

Prengaman, R.D. 2000. Lead product development in the next millennium. In *Lead-Zinc 2000*. Edited by J.E. Dutrizac, D.M. Henke, S.E. James, and A. Siegmund. Warrendale, PA: The Minerals, Metals & Materials Society. pp. 17–22.

Rich, V. 1994. *The International Lead Trade*. Cambridge: Woodhead Publishing.

Richards, G. 2010. The Pyrometallurgical Refining of Lead, Pb-Zn short course: Lead and Zinc Processing. COM 2010. Vancouver, BC.

Richards, G., Aiken, G., and Rioux, D. 2010. The management of continuous drossing furnace accretions. In *Lead-Zinc 2010*. Edited by A. Siegmund, L. Centomo, C. Geenen, N. Piret, G. Richards, R. Stephens. Warrendale, PA: The Minerals, Metals & Materials Society. pp. 1067–1076.

Rioux, D.J., Stephens, R.L., and Richards, G.G. 2008. Teck Cominco's Kivcet furnace operation: The first 10 years. In *Zinc and Lead Metallurgy*. Edited by L. Centomo, M. Collins, J. Harlamovs, and J. Liu. Montreal, QC: Canadian Institute of Mining, Metallurgy and Petroleum. pp. 15–28.

Siegmund, A. 2000a. Modern applied technologies for lead smelting. In *Intensivierung metallurgischer Prozesse, Heft 87 der Schriftenreihe der GDMB*. Clausthal Zellerfeld, Germany. pp. 123–142.

Siegmund, A. 2000b. Primary lead production—A survey of existing smelters. In *Lead-Zinc 2000*. Edited by J.E. Dutrizac, D.M. Henke, S.E. James, and A. Siegmund. Warrendale, PA: The Minerals, Metals & Materials Society. pp. 55–116.

Siegmund. A. 2001. Secondary lead smelting at the beginning of the 21st century. In *EMC 2001, GDMB*. Clausthal-Zellerfeld, Germany. pp. 41–56.

Siegmund, A. 2010. Recycling Lead, Byproduct Metals, Sulfate and Polypropylene in Secondary Smelters, Pb-Zn short course: Lead and Zinc Processing. COM 2010. Vancouver, BC.

Siegmund, A., Püllenberg. R., and Rohkohl, A. 1998. QSL: Stolberg/Germany—Lead smelting in the new millennium. In *Zinc and Lead Processing*. Edited by J.E. Dutrizac, J.A. Gonzalez, G.L. Bolton, and P. Hancock, Montreal, QC: Canadian Institute of Mining, Metallurgy and Petroleum. pp. 825–842.

Stephens, R. 2010. Modern Lead Smelting, Pb-Zn short course: Lead and Zinc Processing. COM 2010. Vancouver, BC.

Stout, M.E. 1993. Secondary lead recovery from spent SLI batteries. In *EPD-Congress Proceedings 1993*. Edited by J.P. Hager. Warrendale, PA: The Minerals, Metals & Materials Society. pp. 967–979.

Young, K., Nener, T., Shannon, T., and Jiao, Q. 2010. Increasing non-concentrate feed to Brunswick smelter's updraft sinter machine. In *Lead-Zinc 2010*. Edited by A. Siegmund, L. Centomo, C. Geenen, N. Piret, G. Richards, and R. Stephens. Warrendale, PA: The Minerals, Metals & Materials Society. pp. 591–602.

CHAPTER 12.20

Lithium

Nicholas J. Welham

Lithium has an atomic number of 3 and occurs in two natural isotopes with atomic masses of 6 and 7; the latter is, by far, the most abundant (92.5%), resulting in an atomic weight of 6.94. Lithium is part of the alkali metal group, along with sodium, potassium, rubidium, and cesium, and occurs naturally as a monovalent cation. It is a soft, silvery metal when pure but rapidly corrodes in moist air to a dull gray. Elemental lithium has a melting point of 181°C (454 K) and a boiling point of 1,330°C (1,603 K). The density at room temperature is 0.534 g/cm^3, the lightest of all metals.

Lithium was first identified in 1817 by Johan August Arfwedson, a Swedish chemist who was unable to identify all components of the mineral petalite. The element was found to have similar properties to sodium and potassium but was discovered from a mineral and initially named "lithion" after *lithos*, the Greek word for *stone*. The same element was also quickly found to be present in spodumene and zinnwaldite. The pure element was first isolated in 1821 by William Thomas Brande, a chemist who electrolyzed lithium oxide. Larger-scale production of lithium was achieved by 1855 when lithium chloride was electrolyzed; this work laid the foundation of modern lithium metal production.

The first industrial-scale production of lithium occurred in Germany in 1923 when a new lead-based antifriction alloy containing 0.04% Li was produced. The alloy was used as bearings in railway trucks by German State Railway until replaced by ball bearings in the 1950s.

PRODUCTION AND USES

The production of lithium is rapidly expanding because of a major increase in the use of lithium ion batteries, mainly for electric vehicles, with an increasing interest in home solar storage. It is predicted that the quantity of lithium produced in 2015 will need to more than double by 2025 to keep up with demand. A large part of the increase in demand for lithium is driven by the numerous estimates for adoption rates of electric vehicles. The increasing rate of uptake seems to be because of recent governmental and car manufacturer announcements regarding emissions restrictions and even the phasing out of combustion engine vehicles. The predicted growth in batteries is 10%–15% per year, whereas the more traditional uses are expected to grow at 3%–5% per year. Figure 1 shows the estimated sector usage of lithium in 2014 and the predicted use in 2025.

Lithium salts, typically the carbonate, are used as a flux to reduce the melting point and viscosity of silica to make glazes with low thermal expansion for use in ovenware. Lithium hydroxide is heated with fatty acids to produce lithium stearate, which can be used to thicken oils to make high-temperature lubricating greases. Lithium carbonate is also added to fluxes during continuous mold casting to increase fluidity and to aid settling. Lithium fluoride is added to the electrolyte in the Hall–Heroult process for producing aluminum metal. The addition reduces the melting temperature and increases electrical resistance. Lithium metal is also alloyed with aluminum to improve chemical resistance and to refine the grain structure. Alloys of lithium with cadmium, copper, and manganese are used to make high-performance aircraft parts.

The red flame color of lithium, first noted by Arfwedson, is utilized in pyrotechnic compositions. Lithium chloride and bromide are hygroscopic and are used as desiccants for gas streams. Lithium hydroxide and peroxide are used in confined spaces, such as spacecraft and submarines, to absorb carbon dioxide. Both reactions lead to the formation of lithium carbonate; the peroxide also releases oxygen. The lithium salts are preferred over other alkaline hydroxides because of their greater performance per unit weight.

Small quantities of lithium fluoride (LiF) are used in specialized optics for infrared, ultraviolet (UV), and vacuum-UV applications. LiF has one of the lowest refractive indices and the widest transmission range in the deep UV of the most common materials. The nonlinear optical properties of lithium niobate find application in more than 60% of mobile phones as resonant crystals. Organolithium compounds are extensively used in organic syntheses, notably as catalysts/initiators in anionic polymerization of unfunctionalized olefins by promoting carbon–carbon bonds. Other lithium compounds used in chemical synthesis include lithium aluminum hydride, lithium triethylborohydride, n-butyllithium, and tert-butyllithium, all of which are commonly used as extremely strong bases.

Nicholas J. Welham, Principal, Welham Metallurgical Services, Perth, Western Australia, Australia

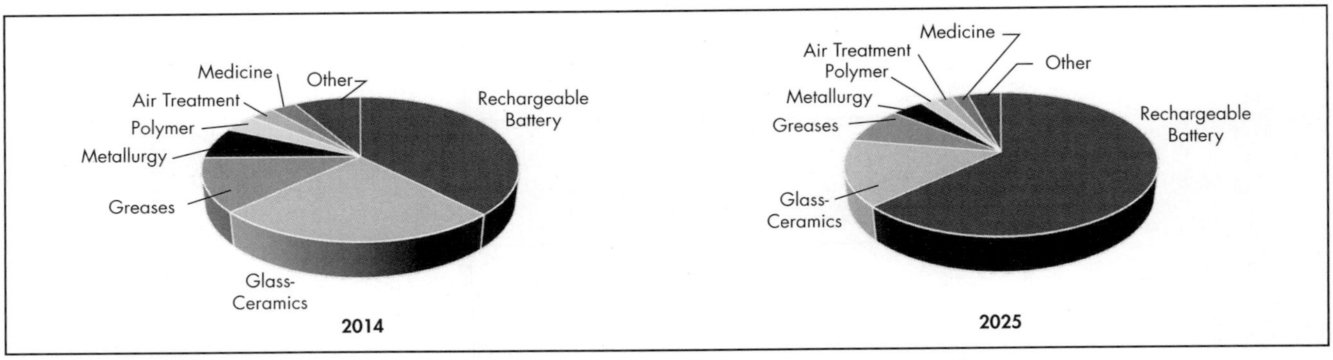

Source: SignumBox 2014

Figure 1 Uses of lithium by sector in 2014 and predicted use in 2025

Lithium metal and lithium aluminum hydride can be used as high-energy additives for rocket propellants; the latter can also be used directly as a solid fuel.

Figure 1 shows that the major expanding use of lithium is in rechargeable lithium ion batteries, which were first produced commercially in 1991. This type of battery has numerous advantages over other rechargeable batteries, including higher energy density, higher cell voltage, a low discharge rate during storage, low hysteresis, high discharge currents, and high charging currents. The battery consists of a graphite electrode, an oxide electrode, and an organic electrolyte containing lithium salts. There are many variations of battery primarily categorized by the composition of the oxide. The main component of the oxide phase can be lithium cobalt oxide, lithium iron phosphate, or lithium manganese oxide; these main oxide structures can be modified with partial substitution by different metals. During charging, lithium is reduced from monovalent to metal, which is stored by intercalation between the sheets of the graphite electrode. On discharge, the lithium metal oxidizes by reacting with the oxide electrode, releasing energy. It is important to avoid safety problems by preventing ingress of moisture when designing and packaging products containing lithium metal. Because it is necessary to dissipate heat during charging and discharging, especially at high currents, electric vehicles typically use numerous relatively small batteries combined into modules to increase the surface area. For example, the Tesla S (85-kW) battery consists of 7,104 separate batteries connected together, whereas the Tesla Model 3 (50-kW) battery has 2,976 individual cells.

Table 1 Abundance of lithium

Environment	Lithium, ppm
Average earth	17–20
Sedimentary rocks	11.5
Igneous rocks	6
Seawater	0.18
Brines	Up to 1,800

Adapted from Garrett 2004

MINERALOGY

The reactivity of lithium is such that it does not occur in elemental form in nature; instead, it forms largely ionic minerals and occurs in many natural waters, including brine and seawater. The approximate abundance of lithium in assorted natural environments is given in Table 1. The average abundance is similar to that of yttrium, cobalt, niobium, and scandium but higher than lead or tin.

There are ~130 minerals where lithium is considered to be a structural requirement and an additional ~100 where lithium is known to substitute for monovalent cations, notably sodium, but is not essential to the structure (Webmineral 2018). However, the vast majority of these are not found in economic quantities. The major economic deposits are largely composed of the minerals shown in Table 2.

Lithium Ore Bodies

The two commercial sources of lithium are hard rock and brines. U.S. Geological Survey data indicate that the majority

Table 2 Major economic lithium minerals

Mineral	Chemical Formula	% Li_2O	Group	Typical % Li_2O in Ore Grade
Spodumene	$LiAlSi_2O_6$	7.99	Pyroxene	1.0–7.7
Petalite	$LiAlSi_4O_{10}$	4.88	Phyllosilicate	3.0–4.7
Amblygonite	$(Li,Na)Al(PO_4)(F,OH)$	7.4–10.2	Phosphate	7.5–9.5
Lepidolite	$K(Li,Al)_3(Si,Al)_4O_{10}(F,OH)_2$	3.0–7.8	Mica	3.0–4.1*
Zinnwaldite†	$KLiFeAl(AlSi_3)O_{10}(F,OH)_2$	4.3–10.7	Mica	2.1–4.3‡
Hectorite	$Na_{0.3}(Mg,Li)_3Si_4O_{10}(OH)_2$	1.17	Clay	0.4–0.7

Adapted from Garrett 2004

*Rb ~0.1%–0.3%; Cs ~0.02%–0.04%.

†Zinnwaldite is no longer recognized as a specific mineral. It is considered to be within a solid solution series between siderophyllite $KFe_2Al(Al_2Si_2)O_{10}(F,OH)_2$ and polylithionite $KLi_2AlSi_4O_{10}(F,OH)_2$.

‡Up to ~0.9% Rb.

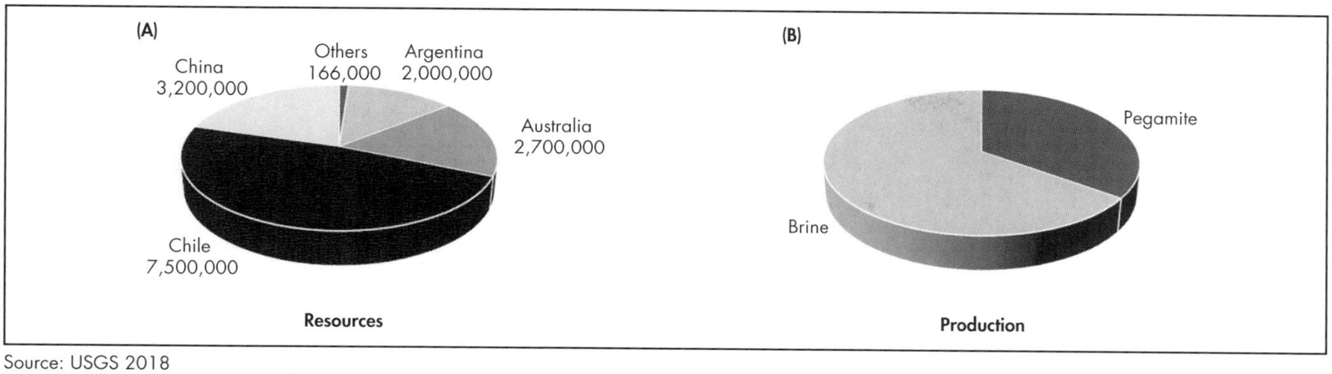

Source: USGS 2018

Figure 2 (A) Location of lithium reserves and (B) reserve type

of reserves (i.e., economically extractable lithium) are in South America (Figure 2A) and that the vast majority of those are brine based (Figure 2B; USGS 2018). Many other potentially significant resources are not considered to be reserves (e.g., Salar de Unuyi in Bolivia, the world's largest salt flat), given the absence of sufficient data. The considerable interest in lithium is also expanding the hard rock resource base around the world.

Figure 3 shows the grade and size of major reported brine and hard rock deposits (in million metric tons [Mt]), the primary difference being much lower lithium grade in the brines.

Brine Deposits

Brine deposits are formed by the ongoing evaporation of water from an enclosed basin into which salty water flows. The inflow of salty water needs to be lower than the outflow of water in order to build up the concentration of salt. Over time, the concentration of salt in the water increases until crystallization of the least-soluble salts occurs, forming a crust on the surface up to several meters in thickness. The lithium salts, being among the most soluble, do not typically crystallize but remain in a concentrated brine solution. The natural processes that formed the lake will have resulted in the deposition of loosely bound sediments ranging in size from gravel to clays. The salty brine occupies the pore space between the particles.

The conditions required to form brine deposits are unusual and occur in only a few places around the world. The host basin needs minimal outflow, the inflow water needs to have a consistent supply of salt, the outflow needs to be extremely low, and the evaporation rate has to be very high.

The highly concentrated brine solution has K^+, Na^+, Li^+, Ca^{2+}, and Mg^{2+} as the main cations and Cl^- and SO_4^{2-} as the main anions. In addition to these, also present will be lower levels of other cations and anions, such as NO_3^-, Br^-, and I^-. The composition of the brine varies across the surface of the lake and with depth.

Table 3 shows the average composition of brines from different salars. The range of elemental concentrations is broad, and deposits are often defined by the theoretical composition based on the relative concentrations of the ions. For extraction purposes, it is the ratios of the elements, especially the (Mg+Ca) to lithium, that are of major importance. Deposits where the (Mg+Ca):Li ratio is greater than 15:1 are harder to process by solar evaporation, as some of the lithium will co-crystallize with the calcium and magnesium salts. The presence of <400 ppm of lithium is also unfavorable for solar evaporation because of the higher volumes of water needing to evaporate to achieve the required lithium concentration for final processing.

Extraction from brine deposits is very different from classical mining methods, being more similar to oil than to ore. The brine solution is extracted from wells, pits, or trenches cut into the brine-loaded sediment layers. The slower the flow of the brine into the collector, the larger the required area. Ultimately, the extraction of the brine is governed by the permeability of the host sediments; a low permeability will result in much lower extraction rates. Thus, when assessing a project, it is critical to determine the maximum pumping rate, as this ultimately determines the production rate. The replenishment rate of the brine within the sediments is also an important factor and can only be assessed by determining the reduction in volume pumped over time. When the permeability and/or the replenishment rate is low, more bores or longer trenches will be required in order to provide the brine flow rate required for production.

Taking all of the factors into account, the level of extraction from brine deposits is significantly lower than from more classical ore deposits, and the method for definition of the resource/reserves is notably different. The CIM Definition Standards for Mineral Resources and Mineral Reserves (Hains 2012) allows for a maximum 33% recovery from brine deposits. This is before a range of different reduction factors are taken into account. In reality, an extraction of 10%–15% is likely to be the realistic maximum for most deposits.

Hard Rock Deposits

The hard rock deposits are composed of granitic pegmatites in which zonation has occurred. The core of the pegmatite is generally barren, while the periphery is lithium rich due to the lower melting point of lithium-rich fluids. Many of the pegmatites also contain other exotic metals, including tantalum/niobium, rubidium/cesium, and tin in separate zones. The occurrence of tin and lithium has led to a new lease on life for mines that were historically exploited for their tin content but now have an inventory of surficial tailings containing lithium and mine workings that have ore-grade lithium remaining from the tin-mining days.

Figure 4 shows the zoning of the pegmatite at the Greenbushes deposit in Western Australia. The main minerals present in the deposits are quartz, feldspar, and mica. Also commonly present in the lithium-rich zones is tourmaline, which presents a minor processing challenge for high-grade

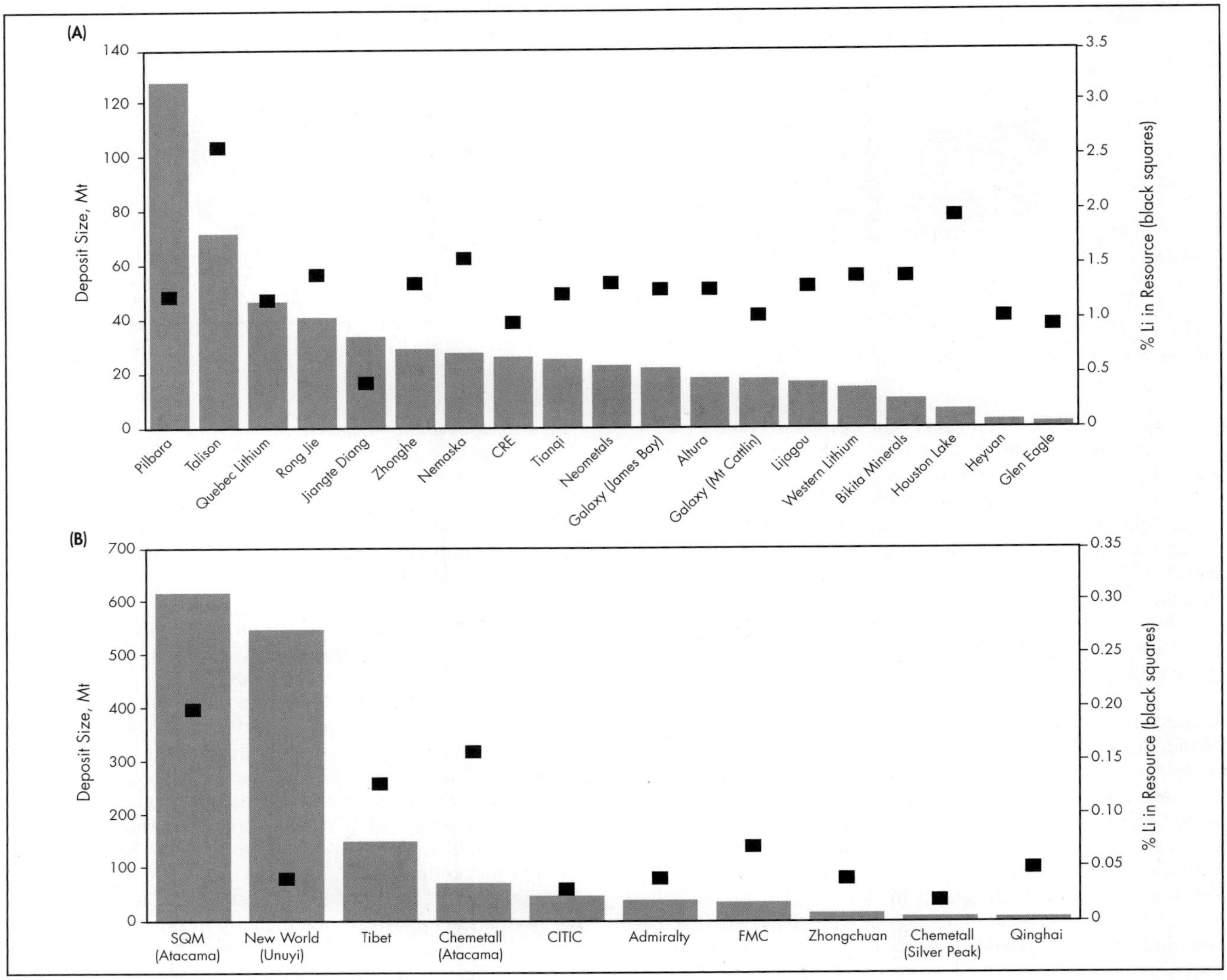

Source: Jephcott 2016

Figure 3 Data for the size and grade of (A) hard rock resources and (B) brine

Table 3 Composition of a range of global brine resources*

Deposit	Type	Li	K	Mg	Ca	SO4	Mg/Li	SO_4/Li	Ca/Li	SO_4/K
Salar de Atacama, Chile	$MgSO_4$-Li_2SO_4-LiCl-$CaCl_2$	1,835	22,626	11,741	379	20,180	6	11	0.2	0.9
Salar de Cauchari, Argentina	Na_2SO_4-K_2SO_4-Li_2SO_4	618	5,127	1,770	476	19,110	3	31	0.8	3.7
Salar de Hombre Muerto, Argentina	Na_2SO_4-K_2SO_4-Li_2SO_4	744	7,404	1,020	636	10,236	1	14	0.9	1.4
Salar de Maricunga, Chile	KCl-LiCl-$CaCl_2$	1,036	8,869	8,247	11,919	1,095	8	1	11.5	0.1
Salar de Olaroz, Argentina	Na_2SO_4-K_2SO_4-Li_2SO_4	774	6,227	2,005	416	18,630	3	24	0.5	3.0
Salar de Rincon, Argentina	$MgSO_4$-Li_2SO_4	397	7,513	3,419	494	12,209	9	31	1.2	1.6
Salar de Unuyi, Bolivia	$MgSO_4$-Li_2SO_4	424	8,719	7,872	557	10,342	19	24	1.3	1.2
Silver Peak, USA	Na_2SO_4-K_2SO_4-Li_2SO_4	245	5,655	352	213	7,576	1	31	0.9	1.3
West Taijinaier, China	$MgSO_4$-Li_2SO_4	256	8,444	15,737	NS†	35,315	61	138	NS	4.2
Zhabuye Salt Lake, China	Li_2CO_3-Na_2SO_4	1,217	17,083	17	NS	38,917	0	32	NS	2.3

Source: Desormeaux 2015

*Elemental concentrations are in parts per million.

†NS = Not stated.

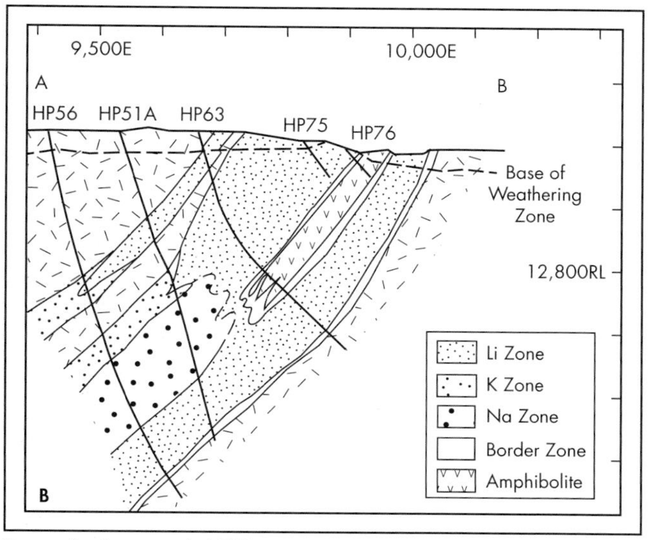

Source: Partington et al. 1995

Figure 4 Cross section of the Greenbushes deposit in Western Australia showing the zonation typical of such pegmatites

products. Extraction of the pegmatites is either via open pit, for those deposits that are close to the surface or at low dip angles, or underground mining for deeper and more steeply dipping deposits.

RECOVERY OF LITHIUM FROM BRINE DEPOSITS

The recovery of lithium from brine deposits is a relatively simple process dominated by the use of solar energy (Garrett 2004). The low initial lithium concentration, high competing ion concentrations, and comparative lack of reactivity of the lithium ion effectively prevent direct recovery of lithium from the brines. Lithium can be solvent extracted from brines as the LiCl neutral ion (Epstein et al. 1981), and further development of this concept may well lead to a commercially viable process. The most recent of these developments has shown promise in the laboratory using a Bateman pulsed column (Lipp 2014).

However, as in many of the new or proposed processes to selectively extract lithium from brine, it is necessary to remove most, if not all, of the calcium and magnesium prior to solvent extraction (SX). In these, the high capital cost of solar evaporation ponds is offset by the increased operational cost of removing calcium and magnesium prior to the lithium.

The low level of lithium is addressed by using solar evaporation to reduce the volume of water, thereby concentrating the lithium and other elements. The ponds cover a large area but are relatively shallow to maximize the surface area to volume ratio to enhance evaporation and therefore have a very high footprint (1,700 ha [SQM 2018]). The ponds need to be leak resistant, and plastic liners similar to those used in heap leaching are employed. The preparation of the ponds is one of the major capital costs.

The significantly higher concentrations of other elements lead to the crystallization of the least-soluble salts in the system as the volume decreases. There is also some variation in phases precipitated due to annual seasonal changes in local atmospheric conditions (Vergara-Edwards and Parada-Frederick 1983). The formation of brine lakes occurs in places that are also excellent for promoting the evaporation of water by a combination of high altitude, low humidity, low rainfall, wind strength, high evaporation rates, and high solar energy density.

The identity and sequence of salts crystallized will vary between deposits and even between brine sources within a single deposit due to the variability in composition outlined in Table 3. Through management of the volume evaporated, the different salts crystallized can be largely separated by allowing the volume to decrease and then transferring the brine to a smaller pond. The sequence for SQM at the Salar de Atacama in Chile is sodium chloride (NaCl), potassium chloride (KCl), and $KMgCl_3 \cdot 6H_2O$ (SQM 2017). The NaCl is discarded, and the KCl and $KMgCl_3 \cdot 6H_2O$ are further processed to pure KCl for sale. As time goes on, the brine becomes increasingly concentrated in lithium until it is ready to transfer to a more intensive recovery process. The solar evaporation process relies on considerable operational experience in knowing when to move the brine onto the next pond to prevent contamination of the product. The typical time required from initial brine to final stages is approximately 18–24 months. This can be considerably longer if there is an unexpectedly high rainfall event onto the ponds.

Solar evaporation is used until it has concentrated the brine to the point where lithium will start to precipitate with other salts. For brines where the (Mg + Ca):lithium ratio is higher than 15, the precipitation of Li + Ca/Mg salts (e.g., $LiMgCl_3 \cdot 7H_2O$) occurs significantly earlier in the evaporation process than for brines where the ratio is lower. Only recently have such high-ratio brines been considered as potential resources due to the increased complexity of lithium extraction compared to the low-ratio brines.

The concentrated brine solution is initially treated to remove boron. This is commonly performed by SX using branched aliphatic alcohols, such as 2-ethylhexanol and 2-butyl-1-n-octanol as extractants. The brine is reduced in pH to <2.5 where the boron is present as neutral boric acid, H_3BO_3, and this is extracted by the alcohol. Recovery of the boron is by contacting with water, after which the solution is evaporated to form crystals of boric acid that can be sold.

Calcium and magnesium are then removed by adding Na_2CO_3 to form insoluble carbonates that are filtered and disposed of. Care needs to be taken to prevent excessive precipitation of Li_2CO_3 by careful monitoring of the solution. Further soda ash is then added to precipitate the lithium carbonate, which is filtered, washed, and dried. The filtrate contains a mixture of sodium and potassium salts that are crystallized, and flotation is used to separate the NaCl and KCl into separate products. The overall flow sheet is shown in Figure 5.

RECOVERY OF LITHIUM FROM HARD ROCK LITHIUM DEPOSITS

The flow sheet for the processing of hard rock lithium ores is largely determined by two factors:

1. **Ore texture.** If the mineral particle is coarse or there is a clear differentiation between particles at crush sizes of >1–2 mm, then physical separation methods can be used to upgrade the lithium minerals. Ore sorting is not presently in use for lithium ores but early test work has indicated that it is possible to upgrade some spodumene

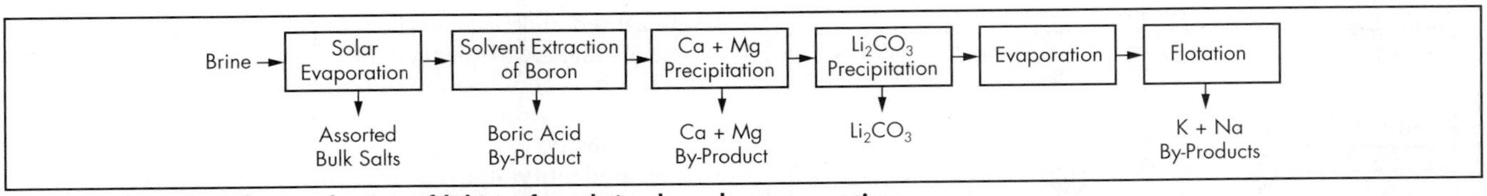

Figure 5 Flow sheet for production of lithium from brine by solar evaporation

ores; mica ores are generally somewhat finer grained and do not separate well at coarse particle sizes. The most common method of physical separation in use is dense medium separation. If the mineral is finely divided, then flotation is the only viable method.

2. **Ore mineralogy.** There is a significant difference between spodumene and mica ores. Spodumene ores are typically coarser and higher grade than mica ores. The higher density of spodumene allows gravity separation to be used, whereas micas are almost exclusively processed by flotation.

Importantly, many pegmatite deposits contain both spodumene and lithium micas in association. The relative ratio of the two will largely determine the processing route, as the subsequent concentrate conversion process will operate on spodumene or, less commonly, mica. The presence of mica in the spodumene concentrate results in a financial penalty, because it increases the operational complexity of the conversion process, most notably due to problems within the initial calcination stage. Lithium from spodumene that is present in a mica ore will not be recovered during the conversion process and is typically not considered when determining the head grade. Ore deposits where the lithium is present at >10% of the minority phase are more economically challenging because of the lower "real" head grade once the unwanted component is eliminated.

Comminution

Crushing is ordinarily undertaken with open-circuit primary and closed-circuit secondary crushing. Tertiary and quaternary crushers are used for ores that are to undergo physical separation. Undersize particles are either discarded, stockpiled for future exploitation, or fed into a grinding circuit in preparation for flotation. Flow sheets that do not use gravity separation and go directly to flotation may use primary crushing followed by semiautogenous milling.

The softness of the micas compared to the other minerals present results in preferential sliming of the micas. For ores where the main lithium mineral is a mica, this can result in high losses when desliming prior to flotation. Unlike most grinding circuits prior to flotation, it is best to avoid using cyclones to size the ground product. The platy mica particles tend to behave according to the largest dimension and report preferentially to the oversize for regrinding. As a consequence, there can be a noticeable concentration of lithium mica in the slimes compared to the flotation feed (e.g., 0.8% Li_2O in the feed and 1.1% Li_2O in the slimes, with >25% loss of Li_2O to slimes). In such situations, an upflow classifier is a more appropriate choice of sizer, as this takes advantage of the platy nature of the mica to separate the particles. To reduce overgrinding of the lithium mica, the classifier is placed prior to the grinding mill. The undersize from the classifier can be passed over Derrick, or similar, screens to remove the slimes prior to flotation.

Table 4 Densities of the major minerals present in pegmatite deposits

Mineral	Density, g/cm^3	Mineral	Density, g/cm^3
Spodumene	3.1–3.2	Quartz	2.65
Feldspar	2.55–2.76	Mica	2.8–3.4
Petalite	2.4	Zinnwaldite	2.9–3.1
Lepidolite	2.8–3.0	Columbite/tantalite	6.3–8.2

Physical Separation

The densities of the major minerals present are given in Table 4. The ranges given depend on the chemical composition of the mineral, which is deposit specific. In order to be separable by simple gravity methods, there needs to be a difference in density > 0.2. This is the case for spodumene and quartz, but the remaining lithium minerals are too close to the gangue to give an effective separation. Thus, gravity methods are only viable for spodumene ores.

Historically, spirals have been used in spodumene plants with some success. Modern plants only use spirals to separate the tantalite from other minerals, as the tantalite may prove to be a significant revenue source in some deposits. Modern lithium plants almost exclusively use dense medium separation (DMS) using cyclones for gravity separation. The required separation density is only achievable using ferrosilicon media. The particle feed size is greater than 0.5 mm, but greater than 2 mm gives a more effective separation. Figure 6 shows a typical single-stage DMS circuit with the ancillary equipment required to recover and make up the ferrosilicon to the required density.

Figure 7 shows two DMS circuit designs. Figure 7A shows a two-stage circuit in which there are primary and secondary dense medium cyclones (DMCs) acting as rougher and cleaner stages. Figure 7B is a three-stage circuit that produces an intermediate density product that is ground and floated.

The inherent complexity of DMS plants precludes the operation of cyclones at different medium densities in order to produce multiple products. The DMS tails may be either sent to tailings, stockpiled, or, more commonly, combined with the screen undersize and ground in preparation for flotation.

DMS is capable of producing a 6% Li_2O concentrate from spodumene ores at good recoveries. This concentration is the current benchmark for conversion to lithium salts. As a consequence, plants may be developed in two stages. Initially, only DMS is used to produce the 6.0% concentrate from higher-quality ore to generate cash flow. The second stage is the installation of grinding and flotation to re-treat the DMS floats and allow higher recoveries at the required grade from lower-quality ores.

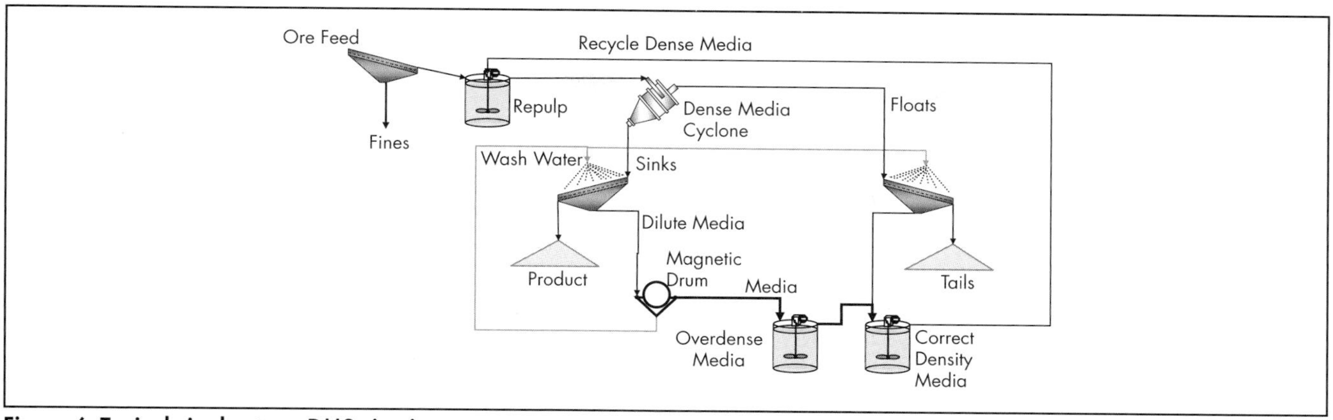

Figure 6 Typical single-stage DMS circuit

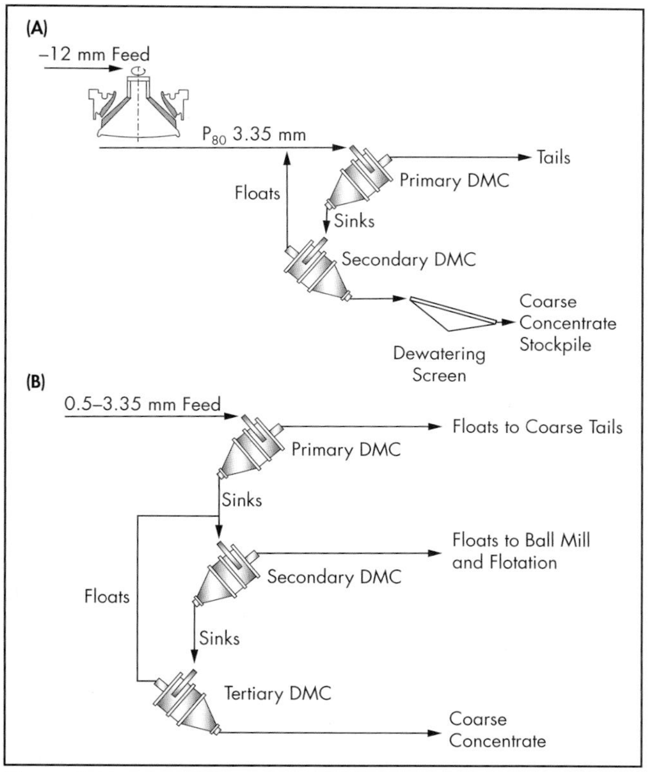

Figure 7 Two- and three-stage DMS circuits

Other Methods

Magnetic Separation

Zinnwaldite is amenable to magnetic separation because of its weakly magnetic nature. Separation from other weakly magnetic minerals, including muscovite, is likely to prove challenging for such deposits. However, no commercial plant currently operates such a process.

Ore Sorting

There is good potential to use ore sorting in upgrading lithium-bearing ores. The potential is greater for spodumene than micas because of the coarser particle size typically present. However, the strong violet/pink color of lepidolite ores can be used to produce a high-grade concentrate from some ores.

Flotation

Spodumene Flotation

The flotation of lithium ores has been a relatively minor market until recently. The classic flotation process using oleic acid as collector has been used since the 1950s and remains the major process used today. The more recent expansion of interest in lithium has led to the development of specialized collectors by manufacturers, though at this stage there is no known plant using reagents other than oleic acid. The flotation of lithium micas is a more recent development, and the typical mica collectors, notably amines, are commonly used. The operating conditions for the process depend on the associated gangue minerals.

Oleic acid is a long-chain fatty acid derived from olive oil. The chemical formula is $CH_3(CH_2)_7CH{=}CH(CH_2)_7COOH$. It has a density when pure of 895 kg/m^3 and is pale yellow or brownish yellow with a lard-like odor. It melts at 13°–14°C and boils at 360°C. Oleic acid is insoluble in water, forming an immiscible layer on top of the water. To increase the solubility of the oleic acid, it is saponified with soda ash (Na_2CO_3) or sodium hydroxide (NaOH) to form water-soluble sodium oleate. The reaction is slow and necessitates a comparatively long conditioning time of 10–30 minutes. The reaction is accelerated by operation at elevated temperature. Typically 500–700 g/t of oleic acid are added in the conditioning stage.

Conditioning is undertaken at 60% slurry density for 10–30 minutes with the addition of 0.5–1.0 kg NaOH/t. The NaOH increases the pH to aid saponification and to help chemically clean the mineral surface. The long conditioning time results in particle attrition. This has two effects: cleaning of the mineral surfaces, which aids adsorption of the collector, and production of slimes, which have to be removed prior to flotation.

The conditioned ore is deslimed and the slimes discarded to tailings. The slurry is diluted to 30%–40% and fed into a bank of rougher cells. The flotation pH is adjusted using sulfuric acid to 6.6–6.9, although ores with more complex mineralogy may use pH 8.0–8.2 to help depress flotation of impurities. No frother is generally required, as the oleic acid also acts as a frother; for low-grade ores, a commercial frother may be added to the second bank of roughers (or scavengers) to ensure recovery.

A typical flow sheet for a flotation plant is shown in Figure 8. The feed is the screen undersize from a previous DMS operation, such as that shown in Figure 7B; the middlings

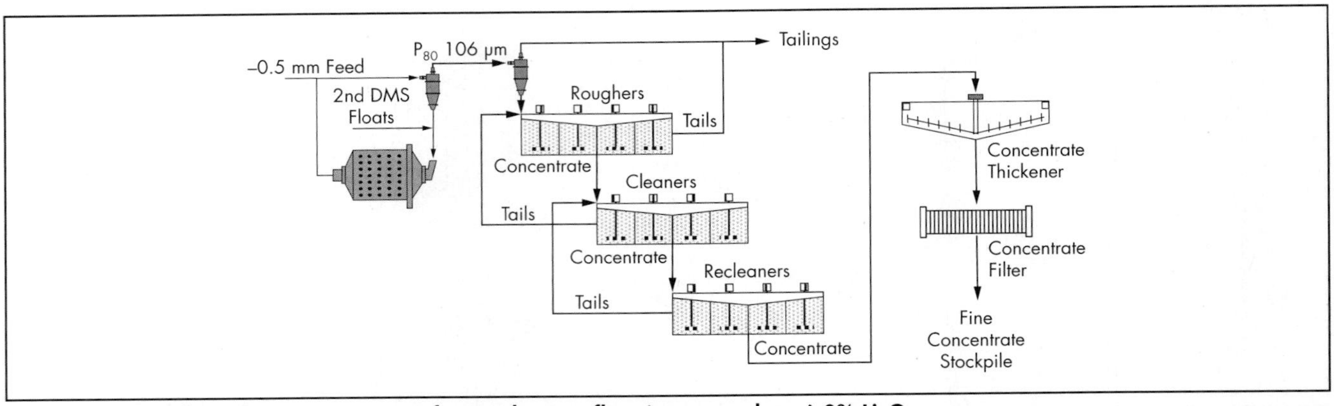

Figure 8 Typical flow-sheet arrangement for spodumene flotation to produce 6.0% Li_2O

from the multistage DMS are also fed into the mill. The ore is ground to a P_{80} of 106 μm in closed circuit with a cyclone; the undersize is deslimed in smaller cyclones and fed into a conditioning tank and rougher bank. The concentrate is diluted to a lower slurry density, cleaned and recleaned to produce a 6.0% Li_2O concentrate; the tails from the cleaners are recycled to the previous stage. The rougher tails may be either scavenged or sent to tailings. The final cleaned concentrate is thickened and filtered to produce a fine concentrate for shipping.

There is a smaller market for higher-grade concentrates than the 6.0% Li_2O destined for conversion to lithium carbonate or hydroxide. To produce grades of up to 7.7% Li_2O, the ore is ground more finely and additional cleaning stages are required. The slurry density in the stages reduces from 35%–40% solids in the roughers to 30%–35% in the cleaners and 25%–30% in the recleaners. The reduced slurry density improves separation primarily by greatly reducing entrainment and increasing flotation time. The froth depth down the bank is adjusted using moveable weirs, with the depth being lowest in the last cells where there is the least floatable material.

For the highest-grade products, the flotation concentrate is thickened and agitated in a tank with sulfuric acid; this reverses the oleic acid saponification reaction, leading to desorption of collector from the mineral surface. The acid also removes small amounts of surficial iron and any acid-soluble minerals (such as apatite), if present. The concentrate undergoes low-intensity magnetic separation to remove strongly magnetic materials and then wet high-intensity magnetic separation to remove weakly magnetic materials, including tourmaline. The final product is belt filtered and then thermally dried before bagging to customer requirements.

Mica Flotation

Flotation of lithium-bearing mica is a relatively new technology and there are no operational plants at present. The standard mica collectors are largely effective, and some newer variations are being marketed as collectors for lithium mica. Performance is strongly dependent on the composition of the lithium mica and the gangue mineralogy. Rubidium-bearing lepidolite is less effectively floated than low-rubidium lepidolite under identical conditions (Bulatovic 2014). The large variation in mica compositions and gangue mineralogy between and within deposits makes ongoing optimization of flotation conditions a challenge for any lithium mica flotation operations.

Among the reagents reported to act as collectors are dodecylamine at pH 2.5–12; hexadecyl amine acetate at pH 3.5; and octadecylamine, which requires sodium silicate and Li_2SO_4 as lepidolite activators. Another effective collector for lepidolite (0.7% Li_2O) from quartz, albite, and calcite is stearyl trimethyl ammonium chloride (STAC) $(CH_3)_3CH_3(CH_2)_{16}NCl$ (Choi et al. 2015). A rougher Li_2O recovery of 95% was reported when using 100 g/t of STAC at pH 9.0. Calcite was removed by lowering the pH to 6.3 and adding 20 g/t of frother; quartz and albite were removed by further reducing the pH to 2, giving a final lithium grade of 2.76% Li_2O at 76.0% recovery. A flow sheet for mica beneficiation is shown in Figure 9.

Flotation of Other Lithium Minerals

The rarity of, and small market for, lithium has limited the amount of work on the flotation of lithium minerals. Petalite has been floated using dodecylamine collector with recovery being possible at pH 2.0–11.0. However, in the presence of a complex pegmatite ore containing feldspar, lepidolite, mica, and albite, selectivity was very poor at pH 2.0–8.0. A somewhat more complex process was patented for an ore containing spodumene, petalite, feldspar, and mica (Jessup and Bulatovic 2000). The mica floated at pH 2 using a cationic amine, and after pH adjustment to 8.5–10, spodumene was floated using oleic acid. The flotation medium was changed to a 10% brine by adding NaCl and KCl, which depressed the feldspar. The brine was reduced to pH 2–3 using a mixture of hydrofluoric acid (HF) and sulfuric acid (H_2SO_4), which cleaned the surface of the petalite, allowing a quaternary ammonium salt to be added at 250–400 g/t as collector. After a 4–10-minute flotation time, a final Li_2O grade of 4.7%–4.8% was achieved, with the concentrate being 96%–98% petalite. A modification of this process is proposed for the Separation Rapids project in Kenora, Ontario, Canada (Aiken et al. 2016).

For ores where there is limited separation possible by flotation of the lithium mineral due to fine particle size or excessive sliming at fine grind sizes, it is possible to undertake reverse flotation and remove the gangue minerals, potentially at a somewhat coarser grind size. This method has been proposed for the removal of calcite from lithium clay deposits to reduce the acid consumption during leaching.

Commercial Aspects

The current lithium market is dominated by the production of the SC6.0 grade with 6.0% Li_2O as this is the most

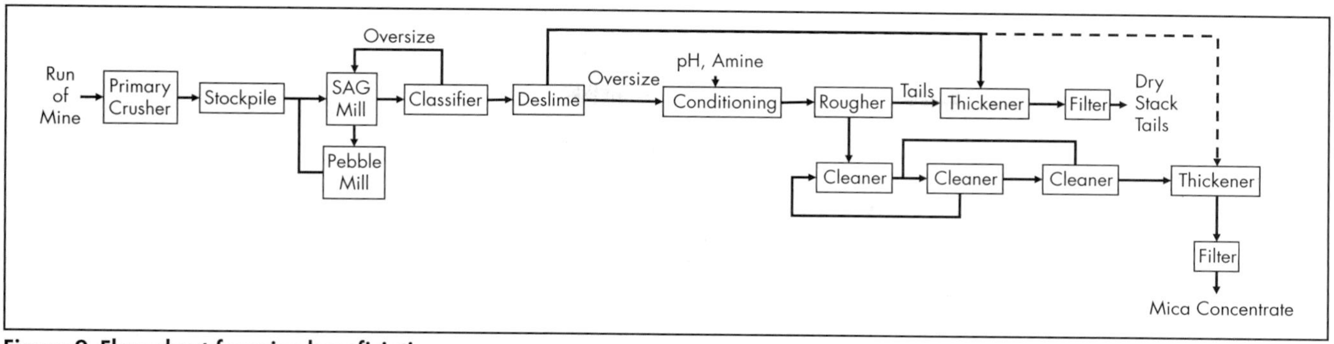

Figure 9 Flow sheet for mica beneficiation

Table 5 Elemental composition of spodumene concentrates

Element	Talison SC6.0	Talison SC7.5	Kings Mountain	Mount Cattlin	Pilgangoora	Pure Spodumene
Li_2O	6.05	7.60	6.00	5.50	6.68	8.03
SiO_2	66.30	64.50	63.00	64.00	64.43	64.57
Fe_2O_3	0.70	0.07	1.90	1.60	1.46	
CaO	0.40	0.05	Not stated	0.84	0.12	
K_2O	0.65	0.015	1.01	1.00	0.47	
Na_2O	0.55	0.08	0.61	0.90	0.39	
Al_2O_3	23.70	26.50	25.40	26.60	24.50	27.40
Totals	98.35	98.92	97.92	100.44	98.05	100.00

Note: Analyses of P_2O_5, TiO_2, MgO, and MnO are also important, especially for the highest Li_2O concentrates.

common feed to conversion plants that make lithium carbonate or hydroxide. Although higher grades are possible, up to 7.7% Li_2O, the end-user market is much smaller and necessitates greater capital expenditures and operating expenditures to achieve. It further needs to be borne in mind that the SC6.0 grade only contains around 75% spodumene—the remaining 25% is an assortment of gangue minerals that are strongly dependent on the ore body being mined.

Table 5 shows the reported composition of a range of different spodumene concentrates and that of pure spodumene. In addition to the elements listed, P_2O_5, TiO_2, MgO, and MnO are also important, especially for the highest Li_2O concentrates. The level of Fe in the Talison products is very noticeably lower than all others; this is due to the unusually low level of impurity minerals in the Greenbushes ore body, leading to a cleaner concentrate.

The only major specification of SC6.0 appears to be a minimum Li_2O content of 6.0%—there do not appear to be any restrictions on the levels of other elements. Such a minimum Li_2O content is only realistically possible for spodumene, because the maximum concentration possible is dictated by the mineralogy, with mica deposits struggling to get >1.5% Li_2O at high recoveries.

One consequence of this is that transportation becomes a significant limiting factor in the value of the concentrate. Table 6 shows the mass of concentrate required to produce 1 t (metric ton) of a range of different lithium products. The table assumes 100% Li recovery during the conversion process. Clearly, a minimum of 6.7 t of SC6.0 spodumene concentrate is required to produce 1 t of Li_2CO_3; for a mica concentrate of ~1.5%, more than 26 t of concentrate are required. As the feed grade of concentrate decreases, the mass required increases

Table 6 Mass of starting material at given Li_2O grade required to produce 1 t of product at 100% recovery

Product	Formula	6.0%	3.0%	1.5%
Metal	Li	35.7	71.3	142.6
Carbonate	Li_2CO_3	6.7	13.4	26.8
Oxide	Li_2O	16.6	33.1	66.3
Hydroxide	LiOH	10.3	20.7	41.3
Chloride	LiCl	5.8	11.7	23.3
Bromide	LiBr	2.8	5.7	11.4
Butyl lithium	C_4H_9Li	3.9	7.7	15.5

significantly, making it economically harder to separate concentrator and conversion plant. It should also be borne in mind that only Li_2O is recovered from the concentrate in the conversion plant, resulting in the vast majority of the mass of the feed reporting to tailings.

Where concentrators use both DMS and flotation, the concentrate will be in the form of a mixture of coarse and fine particles. Conversion plants are typically set up to treat either DMS or flotation products, as the particle size difference changes material handling and affects several areas of the conversion process.

Concentrate Conversion

Spodumene Conversion

Spodumene is not readily soluble in acids unless there are sufficient fluoride ions present to form the stable hexafluorosilicate anion, SiF_6^{2-}. Neither baking with concentrated sulfuric acid at 250°C nor boiling in azeotropic hydrochloric acid (HCl) results in any significant dissolution. To overcome this refractoriness, the spodumene is heated to about 1,050°C for two to three hours. During heating, the spodumene changes crystal form from monoclinic α-spodumene to tetragonal β-spodumene and undergoes a density decrease from 3.1 to 2.4 g/cm^3. The large volume change results in the production of friable particles.

The heating process is generally carried out in a direct-fired rotary kiln, which is designed to give the required residence time for a particular feed particle size range. Blending of the coarse DMS and fine flotation concentrates leads to several process issues. The larger particles are slower to heat up and require a longer residence time but are also more rapidly transported down the kiln by rolling. This necessitates a larger kiln, or installation of lifters and weirs to increase residence time. The finer particles are more readily blown out of the kiln and

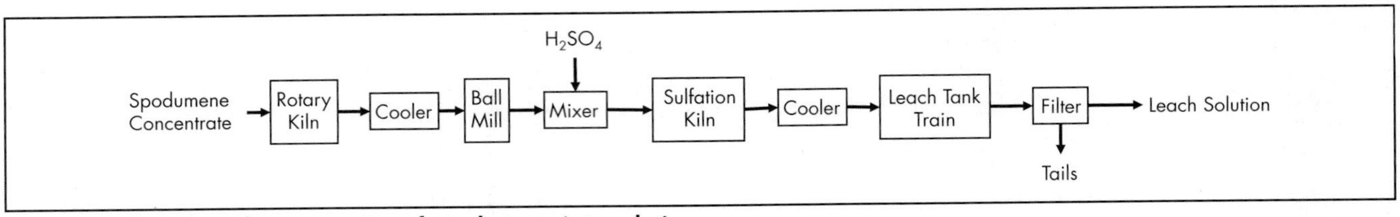

Figure 10 Flow sheet for conversion of spodumene into solution

need to be recovered in a baghouse and returned to the kiln to achieve the required residence time. Typically, 95%–97% of the spodumene is converted into the acid-leachable form in this process.

The presence of impurities in the concentrate (a 6.0% Li_2O concentrate is ~75% spodumene and 25% impurities) becomes a factor in the process at this point. Table 5 shows the composition of a range of different spodumene concentrates. Despite similar Li_2O concentrations, the remaining elements are somewhat different because of the differences in the ore-body mineralogy and processing route.

The impurities have variable behavior during the heating stage. Quartz and feldspar remain largely inert while phases containing iron, phosphorus, potassium, and sodium can melt or form low-melting-point phases by reaction. The formation of liquid phases will bind the particles of spodumene and form larger particles, which will affect the residence time. As it cools, a glassy phase will be somewhat sticky and potentially form barriers within the kiln, leading to blockages and increased downtime for cleaning. The cooled glass-bound spodumene particles will be harder to leach, with <75% Li recovery reported from such particles. Any mica that is present in the concentrate will melt and form glassy phases; thus there is a typical maximum limit to the quantity of mica permissible in the spodumene concentrate.

This heating stage is highly endothermic, requiring a minimum of 4.2 MW·h/t for pure spodumene, and represents the majority of energy costs in the conversion process. The absence of this extremely energy-intensive step is the major reason why brine producers are able to produce Li_2CO_3 at a lower cost.

After conversion to β-spodumene, the solid is cooled in a water-sprayed rotary cooler. After dry grinding to 150 µm, the calcine is mixed with 105%–130% of the stoichiometric amount of concentrated sulfuric acid and heated again to about 250°C. The stoichiometric excess is to ensure complete recovery of lithium and overcome acid consumption by undesirable reactions with the gangue minerals. The main reaction can be summarized as follows:

$$2LiAlSi_2O_6 + H_2SO_4 = Li_2SO_4 + H_2O + Al_2O_3 + 4SiO_2$$

This reaction is highly exothermic and self-heats by ~200°C from the starting temperature, thereby improving the kinetics. After the reaction is complete, the mixture is water-leached at about 50°C, dissolving about 98% of the lithium and other impurities as sulfates, the residue being composed primarily of poorly crystalline phases. The leaching time is kept short to give an initial separation of the fast-dissolving lithium from the slower-dissolving impurities.

It is also possible to leach the β-spodumene directly in hydrochloric acid. The leach requires 6–12 hours of leaching at ~106°C in azeotropic HCl. The resultant solution has higher impurities than the sulfate route, as there is much more time for the impurities to dissolve and chlorides tend to be more soluble than sulfates.

The aluminosilicate residue from the acid leaching is filtered and washed to reduce the level of SO_4^{2-} or Cl^- present. There is potential to use this material as an additive to cement or as a feedstock to other processes, as it is largely amorphous, has a high surface area, and is relatively reactive. A typical flow sheet for solubilization of spodumene is shown in Figure 10.

Of the other lithium minerals, lepidolite and zinnwaldite are directly soluble in sulfuric acid but are often calcined before leaching; petalite is processed similarly to β-spodumene, being heated with H_2SO_4 at 300°C, giving >95% Li extraction; and amblygonite is directly soluble in sulfuric acid.

Mica Conversion

The lithium micas are considerably less refractory than spodumene and require less energy-intensive processes to solubilize the lithium. The lower energy cost is offset by the increased complexity of the process due to the presence of a larger number of impurities. In addition to the impurities in the spodumene process (iron, aluminum, sodium, potassium, cesium, and magnesium), the micas also contain some uncommon impurities such as rubidium, cesium, fluorine, and chlorine, which have to be removed to achieve the required purity.

Several different processes have been proposed for the recovery of lithium from micas; however, only two seem to have been put into large-scale operation. The first mixes ground mica concentrate with concentrated sulfuric acid and heats it to 250°C; the second combines the mica concentrate with potassium sulfate (K_2SO_4) and heats it to 850°C. The result from both routes is much the same: conversion of the lithium into a water-soluble sulfate. However, many impurities also form soluble sulfates, giving high levels of impurities in solution after leaching.

When the mica is roasted with K_2SO_4, there is an additional step to remove K_2SO_4 from solution. This is carried out after the step to precipitate calcium and magnesium using carbonate. The solution is heated and evaporated to the point where it is almost saturated with Li_2SO_4 (35–40 g/L Li), by which time a large proportion of the K_2SO_4 will have crystallized out and can be separated for recycle to the sulfation kiln. A general flow sheet for the processing of mica using K_2SO_4 roasting is shown in Figure 11.

It is also possible to leach lepidolite directly in sulfuric acid using 50 g/L of free acid at 120°C for 12 hours. The final solution is close to Li_2SO_4 saturation with >95% dissolution. The slurry is filtered and the largely amorphous silica residue is either disposed of or re-leached in NaOH to form sodium silicate as a by-product.

The solution at this stage is processed largely the same as that from sulfuric acid leaching of spodumene. The major difference is the presence of ions of rubidium, cesium, and fluorine.

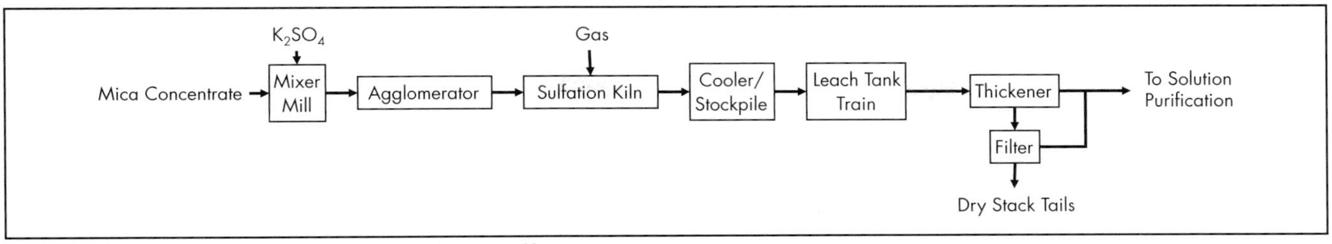

Figure 11 Flow sheet for mica processing using sulfation roasting

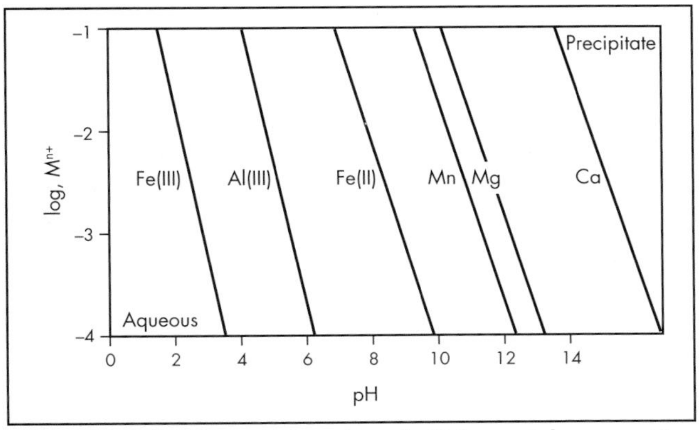

Figure 12 Solubility as a function of pH for metals present in lithium leach solutions

Solution Purification

For all minerals and concentrates, the major ions present in the solution will be highly dependent upon the impurities present in the concentrate. Typically these will include lithium, sodium, potassium, magnesium, calcium, iron, aluminum, and sulfate; lesser amounts of phosphate, fluorine, rubidium, cesium, and SiO_2 may also be present. The solutions from mica leaching will be notably higher in rubidium, cesium, and fluorine than spodumene leach solutions, as these elements are within the mica structure.

Ideally, the lithium ions would be directly and selectively extracted from the other ions. However, lithium is not amenable to standard hydrometallurgical separation processes such as solvent extraction (SX) and ion exchange (IX). Lithium also forms very few highly insoluble salts that are not also equally insoluble for the other metals present, so selective precipitation is not possible. The comparative inertness of the lithium ion is therefore exploited and, instead of selectively removing Li, the other metals are removed.

The principal method is to precipitate the impurities by increasing the solution pH by the addition of bases. Figure 12 shows the solubility of relevant metals as a function of pH. As the pH increases, the solubility of the assorted metals decreases until it reaches the point at which precipitation will occur.

In the purification process for sulfuric acid leaching, the pH increase occurs in three stages. The initial stage, which removes iron and aluminum, uses lime to increase the pH to ~6; this reduces their concentration to <5 ppm. The addition of calcium ions leads to the precipitation of gypsum, which can cause other problems, such as scaling within the circuit and increased tailings volume. Sodium hydroxide can also be used for removing the iron and aluminum, but it adds sodium ions, making the solution predominantly Na_2SO_4. If the plant produces lithium hydroxide (LiOH), that could be used as the alkalizing agent to reduce the levels of impurities, but an ongoing bleed to remove chloride and sulfate would still be required. The precipitate is filtered or centrifuged and washed to reduce lithium losses to around 1%. If some of the iron is present as ferrous ions, hydrogen peroxide can be added to convert it to the less soluble ferric form.

The second step uses sodium carbonate to precipitate magnesium and calcium carbonates, leaving <5 ppm Ca+Mg and a more concentrated Na_2SO_4 solution. This step requires careful process control to prevent substantial losses of lithium as Li_2CO_3. The best control will result in only 1%–2% loss of lithium in this stage. The quantity of lithium in the Mg+Ca carbonate precipitate is insufficient to warrant further treatment to recover the lithium and it is disposed of. If a higher purity of Li_2CO_3 or LiOH is required, the low levels of iron, aluminum, magnesium, and calcium remaining after precipitation can be removed using IX. This is not usually the case for the standard 99.5% Li_2CO_3 battery-grade product. It is also possible to add ethylenediaminetetraacetic acid (EDTA) to complex with the magnesium and calcium ions, preventing them from precipitating during the lithium precipitation. However, the EDTA complexes build up over time, necessitating a bleed that needs to be treated to recover the lithium.

For solutions derived from leaching micas, additional unit operations are required to remove the additional impurities. The rubidium and cesium can be separated by IX, as selectivity for these elements is higher than for potassium, sodium, and lithium—alternatively, it is possible to form insoluble double alums from the leach solution (Jandova et al. 2012). The rubidium is of limited value but the cesium has potential value, if converted into the formate, as a high temperature–high pressure drilling mud in the oil and gas industry. Fluoride is removed using activated alumina, or in higher amounts it will also precipitate as CaF_2 during the initial neutralization of the acid leach solution when using lime or limestone.

The solution at this stage is essentially a mixture of lithium, sodium, and SO_4; the subsequent stages of treatment are dependent upon the required product. A flow sheet for the purification of the leach solution is shown in Figure 13.

Li_2CO_3 Precipitation and Purification

The remaining solution typically contains 35–40 g/L of Li^+, essentially saturated with Li_2SO_4. If the concentration is below 30 g/L, the lithium precipitation efficiency is reduced and the water is generally evaporated to increase the lithium concentration. Figure 14 shows the solubility of lithium as Li_2CO_3 as a function of temperature. From this graph two important points are evident. Li_2CO_3 is unusual in that solubility decreases with increasing temperature, and even the most

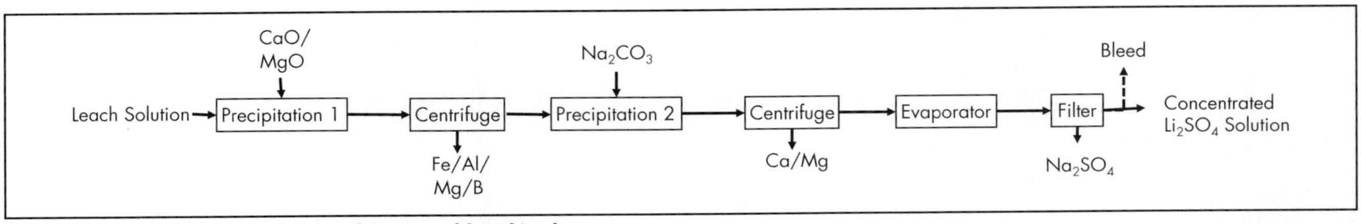

Figure 13 Flow sheet for the purification of leach solutions

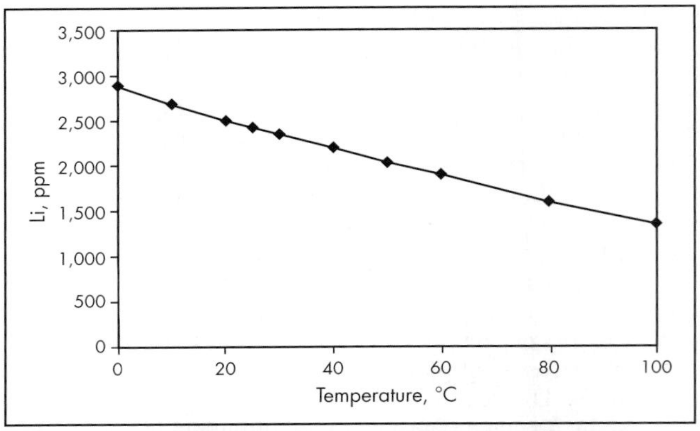

Figure 14 Solubility of lithium in equilibrium with Li_2CO_3 as a function of temperature

efficient precipitation will leave >1.4 g of lithium in solution. The high solubility necessitates recycling of the final solution back into the leaching circuit to minimize lithium losses. In any event, the lithium-bearing solution is heated to 90°C and a solution of Na_2CO_3 mixed in. The precipitation reaction is

$$Li_2SO_4 + Na_2CO_3 \Rightarrow Li_2CO_3 + Na_2SO_4$$

Typically, an excess of around 5% Na_2CO_3 is added to ensure the maximum lithium precipitation.

The Li_2CO_3 product is centrifuged and washed in clean water. At this stage, the major impurities are sodium and SO_4^{2-}, and the overall purity is ~95% Li_2CO_3. To increase the product purity further, a purification stage is undertaken. The impure Li_2CO_3 is slurried in clean water and CO_2 is injected to reduce the solution pH to <8.5, which is sufficient for soluble $LiHCO_3$ to form by the following reaction:

$$Li_2CO_3 + H_2O + CO_2 \Leftrightarrow 2LiHCO_3$$

The reaction is then reversed by heating the solution to precipitate 99.5% Li_2CO_3 with the more soluble impurities remaining in solution. If even higher-purity products are required, the carbonation–decarbonation cycle can be repeated several times. The final solution remains saturated with Li_2CO_3, which can lead to scaling problems and must be recycled within the process to prevent loss of lithium.

The Li_2CO_3 slurry is passed through a high-intensity magnetic separator to remove any magnetic materials present, the majority of which are from the equipment rather than the chemical process. The slurry is filtered, commonly using candle filters, and dried in a rotary kiln prior to micronization and/or bagging to customer requirements. If the product is to be reduced in particle size, a ceramic stirred mill with ceramic media is used to minimize contamination by mill material. Figure 15 shows the flow sheet for the production of lithium carbonate.

LiOH Precipitation and Purification

Direct production of LiOH from the purified solution is based on the following reaction:

$$Li_2SO_4 + 2NaOH \Rightarrow 2LiOH + Na_2SO_4$$

For this process, the solution can be lower in lithium than for carbonate production, with 21–26 g/L Li being typical. A 1.5–1.7-fold excess of NaOH is added to increase the solution pH, which promotes the conversion. The resultant solution is then cooled to –5° to –3°C for around 12 hours, during which time crystals of Glauber's salt, $Na_2SO_4 \cdot 10H_2O$ form and are removed by centrifugation. The crystals are washed and the solution recycled. The salt crystals formed need to be either sold or disposed of. If sold, they are typically converted to anhydrous Na_2SO_4 via additional crystallization steps, centrifuging, and drying.

The remaining solution is then heated to evaporate water until $LiOH \cdot H_2O$ (monohydrate) crystallizes, the evaporation continuing until a 50–60 vol % slurry of $LiOH \cdot H_2O$ is formed. The crystals are centrifuged and washed to remove impurities. The $LiOH \cdot H_2O$ is redissolved at 90°C to give a solution of around 33 g/L Li. The sulfate concentration in this solution should not exceed 15 g/L to prevent it recrystallizing during the next evaporation stage. The evaporation/crystallization of $LiOH \cdot H_2O$ is repeated, with the crystals washed—this time, there is very little sulfate crystallized with the $LiOH \cdot H_2O$ and the product is 99.5% $LiOH \cdot H_2O$. Further stages of dissolution/crystallization can be used to increase purity as required.

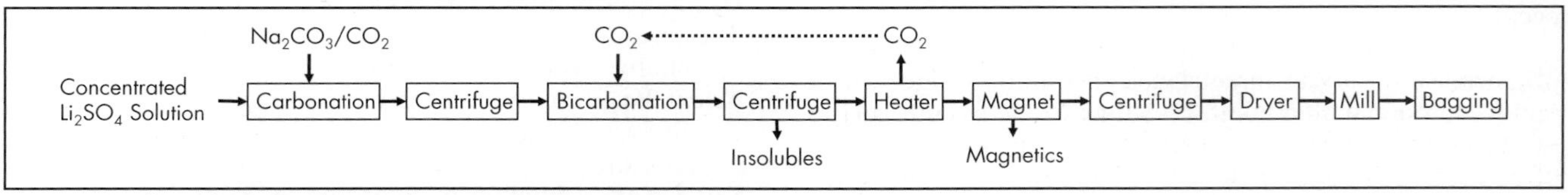

Figure 15 Flow sheet for the production of lithium carbonate

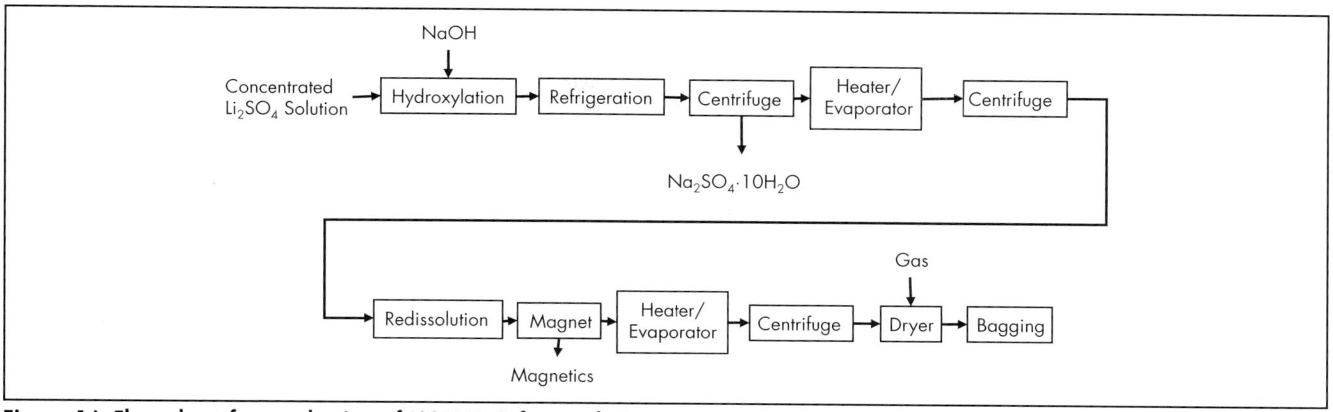

Figure 16 Flow sheet for production of $LiOH \cdot H_2O$ from solution

The final product is dried to remove water and bagged. It is possible to dehydrate the monohydrate to LiOH by heating under vacuum at 180°C. However, the anhydrous hydroxide is extremely hygroscopic and the rehydration generates significant heat, presenting significant difficulties with handling and transporting the anhydrous material. In general, the anhydrous LiOH is made on demand. Figure 16 shows the flow sheet for production of $LiOH \cdot H_2O$ from purified solution.

Li_2CO_3 Conversion to LiOH

It is also possible to convert the carbonate to hydroxide. This used to be the major process route due to the lower demand for hydroxide than carbonate. However, growth in hydroxide demand is expected to be higher than carbonate, so direct production of hydroxide by the preceding route is increasingly attractive.

Carbonate conversion is based on the reaction between lime and lithium carbonate, with the insolubility of the $CaCO_3$ being the main chemical driver:

$$Ca(OH)_2 + Li_2CO_3 = 2LiOH + CaCO_3$$

The starting Li_2CO_3 slurry is made up to around 210 g/L and heated to just below its boiling point; a 5% excess of $Ca(OH)_2$ is added. Higher concentrations of carbonate or higher excesses of $Ca(OH)_2$ lead to $CaCO_3$ in the final product. The slurry is stirred and the reaction occurs with Li_2CO_3 dissolving and $CaCO_3$ precipitating, the final solution containing 48 g/L LiOH. The slurry is filtered to separate the LIOH solution from the $CaCO_3$, and the $CaCO_3$ is washed to recover the lithium and either disposed of or calcined back to CaO for recycle. The LiOH solution is evaporated to crystallize the $LiOH \cdot H_2O$, which is recovered by centrifugation. The impure monohydrate is redissolved in boiling fresh water and impurities removed by adding Na_2CO_3, CaO, and activated carbon. After filtration to remove the solids, further evaporation results in crystallization of $LiOH \cdot H_2O$, which is centrifuged, washed, dried, and bagged.

Conversion of carbonate to hydroxide is a tried and tested process, having been in use for more than 40 years. The hydroxide product is of high quality, mainly because the starting carbonate has already undergone some purification. However, the process is considerably more complex than direct production of hydroxide since both carbonate and hydroxide plants are required.

Residual Li_2CO_3 in Solutions

The constant washing with clean water results in solutions containing an appreciable quantity of lithium as a 1,500-ppm Li solution. In plants there are efforts made to recover this lithium because it may represent a loss of a few percent of total lithium production. The most common method is to recycle the wash waters back through the plant countercurrent to the lithium product and into the water leach step. Ultimately, the constant recycle results in the buildup of impurities above the level where they contaminate the product and a bleed is essential. The bleed can be treated to remove the impurities and the lithium-bearing solution recycled or disposed of; recycling leads to problems with the water balance. There is potential to recover the lithium by precipitation as phosphate, which has a solubility of 0.39 g/L (i.e., 70 ppm Li); however, to achieve this requires the removal of calcium and magnesium, as these also form insoluble phosphates. The phosphate produced can be either dissolved in acid and recycled or, if there is sufficient quantity, sold as an additive for the manufacture of low-expansion enamel glazes (FMC Corporation 2007).

Product Specifications

There are no current international specifications for the purity of lithium salts. A Chinese standard for battery-grade Li_2CO_3 (YS/T582-2013) is the only attempt at codifying the situation. Table 7 shows purity data for a range of different lithium carbonate products from various resources and suppliers. The sample from Olaroz is notably higher in purity than the others, which is indicative of having been through a stage of bicarbonation. The samples from Gangfeng and Tianqi are marketed as being within the specifications of the Chinese standard.

The data in Table 7 show that there is a fairly high variability between products by nominally the same routes. In the absence of an international standard, the acceptable impurity requirements are more likely to be defined by the customer and their requirements. Since the majority of the products are going into battery manufacture, the allowable levels of different impurities will vary between manufacturers because of the different chemistries in use. The rapid development of new battery anodes, cathodes, and electrolytes is likely to see a

Table 7 Reported specifications of battery-grade lithium carbonates

Property	SQM, Chile	FMC Corporation, USA	Orocobre, Australia	Albemarle, USA	Gangfeng, China*	Tianqi, China*	Targray, Canada†
Li_2CO_3, wt %	NS‡	99.5	99.5	99.8	99.57	99.5	99.5
Na, ppm	600	500	100	650	200	250	250
K, ppm	50	NS	10	10	15	10	10
Ca, ppm	100	400	100	160	31	50	50
Zn, ppm	10	5	5	NS	NS	3	10
Mg, ppm	100	NS	60	70	39	80	100
Fe, ppm	10	5	5	10	2	10	20
Ni, ppm	10	6	5	NS	2.9	10	30
Al, ppm	NS	NS	5	NS	22	10	50
Mn, ppm	NS	NS	10	NS	0.9	3	10
Pb, ppm	10	NS	5	NS	1.3	3	10
Cu, ppm	10	5	5	NS	1.2	3	10
Cr, ppm	10	NS	5	NS	NS	NS	NS
B, ppm	NS	NS	10	NS	NS	NS	NS
Si, ppm	NS	NS	10	NS	1	30	50
Cl, ppm	100	100	50	150	17	30	50
SO_4, ppm	300	1,000	300	500	660	800	800
F, ppm	NS	NS	NS	NS	NS	NS	NS
Moisture, wt %	0.2	0.5	0.2	0.35	0.2	0.25	0.4

Adapted from Harman 2018

*The Gangfeng and Tianqi samples are produced from spodumene; the remainder from brines.

†Data from Targray 2011.

‡NS = Not stated.

Table 8 Composition of battery-grade $LiOH{\cdot}H_2O$ (manufacturers' specifications)

Component	FMC Corporation, USA	Leverton, UK	Tianqi, China
LiOH, wt %	56.5	56.5	99.5 $LiOH{\cdot}H_2O$
CO_2, ppm	3,500	2,000	3,000
Cl, ppm	20	30	20
SO_4, ppm	100	300	320
Ca, ppm	15	100	20
Fe, ppm	5	5	7
Na, ppm	20	150	20
Al, ppm	10	NS*	BDL†
Cr, ppm	5	5	NS
Cu, ppm	5	5	BDL
K, ppm	10	100	NS
Ni, ppm	10	NS	BDL
Si, ppm	30	NS	BDL
Zn, ppm	10	5	NS
Heavy metals as Pb, ppm	10	NS	BDL
Acid insolubles, ppm	100	50	20

*NS = Not stated.

†BDL = Below detection limit.

new level of purity being required into the long term to maximize the battery performance.

Table 8 shows the composition of three battery-grade LiOH products. As with the carbonates, there is a wide range of impurity levels within the specifications and different impurities being specified. There is no current standard for this product.

REFERENCES

Aiken, S.R., Gowans, R., Hawton, K.E., Jacobs, C., Pilcher, B., Spooner, J., and Trueman, D.L. 2016. *NI 43-101 Technical Report on the Preliminary Economic Assessment of Lithium Hydroxide Production, Separation Rapids Lithium Project, Kenora, Ontario.* Toronto, ON: Micon International.

Bulatovic, S.M. 2014. *Handbook of Flotation Reagents: Chemistry, Theory and Practice. Vol. 3: Flotation of Industrial Minerals.* Amsterdam: Elsevier.

Choi, J., Hong, J., Park, K., Han, Y., Kim, H., Kim, W., and Kim, S.B. 2015. Lepidolite flotation from low grade ores using a cationic surfactant. SME Preprint No. 15-002. Englewood, CO: SME.

Desormeaux, D. 2015. What to expect from the Chilean lithium industry following the National Commission's advancement plan? Presented at the 7th Lithium Supply and Markets Conference, Shanghai, China, June 16–18.

Epstein, J.A., Feist, E.M., Zmora, J., and Marcus, Y. 1981. Extraction of lithium from the Dead Sea. *Hydrometallurgy* 6(3-4):269–275.

FMC Corporation. 2007. Lithium phosphate data sheet. www.fmclithium.com/Portals/FMCLithium/content/docs/DataSheet/QS-PDS-1038%20r0.pdf. Accessed June 2018.

Garrett, D.G. 2004. *Handbook of Lithium and Natural Calcium Chloride.* Amsterdam: Elsevier Academic Press.

Hains, D.H. 2012. CIM best practice guidelines for resource and reserve estimation for lithium brines. https://mrmr.cim.org/media/1041/best-practice-guidelines-for-reporting-of-lithium-brine-resources-and-reserves.pdf. Accessed June 2018.

Harman, G. 2018. What is battery grade lithium carbonate? Presented at the ALTA Conference, Perth, Australia, May 24–25.

Jandová, J., Dvořák, P., Formánek, J., and Vu, H.N. 2012. Recovery of rubidium and potassium alums from lithium-bearing minerals. *Hydrometallurgy* 119-120:73–76.

Jephcott, B. 2016. *Lithium Industry Analysis 2016.* Beijing: Golden Dragon Capital. www.goldendragoncapital.com/wp-content/uploads/2016/07/Lithium-Industry-Analysis-2016.pdf. Accessed May 2018.

Jessup, T., and Bulatovic, S. 2000. U.S. Patent 6,138,835. Recovery of petalite from ores containing feldspar minerals.

Lipp, J. 2014. LiSX™: A breakthrough in lithium production. Presented at ISEC2014—International Solvent Extraction Conference, Würzburg, Germany, September 7–11.

Partington, G.A., Mcnaughton, N., and Williams, I.S. 1995. Mineralization history of the Li-Sn-Ta Greenbushes pegmatite and granitoids of the southwestern Yilgarn Block, Western Australia. *Econ. Geol. Bull. Soc. Econ. Geol.* 90(3):616–635.

SignumBox. 2014. *Lithium, Batteries and Vehicles—Perspectives and Trends*, 8th ed. Santiago, Chile: SignumBox.

SQM. 2017. SQM Investor Day 2017 [presentation]. Chile: SQM. http://s1.q4cdn.com/793210788/files/doc_news/2017/SQM-Investor-Day-2017_FINAL_Sept6.pdf. Accessed June 2018.

SQM. 2018. Salar brines: Salmueras. Chile: SQM. www.sqm.com/en-us/acercadesqm/recursosnaturales/salmuera.aspx. Accessed June 2018.

Targray. 2011. Product data sheet: Lithium carbonate. https://www.targray.com/content-data/mediafiles/images/documents/pds-lithium-carbonate.pdf. Accessed July 2018.

USGS (U.S. Geological Survey). 2018. Lithium: Statistics and information. In *Mineral Commodity Summaries.* Reston, VA: USGS. pp. 98–99.

Vergara-Edwards, L., and Parada-Frederick, N. 1983. Study of the phase chemistry of the Salar de Atacama brines. In *Proceedings of the Sixth International Symposium on Salt.* Vol. II. Alexandria, VA: Salt Institute. pp. 345–366.

Webmineral. 2018. Mineral species containing lithium (Li). http://webmineral.com/chem/Chem-Li.shtml. Accessed June 2018.

YS/T582-2013. *Nonferrous Metals Industry Standard of the People's Republic of China: Battery Grade Lithium Carbonate.* Beijing: Ministry of Industry and Information Technology of the People's Republic of China.

CHAPTER 12.21

Magnesium Minerals and Metal

David C. Lynch

MAGNESIUM COMPOUNDS OF COMMERCIAL IMPORTANCE

The abundance of magnesium on land, in the seas, and in natural brines leads to its presence in plants and animals, but because of its reactivity, it is not found in elemental form in nature. Magnesium is the eighth most abundant element in the earth's crust and in seawater has a concentration of 1,300 ppmw (parts per million by weight) as Mg^{2+}. The concentration of magnesium in natural brines varies with climate. In the South Arm of the Great Salt Lake, near where US Magnesium processes lake water, the concentration of magnesium has ranged from a low of 0.18 wt % in 1986 when the lake volume was peaking, to approximately 1 wt % in 1963 when the lake was at its historic low (Tripp 2009), which is a water level the lake approached in 2017 (Larsen 2017). The nominal magnesium concentration in the Great Salt Lake is about 0.43 wt % (Tripp 2009), or 3.3 times that of seawater.

Magnesium is found in more than 60 minerals of which 6 are of commercial importance. Those minerals are identified in Table 1. These minerals are found worldwide in large deposits. Magnesium-bearing natural brines represent a resource in billions of metric tons, and there is a virtually unlimited source of magnesium in seawater (USGS 1995–2015).

Magnesite, Magnesia, and Periclase

Worldwide, magnesite is the primary magnesium compound mined, sold, and commercially processed (USGS 1992–2015). Magnesite is obtained from natural and synthetic sources (seawater, natural brines, or deep sea salt beds). In 1992, 75% of world magnesite production took place in six countries (in decreasing level of production): North Korea, China, Czechoslovakia, Turkey, Russia, and Austria. In 2015, more than 82% of production came from China, Turkey, and Russia. The growth of Chinese production of magnesite is presented in Figure 1, which shows that Chinese production grew from 15% of world output in 1992 to more than 69% in 2015. A similar dominance exists in magnesium metal production (covered later in the "Production and Commercial Uses of Magnesium Metal" section).

Table 1 Magnesium minerals of commercial importance

Name	Composition	Mg, wt %
Magnesite	$MgCO_3$	28.8
Brucite	$Mg(OH)_2$	41.7
Dolomite	$CaMg(CO_3)_2$	13.2
Carnallite	$KCl{\cdot}MgCl_2{\cdot}6H_2O$	8.7
Talc	$Mg_3Si_4O_{10}(OH)_2$	19.2
Olivine	$(Mg,Fe)_2SiO_4$	0–34.6

Magnesite is calcined:

$$MgCO_3 \rightarrow MgO + CO_2(g),\ \Delta H_{298K} = 100.9\ \text{kJ} \qquad \textbf{(EQ 1)}$$

at temperatures between 700°–1,500°C (973 K and 1,773 K) in rotary kilns fired with oil or gas (Gill 1980), removing 96%–98% of the carbon as carbon dioxide (CO_2) and producing magnesium oxide (MgO). The oxide product is referred to as *calcined magnesia*, whereas naturally occurring MgO is called *periclase*, except as noted in the following discussion.

Calcined magnesia is classified by the calcine temperature, which impacts its density and reactivity. There are four general classes (Martin Marietta Magnesia Specialties 2018; Grecian Magnesite, n.d.):

1. **Light-burned.** A reactive magnesia is produced at calcine temperatures below 1,000°C (1,273 K) and is referred to as *caustic-calcined magnesia* or CCM. CCM is used in the chemical industry, agricultural supplements, environmental applications, and some cosmetics. The chemical purity of CCM varies from low (90%) MgO content for agricultural uses to the highest grades (~95%) used in the pharmaceutical and food industries.
2. **Hard-burned.** Hard-burned magnesia has low reactivity, having been calcined at temperatures between 1,000° and 1,500°C (1,273–1,773 K). Hard-burned magnesia is typically used in applications where slow degradation of chemical reactivity is required, such as refractories for the steel, cement, and glass industries.

David C. Lynch, Professor Emeritus, University of Arizona, Tucson, Arizona, USA

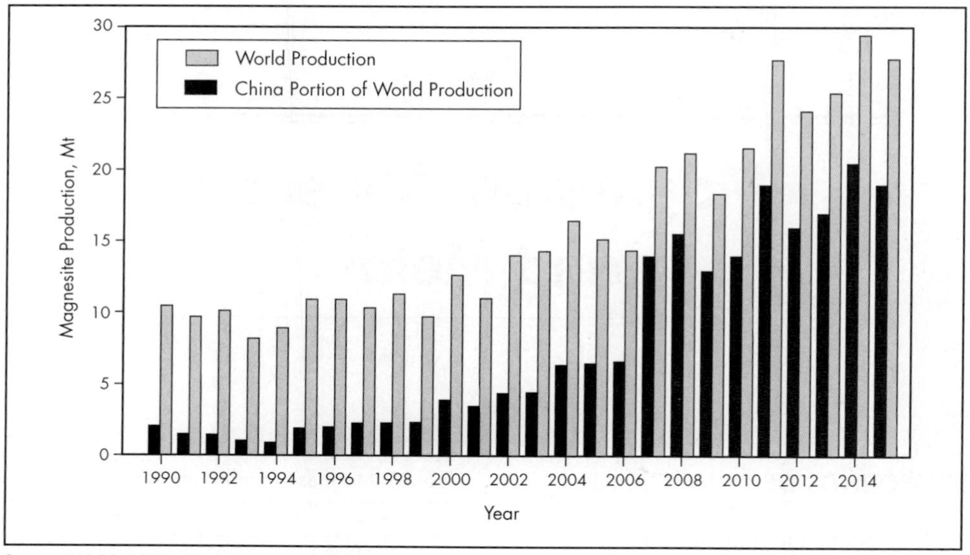

Source: USGS 1992–2015

Figure 1 Annual magnesite production

3. **Dead-burned.** Dead-burned magnesia is formed at calcine temperatures of 1,500°–2,000°C (1,773–2,273 K), resulting in a high-density and low-porosity inert magnesia with very little reactivity. The high temperatures employed produce a near-carbon-free product. Dead-burned magnesia's 50–100-µm crystals have exceptional dimensional stability and strength at high temperatures. These properties make dead-burned magnesia well suited in the manufacture of magnesia-carbon refractories, monolithic gunning refractories, and specialty precast and castable shaped ceramics. Dead-burned magnesia is also referred to as periclase.
4. **Fused magnesia.** High-grade magnesite or calcined magnesia is fused in an arc furnace at temperatures above 2,750°C (3,023 K) over a period of 12 hours with electricity consumption between 3,500 and 4,500 kW·h/t. Fused magnesia crystals greater than 1,000 µm are produced and have near theoretical density of 3.58 g/cm^3.

   Fused magnesia has superior strength, abrasion resistance, and chemical stability and as such finds uses in refractory and electrical insulating markets.

Brucite

Magnesium hydroxide is found in nature as the mineral brucite. It is also precipitated in large quantities from seawater by the addition of calcium hydroxide. Magnesium hydroxide is used as a fire retardant additive, antacid, and laxative. Prior to China's dominance in magnesium metal production (an issue discussed in the "Production of Magnesium Metal" section), magnesium hydroxide was the primary raw material used in the production of magnesium metal.

Dolomite and Dolostone

The term *dolomite* is used in this chapter to designate the near stoichiometric mineral $CaMg(CO_3)_2$, whereas dolostone (or dolomite rock) is a sedimentary rock containing a high concentration of the mineral, usually found with calcite ($CaCO_3$), also known as *limestone*. That association is a result of the process by which dolomite is formed. It is formed by the post-depositional alteration of lime mud and limestone by magnesium-rich groundwater, a process referred to as *dolomization*. Metamorphic transformation of dolostone produces dolomitic marble (King, n.d.).

Dolostone is readily available and has many uses in construction. It is crushed, washed, and sized for use as a base material for roads and railroads, bearing the load and providing drainage that also suppresses vegetation growth. It is used as decorative stone and as aggregate in concrete and asphalt. The larger sized fraction of the crushed stone is used as riprap. Dolostone is also used as a cover barrier to protect underwater tunnels and other structures from accidental anchor strikes and sinking debris. Dolostone is quarried, cut, trimmed, ground, and drilled to desired size and shape for use in construction (King, n.d.; USGS 1995–2015).

Both dolomite and dolostone are used as acid buffers for soil, streams, chemical processes, and feed for livestock. They are used to form basic oxides that decrease the viscosity of slags used in iron and steel production. They are also used as an ingredient in the production of glass, bricks, and ceramics (King, n.d.; USGS 1995–2015).

Dolomite has had a checkered history as a raw material for production of magnesium metal, a subject covered in the "Production of Magnesium Metal" section. Currently, dolomite is the primary ore for production of magnesium metal.

Carnallite

Carnallite ($KCl{\cdot}MgCl_2{\cdot}6H_2O$) is primarily used in producing fertilizers. It is an important source of potash. It is also a source of magnesium chloride and is used in Israel and Russia to produce magnesium metal (Neelameggham 2013). Solution mining is used in recovering carnallite. The process consists of drilling holes into an ore body and pumping heated water into the deposit. The heated water dissolves carnallite and other soluble salts, allowing the dissolved minerals to be pumped out. A portion of the water is removed from the solution through evaporation, leading to precipitation of potassium chloride and many of the other soluble salts. The one exception is magnesium chloride that remains in solution with

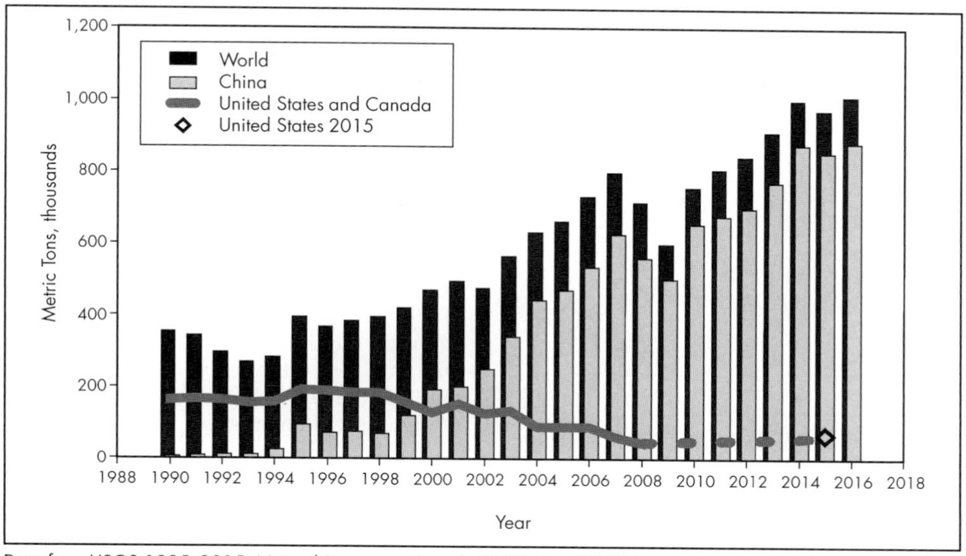

Data from USGS 1995–2015; Natural Resources Canada 2008; Roskill 2016

Figure 2 Primary production of magnesium

the remaining water and can be separated from the precipitates. Evaporation is used to further concentrate the magnesium chloride, leading to its precipitation (Padhy 2017).

Talc

Talc is a clay mineral composed of hydrated magnesium silicate, $Mg_3Si_4O_{10}(OH)_2$. It has a platelet structure with plates held together by weak van der Waals bonds. These two factors contribute to talc's use in plastics, paint, cosmetics, baby powder, and as a high-temtperature lubricant. It is used as filler in paper production and ceramics (whiteware, tile, pottery, and dinnerware). It is also used as a stiffening agent in plastics and rubber. Talc is added to asphaltic roofing material to improve resistance to the sun's ultraviolet radiation (King, n.d.).

Olivine

The International Mineralogical Association does not classify olivine as a mineral species, but as a mineral group with forsterite (Mg_2SiO_4) and fayalite (Fe_2SiO_4) as end members. Fayalite and forsterite create a solid solution, with olivine generally identified as having more forsterite (>69 wt %) than fayalite. "Pure forsterite is uncommon, and pure fayalite is very rare" (King, n.d.).

Norway is the principal producer of olivine. It is mined as a gemstone, and non-gem-quality olivine is used as casting sand in aluminum foundries as it requires less water in holding the mold shape. Olivine is used as a tap plug in iron blast furnace operations, and it substitutes for dolomite in slags used in producing iron and steel.

More recently, olivine is being examined as a low-cost means for carbon dioxide sequestration (Ravilious 2016).

PRODUCTION AND COMMERCIAL USES OF MAGNESIUM METAL

Numbers for world production of primary magnesium are presented in Figure 2. The graph reveals that prior to 1999, the United States and Canada accounted for approximately 50% of world production. From 1999 to 2016, the Chinese share of the world's magnesium market has soared from 28% to 87%, while Canadian and U.S. production declined to 6% in 2008, which was the last year of Canadian production. There has been only one plant producing magnesium in the United States since 2003, operated by the Magnesium Corporation of America. The U.S. Geological Survey (USGS) stopped publishing U.S. production statistics in 1998 for proprietary reasons; however, the Canadians provided numbers for both countries from 2001 through 2007. Primary magnesium production in the United States for 2015 was reported to be 59,000 t (metric tons), which represents about 5.8% of world production (Roskill 2016).

Uses of magnesium metal vary depending on demand and cost. A valuable window on its uses, although limited to U.S. consumption, is provided by USGS for both primary and secondary magnesium. Consumption and use data over a period of 20 years at 5-year intervals is presented in Table 2. In the revised numbers for 2011, the USGS 2012 *Minerals Yearbook* offers the first report of the quantum increase in the use of magnesium as a reducing agent for titanium (USGS 1995–2015). Those who report numbers for magnesium production and use often do not include the magnesium used in the Kroll process for production of titanium, as it involves a production–recycle agreement (Evans 2007).

Table 2 U.S. consumption and uses of primary magnesium

Uses	1995	2000	2005	2010	2015*
Consumption, t	109,000	104,000	82,100	55,700	65,900 (42,900)
Structural products, %†	22	30	45	40	17 (25)
Aluminum alloys, %	55	53	37	43	33 (50)
Iron and steel desulfurization, %	12	12	9	11	11 (17)
Reducing agent, %‡	2	1	1	2	34

* Numbers in parentheses reflect values without the contribution from use as a reducing agent in titanium production.

† Casting and wrought products.

‡ Reducing agent for titanium (2015), zirconium, hafnium, uranium, and beryllium.

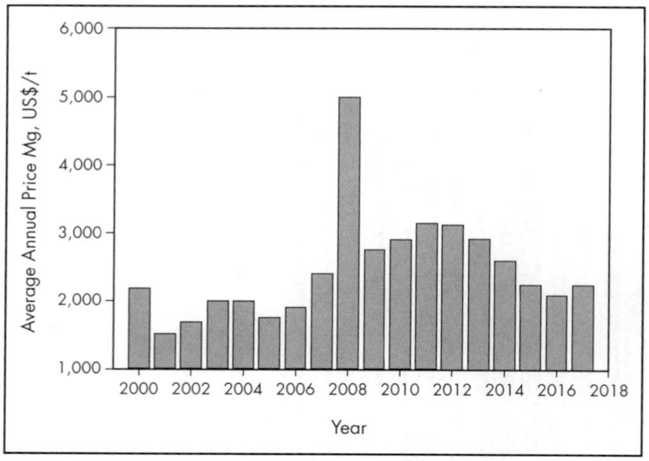

Figure 3 Yearly price history for magnesium metal

In the Kroll process, magnesium is used to reduce titanium tetrachloride, with the resulting magnesium chloride returned as raw material for production of magnesium metal by electrolysis. The numbers in parentheses in the last column of Table 2 are thus a better reflection of the trends observed in the rest of the table, while the other numbers in the last column reflect the magnitude of the effect of the Kroll process on the magnesium metal industry.

Secondary magnesium is recovered from both magnesium- and aluminum-base alloys and used to produce magnesium ingots, magnesium alloy castings, and aluminum alloys. In the United States, recycle and recovery of magnesium has grown and exceeded primary production in 1999. In the United States, secondary magnesium production was 39% of primary production in 1990, 46% in 1995, 198% in 2000, 182% in 2005, 144% (estimated) in 2010, and 136% in 2015. The remelting and casting of secondary magnesium requires between 3.6 and 6 MJ/kg (megajoules per kilogram), whereas production of primary magnesium with the Pidgeon process requires 144 MJ/kg (Johnson and Sullivan 2014).

Magnesium powders have been formed into structural near net shapes by thixocasting and variations of that technique, producing cast parts with 80%–120% greater strength than similar parts produced with conventional casting techniques. In thixocasting a precast billet with a non-dendritic microstructure, produced by intense stirring of the melt as the metal is being cast, is heated to a semisolid state and then injected into hardened steel molds (Neelameggham 2013).

The yearly average price for magnesium metal is presented in Figure 3. The climb in market price from 2005 to 2008 spawned new and revised interest in idled and mothballed plants, and new processes. Those interests faded with the decline in the world's economy at the end of 2008. Currently, there is optimism that the five-year slide in the market price from 2012 to 2016 found a floor at roughly $2,000 per metric ton. By December 2017, the market price had risen 12% over the previous year.

The price for magnesium metal has always been a double-edged sword: use and cost. The low density of magnesium (1.74 g/cm^3) is prized for use in the transportation industry, where energy efficiency is mandated by governments.

Aluminum, with a density of 1.55 times that of magnesium, substitutes for that metal in castings and wrought products and has a price advantage (USGS 2017). In 2008, the prices of magnesium and aluminum peaked at more than $5,000 and $3,000/t, respectively. At the end of 2017, the difference in price between magnesium and aluminum was approximately $250/t.

Zinc also substitutes for magnesium in casings, but not on a weight-reduction basis with a density of 7.14 g/cm^3. Zinc usually has a price advantage, but at the end of 2017 was selling for about $1,000 more per metric ton.

Mineral Processing of Magnesium Ores

There are six commercial magnesium-containing minerals, some in multiple forms. These minerals have multiple uses, and thus the initial treatment of the ore may vary depending on final use. In general, and prior to any thermal processing, lump ore will be crushed, washed, and sized. Some ores will be cleaned using gravitational means, electrostatic separation, magnetic filters, and flotation. Wet or dry attrition scrubbing can be used to remove sharp edges and corners, as well as to remove additional surface material.

Magnesium-containing brines, seawater, and mined carnallite have their water content reduced by evaporation. Brines and seawater also undergo neutralization. Further processing prior to electrolysis is covered in the next section.

Production of Magnesium Metal

Neelameggham (2013) provides a summary of 19th-century efforts in producing magnesium metal. Those efforts involved different electrolytic processes. Late in that century, 1886, the Aluminium und Magnesium Fabrik of Germany electrolyzed molten dehydrated carnallite ($KCl{\cdot}MgCl_2$) in the production of magnesium. Chemische Fabrik Griesheim-Elektron (CFGE) further advanced this technology in 1896, and in the 20th century, CFGE became I.G. Farbenindustrie. Electrolysis of molten carnallite continues in this century (Neelameggham 2013). Prior to World War II, production of magnesium metal was by three primary means (Haughton and Prytherch 1938; Beck 1939; Schambra 1945; Ball 1956; Emeley 1966; Neelameggham 2013):

1. Electrolysis of fused chlorides
2. Electrolysis of magnesium oxide in solution with molten fluorides
3. Carbothermic production of magnesium vapor by direct reduction of magnesium oxide using carbon in an electric arc furnace with a hydrogen atmosphere and subsequent rapid cooling to reduce the extent of reoxidation by carbon monoxide

Silicothermic production of magnesium played an important role in producing magnesium for the Allies during World War II. Its use of thermal energy freed electrical energy for use in other war industries. Today we know this method of producing magnesium as the *Pidgeon process*, named after Lloyd Montgomery Pidgeon of the Canadian National Research Council. However, silicothermic production of magnesium was first developed by the researcher Amati at the University of Padua in Italy. Amati published in his thesis in 1938, a document archived at the university. Immediately afterward, industrial production was established in Bolzano, Italy. The process used externally heated retorts identical in concept to those used by Pidgeon two years later.

"Unlike aluminum, the demand for magnesium took a precipitous drop following World War II, causing" a decline in the variation of process production technology. "Carbothermic and fluoride–melt electrolytic processes exited commercial production. Silico-thermic processes took a backseat, until the mid-1990s, to the electrolytic conversion of magnesium chlorides" (Neelameggham 2013).

A review of magnesium metal production begins with electrolysis of fused chlorides, followed by discussion of the Pidgeon process; the discussion is concluded with review of the magnetherm process. The discussion begins with electrolytic-produced magnesium, not because of its current commercial importance, but as a result of its dominance in the second half of the last century, and because it is still practiced where natural brines are available. The magnetherm process also played a significant role in magnesium production in the second half of the last century, but like electrolytic production, it has precipitously declined with growth of magnesium production by the Pidgeon process in China. Discussion of the magnetherm process follows review of the Pidgeon process, as it is a variant of that process. Reasons for the dramatic growth of magnesium production in China are reviewed in the "Pidgeon Process" section.

Fused Salt Electrolysis

The thermodynamic stability of magnesium compounds precludes recovery of magnesium by aqueous electrolysis, except when the cathode is an amalgam consisting primarily of mercury. Recovery of magnesium in an amalgam leads to recovery of most other metal ions in the aqueous solution. Thus, this approach would require further distillation, which is complicated by a multitude of other metal atoms in the amalgam.

The answer to recovery of magnesium by electrolysis was, as in the case of aluminum production, to turn to halides. Identifying a suitable molten salt for production of molten magnesium by electrolysis begins with the normal melting and boiling point temperatures of the element, which are 649°C (922 K) and 1,090°C (1,363 K), respectively, and identifying a solvent compound of greater thermodynamic stability than the magnesium solute component. The solvents meeting the second restriction are alkali and alkaline earth halides.

The molten salt must exist at temperatures well below the normal boiling point of magnesium to avoid yield reduction through magnesium vapor loss. The operational temperature needs to be above the normal melting temperature of magnesium to prevent magnesium crystal growth on the cathode. The solubility of magnesium oxide in calcium chloride is insignificant, whereas it is highly soluble in calcium fluoride but at temperatures greater than 1,350°C (1,623 K). Magnesium chloride is both soluble in calcium chloride and at temperatures as low as 620°C (893 K), as shown in the phase diagram in Figure 4. The typical fused salt contains, in addition to both magnesium chloride ($MgCl_2$) and calcium chloride ($CaCl_2$), small concentrations of sodium chloride (NaCl), potassium chloride (KCl), and lithium chloride (LiCl).

A flow diagram for electrolytic production of magnesium metal is presented in Figure 5. The diagram includes two options depending on the degree to which water is removed from magnesium chloride. The anhydrous route using $MgCl_2$ is representative of the Norsk Hydro process, whereas use of $MgCl_2{\cdot}H_2O$ is typical of the Dow process.

Magnesium chloride is produced from seawater by producing concentrated brines through evaporation of water in holding ponds. Magnesium cations in the brine are converted to magnesium hydroxide, which has a solubility of only

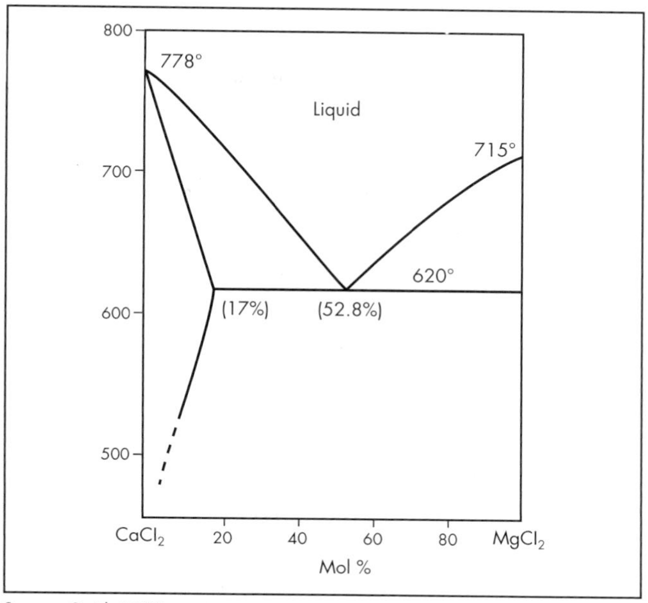

Source: Smith 1983

Figure 4 $CaCl_2$–$MgCl_2$ phase

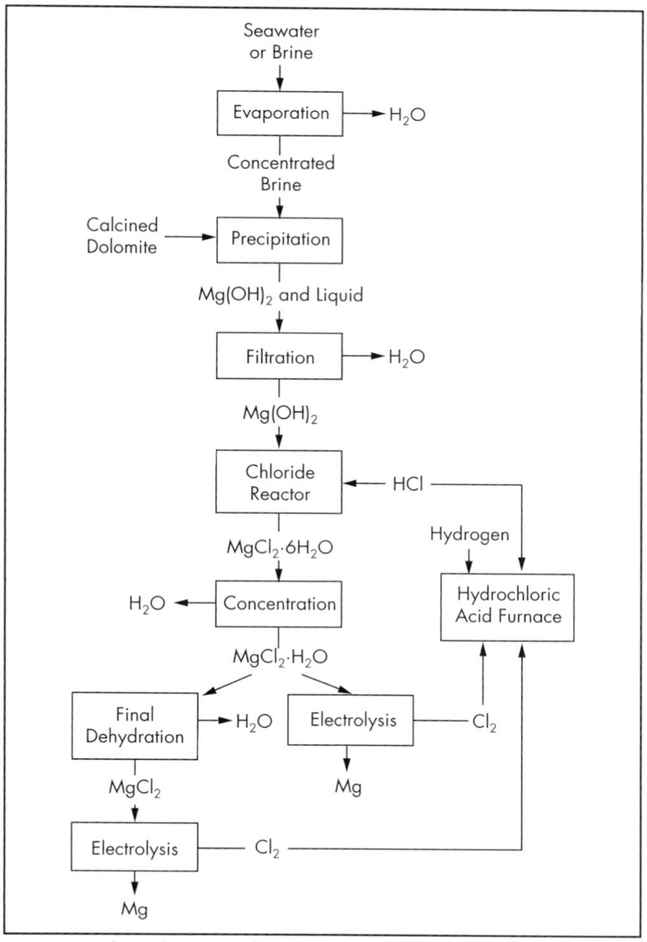

Figure 5 Flow diagram for electrolytic production of magnesium metal

0.00069 g per 100 g of water at 20°C (293 K), by the following reaction:

$$Mg^{2+} + 2OH^- \rightarrow Mg(OH)_2 \quad \textbf{(EQ 2)}$$

The hydroxide ions are supplied by calcium hydroxide, $Ca(OH)_2$, or sodium hydroxide (NaOH) as both are soluble. The magnesium hydroxide is filtered and converted to the chloride by reaction with hydrochloric acid:

$$Mg(OH)_2 + 2HCl \rightarrow MgCl_2 + 2H_2O \quad \textbf{(EQ 3)}$$

The resulting magnesium chloride solution is evaporated and partially dehydrated in the Dow process:

$$\begin{aligned} MgCl_2 \text{ (solution)} &\rightarrow MgCl_2{\cdot}6H_2O \\ &\rightarrow MgCl_2{\cdot}4H_2O \\ &\rightarrow MgCl_2{\cdot}2H_2O \\ &\rightarrow MgCl_2{\cdot}H_2O \end{aligned} \quad \textbf{(EQ 4)}$$

Or complete dehydration is used as in the Norsk Hydro process:

$$MgCl_2{\cdot}H_2O \rightarrow MgCl_2 \quad \textbf{(EQ 5)}$$

It is also possible with natural brines, where extensive solar evaporation ponds are available, to produce "magnesium chloride rich liquor through evaporative crystallization. This involves removal of large quantities of sodium chloride, potassium chloride/sulfate salts by the initial crystallization" (Neelameggham 2013). Near-complete dehydration (Reaction 5) is achieved by spray dryers as practiced by US Magnesium in Utah (United States). A similar process is under development in China at Qarham salt lakes and salt pans.

The resulting magnesium chloride must be purified. The magnesium chloride produced from seawater or natural brines requires removal of sulfates as they decompose in the molten salt, producing oxides that impact removal of other impurities and the electrolytic process.

Boron is a particularly troubling impurity, the source of which is usually seawater. Boron (as borates) in parts per million by weight concentration in the molten salt leads to deposits of borides on the magnesium produced at the cathode, which precludes coalescence of the magnesium droplets, preventing them from joining the pool of molten metal that forms on top of the fused salt (Neelameggham 2013).

Reactions occurring in the electrolytic cell are

$$\text{Cathode: } Mg^{2+}\text{(molten salt)} + 2e^- \rightarrow Mg\text{(liquid)} \quad \textbf{(EQ 6)}$$

$$\text{Anode: } 2Cl^-\text{(molten salt)} \rightarrow Cl_2\text{(gas)} + 2e^- \quad \textbf{(EQ 7)}$$

$$\text{Overall reaction: } MgCl_2 \rightarrow Mg + Cl_2 \quad \textbf{(EQ 8)}$$

The production of Cl_2 gas at the anode and molten Mg metal, which floats on top of the fused salt, at the cathode requires a physical separation between anode and cathode to prevent rechlorination of the magnesium. Considerable effort has been put into improving cell operation; for more in-depth reviews of cell design, readers are directed to Neelameggham (2013) and Evans (2007).

Pidgeon Process

World War II spawned the need for more magnesium production that led to widespread adoption of the Pidgeon process. After the war, economic conditions changed, and most plants using the Pidgeon process closed. Today, and for the first two decades of the 21st century, those conditions that produced a decline in magnesium production by the Pidgeon process have made the process dominant in China. That dominance began around 2000 with many small Chinese producers. Today the magnesium market is dominated by China (Figure 2). According to Abdellatif (2006):

> Chinese dominance may be attributed to several factors, four of which are:
>
> 1. The low cost of raw materials, particularly ferrosilicon
> 2. A cheap and abundant labour force
> 3. Inexpensive coal and energy
> 4. Safety, health, and environmental standards that are lower than those in Europe and the United States

These factors make it possible for the Pidgeon process to succeed in China, in spite of it being a batch and labour-intensive process.

A flow diagram for the Pidgeon process is presented in Figure 6. Dolomite, $CaMg(CO_3)_2$, or a combination of limestone ($CaCO_3$) and magnesite ($MgCO_3$) are mined, crushed, washed, and sized (Step 1 in Figure 6; note that the individual steps are numbered and used in subsequent tables to identify how specific operations contribute to overall energy requirements). The dolomite (or the limestone and magnesite) must be calcined in an operation separate from the reduction of magnesium oxide to remove carbon and oxygen (as CO_2) that otherwise would reoxidize Mg. During calcination, the following reaction occurs:

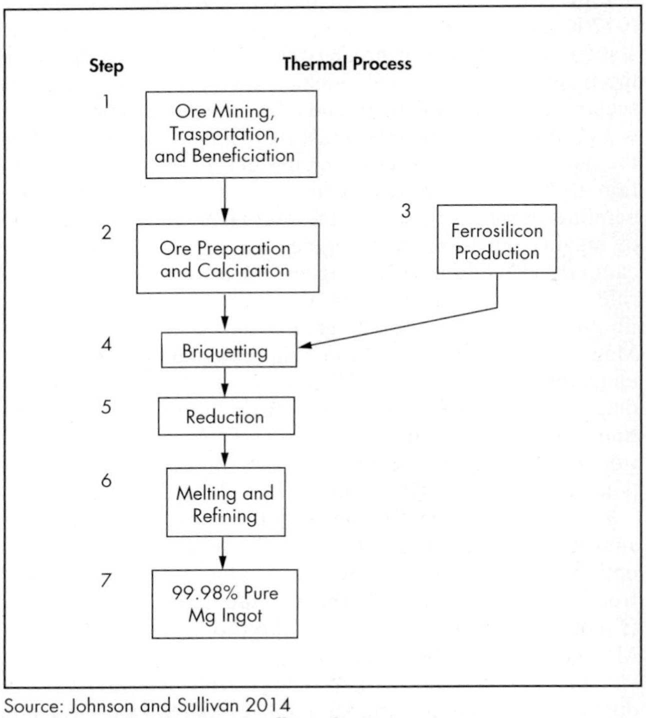

Source: Johnson and Sullivan 2014

Figure 6 Pidgeon process flow diagram

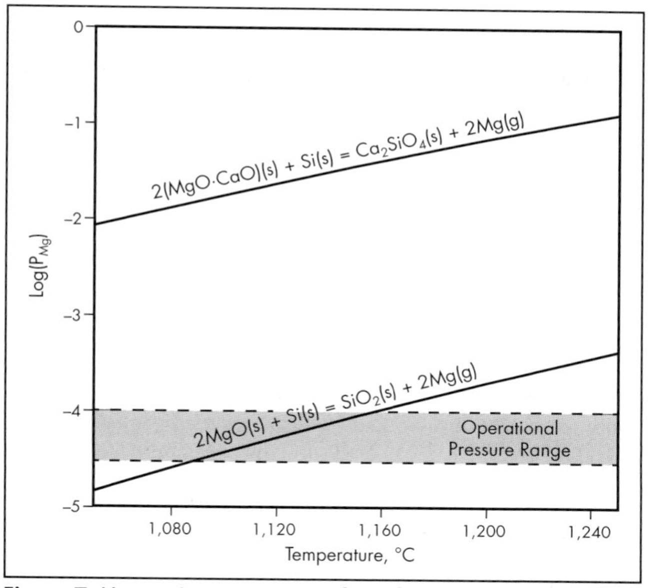

Figure 7 Magnesium pressure in bars for Reactions 10 and 11

$$MgCa(CO_3)_2 = MgO \cdot CaO + 2CO_2(g),\ \Delta H_{900C} = 290\ kJ \qquad \textbf{(EQ 9)}$$

The reaction, carried out at 900°C (1,173 K), is endothermic, and MgO · CaO is a eutectic-type mixture of the two separate oxide phases, often referred to as *dolime* (Step 2).

The Pidgeon process is based on the reduction of magnesium oxide by silicon (Toguri and Pidgeon 1961):

$$2MgO(s) + Si(s) \rightarrow SiO_2(s) + 2Mg(g),\ \Delta H_{1,150C} = 573\ kJ \qquad \textbf{(EQ 10)}$$

"The practical applications of this principal require the reduction of calcined dolomite ore [Step 5] with ferrosilicon at high temperatures. The overall reaction describing the process (Toguri and Pidgeon 1962) is as follows" (Ramakrishnan and Koltun 2004):

$$2(MgO \cdot CaO) + Si(ferrosilicon) = Ca_2SiO_4 + 2Mg(g),\ \Delta H_{1,150C} = 457\ kJ \qquad \textbf{(EQ 11)}$$

Silicon (with a high thermodynamic activity), in combination with calcium oxide (CaO) and in the production of a calcium silicate, serves as a superior reducing agent for MgO. The superiority is presented in Figure 7, where the equilibrium vapor pressure of magnesium by Reaction 11 is compared to that achieved by Reaction 10. The diagram reveals

- That formation of the calcium silicate increases the vapor pressure of the volatilized magnesium over just silica formation;
- It is important to keep a high activity of all the reactants;
- At the temperatures presented, the process must be operated under a vacuum to prevent establishing equilibrium and thus a stop to magnesium vapor production; and
- The vacuum required would need to be substantially greater for Reaction 10 versus Reaction 11.

As a result of the last item, Reaction 10 is no longer considered for further discussion with respect to the Pidgeon process.

Reaction 11 is highly endothermic, requiring a constant heat source. Thus, the reaction rate is subject to heat transfer control. With adequate available heat, the reaction rate initially depends on the degree to which there is physical contact between reactants (Figure 8), and finally limited by solid-state diffusion of silicon as presented in Figure 9. The overall reaction rate is largely limited by solid-state diffusion. Briquetting of fine powders of the calcined dolomite and ferrosilicon is done to improve contact between reactants and to decrease the distance for Si solid-state diffusion.

Retorts used in the Pidgeon process consist of a one-end-closed tube as presented in Figure 10. The open end of the tube has a collar and is sealed with an end plate during production of Mg. The end plate is removed to input the charge, which is pushed all the way to the closed end of the tube. The end plate is then used to seal the retort, and the pressure inside the retort is reduced to approximately 3–10 Pa during Mg production. At the end of production, the end plate is removed and the Mg recovered before the partially spent charge is removed. Retorts are mounted in banks as shown in Figure 11A, with the cooler end of the retort extending outside the furnace.

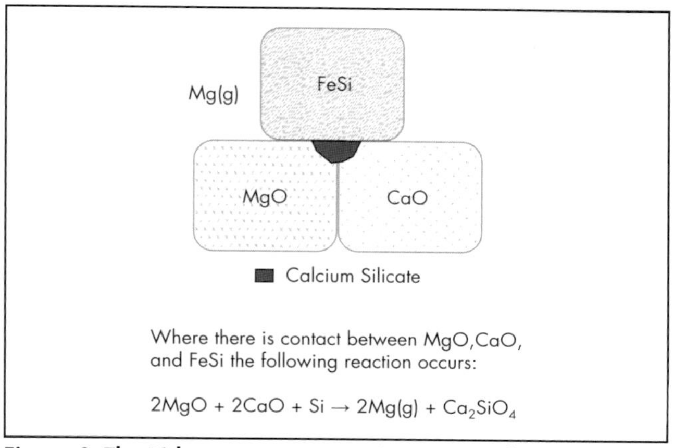

Figure 8 The Pidgeon process requires contact between MgO, CaO, and ferrosilicon to initiate reaction

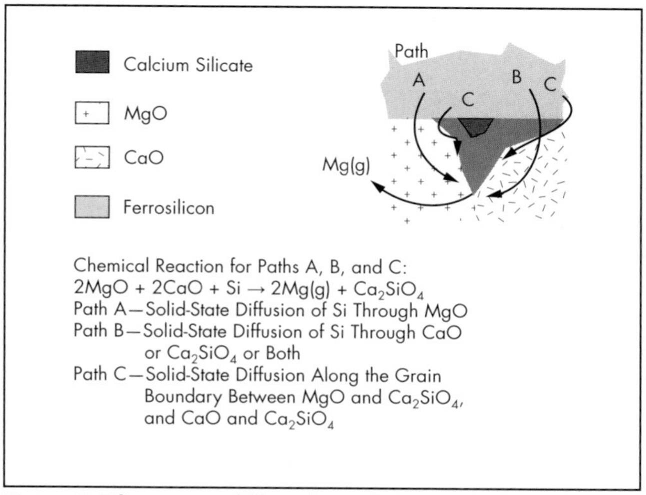

Figure 9 Silicon must diffuse through reactants or Ca_2SiO_4 to the MgO–CaO interface for further Mg(g) production

The furnace temperature is between 1,300°C (1,573 K) and 1,400°C (1,673 K) to achieve the necessary heat transfer for the endothermic reaction, and a charge temperature of 1,150°C (1,423 K) to 1,200°C (1,473 K) inside the retort. The other end of the retort extends outside the furnace and has an internal temperature of 600°C (873 K) to 650°C (923 K). Mg vapor is drawn out of the hotter end of the retort where its equilibrium pressure with the charge exceeds the operational pressure of approximately 3–10 Pa created by condensation of Mg(s) in the cooler end of the retort. Production of Mg continues as long as the rate of production in the hot end of the retort is capable of producing a magnesium vapor pressure greater than that in the cooler end of the retort. The magnesium vapor condenses on an internal sleeve at the cooler end of the retort. The sleeve is attached to the end plate that is used to seal the retort, thus the deposited Mg can be easily removed from the retort without having to extract the retort from the furnace. In China, there have been attempts to create vertical retorts that offer some advantages over the horizontal retorts in automatic loading and unloading of charge and waste product (D. McCallum, personal communication).

Retorts in 1980 were made of nickel chrome steel, were typically 3-m long with inside diameters of 25 cm and 2.5-cm-thick walls (Gill 1980). Today 50-mm-thick boiler steel is used. The retorts lose structural strength after 6 to 9 months with eventual collapse inward under the vacuum (D. McCallum, personal communication). Figure 11B shows that today's retorts are longer and of greater diameter than those used in 1980.

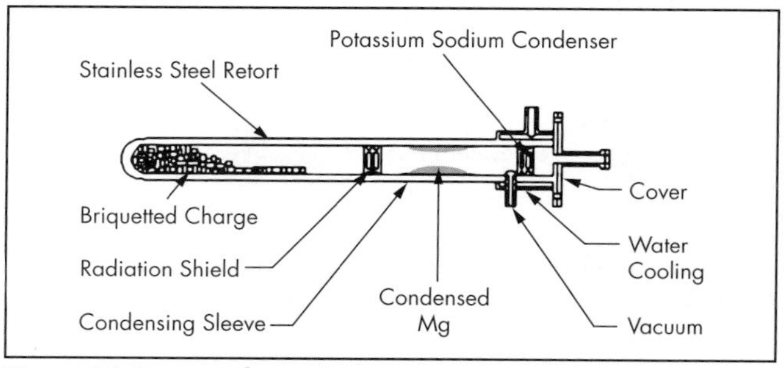

Figure 10 Retort schematic

The Pidgeon process is a batch process and thus labor intensive. The increase in retort size does not impact the chemistry, but does reduce labor cost as more raw material can be charged to a retort.

In 1980, it was reported that each retort produced 14.5 kg of 99.98% pure magnesium from 109 kg of charge in 8 hours. The mass ratio of dolime to ferrosilicon was 5 to 1 (90.8 kg dolime to 18 kg ferrosilicon). The ferrosilicon is typically 75 wt % Si (6 moles Si to each mole of Fe). The dolime and ferrosilicon are ground, mixed, and briquetted. The briquettes are preheated to 700°C (973 K) before being charged to the retort. In 1980, the recovery of Mg in the product stream was about 63% (Gill 1980).

The Pidgeon process dominates commercial production of magnesium metal from ore. Argonne National Laboratory provides an excellent review of both energy requirements and the extent of greenhouse gas emissions from the process (Johnson and Sullivan 2014). Total energy requirements and how that energy is distributed between different energy sources are presented in Table 3, along with total greenhouse gas emissions in kilograms of CO_2 gas equivalent for producing 1 kg of primary Mg ingot for the process steps listed. The numbers in the table are per kilogram of virgin Mg for two different plants using the Pidgeon process. The cost figures in Table 4 have been computed using the unit costs listed in the table and numbers from Table 3. Table 5, from the Argonne National Laboratory report (Johnson and Sullivan 2014), shows how the sources of energy identified in Table 3 for Pidgeon B are distributed to the steps identified in Figure 6.

Using data in Table 5 and the listed heating values, the cumulative energy demand for Pidgeon B in Table 3 exceeds the energy computed by 51 MJ. It is assumed that the difference is associated with the heating value used for the coke oven gas, and that a better value is 27.5 MJ/m^3 rather than 20 MJ/m^3, as the heating value of the gas will depend on the hydrogen content in the coal used to produce the gas.

Using the new value for the coke oven gas, the energy consumption for the reduction step in Figure 6 is computed to be 100.2 MJ/kg of magnesium produced in the Pidgeon process. The theoretical value, assuming the briquettes are charged to retorts at 700°C (973 K), is 13.7 MJ/kg of Mg. The energy efficiency is computed to be 13.7%, a value that compares favorably to the 12% reported in the literature for

Courtesy of Federal Pipe Centrifugal Casting Machine Company

Figure 11 (A) Retorts in operation and (B) new retorts

Table 3 Energy units, cumulative energy demand, and greenhouse gas values

Process, 1 kg Mg	Electricity, kW·h	Gas, m^3	Coal, kg	Diesel, L	Cumulative Energy Demand, MJ/kg	Greenhouse Gas, kg/kg	Included Steps (from Figure 6)
Pidgeon A	12.9	—	10.4*	0.111†	393	37.5	1–7
Pidgeon B	10.16	6.8‡	1.05§	—	256	30.3	1–7

Adapted from Johnson and Sullivan 2014
* Anthracite coal, heating value ~28 MJ/kg.
† Diesel, heating value ~43 MJ/L.
‡ Coke oven gas, heating value ~20 MJ/m^3.
§ Coke, heating value ~31 MJ/kg.

Table 4 Energy cost as a function of energy source

Process, 1 kg Mg	Unit Price						
	Electricity, \$0.10/kW·h	Natural Gas, \$0.106/m^3	Coke Oven Gas*, \$0.053/m^3	Coke, \$0.224/kg	Coal, \$0.108/kg	Diesel, \$0.156/L	Total Energy Cost
Pidgeon A	\$1.29	—	—	—	\$1.123	\$0.017	\$2.43
Pidgeon B	\$1.016	—	\$0.360	\$0.235	—	—	\$1.61

Source: Johnson and Sullivan 2014
* The cost of coke oven gas is assumed to be half that of natural gas, as the heating value is one-half that of natural gas.

Table 5 Inputs to Pidgeon B process, 1 kg of magnesium ingot

Step	Operation	Output	Coke Oven Gas, m^3	Electricity, kW·h	Coke, kg	Other Inputs
1	Mining	10.5 kg dolomite	—	—	—	—
2	Ore preparation and calcining	—	2.6	0.192	—	—
3	Ferrosilicon preparation	1.05 kg $FeSi_6$	—	8.92	1.05	1.19 kg quartz, 0.24 kg iron
4	Briquetting	—	—	0.672	—	0.13 kg CaF_2, 0.1 kg fluxes
5	Reduction	—	3.6	0.320	—	—
6 and 7	Refining and ingot production	1 kg Mg	0.6	0.049	—	—
Total			**6.9**	**10.16**	**1.05**	

Source: Johnson and Sullivan 2014

Chinese production (Ramakrishnan and Koltun 2004). The Pidgeon process is the simplest process, but it is both labor intensive and the least energy efficient process for producing magnesium (Simandl et al. 2007).

In China, the economics of the Pidgeon process has been improved by tailoring the waste from the retorts, which amounts to approximately 80% of the initial charge, to match both the chemistry and mineralogy of portland cement (D. McCallum, personal communication). Composition requirements for both portland cement and waste from the Pidgeon process are presented in Table 6. Numbers for the latter are computed from typical operations data (Gill 1980), assuming the waste from the retort is allowed to be fully oxidized. Portland cement can be produced by increasing the magnesium yield or reducing the amount of MgO charged to a retort, and adding additional CaO, either to the retort or to the oxidized waste. Currently, the prices of portland cement and magnesium metal in the United States are nearly identical on a mass basis. Given that the waste amounts to four times the mass of magnesium metal produced, a magnesium producer may more accurately be viewed as a cement company producing a bit of magnesium (D. McCallum, personal communication). China is the world's leader in production of portland cement.

Table 6 Comparison of portland cement composition to typical waste from Pidgeon process

Composition	Portland Cement, wt %	Pidgeon Process, wt %
CaO	60–66	51
SiO_2	19–23	29
Al_2O_3	2–6	—
Fe_2O_3	1–6	6
MgO	<5	14
CaO:SiO_2 mass ratio	>2	1.8

Magnetherm

Although there are no currently operating magnetherm plants in North America, the process did play an important role in magnesium metal production during the second half of the 20th century. With the uptick in the price of magnesium during the period from 2006 to 2008 (Figure 3), there was optimism that idled magnetherm plants would reopen. That optimism was dashed with the Great Recession in 2009. Plans for a similar process under development by Mintek (Abdellatif and Freeman 2008) failed to gain momentum as well. Magnetherm is a variant of the Pidgeon process, and given its earlier historical importance, its chemistry is worth reviewing.

Christini wrote a Minerals, Metals & Materials Society–award winning paper on the magnetherm process as practiced by Alcoa in Addy, Washington, United States (Christini 1980). The process was originally developed by the French company Pechiney in the 1950s and further adapted by Alcoa in the 1960s. A flow diagram for the process is presented in Figure 12. The magnetherm process, like the Pidgeon process, uses ferrosilicon to reduce magnesia in the production

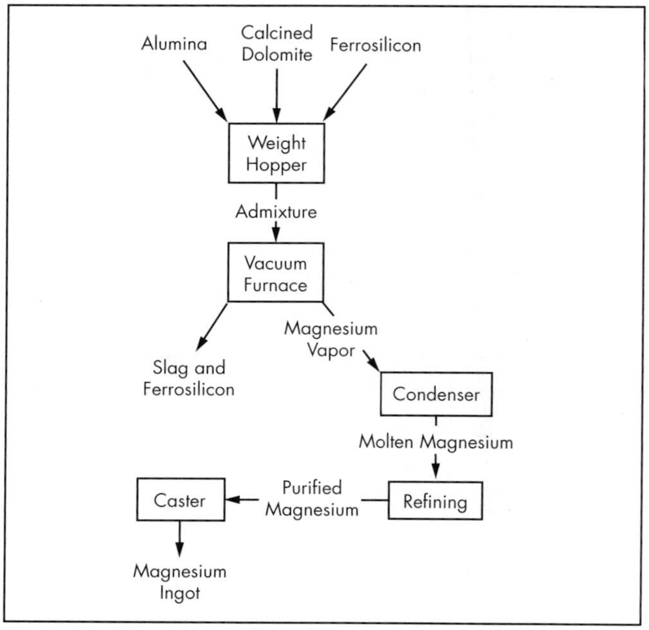

Figure 12 Magnetherm flow diagram

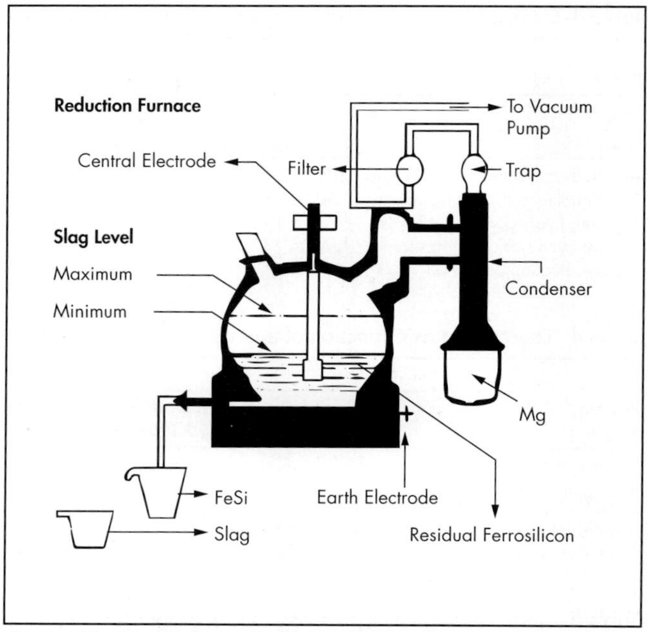

Figure 13 Magnatherm furnace and condenser

of magnesium vapor. However, the operational temperature is 1,550°C (1,823 K) and the magnesia is dissolved in a CaO-SiO_2-Al_2O_3 slag. The magnetherm process was developed to overcome the undesirable aspects of the Pidgeon process. The magnatherm process is less labor intensive, operates semi-continuously at reduced vacuum, and produces magnesium vapor at both higher reaction rate and yield—all aspects that improved the economics of production in the West. The magnetherm process does require a significant initial investment in comparison to the Pidgeon process. That disadvantage and the low labor cost in China are largely responsible for the rapid growth and use of the Pidgeon process in the East and demise of the magnetherm process in the West.

The magnetherm furnace with condenser is represented schematically in Figure 13. Calcined dolomite (dolime), alumina, and ferrosilicon are mixed, weighed, and charged to the furnace intermittently. The addition of alumina to the admix lowers the fusion temperature of the slag that consists of approximately 55 wt % CaO, 25 wt % silicon dioxide (SiO_2), 14 wt % aluminum oxide (Al_2O_3), and 6 wt % MgO. Some of the silica in the slag is produced as a result of the reduction of magnesia with silicon. The operational pressure for the magnetherm process in producing magnesium vapor is approximately 5,000 Pa as compared to the 3–10 Pa for the Pidgeon process. The higher pressure is achieved with the higher operational temperature, producing both molten slag and molten ferrosilicon. With the molten phases and the gaseous product, the process can be run in a quasi-continuous manner with the vacuum in the furnace being broken only twice during a typical 20–24-hour cycle resulting in the production of 3–8 metric tons of Mg. Furnace operation is interrupted when the slag and ferrosilicon rise to the maximum level indicated in the drawing. Tapping of the molten phases produces a minor downtime. A longer downtime occurs when the magnesium crucible is removed, replaced, and slag and molten ferrosilicon tapped.

The molten magnesium is taken to the foundry, where the metal is fluxed with a chloride salt (KCl and $MgCl_2$), and the purified metal is poured into ingot molds (Christini 1980).

The reaction occurring in the furnace

$$2MgO(slag) + Si(l\text{-ferrosilicon}) \rightarrow SiO_2(slag) + 2Mg(g),\ \Delta H_{1,550C} = 514\ \text{kJ} \qquad \textbf{(EQ 12)}$$

requires high activities for MgO and Si and a low activity for SiO_2 to reduce the level of vacuum required to produce Mg vapor. The data in Figure 7 suggests that raising temperature increases the vapor pressure of Mg, and thus reduces the degree to which a vacuum must be applied. From the Fe–Si phase diagram in Figure 14, and at a temperature of 1,550°C (1,823 K), it is possible to deduce that the activity of Si in the molten FeSi (75%) will be high but have a value something less than the ideal value. An activity approaching 0.8 for the silicon is not unexpected.

To achieve the maximum activity for MgO in the slag, the slag composition must be saturated with MgO. That occurs when the slag composition is in contact with the primary phase field for MgO. The phase diagram for the liquidus surface for the CaO-MgO-SiO_2 system presented in Figure 15 reveals that the minimum temperature for the maximum activity of MgO (a_{MgO} = 1) in the slag occurs at a temperature of approximately 1,600°C (1,873 K). That occurs where the primary phase fields for dicalciumsilicate (Ca_2SiO_4), periclase (MgO), and merwinite ($3CaO{\cdot}MgO{\cdot}2SiO_2$) coexist, a point marked by the black diamond on the phase diagram. The high temperature associated with that point would reduce the extent of vacuum required to draw off the magnesium vapor. Higher temperatures pose a serious problem for the long-term structural integrity of the furnace. The exterior of the furnace is "water cooled by a combination of sprays, over flow weirs, and water cooled flanges to maintain a freeze zone of slag within the furnace." The freeze zone is critical to prevent the outer steel shell from being penetrated by the ferrosilicon (Sever and Ballain 2013).

Alumina is added to the admix charged to the furnace to reduce the fusion temperature of the slag. The phase diagram in Figure 16 incorporates the impact of alumina on the liquidus surface presented in Figure 15, but at a constant alumina content of 15 wt %. The temperature at the point in Figure 16 where the primary phase fields for dicalciumsilicate (Ca_2SiO_4), periclase (MgO), and merwinite ($3CaO{\cdot}MgO{\cdot}2SiO_2$) meet is now reduced from 1,600° to 1,485°C (1,873 to 1,758 K).

It is important to remember that the admix charged to the furnace will be rich in MgO and poor in SiO_2, as the former is the source for Mg vapor, and SiO_2 is the product of the reduction Reaction 12. The final desired composition of the slag will be near the boundary separating primary phase MgO (periclase) and Ca_2SiO_4, producing a lower fusion temperature of the slag and maximum thermodynamic activity of MgO in the slag while limiting MgO loss. If a higher temperature was employed in the furnace, an example being 1,900°C (2,173 K), loss of magnesium to the slag would double at MgO saturation.

Although it is essential to have a high value for the activity of MgO in the slag, Reaction 12 indicates that it is also essential to have a low value of the activity of SiO_2 in the same phase. Figure 17 contains the isoactivity curves for SiO_2 in the slag phase at 1,550°C (1,823 K) for the Al_2O_3-CaO-SiO_2 slag. Although MgO is not included in that graph, it has long been established that MgO and CaO behave similarly in slags and, as such, small additions of MgO to a slag with a significant concentration of CaO does not impact the activity of the other components in the slag. Point b in the ternary system presented in Figure 17 has been identified by Christini as being equivalent to the slag composition in the quaternary system (Al_2O_3-CaO-MgO-SiO_2). At that point, the activity of SiO_2 in the slag is 0.001.

With this last missing number, it is possible to compute the equilibrium vapor pressure of Mg in Reaction 12. The standard states for the reactants and products in the reaction are for the pure compound or element in its stable form at 1,550°C (1,823 K) and for magnesium vapor at a pressure of 1 bar. At that temperature and at equilibrium, ΔG° for the

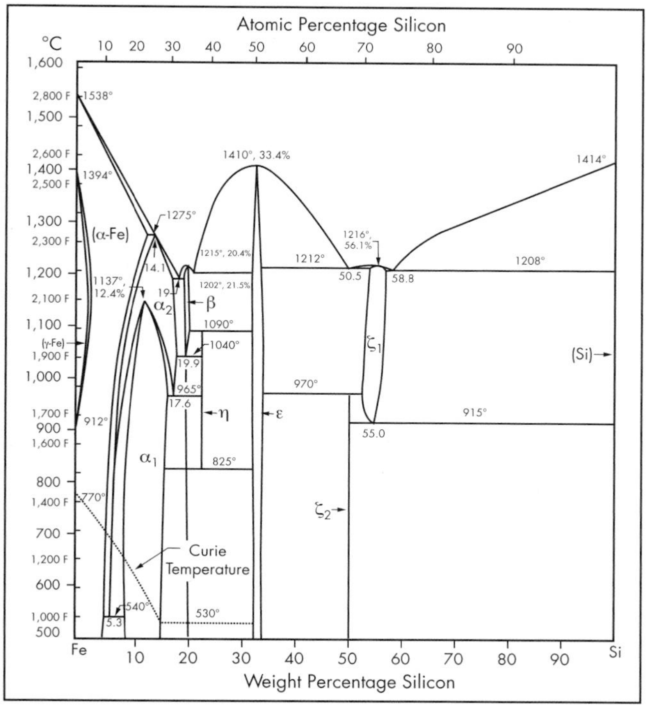

Source: Lyman 1973

Figure 14 Iron–silicon phase diagram

Compounds	
Name	**Composition**
Periclase	MgO
Cristobalite	SiO_2
Tridymite	SiO_2
Lime	CaO
Diopside	CaO : MgO : $2SiO_2$
Pyroxene	Solid solution along section $MgO{\cdot}SiO_2$–diopside
Akermanite	$2CaO{\cdot}MgO{\cdot}2SiO_2$
Monticellite (Mo)	$CaO{\cdot}MgO{\cdot}SiO_2$
Merwinite (Merw)	$3CaO{\cdot}MgO{\cdot}2SiO_2$
Forsterite	$2MgO{\cdot}SiO_2$

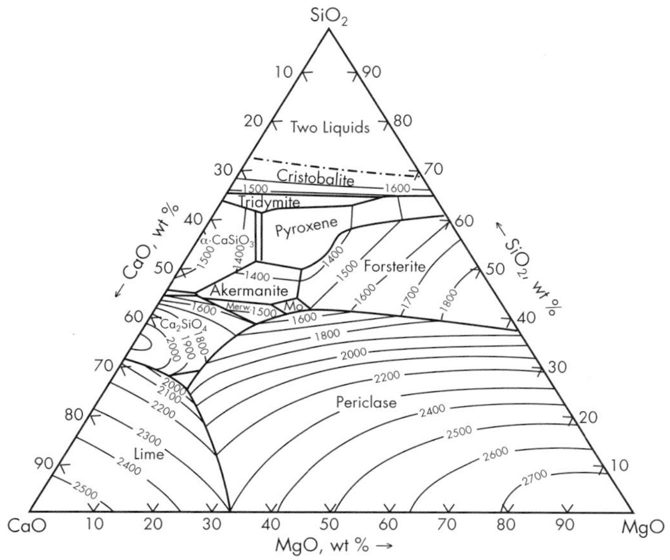

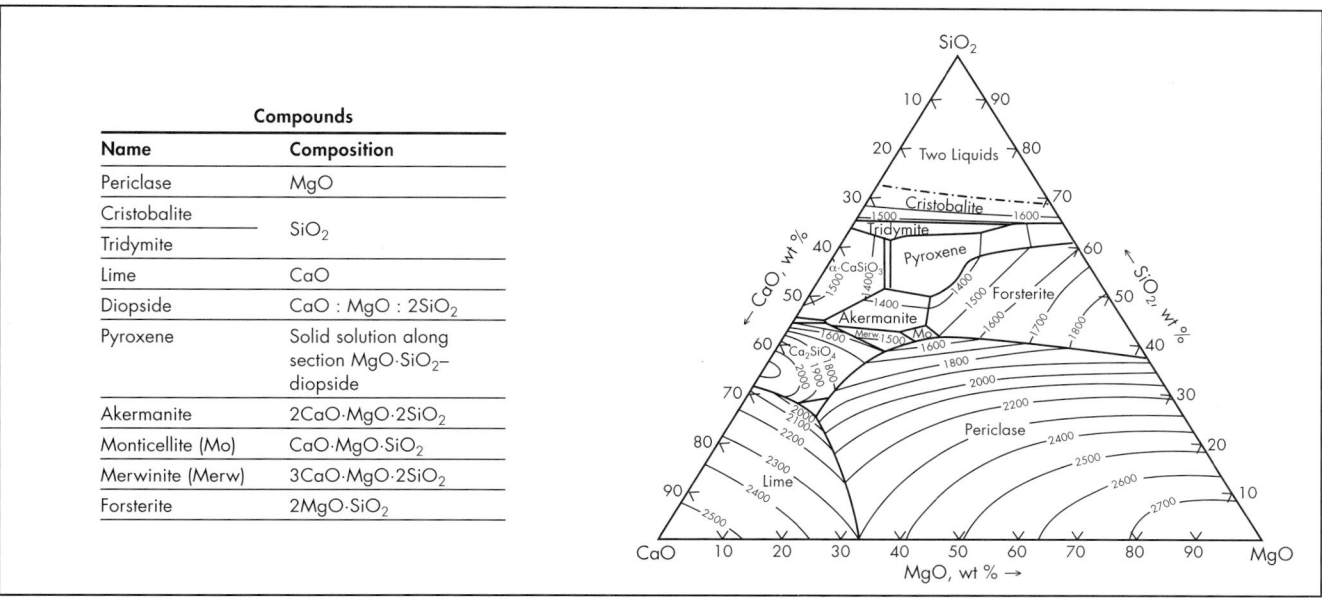

Source: Roth and Vanderah 2005

Figure 15 Liquidus surface of the CaO-MgO-SiO_2 phase diagram

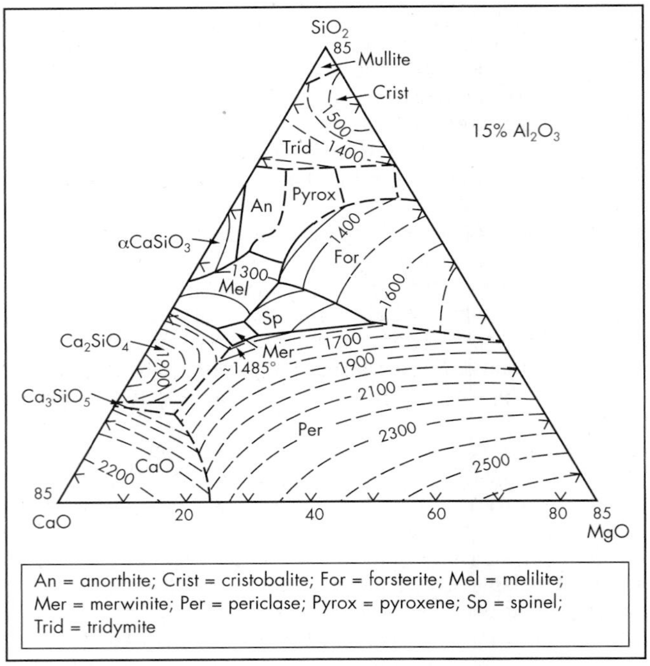

Source: Levin 1969

Figure 16 Iso-section for 15 wt % Al_2O_3 liquidus surface for the Al_2O_3-CaO-MgO-SiO_2 system

reaction is 126.93 kJ and the equilibrium constant, K, has a value of $2.31 \cdot 10^{-4}$. Using the activities just discussed

$$K = 2.31 \cdot 10^{-4} = \frac{a_{SiO_2} \cdot P_{Mg}^2}{a_{MgO}^2 \cdot a_{Si}} = \frac{0.001 \cdot P_{Mg}^2}{1^2 \cdot 0.8} \quad \textbf{(EQ 13)}$$

$$P_{Mg} = 0.43 \text{ bar} \quad \textbf{(EQ 14)}$$

By operating the magnetherm furnace at 0.05 bar, Reaction 12 can never achieve equilibrium. The reaction proceeds as written, and Mg vapor is transformed to liquid in the condenser.

The electrode shown in Figure 13 is submerged in the slag. An AC current passes through the molten slag, providing resistance heating below the upper surface of the slag. That arrangement ensures that the natural convective flow pattern of hot slag rises in the center of the furnace and sweeps vertically away toward the cooler walls of the furnace before sinking and then is drawn back toward the electrode to be heated and rise again.

Christini (1980) has pointed out that the flow pattern is essential to achieve maximum magnesium yield. He has demonstrated that Reaction 12 is thermodynamically favored at the upper surface of the molten slag and that hydrostatic head significantly precludes the reaction from occurring in the bottom of the melt where excess ferrosilicon forms a pool in contact with the molten slag. The flow pattern of the slag helps keep the small particles of ferrosilicon to stay suspended in the slag before sinking to the bottom of the furnace, thereby providing greater time for reaction between MgO in the slag and Si.

Approximately 7 metric tons of material are charged to the magnetherm furnace for every metric ton of Mg metal produced (Yudiarto 2013).

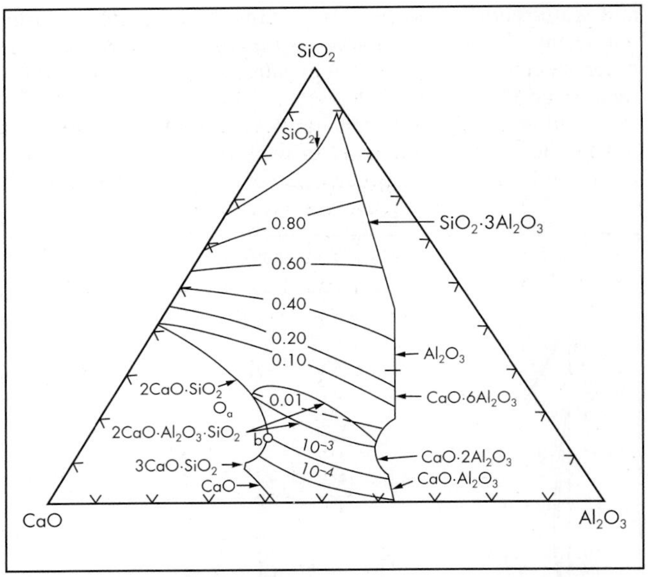

Adapted from Rein and Chipman 1965

Figure 17 Isothermal section of Al_2O_3-CaO-SiO_2 with isoactivity line for SiO_2 in slag phase at 1,550°C (1,823 K)

REFINING OF MAGNESIUM METAL

Magnesium from the Pidgeon process comes in a solid form called *crown*. The crown is fused in a steel crucible along with mixed chloride salt solution containing both $MgCl_2$ and KCl. The flux acts as a heat transfer medium and as a protective layer over the top of the magnesium, preventing oxidation of the metal by isolating it from air. The $MgCl_2$ in the salt also reacts with Ca and K dissolved in the molten magnesium, exchanging the impurities with Mg. In the electrolytic process, this step is excluded as the molten magnesium is removed from the cell along with some of the electrolyte. Fluorspar (CaF_2) and magnesium oxide (a thickening agent) are added to the molten salt to help in the settling of oxide impurities in the molten metal (Neelameggham 2013).

Casting

The "surface of molten magnesium is sensitive to inclusions from oxidation, [therefore] an inert blanket gas is generally used to protect the metal. Alternatively, sulfur powder is spread across the surface of molten magnesium for protection …" (Johnson and Sullivan 2014). Beginning in 1980, sulfur hexafluoride (SF_6), a nontoxic gas, was first used as a preferred blanket gas. By the early 1990s, the gas was determined to be "an extremely potent greenhouse gas, with a global warming potential (GWP) of 23,900·$[CO_2]_{eq}$ per [kilogram] of gas" (Johnson and Sullivan 2014). Alternative blanket gases under consideration are sulfur dioxide (SO_2) with no GWP, but it is corrosive and a known pollutant, and hydrofluorocarbon (HFC)-134a with a GWP of 1,300. Consumption of blanket gases per kilogram of Mg ingot are estimated to be 1.1 g for HFC-134a, 0.58 g for SF_6, and 0.05 kg for S (Johnson and Sullivan 2014).

ADVANCES IN MAGNESIUM METAL PRODUCTION TECHNOLOGY

During the first half of the last century, as previously noted, there was considerable effort in carbothermic production of magnesium vapor by reacting magnesium oxide with carbon. That process produced both magnesium vapor and carbon monoxide. The primary problems were high operational temperatures, T >1,700°C (1,973 K); reoxidation of the magnesium vapor by the CO(g) prior to production of magnesium liquid droplets; and ultimately on further cooling, production of a pyrophoric and explosive powder. Australia's CSIRO (Commonwealth Scientific and Industrial Research Organization) claims to have solved the reversion problem with their MagSonic technology. The new approach still relies on reacting magnesium oxide with carbon to produce Mg(g) and CO(g), but the gases undergo supersonic quenching at four to five times the speed of sound with only about 10% reversion and produce magnesium powder that is 99.9% pure (Prentice et al. 2012).

SAFETY CONSIDERATIONS

Caution is required when producing molten magnesium and magnesium powder. The U.S. Bureau of Mines has listed magnesium powder's relative explosion hazard index as severe. Only metal powders of aluminum-magnesium alloy and atomized aluminum were found to be more dangerous (Jacobson et al. 1964). Molten magnesium when exposed to air can create a dangerous situation. The author of this chapter visited Alcoa's Addy, Washington, facility for the production of magnesium metal by the magnetherm process. During that visit, and shortly after touring the magnesium refining facility, where molten magnesium is covered by a protective molten salt consisting of $MgCl_2$ and KCl, there was a temporary disruption to the process; a loud bang and a half dozen or more small balls of white-hot magnesium rising about 15 m above the refining facility and fanning out a similar distance. The magnesium balls burned in air leaving a trail of white smoke consisting of MgO. The tour guide commented that the incident was not the norm, but not unusual.

REFERENCES

Abdellatif, M. 2006. Mintek thermal magnesium process (MTMP): Theoretical and operational aspects. In *Southern Africa Pyrometallurgy 2006*. Edited by R.T. Jones. Johannesburg: Southern African Institute of Mining and Metallurgy.

Abdellatif, M., and Freeman, M. 2008. Mintek thermal magnesium process: Status and prospective. Presented at the Advanced Metals Initiative, Gold Reef City, Johannesburg, South Africa, October 18–19.

Ball, C.J.P. 1956. The history of magnesium. *J. Inst. Metals*. 84:399.

Beck, A. 1939. *The Technology of Magnesium and Its Alloys* [English translation]. London: F.A. Hughes. pp. 1–19.

Christini, R.A. 1980. Equilibrium among metal, slag, and, gas phases in the magnetherm process. In *Light Metals 1980*. Edited by C.J. McMinn. Warrendale, PA: TMS-AIME. pp. 981–995.

Emeley, E.F. 1966. *Principles of Magnesium Technology*. London: Pergamon.

Evans, J.W. 2007. The evolution of technology for light metals over the last 50 years: Al, Mg, and Li. *JOM*. 59(2):30–38.

Gill, C.B. 1980. *Nonferrous Extractive Metallurgy*. New York: Wiley.

Grecian Magnesite. n.d. Caustic calcined magnesia. www.grecianmagnesite.com/products/caustic-calcined-magnesia. Accessed May 2018.

Haughton, J.L., and Prytherch, W.E. 1938. *Magnesium and Its Alloys*. New York: Chemical Publishing.

Jacobson, M., Austin, R., and Nagy, J., 1964. Explosibility of Metal Powders. Report of Investigations 6516. Washington, DC: U.S. Bureau of Mines.

Johnson, M.C., and Sullivan, J.L. 2014. Lightweight materials for automotive applications: An assessment of material production data for magnesium and carbon fiber. Argonne National Laboratory Report ANL/ESD-14/7. Washington, DC: U.S. Department of Energy, Office of Energy Efficiency and Renewable Energy.

King, H.M. n.d. Dolomite, olivine, talc. Geoscience News and Information. https://geology.com/minerals. Accessed May 2018.

Larsen, L. 2017. Salty lakes are disappearing around the world and we're to blame. *Standard-Examiner*. October 29. http://visuals.standard.net/2017/10/29/saline-lakes-great-salt-lake-utah-are-disappearing-around-the-world-and-were-to-blame/. Accessed May 2018.

Levin, E.M. 1969. *Phase Diagrams for Ceramists: 1969 Supplement*. Westerville, OH: American Ceramics Society.

Lyman, T. 1973. *Metals Handbook. Vol. 8, Metallography, Structure and Phase Diagrams*, 8th ed. Metals Park, OH: American Society for Metals. p. 306.

Martin Marietta Magnesia Specialties. 2018. Magnesia grades. http://magnesiaspecialties.com/magnesia-grades/. Accessed May 2018.

Natural Resources Canada. 2008. *Canadian Minerals Yearbook: 2006 Review and Outlook*. Ottawa, ON: Minerals and Metals Sector, Natural Resources Canada.

Neelameggham, R. 2013. Primary production of magnesium. In *Fundamentals of Magnesium Alloy Metallurgy*. Edited by M. Pekguleryuz, K. Kainer, and A. Kaya. Oxford: Woodhead. pp. 1–32.

Padhy, S. 2017. Why is carnallite valuable? *Potash Investing News*. https://investingnews.com/daily/resource-investing/agriculture-investing/potash-investing/carnallite-a-dual-market-opportunity/. Accessed May 22, 2018.

Prentice, L., Nagle, M., Barton, T., Tassios, S., Kuan, B., Witt, P., and Constanti-Carey, K. 2012. Carbothermal production of magnesium: CSIRO's MagSonic process. In *Essential Readings in Magnesium Technology*. Edited by S.N. Mathaudhu, A.A. Luo, N.R. Neelameggham, E.A. Nyberg, and W.H. Sillekens. Cham, Switzerland: Springer. pp. 127–131.

Ravilious, K. 2016. How much carbon would adding olivine to the oceans remove? *Environmental Research Web*. http://environmentalresearchweb.org/cws/article/news/64739. Accessed May 2018.

Ramakrishnan, S., and Koltun, P. 2004. Global warming impact of the magnesium produced in China using the Pidgeon process. *Resour. Conserv. Recycl*. 42(1):49–64.

Rein, R.H., and Chipman, J. 1965. Activities in the liquid solution SiO_2-CaO-MgO-Al_2O_3 at 1600°C. *Trans. Metall. Soc. AIME*. 233:415–425.

Roskill. 2016. *Magnesium Metal: Global Industry, Markets & Outlook*. London: Roskill Information Services Limited.

Roth, R.S., and Vanderah, T.A. 2005. *ACerS–NIST Phase Equilibria Diagrams*, Vol. 14. Westerville, OH: American Ceramics Society. fig. 11304.

Schambra, W.P. 1945. The Dow magnesium process at Freeport, Texas. *Trans. Am. Inst. Chem. Eng.* 41:35.

Sever, J.C., and Ballain, P.E. 2013. Evolution of the Magnetherm magnesium reduction process. In *Magnesium Technology 2013*. Edited by N. Hort, S.N. Mathaudhu, N.R. Neelameggham, and M. Alderman. Hoboken, NJ: Wiley. pp. 69–74.

Simandl, G.J., Schultes, H., and Paradis, S. 2007. Magnesium—raw materials, metal extraction and economics—global picture. In *Digging Deeper: Proceedings of the Ninth Biennial SGA* [Society for Geology Applied to Mineral Deposits] *Meeting*. Edited by C.J. Andrew. Dublin: Irish Association of Economic Geology. pp. 827–830.

Smith, G. 1983. *Phase Diagrams for Ceramists. Vol. 5, Salts.* Westerville, OH: American Ceramics Society.

Toguri, J.M., and Pidgeon, L.M. 1961. High-temperature studies of metallurgical processes: Part I. The thermal reduction of calcined dolomite with silicon. *Can. J. Chem.* 40(9):1769–1776.

Toguri, J.M., and Pidgeon, L.M. 1962. High-temperature studies of metallurgical processes: Part II. The thermal reduction of magnesium oxide with silicon. *Can. J. Chem.* 39(3):540–547.

Tripp, T.G. 2009. Production of magnesium from Great Salt Lake, Utah USA. *Nat. Resour. Environmen. Issues*. 15(art. 10). https://digitalcommons.usu.edu/nrei/vol15/iss1/10/. Accessed May 2018.

USGS (U.S. Geological Survey). 1992–2015. Magnesium compounds. In *Minerals Yearbook*. Reston, VA: USGS.

USGS (U.S. Geological Survey). 1995–2015. Magnesium. In *Minerals Yearbook*. Reston, VA: USGS.

USGS (U.S. Geological Survey). 2017. *Mineral Commodity Summaries 2017*. Reston, VA: USGS.

Yudiarto, A. 2013. Magnesium extraction by magnetherm process. *Extractive Metallurgy*. http://extractivemetallurgy.blogspot.com/2013/09/magnesium-extraction-by-magnetherm.html. Accessed May 2018.

CHAPTER 12.22

Manganese

William M. Cross and Jon J. Kellar

Manganese, Mn, element 25 in the periodic table, is the 12th most abundant element in the earth's crust, with ~1,000 ppm (~0.1%) concentration and the fourth most-used metal after iron, aluminum, and copper (Gulf Manganese Corporation 2015). As can be seen in Figure 1, manganese metal is not stable in water and therefore is not found in nature in its metallic state. The primary stable manganese oxide/hydroxide minerals are also shown in the figure. The use of manganese minerals by humans dates back at least 22,000 years before present in the "Entrance Gallery" of the Gargas cave in the Hautes-Pyrénées, France. These Paleolithic people mixed manganite (γ-MnOOH) and cryptomelane ((K,Ba) (Mn^{2+}, Mn^{4+})$_8$O$_{16}$·xH_2O) with pyrolusite (MnO_2), calcite, and clay binders, and with hollandite (Ba(Mn^{2+}, Mn^{4+})$_8$O$_{16}$) without binders, to produce slightly different dark colors to adjacent hand prints (Chalmin et al. 2006). This use seems deliberate because, for instance, manganite and hollandite do not generally occur together because of their different thermogeochemical origins.

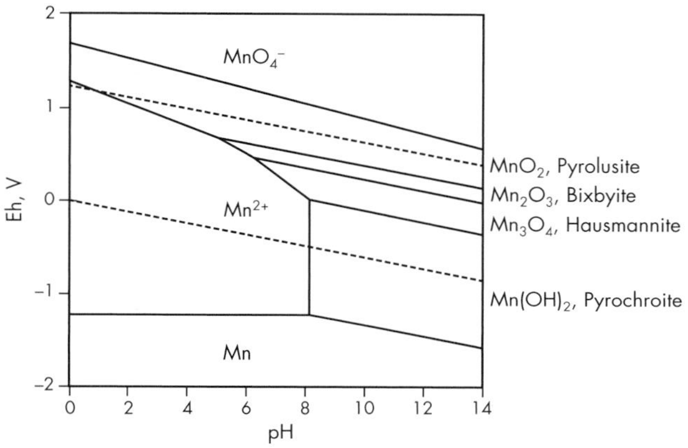

Note: Diagram was generated using Stabcal software (Huang 2012) with the Minteq (Gustafsson 2015) database and assuming a manganese concentration of 2,000 ppm. The dashed lines represent the water stability region.

Figure 1 Pourbaix diagram for the Mn–H_2O system

METAL/MINERAL USAGE

Manganese has several uses. The largest use is in the steel industry, which typically consumes about 90% of the manganese produced each year (Gulf Manganese Corporation 2015a). Other important uses include batteries and electronics, aluminum and copper alloys, fertilizers and micronutrients, water treatment, and as a colorant in materials, including glass, textiles, and plastic (Matos and Corathers 2005).

Steel production typically uses 6–9 kg Mn per metric ton of steel produced (Emsley 2011). About 30% of this total is used in iron ore refining operations prior to making steel (Emsley 2011), although this is decreasing as new steel processing techniques, such as hot metal desulfurization and ladle deoxidation, become more common. The remaining manganese is used to alloy steel. Most common steels have up to 1.0 wt % Mn (Lawcock et al. 2013). Manganese is considered a critical mineral because in its primary uses, particularly steelmaking, manganese cannot be substituted (NRC 2008). The addition of manganese increases the strength and hardenability of the steel (Lawcock et al. 2013; NRC 2008). Manganese serves other purposes in steelmaking, including acting as a deoxidizer, and combining with sulfur, phosphorus, and oxygen, to improve machinability of the steel and to reduce cracking during high-temperature rolling (Lawcock et al. 2013; IMnI 2014). High-strength, low-alloy (HSLA) steels, which contain 1.0–1.65 wt % Mn, have become relatively common in varied applications, with the manganese (and other alloying metals such as niobium and vanadium) used to decrease the carbon content while increasing the strength, weldability, and/or toughness (Davis 2001). In applications requiring high tensile strength, high toughness, and excellent wear resistance, such as gyratory and jaw crushers (IMnI 2014), the manganese content is increased to >11 wt % (Key to Metals AG 2010). In these instances, the steel is referred to as mangalloy. The manganese helps the steelwork harden as

William M. Cross, Professor, Materials & Metallurgical Engineering, South Dakota School of Mines & Technology, Rapid City, South Dakota, USA
Jon J. Kellar, Nucor Professor & Douglas W. Fuerstenau Professor, South Dakota School of Mines & Technology, Rapid City, South Dakota, USA

tensile forces cause a neck to form during elongation, greatly increasing the strength and elongation of such steels. In addition, the manganese inhibits the transformation of austenite to martensite. This causes the austenite phase to be metastable, and the transformation that can occur after impact can increase the surface hardness by up to three times. These properties make mangalloy difficult to machine, limiting its use. Manganese use is also significant in advanced high-strength steels, particularly twinning-induced plasticity (TWIP) steels (Chen et al. 2013). TWIP steels have 18%–30% Mn to stabilize austenite at room temperature. These steels offer up to two times the strength of mild steel, with ~5% lower density and increased energy absorption. This application is expected to lead to the use of considerable manganese in future vehicle manufacture as TWIP and HSLA steels are being used in FutureSteelVehicle construction, and in manganese dioxide cathode batteries in the electric version of the FutureSteelVehicle (Kilic and Ozturk 2014; WorldAutoSteel 2015).

Manganese is an important component of both primary (disposable) and secondary (rechargeable) battery cathodes (Gulf Manganese Corporation 2015a). For primary batteries, manganese dioxide serves as the cathode for the most common types of lithium-ion battery: CR-V3 or Li-Mn (Cadex Electronics 2015). The manganese dioxide cathode can absorb lithium ions and electrons formed at the lithium anode through an organic solvent with dissolved lithium salt to form $LiMnO_2$. Alkaline dry-cell batteries also use manganese dioxide. This type of battery was invented in the late 1950s, and more than 10 billion batteries per year are produced (Olivetti et al. 2011). These dry-cell batteries can also be designed to be rechargeable. A zinc anode and a manganese dioxide cathode are connected by a potassium hydroxide (alkaline) electrolyte.

With respect to aluminum alloys (3xxx series alloys), about 1 wt % Mn is added to wrought aluminum alloys to increase strength and allow strain hardening without decreasing ductility or resistance to corrosion (Woodward 2001). These alloys are typically used for cooking utensils and radiators, and so on, as they retain strength at elevated temperatures. Soda and beer cans typically contain about 1 wt % Mn, with the lids and bottoms having slightly less manganese. Increasing the manganese content creates what are termed *aluminum bronzes*. A representative aluminum bronze incorporating manganese is EN designation $CuAl_{10}Fe_3Mn_2$ with Al: 9–11 wt %; Fe: 2–4 wt %; Ni: <1.0 wt %; Mn: 1.5–3 wt %; Zn: <0.5 wt %; Si: <0.2 wt %; Pb: <0.02 wt %; and the remainder Cu (CDA 2004). Aluminum bronzes are valued for their corrosion resistance, particularly in seawater, leading to their use in several naval systems parts, particularly propellers for Arctic-travelling ships (LeGrand 2011). Other uses for aluminum bronzes include high-end guitar strings (Campus Five 2013), some coins (the Sacagawea dollar has 7 wt % Mn [U.S. Mint 2015]), and dental crowns (Eschler et al. 2003).

Less common uses of manganese include manganese ferrite ($Mn_xFe_{1-x}\ Fe_2O_4$) and manganese-zinc ferrites (Rashad 2006). These materials are often used in transformers and electromagnetic cores, particularly for frequencies <2 MHz (IMA 2011). Manganese compounds have been used to replace lead in unleaded gasoline (Gulson et al. 2006) and are critical nutritional components for aerobic life (Emsley 2011). Manganese-containing superoxide dismutase is present in nearly all eukaryotic mitochondrial cells and in bacteria and is necessary for the catalytic partitioning of toxic superoxides into gaseous oxygen or hydrogen peroxide (HHS 2012).

MINERALS OF ECONOMIC IMPORTANCE

Manganese can take on a variety of valence states, primarily from 2+ to 7+. Because of this speciation, many manganese minerals can form. As shown in Figure 1, water-stable manganese oxide minerals include pyrolusite (MnO_2, 4+ valence), bixbyite (Mn_2O_3, 3+ valence), hausmannite (Mn_3O_4, two 3+ and one 2+ valence), and pyrochroite ($Mn(OH)_2$, 2+ valence) (Hudson Institute of Mineralogy 2015). The prevalence of 2+ and 3+ valence states and their nearness on the periodic table makes iron and manganese similar in behavior, although manganese's higher oxidation states (leading to more minerals) and lower likelihood of forming sulfide minerals lead to separation of manganese from iron primarily into regions of higher solution oxidation potential. Other important manganese minerals include rhodochrosite ($MnCO_3$, 2+ valence), braunite (Mn_7SiO_{12}, six 3+ and one 2+ valence), cryptomelane ($(K,Ba)Mn_8O_{16}\cdot xH_2O$ or $K_2Mn_8O_{16}$, seven 4+ and one 3+ valence), and manganite (MnOOH, 3+ valence) (Hudson Institute of Mineralogy 2015). Rhodochrosite is the primary manganese mineral in about 33% of manganese deposits; braunite is primary in about 25% of deposits. Cryptomelane, manganite, and pyrolusite combine to be the predominant manganese mineral in a little more than 20% of all manganese deposits. Wad or bog manganese was used in ancient times for pigment and by the Romans for coloring glass. Wad is a poorly crystalline mixture of manganese oxides and hydroxides (Hudson Institute of Mineralogy 2015). Nsutite is a hydrated, gray-to-black, opaque manganese oxide of the formula ($Mn^{4+}_{(1-x)}\ Mn^{2+}_x\ O_{2-2x}(OH)_{2x}$, with x = 0.6–0.7 for Ghanaian, Greek, and Mexican nsutite, and x = 0.16 for manganoan nsutite from Ghana) (Zwicker et al. 1962).

GEOLOGICAL SETTING

Land-Based Resources

Figure 2 shows the distribution of the formation of manganese deposits throughout geological time. Volcanogenic manganese occurs in amounts too small to be seen in the figure. Manganese minerals were first deposited in large quantities at the start of the Paleoproterozoic era, coincident with the Great Oxidation Event and are generally associated with the Hamersley–Transvaal-type banded iron formations. The great Kalahari manganese field in South Africa, which contains most of the world's manganese deposits, is thought to be of sedimentary origin, occurring in shallow basins on the continental shelf. Initially, iron precipitated as an oxide or silicate, followed by silica, then manganese, likely as Mn^{4+} oxyhydroxide. Much of the manganese and associated metals transformed to Mn-Fe-Ca carbonates through diagenesis with organic carbon at relatively low Eh values. Two main types of deposit have been recognized, Mamatwan and Wessels. Mamatwan-type deposits make up about 97% of the field, typically have around 38% Mn with an Mn/Fe ratio of up to 10, ~5% SiO_2 (silicon dioxide), ~10% CaO (calcium oxide), and 3% MgO (magnesium oxide). Manganese minerals present in reasonable quantities include rhodochrosite, braunite, and manganite. In addition, the ore contains around 20% calcite, 5% magnesite, and 5% hematite. The remainder of the ore is a higher manganese oxide–type ore called Wessels. The manganese content of the Wessels-type ore is generally in excess of 45% and is mainly braunite, manganite, and hausmannite.

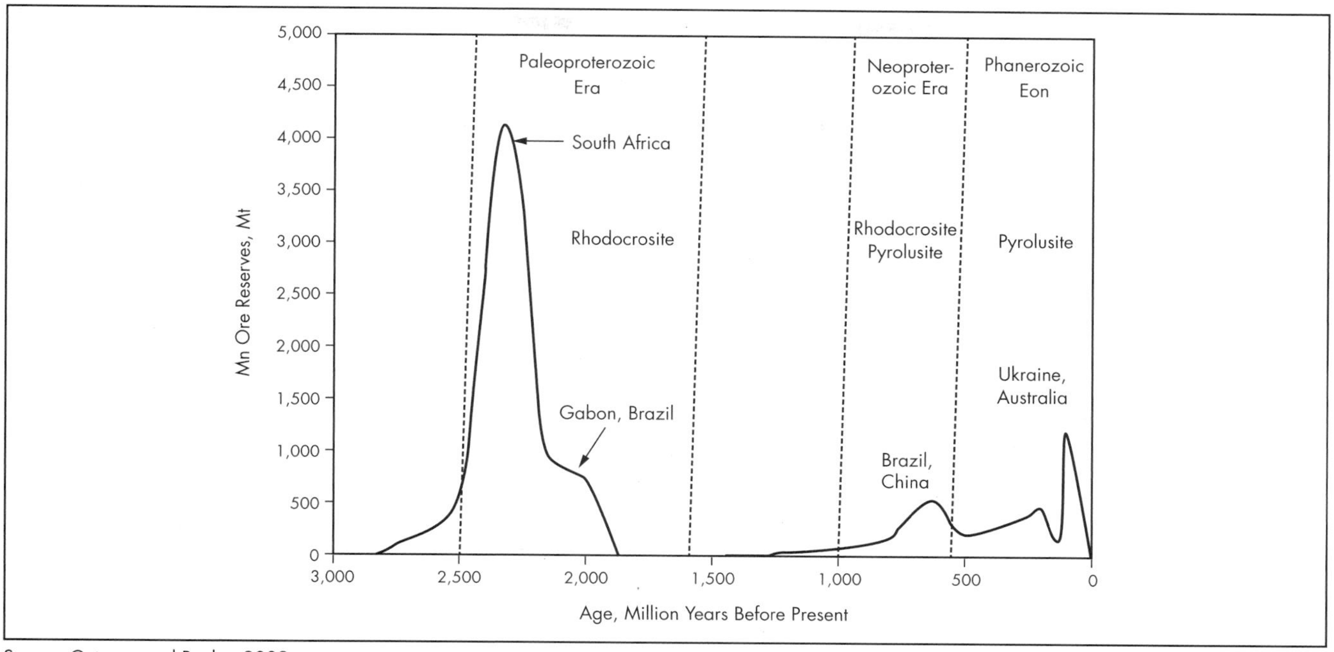

Source: Gutzmer and Beukes 2009

Figure 2 Manganese reserves as a function of time of deposit creation

The Mn/Fe ratio is typically 3–5, lower than in the Mamatwan ore. The calcite content can be 10%. The Wessels ore type is believed to have formed from hydrothermal alteration of the Mamatwan-type ore. Hydrothermal fluids, at 200°–250°C, entered the protore through cracks and leached the carbonate, calcium, magnesium, and silica. This resulted in the observed increase in manganese content and the transformation of the rhodochrosite to hausmannite.

In the late Paleoproterozoic (~2,000 million years ago), manganese deposits generally occurred as manganese-enriched black shale formations and formed as sedimentary deposits with low oxygen levels (<1 mL/L), with rhodochrosite formed by the interaction of organic carbon and Mn^{4+} oxyhydrides. These deposits typically have quartz, various clays, and pyrite as gangue minerals. Black shale–hosted deposits occur mainly in Moanda (Gabon), Azul (Brazil), and various areas of India and China. In addition to rhodochrosite (often calcium enriched), manganese is also present as manganese silicates and lower manganese valence oxides. These black shale–hosted protores have often transitioned to weathered deposits typical of a hot, humid lateritic process, leading to the oxidation of the manganese and loss of carbon dioxide and hydroxide. In these cases, the manganese composition is enriched from <30% Mn in the carbonate protore, some of which is still present at the lowest levels of the deposit, to high-grade oxide and silicate manganese ore deposits like those in Moanda, which in the processed portion contain about 40%–50% Mn as pyrolusite, with some manganite and cryptomelane. Moanda deposits contain about 8% silica, 7% alumina, and 5% iron as gangue.

Little manganese was deposited during the Mesoproterozoic. The deposition in the Neoproterozoic is correlated with the huge glaciation events of the Cryogenian period and decreased with the start of multicellular life in the Edicarian period. Similar to the early Paleoproterozoic Hotazel deposits, the manganese deposition correlates with banded iron formation, in this case, the Rapitan-type formations. For instance, at Mato Grosso do Sul (Brazil), cryptomelane is the most common mineral, followed by hematite, quartz, braunite, and pyrolusite, with some pyrite and a variety of other manganese and iron oxides and hydroxides.

The Phanerozoic eon deposits are of the shallow marine oolitic-type deposits. The primary manganese resources of this type are in Australia at Groote Eylandt and in Nikopol, Ukraine. As with the other manganese deposits, these are sedimentary in origin, usually in continental edge seas with low oxygen content. The manganese was deposited first as oxyhydroxide, was reduced to carbonates, and then weathered to form manganite, then pyrolusite and cryptomelane. For instance, at Groote Eylandt, pyrolusite and cryptomelane are the primary manganese minerals, while the primary gangue minerals are kaolinite, goethite, and quartz (Ostwald 1975).

Land-based deposits are usually mined as open-pit or open-cut operations (Moanda in Gabon; Mamatwan in South Africa; Bootu Creek, Groote Eylandt, and Woodie Woodie in Australia; Azul and Urucum in Brazil) with a few underground operations (Assmang and Wessels in South Africa).

Sea-Based Resources

Polymetallic sea nodules (PSNs) are not yet a productive source, but they have attracted a great deal of attention. The formation process of the nodules is different than the land-based manganese resources, yielding different manganese minerals and different accompanying minerals. PSN minerals form in two ways, hydrogenic and diagenetic precipitation. Initial nodule formation, at >4,000-m depths, is typically hydrogenic nucleation of vernadite (Mn^{4+}, Fe^{3+}, Ca^{2+}, Na^{+})$(O,OH)_2 \cdot n\ H_2O$. Bacteria may also be involved in the nucleation and growth of vernadite, $Mn_{1.4}$, $Fe_{0.6}$, $Ca_{0.1}$, $Na_{0.1}O_{1.5}OH_{0.5} \cdot 1.4H_2O$. Vernadite grows very slowly (1–10 nm/yr). As time passes, the nodules are eventually entrapped in porous sediment and the growth mechanism switches to produce todorokite $(K^+, Ca^{2+}, Na^+)_2(Mn^{4+}, Mn^{3+})_6O_{12}\cdot 3\text{–}4.5H_2O$, and with an empirical formula of $Na_{0.2}Ca_{0.05}K_{0.02}Mn^{4+}{}_4Mn^{3+}{}_2O_{12}\cdot 3H_2O$. Todorokite grows 10–100 times faster than vernadite. In some cases, birnessite $(K^+, Ca^{2+}, Na^+)(Mn^{4+}, Mn^{3+})_2O_4\cdot 1.5H_2O$ can form from todorokite.

PSNs occur throughout the oceans but are concentrated in the Pacific, including the Clarion-Clipperton zone (CCZ) between Mexico and Hawaii (15 kg/m^2 nodule density), the Peru Basin off the west coast of South America (10 kg/m^2 nodule density), near the Cook Islands in the southwestern Pacific Ocean (5 kg/m^2 nodule density), and Indian Ocean (5 kg/m^2 nodule density). The CCZ, Indian Ocean, and Cook Islands nodules have been reasonably well characterized. Manganese is relatively high in the CCZ and Indian Ocean nodules at about 28% and 24%, respectively, with 6%–7% Fe. The Cook Islands nodules have less manganese (~18%) and more iron (~16%) than CCZ and Indian Ocean nodules. The CCZ nodules have appreciable nickel, and the Indian Ocean nodules have nickel and copper, while the Cook Islands nodules are relatively enriched in cobalt and rare earth metals.

In addition to PSNs, ferromanganese sea crusts offer considerable potential as sources for manganese. These crusts form by the precipitation of vernadite and poorly crystalline ferric oxyhydroxide onto sediment-free rock surfaces, sea mounts, knolls, and plateaus. The crust formation in the Pacific Ocean is often similar to hydrogenetic PSN formation via cold water precipitation. In the Atlantic Ocean, crust formation is more often associated with hydrothermal vents. Such vents can reduce the metal concentration in the Atlantic Ocean crusts compared to Pacific Ocean Crusts. Once the crust is formed, other metals are adsorbed onto the crust, which can have specific surface areas up to 325 m^2/cm^3. Ferromanganese crusts are, therefore, enriched in metals, including cobalt, rare earth elements, nickel, and tellurium. The crusts typically contain around 20%–25% Mn regardless of location.

Largest Known Reserves

Current reserve estimates for manganese ores indicate that 150 Mt (million metric tons) of economically recoverable manganese exists in South Africa, primarily in the Mamatwan and Wessels formations. Ukraine has 140 Mt, much of that located in the Nikopol area; 97 Mt in Australia; 54 Mt in Brazil; 52 Mt in India; 44 Mt in China; and 24 Mt in Gabon. Mexico and Kazakhstan round out the list of countries with known significant reserves at 5 Mt each. Most of the world's manganese resources are in South Africa, generally estimated at up to 15 Bt (billion tons) of ore or 75% of the world's total land manganese reserves. Ukraine manganese reserves are estimated at 10% of the world total. In this chapter, the term *reserve* is not meant to denote a geologically proven quantity, and *reserve* and *resource* are used interchangeably, depending on usage in the cited document (Corathers 2015).

The reserves of PSNs are estimated to contain at least 6–7.5 Bt of manganese. Although the number of sea mounts, knolls, and plateaus is not well determined, the tonnage of manganese in ferromanganese crusts is estimated to be about 33% of the total estimated land manganese resources. Figure 3 shows the distribution of manganese reserves throughout the world.

LARGEST PRODUCERS

Manganese ore is processed in several countries, as shown in Figure 3. Production can be divided into large-scale production that is generally available to worldwide markets and small-scale production for local markets.

There are several commercial products. The main products are manganese ore (broken into classes by manganese content: 20%, 36%–39%, 42%, 44%, 46%, and >47% Mn), ferromanganese (broken into classes by carbon content: low carbon [<1% C], medium carbon [1%–2% C, 85% Mn], and high carbon [>2% C, 75% Mn]), silicomanganese, electrolytic manganese dioxide, and chemical manganese dioxide. Permanganate (potassium and sodium), manganese oxides other than dioxide, manganese sulfate, and manganese metal have lesser demand. Ore, from large-scale processing, is generally exported for conversion, usually to ferromanganese

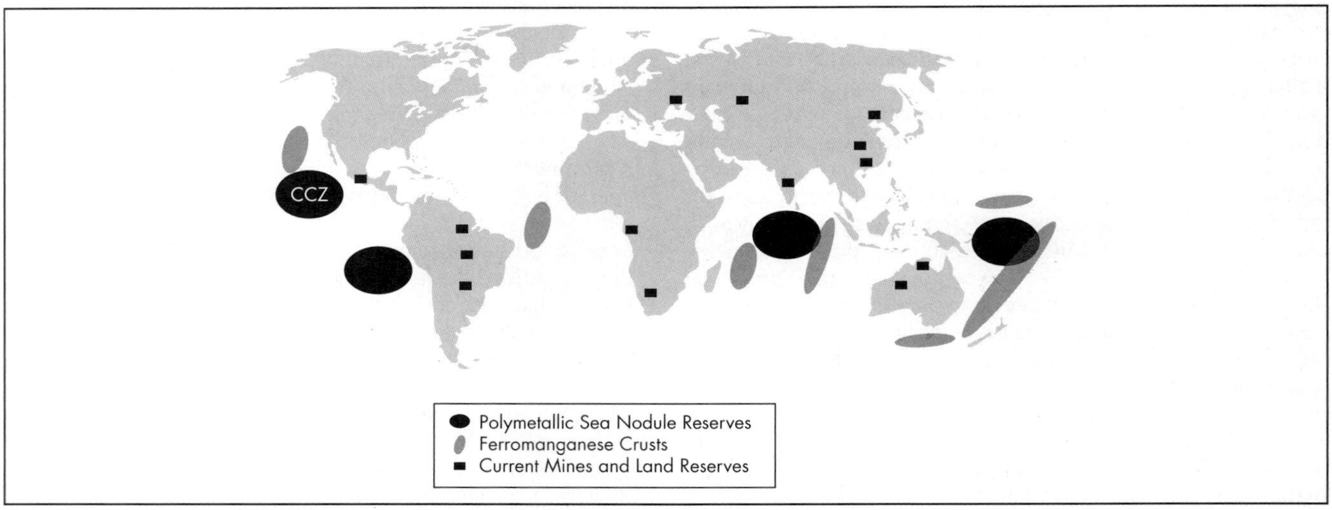

Figure 3 Distribution of world manganese reserves shown by type

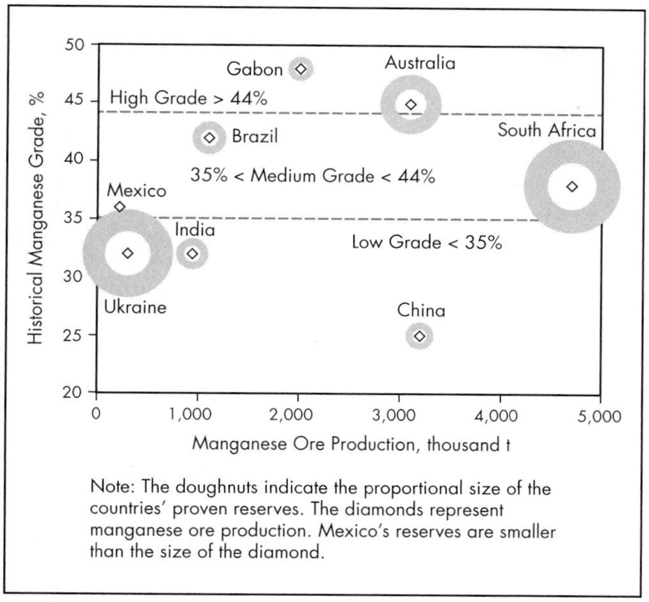

Data from Lawcock et al. 2013; Corathers 2015

Figure 4 Manganese ore production in 2014 by country as a function of historical grade of ore

or silicomanganese for steelmaking. Electrolytic manganese dioxide is used for battery cathodes, and chemical manganese dioxide is used for manganese ferrite production and for agricultural and chemical reaction purposes. Because of the variety of products, the manganese ore producers are discussed as related to the countries in which the mines exist, as shown in Figure 4.

Australia

Groote Eylandt Mining Company (GEMCO) is a subsidiary of South32 (which is currently "de-mergering" from BHP Billiton) and runs the primary processing facility for manganese at Groote Eylandt, Northern Territories, Australia, which opened in 1965 (South32 2017). The GEMCO facility can process 4–5 Mt/yr of ore (Wong 2011). The GEMCO ore is shipped to TEMCO, a wholly owned subsidiary in Tasmania, where the ore is transformed to high-carbon ferromanganese, silicomanganese, and sinter. OM Holdings operates the Bootu Creek mine, which opened in 2006. The ore processing facility is designed to process up to 1 Mt/yr. The Bootu Creek ore is typically shipped to China for ferroalloy (HCFeMn and SiMn) production. Pilbara Manganese, a wholly owned subsidiary of Consmin Ltd., runs the Woodie Woodie mine in Western Australia. The Woodie Woodie operation can process up to 1.5 Mt of ore per year. The processed ore is shipped directly to customers, often for specialty material production, as the ore has low silicon, phosphorus alumina, and iron content. Process Minerals International Pty Ltd. processes about 400,000 t of fines from the Woodie Woodie mine area (Mineral Resources Limited, n.d.).

Brazil

Vale S.A. is the primary manganese producer in Brazil, with ~70% of the Brazilian ore market. Vale product enters the market through a variety of wholly owned subsidiaries, including Vale Manganês S.A., Vale Mina do Azul S.A., and Mineracaõ Corumbaense Reunida S.A. Vale owns the Azul in Para and Urucum in Mato Grosso do Sul and makes ferromanganese (through Vale Manganês S.A.) in Minas Gerais and Bahia states (Gilroy 2014).

China

China produces more manganese than any other country. Most of this production is from imported ore that is used to produce refined manganese products, such as ferromanganese and electrolytic manganese metal. In about 2015, the main producers were Chongqing Tycoon Manganese Company Ltd. and Guangxi Dameng Manganese Industry Company Ltd. China imports manganese ore because the available domestic ore supply is unable to meet demand. Liaoning in Northeast China, Chongqing in Southwest China, Guangxi, Hunan, and Guizhou are the areas of China with the largest manganese reserves (Bloomberg 2015a, 2015b).

Gabon

Comilog (Compagnie minière de l'Ogooué), a subsidiary of Eramet, runs the Moanda mine. This mine produces very high-grade manganese ore. Typically, the ore is shipped after processing to various Eramet operations around the world, including Tinfos Jernverk in Norway for silicomanganese, ferromanganese alloys, and chemical manganese through Erachem Comilog in Belgium, the United States, Mexico, and China. The Moanda mine produced about 4Mt of ore in 2016 (Eramet 2018).

India

India, like China imports considerable manganese ore. MOIL (formerly Manganese Ore India Limited) produced about 51% of the manganese ore in India with 1.14 Mt of ore produced and was seeking to add four mines to its operations and to increase production by ~600,000 t/yr (Gundewar, 2014a, 2014b; Nagpur Pyrolusite 2015).

South Africa

The Mamatwan and Wessels mines (collectively, the Hotazel manganese mines) are primarily (74%) owned by South32 through South Africa Manganese, with the remainder owned by broad-based black economic empowerment investors (Gajigo et al. 2011). The processed ores are smelted by Metalloys at a facility near the mine in South Africa to make high- and medium-carbon ferromanganese and silicomanganese. Assmang Manganese (jointly managed by Assore Manganese and African Rainbow Minerals) also has mines in the Kalahari manganese field. Although much of the ore is exported after processing, Assmang operates four smelters (South32 2015; Kelly and Matos 2014).

HISTORICAL PRICE

The price of manganese ore, in constant-year dollars, tracks world manganese ore production to a certain extent but also has shown years of relative constancy, as shown in Figure 5. This relative constancy seems much less prevalent after 2010.

Mine production has been increasing steadily since the low of 6.4 Mt of manganese in 1999 to 17.4 Mt of manganese in 2014, with an increase of roughly 750,000 t of manganese per year. The price of manganese ore was relatively constant from 1994 through 2007 at about US$3.5/t in 2015 dollars, except for a small price shock in 2005 caused by increased demand for ferroalloys and increased transportation costs

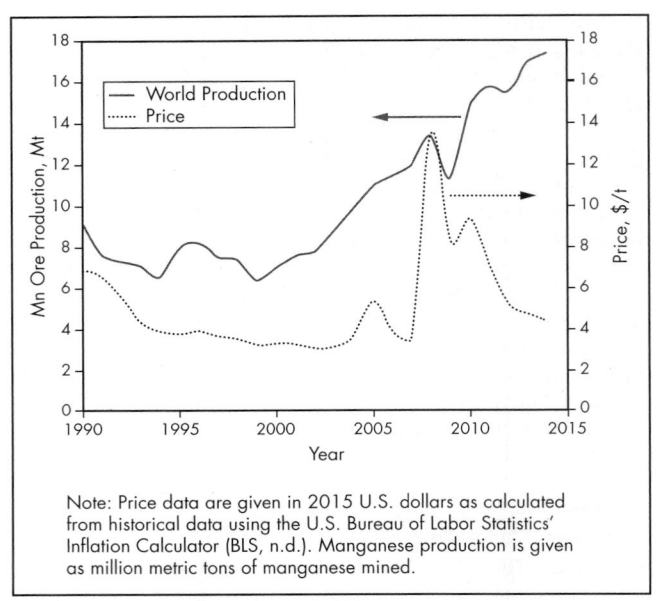

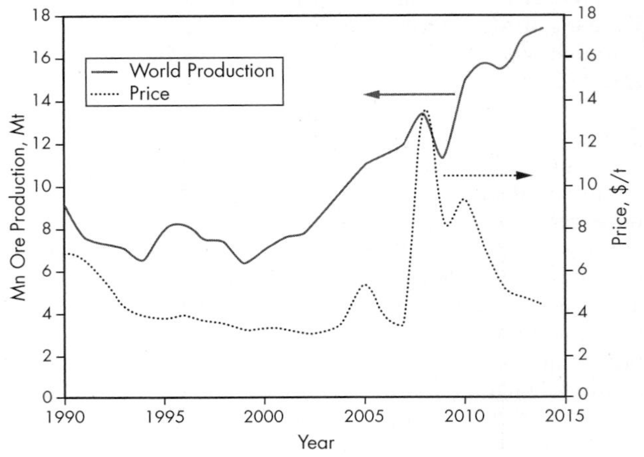

Note: Price data are given in 2015 U.S. dollars as calculated from historical data using the U.S. Bureau of Labor Statistics' Inflation Calculator (BLS, n.d.). Manganese production is given as million metric tons of manganese mined.

Data from Goonan et al. 2015; USGS 2016

Figure 5 Manganese production and price trends since 1990

(Matos and Carothers 2005). The price quadrupled in 2008, primarily because of two main factors, increased global consumption, mainly in China and India, and lower production in Brazil, China, and South Africa (Corathers 2008). Since the Great Recession of 2008–2009, the manganese ore price has begun trending back toward the previous long-term price. In other words, the market participants seem to have adjusted to the previous price fluctuations.

EXTRACTION

Traditionally, pyrometallurgical routes have been used to process manganese minerals to manganese-based materials, primarily ferromanganese and silicomanganese suitable for steelmaking. Beginning in the late-1930s in the United States, hydrometallurgical processing routes, both with and without pyrometallurgical pretreatment, have been investigated and/or commercialized. In this section, pyro- and hydrometallurgical processing routes are described, and recently developed methods for processing low-grade manganese deposits are discussed in more detail.

Beneficiation

A wide variety of manganese mineral beneficiation techniques have been utilized. For high-grade ores, beneficiation is often as simple as separating the fines after crushing and grinding. In this case, the manganese grade is increased by removing clay-type gangue particles (fines) that are often present in the sedimentary manganese deposits. For lower-grade ores, gravity separation, magnetic separation, and/or flotation are commonly performed (Pienaar and Smith 1992). Table 1 shows key separation parameters germane to the beneficiation of the various manganese minerals.

Pyrometallurgical Extraction

Pyrometallurgical extraction often follows a path similar to iron ore extraction. Specifically, the ore is crushed, ground, size fractionated, and beneficiated if necessary. For lateritic ores, the resulting concentrate is typically of a higher oxidation state, 4+, 3+ (Gordon and Nell 2013), including mineral classes such as manganese oxide (pyrolusite, cryptomelane, hausmannite) or silicate (mainly braunite) mineral with associated iron minerals. These types of concentrates are well-suited for preparing manganese for its main use in steelmaking. During smelting, the ore concentrate is mixed with coke (or other carbon form) and heated to at least 1,200°C (often 1,500°C for ferromanganese and 1,600°C for silicomanganese) in a reducing atmosphere. In many cases a flux, such as lime, silica, or calcite, is added to help remove impurities. Under these conditions, the high oxidation state manganese oxides are reduced to MnO (manganosite). A reduction reaction for pyrolusite is given in Equation 1. The MnO formed will then react with the coke to form manganese metal, as shown in Equation 2. Intermediate manganese oxides, such as Mn_2O_3, and Mn_3O_4, can also result.

$$MnO_{2(s)} + CO_{(g)} \rightarrow MnO_{(s)} + CO_{2(g)} \quad \textbf{(EQ 1)}$$

$$2MnO_{(s)} + C_{(s)} \rightarrow 2Mn_{(s)} + CO_{2(g)} \quad \textbf{(EQ 2)}$$

Three primary products are formed for use in steels: silicomanganese (SiMn), high-carbon ferromanganese (HCFeMn), and refined ferromanganese (refined FeMn). The most commonly used (~55%) smelter product is silicomanganese. Silicomanganese contains 65%–70% Mn, 18%–20% Si, and 1%–2% C, with the remainder primarily iron and is usually used to deoxidize steel. The feed material is typically a mix between manganese ore and slag from high-carbon ferromanganese production. Smelting is often performed in a submerged electric arc furnace with the smelting temperature >1,600°C to allow the ore and the ferromanganese slag to melt fully, thereby yielding a high silicon content and low MnO content in the slag. ASTM Standard A483-10 defines silicomanganese compositions. Three grades are distinguished by their carbon and silicon contents. The ASTM standard requires 65%–68% Mn for all grades. For example, Eramet Comilog (2013) specifies P < 0.17% and S < 0.03%. About 75% of the SiMn manufactured goes to steel mini mills. After melting, the SiMn is tapped and cast into billets that are then crushed. The crushed SiMn is typically sold as the bulk crushed alloy. As an example, Eramet Comilog produces SiMn in France and Norway and sells the material in three size ranges (20–80 mm, 10–50 mm, and 3–25 mm) as a bulk material packaged in big bags with other packaging available on request.

For HCFeMn, ASTM 99-03 (2014) indicates that the Mn concentration should be between 74% and 82% depending upon grade, <7.5% C, <1.2% Si, <0.35% P, and <0.05% S. HCFeMn is mainly 85% Mn and is used by integrated steel mills, with the balance used by mini mills. In general, making HCFeMn is similar to making SiMn. Submerged arc furnaces are generally preferred, although blast furnaces can be used. Coke and limestone are added to lower the oxidation potential and to provide a slag former, respectively. The temperature can be a little hotter than that used in silicomanganese production, up to 1,800°C. The feed is typically manganese ore, beneficiated as necessary to achieve >40% Mn, with preheating and oxygen enrichment when blast furnaces are used. Slag characteristics are very important to the performance of the furnaces. Fines should be avoided in submerged electric arc furnaces, as the fines have a tendency to disrupt the flow of carbon monoxide and carbon dioxide (CO and CO_2) through the burden. When fines are to be used, they must be agglomerated and

Table 1 Key separation parameters for manganese minerals

Mineral	Formula	Density, g/cm^3	Reference	Magnetic Susceptibility, 10^{-3} cm^3/g	Reference	Point of Zero Charge pH	Reference
Land-Based Mineral Separation							
Braunite	Mn_7SiO_{12}	4.7–4.8	Hudson Institute of Mineralogy 2015; Drzymala 2007	<1; 0.025–0.5*	Taneja and Garg 1993; Rosenblum and Brownfield 2000	—	—
Cryptomelane	$K_2Mn_8O_{16}$	4.4	Drzymala 2007	0.3–0.6*	Rosenblum and Brownfield 2000	<3, 1.7–2.1	Kosmulski 2009
Hausmannite	Mn_3O_4	4.77–4.85	Todd 2010; Hudson Institute of Mineralogy 2015	0.5–0.76	Drzymala 2007	<5, 6.5, >10	Rosenblum and Brownfield 2000; Drzymala 2007; Tan et al. 2008; Kosmulski 2009
Manganite	MnO(OH)	4.3–4.4	HHS 2012; Drzymala 2007	0.36–0.5	Drzymala 2007	6.3, 7.4	Rosenblum and Brownfield 2000; Kosmulski 2009
Pyrochroite	$Mn(OH)_2$	3.26	Drzymala 2007	—	—	7	Kosmulski 2009
Pyrolusite	MnO_2	4.7, 5.04–5.08	Todd 2010; Hudson Institute of Mineralogy 2015	0.30.48	Drzymala 2007	4–8, 7.2	Kosmulski 2009; Tan et al. 2008
Rhodochrosite	$MnCO_3$	3.7	Hudson Institute of Mineralogy 2015; Drzymala 2007	1.31–1.34	Drzymala 2007	6–10.5	Kosmulski 2009
Sea-Based Mineral Separation							
Birnessite	$(K,Ca,Na)(Mn)_2O_4 \cdot 1.5H_2O$	3.0	Drzymala 2007	—	—	1–1.6	Tan et al. 2008
Todorokite	$Na_{0.2}Ca_{0.05}K_{0.02}Mn_6O_{12} \cdot 3H_2O$	3.7	Drzymala 2007	—	—	3.2, 3.4–4, 3.8	Tan et al. 2008; Kosmulski 2009

*Value given is amperage needed to attract the mineral in a modified Frantz isodynamic separator.

sintered. This can be helpful because this process can be used to pre-reduce the manganese oxides, lowering the coke and temperature requirements.

When SiMn and HCFeMn are made near each other, high-manganese slags are often produced for feed to the SiMn smelter. The high-manganese slag contains 30%–40% Mn, 2%–5% MgO, 10%–30% Al_2O_3 (aluminum oxide), 15% CaO, and 25%–30% SiO_2. Manganese ores containing <40% Mn can also be used, and these are typically smelted to produce low-manganese slag. In these cases, and for most discarded slags, the manganese is less than 20%, with recovery typically >85%. Other slag components include up to 8% MgO, 10% Al_2O_3, 35% CaO, and 30% SiO_2. Again, the HCFeMn is cast, crushed, and sold as the bulk crushed alloy. Eramet Comilog sells this material in the same size and manner as SiMn.

Refined FeMn is also called low- or medium-carbon FeMn. Grades as specified by ASTM A99-03 (2014) are 80%–85% Mn for medium-carbon FeMn and 80%–90% Mn (depends on the grade) for low-carbon FeMn. Medium-carbon FeMn has less than 1.5% C, while low-carbon FeMn contains <0.75% C (depends on grade). The phosphorus and sulfur are slightly lower after refining, <0.3% P in both low- (depends on grade) and medium-carbon FeMn and <0.02% sulfur in both types of refined FeMn. The amount of silicon varies between the different grades of refined FeMn. Refining can be performed by oxidation of some of the carbon in the HCFeMn. Silicothermic reduction can also be used. In silicothermic reduction, SiMn is mixed with manganese ore and lime or a high-manganese slag to reduce the MnO and form Mn. The lime is used to reduce the activity of the silica to force the MnO reduction to produce more Mn. Integrated mills use about 70% of the refined FeMn, with much of the rest procured by specialty steel manufacturers for production of stainless steel and HSLA steels. Eramet Comilog sells this material in the same size and manner as SiMn.

A manganese life-cycle assessment was recently performed by the International Manganese Institute. Typical manganese production methods release about 6 kg CO_2 equivalents per 1 kg of manganese alloy. In addition, 3 g ethane equivalent, 45 g SO_2 equivalents, and 9.6 g particulate matter are released per kilogram HCFeMn. Except for particulate matter, electricity generation is responsible for 60%–80% of these emissions (IMnI 2016).

An example of a modern smelter operation is that proposed by Gulf Minerals in West Timor, Indonesia (Gulf Manganese Corporation 2015b). Indonesia has considerable manganese reserves, both on land and as sea nodules/ferromagnetic crusts (Cunningham 2015); however, little exploitation of these reserves has occurred. The Timor smelter is expected to use high-quality local manganese ore, as Indonesian law limits the exportation of unprocessed ore. The ore to be used in Timor has a low iron content (~2%) as well as low phosphorus and sulfur (see Figure 6). The ore is to be jigged to remove the low-density silica and clay minerals and increase the manganese grade. After jigging, the manganese concentrate, iron ore, metallurgical coal, and limestone will be added to the partially submerged electric arc furnace to make the HCFeMn. Eight furnaces are planned, which will require 64 MW of power, necessitating a dedicated coal-fired power plant at the smelter site. The furnaces will be tapped every 90–100 minutes. When fully completed, 290,000 t of high-manganese ore will be processed per year, leading to 155,000 t of HCFeMn. The smelting requires ~0.944 t of coal, 0.337–0.502 t of limestone, and 0.212 t of iron ore per ton of HCFeMn. For the desired amount of HC FeMn, ~145,000 t of

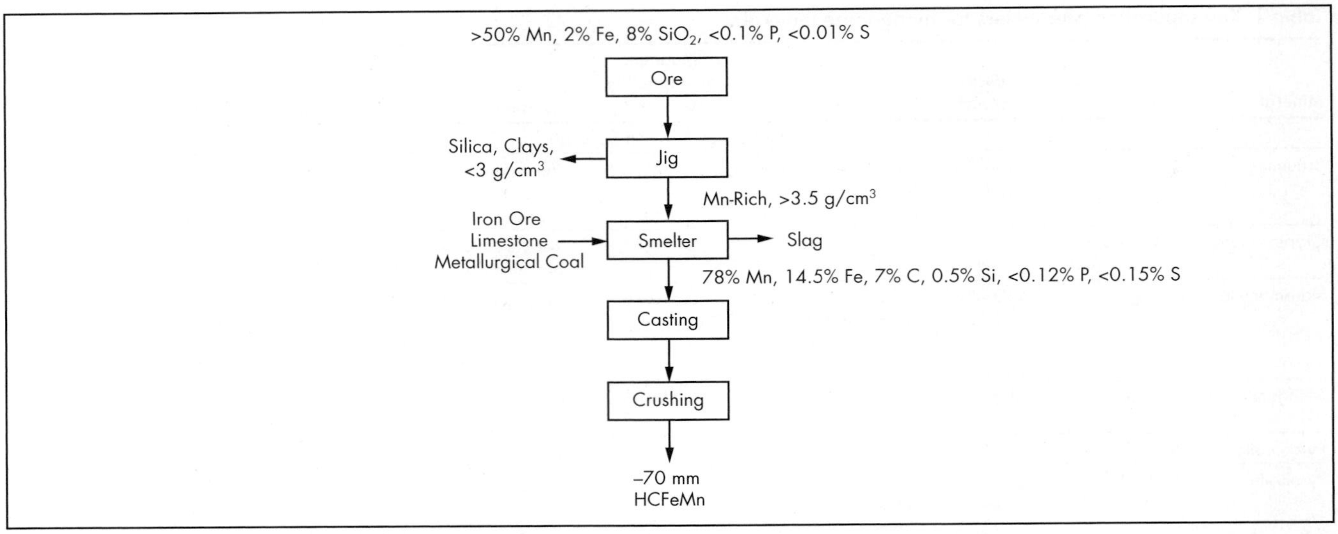

Data from Gulf Manganese Corporation 2015b

Figure 6 Flow sheet for the Gulf Minerals high-carbon ferromanganese plant in West Timor, Indonesia

metallurgical coal are required, along with ~50,000–75,000 t of limestone and 30,000–35,000 t of iron ore. The HCFeMn will be exported, with ~50% to Europe, 25% to Japan, with smaller amounts to Korea and China.

Hydrometallurgical Extraction

The U.S. Bureau of Mines conducted extensive laboratory and pilot-scale investigations in the 1930s and operated a large experimental plant at Boulder City, Nevada (United States), during 1942–1946 and a pilot plant near Oacoma, South Dakota (United States), from 1944–1946. The first commercial hydrometallurgical plant was completed by Electro Manganese Corporation in Knoxville, Tennessee (United States) in 1939. These facilities produced electrolytic manganese metal (EMM) by reduction roasting of pyrolusite to MnO and leaching of the MnO with dilute sulfuric acid. Subsequent installations were built by Union Carbide's Metals Company (later Elkem Metals Company) in Marietta, Ohio; Foote Mineral Company in New Johnsonville, Tennessee; and American Potash Corporation (later Kerr-McGee Chemical Corporation) in Hamilton, Mississippi, and Henderson, Nevada (all in the United States). EMM was initially produced at all of these locations, except Henderson, where only electrolytic manganese dioxide (EMD) was produced as an anode deposit. Eventually, Union Carbide and Foote also produced EMD for use mainly in dry-cell batteries (T.P. McNulty, personal communication).

Flow sheets varied and were periodically upgraded, but reduction roasting was accomplished either in enclosed furnaces or open heaps with natural gas or propane introduced through a supporting grid, producing a reducing environment within the heap. Following solid–liquid separation, the filtrate was treated with sulfide ion or hydrogen sulfide gas to precipitate heavy metals. The clarified filtrate was electrolyzed. Anodic deposition of EMD was on graphite plates that were manually stripped, while water-cooled tubing served as the cathode. Cathodic deposition of EMM was enabled by operating at an elevated electrolyte temperature with perforated anodes to increase the current density on the anode surface and retard deposition of EMD. Diaphragms were used to isolate the electrodes that were not being harvested (e.g., the anodes, in the case of EMM production). The plants previously mentioned had aggregate capacities of about 12,000–30,000 t annually of EMM or EMD. During the early-1980s, EMD producers struggled to make a product that had a 3-year battery shelf life, whereas modern dry-cell batteries typically have a 10-year shelf life. The improvements are attributable to advancements in electrolyte purification, product grinding and classification, process control, and quality assurance and quality control procedures. The original technologies suffered from high labor intensity, largely due to small cathode blanks for EMM and small graphite anodes for EMD. The graphite anodes were fragile and were gradually destroyed by oxidation (from evolved oxygen gas) and by operators being too aggressive with mallets used to dislodge the brittle deposit (T.P. McNulty, personal communication).

Hydrometallurgical processing has been accomplished in two ways: either a hybrid pyro-hydrometallurgical route or a direct hydrometallurgical route can be taken.

Pyro-Hydrometallurgy

In the combined pyro-hydrometallurgy treatments of manganese ores, two main types of pyrometallurgical pretreatments are used. The first of these is sulfation roasting of pyrolusite to form manganese sulfate, and the second is calcination of rhodochrosite to a manganese oxide.

For the sulfation roast, pyrolusite ore is heated in the presence of SO_2 (sulfur dioxide) gas or in the presence of a sulfate chemical such as sulfuric acid or ammonium sulfate. Roasting with SO_2 can be accomplished at relatively low SO_2 partial pressure and moderate temperature. For many manganese ores, too high a partial pressure can lead to increased sulfation of iron oxides, while too high a temperature (>~525°C) leads to transformation of the MnO_2 to bixbyite (Mn_2O_3), which is less amenable to sulfation, thereby increasing the difficulty of leaching. Manganese sulfate is quite soluble in water (>35% at 20°C, decreasing to 10% at 140°C), and the rate and equilibrium extraction can be increased with temperature. Decreasing the pH does not appreciably change the extraction. With sulfuric acid roasting, the roasting can be accomplished at much lower temperatures than roasting with SO_2 gas. Roasting temperatures as low as 150°–200°C can yield >90% extraction of manganese after one hour of ambient temperature leaching (Güler et al. 2008; Zhang et al. 2013; Guo et al. 2009).

In the calcination of rhodochrosite, the ore is heated to drive off the CO_2 gas, typically at temperatures of 800°C or greater (You et al. 2015). At these temperatures in an air atmosphere, the MnO formed oxidizes to MnO_2, which is then cooled and subject to leaching. Thermal reduction of manganic oxides to MnO, as practiced in the United States by Union Carbide and Kerr-McGee Chemical Corporation, was accomplished at moderate temperatures in a reducing atmosphere created by incomplete combustion of natural gas or propane.

Hydrometallurgy

Leaching. The leaching of manganese ores and pyrometallurgically treated concentrates has been studied fairly extensively. As can be seen from examination of Figure 1 and similar Pourbaix diagrams (Veglio et al. 2001), manganese oxides are subject to leaching by lowering the pH and Eh to yield Mn^{2+} ions. Solubility product (K_{sp}) data are available for a few manganese compounds and can be calculated from various databases (Gustafsson 2015; Pourbaix 1974): $MnCO_3$, $K_{sp} = 1.8 \times 10^{-11}$; $Mn(OH)_2$, $K_{sp} = 1.5 \times 10^{-13}$; and MnS, $K_{sp} = 3.7 \times 10^{-15}$ (the preceding calculations were performed with data from the Minteq database [Gustafsson 2015]). Manganese carbonate leaching is assisted by carbonate ions, because at pH <5–6, most of the carbonate will react to form carbon dioxide, reducing the carbonate ion activity and hence causing the manganese activity in solution to be high. Thus, the manganese extraction will be relatively high at these mild pH values. For manganese hydroxide, similar to manganese carbonate, fairly mild pH values allow high manganese ion activity to be achieved. For typical ore constituents, the chemical reactions are slightly more complicated than for the simple carbonate and hydroxide minerals discussed. Also, the manganese solubility is a function of the oxygen fugacity, in addition to the temperature and pH. For pyrolusite, a simple leaching reaction can be written:

$$MnO_{2,s} + 2H^+_{aq} \rightarrow Mn^{2+}_{aq} + H_2O + 0.5O_{2,g} \quad \textbf{(EQ 3)}$$

The equilibrium constant for Equation 3 is 1.02 (calculation performed with data from the Minteq database [Gustafsson 2015]). Therefore, while the reaction proceeds to the right, low pH and/or low oxygen pressures are needed to achieve significant manganese extraction. For instance, when leaching in air, a pH of 1.67 is needed to achieve a manganese concentration of ~55 ppm. Parc et al. (1989) have written similar equations for several manganese minerals, including cryptomelane, lithiophorite, nsutite, birnessite, manganite, and rhodochrosite. A similar reaction to Equation 3 using manganese carbonate rather than manganese oxide has an equilibrium constant of ~2,000 and gives a manganese activity of 0.001 at pH 5.9.

Because of the preceding considerations, more complicated hydrometallurgical methods have been developed. In general, the methods have been primarily aimed at leaching pyrolusite with lixiviants in addition to hydrogen ions. For instance, in methods somewhat similar to sulfation roasting, SO_2 gas or other sulfur-based reductants, such as dithionate or sulfite, are used, usually with sulfuric acid. Sulfur dioxide is oxidized to sulfate and dithionate, while the manganese is reduced and forms Mn^{2+} in aqueous solution. Ferrous sulfate has also been used to leach pyrolusite, with the product being manganese sulfate and various ferric salts, and the composition of these salts depends on the amount of acid available. With enough excess sulfuric acid, ferric sulfate is formed.

For low-manganese-content resources, lower cost methods have been studied to leach manganese (e.g., Zhang and Cheng 2007; Chow et al. 2010, 2012, 2013). Many of these methods revolve around the use of organic, primarily sugar-based, reductants in sulfuric acid solution (Beolchini et al. 2001; Ismail et al. 2004; Furlani et al. 2006; Su et al. 2008, 2009; Lasheen et al. 2009). These organic reductants include glucose, sucrose, lactose, sawdust, waste tea, molasses, glycerin, plant leaves, and several varieties of fungi and microbes. The organic reductants are generally chosen to reduce the cost of leaching with the cost reduction meant to offset the loss in value associated with the lower content of valuable metal. Typically, sulfuric acid is used for leaching, although hydrochloric and nitric can also leach manganese minerals.

Pressure leaching has been explored for nodules and crusts, but this is typically performed to remove non-manganese value, and manganese extraction is low.

Recovery from solution. Once manganese ions have been leached from the ore, the aqueous solution needs to be concentrated and purified, and the manganese recovered in a suitable form. How this is accomplished is dependent on the other ions that are leached along with the manganese ions. For oxide-based manganese ores, the typical gangue minerals include alumina, quartz and hematite, and clays, such as kaolinite. For the carbonate-based ores, calcite, siderite and mixed carbonates, such as dolomite, and mangano calcite can be present, along with quartz and clays. Other metallic cations, such as zinc, arsenic, cobalt, and cadmium, are often present in relatively small amounts in manganese ores but are often sufficient to make these metals the primary recovery target in nodules and crusts. In many cases, useful methods for recovery from solution arise by reversing the leaching processes, as revealed by the use of Pourbaix diagrams (see Figure 1). In these cases, solution purification can be achieved by raising the pH. At 25°C, one expects, thermodynamically, that hydroxides will precipitate from solution in the following order: Fe^{3+}, Al^{3+}, Cu^{2+}, Zn^{2+}, Fe^{2+}, Co^{2+}, Mn^{2+}, Mg^{2+}, and Ca^{2+}. For instance, when testing an Arizona manganese ore for American Manganese, Kemetco Research found that raising the pH of the pregnant leach solution (PLS) from ~1 to ~5.6 removed ~98% of the Al, >94% of the As, >90% of the Cu, 99.7% of the Fe, 89% of the Si, and 57% of the Zn in solution, while the manganese content was reduced ~2% and the cobalt concentration was reduced about 11.5% (Chow et al. 2010, 2012, 2013). The hydroxide-treated PLS was then subject to sulfide precipitation primarily to remove the remaining cobalt and zinc. The thermodynamic order of removal is expected to follow the trend Cu, Cd, Zn, Co, Fe, Mn. Overall, at least 90% of Al, As, Co, Cu, Fe, Si, and Zn were removed sufficiently to yield a suitable feed for further processing of the PLS. The loss of manganese was 2.4%, while K, Ca, Mg, and Na ion concentrations were constant or increased. This type of processing is useful to produce a final leach solution that is primarily manganese in terms of its metal content. In the Kemetco Research data previously discussed, Mn^{2+} makes up about 80% of the metal value after hydroxide and bisulfite precipitations.

Precipitation of manganese carbonate is a common method for recovering manganese solution. Precipitation requires raising the pH of the leach solution and adding carbonate ions. Carbonate ions become the predominant species at pH ~9.5. Sodium or ammonium carbonate can be used, typically, in approximately equal stoichiometry with the manganese ions. Manganese carbonate is typically recovered as a precursor to making sintered-FeO/MnO feed materials for FeMn manufacture.

Another very common precipitation is the formation of MnO_2 as either EMD or chemical manganese dioxide (CMD). To precipitate EMD or CMD, the reaction must be performed with an oxidant (Zhang et al. 2002). The chosen oxidant must be sufficient to account for the potential of the MnO_2 reaction of 1.224 V (Pourbaix 1974). Oxidants that have been used include ozone–oxygen gas mixture, sulfur dioxide–oxygen, chloric acid, and hydrogen ions.

The primary difference between EMD and CMD materials is the purity needed. The EMD material is used in batteries and usually needs a higher purity than the CMD material, which is used for ferrites. EMD is often packaged as powder in 25- or 50-kg net paper or plastic/composite bags from China, or 50-lb bags or 3,000-lb super sacks in the United States by Tronox. CMD is packaged similarly to EMD in 25- or 50-kg woven bags, typically with an inner liner.

Manganese metal is the result of electrowinning PLSs. The electromotive force (emf) for the manganese metal-manganese 2+ ion couple is –1.18 V (Pourbaix 1974). As this emf value is quite low, as evidenced by Figure 1 in which manganese metal is not stable in water, the concentration of all ions having greater emf than –1.18 V should be kept low to keep the current efficiency high. Cobalt has the greatest negative effect, followed by nickel, copper, zinc, silicon, and ferrous iron. American Manganese, in their low-grade manganese process, investigated electrowinning of manganese metal from the purified PLSs. Current efficiencies of >65% were achieved using a Pb-Ag anode and a stainless-steel cathode. The feed solution contained 30–40 g/L Mn, 130–150 g/L ammonium sulfate, 1–2 g/L sodium sulfite, 350 A/cm^2 current density, feed pH 6.2, and a flow rate of 2 mL/min. Deposition was performed for 18–24 hours at 35°C. A small amount of SO_2 gas was used to improve current efficiency, primarily through conversion of γ-Mn to a fine-grained α-Mn. The grade of the plated manganese achieved was greater than 99.5% Mn. One final key to achieving the most efficient electro-deposition is the avoidance of forming MnO_2 at the anode. American Iron and Steel type 316 stainless steel seems to work best for reducing manganese dioxide formation. Temperatures exceeding 36°–38°C also reduce the current efficiency of the manganese electrowinning. These conditions are essentially identical to those developed by the U.S. Bureau of Mines and commercialized as discussed earlier. EMM is often produced as powder or flakes that are sold in 1,000-kg big bags, although smaller bag sizes are often available.

The recovery of manganese from aqueous solution using solvent extraction has also been well studied (Zhang and Cheng 2007). The extraction of manganese has been accomplished using phosphoric acid–based (bis[2-ethylhexyl] hydrogen phosphate, DEHPA, D2EHPA, or HDEHP), phosphinic acid–based (Cyanex 272), phosphonic acid–based (PC-88A/Ionquest 801), and carboxylic acid–based (Versatic 911) extractants. Acids used to alter the solution pH include sulfuric, hydrochloric, and nitric acids. The general solvent extraction equation can be written as follows:

$$\mathrm{mMn^{2+} + nm(H_2A_2)_{organic} \leftrightarrow [\{Mn(H_{2n-2}A_{2n})\}_m]_{organic} + 2mH^+} \qquad \textbf{(EQ 4)}$$

Usually, m = n = 1 is assumed for charge balance and simplicity considerations. However, while in most of the literature m = 1, n has been found typically to have been between 2 and 3. This indicates that some neutral extractant must be part of the extractant complex and that the coordination structure of the complex needs to be considered to better understand the extraction process. Extraction of manganese by DEHPA in sulfuric acid solutions was studied and gave a pH_{50} (the pH at which 50% of the metal ion is extracted) of ~2.7 when performed in the presence of calcium and magnesium. Under these conditions, the manganese is generally not extracted as well as zinc, but it is better than cobalt, copper, and nickel. With PC-88A, manganese extraction varies between sulfate and chloride media. The pH_{50} in sulfate media was found to be 3.2, but shifts to 2.6 in chloride media. With Cyanex 272, the pH_{50} in sulfate media was 4.6. For many systems, the extraction order has been found to be DEHPA > PC-88A > Cyanex 272. Manganese solvent extraction is often performed in the context of separating manganese from zinc, and cobalt, and nickel. Combinations of extractants have become more common recently. This trend is driven by the synergistic nature of the combined extractants.

Low-grade manganese value processing. As the high-grade ores become depleted, new processes for the extraction of low-grade deposits will be of great importance. One example is the process developed by Kemetco Research for American Manganese for their Artillery Peak (Arizona, United States) deposit. Figure 7 shows an outline of the flow sheet developed. Such a hydrometallurgical approach is likely to have considerable cost savings compared to conventional pyrometallurgical processes. One of the keys to this new process is the efficient destruction of dithionate. This method is reasonably versatile, as EMD, CMD, or feed for SiMn and FeMn can be produced with small changes in the process, mostly in the electrowinning circuit (Chow 2010, 2012, 2013).

SUMMARY

Manganese is considered a strategic metal and is essential for making many specialty steels. As >90% of manganese produced is used in steelmaking, the production and consumption of manganese should follow the basic trends expected for steel (Papp et al. 2007), with ~2%–5% annual growth through 2030. New uses related to batteries for electronic devices are also expected to become a greater driver for manganese consumption. Figure 8 shows a flow sheet for manganese materials. Currently, ~94% of manganese ore is processed pyrometallurgically. Of that, 75% is consumed in steel production, 15% ends up as slag from steelmaking, and 10% is used in foundries and welding applications. For the 75% used for steels, 55% goes to silicomanganese production, 35% to high-carbon ferromanganese production, and 9% to refined ferromanganese. The primary consumers of SiMn are mini mills, while integrated steel mills are the main consumers of the ferromanganese alloys. Hydrometallurgically produced manganese is primarily consumed as EMD for batteries and CMD for ferrites, agricultural chemicals, and dietary supplements and pure manganese metal. Approximately 50 new mines or mine

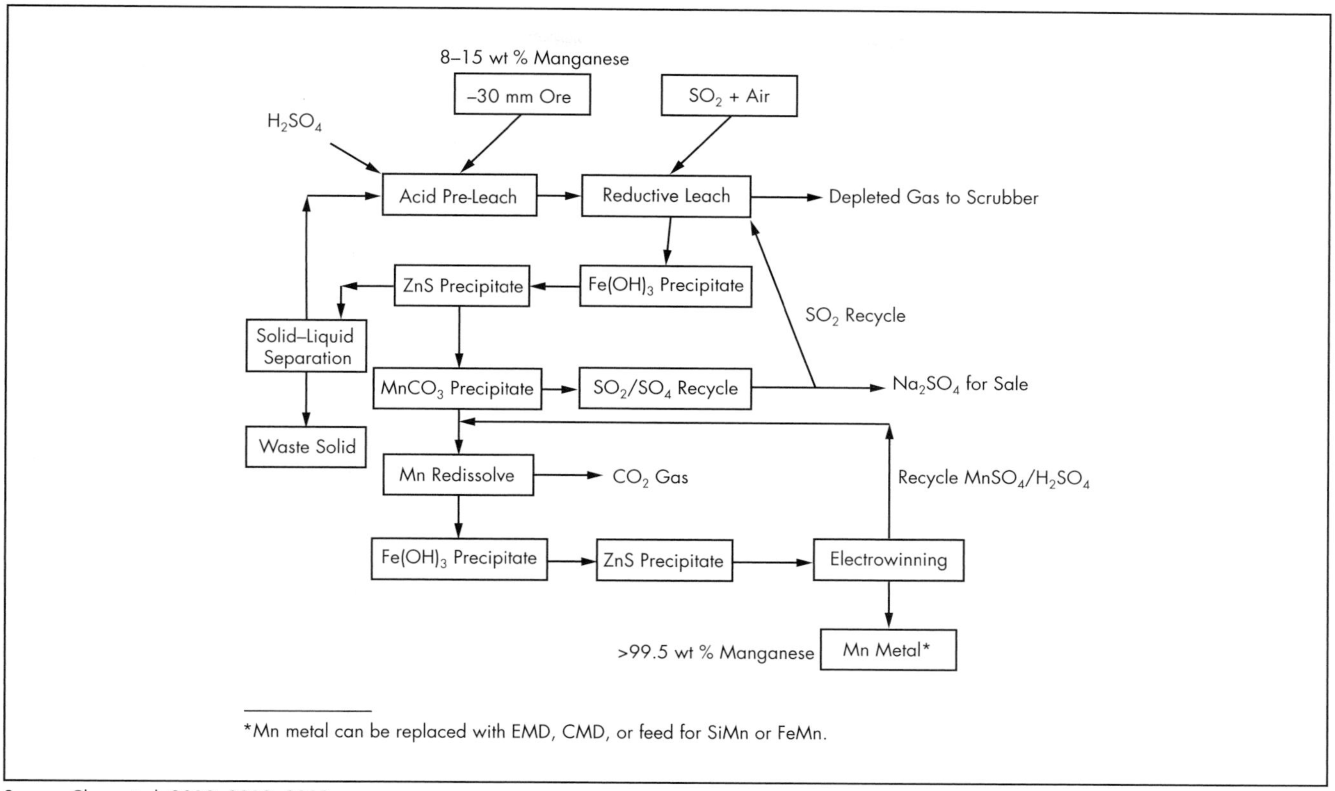

Source: Chow et al. 2010, 2012, 2013

Figure 7 Flow sheet for the Kemetco Research American Manganese/Artillery Peak process

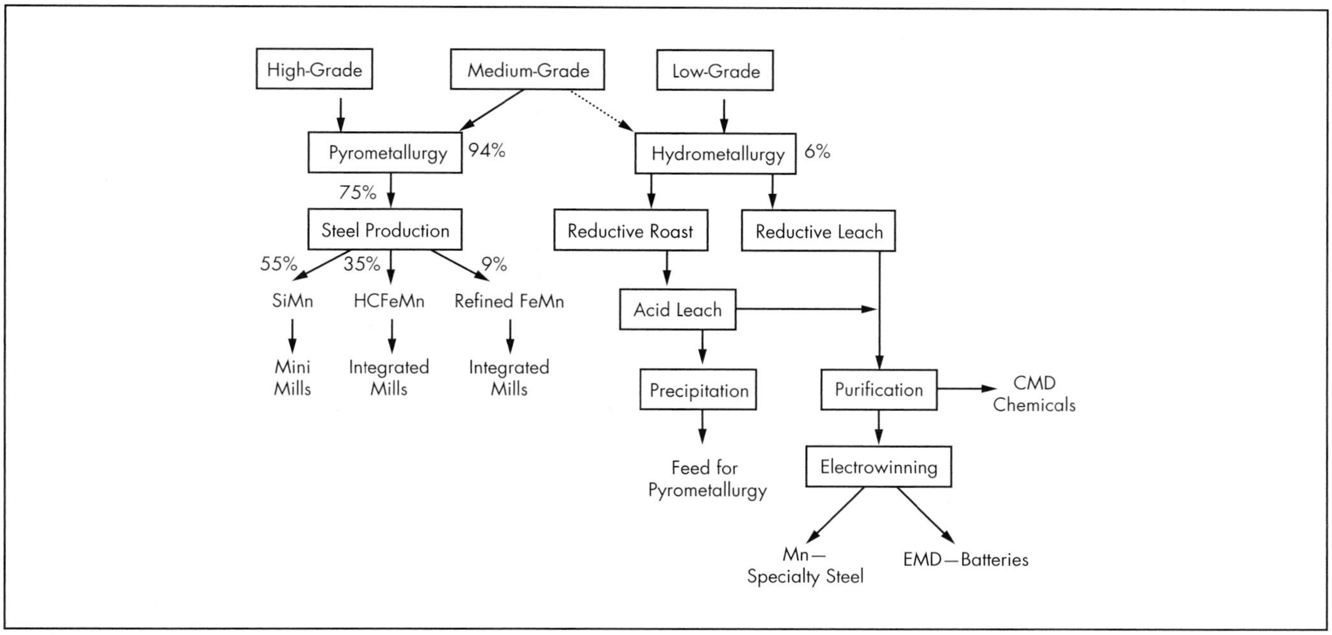

Data from Lawcock et al. 2013; Chow et al. 2012, 2013

Figure 8 Flow sheet for manganese production from ore

expansions were in development at the height of the manganese price bubble from 2009 to 2012 (Corathers 2014). The return of the manganese price to historic levels will likely cause many of the new mines to be placed on standby until most of the high-grade ores are depleted, at which point they may again be commercially viable, depending upon the utilization of sea nodules and ferromanganese crusts.

REFERENCES

ASTM A99-03. 2015. *Standard Specification for Ferromanganese*. West Conshohocken, PA: ASTM International.

ASTM A483-10. 2015. *Standard Specification for Silicomanganese*. West Conshohocken, PA: ASTM International.

Beolchini, F., Papini, M.P., Toro, L., Trifoni, M., and Veglio, F. 2001. Acid leaching of manganiferous ores by sucrose: Kinetic modelling and related statistical analysis. *Miner. Eng.* 14:175–184.

Bloomberg. 2015a. Chongqing Tycoon Manganese Industry Co. Ltd. Company profile. www.bloomberg.com/profiles/companies/0774928D:CH-chongqing-tycoon-manganese-industry-co-ltd. Accessed August 2015.

Bloomberg. 2015b. Metals and Mining: Company overview of Guangxi Dameng Manganese Industry Co., Ltd. www.bloomberg.com/research/stocks/private/snapshot.asp?privcapid=53303562. Accessed August 2015.

BLS (U.S. Bureau of Labor Statistics). n.d. Inflation calculator. www.bls.gov/data/inflation_calculator.htm. Accessed August 2015.

Cadex Electronics. 2015. Battery University—BU-205: Types of lithium-ion. Richmond, BC: Cadex Electronics. www.batteryuniversity.com/learn/article/types_of_lithium_ion. Accessed August 2015.

Campus Five. 2013. Guitar string composition and swing guitar [blog]. Jonathan Stout and his Campus Five. www.campusfive.com/swingguitarblog/2013/5/16/guitar-string-composition-and-swing-guitar.html. Accessed August 2015.

CDA (Copper Development Association). 2004. *Copper and Copper Alloys: Compositions, Applications and Properties*. Publication 120. Hertfordshire, UK: Copper Development Association.

Chalmin, E., Vignaud, C., Salomon, H., Farges, F., Susini, J., and Menu, M. 2006. Minerals discovered in paleolithic black pigments by transmission electron microscopy and micro-X-ray absorption near-edge structure. *Appl. Phys. A* 83:213–218.

Chen, L., Zhao, Y., and Qin, X. 2013. Some aspects of high manganese twinning-induced plasticity (TWIP) steel: A review. *Acta Metall. Sinica* 26:1–15.

Chow, N., Nacu, A., Warkentin, D., Aksenov, I., and Teh, H. 2010. *The Recovery of Manganese from Low Grade Resources: Bench Scale Metallurgical Test Program Completed.* IRAP Project no. 712681. Richmond, VA: Kemetco Research.

Chow, N., Nacu, A., Warkentin, D., The, H., Aksenov, I., and Fisher, J.W. 2012. New developments in the recovery of manganese from lower-grade resources. *Miner. Metall. Process.* 29:61–74.

Chow, N., Nacu, A-C., Warkentin, D., and Fisher, J.W. 2013. Processing of manganous sulphate/dithionate liquors derived from manganese resource material. U.S. Patent 8,460,631 B2.

Corathers, L.A. 2008. Manganese. In *Mineral Commodity Summaries*. Reston, VA: U.S. Geological Survey.

Corathers, L.A. 2014. Ferroalloys. *2012 Minerals Yearbook.* Reston, VA: U.S. Geological Survey.

Corathers, L.A. 2015. Manganese. In *Mineral Commodity Summaries*. Reston, VA: U.S. Geological Survey.

Cunningham, M. 2015. *Review of Manganese Prospects and Deposits in Indonesia.* Denver, CO: SRK Consulting.

Davis, J.R. 2001. High-strength low-alloy steels. In *Alloying: Understanding the Basics*. Materials Park, OH: ASM International. pp. 193–202.

Drzymala, J. 2007. *Mineral Processing: Foundations of Theory and Practice of Minerallurgy*, 1st English ed. Wroclaw: University of Technology.

Emsley, J. 2011. Manganese. In *Nature's Building Blocks: An A-Z Guide to the Elements.* New York: Oxford University Press. pp. 310–315.

Eramet. 2018. Comilog produced 4 million tons of manganese in 2017, Feb. 9. www.eramet.com/en/news/comilog-produced-4-million-tons-manganese-2017. Accessed July 2018.

Eramet Comilog. 2013. Silicomanganese. www.eramet.com/sites/default/files/eramet_silicomanganese_1.pdf. Accessed July 2018.

Eschler, P.Y., Lüthy, H., Reclaru, L., Blatter, A., Loeffel, O., Süsz, C., and Boesch, J. 2003. Copper-aluminium bronze—A substitute material for gold dental alloys? *Eur. Cells Mater.* 5:49–50.

Furlani, G., Pagnanelli, F., and Toro, L. 2006. Reductive acid leaching of manganese dioxide with glucose: Identification of oxidation derivatives of glucose. *Hydrometallurgy* 81:234–240.

Gajigo, O., Mutambatsere, E., and Adjei, E. 2011. *Manganese Industry Analysis: Implications for Project Finance.* Working Paper Series, no. 132. Tunis, Tunisia: African Development Bank.

Gilroy, A. 2014. Vale SA in the manganese and ferroalloys business. http://marketrealist.com/2014/12/vale-sa-manganese-ferroalloys-business/. Accessed December 2014.

Goonan, T.G., Jones, T.S., and Corathers, L.A. 2015. Manganese statistics. *In Historical Statistics for Mineral and Material Commodities in the United States*. USGS Data Series 140. Reston, VA: U.S. Geological Survey.

Gordon, Y., and Nell, J. 2013. Methods of manganese ore thermal-treatment prior to smelting—What to choose? In *The Thirteenth International Ferroalloys Congress: Efficient Technologies in Ferroalloy Industry*, Almaty, Kazakhstan. pp. 5–16.

Güler, E., Seyrankaya, A., and Cöcen, İ. 2008. Effect of Sulfation Roasting on Metal Extraction from Çinkur Zinc Leach Residue. *J. Ore Dress.* 10:1–10.

Gulf Manganese Corporation. 2015a. The elements of manganese. www.gulfmanganese.com/. Accessed August 2015.

Gulf Manganese Corporation. 2015b. *Timor Smelter Study*. South Perth, Western Australia: Gulf Manganese Corporation. http://gulfmanganese.com/wp-content/uploads/2016/07/25-May-2015-GMC-Timor-Smelter-Study-Combined.pdf. Accessed August 2015.

Gulson, B., Mizon, K., Taylor, A., Stauber, J.,Davis, J.M., Louie, H., Wu, M., and Swan, H. 2006. Changes in manganese and lead in the environment and young children associated with the introduction of methylcyclopentadienyl manganese tricarbonyl in gasoline—Preliminary results. *Environ. Res.* 100:100–114.

Gundewar, C.S. 2014a. *Manganese Ore: Vision 2020 and Beyond.* India: Indian Bureau of Mines.

Gundewar, C.S. 2014b. *Market Survey of Manganese Ore.* India: Indian Bureau of Mines.

Guo, X., Li, D., Park, K-H., Tian, Q., and Wu, Z. 2009. Leaching behavior of metals from a limonitic nickel laterite using a sulfation–roasting–leaching process. *Hydrometallurgy* 99:144–150.

Gustafsson, J.P. 2015. Minteq A2 database. http://vminteq.lwr.kth.se/download/. Accessed August 2015.

Gutzmer, J., and Beukes, N.J. 2009. Iron and manganese ore deposits: Mineralogy, geochemistry, and economic geology. In *GEOLOGY: Vol. IV, Encyclopedia of Life Support Systems.* Oxford: EOLSS. pp. 46–69.

Hein, J.R., and Koschinsky, A. 2013. Deep-ocean ferromanganese crusts and nodules. In *Treatise on Geochemistry*, 2nd ed. The Netherlands: Elsevier. pp. 273–291.

HHS (U.S. Department of Health and Human Services). 2012. *Toxicological Profile for Manganese.* Atlanta: HHS, Agency for Toxic Substances and Disease Registry.

Huang, H.-H. 2012. W32-STABCAL (software). Butte, MT: Montana Tech, Metallurgical Engineering.

Hudson Institute of Mineralogy. 2015. Mindat.org database [various minerals]. www.mindat.org/. Accessed December 2015.

IMA (International Magnetics Association). 2011. *Standard Recommendations: Soft Ferrite Cores, A User's Guide.* IMA-STD-100 2011.04. Cleveland, OH: IMA.

IMnI (International Manganese Institute). 2014. Applications—Industrial and metallurgical: Manganese and steelmaking. www.manganese.org/about-mn/applications/. Accessed August 2015.

IMnI (International Manganese Institute). 2016. *The Environmental Profile of Manganese Alloys.* Mississauga, ON: Hatch; Paris: IMnI.

Ismail, A.A., Ali, E.A., Ibrahim, I.A., and Ahmed, M.S. 2004. A comparative study on acid leaching of low grade manganese ore using some industrial wastes as reductants. *Can. J. Chem. Eng.* 82:1296–1300.

Kelly, T.D., and Matos, G.R., compilers. 2014. Manganese statistics. In *Historical Statistics for Mineral and Material Commodities in the United States.* USGS Data Series 140. Reston, VA: U.S. Geological Survey. https://minerals.usgs.gov/minerals/pubs/historical-statistics/ds140-manga.pdf. Accessed July 2018.

Key to Metals AG. 2010. Austenitic manganese steels. http://steel.keytometals.com/Articles/Art69.htm. Accessed August 2015.

Kilic, S., and Ozturk, F. 2014. Advanced high-strength steels in automotive industry to enhance sustainability. Presented at The 16th International Conference on Machine Design and Production, Izmir, Turkey, June.

Kosmulski, M., ed. 2009. *Surface Charging and Points of Zero Charge.* Boca Raton, FL: CRC Press.

Lasheen, T.A., El Hazek, M.N., and Helal, A.S. 2009. Kinetics of reductive leaching of manganese oxide ore with molasses in nitric acid solution. *Hydrometallurgy* 98:314–317.

Lawcock, G., Price, T., Battershill, J., Morgan, D., Brennan-Chong, J., and Bentvelzen, R. 2013. Manganese 101. UBS Investment Research, Australian Resources Weekly, October 23, 2013.

LeGrand, P. 2011. Electrochemical studies of passive film formation and corrosion of friction stir processed nickel aluminum bronze. M.S. thesis, Naval postgraduate school.

Matos, G.R., and Corathers, L.A. 2005. Manganese end-use statistics. Reston, VA: U.S. Geological Survey.

Mineral Resources Limited. n.d. Process Minerals International Pty Ltd. http://minrespublic.powercreations.com.au/entities/pmi.phtml. Accessed July 2018.

Nagpur Pyrolusite. n.d. Executive Summary. http://mpcb.gov.in/notices/pdf/ExeSum_NagpurPyrolusite.pdf. Accessed August 2015.

NRC (National Research Council). 2008. In *Minerals, Critical Minerals, and the U.S. Economy.* Washington, DC: National Academies Press. pp. 154–155.

Olivetti, E., Gregory, J., and Kirchain, R. 2011. *Life Cycle Impacts of Alkaline Batteries with a Focus on End-of-Life: A Study Conducted for the National Electrical Manufacturers Association.* Cambridge, MA: Massachusetts Institute of Technology, Materials Systems Lab.

Ostwald, J. 1975. Mineralogy of manganese oxides from Groote Eylandt. *Miner. Deposita* 10:1–12.

Papp, J.F., Corathers, L.A., Edelstein, D.L., Fenton, M.D., Kuck, P.H., and Magyar, M.J. 2007. Cr, Cu, Mn, Mo, Ni, and steel commodity price influences, Version 1.1. USGS Open-File Report 2007–1257. Reston, VA: U.S. Geological Survey.

Parc, S., Nahon, D., Tanov, Y., and Vieillard, P. 1989. Estimated solubility products and fields of stability for cryptomelane, nsutite, birnessite, and lithiophorite based on natural lateritic weathering sequences. *Am. Miner.* 74:466–475.

Pienaar, P.C., and Smith, W.F.P. 1992. A case study of the production of high-grade manganese sinter from low-grade Mamatwan manganese ore. In *INFACON 6: Proceedings of the 6th International Ferroalloys Congress*, Cape Town, Johannesburg, South Africa. pp. 131–138.

Pourbaix, M. 1974. *Atlas of Electrochemical Equilibria in Aqueous Solutions.* Houston, TX: NACE International. pp. 208, 488.

Rashad, M.M. 2006. Synthesis and magnetic properties of manganese ferrite from low grade manganese ore. *Mater. Sci. Eng. B* 127:23–129.

Rosenblum, S., and Brownfield, I.K. 2000. *Magnetic Susceptibilities of Minerals.* USGS Open-File Report 99-529. Reston, VA: U.S. Geological Survey.

South32. 2015. South Africa: Manganese. www.south32.net/our-operations/south-africa/manganese-south-africa. Accessed August 2015.

South32. 2017. What we do: GEMCO. https://www.south32.net/what-we-do/places-we-work/gemco. Accessed July 2018.

SPC (Secretariat of the Pacific Community). 2013a. *Deep Sea Minerals: Cobalt-rich Ferromanganese Crusts, A Physical, Biological, Environmental, and Technical Review.* Edited by E. Baker and Y.C. Beaudoin. New Caledonia: SPC.

SPC (Secretariat of the Pacific Community). 2013b. *Deep Sea Minerals: Manganese Nodules, A Physical, Biological, Environmental, and Technical Review.* Edited by E. Baker and Y.C. Beaudoin. New Caledonia: SPC.

Su, H., Wen, Y., Wang, F., and Tong, Z. 2009. Leaching of pyrolusite using molasses alcohol wastewater as a reductant. *Miner. Eng.* 22:207–209.

Su, H., Wen, Y., Wang, F., Sun, Y., and Tong, Z. 2008. Reductive leaching of manganese from low-grade manganese ore in H_2SO_4 using cane molasses as reductant. *Hydrometallurgy* 93:136–139.

Tan, W-F., Lu, S-J., Liu, F., Feng, X-H., He, J-Z., and Koopal, L.K. 2008. Determination of the point-of-zero charge of manganese oxides with different methods including an improved salt titration method. *Soil Sci.* 173:277–286.

Taneja, S.P., and Garg, V.K. 1993. Mossbauer and magnetic study of braunite mineral. *Nucl. Instrum. Methods Phys. Res.* B76:239–241.

Todd, M. 2010. Mn ore reduction technologies. Presented at the 7th IMnI EPD China Conference, Nanning, Guangxi, 23 March.

U.S. Mint. 2015. Coin and Medal Programs: Sacagawea golden dollar coin. Washington, DC: U.S. Mint. www.usmint.gov/mint_programs/golden_dollar_coin/?action=sacDesign. Accessed August 2015.

USGS (U.S. Geological Survey). 2016. Manganese. In Mineral Commodity Summaries. Reston, VA: USGS. https://minerals.usgs.gov/minerals/pubs/commodity/manganese/mcs-2016-manga.pdf.

Veglio, F., Trifoni, M., Pagnanelli, F., and Toro, L. 2001. Shrinking core model with variable activation energy: A kinetic model of manganiferrous ore leaching with sulphuric acid and lactose. *Hydrometallurgy* 60:167–179.

Wong, K. 2011. BHP Billiton to hike manganese capacity at Australia's Gemco JV. S&P Global Platts. https://www.spglobal.com/platts/en/market-insights/latest-news/metals/072911-bhp-billiton-to-hike-manganese-capacity-at-australias-gemco-jv. Northern Territory, Australia: Gemco. Accessed August 2015.

Woodward, R. 2001. Aluminum and aluminum alloys: Designations. www.azom.com/article.aspx?ArticleID=310#_Designations_for_Wrought. Accessed August 2015.

WorldAutoSteel. 2015. *FutureSteelVehicle Results and Reports & Cost Model.* Brussels, Belgium: WorldAutoSteel. www.worldautosteel.org/projects/future-steel-vehicle/phase-2-results/. Accessed August 2015.

You, Z., Li, G., Zhang, Y., Peng, Z., and Jiang, T. 2015. Extraction of manganese from iron rich MnO_2 ores via selective sulfation roasting with SO_2 Followed by water leaching. *Hydrometallurgy* 156:225–231.

Zhang, W., and Cheng, C.Y. 2007. Manganese metallurgy review. Part I: Leaching of ores/secondary materials and recovery of electrolytic/chemical manganese dioxide. *Hydrometallurgy* 89:137–159.

Zhang, W., Singh, P., and Muir, D. 2002. Oxidative precipitation of manganese with SO_2/O_2 and separation from cobalt and nickel. *Hydrometallurgy* 63:127–135.

Zhang, Y., You, Z., Li, G., and Jiang, T. 2013. Manganese extraction by sulfur-based reduction roasting–acid leaching from low-grade manganese oxide ores. *Hydrometallurgy* 133:126–132.

Zwicker, W.K., Groeneveld Meijer, W.O.J., and Jaffe, H.W. 1962. Nsutite—A widespread manganese oxide mineral. *Am. Mineral.* 47(99):246–266.

CHAPTER 12.23

Mercury Abatement

Bill Burton

Mercury is a common element in ore bodies containing gold and silver. Alkaline cyanide solution will oxidize and leach mercury along with precious metal, resulting in pregnant solution containing all three metals. As with gold and silver, the mercury–cyanide complex is readily adsorbed by activated carbon, and mercury will be carried into any of the common carbon recovery schemes, including carbon-in-leach, carbon-in-pulp, and carbon-in-column.

The mercury–cyanide reduction potential (–0.33 V) is less negative than the potentials of silver (–0.45 V) and gold (–0.63 V) (Marsden and House 2006), and effectively all mercury in pregnant solution will be recovered along with the precious metal using zinc precipitation (Merrill–Crowe process) or electrowinning. Mercury leached with gold and silver will be present throughout the precious metal recovery operations, specifically the adsorption/elution circuits, and in the electrowinning solution, cathode sludge, zinc precipitate, and carbon regeneration kiln off-gas.

Mercury represents a significant health hazard for both acute and chronic exposure. The Mine Safety and Health Administration's current threshold limit value–time-weighted average (TLV-TWA; 8-hour shift) for airborne mercury is 0.05 mg/m^3 and the American Conference of Governmental Industrial Hygienists' TLV-TWA (8-hour shift) for skin contact with mercury is 0.025 mg/m^3 (MSHA 2015). In addition, mercury presents a persistent environmental hazard with industrial emissions heavily regulated by national and state governments. Mining operations must ensure effective process controls to minimize both employee exposure to mercury (including hygienic considerations and site monitoring) and control emissions to the environment.

Mercury regulation from the Nevada Division of Environmental Protection is widely viewed as a benchmark for the gold and silver mining industry. In reviewing mercury operating permits while working with gold and silver mining clients in Nevada during 2013–2015, the following was found: Mercury emissions (total mercury including airborne particulate and vapor) for process equipment, including retorts, electrowinning cells, smelting furnaces, and carbon regeneration kiln systems, are typically limited to 0.229–0.01145 mg/Nm^3, along with additional operating restrictions and requirements.

MERCURY AND PROCESS EQUIPMENT DESIGN

A review of the select physical properties in Table 1 shows that elemental mercury exists as a liquid with very low vapor pressure at ambient conditions. Condensed mercury will typically be recovered with water and is easily separated by decantation based on immiscibility and the large density difference.

Mercury has a low boiling point and high vapor pressure at elevated temperature compared to precious and base metals and can easily be separated by retorting (vacuum distillation). Retorts operate under reduced atmospheric pressure, typically 150–250 mm Hg (absolute) and temperatures up to 625°C. As an example, heating zinc precipitate or cathode sludge to a temperature of 400°C results in rapid vaporization of mercury (vapor pressure of 1,574 mm Hg) in a retort oven operating at a pressure of 200 mm Hg (absolute).

Heat input for retort oven design, along with operating cycle time, are largely dictated by water content of the zinc precipitate or cathode sludge. The latent heat of vaporization for water is 2,260 J/g compared to 296 J/g for mercury. The difference in latent heat is also a factor in condenser design, where again, the equipment sizing is largely driven by water loading.

Removal of mercury from gas streams in high-temperature systems (retorts and carbon regeneration kiln off-gas) is generally a two-step process. First, the water and mercury vapor

Table 1 Select physical properties of mercury

Vapor pressure, mm Hg/400°C	1,574
Vapor pressure, mm Hg/38°C	0.0052
Melting point, °C	–38.9
Boiling point, °C	356.6
Heat of vaporization, J/g	296
Specific gravity	13.6
Solubility in water	Trace

Source: Weast 1968

Bill Burton, Chief Engineer, Precious Metals Recovery, FLSmidth Salt Lake City Inc., Salt Lake City, Utah, USA are condensed and removed from the process gas, typically in a shell-and-tube-style heat exchanger. The process gas is then passed through a vessel containing sulfur-impregnated carbon (SIC) to adsorb more than 99% of the remaining gas phase mercury. A design strategy is to take advantage of the low mercury vapor pressure by reducing process gas temperature prior to the SIC bed. As shown in Table 1, mercury vapor pressure drops from 1,574 to 0.0052 mm Hg by reducing the process temperature from 400° to 38°C. Reducing process gas temperature to condense as much mercury as possible results in reduced mercury load to the SIC carbon bed (decreasing the operating cost for SIC replacement and maintenance time).

RETORTS AND MERCURY ABATEMENT SYSTEMS

Design and operating details are provided for a typical mercury retort-abatement system used in the gold and silver mining industry to remove mercury from air streams exhausted from carbon regeneration kilns, electrowinning cells, and smelting furnaces.

Retorts

A process flow diagram for a typical retort is shown in Figure 1. The system consists of a retort oven (electric, gas, or diesel fired), condensers, mercury/water trap, mist eliminator, HEPA/coalescing filters, adsorption vessel with SIC, and vacuum pump. Cathode sludge or zinc precipitate is discharged from a filter press into pans (larger retorts use trays and are loaded with a forklift) and loaded into the retort oven. Retorts are operated from a centralized control panel with interlocks and alarms. The vacuum pump is started, and the system is brought to an operating pressure of 150–250 mm Hg (absolute). A minimum system pressure must be maintained (this is often a mercury operating permit requirement) and is interlocked with the oven heating system and an alarm.

The retort oven undergoes a series of ramp and soak temperature increases, operating between 200° and 600°C. Water is evaporated at a lower temperature, followed by mercury vaporization at high temperature and finally cooldown. Retort operating cycles typically run from 18 to 24 hours. Water and mercury vapor leaving the oven pass through a series of two to three water-cooled condensers in series. Cooling water is circulated through the shell side of the condenser, while process gas makes a single pass through the tubes. A refrigerated chiller is typically supplied with the retort system and provides cooling water at 7°C. Condensed water and mercury are collected in a vessel (trap) that is equipped with a decant drain for water and lower discharge for mercury. Provision is made to collect drained mercury in a 1-t (metric ton) container or 34.5-kg flask.

The cooled process gas then passes through a demister vessel and duplex coalescing HEPA filter (one operating and one standby) before entering the adsorption vessel with SIC. The demisting vessel and HEPA filters are key elements to prevent fine aerosols and submicrometer-sized particulate from entering and fouling or passing through the SIC bed. HEPA filter media is specified to resist moisture and provide a collection efficiency of 99.97% at 0.3 μm. The SIC bed is designed to provide a mercury removal efficiency of 99.99% and service life of up to five years before replacement. A final HEPA filter is provided on the discharge of the SIC adsorption vessel to prevent submicrometer carbon fines from discharging from the system to atmosphere.

A liquid ring vacuum pump is used on the retort system with seal water provided from the chiller supply sump. Figure 2 provides an isometric view of a current retort design showing the equipment layout.

Retort systems are designed for regular maintenance and include integral spray systems to wash out particulate that accumulates in the condenser tubes and demister media. The sprays are turned on at the end of the retort cycle to flush out solids that are collected in the mercury/water trap. This system has been developed based on feedback from silver mining operators (Merrill–Crowe process) that experience problems with high solids carryover from the retort oven to the condensers.

Smelting Furnace Systems

Retorts are very effective at removing mercury from cathode sludge and zinc precipitate, typically achieving higher than 99% efficiency. However, residual mercury will be present in the retorted solids that are smelted into doré in the refinery furnace. Dust and fumes from the furnace, doré molds, and slag pots must be contained and captured to prevent exposure to refinery operators and release to the atmosphere.

Figure 3 provides a process flow diagram of a typical furnace system. Fume collection systems must be sized and designed based on furnace type. Gas- and diesel-fired furnaces require much higher extraction rates to accommodate flue gas along with fume and dust compared to electric induction furnaces. In addition, fuel-fired furnaces generate exhaust gas at higher temperatures than equivalent induction furnaces.

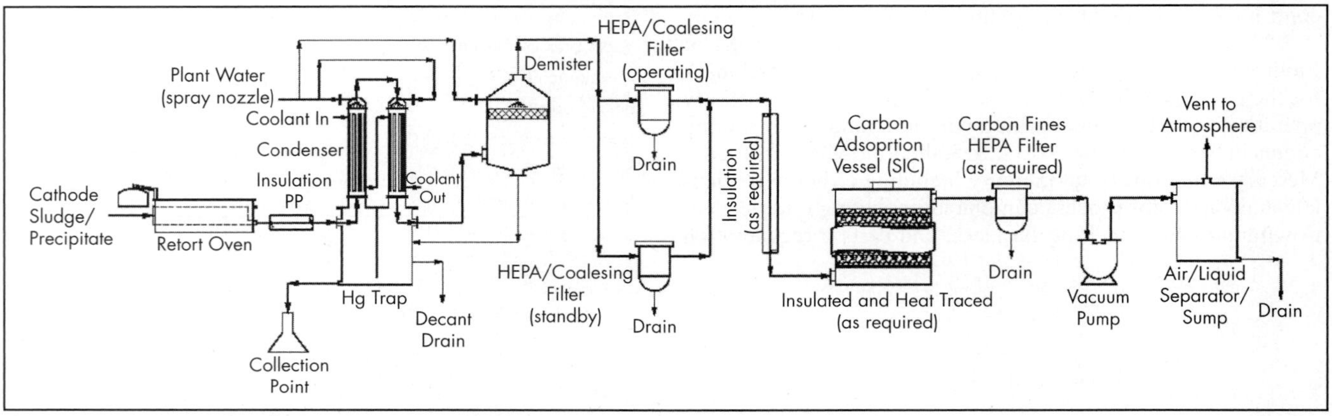

Figure 1 Retort mercury abatement system process

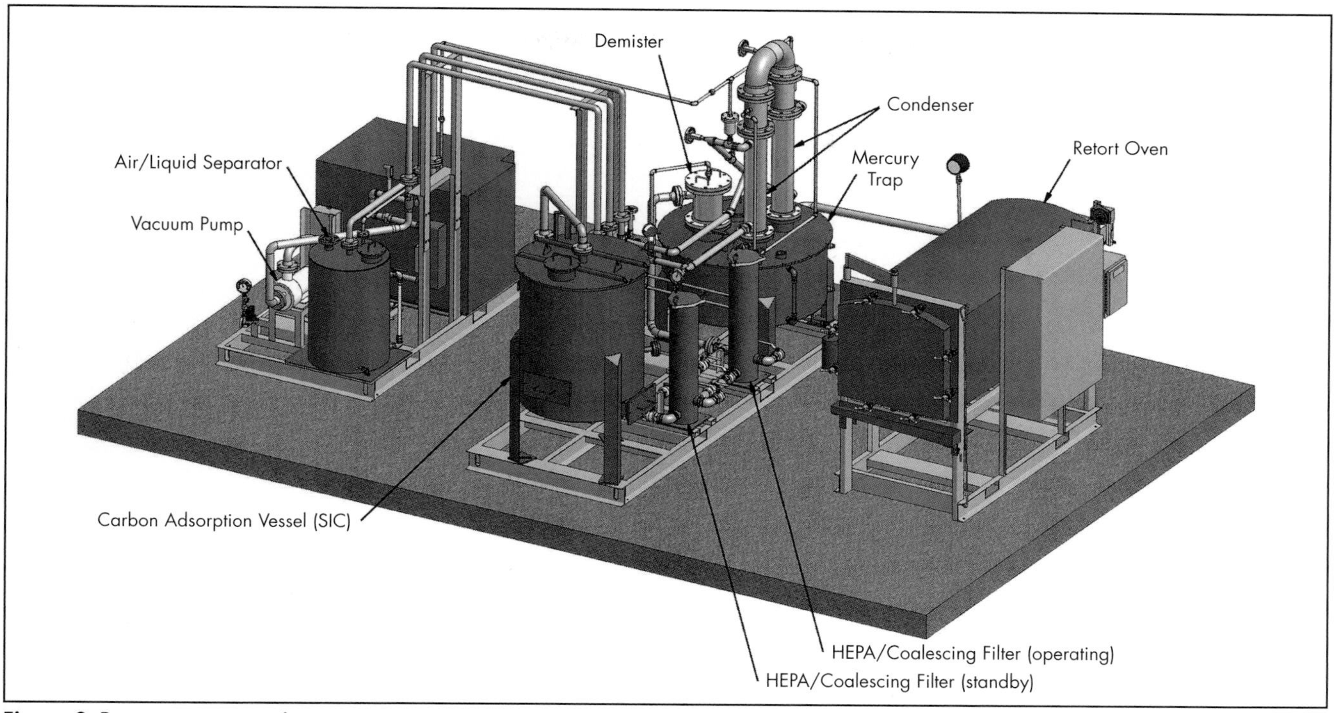

Figure 2 Retort system equipment

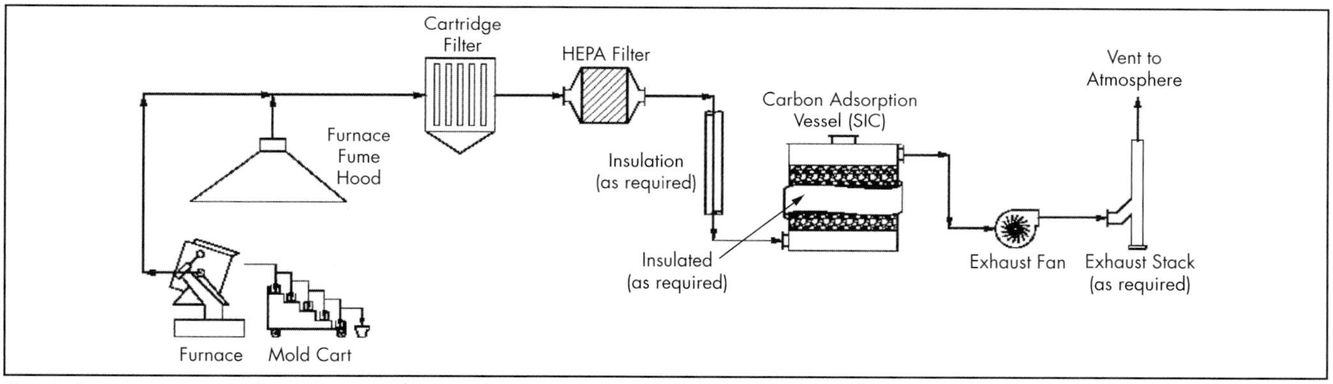

Figure 3 Furnace extraction mercury abatement system process

Primary pollution control for the furnace is a cartridge-style dust collector. Filter media can be specified for moderate- or high-temperature operation at 82° or 138°C, respectively. Following the dust collector is a HEPA filter specified for collection of particulate down to 0.3 μm at an efficiency of 99.97%. An SIC adsorption vessel provides removal of mercury vapor from the furnace off-gas.

An isometric view of a furnace system is shown in Figure 4; depending on the on-site climate, the pollution control equipment will be located inside a refinery building or outdoors. Typically, the exhaust fan for the system will be located outdoors or at a distance from the furnace to reduce noise exposure for operations staff in the refinery.

Electrowinning Cell and Process Solution Tank Ventilation Systems

Ventilation systems for electrowinning cells and process solution tanks must be designed to accommodate saturated gas and condensate. Pregnant solution from an elution circuit will typically enter electrowinning cells and pregnant/barren/return solution tanks at a temperature between 66° and 90°C. Ventilation for electrowinning cells and process solution tanks is critical for removal of gases, including hydrogen and oxygen that are evolved during electrowinning, ammonia that is formed from thermal decomposition of cyanide during elution, and mercury vapor from both pregnant and barren solution that is circulating in an elution/electrowinning circuit. In addition, steam from unventilated electrowinning cells and process solution tanks will create a hot and humid work environment in the refinery.

A process flow diagram for an electrowinning cell ventilation system is shown in Figure 5. Process gas from electrowinning cells and process solution tanks is typically saturated or supersaturated with water vapor and ranges in temperature from 38° to 60°C. The process gas passes through a mist eliminator vessel designed to remove condensate and aerosol

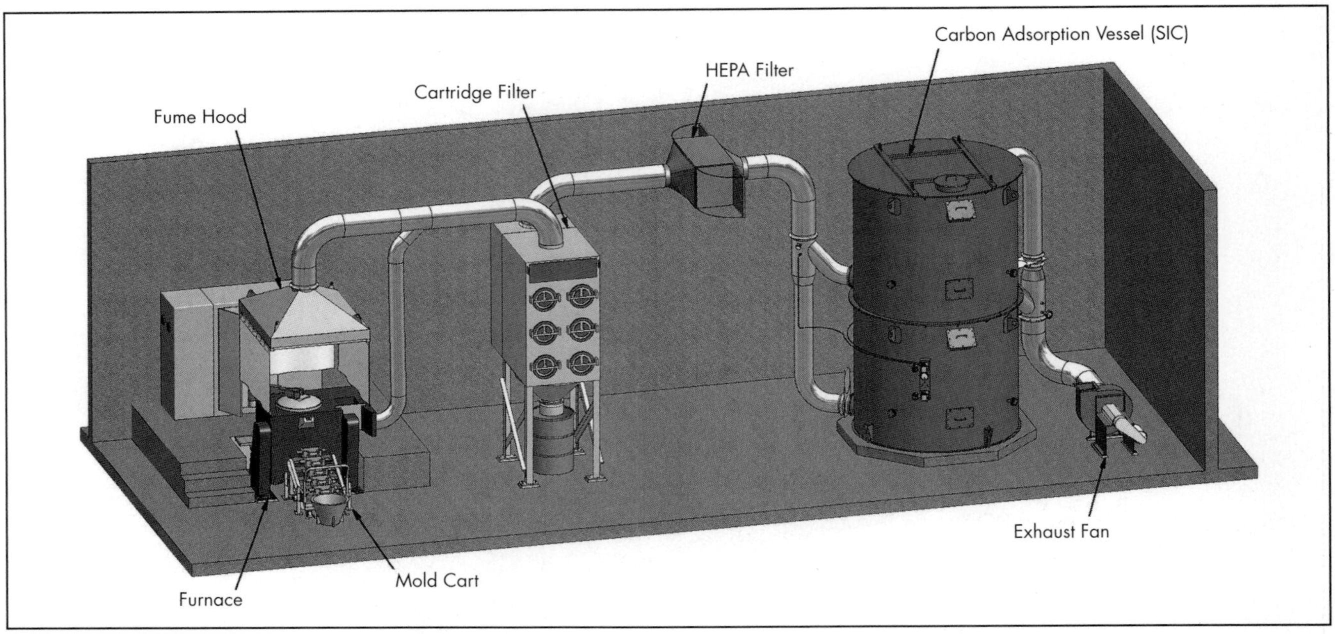

Figure 4 Furnace system equipment

down to 3 μm (98% removal efficiency). The demister vessel is designed with an upper decant drain with a U-bend to discharge collected condensate. A lower drain for condensed mercury is designed for discharge into a 34.5-kg flask and includes a sight glass to monitor accumulation of mercury. A spray nozzle is provided above the mist eliminator media to wash out any accumulated solids buildup.

Process gas exiting the mist eliminator will be saturated at the operating temperature and can condense in the downstream SIC adsorption vessel. An electric duct heater (or fuel-fired unit) is installed immediately after the mist eliminator to heat the process gas and maintain a relative humidity less than 80% through the SIC bed. The temperature of the SIC bed must be maintained at or above the temperature of the process gas, leaving the mist eliminator to prevent condensation and moisture buildup in the SIC. Insulation is typically specified for ductwork leaving the duct heater and for the SIC adsorption vessel, particularly for outdoor installations. An isometric view of a typical electrowinning cell ventilation system is shown in Figure 6.

Carbon Regeneration Kiln Systems

Mercury abatement systems for carbon regeneration kilns are similar to the design for electrowinning cells with the added requirement of addressing particulate. Figure 7 provides a process flow diagram for a kiln system. Kiln off-gas is passed through a coarse particulate filter to remove entrained carbon fines. The filter design is similar to the mist eliminator in the electrowinning cell system as kilns can operate over a range of temperatures and process gas can vary from 80° to 315°C. At lower temperatures, condensate can be present and must be removed with a decant drain and potential for condensed mercury in the lower drain. Construction is all 304 stainless steel to provide corrosion resistance and performance at elevated temperatures. A differential pressure gauge indicates solids buildup, and a spray wash is provided above the filter media to flush out accumulated carbon fines (discharge is typically routed to the plant carbon quench or carbon fines/clarifier tank).

Following the coarse particulate filter, process gas is passed through a water-cooled condenser (shell-and-tube-style heat exchanger). Plant water, if available at a temperature less than 27°C, can be used to cool the process gas; otherwise, a refrigerated chiller is supplied to provide cooling water at 7°C. Condensed water and mercury are collected in and drained from the mist eliminator vessel. The mist eliminator is similar in design to the electrowinning cell system described earlier in the "Electrowinning Cell and Process Solution Tank Ventilation Systems" section.

Following the mist eliminator, process gas is heated through an electric duct heater to prevent condensation in the SIC bed. Installation of a HEPA filter is an option before the SIC bed for kiln systems that generate higher quantities of carbon fines. The HEPA filter is similar in design to the unit described in the "Smelting Furnace Systems" section for furnace systems. The SIC adsorption vessel follows the HEPA filter for removal of final vapor-phase mercury. As with the electrowinning cell ventilation system, the kiln SIC bed must be maintained at a temperature equal to or greater than the gas temperature leaving the demister vessel to prevent condensation and buildup of moisture in the carbon. Insulation on ductwork and the adsorption vessel is a requirement to maintain temperature, particularly on outdoor installations. An isometric view of a typical carbon kiln ventilation system is shown in Figure 8.

SULFUR-IMPREGNATED CARBON AND ADSORPTION VESSEL DESIGN

A key element for all mercury abatement systems is the SIC adsorption bed. After separation of the condensed mercury upstream, the SIC bed is the final pollution control device to remove mercury vapor from process gas before release to the atmosphere. Unimpregnated activated carbon does not adsorb much mercury because its adsorption mechanism is based on

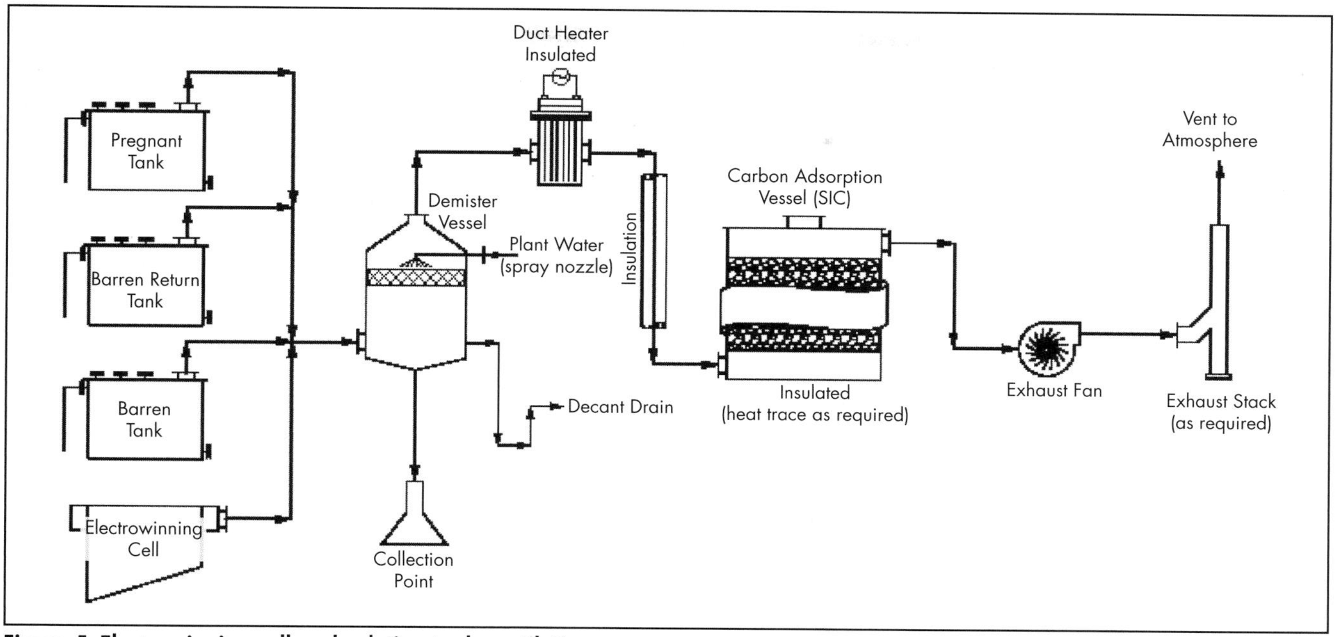

Figure 5 Electrowinning cell and solution tank ventilation process

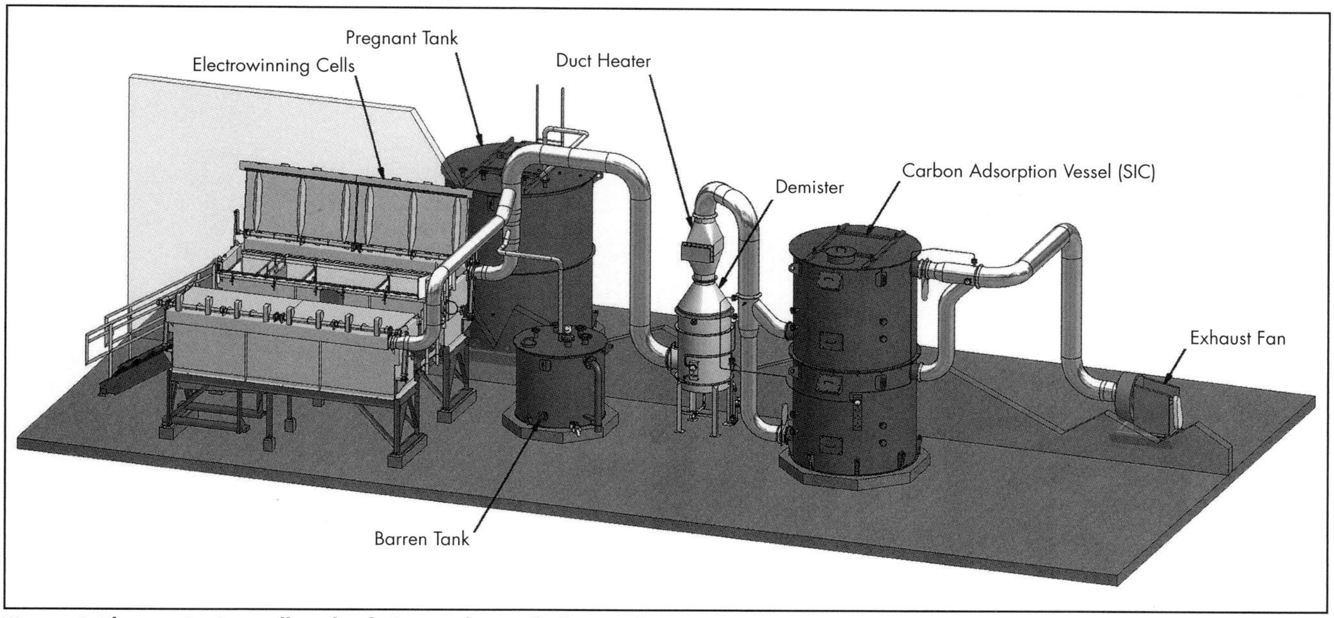

Figure 6 Electrowinning cell and solution tank ventilation equipment

physisorption and depends on intermolecular forces (van der Waals forces) that are relatively weak. SIC is preferred for scrubbing mercury from process gas for two reasons: SIC causes mercury to be strongly adsorbed, and SIC has a high mercury adsorption capacity, which is caused by chemisorption (valence forces) (Perry et al. 1984). Mercury reacts with the elemental sulfur in the carbon, forming mercury sulfide, which is a stable solid that will not revolatilize at operating temperature.

Many variables are important for SIC bed design, specifically process gas temperature, pressure, mercury concentration, flow rate, and moisture concentration (relative humidity). Bed diameter and depth must be sized to provide sufficient residence time for efficient mercury removal, reasonable pressure drop, and service life before replacement. Test data from the SIC manufacturer is key for bed design.

The plotted data in Figure 9 provide a comparison of mercury removal efficiency for varying residence times at an inlet concentration of 32 mg Hg/m^3 in air and velocity of 0.015 m/s. This is a high mercury concentration as air is saturated with mercury at this temperature. Yet mercury removal efficiencies remain at the 99.99% or higher levels for more than a year of operation at a residence time of ~10 seconds.

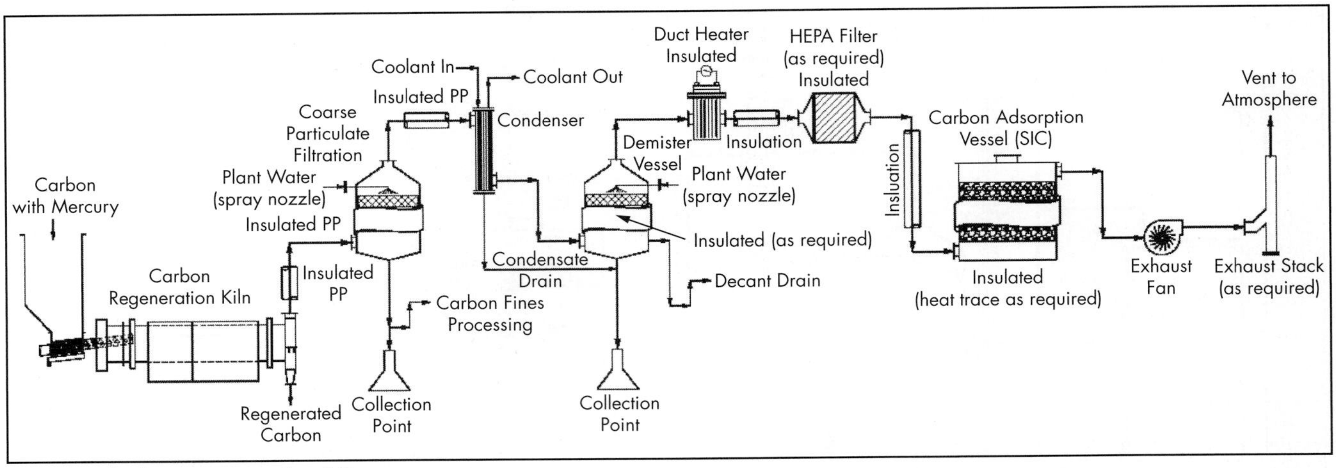

Figure 7 Carbon regeneration kiln process

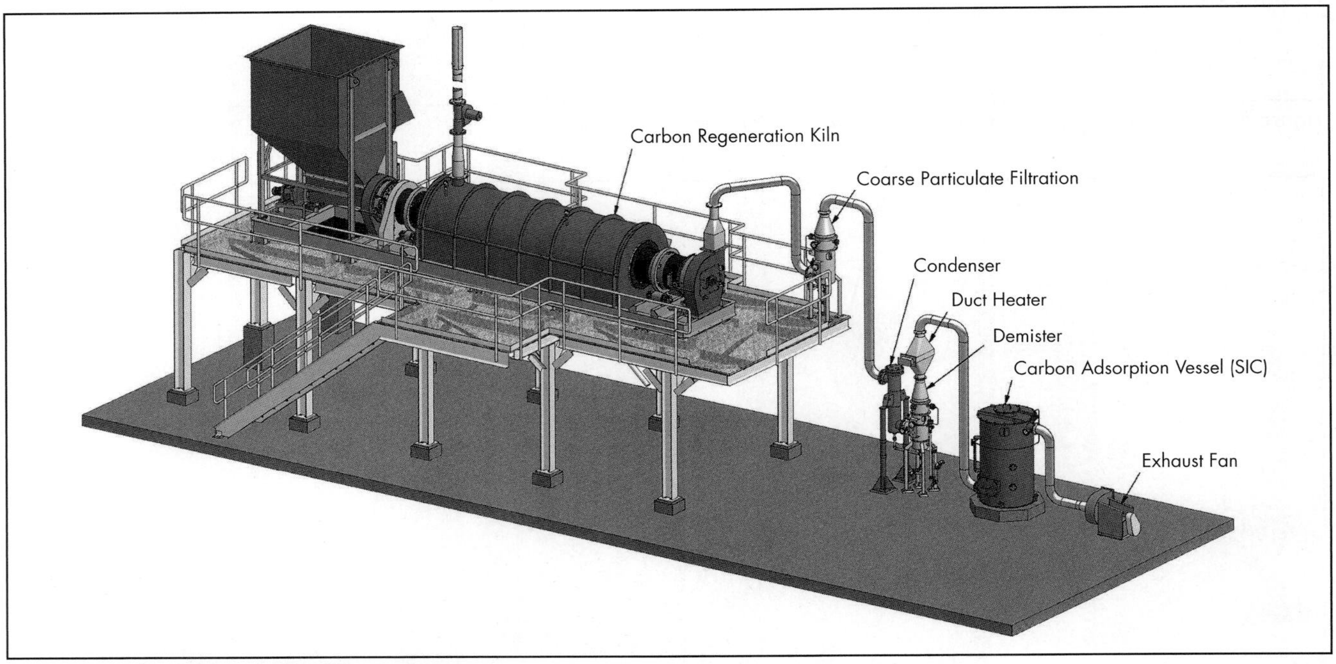

Figure 8 Carbon regeneration kiln equipment

The data were created using Mersorb 3-mm diameter mercury adsorbent pellets. Selective Adsorption Associates Inc. (SAAI) is the exclusive worldwide distributor of Mersorb, which is manufactured by Nucon. Cost-effective and long-wearing SIC adsorption systems have been designed by FLSmidth–Summit Valley Technologies (SV) working in close conjunction with SAAI, and FLSmidth-SV specifies Mersorb HT mercury adsorbent pellets for high-temperature applications.

Most SIC systems operate at temperatures less than 66°C; however, fuel-fired melting furnaces generate combined fume and flue gas that will enter the SIC bed at temperatures up to 135°C. The elemental sulfur impregnant will volatilize from ordinary SIC at elevated temperatures. Another consideration is that all activated carbons can pose a fire risk caused by rapid oxidation when exposed to hot gas and O_2 under certain conditions. Key contributors to rapid oxidation of activated carbons and proven preventive measures are reported in the literature (SAAI 2012).

Figure 10 is a plot showing weight loss of the elemental sulfur impregnant in Mersorb HT mercury adsorbent pellets versus increasing temperature. The impregnant loss remains negligible to 250°C and less than 7% at 300°C. The ignition temperature of Mersorb 3-mm mercury adsorbent pellets is approximately 450°C under the conditions of ASTM test method D3466-06.

SIC adsorption vessel design must include provision for the collection of carbon samples and bed maintenance. The maximum mercury adsorption capacity of SIC is reported as 20-g Hg/100-g adsorbent (16.7 wt % Hg based on weight of used adsorbent) (Nucon 2012). Beds are typically sampled at multiple points and depths to confirm loading. Quarterly testing is often a requirement of the site mercury operating permit. In this author's experience (working with gold and silver

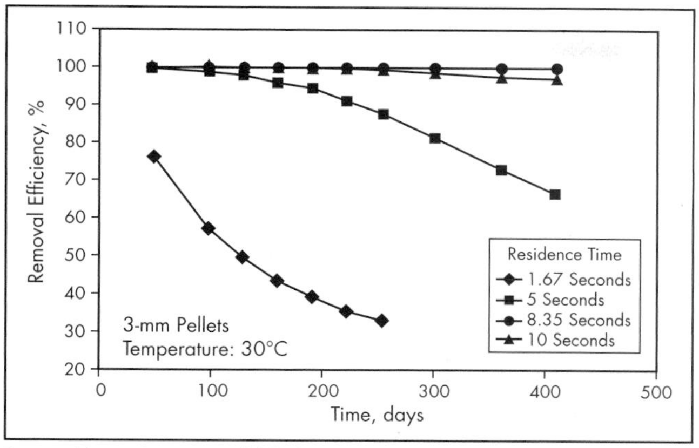

Source: Nucon International 2012

Figure 9 High-inlet-mercury concentration removal efficiency from air

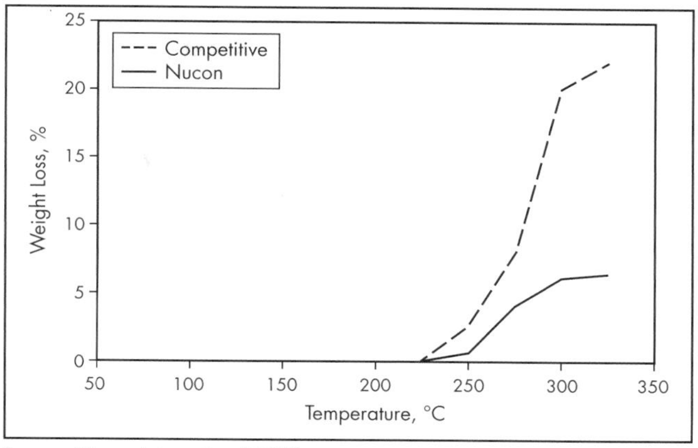

Source: Nucon International 2012

Figure 10 Elemental sulfur impregnant weight loss

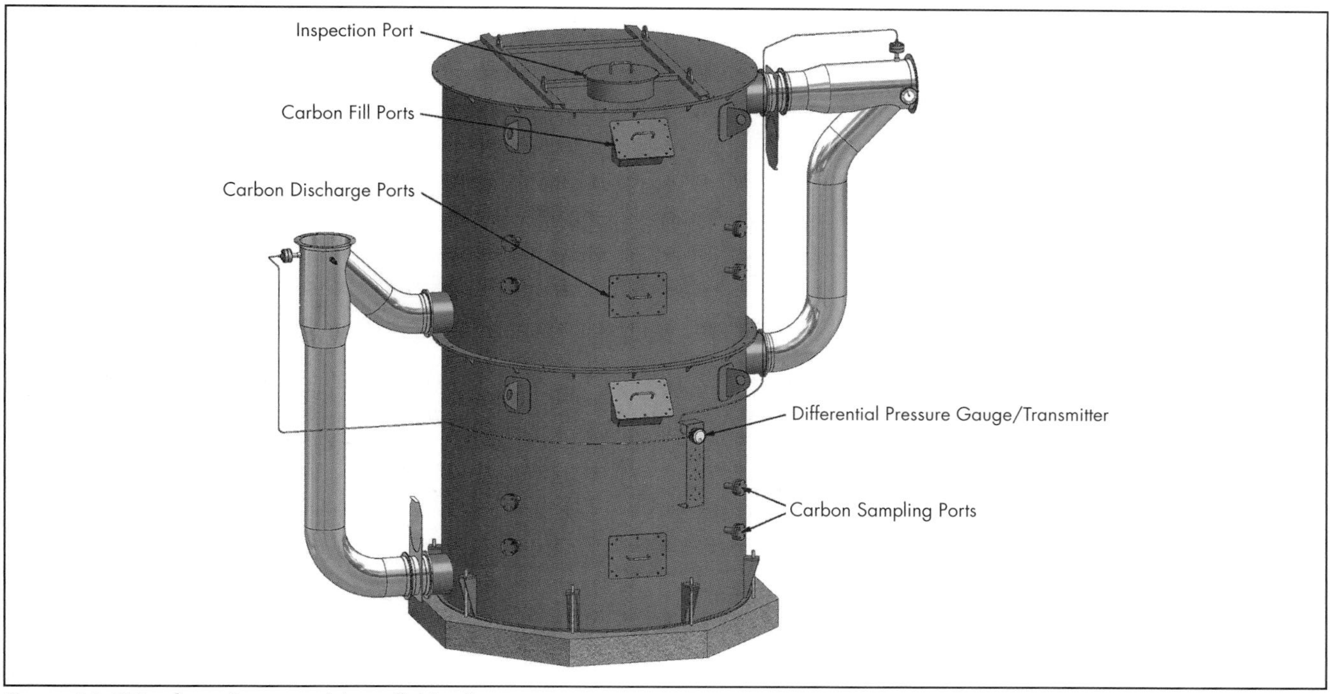

Figure 11 SIC adsorption vessel (parallel bed)

mining clients in Nevada from 2013 to 2015), once 50% of its maximum mercury capacity loading has been confirmed, sampling frequency is increased to monthly, and the carbon bed must be replaced within 30 days of reaching 90% of its maximum mercury capacity loading.

Figure 11 shows a schematic of a typical FLSmidth-SV SIC adsorption vessel. The vessel consists of two SIC beds operating in parallel with the flow split evenly between the upper and lower beds. This is a compact design for higher-process gas flow applications and keeps bed pressure drop below 100-mm H_2O with a reasonable bed diameter. Differential pressure gauges and transmitters are supplied to monitor pressure drop across the SIC beds. Baseline pressures are noted at start-up and are monitored over time to indicate potential problems with moisture or particulate buildup in the beds.

The adsorption vessel is designed with carbon-sampling ports at multiple locations and bed depths and includes an auger tool for convenient collection of samples. Each bed (vessel) is supplied with a lower carbon discharge port to remove spent SIC (typically discharged into a used bulk bag) and an upper fill port. The top inspection port can be used for collection of samples and for filling the vessel. The fill ports accommodate replacement of SIC supplied in bulk bags using a forklift.

MATERIAL HANDLING AND REFINERY HYGIENE CONSIDERATIONS

Refinery design requires consideration of ergonomics and equipment layout for material handling and good hygiene and housekeeping practices. Silver mining operations (Merrill–Crowe process) can generate large volumes of zinc precipitate

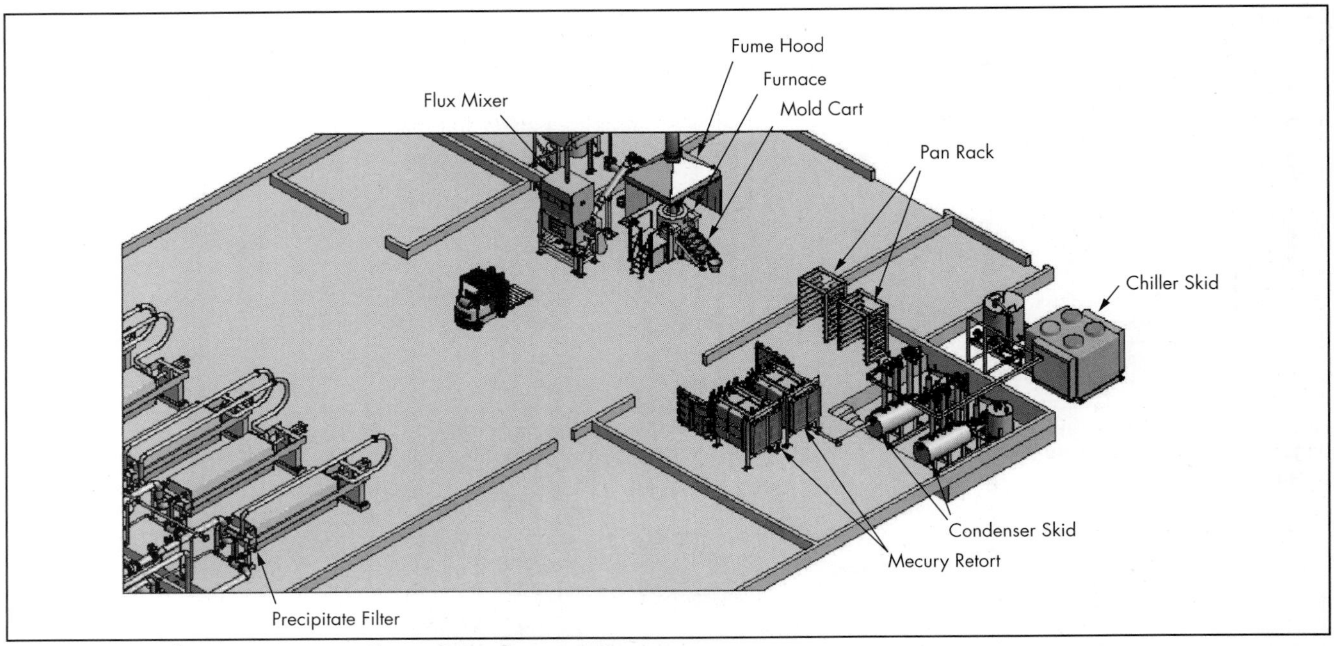

Figure 12 Merrill–Crowe process, silver mine refinery equipment

with mercury concentrations exceeding 20% by weight. Employee exposure to material containing mercury can be significantly reduced with implementation of mechanical handling equipment. Figure 12 shows the equipment layout of a recently commissioned silver mine refinery.

Zinc precipitate is discharged into steel trays below the filter presses. Once filled, the trays are taken by forklift and loaded into the retort ovens. Tray racks are provided for cooling the retorted precipitate and for staging the material prior to smelting. Trays of retorted precipitate are loaded by forklift into the tray-tipper at the top of the flux mixer. The tray-tipper and flux mixer are fully automated and operate under extraction to prevent release of dust. Along with the retorted precipitate, flux is added by tray to the mixer, and the material is fully blended for smelting. The precipitate/flux mix is metered and fed to the furnace using an auger conveyor. Operator contact with material containing mercury is minimized in a refinery with mechanized equipment and safety is improved by significantly reducing repetitive motions such as shoveling or scooping.

The refinery has been configured to provide separation and isolation of the retort room from the zinc precipitate filters and smelting area. For ventilation, the retort room is designed with slightly negative pressure compared to other areas of the refinery to contain any fugitive mercury emissions. The refinery areas are designed with floor sumps to collect and contain process solution and solid materials that may contain mercury or precious metal. Floors and walls are specified to have smooth surfaces with coatings and/or paint to allow frequent wash down with water. The refinery is designed with a locker room for operators to change into work clothing that is contained and laundered separately from regular clothing. Shower and wash facilities adjoin the locker room area to promote operator hygiene.

CONCLUSION

Modern mining technology and techniques enable economic recovery of precious metal from lower-grade ore bodies that often contain deleterious elements such as mercury. Gold and silver recovery processes and equipment must be designed and planned holistically with respect to operator safety, emissions to atmosphere, and efficiency. Environmental regulation will continue to increase and become more stringent, with the expectation that mining owners and operators will employ the best technology currently available. Mining owners will require the support and close collaboration with proactive partners to design processes and equipment for their specific operations.

REFERENCES

ASTM D3466-06. 2018. *Standard Test Method for Ignition Temperature of Granular Activated Carbon*. West Conshohocken, PA: ASTM International.

Marsden, J.O., and House, C.I. 2006. *The Chemistry of Gold Extraction*, 2nd ed. Littleton, CO: SME. p. 371.

MSHA (Mine Safety and Health Administration). 2015. III: Mercury exposure limits. In *Controlling Mercury Hazards in Gold Mining: A Best Practices Toolbox*. Washington, DC: U.S. Department of Labor, Mine Safety and Health Administration. pp. 11–12.

Nucon International. 2012. *Mersorb Mercury Adsorbents*. Bulletin 11B28. Columbus, OH: Nucon International. pp. 7, 9, 15.

Perry, R.H., Green, D.W., and Maloney, J.O., eds. 1984. *Perry's Chemical Engineers' Handbook*, 6th ed. New York: McGraw-Hill. p. 16-5.

SAAI (Selective Adsorption Associates Inc.). 2012. Hot gas, O_2, and activated carbon [technical reference]. Newtown, PA: SAAI.

Weast, R.C., ed. 1968. *Handbook of Chemistry and Physics*, 49th ed. Cleveland, OH: Chemical Rubber Co. pp. B-218, D-33, D-111.

CHAPTER 12.24

Molybdenum

Peter Amelunxen, Christopher Schmitz, Leonard Hill, Nolan Goodweiler, and Josh Andres

Much of the following introductory text and history has been excerpted from *Extractive Metallurgy of Molybdenum*, C.K. Gupta's (1992) excellent and authoritative text on the subject.

"Molybdenum has an atomic number of 42, an average atomic weight of 95.95 [and] belongs to the sixth group of the periodic system of elements. It occurs between chromium and tungsten vertically and niobium and technetium horizontally" in the periodic table (Gupta 1992). The metal, first isolated in 1781 by Peter Jacob Hjelm, does not occur in nature in its metallic (native) form but can be found in a relatively small number of mineral species (12–14) of which only four are of industrial value. By far, the most important mineral for the economic production of molybdenum is molybdenite (MoS_2), which is a soft, metallic-gray mineral whose name is derived from the Greek word *molybdos*, meaning lead. The secondary minerals—powellite ($CaMoO_4$), ferrimolybdite ($Fe_2Mo_3O_{12}{\cdot}8H_2O$), and wulfenite ($PbMoO_4$)—formed during the weathering of molybdenite are the other three molybdenum minerals of potential economic value (Sebenik 2005).

"In its pure state, molybdenum is a lustrous, gray, malleable metal, capable of being filed and polished. It can also be turned and milled without difficulty. Molybdenum is an important refractory metal with a very high melting point of 2,610°C" (only carbon, tungsten, rhenium, tantalum, and osmium possess higher melting points) and a relatively high density at 10.22 g/cm^3 at 20°C (Gupta 1992). Its coefficient of thermal expansion is significantly lower (one-third to one-half) than "that of most steels [and at] elevated temperatures, this low expansion provides dimensional stability and minimizes the danger of cracking" (Gupta 1992). Also, its "low specific heat allows molybdenum to be rapidly heated or cooled, with lower resultant thermal stresses than most other metals" (Gupta 1992). Additionally, molybdenum has one of the highest moduli of elasticity among commercial metals, and the value is not significantly affected by temperature (at 800°C, its modulus of elasticity substantially exceeds that of an ordinary steel at room temperature). Additionally, molybdenum's thermal conductivity exceeds that of all but a handful of elements, and together with its low electrical specific resistance, makes the metal suitable for use in electrical and coating applications.

HISTORY

The first documented mine production of molybdenum was around the end of the 18th century at the Knaben mine in southern Norway (Gupta 1992). However, without a significant commercial use and given the difficulty in extracting the pure metal, the metal remained a laboratory curiosity with little value until 1891, when it was first used in place of tungsten as an alloying element to produce armor-plated steel (Gupta 1992). The French Schneider Electric company was able to demonstrate that molybdenum, when alloyed in small quantities, created a substance that was remarkably tougher than steel alone and highly resistant to heat. Additionally, the material alloyed with molybdenum was lighter than materials created by alloying tungsten, as the atomic weight of molybdenum is approximately 57% that of tungsten.

In the early 20th century, development continued on both the processing of molybdenite ores (most significant were the advances in flotation technology) and the expansion of the commercial use of molybdenum, particularly in the production of high-speed steels and military armament applications. The demand for these alloyed steels increased during World War I because of the high demand for military armaments and shortages of tungsten; interest in molybdenum exploration and mining increased, and annual wartime production of molybdenum metal reached 816 t (metric tons) of contained metal by 1918 (Kelly et al. 2017).

Although the 1918 armistice following World War I almost entirely eliminated the demand for molybdenum, it

Peter Amelunxen, President, Aminpro Peru, Lima, Peru
Christopher Schmitz, Chief Mine Engineer, Climax Molybdenum Company, Leadville, Colorado, USA
Leonard Hill, Technical Services Director, Freeport-McMoRan Inc., Phoenix, Arizona, USA
Nolan Goodweiler, Senior Metallurgist, Climax Molybdenum Company, Parshall, Colorado, USA
Josh Andres, Senior Metallurgist, Freeport-McMoRan Inc., Morenci, Arizona, USA

also led to an unprecedented research program, driven largely by Climax Molybdenum Company, which created new applications and markets for the metal in the postwar period. Over the next 20 years, the valuable physical and mechanical properties that molybdenum imparts to alloys expanded, as it was used both alone and in conjunction with other alloying elements such as chromium, nickel, manganese, silicon, tungsten, copper, and vanadium. There was a notable increase in the production of alloy cast steels, with molybdenum often replacing both nickel and vanadium in many compositions, primarily associated with automotive structural parts and with railway and aircraft steels. Additionally, the use of molybdenum extended to general forgings and castings, and the substitution of molybdenum for tungsten in the tool steels continued to grow.

Before 1927, annual worldwide production of molybdenum never exceeded 1,000 t of contained metal. As research continued into new uses for molybdenum, demand for molybdenum increased, and increased molybdenum production followed. Worldwide production exceeded 10,000 t of contained molybdenum metal for the first time in 1937; 25,000 t first during the wartime production spike in 1942; 50,000 t first in 1966; and 100,000 t first in 1978. Production first exceeded 200,000 t of contained molybdenum in 2007 and is currently (2016) at about 226,000 t annually (Kelly et al. 2017)

A general timeline of world molybdenum production is as follows (based on statistics from Kelly et al. 2017 and Climax Molybdenum Company 1924–2017):

- **Prior to 1925.** World molybdenum production averaged <200 t/yr, except for the aforementioned wartime production spike in 1916 through 1919 that peaked at 816 t of molybdenum in 1918. Most molybdenum production in this period was outside the United States.
- **In 1924.** The Climax mine (Colorado) started producing continuously (there had been some earlier production during World War I). It was the largest U.S. producer for 1924 and provided 26% of world production in its first year of operation.
- **In 1933.** The first documented by-product molybdenite production through differential flotation from copper concentrate was carried out at Greene Cananea Consolidated Copper Company in Sonora, Mexico.
- **In 1936.** By-product molybdenum production was reported from the United States at Utah Copper Company's Bingham Canyon mine and Arizona's Miami Copper Company.
- **From 1924 to 2007.** The United States was the world's leading molybdenum producer.
- **From 1925 to 1981.** The United States produced more than 50% of the world's molybdenum production. From 1925 through 1957 and again from 1960 through 1965, the United States produced more than three-quarters of the world's production.
- **From 1924 to 1946, 1951–1970, and 1974–1977.** The Climax mine provided more than half of U.S. molybdenum production. In 1925 through 1965, the Climax mine alone produced on average 60% of the world's molybdenum production and 71% of the U.S. production.
- **The 1960s and later.** By-product molybdenum production from porphyry copper deposits, pioneered at Cananea, increased in importance as more porphyry copper deposits were brought into production. The shift with time from solution extraction/electrowinning (SX/EW) production from oxidized material to concentrator production from sulfide material also increased the trend toward greater by-product molybdenum production.
- **In 1983.** Climax shuts down for a full year for the first time since it began operating continuously in 1924. From 1984 through 1991, it operated sporadically with minor production relative to its early years before shutting down again in 1992. Molybdenum production restarted in 2012 and continues to the present. In 2014, Climax accounted for only 4% of the world's production—a dramatic drop relative to Climax's early years.
- **In 2008.** China unseated the United States as the world's leading molybdenum producer after 83 years. This was mainly driven by a sharp increase in production of molybdenum in China—from 46,000 t in 2007 to 81,000 t in 2008 and 93,500 t in 2009 (Blossom 2001–2017, 1994–2000).
- **In 2015.** Chile became the second largest molybdenum producer in the world after China, leaving the United States in third place (Blossom 2001–2017).

In summary, historic molybdenum production can be divided into three eras:

1. Prior to 1925, there was a period when molybdenum demand and supply was so low as to be negligible (relative to today's production), and what mining there was focused on small, high-grade deposits with ore sometimes concentrated through hand sorting.
2. From 1925 through 1965, world molybdenum production was dominated by a single mine—the Climax mine in Colorado.
3. From 1966 through the present, molybdenum production has been increasing, and there has been progressive diversification in molybdenum production both by country and by mine. By-product molybdenum production gained sharply in importance during this period.

GLOBAL PRODUCTION

As noted, as of 2016 the global molybdenum production (not including recycled molybdenum) was approximately 226,000 t annually (Kelly et al. 2017). The top producing countries in order of molybdenum production quantities were China, Chile, and the United States. Together they accounted for 77% of the world's production: 40% from China, 23% from Chile, and 14% from the United States (Blossom 2001–2017). The vast majority of the molybdenum produced by these countries comes from porphyry deposits, but there are differences in the details.

For China, detailed production data is scarce, but molybdenum reserves are estimated to be 78% in porphyry molybdenum deposits, with 13% in porphyry-skarns, 5% in skarns, and 4% in veins (Zeng et al. 2013).

Almost all of the molybdenum production in Chile is produced as a by-product from copper porphyries. For 2014, the U.S. Geological Survey (USGS) reported that 84% of Chile's molybdenum production came from the large deposits at Collahuasi, Chuquicamata, El Teniente, Los Bronces, Los Pelambres, and Radomiro Tomic (Wacaster 2017).

Based on the most recent USGS reporting (for 2016) for the United States, production from primary molybdenum mines (the Climax and Henderson mines, both in Colorado and both Climax-type porphyries) accounted for 59% of U.S. molybdenum production. Production of molybdenum as a by-product from seven copper mines accounted for the balance (Blossom 2001–2017; Climax Molybdenum Company 1924–2017). Primary mines are generally more sensitive to variations in the price of molybdenum, and production from these properties can swing significantly as the molybdenum price rises and falls.

GEOLOGY OF MOLYBDENUM ORE BODIES

Mines are developed around ore deposits, and ore is defined as a naturally occurring material which can be mined at a profit. What is ore now—based on present economics and use of current technologies—may not have been ore in the past. Conversely, some material that was mined as ore in the past—when labor was cheap—may not be ore now. And generally, higher grade deposits or deposits that are more economic for other reasons (mineralogy, proximity to the surface, etc.) are mined earlier.

A mineral deposit (a concentration of a particular mineral or element relative to the average concentration in rock) can become an ore deposit based on conditions prevalent at the time the deposit is evaluated for development. The type of ore deposits that were mined early in the 20th century are generally quite different from what we mine as ore deposits now. In the late 1800s and early 1900s, mining was focused on high-grade and generally lower tonnage targets such as veins. The development of mechanized open pit mining, underground block cave operations, and froth flotation in the early 1900s started to change what was regarded as an ore deposit. The development of the first by-product molybdenum recovery circuit at Cananea in Mexico in 1933 put in motion a step change in the economics of molybdenum production.

Metal deposits can be classified in many different ways. One of the most common classifications is based on the economics that make a deposit a mine. A molybdenum-producing mine may be one of the following:

- A primary molybdenum mine in which revenues are driven solely or primarily by molybdenum, possibly with lesser contributions by other commodities
- A by-product mine producing molybdenum where the primary revenue is from another commodity (usually copper in the case of molybdenum) with subordinate revenue provided by the production of molybdenum and possibly other by-products
- A co-product mine, which is uncommon, with the revenue from the production of multiple commodities required to make the mine profitable

Examples of primary molybdenum mines include the Climax and Henderson mines in Colorado, the Questa mine in New Mexico, and the Thompson Creek mine in Idaho (all in the United States). Examples of by-product molybdenum mines include the Bingham Canyon mine in Utah, the Bagdad mine in Arizona, and the Chuquicamata mine in Chile. An example of a co-product mine is the Sierrita mine in Arizona.

Another way to classify deposits is by their geologic *style* of mineralization. The following is a list of the most common types of molybdenum deposits:

- **Vein deposits.** These are generally planar deposits of minerals cutting wall rocks.
- **Pipe deposits.** These are cylindrical/pipe-like deposits of minerals cutting wall rocks.
- **Metasomatic/skarn deposits.** These are characterized by skarn (calc-silicate) minerals that develop due to the interaction of intrusives and (usually) sediments (commonly carbonates).
- **Pegmatites.** Very coarse-grained intrusive rocks, these are usually interpreted as products of very late-stage igneous differentiates with elevated water and incompatible element concentrations.
- **Porphyry deposits.** These are deposits where the minerals of interest are *disseminated* through a large volume of hydrothermally altered rock, usually concentrated along a stockwork of fractures and/or veinlets. Usually in or adjacent to a mass of intrusive rock, some of which commonly has a porphyritic texture, large-scale zoning of hypogene alteration (potassic, phyllic, silicic, argillic, and propylitic) is typical and varies with the type of porphyry deposit. These deposits are often amenable to mining through high-tonnage open pit operations.
- **Sedimentary/metasedimentary-related deposits.** Used here as a catchall category of deposits with elevated molybdenum contents, these deposits include roll-front uranium deposits (Bullock and Parnell 2017), Athabasca-type uranium (Ferguson 1984), Karoo Formation U-Mo deposits (Van der Merwe 1986), black shales (Smedley and Kinniburgh 2017), underwater hydrothermal vent deposits (Smedley and Kinniburgh 2017), and others. Ore-forming processes include the effects of diagenesis and supergene alteration on sedimentary rocks. Molybdenite can be the molybdenum-bearing mineral, particularly in the metasedimentary deposits, but a variety of molybdenum-containing oxide minerals are found in the sedimentary deposits (jordisite, ilsemannite, iron oxides, jarosite, etc.). So far, none of the deposits in this group has gained prominence as an important source of molybdenum.

The preceding list is representative, but not all inclusive—other deposit styles (e.g., greisens) can contain molybdenum mineralization.

Vein deposits and pipe deposits were important in the past, but as stand-alone deposits, they currently account for a minor portion of the world's molybdenum production. There has been some minor molybdenum production from pegmatites in the past. Metasomatic/skarn deposits containing molybdenum are sometimes of a sufficient size to be economic in today's conditions. Skarns sometimes occur in association with, and are mined with, porphyry deposits (e.g., at the Chino mine in New Mexico). Sedimentary- and metasedimentary-related deposits can contain anomalous concentrations of molybdenum, but to the authors' knowledge have not had any substantial molybdenum production to date. The vast majority of molybdenum currently produced, and that produced in the past, comes from porphyry deposits.

Porphyry deposits, which are the major source of molybdenum production, can be further divided by their mineralogy, chemistry, and the genetic models proposed to explain their origins. For molybdenum, the porphyry deposits have generally been divided into four types:

1. **Climax-type deposits.** Also known as high fluorine or alkali-feldspar rhyolite-granite porphyry molybdenum deposits, typical examples are the Climax and Henderson

mines in Colorado, the Mt. Hope deposit in Nevada (United States), and the Questa mine in New Mexico. This type of porphyry has higher molybdenum grades relative to the other two types—usually 0.10%–0.30% Mo. Molybdenum is the primary product. Associated intrusives are more silicic and potassic than the intrusives associated with porphyry coppers or low-F stockwork molybdenum deposits (generally granitic composition rather than monzonites, diorites, or granodiorites). They have high fluorine content (fluorite and/or topaz are common gangue minerals) and often anomalous tin (in cassiterite) and/or tungsten (in wolframite and/or scheelite), sometimes to the extent that tungsten and tin may be produced as by-products. Associated intrusives are generally classed as A-type granites and seem to have been generated when the tectonic regime was transitioning from compressional to extensional (Ludington and Plumlee 2009). The molybdenite concentrate produced from this type of deposit tends to be cleaner with higher Mo content and lower impurities (such as insolubles, which are generally silicate minerals, iron, copper, lead, and others) than that produced from the other two types of porphyries. Higher-value chemical- and lubricant-grade products are often produced from it. The Climax mine, the namesake of this deposit type, has produced more than 450 Mt (million metric tons) of ≥0.20% Mo ore and has reported minable reserves of 170 Mt at 0.15% Mo as of the end of 2016 (Climax Molybdenum Company 1924–2017).

2. **Porphyry copper deposits.** The economic driver for this porphyry type relies on the contained copper, but molybdenum can be present as a by-product. Typical examples include the Bingham Canyon mine in Utah, the Sierrita and Bagdad mines in Arizona, and the Cerro Verde mine in Peru. Molybdenum grades generally range from less than 0.010% (controlled mostly by the economics of the Mo recovery circuit) to 0.075% or so. Associated intrusives mostly range in composition from dioritic through monzonitic, though occasionally late-stage granites and/or rhyolites occur. Typical by-products (beyond the molybdenum) are gold and silver, often recovered during smelting/processing of the copper concentrate. Rhenium may be a by-product of processing of the molybdenum concentrate. Associated intrusives are generally classed as I-type granites and are thought to have been generated above subducting plates in a convergent plate margin boundary or island-arc environment (Berger et al. 2008).
3. **Low-fluorine stockwork molybdenum deposits.** Also known as arc-related porphyry molybdenum deposits and Endako-type deposits, typical examples include the Thompson Creek mine, Endako mine (British Columbia, Canada), and Quartz Hill deposit (Alaska). They are lithologically and tectonically very similar to the porphyry copper deposits. Molybdenum grades generally range from 0.05% to 0.20%, but economic copper is lacking (Ludington et al. 2009; Taylor et al. 2012).
4. **Dabie-type deposits.** These are porphyry molybdenum deposits formed in syn- to post-continental collisional tectonic settings. This is a relatively recent model based on deposits discovered and researched in China. The associated intrusives are high-potassium calc-alkaline to shoshonitic magmas. Examples include the Qian'echong and Yaochong deposits (Wu et al. 2017).

Molybdenum-bearing porphyry deposits in China seem to represent all four types. Shapinggou (Zhang et al. 2014) and Jinduicheng (Stein et al. 1997) appear to be good matches for the Climax-type model, Diyanqinamu and Chalukou are of the low-fluorine stockwork type (Wu et al. 2017), and Qian'echong and Yaochong are of the Dabie type (Wu et al. 2017). By-product molybdenum occurs in deposits in the Yulong porphyry copper belt.

In summary, although there are multiple mineral deposit types that may contain molybdenum, most current and past molybdenum production comes from porphyry deposits. There are at least four major types of molybdenum-producing porphyry deposits, which, although they share enough large-scale similarities to all be classed as porphyry deposits, also have distinct characteristics that permit classification as individual types of porphyries. These characteristics can affect the processing of the ore and the products produced from the deposits.

Molybdenum deposits have been formed throughout geologic time, from the Precambrian through the Cenozoic. Some types of deposits are distinct to certain periods (e.g., Athabasca- and Karoo-type deposits), and some deposit types are typically associated with a specific period in some regions (e.g., Laramide porphyry coppers in western North America).

MINERALOGY

Unlike many other metals, there are no known records of molybdenum usage in ancient times, and this is likely because molybdenum does not occur in nature in its free or native state (Gupta 1992). It is a relatively rare element and is only found in approximately two dozen known natural minerals, the most common of which are summarized next.

Molybdenum Minerals

For almost all production—current and past—the mineral carrying the recovered molybdenum values is molybdenite (molybdenum disulfide, MoS_2). There are other molybdenum-containing mineral species, but the relative abundance of molybdenite along with its amenability to concentration through froth flotation currently make it the predominant ore mineral for molybdenum. Supergene processes can cause the development of secondary molybdenum oxide minerals (e.g., molybdite, ferrimolybdite) and other oxide minerals that may contain molybdenum (iron oxides, jarosite, etc.) in any deposit that has molybdenite as the original molybdenum-containing mineral. In at least one case, a circuit was tested (at Climax, Colorado) for concentration of molybdenum from molybdenum oxide–mineralized material, but with only mixed results (Lane et al. 1972). In the early 20th century, particularly in the southwestern United States, some molybdenum was recovered from the secondary lead molybdate mineral wulfenite (Hess 1924; Kirkemo et al. 1965). Scheelite (calcium tungstate) and powellite (calcium molybdate) are the primary minerals and are usually found in skarns or veins; they form a solid solution series with each other, and tungsten and molybdenum substitute for each other. Scheelite is much more common than the molybdenum-bearing powellite, but scheelite can have some molybdenum content that can sometimes be recovered as a by-product from tungsten deposits. Table 1 shows the most common molybdenum minerals and some of their properties.

Rhenium

Copper-molybdenum porphyry deposits are the world's primary source of rhenium (Re), which occurs as solid substitution for molybdenum atoms on the MoS_2 lattice structure. While many other elements should be able to substitute for Mo, the trigonal arrangement of Mo^{+4} and the coordination number of 6 ensure that only tungsten (W) and Re satisfy the geometric conditions and electrical charge required for substitution, although W's affinity for oxygen means that generally only Re atoms are present with MoS_2 (Mo and Re also have nearly identical atomic radii of 1.39 and 1.37 Å, respectively). Note that Se and Te may also be present, caused by substitution of sulfur (S) (Gupta 1992). All three elements are often recovered from the flue gas of the roaster via hydrolyzation and subsequent extraction with tertiary amine ionic exchange processes.

Table 1 Molybdenite minerals and their properties

Mineral	Molecular Weight, g/mol	Specific Gravity, g/cm³	Molecular Formula
Bamfordite	412.76	3.64	$Fe^{3+}Mo_2O_6(OH)_3 \cdot H_2O$
Biehlite	426.07	5.24	$[(Sb,As)O]_2MoO_4$
Drysdallite	230.41	6.34	$Mo(Se,S)_2$
Ferrimolybdite	717.61	2.99	$Fe^{3+}{}_2(MoO_4)_3 \cdot n(H_2O)$
Hemusite	852.45	4.47	Cu_6SnMoS_8
Kamiokite	527.51	6.02	$Fe_2Mo_3O_8$
Lindgrenite	544.53	4.29	$Cu_3(MoO_4)_2(OH)_2$
Molybdenite	160.07	5.00	MoS_2
Molybdite	143.94	4.75	MoO_3
Monipite	185.60	8.93	MoNiP
Powellite	200.02	4.25	$CaMoO_4$
Tugarinovite	127.94	6.58	MoO_2
Wulfenite	367.14	6.82	$PbMoO_4$

Courtesy of Aminpro

Mineralogical Factors Affecting Recovery

The mineralogy of the deposit is not the only important factor for molybdenite recovery. The grain size of the molybdenite crystals themselves also plays an important role in the selection of the processing equipment and circuit arrangement. For example, the following images in Figure 1 show samples of molybdenite-bearing copper ores from two different deposits. Figure 1A shows drill core from a coarse-grained porphyry deposit in the southwestern United States (Bagdad, Arizona). For this operation, the molybdenite presents primarily as coarse crystals that form as veinlets around fracture sets (average grain size = 35 μm, average grade = 0.02% Mo). Figure 1B is from a fine-grained porphyry in northern Chile (average grain = 6 μm, average grade = 0.04% Mo). For this operation, the molybdenite is finely disseminated and almost invisible to the naked eye, although the average grade is much higher. The coarse, North American operation can achieve higher recoveries at coarser grind sizes, higher pH, and with fewer cleaning stages. The fine-grained South American operation requires finer regrind sizes, much longer retention times, and active pH control to improve the separation kinetics.

In addition to molybdenum mineralogy and grain size, the presence of naturally hydrophobic gangue minerals can also be an important factor in the economics and/or design of the separation process. Many primary and by-product molybdenum deposits contain trace amounts of these minerals, which can include talc, elemental sulfur, carbon, pyrophyllite, or other floatable clays. The geological distribution and association of these minerals can vary greatly between deposits and even within deposits, and therefore significant characterization effort is required prior to process selection, particularly for greenfield projects where little historical information is available to guide the process development team.

ECONOMIC CONSIDERATIONS

Figure 2 shows the significant variations in realized price for technical-grade molybdenum trioxide and ferromolybdenum

A. Coarse-grained molybdenite (Bagdad, AZ)

B. Fine-grained molybdenite (northern Chile)

Courtesy of Freeport-McMoRan

Courtesy of Aminpro

Figure 1 Coarse- and fine-grained molybdenite

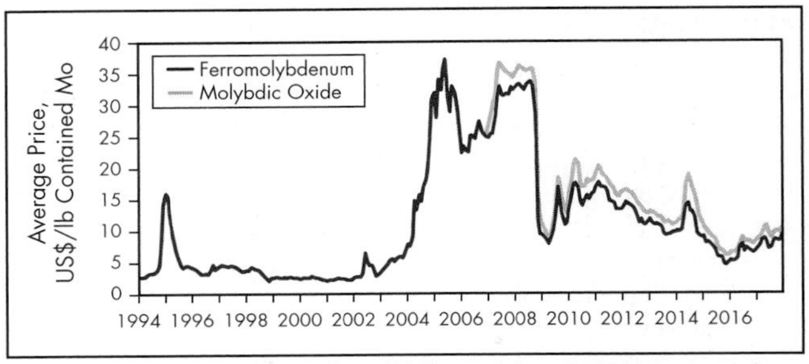

Data from S&P Global Platts, n.d. (molybdic oxide); and CRU International, n.d. (ferromolybdenum)

Figure 2 Molybdic oxide and ferromolybdenum pricing history

over the past 20-plus years. Although the price of molybdenite is primarily driven by molybdenum supply and demand, much of the molybdenum supply is produced as a by-product of copper sulfide recovery processes; as a result, the price, and, therefore, the production rate, can be affected by the supply–demand equilibria of the copper market. For this reason, the supply curve of molybdenum is relatively inelastic with respect to molybdenum price. Other factors further complicate the economics:

- Molybdenum is a much smaller market than copper.
- The requirement of intermediate processing steps (roasting and/or oxidation) can sometimes create artificial or temporary production bottlenecks.
- A significant portion (40%) of the world production originates in China and is susceptible to state export quotas or controls.

The preceding factors cause a relatively high price volatility for molybdenum products, captured by Figure 2, which significantly impacts the sustainability of the primary molybdenum producers, for whom revenue and profitability are much more sensitive to commodity pricing. It is fundamentally for this reason that the primary molybdenum producers are often considered the *swing* producers, where operation is curtailed if market price gets too low.

RECOVERY OF MOLYBDENITE FROM PRIMARY MOLYBDENUM DEPOSITS

Molybdenite is one of the few sulfur-bearing minerals that exhibit natural hydrophobicity. It has been suggested that the native hydrophobicity of molybdenite is primarily caused by the crystal structure, in that a layer of molybdenum atoms is sandwiched between layers of sulfur atoms. The layers are held together by van der Waals bonds that give rise to a well-defined cleavage plane for molybdenite. The surface of molybdenite created by the cleavage plane is highly hydrophobic, and the plane perpendicular to it is only weakly hydrophobic (Chander and Fuerstenau 1972).

Primary molybdenum recovery and concentration relates to the flotation of molybdenite from molybdenum porphyries in which it is the only valuable component (in some cases, particularly in China, tungsten may also be economically recovered from the plant tails). The process is relatively simple and consists of the main processing steps common to most other hydrophobic mineral separation systems. They include comminution, multiple stages of froth flotation, regrind, and concentrate dewatering. In many cases, a chloride leach step is applied to the flotation concentrates to remove lead and other deleterious elements.

Comminution

The comminution circuit generally consists of crushing and grinding equipment (rod milling, autogenous milling, and ball milling), with the purpose of achieving economic liberation of the molybdenite mineral grains. The Henderson primary molybdenum concentrator in Colorado employs primary crushing followed by a single stage of semiautogenous grinding mills operating in closed-circuit configuration with hydrocyclones. The Climax, Endako, and Thompson Creek concentrators all use a combination of autogenous, semiautogenous, and ball milling to achieve liberation. Chinese operations, however, often use a combination of ball mills and rod mills, depending on the age and capacity of the process plant.

Flotation

Once the primary ore has been sufficiently ground, it is conditioned with a series of reagents to improve the selectivity of the separation process. Typical reagents include Calumet oil, pine oil, Syntex, and Nokes. Often, reagent addition occurs during the grinding step, as many of these oils are immiscible with water. After the ore has been properly conditioned, it is directed to a roughing froth flotation stage in which the hydrophobic, predominantly molybdenite minerals are removed. The process of regrinding, conditioning, and cleaner froth flotation is repeated until the desired liberation, concentration, and recovery is achieved.

Flow-Sheet Configuration

Figures 3 and 4 show two typical flow-sheet configurations for primary molybdenum processing (one from a Chinese primary molybdenum processing plant and the other from a U.S. plant). The key difference between the two flow sheets is the use of column flotation cells in the roughing application in the case of the Chinese process configuration. The column cell is typically followed by a mechanical scavenging step with the concentrate advancing (as shown) or returning to the column (not shown). The use of the column cells with wash water in the roughing application eliminates hydraulic entrainment of fine gangue particles, allowing the size of the subsequent cleaning section to be significantly reduced.

Hydrochloric Acid Leaching

Primary concentrate is made into a slurry in a mixture of water and hydrochloric acid (HCl) to maintain approximately 40% solids and 2% acid strength. A surfactant is added to improve the solid/liquid contact and improve the kinetics of impurity removal. The leach tanks are kept at 79°C to support reaction kinetics and keep lead in solution. Leach slurry is filtered through drum filters followed by repulping with hot water to wash and remove the residual acid. The filter cake from the final filtration stage travels through a set of steam dryers to reduce moisture down to 2%–3% prior to storing in the final concentrate bins. Concentrate from the bins is packed and delivered to conversion sites. An example of typical feed and product assays before and after leaching is depicted in Table 2. Customer specifications vary, so lead content in final concentrate can vary.

In general, lead sulfide (PbS) leach kinetics in HCl solutions are quite fast, relative to ferric chloride leaching of copper (which will be discussed in the "Ferric Chloride Leaching"

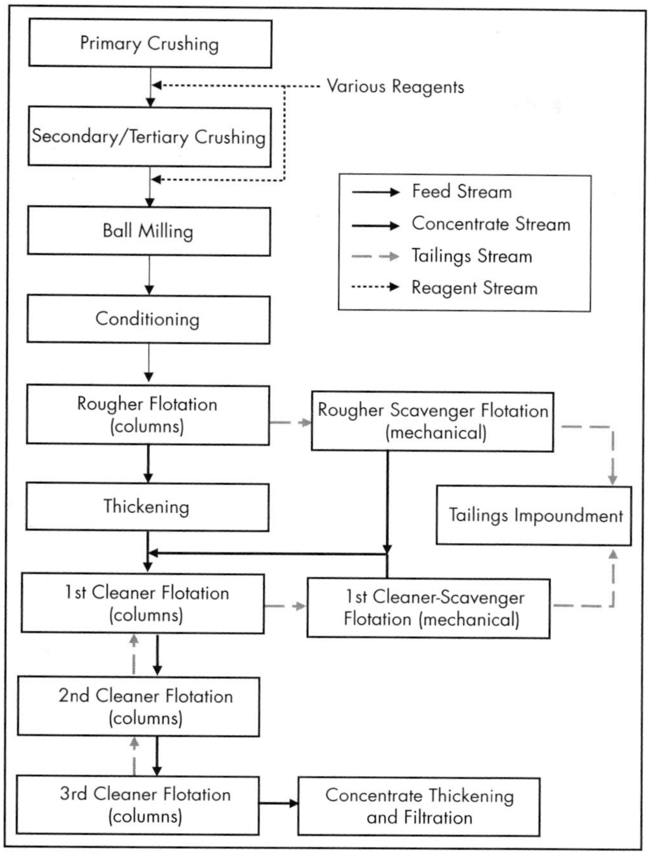

Courtesy of Aminpro

Figure 3 Typical flow sheet for primary molybdenum concentration in China

section). Figure 5 shows laboratory-batch leach kinetics in HCl solutions of various strengths.

RECOVERY OF MOLYBDENITE FROM COPPER PORPHYRY ORES (BULK RECOVERY)

Molybdenite is recoverable from copper ores by froth flotation under much the same conditions as primary copper sulfide. Because molybdenite is naturally hydrophobic, it can be recovered readily without any collectors (although collectors are often used, as discussed in the next section). For economically exploitable molybdenum ore grades, the bulk of the molybdenum reports to the sulfide concentrate, together with the copper minerals, some of the pyrite, and any other hydrophobic minerals that may be present in the ore. The molybdenite is then separated from the bulk sulfides in a subsequent selective flotation processing step. The bulk sulfides, mainly Cu- and Fe-sulfide minerals, are depressed using a depressant—often sodium hydrosulfide (NaHS)—allowing the molybdenite to be recovered and concentrated. After one or more cleaning stages, the concentrated molybdenum is thickened, filtered, (sometimes) dried, and, if the conversion facilities are off-site, packaged and shipped to the customer. The tailings of the molybdenum separation circuit contain the copper concentrates, which are also thickened, filtered, and shipped to the downstream smelting or pressure leaching plant.

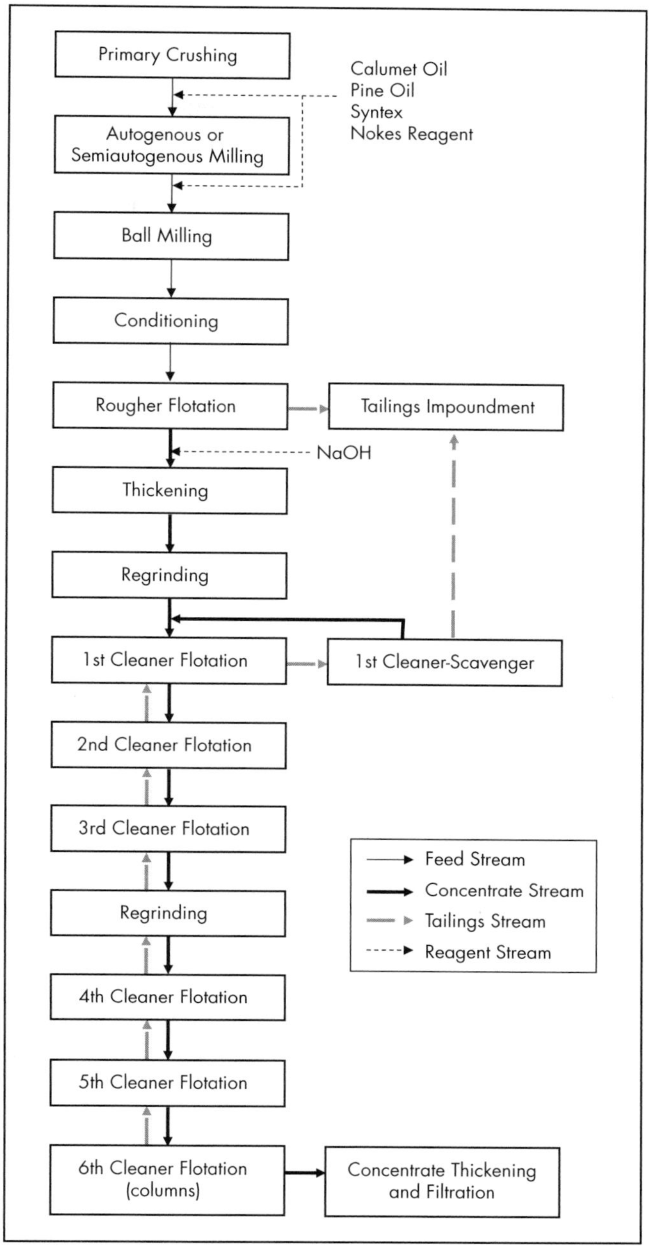

Courtesy of Aminpro

Figure 4 Typical flow sheet for primary molybdenum concentration in the United States

Table 2 Typical feed and product assays

	Slurry Concentrate A Prior to Leach, %	Dry Concentrate B Prior to Leach, %	Dry Concentrate After Leach, %
Molybdenum	58.9	58.6	58.9
Iron	0.15	0.16	0.14
Lead	0.047	0.047	0.010
Insolubles	1.12	1.22	1.26

Courtesy of Aminpro

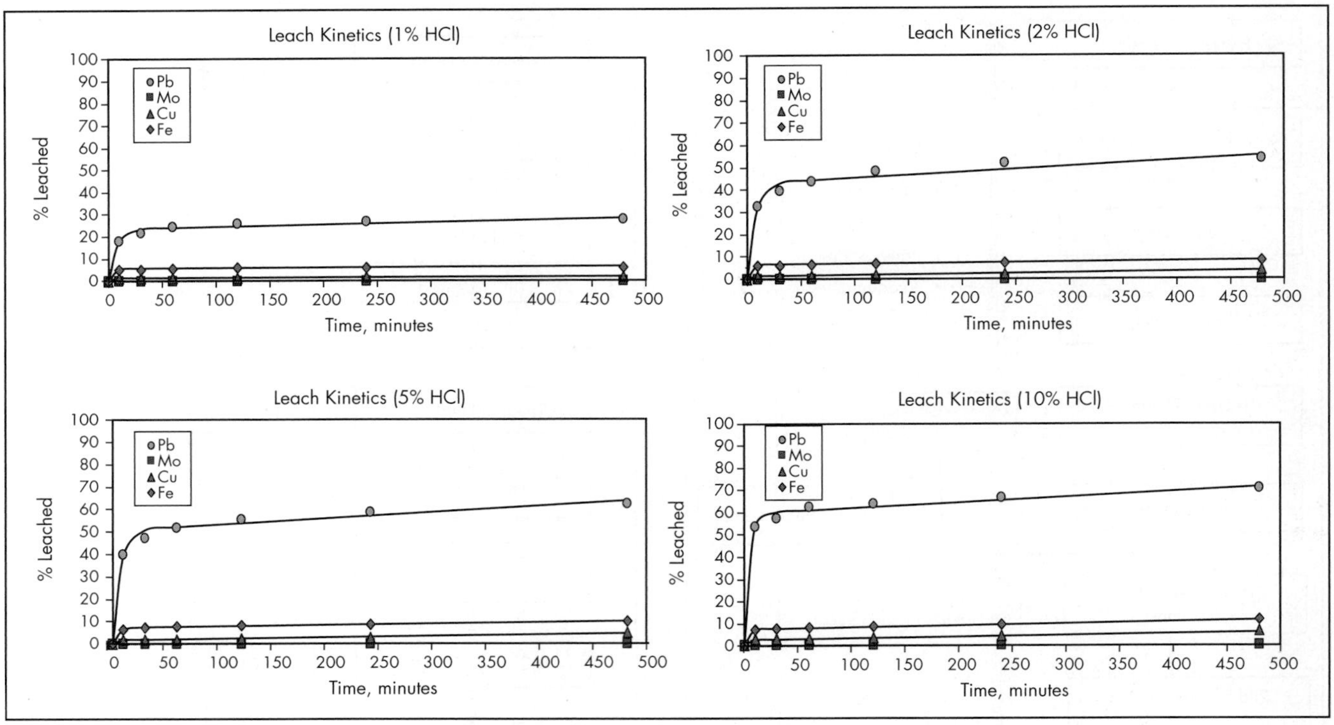

Courtesy of Aminpro

Figure 5 Pb, Mo, Cu, and Fe leach kinetics in HCl leach solutions

In some cases, depending on the ratio of recovered copper to molybdenum, the molybdenum in the bulk concentrates can reach high enough concentrations to actually be considered a smelter penalty element if a molybdenum separation circuit is not implemented prior to smelting.

Factors Affecting Primary or Bulk Recovery

This section describes the four main factors affecting the *primary* or *bulk* molybdenite recovery step (i.e., the bulk sulfide flotation step), including a description of the main factors affecting the *secondary* molybdenite recovery step (i.e., the selective, molybdenite separation step).

Mineralogical Factors

The main mineralogical or geological factors affecting the recoverability of by-product molybdenite are head grade, mineral type, and liberation/association properties, including morphology (Triffett et al. 2008).

For many by-product porphyry properties, the maximum recovery of molybdenite in the bulk sulfide roughing stage decreases rapidly for head grades less than about 0.01% Mo, and this is one of the reasons why economic by-product molybdenum recovery is seldom achieved for lower head grades. Molybdenum recovery data from a South American Cu/Mo porphyry deposit are shown in Figure 6.

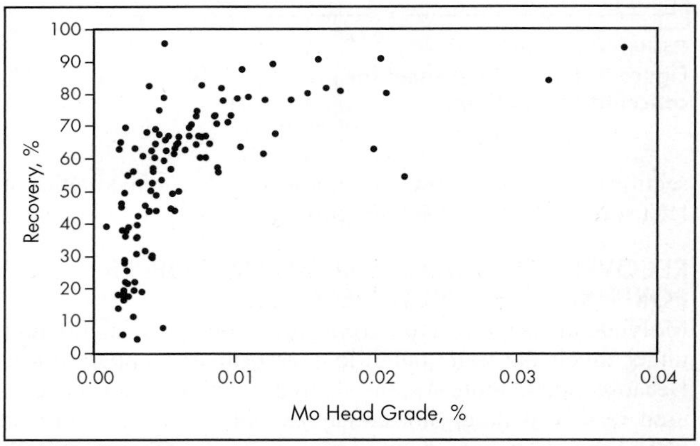

Courtesy of Aminpro

Figure 6 Molybdenum recovery in the bulk sulfide rougher flotation as a function of head grade

Most molybdenite-bearing porphyry ores contain only trace amounts of the nonfloatable molybdate mineral powellite, but in a few cases higher proportions of this mineral have been found. Like molybdenite, powellite is soluble in typical multi-acid solutions used for atomic absorption spectroscopy and is therefore not usually mineralogically identifiable by routine geochemical assaying. Furthermore, it can be difficult to detect using scanning electron microscopy because, when it occurs, it only does so in trace amounts that are often below the detection limits of current technologies. If powellite is detected or suspected in a deposit, the most common measurement approach is to use a separate weak acid digestion (such as dilute sulfuric acid solution) followed by atomic absorption spectroscopy. The results are then compared with molybdenum assayed with the standard multi-acid procedure, and the difference between the two is attributable to powellite (or other nonsulfide molybdenum species).

Pulp Conditions

Like all sulfide minerals, molybdenum flotation is affected by the characteristics of the pulp. In general, the flotation of molybdenite favors lower pH, and recoveries can drop significantly above pH of approximately 11.0. Molybdenite with finer natural grain sizes tend to be more sensitive to higher pH than those with coarser natural grain size. This phenomenon is not clearly understood and may vary from site to site, depending on operating conditions. This sensitivity to higher pH can often lead to very high (relative to copper sulfide) circulating loads of molybdenite in the bulk sulfide cleaning circuit.

Lower pulp density (percent solids) tends to favor molybdenite flotation kinetics. It is thought that the lower pulp viscosities of diluted slurries improve the particle collection kinetics of the molybdenite grains (relative to copper sulfides).

Collectors

Molybdenite is naturally hydrophobic, but kinetics can usually be increased with the use of insoluble nonpolar oily collectors, the most common of which is diesel or fuel oil. Typical collector dosages are between 5 and 25 g/t, although for some high-grade porphyry deposits dosages as high as 100 g/t have been seen in practice. The nature of the activation mechanism is not clearly understood, but two mechanisms have been proposed. In the first, it has been suggested (Born et al. 1976; Cuthebertson 1946) that the oily collectors coat the less hydrophobic grain facets and edges, creating a more compliant hydrophobic surface for bubble attachment and thereby increasing the hydrophobicity of the mineral as a whole. An alternative explanation stems from work done on molybdenite depression using lignosulfonate depressants (Ansari 2006). Lignosulfonate depressants—which work by creating hydrophilic precipitates on the hydrophobic molybdenite faces—can be rendered ineffective if the molybdenite is pretreated with an oily collector. The same phenomenon may be occurring in bulk flotation applications, but instead of preventing lignosulfonate precipitation, it is preventing calcium and other salt precipitates caused by lime saturation (probably calcium hydroxide, carbonate, sulfate, or molybdate).

Flotation Cells

Discussion of flotation equipment is reserved solely to aspects specific to molybdenum. For a broader discussion on flotation technology, see Chapter 7.4, "Froth Management."

The type of flotation cell can also impact the separation efficiency of molybdenite in the bulk circuit. It has been seen that columns, for example, tend to have a lower recovery of molybdenite than expected, compared with copper recovery and/or molybdenite recovery in conventional mechanical cells. It is thought that the reason for this is related to the higher pH at which columns typically operate (by virtue of application in cleaning circuits, where lime is added to depress pyrite). At higher pH, the molybdenite becomes less hydrophobic and the flotation kinetics drop. Froth crowding effects also play a role; that is, when the carry rate ($t/m^2/h$ of concentrate production) approaches the carrying capacity, the least hydrophobic minerals are the first ones that become dislodged in the froth or interface zone. Because copper sulfides are usually more hydrophobic, and because column cells often operate close to the carrying capacity, in this situation molybdenite recovery would suffer. This *crowding* effect has not been observed in molybdenite separation circuits, where there is no strongly hydrophobic specie to displace molybdenite.

SEPARATION OF MOLYBDENITE FROM BULK SULFIDE CONCENTRATES

As noted previously, much of the global supply of molybdenum is produced as a by-product (or a co-product) from copper mining. The molybdenum is recovered in the froth flotation process, together with copper and iron sulfide minerals, and is then separated from the so-called bulk concentrate through a secondary flotation circuit in which the copper and iron sulfides are depressed using chemical depressants, allowing the hydrophobic molybdenite to be selectively recovered and concentrated.

Principles of Molybdenite Separation

The selective froth flotation of molybdenite is based on several key principles. These include the nature and concentration of the depressant, the pH of the pulp, and the impact of hydraulic entrainment (and dispersants).

Depressants

Several depressants have been found to be effective for depressing copper and iron sulfide minerals, including sodium sulfide and sodium hydrosulfide, sodium cyanide, potassium ferri- and ferrocyanide, Nokes reagents, and Anamol-D. This chapter provides a selective overview of the use of sodium hydrosulfide and Nokes reagent, by the far the two most commonly used for by-product molybdenum production (Gupta 1992). Historically, sodium cyanide has been used to depress certain copper sulfide minerals, although this has fallen out of favor because of safety and cost disadvantages.

The Nokes reagent was invented by Charles Nokes at Kennecott (Utah) in the early 1940s. It is a reaction product of phosphorus pentasulfide and sodium hydroxide, which are usually mixed on-site. The reaction is highly exothermic and produces sodium sulfide or hydrogen sulfide (depending on the pH). The reaction equations are as follows (Gupta 1992):

$$6NaOH + P_2S_5 \leftrightarrow 2Na_3PO_2S_2 + 2H_2O + H_2S$$

$$H_2S + NaOH \leftrightarrow NaHS + H_2O$$

$$P_2S_5 + 10NaOH \leftrightarrow Na_3PO_2S_2 + Na_3PO_3S + 2Na_2S + 5H_2O$$

Both the NaHS and the sodium sulfide (Na_2S) act as copper depressants in this chemical system (in solution, Na_2S hydrolyzes and then dissociates, forming HS^-).

The depressing effect of Nokes reagent and NaHS is caused by the formation of hydrosulfide ion, which, at sufficient concentrations, reacts with the surface of the copper and iron sulfide minerals previously coated with cuprous xanthate and other collectors that were used in the upstream sulfide recovery circuit. The collectors desorb, rendering the minerals hydrophilic. Because hydrosulfide also reacts with any dissolved oxygen (or oxidized minerals), it usually has a limited life. Nokes reagent, therefore, has a longer life than NaHS because the complex thiosulfates can act as a reservoir for the sulfide ion, which reduces the oxidation rate and prolongs the time it remains available to play the role of depressant (Gupta 1992).

Another key difference relates to the sodium ion. For sodium hydrosulfide systems, the reaction product is sodium hydroxide. For Nokes systems, the reaction products are water and thiosulfates. Because the natural floatability of

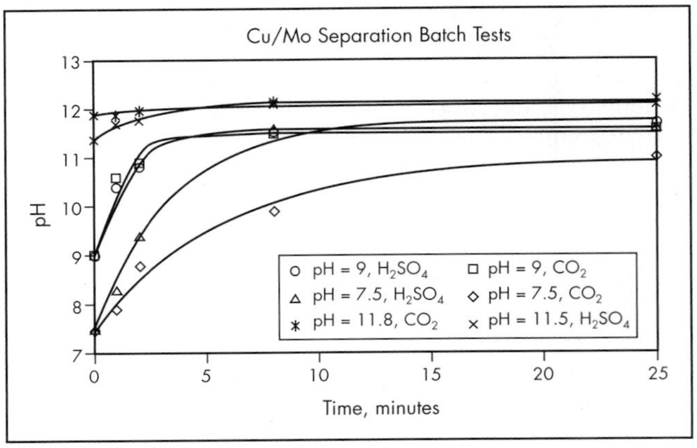

Source: Amelunxen 2009

Figure 7 pH versus time in laboratory-batch flotation

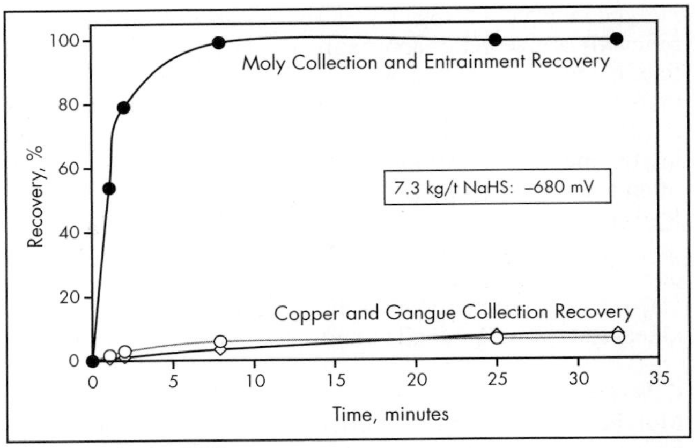

Source: Amelunxen 2009

Figure 8 Collection recovery versus time for Cu/Mo separation

molybdenite is reduced at higher pulp pH levels, it is often necessary to use acidifying agents, such as sulfuric acid or carbon dioxide, to maintain pulp pH in hydrosulfide-based copper/molybdenum separation systems. This is discussed next.

Sodium hydrosulfide (NaHS) is, by far, the most commonly used copper and iron sulfide depressant in by-product molybdenum production. The following two basic reactions occur with NaHS in a Cu/Mo separation circuit:

1. NaHS dissociation to sodium ion and hydroxyl ion (HS^-), reduces the pulp potential and strips the xanthate from the surface of most metal sulfides, rendering them hydrophilic.
2. NaHS oxidation with O_2 to produce elemental polysulfur and sodium hydroxide (NaOH), rapidly driving the pH up. This is illustrated by Figure 7, which shows pH versus time for molybdenum (moly) rougher batch flotation tests using NaHS as the depressant.

The equations are as follows (Morales 1980):

$$NaHS + H_2O \leftrightarrow NaOH + H_2S$$

$$NaOH \leftrightarrow Na^+ + OH^-$$

$$H_2S \leftrightarrow H^+ + HS^-$$

$$HS^- \leftrightarrow H^+ + S^{2-}$$

$$H^+ + OH^- \leftrightarrow H_2O$$

$$\text{Overall: } NaHS + H_2O \leftrightarrow Na^+ + HS^-$$

$$2NaHS \leftrightarrow 2Na^+ + 2HS^-$$

$$2HS^- + O_2(aq) \leftrightarrow OH^- + S_n(s)$$

$$\text{Overall: } 2NaHS + O_2(aq) \leftrightarrow 2Na^+ + 2OH^- + 2S_n(s)$$

NaHS dosage is generally controlled by pulp potential, with moly plants typically operating between –600 mV and –400 mV (using an Ag/Ag-Cl reference electrode). In such an environment, the floatability of MoS_2 is relatively unaffected, but the floatability of copper and iron sulfides is so depressed that their collection recoveries are approximately equal to that of insoluble gangue (Figure 8). Adding NaHS to lower the pulp potential beyond this point does nothing further to improve the separation. Note that NaHS is a surface tension modifier and can act as a defoaming agent in some situations, thereby reducing copper recovery by entrainment (although the authors generally favor commercial defoamers for this application because they are less expensive and do not affect the process pH).

NaHS consumption can be reduced by eliminating process water from the moly plant feed prior to rougher flotation, such as by thickening to 65% and diluting with fresh water. Consumption can be further reduced by using the cleaner tails or copper thickener overflow waters, rather than fresh makeup water. Use of nitrogen and/or enclosed cells for oxygen-depleted air has been demonstrated at many plants to have reduced NaHS by 50% on average.

Table 3 shows some selected Cu/Mo by-product plants and the type of depressant used. Only one plant still uses sodium cyanide and arsenic Nokes, while another uses only Nokes (both are in South America).

Other Depressants

Because of safety concerns associated with the transport and use of NaHS, recent attention has been given to alternative depressants that can be used as a substitute or partial substitute for NaHS.

One such depressant, di-sodium carboxymethyl-tri-thiocarbonate, is marketed by Chevron Phillips under the commercial name Orfom D8 depressant. D8 is effective at depressing copper, iron, and lead sulfides at basic and acid pH ranges. It is applied at varying dosages depending on the application (Chevron Phillips 2000–2018).

Another depressant, calcium polysulfide (Tessenderlo Kerley 2016), marketed by Tessenderlo Kerley under the trade name TKI-330, is reported to provide virtually identical metallurgical recoveries relative to NaHS at similar dosages (Tessenderlo Kerley 2018). Unlike D8, however, TKI-330 still has a hydrogen sulfide (H_2S) vapor pressure, albeit significantly lower than that of NaHS solution. As a result, it reduces safety risks when operating at lower pH, but does not eliminate them. TKI-330 offers an additional advantage of being able to be manufactured from raw materials on-site, eliminating risks associated with the safe transport of the chemical (Tessenderlo Kerley 2018).

Table 3 Copper-molybdenum plants and reagents used

Site	Chalcopyrite, %	Covellite, %	Digenite, %	Chalcocite, %	Bomite, %	Floatable Gangue	Cu Collector	Mo Collector	Frother	Mo in Bulk, %	Bulk Mo Recovery, %
Plant 1, North America	33	33	25	8	<1		Thionocarbamate	MCO	X-133	0.70	70
Plant 2, North America	99						Xanthate	MCO	Nalco 9837/ X-133	3.00	80
Plant 3, North America	97.50	2.50		<1			Xanthate	Fuel oil	X-133	1.50	74
Plant 4, North America	95	1		3	1		Thionocarbamate/xanthate	MolyFlo	Methylisobutyl carbinol (MIBC)	0.70	90
Plant 5, North America	85	<1	<1	<1	14	Talc	Dithiophosphate/monothiophosphate	Fuel oil	X-237	1.40	80
Plant 6, North America	60				40		Xanthate	Fuel oil	Tennafroth 350	1.50	75
Plant 7, North America	100						Xanthate	Fuel oil	MIBC		
Plant 8, North America	100						Thionocarbamate/xanthate	Fuel oil	MIBC (F-2)	0.60	60
Plant 9, South America	98	1			1		Thionocarbamate	Fuel oil	MIBC	0.70	
Plant 10, South America	X	X	X	X	X						
Plant 11, South America	X			X							
Plant 12, South America	X				X						
Plant 13, South America	X				X						
Plant 14, South America	X	X		X	X						
Plant 15, South America	X				X		Dithiophosphate and xanthate	Fuel oil (diesel and kerosene)	MIBC	0.005	0.85

(continues)

Table 3 Copper-molybdenum plants and reagents used (continued)

Site	Mo Rougher, % Mo	Mo Rougher, % Cu	Depressant	Pretreatment	Pretreat Time	Particle Size	Depressant, kg/t	Roughers	Cleaners	pH, Rougher Condenser	eH, Rougher
Plant 1, North America	2.70	37	NaHS	High shear	1.5 hours	30% +325		Air	Air	10, CO_2	
Plant 2, North America	4.50	25	NaHS	High shear	0.5 hour	20% +325	15	N_2	Air	11.5	
Plant 3, North America	5.00	25	NaHS	Conditioning	19 minutes	26% +325	3.5	N_2	N_2	11.5	
Plant 4, North America	1.30	24	NaHS	Conditioning	15 minutes	45 µm	12.5	N_2	Air	10.3, CO_2	
Plant 5, North America	1.40	25	NaHS	Stock and conditioning	24 hours + 20 minutes	75 µm	9	N_2	Air	10.5	
Plant 6, North America	4.00	32	NaHS	Stock and conditioning		50 µm		N_2	N_2	8.8, CO_2	
Plant 7, North America			NaHS					Air, recirculated	Air, recirculated	8.5	
Plant 8, North America	1.20	28	NaHS	Stock and conditioning				N_2	N_2	9.2, H_2SO_4	
Plant 9, South America	1.20	30	NaHS		30 minutes	53 µm	4.4	Air, recirculated	Air	7.5, H_2SO_4 (CO_2 future)	
Plant 10, South America			NaHS				6	N_2	N_2	7.5, H_2SO_4	
Plant 11, South America			NaHS				4	N_2	N_2	10.5	
Plant 12, South America			NaHS				7.4	N_2	N_2		
Plant 13, South America			NaHS		Pipeline		4	Air, recirculated	Air, recirculated	7.5, H_2SO_4 (staged additional)	
Plant 14, South America			NaHS				4.6	N_2	N_2	7.5, H_2SO_4	
Plant 15, South America	0.007	0.0002	NaHS and pentasulfuros	Conditioning with acid	45 minutes	30% +325	4	N_2	First cleaners	9.3–9.5 (H_2SO_4 to pH 8.4 then CO_2 in first cells)	–450 mV

(continues)

Table 3 Copper-molybdenum plants and reagents used (continued)

Site	pH, Cleaners	Open Circuit Cleaner Tails	Cells Covered	Scrubbers	Regrind/Thickener	Cleaning Time, minutes	Column Cells	Mo Concentrate Grade, %	Cu Concentrate Grade, %	Selective Recovery, %	Overall Recovery, %
Plant 1, North America	10, CO_2	No	Yes	Yes	No/No	26	Yes	55	4	95	67
Plant 2, North America		No	Yes (ex cleaners)	No	Yes/No	26	No	48	2.5	95	76
Plant 3, North America		No	No	No	No/No	75	Yes	50	3	95	70
Plant 4, North America	12	No	Yes (ex columns)	No	No/Yes	198	Yes	51	2	91	82
Plant 5, North America	11.8	Sometimes	No	No	Yes/Yes	207	Yes	52	0.40	80	69
Plant 6, North America	9.5, CO_2	No	Yes	Yes	Yes/No		Yes	53	0.70	95	72
Plant 7, North America	10		Yes		Yes		Yes	52	2.50		
Plant 8, North America	11		Yes	No				48	2.50	60	36
Plant 9, South America	9.5, CO_2	No	Yes (ex columns)	No	No/No		Yes	50	3	89	60
Plant 10, South America			No								
Plant 11, South America	11.3		No								
Plant 12, South America			No								
Plant 13, South America	7.5, H_2SO_4 (staged)	No	Yes	Yes					92		
Plant 14, South America		No	Yes	Yes							
Plant 15, South America	9.1 (CO_2) in first cleaners	No	No	No	No/Yes		Yes	48	4 maximum	81	69

Courtesy of Aminpro Benchmarking

Note: Numeric values are from quantitative microscopy. Blank cells indicate that the mineral is not detectable. "X" means that the mineral is present in significant but unknown quantities, either because quantitative mineralogy is not available or because the mineralogy can change significantly between ore types.

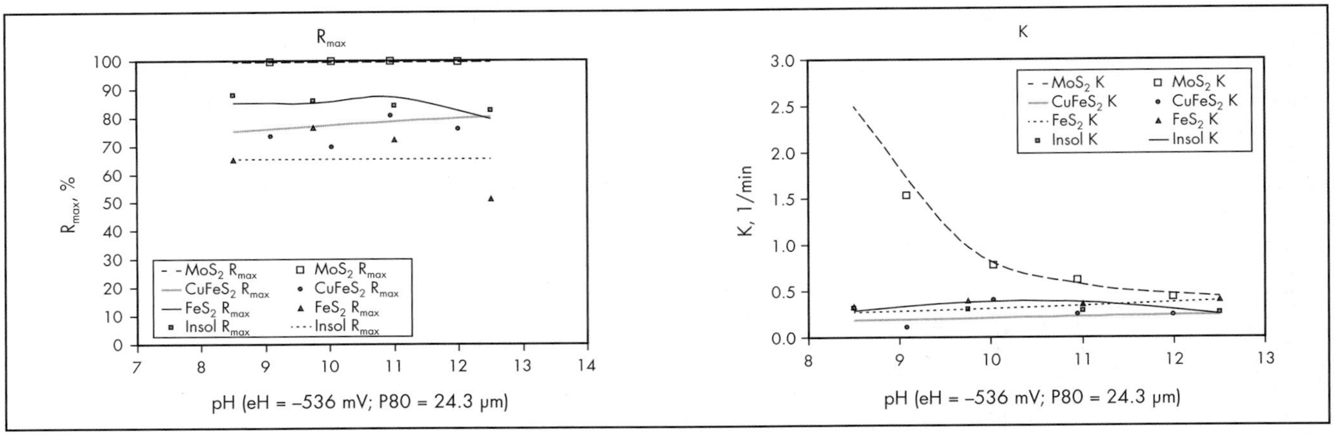

Courtesy of Aminpro

Figure 9 R_{max} and collection zone rate constant as a function of pulp pH for a moly plant cleaner circuit

A third alternative, marketed by Solvay Group (formerly Cytec) under the trade name Aero 7260, has been shown to effectively depress copper sulfides when used as a partial replacement for NaHS (i.e., 50%–90% of the NaHS consumption is replaced). Dosages vary between 1.5 and 2.5 kg/t (Cytec 2013).

Currently, there is not an extensive usage history of these reagents in operating plants, but preliminary results indicate that they show promise. This area of research is expected to see rapid development soon.

pH Modifiers

As part of a previous study performed at a South American moly separation plant, moly cleaner separation kinetics were measured at different pH. The results are shown in Figure 9. It can be seen that at pH of approximately 12.0 and above, the kinetics of moly are virtually indistinguishable from those of chalcopyrite. For the high R_{max} values of the minerals, this means that it is virtually impossible to achieve a good moly separation without extremely high circulating loads. The solution, in this case, is to reduce the pH, thereby moving to the left along the x-axis of the right-hand graph. At pH ≤10, the kinetics become more favorable to moly separation.

Two different approaches to pH control are used in by-product separation. The first, and most common, is passive pH control, in which the NaHS dosage, the use of nitrogen, and the use of natural pH makeup water are carefully controlled to maintain the pH in the appropriate range. In this case, pH is allowed to vary, albeit in a controlled manner. In some cases, with slow-floating molybdenum species or complex circuits (such as those involving talc, pyrophyllite, or other impurities), it may not be possible to control the pH using passive techniques. In these cases, active pH control may be more effective, which considers the use of acidifying agents to control the pH to a target setpoint, using instrumentation and process control strategies.

Sulfuric acid and carbon dioxide (CO_2) gas are the two most common acidifying agents used in active pH control strategies.

1. *Sulfuric acid* donates a hydrogen ion resulting in lower pH. Secondary hydrogen dissociation also occurs, yielding sulfate ion, which can form gypsum with dissolved calcium. The reactions are as follows:

$$H_2SO_4 \leftrightarrow H^+ + HSO_4^-$$

$$HSO_4^- \leftrightarrow H^+ + SO_4^{2-}$$

$$H^+ + OH^- \leftrightarrow H_2O$$

$$SO_4^{2-} + Ca^{2+} \leftrightarrow CaSO_4$$

2. *Carbon dioxide* acidifies the pulp through a series of reactions. The vast majority of dissolved CO_2 is not converted into carbonic acid and therefore is free to react directly with hydroxide ions. The reactions are as follows:

$$CO_2 + 2OH^- \leftrightarrow CO_3^{2-} + H_2O$$

$$CO_3^{2-} + CO_2 + H_2O \leftrightarrow 2HCO_3^-$$

$$2HCO_3^- \leftrightarrow 2H^+ + 2CO_3^{2-}$$

$$2H^+ + 2OH^- \leftrightarrow 2H_2O$$

$$2CO_3^{2-} + 2Ca^{2+} \leftrightarrow 2CaCO_3$$

Thus, each mole of CO_2 neutralizes two moles of hydroxide. The trace amounts of carbonic acid that are formed also reduce the pH:

$$CO_2 + H_2O \leftrightarrow H_2CO_3$$

$$H_2CO_3 \leftrightarrow 2H^+ + CO_3^{2-}$$

$$HCO_3^- \leftrightarrow H^+ + CO_3^{2-}$$

$$3H^+ + 3OH^- \leftrightarrow 3H_2O$$

$$CO_3^{2-} + Ca^{2+} \leftrightarrow CaCO_3$$

It has been found that CO_2 is metallurgically superior to sulfuric acid. For example, Figure 10 shows the relationship between pH and the rate constant (K) for a sample of bulk concentrate from El Teniente. One curve was generated using carbon dioxide as the pH modifier, and the other curve was produced with sulfuric acid as the pH modifier. It can be seen that the molybdenum flotation rate constant is significantly higher when using CO_2, rather than sulfuric acid, as the pH modifier.

It is thought that the improved floatability of molybdenite when using carbon dioxide as the pH modifier, rather than sulfuric acid (Figure 10), is caused by the difference between the

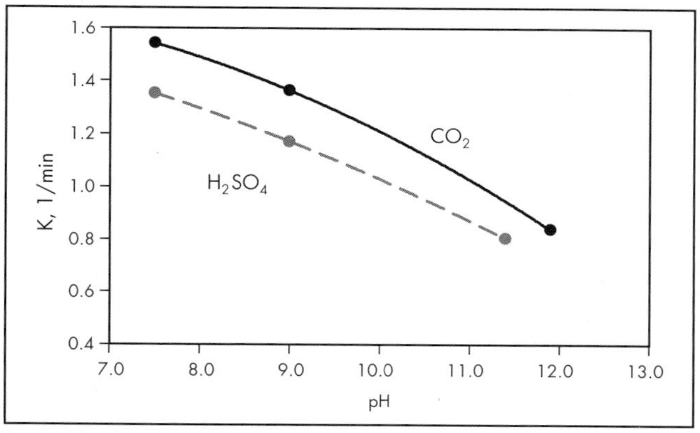

Source: Amelunxen 2009

Figure 10 Molybdenite rate constant versus pulp pH for CO_2 and H_2SO_4 neutralization

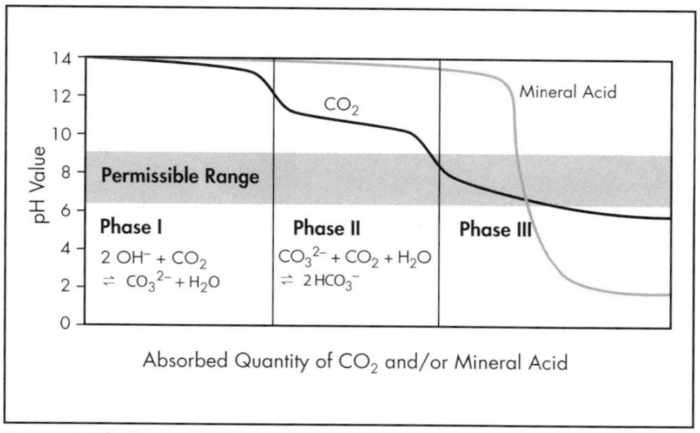

Source: Linde Group 2012

Figure 11 Example of basic neutralization curve for water solutions, mineral acid, and CO_2

equilibrium solubility of calcium carbonate ($CaCO_3$), which is 20 ppm on average at standard temperature and pressure in distilled water, and calcium sulfate (approximately 2,000 ppm) (Green and Maloney 1999). The solubility (Ksp) of $CaCO_3$ is a function of dissolved CO_2 content. At atmospheric CO_2 concentrations, the maximum equilibrium solubility in distilled water is approximately 50 ppm, but this increases as the concentration of carbonic acid in the pulp increases. Note that the presence of other ions will decrease the Ksp.

It has been recognized that calcium ion in solution can depress molybdenite, but the physicochemical explanation is not fully understood. Chander and Fuerstenau (1972) suggest that it forms calcium molybdate precipitate on the surface of MoS_2 crystals. Rhagavan and Hsu (1984) propose a heterocoagulation effect of silicon dioxide (SiO_2) and MoS_2 in the presence of calcium ion (the so-called slimes effect).

In addition to the metallurgical superiority of CO_2, there are other advantages to using CO_2, which include the following:

- It is environmentally benign, and therefore easier, to obtain permits for the transport and use of CO_2.
- It is safer, because there is less risk of mixing with NaHS-containing fluids during spillage. It is also less corrosive than sulfuric acid.
- It has a flatter neutralization curve, as can be seen in Figure 11. This makes it much easier to control the pulp pH and reduces the potential of overshooting.

By-product molybdenum separation plants that are currently employing active pH control include

- Highland Valley, British Columbia (CO_2-based, both roughers and cleaners);
- Antamina, Peru (roughers);
- Cerro Verde (roughers);
- Morenci Metcalf, Arizona (roughers, cleaners);
- Andina, Chile (sulfuric acid in the roughers, carbon dioxide in the cleaners);
- Los Pelambres (sulfuric acid rougher conditioning tank, undergoing CO_2 trials); and
- Los Bronces/Las Tortolas, Chile (sulfuric acid, roughers, and cleaners).

Dispersants

It has been shown that dispersants can be effective at increasing the recovery or separation kinetics of molybdenite in the separation circuit. It is thought that the dispersants help to reduce the heterocoagulation effect; that is, the tendency of excess calcium ion to coagulate with gangue and fine molybdenite particles creates a hydrophilic coating on the molybdenite grains. Water is by far the least expensive dispersant, but many molybdenum separation circuits employ sodium silicate in dosages between 20 and 200 g/t.

Flotation Equipment

Most moly circuits run the roughing section with mechanical cells and run the cleaners with either fully mechanical cells, column cells, or a mixture of both columns and mechanical cells. Column roughers have only been seen in primary moly mines in mainland China.

Upgrading profiles using column flotation in by-product moly plants typically follow the curve shown in Figure 12. The profile indicates that at most five cleaning stages are required to achieve a high-purity product. A similar profile for mechanical cells is shown in Figure 13. For mechanical cells, at least eight stages are required to produce high-grade moly concentrates (i.e., >90% molybdenite in concentrate).

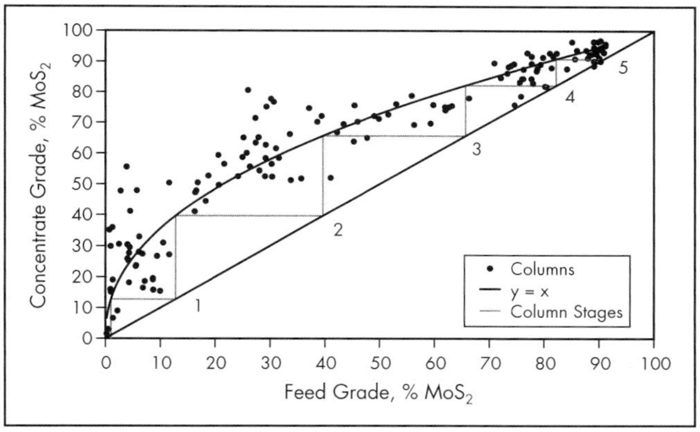

Source: Amelunxen 2009

Figure 12 Upgrading profile for column cells in by-product moly applications

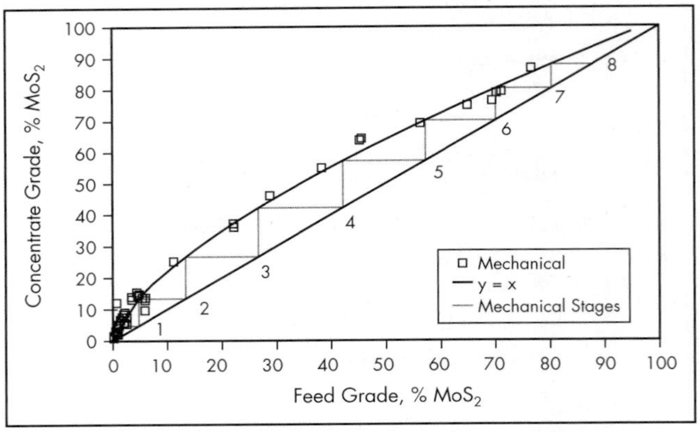

Source: Amelunxen 2009

Figure 13 Upgrading curve for mechanical cells in by-product moly applications

Design engineers generally favor mechanical roughers over column roughers in by-product moly circuits because of column's inability to attain high recoveries. The primary drawback with mechanical cells operating as roughers is that even in a well-operated circuit, at least one-third of the copper floated in the roughers does so through entrainment mechanisms, whereas with the column cell, most of this entrainment is avoided. If the operator is unaware of this, he or she may tend to increase the NaHS dosage when the copper levels are high in the roughers without paying attention to the more relevant parameters of froth bed depth, gas rate, and water recovery. Note that to reduce the entrainment recovery of solids, it is also necessary to reduce the froth recovery of MoS_2, and therefore the total recovery of moly will drop accordingly. It is primarily for this reason that the total rougher retention time is usually much longer in the moly plant than in the bulk circuit, where moly flotation is not affected by pulling rate constraints.

Flow-Sheet Configuration

A typical by-product molybdenum production flow sheet is shown in Figure 14. Depending on mineralogy, association, and molybdenite liberation, different properties have different flow-sheet concepts.

Ferric Chloride Leaching

By-product concentrate feed may contain 45%–48% molybdenum, 3% copper, and 0.3% lead. Leached by-product may contain 50% molybdenum, 0.15% copper, and 0.03% lead. Ferric chloride is often used to leach molybdenum concentrate. Many variables influence leaching: temperature, agitation, particle size, concentration, and oxygen-reduction potential. The chloride ion converts the lead into a form that is soluble. The copper dissolution, which occurs with the chloride ion, is recovered by ferric ion. The ferric ion is then regenerated by chlorination. A ferric digester keeps the total iron at the appropriate level and removes dissolved copper. This is a cementation process that exchanges cuprous for Fe^{+2} using scrap metal. The copper is then recycled at the smelter, and the ferrous solution is sent to the chlorinator to be regenerated to ferric. This reduces reagent costs by minimizing the need to bleed out ferric chloride because of high copper concentrations. An example of a by-product leach process is depicted in Figure 15.

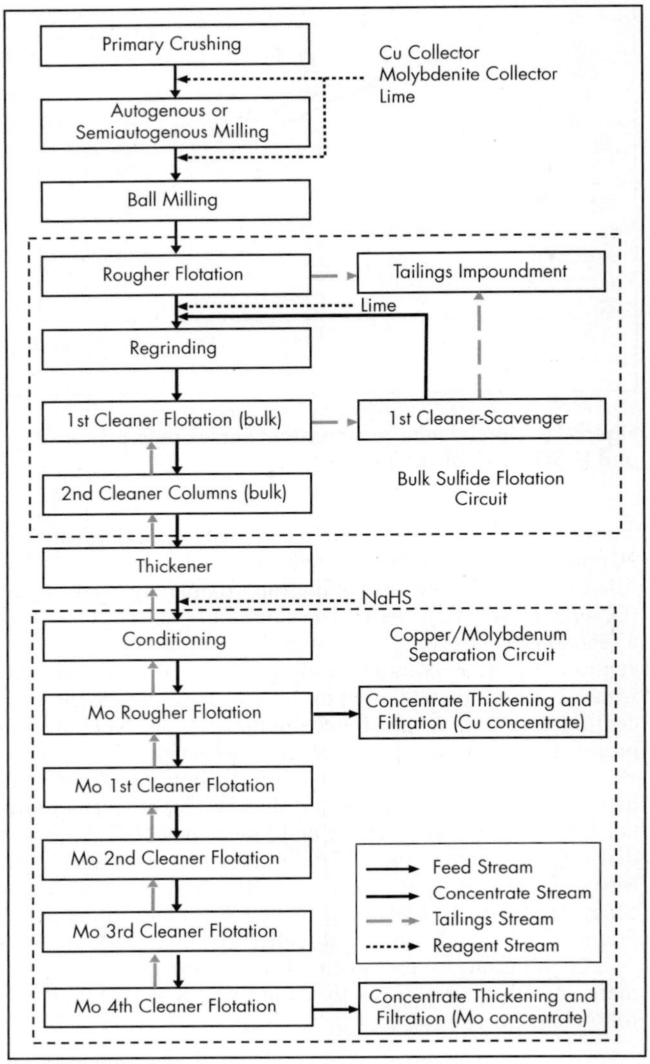

Courtesy of Aminpro

Figure 14 Typical by-product molybdenum flow sheet

Safety Considerations

The use of acidifying agents in combination with sodium hydrochloride can create serious safety concerns caused by the potential to generate lethal concentrations of H_2S gas. H_2S gas is naturally emanated from NaHS solutions in trace amounts, but as the pH drops and the pulp becomes more acidic, the evolution rate begins to increase dramatically. The following stability diagram shown in Figure 16 for HS^- in distilled water shows that below approximately pH 7.2 H_2S becomes the stable species, although in practice, pH of approximately 8.5 can lead to dangerous concentrations in poorly ventilated spaces. Active ventilation systems are therefore suggested for plants operating below pH of approximately 9.5. These systems consist of covered flotation cells that operate under negative pressure, with a NaOH-solution scrubber treating any process off-gas.

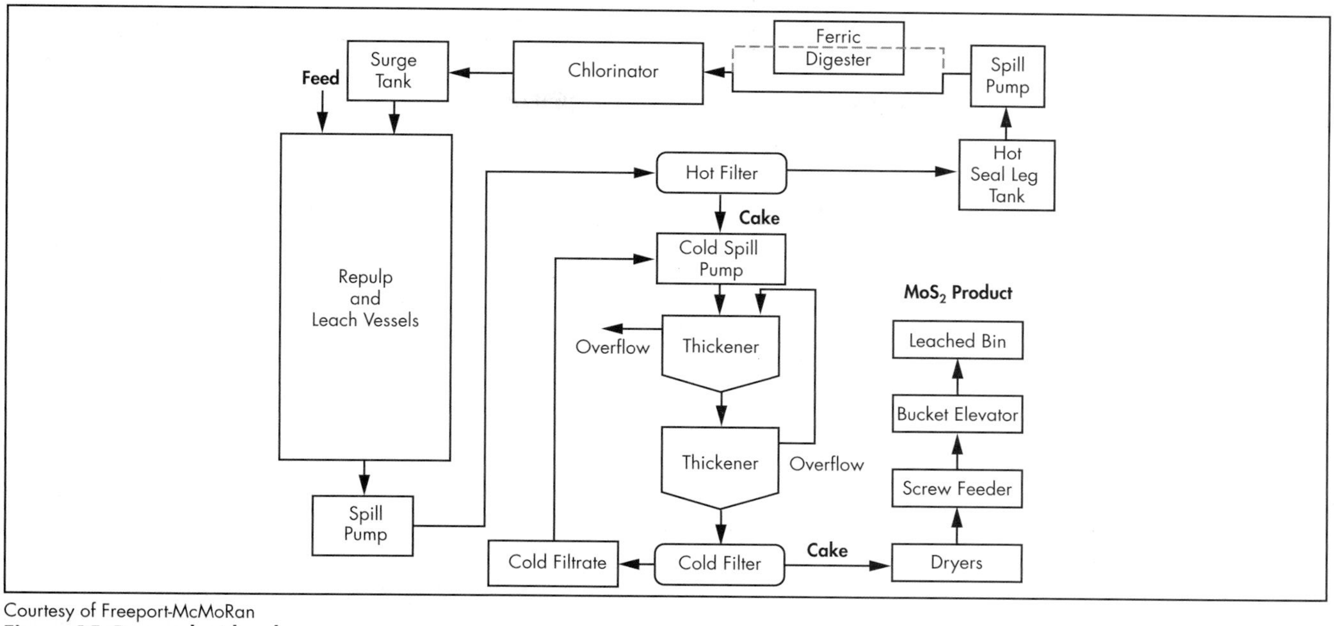

Courtesy of Freeport-McMoRan

Figure 15 By-product leach process

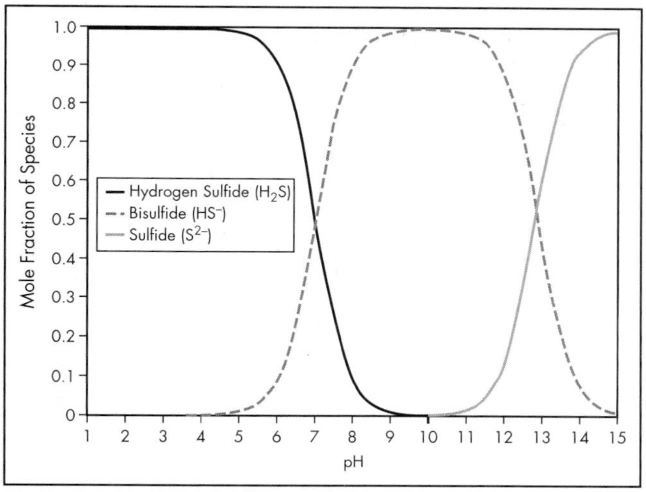

Figure 16 Hydrogen sulfide stability diagram

The use of covered cells is not ideal, as it obstructs visual inspection of the cells and hinders effective operation of the separation circuit. For this reason, it is desirable to maintain the pH above approximately 9.5 whenever possible, and in those cases with marginal separation kinetics, compensate with higher residence time.

Commercial Aspects

Saleable concentrate grades begin at around 44% Mo in the concentrate, although most by-product producers strive for higher grades to avoid penalties associated with copper and other contaminants. The penalties depend on whether the producer converts the molybdenite on-site or the concentrate is *tolled* through a third-party conversion facility. In the former case, the particular conversion facilities and costs dictate the required concentrate quality. In the latter, the tolling or sales contract will dictate the required concentrate quality.

Table 4 Typical metallurgical-grade impurity limits

Impurity	No Penalties	Penalties	Rejected
Cu	<0.4%	0.4% < 3%	>3%
Moisture + oil	<6%	6% < 9%	>9%
Pb	<0.04%	0.04% < 0.1%	>0.1%
Cl	<500 ppm	500 < 2,000 ppm	>2,000 ppm
As	<200 ppm	200 < 1,000 ppm	>1,000 ppm

Courtesy of Aminpro

Most conversion facilities use the pyrometallurgical route (i.e., roasting). Roasters generally charge toll fees ranging from US\$0.44/kg to US\$0.88/kg Mo for toll roasting, US\$0.22/kg to US\$0.44/kg for ferric chloride ($FeCl_3$) leaching, and US\$0.11/kg to US\$0.22/kg for packaging depending on market conditions. This amounts to tolling fees ranging from US\$0.77/kg to US\$1.54/kg, on top of which are added penalties for impurities.

Impurity limits depend on the roaster process and applicable environmental regulations; hence, they are usually roaster specific. Table 4 shows some typical values than can be used as a guideline for design purposes if a roaster contract is unavailable. Penalties generally range from US\$0.11/kg of Mo for relatively clean concentrates with minor amounts of a single impurity to as high as US\$1.10/kg for lower-grade concentrates that exceed all or almost all of the impurity level limits. Following is a brief discussion of some of the impurities.

Copper limits are generally a function of the copper removal process at the roaster or roasters treating the product. The most common options are chloride leach on the roaster feed or sulfuric acid leach of the roaster product, but many variants exist. Depending on the oxide consumer, tech oxide must contain less than 0.1%–0.5% Cu (0.09%–0.45% Cu in the roaster feed after accounting for mass reduction in the roaster); therefore, most commercial roasters treating predominantly by-product concentrate contain some kind of copper removal system. If the intended concentrate buyer does not, it may be necessary to include a copper leach facility (or extra regrind and/or cleaning stages) at the design stage of the moly plant.

Oil and water levels can be controlled with final concentrate dryers—approximately 50% of the typical diesel collector can be volatilized at 250°C and some 90% at 300°C (beyond which SO_2 emission can become a problem). The number one cause of excessive oil content, which refers strictly to hydrocarbon molybdenum collectors, is high moly collector dosage (leaking hydraulic fluid cables being a distant second). Diesel, by far the most common moly collector, would by virtue of its specific gravity float to the top of a flotation cell even if it was not naturally hydrophobic (and it is). Expect a large fraction of the total moly collector to report to the final concentrate.

Lead most often occurs as PbS (galena) and is difficult to depress under normal flotation condition. If Pb levels are excessive, consider HCl leach on the concentrate; the kinetics are fast, and the process is relatively benign from an environmental and safety standpoint.

Arsenic is often locked with copper sulfides, so reducing Cu levels will likely reduce the As levels.

Chlorine levels are only problematic when the concentrate will not be leached after it is sold (such as when the leach circuit is located on-site). It is usually only considered for design purposes when salty or briny water is used in the process.

Rhenium is generally found in solid solution with the MoS_2. Roasters recover anywhere from 50% to 90% but do not generally credit the producer for rhenium content, although this may change in the future as the new pressure oxidation methods (described in the following sections) offer significantly higher rhenium recoveries than traditional roasting. Relatively clean by-product concentrates with unusually high rhenium values (i.e., >800 ppm) may help reduce roaster premiums, but this is not generally seen at the design stage.

The authors have collected a database of primary and by-product concentrate grades; typical industrial impurity levels are shown in Table 5.

Table 5 Typical by-product concentrate quality

Impurity	No. of Mines	Database Statistics		
		Minimum	Average	Maximum
H_2O, %	24	0.1	5.5	12
Oil, %	24	0.1	1.3	4.9
Mo, %	25	46.9	53.0	57.5
Cu, %	25	0.05	0.83	2.06
Pb, %	25	0.005	0.04	0.20
Fe, %	25	0.46	1.67	4.79
Cl, %	23	0.01	0.06	0.42
As, ppm	24	15	215	2,000
Se, ppm	24	100	250	700
Re, ppm	24	10	397	1,100

Courtesy of Aminpro

PLANT DESCRIPTIONS

There have been considerable changes in technology and industry safety culture since the development of the first molybdenum recovery circuit. Engineered safety systems, improved laboratory testing procedures, developments in flotation cell designs, and advancements in froth pumping, material handling, and dewatering systems have all led to significant improvements in plant performance and product quality.

Safety

Safety should be the first and foremost consideration when designing and operating molybdenum plants. In general, molybdenum separation circuits are subject to the same safety risks and mitigation efforts as the rest of the plant (these are not discussed here). Of particular importance with molybdenum separation is the safety risk related to the use of sodium hydrosulfide as a copper depressant, which, as previously noted, has the potential to evolve H_2S gas in lethal quantities. Although safety risks can always be reduced to a negligible residual through application of engineering, equipment, and capital, it is always a good practice to eliminate the root cause rather than apply engineering and equipment fixes. This helps to eliminate risks that may arise as a result of unforeseen process modifications and upgrades. Although there are several promising alternatives to sodium hydrosulfide, these have not yet been proven in industrial applications and therefore, the overwhelming majority of molybdenum separation circuits rely on NaHS as the copper and iron sulfide depressant.

NaHS is usually purchased in the form of a liquid at 40%–45% strength and diluted to 20% strength in on-site solution storage tanks. In fewer cases, NaHS is purchased as a solid and dissolved on-site. The NaHS preparation station and/or offloading stations require special safety isolation equipment and protocols to ensure that concentrated H_2S fumes are safely contained.

If no acidifying agent is used in the molybdenum plant, NaHS solutions naturally trend to higher pH, as the NaOH present in the original NaHS, combined with the NaOH formed during the normal oxidation of solubilized NaHS, tends to increase pH. As a result, these moly plants are relatively safe and only require H_2S sensors in key locations.

If active pH control is used in combination with NaHS, then more safety considerations are required. These include online pH sensors (to ensure that line blockages and process upsets do not lead to lower pH conditions), a good pH meter maintenance and calibration procedure, H_2S gas sensors in key locations, specially designed spillage areas to ensure that spillage does not oxidize or acidify, and special interlock protection to automatically shut off the NaHS addition pumps when the pH is high or H_2S is detected.

In cases with active pH control and particularly difficult separation kinetics, it may be required to operate at quite low pH (i.e., in the 7.5-to-9.5 range). In these cases, in addition to those considerations noted earlier, the cells are likely to be hermetically sealed and operating at negative pressure with a scrubbing system. In addition to these engineered safety controls, it is advisable to install hazardous gas sensors and alarms and implement strict evacuation procedures to maintain a safe working environment and comply with regulatory agencies. Often these plants are required to be isolated from the main separation circuit, and access is strictly controlled.

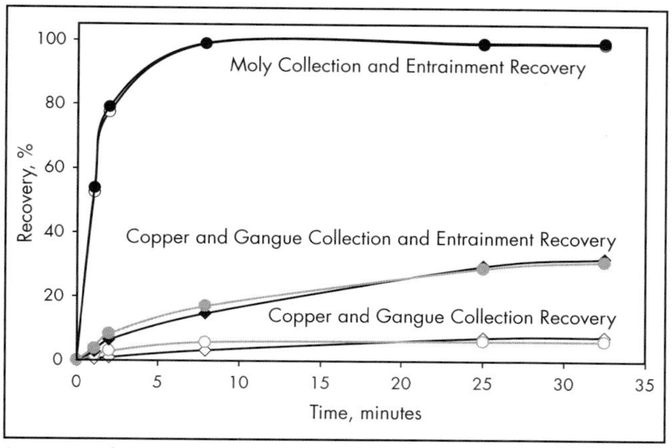

Source: Amelunxen 2009

Figure 17 Collection and entrainment recovery from a batch flotation test

Laboratory Testing

Testing for moly flotation is notoriously difficult for a greenfield project, given the mass of sample required to produce enough bulk concentrate for moly rougher bench-scale flotation tests. As a result, many plants are designed in the absence of test work. This trend is not altogether absurd in that moly plant feed minerals may be divided into two categories: the floatables, consisting of molybdenite; and the non-floatables, made up of the rest of the minerals including copper. All that is needed from the test work is to verify the maximum recovery of molybdenite in the bulk (or Cu-Mo) concentrate and the rate constant in order to size the roughers properly. The remainder of the circuit may be designed using typical kinetics of other operating plants.

When sufficient sample is available for testing, care must be taken to discriminate between collection recovery and entrainment recovery in the batch test. Figure 17 shows the total recovery versus time (collection plus entrainment) for the same batch test shown in Figure 8. Because a batch test is operated with shallow froth and high pulling rates (i.e., high froth recovery), the entrainment recovery is very high, amounting to more than 80% of the total recovery of the non-floatables. Collection and entrainment need to be considered separately within the scale-up models to account for the previously mentioned long residence times and low pulling rates found in typical plant scale moly roughers.

Design Concepts

The process design of bulk sulfide circuits is usually driven by the economics of the primary copper sulfide minerals, although molybdenite considerations can play a role in the cleaning circuit, primarily if column cells are considered. As noted previously, these cells are susceptible to (what are thought to be) froth crowding effects caused by the higher pH and finer molybdenite grain size. As a result, in bulk circuits, column cells often function as inverse flotation stages for molybdenite, with higher molybdenite concentrations in the tails rather than the concentrate. The higher circulating loads are recovered with higher retention times in the cleaner-scavenging stage or stages, thereby recovering the molybdenite from the column tails and returning it to the column feed. The circuit eventually converges on an equilibrium circulating load that yields overall cleaner circuit recoveries slightly below those of copper sulfide minerals (95% for molybdenite vs. typically 97% for copper sulfides). The exception is when insufficient residence time is allowed for in the cleaner-scavengers, or if the column carry rate is already very high (because of undersized columns or high head grades). In these cases, minor process upsets are capable of causing high losses of molybdenum to the cleaner tails.

Unfortunately, there is no validated phenomenological model currently capable of modeling froth crowding and interface phenomena in industrial plants, and issues such as those just discussed are usually either ignored or dealt with through the use of ample safety margins or experience and heuristics.

Brownfields

Brownfield design programs are those in which a molybdenum separation circuit is added to an existing process plant. In this situation, it is usually the case that a large sample of bulk concentrate is available for detailed testing and, perhaps, piloting of the moly separation circuit. Suggested test matrices are shown in Tables 6 and 7 for roughers and cleaners, respectively.

Greenfields

Greenfield design applications—those in which there is no existing process infrastructure—present a significant problem in moly separation circuit design. Because of the typically low molybdenite concentrations in the ore, it is very difficult to procure sufficient sample of bulk concentrate on which to base the design of the molybdenite circuit. For example, each batch kinetics test for a molybdenite rougher separation requires at least 1,000 g of sample (assuming a 2-L cell). To procure this much sample, one would need between 50 and 100 kg of ore, depending on the head grades and mineralogy. This means that to obtain a basic test matrix of nine rougher tests (4 × P80, 3 × pH, 2 × % solids) one would need to convert approximately 1,000 kg of ore into bulk concentrate. This does not include the cleaner tests, of which a minimum of five are required for simulating cleaner circuit performance (3 × pH and 2 × % soils). It is suggested that the cleaner tests be performed with intermediate concentrate grading at least an order of magnitude (10×) higher than that of the bulk concentrate to ensure that any hydrophobic gangue minerals that may be present in trace quantities are sufficiently concentrated for detection. Assuming 10% mass recovery required to achieve a 10:1 upgrade ratio, a bare minimum of approximately 5 t of ore are required to perform a simple rougher and cleaner test matrix. Complex or difficult ores, such as those containing hydrophobic impurities such as talc, pyrophyllite, or carbon, will require much more sample.

It is primarily for this reason that molybdenite separation circuits are often designed with less-than-ideal information, and this is one reason they often require much longer ramp-up times post commissioning. It is not uncommon for molybdenite separation circuits to require up to two or more years, multiple configuration changes, and equipment upgrades to reach design capacity. In the opinion of the authors, modern flotation kinetics test methodologies and phenomenological models are sufficiently advanced and capable of mitigating most of the scale-up risks associated with by-product molybdenum flotation; project development teams are simply unable or unwilling to commit the core sample and funding required

Table 6 Typical test matrix for brownfield plant design—Molybdenum roughers

Test Number	Parameter Studied	% Mo Feed	% Solids	pH	P80	eH	Oily Collector	Dispersant
1	eH	Average/Composite	38	11	Nominal	Low	25	None
2	eH	Average/Composite	38	11	Nominal	Moderate	25	None
3	eH	Average/Composite	38	11	Nominal	High	25	None
4	Collector	Average/Composite	38	11	Nominal	Nominal	0	None
5	Collector	Average/Composite	38	11	Nominal	Nominal	25	None
6	Collector	Average/Composite	38	11	Nominal	Nominal	75	None
7	Collector	Average/Composite	38	11	Nominal	Nominal	150	None
8	pH	Average/Composite	38	8	Nominal	Nominal	Optimum	None
9	pH	Average/Composite	38	9.5	Nominal	Nominal	Optimum	None
10	pH	Average/Composite	38	11	Nominal	Nominal	Optimum	None
11	pH	Average/Composite	38	12.5	Nominal	Nominal	Optimum	Moderate
12	% Solids	Average/Composite	30	11	Nominal	Nominal	Optimum	None
13	% Solids	Average/Composite	38	11	Nominal	Nominal	Optimum	None
14	% Solids	Average/Composite	46	11	Nominal	Nominal	Optimum	None
15	P80	Average/Composite	38	11	2′ regrind	Nominal	Optimum	None
16	P80	Average/Composite	38	11	4′ regrind	Nominal	Optimum	None
17	P80	Average/Composite	38	11	8′ regrind	Nominal	Optimum	None
18	Dispersant	Average/Composite	38	11	Nominal	Nominal	Optimum	Moderate
19	Dispersant	Average/Composite	38	11	Nominal	Nominal	Optimum	High
20	Variability—% Mo feed	Variable	38	11	Nominal	Nominal	Optimum	None
21	Variability—% Mo feed	Variable	38	11	Nominal	Nominal	Optimum	None
22	Variability—% Mo feed	Variable	38	11	Nominal	Nominal	Optimum	None
23	Variability—% Mo feed	Variable	38	11	Nominal	Nominal	Optimum	None
24	Variability—% Mo feed	Variable	38	11	Nominal	Nominal	Optimum	None
25	Variability—% Mo feed	Variable	38	11	Nominal	Nominal	Optimum	None
26	Variability—% Mo feed	Variable	38	11	Nominal	Nominal	Optimum	None
27	Variability—% Mo feed	Variable	38	11	Nominal	Nominal	Optimum	None
28	Variability—% Mo feed	Variable	38	11	Nominal	Nominal	Optimum	None
29	Variability—% Mo feed	Variable	38	11	Nominal	Nominal	Optimum	None

Courtesy of Aminpro

Table 7 Typical test matrix for brownfield plant design—Molybdenum cleaners

Test Number	Parameter Studied	% Mo Feed	% Solids	pH	P80	eH	Oily Collector	Dispersant
1	pH	Average/Composite	38	8	Nominal	Nominal	Optimum	None
2	pH	Average/Composite	38	9.5	Nominal	Nominal	Optimum	None
3	pH	Average/Composite	38	11	Nominal	Nominal	Optimum	None
4	pH	Average/Composite	38	12.5	Nominal	Nominal	Optimum	None
5	% Solids	Average/Composite	30	11	Nominal	Nominal	Optimum	None
6	% Solids	Average/Composite	38	11	Nominal	Nominal	Optimum	None
7	% Solids	Average/Composite	46	11	Nominal	Nominal	Optimum	None
8	P80	Average/Composite	38	11	2′ regrind	Nominal	Optimum	None
9	P80	Average/Composite	38	11	4′ regrind	Nominal	Optimum	None
10	P80	Average/Composite	38	11	8′ regrind	Nominal	Optimum	None
11	Dispersant	Average/Composite	38	11	Nominal	Nominal	Optimum	Moderate
12	Dispersant	Average/Composite	38	11	Nominal	Nominal	Optimum	High

Courtesy of Aminpro

for an adequate test program given the relatively minor portion of the overall revenue stream that molybdenite comprises. Indeed, many greenfield projects prefer to wait until after the primary sulfide concentrator is operational before collecting concentrate samples from the concentrator for subsequent moly separation testing and flow-sheet development.

Pumping and Froth Handling

Although overflowing pump boxes are usually the first indicator of a process upset, poor slurry pumping practice is usually a cause and seldom just a symptom. Moly plant froth can be very tenacious, particularly the concentrate streams, and

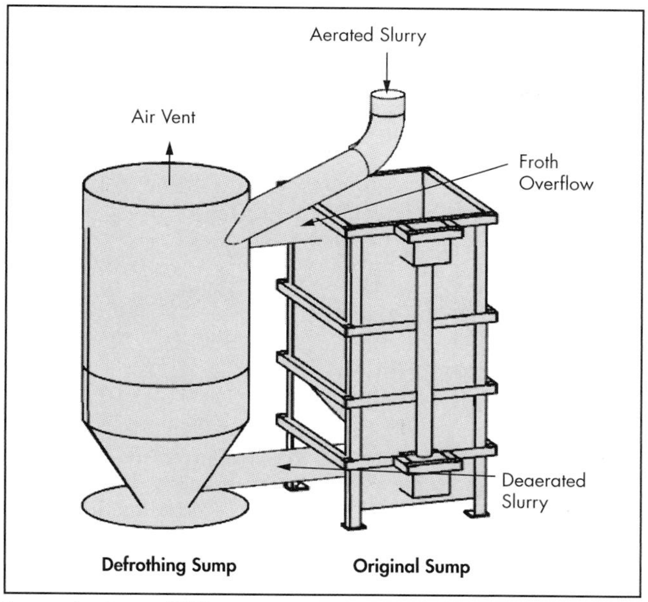

Source: Amelunxen 2009

Figure 18 Retrofitted defrothing sump

care must be taken to consider the implications of the bubbles when selecting the pump and sump configuration. Slurry froth factors can be as low as 1.1 at the pump intake line and as high as 4.5 at the top of the sump. (Slurry froth factor is the ratio of total slurry volume, including air, to slurry volume excluding air.) Any air entering the pump will cause air locking and lead to cavitation and failure of the pump to deliver the required flow. It generally leads the operator to design higher sump heights and add water sprays in an effort to break the froth down. One solution observed at several plants has been to install cyclonic defrothing sumps, such as the retrofit shown in Figure 18, that allow (or promote) the escape of air bubbles before the slurry reaches the pump sump.

Ferric Chloride Leach

In a ferric chloride leach circuit, the lead and copper react with the ferric chloride to form aqueous species. The aqueous species are then removed via filtration, and the ferric chloride is regenerated in a chlorinator by conversion of the ferrous to ferric chlorides using chlorine gas. The reaction equations are

$$CuFeS_2 + 4FeCl_3 \rightarrow CuCl_2 + 5FeCl_2 + 2S^0$$

$$PbS + 2FeCl_3 \rightarrow PbCl_2 + 2FeCl_2 + S^0$$

$$2FeCl_2 + Cl_2 \rightarrow 2FeCl_3$$

Ferric chloride leaching for purifying molybdenite concentrates has been practiced at Urad (Colorado), Henderson, Tonopah (Nevada), Sierrita, Brenda (British Columbia), Andina, Lornex (British Columbia), and other operations on and off for many years.

The impurity in Urad molybdenite concentrates is primarily PbS. The oxidizing ferric chloride leach is required mainly for leaching copper minerals. For leaching PbS from Urad's molybdenite concentrates, the more economic nonoxidative HCl solution is quite adequate. Thus, Urad abandoned ferric chloride leach in favor of HCl leach in the early 1970s. Later, the HCl leach for lead removal was also implemented at Climax's Henderson mine in Colorado. The HCl leach generates hydrogen sulfide.

$$PbS + 2HCl \rightarrow PbCl_2 + H_2S$$

In general, a caustic scrubbing unit is used for removal of the hydrogen sulfide.

The Tonopah plant in Nevada was built and operated by Anaconda Minerals Company in the early 1980s. The ferric chloride leach plant could process 50 t/d of molybdenite concentrates. For purifying its concentrate containing 1.5% Cu as chalcopyrite it used the following leaching conditions: 70°–80°C, 20% pulp density in five agitated tanks for a retention time of 24 hours. The lixiviant typically contains 9% $FeCl_3$ and some cupric chloride ($CuCl_2$). The addition of $CuCl_2$ accelerates the leach kinetics. The basic chemical reactions of chalcopyrite with cupric chloride and regeneration of cuprous to cupric chloride with ferric chloride can be described by the following equations:

$$CuFeS_2 + 3CuCl_2 \rightarrow 4CuCl + FeCl_2 + 2S^0$$

$$CuCl + FeCl_3 \rightarrow CuCl_2 + FeCl_2$$

The lower-temperature operation is to minimize the side reaction of sulfur to sulfate as shown:

$$6FeCl_3 + S^0 + 4H_2O \rightarrow 6FeCl_2 + 6HCl + H_2SO_4$$

At the Sierrita plant in Arizona, continuous atmospheric leaching conditions of molybdenite concentrates through the management of Duval, Cyprus Minerals, Cyprus Amax Minerals, and Phelps Dodge were similar. The plant can process about 100 t/d of concentrates. A 13% $FeCl_3$ solution is used to leach its concentrates containing about 2.5% Cu and 0.3% Pb at 88°C and 30% pulp density in five agitated tanks. After being acquired by Freeport-McMoRan, feed preparation and recycle of $FeCl_3$ leach solution has been optimized, and retention time has been reduced from 20–30 hours to ~10 hours. The improvement of the leaching rate can be attributed to the presence of cupric chloride in the recycled leach liquor. The purified concentrates contain about 0.08% Cu and 0.02% Pb.

Ancillary Operations

Suitably designed upstream and downstream support equipment and processes, proper reagent selection, and automated process controls are important considerations when developing a robust plant design.

Reagent Preparation

NaHS preparation is generally only required if handling NaHS in solid form. In this case, the technician requires protective clothing and an adequate ventilation mask. The NaHS usually arrives in super sacks, which are loaded and dumped in a dissolution tank that feeds a process storage tank.

Nitrogen and CO_2 are received as a liquid. The liquid is vaporized and injected as a gas at the required injection points. Mass flow meters and automatic valves are used to control the flow to each point.

Frothers and defoamers are often used in molybdenum separation circuits. Similar reagent handling protocols are used as in the bulk circuits. The defoamers, when necessary, are usually added to the stage feed (if froth control is

Table 8 Thickeners in separation circuits

Thickener	Unit Area, $m^2/t/d$	Specific Capacity, $t/m^2/d$
Cu/Mo	0.05–0.3	0.14–0.83
Intermediate Mo	0.05–0.15	0.28–0.83
Mo concentrate	0.2–0.5	0.08–0.21
Cu concentrate	0.05–0.3	0.14–0.83

Courtesy of Aminpro

required in the flotation cells) or the launders (if froth control is required to improve hydraulic transport).

Oily collectors are only rarely added in the molybdenite separation circuit, mainly because they are already present in sufficient concentrations because of upstream additions.

Flocculants are often used for dewatering in in-circuit thickeners and/or molybdenum concentrate filters. Similar preparation and injection practices are used as for any concentrate dewatering application. Low-molecular-weight flocculants and/or floccules are usually completely dispersed by a single centrifugal pump, and any surviving floccules are dispersed in the first cell of an agitated mechanical flotation cell. The authors are not aware of any convincing evidence that low-molecular-weight flocculants, in moderate dosages, negatively impact molybdenite flotation.

Dewatering

Molybdenite concentrates and intermediate concentrates are composed of particles that are significantly finer than those in normal copper concentrates, and, as such, higher unit areas are required for sizing thickeners. Flocculants and coagulants are almost always required because of the commonly occurring problem of molybdenum-bearing slimes in the thickener supernatant. Although thickeners should be designed based on settling tests, the ranges shown in Table 8 are typical of the various thickeners employed in moly separation circuits.

Modern molybdenum plants generally favor plate-and-frame pressure filters, although many older plants continue to operate vacuum drum filters. The filtration rate depends on many factors and should be evaluated based on test work.

Instrumentation and Process Control

Instrumentation costs for molybdenum separation circuits are usually between 3% and 5% of the total capital cost of the circuit. The instrumentation requirements are significantly higher for covered cells than for open cells, as they must be operated remotely without visual feedback to the operator. Modern molybdenum roughing stages are usually equipped with similar process control and instrumentation provisions as their counterparts in the bulk sulfide flotation circuit, including froth cameras, automatic level control, online pH meters, and air/nitrogen/CO_2 mass flow meters. Primary slurry streams, such as the rougher tailings, first cleaner tails, first cleaner-scavenger tails, and final concentrates are equipped with flow meters, density gauges, and online X-ray fluorescence analyzers.

MOLYBDENUM CONVERSION

Other than a small quantity that is upgraded for use in lubricant formulations, molybdenum disulfide is converted to technical-grade molybdenum trioxide, which is the starting material for a variety of products for the metallurgical and chemical industries. Typical concentrations in the analysis range of molybdenum disulfide are shown in Table 9. Analytical variations as shown in the table can be possible, depending on the nature and origin of the ore and the beneficiation procedure employed. If the molybdenum disulfide does not meet the specifications for the roasting process, it will have to be leached (as described in earlier sections) to remove impurities. The performance of the roaster component and the nature of the charge as it moves through the roasters are dependent on the impurity levels in addition to other operating parameters.

Table 9 Molybdenum disulfide sample analyses

MoS_2 Analysis	Weight %
Molybdenum	45–59
Sulfur	34.4–37.6
Insolubles	1.6–15.0
Lead	0.002–0.05
Phosphorus	0.002–0.04
Iron	0.3–1.0
Copper	0.006–2.0
Oil	0.5–8.3
Water	0.5–8.4

Courtesy of Freeport-McMoRan

Roasting

The pyrometallurgy conversion practice involves roasting the concentrate in a multiple-hearth roaster. Concentrate is moved from storage by a belt conveyer or other means to the roaster feeder and from there to the top hearth level of the roaster. Rabble arms extending from the central rotary shaft provide plowing action. Drop holes are located at the center or periphery so that the burden is gradually moved toward and through the drop holes on the succeeding lower hearth levels.

As the concentrate is fed at the top hearth level, the flotation oil burns and the concentrate ignites. The reaction, once started, is exothermic. Operating temperatures range between 538° and 816°C. By regulating the air flow to the individual hearths and the overall air inlet and gas outlet flows, these temperatures are maintained throughout the roaster. Most of the oxidation occurs on the upper hearths. Additional heat to complete the reaction is provided by fuel burners at the lower hearth levels. The overall chemical reaction is

$$MoS_2 + \frac{7}{2}O_2 \rightarrow MoO_3 + 2SO_2$$

Off-gas exiting the roaster is cooled either indirectly or by water sprayed directly into the flue gas. Efficient dust collectors, employing cyclones and electrostatic precipitators installed in the flue gas handling system, are utilized for improved product recovery. The off-gas exiting the electrostatic precipitators is treated with a lime slurry or processed in a sulfuric acid plant for sulfur dioxide recovery. For a lengthy and detailed discussion on pyrometallurgy of molybdenum, refer to chapter 4 in Gupta (1992).

Roasting temperatures higher than 815°C can lead to serious operating issues related to sublimation of molybdite (MoO_3) and because of partial melting and softening of the charge. Problems manifest themselves as excessive wear and corrosion of rabble teeth and plugging of drop holes in the upper hearths. Dust losses in the lower hearths can be

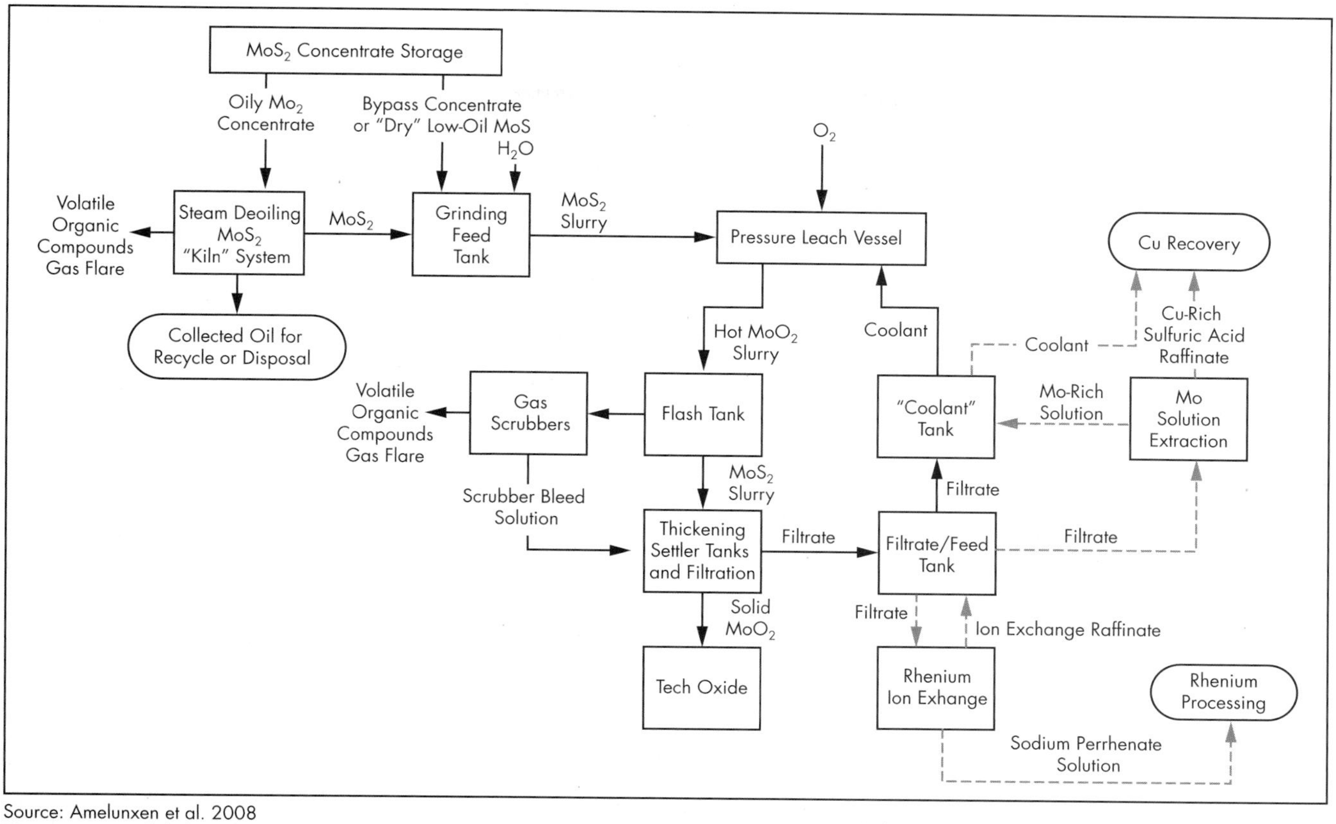

Source: Amelunxen et al. 2008

Figure 19 Molybdenum pressure leach vessel process

experienced because of volatilization as a result of temperatures of about 650°C.

Acid Pressure Oxidation

The roasting process has certain limitations based on the quality of the feed. For example, elevated copper levels can cause material handling issues within the roaster. These issues limit the type of feedstock a roaster can process, or require additional processing steps, such as leaching, prior to roasting. Another method for converting molybdenum disulfide to molybdenum trioxide is through a hydrometallurgical process of oxidation. This process involves the pressure oxidation of an aqueous slurry of molybdenum disulfide to form soluble and insoluble molybdenum species. Within the pressure leach vessel, the molybdenum disulfide is converted to molybdenum trioxide and sulfuric acid by the combination of heat, pressure, and oxygen:

$$2MoS_2 + 9O_2 + 4H_2O \rightarrow 2MoO_3 + 4H_2SO_4$$

The concept of using a pressure leach vessel to convert molybdenum disulfide to molybdenum trioxide has numerous patent holders. The method that is described here was developed, patented, and put into production by Freeport-McMoRan in 2009 and currently produces about 10,000 t of molybdenum trioxide each year. A pressure leach vessel includes a control system to regulate the process conditions including temperature, pressure, and feed rates of molybdenum disulfide concentrate, water, acidic coolant solution, and oxygen, as well as pressure leach vessel discharge and all necessary solution recycle and emission systems. Although some of the molybdenum trioxide enters the aqueous phase within the pressure leach vessel, the molybdenum trioxide compound has a low solubility limit. As such, only a fraction of the molybdenum is leached into the process slurry within the pressure leach vessel; the majority remains as a solid precipitate in the process slurry. Thus, the molybdenum disulfide is converted into partially soluble molybdenum trioxide within the slurry while other metallic elements (e.g., rhenium, copper, iron) that exist in the feedstock slurry are fully leached into solution because of the synthesis of sulfuric acid by the exothermic reaction within the pressure leach vessel. As such, within the pressure leach vessel, water and oxygen combine with the sulfides in the concentrate to generate an aqueous solution consisting largely of (1) sulfuric acid, (2) partially soluble molybdenum trioxide (most of which is a solid precipitate and some of which is in solution), (3) soluble rhenium, (4) soluble iron, and (5) soluble copper. As described next, the pressure leach vessel process incorporates subsystems that allow for the optional capture of these other soluble beneficial metals. When collected by filtration and washed free of the process solution, the solid product is (on a dry basis) generally 98% pure molybdenum trioxide with the remaining impurities composed of insoluble silicate minerals and unreacted molybdenum disulfide. An example of a molybdenum pressure leach vessel process is depicted in Figure 19.

Residual flotation reagents manifest themselves as an oily residue in some molybdenum mining concentrates. These reagent residues can interfere with the efficiency of the process that occurs in the pressure leach vessel. Therefore, concentrates with significant residual reagents are first fed into a de-oiling furnace wherein the agents contained in the flotation/reagent mix are volatilized. The molybdenum disulfide concentrate exiting the de-oiling unit is fed with other low-oil concentrate to a grinding process that ultimately feeds the pressure leach vessel.

As the feed stream is prepared and introduced to the pressure leach vessel, molybdenum disulfide is mixed with process water to create a process slurry. The mixing rates of slurry and oxygen within the pressure leach vessel maintain an oxygen overpressure within the vessel. The overall conditions inside the pressure leach vessel are maintained at a pressure between 3,102 and 3,447 kPa and a temperature of approximately 225°C. Heat in the pressure leach vessel results from exothermic reaction. Addition of water as part of the feed slurry and coolant helps to maintain the optimum operating temperature. The pressure leach vessel's temperature is below the melting or fusion point of the molybdenum concentrates. The chemical oxidation that occurs in the pressure leach vessel effectively converts the molybdenum disulfide contained in the feed stream into molybdenum trioxide, which is generally the same chemical reaction that occurs within a conventional roaster.

Slurry exiting the pressure leach vessel reports to a flash tank where the temperature and pressure of the hot process slurry are reduced to allow the mixture to continue forward into the thickening and filtration stage. At this point, the slurry contains precipitated molybdenum trioxide solids, minor insoluble mineral gangue, and an aqueous solution. This thickened slurry reports to a series of filters that remove solid molybdenum trioxide from the remaining aqueous solution and wash residual solution from it using water, a vacuum belt filter, and a pressure tube filter.

Solute from the pressure leach vessel is collected in a common tank to use for maintaining the process flow and/or additional beneficial metal recovery. Several options exist that may be used simultaneously or in combination. The filtrate can return to the pressure leach vessel as coolant through a coolant tank that also recycles the filtrate's metal and sulfuric acid content back through the process. Filtrate can be sent to a solution extraction (SX) circuit, which recovers soluble molybdenum and rhenium.

In the SX circuit, several extraction and stripping stages occur wherein the aqueous filtrate solution initially is mixed with an organic reagent to extract molybdenum and rhenium from the aqueous phase. Ultimately, after separation and return to an aqueous phase, the molybdenum- and rhenium-bearing solutions from SX are recycled back to the pressure leach vessel through the coolant system. The recycling of this first solution recovers molybdenum and rhenium that are soluble even after processing through the earlier thickener and filter systems. The raffinate solution from the SX contains sulfuric acid and additional metal values such as copper. These values may be recovered by electrowinning, cementation, or by utilizing the acid on a heap leach circuit.

To separate rhenium from the molybdenum an ion exchange (IX) circuit is employed using IX resins that bind selectively with soluble rhenium in the filtrate solution. When the resin is loaded with rhenium, the filtrate flow is stopped, and rhenium is stripped off the resin back into solution as sodium perrhenate. After the rhenium is removed, flow is started again, and the cycle repeats. The IX plant recovers the rhenium from the clear pressure leach vessel solutions, whereas the soluble molybdenum ultimately is returned to the pressure leach vessel for recycle. The rhenium recovery rate from the IX system varies based on the rhenium content present in the molybdenum concentrates that are fed into the pressure leach process. Therefore, the IX plant may operate only when a sufficient concentration of rhenium warrants.

Table 10 Technical-grade molybdenum trioxide sample analyses

MoO_3 Analysis	Weight %
Molybdenum	56–65
Sulfur	<0.1
Insolubles	2.00–11.00
Iron	0.30–1.00
Phosphorus	<0.05

Courtesy of Freeport-McMoRan

Molybdenum Products

Depending on the impurities that are present after conversion, molybdenum trioxide can either be marketed, or it can undergo further processing to meet specific customer or quality requirements.

Technical-Grade Molybdenum Trioxide

Roasted concentrate is discharged from the furnace and either marketed as technical-grade molybdenum trioxide or subjected to additional conversion operations to remove impurities. A typical analysis range of commercially available technical-grade molybdenum trioxide is shown in Table 10.

Technical-grade molybdenum trioxide is marketed either in bulk, drums, bags, or cans containing a predetermined amount of molybdenum. This is convenient to users in the steel industry. When added to molten steel, molybdenum trioxide is reduced to the metallic form and alloyed. The gangue minerals are removed with the slag.

Technical-Grade Molybdenum Trioxide Briquettes

Technical-grade molybdenum trioxide can be formed into briquettes, which facilitates additions of molybdenum in electric and foundry steelmaking processes. In the briquette form, losses experienced in blow-off when molybdenum trioxide is added to the furnace charge are minimized. Storage and handling are also simplified. To prepare the briquettes, technical-grade molybdenum trioxide is mixed with a carbon- or ammonia-based binder and formed into briquettes in a hydraulic press.

Molybdenum Processing for Ammonium Molybdates

Depending on the purity, the molybdenum trioxide feedstock may be leached with hot water and filtered to remove water-soluble impurities. The molybdenum trioxide is further leached in an aqueous-ammonia solution to put the molybdenum into solution as a monomolybdate:

$$MoO_3 + H_2O + 2NH_3 \rightarrow (NH_4)_2MoO_4$$

The ammoniated slurry is filtered to recover the ammoniated molybdate solution. The remaining insoluble material is reprocessed for molybdenum recovery. Solution impurities can be removed by selective precipitation, IX, SX, and/or an impurity bleed from the circuit. The molybdate solution is then evaporated in a continuous crystallizer to form ammonium dimolybdate crystals in equilibrium with ammoniated molybdate solution:

$$2(NH_4)_2MoO_4 \rightarrow (NH_4)_2Mo_2O_7 + H_2O\uparrow + 2NH_3\uparrow$$

The ammonium dimolybdate crystals are allowed to grow within a continuous crystallizer and are recovered by centrifuging. Depending on solution impurities, the liquid centrifuged out of the crystals may be put back into the crystallizer or removed from the crystallization circuit. The ammonium dimolybdate crystals from the centrifuge can either be dried and sold as final product or used as a feedstock for the manufacture of additional products. Ammonium dimolybdate can be completely thermally decomposed to form a calcined molybdenum trioxide or partially thermally decomposed to form ammonium octamolybdate:

$$(NH_4)2Mo_2O_7 \rightarrow 2MoO_3 + 2NH_3\uparrow + H_2O\uparrow$$

$$4(NH_4)2Mo_2O_7 \rightarrow (NH_4)_4Mo_8O_{26} + 4NH_3\uparrow + 2H_2O\uparrow$$

Additionally, calcined molybdenum trioxide that has been dissolved in aqua-ammonia at a 0.86 NH_4:Mo molar ratio can be crystallized to form ammonium heptamolybdate:

$$MoO_3 + NH_4OH + H_2O \rightarrow (NH_4)_6Mo_7O_{24}\cdot 4H_2O$$

Sodium molybdate crystalline is prepared by dissolving molybdenum trioxide in sodium hydroxide followed by filtration to remove gangue elements. After filtration, sodium molybdate is crystallized out of the filtrate:

$$2NaOH + MoO_3 + H_2O \rightarrow Na_2MoO_4\cdot 2H_2O$$

Pure Molybdenum Trioxide Sublimed Process

Sublimed molybdenum trioxide is a pure form of roasted molybdenum disulfide concentrate prepared by purification by sublimation:

$$MoO_3(s) \rightarrow MoO_3(g) \rightarrow MoO_3(s)$$

Sublimation is a thermal process in which molybdenum trioxide undergoes a phase transition directly from a solid to gaseous form, or vapor, without passing through a liquid phase. The pure form of molybdenum trioxide is achieved not by the rejection of impurities as much as by the concentrating of molybdenum trioxide.

ACKNOWLEDGMENT

The authors acknowledge the contributions to this chapter by John Hollow and Ross Edwards.

REFERENCES

Amelunxen, P.A. 2009. Moly plant design considerations. SME Preprint No. 09-136. Littleton, CO: SME.

Ansari, A. 2006. The effect of lignosulfonates on the floatability of molybdenite and chalcopyrite M.Sc. thesis, University of British Columbia, Vancouver, BC.

Berger, B.R., Ayyuso, R.A., Wynn, J.C., and Seals, R.R. 2008. *Preliminary Model of Porphyry Copper Deposits*. Open-File Report 1321. Reston, VA: U.S. Geological Survey.

Blossom, J. 1994–2000. Molybdenum. In *Minerals Yearbook*. Reston, VA: U.S. Geological Survey.

Blossom, J. 2001–2017. Molybdenum. In *Mineral Commodity Summaries*. Reston, VA: U.S. Geological Survey.

Born, C.A., Bender, F.N., and Kiehn, O.A. 1976. Molybdenite flotation reagent development at Climax, Colorado. In *Flotation: A.M. Gaudin Memorial Volume II*. Edited by D.W. Fuerstenau. New York: AIME. pp. 1147–1184.

Bullock, L.A., and Parnell, J. 2017. Selenium and molybdenum enrichment in uranium roll-front deposits of Wyoming and Colorado, USA. *J. Geochem. Explor.* 180(September):101–112.

Chander, S., and Fuerstenau, M. 1972. On the natural floatability of molybdenite. *Trans. AIME* 252:62–69.

Chevron Phillips. 2000–2018. Orfom D8 Depressant. www.cpchem.com/bl/specchem/en-us/Pages/OrfomD8Depressant.aspx. Accessed May 2018.

Climax Molybdenum Company. 1924–2017. Summary of Annual and Life-of-Mine Production from Climax Mine, Phoenix, AZ. Unpublished report.

CRU International. n.d. Ryan's Notes Weekly Commodity Pricing. London, UK. CRU International. www.crugroup.com.

Cuthebertson, R.E. 1946. Unusual reagent combination improves flotation at Climax. *Trans. AIME* 169:505–524.

Cytec. 2013. Aero 7260 HFP depressant: Novel, safe and sustainable alternative to traditional hazardous modifiers—NaSH, Nokes, Na_2S, and cyanide. In *Cytec Solutions for Solvent Extraction, Mineral Processing, and Alumina Processing*. Vol. 17. Woodland Park, NJ: Cytec Industries. www.cytec.com/sites/default/files/files/Cytec%20Solutions%20V17%202013.pdf.

Ferguson, J., ed. 1984. *Proterozoic Unconformity and Stratabound Uranium Deposits*. Report of the Working Group on Uranium Geology Organized by the International Atomic Energy Agency. IAEA-TecDoc-315. Vienna, Austria: International Atomic Energy Agency.

Green, D.W., and Maloney, J.O., eds. 1999. *Perry's Chemical Engineers' Handbook*, 7th ed. New York: McGraw-Hill. pp. 2–122.

Gupta, C.K. 1992. *Extractive Metallurgy of Molybdenum*. Boca Raton, FL: CRC Press.

Hess, F. 1924. *Molybdenum Deposits, A Short Review*. Bulletin 761. Washington, DC: U.S. Geological Survey.

Kelly, T.D., Magyar, M.J., and Polyak, D.E. 2017. Molybdenum. In *Historical Statistics for Mineral and Material Commodities in the United States*. Data series 140. Reston, VA: U.S. Geological Survey.

Kirkemo, H., Anderson, C.A., and Creasey, S.C. 1965. *Investigations of Molybdenum Deposits in the Conterminous United States 1942–1960*. Bulletin 1182-E. Washington, DC: U.S. Geological Survey.

Lane, J.W., Bender, F.N., and Ronzio, R.A. 1972. Recovery of molybdenum from oxidized ore at Climax, Colorado. *Trans. AIME* 255:77–82.

Linde Group. 2012. *When Being Green Matters: Solvocarb Neutralizes Alkaline Waters Quickly and Easily with CO_2*. Murray Hill, NJ: Linde North America. www.lindewatertreatment.com.

Ludington, S., and Plumlee, G.S. 2009. *Climax-Type Porphyry Molybdenum Deposits*. Open-File Report 1215. Reston, VA: U.S. Geological Survey.

Ludington, S., Hammarstrom, J., and Piatak, N. 2009. *Low-Fluorine Stockwork Molybdenite Deposits*. Open-File Report 1211. Reston, VA: U.S. Geological Survey.

Morales, J. 1980. Use of nitrogen in the separation of copper and molybdenum. M.S. thesis, University of Chile, Santiago.

Rhagavan, S., and Hsu, L. 1984. Factors affecting the flotation recovery of molybdenite from porphyry copper ores. *Int. J. Miner. Process*. 12(1-3):145–162.

S&P Global Platts. n.d. Metals (Steel, Copper, Aluminum) Prices, News, and Analysis. London, UK. www.platts.com.

Sebenik, R. 2005. Molybdenum and molybdenum compounds. In *Ulmann's Encyclopedia of Industrial Chemistry*. Weinheim, Germany: Wiley-VCH.

Smedley, P.S., and Kinniburgh, D.G. 2017. Molybdenum in natural waters: A review of occurrence, distributions and controls. *Appl. Geochem*. 84:387–432.

Stein, H.J., Markey, R.J., Morgan, J.W., Du, A., and Sun, Y. 1997. Highly precise and accurate Re-Os ages for molybdenum from the East Quinling molybdenum belt, Shaanxi Province, China. *Econ. Geol*. 92(7):827–835.

Taylor, R.D., Hammarstrom, J.M., Piatak, N.M., and Seal, R.R. II. 2012. Arc-related porphyry molybdenum deposit model. In *Mineral Deposit Models for Resource Assessment*. Scientific Investigations Report 5070-D. Reston, VA: U.S. Geological Survey.

Tessenderlo Kerley. 2016. Safety data sheet: TKI-330. Revised June 10. Phoenix, AZ: Tessenderlo Kerley.

Tessenderlo Kerley. 2018. TKI-330 for Depression of Iron and Copper Sulfide [marketing brochure]. Phoenix, AZ: Tessenderlo Kerly.

Triffett, B., Veloo, C., Adair, B.J.I., and Bradshaw, D. 2008. An investigation of the factors affecting the recovery of molybdenite in the Kennecott Utah Copper bulk flotation circuit. *Miner. Eng*. 21(12-14):832–840.

Van der Merwe, P.J. 1986. *Uranium Prospecting in the Main Karoo Basin in Retrospect* (3 vols.). Pretoria: Atomic Energy Corporation of South Africa.

Wacaster, S. 2017. The mineral industry of Chile. In *2014 Minerals Yearbook: Chile [Advance Release]*. Reston, VA: U.S. Geological Survey. pp. 7.1–7.19.

Wu, Y-S., Chen, Y-J., and Zhou, K-F. 2017. Mo deposits in northwest China: Geology, geochemistry, geochronology, and tectonic setting. *Ore Geol. Rev*. 81(pt 2):641–671.

Zeng, Q.D., Liu, J.M., Qin, K.Z., Fan, H.R., Chu, S.X., Wang, Y.B., and Zhou, L. 2013. Types, characteristics, and time-space distribution of molybdenum deposits in China. *Int. Geol. Rev*. 55(11):1311–1358.

Zhang, H., Li, C-Y., Yang, X-Y., Sun, Y-L., Deng, J-H., Liang, H-Y., Wang, R-L., Wang, B-H., Wang, Y-X., and Sun, W-D. 2014. Shapinggou: The largest Climax-type porphyry Mo deposit in China. *Int. Geol. Rev*. 56(3):313–331.

CHAPTER 12.25

Nickel and Cobalt

Andrew Bamber and Arthur Barnes

NICKEL MARKET

Nickel is a strategic commodity globally, with uses in electronics, energy, aerospace, and manufacturing, the most important application of which is in stainless steel used in the construction, automotive, and petrochemical industries. Although the nickel price is currently suffering an extended period of decline—primarily because of the production of cheap ferronickel (so-called pig nickel) in China from largely imported ores—the market remains significant, with potential for recovery as the world economy normalizes after the global financial crisis of 2008–2011. Total production in 2016 was 1.4 Mt (million metric tons), with the major producing countries being the Philippines, Russia, and Canada (see Figures 1 and 2). Sources are predominantly sulfide deposits (60%) with production from laterite deposits making up the balance. Laterites, in addition to being an important source of nickel, are an increasingly vital source of by-product metals such as cobalt and scandium. Additional potential nickel sources are from nickel-containing manganese nodules, typically containing 0.5%–1.0% Ni and associated cobalt, which occur in vast quantities on the seabed at depths between 3,500 and 4,000 m. Although not currently economically viable solely as a source of nickel, increasing demand for manganese, cobalt, and nickel in battery technologies means these deposits will become an increasingly important source in the future.

Major consumers of nickel—largely as nickel in stainless steel—include Europe, the United States, Japan, and, increasingly, China. Other industrial consumption of nickel by the petrochemical and aerospace industries is in the form of the nonferrous alloys: nickel–copper, nickel–chrome and nickel–molybdenum. An additional and fast-growing use is nickel used in battery technology, driven by growth in renewable energy and battery-electric vehicles. Nickel in lithium-ion batteries is in the form of either NCA (80% Ni, 15% Co, and 5% Al) or NMC (33% Ni, 33% Mn, and 33% Co), demand for which is expected to represent 7% of the total nickel market by 2020 (Padhy 2017).

The most important area of nickel demand, however, remains in stainless steel, with demand from emerging markets being the key driver for growth in this area. Although currently, stainless steel, and therefore nickel, has been experiencing an extended bear market with prices consistently under US$5/lb, in 2016, the stainless market began to recover as a result of increasing demand from China and a cutback in supply mainly from Indonesia in 2014. This recovery was further assisted by cutbacks in the Philippines in 2016 with a moderate price recovery now projected in 2017 and beyond (Figure 3).

COBALT MARKET

Cobalt was originally mined together with silver and arsenic in Saxony and Bohemia in the 15th century, but its use then was only in the oxide form as the colorant called "cobalt blue." Swedish chemist Georg Brandt first isolated metallic cobalt in 1735, and it was subsequently established as an element in 1780 by Torbern Bergman. Despite this progress in physical chemistry, cobalt was still used only in its oxide form until circa 1914. Since then, there has been a steady increase in demand for cobalt as new forms and uses have been developed, including as a metallic or alloying element (Fisher 2011).

Primary cobalt production is relatively small, comprising approximately 5% of world supply. The leading source of mined cobalt, representing approximately 60% of world cobalt mine production and more than 50% of known reserves, is as a by-product from copper mining in the Democratic Republic of the Congo (DRC); see Figure 4. The majority of the balance of cobalt is recovered as a by-product of nickel processing, largely from lateritic nickel operations in the tropics. In 2016, global cobalt mine production decreased from previous years, mainly owing to lower production from these same nickel operations.

China is currently the world's leading refiner of cobalt and thus the leading supplier of refined cobalt to the United States and Europe. Much of China's production was from ore and partially refined cobalt imported from the DRC, with scrap and stocks of recycled cobalt materials also contributing to production. Large increases in refining capacity over the

Andrew Bamber, Principal and Managing Director, Bara Consulting Ltd., London, United Kingdom
Arthur Barnes, President, Metallurgical Process Consultants Ltd., Kamloops, British Columbia, Canada

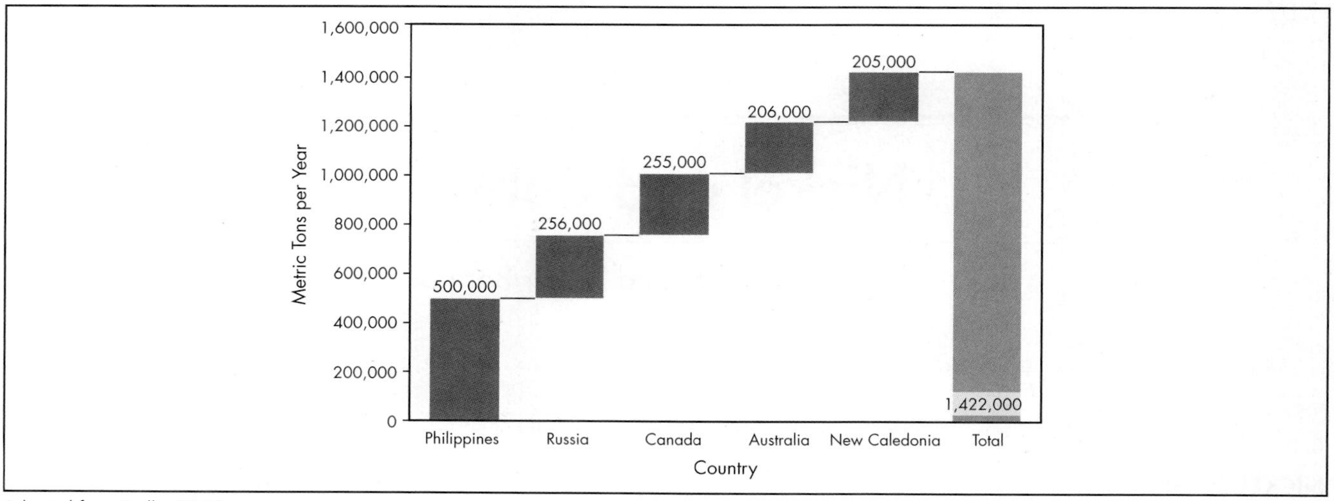

Adapted from Padhy 2017

Figure 1 Major nickel-producing countries

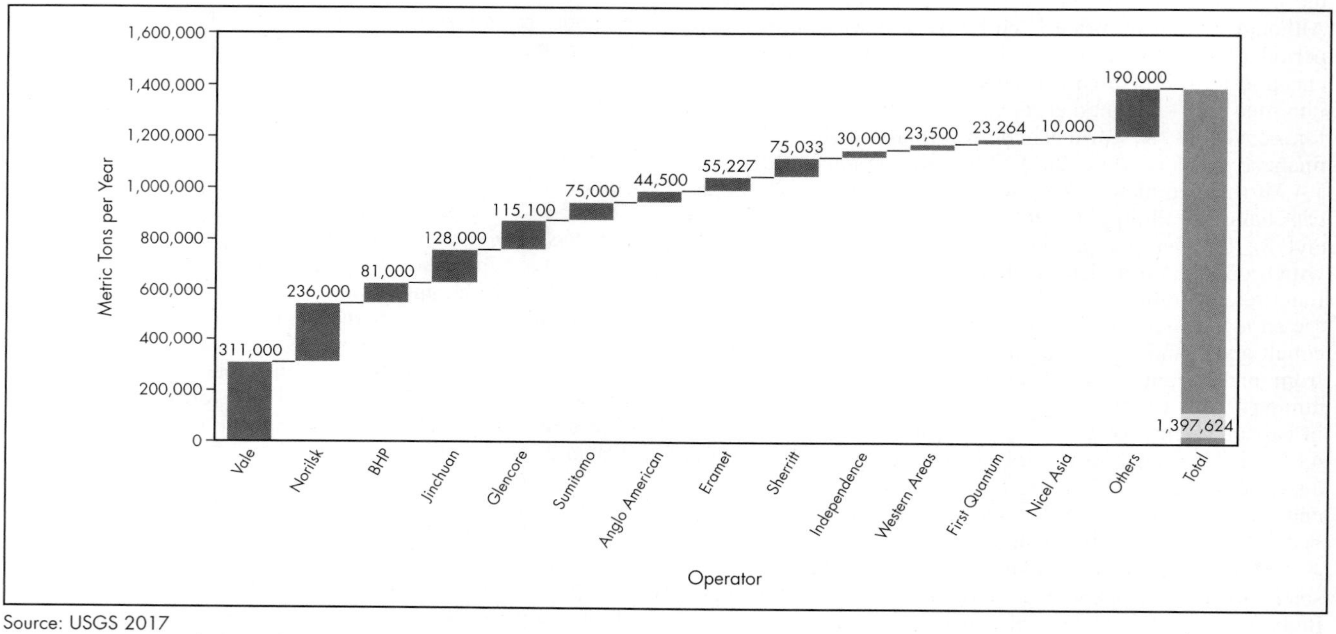

Source: USGS 2017

Figure 2 Major nickel producers

past decade mean that China now accounts for more than 40% of the global cobalt output (USGS 2017).

Operating across these major cobalt-producing regions are several major companies, including cobalt-primary producers in the DRC, such as Freeport (with the Tenke Fungurume mine) and Glencore (with Katanga); and operators producing cobalt as a by-product of nickel or copper operations elsewhere, including Sherritt Gordon (Cuba and Madagascar), Jinchuan Non-Ferrous in China, and Queensland Nickel in Australia (Bell 2018). See Figure 5.

The total value of cobalt consumed globally in 2016 was estimated at more than US$850 million. Growth in world refined cobalt consumption is currently expected to exceed supply, driven mainly by strong growth in the rechargeable battery and aerospace industries. In conjunction with a simultaneous decrease in supply related to decreasing nickel production, the global cobalt market was expected to shift from surplus to deficit in 2017. Rising supplies from ramp-ups at mines such as Ambatovy (Madagascar) and a return to full production at Katanga (Congo) are expected to limit the deficit between now and 2020, but much still depends on copper and nickel prices (Darton Commodities 2009).

In 2016, China was the world's leading consumer of cobalt, with nearly 80% of its consumption being used by the rechargeable nickel–metal hydride (NiMH) and lithium-ion battery industry. The United States is the second-largest consumer of cobalt globally with about US$250 million worth of cobalt either produced or imported, followed by Europe. Approximately 45% of the cobalt consumed in the United States was used in superalloys, mainly in aircraft gas turbine

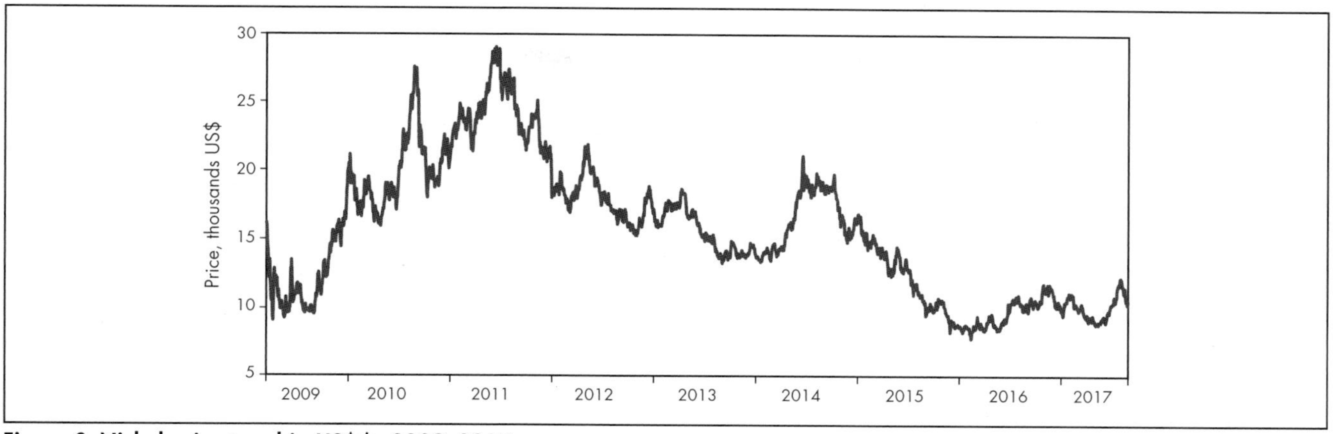

Figure 3 Nickel price trend in US$/t, 2009–2017

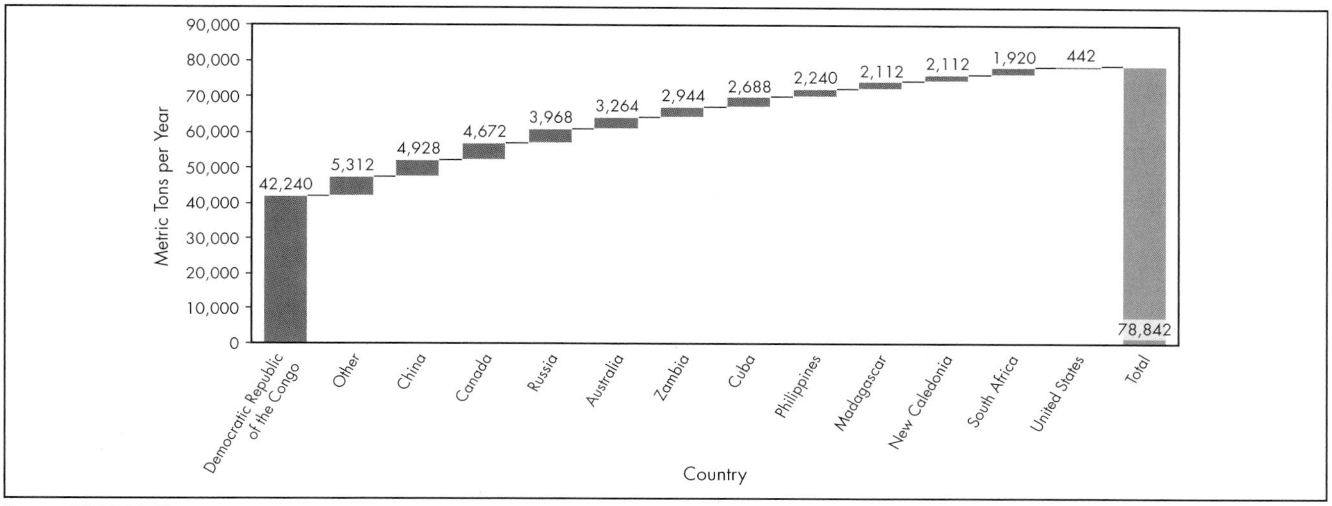

Source: USGS 2017

Figure 4 Cobalt production by country

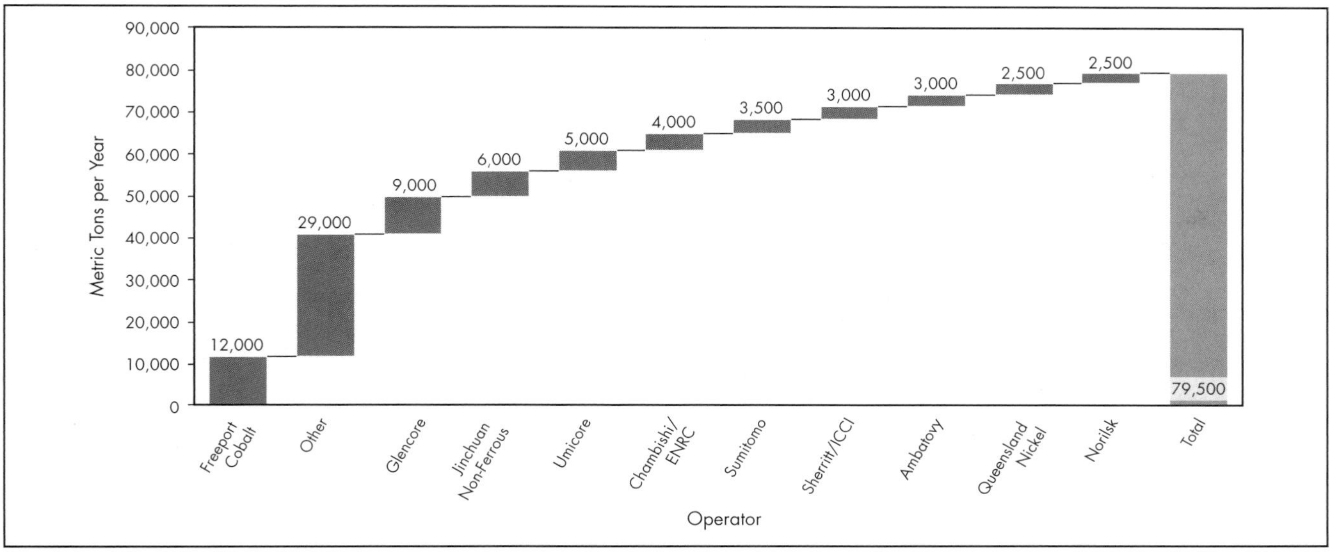

Source: USGS 2017

Figure 5 Major cobalt producers

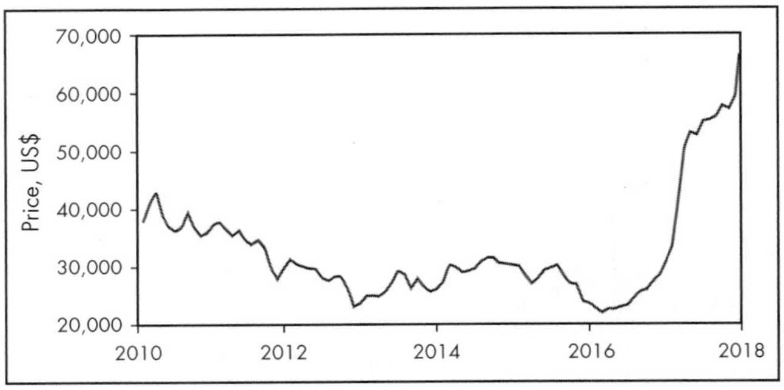

Source: Trading Economics 2018

Figure 6 Cobalt price trend

engines; 8% in cemented carbides for cutting and wear-resistant applications; 16% in various other metallic applications; and 31% in a variety of chemical applications, including rechargeable batteries (Darton Commodities 2016).

Cobalt prices have been largely depressed since the global financial crisis of 2008, reaching a low of US$22,100/t in 2013 (Figure 6). Prices have, however, risen strongly since 2016, because of demand growth and a resulting deficit of almost 7,000 t/yr in supply against demand of 90,000 t/yr. Prices are currently at approximately US$70,000/t and forecast to increase through 2020 against projected demand of approximately 120,000 t/yr (Desai 2016). This growth in demand—as with nickel—is largely from cobalt used in battery technology, driven by growth in renewable energy generation capacity and battery-electric vehicle sales. Cobalt in lithium-ion batteries is in the form of either NCA or NMC, where higher cobalt content generally equates to improved battery performance. Demand growth is primarily driven by increases in electric car sales, which are projected to top 17 million units per year by 2030 from a current base of 540,000/yr (Desai 2016).

NICKEL SULFIDE ORES

Geological and Mining Context

Economic deposits of nickel sulfides globally occur in association with either igneous intrusive or volcanic magmatic complexes. Mineralization can be found concentrated near the margins of localized intrusions, as at Stillwater and Duluth in the United States and Vaaralami in Finland; layered basin-type deposits whether tectonic-related, impact-related, or otherwise, as at Norilsk Nickel in Russia and the Sudbury Basin in Ontario, Canada; komatiitic-type deposits as at Nkomati in South Africa, the Kambalda Dome in Western Australia, and Raglan in Quebec, Canada; or, extensive, stratiform concentrations of disseminated metal sulfide mineralization as in the Bushveld Igneous Complex of South Africa and the Great Dyke of Zimbabwe (Foose et al. 1984). Specific geological, mineralogical setting, and related mining engineering aspects of the main subtypes are discussed in ensuing sections.

The Stillwater Ni–Cu–PGE (platinum group element) deposit is located in the Beartooth Mountains, 130 km southwest of Billings, Montana, United States. It is a localized, syngenetic mafic/ultramafic layered complex in which the layering consists bands of norite, gabbro, and anorthosite. Differential settling of the minerals within the massive melt zone allowed heavy minerals to settle to the base with the lighter siliceous minerals remaining at the top. Gangue minerals include olivine, pyroxenes, plagioclase, and chrome spinels with minor quartz, amphibole, apatite, magnetite, and ilmenite (McCallum 1999). Major ore minerals occur in layers and minor veins averaging 2.5-m thick, comprising predominantly pyrrhotite, chalcopyrite, and pentlandite with sperrylite and minor Pt–Fe alloys.

Access to the mineralized zones is by a 600-m shaft and an internal decline. One of three stoping methods are used: mechanized captive cut-and-fill, ramp-and-fill using electric–hydraulic drilling jumbos and load-haul-dumps (LHDs), and sublevel stoping. Blasted ore is recovered from below using LHDs and the stopes are paste-backfilled for stability. Mucked ore is delivered to an underground crushing station and crushed to –150 mm. The ore is then conveyed to holding bins before loading in the 9-t-capacity skips where it is hoisted to the surface for treatment by grinding and flotation processes.

The Ni–Cu ore deposits of Sudbury are associated with a large body of igneous rock known as the Sudbury Igneous Complex (SIC) or Sudbury Basin (Figure 7). Views on the origins of the complex are mixed; although its location at the junction of the Superior, Southern, and Grenville provinces suggests a tectonic origin, meteor impact sites such as the nearby Lake Wanapitei site have led to widely held impact–origin theories for the complex. The SIC is in the form of a truncated ovoid, with the major axis striking northeast for a distance of 60 km and the minor axis approximately 27 km long, fringed generally by a ring of low hills around the contact area. The main rock units of the SIC range from an outer ring of norite through a transition zone of gabbro to an inner zone of granophyre. Footwall rocks include archean gneisses and granites, as well as a large zone of breccias comprising footwall rocks embedded within igneous material, largely norite (Naldrett 1984).

The SIC is rich in nickel, copper, and PGEs, with nickel-bearing mineralization consisting of zones of massive, net-textured, matrix-textured, and disseminated sulfides associated with brecciated host rocks located in footwall troughs or embayments around the outer norites of the complex. Major sulfides include pyrrhotite, pentlandite, chalcopyrite, and pyrite with minor magnetite. Pentlandite is the main nickel-bearing sulfide with a nickel content of 33–35 wt %, and also containing an average of 1% Co. Nickel can also be mineralogically associated within the pyrrhotites (Charland et al. 2011).

A variety of mining methods are deployed, including mechanized cut-and-fill, sublevel caving, and a variation of sublevel caving called vertical crater retreat mining. Using the Coleman mine near Sudbury (Ontario, Canada) as an example, blasted ore would be mucked by LHD (potentially by remote control) and loaded via ore pass to 20-t or 40-t haul trucks and hauled to the shaft station where it would be crushed to nominally –200 mm and conveyed to the loading pocket for hoisting to surface, typically in 14-t skips. Once on surface, because of the extensive nature of the complex, ore would typically be stockpiled and loaded on to railcars (or road trucks) and hauled to a centralized mill—either Clarabelle or Strathcona depending on the source—for processing by grinding and flotation.

Komatiite-hosted Ni–Cu–PGE deposits, on the other hand, tend to comprise trending zones of disseminated through matrix to laminated massive sulfides and stringers in continuous or semicontinuous zones within a predominantly volcanogenic as opposed to igneous complex. Hanging-wall material is commonly barren olivines or pyroxenes, with a footwall

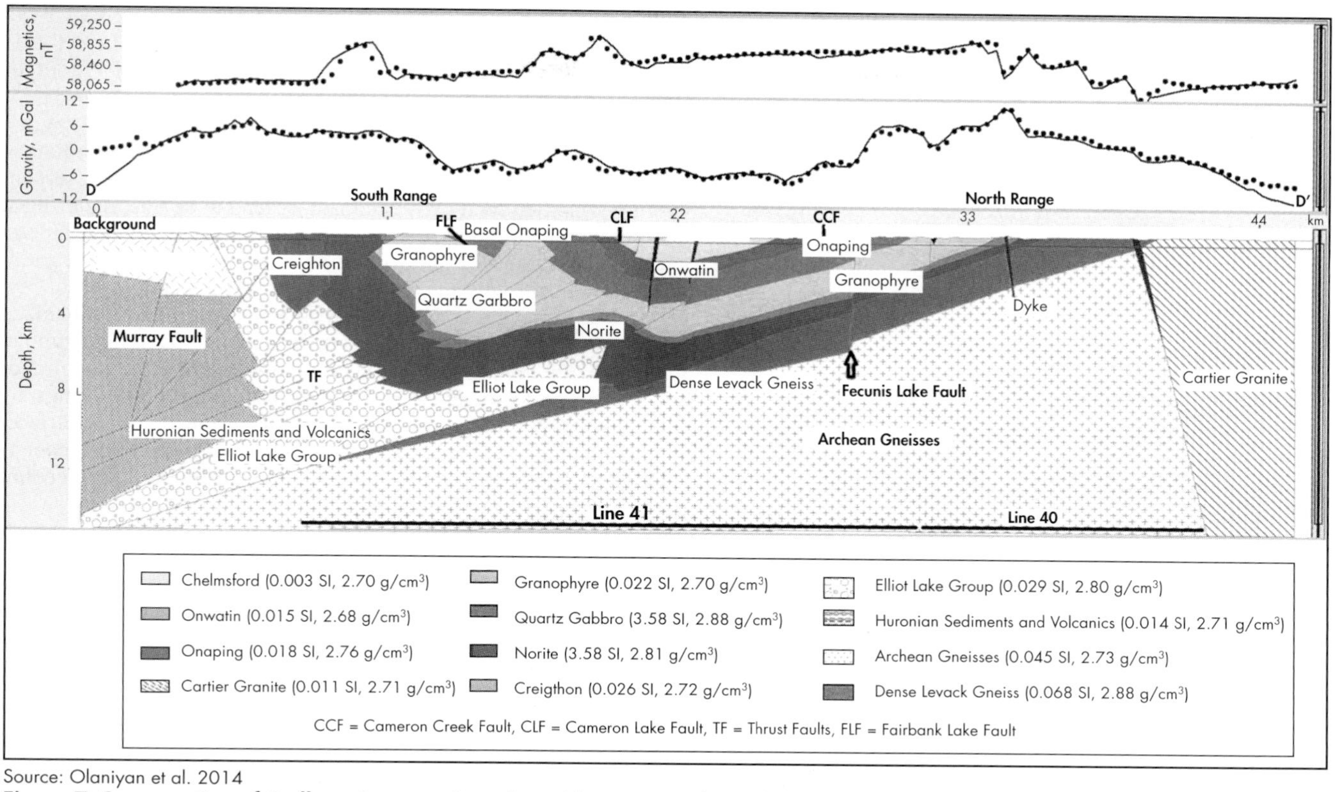

Source: Olaniyan et al. 2014

Figure 7 Cross section of Sudbury Igneous Complex with major geological units

of basalt or similar volcanics (Cowden 1988). Within these trending zones, mineralization styles can vary widely, from net, blebby, and matrix textures to the leopard and "reverse-leopard" textures at Raglan and Voisey's Bay (Labrador, Canada). Although sulfide minerals common to other types of Ni–Cu–PGE deposit, such as pentlandite and chalcopyrite, are the principal ore minerals, weathering, and metamorphic overprinting in some deposits of this type manifest a range of other, more rare minerals such as gaspeite, hellyerite in gossans, and polydymite and violarite in saprolitic zones (Lesher and Keays 2002).

Mining methods are generally underground methods, although minor open pitting has been undertaken at Raglan on account of the shallow, postglaciation situation of that deposit. Jackleg mining of narrow veins with scrapers in steeper zones or small LHDs in shallower zones does occur, although mechanized drift-and-fill, cut-and-fill, and longhole stoping methods are more common. Longwall mining in flat-lying tabular sections can also be found, particularly at Kambalda operations. Ore is typically hoisted or hauled via ramp to surface, and stockpiled prior to treatment via crushing, single-stage semiautogenous grinding (SAG) mill and flotation.

Deposits of the extensive (or elongate) layered intrusive type such as the Bushveld Igneous Complex (the Bushveld) and the Great Dyke of Zimbabwe (the Dyke) differ largely in the wide variety of base metals hosted, from massive and veinous deposits of ferrous minerals such as iron, vanadium, and chrome, to disseminated Ni–Cu–PGE sulfide mineralization associated with massive oxide deposits (e.g., chromite such as the UG2 deposits of the upper Bushveld), flatter lying tabular zones not associated with massive oxides (e.g., the famous Merensky Reef), and extensive amorphous zones of Ni–Cu–PGE mineralization such as the Platreef.

The Great Dyke of Zimbabwe is hosted in archean granitoids, with several magma subchambers occurring along strike of the deposit, and each subchamber hosting disseminated Ni–Cu–PGE mineralization of varying tenors within the ultramafic norites and pyroxenites (Figure 8). Economic deposits tend to be emplaced under or alongside shallow-lying host greenstones. In cross section the Dyke is keel-shaped with layers occurring in several sequences; the main sulfide zone occurs variably at the boundary between the mafics and the ultramafics from the top of the plagioclase and into the base of the websterite, approximately 200–400 m deep (Mwatahwa and Musa 2015).

Mineralization is predominantly pyrrhotite, pentlandite (1%–2%), chalcopyrite, and pyrite, with PGEs occurring at parts-per-million levels as sperrylite, native platinum associated with base metal sulfides, and minor bismuthotellurides. Gangue mineralogy comprises pyroxenes with plagioclase, chlorite, tremolite, and often high levels of talc.

Mining methods are typically mechanized room-and-pillar, with access to the workings by ramp, with haulage by 20–40-t truck, or alternately conveyed to surface. Processing is by crushing, screening, and primary and secondary ball milling with two stages of flotation on account of the typical fine grain size of the sulfides.

The Bushveld is the world's largest layered intrusion, comprising the eastern, western, and northern limbs some 300 km long, 500 km across, and 7–10 km thick (Schoustra et al. 2000). After regional intrusion, the rocks cooled slowly and fractionation occurred according to complex thermodynamics, physics, and chemistry, resulting in light, magnesium silicate–rich minerals in the upper zones; iron-rich phases below; and more dense chromite, nickel, copper, and associated platinum group metals (PGMs) in deeper, so-called critical zones (Figure 9). The Upper Critical Zone, ranging from 200–300 m to more than 1,500 m deep in places, is host to one of the world's largest Ni–Cu resources and the largest known PGE resource in the world. It hosts the UG2 (Ni–Cu–PGE associated with primary chromitite), Merensky (tabular Ni–Cu–PGE), and Platreef (extensive, amorphous Ni–Cu–PGE) economic mineralized zones.

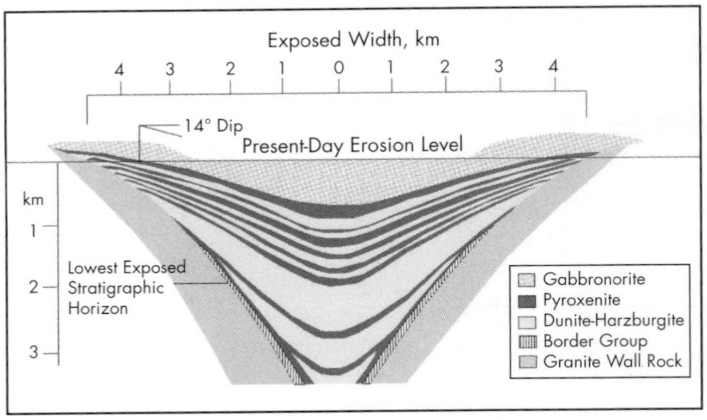

Source: Mwatahwa and Musa 2015

Figure 8 Cross section of the Great Dyke

Ore minerals within these zones include chromite, pyrite, pentlandite, chalcopyrite, pyrrhotite, and sperrylite, with gangue minerals including pyroxene, feldspar, mica, chlorite, serpentines, spinels, and clays. Grain sizes tend to be fine, requiring at least secondary grinding, with some grain sizes noted as fine as –10 µm, requiring tertiary grinding and, more recently, regrinding with stirred mills or other fine grinding technology.

Mining methods include jackleg or airleg, and mechanized low-profile methods in deep sections greater than, for example, 1,000 m requiring shaft access, and mechanized room-and-pillar in shallower sections, with access to the workings by ramp and haulage by 20–40-t truck or conveyor to surface. Processing is by crushing, screening, and primary and secondary ball milling with two or three stages of flotation on account of the increasingly fine grain size of economic sulfides.

Mineral Processing

The low metal content of present-day sulfide ores renders them unsuitable for either direct smelting or direct hydrometallurgical processing. Because the sulfide minerals usually occur as distinct grains in the rock matrix, these ores are amenable to a

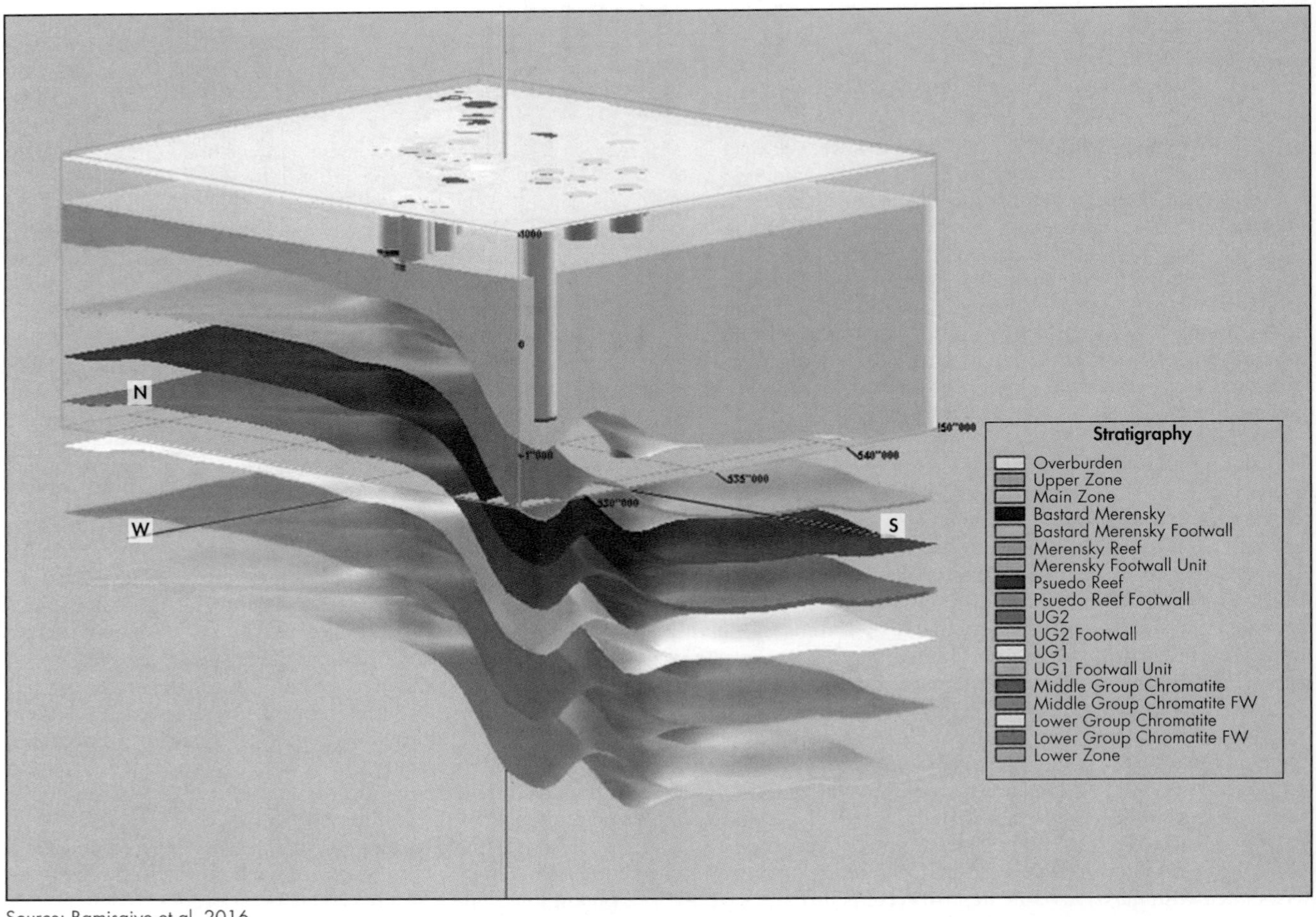

Source: Bamisaiye et al. 2016

Figure 9 Cross section through the western limb of the Bushveld Igneous Complex, near Amandelbult, South Africa, showing ideal sequence

mechanical upgrading in which much of the rock content can be rejected. Beneficiation of nickel sulfide ores is therefore generally by grinding and flotation; however, there are many different flow-sheet styles within this general class, as mineralogy (and mineral associations), texture, grain size, and hardness vary enormously from deposit to deposit.

In principle, it is possible to produce separate nickel (pentlandite), copper (chalcopyrite), and iron (pyrrhotite) concentrates. However, a clean separation of pentlandite from pyrrhotite is difficult in practice since pyrrhotite typically contains intergrown inclusions of pentlandite as well as nickel in solid solution in the pyrrhotite. In fact, pyrrhotite often contains 0.5%–1% Ni that cannot be separated by physical methods. The relatively high nickel content in pyrrhotite makes it difficult to achieve high yields of nickel from the ore unless the pyrrhotite is processed to recover its nickel content; conversely, the presence of nickel makes it difficult to obtain a marketable iron product from the pyrrhotite.

Most nickel producers outside Canada make no attempt to obtain separate nickel, copper, and iron concentrates, preferring a bulk concentrate that contains pentlandite, pyrrhotite, and chalcopyrite as smelter feed. In Canada, both the major producers separate a large part of the pyrrhotite from pentlandite and chalcopyrite, and Vale further separates the pentlandite and chalcopyrite into separate nickel and copper concentrates at its Sudbury operations. Even where the bulk of the pyrrhotite can be separated from pentlandite, residual pyrrhotite in the nickel concentrate still contributes a major part of the sulfide content.

Pyrrhotite can be separated from pentlandite and chalcopyrite either by using its ferromagnetic properties or by flotation. Not all pyrrhotite is ferromagnetic, although in ores from the Sudbury, Ontario, area the magnetic monoclinic form predominates over the paramagnetic or "nonmagnetic" hexagonal form (Toguri 1975). In general, pyrrhotite containing less sulfur than $Fe0.87S$ is nonmagnetic. This material can usually be separated from pentlandite by flotation.

Nickel or Ni–Cu concentrates typically range in grade from 5% to 15% Ni + Cu, depending on the degree of pyrrhotite rejection achieved. An exceptionally high-grade concentrate, containing 28% Ni and consisting largely of pentlandite, is produced in limited amounts at Vale's Thompson, Manitoba, Canada, operation.

One of the more complex nickel concentration process flow sheets is that operated by Vale (previously Inco) at its Sudbury operation. In the two primary concentrators, the Clarabelle mill and the Frood-Stobie mill, 50,000 t/d of ore, grading 1.2% Ni and 1.2% Cu, is ground to a particle size of approximately 200 µm and treated by magnetic separation and flotation with a sodium amyl xanthate reagent to produce 5,000 t/d of a bulk Ni–Cu concentrate and 8,000 t/d of a pyrrhotite concentrate.

The two primary concentrates are further upgraded in the Copper Cliff mill in Ontario, Canada. The bulk Ni–Cu concentrate is separated by flotation into separate nickel and copper concentrates by using lime and sodium cyanide at 30°–35°C to depress pentlandite and pyrrhotite. Under these conditions, about 92% of the copper is recovered to a concentrate grading 29% Cu and 0.9% Ni. About 98% of the nickel and 92% of the pyrrhotite are rejected to the separator tails, which grade 13% Ni and 2.0% Cu.

The Clarabelle rougher pyrrhotite concentrate is treated to separate residual chalcopyrite and pentlandite by grinding to approximately 74 µm followed by magnetic separation and flotation. A magnetic, cleaner pyrrhotite concentrate, a bulk Ni–Cu concentrate, and a nonmagnetic pyrrhotite tail are produced. The nonmagnetic tails stream (1% Cu, 2% Ni, and 30% pyrrhotite) is further treated by regrinding and flotation in the pyrrhotite rejection circuit, for final recovery of pentlandite and chalcopyrite before it is discharged to the tailings pond. Overall, about 80% of the pyrrhotite content of the ore is rejected from smelter feed to the pyrrhotite concentrate and tails.

At its Strathcona mill in the Sudbury Basin, Glencore (previously Falconbridge) treats ore from both the western portion of the basin, containing 1.5% Ni and 1.1% Cu, and high-grade ore from the Nickel Rim South mine on the northeast side of the Sudbury basin, and provides toll milling of numerous ores from other mines. Crushing, grinding, and a flexible flotation flow sheet are used to produce a Ni–Cu concentrate grading around 15%–20% Ni + Cu at an average recovery of 83% Ni and 93% Cu (Hoffman and Kaiura 1985). A large part of the pyrrhotite content of the ore is rejected directly to the tailings pond for environmental (sulfur emission control) reasons. The separation of the nickel and copper in the bulk concentrate is carried out in the matte refining process after smelting.

The ore mined by Vale (Inco) at Thompson consists of pentlandite and pyrrhotite with only a small amount of chalcopyrite (2.7% Ni, 0.2% Cu, 22% pyrrhotite). Since the pyrrhotite contains 78% of the sulfur but only 5.5% of the nickel, a high degree of pyrrhotite rejection is feasible without seriously decreasing the nickel recovery. The ore is ground and floated to recover a Ni–Cu concentrate and reject 90% of the gangue and 60% of the pyrrhotite to tails. The Ni–Cu concentrate is first treated by flotation to separate a copper concentrate and then treated by magnetic separation and flotation to recover a high-grade (28% Ni) pentlandite concentrate. The tailings stream from these steps forms the main nickel concentrate smelter feed (11% Ni, 0.4% Cu). The overall recovery of nickel to the nickel and pentlandite concentrates is about 87%, whereas another 5% of the nickel reports to the copper concentrate. More than 60% of the pyrrhotite is rejected to tailings with only 8% of the nickel.

At the BCL Ni–Cu mine in Botswana, the pyrrhotite/pentlandite ratio in the ore averages about 13:1, and the pyrrhotite contains more than 30% of the nickel either in solid solution or as small pentlandite inclusions. Consequently, high nickel recoveries can only be achieved by recovering the nickel from pyrrhotite. This is achieved by producing a bulk pentlandite–pyrrhotite–chalcopyrite concentrate, with a typical composition of 2.8% Ni and 3.2% Cu (Eliott et al. 1983).

The Norilsk flow sheet in Figure 10 exemplifies the treatment of disseminated pentlandite ore with chalcopyrite comprising crushing followed by grinding in closed circuit with a hydrocyclone to typically 150 µm d_{80}, followed by rougher flotation and two stages of cleaner flotation producing a Ni–Fe–S concentrate at 12% Ni and approximately 80% recovery (Kozyrev et al. 2002).

In the Sudbury context, flow sheets are required to deal with highly variable mixtures of polymetallic ores from various sources, including custom feed, and thus generally accommodate Ni–Cu separation, pyrrhotite rejection, and the generation of nickel primary concentrates plus copper or copper-primary concentrates with precious metal by-products. Primary crushing would typically be done underground, with a –300 mm nominal run-of-mine (ROM) product delivered by

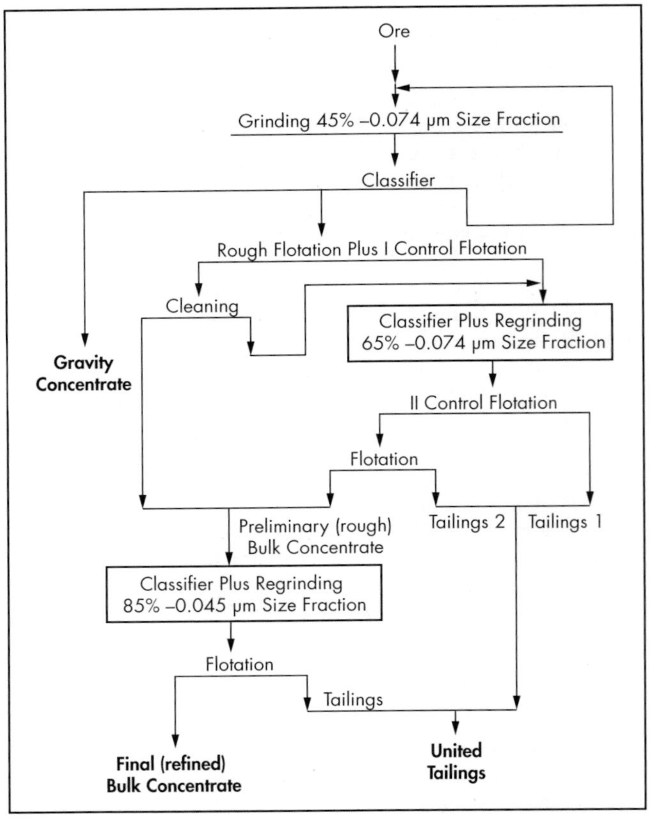

Source: Kozyrev et al. 2002

Figure 10 Grinding and flotation of disseminated nickel ores at Norilsk

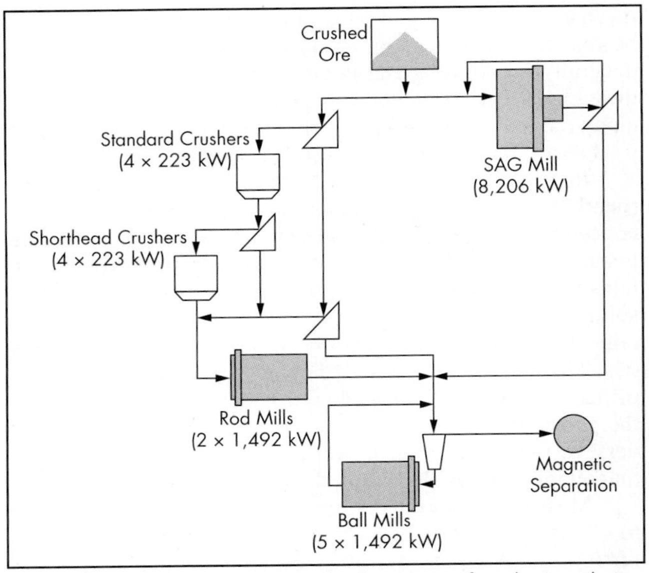

Source: Lawson and Xu 2011, reprinted with permission from the Australasian Institute of Mining and Metallurgy

Figure 11 Clarabelle Mill flow sheet, comminution, and feed preparation

rail to the central mill feed stockpile. Grinding is by SAG/ball/rod mill in combination, followed by magnetic separation at nominally 104 μm on wet high-intensity magnetic separators to separate a predominantly pentlandite- and pyrrhotite-rich stream from a Cu–Ni-rich stream prior to flotation (Figure 11).

The nickel circuit is tuned for pentlandite recovery with depression of the iron-rich pyrrhotite in rougher, cleaner, and scavenger stages, with regrind of the primary concentrate to 38 μm prior to cleaner and scavenger stages (Figure 12). The Cu–Ni circuit is tuned for sequential separation of chalcopyrite and bornite first, followed by pentlandite and, where present, millerite. Combined tailings are thickened and pumped to the tailings dam for deposition and water recovery. Nickel-primary concentrates are combined at around 12% Ni and 7% Cu and delivered to the smelter for processing via fluidized-bed drying followed by flash smelting.

In both the Bushveld and Dyke contexts, ores are lower in grade with respect to both nickel and copper than either their Sudbury or Norilsk counterparts, and finer grained; therefore, flow sheets focus on maximizing concentrate grade as well as recovery within the constraints of the metallurgy of the feed (Figure 13). In these cases the ROM might grade 1% Ni and 1% Cu with perhaps 3 g/t combined platinum, palladium, and gold. Here, so-called MF2 flow sheets are common, particularly on Merensky and Platreef ores with secondary grinding of the rougher tailings to increase liberation and, increasingly, regrinding of the concentrates. Crushing might be done in three stages, scalping of the oversize ROM and crushing in open circuit, followed by secondary crushing in closed circuit with a high-capacity screen, delivering nominally 100% –13 mm to the primary ball mill via a feed silo. Grinding to nominally 120 μm is by ball milling in closed circuit with a cyclone cluster, followed by rougher flotation to a bulk concentrate, with secondary grinding of the rougher tailings to nominally 75 μm and scavenger flotation of the reground ore. Combined concentrates are sent to multiple stages of cleaner flotation, with regrinding of the primary cleaner tails prior to secondary and often ternary cleaning stages to a final concentrate, which is thickened, filtered, and either bagged or tanked and trucked to a centralized smelter for reduction and matte separation.

Several reagents are key in the nickel flotation process. Collectors attach to the sulfide surfaces so the sulfides can more easily attach to air bubbles in the flotation cell. Common collectors include oils and any of the family of xanthates (e.g., potassium amyl xanthate or sodium isobutyl xanthate). Activation of the sulfide surfaces—for example, by copper sulfate ($CuSO_4$), to allow the collector to adsorb onto the pentlandite—may also be required; therefore, reactors for this are often included post–primary grinding ahead of rougher flotation. The most common frothers, used to establish stable air bubbles in the cell, include pine oils, alcohols (e.g., methyl isobutyl carbinol), or glycols. The occurrence of high levels of magnesium is common in these highly ultramafic ores, and depression of talc by carboxy methyl cellulose or soda ash during flotation is also required to achieve desired grades and optimize mass pull to concentrate. pH adjustment by either lime or the addition of acid can also be required, especially in sequential flotation of copper and nickel where chalcopyrite is ideally collected first, followed by pentlandite, requiring a staged adjustment of the acidity of the slurry. Nickel recovery might be 75%–80% to a bulk concentrate grading 7% Ni, 8% Cu, and 200 g/t PGEs.

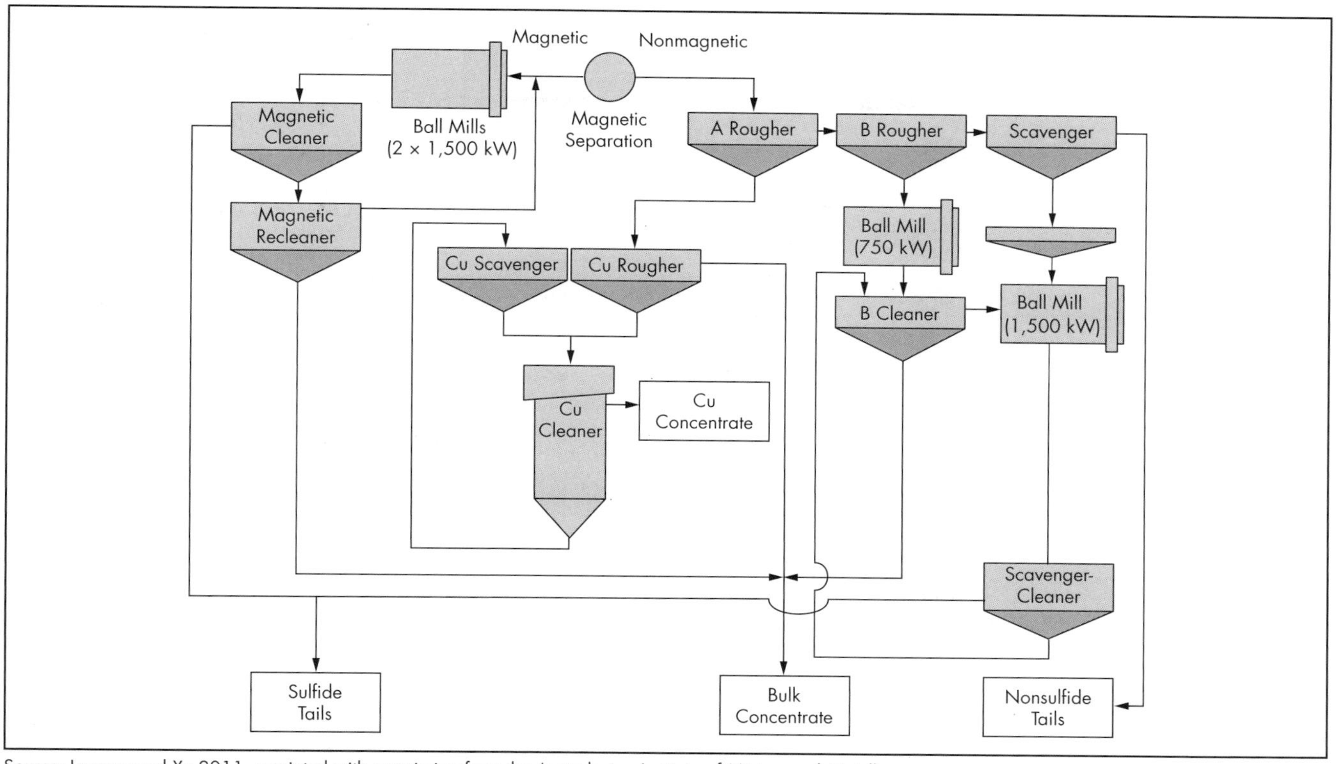

Source: Lawson and Xu 2011, reprinted with permission from the Australasian Institute of Mining and Metallurgy

Figure 12 Clarabelle Mill flow sheet showing Ni and Cu rougher, cleaner, and scavenger circuits

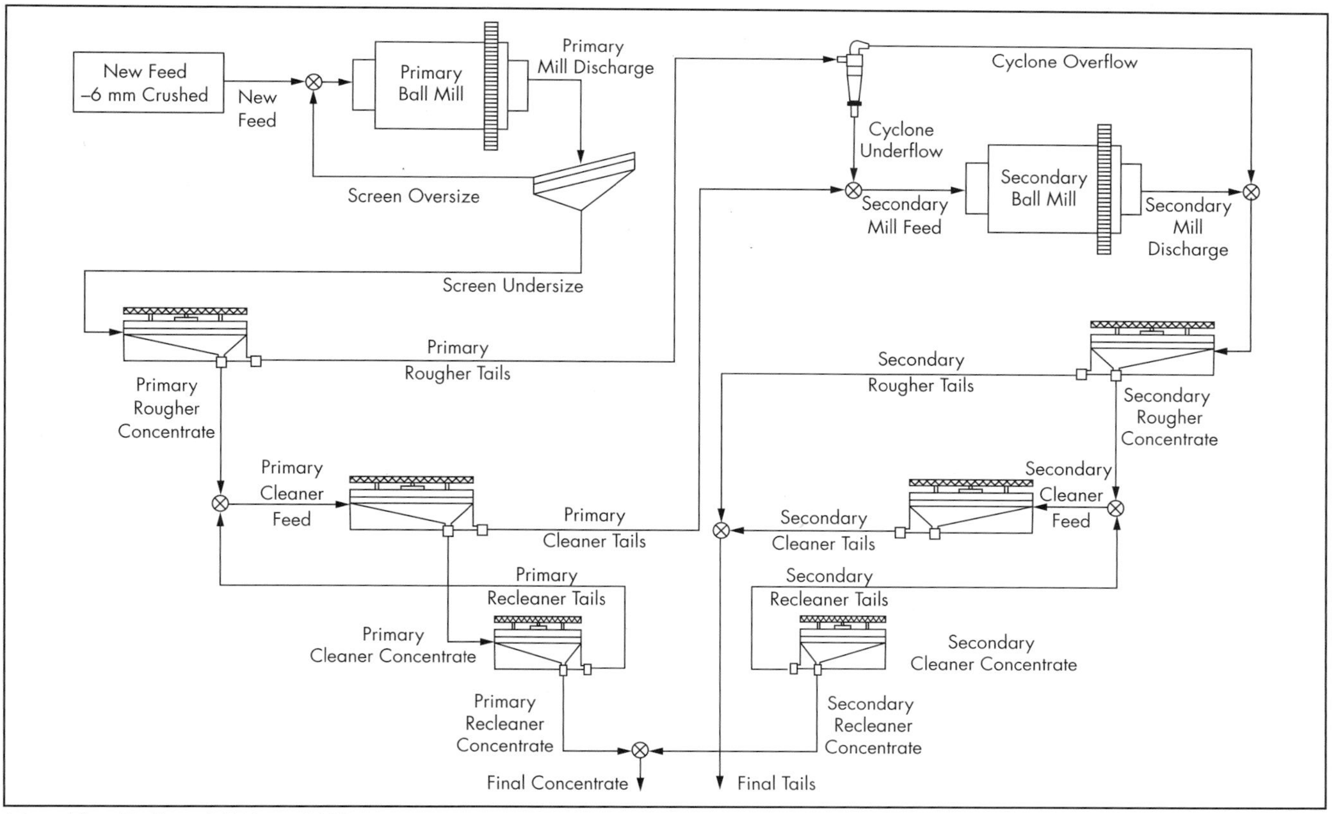

Adapted from Merkle and McKenzie 2002

Figure 13 Typical Bushveld/Dyke MF2 flow sheet with regrind

Source: Bhotoko 2016

Figure 14 Preconcentration of low-grade nickel ores at the Tati nickel mine

Courtesy of Halyard

Figure 15 Chromite removal by heavy mineral spirals: (top right) flotation circuit and (bottom right) spiral circuit

Variations to this flow sheet occur where textures might tend toward massive sulfides but with high dilution to a lower grade (e.g., at the Tati nickel mine in Botswana; see Figure 14), or in the case of the UG2 ores where the ore zones are narrow, the nickel and other metals are associated with chromite, and ROM can be highly diluted with hanging-wall pyroxenites. In the case of low head grades and/or high dilution, preconcentration of ore, by either size classification of coarse, low-grade material, dense media separation, or—increasingly—sensor-based sorting of the ore ahead of grinding and flotation, is common. Here, dense media separation or sorting identifies and rejects the barren or low-grade material at coarse particle size, delivering a reduced mass of higher grade ore into the rougher flotation stage. Additionally, for UG2 ores, where the nickel is associated with massive chromites, gravity separation in large, complex heavy mineral spiral stages after primary grinding is undertaken, which rejects the coarse, heavy chromite and recovers the less dense sulfides ahead of rougher flotation, allowing for a reduced size of flotation circuit and improved response in flotation compared to that of a chromium-rich feed (Figure 15). Further variations to this circuit might be observed where a flash flotation stage might be included, treating the underflow of the primary cyclone to recover coarse precious metals and metal sulfides (Figure 16).

Pyrometallurgy and Hydrometallurgy

Pentlandite is the predominant nickel-bearing mineral in sulfide concentrates and is commonly associated with various iron and copper sulfide minerals including pyrite, pyrrhotite, and chalcopyrite. Millerite and awaruite are less common minerals and require specific flow sheets for concentration. Historically, high-grade ores (in terms of sulfur content), such as those found in the Sudbury area in the early 20th century, were often roasted directly to oxidize the majority of the iron, and then smelted and converted to produce "white matte" Cu–Ni sulfide. Refer to Boldt and Queneau (1967) for details on

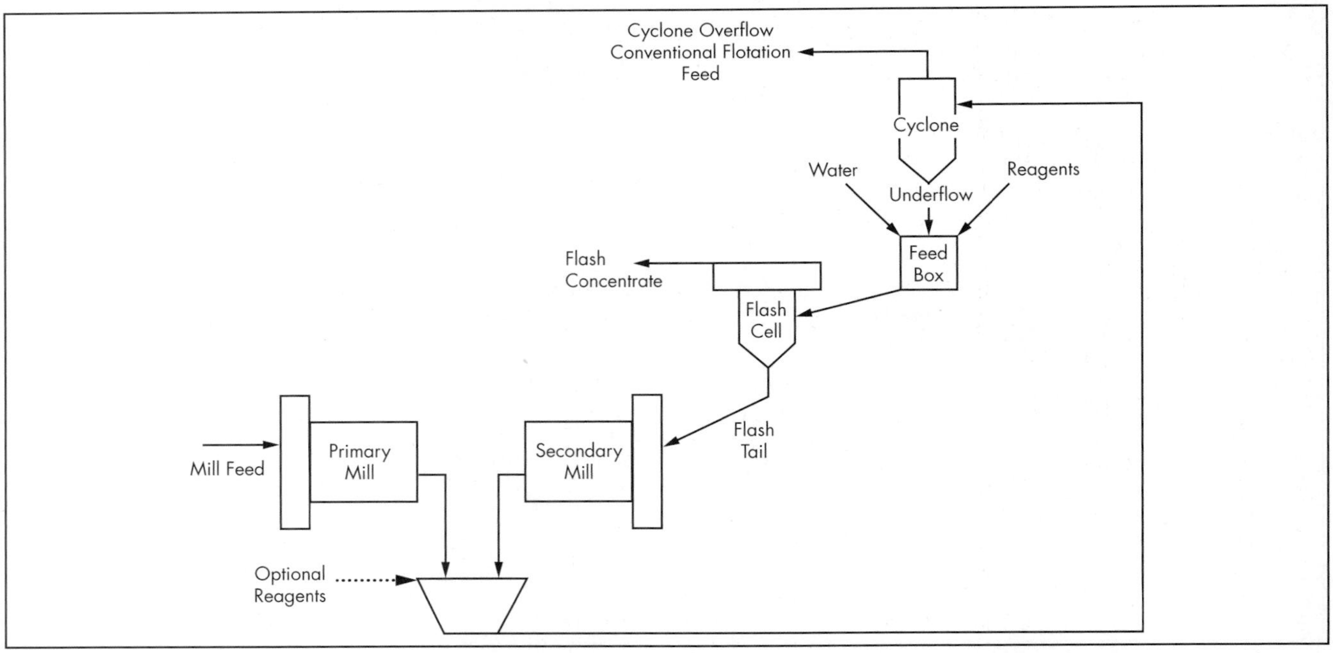

Source: Newcombe et al. 2012

Figure 16 Flash flotation of ball mill cyclone underflow

these methods, now discontinued for obvious environmental reasons. More generically, extraction of nickel from modest grade sulfide ores evolved as a variant of copper extraction: upgrading (beneficiation) of the ores used crushing and grinding followed by froth flotation, initially producing a bulk Fe–Ni–Cu sulfide concentrate. Later improvements permitted production of high-grade copper concentrates based on the better response of chalcopyrite to flotation reagents compared to the more sluggish nickel minerals. Many current circuits include specific pyrrhotite rejection circuits that result in a much higher nickel content in the concentrate and, consequently, a much lower mass of slag generated in subsequent smelting operations.

Concentrates are typically dried, roasted, and smelted or flash smelted (where the roasting heat itself provides the energy for melting the feed), producing a matte (liquid sulfide) containing iron, nickel, copper, cobalt, and sulfur and a discard slag consisting of mainly iron silicates (fayalite) and iron oxides (magnetite). The remaining iron is usually removed by oxidation of the molten matte by converting, usually in a cylindrical (Peirce–Smith) converter. In contrast to copper extraction, converting ceases at or close to the "white metal" stage with less than 4% Fe remaining in the matte. Further oxidation results in cobalt and nickel oxidizing and entering the slag. The white matte is cooled into either flat molds or granulated and becomes feedstock for the refining stages.

Numerous permutations exist for refining, largely determined by the final product desired and market demand, but all result in very high-purity nickel, copper, and cobalt metal (or oxide). In many instances substantial revenue is generated by the recovery of precious metal (Au, Pt, Pd, and Rh) as well. Major evolutionary changes in the extraction process have been driven by environmental concerns, and sulfur capture levels of >97% would be considered the minimum acceptable target for most current operations, in stark contrast to 100 or even 50 years ago when almost all of the sulfur was emitted to atmosphere. Sulfuric acid plants are an indispensable part of nickel sulfide processing, responsible for capturing the major portion of gaseous sulfur emissions. Nickel extraction has a high overall carbon footprint, not easily reduced because most of the processing is of low-grade ores (Bakker et al. 2011).

Pyrometallurgy of Nickel Extraction

More than 90% of the world's nickel sulfide concentrates are treated by pyrometallurgical processes to form nickel-containing mattes. Production capacities and composition of the feed and product for several major nickel smelters are listed in Table 1 (Crundwell et al. 2011).

Table 1 Feed composition, product composition, and production at four major nickel smelters

	Smelter			
	Kalgoorlie (BHP)	Bindura (BCL)	Nadezda (Norilsk)	Gansu (Jinchuan)
Start-up date	1972	1973	1981	1992
Ni in matte, t/yr	100,000	27,000	140,000	65,000
Feed, %				
Ni	15	5	12	9
Cu	0.3	4	5	4
Fe	34	43	38	38
S	32	31	27	27
Product, %				
Ni	47	17	32	29
Cu	1.5	15	15	15
Fe	20	33	23	29
S	27	25	27	23

Source: Crundwell et al. 2011

The pyrometallurgical treatment of nickel concentrates includes three types of unit operation: roasting, smelting, and converting. In the roasting step, sulfur is driven off as sulfur dioxide (SO_2) and part of the iron is oxidized. In smelting, the roaster product is melted with a siliceous flux, which combines with the oxidized iron to produce two immiscible phases: a liquid silicate slag that can be discarded and a solution of molten sulfides that contains the metal values. In converting, iron sulfide is removed from the molten, low-grade furnace matte by intense oxidation and slagging.

Roasting. Roasting is a process in which the nickel sulfide concentrates are heated in an oxygen-containing gas, usually air, to a temperature (600°–700°C) at which oxygen oxidizes sulfide to SO_2 and reacts with the metal values to form solid oxides, termed the *calcine*. The roasting step may also serve to preheat the charge for smelting.

The thermodynamic relationships between the metal sulfides and oxides provide the basis for the separation of iron from nickel, copper, and cobalt contained in the concentrate. The sulfides of iron, nickel, cobalt, and copper are all in the same thermal stability range at the smelting temperatures (1,200°–1,300°C), but, in the presence of oxygen, each sulfide is unstable with respect to its corresponding oxide. Iron has the greatest affinity for oxygen, followed by cobalt, nickel, and copper. Consequently, if the nickel concentrate is roasted with a deficiency of oxygen, the iron is oxidized preferentially, while virtually all of the nickel and copper remain as sulfides. Thus the degree of oxidation of the charge and the degree of elimination of sulfur can be controlled by regulating the supply of air to the roaster (or flash furnace). The degree of sulfur elimination achieved in the roasting step largely determines the grade of the matte produced as well as the losses of nickel and copper to the slag phase in the subsequent smelting operation. All the sulfur present in the calcine in excess of the amount needed to combine with the nickel, copper, and cobalt in the matte combines with iron, which therefore remains in the matte phase and lowers its pay metal (Ni + Cu + Co) grade. The major constraint on achieving a high-matte grade in smelting is the increasing loss of nickel, copper, and cobalt to the slag phase.

Recent modifications to the Glencore Sudbury operation use a much higher degree of roast (>70% S elimination) to produce a sulfur-deficient matte during electric furnace smelting. This is more fully described by Warner et al. (2006, 2007).

Smelting. When a mixture of metal sulfides, iron oxide, gangue, and siliceous flux are melted together, the iron oxide, gangue, and silica form a slag layer that floats on the heavier, molten sulfide (matte) phase. The two immiscible layers are separated and the slag is discarded.

The smelting process for nickel sulfide concentrates is normally carried out in two stages: primary smelting and converting. The oxidizing, slagging, and removal of iron in stages is critical to the efficiency and economics of smelting operations. If most of the iron were to be oxidized in the roaster and then slagged in a single smelting step, a matte high in nickel and low in iron would be produced. However, significant amounts of the nickel and copper sulfides would also be oxidized in the roaster, and these oxides would be lost to the slag in the smelting step. More extensive detail on smelting is provided by Diaz et al. (1988).

Flash smelting. In flash smelting, a finely ground sulfide concentrate is smelted by burning some of its sulfur and iron content while the sulfide particles are suspended in the oxidizing atmosphere. The concentrate is roasted and smelted in a single process step. The pre-dried concentrate and flux materials are injected with preheated oxygen-enriched air or commercial oxygen into the reaction shaft of a specially designed furnace. The exothermic heat of reaction of the iron sulfide with oxygen provides the energy to heat the particles to smelting temperature. The smelting temperature and the grade of matte formed are both controlled by adjusting the ratio of the amount of oxygen supplied to the furnace to the throughput rate of concentrate. More specific details on the evolution of the process are provided by Diaz et al. (1988).

Converting. In the converting step, iron sulfide is removed from the molten, low-grade furnace matte by oxidation and slagging. The slag, which contains high levels of nickel and copper, is in some instances returned to the primary smelting furnace for recovery of metal values. The high-grade, low-iron Ni–Cu matte, which typically contains 20% S, less than 1% Fe, and also any precious metals present in the original concentrates, is the final product of the smelting process.

Converting is a batch operation. The horizontal side blown converter, known as the Peirce–Smith converter in copper smelting, is normally used for the treatment of nickel furnace mattes, although top-blown rotary converters are also used in some plants, particularly for low production rates. Air or oxygen-enriched air is blown through the molten matte to form iron oxides and remove sulfur as SO_2. The iron oxides combine with added silica flux to form an iron silicate slag:

$$2FeS + 3O_2 = 2FeO + 2SO_2 \text{ and}$$
$$2FeO + SiO_2 = 2\ FeSiO_4$$

A substantial amount of magnetite is also formed in the conversion process.

Converter slags contain 20%–35% silica, with the balance being iron oxides. Silica in the slag combines with Fe(II) oxide, thus preventing further oxidation to magnetite. Slags low in silica contain large amounts of magnetite. Typically, up to one-third of the iron may be in the Fe(III) state in a 20% silica slag, whereas a slag with 35% silica contains less than 15% magnetite. Magnetite formation is useful in converting since it can be made to coat the inside of the brick-lined furnace and thus protects it from erosion and corrosion by the slag (Wang et al. 1974).

The oxidation of iron sulfide is strongly exothermic, and much of the heat generated in the conversion process can be used to melt additional ore or concentrate feed or recycled scrap materials. Since the throughput of a converter is almost directly proportional to the amount of oxygen blown through the charge, its capacity can be raised substantially by enriching the air with oxygen. In addition to increasing the converting rate, oxygen enrichment of the air permits the treatment of additional cold charge, and in fact, the addition of cold charge is essential to control the bath temperature when oxygenenriched air is used. More recent developments include the use of shrouded tuyeres to improve refractory life of tuyere blocks that limit campaign life (Kapusta et al. 2012). Engineering of hoods has resulted in improved sulfur capture and cleaner converter aisles.

Electric smelting. The energy required for smelting may be obtained from electrical power as an alternative to burning fossil fuels or sulfides in air or oxygen. The use of an electric furnace for nickel sulfide smelting is favored when the cost of electrical energy is low or when the required smelting temperature is high (e.g., when the concentrate contains high levels of magnesia).

Heating of the bath is achieved by passing a three-phase electrical current through a circuit consisting of carbon electrodes immersed in the slag layer, which has a high electrical resistance (submerged arc technique). Electric furnaces used for Ni–Cu matte smelting are either rectangular with six electrodes in line or circular with three electrodes in triangular configuration. The consumable electrodes are made of carbon and are either prebaked or of the self-baking Soderberg type.

The charge of concentrate (or calcine) and flux is fed through the roof. Thus a layer of unsmelted charge covers the slag, giving it a "cold top." The concentrate or calcine gradually settles into the slag as it melts and then separates into slag and matte layers. The slag and matte are tapped intermittently, as required. Since there is no fuel combustion in an electric furnace, the quantity of off-gas is much less than from a reverberatory furnace, and therefore heat and dust recovery from the off-gas is therefore easier.

Electric furnace smelting of nickel concentrates is practiced by Vale (Inco) at its Thompson, Manitoba, operation; by Glencore (Falconbridge) at Sudbury, Ontario; by the Severonikel Combine in Russia; by Anglo (Rustenburg) Platinum, Impala Platinum, and Lonmin (Western Platinum) in the Republic of South Africa; and by Bindura Nickel in Zimbabwe. The Canadian operations treat partly roasted concentrate calcines, whereas Bindura and the platinum producers treat Ni–Cu concentrates directly. The Severonikel electric furnace smelter treats high-grade ore (3.5% Ni, 3.5% Cu) from Norilsk directly.

Integrated flow sheets. The Vale Ontario operations flow sheet depicted in Figure 17 is an exemplar of an advanced nickel pyrometallurgical flow sheet with refining (Vale Canada Ltd. 2010).

Nickel concentrate from a variety of sources, including custom feeds, is treated at Copper Cliff through a combination of flash smelting, converting, matte preparation, and reconverting, followed by carbonylation and precipitation into a range of products including ferronickel, nickel pellets, and nickel powders including specialized Vale nickel products. By comparison, Glencore (previously Xstrata) operates a simplified flow sheet for nickel production from its own and custom feeds, also in the Sudbury region (Figure 18).

Operations comprise pre-roasting and smelting of nickel concentrates in a submerged electric arc furnace, followed by conversion and matte granulation prior to shipping for refining in the Nikkelverk facility in Finland.

The Outokumpu flash smelting process (Figure 19) is probably the most environmentally compatible of the current commercial nickel smelting processes because it permits the capture of virtually all the SO_2produced prior to the conversion step. The Outokumpu direct oxidation of nickel (DON)

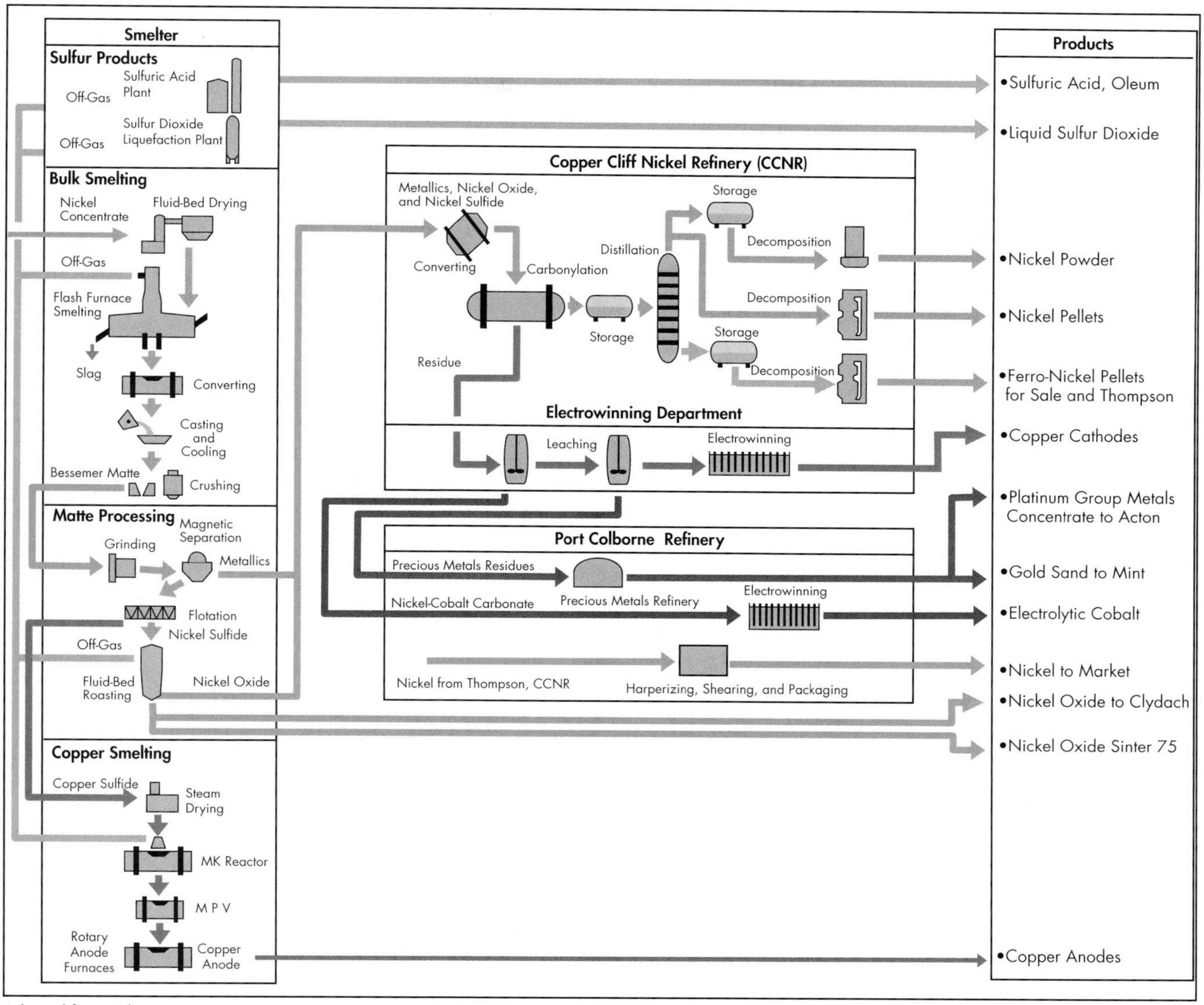

Adapted from Vale

Figure 17 Vale Ontario nickel operations flow sheet

process completely eliminates the separate converting step, instead producing a highly metallized matte that is sent for refining and a highly oxidized slag that is cleaned in an electric furnace to recover pay metal before discarding (Pääkkönen and Mattelmäki 1996).

Hydrometallurgy of Nickel Extraction

Several hydrometallurgical processes are in commercial operation for the treatment of Ni–Cu mattes to produce separate nickel and copper products. In addition, the hydrometallurgical process developed by Sherritt in the early 1950s for the direct treatment of nickel sulfide concentrates, as an alternative to smelting, is still commercially viable and competitive, despite significant improvements in the economics and energy efficiency of nickel smelting technology (Forward 1948).

In a typical hydrometallurgical process, the concentrate or matte is first leached in a sulfate or chloride solution to dissolve nickel, cobalt, and some of the copper, while the sulfide is oxidized to insoluble elemental sulfur or soluble sulfate. Frequently, leaching is carried out in a two-stage countercurrent system so that the matte can be used to partially purify the solution, for example, by precipitating copper by cementation. In this way, a Ni–Cu matte can be treated in a two-stage leach process to produce a copper-free nickel sulfate or nickel chloride solution and a leach residue enriched in copper. The copper-rich residue is treated by pressure leaching, or by roasting and leaching, to solubilize the copper as copper sulfate so that it can be recovered from solution electrolytically as cathode copper. Nickel is recovered from the purified nickel sulfate or chloride solution either electrolytically as a pure nickel cathode or by chemical reduction with hydrogen to give pure nickel powder. Extensive details of the various hydrometallurgical operations are provided by Kerfoot et al. (2011).

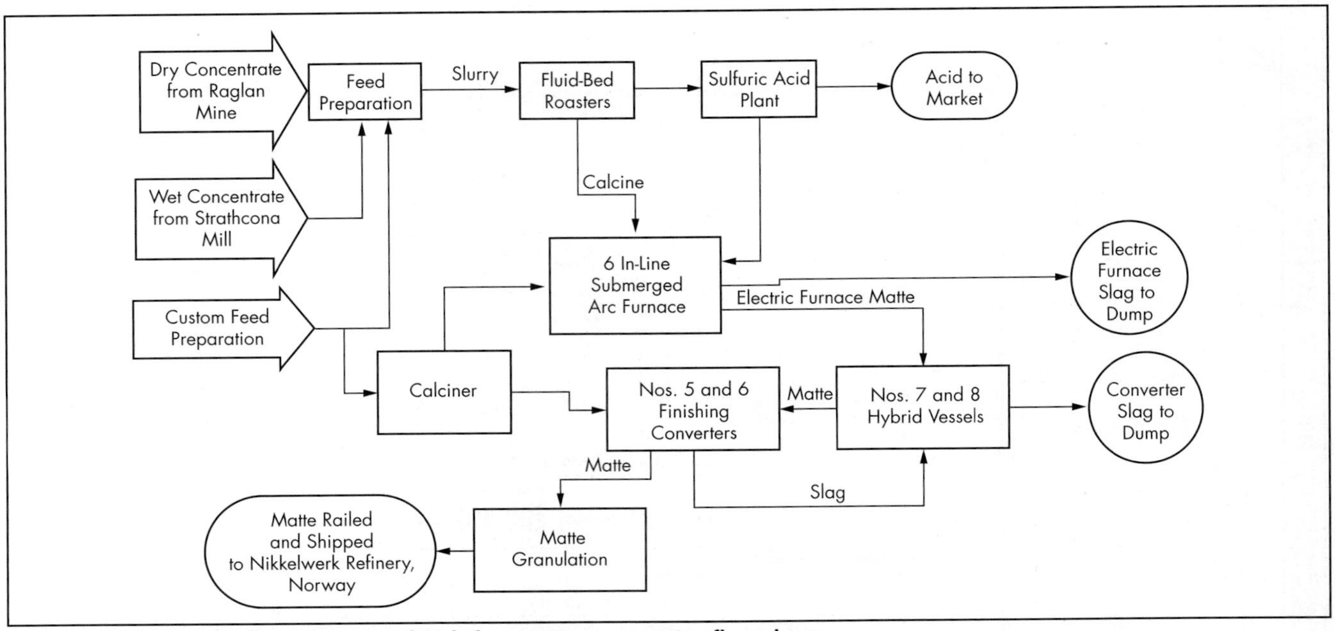

Figure 18 Glencore Sudbury integrated nickel operations extractive flow sheet

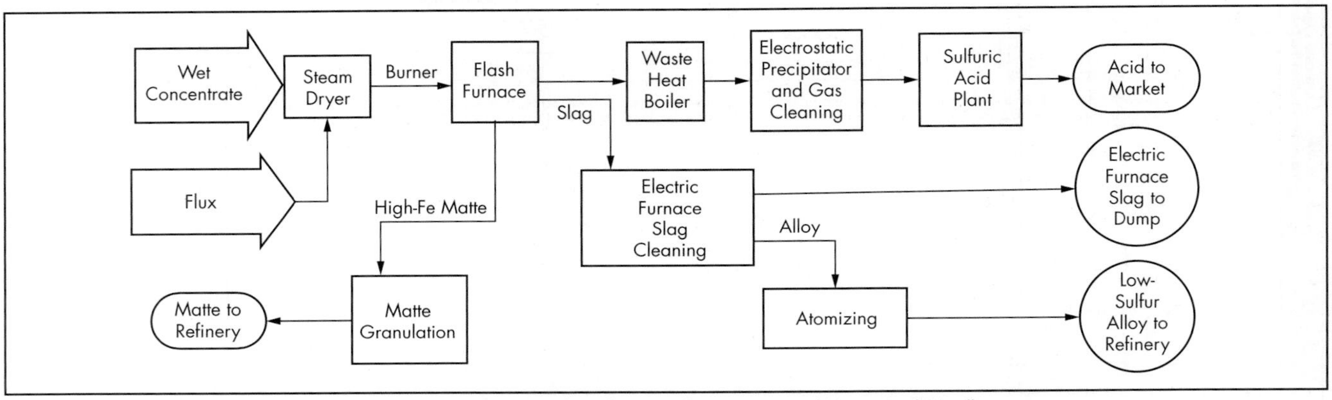

Source: Pääkkönen and Mattelmäki 1996, reprinted with permission from the Australasian Institute of Mining and Metallurgy

Figure 19 Outokumpu DON smelting process

The nickel contents of the mattes treated hydrometallurgically are in the range of 35% to 75%, whereas the copper content varies from 0% to 52%, and sulfur from 6% to 24%. Low-copper mattes (<10% Cu) can be treated by ammoniacal pressure leaching, by ferric chloride leaching, or by direct electrorefining of the matte. High-copper mattes can be treated by high-pressure or atmospheric pressure sulfuric acid leaching or hydrochloric acid leaching. The sulfur contents of the mattes are adjusted during the conversion process in the smelter to the optimal level for the subsequent leaching process. Sulfur levels are typically 20%–23%, except when the matte is to be treated by the Outokumpu atmospheric pressure sulfuric acid leaching process, for which the sulfur content is reduced to about 6%. Advanced hydrometallurgical techniques such as Activox and the Yakabindie process have also been developed specifically to treat high nickel- and magnesium-containing concentrates such as those at Yakabindie and Mount Keith in Western Australia and the Tati nickel mine in Botswana (Becker and Johnson 1996).

Atmospheric acid leaching. The atmospheric acid leach process for the treatment of high-copper, low-sulfur nickel matte was developed by Outokumpu in 1960 to process material from its nickel mine in Finland and was later adopted by the Bindura and Empress nickel refineries in Zimbabwe. In this process, the leaching operations are carried out in trains of air-agitated pachuca tanks, in contrast to the mechanically agitated horizontal autoclaves used in the pressure leaching process.

Acid pressure leaching. The acid pressure leaching process for the treatment of high-copper nickel mattes was developed by Sherritt Gordon during the 1960s and is currently operated by all four platinum producers in the Republic of South Africa: Rustenburg Refiners, Impala Platinum, Western Platinum (now Lonmin), and Barplats (Plaskett and Romanchuk 1978; Brugman and Kerfoot 1986). A similar process was also operated by the Amax nickel refinery in Louisiana, United States, from 1974 to 1986 (Blanco 1978; Llanos et al. 1974).

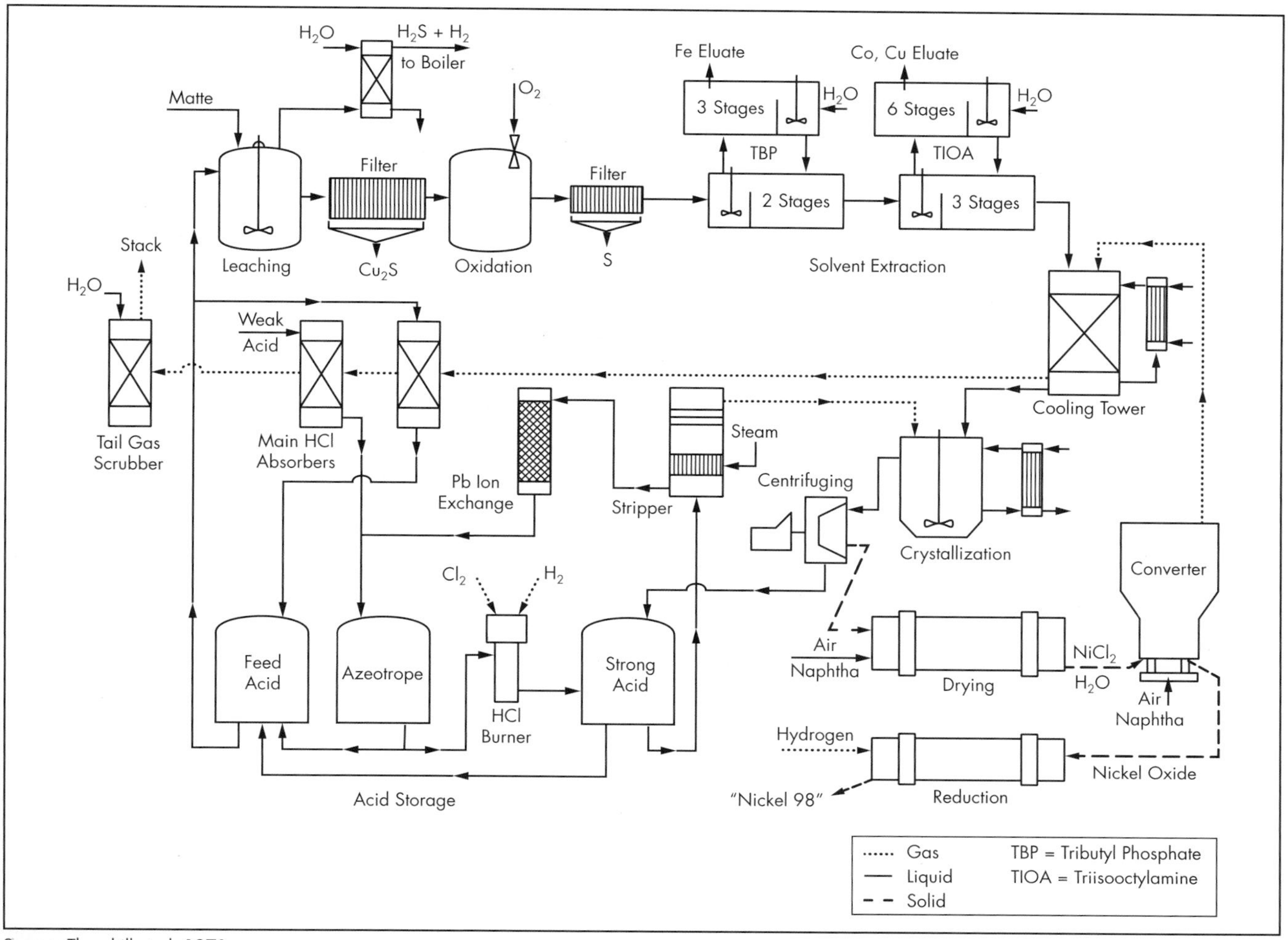

Source: Thornhill et al. 1971

Figure 20 Falconbridge matte leach process

Chloride leach processes. Nickel matte refining processes based on leaching in chloride solution in the presence of chlorine gas have been developed and successfully commercialized by Glencore at Kristiansand in Norway (Stensholdt et al. 1988) and by Société Le Nickel (SLN) at Le Havre-Sandouville in France (Lenior et al. 1981). The flow sheet of the Falconbridge process is shown in Figure 20.

Refining. Nickel differs from copper in that it is largely used as an alloying element in stainless steel and alloy steel production. Thus, although most copper is used in electrical and thermal energy applications, which require a highly refined metal, a high-purity metal is not required for many of nickel's most important markets. Nickel products such as nickel oxide, metallized nickel oxide, and ferronickel, which are designated Class II products, are produced directly by smelting and roasting and are sufficiently pure without refining for many nickel applications such as stainless steel. These Class II products may contain significant levels of cobalt and copper and trace impurities.

In other applications, high-purity nickel is essential and Class I nickel products, which include electrolytic cathode, carbonyl powder, and hydrogen-reduced powder, are made by a variety of refining processes. In general, nickel is refined to decrease to an acceptable level only those impurities that would have an adverse effect on subsequent processing or on the properties of the refined metal product. These elements include antimony, arsenic, bismuth, cobalt, copper, iron, lead, phosphorus, sulfur, tin, and zinc. Cobalt, which closely resembles nickel in physical and chemical properties, is difficult to remove completely, and since it is not often a serious contaminant in most applications for nickel, many Class I products contain quite high levels of cobalt.

The object of the refining process is not only to remove the impurities from the nickel, but also to recover those that have an economic value. Obvious examples are the precious and platinum group metals (platinum, palladium, osmium, rhodium, iridium, and rutheum), cobalt, and copper; sulfur, selenium, and tellurium can sometimes also be recovered economically.

The first successful nickel refining operation was established by the Mond Nickel Company (later absorbed by Inco/Vale) at Clydach, Wales, in 1902, using the carbonyl refining process invented by Mond and Langer in 1889. This refinery is still in operation using a modern version of the carbonyl

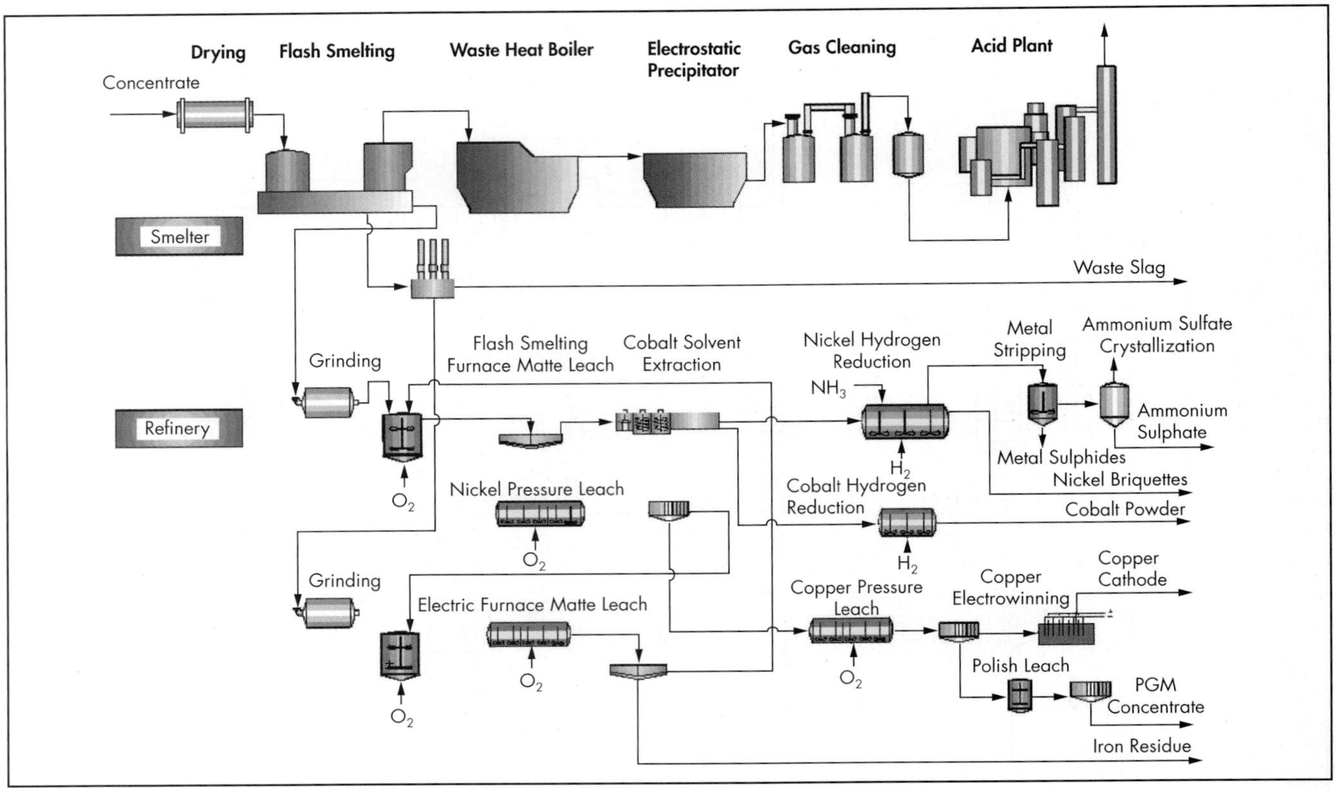

Courtesy of L. Rosenback, as cited in Svens 2013

Figure 21 Outokumpu DON refining process

process. A large refinery that uses a high-pressure version of the carbonyl process was established by Inco at Copper Cliff in Ontario in 1972 (Walker 1989).

The first electrolytic nickel refinery, treating nickel metal anodes, was built by Hybinette in Kristiansand, Norway, in 1910. This plant was acquired by Falconbridge in 1928. The Kristiansand Refinery was modernized in the late 1970s by replacing the Hybinette process with the current chlorine leach process. Inco established its first nickel refinery in Canada at Port Colborne, Ontario, in 1926, using a version of the Hybinette refining process. This refinery operated until 1984. In 1989 the only electrolytic nickel in North America was produced by Inco's (now Vale's) Thompson refinery, which electrorefines matte anodes Boldt and Queneau (1967). This process is also operated in Japan (Inami et al. 1986) and China. The electrorefining of nickel metal anodes, which is effectively obsolete in Western Europe and North America, is still practiced at Norilsk (Renzoni 1943).

Outokumpu, SLN, and Falconbridge in Europe and Bindura, Empress, and Rustenburg Refiners in Southern Africa all recover electrolytic nickel by direct electrowinning from purified solutions produced by the leaching of nickel or Ni–Cu mattes. The Outokumpu process is outlined in Figure 21.

Sherritt Gordon in Canada, Western Mining in Australia, and Impala Platinum in the Republic of South Africa recover refined nickel powder from purified ammoniacal solution by reduction with hydrogen.

Electrolytic refining. Nickel electrorefining is carried out in a tank house containing many electrolytic cells, each filled with a mixed nickel-sulfate–nickel-chloride solution, and containing 30–40 anodes (slabs of metallic nickel or nickel matte) interspersed with cathodes that are thin sheets of pure nickel. The anodes and cathodes in each cell are connected electrically in parallel, and the cells are connected in series. When electrical current is passed through the cells the anodes dissolve, and pure nickel deposits on the cathodes. Cathodes and anodes are replaced periodically to maintain continuity of production.

Electrowinning. Electrowinning differs from electrorefining in that it is used to recover nickel from a leach liquor and can therefore be carried out with an insoluble anode since the only function of the anode is to transfer electrons from the electrolyte to the external circuit. The electrowinning of nickel is practiced commercially with both pure sulfate and pure chloride electrolytes. In the sulfate system the anodes are made of chemical lead or antimonial lead, which have a relatively long life in sulfate solution, although they are soluble in chloride solutions. In the highly corrosive chloride system, dimensionally stable anodes, made by coating a titanium substrate with a platinum group metal oxide, are used. The Outokumpu nickel refinery at Harjavalta in Finland contains 126 electrowinning cells, each having 40 insoluble lead anodes and 39 cathodes. Details are provided in Diaz et al. (2011).

Thermal refining. The reaction of carbon monoxide at atmospheric pressure with active nickel metal at 40°–80°C to form the gaseous nickel tetracarbonyl was discovered by Langer and Mond in 1889 (Walker 1989). The reaction is readily reversible, with nickel tetracarbonyl decomposing to metallic nickel and carbon monoxide at 150°–300°C. Nickel tetracarbonyl is a volatile liquid that melts at –19.3°C and boils at 42.5°C.

Atmospheric pressure carbonyl process. The Inco nickel refinery at Clydach, Wales, began operation in 1902 using the Langer–Mond atmospheric pressure carbonyl process. Originally the plant treated Ni–Cu matte, but now it processes a granular nickel oxide produced by fluidized-bed roasting of nickel sulfide at Vale's Copper Cliff smelter, which typically analyzes 74% Ni, 2.5% Cu, 1.0% Co, 0.3% Fe, and 0.1% S (Walker 1989). The refinery still uses the basic Langer–Mond process, but the operation's efficiency has been greatly increased over the years.

High-pressure carbonyl process. Extraction of nickel with carbon monoxide is also used in processes that operate at high pressure and produce liquid nickel carbonyl. The volume change that occurs in the carbonyl reaction:

$$Ni + 4CO \rightarrow Ni(CO)_4$$

An increase in pressure accelerates the formation of the carbonyl. The increased pressure stabilizes the carbonyl and thus permits the process to be carried out at high temperature, which further increases the rate of reaction. As a result, activation of the nickel feed material is no longer necessary, and a wider range of feeds can be processed. BASF in Germany operated a high-pressure carbonyl nickel refining process from 1932 until 1964.

Inco pressure carbonyl process. Vale's nickel refining complex in Copper Cliff, Ontario, which was commissioned in 1973, has a nominal capacity of 57,000 t/yr of refined nickel. The refinery consists of two operating plants: the converter plant and the pressure carbonyl plant. The converter plant produces granulated metallic nickel with a controlled sulfur content, using top-blown rotary converters. The pressure carbonyl plant produces 45,000 t/yr of nickel pellets, 9,000 t/yr of nickel powder, and 2,200 t/yr of ferronickel by-product from the granulated, sulfided nickel feed (Head et al. 1976; Wiseman et al. 1988).

The principal feed materials supplied to the refinery by the Copper Cliff smelter are the metallics fraction from the matte separation process and the slightly lower purity nickel oxide made by roasting the nickel sulfide concentrate as well as various other nickel-bearing residues and intermediates. The novel feature of this process is the preparation of a nickel feed material that could be carbonylated at moderate pressure without the need for a separate activation step. Full details of the operation can be obtained from Head et al. (1976) and Wiseman et al. (1988).

Hydrogen reduction to nickel powder. Nickel is recovered as a pure metal powder from nickel sulfate leach solutions by chemical reduction with hydrogen under pressure in several hydrometallurgical nickel refineries (Mackiw and Veltman 1980). Precipitation of the metal is carried out with an ammoniacal nickel sulfate solution having a molar ratio of ammonia to nickel of 2:1, so that the acid generated in the reduction reaction is neutralized by reaction with ammonia to form ammonium sulfate:

$$[Ni(NH_3)_2(H_2O)_4]SO_4 + H_2 = Ni + (NH_4)_2\ SO_4 + 4H_2O$$

Full details of the process are provided in Head et al. (1976) and Wiseman et al. (1988).

NICKEL LATERITE ORES

Geological and Mining Context

Although nickel laterite deposits are numerous and abundant, and comprise approximately 70% of the world's nickel resources, very few of them have become producing mines. However, with the global consumption of nickel growing at a rate of 4% per year, permanent new sources of nickel, such as in laterites, need to be found and developed. Production of nickel from laterite ores has occurred for more than 100 years, beginning with processing of garnierite-rich ores from New Caledonia. Since then, there have been major developments in Australasia, Indonesia, and South America with the development of both pyrometallurgically treated laterites (typically saprolite-rich), hydrometallurgically treated laterites (typically limonite rich), and a plethora of direct-shipping operations largely in Indonesia. Development has been challenged by a combination of factors, including the high capital and operating cost intensity of the flow sheet; challenging presentation in typically remote, hilly to mountainous terrain; and material handling challenges related to the sticky and dilatant nature of the saturated, tropical clays. Nevertheless, laterite deposits still have the potential to become the predominant source of nickel—particularly in ferronickel—worldwide (Dalvi et al. 2004).

Nickel laterite deposits occur in present and past zones of the earth that have experienced prolonged tropical weathering of ultramafic rocks containing ferromagnesian minerals. The majority of laterites can be found between the tropics of Cancer and Capricorn: the four main areas in the world for laterite resources are New Caledonia (21%), Australia (20%), the Philippines (17%), and Indonesia (12%). Some deposits do exist outside of this tropical belt, including those found in Oregon, United States, and in the Ural Mountains of Russia.

Lateritic nickel ores are formed by intensive tropical weathering of olivine-rich ultramafic rocks such as dunite, peridotite, and komatiite and their serpentinized derivatives, and serpentinite, which consists largely of the magnesium silicate serpentine and contains approximately 0.3% nickel. Serpentine is the most common product of hydrothermal alteration of olivine in the presence of water at temperatures ranging from 200°C to about 500°C. This initial nickel content is strongly enriched in the course of lateritization (Pelletier 2003). Two kinds of lateritic nickel ore must be distinguished: limonite types and silicate types. Limonite type laterites (or oxide type) are highly enriched in iron because of very strong leaching of magnesium and silica. They consist largely of goethite and contain 1%–2% Ni incorporated in the goethite. Silicate type (or saprolite type) nickel ore formed beneath the limonite zone. It contains generally 1.5%–2.5% Ni and consists largely of magnesium-depleted serpentine in which nickel is incorporated. In pockets and fissures of the serpentinite rock, high nickel contents can occur—mostly 20%–40%—bound in newly formed phyllosilicate minerals. All the nickel in the silicate zone is leached downward (absolute nickel concentration) from the overlying goethite zone (Figure 22).

Typical nickel laterite ore deposits are very large, low-grade deposits located close to the surface, with some examples approaching a billion metric tons of material. Ore deposits of this type are restricted to the weathering mantle developed above ultramafic rocks and tend to be tabular, flat, and extensive, covering many square kilometers. Mining operations generally proceed in the following sequence:

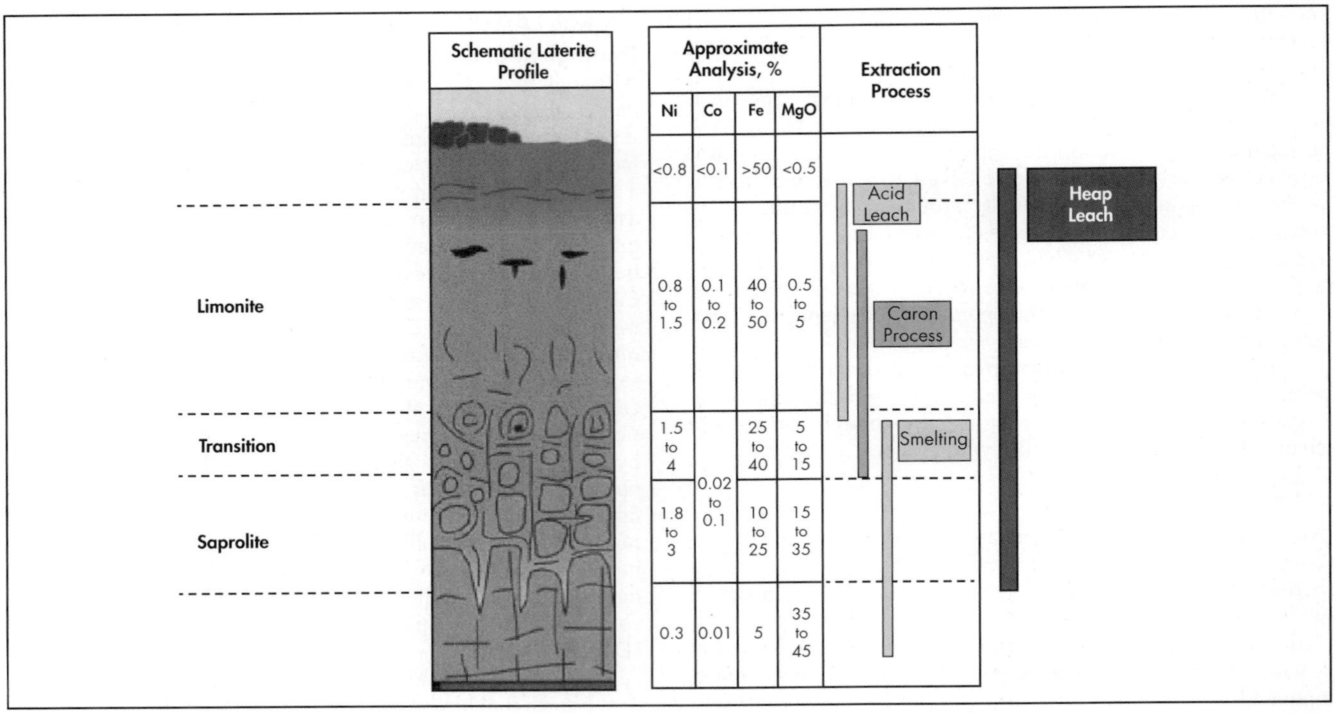

Adapted from Pelletier 2003

Figure 22 Typical laterite cross section

1. Mine top-down from the highest elevation to the lowest within any given mining area.
2. Overburden prestrip by dozers pushing downhill to the temporary roads (typically three more benches or cuts are established below the current operating cut).
3. Load and haul overburden either to permanent dumps or onto adjacent mined-out areas where saprolite or limonite is not exposed.
4. Load with hydraulic backhoes into 40-t capacity articulated or 60-t rigid off-road haul trucks.
5. Haul ore to the ore preparation plant from which upgraded ore is delivered to the processing plant typically via pipeline. Some ore is stockpiled before the slurry plant to "fill in" production during wet weather periods.

Production is maintained under strict geological control because of highly variable weathering profiles, zone thicknesses, and footwall rolls. Operations are typically surveyed daily and are visited at least once per day by the geologist responsible for grade control. Nickel is extracted from the mined ore by a variety of process routes, either pyrometallurgical in nature for primarily saprolitic deposits or hydrometallurgical for primarily limonitic deposits. Many deposits contain significant amounts of both limonite and saprolite material, and therefore close management of blend chemistry is critical. Pyrometallurgical processes are predominantly rotary kiln–electric furnace (RKEF) processes producing a ferronickel or matte product, although several variations to this flow sheet abound and are discussed in the following sections. The Caron process represents a hybrid route with high-temperature calcination of the laterite ore followed by ammoniacal leaching of the calcine to recover nickel and cobalt. Hydrometallurgical processes include high-pressure acid leach (HPAL) and heap leach, both of which are generally followed by solvent extraction and electrowinning (SX-EW) for recovery of nickel and cobalt (Figure 23).

Extraction of Nickel from Oxide Resources

As previously described, nickel oxide ores occur in two forms: a limonitic type that occurs in the upper zone of a lateritic deposit, in which ferric oxide materials are predominant, and a silicate, or garnierite, type (saprolitic), occurring at greater depth in the deposit, which has a lower iron content but is usually richer in nickel than the limonite and contains high levels of magnesia and silica. In nickeliferous limonite [$(Fe,Ni)O(OH){\cdot}nH_2O$] the nickel oxide is mainly present in solid solution with the iron oxides. In nickeliferous silicate, nickel, iron, and cobalt oxides occur in varying proportions, generally replacing part of the magnesium oxide in the serpentinite [$Mg_6Si_4O_{10}(OH)_8$]. Most deposits contain both types of ore. Usually, it is possible to separate the ore by selective mining and screening into an iron-containing limonite fraction and a magnesium silicate fraction enriched in nickel (Golightly et al. 1979). The metallurgy of nickel oxide ores differs from that of sulfides in that oxide ores are not amenable to most of the standard mineral beneficiation methods because of the chemical dissemination of the nickel in the oxide minerals. Screening may be used to reject the oversize, less weathered fragments that contain less nickel, but this provides only a minor degree of upgrading.

Since world resources of nickel are predominantly lateritic (72%) rather than sulfidic (28%), new extraction capacity is largely focused on FeNi production rather than high-purity final products. Of the lateritic material, approximately 70% is extracted by pyrometallurgical techniques and most of the balance by hydrometallurgical methods. More recently, two processes have been developed in which the nickel is first

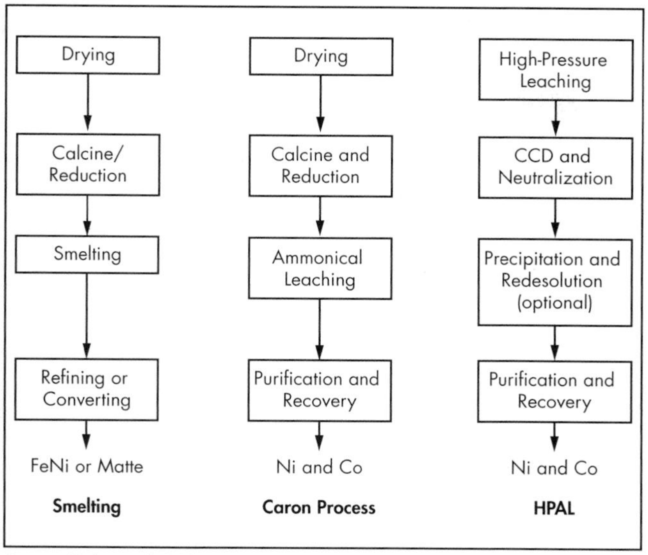

Source: Dalvi et al. 2004

Figure 23 Three main routes for treatment of lateritic ores

converted by high-temperature treatment to a metallic phase, which can then be separated from the bulk of the ore by standard mineral beneficiation techniques. These processes are the Nippon Yakin Oheyama process (which was based on the Krupp-Renn process developed in Germany in the 1930s to produce iron from otherwise uneconomic low-grade iron ores) and the MINPRO-PAMCO segregation process (Watanabe et al. 1985).

Pyrometallurgical Treatment of Laterites

Pyrometallurgical processes are suited for ores containing predominantly saprolite. A typical RKEF flow sheet is shown in Figure 24. Saprolitic ores contain proportionately lower cobalt and iron compared to the limonitic ores and are calcined and smelted to produce a ferronickel matte (Bergman 2003).

Pyrometallurgical processes are energy intensive since ores are typically mined in saturated tropical conditions, where all of the free moisture and interstitial water has to be removed, then the material is calcined to remove water of crystallization and then melted to form metal and slag at about 1,600°C. This requires significant amounts of reductants (coal, oil, or naphtha), fluxes, and electric power. FeNi produced by this route is typically granulated upon tapping, packed, and shipped to steel-producing areas for further treatment.

Variations against the standard RKEF flow sheet include the addition of a derubbling stage to remove coarse, low-grade peridotites or harzburgite to upgrade the ore prior to calcination (Figure 25), or, as at the Koniambo mine in New Caledonia, the addition of a bulk rougher sorting stage based on X-ray fluorescence where bulk ore below a certain grade is diverted to either stockpile or waste to ensure an optimal grade of feed to the calciner.

Nickel Pig Iron

A major (and disruptive) aspect of nickel production from lateritic ores was the widespread use in China of small blast furnaces treating limonitic ores to produce a pig iron containing 4%–6% Ni (Yamada and Hiyama 1985). Since the most popular grade of austenitic stainless steel contains only 8% Ni, 18% Cr, and the balance as iron, these furnaces were able to produce a raw material for stainless steel production that was then blended with cheap ferrochromium. In times of high nickel prices, these plants generated large volumes of nickel pig iron. Environmental pressure and high nickel inventories post-2009 led to the closure of many of these plants or the installation of new RKEF plants in China. Although China has no domestic nickel reserves, imports of raw ore from neighboring countries like Indonesia provided a ready source of feed. Evolving legislation is intended to encourage beneficiation of ores before exporting, and this may in time discourage pig nickel production.

Nippon Yakin Oheyama Process

Nippon Yakin Kogyo produces 12,000 t/yr nickel as low-nickel ferronickel from silicate ores (2.3%–2.6% Ni, 12%–15% Fe), imported from New Caledonia, the Philippines, and Indonesia, at its Oheyama works in Japan. The ore is ground and mixed with anthracite, coal, and flux, and briquetted. The briquettes are fed continuously through a rotary kiln, countercurrent to a flow of hot gas produced by the combustion of coal. As the temperature rises from 700°C to 1,300°C, the charge is successively dried, dehydrated, and reduced (Watanabe et al. 1985). As the charge begins to melt, smelting occurs in the semi-fused state to form coalesced particles of spongy ferronickel (known as luppen, from the Krupp-Renn process) interspersed in a viscous slag phase. The kiln discharge is quenched in water, and the luppen are separated from the slag by grinding, screening, jigging, and magnetic separation. Nickel recovery from ore is typically 95% S. A major advantage of this process is that the ferronickel product is recovered in the form of 2- to 3-mm particles containing 22% Ni + Co, 0.45% S, 0.03% C, and 0.02% P, which can be used directly in the argon-oxygen decarburization steelmaking process, without further upgrading or refining. Although the process is energy intensive, the major energy sources used are cheap anthracite and coal, and no high-cost electrical energy or oil is used for heating. Consequently, energy costs are relatively low.

Caron Process

The Caron process has been utilized for the production of nickel since 1944 with commercial operations at Nicaro, Punta-Gorda, and Tocantins in Brazil and Yabulu in Queensland, Australia (Taylor 2013). In general, the process consists of ore beneficiation and blending, reduction roasting, leaching of roaster product in ammoniacal solvent, separation of nickel and cobalt using solvent extraction, precipitation from solution of basic nickel carbonate (BNC) intermediate product, and thermal decomposition of BNC to NiO and gaseous reduction of NiO to Ni (Figure 26).

The process was developed to treat ores that are a blend of limonitic and saprolitic materials. The mixture is dried to reduce moisture from 35% to approximately 8%, then ground to less than 850 μm before being fed into the roaster. Heavy fuel oil is typically mixed with the feed to provide the high temperature inside the furnace through partial combustion, as well as provide a reducing agent for the process. The maximum temperature of the reduction process is 740°–750°C. This is the critical stage that controls recovery of nickel from the ore. The reduced ore is discharged from the roaster and cooled under a neutral atmosphere to a temperature below 250°C. The cooled product is subsequently

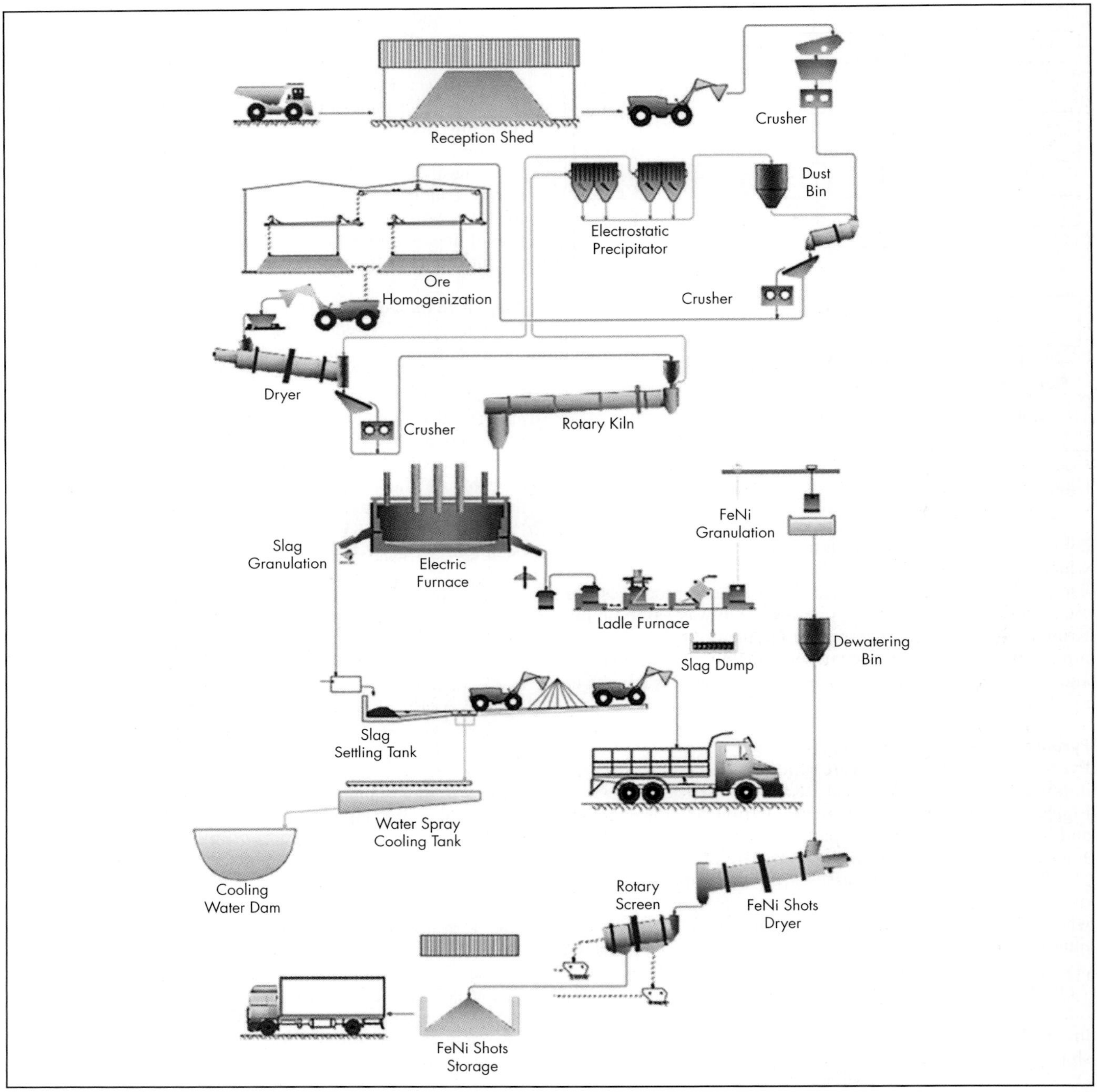

Source: Bergman 2003

Figure 24 Typical RKEF process

leached in an ammonia–ammonium carbonate solution where the nickel is selectively leached from the solid, together with iron and cobalt. The pregnant liquor is then separated from the undissolved solid. The Fe metal complex is unstable and reacts with the oxygen from the atmosphere and precipitates as solid iron hydroxide. Cobalt, on the other hand, is separated from the nickel by an ammonium solvent extraction process and is extracted by hydrometallurgical refining in a separate cobalt processing circuit. The purified liquor is then steam stripped to remove ammonia/ammonium carbonate and to recover the nickel as BNC, a mixed carbonate hydroxide solid with approximately 50% Ni content. The remaining liquid is separated from the solid by filtering process. The BNC is then processed through a series of high-temperature processes involving various gas–solid reactions. The BNC at the Yabulu refinery, for example, is fed to a rotary kiln operated in a slightly reducing atmosphere, fired with naphtha or coal seam gas. This is aimed at partially reducing the BNC to Ni/NiO mixtures. Dehydration and decomposition of BNC into nickel oxide occurs in the kiln, followed by partial reduction

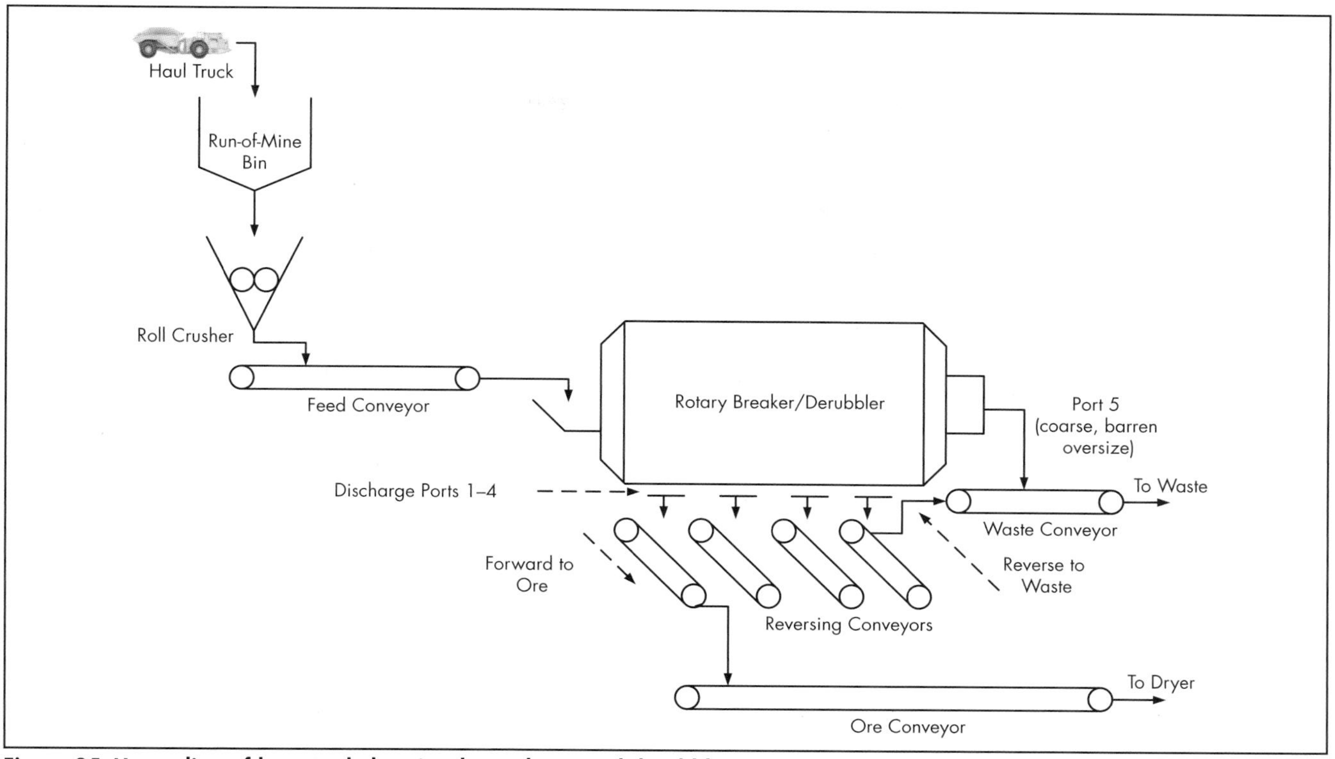

Figure 25 Upgrading of low-grade laterites by scalping and derubbling

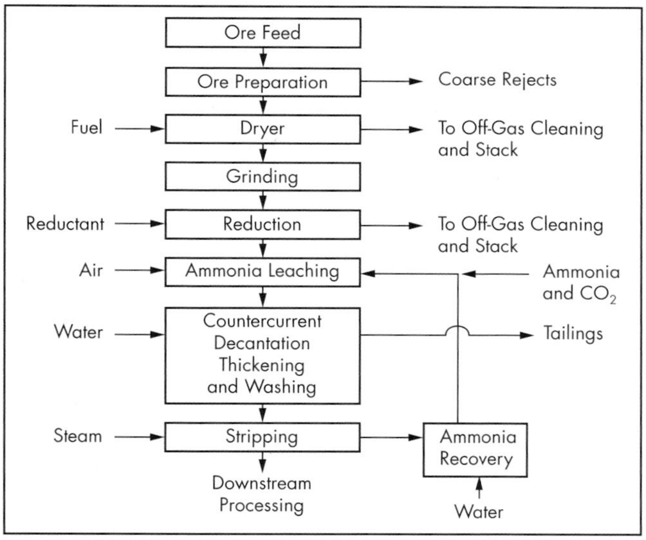

Source: Taylor 2013
Figure 26 Typical Caron process flow sheet

of the nickel oxide, producing a mixture of metal–metal oxide containing 92% Ni (Reid and Barnett 2002). The second stage of the reduction process is carried out in a furnace. The kiln product is crushed and mixed with sawdust to provide sufficient porosity in the bed during the reduction process. The residence time in the furnace is approximately one hour. The furnace carries out reduction at an average temperature of 900°C under an atmosphere containing cracked ammonia, 75% H_2, and 25% N_2. The furnace product contains approximately 97.5% Ni, which is then crushed, blended with stearic acid, and pressed into compacts. The compacts are then fed into a sintering furnace, where a final stage of reduction and sintering of the nickel takes place. The final product of this process is a premium sinter containing >99% Ni.

Caron flow sheets are disadvantaged, despite the premium produced, by high energy consumption, moderate nickel and low cobalt recovery, and high sensitivity to feed composition requiring careful control of blending.

Hydrometallurgical Treatment of Laterites

High-pressure acid leach. HPAL technology enables the recovery of nickel from primarily limonitic nickel oxide ores that traditionally were difficult to process by pyrometallurgical routes. Oxide ores are crushed, ground and pulped, and subjected to high temperature and pressure and reacted under stable conditions with sulfuric acid to produce a nickel-rich refining intermediate (Figure 27). HPAL is employed for two types of nickel laterite ores: (1) ores with a limonitic character, such as the deposits of the Moa District in Cuba and southeast New Caledonia at Goro where nickel is bound in goethite and asbolite, and (2) ores of a predominantly nontronitic character, such as many deposits in Western Australia like the Murrin Murrin Joint Venture, where nickel is bound within clay or secondary silicate substrates in the ores (Reid and Barnett 2002). SMM Mining claims to be the first company in the world to apply HPAL successfully on a commercial scale. Many companies have tried to do so, but some failed because of the complexities of the process, largely relating to rheology of the leach feed slurry and/or the sophisticated materials of construction of equipment needed to treat acidic slurries at high pressure.

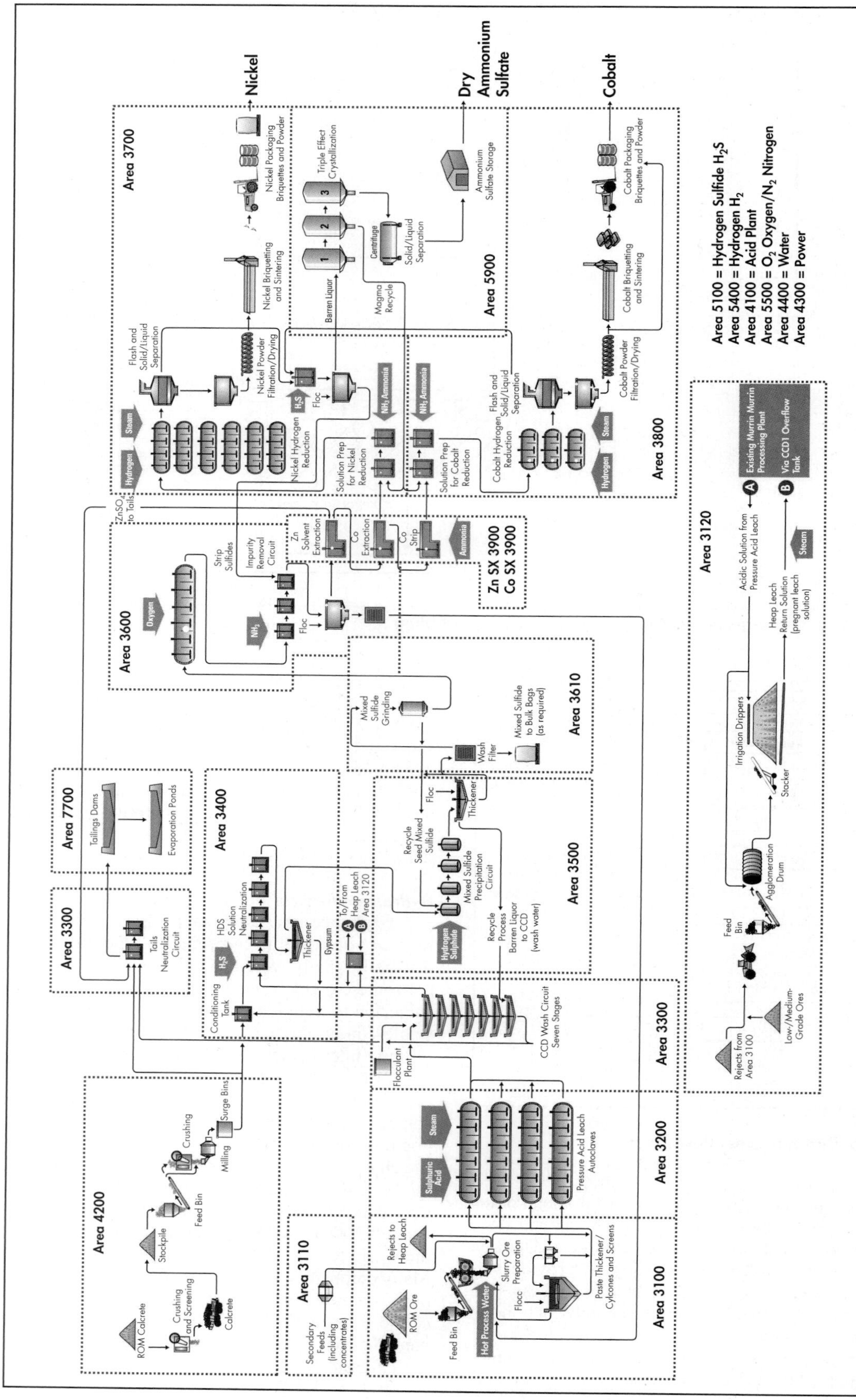

Courtesy of Minara Resources

Figure 27 Typical HPAL flow sheet

Table 2 Index of selected HPAL operations

Plant	Capacity, t Ni/yr	Approximate Cost, US$, millions	Start	Capital, US$/t Ni/yr	Ramp-Up	Comment
Coral Bay Line 1	10,000	180	2004	18,000	Fast	No acid plant, MSP*
Coral Bay Line 2	12,000	370	2008	30,833	Fast	No acid plant, MSP
Ravensthorpe	50,000	3,900	2008	78,000	Sold	Now FQMI, MHP*
Ramu	31,150	2,100	2011	67,416	Steady	MHP only
Goro	60,000	6,000	2012	100,000	Slow	Construction started in 2001
Ambatovy	60,000	7,200	2012	120,000	Steady	Now in operation
Taganito	36,000	1,700	2013	47,222	Fast	No acid plant, MSP

*MSP = mixed sulfide precipitate; MHP = mixed hydroxide precipitate

The advantages of HPAL plants are that they are not as selective about the type of ore minerals, grades, and nature of mineralization. Disadvantages include the high capital cost, the amount of energy required to heat the ore material and acid, the wear and tear of hot acid on plant and equipment leading to high maintenance cost and high downtime, and the complexity in downstream processing of leachates (Taylor 2013). Also, contamination of limonite with magnesium-rich saprolites further increases acid consumption and lowers the grade of the leachate. Projects have been characterized by lengthy ramp-up periods for the majority and total process failure in others (Table 2).

Heap (atmospheric) leach. Heap leach treatment of nickel laterites is primarily applicable to clay-poor, oxide-rich ore types where clay contents are low enough to allow percolation of acid through the heap. Generally, this route of production is much cheaper—up to half the cost of production—because there is no need to heat and pressurize the ore and acid, although recoveries can be low and generally only an intermediate nickel product requiring further treatment by electric furnace is required (Bergman 2003). In heap leach operations, the ore is typically ground, agglomerated, and perhaps mixed with clay-poor rock, to prevent compaction of the clay-like materials and maintain permeability. The ore is stacked on impermeable plastic membranes, and acid is percolated over the heap generally for three to four months, at which stage 60%–70% of the Ni–Co content is liberated into acid solution. That solution is then neutralized with limestone and a Ni–Co hydroxide intermediate precipitate is generated, which is generally sent to a smelter for further treatment. Challenges can include high acid consumption in presence of high magnesium levels, poor heap stability—especially in high rainfall (which is common)—and high levels of impurities carried over into the intermediates, which complicates downstream processing.

COBALT-PRIMARY ORES

Geological and Mining Context

Cobalt naturally occurs in nickel-bearing laterites and nickel–copper sulfide deposits and thus is most often extracted as a by-product of copper and nickel ore. According to the Cobalt Institute, about 48% of cobalt production originates from nickel ores, 37% from copper ores and 15% from primary cobalt production (CI 2017).

Sedimentary rock–hosted stratiform copper–cobalt deposits are the most important source of primary cobalt globally, with the Central African Copperbelt, the world's largest and highest-grade sedimentary copper–cobalt province, hosting the largest reserves of cobalt and more than 15% of the world's copper, the majority of which lies in the Katanga province of the DRC (Figure 28). Economic deposits of cobalt are generally found in the Lower Roan subgroup of the complex on the Zambian side, and the Mines subgroup dolomitic shales on the Congolese side (Broughton 2013). Host rocks are predominantly quartz–carbonate, with minor magnesium and chlorite. Sulfide minerals include vein-controlled, coarse-grained, and disseminated textures of pyrite, chalcopyrite, chalcocite, and cobalt minerals. The main cobalt minerals are cobaltite, erythrite, glaucodot, and skutterudite (Broughton 2013, Jackson et al. 2003).

Mining methods are both open pit and underground, with the Central African Copperbelt host to some of the largest underground mines in the world, including the Konkola and Nchanga mines. Open pit extraction is by truck-shovel, with ore haulage by truck to a primary crusher, from which ore is conveyed to the mill feed stockpile. In underground operations, methods range from open stoping through sublevel caving to vertical crater retreat methods, delivering ore by haul truck to the underground primary crusher, from where ore is conveyed or hoisted to the surface stockpile. Stockpiling of ore on surface is common, prior to crushing, grinding, and flotation stages.

Metallurgical Aspects

Cobalt is usually a by-product of copper and/or nickel production, and the overall process flow sheet will usually feature a leach of some type followed by recovery of the copper first, impurity removal and cobalt recovery second, and nickel recovery last (Figure 29).

Salable cobalt products range from concentrates through various intermediates to high-purity metal and salts. Project stakeholders should decide on the cobalt product they want to make, before the flow sheet can be finalized. Product options could include the following:

- Cobalt sulfide, or mixed Ni–Co sulfide precipitate (MSP)
- Cobalt hydroxide, or mixed Ni–Co hydroxide precipitate (MHP)
- Cobalt oxide or oxy-hydroxide
- Cobalt sulfate
- Cobalt carbonate
- Cobalt powder/briquettes, via hydrogen reduction

Cobalt can also be produced as an electrowon metal, with low-grade at 99.3% Co and high-grade at 99.8% Co.

Historically, cobalt–arsenide ores were typically roasted to remove the majority of the arsenic oxide. The ores were then treated with hydrochloric acid and chlorine, or with sulfuric acid, to create a leach solution that is purified. From

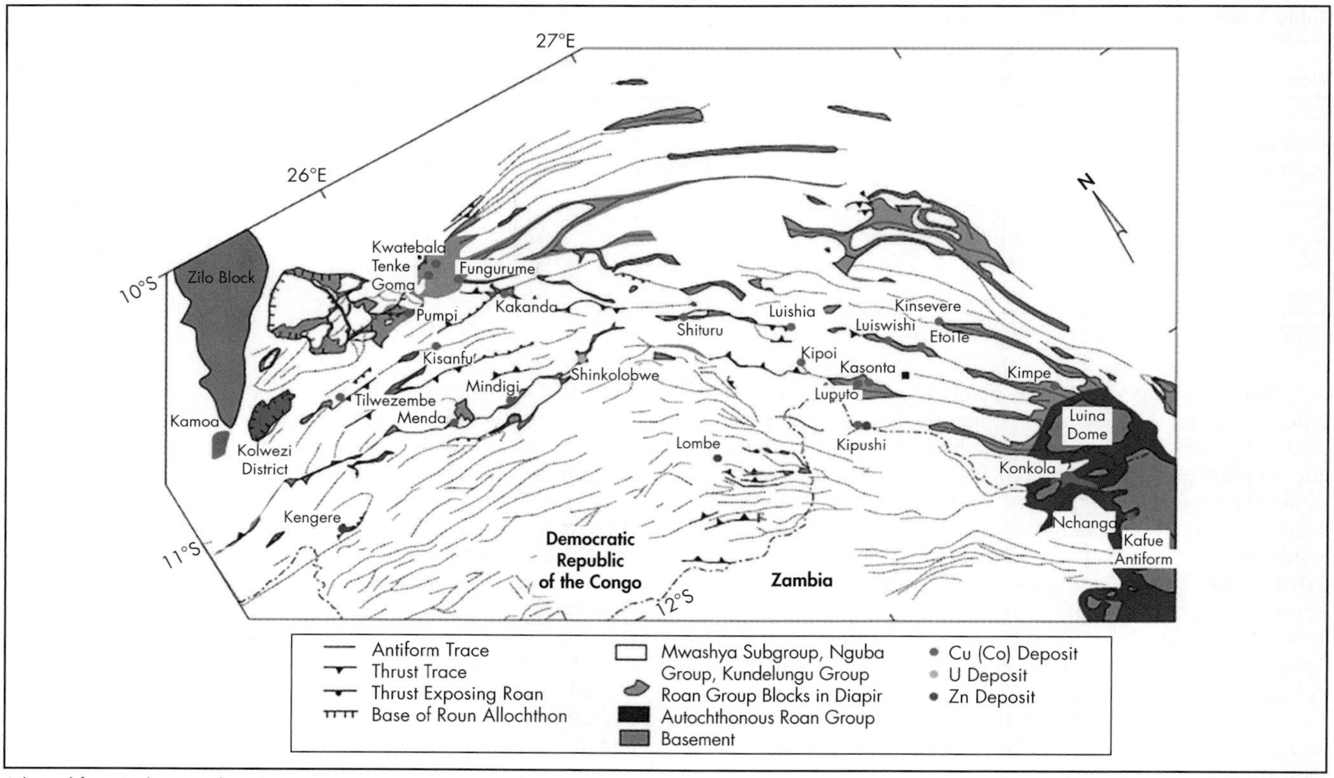

Adapted from Jackson et al. 2003

Figure 28 Sedimentary-hosted deposits of the Congolese copper belt including major producing mines

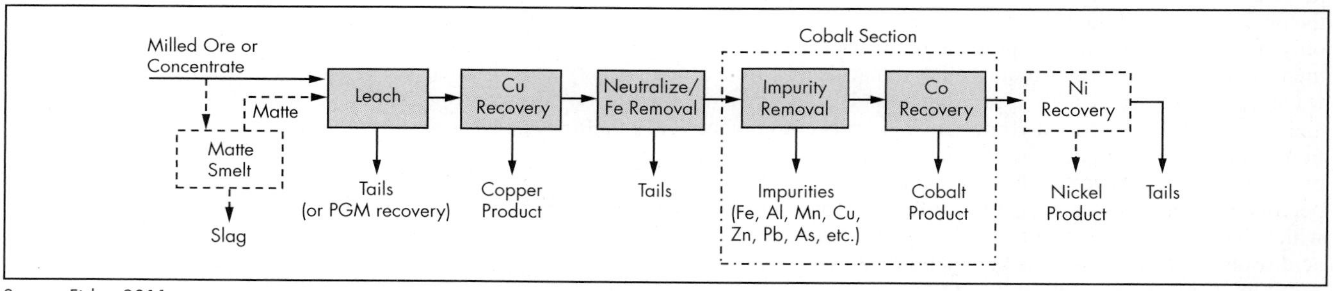

Source: Fisher 2011

Figure 29 Generic Cu–Co–Ni recovery flow sheet

this, cobalt was recovered by electrorefining or carbonate precipitation. A variety of such operations were active during the 20th century, mostly featuring a roasting stage and pressure leach of the roasted concentrate. However, all except the Compagnie de Tifnout Tiranimine operation in Morocco are now closed for obvious environmental reasons. Nevertheless, successful high-temperature/pressure leach test work has been done recently for several cobalt arsenide projects (Ferron 2001). The key to success has been the generation of an environmentally acceptable, stable arsenic residue in the process. As a result, potential exists for future contributions to cobalt production, such as Formation Metals in Idaho, United States (capacity 1,650 t/yr cathode), and Fortune Minerals' NICO Project in northwest Canada (Mezei et al. 2009).

Historically, cobalt from copper ores as in the DRC and Zambia is recovered by selective flotation from the copper-primary ore and then treated by a roast-leach-electrowin process. This would comprise roasting (for the sulfides), sulfuric acid atmospheric leach and direct copper electrowinning, plus impurity removal and cobalt hydroxide precipitation, often followed by releach and cobalt electrowinning. In recent years, apart from a few older plants, direct copper electrowinning has been replaced by copper SX-EW to more easily achieve London Metal Exchange Grade A specification. Many current cobalt projects are employing SX for impurity removal, even to the extent of having three or four SX sections in series, each using a different reagent. Sole et al. (2005) describe several Southern African SX plants and projects, with a particular focus on cobalt. Figure 30 shows a typical flow sheet recently proposed for a DRC cobalt project.

Conditions for cobalt recovery from copper concentrate by flotation at these operations are often suboptimal, with recoveries typically ranging from 40% for oxides to 80% for sulfides. Therefore, a substantial amount of cobalt is delivered

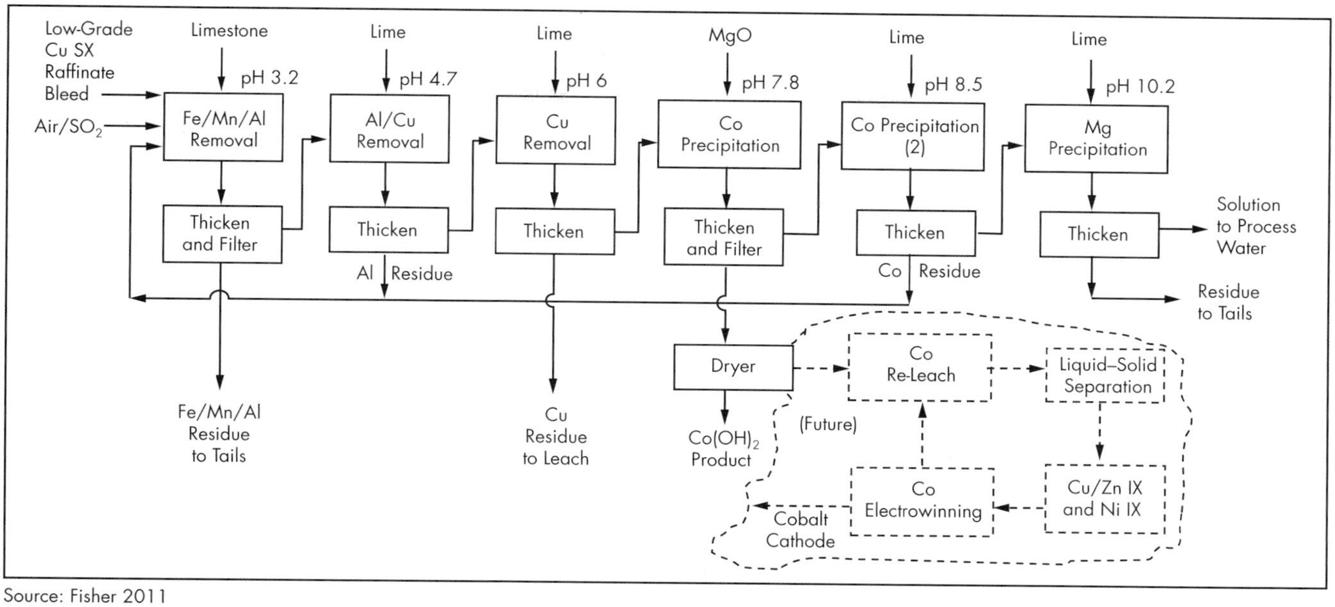

Source: Fisher 2011

Figure 30 Typical DRC cobalt flow sheet

to the copper tailings. An increasing number of operations, particularly Chambishi (Zambia), target these cobalt-rich tailings as a primary source of cobalt; similarly, as cobalt follows iron during smelting, some of the copper smelter operations have generated large slag dumps with a high cobalt content, for example, Nkana (Zambia) containing 0.65% Co (Fisher 2011).

Alternative cobalt production is as a by-product from hydrometallurgical processing of limonitic nickel laterites, either by HPAL or heap (atmospheric) leach. The traditional Caron process employs a reducing roast followed by ammoniacal leach, such as at Yabulu, Australia, which produces cobalt as an oxy-hydroxide cake using steam stripping after nickel is recovered. There are several Caron installations worldwide, but the process is less popular now because of high energy requirements and low cobalt recoveries. Over the years, improvements to autoclave design detail and ancillary equipment have progressively facilitated the treatment of laterites by HPAL in preference to the Caron process or pyrometallurgical processes, as demonstrated in recent Australian projects, albeit with varying degrees of success. Several potentially viable alternatives to the HPAL and Caron processes are currently under development. These include nitric acid leach (e.g., Direct Nickel), atmospheric sulfuric acid leach (e.g., BHP Billiton, Ravensthorpe), hydrochloric acid leach (e.g., Jaguar Nickel, Intec, ARNi), and heap leach. New extractants, including Cyanex 301 and Cyanex 302, have been developed that have the ability to extract cobalt and nickel together at a relatively low pH, leaving behind the impurities manganese, magnesium, and calcium (Fisher 2010). An example is the cobalt circuit employed at Vale's Goro project using Cyanex 301 to produce 5,000 t/yr Co as carbonate (Mihaylov et al. 2000).

Driven by increasing environmental pressures on the smelters, hydrometallurgical alternative processes for nickel sulfide concentrates (as opposed to matte) were seriously developed in the 1990s. Several such processes (acid-oxidative leach) are now possible; practical examples include Vale-Inco's current Voisey's Bay project (capacity 2,450 t/yr Co rounds), which employs a sulfate–chloride pressure leach; and the Activox leach process, which was to be implemented at Tati, Botswana (capacity 640 t/yr Co as carbonate) until it was cancelled in 2008. Others include CESL (chlorine-catalyzed pressure oxidative leach), Intec chloride, bioleach (e.g., BioNIC) and Sherritt Gordon's ammonia leach (Collins et al. 2015).

The advantages and disadvantages of many of the processes summarized earlier are discussed at length in the work by Swartz et al. (2009), which attempts to quantify their relative merits in terms of costs, risks, and benefits. Not surprisingly, there is no preferred process, where the high value but higher risk of cobalt metal production by electrowinning, for example, should be weighed against the simplicity, but lower return, of cobalt hydroxide or other intermediates. The interaction between business and technical considerations have been well summarized by Peek et al. (2009) especially with respect to the hydrometallurgical processing of cobalt salts to metal.

SUSTAINABILITY, ENVIRONMENTAL, AND GREENHOUSE GAS ASPECTS

Trends in Sustainability

The industry continues to move toward mine development and operations more aligned with global trends for more sustainable use of resources, including preventing or reducing adverse environmental impacts (air, water, land); reducing a project's overall environmental footprint (including energy and, again, water and land); adopting safe and environmentally conscious practice among the workforce; and working closely with communities in which companies operate during exploration, development, operations and closure to win and maintain license to operate (Corder and Becker 2017).

Approaches such as SUSOP (SUStainable OPerations), which seek to objectively assess potential impacts of any project across five main areas of capital (natural, human, social, manufactured, and financial), are increasingly adopted to identify risks that could potentially impact the project's viability and rank opportunities to improve the contribution of projects and operations to societal sustainability. Obviously, key areas remain energy, water, and land intensity of operations, and, for sulfide ore in particular, atmospheric emissions of sulfur. However, reducing secondary emissions and eliminating negative impacts on adjacent communities become increasingly essential.

Energy and CO_2 Aspects

Technology development in nickel and cobalt extraction has been inextricably linked with energy cost from the time they were first commercially extracted. For example, the development of the Peirce–Smith converter in France was a result of the high cost of fuel there compared to Welsh smelters, reducing coal consumption per ton of copper from 17 kg/t Cu to about 6 kg/t. One hundred years later, the coal equivalent is down to 1 kg/t Cu.

Nickel and cobalt extraction from laterite ores is very energy intensive, mainly because of the low nickel grade of the ore (typically below 3% Ni). In comparison to copper, energy consumption for nickel extraction in 2009 was 17 kg coal equivalent/t Ni, made up of about 7 kg for the process plant and ~10 kg for power generation (Warner et al. 2006, 2007). Much of the process energy is required for the drying/calcination stage as the saprolitic ore contains a considerable amount of hydroxide.

Specific energy reduction can be achieved in two ways. The first is partial upgrading of the ore based on differential sizing, as was practiced at Falcondo (Dominican Republic) for a short time. Doubling the head grade of feed to the process plant effectively halves the specific energy requirement. The alternative method is partial dissolution of (limonitic) ore in sulfuric acid, followed by precipitation of a mixed hydroxide that contains ~60% Ni. This material can be blended with saprolite feed to the FeNi plant, again substantially raising the head grade. From a technology standpoint this method is proven, but economic factors such as the relative cost and availability of acid and soda ash will likely govern the speed of adoption of these hydro/pyro hybrid process routes.

Sulfur Fixation

During the 1970s and 1980s, the industry responded to increasing public concern over smelter emissions of sulfur dioxide to the environment by replacing outdated process equipment, such as multihearth roasters, sinter machines, blast furnaces, and reverberatory furnaces, by processes such as fluidized-bed roasting, electric furnace smelting, and flash furnace smelting, which are more energy efficient and environmentally cleaner. However, although the technology is available to produce sulfuric acid economically from sulfur dioxide in the high-strength, off-gas streams produced by the modem smelting techniques, there is not always a market for large amounts of acid within an economic shipping distance of the nickel smelter. A process to treat pyrrhotite concentrate separately to recover nickel, iron ore, and sulfuric acid was operated by Inco for many years, but the operation was abandoned in the mid-1980s when it became uneconomic because of increasingly stringent sulfur emission limits.

In response to ever tighter constraints on sulfur dioxide emissions from smelters during the 1970s and 1980s, the Canadian companies have made increased pyrrhotite rejection from smelter feed a major priority, and substantial improvements have been achieved. Glencore at its Sudbury operation rejects more than 55% of the pyrrhotite in its ore and recovers sulfuric acid from the sulfur dioxide in the fluidized-bed roaster off-gas stream. Acid plant tail gas is routed through the electric furnace, further reducing sulfur emissions and overall off-gas volumes, and electric furnace and converter off-gases are vented to atmosphere (Clean Air Sudbury 2016). A pyrrhotite concentrate was treated by Inco (now Vale) for many years for nickel and iron oxide recovery, while Falconbridge (now Glencore) roasted pyrrhotite to produce a feedstock for an FeNi kiln operation (Peek et al. 2009). Environmental constraints on sulfur dioxide emissions and the availability of imported high-grade iron ores resulted in the closure of both plants in the late 1980s. However, with stockpiles of pyrrhotite tailings containing ~0.8% Ni and high in sulfur and iron exceeding 100 Mt in Sudbury, future processes may see a resurgence of interest in this material.

Additional reductions by off-gas cleaning using the best available flue gas desulfurization technologies are possible. Available technologies include wet and dry types. The wet processes involve production of wet slurry wastes as by-products. The flue gas leaving the absorbent is saturated with moisture. Regenerative systems use expensive sorbents that are recovered by stripping sulfur oxides from the scrubbing medium. These produce useful by-products including sulfur, sulfuric acid, and gypsum. Recovery systems have generally higher capital costs than throwaway systems but lower waste disposal requirements and costs. The dry processes produce dry waste material, and the flue gas leaving the absorbent is not saturated with moisture. SO_2 recovery efficiency of dry processes (50%–60%) is considerably less than wet processes (93%–98%). However, more recently, dry processes with more than 90% SO_2 removal efficiency are available (Roy and Sardar 2015). For example, Vale's recently commissioned Clean AER Project is the largest single environmental investment in its history, intended to reduce sulfur dioxide emissions by 85% from current levels and metals and particulate emissions by 35%–40% through additional smelter off-gas wet scrubbing stages, electrostatic precipitation, and recovery of metal particulates from scrubbed gas, and additional catalyzation of unreduced sulfur to recoverable forms.

CONCLUSION

Since 1950, the production of nickel and cobalt has increased on average between 3%–4%/yr. Despite an extended depression in demand since the global financial crisis of 2008, current market trends toward increased use of nickel and cobalt in clean energy and battery-electric applications are expected to continue, suggesting long-term recovery in nickel and cobalt prices from recent local minima is possible.

Historically, nickel production has predominantly come from sulfide deposits and operations. However, a trend toward a higher proportion of nickel production from lateritic deposits by the means described in this chapter is increasingly evident. Similarly, cobalt production has predominantly come from the Central African Copperbelt, with minor contributions from cobalt as a by-product from nickel and copper operations globally. Although cobalt production from cobalt-primary sources in Central Africa is expected to continue and increase, satisfaction of increasing cobalt demand will most likely also come from increasing production of cobalt as a by-product of hydrometallurgical treatment of Ni–Co laterites.

Mining, beneficiation, and extractive metallurgical approaches in nickel and cobalt, which developed massively through the 1970s to 1990s, are by now well established, with incremental improvements in operating cost and selectivity of reagents happening on an ongoing basis. The most recent innovations noted have been in the area of environmental remediation of atmospheric sulfur emissions with an increasing and now sustained focus on emission reduction in many major nickel- and cobalt-producing jurisdictions. Perhaps the next wave of innovations, subsequent to the Paris Climate Accord of 2015, will now be in energy efficiency with bold objectives to reduce the energy intensity of metal production by 40% being adopted specifically in Canada and elsewhere among member countries of the Organisation for Economic Co-operation and Development.

REFERENCES

Bakker, M.L., Nikolic, S., and Mackey, P.J. 2011. ISASMELT TSL—Applications for nickel. *Miner. Eng.* 24(7):610–619.

Bamisaiye, O.A., Eriksson, P.G., Van Rooy, J.L., Brynard, H.M., Foya, S., Billay, A., Nxumalo, V., and Adeola, A.M. 2016. Subsurface geometry of the northwestern Bushveld Complex, South Africa. *Can. J. Trop. Geogr.* 3(1):28–36.

Becker, G.S., and Johnson, G.D. 1996. Nickel metal from sulphide concentrates without smelters: A Yakabinde process and project update. Presented at Nickel '96, Kalgoorlie, Western Australia, November 27–29.

Bell, T. 2018. 10 Biggest Cobalt Producers 2014. *The Balance.* May 31. https://www.thebalance.com/the-10-biggest-cobalt-producers-2014-2339726. Accessed October 2017.

Bergman, R.A. 2003. Nickel production from low-iron laterite ores: Process descriptions. *CIM Bull.* 96(1072):127–138.

Bhotoko, P. 2016. Closure of mines worry Francistown Councilors. *Mmegi Online.* March 8. www.mmegi.bw/index.php?aid=58377&dir=2016/march/08. Accessed July 2017.

Blanco, J.L. 1978. Oxygen flash smelting in a converter. *World Min.* 31(March):80–82.

Boldt, J.R., and Queneau, P.E. 1967. *The Winning of Nickel.* Toronto, ON: Longman's. pp. 464-480.

Broughton, D. 2013. Geology and ore deposits of the Central African Copper Belt. Ph.D. thesis, Colorado School of Mines, Golden, CO.

Brugman, C.F., and Kerfoot, D.G. 1986. Nickel metallurgy, Vol. 1. In *Extraction and Refining of Nickel.* Edited by E. Ozberk and S.W. Marcusen. Montreal, QC: Canadian Institute of Mining, Metallurgy and Petroleum. pp. 512–531.

Charland, A., Kormos, L., and Whitaker, P. 2011. A case study for integrated use of automated mineralogy in plant optimization: The Falconbridge Montcalm smelter. In *Extractive Metallurgy of Nickel, Cobalt and Platinum Group Metals.* Edited by F. Crundwell, M. Moats, V. Ramachandran, T. Robinson, and W.G. Davenport. Amsterdam: Elsevier.

CI (Cobalt Institute). 2017. The production and consumption of cobalt in 2017. In *Proceedings of the 24th Annual Cobalt Conference*, Marrakech, Morocco, May 17–18.

Clean Air Sudbury. 2016. Vale Ontario operations flowsheet. In *Clearing the Air Report No. 3—Air Quality Trends in Greater Sudbury, 2008–2014.* https://d3n8a8pro7vhmx.cloudfront.net/greenconnect/pages/149/attachments/original/1521040916/Clearing_the_Air_2016_Editedfeb28v2.pdf?1521040916. Accessed October 2017.

Collins, M., Yuan, D., Sitter, M., and St. Jeans, S. 2015. Sustainable hydrometallurgical processing at the Ambatovy Nickel operation in Madagascar. Presented at the 2015 Conference of Metallurgists, CIM METSOC, Montreal.

Corder, G.D., and Becker, G. 2017. Sustainability made easy: A business improvement case study. Landmark Paper. Melbourne, Victoria: Australasian Institute of Mining and Metallurgy.

Cowden, A. 1988. Emplacement of komatiite in the Kambalda nickel deposit. In *Geology of the Mineral Deposits of Australia and Papua New Guinea*. Edited by F.E. Hughes. Melbourne, Victoria: Australasian Institute of Mining and Metallurgy. pp. 567–581.

Crundwell, F.K., Moats, M.S., Ramachandran, V., Robinson, T.G., and Davenport, W.G., eds. 2011. *Extractive Metallurgy of Nickel, Cobalt and Platinum Group Metals.* Amsterdam: Elsevier. p. 22.

Dalvi, A.D., Bacon, W.G., and Osborne, R.C. 2004. Past and future of nickel laterites projects. In *International Laterite Nickel Symposium 2004.* Edited by D.M. Lane and W.P. Imrie. Warrendale, PA: The Minerals, Metals & Materials Society. pp. 1–27.

Darton Commodities. 2009. *Cobalt Market Review 2009.* Surrey, England: Darton Commodities.

Darton Commodities. 2016. *Cobalt Market Review 2016.* Surrey, England: Darton Commodities.

Desai, P. 2016. Electric vehicles to power cobalt revival. *Reuters.* June 8. https://www.reuters.com/article/metals-cobalt-demand/electric-vehicles-to-power-cobalt-revival-idUSL8N190219. Accessed October 2017.

Diaz, C.M., Landolt, C.A., Vahed, A., Warner, A.E.M., and Taylor, J.C. 1988. A review of nickel pyrometallurgical operations. *JOM* 40(9):28–33.

Diaz, C., Levac, C., Mackey, P., Marcuson, S., Schonewille, R., and Themelis, N. 2011. Innovation in Nonferrous metals Pyrometallurgy—1961 to 2011. In *The Canadian Metallurgical and Materials Landscape 1960 to 2011.* Edited by J. Kapusta, P.J. Mackey, and N.M. Stubina. Westmount, QC: Canadian Institute of Mining, Metallurgy and Petroleum. pp. 333–360.

Doucet, J., Price, C., Barette, R., and Lawson. V. 2009. Evaluating the impact of operational changes at Vale Inco's Clarabelle Mill. In *Proceedings of the 7th UBC-McGill-UA International Symposium on Fundamentals of Mineral Processing*. Montreal, QC: Canadian Institute of Mining, Metallurgy and Petroleum. pp. 337–347.

Eliott, B.J., Koh, P., and Jorgensen, F. 1983. Expansion project and progress in the Sumitomo metal mining Toyo smelter. In *Advances in Sulfide Smelting.* Vol 2. Edited by H.Y. Sohn. Warrendale, PA: TMS-AIME.

Ferron, J. 2001. Arsenides as a potentially increasing source for the cobalt market. Presented at The Cobalt Conference, Surrey, UK.

Fisher, K.G. 2010. Manganese removal in base metal hydrometallurgical processes. Presented at the Alta 2010 Ni/Co Conference, Perth, Western Australia.

Fisher, K.G. 2011. Cobalt processing improvements. Presented at the 6th Southern African Base Metals Conference, South Africa.

Foose, M.P., Zientek, M.L., and Klein, D.P. 1984. *Magmatic Sulfide Deposits.* Reston, VA: U.S. Geological Survey.

Forward, F.A. 1948. Process for recovering nickel and/or cobalt ammonium sulfate from solutions containing nickel and/or cobalt values. U.S. Patent 2647820 A.

Golightly, J.P. 1979. Nickeliferous laterites: A general description. In *Proceedings of the International Laterite Symposium*. Edited by D.J. Evans, R.S. Shoemaker, and H. Veltman. New York: SME-AIME. pp. 3–23.

Head, M.D., Englesakis, V.A., Pearson, B.C., and Wilkinson, D.H. 1976. Nickel refining by the TBRC smelting and pressure carbonyl route. Presented at the 105th TMS-AIME Annual Meeting, Las Vegas, Nevada.

Hoffman, R.R., and Kaiura, G.H. 1985. Smelting process update at Falconbridge Limited Sudbury Operations. Presented at the 114th TMS-AIME Annual Meeting, New York.

Inami, T., Tsushida, N., Komoda, K., and Shigemura, H. 1986. Studies on anodic dissolution of nickel sulphides in the nickel matte hydrolysis. In *Proceedings of Nickel Metallurgy 1986: The Extraction and Refining of Nickel.* Edited by E. Ozberk and S.W. Marcusen. Montreal, QC: Canadian Institute of Mining, Metallurgy and Petroleum. pp. 486–499.

Jackson, M.P.A., Warin, O.N., Woad, G.M., and Hudec, M.R. 2003. Neoproterozoic allocthonous salt tectonics during the Lufilian orogeny in the Katangan copperbelt, Central Africa. *Geol. Soc. Am. Bull.* 115:314–330.

Kapusta, J.P.T., Mackey, P.J., and Stubina, N.M. 2012. Canadian innovations in metallurgy over the last fifty years. In *The Canadian Metallurgical and Materials Landscape 1960 to 2011.* Edited by J. Kapusta, P.J. Mackey, and N.M. Stubina. Westmount, QC: Canadian Institute of Mining, Metallurgy and Petroleum. pp. 469–481.

Kerfoot, G.E., Collins, M.J., Holloway, P.C., and Schonewille, R.H. 2011. The nickel industry in Canada—1961 to 2011. In *The Canadian Metallurgical and Materials Landscape 1960 to 2011.* Edited by J. Kapusta, P.J. Mackey, and N.M. Stubina. Westmount, QC: Canadian Institute of Mining, Metallurgy and Petroleum. pp. 127–152.

Kozyrev, S.M., Komarova, M.Z., Emelina, L.N., Oleshkevich, O.I., Yakovleva, O.A., Lyalinov, D.V., and Maximov, V.I. 2002. The mineralogy and behaviour of PGM during the processing of the Norilsk Talnaki PGE Cu Ni ore. In *The Geology, Geochemistry, Mineralogy and Beneficiation of Platinum Group Elements, Special Volume 54*. Edited by L.J. Cabri. Montreal, QC: Canadian Institute of Mining, Metallurgy and Petroleum.

Lawson, V., and Xu, M. 2011. Float it, clean it, depress it—Consolidating the separation stages at Clarabelle Mill. In *MetPlant 2011: Metallurgical Plant Design and Operating Strategies*. Melbourne, Victoria: Australasian Institute of Mining and Metallurgy. pp. 589–601.

Lenior, P., Van Peteghem, A., and Feneau, C. 1981. Extractive metallurgy of cobalt. In *Proceedings of the International Conference on Cobalt: Metallurgy and Uses*. Brussels: Benelux Metallurgie. pp. 51–62.

Lesher, C.M., and Keays, R.R. 2002. Komatiite-associated Ni-Cu-(PGE) deposits: Mineralogy, geochemistry, and genesis. In *The Geology, Geochemistry, Mineralogy and Beneficiation of Platinum Group Elements, Special Volume 54*. Edited by L.J. Cabri. Montreal, QC: Canadian Institute of Mining, Metallurgy and Petroleum. pp. 579–617.

Llanos, Z.R., Queneau, P.E., and Rickard, R.S. 1974. Atmospheric leaching of matte at the Port Nickel refinery. *CIM Bull.* 67(742):75–81.

Mackiw, V.N., and Veltman, H. 1980. Recovery of metals by pressure hydrometallurgy—The Sherritt Gordon technology. Presented at the International Mining Exhibition and Conference, Calgary, AB, August 26–28.

Maiera, D., Määttääb, S., Yangb, S., Oberthürc, T., Lahayed, Y., Huhmad, H., and Barnese, S.J. 2015. Composition of the ultramafic–mafic contact interval of the Great Dyke of Zimbabwe at Ngezi mine: Comparisons to the Bushveld Complex and implications for the origin of the PGE reefs. *Lithos* 238:207–222.

McCallum, I.S. 1999. The Stillwater Complex: A review of the geology. In *Layered Intrusions Developments in Petrology, Vol. 15*. Edited by R.G. Cawthorn. Amsterdam: Elsevier Science. pp. 441–483.

Merkle, R.W., and McKenzie, A.D. 2002. The mining and beneficiation of South African PGE ores. In *The Geology, Geochemistry, Mineralogy and Beneficiation of Platinum Group Elements, Special Volume 54*. Edited by L.J. Cabri. Montreal, QC: Canadian Institute of Mining, Metallurgy and Petroleum.

Mezei, A., Ashbury, M., and Tood, I. 2009. Recovery of cobalt from polymetallic concentrates—NICO Deposit, NWT Canada—pilot plant results. Presented at CIM Hydrometallurgy of Nickel and Cobalt 2009, Sudbury, ON.

Mihaylov, I., Krause, E., Colton, D.F., Okita, Y., Duterque, J-P., and Perraud, J. 2000. The development of a novel hydrometallurgical process for nickel and cobalt recovery from Goro laterite ore. Presented at the CIM Congress Mining Millennium 2000, Toronto, ON.

Mwatahwa, C., and Musa, C. 2015. *The PGE Resources of the Great Dyke.* Presentation to the Zimbabwean Branch of the SAIMM, Zimplats Selous Metallurgical Complex, December 2.

Naldrett, A.J. 1984. Mineralogy and composition of the Sudbury ores. In *The Geology and Ore Deposits of the Sudbury Structure*. Edited by E.G. Pye, A.J. Naldrett, and P.E. Giblin. Ontario Geological Survey Special Volume I. Toronto, ON: Ontario Ministry of Natural Resources. pp. 309–325.

Newcombe, B., Bradshaw, D., and Wightman, E. 2012. Flash flotation … and the plight of the coarse particle. *Miner. Eng.* 34(July):1–10.

Olaniyan, O., Smith, R.S., and Lafrance, B. 2014. A constrained potential field data interpretation of the deep geometry of the Sudbury structure. *Can. J. Earth Sci.* 51(7):715–729.

Pääkkönen, E., and Mattelmäki, M. 1996. The direct Outokumpu nickel smelting (DON) process and the Harjavalta expansion. In *Proceedings of Nickel '96: Mineral to Market*. Edited by E.J. Grimsey and I. Neuss. Melbourne, Victoria: Australasian Institute of Mining and Metallurgy. pp. 235–241.

Padhy, S. 2017. 10 top countries for nickel production. *Investing News Network*. August 30. https://investingnews.com/daily/resource-investing/base-metals-investing/nickel-investing/nickel-in-the-philippines/. Accessed October 2017.

Peek, E., Akre, T., and Asselin, E. 2009. Technical and business considerations of cobalt hydrometallurgy. *JOM* 61(10):43–53.

Pelletier, B.C. 2003. Les minerais de nickel de Nouvelle-Calédonie. *Revue Géologues* 138. *Spécial DOM-TOM*. September.

Plaskett, R.P., and Romanchuk, S. 1978. Recovery of nickel and cobalt from high grade matte at Impala Platinum by the Sherritt process. *Hydrometallurgy* 3:135–151.

Reid, J., and Barnett, S. 2002. Nickel laterite hydrometallurgical processing update. Presented at Nickel-Cobalt 8, Alta Metallurgical Services, Perth, Western Australia.

Renzoni, L.S. 1943. The electrorefining of nickel. U.S. Patent 2,394,874.

Roy, P., and Sardar, A. 2015. SO_2 emission control and finding a way out to produce sulphuric acid from industrial SO_2 emission. *J. Chem. Eng. Process Technol.* 6:230.

Schoustra, R.P., Kinloch, E.D. and Lee, C.A. 2000. A short review of the Bushveld Complex. *Platinum Met. Rev.* 44(1):33.

Sole, K.C., Feather, A.M., and Cole, P.M. 2005. Solvent extraction in Southern Africa: An update of some recent hydrometallurgical developments. *Hydrometallurgy* 78:5278.

Stenshold, E.O., Zachariasen, H., Lund, J.H., and Thornhill, P.G. 1988. Recent improvements in the Falconbridge nickel refinery. In *Extractive Metallurgy of Nickel and Cobalt*. Edited by G.P. Tyroler and C.A. Landoldt. Warrendale, PA: TMS-AIME.

Svens, K. 2013. By-product metals from hydrometallurgical processes—An overview. Presented at 3rd International Conference: By-Product Metals in Non-Ferrous Metals Industry, Wroclaw, Poland, May 15–17.

Swartz, B., Donegan, S., and Amos, S. 2009. Processing considerations for cobalt recovery from Congolese Copperbelt ores. In *Hydrometallurgy 2009*. Johannesburg: Southern African Institute of Mining and Metallurgy.

Taylor, A. 2013. Laterites—Still a frontier of nickel process development. In *Ni-Co 2013*. Edited by T. Battle et al. Published by Springer, on behalf of The Minerals, Metals & Materials Society.

Thornhill, P.G., Wigstol, E., and Van Weert, G. 1971. The Falconbridge matte leach process. *J. Miner. Metal. Mater. Soc.* 23(7):13–18.

Toguri, J.M. 1975. A review on the methods of treating nickel-bearing pyrrhotite. *Can. Metall. Q.* 14(4):323–338.

Trading Economics. 2018. Cobalt: 10-year price trend chart. https://www.Tradingeconomics.com/commodity/cobalt. Accessed October 2017.

USGS (U.S. Geological Survey). 2017. Cobalt. In *Mineral Commodity Summaries 2017*. Reston, VA: USGS.

Walker, S. 1989. *Int. Min.* 6(November):18–21.

Wang, S., Santander, N.H., and Toguri, J.M. 1974. The solubility of nickel and cobalt in iron silicate slags. *Met. Trans.* 5(January):261–265.

Warner, A.E.M, Díaz, C.M., Dalvi, A.D., Mackey, P.J., and Tarasov, A.V. *JOM* world non-ferrous smelter survey, part III: Nickel: Laterite. *JOM* 58(4):11–20.

Warner, A.E.M., Díaz, C.M., Dalvi, A.D., Mackey, P.J., Tarasov, A.V., and Jones, R.T. 2007. *JOM* World Nonferrous Smelter Survey—Part IV: Nickel: Sulfide. *JOM* 59(4):58–72.

Watanabe, T., Ono, S., Arai, H., and Matsumori, T. 1985. Direct reduction of garnierite ore for production of ferro-nickel with rotary kiln at Nippon Yakin Kogyo Company, Oheyama Works. Presented at International Seminar on Laterites, Mining and Materials Processing Institute of Japan, Tokyo. pp. 103–115.

Wiseman, L.G., Bale, R.A., Chapman, E.T., and Martin, B. 1988. In *Extractive Metallurgy of Nickel and Cobalt*. Edited by G.P. Tyroler and C.A. Landolt. Warrendale, PA: TMS-AIME. pp. 391–402.

Yamada, K., and Hiyama, T. 1985. A process for stainless steel production directly from nickel laterite and chromium ores. Presented at International Seminar on Laterites, Mining and Materials Processing Institute of Japan, Tokyo. pp. 159–168.

CHAPTER 12.26

Oil Shale

Bruce Peachey

As indicated in other chapters, it is unlikely that use of hydrocarbons, including fossil fuels, will disappear from use any time soon. The energy and hydrocarbon building blocks contained in these resources are too valuable to our modern society. Only limited amounts of carbon are available for recycle through short-term biomass resources, and those resources are often too dispersed and of low quality to meet our energy needs. Impurities in raw biomass also make their use for petrochemicals (plastics, etc.) more difficult and energy intensive to use. Meanwhile, the earth's biosphere has been capturing carbon from the atmosphere and oceans, turning it into organic matter and then processing biomass into higher-quality hydrocarbons for billions of years. Once thermally and/or biologically upgraded, this resource constitutes an extremely large and valuable natural source of energy and carbon. One of the largest bodies of partially upgraded natural hydrocarbon resources are found in oil shales.

DEFINITION

Shales are generically defined as "rock formed by deposition of colloidal particles of clay and mud, and consolidated by pressure" (Schwartz 2016). Shale is fine grained and has a laminated structure, usually containing large amounts of sand, which are colored by metal oxides. Unlike sandstones, however, shales are not usually porous, most shales being hard, slate-like rocks. "Oil shale is a hard shale with veins of greasy solid known as kerogen, which is oil mixed with organic matter. Oil shales are widely distributed in many parts of the world and are regularly distilled in most of the countries of Europe" (Brady and Clauser 1991). The yield varies from 60 to 380 L/t (liters per metric ton).

KEROGEN

The organic material *kerogen* is contained in oil shales. When exposed to heat in a retorting process (sometimes referred to as *destructive distillation* or *pyrolysis*), kerogen yields hydrocarbons that can be processed into synthetic crude oil. Oil shales are often called *immature shales* and differ from tight oil-bearing shales, sands, or carbonate formations from which liquids (formed in *mature* shales) or gaseous petroleum (formed in *overmature* shales) are extracted by drilling, fracturing, and pumping for production of liquids. Once the kerogen has been converted to oil or gas, it can migrate into subsurface geologic traps that have higher permeability and generally comprise *conventional* crude oil or natural gas resources. Recently, other *unconventional* shale oil and gas formations have begun to be developed that contain oil or gas already generated from the kerogen, but which remain trapped in the source rock. Potentially mobile oil found in shales is generally referred to as *shale oil*, while similar formations only containing precursor kerogen and/or bituminous material are referred to as *oil shale*. Exact delineation of these two types of resources is made difficult, as many shale formations contain mature oil or gas, as well as some immature kerogen material.

FORMATION

Oil shale rocks containing organic carbon-rich sediments were deposited with inorganic matter when the rocks were formed by microscopic organisms and algae sinking to the bottom of relatively stagnant bodies of water, along with very fine mineral particles. These types of deposits are still forming in deep waters in the Gulf of Mexico and other water bodies that have poorly mixed, low-oxygen (anaerobic), and low-temperature environments, where decomposition is slowed and the recycle of carbon into the active carbon cycle is prevented. Carbon in these source rocks can only be naturally released by (1) weathering, where tectonic forces push the rocks to the land surface and where rainfall, organisms, or direct oxidation can return the carbon to active circulation over billions of years (slow carbon cycle); (2) thermal conversion to liquid or gaseous hydrocarbons (oil and gas) by heat and pressure from deep burial, through subduction into the earth (millions of years); or (3) biogenesis, which is biological conversion to natural gas through slow, in situ biological activity after burial.

Oil and gas produced by thermogenesis or biogenesis over time may return to the surface and the active carbon cycle by migrating through porous rocks. At least some of those fluids and gases are trapped in shallower geologic formations and become reservoirs of conventional oil and gas, and even more is retained in the nonporous shale as shale oil

Bruce Peachey, President & Senior Consultant, New Paradigm Engineering Ltd., Edmonton, Alberta, Canada and gas. As stated earlier, oil shales are generally considered to be a different resource than the shale oil found in source rocks in geologic petroleum systems. The shale oil found in deeper formations, such as the Bakken and Eagle Ford shales, has formed naturally through subduction and heating but has remained trapped within the shale layers, which must be hydraulically fractured to extract the oil. Conversely, in oil shales, much of the carbon remains in the form of kerogen.

Estonia, China, and Brazil today produce commercial quantities of products derived from oil shales and/or electric power from these types of deposits; and Australian deposits have been, and are continuing to be, developed. Commercial production has yet to be attained in the United States, although the largest delineated oil shale deposits in the world exist in Colorado, Utah, and Wyoming. U.S. Geological Survey estimates of technically recoverable oil shale resources in the United States range as high as 2.6 trillion barrels, while an estimated 1.0 trillion barrels may be technically and economically recoverable (reserves are dependent on oil price), making these oil shale reserves potentially four times larger than remaining conventional oil reserves in Saudi Arabia (Dyni 2006). Any figures on oil shale resources should be considered with caution:

- Many potential oil shale formations around the world have not been evaluated.
- Initial estimates tend to be low to be conservative.
- Proven recovery methods have only been applied in a few specific types of deposit and, thus, may not be technically viable in all situations.
- Economic viability is a function of oil price and recovery costs, which are highly variable over time and geographic regions.

Figure 1 is a comparison of the potential size of technically and economically recoverable oil shale reserves in the United States and the remaining conventional oil reserves in the United States and other major producing countries. Figure 2 shows the location of major oil shale deposits in the United States. As one of the largest consumers of oil in the world, the United States has done considerably more work assessing its oil shale resources than other countries have done, so there is considerable uncertainty in global estimates. Often, few wells have been drilled into shales in some countries, so there is little data on which to estimate potential resources.

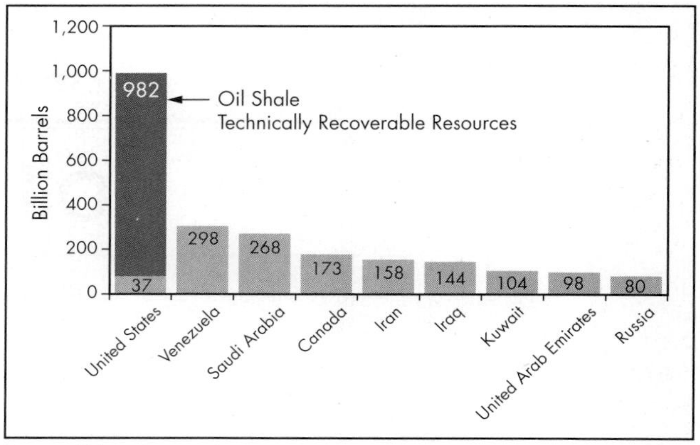

Source: EIA, n.d.

Figure 1 U.S. oil shale resources versus foreign oil reserves

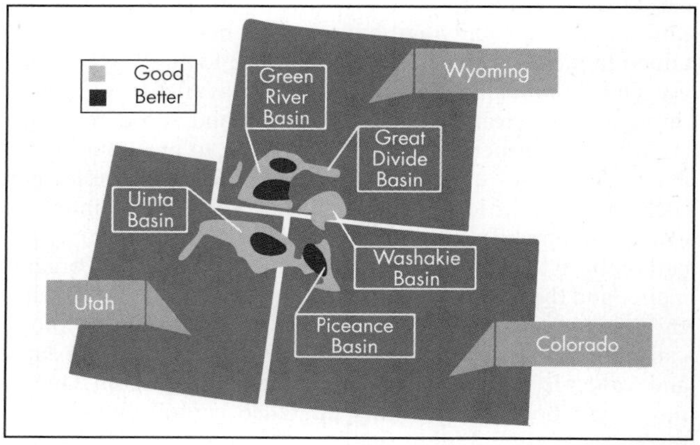

Source: Intek 2011

Figure 2 Major U.S. oil shale resources

KEROGEN TO PETROLEUM AND NATURAL GAS

The kerogen in oil shales is an organic material, whose properties vary depending on where the original organic material was deposited, how the resulting oil shales formed, and what geologic temperatures, pressures, and other conditions they were exposed to over geologic time. The type of kerogen, and the relative composition of the kerogen regarding hydrogen, carbon, and oxygen content, impact the type of petroleum that can be produced from it, either naturally or through processing the oil shale. Recognized generic kerogen types and general properties are shown in Table 1.

Table 1 Kerogen types based on source material

Kerogen Type	Source Material, Composition, and Natural Hydrocarbons Formed	General Environment of Deposition
I	Mainly algae; rich in hydrogen, low in oxygen; oil or gas	Lacustrine (lake)
II	Mainly plankton, some contribution from algae; rich in hydrogen, low in carbon; oil or gas	Marine
II-S	Mainly plankton, some contribution from algae; rich in hydrogen, low in carbon, contains sulfur compounds; sour oil or gas, generation begins earlier (lower temperature)	Marine
III	Mainly higher plants; lower hydrogen, higher oxygen; gas prone	Terrestrial
IV	Reworked, oxidized material; high carbon, hydrogen poor; low potential to generate oil and gas	Varied

Adapted from McCarthy et al. 2011

Natural Maturation Process

Natural thermal maturation of kerogen proceeds through three stages over geologic time as the shales are buried deeper and geothermal temperatures increase. Following is a summary of these stages:

1. **Diagenesis.** At temperatures below 50°C, oxygen, other chemical reactions, and potentially some conversions to methane by methanogenic bacteria gradually convert the organic material to kerogen and smaller amounts of bitumen.

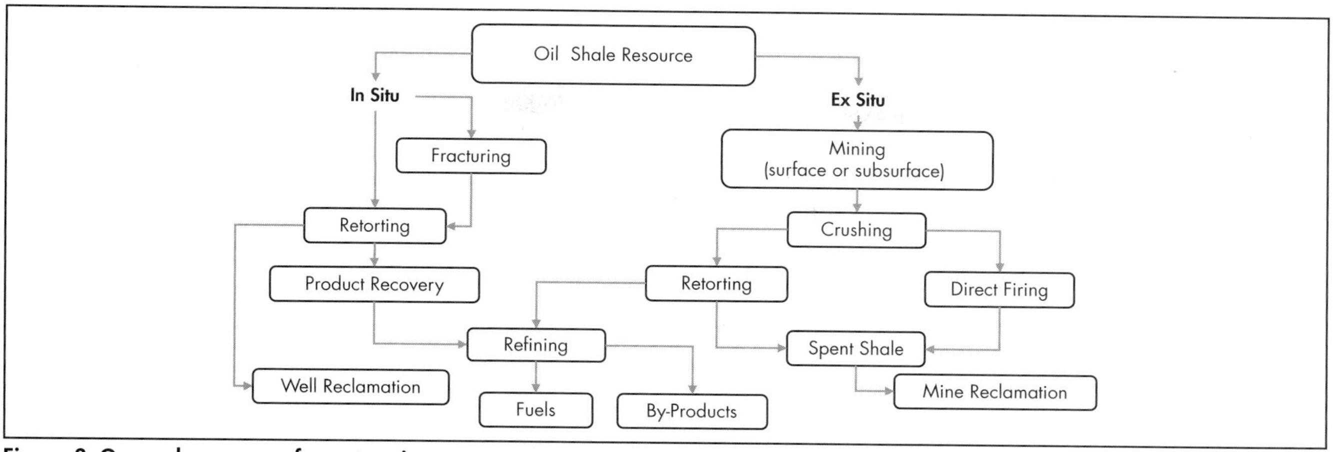

Figure 3 General processes for extracting energy

2. **Catagenesis.** At temperatures between 50° and 150°C, chemical bonds are broken and petroleum is generated in an *oil window*. Further temperature increases and secondary *cracking* of the oil produce wet gas and condensates.
3. **Metagenesis.** Further heating to 150°–200°C converts any remaining kerogen into a solid carbon residue and dry methane, which may also contain carbon dioxide (CO_2), nitrogen (N_2), and hydrogen sulfide (H_2S).

Processing to Energy Streams

Generation of petroleum and gas through natural processes obviously takes a considerable length of time. Even in rocks at temperatures hot enough for conversion to occur, only a small percentage of the kerogen is converted, and it is still difficult to generate enough permeability (through fracturing) for the products to flow into a trap or well. Artificially enhancing the maturation process of the kerogen, either in or ex situ after mining, can accelerate and enhance the natural process and generate higher petroleum yields from a given volume of oil shale. As in nature, processing the kerogen to get liquid, gaseous, and solid fuels is accomplished mainly by heating the oil shale. The heat energy added replaces the natural energy applied to the petroleum source rock deposits over geologic time. The quality and quantity of produced fuels depend not only on the shale properties but also on the retorting process used. Each shale in a given process produces different products and volumes, depending on the type of kerogen it contains, its composition, and the degree of natural maturation that has already occurred. Numerous retorting processes have been developed and categorized into general types based on the method of exposing the raw shale to retorting temperatures. The synthetic crude oil obtained from retorting oil shale is transported to refineries as an acceptable substitute for conventional crudes to produce commercial fuels and petrochemical feedstocks.

The two main categories of oil shale processing are differentiated by how the oil shale is being accessed, with some shallower deposits accessed by mining (surface strip mining or underground mining), but most of the resource requires the use of in situ processing methods if they are too deep and hazardous to mine. Mine-based processes have been the most common methods used to date on oil shales that locally outcrop at surface or near surface, and they have been used for hundreds of years. More recently, a great deal of effort has been devoted by some major oil and gas producers to assess various in situ retorting methods that can be applied to shale oil and oil shale formations after initial depletion of any naturally matured petroleum or gas. Figure 3 shows the difference in the two basic options of in and ex situ extraction. They may proceed to similar end points as refined petroleum products, but they have significantly different technologies, economics, and environmental and sustainability impacts and are applied to oil shale resources at different depths. Ex situ processes are preferred for shallow oil shale resources, while in situ processes tend to be favored for deeper deposits.

EX SITU PROCESSING

Basic extraction of shallow oil shale resources has traditionally focused on mining from either surface (open pit mines) or underground mines. The criteria used to decide on the type of mine operations is similar to that used for coal resources and is a function of ore depth, seam thickness and quality, and the ratio of overburden to ore, which drives mining economics. Historically, oil shale has been mined and used as fuel and/or petrochemicals for centuries, but usually cannot compete economically with lower-cost coal and oil energy resources. Currently, all commercial oil shale operations are based on mining, with ~70% of those operations located in Estonia, with other commercial operations found in China and Brazil, and new large-scale operations actively being considered for Jordan. There have been a range of smaller pilot and demonstration projects in the United States, Australia, and other countries; however, the recent development and rapid growth of hydraulic fracturing to produce shale oil, and the subsequent drop in international oil prices, have reduced interest in oil shales in many locations (Figure 4).

Estonia is unique in that it has few coal, oil, or gas resources of its own, and the low-lying and flat topography of this small country makes hydropower and many other renewable energy sources uneconomic. It does, however, have extensive oil shale deposits that have been exploited since before World War II and have allowed it to be independent of energy imports. In 1997, oil shale provided more than 75% of the country's total primary energy supply, primarily for electrical power generation through direct firing, with smaller volumes of oil shale directed toward production of liquid distillates and

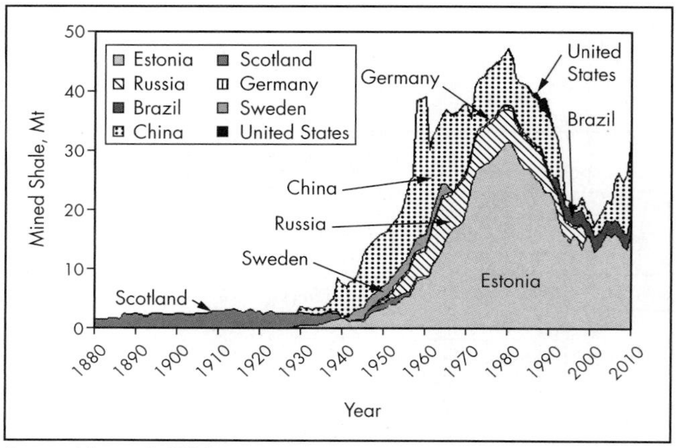

Source: Allix et al. 2010

Figure 4 Oil shale production from 1880 to 2010

Table 2 Size ranges for crushed shale by ex situ process

Process	Crushed Particle Size Range or Description
Oil shale circulating fluidized-bed boiler	Small, 6–12 mm
Kiviter	Coarse oil shale
Fushun	10–75 mm
Petrosix	Lump shale, 12–75 mm
Galoter	Fine, <25 mm
Alberta Taciuk	Fine, <25 mm
Tosco II	<13 mm
Lurgi–Ruhrgas	Fine, 6–13 mm
Chevron staged-turbulent-bed retorting	Small
Union	3–50 mm
Superior multimineral	First stage, <200 mm; second stage, 76 mm for retort feed

petrochemicals through surface retorting processes of various types (U.S. Department of Energy 2003). Like Estonia, China's commercial oil shale operations began in the 1930s for liquid fuels but were largely displaced with the discovery of conventional oil resources from domestic oil fields and imported from the Middle East and other sources. The oil price increases in the 1990s encouraged China to restart oil shale production in 1992, and it is now the largest producer of petroleum from oil shale. Brazil's national oil company, PetroBras, started commercial production of liquid fuels by retorting oil shales in 1992.

Crushing

Crushing of the mined ore is a common feature of ex situ oil shale plants; however, different processes have been designed based on different particle sizes or size ranges, so crushing is a key process consideration. The various oil shale utilization processes are discussed later in this chapter; Table 2 indicates the particle sizes commonly used in those processes.

Power Plants

The largest single use of oil shales globally is for power generation in Estonia, where more than 2,000 MW of capacity of shale oil–fueled power-generation capacity is installed. Originally, the plants were pulverized combustion plants, at medium or high pressure, which used oil shale in a manner similar to coal-fired plants in the rest of Europe. The high mineral content of the shale produced greater volumes of fly ash, some of which was initially used for other purposes, while the high concentration of limestone in the oil shale helped to moderate sulfur oxide (SO_x) emissions. In a drive to increase the economic and environmental performance of power generation, a modified circulating fluidized-bed (CFB) boiler technology was developed for oil shale in Estonia, with commercial operation starting in 2005 (Figure 5).

Retorting Plants

The other commercial processes utilizing oil shale feed are all aboveground retorting operations of various types. These processes often have to be custom designed to match the properties of a specific shale source. Commercial plants currently in operation are classified as either

1. Internal combustion units,
2. External hot gas,
3. Hot recycled solids, or
4. Some hybrid form of those three basic types.

Following are descriptions of the four commercial processes and other systems that are in the demonstration or piloting stage of development.

Internal Combustion

For internal combustion processes, crushed or pulverized feed is fed into the top of a retort and the raw oil shale is heated by the combustion of produced gas and spent residue within the retort at the bottom of the reactor. The vapors generated are condensed, with noncondensable gases fed back into the reactor as fuel. The two commercial oil shale retort processes of this type were developed and first used in the 1920s (Estonia) and 1930s (China), as were some of the first large-scale commercial uses of oil shale.

Kiviter process. Used in Estonia in many retorts, the Kiviter process was previously used for oil shale processing in Russia. The largest unit processes about 960 t/d (metric tons per day) of oil shale (Figure 6).

Fushun process. Similar to the Kiviter process but with a slightly different hot gas flow direction, the Fushun process has some aspects of an external hot gas process. It was developed in China in the 1920s and first commercialized in the 1930s. At one point, there were more than 250 retorts in operation in the 1950s, producing 100–200 t/d of shale oil. Once heavy oil was discovered in China, oil shale production became less economic and was shut down in the 1990s, but was soon restarted and grew to where China is the largest shale oil producer in the world from mined oil shales.

Externally Generated Hot Gas

Externally generated hot gas retort processes are based on generating hot gases at 600°C in an external process unit and injecting these hot gases into the midpoint of the retort vessel for pyrolyzing the oil shale. Another cool gas stream is injected into the bottom of the retort vessel to recover heat from the spent shale before the solids are discharged. Oil and gases are removed from the top of the retort vessel where oil droplets are collected in a precipitator. The gas is used to heat

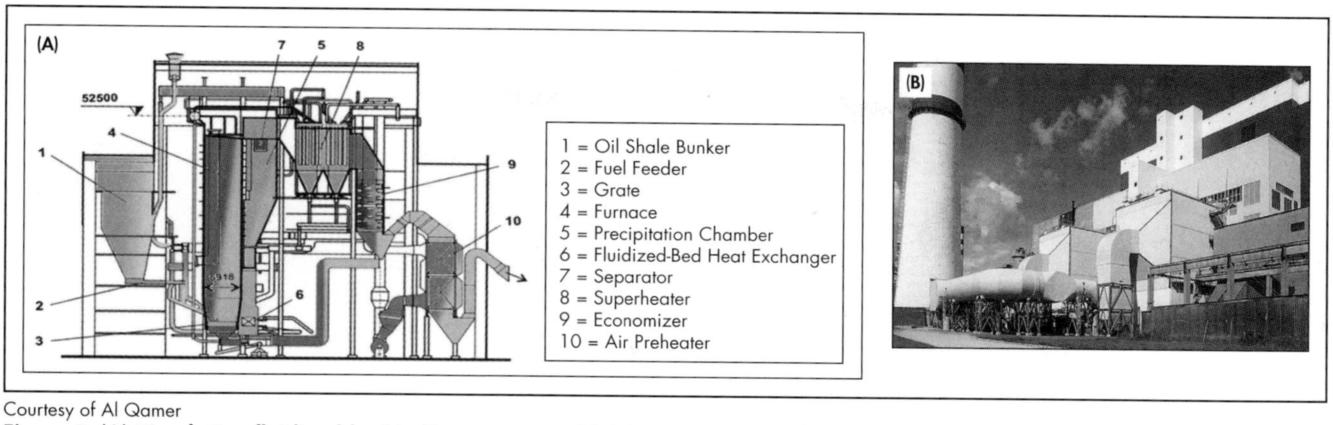

Courtesy of Al Qamer

Figure 5 (A) Circulating fluidized-bed boiler process at (B) AS Narva power plant, Estonia

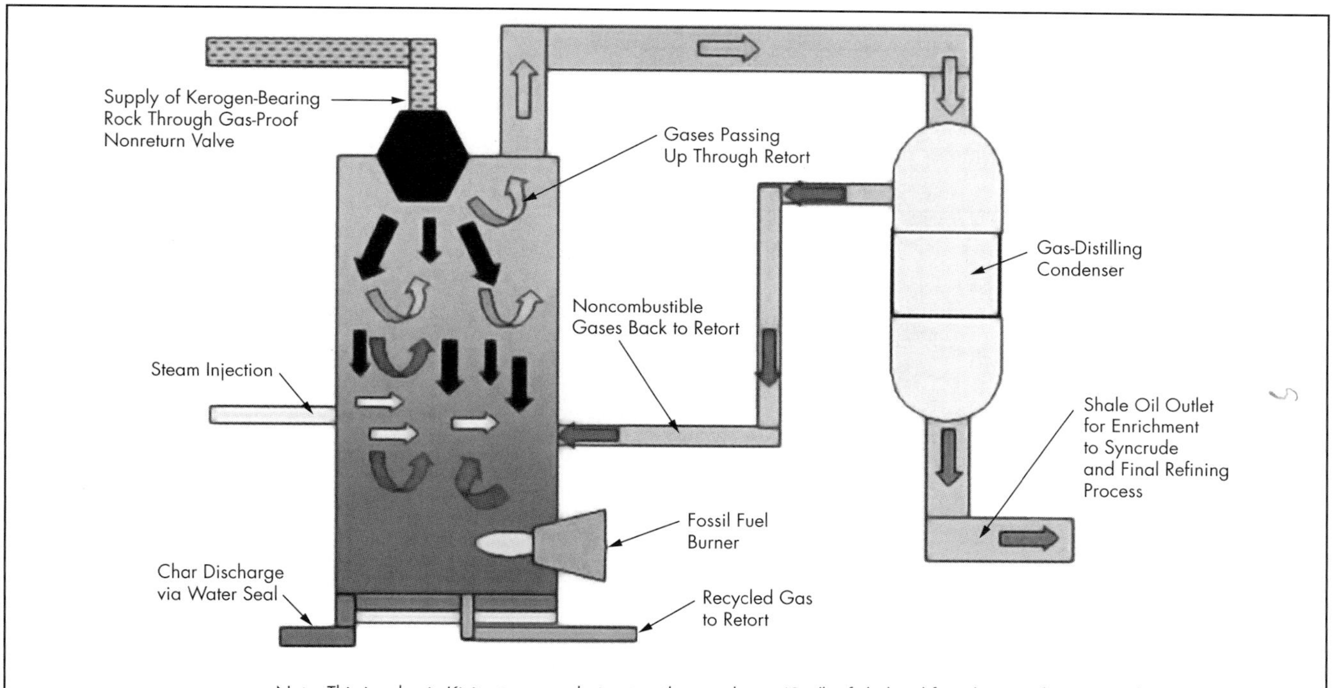

Note: This is a basic Kiviter-type pyrolysis retort that produces 40 t/h of shale oil from kerogen-bearing rock.

Source: Scott 2011

Figure 6 Kiviter process

and cool the retort process, while a side stream of the gas is further cooled to produce naphtha, liquefied petroleum gas, sulfur, and fuel gas for the process. A downside of this process is that the energy in the char in the spent shale is not recovered.

Petrosix, or Petro6, process. Petrobras' Petrosix operation has the world's largest pyrolysis retort vessel with an 11-m-diameter vertical shaft kiln. Work on this process started in the 1950s, with an initial limited commercial operation starting in 1980 and the current large unit brought into commercial operation in 1992. The dual reactors in the plant process 8,500 t/d of shale on a combined basis to produce petroleum products and sulfur.

Hot Recycled Solids

In hot recycled solids processes, much of the energy transfer is accomplished through cross exchange of input ore with hot recycled spent shale. Many processes are based on this concept to increase efficiency and conversion. The commercial operations using this basic method are located in Estonia, but they incorporate some variations and modified processes in the commercial plants.

Galoter or Petroter retort process. The Galoter retort process uses an inclined hot cylindrical rotating kiln as the retort vessel. Solid oil shale particles are separated from the dried crushed feed in a cyclone separator. A mixing chamber combines the feed with hot recycled ash from combustion of the spent shale. Then the mixture is added to the kiln where

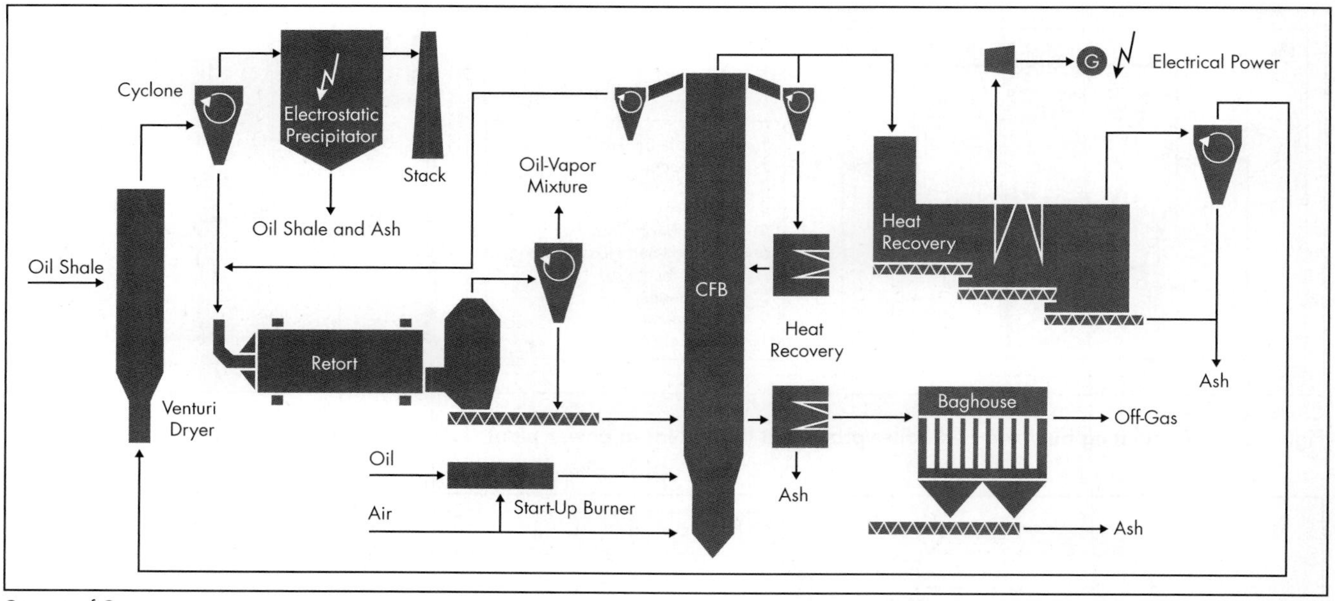

Courtesy of Outotec
Figure 7 Enefit variation of the Galoter process

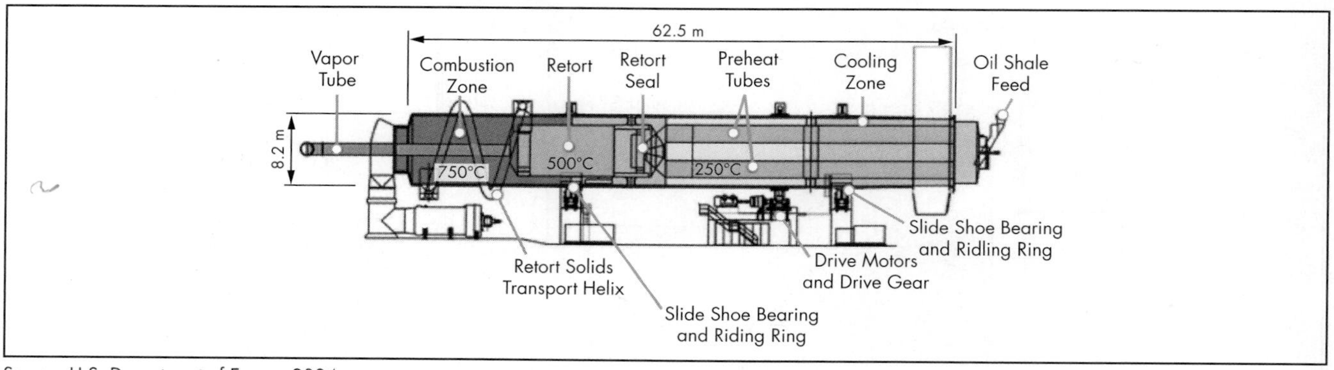

Source: U.S. Department of Energy 2004
Figure 8 Alberta Taciuk processor

the oil and gas vapors are removed and condensed to products. The spent shale is burned in an external furnace with some solids returned to the mixer while the remainder are cooled and sent to disposal. The hot gases from combustion are used to dry and preheat the feed stream. The process has a high thermal efficiency and high oil recovery. Commercial-scale units were first online in the 1950s and 1960s, but were shut down and replaced with two larger 3,000-t/d units in 1980. Between 2009 and 2015, three additional units with slight modifications were built but were called Petroter plants. All units are in Estonia.

Outotec Enefit retort process. The Enefit280 process (Figure 7) was developed based on the Galoter process by combining it with a CFB combustor to replace the spent shale furnace in the Galoter process. One 280-t/d unit is in operation in Auvere, Estonia, which is colocated with the two 140-t/d Galoter units.

Other Noncommercial Retorting Technologies

The three basic retorting processes have been discussed; however, other processes have been proposed, piloted, or demonstrated to attempt to either improve conversion efficiency, reduce costs by reducing the number of units and complexity, or process lower-quality or more variable feeds or end products. Most of these have been piloted or demonstrated at some level and are briefly described in the following paragraphs. The only one that is discussed in a bit more detail is the Alberta Taciuk processor (ATP), which has been widely tested with a range of feedstocks and may be on the verge of being commercialized on a larger scale.

Alberta Taciuk process. Originally developed for minable oil sands in Alberta, Canada, the ATP went through extensive development and testing in the late 1970s. Its first commercial use in the late 1980s was for cleanup of contaminated soils. From 1999 to 2004, it was used at the Stuart Oil Shale Project in Australia, but the plant has since been closed and dismantled. The process has been considered for many other applications in Estonia, China, and Jordan. The ATP is

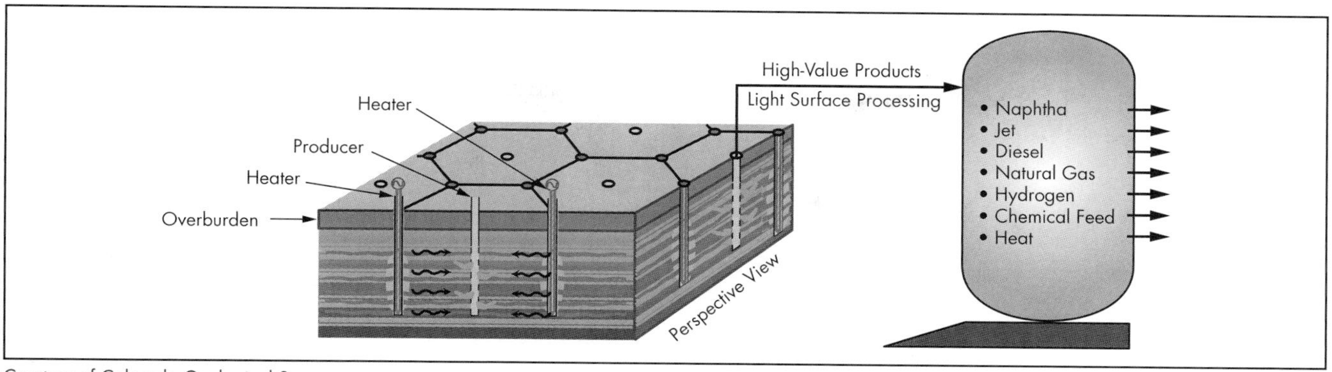

Courtesy of Colorado Geological Survey

Figure 9 In situ conversion process and wells

a hot recycled solids method and its main feature is that most of the process steps occur within a single rotating horizontal retort, through the use of multiple chambers. This provides energy efficiency and high yields with process simplicity and a robust design (Figure 8).

Tosco II process. A hot recycle solids technology, Tosco II uses hot ceramic balls to transfer energy from the heater to the retort vessel, which is a rotating kiln. The process was tested near Parachute, Colorado (United States), in the late 1960s to early 1970s.

Lawrence Livermore National Laboratory (LLNL) hot recycled solids (HRS) process. The LLNL HRS process uses a vertical retort vessel but with hot spent shale and feed mixed in a fluidized-bed mixer. It was tested in Parachute, Colorado, in the mid-1980s.

Lurgi–Ruhrgas process. An HRS technology, the Lurgi–Ruhrgas process is also used for lignite coal liquefaction and was developed in the 1940s. Plants liquefying coal operated in Japan, Germany, the United Kingdom, Argentina, and Yugoslavia, with the plant in Japan also cracking petroleum. Pilot-plant runs using oil shales were done in Frankfurt in the late 1960s and early 1970s.

Kentort II. An HRS concept, the Kentort II was developed by the Center for Applied Energy Research at the University of Kentucky (United States) in the mid-1980s. It is unique in using four cascaded fluidized-bed vessels.

Chevron staged-turbulent-bed (STB) retorting process. An HRS technology, the Chevron STB process uses a vertical fluidized bed as a retort vessel.

Paraho process. With an internal combustion method similar to the Kiviter and Fushun retorts, the Paraho process has different designs for the feed system, gas distribution, and solids discharge. It has direct and indirect heating modes. Pilot operation is located near Rifle, Colorado, and a unit was put into operation in 2011 at the Stuart Oil Shale Project in Australia.

Union process. With a vertical shaft retort, similar to Kiviter and others, the main difference in the Union process is that a *rock pump* is used to add feed shale from the bottom of the retort vessel. It can be heated with direct (Union A) or indirect (Union B) heat. It was developed by Unocal in the 1940s. Union A was tested in the late 1950s, and Union B was tested in Colorado in the late 1980s.

Superior multimineral process. Also known as the *McDowell-Wellman process*, shale oil production is combined with minerals processing to recover aluminum, sodium bicarbonate, and sodium carbonate. Minerals are separated after an initial crushing stage, with the oil shale crushed further prior to the retort vessel. The retort vessel is shaped like a donut with horizontal segments and a traveling grate.

IN SITU PROCESSING

The second major option for accessing and converting oil shales is through in situ (in-place) methods, which are a more recent development, beginning about the 1980s. These methods target deep, thick shale formations that may initially produce shale oil but still contain large amounts of kerogen and shale oil trapped by residual kerogen blocking the fine pores in the shale rock. Directionally, these methods provide access to a much larger volume of oil shale, can be applied incrementally because there is no major investment in mine development, and should result in lower land, air, and water impacts. None of these processes are yet commercial, so it is unclear which process is most likely to proceed. The Shell in situ conversion process (ICP) seems to be furthest along and has been piloted in both oil shale and oil sands formations, with pilot activities still underway. This section provides a basic description of the main processes being considered; however, there may be other methods under development. As with surface processes, all methods require heating the in situ kerogen to temperatures above 300°–350°C to accelerate maturation of the kerogen in the oil shale rock. Temperatures needed are generally lower than surface retorts because residence time and vessel size are not as critical to the economics.

In Situ Conversion Process

The main characteristic of the ICP method is the use of electric heaters in drilled wells, which heat the surrounding oil shale (Figure 9). Initial trials of the process in the Piceance Basin in the United States started in 2004. A unique aspect of the project is the use of subsurface *ice walls* to isolate the test volume from surrounding groundwater aquifers and to allow the shale to be dewatered before heating begins. The ice walls require surface refrigeration systems on relatively close spacing. Vertical wells are then drilled within the ice wall perimeter, and heaters are installed to heat the rock to ~350°C. This slowly converts the kerogen into shale oil and gases, which can then be produced from a pattern of producing wells. With vertical heating and producing wells in close proximity, the initial pilot had a large footprint. ICP requires further development to address issues related to shales, such as low permeability and water impacts.

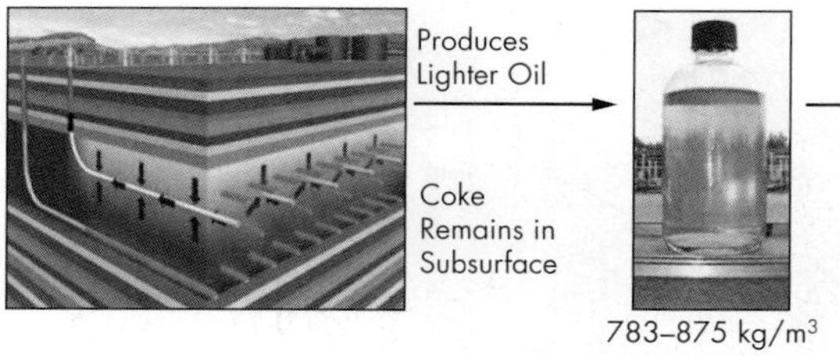

Adapted from Shell Canada Limited 2014

Figure 10 In situ upgrading project

In 2004–2010, Shell tested a similar process at their Viking pilot in the Peace River oil sands in Alberta, Canada, which the company called an "in situ upgrading project." This was followed by a pilot in the Grosmont bitumen-in-carbonate formation, also in Alberta (Shell Canada Limited 2014). In this case, ice walls were not required and heaters were installed in layers of horizontal wells alternating with layers of producing wells (Figure 10). A second project in the Grosmont carbonate bitumen formation has continued work mainly on developing and proving downhole heater technology while upgrading bitumen to lighter fractions. This work, in parallel with the oil shale work, should be complementary but may result in bitumen conversion moving ahead before oil shale conversion.

Electrofrac Process

ExxonMobil's candidate technology for in situ oil shale retorting is the Electrofrac process (Figure 11). The process is designed to heat oil shale by creating hydraulic fractures in the oil shale and propping the fractures open and injecting a proprietary electrically conductive material. Electricity flowing through the conductive medium turns the fracture into a resistive heating element which gradually heats the shale and converts the kerogen into oil and gas that can then be produced by conventional means. This process is supported by ongoing research and technology development for production of shale oil and gas. ExxonMobil indicates that field tests have been initiated to test the concept, but it will be many years before the technology can be shown to be technically, environmentally, and economically feasible.

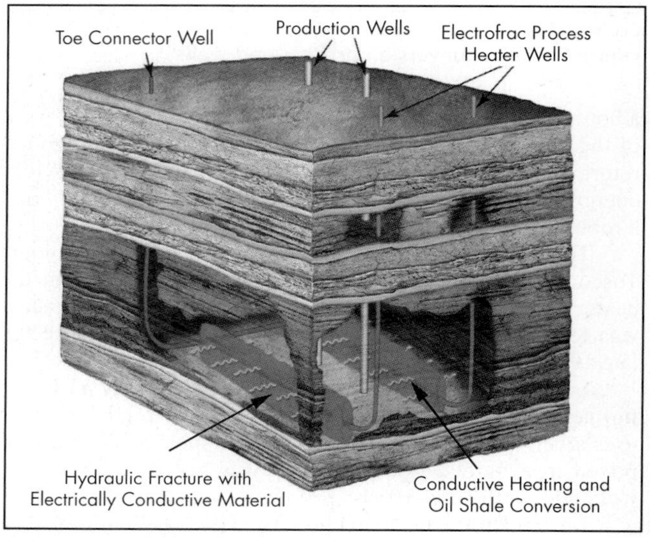

Source: ExxonMobil 2007

Figure 11 Electrofrac process

CRUSH Process

*C*hevron's technology for the *R*ecovery and *U*pgrading of oil from *SH*ale (CRUSH) is based on fracturing, or preferably *rubblizing*, the shale to increase the surface area of oil shale kerogen exposed to flow through the fractures (Figure 12). High-temperature gas, preferably carbon dioxide, is injected through the fractures, then reheated at surface or by being passed through previously depleted portions of the shale zone. There does not appear to be much recent activity on this technology, and Chevron appears to be mainly focused on shale oil developments.

RETORTING BY-PRODUCTS

Some retorting processes can produce by-products from the retorting process that may be of value. All retorting processes produce ash from the combustion of spent shale, which could be used in cement as was done for many years in Estonia. Depending on the composition of the oil shale, other by-products may be natural gas, carbon dioxide, sulfur, aluminum, ammonia, asphalt, or other hydrocarbons.

ENVIRONMENTAL IMPACTS

Production and processing of oil shales can have significant environmental impacts depending on the shale composition, the extraction process used, and the region in which the operation occurs.

Greenhouse Gas Emissions

Greenhouse gas (GHG) emissions vary widely depending on the process. Power generation in Estonia is treated the same as a coal power plant when compared to European emissions. Limestone in the oil shale feed can reduce SO_x emissions but increase CO_2 emissions. One analysis (using oil shale with an average grade of 125 L/t) shows GHG emission ranges, in kilograms of CO_2 per cubic meter of product, for six processes of 264–918 kg/m^3 for combustion; 0–2,214 kg/m^3 as a result of carbonate; and 69–88 kg/m^3 for the pyrolysis step, resulting in an overall total CO_2 range of 352–2,906 kg/m^3 (Burnham and McConaghy 2006). The low end of the range provided is for the Shell ICP process, assuming downhole methane burners are used for heating. This could be lower if electrical heaters were used with power from nuclear or renewable sources. These numbers do not include the life-cycle emissions for

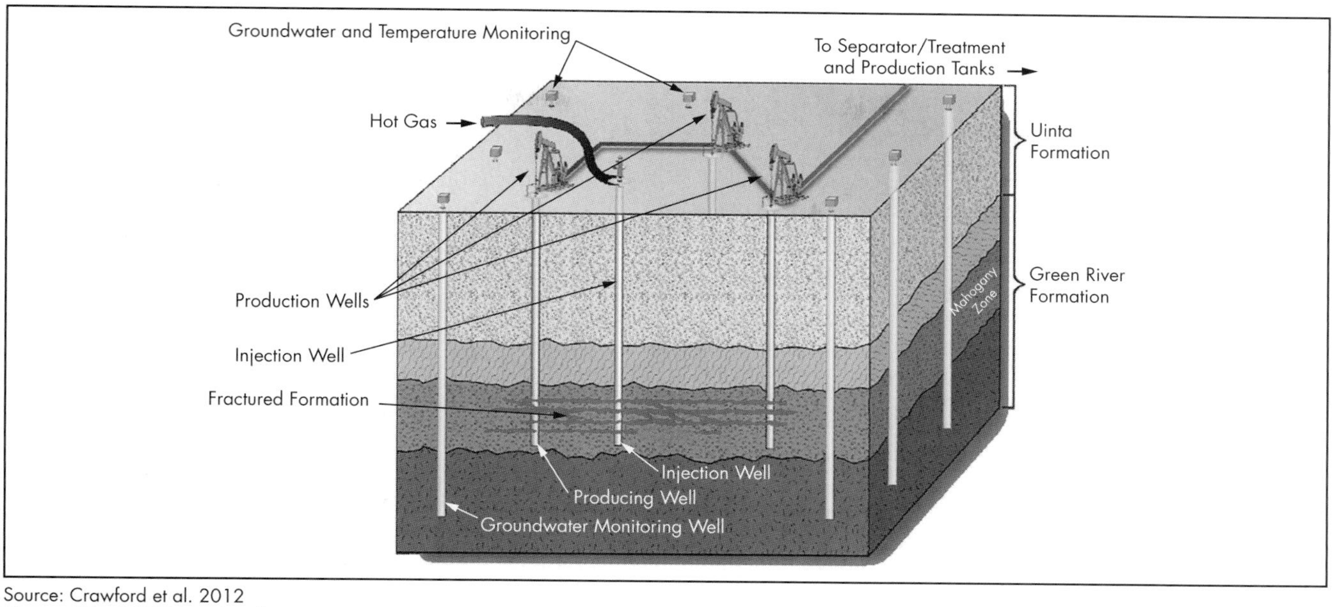

Source: Crawford et al. 2012

Figure 12 CRUSH technology

mining the oil shale or drilling access wells, and thus give an indication of relative GHG emissions only. As with other large-point emission sources, carbon capture technologies might be utilized to mitigate these impacts.

Other Air Emissions

Depending on the shale, processing may produce SO_x emissions, volatile organic compounds, and high-particulate emission levels into the air, which could impact workers, nearby residents, and other industries. These may, in some cases, be similar to coal power plant emissions, other types of mines, refineries, or in situ production operations. They may be better or worse than other energy processes, so there is no clear generic statement that can be made about air emissions from oil shale operations.

Water Use and Contamination

The impact on water use and contamination varies widely across the various shales and extraction processes. All oil shale mining operations tend to impact groundwater and surface drainage. Power-generation water use is likely similar to an equivalent-sized coal-fired plant. Internal combustion retorting plants tend to use a lot of water for quenching the spent shale, and minerals found in the spent shale can also contaminate the water. In contrast, the hot recycled solids processes mostly operate dry, so there is little water consumption, and minerals in the spent shale are less soluble when contacting water at lower temperatures. For in situ conversion processes, the main water concerns are with any water used for hydraulic fracturing, flow-back treatment, and the protection of underground aquifers if the shale is shallow or as a result of potential well failures because of high temperatures and corrosive conditions in the production wells and gathering lines.

Land Impacts

As with other energy resource operations, oil shale operations have land impacts from mining and well development footprints, as well as the processing units, truck haulage, pipelines, and waste disposal operations associated with the chosen extraction method. Land impacts are also affected by the geographic location of the operations. Sensitive landscapes or ecosystems should be considered when looking at potential impacts.

REFERENCES

Allix, P., Burnham, A.K., Fowler, T., Herron, M., Kleinberg, R., and Symington, B. 2010. Coaxing oil from shale. *Oilfield Rev.* 22(4):4–15.

Brady, G.S., and Clauser, H.R. 1991. *Materials Handbook*, 13th ed. New York: McGraw-Hill.

Burnham, A.K., and McConaghy, J.R. 2006. Comparison of the acceptability of various oil shale processes. Presented at the 26th Oil Shale Symposium, Golden, CO, October 16–18.

Crawford, P.M., Dean, C., Stone, J., and Killen, J.C. 2012. Assessment of plans and progress on U.S. Bureau of Land Management oil shale RD&D leases in the United States. www.energy.gov/sites/prod/files/2013/04/f0/BLM_Final .pdf. Accessed September 2018.

Dyni, J.R. 2006. *Geology and Resources of Some World Oil-Shale Deposits.* U.S. Geological Survey Scientific Investigations Report 2005-5294. Reston, VA: U.S. Geological Survey.

EIA (U.S. Energy Information Administration). n.d. International energy statistics. www.eia.gov. Accessed September 2018.

ExxonMobil. 2007. *Financial and Operating Review.* Irving, TX: ExxonMobil.

Intek. 2011. *Secure Fuels from Domestic Resources: Profiles of Companies Engaged in Domestic Oil Shale and Tar Sands Resource and Technology Development*, 5th ed. Report for the U.S. Department of Energy, Office of Petroleum Reserves, Naval Petroleum and Oil Shale Reserves. Arlington, VA: Intek.

McCarthy, K., Rojas, K., Niemann, M., Palmowski, D., Peters, K., and Stankeiwicz, A. 2011. Basic petroleum geochemistry for source rock evaluation. *Oilfield Rev.* 23(2):32–43.

Schwartz, M. 2016. *Encyclopedia and Handbook of Materials, Parts, and Finishes*, 3rd ed. Boca Raton, FL: CRC Press.

Scott, W. 2011. Crude oil from unconventional sources. www.brighthubengineering.com/manufacturing-technology/68900-crude-oil-from-unconventional-sources/. Accessed September 2018.

Shell Canada Limited. 2014. *Field Production and Heater Test Project—Amended Experimental Scheme Approval 11487A Progress Report: Long Horizontal Heater Test (LHT)—2013 Performance Review to AER.* www.aer.ca/documents/oilsands/insitu-presentations/2014AthabascaShellGrosmont11487.pdf. Accessed September 2018.

U.S. Department of Energy. 2003. An energy overview of the Republic of Estonia. www.geni.org/globalenergy/library/national_energy_grid/estonia/EnergyOverviewofEstonia.shtml. Accessed September 2018.

U.S. Department of Energy. 2004. *Strategic Significance of America's Oil Shale Resource.* Vol. II, Oil Shale Resources, Technology and Economics. Washington, DC: U.S. Department of Energy.

CHAPTER 12.27

Oil and Tar Sands

David Harbottle, Jan Czarnecki, Jacob Masliyah, and Zhenghe Xu

Despite great effort in developing alternative and renewable energies to mitigate greenhouse gas emissions, fossil fuels will remain a major mix of our energy strategy for many years to come because of their high-energy intensity and proven utilization with existing infrastructures, in particular considering the rapid increase in population of developing countries and steady improvement in quality of life. Unfortunately, the steady and increased consumption of conventional liquid fossil fuels (oils) will inevitably deplete these attractive high-value energy resources. For the security of future energy demand, we are forced to exploit unconventional fossil fuel resources, most notably shale gas, oil (tar) sands, and gas hydrates. With more than 300 billion barrels of ultimate potential recoverable reserve of heavy oil (bitumen) locked in place in oil sands of northern Alberta (Canada) alone, development of this vast energy resource would alleviate the pressure of depleting conventional liquid fossil fuels for a significant period, buying time to develop new technologies and materials for renewable energies, most notably solar energy. Although the largest and most mature (produced reserves) oil sands reserves are found in Canada, there are several other significant global oil sands deposits in Venezuela, the United States, Trinidad, Russia, Nigeria, Madagascar, and South-Eastern Europe (Romania and Albania). Of the major accumulations, the vast majority of the known initial in-place volumes occurs in Canada and Venezuela. Current estimates of global reserves are not readily available, although orders-of-magnitude assessments can be made from past reviews (Scott 1967; Walters 1974). At present, Canada produces approximately 2.5 million barrels of oil per day from oil sands reserves. This chapter provides an overview of the technologies used to unlock the oil sands energy resources, mainly focusing on minable oil sands where mineral processing technologies are heavily used, starting with an overview on oil sands resources and major properties of bitumen as they are highly pertinent to the processing technologies.

As shown in Figure 1, Alberta oil sands formation is a result of aerobatic biodegradation of conventional petroleum as it migrated updip from deeper sections of the pre-Cretaceous formation of the Western Canada Sedimentary Basin, to the Alberta Basin where it was trapped in the permeable sands of the Fort McMurray formation between the Devonian carbonate below and Cretaceous Clearwater Formation shale above. Based on the environment of sediment deposition, oil sands formation in Fort McMurray can be conveniently divided into three subsections: Lower (continental fluvial), Middle (tidal estuary), and Upper (tidal marine), with increasing content of marine clays in the formation. The biodegradation over millions of years led to the formation of heavy oil known as crude bitumen. On average, bitumen has an in situ viscosity >100 Pa·s and density of about 998 kg/m^3 (an API gravity <10°), which is similar to the density of water. It is important to note a power law dependence of bitumen viscosity to the temperature and roughly a linear decrease in bitumen density with increasing temperature. Such physical properties of bitumen require the bitumen recovery process to operate at elevated temperatures typically greater than 40°C and by air-assisted flotation as described in the following sections.

OVERALL PROCESSING OF MINABLE OIL SANDS

The processing flow sheet of minable oil sands is shown in Figure 2. The unit operations involved in the processing are described in more detail later in this chapter. Here, only the overall bitumen recovery process and connections between the unit operations are described. Oil sands ores typically contain about 10 wt % bitumen and 3–6 wt % of connate (formation) water, with the remaining being a mixture of silica sand and clays including kaolinite, illite, and small amounts of other clay minerals. After removing overburden in the mine, the ore is mined with electric or hydraulic shovels and transported to the slurry preparation unit by trucks. Economy of scale prefers using equipment that is as large as possible.

David Harbottle, Associate Professor, School of Chemical & Process Engineering, University of Leeds, West Yorkshire, United Kingdom
Jan Czarnecki, Adjunct Professor, University of Alberta, Edmonton, Alberta, Canada
Jacob Masliyah, Professor Emeritus, University of Alberta, Edmonton, Alberta, Canada
Zhenghe Xu, Teck Professor (Dean), Univ. of Alberta (Southern Univ. of Science & Technology), Edmonton, Alberta, Canada (Shenzhen, China)

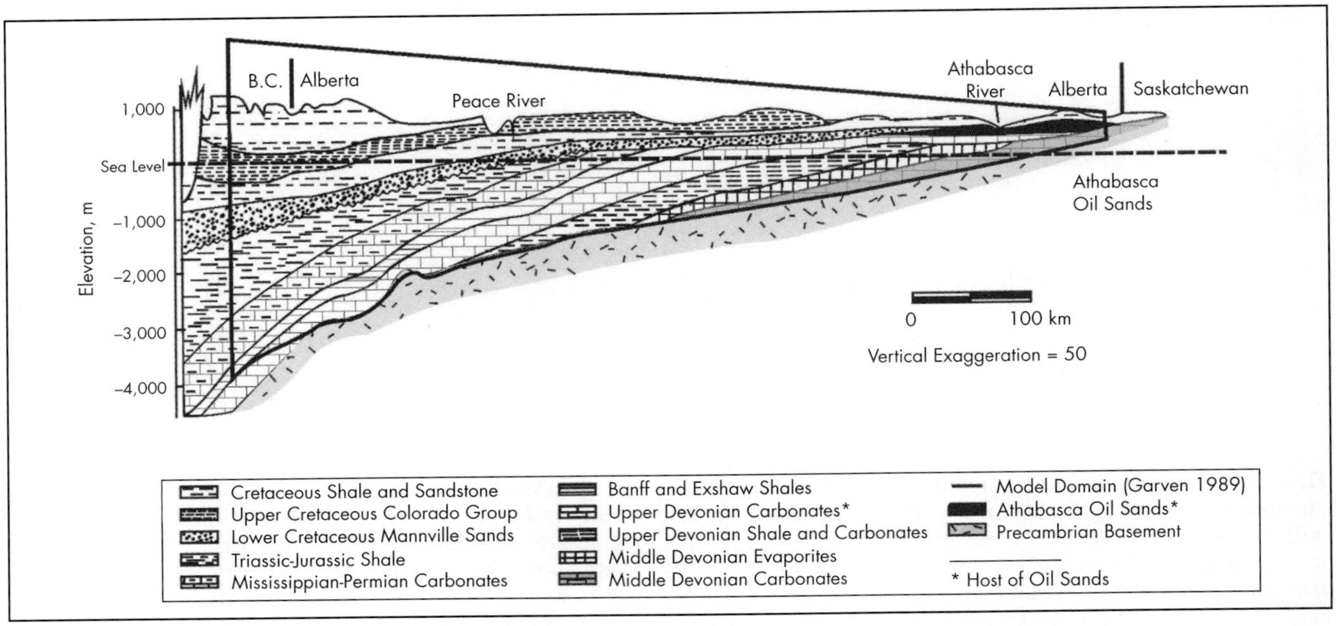

Source: Adams et al. 2004

Figure 1 Geological formation and migration of crude oil

At slurry preparation the ore is crushed, screened, and mixed with hot water to produce slurry that is then fed to a slurry hydrotransport pipeline, which is a major innovation in the oil sands industry.

To enhance bitumen recovery, chemical process aids are often added at this point. Sodium hydroxide (caustic) is the most commonly used chemical to control the slurry pH, although sodium bicarbonate or carbonate can also be used as additives. For example, a mineral named trona and a commercial product called Geosol (mixtures of both sodium carbonate and bicarbonate) have been tested at large pilot-plant scale. In some cases, inorganic dispersants such as sodium silicate (liquid glass) and calcium chelating agents such as sodium citrate have also been tested either in laboratory or commercial operations.

Most of the ore conditioning takes place in the pipeline. *Conditioning* is a term used to cover all elementary processes needed to make bitumen separation feasible. Oil sands ore chunks (lumps) have to be ablated, bitumen washed off the solids (or liberated), and clay minerals dispersed. Bitumen droplets collide with dispersed air bubbles. Under favorable conditions, air bubbles and bitumen droplets form stable aggregates on collision. At temperatures well above 50°C, bitumen spreads over the air, totally engulfing air bubbles. At much lower temperatures, bitumen droplets attach themselves to the bubbles, similar to mineral grain attachment to air bubbles in conventional mineral flotation.

To separate the aerated bitumen from the slurry, the conditioned slurry is fed to primary separation vessels or cells in the extraction plant. The aerated bitumen floats to the top of the separation vessels to form a layer of bituminous froth, which is collected, deareated, and then sent to froth treatment to further remove entrained solids and water. The coarse solids in the slurry settle quickly to the bottom of the primary separation vessels, which are pumped, in most cases, to the tailing ponds. The middle layer, called *middlings*, contains mostly water and dispersed fine solids (clays) with some bitumen in the form of fine droplets that do not have sufficient time to report to the froth. This is usually caused by insufficient aeration, compounded by the attachment of bitumen to some hydrophobic solid particles, making the bitumen-carrying aggregate too heavy to float. To recover this bitumen, the middlings stream is sent to flotation cells, similar to those used in conventional mineral processing. Both column and mechanical flotation cells are used, depending on the configuration of the plant. In flotation, additional air is introduced under a high-shear environment to aerate this portion of bitumen, which is recovered in the form of a secondary froth, usually of much lower quality than the froth formed in primary separation vessels. This secondary froth is typically recycled to primary separation vessels while the underflow from flotation cells is sent to tailings.

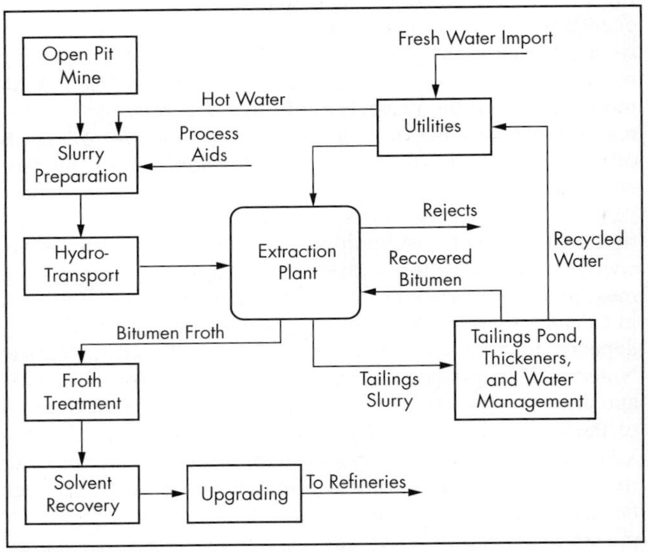

Adapted from Masliyah et al. 2011

Figure 2 Overall flow sheet of minable oil sands processing

The bituminous froth from the primary separation vessels typically contains about 60 wt % bitumen, 30 wt % water, and 10 wt % solids. Water contains dissolved salts, mostly originating from the ore, which contain various amounts of chlorides. Chlorides pose serious corrosion problems in downstream operations. Solids, in particular finely dispersed clays, deactivate upgrading catalysts by clogging catalyst pores. Both fine solids and small amounts of larger solid particles contribute to erosion of pump impellers, valves, and pipes. Before passing bitumen to downstream operations or transporting in pipelines to the market, the entrained solids and water must be removed to meet stringent upgrading or pipeline specifications. Removal of all those impurities takes place in the froth treatment operation.

Since the density of bitumen is very close to the density of water, in the froth treatment process, a light hydrocarbon solvent is added to the froth to provide a density difference between oil and water phases. Solvent addition also lowers the viscosity of the oil phase. Both of these factors enhance the oil–water-phase separation under gravity in inclined plate settlers, centrifuges, or hydrocyclones. The added solvent is then recovered in solvent recovery units by flash distillation and recycled to froth treatment. After stripping the solvent, the bitumen product is then fed to the on-site upgrading plant or sent to off-site refineries by pipelines after adjusting its viscosity either by solvent addition or by heating the bitumen.

Currently, two main technologies are used for froth treatment, depending on the type of solvent used. In naphthenic froth treatment, naphtha from a local upgrading plant is added at a solvent-to-bitumen ratio of about 0.7. In paraffinic froth treatment, a paraffinic solvent imported from outside is added at a solvent-to-bitumen ratio of about 2 for the low-temperature process and a lower value for the high-temperature froth treatment. The paraffinic solvent addition causes partial precipitation of asphaltenes in the diluted bitumen. The precipitating asphaltenes act as flocculant to bind water droplets and fine solids. The oil phase obtained from paraffinic froth treatment is very clean and dry, exceeding the pipelining or refinery specifications. The naphtha-based process produces diluted bitumen that requires addition of polymer demulsifiers and additional centrifugation to meet product specifications of water and solids content.

A utilities plant provides hot water for extraction, steam, and power for the whole integrated plant operations. The extraction plant uses mostly recycled water from tailings operations. Some fresh makeup water has to be imported from outside, usually from the Athabasca River. If there is an upgrading unit on-site, utilities use all low-cost hot gases produced by the upgrader in addition to the imported natural gas. Burning coke is also possible to provide needed energy but difficult to implement because of the high sulfur content of on-site coke.

A crucial part of the overall plant operation is water management. Tailings slurry waste from extraction is typically sent to tailing ponds, where coarse solids settle quickly on the beaches and fine solids settle slowly, forming a semifluid layer of so-called fine tailings toward the bottom of the pond. After several years, the fine tails consolidate only to some degree, forming what the industry calls *mature fine tailings*, or MFT. MFT typically contain approximately 30 wt % solids without further consolidation under natural tailings pond conditions. Nevertheless, sedimentation of solids in the pond leaves a layer of relatively clean water on top. This water is recycled to the extraction plant. Since fine solids take an extended time to settle, the water cools down to ambient temperature. As a result, the recycled water is in most cases cold, requiring reheating to about 80°C before it can be reused in slurry preparation. Therefore, there is a huge incentive to recover and recycle as much water as possible while it is still warm. This could be achieved by passing the whole tailings stream through a hydrocyclone with the warm overflow stream being recycled. Another option is to treat the flotation tailings in a thickener after proper dosing of flocculant to enhance solid–liquid separation that allows overflow warm water to be recycled quickly.

Regardless of the specific plant issues, it is always beneficial to recover as much warm water from the tailings for energy conservation (and thus lower operating costs and greenhouse gas emissions). Also, it is always beneficial to minimize the tailing ponds area. Public perception and regulatory requirements favor dry tailings disposal. Several technologies can be used to achieve this goal. One of them is flocculation of middlings followed by thickening to produce clarified warm water for recycle while generating high-settling-rate flocs. The sediments from the thickener are then mixed with coarse solids from the hydrocyclone to form nonsegregated tailings. Another is known as consolidated (or composite) technology (CT). In the CT process, the MFT is mixed with coarse sands mostly from hydrocycloned extraction tailings to achieve the right mix of fine and coarse solids. Gypsum is added to provide calcium ions needed to flocculate fine solids, allowing them to settle with the coarse sands to produce a nonsegregated sediment. This technology, capable of releasing substantial amounts of water quickly, is currently being used by a few companies, allowing reclamation of the tailings disposal sites a few years after placement. Dosing of gypsum has to be just right to avoid loading the recycle water with calcium, as calcium ions are known to be detrimental to the extraction plant operation.

Following atmospheric and vacuum distillation in the on-site upgrading plant, bitumen is fed to a fluid or delayed coker. Placing a coker at the front end of the upgrading plant allows bitumen feed of a lower quality than that required for a catalytic hydrotreatment unit. If the bitumen is sent off-site for further processing, substantially cleaner product is required. Therefore, plant-specific issues dictate the configuration of the upgrading unit, specifications for the froth treatment product, light hydrocarbon solvent recycle schemes, and many other parameters needed for plant design and operation. Most of the unit operations mentioned earlier in this chapter are discussed in substantially more detail later.

BITUMEN RECOVERY FUNDAMENTALS

For effective bitumen recovery from the oil sand–water slurry, the bitumen has to be *liberated*; that is, the oil sand must be dispersed in water and bitumen must be washed off the sand grains. Individual bitumen droplets formed by liberation collide with dispersed air, forming stable bitumen–air aggregates in a process called *aeration*. In *flotation*, aerated bitumen of sufficient buoyancy reports to the top of the separation vessel to form a layer of froth. Rates of all these elementary steps, including liberation, aeration, and flotation, depend on viscosities of all phases involved (and therefore on temperature) and interfacial tensions between all the phases. Water chemistry, especially pH and divalent ion content, and to a lesser extent the total salt content, also have a profound effect on all the elementary steps and therefore on bitumen recovery. The fundamental science governing these elementary steps involved in bitumen recovery is discussed next.

Bitumen Liberation

Bitumen liberation, or the separation of bitumen from sand grains in water, is the first essential step in bitumen recovery. It proceeds in two consecutive steps: bitumen recession on the sand surface and its separation from sand grains to form small droplets freely flowing in the slurry (Figure 3). The recession step is driven by an energy change relating to the replacement of bitumen–sand interface with sand–water interface, assuming the change in the bitumen–water interface when bitumen forms small droplets can be neglected. The governing formulae of the well-known Young's equation for bitumen on a solid surface in water is given by

$$\gamma_{B/S} - \gamma_{S/W} = \gamma_{B/W}\cos\theta \quad \textbf{(EQ 1)}$$

where
B, *S*, and *W* = bitumen, solid, and water, respectively; that is, $\gamma_{B/S}$, $\gamma_{S/W}$, and $\gamma_{B/W}$ are the bitumen–solid, solid–water, and bitumen–water interfacial tensions, respectively
θ = contact angle measured through the aqueous phase in the bitumen–solid–water system

It is important to distinguish the contact angle of water in two different systems: water–solid–bitumen and water–solid–air. The same contact angle value in these two systems does not indicate the same hydrophobicity of the solids. In other words, for a given solid, the contact angles measured in these two systems are almost always very different. Combining the thermodynamic analysis of the bitumen liberation process with Young's equation (1) yields

$$dG/dA = -\gamma_{B/W}\cos\theta \quad \textbf{(EQ 2)}$$

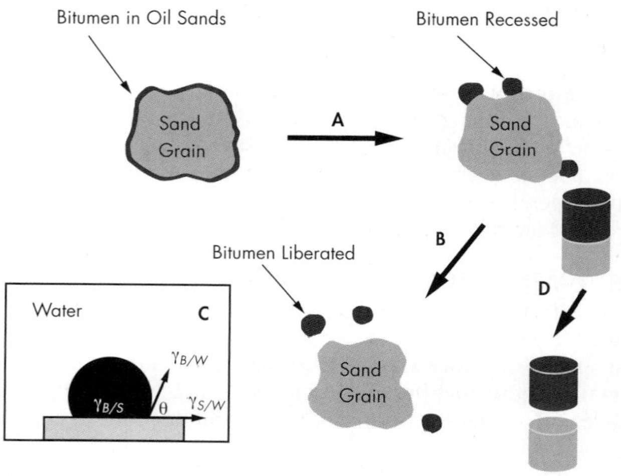

Note: Bitumen recession, A, and bitumen detachment, B (simplified in D), are determined by the aqueous contact angle, θ, of bitumen on a sand grain, C.

Source: Masliyah et al. 2011

Figure 3 Bitumen liberation from a sand grain in two subprocesses

where *dG/dA* represents the energy change during bitumen recession.

Since $\gamma_{B/W}$ is always greater than zero, Equation 2 shows that spontaneous bitumen recession from a sand grain in water can only occur when $\theta < 90°$, pointing out that sand grains have to be hydrophilic for bitumen recession. In other words, the condition of Equation 2 indicates that the interaction between the sand and water must be stronger than the interaction between the sand and bitumen, $\gamma_{B/S} > \gamma_{S/W}$. The smaller the angle θ, the more negative is *dG/dA*, making bitumen easier to recede.

Bitumen recedes to form an approximately spherical droplet still attached to the sand grain via a "neck" unless the contact angle θ defined previously is zero. The neck diameter depends on the contact angle θ. The energy associated with the droplet displacement on solid in water is given by

$$dG/dA = \gamma_{S/W} + \gamma_{B/W} - \gamma_{B/S} \quad \textbf{(EQ 3)}$$

Combining Equation 3 with Young's equation (1) yields

$$dG/dA = \gamma_{B/W}(1 - \cos\theta) \geq 0 \quad \textbf{(EQ 4)}$$

Keeping in mind that $\gamma_{B/W} > 0$, the detachment energy is always greater unless $\theta = 0$. Therefore, for bitumen liberation to occur under static conditions, sand grains must be completely hydrophilic; that is, they must have a contact angle of zero, regardless of the bitumen–water interfacial tension. This is likely never to occur in practice, as even miniscule amounts of impurities adsorbing on sand would make $\theta > 0$. Therefore, bitumen droplets formed by recession have to be washed off the sand grains under the influence of hydrodynamic forces created in the hydrotransport slurry pipelines as practiced. To a lesser extent, the buoyancy of aerated bitumen and inertial forces can also play a role in bitumen detachment.

In practice, a small contact angle is always favorable. It increases the driving force for bitumen recession and makes energy required for detachment lower. Young's equation (1) suggests that to reduce the contact angle, one must reduce the bitumen–water interfacial tension ($\gamma_{B/W}$) and increase $\gamma_{B/S} - \gamma_{S/W}$, that is, increase the bitumen–sand interfacial tension and/or reduce the sand–water interfacial tension. The bitumen–sand interfacial tension cannot be easily changed in operation, whereas the exposure of sand to water on bitumen recession allows for the sand–water interfacial properties to be controlled. Increasing pH, for example, can reduce the sand–water interfacial tension, as it enhances hydrolysis and increases the charge of the sand surface, making the sand more hydrophilic at high pH, which is beneficial for bitumen liberation (Wang et al. 2010). Notably, increasing the temperature from 40°C to 80°C had little effect on contact angle. It is believed that at 40°C, bitumen is already fairly fluidlike so that the transfer of natural surfactant from the bulk bitumen to the bitumen–aqueous interface and bitumen globule formation occurred over the time span of the experiments.

The important role of wettability of sand in bitumen liberation is clearly reflected by difficulties in the processing of weathered ores (Ren et al. 2008). Researchers found that the weathering of oil sands ores led to a significant reduction of water wettability (an increase in the contact angle of water on solids from 42° to 65°) because of the contamination of solids by toluene-insoluble hydrocarbons on solid surfaces. Direct force measurements by atomic force microscopy (AFM) showed an increase in the adhesion force between

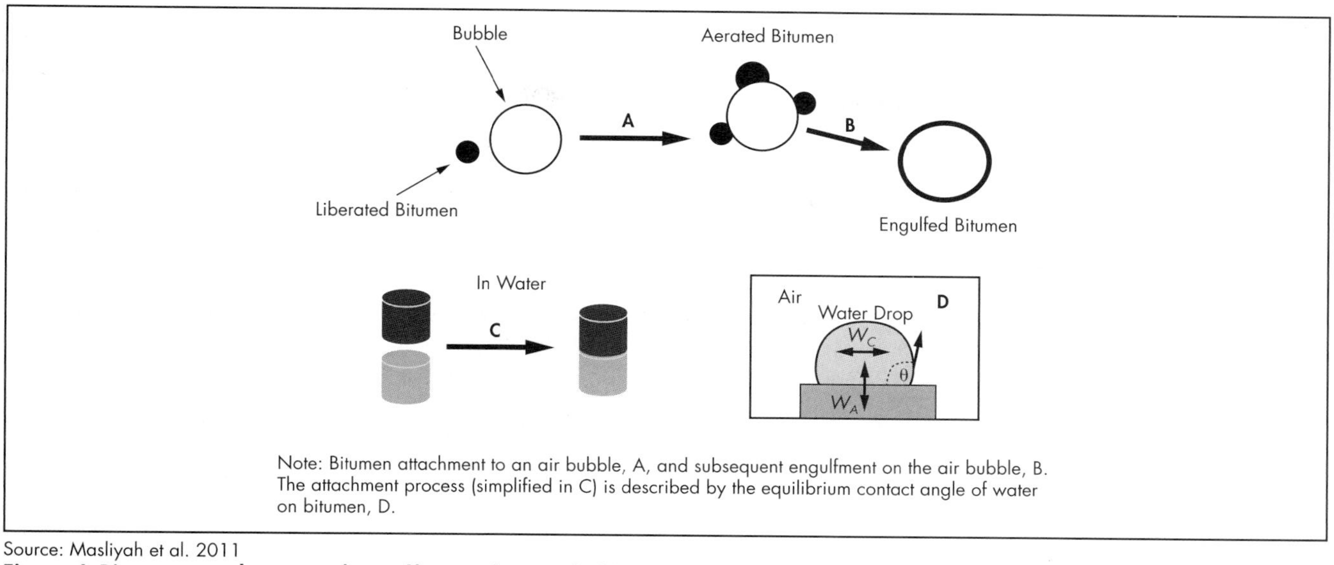
Bubble
Aerated Bitumen
A
B
Liberated Bitumen
Engulfed Bitumen
In Water
C
Air
Water Drop
D
W_C
θ
W_A
Note: Bitumen attachment to an air bubble, A, and subsequent engulfment on the air bubble, B. The attachment process (simplified in C) is described by the equilibrium contact angle of water on bitumen, D.

Source: Masliyah et al. 2011

Figure 4 Bitumen attachment and engulfment of an air bubble

bitumen and silica spheres on weathering of bitumen-coated sand grains. Poor liberation of bitumen from weathered oil sand ores was confirmed experimentally by bitumen recession experiments using bitumen-coated glass slides with and without artificial weathering (Ren et al. 2009). Slow bitumen liberation following artificial weathering is partially responsible for the poor processability of weathered ores, leading to low bitumen recovery and poor froth quality. Note that hot-water washing (at temperatures above 65°C) was able to restore the wettability of contaminated solids and hence improve the processability of weathered ores, further justifying the use of hot water in industrial oil sands extraction operations to ensure a robust processing condition.

In view of the observed significant role of surfactants, water chemistry is expected to influence bitumen liberation, as it affects the soluble surfactant concentration. A recent study using the AFM probe technique showed a significant increase in adhesion force between a silica particle and bitumen in the presence of divalent calcium and magnesium cations (Zhao et al. 2006). The presence of surfactants was found to alleviate the negative impact of divalent cations on bitumen liberation. A further study found that there was strong adhesion between bitumen and silica sand in simulated process water, whereas the addition of bicarbonate to the simulated process water could reduce the adhesion by more than tenfold, suggesting that bitumen liberation is much easier in the presence than in the absence of bicarbonate (Zhao et al. 2009). Bicarbonate is believed to scavenge calcium ions, so that the natural surfactant from bitumen remains effective, reducing the bitumen–water interfacial tension, whereas the sand surface remains hydrophilic by eliminating calcium cation bridging through scavenging of calcium ions with bicarbonate. Sodium bicarbonate addition also increases the pH of the slurry, which could contribute to enhancing bitumen liberation. It should be noted that thermodynamic analysis provides only the direction of liberation. To accurately predict the liberation process, one must consider the hydrodynamic conditions under which the liberation occurs.

Bitumen Aeration

Bitumen aeration is the process of bitumen attachment to air bubbles. It is another key step in bitumen recovery. The density of bitumen is similar to the density of water in that under its own buoyancy the bitumen would not float appreciably for industrial applications. Aeration reduces the apparent density of the bitumen–air mixture so that it is much lower than the density of the separation medium, enabling the aerated bitumen to float within a reasonable residence time to the top of the slurry in the separation vessel, where it can be recovered as bitumen-rich froth. Bitumen aeration can occur in either hydrotransport slurry pipelines or in flotation vessels. In some cases, aeration of bitumen has been found to assist in bitumen liberation (Lelinski et al. 2004).

The strength of the bitumen–air bubble attachment can be analyzed using interfacial energies. In bitumen–air bubble attachment described by A and C in Figure 4, the free energy change associated with air attachment is given by

$$dG/dA = \gamma_{B/A} - (\gamma_{B/W} + \gamma_{A/W}) \quad \textbf{(EQ 5)}$$

Using Young's equation (1) for system D in Figure 4 leads to

$$dG/dA = \gamma_{A/W}(\cos\theta - 1) \quad \textbf{(EQ 6)}$$

Clearly, when the contact angle θ is greater than zero, *dG/dA* is less than zero, that is, the bitumen–air bubble attachment is thermodynamically spontaneous. Equation 6 shows that the larger the contact angle, the more negative the *dG/dA*, and hence the more favorable it is for bitumen to attach to air bubbles. Increasing pH decreases the contact angle of water on bitumen, indicating less favorable attachment of air bubble to bitumen at higher pH, as shown by the less-negative value of *dG/dA* in Equation 6.

To understand how interfacial properties can be optimized for bitumen liberation and aeration, it is instructive to examine Young's equation (1) for bitumen–sand–water and for air–bitumen–water. Table 1 shows the desired trend of various interfacial tensions for this optimization. For practical

Table 1 Desired change of interfacial tensions for better performance of bitumen liberation and aeration*

	cosθ	$\gamma_{B/S}$	$\gamma_{B/W}$	$\gamma_{B/A}$	$\gamma_{S/W}$	$\gamma_{A/W}$
Liberation	↑	↑	↓	—	↓	—
Aeration	↓	—	↑	↓	—	↑

Source: Masliyah et al. 2011
*Subscripts *B*, *S*, *A*, and *W* represent bitumen, solid, air, and water, respectively.

purposes, when the surface tension of bitumen is given, there is little room to change the interfacial tension of bitumen–sand. The only interfacial properties that can be adjusted are those of bitumen–water, sand–water, and air–water. Increasing the pH of the process medium would lead to hydrolysis of sand surfaces and ionization of both sand and bitumen surfaces, making both the bitumen and solids more hydrophilic. Increasing pH therefore would decrease the interfacial tension of the sand–water and bitumen–water interfaces, a condition that is favorable for bitumen liberation.

Increasing the pH of the process water also decreases the air–water interfacial tension, which is undesirable for bitumen aeration. It is evident that high pH is favorable for bitumen liberation, whereas a low pH is favorable for bitumen aeration. There is a clear conflict in the pH requirement, making caustic addition an important process variable. Current warm water–based oil sands extraction processes almost exclusively operate at pH values above 8.0 and below 9.0 as a compromise between bitumen liberation and aeration. Novel ideas are needed either to decouple these two subprocesses through a two-stage processing option or to maximize the air–water interfacial tension for aeration. An effective way to do this would be the use of fresh air bubbles where the air–water interfacial tension is at its maximum. In this respect, producing the bubbles in an area where contact between bitumen droplets and air bubbles could be made immediately would be beneficial. This is the concept behind the aeration technology used in modern flotation cells, such as the Microcell, the Jameson cell, and Canadian Process Technologies' cavitation-based cell.

From the perspective of molecular interaction, the thermodynamic condition of the contact angle of water on bitumen being greater than zero can be thought of as the cohesion of water being stronger than the adhesion of water to bitumen. The work of adhesion (W_A) is given by

$$W_A = \gamma_{A/W} + \gamma_{B/A} + \gamma_{B/W} \quad \text{(EQ 7)}$$

whereas the work of cohesion (W_C) is

$$W_C = 2\gamma_{A/W} \quad \text{(EQ 8)}$$

Inserting Equations 7 and 8 into Young's equation (1) leads to

$$\begin{aligned} dG/dA &= \gamma_{B/A} - (\gamma_{B/W} + \gamma_{A/W}) \\ &= (\gamma_{A/W} + \gamma_{B/A} - \gamma_{B/W}) - 2\gamma_{A/W} = W_A - W_C \end{aligned} \quad \text{(EQ 9)}$$

Equation 9 shows that having $dG/dA < 0$ is equivalent to having $W_A < W_C$, which is identical to the contact angle being greater than zero. A stronger cohesion of water to itself than its adhesion to bitumen is illustrated in Figure 4D, where water molecules pull together to form a water droplet on the bitumen. Anything that weakens the work of adhesion of water on bitumen, such as increasing the hydrophobicity of bitumen, will enhance bitumen–air attachment and improve flotation.

The important role of bitumen surface hydrophobicity (measured by water contact angle on bitumen) in bitumen flotation is illustrated by Zhou et al. (2004). There is a correlation between bitumen recovery and water contact angle on bitumen surfaces: a higher contact angle value corresponds to a higher bitumen recovery. A critical contact angle value of 65° appears to exist. Interestingly, this critical contact angle value for effective bitumen recovery corresponds well with the value required for effective mineral flotation (Fuerstenau 2003).

After attaching to air bubbles, bitumen may spread quickly over the bubble, completely covering the bubble in a bitumen film and resulting in the formation of stable bitumen–air bubble composites. The extent and rate of bitumen spreading on an air bubble depends on the interfacial tension and the temperature of the system. The spreading process, shown in Figure 4B, can be predicted by the spreading coefficients $S_{B|W:A} = \gamma_{A/W} - (\gamma_{B/W} + \gamma_{B/A})$. Applying typical interfacial tension values for the system, the spreading coefficients are calculated to be 32.4 mN/m and 36.5 mN/m for pH 8.5 and 10.5, respectively, indicating the engulfment of bitumen on air bubbles in process water for both pH values. Using Young's equation, the spreading coefficient can be written as

$$S_{B|W:A} = \gamma_{A/W}(1 + \cos\theta) - 2\gamma_{B/A} \quad \text{(EQ 10)}$$

The spreading condition of bitumen on an air bubble is therefore determined by $S_{B|W:A} > 0$, that is,

$$\cos\theta > \frac{2\gamma_{B/A}}{\gamma_{A/W}} - 1 \quad \text{(EQ 11)}$$

Inserting typical values of $\gamma_{B/A}$ = 30 mN/m and $\gamma_{A/W}$ = 68 mN/m for process water into Equation 11 leads to a critical contact angle value of 97°. This value is greater than the contact angle values measured over the entire operating pH range. Therefore, bitumen is likely to spread on the air bubble on attachment in the warm/hot-water extraction process. This finding implies that bitumen–air bubble attachment is a critical step for bitumen flotation, provided that the temperature is sufficiently high so that bitumen viscosity is not a rate-limiting factor.

UNIT OPERATIONS

The process of bitumen extraction begins at the mine face. To expose the oil sands ore, the muskeg and overburden are first removed using shovels and trucks. The excavated land is used for either land reclamation or the construction of tailings dikes. Once exposed, the oil sands ore is extracted and processed to condition the oil sands ore for optimal bitumen recovery. Many of the unit operations are common with other mining activities. Since the first commercial operation, the objectives of oil sands extraction have remained the same, although the methods to satisfy the processing criteria have witnessed several important changes. With increased production rates, oil sands operators became more flexible and adaptable to the receding mine face and the increased distance between the mine and processing facility. Increased flexibility in the mining process allows the opportunity to blend oil sands ores to ensure that the minimum criterion for bitumen recovery is exceeded. Most oil sands producers today operate a shovel-to-truck-to-hydrotransport facility, recognized for its energy efficiency and flexibility with integrated operating systems shown in Figure 2. The unit operations involved in oil sands extraction are shown in Figure 5 and are discussed briefly in the following section.

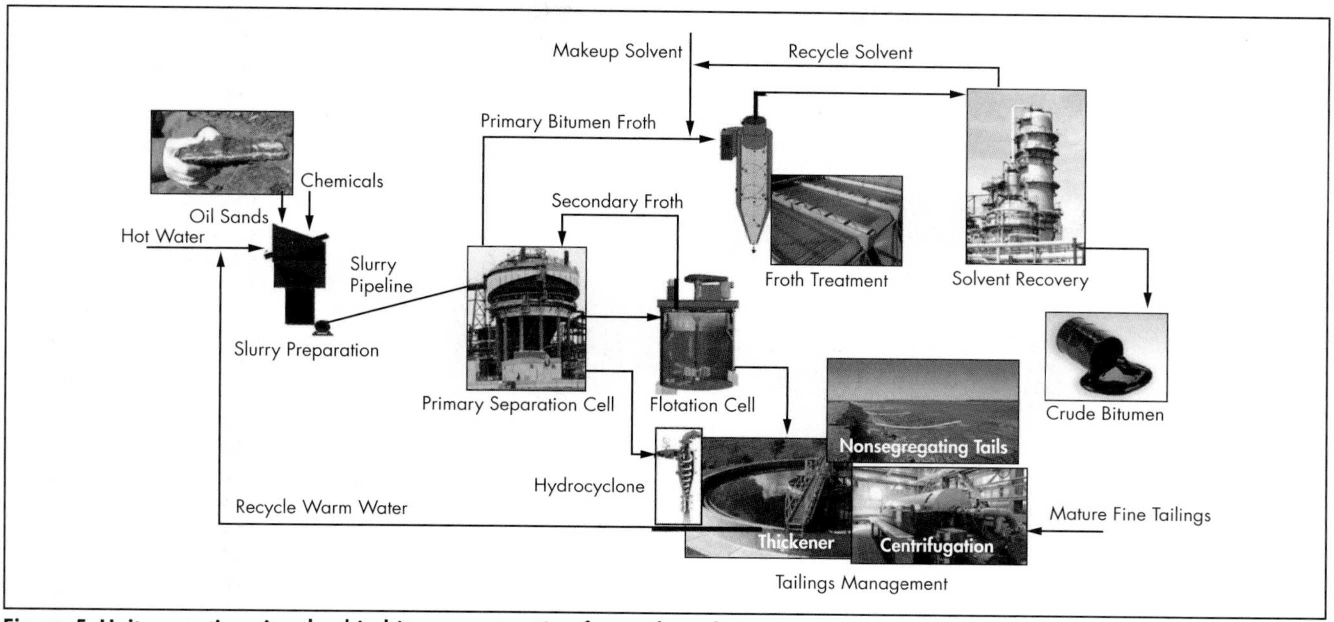

Figure 5 Unit operations involved in bitumen extraction from oil sands

Slurry Preparation

Mining of the oil sands ore typically uses a 100-t (metric-ton) bucket with large ton-capacity trucks used to move the oil sands ore to the preparation facility. The as-received oil sands ore from the mine site frequently contains large rocks and cohesive sediments that must be broken down and crushed to prevent damage to downstream processing equipment. The mined oil sands ore is therefore fed into a crushing plant where double-roll crushers reduce large lumps of oil sands ores to smaller lumps that are conveyed to a storage pile or surge bin, where the storage capacity can often provide approximately 30 minutes of continuous dry feed to the slurry preparation facility.

A typical example of a slurry preparation circuit is shown in Figure 6. The oil sands ore is mixed with process water through a baffled, inclined mixing box. The underflow slurry is then distributed onto a primary vibrating screen where coarse objects are separated from the oil sands slurry. The oversized objects are further reduced using an impact crusher with the product fed to a second screen on a secondary pumpbox and wetted with secondary wash water to improve the screening efficiency. The underflow from the secondary pumpbox is then recycled back to the primary vibrating screen, minimizing the loss of bitumen with oversized objects. Any oversized material from the secondary pumpbox is considered as waste and disposed of. The primary pumpbox underflow, on the other hand, feeds directly into the hydrotransport pipeline, continuing the process of lump ablation and bitumen liberation/aeration.

The oil sands slurry level in the primary pumpbox is maintained by the volumetric flow rate of the hydrotransport pipeline and the addition rate of process feedwater. The slurry preparation circuit produces pumpable oil sands slurry of density typically about 1,580 kg/m^3 and begins the sequence of oil sands lump ablation (size reduction) and bitumen liberation. To promote bitumen liberation, warm process water (normally >80°C) is added with pH-adjusting chemicals. As discussed earlier, warm water addition promotes a reduction in the bitumen viscosity, which improves the breakdown of large oil sands lumps and the flowability of bitumen on solids. Furthermore, operating in basic conditions at pH ≈8.5 creates favorable bitumen–water interfacial tensions and solids wettability to enhance bitumen-receding dynamics.

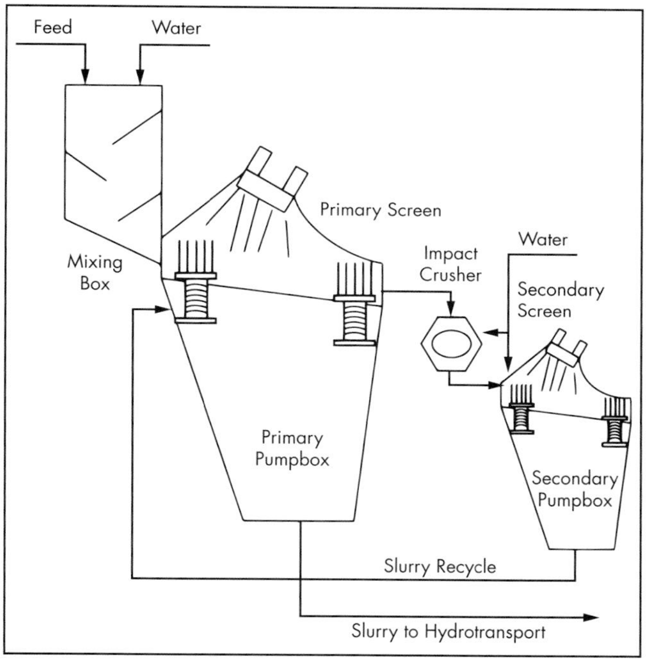

Source: Czarnecki et al. 2013

Figure 6 Slurry preparation circuit

A rotary breaker slurry preparation facility is an alternative to the previously discussed mixing box process. Rotary breakers were introduced by Suncor in the late 1990s at its Steepbank mine (Alberta, Canada). The crushed oil sands ore

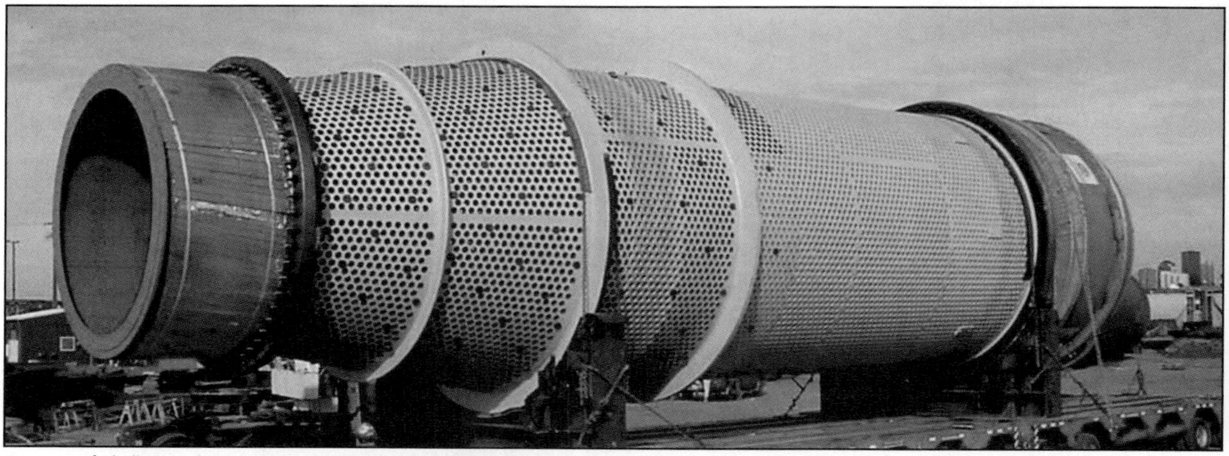

Courtesy of Shell Canada, as cited in *Oil Sands Magazine* 2018
Figure 7 ThyssenKrupp rotary breaker for a typical slurry preparation system

is agitated with warm water, forming a slurry that gradually slides along the inclined perforated drum. The rotary breaker is typically 4 m in diameter and about 20 m long, rotating slowly at ~10 rpm. Under slow rotation, the oil sands lumps are carried by lifters to the top of the drum before falling and impacting on the toe of the rotating drum. The digestion of large oil sands lumps allows slurry drainage into a pumpbox through 5.1- to 7.6-cm-diameter holes in the rotary breaker shell. Often the discharge from the rotary breaker is recycled to increase the slurry residence time and minimize the loss of bitumen with oversized material. An example of a rotary breaker is shown in Figure 7.

Hydrotransport

The oil sands slurry is discharged from the pumpbox and transported to the extraction facility via a hydrotransport pipeline, as shown in Figure 8. The objective of the hydrotransport pipeline is to (1) impart mechanical shear to further ablate lumps of oil sands ore, and (2) provide sufficient residence time that satisfies the requirements of bitumen liberation from the sand grains, bitumen–bitumen coalescence, and attachment of bitumen droplets to entrained air bubbles in the oil sands slurry.

Oil sands operators have to consider many factors when designing an oil sands slurry pipeline, such as the (1) maximum throughput; (2) operating flow conditions to prevent excessive pipe/pump wear and sanding; and (3) slurry flow residence time, which is often influenced by the locations of the mine site and extraction facility, and the operating temperature. Although maximum throughput is aligned to production capacity, the volumetric flow rate to achieve the desired throughput is directly related to the hydrotransport pipe diameter and slurry transport velocity. Predicting the slurry deposition velocity and reducing the potential for sliding bed flow in the pipeline are desirable to avoid excessive pipe wear and potential pipe blockage. The minimum operating velocities for oil sands slurries can be calculated using the two-layer model (Gillies et al. 1991), provided that the fines (<44 μm) content and the carrier fluid (water + fines) viscosity are known. Typical pipeline operating velocities are in the range of 3.0 to 5.0 m/s, and the pipe length is often a few kilometers. Hence, the oil sands slurry residence time is approximately 10–20 minutes.

Courtesy of Syncrude Canada Ltd.
Figure 8 Oil sands hydrotransport pipeline and primary separation vessel at Syncrude Aurora mine

Bitumen Recovery

Bitumen and water have an almost identical density; therefore, the potential for separation under gravity is very low. Attachment of bitumen to air bubbles in the slurry preparation phase, as practiced in mineral flotation, reduces the bitumen apparent density and promotes effective separation of bitumen from water and solids in the primary separation vessel. For most ores the naturally entrained air in the hydrotransport pipeline is sufficient to aerate the liberated bitumen for a recovery of up to 90%; however, operators can add additional air if deemed necessary.

The objectives of the extraction process are to maximize bitumen recovery, produce clean bitumen froth (bitumen-to-solids ratio of 6) with a bitumen recovery >90%, and discard the mined solids to the tailings ponds. To support the processing capacity of an oil sands mine, the primary separation vessels are typically between 20 m and 30 m in diameter and height to ensure a residence time >30 minutes. The cone angle of the primary separation vessel (PSV) is sufficiently steep (>55°) to prevent sand buildup and minimize the potential for plugging the underflow outlet. The bitumen extraction process

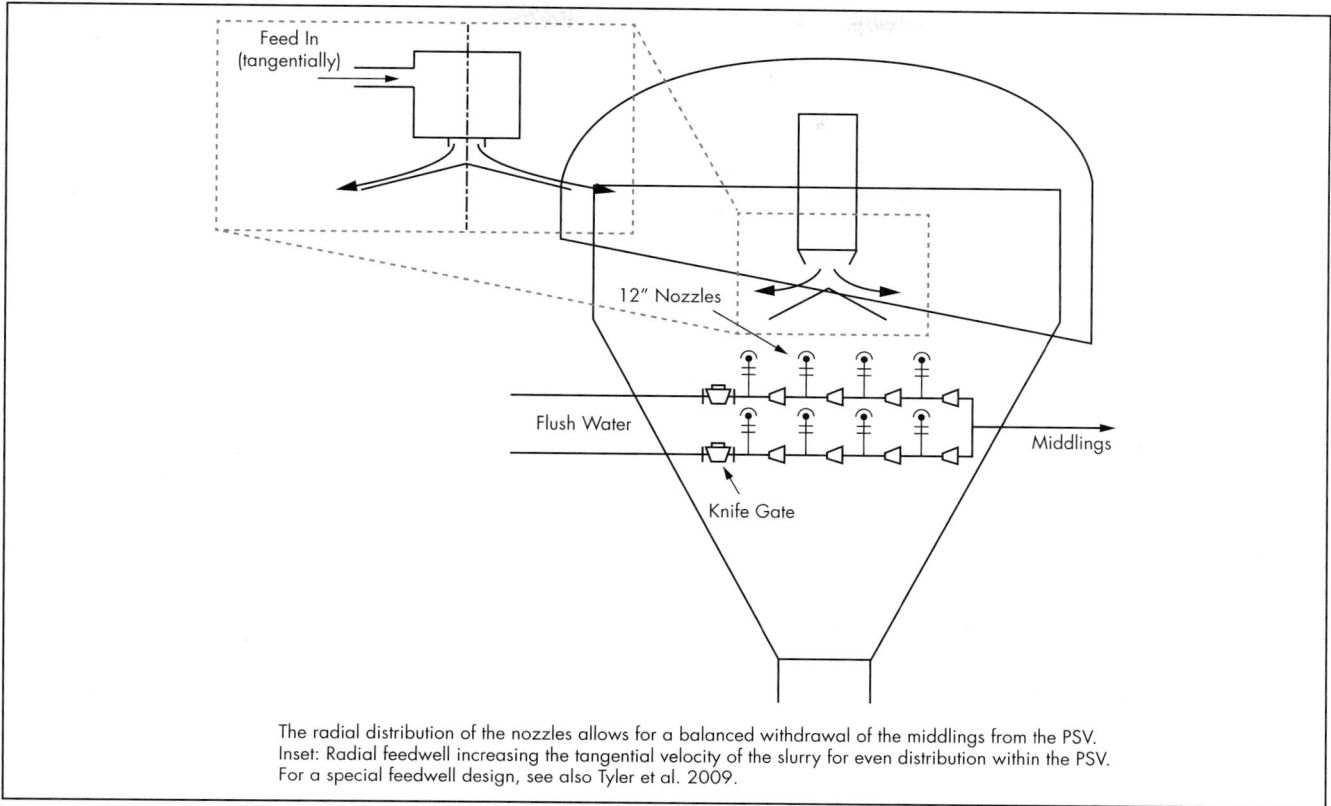

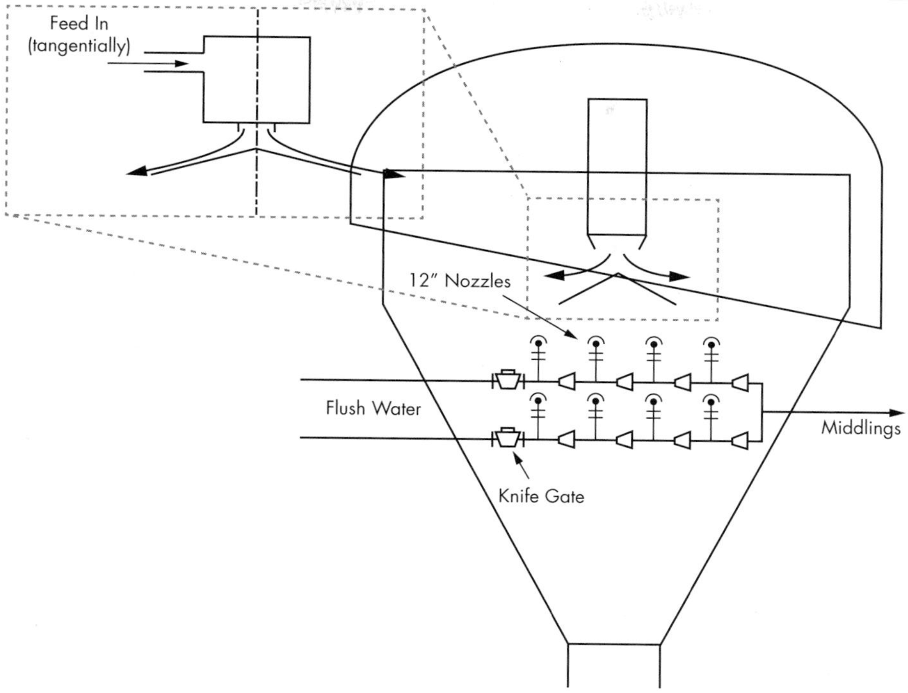

The radial distribution of the nozzles allows for a balanced withdrawal of the middlings from the PSV.
Inset: Radial feedwell increasing the tangential velocity of the slurry for even distribution within the PSV.
For a special feedwell design, see also Tyler et al. 2009.

Adapted from Czarnecki et al. 2013

Figure 9 Middlings withdrawal system positioned a few meters below the froth launder

begins as the oil sands slurry from the hydrotransport pipeline is fed into the PSV. To enhance bitumen separation from the oil sands slurry, process water is added to the oil sands slurry prior to being discharged just below the froth zone in the separation vessel. The oil sands slurry density in the feedwell is controlled to approximately 1,400 kg/m^3. The feedwell as shown in the inset of Figure 9 is designed to introduce the oil sands slurry into the separation vessel with minimal turbulence and disturbance on the separation zone. The tangential feedwell design generates a swirling flow that spreads evenly across the separation vessel as the oil sands slurry is discharged from the feedwell.

Once the oil sands slurry is fed to the separation vessel, the aerated bitumen droplets rise to form primary bitumen froth (1–2 m deep) that flows over the edge of the separation vessel into an insulated froth launder. The primary froth containing up to 40% by volume of air is deaerated with steam, and the deaerated bitumen froth containing 60 wt % bitumen, 30 wt % water, and 10 wt % solids is transferred to the froth treatment stage.

Most of the remaining bitumen is associated with the fines and clay particles that are either too heavy to float or not effectively aerated. This portion of bitumen is suspended in the slurry right below the froth, which is termed as middlings and removed from the separation vessel for further processing (secondary recovery). The middlings withdrawal system as shown in Figure 9 is designed to maintain a low-density suspension (~1,200 kg/m^3) that does not hinder the migration of aerated bitumen to the froth zone, while bitumen-loaded fines and clays are discharged to secondary separation by either flotation column or conventional mechanical flotation cells. In both flotation column and mechanical flotation cells, forced air is added to enhance aeration of fugitive bitumen and hence improve bitumen recovery. The bitumen froth from secondary separation is of much lower quality than the froth from primary separation vessels, containing more fine solids and water. The bitumen froth from the flotation circuit is therefore returned to the primary separation vessel to improve the froth quality. Bitumen recovery by primary separation and flotation is equivalent to a rougher–scavenger closed circuit in mineral processing. Cleaning as a common practice in mineral processing is hardly considered in bitumen extraction, due largely to difficulties in breaking the sticky bitumen froth.

The third exit stream from the primary separation vessel is the underflow consisting of a dense slurry of coarse solids that are transported via pipeline to the tailings pond. The bitumen concentration in the underflow is typically <0.5 wt % and is not currently considered for secondary separation. Hydrophilic particles that remain in the slurry from the flotation vessel are also transported to the tailings pond.

Proper operation of the primary separation vessel is achieved by controlling the withdrawal rate of the underflow stream based on its density, whereas dilution water is used to control the density and withdrawal rate of middlings. By considering the slip velocity of fluid particles, Masliyah and coworkers developed a materials balance model around a source zone that supplies solids to both the underflow and overflow (Nasr-El-Din et al. 1999). The concentration of light (oil or aerated bitumen) and heavy (solid) particles in the source zone is assumed to be uniform. If the downward fluid velocity is greater than the rise velocity of the light particles, the fluid can carry the light particles beyond the lower boundary (underflow). Otherwise, the light particles can only move beyond the upper boundary (overflow). Similar constraints govern the motion of heavy particles in the source zone. A further stipulation of the mathematical model is that light and heavy particles cannot enter the source zone from the lower (light) and upper (heavy) boundaries, respectively. The vertical velocity of the light particles at the upper boundary is given by

$$V_{lo} = V_{fo} + K_{lo} \quad \text{(EQ 12)}$$

where

V_{fo} = vertical velocity of the fluid at the overflow
K_{lo} = generalized slip velocity

Similar equations can be derived for both heavy and light particles at the underflow and overflow boundaries. In addition to the slip velocities, one must define a withdrawal rate from either the underflow or overflow. For the overflow (o), the volumetric withdrawal rate (Q_o) is given by

$$Q_o = -A[V_{lo}\alpha_l + V_{ho}\alpha_h + V_{fo}\alpha_f] \quad \text{(EQ 13)}$$

where downward is taken to be positive and

A = cross-sectional area of the separator
α = volume fraction
l, h, and f = light particle species, heavy particle species, and fluid, respectively

The mathematical model describes the continuous separation of a bi-disperse suspension settling in a vertical settler and neglecting any lateral concentrations gradients. With the feed stream flow rate and composition defined, the model is used to describe the recovery of light and heavy particles as a function of the split ratio (Q_u/Q_F); see Figure 10.

With a high throughput, bitumen recovery in the primary separation vessel can be controlled by adjusting the middlings flow rate, the underflow withdrawal rate, or the flow rates of both streams. The typical residence time in the primary separation vessel is around 30 minutes but is highly dependent on variations in the feed stream flow rate and the quality of the oil sands ore. For low-grade ores, the residence time is often increased to account for the higher dilution ratio (water to oil sands ore), reduced attachment efficiency of bitumen droplets to flotation-sized air bubbles, and the increased hindrance of bitumen droplets navigating through the tortuous middlings layer. Typical operation performance indicators of the primary separation vessel include the bitumen froth–middlings interface level, the interface clarity, and the color of the bitumen froth. Good processing ores typically exhibit a sharp interface and shining dark bitumen froth. Although there is some instrumentation to monitor the performance of separation vessels, such as pressure or density profilers, often the most reliable performance indicator is obtained from the level of the froth–pulp interface imaged via sight glasses on the separation vessel. Sight glasses are installed in the shell of the vessel, providing plant operators with real-time information on the bitumen froth quality, clarity of the bitumen froth–middlings interface, and the depth of the bitumen froth layer. A possible challenge in running primary separation vessels is the gelation of the middlings layer because of high clay content combined with poor water chemistry, such as high divalent cation concentration.

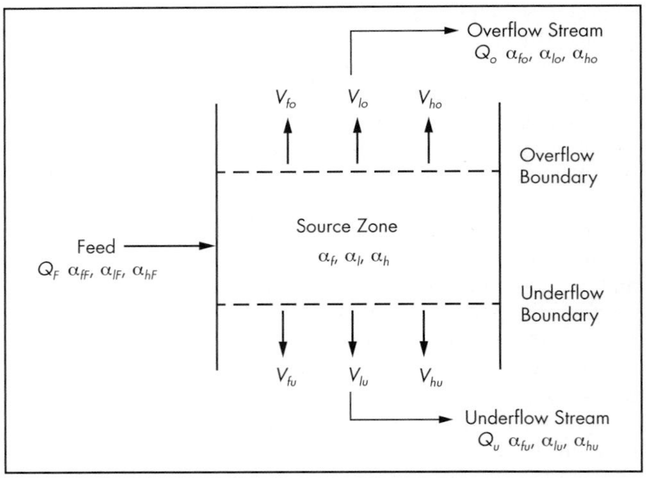

Source: Masliyah et al. 2011

Figure 10 Vertical separation vessel used to construct a mathematic model that describes heavy and light particle separation in the source zone

Bitumen Froth Treatment

The processing objective in froth treatment is to reduce the water and solids content to meet the specifications for upgrading and/or refining of heavy bitumen. The bitumen froth containing typically 40 vol % air is too viscous to be efficiently pumped over long distances. As mentioned earlier, the bitumen froth is first deaerated prior to being fed to the subsequent froth treatment vessel as shown in Figure 11. Reducing the air content reduces the bitumen froth viscosity and increases its flowability. The air bubbles separate from the bitumen froth under buoyant force, with the separation process greatly enhanced by operating at elevated temperatures ~95°C. Through a steam deaerator, the air content in bitumen froth is reduced to approximately 5 vol %. After removal of air (deaeration), bitumen froth from the PSV typically contains 60% bitumen, 30% water, and 10% solids by weight, which remains of little value unless it is cleaned using so-called froth treatment to further reduce the water ~2.0 wt % and solids ~0.5 wt % contents so that the produced bitumen becomes suitable for upgrading and conversion to synthetic crude oil using cokers. However, much lower water and solids content are required when the diluted bitumen is transported for long distances to upgraders.

The separation of water and solids from bitumen is complicated by a weak driving force which results from an almost equivalent density of bitumen and water, and a d_{50} for the suspended water droplets and solids ranging from a few micrometers to hundreds of micrometers. To enhance the rate of separation, deaerated bitumen froth is cooled and diluted with a solvent (light hydrocarbon). The addition of a bitumen-miscible solvent significantly reduces the viscosity and density of the organic phase, assisting the separation of fine water droplets and solids by gravity. Currently, two distinctly different solvents (naphtha and paraffin liquids) are used, leading to two different froth treatment technologies known as naphthenic and paraffinic froth treatment processes.

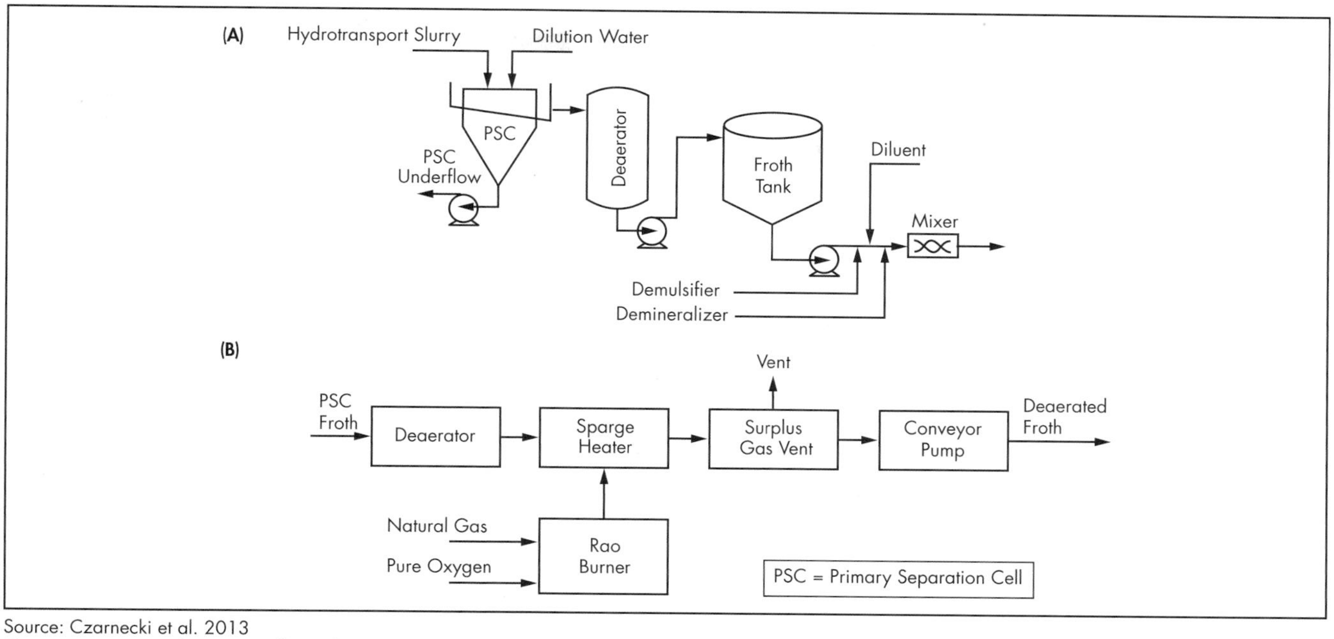

Source: Czarnecki et al. 2013

Figure 11 Typical process flow diagrams applied (A) to treat bitumen froth recovered from the primary separation vessel, and (B) expanding the sequence of steps encountered in the deaeration process via heating

In a typical naphthenic froth treatment process, shown in Figure 12, the operators dilute the bitumen froth with naphtha at a solvent-to-bitumen ratio of about 0.65–0.8 by weight. The source of naphtha is usually from an integrated upgrading plant, where a light hydrocarbon cut can be produced. Although the process provides improved separation of water and solids from bitumen, the naphthenic process requires a series of inclined plate settlers and centrifuges to achieve desirable separation rates and product quality. The complexity of the bitumen separation process is demonstrated by the process flow diagram of a commercial froth treatment plant (Figure 12).

Naphtha-diluted bitumen froth is first processed using scroll centrifuges to remove coarse solids, with a secondary separation of oversized solids achieved by pumping the diluted bitumen through a stack of Cuno filters. The remaining stream of diluted bitumen containing over-specification content of water and fine solids is then fed through a series of disc centrifuges and disc clarifiers to ensure that the water and solids contents are below the specifications of the feed to the upgrading processing plant. The fine solids and water from the disc centrifuges are further treated to recover naphtha in the naphtha recovery unit. To increase plant capacity, inclined plate settlers are installed to rapidly process a fraction of the diluted bitumen froth that immediately satisfies upgrading specifications. The underflow from the inclined plate separators is combined with the feed into the scroll centrifuges to minimize bitumen losses from the operation. The naphthenic froth treatment process is operated at atmospheric pressure and 75°–78°C.

During the naphtha separation process, any agitation resulting from pumping and mixing of the diluted bitumen can cause the formation of "tight emulsions," with micrometer-sized water droplets (~3 μm) stabilized by hydrophobic fines (Figure 13) and asphaltenes. These solids-stabilized interfaces prevent the coalescence of water droplets and inhibit water dropout. The enhanced interfacial film stability is related to the particle wettability. Equation 14 describes the energy (ΔE) needed to detach a colloidal particle from a planar oil–water interface.

$$\Delta E = \pi r^2 \gamma_{\frac{o}{w}} (1 \pm \cos\theta)^2 \qquad \textbf{(EQ 14)}$$

With a constant oil–water interfacial tension (γ) and particle radius (r), the particle detachment energy is maximized when $\cos\theta = 0$ and minimized when $\cos\theta = -1$ and $+1$, depending on the type of emulsion, that is, water in oil (W/O) or oil in water (O/W). The conditions for minimal detachment energy correspond to complete particle wetting in either the oil phase (−1) or water phase (+1), whereas the maximum detachment energy is a consequence of equal partitioning of the particle in both the water and oil phases, as shown in Figure 13B. At this condition, solids-stabilized emulsions exhibit the greatest stability, which results from steric hindrance to compensate for the particle detachment energy.

To promote coalescence of solids-stabilized water droplets, chemical demulsifiers are added to the diluted bitumen at a concentration in the order of parts per million. Chemical demulsifiers are amphiphilic; hence they have a strong affinity to partition at the water–oil interface. The competitive adsorption of the demulsifier molecules at the water–oil interface disrupts the rigid particle networks, allowing the droplets to coalesce on contact. Simultaneously, demulsifier molecules continually adsorb onto the oily-contaminated solids, gradually lowering the particle hydrophobicity and consequently the particle detachment energy into the water phase. Such actions are favorable for increased rates of water separation.

The low-temperature paraffinic froth treatment process uses a paraffinic solvent at about 2:1 mass ratio, causing partial precipitation of asphaltenes. A "rigid" water–oil interface is characteristic in paraffinic froth treatment and is favorable for the sweep flocculation of water droplets by the precipitated

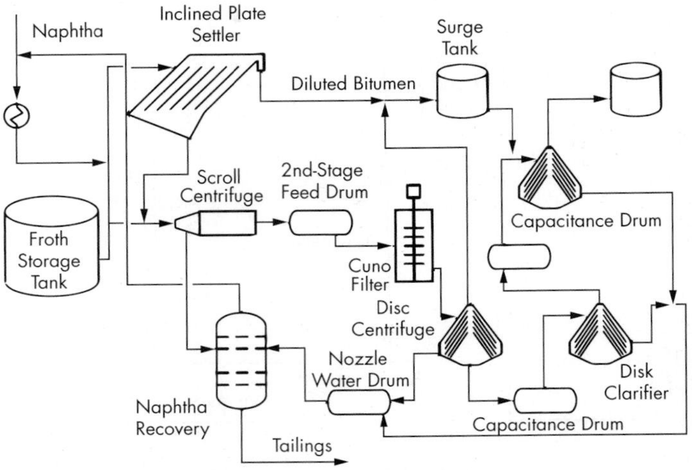

Typical Naphthenic Froth Treatment Plant

Scroll Centrifuge

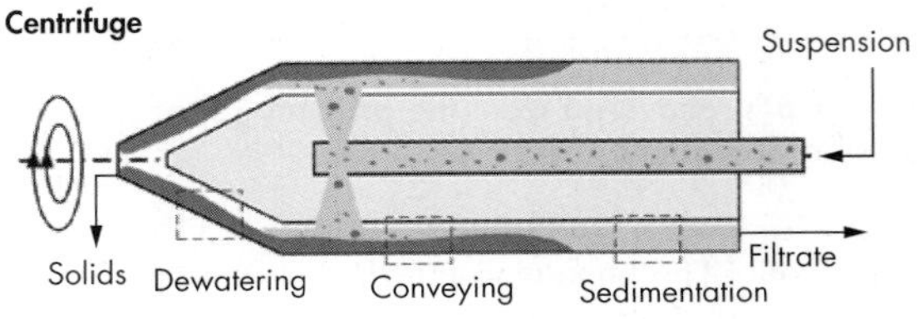

Adapted from Karsten Keller, DuPont

Diluted bitumen froth is fed into the center of a rotating bowl and discharged through a scroll rotating in the same direction. The bowl rotating at ~800 rpm pushes the coarse solids to the bowl wall. The scroll rotating at a slightly lower speed pushes the coarse solids in the opposite direction to the liquid flow. The coarse solids are discharged to a hopper and the diluted bitumen froth recovered over a weir.

Cutaway of an Alfa-Laval Disc Centrifuge

The filtrate from the Cuno filter is centrifuged to separate the naphtha-diluted bitumen from water and fine solids (2–5 µm diameter). The feed enters through a central feed line and is discharged over the disc stack. Diluted bitumen is recovered through the top of the disc centrifuge as the high centrifugal forces (~2,500 g) push the fine solids and water to the outer edge of the disc stack. Water, fine solids, and some hydrocarbon are discharged through nozzles at the edge of the centrifuge. Typical disc centrifuge capacities exceed 100 m^3/h.

Inclined Plate Settler

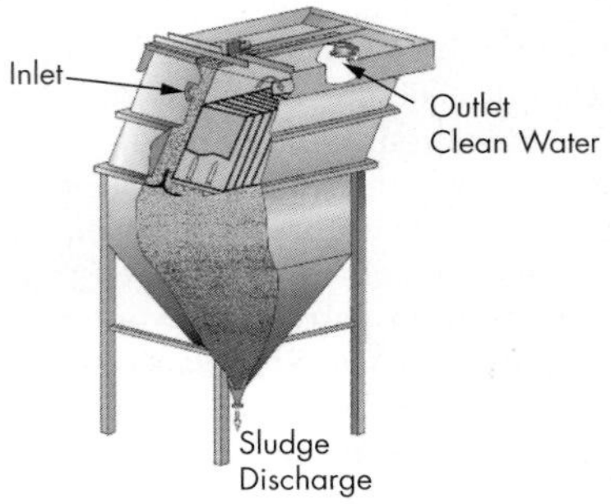

Source: Czarnecki et al. 2013

Diluted bitumen froth flows down the feed chamber and up through the inclined plates. The narrow spacing between the plates reduces the settling distance for the solids, resulting in a higher unit capacity for a given settling area. The settled solids slide down the angled (~55°) plates and are discharged through a cone at the base of the settler along with water and some hydrocarbon emulsions. Clarified diluted bitumen froth is deaerated as the froth is recovered in the overflow launders. The bitumen product typically contains <2 wt % water and 1 wt % solids, suitable for upgrading without further processing.

Figure 12 Simplified process flow diagram of a commercial naphthenic froth treatment plant and an overview of the key processing units

asphaltenes. The rapid settling of flocculated water droplets forms a sludge layer but leads to a virtually water- and solids-free diluted bitumen, producing a better-quality bitumen product than the naphthenic process with the water and solids contents typically <300 ppm and 700 ppm, respectively. The low water content and the partial rejection of asphaltenes mitigate corrosion concerns in downstream processing as well as facilitate higher conversion rates when treating the partially upgraded bitumen with hydrogen. The unit operations involved in a typical paraffinic froth treatment process are shown in Figure 14.

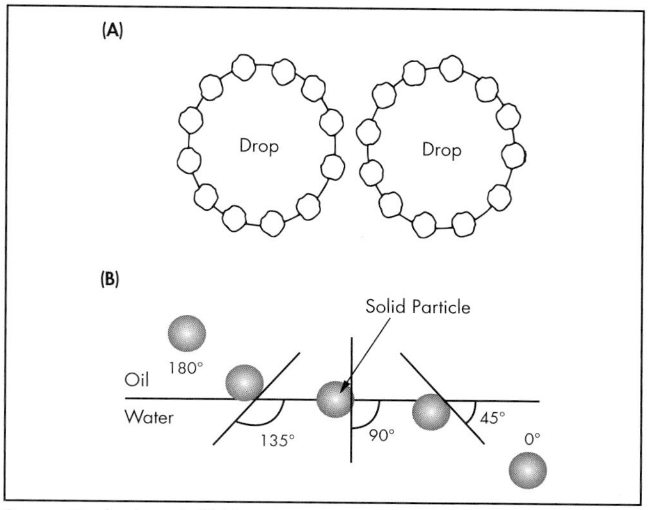

Source: Masliyah et al. 2011

Figure 13 (A) Particle-stabilized film preventing the coalescence of water droplets, and (B) influence of the particle wettability on the particle positioning at the oil–water interface

With enhanced separation driven by asphaltene precipitation, the inclined plate settlers and centrifuges used in the naphthenic froth treatment process can be replaced by large settling vessels. Three settling vessels are operated in a countercurrent flow configuration with fresh solvent blended with the underflow from the second separation vessel. The countercurrent flow ensures that the diluent-to-bitumen ratio increases from the first stage to the third stage. The final diluted bitumen product is removed from the first-stage overflow and the majority of the diluent recovered in a solvent recovery plant. Additional solvent is recovered from the tailings underflow from the third-stage separation vessel. Both solvent streams are recycled back into the third-stage settling vessel. The paraffinic froth treatment process is typically operated at atmospheric pressure and at 30°–32°C, although higher temperature operations are being used at other plants. The latter operations are referred to as high-temperature froth treatment.

Solvent Recovery

With both froth treatment processes requiring significant solvent loading, there is a need to recover and recycle the solvent back into the froth treatment process. The amount of solvent recovered is often a balance between several factors, including economics, availability, environmental impact, and transportability. Solvent recycle is achieved by passing the diluted bitumen (dilbit) through a series of process units that provide dilbit preheating, flash drums to recover the majority of the solvent, and finally fractionation. A typical process flow diagram for a naphtha solvent recovery unit is shown in Figure 15. The dilbit is first heated to a temperature between

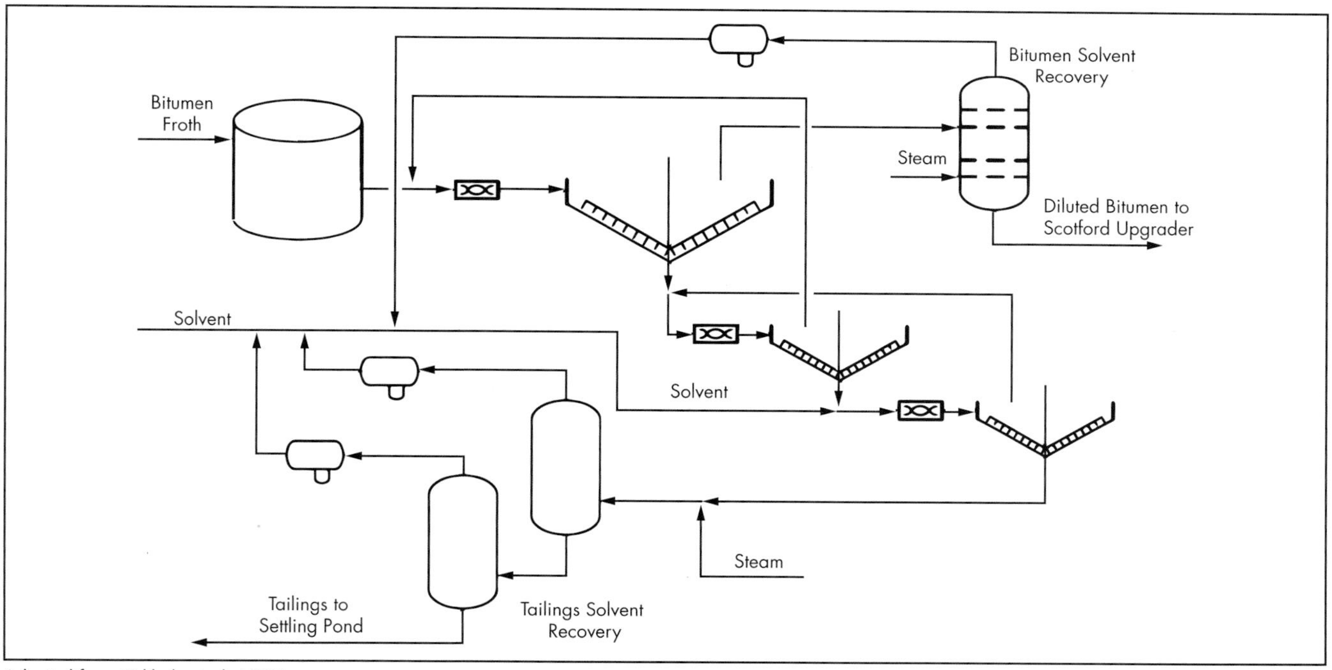

Adapted from Birkholz et al. 1997

Figure 14 Paraffinic froth treatment process flow sheet from Albian Sands Energy

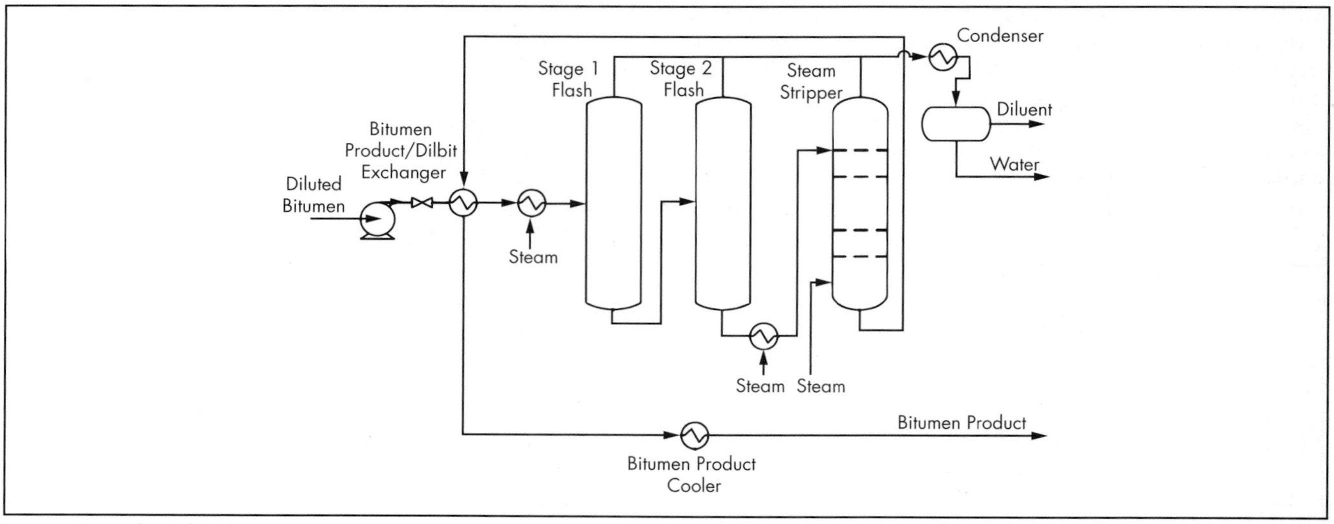

Source: Czarnecki et al. 2013

Figure 15 Naphtha-based diluent recovery unit process flow diagram

80°C and 100°C and then discharged into a flash drum or a series of flash drums. A steam stripper is then used to remove the heaviest fraction of naphtha to produce a dry bitumen product that meets the required specification for upgrading.

In paraffinic froth treatment, the product is already partially upgraded because of the loss of asphaltenes in the froth treatment process. Although solvent is recovered for recycle, the long-term stability of the bitumen is also a requirement, and this is achieved by either partial or complete removal of the paraffinic solvent to prevent further asphaltene precipitation. Since the same solvent is used in the froth treatment and bitumen transportation processes, the solvent recovery process in paraffinic froth treatment operations, as shown in Figure 16, has been designed to recover a fraction of the added solvent. The dilbit entering the solvent recovery unit contains approximately 70 wt % paraffinic (C_5/C_6) solvent. The feed stream is heated from ~33°C to the operating temperature via a sequence of heat exchanges with overhead vapor and diluted bitumen underflow from the solvent recovery unit, and a steam heater. The dilbit is then flashed in the solvent recovery unit column, and the solvent content in the bitumen product is reduced to approximately 25 wt %, suitable for pipeline transportation to the upgrading facility.

Although the majority of the added solvent is removed with the bitumen, a small portion is lost in the tailings underflow (with residual bitumen and solids) of the separation units. To further recover the solvent, the underflow streams are transferred to a tailings solvent recovery unit where the solvent content in the tailings is reduced below limits of no greater than 4 volumes of solvent to every 100 volumes of dry bitumen produced, set by the environmental regulator. A naphtha tailings solvent recovery unit is shown in Figure 17A. The hot tailings are initially diluted with hot water and distributed evenly across the flash vessel to flow down the column of shed decks, which provide countercurrent contact with the rising stripping steam added directly into the tailings pool at the bottom of the flash column. The recovered vapor from the flash vessel is sprayed with water to remove any entrained solids prior to three-phase separation of noncondensable vapors, water, and naphtha. Operating the flash vessel close to 100°C at atmospheric pressure, the solvent recovered is approximately 80% of the feed solvent.

For paraffinic solvent recovery, the flash vessel is specially designed to enable efficient solvent recovery from a feed stream that contains residual bitumen encapsulated by precipitated asphaltene flocs. The solvent recovery process from paraffinic froth treatment tailings (Figure 17B) begins by injecting steam directly into the froth treatment tailings such that the tailings temperature is well above the boiling point of the solvent. The tailings stream is subsequently sprayed into the flash vessel, which operates slightly above atmospheric pressure. Creating a fine mist increases the solvent–vapor surface area to enhance solvent mass transfer to the vapor phase. The recovered vapor separated with the solvent is recycled back to the froth treatment, and residual water and oil returned to the flash vessel. The falling mist is captured in an agitated liquid pool at the bottom of the flash vessel where continuous agitation of the liquid pool provides sufficient shear to break asphaltene flocs and particle aggregates, thus aiding the release of encapsulated bitumen. Further solvent is recovered as the underflow from the first flash vessel is fed to a second flash vessel that operates at a lower temperature and under a slight vacuum pressure. Because of the high asphaltene content, fine solids, and natural surfactants in the bitumen, there is a tendency of building up foams in the first flash vessel. To avoid excessive foaming and the potential carryover of unwanted solids, a secondary vessel (foam-breaking vessel) is operated under vacuum with the addition of chemical defoamers and asphaltene dispersants to collapse the foam before reinjection into the flash vessel. The two-step solvent recovery process as applied to treat the tailings is capable of recovering >99% of the feed solvent with ~95% of the solvent recovered in the first flash vessel.

Tailings Management

In the warm-water extraction process, the production of one barrel of crude bitumen results in approximately 3.3 m^3 of fresh tailings slurry, which is discharged to the tailings ponds. As the rate of oil production increases, industry has focused efforts on reducing the use of fresh water and increasing

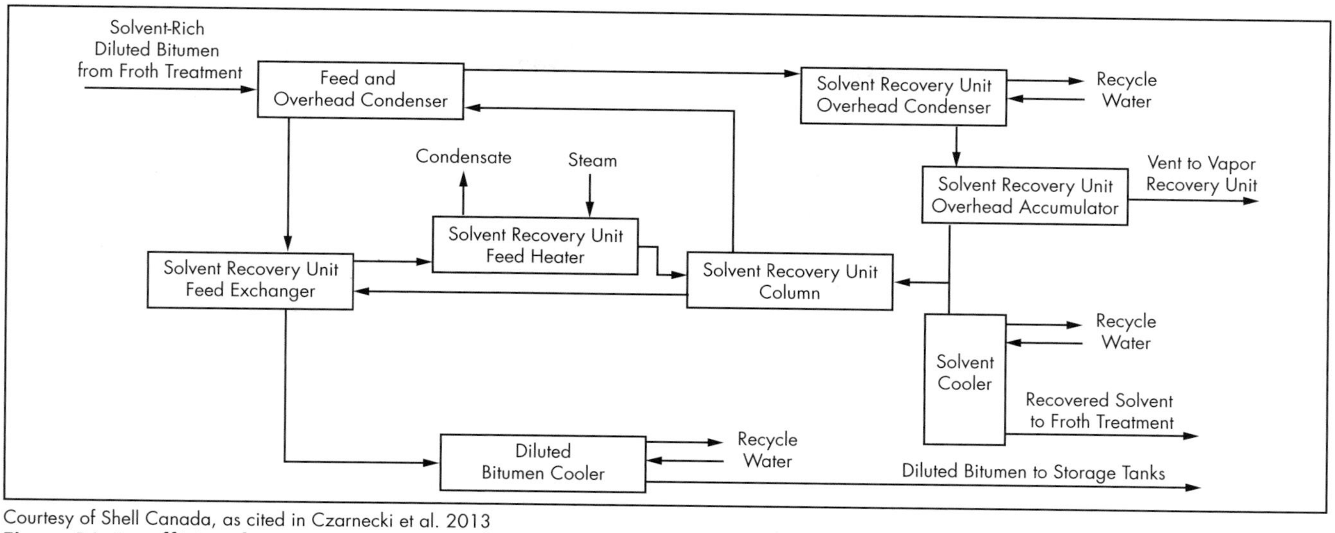

Courtesy of Shell Canada, as cited in Czarnecki et al. 2013

Figure 16 Paraffinic solvent recovery process flow diagram as operated by Albian Sands Energy

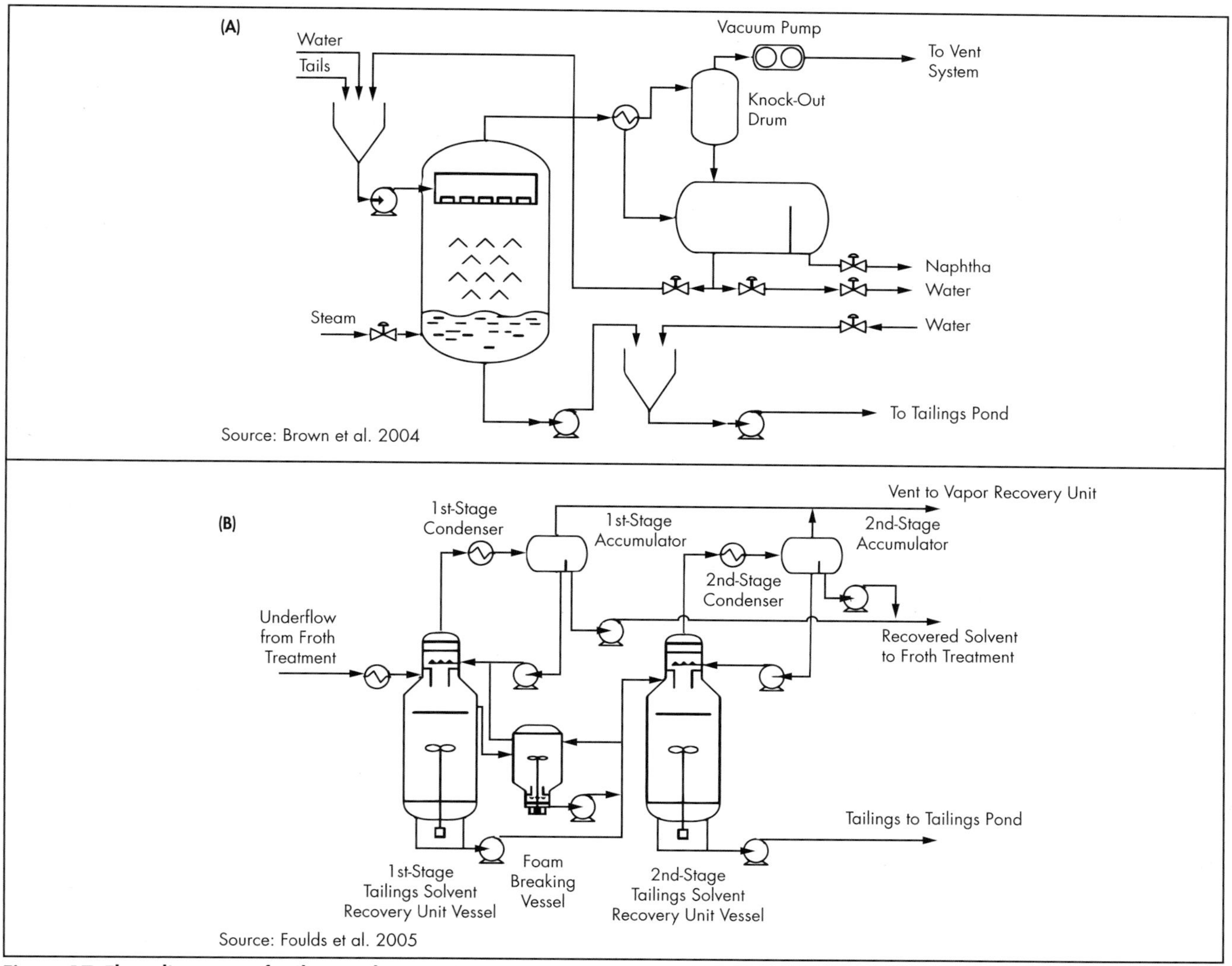

Figure 17 Flow diagrams of tailings solvent recovery processes operated for (A) naphtha froth treatment tailings and (B) paraffinic froth treatment tailings

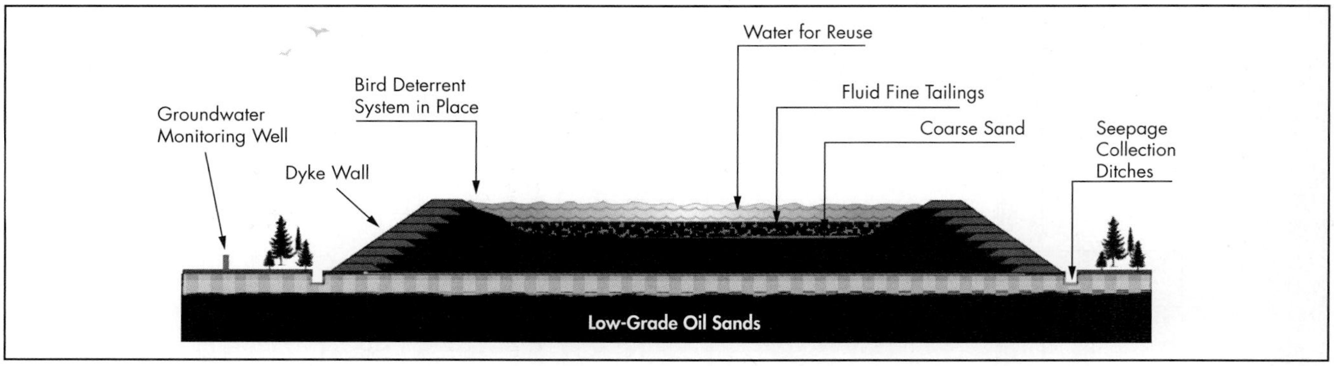

Source: Czarnecki et al. 2013
Figure 18 External tailings management facility

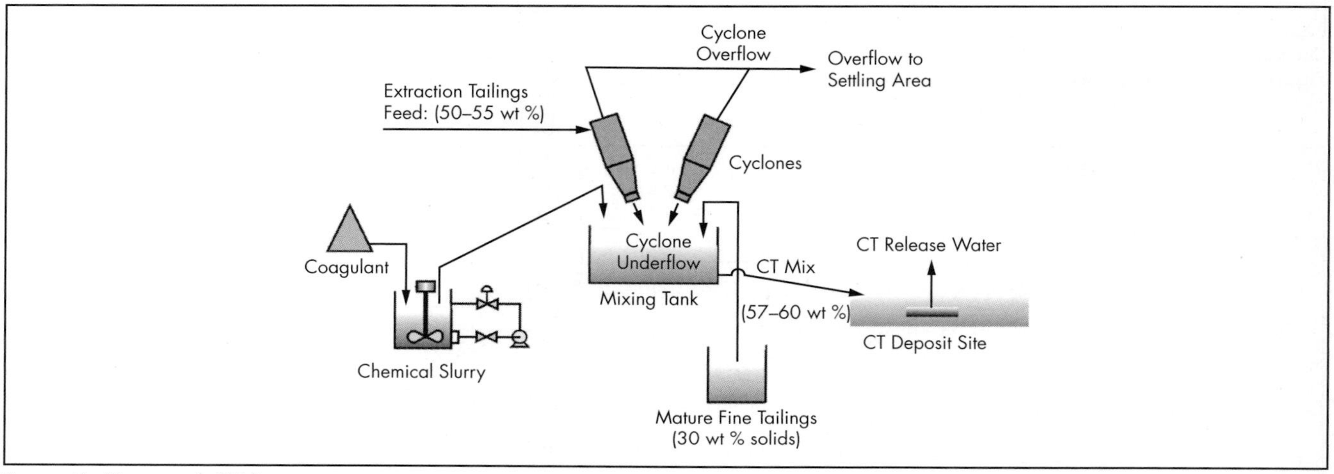

Source: MacKinnon et al. 2010
Figure 19 Typical CT process

energy conservation within the process. Both features rely on the rapid separation and recycle of water from the combined tailings streams received from the bitumen extraction and froth treatment stages.

Tailings ponds are contained by dikes that are formed using the overburden and interburden materials from the mining operation and from the extraction of coarse solids. As the whole tailings is discharged into the pond, the coarse solids (sands) segregate rapidly from the fines and contribute to the formation of sand beaches surrounding the tailings pond. The fine particles (silt and clays) settle along the beach as the slurry flows toward the pond water surface, where the runoff arrives as a thin slurry containing very fine solids. Following a short induction period, the fine particles begin to settle to form a zone of fluid fine tailings (FFT) in the pond. With the expansion of the oil sands industry, Alberta has witnessed the construction of tailings ponds that now occupy a total land area of more than 220 km^2. With the volume of FFT accumulating at a rate more than 20% of the volume of oil sands excavated from the mine, substantial efforts have been directed to enhance the settling rate of FFT and consolidation of MFT. FFT typically contains 85 wt % water, 13 wt % clays, and 2 wt % bitumen, with the FFT zone in the tailings pond often as deep as 40 m. After two or three years of gradual consolidation, MFT of ~30 wt % solids develops with very slow consolidation thereafter because of the relatively high compressive yield strength of the three-dimensional particle network. To avoid disturbance of the FFT during water removal, the tailings pond is usually operated with a 3-m-deep surface water zone. Figure 18 shows a typical tailings pond with the stratified tailings contained between dikes.

Environmental regulations stipulate that tailings ponds should be remediated and reclaimed within 10 years of the end of mine life. The long-term geological stability of the tailings ponds is linked to the compressibility, deformability, strength, and hydraulic conductivity of the tailings deposit. Improving the rate at which the tailings deposit meets geotechnical strength and stability calls for innovation. Tailings management innovation has allowed the industry to rapidly increase oil production during the last decade. The CT process outlined in Figure 19 and developed earlier has been used to enhance settling of FFT and densify MFT. In this process, fresh extraction tailings are first classified by hydrocyclones such that the coarse particles in the underflow of the cyclones are combined with reclaimed MFT from the tailings pond. The MFT feed stream is first treated with a coagulant, typically gypsum at 1,000 g/m^3, to promote the aggregation of the fine clay particles. The coarse particles from the hydrocyclone underflow are then blended with the treated MFT at a sands-to-fines (S/F) ratio of ~4. The mixture is discharged to a CT

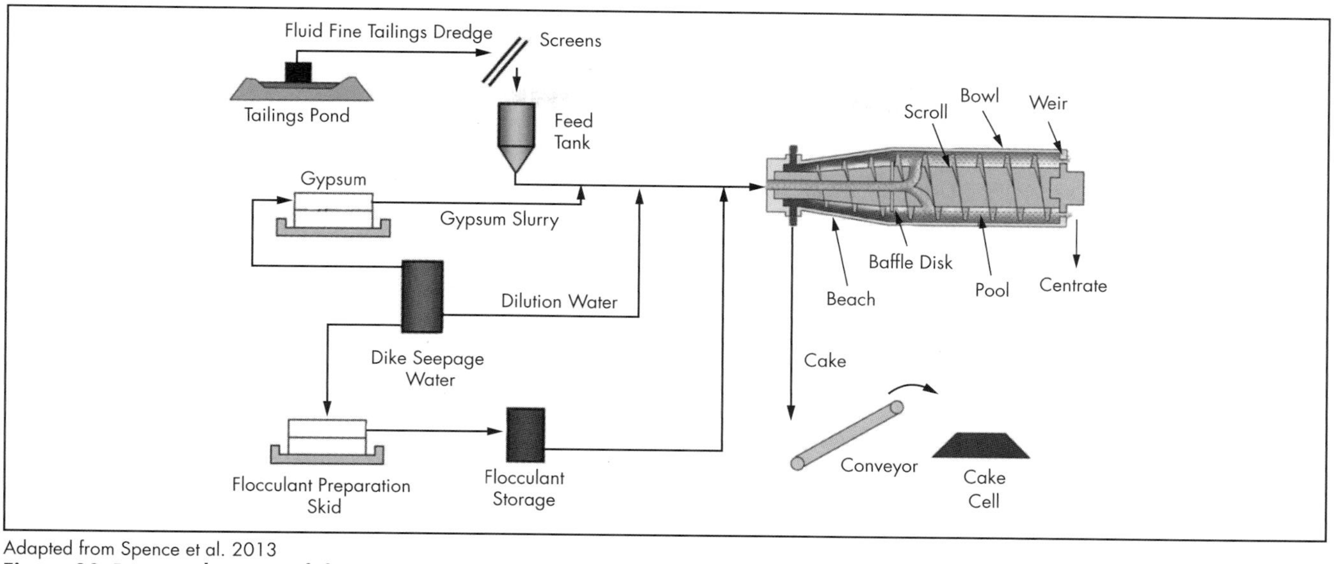

Adapted from Spence et al. 2013

Figure 20 Process diagram of decanter centrifuge for treating fluid fine tailings

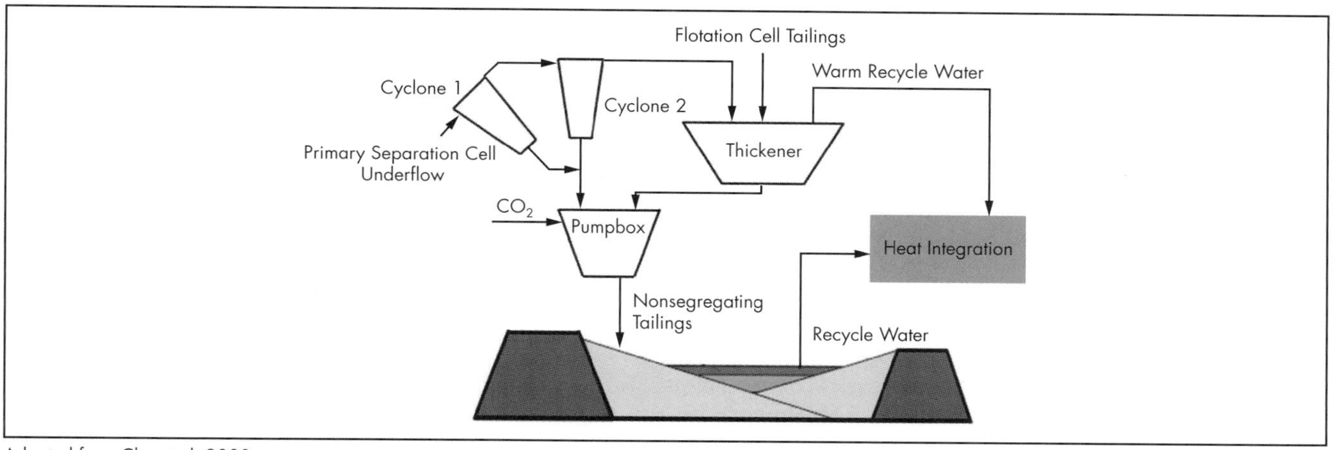

Adapted from Chu et al. 2008

Figure 21 Process diagram of nonsegregating tailings process for treating fresh oil sands extraction tailings

tailings pond where the coagulated fines and the coarse sands settle at equivalent rates to quickly release the clarified water ready for recycle within a few days. After 30 days, the nonsegregating solids form a consolidated sediment of approximately 60 wt % solids, whereas further desiccation occurs over time. One of the downsides to the CT process is the continual buildup of calcium ions in the process water. Unless otherwise treated, a high concentration of calcium ions is detrimental to the processing of certain types of oil sands ores and can cause scaling in pipes and other processing equipment. The demand for coarse sands at an S/F ratio of 4 is another concern that may limit the potential of the CT process in oil sands tailings management.

A recent advance in treating MFT is by flocculation of MFT, followed by centrifugation as schematically shown in Figure 20. The centrifugation process has been commercially demonstrated (Gillies et al. 1991). In this treatment option, MFT is first dredged through a screen to remove coarse sands and debris that could potentially cause severe wear of the downstream centrifuge. The underscreen suspension stream, kept in a holding tank, is mixed with coagulant (gypsum) and further diluted if required prior to flocculant addition. The addition of gypsum is found to increase both the strength of the flocs and the throughput of the centrifuge. The flocculated MFT is sent to a decanter centrifuge to produce a centrifuge cake of solids content >55 wt % with a fines capture of 95%. With a centrifuge of 1-m bowl diameter, it is anticipated to treat 70 t of dry solids per hour (Spence et al. 2015).

There are incentives to treat FFT right after their production to avoid building large tailings ponds and additional materials handling of moving MFT, while utilizing the thermal energy in warm recycle water. To this end, the nonsegregating fresh tailings treatment process shown in Figure 21 represents another industry innovation. In this process, flotation tailings are flocculated using polymer flocculants for enhanced thickening in a specially designed thickener and rapid release of warm water. The underflow stream of primary separation cell is fed to a hydrocyclone to separate fine solids and fugitive bitumen from coarse sands, while fine solids and fugitive bitumen in the hydrocyclone overflow stream are fed to flotation cells for further bitumen recovery. The coarse solids in the underflow stream of the primary separation vessel are blended in the pumpbox with the underflow stream of the thickener. The mixture with added carbon dioxide (CO_2) is discharged as nonsegregating tailings (NST) to the tailings ponds where more water is released for recycle. The addition of CO_2 is to reduce the pH of NST, improve clarity of released water, and accelerate the release of water in the tailings pond for recycle. To form NST, the key is to control the ratio of fines to sand so that they fall in a nonsegregating regime of a solids mix.

Although technology has evolved to treat oil sands FFT, the challenge remains to reduce the cost of treatment and increase operational liability. For this reason, many technologies are currently under investigation or large-scale testing, including thin-lift drying, flocculation-assisted filtration, rim-ditch deposition, in-line filtration, and freeze–thaw cycles, to name just a few. Interested readers are directed to consult the Oil Sands Tailings Road Map (COSIA, n.d.).

REFERENCES

Adams, J.J., Rostron, B.J., and Mendoza, C.A. 2004. Coupled fluid flow, heat and mass transport, and erosion in the Alberta Basin: Implications for the origin of the Athabasca oil sands. *Can. J. Earth Sci.* 41(9):1077–1095.

Birkholz, R.K.O, Power, W.J., and Tipman, N.R. 1997. Method for processing a diluted oil sand froth. Patent CA 2200899A1.

Brown, W., Portwood, M., King, F., Bara, B., and Wagner, M. 2004. Method for recovery of hydrocarbon diluent from tailings. Canadian Patent 2272045.

Chu, A., Paradis, T., and Wallwork, V. 2008. Non-segregating tailings at the Horizon Oil Sands Project. Presented at the First International Oil Sands Tailings Conference, Edmonton, Alberta, December 7–10.

COSIA (Canada's Oil Sands Innovation Allliance). n.d. Oil Sands Tailings Road Map. https://www.cosia.ca/uploads/documents/id9/Tailings%20Roadmap%20Volume%201%20June%202012.pdf. Accessed February 2017.

Czarnecki, J., Masliyah, J., and Xu, Z. 2013. *Handbook on Theory and Practice of Bitumen Recovery from Athabasca Oil Sands*. Vol. II: Industrial Practice. Edmonton, AB: Kingsley Knowledge Publishing Services.

Foulds, G.A., Tipman, N.R., Payne, B., Wamboldt, B.C., and Valgardson, B.E. 2005. Process for removing solvent from an underflow stream from the last separation step in an oil sands froth treatment process—Chevron Canada Resources Limited. Canadian Patent 2353109.

Fuerstenau, M.C. 2003. *Principles of Mineral Processing*. Littleton, CO: SME.

Garven, G. 1989. A hydrogeologic model for the formation of the giant oil sands deposits of the Western Canada sedimentary basin. *Am. J. Sci.* 284:1125–1174.

Gillies, R., Shook, C., and Wilson, K. 1991. An improved two-layer model for horizontal slurry pipeline flow. *Can. J. Chem. Eng.* 69(1):173–178.

Lelinski, D., Drelich, J., Miller, J.D., and Hupka, J. 2004. Rate of bitumen film transfer from a quartz surface to an air bubble as observed by optical microscopy. *Can. J. Chem. Eng.* 82(4):794–800.

MacKinnon, M.D., Matthews, J.G., Shaw, W.H., and Cuddy, R.G. 2001. Water quality issues associated with composite tailings (CT) technology for managing oil sands tailings. *Int. J. Surf. Min. Recl. Environ.* 15(4):235–256.

Masliyah, J., Czarnecki, J., and Xu, Z. 2011. *Handbook on Theory and Practice of Bitumen Recovery from Athabasca Oil Sands*. Vol. I: Theoretical Basis. Edmonton, AB: Kingsley Knowledge Publishing Services.

Nasr-El-Din, H.A., Masliyah, J.H., and Nandakumar, K. 1999. Continuous separation of suspensions containing light and heavy particle species. *Can. J. Chem. Eng.* 77(5):1003–1012.

Oil Sands Magazine. 2018. OPP-Wet: Slurry preparation plants. www.oilsandsmagazine.com/technical/mining/ore-preparation-plant/opp-wet-slurry-preparation. Accessed August 2018.

Ren, S., Trong, D.-V., Zhao, H., Long, J., Xu, Z., and Masliyah, J. 2008. Effect of weathering on surface characteristics of solids and bitumen from oil sands. *Energy Fuels* 23(1):334–341.

Ren, S., Zhao, H., Long, J., Xu, Z., and Masliyah, J. 2009. Understanding weathering of oil sands ores by atomic force microscopy. *AICHE J.* 55(12):3277–3285.

Scott, P. 1967. Major tar sand deposits of the world. Presented at the 7th World Petroleum Congress.

Spence, J., Bara, B., Lorentz, J., Lahaie, R., and Mikula, K. 2013. Development of centrifuge technology for fluid fine tailings. Presented at Tailings and Mine Waste 2013, Banff, Alberta, November 4–6.

Spence, J., Bara, B., Lorentz, J., and Mikula, R. 2015. Development of the centrifuge process for fluid fine tailings treatment at Syncrude Canada Ltd. Presented at the World Heavy Oil Conference, Edmonton, Alberta, March 24–26.

Tyler, J., Kiel, D., Spence, J., Schaan, J., and Larson, G. 2009. The use of physical modeling in the optimisation of a primary separation vessel feedwell. *Can. J. Chem. Eng.* 87(6):821–831.

Walters, E. 1974. *Review of the World's Major Oil Sand Deposits*. The IOCC Monograph Series: Tar Sands. Edited by D. Ball, L.C. Marchant, and A. Goldburg. Calgary, AB: Canadian Society of Petroleum Geologists. pp. 1–11.

Wang, L., Trong, D.-V., Xu, Z., and Masliyah, J.H. 2010. Use of short-chain amine in processing of weathered/oxidized oil sands ores. *Energy Fuels* 24(6):3581–3588.

Zhao, H., Long, J., Masliyah, J.H., and Xu, Z. 2006. Effect of divalent cations and surfactants on silica-bitumen interactions. *Ind. Eng. Chem. Res.* 45(22):7482–7490.

Zhao, H., Dang-Vu, T., Long, J., Xu, Z., and Masliyah, J.H. 2009. Role of bicarbonate ions in oil sands extraction systems with a poor processing ore. *J. Disper. Sci. Technol.* 30(6):809–822.

Zhou, Z., Kasongo, T., Xu, Z., and Masliyah, J. 2004. Assessment of bitumen recovery from the Athabasca oil sands using a laboratory Denver flotation cell. *Can. J. Chem. Eng.* 82(4):696–703.

CHAPTER 12.28

Phosphate Rock

Patrick Zhang

Phosphorus is essential to plant and animal life. It provides the material for skeletal bone structure in animals as well as the material for cell membranes in both plants and animals. The largest and least expensive source of phosphorus is obtained by mining and concentrating the phosphate rock from the numerous phosphate deposits around the world. The principal uses for phosphate rock are in the manufacture of fertilizer ingredients for food crop nutrition and the production of animal feed supplements to promote healthy and quick skeletal growth for meat-producing livestock.

Phosphorus is also one of the most abundant minerals in the human body, second only to calcium. This mineral is essential for the healthy formation of bones and teeth and is necessary for our bodies to process many of the foods that we eat. It is a part of the body's energy storage system and helps with maintaining healthy blood sugar levels. Phosphorus is also found in substantial amounts in the nervous system. The regular contractions of the heart are dependent on phosphorus, as are normal cell growth and repair.

EVOLUTION OF PHOSPHATE FERTILIZERS

The discovery of phosphorus, its role in agriculture, and the early development of phosphate fertilizers were well documented by Vincent Sauchelli (1942, 1951, 1965). Phosphorus was first discovered by a German metallurgist, who made it by heating urine into a solid state. Although as early as 2,000 years ago, Chinese farmers applied calcined or lime-treated bones to their fields to improve crop growth, and ancient farmers from many parts of the world used human wastes and animal manure as the primary source of phosphorus for crops, the essential role of phosphorus in plants' growth was first explained in a scientific manner by Erasmus Darwin in 1799 (Beaton 2010).

The first true phosphate fertilizers were made by acidulating bones with sulfuric acid. This process originated in Europe in the early 1800s. In 1842, John Bennet Lawes of the Rothamsted Experimental Station was granted a patent for the production of superphosphate from bones, and he began manufacturing and selling this fertilizer the same year. This patent was amended in 1848 to include sulfuric acid treatment of phosphate ore. The number of superphosphate plants in England reached 14 by 1853.

Large-scale phosphate fertilizer production was made possible by replacing bones with mined phosphate rock as the phosphate source. Small amounts of phosphate ores were mined in the mid-1840s in England, France, and Spain and in the 1860s in Norway and Germany. Phosphate rock mined in Ontario and Quebec, Canada, was shipped to England for processing between 1863 and 1895. In the United States, phosphate mining began in 1867 in South Carolina, followed by Florida in 1888, Idaho in 1906, Wyoming in 1907, and Montana in 1921.

To date, about 120 phosphate minerals have been identified, with apatite accounting for approximately 95%. Other commercial phosphate minerals include kribergite, guano, and vivianite. There are two primary routes for recovering the phosphate values from phosphate rock. In the thermal process, phosphate is reduced by carbon to elemental phosphorous gas and condensed in water. Today's predominant process is the "wet acid" process in which phosphate rock is acidulated with sulfuric acid.

Superphosphate

A simplified reaction for producing superphosphate is shown in the following equation:

$$2Ca_5(PO_4)_3F + 7H_2SO_4 = 3Ca(H_2PO_4)_2 + 7CaSO_4 + 2HF \qquad \textbf{(EQ 1)}$$

The first patent for this fertilizer was granted in 1842, and production started the same year in England. The product was readily soluble in water compared to the relatively insoluble bones or phosphate ores. This process took off in the early 1850s in the United States and several other countries, initially using bones and later switching to mineral phosphate rock. This fertilizer dominated the world's phosphate fertilizer market for more than 100 years.

Another variation of superphosphate fertilizer is the single superphosphate (SSP), which is still popular in some countries. The molecular formula of SSP is $Ca(H_2PO_4)_2 \cdot H_2O$. SSP is one of the most important fertilizers in Brazil. This

Patrick Zhang, Research Director, Florida Industrial & Phosphate Research Institute, Bartow, Florida, USA phosphorous source is also produced in other regions, such as Australia, China, India, and New Zealand. It accounts for 15% of the phosphate fertilizer used in India.

Triple Superphosphate

Triple superphosphate (TSP) is made based on the following equation:

$$2Ca_5(PO_4)_3F + 12H_3PO_4 + 9H_2O = 9Ca(H_2PO_4)_2 \cdot H_2O + CaF_2 \qquad \textbf{(EQ 2)}$$

This high-analysis phosphate fertilizer was first manufactured in Germany in 1872. Limited triple production began in the United States in 1890, with large-scale production initiated in 1907. Currently, TSP accounts for a few percent of U.S. phosphate fertilizer production, though it accounts for a substantial percentage of worldwide use, representing more than 10% of the total phosphate fertilizer market.

Ammonium Phosphates

Ammonium phosphates are now the most popular phosphate fertilizers. Diammonium phosphate (DAP), $(NH_4)_2HPO_4$, is manufactured by reacting ammonia with phosphoric acid. The nitrogen to phosphate ratio in DAP makes it an excellent product for direct application or one that blends well with other fertilizer materials to produce a variety of NPK (nitrogen, phosphorus, and potassium) fertilizers. Monoammonium phosphate (MAP), $NH_4H_2PO_4$, is also a fertilizer manufactured from phosphoric acid and ammonia. "It is an ideal product for dry bulk blending with other fertilizer materials. MAP is a product-of-choice for manufacturing fluid blends or suspension fertilizers" (Griffith 2010).

Ammonium phosphates were first produced commercially in the United States in 1916. With the development of the dependable Tennessee Valley Authority process for producing granular DAP, DAP became the principal phosphate fertilizer in the United States by late 1960s. Today, DAP represents approximately 40% of phosphate use in the United States. On a worldwide scale, ammonium phosphates accounted for less than 5% of the world's production of phosphate in the early 1960s but have reached roughly 60% today.

Wet Acid Process and Phosphogypsum Production

The term *wet acid process* usually refers to the manufacturing of phosphoric acid by reacting phosphate rock with sulfuric acid. The first phosphoric acid plant was built in Germany in about 1870 and in the United States in 1890 (Beaton 2010). In the past decade, many new phosphoric acid plants have been constructed in major phosphate-producing countries such as Morocco, China, and the Middle East.

The primary chemical reaction in the wet acid process may be expressed in Equation 3 using fluorapatite to represent phosphate rock and sulfuric acid as the reactant (UNIDO and IFDC 1998):

$$Ca_{10}F_2(PO_4)_6 + 10H_2SO_4 + 10nH_2O \rightarrow 10CaSO_4 \cdot nH_2O + 6H_3PO_4 + 2HF \qquad \textbf{(EQ 3)}$$

Depending on the value of n, the process is defined as the dihydrate (n=2) process, hemihydrate (n=1/2) process, or anhydrate process. The term $CaSO_4 \cdot nH_2O$ in the equation is simply referred to as phosphogypsum (PG).

BRIEF HISTORY OF MAJOR PHOSPHATE MINING OPERATIONS

Phosphate rock is the only economical source of phosphorus for the production of phosphate fertilizers and phosphate chemicals. Most of the world's phosphate rock reserves are widely distributed marine phosphorite deposits. In 1847, phosphate rock was first mined in England and used for fertilizer. Between 1863 and 1895, phosphate rock was shipped from deposits in Ontario and Quebec, Canada, to England for processing.

In the United States, phosphate ore was first discovered in South Carolina in 1837, and mining there began in 1867. Extensive exploration of the vast phosphate rock deposits in Florida was carried out during the 1880s, and mining started in 1888. In the Western Phosphate Field covering the Rocky Mountain states, mining of phosphate began in 1906 in Idaho, in 1907 in Wyoming and Utah, and in 1921 in Montana. The North Carolina deposits were the latest to be exploited in the United States; mining started there in the mid-1960s.

Phosphate ore was discovered in Algeria and Tunisia in 1885, and mining initiated in 1889. The world's most abundant and rich phosphate reserve, the Moroccan deposits, was first identified in 1914 and mined in 1921. The major guano phosphate deposits were found in the Pacific Islands in the 1890s. Mining of the rich guano phosphate resource commenced in 1900 on Ocean Island, in 1906 on Nauru, and in 1908 on Makatea.

One of the largest igneous apatite deposits was discovered on the Kola Peninsula in Russia around 1930, and mining started shortly thereafter. Other major igneous deposits were found between 1960 and 1980 in Mexico; Brazil, Peru, and Colombia in South America; Israel, Turkey, Jordan, Iraq, and Saudi Arabia in the Middle East; Angola, South Africa, and Western Sahara in Africa; and in India and Australia. Many of these igneous deposits are being mined in Russia, Brazil, South Africa, Jordan, and Australia, accounting for nearly a quarter of the world's phosphate rock production.

By some estimates, China has the most abundant phosphate resource in the world. However, China was a late starter in phosphate mining, with its first mine initiated in the 1930s and made small progress up to the 1980s. During the past 15 years, China has transformed from the largest phosphate fertilizer importer into a net phosphate exporter. According to the International Fertilizer Association, China produced 51 Mt (million metric tons) of phosphate rock in 2005, against 38 Mt from the United States and 28 Mt from Morocco.

Currently, the leading phosphate-rock-producing countries are the United States, Morocco, China, Russia, Tunisia, Jordan, Brazil, and Saudi Arabia. Other producers of appreciable amounts include Algeria, South Africa, Togo, Israel, Nauru, Senegal, Peru, Australia, and Syria.

South Carolina, United States

Phosphate mining in America started in South Carolina. A phosphate mining company was formed in 1867, 30 years after phosphate was discovered in the state. After about 20 years of profitable mining, the state courts made it easy for anyone to obtain a mining permit. As a result, the profit margin dropped by more than 80%, and most companies shut down their mines there and moved to the new, more profitable source of phosphate in Florida.

Florida, United States

Although phosphate rock mining for fertilizer manufacturing started in England in 1847, the true "Phosphate Rush" originated in central Florida about 40 years later. It began with the river pebble of the Peace River, followed shortly thereafter by the discovery of the hard rock region near Dunnellon, and ultimately the land pebble districts. In 1892, there were 215 mining companies in Florida. By 1900, this number had dwindled to about 50 because of consolidation, and down to less than 10 by early 1990. Today, there are only two companies involved in phosphate mining in Florida. Mining of river pebble peaked in 1893 and ceased in 1908. Hard rock mining ceased in 1965 in the Ocala–Dunnellon region. Pebble mining quickly became the dominant form of mining, and this type of phosphate mining continues today in central and north Florida.

Tennessee, United States

When the "blue rock" was discovered in 1893, Tennessee was a formidable competitor with Florida in phosphate mining operations. But that competition was short lived because the easy-to-access, thick deposits ran out quickly and companies were forced to mine the rather thin seams of about 30 in. of thickness by underground mining. Mining of the brown phosphate rock started around 1896. By 1936, nearly all high-grade deposits in Tennessee were mined out, but phosphate mining continued well into the 1950s because the electric furnace method could process rock as low as 50 BPL (about 23% P_2O_5). *Bone phosphate of lime*, or *BPL*, is a term used by the phosphate industry to express phosphorous content in phosphate rock as percent of tricalcium phosphate, $Ca_3(PO_4)_2$. The most common term to express phosphorous content is *percent* P_2O_5, which can be derived by dividing BPL by 2.185.

Tennessee played a significant role in advancing phosphate mining and processing technologies. It pioneered the use of hydrocyclones for desliming, development of the flotation process, and optimization of the furnace acid technique.

Western United States

The Western Phosphate Field covers a large part of Idaho, a strip of Wyoming, a corner of Utah, and a tip of Montana. This Phosphoria Formation is a 350,000-km^2 area of the northern Rocky Mountains. Phosphate mining began in the early 1900s in southeast Idaho to supply fertilizer to the young agricultural industry. Most of the mines were underground and small. After World War II, technology changed and a growing demand fueled large-scale extraction of phosphate. At this time, open pit mining became the predominant phosphate mining method. Idaho production of phosphates currently constitutes more than 12% of national production. Today there are four open pit operations that produce more than 5 Mt of rock per year.

North Carolina, United States

The North Carolina deposits were the latest to be exploited; mining started in the mid-1960s. Currently, there is one phosphate mine in operation.

Morocco

Shortly after phosphate deposits were discovered in 1914, the Moroccan government established the Office Chérifien des Phosphates Group, commonly known as the OCP Group, and began mining in 1921. Owned solely by the government and managed in nearly the same way as a private company, the OCP Group has held a monopoly on mining, processing, and trading phosphate rock and phosphate fertilizers in Morocco for nearly a century. That monopoly recently showed some flexibility by allowing one private company to mine phosphate and by forming joint ventures with the European Union, India, and Pakistan. Until 1975, the OCP Group's primary business was mining and exporting phosphate rock, fully enjoying the luxury of having the most abundant high-grade ore that required only simple washing and sizing. That business model was revamped during the mid-1970s as accumulation of lower-grade rock made it economically advantageous to produce phosphoric acid or phosphate fertilizers locally. For example, Morocco exported zero phosphate fertilizer up to 1962 but put 75,400 t (metric tons) on the world market in 1976 and passed the one-million-ton mark by 2001. This trend is continuing. Today, Morocco produces well more than 15% of the world's total phosphoric acid. Many of its recent phosphate projects are entirely dedicated to deep processing. The OCP Group's Jorf Lasfar Complex, for example, turns out 2 Mt of P_2O_5 products per year, with 50% exported as phosphoric acid and 50% as fertilizers such as DAP, MAP, and NPK.

Russia

Russia started mining phosphate in the early 1930s and quickly became the world's second largest phosphate rock producer. It maintained that rank during the early 1980s. However, since the fall of the former Soviet Union, phosphate production quickly dropped to roughly 10 Mt a year, making Russia the world's fourth largest producer. Still, approximately one-third of Russia's phosphate rock production is exported.

The Khibiny Complex on the Kola Peninsula is the workhorse of recent phosphate mining in Russia, turning out more than 80% of its total production. This is an igneous deposit. Although the ore grade is not that impressive, the final concentrate analyzes up to 39% P_2O_5.

Russia also has a large sedimentary phosphate deposit in the Karatau Basin with a resource of about 8,000 Mt averaging 21% P_2O_5. In early mining, phosphate ore was crushed, ball milled, and screened to obtain a product of 23% P_2O_5 for elemental phosphorous manufacture. Other major sedimentary deposits are located along the southern margins of the Moscow Basin.

Tunisia

Tunisia is the second largest phosphate producer in Africa, turning out about 8 Mt per year of rock on average in recent years. Phosphate was first discovered in Tunisia in 1885 by an amateur French geologist. Subsequent extensive geological explorations revealed the existence of vast phosphate formations south and north of the so-called Kasserine Island. Like Morocco, the Tunisian government founded the Compagnie des phosphates et de chemins de fer de Gafsa in 1896 to establish a national phosphate industry in the country. The first phosphate was mined in Tunisia in the same year. Rock production reached an impressive level of 200,000 t in 1900.

After these early beginnings, the Compagnie des phosphates et de chemins de fer de Gafsa has gone through several transformations in its structure during the long course of its history before attaining its current status and becoming the Groupe Chimique Tunisien (GCT). GCT is now the world's fifth largest phosphate producer.

For many years, GCT has been gradually developing opencast mines to replace underground capacity to reduce costs. From 1978–1991, five opencast mines were opened. All Tunisian phosphate rock mines are now opencast. This move has resulted in significant savings in labor costs and increased productivity. Over the last decade, GCT has seen its merchant rock output rise from 6 Mt to more than 8 Mt per year.

Togo

Large-scale phosphate mining in Togo started in 1963 but has become the country's most important industry, with an annual production of more than 2 Mt. The phosphate deposit is a relatively thin layer ranging from 2 to 6 m, covered by a thick overburden of sand and clay up to 30 m. The phosphate rock grade, however, is one of the best in the world, analyzing about 36% with a minimal amount of Mg, Al, and Fe.

South Africa

South Africa is blessed with rich mineral resources. Its gold and diamond mining industries are formidable on a global scale. Many people do not realize, however, that the phosphate reserve base in South Africa ranks fourth in the world. The majority of South Africa's phosphate deposits are igneous in nature. The deposits are generally of low to medium grade with high Fe and Al contents.

The major Phalaborwa phosphate deposit was discovered in 1904. This phosphate resource was commercially mined in the early 1930s by South Africa Phosphates Ltd., which mined high-grade apatite pockets and sold them as a fertilizer product, but soon went bankrupt because of the less expensive Moroccan phosphate rock imports. The government later formed the Phosphate Development Corporation Ltd., which eventually became Foskor Ltd. New phosphate mining started in the early 1950s. Gradually, the fertilizer producers adapted their processes to the use of the igneous phosphate rock produced, and in 1976 a significant expansion enabled production of export-quality product. Today, approximately one-third of the phosphate rock is used domestically, one-third is exported as rock, and the remainder is acidulated for phosphoric acid and fertilizer ingredients for export.

China

In China, the Haizhou phosphate deposit (now the Jinping phosphate deposit) was found during the Tongzhi (1862–1874) period of the Qing dynasty. The Kunming phosphate deposit was discovered during a search for refractory clay for a smelting plant in 1931. The Jinping phosphate mine started operations in 1914 and was the first in China.

Before the founding of the People's Republic of China in 1949, the Chinese phosphate fertilizer industry was of little importance. Since then, Chinese geologists have conducted extensive geological reconnaissance, exploration, and evaluation of metamorphic phosphate deposits in the pre-Sinian system. As a result, huge amounts of phosphate deposits were identified in Yunnan, Guizhou, Sichuan, Hunan, and Hubei provinces, earning China the top spot in the world's phosphate reserves.

SSP was first developed in the 1950s, followed by furnace (or fused) calcium–magnesium phosphate (FCMP) fertilizer during the 1960s. High-concentrate phosphate and compound fertilizers such as DAP, MAP, TSP, and NP (nitrogen–phosphorous fertilizers) were developed quickly after the 1980s. During the past 10 years, the output of phosphate fertilizer in China increased in parallel with its gross domestic product (GDP) growth, with an average rate of 13.4% annually. The percentage of high-grade phosphate fertilizer also increased from 1% to more than 20% per year.

Until recently, SSP and FCMP were the primary phosphate fertilizers produced in China, because of the suitability of phosphate rock resources in China. The low-grade, silica- and calcium-containing, semi-self-melting phosphate rock is used for the production of FCMP in Hubei and Hunan. The waste rock of 20% P_2O_5 from the high-grade mines in Yunnan is also used for producing good-quality FCMP. SSP is another significant phosphate fertilizer in China for the benefits of S and Ca.

Although many of the new phosphate mining projects are coupled with producing high-grade fertilizers (DAP, MAP, and NPK), the production of SSP and FCMP will not see much decrease in China for three reasons:

1. These two varieties are relatively inexpensive.
2. Chinese farmers have gained tremendous experience in applying these materials.
3. These fertilizers also supply sulfur, calcium, and magnesium to soils.

Nauru

Phosphate (guano) deposits occupy about 90% of the island of Nauru. These deposits are of the highest grade, with the run-of-mine ore analyzing up to 40% P_2O_5. Nauru once possessed the highest GDP per capita in the world because of its rich phosphate deposits, but they have been nearly depleted.

Nauru was discovered in 1798 by a British sailor. Eleven years later, phosphate was discovered, and it turned out to be profitable for decades to come. In 1906, the Pacific Phosphate Company started mining phosphates with the support of the German government. After World War I, Nauru became a trust territory jointly administered by Australia, Britain, and New Zealand. The British Phosphate Commission was established to run the phosphate mining operations. After a brief pause during World War II, phosphate mining continued at an increasing rate. After the island achieved independence, the government-owned company Nauru Phosphate Corporation (NPC) was formed. NPC took over all phosphate operations in 1970. By the year 2000, the phosphate deposits were nearly depleted, leaving behind some small-scale operations.

Australia

Because of its membership in the Nauru Trust Territory and proximity to the best-quality phosphate rock from Nauru and Christmas Island, Australia had not given any priority to phosphate mining for a long time. Phosphate rock was first discovered in Australia in 1966. Exploration for rock phosphate was at that time encouraged because of the realization that the phosphate deposits at Christmas Island and Nauru were soon to run out. The Phosphate Hill rock phosphate deposit in northwest Queensland was first investigated extensively and targeted for the Queensland Phosphate Project. In 1975, BH South established a phosphate rock mine, a beneficiation plant, rail access, and port facilities. Because of production difficulties and various challenges, the operation was closed in 1978.

In 1980, Western Mining Corporation (WMC) acquired BH South's interests in Phosphate Hill and additional reserves. However, little activity took place during the 1980s and early 1990s. Recognizing that to achieve an acceptable return on investment the phosphate rock would have to be converted into either granulated TSP or ammonium phosphates, WMC decided to invest in a facility for the manufacture of ammonium phosphates. This proved to be a profitable proposition and placed Australia back on the map of the world's major phosphate producers by 2002.

Saudi Arabia

Saudi Arabia is a "sleeping giant" in terms of phosphate mining potential in the Middle East, with phosphate resources estimated up to 3,000 Mt averaging 18%–21% P_2O_5. The largest phosphate formation is the Sirhan-Turayf sedimentary deposit, which extends into Jordan, southern Iraq, and Syria. The best explored and largest deposit is at Al Jalamid, which has measured reserves of 213 Mt averaging 21% P_2O_5 and a stripping ratio of 2.3:1. Indicated resources amount to an additional 187 Mt, with 19.7% P_2O_5 and stripping ratios of 5:1 or less. The second most important deposit is in the Umm Wu'al north area with a total resource of 537 Mt averaging 19.35% P_2O_5 and a stripping ratio of less than 5:1. Saudi Arabia already has a large phosphate operation producing about 3 Mt of phosphate rock, and it is constructing another mining and chemical processing complex touted as the world largest integrated phosphate operation.

Brazil

Geological studies of Brazilian phosphate deposits were initiated as early as 1925, followed by extensive exploration from 1943–1969. However, many of the large phosphate complexes were established during the late 1970s. For example, Bunge's mine at the Araxá Complex began mining in 1977, whereas the largest phosphate mine, Fosfertil's Tapira mine, began production in 1978. Brazil is moving quickly toward self-sufficiency in phosphate fertilizers.

Jordan

Unlike some of its oil-rich neighbors, Jordan has phosphate deposits that are its primary natural resource and a major source of export income. Phosphate deposits cover nearly 60% of Jordan. At the current rate of about 8 Mt of phosphate rock, Jordan's proven reserve can sustain the industry for hundreds of years. Jordan Phosphate Mines Company (JPMC) was one of the earliest phosphate companies. Established in 1935 as a quasi-private company, JPMC has been the only player in phosphate mining and fertilizer production in Jordan. The company started mining in the 1930s from the Ruseifa mine. Mining at the El-Hassa mine was commenced during the 1960s. In the late 1970s, production started from the El-Abiad mine. The newest mine, the Eshidiya mine, was opened in 1988. Jordan is a world-leading phosphate rock exporter.

Canada

Even though phosphate rock from Ontario and Quebec was shipped to England for processing as early as 1863, there had been no major phosphate mining operation in Canada until 1999, when Agrium started its Kapuskasing phosphate operation in Ontario. Phosphate is mined from the Cargill deposit. This deposit was identified as a magnetic anomaly in 1954 and discovered to be a phosphate deposit in 1975. The Cargill deposit is an alkalic igneous complex. Some new deposits were discovered recently in Quebec, with significant amounts of rare earths or niobium. As a result, the Quebec government has recently approved a $750 million phosphate mine project (Topf 2015). The mine will produce phosphate rock product to be shipped from Sept-Îles, Quebec, to Norway, where Yara International will extract the phosphate to produce fertilizer.

Mongolia

Although there is currently no phosphate mining activity in Mongolia, its vast phosphate resources are worth mentioning. There are two phosphate reserve basins in Mongolia. The first area, the Khulosgul Basin, was discovered in the north of the country through geological exploration work from 1963 through 1986. The Zabkhan phosphate basin was discovered in the west of the country from 1987–1997. As a result of broad geological exploration work in Mongolia since 1963, more than 50 deposits have been discovered in the two basins, with an estimated total reserve base of approximately 5.7 billion t. Limited efforts have been made to utilize the phosphate rock by mechanical activation.

GEOLOGY, DEPOSITS, AND RESOURCES

Generally, phosphate minerals are found in igneous, metamorphic, and sedimentary deposits. In sedimentary marine ores, the phosphate is usually a carbonate fluorapatite. In igneous deposits, the phosphates have compositions close to that of fluorapatite. In some deposits, Ca-Fe-Al and Fe-Al phosphates are associated with apatite (McClellan and Germillion 1980). Currently, the apatites recovered from igneous and metamorphic rocks have commercial importance, but such apatites supply only a small fraction of the world market. On the other hand, sedimentary phosphates have been, and no doubt will continue to be, the major source of commercial raw material for the phosphate industry.

Sedimentary Phosphate Deposits

An enlightening description of the phosphate cycle is provided by Bateman (1952):

> The sedimentary cycle of phosphorus is puzzling and fascinating. Dissolved from the rocks, some of it enters the soil from which it is abstracted by plants; from there it passes to the bodies of animals and is returned via their excreta and bones to accumulate into deposits. Those in turn may undergo resolution and reach the sea, and there the phosphorus is accumulated or deposited by sea life, embodied into sediments and returned to the land on uplift, when a new cycle may start. Some of the phosphorus of the sea is absorbed by fish life, and the fish in turn are eaten by birds whose excreta have built up great phosphate deposits on islands of the Pacific.
>
> Commercial beds of phosphate are formed only under marine conditions in the form of phosphorite. The beds have been formed from early to recent geologic periods and extend with remarkable regularity over thousands of square miles. They are sparingly fossiliferous and are interlaminated with marine fossiliferous beds. These features together with their oolitic character indicate a marine origin.
>
> The great phosphate deposits of the world have been formed by the process of sedimentation. Those of Algeria, Tunisia and Morocco together yield the largest production in the world of the greatly desired

> mineral. The Moroccan deposits occur as horizontal beds along with limestone, marls and clays in which there are 3 to 4 beds up to 2.5 meters thick. The western United States deposits underlie parts of Utah, Wyoming, Idaho, and Montana and extend into Canada. The beds are in the Phosphoria formation of Permian age, and the chief bed is about 5 feet thick.... Other sedimentary phosphate deposits are found in Egypt and Russia. The large phosphate deposits of Florida, which supply most of the United States fertilizer and chemical consumption, are not strictly sedimentary beds but are pebble deposits derived from a sedimentary phosphate bed. The guano deposits of the Pacific islands also are not sedimentary.

These sedimentary apatites, because of their widely differing modes of chemical composition, display large variations in physical forms and types of associated gangue. Accordingly, these ores fall into one of the following three principal categories (Lehr and McClellan 1973).

Siliceous Ore

In this type, the major gangue mineral is quartz, chalcedony, or opaline forms of silica. In the United States, the Bone Valley deposit in central Florida, the upper unit of the Hawthorn Formation in north Florida, and the upper zone of the Hawthorn Formation in central Florida belong to this type. Phosphate ores in Australia and Senegal also fall under this category.

Carbonaceous Ore

This type of ore consists of phosphatic limestones or calcareous phosphates in which calcite, dolomite, or ankerite is mixed with the phosphate, or which occurs as intercalated seams. These include the lower zone of the Hawthorn Formation in central Florida; most of the phosphate deposits in Hubei and Guizhou provinces, China; the Karatau Formation in Russia; a majority of the Mongolian deposits; and some of the Moroccan phosphates.

Clayed Phosphate

Clayed phosphate ores are associated with gangue minerals containing mainly clays and hydrous iron and aluminum oxides. These gangue minerals are concentrated mostly in the silt and clay size ranges. Most of the phosphate ores in Morocco and Western Sahara, Tunisia, and Togo contain up to 40% clay, whereas the Florida phosphate matrix is composed of approximately one-third each of clay, phosphate, and sand.

Igneous Phosphate Deposits

Among the more economically important igneous apatite deposits being exploited commercially are the Khibina nepheline syenite massif near Kirovsk in the Kola Peninsula in Russia, the Jacupiranga Alkalic Complex in Brazil, and the Phalaborwa carbonatite in South Africa.

The most important deposit of igneous apatite mined today comes from the Kola Peninsula; the output is more than 9 Mt per year of concentrates with a P_2O_5 grade of 39% from an ore assaying 18% P_2O_5, 23.1% SiO_2, and 13.3% Al_2O_3 in a bonded phase of nephelinic syenites. Other important igneous phosphate ores are present at Phalaborwa in South Africa. The Igneous Complex is comprised of a suite of intrusive alkali rocks, which intruded granite gneiss, to form three coalescing and concentric lobes. The complex is about 7 km long and varies from 2 km to 4 km wide. The outermost zone consists of syenite with the intrusion being a feldspathic pyroxenite. The next zone consists of an apatite-bearing phlogopitic pyroxenite. Foskor operates an open pit for apatite extraction in a corner of this zone. The apatite varies from 0%–30% P_2O_5, averaging 7%, with the principal gangue minerals of diopside and phlogopite. The pyroxenite zone is followed by the foskorite zone, which consists of magnetite, serpentine, and apatite, assaying 6%–11% P_2O_5. Originally within the foskorite zone was a carbonatite core hill, which now has largely been excavated and is the large open pit of the Palabora Mining Company (PMC). In 1979 Foskor and PMC reached an agreement in which PMC would extend its open pit, with Foskor sharing the cost, and in return Foskor would receive certain types of ore mined by PMC. Foskor now produces phosphate rock from three sources: pyroxenite ore, foskorite ore, and PMC tailings (Schmidt 1999).

The Brazilian apatites also represent an important source of igneous phosphate ores. They are included in carbonatites (with simple carbonates as in Jacupiranga or as more complex ones with iron carbonates, phlogopite or vermiculite, titanium minerals, barytine, pyrochlore, etc.). The P_2O_5 contents are comparatively low, from 5% for Jacupiranga to 15% for Araxá.

World Reserves and Resources

How to classify mineral resources remains a topic of debate. This author believes in the simple way of defining phosphate deposits as reserves and resources, which also seems to be the dominant view in the United States. *Reserves* are the deposits that could be economically mined and processed to produce salable products under current economic conditions using available technologies. *Resources* are identified deposits that meet specified minimum physical and chemical criteria related to current mining and production practices, including those for grade, quality, thickness, and depth. Resources encompass reserves.

The International Fertilizer Development Center (IFDC) recently conducted a detailed study of world phosphate resources (Kauwenbergh 2010). Based on this study, the world has total phosphate reserves of 60 billion t, on the basis of 30% P_2O_5, and 290 billion t of resources. These numbers are close to U.S. Geological Survey (USGS) newly updated estimates of 67 billion t of reserves and 300 billion t of resources (Table 1). Table 2 lists reserves and resources of major phosphate-hosting countries based on the IFDC study.

UTILIZATION OF PHOSPHATE ROCK

World phosphate rock production reached nearly 200 million t by the year 2012 and has maintained at that level since. Table 3 shows phosphate rock production data for the major producing countries.

Figure 1 shows the distribution of various uses for phosphate rock in 2005. Wet-acid-based products account for about 71%, followed by SSP (14%), other non–wet-acid-based products (about 10%), elemental phosphorus (5%), and direct application.

The majority of phosphate rock requires deep beneficiation involving flotation prior to chemical processing. Figure 2 shows a simple classification of phosphate rock based on the degree of beneficiation difficulty. Figure 2 illustrates the importance of beneficiation/flotation in the processing of phosphate. All new mines being constructed or considered will include flotation, with the exception of the phosphate mine in Peru. More importantly, about three-quarters of the phosphate deposits are contaminated by carbonaceous minerals, which makes beneficiation more challenging.

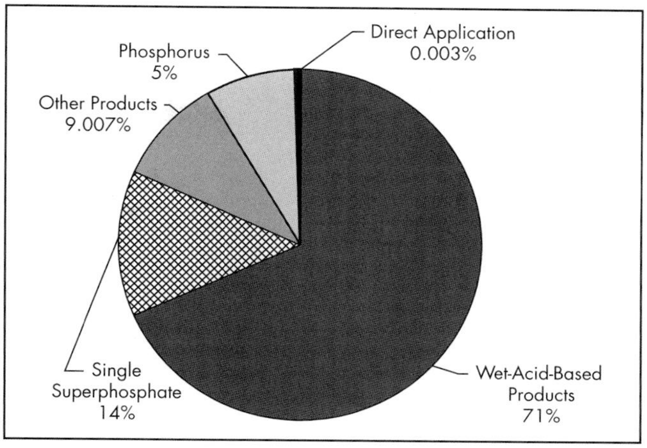

Figure 1 Distribution of various uses for phosphate rock

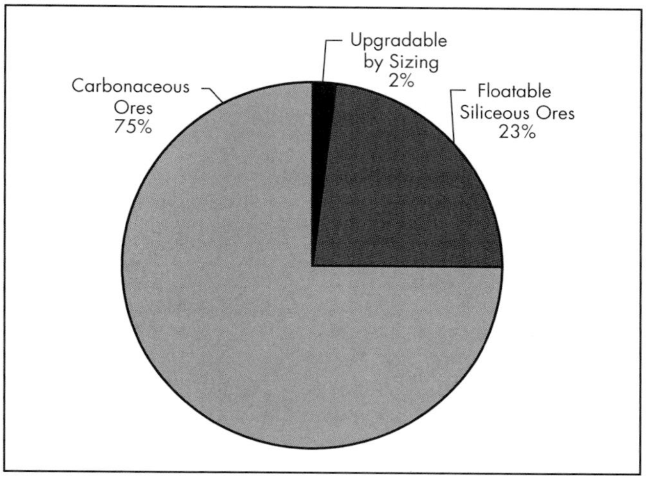

Figure 2 Classification of phosphate ores by beneficiation readiness

Table 1 Updated world phosphate reserves and resources

Data Source	Reserves, billion t	Resources, billion t
International Fertilizer Development Center	60	290
U.S. Geological Survey	67	>300

Table 2 Major phosphate reserves and resources in the world

Country	Reserves, Mt	Resources, Mt
Australia	82	3,500
Brazil	400	2,800
Canada	5	130
China	3,700	16,800
Egypt	51	3,400
Israel	220	1,600
Jordan	900	1,800
Morocco	51,000	170,000
Russia	500	4,300
Senegal	50	250
South Africa	230	7,700
Syria	250	2,000
Togo	34	1000
Tunisia	85	1,200
United States	1,800	49,000
Other countries	600	22,000
World total	**60,000**	**290,000**

Source: USGS 2010–2016.

Table 3 Phosphate rock production, thousand tons

Country	2010	2011	2012	2013	2014	2015 (estimated)
Australia	2,095	2,519	2,000	1,919	2,600	2,600
Brazil	5,693	6,094	6,094	5,939	6,040	6,700
China	69,100	73,000	78,500	77,000	100,000	100,000
Egypt	3,435	4,746	5,738	5,315	5,500	5,500
India	2,000	1,910	1,154	1,211	1,110	1,100
Israel	3,078	3,118	2,871	3,437	3,360	3,300
Jordan	6,529	7,589	6,382	5,399	7,140	7,500
Morocco	25,655	27,821	26,844	25,489	30,000	30,000
Peru	791	2,544	3,209	3,547	3,800	4,000
Russia	10,844	10,304	10,282	10,743	11,000	12,500
Saudi Arabia	121	1,068	3,097	3,055	3,000	3,300
South Africa	2,499	2,468	2,065	1,925	2,160	2,200
Syria	3,765	3,542	1,780	898	1,230	750
Tunisia	8,132	2,510	2,607	3,284	3,780	4,000
United States	25,244	27,619	29,475	30,548	25,300	27,600
Others	15,629	14,104	14,578	13,584	11,980	11,950
Total	**182,111**	**190,957**	**196,676**	**193,293**	**218,000**	**223,000**

Source: USGS 2010–2016.

PHOSPHATE MINING AND BENEFICIATION TECHNOLOGY

During the late 19th century, phosphate mining operations in Florida involved wheelbarrows, picks, and shovels, followed by mule-drawn scrapers. Next came steam shovels and centrifugal pumps mounted on barges. Large-scale, more efficient mining was made possible only when draglines were introduced in the late 1920s.

The following numbers demonstrate how draglines increased mining efficiency. In the 1900s, it took 3–4 years to mine 15 acres with picks and shovels. Even a small dragline in the early days could mine about 5 acres a year. As draglines became larger, companies were able to mine 500–600 acres a year. Today, a dragline with an 80-yd bucket can mine up to 20 acres a month.

Phosphate ores usually require upgrading, that is, beneficiation, before they can be used for producing fertilizers (e.g., SSP, TSP, and FCMP) or products via the wet acid process. Early beneficiation methods included simple washing, screening, and crushing/grinding. A small amount of deposit could even be used directly.

In the 1920s and 1930s, separation advancements focused on improving washing and screening reduced the amount of phosphate rock discarded. The most dramatic change was the introduction of flotation technology in about 1927. Flotation of phosphate initially used oil as reagent and then changed to fatty acid flotation with oil as an extender. In Florida, the Crago "double float" process truly revolutionized phosphate beneficiation, in terms of both increasing P_2O_5 recovery and reducing silica content in the final concentrate. This process was adopted during the 1940s and is now the predominant beneficiation practice in Florida.

Study of the magnetic susceptibility of minerals dates to the late 1700s, and magnetic separation was used for nearly 200 years in upgrading iron ores. This separation technology finds its use in processing igneous phosphate ores in Brazil, Canada, and South Africa. Both high-intensity and low-intensity separators are used.

Limited gravity separation circuits are found in Florida beneficiation plants. Spiral separators are used to remove sand from the ultra-coarse (14 × 28 mesh) fraction. Heavy media separation was used to remove dolomite in the pebble fraction from the late 1980s to early 2000s, but these machines are no longer in use because of low separation efficiency and high operation costs.

The calcination process was employed in the western United States to remove carbonaceous materials from phosphate rock, and it is still used in Morocco to eliminate organic gangues.

Mining

Surface Mining with Dragline

Today, most phosphate rock is mined using large-scale surface methods. In the past, underground mining methods played a greater role, but their contribution to world production has declined. For example, in Idaho, phosphate mining has been transformed from 100% underground during the 1930s to nearly all surface mining today. Surface phosphate rock mining operations can vary greatly in size. Extraction may range from several thousand to more than 10 million tons of ore per year. In many cases, operations supply concentrate to a nearby fertilizer processing complex for the production of fertilizer products. The land area affected by the surface operations can vary widely, depending on the ore-body geometry and thickness. At similar extraction rates, mining of flat-lying thin ore bodies as found in Florida (United States) will affect a far wider area of land than the mining of thicker ore bodies as found in Brazil and Idaho. The depth of excavations may range from a few meters to more than 100 m. Currently, most phosphate rock production worldwide is extracted using opencast dragline or open pit shovel/excavator mining methods. This method is widely employed in parts of the United States, Morocco, and Russia. Underground mining methods are currently used in Tunisia, Morocco, Mexico, and India. Figure 3 shows a large dragline used for mining phosphate in Florida.

Figure 3 Surface mining of phosphate rock with dragline in Florida, United States

Surface Mining with Drilling and Blasting

In Khouribga, Morocco, the overburden in thickness of 16–17 m is drilled, blasted, and removed by dragline to the side of the mining area for subsequent reclamation. Small draglines, electric shovels, and bulldozers are used to recover the upper ore body. The intercalating limestone layer is blasted and removed to expose the Bed I phosphate, and then it is loaded onto trucks. The process is repeated to mine the Bed II phosphate deposit. At Vernal, Utah, mining begins with removal of topsoil and temporary storage for reclamation. Overburden is drilled and blasted with ANFO and shoved into mined-out areas. The ore is then drilled and blasted and loaded by shovel into trucks for transport to a feeder-breaker. The broken product at 25 cm is belt-conveyed to a 10,000-t live storage pile. Overburden in mined areas is contoured, covered with topsoil, and seeded.

Dredge Mining

Consideration of mining systems for the Santo Domingo, Mexico, project included bucketwheel excavators, draglines, scrapers, shovels, and dredges. A dredge-based system was favored primarily because the other systems could not operate effectively below sea level. Low operating and maintenance costs prompted the final selection of floating hydraulic cutter suction dredges with 27-in.-diameter suction heads. Two Ellicott dredges were used, each capable of pumping about 2,000 t/h of slurried solids to the floating primary beneficiation

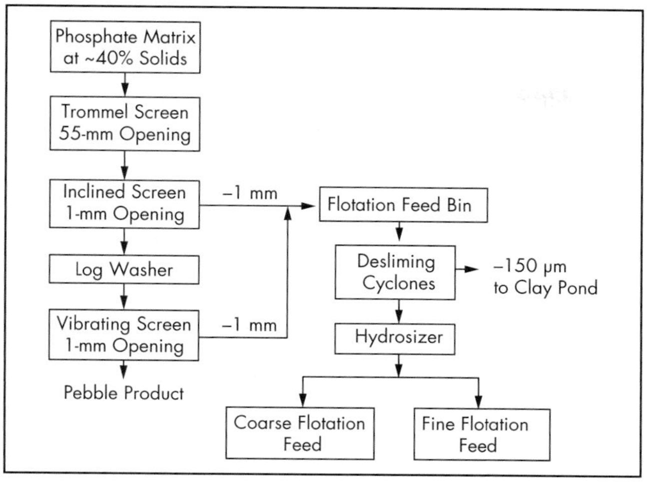

Figure 4 Simplified flow sheet of washing and sizing of Florida phosphate ores

plant. Dredge mining is also practiced at the Wingate Creek Mine in central Florida.

Materials Haulage

The three major methods of hauling materials in phosphate mining are pumping, trucking, and transporting using conveyer belts. In the United States, pumping is the predominant means of moving ore and tailings, with the longest pipeline running 87 mi. The J.R. Simplot Company transports phosphate rock slurry from its Smoky Canyon Mine in Wyoming to Pocatello, Idaho. OCP Group's recently constructed pipeline in Morocco is equally long.

PHOSPHATE MINERAL PROCESSING

According to the beneficiation difficulty, phosphate minerals may be classified into three categories: siliceous, carbonaceous, and magnetic. Typical siliceous phosphate ores are found in Florida, Morocco, and Jordan. Carbonaceous phosphate deposits are found all over the world (China, the western United States, the lower zone in central Florida, Saudi Arabia, and some parts of Morocco) and will constitute the source of most future phosphate rock production.

Beneficiation of Siliceous Phosphates

Washing and Sizing

Washing and sizing are commonly used on clayey ores as pretreatment steps prior to flotation. This operation is of paramount importance in Florida, because the phosphate ore as mined, known locally as the matrix, contains a high percentage of clay or clay-sized minerals. Figure 4 shows a simplified processing flow sheet. Details may vary significantly from mine to mine, particularly in sizing equipment and flotation feed size fractions.

The coarsest portion of the disaggregated matrix (+50 mm) is removed as waste by trommel screens because it usually represents lower phosphate grade and is likely to contain significant dolomite-cemented conglomerates. The portion of phosphate rock processed in the washer by disaggregation and simple screening is expected to diminish from the roughly 50% for the Bone Valley to about 20% as the mining moves to the south or west of the current Florida mining district. Because pebble production is relatively inexpensive, the reduction of this production has an unfavorable impact on costs. The disaggregation, which largely takes place in the pumps and transport lines, is completed by log washing at high percent solids to scrub residual clay from the mineral surface and is countercurrently washed away from the rock. Vibrating, plastic-coated screen decks permit the final removal of the pebble (1 × 55 mm).

For simplicity, Figure 4 does not show the split washing and sizing for the coarse (20 × 55 mm) and fine (1 × 20 mm) pebbles. The coarser washer pebbles are discarded when the dolomite content is too high, which is quite common when the draglines dig deeper.

The slurry containing flotation feed and clays, which represents the combination of pebble screen undersize, log washer overflow, and most other slurry streams created in the washer, is pumped through two stages of hydrocyclones with recycled water added to the first hydrocyclone underflow for dilution. This removes roughly 98% of the fine suspended clay particles from the flotation feed and creates a clay slurry waste containing 3% solids on average, which by themselves settle very slowly in clay ponds. Because roughly 30%–40% of the mined material weight is clay, these ponds represent a substantial capital investment in terms of both construction and reclamation when filled. These ponds also act as reservoirs for water recovery and recirculation. Most phosphate flotation plants separate the flotation feed (−1 mm to + 105 μm) into at least two size fractions with the split point at about 420 μm or 35 Tyler mesh.

In many other operations, washing and sizing produce the finished product. WMC in Australia uses water scrubbing to remove fine clays before shipping the low-grade rock (23.5% P_2O_5) to its chemical plant.

In Togo operations, the phosphate ore is first scrubbed with seawater and then wet screened at between 0.8 and 3 mm, depending on the quality of the ore. The clay is then removed using hydrocyclones. The high-ferrous product is further dried and upgraded using electromagnetic separation technology.

A large portion of the currently mined phosphate rock in Jordan only requires crushing and screening to become a salable product. In most operations, phosphate ore is first crushed and screened to reject the + 12.7 mm material. When the −12.7 mm fraction makes the grade of 66%–68% BPL, it is directly fed to rotary cascading dryers to produce a final product. The lower-grade ore requires beneficiation, which involves sizing and desliming at about 200 mesh using hydrocyclones.

Until recently, beneficiation of phosphate in Morocco has involved only crushing and sizing, except for the "brown" Youssoufia phosphate, which is calcined above 700°C to eliminate organic matter. At the Benguérir mine, located 70 km north of Marrakesh, the first treatment step is removing + 100 mm material, followed by crushing and wet screening at 10 mm, producing nearly 4 Mt per year of wet phosphate rock of −10 mm. Processing of phosphate ore at the Bou Craâ mine is almost as simple as it gets: ore is crushed and slurried with seawater, followed by sizing and desliming to remove unwanted materials. At the Khouribga mine, beneficiation is more complicated. High-grade ore is screened and stored, low-grade ore is treated using log washers and hydrocyclones, and lower-grade ore is calcined.

Crago "Double Float" Process

The Crago "double float" process shown in Figure 5 has been practiced in Florida for more than half a century. Modified versions or parts of the process are also practiced in other parts of the world. This process is particularly suitable for processing siliceous phosphates, where low-silica product is required. The flotation plant consists of several important processing steps:

1. Remove the flotation feed at a controlled rate from partially consolidated storage bins.
2. High solids condition the solids (65%–70%) with a pH modifier and anionic flotation reagent (namely, a vegetable-derived fatty acid) as a collector, a petroleum extender, and sometimes surfactants to impart desirable selectivity or frothing characteristics.
3. Dilute with clean, recirculated water to roughly 20% solids for the rougher flotation step of the bulk of the phosphate mineral from the sand.
4. Collect the froth concentrate and scrub the solids with sufficient sulfuric acid to free the fatty acid and fuel oil coating, and then rinse and dewater with clean, recycled water to provide a clean mineral surface for further separation.
5. Dilute with either fresh or very clean recycled water, adjust the pH to approximately neutral, and add a cationic flotation reagent (an amine, a light petroleum extender, and sometimes a surfactant) into a mixing tank or the feedbox of the flotation cells. (This cleaning flotation step is aimed at removing any fine sand inadvertently carried into the rougher concentrate by the relatively high reagent levels required to float the phosphate mineral.)
6. Partially dewater the cleaner cell underflow concentrate and hold it in bins until it is routed to the appropriate storage pile locations when chemical assays are available.
7. Combine the cell underflow tailing product from rougher flotation and the froth tailing product from cleaner flotation to create a combined tailing that is pumped to reclamation areas for land filling in the mine.

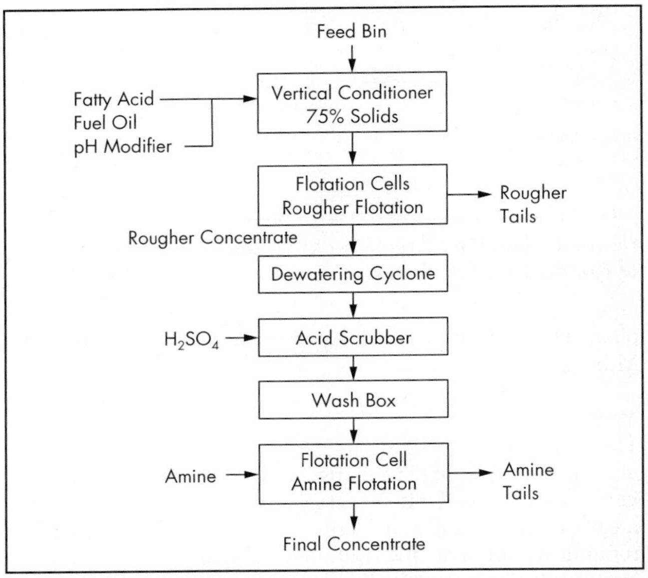

Figure 5 Simplified flow sheet of the "double float" process

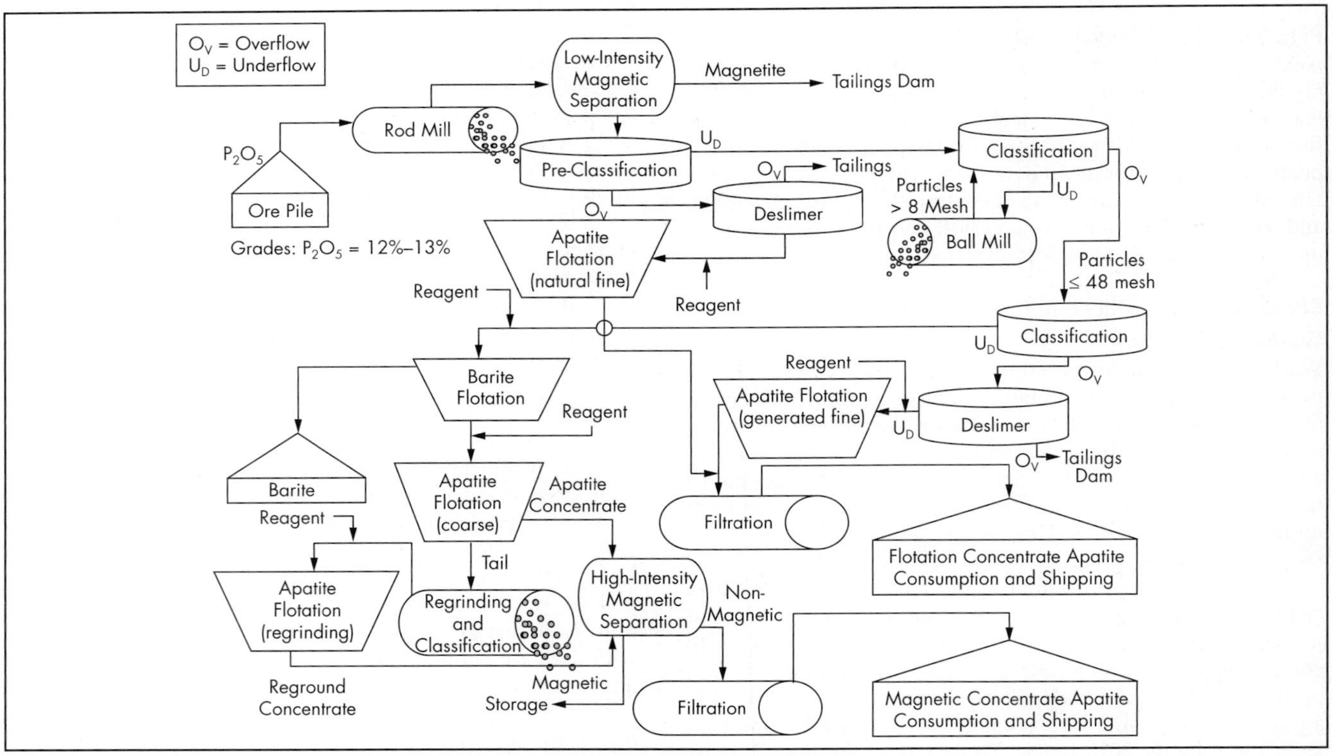

Courtesy of Bunge

Figure 6 Industrial flow sheet of phosphate beneficiation in Brazil

Anionic Flotation Only

Outside Florida, direct flotation of phosphate using anionic collectors is the most prevalent practice. In Senegal phosphate operations, after desliming and sizing, the phosphate is upgraded using a tall oil fatty acid. The silica and the iron and aluminum oxides are depressed, and a high-grade concentrate analyzing up to 80% BPL is produced.

Foskor Ltd. of South Africa produces phosphate rock from three sources: pyroxenite ore, foskorite ore, and PMC tailings. The pyroxenite ore is crushed and milled to 15% + 425 µm and 20% –74 µm. The mill product is pumped to conditioners from which it flows to the flotation circuit, where apatite is floated using Wemco flotation cells. Flotation is done in a four-stage flotation circuit comprising rougher, scavenger, cleaner, and re-cleaner stages, with the recirculation of middlings. The pyroxenite process with only two reagents in the system is relatively simple and relatively trouble-free. A straight-chain petroleum sulfonate or sulfonic acid and a tall oil fatty acid are added to the pulp at the conditioners. From ore with a head grade of 7% P_2O_5, a concentrate with a P_2O_5 content of 39.6% is produced with a recovery of 70%. In the circuit for foskorite, flotation is also done in a four-stage flotation circuit comprising rougher, scavenger, cleaner, and re-cleaner stages, with recirculation of middlings, using Wemco flotation cells. Three reagents are used: sodium silicate ($Na_2O{\cdot}SiO_2$) as a dispersant, nonyl phenyl tetraglycol ether as a modifier and depressant, and a tall oil fatty acid as a collector. The average head grade of the foskorite is 7.5% P_2O_5, whereas the concentrate grade is 38.5% P_2O_5 with a recovery of 67%.

In Russia, a majority of the phosphate rock is turned out from the phosphate mines at the central Kola Peninsula. The mines are about half open pit and half underground. Phosphate ore is first crushed to –20 mm before being sent to the beneficiation plant. In the beneficiation plant, the feed is ground to approximately 55% passing 74 µm. Anionic rougher flotation is followed by several stages of cleaning and scavenging flotation.

Some low-grade phosphate ores in Jordan are treated by direct flotation using an aqueous blend of tall oil and diesel oil as phosphate collector, and sodium silicate as a dispersant for both clay and silica depressant for both fine and coarse fractions, with several rougher and cleaner stages.

Beneficiation of Igneous Phosphates with Iron

Brazilian igneous phosphate ores require numerous beneficiation steps using both magnetic separation and flotation. Flotation is typically carried out after a double desliming, generally with a roughing operation followed by one or two scavenging operations, with the froth being cleaned two or three times. The depression of carbonates and iron oxides is achieved at a pH of about 10 with causticized starch, and the flotation is carried out with fatty acids (tall oil). Concentrates of about 35% P_2O_5 with recovery ranging from 45% to 78% are obtained from different localities. The flow sheet shown in Figure 6 demonstrates the complexity of phosphate beneficiation operations in Brazil.

Another phosphate beneficiation flow sheet in Brazil is shown in Figure 7. The Brazilian phosphate industry makes extra efforts to recover phosphate from the "slime." Up to 35%

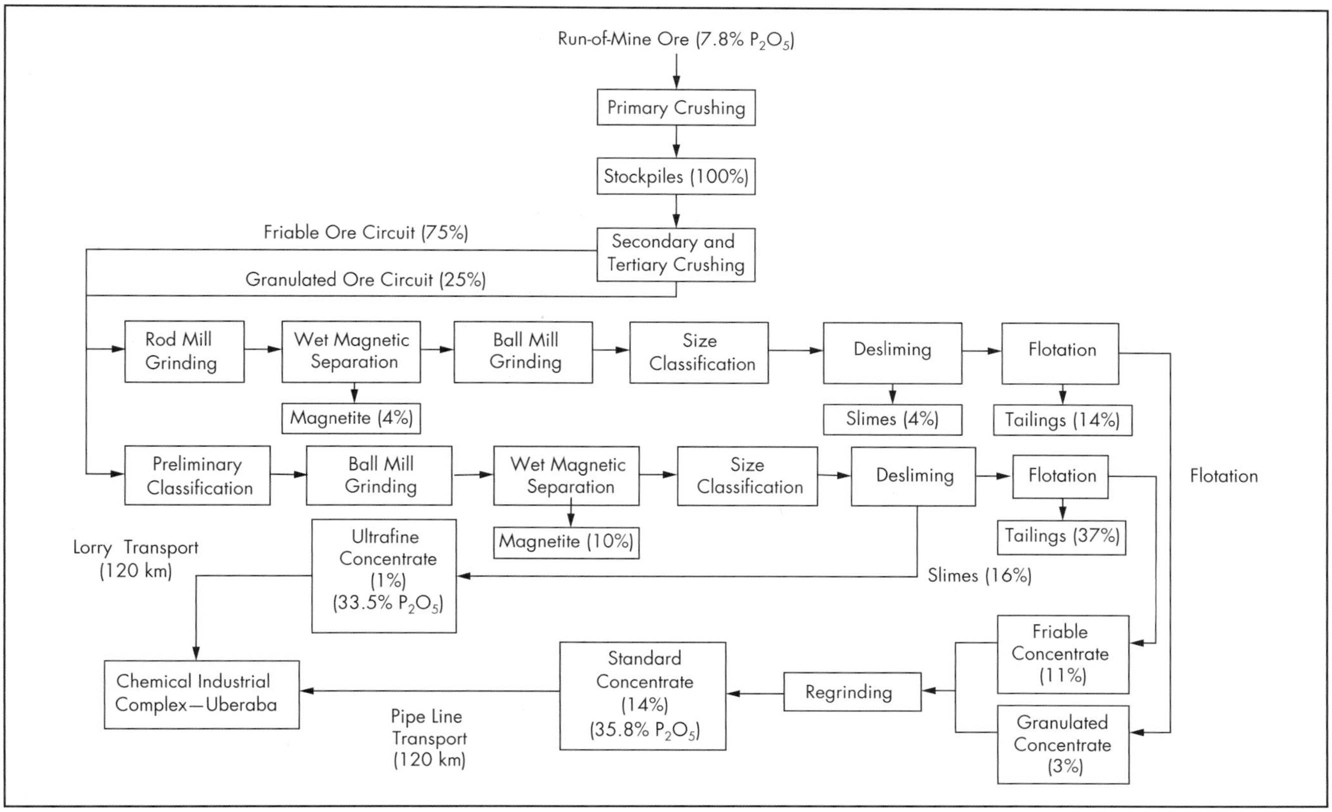

Courtesy of Fosfertil

Figure 7 Processing flow sheet of a Brazilian plant

of the product P_2O_5 is derived from the extremely fine fraction. Small hydrocyclones (2 in. diameter) are used to deslime, and fine flotation is carried out on the cyclone underflow.

At the newly opened Kapuskasing mine, the only phosphate mining operation in Canada, beneficiation is quite challenging, particularly at the beginning. Phosphate minerals are mainly fluorapatite with a small amount of crandallite. Gangue minerals are associated with iron, including iron sulfides, iron and titanium oxides, iron hydroxides, and iron and magnesium carbonates. Silica (mostly quartz) content ranges from 1% to 50%. After the strongly magnetic iron minerals are rejected using low-intensity magnetic separation (LIMS) and the weakly to moderately magnetic iron and titanium minerals are rejected using high-intensity magnetic separation (HIMS), the flotation feed goes through three stages of conditioning. Caustic starch is added into the first conditioner to both depress iron and adjust pH. In the second conditioner, a saponified tall oil is added as apatite collector. An optional third conditioner is used to further modify pH using NaOH. Flotation is conducted in mechanical cells. Flotation concentrate is dewatered with cyclones to 70% solids, followed by dewatering with belt filters to 85%–90% solids and drying with fluidized beds to 99.9% solids.

For Finnish igneous phosphates (which contain about 10% apatite, 22% calcite and dolomite, 65% phlogopite, and 3% amphibole and other silicates), the flotation process is carried out using substituted N sarcrosine (an amphoteric compound). Flotation is performed at a basic pH (8–11). The flow sheet adopted is grinding to 38.5% –74 µm, one roughing flotation stage, and five cleaning flotation stages. This flotation scheme is applied in the Siilinjarvi plant, achieving a concentrate of 33.7% P_2O_5 with a recovery of 85%.

Anionic rougher-cleaner flotation has been practiced in Mexico for many years. The flotation collector is a fatty acid emulsified with petroleum sulfonate and diesel oil.

Magnetic Separation

Magnetite, pyrite, hematite, and other iron-containing minerals are often associated with igneous phosphate ores. These minerals are usually removed using magnetic separation. Both high-intensity and low-intensity separation techniques are used. At the Phalaborwa Igneous Complex in South Africa, the pyroxenite zone is followed by the foskorite zone. This part of the ore body consists mainly of magnetite, serpentine, and apatite. The P_2O_5 content in this zone varies from 6%–11%. The magnetite must be removed using magnetic separation to upgrade the ore to commercial-grade rock. In the foskorite circuit, the crushed ore is conveyed to a 60-kt stockpile, reclaimed by three plough reclaimers, and conveyed to the milling plant. After copper flotation, Sala low-intensity magnets are employed to remove the magnetite after 750-mm and 100-mm Multotec cyclones remove the slimes. Magnetic separation is also conducted on the PMC tailings before phosphate flotation.

Magnetic separation as shown in Figures 6 and 7 is used in treating most Brazilian phosphate ores. The uniqueness of the Brazilian practice is that magnetic separation is used in both the front end and final stage of beneficiation.

At the Kapuskasing Phosphate Operations in Canada, magnetic separation is carried out on the final flotation product. The re-cleaner flotation concentrate is pumped to a reconditioned high-gradient magnetic separator (HGMS) purchased from the Iron Ore Company of Canada. The HGMS reduces the iron content from approximately 5% to less than 2%. HGMS product is pumped to a concentrate thickener.

At an Egyptian mine, phosphate ore is crushed and screened at 60 mm. The resultant oversize is discarded because of its high impurities. The –60 mm fraction is delivered to the beneficiation plant for further treatment. The first step of beneficiation is scrubbing and sizing to remove the +2 mm, a high-dolomite fraction. With further scrubbing, washing, sizing, and desliming, a 0.2 × 2 mm product is obtained. The –0.2 mm fraction is treated using a HIMS to obtain the fine concentrate. Magnetic separation is also practiced in Togo and Mexico.

Calcination

Calcining is used to treat phosphate rock to achieve one or more of the following objectives:

- Remove carbonaceous materials, dolomite, or calcite
- Remove organic matters
- Improve the reactivity of the rock
- Make low-grade, slow-release fertilizers

Calcination (at higher than 700°C) is widely used in Morocco to eliminate organic materials from the so-called brown or black phosphate rock.

During the last decade, calcination was studied and recommended to treat phosphate ores in Saudi Arabia, where the water supply is limited and energy is inexpensive. In the study, a Saudi phosphate ore containing 40%–50% carbonate and 16%–25% P_2O_5 was treated by calcination at 850°C for about an hour, followed by quenching with 5% ammonium nitrate (NH_4NO_3), 5% ammonium chloride (NH_4Cl), or water. Under the best test conditions, a concentrate containing 38% P_2O_5 was obtained.

Two low-grade Indian carbonaceous ores were successfully upgraded using a continuous-flow calcination process. Phosphate recoveries ranged from 63% to 84.6%, with concentrate grades of 31.3%–38.5% P_2O_5. The roasting temperature was 900°C.

Although the process of making the popular FCMP in China is not defined as calcination, it is a thermal process. Some of the chemical reactions taking place in the FCMP process also occur in calcination kilns.

Beneficiation of Carbonaceous Phosphates

Sedimentary carbonaceous ore is, by far, the most widely present form of phosphate in the world and constitutes roughly two-thirds of present-day reserves. Some carbonaceous phosphates are beneficiated using a calcination process followed by elimination of CaO fines. Because of ever-increasing energy costs, more economical means are becoming increasingly attractive, including the froth flotation process. The difficulty, however, arises from the fact that the physicochemical properties of phosphatic minerals and carbonates are very similar. In the last three decades, numerous studies have been carried out to separate carbonaceous gangue from sedimentary phosphate ores. These processes include direct flotation of phosphate with depression of the carbonate gangue and reverse flotation of the carbonate gangue with depression of the phosphates.

China is quite successful in processing sedimentary carbonaceous phosphates using flotation technology, perhaps because most of its phosphate deposits (about 80%) belong to this category. The Chinese phosphate resources have three distinct characteristics: old geological age, a high content of

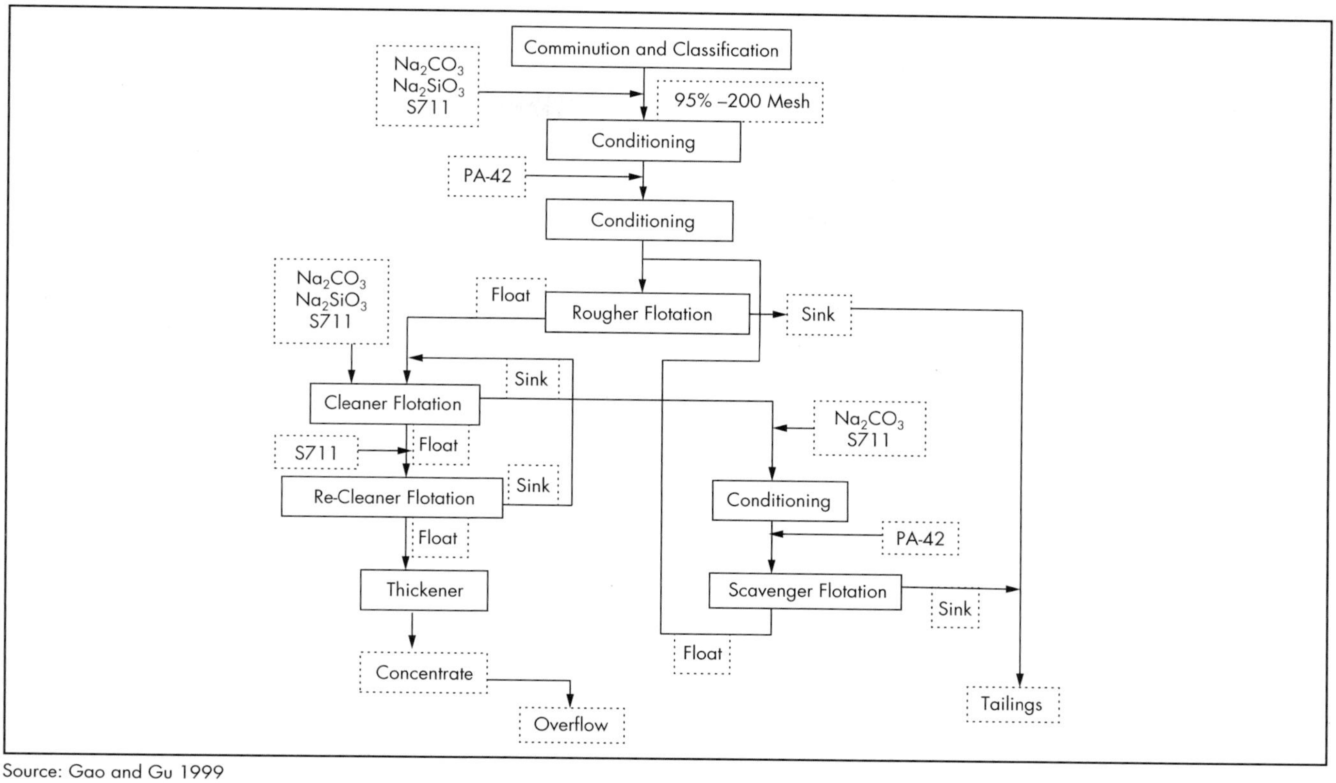

Source: Gao and Gu 1999

Figure 8 Processing flow sheet for Dayukou carbonaceous phosphate ore

dolomite and other impurities, and fine dissemination. Before beneficiation, fine grinding is usually required to liberate impurity minerals from phosphate.

Figure 8 shows the processing steps at the Dayukou phosphate mine in Hubei Province, China. The phosphate ore contains 17%–18% P_2O_5 and 4%–5% MgO. Phosphate, dolomite, and silicate in the ore were intergrown as extremely fine particles. To achieve the desired degree of liberation of phosphate from gangue minerals, the ore must be ground finely to 90% passing 200 mesh. For such fine particle sizes, it was once considered almost impossible to separate dolomite from phosphate by the flotation method. This beneficiation plant was put into operation in 1996. The success of the "direct flotation" process was attributed to the S series depressants developed by China Lianyungang Design and Research Institute (CLDRI). These depressants are effective for both carbonates and silicates. The S series depressants were derived from a by-product of the petroleum industry. The phosphate collector PA-42 is derived from waste material from vegetable oil processing. The flotation section of the process is in a closed circuit. The underflow of first cleaning flotation is subject to a scavenging operation, which is a key step for effective rejection of MgO impurity.

Figure 9 shows a "reverse flotation" process for processing carbonaceous phosphates in China. The Wengfu phosphate project in Guizhou Province, China, is currently the largest integrated phosphate fertilizer complex in China, producing 2.5 million t/yr of rock, 800,000 t/yr of TSP, 800,000 t/yr of sulfuric acid, and 300,000 t/yr of P_2O_5 as phosphoric acid. The run-of-mine ore containing about 30% P_2O_5 and 4%–4.5% MgO is crushed and then ground to about 60% passing 200 mesh for mineral liberation. The feed slurry of 40% solids is conditioned with H_2SO_4 as a pH modifier and PA-31 as dolomite collector, and is then subject to carbonate flotation to remove dolomite as a froth product. The rougher sink is cleaned and refloated to further reject dolomite. The concentrate analyzes 36% P_2O_5 and 0.95% MgO at the overall P_2O_5 recovery of 95%.

The Vernal, Utah, phosphate operation recently reached a capacity of 1.4 million t/yr. Although the ore body (4.5–6 m in thickness) is overlaid by 29 m of overburden, and beneficiation involves fine grinding and complex flotation steps, Vernal boasts the lowest cost phosphate production facility in North America. Primary grinding using the semiautogenous grinding mill brings the ore down to 35 mesh. The + 35 mesh fraction is sent to the secondary grinding circuit with ball mill. The –35 mesh (–420 μm) material from the secondary grinding is the flotation feed. The flotation feed is deslimed using hydrocyclones to remove the 400 mesh (37 μm) primary slime. The cyclone underflow is sized at 200 mesh using hydrosizers, generating a primary flotation feed (35 × 200 mesh) and a secondary slime (200 × 400 mesh). The flotation includes many steps. In the primary step, rougher flotation feed is conditioned in a three-stage tank system at a natural pH ranging from 7.5 to 7.9. This product is cleaned in a single four-cell flotation bank identical to those used in the rougher circuit. Final concentrate grade averages 30.8% P_2O_5 with an MgO content of 0.8%. Two-thirds of the phosphate production comes from the primary circuit. Primary rougher and cleaner tails are combined as the scavenger flotation feed for recovering the P_2O_5 loss in the primary tails. This feed is also conditioned at a natural pH ranging from 7.5 to 7.9 and

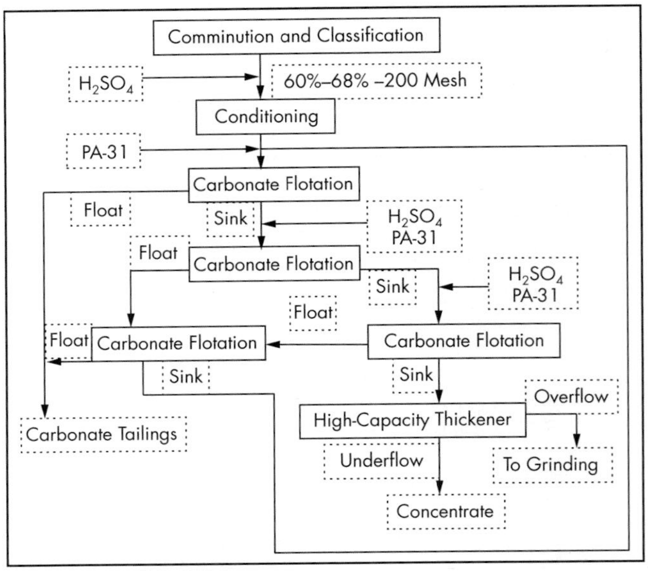

Source: Gao and Gu 1999

Figure 9 "Reverse flotation" process at Wengfu, China

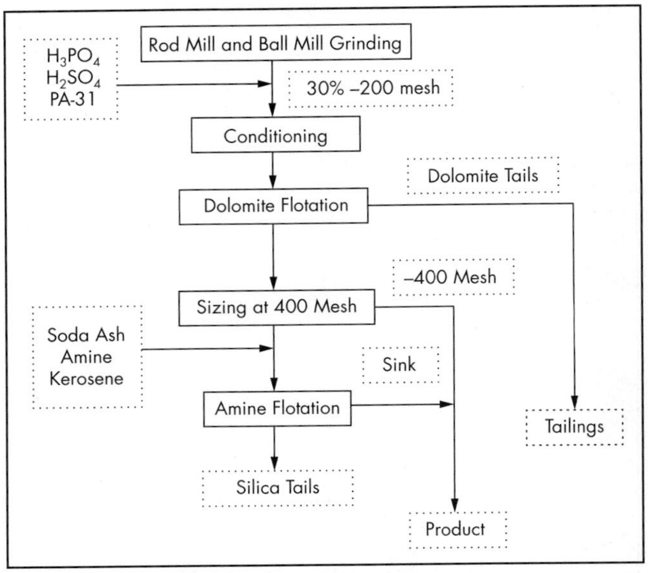

Figure 10 CLDRI fine flotation process for Florida dolomitic phosphate pebbles

70% solids with a fatty acid collector, a petroleum sulfonate, a diesel fuel, and a frother. The scavenger cleaner concentrate averaging 26.8% P_2O_5 and 2.2% MgO is acid-scrubbed prior to dolomite flotation using a dolomite collector. A frother is required to stabilize the carbonate flotation circuit.

CLDRI Process for Florida Dolomitic Pebbles

Under Florida Industrial and Phosphate Research (FIPR) Institute funding, IMC Phosphates conducted two major projects to develop an economically feasible process for high-dolomite phosphate pebbles in Florida. The CLDRI worked with IMC as the primary subcontractor for laboratory development and pilot-scale evaluation of flotation processes for dolomitic pebbles. As a result, the CLDRI process was developed. Figure 10 shows a typical flow sheet of the CLDRI process, which involves grinding, dolomite flotation, sizing, and silica flotation. Unlike most previous processes, the CLDRI process does not require the desliming step after grinding, thus reducing phosphate loss. Pilot-plant testing on two Florida samples achieved encouraging results similar to or better than those obtained in laboratory tests. An economic analysis was conducted based on the pilot testing results for a 272-t/h (300-stph) battery limit plant. The estimated variable (production rate dependent) costs (including raw materials, electric power, reagents, consumables, severance tax, dam building, services, and contract maintenance) add up to \$13.80/t (\$12.52/st) of concentrate. Fixed costs (including labor, overhead, depreciation, supplies, taxes, and insurance) total an estimated \$3.42/t (\$3.10/st) of concentrate. Total operating costs, which are the sum of variable and fixed costs, are estimated at \$17.22/t (\$15.62/st) of concentrate. These numbers suggest that the CLDRI process is economically feasible under the current conditions. The battery limit plant includes silica flotation, which reduces the concentrate acid insolubles from 19%–6%.

In the CLDRI fine flotation process, the H_3PO_4-H_2SO_4 mixture serves dual purposes as a pH modifier and a phosphate depressant. Dolomite flotation is conducted at pH about 5. The

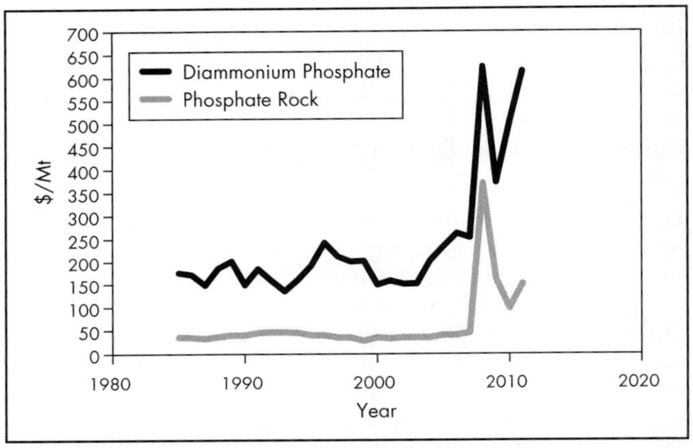

Figure 11 Historic phosphate prices

dolomite collector PA-31 is a saponified fatty acid containing a plasticizer to improve dispersion of the collector.

PHOSPHATE MARKET

Phosphate is mainly traded as finished products such as fertilizers (DAP, MAP, SSP, etc.), phosphoric acid, animal feeds, and industrial chemicals. However, Morocco, Jordan, Tunisia, and Peru export large amounts of phosphate rock. Like the price for other mineral commodities, the price for phosphate rock varies cyclically, as demonstrated in Figure 11.

CURRENT PROBLEMS AND FUTURE DEVELOPMENTS

World production of phosphate rock is exceeding 200 million t/yr. Regardless of the source of the ore, the quality of the rock concentrates produced is a major criterion for its subsequent use and/or processing operations. For instance, in acidulation processes for production of phosphoric acid, a set of physical and chemical quality factors is highly important and must be met before using the produced phosphate rock. The contents of iron, aluminum and magnesium oxides, silica, and chlorine, for example, should not exceed certain limits (Lehr and McClellan 1973). From the chemical point of view, the iron and aluminum content (expressed as R_2O_3) is the principal cause of sludge and postprecipitation in phosphoric acid production (Lehr and McClellan 1973). The ratio R_2O_3/P_2O_5 should be less than 0.095 if solid fertilizer products meeting specifications are to be produced (Lawver et al. 1978). Magnesium is the most undesirable impurity in phosphate ore. It precipitates fluorine as colloidal particles that blind the gypsum filters. It also raises the viscosity of superphosphoric acid (Lehr and McClellan 1973). Consequently, magnesium should be removed from phosphate rock feed to phosphoric acid production plants. The MgO content in acid plant feed should not be higher than 1.0% (McClellan and Germillion 1980). Chlorine is also a particularly harmful impurity in phosphate ores because of its corrosive action on plant equipment. Generally, chlorine contents higher than 0.1%–0.2% cannot be tolerated in acid-grade concentrates except in SSP manufacture where a higher level of chlorine may be used (Everhart 1971).

It is also important to maintain the correct balance of Si to F during the production of phosphoric acid to ensure that the fluorine is removed as silicon tetrafluoride, which is less corrosive than hydrogen fluoride (Frazier and Lee 1972). Also, the weight ratio of CaO/P_2O_5 should not exceed 1.6 to avoid high sulfuric acid consumption during the acidulation (Frazier and Lee 1972).

On the other hand, the minor elements such as Cu, Mn, and Zn contribute to postprecipitation of insoluble phosphate. Other elements, such as Cd, Pb, Cr, As, Hg, Se, and V, are either toxic or harmful in agricultural products

In its recently updated strategic plan for 2012–2015, the FIPR Institute assigned high priority to research on two of the major waste products from phosphate mining and processing, phosphatic clay and PG. Phosphatic clays are a major problem. They occupy large land areas, which are difficult to reclaim for either economic or environmental purposes. They also waste phosphate and clay minerals. PG is also wasted resource. Pond water associated with PG stacks poses a significant environmental hazard.

Because of limited space, this chapter focuses only on the metallurgical problems as related to beneficiation and chemical processing of Florida phosphates. These may be summarized as follows:

- Developing methods for reducing or eliminating clay settling ponds
- Recovering phosphate from phosphatic clays
- Developing technologies for solving the dolomite problem
- Reducing the magnitude of the process water problem
- Reducing the accumulation of phosphogypsum in the stacks
- Developing an alternative to the Crago process
- Extracting all useful elements from phosphate

Developing Methods for Reducing or Eliminating Clay Settling Ponds

The Florida phosphate matrix (ore) is composed of roughly one-third each of phosphate, clay, and sand. The clay must be removed before upgrading phosphate using flotation. Therefore, approximately 1 st ton of clay waste (phosphatic clay) is generated for each ton of phosphate rock product. This translates to nearly 100,000 stpd of waste clay produced in Florida. Under current practice, phosphate clay slurry with an average solids content of about 3% is pumped through pipelines to clay storage ponds where the clay slowly settles.

Although impounding is the most economical method of waste clay disposal, it has several major disadvantages. Clay settling ponds occupy about 40% of mined lands and generally have limited use after reclamation, causing adverse economic impacts. The waste clay not only ties up a large amount of water but a significant amount of water is also lost through evaporation over the clay settling areas that can occupy up to 800 acres each. The impacts of clay settling ponds on the hydrological systems must be better defined. Wetlands created on clay ponds are not accepted by the Florida Department of Environmental Protection as suitable for replacing wetlands destroyed during mining.

A near-term approach to achieve this goal is to develop rapid dewatering methods for instantaneous reclamation of mine cuts. The long-term solution is to develop alternative minerals processing technologies that would not create waste clay in the first place. For example, direct leaching of the phosphate ore (matrix) would not produce waste clay. This may be accomplished by continuous-circuit, vat, heap, or in situ leaching. The former U.S. Bureau of Mines (USBM) conducted a pilot-scale evaluation of continuous-circuit leaching of phosphate matrix in a five-tank reactor (White at al. 1978). The results from that study may be summarized as follows:

1. Phosphate recovery from the matrix ranged from 80% to 90%.
2. Sulfuric acid consumption based on P_2O_5 content was 10% higher than in commercial plants processing concentrates.
3. Total impurity (Al and Fe) content in product phosphoric acid was 60% higher than in commercial filter acid.

Another problem with this approach is the total operating cost, which was not analyzed but should be much higher than processing concentrate. Today, the industry spends about $20 on beneficiation and $70 on chemical processing per ton of phosphate rock in the near term.

Biohydrometallurgy may also offer some solutions. It has been demonstrated that phosphate rock can be leached by bacterial solutions to produce either phosphoric acid or elemental phosphorus. The practical and economic feasibility of bioprocessing for phosphate remain to be determined, but preliminary studies indicate they are not likely to be feasible or practical.

Whether traditional leaching or bioleaching was used, the resultant phosphoric acid would be contaminated by Fe, Al, and Mg. Acid purification steps would be necessary if the acid were used to make high-grade DAP. Hydrometallurgical approaches to purify the acid, such as solvent extraction and ion exchange techniques, would be needed.

Recovering Phosphate from Phosphatic Clays

Phosphate mining in central Florida generates approximately 1 st of phosphatic waste clays for each ton of phosphate rock produced. Based on this ratio, approximately 1.5 billion st of phosphatic clays have been generated by the Florida phosphate industry over the years. In many cases, phosphate content in the phosphatic clays is higher than in the matrix. About one-third of the phosphate mined has ended up in the clays.

Table 4 Concentration of P_2O_5 in phosphatic clays

Year	Author/Organization	% P_2O_5 in Clays
1954	Tyler and Waggaman, National Academy of Science	20.0
1966	Stanczyk and Feld, U.S. Bureau of Mines	12.5
1969	Davenport and Watkins, Tennessee Valley Authority	17.7
1982	Ardaman and Associates	4.6–17.9
1987	Davis et al., U.S. Bureau of Mines	6–15.9
2001	Zhang et al., Florida Industrial and Phosphate Research Institute	9.0

Many investigators have verified this enormous loss of phosphate (Table 4.)

Phosphatic clays are extremely fine, with more than 50% a few micrometers in size. Therefore, traditional beneficiation techniques, such as flotation and gravity separation, are not suitable for upgrading the phosphate component. Again, leaching or other hydrometallurgical processes may be needed for recovering the phosphate values from the clays.

Developing Technologies for Solving the Dolomite Problem

Among the deleterious materials (Fe_2O_3, Al_2O_3, CaO, MgO, F) in phosphate rock feed for phosphoric acid production, MgO is the most common and the one that causes the most problems. Generally, the phosphate rock acidulation process requires a feed of less than 1% MgO. This feed requirement has not been achieved in Florida at a commercial scale by removing magnesium from rock containing more than 2% MgO. Phosphate rock that does not meet the requirement consumes too much sulfuric acid per ton of P_2O_5 produced in the wet phosphoric acid process. In addition, MgO in the wet acid reduces the filtering capacity and ties up an equivalent amount of P_2O_5 when acidulated. MgO also causes problems in the production of DAP, which has a strict requirement for the minor element ratio, of which Mg is a constituent.

With the depletion of the higher-grade, easy-to-process Bone Valley deposits, the central Florida phosphate industry has moved into the lower-grade, more contaminated ore bodies from the southern extension of the Hawthorn Formation. Phosphate deposits in the southern extension may be divided into two zones: upper and lower. Rock in the upper zone can be processed readily using the current technology; however, the lower zone is highly contaminated by the magnesium-containing dolomite. Geological and mineralogical statistics show that nearly 50% of the phosphate resource would be wasted if the high-dolomite deposits were bypassed in mining.

Reducing the Magnitude of the Process Water Problem

Process water is essential to the operation of phosphoric acid plants. It is used as makeup water for phosphoric acid manufacture, to convey the PG to the stack, to remove the heat generated during phosphoric acid manufacture, and to scrub all the exhaust gases within the chemical complex to prevent the discharge of undesirable gases to the atmosphere. In the course of serving all these functions, the process water becomes acidic and reaches equilibrium for all the materials found in both the phosphate rock and the sulfuric acid. While the plants are designed to operate with a negative water balance, periods of excessive rainfall may make it necessary to treat and discharge surplus water. Treatment before discharge involves a two-stage neutralization with lime to remove fluoride and phosphate, followed by aeration to remove ammonia, and then the addition of sulfuric acid to reduce the pH to 6–7. Even after treatment, the conductivity of the treated water is higher than allowed and fresh water must be added before process water can be discharged to the surface waters of the state. Ammonia concentrations are also too high. Finding or developing a technology that would significantly remove or reduce any or all of the soluble salts, or raise the pH of the process water economically, would be desirable.

Double liming is also expensive, at about $20 per 1,000 gallons of process water. Reverse osmosis (RO) is now being used to obtain higher quality for discharge water. However, the major contaminants in the process water, particularly ammonia and phosphate, reduce RO efficiency significantly. Selective precipitation, ion exchange, and solvent extraction could play a role in treating the process water.

Reducing the Accumulation of Phosphogypsum in the Stacks

PG has the potential to become a major asset to the state of Florida. PG is a by-product of phosphoric acid manufacture. For every ton of phosphoric acid produced, approximately 5 tons of PG are produced. If acceptable uses can be found for PG, it may have the benefit of improving the environment, economy, and welfare of the citizens of the state. Research has demonstrated that using PG in both road building and agriculture offers substantial economic benefits without creating excessive risks. Other uses that are environmentally and/or economically desirable are also possible. The major barrier to using PG is concern about the low level of radiation associated with it. Since it is highly unlikely that methods can be developed that would use all PG production, it is desirable to find methods that will reduce its rate of production.

Some efforts have been made in reducing PG production by manipulating the chemistry, such as the MonoCal and DiCal processes. However, dramatic PG reduction requires entirely different hydrometallurgical processes.

Developing an Alternative to the Crago Process

The Florida phosphate reserves that can be economically processed with current technology may last for only about 20–30 years at the present mining rate. While development of a viable dolomite separation process is critical to extending the life of Florida's phosphate reserve, improvement in P_2O_5 recovery from the currently mined siliceous phosphates is equally important. Phosphate recoveries from the flotation feeds in most plants in Florida do not exceed 85%, with <80% being more common. Assuming that Florida's total production of flotation concentrate is about 20 Mt/yr at 32% P_2O_5 from feeds averaging 7% P_2O_5, an improvement of 1.8% P_2O_5 recovery translates to an additional 1 Mt of phosphate rock.

In the Crago process, sized flotation feed is dewatered and conditioned at about 70% or higher solids with fatty acid/fuel oil at pH ~9 for 3 minutes. The phosphate is then floated to produce a rougher concentrate and a sand tailing. A significant amount (30%–40%) of silica is also floated in this step. The rougher concentrate goes through a dewatering cyclone, an acid scrubber, and a wash box to remove the reagents from the phosphate surfaces. After rinsing, the de-oiled rougher concentrate is transported into flotation cells where amine is added. The silica is finally floated at neutral pH.

In this conventional process, about 30%–40% by weight of the sands in the feed are floated twice, first by fatty acid and then by amine. The Crago process is, therefore, inefficient in terms of collector utilization. One of the major drawbacks of this process is the de-oiling process. De-oiling consumes a significant amount of sulfuric acid, which calls for special safety cautions and equipment maintenance. Insufficient de-oiling, which is a frequent phenomenon, often causes loss of phosphate and poor concentrate grade. De-oiling also causes loss of fine phosphate particles, amounting to more than 1% phosphate recovery in most operations. Another problem with the Crago process relates to the amine flotation step. Not only are amines more expensive than fatty acids, but they are also very sensitive to water quality, particularly the clay content in water.

Direct flotation using anionic reagents was practiced on high-grade phosphate flotation feeds in some U.S. plants and is still being used in a Mexican plant. Research efforts have also been directed at developing an anionic flotation process for phosphate, but the interest has faded because of the stringent requirement for Insol (at 4%–5%) content in beneficiated phosphate rock product in the past. However, that requirement has recently been relaxed, with chemical plants now accepting concentrate analyzing as high as 10% Insol, which an anionic rougher-cleaner process could achieve without sacrificing much flotation recovery.

Another driving force for an anionic flotation process is its environmental friendliness. The environmental benefits of such a process are evident: eliminating the use of sulfuric acid and amines for phosphate beneficiation, reducing water and energy usage, and curtailing the total discharge of chemicals to the environment. An anionic flotation process also offers improved phosphate recovery. With companies having difficulty getting permits for new mines or for expansion of existing mines, there is a growing interest in extracting as much phosphate as possible from each acre mined.

Extracting All Useful Elements from Phosphate

It is well known that phosphate is a nonrenewable resource essential for plant growth and crop production, and it is therefore vital to feeding the fast-growing population of the world. But it is not widely known that there are many other valuable elements in phosphate ore that may play significant roles in the development of future energy, particularly green energy, high-tech equipment, and advancement of various key technologies. These elements include rare earths, uranium, and thorium. Uranium in phosphate accounts for more than 80% of the world's unconventional uranium resources, whereas rare earth elements (REEs) in the world's annual production of phosphate rock (about 200 Mt) total about 100,000 t, assuming an average REE concentration of 500 ppm. If uranium is not recovered during phosphate mineral processing and phosphoric acid manufacturing, it mostly ends up in fertilizers and eventually is spread on farm lands, making it impossible to ever recover it. Except for the two "waves" of uranium recovery from phosphoric acid when the uranium price was high, millions of tons of these critical elements have been discarded with fertilizers and other processing streams.

Uranium content in the world's phosphate rock products ranges from 50 to 200 ppm, averaging about 100 ppm, as shown in Table 5 (Menzel 1968; El-Shall and Bogan 1994). Worldwide, unrecovered uranium from phosphate totals about 20 million kg (based on 200 Mt rock production) per year. Using the same average uranium concentration of 100 ppm, total uranium in the world's 290 billion t of phosphate resources amounts to 29 billion kg. Florida phosphate ranks on the top in terms of uranium content, which is why in the past, most uranium extraction plants from phosphoric acid were built in Florida. The next-generation uranium extraction plants may also start in Florida.

A majority (about 65%) of the uranium in sedimentary phosphate exists in the tetravalent state in isomorphous substitution with calcium ion, because the ionic radius of U^{4+} is 0.97Å, very close to that of Ca^{2+}, which is 0.99Å. Uranium in hexavalent (+6) state is held by chemisorption on the phosphate surface as $UO_2 \cdot HPO_4$. Some uranium in phosphate, particularly that in in igneous phosphate, is associated with REEs and thorium.

Rare Earth Elements in Phosphate

Although there are more than 200 minerals known to contain appreciable amounts of REEs, only three of them are economically significant: bastnaesite, monazite, and xenotime, with bastnaesite and monazite accounting for about 95% of the current sources for light rare earths. Some rare-earth-bearing clays are also significant sources for REEs. Xenotime is the primary mineral for heavy REE and yttrium.

REEs may also be extracted as a by-product from processing of minerals, such as copper, gold, uranium, and phosphate ores, with phosphate having a great potential. Certain phosphate deposits, specifically the fluorapatite ores, contain significant amounts of REEs (Jorjani et al. 2008; Becker 1983; Preston at al. 1996). Table 6 shows the lanthanide content in some phosphate rock (Altschuler et al. 1967; Bliskoyskii et al. 1969).

Some phosphate ores or processing streams contain much greater amounts of REEs than shown in Table 6. A Canadian phosphate deposit near Quebec, for example, contains about 1,800 ppm of REEs. In recently discovered phosphate deposits in northern China (Xia 2011), REE concentration (total R_2O_3) ranges from 1.5% to ~6.41%.

Table 5 Analysis of uranium and thorium in phosphate rock from different sources

Sample	Number of Samples	Median Content of Element, ppm	
		Uranium	Thorium
Florida, United States, pebble, 1926–1935	11	208	14
Florida, United States, pebble, 1946–1955	14	148	13
Florida, United States, pebble, 1959–1964	12	127	17
Florida, United States, pebble, 1994	3	95	—
North Carolina, United States, 1957–1964	3	79	9
Utah, United States, 1936–1961	9	128	7
Idaho, United States	5	151	8
Peru, washed rock, 1961–1964	7	106	8
Morocco, 1937–1943	5	141	8
Tunisia, 1927–1955	6	48	23
Jordan, 1956–1963	6	48	0
Egypt, 1936–1937	6	122	6
Senegal	6	107	17

Source: Menzel 1968; El-Shall and Bogan 1994

Table 6 Lanthanide content in selected phosphate rock

Phosphate Rock Source	% Ln_2O_3
Kola, Russia	0.8–1.0
Florida, United States	0.06–0.29
Algeria	0.13–0.18
Morocco	0.14–0.16
Tunisia	0.14

Table 7 Total REE content in two sets of Florida samples

	Total Rare Earth Elements	
Sample Identification	Plant A	Plant B
Final concentrate	608	901
Pebble product	163	262
Flotation feed	160	209
Cleaner tails	153	335
Rougher tails	40	20
Waste clay	102	346
Phosphogypsum	118	112

Fate of Uranium in the Acidulation Process

In phosphoric acid manufacturing using the dihydrate process, about 90% of the uranium in the phosphate rock feed reports to phosphoric acid and 10% ends up in PG, versus 80% and 20%, respectively, using the hemihydrate process. Therefore, the dihydrate process favors uranium recovery. Past uranium recovery plants all used phosphoric acid produced from dihydrate-based acid plants.

Rare Earth Elements in Florida Phosphate

The Florida phosphate ore (matrix) is mined in open pits using large draglines. Phosphate matrix is first transported to the beneficiation plant and, after several washing and separation steps, is turned onto four streams: pebble product, flotation concentrate product, sand tailings, and waste clay. In the chemical processing plant, the combined pebble/concentrate rock product is reacted with sulfuric acid producing a relatively concentrated phosphoric acid and PG by-product.

Decades-old published analyses of trace elements in Florida phosphate rock showed that many of the vital REEs are present. Though only present in trace amounts, because of the tonnage of phosphate produced, these elements are still significant in aggregate mass and have not been recovered.

A comprehensive investigation of REEs in Florida phosphate was conducted by Kremer and Chokshi (1989) of the Mobil Research and Development Corporation. The total REEs in Florida phosphate matrix analyzed 282 ppm (88 ppm neodymium, 68 ppm cerium, 57 ppm yttrium, and 49 ppm lanthanum, accounting for 90%). Distributions of REEs in the mining and chemical processing streams were also determined, showing 40% in waste clay, 37.5% in PG, 12.5% in phosphoric acid, and 10% in sand tailings. Data further indicated that REEs were concentrated in fine phosphate particles, as the pebble product analyzed 284 ppm REEs versus 575 ppm in the flotation concentrate. The high REE concentration (336 ppm) in waste clay is evidence of REEs concentrating in fine phosphate particles.

Another set of data by the USGS (Altschuler et al. 1967) showed that Florida phosphate rock contained about 500 ppm REEs, with 150 ppm lanthanum, 120 ppm cerium, and 110 ppm yttrium. According to a report by the former USBM (May and Sweeney 1983), Florida PG contained 300 ppm REEs, with 130–170 ppm gadolinium, 49 ppm cerium, and 39 ppm lanthanum.

The most recent characterization of REEs in Florida phosphate was conducted by the FIPR Institute (Zhang 2012). In this study, two sets of samples were collected from two operating mines, with each set consisting of eight samples, including fine flotation feed, coarse flotation feed, flotation concentrate, pebble product, fatty acid flotation tails, amine flotation tails, waste clay (primary slime), and PG. Total REE concentrations in various samples are listed in Table 7. REE mass balance was calculated based on chemical analysis and plant mass flow rates. For Plant A, the mass balance analysis shows the following REE distributions: 61.9% in rock product, 19.6% in waste clay, and 18.2% in flotation tailings. These distributions add up to 99.7%. For the Plant B samples, mass balance analysis shows the following REE distributions: 53.19% in rock product, 39.83% in waste clay, and 8.5% in flotation tailings. These distributions add up to 101.25%.

Other Elements of Interest

Recovery of iodine, fluorine, and magnesium have been commercialized. Some recently discovered phosphate deposits in Canada contain economically recoverable niobium.

REFERENCES

Altschuler, Z.S., Berman, S., and Cuttitta, F. 1967. Rare earths in elements—Geochemistry and potential recovery. In *Intermountain Association of Petroleum Geologists Annual Conference Guidebook No. 15*. Salt Lake City: Intermountain Association of Petroleum Geologists. pp. 125–135.

Ardaman and Associates. 1982. *Evaluation of Phosphatic Clay Disposal and Reclamation Methods. Volume 1: Index Properties of Phosphatic Clays*. FIPR Publication 02-002-003. Bartow, FL: Florida Industrial and Phosphate Research Institute. p. 68.

Bateman, A. 1952. The formation of mineral deposits. *Science* 115(2990):442–443.

Beaton, J. 2010. History of fertilizer. In *Efficient Fertilizer Manual*. Mosaic. www.back-to-basics.net/efu/pdfs/HistoryofFertilizers.pdf. Accessed May 2012.

Becker, P. 1983. *Phosphates and Phosphoric Acid: Raw Materials, Technology, and Economics of the Wet Process*. New York: Marcel Dekker.

Bliskoyskii, V.Z., Mineev, D.A., and Kholodov, V.N. 1969. Composition of lanthanides in phosphorites. *Chem. Abstr.* 72:34390-d.

Davenport, J.E., and Watkins, S.C. 1969. Beneficiation of Florida pebble phosphate slime. Size distribution of constituent minerals. *Ind. Eng. Chem. Process Des. Dev.* 8(4):533–539.

Davis, B.E., Davis, E.G., and Llewellyn, T.O. 1987. *Recovery of Phosphate from Florida Phosphate Slimes*. Report of Investigation 9110. Washington, DC: U.S. Bureau of Mines.

El-Shall, H., and Bogan, M. 1994. *Characterization of Future Florida Phosphate Resources, Final Report*. FIPR Publication No. 02-082-105. Bartow, FL: Florida Industrial and Phosphate Research Institute.

Everhart, D.L. 1971. Evaluation of phosphate rock deposits. Presented at the Society of Economic Geologists Meeting, New York, March 1971.

Frazier, A.W., and Lee, R. G. 1972. Stabilization of polyphosphate fertilizer solutions. *Fert. Solut.* 16(4):32–43.

Gao, Z., and Gu, Z. 1999. Plant practices of phosphate. In *Beneficiation of Phosphates: Advances in Research and Practice*. Edited by P. Zhang, H. El-Shall, and R. Wiegel. Littleton, CO: SME. pp. 289–301.

Griffith, B. 2010. Phosphorus. In *Efficient Fertilizer Use Manual*. Plymouth, MN: Mosaic.

Jorjani, E., Bagherieh, A.H., Mesroghli, S., and Chehreh Chelgani, S. 2008. Prediction of yttrium, lanthanum, cerium, and neodymium leaching recovery from apatite concentrate using artificial neural networks. *J. Univ. Sci. Technol. Beijing* 15(4):367–374.

Kauwenbergh, S.J. 2010. *World Phosphate Rock Reserves and Resources*. Muscle Shoals, AL: International Fertilizer Development Center.

Kremer, R.A., and Chokshi, J.C. 1989. *Fate of Rare Earth Elements in Mining/Beneficiation of Florida Phosphate Rock and Conversion to DAP Fertilizer*. Research report. Nichols, FL: Mobil Mining and Minerals.

Lawver, J.E., McClintock, W.O., and Snow, R.E. 1978. Beneficiation of phosphate rock: A state of the art review. *Miner. Sci. Eng.* 10(4):278–294.

Lehr, J.R., and McClellan, G.H. 1973. Phosphate rocks: Important factors in their economic and technical evaluation. Presented at the CENTO Symposium on Mining and Beneficiation of Fertilizer Materials, Istanbul, Turkey, November.

May, A., and Sweeney, J.W. 1983. *Evaluation of Radium and Toxic Element Leaching Characteristics of Florida Phosphogypsum Stockpiles*. Report of Investigation No. 8776. Washington, DC: U.S. Bureau of Mines.

McClellan, G.H., and Germillion, L.R. 1980. The role of phosphorous in agriculture. *Proceedings of Symposium at the National Fertilizer Development Center, Tennessee Valley Authority*, Muscle Shoals, AL, June 1–3, 1976.

Menzel, R. 1968. Uranium, radium, and thorium content in phosphate rocks and their possible radiation hazard. *J. Agric. Food Chem.* 16(2):231–234.

Preston, J.S., Cole, P.M., Craig W.M., and Feather, A.M. 1996. The recovery of rare earth oxides from a phosphoric acid by-product. Part 1: Leaching of rare earth values and recovery of a mixed rare earth oxide by solvent extraction. *Hydrometallurgy* 41:1–19.

Sauchelli, V. 1942. *Manual on Phosphates in Agriculture*. Baltimore: Davison Chemical.

Sauchelli, V. 1951. *Manual on Phosphates in Agriculture*. Baltimore: Davison Chemical.

Sauchelli, V. 1965. *Phosphates in Agriculture*. New York: Reinhold Publishing.

Schmidt, C. 1999. Foskor phosphate beneficiation. In *Beneficiation of Phosphates: Advances in Research and Practice*. Edited by P. Zhang, H. El-Shall, and R. Wiegel. Littleton, CO: SME. pp. 325–334.

Stanczyk, M.H., and Feld, I.L. 1966. *Chemical Processing of Florida Phosphate Rock Slime*. Report of Investigation 6844. Washington, DC: U.S. Bureau of Mines.

Topf, A. 2015. Quebec okays $750M phosphate mine. *Mining.com*. March 19. www.mining.com/quebec-okays-750m-phosphate-mine/. Accessed November 2015.

Tyler, P.M., and Waggaman, W.H. 1954. Phosphatic slime—A potential mineral asset. *Ind. Eng. Chem.* 46:1049–1956.

UNIDO and IFDC (United Nations Industrial Development Organization and International Fertilizer Development Center), eds. 1998. *Fertilizer Manual*. Dordrecht, The Netherlands: Kluwer Academic Publishers.

USGS (U.S. Geological Survey). 2010–2016. Phosphate rock. In *Mineral Commodity Summaries [various years]*. Reston, VA: USGS.

White, J.C., Goff, T.N., and Good, P.C. 1978. *Continuous-Circuit Preparation of Phosphoric Acid from Florida Phosphate Matrix*. Report of Investigation 8326. Washington, DC: U.S. Bureau of Mines.

Xia, X.H. 2011. The distribution characteristics and potential of endogenetic phosphate rock in north China. Presented at the Beneficiation of Phosphates VI Conference, Kunming, China, March 6–11.

Zhang, P. 2012. Recovery of critical elements from Florida phosphate: Phase I. Characterization of rare earths. Presented at the ECI International Conference: Rare Earth Minerals/Metals—Sustainable Technologies for the Future, San Diego, CA, August 12–17.

Zhang, P., Snow, R., Yu, Y., and Bogan, M. 2001. *Recovery of Phosphate from Florida Phosphatic Clay*. FIPR Publication 02-096-179. Bartow, FL: Florida Industrial and Phosphate Research Institute. p. 67.

CHAPTER 12.29

Platinum Group Metals

Frank Crundwell

The platinum group metals (PGMs) consist of six elements: platinum, palladium, rhodium, ruthenium, iridium, and osmium, which are used as jewelry, catalysts, and stores of wealth. These elements are all located in the same region of the periodic table. Consequently, they all exhibit similar chemistries, which explains why they are found together in the earth's crust, and why their production and refining pipeline is much longer than that of other metals. The uses, mining sources, market supply, primary extraction, and recycling of these metals are described in this chapter.

USES OF PLATINUM GROUP METALS

PGMs are used in jewelry, coinage, and investment bars, and as powerful catalysts for industrial reactions. Over the long term, the requirement for lower emissions from cars and trucks has driven an increase in demand for these metals. In the United States, legislation was first introduced in 1975 requiring light vehicles to be fitted with autocatalysts. Other countries followed this lead, and it is estimated that at least 95% of all new vehicles are fitted with autocatalysts. Other uses of the PGMs are given in Table 1.

Table 1 Demand by application for 2013*

Uses	Platinum (Pt)	Palladium (Pd)	Rhodium (Rh)	Iridium (Ir)	Ruthenium (Ru)
Autocatalyst	3,125	6,970	801	—†	—
Chemical	540	530	79	20	104
Dental	—	510	—	—	—
Electrical	205	1,055	7	36	531
Electrochemical	—	—	—	59	125
Glass	235	0	40	—	—
Investment	765	75	—	—	—
Jeweler	2,740	390	—	—	—
Medical and biomedical	235	—	—	—	—
Petroleum	155	—	—	—	—
Other	420	100	89	83	68
Total demand	**8,420**	**9,630**	**1,016**	**198**	**828**

Adapted from Johnson Matthey 2003–2015b
*Units per 1,000 troy ounces.
†A dash indicates minimal usage in that field or usage is unquantified.

SOURCES AND SUPPLY

Production

PGMs are produced from primary sources in only five countries of the world, as shown in Table 2. Most platinum is produced in South Africa, while most palladium is produced in Russia. The annual global production of platinum, palladium, and rhodium are shown in Figures 1 and 2. The global demand for iridium and ruthenium is shown in Figure 3.

Table 2 Supply of platinum group metals by country for 2013*

<table>
<tr><th>Country</th><th>Platinum (Pt)</th><th>Palladium (Pd)</th><th>Rhodium (Rh)</th></tr>
<tr><td>South Africa</td><td>4,120</td><td>2,350</td><td>574</td></tr>
<tr><td>Russia</td><td>780</td><td>2700</td><td>85</td></tr>
<tr><td>Canada</td><td>209</td><td>541</td><td rowspan="2">24</td></tr>
<tr><td>United States</td><td>106</td><td>389</td></tr>
<tr><td>Zimbabwe</td><td>400</td><td>310</td><td>33</td></tr>
<tr><td>Others</td><td>125</td><td>140</td><td>5</td></tr>
<tr><td>Total supply</td><td>1,620</td><td>4,080</td><td>147</td></tr>
</table>

Adapted from Johnson Matthey 2003–2015b
*Units per 1,000 troy ounces.

Occurrence

The economically viable sources of platinum are concentrated in South Africa and Russia. The platinum-bearing ores of South Africa occur in a massive igneous intrusion referred to as the Bushveld Igneous Complex. The main mining region in Russia is the area around Norilsk and Talnakh, which produces large quantities of nickel and copper in addition to PGMs. Small amounts of PGMs are produced from alluvial deposits at Kondyor and Koryak in the Urals. PGMs are also mined at Stillwater and East Boulder, in Montana, United States, and from the Great Dyke area of Zimbabwe.

A significant amount of PGMs are produced as by-products of nickel production. All the Canadian production and the

Frank Crundwell, Director, CM Solutions Pty Ltd., Johannesburg, South Africa

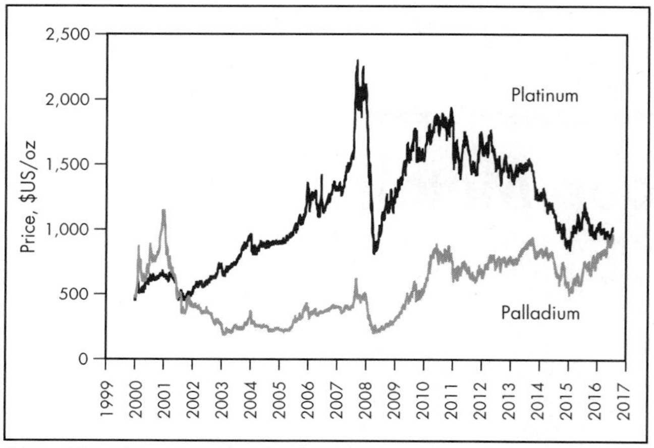

Adapted from Johnson Matthey 2003–2015a

Figure 1 Global production of platinum and palladium

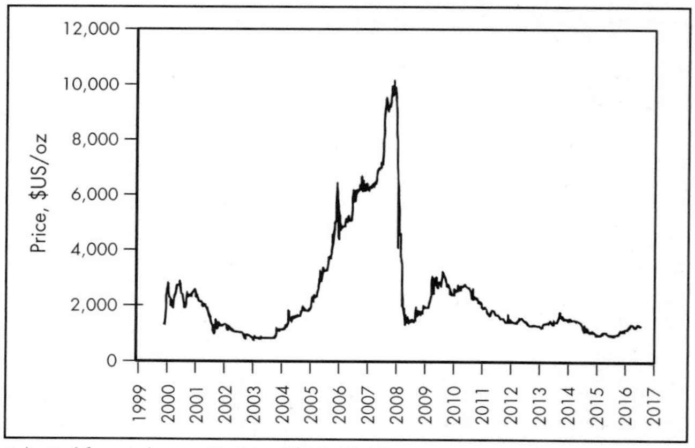

Adapted from Johnson Matthey 2003–2015a

Figure 2 Global production of rhodium

production from countries that toll refine nickel-bearing ores, such as Norway, falls into this category. The locations of primary mining, smelting, and refining facilities are given in Table 3 and Figure 4.

PRICES AND MARKETING

The prices of PGMs, from January 1, 2004, to August 20, 2015, are shown in Figures 5 to 7. Prices dropped dramatically during the global credit crisis of August 2008, recovered over the following two years, and have been in decline since 2011.

PRIMARY EXTRACTION

The primary extraction of PGMs from the Bushveld Igneous Complex in South Africa is the most complex with the longest processing pipeline of all metals. The mined ore is milled and the sulfide fraction is removed by flotation. This sulfide concentrate is smelted to remove most of the silicate material, and the resulting matte is converted to reduce the sulfur and iron content. The converter matte can be treated in a base metals refinery, or it can be treated by slow cooling and magnetic separation. After either the base metals refinery or the magnetic separation plant, the platinum-containing fraction is then transferred to the precious metals refinery, where each of the metals is separated into their pure forms.

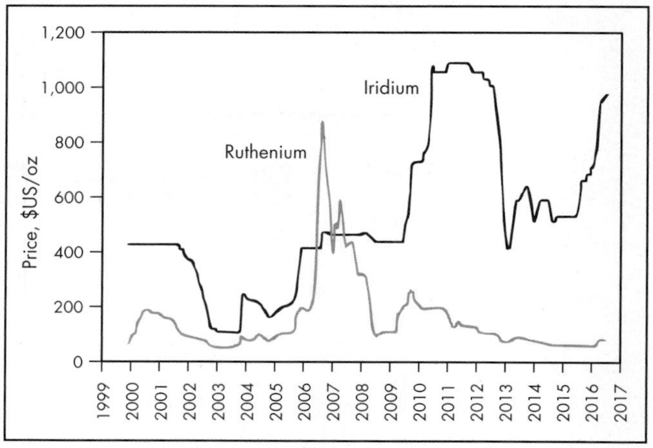

Adapted from Johnson Matthey 2003–2015a

Figure 3 Global demand for iridium and ruthenium

Table 3 Location of mining, smelter, and refining facilities for the production of platinum group metals

	Location	Mine	Concentrator	Smelter	Base Metals Refinery	Precious Metals Refinery
1	Bushveld Igneous Complex, South Africa	~20	20	6	4	3
2	Sudbury Basin, Canada	~10	2	2	1	*
3	Norilsk/Talnakh, Russia	7	2	3	1	*
4	Stillwater, United States	2	1	1	1	*
5	Lac des Iles, Canada	1	1	*	*	*
6	Selous, Zimbabwe	1	1	1	*	*
7	Kola Peninsula, Russia	1	1	1	1	*
8	Kondyor, Russia	1	*	*	*	*
9	Koryak, Russia	1	*	*	*	*
10	Port Elizabeth, South Africa	*	*	*	*	1
11	Hanau, Germany	*	*	1	1	1
12	Hoboken, Belgium	*	*	1	1	1
13	Acton, England	*	*	*	*	1
14	Royston, England	*	*	*	*	1
15	Kristiansand, Norway	*	*	*	1	1
16	Prioksk, Russia	*	*	*	*	1
17	Yekaterinburg, Russia	*	*	*	*	1
18	Novosibirsk, Russia	*	*	*	*	1
19	Krasnoyarsk, Russia	*	*	*	*	1
20	Niihama, Japan	*	*	*	*	1

* Indicates that no production of that type is undertaken at that site.

PGMs are extracted from Southern African sources primarily for their precious metals values, even though nickel and copper are produced as by-products. However, PGMs are produced from Russian sources as a by-product of nickel

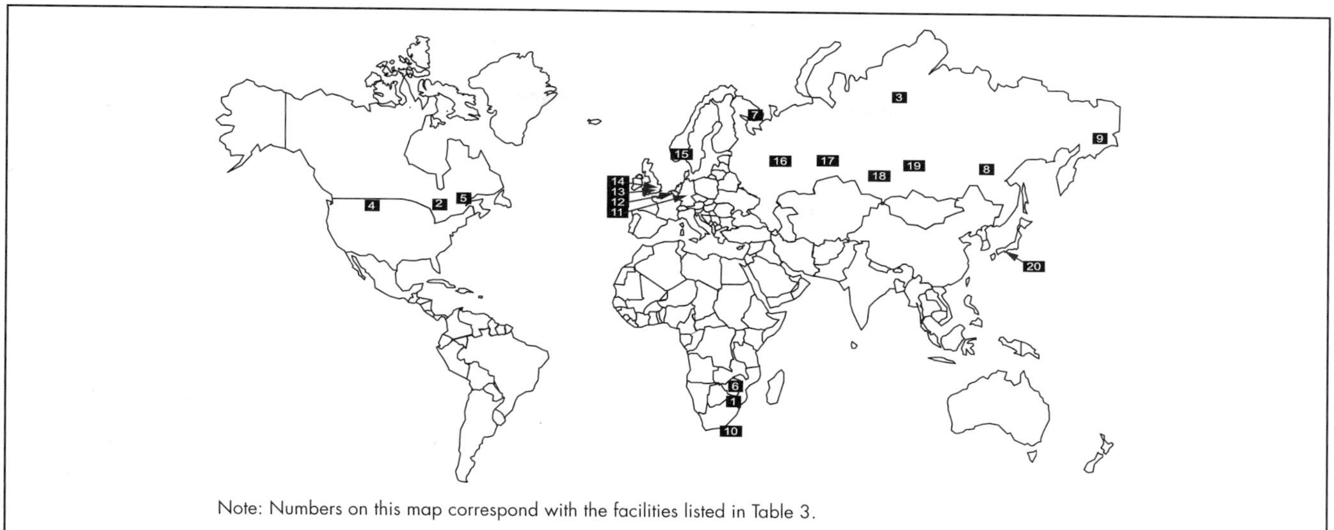

Figure 4 Locations of the major platinum group metals mines, smelters, concentrators, and refineries around the world

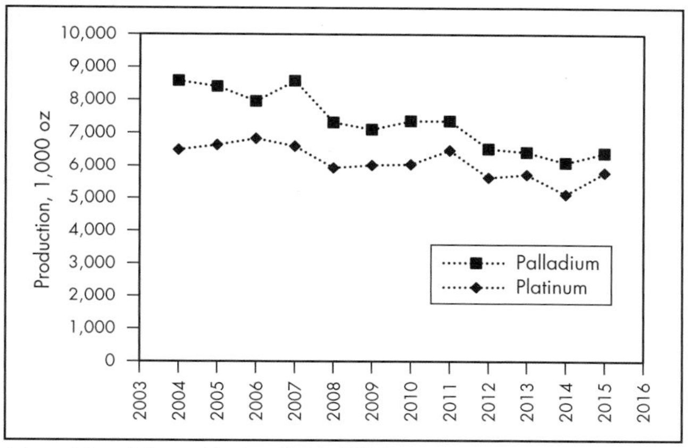

Adapted from Johnson Matthey 2003–2015a

Figure 5 Daily price for platinum and palladium from 2004 to 2015

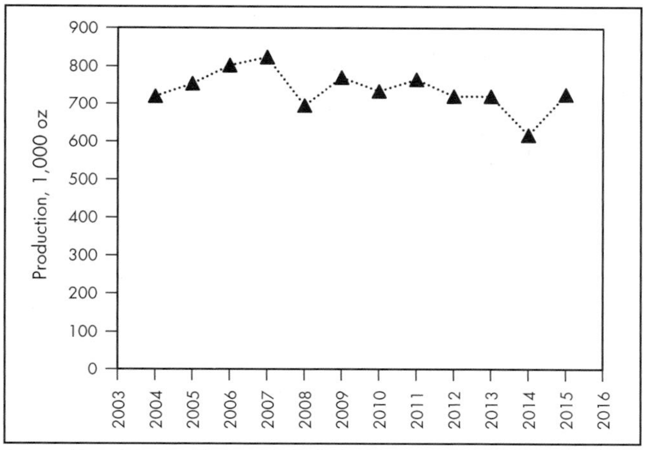

Adapted from Johnson Matthey 2003–2015a

Figure 6 Price of rhodium from 2004 to 2015

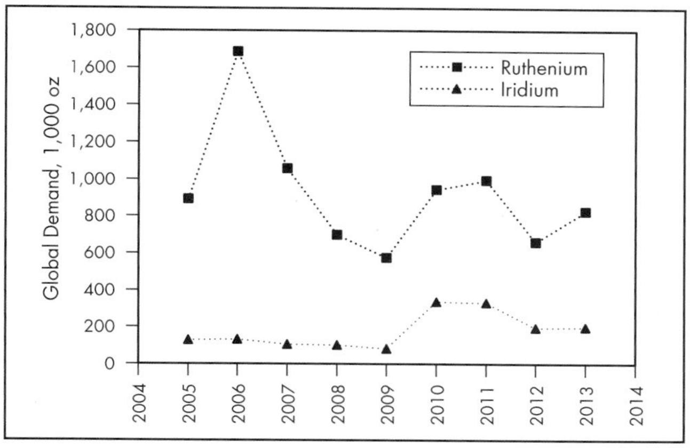

Adapted from Johnson Matthey 2003–2015a

Figure 7 Prices of iridium and ruthenium for 2005 to 2013

production. Thus, a flotation concentrate is produced from the mined ore, which is smelted; the matte is cast into anodes used in electrorefining to produce nickel metal. The insoluble precious metals report to the anode slime, which is refined in the precious metals refinery.

PGM extraction from both South African and Russian sources is discussed in further detail in the following sections.

Primary Extraction from South African Ores

The overall process for the extraction of PGMs from South African sources is shown in Figure 8. The mined ore typically contains between 1 and 10 g/t of PGMs in either metallic form or as sulfides or tellurides. These PGMs are associated with the sulfide of copper and nickel.

Flotation

The mined ore is ground and floated to produce a nickel-copper concentrate. The PGM concentration is between 100 and 200 g/t (Cramer 2008), which means that the concentration has been upgraded by a factor of approximately 20. However,

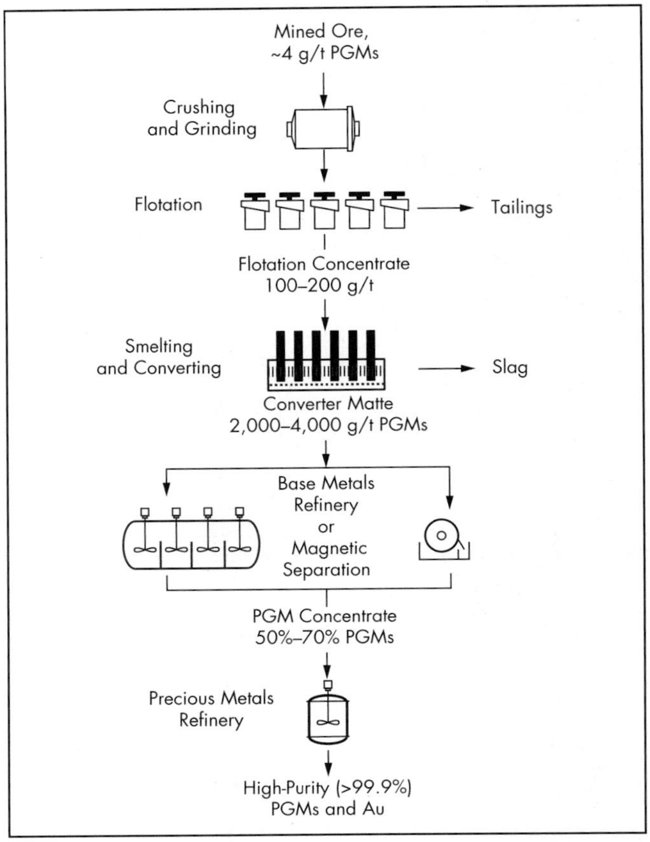

Source: Crundwell et al. 2011

Figure 8 Overview of the processing of PGMs from South African ores, showing the major steps of flotation, smelting, base metal refining, and precious metal refining

the recovery of PGMs is typically between 75% and 85%, which means that this step represents the biggest loss of value in the process.

The ores of the Merensky Reef have been typically treated by conventional sulfide flotation, in a rougher, cleaner, and re-cleaner circuit configuration. The rougher tailings are discarded, while the cleaner tailings are recycled to the rougher feed. The conventional flow sheet is given in Figure 9.

The ores of the Upper Group 2 Reef are generally more challenging, because these ores have a higher content of chromite (between 20% and 60% chromium oxide [Cr_2O_3]), which is deleterious in the downstream smelting process. Therefore, reducing the content of chromite in the flotation product to less than 3% is important. The main mechanism by which chromite reports to the flotation concentrate is by entrainment of fine particles of chromite between rising bubbles (rather than by true flotation) (Knights and Bryson 2009). The effective control of grinding is critical in reducing such entrainment. In addition, the circuit is configured in an MF2 (mill-float-mill-float) configuration, shown in Figure 10, in which the tailings from the rougher stages are reground and refloated. The objective is to achieve a relatively coarse grind (35% passing 75 µm) in the first milling stage, which reduces the chromite entrainment while managing to recover the fast-floating fraction. The rougher tailings are reground and refloated to recover the slow-floating fraction.

Because the density of the platinum minerals is higher than that of the host rock, separation techniques based on density can be effective. A gravity concentrate can be produced by cycloning the primary mill fines. This concentrate can be of a sufficiently high concentration to be sent directly to the precious metals refinery.

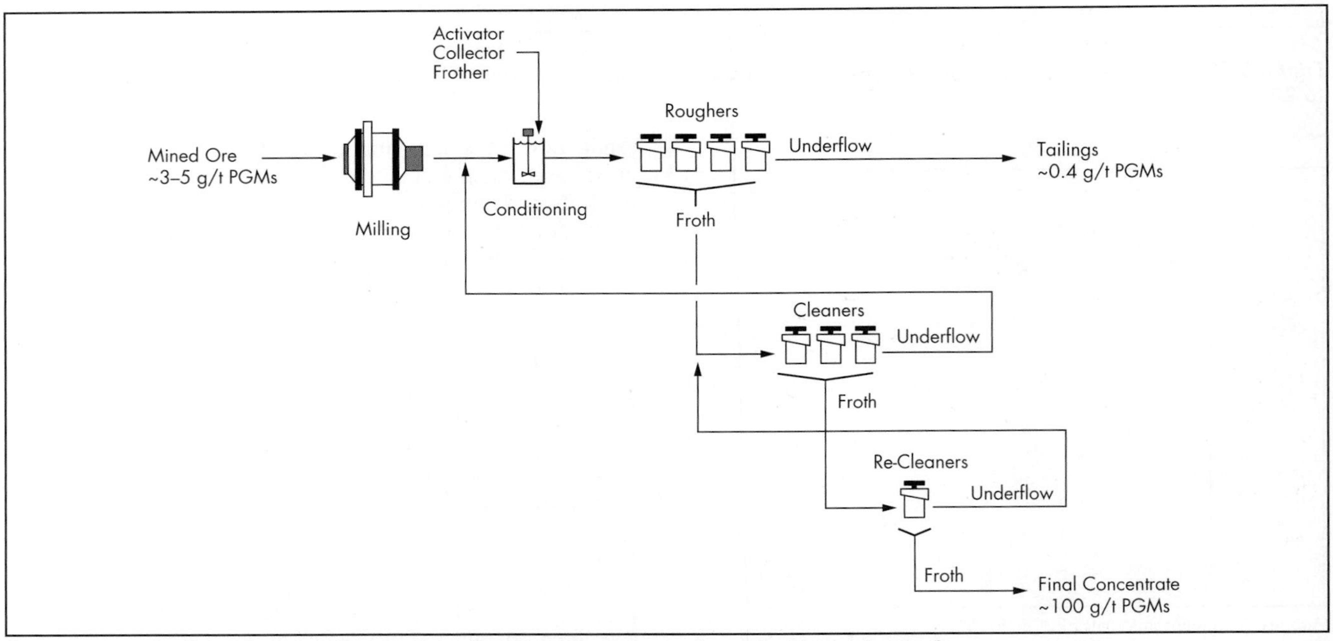

Figure 9 Process flow sheet used to produce flotation concentrates from the Merensky Reef

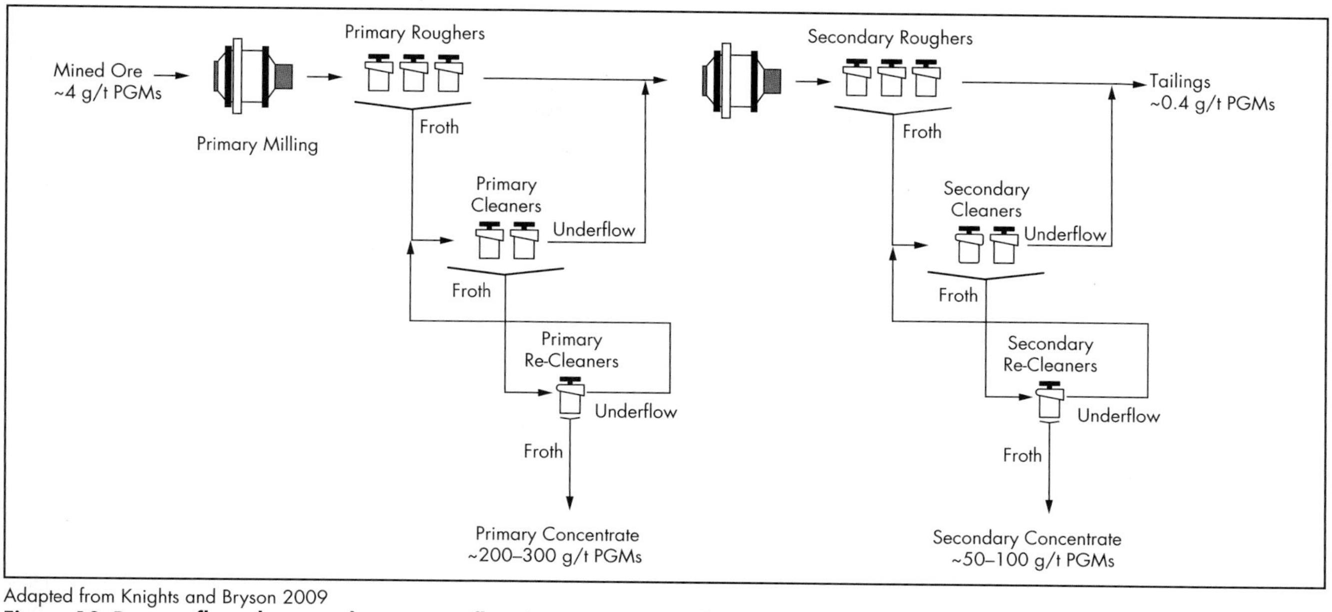

Adapted from Knights and Bryson 2009

Figure 10 Process flow sheet used to recover flotation concentrates from ores containing increased concentrations of chromite

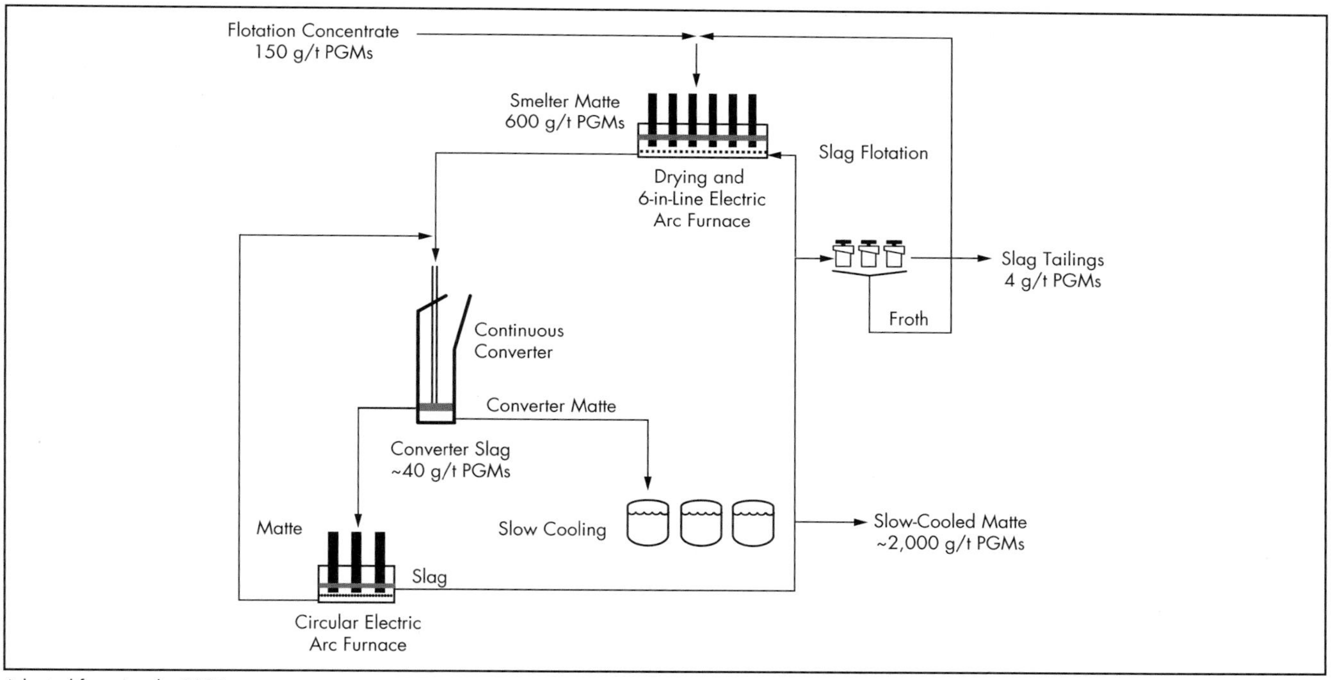

Adapted from Jacobs 2006

Figure 11 Smelting of platinum-containing sulfide concentrates at Anglo American Platinum's Waterval smelter

Smelting

The flow sheet for the smelter circuit at Anglo American Platinum is shown in Figure 11. Moist or wet concentrates are first dried in flash dryers or spray dryers to a moisture content of less than the requirement of 0.5% (Jacobs 2006). The dried concentrate is fed to the smelter, where the objective is to separate the PGM matte phase from a slag phase containing the gangue minerals. The concentrate is heated to approximately 1,600°C in an electric arc furnace. This temperature is about 250°C hotter than that of typical nickel-copper smelters because of the higher melting points of magnesium oxide (MgO) and Cr_2O_3 slags associated with these ores. The melt separates by gravity into two phases: namely, the denser matte phase at the bottom of the furnace and the less dense slag phase above it. The furnace is operated semicontinuously, with matte and slag phases being tapped off at regular intervals.

The minerals containing the PGMs are small, typically less than 10 µm in size. Because of their size, droplets of these minerals would take a long time to settle through the slag. It is thought that the larger droplets of sulfide collect these platinum-containing minerals as they descend. In this manner, PGMs are collected by the "rain" of sulfide minerals falling through the slag (Nelson et al. 2005). This means that it can be more difficult to recover PGMs in ores in which the content of sulfides is particularly low.

The smelter matte is fed to the converter, which can be of the Peirce–Smith, top-blown horizontal, or vertical-blown lance type. The iron content of the feed to the converter is typically 40%, and this is reduced to between 0.6% and 3% in the converter. The removal of the majority of the iron in the converter greatly simplifies the downstream hydrometallurgical processing.

Slow-Cooled Matte and Magnetic Separation

Most producers prefer to send the converter matte directly to the base metals refinery. However, one of the producers (Anglo American Platinum) employs the slow-cooled process, in which molten matte is poured into large molds that are covered to reduce the rate of heat loss. The matte in these molds cools over several days, during which time, crystals of heazlewoodite (Ni_3S_2) and chalcocite (Cu_2S) grow to a size of approximately 100 µm. The converter matte contains less than the stoichiometric amount of sulfur for the formation of these two sulfide phases. As a result, an alloy phase also forms. It is this alloy phase that collects the PGMs.

The slow-cooled matte is crushed and milled, and passed through a series of magnetic separators. The magnetic concentrate, which contains virtually all the PGMs, is transferred to the precious metals refinery, while the nonmagnetic fraction is sent to the base metals refinery.

Table 4 Products from different base metal refineries

Process	Nickel Product	Sulfur Product	Producer
1	Nickel sulfate	Nickel sulfate	Lonmin Platinum, Northam Platinum, Stillwater
2	Nickel powder	Ammonium sulfate	Impala Platinum
3	Nickel cathode	Sodium sulfate	Anglo America Platinum

Base Metals Refining

As mentioned, most producers send the converter matte to the base metals refinery, where it is milled and leached. In this case, the feed to the refinery is the converter matte, which contains typically 46% Ni, 33% Cu, and 19% S. The first two stages are leaching steps, and the leach residue from these steps (after some further batch dissolution steps to remove base metals and amphoteric elements) forms the feed to the precious metals refinery.

There are different processes for the refining of the base metals, and these processes might be classified on the basis of the form in which the nickel and sulfur are produced. Different options have been selected by different producers, as shown in Table 4.

Flow sheets for the three processes shown in Table 4 are given in Figures 12 to 14. The least complex of these flow sheets is the nickel sulfate process shown in Figure 12.

Nickel sulfate process. This process is implemented by Lonmin Platinum and Northam Platinum, both in South Africa, and by Stillwater in Montana, United States. The converter matte is contacted with copper spent in the first-stage leach at 85°C, which consists of five tanks in series. Oxygen is supplied to the first three tanks. The Ni_3S_2 phase in the converter matte reacts as follows:

$$Ni_3S_2(s) + 2\,CuSO_4(aq) \rightarrow 2NiSO_4(aq) + Cu_2S(s) + NiS(s) \qquad \textbf{(EQ 1)}$$

$$Ni_3S_2(s) + H_2SO_4(aq) + 0.5O_2(g) \rightarrow NiSO_4(aq) + 2NiS(s) + H_2O(l) \qquad \textbf{(EQ 2)}$$

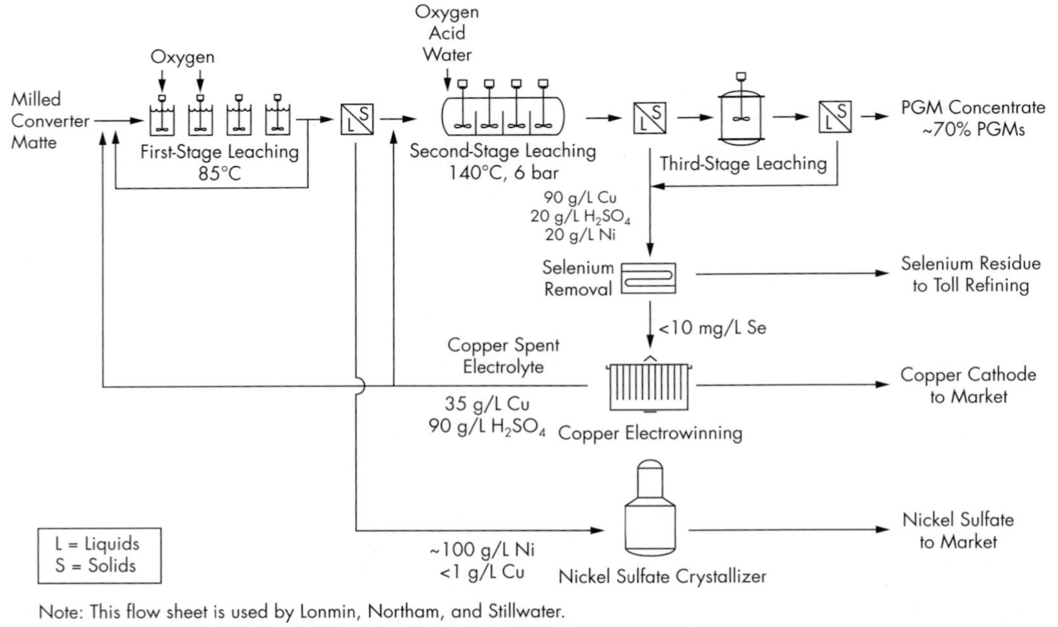

Figure 12 Process flow sheet for the base metal refinery where nickel is produced as nickel sulfate

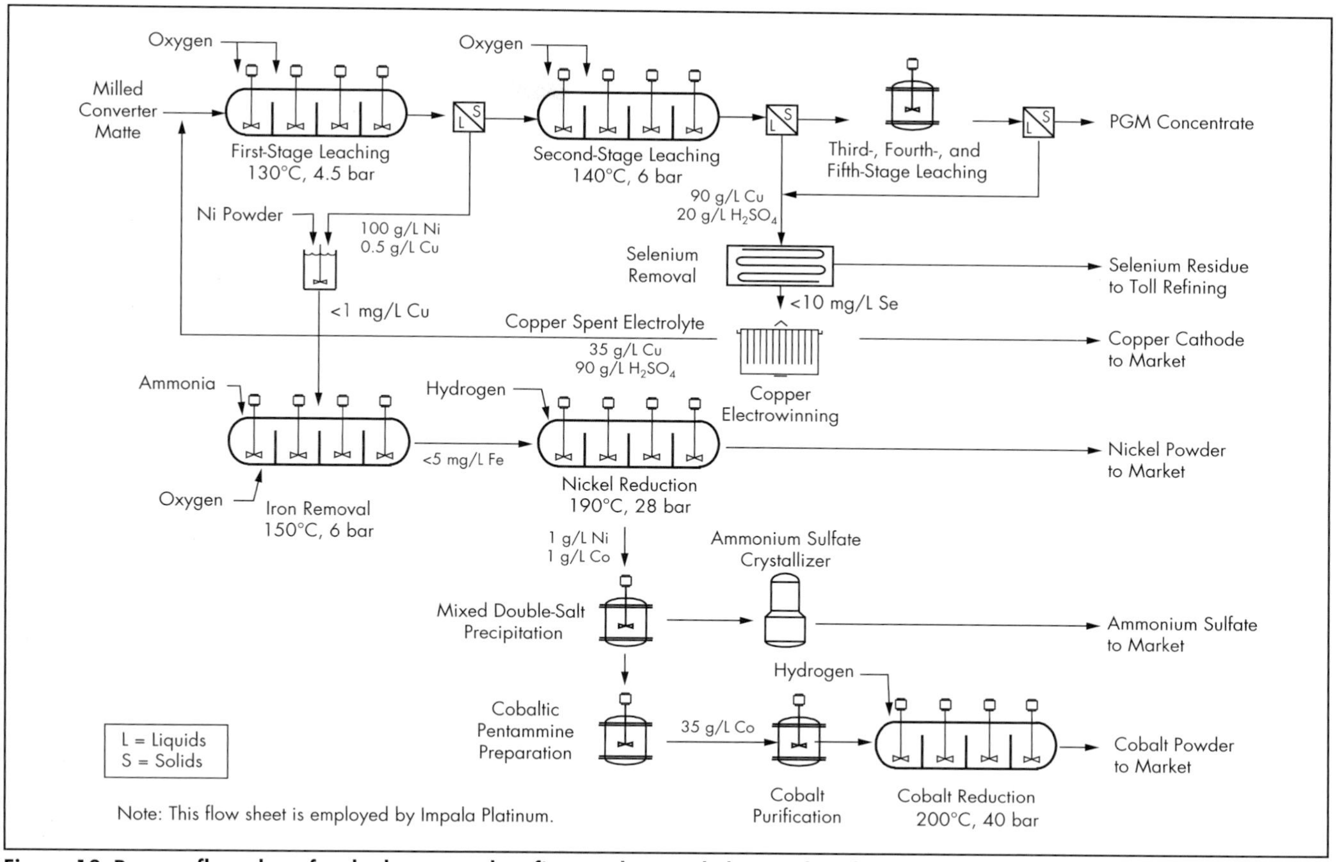

Figure 13 Process flow sheet for the base metals refinery where nickel is produced as nickel powder

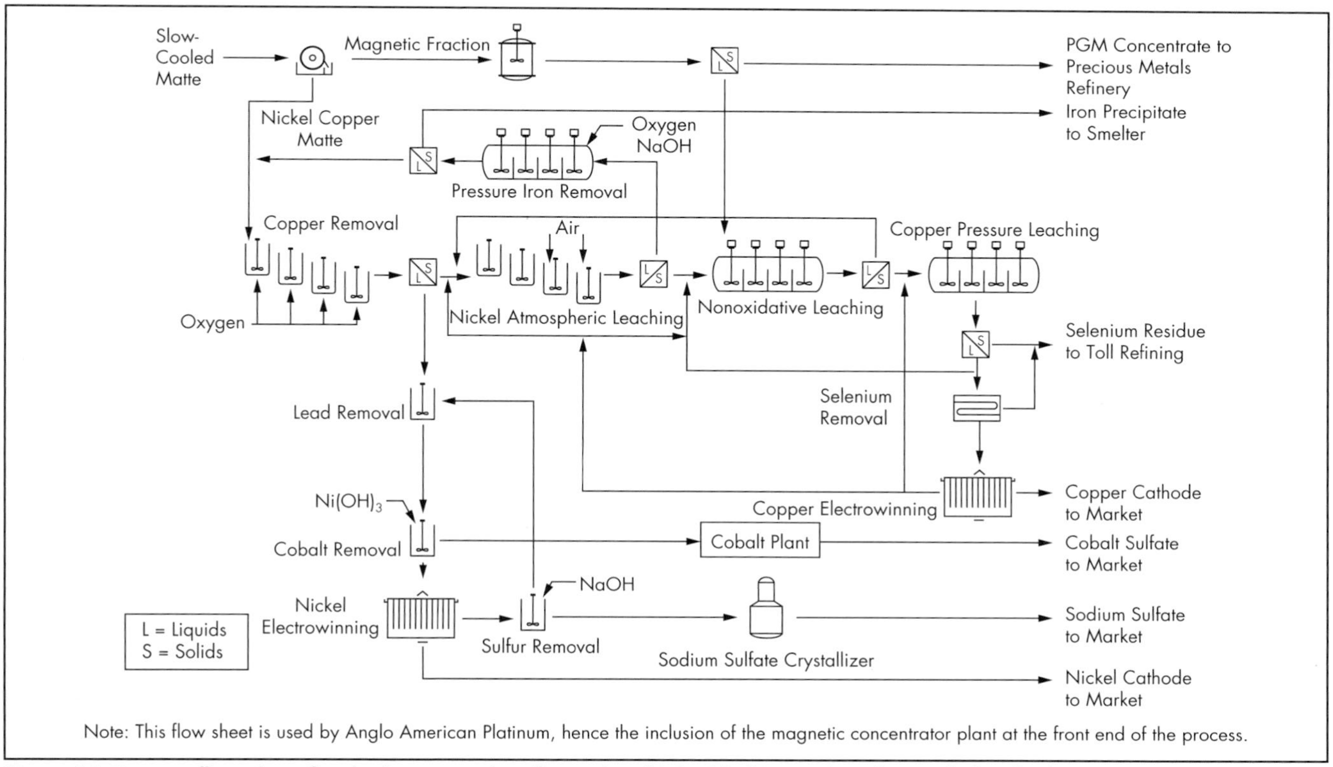

Figure 14 Process flow sheet for the base metals refinery in which nickel is produced as nickel cathode

Reaction 1 is the primary method used for separating copper from nickel in these base metals refineries. The consequence of these two reactions is that approximately 75% of the nickel dissolves, reaching about 90 g/L. Copper precipitates to less than 0.5 g/L, and acid is consumed to the point where its concentration is less than 1 g/L.

The reaction conditions are intensified in the second-stage leach by increasing the temperature and the oxygen partial pressure. The reactions are carried out in a horizontal autoclave, which is separated into various compartments so that internally it forms a series of continuous stirred-tank reactors (CSTRs). The reactions that occur are the following:

$$NiS(s) + 2O_2(g) \rightarrow NiSO_4(aq) \qquad \textbf{(EQ 3)}$$

$$Cu_2S(s) + H_2SO_4(aq) + 0.5O_2(g) \rightarrow CuSO_4(aq) + CuS(s) + H_2O(l) \qquad \textbf{(EQ 4)}$$

$$CuS(s) + 2O_2(g) \rightarrow CuSO_4(aq) \qquad \textbf{(EQ 5)}$$

As a result of these reactions, all the remaining nickel and copper dissolves, leaving a residue rich in PGMs. This residue is leached in formic acid to remove additional copper and nickel. In another step, it is leached in caustic soda with oxygen to remove sulfur, selenium, and tellurium. The remaining residue is transferred to the precious metals refinery, where the individual PGMs are separated.

The solution from second-stage leaching is purified by reducing the selenium and tellurium using sulfurous acid. The reactions are as follows:

$$2CuSO_4(aq) + H_2SO_3(aq) + H_2O(l) \rightarrow Cu_2SO_4(aq) + 2H_2SO_4(aq) \qquad \textbf{(EQ 6)}$$

$$4Cu2SO_4(aq) + H_2SeO_3(aq) + 2H_2SO_4(aq) \rightarrow Cu_2Se(s) + 6CuSO_4(aq) + 3H_2O(l) \qquad \textbf{(EQ 7)}$$

This reaction occurs at approximately 90°C. The initial contacting is in a pipe reactor, after which the reactions continue in two tanks with residence times of the order of 24 hours. The resulting slurry is filtered, and the filtrate, containing between 60 and 75 g/L Cu and 15 g/L H_2SO_4, is sent to copper electrowinning.

Copper is plated onto stainless-steel cathodes (or copper starter sheets), and oxygen evolves at the lead anode. The solution discharge from copper electrowinning, referred to as *copper spent*, contains 25–35 g/L Cu and 80–100 g/L sulfuric acid (H_2SO_4). This copper spent is recycled to all the leaching stages.

The solution from the first-stage leach, containing 100 g/L nickel sulfate along with impurities of copper, iron, and acid, is sent to the nickel crystallizer, where nickel is crystallized as nickel sulfate heptahydrate. Because the feed to the crystallizer is impure, the nickel sulfate product is not a pure product. If a purer nickel product is desired, significantly more processing steps are required, as is evident from a comparison of the nickel sulfate circuit with the nickel powder circuit shown in Figure 13, which is discussed next.

Nickel powder process. This process, shown in Figure 13, is implemented at Impala Platinum. The first-stage leach is performed in an autoclave at 130°C. Oxygen is added to the first two compartments of the autoclave, and none to the last two. As a result, the last two compartments are referred to as *nonoxidizing*. The final conditions are similar to those for the nickel sulfate process, that is, approximately 0.5 g/L Cu and 15 g/L H_2SO_4. The reactions that occur in the first stage are similar to those of the first stage in the nickel sulfate process, given by Reactions 1 and 2, but because of the harsher conditions in the first-stage leach, two additional reactions occur:

$$NiS(s) + CuSO_4(aq) \rightarrow CuS(s) + NiSO_4(aq) \qquad \textbf{(EQ 8)}$$

$$NiS(s) + 4O_2(g) \rightarrow NiSO_4(aq) \qquad \textbf{(EQ 9)}$$

As a result, the nickel extraction reaches typical values of 85%, which is higher than that achieved in the nickel sulfate process.

Slurry from the thickener underflow from first-stage leach is diluted with process water and fed to the second-stage leach. The reactions occurring in this section are similar to those in the nickel sulfate process, that is, Reactions 3 to 5. The autoclave is operated at a temperature between 120° and 140°C and at a pressure of 9.5 bar gauge. Over the last five years, the operating temperature has been lowered to reduce the dissolution of PGMs in this stage.

The leach residue from the second stage is leached in a series of batch stages to remove base metals and amphoteric elements. The amphoteric elements of interest are selenium, tellurium, sulfur, and arsenic. The residue from these leaching steps is the feed to the precious metals refinery.

The solution from the second-stage leach contains approximately 90 g/L Cu and 20 g/L H_2SO_4. This solution is purified in the selenium-removal section in a manner similar to that described previously, and sent to copper electrowinning. The spent solution from copper electrowinning goes to the first-stage leaching section.

The solution from the first-stage leach, containing 100 g/L Ni, 0.5 g/L Cu, and 1.5 g/L H_2SO_4, is purified by removing copper and iron in separate steps. Copper is removed by cementation with nickel powder. The resulting solution contains less than 1 mg/L Cu. Iron is removed from this solution in an autoclave operated at a temperature of 150°C and a pressure of 7.5 bar gauge using oxygen. Ammonia is added to precipitate the iron as ammonium jarosite, $NH_4Fe_3(SO_4)_2(OH)_6$. Lead and arsenic are also removed in this autoclave. Lead precipitates in the form of plumbojarosite, $Pb_{0.5}Fe_3(SO_4)_2(OH)_6$, and arsenic in the form of scorodite, $FeAsO_4 \cdot 2H_2O$. The solution, after filtering, is pumped to nickel reduction.

The solution is diluted to 60 g/L Ni using a solution containing ammonium sulfate. Ammonia is added so the molar ratio of nickel to free ammonia is approximately 1.9. This solution is fed to a batch autoclave that operates at a temperature of 190°C and a pressure of 28 bar gauge. Nickel sulfate is reduced to nickel metal using hydrogen gas. The operation is as follows: Solution is pumped into the reactor with seed crystals, heated and pressurized until the concentration of nickel is reduced to between 2 and 3 g/L. The reactor is depressurized to about 2 bar gauge, and the solution is drained while leaving the nickel powder in the reactor. Fresh solution is then pumped in. This cycle is repeated about 60 times, referred to as *densification*. After approximately 20–50 t (metric tons) of nickel powder is produced, the reactor is depressurized, flushed with nitrogen, and the slurry of nickel powder is drained from the reactor. The powder is filtered, washed, and dried. This dried powder can be mixed with a binding agent and briquetted, which is the final product.

The solution from nickel reduction contains 2–3 g/L nickel and about 0.5 g/L cobalt. The nickel and cobalt is precipitated from solution as a mixed double salt of nickel diammine disulfate hexahydrate, $Ni(NH_4)_2(SO_4)_2 \cdot 6H_2O$, and cobalt diammine disulfate hexahydrate, $Co(NH_4)_2(SO_4)_2 \cdot 6H_2O$. The mixed double salt is filtered and dissolved in water so that the concentration of nickel is 85 g/L and that of cobalt is 20 g/L. In the next step, ammonia is added to the solution. This ammonia forms stable cobaltic pentammine complexes with the cobalt. The solution is cooled and acidified so that nickel precipitates as $Ni(NH_4)_2(SO_4)_2 \cdot 6H_2O$. This leaves a solution containing about 35 g/L of cobalt.

The reduction of cobalt is achieved using hydrogen reduction in a similar manner to that of nickel, except at higher temperature and pressure. The temperature is 200°C and the pressure 40 bar gauge. After washing and drying, the cobalt powder is briquetted.

The solution from the reduction of cobalt and from mixed double-salt precipitation contains about 450 g/L ammonium sulfate. This ammonium sulfate is crystallized in a crystallizer, and the crystals are sold into the fertilizer market.

Nickel cathode process. This process, shown in Figure 14, is implemented at Anglo American Platinum. It is significantly more complex than the other two processes because of the five leaching stages arranged in a countercurrent configuration, although the main separation techniques are similar. The implementation is also different from that of the other producers in that the slow-cooled matte and magnetic separation process lies between that of the converter matte and the base metals refinery.

The slow-cooled converter matte is crushed, milled, and separated by magnetic concentration in three stages. The mass pull of the magnetic concentrate is between 13% and 17% of the mass of the slow-cooled matte. The nonmagnetic fraction, referred as the *nickel-copper matte*, forms the feed to the base metals refinery.

The magnetic concentrate is leached in three different solutions to remove nickel, copper, and iron. The first of these three leaching steps is performed at 160°C and 9 bar gauge using acid and oxygen to remove copper. The second leaching step is performed at atmospheric pressure using formic acid to remove iron from spinel-type minerals such as trevorite ($NiFe_2O_4$). The objective of the third leaching step is to dissolve any remaining copper, nickel, or iron. This leaching stage operates under conditions similar to those of the first-step leach.

The residue from these leaching steps contains about 50% PGMs and forms the primary feed to the precious metals refinery. The filtrate, containing approximately 100 g/L Ni and 30 g/L Cu, is pumped to the base metals refinery.

The first step in the base metals refinery is referred as *copper removal*. In this step, the nickel-copper matte from the magnetic separation plant is contacted with solution from iron removal in a series of four CSTRs. The solution contains about 70 g/L Ni^{2+} and 12 g/L Cu^{2+} at a pH value of 2. Copper precipitates as antlerite, $Cu(OH)_2 \cdot CuSO_4$, so that the concentration of copper solution is less than 0.5 g/L by the following reaction:

$$3CuSO_4(aq) + 2H_2O(l) \rightarrow Cu(OH)_2CuSO_4(s) + 2H_2SO_4(aq) \quad \textbf{(EQ 10)}$$

At the same time, acid is consumed by the partial leaching of Ni_3S_2:

$$Ni_3S_2(s) + H_2SO_4(aq) + 0.5O_2(g) \rightarrow NiSO_4(aq) + 2NiS(s) + H_2O(l) \quad \textbf{(EQ 11)}$$

The precipitation reaction is at equilibrium, which means that the concentration of copper is dependent only on the temperature and the pH. However, if the pH rises above a value of 6, nickel precipitates from solution as nickel basic sulfate, $Ni(OH)_2 \cdot NiSO_4$, so careful control of pH is needed. The slurry is thickened, and the overflow solution is pumped to lead removal, while the underflow solids are pumped to nickel atmospheric leaching.

In the nickel atmospheric leaching step, the slurry from copper removal is mixed with copper spent, which contains about 30 g/L Cu and 90 g/L H_2SO_4, and solution from the nonoxidizing leaching, which contains about 20 g/L Cu and 35 g/L H_2SO_4. The equipment consists of four CSTRs in series, and air is injected to some of the tanks. The concentration of copper in solution is reduced by about half in this step, while all the remaining Ni_3S_2 dissolves. The reactions that occur are the metathesis leaching of Ni_3S_2:

$$Ni_3S_2(s) + 2CuSO_4(aq) \rightarrow 2NiSO_4(aq) + Cu_2S(s) + NiS(s) \quad \textbf{(EQ 12)}$$

and the dissolution of Ni_3S_2 by oxygen:

$$Ni_3S_2(s) + H_2SO_4(aq) + 0.5O_2(g) \rightarrow NiSO_4(aq) + 2NiS(s) + H_2O(l) \quad \textbf{(EQ 13)}$$

In addition, all of the antlerite that is precipitated in copper removal redissolves in nickel atmospheric leaching. The slurry is thickened, the overflow is pumped to iron removal, and the underflow solids are pumped to the nonoxidative leaching stage.

The next step is the nonoxidative leaching stage, in which the slurry from nickel atmospheric leaching is combined with solution from the magnetic concentrator plant and with solution obtained from the feed to copper electrowinning, referred to as *copper advance*. A four-compartment autoclave operated at 150°C and 6 bar gauge is used to dissolve as much of the nickel as possible. The reactions that occur in this reactor are the metathesis of millerite (NiS) in the following two reactions:

$$NiS(s) + CuSO_4(aq) \rightarrow NiSO_4(aq) + CuS(s) \quad \textbf{(EQ 14)}$$

$$6NiS(s) + 9CuSO_4(aq) + 4H_2O(l) \rightarrow 6NiSO_4(aq) + 5Cu_{18}S(s) + 4H_2SO_4(aq) \quad \textbf{(EQ 15)}$$

Because this last reaction produces acid, the concentration of acid rises across nonoxidative leaching. The solution leaving nonoxidative leaching contains about 20 g/L Cu^{2+} and 30 g/L H_2SO_4. The leach slurry is thickened, the overflow is pumped to nickel atmospheric leaching, and the underflow slurry is pumped to copper pressure leaching.

The next step is copper pressure leaching, in which the slurry from the nonoxidizing leach step is contacted with copper spent from electrowinning in an autoclave operating at 140°C and 6 bar gauge. Oxygen is sparged into each of the four compartments of the autoclave. The following reactions occur under these conditions:

$$NiS(s) + 2O_2(g) \rightarrow NiSO_4(aq) \quad \textbf{(EQ 16)}$$

$$Cu_2S(s) + H_2SO_4(aq) + 0.5O_2(g) \rightarrow CuS(s) + H_2O(l) \quad \textbf{(EQ 17)}$$

$$CuS(s) + 2O_2(g) \rightarrow CuSO_4(aq) \quad \textbf{(EQ 18)}$$

The solution leaving the copper pressure leaching autoclave contains about 90 g/L Cu, 40 g/L Ni, and 15 g/L H_2SO_4. The slurry is filtered, the filter cake is collected for toll refining to recover the value of residual PGMs, and the filtrate is pumped to selenium and tellurium removal.

Selenium and tellurium removal and copper electrowinning are similar in nature to these same steps for the nickel sulfate and nickel powder processes discussed earlier.

The solution from copper removal is purified further by the removal of lead and cobalt from solution. Lead is removed by precipitation with barium hydroxide. The cobalt in this product stream is oxidized to the cobaltic state using nickelic hydroxide. This causes cobaltic hydroxide to precipitate. The precipitate is dissolved, and a pure cobalt stream is obtained by solvent extraction using D2EPHA as the extractant. The cobalt-lean solution from cobaltic precipitation is fed to the nickel electrowinning tank house.

The feed solution to nickel electrowinning contains about 90 g/L Ni and 140 g/L sodium sulfate (Na_2SO_4) at a pH of 3.5. The solution is fed individually to the cathode compartment of a cell, which is divided into compartments using a cathode bag. Solution flows from the cathode compartment into the anode compartment, from which it exits the cell. The flow through the cathode bag opposes the diffusion of H^+ ions from the anode compartment to the cathode compartment. The nickel plates onto titanium cathodes, and once the plate is sufficiently thick, the cathodes are harvested and stripped. The nickel cathode is washed, strapped into bales, and shipped to market. The solution from nickel electrowinning goes to sulfur removal.

In sulfur removal, the flow from nickel electrowinning is split into two streams. In one stream, all the contained nickel is precipitated as nickel hydroxide and filtered. The filtrate goes to the sodium sulfate crystallizer. The cake is dissolved in the fraction of the stream from nickel electrowinning that was not subject to precipitation.

The nickel-free solution from sulfur removal is transferred to the sodium sulfate crystallizer. The product crystals of sodium sulfate are sold as filler in detergents.

Comparison of different operations. A comparison of the different processes is presented in Table 5.

Precious Metals Refinery

The typical feed composition to a precious metals refinery is shown in Table 6. From this feed, PGMs must be produced at a purity of equal to or greater than 99.9% with as high a recovery as possible. The flow sheet of platinum metals refineries generally follow a pattern in which there is a sequence of primary separations, in which each of the PGMs is removed sequentially from the feed, secondary purification of each the products is made from the primary separations, and finally reduction of the salt to metallic form suitable for sale to the market is performed. The *sharpness* of the separation processes used affects the *first-pass yield*, which is a measure of the fraction of PGMs in the feed that emerges from the refinery without having to be reprocessed.

There are three main types of modern refineries, classified by the methods used in the primary separations, particularly the separation of platinum and palladium: (1) those based on precipitation, (2) those based on solvent extraction, and (3) those based on ion exchange. Each of these will be discussed in turn.

Precipitation-based refining. An overview of the precipitation process as implemented at Lonmin is shown in Figure 15. The platinum-metal concentrate as received from the base metals refinery contains 68%–70% PGMs. This material is digested in a solution of 6 M HCl at 65°C for about 6 hours. Chlorine is continuously sparged through the glass-lined reactor while the slurry is agitated using a pitched blade turbine.

Once the digestion is complete, air is blown through the solution to remove the dissolved chlorine. The slurry is filtered, and the remaining solids are redissolved using similar conditions. The dissolution efficiency of the platinum and palladium is in excess of 99%. The dissolution efficiency of the other metals is in excess of 95%, except that occasionally the extraction of iridium may drop to 85%.

Gold, which is in solution following the digestion step, is precipitated using hydrazine. Following this, the pH of the filtrate is adjusted to a value of 1 by the addition of sodium hydroxide. This results in the precipitation of approximately 40% of the base metals in this solution.

In the next step, ruthenium is removed from solution by distillation. Sodium chlorate and sodium bromate are added to the solution to oxidize the ruthenium and osmium to RuO_4 and OsO_4. In this form, ruthenium and osmium are volatile, and are easily stripped from solution into the gas phase. This stripping is achieved by passing a stream of air through the vessel. This gaseous stream, now laden with RuO_4 and OsO_4, flows through a packed column where it is contacted with a solution containing hydrochloric acid. The solution causes the reduction of the gaseous ruthenium, which then reports to the aqueous phase. The gaseous phase passes through a second column where it is contacted with a solution containing caustic soda, which scrubs chlorine and osmium from the gas phase.

The pH of the solution remaining after ruthenium distillation is increased to about a value of 5, resulting in the precipitation of rhodium and iridium hydroxide. These salts are filtered from the solution and transferred to this sidestream.

The solution remaining after pH 5 contains mainly platinum and palladium. These two elements are separated from one another by the addition of hydrochloric acid and hydrogen peroxide, which causes the precipitation of ammonium hexachloroplatinate, $(NH_4)_2PtCl_6$. This salt is filtered from solution and ignited to produce an impure platinum sponge of between 92% and 97% platinum. The main impurity in the platinum sponge is palladium. The sponge is purified in the platinum sidestream.

The pH of the solution remaining after platinum precipitation is increased to 0.9 by adding sodium hydroxide. The solution is boiled, during which time the pH increases to about a value of 3. Ammonium acetate is added and the solution boiled until the pH reaches a value of 4.2. When this solution is cooled, palladium precipitates as diammino-palladous dichloride, $(NH_3)_2PdCl_2$. This precipitate is filtered off and transferred to the palladium purification sidestream.

Table 5 Comparison of the different base metals refineries

	Nickel Sulfate Process	Nickel Powder Process	Nickel Cathode Process
Operation	Lonmin Platinum	Impala Platinum	Anglo American Platinum
Year of first production	1971	1969	1954
Converter matte capacity, t/a (metric tons per annum)	10,000	30,000	45,000
Nickel Leaching			
Temperature	85°C	150°C	NAL*: 85–90; NOX†: 150°C
Pressure	Atmospheric	4.5 bar	NAL: ambient; NOX: 6 bar
Nickel extraction	75%	85%	80%–85%
Copper Leaching Stage			
Temperature,	140°C	130°–140°C	140°C
Pressure	6 bar	6 bar	6 bar
Copper extraction	90%	>95%	>95%
Iron Removal Step			
Product	Ferric hydroxide	Ammonium jarosite	Sodium jarosite
Temperature	85°C	140°C	140°C
Cobalt Product			
Purification method	None	Cobaltic pentammine	Cobaltic hydroxide
Product	None	Cobalt powder or briquette	$CoSO_4{\cdot}6H_2O$
Nickel Product			
Production capacity, t/a	3,500	16,000	33,000
Process route	Nickel crystallizer	Hydrogen reduction	Electrowinning
Product type	$NiSO_4{\cdot}6H_2O$	Nickel powder or briquette	Nickel cathode
Copper Product			
Electrowinning capacity, t/a	2,500	9,000	18,000
Platinum Group Metal Product			
Composition of PGMs	68%–70%	50%–60%	50%

Source: Crundwell et al. 2011
*NAL = nickel atmospheric leaching
†NOX = nonoxidative leaching

This completes the description of the primary separation step. The most important sidestreams are those of platinum and palladium, which are discussed next.

The feed to the platinum sidestream is the impure platinum sponge. This sponge is digested in hydrochloric acid using chlorine gas. The solution is heated, the pH is increased to a value of 3, and sodium chlorate and sodium bromate are added. The pH is increased to a value of 5, which results in the precipitation of iridium and rhodium hydroxides. The pH of the solution remaining after this treatment is increased to 8, and any precipitated base metals are filtered off. After this step, dimethylglyoxime is added, which results in the precipitation of palladium from solution. Following these steps, the solution should contain only platinum. This solution is conditioned by adding hydrochloric acid and hydrogen peroxide, and then ammonium chloride is added, which causes $(NH_4)_2PtCl_6$ to precipitate. This salt is filtered off, redissolved in water, and the dissolved platinum reduced to metal using hydrazine. The metal is calcined to produce a platinum sponge with a purity of 99.99% platinum.

The feed to the palladium sidestream is the impure palladium salt. The main impurity in this salt is platinum. The salt is dissolved is a solution of pH 9, and the redox potential is adjusted using hydrogen peroxide. Under these conditions, the

Table 6 Composition of the feed to a South African PGM refinery

Element	%
Pt	24.6
Pd	13.3
Au	1.12
Rh	4.12
Ru	5.03
Ir	1.49
Ag	2.20
Ni	1.20
Cu	3.00
Fe	2.60
Pb	1.90
SiO_2	12.9
As	0.43
Se	1.20
Te	2.40
Ba	0.20
Na	1.70
O, H	~20.00

Source: Asamoah-Bekoe 1998

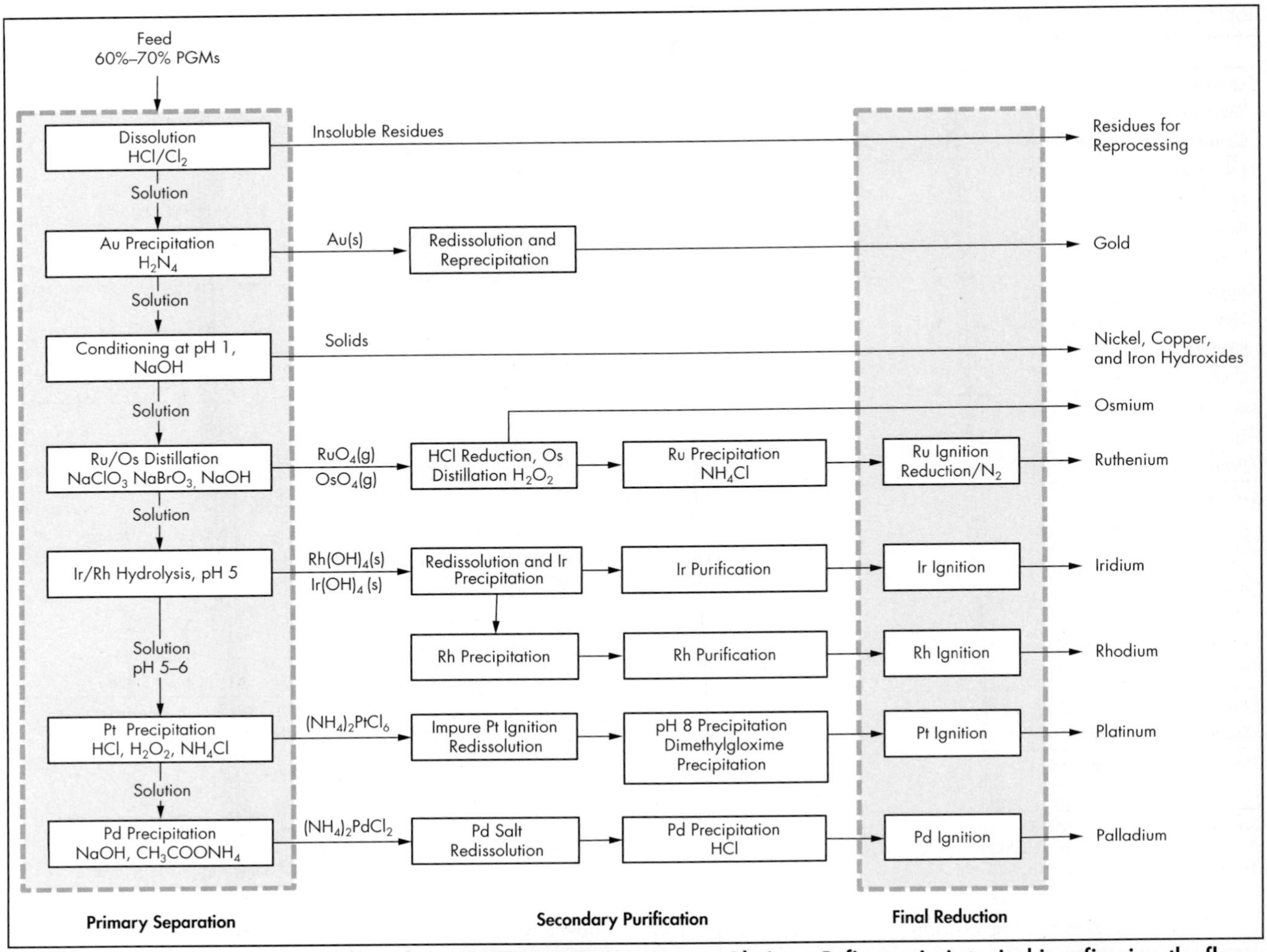

Figure 15 Process flow sheet used for refining PGMs at Lonmin's Western Platinum Refinery. As is typical in refineries, the flow sheet consists of primary separation steps, secondary purification steps (also called *sidestreams*), and final metal production or reduction steps.

palladium salt dissolves, but the platinum in the salt does not. After filtering, the solution contains primarily palladium. The solution is acidified to a pH of 0.6, which causes $(NH_3)_2PdCl_2$ to precipitate. This precipitate, ignited in air, produces palladium metal with a purity of between 99.98% and 99.99% palladium.

Solvent-extraction-based refining. An overview of the solvent-extraction-based process implemented at Vale's refinery at Acton in England is shown in Figure 16. The plant treats anode slimes and secondary materials in addition to primary concentrates (Barnes and Edwards 1982). The first step in the process at Acton is the smelting of the feed with iron, which collects PGMs. This iron is dissolved from the smelted solids, which is the feed to the rest of the process.

The residue from iron dissolution is dissolved in hydrochloric acid using chlorine gas at 90°C and at ambient pressure. The residues from this digestion step contain silver and lead that are treated with nitric acid. After this nitric acid treatment, the residues are fused with sodium hydroxide at temperatures between 500° and 600°C, and then subject to a second dissolution step with hydrochloric acid and chlorine gas. The resulting solution is neutralized using sodium hydroxide.

In the next step, ruthenium is removed by distillation. Sodium bromate is added to oxidize ruthenium in solution to ruthenium tetroxide at a temperature of about 90°C. The ruthenium tetroxide is stripped from solution by passing a stream of air through the solution. The ruthenium tetroxide in the gas phase is reduced by contacting the gas with a solution containing hydrochloric acid. Ammonium chloride is added to this solution, resulting in the precipitation of ammonium ruthenium chloride, $(NH_4)_3RuCl_6$. This salt is heated and reduced using hydrogen gas with a purity of 99.9% ruthenium.

The pH of the solution remaining after ruthenium distillation is increased using sodium hydroxide, resulting in the precipitation of base metals, such as copper and nickel.

Hydrochloric acid is added to the solution remaining after base metal precipitation so that the concentration is about 3 M HCl. This solution is fed to the gold–solvent–extraction circuit, where it is extracted by dibutyl carbitol in two stages: The loaded organic is washed with a solution of hydrochloric acid and stripped from the loaded organic using oxalic acid. The oxalic acid reduces the gold to the metallic state. This gold powder is melted into grains with a purity of 99.99% gold.

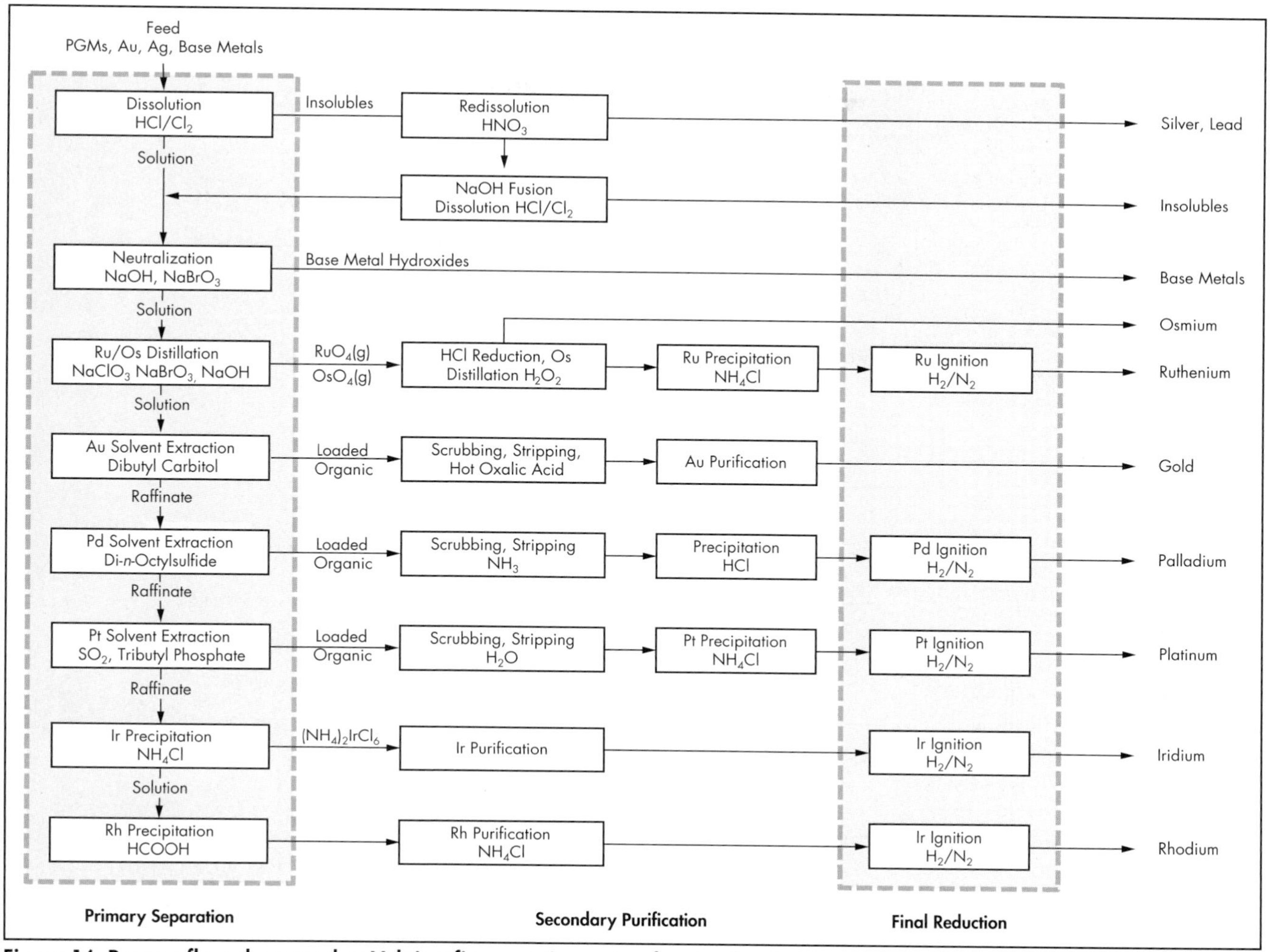

Figure 16 Process flow sheet used at Vale's refinery at Acton, London, England. This flow sheet is typical of solvent-extraction-based processes.

The raffinate from gold extraction is fed to palladium solvent extraction, where the palladium is extracted from the solution using di-*n*-octyl sulfide. The raffinate contains less than 1 mg/L. The loaded organic is scrubbed with hydrochloric acid and stripped from solution, using ammonium hydroxide. Palladium is precipitated as $(NH_3)_2PdCl_2$ from the strip solution using hydrochloric acid. This salt is reduced at 1,000°C by hydrogen. The final metal sponge has a purity of greater than 99.95% palladium.

The raffinate from palladium extraction forms the feed to platinum solvent extraction. The solution is conditioned using SO_2 to reduce iridium from the tetravalent state to the trivalent state. The platinum in this solution is extracted using tri-*n*-butyl phosphate in four stages. The loaded organic is scrubbed with a solution of 5 M hydrochloric acid and stripped with water. Platinum is recovered from the strip solution by adding ammonium chloride, which causes $(NH_4)_2PtCl_6$. This salt is ignited in air to produce a sponge with a purity of greater than 99.95% platinum.

The raffinate from platinum solvent extraction is conditioned so that the iridium is oxidized from the trivalent state to the tetravalent state. Ammonium chloride is added to the solution, causing ammonium hexachloroiridate, $(NH_4)_2IrCl_6$, to precipitate. This salt is heated in the presence of hydrogen gas, which results in iridium metal with a purity of greater than 99.9% iridium.

Rhodium is precipitated from the solution remaining after iridium precipitation by formic acid. This results in *rhodium black*, which is heated to form a powder with a purity of greater than 99.9% rhodium.

Ion-exchange-based refining. An overview of an ion-exchange-based process implemented at the Impala Platinum refinery at Springs in South Africa is shown in Figure 17. The refinery treats primary material from the base metals refinery. Impala Platinum reprocesses a significant amount of materials that contain PGMs, but these materials are added to the smelter rather than as direct feed to the precious metals refinery. The feed to the refinery contains about 65% PGMs.

The initial step is the digestion of the feed in hydrochloric acid using chlorine gas at 70°C. Digestion is carried out in two steps to ensure complete conversion. Gold is extracted from this solution using an ion-exchange resin. The columns are arranged in a lead-middle-trail configuration. The gold is stripped from the resin and precipitated as gold metal using a reductant. The final gold metal has a purity of 99.95% gold.

The raffinate from gold ion exchange is fed to the palladium ion-exchange step. The media that is used, SuperLig 2,

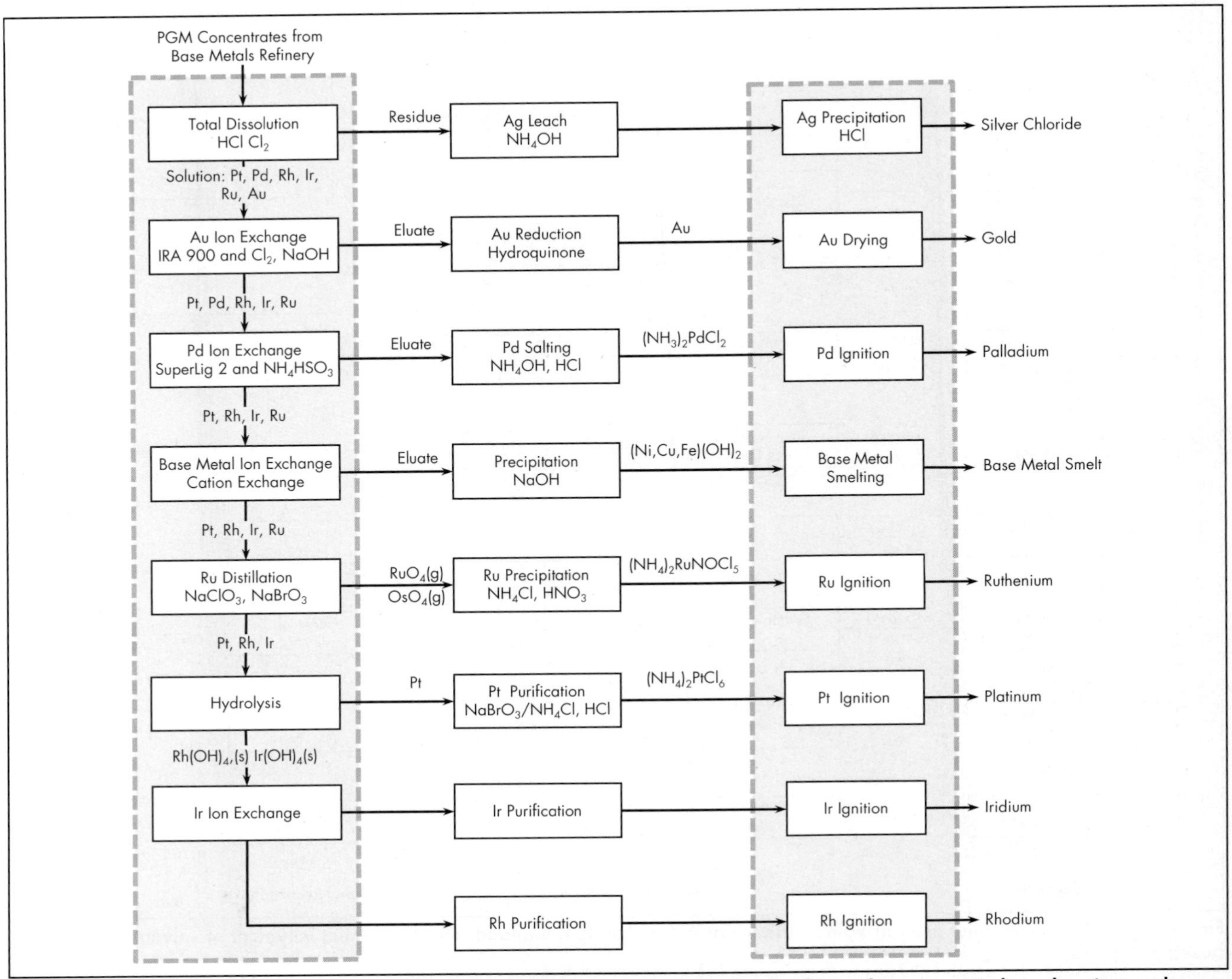

Figure 17 Process flow sheet used at Impala Platinum's refinery in Springs, South Africa. This process is based on ion-exchange-type separations.

is a molecular-recognition technology compound bound to a silica-based substrate. This step is particularly effective: Less than 1% of the platinum in the feed is co-extracted with the palladium, and less than 1 mg/L palladium remains in the solution. Loaded solution is eluted using ammonium bisulfite. Palladium is precipitated as $(NH_3)_2PdCl_2$ from the eluate by the addition of ammonium hydroxide and hydrochloric acid. The palladium salt is ignited to produce palladium sponge.

The raffinate from palladium ion exchange is boiled down to increase the concentrations of the metals in solution. Following boil-down, base metals are removed from the solution using ion exchange.

This solution is fed to the ruthenium distillation section, where ruthenium is oxidized to ruthenium tetroxide by the addition of sodium chlorate and sodium bromate. The volatile RuO_4 is stripped from the solution into the gas phase by passing a stream of air through the solution. The ruthenium is removed from the gas phase by contacting it with a solution of hydrochloric acid. The ruthenium in the resulting solution is precipitated as a ruthenium nitrosyl salt, $(NH_4)_2RuNOCl_5$, by the addition of nitric acid and ammonium chloride. The ruthenium nitrosyl salt is ignited to produce pure ruthenium sponge.

The pH of the solution left behind after ruthenium distillation is increased to a value greater than 5 by the addition of sodium bicarbonate to precipitate rhodium and iridium as hydroxides. These hydroxides are filtered from solution and treated in the rhodium–iridium circuit. The filtrate is transferred to the platinum recovery section.

The pH 5 hydrolysis is repeated to further remove rhodium and iridium from the solution. Finally, the solution is boiled to dryness in a glass-lined vessel, the dry contents resolubilized with demineralized water, and the pH adjusted to a value of about 8. After filtration, ammonium chloride is added to the solution so that $(NH_4)_2PtCl_6$ precipitates. The salt is ignited to produce platinum sponge with a purity of greater than 99.95% platinum.

Primary Extraction from Russian Ores

The source of PGMs of the Russian deposits is mainly from ores mined primarily for their nickel content. These ores

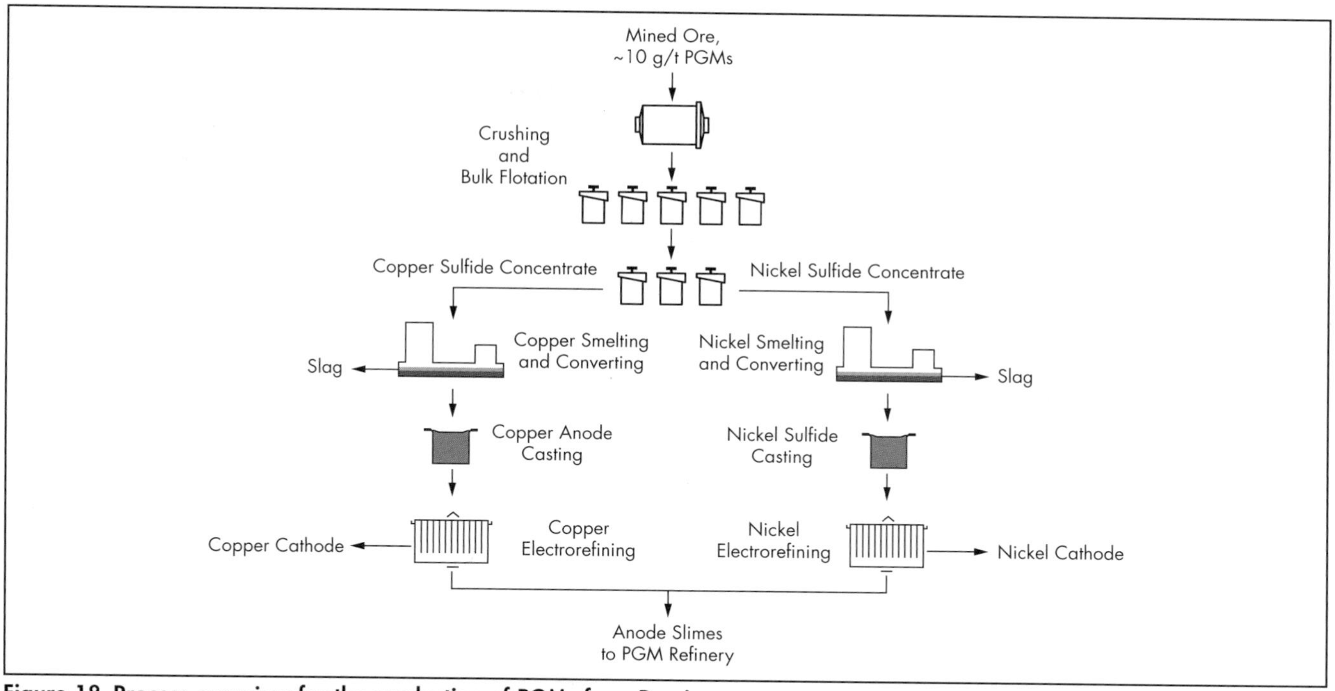

Figure 18 Process overview for the production of PGMs from Russian ores

contain perhaps 50 times more nickel, copper, and cobalt than ores mined in South Africa for their PGMs. Another advantage is that they are more concentrated in PGMs than the South African ores. However, the South African ores have a higher ratio of platinum to palladium, which is advantageous because platinum usually yields a higher price than palladium.

An overview of the extractive metallurgy of these ores is shown in Figure 18. The ore is mined, and after a bulk concentration step, a copper-rich concentrate and nickel-rich concentrate are prepared. Each of these concentrates is smelted separately and cast into anodes. The anodes are purified in an electrorefining step. Precious metals report to the anode slimes, which are then the feed to the platinum refinery.

An overview of the Krastsvetmet refinery at Krasnoyarsk is given in Figure 19. The feed material is sampled and separately routed. Feed rich in platinum and palladium is routed to the platinum–palladium circuit, while that rich in the other PGMs or of low grade is routed to the smelting circuit.

The first step in the platinum–palladium circuit is the digestion in a solution of hydrochloric acid using chlorine gas. From the resulting solution, gold is precipitated using a small quantity of fresh feed. The precipitated gold is redissolved in aqua regia and precipitated by reduction using sucrose, which results in gold with a purity of 99.99% gold.

Ammonium chloride is added to the solution remaining after gold precipitation. This causes platinum to precipitate as $(NH_4)_2PtCl_6$. An impure sponge is formed by igniting this salt, which forms the feed to the platinum purification section.

Ammonium hydroxide and hydrochloric acid are added to the solution remaining after platinum precipitation. This results in the precipitation of $(NH_3)_2PdCl_2$. This salt is transferred to the palladium purification section.

Feeds that are low in platinum and palladium and all internal recycle streams are smelted. The alloy from the smelting step is dissolved in a solution of hydrochloric acid and chlorine gas. Ammonium nitrite is added to the solution to precipitate rhodium in the form of ammonium rhodium nitrite, $(NH_4)_3Rh(NO_2)_6$. This salt is redissolved and the solution is fed to a solvent extraction step in which tri-*n*-butyl phosphate is used the remove impurities. Rhodium metal is produced by electrowinning from the raffinate.

Solution remaining after the precipitation of rhodium nitrite is fed to the ruthenium circuit. Platinum and palladium are extracted from this solution using tri-*n*-butyl phosphate, and the strip solution is transferred to the platinum and palladium circuit. Ruthenium is precipitated from the raffinate by the addition of sodium chlorate.

The residue from the alloy dissolution step contains iridium. This residue is alloyed with barium peroxide and then dissolved in a solution of hydrochloric acid using chlorine gas. Barium is removed by precipitation as the sulfate. Impure ammonium chloroiridate is precipitated by the addition of ammonium chloride. After several redissolution and precipitation steps, this salt is ignited to form iridium with a purity greater than 99.9% iridium.

Comparison of Refineries

A comparison of the major features of some precious metals refineries is given in Table 7.

SECONDARY PRODUCTION

It is estimated that approximately 25% of the consumption of PGMs is supplied from recycled sources (Jollie 2010). The supply of secondary production of PGMs into the market from different sources is given in Table 8. The largest source of the recycling in the PGM market is that of autocatalysts (or catalytic converters) used in trucks and cars to catalyze the conversion of hydrocarbons, carbon monoxide, nitrous oxide, carbon dioxide, water, and nitrogen.

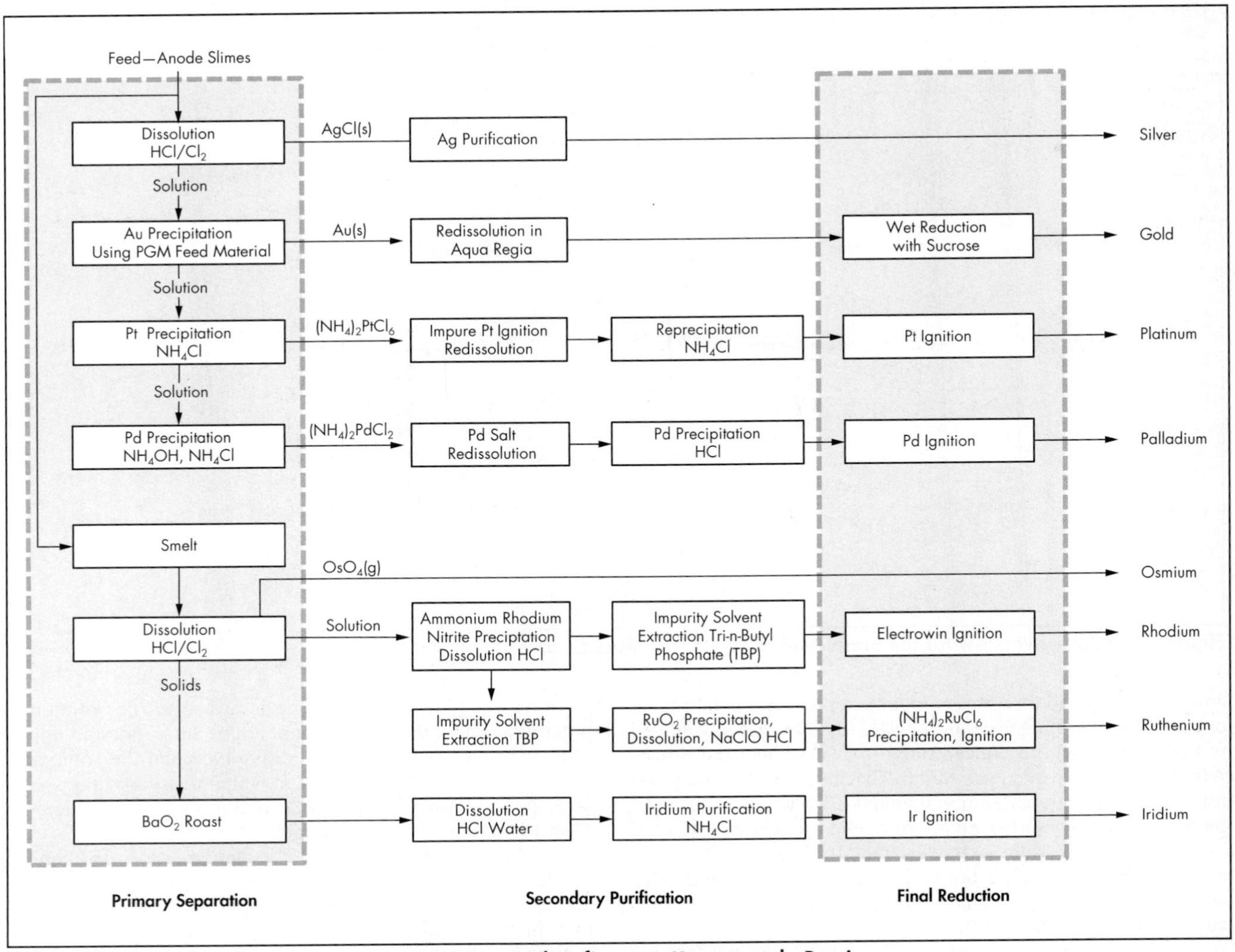

Figure 19 Overview of the Krastsvetmet's precious metals refinery at Krasnoyarsk, Russia

Table 7 Comparison of the major features of several refineries

Feature	Lonmin	Krastsvetmet	Vale (Acton)	Anglo American Platinum	Impala Platinum
Capacity	32,000 kg 1 million oz Pt		32,000 kg 1 million oz Pt	112,000 kg 3.5 million oz Pt	71,000 kg 2.2 million oz Pt
Year started	1975/1987	1943	1975	1989	1998
Production	Primary	Primary	Mainly toll refining	Primary	Primary with large-scale toll refining
Technology	Precipitation	Precipitation	Solvent extraction	Solvent extraction	Ion exchange
Feed quality	65%–75% platinum group metals (PGMs)		<15% PGMs	30%–50% PGMs	60%–65% PGMs
Dissolution	HCl/Cl_2 at 65°C and atm pressure	HCl/Cl_2 at 70°C, 3 bar pressure	HCl/Cl_2 at 65°C and atm pressure	HCl/Cl_2	HCl/Cl_2 at 85°C, 1 bar pressure
Order of extraction	Au, Ru, Pt, Pd, Rh, and Ir separated after Ru distillation (pH 5 cake)	Ag, Au, Pt, Pd, Rh, and Ir separated after smelting, dissolution	Ru, Au, Pd, Pt, Ir, Rh	Au, Pd, Pt, Ru, Rh, and Ir separated after Ru distillation (pH 5 cake)	Au, Pd, Ru, Pt, Rh, and Ir separated after Ru distillation (pH 5 cake)

Source: Crundwell et al. 2011

Table 8 Secondary production of platinum group metals for 2013*

Source	Platinum	Palladium	Rhodium
Autocatalyst	1,275	1,860	281
Electrical	25	420	0
Jewelry	775	180	0
Total recycling	**2,075**	**2,460**	**281**

Adapted from Johnson Matthey 2003–2015b
*Units per 1,000 troy ounces.

Table 9 Recycle smelters and recycle refineries around the world

Smelters	Description
Anniston, United States (Multimetco)	Catalyst recycle smelter
Hanau, Germany (Heraeus)	Precious metals smelter
Hoboken, Belgium (Umicore)	Base and precious metals smelter
Kosaka, Japan (Dowa Mining)	Catalyst recycle smelter
Niihama, Japan (Sumitomo)	Copper smelter and refinery, nickel-cobalt refinery, precious metals refinery
Rustenburg, South Africa (Impala Platinum)	Primary platinum group metal (PGM) smelter with catalyst recycle
Stillwater, United States (Norilsk)	Primary PGM smelter with catalyst recycle
Williston, United States (Sabin)	Catalyst recycle smelter
Refineries That Produce PGMs and/or Chemicals	
Acton, England (Vale Limited)	Primary and recycle refinery
Badia Al Pino, Italy (Chimet)	Recycle refinery
Balerna, Switzerland (Valcambi)	Recycle refinery
Castel San Pietro, Switzerland (PAMP)	Recycle refinery
Chikusei, Ibaraki, Japan (Furuya)	Recycle refinery
Hanau, Germany (Heraeus)	Feed from Hanau smelter
Hoboken, Belgium (Umicore)	Feed from Hoboken smelter
Hong Kong (Heraeus)	Recycle refinery
Los Angeles, United States (Heraeus)	Recycle refinery
Mendrisio, Switzerland (Argor-Heraeus)	Recycle refinery
Neuchâtel, Switzerland (Metalor)	Recycle refinery
Newark, United States (Heraeus)	Recycle refinery
Niihama, Japan (Sumitomo)	Feed from Niihama base metals plants
Rochester, United States (Sabin)	Feed from Williston smelter
Rome, Italy (BASF)	Recycle refinery
Royston, England (Johnson Matthey)	Primary and recycle refinery
Sao Paulo, Brazil (Umicore)	Recycle refinery
Seneca, United States (BASF)	Recycle refinery
Sodegaura, Chiba, Japan (Tanaka K.K.)	Recycle refinery
Springs, South Africa (Impala Platinum)	Feed is from Impala Rustenburg smelter
Tainan, Taiwan (Solar Applied Technology)	Recycle refinery
Wayne, United States (Johnson Matthey)	Recycle refinery

Source: Crundwell et al. 2011

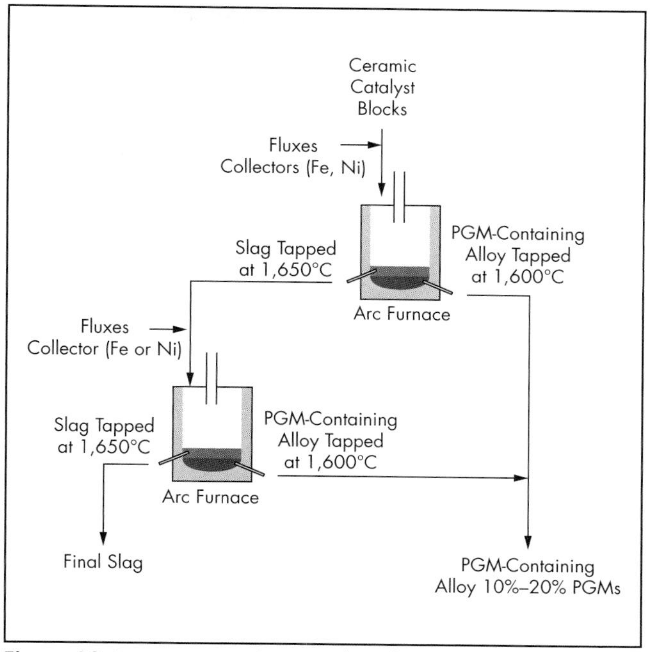

Figure 20 Process overview employed at many recycling smelters for the collection of PGMs as an alloy of either iron or copper

PGMs are deposited on the internal surfaces of a porous ceramic block, which contains between 0.1% and 0.3% PGMs, within the autocatalyst. This ceramic block is encased in a steel container. Recycling of the autocatalyst requires the steel container to be opened and the ceramic block to be extracted. The crushed ceramic blocks can either be fed to the smelter of a primary PGM producer, such as Impala Platinum, or they can be processed in a recycle smelter. A list of some recycle smelters and refineries is given in Table 9.

The typical process followed in recycle smelting is shown in Figure 20. Crushed ceramic blocks, together with a collector and fluxes, are fed to an arc furnace. Iron is usually used as a collector, but copper and nickel might also be used. The furnace is operated at temperatures as high as 1,700°C. The alloy phase and the slag phase are tapped intermittently. The slag is crushed and resmelted in a slag-cleaning furnace. Dusts from the furnaces are also collected. The alloy phase, containing 10%–20% PGMs, and the dusts are sent to the precious metals refinery.

REFERENCES

Asamoah-Bekoe, Y. 1998. Investigation of the leaching of the platinum-group metal concentrate in hydrochloric acid by chlorine. M.Sc. thesis, University of the Witwatersrand, Johannesburg, South Africa.

Barnes, J.E., and Edwards, J.D. 1982. Solvent extraction at Inco's Acton precious metal refinery. *Chem. Ind.* 5:151–155.

Cramer, L.A. 2008. What is your PGM concentrate worth? In *Proceedings of Platinum in Transformation, Third International Platinum Conference*. Edited by M. Rogers. Johannesburg: Southern African Institute of Mining and Metallurgy. pp. 387–394.

Crundwell, F.K., Moats, M.S., Ramachandran, V., Robinson, T.G., and Davenport, W.G. 2011. *Extractive Metallurgy of Nickel, Cobalt and Platinum-Group Metals*. Oxford: Elsevier.

Jacobs, M. 2006. Process description and abbreviated history of Anglo Platinum's Waterval smelter. In *Southern African Pyrometallurgy 2006 International Conference*. Edited by R. Jones. Johannesburg: Southern African Institute of Mining and Metallurgy. pp. 17–28.

Johnson Matthey. 2003–2015a. Various price data charts. www.platinum.matthey.com/prices/price-charts. Royston, Hertfordshire: Johnson Matthey.

Johnson Matthey. 2003–2015b. Various market data tables. www.platinum.matthey.com/services/market-research/market-data-tables. Royston, Hertfordshire: Johnson Matthey.

Jollie, D. 2010. *Platinum 2010*. Hertfordshire, UK: Johnston Matthey.

Knights, B.D.H., and Bryson, M.A.W. 2009. Current challenges in PGM flotation of South African ores. In *Advances in Minerals Processing*. Edited by C.O. Gomez, J.E. Nesset, and S.R. Rao. Montreal, QC: Canadian Institute of Mining, Metallurgy and Petroleum. pp. 285–396.

Nelson, L.R., Stober, F., Ndlovu, J., de Villiers, L.P.S., and Wanblad, D. 2005. Role of technical innovation on production delivery at the Polokwane smelter. In *Nickel and Cobalt 2005: Challenges in Extraction and Production: Proceedings of the International Symposium on Nickel and Cobalt*. Edited by J. Donald and R. Schonewille. Montreal, QC: Canadian Institute of Mining, Metallurgy and Petroleum. pp. 91–116.

CHAPTER 12.30

Portland and Other Cements

Rackel San Nicolas, Brant Walkley, and Jannie S.J. van Deventer

The Roman term *opus caementicium* describing a combination of volcanic ash or pulverized brick plus burnt lime as a binder is deemed to be the origin of the word *cement*. A general definition of cement may be "a substance used in construction that adheres to other materials, binding them together when it sets and hardens." Cement is usually mixed with fine aggregate to produce mortar for masonry, or mixed with sand and gravel aggregates to produce concrete. Cements can be classified as either hydraulic or non-hydraulic, depending upon the ability of the cement to set in the presence of water. Hydraulic cements (e.g., portland cement, or PC) set and become adhesive because of a chemical reaction between the dry ingredients and water, resulting in mineral hydrates with low solubility in water and resistance to chemical attack.

TRENDS IN CEMENT CONSUMPTION

Global cement production in 2015 was approximately 4 Bt/yr (billion metric tons per year; Beyond Zero Emissions 2017), making it the second most widely used commodity besides water. Global construction output is expected to exceed US$15 trillion by 2030, with China, the United States, and India accounting for 57% of all growth. Global construction growth is also expected to outpace that of global gross domestic product (GDP) by over one percentage point (Global Construction Perspectives and Oxford Economics 2015), so with construction contributing about 10% of global GDP, demand for cement will continue to grow.

As outlined later in this chapter, production of PC contributes about 8% of global carbon emissions—more than the total emissions from all global road transport. Even if the global cement industry achieves its own emissions targets, cement-related emissions will contribute an alarming 20%–26% of global carbon emissions by 2050 (Beyond Zero Emissions 2017). Clearly, there is a need to change cement production practice to reduce carbon emissions. This chapter aims to equip the mining professional with the ability to identify mineral deposits that will result in new opportunities for making cements and to reduce the consumption of PC.

CEMENTITIOUS MATERIALS

The main source materials used in the production of PC are limestone and clays. Supplementary cementitious materials (SCMs), such as fly ash, ground granulated blast-furnace slag (GGBS) from ironmaking, natural pozzolans, volcanic rock, calcined clays, fine limestone, and silica fume, are added to PC to reduce carbon emissions, reduce cost, and improve durability of the concrete. Figure 1 shows the compositional relationship between these materials. PC with addition of approximately 20–50 mass % SCMs are called "blended cements." It is possible to make a cement containing no PC from SCMs and an alkali activator, which is called "alkali-activated cement." These cements have been demonstrated in many projects but are not yet widely available commercially. In view of the dominance of portland and blended cements, and because most of the existing standards pertain to such cements, the following discussion focuses primarily on these cements.

CEMENT TERMINOLOGY

Before describing cement chemistry, it is necessary to introduce its terminology, aimed at simplifying compositions by leaving out the oxygen and abbreviating elemental symbols, as shown in Table 1. Examples of typical cement minerals and products are given in Table 2, with Figure 2 showing the compositional relationship between some of these binding phases.

About 90%–95% of PC is composed of four minerals (C_3S, C_2S, C_3A, and C_4AF), with the remainder consisting of calcium sulfate, alkali sulfates, unreacted (free) lime (CaO), magnesia (MgO), and other minor constituents. The C_3S and the C_2S contribute virtually all of the beneficial properties by generating the main hydration products C-S-H and CH. In blended PCs, the reaction of SCMs such as fly ash, GGBS, and calcined clay result in a higher aluminum content of the C-S-H gel, and often a decrease in the Ca/Si ratio, with these gels called C-A-S-H. In alkali-activated cements, some alkali cations like sodium are included in the gel, so that the gel is called C-(N,K)-A-S-H. A high alkali content in

Rackel San Nicolas, Lecturer, Department of Infrastructure Engineering, The University of Melbourne, Parkville, Victoria, Australia
Brant Walkley, Research Associate, The University of Sheffield, Sheffield, United Kingdom
Jannie S.J. van Deventer, Chief Executive Officer, Zeobond Pty Ltd., Melbourne, Victoria, Australia

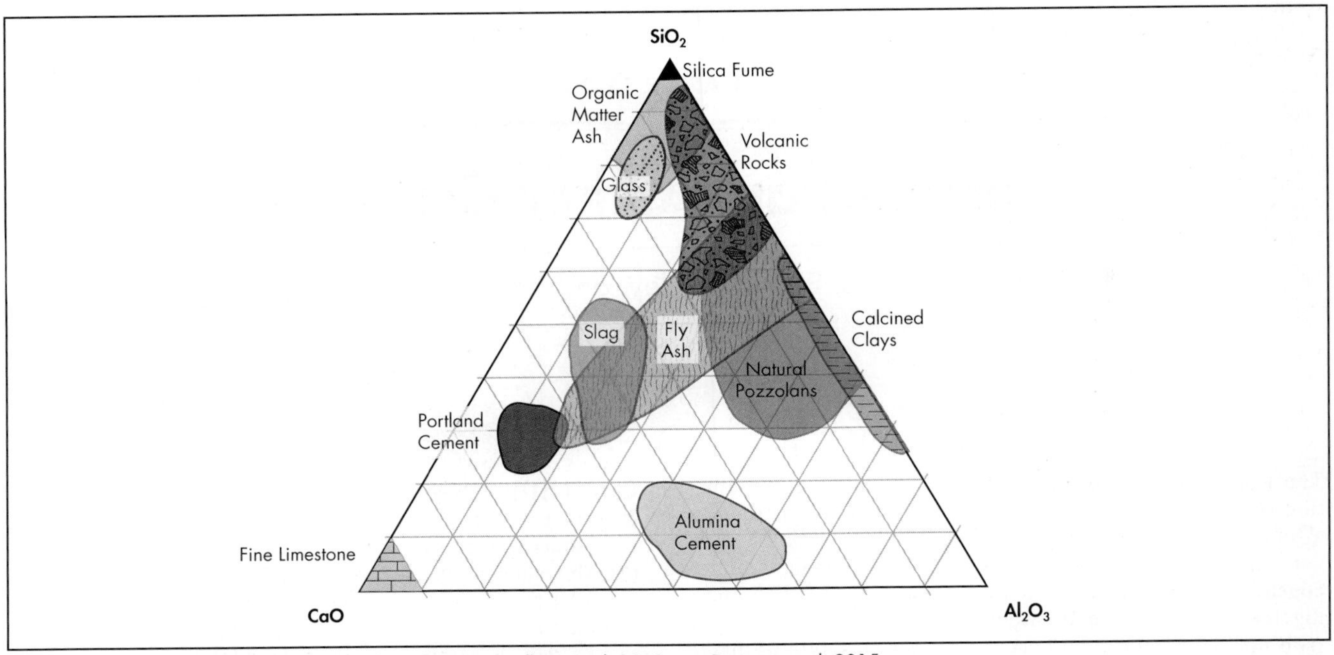

Adapted from Glasser et al. 1987, Lothenbach et al. 2011, Snellings et al. 2012, van Deventer et al. 2015

Figure 1 Ternary CaO–SiO$_2$–Al$_2$O$_3$ diagram (mass % based) depicting the chemical composition of major cementitious materials

Table 1 Cement chemistry notation based on oxide

Oxide Form	Notation	Oxide Form	Notation
CaO	C	SO_3	$\overline{S}$
SiO_2	S	H_2O	H
Al_2O_3	A	MgO	M
Fe_2O_3	F	Na_2O	N

alkali-activated cements may result in the formation of (N,K)-A-S-H gel as well.

TYPES OF PORTLAND CEMENT

There are five main types of PC in the United States, according to standard ASTM C150/C150M, with some of them subdivided, giving 10 types of cements (Table 3). These different types of PC need to satisfy many physical and chemical requirements given in ASTM C150/C150M. The differences between these cement types are rather subtle. All five types contain about 75 mass % calcium silicate minerals, and the properties of mature concretes made with all five are quite similar.

Type I cement is conventional PC. Types II and V PC are resistant to damaging sulfate attack resulting from expansive ettringite crystals formed by the reaction of hydration products of C_3A with sulfate ions entering the concrete pores. To control this reaction, Type II and Type V cements focus on keeping the C_3A content low. However, the most effective way to prevent sulfate attack is to prevent the sulfate ions from entering the concrete in the first place by using a low water/cement (w/c) ratio. Alkali-activated concrete has a lower susceptibility than PC to sulfate attack.

Type III cement develops early strength more quickly than Type I cement, which is useful for a rapid pace of construction, since it allows cast-in-place concrete to bear loads sooner. These advantages are important in cold weather, which reduces the rate of hydration. The disadvantage of rapid-reacting cements is a shorter period of workability, greater heat of hydration, and a slightly lower ultimate strength.

Type IV cement is designed to release heat more slowly than Type I cement, meaning that it also gains strength more slowly. A slower rate of heat release limits the increase in the core temperature of a concrete element. The maximum temperature scales with the size of the structure, and Type IV cement was developed to prevent excessive temperatures in large concrete structures such as dams. Type IV cement is rarely used today, because similar properties can be obtained by using a blended PC.

SOURCE MATERIALS

Cementitious materials comprise primarily of lime (CaO), silica (SiO_2), alumina (Al_2O_3), and ferrite (Fe_2O_3) sourced from a wide range of materials (Figure 1), including limestone, clays, metallurgical slags, and ashes. Small quantities of gypsum and other admixtures are added during clinker grinding to manipulate setting.

Limestone

Limestone ($CaCO_3$) for PC production is quarried and transported for crushing and grinding prior to delivery into a clinker plant. Frequently the limestone is impure and can contain as low as 80% $CaCO_3$ (Moir 2003). ASTM C1797 gives the requirements for the different types of limestone, which can be used for PC production, as SCM, interground with clinker and gypsum, or calcined to produce lime. Furthermore, alkali-activated binders based on limestone and slag (containing up to 68% granular limestone) have been reported (Moseson et al. 2012). Limestone can also be blended with calcined clays and used as an SCM, partially replacing PC clinker (limestone calcined clay cement, LC^3) (Sánchez Berriel et al. 2016; Cancio Díaz et al. 2017). With overall limestone sources widespread and abundant, future supply is secure.

Table 2 Examples of cement minerals and products

Chemical Name	Chemical Formula	Oxide Formula	Cement Notation	Mineral Name
Tricalcium silicate	Ca_3SiO_5	$3CaO{\cdot}SiO_2$	C_3S	Alite
Dicalcium silicate	Ca_2SiO_4	$2CaO{\cdot}SiO_2$	C_2S	Belite
Tricalcium aluminate	$Ca_3Al_2O_6$	$3CaO{\cdot}Al_2O_3$	C_3A	Aluminate
Tetracalcium aluminoferrite	Ca_2AlFeO_5	$4CaO{\cdot}Al_2O_3{\cdot}Fe_2O_3$	C_4AF	Ferrite
Calcium hydroxide	$Ca(OH)_2$	$CaO{\cdot}H_2O$	CH	Portlandite
Calcium sulfate dihydrate	$CaSO_4{\cdot}2H_2O$	$CaO{\cdot}SO_3{\cdot}2H_2O$	$C\bar{S}H_2$	Gypsum
Calcium oxide	CaO	CaO	C	Lime
Alumina, ferric oxide, trisulfate	—	$Ca_3(Al,Fe)(OH)_6{\cdot}12H_2O]_2{\cdot}X_3{\cdot}nH_2O$ where X = a doubly charged anion or, sometimes, two singly charged anions	AFt	Example: ettringite, where X = SO_4^{2-}
Alumina, ferric oxide, monosulfate	—	$Ca_2(Al,Fe)(OH)_6)]{\cdot}X{\cdot}nH_2O$ where X = a singly charged anion or "half" a doubly charged anion. X = OH^-, SO_4^{2-}, and CO_3^{2-}.	AFm	Examples: hemicarboaluminate, monocarboaluminate, monosulfoaluminate, gehlenite hydrate, strätlingite
Calcium-silicate-hydrate gel	—	$CaO–SiO_2–H_2O$ gel	C-S-H gel	Structural analogue: tobermorite or jennite
Calcium-alumina-silicate-hydrate gel	—	$CaO–Al_2O_3–SiO_2–H_2O$ gel	C-A-S-H gel	Structural analogue: tobermorite or jennite
Sodium-alumina-silicate gel	—	$Na_2O–Al_2O_3–SiO_2–H_2O$	N-A-S-H gel	Structural precursor to zeolites

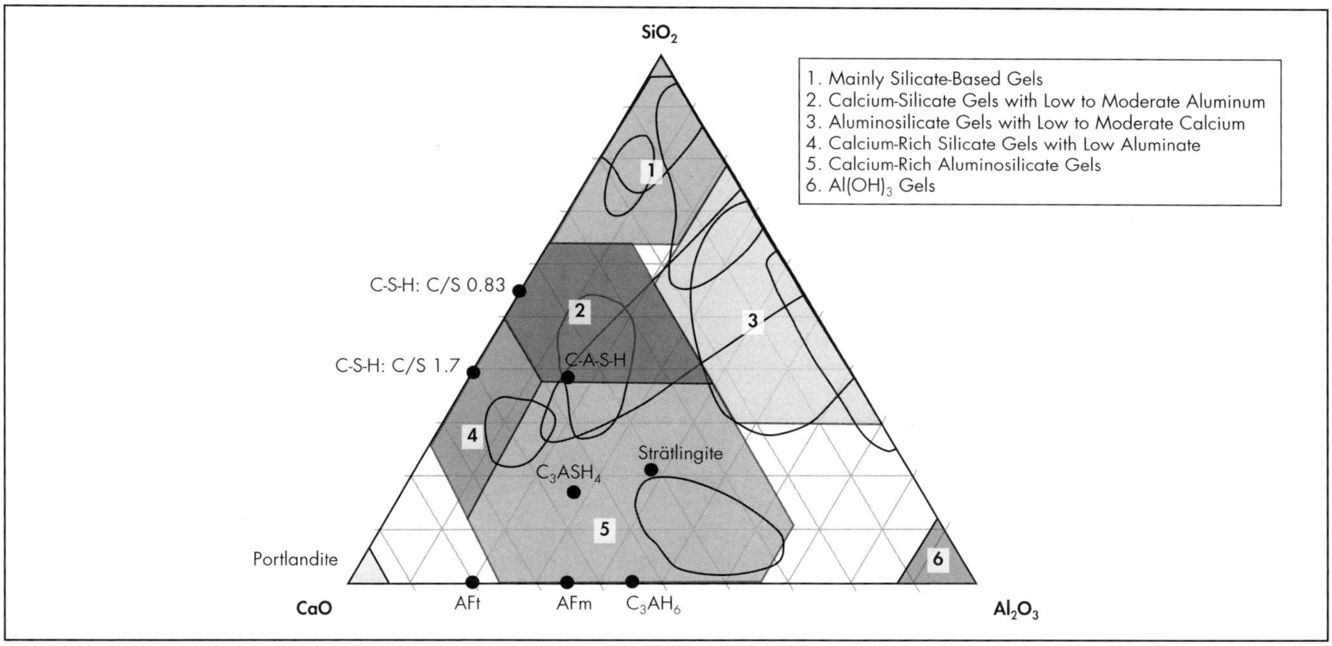

Adapted from Durdiński et al. 2016, van Deventer et al. 2015, Lothenbach et al. 2011

Figure 2 Ternary $CaO–SiO_2–Al_2O_3$ diagram (mass % based) depicting the chemical composition of cementitious binders

Clays

Clay minerals used in cement are formed primarily by weathering of silicate minerals in rocks (e.g., feldspar, micas, amphiboles, and pyroxenes); however, clay minerals in waste products from sand quarries and clayey mines may also be used if calcined properly. Clays comprise silica tetrahedra and aluminum octahedral sheets arranged in layers that are stacked in a variety of arrangements. Clay minerals that are naturally abundant and of interest for cement production include kaolinite, illite, and montmorillonite.

Clays are not inherently cementitious and must be heated to remove structural water and to form a metastable/disordered structure with pozzolanic reactivity. Metakaolin exhibits the greatest degree of disorder after calcination at low temperatures (500°–800°C), while montmorillonite and illite retain their structure after calcination at these temperatures. Consequently, metakaolin exhibits the highest pozzolanic activity (Wild and Khatib 1997; Sabir et al. 2001). Calcined calcium-montmorillonite exhibits a slightly lower degree of pozzolanic reactivity than metakaolin, whereas illite and other thermally activated clay minerals exhibit relatively low pozzolanic reactivity.

Montmorillonite is the most abundant of these clay minerals and has much potential for use in cements; however,

Table 3 Types of portland cement according to ASTM C150/C150M

Types	Classification	General Characteristics	Applications
Type I and Type IA*	Special properties are not required	Fairly high C_3S content for good early-strength development	Most buildings, bridges, pavements, precast units, etc.
Type II and Type IIA*	Moderate sulfate resistance is desired	Low C_3A content (<8%)	Structures exposed to soil or water containing sulfate ions
Type II(MH†) and Type II(MH)A*	Moderate heat of hydration and moderate sulfate resistance are desired	Low C_3A content (<8%)	Structures exposed to soil or water containing sulfate ions
Type III and Type IIIA*	High early strength is desired	Ground more finely, may have slightly more C_3S	Rapid construction, cold-weather concreting
Type IV	Low heat of hydration is desired	Low content of C_3S (<50%) and C_3A	Massive structures such as dams (now rare)
Type V	High sulfate resistance is desired	Very low C_3A content (<5%)	Structures exposed to high levels of sulfate ions

Adapted from ASTM C150/C150M
*Air-entraining cement where air entrainment is desired.
†MH = Moderate heat.

there has been limited work done to enhance pozzolanic reactivity of this clay. Sodium-montmorillonite in particular has received limited attention due to the high sodium content being undesirable in PC, whereas in alkali-activated systems, a high sodium content can be accommodated. Metakaolin is the most widely used calcined clay in cementitious materials and exhibits pozzolanic reactivity when mixed with lime or PC (Ambroise et al. 1985; He et al. 1995) and drastically improves the mechanical properties and durability of concrete (Zhang and Malhotra 1995; Singh and Garg 2006).

Slags

Slag is a glassy by-product of both ferrous and nonferrous smelting processes, and comprises primarily metal oxides, silicon dioxide, and sulfides. The type of slag and its characteristics are dependent on processing conditions during production. Common examples are air-cooled blast-furnace slag, air-cooled basic oxygen steel slag, air-cooled high-alloy electric arc furnace slag, stainless-steel slag, electric arc furnace carbon steel slag, granulated blast-furnace slag, and GGBS.

GGBS is a latent hydraulic material consisting primarily of depolymerized calcium silicate glasses, as well as small amounts of poorly crystalline phases within the melilite group (Shi et al. 2006; Duxson and Provis 2008; Provis and van Deventer 2009; Le Saoût et al. 2011; Li et al. 2011). During granulation, slag obtained from blast-furnace smelting of iron ore is cooled quickly to facilitate the formation of glassy phases, enabling it to exhibit cementitious properties. This cooling process can cause GGBS to contain up to 30% water (particularly older slags produced in China and Russia) (Shi et al. 2006), although modern slags are almost completely dry. After granulation, the slag must be ground for use as a cementitious material. GGBS is particularly useful in cement because of its composition, which contains high levels of CaO, SiO_2, and Al_2O_3, and lower levels of Fe_2O_3, relative to other slags and exhibits high pozzolanic reactivity.

It is estimated that global iron slag production in 2016 was between 300 and 360 Mt (million metric tons) and global steel slag production in 2016 was 160–240 Mt (USGS 2017). It is estimated that global slag production is approximately one-half to two-thirds of coal fly ash production (Fernández-Jiménez and Palomo 2005) and far below that of the available natural clay mineral deposits. Consequently, the amount of GGBS available for use as a PC substitute is significantly reduced compared to that of fly ash. The composition of GGBS varies across different locations, with CaO, Al_2O_3, SiO_2, and MgO typically varying between 35 and 42 mass %, 7 and 13 mass %, 34 and 36 mass %, and 6 and 15 mass %, respectively (Shi et al. 2006; Duxson and Provis 2008; Provis and van Deventer 2009; Ben Haha et al. 2011, 2012; Bernal et al. 2014). The reactivity of GGBS is dependent on the level of depolymerization of the aluminosilicate framework and on the particle size distribution (PSD). The chemical requirements for slag used in cement are specified by ASTM C989. Other metallurgical slags can be modified while hot and in the liquid phase (hot-stage engineering) to tailor the properties of the final cooled slag for use in cement (Engström et al. 2011).

Ashes

About 80% of ash leaves the coal-fired boiler as fly ash entrained in the flue gas, while the remainder is collected as bottom ash. Bottom ash is cooled naturally and consequently is coarse, so size reduction is required if used in cement. Also, bottom ash often contains high levels of unburned carbon that must be removed. Fly ash particles are roughly spherical in shape and can be completely or partially hollow, containing primarily Al_2O_3, SiO_2, and Fe_2O_3.

Fly ash is collected by electrostatic precipitators, after which it is either collected dry and dispatched unclassified, or collected dry and classified with the fines (approximately one-third of the fly ash), then separated and dispatched for use in cement. If the fly ash is deemed to be unsuitable for use, it is collected wet and pumped into disposal ponds. In 2016, worldwide fly ash production was estimated to be 0.7–1.1 Bt/yr, with approximately 30% being utilized (Scrivener et al. 2016; Snellings 2016), which makes it a low-cost partial PC substitute.

The composition of fly ash varies significantly from one source to the next. The CaO content of fly ash can be as high as 15 mass %; however, many fly ashes contain less than 5% CaO. Fly ash is categorized based on the CaO content, with Class F fly ash specified to contain at least 70% combined silica, alumina, and iron oxide and exhibiting pozzolanic reactivity, and Class C fly ash typically containing more than 20 mass % CaO and exhibiting both pozzolanic reactivity and some self-cementing properties (ASTM C618). Fly ash also contains crystalline minerals such as quartz, mullite, and ferrite spinels; kaolinite and other layered silicates present in the coal melt upon combustion to form the glassy phase in fly ash, and consequently are mostly absent (Provis and van Deventer 2009). ASTM C618 details the chemical and physical requirements of coal fly ash for use in concrete.

Coal can also be burnt via fluidized-bed combustion (FBC), which enables high-efficiency combustion of low-grade fuels. Limestone is often added to react with SO_2 in high-sulfur coal to minimize SO_x (sulfur oxides) emissions; consequently, ashes from FBC can contain gypsum ($CaSO_4 \cdot 2H_2O$) and free lime (CaO). Currently, the vast majority of FBC ashes are deemed unsuitable for use as a pozzolanic material in blended cements and are disposed of in ponds; however, alkali-activated FBC ashes can exhibit mild strength development (Yliniemi et al. 2015, 2017; Pesonen et al. 2016).

Natural Pozzolans

Natural pozzolans and volcanic ashes can also be used in PC blends with up to 50 mass % replacement achievable (Celik et al. 2014), as well as in alkali-activated cements (Lemougna et al. 2011, 2013; Takeda et al. 2014). Volcanic ash is often a locally available material and comprises primarily silica at varying content, depending on the ash source, and lower amounts of Al_2O_3, Fe_2O_3, CaO, MgO, and sodium oxide (Na_2O). ASTM C618 details the chemical and physical requirements of natural pozzolans for use in concrete. They may need to be dried or calcined and ground prior to use in cement. In 2016, global annual production of natural pozzolans was estimated to be approximately 75 Mt (Snellings 2016).

Silica Fume

Silica fume (also referred to as condensed silica fume and microsilica) is an amorphous polymorph of silica (SiO_2) produced as a by-product in the production of silicon metal and ferrosilicon alloys. Silica fume particles are spherical and typically have a diameter of approximately 150 nm. In general, it comprises 85–99 mass % SiO_2, with oxides Fe_2O_3, Al_2O_3, CaO, MgO, Na_2O, and potassium oxide (K_2O) accounting for less than 1 mass % and loss on ignition approximately 1–2 mass % (Snellings et al. 2012). It is a commonly used supplementary cementitious material with high pozzolanic reactivity and consequent rapid rate of reaction when blended with PC and hydrated lime, due to the very small particle size and spherical morphology (Lothenbach et al. 2011). The chemical requirements for silica fume used in cement are specified by ASTM C1240. In 2016, global annual production of silica fume was estimated to be approximately 1–2.5 Mt (Snellings 2016).

TRANSFORMATIONS

Raw materials such as limestone, clays, sand, and aluminates have to be transformed to become cementitious materials. This section outlines the processing equipment, energy requirements, and physical and chemical processing steps required to transform raw materials into PC and calcined clay.

Manufacturing of Clinker and PC

Figure 3 depicts the general flow diagram of PC production. The first step in making PC is to combine a mixture of quarried materials (raw materials), composed mainly of limestone mixed in accurate proportions (step 7) with silica and alumina (from sands, clays, or shales), and iron (from iron ore or other iron-bearing minerals) to form a "raw mix." Limestone provides the required calcium oxide and some of the other oxides, whereas clay, shale, and other materials provide most of the silicon, aluminum, and iron oxides required for

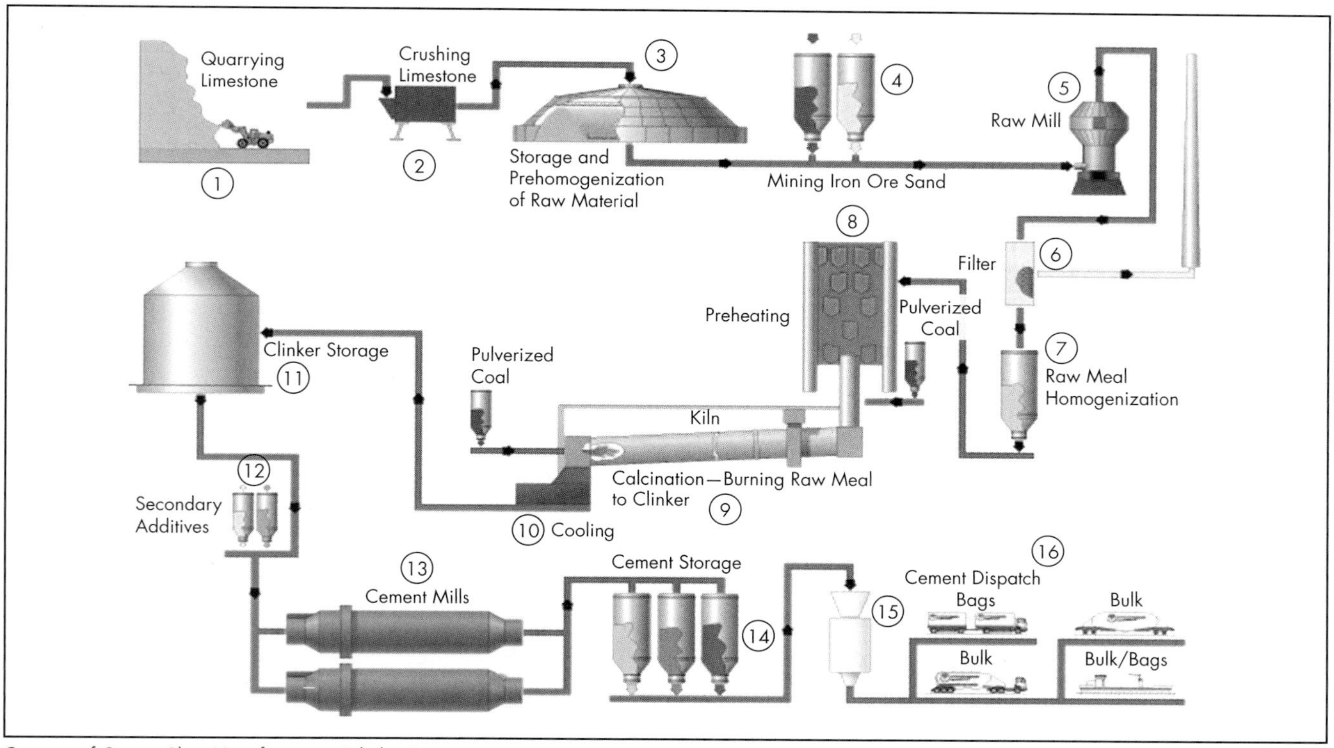

Courtesy of Cement Plant Manufacturers, Ashoka Group

Figure 3 Flow diagram of the PC manufacturing process

the manufacture of PC. The raw mix is then ground to a fine powder called "raw meal" (steps 2 to 4). The size reduction of the raw mix can be done by the "dry process" or the "wet process," as explained below. The raw meal is then chemically transformed by intensive heating to form an intermediate product called clinker (steps 8 to 10), which is then ground with gypsum and small proportions of additional materials to produce PC (steps 12 and 13). This cement is often blended with SCMs such as fly ash, slag, metakaolin, fine limestone, and fillers to make blended cements (step 15). The following subsections provide a more detailed description of the processing steps to transform raw materials to PC.

Quarrying

PC plants usually operate their own quarries for limestone (step 1 in Figure 3), while other materials, such as clays, iron, and sand, may be sourced externally. More than 1.5 t of raw materials are required to produce 1 t of PC (Alsop 2007).

Crushing and Homogenization

The PSD of the kiln feed has a direct effect upon the burnability, time, and cost of the process. Hence, multiple screening steps are performed at different stages of the process and retained particles are recycled to the mill (step 6). The raw mix is reduced to a powder with sizes smaller than 100–150 µm.

The limestone is first crushed by a jaw crusher (step 2) and then fed into the second crusher, usually an impact crusher or gyratory jaw crusher to reduce particle size below 50 mm. The clay and sand portion of the raw mix can be reduced in size using the same principle. All the raw materials are then subjected to a second size reduction with rollers or hammer mills (step 5). This size reduction can be with the addition of water, called the wet process, or without water, called the dry process.

In the wet process, water is usually added to the limestone and the clay separately, because of differences in the handling of moist clay and limestone. The wet process was preferred in the 1950s because of its ease of use, but it requires a long wet kiln, which has a very poor fuel efficiency. In the wet process, once materials are milled and homogenized in the desired proportion of the raw mix (limestone 70%, clays 30%), the slurry is fed to the kiln. The slurry is fed into the kiln at the upper end and pulverized coal is blown in by an airblast at the lower end of kiln, where the temperature is about 1,450°C. This type of kiln has to operate continuously to ensure a steady regime. A large kiln in a wet process plant typically produces 3,600 t of clinker per day.

In the dry and semidry processes, the raw materials are crushed and fed in the correct proportions into a grinding mill, where the raw materials are dried and reduced in size to a fine powder. Although ball mills have been used in the past for raw meal grinding, vertical roller mills (VRMs) have become the industry standard in more than 80% of new plants (Harder 2010). These powdered minerals are mixed by compressed air to get the dry raw mix, which is then stored in silos before being sent into the rotary kiln.

In the semidry process, the blended raw meal is then sieved and fed into a rotating dish where water (about 12% of the meal) is added at the same time. In this manner, hard pellets about 15 mm in diameter are formed. This is necessary, as cold powder fed directly into kiln would not permit the airflow and exchange of heat necessary for the formation of clinker. The pellets are heated in a preheating grate by hot gases from the kiln and then enter the kiln heated at about 1,450°C. The dry material undergoes reactions in the hottest part of the kiln, with 20%–30% of the material becoming liquid. The consumption of coal in this dry method is only about 100 kg/t cement compared to about 220 kg/t cement in the wet process. The dry process uses smaller equipment, is more economical, and gives better control over clinker quality.

Preheating

Nearly all new kilns have precalciners or preheaters (step 8) in the dry process. If a long dry rotary kiln is used, the preheating step is not necessary. The long dry kiln is similar to the long wet kiln, however less fuel is consumed. Modern kilns contain ceramic heat exchangers, allowing the kiln to be split into three or four areas, dividing the feed and gas flow, improving heat transfer, and resulting in energy savings.

After preheating, the raw meal is fed to a shorter rotary kiln (length/diameter = 15) where it enters at approximately 800°C, so decarbonation is almost complete and only clinkerization occurs in the rotary kiln. Since the decarbonation reaction is highly endothermic (~1.88 MJ/kg), this involves introducing about 60% of the total fuel required by the kiln system into the preheater.

Calcination/Pyroprocessing

The burning process is carried out in the rotary kiln (step 9) while the raw materials are rotated at 1–2 rpm along the longitudinal axis of the kiln. The rotary kiln has a diameter of approximately 2.5–3.0 m and a length of 90–120 m and comprises steel tubes. The inside of the kiln is lined with refractory bricks. The kiln rests on roller bearings at a gradient of 1 in 25 to 1 in 30, with material moving at 15 m/h. The raw mix from the dry process or the slurry from the wet process is injected into the kiln from the upper end. The kiln is heated by powdered coal, oil, or hot gases from the lower end of the kiln so that a long, hot flame is produced.

Figure 4 shows the chemical changes as the raw mix is converted into clinker:

- Water is evaporated during a drying step below 750°C. Unbound water evaporates between 20°C and 100°C, adsorbed water evaporates between 100°C and 300°C, and chemically bound water evaporates between 400°C and 750°C. Clays are calcined during this stage (e.g., kaolinite is converted to metakaolin via $Al_4Si_4O_{10}(OH)_8 \rightarrow 2(Al_2O_3{\cdot}2SiO_2) + 4H_2O$).
- Decomposition reactions occur between 750°C and 950°C.
- Calcined clays are converted to aluminates and silica via $Al_2O_3{\cdot}2SiO_2 \rightarrow Al_2O_3 + 2SiO_2$.
- Limestone is converted to free lime, releasing carbon dioxide (CO_2), via $CaCO_3 \rightarrow CaO + (CO_2)_{gas}$.
- Between 950°C and 1,350°C, calcium, silicon, and aluminum oxides react to form calcium aluminates, or CA ($CaO{\cdot}Al_2O_3$), and calcium silicates, or CS ($CaO{\cdot}SiO_2$), via $3CaO + 2SiO_2 + Al_2O_3 \rightarrow 2(CaO{\cdot}SiO_2) + CaO{\cdot}Al_2O_3$.
- Lime reacts with CS to form belite (C_2S) via $CS + C \rightarrow C_2S$ as well as $2C + S \rightarrow C_2S$.
- Between 1,350°C and 1,450°C, lime reacts with belite (C_2S) to form alite (C_3S) in a process called *clinkerization* via $C_2S + C \rightarrow C_3S_{(solid)}$. Lime also reacts with CA and ferrite (F) to form tricalcium aluminate and tetracalcium

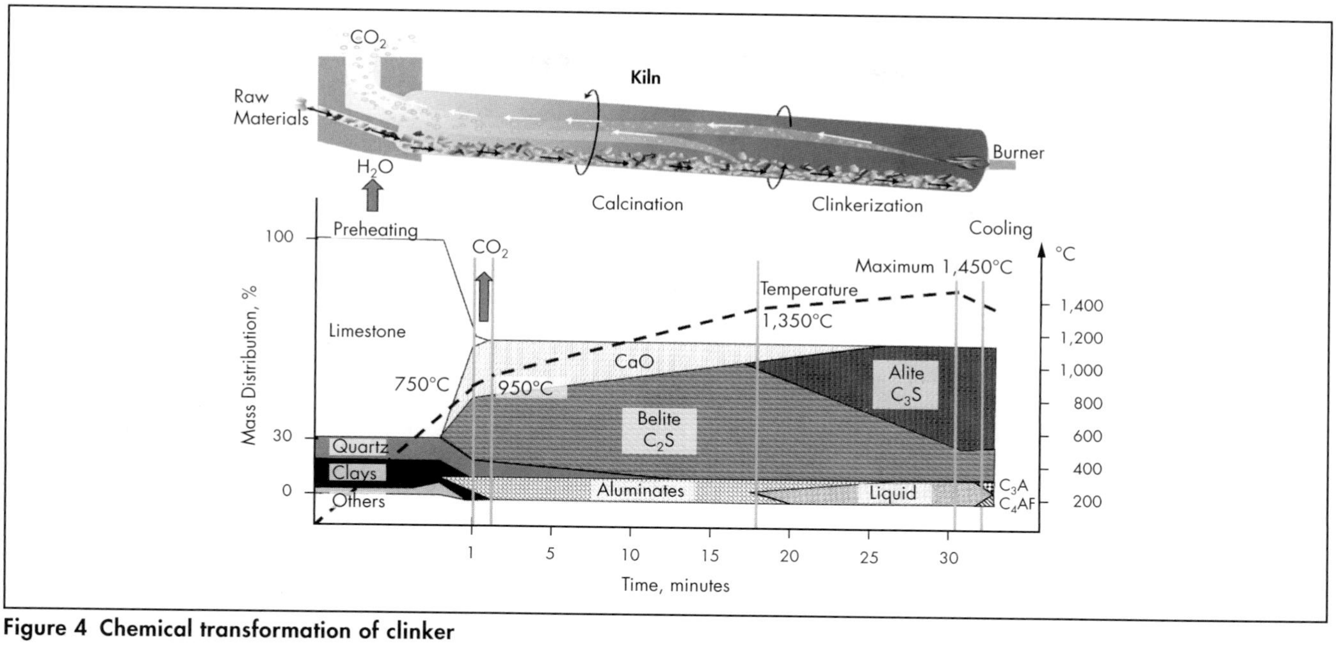

Figure 4 Chemical transformation of clinker

aluminoferrite via CA + 2C → $C_3A_{(liquid)}$ and CA + 3C + F → $C_4AF_{(liquid)}$.

- Cooling then occurs, resulting in solidification of tricalcium aluminate (C_3A) and tetracalcium aluminoferrite (C_4AF).

Cooling and Storage of Clinker

After exiting the kiln (steps 10 and 11), the clinker falls onto a grate and is cooled by large air fans. The heat recovered from the cooling process is recirculated back to the kiln or preheater. The cooling of clinker has a significant effect on the quality of cement, as quenching of the clinker (~50-mm diameter) preserves the reactive metastable high-temperature minerals and maximizes its strength potential.

Grinding of Clinker to Cement

Clinker is transported to the finishing mill (step 13) by a conveyor belt, where a small amount of gypsum is added to control setting of the cement. Blended cements and masonry cements may include large additions of natural pozzolans, fly ash, slag, limestone, silica fume, or metakaolin. The rate of the initial reaction of cement is dependent on the PSD; therefore, the grinding process is closely controlled to obtain a specific surface area, called the Blaine fineness, which usually ranges from 300 to 500 m^2/kg.

Traditionally, closed-circuit ball mills have been used for clinker grinding, as they are simple to operate and can deal with variations in mill feed quality and quantity. Harder (2010) stated that ball mills are used in <50% of plants, while VRMs are used in about 40% of plants for clinker grinding. The maintenance costs for a VRM and a ball mill are comparable; while the wear rates for a ball mill are higher than for a VRM, the cost of the replacement parts is lower. Because of its high internal circulation and high separation efficiency, a VRM tends to produce a narrow PSD, which is not always desirable in cement. Usually, a compromise is found between the PSD and energy consumption. Although the capital cost of an installed VRM may be 20% more than for a ball mill, the lower energy consumption of a VRM is a determining factor for increased use of VRMs.

Storage of Cement

The ground cement is conveyed by belt or powder pump into a silo for storage, where it is ready for bagging and/or shipping in bulk by trucks, rail, or barge. The capital necessary for cement storage represents usually about 10% of the total capital cost of the cement plant. The principal function of the cement storage silos is to provide surge capacity between the grinding mills and the shipping media. There are usually several separated storage silos for the different cements produced. The storage silos must be waterproof to prevent early hydration of the cement. Figure 5 depicts the complete flow sheet for a dry PC plant.

Metakaolin and Other Calcined Clays

Metakaolin and calcined clays are produced by burning kaolinite clay or other clays (e.g., montmorillonite or illite) at temperatures between 700°C and 850°C, which induces dehydroxylation and produces a disordered metastable state. Calcined clays can be used as SCMs and as precursors for alkali-activated cement. Unlike the kaolin used in paper and ceramics which require further processing, clay used as a SCM can contain up to 50% sand and 10% limestone.

Rotary Kiln Calcination

Clays are usually calcined in a rotary kiln by both dry and wet processes, similar to the kilns used for cement production. The dry process is simpler but produces a lower-quality product than the wet process. In the dry process, the raw material is crushed to the desired size, dried in rotary dryers, pulverized, air-floated to remove coarse grit, and then fed to the kiln. The calcination process, between 650°C and 700°C, which lasts about 3–5 hours, drives the particles of metakaolin to agglomerate in pellets of 5–10 cm in diameter, which are

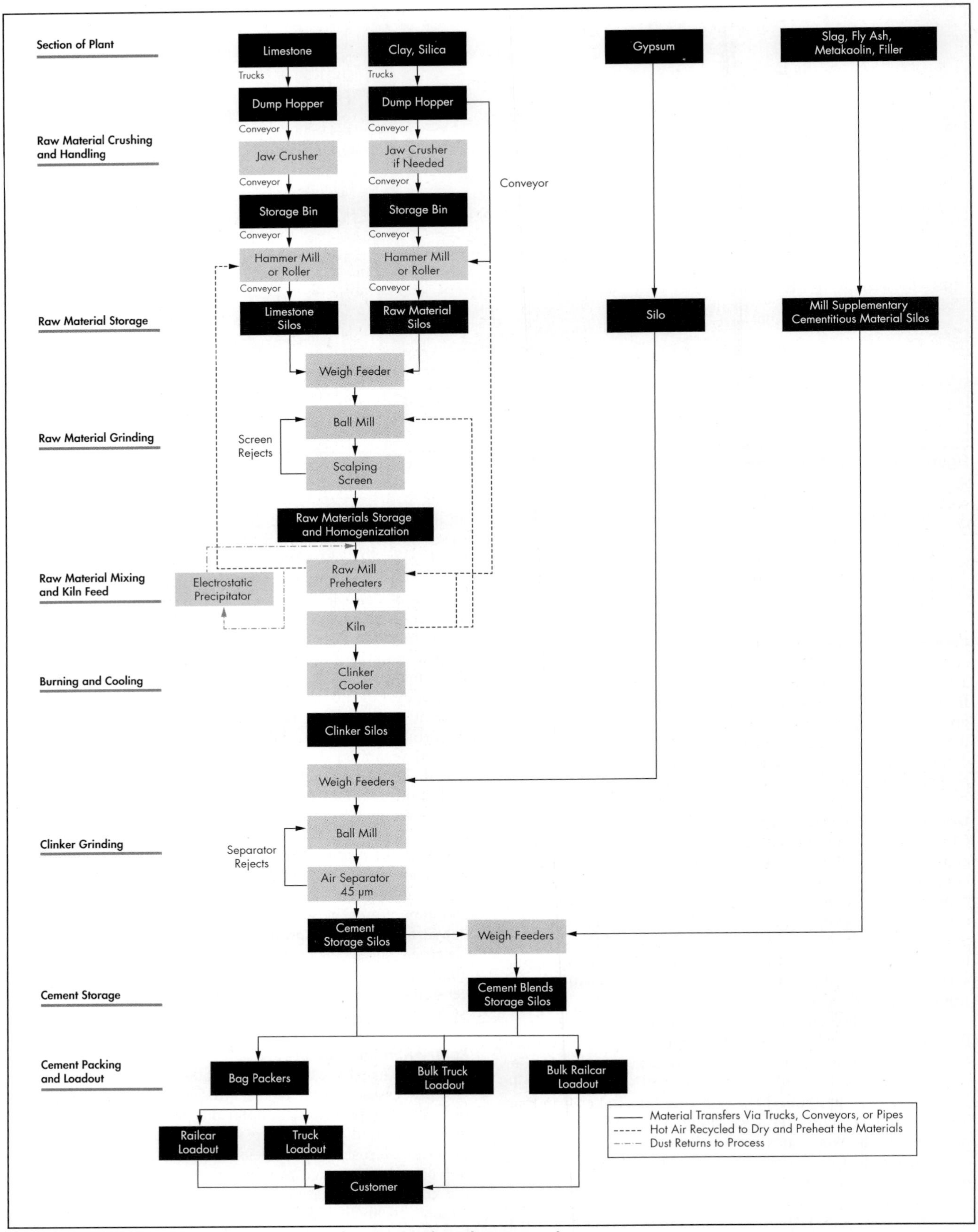

Figure 5 Flow sheet of typical dry process of cement with preheating cyclones

then ground by a ball mill to the desired fineness. This method results in flaky, irregular morphology and high specific surface area particles.

Flash Calcination

More recently, flash calcination has been used in metakaolin production, which involves rapid heating and calcining for a tenth of a second, then cooling at a rate of 103°–105°C/s, with insufficient time for agglomeration of the particles, so no subsequent grinding is required as in rotary calcination. This type of installation consumes 2.2 MJ/t of metakaolin, which is 80% less than the energy consumed during cement production (San Nicolas 2011). As such, flash-calcined metakaolin particles are usually spherical, whereas rotary kiln–calcined metakaolin particles are generally flat. Flash calcination therefore results in metakaolin with drastically improved workability, water demand, and reactivity when compared to rotary kiln–calcined metakaolin (San Nicolas et al. 2013).

BLENDED CEMENTS

The composition, structure, and performance of PC, blended cement consisting of PC and SCMs, and alkali-activated cements are discussed in this section.

PC Reactions

Only the principal reactions and phases in PC hydration are described here; the comprehensive texts by Hewlett (2018) and Taylor (1997) may be consulted for further details. PC hydration can be separated into five stages, as determined by calorimetry (Figure 6):

1. **Pre-induction period.** Immediately after contact with water, an intense, short hydration of PC occurs. The hydration rate increases rapidly and subsequently decreases rapidly to very low values (Taylor et al. 1984). An intense liberation of heat may be observed in this stage of hydration. The duration of this period is typically no more than a few minutes, represented by the first very sharp peak at the beginning of the curve in Figure 6.
2. **Induction period or dormant period.** The rate of reaction slows down significantly, while simultaneously the liberation of heat of hydration is also significantly reduced. This stage typically lasts a few hours. This stage is very important because it is the period over which concrete can be handled, transported, and placed with ease.
3. **Acceleration (post-induction) period.** After several hours, the rate of hydration accelerates suddenly, reaching a maximum within about 5 hours from the commencement of acceleration. This phenomenon is explained in detail in the literature (Regourd et al. 1980; Brown et al. 1984; Taylor et al. 1984; Jennings 1986; Grutzeck and Ramachandran 1987).
4. **Deceleration period.** After reaching a maximum, the rate of hydration starts to slow down gradually.
5. **Diffusion limited reaction period.** After the deceleration period, the hydration reactions continue, however at a greatly reduced rate, and are measurable after months of curing.

It is the combination of PC compounds that determines the shape of the reaction kinetic curves. Each of the four main components of PC (C_3S, C_2S, C_3A, C_4AF) dissolves and diffuses in water and then precipitates in different phases that form the binder; these reactions are described below in more detail.

Tricalcium silicate ($3CaO{\cdot}SiO_2$, abbreviated C_3S), called alite, is the main constituent of PC, and largely controls its setting and hardening (Figure 6). The hydration of C_3S at ambient temperature forms a calcium-silicate-hydrate (C-S-H) phase with a CaO/SiO_2 molar ratio of <3.0, and calcium hydroxide (CH). C_3S hydration is summarized by the following equation:

tricalcium silicate + water →
calcium-silicate-hydrate + lime + heat

$$2C_3S + 6H \rightarrow C_3S_2H_3 + 3CH \qquad \Delta H = 502.8 \text{ kJ/kg}$$

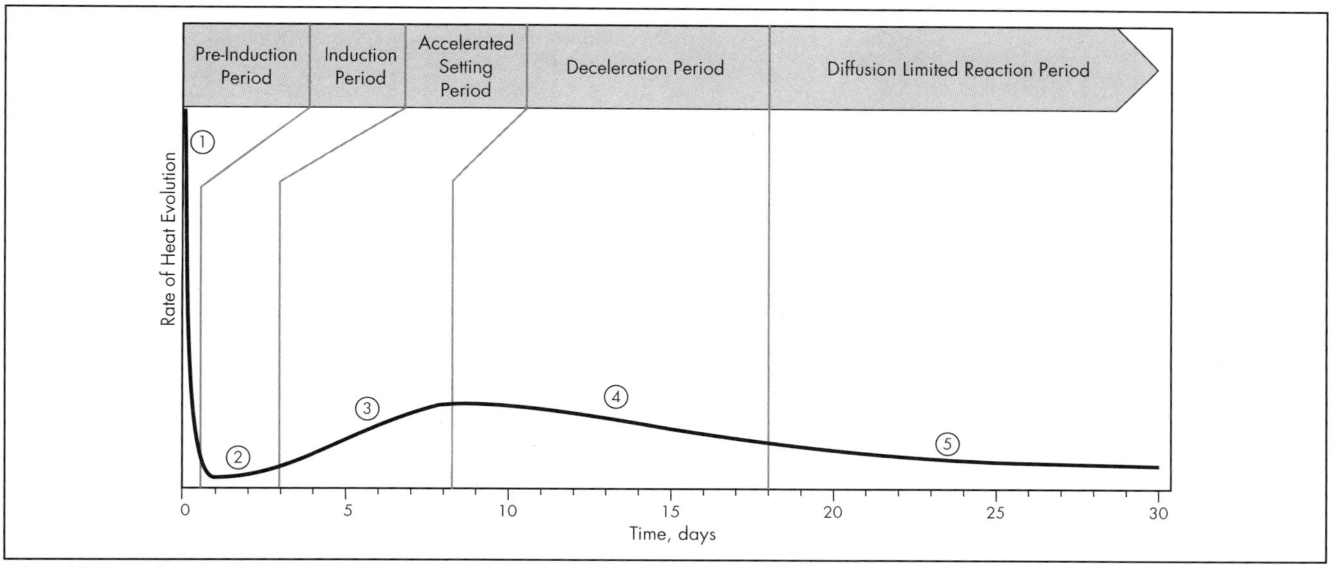

Adapted from Hewlett 2018

Figure 6 Rate of heat of hydration over time

Initially, the dissolution rate of C_3S is higher than the diffusion rate, which allows the movement of dissolved ions away from the surface. This results in a concentration gradient on the surface of C_3S particles; very close to the surface, the liquid phase becomes oversaturated with respect to calcium-silicate-hydrate and a layer of C-S-H precipitates. This precipitated C-S-H is responsible for the large reduction in hydration rate within the first few minutes of C_3S hydration and subsequent acceleration of the reaction after a dormant period. The hydration rate accelerates with increasing water/solid (w/s) ratio. The induction period becomes shortened and at very high w/s ratios may be absent altogether. Also, rapid cooling in the course of C_3S synthesis may increase the reactivity of C_3S.

Dicalcium silicate ($2CaO{\cdot}SiO_2$, abbreviated C_2S), called belite, is another important constituent of PC. Similar to C_3S, hydration of C_2S at ambient temperature forms an amorphous C-S-H phase and portlandite. The hydration of C_2S contributes to the strength of the cement paste (Figure 6) and can be summarized by the following equation:

dicalcium silicate + water →
calcium-silicate-hydrate + lime

$$C_2S + 4H \rightarrow C_3S_2H_3 + CH \qquad \Delta H = 259.4 \text{ kJ/kg}$$

This reaction generates less heat and proceeds at a slower rate, meaning that the contribution of C_2S to the strength of the cement paste will be slow initially. This compound is, however, responsible for the long-term strength of PC concrete. Similar to C_3S, hydration of C_2S may be accelerated by increasing the fineness of grinding, by increasing the temperature of hydration, or by increasing the w/s ratio.

Tricalcium aluminate ($3CaO{\cdot}Al_2O_3$, abbreviated C_3A) is usually present in PC in a much lower quantity than C_3S or C_2S. In the presence of calcium sulfate or gypsum, C_3A reacts according to the following equation:

tricalcium aluminate + gypsum + water →
ettringite + heat

$$C_3A + 3C\overline{S}H_2 + 26H \rightarrow C_6A\overline{S}_3H_{32} \text{ (AFt)} \qquad \Delta H = 866.1 \text{ kJ/kg}$$

Ettringite (abbreviated AFt) consists of long crystals that are only stable in solution with gypsum and does not contribute to the strength of the cement. Hydration of C_3A and gypsum is accompanied by a significant liberation of heat. After a rapid initial reaction, the hydration rate slows down distinctly. The length of this dormant period may vary and escalates with increasing amounts of calcium sulfate in the original paste. A faster hydration, associated with a second heat release maximum, occurs after all the available calcium sulfate has been consumed. Under these conditions, the ettringite formed initially reacts with additional amounts of tricalcium aluminate, yielding calcium aluminate monosulfate hydrate (abbreviated AFm, often called monosulfate) as the product of the following reaction:

tricalcium aluminate + ettringite + water →
monosulfate aluminate hydrate

$$2C_3A + 3\,C_6A\overline{S}_3H_{32} \text{ (or AFt)} + 22H \rightarrow 3C_4ASH_{18} \text{ (AFm)}$$

The monosulfate crystals are only stable in a sulfate-deficient solution. In the presence of sulfates, the monosulfate is converted to ettringite, whose crystals occupy 2.5 times the volume of those of monosulfate. In a paste hydration at ambient temperature, nearly complete hydration of C_3A is attained within several months. The actual reaction rate depends on a variety of factors, such as the w/s ratio and temperature. The kinetics of hydration of C_3A are also used to explain the induction period. The buildup of ettringite at the surface of C_3A may slow down hydration, but this has been disputed by Mehta and Monteiro (2006).

Calcium aluminoferrite (C_4AF) phases, usually present in PC in low amounts, react initially with gypsum and then with ettringite, according to the following equations:

ferrite + gypsum + water →
ettringite + ferric aluminum hydroxide + lime

$$C_4AF + 3C\overline{S}H_2 + 3H \rightarrow C_6(A,F)\overline{S}_3H_{32} \text{ (or AFt)} + (A,F)H_3 + CH$$

ferrite + ettringite + lime + water → hydrogarnets

$$C_4AF + C_6(A,F)\overline{S}_3H_{32} \text{ (or AFt)} + 2CH + 23H \rightarrow 3C_4(A,F)\overline{S}H_{18} \text{ (AFm)} + (A,F)H_3$$

The hydrogarnets assume a space-filling role and do not contribute to the strength of the cement paste (Figure 6). The assemblage of these phases in concrete, dictating its strength and durability, depends on the composition of the PC and SCMs (Table 2 and Figure 2).

Blended Cements and Supplementary Cementitious Materials

Blended cements consist of PC blended with one or more SCMs, which contribute to the strength-gaining properties of the cement (ASTM C219). Blended cements account for the majority of cementitious binders (Schneider et al. 2011). The increased use of blended cement is due to the growing demand for concrete, the aim to reduce cost and CO_2 emissions, and to provide enhanced durability.

Most standards allow substitution of PC with different SCMs in specified proportions (ASTM C1697). All SCMs must also fulfill requirements of other relevant standards. For example, slag cement (Type S) needs to conform to ASTM C989/C989M, natural pozzolans (Type N) and coal fly ash (Types F and C) need to conform to ASTM C618, and silica fume (Type SF) to ASTM C1240. In Europe, limestone filler and calcined clays (such as metakaolin) are also considered SCM. There are no chemical requirements for final blended cement; however, there are composition requirements for the individual constituents. There are physical requirements for the final blended cement, with further optional requirements listed in ASTM C1697. ASTM standards specify the proportions of SCMs in blended cements. Ternary blended cement cannot have more than 15% by mass of limestone, no more than 40% of pozzolan content, and slag shall be less than 70% by mass of the blended cement. Selected blends are described in the following subsections.

Limestone Cement

In Europe, limestone content can be as high as 35%. Such practice is not permitted by ASTM standards; however, limestone is permitted in high–early strength cement (ASTM cement Type III). The limestone content allowed by ASTM standards is more than 5% but less than or equal to 15% by mass of the blended cement. Blending finely ground limestone to PC has several effects on the properties of concrete. The "filler effect" tends to accelerate PC hydration rates, as replacing PC by limestone increases the w/c ratio and surface area for nucleation, with increase in early strength (Gutteridge and Dalziel 1990; Berodier and Scrivener 2014).

Binders with limestone tend to be denser, which can result in improved durability properties, including enhanced sulfate resistance, and decreased drying shrinkage and carbonation depths (Baron 1988; Barker and Matthews 1994). Even if the limestone does not have pozzolanic properties, it reacts with the alumina phases (C_3A) of cement to form an AFm phase (calcium monocarboaluminate hydrate) with no significant changes in the strength of blended cement (Bonavetti et al. 1999).

Pozzolan Cements

According to ASTM C219, PC-pozzolan cement (Class N) consists of an intimate and uniform blend of PC or PC blast-furnace slag cement and fine pozzolan produced by blending PC or PC blast-furnace slag cement and finely divided pozzolans within specified limits. *PC-pozzolan cement* is defined as a hydraulic cement in which the pozzolan constituent is up to 40 mass %. Several materials qualify as pozzolans, including slag, calcined clay, fly ash, natural pozzolans (volcanic ash), silica fume, and rice husk ash. These materials all comprise amorphous silica to some extent. When a pozzolan is blended with PC and hydrated, the silica within the pozzolan reacts with CH and forms additional C-S-H. This C-S-H has a lower Ca/Si ratio (around 1.55, cf 1.7 for hydrated PC) and is consequently called secondary C-S-H, which fills up large capillary pores, thus improving the strength and decreasing the permeability of the system (Mehta and Monteiro 2006). The general pozzolanic reaction can by expressed as follows:

silica component of pozzolan (S) + portlandite (CH) → secondary C-S-H

$$S + 1.5CH + yH \rightarrow C_{1.15}\text{-}S\text{-}H_{y+1.15}$$

Calcined clays do not have a specific ASTM standard specifying these requirements, so the chemistry of hydrated calcined clay/PC blends is discussed here. Following calcination, kaolin and some montmorillonites have the highest pozzolanic activity compared to other clays, as measured according to ASTM C311/C311M (He et al. 1995). Amorphous aluminosilicate phases (ASx) in calcined clays react with CH in the presence of water, producing a cementing compound like C-S-H, C-A-S-H, or C-A-H according to the following equations:

$$A + 3CH + 3H \rightarrow C_3AH_6$$
$$A + S + 2CH + 6H \rightarrow C_2ASH_8$$
$$A + xS + 3CH + (y - -3)\ H \rightarrow C_3AS_xH_y$$

In PC-calcined clay blends, the clay content is often less than 40% due to strong reduction in the workability of the blended concrete, usually associated with higher water demand (Badogiannis et al. 2005). However, water demand can be reduced by use of flash calcination instead of rotary kiln calcination, allowing higher amounts of PC to be substituted. PC-calcined clay blends also exhibit enhanced mechanical properties. In the case of metakaolin, substitution between 10% and 20% seems to be optimal regarding mechanical properties (Badogiannis et al. 2002).

The pore size of PC-calcined clay cements is often reduced compared to PC due to formation of secondary C-S-H, C-A-H, and C-A-S-H, which reduces permeability and mass transfer (Sabir et al. 2001) and enhances the resistance to chloride migration, sulfate attack, and acid attack (Courard et al. 2003; Al-Akhras 2006). Accelerated carbonation tests, however, showed that increasing the metakaolin content tends to increase the carbonation depth of concretes due to the consumption of portlandite by the pozzolanic reaction (Courard et al. 2003; San Nicolas et al. 2014). However, the carbonation was predominantly within the range of carbonation depths found in commercially available blended cements with proven industrial performance. The combination of metakaolin and limestone filler behaves well with respect to carbonation, similarly to PC on its own. The modeling of CO_2 ingress into concrete shows that, although metakaolin increases carbonation, the carbonation depth does not exceed 30 mm after 5 years, which is acceptable.

Fly Ash Cements

Fly ashes are pozzolanic constituents that are used to partially substitute for PC in Class F and Class C cements. When PC-fly ash blends are hydrated, a pozzolanic reaction occurs, generating low Ca/Si C-S-H and increasing the density and durability of the concrete via two mechanisms:

1. Fly ash substitution for PC results in a binder with a higher w/c ratio, improving the hydration of PC grains (Xu et al. 1993).
2. Fly ash particles tend to absorb Ca^{2+} and provide precipitation sites that accelerate hydration (Lawrence et al. 2003; Cyr et al. 2005, 2006).

Because of the presence of significant quantities of aluminum and silicon in fly ash, additional reactions also occur (Marsh and Day 1988; Papadakis 1999; Zeng et al. 2012):

- $1.5CH + S \rightarrow C_{1.5}\ SH_{1.5}$
- $A + 4CH + 9H \rightarrow C_4AH_{13}$
- $A + C\bar{S}H_2 + 3CH + 7H \rightarrow C_4A\bar{S}H_{12}$

Fly ash reacts slower than other pozzolans, which explains the small increase in setting time for fly ash–based concretes compared with PC on its own, especially if Type III and not Type I is used (Carette et al. 1993; Bilodeau and Malhotral 1995). Usually, even a high fly ash substitution shows acceptable setting time and adequate early strength (Dunstan and Hicks 2003). In the fresh state, the substitution of PC by fly ash increases the workability of the concrete at the same w/s ratio, resulting from the spherical shape of fly ash particles and the wider size distribution that decreases yield stress; consequently, the w/c ratio can be reduced. In general, the mechanical properties of fly ash concrete are excellent, due to lower water content and denser microstructure. High-volume fly ash concrete has low chloride penetration, low permeability, high sulfate resistance, and good freeze–thaw resistance, but it has inferior deicing salt scaling resistance compared with PC on its own (Bisaillon and Malhotra 1988).

Silica Fume Cements

Silica fume in Class SF cement is a highly reactive pozzolan, typically with a much smaller particle size than PC, which gives an added filler effect by accelerating the hydration of C_3S (Cheng-yi and Feldman 1985a, 1985b) and provides more nucleation sites. The hydrates obtained in this system are primarily C-S-H, ettringite, and AFm phases and a reduced quantity of portlandite. The blending of PC with ≥24 mass % silica fume results in complete consumption of portlandite after longer hydration times, with ettringite and C-S-H of a reduced Ca/Si ratio as the only hydrate phases (Cheng-yi and Feldman 1985a, 1985b; Calvo et al. 2010; Lothenbach 2010; Rossen et al. 2015). The substitution of PC in Class SF cements is usually between 5% and 10%, which gives reduced bleeding, enhances mechanical properties, and improves durability of the concrete. Higher additions of silica fume (up to 50%) can also be used for low-pH materials (Lothenbach et al. 2012).

Slag Cements

The use of GGBS as a PC replacement in Class S cements is widely practiced due to technological, economic, and environmental benefits. According to ASTM C219, slag cement is a uniform blend of PC and GGBS, where the content of GGBS has to be kept below 95% by mass of the blended cement. GGBS shows both cementitious behavior and some pozzolanic characteristics (reaction with portlandite). The main hydration products of the slag-cement are C-A-S-H, $Ca(OH)_2$, the sulfoaluminate hydrate phases AFt and AFm, and a Mg, Al-rich hydroxide phase. The phases formed are similar to those in a PC-only binder; however, the chemistry is somehow different. For example, the morphology and composition of the C-A-S-H gel may be modified by partial accommodation of Mg^{2+} and Al^{3+} within the micro- or nanostructure, and their Ca/Si ratio becomes lower (e.g., 1.55). A hydrotalcite-like phase with approximate composition $Mg_6Al_2(OH)_{16}(CO_3){\cdot}4H_2O$ is formed from the MgO content of the GGBS, typically 5%–9% (Chen 2006).

One of the principal reasons for the use of GGBS in blended cements is the reduction of the heat released during the hydration process, which reduces thermal stresses in concrete (Sakai et al. 1992; Binici et al. 2007) and improves long-term strength and modulus of elasticity, reduces its long-term shrinkage and creep, and significantly decreases its permeability at later ages. The long-term durability properties of PC-slag blends are enhanced compared with neat PC, which increases resistance to the penetration of chloride ions (Escalante et al. 2001; Kolani et al. 2012), with little impact on freeze–thaw resistance. Concrete incorporating GGBS can cost less than concrete made with PC only; costs are dependent on the availability of GGBS in a market, transportation and handling costs, and who controls that market.

Ternary Blended Cements

Ternary blended cements can be developed by combining the different positive attributes of SCMs with the PC system, such as the filler effect, pozzolanic effect, and chemical balancing. LC^3 (limestone calcined clay cement) has recently been promoted strongly in Europe and the United States and may meet ASTM guidelines. Parallel chemical reactions occur in these blends, with the cement reacting with water in the same way as in pure PC, and the limestone reacting with alumina from the metakaolin to form supplementary AFm phases and stabilized ettringite (Antoni et al. 2012). The pore size of LC^3 concrete appears to be smaller than in PC concrete, improving durability performance (e.g. reducing chloride migration; Antoni 2013).

Alkali-Activated Cements

It is possible to make a cement without PC, containing just SCMs and alkali activator, which is called alkali-activated cement. These cements have been demonstrated in many projects but are not yet widely available commercially. As more countries have standards based on performance instead of being prescriptive, these materials can be used and have been demonstrated to perform well. The following paragraphs describe the composition of the binder and general properties of concretes made from these alkali-activated binders. For further details, please see Pacheco-Torgal et al. (2014) and Provis and van Deventer (2014).

The main reaction product of alkali activation of silica-rich precursors such as fly ash or metakaolin is a three-dimensional alkali aluminosilicate hydrate gel network consisting of cross-linked AlO_4^- and SiO_4 tetrahedra linked via shared oxygen atoms denoted as "bridging oxygen," with terminal hydroxyl groups forming at the gel surface, or (N,K)-A-S-H (Davidovits 1994; Duxson et al. 2007; Bernal et al. 2013; Provis and van Deventer 2014). The level of network-modifying cations present affects the compressive strength of the geopolymer binder formed upon alkali activation, with strength generally increasing with increasing network-modifying cation content. Alkali activation of GGBS (generally with sodium silicate solution, although other alkali solutions can be used) produces an aluminum-substituted calcium-silicate-hydrate (C-(N)-A-S-H) gel plus possibly portlandite, calcium monosulfoaluminate (and similar hydrocalumite-like "alumina, ferric oxide, monosulfate [AFm]" phases with carbonate, aluminosilicate, or hydroxide anion substitution), the "third aluminate hydrate," Mg-Al layered double-hydroxide phases such as hydrotalcite and M_4AH_{13}, ettringite, and hydrogarnet. These additional reaction products are intimately mixed with the C-(N)-A-S-H gel framework because of differences in bulk composition stability (García-Lodeiro et al. 2011; Lothenbach et al. 2011) and can significantly influence gel-phase evolution and nanostructural development.

Coexistence of N-A-S-H and C-(N)-A-S-H gel frameworks has been observed in binders based on blends of low-calcium and high-calcium precursors (Yip et al. 2005; Bernal et al. 2011b, 2013; Ismail et al. 2014); however, the stability of the gel coexistence remains unclear. Blends of different precursors can often show advantages over pure systems (Bernal et al. 2011a).

ENERGY CONSUMPTION AND EMISSIONS

The manufacturing of PC is energy intensive, and with the associated emission of carbon dioxide resulting from the decomposition of limestone, the PC industry globally is one of the highest emitters of CO_2. This section analyzes the sources of energy consumption and emissions for PC and explains how CO_2 emissions can be reduced by blending with SCMs and/or alkali activation.

Energy Consumption of PC Manufacturing

The energy consumption in a PC plant is shared between thermal and electrical energy. Assessments of energy consumption conducted on more than 100 PC plants in Europe and the United States estimated the thermal energy share at around 90% against 10% for the electrical energy share. The kiln alone is responsible for 99% of this thermal energy consumption plus 25% of the electrical energy consumption (Schneider et al. 2011). The rest of the electrical energy is associated with grinding and drying of the raw mix (60%), and finally, 15% is associated with handling and carrying materials (Afkhami et al. 2015).

Much effort has been dedicated to reducing the energy consumption of PC plants. It has been demonstrated that if the efficiency of the kiln is low, the energy loss is around 40% (Atmaca and Yumrutaş 2014). To date, the most economical and practical option to improve the efficiency of the kiln is to add heat exchangers, which recycle heat at different stages of PC manufacturing. Improving the insulation and the combustion efficiency of the kiln also improves the general efficiency of the kiln system. The optimum size of kiln in regard to energy efficiency is dependent on the moisture content of the raw materials.

PC plants have actively reduced their energy consumption over the past 40 years, especially with the increased use of alternative and waste fuel such as tires, animal residues, sewage sludge, waste oil, and lumpy materials. In most plants, these alternative fuels are replacing all of the fuel used in the precalcination stage of the PC process; however, their use for clinker kilns is still progressing. Importantly, the combustion process needs to be adapted to the fuel employed; consequently, it is common to find modern multichannel burners designed for the use of alternative fuels and thermograph systems allowing control of the flame shape to optimize the burning behavior of the different fuels and burning conditions. The use of most alternative fuels does not have any effect on the final clinker. Nevertheless, phosphorous-rich fuel such as meat and bone meal or sewage sludge retards setting.

Emissions of PC

The CO_2 emissions associated with PC production can be classified into two groups: direct emissions as a result of the intrinsic chemical reaction occurring during thermal processing (calcination) of limestone ($CaCO_3$) to produce cement clinker (55%), and indirect emissions associated with energy requirements for thermal processing, crushing, grinding, and transport (45%).

The use of limestone as a raw feedstock for clinker production results in a total CO_2 embodiment of 0.73–0.99 t CO_2/t PC, with the variation primarily due to the kiln efficiency during PC clinker production (Bernstein et al. 2007). It has been estimated that by 2050, CO_2 emissions associated with PC production will account for 20%–26% of the total global anthropological CO_2 emissions (Beyond Zero Emissions 2017). Therefore, there is ample justification for seeking an alternative binder to high-PC blends.

The manufacturing of PC also causes emissions other than just CO_2 and has global environmental impacts, including abiotic depletion, ozone layer depletion, human toxicity, fresh-water aquatic effects, ecotoxicity, photochemical oxidation, acidification, and eutrophication (Oss and Padovani 2003; Kumar et al. 2008; Huntzinger and Eatmon 2009; Habert et al. 2011).

Reduction of Energy Consumption and Emissions of Cement by Blending and Alkali Activation

The energy consumption and emissions of PC can be reduced by blending with SCM. All the SCMs have a lower energy consumption compared to PC; hence, the more SCMs used in a blend, the lower the net energy consumption. The hierarchy of energy consumption of SCMs is: calcined clays > slag, limestone > fly ash, silica fume.

Clay calcined at about 700°C is the SCM with the highest energy consumption, even when its production uses 50%–80% less energy than PC production. Similar to clinker kilns, heat exchangers and plant modifications can reduce the energy needs of calciners. Rotary kiln calcining requires higher energy than flash calcining, not only regarding thermal energy used for calcination but also concerning the electrical energy required to grind the calcined product from a rotary kiln, which is not required in flash calcination (Salvador 1995; San Nicolas 2011; San Nicolas et al. 2013).

Slag and limestone as SCMs require less energy than calcined clay but still need to be ground finely. The power consumption for slag grinding in a VRM is about 30 kW·h/t compared with the 40 kW·h/t for a ball mill. Silica fume and fly ash (requires classification) do not need energy for grinding because they are both fine products and hence are the SCMs with the lowest energy consumption. Even when the processing of SCMs used in blended cement involves some energy consumption, it is much less than when only PC is used (Schneider et al. 2011). The embedded CO_2 emissions of the SCMs are very low, so the higher the SCM content of a blended cement, the lower its CO_2 emissions.

Many SCMs can be alkali activated, which gives further reduction in CO_2 emissions as a result of the elimination of PC with high CO_2 emissions. The production of alkali activators requires energy and generates CO_2 emissions, which vary vastly in intensity from one activator to the next. For example, some alkali activators may be waste products from other industries with no inherent emission penalty (Pacheco-Torgal et al. 2014; Phetchuay et al. 2014). A commonly used activator is sodium silicate, which has inherent emissions from the smelting of sodium carbonate with silica sand, where the emissions of sodium carbonate may vary vastly, depending on whether it is mined or chemically synthesized. Activators also have environmental impacts other than CO_2 emissions (Habert et al. 2011). Alkali-activation technology can reduce CO_2 emissions by about 80% in comparison with blended PC available in Australia (van Deventer et al. 2012).

TECHNO-ECONOMIC PERSPECTIVES

Cement production cost and market pricing are complex functions of the cost of raw material, energy, labor, transportation and capital, the scale of operations, market positioning, and competition. These factors are discussed with regard to changes in the market, partly due to climate change, and technological developments.

Production Costs and Cement Pricing

A survey of PC producers in the United States gives the following breakdown of production cost of PC as a percentage of the sales price (Pabon 2014):

- Power and fuel: 30%
- Limestone and other raw materials: 30%–40%
- Transportation: 10%
- Maintenance, labor, administration: 15%–20%

As explained earlier in this chapter, the production of clinker is energy intensive, so the cost of power and fuel (natural gas, fuel oil, coal, diesel, pet coke, lignite, used tires, and other waste) has a major impact on the operating expenditure of a PC plant. According to the survey by Pabon (2014), the energy consumption is 130–150 kW·h/t of clinker.

The location of a PC plant is largely dependent on the proximity of sources of raw materials and the market, and hence a function of the cost of transportation. A PC plant is usually located near a limestone quarry, but there are cases where low-cost water transport justifies a location closer to the market. Road and rail contribute more than 90% of the transportation of PC product.

Modern PC plants rarely have a capacity of less than 300,000 t clinker per year, whereas large plants in China frequently produce in excess of 10 Mt clinker per year. The capital and operating costs of clinker production are strongly dependent on scale, so there is a trend for clinker to be produced in very large plants close to limestone quarries, not just for the local market but also for export. Countries like China, Thailand, United Arab Emirates, Turkey, Vietnam, Japan, and India export a sizable portion of their PC production, usually at a lower landed cost than the production cost in the destination country. To protect local PC production, some countries impose tariffs on imported PC, but usually not on imported clinker. Therefore, clinker is often shipped from the low-cost countries to remote locations with smaller markets where it is ground to become PC. These grinding plants may also source slag granulates from remote suppliers and grind it to become GGBS, which they may blend with the PC. Similarly, fly ash from local coal-fired power stations may also be used for blending with PC.

Cement prices vary by a factor of 3 or more from one location to another. Evidently, large local markets near deposits of limestone result in low transportation costs and, associated with low labor costs, give low PC prices, especially in the presence of competitive market action. In contrast, locations where PC is most expensive are characterized by a small market size, a high cost of transportation, restricted access to importation infrastructure, and limited or ineffective competition.

The availability of local SCMs such as fly ash or slag that can be ground (to become GGBS) to blend with PC should, in theory, reduce the price of blended cement in the market. In practice, such price reduction does not always happen, especially when fly ash or GGBS is in short supply or when cement companies also control the supply chain of these SCMs. For example, unclassified ash may be more widely available, but cement companies may have exclusive supply arrangements with coal-fired power stations and control classification plants, and hence determine the price of classified fly ash. Unground slag granulates may be available in a market, but often there are only a few independent facilities for the grinding of slag, apart from the cement companies purchasing and grinding slag for blending with PC. The result is that the prices for GGBS and classified fly ash in a market may be on par with or slightly less than the PC price. In such markets there is little incentive for new entrants to introduce high-SCM-blended cement or alkali-activated cement, unless an independent supply chain can be established to reduce the cost of SCM.

Developments in Grinding Technology

Currently, cement and slag grinding plants are large ball mills or VRMs that are capital intensive and often located in association with PC operations. It is not practical to use such equipment in smaller markets to exploit waste streams such as metallurgical or steel slag, or virgin materials like volcanic ash or rock that can be used as SCM.

Loesche (2017) in Germany has recognized this limitation and now has a compact VRM available for remote locations where a customer will grind PC clinker and/or slag granules in smaller volumes. The Universal Vortex acoustic grinder produces a shock wave generated in a precisely designed geometry to shatter brittle materials and has been used at demonstration scale for clinker and slag granules (van Deventer 2017). The advantage of the Universal Vortex is that it fits into a shipping container, so it is mobile and can be deployed easily in remote locations for the grinding of smaller volumes of material and also for one-off construction projects.

Potential Impact of Alkali Activation and Modularized Grinding on Cement Value Chain

Alkali activation has the potential to expand the scope of SCMs used in cement, especially in hybrid cements incorporating clinker as a minor component. The result will be the valorization of waste streams currently not utilized as cementitious materials, which should reduce the cost of cement in remote locations or smaller markets. Mining often takes place in locations remote from established markets and PC supply chains. Consequently, PC used in either backfill or shotcrete may be a substantial part of the operating cost of a mine. If indigenous mine waste such as clays or volcanic material could be used as SCM, even when alkali activated, less PC will be required, with substantial savings in transportation cost. Usually, the beneficiation of mine waste as a SCM will require dry fine-grinding in addition to calcining if clays are used.

With the development of modularized grinding technology, it becomes possible to grind smaller amounts of SCMs and even clinker. The possibility to decentralize grinding gives a lower barrier of entry to the cement market that may change the cement value chain. An associated benefit of modularized grinding is that it becomes economical to transport smaller volumes of clinker or slag granulates in bulk carriers (shipping, road, or rail) to remote locations, which is cheaper than transporting cement powder in pneumatic tanks.

Performance-Based Versus Prescriptive Standards

The regulatory framework for cement and concrete in most jurisdictions has been prescriptive, even when it is claimed superficially that such standards are performance based. Standards frameworks often restrict the definition of SCMs to GGBS and fly ash, usually place an upper limit on the SCM content, and hence prescribe concrete compositions. To enable the wide adoption of advanced concrete technologies, including alkali activation, it is essential that performance-based standards be adopted instead of the current prescriptive system. There should be no restriction on the type of components used, either as cementitious material, aggregates, or admixtures/activators. The amount of cementitious material and the w/c ratio should also not be restricted.

It is important to decide which performance and durability testing methods should be used to specify performance criteria. In a critical review of performance-based approaches (Alexander and Thomas 2015), it was explained that it is possible to relate service-life prediction models to durability testing, even when it is known that the diffusion parameters in concrete are complicated by several factors, including interaction between the diffusing species and the matrix, and the reduction of diffusion coefficients with age.

REFERENCES

Afkhami, B., Akbarian, B., Beheshti, N., Kakaee, A., and Shabani, B. 2015. Energy consumption assessment in a cement production plant. *Sustainable Energy Technol. Assess.* 10:84–89.

Al-Akhras, N.M. 2006. Durability of metakaolin concrete to sulfate attack. *Cem. Concr. Res.* 36(9):1727–1734.

Alexander, M., and Thomas, M. 2015. Service life prediction and performance testing—Current developments and practical applications. *Cem. Concr. Res.* 78(Part A):155–164.

Alsop, P.A. 2007. *Cement Plant Operations Handbook: For Dry Process Plants*. Surrey, UK: Tradeship Publications.

Ambroise, J., Murat, M., and Pera, J. 1985. Hydration reaction and hardening of calcined clays and related minerals V: Extension of the research and general conclusions. *Cem. Concr. Res.* 15(2):261–268.

Antoni, M. 2013. Investigation of cement substitution by blends of calcined clays and limestone. Doctoral thesis, École Polytechnique Fédérale de Lausanne, Switzerland.

Antoni, M., Rossen, J., Martirena, F., and Scrivener, K. 2012. Cement substitution by a combination of metakaolin and limestone. *Cem. Concr. Res.* 42(12):1579–1589.

ASTM C150/C150M. 2018. *Standard Specification for Portland Cement.* West Conshohocken, PA: ASTM International.

ASTM C219. 2014. *Standard Terminology Relating to Hydraulic Cement.* West Conshohocken, PA: ASTM International.

ASTM C311/C311M. 2017. *Standard Test Methods for Sampling and Testing Fly Ash or Natural Pozzolans for Use in Portland-Cement Concrete.* West Conshohocken, PA: ASTM International.

ASTM C618. 2015. *Standard Specification for Coal Fly Ash and Raw or Calcined Natural Pozzolan for Use in Concrete.* West Conshohocken, PA: ASTM International.

ASTM C989. 2016. *Standard Specification for Slag Cement for Use in Concrete and Mortars.* West Conshohocken, PA: ASTM International.

ASTM C1240. 2015. *Standard Specification for Silica Fume Used in Cementitious Mixtures.* West Conshohocken, PA: ASTM International.

ASTM C1697. 2016. *Standard Specification for Blended Cementitious Materials.* West Conshohocken, PA: ASTM International.

ASTM C1797. 2017. *Standard Specification for Ground Calcium Carbonate and Aggregate Mineral Fillers for Use in Hydraulic Cement Concrete.* West Conshohocken, PA: ASTM International.

Atmaca, A., and Yumrutaş, R. 2014. Analysis of the parameters affecting energy consumption of a rotary kiln in cement industry. *Appl. Therm. Eng.* 66(1):435–444.

Badogiannis, E., Tsivilis, S., Kakali, G., Papadakis, V., and Chaniotakis, E. 2002. The effect of metakaolin on concrete properties. In *Innovations and Developments in Concrete Materials and Construction: Proceedings of the International Congress Held at the University of Dundee, Scotland, UK.* Edited by R.K. Dhir, P.C. Hewlett, and L.J. Cetenyi. London: Thomas Telford. pp. 81–89.

Badogiannis, E., Kakali, G., Dimopoulou, G., Chaniotakis, E., and Tsivilis, S. 2005. Metakaolin as a main cement constituent: Exploitation of poor Greek kaolins. *Cem. Concr. Compos.* 27(2):197–203.

Barker, A., and Matthews, J. 1994. Concrete durability specification by water/cement or compressive strength for European cement types. *ACI Spec. Publ.* 145:1135–1160.

Baron, P. 1988. The durability of limestone composite cements in the context of the French specifications. In *Durability of Concrete: Aspects of Admixtures and Industrial Byproducts.* Edited by L-O. Nilsson. Swedish Council for Building Research. pp. 115–122.

Ben Haha, M., Lothenbach, B., Le Saoût, G., and Winnefeld, F. 2011. Influence of slag chemistry on the hydration of alkali-activated blast-furnace slag—Part I: Effect of MgO. *Cem. Concr. Res.* 41(9):955–963.

Ben Haha, M., Lothenbach, B., Le Saoût, G., and Winnefeld, F. 2012. Influence of slag chemistry on the hydration of alkali-activated blast-furnace slag—Part II: Effect of Al_2O_3. *Cem. Concr. Res.* 42(1):74–83.

Bernal, S.A., Provis, J.L., Rose, V., and de Gutierrez, R.M. 2011a. Evolution of binder structure in sodium silicate-activated slag-metakaolin blends. *Cem. Concr. Compos.* 33(1):46–54.

Bernal, S.A., Rodriguez, E.D., de Gutierrez, R.M., Gordillo, M., and Provis, J.L. 2011b. Mechanical and thermal characterisation of geopolymers based on silicate-activated metakaolin/slag blends. *J. Mater. Sci.* 46(16):5477–5486.

Bernal, S.A., Provis, J.L., Walkley, B., San Nicolas, R., Gehman, J.D., Brice, D.G., Kilcullen, A.R., Duxson, P., and van Deventer, J.S.J. 2013. Gel nanostructure in alkali-activated binders based on slag and fly ash, and effects of accelerated carbonation. *Cem. Concr. Res.* 53:127–144.

Bernal, S.A., San Nicolas, R., Myers, R.J., Mejía de Gutiérrez, R., Puertas, F., van Deventer, J.S.J., and Provis, J.L. 2014. MgO content of slag controls phase evolution and structural changes induced by accelerated carbonation in alkali-activated binders. *Cem. Concr. Res.* 57:33–43.

Bernstein, L., Roy, J., Delhotal, K.C., Harnisch, J., Matsuhashi, R., Price, L., Tanaka, K., Worrell, E., Yamba, F., and Fengqi, Z. 2007. Industry. In *Climate Change 2007: Mitigation Contribution of Working Group III to the Fourth Assessment Report of the Intergovernmental Panel on Climate Change.* Edited by B. Metz, O.R. Davidson, and P.R. Bosch. Cambridge, UK: Cambridge University Press.

Berodier, E., and Scrivener, K. 2014. Understanding the filler effect on the nucleation and growth of C-S-H. *J. Am. Ceram. Soc.* 97(12):3764–3773.

Beyond Zero Emissions. 2017. *Rethinking Cement, Zero Carbon Industry Plan.* Fitzroy, Victoria: Beyond Zero Emissions.

Bilodeau, A., and Malhotra, V.M. 1995. Properties of high-volume fly ash concrete made with high early-strength ASTM Type III cement. *ACI Spec. Publ.* 153:1–24.

Binici, H., Temiz, H., and Köse, M.M. 2007. The effect of fineness on the properties of the blended cements incorporating ground granulated blast furnace slag and ground basaltic pumice. *Constr. Build. Mater.* 21(5):1122–1128.

Bisaillon, A., and Malhotra, V.M. 1988. Permeability of concrete using a uniaxial water-flow method. *ACI Spec. Publ.* 108:175–194.

Bonavetti, V., Donza, H., Rahhal, V., and Irassar, E. 1999. High-strength concrete with limestone filler cements. *ACI Spec. Publ.* 186:567–580.

Brown, P.W., Franz, E., Frohnsdorff, G., and Taylor, H. 1984. Analyses of the aqueous phase during early C3S hydration. *Cem. Concr. Res.* 14(2):257–262.

Calvo, J., Hidalgo, A., Alonso, C., and Luco, L. 2010. Development of low-pH cementitious materials for HLRW repositories: Resistance against ground waters aggression. *Cem. Concr. Res.* 40(8):1290–1297.

Cancio Díaz, Y., Sánchez Berriel, S., Heierli, U., Favier, A.R., Sánchez Machado, I.R., Scrivener, K.L., Martirena Hernández, J.F., and Habert, G. 2017. Limestone calcined clay cement as a low-carbon solution to meet expanding cement demand in emerging economies. *Dev. Eng.* 2:82–91.

Carette, G., Bilodeau, A., Chevrier, R.L., and Malhotra, V. 1993. Mechanical properties of concrete incorporating high volumes of fly ash from sources in the US. *ACI Mater. J.* 90(6):535–544.

Celik, K., Jackson, M.D., Mancio, M., Meral, C., Emwas, A.H., Mehta, P.K., and Monteiro, P.J.M. 2014. High-volume natural volcanic pozzolan and limestone powder as partial replacements for portland cement in self-compacting and sustainable concrete. *Cem. Concr. Compos.* 45(Suppl. C):136–147.

Chen, W. 2006. Hydration of slag cement: Theory, modeling and application. Ph.D. thesis, Enschede, University of Twente.

Cheng-yi, H., and Feldman, R.F. 1985a. Influence of silica fume on the microstructural development in cement mortars. *Cem. Concr. Res.* 15(2):285–294.

Cheng-yi, H., and Feldman, R.F. 1985b. Hydration reactions in portland cement-silica fume blends. *Cem. Concr. Res.* 15(4):585–592.

Courard, L., Darimont, A., Schouterden, M., Ferauche, F., Willem, X., and Degeimbre, R. 2003. Durability of mortars modified with metakaolin. *Cem. Concr. Res.* 33(9):1473–1479.

Cyr, M., Lawrence, P., and Ringot, E. 2005. Mineral admixtures in mortars: Quantification of the physical effects of inert materials on short-term hydration. *Cem. Concr. Res.* 35(4):719–730.

Cyr, M., Lawrence, P., and Ringot, E. 2006. Efficiency of mineral admixtures in mortars: quantification of the physical and chemical effects of fine admixtures in relation with compressive strength. *Cem. Concr. Res.* 36(2):264–277.

Davidovits, J. 1994. Properties of geopolymer cements. In *Proceedings of the First International Conference on Alkaline Cements and Concretes.* Saint-Quentin, France: Geopolymer Institute. pp. 131–149.

Dunstan, M.R.H., and Hicks, C.H. 2003. Adaptive construction methodology for the Ghatgher Saddle dam, India's first RCC dam. In *Roller Compacted Concrete Dams.* Rotterdam: A.A. Balkema. pp. 239–247.

Durdziński, P.T., Dunant, C.F., Haha, M.B., and Scrivener, K.L. 2015. A new quantification method based on SEM-EDS to assess fly ash composition and study the reaction of its individual components in hydrating cement paste. *Cem. Concr. Res.* 73:111–122.

Duxson, P., and Provis, J.L. 2008. Designing precursors for geopolymer cements. *J. Am. Ceram. Soc.* 91(12):3864–3869.

Duxson, P., Mallicoat, S., Lukey, G., Kriven, W., and van Deventer, J. 2007. The effect of alkali and Si/Al ratio on the development of mechanical properties of metakaolin-based geopolymers. *Colloids Surf. A* 292(1):8–20.

Engström, F., Pontikes, Y., Geysen, D., Jones, P.T., Björkman, B., and Blanpain, B. 2011. Review: Hot stage engineering to improve slag valorisation options. In *Proceedings of the Second International Slag Valorisation Symposium.* Leuven, Belgium: Katholieke Universiteit Leuven.

Escalante, J., Gomez, L., Johal, K., Mendoza, G., Mancha, H., and Mendez, J. 2001. Reactivity of blast-furnace slag in portland cement blends hydrated under different conditions. *Cem. Concr. Res.* 31(10):1403–1409.

Fernández-Jiménez, A., and Palomo, A. 2005. Composition and microstructure of alkali activated fly ash binder: Effect of the activator. *Cem. Concr. Res.* 35(10):1984–1992.

García-Lodeiro, I., Palomo, A., Fernández-Jiménez, A., and Macphee, D.E. 2011. Compatibility studies between N-A-S-H and C-A-S-H gels: Study in the ternary diagram Na_2O–CaO–Al_2O_3–SiO_2–H_2O. *Cem. Concr. Res.* 41(9):923–931.

Glasser, F.P., Lachowski, E.E., and Macphee, D.E. 1987. Compositional model for calcium silicate hydrate (C-S-H) gels, their solubilities, and free energies of formation. *J. Am. Ceram. Soc.* 70(7):481–485.

Global Construction Perspectives and Oxford Economics. 2015. *Global Construction 2030: A Global Forecast for the Construction Industry to 2030.* London: Global Construction Perspectives and Oxford Economics.

Grutzeck, M.W., and Ramachandran, A. 1987. An integration of tricalcium silicate hydration models in light of recent data. *Cem. Concr. Res.* 17(1):164–170.

Gutteridge, W.A., and Dalziel, J.A. 1990. Filler cement: the effect of the secondary component on the hydration of portland cement: Part I. A fine non-hydraulic filler. *Cem. Concr. Res.* 20(5):778–782.

Habert, G., De Lacaillerie, J.D.E., and Roussel, N. 2011. An environmental evaluation of geopolymer based concrete production: Reviewing current research trends. *J. Cleaner Prod.* 19(11):1229–1238.

Harder, J. 2010. Grinding trends in the cement industry. *ZKG (Zement Kalk Gips) Int.* 63(4):46–58.

He, C., Osbaeck, B., and Makovicky, E. 1995. Pozzolanic reactions of six principal clay minerals: Activation, reactivity assessments and technological effects. *Cem. Concr. Res.* 25(8):1691–1702.

Hewlett, P. 2018. *Lea's Chemistry of Cement and Concrete.* Oxford: Elsevier Science.

Huntzinger, D.N., and Eatmon, T.D. 2009. A life-cycle assessment of portland cement manufacturing: comparing the traditional process with alternative technologies. *J. Cleaner Prod.* 17(7):668–675.

Ismail, I., Bernal, S.A., Provis, J.L., San Nicolas, R., Hamdan, S., and van Deventer, J.S.J. 2014. Modification of phase evolution in alkali-activated blast furnace slag by the incorporation of fly ash. *Cem. Concr. Compos.* 45:125–135.

Jennings, H.M. 1986. Aqueous solubility relationships for two types of calcium silicate hydrate. *J. Am. Ceram. Soc.* 69(8):614–618.

Kolani, B., Buffo-Lacarrière, L., Sellier, A., Escadeillas, G., Boutillon, L., and Linger, L. 2012. Hydration of slag-blended cements. *Cem. Concr. Compos.* 34(9):1009–1018.

Kumar, S.S., Singh, N.A., Kumar, V., Sunisha, B., Preeti, S., Deepali, S., and Nath, S.R. 2008. Impact of dust emission on plant vegetation in the vicinity of cement plant. *Environ. Eng. Manage. J.* 7(1):31–35.

Lawrence, P., Cyr, M., and Ringot, E. 2003. Mineral admixtures in mortars: Effect of inert materials on short-term hydration. *Cem. Concr. Res.* 33(12):1939–1947.

Le Saoût, G., Ben Haha, M., Winnefeld, F., Lothenbach, B., and Jantzen, C. 2011. Hydration degree of alkali-activated slags: A ^{29}Si NMR study. *J. Am. Ceram. Soc.* 94(12):4541–4547.

Lemougna, P.N., MacKenzie, K.J., and Melo, U.C. 2011. Synthesis and thermal properties of inorganic polymers (geopolymers) for structural and refractory applications from volcanic ash. *Ceram. Int.* 37(8):3011–3018.

Lemougna, P.N., MacKenzie, K.J., Jameson, G.N., Rahier, H., and Melo, U.C. 2013. The role of iron in the formation of inorganic polymers (geopolymers) from volcanic ash: A 57Fe Mössbauer spectroscopy study. *J. Mater. Sci.* 48(15):5280–5286.

Li, Y., Liu, X., Sun, H., and Cang, D. 2011. Mechanism of phase separation in BFS (blast furnace slag) glass phase. *Sci. China Technol. Sci.* 54(1):105–109.

Loesche. 2017. Plant engineering solutions for any mission. https://www.loesche.com/. Accessed October 2017.

Lothenbach, B. 2010. Thermodynamic equilibrium calculations in cementitious systems. *Mater. Struct.* 43(10):1413–1433.

Lothenbach, B., Scrivener, K., and Hooton, R.D. 2011. Supplementary cementitious materials. *Cem. Concr. Res.* 41(12):1244–1256.

Lothenbach, B., Le Saout, G., Haha, M.B., Figi, R., and Wieland, E. 2012. Hydration of a low-alkali CEM III/B–SiO_2 cement (LAC). *Cem. Concr. Res.* 42(2):410–423.

Marsh, B.K., and Day, R.L. 1988. Pozzolanic and cementitious reactions of fly ash in blended cement pastes. *Cem. Concr. Res.* 18(2):301–310.

Mehta, P., and Monteiro, P.J.M. 2006. *Concrete: Microstructure, Properties, and Materials.* New York: McGraw-Hill.

Moir, G. 2003. Cements. In *Advanced Concrete Technology.* Edited by B.S. Choo and J. Newman. Oxford: Butterworth-Heinemann. pp. 3–45.

Moseson, A.J., Moseson, D.E., and Barsoum, M.W. 2012. High volume limestone alkali-activated cement developed by design of experiment. *Cem. Concr. Compos.* 34(3):328–336.

Oss, H.G., and Padovani, A.C. 2003. Cement manufacture and the environment, part II: Environmental challenges and opportunities. *J. Ind. Ecol.* 7(1):93–126.

Pabon, A. 2014. The cost elements of cement. http://marketrealist.com/2014/08/business-overview-cement-industry/. Accessed October 2017.

Pacheco-Torgal, F., Labrincha, J., Leonelli, C., Palomo, A., and Chindaprasit, P. 2014. *Handbook of Alkali-Activated Cements, Mortars and Concretes.* Cambridge: Elsevier.

Papadakis, V.G. 1999. Effect of fly ash on portland cement systems: Part I. Low-calcium fly ash. *Cem. Concr. Res.* 29(11):1727–1736.

Pesonen, J., Yliniemi, J., Illikainen, M., Kuokkanen, T., and Lassi, U. 2016. Stabilization/solidification of fly ash from fluidized bed combustion of recovered fuel and biofuel using alkali activation and cement addition. *J. Environ. Chem. Eng.* 4(2):1759–1768.

Phetchuay, C., Horpibulsuk, S., Suksiripattanapong, C., Chinkulkijniwat, A., Arulrajah, A., and Disfani, M.M. 2014. Calcium carbide residue: Alkaline activator for clay–fly ash geopolymer. *Constr. Build. Mater.* 69:285–294.

Provis, J.L., and van Deventer, J.S.J. 2009. *Geopolymers—Structure, Processing, Properties and Industrial Applications.* Cambridge: Woodhead Publishing.

Provis, J., and van Deventer, J. 2014. *Alkali Activated Materials.* Berlin: Springer.

Regourd, M., Thomassin, J.H., Baillif, P., and Touray, J.C. 1980. Study of the early hydration of Ca_3SiO_5 by X-ray photoelectron spectrometry. *Cem. Concr. Res.* 10(2):223–230.

Rossen, J.E., Lothenbach, B., and Scrivener, K.L. 2015. Composition of C-S-H in pastes with increasing levels of silica fume addition. *Cem. Concr. Res.* 75:14–22.

Sabir, B., Wild, S., and Bai, J. 2001. Metakaolin and calcined clays as pozzolans for concrete: a review. *Cem. Concr. Compos.* 23(6):441–454.

Sakai, K., Watanabe, H., Suzuki, M., and Hamazaki, K. 1992. Properties of granulated blast-furnace slag cement concrete. *ACI Spec. Publ.* 132:1367–1384.

Salvador, S. 1995. Pozzolanic properties of flash-calcined kaolinite: A comparative study with soak-calcined products. *Cem. Concr. Res.* 25(1):102–112.

San Nicolas, R. 2011. Performance-based approach for concrete containing metakaolin obtained by flash calcination. PhD thesis (in French), Université de Toulouse, Toulouse.

San Nicolas, R., Cyr, M., and Escadeillas, G. 2013. Characteristics and applications of flash metakaolins. *Appl. Clay Sci.* 83:253–262.

San Nicolas, R., Cyr, M., and Escadeillas, G. 2014. Performance-based approach to durability of concrete containing flash-calcined metakaolin as cement replacement. *Constr. Build. Mater.* 55:313–322.

Sánchez Berriel, S., Favier, A., Rosa Domínguez, E., Sánchez Machado, I.R., Heierli, U., Scrivener, K., Martirena Hernández, F., and Habert, G. 2016. Assessing the environmental and economic potential of limestone calcined clay cement in Cuba. *J. Cleaner Prod.* 124:361–369.

Schneider, M., Romer, M., Tschudin, M., and Bolio, H. 2011. Sustainable cement production—Present and future. *Cem. Concr. Res.* 41(7):642–650.

Scrivener, K.L., John, V.M., and Gartner, E.M. 2016. Eco-efficient cements: Potential, economically viable solutions for a low-CO_2, cement-based materials industry. Paris: United Nations Environment Programme.

Shi, C.J., Krivenko, P.V., and Roy, D.M. 2006. *Alkali Activated Cements and Concretes.* New York: Taylor and Francis.

Singh, M., and Garg, M. 2006. Reactive pozzolana from Indian clays—Their use in cement mortars. *Cem. Concr. Res.* 36(10):1903–1907.

Snellings, R. 2016. Assessing, understanding and unlocking supplementary cementitious materials. *RILEM Tech. Lett.* 1:50–55.

Snellings, R., Mertens, G., and Elsen, J. 2012. Supplementary cementitious materials. *Rev. Mineral. Geochem.* 74(1):211–278.

Takeda, H., Hashimoto, S., Kanie, H., Honda, S., and Iwamoto, Y. 2014. Fabrication and characterization of hardened bodies from Japanese volcanic ash using geopolymerization. *Ceram. Int.* 40(3):4071–4076.

Taylor, H.F.W. 1997. *Cement Chemistry.* London: Thomas Telford.

Taylor, H., Barret, P., Brown, P., Double, D., Frohnsdorff, G., Johansen, V., Ménétrier-Sorrentino, D., Odler, I., Parrott, L., and Pommersheim J. 1984. The hydration of tricalcium silicate. *Matériaux et Construction* 17(6):457–468.

USGS (U.S. Geological Survey). 2017. Iron and steel slag. In *Mineral Commodity Summaries.* Reston, VA: USGS.

van Deventer, J.S.J. 2017. Valorisation of slag in construction: So much more than just technology. In *Proceedings of the Fifth International Slag Valorisation Symposium: From Fundamentals to Applications*. Leuven, Belgium: Katholieke Universiteit Leuven.

van Deventer, J.S.J., Provis, J.L., and Duxson, P. 2012. Technical and commercial progress in the adoption of geopolymer cement. *Miner. Eng.* 29:89–104.

van Deventer, J.S.J., San Nicolas, R., Ismail, I., Bernal, S.A., Brice D.G., and Provis, J.L. 2015. Microstructure and durability of alkali-activated materials as key parameters for standardization. *J. Sustainable Cem. Based Mater.* 4(2):116–128.

Wild, S., and Khatib, J. 1997. Portlandite consumption in metakaolin cement pastes and mortars. *Cem. Concr. Res.* 27(1):137–146.

Xu, A., Sarkar, S., and Nilsson, L-O. 1993. Effect of fly ash on the microstructure of cement mortar. *Mater. Struct.* 26(7):414–424.

Yip, C.K., Lukey, G.C., and van Deventer, J.S.J. 2005. The coexistence of geopolymeric gel and calcium silicate hydrate at the early stage of alkaline activation. *Cem. Concr. Res.* 35(9):1688–1697.

Yliniemi, J., Pesonen, J., Tiainen, M., and Illikainen, M. 2015. Alkali activation of recovered fuel–biofuel fly ash from fluidised-bed combustion: Stabilisation/solidification of heavy metals. *Waste Manage.* 43(Suppl. C):273–282.

Yliniemi, J., Pesonen, J., Tanskanen, P., Peltosaari, O., Tiainen, M., Nugteren, H., and Illikainen, M. 2017. Alkali activation–Granulation of hazardous fluidized bed combustion fly ashes. *Waste Biomass Valor.* 8(2):339–348.

Zeng, Q., Li, K., Fen-chong, T., and Dangla, P. 2012. Determination of cement hydration and pozzolanic reaction extents for fly-ash cement pastes. *Constr. Build. Mater.* 27(1):560–569.

Zhang, M., and Malhotra, V.M. 1995. Characteristics of a thermally activated alumino-silicate pozzolanic material and its use in concrete. *Cem. Concr. Res.* 25(8):1713–1725.

CHAPTER 12.31

Potash

John G. Mansanti

The term *potash* originated from the practice of burning seacoast plants or hardwoods to ashes in iron pots and then leaching the potassium salts, primarily potassium carbonate, from the residue to provide a usable form of potassium. Until about the 1950s, this term referred to potassium carbonate, but today potash is a generic term used to describe various potassium salts used in agriculture and industry. Potassium chloride, potassium sulfate, potassium-magnesium sulfate, and potassium nitrate are the most common of these salts. As potassium is usually the element of interest, potash is measured as the oxide of potassium, K_2O, and the potassium content of ores and products is stated as the weight percent of K_2O present. Its salts are commonly referred to as muriate of potash (MOP), sulfate of potash (SOP), nitrate of potash (NOP), and so on.

HISTORY

Early uses for potash included soap making, glass making, dyeing, baking, and manufacturing of gunpowder. Prior to the late 1800s, the majority of potash was produced by burning hardwood trees and by leaching kelp, with the first U.S. patent issued in 1790 for producing potash from the ash of hardwood trees. In 1840 Justus von Liebig, a German chemist, determined that the addition of minerals to soils was necessary to provide healthy plant life and could significantly improve crop yields. This recognition triggered a surge in demand for potassium. Potash ore was initially discovered in 1857 in Germany, with additional discoveries found in German-controlled Alsace in the early 1900s. The first potash mine and refinery began operating in Strassfurt, Germany, producing both crude salts and chemically refined potash from carnallite ores.

Germany and France supplied nearly the full world demand, with Germany dominating the market until almost World War II. During the potash shortages of World War I, the United States produced potash from lake water in Nebraska, brines in California, kelp beds on the Pacific Coast, waste from molasses distilleries, and alunite ores in Utah. After the war, many nations recognized the strategic significance of potash, and potash mining and processing began in Spain, Russia, and Palestine, with the United States starting mine production in 1931. Prior to 1931, the U.S. potash production was limited to production from Searles Lake in California and other small brine operations, but by 1941, domestic self-sufficiency was achieved when production from underground mines in New Mexico began to exceed all the U.S. potash consumption.

Increased demand, driven by the benefits of higher fertilization, was supported by the addition of production from Russia and Belarusia and, most notably, new Canadian production in the 1960s. Potash production rose steadily from just under 3.0 Mt (million metric tons) of K_2O just after World War II to 31.8 Mt K_2O in 1988. With the collapse of the former Union of Soviet Socialist Republics, Soviet potash entered the global market and created market destruction that extended into the following years. Global potash production ultimately decreased due to oversupply, and after finding a new bottom in 1993, it resumed a slightly lower average annual growth rate of approximately 4% over the next two decades, reaching a value of 40.7 Mt K_2O in 2015 (Figure 1).

OCCURRENCE

Potassium is the seventh most abundant element in the earth's crust and in seawater. It is estimated to constitute approximately 2.6% of the lithosphere and is found in soluble forms in our oceans, lakes, rivers, and subsurface waters. Potassium is also an important constituent of plant and animal life and the seventh most abundant element in the human body. Potash belongs to a family of minerals characterized as evaporites, and its origin is similar to that of salt, gypsum, and soda ash (Warren 2006). Commercially significant accumulations of potassium salts occur in underground deposits formed in past geologic times by the evaporation of ancient bodies of saltwater, inland drainage basins, and shallow sea lagoons. The concentrations of potassium salts in the natural brines of some lakes in the arid and semiarid regions of the continents are also important as sources of potash, for example, the brines of the Dead Sea in Asia Minor, the Great Salt Lake in Utah (United States), and Qarhan (Chaerhan) Lake in China.

The U.S. Geological Survey (USGS) completed a Global Mineral Resource Assessment for potash identifying more than 980 sites with evaporite-related potash mineral occurrences or potassium-enriched brines (Orris et al. 2014). Sites were

John G. Mansanti, President and CEO, Crystal Peak Minerals Inc., Salt Lake City, Utah, USA

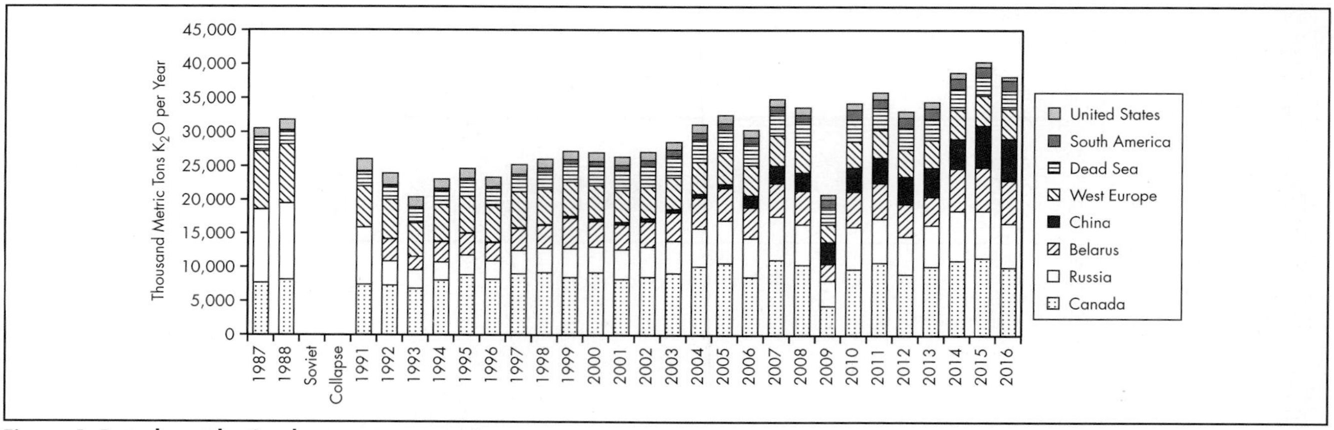

Figure 1 Potash production by country or region

classified by the nature of their occurrence, and the majority of the sites were identified as either stratabound deposits or halokinetic deposits. These two types of deposits represented about 40% each of the total sites inventoried. Stratabound deposits are described as thick, laterally continuous strata with minimal if any deformity. Seventy-five percent of the world potash production comes from this type of deposit. The Elk Point Basin in Canada, the Pripyat Basin in Belarusia, the Solikamsk Basin in Russia, and the Permian Basin in New Mexico are typical of this type of deposit. Halokinetic deposits result from differential loading of stratabound deposits or post-depositional folding, creating complex deformed or altered strata or beds. The halokinetic deposits are made up of salt structures such as anticlines, domes, and diapirs that were created from stratabound deposits. Between 10% and 15% of the world's production comes from this type of deposit. The Carpathian Subbasin of Poland and Ukraine and the Paradox Basin of Utah are typical of this type of deposit. Surface brines represent 7% of the sites, and the balance of the remaining deposits was classified as either a mixture of stratabound and halokinetic or as unclassified.

In general, the stratabound deposits are much larger deposits, typically an order of magnitude larger than halokinetic type deposits and much more uniform. Therefore, stratabound deposits are more conducive to less-expensive bulk mining methods. Where the grades of stratabound deposits vary regionally and may range from 11% to 25% K_2O, the grades of halokinetic deposits are highly variable and contain less than 20% K_2O. The vertical heights of the two deposit types vary significantly, again influencing the mining methods selected for extraction. The stratabound deposits may vary from one to several meters in seam thickness, whereas the potash seam thickness for halokinetic deposits can range from less than a meter to several tens of meters.

In addition to cataloging the location of known potash resources, the USGS highlighted 19 potash-producing areas, or tracts, including surface brine production, from fewer than 20 producing countries; they are listed in Table 1.

POTASH MINERALS

Because of its high chemical reactivity, potassium is never found in nature in its elemental form, and all potash deposits contain a variety of potassium compounds in both soluble and insoluble forms. The most common potash mineralizations are chloride salts, sylvite, and carnallite; the most common sulfate-containing potash mineral is polyhalite. The temporal differences in the chemistry of pre-depositional seawater, primarily magnesium sulfate content, influenced the nature of potash minerals contained in marine deposits (Hardie 1996). Seawater rich in magnesium sulfate resulted in sulfate-enriched deposits typically composed of halite, carnallite, and lesser sylvite, with varied amounts of polyhalite, kieserite, kainite, and langbeinite. Seawaters richer in magnesium sulfate resulted in sulfate-enriched deposits with little to no sylvite composed of halite, polyhalite, kieserite, and carnallite. Seawater low in magnesium sulfate resulted in sulfate-depleted deposits composed of halite, sylvite, and carnallite. More than 60% of potash production comes from the sulfate-depleted deposits (Hardie 1991). A list of common potash and potash-associated minerals is provided in Table 2.

POTASH ORES AND WORLD RESERVES

The principal sources of potash production are from potash ores found as bedded underground deposits. The most important type of ore is sylvinite, a mixture of sylvite and halite. Another important type of ore is carnallitite, a mixture of carnallite and halite. These two types of ores may contain varied amounts of the other primary potassium salts, carnallite and sylvite, respectively, and exist concurrently in some deposits. These ores are typically lower in sulfates but, in addition to clays, may contain minor amounts of anhydrite and occasionally potassium complexes such as polyhalite, kainite, kieserite, langbeinite, and leonite. Frequently, iron oxides occluded within the crystals tend to color sylvinite and carnallitite ores red, while clays contribute to gray or yellow colorations. Sylvinite ores are noted for their relative abundance, high potassium chloride (KCl) content, and ease of beneficiation. The high production rates from Canada, Russia, and Belarusia are predominantly from such ores. Carnallitite ores, however, are less abundant than sylvinite ores, contain less K_2O, and are more involved to beneficiate. Carnallitite ores have been mined since the discovery of potash ores in Germany. They are also typically the primary ore associated with the solar evaporation of some natural brines, such as the brines from the Dead Sea and Qarhan Lake.

The hard salts make up another type of ore, which contains increased sulfate mineralization. *Hartsalz*, a term used in Germany, applies to a mixture which is unique to Germany of

Table 1 Known resources exceeding 0.5 billion t of potash ore

Countries	Potash Tract	Resources, billion t	Deposit Type	Nature of Production*
Canada	Elk Point	>5	Stratabound	Active
Canada	Maritimes	1–5	Mixed	Active
United States	Permian Salado	1–5	Stratabound	Active
United States	Permian Rustler	0.5–1	Halokinetic	No
United States	Bonneville	Brine	Brine	Active
United States	Paradox	1–5	Halokinetic	Active
United States	Holbrook	0.5–1	Stratabound	No
United States	Michigan†	—	Stratabound	Active
Brazil	Sergipe	1–5	Stratabound	Active
Brazil	Amazonas	1–5	Stratabound	No
Argentina	Neuquén	1–5	Stratabound	Suspended
Chile	Atacama	Brine	Brine	Active
United Kingdom	Zechstein, UK	0.5–1	Stratabound	Active
Germany	Zechstein	1–5	Mixed	Active
Spain	South Pyrenean	0.5–1	Mixed	Active
Belarus, Ukraine	Pripyat	1–5	Stratabound	Active
Russia	Solikamsk	>5	Stratabound	Active
Russia	East Siberian	1–5	Stratabound	No
Ukraine, Poland, Romania	Carpathian	1–5	Halokinetic	Active
Kazakhstan, Russia	Cisuralian South	1–5	Halokinetic	No
Kazakhstan, Russia	Pricaspian‡	>5	Halokinetic	Yes
Uzbekistan, Afghanistan Tajikistan, Turkmenistan	Gissar	>5	Mixed	Active
Israel, Jordan	Dead Sea	Brine	Brine	Active
China	Tarim	Brine	Brine	Active
China	Qaidam	0.5–1	Brine	Active
China, Laos	Simoa	—	Halokinetic	Active
Laos, Vietnam	Lao	0.5–1	Mixed	Development
Thailand	Sakon Nakhon	1–5	Mixed	Development
Thailand	Khorat	1–5	Mixed	No
India, Pakistan	Nagaur	1–5	Mixed	No
Eritrea, Ethiopia	Danakil	1–5	Stratabound	Development
Congo, Angola Gabon, Zaire	Lower Congo	0.5–1	Stratabound	No

Adapted from Orris et al. 2014
*Assessment through 2012.
†Classified as active in report, no longer in production.
‡Not highlighted as active in report but listed as mining.

sylvite, halite, anhydrite, kieserite, and sometimes carnallite. Polyhalite, a polymineral, is currently experiencing renewed interest as a source of fertilizer with several new projects in development. Another polymineral, langbeinite, has notable mineralization located in the Carpathian Mountains of Ukraine and in New Mexico and is found in horizontal beds mixed with halite and varying amounts of sylvite. Kainite ore is another type of mixed salt that was commercially exploited in Sicily and Ethiopia and is being reconsidered in the Danakil Depression of Eritrea and Ethiopia.

Known world reserves of potash are estimated to be in excess of 4.3 billion t of K_2O as detailed in Table 3 (Jasinski 2017). The majority of the reserves are located in Canada, Russia, and Belarusia. At the current global consumption rate, more than a century of known potash reserves is identified. The USGS estimates resources at 250 billion t of K_2O.

POTASH PRODUCTS AND THEIR USES

The primary products of the potash industry are MOP, SOP, and sulfate of potassium magnesia recovered from sylvite, carnallite, langbeinite, kieserite, and kainite ores. Muriate of potash makes up 90%–95% of the global potash products and 60% of the domestic potash products. The final products are compatible with current agricultural products, namely granular, standard, fine standard, and the soluble (fine white product for liquid application) grades. The potassium and potassium-magnesium sulfates are also available to the agricultural sector as granular, standard, and fine standard grades.

Agricultural fertilizers account for 85% of the domestic potash consumption and 90%–95% of the global potash consumption. Potash is generally applied to the soil as muriate mixed with phosphorus and nitrogen materials in specific proportions, depending on the needs of the soil. Some crops, such

Table 2 Potash-related minerals and ores

Mineral or Ore	Composition	Percent K_2O	Specific Gravity
Primary Potash Minerals			
Sylvite	KCl	63.2	2.00
Langbeinite	$K_2SO_4 \cdot 2MgSO_4$	22.7	2.83
Kainite	$KCl \cdot MgSO_4 \cdot 3H_2O$	19.3	2.10
Carnallite	$KCl \cdot MgCl_2 \cdot 6H_2O$	16.9	1.60
Polyhalite	$K_2SO_4 \cdot 2CaSO_4 \cdot MgSO_4 \cdot 2H_2O$	15.6	2.77
Accessory Potash Minerals			
Arcanite	K_2SO_4	54.1	2.66
Niter (saltpeter)	KNO_3	44.0	2.10
Aphthitalite (glaserite)	$(K,Na)_3Na(SO_4)_2$	42.5	2.69
Leonite	$K_2SO_4 \cdot MgSO_4 \cdot 4H_2O$	25.7	2.20
Picromerite (schoenite)	$K_2SO_4 \cdot MgSO_4 \cdot 6H_2O$	23.4	2.03
Accessory Non-Potash Minerals			
Anhydrite	$CaSO_4$	0.0	2.98
Gypsum	$CaSO_4 \cdot 2H_2O$	0.0	2.30
Halite	NaCl	0.0	2.17
Kieserite	$MgSO_4 \cdot H_2O$	0.0	2.57
Primary Potash Ores			
Brine	Potassium-enriched brine	>0.5	—
Carnallitite	Halite and carnallite	<15	—
Hartsalz	Halite, sylvite, anhydrite, and kieserite	Typically <15	—
Sylvinitite	Halite and sylvite	Typically <25	—

Adapted from Orris at al. 2014

as tobacco, citrus fruit, coffee, and nuts, have low chloride tolerances and require other forms of potassium, such as potassium sulfate, potassium-magnesium sulfate, or potassium nitrate. Potassium chloride is also used in the conditioning of drilling muds and completion fluids, producing potassium hydroxide, refining aluminum, heat treating steel, electroplating, melting snow and ice, and treating water. The secondary potassium products are potassium hydroxide, potassium carbonate, potassium nitrate, and other miscellaneous potassium salts that are produced in small quantities for special agricultural or industrial uses. The MOP is converted to potassium hydroxide by electrolysis, then to potassium carbonate and a variety of other salts. Potassium carbonate is used in animal food supplements, cement, fire extinguishers, pharmaceuticals, textiles, and high-quality glass, and as a catalyst for synthetic rubber manufacturing.

WORLD PRODUCTION AND CONSUMPTION

Potash production and consumption are generally reported as either metric tons of K_2O or equivalent metric tons of MOP. Generally, potash consumption increases with population growth or, more importantly, the affordability of the growing population (Williams-Stroud et al. 1994). Until recently, potash sales volumes and pricing were closely controlled through two major cartels and the oligarchy of potash producers, but with the breakup of the Belarusian Potash Company in 2013, Canpotex (headquartered in Canada) remains the sole potash cartel. In response to demand, potash production typically varied from 70% to 80% of installed productive capacity. Global potash production was 39.0 Mt K_2O in 2016. This equals about 64.5 Mt MOP. The world potash production statistics for the last 10 years are displayed by country in Table 4. Increased levels of production from China, Belarus, Russia, and Chile are evident in the production trends, while the levels of production from the Dead Sea, Europe, and the United States are declining. Canada remains the top potash producing country at 29% of the total world production, followed by Russia, 17%, Belarusia, 17%, and China, 13% (Natural Resources Canada 2016). These four countries represent 76% of world potash production. The contributions of the major potash producing companies are displayed in Figure 2.

Table 3 World potash reserves, in million metric tons of K_2O

Country	2016	%
Belarus	750	17.4
Brazil	13	0.3
Canada	1,000	23.3
Chile	150	3.5
China	360	8.4
Germany	150	3.5
Israel	270	6.3
Jordan	270	6.3
Russia	860	20.0
Spain	20	0.5
United Kingdom	70	1.6
United States	270	6.3
Other countries	90	2.1
Total	4,300	100.0

Source: Jasinski 2017

Because of capital investments and increased productive capacity, potash is currently in an oversupply situation. The Food and Agricultural Organization of the United Nations (FAO 2015) estimated 2014 potash demand at 34.87 Mt K_2O, about 58.12 Mt MOP. The four primary potash consuming nations are China, Brazil, the United States, and India, representing 65% of the world consumption. Significant capital was invested to increase productive capacity over the last decade, with major projects in Canada, Russia, China, and Belarusia; many of these projects are still in progress. The FAO forecasted a compound annual rate of growth (CARG) of 4.6% for potash capacity between 2014 and 2018 because of these investments. This compares to a projected CARG of 2.6% for potash consumption over the same period. The projected volume of demand over this period is distributed as follows: 56% Asia (23% China, 17% India), 27% Americas (18% Brazil), 11% Europe, 6% Africa, and 0.4% Oceania. The current oversupply situation is expected to be exacerbated by these relative projections.

MINING

As described earlier, many potash deposits are found in nearly horizontal strata of various thicknesses and at depths ranging from 300 m to 3,000 m below the surface. In some localities, however, the beds or portions of beds may be folded, faulted, or tilted. The predominant mining methods for potash

Table 4 World potash production, thousand metric tons of K_2O

Country	2007	2008	2009	2010	2011	2012	2013	2014	2015	2016 (estimated)	2016 vs. 2007	Increase (Decrease)
Belarus	4,972	4,968	2,485	5,223	5,306	4,840	4,243	6,290	6,470	6,400	127%	1,318
Brazil	424	383	453	448	424	425	430	311	293	300	73%	(113)
Canada	11,085	10,455	4,297	9,700	10,686	8,976	10,140	11,000	11,400	10,000	99%	(85)
Chile	515	559	691	964	861	1,053	1,050	1,200	1,200	1,200	233%	685
China	2,600	2,750	3,200	3,600	3,800	4,100	4,300	4,400	6,200	6,200	169%	1,800
Germany	3,637	3,280	1,825	3,024	3,215	3,149	3,200	3,000	3,100	3,100	82%	(637)
Israel	2,182	2,170	1,900	2,080	1,820	2,100	2,100	1,770	1,260	1,260	81%	(412)
Jordan	1,096	1,223	683	1,185	1,378	1,092	1,080	1,260	1,410	1,400	115%	164
Russia	6,429	5,992	3,727	6,283	6,498	5,563	6,100	7,380	6,990	6,500	115%	951
Spain	427	435	400	415	420	420	420	715	690	700	167%	288
United Kingdom	427	411	411	427	470	470	470	610	610	600	143%	183
United States	1,100	1,100	715	930	1,000	900	960	850	740	520	77%	(250)
Total	34,900	33,700	20,800	34,300	35,900	33,100	34,500	38,800	40,700	39,000	111%	3,900

Adapted from Jasinski 2012–2017

Note: World totals, U.S. data, and estimated data are rounded to no more than three significant digits; may not add to totals shown.

have been adopted from the coal mining industry; the modified room-and-pillar and the longwall methods. Some potash mines employ conventional coal mining sequences of undercutting, drilling, blasting, mucking, and transport to hoisting stations. Most modern mines are highly mechanized and use heavy-duty mechanical cutting and boring machines, automatic belt conveyors for ore transport, and high-capacity automated skips for hoisting. Conventional mining depths are limited by the plasticity of salt, and the deepest conventional potash mining occurs at depths of 1,300–1,500 m in Germany (e.g., Niedersachsen-Riedel, Sigmundshall, and Zielitz) and to 1,400 m in Great Britain (e.g., Boulby). Conventional mines in Saskatchewan (Canada) are operating at depths of approximately 1,000–1,100 m, and due to the characteristics of the salts, solution mining is preferred for deposits with depths greater than 1,100 m.

Potash can be extracted from deeper deposits by utilizing solution mining techniques. Typically, solution mines extract potash from depths beyond the reach of practical conventional salt mining, such as Mosaic's Belle Plaine mine (Saskatchewan) and Intrepid Potash's Moab (Utah) mine at 1,600 and 1,400 m deep, respectively. Potash is also extracted when unintentional or intentional flooding of prior mine workings takes place, which happened at Potash Corporation of Saskatchewan's (PCS's) Patience Lake (Saskatchewan) and Intrepid's Moab and HB (Utah and New Mexico, respectively) mines. Solution mining of carnallite occurs at shallower depths in Germany. Solution mining of kainite is being developed in east Africa at depths of 1,400–1,800 m.

Lake deposits may exist as brine, crystallized salts, or a mixture of the two phases. The extraction of these materials is usually accomplished by dredging and pumping or by draining ponds and mechanically harvesting the crystals with scrapers or similar earthmoving equipment. The Dead Sea operators, Israeli Chemicals Ltd. (ICL) and Arab Potash Company, Qinghai Salt Lake Potash Company Ltd. (QSL) in China, and Sociedad Quimica y Minera (SQM) in Chile represent the larger producers of potash from brines. Domestically, Compass Minerals and Intrepid Potash extract potash from the Great Salt Lake and Bonneville (Utah), respectively.

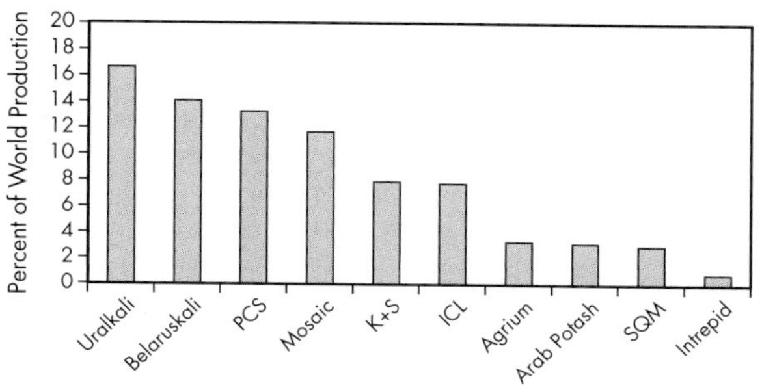

Note: Does not include China at 17% combined. PCS production does not include ICL, Arab Potash, or SQM production.

Figure 2 Potash production by company

BENEFICIATION

Every mineral has distinct physical and chemical properties that may be exploited to separate it from other minerals or substances. A key difference for many evaporites is mineral solubility. High solubility in water presents both opportunities and challenges. The increased solubility of sylvite relative to halite as a function of temperature provides a significant opportunity to separate the two minerals. Leaching with single saturated brines allows the removal of halite from sylvite and an upgrade of sylvite concentrates with minimal sylvite losses. However, all wet processing must be conducted in at least a doubly saturated brine, sodium chloride and potassium chloride, to avoid the loss of sylvite. The high solubilities of potash minerals create a series of challenges less common in the processing of most other minerals. Some such challenges are as follows:

- The higher specific gravity and increased viscosities of brines relative to fresh or mill water slurry systems have different impacts on hydraulic separations, sedimentation, bubble formation, and so on.
- The pseudo-phase equilibrium with normal temperature fluctuations can result in dissolution and recrystallization

of fine crystals dissolving targeted minerals or changing the surface chemistry of mineral particles.
- The unintended salting of screen openings, orifices, venturis, and other air–brine interfaces creates operational issues.
- The hygroscopic nature of sylvite may result in the development of salt bridges at higher relative humidity and create screen blinding, baghouse blinding, and materials handling difficulties.
- Covered storage is required due to product solubility, and a lower relative humidity is important to sustain the quality of inventoried potash products.
- The high ionic strength of the brines impacts the selection and effectiveness of surface modifying reagents, and the high ionic strength creates an environment where fine halite and sylvite will self-agglomerate due to opposite surface charges and make the recovery of minus 0.1 mm particles impractical.
- The high ionic strength of the brines also creates a corrosive process environment due to the higher chloride content of the brines, thus the sustaining capital requirements for a salt plant are higher than for other conventional mineral processing plants.

Sylvinite Ores

Most potash production comes from sylvinite ores, primarily in Canada, Russia, and Belarusia. Flotation and crystallization are the two principal processes for the beneficiation of sylvinite ores. Flotation is the lower-cost alternative due to the higher energy and maintenance costs associated with hot dissolution of sylvite prior to crystallization. Dense media separation, a third process, is minimally used to scalp coarse potash ahead of flotation and has been limited to ores with coarser liberation sizes.

The flotation process for the recovery of KCl from sylvinite ores was first commercially implemented at Carlsbad, New Mexico, in the early 1930s. A fatty acid flotation of halite from sylvite using a lead chloride activator was initially used. The differential flotation of the sylvite in preference to halite was initially used at Carlsbad in 1940 and is the universal practice in processing of sylvinite ores today. The development and use of the primary aliphatic amines as KCl collectors was critical to the development of this process. The larger potash-producing plants use flotation, several plants between 2 to 3 Mt MOP/yr. Brownfield expansions at Saskatchewan (Canada) plants Rocanville and Esterhazy project annual average production of 6.0 Mt and 7.1 Mt MOP/yr, respectively.

Hot leach/crystallization was the first commercial process for refining potash; it was developed in Germany and became the standard method for early European producers. Crystallization was adopted in 1916 for commercial production of KCl from the complex brines of Searles Lake, California, and the first vacuum cooling was introduced in 1918 (Schultz et al. 2005). The second crystallization plant in North America was erected in 1931 near Carlsbad, New Mexico. Crystallization of KCl yields a white, higher purity product with a more uniform particle-size distribution than that obtained by flotation. The process is normally selected for recovery of KCl when elevated levels of insoluble materials prevent reasonable recovery by desliming and flotation or when higher grade, 62% K_2O (98% KCl), product is preferred for market reasons. Many operators may use small crystallization plants to augment production or to produce quality potash from low-grade dusts, fines, or brines. Stand-alone crystallization plants are less common for potash recovery; however, Mosaic's Belle Plaine, Belaruskali's Soligorsk No. 4 (Belarus), and Uralkali's Solikamsk No. 1 and Berezniki No. 4 (Russia) are such examples. These plants produce from 1–2 Mt MOP/yr.

Plant recoveries usually are 85%–90% for flotation, with the losses of KCl occurring in separated slimes, flotation tailings, and necessary makeup of process brine to compensate for the brine lost as moisture in filtration and drying. Some plants combine flotation and crystallization to overcome variations in liberation characteristics of the ore. These plants diversify their muriate sales by marketing agricultural, soluble, and chemical grades of KCl. The major component of plant production, however, is a flotation concentrate that is sized and converted into granular and standard muriates.

Crushing and Grinding

Potash ores are friable and require careful selection of size-reduction equipment to minimize production of undesirable fines. Depending on the process and the market preference for final products, sizing is as important as size reduction in flotation or gravity plants. The first stage of crushing takes place underground usually by portable breakers near the mine face. The maximum size of ore hoisted is 100–125 mm. A somewhat recent trend in North America is to improve skip factors and feed sizing by utilizing underground roll crushers to deliver –50-mm material to the skips. Handling of the hoisted ore on the surface requires secondary crushing, usually a hammer-type rotary crusher preceded by a suitable grizzly typically delivering a –9.5-mm product to the final stage of size reduction, either wet milling using a KCl-saturated brine or tertiary dry crushing. Conventional rod mills in closed circuit with screens or cyclones are used for the wet milling option. In North America, many of the rod mills have been replaced by cage mills for this stage of size reduction. Most rod mills used for the regrind of rougher tails have also been replaced with cage mills to better control the generation of fines prior to flotation. The tertiary dry crushing stage, when employed, is accomplished by impact crushers in closed circuit with vibrating screens.

KCl crystals larger than 6 mesh cannot be floated efficiently. Therefore, the largest particle size of flotation feed is –6 mesh even for ores that have coarse liberation sizes. Potash particles float best between 0.1 and 1.2 mm; Canadian ores are typically reduced to passing 6–8 mesh, 3.35–2.36 mm (Strathdee et al. 2007). New Mexico limits ores to the maximum size of 2.4 mm and usually less than 1 mm, while German sylvinite ores are ground to a maximum particle size of 0.8–1.0 mm (Schultz et al. 2005). For crystallization, ore is crushed to 4 mesh, 4.75 mm, or finer.

Scrubbing

The slurried mill feed is subjected to a vigorous scrubbing, in either agitated tanks or rotating tumblers to remove insoluble particles from the crystal surfaces, to break up amorphous clay agglomerates, and to disperse freed clay slime in the brine medium. The ores are usually scrubbed in stages, with each stage followed by a screening step to remove the slime-laden brine from the scrubbing circuit. The screen oversize is repulped with clear brine at each succeeding stage. Pulp densities are critical in scrubbing, and densities of 60%–70% solids are preferred.

Mechanical desliming is the most common method of desliming. The usual practice is to separate the scrubbed ore into the coarse (+20 mesh, 0.85 mm) and fine (–20 mesh) portions and deslime these separately. Sieve-bend screens, cyclones, or mechanical classifiers serve equally well in desliming coarse fractions. Hydraulic classifiers (e.g., the CrossFlow classifier) have recently gained popularity in desliming and classifying coarser sized feeds prior to coarse flotation or in the reprocessing of rougher tails. Good desliming of the fine fractions requires efficient removal of –270 mesh, 0.05-mm solids. Mechanical classifiers, hydroseparators, cyclones, or combinations thereof are used. Desliming can be aided by the use of dispersants. A typical flow sheet of a scrubbing and mechanical desliming circuit is displayed in Figure 3. Desliming typically reduces insoluble levels in flotation feed to 0.8%–1.5% insoluble residue. Clay depressants generally work best between 1.5% and 2.0% insoluble residue.

At insoluble concentrations >4%–5%, mechanical desliming becomes less effective and potash recoveries suffer. Flocculation and insoluble flotation or hot leach and crystallization are used for ores with higher insoluble contents. For flotation, two stages of flotation are usually utilized with reagent conditioning prior to each stage as the agitation in flotation liberates additional insolubles. Typically, an ether amine is utilized as a collector for the flocculated insoluble minerals. The Russian plants in Berezniki and Solikamsk, which may contain as much as 8% insoluble, utilize this process as did the original Vanscoy and Rocanville plants in Canada.

The presence of carnallite in sylvinite ores produces changes in scrubbed ore and plant brine composition. Carnallite is dissolved, introducing magnesium chloride to the brine and causing partial crystallization of KCl and sodium chloride (NaCl) from the brine. At magnesium chloride ($MgCl_2$) concentrations as low as 1%, sylvite recoveries begin to be adversely affected, especially for particles larger than 0.5 mm or at higher temperatures (Konoplev et al. 2014). Because of increased brine viscosity, increased reagent consumption, and decreased recoveries, magnesium concentrations are limited to <2.5% magnesium chloride in the brine. The negative impact of insoluble materials is exacerbated by increasing $MgCl_2$ concentrations. Due to a combination of geotechnical issues underground and increased magnesium concentrations in plant brines, ore is usually limited to less than 6% K_2O as carnallite.

Potash Flotation Reagents

A significant body of work has been generated in an attempt to understand the fundamentals of potash and soluble salt flotation. Summaries of the evolution of potash flotation can be found in Miller et al. (2007), Laskowski (2013), and Ozdemir et al. (2011).

Initial potash collectors were primary aliphatic amines produced from beef tallow, either saturated or unsaturated, containing predominantly C_{16} to C_{18} amines. The secondary, tertiary, or quaternary amines do not support hydrogen bonding at the mineral surface and therefore do not float sylvite. In response to coarser flotation and temperature variations, the inclusion of fish-based amines and other longer chain amines may make collector blends ranging from C_{16} to C_{26} amines. Sylvite flotation is affected by temperature, and it is typical

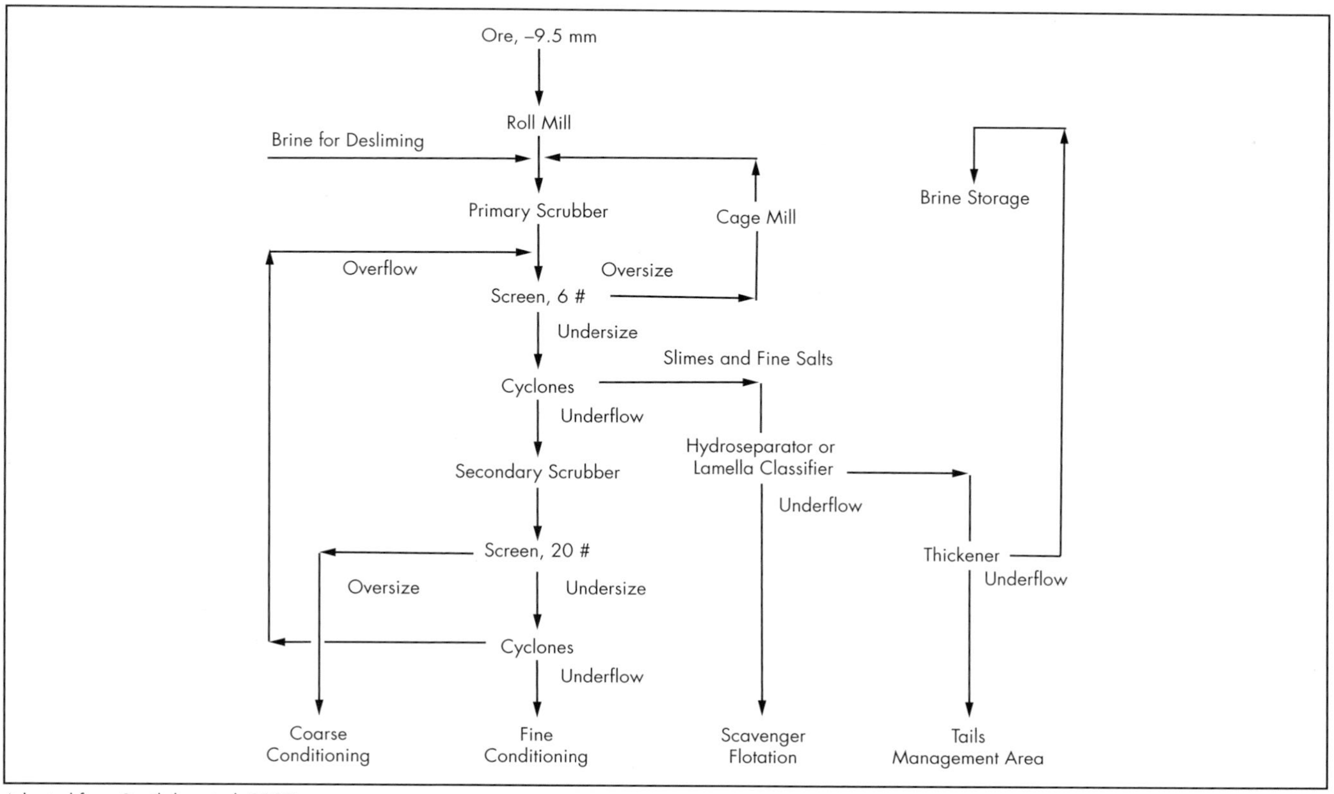

Adapted from Strathdee et al. 2007

Figure 3 Typical desliming flow sheet

Table 5 Reagent consumption ranges, in grams per ton processed

For Potash Ores	Strathdee (1982)		Zandon (1985)		Schultz (2005)		Miller (2007)	
	Low	High	Low	High	Low	High	Low	High
Depressant, starches	—	510	500	900	—	—	—	—
Depressant, others	20	70	75	125	—	—	200	300
Collectors, amines	40	100	50	150	40	80	50	60
Flotation aids, oils	10	85	75	200	40	—	30	40
Flotation aids, others	—	—	20	50	—	—	—	—
Frothers	20	23	25	50	20	—	20	0
Flocculants, brine recovery	—	—	40	75	—	—	—	—
Flocculants, filters	—	—	50	125	—	—	—	—
Anticaking, amines, etc.	30	100	15	60	12	75	—	—
Dust control	150	400	150	375	75	1,250	—	—

Source: Strathdee et al. 1982; Zandon 1985; Schultz et al. 2005; Miller and Nalaskowski 2007

for operators to alternate to longer chain amines during the warmer summer months. The flotation of coarse particles is difficult and is improved with the support of a promoter such as extender oil, especially when the oil is emulsified with the amine.

Slime depressants include starches, guar gum, dextrins, and synthetic polymers, such as polyglycols and polyacrylamides. Urea formaldehyde resin is also used as an effective cost-competitive alternative to guar, and a reduction of 25%–40% in amine consumption may result with its use (Titkov et al. 2005).

Flotation amines have frothing characteristics, and as a result, frother reagents often act more as a dispersant for the collector than as a frother. Methylisobutylcarbinol (MIBC) is employed mainly as a froth modifier in potash flotation, inhibiting the formation of excessive amine froth. In consideration of health and safety issues, hexanol or polyglycols may be preferred to manage froths. Glycol ethers, pine oil, and dioxane alcohols are used in Russia with a preference for higher dosing of the more dispersive type frothers, the glycol ethers. Guar and polyacrylamides are employed as flocculants in slime thickeners, in clarification of liquors for crystallization, and as internal flocculants in slime centrifuges.

The application of reagents in potash processing merits special attention. A slimes depressant, collector, and modifier are added to the slurry prior to flotation and preferably in the order given. Emulsifying the amine, modifier, and some frother prior to addition to conditioning improves flotation and reagent dosage. The fines require less collector but more depressant, and there is considerable advantage in separate conditioning systems for the coarse and fine fractions of the ore. Conditioning at high density is favored. Conditioning in drums, screw conveyor mixers, or launders is suggested for coarse flotation feed. Ranges of reagent usages in potash flotation plants are reported in Table 5.

Flotation Systems

Flotation kinetics are inversely related to particle size. Potash flotation systems generally consist of roughing, cleaning, and scavenger circuits. Some ores require two to three stages of cleaning. There is an emphasis on the capability of the circuit to float and recover coarse KCl. General experience is that only the intermediate and fine fractions of the rougher concentrate are readily refloated in the cleaner. The coarse KCl crystals, 6 × 14 mesh, fail to float efficiently for a second time unless appropriately debrined and conditioned anew. These particles are recovered after classification, utilizing either regrind or coarse fraction flotation (Figures 4 and 5). Recent developments with teeter bed, or hydrodynamic flow-assisted, flotation provides a highly efficient alternative to successfully recover coarse particles from either a rougher feed or rougher tail.

Ores containing sizable amounts of free coarse KCl may have their coarse and fine fractions floated separately as in Figure 5. The coarse circuit may be augmented with scavenging flotation where the rougher tail is screened and the oversize is conditioned and floated. The scavenger concentrate may be combined with the rougher concentrate or used as crystallizer feed while the scavenger tails report to tails. Scavenger flotation is also used for deslimed fines or centrate from centrifuges generating a dirty concentrate for brine makeup or crystallizer feed.

Final flotation concentrates contain minor amounts of middlings, fine NaCl, and traces of insolubles; grades of 90%–92% KCl are common.

Flotation Equipment

The Denver DR 300 has been a workhorse in the industry because of its ability to recover coarser potash particles. Until recently, it was the preferred machine for coarse flotation and is still the preferred equipment for coarser mixed flotation. In Canada, Fairbanks Morse and Gallagher cells are also used. Agitair and Mechanobr cells are commonly used in eastern Europe (Schultz et al. 2005). When a coarser, more uniform process stream can be created by preclassification, especially hydraulic preclassification, Eriez's HydroFloat has been used to recover coarser potash particles, 1–4 mm, from plant feed or rougher tails.

Mechanical or column cells are used for the flotation of intermediate-sized and fine-sized particles, 0.1–1.2 mm, and newer facilities or modernized plants are using column cells in the cleaning and recleaning circuits. Pneumatic flotation cells, such as the Imhof cell, have become common for cleaner flotation in Belarusia and are being used in Germany (Imhof et al. 2002; Fertilizer International 2010). The Jameson cell has successfully replaced mechanical cells for slimes and cleaner flotation, initially at Boulby (United Kingdom) and sequentially at the Dead Sea Works (Israel) and in Canada (Hall and Harrison 1995).

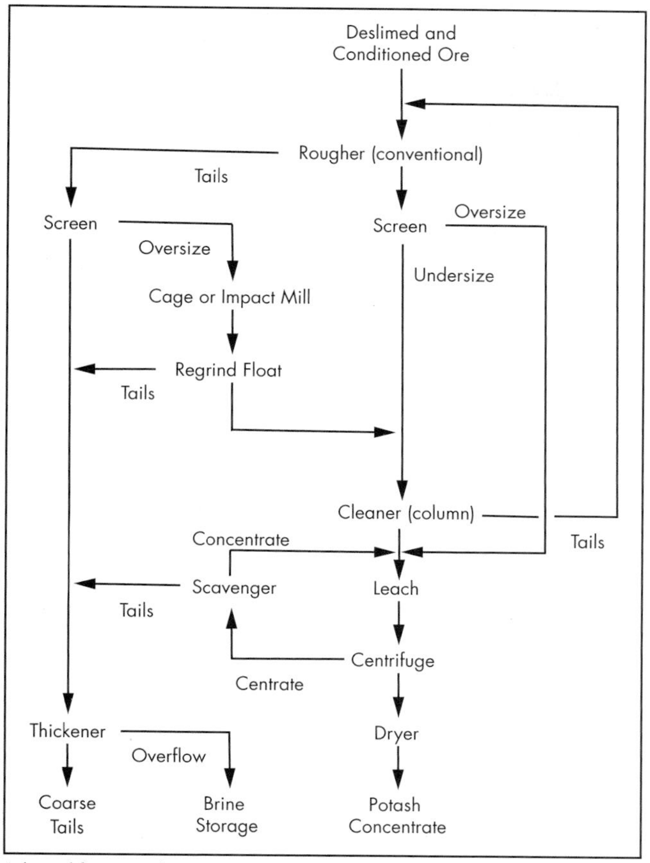

Adapted from Hatch 2012

Figure 4 Typical flotation option

Flotation–Crystallization System

As was noted previously, many potash plants find it necessary to combine flotation and crystallization methods in treatment of their ores for optimum KCl recoveries and acceptable grades of all products. In some cases, coarse deslimed ore reports to flotation and the fines report directly to crystallization. Crystallizer feed may vary (e.g., hot rinse brines from tails and slimes, scavenger flotation concentrates, flotation concentrates from product centrate brines, potash dusts, and product losses from materials handling losses).

Crystallization

The crystallization of KCl from hot saturated brines is governed by equilibrium solubilities of KCl, NaCl, and other brine components, if any, in cooled solutions. The principle is demonstrated by the solubility–temperature relationship for a simple KCl-NaCl-H_2O system (Figure 6). The solubility of KCl increases rapidly with the increase in temperature, while that of NaCl decreases. Conversely, the solubility of KCl decreases sharply as the temperature is lowered, whereas the concentration of NaCl changes very little. This differential solubility of the two chlorides at lower temperatures produces supersaturation of the solution with respect to KCl which allows its separation and recovery by fractional crystallization.

The presence of other salts in the system alters the solubility of KCl and NaCl and therefore the equilibrium solubilities of all the salts present in the feed brine to crystallizers. The rate of crystallization of KCl from the complex mixture of salts must be considered in the design and operation of the crystallizer plant. Solubility data for a wide range of concentrations of KCl, NaCl, Mg, SO_4, and Ca are available in reference literature. A classic solubility reference for many salt systems has been written by D'Ans (1933).

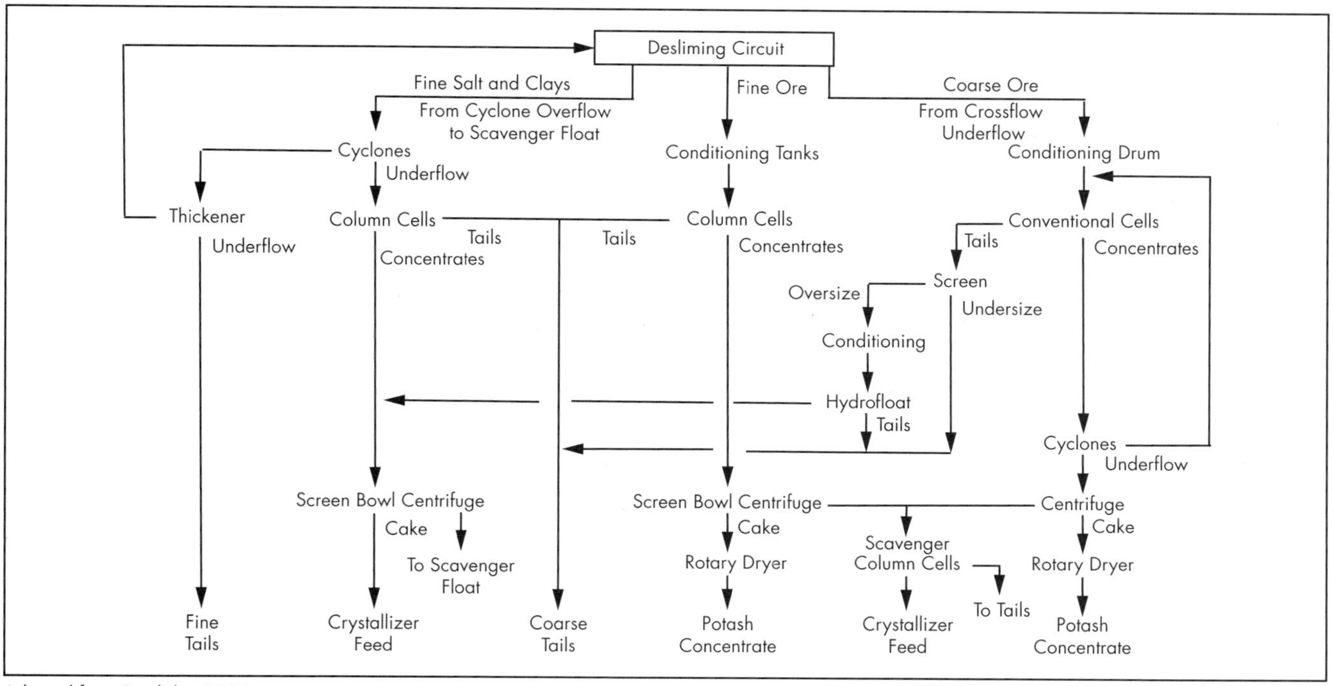

Adapted from Strathdee 2007

Figure 5 Potash flotation with separate coarse and fine flotation

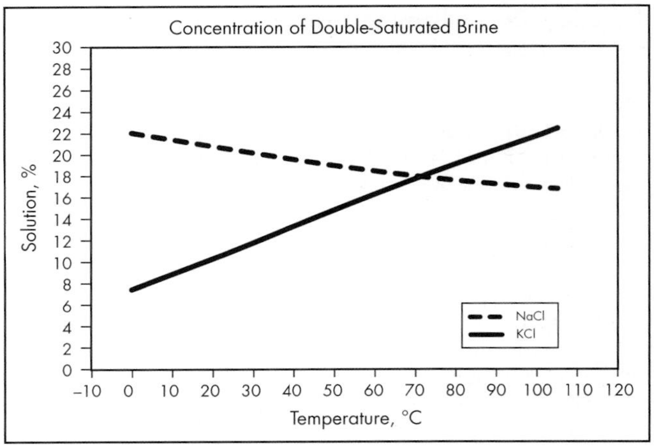

Source: Garrett 1996; Siedel 1919

Figure 6 Solubilities in a NaCl–KCl system

Crystallizers

The majority of modern potash operations use continuous vacuum crystallizers, which develop the needed degree of supersaturation by cooling the hot saturated brines. The cooling is produced by the evaporation of water from the hot brines and is governed by a vacuum in the crystallizing vessel. The decrease in brine temperature reduces the solubility of KCl, and its excess crystallizes out as fine nuclei that subsequently grow to the crystal size desired. After crystallization, the brine reverts to a saturated state, and the cycle is repeated with fresh hot feed. Uncontrollable formation of fine seed crystals or spontaneous nucleation takes place when supersaturation is maintained at too high a level, reaching into the labile field of supersaturation. The resulting crystal crop will be confined to extremely fine sizes, which are undesirable. Production of larger crystals requires operation of the crystallizer at a lower level of supersaturation, the metastable field in which a limited formation of new nuclei occurs and conditions are favorable to the growth of existing crystals.

The continuous crystallizing plants contain several cooling stages in series, usually between four to eight stages. The last stage will have the highest vacuum and the coolest brine. In some designs the entire crystallizer body may be under vacuum; other designs favor a separate vacuum or "flash" vessel directly above the crystal suspension chamber of the crystallizer body proper. Such vaporizer discharges through a seal leg into the lower chamber, which is at atmospheric pressure. The vapor released in the crystallizing vessels is condensed in either barometric or surface condensers; in either case a significant quantity of heat is recovered when recycle brine is used as a coolant prior to its return to the dissolvers.

Typical Crystallizer System

A simplified flow sheet for a crystallization circuit for the basic case of a sylvinite ore is shown in Figure 7. Hot recycle brine between 95° and 105°C is used to dissolve the solid KCl into a liquor that is clarified and reports to crystallization. The undissolved solids in the clarifier underflow, primarily NaCl and insoluble materials, are debrined and discarded. Depending on the level of insolubles in the ore, multiple stages of thickening and countercurrent decantation may be utilized to minimize soluble KCl losses to the tails. The filtrates from both debrining operations are recycled to the system. Recycling of the brine may lead to a buildup in the concentration of undesirable ions in solution, which can be detrimental to crystallization. General plant practice is to minimize the buildup of harmful impurities by a controlled bleeding of brine from the system.

When chemical-grade KCl is desired, the concentration of NaCl and other impurities in the brine must be limited. This is usually accomplished by redissolving the commercial grade of KCl in water to remove the objectionable impurities and recrystallizing the purified brine in a separate circuit

Hot leaching and crystallization is employed with sylvinitic, carnallitic, and hartsalz ores. Operational leach and brine management practices are significantly different between the different ore types as magnesium concentrations must be closely managed in the case of carnallite and hartsalz ores. Schultz et al. (2005) provide a good reference on crystallization with a relatively detailed description of the impacts of other ions, especially magnesium. Crystallization is also utilized to recover other potassium salts, such as potassium sulfate, from schoenite and kieserite.

Other Methods of Concentration

The dense media process has been successfully employed to recover coarser KCl crystals less suited to flotation, such as at Mosaic's Esterhazy operation where the ores are known for their relatively coarser liberation size compared to the other Saskatchewan ore zones, 9.5 mm versus 1.5 mm, respectively (Perucca 2003).

The separation is favored by the difference in specific gravities of the ore components (i.e., 2.17 for halite and 1.98 for sylvite). The crushed middlings and –10 mesh product from the screen undersize are combined and proceed to flotation.

Several other methods were used or considered for the concentration of potash ores, such as tabling and electrostatic separation (LeBaron and Knopf 1958).

Carnallitic Ores

Carnallite is considered a minor impurity in North American sylvinite ores, but it is a major potassium-bearing mineral in several European deposits, notably in Germany and Solikamsk in Russia, and evaporative products from lake brines of the Dead Sea and the Qarhan Lake in China's Qaidam Basin. ICL's Dead Sea Works has been extracting potash from carnallite salts since 1955 and currently produces around 2 Mt MOP/yr. APC (Jordan) produces at almost the same level of production.

Carnallite was the first potash ore refined at Strassfurt, Germany, in 1861. Following hand sorting as the primary form of beneficiation, leaching and crystallization were used to obtain a purer product, initially up to 80% KCl. Commercial flotation of carnallite was not introduced until the 1980s. Today several processes are combined in the recovery of KCl from carnallitic ores or raw salts. Generally, some combination of selective decomposition of carnallite, hot leach and crystallization, cold crystallization, sylvite flotation with alkylamines, and reverse flotation of halite with alkylmorpholines is used.

Elevated magnesium levels in the flotation brine activate insoluble materials relative to alkylmorpholines, and therefore the Russian operations in Perm favor desliming by flotation where both halite and insolubles can be removed from carnallite (Titkov et al. 2014).

The known German practice for processing carnallite–sylvite ores employs hot leaching of the ore and recovery of the low-grade KCl product (80% KCl or lower) by vacuum

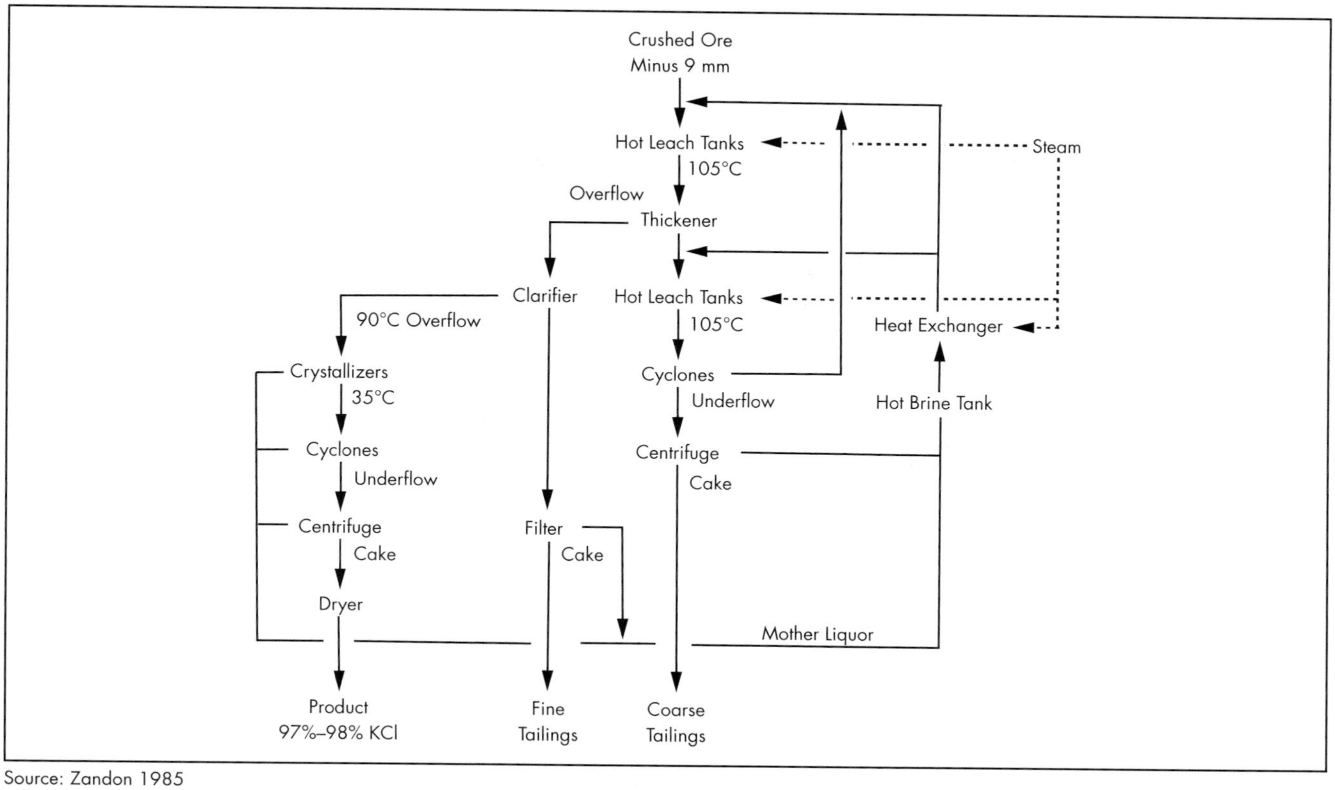

Source: Zandon 1985

Figure 7 Hot leach/crystallization of sylvinite ores

crystallization. The essential feature of this particular process is the control of $MgCl_2$ concentration of the brine below the point at which crystallization of carnallite might occur. Such control is usually achieved by a bleed of the lean circuit brine containing $MgCl_2$ relative to the amount of carnallite in the ore.

Raw carnallite salts are recovered from the solar evaporation of Dead Sea brines. Israel began recovering potash from these carnallite salts by utilizing selective decomposition of carnallite followed by hot leaching and crystallization of sylvite to remove the potash from halite. This became the standard approach for recovering potash in the region until the development of the cold crystallization process (Alamer 2009; Garrett 1996; Wisniak 2002; Zbranek 2013). In hot crystallization, raw carnallite salts containing about 16% halite are recovered from the solar ponds, debrined, and mixed with weak brine. The more soluble $MgCl_2$ is selectively dissolved from the carnallite double salt while undissolved potassium and sodium salts remain or recrystallize as sylvinite. The washed sylvinite is hot leached at 105°C with mother liquor from crystallization and thickened. Sylvite is recovered from the thickener overflow in multiple stages of crystallization. Including pond losses, regionally about 50% of the KCl that is introduced to the solar ponds is recovered as product (Zbranek 2013).

A cold crystallization process was developed by the Israelis to lower costs. The carnallite crystals from the evaporation of Dead Sea brines are coarser than the halite crystals. By exploiting this fact, about 25% of the raw carnallite is removed as a high-grade carnallite oversize and bypasses halite flotation prior to cold crystallization, as detailed in Figure 8. A sylvite cake containing less than 0.7% $MgCl_2$ is produced from the process.

Almost all MOP production in China is extracted from carnallite salts evaporated from lake brines at Lake Qarhan in the Qinghai Province. These carnallite crystals are smaller than the carnallite crystals produced from Dead Sea brines and therefore influence the Chinese processing options. Two main methods are used to recover muriate from these carnallite salts: selective decomposition–flotation and reverse flotation–selective decomposition. Due to better product quality, coarser product size, improved debrining properties, and higher recoveries, the industry is migrating from the older selective decomposition–flotation process to reverse flotation-selective decomposition. QSL is the largest producer and uses the reverse flotation process (Wang et al. 2014; Zhang et al. 2009). Although there may be as many as 50 smaller operators producing small quantities of MOP from the lake (Zhang et al. 2009), many of these smaller operators use cold decomposition–flotation. A comparison of the two processes is provided in Table 6.

Hard Salts and Polymineral Ores

The production of potash from hartsalz represents some of the earliest potash production techniques. The product mix from hartsalz can vary depending on the composition of the ore and the selection of the processing methods. Potash products are generally MOP, SOP, kieserite, and epsomite. Since the first commercial introduction of flotation in 1953 and electrostatic separation in 1974, crystallization, electrostatic separation, and flotation typically have been used in the recovery of products. For ore containing greater than 15%–20% carnallite,

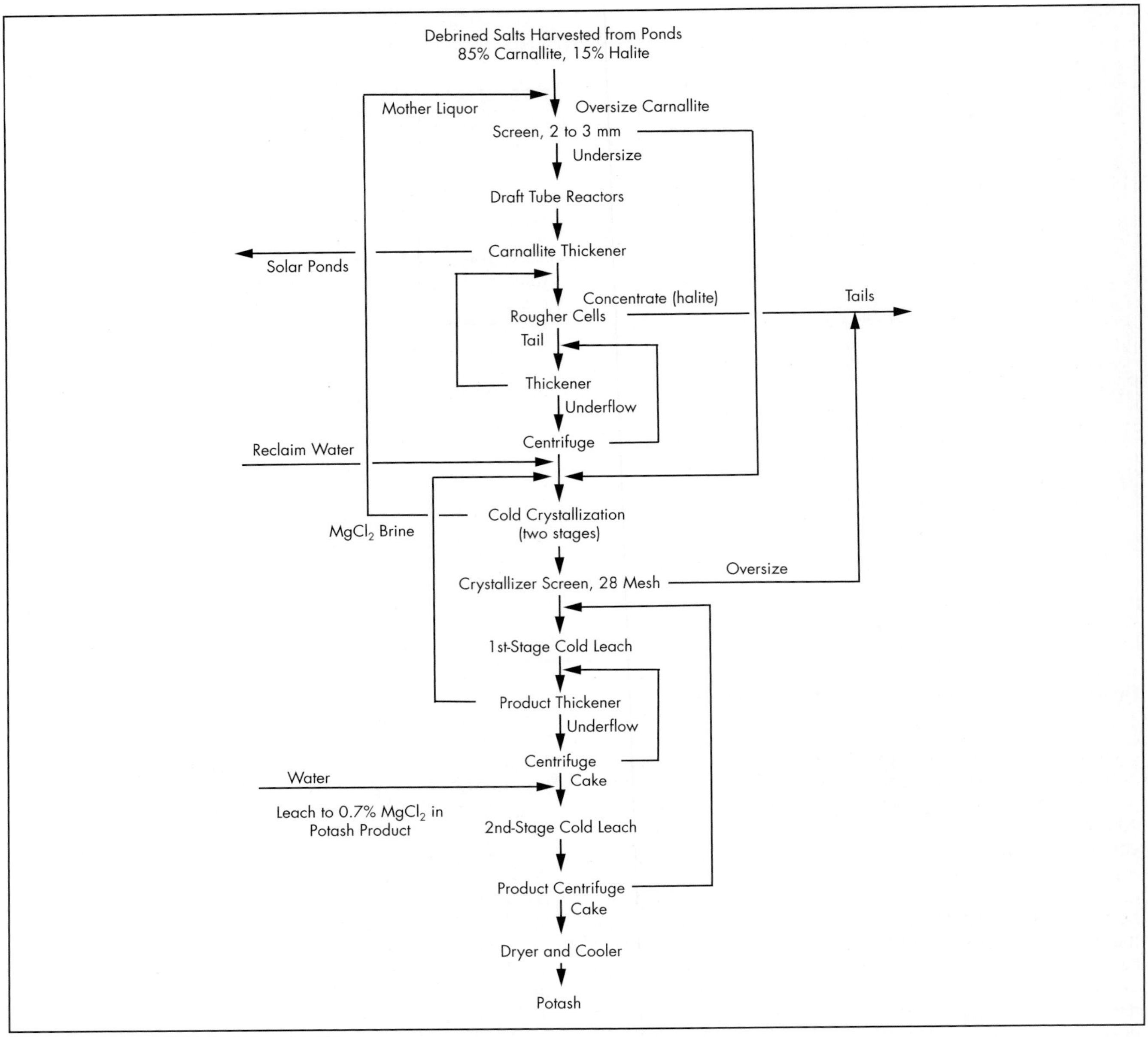

Adapted from Alamer 2009 and Zbranek 2013

Figure 8 Cold crystallization for Dead Sea salts

Table 6 Comparison of Chinese process options at Qinghai Province

Production Attributes	Decomposition and Sylvite Flotation	Halite Flotation and Decomposition
Product grade, % KCl	86–92	90–95
Overall recovery, %	40–50	59–62
Particle size distribution, mm	P_{80} of 0.088	D_{50} of 0.2
Temperature sensitivity	High	Low
Collector dosage, g/t	450–500	50
Product moisture, %	6	3
Production, t/yr in mm	0.5	2.5

Adapted from Wang et al. 2014 and Zhang et al. 2009

cold decomposition is typically used to initially remove higher levels of magnesium.

The hot leach crystallization process for ores containing carnallite requires close control of magnesium content in the mother liquor and yet still produces lower quality MOP. By using a quick hot leach, the solids to liquid ratio, and coarser particle sizes, sylvite and carnallite can be selectively leached from the slower leaching kieserite. A liquor bleed is used to control $MgCl_2$ concentrations, or the production of a 60% K_2O product is not possible without excessive leaching. The flotation of kieserite from the hot leach residue results in a final halite residue.

Electrostatic separation is successfully used at particle sizes >0.1 mm. One processing sequence is to remove the halite and then separate and upgrade kieserite from the potassium salts. A variety of other mineral separation sequences are utilized depending on the product preference. Schultz et al. (2005) provide a more detailed reference for the processing of hartsalz using electrostatic separation. MOP, kieserite, epsomite, KCl brine, and $MgCl_2$ brines are produced for sales or the creation of other products. For instance, epsomite or kieserite are reacted with a KCl brine to form intermediates leonite ($K_2SO_4 \cdot MgSO_4 \cdot 4H_2O$) or schoenite ($K_2SO_4 \cdot MgSO_4 \cdot 6H_2O$), respectively, and $MgCl_2$ brine. These intermediates are further reacted with KCl brine to form SOP and additional $MgCl_2$ brine.

Schoenite produced as raw salts from evaporation ponds is reacted in a similar fashion. Halite has a negative impact on this conversion and therefore is generally removed by flotation ahead of schoenite decomposition. Schoenite can also be floated from halite.

Kainite ($KCl \cdot MgSO_4 \cdot 3H_2O$) often appears as minor quantities in potash ores. Kainite is usually upgraded by flotation, reacted with excess sulfate brines to form a crystalline schoenite intermediate and $MgCl_2$ brine. The schoenite is further converted to form SOP and additional $MgCl_2$brine. Raw salts with higher levels of kainite are reacted with sulfate brines to convert the kainite to schoenite prior to flotation.

Langbeinite is another polymineral, a twin salt ($K_2SO_4 \cdot 2MgSO_4$) that is currently produced and marketed as sulfate of potash magnesia by Mosaic and Intrepid Potash. The ore is crushed, ground, and screened and the denser langbeinite sink is removed from a lighter, predominantly halite, float using dense media cyclones (Harrison 1972; Garrett 1996). The screen undersize is water leached to remove the fine salt from the fine langbeinite. The recovery of langbeinite in the dense media circuit is high (90%–95% langbeinite), while the overall recovery is 70%–80%.

Another polymineral, polyhalite, with a lower potassium content, 15.6% K_2O, is used as a lower grade fertilizer or soil amendment. Cleveland's Boulby mine in the United Kingdom, owned by ICL mines, crushes, screens, and direct sells the unrefined ore as a fertilizer with two products sizes, granular or powder. Two greenfield projects, one in New Mexico and the other near the Boulby mine, are in the financing phase.

About half of the SOP production is obtained from sulfate containing ores or brines. The other half of production is as a secondary product created by reacting MOP with sulfuric acid or sulfate brines. The majority of secondary SOP is produced in the Mannheim process in which KCl and sulfuric acid are reacted in Mannheim furnaces at 600°C to produce SOP and hydrochloric acid. A less common method is to react sodium sulfate with MOP to produce glaserite ($Na_2SO_4 \cdot 3K_2SO_4$). The glaserite is further reacted with MOP to produce SOP and salt brine.

Alunite, an aluminum potash sulfate mineral ($KAl_3(SO_4)_2(OH)_6$), has also been processed to produce SOP, primarily from ores in Russia, and historically from ores in Italy and for a brief period during World War I from ores in Utah. Processes vary, but after either a soda roast or a reduction roast, the calcine is hot leached and SOP is crystallized from resulting liquors.

Nitrate Ores

No natural form of potassium nitrate, NOP, is mined or harvested from brines, and therefore its production is all secondary. Sodium nitrate is mined from natural sources, caliche ores, in Chile and reacted with MOP from solar brine operations to produce NOP and NaCl brine.

BRINE SOURCES AND SOLUTION MINING

Between 20% and 25% of the global potash production originates from brine sources. As these brine sources are predominantly located in arid environments, the potash minerals are recovered from the brines through solar evaporation in the form of raw salts creating a synthetic ore. The nature of the evaporated raw salts depends on the composition of the brine, and the management of brine chemistry by advancing concentrated brines through successive ponds to selectively deposit different salts (Table 7). Low-sulfate brines will deposit halite first and sylvinite next, for example, Bonneville, Utah, and the Salar de Atacama, Chile. The Atacama brines contain higher sulfate levels, and schoenite and kainite will deposit later in the sequence; therefore the production of MOP and SOP are possible from the same region. For low-sulfate brines with increased magnesium levels, the raw salts are carnallitic with little if any sylvinite, for example, the Dead Sea (Israel) and the Qarhan Salt Lake (China). The raw salts from brines with increased magnesium and sulfate levels contain halite, schoenite, and kainite (e.g., Lop Nur Salt Lake in China and Great Salt Lake [Utah]). Processing of sylvinitic, carnallitic, and potassium sulfate raw salts was described earlier.

Solution mining of sylvite beds was first used in 1964 at Belle Plaine in Saskatchewan; it is the largest solution mining operation and is still producing at about 2 Mt MOP/yr. K+S produces MOP from the Bethune solution mine in Saskatchewan. At three sites in North America (Patience Lake, Moab, and Carlsbad HB), brines are injected and extracted from old mine workings. Patience Lake is operated seasonally, taking advantage of the cold winters with sylvite depositing in the cooling ponds. At Moab and HB, a raw sylvinitic salt is deposited from the extraction brine in solar evaporation ponds. Direct flotation of potash from the harvested salts produces the final MOP product. MOP and $MgCl_2$are recovered from carnallite brines at DEUSA International's Bleicherode, Germany, solution mine by the careful management of brine temperatures and brine magnesium levels. In all cases the processing of brines or salts from solution mining is simplified due to the absence of insoluble materials, which remain underground.

PREPARATION FOR MARKET

The fertilizer market is diverse, and the application of product will vary with crop type and location. As potassium products are generally blended with nitrogen and phosphate, the nature and quantity of contained impurities and the size of the final product are targeted to meet the needs of specific markets. Value-added steps are required to prepare potash concentrates for the diverse needs of the market.

Debrining and Drying

Whether recovered by flotation or crystallization, the recovered potash is a suspension of solids in brine. It must be debrined, dried, sized, stored, and, finally, shipped to the customers. Potash concentrates can be debrined either by centrifuges or vacuum filters. Centrifuges are favored because of their ability to lower the concentrate moisture to a level of 2%–4% versus the range of 6% to 8% moisture obtainable with vacuum filters. Lower cake moistures are preferred as the

Table 7 Chemical composition of commercially extracted brines

	Concentration, g/L					Mass Ratios		
Brine Source	Na	K	Mg	Cl	SO_4	K/Mg	SO_4/(Cl+SO_4)	Potash Salts
Dead Sea	35	7	35	190	<1	0.2	0.01	Carnallite
Qarhan Salt Lake	14	22	75	250	5	0.3	0.02	Carnallite
Bonneville	118	7	5	202	4	1.4	0.02	Sylvite
Salar de Atacama	90	22	12	190	23	1.8	0.11	Sylvite, schoenite, kainite
Great Salt Lake	95	7	10	155	20	0.7	0.11	Schoenite, kainite
Lop Nur Salt Lake	—	10	17	180	45	0.6	0.20	Schoenite, kainite

Adapted from Garrett 1996; Fertilizer International 2015

crystallization of soluble salts from the entrained brine will decrease the K_2O grade of the dried product.

Drying of the filtered concentrates is conducted at relatively low temperatures ranging from 150° to 200°C. Product moisture levels are generally less than 0.1% H_2O. Product leaching, brine management, and cake washing are used to control the final product grade. An elevated presence of $MgCl_2$ in brine leads to some deterioration in the free-flowing quality of the dried product. Potash operations use both direct-flow rotary dryers and fluid bed dryers.

Product Sizing and Compaction

Prior to 1957, MOP was sold on the basis of K_2O content only. The fertilizer trade adopted sophisticated practices in the formulation of complete plant foods during this time. With the majority of MOP usage as fertilizer, either applied separately or blended with nitrogen or phosphate products, a standard uniform size was preferred. The product, known as granular muriate, has a firmly established position in bulk blending or dry mixing of nitrogen, phosphorus, and potassium (NPK) fertilizers for direct application to soils.

Because of plasticity, fine particles of potash can be fused together with the application of pressure and without the use of binders and then subsequently broken into granules. Compaction can be an important means of salvaging noncontaminated dusts and fine products by converting them to a readily saleable granular product. Granulation by compaction started in the 1950s and has proven to be an economic size-enlargement process for MOP and other potassium salts (Figure 9). Granular sales in the United States have grown from 8% of total, or 300,000 t, in 1962 to 70%, almost 7,200,000 t in 2013. Granular consumption is greatest in the more mature fertilizer markets such as in the United States and Brazil, and overall represents about 45% of the global market.

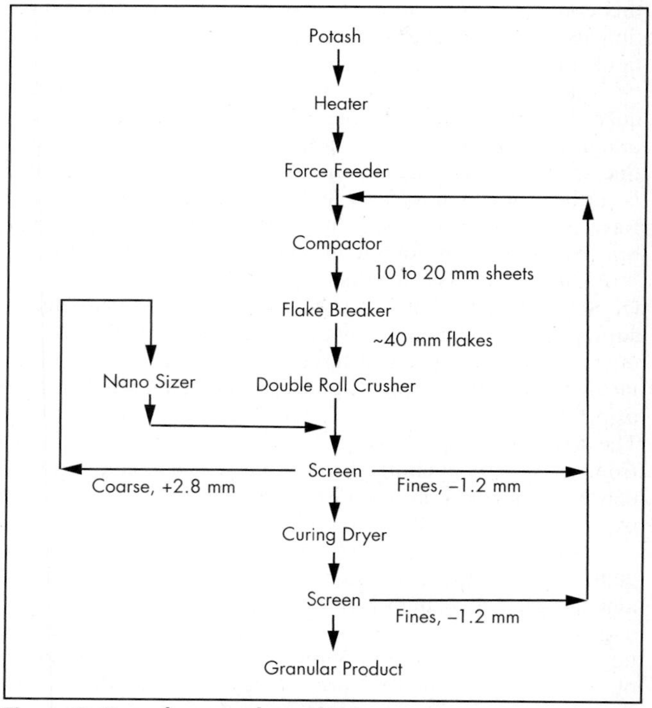

Figure 9 Granular potash production

Storage and Loadout

Since 90%–95% of all potash production is used in agricultural applications, the peak sales and shipping demands occur during the spring and fall crop-planting seasons. The logistics involved in production and movement of millions of tons of potash from mines to market under such conditions suggest the need for large storage facilities at the plant sites. In practice, however, the product storage capacity is limited to about 10%–30% of the annual production by individual plants. The practice is supported with long-term sales contracts calling for steady shipments during the year, off-season export sales, seasonally adjusted price schedules stimulating extra domestic sales in slack periods, and rentals of warehouse space at some marine terminals and strategic inland locations. Fertilizer application rates have dramatically increased due the modernization of farm equipment, and therefore application seasons have been compressed. The establishment of large wholesaler-owned storage and load-out facilities at transportation hubs and warehouse contracts are used to meet just-in-time farm demands.

Normal warehousing practice is to store each product separately, but some producers store several grades in one large structure. Warehouse storage capacity can be comprised of a single structure or multiple structures. With the average size of current operations, most producers have at least 100,000 t of on-site storage. Individual warehouses can range from 60,000 t to 300,000 t. The 526,000-t product warehouse at PCS Rocanville's Scissors Creek is one of the largest reported warehouses. Stored materials can be reclaimed through multiple draw points in the floor slab, partly by gravity, but mostly with the aid of slushers, front-end loaders, or bulldozers. Newer warehouses have eliminated reclaim tunnels and utilize stacker/reclaimers to manage warehouse inventories.

SUPPORT SYSTEMS

In addition to the processing of potash ores, other infrastructural needs and support systems are required in the production of potash. The materials handling and deposition of the non-potash materials must be appropriately managed, and water, brine, and energy systems must also be managed to minimize costs and to optimize stewardship.

Potash Tailings

In practice, potash tailings are managed in three forms: coarse slurry, fine slurry, and excess brine. The composition of slurries varies depending on the ore type processed and the recovery processes selected. Local regulations, proximity to communities, and climate are major factors that influence the choice in management of tails. Tailings may be slurried to impounded piles, dry stacked in heaps, backfilled in underground cavities, or slurried with fresh or seawater prior to riverine or marine deposition. Excess brines are injected through deep wells into deeper underlying strata or in some cases pumped into rivers or seas. The coarse and intermediate fractions are usually debrined by centrifuges, horizontal rotary filters, or belt filters.

In Saskatchewan, coarse and fine tails are dewatered separately, slurried, and pumped onto separate tailings piles. Slurry transport and deposition is preferred to conveyors. Well injection is used for excess brines that predominantly contain NaCl and $MgCl_2$. German operations have been backfilling underground cavities since 1908. Surface storage of tails is also used in Germany; filtered tails are combined with the dry tails from electrostatic separation and dry stacked in piles using conveyors. Increased pile heights are possible with dry stacking, and Monte Kali, the largest of the salt piles at the time of writing, produced by K+S, is approximately 240 m high. The Belarusian operations also use conveyors to dry stack tails. To manage subsidence, Russian operations backfill tails as slurry; dry stacking of tails is also used. The Boulby mine in England uses a combination of methods for tailing management. Centrifuged coarse tails are repulped with seawater and discharged in the North Sea; the finer tails are slurry backfilled underground.

Brine Systems

Unwarranted losses of brine are detrimental to efficiency of the plant, and the optimum recovery of brine for reuse is of paramount importance. Thickeners or lamella-type settlers are utilized in the recovery and clarification of brine. Producing a high clarity overflow is one of the most important functions in potash operation as a flotation circuit may experience 2%–5% lower recoveries by using brines with higher suspended solids. The thickener underflow usually contains most of the clay, a high proportion of fine NaCl, some fine KCl, and a substantial volume of brine.

Temperature is an important factor in the brine system; cooling of the brine produces partial recrystallization of its components, whereas at high temperatures the brine dissolves additional minerals, decreases flotation efficiency, and increases consumption of reagents. Several methods are used to regulate brine temperatures, such as enclosure in heated buildings, cooling towers, and crystallization–refrigeration systems.

Process Control

Process control and control variables for potash flotation plants are common to other branches of the mineral processing industry. The soluble nature of the ore and its sensitivity to temperature makes online density and mass measurements difficult and complicates accurate mass balances. On-line ^{40}K analyzers are successfully used to determine the potassium content of process streams and readily integrated into process control strategies. In Russia, laser-induced breakdown spectroscopy has been successfully utilized to measure the grade of final product and control product leaching (Groisman and Gaft 2010). Flotation controls such as optical froth monitoring systems to identify bubble properties and mass pull rates are finding a home in several potash operations. Special design considerations should be given to the location, installation and long-term operation and serviceability of belt weightometers as accurate measurements will be required to complete mass balances and to manage product inventories. Process control automation, integrated with recording, reporting, data analysis and data storage has found wide application in potash plants.

Energy

Several energy studies reviewed Canadian potash operations such as Natural Resources Canada through the Canadian Industry Program for Energy Conservation (CIPEC) (Natural Resources Canada 2003) and Canadian Industry Energy End-Use Data and Analysis Center (Nyboer and Bennett 2015). Total energy consumption ranges from 1.1 to 2.3 GJ/t MOP product depending on the scale of the operation, the nature of the process, the age of the plant, and the plant utilization. Natural gas generally makes up about 75% of the energy used. Russian operations report similar rates and distributions. CIPEC reported that nearly 70% of the energy is used in the surface operation, whereas 30% is used underground. Solution mines and plants with thermal leach crystallization are about 2.5–3.5 times more energy intensive and range from 3.0–6.0 GJ/t.

Water

The water requirement for a potash plant also varies, primarily based on the ore mineralogy and process selection. For sylvinite ores, water consumption will increase with carnallite and insoluble content. The efficiency of the brine recovery system and water management practices will also influence water requirements. In a conventional operation the total water requirement is in the range of 2.0–6.0 m^3/t MOP, typically between 3.4 and 4.0 m^3/t. Most operations reclaim the clarified tailings pond brine from transporting slurried tailings, from process use, and from plant housekeeping purposes. Such practices reduce fresh water makeup requirements to 1.5–2.0 m^3/t.

CAPITAL AND OPERATING COSTS

Potash is a capital-intensive business, and the general rule of thumb states that it requires between US$500 and US$1,000 of capitalization per metric ton of annual MOP production to develop brownfields and approximately US$2,000/t for greenfield site costs. PCS regularly reports that a 2.0 Mt MOP/yr greenfield project will require from US$4–$6 billion of capitalization for site costs and support infrastructure (Potash Corporation of Saskatchewan 2014; Fortney 2013). Operating costs are reported by most potash producers, potash analysts, or potash development projects. Significant levels of capitalization were invested over the last decade, significantly reducing operating costs. As with any mineral operation, operating costs vary with ore grades, the scale and nature

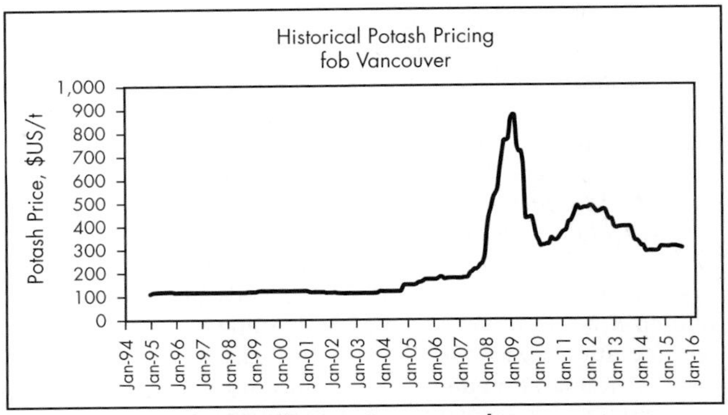

Figure 10 Price for North American potash

of the operation, and by-product credits. Cash operating costs, including royalties and selling, general, and administrative expenses, currently range, in U.S. dollars, from $55/t to $200/t MOP product for approximately 90% of the market. The cash costs for several smaller producers ranges from $200–$330/t.

PRODUCT PRICING

A 10-year pricing trend is presented in Figure 10. Potash products are sold by the ton, either short ton or metric ton, depending on the location of the sale. Sale prices for muriate, sulfates, and sulfates magnesia will depend on the product grade and size. Per-ton pricing is tied to the producers' product specification. Generally, an increased price is commanded for coarser products, such as granular or pellet products, and for increased potassium content. Prices are quoted per short ton or metric ton of product based on a minimum potassium content. This minimum is 60% K_2O for muriate and 50% K_2O for potassium sulfate. Price quotations are usually free-on-board plant sites for domestic shipments or specific ports for export sales. Large volume pricing contracts are annually negotiated with major potash consumers such as China, Brazil, and India, setting the tone for pricing in the rest of the market.

Average potash prices are published in trade publications such as Green Markets, Fertecon, and Argus. The individual potash producers in North America publish their own price schedules and distribute them to their customers. These schedules usually cover the so-called fertilizer year, a period beginning on or about July 1 of one year and ending on June 30 of the following year. Schedules state price differentials for various grades available and include seasonal price adjustments.

REFERENCES

Alamer, A. 2009. Carnallite froth flotation optimization and flotation cells efficiency in the Arab Potash Company. Presented at the Jordan 22nd AFA International Fertilizers Technology Conference and Exhibition, Marrakech, Morocco.

D'Ans, J. 1933. Die Lösungsgleichgewichte der Systeme der Salze ozeanischer Salzablagerungen. Verlagsgesellschaft für Ackerbau, Berlin.

Fertilizer International. 2010. Improving the technology, improving the product quality. *Fertilizer Inst.* 43:42–46.

Fertilizer International. 2015. SOP production from brine. *Fertilizer Inst.* 468:50–51.

FAO (Food and Agriculture Organization of the United Nations). 2015. *World Fertilizer Trends and Outlook Through 2018.* Rome, Italy: FAO.

Fortney, S. 2013. Potash Industry Developments. www.saskmining.ca/uploads/news.../steve-fortney-potashcorp.pdf. Accessed June 2015.

Garrett, D.E. 1996. *Potash: Deposits, Processing, Properties and Uses.* Berlin, Germany: Springer Science and Business Media.

Groisman, Y., and Gaft, M. 2010. Online analysis of potassium fertilizers by laser-induced breakdown spectroscopy. *Spectrochim. Acta B* 65(8):744–749.

Hall, S., and Harrison, M. 1995. New Jameson cell flotation of industrial minerals. *Ind. Min.* 333:61–67.

Hardie, L.A. 1991. On the significance of evaporites. *Annu. Rev. Earth. Planet. Sci.* 19:131.

Hardie, L.A. 1996. Secular variation in seawater chemistry: An explanation for the coupled secular variation in the mineralogies of marine limestones and potash evaporites over the past 600 my. *Geology* 24(3):279–283.

Harrison, M.H. 1972. International Minerals and Chemical Corporation. Method for treatment of heavy media. U.S. Patent 3,638,791.

Hatch Ltd. 2012. *Potash Mining Supply Chain Requirement Guide for Greenfield Mine Lifecycle Costs.* Report to Ministry of the Economy H341318-0000-00-236-0001 Rev. 0. Saskatoon, SK: Saskatchewan Ministry of the Economy.

Imhof, R.M., Battersby, M.J.G., and Ciernioch, G. 2002. Pneumatic flotation: An innovative application in the processing of potash salts. SME Preprint No. 02-162. Littleton, CO: SME.

Jasinski, S.M. 2012–2016. Potash. In [various years] *2010–2014 Minerals Yearbook.* Reston, VA: U.S. Geological Survey.

Jasinski, S.M. 2017. Potash. In *Mineral Commodity Summaries 2017.* Reston, VA: U.S. Geological Survey. pp. 128–129.

Konoplev, E.V., Titkov, S.N., Gurkova, T.G., and Chumakova, T.G. 2014. Sylvite flotation's activation from carnallite potash ores. Presented at the International Mineral Processing Congress XXVII, Santiago, Chile.

Laskowski, J.S. 2013. From amine molecules adsorption to amine precipitate transport by bubbles: A potash ore flotation mechanism. *Miner. Eng.* 45:170–179.

LeBaron, I.M., and Knopf, W.C. 1958. Application of electrostatics to potash beneficiation. *Min. Eng.* 1(10):1081–1083.

Miller, J.D., and Nalaskowski, J. 2007. Soluble salts. In *Froth Flotation: A Century of Innovation.* Edited by M.C. Fuerstenau, G.J. Jameson, and R.H. Yoon. Littleton CO: SME. pp. 529–533.

Miller, J.D., Khalek, N.A., Basilio, C., El-Shall, H., Forsberg, K.S.E, Fuerstenau, M.C., Mathur, S., Nalaskowski, J., Rao, K.H., Somasundaran, P., Wang, X., and Zhang, P. 2007. Flotation chemistry and technology of nonsulfide minerals. In *Froth Flotation: A Century of Innovation.* Edited by M.C. Fuerstenau, G.J. Jameson, and R.H. Yoon. Littleton CO: SME. pp. 533–543.

Natural Resources Canada. 2003. *Energy Benchmarking: Canadian Potash Production Facilities.* Ottawa, ON: Natural Resources Canada Office of Energy Efficiency.

Natural Resources Canada. 2016. *Minerals and Metals Factbook 2016.* https://www.nrcan.gc.ca/sites/www.nrcan.gc.ca/files/mineralsmetals/pdf/mms-smm/Minerals%20and%20Metals_factbook_En.pdf. Accessed September 2016.

Nyboer, J., and Bennett, M. 2015. *Energy Use and Related Data: Canadian Mining and Metal Smelting and Refining Industries 1990 to 2013.* For Mining Association of Canada. Burnaby, BC: Simon Fraser University.

Orris, G.J., Cocker, M.D., Dunlap, P., Wynn, J.C., Spanski, G.T., Briggs, D.A., Gass, L., Bolm, K., Yang, C., Lipin, B.R., Ludington, S., Miller, R.J., and Slowakiewicz, M. 2014. Potash: A global overview of evaporate-related potash resources, including spatial databases of deposits, occurrences, and permissive tracts. In *Global Mineral Resource Assessment.* Scientific Investigations Report No. 2010-5090-S. Reston, VA: U.S. Geological Survey.

Ozdemir, O., Du, H., Karakashev, S.I., Nguyen, A.V., Celik, M.S., and Miller, J.D. 2011. Understanding the role of ion interactions in soluble salt flotation with alkylammonium and alkylsulfate collectors. *Adv. Colloid Interface Sci.* 163(1):1–22.

Perucca, C.F. 2003. Potash processing in Saskatchewan: A review of process technologies. *CIM Bull.* 96(1070):61–65.

Potash Corporation of Saskatechewan. 2014. Overview of PotashCorp and its industry. www.potashcorp.com/overview/advantages/potash/saskatchewan-potash-capacity-costs. Accessed May 2016.

Schultz, H., Bauer, G., Schachl, E., Hagedorn, E., and Schmittinger, P. 2005. Potassium compounds. In *Ullmann's Encyclopedia of Industrial Chemistry.* Weinheim, Germany: Wiley-VCH.

Siedel, A. 1919. *Solubilities of Organic and Inorganic Substances.* New York: Van Nostrand.

Strathdee, G.G., Haryett, C.R., Douglas, C.A., Senior, M.V., and Mitchell, J. 1982. The processing of potash ore by PCS. *CIM Bull.* 75(845):82.

Strathdee, G., Gotts, B., and McEachern, R. 2007. Potassium flotation practice. In *Froth Flotation: A Century of Innovation.* Edited by M.C. Fuerstenau, G.J. Jameson, and R.H. Yoon. Littleton CO: SME. pp. 533–543.

Titkov, S., Buksha, J., Panteleeva, N., and Afonina, N. 2014. Effect of electrolyte composition on flotation of water-soluble minerals. Presented at the International Minerals Processing Congress XXVII, Santiago, Chile.

Titkov, S., Sabirov, R., and Novoslev, V. 2005. The new flotation of water-soluble minerals: Peculiarities and new technologies. Presented at the Century of Flotation Symposium, International Minerals Processing Congress, Brisbane, Australia, June 6–9.

Wang, X., Miller, J.D., Cheng, F., and Cheng, H. 2014. Potash flotation practice for carnallite resources in the Qinghai Province, PRC. *Miner. Eng.* 66:33–39.

Warren, J.K. 2006. Evaporites as exploited minerals. In *Evaporites: Sediments, Resources and Hydrocarbons.* Berlin, Germany: Springer Science and Business Media.

Williams-Stroud, S., Searls, J.P., and Hite, R.J. 1994. Potash resources. In *Industrial Minerals and Rocks: Commodities, Markets and Uses.* 7th ed. Edited by J.E. Kogel, N.C. Trivedi, J.M. Barker, and S.T. Kruskowski. Littleton, CO: SME. p. 797.

Wisniak, J. 2002. The Dead Sea: A live pool of chemicals. *Indian J. of Chem. Tech.* 9(1):79–87.

World Bank. 2018. Data Bank: Global Economic Monitor (GEM) Commodities. http://databank.worldbank.org/data/reports.aspx?source=global-economic-monitor-commodities. Accessed January 2018.

Zandon, V.A. 1985. Potash. In *SME Mineral Processing Handbook.* Vol 2. Edited by N.L. Weiss. Littleton, CO: SME-AIME. pp. 22-1–22-25.

Zbranek, V. 2013. Red Sea–Dead Sea water conveyance study program reports: Chemical industry analysis study. http://siteresources. worldbank.org/INTREDSEADEADSEA/Resources/Chemical_Industry_Analysis_Study_Final_Report.pdf. Accessed January 2015.

Zhang, F., Zhang, W., and Ma, W. 2009. *The Chemical Fertilizer Industry in China: A Review and Its Outlook.* Paris: International Fertilizer Industry Association.

CHAPTER 12.32

Rare Earth Elements

J.R. Goode

The International Union of Pure and Applied Chemistry (IUPAC) defines the lanthanoids as the 15 elements with atomic numbers of 57–71 inclusive: lanthanum (La), cerium (Ce), praseodymium (Pr), neodymium (Nd), promethium (Pm), samarium (Sm), europium (Eu), gadolinium (Gd), terbium (Tb), dysprosium (Dy), holmium (Ho), erbium (Er), thulium (Tm), ytterbium (Yb), and lutetium (Lu). The rare earth elements (REE) are defined as the lanthanoids plus yttrium (Y) and scandium (Sc) (Connelly et al. 2005). Pm is an unstable element and not found in nature, so in practical terms there are 16 REE.

The abbreviation Ln is sometimes used to signify the lanthanoids or the entire group of REE. Didymium, sometimes abbreviated as Dd or NdPr, is an unseparated, and therefore lower cost, mixture of Pr and Nd, usually at the natural Pr:Nd ratio of approximately 1:3.4 and often used in NdFeB magnets. The REE industry uses other abbreviations as follows:

- LREE = light REE
- MREE = medium REE
- HREE = heavy REE
- TREE = total REE
- CREE = critical REE
- SEG = Sm, Eu, and Gd

IUPAC does not refer to LREE, MREE, and HREE, but the terms are in common use although defined in different ways by different parties. The U.S. government agencies only use LREE and HREE and define LREE as the elements from La to Gd inclusive, all of which have unpaired electrons in the 4f electron shell. The HREE, with paired 4f electrons, are the elements from Tb to Lu plus Y. Y usually behaves like an HREE with properties between those of Ho and Er. IUPAC includes Sc as a REE, but this element does not fit comfortably into either the LREE or HREE group.

Actinium, particularly ^{227}Ac, which is a radioactive decay product of ^{235}U and found in most REE deposits, behaves like La and tends to report to separated La products where it can concern end users.

The U.S. Department of Energy (U.S. DOE 2010) and European Union (European Commission 2014) have defined the critical rare earth elements (CREE) as those for which there is a serious risk of assured supply and that are important to the present way of life. The DOE report identified Dy, Nd, Tb, Y, and Eu as critical through the year 2025. The European Union defined the HREE (Gd to Lu plus Y) as being particularly critical, the LREE (La to Eu) as being somewhat less critical, and Sc being noncritical. The criticality matrices are changing as REE uses and production levels change.

When rare earth oxides are being referenced, common with product prices and analytical data, the abbreviation REO is used as well as LREO, MREO, HREO, TREO, and CREO. REM is occasionally used to signify the metal.

The REE are usually encountered in the trivalent state, so most of the oxides are written as La_2O_3, Y_2O_3, and so on. However, Ce, Pr, and Tb do not ignite to trivalent oxides and their oxides are CeO_2, Pr_6O_{11}, and Tb_4O_7, respectively.

Certain fundamental data and oxide–metal conversion factors for the REE are presented in Table 1. Selected properties of the REE, which govern the way in which the REE behave in minerals and in solution, are presented in the following sections.

VALENCY

The REE are generally trivalent. However, Nd, Sm, Eu, Dy, Tm, and Yb can be reduced to the divalent state and Ce, Pr, and Tb can be oxidized to the quadrivalent state.

IONIC RADIUS

The ionic radii of the trivalent REE get smaller as the atomic mass increases (the "lanthanoid contraction"), except for those of Y and Sc. The Y ionic radius falls between that of Ho and Er and, consequently, the solution chemistry of Y is often intermediate to that of Ho and Er. The lightest REE, Sc, has the smallest ionic radius and often, but not always, behaves as an HREE. Ac, a progeny of ^{235}U and ^{232}Th, has an ionic radius similar to that of La and behaves in a similar way to La. The ionic radii of the REE, expressed in picometers (1 pm = 10^{-12} m) are plotted in Figure 1.

J.R. Goode, Principal & Consulting Metallurgist, J.R. Goode & Associates, Toronto, Ontario, Canada

Table 1 Fundamental REE data and oxide–metal conversion factors

	Light REE							Heavy REE
Element	La	Ce	Pr	Nd	Sm	Eu	Gd	Tb
Atomic number	57	58	59	60	62	63	64	65
Atomic mass	138.905	140.116	140.907	144.242	150.360	151.964	157.250	158.925
Oxide formula	La_2O_3	CeO_2	Pr_6O_{11}	Nd_2O_3	Sm_2O_3	Eu_2O_3	Gd_2O_3	Tb_4O_7
Conversion Factors								
Element to oxide	1.17277	1.22837	1.20816	1.16638	1.15961	1.15792	1.15261	1.17617
Oxide to element	0.85268	0.81408	0.82770	0.85735	0.86235	0.86361	0.86759	0.85021

	Heavy REE							–
Element	Dy	Ho	Er	Tm	Yb	Lu	Y	Sc
Atomic number	66	67	68	69	70	71	39	21
Atomic mass	162.500	164.930	167.259	168.934	173.054	174.966	88.906	44.956
Oxide formula	Dy_2O_3	Ho_2O_3	Er_2O_3	Tm_2O_3	Yb_2O_3	Lu_2O_3	Y_2O_3	Sc_2O_3
Conversion Factors								
Element to oxide	1.14768	1.14551	1.14348	1.14206	1.13867	1.13716	1.26993	1.53383
Oxide to element	0.87131	0.87297	0.87451	0.87560	0.87820	0.87938	0.78743	0.65196

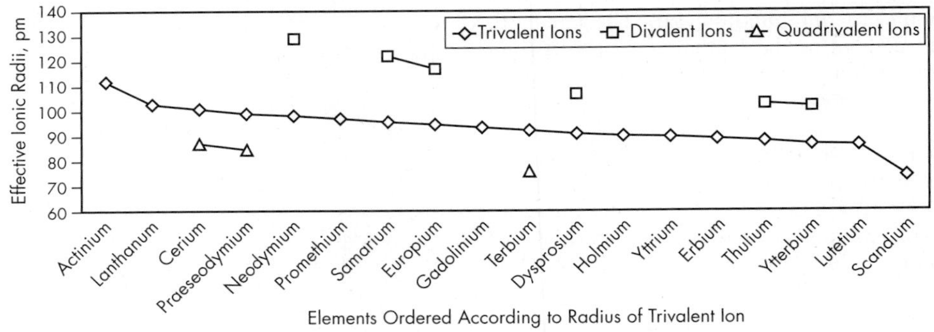

Data from Shannon 1976

Figure 1 Effective ionic radii of the REE and Actinium

DISCOVERY AND DEVELOPMENT

Swedish Army Lieutenant C.A. Arrhenius discovered an unusual mineral near Ytterby, Sweden, in 1787. Seven years later, Johan Gadolin, a chemist at the University of Abo (now Turku) in Finland, separated a previously unknown metal oxide from a sample of the mineral. The mineral was later dubbed gadolinite and the oxide named yttria after the discovery location.

In 1803, Jöns Jacob Berzelius and Wilhelm Hisinger in Sweden studied a heavy mineral discovered at Bastnas in Sweden and extracted a metal oxide that they named ceria. Martin Klaproth independently separated ceria at about the same time in Germany.

The crude metal oxides, yttria and ceria, isolated from these minerals were studied, and it became evident that both were mixed oxides of other elements. A Swedish chemist, Carl Mosander, extracted La_2O_3 and didymia, a mixture of Pr and Nd oxides, from ceria in the late 1830s. A couple of years later, he separated oxides of Y, Tb, and Er from yttria.

Between 1869 and 1900, work by Paul-Émile Lecoq de Boisbaudran, Eugène-Anatole Demarçay, and Carl Auer von Welsbach had produced compounds of Sm, Eu, Gd, Pr, and Nd from ceria. Several chemists contributed to the further separation of the HREE in yttria after Mosander's pioneering work. The Swiss chemist Jean Charles Galissard de Marignac isolated Gd, Tb, and Yb in the late 1870s. Lars Nilson of Sweden discovered Sc in 1879, and his countryman, Per Cleve, identified Ho and Tm the same year. Georges Urbain, Charles James, and Welsbach independently discovered Lu in 1907.

Separation of the REE during the pioneering years was achieved using oxidation–reduction reactions, double salt formation, and fractional crystallization or precipitation procedures. As an illustration of the time-consuming nature of these methods, Urbain performed 15,000 sequential nitrate crystallization operations to separate Lu from a crude Yb product.

The earlier separation processes satisfied the low demand for individual REE through the first half of the 20th century. During the 1950s, separation by ion exchange (IX), with chromatographic elution, was developed and commercial plants were constructed in the United States, United Kingdom, former Union of Soviet Socialist Republics, and France (Spedding and Powell 1954). These plants operated until the 1960s when they were replaced by more efficient solvent extraction systems similar to those still used today (Kaczmarek 1981).

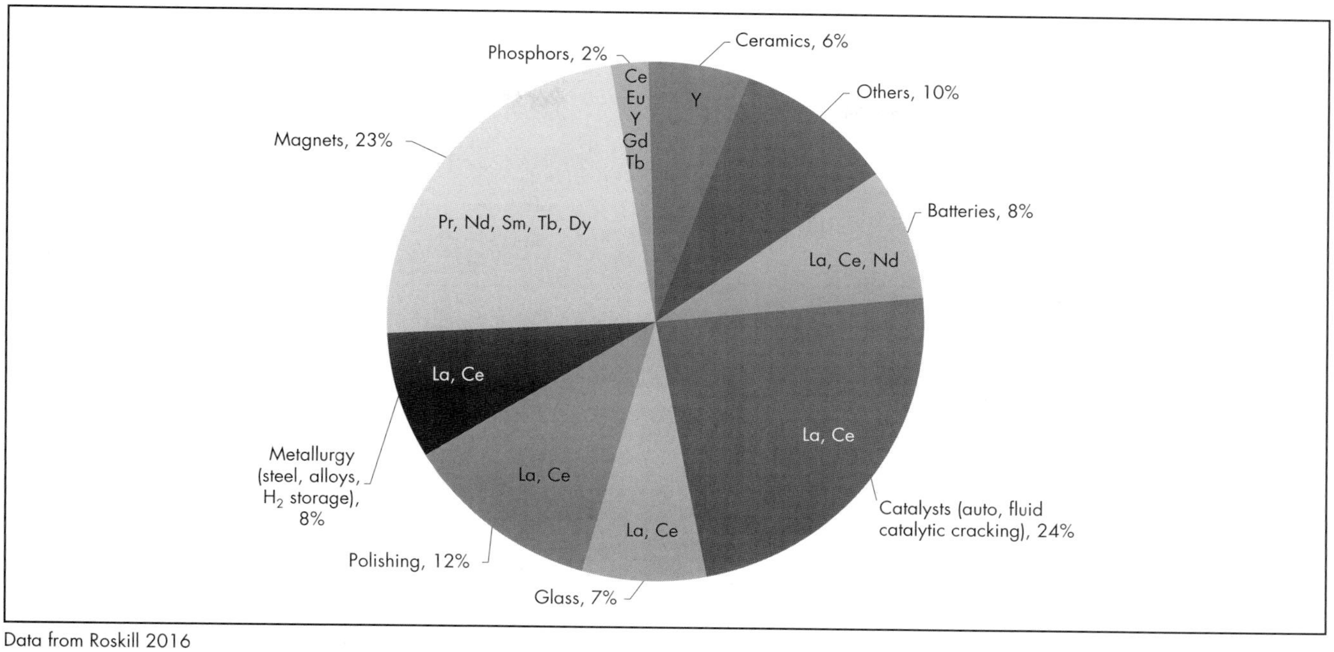

Data from Roskill 2016

Figure 2 Distribution of REE use

USES

In the 1880s, Welsbach discovered that the REOs emit a brilliant light when heated in a gas flame, and he went on to develop a lighting system using gas mantles incorporating La and Y (Habashi 2012). A much-improved formulation using ThO_2 with about 1% CeO_2 was extensively employed for street lighting and in kerosene-burning lanterns known as Petromax or Tilley lamps in much of the world and as Coleman lanterns in North America. Welsbach factories in Germany and France produced Th from monazite and made a bulk REE by-product. In 1903, seeking a market for the REE, Welsbach developed a cerium–iron alloy, mischmetal, that sparked when struck. It became widely used as lighter flints and is still used today.

After these earliest commercial applications, further REE uses were developed for mixed and separated REE, including glass polishing, ultraviolet (UV) cut glass (which reduces UV light transmission), Al and Mg alloying, high refractive-index lenses, and catalysts. The demand for REE increased dramatically after 1955 with the use of red phosphors incorporating Eu and Y in television cathode ray tubes. Permanent magnets comprising an alloy of Sm and Co with a formula of $SmCo_5$ (about 34% Sm by mass) were developed in about 1970, followed by NdFeB magnets in 1982. The latter are nominally $Nd_2Fe_{14}B$ alloys, but in addition to about 30% Nd they also contain between 2% and 9% Dy and Tb. A mixture of Pr and Nd, which is less expensive since the difficult Pr–Nd separation is avoided, can be used instead of pure Nd. NdFeB magnets remain the most important end use for REE and largely control the present demand and production of REE. The distribution of REE uses is illustrated in Figure 2.

REE deposits contain each of the REE in certain proportions and, as a general rule, the percentage recovery of each REE is similar. Therefore, to produce a given mass of a particular REE, say Nd, a given mine will incidentally produce certain quantities of the other REE in the deposit. There are exceptions to this depending on exact mineralogy and process methods. However, it is generally true that some high-demand REE are relatively valuable, whereas others, for which supply exceeds demand, command low prices or are not marketable.

The history of REE demand has changed as new uses are developed and older technologies discarded. A recent example is a reduction in demand for Y and especially Eu and Tb as compact fluorescent lamps are replaced by longer lasting, more energy-efficient LED lights.

MINERAL DEPOSITS

The most common REE (Ce, La, Y, and Nd) are more abundant than Li, Cu, Mo, Sn, W, Pb, and U. Furthermore, REE minerals can occur in mineable deposits with grades as high as 50% REE—far higher than most other elements. REE are usually found in deposits in which they are the most valuable elements, although they are also found in deposits that are primarily deposits of Fe, niobium (Nb), U, Cu, or other elements.

The U.S. Geological Survey (USGS) divides REE deposits into nine types: peralkaline igneous rocks, carbonatites, iron oxide copper–gold, pegmatites, porphyry molybdenum, metamorphic, stratiform phosphate residual, paleoplacer, and placer. The USGS identifies various REE deposit types within each association and provides examples (Long et al. 2010).

The British Geological Survey (2017) classifies REE deposits into four primary deposit types: carbonatite associated, associated with alkaline igneous rocks, iron–REE deposits, and hydrothermal deposits; and five secondary deposit types: marine placers, alluvial placers, paleoplacers, lateritic deposits, and ionic clays, for a total of nine types.

Table 2 lists the main deposit types, typical examples, and general characteristics of each deposit type. Figure 3 presents grade and tonnage data for REE deposits and identifies important deposits.

Currently, almost all primary REE production comes from four mining areas (Goode 2016c):

Table 2 Main REE deposit types

Deposit Type	Examples	Typical REE-Bearing Minerals	Typical Non-REE Minerals	Average in Situ REO Distribution	By-Products or Co-Products
Carbonatites	Mountain Pass* and Bear Lodge, USA Bayan Obo† and Mianning†, China Niobec and Ashram, Canada	Bastnaesite, parisite, synchysite, monazite, xenotime, aeschynite	Calcite, dolomite, ankerite, aegerine, riebeckite, biotite, magnetite, hematite, fluorite, barite	HREO 2.8% CREO 23.6%	Fe, Nb, CaF_2
Weathered carbonatites (laterites)	Mt. Weld, Australia† Ngualla, Tanzania Kangankunde, Malawi Morro de Ferro, Brazil	Bastnaesite, apatite, monazite, synchysite, cerianite, churchite	Goethite, gibbsite, magnetite, mica, kaolinite	HREO 4.4% CREO 24.9%	None
Intrusive peralkaline complexes (nephaline syenites)	Bokan Mountain and Pajarito Mountain, USA Weishan, China† Lovozero, Russia† Strange Lake, Thor Lake, Kipawa, Canada Norra Karr, Sweden Dubbo, Australia Illimaussaq, Greenland	Eudialyte, loparite, bastnaesite, steenstrupine, synchysite, allanite, fergusonite, gadolinite, mosandrite, britholite, monazite, xenotime	Feldspars, nepheline, sodalite, pyroxene, amphibole, quartz, arfvedsonite	HREO 26.5% CREO 38.4%	Nb, Ta, Zr, U, Zn, phosphates
Hydrothermal deposits	Lehmi Pass, USA Steenkampskrall, South Africa* Hoidas Lake, Canada Lofdal, Namibia Browns Range, Australia	Monazite, xenotime		HREO 27.3% CREO 41.8%	None
Volcanic	Foxtrot, Canada	Allanite, chevkinite, fergusonite, monazite, zircon	Quartz, K-feldspar, amphibole/pyroxene, plagioclase, Fe-oxides, biotite	Only two examples, so average is meaningless	None
	Brockman, Australia	Bertrandite, bastnaesite, parisite, synchysite	Quartz, mica, albite		
Iron-REE deposits	Olympic Dam, Australia‡ Pea Ridge and Minesville, USA Kiruna, Sweden‡	Monazite, apatite		Limited data but generally dominantly LREO	Cu, U, Fe
Placers	Chavara, India Richards Bay, South Africa‡ Aksu Diamas, Turkey	Monazite, zircon, allanite, chevkinite		HREO 12.9% CREO 30.4%	Ilmenite, zircon, garnet, magnetite, Nb
Paleoplacers	Bald Mountain, USA Elliot Lake, Canada*	Monazite		Limited data but generally dominantly LREO	U, Th, pyrite
Ionic clay deposits	Longnan†, Xunwu, etc. China Serra Verde, Brazil Tantalus, Madagascar	REE on clays, residual primary monazite, bastnaesite, etc.	Clay, granites	Variable but Ce in deposits usually relatively insoluble, so low Ce in product	None

Note: Distribution by author. Numerical assay data from company releases collated from public sources in TMR 2015 with analysis by author.
*Past REE producer.
†Current REE producer.
‡Currently producing but not REE.

1. **Bayan Obo, Inner Mongolia Autonomous Region, People's Republic of China.** Carbonatite with bastnaesite and monazite, three large open pits, beneficiation by magnetic and flotation methods with REE recovered as a co-product of iron concentrate, acid bake to process REE concentrate.
2. **Mianning–Dechang District, Sichuan Province, People's Republic of China.** Carbonatite with bastnaesite; several widely spaced open pits; beneficiation by gravity, magnetic, and flotation methods; roast and leach or acid bake to process concentrate.
3. **Southern China: Jiangxi, Guangdong, Fujian, and Hunan provinces and Guangxi Zhuang Autonomous Region, People's Republic of China.** Ionic clay deposits; heap or in situ leaching with weak electrolyte, such as $(NH_4)_2SO_4$.
4. **Mt. Weld, Western Australia.** Weathered carbonatite, single open pit, beneficiation by flotation, acid bake to process concentrate.

Originally operated in about 1950, Molycorp's Mountain Pass rare earth mine in California (United States) included an open pit mine, and flotation and hydrometallurgical plants. Molycorp closed the operation in 1998 because it was not profitable at the prevailing prices and because of environmental issues. With improving REE prices in about 2010, and under new ownership and management, the plant was redesigned and refurbished and restarted in 2013. However, prices for the REOs produced by Molycorp were far lower than projected and there were severe technical issues limiting REO production that together rendered the operation unprofitable. Molycorp declared bankruptcy in June 2015 and placed the plant on a care-and-maintenance basis.

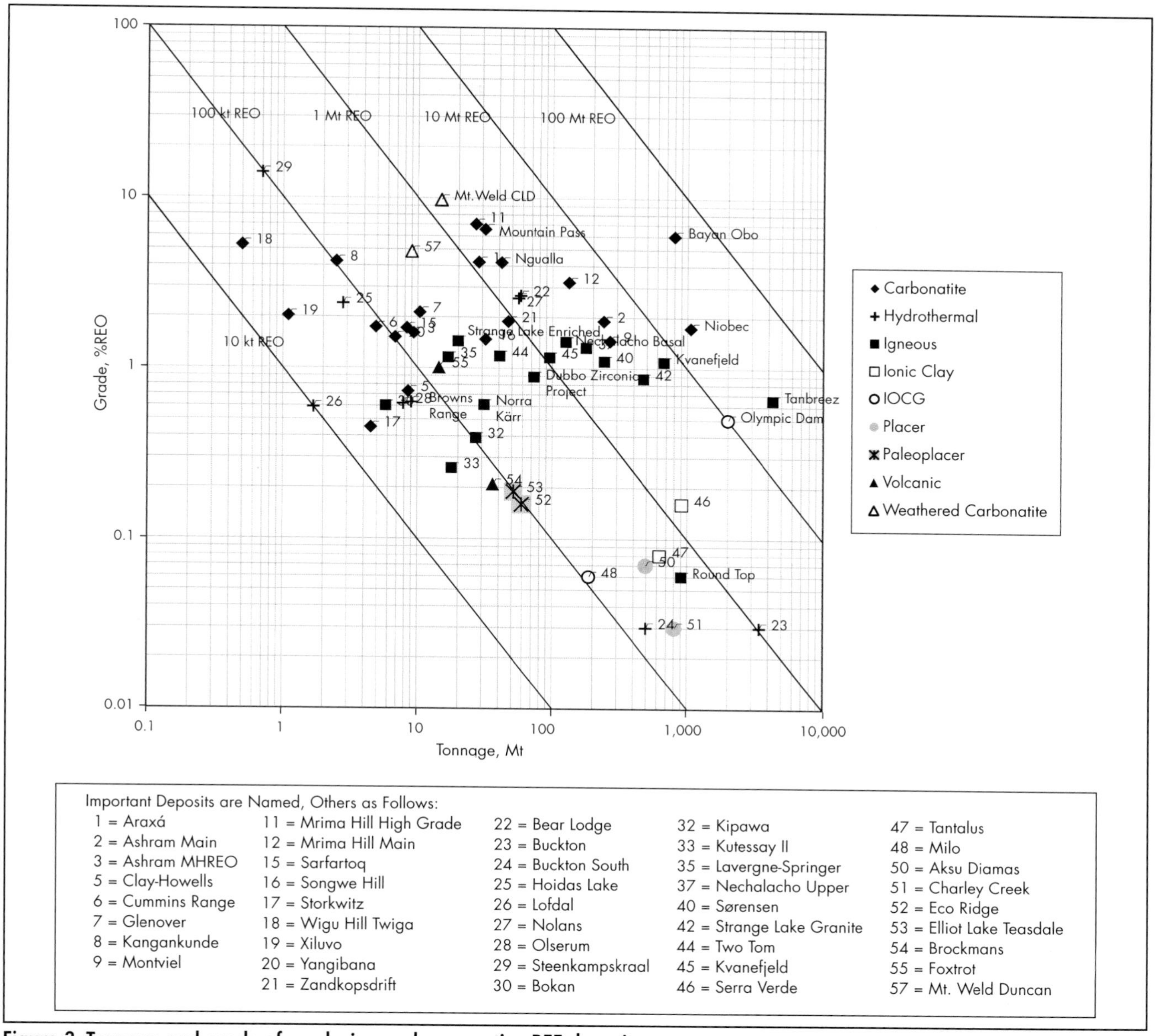

Figure 3 Tonnage and grade of producing and prospective REE deposits

The elemental distribution of the REE within each deposit varies widely according to the deposit type and within the deposit type. The minerals in the carbonatite deposits are frequently bastnaesite, monazite, and other LREE-bearing minerals, so the carbonatites and weathered carbonatites are generally very low in HREE content. The placer and paleoplacer deposits also tend to be poor in HREE.

In contrast, the intrusive peralkaline deposits tend to be relatively rich in HREE. The products from the ionic clay deposits also tend to be rich in HREE largely because the weathering and ion exchange processes cause the Ce to be oxidized and precipitated as cerianite, $Ce(Th)O_2$, a refractory mineral, rather than being sorbed on the clay. Thus, the product obtained by elution of REE from the clays with a weak electrolyte generally contains little Ce, leading to a high-HREE product.

The distribution of REE within selected deposits, and available Sc, U, and Th assays are presented in Table 3. This table also includes an indicative value for the mixed oxide that would be produced from each deposit using projected 2020 REO prices (Roskill 2016). This way of calculating relative value is deceptive since several REE, including La and Ce, cost more to separate than they can be sold for. The relatively low value of the carbonatite deposits is evident.

MINERALOGY

There are several hundred REE-bearing minerals. The REE minerals that are most commonly encountered and typical gangue minerals are listed with formulae and important properties in Table 4.

Table 3 REO distribution; REO, Sc, U, and Th content; and nominal product value for selected deposits

	Oxide	Price, US$/kg oxide[*]	Carbonatite				Ionic Clay (distribution in products, not ore)			Intrusive Peralkaline	
			Bayan Obo	Mianning	Mountain Pass	Mt. Weld	Longnan	Xinfeng	Serra Verde	Dubbo	Strange Lake
			China	China	USA	Australia	China	China	Brazil	Australia	Canada
REO Distribution as %TREO in Carbonatite and Intrusive Alkaline Ore in Product from Ionic Clay	La_2O_3	1	25.6	29.5	32.9	25.6	2.5	26.2	33.7	19.6	10.4
	CeO_2	1	48.5	47.6	49.7	46.9	0.5	1.9	4.2	36.9	25.0
	Pr_6O_{11}	65	5.7	4.4	4.5	5.3	1.0	6.0	6.3	4.0	2.7
	Nd_2O_3	71	16.8	15.2	12.0	18.6	5.1	21.1	20.4	14.1	9.7
	Sm_2O_3	1	1.6	1.2	0.6	2.4	3.9	4.5	3.3	2.2	2.5
	Eu_2O_3	48	0.12	0.23	0.10	0.44	0.30	0.71	0.20	0.08	0.14
	Gd_2O_3	10	0.31	0.65	0.03	0.10	6.62	4.80	3.20	2.17	2.71
	Tb_4O_7	377	0.20	0.12	0.02	0.07	1.34	0.77	0.50	0.31	0.63
	Dy_2O_3	205	0.19	0.21	0.02	0.12	8.83	4.10	3.00	2.02	4.58
	Ho_2O_3	0	0.02	0.05	0.02	0.07	1.60	0.80	0.60	0.39	1.04
	Y_2O_3	4	0.7	0.7	0.0	0.1	58.3	25.1	20.7	15.8	32.6
	Er_2O_3	21	0.02	0.06	0.02	0.07	5.10	2.00	1.90	1.16	3.40
	Tm_2O_3	0	0.10	0.04	0.02	0.07	0.66	0.20	0.30	0.16	0.56
	Yb_2O_3	0	0.01	0.05	0.02	0.07	3.94	1.60	1.60	1.01	3.54
	Lu_2O_3	0	0.13	0.01	0.02	0.07	0.51	0.20	0.20	0.16	0.49
Ore Assay[†]	REO, %	—	5.5	7	8.5	15	0.06	—	0.148	0.9	1.1
	Sc, %	—	—	—	—	—	—	—	~0.0003	—	—
	U, %	—	—	—	0.002	0.003	—	—	0.001	0.012	0.007
	Th, %	—	0.04	—	0.02	0.069	—	—	0.011	0.038	0.037
Value of Mixed Oxide, $/kg REO			17.4	15.3	12.3	17.9	31.6	32.6	28.5	19.5	23.0

*Prices from Roskill 2016 for 2020, allowing for inflation. Prices for individual REE can and will change, and relative values might also change.
†Ore assays for Sc, U, and Th are generally not available.

Table 4 Common REE-bearing minerals and associated gangue minerals

Mineral[*]	Ore–Gangue[†]	Class[‡]	Formula[§]	Typical REE%	Density, g/mL	Hardness, Mohs scale	Electrostatic Response[**]	Magnetic Response (USGS)[††]	Acid Leach Response[‡‡]	Caustic Crack Response[§§]
Bastnaesite	O	C	$(Ce,La)CO_3F$	64	5.0	4–5	N	0.7–0.9	2, 3	2, 3
Huanghoite	O	C	$BaCe(CO_3)_2F$	34	4.6	5	—	0.7–0.8	3	2
Parisite	O	C	$Ca(Ce,La)_2(CO_3)_3F_2$	52	4.4	4–5	—	0.7	3	2
Synchysite	O	C	$Ca(Ce,La)(CO_3)_2F$	44	4.0	4–5	—	—	2, 3	2,3
Ankerite	G	C	$Ca(Fe^{2+},Mg,Mn)(CO_3)_2$	0	3.1	3–4	N	0.4–0.5	1	—
Calcite	G	C	$CaCO_3$	0	2.7	3	N	>1.7	1	—
Dolomite	G	C	$CaMg(CO_3)_2$	0	2.8	3–4	N	>1.7	1	—
Siderite	G	C	$Fe^{2+}CO3$	0	4.0	3.5	N	0.2–0.3	1	—
Fluocerite	O	F	$(Ce,La)F_3$	71	6.1	4–5	—	0.6–0.7	2, 3	2
Fluorite	G	F	CaF_2	0	3.1	4	N	>1.7	2, 3	2
Fergusonite	O	N	$YNbO_4$	36	5.1	5–6	—	0.4–0.6	3	3
Pyrochlore	O	N	$(Na,Ca)_2Nb_2O_6(OH,F)$	0	5.3	5–6	C	0.9–>1.7	3	2
Aeschynite	O	O	$(Ce,Ca,Fe,Th)(Ti,Nb)_2(O,OH)_6$	20–28	4.8–5.2	5–6	—	0.7–0.8	3	—
Cerianite	O	O	$(Ce^{4+},Th)O_2$	53	7.2	—	—	—	2	—
Euxenite	O	O	$(Y,Ca,Ce)(Nb,Ta,Ti)2O6$	19	4.8	6–7	C	0.5–0.6	—	2
Loparite	O	O	$(Ce,Na,Ca)_2(Ti,Nb)_2O_6$	25	4.8	5–6	—	—	—	—
Uraninite	O	O	UO_2	Varies	8.7	5–6	N	0.5–>1.7	2	—
Goethite	G	O	$Fe^{3+}O(OH)$	0	3.8	5–6	N	0.4–0.5	1	—
Hematite	G	O	Fe_2O_3	0	5.3	6–7	C	0.1–0.3	2	—
Ilmenite	G	O	$Fe^{2+}TiO_3$	0	4.7	5.6	C	0.2–0.3	—	—

(continues)

Table 4 Common REE-bearing minerals and associated gangue minerals (continued)

Mineral*	Ore–Gangue†	Class‡	Formula§	Typical REE%	Density, g/mL	Hardness, Mohs scale	Electrostatic Response**	Magnetic Response (USGS)††	Acid Leach Response‡‡	Caustic Crack Response§§
Magnetite***	G	O	$Fe^{2+}Fe^{3+}{}_2O_4$	0	5.2	5–6	C	0.01	2	—
Muscovite	G	O	$KAl_2(Si_3Al)O_{10}(OH,F)_2$	0	2.8	2–3	N	0.6>1.7	—	—
Quartz†††	G	O	SiO_2	0	2.6	7	N	>1.7	4	2
Brockite	O	P	$(Ca,Th,Ce)PO_4 \cdot H_2O$	6	3.9	3–4	—	—	—	2
Churchite	O	P	$(Nd,Y)PO_4 \cdot 2(H_2O)$	40–54	3.3	3–4	—	—	3	2
Monazite	O	P	$(Ce,La,Nd,Th)PO_4$	55	5.2	5–6	N	0.5–0.8	2, 3	2, 3
Steenstrupine	O	P	$Na_{14}Ce_6Mn^{2+}Mn^{3+}Fe^{2+}{}_2(Zr,Th)(Si_6O_{18})_2(PO_4)_7 \cdot 3(H_2O)$	27	3.4	4	—	—	—	2
Xenotime	O	P	YPO_4	48–65	4.8	4–5	N	0.4–0.5	2, 3	2, 3
Apatite	G	P	$Ca_5(PO_4)_3(OH,F,Cl)$	0	3.1	5	N	>1.7	2, 3	—
Allanite	O	S	$(Ce,Ca,Y,La)_2(Al,Fe^{3+})_3(SiO_4)_3(OH)$	18–32	3.8–3.9	5–6	—	0.4–0.5	2, 3	2, 3
Britholite	O	S	$Ca_2(Ce,Ca)_3(SiO_4,PO_4)_3(OH,F)$	28–41	4.1–4.5	5–6	—	0.4	—	—
Cerite	O	S	$(Ce,La,Ca)_9(Mg,Fe^{3+})(SiO_4)_6(SiO_3OH)(OH)_3$	55–62	4.7–4.9	5–6	—	0.6–0.7	—	2
Chevkinite	O	S	$(Ce,La,Ca,Th)_4(Fe^{2+},Mg)_2(Ti,Fe^{3+})_3Si_4O_{22}$	36	4.5	5–6	—	0.4–0.5	3	—
Eudialyte	O	S	$Na_{15}Ca_6(Fe^{2+},Mn^{2+})_3Zr_3SiO(O,OH,H_2O)_3(Si_3O_9)_2(Si_9O_{27})_2(OH,Cl)_2$	8	2.9	5–6	—	—	1, 2	—
Gadolinite	O	S	$(Ce,La,Nd,Y)_2FeBe_2Si_2O_{10}$	38–50	4.2	6–7	—	0.2–0.4	2	3
Mosandrite	O	S	$Na(Na,Ca)_2(Ca,Ce,Y)_4(Ti,Nb,Zr)(Si_2O_7)_2(O,F)_2F_3$	28	3.3	4	—	—	2	—
Stillwellite	O	S	$(Ce,La,Ca)BSiO_5$	51	4.6	6–7	—	—	—	—
Zircon	O	S	$ZrSiO_4$	4	4.7	7.5	N	>1.7	3	3
Kainosite	O	S	$Ca_2(Y,Ce)_2Si_4O_{12}(CO_3) \cdot (H_2O)$	30	3.5	5.5	—	—	2	—
Albite	G	S	$NaAlSi_3O_8$	0	2.6	7	N	>1.7	—	—
Augite	G	S	$(Ca,Na)(Mg,Fe,Al,Ti)(Si,Al)_2O_6$	0	3.4	5–7	V	0.4–0.9	—	—
Biotite	G	S	$K(Mg,Fe^{2+})_3[AlSi_3O_{10}(OH,F)_2$	0	3.1	2–3	N	0.3–0.8	3	—
Chlorite	G	S	$(Mg,Fe)_3(Si,Al)_4O_{10}(OH)_2(Mg,Fe)_3(OH)_6$	0	2.6–3.3	2–3	N	0.2–0.5	—	—
Diopside	G	S	$CaMgSi_2O_6$	0	3.4	6	N	0.4–0.6	—	—
Microcline	G	S	$KAlSi_3O_8$	0	2.6	6	N	>1.7	—	—
Orthoclase	G	S	$KAlSi_3O_8$	0	2.6	6	N	>1.7	—	—
Titanite	G	S	$CaTiSiO_5$	4	3.5	5–6	—	0.8–>1.7	4	4
Barite	G	Su	$BaSO_4$	0	4.5	3–4	N	>1.7	4	4
Pyrite	G	Su	FeS_2	0	5	6–7	C	>1.7	4	1
Wakefieldite	O	V	$(La,Ce,Nd,Y)VO_4$	30–56	4.2–4.8	4–5	—	—	—	—

*Mineral characteristics and response to acid or caustic attack varies depending on exact composition of mineral and conditions of attack.
†Ore–gangue abbreviations: O = ore, G = gangue.
‡Class abbreviations: C = carbonate, F = fluoride, N = niobate, O = oxide, P = phosphate, S = silicate, Su = sulfides/sulfate, V = vanadate.
§Formulae and mineral properties generally from Webmineral (2016) and Outokumpu Technology (2006).
**Electrostatic abbreviations: N = nonconductor; C = conductor; V = variable; — = no data.
††Magnetic response data from Rosenblum and Brownfield (1999) of the USGS and showing best range amperage data determined using a Frantz isodynamic magnetic separators, Model L-1.
‡‡Leach response scale:
1 – Readily attacked by dilute acid at ambient temperature
2 – Attacked by strong acids at elevated temperature
3 – Responds to concentrated sulfuric acid baking
4 – Not attacked by simple mineral acids
§§Alkali treatment scale:
1 – Attacked by weak caustic solution
2 – Attacked by strong (>50%) hot (>100°C) caustic solution
3 – Responds to caustic fusion or alkali fusion
4 – Not attacked by strong caustic solution
***Magnetite is ferromagnetic and gives a positive response in the Frantz separator at 0.01 A.
†††Quartz gives no response, even at Frantz separator currents exceeding 1.7 A.

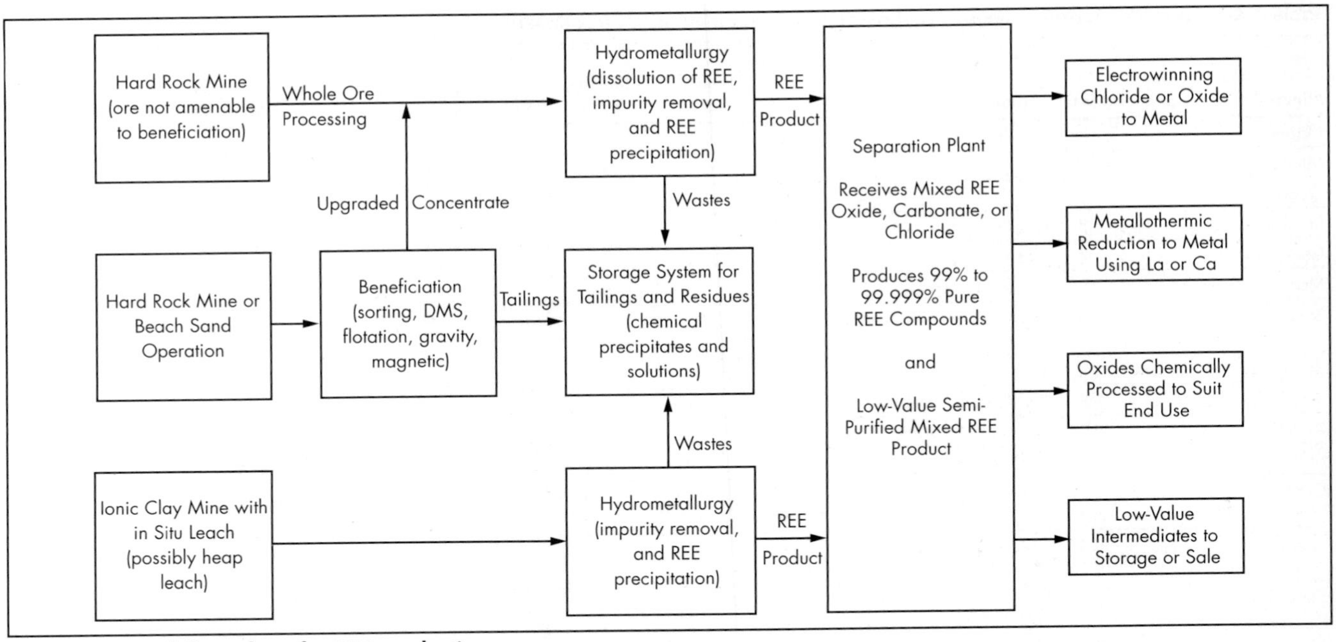

Figure 4 Overall flow sheet for REE production

PROCESSING

Process details for operating REE facilities are usually not published. Although companies are often legally required to state ore reserves, there are no requirements to provide details of how the ore is being processed. Furthermore, the REE industry is competitive and there is nothing like the sharing of technical information between producers of gold, copper, and other REE. Often, there is copious information on what has been tested in the laboratory but no reliable information on plant practice.

More information is available regarding proposed operations, especially for companies listed on stock exchanges in Canada, Australia, and the United States. In these cases, information can be obtained from company websites and also from SEDAR (System for Electronic Document Analysis and Retrieval, www.sedar.com), EDGAR (Electronic Data Gathering Analysis and Retrieval, www.sec.gov/edgar), or the ASX website (www.asx.com.au).

Overall Process Flow Sheet

REE ore deposits contain between a few hundred parts per million to as high as 50% REE. The ultimate end uses of REE typically require high-purity separated products. The general path from ore to end product is illustrated in Figure 4.

The beach sands and hard rock ores that are presently being processed are beneficiated using gravity, flotation, and/or other beneficiation methods to reject low-grade or waste material. The upgraded concentrate, usually containing between 40% and 60% REO, is then processed hydrometallurgically to "crack" the REE minerals, and the REE are then dissolved. Leach solutions are treated for the removal of impurities, including specific radionuclides, and processed to generate a purified bulk rare earth product. This product is often a solid oxide, carbonate, or chloride that is low in impurities and suitable for separation into the individual REE. If the hydrometallurgical plant and separation plant are closely coupled, then the product of the hydrometallurgical plant could be a purified solution suitable for separation. Some REE ores now under investigation are not amenable to beneficiation, and the developers are proposing to hydrometallurgically process the whole ore.

The ionic clay deposits in China are the dominant source of HREE, and although heap and vat leaching have been used, they are generally exploited by in situ leaching. Leach solutions are purified and a high-grade REE product recovered by precipitation and sent to separation plants.

Regardless of the ore type and manner of hydrometallurgical processing, the mixed REE product of the hydrometallurgical plant requires separation into the REE products of interest. The separated products will then be further processed (e.g., reduced to metal) before being used.

Less than 5% of the REE consumed is recycled and most of the mined REE contained in scrap fluorescent lamps, electronics, and automobiles end up in landfills. When reprocessing is practiced, it typically starts with physical upgrading, sometimes a thermal treatment, followed by hydrometallurgical processing to remove impurities or other products such as Hg, Ni, or Co, to yield a pure solid or solution containing REE ready for separation or for direct recycle as a mixed REE product. At least one proposed mining operation intends to incorporate scrap phosphor in the feed to its hydrometallurgical plant.

BENEFICIATION

Table 4 presents data on the density, magnetic response, and conductivity of common REE minerals and associated gangue. The density of most of the REE minerals is substantially higher than most of the gangue minerals with averages of 4.7 for the REE minerals (range 2.9–7.2) and an average of 3.5 for the gangue minerals, although the more common gangue minerals have densities of less than 3. Gravity separation can be an effective means of beneficiating REE ores and is used at some operations and proposed for some projects under development.

Several of the REE-bearing minerals respond well to magnetic separation, and this process is used in, or proposed for, several REE beneficiation plants. Some REE minerals have different conductivities to other REE minerals or accompanying gangue minerals, and electrostatic separation is used at beach and riverine sand processing operations.

The present REE producers and most potential producers are using some form of beneficiation to upgrade the ore before hydrometallurgical processing. However, some ores cannot be effectively beneficiated because the REE minerals are too fine, poorly liberated, or simply not amenable. In such cases, the whole ore is subjected to hydrometallurgical processing. Examples of such proposed projects include Alkane Resources' Dubbo Zr-REE project; Search Minerals' Foxtrot project; Frontier Rare Earths' Zandkopsdrift project; the Chuktukon and Tomtor REE projects in Russia; Texas Rare Earth Resources Corp.'s Round Top project; and all ionic clay deposits.

Beneficiation is discussed in more detail in the following section. Additional information can be obtained from the literature (Zhang and Edwards 2012; Jordens et al. 2013; Goode 2016c; Li and Yang 2014).

Coarse Ore Beneficiation

Ore sorting and dense media separation (DMS) are coarse ore beneficiation processes that operate on material in the approximate size ranges of 300 mm down to 20 mm, and 150 mm down to 0.5 mm, respectively. Both techniques have been, and continue to be, widely used in industries other than the REE industry.

Automated ore sorting has been used since the 1950s, and there are presently hundreds of machines sorting industrial minerals and ores of gold, uranium, and base metals. Many more machines are used for sorting glass cullet, metallic scrap, and seeds are used in other nonmining applications. Sorting machines can process rock particles as large as 300 mm and as small as 5 mm. However, the capacity of sorting machines decrease as the feed size becomes finer, and the sorting of particles finer than 20 mm is generally not economical. DMS is widely used for processing coal and base metal ore.

A 2016 review of the application of coarse ore beneficiation to REE ores prepared for Natural Resources Canada (Goode 2016a) indicated that there were no REE production facilities using ore sorting or DMS. The review also showed that few potential REE mine developers had examined the possibility of coarse ore beneficiation. However, three junior companies, Ucore Rare Metals, Quest Rare Minerals, and Namibia Rare Earths, had conducted tests and were planning to include dual energy X-ray transmission (DEXRT) ore sorting in their flow sheets. With more favorable ores, 50% mass rejection and >95% REE recovery were indicated.

The 2016 review showed that Great Western Minerals Group, Arafura Resources, and Northern Minerals had all tested DMS on their REE ores, but all three had decided not to include this process in their flow sheets.

The 2016 sorting study examined economics and concluded that coarse ore beneficiation could offer significant advantages on amenable ores (Goode 2016a). This was particularly true for remote processing sites using relatively expensive diesel electric power. Sorting or DMS substantially reduced the amount of material to be ground ahead of fine ore beneficiation (flotation, etc.) and thereby reduced power demand and costs.

Coarse ore beneficiation tests are simple to perform, and this option should be examined by anyone considering development of a REE deposit.

Flotation

Except for the ionic clay and beach/riverine sand operations, all five current and recent REE recovery facilities include flotation upgrading and exhibit the following common features.

- Deposit types: Carbonatite; Mt. Weld is weathered carbonatite.
- REE minerals: Dominantly bastnaesite (with associated monazite at Bayan Obo and Mountain Pass) and other minor REE minerals.
- Non-REE minerals: Carbonates, iron oxides, barite, fluorite.
- Grind—Relatively fine: 95% passing 75 μm (Bayan Obo), 80% passing 43 μm (Sichuan), 75% passing 74 μm (Shandong), and 38 μm at Mt. Weld.
- Beneficiation flow sheet:
  - Bayan Obo: Grind, LIMS (low-intensity magnetic separation) for magnetite recovery, WHIMS (wet high-intensity magnetic separation) for hematite+REE recovery, then a lower intensity WHIMS to separate hematite and REE, reverse flotation to upgrade magnetite+hematite concentrate, flotation upgrading of REE, concentrate transport to Baotou process plant.
  - Sichuan: Grind, gravity, flotation, concentrate transport to process plant.
  - Mt. Weld: Grind, float, concentrate bagged and containerized, ship to process plant in Malaysia.
  - Mountain Pass: Grind, float, leach to dissolve carbonates, concentrate to on-site process plant.
- REE collectors: Hydroxamates, except Mountain Pass, which used fatty acid.
- Depressants: Sodium silicate, except Mountain Pass, which used sodium lignin sulfonate.
- Flotation temperature: Typically elevated (near 90°C at Mountain Pass), except Sichuan.
- Overall metallurgical performance: As shown in Table 5.

All plants target high-grade REE concentrates at the expense of REE recovery. The reason is the high operating cost of the typical hydrometallurgical process, and because mineral resources are generally large, and mining and beneficiation costs are relatively low.

Most of the REE projects under development include beneficiation by flotation. With few exceptions, the grade of concentrate achieved is far lower than the 40% REO or greater concentrate grade of the producing operations. These low-concentrate grades are a detriment to good economics.

Gravity Concentration

The REE minerals generally have high density values and, if the mineralized particles are coarse enough (say, >10 μm) and liberated, could be separated from lighter gangue minerals by density-based processes. Beach and riverine sands are processed for REE recovery by a combination of gravity, magnetic, and electrostatic methods. Spirals, shaking tables, and jigs are variously used for the gravity separation step. A few hard rock REE beneficiation plants in Sichuan and Shandong in China use gravity concentration on shaking tables either alone or in combination with flotation or magnetic separation.

Table 5 Overall performance of REE beneficiation plants

		Non-REO Streams*		REO Concentrates		
Operation	Feed Grade, % REO	Grade, % REO	REO Distribution, % Feed	Mass, % Feed	Grade, % REO	REO Recovery, % Feed
Bayan Obo	5.5	3.7	64	4	49	36
Mianning	7.0	1.7	22	11	61	78
Mountain Pass	8.5	3.2	35	8	68	65
Mt. Weld	15.0	6.1	30	26	40	70

*Non-REO streams for Bayan Obo include Fe concentrate (44% mass, grading 61% Fe at 80% Fe recovery) and tailings. REO concentrate is blend of 61% REO and 40% REO concentrates.

Several potential REE projects under investigation have included gravity separation at various stages of the development process. Avalon Advanced Materials proposed to use a Mozley multi-gravity separator, essentially a centrifugal shaking table, in its flotation circuit. Arafura Resources is proposing gravity separation in the beneficiation of its Nolans REE-U-P project. AMR Minerals Metals proposes to employ gravity concentration using spirals followed by magnetic separation to eliminate magnetite, then further upgrading on shaking tables ahead of flotation at its Aksu Diamas REE project in Turkey. Rare Element Resources has included gravity concentration with magnetic separation for certain ore types from the Bear Lodge deposit.

Magnetic Separation

The Bayan Obo plant in China uses LIMS to extract magnetite from the ground ore, then two stages of WHIMS to first recover hematite and REE minerals (bastnaesite, monazite, huanghoite, etc.) and then separate the hematite and REE minerals, followed by flotation upgrading of the separated hematite and REE streams. Some of the Sichuan operations also use LIMS and WHIMS as part of their beneficiation processes.

Regarding potential REE operations, Leading Edge Materials Corp. is considering development of the Norra Kärr project in Sweden in which eudialyte is the dominant REE mineral. Matamec Explorations is similarly considering development of the Kipawa project in Canada, which contains eudialyte, mosandrite, and britholite. Both companies propose to use LIMS to eliminate magnetite, and similar minerals, ahead of WHIMS for the recovery of the REE minerals. Tanbreez Mining Greenland A/S is proposing to use magnetic separation to upgrade its eudialyte-bearing ore in Greenland. Commerce Resources Corp. is studying the Ashram REE deposit in Quebec and proposes to make a flotation concentrate, leach it with hydrochloric acid (HCl) to remove carbonates, regenerate the HCl, and use WHIMS to recover the monazite, bastnaesite, and xenotime from the leach residue. The nonmagnetic fraction could be marketed as metallurgical-grade fluorspar and the REE concentrate subjected to sulfuric acid baking (Commerce Resources Corp. 2014).

Electrostatic Separation

Mineral sand deposits in Australia, India, and elsewhere are typically wet processed using spirals, or Reichert cones and other gravity devices to produce a heavy mineral concentrate (HMC) and tailings. The HMC is usually dried and separated into the constituent minerals. Commonly, a first electrostatic separation is used to separate a rutile/ilmenite/leucoxene conducting concentrate from a zircon/monazite/garnet/sillimanite nonconducting concentrate. The two concentrates are further separated using magnetic, additional electrostatic, and dry gravity methods.

MINERAL "CRACKING" AND LEACHING

Most REE minerals are refractory (inert to chemical attack) and need "cracking" using aggressive methods, such as baking with strong sulfuric acid or treatment with strong alkali. These processes can be capital intensive and expensive to operate but are viable with certain ores. The ionic clay deposits are more readily processed since they only need contacting with a weakly acidic electrolyte to displace the REE from the exchange sites on the clay. REE ore bodies dominated by silicate minerals, such as allanite, eudialyte, mosandrite, and britholite, are also an exception since the silicates are generally easy to leach. Reference material includes the various company websites, Verbaan et al. (2015), Goode (2016c), Buchanan et al. (2014), Gupta and Krishnamurthy (2005), Sadri et al. (2017), Zhang and Edwards (2013), and publicly available company reports from SEDAR, EDGAR, and ASX.

REE deposits usually contain more than one REE-bearing mineral, and occasionally one leach or cracking system needs to be followed by a second cracking operation to extract all of the REE from the ore. In other cases, the main mineral can be cracked, but part of the REE might be re-precipitated requiring a second crack/leach stage for full recovery. Examples include Molycorp's Mountain Pass operation where fluoride re-precipitation occurs (SRK Consulting 2010) and Greenland Minerals and Energy's Kvanefjeld project where a sulfuric acid leach of a high-sodium material leads to the formation of an insoluble double sulfate (Krebs 2014). Both issues are solved using caustic metathesis and releaching.

Low-Temperature Acid Leaching

Some REE minerals are amenable to low-temperature (<100°C) acid leaching. If all the REE in a given material are contained entirely in such minerals, then low-temperature acid leaching can be effective.

Fluorocarbonate Minerals

Dilute mineral acids readily attack bastnaesite and synchysite, but these minerals contain fluorine and the REE fluorides are insoluble. Accordingly, low-temperature acid processing of ores and concentrates containing bastnaesite and/or synchysite can be challenging because the REE can enter solution but immediately re-precipitate as fluorides.

In its early operations, Mountain Pass produced a flotation concentrate that was leached in weak HCl to remove calcite and strontianite and produce a 60%–70% REO grade concentrate dominantly containing bastnaesite. In the pre-2010 flow sheet, the concentrate was roasted at 650°C to crack the bastnaesite, volatilize the contained fluorine, and oxidize the Ce to Ce(IV). The resulting calcine was cooled and leached in cold, weak HCl, which dissolved all the REE except for the Ce(IV). The leach residue was further processed and marketed as a glass-polishing agent with some upgraded to high-grade CeO_2.

The Mountain Pass flow sheet operated after 2010 subjected the preleached bastnaesite to a strong HCl leach that extracted most of the REE, except for about 30%, which combined with the F released from the bastnaesite and re-precipitated as fluorides (SRK Consulting 2010). To get high overall leach extraction, the leach residue was metathesized with NaOH and releached in the pregnant leach solution (PLS) obtained in the first leach in a process similar to that operated in the 1950s by Molycorp at its York, Pennsylvania (United States) facility. However, severe difficulties were encountered with the new Mountain Pass leach system that, with issues elsewhere in the operation including the chlor-alkali plant for reagent regeneration, resulted in unprofitable operations and closure of the plant in 2015 (Thompson Reuters 2015).

Chinese researchers have investigated ways of avoiding the F precipitation effect and otherwise improving the environmental aspects of its REE operations processing F-bearing concentrates. Zhou and Liu (2015) describe pelletizing a mixture of Baotou flotation concentrate and soda ash, roasting at 600°–700°C, washing the calcine with water or weakly acidified water to remove sodium fluoride and other contaminants, then HNO_3 leaching to extract the REE and Th. The process avoids the loss of fluorine and/or the adverse effects of F on REE dissolution. It also oxidizes Ce(III) to Ce(IV), which is extracted with a solvent along with Th, leaving the trivalent REE free of Ce and Th and ready for separation. In 2000, Zhu et al. (2000) described a similar process to remove F from a Sichuan bastnaesite ore using a Na_2CO_3 roast for 30 minutes at 650°C followed by chlorination roasting using NH_4Cl for 1 hour at 325°C to render the REE water soluble.

Peak Resources is developing the Ngualla REE project in Tanzania, which will produce a flotation concentrate containing bastnaesite and assaying about 40% REO. This will be treated by a selective alkaline roast with soda ash at 700°C followed by a water wash to remove the solubilized fluoride and avoid REE re-precipitation. The water wash is followed by a weak (<1%) HCl leach at 80°C to dissolve the REE. The Ce is oxidized to Ce(IV) during roasting such that Ce leach extraction is low. This is seen as an advantage given low Ce prices (Peak Resources 2016).

Silicate Minerals

Eudialyte, a silicate mineral, is readily leached with dilute sulfuric acid, and potential producers, including Leading Edge Minerals Corp. (Norra Kärr project) and Matamec (Kipawa Zeus project), are proposing this leach system. Although much of the REE is dissolved from the silicate mineral, a significant amount of silica also enters the solution and can cause filtration and SX issues.

Leading Edge Minerals determined that its eudialyte concentrate could be effectively leached at pH 1, 30°C, a low pulp density of 15% to avoid silica precipitation, and a 2-hour leach time to give 90% REE extraction. Silica in solution is later removed by raising the pH and filtration followed by reheating the solution and re-acidification ahead of solvent extraction (SX). Matamec found that with a 150-µm grind, a 5-hour leach with 20 g/L free acid, ambient temperature, and 35% solids, its eudialyte-bearing concentrate would yield REE extractions of about 90% and a leached slurry that could be handled through normal liquid–solid separation equipment.

Ucore proposes to upgrade its Bokan Mountain ore using DEXRT sorting and magnetic concentration to produce a 25% mass concentrate containing the silicate minerals allanite and kainosite, as well as bastnaesite. The concentrate is ground to 100% passing 40 µm and then leached for 8 hours at 90°C in two stages using 30% nitric acid. Ucore proposes to recover excess HNO_3 from the PLS using diffusion dialysis.

Sulfuric Acid Baking

Almost all the REE-bearing minerals respond well to baking with concentrated sulfuric acid at temperatures greater than 200°C. An exception is zircon, which is generally not affected by acid baking and can carry significant levels of HREE. In this case, the residue from an acid bake can be caustic cracked to access the REE and Zr in the zircon. Fergusonite, samarskite, and other Nb-bearing REE minerals do not respond well to low-temperature acid baking.

A simple acid bake of REE minerals using excess sulfuric acid at 200°–300°C leads to the following reactions (Xu 1995) for REE minerals and common accessory minerals, using La as a surrogate for all the REE.

Bastnaesite:

$$2La(CO_3)F + 3H_2SO_4 \rightarrow La_2(SO_4)_3 + 2HF\uparrow + 2H_2O\uparrow \quad \textbf{(EQ 1)}$$

Monazite:

$$2LaPO_4 + 3H_2SO_4 \rightarrow La_2(SO_4)_3 + 2H_3PO_4 \quad \textbf{(EQ 2)}$$

and $Th(SO_4)_2$

Hematite:

$$Fe_2O_3 + 3H_2SO_4 \rightarrow Fe_2(SO_4)_3 + 3H_2O\uparrow \quad \textbf{(EQ 3)}$$

Fluorspar:

$$CaF_2 + H_2SO_4 \rightarrow CaSO_4 + 3HF\uparrow \quad \textbf{(EQ 4)}$$

Calcite:

$$CaCO_3 + H_2SO_4 \rightarrow CaSO_4 + CO_2\uparrow + H_2O\uparrow \quad \textbf{(EQ 5)}$$

Silica:

$$SiO_2 + 6HF \rightarrow (H_3O)_2SiF_6\uparrow \quad \textbf{(EQ 6)}$$

and $SiO_2 + 4HF \rightarrow SiF_4\uparrow + 2H_2O\uparrow$

At progressively higher temperatures through the range of 300° to 800°C, other reactions take place, including

$$2H_3PO_4 \rightarrow H_4P_2O_7 + H_2O\uparrow \quad \textbf{(EQ 7)}$$

$$Th(SO_4)_2 + H_4P_2O_7 \rightarrow ThP_2O_7 + H_2SO_4 \quad \textbf{(EQ 8)}$$

$$H_2SO_4 \rightarrow SO_3\uparrow + H_2O\uparrow \quad \textbf{(EQ 9)}$$

$$Fe_2(SO_4)_3 \rightarrow Fe_2O(SO_4)_2 + SO_3\uparrow \quad \textbf{(EQ 10)}$$

$$H_4P_2O_7 \rightarrow 2HPO_3\uparrow \quad \textbf{(EQ 11)}$$

$$Fe_2O(SO_4)_2 \rightarrow Fe_2O_3 + 2SO_3\uparrow \quad \textbf{(EQ 12)}$$

$$La_2(SO_4)_3 \rightarrow La_2O(SO_4)_2 + SO_3\uparrow \quad \textbf{(EQ 13)}$$

$$La_2O(SO_4)_2 \rightarrow La_2O_3 + 2SO_3\uparrow \quad \textbf{(EQ 14)}$$

A low-temperature acid bake, say at 220°C, will convert most of the REE into water-soluble forms, allowing a simple water leach of the baked material to dissolve the REE. Th, Fe species, and other gangue components are also largely or completely sulfated. These reactions consume acid and solubilize Th and Fe, resulting in a PLS containing high impurity levels.

A high-temperature bake, or a high-temperature second-stage bake of a low-temperature baked residue, will, depending on the temperature, cause Reactions 7 to 14 to occur, with Sc sulfate decomposition starting at 700°C and the other REE at about 800°C (Borra et al. 2016). Thorium pyrophosphate and Fe_2O_3 are sparingly soluble, so baking at a higher temperature leads to a PLS that is low in impurities. Depending on the minerals in the feed, a high-temperature bake (>500°C) can also lead to reduced extraction of Zr, Nb, and Al. Finally, much of the acid consumed in sulfation reactions at the lower temperature is released as sulfur oxides at higher temperatures, and these can be used to reconstitute sulfuric acid from the off-gas stream.

Although there are positive aspects to the high-temperature operation, REE recovery can suffer at the higher temperatures. Additionally, the thermal demand of the system is increased and the design and installation of the high-temperature baking equipment, its refractory lining, and the peripheral equipment all become more complex and expensive.

Acid baking is typically achieved by mixing concentrated acid and REE-bearing feed and then placing the mixture into a horizontal rotating kiln for heating to the desired temperature and holding for the necessary time—often between 2 and 4 hours. The high-temperature (>500°C) acid baking process is used to process concentrates obtained from Bayan Obo, Mianning-Dechang, and Mt. Weld. Kilns are typically direct fired and lined with acid-proof refractory. Accretion can be an issue, and scrapers, calcine recycle, and other devices and strategies are used or proposed to reduce such problems. Off-gases need to be scrubbed for the capture of fluorine and sulfur compounds.

Acid baking in rotary kilns is common in Chinese REE recovery plants where the kilns have been operating trouble-free for many years. Initial operations were low tonnage and kilns were small. For example, a processing rate of 3 t/d concentrate could be handled by a kiln that was 0.7 m internal diameter by 11 m long (Xu 1995). As demand and production increased, kilns as large as 1.3 m diameter by 25 m long, and greater, were in use and capable of processing 15 t/d to produce about 2,500 t/yr of REO. Mixing of acid and concentrate, kiln feeding, kiln detailed design, off-gas handling, and so on, are all well established.

Large-scale acid baking of REE concentrates has only recently been applied outside of China. Lynas Corporation's LAMP (Lynas Advanced Materials Plant) operation at Kuantan in Malaysia is designed to process 66,000 t/yr of flotation concentrate containing 40% REO produced at Mt. Weld in Australia. Four 60-m-long kilns have been installed at the LAMP site, each nominally processing about 55 t/d of concentrate at 600°C. The Lynas kilns, and associated mixing and feeding systems, are far larger than the equivalent plants in China. First feed to two Phase 1 kilns was made late in 2012, but serious issues in the baking systems took time to resolve and full design production was only attained in September 2017 (Lynas Corporation 2017). Clearly, high-capacity acid mixing–transfer–bake systems are difficult to design correctly.

Outotec has proposed the use of a high-intensity mixer to contact concentrated sulfuric acid and REE concentrate to form pellets in the 0.1–2 mm size range. Although sulfation in a rotary kiln frequently needs 2–4 hours, Outotec claims 15 minutes is sufficient for complete sulfation in a fluidized bed roaster operating at about 250°–300°C. The sulfated product can then be transferred into a second fluidized bed roaster operating at about 700°C for the selective decomposition of the Fe, Al, and other sulfates (Guentner et al. 2016). Off-gas from the two-stage roasting system can be processed for the regeneration of sulfuric acid, capture of fluoride species, and so on. The Outotec two-stage roasting system forms an integral part of Frontier's proposed Zandkopsdrift project (Venmyn Deloitte 2015).

Several proposed REE projects would include sulfuric acid baking, including Strange Lake in Quebec, Canada (REE silicates); Ashram (monazite); Steenkampskraal in South Africa (monazite); Browns Range in Western Australia (xenotime); and Eco Ridge in Ontario, Canada (monazite). Sulfuric acid baking at high temperature has been proposed and tested for recovering REE from bauxite residue (Borra et al. 2016). In some cases, a preleach, often using HCl, is proposed ahead of the acid bake to eliminate undesirable elements or reduce the feed mass going to the kiln feed.

Agitated Leaching with Concentrated Sulfuric Acid

Concentrated sulfuric acid can be used in an agitated leach system at temperatures up to the boiling point of concentrated acid, which is 287°C for 93% acid. This form of leaching has been referred to as SAAB (strong acid agitated bake) or concentrated acid "pot digestion." Test results have been reported by Commerce Resources (2014) and Rainbow Rare Earths (2016).

The strong acid agitated leach approach typically gives about the same REE extraction as acid baking, but it avoids the difficulties associated with kilns. However, the high-acid addition of the strong acid agitated leach necessitates the recovery of the excess acid, which presents difficult process challenges that have yet to be solved on an industrial scale.

Caustic Cracking

Fusion with pure alkali (NaOH, KOH, etc.) in the absence of water or agitation with strong caustic solutions (e.g., 50% NaOH or greater) at elevated temperatures will crack most REE-bearing minerals, leading to the formation of hydroxide-like compounds that can be leached with a weak-acid solution. Typical reactions for selected REE minerals and typical gangue minerals include the following:

Bastnaesite:

$$La(CO_3)F + 3NaOH \rightarrow La(OH)_3 + NaF + Na_2CO_3 \quad \textbf{(EQ 15)}$$

Monazite:

$$LaPO_4 + 3NaOH \rightarrow La(OH)_3 + Na_3PO_4 \quad \textbf{(EQ 16)}$$

and $Th(OH)_4$

Fergusonite:

$$YNbO_4 + 3NaOH \rightarrow Y(OH)_3 + Na_3NbO_4 \quad \textbf{(EQ 17)}$$

Apatite:

$$Ca_{10}(PO_4)_6F_2 + 20NaOH \rightarrow 10Ca(OH)_2 + 20Na^+ + 6PO_4^{3-} + 2F^- \quad \textbf{(EQ 18)}$$

Feldspar:

$$KAlSi_3O_8 + 6NaOH \rightarrow K^+ + 6Na^+ + Al(OH)_4^- + 3SiO_3^{2-} + H_2O \qquad \textbf{(EQ 19)}$$

Fluorite:

$$CaF_2 + 2NaOH \rightarrow Ca(OH)_2 + 2Na^+ + 2F^- \qquad \textbf{(EQ 20)}$$

Silica:

$$SiO_2 + 2NaOH \rightarrow Na_2SiO_3 + H_2O \qquad \textbf{(EQ 21)}$$

The preceding reactions show the use of NaOH, but KOH and mixtures can also be effective, although there are differences between NaOH and KOH. For example, the fusion temperature of NaOH is 318°C, whereas that of KOH is 360°C. Another example concerns the solubility of the reaction products. If niobium is present, it could be important that Na_3NbO_4 is soluble in water but insoluble in strong NaOH solution, whereas the K equivalent is soluble in KOH solutions.

Fusion cracking has been, and is, used industrially for the cracking of tungsten concentrates (NaOH, or Na_2CO_3), niobium concentrates (NaOH, with soda ash; or KOH), and zircon (NaOH). Fusion cracking of REE minerals has not been practiced, although development companies have investigated the process in the laboratory. Typically, fusion involves mixing excess alkali with the mineral of interest, heating the mixture on a batch basis to above the fusion point, and holding at that temperature for a required period, often 2–4 hours. When fusion is complete, the melt is allowed to cool or else granulated, then mixed with water, the excess alkali solution filtered off, and the hydroxides processed for metal recovery.

Caustic cracking with strong NaOH solutions (60%–70%) was commonly used for cracking monazite. The process was employed in Brazil, by Rhone Poulenc (now Solvay) at La Rochelle in France, and by REE and Th producers in India. The monazite was generally ground to –40 µm, mixed with an excess of NaOH as a ~50% solution, and held with agitation at 140°C or greater for several hours. At the end of the cracking period, the slurry was diluted with water and agitated, solution was recovered by filtration, the filter cake washed to remove residual caustic solution, the solution cooled for crystallization and recovery of trisodium phosphate (TSP) recovery and then recycled, and the hydroxide-bearing filter cake leached for REE and Th recovery. This was usually done by an initial HCl leach to a pH of about 3.4 to dissolve the REE hydroxides followed by liquid–solid separation, then a strong HCl leach to dissolve Th (Krumholz 1957; Kaczmarek 1981).

One of the key aspects of the caustic cracking process is the removal of excess NaOH and its recovery for reuse, and elimination of soluble cracking products, such as TSP, aluminates, and silicates, without crystallization of the TSP. The solubility of TSP is a strong function of NaOH strength and temperature (Youtz et al. 1950), and test and plant designs need to recognize this issue.

Caustic cracking is not now being used. Of 18 REE development projects summarized by Verbaan et al. (2015), only three are proposed with caustic cracking. Along with Mountain Pass, discussed earlier, the proposed REE caustic crack operations are as follows:

- The Kvanefjeld project in Greenland will use caustic cracking to treat a preleach residue containing REE-Na double sulfates. Details of the cracking step are not provided.
- Lofdal in Namibia will batch process a preleached concentrate containing xenotime. A 50% NaOH solution, 20% solids, and 140°C for 4 hours are proposed followed by hot water washing, then HCl leaching.
- Mkango's Songwe Hill deposit in Malawi contains REE in apatite and synchysite with calcite as the dominant gangue mineral. The proposed flow sheet includes a pH 3.5 HCl preleach to remove calcite with HCl regeneration from the preleach PLS. The preleached concentrate is then leached at ambient temperature with 30% HCl for 0.5 hours, which gives about 80% extraction of LREE and >90% extraction of HREE. A caustic crack applied to the REE leach residue (50% NaOH, 100°C for 4 hours) releases REE from fluoride precipitates and any minerals not cracked by the 30% HCl leach. The NaOH crack step is followed by a 5% HCl releach, which increases overall recoveries for all REE (excluding Ce, Yb, and Lu) to >92% (MSA Group 2015).
- Mountain Pass included caustic cracking of an HCl leach residue to recover insoluble fluorides formed during HCl leaching of the bastnaesite in the preleached concentrate.

Other Cracking Systems

Numerous alternative roasting and baking systems for REE ores and concentrates have been proposed, and some have been used commercially. The roasting of bastnaesite concentrate with a sodium salt ahead of hydrochloric acid leaching has been mentioned earlier as a means of allowing the elimination of fluorine and oxidizing Ce.

The Kashka REE plant in Kyrgyzstan processed a complex ore, high in HREE content, from the Kutessay II mine. Concentrate containing 6%–7% REE was cracked by reaction with Na_2CO_3 at elevated temperature in a rotary kiln. The calcine was ground in a ball mill and dewatered; then the REE was dissolved in nitric acid, purified, and separated using tri-n-butyl phosphate (TBP) SX and IX (Stans Energy 2017; Danilov 2011).

Roasting at 700°C for 1 hour with 15% CaO and 15% $NaCl\text{-}CaCl_2$ has been proposed as a means of decomposing the bastnaesite in a Mianning REE concentrate while preventing HF formation (Bian et al. 2006). The bastnaesite was 94% decomposed during the roast. The calcine from this process would be acid leached to dissolve the REE compounds formed during roasting. Similar experiments on a sample of Bayan Obo concentrate, containing bastnaesite and monazite, gave 70% decomposition with CaO alone and 90% decomposition, and almost total retention of F, with $CaO\text{-}NaCl\text{-}CaCl_2$ mixture (Bian et al. 2010). The $NaCl\text{-}CaCl_2$ becomes molten at about 520°C and facilitates the reactions.

Investigators have studied fixing fluorine by mixing bastnaesite-bearing concentrate containing 67% REO and 7% F with MgO (0.1 to 0.3 t/t concentrate) followed by chlorination roasting at ~325°C for 1 hour using about 1:1 mass ratio of NH_4Cl (Chi et al. 2004).

Chinese investigators working with Baotou concentrate and Peak Resources propose to add sodium carbonate to bastnaesite concentrate to promote the formation of NaF during roasting, which can then be water-washed from the calcine to eliminate fluoride precipitation issues.

Avalon Advanced Materials obtained promising results from a mixed alkali cracking pilot plant in which REE concentrate was pelletized with Na_2CO_3 and $CaCO_3$ and then fed to a rotary kiln, the product water leached to remove excess alkali and then acid leached (Avalon Advanced Materials 2013).

Chlorination

Up to 13,000 t/yr of loparite concentrate recovered from mines in the Kola Peninsula are processed by chlorination at the Solikamsk Magnesium Works in the Urals. Loparite concentrate contains about 9% Nb+Ta oxides, 36% TiO_2, and 29% REO; it is processed to yield mixed REE carbonates and oxides, and Nb and Ta products. In 2015, Solikamsk processed 8,509 t loparite and produced 2,312 t REO, 2,014 t Ti, 628 t Nb_2O_5, and 32 t Ta_2O_5 (JSC Solikamsk Magnesium Works 2016).

Zelikman et al. (1964) described chlorination in which loparite was mixed with coke and a binder, briquetted, "coked" at 700°C, then chlorinated at about 800°C in an electric shaft furnace, 4.5–5 m internal diameter by up to 8-m-high partially filled with carbon cylinders as the resistive heating elements. The off-gas was fractionally condensed with a solid containing $NbOCl_3$, $NbCl_5$, $TaCl_5$, and $FeCl_3$ collected first and further processed for Nb and Ta recovery. $TiCl_4$ was then recovered in a packed tower irrigated with cooled liquid $TiCl_4$. REE, Ca, and Na chlorides were not volatilized at the furnace operating temperatures and were removed as a molten salt and refined to yield mixed REE carbonates and oxides.

The Australian Mineral Development Laboratory described the chlorination of briquettes made from a carbon–monazite mixture at temperatures greater than 900°C. Thorium and phosphate were volatilized, while the REE chlorides were collected as molten anhydrous chloride with a recovery exceeding 85% (Hartley 1952).

Th. Goldschmidt AG also developed a process in which finely ground concentrate was mixed with carbon and a binder such as starch or sugar, briquetted, dried, then chlorinated using chlorine gas at 1,000–1,200°C. A special furnace was designed for the aggressive system with heat generated by passing electricity through a bed of coke upon which the carbon–REE concentrates were supported (Brugger and Greinacher 1967).

If the REE concentrate is a phosphate or niobate, the REE are converted to chlorides that are removed as a molten salt mixture, and many of the other chlorides, including $ThCl_4$, are volatilized and leave with the off-gas. If the REE concentrate is bastnaesite, at least part of the REE report as fluorides and part of the Th and U will similarly be converted to fluorides and not volatilized.

Several researchers have developed solutions for the bastnaesite chlorination problem by the addition of fluorine complexing reagents. Wang and Zhang (2002) propose a first bake of the carbon–concentrate mixture with the addition of a silicon, phosphorous, or boron-containing material at between 400° and 700°C to remove F as well as Fe and P. A second-stage roast with Cl at 700°–900°C removes U and Th and chlorinates the REE, which are separated as molten salts from solid residues by filtration at 700°–1,000°C to yield a pure REE chloride melt. The work is described in detail by Zhang et al. (2004). The preferred agent for F removal is revealed as being $SiCl_4$. $AlCl_3$ is used to enhance Th removal.

In 2015, Niobec was granted a patent covering the chlorination of bastnaesite (Bergeron et al. 2015). The patent discusses the fluoride issue and identifies boron-containing Lewis acids, such as $B(OH)_3$, BF_3, BCI_3, B_2O_3, and $Na_2B_4O_7$, as being useful for eliminating the adverse effect of fluorine. Using La to represent the REE, the reactions are

$$LaFCO_3 \rightarrow LaFO + CO_2 \quad \textbf{(EQ 22)}$$

$$3LaFO + \text{Lewis acid-}Cl_3 \rightarrow 3LaClO + \text{Lewis acid-}F_3 \quad \textbf{(EQ 23)}$$

In the proposal, bastnaesite concentrate is mixed with a carbonaceous reductant, the Lewis acid, and a binder, pelletized to a 10-mm top size, screened at 1 mm, and then dried before being chlorinated. Off-gases are fractionated to remove volatilized chlorides of Fe, Al, P, Ti, Si, Nb, Zr, and Hf, and excess Cl and BF_3 are recycled. Th and U are eliminated by distillation from the molten chloride at 900°C. Gaseous reductants, such as CO, can be used instead of solid reductants.

Elution of Ionic Clay

The ionic clay deposits of Southern China are the main source of HREE (Goode 2016c). Undeveloped deposits are found in Madagascar, Africa, Brazil, and Chile. The genesis, prevalence, mineralogy, and exploitation of such deposits have been discussed by Chi and Tian (2008); Rocha et al. (2014); Moldoveanu and Papangelakis (2013); Goode (2016c); and Vossenkaul et al. (2015).

The ionic clay deposits, also known as "weathered crust elution-deposited" and "saprolite" deposits, that are in production are located in China's Jiangxi, Fujian, Guangdong, Yunnan, Hunan, and Zhejiang provinces and Guangxi Zhuang Autonomous Region. They are low grade, typically 0.1% REO, but are economical to process and usually deliver products that are low in Ce. Typical REE distributions and in situ values are shown in Figure 5 with corresponding data for selected operating hard rock deposits.

The ionic clay deposits were formed by intensive weathering in which REE-bearing bedrock was altered into clay-rich saprolite, often for tens of meters from present surface. The original REE minerals, typically xenotime and monazite, were decomposed to an extent between 50% and almost 100%. The dissolved REE were then sorbed through IX processes by the clays of the saprolite. The extraction of the REE sorbed on the clay is readily achieved by contacting the ore with an electrolyte solution such as 0.5 M NaCl or 0.2 M $(NH_4)_2SO_4$.

During deposit formation, cerium was usually oxidized to Ce^{4+} and precipitated with any dissolved Th as cerianite $(Ce^{4+},Th^{4+})O_2$. Cerianite is not dissolved by the leach solutions employed, hence the low Ce of the products.

When ionic clay deposits were first exploited in 1969, REE were extracted from the ore by mining and vat leaching or by heap leaching. However, environmental issues were evident in the form of massive destruction of forests and land, landslides because of the removal of vegetation, and surface water and agricultural land polluted by leach reagents. Eventually, in situ methods of recovery were adopted, particularly for the hilltop deposits that are typical of the area, and are now mandated for new operations.

In situ leach sites in China are often small and might produce just a few hundred tons annually of REO over a couple of years. Larger operating complexes might produce 1,000 or 2,000 t/yr. Preparation for in situ leaching is minimal with vegetation left in place, injection wells drilled on roughly 10-m center around the hilltop, and a low-cost plastic pipe distribution system installed. PLS is collected in tunnels or

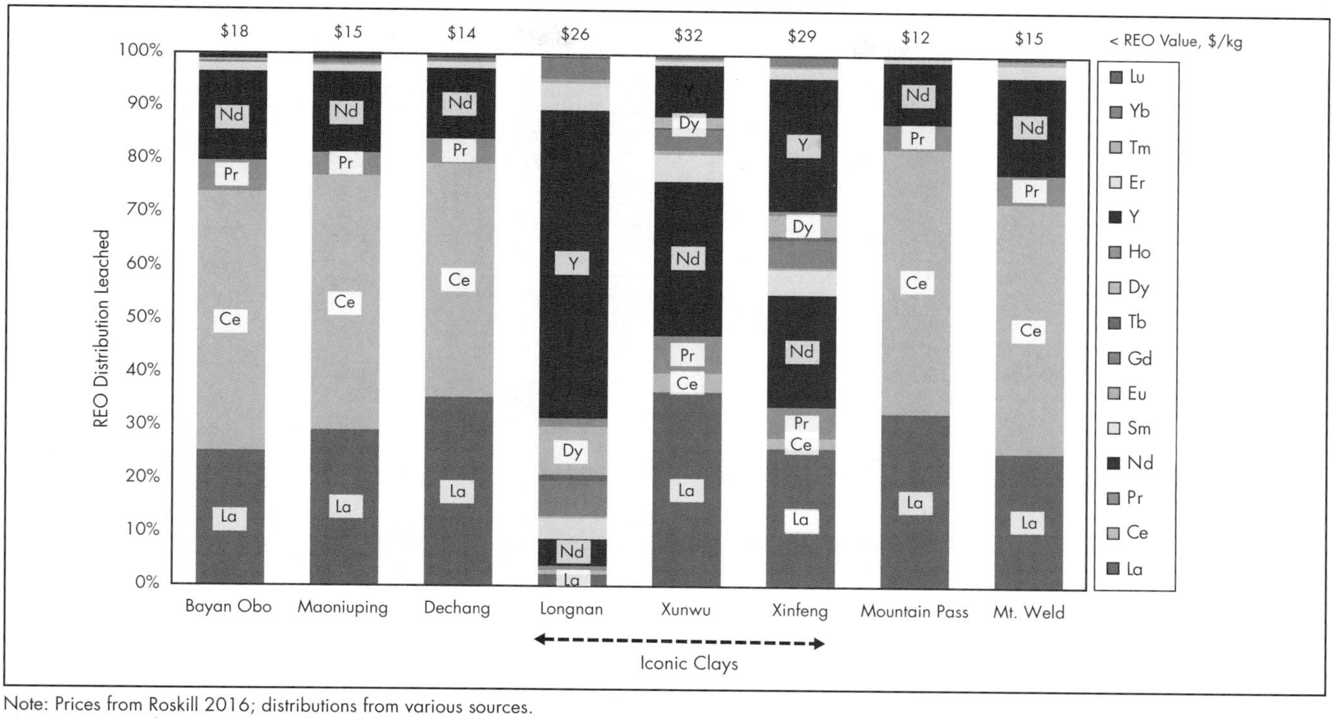

Note: Prices from Roskill 2016; distributions from various sources.

Figure 5 Distribution and basket value of in situ hard rock and recovered ionic clays

ditches at the base of the hill. Some recirculation of solution can be practiced to increase grades, but eventually the solution is processed by batch neutralization to pH 5.3 and filtration to remove impurities, then batch precipitation of REE using either oxalic acid or ammonium bicarbonate, followed by filtration and ignition of the precipitate to yield a mixed REO product. Equipment at the smaller operations is rudimentary, with in-ground excavations lined with plastic sheet used as precipitation tanks and plate-and-frame filters for liquid–solid separation.

Some are concerned about the in situ process since lixiviant or PLS can be lost to bedrock fractures or inadequate solution retrieval systems, leading to contaminated aquifers. The recovery of REE from the resource can be low if there is a loss of lixiviant or PLS or if there are extensive areas of low permeability in the deposit. The injection of leach solution into the clay and subsequent rinse water injection and/or rainfall can cause swelling of the clays, leading to landslides.

Much has been said about the illegal production of REE in China, perhaps 40,000 t/yr REO in recent years (China State Council 2012; Roskill 2016). Small, in situ, ionic clay operations are inconspicuous; would readily escape detection; are cheap to build and operate; and are undoubtedly a significant source of illegal material. Illegal REE mining is seen as a cause of environmental damage, health issues, and depressed prices, and the Chinese government is attempting to eliminate such activity.

Development companies with ionic clay deposits in Brazil, Chile, and Madagascar have proposed large-scale operations using mining and either heap or agitation leaching. Although the developers occasionally mention in situ leaching, it does not appear to be a high-priority option. This might be because of the difficulty of demonstrating bedrock impermeability to the satisfaction of the regulators.

HYDROMETALLURGICAL PROCESSING OF REE-BEARING PREGNANT SOLUTION

The cracking and leach processes produce a PLS containing REE and impurity elements such as Fe, Al, Th, and U. The PLS must be treated to eliminate the impurities and produce the REE in an intermediate solution or solid amenable to separation into individual REE products that are typically 99.9% pure or greater. Non-REE impurities complicate the task of the separation plant, and a typical specification for a mixed REE feed to a separation plant will require that it contain very low levels of U, Th, Al, Ca, Fe, Al, Zn, Cd, Pb, SO_4, and U and Th progeny.

Specifications will generally be agreed upon between the producer of the intermediate REE product and the separation plant. Published standards are generally not freely available; however, a Chinese Standard, GB/T 20169–2015, gives standards for "mixed rare earth oxide of ion-absorbed type rare earth ore."

Impurity Removal

Various techniques are being used or have been proposed for the removal of impurities from REE pregnant solutions (Lucas et al. 2015; Verbaan et al. 2015; Yahorava et al. 2013; Zhu et al. 2015; Zhang and Edwards 2013).

Selective Dissolution

Caustic cracking produces a mixture of the hydroxides of those metals that were released during the mineral cracking process. Thus, the solid residue produced by cracking monazite will comprise hydroxides and similar compounds of the REE, Th, Fe, U, and whatever other metals might be converted during the cracking process. Just as Fe and Th hydroxides are precipitated earlier than the REE during selective hydrolysis, so too can REE be dissolved from a mixed hydroxide while leaving

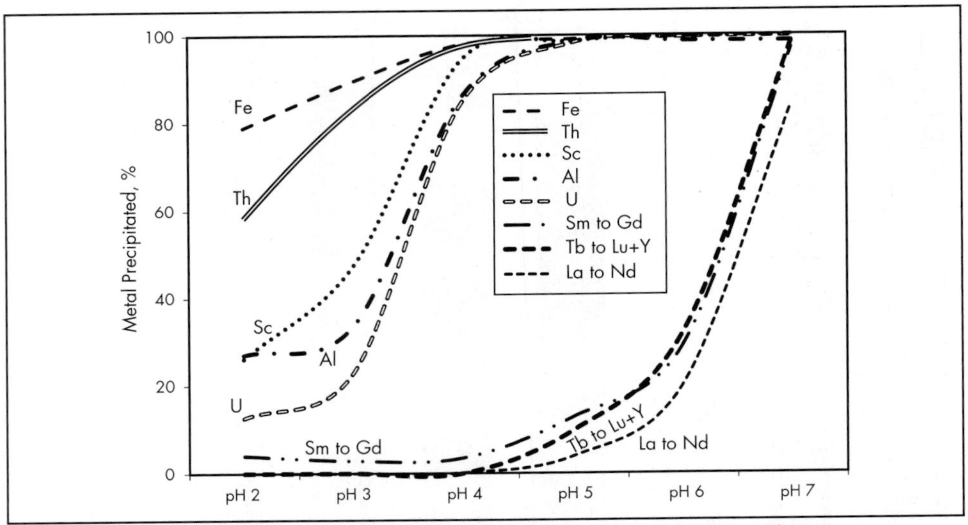

Figure 6 Precipitation of impurities and REE by hydrolysis from chloride solution

the Fe and Th in the residue. Selective dissolution by adding HCl to a pH no lower than 3.5 has long been part of the process for recovering REE from monazite by caustic cracking. When practiced on Brazilian monazite containing 64% REO and 6% ThO_2, the residue from selective leaching contained 60% ThO_2 and 4% REO (Krumholz 1957). Rhone Poulenc practiced selective dissolution from cracked monazite at its La Rochelle plant (Kaczmarek 1981) as did the Indian monazite processors (Pillai 2008; Swaminathan et al. 1988).

Selective dissolution can also be employed if an intermediate REE product is precipitated as a carbonate or hydroxide and then redissolved for further processing.

Fe, Al, and Th Removal by Selective Hydrolysis

PLSs commonly contain Fe, Al, and Th, and these elements can be removed as hydroxides by a careful partial neutralization process in which the pH is raised to between 3 and 5. At the optimum pH, the REE remain in solution, but Fe^{3+}, Al, and Th are precipitated as indicated in Figure 6 for chloride solution. Depending on the starting solution analysis and the operation details, more than 95% of many contaminants can be precipitated with REE losses of well under 5%. Fe^{2+} is only hydrolyzed at about pH 6, so removal of iron will only be complete if the iron has been oxidized.

The precipitation trends shown in Figure 6 do not radically change with the choice of neutralizing agent (e.g., NaOH, MgO, NH_4OH, etc.) except in the case of carbonates (e.g., $CaCO_3$, Na_2CO_3, etc.), in which case the precipitation curves are displaced by about 2 pH units lower and the sharpness of separation is considerably reduced.

The use of CaO or $CaCO_3$ to precipitate impurities from sulfate solutions causes the precipitation of gypsum as well as the impurities. The gypsum precipitate tends to carry down some REE, and clean separations with low-value losses are difficult to achieve. The use of NaOH or Na_2CO_3 as an impurity precipitant from sulfate solutions is not appropriate, especially at higher solution grades, since a part of the LREE will likely be precipitated as the double sulfate salts, typically $La_2(SO_4)_3{\cdot}Na_2SO_4{\cdot}2H_2O$. For the partial neutralization of sulfate solutions, with low REE losses, the preferred reagent is either MgO or $MgCO_3$.

Kim et al. (2014) present data on the speciation of REE and common impurity elements across a range of pH values, which is very much in accordance with the aforementioned observations.

In work conducted for Natural Resources Canada, SGS Canada demonstrated that a phosphate addition equimolar to Fe to a sulfate solution led to 99% elimination of Th at pH 2 (and enhanced Fe and Sc removal), whereas without phosphate, the same level of elimination was not obtained until a pH of 4.5 (SGS Canada 2017).

Although Cu, Zn, and Pb are not commonly encountered in REE solutions, they can be eliminated by sulfide addition. This approach has been used to eliminate radiogenic Pb from PLS. Similarly, sequential additions of $BaCl_2$ and Na_2SO_4 will precipitate $BaSO_4$, which will carry down Ra (Pillai 2008). A simple sulfate addition to a Ba-containing HCl-based PLS will precipitate $BaSO_4$, which will also precipitate Ra.

Solvent Extraction and Ion Exchange

SX or IX is useful for the extraction of U and Th away from REE, and this has been reviewed by Zhu et al. (2015) and Xie et al. (2014).

Li et al. (2013) operated a pilot-plant campaign using a primary amine SX circuit on a solution containing 30–50 g/L of REE and 0.1–0.2 g/L Th obtained by a sulfuric acid bake operation on Bayan Obo concentrates. Using seven loading, eight scrubbing, and eight HNO_3 stripping stages, they achieved REE-bearing raffinates containing about 0.0001 g/L Th and made a Th product that was 98% pure. The raffinate from this operation was contacted with a phosphonic acid extractant (P507), which extracted the REE away from Zn, Mg, Ca, and SO_4, yielding a 223-g/L chloride strip stream immediately ready for separation.

The Kvanefjeld project in Greenland will process ore containing about 1% REO and about 300 ppm of U. The ore will be beneficiated and the concentrate will then be sulfuric acid leached to recover most of the uranium, with the REE reporting to the residue as double sulfates. The leach residue will be metathesized to destroy the double sulfates, then HCl leached to take the REE and remaining U into solution. Selective hydrolysis, $BaCl_2$, and NaHS will be used to eliminate most impurities from the chloride leach solution, followed by IX to remove U immediately ahead of REE precipitation using Na_2CO_3 (Krebs 2014).

A mixture of a tertiary amine (Alamine-336) and a primary amine (Primene-JMT) has been shown to be effective at simultaneously extracting U and Th from sulfate leach solution (SGS Canada 2017).

An industrially used approach is to extract the REE, Th, and U using a mixture of di-(2-ethylhexyl)phosphoric acid (DEHPA) and TBP, and sequentially and selectively strip the elements. This approach was used to recover REE from the barren solution of uranium plants in the Elliot Lake mining camp (Goode 2013).

Double Sulfates

The addition of Group 1 ions, such as Li, Na, K, or the ammonium ion to a sulfate solution of the REE causes the formation of variously soluble double sulfate salts. The formulae for these are commonly written as $La_2(SO_4)_3{\cdot}Na_2SO_4{\cdot}2H_2O$ (Kul et al. 2008), but Bickerdike (1937) recorded a wide variety of compositions.

Kul et al. (2008), working with a solution containing about 26 g/L of mostly LREE, showed that precipitation efficiencies greater than 95% could be obtained for the LREE at about 1.5 times the stoichiometric demand for the REE. However, the HREE and Y are only partially precipitated, perhaps 75%. Temperatures up to about 55°C promote the formation of double sulfates, but higher temperatures provided little benefit. Thorium precipitation from the solution, which contained 0.3 g/L Th, increased with the Na addition from 30% at 1.25 times REE stoichiometric to 64% at 3 times stoichiometric demand.

Abreu and Morais (2010) showed that the yield of double sulfates was essentially the same whether NaOH or Na_2SO_4 was the source of Na and that excellent Fe elimination was obtained unless NaOH was used. LREE recovery of 98% was possible at about 2 times stoichiometric Na addition, but Y recovery was only 31%. Y recovery of 99% was obtained at a Na addition of 8.5 times stoichiometric. The authors suggested an optimum temperature of 70°C for maximum Y precipitation and reaction times of 2 hours. They reported the double sulfates as having formulae such as $LaNa(SO_4)_2{\cdot}H_2O$ and containing about 44% REO.

Formation of the double sulfate is a useful way of getting a high yield of reasonably pure REE from solution, especially if HREE are not at high levels, but the double sulfates need to be converted to other forms to become commercial commodities or further processed. This is usually done by metathesis using NaOH, which results in the following reaction:

$$LaNa(SO_4)_2{\cdot}H_2O + 3NaOH \rightarrow La(OH)_3{\downarrow} + 2Na_2SO_4 + H_2O \qquad \textbf{(EQ 24)}$$

After filtration, the hydroxide can be dissolved in a weak acid as a reasonable pure and high-strength REE solution and further processed. The sodium sulfate solution can be recycled back to the double sulfate precipitation step. Abreu and Morais (2010) showed that metathesis was best achieved at 70°C with 1.1 times stoichiometric demand.

Abreu and Morais (2010) showed how, after metathesis and dissolution of double sulfate, the Ce could be oxidized from Ce(III) to Ce(IV) by adding about 30% excess permanganate ion, which leads to the precipitation of Ce(IV) hydroxide at a pH of about 3. The precipitate contained Mn, which was removed by releaching the hydroxide in HCl, reprecipitating the Ce with oxalic acid, and calcining it to give a 99.5% CeO_2 product.

Double sulfate precipitation from REE solutions containing low-HREE values can be a useful and economical way of processing since it allows high recovery of the REE from an acidic solution to a high-grade, reasonably pure precipitate.

Precipitation

After the REE have been obtained in a relatively pure solution through processes of mineral cracking, leaching, and impurity removal, it is generally necessary to precipitate the REE from solution to facilitate downstream processes such as separation of the individual REE.

Hydroxides and Carbonates

The REE are readily precipitated from purified solutions as hydroxides by neutralizing to about pH 8 using a suitable base such as CaO, MgO, NH_3, or NaOH. With chloride solutions, any base should be considered. However, if precipitating a REE hydroxide from a sulfate solution is required, MgO or ammonia could be the preferred base since gypsum precipitation and double sulfate salt formation are avoided.

CaO, MgO, NaOH, and NH_3 are relatively expensive, hydroxide precipitates generally do not filter well, and neutralization is not very selective. For these reasons, hydroxide precipitation of the REE is not practiced to any significant degree, although the process has been proposed by some development groups. Carbonate precipitants such as $MgCO_3$, Na_2CO_3, $(NH_4)_2CO_3$, or NH_4HCO_3 are relatively inexpensive and have been used in REE operations and proposed by several project developers. Luo et al. (2013) present data for the ammonium bicarbonate precipitation of REE from low-grade solutions obtained from ionic clay solutions. The work showed that a 3.5:1 molar ratio of NH_4HCO_3 to REE was required and that crystal size and settling rates improved with aging.

Oxalate Precipitation

The REE oxalates are insoluble, oxalate precipitation is quite selective for the REE, and REE oxalates can be calcined at about 900°C to give oxides. For these reasons, oxalic acid is commonly employed in the REE industry for both mixed REE solutions and the highly refined REE solutions produced in separation plants. The reaction is as follows, using La as an example:

$$2La^{3+} + 3H_2C_2O_4 + 10H_2O \rightarrow La_2(C_2O_4)_3{\cdot}10H_2O{\downarrow} + 6H^+ \qquad \textbf{(EQ 25)}$$

Chi and Xu (1999) provide information on the efficiency of precipitation from solutions produced by leaching ionic clay and containing about 4 g/L REE, 0.1 g/L Al, and 0.3 g/L Fe. The oxalic acid addition for 95% REE recovery was 2.5–3 mol/mol REE or 170%–200% stoichiometric. A residence time of 1 hour was satisfactory and a pH of <2 avoided the precipitation of impurities.

With very pure, higher-grade solutions, such as those produced in separation plants, an oxalate excess of about 10% over stoichiometric is normal for complete precipitation (Liao et al. 2013).

Oxalate precipitation is very selective for REE and against many impurities, but Th co-precipitation is nearly complete and Ca, Sr, and Ba oxalates are partially precipitated along with the REE and consume oxalic acid.

Normally, oxalic acid is added as a solution containing 10% oxalic acid. The precipitated REE oxalates are then recovered by filtration or centrifugation. The high-grade products (>99.9% pure) made in separation plants are washed with copious amounts of water to eliminate dissolved impurities before being calcined. Calcining can be done using a rotary dryer followed by a rotary calciner and cooler, as at Molycorp, or using crucibles that are conveyed through a tunnel furnace, as used by Lynas and most of the Chinese producers.

As an alternative to calcining REE oxalate, the precipitate can be metathesized using NaOH or Na_2CO_3, the resulting $Na_2C_2O_4$ recycled back to precipitation and the REE product calcined.

Early Separation of Cerium as Ce(IV)

Ce makes up about 40% of the REE in the mixed product of most deposits. Ce does not command a high price, yet it consumes essentially the same mass of chemicals as more valuable REE, such as Nd or Dy, in the separation plant. Hence there has been much interest in removing Ce from the main REE stream before the separation plant is reached. Ce in solution is normally trivalent and therefore behaves much as the other REE. However, Ce(III) can be oxidized to the quadrivalent state, Ce(IV), after which it is readily separated from the other REE through selective dissolution or precipitation or SX.

Molycorp roasted its concentrate at 650°C in a multiple-hearth roaster that drove off fluorine and oxidized Ce(III) to Ce(IV). A cold dilute HCl leach solubilized all the REE except for the Ce(IV), which reported to the leach residue. The leach residue was sold as a glass polishing powder, although a portion was purified to give a carbonate of 96% CeO_2 purity. Leach recovery of the trivalent REE was not complete, and the Ce(IV) product, which contained about 60% Ce, did carry about 10% non-Ce REE. The Ce-depleted HCl leach solution went on to impurity removal and the separation of individual REE (Molycorp 1996).

When Molycorp upgraded its plant, it replaced roasting with a process in which all the REE were dissolved in HCl, impurities removed by hydrolysis and sulfide precipitation, and the Ce(III) chemically oxidized to Ce(IV), which precipitated from solution. Data are not given, but it is reasonable to assume that hypochlorite was used as the oxidant (SRK Consulting 2010; Goode 2016c).

Ho et al. (2014) researched the effect of dose, temperature, and pH on the oxidation of Ce using hypochlorite. In a mixed REE solution, 99% oxidation was obtained at 2.5 times stoichiometric hypochlorite addition. A pH of 4 appeared to give best separation of Ce from other REE with poor selectivity at higher pH values. A temperature of 60°C was optimal. Significant co-precipitation of other REE, especially the heavier REE, occurred at all levels investigated. For example, with 1.9 times stoichiometric NaOCl addition, 60°C, and pH 4, levels of co-precipitation were about 30% HREE, 20% medium REE, 12% Nd, 9% Pr, and 5% La.

In 1968, Bauer and Lindstrom reported on the separation of Ce by ozone oxidation from sulfate, nitrate, and chloride solutions containing REE in the proportions of 50% CeO_2, 36% La_2O_3, 10% Nd_2O_3, and 4% Pr_6O_{11} and others. High recovery of high-purity Ce precipitate was obtained from all three solutions at ambient temperature. A pH range of 4 to 5 was recommended. Variations on the basic chemistry were explored. The reaction was

$$2CeCl_3 + O_3 + 3H_2O \rightarrow 2CeO_2\downarrow + 6HCl + O_2 \quad \textbf{(EQ 26)}$$

Bauer reported co-precipitation of other REE, especially the heavier REE, and investigated a process to redissolve and reprecipitate Ce(IV) to purify Ce and isolate a pure La stream.

Zou et al. (2014) developed a wet oxidation process for Ce starting with a mixed REE hydroxide precipitated from a nitrate medium. They obtained 97% oxidation of Ce to Ce(IV) by first precipitating the mixed hydroxides and then blowing air through the hydroxide slurry at 80°C at pH 13. It was determined that F had little effect on the extent of oxidation, but PO_4^{3-} had an adverse effect on oxidation. The oxidation reaction was given as

$$2Ce(OH)_3 + 0.5O_2 + H_2O \rightarrow 2Ce(OH)_4\downarrow \quad \textbf{(EQ 27)}$$

Although not detailed in the work by Zou et al., dissolution of the oxidized hydroxides in a weak acid would lead to dissolution of the REE except for $Ce(OH)_4$. Possibly there would be some residual non-Ce hydroxides left with the $Ce(OH)_4$ leading to some losses of the other REE, as experienced by Molycorp and indicated by researchers.

Abreu and Morais (2010) oxidized Ce(III) in a pH 3 solution using potassium permanganate.

Electrolysis of chloride solutions leads to the formation of hypochlorous acid (HClO), chlorite ion (ClO^-), and chlorate ion (ClO_3^-). Vasudevan et al. (2005) oxidized Ce(III) through the in situ electrolytic generation of such oxidants. A temperature of about 30°C, a pH of about 6.7, and concentration of 130 g/L $CeCl_3$ were found to be optimum. The cathode of the electrolytic cell was rotated to prevent the formation of a coating. Using a solution containing a typical LREE mixture, cell electrochemical efficiency was reported as 60% and the purity of the precipitated $Ce(OH)_4$ was 95%. The impurities appear to be other REE, which implies significant losses.

Martin and Rollat (1986) patented a system in which Ce(III) in a mixed REE nitrate solution was electrolytically oxidized to Ce(IV) in the anode compartment of a cell. Ce(IV) was then selectively extracted away from the other REE by, preferably, TBP. The loaded organic was then stripped by placing it, with nitric acid, in the cathode compartment of the same electrolytic cell where it was reduced to Ce(III) and simultaneously stripped from the solvent. In an alternative stripping process, nitric acid was reduced in the cathode compartment to form NO and NO_2 gases, which were mixed with the loaded solvent to form Ce(III), which was then water stripped.

Oxidation of Ce can be achieved through the drying of hydroxide precipitates produced by caustic cracking monazite. Using hydroxide containing about equal proportions of Ce and other REE, Swaminathan et al. (1988) showed that a drying temperature of 160°C was optimal with higher temperatures leading to difficulty in subsequent HCl leaching. The Ce in a 5-mm-thick layer of hydroxide attained 95% oxidation to Ce(IV) after 6 hours and 99% after 12 hours. The dried and oxidized hydroxide was leached with HCl at 50% solids at pH 3 and produced a solution containing 95% of the trivalent REE and CeO_2/REO of about 7%. The CeO_2 in the leach residue was selectively dissolved away from the Th and gangue materials using HCl at pH 1.5 with a sulfite reductant leading to a product containing up to 85% CeO_2.

Ce(IV) in acidic solution is preferentially extracted by solvents away from other REE. TBP and Cyanex 923 are effective from nitrate solutions, carboxylic and organophosphate solvents from sulfate, or chloride solutions (Xie et al. 2014; Lucas et al. 2015; Zhou and Liu 2015).

SEPARATION OF INDIVIDUAL RARE EARTH ELEMENTS

The REE recovery operations described earlier produce a mixed REE solution or solid. The REE distribution in these intermediate products is similar to those indicated in Table 3. However, the manufacturers of magnets, phosphors, and so on, require separated products of individual elements with little or no adjoining REE and with purities ranging from 99% to as high as 99.9999%; therefore, separation plants are necessary.

Large-scale separation was performed using IX until the advent of SX methods in the mid-1960s (Kaczmarek 1981). A typical SX-based production plant separating all the REE will include up to 1,500 mixer-settlers, extensive product precipitation, liquid–solid separation and washing systems, and calcining equipment. Such plants are expensive to build and operate. Capital and operating costs for any separation system can be reduced by not separating all the individual REE. The following can be considered:

- La, Ce, and Y are often high-mass components of the separation plant feed, but they can be removed from the feed (or partway through the separation process) using classic chemistry procedures or more advanced methods. Ce can be oxidized to the Ce(IV) state and isolated as noted earlier. Alternatively, La, Ce, and Y can be extracted away from the other REE using certain solvents as explained later. In another option, the LREE and HREE can be separated through double sulfate precipitation.
- NdFeB magnets can be effectively made using a NdPr mixture (didymium) instead of pure Nd. The Lynas LAMP operation produces much of its output as a NdPr product, thus avoiding the costs of separating these elements.
- Several REE are not in high demand, or their content in the intermediate REE product is low, so they are not worth separating but best made as a bulk product. Bulk products might comprise Sm+Eu+Gd, or Ho+Er+Tm+Yb, or the entire HREE suite (Sm to Lu, including Y).

These strategies can reduce the capital and operating costs of a separation plant by as much as 50%.

Chloride-Based SX Separation Systems

All commercial REE separation plants currently use SX and almost entirely from a chloride-based aqueous phase. Prior to 2015, there were about 100 REE separation plants in China with a combined capacity of more than 400,000 t/yr of REO (Roskill 2016). Most of these are now closed due to market conditions and government regulations intended to consolidate the industry and reduce illegal production. The Chinese separation plants are generally based on the use of a phosphonic acid extractant, H(EH)EHP, at 1.5 molar concentration (50% v/v) in an aliphatic diluent. The reagent is known as P507 in China and as PC88A and Ionquest 801 elsewhere. Separation plants in Japan and Vietnam also use chloride chemistry and H(EH)EHP.

The plants that process Bayan Obo flotation concentrates and the Lynas LAMP operation use sulfuric acid baking to crack the REE minerals and therefore generate a sulfate-based PLS. This is partially neutralized and filtered, and then contacted with DEHPA to extract the REE. HCl is used to strip the REE, which can then be separated using P507.

Although P507 or equivalent is the main extractant in most chloride-based separation plants, certain plant sections might use other extractants, such as naphthenic acid or IX, for special tasks (Li et al. 2013; Huang et al. 2015). Europium is readily reduced to the divalent state, and this fact is exploited in many separation plants to separate Eu from the trivalent REE.

The separation factors (abbreviated as β) between adjacent rare earths are generally small, meaning that many sequential separations are required to obtain a high-purity product from a mixture of rare earths. If all 15 REE are separated to high-purity products (>99.9% pure), the separation plant might contain 1,500 or more mixer-settler units arranged in approximately 15 different circuits. Each circuit would accept a feed solution containing mixed REE and generate two or three outlet streams containing partially or fully refined individual REE. These circuits would be cascaded with partially separated streams directed to other circuits for separation.

A simplified block diagram for a P507-based separation plant producing separated LREE, Sm, Eu, Gd, Tb and Dy, and a bulk concentrate of the heavier REE is provided in Figure 7.

A single circuit, say for the separation of the Sm-Dy group from the heavier REE (Ho-Lu including Y), might contain 50 extraction stages, 30 scrub stages, 15 strip stages, and 5 stages for water washing and saponification for a total of 100 mixer-settlers.

The number of stages needed in a separation circuit is a function of the composition of the feed, the desired REE purity in the scrubbed solvent and the raffinate (aqueous phase leaving the extraction section), and the β for the two REE at which a split is being made. The β for P507 varies from 6.8 for the separation of Ce and La to as low as 1.4 for the separation of Er and Y. A given Er–Y separation will need at least five times as many stages as the equivalent Ce–La separation because of the differences in β. Much research effort is being expended on searching for solvent systems offering improved β.

In many situations, for example, SX using P507, Y acts as an HREE lying between Ho and Er. However, under some circumstances, Y can behave as an LREE (Collier 2012). Figure 8 shows how Y behaves with different SX reagents. Naphthenic acid is commonly used for the separation of Y from the other HREE after a Nd–Sm separation but is also practiced or proposed for the separation of Y (and La+Ce) from other REE before the main separation circuits (Li et al. 1994; Griffith et al. 2016).

Separation plants using an extractant such as P507 typically use HCl to dissolve the incoming oxide feed and to scrub and strip the REE from the loaded solvent. NaOH is used to regulate pH by partially converting the P507 to the sodium form in the saponification step. The use of HCl and NaOH leads to the production of a NaCl brine solution as a separation plant effluent. A conventional P507 separation plant uses about 3 t HCl (100% basis), about 2.5 t NaOH (100% basis), and up to 20 t water, much of it for washing the REE products, to fully separate 1 t mixed REO into purified oxalates (Liao et al. 2013). This water, and the contained NaCl, forms the main plant effluent.

The Chinese have pioneered new plant configurations, such as hyperlinking, that reduce the reagent consumption

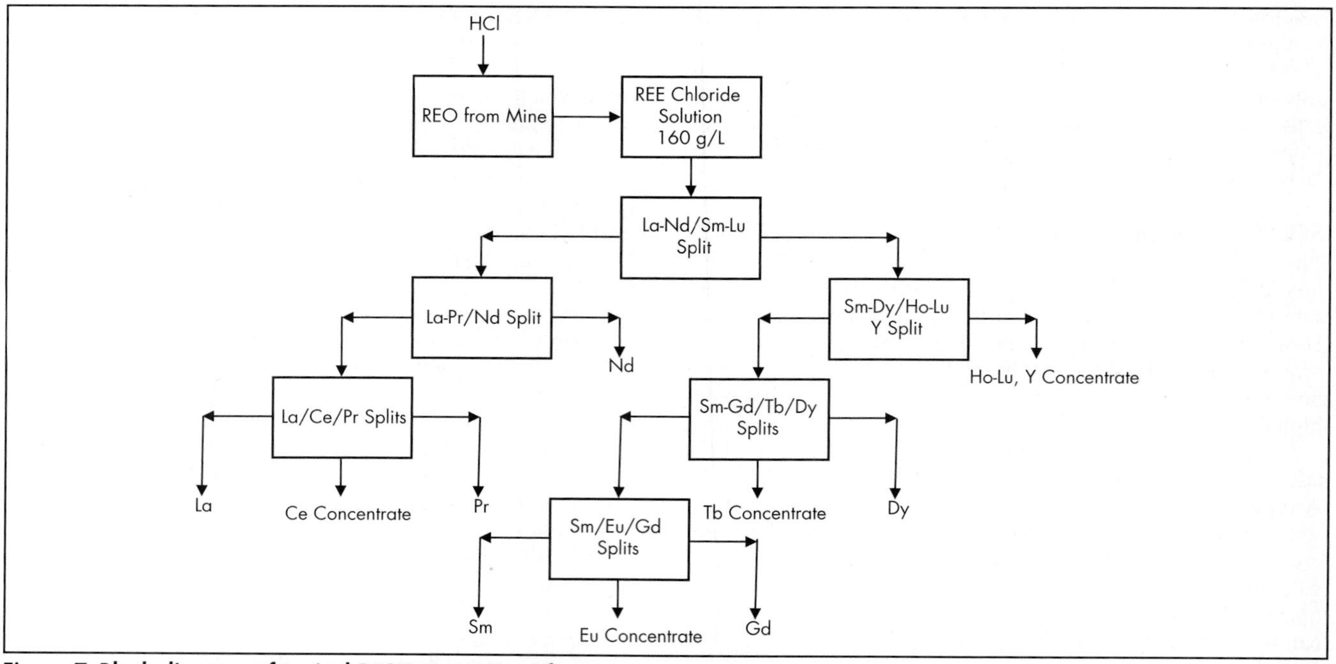

Figure 7 Block diagram of typical P507 separation plant

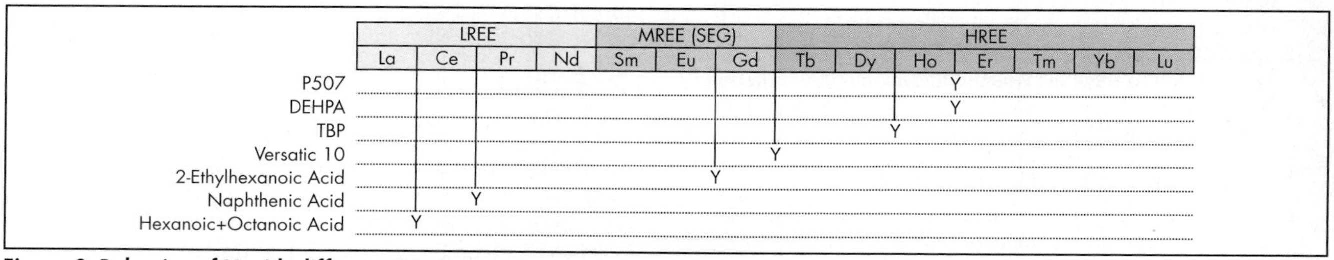

Figure 8 Behavior of Y with different SX reagents

by at least 30% (Liao et al. 2013; Huang et al. 2015). This process, which uses REE in the saponification step instead of NaOH, has been widely adopted in China and is mandated for new plants as a means of reducing environmental impact. The Chinese have also developed saponification processes using slurries of $Ca(OH)_2$, or MgO, or REE carbonates, all aimed at reducing costs and mitigating environmental issues (Huang et al. 2015).

An alternative means of reducing new reagent demand is to operate a chlor-alkali plant on the brine waste stream to regenerate HCl and NaOH. Given low power costs, a chlor-alkali plant might economically regenerate the reagents while also eliminating the costs associated with disposing of the waste brine effluent. Molycorp attempted to operate a chlor-alkali plant at its refurbished Mountain Pass plant, but it appears that the preparation of suitable feed brine from the plant effluents was problematic. The company was forced to buy reagents at a higher-than-projected cost and had to dispose of wastewater at a significant cost, both contributing to its bankruptcy (Molycorp 2015).

Another alternative for the reduction of acid and alkali demand is the use of SX reagents such as Cyanex 272, a phosphinic acid, or Cyanex 572, a chelating extractant. These extractants are weaker than P507, so they extract at a slightly higher pH and are easier to strip, leading to acid savings of about 30% (Soderstrom et al. 2014). REE separation factors are about the same as for P507, so these new reagents do not reduce stage requirements.

In China, a great deal of development activity is aimed at better modeling and optimizing separation circuits; improving separation factors through different reagents; and reducing the environmental impact and operating costs through hyperlinking, online analysis, and other strategies. Chinese law limits the export of technology concerning advanced REE separation plants.

Nitrate-Based SX Separation Systems

Solvay's La Rochelle separation plant uses nitric acid dissolution and TBP for extraction and separation of the LREE (Kaczmarek 1981). TBP is effective for the separation of the dominantly LREE in monazite but is not suitable for separating the HREE. Solvay uses several other solvent extractants, including a quaternary amine such as Aliquat 336, for separating the HREE (Lucas et al. 2015).

Molycorp's Silmet plant in Estonia, which was originally designed and constructed by the Soviets for processing a

LREE-bearing loparite concentrate from the Kola Peninsula, also uses a nitrate-TBP system. Similarly, the Kashka REE plant in Kyrgyzstan used TBP to separate LREE into finished products. The HREE were separated using IX with chromatographic elution (Stans Energy 2017; Danilov 2011).

The plants using nitrate-TBP systems typically dissolve a REE-bearing material in nitric acid; concentrate the solution to typically 200–300 g/L REE; add a "salting-out" reagent such as ammonium, calcium, or sodium nitrate; and then extract the REE with TBP in mixer-settlers arranged in a load-scrub-strip configuration. One such SX battery makes a single separation just as in the chloride systems. Stripping is done with water rather than acid. Although this reduces reagent demand, the dilute strip solution, perhaps containing 100 g/L REE, must be evaporated to a higher concentration before it can be fed to the next separation battery.

TBP circuits consume relatively small amounts of HNO_3 and NH_3 and generate an ammonium nitrate solution as the effluent. After purification, this material can be crystallized and sold as a fertilizer. The reagent consumption in a nitrate-TBP separation plant is less than that of a chloride-P507 plant; however, energy costs are higher because of the evaporation required between SX batteries. Separation factors in the nitrate separation system for the LREE are similar to those in the chloride system, so the number of stages required for a given set of separations is similar. With appropriate extractants for the HREE, separation factors similar to those for P507 can be obtained, so stage requirements are similar.

Alternative Separation Systems

A 2016 review (Goode 2016b) of new separation systems examined the development of non-SX systems and improved SX systems. The SX developments fell into two general categories: (1) replacement of the conventional mixer-settler with more compact devices and (2) new SX chemistries. The new devices include microfluidic contactors, centrifugal contactors, and membranes. The new SX chemistries include the use of novel extractants, precipitative stripping, and ionic liquids.

The non-SX options under investigation included molecular recognition technology by Ucore Rare Metals; free-flow electrophoresis by Geomega Ressources; high-precision liquid chromatography by REEtec AS; electrolytic systems developed by Rare Earth Salts; and new IX systems, for example, those developed by K-Technologies.

In the absence of any large-scale operation using such new systems, the review concluded that several developments held promise but only conventional SX separation could be financed.

POST-SEPARATION PROCESSING

Product Handling

Separation systems produce solutions of highly refined or partially refined individual REE or mixtures of REE. These are precipitated, usually as oxalates but frequently as carbonates, or evaporated to give chloride salts. Precipitates are separated from mother solutions and washed using centrifuges or filters.

REE oxalates are calcined to oxides at temperatures of about 1,000°C in rotary kilns or tunnel furnaces. Carbonates can be marketed as is or calcined to oxide depending on the intended use. Co-precipitates (e.g., of Y and Eu) are frequently produced for special applications. These and other products are often needed with specific physical properties (particle size, etc.) and therefore require carefully controlled precipitation and calcining conditions.

Metal Production

Magnet makers require metallic REE, such as NdPr, Pr, Nd, Sm, Dy, and Tb. Metals are used in the production of nodularized iron, magnesium alloys, and aluminum alloys. Mischmetal, a mixture of the LREE metals, is used in alloying but also in lighter flints. Nickel–metal hydride (NiMH) batteries include an intermetallic negative electrode incorporating LREE with Ni, Co, and Al.

REE metals are produced by electrowinning or by metallothermic processing using calcium, magnesium, or lanthanum metal as the reductant. Electrowinning can be from molten NaCl-REE chloride at about 1,000°C, leading to the production of molten metal at the cathode and Cl_2 at the graphite anode. Alternatively, an REO, such as La_2O_3, can be fed into a molten La/Li/Ca/Ba fluoride electrolyte with molten metal formed at the cathode and O_2 at the graphite anode. Electrowinning is used to produce several pure REE metals and, through co-reduction, various REE-REE, REE-Fe, REE-Cu, and REE-Mg alloys.

Some of the REE metals and their halide salts, particularly Sm, Eu, Tm and Yb, have high vapor pressures at the temperatures used for electrowinning. These have been recovered by metallothermic reduction using La or Ce metal with simultaneous sublimation of the metal. These reactions are highly exothermic, and operating temperatures of 1,200°–1,500°C are common (Songina 1964; Riedemann 2011; Lucas et al. 2015).

ENVIRONMENTAL ISSUES

The rapid development of the REE industry has led to some environmentally unsatisfactory situations (Keith-Roach et al. 2014; Wall et al. 2017). Molycorp's earlier operations at Mountain Pass were curtailed because of environmental issues. Concerns over the massive tailings storage facility near Baotou led to new milling and tailings facilities at the Bayan Obo mine. Concerns remain over radionuclide storage. The earlier ionic clay development projects in China were carried out, often illegally, with little regard for the environment, leading to long-term problems with high projected remediation costs.

The radionuclides that generally occur in REE deposits not only have caused issues at Mountain Pass and Baotou, but have also triggered strong opposition to future REE developments. Concern over the fate of thorium contained in the concentrate processed at the Lynas plant in Malaysia resulted in a commissioning delay of about a year while the International Atomic Energy Agency (IAEA) undertook a third-party review. Local concerns about the Lynas plant were sparked by major Th-based radiological problems resulting from earlier REE operations in Malaysia.

Radionuclides

The regulations covering working with and transporting radionuclides, their incidence in REE processing, concerns, and methods of management have been discussed by Feasby et al. (2013). Concerns include Th and U and their progeny but also the natural radioactivity of some of the REE. ^{147}Sm is particularly radioactive such that the transport of natural Sm can exceed IAEA transport regulation exemption limits. Feasby et al. concluded that radionuclides can be safely handled within process plants and during transportation through proper process and repository design supported by test work to demonstrate environmental stability.

The ways in which radionuclides deport during processing are discussed by several authors (Feasby et al. 2013; Pillai 2008; Anvia 2015).

Overall Environmental Footprint

Haque et al. (2014) discuss various aspects of REE production, sustainability, and environmental impact. The authors reported on the greenhouse gas (GHG; expressed as CO_2) during the production of a mixed REE oxide from a deposit similar to the Bayan Obo operation in China. The total was estimated at 1.4 kg CO_2 equivalent/kg of REO with the main contributors being due to HCl (38%) and steam (32%) usage. The GHG generated during the production of separated products ranged from about 9 kg CO_2 equivalent/kg of REO for La to about 56 kg CO_2 equivalent/kg of REO for a mixed Sm-Eu-Gd (SEG) product. Even higher emissions would be expected for the heavier REE.

In contrast to the GHG data for REE production, Rankin (2012) reported that copper generates between 3 and 6 kg CO_2/kg metal, nickel 11–16 kg CO_2/kg metal, and aluminum 22 kg CO_2/kg metal.

Haque et al. (2014) noted the high water consumption incurred during the production of the REE. Values range from 330 L/kg oxide for La to 1,750 L/kg oxide for SEG. These can be compared with values of 38 L/kg Cu metal for heap leaching, SX, and electrowinning; 110 L/kg for Ti metal; and 377 L/kg for Ni recovery by pressure acid leaching and SX-electrowinning (Norgate and Lovel 2006).

Koltun and Tharumarajah (2014) have analyzed the life-cycle impact of the REE and present data for the consumption of electrical and thermal energy and water, and the GHG emissions for each of the individual REOs. The authors used a price-based method of distributing consumption across the REE production chain. A large range is evident for each parameter using this method. For example, water consumption ranges from 11 m^3/kg REO for CeO_2 to 7,068 m^3/kg Sc_2O_3; electrical energy ranges from 24 mJ/kg REO for CeO_2 to 11,630 mJ/kg Sc_2O_3.

Wall et al. (2017) also published a paper on the responsible sourcing of critical minerals with an emphasis on REE.

Recycling

The level of REE recycling is not well documented but is undoubtedly low, probably less than 5%. Noteworthy recycling efforts include the Showa Denko operation in Vietnam (800 t/yr of REE from magnets), Solvay's recovery of REO from scrap fluorescent lamps (1,000 t/yr capacity), and efforts in Japan to recycle magnet material from scrap motors (e.g., Mitsubishi and Panasonic) and from NiMH batteries (e.g., Toyota) (Roskill 2016).

Although the level of REE recycling was low in 2017, there are signs that recycling will become more important in the future. Research is being conducted in Europe, China, Japan, Korea, and the United States (Binnemans et al. 2013; Bedrossian et al. 2016; Dutta et al. 2016).

CHARACTERIZATION, TESTING, AND ANALYSIS

Because REE ores are typically highly complex, the development of a workable process flow sheet for a new REE deposit can be a long, drawn-out, and expensive affair. An example is the Strange Lake deposit that was discovered in 1980 and subjected to metallurgical test work starting in 1982, for which a final flow sheet has yet to be selected. Allowing for periods of inactivity, test work has probably taken up at least 10 years.

Potential REE by-products, such as Zr, Nb, and Sc, can be a distraction and must be considered pragmatically. Understanding and controlling the deportment of radionuclides is usually more important. The logical steps to developing a viable metallurgical process are described in the following sections.

Geology of the Deposit

The geology of the deposit must be understood in terms of ore types, REE grades, non-REE grades, and mineralogy of REE and gangue. The variability of the analytical grades and suite of minerals must be understood.

Mineralogy

Exploration and metallurgical samples should be thoroughly characterized using QEMSCAN (Quantitative Evaluation of Minerals by Scanning Electron Microscopy) and similar techniques. EMPA (electron probe microanalysis), laser ablation ICP-MS (inductively coupled plasma mass spectrometry), and other techniques can be used to determine the elemental composition of the minerals (Grammatikopoulos et al. 2016).

Composite and Variability Sample Selection

The project team, which should include geologists, mining engineers, and metallurgists, needs to select material suitable for the preparation of one or more master composites representing important ore types, locations, or production periods. Several variability samples covering different locations, lithologies, and grades should also be prepared and tested after the main processing route has come into focus.

Comminution

Crushing and grinding tests must be conducted on appropriate samples using basic methods such as the Bond comminution tests. Bond ball mill tests should be performed using a closing screen size that matches the intended grind size. Tests aimed at determining design parameters for semiautogenous grinding and high-pressure grinding rolls will probably be needed and should be undertaken by experienced testing laboratories.

Beneficiation

Tests should examine the response of ore samples to coarse ore sorting, DMS, and the full range of beneficiation processes. The test program can be guided by geological and mineralogical information. Testing of coarse ore beneficiation processes, such as ore sorting and DMS, requires samples of coarse ore for thorough testing

Hydrometallurgy

Tests should be guided by the mineralogy of the REE mineral and could include roasting, with or without reagent addition, followed by acid dissolution, sulfuric acid baking, caustic cracking followed by acid dissolution, and direct acid attack. These processes will produce an acidic solution containing the REE along with impurities such as Th, U, Fe, and Al. Tests aimed at removing impurities could include partial hydrolysis, SX, IX, and other methods.

The liquid–solid separation characteristics of REE-based slurries are often challenging, and the preferred process flow sheet is often driven by the thickening and filtration characteristics of different process options. Liquid–solid separation characteristics should be monitored at all stages of the test work.

Acid regeneration or recovery is a common requirement in REE circuits. Nanofiltration, reverse osmosis, acid springing (e.g., HCl regeneration through H_2SO_4 addition), electrodialysis, pyrohydrolysis, distillation, and other techniques will need to be examined as appropriate.

Environment

The various waste streams from the REE operation will require characterization to allow effective environmental systems to be designed. Waste characterization is typically obtained through comprehensive elemental analysis using low detection system packages, X-ray diffraction studies, and radionuclide analysis. Environment-specific tests including toxicity characteristic leaching procedure, acid–base accounting, net acid production and net acid generating, shake flask extraction, humidity cell tests, and others will probably be needed.

Geotechnical tests will be needed to allow proper design of tailings facilities. These could include settling tests, consolidation/percolation tests, and others.

The ore likely will include radionuclides, and the way that these are handled must be carefully considered. Ideally, the radionuclides will be produced in a chemically inert form and combined with the main tailings stream. Alternatively, the radionuclide stream would require isolation and special handling.

Pilot-Plant Operations

After bench-scale test work has determined a series of process steps needed to recover a commercially acceptable intermediate mixed REE product, it is essential to combine the operations in an integrated pilot plant. The pilot plant must operate at a scale, and for a duration, that will demonstrate that the selected process will work, confirm its operability, examine materials of construction, and generate or confirm process design criteria needed for engineering design.

The pilot plant will also serve to generate large volumes of slurry needed to allow effective liquid–solid separation testing. It is common to have liquid–solid separation equipment specialists or vendors attend the pilot plant to independently generate reliable data for plant design. Pilot plants are also useful to produce enough product for evaluation of intermediate products by potential toll processors or customers.

The failure to adequately demonstrate a process frequently results in a catastrophic failure of a production plant to attain design throughput, recoveries, and operating costs (McNulty 2004; Verbaan et al. 2014).

Separation

It is possible to demonstrate in a pilot operation that a given intermediate REE product is amenable to separation. This might be useful for the investment community, but in the author's view, this is not necessary since the separation factors between adjacent REE for a given separation system are totally independent of the feed source.

A separation pilot plant is an expensive undertaking. Each separation circuit will require several kilograms of REE to fill and then operate for the several weeks needed to reach equilibrium and deliver the required purities—and there are several separation circuits. To reach equilibrium faster, the circuit could be preloaded with organic and aqueous solutions containing estimated equilibrium compositions, but the validity of the demonstration is then questionable.

The demonstration of a single circuit of a conventional SX separation plant will add about \$1 million in costs and significant additional time to the development of a project. The newer separation processes under development might not need such long times and mass of feed material to come to equilibrium. However, as of the time of publication, their industrial-scale viability has not been demonstrated.

Analysis

In the evaluation of a REE deposit, a complete analytical suite should be commissioned, particularly in the initial phase of evaluation. A whole rock assay (WRA = Al_2O_3, CaO, Cr_2O_3, K_2O, MgO, MnO, Na_2O, P_2O_5, Fe_2O_3, SiO_2, TiO_2, and V_2O_5) is usually obtained by X-ray fluorescence (XRF) and is economical and essential. Analytes that can be important and should not be overlooked include Sr, Ba, Sc, U, Th, Zr, Nb, F, $S_{sulfide}$, S_{total}, and CO_2.

Some of the analytes can be obtained as part of an inductively coupled plasma optical emission spectrometry scan, others using Leco-based methods.

The individual REE, including Y, Sc, Th, and U, can be analyzed by ICP-MS REE scan. For mixed REE samples with a REE content of >80%, the determination of La, Ce, Nd, Sm, and Y is best performed using XRF rather than the normal ICP-MS method.

Analysis of the radionuclides will be necessary and requires the services of specialized analytical laboratories.

ACKNOWLEDGMENTS

The author acknowledges mentors and colleagues, many no longer with us, associated with REE-Th operations at Rio Algom and Denison; good friends from Molycorp, in China, at various laboratories around the world; and development companies in Canada, Brazil, and Russia. Massoud Aghamirian, Kevin Bradley, D. Grant Feasby, Mike Robart, and Niels Verbaan are thanked for reviewing a draft. Parts of Goode 2016b are included in the section titled "Separation of Individual Rare Earth Elements" and are reproduced with permission from the Canadian Institute of Mining, Metallurgy and Petroleum. Special thanks to Ev King for her support and patience.

REFERENCES

Abreu, R.D., and Morais, C.A. 2010. Purification of rare earth elements from monazite sulphuric acid leach liquor and the production of high-purity ceric oxide. *Miner. Eng.* 23(2010): 536–540.

Anvia, M. 2015. Radionuclide deportment in rare earth processing from monazite and bastnasite using conventional and alternative processing routes. Ph.D. thesis, University of Sydney, Australia.

Avalon Advanced Materials. 2013. Form 40-F. Filed 12/02/13 for the Period Ending 08/31/13. www.otcmarkets.com/edgar/GetFilingPdf?FilingID=9639447. Accessed March 2018.

Bauer, D.J., and Lindstrom, R.E. 1968. *Recovery of Cerium and Lanthanum by Ozonation of Lanthanide Solutions*. Report of Investigations 7123. Washington, DC: U.S. Department of the Interior, Bureau of Mines.

Bedrossian, S., Gonzales-Calienes, G., and Baxter, C. 2016. *The State of Global Rare Earths Industry: A Review of Market, Production, Processing and Associated Environmental Issues.* National Research Council Canada.

Bergeron, M., Langlais, A., Ourriban, M., and Pelletier, P. 2015. Dry chlorination process to produce anhydrous rare earth chlorides. U.S. Patent 20160298211 A1.

Bian, X., Wu, W., Chang, H., Peng, K., Xu, J., Sun, S., and Tu, G. 2006. Study on CaO decomposing bastnaesite in assistant of NaCl-$CaCl_2$. Accessed March 2018. https://wenku.baidu.com/view/0f7e7105e87101f69e31953b.html.

Bian, X., Chen, J., Zhao, Z., Yin, S., Luo, Y., Zhang, F., and Wu, W. 2010. Kinetics of mixed rare earths minerals decomposed by CaO with NaCl-$CaCl_2$ melting salt. *J. Rare Earths* 28(December):86–90.

Bickerdike, E.L. 1937. A study of the double sulphates of some rare-earth elements with sodium, potassium, ammonium and thallium. PhD dissertation. University of Southern California.

Binnemans, K., Jones, P.T., Blanpain, B., Van Gerven, T., Yang, Y., Walton, A., and Buchert, M. 2013. Recycling of rare earths: A critical review. *J. Clean. Prod.* 51(July): 1–22.

Borra, C.R., Mermans, J., Blanpain, B., Pontikes, Y., Binnemans, K., and Van Gerven, T. 2016. Selective recovery of rare earths from bauxite residue by combination of sulfation, roasting and leaching. *Miner. Eng.* 92(June):151–159.

British Geological Survey. 2017. Rare earth elements. www.bgs.ac.uk/research/highlights/2010/rare_earth_elements.html. Accessed November 2017.

Brugger, W., and Greinacher, E. 1967. A process for direct chlorination of rare earth ores at high temperatures on a production scale. *J. Met.* 19(12):32–35.

Buchanan, G., Reveley, S., Forrester, K., and Cummings, A. 2014. Review of current rare earth processing practice. Presented at ALTA 2014 Uranium-REE conference, Perth, Western Australia, May 2014.

Chi, R., and Tian, J. 2008. *Weathered Crust Elution-Deposited Rare Earth Ores*. New York: Nova Science Publishers.

Chi, R., and Xu, Z. 1999. A solution chemistry approach to the study of rare earth element precipitation by oxalic acid. *Metall. Mater. Trans. B* 30B(April):189–195.

Chi, R., Zhang, X., Zhu, G., Zhou, Z.A., Wu, Y., Wang, C., and Yu, F. 2004. Recovery of rare earth from bastnasite by ammonium chloride roasting with fluorine deactivation. *Miner. Eng.* 17(2004):1037–1043.

China State Council. 2012. Situation and Policies of China's Rare Earth Industry. Beijing: Information Office of the State Council, The People's Republic of China.

Collier, D.C. 2012. The itinerant position of yttrium as evidenced by carboxylic acid extractions. PhD dissertation, University of Tennessee.

Commerce Resources Corp. 2014. Commerce Resources Corp. updates metallurgical flowsheet and outlines next steps for the Ashram Rare Earth Deposit, Northern Quebec. Press release.

Connelly, N.G., Damhus, T., Hartshorn, R.M., and Hutton, A.T. 2005. Nomenclature of Inorganic Chemistry: IUPAC Recommendations 2005. Cambridge, UK: Royal Society of Chemistry.

Danilov, V.V. 2011. *Technical Report on the Kutessay II Rare Earth Property, Kemin District, Kyrgyzstan, with Rare Earth Resource Estimate.* www.sedar.com/DisplayCompanyDocuments.do?lang=EN&issuerNo=00027886. Accessed March 2018.

Dutta, T., Kim, K-H., Uchimiya, M., Kwon, E.E., Jeon, B-J., Deep, A., and Yun, S-T. 2016. Global demand for rare earth resources and strategies for green mining. *Environ. Res.* 150(October):182–190.

European Commission. 2014. *Report on Critical Raw Materials for the EU.* http://ec.europa.eu/growth/sectors/raw-materials/specific-interest/critical/indexen.htm. Accessed May 2016.

Feasby, D.G., Chambers, D.B., and Lowe, L.M. 2013. Assessment and management of radioactivity in rare earth element production. In *Rare Earth Elements: Proceedings of the 52nd Conference of Metallurgists.* Edited by I.M. London, J.R. Goode, G. Moldoveanu, and M.S. Rayat. Westmount, QC: Canadian Institute of Mining, Metallurgy and Petroleum.

GB/T 20169-2015. *Mixed Rare Earth Oxide of Ion-Absorbed Type Rare Earth Ore.* China: General Administration of Quality Supervision, Inspection and Quarantine of the People's Republic of China; Standardization Administration of the People's Republic of China.

Goode, J.R. 2013. Thorium and rare earth recovery in Canada: The first 30 years. *Can. Metall. Q.* 52(3):234–242.

Goode, J.R. 2016a. A study of coarse ore pre-concentration and potential applicability to Canadian rare earth deposits. Report to Natural Resources Canada.

Goode, J.R. 2016b. Options for the separation of rare earth elements. Presented at IMPC 2016/COM 2016, Quebec City, QC, September.

Goode, J.R. 2016c. Where did that earbud come from? Current rare earth production facilities. Presented at 48th Annual Canadian Mineral Processors Operators Conference, Ottawa, ON, January 2016.

Grammatikopoulos, T., Jordens, A., and Gunning, C. 2016. Advanced characterization of the REE mineralogy and implications in mineral processing. Presented at IMPC 2016/COM 2016, Quebec City, QC, September.

Griffith, C.S., Fainerman-Melnikoval, M., Manis, A., Safinski, T.A., Soldenhoff, K.H., Catovic, E., Hadley, T., and Jones, R. 2016. Northern Minerals Ltd. Browns Range Project: Process flowsheet improvements. Presented at IMPC2016/COM2016. Quebec City, QC, September.

Guentner, J., Kolehmainen, E., Tiihonen, M., Karonen, J., Haavanlammi, L., and Hammerschmidt, J. 2016. Recovery of rare earth elements from complex raw materials by Outotec's solutions. Presented at IMPC 2016/COM 2016, Quebec City, QC, September.

Gupta, G.K., and Krishnamurthy, K. 2005. *Extractive Metallurgy of Rare Earths*. Boca Raton, FL: CRC Press.

Habashi, F. 2012. Carl Auer and the beginning of the rare earths industry. In *Rare Earths 2012 Elements, Proceedings of the 52nd Conference of Metallurgists (COM)*. Edited by J.R. Goode, G. Moldoveanu, and M.S. Rayat. Westmount, QC: Canadian Institute of Mining, Metallurgy and Petroleum.

Haque, N., Hughes, A., Lim, S., and Vernon, C. 2014. Rare earth elements: Overview of mining, mineralogy, uses, sustainability and environmental impact. *Resources* 3:614–635.

Hartley, F.R. 1952. The preparation of anhydrous lanthanon chlorides by high-temperature chlorination of monazite. *J. Appl. Chem.* 2(January):24–31.

Ho, E., Wilkins, D., and Soldenhoff, K. 2014. Recovery of cerium from chloride solution by oxidation with sodium hypochlorite. Presented at the Seventh International Symposium on Hydrometallurgy 2014, Victoria, BC.

Huang, X.W., Long, Z.Q., Wang, L.S., and Feng, Z.Y. 2015. Technology development for rare earth cleaner hydrometallurgy. *Rare Metals* 34(4):215–222.

Jordens, A., Cheng, Y.P., and Waters, K.E. 2013. A review of the beneficiation of rare earth element bearing minerals. *Miner. Eng.* 41(February):97–114.

JSC Solikamsk Magnesium Works. 2016. *Annual Report 2015*. http://смз.рф/raport/2016/2015_Solikamsk_Magnesium_Works_Annual_Report_web_v.pdf. Accessed March 2018.

Kaczmarek, J. 1981. Discovery and commercial separations. In *Industrial Applications of Rare Earth Elements*. ACS Symposium Series, Vol. 164. Edited by K.A. Gschneidner Jr. Washington, DC: American Chemical Society.

Keith-Roach, M., Grundfelt, B., Kousa, A., Pohjolainen, E., Magistrati, P., Aggelatou, V., Olivieri, N., and Ferrari, A. 2014. Past experience of environmental, health and safety issues in REE mining and processing industries and an evaluation of related EU and international standards and regulations. www.eurare.eu/publications.html. Accessed March 2018.

Kim, R., Cho, H.C., and Han, K.N. 2014. Behavior of anions in association with metal ions under hydrometallurgical environments: Part I: OH^- effect on various cations. *Miner. Metall. Proc.* 31(1):34–39.

Koltun, P., and Tharumarajah, A. 2014. Life cycle impact of rare earth elements. *ISRN Metall.* 2014, Article ID 907536.

Krebs, D. 2014. Separation of rare earths from uranium and thorium. Presented at IAEA, June 23–27. Vienna, Austria: International Atomic Energy Agency.

Krumholz, P. 1957. Brazilian practice for monazite treatment. In *Symposium on Rare Metals Held on 1st, 2nd, 3rd December 1957*. Trombay, Bombay, India: Atomic Energy Establishment.

Kul, M., Topkaya, Y., and Karakaya, I. 2008. Rare earth double sulfates from pre-concentrated bastnasite. *Hydrometallurgy* 93:129–135.

Li, D.Q., Wang, Z.H., Song, W.Z., Meng, S.L., and Ma, G.X. 1994. Recommended separation processes for ion-absorbed rare earth minerals. *Hydrometallurgy* 94:627–634.

Li, D.Q., Bai, Y., Sun, X.Q., and Liu, S.Z. 2013. Advances in solvent extraction and separation of rare earths. In *Rare Earth Elements: Proceedings of the 52nd Conference of Metallurgists*. Edited by I.M. London, J.R. Goode, G. Moldoveanu, and M.S. Rayat. Westmount, QC: Canadian Institute of Mining, Metallurgy and Petroleum.

Li, Z.L., and Yang, X.S. 2014. China's rare earth ore deposits and beneficiation techniques. Presented at ERES2014: 1st European Rare Earth Resources Conference.

Liao, C.S., Wu, S., Cheng, F.X., Wang, S.L., Liu, Y., Zhang, B., and Yan, C.H. 2013. Clean separation technologies of rare earth resources in China. *J. Rare Earths* 31(4):331–336.

Long, K.R., Van Gosen, B.S., Foley, N.K., and Cordier, D. 2010. *The Principal Rare Earth Elements Deposits of the United States—A Summary of Domestic Deposits and a Global Perspective*. Scientific Investigations Report 2010–5220. Reston, VA: U.S. Geological Survey.

Lucas, J., Lucas, P., Le Mercier, T., Rollat, A. and Davenport, W. 2015. *Rare Earths: Science, Technology, Production and Use*. Amsterdam: Elsevier.

Luo, X.P., Qian, Y.J., Chen, X.M., and Liang, C.L. 2013. Crystallization of rare earth solution by ammonium bicarbonate. In *Rare Earth Elements: Proceedings of the 52nd Conference of Metallurgists*. Edited by I.M. London, J.R. Goode, G. Moldoveanu, and M.S. Rayat. Westmount, QC: Canadian Institute of Mining, Metallurgy and Petroleum.

Lynas Corporation. 2017. Investor presentation, September 2017. https://www.lynascorp.com/Pages/Announcements-and-media-releases.aspx. Accessed March 2018.

Martin, M., and Rollat, A. 1986. Electrolytic separation of cerium/rare earth values. U.S. Patent 4,676,957.

McNulty, T. 2004. Minimization of delays in plant startups. In *Plant Operators Forum*. Edited by E.C. Dowling and J.O. Marsden. Littleton, CO: SME. pp. 113–120.

Moldoveanu, G.A., and Papangelakis, V.G. 2013. Recovery of rare earth elements adsorbed on clay minerals: II. Leaching with ammonium sulfate. *Hydrometallurgy* 131–132:158–166.

Molycorp. 1996. *Molycorp Mountain Pass Mine Expansion Project: Mountain Pass, California*. Draft Environmental Impact Report SCH 92092040. Camarillo, CA: ENSR Consulting and Engineering.

Molycorp. 2015. *Form 10-Q: Quarterly Report Pursuant to Section 13 or 15(d) of the Securities Exchange Act of 1934 for the Quarterly Period Ended March 31, 2015*.

MSA Group. 2015. *Mkango Resources Limited Songwe REE Project Malawi NI 43-101 Pre-Feasibility Report*.

Norgate, T.E., and Lovel, R.R. 2006. Sustainable water use in minerals and metal production. In *Water in Mining 2006*. Melbourne, Victoria: Australasium Institute of Mining and Metallurgy. pp. 331–339.

Outokumpu Technology. 2006. Physical characteristics of select minerals and materials. www.outokumputechnology.com/. Accessed January 2008.

Peak Resources. 2016. Ngualla project study delivers substantial capex and opex savings. ASX announcement, March 16. http://investorintel.com/technology-metals-news/peak-resources-limited-ngualla-study-delivers-substantial-capex-and-opex-savings/. Accessed March 2016.

Pillai, P.M.B. 2008. Naturally occurring radioactive materials (NORM) in the extraction and processing of rare earths. In *Proceedings of the Fifth International Symposium on Naturally Occurring Radioactive Material*. Vienna, Austria: International Atomic Energy Agency.

Rainbow Rare Earths. 2016. African Exploration Showcase, November 11. www.gssaconferences.co.za/African%20Explorattion%20Showcase%20Presos/African%20Exp.%20Session%203/Morelli%20Cesare.pdf. Accessed March 2018.

Rankin, J. 2012. Energy use in metal production. Presented at the High Temperature Processing Symposium 2012, Melbourne, Australia. https://publications.csiro.au/rpr/download?pid=csiro:EP12183&dsid=DS3. Accessed March 2018.

Riedemann, E.I. 2011. High purity rare earth metals preparation. Washington, DC: Materials Preparation Center, U.S. Department of Energy.

Rocha, A., Coutinho, D., Soares, E., de Tarso, P., Sprecher, A., Ferreira, J., and Goode, J.R. 2014. The Serra Verde rare earth deposit, Goias State, Brazil: Advancing to production. In *Proceedings of the 53rd Conference of Metallurgists*. Westmount, QC: Canadian Institute of Mining, Metallurgy and Petroleum.

Rosenblum, S., and Brownfield, I.K. 1999. *Magnetic Susceptibilities of Minerals*. Open-File Report 99-529. Reston, VA: U.S. Geological Survey.

Roskill 2016. *Rare Earths: Global Industry, Markets and Outlook to 2026*, 16th ed. London: Roskill Information Services Limited.

Sadri, F., Nazari, A.M., and Ghahreman, A. 2017. A review on the cracking, baking and leaching processes of rare earth element concentrates. *J. Rare Earths* 35(8):739–752.

SGS Canada. 2017. *Impurity Control for Rare Earth Element (REE) Processing—Phase 3—Impurity Removal from PLS*. Project 15484-004, Final Report 3, July 24. Report to Natural Resources Canada.

Shannon, R.D. 1976. Revised effective ionic radii and systematic studies of interatomic distances in halides and chalcogenides. *Acta Crystall. A-Crys.* 32(5):751–767.

Soderstrom, M.D., McCallum, T., Jakovljevic, B., and Quilodrán, A.J. 2014. Solvent extraction of rare earth elements using Cyanex® 572. In *Proceedings of the 53rd Conference of Metallurgists*. Westmount, QC: Canadian Institute of Mining, Metallurgy and Petroleum.

Songina, O.A. 1964. *Rare Metals (Redkie Metally)*. Translated from Russian by Israel Program for Scientific Translations, Jerusalem, 1970. Jerusalem: Keter Press.

Spedding, F.H., and Powell, J.E. 1954. Methods for separating rare-earth elements in quantity as developed at Iowa State College. *Trans. AIME* 200:1131–1135.

SRK Consulting (U.S.). 2010. *Alternative Technical Economic Model for the Mountain Pass Re-Start Project. Report to Molycorp filed with U.S. Securities Exchange Commission*. www.sec.gov/Archives/edgar/data/1489137/000095012310065239/d74323fwfwp.htm. Accessed March 2018.

Stans Energy. 2017. Home page. www.stansenergy.com. Accessed October 2017.

Swaminathan, T.V., Nair, V.R., and John, C.V. 1988. *Stepwise Hydrochloric Acid Extraction of Monazite Hydroxides for the Recovery of Cerium Lean Rare Earths, Cerium, Uranium and Thorium. Rare Earths, Extraction, Preparation and Applications*. Edited by R.G. Bautista and M.M. Wong. Warrendale, PA: TMS-AIME.

Thompson Reuters. 2015. Edited transcript MCP–Q4 2014 Molycorp Inc. Earnings Call. Event date March 17, 2015. www.molycorp.com/investors/. Accessed June 2016.

TMR (Technology Metals Research). 2015. TMR Advanced Rare-Earth Projects Index. www.techmetalsresearch.com/metrics-indices/tmr-advanced-rare-earth-projects-index/. Accessed June 2016.

U.S. DOE (U.S. Department of Energy). 2010. Critical Materials Strategy. http://energy.gov/sites/prod/files/DOECMS2011FINALFull.pdf. Accessed June 2016.

Vasudevan, S., Sozhan, G., Mohan, S., and Pushpavanam, S. 2005. An electrochemical process for the separation of cerium from rare earths. *Hydrometallurgy* 76:115–121.

Venmyn Deloitte. 2015. *National Instrument 43-101 Independent Technical Report on the Results of a Preliminary Feasibility Study on the Zandkopsdrift Rare Earth Element and Manganese By-product Project in the Northern Cape Province of South Africa for Frontier Rare Earths Limited*. www.sedar.com/DisplayCompanyDocuments.do?lang=EN&issuerNo=00030507. Accessed March 2018.

Verbaan, N., Yu, M., Brown, J., and Mezei, A. 2014. Are REE exploration companies doing enough and/or the right testwork to avoid plant failure? In *COM 2014: Conference of Metallurgists Proceedings*. Westmount, QC: Canadian Institute of Mining, Metallurgy and Petroleum.

Verbaan, N., Bradley, K., Brown, J., and Mackie, S. 2015. A review of hydrometallurgical flowsheets considered in current REE projects. In *Symposium on Strategic and Critical Materials Proceedings*. British Columbia Geological Survey Paper 2015-3. Edited by G.J. Simandl and M. Neetz. Victoria, BC: British Columbia Ministry of Energy and Mines. pp. 147–162.

Vossenkaul, D., Stoltz, N.B., Meyer, F.M., and Friedrich, B. 2015. Extraction of rare earth elements from non-Chinese ion adsorption clays. Presented at EMC 2015, Dresden, Germany.

Wall, F., Rollat, A., and Pell, R.S. 2017. Responsible sourcing of critical metals. *Elements* 13:313–318.

Wang, Z.C., and Zhang, L.Q. 2002. Carburizing chlorination process for extracting and separating cerium and non-Ce rare earth from rare-earth ore. Chinese Patent CN 1373233 A.

Webmineral. 2016. Mineralogical database. www.webmineral.com/. Accessed June 2016.

Xie, F., Zhang, T.A., Dreisinger, D., and Doyle, F. 2014. A critical review on solvent extraction of rare earths from aqueous solutions. *Miner. Eng.* 56:10–28.

Xu, G.X. (ed.) 1995. *Rare Earths*. 2nd ed. Beijing: Metallurgy Industry Press.

Yahorava, V., du Preez, A.C., and Kotze, M.H. 2013. Some processing options for the removal of critical impurities from multi-source concentrates feeding a rare earths refinery. In *Rare Earth Elements: Proceedings of the 52nd Conference of Metallurgists*. Edited by I.M. London, J.R. Goode, G. Moldoveanu, and M.S. Rayat. Westmount, QC: Canadian Institute of Mining, Metallurgy and Petroleum.

Youtz, M.A., Warren, F.A., and Bearse, A.E. 1950. *Topical Report on the Solubility of Sodium Phosphate in Sodium Hydroxide Solutions*. Columbus, OH: Battelle Memorial Institute.

Zelikman, A.N., Krein, O.E., and Samsonov, G.V. 1964. *Metallurgiya redkikh metallov*. Moscow, Russia: Izdatel'stvo Metallurgiya Moskva. Translated from Russian by Israel Program for Scientific Translations, Jerusalem, 1966.

Zhang, J., and Edwards, C. 2012. A review of rare earth mineral processing technology. Presented at the 44th Annual Canadian Mineral Processors Operators Conference, Ottawa, ON, January 17–19.

Zhang, J., and Edwards, C. 2013. Mineral decomposition and leaching processes for treating rare earth ore concentrates. *Can. Metall. Q.* 52(3):243–248.

Zhang, L.Q., Wang, Z.C, Tong, S.X., Lei, P.X., and Zou, W. 2004. Rare earth extraction from bastnaesite concentrate by stepwise carbochlorination-chemical vapor transport-oxidation. *Metall. Mater. Trans. B* 35b.

Zhou, J.M., and Liu, H.Z. 2015. A method for clean separation of Baotou Rare Earth Concentrate. Chinese Patent 103540740 B.

Zhu, G.C., Qiu, X., Chi, R.A., Xu, S.M., Zhang, Z.G., and Tian, J. 2000. Recovering RE with NH_4Cl Roasting from Bastnasite Ore. Presented at MINPREX 2000, Melbourne, Australia, September 11–13.

Zhu, Z., Pranolo, Y., and Cheng, C.Y. 2015. Separation of uranium and thorium from rare earths for rare earth production—A review. *Miner. Eng.* 77:185–196.

Zou, D., Chen, J., Cui, H., Liu, Y., and Li, D.Q. 2014. Wet Air oxidation and kinetics of cerium(III) of rare earth hydroxides. *Ind. Eng. Chem. Res.* 53(35):13790–13796.

CHAPTER 12.33

Silica, Quartz, and Silicon

David C. Lynch and Courtney A. Young

SILICA AND QUARTZ

Silica, also known as quartz (SiO_2), is categorized by statistical historians as sand and gravel. It is divided into two categories of commodities: construction-grade and industrial-grade sand and gravel.

Industrial-grade sand and gravel generally has silica content that is much higher than that of construction grade in some instances, such as in glass production where the silica content will be as high as 99.9% SiO_2. Production levels of silica are relatively unchecked on a world basis; however, U.S. production of the two grades in 2014 was 911,000 t (metric tons) and 75,000,000 t of construction-grade and industrial-grade silica, respectively. Industrial-grade silica is used in the production of glass (39% of the total), foundry sand (22%), and abrasives (5%), and the remaining 34% is used in a multitude of products, which include fracking sand in the production of petroleum and lump ore used in the production of ferrosilicon and metallurgical-grade silicon (MG-Si).

Construction-grade sand and gravel is even less specific in that 50% of the products are unspecified while the second half is generally categorized as being concrete for road construction, additives for asphalt, and the remaining used in concrete blocks, bricks, pipe, and roofing shingles.

COMMERCIAL USES OF SILICON METAL

Silicon is a metalloid, and its primary use over the past century has been as an alloying agent. Thus, it has typically been referred to as a metal. Silicon as an element begins to take on metallic properties at temperatures above 1,000 K (727°C). Silicon metal (Si) is produced most commonly in an electric arc furnace, reducing SiO_2 to Si metal (covered in detail in the "Production of Metallurgical-Grade Silicon" section later in this chapter). The products from this industry are dependent on the demand for the various end uses, typically silicon ferroalloy and silicon metal. Silicon ferroalloy, or ferrosilicon, is produced for alloying in cast iron, cast steel, and steel. Silicon metal, or MG-Si, is used to alloy with aluminum and in the production of chemicals, particularly silicones, and, to a lesser degree, in the production of high-purity silicon for the photovoltaic industry (i.e., solar-grade silicon), and in the production of ultra-high-purity silicon in the semiconductor industry (i.e., electronic-grade silicon).

Silicones Production

Silicones, or organosilicon compounds, are produced throughout the world using direct synthesis, the Rochow process, and/or the Müller–Rochow process, typically as a common technology on an industrial scale. The process can be completed using alkyl halides in a copper-catalyzed reaction with silicon metal. Processing is conducted in a fluidized bed reactor. The best result in terms of both selectivity and yield occurs with use of methyl chloride. Typical conditions for the production to occur are at temperatures of 573 K (300°C) at 2–5 bars of pressure. These conditions allow for 90%–98% conversion for silicon and 30%–90% for chloromethane. Approximately 1.4 million t of dimethyldichlorosilane (Me_2SiCl_2, where Me represents CH_3) is produced annually using this process (Elschenbroich 1992). The generalized chemical reaction for the production of organosilicon compounds is as follows:

$$xMeCl + Si \rightarrow H_2MeSiCl,\ Me_2SiCl_2,\ MeSiCl_3,\ \text{and many other products} \qquad \textbf{(EQ 1)}$$

Ultra-high-purity silicon metal for use in electronics and high-purity silicon metal for use in photovoltaics are presented in detail in the "Metallurgical-Grade Silicon Applications" section later in this chapter. Each of these products are produced using chlorine-based chemistry and are completed utilizing hydrogenation and distillation of trichlorosilane to produce the base chemicals, which are further refined through thermal decomposition to produce polycrystalline silicon (poly-Si) of very high purity. Further treatment of the poly-Si is required to produce the electronic-grade single-crystal silicon during this refining process. Poly-Si is produced in rod or granular form depending on the technology employed. Poly-Si rods, having diameters of 60–300 mm, are produced in Siemens reactors. Poly-Si is sold in rod form or the rods are broken into smaller particles (i.e., minus 10 cm and sold in a chunk form). Typically, chuck is sold in 5-kg bags. Granular

David C. Lynch, Professor Emeritus, University of Arizona, Tucson, Arizona, USA
Courtney A. Young, Department Head & Professor, Metallurgical & Materials Engineering, Montana Tech, Butte, Montana, USA

silicon, generally the size of rock salt and 5–10 mm in diameter, is produced using fluid bed reactors. Granular poly is sold in bags from 5 kg up to 1,500-kg super sacks.

SOURCES OF QUARTZ AND SILICA FOR PRODUCTION OF SILICON

Quartz is the second most common mineral in the earth's crust. It is found in all three of the earth's rock types: igneous, metamorphic, and sedimentary. Quartz is colorless but is frequently colored by impurities, such as iron, and may then be almost any color. It may be transparent to translucent, hence its use in glassmaking, and has a vitreous luster. Quartz is a hard mineral, rating 7 out of 10 on the Mohs scale of mineral hardness, owing to the strength of the bonds between the atoms. It will scratch glass. It is also relatively chemically inert and does not react with dilute acids. These are desirable qualities in various industrial uses. Figure 1 shows Brazilian quartz crystals.

Figure 1 Brazilian quartz crystals

Quartz is particularly prevalent in sedimentary rocks since it is extremely resistant to chemical breakdown via the weathering process. Physical weathering—such as erosion as a result of freeze and thaw, rain splash, and surface runoff resulting in abrasion—produces silica sands and sandstones that are of great use to the industry in many areas, such as construction, landscaping, and fiberglass production. However, most of the quartz/silica that is used by the industry for the production of silicon metal is the metamorphic form of quartz or the mineral quartzite (Schei et al. 1998). Sandstone is converted into quartzite through heating and pressure usually related to tectonic compression within orogenic belts. When sandstone is cemented to quartzite, the individual quartz grains recrystallize along with the former cementing material to form an interlocking mosaic of quartz crystals. Most or all of the original texture and sedimentary structures of the sandstone are erased by the metamorphism. The grainy, sandpaper-like surface becomes glassy in appearance. Minor amounts of former cementing materials, iron oxide, silica, carbonate, and clay often migrate during recrystallization and metamorphosis. This causes streaks and lenses to form within the quartzite.

Pure quartzite is usually white to gray, though quartzite often occurs in various shades of red because of varying amounts of iron oxides (FeO, Fe_2O_3, and Fe_3O_4). Other colors of quartzite include yellow, green, blue, and orange, which occurs because of other chemical or mineral impurities. High-grade silica is found in unconsolidated deposits below thin layers of overburden or in metamorphic deposits that sometimes are millions of metric tons (i.e., deposits that result in mountains).

For industrial use, high-purity deposits of silica capable of yielding products of at least 95% SiO_2 are required. Often, even higher purity values are needed. Silica is produced by strip mining, quarrying methods, and open pit mining; it is rare that extraction is by underground mining methods. Additionally, very little or no physical processing is normally required to produce high-grade silica products, due to the high-grade deposits that are currently found in the world. Often, the only processing is by human interaction in the course of sorting, visual observation, and physical removal of undesired rock types and other waste products (i.e., wood or other organic matter).

The preferred quartzite deposits, which are used by smelters in the reduction of silica to silicon metal, are those that produce "lump" silica. The size is variable but normally consists of 80% in the range of 75 to 150 mm lump, with the other 20% being smaller in size. The lump quartzite retards compression in the shaft of the electric arc furnace because of the toughness of the mineral, providing the needed voids in the furnace charge and allowing the hot gases of the furnace to easily move upward through the burden. More on the dynamics of the electric arc furnace is covered later in this chapter.

Depending on how the silica deposit was formed, quartz grains may be sharp and angular, subangular, subrounded, or rounded. Foundry and filtration applications require subrounded or rounded grains for best performance.

Historical pricing of silicon metal products is shown in the following list for the period of 1995 through 2014. The pricing of these products has fluctuated significantly with pricing in the five-year period from 2000 through 2005 having some lower pricing than that which occurred during the previous five-year period (i.e., 1995 through 1999). The source of pricing information is the U.S. Geological Survey's *Minerals Yearbook* (USGS 1996–2015).

- Ferrosilicon (50% Si): \$0.41–\$1.16 per lb
- Ferrosilicon (75% Si): \$0.32–\$1.09 per lb
- Metallurgical Si (>96.5% Si): \$0.51–\$1.62 per lb

Both the construction industry and other commodity pricing are variable and the number of commodities that exist does not allow for listing. Resources for pricing should be checked by the investigator at the time of need.

MINERAL PROCESSING OF SILICA AND QUARTZ

Impurities in quartz play a significant role in ore selection and the degree of mineral processing required for its use in production of silicon, optics, optical fiber, glass, plastics, and high-temperature refractories. The shape, size, surface texture (rounded versus sharp edges), and crystal structure of granular quartz has a significant effect on product choices for foundry and fracking sands, concrete, bricks, sand traps (golf), sandboxes, and playgrounds.

Mine-site selection and the degree of mineral processing of quartz is dictated by product specifications, ore mineralogy, and production method. An example of the latter is the production of silicon in the silicon-submerged arc furnace where quartz lumps from mines, or river rocks, are charged to the

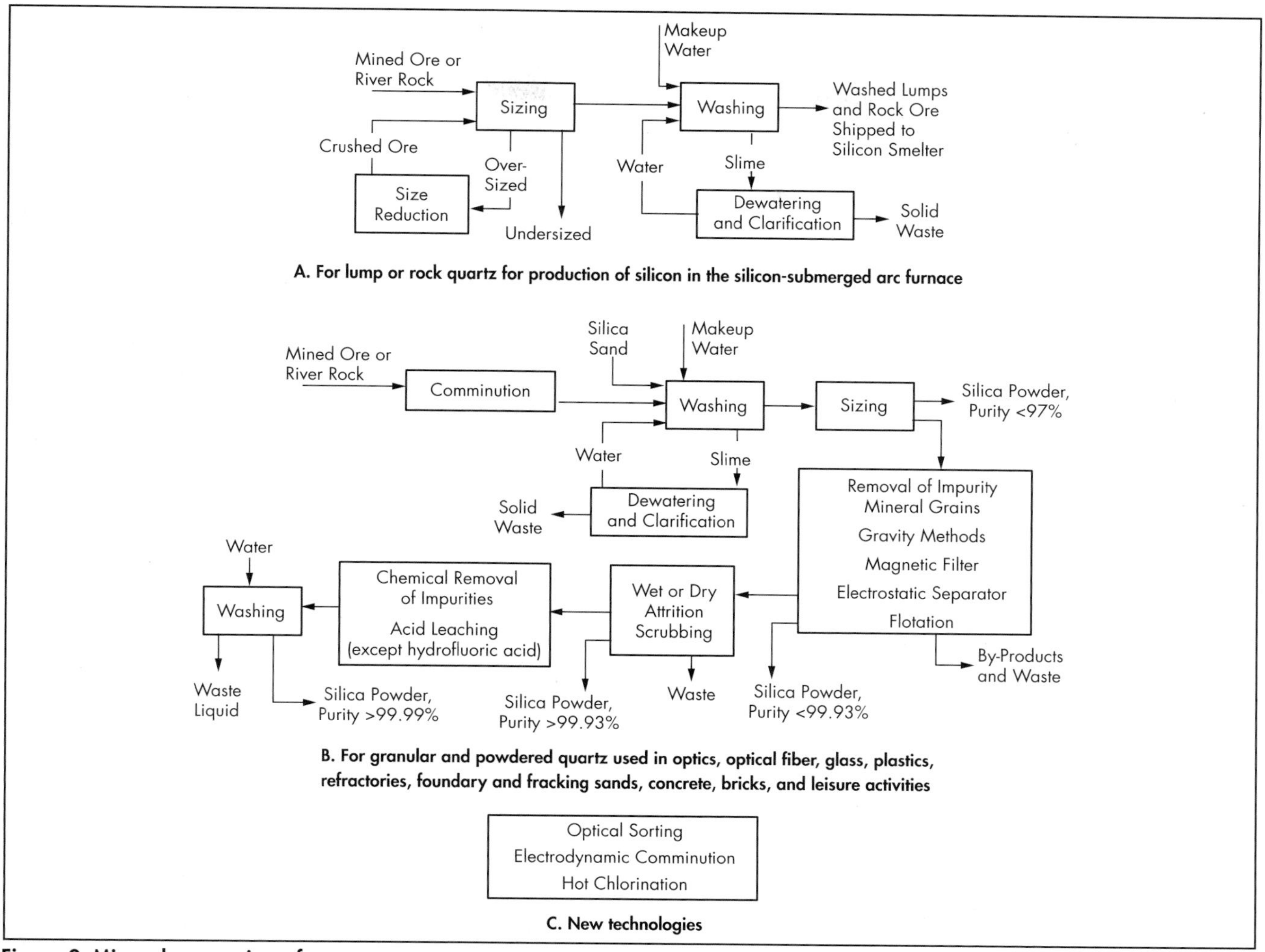

Figure 2 Mineral processing of quartz

furnace. The control of impurity content rests largely with ore selection, as impurities remain locked in the lumps and rocks, and thus mineral processing amounts to only sizing and washing of the ore, as shown in Figure 2A.

Meeting product specifications with granular and powdered quartz often requires greater mineral processing, as presented in Figure 2B. Comminuted ore consists of five classes of impurities:

1. Separate grains of other minerals, either present in quartz sand or liberated upon crushing and grinding of lump ore
2. Impurity minerals attached to quartz grains, or impurity atoms that have diffused into the quartz but are concentrated in a thin skin at the surface
3. Impurity minerals that extend from the surface and into the grain itself
4. Impurity minerals contained within the grain of quartz
5. Impurity atoms that substitute for either Si or O atoms throughout the grains

Comminuted ore provides more opportunities for removal of impurities than lump and rock ore used to produce silicon, as indicated in Figure 2B. The purities indicated in the figure are approximate numbers; purity <97% is used for playgrounds, sandboxes, sand traps; purity <99.93% is for foundries, fracking, glass, and bricks; purity >99.93% is for specialty glass, high-temperature refractories, and plastics; purity >99.99% is for optics and optical fiber. Individual grains of other minerals can be removed after washing and sizing by any or all of several methods: electrostatic separator, flotation, magnetic filter, and gravity methods. The latter includes jigs, shaker tables, spirals, and so forth. Attrition scrubbers produce a high-shear environment in which particles scrub against themselves to scour their surface, removing both contaminated surface silica and deleterious minerals bonded or physically attached to the surface of the quartz grains. If impurity minerals extend from the surface into the body of the grain, acid leaching can be used in their removal. Impurity minerals are immune to acid leaching if they are encapsulated by silica. Thus, the success of acid leaching is dependent on the degree of comminution. Unfortunately, impurity atoms (Al, B, P, and S) that substitute for either Si or O in the quartz remain with the quartz.

Newer mineral processing techniques identified in Figure 2C are all being employed by Anzaplan GmbH. The coupling of optical images and high-speed computer to operation of mechanical devices makes optical sorting on the basis of color, transparency, and shape a means for improved quality control of quartz products for the glass, optics, foundry, and cement industries. Electrodynamic comminution uses electrical-induced shock waves that fracture minerals along their weaker natural boundaries. Those boundaries exist between different minerals in composite rock. In processing of quartz, electrodynamic comminution exposes impurity minerals, making their removal by acid leaching possible. Hot chlorination can replace the use of acid. This technique relies on the volatility of metal chlorides. Passing granular or powdered quartz through a hot gas (~1,300 K [1,027°C]) containing chlorine or hydrogen chloride will remove most metallic minerals from quartz, provided the impurities are not isolated from the gas.

The degree of mineral processing of quartz is dictated by product specifications and production cost; those factors are linked by ore selection and its mineralogy. Successful development of a quartz deposit requires careful evaluation of all these factors.

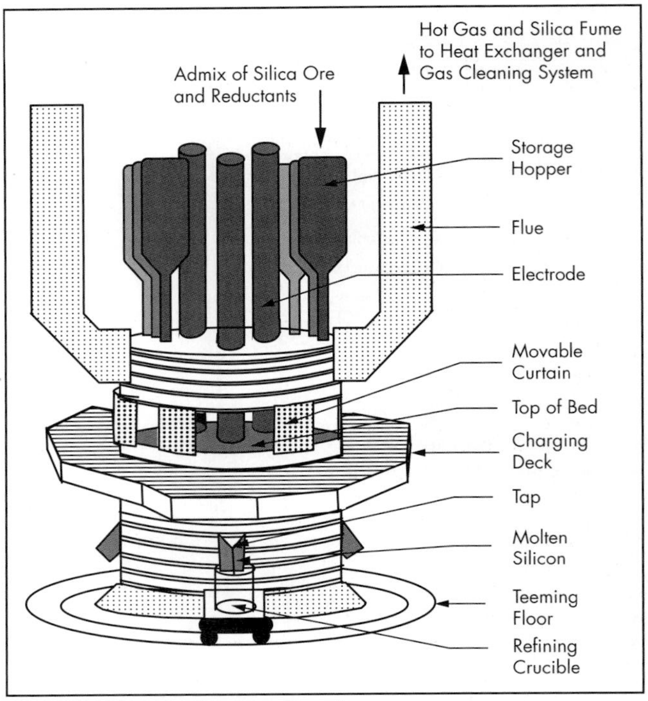

Figure 3 Silicon-submerged arc furnace

PRODUCTION OF METALLURGICAL-GRADE SILICON

Silicon-Submerged Arc Furnace

Metallurgical-grade silicon (MG-Si) and ferrosilicon are produced in the silicon-submerged arc furnace, where ore and reductants are processed at temperatures as high as 2,400 K (2,127°C). The arc furnace, as shown in Figure 3, is the centerpiece to a modern silicon plant. The raw materials (silica ore, coke or coal, and wood chips) are delivered to the plant and stored in covered sheds. Operators, using sophisticated computer programs and automated controls, draw on the stored materials, mix them, and transport the furnace feed to hoppers above the arc furnace. When needed, the operator releases the furnace feed so that it falls onto the bed material in the furnace. Movable curtains allow workers on the charging deck easy access to the furnace to use mechanized rakes to smooth out the piles of new furnace feed, maintaining bed height and filling in blowholes formed since the previous addition of furnace feed.

Countercurrent flow exists in the furnace; gases generated in the hottest portion of the furnace pass up through the bed of solids that move downward. The latter is consumed in the production of silicon and in generation of gases. Silicon is extracted from the furnace through tap holes and allowed to collect in the refining crucible. Gases emanating from the top surface of the bed are drawn off, along with air, to the heat exchanger and gas cleaning system. Silica fume is collected from the gas cleaning system. The furnace can be operated to produce commercial quantities of both silicon and silica fume.

The hearth of the furnace is rotated around the fixed electrodes, resulting in a cavity forming around each electrode, as shown in Figure 4. The cavity serves to promote radiant heat transfer to satisfy energy requirements for the endothermic

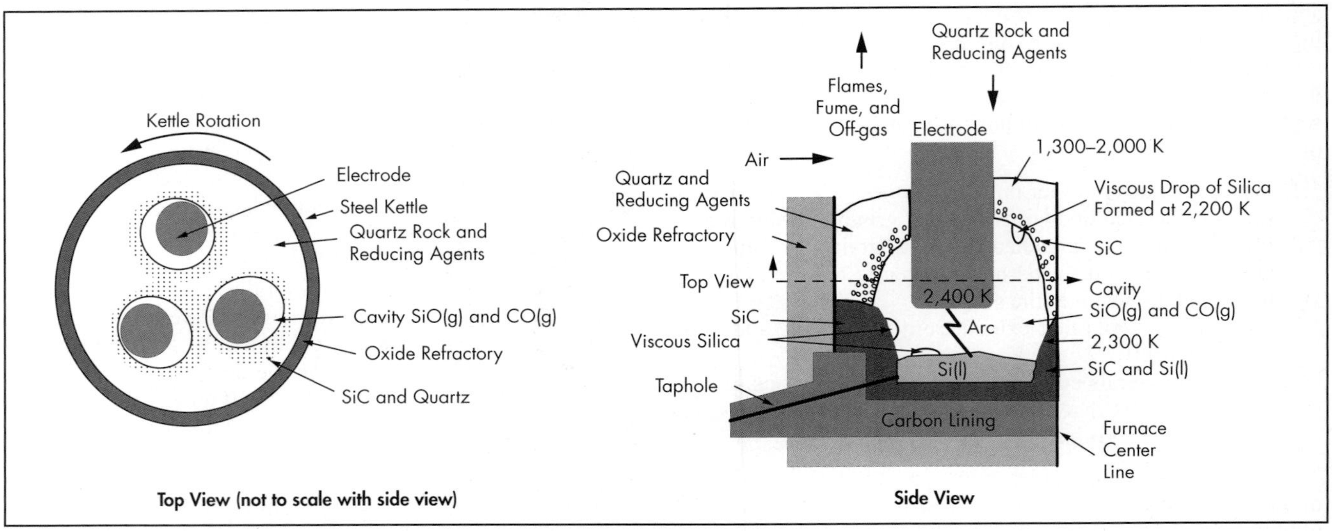

Adapted from Schei et al. 1998

Figure 4 Interior of the silicon-submerged arc furnace

reactions occurring in and around the cavity. Newly charged material caves in behind the electrode as the kettle is rotated, renewing the porosity of the bed, while on the other side of the electrode, older furnace feed with less porosity is consumed. The drawing also shows that SiC builds up around the hearth, protecting it from thermal and physiochemical corrosion. Furnace linings typically last two years, and the lining in a well-run furnace will last three years.

Chemistry in the Arc Furnace

The overall reactions in producing silicon are

$$SiO_2 + 2C \rightarrow Si(l) + 2CO(g) \quad \textbf{(EQ 2)}$$
(composite reaction occurring within the bed)

$$2CO(g) + O_2(g) \rightarrow 2CO_2(g) \quad \textbf{(EQ 3)}$$
(occurs as off-gas exits bed)

$$2SiO(g) + O_2(g) \rightarrow 2SiO_2 \quad \textbf{(EQ 4)}$$
(silica fume produced as off-gas exits bed)

The composite Reaction 2 is composed of several reactions occurring in the furnace. While there have been many studies of the silicon-submerged arc furnace (reviewed in Schei et al. 1998), it was Anders Schei in 1967 who published the classical equilibrium diagram used, along with the Si-C phase diagram, to explain reactions occurring in the arc furnace. Through use of the equilibrium diagram and the Si-C phase diagram, the bed within the furnace is viewed as consisting of inner and outer reaction zones. The demarcation between zones is the transformation of all carbon in the reductants to SiC, which in well-run furnaces occurs at a temperature of approximately 2,050 K (1,777°C). The phase diagram in Figure 5 reveals that silicon cannot coexist in equilibrium with carbon at temperatures below 3,100 K (2,827°C), thus the necessity that all carbon be converted to SiC.

Schei's equilibrium diagram (with modifications in Lynch 2002) is presented in Figure 6. Using the phase rule, Schei (1967) demonstrated that any equilibrium condition within the arc furnace can only involve two condensed phases. The equilibrium diagram, which is presented in two graphs, represents all possible equilibrium reactions. Figure 6A represents conditions in the upper bed (outer reaction zone) where it is essential to reduce the SiO content in the gas to ensure significant Si yield for the overall process. Figure 6B reflects the conditions in the inner reaction zone where molten Si is produced. The diagram reflects the high temperatures and significant partial pressure of SiO(g) necessary to produce Si.

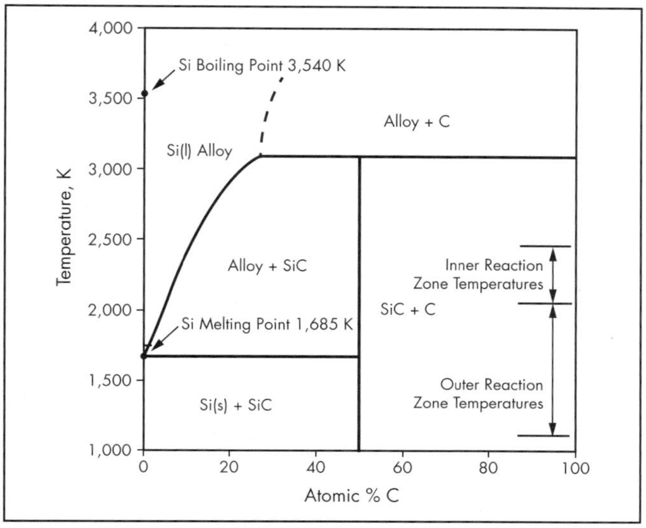

Adapted from Schei et al. 1998
Figure 5 Si-C phase diagram

Analysis of reactions occurring in the arc furnace begins with the outer reaction zone and with knowledge that the gas in the inner reaction zone has significant concentration of both CO and SiO, while the gas leaving the furnace is primarily CO. Quartz ore, coal (or coke), and wood chips are dropped from the storage hopper onto the bed in the furnace, where the temperature is 1,300 K (1,027°C) and higher. The heat and the reducing nature of the gas exiting the bed convert the wood chips to charcoal. That charcoal, while a source of impurity elements, is essential to maintaining a porous bed through which gases can flow from the inner reaction zone and through the bed material in the outer zone. The SiO content in the gas as it moves through the outer zone declines owing to any of the following reactions:

$$SiO(g) + 2C \rightarrow SiC + CO(g),\ \Delta H_{1473K} = -78.3\ \text{kJ},\ T \geq 1{,}473\ \text{K} \quad \textbf{(EQ 5)}$$

$$2SiO(g) \rightarrow SiO_2 + Si,\ \Delta H_{1273K} = -688.9\ \text{kJ},\ T \geq 1{,}273\ \text{K} \quad \textbf{(EQ 6)}$$

$$3SiO(g) + CO(g) \rightarrow SiC + 2SiO_2,\ \Delta H_{1785K} = -1{,}415.7\ \text{kJ},\ T \geq 1{,}785\ \text{K} \quad \textbf{(EQ 7)}$$

$$SiO(g) + CO(g) \rightarrow C + SiO_2,\ \Delta H_{1785K} = -668.5\ \text{kJ},\ T \leq 1{,}785\ \text{K} \quad \textbf{(EQ 8)}$$

The equilibrium conditions for these reactions are presented in Figure 6A, some written in the reverse direction. The enthalpy of these reactions reveals that consumption of SiO(g) moves heat from the cavity into the upper bed. That evidence—light brown deposits or white deposits with brown spots—from excavation of cooled furnaces reveals that Reaction 6 is the most important reaction in moving heat into the upper bed (Schei et al. 1998). Kinetic issues play important roles in determining which reactions occur at meaningful rates. The temperature listed with each reaction represents conditions where they occur. Reaction 8 is not thought to be significant much below the temperature indicated.

The movement of heat to the upper bed plays an important role in converting carbon to SiC. The equilibrium line for Reaction 6 lies in the unstable gas region (defined by the line connecting points i, b, and g in Figure 6A, and with the line connecting points g, h, and i in Figure 6B, indicating that the rate of Reactions 7 and 8 are slow in comparison and that a pseudo equilibrium is established by Reaction 6. At temperatures below 1,473 K (1,200°C), Reaction 6 dominates and SiO(g) decomposes, depositing SiO_2 and Si on the ore and reductants, and raising their temperature. At 1,473 K (1,200°C), SiO(g) concentration is not so depleted by Reaction 6 as to prevent reaction between the gas and C to produce SiC as per Reaction 5. At temperatures higher than that at point b in Figure 6A, Reaction 5 continues to reduce the SiO(g) content in the gas passing up and out of the inner reaction zone. Production of SiC continues until all the carbon is consumed. The equilibrium line for Reaction 5 in Figure 6A extends into Figure 6B, which is not shown in the latter figure

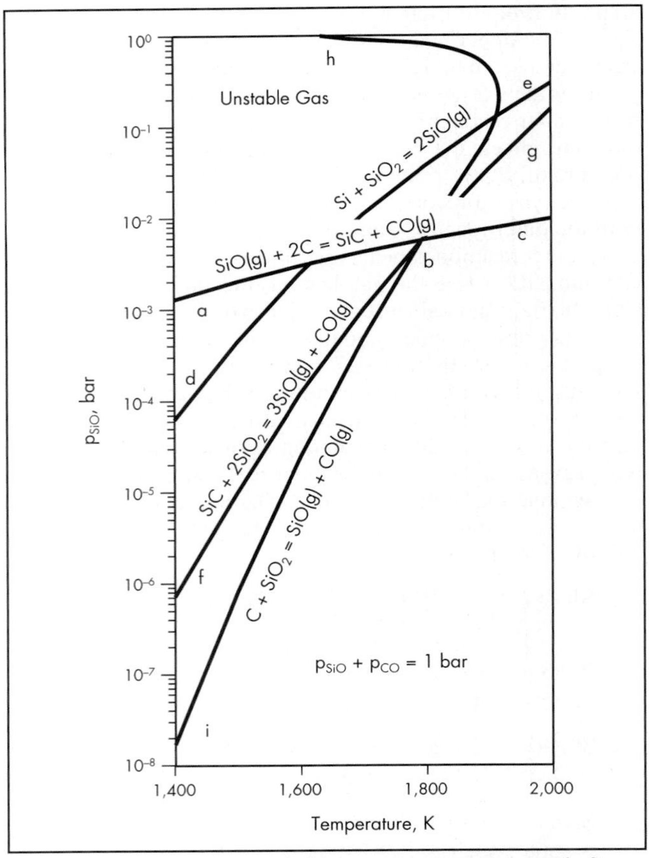

A. Lower temperatures are typical of the outer reaction zone (upper bed)

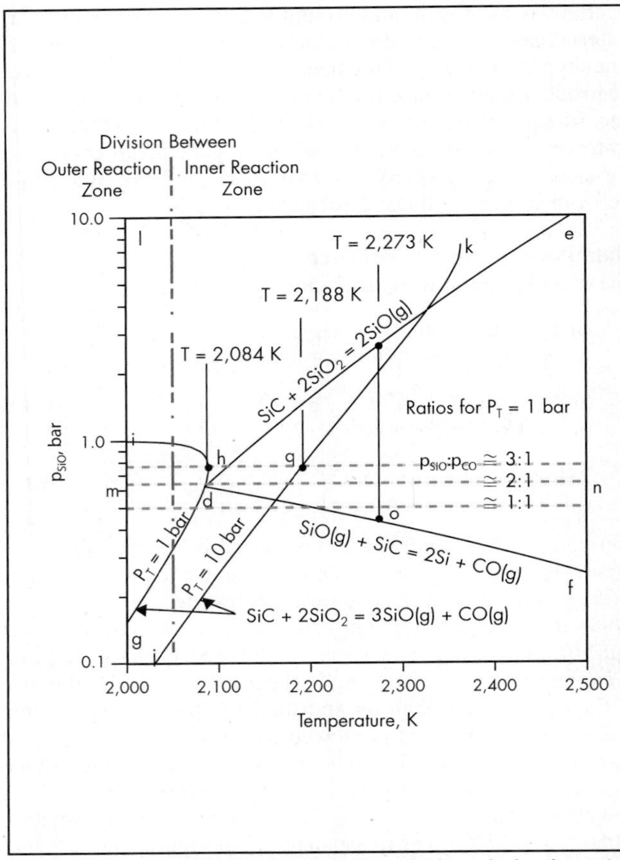

B. Represents conditions in the inner reaction zone (lower bed and cavity)

Adapted from Schei et al. 1998

Figure 6 Stability diagrams for reactions occurring in the silicon-submerged arc furnace

because the equilibrium vapor pressure associated with the reaction is lower than that plotted.

The production of Si requires that all carbon entering the inner reduction zone be converted to SiC. While Si and C cannot coexist in the arc furnace, Si and SiC can, as shown in the Si-C phase diagram (Figure 5). Too much C charged to the furnace will cause SiC to build up in the bottom of the hearth. That buildup will push the electrodes up and out of the furnace. Charging too little C will cause the Si yield to drop as silica fume is formed when SiO(g) leaves the furnace.

There is a balance between SiC entering the inner reaction zone as well as its consumption in the production of Si. Reactions occurring in this zone are represented by the equilibrium lines presented in Figure 6B. Silicon is produced by the following reaction:

$$SiO(g) + SiC \rightarrow 2Si(l) + CO(g),\ \Delta H_{2084K} = 166.6\ \text{kJ},\ T \geq 2{,}084\ \text{K} \qquad \textbf{(EQ 9)}$$

That reaction requires SiO(g) that is produced by both

$$SiO_2(l) + Si \rightarrow 2SiO(g),\ \Delta H_{2084K} = 606.1\ \text{kJ},\ T \geq 2{,}084\ \text{K} \qquad \textbf{(EQ 10)}$$

and

$$SiC + 2SiO_2(l) \rightarrow 3SiO(g) + CO(g),\ \Delta H_{2084K} = 1378.7\ \text{kJ},\ T \geq 2{,}084\ \text{K} \qquad \textbf{(EQ 11)}$$

All three reactions are endothermic, particularly Reactions 10 and 11.

Production of SiO(g) is necessary for producing silicon. In Figure 4, viscous drops of SiO_2 fall into the pool of molten silicon at a temperature approaching 2,300 K (2,027°C). Examination of Figure 6B reveals that the equilibrium vapor pressure of SiO(g) at 2,300 K (2,027°C) for Reaction 10 is about 2 bars. Since the total pressure in the arc furnace is 1 bar, Reaction 10 can never achieve equilibrium. There is a similar situation for Reaction 11. Equilibrium conditions for that reaction are plotted for total pressures of 1 and 10 bars. The location of those two equilibrium lines indicates that any contact of SiC and SiO_2 at temperatures above 2,084 K (1,811°C) can only achieve equilibrium at total pressures >1 bar. At 2,300 K (2,027°C), the equilibrium vapor pressure is well above 10 bars. In the inner reaction zone at the higher temperatures, SiO(g) production by Reactions 10 and 11 cannot be stopped. The SiO(g) is either consumed in the production of Si by Reaction 9, or the gas passes up through the bed where most of it reacts with C in production of SiC, or it condenses as per Reaction 6.

Kinetic factors play important roles in the composition of the gas leaving the inner reaction zone. Widespread condensation of SiO(g) by Reaction 6 on reductant and SiC ensure good contact between SiC and the SiO_2 required for Reaction 11. The energy requirement is substantial, and the source of that energy is likely photon transfer and convective heat transfer in the cavity. Similarly, the intimate contact between SiO_2 and Si produced by Reaction 6 is reversed as the material enters the cavity; again, heat transfer is the same as for Reaction 11. Reaction 10, as noted previously, also occurs when viscous drops of SiO_2 fall into the pool of molten Si, which acts as a source of energy. Silicon above 1,000 K (727°C) takes on metallic characteristics; thus the molten Si in the arc furnace is a good conductor of thermal energy, which is further enhanced by the turbulent conditions created by the arc striking the pool of Si. The area of contact between the drops and Si is limited compared to the contact created when SiO(g) is precipitated as silica and Si in the outer reaction zone. Reactions 10 and 11 feed SiO(g) for production of Si by Reaction 9. Consumption of SiO(g) by the reaction is considered to be faster than the rate of its production. Thus, gas composition in the cavity around the electrodes is established by Reaction 9. The equilibrium line for the reaction has negative slope in Figure 6B, and, thus, higher temperatures in the cavity are preferred to reduce the concentration of SiO(g) rising through the bed.

Impurities

Si refining is dependent on its final use; minimum refining for the steel and aluminum industries and maximum refining for the electronics industry are covered in the next section. A significant fraction of impurities entering the arc furnace in the ore as well as reductants are retained in the Si leaving the furnace. Myrhaug and Tveit (2000) performed an extensive mass balance on the retention of impurities in ferrosilicon leaving the silicon-submerged arc furnace, coming up with the boiling-point model. Some of their results are presented in Figure 7, where elements with boiling-point temperature higher than the cavity temperature are expected to dissolve in the Si product. Elements with boiling-point temperatures below that of the top bed material should leave the furnace with either the silica fume or the off-gas. The one anomaly is phosphorus, which should be volatilized but primarily enters the final product. Unfortunately, P as well as B are the two most difficult elements to remove from Si and require critical control in Si used by both the electronic and photovoltaic industries.

With large arc furnaces, tapping is continuous. SiO_2-rich flux can be added to the MG-Si flowing into a refining crucible to create a slag. The silica in the flux is reduced by Al and Ca dissolved in the Si, producing a slag consisting primarily of SiO_2, Al_2O_3, and CaO. Depending on the combined Al_2O_3 and CaO content of the slag, it either floats on top of the molten

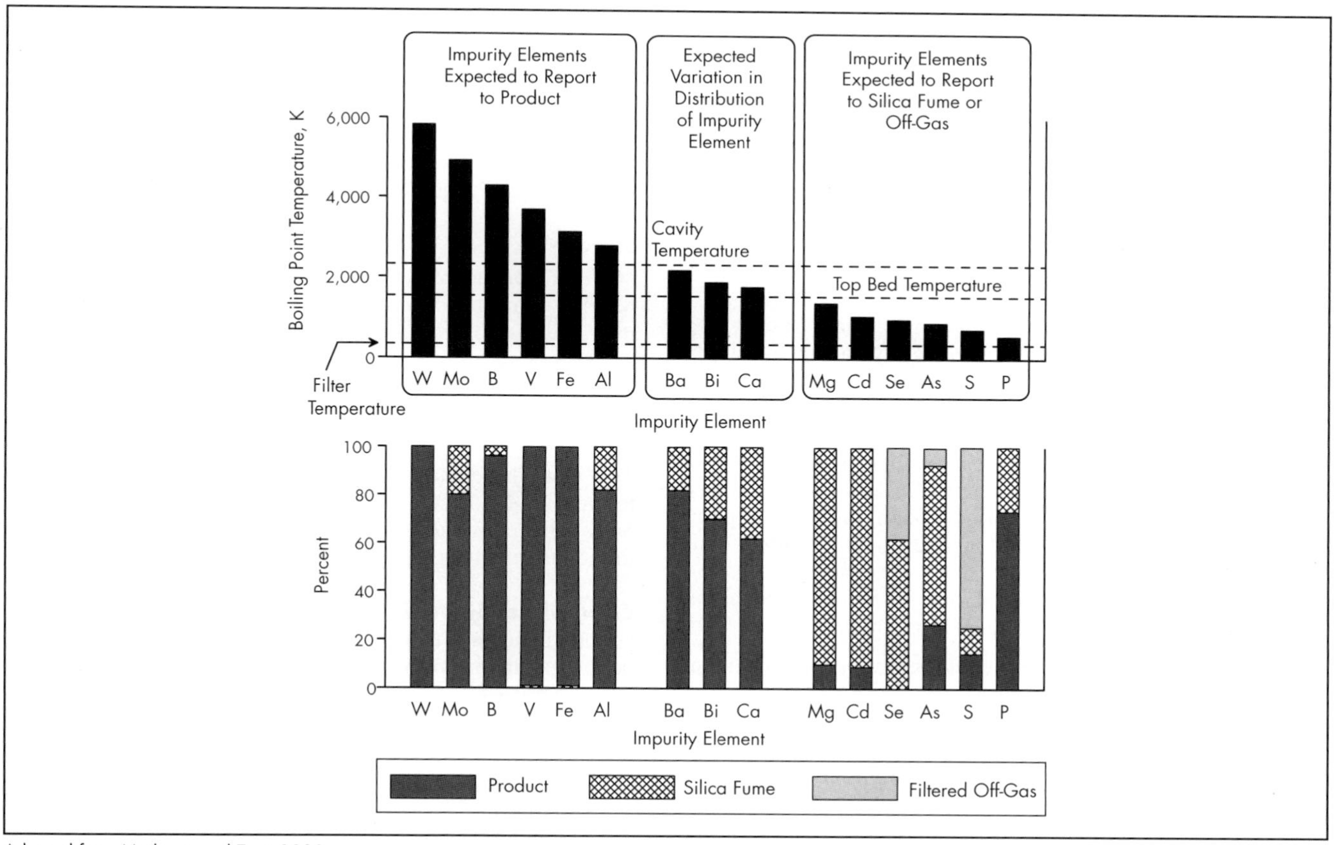

Adapted from Myrhaug and Tveit 2000

Figure 7 Myrhaug and Tveit's boiling-point model and test results

Table 1 Annual production for silicon, in thousands of metric tons

Product	Circa 1997	2008	2009	2014
Silicon metal	980	1,680* + 210†	1,800* + 160†	2,600* + 200†
Ferrosilicon	2,600	~4,270	~4,350	~5,310
Total silicon	3,580	6,160	6,310	8,110

Data from Schei et al. 1998; USGS Mineral Commodity Summaries 2010, 2011, 2016; USGS Minerals Yearbook 2011, 2014
*Does not include U.S. production.
†Authors' estimate of U.S. production.

Si or sinks. Droplets of slag passing through the molten Si improve the refining action and also trap SiC particles that are likewise slowly sinking through the silicon. After sufficient refining time, the Si is cast in shallow trays.

METALLURGICAL-GRADE SILICON APPLICATIONS

The world production figures for silicon metal, and silicon in ferrosilicon, are presented in Table 1. In 2014, total domestic consumption from all sources, domestic and foreign, of silicon metal of purity <99.6% was 187,000 t, of which 15% was used primarily in producing aluminum alloys and the remainder used to produce silicones and other chemicals. Similarly that same year, 164,000 t of ferrosilicon was used in steel production and 23,000 t used in producing cast iron (USGS *Minerals Yearbook* 2014).

Worldwide production of high-purity silicon for electronics and photovoltaics amounted to 74,000 t in 2009. In 2010 to 2014, it is estimated that less than 5% of all silicon metal production was used in this capacity (USGS *Minerals Yearbook*, 2009–2014). While the percentage mass of silicon metal used in electronics and photovoltaics is small in comparison to other uses, its value on a per-mass basis far exceeds that of silicon used in other applications. Thus, much of this section is devoted to the purification of silicon.

Aluminum

Adding silicon to aluminum or aluminum alloys results in a strong, light alloy. As a result, these alloys are increasingly used in the automotive industry to replace heavier cast-iron components. This allows weight reductions, a reduction in fuel consumption, increased efficiencies, and reducing gas emissions. The Al-Si binary phase diagram shows that a eutectic point exists at 850 K (577°C) and 12.6% silicon (Figure 8). At this temperature and Al-Si composition, the melting point is lower than the melting point for pure silicon or pure aluminum (ASM International 1992).

There are three classes of aluminum foundry alloys that are based on their silicon content (Sigworth 2014):

1. **Hypoeutectic alloys.** These alloys have a silicon content less than the eutectic composition. Most of the common hypoeutectic alloys have between 5% and 10% silicon. Some examples: C355, 356, 357, and 359. These alloys are designed primarily for high-strength applications where good ductility is also required.
2. **Eutectic alloys.** These alloys have between 10% and 13% silicon and consist mainly of Al-Si eutectic in the cast structure. They have a narrow freezing range, excellent fluidity, and are easy to cast. They also have good wear resistance and are quite ductile when not alloyed and heat treated to high strength. Eutectic alloys containing Cu, Mg, and sometimes Ni are used extensively for pistons.
3. **Hypereutectic alloys.** These alloys have between 15% and 20% silicon, so their cast structure is composed of primary silicon particles embedded in a matrix of Al-Si eutectic. These materials have remarkable wear resistance and are used where this characteristic is desired: for pistons, liner-less engine blocks, and compressor parts. They also have good high-temperature strength but are difficult to machine. Diamond tools are necessary.

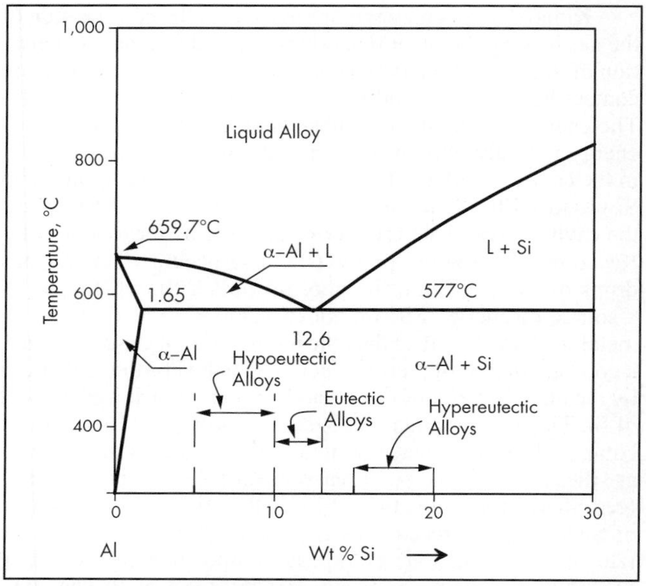

Adapted from ASM International 1992
Figure 8 Al-Si phase diagram

Silicone

Methyl Chloride Production

Methyl chloride is produced by condensing methanol with hydrochloric acid:

$$\underset{\text{methanol}}{CH_3OH} + \underset{\text{hydrochloric acid}}{HCl} \rightarrow \underset{\text{methyl chloride}}{CH_3Cl} + \underset{\text{water}}{H_2O} \qquad \textbf{(EQ 12)}$$

Two methyl groups (CH_3) must attach to each silicon atom to produce one molecule of dimethyldichlorosilane, which is the basic silicone building block. Methanol is a major ingredient in the manufacturing of silicone.

Chlorosilanes Synthesis

Chlorosilanes are synthesized in a fluid bed reactor at temperatures ranging from 523 K to 573 K (250°C to 300°C) and at pressures of 1–5 bars. The methyl chloride (CH_3Cl) flows through a fluidized bed of silicon metal powder. The reaction, which is catalyzed by a copper-based catalyst, generates a complex mix of methyl chlorosilanes:

$$xSi + yCH_3Cl \rightarrow \text{methyl chlorosilanes} \qquad \textbf{(EQ 13)}$$

The methyl chlorosilane mix contains (Me = CH_3)

- A large amount of dimethyldichlorosilane (Me_2SiCl_2)—the primary building block,
- A moderate amount of methyltrichlorosilane ($MeSiCl_3$),

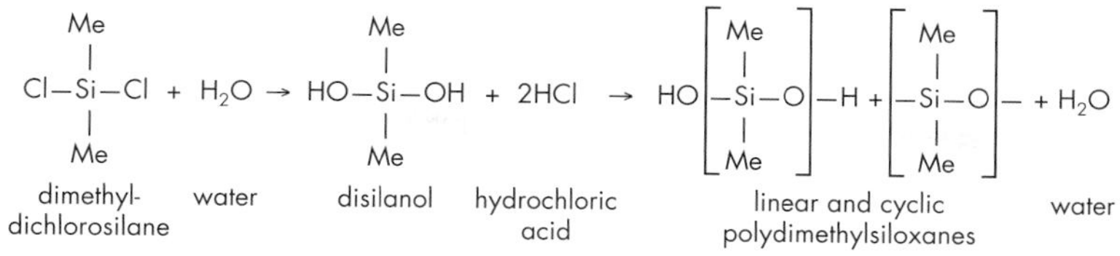

$$
\begin{array}{c}
\underset{\substack{\text{Me} \\ | \\ \text{Cl—Si—Cl} \\ | \\ \text{Me}}}{} + H_2O \rightarrow \underset{\substack{\text{Me} \\ | \\ \text{HO—Si—OH} \\ | \\ \text{Me}}}{} + 2HCl \rightarrow \text{HO}\left[\begin{array}{c}\text{Me} \\ | \\ \text{—Si—O—} \\ | \\ \text{Me}\end{array}\right]\text{—H} + \left[\begin{array}{c}\text{Me} \\ | \\ \text{—Si—O—} \\ | \\ \text{Me}\end{array}\right] + H_2O
\end{array}
\qquad \textbf{(EQ 14)}
$$

dimethyl- water disilanol hydrochloric linear and cyclic water
dichlorosilane acid polydimethylsiloxanes

- Some trimethylchlorosilane (Me_3SiCl),
- A small amount of methyldichlorosilane ($MeHSiCl_2$), plus
- Smaller amounts of other silanes.

The chlorosilanes are then separated from one another through the process of distillation.

Chlorosilane Distillation

Distillation is not a reaction. It is a liquid \rightarrow gas \rightarrow liquid phase change. Each of the chlorosilanes produced during the synthesis stage has a unique boiling point. These boiling points are used to distill the chlorosilanes from one another based on the number of chlorine atoms attached to the molecule. Once separated, the chlorosilanes are ready to be turned into useful siloxanes through the process of hydrolysis and condensation.

Chlorosilane Hydrolysis and Condensation

When water is added to dimethyldichlorosilane (the principal chlorosilane), the two react to form disilanol and hydrochloric acid. This occurs because oxygen "likes" silicon more than chlorine, and chlorine "likes" hydrogen more than oxygen (see Equation 14).

The disilanols are unstable and strongly attracted to one another. Catalyzed by the hydrochloric acid, they condense into polydimethylsiloxanes—molecules containing a backbone of silicon atoms bonded to oxygen atoms. Hydrolysis and condensation occur spontaneously, simultaneously, and very quickly. These reactions are a veritable "perpetual chemistry machine," resulting in a mixture of

- Cyclic rings (with 3–6 repeating SiO units) and
- Linear chains (with 30–50 repeating SiO units).

The cyclic and linear oligomers (mini-polymers) are separated from one another and distilled into cuts based on the number of SiO units in the chain. Some are used as they are (cyclosiloxane and low-molecular-weight silicone fluids, for example). But the majority are further polymerized and finished to create a wide array of materials with an amazing range of performance capabilities (Dow Corning Corporation 2015).

Fumed Silica

Fumed silica is produced by hydrolyzing silicon tetrachloride ($SiCl_4$) in an oxygen-hydrogen flame at high temperature, resulting in fine SiO_2 primary particles (10–40 nm). The primary particles accumulate and collide in the flame and agglomerate. The agglomerates and the waste gas (hydrochloric acid vapor, $HCl(g)$) are placed in a high-efficiency gas–solid separator column. The waste gas exits from the top of the separator and goes to a tail gas treatment system. The fumed silica (SiO_2) is removed from the bottom of the separator, into a container, and subsequently packaged. The reaction formula is

$$SiCl_4 + 2H_2(g) + O_2(g) \rightarrow SiO_2 + 4HCl(g) \qquad \textbf{(EQ 15)}$$

Fumed silica is used as an additive in silicone rubbers to increase the strength, as a thickening agent for paints, and an as a desiccant. It is also used in cosmetics, abrasives, and coatings.

Functional Silanes

This generic term covers a broad range of products built on silane molecules, in which an atom of hydrogen or of chlorine is substituted with an organic radical bearing a functional group (e.g., amine, acid, ester, alcohol). The general formula of functional silanes is

$$X_{4-(n+p+q)} SiRnR'pR''_q \qquad \textbf{(EQ 16)}$$

R, R', and R'' are the organic radicals, and X is Cl or OH. The major application for functional silanes is a coupling agent between inorganic and organic compounds, such as inorganic fillers (glass, silica, clays, etc.) in organic matrices (epoxy, polyester, etc.) (Luque and Hegedus 2003).

Polycrystalline Silicon Production

Silicon is a semiconductor and cannot conduct electricity when in a pure (minimal impurities) form, because there are very few free electrons available to carry a current:

	Resistivity, Ω-cm
Conductors	10^{-8} to 10^{-6}
Semiconductors	10^{-6} to 10^{-2}
Insulators	10^{7} to 10^{22}

The electrical conduction properties of silicon can be increased through either of the following methods.

Intrinsic Conduction

Intrinsic conduction is possible through physically heating the silicon. At any temperature above 0 K ($-273°C$), some of the bonds linking the silicon atoms are broken, resulting in a free electron and an absence of an electron (hole). These free electrons can now conduct electricity and the holes are free to move within the lattice. The magnitude of the number of free electrons based on temperature is as follows:

- At 300 K (27°C), the number of free electrons $= 1.6 \times 10^{10}/cm^3$, and the resistivity $= 230,000$ ohm·cm
- At 673 K (400°C), the number of free electrons $= 1 \times 10^{17}/cm^3$, and the resistivity $= 0.1$ ohm·cm

Extrinsic Conduction

Extrinsic conduction is feasible by the addition of trace impurities or "doping" to the silicon. Silicon has four electrons in the valence shell. Group VA (phosphorus, arsenic, and antimony) have five electrons in the valence shell and replace a silicon atom in the crystalline lattice (substitutional), resulting in an extra electron free to conduct electrical current considered "donors" or "n-type" impurities. Elements from group IIIA of the periodic table of elements (boron, aluminum, and gallium) have three electrons in the valence shell and replace

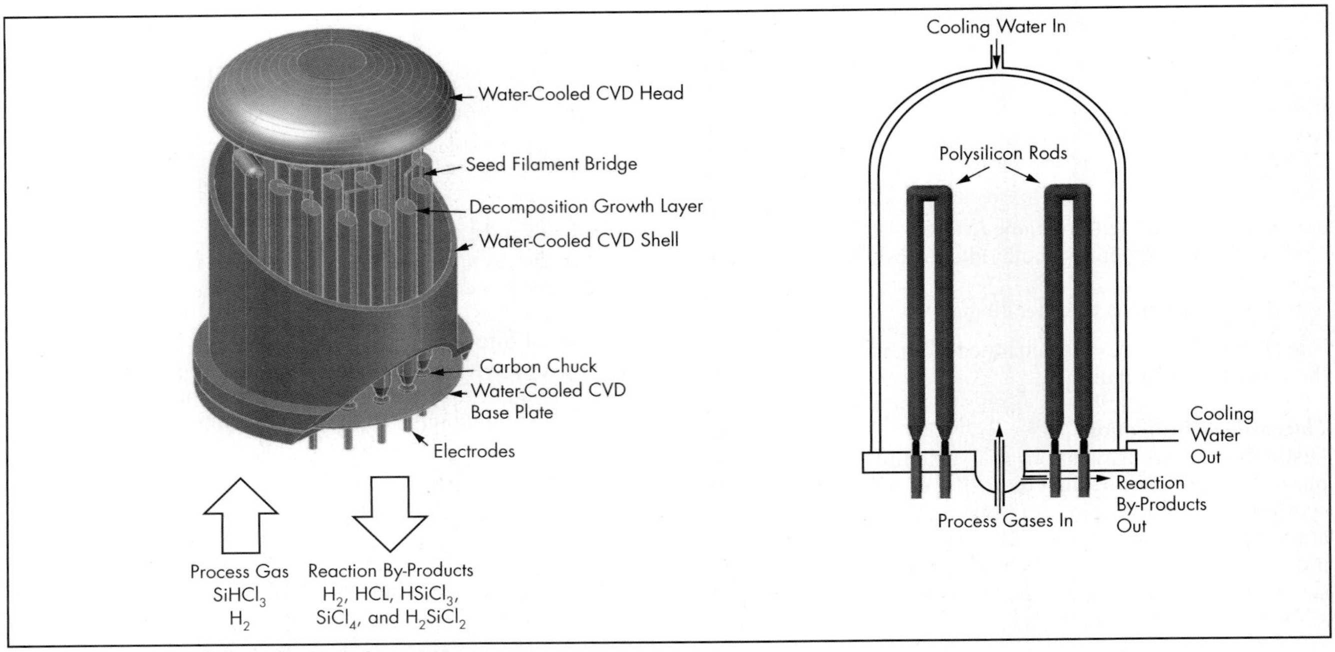

Figure 9 Siemens chemical vapor deposition reactor

a silicon atom in the crystalline lattice (substitutional). Group IIIA elements require an electron to form a bond considered "acceptors" or "p-type." Extrinsic conduction, using acceptors and donors, is the foundation for the creation of silicon-based devices such as computer chips and solar panels.

Many processes developed to produce poly-Si and have been tested, patented, and operated for many years. Three commercial processes make the bulk of the world's current poly production: Siemens, Union Carbide, and Ethyl Corporation. These processes are explained in the following sections.

The Siemens process. The main process steps are as follows:

1. Silicon tetrachloride (STC), hydrogen gas (H_2), and metallurgical silicon (Si) are fed into a fluidized bed reactor. The hydrochlorination process to produce trichlorosilane (TCS) is represented by the following equation:

   $$SiCl_4 + H_2 \rightarrow HSiCl_3 + HCl \quad \textbf{(EQ 17)}$$

   This is followed by the direct chlorination reaction where HCl reacts with raw MG-Si:

   $$3HCl + Si \rightarrow HSiCl_3 + H_2 \quad \textbf{(EQ 18)}$$

   The hydrochlorination reaction results in approximately 20% TCS conversion at 773 K (500°C), 35 bar, with a 1:1 ratio of STC to H_2 in one pass through a fluidized bed of MG-Si.
2. The TCS produced is then subjected to a double purification through fractional distillation. The first phase is removing the heaviest components resulting from the direct synthesis.
3. The second phase is eliminating the components lighter than TCS, also called volatiles.
4. High-purity TCS is then vaporized, diluted with high-purity hydrogen, and then directed into a chemical vapor deposition (CVD) reactor where the main reaction takes place (Figure 9). The gas is decomposed onto the surface of heated silicon seed rods, electrically heated to about 1,100°C, and growing large rods of ultra-high-purity silicon. The reactions within the CVD reactor include the following:

   $$2HSiCl_3 \rightarrow H_2SiCl_2 + SiCl_4 \quad \textbf{(EQ 19)}$$

   $$H_2SiCl_2 \rightarrow Si + 2HCl \quad \textbf{(EQ 20)}$$

   $$H_2 + HSiCl_3 \rightarrow Si + 3HCl \quad \textbf{(EQ 21)}$$

   $$HCl + HSiCl_3 \rightarrow SiCl_4 + H_2 \quad \textbf{(EQ 22)}$$
5. The stream of reaction by-products, which leaves the reactor, contains H_2, HCl, $HSiCl_3$, $SiCl_4$, and H_2SiCl_2. These products are recycled back to the front of the hydrochlorination process for further treatment and capture of the silicon.
6. The polysilicon rod products from the CVD reactor are transported to a clean room to be sized, or broken into chunk, and sealed in virgin polyethylene bags that will be sent to silicon ingot producers. The bags contain a precisely weighed amount of polysilicon, typically 5–10 kg product weight (Hunt 1990; Ceccaroli and Lohne 2003).

The Union Carbide process. The main process steps are as follows:

1. The hydrogenation of STC, mixed with H_2 gas through a mass bed of MG-Si, is carried out in a fluidized bed reactor through the following reaction:

   $$3SiCl_4 + 2H_2 + Si \rightarrow 4HSiCl_3 \quad \textbf{(EQ 23)}$$
2. The TCS is separated by distillation while the unreacted STC is recycled back to the hydrogenation reactor.
3. The purified TCS is then redistributed in two separate stages through fixed bed columns filled with ion exchange

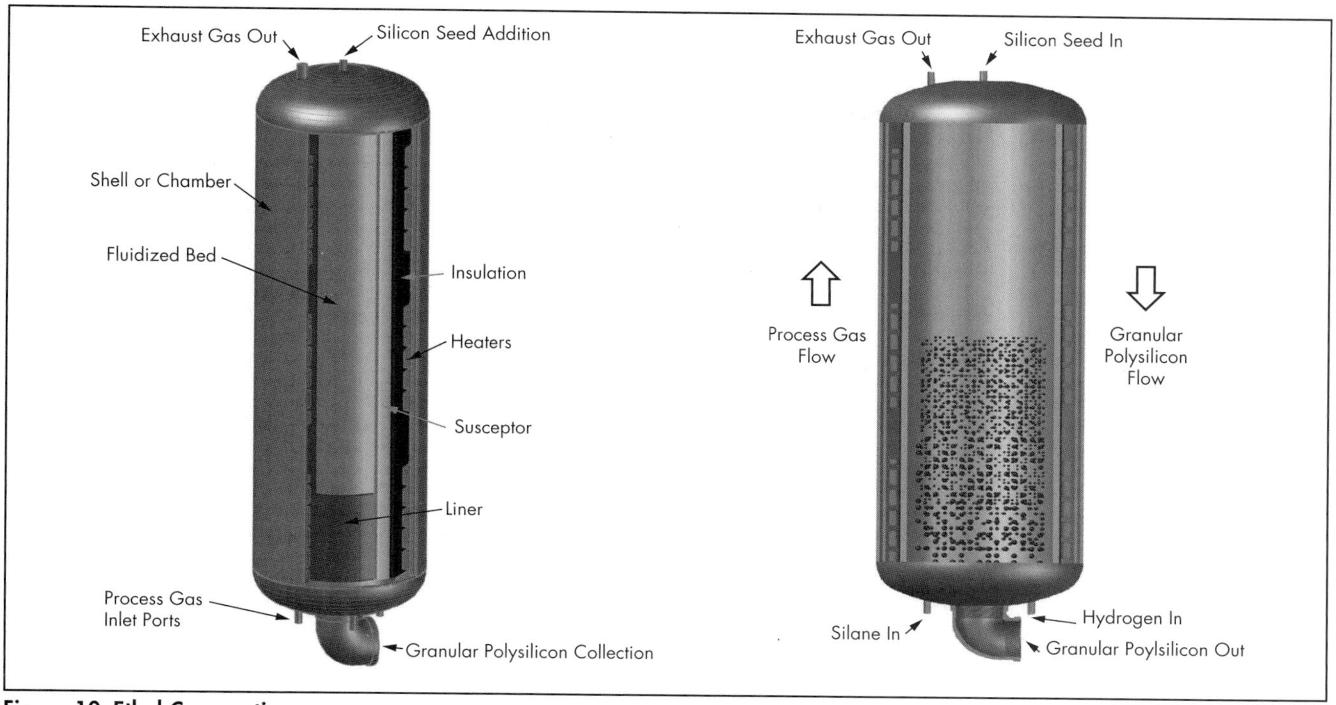

Figure 10 Ethyl Corporation process

resins acting as a catalyst to the following two redistribution reactions:

$$2HSiCl_3 \rightarrow H_2SiCl_2 + SiCl_4 \quad \textbf{(EQ 24)}$$

$$3H_2SiCl_2 \rightarrow SiH_4 + 2HSiCl_3 \quad \textbf{(EQ 25)}$$

4. The products of the two preceding reactions are separate distillations. STC and TCS are recycled back to the hydrogenation reactor and the first redistribution stage, respectively.
5. Silane (SiH_4) is further purified by distillation to ultra-high-purity levels. The silane is then pyrolyzed to produce polysilicon in a Siemens-style CVD reactor by the following reaction:

$$SiH_4 \rightarrow 2H_2 + Si \quad \textbf{(EQ 26)}$$

6. This Union Carbide method produces a measurable amount of amorphous silicon powder due to homogeneous decomposition of silane, and this powder typically adheres to the cooler reactor walls.
7. STC and TCS are recycled and purified many times before conversion to silane, and the results are extremely high-purity silane and subsequent polysilicon.
8. Polysilicon rods produced by the Union Carbide process are strong and typically void-free and are particularly suitable for single-crystal manufacturing by the floating zone method described in a later section. The majority of the product is sized similar to the Siemens process in 5-kg bags (Luque and Hegedus 2003).

The Ethyl Corporation process. The main process steps are as follows:

1. High-purity silicon is produced in the fluidized bed reactor by decomposition of silane in a hydrogen atmosphere by the following reaction:

$$SiH_4 \rightarrow 2H_2 + Si \quad \textbf{(EQ 27)}$$

2. This reaction is exothermic, but additional heat must be provided to balance the thermal losses. Recycled purified hydrogen and silane are introduced continuously through the bottom of the reactor, while high-purity silicon seed particles are introduced through the top of the reactor (Figure 10). The granular polysilicon finished product is cooled and collected through the bottom of the reactor. The hydrogen-rich off-gas is cooled, and the fines are removed, compressed, purified, and recycled back to the reactor. The equipment contact surfaces are typically constructed from high-purity silicon or are silicon coated to maintain purity of the granular silicon.
3. Operating conditions for the reactor are stated as 600–800°C and 5–15 psig. The average particle size for the granular polysilicon is 700–1,100 μm. As with the Union Carbide process, this process produces a measurable amount of amorphous silicon powder due to homogeneous decomposition of the silane (Luque and Hegedus 2003).

Polycrystalline silicon impurities. The purity of silicon produced by the refining operations mentioned previously dictate how it can be used, as photovoltaics, semiconductors, or waste for recycle. Required purity levels are presented in Table 2. Recent efforts to produce photovoltaics from upgraded MG-Si have produced good results from less restrictive purity requirements and at less cost (Søiland et al. 2012).

Table 2 Polycrystalline silicon impurities

Attribute		Grade: Photovoltaic	Grade: Semiconductor
Purity	Boron, B	≤3 ppba	≤0.1 ppba
	Donors: P + As	≤5 ppba	≤0.2 ppba
	Carbon	≤1 ppma	≤0.2 ppma
Surface metals	Aluminum, Al	<5 ppbw	<0.6 ppbw
	Cobalt, Co	—	<0.6 ppbw
	Copper, Cu	<2 ppbw	<0.1 ppbw
	Chromium, Cr	<2 ppbw	<0.2 ppbw
	Iron, Fe	<10 ppbw	<0.8 ppbw
	Nickel, Ni	<2 ppbw	<0.2 ppbw
	Potassium, K	—	<0.5 ppbw
	Sodium, Na	<5 ppbw	<0.5 ppbw
Bulk metals	Total	<30 ppbw	<2 ppbw
Packaging	Virgin polyethylene bags	5 or 10 kg ±0.5%	5 kg ±0.5%

SILICON INGOT PRODUCTION METHODS

To produce high-quality silicon ingots, the distribution of impurities must be determined. The equilibrium segregation coefficient (k_o) is the ratio of the concentration of the impurity in the solid silicon (C_S) and the bulk liquid silicon (C_L):

$$k_o = C_S / C_L \qquad \textbf{(EQ 28)}$$

Table 3 contains the equilibrium segregation coefficient for various impurities in silicon. Equilibrium segregation coefficients are only pertinent if the ingot growth rate is very low, such as with the directional solidification system.

If the equilibrium segregation coefficient is much lower than 1, then the impurity will most likely stay in the molten silicon and end up in the top layer of the ingot, or will be the last section to solidify.

For processes that are carried out at a rate which would not allow the pickup of impurities under true equilibrium conditions, the distribution of impurities in the liquid and the solid would depend on the effective segregation coefficients (k_{eff}).

Table 3 The equilibrium segregation coefficient for various impurities in silicon

Impurity	Ta	Ag	Fe	Co	Ti	Mn
Equilibrium segregation coefficient	1×10^{-7}	1×10^{-6}	8×10^{-6}	8×10^{-6}	2×10^{-6}	4.5×10^{-5}
Impurity	**Ni**	**Au**	**Zn**	**S**	**Cu**	**In**
Equilibrium segregation coefficient	3×10^{-5}	2.5×10^{-5}	1×10^{-5}	1×10^{-5}	4×10^{-4}	4×10^{-4}
Impurity	**Al**	**Ga**	**Sn**	**Sb**	**C**	**Ge**
Equilibrium segregation coefficient	0.002	8×10^{-3}	0.016	0.023	0.08	0.33
Impurity	**P**	**B**	**O**			
Equilibrium segregation coefficient	0.35	0.8	1.2			

Source: Davis et al. 1980; Trumbore 1960; Shimura 1989

The following equations are used for calculating the effective segregation coefficient (Shimura 1989):

$$k_{eff} = k_o / [k_o + (1 - k_o) \exp(-G_s\sigma/D)] \qquad \textbf{(EQ 29)}$$

$$\sigma = 1.6D^{1/3}\, v_k^{1/6}\, \omega^{-1/2} \qquad \textbf{(EQ 30)}$$

where

k_{eff} = effective segregation coefficient
k_o = equilibrium segregation coefficient
G_s = solidification rate
σ = diffusion layer thickness
D = diffusion coefficient of impurity in the melt
v_k = kinematic viscosity of liquid
ω = relative rotation of the crystal

The effective segregation coefficient is usually determined experimentally, due to the difficulty determining the diffusion layer thickness (Schei et al. 1998).

Directional Solidification System for Multicrystalline Ingots

Multicrystalline ingots are produced via directional solidification of silicon in quartz crucibles:

1. Electronic-grade polysilicon is melted in a quartz crucible at a temperature >1,673 K (>1,400°C) in an argon gas atmosphere by a resistance or induction heater. The silicon melt temperature is kept above the silicon melting point.
2. The silicon seed product is placed in the bottom of the crucible during loading and acts as a starting point for the large grain formation.
3. The ingot is solidified from the bottom to the top by moving the crucible down through the heater, or moving the heater up past the crucible, or using induction heating and turning of power sequentially from the bottom coils to the top coils.
4. As the process continues, all silicon in the crucible freezes, and the impurities are collected on the top surface (see Figure 11).

Czochralski Zone Puller for Single-Crystal Rods

The Czochralski method is a technique to pull a single crystal with the same crystallographic orientation of a small monocrystalline seed crystal out of melted poly-Si (Figure 12).

1. Electronic-grade poly-Si is melted in a quartz crucible at a temperature >1,673 K (>1,400°C) in an argon gas atmosphere by a graphite heater. The temperature of the silicon melt is kept above the silicon melting point.
2. A monocrystalline silicon seed crystal is immersed into the melt and acts as a starting point for the crystal formation. A thin "dashed neck" is established between the seed and the ingot to reduce dislocations from freezing.
3. The seed crystal is slowly pulled out of the melt, where the pull speed determines the crystal diameter. During crystal growth, the crystal as well as the crucible counter-rotate to improve the homogeneity of the crystal.
4. Before the crystal growth is finished, a continuous increase of the pull speed reduces the crystal diameter toward zero. This helps prevent thermal stress in the ingot that could happen by an abrupt lifting out of the melt, rapidly reducing the ingot temperature, which can destroy the crystal.

Float Zone Method

The float zone (FZ) method is a technique to pull a single crystal with the same crystallographic orientation of a small monocrystalline seed crystal out of melted polysilicon rod using a radio-frequency coil as the heat source.

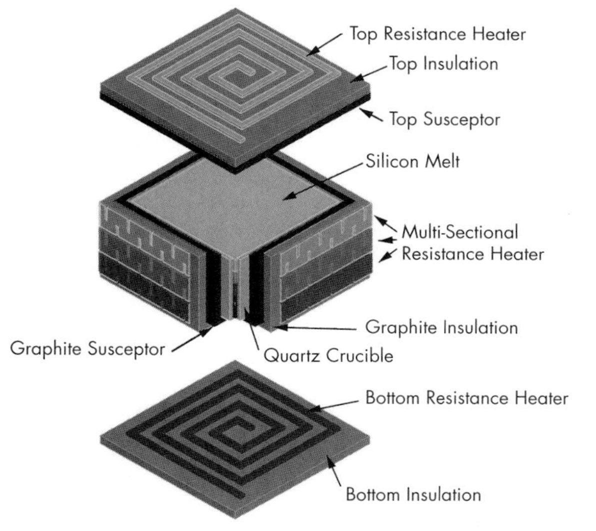
Top Resistance Heater
Top Insulation
Top Susceptor
Silicon Melt
Multi-Sectional
Resistance Heater
Graphite Insulation
Graphite Susceptor
Quartz Crucible
Bottom Resistance Heater
Bottom Insulation

Figure 11 Directional solidification system

1. The FZ method starts with a high-purity polycrystalline rod and a monocrystalline seed crystal that are held face to face in a vertical position and are rotated countercurrently (see Figure 13).
2. A radio-frequency coil is placed around the poly-Si rod and a molten area is established. The seed is brought up from below to contact the drop of melt formed at the tip of the polysilicon rod. A necking process is carried out to establish a dislocation-free crystal before the neck is allowed to increase in diameter to form a taper and reach the desired diameter for steady-state growth.
3. As the molten zone is moved along the poly-Si rod, the molten silicon solidifies into a single-crystal ingot. FZ crystals are doped by adding the doping gas phosphine (PH_3) or diborane (B_2H_6) to the inert gas for n- and p-type, respectively.

Finished silicon ingots are shipped to wafer fabricators and device manufacturers for further processing. Multicrystalline ingots are used exclusively for solar cells while the main end user for single-crystal ingots is the semiconductor industry.

SAFETY CONSIDERATIONS

Caution is required when producing the chemicals and metals that are described in this chapter. Chlorine chemistry is the basis of the reaction chemistry, beginning with the feedstock that is used. Silicon tetrachloride (STC, $SiCl_4$) is highly toxic, producing hydrochloric acid vapor (HCl) when it is exposed to air. HCl vapor is lethal to most life forms. Intermediate products that are produced during the production of silicones

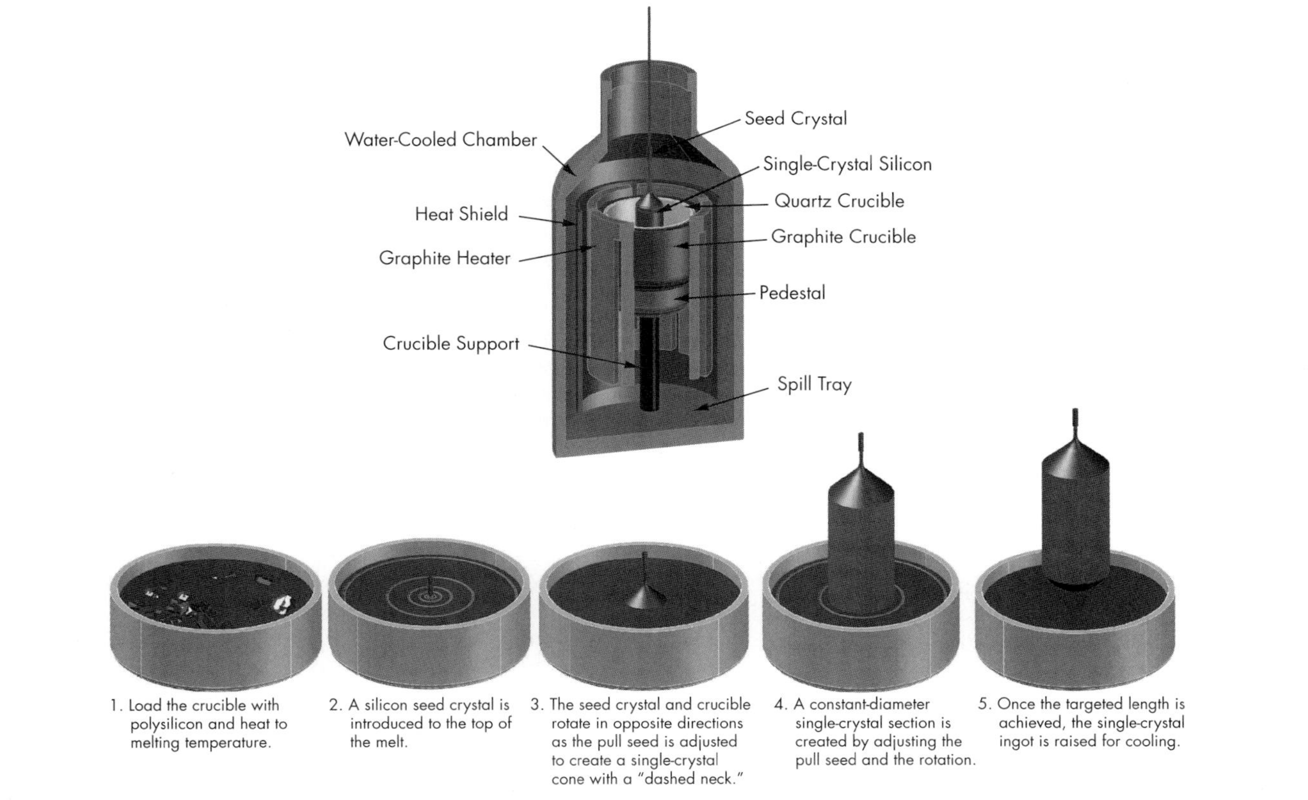
Water-Cooled Chamber
Seed Crystal
Single-Crystal Silicon
Quartz Crucible
Heat Shield
Graphite Crucible
Graphite Heater
Pedestal
Crucible Support
Spill Tray
1. Load the crucible with polysilicon and heat to melting temperature.
2. A silicon seed crystal is introduced to the top of the melt.
3. The seed crystal and crucible rotate in opposite directions as the pull seed is adjusted to create a single-crystal cone with a "dashed neck."
4. A constant-diameter single-crystal section is created by adjusting the pull seed and the rotation.
5. Once the targeted length is achieved, the single-crystal ingot is raised for cooling.

Figure 12 Czochralski zone puller for single-crystal ingots

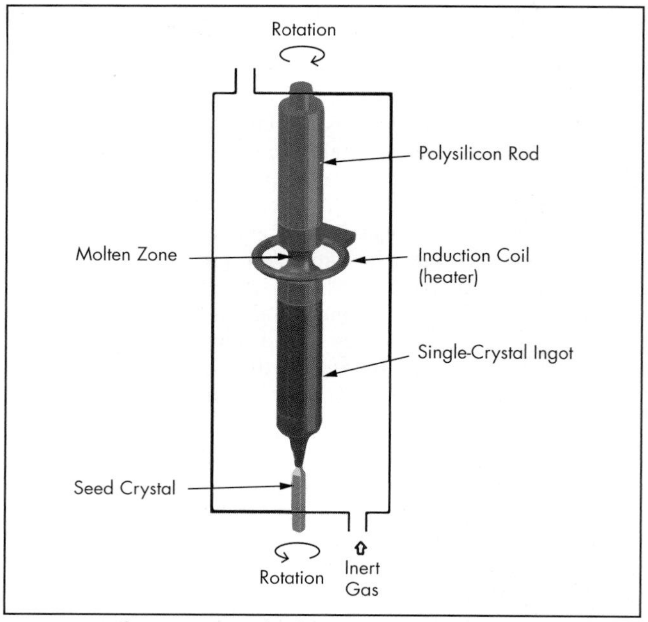

Figure 13 Float zone method

and high- and ultra-high-purity silicon are chlorine-based chemicals—that is, trichlorosilane (TCS, $HSiCl_3$), dichlorosilane (DCS, H_2SiCl_2) and monochlorosilane (MCS, H_3SiCl)—and these chemicals will also produce hydrochloric acid upon exposure to air. Care must be used in handling any of these chemicals.

Silane (SiH_4) can autoignite at temperatures under 327 K (54°C). Regrettably, it has been the cause of injuries and deaths. Extreme care is essential in handling of the gas.

Additionally, when designing processes to treat these chemicals, proper material selection is required to contain the chemicals. Intense mechanical integrity programs using sophisticated nondestructive examination techniques must be applied to ensure that the processes remain safe after processing has been intimated.

Safety concerns are also associated with silicon metal dust that is part of the feed used in the fluid bed reactors that produce silicon-based chemicals. Three processes that are commonly used in the preparation of silicon metal for fluid bed reactors are crushing, to produce the required sizes for use in the fluid bed reactors; pneumatic conveyance, for moving the sized product to storage from the crusher and then from the storage bins to the various reactors; and dust handling, to capture the small particles of dust in baghouses, cyclones, or dry scrubbers. If the following conditions exist at the same time, the metal dust can be highly explosive:

1. Correct particle sizing (<60 μm)
2. Correct concentration of silicon metal dust and air
3. Oxygen levels greater than 8%
4. An energy source of sufficient magnitude

Explosions in facilities that contain these processes have been damaged beyond repair and, unfortunately, human lives have been lost. Removal of any one of the preceding four conditions eliminates the potential occurrence of an explosive condition. Therefore, nitrogen gas is typically used to inert operations and maintain very low levels of oxygen to keep these operations in the reduced state and safe.

Finally, pressure relief devices, also referred to as safety valves, are required and used on any and all vessels that are used to contain, store, and feed the chemical plants that produce these products. The utmost of care must be used while handling silicon metal products and the chemicals produced in the manufacture of these products.

REFERENCES

ASM International. 1992. *ASM Handbook: Volume 03, Alloy Phase Diagrams.* Materials Park, OH: ASM International.

Ceccaroli, B., and Lohne, O. 2003. Solar grade silicon feedstock. In *Handbook of Photovoltaic Science and Engineering*. Edited by A. Luque and S. Hegedus. Chichester, West Sussex, England: John Wiley and Sons.

Davis, J.R., Rohatgi, A., Hopkins, R.H., Blais, P.D., Choudhury, P.R., McCormick, J.R., and Mollenkopf, A.C. 1980. Impurities in silicon cells. In *IEEE Transactions on Electron Devices*, Vol. ED-27. New York: IEEE. pp. 677–687.

Dow Corning Corporation. 2015. Silicone Manufacturing, Discovery Center. Auburn, MI: Dow Corning Corporation.

Elschenbroich, C. 1992. *Organometallics*. Weinheim: Wiley-VCH.

Hunt, L. 1990. Silicon precursors: Their manufacture and properties. In *Handbook of Semiconductor Silicon Technology.* Park Ridge, NJ: Noyes Publications.

Luque, A., and Hegedus, S. 2003. *Handbook of Photovoltaic Science and Engineering.* Chichester, West Sussex, England: John Wiley and Sons.

Lynch, D.C. 2002. *Silicon for the Chemical Industry VI.* Trondheim, Norway: Department of Chemistry, Norwegian University of Science and Technology. pp. 73–91.

Myrhaug, E.H., and Tveit, H. 2000. Material balances of trace elements in the ferrosilicon and silicon processes. In *58th Electric Furnace Conference and 17th Process Technology Conference Proceedings*. Warrendale, PA: Iron and Steel Society. pp. 591–604.

Schei, A. 1967. On the chemistry of ferrosilicon production. *Tidsskrift for Kjem, Bergvesen, og Metallurgi*, 27(8-9):152–158.

Schei, A., Tuset, J.K., and Tveit, H. 1998. *Production of High Silicon Alloys.* Trondheim, Norway: Tapir Forlag.

Shimura, F. 1989. *Semiconductor Silicon Crystal Technology*. San Diego, CA: Academic Press. pp. 139–150.

Sigworth, G.K. 2014. Fundamentals of solidification in aluminum castings. *Int. J. Metalcast.* 8(1).

Søiland, A.K., Odden, J.O., Sandberg, B., Friestad, K., Håkedal, J., Enebakk, E., and Braathen, S. 2012. Solar silicon from a metallurgical route by ELKEM Solar—A viable alternative to virgin polysilicon. In *Proceedings of the 6th International Workshop on Crystalline Silicon for Solar Cells*, Aix Les Bains, France.

Trumbore, F.A. 1960. Solid solubility's of impurity elements in germanium and silicon. In *Bell System Technical Journal*, Vol. 39. pp. 205–233.

USGS (U.S. Geological Survey). 1996–2015. *Minerals Yearbook, Silicon.* Reston, VA: USGS.

USGS (U.S. Geological Survey). 2010, 2011, and 2016. *Mineral Commodity Summaries, Silicon.* Reston, VA: USGS.

CHAPTER 12.34

Steelmaking

P. Chris Pistorius, Xu Gao, Lauri Holappa,
Hiroyuki Matsuura, Joohyun Park, and Muxing Guo

Steel is an iron-based engineering material, often containing carbon as the principal strengthening element, and is the most-produced metallic material; more than 1.6 billion t (metric tons) of crude steel was produced in 2016. The many different grades of steel provide a wide range of properties, ranging from soft and ductile interstitial-free steel to tool steels that can retain their strength even when red hot; stainless steels contain more than 12% chromium for improved corrosion resistance. Steels that combine high strength and high ductility at reasonable cost are currently being developed and improved for use in automobiles.

The properties of steel depend on the chemical composition and physical processing of the material. In this overview, the focus is on production of crude steel from raw materials. Detail on processing routes and reactions can be found in comprehensive reviews (Fruehan 1998; Remus et al. 2013; Seetharaman et al. 2014).

Steelmaking is the conversion of raw materials to liquid steel—with the correct chemical composition and temperature—for subsequent solidification into intermediate products such as slabs, blooms, billets, beam blanks, or ingots. Simplified flow sheets are shown in Figure 1.

The main metallic raw materials are hot metal (the carbon-rich product of blast furnace [BF] ironmaking), scrap (recycled steel), and other metallics such as direct reduced iron (DRI). The step labeled "Steelmaking" in Figure 1 involves several other steps. As discussed in subsequent sections, hot metal is normally processed by converter steelmaking, often after an initial pretreatment step to remove phosphorus and sulfur. Electric furnace steelmaking mainly uses scrap and DRI. These initial operations (primary steelmaking) remove oxidizable impurities by injecting oxygen: Carbon is removed in gaseous form (as CO); phosphorus and silicon are converted to oxides and captured in slag. Slag is the molten (or largely molten) mixture of oxides that floats on top of the steel; slag originates from elements and compounds in the raw materials and fluxing additives (such as limestone). Slag–metal reactions are essential to many of the refining reactions and can be manipulated by changing the overall slag composition, operating the process under relatively oxidizing or reducing conditions, and by stirring. Slag is a major by-product of steelmaking.

Following the primary steelmaking step, dissolved oxygen is normally removed from the crude steel by the addition of deoxidizers. Subsequent processing (*secondary metallurgy*) involves control of dissolved impurities (such as sulfur, hydrogen, and nitrogen) and of second-phase impurities (nonmetallic inclusions).

Quality control of steelmaking relies on discrete measurements of composition (typically using optical emission spectroscopy and inert-gas fusion of *lollipop* samples taken from liquid steel), temperature (taken with immersion thermocouples), and activity of dissolved oxygen (measured with electrochemical probes). Some variables are or could be continuously monitored during processing, including off-gas composition, cooling-water flow rate and temperature, and steelmaking vessel acceleration.

Steelmaking involves physical and chemical processing, manipulated by stirring (electromagnetically or by gas bubbling); physical separation of metal, slag, and gas; and control of reaction equilibria by temperature, slag composition, vacuum, and redox conditions. This rich variety of processing options allow tight and economical control of steel compositions.

Steelmaking costs vary by geographical location but are typically dominated by the cost of raw materials. As indicated by Figure 2, approximately two-thirds of the variable cost is scrap (for electric arc furnace [EAF] steelmaking) or coal (for coke production) and iron ore (for BF/basic oxygen furnace [BOF] steelmaking). The estimates are based on hypothetical

P. Chris Pistorius, POSCO Professor of Materials Science & Engineering, Carnegie Mellon University, Pittsburgh, Pennsylvania, USA
Xu Gao, Assistant Professor, Institute of Multidisciplinary Research for Advanced Materials, Tohoku University, Sendai, Miyagi, Japan
Lauri Holappa, Emeritus Professor, Aalto University, Espoo, Finland
Hiroyuki Matsuura, Associate Professor, Department of Materials Engineering, University of Tokyo, Tokyo, Japan
Joohyun Park, Professor, Department of Materials & Chemical Engineering, Hanyang University, Ansan, Korea
Muxing Guo, Senior Researcher, Department of Materials Engineering, Katholieke Universiteit Leuven, Leuven, Belgium

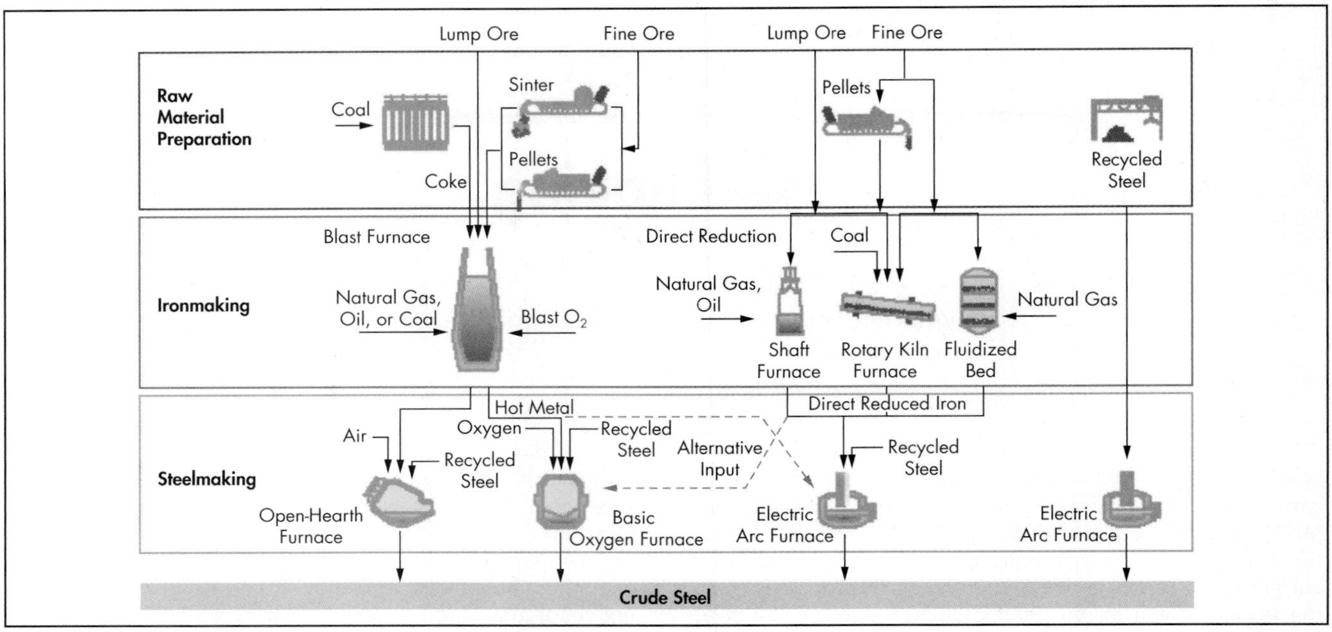

Source: World Steel Association 2018

Figure 1 Steel production routes for basic oxygen furnace and electric arc furnace

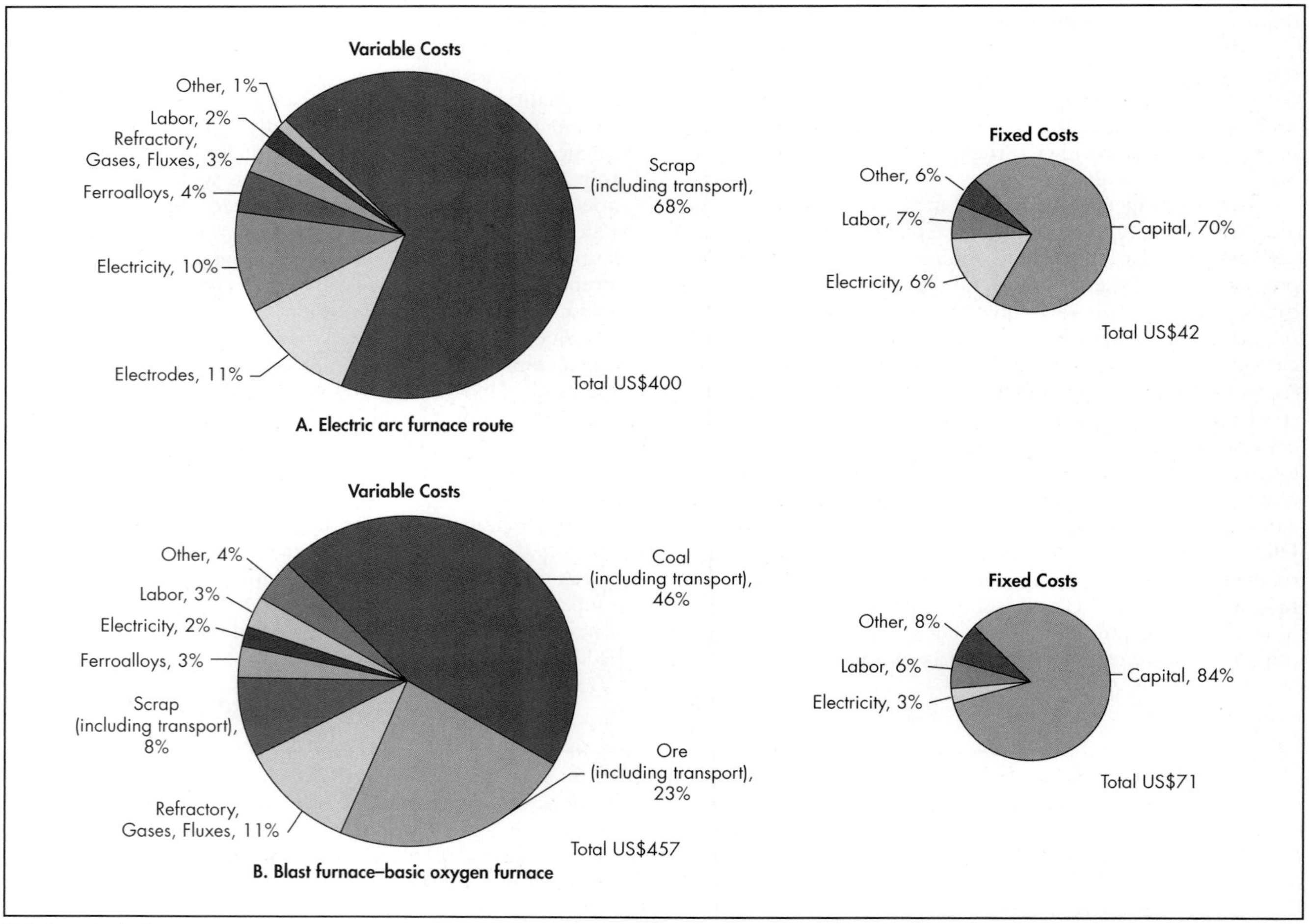

Data from steelonthenet.com

Figure 2 Indicative costs of (A) electric furnace steelmaking and (B) integrated steelmaking

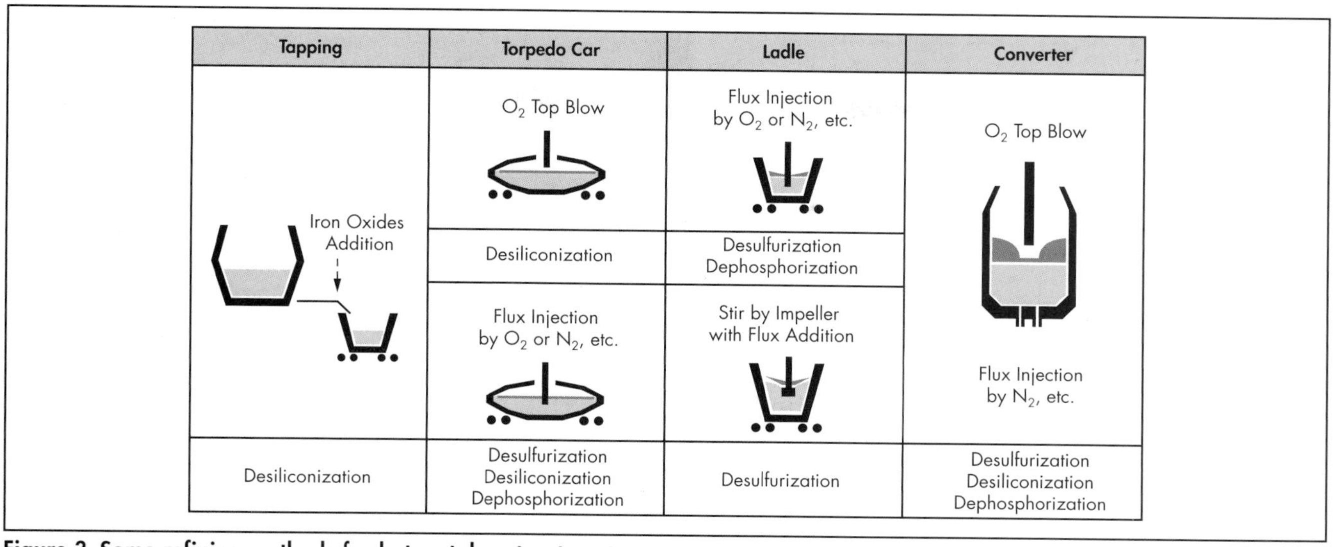

Figure 3 Some refining methods for hot metal pretreatment

coastal Japanese plants in 2017. Credits for by-products and export gas are not included in the costs.

Steelmaking occurs by batch processing; that is, a quantity (*heat*) of material is processed until it achieves the required chemical composition and temperature. In contrast, ironmaking (in BFs and direct reduction) is typically a continuous process.

HOT METAL PRETREATMENT

Hot metal pretreatment is a typical first step in integrated steelmaking, that is, the process that uses hot metal from the BF for oxygen steelmaking. The aim is to remove most of the sulfur and phosphorus when their concentrations in hot metal are much higher than the requirement of the final product. Hot metal pretreatment is normally conducted during the transportation period after the tapping of hot metal from the BF and before charging into the converter for decarburization. A torpedo car, ladle, or converter is used as a refining vessel. Several examples of refining methods and their functions for hot metal pretreatment are compared in Figure 3 (Fuwa et al. 1984; ISIJ 2006; Kitamura et al. 1990; JFE 21st Century Foundation 2003; Kawabata et al. 2002; Kanbara et al. 1972; Kraemer et al. 1968; Mukawa and Mizukami 1994; Ogawa et al. 2001; Sasaki et al. 2013). In the following sections, some thermodynamics and kinetics considerations for desulfurization, desiliconization, and dephosphorization are introduced.

Desulfurization

Desulfurization of hot metal by slag can be expressed as Equation 1 when using solid calcium oxide (CaO) as reactant. In principle, the desulfurization reaction can be promoted by using basic slags, such as soda ash (Na_2CO_3), and CaO-based slags, and by a low oxygen activity in the metal. Desulfurization is commonly not based on slag reactions, however, but on injection of solid calcium carbide or co-injection of magnesium and calcium oxide (Fruehan 1998).

$$\mathrm{CaO(s)} + [\mathrm{S}]_{\mathrm{metal}} = \mathrm{CaS(s)} + [\mathrm{O}]_{\mathrm{metal}} \qquad \textbf{(EQ 1)}$$

Under thermodynamically favorable conditions for desulfurization, mass transfer of sulfur in hot metal is the rate-controlling step. Therefore, the shape and size of refining vessel, size of added flux and addition method, and stirring intensity are important factors for designing the desulfurization process. Flux injection and mechanical stirring as shown in Figure 3 are commonly used to improve kinetics. Rate control by mass transfer of sulfur in the hot metal implies first-order kinetics. This leads to a semilogarithmic relationship between the extent of desulfurization and the amount of reagent needed (Delhey et al. 1989).

Desiliconization

Both silicon and phosphorus are removed from hot metal by oxidation—the opposite of desulfurization, which requires reduction. BF hot metal typically contains 0.5%–1% Si by mass. Traditionally, desiliconization was often adopted to decrease the CaO consumption and shorten the refining time in the following processes, such as dephosphorization and decarburization. However, because low-basicity slag can be used for dephosphorization (Mukawa and Mizukami 1994; Ogawa et al. 2001), today, desiliconization is not conducted separately in some steelworks but rather occurs as a first step before dephosphorization.

Oxidizers such as iron oxides (mill scale, sinter ore, etc.) and gaseous O_2 are used for desiliconization. Iron oxides are added during tapping or injected into the torpedo car, and O_2 is introduced by top blowing or injection, as shown in Figure 3. Desiliconization changes the hot metal temperature: O_2 injection increases hot metal temperature, whereas use of iron oxides decreases it (Figure 4).

Dephosphorization

Dephosphorization of hot metal pretreatment requires oxidation; see Equation 2. Use of basic slags, a higher oxygen activity, and lower temperatures all promote dephosphorization. Dephosphorization typically involves a basic iron oxide (FeO)-containing slag reacting with hot metal that contains carbon. This slag and metal cannot coexist at equilibrium: Some carbon is oxidized in parallel with oxidation of phosphorus. The relative extents of these competing reactions depend on both the thermodynamics (temperature and slag

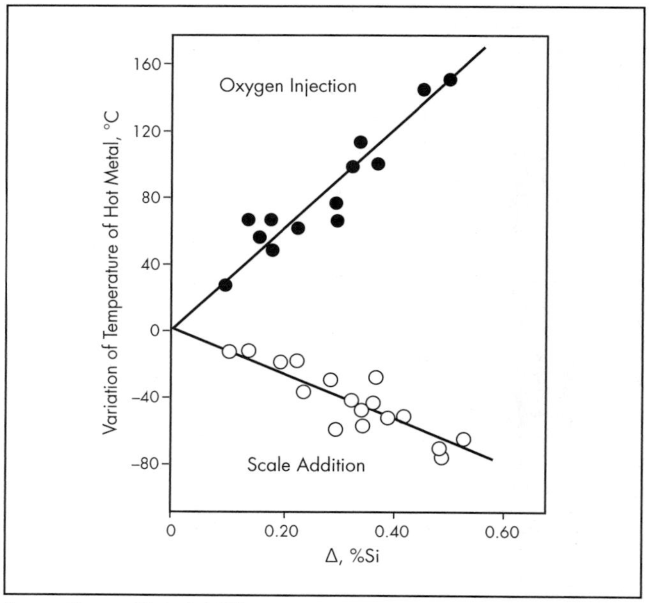

Source: Kawauchi et al. 1983

Figure 4 Oxygen injection and scale addition affect temperature

composition) and kinetics. An empirical equation has been proposed to quantify the joint effects of temperature, slag, and carbon in hot metal (Ogawa et al. 2001), and kinetic models have been used to quantify the competitive reactions (Kitamura et al. 2014).

$$[P]_{metal} + \frac{5}{4}O_2 + \frac{3}{2}(O^{2-})_{slag} = \left(PO_4^{3-}\right)_{slag} \qquad \textbf{(EQ 2)}$$

In Equation 2, [P] is phosphorus dissolved in the metal, O_2 is gaseous oxygen, O^{2-} is anionic oxygen in the slag, and PO_4^{3-} refers to phosphate ions in the slag.

CaO-based slag is often used for the dephosphorization process, either of high CaO-to-silicon dioxide (SiO_2) ratio with low FeO content or low CaO/SiO_2 ratio with high FeO content. During the dissolution of CaO into slag, $2CaO{\cdot}SiO_2$ solid solution forms a layer surrounding the CaO particle (Hamano et al. 2006), and $2CaO{\cdot}SiO_2$-$n3CaO{\cdot}P_2O_5$ solid solution also forms within and near this layer (Saito et al. 2009). This phenomenon concentrates phosphorus oxide (P_2O_5) from liquid phase to solid phase, maintains a low P_2O_5 concentration in the liquid slag, and therefore promotes the transfer of phosphorus from metal to slag.

CONVERTER STEELMAKING

The converter process is the main steelmaking method, responsible for nearly 75% of world steel production. The modern oxygen converter evolved from air-blown Bessemer and Thomas converters developed in the second half of the 19th century. These processes had high productivity but also quality problems caused by high nitrogen content from air. As blowing was through the vessel bottom, the technique could not use pure oxygen because of overheating and catastrophic lining wear. The problems of oxygen injection were solved by top blowing with a water-cooled lance, first realized in Linz, Austria, in 1951. The process is known as LD (Linz-Donawitz) or basic oxygen process (BOP)/BOF steelmaking. The new method emerged rapidly during the 1960s and became the dominant steelmaking process (Michaelis 1979; Jalkanen and Holappa 2014).

Oxygen bottom blowing was successfully resolved by using annular nozzles, in which O_2 is blown through the inner tube and a shrouding gas (hydrocarbons [C_xH_y], Ar) blown through the outer annulus. The first OBM/Q-BOP (oxygen-bottom-Maxhütte/quick, quiet, quality BOP) was commissioned in 1967. Since then it has spread and currently accounts for about 10% of the overall converter capacity (Brotzmann 1979). Experiences from the OBM process clearly revealed the principal drawback of top blowing: the relatively weak stirring effect of the oxygen jet, resulting in over-oxidation of the slag and of the metal bath.

The benefits of both processes were combined into one process; numerous combined-blowing processes were developed since the middle 1970s, most based on the LD process with inert-gas bottom blowing. A great majority of the converters today are these *hybrid* processes. The advantages compared to pure top blowing are reported to be increased blowing efficiency owing to strongly intensified melt stirring, lower FeO in the slag, increased refractory lining life, better accuracy in target end composition and temperature because of better homogeneity, as well as less splashing and spitting of slag (Jalkanen and Holappa 2014). The following section describes the features of the modern oxygen top blowing plus inert-gas bottom stirring practice.

Process Description

Hot metal from BFs typically contains about 94.5% Fe, 4.5% C, 0.3%–0.4% Si, 0.1%–0.3% Mn, and 0.03%–0.05% S plus varying amounts of other impurities (such as P, Ti, and V), depending on the raw materials for sinter, pellets, coke, and coal. High-phosphorus ores used to be more common but were discarded during the last 50 years. However, two categories still exist. In Asian plants, typical hot metal has 0.08%–0.13% P, whereas in Europe and the United States, the typical level is 0.05%–0.07% P (Jalkanen and Holappa 2014). For these reasons, special processes for hot metal dephosphorization are in use in Japan, Korea, and China (Kitamura 2014). For low-phosphorus hot metal, external dephosphorization is not needed. Sometimes, iron ore contains significant concentrations of other impurities, such as vanadium, resulting in up to 0.1% V in BF hot metal and requiring special attention in the BOF process. This section focuses on decarburization, which is the main goal of the converter process.

Figure 5 shows the main features (reactions and zones) of a BOF vessel and the physical and chemical phenomena that occur during converting (Jalkanen and Holappa 2014). A heat starts with charging of steel scrap and liquid hot metal. The permissible amount of scrap depends on the thermal balance of the process, considering physical and chemical heats, as well as input and output temperatures of the system phases. Most of the heat generation comes from oxidation of carbon and a smaller share from oxidation of silicon and iron and other reactions. Typical scrap rates are 15%–25%. Even higher fractions are possible, for example, by utilizing post-combustion or addition of extra fuel (such as coke or ferrosilicon) (Dubois et al. 2003).

As seen, carbon oxidation occurs not only in the impact area of the oxygen jet but also in the slag and at the slag–metal interface. The slag has other refining roles: It binds the condensed-state oxidation products (mainly SiO_2, MnO, FeO, and P_2O_5) and protects the basic refractory lining. Lime (CaO)

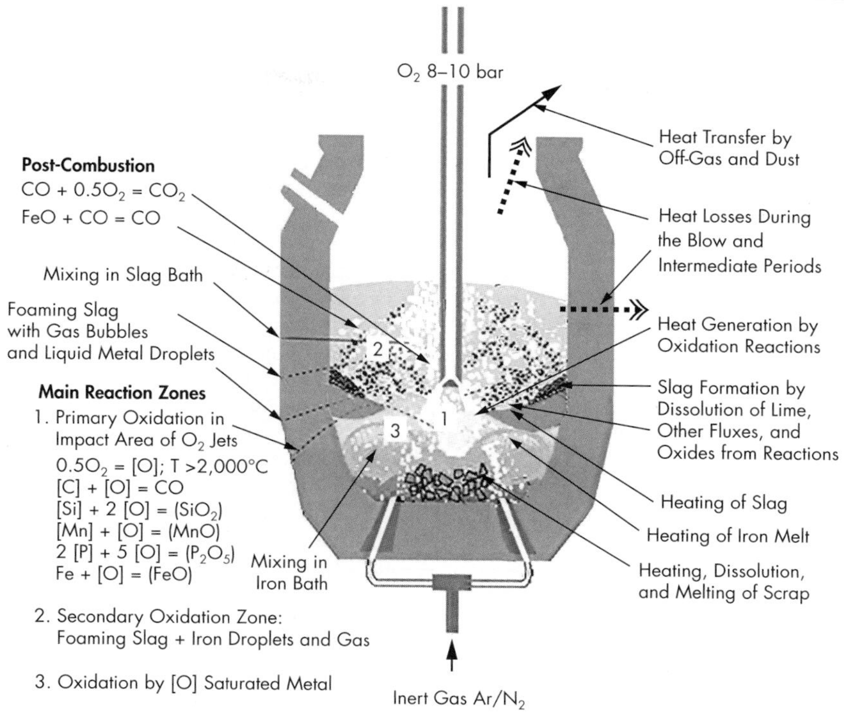

Source: Jalkanen and Holappa 2014

Figure 5 Basic oxygen furnace converter with oxygen lance and bottom-blown inert gas

is added before the start of the blow and during blowing to attain a high slag basicity value (the CaO/SiO_2 mass ratio is often larger than 3), which is favorable for dephosphorization and lining protection. Typical composition evolution in the metal bath and slag during the oxygen blow period is shown in Figure 6. For more information about the phenomena behind the features of the curves, see Jalkanen and Holappa (2014). The lines were drawn as trend lines based on the experimental data (Cicutti et al. 2000) and generally known phenomena taking place in a typical process.

During blowing, the temperature increases from the hot metal temperature, typically 1,300°–1,350°C, to the steel target temperature in the 1,650°–1,700°C range. The heat generation from reactions (Figure 5) and the amounts of cooling materials, scrap and other additions, lime, doloma lime, iron ore, as well as heat losses, define the end temperature. The main targets of process control are that the carbon concentration should be less than a prescribed maximum and the temperature higher than a set minimum; often a maximum phosphorus concentration is an additional target. If all targets are not attained, re-blow is needed. The carbon target depends on the final C content of the steel to be produced: For high–C grades $[C]_{max}$ can be up to 0.7% C, and for ultra-low-carbon (ULC) grades, for example, 0.04% C. The target temperature depends on the steel grade, but also on the subsequent process stages, which often include the possibility for heating in a ladle furnace. In such a case, the optimum temperature is related to the requirements of both converter process and ladle metallurgy. Process control systems are generally based on a static model with mass and heat balance calculations, which calculates the amount of oxygen, scrap, and other additions.

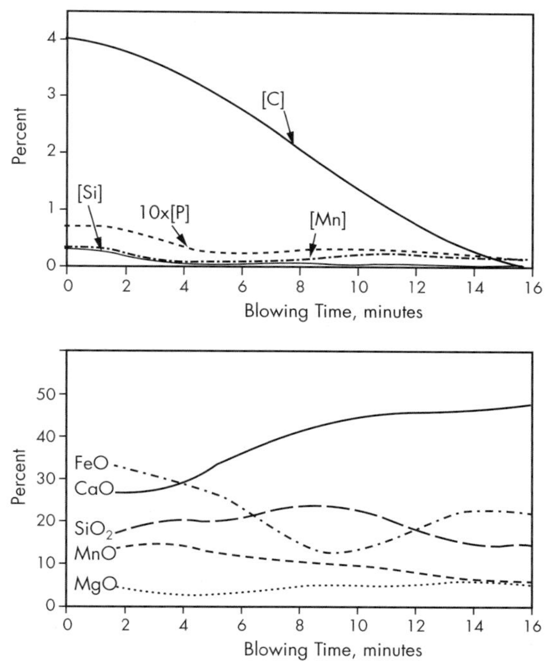

Data from Cicutti et al. 2000

Figure 6 Evolution of composition of iron bath and slag during oxygen blowing

The model starts from measured input values (analyses of hot metal, scrap, etc., and temperatures) and output values (target C, T, P), then calculates reactions in the metal bath and in the slag phase. The target values (C, T) are then attained via iteration. The hit rate can be improved by adaptive control using experimental data from previous heats. Today, dynamic control systems are common in large converters, often utilizing a sub-lance device for direct measurement of T, [C] and/or $a_{[O]}$ with probes. The measurement is performed 1–2 minutes before the predicted blow end to have enough time for corrective actions. Another method for carbon control is a cumulative calculation of carbon oxidation by applying continuous off-gas analysis and flow measurement. Strict control improves productivity by shortening average time from converter tapping to next tapping. Typical tap-to-tap times are 35–40 minutes, which is compatible with other downstream unit processes.

ELECTRIC ARC FURNACE STEELMAKING

Increased steel stock worldwide results in the generation of steel scrap, and thus a suitable recycling process is required. The major steel recycling process is EAF steelmaking, in which steel scrap is melted and refined. EAF steelmaking is also a major process for the production of special grades of steel, for example, stainless steel or highly alloyed steel.

EAF steelmaking has also become a major steelmaking process in newly developed countries because of its relatively low initial investment and lesser technical and metallurgical difficulty compared with integrated ironmaking and steelmaking (that uses iron ore and reducing agents). Today, some steel mills use large-production-capacity EAFs with tap sizes of up to 400 t.

Furnace Type

The majority of EAFs use alternating current (AC), converting electrical energy to thermal energy by using electric arcs produced by three-phase AC. The energy is used for melting and refining the steel. Direct current arc heating is also utilized in some furnaces. There are several systems to improve the energy efficiency of operation. Because much of the energy input into the furnace is lost as sensible heat in the off-gas (Table 1), systems are available to use this heat to preheat steel scrap before charging to the EAF furnace. These include conveyor and shaft systems, in which scrap is continuously preheated by off-gas from the EAF. However, there are various technical challenges; therefore, preheating systems are still under development and not widely utilized in commercial operating furnaces.

Raw Materials

The main raw material for EAF steelmaking is various grades of steel scrap. Selection of steel scrap grade depends on the quality and type of steel to be produced. Heavy steel scrap has a known composition and fewer impurities, while some steel scrap, such as waste household appliances or automobile shredder residue, contain various kinds of metallic impurities. During oxidizing refining of molten steel, some elements cannot be removed, including copper, tin, and nickel. Contamination of steel products by copper has become a serious problem. Although some special refining processes have been proposed to remove copper, these are impractical, and other economically feasible countermeasures, such as scrap sorting, are required.

In some regions, the supply of steel scrap cannot satisfy the demand of EAF steel producers. In such a case, steel scrap is imported from other countries, or other iron resources, such as cold pig iron, DRI or hot briquetted iron, are used as raw materials.

Typical Energy and Electricity Consumption

A typical thermal energy balance for a Japanese EAF is summarized in Table 1. Modern operations utilize a large proportion of chemical reaction energy to increase productivity and reduce electricity consumption. In addition, various lances and injectors are used in modern EAFs. Oxygen or oxygen and gas coherent burners are used for scrap cutting, steel refining, and supplying supplementary energy. Immersed lances are used for coal and coke injection into slag, mainly for slag foaming. The other essential technology for energy utilization is post-combustion above the furnace melt, where oxygen gas is injected to burn CO gas generated through slag–carbon reaction and the combustion energy provides further heating.

Efficient energy utilization, less thermal damage to the furnace body, and stabilized arcing are promoted by covering the electric arc with foamy slag.

Operation

EAF operations are divided into four steps: (1) raw material charging, (2) melting, (3) refining, and (4) tapping. Except for furnaces equipped with a scrap preheating system, raw materials, such as steel scrap, DRI, or cold pig iron, are put into a charging bucket according to a prescribed recipe. The recipe prescribes not only the chemical composition but also the layering order by scrap density, size, and other feedstocks, for smooth melting and furnace protection. The charging bucket is transferred onto the furnace body and the bucket bottom is opened for charging. The furnace is subsequently covered by a water-cooled roof, and the graphite electrodes (three electrodes for an AC furnace) are lowered.

The melting process, during which the melted and unmelted materials coexist, should be completed as quickly as possible; this requires a well-considered procedure depending on type of raw materials and their mixing ratio, furnace size, electric power input, auxiliary materials, other inputs, and use of a hot heel (molten material retained from the previous heat). Some furnaces that consume a large proportion of DRI continuously feed this material (through one or more ports in the furnace roof) during arcing.

Table 1 Typical thermal energy balance of a Japanese electric arc furnace

Input	kW·h/t	Output	kW·h/t
Electricity	313	Molten steel	396
Metal heel	41	Slag sensible heat	54
Slag heel	11	Remaining metal heel	40
Scrap sensible heat	170	Remaining slag heel	11
Fuel burning	20	Electric energy loss	11
Oxidation of electrode	12	Cooling water	58
Oxidation of auxiliary materials	192	Off-gas	189
Slag formation heat	5	Furnace radiation heat	9
		Others	–4
Total	**764**	**Total**	**764**

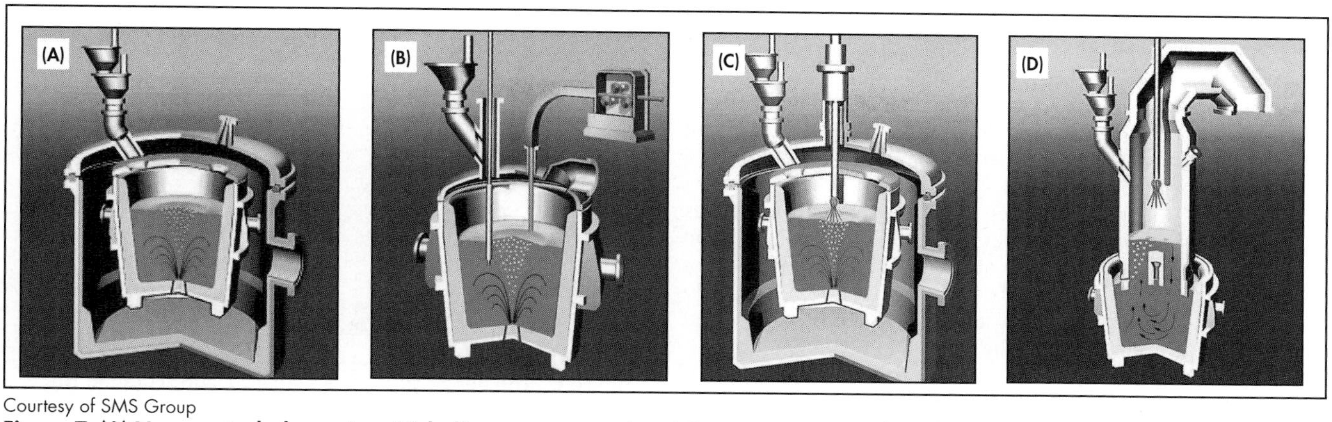

Courtesy of SMS Group

Figure 7 (A) Vacuum tank degassing, (B) ladle treatment station, (C) vacuum oxygen decarburizing, and (D) Ruhrstahl Heraeus degasser with oxygen blowing units for secondary refining

The conventional EAF operation also refines the steel. Decarburization and dephosphorization reactions proceed during the initial refining stage in oxidation mode. The generated slag is continuously removed from the furnace as a foamy slag. During an optional last stage of refining, desulfurization, iron or chromium recovery, and temperature tuning can be achieved under reducing conditions. Commonly, secondary refining processes are available in the steel plant, in which case it is not necessary to refine the steel in the EAF under reducing conditions.

Tapping of molten steel is conventionally conducted by tilting the furnace. Modern EAFs have a bottom tapping hole (eccentric bottom tapping) to remove molten steel without moving the furnace, promoting quick tapping and avoiding entrapment of slag. A portion of the molten steel is retained in the furnace as a *hot heel* to smoothly melt steel scrap in the next heat.

SECONDARY STEELMAKING

Decarburization and (to some extent) dephosphorization are carried out during primary steelmaking. Final adjustments of composition, cleanliness, and temperature are performed in a separate vessel after primary steelmaking. Generally, this step is referred to as *secondary steelmaking* (refining) or *ladle metallurgy*. Ladle metallurgy is important in the production of high-quality steel.

Alloying

Usually basic alloying is performed during tapping, which is the process of transferring liquid steel from the BOF/EAF furnaces to a ladle; the kinetic energy of the tapping stream is utilized to intensify alloy melting and promote homogenization in the ladle. Secondary additions are then made during ladle treatments. For small additions, such as trimming of major components and micro-alloying, a special technique, wire injection, is often used. The alloying system in vacuum tank degassing and wire feeding at a ladle treatment station are schematically shown in Figures 7A and 7B.

Gas Stirring

Inert gas (argon) is generally used to stir the steel in the ladle. The gas rising through the liquid steel induces a turbulent recirculatory motion that provides the necessary bath agitation for increasing the rates of various heat- and mass transfer–controlled processes (such as melting of deoxidizer and alloying additions and their dissolution and dispersion). As the injected gas escapes to the surroundings, the redirected bulk flow from the spout region (plume eye) pushes the slag layer radially outward, exposing the melt surface. This plume eye is a potential site for reoxidation, nitrogen pickup, and slag entrainment phenomena and hence can influence quality of steel profoundly. Therefore, during the final stage of ladle refining and immediately prior to continuous casting, it is customary to practice gentle stirring to ensure a small plume eye area.

Temperature and Composition Control

Temperature and composition control are routinely carried out in a ladle furnace following furnace tapping. A top cover is placed on the ladle and graphite electrodes are lowered into the ladle through holes in the cover. An arc is struck between the electrodes and steel, transferring heat to the melt by convection and radiation. Convection generated by argon bubbling helps to distribute heat within the melt; relatively large argon flow rates are employed. The top cover also provides significant protection from atmospheric oxidation and radiation heat losses.

Degassing

Degassing is the removal of carbon, nitrogen, and hydrogen. The carbon content cannot be reduced to very low levels in the BOF because oxidation of iron increases strongly when [C] decreases below approximately 0.03%. The maximum carbon requirements for ULC steel can be as low as 20–30 ppm (by mass). This is achieved by vacuum degassing. As subsequent carbon contamination (pickup) cannot be fully avoided, the aim value after vacuum treatment is 10–20 ppm. In certain processes, oxygen is blown while placing the steel under vacuum to promote deep decarburization (Figures 7C and 7D).

Hydrogen in steel originates from ambient moisture (in air), either directly or indirectly from added lime and other hygroscopic slag formers, and from other sources such as alloying materials and moisture in refractory linings. Typical [H] contents in steel produced by EAF and BOF processes are in the range of 4 to 6 ppm and 2 to 4 ppm, respectively. Hydrogen removal is done by vacuum degassing; [H] contents less than 2 ppm are readily attained. For some critical grades, the aim value after hydrogen removal can be less than 1 ppm, requiring a long and efficient vacuum treatment. For efficient dehydrogenation, the pressure in the vacuum chamber should be below 1 mbar.

Nitrogen in steel originates from N_2 in air dissolving as nitrogen atoms. In the BF process, some nitrogen is dissolved in iron, and typical values are about 50 ppm (or less) in hot metal. In the BOF converter, the CO gas evolution efficiently flushes the liquid steel, and [N] contents down to 10–20 ppm are attained. During tapping, the steel is contaminated by air, causing some [N] pickup. Because of the high solubility of nitrogen in steel and kinetic limitations, only limited nitrogen removal can be achieved by vacuum degassing.

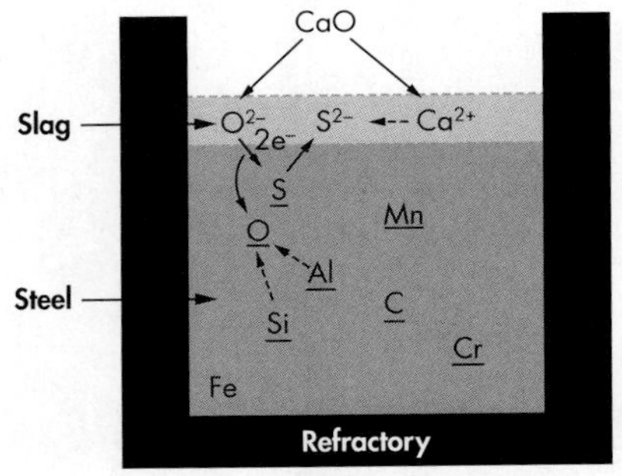

Figure 8 Transfer of sulfur from steel to slag, coupled to transfer of oxygen from slag to steel

Desulfurization

Desulfurization can be understood as an electrochemical exchange reaction, as shown in Figure 8, where dissolved elemental sulfur (in the steel) receives two electrons and turns into a sulfide anion that dissolves in the slag. The electrons are obtained from free oxygen ions in the slag; [O] atoms are formed and transferred to steel as dissolved oxygen, which can further react with deoxidizing elements such as silicon and aluminum.

The sulfide capacity quantifies the partitioning of sulfur from steel to slag. The sulfide capacity strongly increases as the composition approaches CaO saturation, indicating better thermodynamic conditions for desulfurization.

It has been commonly concluded that the overall desulfurization rate is controlled by mass transfer of sulfur in molten steel, which is promoted by strong argon stirring (Roy et al. 2013). However, mixed control (by mass transfer in both steel and slag) may arise in some cases: For less-basic slags with a viscosity greater than about 1.1 dPa·s (poise) and an equilibrium S partition ratio L_S lower than about 400, the overall mass transfer coefficient is affected by a change in the physicochemical properties of the slag (Kang et al. 2017).

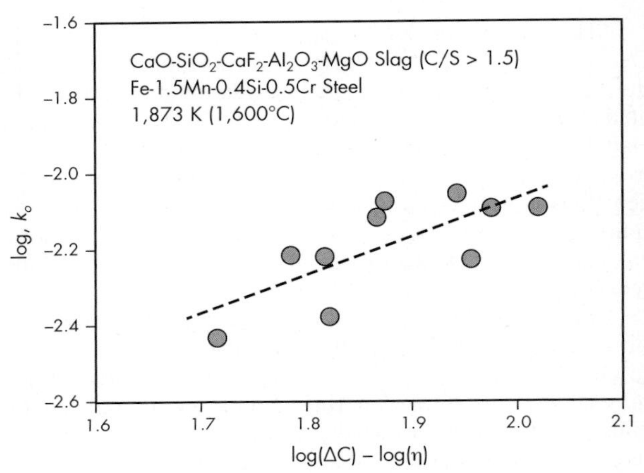

Source: Park and Park 2016

Figure 9 Apparent rate constant of inclusion removal reaction and slag parameter relationship

Deoxidation and Inclusion Control

After the BOF converter process or scrap melting and oxygen blowing in EAF steelmaking, the crude steel has a high concentration of dissolved oxygen (typically in the range of 200 to 800 ppm). Currently, most steelmaking aims at complete *killing* (lowering dissolved oxygen to around 50 ppm or less) by addition of strong oxide-forming elements (deoxidants) such as aluminum and silicon.

When deoxidants are added into liquid steel, intensive nucleation results in many small inclusions. After initial nucleation, the inclusions grow by diffusion, coalescence, and collisions; collisions are regarded as the main growth mechanism (Linder 1974). Because of the rapid initial deoxidation reaction, the dissolved oxygen content decreases to a value close to the equilibrium shortly after deoxidant addition and can be regarded as almost constant until steel casting (Patil and Pal 1987). However, the total oxygen content, which is the sum of the dissolved oxygen and the oxygen bound in inclusions (deoxidation products), decreases relatively slowly during the subsequent ladle metallurgical operations. The decrease in total oxygen directly reflects removal of inclusions from the steel to the top slag. It has been suggested that the dissolution rate of inclusions into the slag phase is proportional to the ratio of the driving force for inclusion dissolution (expressed as a concentration difference, ΔC) to slag viscosity (η) (Valdez et al. 2006). This implies that the inclusion removal rate would strongly depend on the physicochemical slag properties as shown in Figure 9, where k_{ox} represents the apparent rate constant of inclusion removal (Park and Park 2016), including the driving force of the chemical dissolution of alumina (Al_2O_3) and the viscosity of the slag.

ENVIRONMENTAL ISSUES

Efficient scrap recycling, valorization of the steelmaking by-products (such as slag, dust, and sludge), and innovative process development for reducing carbon intensity of steelmaking contribute to lowering the environmental effects of steelmaking.

Scrap Recycling

The steel industry has been recycling steel scrap for more than 150 years. Increased consumption of scrap reduces the needs for additional resource extraction and hence the environmental impact.

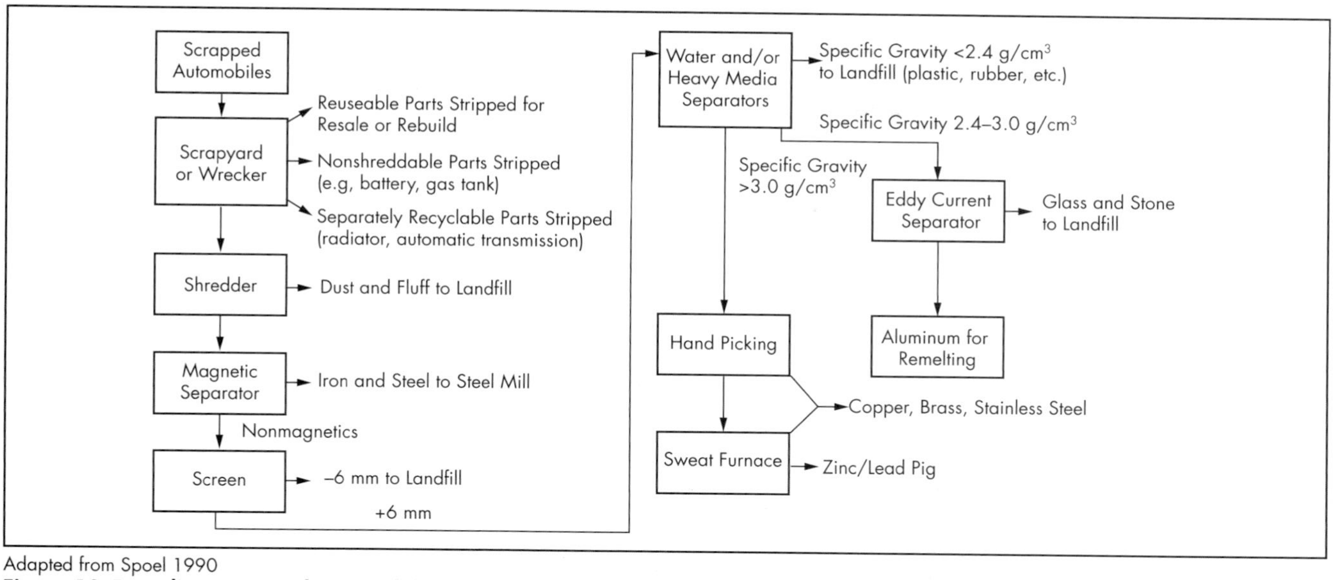

Adapted from Spoel 1990

Figure 10 Recycling scrapped automobiles

Scrap Source

The three main sources of steel scrap are home scrap, prompt scrap, and obsolete scrap. Home scrap is generated within the steel mill during production of iron and steel. It presents no major problems of collection, transportation, and sorting. This scrap has a known composition and is easily recycled to steelmaking. Prompt scrap is generated from the manufacturing process of steel products. It is recycled as part of normal industrial housekeeping as well as for economic reasons. Obsolete scrap consists of obsolete machinery, industrial equipment, agricultural equipment, household appliances, and automotive steel. Because of the wide variety of chemical and physical characteristics, old scrap often requires significant preparation for its recycling, such as sorting, de-tinning, and de-zincing, prior to consumption in mills.

Scrap Processing

Using a variety of equipment, scrap dealers collect and process scrap into a physical form and chemical composition that steel mill furnaces can consume. Figure 10 shows a flow-sheet example for recycling of scrapped automobiles (Spoel 1990). The largest and most expensive equipment is the shredder. The shredder can fragment vehicles into fist-sized pieces of various metals, glass, rubber, and plastic. Hydraulic shears, which have cutting knives of chromium-nickel-molybdenum alloy steel for hardness, slice heavy pieces of ship plate, railroad car sides, and structural steel into chargeable pieces. Various separation processes are applied for sorting and preparation of steel scrap, including magnetic, eddy current, and heavy medium methods; separation by other physical and chemical characteristics; and incineration. Recently developed scrap-sorting technologies include portable optical emission spectrometers, color sorting, and laser-induced breakdown spectroscopy (Yellishetty et al. 2011). Baling presses are used to compact scrap into manageable bundles, thereby reducing scrap volume and shipping costs.

Scrap in Steelmaking

Steel plants melt scrap in BOFs, EAFs, and, to a minor extent, BFs. The proportion of scrap in the charge of a BOF is limited to less than 30%, whereas that in an EAF can be up to 100%. Steel and iron foundries use scrap in EAFs and cupola furnaces. Scrap recycling in steelmaking allows the BF to be bypassed, eliminating the cost of using expensive equipment, labor, energy, iron ore, and coke, resulting in significant primary resource savings and reductions of CO_2 emission and waste generation. One published estimate is that, on average, recovery of 1 t of steel from scrap conserves 1,030 kg of iron ore, 580 kg of coal, and 50 kg of limestone (Fenton 2003). Another estimate can be made: 1.1 t of scrap is needed to produce 1 t of EAF steel, avoiding use of 1.5 t of ore to produce 1 t of BOF steel (steelonthenet.com, n.d.). This means that 1 t of scrap conserves approximately 1.4 t of iron ore.

A major issue of the scrap recycling in steelmaking arises from tramp elements (Cu, Sn, Zn, Pb, Mo) in scrap. For example, automotive scrap contains copper and tin from the electrical systems. These tramp elements are hard to remove in steelmaking, degrading the steel quality, complicating the slag chemistry, and causing refractory degradation (e.g., lead oxide can penetrate refractory bricks and damage the furnace).

Carbon Intensity of Steelmaking

Currently, iron and steel production is the largest industrial source of CO_2 emissions, at 6%–7% of global anthropogenic CO_2 emissions. This is attributed to using carbon-based fuels and reductants and the large volume of steel produced (1,630 Mt [million metric tons] in 2016) (World Steel Association 2017).

CO_2 Emissions in Steel Production

The main CO_2 emissions in steel production are from BFs where iron ore reacts with a reducing agent, like coke, ultimately producing large volumes of CO_2. The carbon intensity of iron and steel production varies considerably between

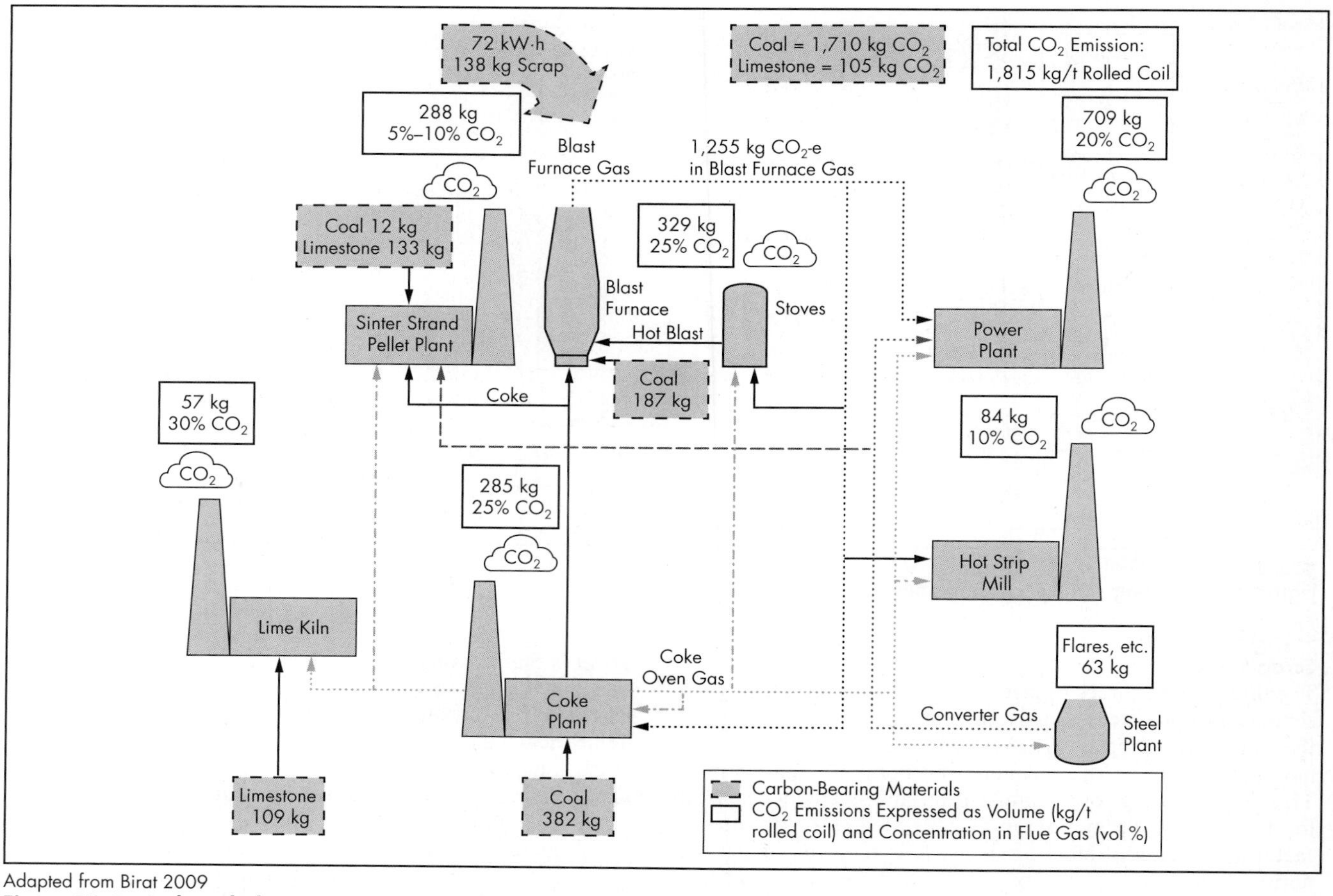

Adapted from Birat 2009

Figure 11 Typical steel plant CO_2 emission

the production routes, ranging from around 0.4 t CO_2/t crude steel for scrap EAFs, to 1.7–1.8 t CO_2/t crude steel for the integrated BF/BOF, and 2.5 t CO_2/t crude steel for coal-based DRI processes (Carpenter 2012). Figure 11 shows the principal CO_2 sources for a typical integrated steel plant producing hot rolled coil. The figure gives a simplified carbon balance, indicating the major entry sources (coal and limestone) and the stack emissions, in volume (kilogram per ton of hot rolled coil) and concentration of CO_2 (volume percent). The BF (69% of emissions) is the largest emitter of CO_2, followed by the sinter/pellet and coking plants.

Reducing Carbon from Steelmaking

Available technologies for reducing CO_2 emissions from iron- and steelmaking processes include (1) minimizing energy consumption and improving the energy efficiency of the process; (2) using clean energy and low-carbon materials in the process; and (3) capturing the CO_2 and storing it.

CO_2 emissions can be reduced by recovering heat from the various gaseous steams and solid and liquid products. The calorific value of off-gases can be recovered before these are emitted to the atmosphere. The gases can also be utilized as a fuel, to produce steam for internal use or to generate electricity. Condensed-phase streams suitable for heat recovery include liquid slag (by dry granulation) and hot coke (by dry quenching; recovering around 80% of the sensible heat of coke and generating around 160 kW·h/t coke [Carpenter 2012]).

CO_2 emissions would be significantly cut if steel plants could switch to clean energy and low-carbon materials, for example, scrap as charge materials (in EAF and BOF), hydrogen (or biomass) as a reducing agent, and using electricity generated in hydroelectric or nuclear power plants or from renewable energy.

Finally, CO_2 emissions could be cut by equipping steel plants with carbon capture and storage. In-process CO_2 capture with oxygen operation could also reduce CO_2 emissions from BFs. In addition, mineral sequestration has been proposed for storage of CO_2: Some minerals (e.g., in magnesium-rich ultramafic rocks) can react with CO_2 to form stable carbonates. However, application of the mineral sequestration is limited by the reaction kinetics.

Steelmaking By-Products

Depending on technological level, production route, and charging materials, approximately one ton of solid wastes or by-product is generated per ton of steel produced from steel industries. These solid wastes (Table 2) or by-products include slag, sludge, dust, and mill scale, and contain valuable metals and minerals.

Ironmaking Slag

Ironmaking slag (approximately 300–400 kg/t of pig iron) is produced during BF processes, from the gangue in the raw materials and fluxing additions. The slag contains mainly

Table 2 Solid and liquid waste generated from steel plants

Solid and Liquid Wastes	Mass Relative to Hot Metal, kg/t	Source of Generation
Coke breeze	—	Coke oven
Nut coke	—	Coke oven
Coke dust and sludge	—	Coke oven
Blast furnace slag	300–400	Blast furnace
Blast furnace dust and sludge	28	Blast furnace
Sintering plant sludge	—	Sinter plant
Basic oxygen furnace slag	100–200	Steel melting shop
Lime fines	15–16	Steel melting shop
Electric arc furnace slag	100–200	Steel melting shop
Electric arc furnace dust	11–20	Steel melting shop
Gas-cleaning-plantsludge	—	Steel melting shop
Carbide sludge	—	Acetylene plant
Mill scale	22	Mills
Mill sludge	12	Rolling mills
Refractory, bricks	11.6	Steel melting shop, mills, etc.
Sludges and scales	—	Water treatment plant
Fly ash	—	Power plant

Source: Das et al. 2007

inorganic constituents such as 30%–35% SiO_2, 28%–35% CaO, 5%–10% MgO, 10% Al_2O_3, 1% iron oxide (expressed as Fe_2O_3). Because BF slag contains little iron and is high in CaO, it is used in cement production. The slag is either air-cooled or granulated. Air-cooled slag is used as aggregate in road construction, and granulated slag is used for cement making. Almost 100% of BF slag is utilized in many areas, such as cement production, road construction, civil engineering work, fertilizer production, landfill daily cover, soil reclamation, and others. Recovery of metals from BF slag is not important because of its low iron content (Fe <2%). In leaching studies, the concentrations of hazardous elements (including As, Ba, Cd, Cr, Pb, Hg, Se, and Ag) were much lower than the toxicity characteristic leaching procedure criterion, indicating that the elements are very tightly bound and not released from the matrix (Proctor et al. 2000).

Steel Slag

Steel slag (100–200 kg/t of liquid steel) is formed in steelmaking from oxidation of Si, P, Mn, and Fe in BOF and EAF steelmaking; other impurities; and addition of fluxes. The main components in steel slags are CaO, Fe, SiO_2, MgO, and MnO. Iron is present as metallics (7%–10%), iron oxide, and iron-bearing minerals. Figure 12 summarizes the recycling of steel slag (Okumura 1993). The metallic iron can be separated by mineral processing technology and recycled by sintering, BF ironmaking, and steelmaking. After metal recovery, the high content of 30%–50% CaO, 3%–10% MgO, and MnO in steel slags can be used to substitute for a part of limestone, dolomite, and manganese ore in steelmaking. However, 50%–90% of the total steel slag (Das et al. 2007) has higher concentrations of phosphorus and sulfur, limiting direct recycling to the ironmaking and steelmaking. Such slag is usually subjected to metal recovery and then applied outside ironmaking and steelmaking.

Difficulties in steel slag reuse include poor volume stability (with BOF slag, hydration of free CaO and MgO; with argon oxygen decarburization slag, C_2S disintegration), high iron oxide content (unsuitable for cement making), high alkalinity (with BOF slag, pH of 10–12.5), and heavy metal (e.g., Cr in EAF stainless-steel slag) (Horii et al. 2015). Hot-stage slag engineering can improve slag properties. Approaches include (1) additions to molten slag for reduction and separation of a metallic phase and/or stabilization of minerals (e.g., C_2S, free CaO and MgO); and (2) selection of appropriate cooling paths to obtain a desirable microstructure. Currently in most steel plants, however, liquid slag is simply cooled slowly to ambient temperatures in slag yards; minimal additions take place, and granulation is not widely practiced.

Steel slag has been recycled in different applications, including aggregate for road construction, metallurgical powder for desulfurization of steel, cement and concrete for preparing building materials, phosphorus-rich fertilizers in agricultural application, soil stabilizer and soil conditioner, use for wastewater treatment, application in CO_2 capture, and flue gas desulfurization (Chand et al. 2016).

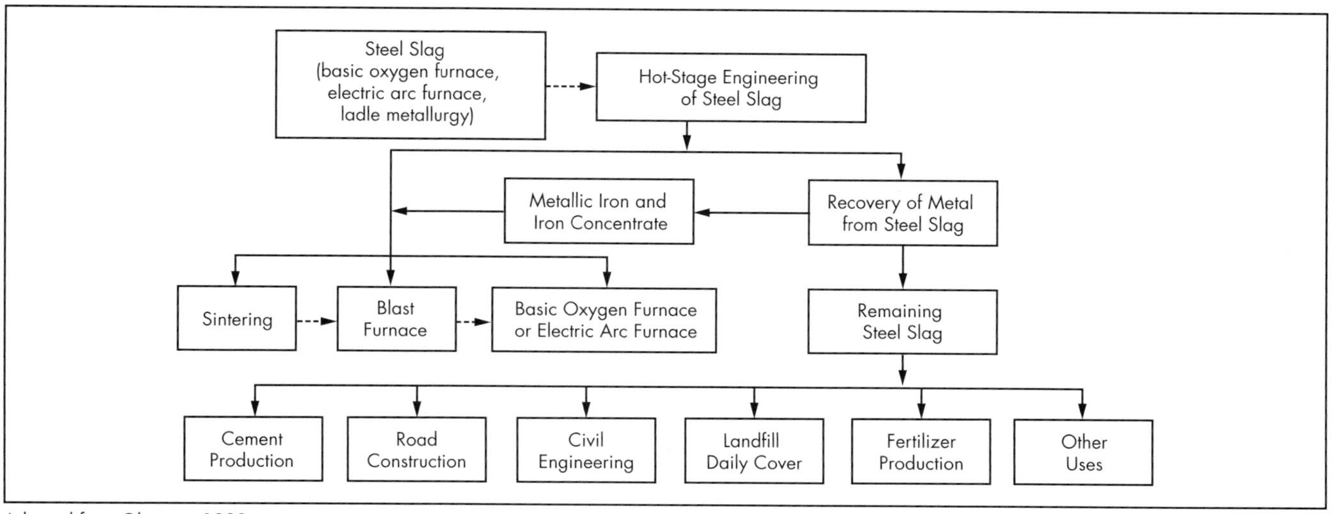

Adapted from Okumura 1993

Figure 12 Recycling of steel slag

Dust and Sludge

Dust and sludge are generated during iron and steelmaking from the top of the furnaces (BF, EAF, and BOF). The BF flue dust is a mixture of iron oxides (50% Fe_2O_3) and coke fines (30% C) (Das et al. 2007). It also contains SiO_2, CaO, MgO, and trace elemental oxides. Direct recycling of the BF dust is not possible because of undesirable elements (like Zn, Pb, and alkali metals) causing BF operational difficulties. Sometimes the dust contains toxic elements (Cd, Cr, and As), and a suitable beneficiation method is needed to recycle it within the plant, avoiding landfill.

During EAF steelmaking, around 11–20 kg of EAF dust is formed per ton of steel; the EAF dust is rich in oxides of iron (20%–45%) and zinc (14%–35%) and has been classified as a hazardous waste by the Environmental Protection Agency because it contains leachable cadmium, lead, and chromium (Suetens et al. 2014). Currently, 80% of EAF dust is recycled to recover iron and zinc by using the Waelz kiln process.

BOF sludge consists of the fine particles recovered after wet cleaning of the off-gas from BOF converters, with high levels of FeO (61%–64%), CaO (10%), and some zinc and lead, depending on the type of limestone and chemistry of scrap used in BOF steelmaking. The high moisture content of the BOF sludge (35%–40%) is a major obstacle in its recycling to the sinter plant. Because it becomes sticky and forms agglomerates after long exposure to the atmosphere, the sludge must be dried before recycling. Mill scale (from the rolling process) contains more than 70% Fe; 70%–100% of mill scale is recycled by briquetting or sintering. Technologies to recycle steel sludge and dusts range from simple agglomeration to thermal, hydrometallurgical, and physical beneficiation methods. An example is a briquetting plant, commissioned at the then National Steel Corporation, Great Lakes, to recycle 300,000 t of waste materials such as BOF sludge, BF dust and sludge, mill scale, and other materials (Landow et al. 1997).

REFERENCES

Birat, J.P. 2009. Steel and CO_2—The ULCOS program: CCS and mineral carbonation using steelmaking slag. In *Proceedings of First International Slag Valorisation Symposium*. Leuven: ACCO. pp. 15–25.

Brotzmann, K. 1979. The development of the OBM/Q-BOP process. In *Basic Oxygen Steelmaking—A New Technology Emerges?* London: Metals Society. pp. 41–45.

Carpenter, A. 2012. *CO_2 Abatement in the Iron and Steel Industry*. Publication No. CCC/193. International Energy Association (IEA) Clean Coal Centre. www.usea.org. Accessed July 2018.

Chand, S., Paul, B., and Kumar, M. 2016. Sustainable approaches for LD slag waste management in steel industries: A review. *Metallurgist* 60(1-2):116–128.

Cicutti, C., Valdez, M., Pérez, T., Petroni, J., Gómez, A., Donayo, R., and Ferro, L. 2000. Study of slag-metal reactions in an LD-LBE converter. In *Proceedings of the Sixth International Conference on Molten Slags, Fluxes and Salts*, Stockholm-Helsinki. Stockholm, Sweden: Department of Materials Science and Engineering, Kungliga Tekniska Högskolan. p. 17.

Das, B., Prakash, S., Reddy, P.S.R., and Misra, V.N. 2007. An overview of utilization of slag and sludge from steel industries. *Resour. Conserv. Recycl.* 50:40–57.

Delhey, H.M., Schürmann, E., Fix, W., and Fiege, L. 1989. Lime and calcium carbide treatment of hot metal according to the immersed lance process. *Stahl und Eisen* 109:1207–1214.

Dubois, C.-A., Leclercq, F., Schutz, R., and Huber, J.-C. 2003. Development of the scrap ratio at the Sollac Lorraine steelshop. In *Proceedings of the 4th European Oxygen Steelmaking Conference*. Essen, Germany: Verlag Glückauf. pp. 341–348.

Fenton, M. 2003. *Iron and Steel Recycling in the United States in 1998*. Open-File Report 01-224. Reston, VA: U.S. Geological Survey.

Fruehan, R.J., ed. 1998. *The Making, Shaping and Treating of Steel*, 11th ed. Steelmaking and Refining Volume. Pittsburgh, PA: Association of Iron and Steel Engineers.

Fuwa, T., Mizoguchi, S., and Harashima, K. 1984. Fundamental aspects of hot metal treatment. *Steelmaking Proc.* 67:257–268.

Hamano, S., Fukagai, S., and Tsukihashi, F. 2006. Reaction mechanism between solid CaO and FeO_x–CaO–SiO_2–P_2O_5 slag at 1 573 K. *ISIJ Int.* 46:490–495.

Horii, K., Tsutsumi, N., Kato, T., Kitano, Y. and Sugahara, K. 2015. Overview of iron/steel slag application and development of new utilization technologies. *Nippon Steel Sumitomo Met. Tech. Rep.* 109:5–11.

ISIJ (Iron and Steel Institute of Japan). 2006. *Physical and Chemical Data Book for Iron- and Steelmaking*. Tokyo, Japan: Nissei Company Press.

Jalkanen, H., and Holappa, L. 2014. Converter steelmaking. In *Treatise on Process Metallurgy.* Vol. 3, Industrial Processes: Part A. Edited by S. Seetharaman, A. McLean, R. Guthrie, and S. Sridhar. Oxford: Elsevier. pp. 223–270.

JFE 21st Century Foundation. 2003. Smelting, refining and continuous casting. In *An Introduction to Iron and Steel Processing*. www.jfe-21st-cf.or.jp/index2.html. Accessed July 2018.

Kanbara, K., Nisugi, T., Shiraishi, O., and Hatakeyama, T. 1972. Impeller stirring type desulfurization (in Japanese). *Tetsu to Hagané* 58:S34.

Kang, J.G., Shin, J.H., Chung, H., and Park, J.H. 2017. Effect of slag chemistry on the desulfurization kinetics in secondary refining processes. *Metall. Mater. Trans. B* 48B:2123–2122.

Kawabata, R., Kohira, S., Watanabe, A., Kawashima, H., Isawa, T., Matsuno, H., and Kikuchi, Y. 2002. Development of environmentally-friendly steelmaking process by zero-slag operation in BOF. *NKK Tech. Rep.* 178:1–5.

Kawauchi Y., Maede, H., Kamisaka, E., Satoh, S., Inoue, T., and Naki, M. 1983. Metallurgical characteristics of hot metal desiliconization by injecting gaseous oxygen. *Tetsu to Hagané* 69:1730–1737.

Kitamura, S. 2014. Hot metal pretreatment. In *Treatise on Process Metallurgy.* Vol. 3, Industrial Processes: Part A. Edited by S. Seetharaman, A. McLean, R. Guthrie, and S. Sridhar. Oxford: Elsevier. pp. 177–221.

Kitamura, S., Mizukami, Y., Kaneko, T., Yamamoto, T., Sakomura, R., Aida, E., and Onoyama, S. 1990. Establishment of steel making process with high productivity and high efficiency by use of hot metal pretreatment. *Tetsu to Hagané* 76:1801–1808.

Kitamura, S., Ito, K., Pahlevani, F., and Mori, M. 2014. Development of simulation model for hot metal dephosphorization process. *Tetsu to Hagané* 100:491–499.

Kraemer, F., Mots, J., and Rohrig, K. 1968. Metallurgical possibilities in application of agitation devices in foundry plants (in German). *Giesserei* 55:145–149.

Landow, M.P., Martinez, M., and Barnett, T. 1997. The recycling of waste oxides at Great Lakes Division, National Steel Corporation. *Ironmaking Conf. Proc.* 56:13–16.

Linder, S. 1974. Hydrodynamics and collisions of small particles in a turbulent metallic melt with special reference to deoxidation of steel. *Scand. J. Metall.* 3:137–150.

Michaelis, E.M. 1979. A review of progress from Bessemer and Thomas to LD, LD-AC, OLP, Graef-Rotor, Kaldo, OBM/Q-BOP, and LWS. In *Basic Oxygen Steelmaking—A New Technology Emerges?* London: Metals Society. pp. 1–10.

Mukawa, S., and Mizukami, Y. 1994. The effect of stirring energy and the rate of oxygen supply on the rate of hot metal dephosphorization. *Tetsu to Hagané* 80:207–212.

Ogawa, Y., Yano, M., Kitamura, S., and Hirata, H. 2001. Development of the continuous dephosphorization and decarburization process using BOF. *Tetsu to Hagané* 87:21–28.

Okumura, H. 1993. Recycling of iron- and steelmaking slags in Japan. In *Proceedings of the First International Conference on Processing Materials for Properties*. Warrendale, PA: The Minerals, Metals & Materials Society. pp. 803–806.

Park, J.S., and Park, J.H. 2016. Effect of physicochemical properties of slag and flux on the removal rate of oxide inclusion from molten steel. *Metall. Mater. Trans. B* 47B:3225–3230.

Patil, B.V., and Pal, U.B. 1987. Fundamental studies on the deoxidation of low carbon steels with ferrosilicon. *Metall. Trans. B* 18B:583–589.

Proctor, D.M., Fehling, K.A., and Shay, E.C. 2000. Physical and chemical characteristics of blast furnace, basic oxygen furnace, and electric arc furnace steel industry slags. *Environ. Sci. Technol.* 34:1576–1582.

Remus, R., Aguado Monsonet, M., Roudier, S., and Delgado Sancho, L. 2013. *Best Available Techniques (BAT) Reference Document for Iron and Steel Production*. Luxembourg: Publications Office of the European Union.

Roy, D., Pistorius, P.C., and Fruehan, R.J. 2013. Effect of silicon on the desulfurization of Al-killed steels: Part I. Mathematical model. *Metall. Mater. Trans. B* 44B:1086–1094.

Saito, R., Matsuura, H., Nakase, K., Yang, X., and Tsukihashi, F. 2009. Microscopic formation mechanisms of P_2O_5-containing phase at the interface between solid CaO and molten slag. *Tetsu to Hagané* 95:258–267.

Sasaki, N., Ogawa, Y., Mukawa, S., and Miyamoto, S. 2013. Improvement in hot-metal dephosphorization. *Nippon Steel Tech. Rep.* 104:26–32.

Seetharaman, S., McLean, A., Guthrie, R., and Sridhar, S., eds. 2014. *Treatise on Process Metallurgy.* Vol. 3, Industrial Processes: Part A. Oxford: Elsevier.

SMS Group. n.d. Steelmaking and secondary metallurgy. www.innse.com/steel/ne_produ.htm. Accessed July 2018.

Spoel, H. 1990. The current status of scrap metal recycling. *JOM* (April): 38–41.

Steelonthenet.com. n.d. Steelmaking input costs. www.steelonthenet.com/cost-bof.html. Accessed January 2018.

Suetens, T., Klaasen, B., Van Acker, K.V., and Blanpain, B. 2014. Comparison of electric arc furnace dust treatment technologies using exergy efficiency. *J. Cleaner Prod.* 65:152–167.

Valdez, M., Shannon, G.S., and Sridhar, S. 2006. The ability of slags to absorb solid oxide inclusions. *ISIJ Int.* 46:450–457.

World Steel Association. 2017. *World Steel in Figures 2017*. Brussels, Belgium: World Steel Association.

World Steel Association. 2018. Fact sheet: Steel and raw materials. www.worldsteel.org/en/dam/jcr:16ad9bcd-dbf5-449f-b42c-b220952767bf/fact_raw%2520materials_2018.pdf. Accessed July 2018.

Yellishetty, M., Mudd, G.M., Ranjith, P.G., and Tharumarajah, A. 2011. Environmental life-cycle comparisons of steel production and recycling: sustainability issues, problems and prospects. *Environ. Sci. Policy* 14:650–663.

CHAPTER 12.35

Sulfur

David B. George

Sulfur is an abundant, multivalent, nonmetal chemical element (symbol S and atomic number 16 on the periodic table). Under normal conditions, sulfur atoms form cyclic octatomic molecules with the chemical formula S_8. At room temperature, elemental sulfur presents as a bright-yellow crystalline solid. Chemically, sulfur reacts as either an oxidizing or reducing agent. Sulfur oxidizes most metals and several nonmetals, including carbon. This leads to its negative charge in most organosulfur compounds. Used as a reducing agent, sulfur reduces several oxidants including oxygen and fluorine.

Sulfur occurs naturally as the pure element (native sulfur); as sulfide and sulfate minerals; and in volcanic gases as sulfur dioxide (SO_2), hydrogen sulfide (H_2S), and sometimes sulfur vapor. Plentiful in native form, sulfur and its uses were known in ancient times in China, Egypt, India, and Greece. Fumes from burning sulfur, which emits a blue flame, were used as bleaches and fumigants, and sulfur-containing medicinal mixtures were used as balms and antiparasitics. Sulfur is referred to in the Bible as *brimstone* (*burn stone* in English) and the name is still in use today, although archaic. In 1777, the French chemist Antoine Lavoisier argued and convinced his scientific community that sulfur was a basic rather than compound element.

Historically, elemental sulfur was extracted from volcanic deposits as native sulfur where it occurred in lumps or veins and occasionally in beautifully crystalline forms. More recently, sulfur has been extracted from salt domes, where it sometimes occurs in nearly pure form, using the Frasch (superheated water) process (Tuller 1954) or it is physically mined as lump sulfur from Poland. Physically mining sulfur, however, has become obsolete because of the high cost of extraction and lower-cost alternative sources. Today, almost all elemental sulfur is produced as a by-product of removing sulfur from natural gas and crude oil.

Sulfur's largest commercial use (mostly after conversion to sulfuric acid) is to produce sulfate and phosphate fertilizers, because of the high requirement of all plant life for sulfur and phosphorus (USGS 1999). Sulfuric acid is also a primary industrial chemical. About 10% of sulfuric acid is produced as the by-product of nonferrous metal smelting, mainly copper, nickel, and, to a lesser extent, lead and zinc. An additional 10% of sulfuric acid production is in the form of recycled/recovered acid from petroleum alkylation units or chemicals production, such as methyl methacrylate (MMA).

Other well-known uses for sulfur are insecticides and fungicides. And although matches are associated with sulfur, they have been almost entirely replaced with butane lighters. Many sulfur compounds are odoriferous, and the smell of odorized natural gas, skunk scent, grapefruit, and garlic is caused by the presence of organosulfur compounds. Hydrogen sulfide produced by living organisms imparts the distinctive odor to rotting eggs and other biological processes.

As an essential element for life, sulfur is extensively used in biochemical processes. Sulfur compounds serve as both fuels (electron donors for CO_2 fixation; Brune 1995) and respiratory (anaerobic) materials in metabolic reactions (electron acceptors; Zhang et al. 2018). Sulfur in organic form is present in the vitamins biotin and *thiamine*, the latter named for the Greek word for *sulfur*. Organically bonded sulfur is a component of all proteins, such as the amino acids cysteine and methionine.

The information presented in this chapter provides an overview of sulfur and its uses, trade, and resources. Almost all sulfur is now the by-product of natural gas and oil production and refining. Large-scale mining of sulfur is almost nonexistent and practiced only by small artisanal miners and in a few legacy locations.

Excellent and up-to-date information is available from the U.S. Geological Survey (USGS), Department of the Interior, which publishes a comprehensive annual summary of sulfur. The *Mineral Commodity Summaries* are updated annually and currently include data through 2015. The *Minerals Yearbook* is currently updated to 2013. Many of the consumption and use figures presented in this chapter are derived from these open USGS sources (1932–2018). In addition, there are other online resources that offer useful references and color images.

PHYSICAL PROPERTIES

Sulfur forms polyatomic molecules with different chemical formulas. The best-known allotrope is octasulfur, cyclo-S_8.

David B. George, Principal Consultant, David B. George and Associates LLC, Heber City, Utah, USA

Octasulfur is a soft, bright-yellow solid with only a faint odor, similar to that of matches. It melts at 115.2°C, boils at 444.6°C, and sublimes easily (Lumen Learning, n.d.). Amorphous or *plastic* sulfur is produced by rapid cooling of molten sulfur—for example, by pouring it into cold water. At 95.2°C, cyclo-octasulfur changes from α-octasulfur to the β-polymorph. The structure of the S_8 ring is virtually unchanged by this phase change. Between its melting and boiling temperatures, octasulfur changes its allotrope again, turning from β-octasulfur to γ-sulfur, again accompanied by a lower density but increased viscosity because of the formation of polymers. At even higher temperatures, however, the viscosity decreases as depolymerization occurs. Molten sulfur becomes dark red at temperatures higher than 200°C. The density of sulfur is approximately 2 $g \cdot cm^{-3}$, depending on the allotrope; all of its stable allotropes are excellent electrical insulators (Lumen Learning, n.d.).

Liquid sulfur is always handled between 130° and 140°C with a viscosity of 7–10 cP. Below this temperature, the viscosity increases gradually; above 155°C, viscosity rapidly increases and the sulfur becomes unpumpable.

CHEMICAL PROPERTIES

Notable for its peculiar stifling odor, sulfur dioxide is formed from vaporized sulfur. Sulfur vaporizes at 445°C and autoignites at approximately 235°C. Insoluble in water, sulfur is soluble in carbon disulfide and, to a lesser extent, in other nonpolar organic solvents, such as benzene and toluene. The first and the second ionization energies of sulfur are 999.6 and 2,252 $kJ \cdot mol^{-1}$, respectively. The +2 oxidation state is rare, however, with +4 and +6 being more common. Because of electron transfer between orbitals, the fourth and sixth ionization energies are 4,556 and 8,495.8 $kJ \cdot mol^{-1}$, and these states are only stable with strong oxidants such as fluorine, oxygen, and chlorine.

Both sulfur dioxide and sulfuric acid are corrosive to many metals and materials. Sulfuric acid, however, has the unusual property of being significantly more corrosive when diluted with water than when nearly pure. Although dilute sulfuric acid requires exotic alloy steels for its containment, concentrated sulfuric acid can be safely stored in carbon steel tanks. This is because sulfuric acid forms a stable iron sulfate film on the carbon steel, and this restricts corrosion.

Polymer-lined tank trucks or railcars, often in unit trains, are commonly used for handling and transporting sulfuric acid. Barges, ships, and occasionally pipelines are used to transport large shipments. To prevent serious chemical burns from contact with the acid, special precautions and protective suits are required when handling sulfuric acid.

NATURAL OCCURRENCES

Sulfur is created inside massive stars, at a depth where the temperature exceeds 2.5×10^9 K, by the fusion of one nucleus of silicon plus one nucleus of helium (Cameron 2013). This synthesis, the alpha process, produces elements in abundance. Sulfur is the 10th most common element in the universe.

According to Harraz (2015), "On Earth, elemental sulfur can be found near hot springs and volcanic regions in many parts of the world, especially along the Pacific Ring of Fire; such volcanic deposits are currently mined in Indonesia, Chile, and Japan," but are of little commercial importance. "Historically, Sicily was a large source of sulfur during the Industrial Revolution."

Native sulfur is geologically produced from sulfate minerals, such as gypsum (calcium sulfate), by anaerobic bacteria in salt domes in the presence of methane and other hydrocarbons. "Significant deposits in salt domes occur along the coast of the Gulf of Mexico, and in evaporative deposits in Eastern Europe and Western Asia" (Harraz 2015). Until recently, sulfur deposits from salt domes were the basis for most commercial production in the United States, Russia, Turkmenistan, and Ukraine. Such sources are now of little commercial importance, and most are no longer worked.

Common naturally occurring sulfur compounds include the sulfide minerals, such as pyrite (iron sulfide), chalcopyrite (copper-iron sulfide), nickel sulfides, cinnabar (mercury sulfide), galena (lead sulfide), sphalerite (zinc sulfide), and stibnite (antimony sulfide); and the sulfates, such as gypsum (calcium sulfate), alunite (potassium aluminum sulfate), and barite (barium sulfate) (Harraz 2015). In some organic-rich environments, sulfates and iron compounds are reduced to iron sulfide by bacterial action yielding small pyrite or marcasite nodules. These are common in the southern United States but are only a curiosity.

SULFUR EMISSIONS AND THE SULFUR CYCLE

The discharge of sulfur compounds and their fate in the atmosphere are important to the world's climate. Many views on the sulfur balance in the atmosphere exist, but the consensus is that anthropogenic emissions of sulfur range between 60 and 100 Mt/yr (million metric tons per year), and these account for approximately 70% of all sulfur emissions. Approximately 65% of the discharged sulfur is from power generation. Organic sulfides from oceanic plankton are estimated to emit between 15 and 40 Mt/yr of sulfur equivalent. Smaller sulfur emissions come from natural fires and biological processes in soils and wetlands. Volcanos discharge between 5 and more than 20 Mt/yr of sulfur, mainly as sulfur dioxide. Volcanic emissions of sulfur dioxide and hydrogen sulfide from an individual eruption can be significant. It is well known that large volcanic eruptions, for example, the 1991 Pinatubo eruption in the Philippines, cause short-term lowering of global temperatures.

Sulfur dioxide has a short life in the atmosphere, with some estimates calculating that it oxidizes to sulfur trioxide (SO_3) in a day. The sulfur trioxide reacts with water vapor to form sulfuric acid mist, which in turn is removed from the atmosphere through precipitation (acid rain) or by absorption on dust particles and sedimentation. Volcanic sulfur emissions may be more persistent than most anthropogenic emissions because of their release at higher altitudes. Hydrogen sulfide emissions are similarly oxidized in the atmosphere and removed.

SULFUR PRODUCTION

Sulfur may be found in its elemental form, not associated with other elements, and historically was obtained in this way. Natural sulfide minerals—metal sulfides—have been sources of sulfur via their oxidation to produce sulfur dioxide followed by conversion to sulfuric acid. Today, virtually all sulfur production is as a by-product of other industrial processes such as oil refining and natural gas production. In these processes, sulfur often occurs as undesired or detrimental sulfide compounds that must be removed, mainly as hydrogen sulfide. The hydrogen sulfide is then converted to elemental sulfur using the Claus process.

The Frasch process has been used for more than a century to extract elemental sulfur from sulfur-bearing porous limestones in the salt dome caprocks of Texas and Louisiana (United States). The caprocks are calcite (limestone) and the basement is halite (salt) overlain by anhydrite ($CaSO_4$). Typically, a 150-mm-diameter hole is drilled from the surface to a depth of approximately 300–900 m, and a string of three concentric pipes is lowered into the hole. The 150-mm-diameter outer pipe is the casing, and the inner pipes are ~75 mm and 25 mm in diameter, respectively. Water heated to approximately 165°C is pumped down the annulus at high pressure between the two larger pipes. The casing rests on anhydrite and is perforated near the bottom and again at some distance up into the sulfur-bearing calcite bed. An annular collar seals the space between the two larger pipes and is positioned between the sets of perforations. Hot water emerging from the upper perforations melts the sulfur. The molten sulfur forms a pool that submerges the lower set of perforations, allowing the liquid sulfur to enter the perforations and rise through the inner annulus. Hydrostatic pressure exerted by the column of hot water causes the liquid sulfur to rise in the annulus. Because sulfur is denser than water, it will only rise about halfway. Compressed air is discharged from the bottom of the 25-mm pipe and mixes with the molten sulfur, sufficiently lowering the density of the two-phase fluid so that it reaches the surface, where it is deaerated and collected.

In another method, superheated water is pumped into a native sulfur deposit to melt the sulfur, and then compressed air is injected to return the 99.5% pure melted product to the surface. Throughout the 20th century, this procedure produced elemental sulfur that required no further purification. However, because of a limited number of such sulfur deposits and the high operating cost, this process has not been employed in a major way anywhere in the world since 2002 (Harraz 2015).

Today, most sulfur is produced by its removal from petroleum, natural gas, and related fossil resources, mainly as hydrogen sulfide but also as more complex organosulfur compounds. Organosulfur compounds, which are undesirable impurities in petroleum, may be upgraded by subjecting them to hydrodesulfurization. This action cleaves the C–S bonds as shown in the following reaction (Harraz 2015):

$$R\text{–}S\text{–}R + 2H_2 \rightarrow 2RH + H_2S \quad \textbf{(EQ 1)}$$

This process forms hydrogen sulfide, which also occurs in natural gas, and it is converted into elemental sulfur by the Claus process. Half of the hydrogen sulfide is oxidized to sulfur dioxide in the Claus process. The two are then reacted at high temperature so that comproportionation occurs:

$$3O_2 + 2H_2S \rightarrow 2SO_2 + 2H_2O \quad \textbf{(EQ 2)}$$

$$SO_2 + 2H_2S \rightarrow 3S + 2H_2O \quad \textbf{(EQ 3)}$$

Large stockpiles of prospective elemental sulfur exist within the Athabasca oil sands in western Canada, owing to the high sulfur content of the sands. Marketing this sulfur, however, is not currently economic because of high shipping costs.

BY-PRODUCT SULFUR AND MINING

In the 1930s, the first sulfuric acid plant treating smelter gases was installed at the Kennecott Utah Copper smelter near Salt Lake City, Utah (United States). This small plant produced only 50 t/d of sulfuric acid, which was sold to the local explosives industry. By the 1950s, the nonferrous industry was installing sulfuric acid plants to reduce emissions—although the impact was small—and produced a useful by-product. For years, sulfide smelting industry emitted most of the sulfur dioxide into the atmosphere through highly visible smoke stacks. Figure 1 shows a typical nonferrous smelter up until the 1980s.

© David B. George

Figure 1 Large U.S. copper smelter, circa 1978

Although the United States had promulgated regulations to reduce emissions from industrial sources for many years, it was not until the 1970 amendment to the Clean Air Act of 1963 that companies were forced to seriously consider how to reduce sulfur and other air pollutant emissions. By the late 1980s, most U.S.-based smelters had either modernized or shut down. The nonferrous industry in the United States went into a major decline starting around 1970. Many mines and smelters closed, some because of the end of economic mine production and others because of the high cost of meeting ever more stringent air quality regulations. In the United States in 1970, there were more than 30 copper smelters; today, there are only 3.

Worldwide, the recovery of sulfuric acid from nonferrous smelters has dramatically increased over the last 40 years. Starting in the 1970s, several new smelters were constructed in Japan and Europe as part of the post–World War II industrialization. These smelters all achieved high sulfur capture using sulfuric acid plants and more modern smelting technologies than was practiced in the United States and elsewhere. Sulfur capture was greater than 90% and approached 98% at some smelters. The by-product acid found a ready local market.

In the United States, sulfuric acid production at copper, nickel, lead, molybdenum, and zinc roasters and smelters accounted for approximately 7% of total domestic production of sulfur in all forms and totaled the equivalent of 620,000 t of elemental sulfur in 2010 (Apodaca 2015–2017). Three acid plants operated in conjunction with copper smelters, and several more were by-product operations at lead, molybdenum, and zinc smelting and roasting operations. The three largest by-product sulfuric acid plants, in size and capacity, were associated with copper smelters and accounted for 86% of the by-product sulfuric acid output (Apodaca 2015–2017). The copper producers Asarco, Rio Tinto Kennecott Utah, and Freeport-McMoRan Copper and Gold each operated a sulfuric acid plant at their primary copper smelters.

Except in some countries such as Chile, Russia, and parts of Africa, most nonferrous smelters around the world are now achieving greater than 90% sulfur capture. The latest copper smelting technology, for example, can capture more than 99.9% of the input sulfur, making copper smelting a minor source of sulfur dioxide emissions in almost every developed location. In contrast, some smelters in Russia have almost no sulfur capture because of the enormous problems of sulfuric acid transport in the Arctic north (e.g., at Norilsk in Siberia). Some of these smelting complexes emit more than 2 Mt/yr of sulfur dioxide compared to a few hundred metric tons per year at the most advanced smelters. Chile, the world's largest copper producer, has lagged in sulfur emissions control, and many smelters are still achieving less than 80% sulfur capture, making these among the dirtiest smelters in the developed world. China, in contrast, has embraced environmental control in their newest smelters and achieves world-class performance. But many small and dirty smelters remain in China, seemingly immune to normal market and regulatory forces. Today, China treats more than 40% of the copper concentrate produced at the world's mines.

Although production of sulfuric acid from smelter and roaster off-gas is relatively easy, the capital cost is significant, and the product is difficult to ship for long distances unless there is ready access to barges or ships. Generally, the production of sulfuric acid at smelters is not a money-making proposition, although there are many exceptions to this generality. In the United States, the acid produced by the three copper smelters in the West is consumed locally, often for ore leaching, but some is shipped as far as Florida for use in phosphate production.

In Japan, where all the copper smelters achieve very high sulfur capture, the acid produced now exceeds domestic requirements. This has necessitated shipping sulfuric acid to Chile or Peru, where much of the copper concentrate that is smelted in Japan originates, where it is used in oxide copper leaching. It is anticipated that China will soon have a surplus of sulfuric acid, which could further depress sulfuric acid demand and prices.

U.S. CONSUMPTION

Apparent domestic consumption of sulfur in all forms has remained nearly stable with growth in the 2%–4% range, depending on fertilizer demand, which is often cyclical. Of the sulfur consumed, 65% is obtained from domestic sources as elemental sulfur (60%) and by-product acid (5%). The remaining 35% is supplied by imports of recovered elemental sulfur (26%) and sulfuric acid (9%) (Apodaca 2015–2017).

Sulfur differs from most other major mineral commodities in that its primary use is as a chemical reagent rather than as a component of a finished product. This use generally requires that it be converted to an intermediate chemical product prior to its initial use by industry. The leading sulfur end use, sulfuric acid, represents 66% of reported consumption with an identified end use. Although reported as elemental sulfur in many sources, it is reasonable to assume that nearly all the sulfur consumption reportedly used in petroleum refining was first converted to sulfuric acid, bringing sulfur used to produce sulfuric acid to 85% of the total sulfur consumption (Apodaca 2015–2017).

Practically all of the sulfuric acid production in the world is through the contact process, where sulfur dioxide (SO_2) is oxidized to sulfur trioxide (SO_3) over a vanadium-based catalyst and absorbed into circulating strong sulfuric acid. The source of SO_2 gas is either off-gas from a nonferrous smelter or generated by burning sulfur (or a sulfur-containing compound) in a furnace. Since the mid-1970s, most new acid plants have been built in the double-contact, double-absorption configuration to achieve reduced tail gas emissions of SO_2.

Because of its desirable properties, sulfuric acid retains its position as the most universally used mineral acid and the most produced and consumed inorganic chemical, by volume. Data based on USGS surveys of sulfur and sulfuric acid producers show that reported U.S. consumption of sulfur in sulfuric acid (100% basis) between 2012 and 2013 increased by 8%, and total reported sulfur consumption increased by 7% (Apodaca 2015–2017). The reported increase in sulfuric acid consumption can be attributed to a threefold increase in sulfuric acid use in petroleum refining and other petroleum and coal products. Some inconsistencies exist in reported sulfur consumption and use figures, so this information is not considered particularly accurate.

Agriculture is the leading sulfur-consuming industry, mainly for the production of fertilizers, primarily phosphate fertilizer. Approximately 45% of domestic phosphate fertilizer production is exported. The second-ranked end use for sulfur is in petroleum refining and other petroleum and coal processing. Demand for sulfuric acid in copper ore leaching, which is the third-ranked end use, has decreased in recent years.

Adding to the problem of quantifying the production of sulfur and sulfur chemicals is the practice of recycling some sulfuric acid, mainly that used in petroleum refining and explosives and chemical production. Some acid is recycled internally by companies and some is sent to recyclers who regenerate sulfuric acid. Data on this part of the industry is very hard to obtain and verify and could comprise 10% of the total sulfuric acid production.

PRICES

The price of sulfur is highly influenced by location and cyclical demand for fertilizer. Prices vary greatly on a regional and international basis, and it is recommended that previously mentioned USGS publications be referred to in any effort to understand current trends.

During periods of poor agricultural commodity prices, the agribusiness sector and small farmers elect to not use fertilizers. This can be sustained for a few years, but the drop in yield and recovering commodity prices typically spurs a demand in fertilizer. Usually, the highest reported prices in the United States are from Tampa, Florida, because of the large demand for sulfur for fertilizer production using local phosphate deposits in the central Florida area.

At year-end 2013, U.S. West Coast prices ranged from US$93 to $98/t (Apodaca 2015–2017). Nearly all the elemental sulfur produced in some regions, such as the West Coast, is processed at sulfur-forming plants to form sulfur prills, which are necessary to make solid sulfur acceptable for shipping overseas. The cost for prilling is significant. Recently, world sulfur prices generally were higher than domestic prices. This is evidenced by the Abu Dhabi National Oil Company (ADNOC) price. Although prices can vary according to locations, providers, and types of sulfur produced, the ADNOC contract price is recognized as an indicator of world sulfur price trends. The ADNOC price averaged approximately US$120/t in 2013, with price ranging from a low of US$70/t in August and as high of US$160/t in April (Apodaca 2015–2017). As a rule, the sulfuric acid price is typically one-third the sulfur price.

REFERENCES

Apodaca, L.E. 2015–2017. Sulfur. In *Minerals Yearbook* [various years]. Reston, VA: U.S. Geological Survey.

Brune, D.C. 1995. Sulfur compounds as photosynthetic electron donors. In *Anoxygenic Photosynthetic Bacteria*. Advances in Photosynthesis and Respiration. Vol 2. Edited by R.E. Blankenship, M.T. Madigan, and C.E. Bauer. Dordrecht: Springer.

Cameron, A.G.W. 2013. *Stellar Evolution, Nuclear Astrophysics, and Nucleogenesis*. Mineola, NY: Dover.

Harraz, H.Z. 2015. Topic 7: Sulfur ore deposits [slide presentation]. www.slideshare.net/hzharraz/sulfur-ore-deposits. Accessed September 2018.

Lumen Learning. n.d. Properties of sulfur. In *Introduction to Chemistry*. Portland, OR: Lumen Learning.

Tuller, W.N., ed. 1954. *The Sulphur Data Book*. New York: McGraw-Hill.

USGS (U.S. Geological Survey). 1932–2018. Sulfur: Statistics and information. http://minerals.usgs.gov/minerals/pubs/commodity/sulfur/. Reston, VA: USGS.

USGS (U.S. Geological Survey). 1999. *Fertilizers: Sustaining Global Food Supplies*. Fact Sheet 155-99. Reston, VA: USGS.

Zhang, L., Xiaojuan, L., Zhang, Z., Chen, G-H., C., and Jiang, F. 2018. Elemental sulfur as an electron acceptor for organic matter removal in a new high-rate anaerobic biological wastewater treatment process. *Chem. Eng. J.* 331:16–22.

CHAPTER 12.36

Talc

Edward F. McCarthy and Jorge L. Yordan Hernandez

INTRODUCTION

The objective of this chapter is twofold: (1) to review the traditional mineral processing technology used by the talc industry and (2) to introduce and discuss the new technologies that are being implemented by talc producers in response to a changed marketplace.

Fifty years ago or so, talc processing was an easy manufacturing process because of an abundance of high-quality talc ores and relatively simple end-user applications. For instance, talc products were used as fillers in unsophisticated ceramic articles or in cosmetics and body powder. Nevertheless, the talc industry was reasonably profitable.

Today, the markets for talc have completely changed. Additionally, the good-quality ores are being depleted around the world, especially in China. Many simple end uses for talc are gone and have been replaced by applications that require the talc products they buy to be produced using state-of-the-art quality control programs and to comply with difficult-to-meet specifications. The talc industry worldwide continues to consolidate. The profitability of talc companies will be poor unless a substantial portion of their total revenues comes from value-added talc products that are targeted to enhance important properties of sophisticated end products manufactured by the plastics, specialty technical ceramics, and coatings industries.

Even with major new world-class talc deposits being developed in Afghanistan and Pakistan in the last decade, it is challenging for talc producers to find talc ores that—when beneficiated the traditional way—would yield end products that meet the new market requirements. Consequently, talc producers have made significant investments to upgrade beneficiation, micronizing, and compaction capability to fulfill the market demands and also to alleviate the pressure of having to high-grade mines. These separation technologies, together with advances in wet grinding, fine particle sizing, surface coatings of talc particles, and talc synthesis, are discussed in this chapter.

The mining, processing, and applications of value-added talc products is a sophisticated process involving geologists, mining engineers, mineral processing engineers, chemical engineers, surface chemists, applications specialists, and a solid commercial group. In some instances, the applications are unique to talc and even to specific talc deposits. Indeed, for the most part, the market for talc is heavily dependent on the scientific research knowledge and technical skills of the producer. Without such application development efforts, new markets are very difficult to penetrate. This chapter borrows some content from the "Talc" chapter written by McCarthy et al. (2006) in *Industrial Minerals and Rocks* published by the Society for Mining, Metallurgy & Exploration (SME).

The Mineral Talc

Talc is a crystalline hydrated magnesium silicate of the general formula $3MgO\text{-}4SiO_2\text{-}H_2O$ (see Figure 1), which theoretically is 31.7% MgO, 64.5% SiO_2, and 4.8% H_2O. It is one of the layer silicates such as kaolin and mica. Talc ores can occur as a relatively pure mineral or as a mixture with other minerals. In its pure form, talc is colorless or appears white, but when the ore contains other minerals, it can also appear green, light green, yellowish, pink, or even black. Accompanying minerals include magnesite, dolomite, chlorite, calcite, and quartz.

The adjacent layers of silica are very weakly bonded with only van der Waals forces, and this allows talc to be easily sheared along this plane. This gives talc its natural slippery

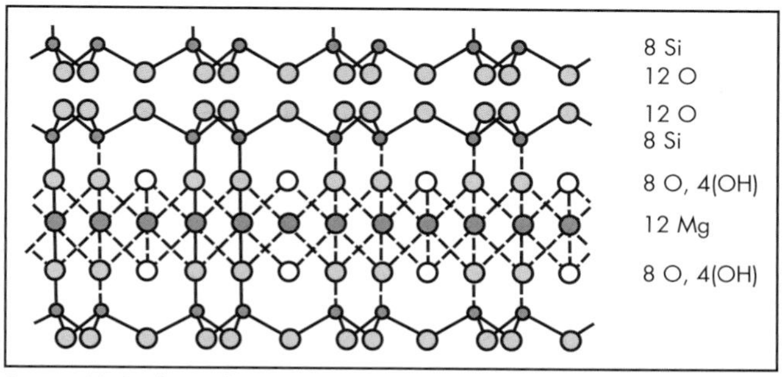

Source: McCarthy et al. 2006

Figure 1 Talc crystal structure

Edward F. McCarthy, Manager, Performance Minerals LLC, Morgan Hill, California, USA
Jorge L. Yordan Hernandez, Retired, Formerly with Rio Tinto Minerals, Parker, Colorado, USA

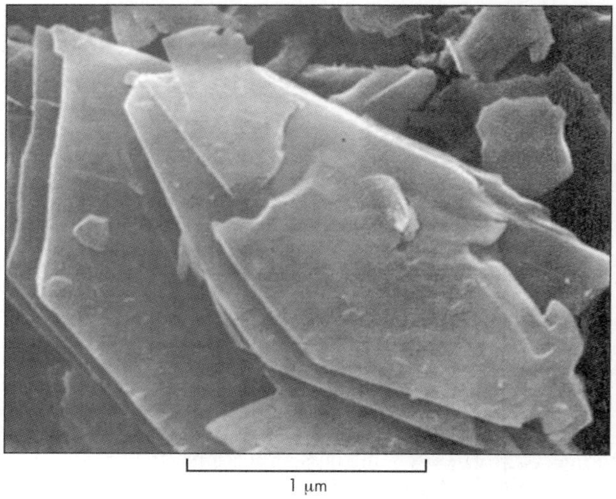

Source: McCarthy et al. 2006

Figure 2 Scanning electron micrograph of talc

feel as well as its platyness and softness (Figure 2). Talc is the softest mineral, with a Mohs scale hardness of 1. Additionally, talc is relatively inert and almost insoluble in conventional acids and bases but soluble in aqua regia and hydrofluoric acid. Talc has a pH in water of 9.0 to 9.5 and very little buffering capacity due to its insolubility. Regulatory authorities consider it a nuisance dust, but it has GRAS (generally recognized as safe) status as a cosmetics and food ingredient.

Talc has a specific gravity of 2.78, a refractive index of 1.58, a specific heat of 0.208 cal/g/°C, a thermal conductivity of 5 cal/g/s/°C, and a surface-free energy of 30–40 mJ/m^2 (Yordan et al. 2002) that makes talc a hydrophobic or water-hating mineral. Interestingly, when talc is split along the silica surface, the surface created is hydrophobic. When the talc is fractured across the silica–brucite–silica layers, the surface created is ionic (electrically charged) and hydrophilic. Materials that have both hydrophobic and hydrophilic components are surface active, and this gives talc its unique surface-active properties. Recently, Yan et al. (2011) confirmed and quantified the hydrophobicity of the planar talc surface as well as the hydrophilicity of the talc edges using a modified atomic force microscope. In water, the talc particles have a negative zeta potential with a maximum of –40 mV at a pH of 9. Its point of zero charge is around pH 2.5–3.0.

Geology of Talc Deposits

Talc can be formed in different geological environments or processes. The most commercially important deposits are metasedimentary-type deposits that are formed from the reaction of dolomite or magnesite host rocks and silica-containing fluids. Such deposits are common in the United States, China, and South Asia. They are amorphous when first formed but become progressively more crystalline when they are sheared into thinner lenses by geological tectonic movements. Such deposits will have carbonates or chlorite as the major co-minerals.

Also of significant commercial importance are serpentine-derived, ultramafic talc deposits typical of eastern Canada (Ontario) and Finland. These deposits are formed from the reaction of CO_2 with a host serpentine. Such deposits will always have high levels of divalent iron substitution in the talc lattice along with a magnesium–iron carbonate isomorph as the major co-minerals. Also, a second type of metasedimentary deposit is found in France, where magnesium-rich fluids-altered mica schists to form chlorite, and the silica released during this reaction altered overlying dolomite to talc. These deposits have 25% to 75% chlorite along with the talc and are often referred to as chloritic talc. Table 1 summarizes the typical mineralogy of the aforementioned ores.

Table 1 Typical mineralogy of North American and international talc ores

Ore Location	Talc, %	Carbonates, %	Chlorite, %	Quartz, %
Montana, United States	95	2	1	0.5
Ontario, Canada	45	50	5	Trace
Finland	55	40	1	Trace
France	55	2	40	1
China	93	3	3	0.5

Adapted from McCarthy et al. 2006

The value of a talc deposit is a function of its purity, color, and location relative to the market. The purity issue is not related to the absolute percentage of talc present but more a function of specific mineral or metal impurities. Asbestiform minerals and metals such as arsenic and lead are particularly deleterious even at parts-per-million (ppm) levels, and quartz, serpentine, chromium, and pyrite are harmful at levels greater than 1,000 ppm. However, carbonates and chlorite (magnesium–aluminum silicate) can be beneficial at levels up to 50%, especially when they are white and the quality is consistent.

The profitability of a modern talc operation mining a world-class talc ore such as the ones found in Montana (United States), Ontario (Canada), Finland, France, and China depends on minimizing the mining, beneficiation, and processing costs and maximizing the production of value-added talc products.

Exploration

Despite the availability of many modern geophysical techniques, the most effective talc exploration techniques are traditional field diagnostic, sampling, and mapping methods. Although talc is physically weak, it is chemically very stable, and the presence of talc in outcrops and in the overlying soils is the best diagnostic tool. After the deposits are sampled and mapped, if the quality is acceptable, the deposit is drilled, the quality is confirmed, and the reserve is estimated. The value of any talc deposit is a function of its color, purity, and location, along with the availability of appropriate markets. Nowhere is this more true than in Finland, where relatively low-purity and low-color ultramafic deposits were developed successfully to serve the local paper market and to replace expensive imported kaolin.

Mining

Most talc is mined today by conventional open pit, drill-and-blast, and shovel-and-truck techniques. The major difference from conventional technology is that blasting is minimized to avoid breakage of the soft talc ore, and all shovel work is accompanied by a high level of selectivity to minimize cross-contamination of high- and low-grade material. Many producers analyze blast-hole cuttings to delineate ore grades and select blast patterns accordingly. Experienced shovel operators can select ore or waste by color or texture for different grades. Equipment such as 10-m^3 shovels and 150-t trucks are common in North America and Europe, whereas 1-m^3 shovels, 20-t trucks, and even hand-carried baskets are used in Asia.

A typical western talc deposit is drilled on 30- to 50-m spacing, and the ore body is characterized by mineralogy, color, and chemistry. From this, the ore body is defined and a computer-aided mine plan is generated. Overburden is removed and stored elsewhere for eventual use in mine reclamation. Mining is typically done on 5- to 10-m benches. Before each blast, drill holes are analyzed and, if necessary, the shot is reconfigured to remove potential waste or segregate a better quality of ore. After the blast, the shovel operator will selectively scoop the rock into haul trucks and designate it as waste or as a low or high grade of ore.

Waste-to-ore ratios are often quite high, especially for massive steatite deposits. Typical values range from 5:15 for massive ores and 1:3 for the lower-purity ultramafic grades. Underground mining is now less common and is declining rapidly in importance. Pure talc is a very soft, noncompetent rock, which is often highly fractured, of varying thickness, and in steeply dipping bodies. Where veins are thin and the surrounding rock is competent, overhand mining is acceptable with limited timbering. In thicker veins, underhand mining is necessary; where the veins dip steeply, shrinkage stoping is used. Some low-grade deposits can be mined by room-and-pillar methods, for which equipment is much smaller, blasting and mucking much more selective, and waste-to-ore ratios much lower than in open pit mining. Continuous miners can be used on softer high-grade veins.

Reclamation

In contrast to past practice, reclamation is now an integral part of all talc mining in developed countries. Waste piles are graded, covered with topsoil, seeded, and monitored for a return to natural vegetation. For both surface and underground mining, both surface and subsurface water flow is measured, monitored, analyzed, and, if necessary, treated to meet local and national discharge standards. Some pits are backfilled or converted to a beneficial long-term use, such as recreation.

Manufacturing of Value-Added Talc Products

The manufacturing of value-added talc products from talc ores can be viewed as a two-stage process: a beneficiation stage and a finishing stage. The goal of the first stage is to remove, using mineral processing technology, as much of the deleterious mineral impurities from the ore as possible to produce a high-quality talc concentrate. The goal of the second stage is to manufacture high-quality products by properly milling or micronizing the talc concentrate, removing additional impurities, sizing, compacting, deaerating, and packaging and shipping the finished goods.

BENEFICIATION STAGE

The beneficiation stage can be a dry or a slurry process. If dry, the predominant technology is sorting; if it is carried out wet, it involves unit operations such as froth flotation and wet high-intensity magnetic separator (WHIMS).

Sorting

In reality, the sorting of the talc ore starts at the mining front; after the blast, the shovel operator will selectively scoop the rock into haul trucks and designate it as waste or as a low or high grade of ore.

Typically, the ore selected to be processed any given day is crushed with a jaw crusher, and the crushed ore is conveyed to the sorting plant. The sorting technology used to separate talc from dolomite and other impurities includes hand sorting of lump ore, mechanized optical sorting of pebble-sized ore, and slide or friction sorting of lump ore.

Hand sorting usually takes place in sheds at process plants. Screened ore (5–25 cm in size) is run across belts, and if low grade, talc is picked from the ore; if high grade, waste is picked from the talc.

Optical sorting processes usually require talc to be washed and screened into narrow pebble-sized (1–2 cm) range. A camera scans each particle as it falls through a testing zone, and then jets of air remove either ore (for low grade) or waste (for high grade). This type of sorting is suitable for higher-value products such as those required for cosmetics or plastics applications.

This technology has now been extended to mineral-based separations based on using near-infrared and X-ray transmission sensor technologies instead of color meters (Bergmann 2014).

A slide sorting process conducted on friction plates and chutes to replace hand sorting was patented (Nichols et al. 1991). It is based on the fact that a talc-rich rock has a much lower coefficient of friction than dolomite or quartz rocks, so when dropped on a rotating plate, the talc rocks will quickly slide off the plate, whereas the dolomite or quartz rocks will be carried around and removed by a scraper. This type of sorting is particularly useful for high-grade talc that occurs in narrow (<1 m) veins in a carbonate host rock.

Froth Flotation

Talc concentration by froth flotation is usable for almost any talc ore, but it is particularly useful for ores where the talc is intercalated with the waste, such as talc ores of ultramafic origin (talc–carbonate ores). Ore is crushed and then milled to liberation size, typically about 50% passing 325 mesh. It is slurried in water at about 25% solids, and a frother, such as methyl isobutyl carbinol, is added at very low levels (<100 ppm). It is then passed through two to five stages of conventional flotation, where the talc content is upgraded to >95% purity. Carbonates, quartz, chlorite, and serpentine are selectively removed by this process. Most operations use traditional agitated float cell technology; flotation columns have not found extensive application, even when they are used as cleaner devices.

After flotation, the flotation concentrate is usually gravity thickened and then filtered to dewater to <20% moisture on a rotary vacuum filter. A Finnish producer uses a continuous Larox belt press to minimize the residual moisture. The cake is then flash-dried with hot air, sometimes in combination with milling. If the cake is too difficult to feed, it is premixed with some dried powder to improve its handling. The drying is usually done in a doughnut-type fluid mill at lower hot air pressure; machines such as the Bepex Pulvacron mill or the Hosokawa pin mill can also be used, as can more sophisticated fluid energy mills.

Wet High-Intensity Magnetic Separators

If a problematic level of dark, magnetic minerals is still present, WHIMS can be used to remove these from the floated product. Even stronger cryogenic magnets developed for the kaolin industry are being used in Finland. By using flotation and WHIMS together, it is possible to produce a higher-quality talc concentrate but also to use lower-quality talc ores that normally would be rejected as waste.

Leaching and Acid Washing

Leaching of floated talc for removal of yellow iron stains found on the surface of the talc particles was practiced in the past in the United States as a way to further increase the whiteness of floated talc concentrates. The leached talc products were mainly used in cosmetics. However, as the cosmetic talc market became smaller, this costly and environmentally problematic technology was eliminated and the talc requirements for cosmetic applications were replaced by very white, double-sorted, imported talc ores.

Acid washing to remove both carbonate and sulfide impurities for products used in chewing gum or pharmaceuticals was also used, but this has also been replaced by more environmentally acceptable separation technology such as slide sorting.

FINISHING STAGE

This stage includes milling the talc concentrate, additional magnetic impurities removal, sizing, compacting into pellets or powder densification, and packaging and shipping of finished products. The flow scheme of a typical multiproduct talc plant is shown in Figure 3.

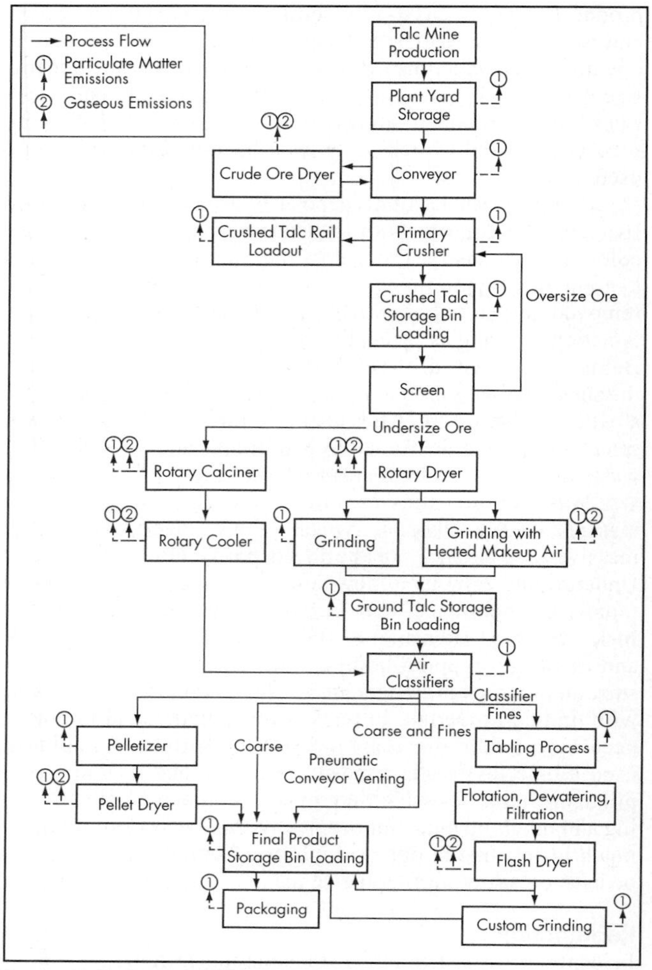

Source: McCarthy et al. 2006

Figure 3 Typical flowchart of a talc operation manufacturing multiple products

Analytical Techniques and Measurements Nomenclature

In describing the physical and chemical properties of milled products, a substantial number of analytical techniques and measurements with a unique nomenclature are used. In this section, the techniques and measurements are described and the nomenclature is defined.

Among the most notable measurements are the median particle size and the top size. Median particle size is most often measured by sedimentation techniques using an instrument called a sedigraph. The sedimentation technique assumes that all particles are spherical; talc particles are not. A sedimenting talc platelet will behave like a much smaller sphere, skewing the reported distribution toward the fine end. The technique is generally used only for subsieve particles (100% finer than 100 mesh [150 µm]). The technique will give a full distribution and is accurate down to approximately 0.4 µm. The other major subsieve technique is laser particle size analysis with instruments made by several companies, including Malvern and Horiba. Unlike the sedimentation technique, the laser analyzer measures the long dimension of the talc platelet, and thus the distribution reported by the laser will be different and larger than that reported by the sedigraph.

For top-size measurements, sieving and Hegman drawdowns are used. Sieves from 40 to 325 mesh are common. The Hegman bar, also called the fineness-of-grind gauge, is a technique borrowed from the paint industry. A small quantity of talc powder is dispersed under high shear in linseed oil. A small portion of the talc in oil dispersion is placed on the Hegman bar, which has a very accurately machined groove with a depth of 100 µm at the top end and 0 µm at the bottom. When the talc in oil dispersion is drawn down through the groove, the coarsest particles will show somewhere in between, and the material is given a Hegman rating based on the largest particles in the sample. For instance, particles of 70 µm would give a 3 Hegman and particles of 25 µm would give a 6 Hegman.

Other critical properties that are measured include surface area, bulk density, and color. Surface area is measured by the Brunauer, Emmett, and Teller (BET) method. It consists of adsorbing a monolayer of nitrogen on the surface of the sample and weighing it. Values vary from 1 m^2/g for the coarsest macrocrystalline products to 20 m^2/g for the finest microcrystalline products. Bulk density is reported as either loose bulk density (LBD) or packed bulk density in grams per cubic centimeter. LBD is measured in a very simple piece of equipment called a Scott volumeter in which the powder is cascaded over a series of glass plates into a 1-in. cube. After the cube fills, it is smoothed off and weighed. LBD values range from 1 to 1.2 g/cc for coarse roofing products to <0.15 g/cc for the finest paint and plastic grades. LBD strongly correlates with median particle size and in many plants is used as a primary control parameter because it is quicker to run than a median particle size.

The science of color measurement as it affects talc is quite complex and constantly evolving. Most producers measure powder color, which is achieved by pressing the powder into a pellet at standard conditions (3 bar pressure) and then measuring the color and brightness on the surface. A wide variety of measurement systems are available. The most commonly used today is the Y reflectance system, whereas the General Electric brightness (GEB) method is used only in the paper industry. Y reflectance and GEB are quite close in value and are measured on a simple scale of 0 to 100 on a brightness tester. Products vary from about 60 for coarse products with low talc content to 96 for the finest, whitest talc products. Almost always, as talc products are milled more finely, dry brightness values increase.

The CIELAB scale measures color on three scales: *L* ranges from 0 to 100 on a white or black scale; *a* ranges from red to green on a similar scale; and *b* ranges from blue to yellow. A typical talc product with a GEB of 80 might have an *L* value of 86, an *a* value of 0.5, and a *b* value of 3. The *b* value, sometimes called *tint*, is the most critical, as it is often reflective of iron staining. A lower *b* value is almost always better.

The powder brightness does not always reflect how the product performs in an end use. In plastics and bar soaps, for instance, the brightness of talc is typically lowered significantly as the talc surface is wet out by resin, which changes the optical image. To measure this, the powder is wet out in mineral oil and the color is measured on the paste.

The other major analyses involve the mineralogy and chemistry of the ore and products. In its simplest form, talc mineralogy is estimated from loss on ignition (LOI) data. The pure mineral talc loses its (4.75%) water of crystallization at 960°C. Thus, the closer the 1,000°C LOI is to this, the purer the talc. If the major impurity such as dolomite or magnesite is known, the purity can be accurately estimated. Mineralogy is measured more accurately by X-ray diffraction (XRD) techniques. An X-ray beam is sent through a pressed sample of the powder, and the beam is scattered at a specific angle by each mineral present. The location and intensity of that scattered beam are characteristic of the mineral and its level in the sample. Talc, chlorite, carbonates, amphiboles, and quartz are readily identified and quantified down to concentrations of approximately 0.1% (1,000 ppm) using this technique. Asbestiform minerals can be measured down to <10 ppm sensitivity using transmission electron microscope analysis (Pier 2016).

A method to quantify the crystalline nature of talc was proposed by Holland and Murtagh (2000). They measured the intensity of the 004 and 020 XRD peaks and combined them in the following equation:

$$MI = I_{004}/(I_{004} + 2I_{020}) \quad \textbf{(EQ 1)}$$

where

MI = morphology index

I = beam intensity

Their MI varied from 0.98 for the more macrocrystalline and platy Ontario talc to 0.45 for the microcrystalline Montana ores.

Chemistry is determined by either X-ray fluorescence (XRF) techniques or by more detailed wet chemistry techniques, such as inductively coupled plasma (ICP) or atomic adsorption (AA). XRF is done on a pressed powder sample and is accurate for silica, magnesium, calcium, iron, and aluminum down to perhaps 0.2%. The wet chemistry techniques are much more accurate but require that the samples be digested in a very strong acid, such as aqua regia or hydrofluoric acid, before they are run through the ICP or AA instruments. This makes the technique much more expensive and time-consuming.

Talc Milling

Talc is an extremely soft mineral and generally very easy to mill or to reduce in particle size, although the aspect ratio or platy nature of the talc can complicate matters. Preserving that aspect ratio is difficult in some milling processes, and a fine talc platelet is very difficult to handle and "manage" in an airstream, where almost all talc milling occurs. Very little wet milling is used. But it is slowly becoming more popular, as it is one of a few milling techniques that can increase the aspect ratio of the talc being milled. If the ore is wet, the process will typically start with drying. Direct-fired rotary dryers are the most common. The coarsest talc products can be produced by crushers run in closed circuit with vibrating screens. Roofing products are produced in this fashion by running crusher discharge over a 35-mesh screen.

A tiny fraction of talc is sold in lump form for carving, machining, or decorative applications. A significant volume of Chinese and South Asian production is exported in lump form to the consuming countries. However, the vast majority of the ore must be reduced to a powder for end use. The size needed ranges all the way from −10 mesh (top size of 2.5 mm or 2,500 µm) to 1,250 mesh (top size of 10 µm). Obviously, in a milled talc, all the particles are not the same size; they have a distribution of sizes—sometimes called the granulometry—which also has end-use importance because some applications require a narrow particle size distribution and some require a broad distribution. Finally, the particle shape is critical in some applications and is affected by the grinding method as well.

The most common milling machine in the talc industry is the ring roller mill, with the Raymond roller mill being the most popular type (Figure 4). In this mill, ore up to 50 mm in size is swept into a grinding zone between a set of rotating rolls that grind the talc between them and a horizontal ring. The fine product is swept out of the milling zone up to an air separation zone. Here, a set of horizontal blades, called a whizzer, creates an airflow in a horizontal direction to the sweep air and forces the coarse particles out to the side and back down the circumference of the cylinder into the grinding zone. The fine product continues with the sweep air and is removed by centrifugal action in a cyclone. The air is recirculated with a large fan back into the grinding zone. A slipstream of air is pulled off after the fan into a dust collector. This keeps the whole system at a slight negative pressure and removes particles that are too fine for the cyclone to capture.

Roller mills are effective devices for milling softer (up to 3.5 Mohs hardness) materials down to 325 mesh top size (44 µm). They appear to preserve the aspect ratio and, if heat is applied, have the capability to do limited drying of wet feed. The primary control mechanism on the mill is the whizzer speed, with a higher speed giving a finer top size and a finer median particle size.

Newer mills use the more sophisticated cage classifiers in place of the whizzer, which is not an efficient separation device. Cage classifiers require a higher pressure drop and usually a bigger fan but will enhance both throughput and

Courtesy of Arvos Raymond Bartlett Snow LLC

Figure 4 Raymond roller mill

top-size control. Products as fine as 5 Hegman can be made with enhanced classifier systems.

Producers use ball mills in closed circuit with air classifiers, especially for harder chloritic talc or ore with a high carbonate content. In South Asia, it is common to use less expensive hammer mills for coarse milling, sometimes in open circuit with air classification.

To mill from a 325 mesh to a 6 Hegman, the most efficient machines are hammer mills with built-in classifiers, such as air classifying mills (ACMs). The ACM is a compact hammer mill with an integral, highly efficient cage classifier. The hammers are mounted on a plate in a cylindrical housing, and the rotation forces the particles to impact a serrated liner. They are then air-swept into the cage classifier zone, where the coarse particles are knocked back and the fine particles are sent on to collection in a baghouse. The ACM can accept feed up to approximately 6 mm. The primary control mechanism in the ACM is classifier speed, with a higher speed giving a finer top size and a smaller median particle size.

For ultrafine grinding, often called micronizing, fluid energy mills, alternatively called jet mills, are used. In this process, high-pressure (>10 bar) air or steam is expanded through nozzles into a grinding chamber. In some of these, the grinding chamber is a doughnut-shaped system, and the high-speed air flowing around the chamber in a centrifugal motion causes the powder to grind against itself. In the Aljet type, the fine product is taken off on a slipstream on the inside of the doughnut. In the Alpine type, the grinding is done by streams of opposing nozzles that inject pressurized air and powder into the base of a cylinder (Figure 5). The fines are taken by the milling air into the classification zone on the top of the cylinder where multiple cage classifiers reject coarse particles back into the mill. Product from all these systems is collected in baghouse filters, with care being taken to ensure that any cooling does not approach the dew point of the steam effluent.

Fluid energy mills are effective for milling from approximately 4 down to 7 Hegman top-size products. At the same top size, they will give a much finer median particle size than the ACM. Furthermore, centrifugal-type mills are believed to be better delaminating devices than the ACM.

The primary control mechanism in fluid mills is the integral or external classifier speed, with a higher speed giving a finer product. For mills without a classifier, the product can be made finer with a higher fluid-to-talc ratio or a higher fluid pressure.

Mills are normally hardened steel construction, but grinding zones can be lined with urethane, ceramic, or ultra-high molecular-weight polyethylene for critical end uses (such as pharmaceuticals) where contamination, wear, and minimal color loss are important.

Size Classification of Talc Powders

Screens are excellent absolute-size separation devices, but their capacity and efficiency decline rapidly with finer meshes, and they are subject to rapid blinding and high maintenance

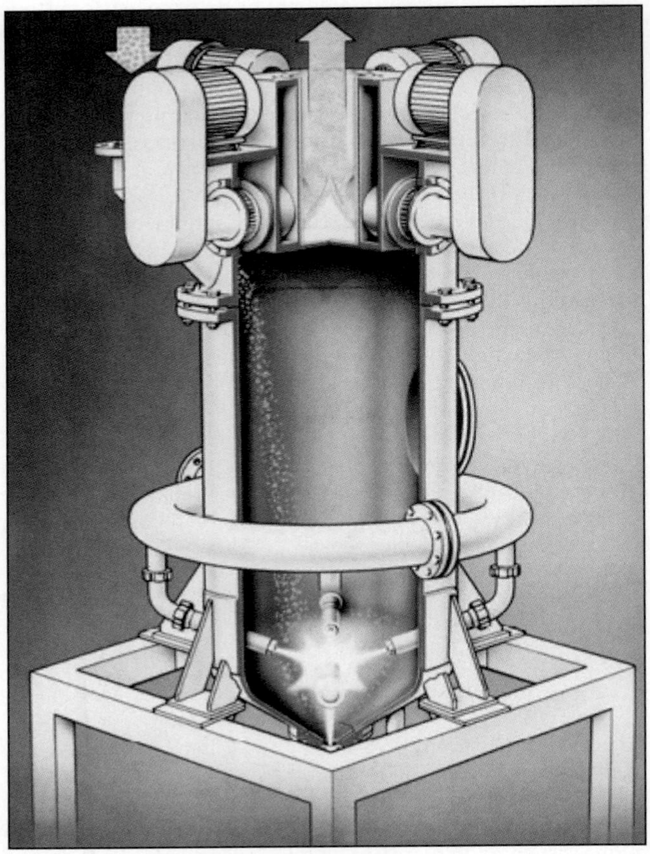

Courtesy of Hosokawa Micron Corporation

Figure 5 Hosokawa jet mill

with fine powders, platy materials, and agglomerates. In general, conventional screens are effective for separation of talc powders down to about 60 mesh. Below this, the powders agglomerate and blind the screens. For scalping-type applications (<0.5% oversize), air-swept screens can be used for powders in sizes down to 100 μm.

Air classifiers are devices in which a stream of air that contains fine particles (usually <100 mesh) is subjected to a velocity perpendicular to the direction of flow. This will cause the coarse particle to be deflected, allowing the fine particles to pass through. Two types of devices are used to create the perpendicular flow: blades (sometimes called whizzers) and cages that have a fan-like design. The latter are much more efficient, especially at fine particle sizes, but have higher pressure drop, lower capacity, and more potential problems. Typically, the flow is upward, and the whizzer is mounted vertically. When the separation points get finer (<20 μm), however, it is often advantageous to mount the unit horizontally.

Classifiers do not make a clean cut, and a platy particle behaves like a much smaller sphere. In addition, there are problems with bypassing, dispersion, and airflow control. In the most sophisticated units, all the accepted (fine) particles are forced through the vanes of the cage and exit through a hollow shaft on which the cage rotates.

The primary control parameter is the speed of the whizzer or cage, but other variables such as airflow, the air-to-solids ratio, and particle surface energy have significant effects. For any classifier, the higher the air-to-solids ratio, the better the classifier will work. But the operating cost is higher and the capacity is proportionally lower. Higher surface energy will have a detrimental effect in encouraging agglomeration, and this energy is an inevitable by-product of the milling process.

Classified products with fines removed are essential for use in solvent-based coatings today. All of these coatings have reduced levels of solvents to comply with volatile organic carbon emission regulations; thus, a pigment used in the formulation must minimize the viscosity build, and that is best done with talc by removing its fine particles.

Densification and Compaction

The fluffy nature and low bulk density of fine talc products make packaging and shipping difficult and expensive. To overcome this, producers use deaeration and agglomeration technology to increase the bulk density of fine products.

The standard deaeration technology uses a horizontal screw densifier to pull air out of powders. The powder is transferred through a horizontal screw that is surrounded by a filter cloth. A vacuum is pulled on the outside of the cloth, increasing the bulk density of a fine product by more than 100%, from 0.15 to 0.3 g/cc. The process has little effect on the talc itself, as the powder reaerates with minimal agitation. It does allow the product to be packaged in smaller sacks or to get 100% more talc in a regular sack. Such machines are manufactured by Carman in the United States, Alexanderwerk in Europe, and Kurimoto in Japan.

In compaction, the fine talc is wetted with about 20% water in a continuous mixer, and this paste is forced through a circular die to form pellets. Kahl and California Pellet mills are the most commonly used; some pellets are shipped wet, but most pellets are dried on a continuous tray or vibrating conveyor dryers. Typically, approximately 2% moisture is preferred for paper applications and <0.5% moisture is required for plastic and rubber. Bulk density is increased from 0.15 to 1 g/cc for fine products.

Dry compaction equipment is also available from Kurimoto, Bepex, and Hosokawa. In dry compaction, the fine powder is forced between two rolls and subjected to a high local pressure that agglomerates the fine particles. This process has the potential of making the agglomerates too hard to redisperse. Uetake (2002) patented a process that controlled the density between 0.7 and 0.9 g/cc to optimize density improvement and dispersion in polymer.

Densified and compacted products are now the standard product form in the plastics, paper, and rubber markets. Along with savings in transportation, the compacted grades are easier and faster to compound into polymers and disperse into water.

Packaging and Shipping

Talc products are shipped mainly in super sacks (400–1,200 kg) and, to an increasingly lesser degree, in paper bags (usually 25 kg) stacked on wooden pallets. Nowadays, packaging systems are increasingly automated and use robotics.

Transportation to the customer is a significant portion of the total cost of talc products, averaging 25% to 30% of the cost and, in some cases, more than 50%. Most talc products are now shipped by truck. The combination of faster, more reliable delivery and lower costs for most customers continues to favor this mode. Rail is used for larger-volume bulk customers when the haul distance is more than 500 km. Most trucks will deliver coast to coast in 4 or 5 days, whereas rail delivery will typically take 2 to 3 weeks. Bulk trucks are popular with customers because the truck driver unloads the material with a compressor on the truck, in contrast to bulk railcars where the customer has to supply the compressor and unload the material. Most product is shipped directly to customers. Smaller (less than truckload) customers typically purchase through chemical distributors who purchase from the producer and warehouse talc products locally.

Other Processing Technologies

Wet Milling

Wet milling in stirred media mills is a patented technology by Fourty et al. (2002) for the manufacture of highly delaminated talc products. Water is the liquid phase that is most commonly used. The milling media are typically small steel or ceramic balls with a narrow size distribution in the 1- to 5-mm range.

The main control variables in stirred media mills are the type of talc feed; the talc-to-liquid-media ratio; the type, shape, size, and amount of the milling media; and the stirring speed, stirring energy input, and milling time. After milling, the wet slurry is screened to remove oversize and broken media and then spray-dried; alternatively, it can be centrifuged or filtered to dewater and then flash-dried. Highly delaminated talc products have had commercial success in the marketplace being used for improved reinforcing in thermoplastic olefins and for reduced permeability in rubber tire inner liners.

Submicrometer or nanosized talc products can be manufactured with a dry-stirred media mill, followed by wetting the talc with water to make products with a BET surface area in the range of 300–600 m^2/g. Although the process has been patented, the products have not been commercially successful, as the application performance is quite poor in markets such as plastics and coatings. There are also patents (Martin et al. 2012) issued on a process to make synthetic nano-talc via the high-pressure, moderate-temperature, water-based reaction of sodium silicate with magnesium chloride.

Surface Treatments

A limited amount of talc is surface treated to improve its performance in specific applications such as plastics, rubber, and cosmetics. Aminosilane, polyethylene glycol, and waxes are some of the materials used. In general, the level of coating on talc products is relatively low—in most cases, less than 1%. This is in contrast to carbonates, where stearate coatings reach 3 wt % for ultrafine products.

A wide variety of equipment is used to apply the coatings. When the volume is small, as it typically is in cosmetic applications, batch technology is used. Horizontal drum mixers with slow macromixing and high-speed micromixing are common, as are V blenders, Nauta-type mixers, and rotating drums. For larger-volume applications, continuous mixers such as Littleford or Hosokawa turbulizers are used. They provide a modest (5–30 s) residence time along with high-shear mixing. The additive is usually added via a spray nozzle.

Talc for use in cosmetic, food, or pharmaceutical applications must be bacterially controlled to <100 cfu/g. This is generally done in a heat-treating circuit, where the talc is heated to >170°C for 1 minute; this can be done in an air-conveyed system or in a conveying heat exchanger.

Talc Slurries

A major development, especially in northern Europe, is the delivery of fine talc to paper mills in slurry form for use in paper coating. Fine talc, which has low surface energy and hydrophobicity and a platy structure, is difficult to wet in water and extremely difficult to make into a stable, high-solids slurry. Nevertheless, close to 250,000 t/a was shipped in this form in Europe, but the paper coating market has been in decline. The talc is dry-milled and then made into slurry using high-energy mixers and a proprietary package of dispersants and stabilizers. The slurries have a solids content of 61% to 65% and a viscosity of approximately 200 cP. A limited amount of talc is also shipped for pitch-control applications in slurry form, but the solids content is much lower and no chemicals are used.

Other Innovations

There are applications where some of the properties of talc are detrimental (such as the viscosity build in solvent-based coatings). To address this, producers in Europe and the United States are selling blends of talc and calcium carbonate for the coatings and putty industries, and another producer is selling blends of talc and feldspar for film antiblock.

CONCLUSION

Talc producers are well aware of the changes in the end uses of talc as well as the significant reduction of high-quality talc ore reserve. Consequently, they are investing in mineral processing technologies that will allow them to manufacture value-added talc products from lower grades of ores. These new products will have to comply with very strict product specifications and provide the expected functional performance to the end user.

REFERENCES

Bergmann, J.-M. 2014. Innovative water free mineral processing. Presented at the 22nd Industrial Mineral Conference, Vancouver, BC, Canada.

Fourty, G., Jouffret, F., and Monnot, P. 2002. Lamellar filler process for the treatment of polymers. U.S. Patent 6,348,536.

Holland, H.J., and Murtagh, M. 2000. An XRD morphology index for talcs: The effect of particle size and morphology on the specific surface area. *Adv. X-Ray Anal.* 42:421–428.

Martin, F., Ferret, J., Lebre, C., Petit, S., Granby, O., Bonino, J.P., Arseguel, D., Decarreau, A., and Ferrage, E. 2012. Method for preparing a synthetic talc composition from kerolite composition. U.S. Patent 8,202,901.

McCarthy, E.F., Reade, E., and Genco, N.A. 2006. Talc. In *Industrial Minerals and Rocks*, 7th ed. Edited by J.E. Kogel, N.C. Trivedi, J.M. Barker, and S.T. Krukowski. Littleton, CO: SME.

Nichols, C.W., Lorang, M.J., Wold, M.O., and Rayfield, J.W. 1991. Method and apparatus for friction sorting of particulate materials. U.S. Patent 5,069,346.

Pier, J. 2016. Development of a new ASTM method to detect asbestiform minerals in talc. Presented at the 2016 SME Annual Conference and Expo, Phoenix, February 21–24.

Uetake, T. 2002. Manufacturing methods of polypropylene resin composition with talc content. Japanese Patent 3,241,746.

Yan, L.A., Masliyah, J.H., and Xu, Z. 2011. Determination of anisotropic surface characteristics of different phyllosilicates by direct force measurements. *Langmuir* 27(21):12996–13007.

Yordan, J., Kortmeyer, J., Yildirim, I., and Yoon, R.H. 2002. Studies of the interactions between talc and stickies during paper recycling. SME Preprint No. 02-152. Littleton, CO: SME.

CHAPTER 12.37

Titanium Dioxide Pigment

Thomas P. Battle

The dominant color used in coatings around the world is white. Titanium dioxide white pigment alone was 69% of all inorganic pigment tonnage in 2000 (Buxbaum and Pfaff 2005). This is not only because of the many objects that actually need to be painted white, but part of the value of a quality white pigment is being an opacifier—coating whatever is below such that it can no longer be seen. In fact, the major cost differential in paints that are advertised as needing only one coat is because of the amount and quality of the white pigment used to increase the opacity. Specific advantages of titanium dioxide (TiO_2) pigment are (Elsner 2010)

- High refractive index of 2.55 or 2.70, for high opacity (covering power);
- Good reflectivity, resulting in luminosity and whiteness;
- Inertness and chemical stability (insoluble in acids, bases, organic solvents, and air pollutants);
- High resistance to ultraviolet (UV) light degradation (color retention);
- Nontoxicity; and
- High thermal stability.

It is the "strongest, most brilliant white available to artists in the entire history of art. Its chemical stability is likewise outstanding" (Douma 2008).

White pigments have been used for centuries, but use of TiO_2 in this context is comparatively recent. The white characteristics of TiO_2 were discovered as early as 1821, but it was another 100 years before an economically viable technology was developed to convert a low-grade ilmenite ore to a quality pigment that could compete with white lead, zinc white, and lithopone, a compound of zinc sulfide and barium sulfate (Douma 2008). Lithopone had been developed for use in pigmentary applications in the 1870s and had grown to a total production of more than 180,000 t/yr (metric tons per year) in the United States in 1928 (Hounshell and Smith 1988). By that time, however, at least two companies had developed the technology to make quality TiO_2 pigment. A U.S. company subsequently bought the rights to the sulfate process developed in Norway (Korneliussen et al. 2000) and commercialized it in 1920 under the ownership of the National Lead Corporation (Hounshell and Smith 1988; BNP Media 2004).

The development of cost-effective TiO_2 pigments was monitored by the major U.S. producer of lithopone, DuPont, but no concern was expressed until familiarity with the sulfate process allowed National Lead to significantly lower the cost of TiO_2 in 1929. When DuPont realized that the lithopone market might be overtaken by TiO_2 (lithopone sales dropped 20% in 1930), it began a frantic program to develop its own version of the technology (Hounshell and Smith 1988). DuPont soon realized that this was unlikely in the time frame of concern, and it eventually purchased Commercial Pigments Corporation, which owned some patents in the field, and signed a patent- and process-sharing agreement with National Lead. As the market grew, the first DuPont sulfate plant, Edge Moor, was built in Delaware (United States).

The market for pigment continued to grow, and DuPont continued to look for ways to differentiate its technology from the competition. First, DuPont discovered the added 20% hiding power of rutile TiO_2 pigment (compared to the standard anatase) and a cost-effective way to make rutile pigment from the sulfate process, which makes anatase by default. Then DuPont started developing an entirely new production technology, using chlorine gas instead of sulfuric acid liquid as the main reactant. The chloride process was commercialized in the early 1950s, again at Edge Moor. This technology has subsequently spread over the world.

Figure 1 shows the evolution in production of TiO_2 worldwide, by technology, over the past 45 years. Sulfate technology dominated until about 1990, when chloride took the lead, which it held until about 2010, when the large number of new plants in China (virtually all sulfate) led to sulfate pigment taking the lead again (Iluka 2014).

Using the new chloride technology, DuPont was able to regain its share of the U.S. market, but it took close to 50 years to get back to the level of dominance (37%) it had held in the lithopone market. (In 2015, DuPont spun off its performance chemicals division and the new company was rebranded as Chemours [Sanati 2016]; for simplicity, the DuPont name will

Thomas P. Battle, Extractive Metallurgy Consultant, Self-Employed, Charlotte, North Carolina, USA

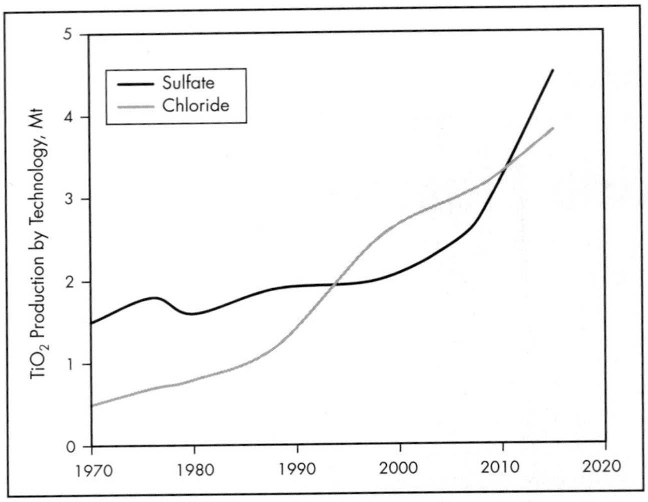

Adapted from Adams 2015

Figure 1 Titanium dioxide production by technology

continue to be used for the rest of this chapter.) DuPont is still the largest individual producer of TiO_2, although the cumulative production in China now exceeds DuPont's production, as shown in Table 1 (Iluka 2014) and Figure 2 (USGS 1996–2017). The dominance of production in the United States over this period is overshadowed by the immense growth in China. Only minor changes in production occurred in the other four major TiO_2-producing countries: Germany, Japan, United Kingdom, and Australia.

Whereas the sulfate process operates by liquid–solid interactions of titanium ores with various sulfuric acid or aqueous solutions, the chloride process operates by gas–solid interactions of the titanium ores with chlorine. The two processes differ in the method of separation of titanium from impurities. In the chloride process, gaseous metallic species are separated because of their differing melting and boiling points; in sulfate liquid-phase processing, separation occurs through the differing solubilities of metal sulfate species with temperature and pH.

In the following sections, after a brief consideration of feedstocks, the sulfate and chloride routes to make TiO_2 are discussed. This is followed by a brief consideration of TiO_2 pigment properties, applications, and future prospects.

FEEDSTOCKS

Titanium ores and mineralogy are discussed in detail in Chapter 12.38, "Titanium Ores." Because the most common titanium-rich ore is ilmenite (nominally 50% TiO_2), rather than rutile (90%–97% TiO_2), an entire industry has developed to upgrade ilmenite to a higher-grade product, either titanium slag or synthetic rutile. Because the sulfate and chloride processes have different requirements for their feeds, a differentiation has been made between *sulfatable* and *chlorinatable* ilmenites and similarly with titanium slag. The feeds available to a pigment producer are thus wide and varied. In 2006, of the three feedstock classes (rutiles, slags, and ilmenite), the largest fraction was the titanium slags (990,000 t chloride slag; 780,000 t sulfate; 170,000 t upgraded slag), followed almost equally by the ilmenites and natural and synthetic rutile.

Although, theoretically, a given pigment plant can use many of the options shown in Table 6 of Chapter 12.38, "Titanium Ores," in widely varying fractions, this is not really true in practice (McCoy et al. 2011):

> Significantly changing a feedstock, or blend of feedstocks, is not a short-term option for most pigment producers. Once a pigment plant has been optimised for a known feedstock blend, slight variations can have a dramatic effect on the plant performance and final pigment quality.

Table 1 Major producers of TiO_2 pigment, 2011

Company	Percentage of 5.6 Mt
DuPont	19
Cristal	13
Huntsman	9
Kronos	9
Tronox	6
Sachtleben (has since been purchased by Huntsman)	4
Ishihara	3
Other	37

Adapted from Iluka 2014

More importantly, most of the DuPont TiO_2 plants were built to accommodate use of at least some fraction of ilmenite ores, rather than the high-grade ores of its competitors. Because much of the upstream equipment in a pigment plant must be sized to handle the impurities in the ore, not just the titanium fraction, the DuPont plants are much larger in those areas than competitors' plants; and even if the competitors wanted to switch to lower-grade feedstocks, their overall production rate would suffer (Minkler and Baroch 1981; Charlier et al. 2010).

That said, as in other industries, having flexibility in feedstock capabilities is a desirable attribute. The company or plant with this flexibility can change ore blends or particular ores over a wide range, which can allow them to take advantage of short-term opportunities, increase or decrease production rates while still utilizing plant facilities efficiently, and so on (McCoy et al. 2011; Iluka 2014). Many TiO_2 producers have some level of vertical integration, including Cristal, Tronox, DuPont, Pangang, and some other Chinese producers.

SULFATE PROCESS

In the sulfate process, the titanium ore is reacted with concentrated sulfuric acid to form a solution rich in metal sulfates. The main reaction (as defined for a pure ilmenite feed) is given by Equation 1 (Elsner 2010; Buxbaum and Pfaff 2005); a process flow sheet can be seen in Figure 3 (McNulty 2008). This reaction generally takes place in batch mode in a digester that is fed ore sized finer than 40 μm and sulfuric acid of a strength between 80% and 95%. During the course of the reaction, the liquor heats up to 200°C or more during the 1- to 12-hour reaction time (Buxbaum and Pfaff 2005).

$$FeTiO_3 + 2H_2SO_4 \rightarrow FeSO_4 + TiOSO_4 + 2H_2O \qquad \textbf{(EQ 1)}$$

The cake produced in the digester is often redigested in water or recycled weak acid; if any ferric ion is present, it is converted to ferrous, using scrap iron, as ferric sulfate species are very difficult to remove from the titanium product (Buxbaum and Pfaff 2005). With digestion complete, unreacted solids are

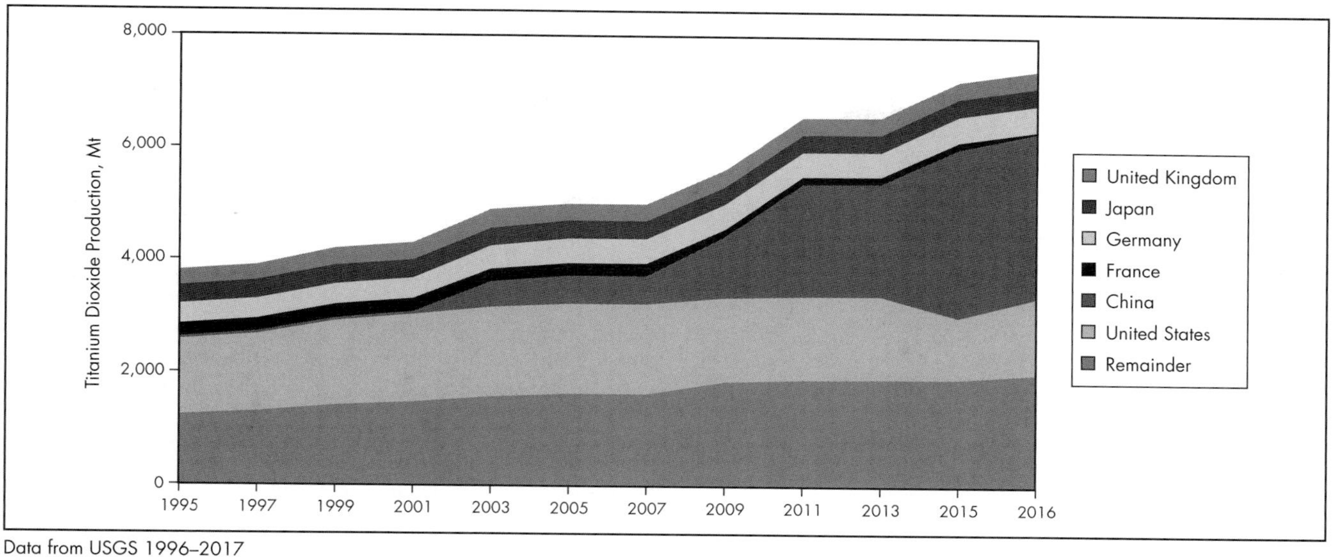

Data from USGS 1996–2017

Figure 2 Major producers of titanium dioxide pigment, by country

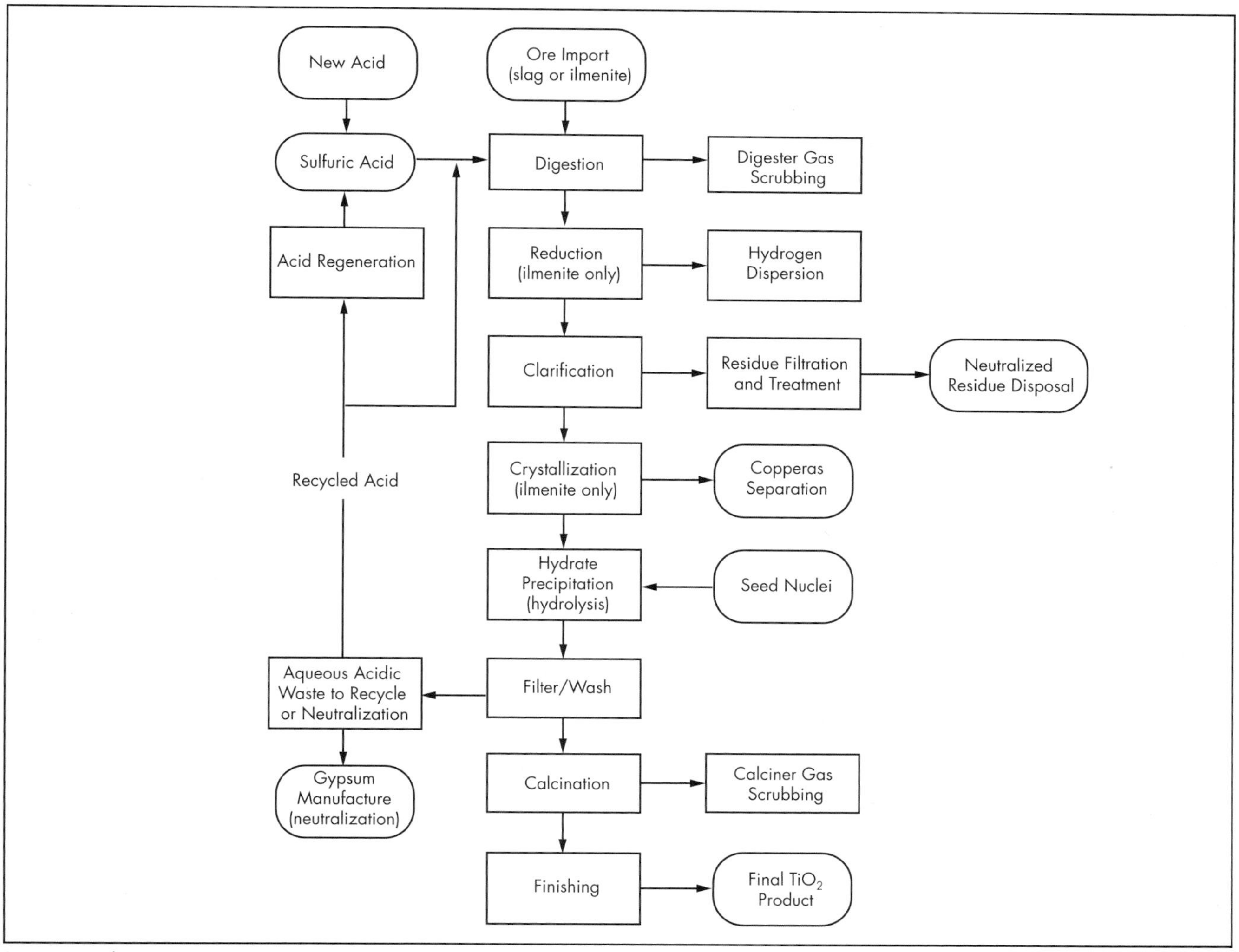

Source: McNulty 2008

Figure 3 Titanium dioxide (TiO_2) sulfate process

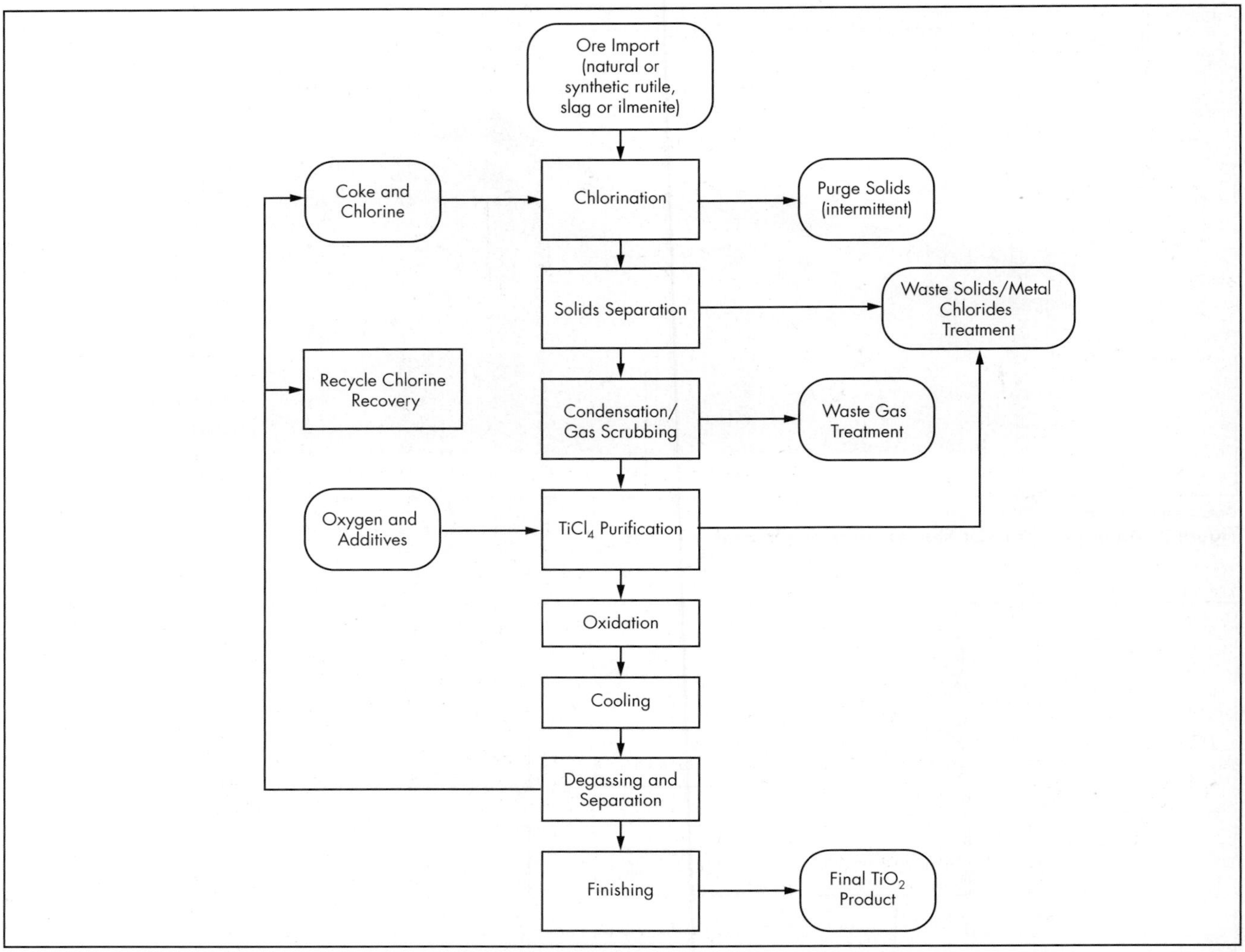

Source: McNulty 2008

Figure 4 Titanium dioxide chloride process

removed using a clarifier and a filter press. After this, the temperature in the bath is lowered to crystallize much of the iron as copperas, $FeSO_4{\cdot}7H_2O$. Steam is then added to bring the titanium out of solution as an oxyhydroxide (Equation 2). This titanium intermediate is often doped with additives (depending on the exact pigment grade to be produced) and then calcined in a rotary kiln at 800°–1,100°C for 7–20 hours. The final product is ground and coated as necessary.

$$TiOSO_4 + 2H_2O \rightarrow TiO(OH)_2 + H_2SO_4 \quad \textbf{(EQ 2)}$$

Depending on the titanium source used, sulfuric acid consumption in the sulfate process is 2.4–3.5 t of concentrated acid per metric ton of TiO_2 produced (Buxbaum and Pfaff 2005). In the early days of the technology, this was virtually all lost, as copperas was landfilled and the weak acid product (with impurity salts) was dumped into rivers or the ocean (Elsner 2010; Buxbaum and Pfaff 2005; Battle et al. 1993). Since then, regulations have been tightened, and producers have altered their processes to convert former waste products to coproducts or stable wastes that can be landfilled or disposed of as liquids. In some cases, copperas and ferrous sulfate monohydrate products are produced, with much of the recovered sulfuric acid strong enough to be used as recycle in the process. The remaining weak acid is neutralized with lime or limestone to make a mixture of calcium sulfate, iron, and calcium oxides known as *red gypsum* (Gázquez et al. 2014). Titanium recovery from feed to product is typically 80% (Henry et al. 1987).

CHLORIDE PROCESS

A typical flow sheet for the chloride process can be seen in Figure 4. The core reactor is a fluidized bed. Fine titanium ore and coke are injected into the middle of a large refractory-lined vessel that operates near 1,000°C. These particles are fluidized by chlorine in the presence of coke (major reactions given as Equations 3 through 5) and immediately begin reacting. Most of the species in the ore will form gaseous metal chlorides at these temperatures, predominantly titanium tetrachloride ($TiCl_4$), ferric chloride ($FeCl_3$), and ferrous chloride ($FeCl_2$); melting and boiling points for species of interest can

Table 2 Phase transition temperatures for relevant chlorides

Species	Melting Point, °C	Boiling Point, °C
$AlCl_3$	193	180
$AsCl_3$	−16	130
$CaCl_2$	775	1,935
HCl	−114	−85
$FeCl_2$	677	1,023
$FeCl_3$	308	316
$MgCl_2$	714	1,412
$MnCl_2$	650	1,190
$NbCl_5$	206	247
PCl_5	167	160
$POCl_3$	1	106
$SbCl_5$	3	92
$SiCl_4$	−70	58
$SnCl_4$	−33	114
$ThCl_4$	770	921
$TiCl_4$	−24	136
VCl_4	−28	149
$VOCl_3$	−79	127
$VOCl_2$	380	Unknown
$ZrCl_4$	437	331

Source: Grant et al. 2004; Helberg 2011; Lynch 2002

be seen in Table 2. These gaseous species (along with some fines carryover) exit the top of the chlorinator into a long piping system. Left behind in the bed are less-reactive species, typically some Si-containing compounds, as well as Zr, U, and Th (Gázquez et al. 2014; McCoy et al. 2011). Because liquids can cause loss of fluidization in this type of reactor, it is clear why operators prefer low levels of species such as Mn, Mg, and Ca in their ore feeds.

$$2TiO_2 + 3C + 4Cl_2 \rightarrow 2TiCl_4 + 2CO + CO_2 \quad \textbf{(EQ 3)}$$

$$FeO + C + Cl_2 \rightarrow FeCl_2 + CO \quad \textbf{(EQ 4)}$$

$$2FeO + 2C + 3Cl_2 \rightarrow 2FeCl_3 + 2CO \quad \textbf{(EQ 5)}$$

Cooling operations begin just downstream of the chlorinator, where recycled liquid $TiCl_4$ begins to cool the hot off-gases toward the $TiCl_4$ condensation temperature, 136°C. Along the way are one or more separation points for blowover solids from the chlorinator and condensed metal chlorides. By the time it is cool enough for $TiCl_4$ to condense, there is little else in the gas but noncondensable species such as HCl, Cl_2, CO_2, N_2, and COS. However, the condensed $TiCl_4$ product contains enough impurity chlorides that it is known as *crude* $TiCl_4$. Crude $TiCl_4$ must be purified to remove species such as arsenic, phosphorus, and tin, but mainly vanadium (V), which would lead to discolored pigment if not reduced to low limits (Lynch 2002). These impurities are removed by the addition of species that preferentially react with them, such as copper, hydrogen sulfide, or fatty acids (Helberg 2011), preferably without removing titanium from the liquid as well. The result is *pure* $TiCl_4$ (Henry et al. 1987). Crude $TiCl_4$ can contain up to 2,000 ppm V (Lynch 2002), with the product often below 5 ppm (Buxbaum and Pfaff 2005). Approximately 1% of the $TiCl_4$ is lost with the vanadium in the fatty acid stream (Lynch 2002).

Pure $TiCl_4$ is reacted with pure oxygen at high temperature to create fine TiO_2, with Cl_2 recycled back to the chlorinator (Equation 6). Different pigment grades can be made by controlling how the oxidation occurs.

$$TiCl_4 + O_2 \rightarrow TiO_2 + 2Cl_2 \quad \textbf{(EQ 6)}$$

The waste produced in the chloride process is less than that of the sulfate process; how much less depends on the ore used. Chlorine consumption per metric ton of product ranges from 0.1 t for some high-grade ores to 0.9 t for ilmenite; impurity production is 0.7 t (mainly iron chlorides) for high-grade ores. TiO_2 recovery is typically around 89% (Henry et al. 1987). Purification yields 0.4–0.9 t of by-products.

FINISHING STEPS

Equation 6 defines the main reaction in the oxidation of titanium tetrachloride, converting it to titanium dioxide and chlorine, which can be recycled. However, pure TiO_2 is not used in most applications, and the particle size of the product is critical. Alterations to pure TiO_2 take place during the oxidation (chloride process) and calcination (sulfate process) steps to prepare the product for the last production stage, often called *finishing*. Conditions are adjusted before finishing to make sure the desired phase structure is present (e.g., aluminum chloride can be added during oxidation to ensure formation of rutile, not anatase, in the chloride process; calcining temperature does the same for sulfate product [Tian 2016]). Finishing operations ensure the development of the desired particle size distribution, which may include fine grinding (micronizing). Additives to coat the surface of the pigment particles can also be included to impart desired physical properties. When processing is complete, TiO_2 is often shipped in bulk bags, typically containing 1 or 2 t. A common product, particularly in the paper industry, is a slurry containing up to 75% TiO_2, which is stabilized to minimize settling, even over significant periods of time (Elsner 2010).

APPLICATIONS

As mentioned earlier, properties of treated TiO_2 pigments include a high refractive index, good thermal stability, resistance to chemical attack, and resistance to degradation by UV light (Gázquez et al. 2014; Elsner 2010). These properties allow for its use in a myriad of applications (as discussed in detail by Iluka 2014; Gázquez et al. 2014). The major application has always been coatings, followed by plastics and paper; sales in these areas trend with a country's gross domestic product. Major uses of titanium dioxide in the United States over time can be seen in Figure 5 (USGS 1975–2006). Minor applications include inks, enamel, and rubber, with lesser applications including "catalysts, ceramics, coated fabrics and textiles, floor coverings, printing ink, and roofing granules" (USGS 1975–2006).

> Approximately 80% of end applications can use either chloride or sulphate pigment. Some of the applications that need chloride pigment specifically are specialty heat treated coatings, including automotive paint and some industrial coatings, while fibres and cosmetic applications need sulphate pigment. (Iluka 2014)

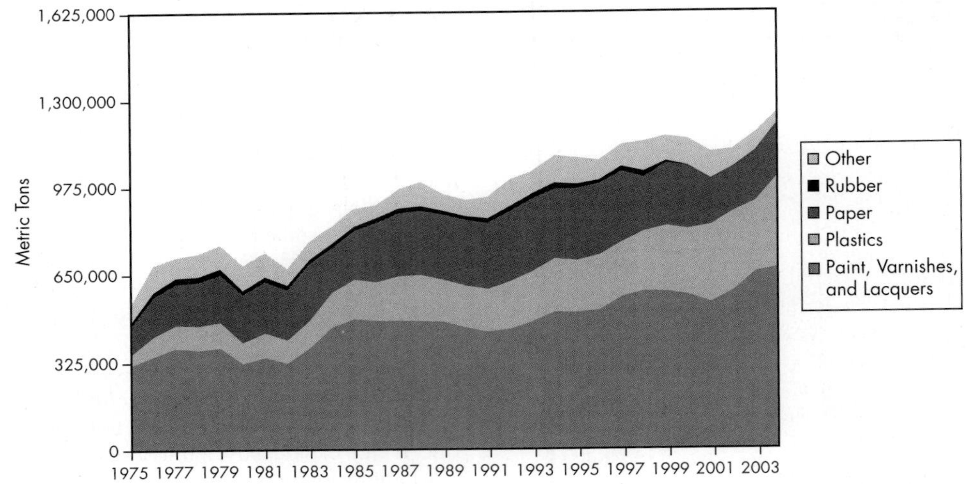

Data from USGS 1975–2006

Figure 5 Major uses of titanium dioxide pigment in the United States, 1975–2005

FUTURE

Work continues on the processing of titanium ores to make high-quality pigment process feeds and for the pigment processes to handle a wider variety of ores, yet still produce quality product with reduced waste streams (Battle et al. 1993). Relatively little research has been conducted on new TiO_2 production technologies. But every few years, a new technology is unveiled, most recently by Argex Titanium (n.d.).

The biggest changes in the industry pertain to the location of pigment plants and the potential markets for it. Figure 2 clearly shows that the center of pigment production has shifted from North America to China, whose pigment quality has been increasing such that many grades are considered roughly equal to the traditional suppliers (Barlow 2015), and new pigment grades continue to be developed (Cappelle and de Backer 2015).

Perhaps the biggest potential change in the industry is the addition of new major applications for TiO_2. This includes possible use in solar cells, water treatment, cancer treatments, cement with self-cleaning properties, catalysis, and air purification (Gázquez et al. 2014). Along with increasing needs in developing nations for paint, plastics, and paper that require white pigment for their functionality, the TiO_2 market is expected to continue expanding.

REFERENCES

Adams, R. 2015. A few more shakes of the TiO_2 kaleidoscope. Presented at Mineral Sands Conference, Melbourne, March 24–25.

Argex Titanium. n.d. Our company. http://argex.ca/en/our-company. Laval, QC.

Barlow, E. 2015. A review of global supply and demand for titanium dioxide. *PCI Magazine: Paint Coat. Ind.* (February 6). www.pcimag.com/articles/100036-a-review-of-global-supply-and-demand-for-titanium-dioxide.

Battle, T.P., Nguyen, D., and Reeves, J.W. 1993. The processing of titanium-containing ores. In *The Paul E. Queneau International Symposium on Extractive Metallurgy of Copper, Nickel and Cobalt*. Vol. I, Fundamental Aspects. Edited by R.G. Reddy and R.N. Weizenbach. Warrendale, PA: Minerals, Metals and Materials Society.

BNP Media. 2004. 90 years with PCI: A retrospective in the 1910s. *PCI Magazine: Paint Coatings Ind.* (January 1). www.pcimag.com/articles/86686-90-years-with-pci-a-retrospective-in-the-1910s. Accessed March 2018.

Buxbaum, G., and Pfaff, G. 2005. *Industrial Inorganic Pigments*, 3rd ed. Weinheim, Germany: Wiley-VCH.

Cappelle, S., and De Backer, S. 2015. Specialised TiO2 grade for flat paints. *Polym. Paint Color J.* (March):42–44.

Charlier, B., Namur, O., Malpas, S., de Marneffe, C., Duchesne, J-C., Vander Auwera, J., and Bolle, O. 2010. Origin of the giant Allard Lake ilmenite ore deposit (Canada) by fractional crystallization, multiple magma pulses and mixing. *Lithos* 117(1-4):119–134.

Douma, M., curator. 2008. Titanium dioxide whites. In Pigments through the Ages. www.webexhibits.org/pigments/indiv/history/tiwhite.html. Accessed March 2018.

Elsner, H. 2010. *Heavy Minerals of Economic Importance: Assessment Manual*. Hannover, Germany: Bundesanstalt für Geowissenschaften und Rohstoffe (BGR) [Federal Institute of Geosciences and Natural Resources].

Gázquez, M.J., Bolívar, J.P., Garcia-Tenorio, R., and Vaca, F. 2014. A review of the production cycle of titanium dioxide pigment. *Mater. Sci. Appl.* 5(7):441–458.

Grant, S., Freer, A.A., Winfield, J.M., Gray, C., Overton, T.L., and Lennon, D. 2004. An undergraduate teaching exercise that explores contemporary issues in the manufacture of titanium dioxide on the industrial scale. *Green Chem.* 6(1):25–32.

Helberg, L. 2011. The purification of titanium tetrachloride: A history. In *EPD Congress 2011*. Edited by S.N. Monteiro, D.E. Verhulst, P.N. Anyalebechi, and J.A. Pomykala. Hoboken, NJ: Wiley.

Henry, J.L., Stephens, W.W., Blue, D.D., and Maysilles, J.H. 1987. *Bureau of Mines Development of Titanium Production Technology*. Bulletin 690. Washington, DC: U.S. Bureau of Mines.

Hounshell, D.A., and Smith, J.K. 1988. Development power: The case of titanium dioxide pigments. In *Science and Corporate Strategy: DuPont R and D, 1902–1980*. Studies in Economic History and Policy: USA in the Twentieth Century. New York: Cambridge University Press.

Iluka. 2014. *Mineral Sands Industry: Fact Book*. Perth, Western Australia. www.iluka.com/docs/default-source/industry-company-information/the-mineral-sands-industry-factbook-(feb-2014).

Korneliussen, A., McEnroe, S.A., Nilsson, L.P., Schiellerup, H., Gautneb, H., Meyer, G.B., and Storseth, L.R. 2000. *An Overview of Titanium Deposits in Norway*. Bulletin 436. Trondheim, Norway: Norges Geologiske Undersøkelse (Geological Survey of Norway [NGU]). p. 27.

Lynch, D.C. 2002. Conversion of $VOCl_3$ to $VOCl_2$ in liquid $TiCl_4$. *Metall. Mater. Trans. B* 33(1):142–146.

McCoy, D., Coetzee, B., Keegel, M., and Bender, E. 2011. Titanium feedstocks—Opaque quality requirements. In *Eighth International Heavy Minerals Conference*. Melbourne, Victoria: Australasian Institute of Mining and Metallurgy.

McNulty, G.S. 2008. Production of titanium dioxide. In *Naturally Occurring Radioactive Material (NORM V): Proceedings of the Fifth International Symposium on Naturally Occurring Radioactive Material*. Vol. 136. Vienna, Austria: International Atomic Energy Agency. pp. 169–188.

Minkler, W.M., and Baroch, E.F. 1981. The production of titanium, zirconium, and hafnium. In *Metallurgical Treatises*. Edited by J.K. Tien and J.F. Elliott. Warrendale, PA: TME-AIME.

Sanati, C. 2016. How DuPont spinoff Chemours came back from the brink. Fortune. (May 18). http://fortune.com/2016/05/18/how-dupont-spinoff-chemours-came-back-from-the-brink/. Accessed May 2018.

Tian, C. 2016. Calcination intensity on rutile white pigment production via short sulfate process. *Dyes Pigm.* 133(October):60–64.

USGS (U.S. Geological Survey). 1975–2006. Titanium dioxide pigment end-use statistics. In *Historical Statistics for Mineral and Material Commodities in the United States*. Compiled by T.D. Kelly and G.R. Matos. Reston, VA: U.S. Geological Survey.

USGS (U.S. Geological Survey). 1996–2017. Titanium and titanium dioxide. In *Mineral Commodity Summaries*. Reston, VA: U.S. Geological Survey.

CHAPTER 12.38

Titanium Ores

Thomas P. Battle

The only chemical elements formed in the early years of the universe were hydrogen and helium, with a trace of lithium. All the other elements were created much later, in the cores of massive stars toward the end of their lifetimes.

Stars, such as the Sun, are powered by the conversion of mass into energy as four hydrogen atoms react to form one helium atom. This requires such high temperatures and pressures that these reactions only occur in the centers of stars and provide enough power to light the star for millions, often billions, of years. For small stars, nucleosynthesis ends here—the core of the star never grows hot enough to fuse helium into larger nuclei. The larger the star, the higher the temperature and pressure of its core; stars with an initial mass around that of the Sun will ultimately reach the conditions required for the next stage of elemental development, the triple alpha process, where three ionized helium atoms (alpha particles) fuse to form carbon 12 (Gribbin 2000). At yet higher temperatures and pressures, alpha particles will react with carbon to form oxygen; for stars with masses similar to the Sun, that is as far as it goes. For stars with initial masses larger than approximately eight times that of our Sun, however, even higher temperatures and pressures permit fusion to higher mass nuclei. The most common sequence of events leads to nuclei formed by the addition of alpha particles to carbon or oxygen atoms. These form, in turn, neon, magnesium, silicon, sulfur, argon, calcium, titanium, chromium, and iron (heavier elements are formed only during the last brief moments of a massive star's life, before it becomes a supernova). These elements are dispersed to the interstellar medium at the end of the star's lifetime and become available for the next generation of stars (Gribbin 2000).

It was in the cold of space that the various elements formed solids, and the first minerals were created. It is believed that the earliest mineral formed was diamond (Hazen et al. 2008), followed by oxides (first aluminum oxides, then by Ca, Ti, Ni, Fe, Mg, and Si), and then silicates (Gribbin 2000).

According to Chown (2001):

> Nobody knows for sure how many times this cycle of cosmic death and rebirth was repeated before a cold cloud of gas and dust on the outskirts of the Milky Way began to slowly shrink under its own gravity 5 billion years ago. The cloud had hung in space, inky black against the background stars, for tens of millions, perhaps hundreds of millions, of years. What set it in motion was not certain. Perhaps it was the impact of debris from a nearby supernovae. Once it began shrinking, however, there was no stopping it. In time, the heavy elements in the cloud became incorporated into the newborn sun and its entourage of planets. They became incorporated into the earth and its rocks and ultimately into the first primitive living cells.

Thus, each generation of stars will have slightly higher contents of *metals* (meaning, in astronomical terms, any element beyond helium on the periodic table), depending on the composition of the gas it formed from. The Sun is considered to be at least a third-generation star, since it has a relatively high content of metals, but the Sun is still dominated by hydrogen and helium (Chown 2001). For every 1 million atoms of hydrogen in the Sun, there are 98,000 of helium, 850 of oxygen, 360 of carbon, 120 of neon, 110 of nitrogen, 40 of magnesium, 35 of iron, and 35 of silicon (Choi 2014). On this scale, the number of titanium atoms would only be one or two (Amethyst Galleries 2015). Titanium is relatively more abundant in the earth's crust because of the absence of elements that vaporized away from the forming earth (such as He and Ne), or historically have ended up in the liquid or gaseous state (H, C, N).

MINERALOGY AND MINING OPERATIONS

At approximately 4,400 ppm, titanium is the ninth most common element in the earth's crust (Emsley 2002). Titanium mineralogy is marked by its strong attraction to oxygen, such that titanium is virtually never found in nature in the elemental state. Instead, economic concentrations of titanium are found either in the form of simple oxide ores (based on rutile, nominally titanium dioxide), or combined iron titanium oxides (the basic mineral being ilmenite, $FeO{\cdot}TiO_2$). Both of

Thomas P. Battle, Extractive Metallurgy Consultant, Self-Employed, Charlotte, North Carolina, USA these minerals have been identified in meteorites, and thus are among the 60 or so minerals to be present since the formation of the earth (Hazen et al. 2008). The extractive metallurgy of titanium has been defined by the fact that ilmenite ores, of one type or another, are far more common than rutile ores, forcing metallurgists to remove not just oxygen, but a significant amount of iron to make a quality titanium product.

The major minerals of titanium are discussed first, followed by the key structural feature that defines the type of mining operation required, that is, whether the ore deposits in a given situation happen to be in rock or in sand. Finally, typical mining operations are discussed, along with the major producing countries and companies.

Minerals of Titanium

An unusual feature of titanium is that "in many rocks, common silicate minerals rather than oxides contain most of the TiO_2 in the rock. Because only oxide minerals are economically useful at present, the economic geology of titanium is as much a function of mineralogy as it is of chemistry" (Force 1976). Thus, it is not just a matter of looking for regions of overall high titanium content, because that may have nothing to do with economic concentrations of ore.

Rutile

The name *rutile* comes from the Latin *rutilis*, or red (Mindat .org, n.d.). Its nominal composition is titanium dioxide (TiO_2), and it is isomorphous with two other fairly common titanium minerals, brookite and anatase (Pellant 2002). It is formed as an accessory mineral in igneous rocks and either remains in the rocks or is removed from the structure by weathering away of the other components. It is the "most stable of all soil components: They [TiO_2 and titanates] are so resistant to weathering that very little titanium is leached into rivers" (Emsley 2002). Consequently, it often forms a part of a heavy mineral suite in sand deposits. Because of substitutions and intergrowths, the actual composition of rutile is rarely 100% TiO_2; Fe, Ta, Nb, Cr, V, and Sn are common impurities. Typically, there are 3%–5% impurities in natural rutile ores (Iluka 2014; Stanaway 1994; Rozendaal et al. 2010).

Ilmenite

Ilmenite was first identified in rock from the Ilmen Mountains in Russia (Amethyst Galleries 2015). The nominal composition of ilmenite is $FeTiO_3$, but as used in the titanium ore literature, the word *ilmenite* expresses a fairly large range of composition compared to the pure mineral, which is 52.7 wt % TiO_2, 47.3% FeO. In the mining literature, the high TiO_2 composition in ilmenite can reach 67% or even more, because of enrichment during weathering and transport. In fact, at higher TiO_2 contents, "much of this material actually consists of alteration products pseudomorphous after ilmenite" (Force 1976). Examples of minerals commonly defined at higher TiO_2 contents are leucoxene and pseudorutile. Interestingly, although the word *leucoxene* is commonly used in the industry, it is not officially considered to be a mineral; instead it is "a loose term for the end products of this alteration of ilmenite. The titanium is contained in fine-grained aggregates of rutile, brookite, anatase, hematite, and (or) sphene" (Force 1976). Pseudorutile ($Fe_2Ti_3O_9$), however, is an alteration product "in which all the iron has been oxidized to the trivalent state and one-third of it has been leached out" (Grey and Reid 1975). Many of the titanium minerals and pseudo-minerals are listed in Table 1. The following, written more than 40 years ago by Garnar (1972), still seems to apply:

> Because present nomenclature is well established with pigments-oriented geologists, I believe that future workers will continue to call the black, opaque, para-magnetic grains containing less than 65% TiO_2 by the name "ilmenite"; the less magnetic, lighter colored, alteration products containing from 85% TiO_2 to 98% TiO_2 will continue to be called "leucoxene." The primary unaltered rutile grains associated with heavy minerals have always been called rutile—even though some of these grains may be anatase or brookite.

The titanium content of ilmenite ore can also go lower than the pure mineral, because of intergrowths or solid solutions with hematite and magnetite. At the low end of composition, some Fe-Ti compounds grade toward pure magnetite, becoming known as *titaniferous magnetites*. In fact, at the lower end, titanium is a nuisance impurity in ores being mined for their iron, or perhaps vanadium, content (Taylor et al. 2006). More specifically, some sources define a titanomagnetite as a mineral in the solid–solution series between magnetite and ulvöspinel.

Variability in composition of ilmenite is not limited to the wide range in TiO_2 contents; elemental substitutions into the main structure of the ilmenite molecule can occur, either substituting for the iron constituent, or the titanium. Such impurities include manganese, magnesium, and vanadium (Mindat .org, n.d.). Chromium may also be present, sometimes in the form of chrome spinels that are hard to separate from ilmenite using physical beneficiation (Ahmad et al. 2015). In addition, many titanium deposits also contain measurable amounts of

Table 1 Titanium minerals

Mineral	Formula or Range	Comments
Anatase	TiO_2	Isomorphous with rutile
Anosovite	Ti_3O_5	Discredited in 1988 (Mindat.org, n.d.)
Brookite	TiO_2	Isomorphous with rutile
Ilmenite	$FeTiO_3$	Major ore mineral
Leucoxene	$FeTiO_3 \cdot TiO_2$	Not an official mineral, but a composition range*
Perovskite	$CaTiO_3$	—
Pseudobrookite	Fe_2TiO_5	As solid solution in slags, generically given as M_3O_5
Pseudorutile	$Fe_2Ti_3O_9$	Ilmenite weathering product
Rutile	TiO_2	Major ore mineral
Sphene	$CaTiSiO_5$	Also known as titanite
Titanohematite	$Fe_{2-y}Ti_yO_3$	Rhombohedral (Park and Ostrovski 2003)
Titanomagnetite	$Fe_{3-x}Ti_xO_4$	The solid solution mentioned earlier (Hu et al. 2013; Park and Ostrovski 2003); also used as a more general term
Ulvöspinel	Fe_2TiO_4 (or $2FeO \cdot TiO_2$)	Forms solid solution with magnetite

* This definition is from Jones (2009).

Table 2 Analyses of typical titanium ores

	Green Cove Springs Leucoxene	Trail Ridge	Quilon	Capel	Kuranakh	Tellnes	Panzhihua	Allard Lake
Location	United States	United States	India	Australia	Russia	Norway	China	Canada
Type	Sand	Sand	Sand	Sand	Rock	Rock	Rock	Rock
TiO_2, %	76	66	60	54	50	45	38	36
Fe_2O_3, %	21	27	26	18	—*	11	6	28
FeO, %	—	2	10	24	38	35	30	28
CaO, %	—	0.00	—	0.00	—	0.00	0.04	0.00
MgO, %	—	0.003	0.005	0.002	—	0.05	0.06	0.04
SiO_2, %	1.2	0.005	0.008	0.005	0.02	0.02	0.10	0.02
Al_2O_3, %	—	0.014	0.016	0.009	0.03	—	0.02	0.02

Source: Battle et al. 1993

* Dash indicates analyses not supplied, not measured, or below detection.

thorium and radium. Miners and consumers of these ilmenites must watch for concentration of the radioactive species in waste product streams or pipe deposits (McNulty 2007), even if the starting mineral suite is fairly low in thorium and uranium (a common source of radioactivity in mineral sands is the mineral monazite).

A primary formation mechanism for ilmenite is magmatic segregation. Because of the comparatively high levels of titanium, iron, and oxygen in some magmas, ilmenite precipitates early in the solidification process, and because the density of ilmenite is relatively high, crystals will sink to the bottom of the magma chamber in layers (Amethyst Galleries 2015). The titanium component of ilmenite does not weather as rapidly or completely as other minerals and is often left behind to form a major impurity in many silica sands worldwide. Ilmenite and its weathered products account for roughly 92% of all the titanium minerals consumed on an annual basis (Gribbin 2000).

Other Titanium Minerals of Interest

Virtually all titanium ore deposits are based on rutile, ilmenite, or one of the ilmenite alteration products. Other titanium minerals exist, and many have been explored for possible future use. This includes the titanium silicate sphene (also known as *titanite*, $CaTiSiO_5$), the rutile isomorph anatase, *perovskite* ($CaTiO_3$), and *eclogite* (Sandvik et al. 2010; Korneliussen et al. 2000). There are very large deposits of anatase in different regions of the world, for example, in Brazil; but none have been commercialized at scale because of the large amounts of ilmenite available and the presence of deleterious impurities such as calcium, phosphate, and radioactive elements (Stanaway 1994).

Economic Ores and Mining Processes

Two types of titanium mineral formations are being used commercially—rock and sand deposits—which are mined quite differently. The only rock ores mined in quantity are the Allard Lake deposit in Canada and Tellnes in Norway, both formed in anorthosite massifs as discussed in the literature (Morrisset 2008; Charlier et al. 2010). The other major mines are in mineral sands deposits, where rutile and primary and secondary ilmenite (along with other minerals) wash down from their region of exposure and mix with the dominant silica sand. Typically, these *heavy minerals* (defined as having a density greater than 2.9 kg/m^3) make up about 0.5% of the overall sand (Benjamin 2011). However, the action of waves and currents can lead to economic levels of heavy minerals, sometimes as high as 15%–30%, or even more. These ore deposits are found along many current or relict shorelines worldwide. Although mineral sands are much easier to mine than rock ores, they typically have higher costs of mineral processing, as the products more often go directly to the final product manufacturer, whereas the rock ores go into a smelting operation to increase their titanium content.

Rock Ores

For the huge open pit ilmenite mining operation in Quebec (Canada)—the Rio Tinto Fer et Titane (RTFT) Tio mine (also called the Allard Lake mine)—the ore is removed from the rock face by blasting, followed by crushing to –7.5 cm. The rock is then shipped to the smelter by train and ship (Rio Tinto, n.d.). Like other rock ores, the titanium content of the product after physical beneficiation is comparatively low (Table 2; Battle et al. 1993). No other rock ilmenite operation is nearly the size of Allard Lake. The only other deposit about which much is known is the Tellnes deposit in Norway (Korneliussen et al. 2000). Some rock ores are being mined in China (Hazen et al. 2008) and Russia (Makarov 2013).

Mineral Sands

The nature of mineral sands deposits lends itself to a unique form of mining. In wet dredging, the visible mining operation only affects a relatively small area at any given time, and it is difficult to visually detect that mining had taken place only a few years later. In this process, sand is taken up from a dune face using a scoop and fed to a wet concentrator (Minkler and Baroch 1981), typically containing spiral classifiers. This type of classifier can easily separate silica from heavier minerals. The silica is rejected into the pond behind the dredge, while the heavy mineral concentrate is taken ashore for further processing. Eventually the pond behind the dredge fills in, and the sand is planted in dune grasses (Geoscience Australia 2015). Dry dredging is also used in some mineral sands deposits where water is scarce or the characteristics of the mineral suite demand it (Jones 2009).

The details of a mineral separation plant (or dry mill) depends on the particular minerals in the heavy mineral suite, their characteristics in the specific locality, and the amount of those minerals that have some value in the open market. Generally at least two titanium minerals are separated (rutile and ilmenite), and a leucoxene product may be sold. Zircon is

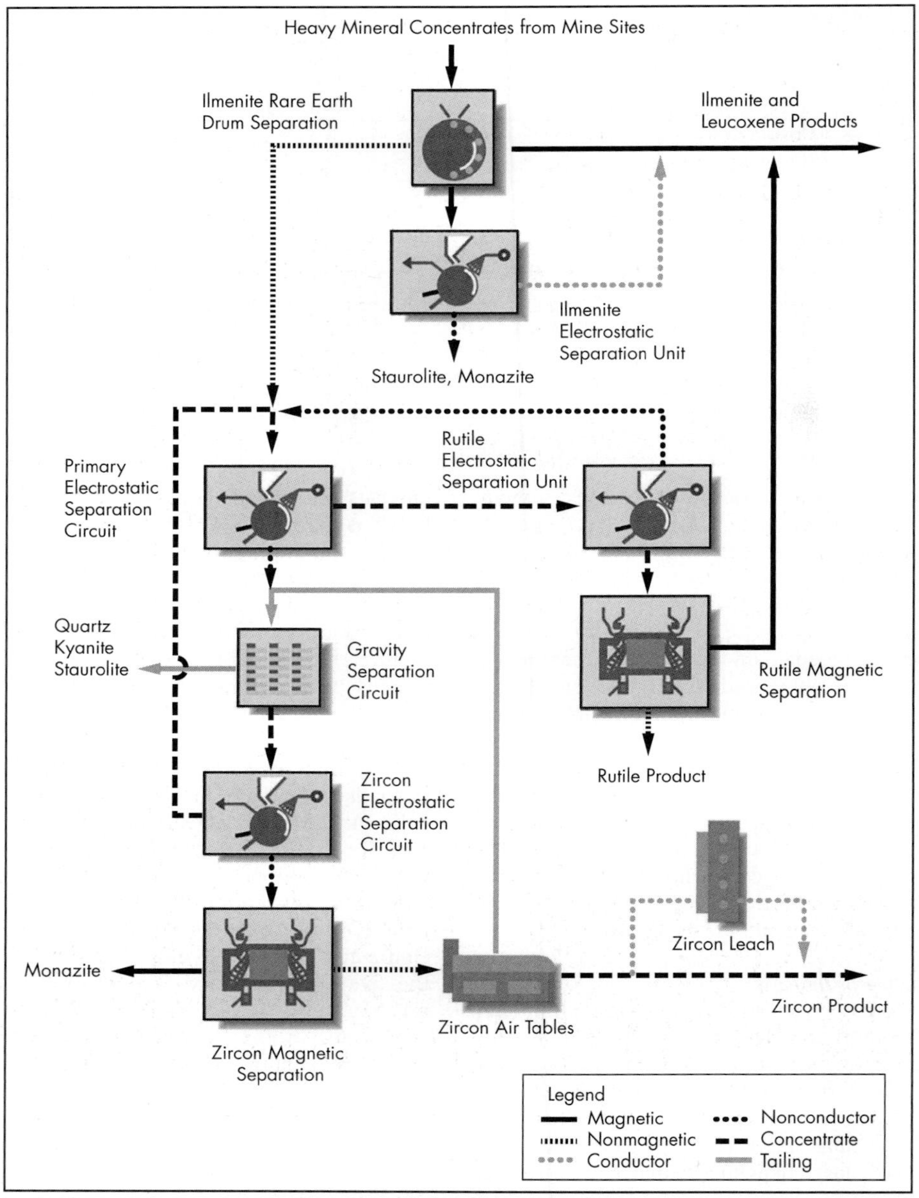

Source: Jones 2009, reprinted with permission from the Australasian Institute of Mining and Metallurgy

Figure 1 Typical mineral sands dry mill

the most valuable coproduct and, at times, is more valuable than the titanium minerals. Some other minerals are also sold from mineral sands, including staurolite and foundry sands (Elsner 2010), but heavy mineral suites can also include garnet, tourmaline, epidote, monazite, xenotime, and aluminosilicates (Garnar 1972).

A typical dry mill flow sheet can be seen in Figure 1 (Jones 2009). Ilmenite can be separated from other heavy minerals (except magnetite) because it is slightly magnetic, whereas the other minerals are not. Rutile and zircon can be separated from each other by their differences in conductivity, with residual gangue removed by density differences (Williams and Steenkamp 2006). The dry mill at the new mineral sands mine in Georgia includes electrostatic and magnetic rollers, spirals, and shaker tables in its flow sheet. The result is four products: zircon, rutile, ilmenite, and leucoxene (Dunn et al. 2015).

TITANIUM ORES AND INTERMEDIATES

South Africa and Canada are the leading ilmenite-producing countries. Australia continues to produce the most rutile, with Sierra Leone not far behind.

Commercial Ores and Locations

Table 3 shows the major individual titanium ore suppliers in 2013 (Bedinger 2018; Murty et al. 2007), while Figures 2 and 3 show the production of ilmenite/leucoxene and rutile, respectively, by country over the last 20 years (USGS 1993–2016). In that period, the amount of ilmenite and rutile mined by year has roughly doubled, with the amount of contained TiO_2 in

Table 3 Major titanium ore suppliers, 2013

Mine	Company	Country	Ore TiO_2 Concentration, %	Production of Contained TiO_2, t
Ilmenite				
Richards Bay Minerals (RBM)	Rio Tinto/BHP Billiton	South Africa	49	900,000
Rio Tinto Fer et Titane	Rio Tinto	Canada	35	770,000
Tellnes	TiZir	Norway	45	498,000
Moma Titanium Minerals Mine	Kenmare Resources	Mozambique	—*	430,000
Pangang	Pangang	China	48	384,000
Cooljarloo	Tronox	Australia	61	274,500
QIT Madagascar Minerals	Rio Tinto	Madagascar	—	264,000
Rutile				
Murray Basin	Iluka	Australia	—	170,000
Sierra Rutile	Iluka	Sierra Leone	96	125,000
Richards Bay Minerals	Rio Tinto/BHP Billiton	South Africa	—	100,000
Stradbroke Island	Sibelco/Unimin	Australia	—	70,000
Multiple	Group DF	Ukraine	—	60,000

Source: Bedinger 2018; Murty et al. 2007

*A dash indicates that the available references do not list the average TiO_2 content of the rutile or ilmenite mined.

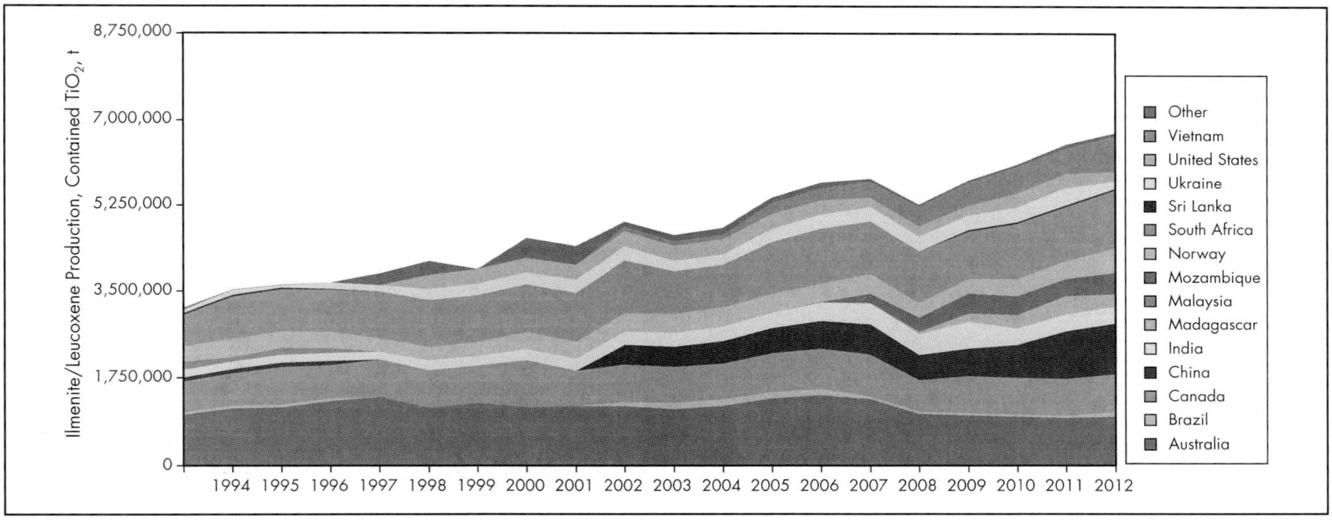

Adapted from USGS 1993–2016

Figure 2 Production of ilmenite and leucoxene by country

the mined ilmenite 9–10 times that of the rutile. Before further discussion, however, note that these numbers are not exact nor always self-consistent, for several reasons. The U.S. numbers are often not included because the U.S. Geological Survey (USGS) protects the confidentiality of individual companies. The numbers found in the literature were generally quoted as per metric ton contained TiO_2, but sometimes they are given as metric tons of ilmenite, without including the titanium content of the ilmenite. And sometimes, numbers are even given per metric ton of heavy mineral concentrate. Care must also be taken to minimize (or at least document) double counting those ores used to make slag or synthetic rutile.

In 1994, the dominant countries for ilmenite production were Australia, South Africa, Canada, and Norway. All four are currently among the leaders in production, but they have been joined by many other countries that were minor or nonexistent producers 20 years ago. In fact, of the four countries dominant in 1994, only South Africa has significantly increased the absolute amount of ilmenite produced over the last 20 years. Newer countries in the mix for ilmenite production are China, India, Ukraine, and Vietnam. In 2014, 69% of the ilmenite was produced by five countries: 18% in South Africa, 15% China, 14% Australia, and 11% each Canada and Vietnam. As for the companies involved in ilmenite production, the leaders are Rio Tinto and Iluka. Rio Tinto controls the former Quebec Iron and Titanium (QIT) property in Quebec, Canada (now called RTFT), and the Richards Bay Minerals (RBM) operation in South Africa, along with a new mine in Madagascar (the operation is called QIT Madagascar Minerals). Iluka owns or markets much of the ilmenite in Australia, along with a deposit in the United States. Norway is entirely represented by the Tellnes deposit, now owned by TiZir Limited. There are many newer individual producers, particularly in China, Ukraine, and Vietnam, along with India, which has always

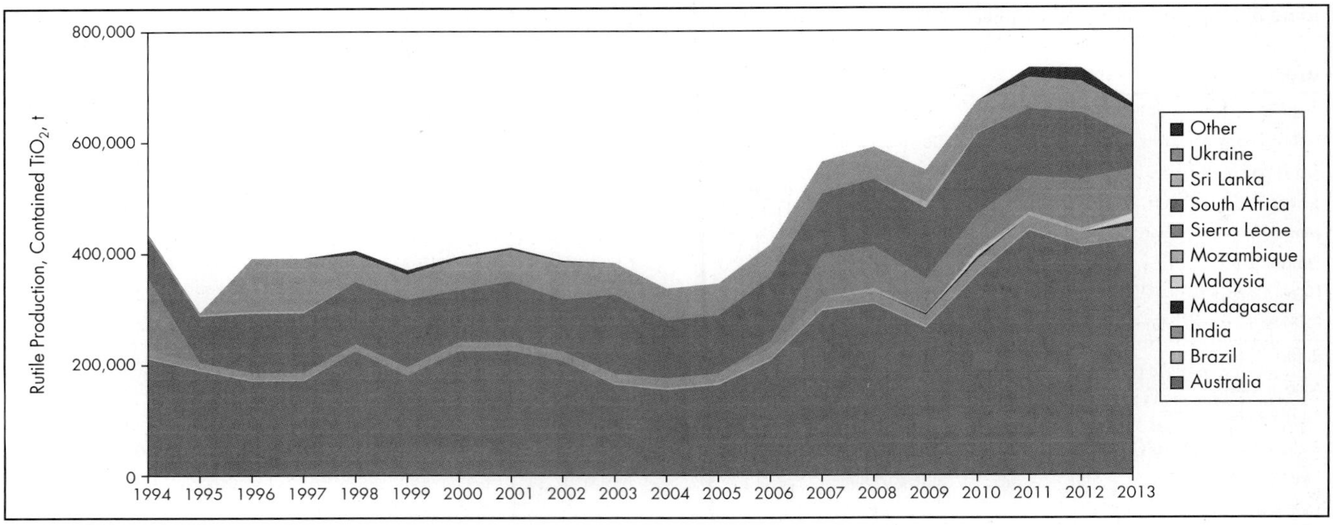

Source: USGS 1993–2016

Figure 3 Production of rutile by country

supplied some level of ilmenite to the world market but was never one of the leading suppliers.

Overall year-to-year rutile production is much more variable, with production again approaching twice the level in 2013 as in 1994. Australia remains the dominant rutile supplier, and unlike the situation for ilmenite, its share has increased over time, from 49% to 63% of the total. South Africa has been steadily in second place, with lesser, but consistent amounts from Ukraine and India (the United States is not included in this figure). The major variable has been the production from Sierra Leone. It supplied 30% of world rutile in 1994, but from 1995 to 2005, Sierra Leone produced no rutile at all. The mine started up again in 2006 and supplied 12% of the world's rutile production in 2013. As for companies producing rutile, Iluka has the largest amount of the supply from Australia. The entire supply from Sierra Leone is the resurgent Sierra Rutile (now part of Iluka). Several different companies produce rutile in South Africa, India, and Ukraine.

Background and the Need for Chemical Beneficiation

The most economical titanium-containing feed to a production process is dependent on the final product to be made and the technology used to make that product. The vast majority of titanium ores are used to make titanium dioxide white pigment. TiO_2 is made by two methods. In the earlier technology development, the titanium minerals are dissolved in sulfuric acid, and later titanium is precipitated as a titanium oxysulfate, which is purified and converted to titanium dioxide. In the chloride process, the ore is reacted with coke and chlorine in a fluidized bed, and virtually all minerals are converted to gaseous chlorides. Titanium tetrachloride ($TiCl_4$) is then isolated from most other chlorides and further purified. To make TiO_2, pure $TiCl_4$ is reacted with oxygen at high temperatures, and the chlorine gas is recycled. In the major production process for titanium metal—the Kroll process—high-purity $TiCl_4$ is reacted with elemental magnesium to make titanium metal sponge and recyclable magnesium chloride.

Each of these production processes place their own demands on ore supplies. Both sulfate and chloride pigment processes are capable of using ilmenite directly, but not all ilmenite ores are suitable for both processes, hence the subcategories of *sulfatable* and *chlorinatable* ilmenite. In theory, ores used to make titanium metal could use chlorinatable ilmenite directly, but the plants are all so small that it is not considered economically viable to add the extra infrastructure to chlorinate, separate, and dispose of the non-titanium chlorides. Instead, their sources of titanium are natural rutile or artificial products made by chemical processing of ilmenite.

The three *natural* products (those made by mining and physical beneficiation only) are discussed first: sulfatable ilmenite, chlorinatable ilmenite, and natural rutile. The desirability of chemical beneficiation is then considered, along with the technologies that are being used to do the upgrading. Finally the entire suite of products available to a TiO_2 pigment manufacturer or titanium metal producer are summarized, along with expectations for the future.

Sulfatable Ilmenite

The most important characteristic of sulfatable ilmenite is that the ore is easily dissolved in sulfuric acid. This means that weathered ilmenite containing large amounts of ferric iron must be avoided, and tetravalent titanium is preferred over trivalent (McCoy et al. 2011). Impurities that do not dissolve in sulfuric acid, such as silica and zircon, can be easily separated from the titanium-rich solution, so that is not a limitation. However, the intermediate solid, titanium oxygen sulfate ($TiOSO_4$), is hard to purify, so several species that tend to follow titanium into solution (such as Cu, Ni, Mn, Cr, V, and Nb) should be avoided in the feed (McCoy et al. 2011). Ore size is not critical; the feed is generally ground to –45 µm.

Chlorinatable Ilmenite

Because the chloride process involves the use of a fluidized bed to react titanium ore with coke and chlorine, there are strict physical constraints on the ore feed size and density. The ore particles must be suspended in the stream of chlorine gas for long enough that the constituents of the ore can react with coke and chlorine to convert to gaseous chlorides or oxychlorides. Particles too large and dense cannot be supported by the gas stream and result in defluidization of the bed. Fine particles can elutriate from the bed before reactions are completed. At typical densities of ilmenite or rutile, the feed size needs to be in the range of 50 to 150 μm.

Chemically, the key issue for successful operation of the chloride-process fluidized bed is to avoid the production of significant quantities of liquids. The bed is designed for rapid conversion of fine solids to gases; liquid can lead to clumping of solids and defluidization of the bed. The main culprits at the temperatures and pressures of typical chloride-process operations are calcium, magnesium, and to a lesser extent, manganese. This greatly restricts use of some of the rock ilmenites that often contain significant concentrations of these impurities. Other impurities are mainly an issue because they raise or lower operating costs. The presence of trivalent titanium, rather than tetravalent, for example, reduces coke consumption. High levels of sulfur increase downstream gas-handling costs. Various impurities can be removed within the chloride process, such as Si, Zr, Nb, and V. This topic is discussed in more detail in Chapter 12.37, "Titanium Dioxide Pigment."

Natural Rutile

Natural rutile is a desirable feed for both titanium metal and pigment manufacture, because of the high TiO_2 and concomitantly low impurity content. However, there is not enough rutile to supply both customers, especially with the low cost of ilmenite relative to rutile. Ilmenite is inexpensive enough that it can be upgraded to various *synthetic* versions of rutile that can compete with the natural mineral in many applications.

Chemical Beneficiation of Ilmenite

Chemical analyses of eight natural ilmenites are given in Table 2. The titanium content of the different ilmenites can be compared to that of typical rutile TiO_2 contents of 90%–95%, and relative prices in Figure 4, which represents the period from 1992 to 2015 (Elsner 2010; USGS 1993–2016). The cost of ilmenite is roughly a fifth that of rutile, but of course, the value-in-use is much less as well, as the end user must now cope with increasing quantities of iron and possibly other impurities as well. Given the lack of enough natural rutile to supply all downstream needs, the possibility of chemically upgrading ilmenite to make a higher TiO_2 product is appealing both for the customer and for certain suppliers, especially those possessing ilmenite deposits of very low grade. The main question is whether this upgrading can be accomplished while keeping product cost competitive with that of rutile.

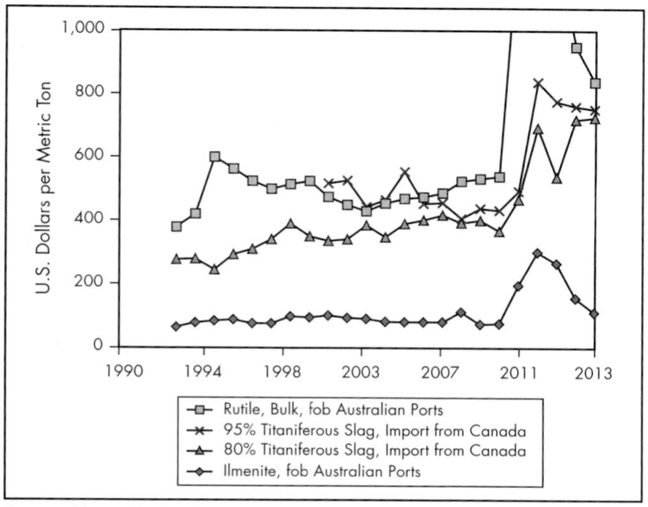

Source: Elsner 2010; USGS 1993–2016

Figure 4 Ilmenite, rutile, and slag prices, 1992–2013

In the 1940s, a huge deposit of rock ilmenite was discovered in Quebec, not far inland from the St. Lawrence River near Allard Lake. However, the percentage of titanium obtainable by standard physical beneficiation was too low for any application at the time. A technology was needed to upgrade the titanium content to an acceptable level, while ideally getting value for the iron impurity. The technology developed was carbothermic reduction in an electric arc furnace at elevated temperature. Under these conditions, the carbon in the feed preferentially reacts with the iron component of the ore to make metallic iron. Because metallic iron is much heavier than the slag phase that contains the titanium, the iron sinks to the bottom of the reaction vessel and is easily separated from the enriched titanium above. The process was successfully implemented in Sorel, Quebec, in the mid-1950s, and Sorelslag has been part of the titanium feedstock equation ever since (Habashi 1986, 2010).

Since that time, many other ilmenite smelting plants have been developed, all producing a high-TiO_2 slag product. Because of the high operating costs and operational difficulties with the smelting process, many other ilmenite upgrading technologies have been developed (Battle et al. 1993). Most successful have been the solid-state reduction technologies that produce synthetic rutile, and many of these plants are currently operating, mainly in Australia and India. Smelting and commercial synthetic rutile manufacturing technologies and practitioners are discussed in the following section, with a brief consideration of upgrading technologies not yet commercialized.

Ilmenite Smelting

Not every low-grade titanium feedstock can be used to make a high-quality titania slag: The chemical composition must be within a particular range. Ilmenite smelting does not separate all of the impurities from the titanium fraction of an ilmenite ore. It removes most of the iron, indeed, but removal of other impurities depends on system temperature and pressure, and the thermodynamics and kinetics of elemental partitioning between molten iron and the liquid slag phase dominated by titanium oxides. Those considering ilmenite smelting of a particular ore must consider the actual partitioning of elements in slag versus metal, and they must determine if any particular component is likely to exceed desired limits in either the slag product or the metallic iron.

If modeling and experiments indicate that the predicted slag is likely to have an unfavorable composition, decisions will be necessary to continue the project. Changes could include the following:

- Use a different ore source or blend.
- Modify the ore before it reaches the smelter.
- Modify the operating conditions of the smelter to optimize operation with that particular feed.
- Modify the product slag after it is created.
- Live with the reduced value of a product with excessive levels of one or more impurities.

All of these options have been explored at one time or another by smelter operators.

For example, it was realized early in the development of the RBM smelting operation that the ores of choice were relatively high in chromium, which would produce a slag with chromium oxide levels unacceptable to customers. Their solution was to subject the concentrate to an oxidizing roast, which converts the ilmenite fraction of the ore to a more magnetic state allowing for separation from the high-Cr minerals (Steenkamp et al. 2005). The same approach does not necessarily work on all ilmenites; in Australia, a sulfidizing pretreatment might be more appropriate (Ahmad et al. 2015).

Although carbothermic reduction of metal oxides is a well-established technology, there are certain unique features making ilmenite processing more difficult than for most other metal oxides. One problem is that the slag product is actually more valuable than the iron. In normal metal smelting, the slag has the critical role of absorbing impurities that would otherwise enter the valuable metal phase and helps to control heat transfer, the liquid state of the system, and its flow properties. Often, fluxes are added to the slag to lower its melting point and make operations safer and lower cost. That cannot be done here, as any additions to the slag would dilute product quality. Thus, ilmenite smelters must operate at temperatures in excess of 1,750°C for the high-TiO_2 slag to have the right fluidity for feed materials to rapidly fall through the slag to the metal phase and for adequate product flow exiting the system when tapped.

Another problem is that no one has yet found a refractory material for lining smelters at these temperatures that can hold up to the corrosive and erosive effects of high titania slags. The ingenious solution has been to water-cool the outer shell of the reactor such that a layer of slag freezes against the inner shell, thus protecting the refractory from further attack. The downside of this is that tight thermal control of the reactor is required to keep this frozen shell intact.

It is also not economically realistic to remove all of the iron oxide from the slag; typical operations lead to a slag containing roughly 10% FeO. This happens because as the iron content of the slag is reduced, its activity decreases while that of the titanium increases. At a certain time, therefore, the reducing agent acts to reduce tetravalent titanium to trivalent, with less and less of the carbon being used to reduce ferrous to metallic iron. The increased value of slightly more pure slag is generally outweighed by the extra energy costs to get to the lower iron (Burger et al. 2009); in addition, not all customers are happy to get a slag with high Ti_2O_3 content as opposed to TiO_2.

The original ilmenite smelting operation in Quebec now has nine smelting furnaces, all more than 40 years old, but modernized over time. The technology was transferred to Rio Tinto's RBM operation in South Africa, which makes similar production levels from "... four of the world's largest six-in-line electric arc furnaces. The furnaces are rectangular in shape, being 18 metres long and eight metres wide, and have six electrodes in line" (Williams and Steenkamp 2006). Although there are several other ilmenite smelting operations now in operation around the world, none approach the size of the two Rio Tinto operations.

The quality of the titania slag is directly related to the quality of the ore feed. This is quite obvious with a comparison between Sorelslag and the slag created using the sands from Richards Bay. Because the rock ilmenite contains much higher levels of impurities (particularly MgO) that tend to stay in the slag and not be reduced, the final TiO_2 content of rock-ilmenite slag is considerably lower than Richards Bay slag. And it gets worse, because the impurities accentuated in the Sorelslag are the most problematic impurities for the operation of the chloride process to make TiO_2 or the Kroll process to make titanium metal. At the temperatures of operation of the fluid-bed chlorinator in these processes, both magnesium chloride and calcium chloride are liquid, which can lead to severe operating problems if not carefully controlled and monitored. No chloride operation can operate on high percentages of Sorelslag. This became a particular problem in the 1990s as the number and size of chloride plants increased (Rio Tinto, n.d.). Rio Tinto's solution was to develop new technology to remove impurities from the Sorelslag to make an upgraded slag (UGS). This process involves oxidation of the slag to convert all reduced titania to insoluble TiO_2, followed by reduction to convert all iron to ferrous, then an acid leach to dissolve the iron and other impurities, leading to a product containing 94.5% TiO_2, suitable for all titanium applications. Since that time, Rio Tinto has marketed Sorelslag to sulfate pigment producers and UGS to others. Most recently, they have begun making higher quality slag by another route, which is shipping high-grade ilmenite from QIT Madagascar Minerals to the smelters in Quebec.

Known smelting operations are listed in Table 4 (Elsner 2010). Prominent are the multiple-smelter operations of Rio Tinto in Quebec, Canada, and Richards Bay, South Africa. In the last 20 years, two smaller operations have started up in South Africa—Namakwa Sands and KZN Sands. The Tinfos smelter in Norway began operation in 1986 and soon encountered the same problems as Rio Tinto, as their local rock ilmenite led to slags that could be sold only to sulfate pigment producers. They were the first to try and make a high-grade slag by importing higher-grade ilmenite, but the initiative did not last long (Gambogi 1997, 1999). Now it appears they will try that approach again, this time with a new high-quality ilmenite sand from Senegal (Bedinger 2015).

Little is known about the other slag producers listed in Table 4, other than their existence. There is one article summarizing titanium slag production in Russia and Kazakhstan (Toembaeva 2015).

Hydrometallurgical Upgrading

Even with the success of the smelting operation in Quebec, it was not obvious to other ilmenite producers that this was the technology to emulate for upgrading their own ores. Operation, particularly in the early years, was very tricky (Rio Tinto, n.d.). Also, energy costs are inherently very high, with significant capital costs required for the smelter and attendant equipment. In Australia in the 1960s, a much lower-cost technology was developed—the Becher process, or as it is sometimes known, the Lurgi–Becher process.

The flow sheet for the Becher process can be seen in Figure 5 (Iluka 2012). The ore is pre-oxidized, if necessary, then reduced in a rotary kiln in the solid state to make metallic iron. After magnetic separation of the reduced ilmenite from char and ash, it is leached for several hours in an aerated aqueous solution that contains ammonium chloride as a catalyst (Battle et al. 1993). The iron is converted to oxides that can be washed off the TiO_2, resulting in what became known as *synthetic rutile* (SR). This process has low capital and operating costs but does not result in complete removal of the iron,

Table 4 High-TiO_2 slag producers

Company	Location	Initial Operation	Number and Type of Furnaces*	Overall Capacity, t/yr slag	TiO_2 Content of Ore, %	TiO2 Content of Slag, %
Rio Tinto	Quebec, Canada	1950	9 AC	1,000,000	34	80
Rio Tinto	Richards Bay, South Africa	1977	4 AC	1,000,000	50	85
TiZir	Tyssedal, Norway	1986	1 AC	200,000	44	80
Tronox	Namakwa Sands, South Africa	1995	2 DC	200,000	52	86
Tronox	KZN Sands, South Africa	2003	2 DC	250,000	47	85
Pangang	Pangang, China	2006	Unknown	100,000	48	Unknown
Various	Vietnam	2010–2014	5	64,000	Unknown	Unknown
Xinli	Kunming, China	2009	DC	80,000	Unknown	80–90

Source: Elsner 2010
*AC = alternating current; DC = direct current.

and the iron removed must be disposed of, rather than being a valuable coproduct as in smelting. Besides reducing the iron content of the feed, manganese can be removed by adding sulfur to the reduction kiln, then acid washing the manganese sulfide from the product; something similar might be done to remove chromium residues as well (Ahmad et al. 2015).

Commercial Becher operations are listed in Table 5. All operations are located in Western Australia, operated by either Iluka or Tronox. Tronox markets half of their SR and uses the remainder in their pigment plant in Kwinana. Iluka makes several different products in their SR plants: basic SR; premium SR (removal of Mn and some Fe as sulfides, mentioned earlier); synthetic rutile enhanced product, which uses a complicated flow sheet specifically to lower the thorium and uranium contents of the ore; and a low-grade SR85, made from their Murray Basin ilmenite (Iluka 2012).

No other SR process has been as successful as the Becher process. The only other technology known to have multiple plants is the Benelite process, where reduced ilmenite is leached in 20% HCl at elevated temperatures and pressures. This can be followed by acid regeneration to reform the HCl. The Wah Chang process is similar but uses stronger acid (Battle et al. 1993). Plants using these technologies have predominantly been built in India (Indian Bureau of Mines 2015) and are also listed in Table 5.

The Ishihara process used the waste sulfuric acid from the company's sulfate-route pigment plant in Japan. The acid was used to leach ilmenite after partial reduction of the iron. Colloidal hydrated TiO_2 was used in the leaching step as a reaction accelerator. The waste acid was used in the production of ammonium sulfate. The original ore particle size was retained during processing (Battle et al. 1993). This process was only used by Ishihara in its Osaka facility; the plant was shut down in 1994.

Other Ilmenite Upgrading Technologies

Research continues to improve or replace current ilmenite upgrading technologies; many of these were summarized 20 years ago (Battle et al. 1993). Only a few more recent technologies can be summarized here.

One of the major goals has been to unite the great advantages of the smelting and SR routes—to run low-temperature, relatively simple, energy-saving unit operations—yet achieve a high removal percentage of iron and create a valuable iron coproduct. The world's largest single consumer of titanium raw materials, DuPont (recently separated from the parent company as part of Chemours) has made several attempts at this type of technology improvement over the years (Battle et al. 1996; Shekiro et al. 2011).

Quite a number of newer technologies are focused on upgrading Ti-rich slags to make the product more saleable (in the case of some, converting a slag of little or no value to high value). More than 20 years ago, six options were identified, as given in the following list (Battle et al. 1993). Modern-day flow sheets still seem to fit into this list, with perhaps the acid changed in leaching operations, use of new power sources such as microwaves, or additional unit operations included (Guo et al. 2011; Qian et al. 2011; Lakshmanan et al. 2011; Mehdilo and Irannajad 2012; Nell 2007; Abdou et al. 2005; Lei et al. 2011):

1. **Extended reduction.** The iron content is reduced to 1%–2%. Much of the Si, Mn, Cr, and V is removed. Essentially, all of the titanium is reduced to the +3 valence state.
2. **Selective chlorination.** Residual Fe, as well as Cr, V, and some Mg are removed.
3. **Sulfidation and acid leach.** Fe, Cr, and V are removed.
4. **Sulfation roast and leaching.** Mg and Ca are converted to soluble sulfates. Iron must be removed first, perhaps by Na_2S. Silica is also removed.
5. **Reduction of Ti to Ti(O,C).** This species chlorinates at low temperatures, before remaining impurities.
6. **Absorption of impurities in coke.** The *Scoke* process creates micropellets from slag and coke. The liquid chlorides soak into residual coke, which may prevent the fluidization problems that they would otherwise cause.

Iluka is working on developing acid-soluble SR, to open the market for this product to sulfate pigment producers (Iluka 2012). Other concepts include separating different Ti-containing phases in low-grade ilmenite ores by magnetic separation, then treating the two streams separately to optimize titanium separation and yield (Jha et al. 2011) and using microwave energy to reduce costs (Shaohua et al. 2012).

Technologies are also being developed to support downstream technological innovations in the titanium field, for example, modifying the Becher process to make titanium oxycarbide instead of titanium dioxide, to be used as the feed to an electrochemical process to make titanium metal (Fatollahi-Fard and Pistorius 2015).

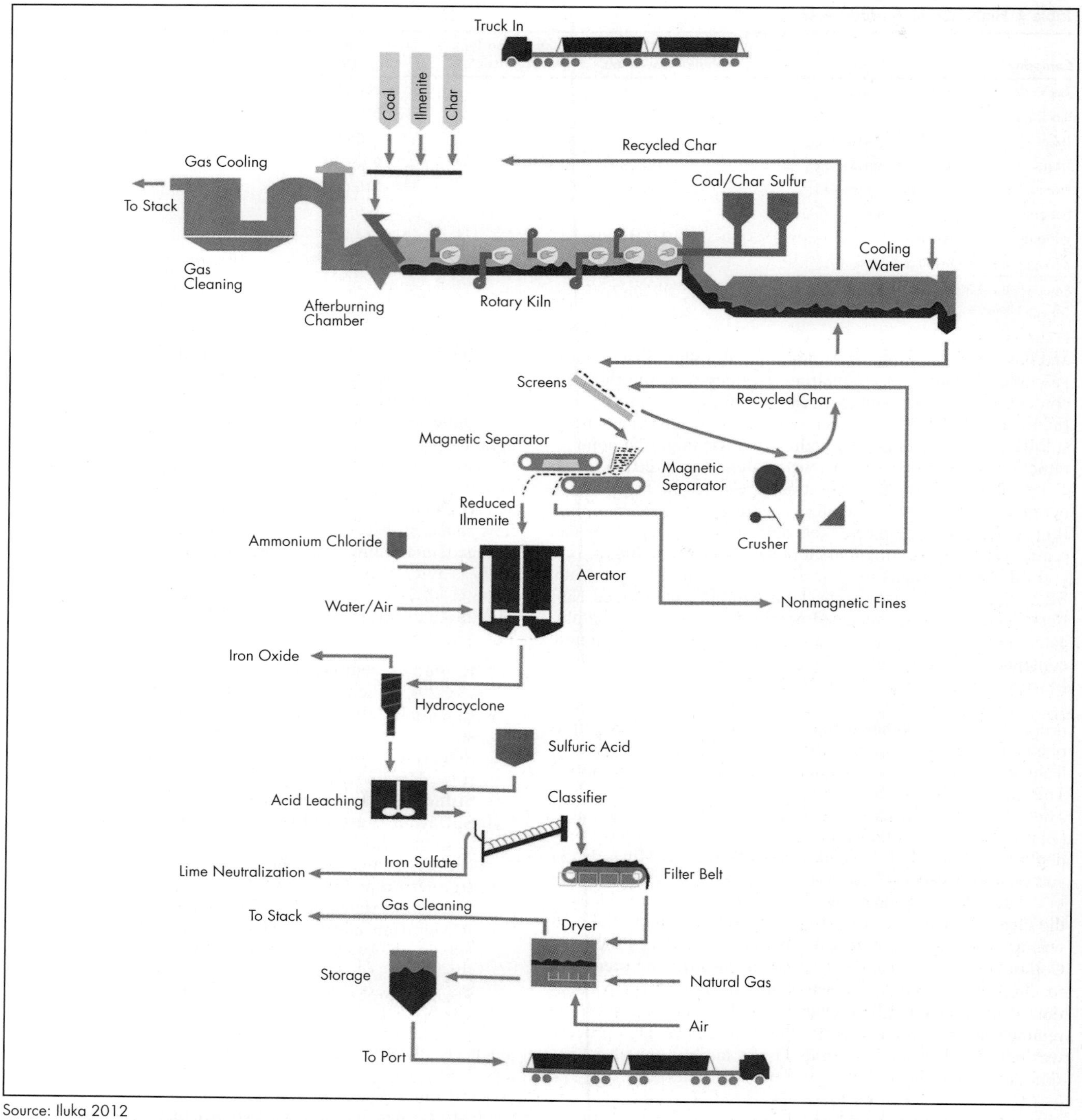

Source: Iluka 2012

Figure 5 Flow sheet for the improved Becher synthetic rutile process

TITANIUM-RICH FEEDSTOCKS FOR TITANIUM DIOXIDE AND TITANIUM METAL PRODUCTION

In previous sections, the titanium minerals processing business has been discussed, from the prevalence of ilmenite over rutile to the desire by customers for feeds at ilmenite prices but the purity of rutile. The result is a plethora of products available for use by the TiO_2-pigment and Ti-metal industries. These products are listed in Table 6 (Iluka 2014; Nell 2007; USGS 1993–2016), along with approximate amounts produced in 1990 and 2006, and prices.

Overall production of titanium feedstocks has increased appreciably over the past 20 years, although not across the board. High-grade feeds have increased (particularly those for the chloride process) at the expense of lower-grade and sulfate feedstocks.

Any snapshot of pricing, especially in the environment of the past several years, should be examined with caution.

Table 5 Commercial synthetic rutile production

Company	Location	Capacity, t/yr	Ilmenite TiO_2 Content, %	SR TiO_2 Content, %
Becher Process				
Iluka		500–550,000	–*	–
	Capel, Australia	–	–	–
	Narngulu, Australia	–	–	–
Tronox	Chandala, Australia	220,000	61	92
Benelite Process				
Indian Rare Earths Limited (Orissa Sands Complex)	Chatrapur, India	100,000†	50	91
Kerala Minerals and Metals Limited	Chavara, India	50,000	60	93
Cochin Minerals and Rutile Limited	Kerala, India	45,000	–	–
Tor Minerals	Ipoh, Malaysia	–	–	–
Kerr-McGee	Alabama, United States	Operated 1976–2003†		
Wah Chang Process				
DCW	Sahupuram, India	48,000	–	95
Ishihara Process				
Ishihara	Ishihara, Japan	Operated 1971–1994†		

*A dash indicates that the available references do not list the average TiO_2 content of the rutile or ilmenite mined.
†Not operating.

Table 6 Titanium feedstocks

Feedstock*	TiO_2 Content, %	Total Production of Contained TiO_2 in 1990, t	Total Production of Contained TiO_2 in 2006, t	Price in 2013, $/t
Natural rutile	95–97	492,000	390,000	1,250
Synthetic rutile	88–95	397,000	810,000	1,150
Upgraded slag	95	0	170,000	No information
Chloride slag	85–90	531,000	990,000	777
Sulfate slag	80–85	1,050,000	780,000	538
Chloride ilmenite	58–62	487,000	470,000	No information
Sulfate ilmenite	52–54	1,205,000	890,000	265

Source: Iluka 2014; Nell 2007; USGS 1993–2016
*These feedstocks are used to produce TiO_2 pigment and Ti metal.

From 1992 to 2015 (Elsner 2010; USGS 1996–2016), ilmenite prices were steady at just less than US$100/t, whereas rutile prices increased from US$400/t to $500/t and became steady at that level. This compares with numbers quoted at US$60–$80 for ilmenite in India in 2006, SR at US$360–$440, and slag at US$370–$410 (Murty et al. 2007). Prices skyrocketed in 2010, however, and have only just begun to settle down. According to the USGS (Bedinger 2018), this was caused by a shortage of concentrate, or more picturesquely, "shortages in 2010 and 2011 led to a price spike and super profits before both collapsed under a mountain of inventories in 2012" (Barlow 2015). The feedstock market is improving, although the market has to adjust to China moving from a large importer of ilmenite to a country exporting low-cost ilmenite that is a by-product of iron ore mining: "Current factory gate prices of around US$100 are a symptom of the massive stockpiles that have accrued" (Barlow 2015).

In their detailed discussion of mineral sands pricing, Iluka describes former "constrained pricing" because of long-term contracts (Iluka 2014). Iluka ended legacy contracts in 2010, and now contracts are renegotiated quarterly or half yearly, with some spot sales.

The overall changes in the market in the past few years are summarized by Barlow (2015):

> Titanium dioxide is never more than two or three feet away from you at any given time. The ubiquity of its application means that global demand trends run closely with gross world economic product. This should mean that producers can rely on a fairly predictable set of metrics by which to forecast their market. However, the last 10 years have been anything but predictable. We have witnessed the departure of synchronicity between feedstock mining and pigment production, exacerbated by the global financial crisis and the explosion of Chinese capacity, which saw that country turn from a key net importer to an exporting behemoth in a few short years.

Both titanium dioxide and titanium metal markets are expected to continue to grow in upcoming years. There are few cost-effective alternatives to titanium dioxide in many of its applications, and titanium metal's unique combination of physical and chemical properties will make it more popular, particularly if the price of the finished product can be lowered. So there will be demand for more titanium ore, although new ores will need to meet ever more stringent demands in product quality and environmental responsibility. Fortunately, titanium ores, particularly ilmenites, are in great supply and

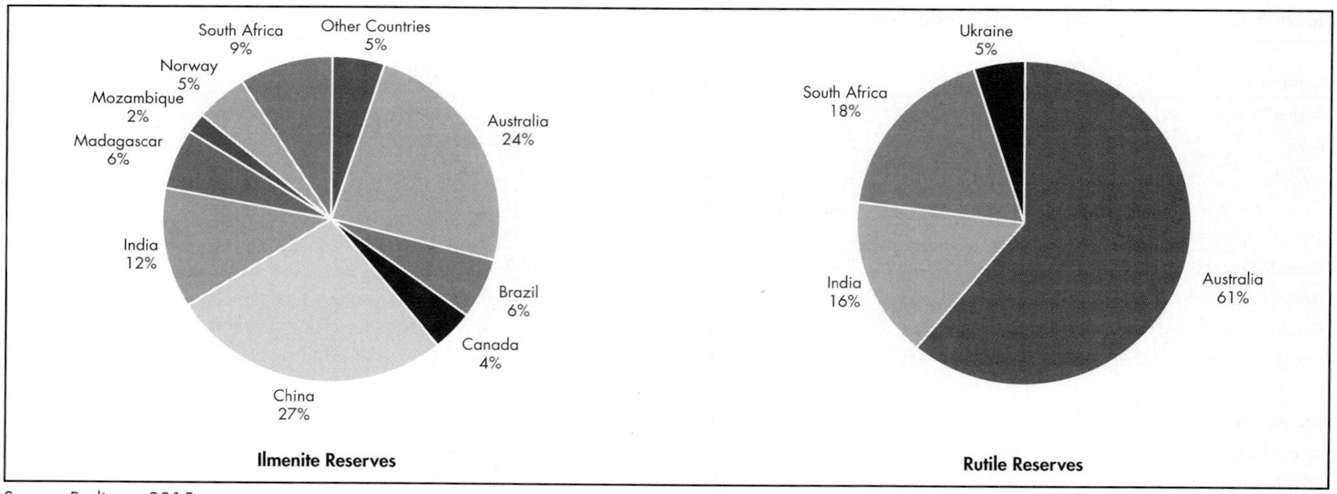

Source: Bedinger 2015

Figure 6 Current world reserves of titanium minerals

should not run out anytime soon. As shown in the pie charts of Figure 6 (Bedinger 2015), there are roughly 720 Mt (million metric tons) of ilmenite reserves (more than a 100-year supply for all titanium units needed) and 47 Mt of rutile and leucoxene (over a 50-year supply). This likely does not include the great supply of titanium in the titaniferous ores that are currently being mined almost exclusively for their iron content. Work continues on ways to economically use the titanium content of these ores (Taylor et al. 2006).

REFERENCES

Abdou, A., Manaa, E.A., and Zaki, S.A. 2005. Synthetic rutile preparation from Egyptian ilmenite using hydrochloric acid in the presence of cellulose as reducing agent. *Int. J. Multidiscip. Res. Dev.* 2(1):443–451.

Ahmad, S., Rhamdhani, M.A., Pownceby, M.I., and Bruckard, W.J. 2015. Analysis of sulfidation routes for processing weathered ilmenite concentrates containing impurities. In *6th International Symposium on High-Temperature Metallurgical Processing*. Edited by T. Jiang, J-Y. Hwang, G.A.F. Alvear, O. Yucel, X. Mao, H.Y. Sohn, N. Ma, P.J. Mackey, and T.P. Battle. Hoboken, NJ: Wiley.

Amethyst Galleries. 2015. Mineral gallery: The mineral ilmenite. www.galleries.com/Ilmenite. Accessed February 2018.

Barlow, E. 2015. A review of global supply and demand for titanium. *Paintings and Coatings Ind. Mag*. (February 6). www.pcimag.com/articles/100036-a-review-of-global-supply-and-demand-for-titanium-dioxide.

Battle, T.P., Nguyen, D., and Reeves, J.W. 1993. The processing of titanium-containing ores. In *The Paul E. Queneau International Symposium on Extractive Metallurgy of Copper, Nickel and Cobalt*. Vol. 1, Fundamental Aspects. Edited by R.G. Reddy and R.N. Weizenbach. Warrendale, PA: Minerals, Metals and Materials Society.

Battle, T.P., Leary, K.J., Reeves, J.W., Ericson, A.S., and Zander, B. 1996. Reduction-segregation of ilmenite: Iron and titanium co-products from solid-state processing. In *Second International Symposium on Extraction and Processing for the Treatment and Minimization of Wastes*. Edited by V. Ramachandran and C.C. Nesbitt. Warrendale, PA: Minerals, Metals and Materials Society.

Bedinger, G.M. 2015. Titanium. In *USGS Minerals Yearbook 2015 (Advance Release)*. Reston, VA: U.S. Geological Survey.

Bedinger, G.M. 2018. Titanium mineral concentrates. In *Mineral Commodity Summaries 2018*. Reston, VA: U.S. Geological Survey.

Benjamin, P. 2011. Mineral sands geology and exploration. Technical presentation. http://iluka.com/docs/company-presentations/mineral-sands-technical-presentation-sydney-3-may-2011.pdf. Accessed February 2018.

Burger, H., Bessinger, D., and Moodley, S. 2009. Technical considerations and viability of higher titania slag feedstock for the chloride process. Presented at the Southern African Institute of Mining and Metallurgy 7th International Heavy Minerals Conference, Johannesburg, South Africa.

Charlier, B., Namur, O., Malpas, S., de Marneffe, C., Duchesne, J.-C., Vander Auwera, J., and Bolle, O. 2010. Origin of the giant Allard Lake ilmenite ore deposit (Canada) by fractional crystallization, multiple magma pulses and mixing. *Lithos* 117(1-4):119–134.

Choi, C.Q. 2014. Earth's sun: Facts about the sun's age, size and history. Space.com. www.space.com/58-the-sun-formation-facts-and-characteristics.html.

Chown, M. 2001. *The Magic Furnace*. New York: Oxford University Press.

Dunn, P.L., Rose R., and Settles, D. 2015. Development of the SII "Mission" mineral sands project. In *Proceedings 2015, 47th Annual Meeting of the Canadian Mineral Processors*. Edited by P. Blatter. Westmount, QC: Canadian Institute of Mining, Metallurgy and Petroleum. pp. 106–117.

Elsner, H. 2010. *Heavy Minerals of Economic Importance: Assessment Manual*. Hannover, Germany: Bundesanstalt für Geowissenschaften und Rohstoffe (BGR) [Federal Institute of Geosciences and Natural Resources].

Emsley, J. 2002. *Nature's Building Blocks: An A-Z Guide to the Elements*. New York: Oxford University Press.

Fatollahi-Fard, F., and Pistorius, P.C. 2015. Production of titanium oxycarbide from titania-rich mineral sands. In *EPD Congress 2015*. Edited by J. Yurko, A. Allanore, L. Bartlett, J. Lee, L. Zhang, G. Tranell, Y. Meteleva-Fischer, S. Ikhmayies, A.S. Budiman, P. Tripathy, and G. Fredrickson. Hoboken, NJ: Wiley.

Force, E.R. 1976. Titanium contents and titanium partitioning in rocks. Paper 959-A. In Geology and Resources of Titanium. U.S. Geological Survey Professional Paper[s] 959-A–F. Washington, DC: U.S. Government Printing Office.

Gambogi, J. 1997. Titanium. In *Minerals Yearbook 1997*. Reston, VA: U.S. Geological Survey. https://minerals.usgs.gov/minerals/pubs/commodity/titanium/670497.pdf.

Gambogi, J. 1999. Titanium. In *Minerals Yearbook 1999*. Reston, VA: U.S. Geological Survey. https://minerals.usgs.gov/minerals/pubs/commodity/titanium/670499.pdf.

Garnar, T.E., Jr. 1972. Economic geology of Florida heavy mineral deposits. In *Proceedings of the 7th Forum on Geology of Industrial Minerals*. Florida Bureau of Geology, Special Publication 17. Edited by H.S. Puri. pp. 17–21.

Geoscience Australia. 2015. Australian Atlas of Minerals Resources, Mines and Processing Centres: Titanium fact sheet. Symonston, ACT: Geoscience Australia. www.australianminesatlas.gov.au/education/fact_sheets/titanium.html.

Grey, I.E., and Reid, A.F. 1975. The structure of pseudorutile and its role in the natural alteration of ilmenite. *Am. Mineral.* 60:898–906.

Gribbin, J. 2000. *Stardust*. New York: Penguin Books.

Guo, Y-F., Liu, S-S., Jiang, T., and Qiu, G-Z. 2011. Study on preparation of high-quality synthetic rutile from titanium slag by activation roasting followed by acid leaching. In *2nd International Symposium on High-Temperature Metallurgical Processing*. Edited by J-Y. Hwang, J. Drelich, J. Downey, T. Jiang, and M. Cooksey. Warrendale, PA: Minerals, Metals and Materials Society.

Habashi, F. 1986. *Principles of Extractive Metallurgy*. Vol. 3, Pyrometallurgy. New York: Gordon and Breach.

Habashi, F. 2010. The beginnings of mineral processing research in Canada: A short account. In *XXV International Mineral Processing Congress Proceedings*. Melbourne, Victoria: Australasian Institute of Mining and Metallurgy.

Hazen, R.M., Papineau, D., Bleeker, W., Downs, R.T., Ferry, J., McCoy, T., Sverjensky, D., and Yang, H. 2008. Mineral evolution. *Am. Mineral.* 93:1693–1720.

Hu, T., Lv, X., Bai, C., Zun, Z., and Qiu, G. 2013. Reduction behavior of Panzhihua titanomagnetite concentrates with coal. *Metall. Mater. Trans. B* 44(2):B252–260.

Iluka. 2012. *Iluka's Synthetic Rutile Production*. Perth, Western Australia. www.iluka.com/docs/default-source/mineral-sands-briefing-papers/iluka's-synthetic-rutile-production-june-2012.

Iluka. 2014. *Mineral Sands Industry: Fact Book*. Perth, Western Australia. www.iluka.com/docs/default-source/industry-company-information/the-mineral-sands-industry-factbook-(feb-2014).

Indian Bureau of Mines. 2015. Ilmenite and rutile. In *Indian Minerals Yearbook 2013*. Vol. 3, Mineral Reviews. Nagpur, India: Indian Bureau of Mines.

Jha, A., Cooke, G., Lahiri, A., and Kumari, E.J. 2011. The alkali roasting and leaching of ilmenite minerals for the extraction of high purity synthetic rutile and rare-earth oxides. In *Energy Technology 2011: Carbon Dioxide and Other Greenhouse Gas Reduction Metallurgy and Waste Heat Recovery*. Edited by N.R. Neelameggham, C.K. Belt, M. Jolly, R.G. Reddy, J.A. Yurko. Hoboken, NJ: Wiley.

Jones, G. 2009. Mineral sands: An overview of the industry. In *Seventh International Mining Geology Conference*. Melbourne, Victoria: Australasian Institute of Mining and Metallurgy. p. 213.

Korneliussen, A., McEnroe, S.A., Nilsson, L.P., Schiellerup, H., Gautneb, H., Meyer, G.B., and Storseth, L.R. 2000. *An Overview of Titanium Deposits in Norway*. Bulletin 436. Trondheim, Norway: Norges Geologiske Undersøkelse (Geological Survey of Norway [NGU]). p. 27.

Lakshmanan, V.I., Sridhar, R., and Roy, R. 2011. Synthesis of TiO_2 by an innovative atmospheric mixed–chloride leach process. In EPD Congress 2011. Edited by S.N. Monteiro, D.E. Verhulst, P.N. Anyalebechi, and J.A. Pomykala. Hoboken, NJ: Wiley.

Lei, Y., Li, Y., Peng, J., Zhang, L., Guo, S., and Li, W. 2011. Research on the ball milling and followed by microwave reduction of Panzhihua low-grade ilmenite concentrate. In *2nd International Symposium on High-Temperature Metallurgical Processing*. Edited by J-Y. Hwang, J. Drelich, J. Downey, T. Jiang, and M. Cooksey. Hoboken, NJ: Wiley.

Makarov, Y. 2013. *Bolshoi Seym Deposit: New Prospects for Russian Ilmenite*. Hong Kong, China: IRC Company Limited. www.ircgroup.com.hk/attachment/2013111911581217_en.pdf.

McCoy, D., Coetzee, B., Keegel, M., and Bender, E. 2011. Titanium feedstocks—Opaque quality requirements. In *Eighth International Heavy Minerals Conference*. Melbourne, Victoria: Australasian Institute of Mining and Metallurgy.

McNulty, G.S. 2007. Production of titanium dioxide. Presented at the NORM [naturally occurring radioactive material] V Sevilla EU-NORM Symposium, Spain.

Mehdilo, A., and Irannajad, M. 2012. Iron removing from titanium slag for synthetic rutile production. *Physicochem. Probl. Miner. Process*. 48(2):425–439.

Mindat.org. n.d. Database [Rutile; ilmenite; anosovite]. www.mindat.org. Accessed April 2018.

Minkler, W.M., and Baroch, E.F. 1981. The production of titanium, zirconium, and hafnium. In *Metallurgical Treatises*. Edited by J.K. Tien and J.F. Elliott. Warrendale, PA: TMS.

Morrisset, C-E. 2008. Origin of rutile-bearing ilmenite Fe-Ti deposits in proterozoic anorthosite massifs of the Grenville province. Ph.D. thesis, University of British Columbia, Vancouver, BC.

Murty, CVGK, Upadhyay, R., and Asokan, S. 2007. Electro smelting of ilmenite for production of TiO_2 slag—Potential of India as a global player. Presented at the INFACON: International Ferro-Alloys Congress 11, New Delhi, India, February 18–21.

Nell, J. 2007. Upgrading of titania slag: A mineralogical perspective. www.mintek.co.za/Mintek75/Proceedings/E03-Nell.pdf. Accessed April 2018.

Park, E., and Ostrovski, O. 2003. Reduction of titania–ferrous ore by carbon monoxide. *ISIJ Int.* 43(9):1316–1325.

Pellant, C. 2002. *Smithsonian Handbook: Rocks and Minerals*. New York: Dorling Kindersley.

Qian, X., Yjun, F., Lin, S., Jianfeng, J., and Jihong, D. 2011. Investigations in the mechanism of acid leaching of alkali-activated ilmenite concentrate and titanium-rich slag. In *Supplemental Proceedings: Materials Processing and Energy Materials*, vol. 1. Hoboken, NJ: Wiley.

Rio Tinto. n.d. About Rio Tinto [History; locations]. www.riotinto.com. Accessed April 2018.

Rozendaal, A., Philander, C., and Carelse, C. 2010. Characteristics, recovery and provenance of rutile from the Namakwa Sands heavy mineral deposit, South Africa. *J. S. Afr. Inst. Min. Metall.* 110(2):67–74.

Sandvik, K., Norkyn, P., and Malvik, T. 2010. Rutile beach sand from hard rock. In *XXV International Mineral Processing Congress Proceedings*. Melbourne, Victoria: Australasian Institute of Mining and Metallurgy.

Shaohua, J., Libo, Z., Jin-hui, P., Sheng-hui, G., Lei, X., Xin, W., and Meng-yang, H. 2012. Extraction impurities such as Fe, Ca and Mg from a titanium material in an acid chloride system with microwave energy leaching. In *T.T. Chen Honorary Symposium on Hydrometallurgy, Electrometallurgy and Materials Characterization*. Edited by S. Wang, J. Dutrizac, M.L. Free, J.Y. Hwang, and D. Kim. Hoboken, NJ: Wiley.

Shekiro, J.M., Barnes, J.J., Hill, P., Liu, G., Lyke, S.E., and Nguyen, D. 2011. TiRON—A new process for the production of high-grade feedstock for TiO_2 pigment manufacture. In *Eighth International Heavy Minerals Conference*. Melbourne, Victoria: Australasian Institute of Mining and Metallurgy.

Stanaway, K.J. 1994. Overview of titanium dioxide feedstocks. *Min. Eng.* 12:1367–1370.

Steenkamp, J.D., Pistorius, P.C., and Allen, N.R. 2005. Reflections on ilmenite roasting and magnetic separation. In *Heavy Minerals 2005*. Edited by Mustafa Akser and John Elder. Littleton, CO: SME. pp. 133–141.

Taylor, P.R., Shuey, S.A., Vidal, E.E., and Gomez, J.C. 2006. Extractive metallurgy of vanadium-containing titaniferous magnetite ores: A review. *Miner. Metall. Process.* 23(2):80–86.

Toembaeva, A. 2015. 50 years of UKTMC JSC: Production achievements and development prospects. Trans. Google Translate. *Kazakhstanskaya Pravda* [newspaper], April 1.

USGS (U.S. Geological Survey). 1993–2016. Titanium mineral concentrates. In *Mineral Commodity Summaries*. Reston, VA: U.S. Geological Survey.

Williams, G.E., and Steenkamp, J.D. 2006. Heavy mineral processing at Richards Bay Minerals. In *Southern African Pyrometallurgy 2006*. Edited by R.T. Jones. Johannesburg: Southern African Institute of Mining and Metallurgy.

CHAPTER 12.39

Titanium and Titanium Alloys

M. Ashraf Imam and Francis H. (Sam) Froes

The characteristics of titanium are shown in Figure 1. The conventional processing of titanium metal occurs in four major steps: reduction of titanium ore into *sponge*, a porous form; melting of sponge, or sponge plus a master alloy, to form an ingot; primary fabrication, where an ingot is converted into general mill products, such as a billet, bar, plate, sheet, strip, and tube; and secondary fabrication of finished shapes from mill products. An alternate fabrication route is by powder metallurgy (including additive manufacturing), which can have economic advantages compared to the conventional processing route.

Titanium has high passivity; therefore, it exhibits high levels of corrosion resistance to most mineral acids and chlorides. It is also nontoxic and biologically compatible with human tissue and bone, making it an ideal material for medical implant products (Housley 2007; Imam et al. 2010; Froes et al. 1985).

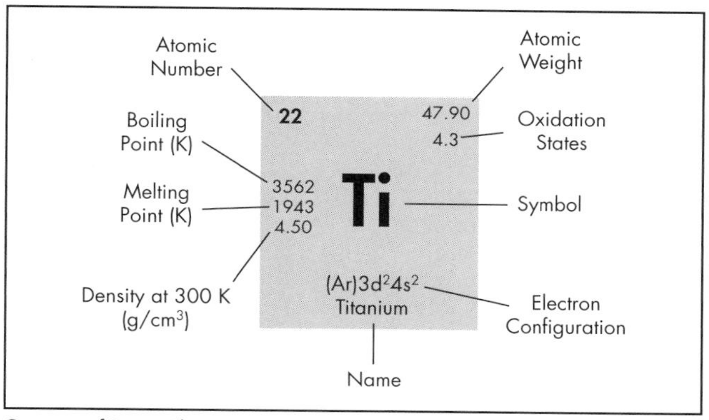

Courtesy of Rector-Elements

Figure 1 Characteristics of titanium

USES

Titanium is used extensively in aerospace applications (both engines and airframes). It is also commonly used in implant components, automobiles (particularly in rotating and reciprocating parts of the internal combustion engine), the oil and gas industries, chemical processing equipment, sporting equipment, and consumer products.

HISTORICAL PRICE

The historical price of titanium sponge is shown in Figure 2. As this sponge is melted to form an ingot and subsequently fabricated to various mill products, the price increases, which is shown in Figure 3.

OCCURRENCE

Titanium occurs in nature in many forms, most notably as ilmenite ($FeTiO_3$) and rutile (tetragonal TiO_2). In the production of titanium metal, only high-grade titanium ores are utilized, such as natural rutile, ilmenite processed to make synthetic rutile, or high-TiO_2 slags. Titanium ores and their processing are discussed in another chapter of this handbook.

PROPERTIES

Titanium alloys may be divided into two principal categories: corrosion-resistant and structural (Housley 2007; Imam et al. 2010; Froes et al. 1985). The corrosion-resistant alloys are generally based on a single-phase (alpha) microstructure with dilute additions of solid solution strengthening and alpha-stabilizing elements, such as oxygen (interstitial), palladium, ruthenium, and aluminum (substitutional). These alloys are used in the chemical, energy, paper, and food processing industries to produce highly corrosion-resistant tubing, heat exchangers, valve housings, and containers. The single-phase alpha alloys provide excellent corrosion resistance, good weldability, and easy processing and fabrication, but a relatively low strength.

In markets served by major U.S. titanium producers, corrosion-resistant alloys comprise approximately 25% of the total output; Ti-6Al-4V, 60%; and all other structural alloys, the remaining 15%.

MANUFACTURING

The commercial production of titanium metal, by the Kroll process (Figure 4), involves the chlorination of rutile (TiO_2)

M. Ashraf Imam, Faculty/Research Professor, George Washington University, Washington, D.C., USA
Francis H. (Sam) Froes, Consultant, Tacoma, Washington, USA

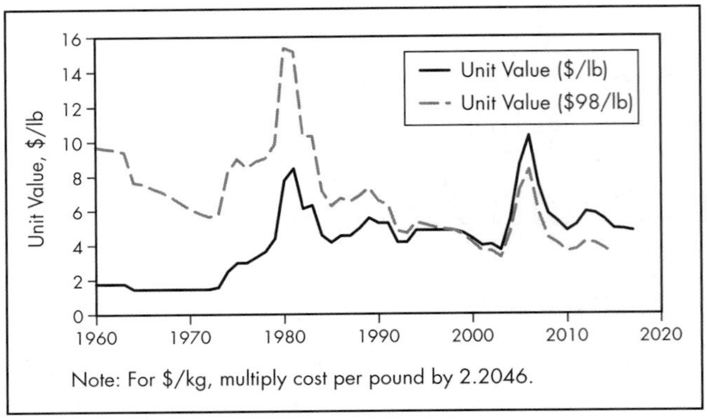

Data from USGS 1960–2017

Figure 2 Titanium sponge price

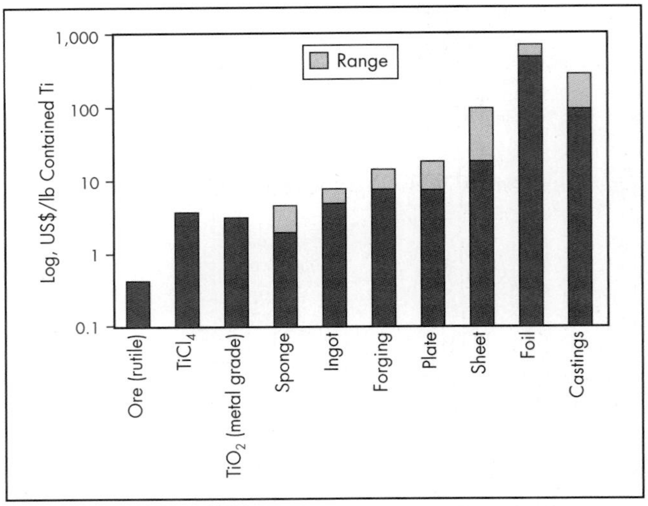

Source: Dutta and Froes 2016

Figure 3 Price of titanium product forms, in 2014 U.S. dollars

in the presence of coke or another form of carbon (Imam et al. 2010; Froes et al. 1985). The resulting titanium tetrachloride ($TiCl_4$; *tickle*) is purified by distillation and chemical treatments and subsequently reduced to titanium sponge using either the Kroll process (Mg) or Hunter process (Na). Titanium tetrachloride for metal production must be of very high purity.

Nearly all sponge is produced by the Kroll magnesium reduction process:

$$TiCl_4(g) + 2Mg(l) \rightarrow Ti(s) + 2MgCl_2(l)$$
$$\Delta G_{900°C} = -301 \text{ kJ } (-72 \text{ kcal})$$

$TiCl_4$ gas (g) is metered into a carbon-steel or 304 stainless-steel reaction vessel that contains liquid (l) magnesium. An excess of 25% magnesium over the stoichiometric amount ensures that the lower chlorides of titanium solids (s) ($TiCl_2$ and $TiCl_3$) are reduced to metal. The highly exothermic reaction [$DH_{900°C}$ = −420 kJ/mol (−100 kcal/mol)] is controlled by the feed rate of $TiCl_4$ at approximately 900°C. The reaction atmosphere is helium or argon. Molten magnesium chloride is tapped from the reactor bottom and recycled using conventional magnesium reduction methods. The production is in batches up to 10 t (metric tons) of titanium. The product, the so-called sponge, is further processed to remove the unreacted titanium chlorides, magnesium, and residual magnesium chlorides. These impurities, which can be as much as 30 wt %, are removed by either acid leaching in dilute nitric and hydrochloric acids at a low energy requirement of approximately 0.3 kW·h/kg of sponge but effluent production of 8 L/kg of sponge; vacuum distillation at 960–1,020°C for as much as 60 hours; or the argon sweep at 1,000°C used by the Oregon Metallurgical Company (Oremet), one of the subsidiaries of Allegheny Technologies Incorporated (ATI). After purification, the sponge is crushed, screened, dried, and placed in airtight, 23-kg drums to await consolidation. The energy required to convert $TiCl_4$ to sponge, which is ready for further processing by the leaching routes, is approximately 37 kW·h/kg of sponge, of which approximately 97% is required for magnesium production. Figure 5 shows the sponge produced by magnesium reduction at ATI Wah Chang (Albany, Oregon, United States).

The sodium-reduction process was employed in Japan, the United States, and England for several years as an alternative to magnesium reduction. The last large production plant was closed in the early 1990s. Although the process was more costly than magnesium reduction, the product contained fewer metallic impurities, that is, iron, chromium, and nickle. Desirable for the electronics and computer industries, high-purity titanium is produced by the Alta Group in Fombell, Pennsylvania (United States), which is a subsidiary of Honeywell Electronic Materials.

Hardness, indicating the degree of purity, is affected both by the interstitial impurities (i.e., oxygen, nitrogen, and carbon) and by the non-interstitial impurity iron. Hardness numbers range from 80 to 150 BHN (Brinell hardness number) units; typical commercial sponge is characterized by 110–120 BHN units. Some developmental processes, such as electrolysis reduction, produce sponge having 60–90 BHN units. Iron impurities in Kroll sponge are difficult to control because of diffusion into the sponge from the reactor wall. In the sodium-reduction process, the sponge is protected from the wall by sodium chloride. The other impurities originate from tetrachloride, residual gases in the reactor, helium or argon impurities, and magnesium or sodium residues.

Sponge Consolidation

Conventionally (the ingot metallurgy approach), the next step is the consolidation of the sponge into ingot. The crushed sponge is blended with alloying elements and cleaned revert (scrap). Consumable electrodes are produced by welding 45–90-kg sponge compactions (electrode compacts) in an inert atmosphere, and then double–vacuum arc remelted (VAR). A portion of the elemental sponge compacts are often replaced with bulk (cleaned) scrap. The ingots are approximately 71–91-cm diameter and long enough to weigh 4.5–9.0 t. The double melt, included in aerospace specifications, is required for thorough mixing of alloying elements, scrap, and titanium sponge, and for improving yields because vaporization of volatiles during the first melt leaves a rough, porous surface. The double melt removes residual volatiles such as Mg, magnesium chloride ($MgCl_2$), Cl_2, and H_2. Triple melts are specified for critical applications such as rotating components in gas turbine engines. The third melt allows more time to dissolve high melting point inclusions that infrequently occur. This is often referred to as *rotating quality titanium*.

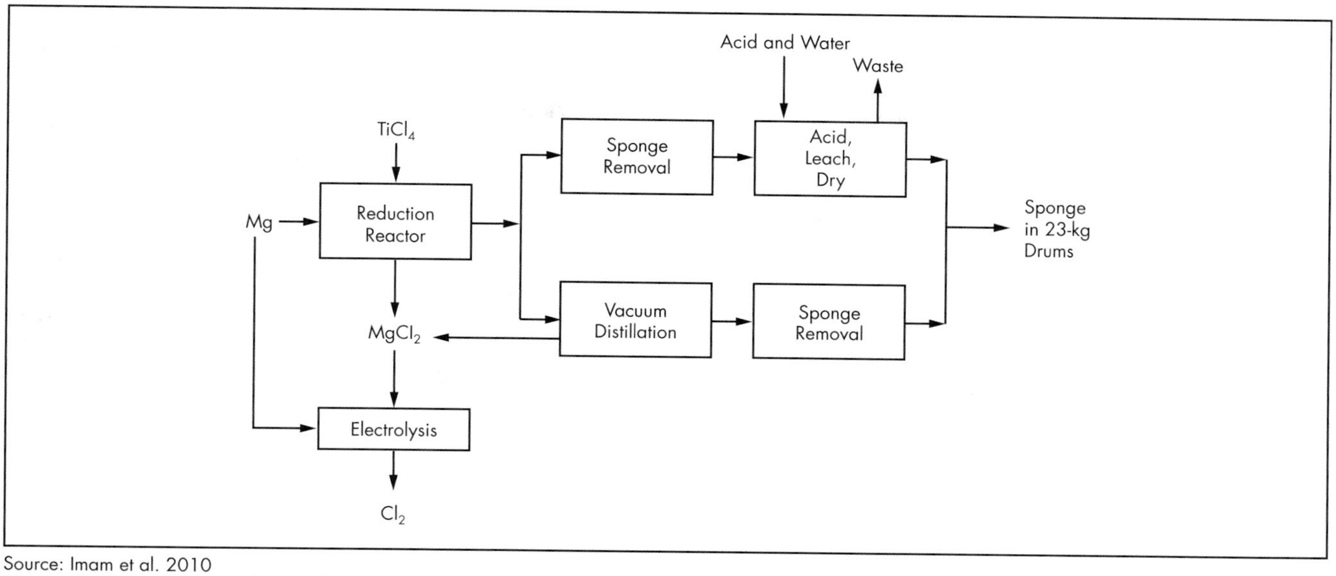

Source: Imam et al. 2010
Figure 4 Flow diagram for Kroll titanium sponge production

Source: Imam et al. 2010
Figure 5 Vacuum-distilled titanium sponge

A two-station VAR furnace for double melting has an annual production capacity of approximately 1,400–3,000 t, depending on product mix, that is, the alloy and the number of remelts. The energy requirement is approximately 1.1 kW·h/kg per single melt. Plasma cold hearth melting (also called plasma arc melting [PAM]) and electron beam cold hearth melting (Figure 6) have been employed more recently for both consolidation and final melting. The hearth processes are well suited for utilizing scrap in various shapes and forms and for avoiding the costly electrode fabrication inherent in consumable vacuum arc melting. In addition, these processes can produce cast metal into shapes such as slabs. For many industrial applications, a single hearth melt is acceptable. The hearth process is low cost and can be designed to trap high-density inclusions, such as carbide tool bits, and oxynitride-rich (type I) inclusions in the hearth skull.

Development of New Production and Processing Technologies

One of the problems with the Kroll process is that it is not continuous, as is steel production, but is a slow batch process, making it very expensive. Kroll himself considered a new process to be essential and eventually achievable, predicting in 1959 that an electrolytic process might be possible within 5–10 years. In the 1950s, at least five companies—DuPont, Horizons, Kennecott Copper, Monsanto, and National Research Corporation—invested heavily and futilely in developing new processes. Subsequent attempts by Dow-Howmet, RTI, and other companies led to incremental improvements in the competing processes, but nothing has ever replaced the Kroll process. Besides the metallurgical difficulties, the extreme volatility of the economic market for titanium made heavy investment in experimental technologies difficult.

Although part of the expense of titanium is caused by the batch sponge process, this actually accounts for less than 40% of the final mill product cost. Pressing and/or forging is the next step. These are time-consuming processes, which explain why a replacement for the batch process alone will not sufficiently bring the price down to make the use of titanium commonplace. It also explains the interest in titanium powder metallurgy (PM) technology (Dutta and Froes 2016; Sun et al. 2016) that would lower the final cost by greatly reducing the number of processing steps, opening the metal to a wider range of applications, for example, military land vehicles and artillery, which need to be both lightweight and strong so that they can be airlifted yet can also withstand combat. PM does not require VAR, electron beam melting, or PAM, producing near-net-shape parts directly. Figure 7 demonstrates the advantage of the PM approach using the Armstrong process. There is a significant reduction in processing steps with PM, which will lower the final product cost, compared to the conventional ingot metallurgy technique.

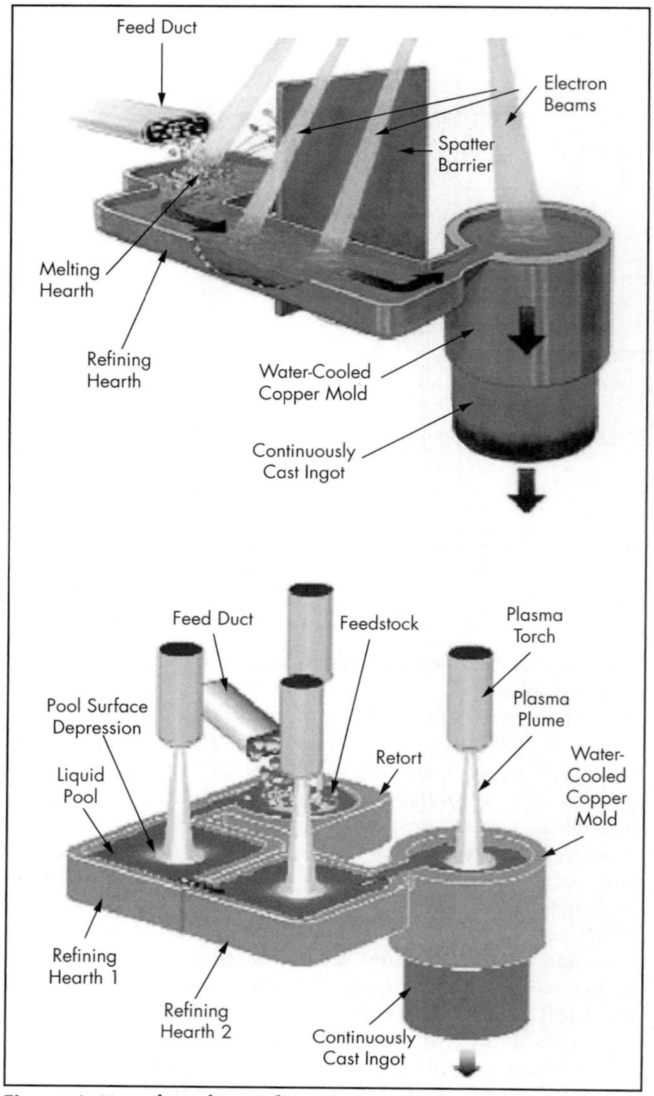

Figure 6 Hearth melting of titanium

Since the 1990s, research has increased on finding replacements for, or modifications to, the Kroll process, with the greatest concentration being on PM. A non-melt consolidation process being developed by Oak Ridge National Laboratory in Oak Ridge, Tennessee (United States), and industry partners may reduce the energy required and the cost to make titanium parts from powders by up to 50%. The process includes roll compaction for directly fabricating sheets from powder and press and sinter techniques to produce net shape components and extrusions. In Japan, research has concentrated on a high-speed continuous process using subhalide reduction in a titanium container.

Because Australia has some of the largest mineral sand deposits in the world, the Commonwealth Scientific and Industrial Research Organisation has developed a technology for producing, it is claimed, cost-effective commercially pure (CP) titanium using a continuous fluidized bed in which titanium tetrachloride is reacted with magnesium (the TiRO process; Froes 2013). Continuous production of a wide range of alloys including aluminides, Ti-6Al-4V, and many Ti-Al alloys has been demonstrated on a large laboratory scale. The powder produced has been used to fabricate extrusions, thin sheet by continuous roll consolidation, and cold sprayed complex shapes including ball valves and seamless tubing. Commercialization of the process is now in the planning stage with a decision to proceed to the pilot plant stage likely to be taken soon.

In the United Kingdom, scientists at Cambridge University have developed a unique approach that uses titanium as an efficient cathode, made possible by the removal of oxygen from the cathode (Froes 2013). The Fray–Farthing–Chen Cambridge process extracts metals from their solid oxides by molten salt electrolysis (generally with a calcium chloride [$CaCl_2$] electrolyte). Using titanium dioxide as an example, the sequence is as follows: TiO_2 (solid, cathode) → molten salt electrolysis → Ti (cathode) + O_2 (anode). The process is being commercialized by Metalysis, located in South Yorkshire in the United Kingdom.

In the United States, several initiatives for industrially scalable methodologies were funded by the Defense Advanced Research Projects Agency (DARPA). In 2003, DARPA listed as the first goal of its Titanium Initiative the "establishment of a U.S.-based, high-volume, low-cost, environmentally benign production capability" (Froes 2013). Its second goal was the development and demonstration of "unique, previously unattainable titanium alloys, microstructures, and properties that enable new high-performance applications" (Froes 2013). Raising the bar significantly higher, DARPA set as its goal the production of high-quality titanium less than US$9/kg for billet with no more than 500 ppm of oxygen. (Generally oxygen in CP titanium is approximately 2,000 ppm.)

One of the projects that received DARPA funding was the Armstrong process as developed by International Titanium Powder (ITP; Froes 2013). In 2008, ITP was acquired by Cristal US (a wholly owned subsidiary of Cristal Global). Cristal Global has extensive experience in the manufacture of titanium tetrachloride and has been supplying the chemical to titanium sponge producers for more than 20 years. The Armstrong process, named after its development director Donald Armstrong (Froes 2013), is continuous and uses molten sodium for reduction of titanium tetrachloride (the Hunter process), which is injected into it as a vapor. Because the sodium exceeds the requirements for the reduction, it cools the reaction products and carries them to separation stages, at which point the excess is removed along with salt. As a result, the powder does not need further purification and can be used in traditional melt-to-billet processes, allowing use of the product in the current value chain. However, the powder is better utilized in direct non-melt consolidation to final end product, eliminating the need to process sponge, thereby reducing supply chain cycle time, energy consumption, manufacturing costs, and environmental impact (Froes 2013). One apparent advantage of the Armstrong process is its ability to produce a range of alloys (including Ti-6Al-4V) as high-quality homogeneous powders suitable for many applications. ITP currently operates a research and design facility and pilot plant in Lockport, Illinois (United States). This site provides limited quantities of materials necessary to successfully complete solid-state consolidation development activities. Additionally, ITP broke ground on a 4-million-t/yr facility in Ottawa, Illinois, in 2010, which was designed to produce both CP titanium and Ti-6Al-4V alloy powder. Independent

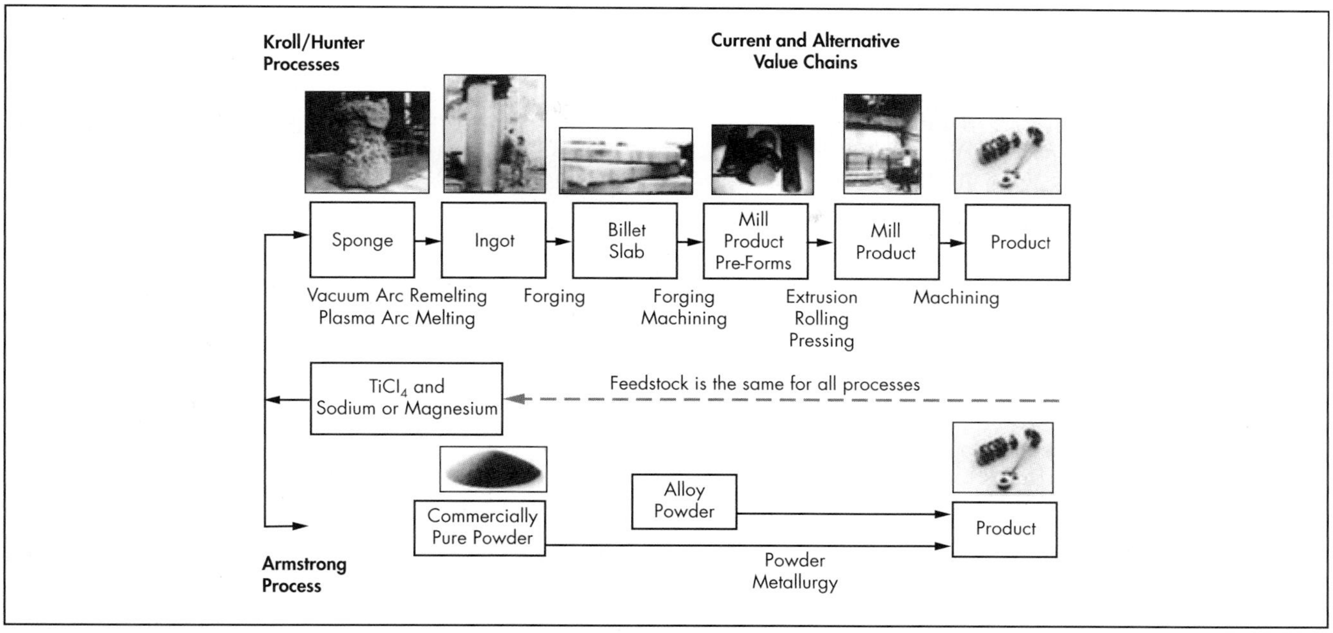

Source: Froes 2013

Figure 7 Titanium powder metallurgy technology

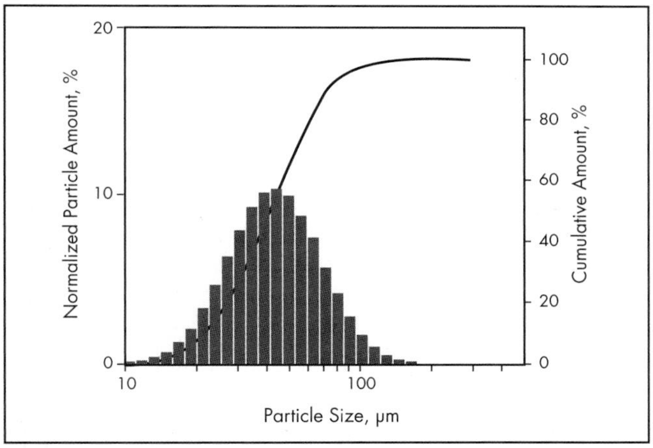

Adapted from Sun et al. 2016

Figure 8 Particle size distribution of hydrogen-assisted reduction of TiO_2 with magnesium

research on the process has found that the use of low-grade $TiCl_4$ leads to higher parts density because of the difference in sintering behavior that affects the entrapment of volatiles within the sintered body. This full-scale facility, however, is still not in production in 2017.

Another process is being developed by the Materials and Electrochemical Research Corporation and DuPont. An electrolytic powder process, it uses a composite anode of TiO_2, a reducing agent, and an electrolyte mixed with fused halides (Froes 2013). The titanium is deposited at the cathode. Projections are for titanium production at significantly lower cost than the Kroll process.

A further emerging powder process is being developed by Advanced Materials Products, which also operates under the name ADMA Products (Froes 2013), in which sponge titanium is cooled in a hydrogen atmosphere rather than the conventional inert gas. The hydrogenated sponge is then easily crushed and in the hydrogenated condition can be compacted to a higher density than conventional low-hydrogen sponge. The remnant chloride content of the hydrogenated sponge is reported to be at low levels (helping to avoid porosity and enhance weldability).

Further development of the use of hydrogen-assisted reduction of titanium oxide with magnesium and production of low-cost spherical meltless titanium powder has been conducted by Sun and coworkers (Sun et al. 2016). Figure 8 shows the size distribution of this powder, which features a reduction to <0.2 wt % O_2 at 50% lower cost than with a conventional hydride-dehydride powder.

The hydrided Ti-6Al-4V or blended elemental material is prepared using the following steps:

1. It is ball-milled in a solvent to reduce the particle size to <10 μm.
2. A thermoplastic binder is added to facilitate formation of granules.
3. The product is spray dried with Ar gas to form spherical granules of Ti-6Al-4V hydride.
4. It is thermally debinded, using an inorganic separator to prevent granules sintering together.
5. Finally, the sintered spherical Ti-6Al-4V powder is deoxidized with calcium or magnesium (with an O_2 content tailored in the range 0.08–0.20 wt %).

ECONOMIC ASPECTS

Before 1970, more than 90% of the titanium produced was used for aerospace applications. By 1982, the percentage decreased to approximately 70%–80%. Military use continually decreased from nearly 100% in the early 1950s to 20% in the late 1990s (Housley 2007; Imam et al. 2010; Froes et al. 1985). However, industrial demand for titanium continued to increase through the late 1990s and into the new millennium. Although the destruction of the World Trade Center in New York City on September 11, 2001, had a marked negative effect on the commercial aerospace industry, Boeing and Airbus began to receive orders for their new planes—the 787 Dreamliner and the A380, respectively—both of which contain between 130 and 150 t of titanium components per unit. Also in 2003, the U.S. Air Force's F-22 Raptor went into production. Simultaneously, there was a price surge in titanium. Between 2003 and 2006, the price of titanium more than doubled. However, the recession that began in 2008 reversed the cycle, with cancellations outnumbering orders of new planes. Production of the F-22 came to an end, replaced by production of the F-35 Lightning II.

With the breakup of the former Soviet Union, the internal consumption of titanium was a modest fraction of its former capacity, thus leaving a large capacity available for export. Throughout the 1990s, the Russian company VSMPO-Avisma upgraded their quality standards so that their products could be used in aerospace applications; it completed its transformation by 2000. By 2009, the giant company was exporting 70% of its products, with the remainder supplied to the domestic market, including large contracts with Boeing and Airbus, as well as Safran Aircraft Engines (previously SNECMA), General Electric, Rolls-Royce, and Pratt and Whitney.

REFERENCES

Dutta, B., and Froes, F.H. 2016. *Additive Manufacturing of Titanium Alloys: State of the Art, Challenges and Opportunities*. Oxford: Butterworth-Heinemann.

Froes, F.H. 2013. Titanium powder metallurgy: Developments and opportunities in a sector poised for growth. *Powder Metall. Rev.* 2(4):29–43.

Froes, F.H., Eylon, D., and Bomberger, H.B., eds. 1985. *Titanium Technology: Present Status and Future Trends*. Boulder, CO: Titanium Development Association.

Housley, K.L. 2007. *Black Sand: The History of Titanium*. Hartford, CT: Metal Management.

Imam, M.A., Froes, F.H., and Housley K.L. 2010. Titanium and titanium alloys. In *Kirk-Othmer Encyclopedia of Chemical Technology*. Hoboken, NJ: Wiley. pp. 1–41.

Sun, P., Fang, Z.Z., Xia, Y., Zhang, Y., and Zou, C. 2016. A novel method for production of spherical Ti-6Al-4V powder for additive manufacturing. *Powder Technol.* 301(November):331–335.

USGS (U.S. Geological Survey). 1960–2017. Titanium. In *Mineral Commodity Summaries*. Reston, VA: U.S. Geological Survey.

CHAPTER 12.40

Tungsten

Harold E. Kelley

PROPERTIES

Tungsten (atomic number 74) is a high-density metal with an extremely high melting point. It was first noted during the 16th century in the tin mines of Germany and later in the tin mines in Cornwall, England. At first, it was said to consume the tin like a wolf and was given the German name *wolfram*. Hence, this is why the atomic symbol is W versus T or Tu.

In the late 1700s, Swedish chemist Carl Scheele identified a compound with calcium and an unknown acid. The acid-forming element was named *tungsten* by Axel Fredik Cronstedt. In Swedish, *tung* means "heavy" and *sten* means "stone." A short time later, the brothers Juan José and Fausto de Elhujar discovered wolframite. Because iron and manganese mainly occur together, and because wolframite contains both iron and manganese ($FeMnWO_4$), it is thus named. The mineral is called ferberite ($FeWO_4$) if it contains only iron. If it contains only manganese, the mineral is called huebnerite ($MnWO_4$). The calcium tungstate that Scheele identified was eventually named after him as *scheelite* ($CaWO_4$). These two minerals—wolframite and scheelite—are the two economically important ores of tungsten and both are mined around the world.

In the mid-1800s, Robert Oxland developed a process to produce pure tungsten metal and iron–tungsten alloys. It would not be until the early 1900s that a practical use for tungsten would be found in filaments for incandescent lightbulbs. About this same time, it was discovered that tungsten could be alloyed with iron to form tool steel. This first became important during World War I when it was misconstrued that Germany would quickly use its supply of ammunition. Germany's use of tungsten–iron alloy as tool steels and cutters produced much more ammunition than thought possible. After the war, it was learned that much of the tungsten had come from the tin mines in Cornwall, England.

Although tungsten carbide had been known since the 1890s, it only became practical to use in 1927 when the Krupp Company in Germany developed a process to cement the carbide together with a binder. Once again, Germany used this new technology to help arm itself for World War II, even with limited natural resources. After the war, tungsten carbide development outpaced tungsten, although both products are extremely important in today's manufacturing.

ORES

Tungsten ores are located around the world including the United States, Russia, Southeast Asia, Portugal, Bolivia, Brazil, Canada, Africa, Korea, and China. China holds about 70% of the world reserves and is the largest producer. Tungsten deposits are all formed from igneous activity, mostly following a magmatic-hydrothermal model. Although the commercial ores are scheelite and wolframite, tungsten does form several other minerals, including the following (Yih and Wang 1979):

- Tungstite or meymacite ($WO_3 \cdot H_2O$)
- Ferritungstite [$Ca_2Fe_2^{2+}Fe_2^{3+}(WO_4)_7 \cdot 9H_2O$]
- Raspite (tetragonal) and stolzite (monoclinic) ($PbWO_4$)
- Russelite (Bi_2WO_6)
- Tungstenite (WS_2)
- Sanmartinite ($ZnWO_4$)
- Cuprotungstite [$Cu_2(WO_4)(OH)_2$]
- Anthoinite [$Al(WO_4)(OH) \cdot H_2O$]
- Powellite [$Ca(MoW)O_4$]

The main ore deposits are classified as pegmatite, greisen, and contact metasomatic. Several minor deposits include hot springs, replacement, pnuematolytic, secondary enrichment, and placer. The mining of these deposits follow three traditional methods depending on the type: underground, open pit, and placer. Underground is the most common.

Ore Valuation and Trends

The ore price is based on tungsten content and the market value of ammonium paratungstate (APT). The unit of measurement is the metric ton unit (MTU), which is equal to 10 kg. During the 1980s and 1990s, the price was between US$25 and US$30/MTU. This was mostly because of the large amounts of Chinese ore that were available. Because of this low-cost ore, most Western producers were not profitable, and many mines closed, including all the mines in the United States and Canada. At the turn of the century, the Chinese

Harold E. Kelley, Director of Technical Services, TungsteMet, Oak Hill, West Virginia, USA stopped all ore exports, and the price gradually increased to nearly US$500/MTU in 2008. With the economic downturn, the price slowly dropped to US$150/MTU and has currently rebounded to US$282/MTU. The first new and only mine in the United States is scheduled to open on the Utah–Nevada border, south of Wendover, Nevada, when the price of tungsten rebounds.

MINERAL PROCESSING

Once the material is removed from the ground, the beneficiation steps required are crushing, grinding, sizing, and concentrating; the steps are supplemented with leaching, roasting, or magnetic separation if needed. The crushing and grinding is typically accomplished with cone crushers, roll crushers, jaw crushers, ball mills, and rod mills. Sizing is mainly done with screens and classifying. The typical size of finished ore is 1–4 mm (sand or finer). Wolframite is usually coarser than scheelite. Concentrating is typically performed with gravity tables and/or spirals or flotation. Flotation is most often used when treating scheelite, as scheelite produces more fines during crushing. This concentrating purity of tungsten oxide (WO) is around 60%–70%, with some processes requiring at least 65% tungsten trioxide (WO_3) content.

Extraction Processes

At this point, some of the ores need to be processed further to be used. Many of the ores (scheelite and wolframite) are roasted to remove sulfur. These ores are then further treated with chemicals. The main process is to add hydrochloric acid (HCl) and carbon to form tungstic acid. The tungstic acid is then processed into APT by digesting the tungstic acid with ammonia. To help purify the APT, additives such as magnesium oxide and carbon (to help remove phosphorus, arsenic, iron, and silica) are added before filtration. The purified APT solution next goes through an evaporation process to produce APT crystals. These APT crystals are converted to oxides. The three main oxides produced are WO_3 (yellow), tungsten dioxide WO_2 (brown), and an intermediate oxide W_4O_{11} (blue). These oxides are produced by heating the APT crystals in purified air. The blue oxide is done with a slightly reducing atmosphere. Both of these can be done with either stationary or rotary furnaces.

The oxides now need to be further reduced with hydrogen. The same types of furnaces as before can be used. This is accomplished by using hydrogen and a temperature of ~850°C. The end product is pure tungsten metal powder, which can now be further processed into a variety of useful products.

Tungsten and Ferrotungsten Powder Production

Although there are several other methods (reduction with carbon, aluminothermic reduction, silicothermic reduction, reduction with calcium, and reduction with other metals such as zinc) of producing tungsten and ferrotungsten, the major method is to use powder metallurgy. The tungsten metal powder can be made at different micrometer levels. Mainly, these range from 1 to 15 µm but can be made as large as 500 µm with doping agents added. These tungsten metal powders can be mixed with iron at a 35:65 ratio to form ferrotungsten. The same tungsten particles can be milled with lube or wax for green strength and then pressed into shapes. The two main ways of pressing are mechanical and isostatic.

With mechanical pressing, a die cavity is filled with the powder. Two rams compress the powder at high pressure (typically, mechanical presses range from 50 to 500 t). The isostatic pressing occurs in a chamber filled with a liquid, usually water. The powder is placed in a tube or boot made from rubber or plastic. The tube is sealed and placed in the chamber then pressed with an average of 172 kPa. The boot is withdrawn from the chamber and dried. The slug or billet is removed from the boot. This slug has green strength from the wax or binder and can either be sintered as is or green machined into shapes, then sintered.

Tungsten Production

The billets or parts can be sintered at high temperatures. This process solidifies and produces a dense metallurgical bond. These large billets can be sintered using induction. The smaller parts can use more conventional furnaces. Temperatures of 2,300°–2,400°C are needed during the final cycle to sinter the product. This process is also done in a reducing environment, most commonly using hydrogen.

The sintered parts can be finished machined, ground, or polished. The billets are usually extruded, forged, rolled, or swaged. The process used is dependent on the end product desired and the crystal structure of the grains. The majority are extruded to form bars or rods. The rods are further processed to form wire that will be used in lamp filaments, heater filaments, and heat elements. Other products include sheets, bars, and tubes. These materials will become rocket nozzles, electrodes, supports, seals and contacts, to name a few. For most of its history, filaments for lamps and furnaces were the major use of pure tungsten.

Other Tungsten Products and Applications

Another industrial application uses the same milling process to mix the tungsten with nickel, iron, and/or copper to produce what is known as *heavy* metal. This material is mainly used as radiation shielding, heat sinks, and counterweights. These products are used in electronics, medical equipment, and aerospace industries. Powder metallurgy is used to blend powders of tungsten and the alloying ingredients. Then they are milled, pressed, or extruded and sintered. These parts can be machined or ground to finish quality. Medical equipment that performs radiography, magnetic resonance imaging, and computed tomography would not be available without these alloys.

The original use of tungsten was to make steels, especially tool steels. The high-strength steels produced by adding tungsten began the modern machining era during the first quarter of the 20th century. Later, high-temperature and wear-resistant steels were developed. The wear-resistant steels mixed with cobalt and chrome are trademarked as Stellite alloys, which are used in high-temperature applications that also require wear and corrosion resistance.

Some magnetic steels, stainless steels, and specialty steels are also made with tungsten. Again, many of these are used in specific applications. Pumps, turbines, nozzles, tubing, food industry knives, as well as space vehicles, all use these types of steels.

Tungsten Carbide

Tungsten carbide (WC) became important before World War II, critical during the war, and since has become the major use of tungsten. Tungsten forms two carbide phases: The stable completely carburized WC and W_2C. Macrocrystalline carbide, cast carbide, and conventionally carburized tungsten carbide are the three main types of tungsten carbide. Macrocrystalline

WC is produced directly from the ores through an alumino-thermic process. Macrocrystalline WC is single-crystal WC that has a stoichiometric carbon content of 6.13%. This material is used in hard-facing welding rods; plasma tungsten arc welding powders; hot-press powders; cemented carbide powders and parts; and matrix powders, which are infiltrated with copper alloys to produce polycrystalline diamond (PCD) oil and gas drill bits.

Cast carbide is the near eutectic of the two phases of tungsten carbide (WC/W_2C). The carbon content varies between 3.8% and 4.2%. It is produced by arc melting tungsten and carbon and pouring it into chilled molds. The quicker the material cools, the better the microstructure. The material is then crushed into various screen sizes. The best microstructure is a finely layered or feathery structure of the WC and W_2C phases.

Over the last decade, a spherical cast carbide has been produced by melting normal cast carbide through a plasma torch and cooling immediately with an inert atmosphere. This material is even harder and more erosion resistant. Spherical cast carbide is expensive to produce and therefore used in only specific high-wear applications. Cast carbide is used in most of the same areas as macrocrystalline WC.

Conventionally carburized WC is the most abundantly used tungsten carbide. This WC is made by adding tungsten and carbon into boats. They are then placed through a furnace to cause a diffusion of the carbon into the tungsten. The carbon level has to be controlled to produce a near stoichiometric product. Most of the WC is from submicrometer to 30 μm in size, but with special treatment, it can be made to 60 mesh. This process produces a low-impurity carbide with specific micrometer sizes. It is used to make cemented carbide powders and sintered parts.

Cemented Tungsten Carbide

Usually, cemented carbide is what most people mean when talking about tungsten carbide. Tungsten carbide is typically cemented together with a binder of cobalt. Nickel and iron can be used with other smaller fractions of materials such as titanium carbide, tantalum carbide, niobium carbide, vanadium carbide, or chromium carbide. The amount of cobalt varies from 6% to 30%, depending on the properties desired. The lower the cobalt content, the more wear resistant. The higher the cobalt content, the tougher the grade.

Another factor that becomes important is that the size of the carbide used has an effect on the toughness. Similar to producing tungsten powders, cemented carbides are milled with a wax or lubricant and a milling liquid. The powders are then either spray dried or tumble dried. These powders are then pressed into green parts. Like tungsten, they can be mechanically pressed, cold isostatically pressed (CIP), or extruded. Most parts are mechanically pressed using single- or dual-axial presses. The cavity of the press is filled with powder and then loaded from the top, the top and bottom, or even with a floating die. The pressed parts are placed on trays, which are loaded into furnaces for sintering.

The CIP method is used for bigger parts that are usually green machined before sintering. The sintering process is similar to tungsten with vacuum furnaces being cycled through a ramp, de-lube, ramp, and then held at sintering temperature before entering a slow cooling section.

Many tungsten cemented carbide parts are processed using sinter-HIP (hot isostatic pressure). This is the same as sintering, except once the sintering temperature is reached, the furnace is filled with an inert gas to produce a high pressure. Using sinter-HIP helps eliminate very fine porosity for maximum density. Depending on the type of parts, after sintering, they can be machined, ground, polished, etched, or finished using electrical discharge machining (EDM). They can also be coated by chemical vapor deposition or physical vapor deposition for enhanced properties. These cemented carbide parts are used for multiple applications.

Applications. Most cemented carbide parts are produced for metalworking, including cutting tool inserts, bores, drills, end mills, taps, and milling tools. In the oil and gas industry, carbide is found in compacts, bit nozzles, stabilizers, mud motor bearings, balls and seats, shafts, substrates for PCD cutters, anvils for the cubic presses to produce PCD cutters and synthetic diamonds, fracking nozzles, flow control parts, and disk stacks. Mining and construction applications include inserts for mining bits, picks for road planning and coal mining, roof bits, trenching tools, snowplow blades, skid plates, and ground-engagement tools. In general, industry carbide is supplied for rollers, pumps, centrifuge nozzles, slurry nozzles, mixer paddles, crusher teeth, seals, hammers, EDM blanks, and saw teeth.

Hard-facing applications. Hard-facing, thermal spray, and cladding are also major uses of tungsten carbide. As discussed earlier, plasma tungsten arc powders use WC mixed with nickel chrome alloys and are applied to all kinds of wear parts. Cladding is used for extrusion barrels and screws for plastics and pet foods, pump parts, turbines, fan blades, boiler tubes, screens, shafts, and bearings.

Thermal spray powders are made from sintered spherical spray dried or sintered crushed cemented carbides. These are used in plasma and spray guns to spray jet engine turbines, shafts, molds, downhole, and other parts needing a thin uniform coating.

Hard-facing wires, stick electrodes, and tube-filled rods are made containing macrocrystalline WC, cast carbide, cemented carbide pellets, and crushed recycled cemented carbide. These are deposited with oxyacetylene gas, arc, tungsten inert gas or metal inert gas welding. Hard-faced items include everything from tri-cone mill tooth drill bits, drums, agriculture and mining wear parts to flour mill hammers. Because of the extensive industrial applications, tungsten is one of the most critical strategic metals.

REFERENCE

Yih, S.W.H., and Wang, C.T. 1979. *Tungsten: Sources, Metallurgy, Properties, and Applications*. New York: Plenum Press.

CHAPTER 12.41

Uranium

Henry Schnell

Uranium is a naturally occurring element with an average concentration of 2.8 ppm in the earth's crust. Traces of it occur almost everywhere. It is more abundant than gold, silver, or mercury, with about the same abundance as tin but slightly less plentiful than cobalt, lead, or molybdenum. Vast amounts of uranium also occur in the world's oceans, but in very low concentrations.

Miners had noticed uranium minerals for a long time prior to the discovery of uranium in 1789. The uranium mineral pitchblende, also known as uraninite, was reported from the Krušnéhory (Ore Mountains), Germany, as early as 1565. Other early reports of pitchblende date from 1727 in Jáchymov (Czech Republic) and 1763 in Schwarzwald, or the Black Forest (Germany). Eugene M. Péligot isolated the element in 1841. Antoine H. Becquerel discovered its radioactivity in 1896. Before the discovery of nuclear fission by Otto Hahn and Fritz Strassmann in 1938, the principal use of uranium (chiefly as the oxides) was in pigments, ceramic glazes, and a yellow-green fluorescent glass, and as a source of radium for medical purposes. It has also been added to steels to increase their strength and toughness. However, because of the high toxicity (both chemical and radiological) of uranium and its compounds, and because of their importance as nuclear fuel, these earlier uses have been curtailed.

Today, the main market for uranium is for nuclear fuels and medical requirements for some of the associated isotopes. Uranium differs from many metal commodities, as its markets are energy demand based. The uranium market has had many ups and downs—affected by the energy crises of the 1970s and nuclear accidents such as Three Mile Island, Chernobyl, and, more recently, Fukushima—that continue to fuel the nuclear debate. However, demand for uranium continues to grow to supply the existing 447 worldwide nuclear reactors and 61 new reactors under construction in 15 countries.

Uranium minerals are very commonly distributed geographically and in a very wide range of mineral species and deposit grades. In the past 25 years, this diversity has been evident, with production in more than 20 countries and many more countries preparing for production to support their nuclear power industries. In 2014, approximately 54% of world production came from just 10 mines in six countries, with these six countries providing 85% of the world's mined uranium.

With uranium found in a large variety of settings, uranium processing must be adaptive to these settings and the large variety of mineralization. Historically, flow sheets were based on a tank leach process that tended to have very similar flow sheets that are referred to as "conventional." There has been a trending away from such a conventional flow sheet, or at least the technology within each unit operation, as high-grade deposits were exploited in Canada or as extreme low grades were considered in places such as Namibia.

The large diversity of mineral species and ore grades has resulted in an increased diversity of processing options and interesting adaptations of hydrometallurgical processes. The high grades found in Canadian ores have seen remote mining methods and special plant designs to address the challenges of these special ores. Kazakhstan has advanced the in situ leach/leaching (ISL) technologies, making it currently the largest producer. The low grades found in Namibia have resulted in higher tonnage operations (up to 100,000 t/d of ore) and significant technology advances of preconcentration, ion exchange (IX) adaptations, and other innovations. The advancement of resins has resulted in the reemergence of IX options. The ores of Niger similarly needed adaptation to deal with high clays found in these ores.

As concern about environmental impacts and climate change has grown, a number of high-profile environmentalists have decided that this is a more serious problem than their previous concerns with nuclear power. They have, to varying degrees, either changed their public stance or conceded that because nuclear power is virtually emission free regarding carbon dioxide (CO_2), it merits at least grudging support as part of the response to increasing atmospheric CO_2 levels and the depletion of fossil fuels. But public education and the mystique around radiation continue to play an important part in advancing uranium and nuclear projects.

Uranium will continue to lead the way in technology advances as by-product production gains importance. The environmental advances for these mines will result in

Henry Schnell, Retired, Formerly with AREVA (Paris, France), Eagle Bay, British Columbia, Canada adaptations for other metal resources. Hydrometallurgy will continue to provide new adaptations, and these advances will also flow into other metallurgical areas.

GEOLOGY, MINERALOGY, AND URANIUM DEPOSIT TYPES

History of Uranium Use

Uranium was discovered by H.M. Klaproth in 1789 in samples coming from the silver mines of Bohemia but was first isolated as a metal by E.M. Peligot, a chemist from the Baccarat glass factory in France. Uranium was initially used essentially because of the bright yellow color of the uranyl ions and hexavalent uranium minerals to color the glasses (autunite) until the early 1980s and was used for body paint by the early Navajos and Ute in Colorado (carnotite).

Exploration for uranium ores began after the French chemist Marie Curie discovered radium in 1898. About 1 g of radium was produced for every 8 t (metric tons) of uranium ore. The total amount extracted until 1939 did not exceed about 6,000 t U. Uranium ores were mined mainly from the former Belgian Congo (Shinkolobwe in the Democratic Republic of the Congo), Bohemia (Jáchymov), the Colorado Plateau in the United States, and Port Radium in Canada. Radium had a large variety of applications but was mainly used for radiotherapy.

A second rush for uranium exploration started at the end of the 1940s, essentially for military use after the discovery of atomic fission by Hahn and Strassmann in 1938 and development by the U.S. Atomic Energy Commission of the first atomic bombs by the end of World War II. For these bombs, the uranium was mainly from Shinkolobwe, a uranium inventory present on a Belgian ship rerouted to New York Harbor when the Germans invaded Belgium in 1939. Until the end of the 1960s, uranium was mined without economic considerations. In 1959, production reached a peak of 50,000 t U/yr.

The third and strongest rush occurred in the 1970s as a consequence of the huge increase of oil prices in 1973. A vast program of nuclear reactor construction started all over the world. Before 1973, uranium prices, which were stable and very low (~\$6/lb U_3O_8 [uranium oxide]), peaked in 1979 at \$44/lb U_3O_8. Another production peak was reached in 1980 at 66,000 t U/yr. World uranium production largely exceeded military and civil needs of about 20,000 t/yr until 1989. The prices declined to about \$10/lb U_3O_8 from 1990 to 2005, mainly because of the accumulation of huge uranium inventories and the 1993 Megatons to Megawatts Program agreement. This agreement, between Russia, RWE Nukem, AREVA, and Cameco for the dilution of military highly enriched uranium, had delivered the equivalent of 9,000 t of natural uranium per year on the market, until 2013. The decline of production created a strong unbalance (about 20,000 t/yr) with respect to the need of the nuclear reactors, during 24 years. After the oil crisis in 2004 and 2008, uranium production and uranium prices started again to increase, with a short peak at \$137/lb U_3O_8. Then the uranium price declined again because of stabilization followed by a strong decline in oil prices resulting from a world financial crisis, a strong increase of shale gas production in the United States, and the Fukushima nuclear accident. At a price of about \$40/lb U_3O_8, most uranium deposits cannot be mined in economic conditions.

Uranium Geochemistry

The initial uranium concentration for primitive Earth of 7 ppb has been progressively enriched through time, first in the mantle after the segregation of its Ni-Fe core and then in the continental crust by partial melting of the mantle and fractional crystallization. Average uranium concentration in the continental crust is 1.7 ppm and reaches 2.7 ppm in its upper part. Uranium tends to be enriched in the most silicic magmatic rocks (up to several tens to hundreds of parts per million), such as granites and rhyolites and in sedimentary rocks rich in organic carbon (black shale, lignite) or phosphorus (phosphorites). Uranium concentration in seawater is 3.3 ppb and only about one-tenth of parts per billion in lake and river waters.

Uranium has three natural radioactive isotopes: ^{238}U, ^{235}U, and ^{234}U with relative abundances of 99.274%, 0.720%, and 0.0055%, and half-lives of 4.469×10^9, 0.7038×10^9, and 246,000 years, respectively. The present $^{238}U/^{235}U$ ratio is 137.88 everywhere on the earth, except in some ore bodies at Oklo (Gabon, in Africa) where natural fission reactions have occurred. The decays of ^{235}U and ^{238}U are complex, with many intermediate daughter products until the final stable daughter isotope ^{207}Pb in the ^{235}U chain and ^{206}Pb in the ^{238}U decay chain are reached. These daughter elements have variable geochemical properties and thus can be differentially mobilized and dispersed into the environment, resulting in disequilibrium in the radioactivity of isotopes in the decay schemes. This disequilibrium must be assessed for using radiometric measurements in the evaluation of uranium ore grades in mining.

Uranium oxide has the lowest solubility in water in reducing conditions at neutral pH and low temperature. The solubility of uranium minerals in dilute waters in oxidizing conditions is controlled primarily by complexes of the uranyl cation UO_2^{2+}. Its solubility increases markedly in the presence of anions such as F^-, Cl^-, CO_3^{2-}, SO_4^{2-}, and PO_4^{3-}, and with increasing acidity or alkalinity of the waters (Langmuir 1978). For example, uranium is extremely soluble in the strongly acid, oxidizing water commonly associated with acid mine drainage because of the formation of uranyl sulfate complexes.

Uranium Deposit Types

The most recent and most widely used classification of uranium deposits is the one provided by the International Atomic Energy Agency (IAEA) in the last edition of the "*Red Book*" (NEA and IAEA 2016), mostly based on the work of Dahlkamp (2009), with 15 main types and 50 subtypes and classes. The classification is based on deposits listed in the IAEA World Distribution of Uranium Deposits (UDEPO) database (IAEA 2009) according to the nature of host rock lithologies. The 15 types of deposits are listed here from high temperature to surficial environments:

1. Intrusive-type deposits with two subtypes: (1.1) *intrusive anatectic* related to partial melting processes and hosted in granite-pegmatite from very high-grade metamorphic domains (e.g., Rössing and Husab in Namibia); and (1.2) *intrusive plutonic* related to magmatic differentiation processes, subdivided into three classes: (1.2.1) *quartz monzonites* associated with porphyry Cu deposits (Bingham, United States); (1.2.2) *peralkaline complexes* (Kvanefjeld, Greenland); and (1.2.3)

carbonatites (Palabora, Republic of South Africa). Resources are generally large, but the grades are generally low.

2. Granite-related deposits include veins hosted in granite or adjacent rocks and disseminated mineralization in episyenite bodies where quartz has been hydrothermally dissolved. In the Hercynian belt of Europe, these deposits are generally associated with large, peraluminous, two-mica granites. Two subtypes are defined: (2.1) *endogranitic*, occurring within the granitic pluton; and (2.2) *perigranitic*, occurring within the country rocks. Resources are small to large with low to high grades.
3. Polymetallic iron-oxide breccia complex deposits belong to the iron-oxide-copper-gold (IOCG) category. Olympic Dam (Australia) is the only IOCG deposit where uranium is extracted as a by-product of copper and gold. It is the world's largest uranium resource with more than 2 Mt (million metric tons) U at 230 ppm. These deposits occur as huge granite (Olympic Dam) or metasedimentary-metavolcanic (Salobo, Brazil) breccias associated with large quantities of magnetite or hematite. Uranium is disseminated as uraninite, coffinite, and U-Ti oxides.
4. Volcanic-related deposits occur within and near volcanic calderas filled with mafic to felsic volcanic lavas or pyroclastic rocks and interlayers of clastic sediments with (4.1) *structure-bound deposits* where uranium mineralization is controlled by veins and stockworks that may extend deeply into the basement; (4.2) *stratabound deposits* as disseminations in clastic sediments; the largest district of these subtypes is the Streltsovska caldera in Russia; and (4.3) *volcano-sedimentary deposits* where uranium mineralization is nearly concordant, low grade, in carbonaceous lacustrine sediments, with an important tuffaceous component (Anderson mine, United States). The association of U-Mo-F is quite common.
5. Metasomatite deposits occur in rocks affected by metasomatic alteration: (5.1) *Na-metasomatite subtype* (e.g., Novokonstantinovskoye deposit in Central Ukraine); (5.2) *K-metasomatite subtype* (only known in the Elkon District in Russia) that produced albitized or illitized rocks, respectively, along crustal-scale ductile to brittle structures; the resources range from medium to very large with average grades of about 0.1 wt % uranium; (5.3) *skarn subtype* is developed at the contact between granites and enclosing carbonate rocks (e.g., Mary Kathleen, Australia).
6. Metamorphite deposits occur are disseminations, veins, and shear zones within regionally metamorphosed rocks having no known relation to granitic intrusions, with three subtypes: (6.1) rare *stratabound* (e.g., Forstau, Austria); (6.2) *structure-bound* (e.g., Schwartzwalder, United States); and (6.3) *marble-hosted phosphate* (e.g., Itataia in Brazil). These deposits are highly variable in size and grade.
7. Proterozoic unconformity deposits occur below, along, or above the contact between an Archean to Paleoproterozoic crystalline basement and Proterozoic oxidized siliciclastic sediments. Deposits consist of pods or veins essentially consisting of massive uranium oxide. Strong quartz dissolution is generally associated with them. Three subtypes are distinguished: (7.1) *unconformity-contact* (e.g., Cigar Lake, Athabasca Basin, Canada); (7.2) *basement-hosted* (e.g., Ranger in Australia); and (7.3) *stratiform structure-controlled* (e.g., only at Chitral and Lambapur in India). The resources range from medium to large with very high grades in Canadian deposits but low grades in Indian deposits.
8. Collapse breccia pipe deposits are made of down-dropped fragments from overlying lithological sedimentary units filling large karst dissolution cavities developed in carbonate layers. Uranium oxides occur in the breccia matrix with sulfides or oxides of Cu, Fe, Zn, Pb, Ag, Mo, Ni, and Co. They are only known in the United States (Grand Canyon area). Resources are small to medium, but grades average 0.5% U.
9. Sandstone deposits occur in sandstones deposited in continental fluvial or marginal marine environments. They represent more than 600 deposits over the 1,580 registered in the UDEPO database. Five main subtypes are distinguished: (9.1) *basal channel deposits* occur within elongated channels filled with alluvial-fluvial sediments, and uranium is dominantly associated with plant debris (e.g., Khiagdinskoye, Vitim District, Russia); resources are large to medium, at grades ranging from 0.01% to 3% U; (9.2) *tabular deposits* are elongated parallel to the stratigraphy within reduced sediments (e.g., Akouta, Niger; and the Colorado Plateau, United States); individual deposits are small to large, at grades ranging from 0.05% to 0.5% U; (9.3) *roll-front deposits* with a crescent shape ore-body morphology (e.g., Inkai, Kazakhstan; and Crow Butte, United States); resources range from small to very large, with grades varying from 0.05% to 0.25% U; (9.4) *tectonic-lithologic deposits* occur along fault zones and adjacent sandstone beds where uranium is associated to hydrocarbons or detrital organic matter (e.g., Lodève District, France; Franceville Basin, Gabon); deposits are small to medium, and grades vary from 0.1 to 0.5% U; and (9.5) uranium mineralization associated to mafic dykes and sills in Proterozoic sandstones (Matoush, Otish Basin, Canada); small to medium deposits with low to medium grades (0.05%–0.40%).
10. Paleo quartz-pebble conglomerate deposits are characterized by uranium oxides associated with pyrite and detrital minerals, including gold, Fe-Ti oxides, and other sulfides in monomictic conglomerates, older than 2.4 Ga (billion years), deposited as basal (Elliot Lake District, Canada) or intraformational units (Witwatersrand Basin, Republic of South Africa) of fluvial to lacustrine braided streams. Hydrothermal remobilization may be important. Two subtypes are recognized: (10.1) *uranium dominant* associated with rare earth elements and Th (Elliot Lake), and (10.2) *Au-dominant* where uranium is a by-product (Witwatersrand).
11. Surficial deposits occur in near-surface sediments and soils, with the largest ones in calcretes. They occur (11.1) in *fluvial valleys* along drainage channels (e.g., Langer Heinrich, Namibia); (11.2) in *lacustrine-playa sediments* (e.g., Lake Way, Australia), where uranium always occurs as vanadates; (11.3) in *peat bogs* (e.g., Kamushanovskoye, Kyrgyzstan; (11.4) in *karst caverns* (Tyuya-Muyun, Kyrgyzstan); and (11.5) as *pedogenic/fracture-filling* (e.g., Beslet, Bulgaria).
12. Lignite-coal deposits correspond to elevated uranium contents in lignite or coal mixed with detrital minerals (silt, clay) adjacent to carbonaceous mud and silt or sandstone beds, with two subtypes: (12.1) *stratiform*

deposits (the Dakotas, United States), and (12.2) *fractured-controlled deposits* (e.g., Freital, Germany).

13. Carbonate deposits are hosted in limestones or dolostones. Mineralization can be (13.1) *syngenetic stratabound* in carbonate (e.g., Tumalapalle, India), (13.2) in *cataclastic carbonate* (e.g., Todilto Formation, United States), and (13.3) in *karst* (e.g., Pryor Mountains, United States).
14. Phosphate deposits with three subtypes: (14.1) *minero-chemical phosphorite* in marine continental-shelf origin containing synsedimentary, stratiform, disseminated uranium in the structure of fine-grained apatite; these deposits represent millions of tons of uranium but at very low grade (0.005%–0.015% U); uranium is a by-product of phosphoric acid production (Phosphoria Formation, Idaho and Montana, United States; Gantour Basin, Morocco); (14.2) *organic phosphorite* in argillaceous marine sediments with uranium-rich fish remains (e.g., Melovoe District, Kazakhstan); and (14.3) *continental phosphate* (e.g., Bakouma, Central African Republic).
15. Black shale deposits include (15.1) *stratiform subtype* in organic-rich marine shale and coal-rich pyritic shale, with synsedimentary uranium adsorbed onto organic material and clay minerals (e.g., Haggan alum shale, Sweden; and (15.2) *stockwork sun-type* (e.g., Gera–Ronneburg District, Germany).

An alternative genetic classification has been proposed by Cuney (2009, 2011) and is based on the dominant process involved in concentrating the uranium in a deposit.

MINING OPTIONS

Underground mining—including the Canadian ores (McArthur River, Cigar Lake, and Rabbit Lake), Cominak (Compagnie minière d'Akokan) in Niger, and other smaller producers—represents a large source of uranium. An interesting development is the underground mining methods used for the very-high-grade ores in Canada, with the McArthur River mine using raise boring, eliminating the need for crushing, and the Cigar Lake mine using jet boring with water to recover the ore directly into a coarse slurry form.

The by-product production segment in Table 1 is primarily from the Olympic Dam operation in Australia that is a copper producer foremost, with significant uranium as a by-product or co-product. In recent years, ISL or in situ recovery has become a very significant source of uranium, most notably from Kazakhstan (Table 2). Uranium is produced not only by various methods but also in a wide variety of geographical locations, with Kazakhstan currently the largest producer.

DEMAND AND MARKET CONSIDERATIONS

Historically, uranium production in developing countries has been dominated by underground or open pit mining and conventional milling. However, experimental ISL of uranium was begun as early as 1961 in Ukraine, and today ISL is the cornerstone of the uranium production industries of Kazakhstan and Uzbekistan. In addition, several other developing countries have either begun ISL operations or are assessing the amenability of their uranium resources to ISL extraction (WNA 2015).

Uranium processing tends to follow basic unit processes, as illustrated in Figure 1. The figure shows the relationships of various mining options to what referred to as "conventional processing."

Table 1 2016 Mining methods for uranium production

Method	Uranium, t	Percent
In situ leach	30,062	48
Underground and open pit (except Olympic Dam)*	29,030	47
By-product*	3,274	5

Source: WNA 2017

*Considering Olympic Dam as by-product rather than in underground category.

Table 2 2016 world distribution of uranium production

Country	Uranium, t	Percent
Kazakhstan	24,575	39
Canada	14,039	22
Australia	6,315	10
Namibia	3,654	6
Niger	3,479	5
Russia	3,004	5
Uzbekistan*	2,404	4
China*	1,616	3
United States	1,125	2
Ukraine*	1,005	2
Other	1,150	2
Total	**62,366**	**100**

Source: WNA 2017

*Estimates.

Although ISL has emerged as a prominent production alternative, there is a recent reemergence of heap leaching, which is different from historical uranium heap operations that used "dump leaching."

Markets and Prices

The market for uranium during the period 1945–1970 was primarily for nuclear weapons through various governments. Since that time, the market has consisted solely of nuclear fuel for nuclear power plants through both government and private entities. It is a well-known and easily calculated market given that the number and character of nuclear plants worldwide is well documented. As of mid-2015, there were 447 operable nuclear power plants in 30 countries. Uranium demand in 2015 was estimated to be 63,000 t (WNA 2017). Historical uranium production is illustrated in Figure 2.

These uranium requirements are filled from two sources: primary production and secondary supply. Primary production consists of newly mined uranium. Secondary supply consists of previously mined uranium in a variety of forms, including inventories, blended-down nuclear weapons, tailings from the uranium enrichment process, and reprocessed spent fuel. Secondary supply currently amounts to about 17,000 t U/yr and is expected to remain relatively constant at this level for the foreseeable future. Thus, its market share is expected to decline. Future uranium requirements, as projected by the IAEA, are set forth in Figure 3.

Sales of uranium to owners of nuclear power plants are made mainly under long-term (5- to 10-year) contracts, in which future delivered prices might be (1) base price escalated, (2) market based, or (3) some combination of the two. Deliveries of uranium under long-term contracts typically

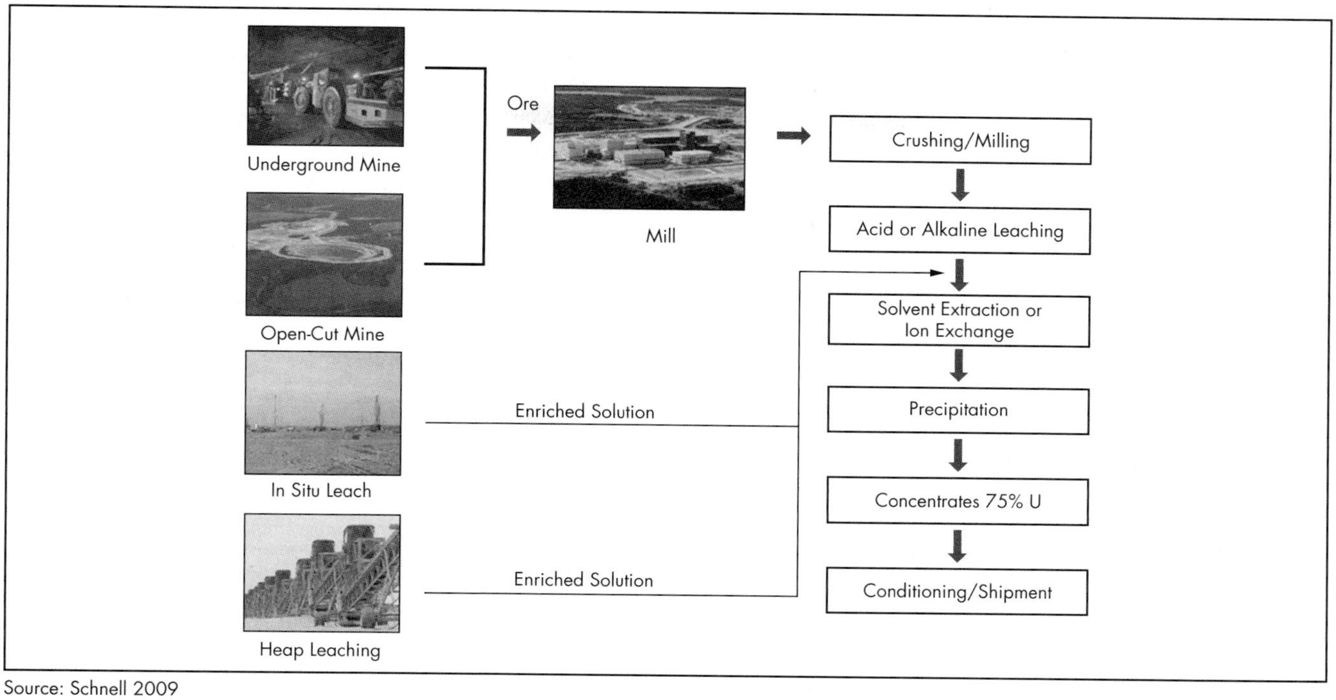

Source: Schnell 2009

Figure 1 General uranium flow sheet

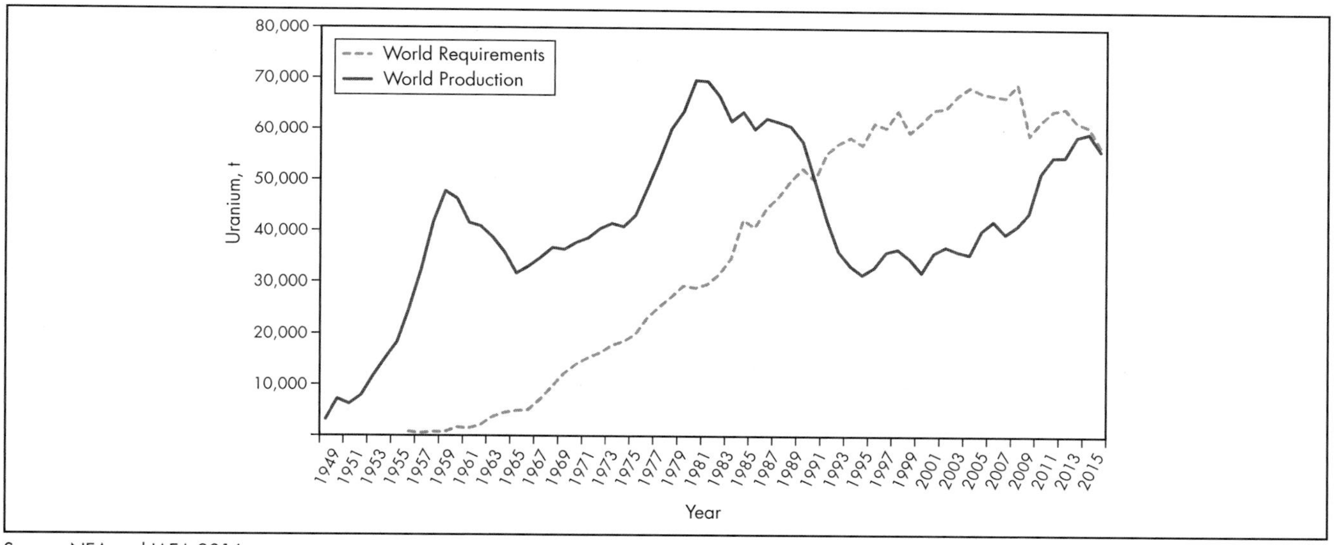

Source: NEA and IAEA 2016

Figure 2 Annual uranium production and demand

cover about 75% of the market, but pricing terms are rarely made public. Prices in the spot market, however, are widely available and form the basis for most analyses. World production of uranium is expressed in metric terms as tons or kilograms of uranium although some producers will use U_3O_8 to exaggerate reserves or production. The spot price, a historical carryover, is expressed as pounds of U_3O_8 (typically in U.S. dollars). (For reference, 1 kg U = 2.6 lb U_3O_8.) Historical spot market prices are shown in Figure 4.

URANIUM UNIT OPERATIONS

Uranium is a relatively common element in the crust of the earth (very much more than in the mantle). It is a metal approximately as common as tin or zinc, and it is a constituent of most rocks and even of the sea. Some typical concentrations are given in Table 3.

Uranium is found in a large variety of settings; consequently, uranium processing must be adaptive to these settings. Historically, flow sheets based on a tank leach process tended to be very similar and are referred to as conventional. There

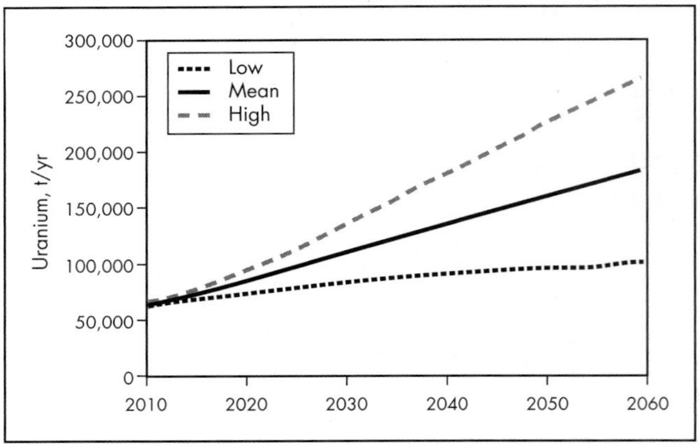

Source: Pool 2013
Figure 3 Future nuclear reactor uranium requirements

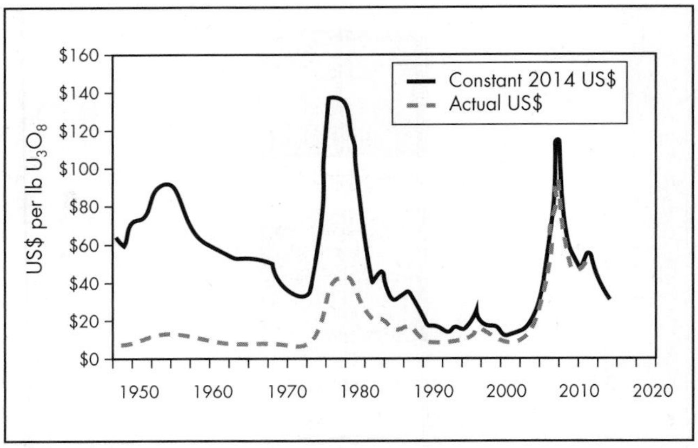

Source: Pool 2013
Figure 4 Historical uranium price

is a trending away from such conventional flow sheets, or at least the technology within each unit operation is changing. The range of processing now varies from the extreme high-grade deposits in Canada to the other low-grade extremes in places such as Namibia.

When determining a flow sheet for a particular ore, many factors need to be considered: mining method, deposit type, deposit size, mineralogy, uranium grade, geographical location, water balance, production capacity, costs (capital and operating) schedule, and other considerations. Most notably, costs are a result, although the deposit size and production target play an important role. Many of the unit processes depend upon the host rock or gangue characteristics rather than uranium mineralogy.

Mineral Variability

The known types of uranium mineralization are overwhelming because of the large uranium atom with four electrons loosely bound to the nucleus. Uranium is therefore found in a large range of minerals—vanadates, carbonates, oxides, phosphates, silicates, arsenates, sulfates, and molybdates. These minerals are all potential uranium sources, if found in sufficient abundance. The principal minerals exploited today are uraninite (pitchblende), carnotite, coffinite, brannerite, torbernite, and wulfenite.

Process Choice

There are three basic process alternatives—conventional, ISL (or in situ recovery), and heap leach. Although one speaks of conventional, there are no two processing plants that are the same today in the world. Most similar, but still distinct, are the Somair and Cominak operations in Niger. Reviewing the three sets of unit operations, as shown in Figure 5, heap leach has the advantage of not requiring grinding and eliminating solid–liquid separation. In ISL, the mine is eliminated, but it is now the leach section of the plant with less control.

It is commonly thought that ISL and heap leach have much lower costs, both capital expenditures (CAPEX) and operating expenditures (OPEX), but with the complex back end of the process costing about 50% of investment, cost differences of the three basic flow sheets are not huge. Certainly, ISL is the lowest cost option for uranium production, with conventional agitated leach being the highest cost, and the choice must be balanced against risk and recovery.

A comparison of the main variables for each process option is shown in Table 4 and Figure 6. The other considerations are costs, with the key OPEX and CAPEX element proportions surprisingly similar, regardless of location. The highest CAPEX components are the leaching and solid–liquid separation areas. The highest OPEX components are power, labor, and reagents in similar proportions.

Table 3 Natural uranium concentration occurrences

Very high-grade ore (Canada), 20% U	200,000 ppm U
High-grade ore, 2% U	20,000 ppm U
Low-grade ore, 0.1% U	1,000 ppm U
Very low-grade ore (Namibia), 0.01% U	100 ppm U
Granite	3–5 ppm U
Sedimentary rock	2–3 ppm U
Earth's continental crust (average)	2.8 ppm U
Seawater	0.003 ppm U

Source: WNA 2016

Conventional Uranium Process

The following subsections discuss each unit process, with leaching separated into conventional, heap leach, and ISL. Further technical details can be found in other chapters of this handbook.

Crushing

Uranium most commonly occurs along the fissures or cracks within the ore or gangue matrix, and crushing is preferential at these fissures and cracks in the ore matrix, thereby exposing the uranium minerals. Gyratory crusher installations are no longer typical because costs and uranium throughput tonnages are generally small. Jaw crushers are still common and are used at the Langer Heinrich operation, but there is a trend toward the use of sizers for low-grade operations such as Trekkopje in Namibia or Kayelekera in Malawi due to their lower capital cost.

Sizers require testing with the specific ores and have limited application for hard-rock ores. Under development, with

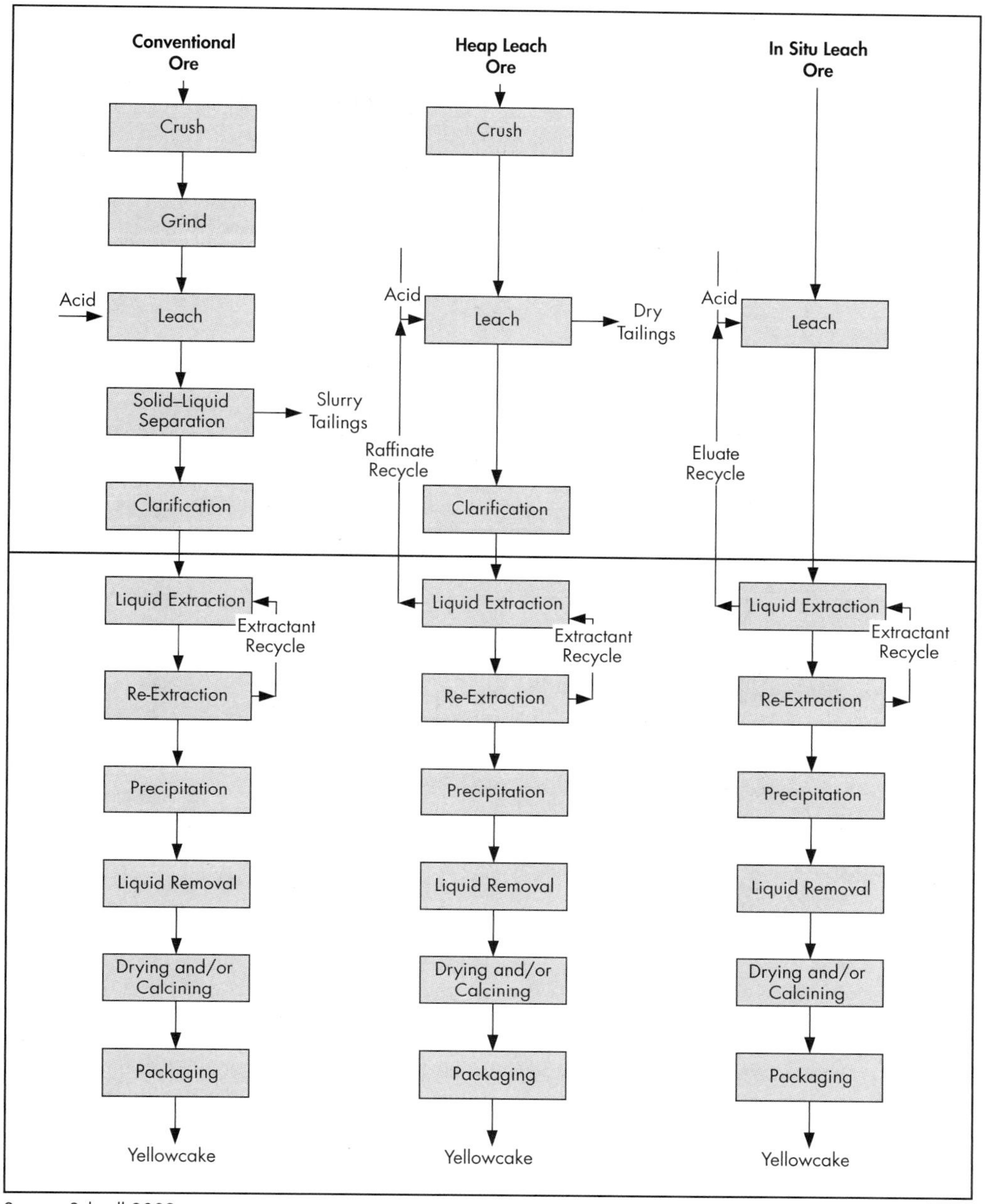

Source: Schnell 2009

Figure 5 Principle uranium process flow sheets

uses in gold and coal, are high-pressure grinding rolls that pass the ore through a set of opposite rotation rolls under high pressure. This has been considered for uranium, but to date, there are no commercial uranium-based installations.

Crushing is generally limited down to the 10-to-15-mm size, which is usually not sufficient to expose all the uranium minerals.

Ore Stockpile

A stockpile is generally used to separate the crushing operation from the mine and reduce the downstream plant size. Ore stockpiles have been open air using either a reclaim system under the stockpile or a set of feeders, but best practice today is for a covered stockpile requiring consideration for radon buildup in such enclosures while providing personnel access.

Preconcentration

Historically, radiometric sorting has been used to reduce the tonnage to the main processing plant, and this option is still studied, with several pilot ore sorters installed, at the Ranger operation in Australia (ERA 2008) and at Rössing in Namibia (Rössing Uranium Ltd. 2007). However, neither of these installations have been commercially accepted.

The preconcentration step is employed at the Langer Heinrich operation in Namibia, with uranium as carnotite hosted in gypcrete ore. Lunt et al. (2007) use an attrition grind

Table 4 Leach option comparison

Conventional	Heap Leach	In Situ Leach
• Conventional mining	• Conventional mining	• No mining
• Higher capital cost	• Reduced capital cost	• Plant capital cost similar to heap leach; reduced overall capital cost
• Higher operating cost	– No grinding	• Operating costs reduced (no mining)
• Best operating control	– No solid–liquid separation	• Long leach times
• Short leach times	• Reduced operating cost	• Reduced reagent consumption
• Best recovery, +95%	• Long leach times	• No real leach control
• Adaptable to most ores	• Reduced reagent consumption	• Recovery questionable
	• Reduced leach control	• Only applicable to specific ore/ geological settings
	• Reduced recovery, ~70%	
	• Not applicable to all ore types	

Source: Schnell 2009

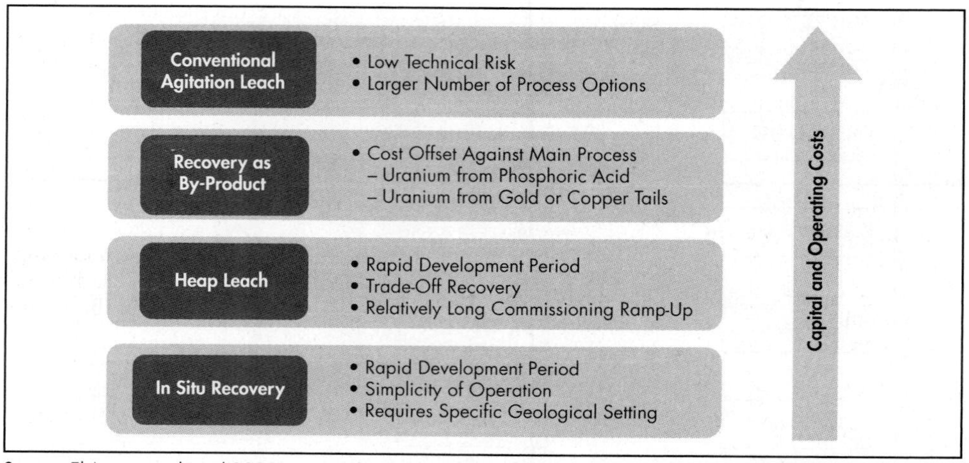

Source: El-Ansary and Lund 2008, reprinted with permission from the Australasian Institute of Mining and Metallurgy

Figure 6 Capital and operating cost considerations for process options

followed by sizing of the various ore fractions to reduce ore tonnage.

Gravity separation has been applied to uranium ores in the past (Schnell 2011) with some success but is associated with serious radiation issues and, consequently, is currently not used. One exception is the use of a Knelson concentrator in the ore receiving circuit of Key Lake to help reduce cement carryover coming from the McArthur River underground mine backfill (Cameco Corporation 2009).

Grinding

Ore sizes of 10–15 mm are then further reduced with the use of grinding mills, but final grind size is coarse (100–500 µm) compared to other metals operations, given that only mineral exposure is required, rather than mineral liberation.

Most grinding operations are wet and then use either coarse ore to grind the fine ore (semiautogenous grinding, or SAG) or steel rods or balls in more conventional grinding. SAG milling is only applicable to specific hard ores after suitable testing. An exception is dry grinding found in the Somair and Cominak operations in Niger. Dry grinding is specific to the Niger ores, which require a very intense leach reagent concentration to liberate the uranium minerals from an expanding clay matrix.

Grinding for high-grade deposits generally uses small autogenous mills, as in the case of McClean Lake in Canada. McArthur River and Cigar Lake operations in Canada have small autogenous mills placed underground to avoid high-grade ore hoisting. In these cases, the ore is transported to surface as a ground slurry with positive displacement pumps. Both operations then transport the ground slurry by truck in special dual-containment containers to mills more than 80 km away.

Closed-circuit operations using cyclones are preferred so that the higher specific gravity uranium minerals will be ground finer than the gangue, improving leaching.

Conventional Leach Circuits

Following comminution, the uranium minerals are exposed and available for dissolution of uranium in a hydrometallurgical process. There are two basic uranium leaching alternatives—acid or alkaline process, with U^{6+} dissolving as either a sulfate or a carbonate.

The gangue of the ore will define whether to use acid or alkaline leaching, primarily depending upon reagent consumption costs. Conventional leaching is most common with sulfuric acid (25–200 kg/t acid) and typically at about 55°C. Higher temperatures, as used in pressure leaching, will result in silica dissolution, making the subsequent unit operations more difficult.

Leaching can be carried out in a large variety of equipment types—heap, vat, agitated, autoclave, and so on—all dependent upon many variables and determined by laboratory testing. The interdependence of time/temperature/pressure results in the increase of one variable while decreasing the other variables. Other variables are similar, but their effects are perhaps not as dramatic.

Table 5 Typical conventional uranium leach parameters

	Typical Leach Parameters	
Parameter	**Acid Leach**	**Alkaline Leach**
Reagent	25–200 kg H_2SO_4/t	$Na_2CO_3/NaHCO_3$
Time	8–12 hours	48–96 hours
Pressure	Atmospheric	Pachuca or autoclave
Oxidant	H_2O_2, O_2, MnO_2	O_2
Temperature	50°–60°C	85°–90°C
Recovery	93%–97%	85%–90%

The basic acid leach chemistry dissolves the U^{6+} directly, while the U^{4+} species is oxidized. Typical oxidants in use are sodium chlorate, manganese dioxide, hydrogen peroxide, and oxygen. Oxygen has become most desirable because of low-cost oxygen production using vacuum swing adsorption technology. There is a general move away from the use of chlorates due to environmental and corrosion issues. Pyrolucite (MnO_2) is still common where it is economically available. Traditionally, the oxidation process found in most textbooks depends upon first oxidizing a ferrous iron to ferric and then the ferric oxidizing the uranium, although there are ores where direct oxidation takes place. The simplified chemistry of alkaline leaching is oxidation followed by formation of a uranyl–carbonate complex.

Acid leach reactions are as follows:

$$UO_2^{2+} + SO_4^{2-} \leftrightarrow UO_2SO_4$$

$$UO_2SO_4 + SO_4^{2-} \leftrightarrow [UO_2(SO_4)_2]^{3-}$$

$$[UO_2(SO_4)_2]^{3-} + SO_4^{2-} \leftrightarrow [UO_2(SO_4)_3]^{4-}$$

$$6Fe^{2+} + NaClO_3 + 6H^+ \leftrightarrow 6Fe^{3+} + NaCl + 3H_2O$$

$$UO_2 + 2Fe^{3+} \leftrightarrow UO_2^{2+} + 2Fe^{2+}$$

Alkaline leach reactions are as follows:

$$2UO_2 + O_2 \leftrightarrow 2UO_3$$

$$UO_3 + Na_2CO_3 + 2NaHCO_3 \leftrightarrow Na_4UO_2(CO_3)_3 + H_2O$$

Continuous tank leaching typically uses an agitated system of no fewer than five or six reactors in series to eliminate short-circuiting. Typical acid leach times are 8–12 hours in a 35%–50% solids slurry at 50°–60°C. Most leach circuits are arranged to advance the slurry by gravity. Some leach circuits use a two-stage leach so that the first stage uses the discharge solution from the second stage to help reduce acid consumption, providing a pre-leach. Typical parameters for conventional leaching are given in Table 5.

An alternative to the typical agitated tank leach are the Somair and Cominak operations that use a pug leach process. Using this method, the dry ore is combined with concentrated sulfuric acid and nitric acid in a drum and then the moist mixture is cured using a slow-moving conveyor for 2 hours before entering agitated tanks and diluted before solid–liquid separation (Thiry and Roche 2000).

Carbonate leaching has reemerged with the start-up of the Langer Heinrich operation in Namibia, and Trekkopje will be the first application using carbonate in a heap leach operation (Schnell 2010).

Leaching equipment ranges from pressure leach (Key Lake and McClean Lake) to agitated tank leaching (Langer Heinrich, Rössing, etc.) and a recent return to pachuca leaching (Cameco Corporation 2012).

Leaching of uranium would not be complete without considering the pugging-and-curing flow sheet used at the Somair and Cominak operations in Niger. These ores in Niger contain the mineral structure of siliceous clay that requires a high-acid curing step to achieve good metal recovery.

Heap Leach

Although some heap leach work was carried out in the 1970s and 1980s, uranium heap and dump leaching did not develop as it did in gold and copper, primarily because of market conditions and the availability of higher-grade deposits. Recent increased interest in lower-grade deposits and the efforts of countries to secure internal uranium production has resulted in heap leach application in uranium production. Heap leaching based on the fine crushed, agglomerated technology for copper operations in Chile with sulfuric acid has been applied for many years at the Caetité operation in Brazil (Gomiero et al. 2010) and more recently in Niger at Somair for the processing of mineralized waste (Durupt et al. 2013). In addition, the large Imouraren project in Niger will also be using crushed agglomerated heap leach (Thiry and Bustos 2014). Many junior companies have been looking at acid heap leaching to exploit low-grade deposits, and more heap leach uranium operations will be seen in the future as the application of this technology to uranium matures. The Trekkopje project is unique in applying crushed agglomerated heap leach in carbonate (alkaline) leach chemistry.

The objectives of a good heap leach operation are

- Recoverable metal to solution of 80% plus,
- Leach times of <300 days,
- Results that are consistent and independent of location within the heap,
- Virtually no solids in the pregnant leach solution, or PLS (no requirement for solid–liquid separation step),
- Little sensitivity to head grade, and
- Heap stability.

The most difficult objective is achieving the "consistent and independent of location" result within a heap. This challenge requires that leach conditions are the same everywhere on the heap and the ore is also equally responsive.

When considering the heap leach process option, the following must be true:

- Months (years) of careful column test work have confirmed good results in all areas of the ore body.
- The ore body has been carefully sampled for variability.
- Initial test work is generally performed on a blended sample, but variability testing is required to better predict actual performance.

The general operating conditions for uranium heap leach are as presented in Table 6.

Heap leaching seems simple, but there are some pitfalls, such as poor heap permeability due to compaction or jarosite formation. Compaction can cause possible increased phreatic head, resulting in heap failures; heap stability and extraordinary measures are required to correct such problems. Solution flow considerations, including rain events, require a design that considers the inherent solution energy and avoids solution discharges into the environment. Figure 7 presents a typical acid heap leach flow sheet.

Table 6 General uranium heap leach operating conditions

Description	Acid Heap Leach	Alkaline Heap Leach
Typical uranium grade	200–500 ppm	150–500 ppm
Crush size	10–25 mm	10–25 mm
Clay content	<35%	<35%
Agglomeration time	45–90 seconds	45–90 seconds
Agglomeration moisture	7%–10%	7%–10%
Heap height	6–9 m	7–9 m
Irrigation rate	4–10 L/min/m^2	5–10 L/min/m^2
Leach reagent	10 g/L H_2SO_4	40 g/L CO_3 and 10 g/L HCO_3
Leach cycle	1 or 2 countercurrent	3–4 countercurrent
Leach time	30–90 days	120–180 days
Recovery	80%–85%	70%–80%

In Situ Leach

The in situ mining method for uranium is in sandstone deposits with typical low-grade uranium concentrations of (0.06%–0.2% U). ISL is currently responsible for more than 40% of world uranium production. Mining and uranium leaching take place in the deposit.

In order to consider an ISL option, the deposits must be located in a permeable formation, water should normally be over the formation, and the deposit must be located between confining layers to be able to contain the solutions. Figure 8 shows a conceptual model of a typical ISL formation as found in Kazakhstan, Australia, and the United States.

The leach process is conducted within the ore formation, and the leach chemistry is either with acid or carbonate solutions, similar to conventional uranium leach procedures but with much lower reagent concentrations and requiring very long leach times, in the three-to-five-year time frame. Specifically, there are two leach options: (1) Oxygen, carbon dioxide, and sodium bicarbonate are added to the native groundwater, or (2) weak sulfuric acid is pumped into formation, and at times with use of an oxidant. Recoveries should normally be in the +75% range.

The PLS from the well fields will have also leached other contaminants, mostly iron in the case of acid leach since the leach solution is low grade. Ion exchange is normally used to fix the uranium to a resin with the PLS feeding IX columns. The uranium is complexed with an organic radical and then is absorbed onto styrene beads. Most operations in Australia and the United States use fixed columns, while pinned-bed columns are used in Kazakhstan. The uranium is then eluted from the resin with a carbonate of acid solution. This elute maybe then be precipitated directly (alkaline leach) or forwarded to solvent extraction (SX) for iron removal or a double precipitation (pH ~3.4 to precipitate iron, followed by pH +8 for uranium precipitation) may be required prior to final yellowcake product precipitation. The water from ISL needs to be treated before being recycled back into the aquifer, or some bleed may need to be discharged into the environment. The available alternatives for water treatment are deep well disposal, reverse osmosis treatment, or chemical treatment.

The environmental considerations for ISL can be onerous. In the case of alkaline leach, the well field needs to be washed or rinsed after leaching is complete to reduce contamination. Reverse osmosis is used to recycle this rinse solution over a period of several years. For acid ISL, the dissolved iron will precipitate as the leach solution advances out of the well field, creating a natural bearer for solution contamination. When the groundwater remediation is complete, piping is removed, topsoil is replaced, and the area is revegetated.

The advantages of ISL are that the employees have minimal radiation exposure, ground surfaces have little permanent disturbance, capital and operating costs are low, and low-grade ore bodies can be economically mined. But one must keep in mind that this method is only applicable to select types of deposits contained in permeable sandstone.

Solid–Liquid Separation

After leaching the uranium into a soluble complex, it must be separated from the residue with a physical separation using either decantation or filtration. The solid–liquid separation choices for uranium are dependent on the characteristics of the leached pulp—basically, how the pulp filters or decants. The options used are numerous, such as filtration (belt, drum, pressure), countercurrent decantation (CCD), countercurrent cyclones, countercurrent classifiers, or resin-in-pulp (RIP).

RIP has recently been applied in the Kayelekera mine in Malawi, with the development of modern resins where the uranium complex is adsorbed or complexed with a resin and the pulp is then screened to separate the resin bead (larger in diameter than the ore solids) by screening (Figure 9).

Solution Clarification

Uranium operations generally have low throughput compared to base metal operations, and, subsequently, clarification is a viable option in most conventional and heap leach operations. The leach solutions (PLS) will commonly contain fine solids that needs to be clarified prior to solution purification. Conventional clarifiers, sand filters, and sand filtration are typical. Some continuous sand filtration types have shown poor availability, but the use of pinned-bed clarifiers is being more recently applied for high-PLS flows (Thiry and Bustos 2014).

Thickening

Settling of solids from solution typically uses a thickener or decanter or settler (different terms are often used for the same equipment) with the solids raked into the center and with the uranium-bearing solution overflowing. The decantation circuits operate as CCD units, typically using five to eight thickeners in series with a water consumption of 1–2 m^3/t. A recent variation has been applied at the McClean Lake operation in Canada with the use of "high rate" thickeners where the pulp bed is partially used as a filter to improve solid–liquid separation, thereby reducing thickener diameters.

Filtration

Filtration is another option to separate the uranium solution from the pulp and wash the pulp to reduce uranium losses. A wide variety of filtration equipment is available. Most commonly used in uranium processing are horizontal belt filters as used at the Somair and Cominak operations in Niger. The solids from a belt filter are typically discharged directly from the filter onto a conveyor belt to send the washed pulp to a dry tailings disposal area.

Solvent Extraction

Various other metals are dissolved during leaching, requiring purification of the uranium-bearing solutions. The purity requirements of the final uranium precipitate are becoming

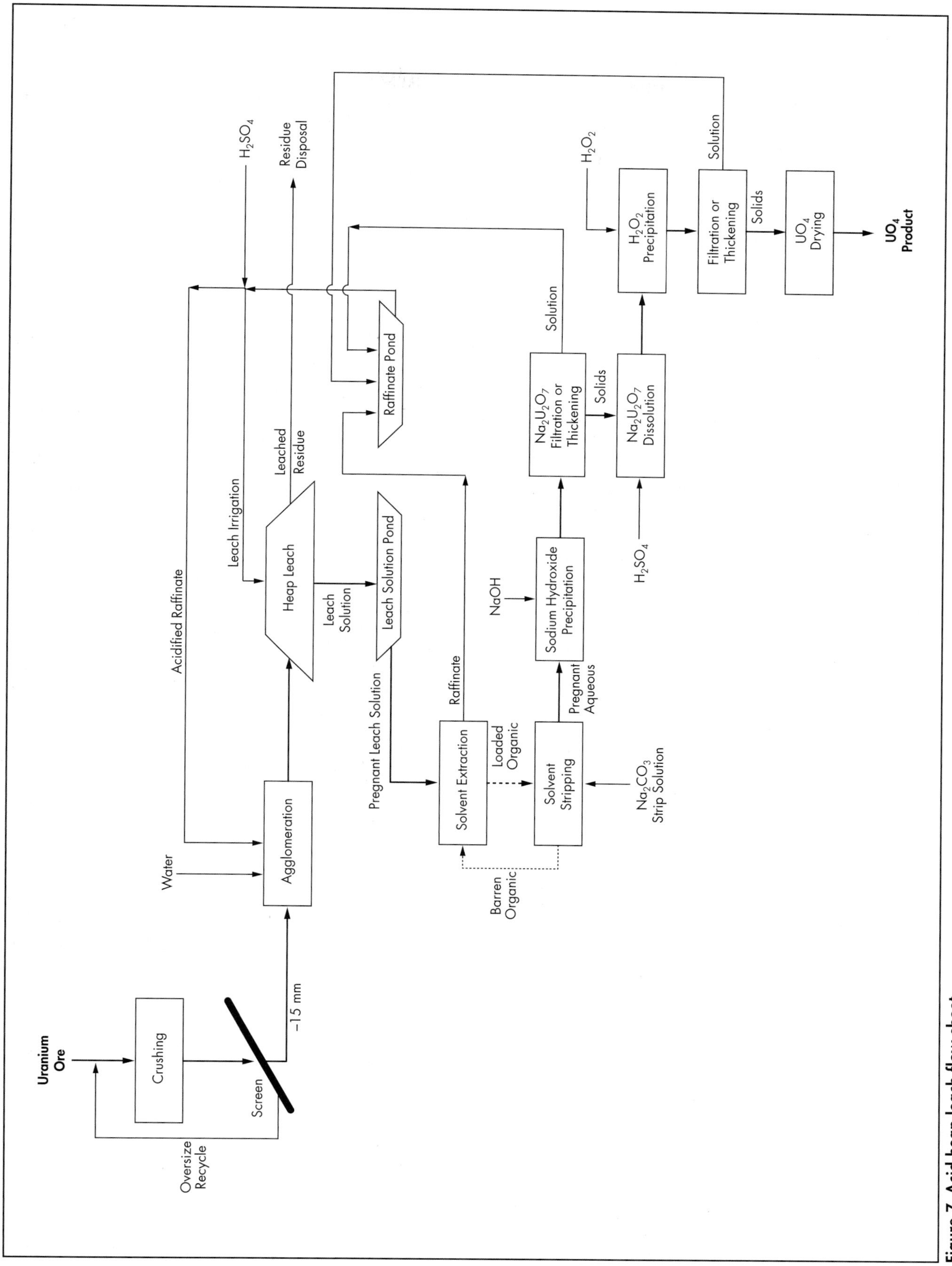

Figure 7 Acid heap leach flow sheet

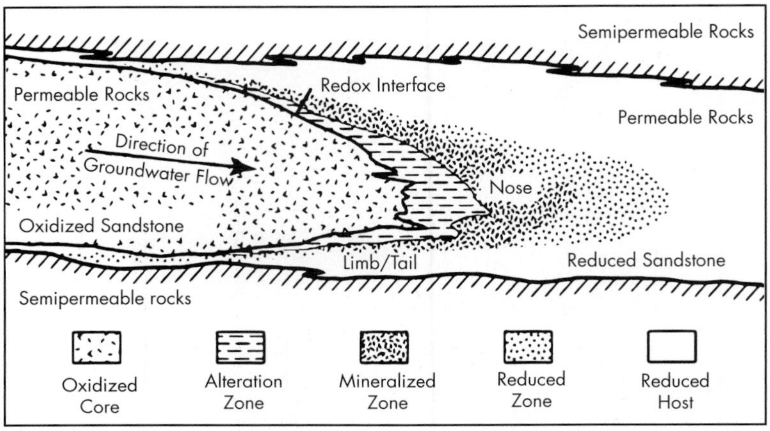

Source: Cyclone Uranium Corporation 2012

Figure 8 Conceptual model of a roll-front deposit

Source: Kemix Pty Ltd. 2015

Figure 9 Resin-in-pulp example

stricter, and most uranium conversion plants require a precipitate quality better than the ASTM International standard. The most common impurities associated with uranium are molybdenum and vanadium, but care must be taken to also consider other metal contaminants, most notably zirconium, arsenic, copper, and nickel.

Solvent extraction is typical for uranium, but little progress has been made for extractants and modifiers and the tertiary amines predominate with some use of the organophosphorus compound DEPHA. Equipment is usually conventional mixer-settlers or Krebs mixer-settlers, as used in Canada (McClean Lake and Key Lake; see Figure 10). At Olympic Dam in Australia, Bateman columns have been applied (Bateman Litwin 2012). Future projects are also considering vertical smooth flow technology (Paatero et al. 2010).

The uranium stripping in solvent extraction is with ammonium sulfate (Key Lake, McClean Lake), sodium carbonate (Somair), or strong acid (Rabbit Lake). In one interesting alternative, the uranium is precipitated directly in the organic extractant. Extractants are generally tertiary amines with typical concentrations in the 5% amine/g/L U levels, although for higher-grade solutions, extractant concentrations can be in the 10% amine/g/L U. Stripping generally produces 35 g/L U concentrations but can be up to 100 g/L for high-grade operations. Figure 11 depicts a typical solvent extraction pulse column.

Ion Exchange

The IX process has progressed both in the resin and equipment design areas and in the recovery of uranium from lower-grade deposits. ISL operations with lower-grade solutions make ion exchange the preferred choice.

Normally, ion exchange is chosen for uranium solution grades at the <1.0-g/L range; however, slightly lower solution grades may use the SX option. In the case of ion exchange, additional solution purification may be needed, which usually adds a small SX circuit, a process known as the Eluex process

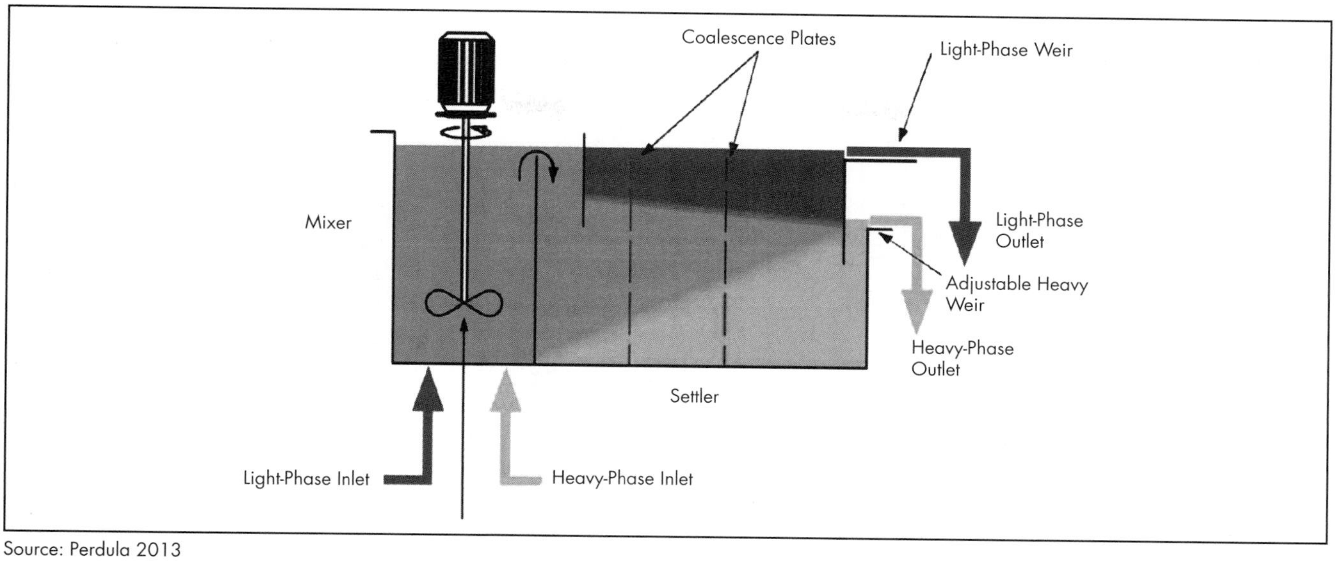

Source: Perdula 2013

Figure 10 Typical solvent extraction mixer-settler

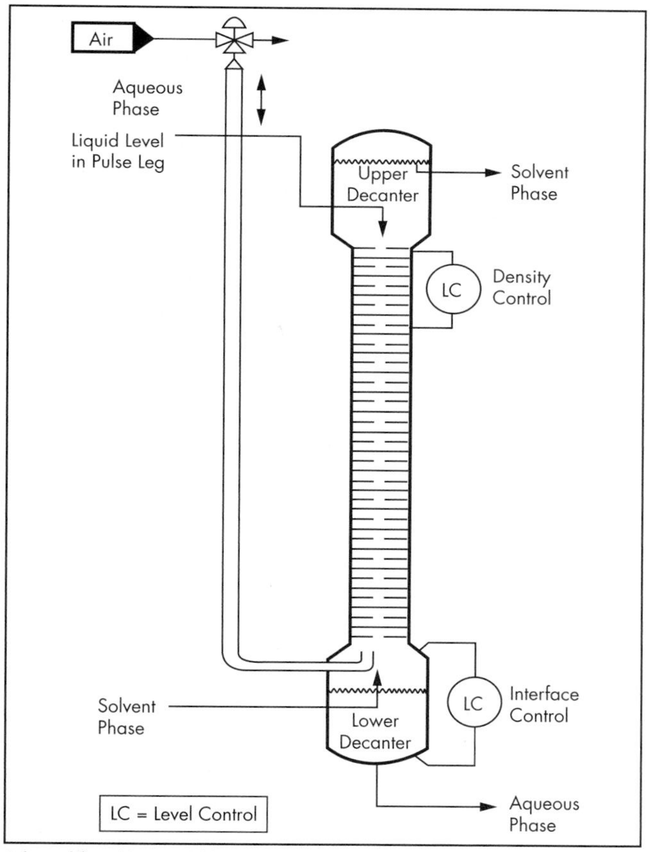

Adapted from Vancas 2003

Figure 11 Solvent extraction pulse column

(Merritt 1971). The most notable example of an Eluex circuit is the Rössing operation in Namibia (Griebel 2000). Basic IX and SX reactions are shown in Figure 12.

Three basic technologies are used for uranium ion exchange: fixed-bed columns, pinned-bed columns, and

Anionic Exchange	$R^+X^- + A^- \Leftrightarrow R^+A^- + X^-$
Cationic Exchange	$R^-Y^+ + B^+ \Leftrightarrow R^-B^+ + Y^+$

Figure 12 Basic ion exchange and solvent extraction reactions

fluid-bed columns (NIMCIX [National Institute for Metallurgy Countercurrent Ion eXchange]). Fixed-bed columns have been applied in the Langer Heinrich operation, with pinned-bed predominant in the Kazakhstan ISL operations, and NIMCIX has been installed at Trekkopje and the Langer Heinrich operations in Namibia. NIMCIX offers the advantage of handling high suspended solids.

Precipitation

After the uranium solution has been purified and concentrated, the uranium can be precipitated. The process choice for final uranium precipitation is dependent upon solution chemistry, environmental requirements, reagent availability, and converter requirements. The most common reagents used are ammonia, hydrogen peroxide, sodium hydroxide, and magnesium oxide. However, most converters prefer oxide precipitates rather than an alkali precipitate because the latter results in additional environmental concerns.

Equipment choice most commonly includes a series of stirred tanks, but recent innovation using a fluid bed (Courtaud et al. 2011) produces an improved product with less dusting. One unusual case involves precipitating the uranium directly in SX organic (Kazakhstan).

The various chemistries of precipitation are as follows:

- Magnesia precipitation

$$UO_2SO_4 + 2Mg(OH)_2 \rightleftharpoons MgUO_4\downarrow + MgSO_4 + 2H_2O$$

$$UO_2SO_4 + Mg(OH)_2 \rightleftharpoons UO_2(OH)_2\downarrow + MgSO_4$$

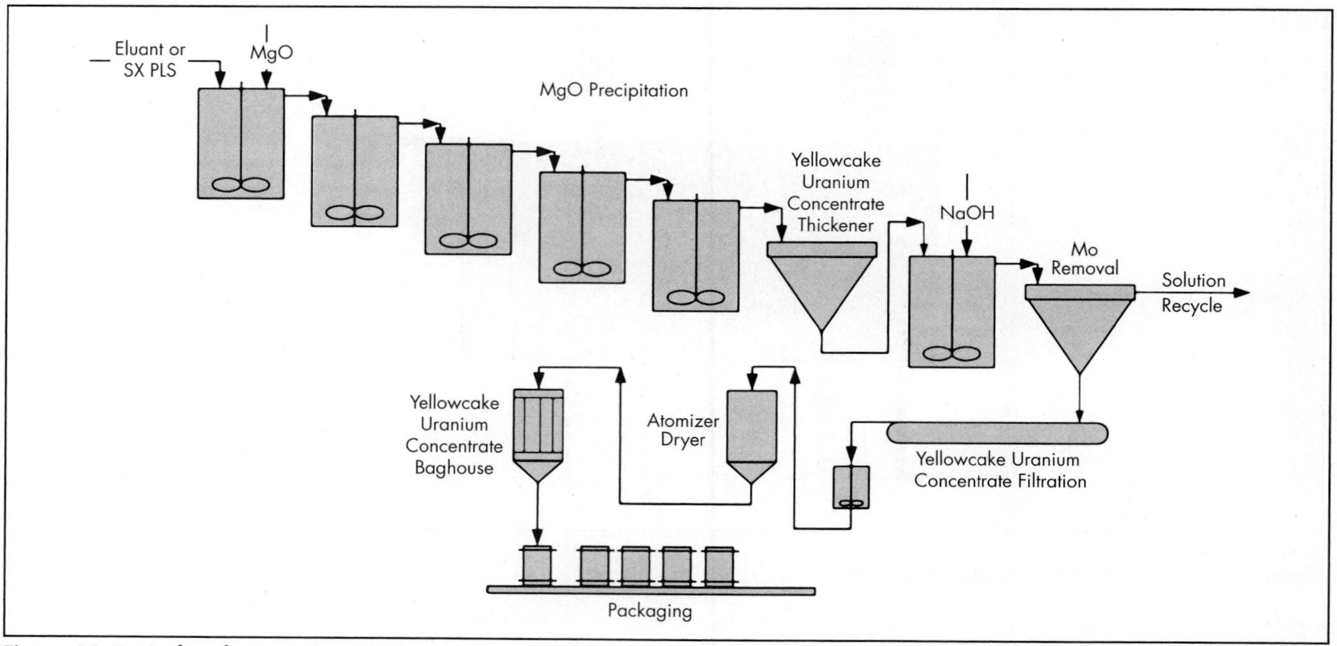

Figure 13 Typical tank precipitation circuit

- Sodium precipitation

$$2Na_4UO_2(CO_3)_3 + 6NaOH \rightleftharpoons Na_2U_3O_7\downarrow + 6Na_2CO_3 + 3H_2O$$

$$Na_4UO_2(CO_3)_3 + NaOH \rightleftharpoons UO_2(OH)_2\downarrow + 3Na_2CO_3$$

- Hydrogen peroxide precipitation

$$UO_2SO_4 + H_2O_2 + 2\ NaOH \rightleftharpoons UO_4{\cdot}2H_2O\downarrow + Na_2SO_4$$

- Ammonia precipitation

$$2UO_2SO_4 + 6NH_4OH \rightleftharpoons (NH_4)_2U_2O_7\downarrow + 2(NH_4)_2SO_4 + 3H_2O$$

$$UO_2SO_4 + 2NH_4OH \rightleftharpoons UO_2(OH)_2\downarrow + (NH_4)_2SO_4$$

Illustrated in Figure 13 is a typical tank precipitation circuit with thickener operation to recycle precipitate and feed a filter before final precipitate drying. Figure 14 shows a schematic of fluid-bed precipitation.

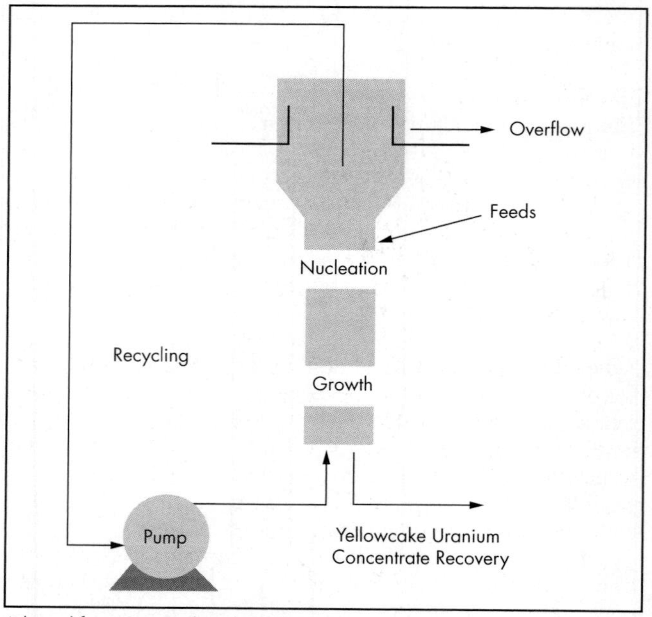

Adapted from Courtaud et al. 2011
Figure 14 Fluid-bed precipitation

Drying and Packaging

The precipitate is generally dewatered before drying with use of a centrifuge or filter (horizontal filter or pressure filter). The final dewatered precipitate is then dried or calcined with the final product and is referred to as "yellowcake." Commercially it is then marketed as U_3O_8, although there is a general attempt to express everything as uranium rather than U_3O_8. The color of the precipitated uranium (from ammonia, magnesia, or peroxide precipitation) is dependent upon the final drying temperature—yellow at 100°C, orange at 250°C, and very dark green (black) at higher than 800°C, as illustrated in Figure 15 for ammonium diuranate. The final product has a uranium concentration above 75% U.

Product drying and/or calcining equipment ranges from multi-hearth furnaces (Key Lake and McClean Lake), to rotary tube dryers and calciners (Kazakhstan), to hollow Flyte dryers (Rabbit Lake and Langer Heinrich), to spray dryers (Cominak). The choice of calcining or drying to a lower temperature is dependent upon economics and the destination of final product.

In all cases, the final precipitate is packed into standard 170-L steel drums for shipping. Packaging equipment and installations are being upgraded to reduce personnel exposures to yellowcake dust. With these upgrades, facilities are generally located in a separate enclosed space with remote operation and drum lids added with glove-box type

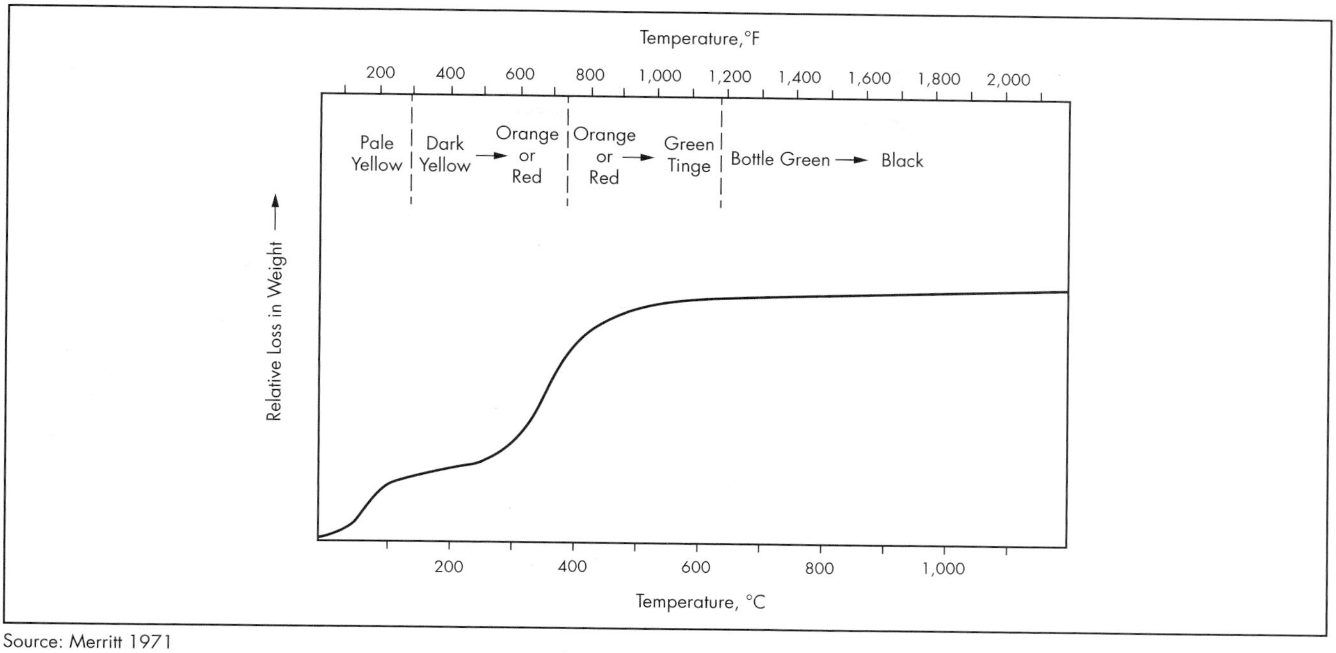

Source: Merritt 1971

Figure 15 Effect of temperature on the decomposition of ammonium diuranate

of equipment. Table 7 lists comparative converter yellowcake quality requirements.

Effluent Treatment

The plant product is yellowcake, but the wastes from the process must also be considered. There are basically two waste streams to consider: waste/process water and tailings from leaching. Both of these streams, and other less important streams, require environmentally acceptable treatment and disposal. Some of these waste streams also have value that must be considered. For example, all solutions in alkaline leach contain reagent values that can be recycled.

Liquid wastes must be treated to neutralize them, remove radioactive species such as radium, and eliminate other metals dissolved in the process. The tailings pulp needs similar treatment and neutralization. In Canada, waste solutions are treated and temporarily stored, then analyzed to ensure environmental compliance before the water is released into the environment. These streams are heavily monitored to confirm compliance and minimize environmental effects.

In warmer climates with high evaporation rates, evaporation ponds are typically used to treat liquid wastes. The residues in these ponds then need recovery and suitable long-term disposal. Ponds have a risk of leakage and are typically double lined with a leak detection system.

Tailings Solids

The tailings solids require final disposal after treatment. Traditionally, dammed structures have been used for tailings disposal and final dewatering. These structures have associated risks, and, more recently, an in-pit disposal or lined cell disposal has become a requirement. The schematic of an in-pit tailings disposal as practiced in the most recent operations (Rabbit, McClean, and Key Lakes) is shown in Figure 16.

URANIUM TESTING AND FLOW-SHEET DEVELOPMENT

Several key steps are required in establishing the processing route for a new uranium project. Before preliminary laboratory test work is undertaken to recover the uranium, a thorough understanding of the deposit's mineralogy is necessary, including both the uranium and gangue composition. The uranium grade, liberation characteristics, gangue associations, and grain size will all guide selection of the required leach conditions and leaching method (e.g., tank versus ISL).

It is important to also understand how the uranium and gangue mineralogy may vary through the deposit, as this will influence the sample selection for a test-work program. The leach test conditions should be optimized with consideration given to the variations in a deposit; therefore, testing should include samples representing this variability to determine the response of the leaching process to different feeds. Comminution studies play a crucial role in further understanding the physical characteristics of an ore body, including the potential to upgrade an ore through rejection of gangue minerals by scrubbing. It may also be possible to concentrate the uranium through methods such as radiometric sorting and gravity separation, or potentially problematic phases may be able to be rejected through processes such as flotation.

In most instances, the uranium and gangue mineralogy will define whether a sulfuric acid or carbonate leach is most suitable for uranium extraction; however, it is prudent to test and assess both systems in the first instance. Laboratory test work is then required to test the suitability of the chosen lixiviant, determine the maximum uranium extractable, and establish the preliminary leach parameters. These would include pH, temperature, particle size, leach duration, and oxidation–reduction potential. The composition of the expected site water and its impacts on leaching and reagent consumption, particularly for alkaline leaching, should also be investigated.

Table 7 Comparative converter yellowcake quality requirements

Element	Unit	ASTM Standard		Converter A		Converter B		Converter C	
U	% dry	>65%		>60%		>65%		>65%	
Isotope	%	0.7105–0.7115		0.7105–0.7115		0.7105–0.7115		0.7105–0.7115	
U234	ppm/U	56	62	56	62	58	62	62	
Particle size	mm	<6.35		<6		<6.35		—	
U insoluble	%	<0.10		<0.10		<0.10		—	
Limits	**Unit**	**Penalty**	**Reject**	**Penalty**	**Reject**	**Penalty**	**Reject**	**Penalty**	**Reject**
Moisture	%	2.00	5.00		10.00	5.00	10.00		5.00
As	%/U	0.05	0.10	1.00	2.50	0.05	0.15	0.01	0.40
B	%/U	0.005	0.10		0.20	0.01	0.15	0.01	0.10
Ca	%/U	0.05	1.00	1.00	5.00	3.00	4.00		1.00
CO_3	%/U	0.20	0.50	2.00	3.00	2.00	4.00	0.20	0.50
F	%/U	0.01	0.10	0.15	0.30	0.10	0.15	0.01	0.10
Halogen	%/U	0.05	0.10	0.15	0.25	0.10	0.20	0.05	0.10
Fe	%/U	0.15	1.00			1.00	2.00	0.15	1.00
Mg	%/U	0.02	0.50			3.00	4.00	0.02	0.50
Mo	%/U	0.10	0.30	0.10	0.30	0.10	0.30	0.10	0.30
P	%/U	0.10	0.70		0.32	0.50			0.20
PO_4	%/U	*~ 0.30*	*2.15*		1.00			0.10	1.00
K	%/U	0.20	3.00			+Na=1	+Na=2	0.20	3.00
SiO_2	%/U	0.50	2.50	0.50	2.50	0.50	2.00	0.50	2.50
Na	%/U	0.50	7.50	1.00	7.50	+K=1	+K=2	0.50	7.50
S	%/U	1.00	4.00	1.00	3.33	1.00	3.50		
SO_4	%/U			3.00	10.00			3.00	12.00
Th	%/U	1.00	2.50	1.00	4.00	0.50	2.00		
Ti	%/U	0.01	0.05			0.05	0.10	0.01	0.05
V	%/U	0.06	0.30		0.08	0.10	0.50		
V_2O_5	%/U	*~0.21*	*1.07*		0.30			0.10	0.75
Zr	%/U	0.01	0.10	0.20	2.00	0.10	0.50	0.01	0.50

Note: Numbers in italic represent preferred levels.

Typically, laboratory test work would commence as bench-scale tank leaches, which would be used to optimize the leach conditions and establish typical reagent consumptions. The results from these initial tests would also guide the anticipated leaching method. Although tank leaching is the most typical mode of leaching, heap leaching may be feasible for lower-grade ores with low acid consumption, with mineralogy that allows sufficiently porous heaps to be stacked. The amenability of a deposit to an ISL operation is dependent on the geology of the deposit, with smaller, porous sandstone deposits most suitable. If a heap leach or ISL process option was under consideration, a series of bottle roll, or preferably column leach tests, would be necessary. If more aggressive leach conditions are required, autoclaves may be used to investigate pressure leaching.

The fundamental purpose of laboratory-scale test work is to determine optimal leach conditions, maximum recoveries, and expected reagent consumptions. All this information is critical to understanding the viability of a uranium project. Depending on the results from preliminary leach test work, it may be necessary to investigate methods such as resin-in-pulp or resin-in-leach to counteract preg-robbing of uranium from liquors by ores containing clays or organic materials.

At this stage, it is necessary to consider the choice of a suitable oxidant, if one is mandatory. This will depend on the amount required to obtain an acceptable uranium extraction, as shown by the leach tests, the cost, and the proximity of the source of a supplier. For example, if a suitable manganese mine is close to the operation, then it is likely that the use of pyrolusite, MnO_2, may be feasible. If only a small amount of oxidant is required, then it may be reasonable to use a more-expensive, faster-reacting reagent. If initial leach tests show low iron dissolution, then it may also be necessary to investigate the addition of ferric iron to leach liquors to promote

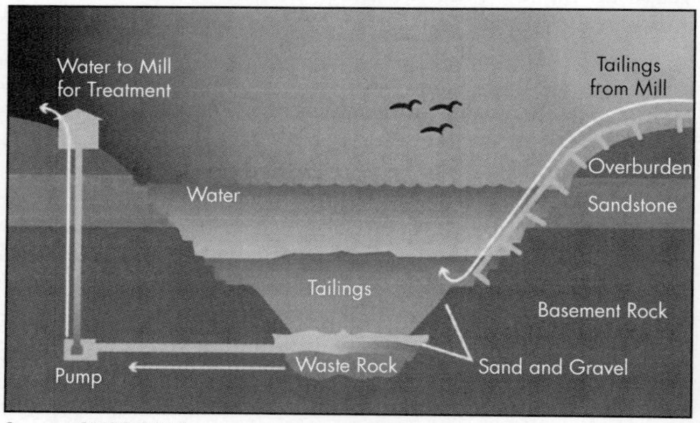

Source: CNSC 2015

Figure 16 Example of in-pit tailings disposal

U^{4+} oxidation. The concentration of impurities in the leach liquors, and their potential effects in building up in concentration, need to be evaluated, with the subsequent requirement of bleed streams.

Another important factor to consider is the material handling characteristics of ores and slurries. Rheological studies at varying solids density should be performed on leach feeds and residues to identify potential issues with agitation and pumping of slurries, which can occur in ores with high clay content. In addition, the amenability of the slurry to solid–liquid separation, and the suitability of filtration or CCD, should be determined from filtration and settling tests.

Once the appropriate leaching and solid–liquid separation routes have been determined, concentration and subsequent precipitation of the uranium can then be tested using standard methods. The composition of the leach liquor will influence the selection of the concentration method. Ion exchange is typically favored for liquors with lower uranium tenor; however, it is sensitive to species that also load onto the resin, such as ferric iron. Liquors with high chloride concentrations are also generally not amenable to ion exchange. Solvent extraction is the most typical concentration method employed and is a robust process, although problems can arise for liquors with high silicon concentrations, with colloidal silica causing the formation of stable emulsions. Elements such as molybdenum, vanadium, and zirconium in leach liquors can also present challenges. Both ion exchange and solvent extraction are tested by generating adsorption isotherms to identify the impact of liquor composition on extraction and to identify the number of stages required. Equilibrium data generated for ion exchange can then be fed into appropriate models. Loading and elution kinetics must also be determined. The eluent or stripping reagent to recover the uranium from the chosen process must be selective and also not strip species that will be deleterious to the precipitated uranium product.

The final stage of processing—precipitation of the uranium as uranyl peroxide, sodium diuranate, or ammonium diuranate—is a standard process. If selection of the concentration method and stripping reagent is appropriate, then uranium can typically be effectively recovered without excessive levels of undesired impurities. Alternatively, it may be possible to selectively precipitate uranyl peroxide directly from the leach liquor without a concentration step. Uranium precipitates should be assayed to ensure that the product meets the specifications required for the converter to which the product would be sold.

Following evaluation of the preceding bench-scale batch work, it is then necessary in the next phase to move to miniplant work to test and refine the process and identify any unforeseen challenges.

ACID LEACH OPERATIONS

Key Lake Mill for McArthur River Mine Ore

The Key Lake mill in Northern Saskatchewan's Athabasca Basin is operated by Cameco Corporation. Ownership is by Cameco Corporation with 83.33% and AREVA Resources Canada Inc. with 16.67%. The licensed annual production capacity is 10,000 t U. Since production start-up in 1983, Key Lake's lifetime production at the end of 2014 was 185,000 t U—about 210,000 t U by the end of 2017 (Edwards 1992; Cameco Corporation 2015).

The Key Lake mill process was designed to deal with the relatively high uranium ore grade and the presence of significant levels of arsenic and nickel sulfides (Neven and Gormley 1982; Neven et al. 1985). Particular attention was paid to effectively sequestering arsenic by ensuring a molar Fe/As ratio of 4 or more and maintaining a proper pH profile through contaminant precipitation operations (Krause and Ettel 1987).

Currently, McArthur River mine ore is processed in the Key Lake mill. A simplified flow sheet is shown in Figure 17. Run-of-mine (ROM) ore slurry is received by B-train truck from the mine. On arrival at Key Lake, the high-grade ore slurry is removed from the truck containers and mixed with a low-grade gravity concentrate from ground special waste rock. This dilutes the McArthur River ore slurry, grading on average 15% U_3O_8 down to 5% U_3O_8 for mill feed. This blended mill feed is transferred to storage pachucas in the mill and then pumped to leaching.

Key Lake mill benefited from the January 2002 replacement of the McArthur River grinding circuit classification screens with cyclones. With cyclones, the pitchblende in the ore is preferentially ground finer than 75 μm. In addition to improvements in ore slurry handling, shipping and unloading, and radiation safety, uranium recovery at Key Lake increased by approximately 1% (Edwards and Dyck 2004).

Sulfuric acid leaching is carried out in pachucas followed by continuous stirred-tank reactors, with oxygen as oxidant. Steam injection is required to maintain leach temperature. Two autoclaves, previously used for leaching, are on standby if extra leaching capacity is needed. Leach circuit discharge is pumped to the CCD circuit. Final CCD underflow slurry is pumped to tailings neutralization. Final CCD overflow is pumped through sand filters for clarification. The clarified solution is SX feed. The SX plant uses tertiary amine in a kerosene solvent with isodecanol modifier and ammonia stripping in a series of mixer-settlers. Pregnant strip solution from the uranium SX plant is fed to a smaller SX plant to remove molybdenum. This plant uses LIX 63 in a kerosene solvent to extract molybdenum (U.S. Patent No. 5,229,086; Lam et al. 1993). Uranium is precipitated from the cleaned pregnant strip solution by ammonia addition, forming ammonium diuranate. Washed and dewatered precipitate is calcined and packaged in standard 210-L steel drums.

To prevent ammonia contamination of mill effluent, a crystallization plant removes ammonium sulfate from the barren strip solution. The crystallization plant produces a quality product that meets the specifications for sale as ammonium sulfate fertilizer.

SX raffinate is neutralized with slaked lime. Overflow solution from the bulk neutralization thickener is treated for radium removal with barium chloride. Bulk neutralization and radium removal precipitates, collected as thickener underflow slurry, are disposed in the Deilmann in-pit tailings management facility via the tailings neutralization circuit. Overflow from the radium removal thickener flows to a series of monitoring ponds. These ponds hold effluent until assays on the contents of each pond confirm that effluent specifications are met, and only then are the ponds individually discharged (Neven et al. 1986; Rodgers and Sarion 1998; Rodgers 1996, 2000; Bharadwaj et al. 2010; Lien 2010).

McClean Lake Mill

The McClean Lake mill, referred to as JEB Mill in Northern Saskatchewan's Athabasca Basin, is operated by AREVA Resources Canada Inc. Ownership is by AREVA Resources

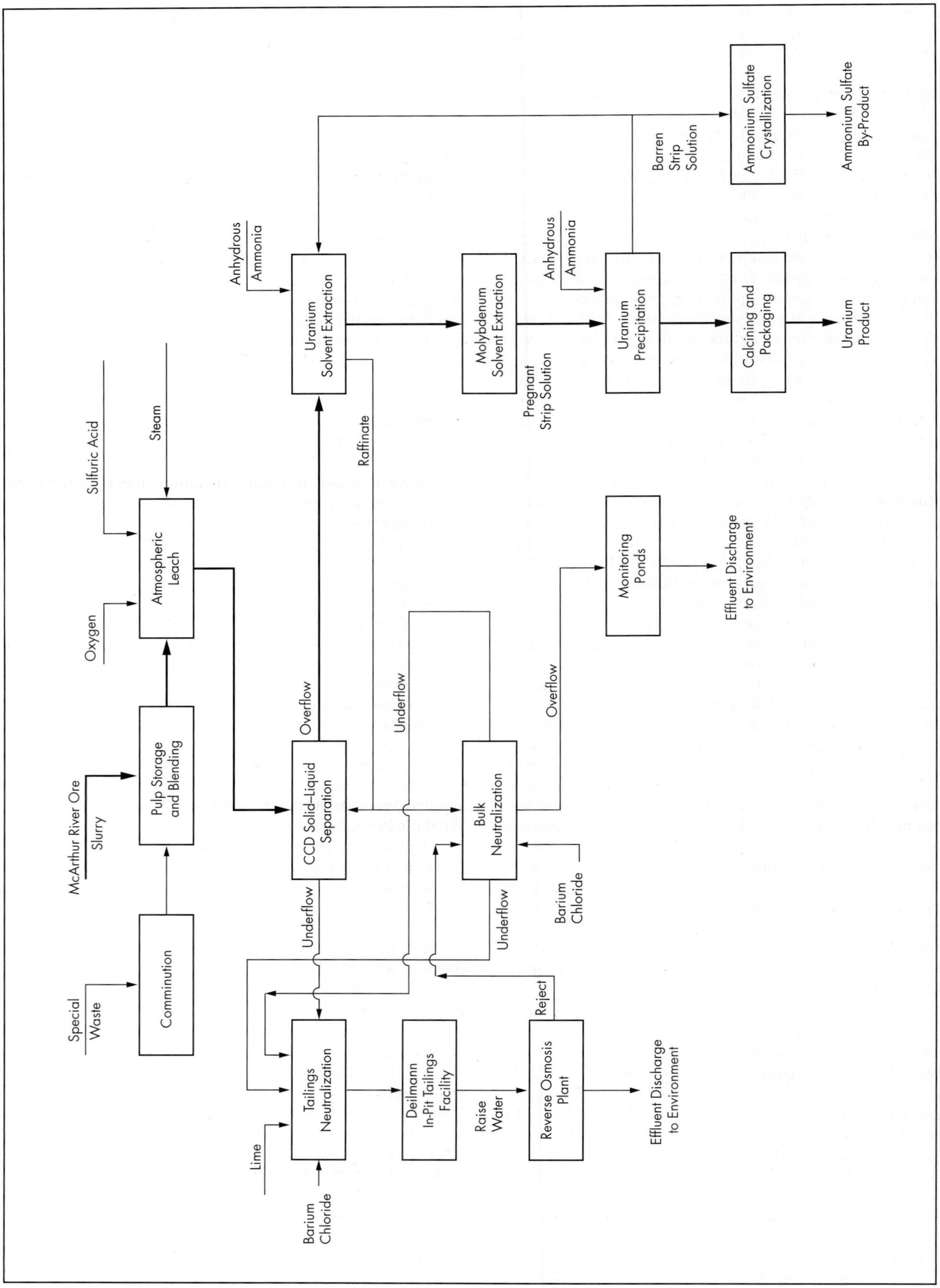

Courtesy of Cameco Corporation

Figure 17 Simplified flow sheet for Key Lake mill processing McArthur River ore

Canada Inc. with 70%, Denison Mines Inc. with 22.5%, and OURD (Canada) Co. Ltd. with 7.5%. The licensed annual production capacity is 5,000 t U. Since production start-up in 1999, the JEB Mill lifetime production at the end of 2016 was approximately 30,000 t U (Edwards 1992; Pacquet 2017; CNSC 2012; Denison Mines Corporation 2015) before the Cigar Lake mine commenced shipment of ore to McClean Lake.

Although the McClean Lake mill is based on the same chemistry as the Key Lake mill, McClean Lake's design is not a clone of Key Lake. One major difference is that the radiation protection installations at Key Lake limit maximum mill feed grade to 5% U_3O_8, yet the McClean Lake mill was designed for feed grades as high as 35% U_3O_8 (30% U), in anticipation of processing Cigar Lake ore (Badea and Schwartz 2000; Carino 2001; McKinnon 2000; St-Pierre and Huffman 2000).

Cigar Lake mine ore is currently processed in the McClean Lake mill (Figure 18). ROM ore slurry is received by B-train truck from the mine. The average ore grade is 18% U_3O_8. Sulfuric acid leaching is carried out in continuous stirred-tank reactors, with hydrogen peroxide as oxidant. With higher uranium and arsenic contaminant grades in the feed, cooling is required to maintain leach temperature. Leach circuit discharge is pumped to the CCD circuit. Final CCD underflow slurry is pumped to tailings neutralization. Final CCD overflow is pumped through sand filters for clarification. The clarified solution is SX feed. The SX plant uses tertiary amine in a kerosene solvent with isodecanol modifier and ammonia stripping in a series of mixer-settlers. Impurity molybdenum is removed from pregnant strip solution using activated carbon columns. Uranium precipitation is by ammonia, forming ammonium diuranate. Washed and dewatered precipitate is calcined and packaged in standard 210-L steel drums.

To prevent ammonia contamination of mill effluent, a crystallization plant removes ammonium sulfate from barren strip solution. The crystallization plant produces a quality product that meets the specifications for sale as ammonium sulfate fertilizer.

SX raffinate is pumped to tailings neutralization. Neutralization is with slaked lime. Tailings neutralization precipitates are disposed in the JEB in-pit tailings management facility. Water recovered from tailings slurry densification is pumped to the water treatment circuit. Ferric sulfate is added to remove dissolved heavy metals. Dissolved radium is precipitated with barium chloride. Neutralization is with slaked lime. Initially, treated water was pumped through sand filters prior to entering the monitoring ponds, a practice first used at Rabbit Lake (Edwards 1987, 2002b; Milde 1989). Currently, the effluent sand filters are bypassed. The monitoring ponds hold effluent until assays on the contents of each pond confirm effluent specifications are met, and only then are the ponds individually discharged (Edwards 1997, 2002a; Edwards and Schnell 2000; Remple and Schnell 2000; Stewart 2010).

Kayelekera Mine—Crush/Grind/Leach/RIP/Double-Precipitation Operations

The Kayelekera mine is located in Malawi about 52 km west of Karonga on the northern tip of Lake Malawi. Uranium mineralization was first identified in the North Rukuru River valley in Malawi in 1957, but it was only following an airborne radiometric survey in 1977 that Agip located several radiometric anomalies in the area. In 1983, the United Kingdom's Central Electricity Generating Board (CEGB) was granted a reconnaissance license, and a year later, when they realized that there could be a significant deposit of uranium at Kayelekera, they applied for a three-year Exclusive Prospecting Licence.

The Karoo rocks of the North Rukuru Basin consist of a succession of clastic sediments, overlain by a series of thick arkosic sandstones and mudstones. The succession is indicative of cyclic sedimentation within a broad, shallow, intermittently subsiding basin. Each cyclotherm passes upward from coarse reduced-facies arkose, through oxide-facies mudstone, to carbonaceous silty mudstone with coal-rich horizons. The arkose units, on average, are 8 m thick. They consist of coarse-grained quartz and fresh feldspar (albite) together with biotite and muscovite. The carbonaceous mudstone units contain discrete quartz grains with calcite veining and laminated carbonaceous pyritic black shale.

Most of the uranium mineralization is contained in the arkose units, but some secondary mineralization occurs in the mudstones. Coffinite is the main primary uranium mineral in the reduced zone; it is often associated with organic debris and pyrite and it is finely intergrown with chlorite/clay that fills the interstices. Extremely fine-grained uranium oxide and a uranium-titanium mineral have been identified in minor quantities in the transitional zone. In the oxidized zone, several secondary uranium minerals, including meta-autunite, boltwoodite, and uranophane, are the result of oxidative weathering of the primary uranium minerals.

A prefeasibility study, including preliminary laboratory testing, confirmed the potential of the deposit, and a full feasibility study was completed by Wright Engineers in 1991 for the CEGB. However, the low uranium price and the privatization of the CEGB at that time led to the project being placed on a "minimum cost" basis. Paladin Resources purchased a controlling interest in the deposit in 1998. It updated the earlier feasibility study and undertook a bankable feasibility study that led to a development agreement with the government of Malawi. The environmental impact assessment was approved and a mining license was granted in 2007.

The study was based on a conventional acid leach, CCD, SX, and ammonium diuranate flow sheet. However, following test work at ANSTO (Australian Nuclear Science and Technology Organisation) and an optimization review, it was found that a process using resin-in-pulp (RIP) and hydrogen peroxide precipitation offered lower capital and operating costs; better plant operability; and reduced safety, health, and environmental concerns. This innovative flow sheet represented the first RIP plant to be built in more than 50 years and the first ever in the Western world to treat high-density slurry (over 30% solids). It also included the use of hydrogen peroxide as the leach oxidant with multistage addition of acid and oxidant.

In 2009, a plant with a capacity of 1.3 t/yr U was commissioned. It was designed to treat 1.5 Mt/yr of ore at a grade of ~1,100 ppm U_3O_8. After open pit mining, the ore is delivered to dedicated stockpiles classified by ore type and grade. The ore is then blended to control the plant feed grade and to maintain the mudstone component to between 10% and 20% of the ROM feed. A front-end loader transports the ore from each stockpile to a dump hopper fitted with a grizzly. Ore is drawn from the hopper using a variable-speed apron feeder that discharges into a rotary tooth mineral sizer. The mineral sizer was subsequently supplemented with a jaw crusher to provide greater crushing flexibility. The crushed ore, cyclone

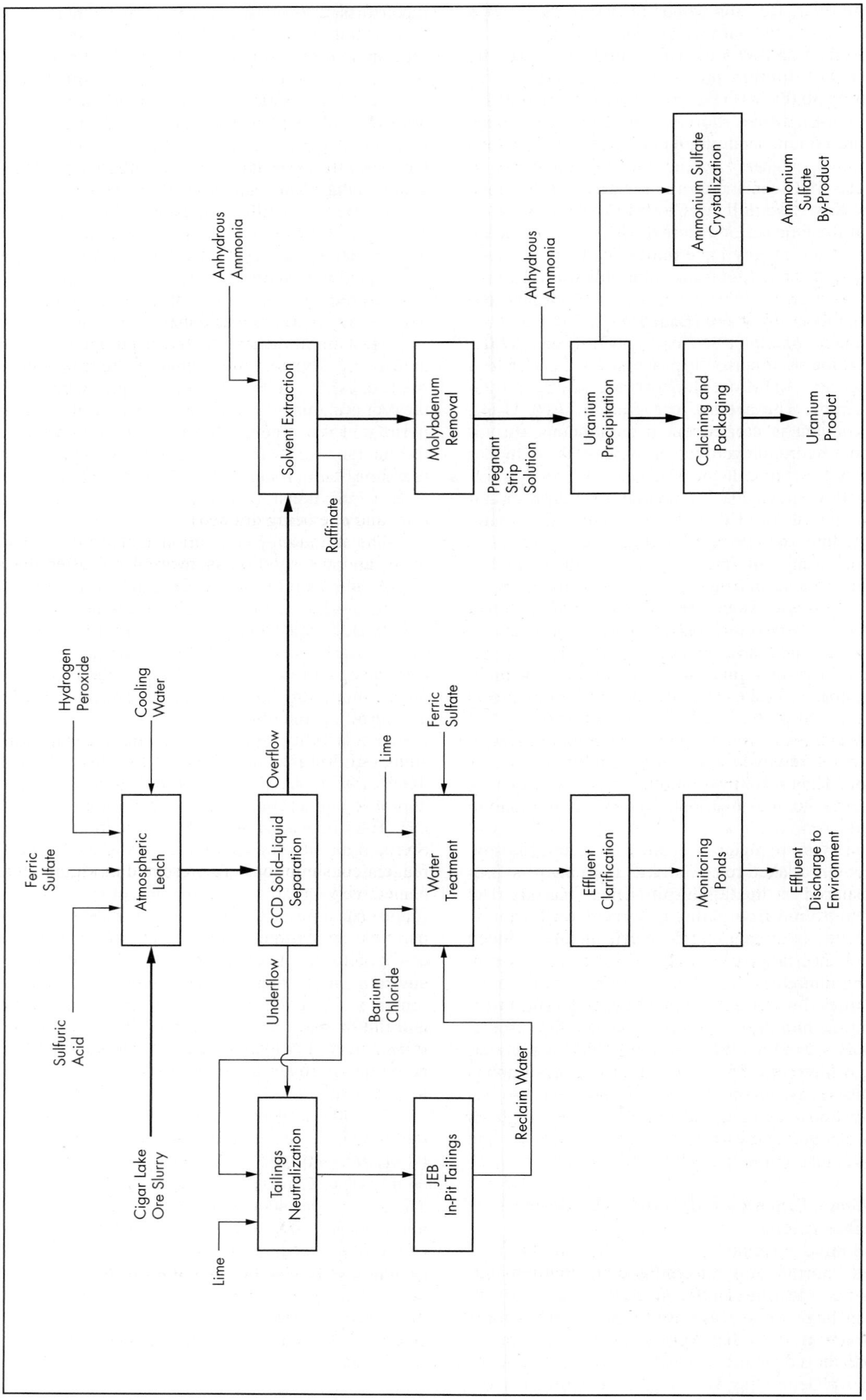

Courtesy of Cameco Corporation

Figure 18 Simplified flow sheet for McClean Lake mill processing Cigar Lake ore

underflow, and water are fed to a ball mill with grinding media added to the mill feed conveyor as required.

The mill discharge, after removal of scats and grinding media on a trommel, is pumped to a hydrocyclone cluster operating in closed circuit with the mill. The target grind size is P_{100} of 300 μm (P_{80} of 180 μm). Water is added to the mill discharge to control the cyclone feed density. The cyclone overflow is directed to a trash screen and then gravitates to the pre-leach thickener where flocculant is added to the feed launder. The thickener overflow gravitates to a process water tank for reuse in the mill, and the underflow is pumped to the leaching circuit.

Slurry feed from the pre-leach area, liquor from the tailings thickener overflow, and leach spillage all flow into the leach feed dilution box. From this box, slurry is fed to a series of six mechanically agitated leach tanks (with bypass launders). Sulfuric acid can be added to the first four leach tanks and hydrogen peroxide (H_2O_2) to tanks 2, 3, and 4 to control both the pH and oxidation–reduction potential. Slurry leaving the leach gravitates onto a linear screen to protect the RIP plant by removing any particles coarser than 400 μm. Screened slurry is pumped to the first RIP contactor.

The RIP tanks are laid out for carousel operation so that any tank can be used as the feed or tail contactor and any tank can be taken off-line and drained for recovery of the resin. The RIP tanks are each equipped with a pump-cell mechanism. This device consists of a down-pumping agitator to keep the pulp and resin in suspension, a wedge-wire screen to retain the resin beads in the contactor, and an up-pumping impellor inside the screen basket to develop sufficient head for the pulp to flow to the next contactor via an open volute. The RIP plant comprises 10 such contactors with 8 online at any one time extracting uranium from the leached slurry. The other two contactors are off-line, one of which transfers loaded resin to the resin recovery circuit while the second is on standby or receiving stripped resin from the elution circuit.

Slurry that has passed through the series of adsorption contactors flows from the final RIP tank to another linear screen to ensure that any resin that may have escaped the interstage screens is recaptured. The RIP tail slurry is then pumped, via a sampler, to the tailings thickener feed box. The overflow from this thickener is returned to the leach feed box, and the underflow is pumped to two mechanically agitated tanks, where it is neutralized with slaked lime before being pumped to the tailings dam.

On a fixed cycle or when the resin in the lead RIP contactor is sufficiently loaded, the lead RIP contactor is taken off-line, whereby the second contactor in the sequence becomes the lead contactor and the tank containing freshly stripped resin is added to the end of the adsorption train. The tank containing the loaded resin and pulp is drained in its entirety and pumped over a linear screen with water sprays to clean and recover the resin. The screen undersize is returned to the RIP feed sump, and the recovered resin is subjected to additional cleaning using spirals and a secondary screen to remove near-size grit.

The clean loaded resin is fed to the last contactor in the elution circuit, which operates in a similar countercurrent carousel fashion to the adsorption circuit. Fresh eluent (1.1 M sulfuric acid) is pumped to the lead elution contactor and then passes by gravity flow through the other contactors in sequence before the concentrated eluate is discharged from the last elution vessel. The elution vessels are air-agitated conical bottom tanks with wedge-wire "spiders" to retain the resin. The design provided for resin scrubbing and regeneration; however, this part of the process has not been required.

In 2012, another pioneering innovation was introduced at Kayelekera (patent pending). To recover sulfuric acid from the eluate and reduce the magnesia demand for neutralization of the concentrated eluate, the concentrated eluate is first fed to a membrane plant where ultrafiltration followed by nanofiltration recovers a significant fraction of the acid in the eluate as permeate (which is recycled to the leach, offsetting the new acid demand for leaching) and simultaneously increases the uranium concentration in the retentate. The retentate is then advanced to the neutralization section, instead of untreated concentrate eluate as was done in the original circuit.

The nanofiltration retentate is pumped through a series of neutralization tanks, and magnesia slurry is added to raise the pH to 2.5–3.0. Some of the iron in the eluate precipitates, and this precipitate, along with impurities introduced by the magnesia, is captured on a filter press. The filter cake is recycled to the leach or RIP feed to recover any co-precipitated uranium, and the filtrate is pumped to the batch uranium peroxide precipitation tanks (one being filled, one being used for precipitation, and one being drained).

The uranium precipitation tanks are mechanically agitated, and the solution is recirculated after the addition of hydrogen peroxide and sodium hydroxide (to control the pH) as a batch process before being transferred in batches to the uranium precipitation thickener. The underflow from this thickener is washed in two countercurrent wash thickeners and the overflow returned to the pre-leach thickener. The underflow is dewatered in a centrifuge before reporting to the drying and packaging plant.

Solids from the centrifuge are blended with a recycle stream of dried yellowcake (recycled from the dryer) and fed to the dryer with a screw feeder. The drying is done indirectly with hot oil that flows through the screws. Dried product gravitates into a product hopper and is then fed into 210-L drums using a rotary feeder. A schematic flow diagram and the reagent consumptions are presented in Figure 19 and Table 8, respectively.

In April 2014, the mine and plant was placed on care and maintenance because the uranium price was insufficient to justify continuing operation. However, from a technical perspective, the plant was performing close to design expectations with uranium extraction in the leach of around 88% and more than 98% uranium recovery by RIP. The ore is variable in its acid and oxidant demand, and, in the future, this aspect could be improved by better management of stockpile blending and the application of radiometric sorting.

ALKALINE LEACH OPERATIONS

Langer Heinrich—Attrition/Tank Leach/CCD/IX/H_2O_2 Precipitation

The Langer Heinrich mine (LHM) is located in Namibia about 80 km inland from a deep-water port at Walvis Bay. When it was commissioned in early 2007, it was the first conventional mining and processing operation to be brought into production for more than a decade and the first one to use alkaline leaching since 1981. Remarkably, the head grade is less than one-third that of previous alkaline leach plants.

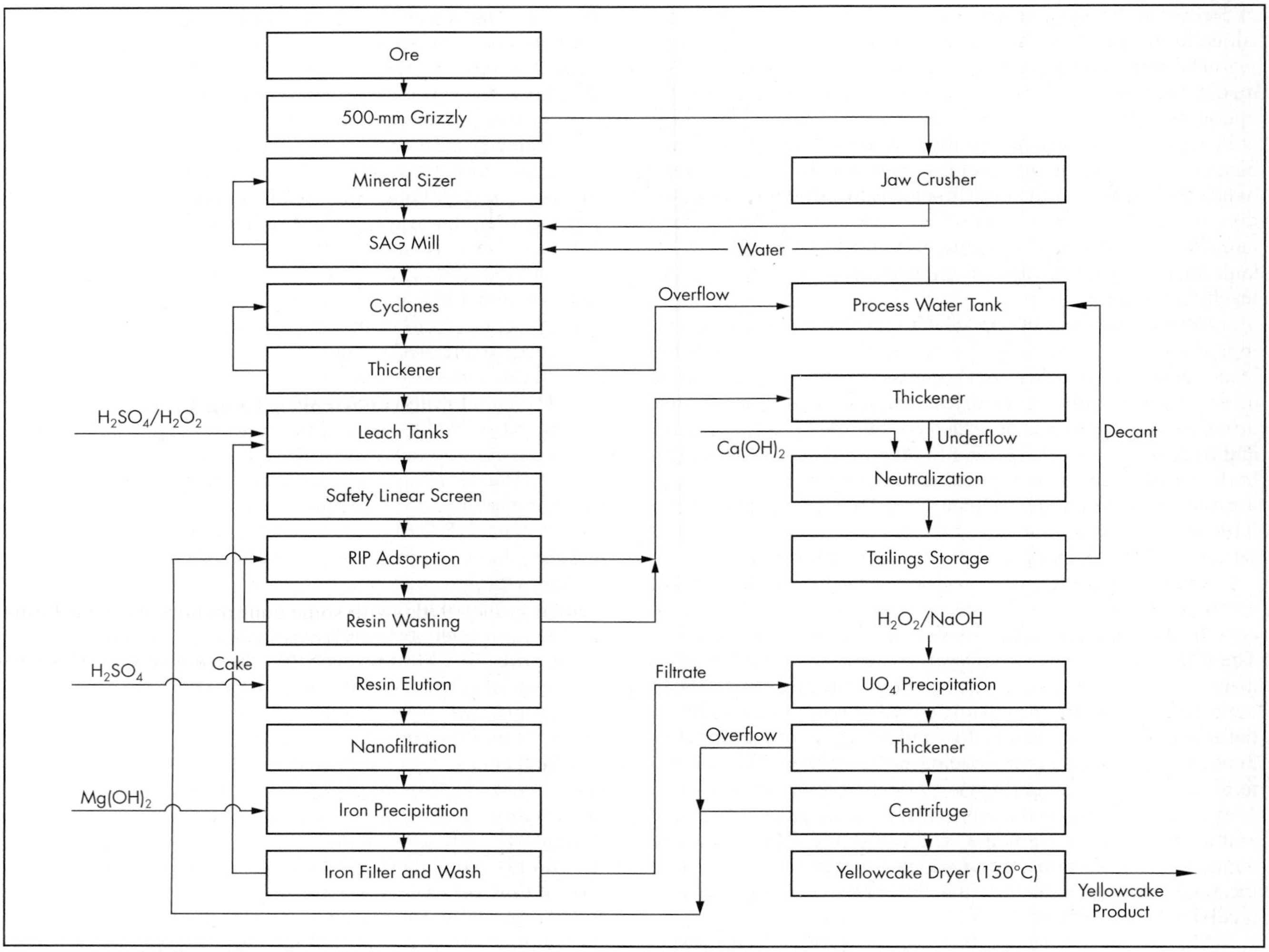

Figure 19 Kayelekera plant flow sheet

Gencor (now BHP) discovered the deposit in the 1970s and conducted an extensive project evaluation over a period of eight years. In 1998, the project was sold to Acclaim Uranium and then in 2002, to Paladin Energy. Paladin completed a bankable feasibility study in April 2005 and, based on a mill throughput of 1.5 Mt/yr at a feed grade of 740 g/t U, concluded that a production rate of 1 million kg U/yr could be sustained for more than 11 years. Site works began in September 2005, and the mine was officially opened in March 2007. Nameplate production was achieved in December 2007.

Uranium mineralization at LHM is associated with the calcretization of fluvial sediments in an extensive paleo-drainage system that extends more than 15 km. It is between 50 and 1,100 m wide and occurs at the depth of between 1 and 30 m below the surface. The uranium occurs as carnotite (an oxidized secondary uranium/vanadium mineral) as thin films lining cavities and fracture planes and as grain coatings.

In May 2008, work on the Stage 2 expansion of production to 2.6 Mt/yr of ore and 1.4 million kg U/yr was started. This extension was commissioned in early 2009, and, before the end of the year, the design and engineering of Stage 3 expansion to 2 million kg U/yr was initiated, and nameplate production was achieved in early 2012.

Table 8 Kayelekera key reagent consumptions

Reagent	Units	Value
Fresh water	kg/t ore	540
Grinding media	kg/t ore	1.5
Flocculant	kg/t ore	0.2
Sulfuric acid	kg/t ore	55
Hydrogen peroxide—leach	kg/t ore	2.5
Lime	kg/t ore	10
Resin	L/t ore	0.25
Magnesia	kg/kg U_3O_8	3.0
Hydrogen peroxide—precipitation	kg/kg U_3O_8	0.2
Sodium hydroxide	kg/kg U_3O_8	0.4
Sulfur	kg/kg acid	0.33
Electricity	kW·h/kg U_3O_8	36
Diesel fuel for power	L/kW·h	0.26

The LHM process is possibly unique in that after crushing to a nominal top size of 100 mm, the ore is fed to an autogenous rotary scrubber where the agglomerates are broken down into individual grains and the uranium minerals are scrubbed from the grain surfaces. Size classification is used to separate the barren sands from the rich fines. This size classification is done with a combination of cyclones and screens, permitting around 40% of the mass to be rejected (with less than 6% of the uranium) and the leach feed to be upgraded to 1,000 ppm U.

Leaching is done using a mixture of sodium carbonate and sodium bicarbonate in a series of mechanically agitated tanks at elevated temperature. Two sets of spiral heat exchangers are used. The first set recovers heat from the leached pulp and the second set uses hot water (recycled through electric and diesel boilers) to raise the temperature in the leach feed to 75°C. In addition, inter-tank hot-water-immersion heating bundles are installed to compensate for the heat losses during the leach.

The leached solids are then washed in a six-stage CCD circuit and the tailings are pumped to a tailings storage facility. The decant liquor is recycled to the plant to conserve water and reagents. Part of the overflow from CCD2 is recirculated to the leach conditioning tank to control the leach density (30% solids), and the overflow from CCD1 is clarified in preparation for uranium recovery by ion exchange. The pH of the PLS is increased to about 11.2 using lime and sodium hydroxide to improve the selectivity of the resin for uranium over vanadium.

In the very few instances where ion exchange has been used for uranium recovery from an alkaline PLS, sodium chloride has been used as the eluent. However, in the LHM context, situated as it is in a national park in the Namib Desert, with no ability to bleed process water, the chlorides would build up in the process water to levels that would severely depress the uranium loading on the resin. Sodium bicarbonate was found to be a suitable alternative eluent and although it is not as efficient an eluent as sodium chloride, it avoids adding chloride to the system and indirectly satisfies the leach reagent demand (with no other sodium carbonate or sodium bicarbonate being required).

The concentrated eluent is fed to the continuous SDU (sodium diuranate) precipitation plant where sufficient sodium hydroxide is added to react with the remaining sodium bicarbonate and uranium as well as leaving 10 g/L residual NaOH (sodium hydroxide). The final liquor is thickened and the underflow is repulped with fresh water to wash out sodium carbonate and caustic soda.

The thickened SDU slurry is then pumped to one of three batch tanks, where, after decanting excess liquor, sulfuric acid is added to dissolve the SDU, decompose any carbonates, and adjust the pH to 3.5. The next step uses hydrogen peroxide to precipitate a pure uranyl peroxide, while the pH is controlled using sodium hydroxide. The UO_4 precipitate is thickened, repulped with fresh water, and dewatered with a centrifuge. The centrifuge product is then sent to a dryer where the remaining moisture is removed at low temperatures (115°C). The dried product discharges to a bin and is packaged in tightly sealed 210-L drums.

Plant expansions, as noted previously, were the inclusion of recycle crushing of coarse scrubbed rocks and a second larger pre-leach thickener. The next expansion added a second primary crushing circuit and a tertiary crushing plant, plus a third larger scrubber and classification equipment. The expansions also added a second leaching train, with the associated heat exchangers. The original six 25-m-diameter CCD thickeners were rearranged into two trains of three and supplemented by three additional 35-m-diameter thickeners, and then the original six thickeners were reconfigured as three trains of two thickeners and, with two additional 35-m-diameter thickeners, created a seven-stage CCD circuit.

The original IX plant consisted of 16 downflow fixed-bed columns (3.1 m-diameter), each containing 22.4 m^3 of strong base anion resin in lead, lag, or elution mode. The number of columns was initially increased to 24. However, the resin bed clogged up with suspended solids and frequent backwashing was required, introducing inefficiencies to the system. Finally it was decided to use NIMCIX multistage fluidized-bed contactors and two parallel modules of adsorption, and elution columns were installed with a nominal capacity of 600 m^3/h PLS. A schematic flow diagram and the reagent consumptions are presented in Figure 20 and Table 9, respectively.

Trekkopje, Namibia—Alkaline Heap Leach

Trekkopje is a shallow, high-tonnage, low-grade, calcrete surface deposit. The principal mineralization extends over an area approximately 14 km long and 3 km wide. It will be exploited with an open pit mine. Owing to the very low grade of the ore (~120 ppm U) and the method used to process it, exploiting this deposit presents a technical challenge.

The Trekkopje Project uranium mineralization was first drilled in the 1970s, with some confirmatory work performed during the 1990s, and was reestablished as a uranium development project by UraMin in 2005. The project was then taken over by AREVA in 2007. The project was advanced through to the feasibility study, incorporating a MINI pilot plant of 30,000 t and then to a demonstration MIDI-phase plant with the first uranium concentrate (SDU) from the Trekkopje deposit produced in January 2011. During this demonstration-plant phase, ending in mid-2013, almost 435 t of SDU was produced, confirming the technical aspects of this large-scale alkaline leaching operation. The installations for the full-scale production phase, called MAXI, were mothballed in 2013 waiting for improved uranium market conditions.

Heap leaching was selected as the preferred process for the Trekkopje Project. Given a total leach cycle time of 120 days (plus washing and rinsing cycles), an irrigation rate of 5–10 L/h/m^2, a crush size of 38 mm, and a carbonate concentration of ~20 g/L carbonate and ~10 g/L bicarbonate, an extraction of +75% of the uranium was achieved.

The performance of a multistage countercurrent fluidized bed IX system was modeled based on the batch equilibrium loading and loading rate data. To optimize uranium recovery, the IX plant needs a total resin inventory (for a PLS of 2,300 m^3/h containing 300 mg/L U) of between 1,000 and 1,300 m^3.

The crushing plant was designed to crush ore at a rate of 7,000 t/h to a product size of 100% passing 38 mm. The crushing plant consists of two primary movable mineral sizer crushers running in parallel and two secondary crushers, also running in parallel, each with a throughput of 3,500 t/h. All ore is transported from the primary crusher by overland conveyors to the stationary secondary and tertiary crusher installation.

Both sets of secondary crushing circuits deliver crushed ore into one of two storage silos. Material is reclaimed from these silos by means of apron feeders feeding onto the agglomeration drum feed conveyors that discharge material directly into the agglomeration drums.

Crushed ore is fed to the agglomeration drum at 100,000 t/d, with water added to bring the moisture to ~8%. The agglomerated ore is then stacked with a RAHCO

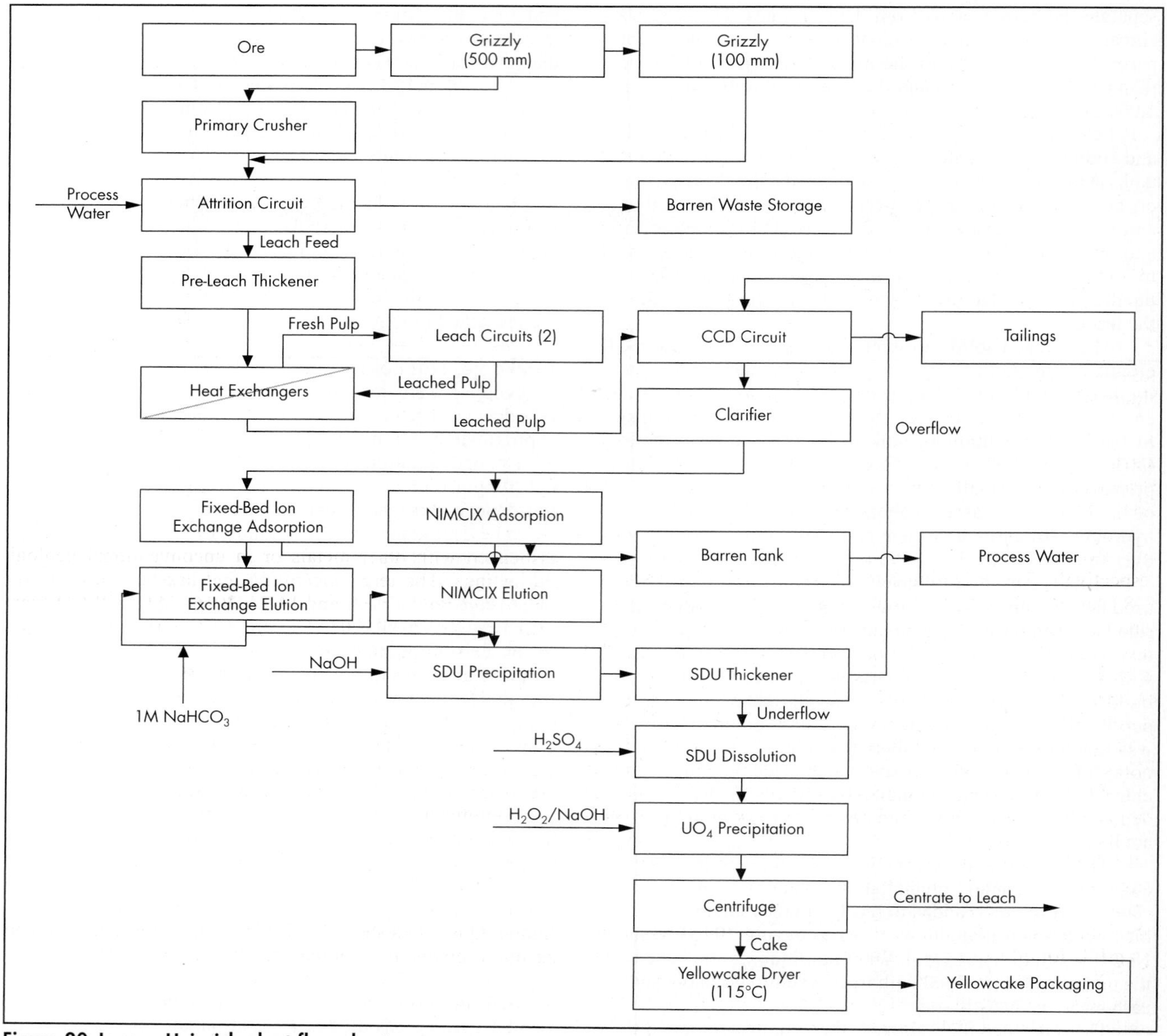

Figure 20 Langer Heinrich plant flow sheet

Table 9 Langer Heinrich key consumption data

Reagent	Units	Value
Fresh water	kg/t leach feed	830
Steam	kg/t leach feed	90
Flocculant	kg/t leach feed	0.21
Sodium bicarbonate	kg/kg U_3O_8	30
Sodium hydroxide	kg/kg U_3O_8	13
Hydrogen peroxide	kg/kg U_3O_8	0.25
Hydrated lime	kg/kg U_3O_8	2
Sulfuric acid	kg/kg U_3O_8	0.80
Electricity	kW·h/kg U_3O_8	30

conveyor/stacker in 1-million-t cells (~900 m long × ~100 m wide × 9 m high).

The Trekkopje ore deposit contain chlorides that must undergo washing with fresh water to remove the chlorides. The chlorides compete with uranium for loading on the resin, and if the concentration in the PLS is in appreciable amounts (>1.6 g/L), it can lead to poor uranium loading on the resin, affecting the performance of the entire circuit downstream.

Dripper layout design is such that the pad can be irrigated at 5–8 L/m^2/h, with the cell area at 75,000 m^2, giving a total flow rate to each cell around 600 m^3/h.

The wash is comprised of two cells operating in series, with the discharge of the first cell feeding the next. The

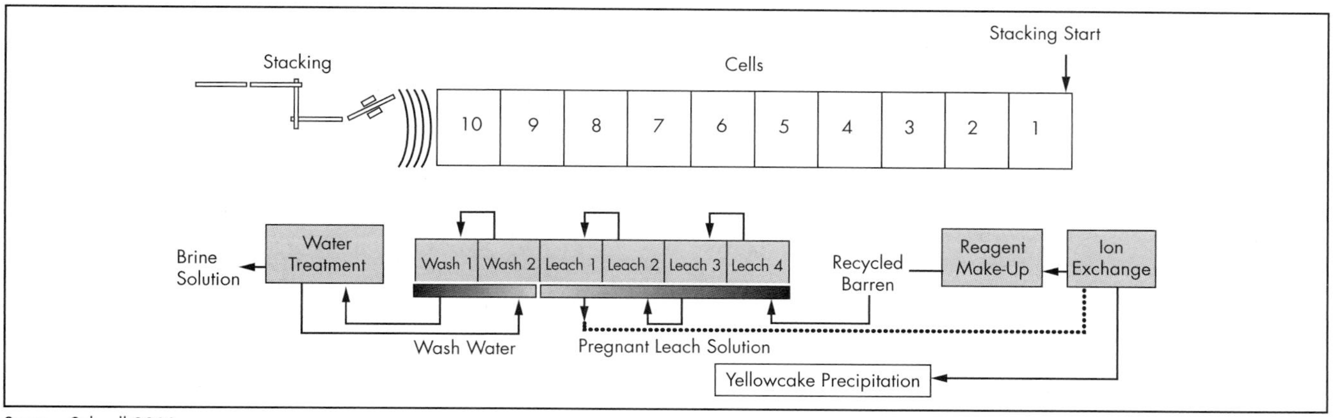

Source: Schnell 2011

Figure 21 Alkaline heap leach basic flow sheet

discharge of the second cell reports to the high-chloride water treatment plant where reverse-osmosis technology is used to bring the chloride down to about 600 ppm.

Leaching is carried out with a sodium carbonate and sodium bicarbonate solution at 40 g/L and 8–10 g/L, respectively. The irrigation rate during the leach cycle is at 5–8 $L/m^2/h$ with a 40-day leach cycle and a three-cycle countercurrent leach for ~120 days of total leach time.

The average uranium grade of the stacked ore is expected to be around 106 g/t, and the sulfate, 0.233%. Sodium carbonate consumption due to the ore is about 14 $kg/t/\%SO_4$, which is equivalent to a consumption of 3.2 kg/t. The sodium bicarbonate consumption for leach due to the ore is ~1 kg/t. The PLS grade of three cells in series reaches a peak of 150 ppm U.

The PLS is fed through a series of ponds to the NIMCIX IX columns with solution characteristics as shown in Figure 21, which provides a basic flow sheet for an alkaline heap leach facility. These columns use resin to adsorb the uranium in a multistage fluidized column. The uranium is eluted from the resin using 1 M sodium bicarbonate solution. The uranium grade of the solution leaving the elution columns is <4 g/L U for effective precipitation.

SDU is produced by direct precipitation with caustic soda added to the eluate. Before the precipitate can form, the sodium bicarbonate in the eluate must first be oxidized by the caustic according to the following reaction:

$$NaOH + NaHCO_3 \rightarrow Na_2CO_3 + H_2O$$

Excess caustic in the region of 5–7 g/L of eluate is required for effective precipitation, and once the required amount of excess caustic is achieved, SDU precipitate is formed with a precipitation efficiency of 98%, with eluate at 4.6 g/L. Precipitated SDU is sent to a thickener for dewatering to increase the slurry density from 1.07 to around 1.2. The dewatered SDU is washed with water to remove excess caustic.

Sulfuric acid is then added to the SDU slurry to dissolve the SDU. Caustic and hydrogen peroxide are then introduced to increase the pH of the solution, and thereby precipitate UO_4. The precipitate is filtered, dried, and packaged into standard 170-L drums ready for shipment.

BY-PRODUCTS AND CO-PRODUCTS FROM URANIUM OPERATIONS

Uranium is commonly found in many geological and background settings, and there are also many examples of uranium associated with other metals or in unconventional geological settings. The term *unconventional* has been used in the IAEA *Red Book* (NEA and IAEA 2016: IAEA/OECD biannual Uranium Resources, Production and Demand Review) for many years to suggest geological settings that have not been generally associated with uranium production.

In this section various options are considered where uranium can be produced from unconventional resources or as a by-product with other production. The *Red Book* defines a series of unconventional resources—phosphates, nonferrous ores, carbonatite, black schists, and lignite.

Significant quantities of uranium are found in many unconventional resources (NEA and IAEA 2016). These are resources from which uranium is only recoverable as a minor by-product, such as phosphate rocks, nonferrous ores, peralkaline intrusions and carbonatite, and black shale and coal-lignite. Quantities shown in Table 10 are largely incomplete, as many potential resources are not assessed properly and reported. Apart from these deposit types, reprocessing of previous tailings, wastewater, and residues (such as coal ash) can also be a source of unconventional uranium. Historically, significant quantities of uranium have been produced from phosphates as a by-product of fertilizer production (IAEA 2013b). About 17,150 t U were recovered in the past from phosphate rocks in Florida (United States; IAEA 1989). Gold tailing projects in South Africa are another source that contributed uranium production considerably in the past, which continues to date, though at reduced levels (IAEA 1985).

By-Product Production Classification

Uranium will be produced from a variety of unconventional resources in the future, and there are some examples already close to or in production. A possible alternative unconventional classification is to use the process options available for these resources:

- **Category 1—By-production.** By-production is used for resources when a primary product also contains a

Table 10 Unconventional resources of uranium (metric tons of uranium)

Country	Phosphate Rocks	Nonferrous Ores	Carbonatite	Black Schist/Shales, Lignite
Brazil	84,500	2,000	13,000	—
Chile	400	800	—	—
Columbia	20,000–60,000	—	—	—
Egypt	35,000–100,000	—	—	—
Finland	1,000	—	2,500	35,000
Greece	500	—	—	—
India	1,700–2,500	6,600–22,900	—	4,000
Iraq	19,000–42,800	—	—	—
Jordan	60,000	—	—	—
Kazakhstan	29,000	—	—	—
Mexico	240,000	1,000	—	—
Morocco	6,526,000	—	—	—
Peru	41,600	140–1,410	—	—
South Africa	180,000	—	—	70,700
Sweden	42,300	—	—	1,012,000
Syria	60,000–80,000	—	—	—
Thailand	500–1,500	—	—	—
United States	14,000–33,000	1,800	—	—
Venezuela	42,000	—	—	500

Adapted from NEA and IAEA 2016

secondary product or more than one other product. Often, the secondary product is not economic by itself, but perhaps as an add-on it can take advantage of the primary commodity production process to be available for recovery.

An example is uranium from phosphoric acid. The uranium is already soluble in the phosphoric acid and could be extracted, thereby reducing environmental concerns of uranium in fertilizers and producing a higher-quality phosphoric acid. Considerable work is underway to make the uranium economically attractive. Another example would be uranium in copper heap leach operations with uranium leached during the copper process.

- **Category 2—Co-production.** When more than one metal has economic values, co-production is employed. This is common in base metals when, typically, copper and nickel or zinc and lead are produced from a single ore source.

  The most notable example for uranium is the Olympic Dam operation in Australia, with economic uranium production from a copper producer. This places the Olympic Dam operation in the co-production category given that the primary source of uranium is from the copper tailings. Considering the high value of the uranium, this is normally considered a co-product relationship. Talvivaara in Finland is an interesting case of multiple products: the Talvivaara operation has copper, zinc, nickel/cobalt, and uranium as products using a heap leach flow sheet in a cold climate. This operation is a good example of "comprehensive extraction."

- **Category 3—Tailings reprocessing.** This option is used for tailings or mine waste reprocessing. Generally, this category would be for reprocessing previously placed tailings.

  The best examples are the tailings reprocessing operations for gold in South Africa. The tailings are hydraulically mined with high-pressure water, then fed to a leach plant for gold recovery. Some of these tailings also contain significant uranium values, and in some cases, this uranium is also recovered (Mine Waste Solutions plant of First Uranium Corporation). Similarly, a recent project at the Somair operation in Niger is using heap leach on previous mine waste with previous uneconomic uranium values to produce 400–800 t U/yr. As metal prices change with environmental considerations, reprocessing of old tailings will continue. The better option, of course, is to extract the most value initially.

- **Category 4—Unusual treatment.** This option would be used for new, innovative processes on previously uneconomic or unconventional resources.

  For example, the Trekkopje deposit in Namibia has extremely low grades (120 ppm U) that have been considered too low, but applying economics of scale combined with heap leach technology and new application of alkaline chemistry to heap leach, this deposit has potential for significant uranium production in the future. Another unusual resource is the Bakouma deposit in Central African Republic. The difficulty of this deposit is the high apatite content that requires very high-acid consumptions and subsequent neutralization that makes even a good-grade uranium deposit uneconomic. Also, this deposit has an unfavorable geographic location. It will be interesting to see how eventually these difficulties will be overcome for these projects.

Uranium has been produced as a by- or co-product of other metals and minerals, most notably: gold, copper, vanadium, and phosphate. Historically, by- and co-product uranium has accounted for approximately 12% of world production.

By-Product Operation Examples

Gold

Gold/uranium ores of South Africa's Witwatersrand Basin consist of a quartz-pebble conglomerate of Precambrian age containing low to moderate concentrations (0.02–0.06% U) of uraninite- and brannerite-type minerals. Uranium production from these ores has been ongoing since 1952, from both primary ore and historical tailings (slimes). Uranium extraction is by sulfuric acid leach of slurry either pre- or post-gold recovery. Pre-gold recovery leaching has been employed to enhance subsequent gold recovery. Uranium is recovered from the leach solution by ion exchange, concentrated by solvent extraction, and precipitated with ammonia. Overall uranium recovery is quite low at about 30%–40%.

Copper

Uranium has been recovered as a by- or co-product of copper in several ways, including (1) direct sulfuric acid leaching of initial copper ore slurry (Twin Buttes, United States); (2) IX extraction from dump leach solutions (Bingham Canyon, United States); (3) as a gravity concentrate (Palabora in South Africa; and Rakha, Surda, and Mosaboni in India); and (4) sulfuric acid leaching of copper flotation concentrates and tailings (Olympic Dam, Australia). More specific details for some operations are given here:

- **India (1970s–2000).** Small quantities of uranium were produced as a by-product of copper in India at three uranium recovery plants: Rakha, Surda, and Mosaboni. These plants treated flotation tails (~0.015% U) from copper concentrators by means of Wilfley tables to produce uranium-bearing heavy mineral concentrates (0.13% U). Concentrates were shipped by truck to the Jaduguda uranium mill for final recovery.
- **Palabora (1971–2002).** Uranium at Palabora occurs as uranothorianite at an average concentration of 35 ppm U in close association with copper sulfide mineralization in a carbonatite. Uranium recovery was from copper flotation tailings through a complex process of magnetic separation, gravity concentration, nitric acid leaching, and solvent extraction. The average recovered uranium grade was about 6 ppm.
- **Bingham Canyon (1978–1985).** The Bingham Canyon uranium plant ran four Higgins Loop IX columns operating in parallel to process approximately 1,800 m^3/h of copper–barren dump leach solution. Uranium concentration in the feed solution was typically in the range of 4 to 6 ppm U. Uranium was eluted with sulfuric acid, concentrated by solvent extraction, and precipitated with ammonia.
- **Twin Buttes (1980–1985).** At Twin Buttes, copper oxide ore containing about 30 ppm U was crushed and wet-ground to liberate both copper and uranium. Ore pulp was acidified with sulfuric acid to dissolve copper and uranium. Uranium was extracted by ion exchange, concentrated by solvent extraction, and precipitated with ammonia.
- **Olympic Dam (1988–present).** Sulfide mineralization at Olympic Dam averages 0.9% Cu and 0.04% U with minor amounts of gold and silver. Copper concentrates are recovered by flotation, and both concentrates and tailings are then leached with sulfuric acid to dissolve uranium. Uranium is recovered by solvent extraction, stripped with ammonium sulfate, and precipitated with ammonia.

Vanadium

Uranium/vanadium deposits of the Colorado Plateau in the western United States were first mined for radium in the late 1800s and then for vanadium beginning in the 1920s. Uranium for the Manhattan Project (the first atom bombs) was recovered from these tailings during World War II. Following the war, uranium and vanadium were produced as co-products, a practice that continues sporadically to the present.

Ores with a vanadium-oxide-to-uranium-oxide ratio of about 6:1 occur in sedimentary rocks of the Colorado Plateau district. Uranium and vanadium grades average 0.19% U and 1.23% V_2O_5 (vanadium oxide), respectively. Current milling operations incorporate a sulfuric acid leach with an SX process containing separate recovery circuits for uranium and vanadium extraction. Uranium is extracted first, stripped with sodium chloride, and precipitated with ammonia. Vanadium is stripped with soda ash, batch precipitated, dried, and fused to V_2O_5.

Phosphate

Uranium occurs in low concentrations (typically 50–150 ppm U) in most marine phosphate deposits. Phosphate rock is beneficiated, concentrated by flotation, and acidified with sulfuric acid to produce phosphoric acid, a basic building block for phosphate fertilizers. Phosphoric acid typically contains about 0.5 kg U/t of acid. During the period 1977–1999, several uranium by-products from phosphoric acid facilities operated in the United States, Canada, and Belgium. All used similar SX processes and all have now been decommissioned.

Typically, a uranium recovery plant consisted of two SX circuits. The primary circuit was designed to remove uranium from the large volume of feed acid and concentrate it (by a factor of 50). Concentrated uranium was then sent to the secondary circuit where uranium was further concentrated, stripped with ammonium carbonate, and then steam stripped to remove ammonia and carbon dioxide, which caused uranium to precipitate.

Future Technologies and Challenges

Efforts are underway on several fronts to develop new technologies for by- and co-product uranium recovery and to recover uranium from polymetallic and multielement deposits. Uranium from phosphate deposits are advancing with various levels of research and some pilot-plant operations. The Black Shale deposits, predominantly found in Sweden, have one new operation underway and research is continuing for these deposits.

Rare earth deposits typically contain some uranium that may be recoverable as a by-product if economics permit. The proposed Round Top project in Texas (United States), for example, exhibits concentrations of rare earths on the order of 600 ppm and potentially recoverable quantities of niobium, hafnium, tantalum, and tin, along with 45 ppm U. Extraction would utilize a sulfuric acid heap leach.

Good ore deposits of high grade and without high impurities are rare and exceptional. As a result, future challenges for uranium production are to produce uranium as a by-product and to produce by-products from uranium ore deposits. Many other challenges also exist: to keep costs low, reduce environmental problems, provide safe working conditions, reduce worker exposure, and consider sustainable development.

BEST PRACTICES FOR OPERATIONS AND DESIGN

Recognition and adoption of best practice principles are considered fundamental cornerstones of sustainable development for the uranium industry. This covers the social, environmental, and economic aspects of an operation and includes the active search, documentation, and implementation of those practices and principles that are most effective in improving the social, environmental, and economic performance of an operation (IAEA 2010).

The guiding principle for the peaceful use of nuclear energy defines that "any use of nuclear energy should be beneficial, responsible and sustainable, with due regard to the protection of people and the environment, non-proliferation, and security" (IAEA 2008). Based on this principle, the criteria for uranium are well defined under (1) beneficial, (2) responsible, and (3) sustainable uses (IAEA 2013a, 2013b). Sustainability of uranium production depends on improved recovery techniques and continued exploration and development of new ore bodies as older ones are depleted. At the end of mine life, decommissioning and remediation of facilities are also required (IAEA 2009).

The International Council on Mining and Metals (ICMM) has developed 10 principles for sustainable development performance (ICMM 2003). The World Nuclear Association (WNA) has published *Sustaining Global Best Practices in Uranium Mining and Processing*, which sets down a corresponding set of principles applicable to the worldwide uranium production industry (WNA 2008).

The application of best practice principles for a project starts at the conceptual phase and continues throughout all of the stages of a project, including conceptual design and/or exploration, feasibility studies, construction, operation, remediation, and closure and post-closure stewardship. Best practice principles are continually developed and improved upon for any project as it moves through the stages of the cycle and as more information is collected and a better understanding gained.

Generally, the first step is baseline data collection. Baseline information is required to characterize both the physical and social environment prior to project development. Typically, baseline studies are required to understand the pre-development conditions and integrate information into project supporting documents. Public and stakeholder consultation processes should commence and be managed in parallel with the baseline data collection program during the exploration or conceptual design stage. Recognition and response to stakeholder concerns and expectations can minimize the potential for conflict and be of mutual benefit to the communities and the operators.

The impact assessment process identifies potential adverse impacts of the project. Impact assessment is a process of identification, communication, prediction, and interpretation of information to identify potential (both adverse and beneficial) impacts through the life of a project and determine measures to manage these impacts. Impacts are predicted based on the comparison of baseline information and anticipated future conditions, both with and without the project occurring.

Undertaking a formal risk analysis is a fundamental component of the decision-making process for the sustainable operation of projects. Risk assessment can be used to determine the existing level of risk to the social, environmental, and economic aspects of a project. Risk assessment can also be used to evaluate the relative risk reduction achieved by various risk management options and is site or project specific. Sensitivity analysis is used to examine how robust an alternative is to changes in the information or assumptions used in the original analysis.

A great challenge to operational design is that standards continue to evolve over time. Any design and subsequent reclamation strategy that does not anticipate and allow for changes in legislative and regulatory requirements, as well as evolving community expectations, carries a high risk of failure.

Uranium projects need to incorporate an environmental management system (EMS) into the operation. Several EMS standards have been developed for operations to ensure they are designed and operated to meet the objectives and specific needs of the project. Two series of International Organization for Standardization (ISO) standards that are particularly relevant in this regard are the ISO 9000 and 14000 series. The ISO 9000 series focuses on quality, while the ISO 14000 series defines an EMS based on a commitment to continuous improvement. Another operational management tool is the use of key performance indicators (KPIs). KPIs are targets that may be either quantitative or qualitative, and, in contrast to ISO 14000, are used to measure performance against specific objectives or set values.

Management systems required for waste products associated with a mine or processing facility are site specific and in some rare instances region specific. Typically, regional optimization of waste management is not feasible, and each site must manage their own waste streams, which generally include water, waste rock, process residues, and radiologically and chemically contaminated equipment.

Continued stewardship of an operation post-closure is required to meet the best practice of sustainable development. This may consist of, but not be limited to, ongoing monitoring, collection and treatment of contaminated water, management and storage of water treatment sludges, and maintenance of facilities such as water diversion structures.

ENVIRONMENTAL IMPACTS

Documented environmental impacts from uranium mining and processing include elevated concentrations of trace metals, arsenic, and uranium in water; localized reduction of groundwater levels; and exposures of populations of aquatic and terrestrial biota to elevated levels of radionuclides and other hazardous substances. Such impacts have mostly been observed at mining facilities that operated at standards of practice that are not acceptable today. Designing, constructing, and operating uranium mining, processing, and reclamation activities based on modern international best practices has the potential to substantially reduce near- to moderate-term environmental effects. The exact nature of any adverse impacts from uranium mining and processing would depend on site-specific conditions and on the nature of efforts made to mitigate and control these effects.

Uranium tailings present a significant potential source of radioactive contamination for thousands of years into the future and, therefore, must be controlled and stored carefully. Over the past few decades, improvements have been made to tailings management systems to isolate tailings from the environment, and below-grade disposal practices have been developed specifically to address concerns regarding tailings dam failures. Modern tailings management sites are designed so that the tailings remain segregated from the water cycle to control mobility of metals and radioactive contaminants for at least 200 years, and possibly up to 1,000 years. However, because monitoring of tailings management sites has only been carried out for a short period, monitoring data are insufficient to assess the long-term effectiveness of tailings management facilities designed and constructed according to modern best practices (NRC 2012).

The precise impacts of any uranium mining and processing operation would depend on a range of specific factors for a particular site. Therefore, a thorough site characterization, supplemented by air quality and hydrological modeling, is essential for estimating any potential environmental impacts and for designing facilities to mitigate potential impacts. Additionally, until comprehensive site-specific risk assessments are conducted, including accident and failure analyses, the short-term risks associated with natural disasters, accidents, and spills remain poorly defined.

A well-designed and executed monitoring plan is essential for gauging the performance of best practices to limit environmental impacts, determining and demonstrating compliance with regulations, and triggering corrective actions if needed. Making the monitoring plan available to the public helps foster transparency and public participation. Regular updates to the monitoring plan, along with independent reviews, allows the incorporation of new knowledge and insights gained from analysis of monitoring data. In addition, best practice is to undertake an assessment of the appropriate mitigation and remediation options that are required to minimize predicted environmental impacts, such as acid mine drainage control, and tailings and waste management.

HEALTH AND SAFETY

Any mining operation or other industrial activity involving a mineral or raw material has the potential to increase the effective dose received by individuals from natural sources, resulting from exposure to radionuclides of natural origin contained in or released from such material. Where this increase in dose is significant, radiation protection measures may be needed to protect workers or members of the public.

For many of its aspects, the potential adverse health effects associated with uranium mining are no different than the risks identified in other types of non-radiation-related mining activities (Laurence 2011). Uranium mining, however, adds another dimension of risk because of the potential for exposure to elevated concentrations of radionuclides. Internal exposure to radioactive materials during uranium mining and processing can take place through inhalation, ingestion, or absorption through an open cut or wound. External radiation exposure from beta particles or gamma rays can also present a health risk.

All minerals and raw materials contain radionuclides of natural, terrestrial origin—these are commonly referred to as primordial radionuclides. The ^{238}U and ^{232}Th decay series and ^{40}K are the main radionuclides of interest. The activity concentrations of these radionuclides in normal rocks and soil are variable but generally low. However, certain minerals, including some that are commercially exploited, contain uranium and/or thorium series radionuclides at significantly elevated activity concentrations. Furthermore, during the extraction of minerals from the earth's crust and subsequent physical and/or chemical processing, the radionuclides may become unevenly distributed between the various materials arising from the process, and selective mobilization of radionuclides can disrupt the original decay-chain equilibrium. Consequently, radionuclide concentrations in materials arising from a process may exceed those in the original mineral or raw material, sometimes by orders of magnitude.

External exposure from ionizing radiation (beta and gamma) is usually not a major problem in uranium mining. Where this is a potential problem, such as the high-grade operations in Canada, extra precautions are put in place. The high-grade Canadian operations use remote ore handling methods, add shielding and additional worker controls. Lower-grade uranium (<1% U) ores, process solutions, and concentrates do not emit large quantities of penetrating radiation.

The most common radiation hazard encountered in the milling of uranium ore is airborne radioactive material. This airborne material may consist of uranium together with all its radioactive decay products in the case of ore dust, or uranium dust alone in the case of yellowcake. The other airborne hazard is radon gas and the associated radon daughter decay products, particularly during crushing, grinding, and leaching unit processes. Ventilation of process equipment and buildings is essential. Many precautions are taken at a uranium mine to protect the health of workers:

- Dust is controlled, to minimize inhalation of gamma- or alpha-emitting minerals. In practice, dust is the main source of radiation exposure in an open-cut uranium mine and in the mill area.
- Radiation exposure of workers in the mine, plant, and tailings areas is limited. In practice, radiation levels from the ore and tailings are usually very low.
- Radon daughter exposure is minimal in an open-cut mine because there is sufficient natural ventilation to remove the radon gas. In an underground mine, a good forced ventilation system is required. For example, at Olympic Dam in Australia, radiation doses in the mine from radon daughters are kept very low, with an average of less than about 1 mSv/yr. Canadian doses (in mines with high-grade ore) average about 3 mSv/yr.
- Strict hygiene standards are imposed on workers handling the uranium oxide concentrate. If it is ingested, it has a chemical toxicity similar to lead oxide. (Both lead and uranium are toxic and affect the kidney. The body progressively eliminates most lead or uranium via the urine.) In effect, the same precautions are taken as in a lead smelter, with use of respiratory protection in particular areas identified by air monitoring.

Health, Safety, and Environmental Management

International good practice for sound health, safety, and environmental (HSE) management of a uranium mining and processing site has been accumulated through long-term development of mining and processing of uranium and other minerals.

An integrated approach is one good and important practice in tackling HSE management of a uranium mining and processing site. This method considers radiological and non-radiological issues of HSE in a single, comprehensive system, which means that observing one aspect of HSE good practice will not compromise any other. Many HSE elements share a common engineering approach of sound control of radiological and non-radiological risk in HSE management, such as control of worker safety, surface run-off, and stability of the tailings dam. IAEA standards and guidelines are implemented around the world for the control of radiological risk related to uranium mining and processing sites and for protection against ionizing radiation. In particular, the IAEA safety and security standards provide a framework of international standards (including safety fundamentals, safety requirements, safety guides, safety reports, and technical documents) that address radiation, environmental, waste, and transport safety and security for uranium mining and processing sites and their related activities.

While IAEA safety standards and security series are more focused on radiological concern, the importance of an integrated approach in HSE management is addressed and encouraged in the planning and operation of radiological-related facilities and activities. The WNA policy document *Sustaining Global Best Practices in Uranium Mining and Processing: Principles for Managing Radiation, Health and Safety, Waste and the Environment* (WNA 2008) is another good reference that proportionally positions radiation risk among the diverse HSE risks that need to be addressed in uranium mining and processing.

In practice, some non-radiological HSE elements associated with any mining project may result in acute consequences, including injuries and fatalities, which rank high in the HSE inventory. Issues such as handling explosives, ground support for open pit and underground mines, operating of large vehicles and equipment, and so on, each bear significant safety risks that require careful design, practice, and management.

Integration of HSE aspects as part of design and operations tends to lead to greater efficiency and performance. This is an important evolution that took place over the last decade or so. The management of HSE aspects in isolation from each other and from design and operations is of a past era. Moreover, for greater efficiency, HSE issues are managed as part of an integrated management system such as ISO 9000, 14000, and 18000 series or an Occupational Safety and Health Administration equivalent.

As part of integrated management, HSE issues are addressed to an increasing level of depth in successive project development phases from prefeasibility, feasibility, detailed feasibility, and preparation for commissioning and operations.

There is a wealth of relevant international experience for a sound occupational health and safety program, based on many decades of experience in mining, which is also directly applicable to uranium mining. Among others, this experience pays particular attention to the risk of incidents (near misses), accidents, injuries, and fatalities due to "conventional" accidents (i.e., those with physical causes). For uranium mining, experience also includes specific radiation protection measures for the control and monitoring of radiation levels and of exposures to external gamma radiation, radon gas, and long-lived radioactive dust. In addition to the main items already mentioned earlier, extra radiation protection measures include the control and monitoring of possible incidental/accidental intakes of uranium by site personnel (e.g., uranium in urine program) as well as the use of personal protective equipment for both conventional and radiation health risks.

Radiation Safety Basis

The following basis of radiation protection standards is from the WNA (2014):

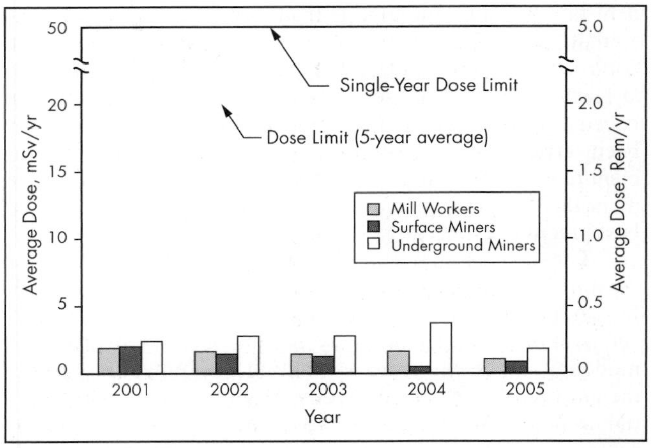

Source: Chambers 2011

Figure 22 Evolution of radiation dose in Canadian mining operations

In practice, radiation protection is based on the understanding that small increases over natural levels of exposure are not likely to be harmful but should be kept to a minimum [Figure 22]. To put this into practice the International Commission [on] Radiological Protection (ICRP) has established recommended standards [ICRP 1991] of protection (both for members of the public and radiation workers) based on three basic principles:

- Justification. No practice involving exposure to radiation should be adopted unless it produces a net benefit to those exposed or to society generally.
- Optimisation. Radiation doses and risks should be kept as low as reasonably achievable (ALARA), economic and social factors being taken into account.
- Limitation. The exposure of individuals should be subject to dose or risk limits above which the radiation risk would be deemed unacceptable.

These principles apply to the potential for accidental exposures as well as predictable normal exposures.

Underlying these is the application of the "linear hypothesis" based on the idea that any level of radiation dose, no matter how low, involves the possibility of risk to human health. This assumption enables "risk factors" derived from studies of high radiation dose to populations (eg from Japanese bomb survivors) to be used in determining the risk to an individual from low doses (ICRP Publication 60 [ICRP 1991]). However the weight of scientific evidence does not indicate any cancer risk or immediate effects at doses below about 50 millisievert (mSv) per year.

Based on these conservative principles, ICRP recommends that the additional dose above natural background and excluding medical exposure should be limited to prescribed levels. These are: one millisievert per year for members of the public, and 20 mSv per year averaged over 5 years for radiation workers who are required to work under closely-monitored conditions.

The frameworks of radiation safety in countries where most uranium is mined are based on the full adoption of international recommendations. This is not the case in all parts of the world. Even the 1977 Recommendation of the ICRP has not been universally adopted.

The safety record of the uranium mining industry is good. Radiation dose records compiled by mining companies under the scrutiny of regulatory authorities have shown consistently that mining company employees are not exposed to radiation doses in excess of the limits. The maximum dose received is about half of the 20 mSv/yr limit and the average is about one tenth of it. (This compares with natural doses of up to 50 mSv/yr for some places in India and Europe, without any adverse effects being evident, and mean exposures of 750 mSv/yr in some East German mines from 1946 to 1954, resulting in thousands of cases of lung cancer.)

ACKNOWLEDGMENTS

The author thanks the following individuals for their contributions to various sections of this chapter: Michel Cuney (French National Centre for Scientific Research) for the "Geology, Mineralogy, and Uranium Deposit Types" section, Thomas Poole (International Nuclear, Inc.) for details on historical uranium production, Bob Ring (ANSTO Minerals) for the "Uranium Testing and Flow-Sheet Development" section, Harikrishnan "Hari" Tulsidas (United Nations Economic Commission for Europe) for the "Best Practices" section, Chuck Edwards (Amec-Foster Wheeler) for the "Key Lake Mill" section, and Merrill A. Ford (Paladin Energy Ltd.) for the "Kayelekera Mine" and "Langer Heinrich" sections.

REFERENCES

Badea, A.L., and Schwartz, L. 2000. Uranium milling operations at McClean Lake. In *Uranium 2000: Proceedings of the International Symposium on the Process Metallurgy of Uranium*. Montreal, QC: Canadian Institute of Mining, Metallurgy and Petroleum.

Bateman Litwin. 2012. *From Concept to Turnkey: Solvent Extraction*. Corporate brochure. Amsterdam: Bateman Litwin.

Bharadwaj, B., Moldovan, B., Yesnik, L., Grant, S., and Piche, J. 2010. Cameco Corporation—The Key Lake uranium mill: Current status and vision for the future. In *Uranium 2010: Proceedings of the 3rd Annual International Conference on Uranium*. Montreal, QC: Canadian Institute of Mining, Metallurgy and Petroleum.

Cameco Corporation. 2009. *Cameco 2009 Annual Report*. Saskatoon, SK: Cameco.

Cameco Corporation. 2012. Revitalization Plan, Corporate Quarterly report 2012. Saskatoon, SK: Cameco.

Cameco Corporation. 2015. *Cameco 2015 Annual Report*. Saskatoon, SK: Cameco.

Carino, A. 2001. The McClean Lake operation: Jeb Mill process description and commissioning. In *33rd Annual Operator's Conference of the Canadian Mineral Processors*. Montreal, QC: Canadian Institute of Mining, Metallurgy and Petroleum.

Chambers, D. 2011. Overview of conventional uranium mining and milling technologies. Presented to Institute for Advanced Learning and Research, Virginia Tech, Blacksburg, VA.

CNSC (Canadian Nuclear Safety Commission). 2012. *Record of Proceedings, Including Reasons for Decision in the Matter of AREVA Resources Canada Inc.* Ottawa, ON: CNSC.

CNSC (Canadian Nuclear Safety Commission). 2015. Radioactive waste: Uranium mines and mills waste. Ottawa, ON: CNSC.

Courtaud, B., Auger, F., and Thiry, J. 2011. Method for preparing uranium concentrates by fluidized bed precipitation, and preparation of UO_3 and U_3O_8 by Drying/Calcining Said Concentrates. U.S. Patent Application 2011/0212005 A1.

Cuney, M. 2009. The extreme diversity of uranium deposits. *Miner. Dep.* 44:3–9.

Cuney, M. 2011. Uranium and thorium: The extreme diversity of the resources of the world's energy minerals. In *Non-Renewable Resource Issues: Geoscientific and Societal Challenges*. Edited by R. Sinding-Larsen and F.-W. Wellmer. Dordrecht: Springer. pp. 91–129.

Cyclone Uranium Corporation. 2012. Idealized cross-section of a sandstone-hosted roll front uranium deposit.

Dahlkamp, F.J. 2009. *Uranium Deposits of the World–Asia*. Berlin: Springer-Verlag.

Denison Mines Corporation. 2015. Exploration and Development: McClean Lake and the McClean Lake Mill. Toronto, ON: Denison Mines Corporation.

Durupt, N., Blanvillain, J.J., Jourde, J., and Sanoussi M. 2013. Heap leaching treatment of clayey ore at Somair. Presented at Heap Leach Solutions 2013 Conference, Vancouver, BC, Canada.

Edwards, C.R. 1987. Hydrogen peroxide precipitation of uranium in the Rabbit Lake mill. In *Crystallization and Precipitation: Proceedings of the International Symposium*. New York: Pergamon Press. pp. 149–157.

Edwards, C.R. 1992. Uranium extraction process alternatives. *CIM Bull.* 85:112–136.

Edwards, C.R. 1997. Ore handling for the McArthur River and Cigar Lake projects. Presented at Canadian Mineral Processors Division Conference of Canadian Institute of Mining, Metallurgy and Petroleum, Ottawa, ON.

Edwards, C.R. 2002a. The Cigar Lake project—Mining, ore handling and milling. Presented at Canadian Mineral Processors Division Conference of Canadian Institute of Mining, Metallurgy and Petroleum, Ottawa, ON.

Edwards, C.R. 2002b. Methods of evaluating ore processing and effluent treatment for Cigar Lake ore at the Rabbit Lake Mill. In *Technologies for the Treatment of Effluents from Uranium Mines, Mills and Tailings*. IAEA-TECDOC-1296. Vienna: International Atomic Energy Agency. pp. 5–14.

Edwards, C.R., and Dyck, K. 2004. Replacement of classification screens with cyclones in the McArthur River uranium mine grinding circuit. Presented at Canadian Mineral Processors Division Conference of Canadian Institute of Mining, Metallurgy and Petroleum, Ottawa, ON.

Edwards, C.R., and Schnell, H. 2000. Process design for handling, transporting and milling Cigar Lake ore. In *Uranium 2000: Proceedings of the International Symposium on the Process Metallurgy of Uranium.* Montreal, QC: Canadian Institute of Mining, Metallurgy and Petroleum.

El-Ansary, Z., and Lund, D. 2008. Tailoring process selection to uranium mineralogy and ore type. Presented at AusIMM International Uranium Conference, Adelaide, South Australia, June 18–19.

ERA (Energy Resources of Australia). 2008. *Energy Resources of Australia–Rio Tinto, Sustainable Development Report.* Darwin, Northern Territory: ERA.

Gomiero, L.A., Lima, H.A., and Morais, C.A. 2010. Evaluation of a new milling process for the Caetite. In *Uranium 2010: Proceedings of the 3rd Annual International Conference on Uranium.* Montreal, QC: Canadian Institute of Mining, Metallurgy and Petroleum.

Griebel, E.G. 2000. Rössing uranium update. In *Uranium 2000: Proceedings of the International Symposium on the Process Metallurgy of Uranium.* Montreal, QC: Canadian Institute of Mining, Metallurgy and Petroleum.

IAEA (International Atomic Energy Agency). 1985. *Advances in Uranium Ore Processing and Recovery from Non-Conventional Resources.* Panel Proceedings Series. STI/PUB/689. Vienna: IAEA.

IAEA (International Atomic Energy Agency). 1989. *The Recovery of Uranium from Phosphoric Acid.* IAEA-TECDOC-533. Vienna: IAEA.

IAEA (International Atomic Energy Agency). 2008. *Nuclear Energy Basic Principles.* IAEA Nuclear Energy Series No. NE-BP. Vienna: IAEA.

IAEA (International Atomic Energy Agency). 2009. *Establishment of Uranium Mining and Processing Operations in the Context of Sustainable Development.* IAEA NES NF-T-1.1. Vienna: IAEA.

IAEA (International Atomic Energy Agency). 2010. *Best Practice in Environmental Management of Uranium Mining.* IAEA NES NF-T-1.2. Vienna: IAEA.

IAEA (International Atomic Energy Agency). 2013a. *Nuclear Fuel Cycle Objectives.* IAEA Nuclear Energy Series No. NF-O. Vienna: IAEA.

IAEA (International Atomic Energy Agency). 2013b. *Radiation Protection and Management of NORM Residues in the Phosphate Industry.* Safety Reports Series No. 78. Vienna: IAEA.

ICMM (International Council on Mining and Metals). 2003. *Sustainable Development Framework: 10 Principles.* London: ICMM.

ICRP (International Commission on Radiological Protection). 1977. *Annals of the ICRP: Recommendations of the International Commission on Radiological Protection.* Vol. 1, No. 3. ICRP Publication 26 (superseded by Publication 103). New York: Pergamon Press.

ICRP (International Commission on Radiological Protection). 1991. *Annals of the ICRP: 1990 Recommendations of the International Commission on Radiological Protection.* ICRP Publication 60 (superseded by Publication 103). New York: Pergamon Press.

Kemix Pty Ltd. 2015. *AAC Pumpcell.* Corporate brochure. Midrand, Gauteng Province, South Africa: Kemix.

Krause, E., and Ettel, V.A. 1987. Solubilities and stabilities of ferric arsenates. In *Proceedings of the International Symposium on Crystallization and Precipitation.* Toronto, ON: Pergamon Press. pp. 195–210.

Lam, E.K., Neven, M.G., Steane, R.A., and Ko, K.F. 1993. Removal of molybdenum from uranium-bearing solutions. U.S. Patent No. 5,229,086.

Langmuir, D. 1978. Uranium solution-mineral equilibria at low temperatures with applications to sedimentary ore deposits. *Geochim. Cosmochim. Acta* 42:547–569.

Laurence, D. 2011. Mine safety. In *SME Mining Engineering Handbook*, 3rd ed. Edited by P. Darling. Englewood, CO: SME.

Lien, L. 2010. The use of engineered membrane separation technology "EMS" for radionuclide separations. In *Uranium 2010*: *Proceedings of the 3rd International Conference on Uranium, 40th Hydrometallurgical Meeting*, Saskatoon, SK.

Lunt, D., Boshoff, P., Boylett, M., and El-Ansary, Z. 2007. Uranium extraction: The key process drivers. *J. S. Afr. Inst. Min. Metall.* 107:419–426.

McKinnon, C. 2000. Water treatment at McClean Lake. In *Uranium 2000: Proceedings of the International Symposium on the Process Metallurgy of Uranium.* Montreal, QC: Canadian Institute of Mining, Metallurgy and Petroleum.

Merritt, R.C. 1971. *The Extractive Metallurgy of Uranium.* Golden, CO: Colorado School of Mines Research Institute.

Milde, W.W. 1989. Rabbit Lake project: Milling and metallurgy. *CIM Bull.* 82(932):69–75.

NEA and IAEA (Organisation for Economic Co-Operation and Development/Nuclear Energy Agency and International Atomic Energy Agency). 2016. *Uranium 2016: Resources, Production and Demand ("Red Book").* NEA No. 7301. Paris: OECD/NEA.

Neven, M., and Gormley, L. 1982. Design of the Key Lake leaching process. Presented at Uranium '82: 12th Annual Canadian Hydrometallurgical Conference, Toronto, ON.

Neven, M., Steane, R., and Becker, J. 1985. Arsenic management and control at Key Lake. In *Impurity Control and Disposal, Proceedings of the 15th Annual Meeting of the Hydrometallurgy Section of the Metallurgical Society.* Montreal, QC: Canadian Institute of Mining, Metallurgy and Petroleum.

Neven, M., Steane, R.A., and Lam, E. 1986. Design and operation of the Key Lake uranium mill. In *Proceedings of the Canadian Mineral Processors Division Conference of CIM.* Montreal, QC: Canadian Institute of Mining, Metallurgy and Petroleum. pp. 352–376.

NRC (National Research Council). 2012. *Uranium Mining in Virginia: Scientific, Technical, Environmental, Human Health and Safety, and Regulatory Aspects of Uranium Mining and Processing in Virginia.* Washington, DC: National Academies Press.

Paatero, E., et al. 2010. Recovery of molybdenum and uranium as by-products with solvent extraction. Metals and Materials Recycling Workshop, Santiago, Chile.

Pacquet, E. 2017. Vice President AREVA Resources Canada—Operations and Projects Update. Presentation to Saskatchewan Mining Supply Chain Forum, Saskatchewan Mining Association, April 5.

Perdula. 2013. Mixer-settler schema. https://commons.wikimedia.org/wiki/File%3AMixer-settler_schema.jpg. CC BY-SA 3.0. https://creativecommons.org/licenses/by-sa/3.0, via Wikimedia Commons.

Pool, T. 2013. Uranium market update. Presented at IAEA Technical Meeting on Uranium Production Cycle Pre-Feasibility and Feasibility Assessment, Vienna, October 7–10.

Remple, G., and Schnell, H. 2000. Processing of Cigar Lake Ore in the McLean Lake Mill. In *Uranium 2000: Proceedings of the International Symposium on the Process Metallurgy of Uranium.* Montreal, QC: Canadian Institute of Mining, Metallurgy and Petroleum.

Rodgers, C. 1996. Molybdenum extraction from uranium rich strip solution. In *Proceedings of the Canadian Mineral Processors Division Conference of CIM.* Montreal, QC: Canadian Institute of Mining, Metallurgy and Petroleum. pp. 3–22.

Rodgers, C. 2000. Converting the Key Lake Mill to Process McArthur River Ore. In *Uranium 2000: Proceedings of the International Symposium on the Process Metallurgy of Uranium.* Montreal, QC: Canadian Institute of Mining, Metallurgy and Petroleum.

Rodgers, C., and Sarion, A. 1998. Modifications of the CCD Circuit at Cameco's Key Lake Uranium Circuit. In *Proceedings of the Canadian Mineral Processors Division Conference of CIM.* Montreal, QC: Canadian Institute of Mining, Metallurgy and Petroleum. pp. 309–321.

Rössing Uranium Ltd. 2007. *Socio-Economic Component of the Social and Environmental Impact Assessment Report for the Rössing Uranium Mine Expansion Project: Socio-Economic Impact Assessment.* Public Information Document.

Schnell, H. 2009. Uranium from unconventional sources. Presented at International Atomic Energy Agency Technical Meeting on Uranium from Unconventional Resources, Vienna, Austria, November 4–6.

Schnell, H. 2010. Trekkopje. Presented at International Atomic Energy Agency Technical Meeting on Low Grade Uranium Ore, Vienna, Austria, March 29–31.

Schnell, H. 2011. Uranium processing, extremes of ore grade. Presented at SME Annual Conference and Exposition.

Stewart, K. 2010. Unique radiation protection design for processing high-grade uranium ores at the McClean Lake mill. *Uranium 2010: Proceedings of the 3rd Annual International Conference on Uranium.* Montreal, QC: Canadian Institute of Mining, Metallurgy and Petroleum.

St-Pierre, S., and Huffman, D. 2000. Unique radiation protection design for processing high grade uranium ores at the McClean Lake mill. In *Uranium 2000: Proceedings of the International Symposium on the Process Metallurgy of Uranium.* Montreal, QC: Canadian Institute of Mining, Metallurgy and Petroleum.

Thiry, J., and Bustos, S. 2014. Heap leaching technology—Moving the frontier for treatment: Applications in Niger and Namibia. Presented at IAEA-URAM Conference, Vienna, Austria, June 28.

Thiry, J., and Roche, M. 2000. Recent developments at COMINAK. In *Uranium 2000: Proceedings of the International Symposium on the Process Metallurgy of Uranium.* Montreal, QC: Canadian Institute of Mining, Metallurgy and Petroleum.

Vancas, M.F. 2003. Pulsed column and mixer-settler applications in solvent extraction. *JOM* 55(7):43–45.

WNA (World Nuclear Association). 2008. *Sustaining Global Best Practices in Uranium Mining and Processing: Principles for Managing Radiation, Health and Safety, Waste and the Environment.* WNA Policy Document. London: WNA.

WNA (World Nuclear Association). 2014. Occupational safety in uranium mining. London: WNA. www.world-nuclear.org/information-library/safety-and-security/radiation-and-health/occupational-safety-in-uranium-mining.aspx. Accessed March 2018.

WNA (World Nuclear Association). 2015. World uranium mining production. London: WNA. www.world-nuclear.org/info/Nuclear-Fuel-Cycle/Mining-of-Uranium/World-Uranium-Mining-Production. Accessed March 2017.

WNA (World Nuclear Association). 2016. Supply of uranium. August. London: WNA. www.world-nuclear.org/information-library/nuclear-fuel-cycle/uranium-resources/supply-of-uranium.aspx. Accessed February 2018.

WNA (World Nuclear Association). 2017. World uranium mining production. London: WNA. www.world-nuclear.org/information-library/nuclear-fuel-cycle/mining-of-uranium/world-uranium-mining-production.aspx. Accessed March 2018.

CHAPTER 12.42

Vanadium

Sai Wei Lam and Daniel Harris

Vanadium is commercially produced around the world from a variety of feedstocks. These include vanadium ores, vanadium slag concentrates, boiler and fly-ash residues from power-generation plants, vanadium cokes from oil refining, and spent catalysts from chemical processing and oil refining. Approximately 85% of worldwide vanadium production is from titaniferous magnetite ore, either as primary production from magnetic concentrate recovered from the ore, or as co-production from vanadium-bearing steel slag produced during steelmaking operations utilizing the ore. Approximately 91% of worldwide vanadium production is consumed by the steelmaking industry to achieve strength and hardness requirements in various steel grades. The remaining 10% is consumed in the production of high-strength titanium alloys, catalysts for manufacturing chemicals, catalysts for the removal of sulfur dioxide (SO_2) and nitrogen oxide (NO_x), and energy storage batteries.

The processing of vanadium, regardless of feedstock source, utilizes common industry metallurgical extraction and chemical unit operations. These include comminution; physical concentration, involving magnetic or gravitational separation; vanadium dissolution by common alkaline or acid fluxes or leachates; chemical separation for purification by selective precipitation and ion exchange or solvent extraction; vanadium precipitation and crystallization; and filtration, drying, and calcining.

Commodity-grade vanadium pentoxide (V_2O_5), used to produce ferrovanadium (FeV) for the steel industry, has a purity level of +96% V_2O_5, whereas chemical-grade material has a purity level of +99.5% V_2O_5. High-purity V_2O_5 (+99.5%) production generally requires solvent extraction or ion exchange, followed by crystallization to achieve the purity requirements, although commodity-grade material can be produced using only selective precipitation.

This chapter primarily focuses on the processing of commodity-grade vanadium, predominantly from titaniferous magnetite ores and steel slag concentrates.

EXTRACTION VIA SALT ROAST–WATER LEACH

Regardless of whether the raw material is slag or vanadium-containing ore, the basic extraction process for recovering vanadium is similar, involving a salt roast–water leach route (McRae et al. 1969; Gupta and Krishnamurthy 1992; Goddard and Fox 1980). Typically, vanadium in titaniferous magnetite concentrate is in a reduced state. To improve its solubility, the vanadium needs to be oxidized to the pentavalent state to form sodium metavanadate. The oxidation is performed in a rotary kiln.

The key to successful vanadium recovery from titaniferous magnetite involves not only the oxidation of vanadium but also the oxidation of magnetite to hematite. When magnetite is oxidized to hematite, the accessibility of oxygen and sodium salt to react with vanadium in the titaniferous magnetite concentrate improves because of the increase in porosity. This process liberates vanadium from the spinel structure.

Sodium chloride, sodium carbonate, sodium sulfate, and/or a mixture of these have been widely applied as flux. Sodium oxalate has been used as flux at the Windimurra operation in Australia. In Russia, lime has been used as flux in the roasting reaction. With sodium salts, the roasted product is leached in water. Lime, however, requires acid leaching to promote the solubility of vanadium.

Sodium chloride fluxing requires the presence of water vapor or oxygen to drive the formation of sodium metavanadate. The reaction is represented by the following:

$$2NaCl + O_2 + H_2O + V_2O_3 = 2NaVO_3 + 2HCl(g) \quad \textbf{(EQ 1)}$$

or

$$2NaCl + 3/2O_2 + V_2O_3 = 2NaVO_3 + Cl_2(g) \quad \textbf{(EQ 2)}$$

If the vanadium is in the pentavalent state, the reaction is

$$2NaCl + V_2O_5 + H_2O = 2NaVO_3 + 2HCl(g) \quad \textbf{(EQ 3)}$$

Sai Wei Lam, Senior Process Engineer, Wood, Perth, Western Australia, Australia
Daniel Harris, Consultant & Professional Board Director, EnergySource LLC, Reno, Nevada, USA

Under prolonged roasting, sodium pyrovanadate ($Na_4V_2O_7$) can form (Goddard and Fox 1980). Sodium pyrovanadate is relatively water soluble, but its formation is unnecessary as it consumes more salt. The mechanisms of sodium pyrovanadate formation are as follows:

$$2NaCl + 3/2O_2(g) + V_2O_3 = 2NaVO_3 + Cl_2(g) \quad \textbf{(EQ 4)}$$

$$2NaCl + H_2O(g) + 2NaVO_3 = Na_4V_2O_7 + 2HCl(g) \quad \textbf{(EQ 5)}$$

Insufficient salt and/or water vapor in the reaction with sodium chloride can lead to water-insoluble vanadates called *bronzes*, with a sodium-to-vanadium mole ratio of less than 1.

Use of sodium sulfate flux is reported to produce superior vanadium recovery over sodium chloride. As opposed to sodium chloride, the reaction requires no water vapor or oxygen to trigger the formation of sodium metavanadate. In fact, sulfur trioxide (SO_3), an oxygen supplier, is generated when sodium sulfate decomposes at high temperature through the equilibrium reaction of $SO_3 = SO_2 + ½O_2$.

Sodium carbonate roasting has been repeatedly reported to generate the best vanadium water leach recoveries from titaniferous magnetite when compared with sodium sulfate or sodium chloride. This is mainly related to the alkaline character of sodium carbonate, which has been reported to be more aggressive in the attack of the magnetite spinel as compared to sodium chloride or sodium sulfate. However, for this same reason, it has been found to be less selective than sodium chloride or sodium sulfate. When sodium carbonate is used, the presence of an oxidizing environment is critical. Free oxygen content of 4% is required to maintain a good oxidizing environment.

Sodium oxalate found its application in salt roasting at the Windimurra operation because of the availability of a low-cost waste resource. Sodium oxalate decomposes to sodium carbonate on heat treatment. The reaction mechanism with sodium oxalate is shown in the following equations:

$$Na_2C_2O_4 = 2Na_2CO_3 + CO \quad \textbf{(EQ 6)}$$

$$V_2O_5 + Na_2CO_3 = 2NaVO_3 + CO_2 \quad \textbf{(EQ 7)}$$

The proportion of vanadium recoverable from a salt roasting operation usually ranges from 70% to 90%. Most of the problems in the recovery arise from side reactions of sodium compounds with other constituents in the ore. These constituents include silica and silicates; claylike complexes of calcium, aluminum silicate, gypsum, calcium carbonates, iron oxides; and also organic matter.

The high silica and lime content in vanadium-bearing materials is the main cause of unproductive side reactions in salt roasting. The detrimental effect of siliceous materials is caused by the formation of low-melting sodium iron silicate-acmite ($Na_2O{\cdot}Fe_2O_3{\cdot}4SiO_2$) that retains vanadium oxide in a water-insoluble solid solution. Its formation can retard magnetite oxidation by fusing. Goddard and Fox (1980) found that vanadium extraction drops below 20% from magnetite with 7.5% silicon dioxide (SiO_2) or more. At Windimurra, the vanadium extraction dropped to 70% when the titaniferous magnetite concentrate contained approximately 4.6% SiO_2, but increased to 88% as the silica content reduced to 2.7% SiO_2.

Silica is a sodium salt consumer. Silica is typically rejected by magnetic separation in the concentrator plant. Alternatively, alumina can be added to form sodium aluminum silicates with higher melting temperature. This suppresses the competing effect from silica for sodium salt in the roasting operation. However, its commercial application is unknown. Typically, the silica content should not be greater than 3% SiO_2. The rule of thumb is that the alumina-to-silica ratio should be greater than 1, or the combined silica and alumina content should be less than 5%.

The presence of lime can lead to the formation of water-insoluble calcium vanadate. This is particularly true when roasting with neutral salts such as sodium sulfate. Alkaline salts, such as sodium carbonate, tend to yield higher extraction of vanadium for ore containing high lime levels. Controlling the gangue minerals content along with the roasting conditions is important to maximize vanadium extraction.

PRODUCTION AND PROCESSING HISTORY

Vanadium was first discovered in 1801 by Andrés Manuel del Río, a Spanish mineralogist, in a brown lead sample collected from a mine near Hidalgo, Mexico, but it was mistaken as a form of chromium (Hilliard 1997). In 1830, Nils Gabriel Sefström, a Swedish chemist, rediscovered vanadium in a converter slag produced from an iron ore sample in Taberg, Sweden, and named it vanadium, in honor of the Scandinavian goddess of beauty, Vanadis. Then in 1868, Henry Roscoe successfully produced vanadium metal from vanadium chloride by reducing it in hydrogen.

In the early 1870s, the high cost and scarcity of vanadium restricted its commercial potential. Nevertheless, extensive studies were carried out to investigate the suitability of vanadium as an alloying agent for steel (Duke 1983). Commercial application of vanadium as an alloying agent for steel started in 1905 when Henry Ford recognized its benefits for use in the automobile industry. Later, in 1960, vanadium found its application in titanium alloys for jet engines and airframes.

China, South Africa, and Russia are the largest global producers, supplying approximately 90% of the global vanadium demand in 2014. It is estimated that China supplies 55% of the global vanadium demand, and South Africa and Russia contribute 21% and 14%, respectively. World vanadium resources exceed 63 Mt (million metric tons) (Polyak 2013). Vanadium occurs in combination with more than 65 different minerals in deposits such as phosphate rock, titaniferous magnetite, and uraniferous sandstone and siltstone. Vanadium is also found in some crude oils in the form of organic complexes. This vanadium is concentrated in boiler slag and ash generated from the burning of oils in power stations. Vanadium is also concentrated in spent catalysts from oil refinery processing. These vanadium raw materials are classified as secondary material. The extraction of vanadium from secondary material equals 9% of global production (TZMI 2015).

Titaniferous magnetite ore accounts for ~85% of global vanadium resources (Vanitec 2014), making it the most important source of vanadium. In titaniferous magnetite ore, the titanium is mainly present as a solid solution in the titanium-rich magnetite phase (ulvöspinel [Fe_2TiO_4]) and to a lesser degree as ilmenite (Fischer 1975). The vanadium in the ore occurs as a solid solution within the magnetite-ulvöspinel, where V^{3+} substitutes Fe^{3+} in the magnetite crystal lattice.

Early Production

The change in the supply and commercial value of vanadium started with the discovery of vanadium-bearing asphaltic deposits at Minas Ragra in Peru in 1905. American Vanadium Company began mining the deposits in 1907 (Anon. 1920; Hammond Swayne LLC 2015). The Minas Ragra deposit was the most productive single deposit of vanadium in history.

The vanadium mineralization was hosted within patronite, containing up to 24.8% V_2O_5. Maximum production of 2,074 t (metric tons) of V_2O_5 was recorded in 1920. By 1929, the high-grade vanadium ore was nearly exhausted. The mine was subsequently closed in 1930 but reopened in 1934 after a new process was developed to extract vanadium from low-grade ore (Gomi and Whetham 2012). It involved salt roasting and sulfuric acid leaching. The mine continued to operate until 1955 and produced a total of 39,009 t of V_2O_5 between 1907 and 1955 (Crabtree and Seidel 1961).

Significant mining activities for vanadium also took place in southwestern Colorado and southeastern Utah (United States). Roscoelite ore was mined exclusively for vanadium in Colorado near Placerville from 1910 to 1920 and then at Rifle Creek from 1925 to 1954 (Hammond 2013). Vanadium was also recovered from carnotite ore (containing both uranium and vanadium) in the Colorado Plateau, spanning Colorado, Utah, Arizona, and New Mexico in the United States. From 1946 to 1960, the Colorado Plateau supplied 55% of the world's recoverable vanadium pentoxide, producing a total of 47,428 t of V_2O_5 (Crabtree and Seidel 1961).

Salt roasting has been practiced in Colorado since 1910 (Crabtree and Seidel 1961). Carnotite ore is roasted with sodium chloride. Sodium chloride is added at a flux-to-ore ratio of 6–10 wt %. The mixture is roasted at 820°–850°C for 1–3 hours in a multiple-hearth roaster or rotary kiln. Vanadium extraction typically ranges from 70% to 80%. However, if the carnotite ore contains lime in excess of 6%, vanadium extraction can drop below 60% because of the formation of water-insoluble calcium vanadate. The roasted product is subsequently subjected to either alkaline or water leaching for vanadium recovery. When leached in water, the vanadium is selectively dissolved in solution and recovered by precipitation.

At the Newmire operation in Colorado, vanadium was precipitated as iron vanadate using ferrous sulfate. The iron vanadate was then reduced to ferrovanadium in an aluminothermic reactor (Crabtree and Seidel 1961). At the Rifle operation, the vanadium in solution was precipitated as sodium hexavanadate, commonly known as *red cake*.

The vanadium in solution is heated to 80°–90°C, and sulfuric acid is added to yield an optimum pH of 2.5–3.0 for the red cake formation. The reaction mechanism is as follows:

$$6NaVO_3 + 2H_2SO_4 = Na_2H_2V_6O_{17} + 2Na_2SO_4 \quad \textbf{(EQ 8)}$$

The red cake is then fused into vanadium oxide. Typically, the fused oxide produced directly from red cake contains 88%–92% V_2O_5 with ~8% sodium oxide (Na_2O) as the principal contaminant.

When alkaline leaching is employed, both uranium and vanadium dissolve, requiring further treatment to separate uranium from vanadium. The leaching is performed using sodium carbonate, sodium bicarbonate, or sodium hydroxide solution as the lixiviant (Gupta and Krishnamurthy 1992). Alkaline leaching is attractive for feed material containing high amounts of lime, as insoluble calcium vanadate is converted to soluble sodium vanadate.

When acid or alkaline leaching is employed, the vanadium in solution is purified via either chemical separation, ion exchange, or solvent extraction. The chemical separation process is highly complex and was gradually phased out and replaced by ion exchange and/or solvent extraction. Solvent extraction and ion exchange are employed to recover vanadium from acidic solutions (Gupta and Krishnamurthy 1992; Crabtree and Seidel 1961).

At the Uravan operation in Colorado, Union Carbide Nuclear Company employed column ion exchange technology to separate uranium from vanadium. Any pentavalent vanadium in acid leach liquors was initially reduced to the quadrivalent state with sulfur dioxide. The uranium was loaded to the resin, while the vanadium passed through the columns. Vanadium recovered from ion exchange effluent was recovered by precipitation.

Solvent extraction technology was applied at the Grand Junction and Rifle operations for separating uranium from vanadium in acid leach liquors. The uranium in the acid leach liquors was loaded onto di-2-ethylhexyl phosphoric acid (DEHPA). The extractant loaded with uranium was stripped with 10% sodium carbonate solution. The vanadium in the raffinate was recovered by precipitation. At the Rifle operation, Union Carbide Nuclear Company adopted a similar solvent extraction technology to that used by Climax Uranium Company. However, the vanadium in raffinate, recovered from the uranium solvent extraction circuit, was subjected to purification by solvent extraction using DEHPA. The extractant loaded with vanadium was stripped with sulfuric acid. The generic processing route for vanadium extraction from carnotite ore is presented in Figure 1.

Apart from patronite, roscoelite, and carnotite, vanadium has also been recovered from descloizite ores as the principal metal. Descloizite ore was mined at Otavi Mountainland in Namibia. This mine was in operation from 1925 until 1978.

Vanadium Extraction from Titaniferous Magnetite

In the 1970s, titaniferous magnetite ore emerged as the most important source for vanadium, replacing carnotite ore. It was discovered that ~98% of globally known vanadium reserves are contained in titaniferous magnetite ore. Mining titaniferous magnetite ore for vanadium on a commercial scale began in Finland around 1953, and then in South Africa by 1957. Titaniferous magnetite ore was mined in the former Soviet Republic in the early 1960s. Then in the early 1970s, it was mined in large quantities in China.

This raw material can be processed directly for vanadium in primary production, or it can be recovered as a by-product from vanadium-bearing slag from the steelmaking industry. The latter is the most common processing method used by the key global vanadium producers. The generic process flow sheet for recovering vanadium from titaniferous magnetite ore is presented in Figure 2.

Finland

In Finland, titaniferous magnetite ore was sourced from the mines at Otanmaki and Mustavaara (Gupta and Krishnamurthy 1992). The operation at the Otanmaki mine started in 1953. The as-mined ore comprised a magnetite-ilmenite deposit hosting about 0.4% V_2O_5. The ore was physically beneficiated using magnetic separation, producing ilmenite and magnetite concentrates. The ilmenite concentrate was sold as a marketable titanium concentrate. The magnetite concentrate, which contained 1% V_2O_5, 60% Fe, and 2.2% titanium dioxide

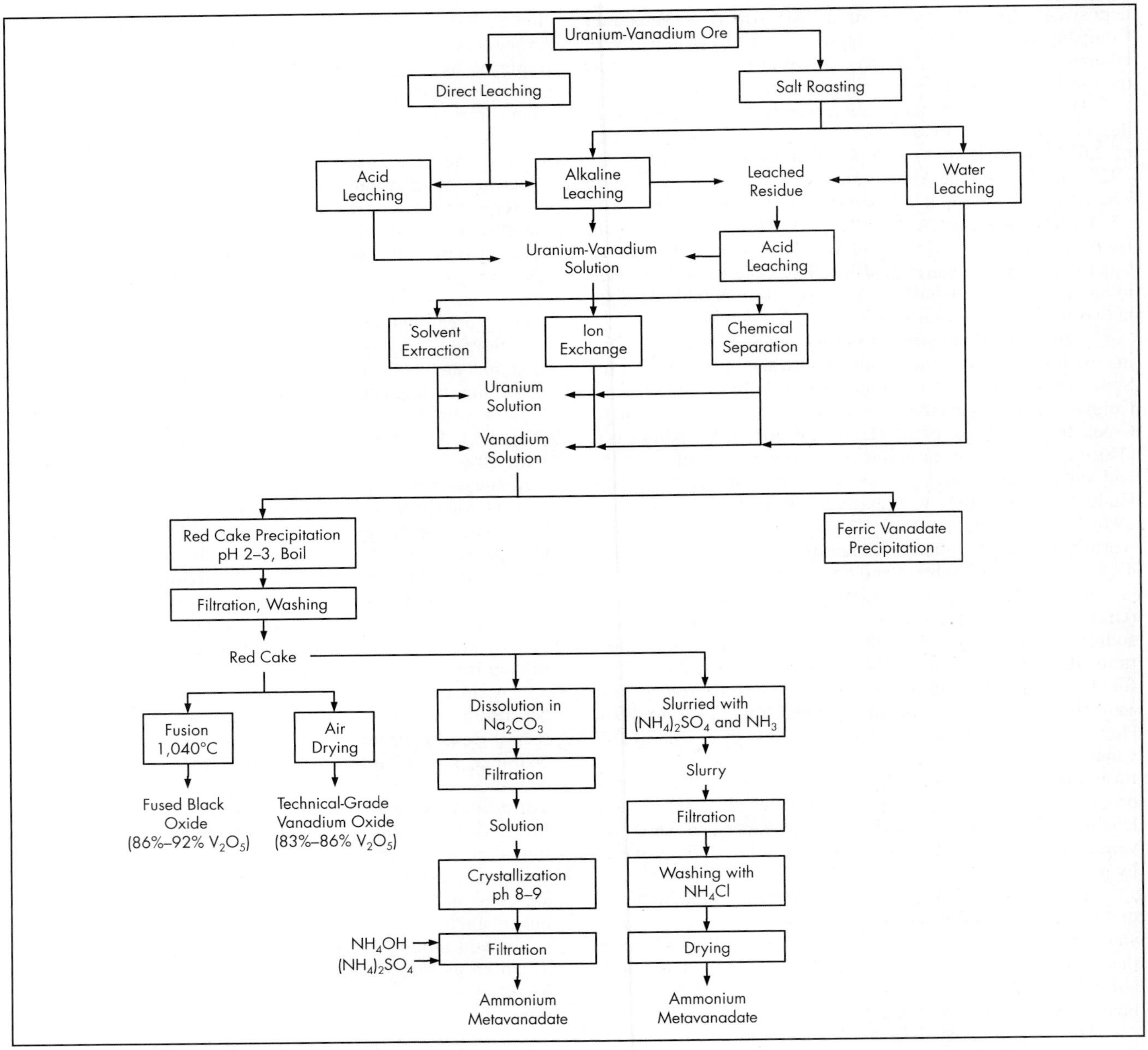

Source: Gupta and Krishnamurthy 1992

Figure 1 Extracting vanadium from uranium-vanadium ore

(TiO_2), was processed through a salt roast–water leach route for vanadium recovery.

Sodium sulfate was originally used as a sodium flux but was later replaced with sodium carbonate. The magnetite concentrate is mixed with sodium carbonate and pelletized in balling drums. The pellets are heated to 1,200°C in two parallel oil-fired shaft furnaces. The vanadium in the concentrate reacts with the sodium to form sodium metavanadate. Hot water is used to leach vanadium from the roasted pellets. Vanadium is precipitated from the liquor as ammonium polyvanadate. The ammonium polyvanadate is filtered, dried, decomposed, fused, and flaked to vanadium pentoxide. The leached pellets can be sold to steel producers as a high-grade concentrate.

The operation at Mustavaara mine began in 1976. The deposit was similar to the Otanmaki deposit except that the magnetite and ilmenite were finely disseminated. This made the separation of magnetite from ilmenite impossible. The vanadium grade in the feed ore was approximately 1.6% V_2O_5. Vanadium was recovered using the same salt roast–water leach route as that used at the Otanmaki operation, but it required a desilication process for silicate removal prior to precipitation. The leached residue contained high TiO_2 content and was unsuitable for smelting in a conventional blast furnace for steelmaking. The leached residue was therefore discarded into a tailings storage facility. The mine was closed in 1986. Since then, there has been no reported vanadium production in Finland.

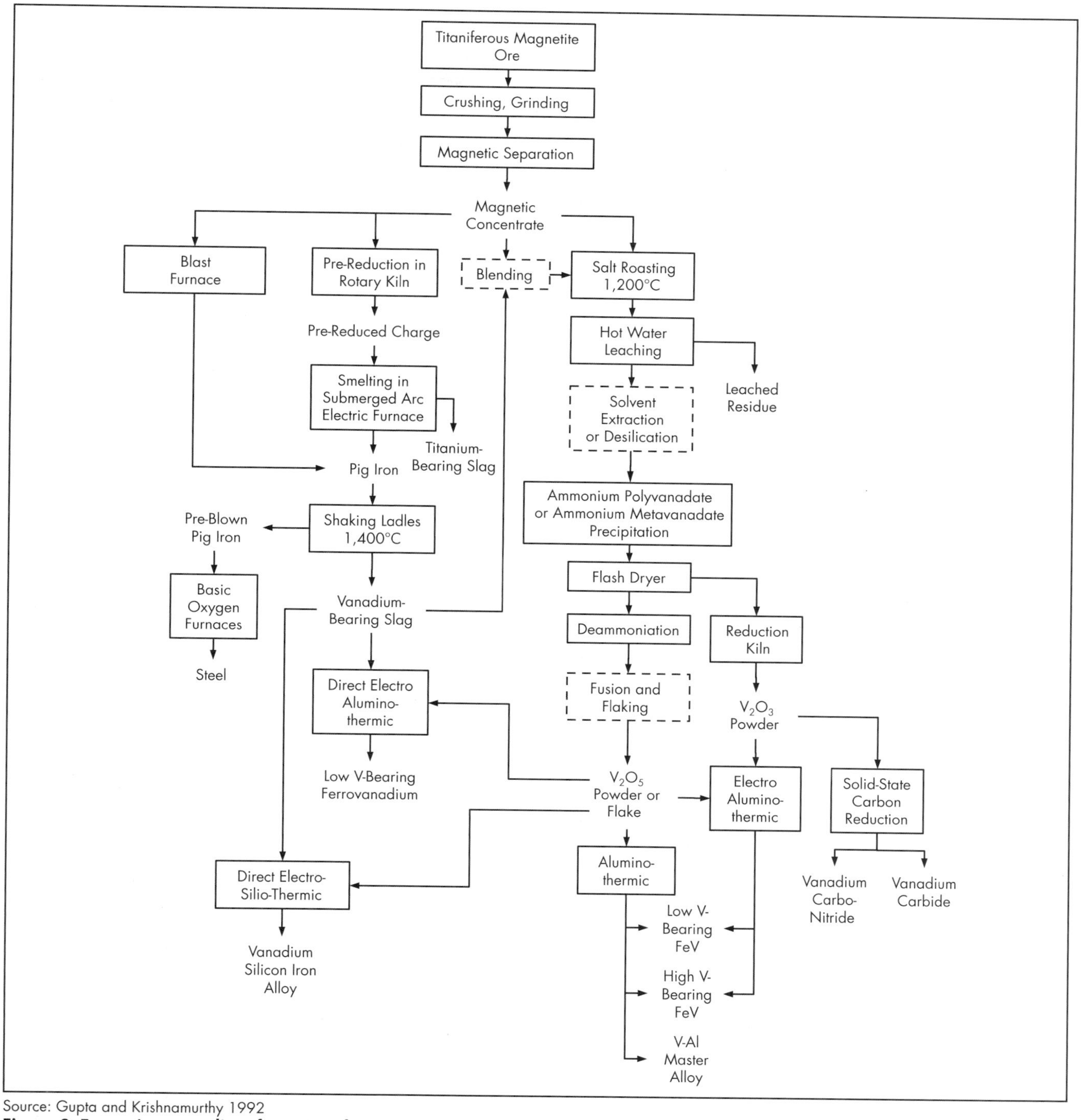

Source: Gupta and Krishnamurthy 1992

Figure 2 Extracting vanadium from titaniferous magnetite ore

South Africa

Multiple vanadium production operations exist in South Africa.

Vantra. Minerals Engineering Corporation of South Africa, a subsidiary of Minerals Engineering of Colorado, commissioned a processing plant in 1957 at Kennedy Vale in Eastern Transvaal (Rohrmann 1985). Titaniferous magnetite ore from the Bushveld Complex was processed using a salt roast–water leach route.

In 1966, Anglo American acquired the company and it became known as the Vantra division of Highveld Steel and Vanadium Corporation. Kennedy Vale mine was the main ore supplier to Vantra up to 1972 (Rohrmann 1985). Thereafter, Mapochs mine became the principal supply source. The ore was processed through wet milling and upgraded by magnetic separation for rejection of gangue minerals. The magnetic concentrate contained 1.65% V_2O_5, 56.4% Fe, 14.1% TiO_2, 1.2% SiO_2, and 3.1% aluminum oxide (Al_2O_3).

The ore, mixed with sodium flux, was roasted either in multiple-hearth Skinner roasters or rotary kilns. The processing plant was comprised of four identical 10-hearth Skinner roasters (6.1 m in diameter), three small rotary kilns (1.52 m in diameter × 18.3 m long), and one larger rotary kiln (2.6 m in diameter × 36.5 m long).

Roasting was initially carried out in the Skinner roaster using sodium chloride. Sodium chloride is used because of its ability to selectively attack vanadium and its comparatively lower operating temperature of 800°–900°C. This process generates large amounts of hydrochloric acid. The hydrochloric acid is captured and converted to ammonium chloride in an ammonia solution. The process is inefficient and presents major pollution and corrosion problems, and it delivers lower vanadium conversion compared to roasting with sodium carbonate or sodium sulfate. After 1974, sodium carbonate, sodium sulfate, or a mixture of both, replaced sodium chloride.

The reaction with sodium sulfate was found to be highly selective in its attack on vanadium. It produces a higher-purity pregnant solution compared to sodium carbonate. Co-dissolution of sodium aluminate, chromate, and silicate, which was encountered in the roasting with sodium carbonate, was not observed.

Roasting with sodium carbonate requires a roasting temperature of 900°–1,200°C. For sodium sulfate, the effective roasting temperature needs to be higher than 1,200°C. At Vantra, the upper temperature limit of the Skinner roasters was lower than 1,200°C; thus, the Skinner roasters were unable to deliver the required temperature for maximum recovery of vanadium. This subsequently led to the installation of rotary kilns.

The hot calcined ore is discharged onto drag conveyors for transport to quench tanks. The leach liquor is concentrated to 50–60 g/L V_2O_5 and pumped to precipitation tanks. After several displacement washes, the leached calcine is removed from the leaching vats and discarded into the tailings dump.

The concentrated pregnant solution is initially precipitated as ammonium metavanadate (AMV). Excess ammonium chloride is added as a precipitating agent to the continuous flow of pregnant solution inside an air-agitated reactor. The slurry is then thickened and pumped into filter boxes. The thickener overflow, containing barren solution, is pumped to coal-fired flash evaporators where the ammonium chloride solution is concentrated and returned to the precipitation plant for reuse.

After filtration and washing, the AMV precipitates are discharged from the filter boxes and fed to externally heated tube spiral conveyor-type deammoniators. The vanadium pentoxide powder is then melted in furnaces heated with glow bars at around 850°C for the production of vanadium pentoxide flakes.

The production of AMV was stopped in 1974 and replaced with the ammonium polyvanadate (APV) precipitation route. Sulfuric acid is used for pH adjustment. The pregnant solution is initially adjusted to pH 5.5 before the required quantity of ammonium sulfate is added. The solution is then adjusted to pH 2 with further additions of sulfuric acid to promote APV precipitation.

The coarse bright-orange APV precipitate is pumped into filters for washing and dewatering. Deammoniation and fusion follow thereafter and are carried out using the same reactor setup as in the AMV route. Reactors were later replaced with a rotary dryer to improve efficiency. The fused vanadium pentoxide assays at 99.5% V_2O_5, 3.5% V_2O_4, 0.25% Na_2O, 0.15% Fe, 0.006% S, and 0.002% P.

The filtrate, containing the barren solution from the APV plant, is evaporated in a two-stage vacuum evaporator. The condensate is sent to the leach dams. The sodium sulfate salt, produced after flash drying, is fed back to the roasting kilns for reuse as sodium flux.

Evraz Highveld Steel and Vanadium. In 1960, Anglo American Corporation of South Africa established Highveld Development Company. The purpose was to study the viability of producing pig iron and vanadium-bearing slag from titaniferous magnetite ore and to conduct a market survey of these new products. The study demonstrated that the Bushveld Complex ore could be melted in a submerged arc furnace using the addition of carbon to selectively separate vanadium and iron from titania. The vanadium and iron were recovered as liquid pig iron with most of the titania being absorbed in the slag. The liquid pig iron was subsequently processed in a side-blown air converter to produce vanadium-bearing slag and pig iron.

The Highveld process was developed to process ore from the Mapochs mine, which produces lump and fines iron ore. The ore treatment plant at the Mapochs mine is illustrated in Figure 3. Lump iron ore is processed through the Highveld steelmaking plant at Witbank (the city has since been renamed Emalahleni), and fine ore is sold to Vanchem Vanadium Products.

In 1965, Highveld Development Company changed its name to Highveld Steel and Vanadium Corporation (Highveld). By 1966, Highveld emerged as the global leader in vanadium production. In 2007, Evraz acquired an 85.1% stake of Highveld and then changed its name to Evraz Highveld Steel and Vanadium (Evraz Highveld) in 2010.

The Evraz Highveld processing plant comprises an iron production and steel facility with casters, a vanadium slag-crushing plant, a structural products rolling complex, and a flat products rolling complex. The iron plant treats the titaniferous magnetite ore in a pre-reduction process in 13 rotary kilns. Once reduced, liquid pig iron is produced in four open slag bath furnaces and two submerged arc furnaces with a combined annual capacity of 1,000,000 t of liquid pig iron (Evraz 2014). Vanadium slag is extracted at four shaking ladle emplacements in the steel plant, and the remaining liquid iron is processed into liquid steel in three basic oxygen furnaces and two ladle furnaces. The liquid steel is then cast into blooms and slabs for hot rolling, and the resulting billets are sold as semifinished product. In 2014, Evraz Highveld produced 7,130 t of contained vanadium in slag (Evraz 2015). The Evraz Highveld process flow sheet is shown in Figure 4.

The 13 refractory-lined rotary kilns in the iron ore plant are each 60 m long (Rohrmann 1985; Steinberg et al. 2011). The rotary kilns are co-currently fired using pulverized coal. Combustion air is introduced into each rotary kiln at the feed end. This design prevents the particles from fusing together by allowing the charge to absorb heat. The temperature profile in the rotary kiln is regulated via the supply of air and pulverized coal and is maintained close to a maximum temperature of 1,140°C.

The titaniferous magnetite ore is pre-reduced to 50%–70% metallic iron. The pre-reduction step provides hot charge, thereby reducing power consumption in the arc furnace. Dolomite and quartz are added to flux the titania into slag. The hot kiln discharge is collected into hoppers from which it is transferred to skip bins before feeding into a submerged arc furnace.

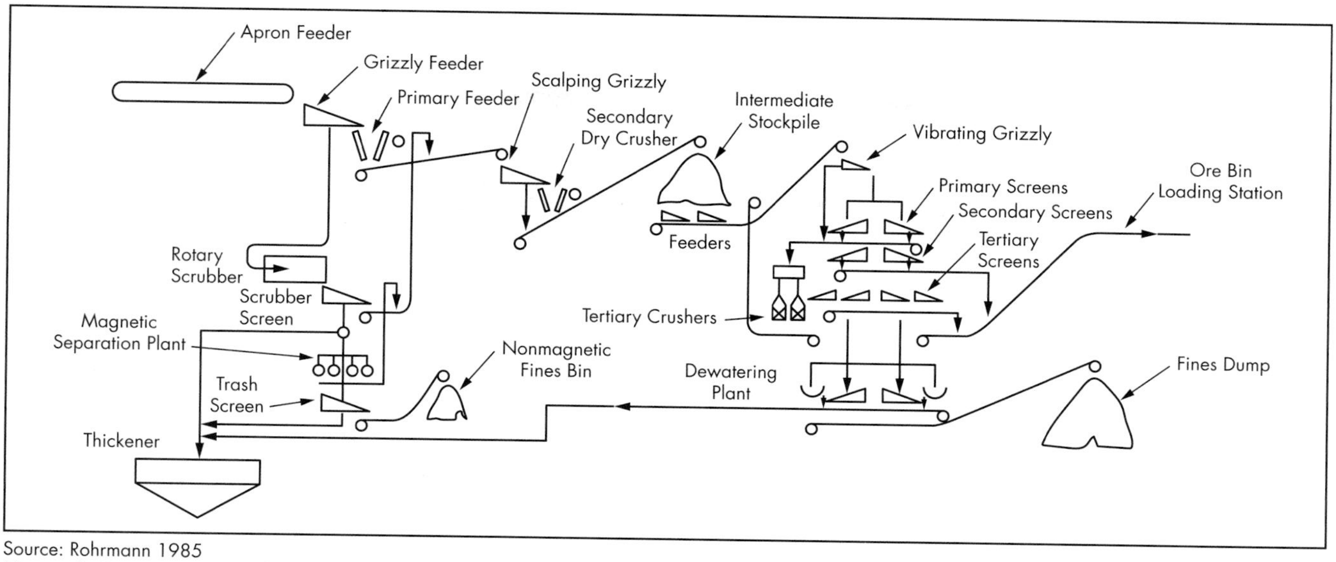

Source: Rohrmann 1985

Figure 3 Ore treatment plant at Mapochs mine

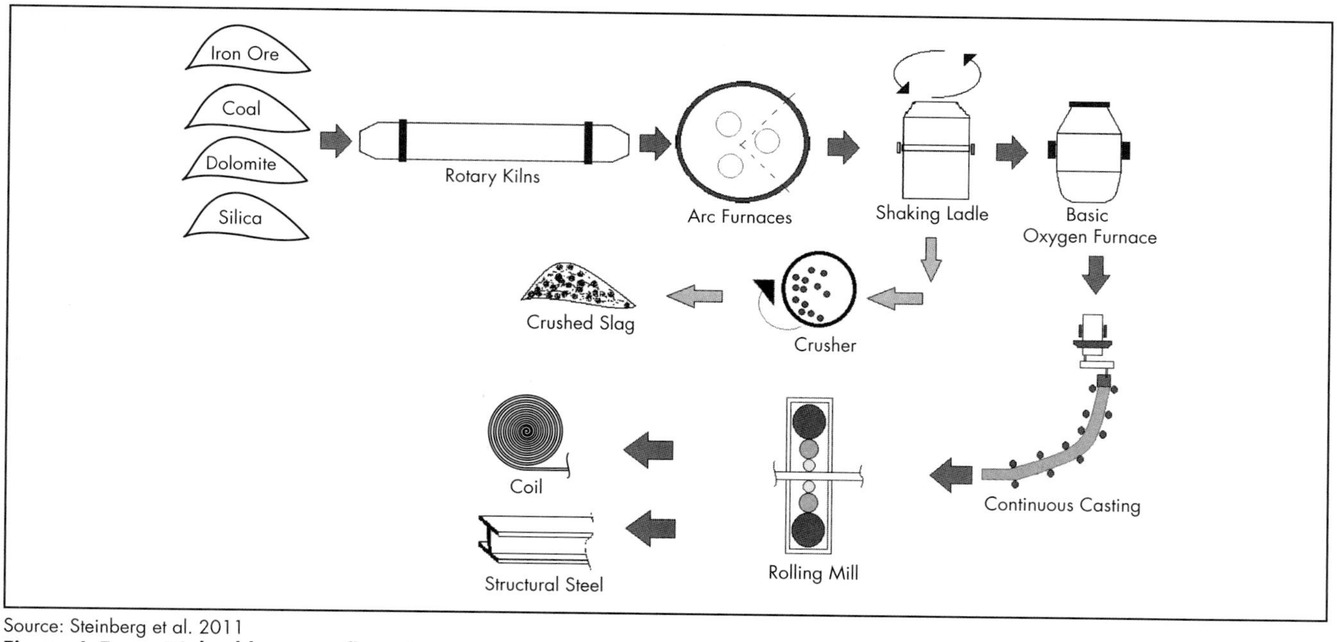

Source: Steinberg et al. 2011

Figure 4 Evraz Highveld process flow sheet

In the electric submerged arc furnace, liquid pig iron, titania slag, and carbon monoxide (CO) gas are produced. The submerged arc Elkem furnaces used are of closed-top design. The shell, with a diameter of 14 m, is equipped with three 1.6-m-diameter Soderberg electrodes. The temperature is maintained at ~1,350°C.

Each furnace is equipped with two tapholes, which are alternately used. A weir placed in the tapping launder separates the liquid pig iron and slag. The hot metal flows into railroad transfer ladles, and the slag is collected into slag pots. The titania slag is then crushed. Any entrained metal particles are magnetically removed and recycled through the process.

From the arc furnaces, the pig iron is transferred to the steel plant. The steel plant consists of four shaking ladles, three 70-t basic oxygen furnaces, two ladle furnaces, and four continuous casting machines. The steel plant can produce slab, blooms, or billets at a production rate of 2,600 t/d.

Vanadium is transferred from the liquid pig iron to slag using a shaking ladle process. The shaking ladle follows soft oxygen blowing. The oxygen blows into a ladle to form iron oxide (FeO) and then reacts with vanadium to form slag enriched with vanadium oxide. The reactions of this mechanism are as follows:

$$2V + 3/2O_2 = V_2O_3 \text{ (in slag)} \quad \textbf{(EQ 9)}$$

$$Fe + ½O_2 = FeO \text{ (in slag)} \quad \textbf{(EQ 10)}$$

$$2V + 3FeO = 3Fe + V_2O_3 \text{ (in slag)} \quad \textbf{(EQ 11)}$$

Scrap and iron ore are used as a coolant. The bath temperature is controlled to below 1,400°C such that a high rate of vanadium oxidation and a low rate of carbon oxidation can be achieved. The addition of iron ore improves the FeO activity in the slag, which in turn enhances the oxidation of vanadium from the metal.

During blowing, most of the carbon is oxidized to CO or carbon dioxide (CO_2). Anthracite is used to replenish carbon. The blowing rate and anthracite addition need to be carefully controlled to prevent the anthracite from reducing the vanadium oxide back to metal. The metal is poured into a basic oxygen furnace. The slag is retained and transferred to the crushing and packaging plant. The vanadium slag typically contains 12%–24% V_2O_5. It is used as feedstock for vanadium oxide and ferrovanadium production.

Vanadium slag is crushed, milled, and packaged at the vanadium slag-crushing plant. It is then sold to Hochvanadium Handels (a wholly owned subsidiary of Evraz Highveld), which supplies it to Treibacher Industrie. In turn, Treibacher Industrie processes it further into vanadium products, predominantly ferrovanadium, for sale in the European market.

Vametco. Vanadium slag is also supplied to Bushveld Vametco Alloys, situated near Brits in the North West Province of South Africa, for toll conversion into products such as modified vanadium oxide and Nitrovan. These products are sold on global markets through East Metals. Vametco's primary production is from magnetite ore from its mine located near the processing plant. Vametco has an annual production capacity of 3,600 t of contained vanadium as Nitrovan.

Vanchem Vanadium Products. A small amount of the vanadium-enriched slag is also sold to Vanchem Vanadium Products (VVP), a subsidiary of Duferco of Switzerland. VVP also purchased fines iron ore from the Evraz Highveld Mapochs mine. VVP owns and operates a processing plant at Witbank, located ~10 km from the Evraz Highveld's steelmaking plant. The processing plant at Witbank can produce ferrovanadium, vanadium chemicals, and fused V_2O_5 flakes. The rated production capacity is approximately 5,400 t/yr of contained vanadium. The estimated production in 2014 was 4,000 t/yr of contained vanadium (TZMI 2015). VVP also holds mineral rights for vanadium-containing iron ore reserves in the Steelpoort area through Rakhoma Mining Resources. In addition, VVP owns and manages 50% of South Africa Japan Vanadium (SAJV), a joint venture between VVP and Nippon Denko Company, which produces ferrovanadium for the Japanese market.

Glencore Rhovan. The Glencore Rhovan operation is another major vanadium producer in South Africa, located near Brits in the North West Province. Glencore acquired the Xstrata Rhovan operation when Glencore and Xstrata merged as corporate entities. The Ba-Mogopa mine, in operation since 1989, supplies titaniferous magnetite ore to the Rhovan operation.

Vanadium is extracted from the magnetite ore via primary production. The ore is crushed, milled, and then upgraded by magnetic separation to produce a concentrate, which is then processed using a salt roast–water leach route. A mixture of concentrate and sodium flux is processed in a coal-fired rotary kiln. The leaching is performed in leach dams to recover the vanadium-bearing solution. Silica in the pregnant solution is removed via precipitation with aluminum sulfate and filtration. The purified pregnant solution, recovered from the filtration, is reacted with ammonium sulfate to form AMV. The AMV is filtered, dried and deammoniated, and fused into V_2O_5 flakes. A portion of the AMV is also converted in reactors to produce V_2O_3 or V_2O_5 powder.

The processing plant can produce the equivalent of 10,000 t/yr V_2O_5. Glencore Rhovan also has a conversion capacity to produce 7,800 t/yr FeV. In 2014, Glencore Rhovan produced a total of 9,435 t V_2O_5 (Glencore 2015). Figure 5 shows the Glencore Rhovan process flow sheet.

New Zealand

In New Zealand, vanadium is extracted as a by-product from the steelmaking industry (Robson 1973; Hitching and Kelly 1982). The steelmaking plant, located at Glenbrook, ~40 km southwest of Auckland, employs similar technology to that used in South Africa by Evraz Highveld.

New Zealand Steel produces ~15,000 t/yr vanadium slag with 15% V_2O_5 concentration, which is sold to Chinese convertors, to be processed into V_2O_5 and FeV. New Zealand Steel also exports ~750,000 t/yr of its iron sands concentrates to Tianjin in northern China as feedstock for steel and vanadium slag production at Chengde.

China

As discussed previously, blast furnace operation is unsuited for treating titaniferous magnetite ore. Nonetheless, a blast furnace operation using titaniferous magnetite concentrate has been successfully implemented in China (Zhou 2000; Taylor et al. 2006; Vermeulen 2009). It has been reported that the concentrate is mixed with high-quality iron ore sourced to dilute the TiO_2. This reduces the risk of reducing the high-melting-temperature titanium carbides and titanium nitrides into the hot metal. The vanadium-bearing concentrate is first reduced to hot metal in a blast furnace, and then the vanadium is oxidized into a slag phase by blowing oxygen into the hot metal in the converter. Taylor et al. (2006) reported that Panzhihua controls the slag basicity (the ratio of calcium oxide [CaO] and SiO_2) to suppress the reduction of TiO_2. Du (1989) observed that the silicon-to-titanium ratio should be lowered to <0.8 in the blast furnace to suppress reduction of TiO_2.

The titaniferous magnetite resources are located in Ma'anshan in Anhui Province, Chengde in Hebei Province, and Panzhihua in Sichuan Province. Panzhihua New Steel and Vanadium Company, a subsidiary of Panzhihua Iron and Steel Group Company (Pangang), is the largest vanadium slag producer in China. Chengde Xinxin Vanadium and Titanium Company is the second largest vanadium slag producer in China. Together, they constitute approximately two-thirds of the total vanadium supply in China (TZMI 2015). Table 1 shows the production capacity of the various vanadium producers in China (TZMI 2015; Panzhihua 2013).

Slag in China typically contains 14%–22% V_2O_5. The slag is then roasted with sodium or calcium salts. China produces ferrovanadium containing 50% and 80% vanadium. Apart from that, the vanadium extracted from the slag is also processed into V_2O_5 flake, V_2O_5 powder, vanadium nitride, and vanadium carbonitride.

Qin et al. (2015) reported that Pangang invested 290 million yuan and constructed a pilot plant in 2009 to investigate the feasibility of rotary hearth furnace direct reduction and electric arc furnace for extracting vanadium from titaniferous magnetite ore. The pilot plant has a production capacity of 52,000 t of pig iron with low carbon content, 5,900 t of vanadium slag, and 24,000 t of titanium slag.

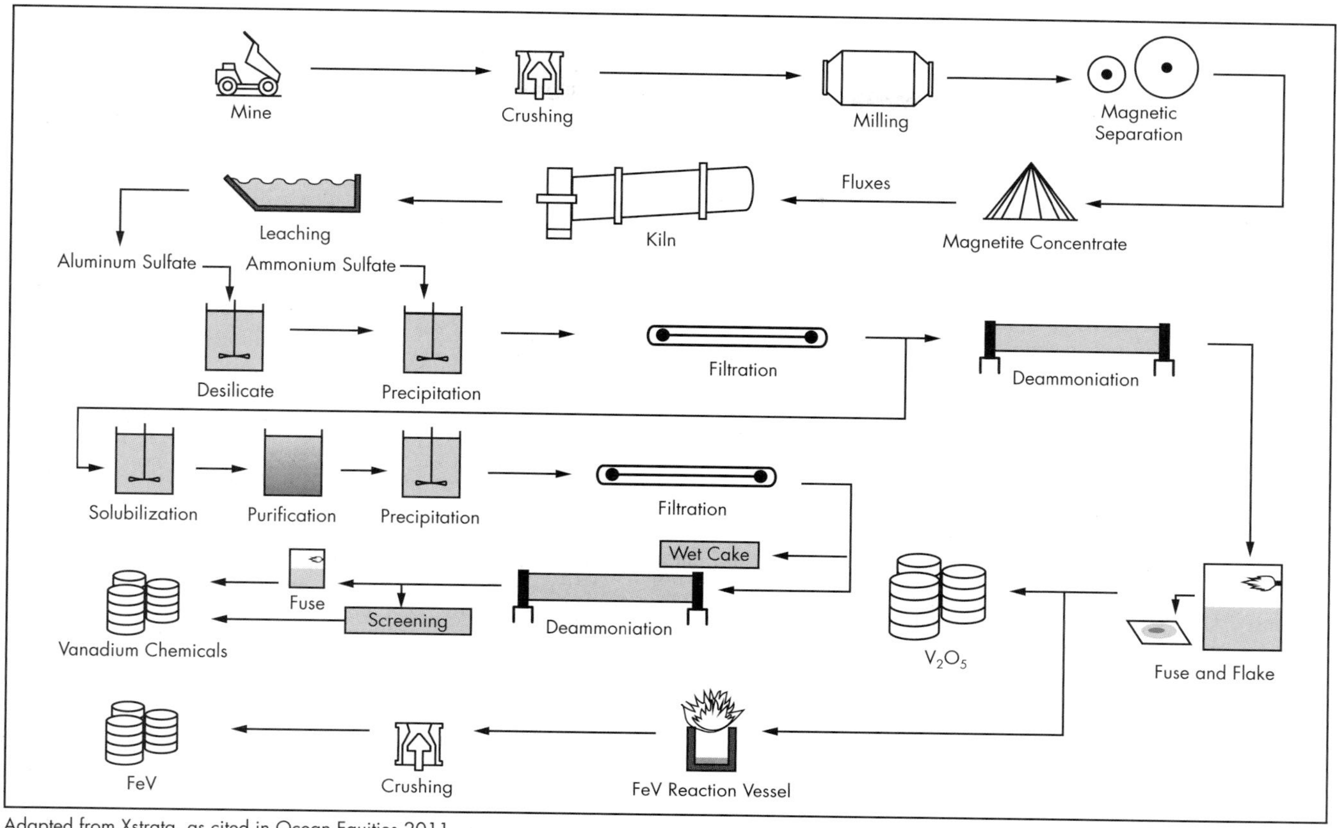

Adapted from Xstrata, as cited in Ocean Equities 2011

Figure 5 Glencore Rhovan process flow sheet

The titaniferous magnetite concentrate is first pelletized and then charged into the rotary hearth furnace. The pellets are reduced to 85% metallics, then transferred into an electric arc furnace for deep reduction. In the furnace, the titanium slag is separated from vanadium-bearing hot metal. The titanium slag contains about 45% TiO_2. After desulfurization, the vanadium-bearing hot metal is transferred to a vanadium recovery station. At the recovery station, oxygen is injected from the surface of hot metal with nitrogen being supplied at the bottom of the ladle. The vanadium in the slag is approximately 12% V_2O_5. Approximately 80% of the vanadium reports to slag. The vanadium in slag is then recovered using salt roasting. With the titanium slag, a sulfuric acid process is applied to produce a titanium dioxide product. Titanium recovery of 80% is achieved.

Table 1 Estimated vanadium slag production capacity in China

Company	Region	Product	Capacity, t/yr V_2O_5
Panzhihua New Steel and Vanadium	Sichuan	Vanadium slag	47,000
Chengde Xinxin Vanadium and Titanium	Hebei	Vanadium slag	20,000
Jianlong Steel Industrial	Heilongjiang	Vanadium slag	10,000
Kunming Iron and Steel	Yunnan	Vanadium slag	2,000
Chuanwei Steel Group	Sichuan	Vanadium slag	10,000
Other		Imported, vanadium slag	8,000

Source: Panzhihua 2013

Russia

Kachkanarsky mining and dressing complex (KGOK), a subsidiary of the Evraz Group, mines titaniferous magnetite ore from an open pit at Kachkanarskoye in western Russia. The beneficiation plant consists of dry magnetic separation, two-stage milling, and three-stage wet magnetic separation, followed by dehydration (Evraz 2010; Moskalyk and Alfantazi 2003; Taylor et al. 2006).

It produces a concentrate assaying at 0.57%–0.66% V_2O_5. The KGOK facility can produce 10.9 Mt/yr concentrate, which is further processed to 4.0 Mt/yr sinter and 6.9 Mt/yr pellet. KGOK supplies 70% of its production to Evraz Nizhny Tagil Iron and Steel Works (NTMK). The remaining is supplied to Chusovoy Metallurgical Works (CMW), Magnitogorsk Iron and Steel Works, and Novokuznetsk Iron and Steel Works.

NTMK processes material from KGOK into vanadium-bearing slag and microalloyed steels using two-stage basic oxygen converters. The first stage involves the production of vanadium slag (16%–28% V_2O_5) and a carbon-bearing intermediate product (3% C, 0.03% V). In the second stage, the intermediate product is processed into various grades of steel. The process recovers ~80% of vanadium into slag.

Most of the vanadium slag from NTMK is sold to Evraz Vanady-Tula. Vanady-Tula, established in the 1970s, joined the Evraz Group in 2009. Vanady-Tula employs calcium roasting technology followed by hot acid leach to produce vanadium red cake, which is fused to V_2O_5 flake. Vanady-Tula converts a portion of its V_2O_5 production into 50% FeV and 80% FeV to sell in the Russian and Commonwealth of Independent States markets. It has the capacity to produce 7,500 t of contained vanadium as V_2O_5 and 5,000 t of contained vanadium as FeV. In 2014, it produced 7,309 t V of V_2O_5, of which, 2,755 t was converted into FeV at Vanady-Tula, 3,538 t was consumed by Evraz Nikom, and the remaining V_2O_5 was sold to third parties (Evraz 2015).

NTMK also supplies CMW with vanadium slag. CMW employs a silica-thermic process to produce ferrovanadium. The mixture of vanadium slag, V_2O_5, lime, and fluoride is fused and then reduced with silicon and aluminum powder. The first stage produces metal assaying at 35%–40% V and 9%–12% Si. The metal is refined in the second stage by successive additions of V_2O_5 and lime until the residual silicon content is less than 2%.

Australia

The development of the Windimurra project led to the first commercial production of vanadium in Australia. The Windimurra vanadium deposit was discovered in 1961, but construction of the processing plant only began in 1998. Precious Metals Australia (PMA) formed a joint venture with Xstrata Alloys (now Glencore) in 1998 to develop the Windimurra project, and Xstrata Alloys acquired 51% interest in the project. Mining commenced in December 1999 with the first shipment of vanadium taking place in February 2000.

The project comprised an open cut mine followed by a processing plant employing primary crushing, single-stage semiautogenous milling, and magnetic separation. The concentrate was subjected to grinding in a Vertimill before a final polishing step with magnetic separation to produce a concentrate for the downstream refinery processing plant. The vanadium in concentrate was extracted via salt roast–water leach route using sodium oxalate. The sodium oxalate was a by-product of Alcoa's alumina refining operations at Wagerup and Kwinana and therefore offered an advantage regarding reagent cost. The pregnant solution was subjected to desilication before precipitation. It was then subjected to deammoniation, fusing, and flaking for the production of vanadium pentoxide flakes with purity of 98%.

In June 2000, after six months in production, the mine produced 2,722 t V_2O_5; in 2001 it produced 4,763 t/yr V_2O_5; and in 2002, it produced 5,443 t/yr. The mine reached its operating target in August 2001, producing 7,257 t/yr V_2O_5 before the production level was reduced (Economics and Industry Standing Committee 2004). The processing plant processed predominantly weathered ore. Milling the soft weathered ore created excessive fines and therefore hindered magnetic recovery.

In February 2003, the mine was placed on care and maintenance with its permanent closure announced in May 2004. The decision to suspend production was caused by a prolonged period of low vanadium prices. The processing plant was partially dismantled with significant components sold and removed from the site.

In 2005, PMA regained 100% ownership of the Windimurra project and formed a subsidiary company—Windimurra Vanadium Australia (WVA)—to develop the project. PMA adopted a new process flow sheet involving three-stage crushing followed by high-pressure grinding rolls. The crushed ore was then upgraded by rougher-scavenger magnetic separation. The concentrate was reground in a ball mill before final polishing with cleaner-scavenger magnetic separation. The vanadium processing facility was similar to the historical operation by Xstrata-PMA. The new plant also incorporated a new ferrovanadium production facility. Sodium carbonate was used to replace sodium oxalate.

In 2009, PMA was unable to raise sufficient funds to complete the project, and hence WVA was placed in receivership. In 2010, Atlantic acquired the Windimurra project and completed the construction of the remaining processing facility. At the time of the purchase, the construction of the concentrator plant had almost been completed. Atlantic formed a subsidiary company, Midwest Vanadium Pty Ltd. (MVPL), to operate the Windimurra project. Atlantic completed the construction in 2011. Mining commenced in January 2012 with the first shipment of ferrovanadium reported in June 2012. The concentrator processing plant was unable to reach its nameplate capacity, mainly because of the incorrect selection and/or sizing of equipment in the crushing, milling, and beneficiation area.

Production ceased in February 2014 because of a hot work accident that caused a fire in the beneficiation facility.

Brazil

Largo Resources (Largo) emerged in Brazil as a new vanadium producer following the construction of the Maracas Menchen mine in 2012. This mine is located ~250 km southwest of Salvador (Arsenault et al. 2013). The mine phase I nameplate annual production capacity is 9,600 t of V_2O_5. Ramp-up to full production was achieved 2017.

The processing facility comprises three-stage crushing with an intermediate stockpile to provide continuous feed to a ball mill. The crushed ore is then milled in the ball mill before it is upgraded by magnetic separation. The concentrate is roasted in a rotary kiln using sodium carbonate and sodium sulfate. The hot calcine is conveyed to a regrind mill where the lump is broken down. Leaching is conducted in agitated tanks. The slurry is filtered using a vacuum belt filter. The pregnant solution is pumped to desilication tanks, where it reacts with aluminum sulfate for silica removal. Sulfuric acid is added to control the solution pH. The slurry is then filtered to separate the solid enriched with silica. The purified pregnant liquor is precipitated as AMV using ammonium sulfate. The precipitate is filtered and fed to a flash dryer and deammoniator to produce V_2O_5. The V_2O_5 is then melted and cast into flakes for sale. The barren solution is treated in evaporators to recover sodium sulfate salt for reuse in the rotary kiln. Figure 6 shows the process flow sheet of the Maracas mine.

OUTLOOK AND MARKET

The near-term outlook for vanadium production is for a very tight market to continue, as many vanadium producers have stopped production or gone out of business. In April 2015, Evraz Highveld went into business rescue (Marsden and Terblanche 2015), and much of the iron plant has been dismantled. The future of the steelmaking plant at Witbank and the Mapochs mine remains unknown. With suspended production at Mapochs mine, VVP was forced to stop production because of lack in supply of vanadium raw material.

In February 2015, MVPL was placed into receivership. The Windimurra mine and mill has not operated since February 2014, and the future of the project remains uncertain. Longer

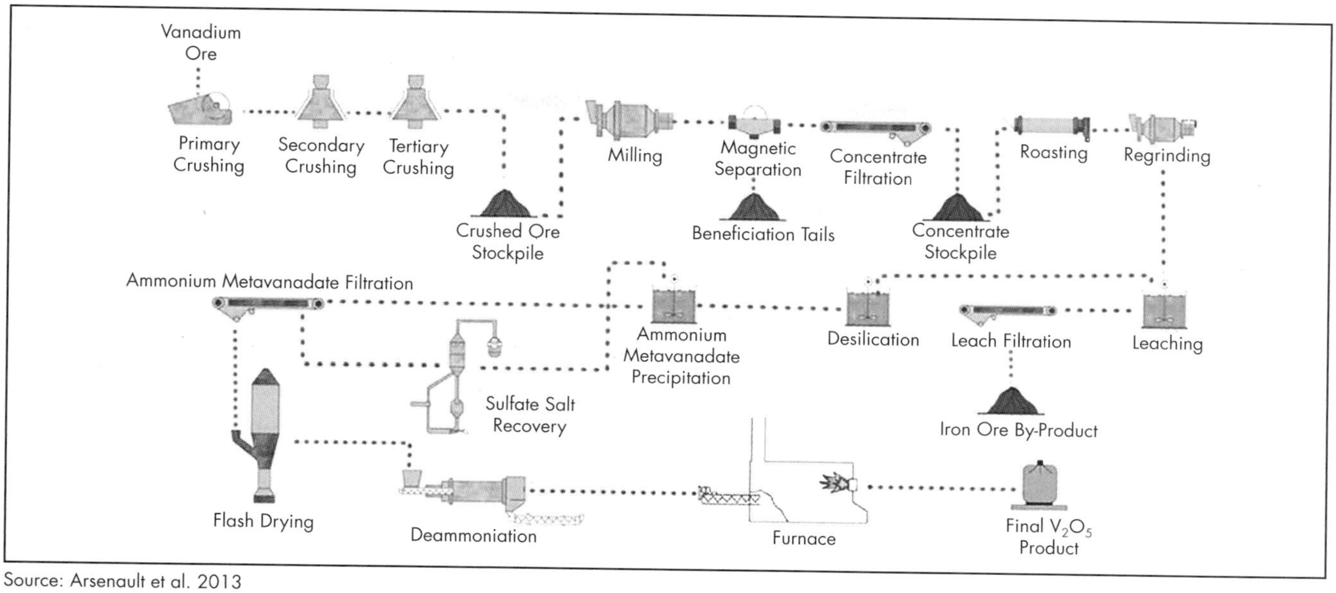

Source: Arsenault et al. 2013

Figure 6 Maracas process flow sheet

term, vanadium demand is projected to grow at its historic compound annual growth rate of 6.6% (Perles 2015; TZMI 2015). Growth in global steel production and the increase in specific vanadium consumption rates within the steel industry continue to drive vanadium growth. Vanadium intensity of use in North America is almost double that in China today. As China increases the percentage of alloyed steel production, the growth in demand for vanadium is expected to exceed that of growth in steel production. Growth in vanadium demand has exceeded vanadium production capacity, and all excess inventory has been worked out of the system. Additional new production is required to be brought on line.

Numerous vanadium projects around the world can be brought into production over time. Many of these are based on titaniferous magnetite ore bodies that require significant capital investment to be brought online. The projected cost to bring a greenfield mine and mill into production is US$300–$500 million, with an anticipated project timeline of four years.

Today, vanadium is used not only as an alloying component in steel and titanium alloys, but also as a catalyst for chemical reactions (Perles 2010, 2012). In 2014, world vanadium production amounted to 89,000 t of contained vanadium (Vanitec 2014). Approximately 91% of vanadium is consumed in the steelmaking industry (Roskill 2015). An alloy called *ferrovanadium* is used to produce higher strength steels for use in the construction and automotive and transportation industries and in manufacturing high-speed tools. The application of vanadium in the production of titanium alloys for aerospace and industrial purposes represents 4.5% of global demand. Approximately 3.5% of vanadium consumed in 2014 was used in petrochemical, catalyst, and pollution-control applications and in production of consumer products, such as ceramic pigment, special glasses, and so forth. The remaining 1% was used in energy storage applications.

As the world turns more and more to green energy power production, utilizing solar and wind power generation, there is much discussion about vanadium storage batteries and the significant potential demand from this area. Substantial

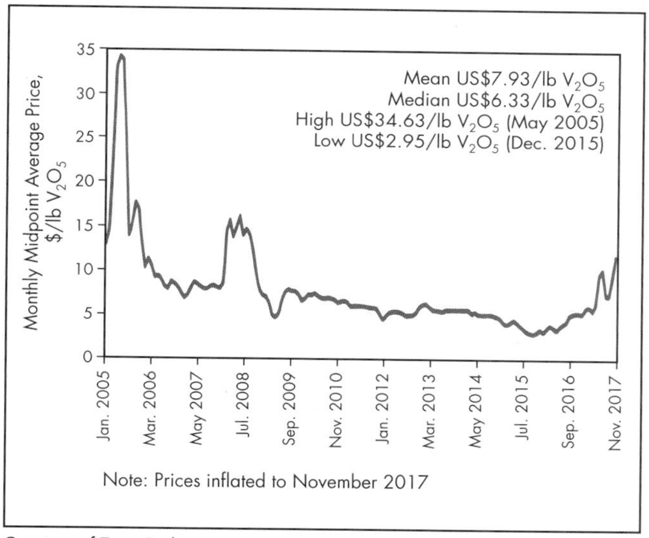

Courtesy of Terry Perles

Figure 7 V_2O_5 monthly midpoint average price

investment has been made in this area as a result of improvements in cell technology and electrolyte chemistry. Large megawatt batteries are now being produced and placed into service. With continued growth, this application is expected to be a major consumer of vanadium in the future.

The vanadium price is reported by the London Metal Bulletin (LMB) for European and Asian market pricing and by Ryan's Notes for North American pricing. V_2O_5 is typically priced per pound V_2O_5 contained, and FeV is quoted per kilogram vanadium by LMB and per pound vanadium by Ryan's Notes. Pricing is driven by supply and demand and has been quite volatile through history, with large swings in pricing occurring based on the balance between supply and demand. V_2O_5 pricing as quoted by LMB from 2005 through January 2018 is shown in Figure 7.

REFERENCES

Anon. 1920. Vanadium: Peru's large alloy deposits. *South Australian Register*. January 20. https://trove.nla.gov.au/newspaper/article/65025922. Accessed October 2018.

Arsenault, D., Hennessey, B.T., Oliveira, H.L., Spooner, J., Tanas, K., and Weston, S. 2013. *Preliminary Economic Assessment of the Maracas Vanadium Project*. Technical Report NI 43-101F1. Toronto, ON: Largo Resources Ltd.

Crabtree, E.H., and Seidel, D.C. 1961. *Colorado Vanadium: A Composite Study*. Denver, CO: State of Colorado Metal Mining Board.

Du, H. 1989. Titanium transfer in slag/metal systems for ironmaking. *Chin. J. Met. Sci. Technol.* 5:325–327.

Duke, V.W.A. 1983. *Vanadium: A Mineral Commodity Review*. Minerals Bureau Report No. 4/82. Johannesburg: Department of Mineral and Energy Affairs, Republic of South Africa. pp. 1–4.

Economics and Industry Standing Committee. 2004. *Inquiry into Vanadium Resources at Windimurra*. Legislative Assembly Report No. 10. Perth: Parliament of Western Australia. pp. 1–95.

Evraz. 2010. OJSC Nizhniy Tagil Iron and Steel Works. www.evraz.com/products/business/constructional/nizhny_tagil/. Accessed October 2018.

Evraz. 2014. *Highveld Steel and Vanadium: Integrated Annual Report 2013*. Mpumalanga, South Africa: Highveld Steel and Vanadium Corporation Ltd.

Evraz. 2015. *Annual Report and Accounts 2014*. Chicago, IL: Evraz.

Fischer, R.P. 1975. *Vanadium Resources in Titaniferous Magnetite Deposits: Geology and Resources of Vanadium Deposits*. Washington, DC: U.S. Government Printing Office.

Glencore. 2015. Corporate update and production report for the 12 months ended 31 December 2014 [news release]. Baar, Switzerland: Glencore International AG.

Goddard, J.B., and Fox, J.S. 1980. Salt roasting of vanadium ores. In *Extractive Metallurgy of Refractory Metals*. Edited by H.Y. Sohn, O.N. Carlson, and J.T. Smith. Warrendale, PA: TMS-AIME. pp. 127–154.

Gomi, A., and Whetham, R.D. 2012. The rocks at the top of the world. *Distillations* 30(2).

Gupta, C.K., and Krishnamurthy, N. 1992. *Extractive Metallurgy of Vanadium: Process Metallurgy 8*. Amsterdam: Elsevier Science. pp. 203–380.

Hammond, A.D. 2013. Vanadium, the new green metal: Mineral deposits in the Colorado Plateau. *Min. Eng.* (December):1–8.

Hammond Swayne LLC. 2015. Report released on vanadium resources contained in the asphaltite deposits of Central Peru. Chicago, IL: PR Newswire.

Hilliard, H.E. 1997. Vanadium and vanadium compounds, materials flow in the United States. In *Encyclopedia of Chemical Processing and Design: Vacuum System Design to Velocity*. Vol. 61. Edited by J.J. McKetta. New York: Marcel Dekker. pp. 271–287.

Hitching, K.D., and Kelly, E.G. 1982. Extraction of vanadium from N.Z. steel slags using a salt-roast/leach/liquid ion exchange process. *Proc. Aust. Inst. Min. Metall.* 283:37–42.

Marsden, P., and Terblanche, D. 2015. *Evraz Highveld Steel and Vanadium Limited (in Business Rescue)*. Registration No. 1960/001900/06. Johannesburg: Matuson and Associates.

McRae, L.B., Pothas, E., Jochens, P.R., and Howat, D.D. 1969. Physico-chemical properties of titaniferous slags. *J. S. Afr. Inst. Min. Metall.* (June):45–49.

Moskalyk, R.R., and Alfantazi, A.M. 2003. Processing of vanadium: A review. *Miner. Eng.* 16:793–805.

Ocean Equities. 2011. Vanadium sector review. London: Ocean Equities.

Panzhihua. 2013. Panzhihua I&S current situation and development suggestion. www.pzhsteel.com.cn/. Accessed October 2018.

Perles, T. 2010. Vanadium supply chain. Presented at the International Flow Battery Forum, Vienna, Austria, June 15–16.

Perles, T. 2012. Vanadium market fundamentals and implications. Presented at the Metal Bulletin 28th International Ferroalloys Conference, Berlin, Germany.

Perles, T. 2015. *Vanadium Market Report*. Pittsburgh, PA: TTP Squared.

Polyak, D.E. 2013. Vanadium. In *Mineral Commodity Summaries*. Reston, VA: U.S. Geological Survey. pp. 178–179.

Qin, J., Lin, G., Li, Z.J., and Qi, J.L. 2015. A study on the comparison of the several typical processes for dealing with vanadium titanium magnetite resources. In *Energy and Environmental Engineering: Proceedings of the 2014 International Conference on Energy and Environmental Engineering*. Edited by Y. Wu. Boca Raton, FL: CRC Press. pp. 145–148.

Robson, A.H. 1973. *Recovery of Vanadium from N.Z. Steel Slag Using a Salt Roast/Leach/Liquid Ion Exchange Process*. Report No. CME 33. Auckland, New Zealand: School of Engineering, University of Auckland.

Rohrmann, B. 1985. Vanadium in South Africa. *J. S. Afr. Inst. Min. Metall.* 85(5):141–150.

Roskill. 2015. *Market Outlook for Vanadium to 2025*. London: Roskill Information Services.

Steinberg, W.S., Geyser, W., and Nell, J. 2011. The history and development of pyrometallurgical processes at Evraz Highveld Steel and Vanadium. *J. S. Afr. Inst. Min. Metall.* 111:705–710.

Taylor, P.R., Shauey, S.A, Vidal, E.E, and Gomez, J.C. 2006. Extractive metallurgy of vanadium containing titaniferous magnetite ores: A review. SME Preprint No. 05-025. Littleton, CO: SME.

TZMI (TZ Minerals International). 2015. Global vanadium market review. Burswood, Western Australia: TZMI.

Vanitec. 2014. Production and consumption: 2011–2017 Vanadium statistics. http://vanitec.org/vanadium/production-consumption. Accessed October 2018.

Vermeulen, P. 2009. Uncovering the potential of ultra-low cost steel making using titano-magnetite ores in blast furnace-based mills. Presented at the SBB Steel Focus China.

Zhou, J. 2000. Development of China vanadium industry. In *Proceedings of the Vanitec Symposium*. Kent, UK: Vanitec Ltd. pp. 7–24.

CHAPTER 12.43

Zirconium

Steven C. Evans

Zirconium is a silver-gray metal with a close-packed hexagonal crystal structure at room temperature. At 862°C, the crystal structure changes to a body-centered cubic structure. Both structures are very ductile, and the metal is easily machined, rolled, and extruded using conventional methods. Its principal alloys have outstanding resistance to corrosion in water and steam at high temperatures. Zirconium has a very low-capture cross section for thermal neutrons and is commonly used for fuel rod cladding and structural materials in nuclear power generation.

Zirconium naturally occurs as zircon ($ZrSiO_4$). It is found in zircon ore bodies and is a significant impurity in the titanium ore rutile. Zirconium ore contains between 1% and 4% hafnium as an impurity. Hafnium is highly resistant to hot water corrosion but exhibits a high-capture cross section for thermal neutrons and is commonly used for control rods in nuclear power generation.

Zirconium metal has two distinct industrial applications. Zirconium is exceptionally resistant to many common acids, alkalis, and seawater. Zirconium, used for its high corrosion resistance in nonnuclear applications, is manufactured with 1%–4% naturally occurring hafnium content, depending on the ore body. Nuclear-grade zirconium alloys (zircaloys) require that the hafnium content is reduced to less than 100 ppm from the naturally occurring ore levels. The cost of producing zirconium is higher than most common metals, limiting its use to specific applications.

PRODUCTION AND RESOURCES

Australia is the largest producer of zirconium concentrate, followed by South Africa and then China. This trend has not changed for the last 10 years. World production trends for zirconium concentrates are shown in Figure 1. Production statistics by country for 2017 are shown in Table 1. Current zirconium reserves are around 74 Mt (million metric tons). The largest reserves at 47 Mt are in Australia, followed by South Africa at 14 Mt.

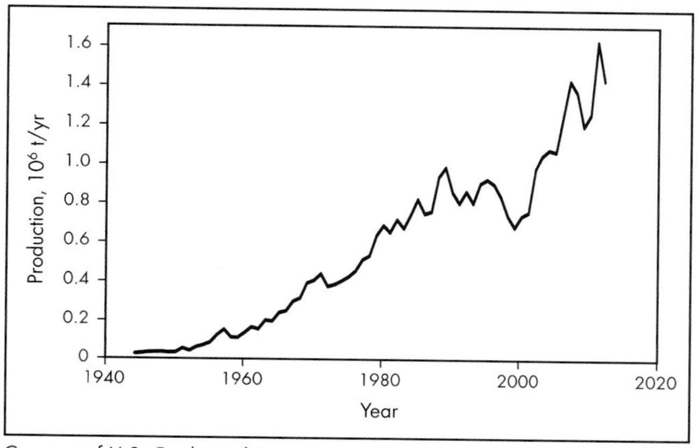

Courtesy of U.S. Geological Survey

Figure 1 Zircon concentrate production trend

PRICE OF ZIRCON CONCENTRATES

The price trend for zircon concentrates for the 10-year period from 2007 to 2017 are shown in Table 2. The prices of zirconium concentrates rose dramatically during 2011; this was mainly caused by booming Chinese demand for zircon to meet the requirements of the ceramic and zirconium chemical sectors (Adams 2012). The subsequent decrease in world production and therefore price of zirconium mineral concentrates in 2013 was in response to weakened demand, particularly in China, because of a slowdown in housing construction.

ORE DEPOSITS

Zircon and the less common mineral, baddeleyite (ZrO_2), are found as accessory minerals in igneous rocks but rarely in economic concentrations. The world's largest primary deposits of zirconium are associated with alkaline igneous rocks. Baddeleyite is produced as a by-product from apatite and magnetite mining in Russia, Brazil, and South Africa.

Steven C. Evans, Principal Engineer (Retired), Westinghouse Electric Company, Western Zirconium Plant, Ogden, Utah, USA

Table 1 Production and reserve statistics

Country	2017 Zirconium Concentrates, t	Zirconium Reserves, Mt ZrO_2
Australia	600,000	47
South Africa	400,000	14
China	140,000	0.5
Indonesia	120,000	Not applicable
Mozambique	75,000	1.8
Senegal	60,000	Not applicable
United States	50,000	0.5
India	40,000	3.4
Other countries	110,000	7.2
Total	**1,600,000**	**74**

Source: USGS 2007–2017

Table 2 Price of zircon concentrates*

Year	U.S. Domestic, \$/t	U.S. Imported, \$/t
2007	763	872
2008	788	773
2009	830	850
2010	850	860
2011	2,650	2,122
2012	2,650	2,533
2013	1,050	996
2014	1,050	1,106
2015	1,050	1,052
2016	1,025	877
2017	1,025	881

Source: USGS 2007–2017
*Prices are for domestic standard-grade bulk.

The main worldwide zircon ore deposits are heavy-mineral sands produced by the weathering and erosion of preexisting rocks and the concentration of zircon and other economically important heavy minerals, particularly in coastal environments. In coastal deposits, heavy-mineral enrichment occurs where sediment is repeatedly reworked by wind, waves, currents, and tidal processes. The resulting heavy-mineral sand deposits, called *placers* or *paleoplacers*, preferentially form at relatively low latitudes on passive continental margins and supply nearly all of the world's zircon.

Zircon makes up a relatively small percentage of the economic heavy minerals in most deposits and is produced primarily as a by-product of heavy-mineral sand mining for titanium minerals. Two styles of mineral sand deposits are found:

1. Strandline deposits are characterized by their relatively linear geometry.
2. Deposits found in the Wimmera region (WIM-style) of western Victoria, Australia, have a sheetlike geometry.

The strandline deposits tend to be coarse grained (>100 μm grain size), relatively rich, with grades in the range of 5%–20% heavy-mineral sands but are of relatively low tonnage. The WIM deposits are relatively fine grained (<100 μm grain size) with lower grades, in the range of 2%–5% heavy-mineral sands, but tonnages are generally at least one order of magnitude greater than that of the strandline deposits.

MINING

Mineral sands are mined using either wet or dry methods. Wet mining is by a dredge floating on the surface of a pond that is used to undercut the sand and pump the slurried sand to the wet processing plant usually situated in the pond behind the dredge. This method is generally used where access to water is not an issue. Dry mining uses large-scale earthmoving equipment (self-elevating scrapers, bulldozer traps, and front-end loaders) to excavate and transport the sand to the wet processing plant.

MINERAL PROCESSING

After mining, the ore is subjected to beneficiation process, which involves a wet primary concentration and a dry mineral separation. The primary concentration plant upgrades the run-of-mine ore by rejecting light gangue materials, such as quartz and shells, to produce a heavy-mineral concentrate (HMC) containing greater than 95% heavy mineral with minimum losses of valuable heavy minerals, such as ilmenite, rutile, and zircon. After separation of ilmenite and rutile by magnetic and high-tension (HT) separators, the rest of the HMC is subjected to zircon recovery processes.

In the wet processing plant, the heavy minerals are separated from the lighter gangue mineral, mainly quartz, by gravity methods, usually spirals and shaking tables. The light, rejected minerals in the processing plant (e.g., quartz, feldspar, mica, and clay minerals) are returned to the mined-out area of the mine, usually behind the dredge. The resulting HMC, containing up to 85%–98% heavy minerals, is composed mainly of ilmenite, zircon, rutile, and leucoxene, with smaller amounts of monazite, xenotime, and aluminum silicate minerals, such as garnet, kyanite, sillimanite, and staurolite.

In spiral separations, the slimes can affect the rheology of the slurry, preventing a clean separation of valuable mineral into the concentrate, thus increasing the loss of fine heavy-mineral particles into the tailings. As well, the mineral grains are often coated with iron oxides or clay minerals that can affect their magnetic and electrostatic properties, and so, the subsequent separations of the minerals. Desliming of the ore by attritioning and removing the fines with hydrocyclones help overcome these particular problems (Pownceby et al. 2105).

A clean separation of the heavy minerals is not achieved in a single spiral separation; consequently, several stages of separation (e.g., rougher, middlings, cleaner, recleaner, and scavenger stages) are required to obtain a final acceptable HMC. Developments in spiral separators have resulted in improved metallurgical performance, and changes in construction materials have increased unit capacities, simplified operation, and increased the recovery of fine heavy minerals. However, for the fine-grained WIM-style deposits, the heavy minerals may not be recovered with acceptable recoveries and grades using conventional processing equipment.

The individual heavy minerals in the HMC are separated in the dry processing plant. This is achieved by a combination of magnetic separations to recover ilmenite and associated magnetic products; and electrostatic separations to separate rutile, a conductor, from zircon, a nonconductor (Pownceby et al. 2105).

The advent of rare earth magnets has provided higher-strength permanent magnetic separators that are widely applied in the mineral sands industry. The performance of these separators depends on their field strength (gauss level) and their shape (field gradient). Several types of rare earth magnetic separators are used, including rare earth roll magnetic separators; rare earth drum magnetic separators, for both wet and dry separations; and matrix-type separators, such as wet high-intensity magnetic separators.

These separators are applied to concentrate and separate heavy minerals in different sections of the processing circuit. As an example, low-intensity magnetic separation and wet high-intensity magnetic separation may be used for separation of ilmenite from an HMC. High-intensity magnetic separations, however, are required to remove residual magnetic minerals from the resulting nonmagnetic product containing rutile and zircon. Wet magnetic separators are usually considered to be efficient for minerals with particle sizes larger than 75 µm.

Traditional electrostatic separations in mineral sand operations use combinations of HT roll and electrostatic plate separators. Heating the concentrate to 110°–130°C before the electrostatic separation improves the separation of zircon from other minerals. Substantial middling streams of conductors and nonconductors can be generated that require recirculation to different parts of the circuit (Pownceby et al. 2105). Flotation has been used for processing environmentally sensitive minerals, such as monazite, that are a health hazard when present as airborne dust during dry processing (Dunkin 1953; Subramanya 1960; Senior et al. 1996; Bruckard et al. 1999).

Impurity Removal from Zircon Concentrates

Zircon collected in the nonconductors fraction from electrostatic separations usually still contains small amounts of other nonconducting impurity minerals, such as monazite and aluminum silicates (e.g., kyanite), that must be removed. Monazite has a higher magnetic susceptibility than zircon and can be removed with a magnetic separation and centrifugal-type gravity concentrators, such as a Kelsey centrifugal jig (Jones and Foster 2010; Van der Westhuyzen et al. 2011; Capps and Waldram 1986).

Radioactivity in zircon concentrates is usually caused by uranium and thorium and/or their decay products contained in the crystal lattice, in mineral or fluid inclusions, in cracks and fissures in the zircon grains, or in other radioactive minerals such as monazite, xenotime, allanite, tantalite, betafite, thorianite, thorite that are present in the zircon concentrate. A global limit of 500 ppm U + Th or U + (0.4 × Th) <100 ppm usually applies to zircon concentrates. However, there are some countries that have different limits. The lower the level of uranium and thorium, the higher the marketing potential. In many treatment plants, the radioactive zircons, which are often weakly magnetic, are separated out during processing.

In many instances, the final zircon concentrate may contain zircon grains that are discolored and stained with coatings of iron oxide or clay minerals that prevent its use in many applications. Attritioning in water, acid, or alkaline solutions has been used to dissolve these impurities. However, more vigorous chemical treatments are required in many situations to meet market specifications. Table 3 provides an example of the quality specifications of a premium-grade zircon concentrate. Methods that are generally used include the following:

- In the hot acid leach process, hot zircon is treated with moderately concentrated sulfuric acid and reacted in a kiln, after which the product is attritioned, neutralized, and dried. A variant of this process employs a plug flow reactor in place of the rotary kiln.
- In the zircon upgrading process, the zircon is heated to ~400°C, after which it is quenched in dilute sulfuric acid.

Table 3 Premium-grade quality specifications

Constituent	Weight %
$ZrO_2 + HfO_2$	Minimum 66
TiO_2	Maximum 0.15
Al_2O_3	Maximum 0.45
Fe_2O_3	Maximum 0.10
SiO_2	32.80
U + Th, ppm	Maximum 500

MARKETING

Approximately 54% of zircon is used in finely ground form in the ceramics industry, 14% goes into the foundry industry, and 11%–14% into refractory applications. A growing portion of the zircon mined worldwide is used for the extraction of the metals zirconium and hafnium or their further processing to produce zirconium chemicals.

Zircon used in the ceramics industry has a wide range of applications. It is used as zircon sand directly from the mine and as a milled product in the form of zircon flour. Many milled zircon products, which vary in grain size and purity, are available commercially. The following are the most commonly used grades of opacifier used in the ceramics industry:

- Grade 9 µm (d_{50} = 1.5–2.0 µm) is widely use in ceramic glazes.
- Grade 6 µm (d_{50} = 1.2–1.5 µm) is used in ceramic glazes, as a cheaper substitute for the 5-µm material in sanitary ware, and for fully vitrified unglazed tiles.
- Grade 5 µm (d_{50} = 0.8–1.2 µm) is the highest-use product in fully vitrified unglazed tiles and sanitary ware. This grade tends to be the benchmark because of its high value-added-to-volume ratio of 4:5.

The application of zircon is determined by the particle size of the ground zircon. In addition to particle size, other physical properties are important in marketing zirconium concentrates. These are shown in Table 4.

REFINING

Unlike most metals that utilize smelting to refine the raw ore, zirconium ore refining is accomplished using chemical processes. Zirconium ore contains significant quantities of silicon, hafnium, iron, and aluminum.

Separation

Raw zirconium ore is dissolved and chemically processed to remove the silicon, iron, and aluminum, resulting in an aqueous solution of zirconium. Zirconium is then separated from the hafnium using a liquid–liquid separation process, where the zirconium resides in the aqueous phase and the hafnium in the organic phase that floats atop the aqueous phase. The resultant zirconium solution is precipitated as a zirconium sulfate, which is subsequently roasted in a rotary kiln to convert the sulfate to an oxide. The result is a very pure, finely divided ZrO_2 powder.

Table 4 Physical properties and applications

Physical Properties	Ceramics	Abrasives	Refractory	Foundry
	Application			
AFS*				X
Grain size distribution		X	X	X
Low loss on ignition	X	X	X	X
pH	X		X	X
Apparent density			X	
Low demand for acid				X
Color	X			X
Grain shape		X	X	X
Calcination	X			X
High tensile strength				X
High melting point	X		X	X
Great hardness		X		

Source: Pirkle and Podmeyer 1992
* American Foundry Society value for characterizing the grain fineness of foundry sand.

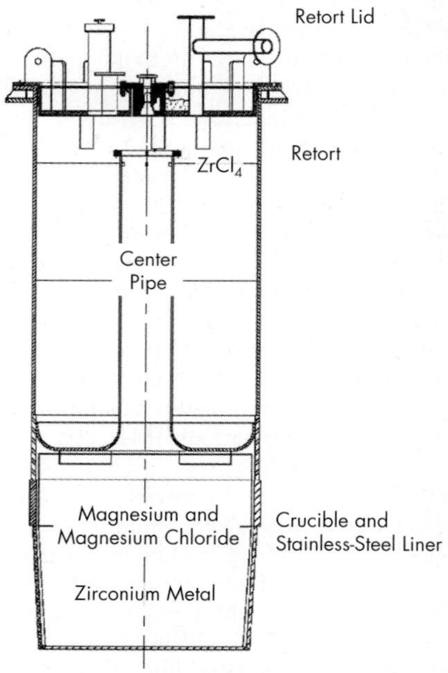

Courtesy of Westinghouse Electric Company
Figure 2 Zirconium reduction retort

Chlorination

The manufacture of zirconium from ore is essentially identical to the manufacture of titanium. The commercial process predominantly used is the Kroll process, developed by William J. Kroll for manufacture of zirconium through carbochlorination of the ZrO_2 followed by subsequent reduction of the resultant zirconium tetrachloride ($ZrCl_4$) with magnesium to form pure zirconium metal and magnesium chloride salt. ZrO_2 powder is combined with a fine powdered petroleum coke (essentially pure carbon) and heated in a fluidized-bed reactor with pure chlorine as the fluidizing gas. $ZrCl_4$ vapor is condensed to pure zirconium tetrachloride powder. Unlike titanium, zirconium tetrachloride (and hafnium tetrachloride, as well) does not exhibit a stable liquid phase at ambient conditions and condenses from the vapor phase directly to a solid particle, as shown in the following reaction.

$$ZrO_2(s) + C(s) + 2Cl_2(g) + heat = ZrCl_4(g) + CO(g) + CO_2(g) \quad \textbf{(EQ 1)}$$

Reduction

Processing of zirconium from a tetrachloride to a metal is similar to titanium and hafnium processing. Zirconium tetrachloride vapor is reacted with pure magnesium to form pure zirconium metal and magnesium chloride salt ($MgCl_2$), following the reaction:

$$ZrCl_4(g) + 2Mg(l) = Zr(s) + 2MgCl_2(l) + heat \quad \textbf{(EQ 2)}$$

The processing of zirconium ore through the $ZrCl_4$ stage is typically accomplished in air, and the materials routinely are in contact with air. $ZrCl_4$ is hygroscopic, absorbing moisture (humidity) and creating a dilute hydrochloric acid (HCl) on its surface. Although the water is not necessarily a quality issue, the bound water can create a safety issue in subsequent processing.

The zirconium tetrachloride is loaded into a large steel container called a retort. The retort is typically a very-large-diameter cylinder with a large-diameter center pipe protruding through the retort bottom and extending to near the retort top. Zirconium forms a eutectic compound with iron at the Zr/Mg reaction temperature, preventing the use of mild steel containment for the reaction products. Pure magnesium metal pigs are therefore loaded into a liquid-tight, cylindrical, stainless-steel liner inserted within a mild steel crucible, which is then welded to the bottom of the retort. A welded pressure-tight joint between the crucible and retort is required because the retort/crucible assembly is heated to 1,000°C and held for approximately 48 hours. A lid is then affixed to the top of the retort. The entire assembly is checked to ensure that it is leak tight. The entire assembly, as depicted in Figure 2, is then evacuated to remove all air and then backfilled to atmosphere with high-purity argon.

Magnesium and zirconium both readily react to form oxides and nitrides, thereby requiring an atmosphere with no residual air. Because the magnesium and $ZrCl_4$ have been in contact with air and the $ZrCl_4$ is hygroscopic, the feed materials hold a quantity of trapped moisture. While under vacuum, the assembly is heated to just below the $ZrCl_4$ sublimation temperature of 331°C to drive off any absorbed water vapor. Any water vapor allowed to remain in the assembly could react with the magnesium as it melts, creating magnesium oxide (MgO) and hydrogen gas. As the temperature increases, the vapor expands, building pressure within the assembly, which must be vented. The vented hydrogen vapors can react explosively with air. Therefore, all residual water must be removed along with the air.

The temperature of the top zone (top half) of the reduction furnace is controlled to sublime the $ZrCl_4$, typically no greater than 500°C. The bottom zone (bottom half) is heated above the melting point of magnesium (649°C) and held until most of the magnesium is molten and a vapor of magnesium is present. The sublimed $ZrCl_4$ vapor travels from the retort, down the center pipe, reacting with the molten magnesium and magnesium vapor, forming very finely divided zirconium particles ~10 μm in size in a bath of molten magnesium and magnesium chloride salt. Fine zirconium particles settle through the molten magnesium/magnesium chloride bath to the bottom of the crucible. The density of the magnesium is slightly lower than the molten magnesium salt, and the magnesium floats on top of the magnesium chloride salt, which continues the reaction.

The reaction of $ZrCl_4$ and magnesium is exothermic. The excess heat of reaction, within the retort and crucible assembly, can result in an acceleration of the sublimation of the $ZrCl_4$, causing a rapid increase in reaction rate. Reaction kinetics must be controlled to prevent this runaway reaction, resulting in a relatively lengthy reaction time.

After the complete consumption of the zirconium tetrachloride, the entire run is allowed to cool to room temperature at 1 atm of argon pressure. Because nitrogen reacts with hot zirconium, the only acceptable inert gases are helium and argon. When the entire run has cooled, the weld connection between the retort (top) and the crucible (bottom) is milled apart. The empty retort is washed and prepared for subsequent loading. The crucible now contains a regulus of zirconium metal on the bottom and solid magnesium chloride salt on top. The magnesium chloride is removed and processed as a coproduct. The zirconium metal regulus is a semisolid mass of very fine zirconium particles entrained in solidified magnesium metal and magnesium chloride salt mixture. Because magnesium chloride is highly hygroscopic, exposure of the zirconium regulus to the humidity in the air must be limited.

Vacuum Distillation

After removal of the zirconium sponge mass from the crucible, multiple zirconium reguluses are loaded into a vacuum furnace for vacuum distillation. A high vacuum is maintained while the zirconium reguluses are heated to temperatures above the vapor transition temperature for both magnesium and magnesium chloride. A water-cooled section of the furnace recondenses the vaporized magnesium and magnesium vapors. Figure 3 shows a schematic of a vacuum distillation vessel.

The size and density of the zirconium regulus limits the movement of the vapors from the interior of the regulus to the surface. As a result, the vacuum distillation process requires extended time at temperature. At the end of the distillation process, the temperature within the distillation furnace is raised to sinter the fine particles of zirconium into a semisolid mass. The zirconium reguluses are subsequently cooled to room temperature in an inert atmosphere. After distillation, the pores in the metal that previously contained the magnesium and magnesium chloride are now voids. The zirconium regulus now has the appearance of a metal sponge.

Newly formed zirconium metal has a high affinity for oxygen and, to a lesser extent, nitrogen. When freshly produced zirconium is initially exposed to air, the surface zirconium particles react rapidly with the air. The reaction is very exothermic, and extreme care must be exercised during air exposure and initial handling of the zirconium regulus to prevent ignition.

Zirconium burns extremely hot. Typically, a zirconium fire brightly glows, emitting very little smoke. Burning zirconium is highly reactive with water, forming $ZrO_2(s)$ and hydrogen gas. Therefore, zirconium fires should never be attacked with water or water-based extinguishers. Dry sand or fine table salt should be used. The sand or salt forms a crust that limits the exposure to air, eliminating the oxidizer. Zirconium fires should be buried and allowed to cool completely prior to cleanup.

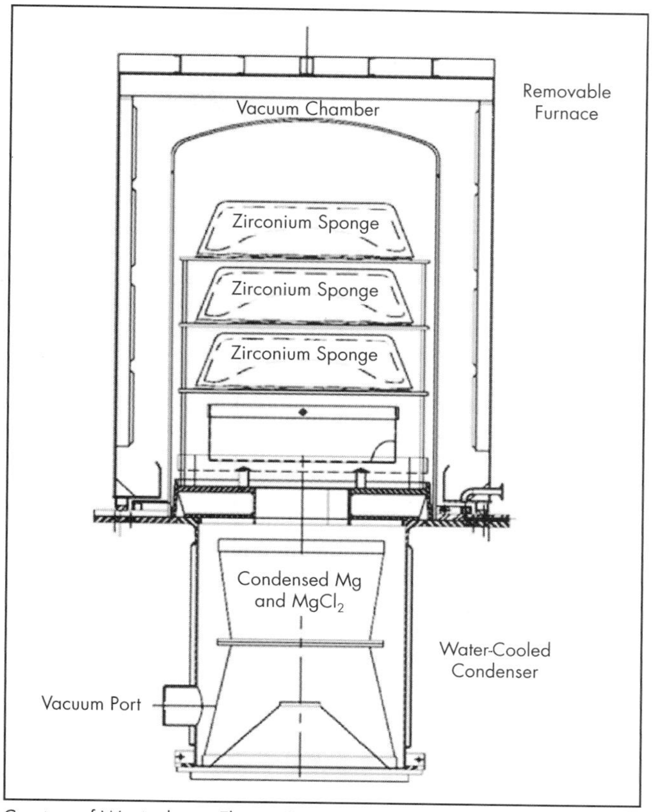

Courtesy of Westinghouse Electric Company

Figure 3 Vacuum distillation furnace

Before further processing, the entire surface of the regulus is visually inspected for rejectable materials. Rejectable materials are, for example, excessive residual metal chlorides and burned material. A mechanical press is usually used to break the large sponge mass into smaller pieces. These pieces are further shredded into sizes compatible with subsequent use (Figure 4).

To prevent fires, the mechanical shredding is usually performed in a sealed unit under a cover of argon gas. After shredding, the entire quantity of sponge is 100% visually inspected to remove any rejectable materials. The most common reject is burned sponge. Burned sponge contains a high proportion of ZrO_2 and zirconium nitride (ZrN) compounds. The ZrO_2 can result in elevated oxygen in the sponge, limiting its use. The presence of ZrN is a significant quality detractor because it has a very high melting point and limited solubility in molten zirconium. ZrN particles in sponge can result in hard nitride defects in products, causing failed final components.

After crushing and visual inspection, a sample is melted into a small ingot for chemical analysis. Because the sponge regulus is not uniform in chemistry across its entire volume, it is imperative that a representative sample of the entire crushed sponge mass is used for the evaluation ingot. Zirconium sponge has no direct applications as a finished product. Zirconium sponge can be used as an alloy addition in other metal systems such as aluminum, titanium, or steel.

MELTING

Melting of zirconium must be performed in furnaces isolated from exposure to air because zirconium reacts rapidly with air

Courtesy of Westinghouse Electric Company

Figure 4 Zirconium sponge

to form ZrO_2. Zirconium readily dissolves or is contaminated by all refractory materials. Consequently, zirconium melting is typically performed in water-cooled copper containment. Four typical furnace types are used to melt zirconium:

1. Vacuum arc remelting (VAR)
2. Electron beam melting (EBM)
3. Plasma arc melting (PAM)
4. Vacuum induction melting (VIM)

VAR and EBM are performed under a high vacuum, whereas PAM is performed in an atmosphere of argon and/or helium. Furnace type is selected for a specific material outcome.

Vacuum Arc Remelting

VAR is a consumable electrode melt that is very similar to arc welding. An electrode of alloyed zirconium is loaded into a water-cooled copper crucible with a gap between the outside of the electrode and the inside of the crucible. An arc is struck between the electrode and copper crucible bottom. A high-current, low-voltage arc melts the bottom of the electrode, which drips into the copper crucible, forming a solidifying ingot. At the end of the melt, the electrode has been consumed, creating a solidified ingot.

The entire electrode/ingot is never molten at the same time; therefore, the homogeneity of the input material along the entire length of the electrode is critical to produce a homogeneous final ingot. VAR is a high-throughput melt technique resulting in ingots up to 86 cm in diameter and weighing up to 9,072 kg.

The melting technique provides for exposure of the molten metal droplets to high vacuum, causing vacuum purification of the metal. The high power melt also results in a deep molten pool that mixes the alloy, resulting in good composition control. The melt rate is typically 14–41 kg/min. Because of the rapid melt rate, the purification from a single melt can be limited. Standard nuclear alloys require three consecutive melts. The primary melt electrode is very long but small in diameter. Each subsequent melt becomes shorter but increases in diameter. VAR melting is the predominant method for most zirconium manufacture.

Electron Beam Melting

EBM furnaces are of two types: drip melt and hearth melt. EBM takes place in a high-vacuum chamber. Under a high vacuum, high-energy beams of electrons are directed onto the zirconium metal. The beams are created by stripping electrons from a filament within an electron beam generator.

The electron beam operates at very high voltage and low current. EBM furnaces generate a significant quantity of high-energy X-rays. The furnaces typically are double walled with viewing systems of leaded glass, or they utilize cameras to prevent X-ray exposure.

In a drip melt furnace, the melt stock is suspended directly over a copper crucible. The electron beams impinge on the melt stock and the molten pool simultaneously, dripping the melt stock into the molten pool. The copper crucible has a moveable copper puller inserted into the bottom of the crucible. As metal is dripped in the top, the puller is slowly removed from the bottom, withdrawing a solidified ingot.

In the hearth melt furnace, solid or granular loose material is fed into a water-cooled hearth. The electron beams impinge on the metal in the hearth, melting the new material into the existing molten hearth. Once the hearth is adequately filled, the metal is allowed to overflow into a copper crucible with a puller. The electron beams heat the top of the withdrawing ingot to allow for a continuous ingot withdrawal.

EBM is much slower than VAR. The molten pool is relatively shallow, and the homogeneity of the electron beam melts is not as good as the VAR technique. The extremely high energy of the beam, the high vacuum, and the slow melt rate evaporate most of the highly volatile elements from the molten pool, which then deposit on the EBM chamber walls. EBM is typically used for purification melts. Alloy control is very difficult in EBM furnaces because large quantities of alloy must be added to compensate for the evaporation. The shape of the crucible is essentially unlimited in two dimensions. A semifinished shape can be made with the hearth/withdrawal system.

Plasma Arc Melting

PAM takes place in an inert atmosphere. The PAM furnace is typically a vacuum chamber with an inert gas system. When the furnace is closed, the vacuum system evacuates all of the air. The furnace is then back filled with either argon or helium. The hearth melting takes place exactly like EBM, with the exception of the heat source. In PAM, high-throughput plasma torches provide the energy for melting. Input material is fed into the hearth; the metal is melted and cast into an ingot by the hearth overflow system.

PAM is performed with a partial or full atmospheric pressure, significantly reducing the alloy loss caused by evaporation. With a PAM system, chemical composition control is much better than with an EBM system. As in EBM, the molten pool is relatively shallow. Additionally, because the melt takes place at a higher pressure, the refinement of the zirconium is limited in PAM.

Zirconium sponge contains small quantities of residual Mg and $MgCl_2$. VAR and EBM eliminate the high vapor pressure material through evaporation; PAM does not appreciably remove the volatile materials. Residual magnesium chloride in the final cast ingot results in microscopic longitudinal defects, known as *stringers*, in the fabricated product, which are detrimental to performance of the final components. The PAM furnace also uses a large quantity of inert gas, which can be very expensive. Expensive gas recovery systems are typical in PAM furnace operations.

Other Melting Methods

Most applications for zirconium are fabricated products. Several other techniques are infrequently used to melt zirconium that may or may not be fabricated.

Vacuum Induction Melting

VIM can be used to initially melt input materials of zirconium sponge, recycle, and alloy to form a primary melt ingot for subsequent VAR. VIM can also be utilized to create a volume of molten alloyed zirconium that can be used to pour castings.

VIM is performed by melting a volume of zirconium inputs in a water-cooled, segmented copper crucible. Zirconium readily dissolves all common refractories. A solidified layer of zirconium (called a *skull*) forms on the surface of the water-cooled copper. The balance of the molten material is therefore contained within a zirconium shell of the same composition. When the entire charge of input material is molten, the crucible is tilted, pouring the metal into the previously placed mold. The crucible and mold are contained within a vacuum chamber, preventing contamination with air.

Vacuum Arc Skull Melting

Vacuum arc skull melting is a hybrid of the VAR process where an electrode is melted in a water-cooled copper crucible to form a molten pool. When the appropriate amount of metal has been melted, the crucible is tilted to pour into a previously placed mold. The quantity of metal is relatively small because the arc-melted material is constantly being cooled by the water-cooled crucible during the melting process.

APPLICATIONS

Typically, zirconium metal is melted into ingots that can be fabricated into forms that can be used in industrial applications. The standard commercial alloy compositions are unalloyed zirconium ASTM grade 702 and ASTM grade 705, which is a Zr-2.5% Nb alloy (ASTM B350/B350M-11 [2016]). Two ASTM nuclear grades are common:

1. Zr-2 (Zr-1% Sn, 0.15% Fe, 0.1% Cr, 0.07% Ni)
2. Zr-4 (Zr-1% Sn, 0.2% Fe, 0.1% Cr)

Several other trademarked alloys are gaining wide nuclear industry use as Zr-Sn-Nb-Fe alloys.

CONCLUSION

Zirconium is a metal with several desirable attributes of water corrosion resistance and moderate ease of fabrication. However, as discussed, the costs associated with its production limits zirconium to specific niche applications. The production of zirconium metal from raw ore is expensive and requires significant care to produce an acceptable final product. Although zirconium is more expensive and difficult to fabricate, it is the best metal for certain highly critical applications.

REFERENCES

Adams, R. 2012. 2011 Review of zirconium. *Min. Eng.* (July):57–60.

ASTM B350/B350M-11. 2016. *Standard Specification for Zirconium and Zirconium Alloy Ingots for Nuclear Application.* West Conshohocken, PA: ASTM International.

Bruckard, W.J., Creed, M.D., Guy, P.J., and Heyes, G.W. 1999. Beneficiation of Australian mineral sands deposits using flotation. In *Murray Basin Mineral Sands, Extended Abstracts*. Bulletin No. 26. Edited by R. Stewart. Crows Nest, New South Wales: Australian Institute of Geoscientists. pp. 111–115.

Capps, P.G., and Waldram, J.T. 1986. Concentration of fine grained heavy minerals using the Kelsey centrifugal jig. In *Australia: A World Source of Ilmenite, Rutile, Monazite and Zircon*. Melbourne, Victoria: Australasian Institute of Mining and Metallurgy. pp. 99–106.

Dunkin, H.H. 1953. Concentration of zircon, rutile beach sands. In *Proceedings of the 5th Empire Milling and Metallurgical Congress on "Ore Dressing Methods in Australia and Adjacent Territories, Australia and New Zealand."* Vol. III. Edited by H.H. Dunkin and M.R. McKeown. Melbourne, Victoria: Australasian Institute of Mining and Metallurgy. pp. 230–274.

Jones, T., and Foster, A. 2010. Smarter gravity separation using the Kelsey centrifugal jig. In *Proceedings of the 25th International Mineral Processing Congress 2010*. Melbourne, Victoria: Australasian Institute of Mining and Metallurgy. pp. 859–869.

Pirkle, F.L., and Podmeyer, D.A. 1992. Zircon: Origin and uses. *Trans. SME* 292:1872–1880.

Pownceby, M.I., Sparrow, G.J., Aral, H., Smith, L.K., and Bruckard, W.J. 2015. Recovery and processing of zircon from Murray Basin mineral sand deposits. *Trans. Inst. Min. Metall. Sect. C* 124(4):240.

Senior, G.D., Trahar, W.J., and Creed, M.D. 1996. Processing of mineral deposits. Australian Patent 668807.

Subramanya, G.V. 1960. Selective flotation of zircon from beach sands. *J. Min. Met. Fuels* 7:47–48.

USGS (U.S. Geological Survey). 2007–2017. Zirconium and hafnium. In *Mineral Commodity Summaries*. Reston, VA: USGS.

Van der Westhuyzen, P.V.A., Jones, T.A., and Malonde, W. 2011. Kelsey centrifugal jig implementation at Namakwa Sands for zircon recovery from quartz rejects. In *Proceedings of the 8th International Heavy Minerals Conference*. Melbourne, Victoria: Australasian Institute of Mining and Metallurgy. pp. 311–321.

Index

Note: *f.* indicates figure; *t.* indicates table.
Pages 1–1176 are found in Volume 1; pages 1177–2204 are found in Volume 2.

SME
MINERAL PROCESSING & EXTRACTIVE METALLURGY HANDBOOK

VOLUME ONE

SME MINERAL PROCESSING & EXTRACTIVE METALLURGY HANDBOOK

VOLUME ONE

MANAGING EDITOR: ROBERT C. DUNNE
EDITORS: S. KOMAR KAWATRA & COURTNEY A. YOUNG

PUBLISHED BY THE
SOCIETY FOR MINING, METALLURGY & EXPLORATION

Society for Mining, Metallurgy & Exploration (SME)
12999 E. Adam Aircraft Circle
Englewood, Colorado, USA 80112
(303) 948-4200 / (800) 763-3132
www.smenet.org

The Society for Mining, Metallurgy & Exploration (SME) is a professional society whose more than 15,000 members represent all professionals serving the minerals industry in more than 100 countries. SME members include engineers, geologists, metallurgists, educators, students, and researchers. SME advances the worldwide mining and underground construction community through information exchange and professional development.

ISBN 978-0-87335-385-4
Ebook 978-0-87335-386-1

Library of Congress Cataloging in Publication Control Number: 2018054404

Contents

PART 4: CLASSIFICATION AND WASHING

PART 5: TRANSPORT AND STORAGE

PART 6: PHYSICAL SEPARATIONS

PART 7: FLOTATION

PART 8: SOLID AND LIQUID SEPARATION

PART 9: DISPOSAL

VOLUME TWO

PART 10: HYDROMETALLURGY

Foreword

Mineral processing and extractive metallurgy are atypical disciplines, requiring a combination of knowledge, experience, and art. This was recognized by Peter Ritter von Rittinger in the aptly titled *Lehrbuch der Aufbereitungskunde*, or *Textbook of Processing Art*, that was published in Berlin in 1867. Subsequent handbooks continued the pattern, combining technical analyses of processes with numerous examples and detailed descriptions of successfully operating process plants. Arthur F. Taggart's 1927 *Handbook of Ore Dressing* was, in the author's words, intended to "serve as a reference handbook for engineers … and also as a text-book for students." Its compact size made it a true handbook—one that could be easily carried in an engineer's knapsack to a mill site in Chuquicamata, Johannesburg, Butte, or Kalgoorlie. With a second edition appearing in 1945, it was the only book on the shelf in many mills.

The *SME Mineral Processing Handbook* published in 1985 rapidly became a standard reference, providing comprehensive coverage of all the unit operations in mineral processing, descriptions of process plants for more than 26 minerals and materials, and small sections on hydrometallurgy and pyrometallurgy. The authors of its many chapters contributed their collective expertise unselfishly to provide a handbook that was truly useful to all the practitioners of mineral processing—students, engineers, mill managers, and operators.

This new *SME Mineral Processing and Extractive Metallurgy Handbook* provides up-to-date coverage of all mineral processing unit operations, but it also has much larger sections on hydrometallurgy and pyrometallurgy as well as a section on management and reporting, which discusses such topics as health and safety, community and social issues, and project management. Once again, each chapter is authored by an acknowledged expert. These selfless experts, recruited by the editors, have each made an invaluable contribution. The chapters were skillfully organized and refined by the managing editor, with the incomparable assistance of SME's book publishing team. The new SME handbook is truly a timely document, addressing the new technologies and important cultural and social issues that are important to today's minerals community.

With the release of the *SME Mineral Processing and Extractive Metallurgy Handbook*, SME continues to fulfill its mission as the world's premier publisher and distributor of technical information related to the mineral industries. I am proud to have had some association with the society's continuing efforts and ongoing success in that endeavor.

Michael G. Nelson
Professor and Chair, Mining Engineering Department
University of Utah, Salt Lake City, Utah

Preface

A *handbook* is a concise reference book intended to provide ready reference material such as facts on a particular subject; details regarding how equipment works, how it operates, and how it is controlled; and guidance for problem solving. Such a text should be easy to consult and should provide quick answers where possible. This new *SME Mineral Processing and Extractive Metallurgy Handbook* has been designed and written to meet these criteria and is intended for those interested about, and working in, these industries in the 21st century.

The handbook contains the most current information on unit operations used in mineral processing, hydrometallurgy, and pyrometallurgy. Integrated optimization and value-chain processes that have gained importance during the last 20 years, such as mine-to mill, mine-to-market, and mine-to-metal, are explained, and examples of mine-to-market and mine-to-metal applications are inherent in the commodity chapters. An important input to these processes depends on proper and quantitative ore characterization. The latest available analytical methods and automated mineral identification systems that are used for ore characterization are described in separate chapters and will help metallurgists select the most appropriate procedures for their particular situations. The important design and equipment requirements for processes such as carbon-in-pulp, carbon-in-leach, bacterial tank leaching, heap and in situ leaching, and pressure oxidation will help both those looking at these options for a new project and operators of these processes. How best to model and control these and other methods is an important aspect when optimizing metallurgical operation performance. Chapters on modeling and simulation as well as process control and operational intelligence provide excellent overviews of what techniques are available and what works. The importance of proper and appropriate health and safety programs, environmental stewardship, and attention to community and social engagement are also covered in these pages. Those interested in recycling, graphene, and "battery" minerals will also find chapters appropriate to these topics. Overall, the handbook contains a wealth of valuable information.

The search for an editor to take on the task of producing an updated handbook began in 2012, as the previous handbook went out of print in 2003 and there was a steady and undeniable demand for a new version. Professors S. Komar Kawatra and Courtney A. Young from Michigan Technological University and Montana Tech, respectively, submitted a joint handbook proposal that incorporated both mineral processing and extractive metallurgy. The proposal was unanimously approved by the SME Information Publishing Committee in 2013. After six years of intense work by the authors, reviewers, editors, and SME book publishing team, the new *SME Mineral Processing and Extractive Metallurgy Handbook* is now here, comprising two volumes of 2,312 pages, with 128 chapters.

A project of this magnitude owes much appreciation to the commitment of the many people who brought it to fruition, including 192 authors and co-authors, and 102 technical reviewers, to realize a product of technical and practical worth. I applaud Kawatra and Young for taking on the challenge of such a large, complex, and diverse project. An unexpected obstacle that hindered the long-term progress of the task was the turbulent time in the mining industry during 2014–2017, created by the global recession in the mining industry. It was an extremely stressful and unsettling stretch for all those associated with the handbook. Many of the authors who agreed to contribute were retrenched or relocated, which meant that many were unable to fulfil their commitment and new authors had to be found. During this period, I was asked to join the technical team and assist with locating new contributors and provide technical management of the project.

Another issue that reared its head and is aptly described by Arthur F. Taggart is that "No one who has not tried to write a technical chapter in a handbook, or a part thereof, realizes how little he or any other one person knows about the subject that he considers his specialty." Consequently, many authors needed to spend much more of their personal and precious time either late at night and/or over weekends to source the appropriate information and complete the development of their chapters. On behalf of Kawatra, Young, and I, we extend our appreciation and sincere thanks to all the authors, co-authors, and technical reviewers for their dedication and commitment to make the handbook a reality. My special and heartfelt thanks goes to Michael G. Nelson and Thomas Battle who were stalwarts willing to give a lot more of their valuable time to take on additional authorship and reviewing tasks at short notice. Undoubtably, owners of the new handbook will greatly appreciate everyone's efforts in delivering these two volumes of technical and practical excellence.

The five professional and dedicated individuals who supported SME efforts and were intimately associated with the production of the handbook require special mention and a sincere thank-you for a job well done: Jane Olivier, manager of book publishing, who initiated the project; Diane Serafin, senior technical editor; Terese Platten, technical editor; Karen Ehrmann, permissions editor; and Stephen Adams who composed the pages. Readers of the handbook will appreciate the enormity of the overall editing task that included obtaining thousands of permissions and redrawing or reworking almost 3,000 figures to produce this high-quality handbook.

Finally, SME's management and Mineral & Metallurgical Processing Division members are thanked for their support and encouragement throughout the project.

Robert C. Dunne
Managing Editor
Gooseberry Hill, Western Australia

About the Editors

Robert C. Dunne

Robert C. Dunne, now retired, has worked in the mining industry for 41 years and is currently an adjunct professor at both Curtin University (Gold Technology Group) and the University of Queensland (Julius Kruttschnitt Mineral Research Centre [JKMRC]) in Australia. He graduated from the University of Witwatersrand in Johannesburg, South Africa, with a BScEng degree in metallurgy.

During his career, Dunne held group executive positions in metallurgical development and technology and also in innovation, and he spent time at Newcrest Mining, Anglovaal, and Anglo American. He was the assistant director of ore dressing at Mintek for 14 years, and during this time, he visited most of the gold and base metal mines in South Africa and initiated mineral processing research and development at the chemical engineering departments of Cape Town and Stellenbosch universities. He immigrated to Australia in 1986 and spent three years at the Western Australian School of Mines as senior lecturer in the Minerals Engineering and Extractive Metallurgy Department and oversaw the new Research and Development Department. At the time of his retirement in 2013, he was the Fellow Metallurgy at Newmont Mining Corporation in Denver, Colorado.

In 2016, Dunne received the Robert H. Richards Award from the American Institute of Mining, Metallurgical, and Petroleum Engineers for pioneering work in the application of semi-autogenous milling, flotation, and gravity concentration for gold and copper recovery. He was a Society for Mining, Metallurgy & Exploration (SME) Henry Krumb Lecturer in 2012 with his presentation titled Water in Mining. In 2017, Dunne was nominated as the Australasian Institute of Mining and Metallurgy's (AusIMM's) Metallurgical Society G.D. Delprat Distinguished Lecturer, and his presentation titled Water: Facts, Perceptions and Conflicts was delivered to local AusIMM branches throughout Australia. As an invited conference speaker, Dunne has given discourses on refractory gold, gravity separation, gold flotation, and mine-to-mill application. He has also been a contributor to numerous short courses on gold processing and flotation. Dunne has authored and co-authored more than 80 technical papers in conference proceedings and professional journals and has contributed to numerous chapters on gold and precious metal flotation.

Dunne is a member of SME, AusIMM, and the Southern African Institute of Mining and Metallurgy. He has been a reviewer for *Minerals Engineering*; *Mineral Processing and Extractive Metallurgy: Transactions of the Institutions of Mining and Metallurgy*; and for research applications for the Natural Sciences and Engineering Research Council of Canada. He has also represented the mining industry on several committees, including the Chamber of Minerals and Energy of Western Australia, and as a board member of JKMRC and a panel member of the Gold Advisory Group at the A.J. Parker Cooperative Research Centre for Hydrometallurgy at Murdoch University in Western Australia.

Although retired, Dunne provides consulting services through his metallurgical consulting company and enjoys mentoring young engineers and advancing new ideas and technologies within the mining industry. Spending more time with family, sons Kevin and James and daughter Tracy, and his five grandchildren, is a favorite pastime, as is traveling the world with his wife Deidre.

S. Komar Kawatra

S. Komar Kawatra is a professor of chemical engineering at Michigan Technological University in the United States. Kawatra's primary research focuses on analysis, control, and optimization of mineral processing operations and the treatment and utilization of industrial waste materials. He obtained an MS degree in physics from the former University of Poona in India and a PhD in metallurgy from the University of Queensland in Brisbane, Queensland, Australia. His doctoral work was carried out at Mount Isa Mines in Mount Isa, Australia.

Kawatra's research experience includes work at the Bhabha Atomic Research Centre in Trombay, Bombay, India; the Julius Kruttschnitt Mineral Research Centre in Brisbane; and the Canada Centre for Mineral and Energy Technology in Ottawa, Ontario, Canada. He was also a research associate at the University of Alberta in Edmonton, Canada. He moved to Michigan Technological University in 1977 and became chair of the Department of Chemical Engineering from 2007 to 2017 and, during this time, established several endowments. Kawatra has been continuously active in research projects of academic, industrial, and governmental interest and has received several patents for technologies stemming from his research.

Since 1997, Kawatra has been editor in chief of the Society for Mining, Metallurgy & Exploration's (SME's) *Minerals & Metallurgical Processing* journal. Since 2001, he has also been editor in chief of *Mineral Processing and Extractive Metallurgy Review*, an international journal published by Taylor and Francis. He has been invited to give many keynote addresses and plenary lectures on the future of mineral processing and extraction, and has also authored or edited nine books and more than 200 peer-reviewed publications.

Kawatra has received several SME awards, including the Distinguished Member Award, the Arthur F. Taggart Award for advances in coal flotation technology, and the Antoine M. Gaudin Award for sustainable by-product management treatment and utilization. He has also received an SME Presidential Citation in honor of his outstanding contributions to SME.

Kawatra has received prestigious awards from several other organizations, including the Robert H. Richards Award and the Frank F. Aplan Award from the American Institute of Mining, Metallurgical, and Petroleum Engineers; the Michigan Technological University Research Award and Graduate Student Mentor Award; and the IEEE Certificate of Service from the Institute of Electrical and Electronics Engineers. He has also received the Michigan Association of Governing Boards Distinguished Faculty Member Award for extraordinary contribution to Michigan higher education. Kawatra is a Fulbright Scholar, having completed the Fulbright Scholarship Program in 2012.

Kawatra is grateful to his graduate students and Leeyana Gupta for their assistance during the course of his career.

Courtney A. Young

Courtney A. Young is the Department Head and Lewis S. Prater Distinguished Professor of Metallurgical and Materials Engineering at Montana Technological University (Montana Tech). As a faculty member there for the past 25 years, Young has taught various topics centered around mineral processing and extractive metallurgy to more than 2,000 students. He has also worked with numerous local and international companies doing research in these areas, focusing on mining sustainability issues such as water remediation, slag recycling, tailings repurposing, and waste treatment.

Young is a graduate of three premier mineral processing and extractive metallurgy institutions, having obtained his BS degree in mineral processing engineering from the former Montana College of Mineral Science and Technology in 1984, his MS degree in mining and minerals engineering from Virginia Polytechnic Institute and State University in 1987, and his PhD degree in metallurgical engineering from the University of Utah's College of Mines and Earth Sciences in 1994.

Young is extremely active in the Society for Mining, Metallurgy & Exploration (SME), having served as a co-advisor to its local student chapter since 1995 and on at least 30 committees, including 18 as chair or co-chair. He has organized and edited seven symposia and their corresponding proceedings, chaired and organized 15 sessions at SME meetings, and presented more than 30 papers, with about half being published in SME proceedings or the *Minerals & Metallurgical Processing* journal. He has co-taught an extremely popular short course in mineral processing and extractive metallurgy with his good friend, Corby Anderson. The course will continue to be offered annually at the SME Annual Conference and Expo.

Young has authored and co-authored 213 publications and presentations. Because of his devoted efforts to this handbook, SME, and the mining industry in general, Young has been accordingly recognized with plenary lectures and has received many awards, including the Outstanding Young Engineer Award and Millman of Distinction Award from the Mineral & Metallurgical Processing Division of SME; the Frank F. Aplan Award and Mineral Industry Education Award from the American Institute of Mining, Metallurgical, and Petroleum Engineers; and two Presidential Citations and the Distinguished Member Award from SME. His and others' research at Montana Tech have led to the establishment of a Materials Science PhD program, a first for Montana Tech, with materials being specifically defined to include minerals and metals.

Young acknowledges that none of this would have happened without the support of his wife, Miriam, of 30 years and their daughters, Jessica and Jamie.

Contributing Authors

Russell D. Alley
President
Minerals Advisory Group LLC
Tucson, Arizona, USA

Peter Amelunxen
President
Aminpro
Lima, Peru

Basak Anameric
Chief Scientist
Basak Anameric Consulting
Grand Rapids, Minnesota, USA

Corby G. Anderson
Harrison Western Professor
Dept. of Metallurgical & Materials Engineering
Colorado School of Mines
Golden, Colorado, USA

Josh Andres
Senior Metallurgist
Freeport-McMoRan Inc.
Morenci, Arizona, USA

Ian Arbuthnot
Retired
Outotec Pty Ltd.
Perth, Western Australia, Australia

Esau Arinaitwe
Group Leader
Solvay Technology Solutions
Stamford, Connecticut, USA

Jari Aromaa
Senior Lecturer
Aalto University
Espoo, Finland

Kevin Ausburn
Chief Mineralogist
FLSmidth USA Inc.
Midvale, Utah, USA

Mark G. Aylmore
Senior Research Fellow
John de Laeter Centre, Faculty of Science & Engineering
Curtin University
Perth, Western Australia, Australia

Manfred Bach
Senior Sales Manager Alumina
FLSmidth Wiesbaden GmbH
Walluf, Hessen, Germany

Andrew Bamber
Principal and Managing Director
Bara Consulting Ltd.
London, United Kingdom

Arthur Barnes
President
Metallurgical Process Consultants Ltd.
Kamloops, British Columbia, Canada

Osvaldo Bascur
Principal, Academia–Industry Innovation
OSIsoft LLC
Houston, Texas, USA

Thomas P. Battle
Extractive Metallurgy Consultant
Self-Employed
Charlotte, North Carolina, USA

Wolfgang Baum
Managing Director
Ore & Plant Mineralogy LLC
San Diego, California, USA

Richard (Ted) Bearman
Director
Bear Rock Solutions Pty Ltd.
Attadale, Western Australia, Australia

Megan Becker
Associate Professor
Dept. of Chemical Engineering, University of Cape Town
Cape Town, South Africa

Jose A. Botin
Retired Professor
Universidad Politecnica de Madrid
Madrid, Spain

Sylvie C. Bouffard
Manager, Studies Optimization
BHP
Saskatoon, Saskatchewan, Canada

David J. Bowman
Principal Engineer
Bear Rock Solutions Pty Ltd.
Attadale, Western Australia, Australia

Dee Bradshaw
Emeritus Professor
Dept. of Chemical Engineering, University of Cape Town
Cape Town, South Africa

Corale L. Brierley
President/Principal
Brierley Consultancy LLC
Highlands Ranch, Colorado, USA

James A. Brierley
Biohydrometallurgy Consultant
Brierley Consultancy LLC
Highlands Ranch, Colorado, USA

Bill Burton
Chief Engineer, Precious Metals Recovery
FLSmidth Salt Lake City Inc.
Salt Lake City, Utah, USA

Frank Cappuccitti
President
Flottec LLC
Boonton, New Jersey, USA

Jannette L. Chorney
Adjunct Professor
Montana Tech
Butte, Montana, USA

William M. Cross
Professor
Materials & Metallurgical Engineering
South Dakota School of Mines & Technology
Rapid City, South Dakota, USA

Frank Crundwell
Director
CM Solutions Pty Ltd.
Johannesburg, South Africa

Jan Czarnecki
Adjunct Professor
University of Alberta
Edmonton, Alberta, Canada

Michael Daniel
Consulting Metallurgist & Director
CMD Consulting Pty Ltd.
Brisbane, Queensland, Australia

Avimanyu Das
Associate Professor, Metallurgical & Materials Engineering
Montana Tech
Butte, Montana, USA

Dean M. David
Technical Director–Process
Mining & Minerals Australia Division, Wood
Perth, Western Australia, Australia

Alex de Andrade
General Manager Equipment & Technology
Mineral Technologies Pty Ltd.
Gold Coast, Queensland, Australia

Jerome P. Downey
Professor of Extractive Metallurgy
Montana Tech
Butte, Montana, USA

Michael J. Dry
Owner
Arithmetek Inc.
Peterborough, Ontario, Canada

Martin du Plessis
Lead Mechanical Engineer
Fluor
Perth, Western Australia, Australia

Kristy-Ann Duffy
Process Consultant
Consulting & Technology, Mining & Minerals Processing
Hatch
Brisbane, Queensland, Australia

Robert C. Dunne
Principal
Robert Dunne Consulting
Gooseberry Hill, Western Australia, Australia

Timothy C. Eisele
Assistant Professor, Department of Chemical Engineering
Michigan Technological University
Houghton, Michigan, USA

Cathy Evans
Senior Research Fellow
Julius Kruttschnitt Mineral Research Centre
Brisbane, Queensland, Australia

Daryl Evans
Director of Metallurgy
IMO Pty Ltd.
Perth, Western Australia, Australia

Ken Evans
Formerly with Rio Tinto Alcan (Retired)
Evans Technical Consulting Limited
Chalfont St. Peter, United Kingdom

Steven C. Evans
Principal Engineer (Retired)
Westinghouse Electric Company, Western Zirconium Plant
Ogden, Utah, USA

Raymond S. Farinato
Senior Research Fellow
Solvay Technology Solutions
Stamford, Connecticut, USA

Brian Flintoff
Consultant
Sigmation Inc.
Kelowna, British Columbia, Canada

Olof Forsén
Professor Emeritus
Aalto University
Espoo, Finland

Kevin S. Fraser
Co-Director & Principal Metallurgist
High-Pressure Metallurgy, Metals, Autoclave Technology
Hatch Ltd.
Mississauga, Ontario, Canada

Michael L. Free
Professor
University of Utah
Salt Lake City, Utah, USA

Rudi Frischmuth
Senior Metallurgist
High-Pressure Metallurgy, Metals, Autoclave Technology
Hatch Ltd.
Mississauga, Ontario, Canada

Francis H. (Sam) Froes
Consultant
Tacoma, Washington, USA

Xu Gao
Assistant Professor
Institute of Multidisciplinary Research for Adv. Materials
Tohoku University
Sendai, Miyagi, Japan

Mike Garska
Senior Process Engineer
Hudson Ranch Energy Service
Calipatria, California, USA

David B. George
Principal Consultant
David B. George and Associates LLC
Heber City, Utah, USA

David George-Kennedy
Chief Advisor, Smelting & Refining
Rio Tinto Copper & Diamonds
Salt Lake City, Utah, USA

Mike Germain
Process Metallurgist
Mineral Technologies Pty Ltd.
Gold Coast, Queensland, Australia

Andrea R. Gerson
Professor & Managing Director
Blue Minerals Consultancy
Middleton, South Australia, Australia

Aidan Giblett
Senior Technical Advisor—Mineral Processing
Newmont Mining Services
Subiaco, Western Australia, Australia

J.R. Goode
Principal & Consulting Metallurgist
J.R. Goode & Associates
Toronto, Ontario, Canada

Nolan Goodweiler
Senior Metallurgist
Climax Molybdenum Company
Parshall, Colorado, USA

Muxing Guo
Senior Researcher
Department of Materials Engineering
Katholieke Universiteit Leuven
Leuven, Belgium

Kenneth N. Han
Professor Emeritus
South Dakota School of Mines & Technology
Rapid City, South Dakota, USA

Greg Harbort
Technical Director—Process
Wood plc
Brisbane, Queensland, Australia

David Harbottle
Associate Professor
School of Chemical & Process Engineering
University of Leeds
West Yorkshire, United Kingdom

Daniel Harris
Consultant & Professional Board Director
EnergySource LLC
Reno, Nevada, USA

Brian R. Hart
Adjunct Professor & Senior Research Scientist
Univ. of Western Ontario & Surface Science Western
London, Ontario, Canada

Howard Haselhuhn
Applications Engineer–Industrial Minerals
Solvay Technology Solutions
Stamford, Connecticut, USA

Nick Hazen
President
Hazen Research Inc.
Golden, Colorado, USA

Jason Heath
Senior Metallurgist
Talison Lithium
Greenbushes, Western Australia, Australia

Paul J. Henkels
Senior Technical Manager
United States Gypsum Company
Chicago, Illinois, USA

Brandt Henriksson
Vice President Thickeners & Clarifiers
Outotec OY
Espoo, Finland

Leonard Hill
Technical Services Director
Freeport-McMoRan Inc.
Phoenix, Arizona, USA

J. Brent Hiskey
Professor Emeritus, Mining & Geological Engineering
University of Arizona
Tucson, Arizona, USA

Lauri Holappa
Emeritus Professor
Aalto University
Espoo, Finland

John Michael Holden
Materials Handling Engineer
Retired
Littleton, Colorado, USA

John Hollow
Owner
CRCW Consulting & Management Services
Castle Rock, Colorado, USA

Rick Q. Honaker
Professor
Mining Engineering, University of Kentucky
Lexington, Kentucky, USA

Hsin-Hsiung Huang
Professor
Department of Metallurgical & Materials Engineering
Montana Tech
Butte, Montana, USA

Peter Huxtable
Principal
Huxtable Associates
Calver, Derbyshire, United Kingdom

M. Ashraf Imam
Faculty/Research Professor
George Washington University
Washington, D.C., USA

Dusty Jacobson
Senior Cushing & Screening Specialist
Metso Minerals Industries Inc.
Waukesha, Wisconsin, USA

Alex Jankovic
Principal Consultant
Consulting & Technology, Mining & Minerals Processing
Hatch
Brisbane, Queensland, Australia

Matthew Jeffrey
Process Manager
Newmont USA Limited
Englewood, Colorado, USA

Bill Johnson
Principal Consultant & Adjunct Professor
Mineralis Consultants Pty Ltd. & University of Queensland
Brisbane, Queensland, Australia

Adam Johnston
Chief Metallurgist
Transmin Metallurgical Consultants
Lima, Peru

Myungwon Jung
Postdoctoral Researcher
Metal Processing Institute, Worcester Polytechnic Institute
Worcester, Massachusetts, USA

Sarma S. Kanchibotla
Professor
JKMRC, Sustainable Minerals Institute
University of Queensland
Brisbane, Queensland, Australia

S. Komar Kawatra
Professor & Chair, Department of Chemical Engineering
Michigan Technological University
Houghton, Michigan, USA

Jon J. Kellar
Nucor Professor & Douglas W. Fuerstenau Professor
South Dakota School of Mines & Technology
Rapid City, South Dakota, USA

Harold E. Kelley
Director of Technical Services
TungsteMet
Oak Hill, West Virginia, USA

Brandon Kern
Technical Service Specialist, Ion Exchange
The Dow Chemical Company
Midland, Michigan, USA

Bern Klein
Professor, Norman B. Keevil Institute of Mining Engineering
University of British Columbia
Vancouver, British Columbia, Canada

Brian Knorr
Director, Research & Development
Metso Minerals Industries Inc.
York, Pennsylvania, USA

Jaisen N. Kohmuench
Managing Director
Eriez Manufacturing Pty Ltd.
Melbourne, Victoria, Australia

Stacy Kramer
Vice President Global H&S
Freeport-McMoRan Inc.
Phoenix, Arizona, USA

Tom Krumins
Process Engineer
High-Pressure Metallurgy, Metals, Autoclave Technology
Hatch Ltd.
Mississauga, Ontario, Canada

Martin C. Kuhn
Chief Executive Officer
Minerals Advisory Group LLC
Tucson, Arizona, USA

Halvor Kvande
Professor Emeritus
Norwegian University of Science & Technology
Trondheim, Norway

Richard LaDouceur
Postdoctoral Research Scholar
Montana Tech
Butte, Montana, USA

Sai Wei Lam
Senior Process Engineer
Wood
Perth, Western Australia, Australia

Tim Laros
Owner
Filtration Technologies LLC
Park City, Utah, USA

Mike Larson
Applications Engineering Manager
Moly-Cop USA
Ewen, Michigan, USA

Marko Latva-Kokko
R&D Manager—Hydromet Equipment
Outotec
Pori, Finland

Dariusz Lelinski
Global Product Manager of Flotation
FLSmidth Minerals
Salt Lake City, Utah, USA

Meg Dietrich LeVier
Consulting Analytical Chemist
K. Marc LeVier & Associates Inc.
Highlands Ranch, Colorado, USA

Alison Lewis
Dean, Faculty of Engineering & Built Environment
Dept. of Chemical Engineering, University of Cape Town
Cape Town, South Africa

Leendert (Leon) Lorenzen
Managing Director (Professor)
Lorenzen Consultants, Mintrex (Stellenbosch Univ.)
Perth, Australia (Stellenbosch, South Africa)

Norman O. Lotter
President
Flowsheets Metallurgical Consulting Inc.
Sudbury, Ontario, Canada

David Love
Principal Consultant & Director
Water Logics
Adelaide, Australia

Mari Lundström
Professor
Aalto University
Espoo, Finland

John Lupo
Senior Director Geotechnical and Hydrology
Newport Mining Corporation
Denver, Colorado, USA

Gerald H. Luttrell
E. Morgan Massey Professor
Mining & Minerals Engineering, Virginia Tech
Blacksburg, Virginia, USA

David C. Lynch
Professor Emeritus
University of Arizona
Tucson, Arizona, USA

Phillip J. Mackey
President
P.J. Mackey Technology Inc.
Kirkland, Quebec, Canada

Deepak Malhotra
President
Resource Development Inc.
Wheat Ridge, Colorado, USA

Michael J. Mankosa
Executive Vice President of Global Technology
Eriez Manufacturing
Erie, Pennsylvania, USA

John G. Mansanti
President and CEO
Crystal Peak Minerals Inc.
Salt Lake City, Utah, USA

John O. Marsden
President & Consultant
Metallurgium
Phoenix, Arizona, USA

Jacob Masliyah
Professor Emeritus
University of Alberta
Edmonton, Alberta, Canada

Hiroyuki Matsuura
Associate Professor
Department of Materials Engineering, University of Tokyo
Tokyo, Japan

Edward F. McCarthy
Manager
Performance Minerals LLC
Morgan Hill, California, USA

Ed McGowan
Safety Manager
Kinross Gold Mining (Bald Mountain) Inc.
Elko, Nevada, USA

Terry McNulty
President
T.P. McNulty and Associates Inc.
Tucson, Arizona, USA

David G. Meadows
Manager of Global Process Technology
Bechtel Mining and Metals
Phoenix, Arizona, USA

Glenn C. Miller
Professor
Dept. of Natural Resources & Environmental Science
University of Nevada–Reno
Reno, Nevada, USA

Paul Miller
Managing Director
Sulphide Resource Processing Pty Ltd.
Perth, Western Australia, Australia

Brajendra Mishra
Director of Metal Processing Institute
Mechanical Engineering, Worcester Polytechnic Institute
Worcester, Massachusetts, USA

Michael S. Moats
Associate Professor
Missouri University of Science & Technology
Rolla, Missouri, USA

Stephen Morrell
President
SMC Testing Pty Ltd.
Brisbane, Queensland, Australia

Robert D. Morrison
Former Chief Technologist
Julius Kruttschnitt Mineral Research Centre
University of Queensland
Brisbane, Queensland, Australia

Trond Muri
Senior Project Engineer
PASSER
Vear, Vestfold, Norway

D.R. Nagaraj
Principal Research Fellow
Solvay Technology Solutions
Stamford, Connecticut, USA

Michael G. Nelson
Professor & Chair, Mining Engineering Department
University of Utah
Salt Lake City, Utah, USA

Khosrow Nikkhah
Senior Process Engineer
AMEC Foster Wheeler
Vancouver, British Columbia, Canada

Aaron Noble
Associate Professor
Mining & Minerals Engineering, Virginia Tech
Blacksburg, Virginia, USA

Stefan Norgaard
Operations Manager
Australian Laboratory Services
Australasian Institute of Mining and Metallurgy
Perth, Western Australia, Australia

Daniel A. Norrgran
Former Manager Minerals & Materials Processing Division
Eriez Manufacturing
Erie, Pennsylvania, USA

Ricardo Maerschner Ogawa
Product Manager, Mining Screens
Metso Brasil Indústria e Comércio
Sorocaba, Brazil

Joohyun Park
Professor
Department of Materials & Chemical Engineering
Hanyang University
Ansan, Korea

Bruce Peachey
President & Senior Consultant
New Paradigm Engineering Ltd.
Edmonton, Alberta, Canada

Murray Pearson
Director, Technology Development
High-Pressure Metallurgy, Metals, Autoclave Technology
Hatch Ltd.
Mississauga, Ontario, Canada

P. Chris Pistorius
POSCO Professor of Materials Science & Engineering
Carnegie Mellon University
Pittsburgh, Pennsylvania, USA

Graham Popplewell
Technical Director & Senior Fellow, Diamond Processing
Fluor Corporation
Perth, Western Australia, Australia

Joe Poveromo
Raw Materials & Ironmaking Global Consulting
Self-Employed
Bethlehem, Pennsylvania, USA

Robert J. Pruett
Formerly with Imersys Oilfield Solutions
Sandersville, Georgia, USA

Brian Putland
Principal Metallurgist
Orway Mineral Consultants
Perth, Western Australia, Australia

Randall Pyper
General Manager
Kappes, Cassiday & Associates Australia Pty Ltd.
Perth, Western Australia, Australia

Benny E. Raahauge
General Manager, Alumina & Pyro Technology
FLSmidth
Copenhagen, Denmark

Shashi Rao
Metallurgical Engineer
Natural Resources Research Institute, Coleraine Labs
University of Minnesota
Coleraine, Minnesota, USA

Amy J. Richins
Research Associate, Mining Engineering Department
University of Utah
Salt Lake City, Utah, USA

Tiago Ramos Riebeiro
Researcher
Laboratory of Metallurgical Processes
IPT (Institute for Technological Research)
São Paulo, Brazil

S. Jayson Ripke
Manager–R&D
Midrex Technologies Inc.
Pineville, North Carolina, USA

Joanna Robertson
Director of Metal Recovery
Freeport-McMoRan Technology Center
Tucson, Arizona, USA

David Rohaus
(Retired)
United States Steel Corporation
Pittsburgh, Pennsylvania, USA

Kym Runge
Program Leader, Separation
JKMRC, Sustainable Minerals Institute
University of Queensland
Brisbane, Queensland, Australia

Rackel San Nicolas
Lecturer
Department of Infrastructure Engineering
The University of Melbourne
Parkville, Victoria, Australia

Vicki J. Scharnhorst
Operations Manager
Tetra Tech Inc.
Denver, Colorado, USA

Mark E. Schlesinger
Professor of Metallurgical Engineering
Missouri University of Science & Technology
Rolla, Missouri, USA

Christopher Schmitz
Chief Mine Engineer
Climax Molybdenum Company
Leadville, Colorado, USA

Henry Schnell
Retired
Formerly with AREVA (Paris, France)
Eagle Bay, British Columbia, Canada

Fred Schoenbrunn
Director for Thickeners
FLSmidth
Salt Lake City, Utah, USA

Rebecca Sciberras
Lead Metallurgist
Orway Mineral Consultants
Perth, Western Australia, Australia

Andrew Scogings
Principal Consultant
CSA Global Pty Ltd.
Perth, Western Australia, Australia

Thom Seal
Director of the Institute for Mineral Resource Studies
Mining & Metallurgical Eng. Dept.
University of Nevada–Reno
Reno, Nevada, USA

Roger D. Sharpe
Director, GeoTechnical & Mining Services
United States Gypsum Company
Chicago, Illinois, USA

Andreas Siegmund
Principal
LanMetCon LLC
Lantana, Texas, USA

Roger St.C. Smart
Emeritus Professor & Senior Consultant
Univ. of South Australia & Blue Minerals Consultancy
Mawson Lakes, South Australia, Australia

Stuart Smith
Technical Director
Metifex Pty Ltd.
Brisbane, Queensland, Australia

Matthew Soderstrom
Business Director Metal Extraction Products
Solvay S.A.
Tempe, Arizona, USA

Peter-Hans ter Weer
Director
TWS Services and Advice Bauxite & Alumina Consultancy
Huizen, The Netherlands

Daniel Thomas
Principal Consultant, Alumina
Advisian, WorleyParsons Group
Brisbane, Queensland, Australia

Philip Thompson
Senior Minerals Processing Advisor
FLSmidth
Salt Lake City, Utah, USA

Larry G. Twidwell
(Retired) Emeritus Professor of Metallurgical Engineering
Metallurgy/Materials Engineering Dept., Montana Tech
Butte, Montana, USA

John L. Uhrie
President Consulting Services–Americas
RPM Global USA Inc.
Greenwood Village, Colorado, USA

Walter Valery
Global Director
Consulting & Technology, Mining & Minerals Processing
Hatch
Brisbane, Queensland, Australia

Jannie S. J. van Deventer
Chief Executive Officer
Zeobond Pty Ltd.
Melbourne, Victoria, Australia

Edgar E. Vidal
Director of Business Dev. for Defense, Nuclear & Science
Materion Beryllium & Composites
Mayfield Heights, Ohio, USA

Judith C. Vidal
Building Energy Science Group Manager
National Renewable Energy Laboratory
Golden, Colorado, USA

Brant Walkley
Research Associate
The University of Sheffield
Sheffield, United Kingdom

Henry Walqui
Director, Field Pilot Plant Programs
CiDRA Minerals Processing
Windsor, Connecticut, USA

Chengtie Wang
PhD Candidate
University of British Columbia
Vancouver, British Columbia, Canada

Barry Welch
Professor Emeritus
University of Auckland
Auckland, New Zealand

Nicholas J. Welham
Principal
Welham Metallurgical Services
Perth, Western Australia, Australia

Jessica M. Wempen
Assistant Professor of Mining Engineering
University of Utah
Salt Lake City, Utah, USA

Greg Wilkie
Program Coordinator
CRC Ore
Brisbane, Queensland, Australia

Richard Williams
Regional Manager Western Australia
McLanahan Corporation
Perth, Western Australia, Australia

Gary R. Wilson
Graduate Student
Department of Metallurgical & Materials Engineering
Montana Tech
Butte, Montana, USA

David Wiseman
Principal
David Wiseman Pty Ltd.
Blackwood, South Australia, Australia

Zhenghe Xu
Tech Professor (Dean)
Univ. of Alberta (Southern Univ. of Science & Technology)
Edmonton, Alberta, Canada (Shenzhen, China)

Jorge L. Yordan Hernandez
Retired
Formerly with Rio Tinto Minerals
Parker, Colorado, USA

Courtney A. Young
Department Head & Professor
Metallurgical & Materials Engineering
Montana Tech
Butte, Montana, USA

Patrick Zhang
Research Director
Florida Industrial & Phosphate Research Institute
Bartow, Florida, USA

Technical Reviewers

Jeffrey F. Adams
Specialist Process Engineer, Hydrometallurgy
Hatch Ltd.
Mississauga, Ontario, Canada

John E. Angove
Principal
AFT Metallurgy
North Beach, Western Australia, Australia

Dan Apai
Senior Civil Engineer
Fluor Canada
Vancouver, British Columbia, Canada

Tony Bagshaw
Retired
Formerly with Minerals Research Institute of WA
Nedlands, Western Australia, Australia

Amanda S. Barnard
Chief Research Scientist, Data61
CSIRO
Docklands, Victoria, Australia

Thomas Battle
Ext. Metallurgy Consultant & Senior Consulting Engineer
Kingston Process Metallurgy
Charlotte, North Carolina, USA

Richard (Ted) Bearman
Director
Bear Rock Solutions Pty Ltd.
Attadale, Western Australia, Australia

Gavin Becker
Principal
Gavin Becker & Associates
Ormiston, Queensland, Australia

Sylvie C. Bouffard
Manager, Studies Optimization
BHP
Saskatoon, Saskatchewan, Canada

Quentin Campbell
Professor, School of Chemical and Minerals Engineering
North-West University
Potchefstroom, South Africa

John A. Cole
Senior Director of Process Development
Newmont Mining Corporation
Englewood, Colorado, USA

Rob Coleman
Account Director
Outotec
Brisbane, Queensland, Australia

Avimanyu Das
Associate Professor, Metallurgical & Materials Engineering
Montana Tech
Butte, Montana, USA

Dean M. David
Technical Director–Process
Mining and Minerals Australia Division, Wood
Perth, Western Australia, Australia

Emmanuel De Moor
Assistant Professor
Department of Metallurgical & Materials Engineering
Colorado School of Mines
Golden, Colorado, USA

Carl E. Defilippi
Engineering Manager
Kappes, Cassiday & Associates
Reno, Nevada, USA

Jaroslaw W. Drelich
Professor, Department of Materials Science & Engineering
Michigan Technological University
Houghton, Michigan, USA

David C. Duffy
Mining Engineer
ACG Materials
Norman, Oklahoma, USA

Robert C. Dunne
Principal
Robert Dunne Consulting
Gooseberry Hill, Western Australia, Australia

Akbar Farzanegan
Professor, School of Mining, College of Engineering
University of Tehran
Tehran, Iran

Brian Flinthoff
Consultant
Sigmation Inc.
Kelowna, British Columbia, Canada

Robert J. Fraser
Global Director, Hydrometallurgy
Hatch Ltd.
Mississauga, Ontario, Canada

Mike Garska
Senior Process Engineer
Hudson Ranch Energy Service
Calipatria, California, USA

Andrea R. Gerson
Professor & Managing Director
Blue Minerals Consultancy
Middleton, South Australia, Australia

Aidan Giblett
Senior Technical Advisor—Mineral Processing
Newmont Mining Services
Subiaco, Western Australia, Australia

Rick S. Gilbert
Vice President Process Technology
Freeport-McMoRan Inc.
Phoenix, Arizona, USA

Eric J. Grimsey
Emeritus Professor, Minerals Engineering
Curtin University
Perth, Western Australia, Australia

Fathi Habashi
Professor Emeritus of Extractive Metallurgy
Laval University
Quebec City, Quebec, Canada

Ian Harmsworth
Engineering Manager
Nexus Engineering Services Pty Ltd.
Brisbane, Queensland, Australia

Steven D. Hart
Principal Advisor Processing
Newmont Mining Corporation
Perth, Western Australia, Australia

Steve Hearn
Senior Staff Engineer
Huntsman
Centennial, Colorado, USA

J. Brent Hiskey
Professor Emeritus
Department of Mining & Geological Engineering
University of Arizona
Tucson, Arizona, USA

Kirsty Hollis
Process Manager
Phu Bia Mining
Vientiane, Laos

Ralph Holmes
Honorary Fellow
CSIRO Mineral Resources Flagship
Melbourne, Victoria, Australia

Matthew Jeffrey
Process Manager
Newmont USA Limited
Englewood, Colorado, USA

Bill Johnson
Principal Consultant & Adjunct Professor
Mineralis Consultants Pty Ltd. & University of Queensland
Brisbane, Queensland, Australia

Adam Johnston
Chief Metallurgist
Transmin Metallurgical Consultants
Lima, Peru

Richard H. Johnson
Principal
JohnsonMCS LLC
Bellaire, Texas, USA

Ronel Kappes
Process Manager
Newmont Mining Corporation
Englewood, Colorado, USA

S. Komar Kawatra
Professor & Chair, Department of Chemical Engineering
Michigan Technological University
Houghton, Michigan, USA

Jon J. Kellar
Nucor Professor & Douglas W. Fuerstenau Professor
South Dakota School of Mines & Technology
Rapid City, South Dakota, USA

Harold E. Kelley
Director of Technical Services
TungsteMet
Oak Hill, West Virginia, USA

Bern Klein
Professor, Norman B. Keevil Institute of Mining Engineering
University of British Columbia
Vancouver, British Columbia, Canada

Gary Kordosky
(Retired) Consultant in Hydrometallurgy & Solvent Extraction
Formerly with Cognis
Tucson, Arizona, USA

Halvor Kvande
Professor Emeritus
Norwegian University of Science & Technology
Trondheim, Norway

James Kyle
Professor
Murdoch University
Murdoch, Western Australia, Australia

Luis M. La Torre
Principal Metallurgist
Transmin Metallurgical Consultants
Lima, Peru

Jaeheon Lee
Associate Professor
Department of Mining & Geological Engineering
University of Arizona
Tucson, Arizona, USA

Alison Lewis
Dean, Faculty of Engineering & Built Environment
Dept. of Chemical Engineering, University of Cape Town
Cape Town, South Africa

Jens Lichter
Head of Comminution
Anglo American
Denver, Colorado, USA

David McCallum
Senior Experimental Scientist
CSIRO Mineral Resources
Clayton, Victoria, Australia

Robbie McDonald
Sr. Research Scientist, Pressure Hydrometallurgy
Precious & Base Metal Hydrometallurgy
CSIRO Mineral Resources
Karawara, Western Australia, Australia

Thomas F. McIntyre
Consulting Engineer
REC Silicon
Butte, Montana, USA

Terry McNulty
President
T.P. McNulty and Associates Inc.
Tucson, Arizona, USA

James P. Metsa
Director of Sales & Technical Support
Weir Minerals
Madison, Wisconsin, USA

Paul Miller
Managing Director
Sulphide Resource Processing Pty Ltd.
Perth, Western Australia, Australia

Michael S. Moats
Associate Professor
Missouri University of Science & Technology
Rolla, Missouri, USA

Robert D. Morrison
Former Chief Technologist
Julius Kruttschnitt Mineral Research Centre
University of Queensland
Brisbane, Queensland, Australia

Brij Moudgil
Distinguished Professor & Alumni Professor
Materials Science & Engineering
University of Florida
Gainesville, Florida, USA

D.R. Nagaraj
Principal Research Fellow
Solvay Technology Solutions
Stamford, Connecticut, USA

Andrew Neale
Executive Director–Technology
PT Merdeka Copper and Gold
Jakarta, Indonesia

Michael G. Nelson
Professor & Chair, Mining Engineering Department
University of Utah
Salt Lake City, Utah, USA

Aaron Noble
Associate Professor
Mining & Minerals Engineering, Virginia Tech
Blacksburg, Virginia, USA

Jim Orlich
Senior Director Metallurgical Technology
Newmont Mining Corporation
Englewood, Colorado, USA

Jaye Pickarts
Consultant
Rare Element Resources Inc.
Littleton, Colorado, USA

P. Chris Pistorius
POSCO Professor of Materials Science & Engineering
Carnegie Mellon University
Pittsburgh, Pennsylvania, USA

Holger Plath
Vice President, Minerals
ThyssenKrupp Industrial Solutions (USA)
Atlanta, Georgia, USA

Malcolm Powell
Professorial Research Fellow
JKMRC, Sustainable Minerals Institute
University of Queensland
Brisbane, Queensland, Australia

Randall Pyper
General Manager
Kappes, Cassiday & Associates Australia Pty Ltd.
Perth, Western Australia, Australia

W. John Rankin
Honorary Fellow
CSIRO Mineral Resources
Clayton, Victoria, Australia

Samira Rashidi
Process Engineer
ThyssenKrupp Industrial Solutions (USA)
Atlanta, Georgia, USA

Timothy Reeves
Principal Hydrologist
Tetra Tech
Golden, Colorado, USA

S. Jayson Ripke
Senior Applications Engineer
Solvay
York, South Carolina, USA

Andrew Robertson
Principal
Andrew Robertson & Associates
East Freemantle, Western Australia, Australia

Dave Robinson
Technology Leader, In Situ Recovery & Processing
Mining3
Perth, Western Australia, Australia

Angus A. Rockett
Professor
Department of Metallurgical & Materials Engineering
Colorado School of Mines
Golden, Colorado, USA

Greg Roset
VP & General Manager of Recycling Operations
Stillwater Mining Company
South Columbus, Montana, USA

Kym Runge
Program Leader, Separation
JKMRC, Sustainable Minerals Institute
University of Queensland
Brisbane, Queensland, Australia

Axel Schippers
Professor, Resource Geochemistry
Federal Institute for Geosciences & Natural Resources
Hannover, Germany

Mark E. Schleisinger
Professor of Metallurgical Engineering
Missouri University of Science & Technology
Rolla, Missouri, USA

Henry Schnell
Retired
Formerly with AREVA (Paris, France)
Eagle Bay, British Columbia, Canada

Fred Schoenbrunn
Director for Thickeners
FLSmidth
Salt Lake City, Utah, USA

Scott Schuey
Process Manager, Sr. Technical Advisor
Refractory Hydrometallurgical Technology
Newmont Mining Corporation
Englewood, Colorado, USA

Robert A. Seitz
Consultant
Seitz Solutions LLC
Phoenix, Arizona, USA

Nicholaos Serpentzis
Project Director
Fluor Australia Pty Ltd.
Perth, Western Australia, Australia

Roger St.C. Smart
Emeritus Professor & Senior Consultant
Univ. of South Australia & Blue Minerals Consultancy
Mawson Lakes, South Australia, Australia

Stuart Smith
Technical Director
Metifex Pty Ltd.
Brisbane, Queensland, Australia

James G. Speight
Consultant
CD&W Inc.
Laramie, Wyoming, USA

D. Erik Spiller
Research Professor, Kroll Institute for Extractive Metallurgy
Colorado School of Mines
Golden, Colorado, USA

Regis R. Stana
President
R Squared S Inc.
Lakeland, Florida, USA

Patrick R. Taylor
Director, Kroll Institute for Extractive Metallurgy
Colorado School of Mines
Golden, Colorado, USA

Frank Trask
Technical Director
Mining & Process Solutions Pty Ltd.
Darch, Western Australia, Australia

Jack Tryon
Health, Safety & Compliance Specialist
Newmont Mining Corporation
Englewood, Colorado, USA

Larry G. Twidwell
(Retired) Professor Emeritus of Metallurgical Engineering
Metallurgy/Materials Engineering Dept., Montana Tech
Butte, Montana, USA

Maureen T. Upton
Principal
Maureen Upton Ltd.
Denver, Colorado, USA

André Vien
Senior Manager Metallurgy & Process Development
Freeport-McMoRan Mining Company
Phoenix, Arizona, USA

David Wiseman
Principal
David Wiseman Pty Ltd.
Blackwood, South Australia, Australia

Zhenghe Xu
Tech Professor (Dean)
Univ. of Alberta (Southern Univ. of Science & Technology)
Edmonton, Alberta, Canada (Shenzhen, China)

Jorge L. Yordan Hernandez
Retired
Formerly with Rio Tinto Minerals
Parker, Colorado, USA

Courtney A. Young
Department Head & Professor
Metallurgical & Materials Engineering
Montana Tech
Butte, Montana, USA

Patrick Zhang
Research Director
Florida Industrial & Phosphate Research Institute
Bartow, Florida, USA

Xia Zhang
Metallurgist
Freeport-McMoRan Copper & Gold Inc.
Tucson, Arizona, USA

PART 1

Mineral Characterization and Analysis

CHAPTER 1.1

Mineral Properties and Processing

Avimanyu Das and Courtney A. Young

Critical players employed in the mining industry include scientists and engineers involved in geology, mining, mineral processing, extractive metallurgy, environmental stewardship, and occupational safety and health. All these professionals should have overlapping knowledge about minerals and how their properties are used for both identification and separation purposes. This promotes cross-communication among all participants as required in ore-to-product optimization, a concept that is also termed *mine-to-mill* and *pit-to-plant*. By optimizing all facets of the mining industry, ores will be mapped, mined, and processed to their highest efficiency, ultimately resulting in maximum metallurgical performance and therefore recovery and concentrate grade of the valuable minerals. The more professionals know about minerals and the better they communicate with others, the greater their success will be. This is particularly true for mineral processors and extractive metallurgists.

Mineral processors and extractive metallurgists work with geologists throughout most of the exploration phase to help identify the type and amount of minerals present and how they are associated with one another. Associations can lead to the identification of other minerals, particularly if those minerals are present in amounts that are difficult to detect. To do this, mineral processors and extractive metallurgists will collect and examine hand specimens, chip samples, and drill core and cuttings from all over the mining property. Mineral identification will initially be accomplished through visual examination in the field with the naked eye, magnifying lenses, and portable light microscopes and then progress to the laboratory where high-power, ore microscopy, and various spectroscopic analyses are conducted. Accuracy normally increases with the cost of the identification technique, not only because of the analytical devices being used but also because of the equipment needed to properly prepare the materials for examination.

After the minerals are identified and quantified, mineral processors and extractive metallurgists will catalog their properties and evaluate differences to determine which can be exploited to provide the best separation. They will then study and optimize those separations in the laboratory, gather necessary data for scaling up to pilot plant or industrial-size equipment, estimate capital and operating costs, and ultimately recommend the best, most economical strategy. These studies also require mineral processors and extractive metallurgists to consider ore hardness and liberation patterns to determine the best comminution practices to minimize energy consumption and match the resulting product size to the recommended separation strategy. The recommended strategy might be the reverse of the process that formed the ore in the first place. Therefore, understanding ore genesis can aid process development, and it is another example of why communicating with geologists is critical.

While the mineral and metallurgical testing is being conducted, geologists will be mapping the ore body and working with mining engineers to determine the best way to mine and haul the ore to the process plant. Mining engineers will also be working with the mineral processors and extractive metallurgists to determine the best method for blasting so that the optimal feed size is delivered to the first stage in comminution. Ore-to-product optimization in subsequent stages will lead to appropriate particle sizes being fed to the separation process and therefore obtain maximum metallurgical performance.

Clearly, mineral identification must be established beyond a doubt so that a strategy can be developed for concentrating valuable minerals or extracting valuable metals. However, the final product also must meet the quality specified by the end user. While this implies that the primary goal is to keep waste minerals, referred to as gangue, away from valuable minerals reporting to the concentrate, it is also important to get as much valuable mineral to report to the concentrate as possible. Both actions help maximize concentrate grade and recovery and therefore lead to optimal metallurgical performance, as already mentioned. A good understanding of how differentially the valuable and gangue minerals behave in the separation process is important. However, given the set of machinery in the processing plant, it will be difficult for mineral processors and extractive metallurgists to meet the specifications

Avimanyu Das, Associate Professor, Metallurgical & Materials Engineering, Montana Tech, Butte, Montana, USA
Courtney A. Young, Department Head & Professor, Metallurgical & Materials Engineering, Montana Tech, Butte, Montana, USA

without having a thorough knowledge of minerals and their properties and being able to communicate with one another.

KEY DEFINITIONS

Minerals

Formed through biogeological processes, *minerals* are naturally occurring, predominantly inorganic solids that have specific physical and chemical properties, highly ordered atomic/crystal structures, and characteristic chemical compositions usually of high purity and homogeneity, ranging from simple elements to complex solid solutions. By comparison, *rock* is an aggregate of minerals and may be igneous, sedimentary, metamorphic, or any combination thereof. A *deposit* is an aggregate of rock. An *ore* is a deposit or a system of deposits containing at least one valuable mineral that can be separated from the other minerals and marketed for a profit.

Properties

Polymorphs are minerals that have the same chemical composition but possess different crystal structures. Because structure determines properties, the physical and chemical properties of the polymorphs will vary, albeit usually minimally. Of course, changing the chemical composition will also change the properties, but in this case the changes are likely to be significant. Measuring the various physical and chemical properties of a mineral can lead to its identification. Furthermore, if there are differences in those properties, even if those differences are small, the properties can be exploited to separate the minerals.

Processing

Processing is a series of actions taken to achieve mineral separations. Valuable minerals report to a concentrate, or con, and gangue minerals report to tailings, or tail. To achieve separation, differences among the physical or chemical properties of minerals must be exploited using physical or chemical processes. In general, *mineral processing* is the field of engineering concerned with the separation of valuable minerals from ores into concentrates usually without chemical change, while *extractive metallurgy* is the field of engineering concerned with the extraction of valuable metals from ores or concentrates usually with chemical change.

MINERAL IDENTIFICATION

Although many physical and chemical properties are used to identify minerals, typically several properties are needed to identify them definitively. The properties most commonly used include chemical composition, crystal structure, hardness, tenacity, cleavage, fracture, streak, color, luster, transparency, and refractivity. Except for chemical composition, all are physical properties typically quantified with the naked eye, a magnifying lens, or an ore microscope. Because there are more than 4,000 minerals, all these properties are important for identification purposes, but not in all cases. The properties are briefly described in the following sections. Table 1 includes a quantitative compilation for a few select minerals using information from Hurlbut and Klein (1977), Craig and Vaughan (1981), Bolles and McCullough (1985), and Thomas (2010). Other important properties are also discussed but are not included in the table.

Chemical Composition

All minerals have definite chemical compositions with specific stoichiometric ratios. A few are native elements, such as Fe, Ni, Cu, Ag, Au, S, and C as graphite and diamond; and some are alloys, including electrum (AuAg) and tetrataenite (FeNi). However, most are molecular, ranging from simple compounds to complex solid solutions. Compositions indicate the chemical analyses of the minerals and, thereby, the anions and cations that comprise the minerals. Because an anion is usually significantly larger than a cation, the anion plays a dominant role in crystal structure and which characteristics the minerals are likely to exhibit. Consequently, mineral classification systems such as Dana or Nickel–Strunz are based on crystal structure and type of anion. Because of recent modification by Mills et al. (2009), mineralogists tend to favor the latter.

Crystal Structure

Minerals have a well-defined crystal structure. However, that structure can be modified by a mineral's association with other minerals, impurities, and inclusions. Because it is not always possible to establish a mineral's crystal structure by visual examination alone, techniques such as X-ray diffraction are used. Some minerals may have poorly organized internal arrangement and will thus exhibit near-amorphous behavior rather than crystallinity. There are six major crystal structures: isometric or cubic, tetragonal, orthorhombic, monoclinic, triclinic, and hexagonal. Some classify trigonal as a seventh structure, but it is usually recognized as a division under the hexagonal structure. Depending on how the crystal structure grows, minerals generally take on one of 11 forms (i.e., monohedron, parallelohedron, dihedron, disphenoid, prism, pyramid, dipyramid, trapezohedron, scalenohedron, rhombohedron, and tetrahedron), but numerous subcategories have also been defined.

Mohs Hardness

This property is a measure of resistance to scratching abrasion. Because talc is the softest known mineral, it cannot scratch other minerals. By comparison, diamond is the hardest known mineral and cannot be scratched. Using these as extreme cases, Mohs developed a hardness classification system on a 10-point scale. The higher the number, the harder the mineral:

1	Talc	6	Orthoclase feldspar
2	Gypsum	7	Quartz
3	Calcite	8	Topaz
4	Fluorite	9	Corundum
5	Apatite	10	Diamond

While kits are available that contain these minerals, the following common objects can also be used:

2.5	Thumbnail
3	Copper penny
4.5	Iron nail
5.5	Knife blade and window glass
6.5	Steel file
7	Flint
8.5	Emery wheel/paper
9.5	Carborundum wheel/paper

Tenacity

Tenacity refers to the cohesiveness of a mineral and its resistance to breaking, bending, and deforming. A mineral is sectile if it can be cut with a knife. *Tough* means a mineral resists hammering. *Malleable* indicates it flattens when hammered. *Brittle* or *fragile* means it breaks when hammered. If much smaller forces than hammering are used and the mineral breaks into pieces, it is *friable*; however, if it breaks into powder, it is *pulverulent*. *Flexible* means it can be bent but stays bent. *Elastic* indicates the mineral returns to its original state after being bent. These and other terms are also used to describe metals when they are worked.

Cleavage

Cleavage occurs when minerals break preferentially along crystallographic planes as a result of relatively weak bonds across the planes where particular ions are located. Most classification systems have at least five categories of cleavage quality: perfect, good, poor, indistinct, and nonexistent. *Perfect cleavage* occurs when breakage is with ease; continues to be parallel to crystallographic planes; and reveals smooth, lustrous surfaces. The crystal structures and planes involved define the types of cleavage that occur. Types of crystal structures and planes include basal, cubic, octahedral, pinicoidal, rhombohedral, and prismatic.

Fracture

Fracture refers to a breakage that does not take place along defined crystal planes. There are seven types of fracture:

1. Even results in smooth, straight surfaces.
2. Conchoidal produces smooth, curved surfaces.
3. Subconchoidal yields smooth but irregular curved surfaces.
4. Uneven generates rough, irregular surfaces.
5. Hackly indicates sharp, jagged surfaces.
6. Splintery forms elongated slivers.
7. Earthy makes small pieces, often referred to as crumbles.

Streak

Streak is the color of a powdered mineral. The powder should be as fine as possible. It is made by crushing the mineral with a mortar and pestle or, more commonly, by swiping the mineral across a streak plate usually composed of unglazed porcelain; however, clean whetstone and fine-cut files can also be used. This technique requires the mineral to be softer than the streak plate or mortar and pestle. On the Mohs hardness scale, the streak plate has a hardness of about 6.5.

Color

In some instances, the *color* of the mineral is definite and clearly helps in its identification. Minerals that have a metallic luster are good examples. However, many minerals exist in a variety of colors. Variations can be caused by defects in the lattice or by the presence of impurities, leading to polymorphism and isomorphism. Some minerals possess prismatic effects in light, including play in color, opalescence, asterism, and iridescence. The degree of these effects often depends on the mineral's amount of polish and tarnish. Consequently, the same mineral can possess a range of colors.

Luster

Like color, *luster* is an optical property of a mineral. It describes how a mineral appears to reflect light and how bright the reflection is. If a mineral is transparent, it will change with its refractivity. There are 11 categories of luster: metallic, submetallic, vitreous, adamantine, resinous, silky, pearly, greasy, pitchy, waxy, and dull. Metallic lusters are opaque with high mirror-like reflectivity, whereas submetallic lusters are nearly opaque and slightly translucent. Approximately 70% of minerals have vitreous luster, which makes them look like broken glass. Adamantine minerals may be translucent but normally have a high refractivity. They display a high brilliance and shine similar to that of diamonds. Resinous minerals are also highly refractive but are dull in color, often yellow-to-brown like honey. Silky lusters are usually caused by fibrous minerals. Pearly typically refers to reflections that come from not only the surface but also the layers below. Greasy, pitchy, and waxy lusters have the appearance and often the feel of organic coatings of grease, tar, and wax, respectively. Dull lusters have little reflectivity, which is usually caused by the mineral's rough or porous surfaces.

Diaphaneity

Diaphaneity refers to a mineral's degree of transparency or the percentage of light that is transmitted through it. There are three cases of diaphaneity: transparent, translucent, and opaque. If objects can be clearly seen through a mineral, then it has *transparent diaphaneity*. *Translucent diaphaneity* occurs when objects cannot be clearly seen through a mineral. *Opaque diaphaneity* means the objects cannot be seen at all.

Refractivity

Refractivity is an optical property of a transparent mineral. It can be quantified by measuring its refractive index (n), which is defined as the ratio of the velocity of light in a vacuum (c) to the velocity of light in the mineral (v). If c is taken as unity, then n is the reciprocal of v. Thus, as n increases, v decreases. In addition, Snell's law states that the refractive index (n) is also equivalent to the ratio of the sine of the angle of incidence (θ) to the sine of the angle of refraction (α).

Other Properties

Many other properties are used to identify minerals. While not included in Table 1 for simplicity reasons, they include a variety of optical and hardness properties other than those already discussed in this chapter.

Optical

Ore microscopy can also be used in reflection mode to observe several optical properties, including but not limited to pleochroism, bireflectance, polarization colors, anisotropy, internal reflection, and, of course, reflectance. These properties are differentiated by rotating a polished mineral in the absence or presence of polarized light with or without crossed polarizers. The techniques are predominantly used on opaque minerals, but they can also be used to examine translucent minerals.

Pleochroism. With pleochroism, a mineral changes colors as it is rotated while being illuminated with plane polarized light. The polarizers are not crossed.

Table 1 Select common minerals and their major properties used for identification

Name/Chemical Composition	Mohs Hardness	Color	Luster	Streak	Crystal Structure	Refractive Index (589 nm)
Apatite $Ca_5(PO_4)_3(F,Cl,OH)$	5.0	Seagreen	Vitreous to resinous	White	Hexagonal dipyramidal	1.634–1.638
Beryl $Be_3Al_2Si_6O_{18}$	7.5–8.0	White, bluish green, greenish yellow, etc.	Vitreous to resinous	White	Hexagonal dihexagonal dipyramidal	1.564–1.602
Borax $Na_2B_4O_7{\cdot}10H_2O$	2.3	Colorless, white, grayish, bluish, greenish	Vitreous to resinous	White	Monoclinic prismatic	1.45
Calcite $CaCO_3$	3.0	White or colorless	Vitreous to pearly	White to grayish	Hexagonal hexagonal-scalenohedral	1.640–1.660, 1.486
Chalcopyrite $CuFeS_2$	3.5–4.0	Brass yellow	Metallic	Greenish black	Tetragonal scalenohedral	—
Copper Cu	2.7	Copper-red, tarnishes to brown, red, black, green	Metallic	Copper-red metallic	Isometric hexoctahedral	—
Dolomite $CaMg(CO_3)_2$	3.5–4.0	Colorless or white gray	Vitreous to pearly	White	Hexagonal rhombohedral	1.679–1.681, 1.500
Galena PbS	2.5	Silvery	Metallic	Lead-gray	Isometric hexoctahedral	—
Gibbsite $Al(OH)_3$	2.5–3.0	White, greenish, blue, gray	Vitreous to dull		Monoclinic prismatic	1.58
Gypsum $CaSO_4{\cdot}2H_2O$	2.0	Colorless to white, gray, yellowish	Vitreous to silky	White	Monoclinic prismatic	1.519–1.530
Ilmenite $FeO{\cdot}TiO_2$	5.0–6.0	Iron-black, brown	Metallic to submetallic	Black	Hexagonal rhombohedral	—
Limonite $FeO(OH){\cdot}nH_2O$	4.0–5.5	Dark brown to yellow	Vitreous to dull	Yellowish brown	Amorphous nature	2.27–2.28
Marcasite FeS_2	6.5	Pale bronze-yellow	Metallic	Grayish to brownish black	Orthorhombic dipyramidal	—
Molybdenite MoS_2	1.0–1.5	Black, silvery gray	Metallic	Greenish/bluish	Hexagonal dihexagonal dipyramidal	—
Niccolite NiAs	5.0–5.5	Pale red, white with yellowish pink	Metallic	Pale brownish	Hexagonal dihexagonal dipyramidal	—
Pyrite FeS_2	6.0–6.5	Brass-yellow	Metallic	Greenish or brownish	Isometric diploidal	—
Rutile TiO_2	6.0–6.5	Red, brown, pale yellow or blue	Metallic to adamantine	Pale brown to yellowish	Tetragonal ditetragonal dipyramidal	2.613, 2.909
Scheelite $CaWO_4$	4.5–5.0	Colorless, white, gray, brown	Vitreous to adamantine	White	Tetragonal dipyramidal	1.918–1.938
Sphalerite ZnS	3.5–4.0	Brown, yellow, red, green, black	Adamantine to resinous	Brownish to light yellow	Isometric hextetrahedral	2.369
Sylvite KCl	2.0	Colorless to white, pale gray or blue, reddish	Vitreous	White	Isometric hexoctahedral	1.4903
Talc $Mg_3Si_4O_{10}(OH)_2$	1.0	Green, brown, white gray	Pearly to greasy	White	Monoclinic triclinic	1.538–1.600
Uraninite UO_2	5.5	Steel black, brownish, pale gray	Submetallic to greasy	Brownish black	Isometric hexoctahedral	—
Wolframite $(Fe,Mn)WO_4$	4.0–4.5	Grayish to brownish black	Submetallic to resinous	Reddish brown	Monoclinic prismatic	—
Zircon $ZrSiO_4$	7.5	Reddish brown, yellow, green	Vitreous to adamantine	Uncolored	Tetragonal ditetragonal dipyramidal	1.925–2.015

Adapted from Bolles and McCullough 1985

Table 1 Select common minerals and their major properties used for identification (continued)

Tenacity	Cleavage	Fracture	Diaphaneity	Occurrence	Common Names	Special Feature
Brittle	Imperfect	Conchoidal/ uneven	Transparent to opaque	Metamorphic crystalline rocks, associated with beds of iron ore	Asparagus stone, cellophane	May be confused with beryl
Brittle	Imperfect	Conchoidal/ uneven	Transparent to subtranslucent	Granite rocks and pegmatites	Aquamarine, emerald, goshenite	May be confused with apatite
Rather brittle	Perfect	Conchoidal	Translucent to opaque	Saline lakes, beds due to evaporation of such lakes	Tincal	Co-complexing with others
Brittle	Highly perfect	Conchoidal	Transparent to opaque	Widespread constituent of sedimentary rocks and minor constituent of igneous rocks	Iceland spar, limestone	Phosphorescent
Brittle	Distinct	Uneven	Opaque	Primary veins or disseminated often with pyrite, quartz	Copper pyrites, cupropyrite	Turns magnetic on heating
Malleable and ductile	None	Hackly	Opaque	Secondary, with copper minerals near igneous rocks	Native copper	—
Brittle	Perfect	Subconchoidal	Transparent to translucent	Vein mineral or altered limestone	Pearl spar, rhomb spar, bitter spar	—
Brittle	Cubic	Even	Opaque	Veins often with pyrite, sphalerite, chalcopyrite, intrusive replacement	Galenite, lead glance, plumbago	Semiconductor
Tough	Eminent	—	Translucent	Usually with bauxite	—	—
Flexible or brittle	Eminent	Conchoidal	Transparent to opaque	Forms extensive sedimentary beds	Satin spar, alabaster, selenite	Retrograde solubility
Brittle	—	Conchoidal	Opaque	Veins near igneous rocks	Titanic iron ore, menaccanite	Paramagnetic
Brittle	None	Uneven	Opaque	Secondary iron mineral	Brown ocher, bog iron ore	Mix of hydrated iron oxides
Brittle	Poor	Uneven	Opaque	Formed near surface with galena, sphalerite, calcite, dolomite	White iron pyrites, cockscomb	Exhibits strong anisotropism
Flexible, sectile	Perfect	—	Opaque	Veins often with quartz and copper sulfides	Moly, molybdena	Feels greasy
Brittle	None	Uneven	Opaque	With sulfides and silver-arsenic minerals	Copper nickel, nickeline	Garlic odor when hot
Brittle	Indistinct	Uneven	Opaque	Primary, veins or disseminated, usually crystalline	Fool's gold, iron pyrites, mundic	Semiconductor
Brittle	Distinct	Uneven	Transparent to opaque	Frequently secondary in micas or igneous rocks; black sands	Edisonite, titanite	Among highest refractive index
Brittle	Distinct	Uneven	Transparent to translucent	Pegmatite veins or in veins associated with granite/gneiss	Tungstein, schellspath	Fluorescent under UV light
Brittle	Perfect	Conchoidal	Translucent	Often in limestone with other sulfides	Zinc blende, ruby zinc, black jack	Fluorescent and triboluminescent
Brittle	Cubic perfect	Uneven	Transparent to translucent	An evaporite	Muriate of potash, hoevelite	Optically isotropic
Sectile	Perfect	—	Subtransparent to translucent	Secondary mineral formed by alteration of nonaluminous magnesium silicates	Steatite, soapstone	Fluorescent under UV light
Brittle	—	Uneven	Opaque	Granitic pegmatites, or with ores of silver, lead, copper	Pitchblende, ulrichite	Radioactive
Brittle	Very perfect	Uneven	Opaque	In granite and pegmatite veins	Wolfram, mock-lead	Strategic mineral
Brittle	Imperfect	Concoidal	Transparent to opaque	Accessory mineral in igneous rocks	Hyacinth, azurite	Fluorescent and radioactive

Bireflectance. Bireflectance is similar to pleochroism. In this case, changes in intensity are observed as the mineral is rotated while being illuminated with plane-polarized light. The polarizers are not crossed.

Polarization colors. With polarization colors, a mineral changes colors as it is rotated while being illuminated with plane-polarized light and observed with crossed polarizers.

Anisotropy. Like bireflectance, anisotropy occurs when changes in intensity are observed as the mineral is rotated while being illuminated with plane-polarized light and observed with crossed polarizers.

Internal reflection. Internal reflections are observed when translucent minerals allow light to penetrate below the surface and reflect to the observer from cracks, crystal boundaries, cleavages, and other flaws within the crystal. The internal reflections are usually visible with plane-polarized light but are best observed with crossed polarizers.

Reflectance. Reflectance is the ratio of reflected light intensity to incident light intensity as measured with a photometer. Because some portions of the light are absorbed or transmitted, reflectance is never 100% but approaches this value with opaque minerals that have metallic lusters. Silver has the highest reflectance, near 95%. A common reference is pyrite, which has a reflectance of 55%. Reflectance can also be expressed by measuring the reflectance of a mineral, dividing by the value measured for pyrite under the same conditions, and then multiplying by 55%. If quartz (5%), magnetite (20%), galena (43%), or other common minerals with known reflectances are also present, reflectances can be simply estimated by visual inspection and comparison.

Hardness

In ore microscopy, there are three types of hardness: scratch, polishing, and micro-indentation. While they are not equivalent because they are responses to different types of forces, they are used for mineral identification.

Scratch hardness. This is Mohs hardness; however, scratches made intentionally or left behind from polishing during the sample preparation process might be observed in some minerals and not others.

Polishing hardness. During polishing, soft minerals are abraded away faster than hard minerals, producing a relief in which the harder minerals lie above the softer minerals. The Kalb-line test can be used to determine which mineral is softer if the relief is appreciable.

Micro-indentation hardness. Micro-indentation is favored over the other two methods for measuring hardness because it is more quantitative and yields linear responses to Mohs hardness in a log-log plot. Although the Vickers and Knoop techniques yield similar results, Vickers has been used more widely. When micro-indentation hardness and reflectance at a defined wavelength are plotted against one another, an identification map results. Although the map is crowded at moderately low hardness and reflectance, it is a useful tool in the identification process despite the ranges that some minerals have for both properties.

MINERAL PROCESSING

Properties that are relevant for mineral identification might not be very useful to a process engineer when the main goal is to separate the valuable mineral from the gangue. In fact, only a handful of properties can be exploited to accomplish separation on a commercial scale: specific gravity, magnetic susceptibility, electrical conductivity, and hydrophobicity. These properties are briefly described in the following sections and are compiled in Table 2 for 150 of the more common minerals using information from Carpco (2000). These properties are also used for identification purposes but were excluded from the previous discussion in this chapter because of their criticality here. However, they are discussed in detail later throughout Volume 1 of this handbook.

Specific Gravity

Density is the weight per unit volume of a substance. Specific gravity is the density ratio of the substance to water. A specific gravity of 7.5 implies that the substance is 7.5 times heavier than water of the same volume. The pycnometer is the tool most commonly used for measuring specific gravity. Its accuracy is dependent on weighing precisely and removing air, usually in the form of bubbles but sometimes in pores. The easiest way to remove air is by pulling a weak vacuum over the system. Most minerals have a definite specific gravity, but many have a range because of substitutions, impurities, and inclusions. Gravity separations are very common in mineral processing and are used to separate denser minerals from less dense minerals.

Magnetic Susceptibility

Depending on how minerals are affected by an applied magnetic field, they are generally classified as diamagnetic, paramagnetic, or ferromagnetic. When the field repels a mineral, it has a negative magnetic susceptibility and is therefore classified as diamagnetic. However, the repulsion is negligible and unlikely to be noticed. Paramagnetic minerals behave oppositely; they have positive susceptibility and thus are attracted to the magnetic field. If a paramagnetic mineral retains magnetic properties after the applied field is turned off, it is called ferromagnetic. All minerals have a magnetic susceptibility, but the applied field must be strong enough to invoke a response. Minerals that do not respond traditionally have been called nonmagnetic, although this is a misnomer. This has been corrected in Table 2 as best as possible; however, some minerals might not be diamagnetic because of substitutions, impurities, and inclusions that slightly change their composition or crystal structure. Such minerals would be very weakly paramagnetic and, similar to diamagnetic minerals, unlikely to respond to an applied field. Many minerals are paramagnetic and are normally classified as weak to strong, but only a few are ferromagnetic (e.g., magnetite and pyrrhotite).

Magnetic separations are becoming more popular in mineral processing because of the somewhat recent advent of rare earth magnets and supercooled electromagnets. Prior to approximately 1990, magnetic separations were primarily conducted with permanent magnets and electromagnets. In typical separations, the paramagnetic minerals are separated into the "mag" fraction away from the diamagnetic minerals that then report to the "nonmag" fraction. There are two other classifications of magnetic minerals: ferrimagnetic and antiferromagnetic; however, as always, they separate with the paramagnetic minerals into the mag fraction. To avoid confusion, Table 2 was not corrected in this regard.

Electrical Conductivity

Minerals are either conductors or insulators depending on their ability to transport electrical charge. When minerals are placed on a metal surface and charged, usually by passing

Table 2 Common minerals and their major properties used for identification and separation

Mineral Name	Chemical Composition	Specific Gravity	Magnetic Property*	Electrostatic Property†	Naturally Hydrophobic
Actinolite	$Ca_2(Mg,Fe)_5(Si_4O_{11})_2(OH)_2$	3.0–3.2	Paramagnetic	Insulator	No
Albite	$Na(AlSi_3O_8)$	2.6	Diamagnetic	Insulator	No
Almandine	$Fe_3Al_2(SiO_4)_3$	4.3	Paramagnetic	Insulator	No
Amphibole	$(Ca,Mg,Fe)_x(SiO_3)$	2.9–3.5	Paramagnetic	Insulator	No
Anatase	TiO_2	3.9	Diamagnetic	Conductor	No
Andalusite	Al_2SiO_5	3.2	Diamagnetic	Insulator	No
Andradite	$3CaO \cdot Fe_2O_3 \cdot 3SiO_3$	3.8	Paramagnetic	Insulator[4]	No
Anhydrite	$CaSO_4$	3.0	Diamagnetic	Insulator	No
Ankerite	$Ca(Mg,Fe)(CO_3)_2$	2.9–3.1	Paramagnetic	Insulator	No
Apatite	$Ca_5(PO_4)_3(F,Cl,OH)$	3.2	Diamagnetic	Insulator	No
Aragonite	$CaCO_3$	3.0	Diamagnetic	Insulator	No
Arsenopyrite	FeAsS	5.9–6.2	Paramagnetic[1]	Conductor	Weak
Asbestos	$Mg_3(Si_2O_5)(OH)_4$	2.4–2.5	Diamagnetic	Insulator	No
Augite	$Ca(Mg,Fe,Al)[(Si,Al)_2O_6]$	3.2–3.5	Paramagnetic	Conductor[5]	No
Azurite	$Cu_3(CO_3)_2(OH)_2$	3.8	Diamagnetic	Insulator	No
Baddeleyite	ZrO_2	5.6	Diamagnetic	Insulator	No
Barite	$BaSO_4$	4.5	Diamagnetic	Insulator	No
Bastnaesite	$(Ce,La,F)CO_3$	5.0	Paramagnetic	Insulator	No
Bauxite	$Al_2O_3 \cdot 2H_2O$	2.6	Diamagnetic	Insulator	No
Beryl	$Be_3Al_2(Si_6O_{18})$	2.7–2.8	Diamagnetic	Insulator	No
Biotite	$K(Mg,Fe)_3(Si_3AlO_{10})(OH,F)_2$	3.0–3.1	Paramagnetic	Insulator	No
Bismuth	Bi	9.8	Diamagnetic	Conductor	No
Borax	$Na_2B_4O_7 \cdot 10H_2O$	1.7	Diamagnetic	Insulator	No
Bornite	Cu_5FeS_4	4.9–5.0	Diamagnetic[2]	Conductor	Weak
Brannerite	$(UO,TiO,UO_2)(TiO_3)$	4.5-5.4	Paramagnetic	Conductor	No
Brookite	TiO_2	4.1	Diamagnetic	Conductor	No
Calcite	$CaCO_3$	2.7	Diamagnetic	Insulator	No
Carnotite	$K_2(UO_2)_2(V_2O_8) \cdot 2H_2O$	5.0	Diamagnetic	Insulator[4]	No
Cassiterite	SnO_2	7.0	Diamagnetic	Conductor	No
Celestite	$SrSO_4$	4.0	Diamagnetic	Insulator	No
Cerussite	$PbCO_3$	6.6	Diamagnetic	Insulator[4]	No
Chalcocite	Cu_2S	5.5–5.8	Diamagnetic	Conductor	Weak
Chalcopyrite	$CuFeS_2$	4.1–4.3	Diamagnetic[2]	Conductor	Weak
Chlorite	$(Mg,Al,Fe)_{12}[(Si,Al)_8O_{20}](OH)_{16}$	2.6–3.2	Paramagnetic	Insulator	No
Chromite	$(Mg,Fe)(Cr,Al)_2O_4$	4.6	Paramagnetic	Conductor	No
Chrysocolla	$CuSiO_3 \cdot nH_2O$	2.0–2.3	Diamagnetic	Insulator	No
Cinnabar	HgS	8.1	Diamagnetic	Insulator	Yes
Cobaltite	(Co,Fe)AsS	6.0–6.3	Paramagnetic	Conductor	Weak
Colemanite	$Ca_2B_6O_{11} \cdot 5H_2O$	2.4	Diamagnetic	Insulator	No
Collophanite	$Ca_3P_2O_8 \cdot H_2O$	2.6–2.9	Diamagnetic	Insulator	No
Columbite	$(Mn,Fe)(Ta,Nb)_2O_6$	5.2–8.2	Paramagnetic	Conductor	No
Copper	Cu	8.9	Diamagnetic	Conductor	No
Corundum	Al_2O_3	3.9–4.1	Diamagnetic	Insulator	No
Covellite	CuS	4.7	Diamagnetic	Conductor	Weak
Cryolite	Na_3AlF_6	3.0	Diamagnetic	Insulator[4]	No
Cuprite	Cu_2O	5.8–6.2	Diamagnetic	Insulator	No
Diamond (nat)	C	3.5	Diamagnetic	Insulator	Yes
Diamond (syn)	C	3.5	Paramagnetic	Insulator	No
Diopside	$CaMg(Si_2O_6)$	3.3–3.4	Paramagnetic[1]	Insulator	No
Dolomite	$CaMg(CO_3)_2$	1.8–2.9	Diamagnetic	Insulator	No
Epidote	$Ca_2(Al,Fe)_3(Si_3O_{12})(OH)$	3.4	Paramagnetic	Insulator	No
Euxenite	$(Y,Er,Ce,La,U)(Nb,Ti,Ta)_2(O,OH)_6U_3O_8$	4.7–5.2	Paramagnetic	Conductor	No

(continues)

Table 2 Common minerals and their major properties used for identification and separation (continued)

Mineral Name	Chemical Composition	Specific Gravity	Magnetic Property*	Electrostatic Property†	Naturally Hydrophobic
Feldspar	$(K,Na,Ca...)_x(AlSi)_3O_8$	2.6–2.8	Diamagnetic	Insulator	No
Ferberite	$FeWO_4$	7.5	Paramagnetic[3]	Conductor	No
Flint	SiO_2	2.6	Diamagnetic	Insulator	No
Fluorite	CaF_2	3.2	Diamagnetic	Insulator	No
Franklinite	$(Zn,Mn)Fe_2O_4$	5.1–5.2	Ferromagnetic	Conductor	No
Gahnite	$ZnAl_2O_4$	4.6	Diamagnetic	Insulator	No
Galena	PbS	7.5	Diamagnetic	Conductor	Weak
Garnet	Complex Ca,Mg,Fe,Mn silicates	3.4–4.3	Paramagnetic[1]	Insulator[4]	No
Gibbsite	$Al(OH)_3$	2.4	Diamagnetic	Insulator	No
Goethite	$HFeO_2$	4.3	Paramagnetic	Insulator[4]	No
Gold	Au	15.6–19.3	Diamagnetic	Conductor	No
Graphite	C	2.1–2.2	Diamagnetic	Conductor	Yes
Grossularite	$Ca_3Al_2(SiO_4)_3$	3.5	Diamagnetic	Insulator[4]	No
Gypsum	$CaSO_4 \cdot 2H_2O$	2.3	Diamagnetic	Insulator	No
Halite	NaCl	2.5	Diamagnetic	Insulator[4]	No
Hematite	Fe_2O_3	5.2	Paramagnetic	Conductor	No
Hornblende	$Ca_2Na(Mg,Fe^{2+})_4(Al,Fe^{3+})[(Si,Al)_4O_{11}](OH)_2$	3.1–3.3	Paramagnetic	Insulator[4]	No
Hubnerite	$MnWO_4$	6.7–7.5	Paramagnetic[1]	Conductor	No
Hypersthene	$(Mg,Fe)SiO_3$	3.4	Paramagnetic	Insulator	No
Ilmenite	$FeTiO_3$	4.7	Paramagnetic	Conductor	No
Ilmenorutile	$(Nb_2O_5,Ta_2O_5)_xTiO_2$	5.1	Paramagnetic	Conductor	No
Ilvaite	$CaFe_2(FeOH)(SiO_4)_2$	4.0	Paramagnetic	Conductor[5]	No
Kaolinite	$Al_2Si_2O_5(OH)_4$	2.6	Diamagnetic	Insulator	No
Kyanite	$Al_2O(SiO_4)$	3.6–3.7	Diamagnetic	Insulator	No
Lepidolite	$KLiAl_2(Si_3O_{10})(OH,F)_2$	2.8–2.9	Diamagnetic	Insulator	No
Leucoxene	$FeTiO_3 \rightarrow TiO_2$ (Alteration product)	3.6–4.3	Paramagnetic[1]	Conductor	No
Limonite	$HFeO_2 \cdot nH_2O$	2.2–2.4	Paramagnetic[1]	Insulator[4]	No
Magnesite	$MgCO_3$	3.0	Diamagnetic	Insulator	No
Magnetite	Fe_3O_4	5.2	Ferromagnetic	Conductor	No
Malachite	$Cu_2CO_3(OH)_2$	4.0	Diamagnetic	Insulator	No
Manganite	$MnO(OH)$	4.3	Paramagnetic[1]	Conductor	No
Marcasite	FeS_2	4.6–4.9	Diamagnetic	Conductor	Weak
Martite	Fe_2O_3	5.2	Paramagnetic	Conductor	No
Microcline	$KAlSi_3O_8$	2.6	Diamagnetic	Insulator	No
Microlite	$Ca_2Ta_2O_7$ (Pyrochlore group)	5.5	Diamagnetic	Insulator	No
Millerite	NiS	5.2–5.6	Paramagnetic	Conductor	Weak
Molybdenite	MoS_2	4.7–5.0	Diamagnetic	Conductor	Yes
Monazite	$(Ce,La,Y,Th)PO_4$	4.9–5.5	Paramagnetic	Insulator	No
Mullite	$Al_6Si_2O_{13}$	3.2	Diamagnetic	Insulator	No
Muscovite	$KAl_2(AlSi_3O_{10})(F,OH)_2$	2.8–3.0	Diamagnetic	Insulator	No
Nahcolite	$NaHCO_3$	2.2	Diamagnetic	Insulator	No
Nepheline Syenite	$(Na,K)(Al,Si)_2O_4$	2.6	Diamagnetic	Insulator	No
Niccolite	NiAs	7.6–7.8	Paramagnetic	Conductor	No
Olivine	$(Mg,Fe)_2(SiO_4)$	3.3–3.5	Paramagnetic	Insulator	No
Orpiment	As_2S_3	3.4–3.5	Diamagnetic	Conductor	Weak
Orthoclase	$K(Al,Si_3O_8)$	2.5–2.6	Diamagnetic	Insulator	No
Periclase	MgO	3.6	Diamagnetic	Insulator	No
Perovskite	$CaTiO_3$	4.0	Diamagnetic	Insulator	No
Petalite	$LiAl(Si_2O_5)_2$	2.4	Diamagnetic	Insulator	No
Phosphate (pebble)	$Ca_3P_2O_8 \cdot H_2O$	2.6–2.9	Diamagnetic	Insulator	No
Platinum	Pt	14.0–21.5	Diamagnetic[2]	Conductor	No
Pyrite	FeS_2	5.0	Diamagnetic[2]	Conductor	Weak

(continues)

Table 2 Common minerals and their major properties used for identification and separation (continued)

Mineral Name	Chemical Composition	Specific Gravity	Magnetic Property*	Electrostatic Property†	Naturally Hydrophobic
Pyrochlore	$(Na,Ca...)_2(Nb,Ta...)_2O_6(F,OH)$	4.2–4.4	Diamagnetic	Conductor	No
Pyrolusite	MnO_2	4.7–5.0	Diamagnetic[2]	Insulator	No
Pyrope	$Mg_3Al_2(SiO_4)_3$	3.5	Diamagnetic	Insulator[4]	No
Pyroxene	$(Ca,Mg,Fe,Al)_2Si_2O_6$	3.1–3.6	Paramagnetic[1]	Insulator[4]	No
Pyrrohotite	$Fe_{x-1}S_x$	4.6–4.7	Ferromagnetic	Conductor	Weak
Quartz	SiO_2	2.7	Diamagnetic	Insulator	No
Realgar	AsS	3.6	Diamagnetic	Conductor	Weak
Rhodochrosite	$MnCO_3$	3.7	Diamagnetic	Insulator[4]	No
Rhodonite	$MnSiO_3$	3.6–3.7	Diamagnetic	Insulator[4]	No
Rutile	TiO_2	4.2–4.3	Diamagnetic	Conductor	No
Samarskite	$(Y,Er...)_4[(Nb,Ta)_2O_7]_3$	5.6–5.8	Paramagnetic[3]	Conductor	No
Scheelite	$CaWO_4$	6.1	Diamagnetic	Insulator	No
Serpentine	$Mg_6(Si_4O_{10})(OH)_8$	2.5–2.7	Paramagnetic	Insulator	No
Siderite	$FeCO_3$	3.9	Paramagnetic	Insulator[4]	No
Sillimanite	$Al_2O(SiO_4)$	3.2	Diamagnetic	Insulator	No
Silver	Ag	10.1–11.1	Diamagnetic	Conductor	No
Smithsonite	$ZnCO_3$	4.1–4.5	Diamagnetic	Insulator	No
Sodalite	$Na_8(Al_6Si_6O_{24})Cl_2$	2.1–2.3	Diamagnetic	Insulator	No
Spessartine	$Mn_3Al_2(SiO_4)_3$	4.2	Diamagnetic	Insulator	No
Sphalerite	ZnS	3.9–4.0	Paramagnetic[1]	Conductor[5]	Weak
Sphene	$CaTi(SiO_4)(F,OH)$	3.3–3.6	Diamagnetic	Insulator[4]	No
Spinel	$MgAl_2O_4$	3.6	Diamagnetic[2]	Conductor	No
Spodumene	$LiAl(SiO_3)_2$	3.1–3.2	Diamagnetic	Insulator	No
Stannite	Cu_2FeSnS_4	4.3–4.5	Diamagnetic	Conductor	No
Staurolite	$Fe^{2+}Al_4(Si_4O_{11})_2O_2(OH)_2$	3.6–3.8	Paramagnetic	Insulator[4]	No
Stibnite	Sb_2S_3	4.6	Diamagnetic	Conductor	Weak
Struverite	$(Ta_2O_5,Nb_2O_5)_xTiO_2$	5.1	Paramagnetic	Conductor	No
Sulphur	S	2.1	Diamagnetic	Insulator	Yes
Sylvite	KCl	2.0	Diamagnetic	Insulator	No
Talc	$Mg_3(Si_4O_{10})(OH)_2$	2.7–2.8	Diamagnetic	Insulator	Yes
Tantalite	$(Fe,Mn)(Ta,Nb)_2O_6$	5.2–8.2	Paramagnetic	Conductor	No
Tapiolite	$Fe(Ta,Nb)_2O_6$	7.3–7.8	Paramagnetic	Conductor	No
Tetrahedrite	$(Cu,Fe)_{12}Sb_4S_{13}$	5.0	Paramagnetic	Conductor	No
Thorianite	ThO_2	9.7	Diamagnetic	Insulator	No
Thorite	$ThSiO_4$	4.5–5.4	Diamagnetic	Insulator	No
Topaz	$Al_2SiO_4(F,OH)_2$	3.5–3.6	Diamagnetic	Insulator	No
Tourmaline	$(Na,Ca)(Mg,Fe^{2+},Fe^{3+},Al,Li)_3Al_6(BO_3)_3Si_6O_{18}(OH)_2$	2.9–3.2	Paramagnetic[1]	Insulator[4,6]	No
Uraninite	UO_2	11.0	Paramagnetic	Insulator	No
Vermiculite	$Mg_3(AlSi_3O_{10})(OH)_2.nH_2O$	2.4–2.7	Diamagnetic	Insulator	No
Wolframite	$(Fe,Mn)WO_4$	6.7–7.5	Paramagnetic	Conductor	No
Wollastonite	$CaSiO_3$	2.8–2.9	Diamagnetic	Insulator	No
Wulfenite	$PbMoO_4$	6.7–7.0	Diamagnetic	Conductor	No
Xenotime	YPO_4	4.4-5.1	Paramagnetic	Insulator	No
Zeolite	Hydrated alumino-silicate of Ca and Na	2.0–2.5	Diamagnetic	Insulator	No
Zincite	ZnO	5.7	Diamagnetic	Insulator	No
Zircon	$ZrSiO_4$	4.7	Diamagnetic	Insulator	No

Adapted from Carpco Inc. 2000

* Magnetic property will change subject to exact composition of the mineral from various locations. The most common state is shown in the table. 1: Magnetic property may change from paramagnetic to diamagnetic. 2: Magnetic property may change from diamagnetic to paramagnetic. 3: Magnetic property may change from paramagnetic to ferromagnetic.

† Electrical conductivity will change subject to exact composition and treatment of temperature on the mineral from various locations. The most common state is shown in the table. 4: Transformation may occur from insulator to conductor due to temperature. 5: Transformation may occur from conductor to insulator due to composition. 6: Transformation may occur from insulator to conductor due to composition.

through an electrical field or corona, they become pinned to the metal surface. However, conductors dissipate that charge quickly and no longer remain pinned, whereas insulators retain their charge and remain pinned. Other forces, such as momentum or gravity, are then used to separate the minerals. Such practices are rare in mineral processing with the exception of beach sands, where differences among insulating minerals are significant compared to differences among conducting minerals.

Hydrophobicity

Hydrophobicity is the property exploited in froth flotation separations. If a water droplet spreads across the surface completely, the surface is wetted and referred to as hydrophilic. If the droplet stays beaded up, even partially, it is hydrophobic. Bubbles generated in a slurry of mineral particles will attach to the hydrophobic particles and float them to the top where a froth forms and is scraped off. Because hydrophilic particles stay in the slurry, a separation is made. Coal, diamond, talc, and sulfur are known to possess strong natural hydrophobicity. Several sulfide minerals do as well, but most are naturally weak. However, in many instances, the sulfide can be made more hydrophobic by slight oxidation of its surface forming metal-deficient surfaces, sulfide-rich surfaces, and even elemental sulfur. Most minerals are not naturally hydrophobic and have to be treated with surface-active reagents called collectors to make them hydrophobic. Many other reagents are used to enhance the process, including frothers to stabilize the froth, activators to promote collector adsorption and help make particles hydrophobic, depressants to prevent collector adsorption and help keep other particles hydrophilic, and modifiers to act as depressants for one mineral and activators for another. Most modifiers are used to adjust the pH.

From the perspective of flotation reagents, minerals/ores are classified into two distinct groups: (1) sulfide minerals/ores and (2) nonsulfide minerals/ores. The rationale for this classification is that separation schemes and reagent selection and usage for nonsulfide minerals are distinctly different from those for base metal sulfide minerals. Tables 3 and 4 list the reagents used for common sulfide and nonsulfide minerals, respectively, using information from two major flotation reagent manufacturers: Cytec and Clariant (Thomas 2010; Clariant Mining 2013).

Natural resources are nonrenewable. It is indisputable that richer ores are becoming less abundant and poor-quality resources are being processed. Not only are the valuable contents in these ores decreasing, but the association of the gangue is also becoming increasingly intricate with the valuables. This calls for very fine grinding to achieve an adequate degree of liberation. Most unit operations in mineral processing fail to produce good separation in such fine ranges. Therefore, froth flotation is increasingly used in ore beneficiation. Experts estimate that more than 50% of the world's natural resources are now processed by flotation.

EXTRACTIVE METALLURGY

In extractive metallurgy, hydrometallurgical and pyrometallurgical processes are used in the presence and absence of liquid water, respectively. The processes involve extracting the valuable metal content in concentrates and ores through chemical change and then recovering the metal value. To do both, chemical properties are primarily exploited and are greatly dependent on the final metal product being made. However, only a few of these chemical properties are used to identify minerals (e.g., if dilute acid is added to the mineral and it effervesces, it is likely calcite; if a slag cools at 800°C as opposed to 1,000°C, it may be because of the presence of Na-rich plagioclase). The primary property being exploited in hydrometallurgy is solubility that, in turn, is affected by system factors such as temperature, pressure, and chemistry (e.g., pH, lixiviant, concentration, aqueous/gaseous oxidant, presence of contaminants). Likewise, for pyrometallurgy, the property being exploited is phase change (e.g., melting or sublimation points) as determined by the same system factors although the chemistry is slightly different (e.g., partial pressures, reactants, solid/gaseous oxidant, and presence of contaminants). These system factors can be thermodynamically modeled to explain how a process works or to determine whether a proposed process is feasible. The details are discussed in Part 12 of this handbook.

REFERENCES

Bolles, J.L., and McCullough, E.J. 1985. Minerals and their properties. In *SME Mineral Processing Handbook*. Vol. 1. Edited by N.L. Weiss. Littleton, CO: SME-AIME.

Carpco Inc. 2000. *Gravity, Electrostatic and Magnetic Characteristics of Selected Minerals*. Jacksonville, FL: Carpco.

Clariant Mining. 2013. *Beneficiation of Frequently Occurring Minerals*. The Woodlands, TX: Clariant Mining.

Craig, J.R., and Vaughan, D.J. 1981. *Ore Microscopy and Ore Petrography*. New York: John Wiley and Sons.

Hurlbut, C.S., Jr., and Klein, C. 1977. *Manual of Mineralogy*, 19th ed. New York: John Wiley and Sons.

Mills, S.J., Hatert, F., Nickel, E.H., and Ferraris, G. 2009. The standardization of mineral group hierarchies: Application to recent nomenclature proposals. *Eur. J. Mineral.* 21:1073–1080.

Thomas, W., ed. 2010. *Mining Chemicals Handbook*. West Patterson, NJ: Cytec Industries.

Table 3 Reagent selection guidelines for the flotation of common base metal sulfide minerals*

Valuable	Major Minerals	Common Ore Concentration	Ore Types	Collector†	Modifier‡	Depressant or Dispersant‡§	Activator‡	pH Range
Ag (silver)	Argentite Ag_2S Pyrargyrite Ag_3SbS_3 Proustite Ag_3AsS_3 Acanthite Ag_2S Freibergite $(Ag,Cu,Fe)_{12}(Sb,As)_4S_{13}$	0.1–100 ppm	Ag values found in polymetallic ores (e.g., Pb-Zn, Cu-Pb-Zn, and Cu-Zn)	Dithiophosphinates, xanthate, alkyl/aryl dithiophosphates, mercaptobenzothioazoles, dithiocarbamates	Na_2CO_3, lime, H_2SO_4	$ZnSO_4$, dextrin, nigrosine, NaCN, $Zn(CN)_2$, tannin, sodium silicate, sulfoxy compounds	—	5–10.5
	Native silver Ag-Au alloys (electrum)		Ag values in primary Au ores	Xanthate, dithiocarbamates, mercaptobenzothioazoles, dithiophosphinates, monothiophosphates	H_2SO_4, lime	—	$CuSO_4$, $Pb(NO_3)_2$	5–9.5
Au (gold)	Native gold Au-Ag alloys (electrum) Au-Cu alloys	0.05–10 ppm	Bulk flotation of tellurides and alloys	Xanthates, dithiophosphinates, mercaptobenzothioazoles, dithiocarbamates, alkyl/aryl mono/dithiophosphates	H_2SO_4, lime	Na_2CO_3, sodium silicate, polyphosphates, synthetic polymeric modifiers	—	5–10
	Krennerite $(Au,Ag)Te_2$ Calaverite $AuTe_2$ Petzite Ag_3AuTe_2 Sylvanite $(AuAg)_2Te_4$							
	Pyrite FeS_2 Marcasite FeS_2 Arsenopyrite FeAsS, Enargite Cu_3AsS_4, Tennantite $Cu_{12}As_4S_{13}$ Tetrahedrite $Cu_{12}Sb_4S_{13}$ Chalcopyrite $CuFeS_2$, Pyrrhotite $Fe_{(1-x)}S_x$	0.05–10 ppm	Flotation of Au/Ag values associated and disseminated in these minerals	Xanthates, dithiocarbamates, mercaptobenzothiazoles, alkyl/aryl mono/dithiophosphates	H_2SO_4, lime	Na_2CO_3, sodium silicate, polyphosphates, synthetic polymeric modifiers	$CuSO_4$, $Pb(NO_3)_2$	5–9
Co (cobalt)	Cobaltite CoAsS Heterogenite CoO(OH) Safflorite $(CoAs_2)$ Glaucodot (Co,Fe)AsS Skutterudite $(Co,Ni,Fe)As_3$	0.2%–1%	Cu-Co ores (Zambian Cu belt)	Dithiophosphates, xanthates, xanthate ester thionocarbamates	Na_2CO_3	Na_2CO_3	Na_2S NaSH	8–10
Cu (copper)	Azurite $Cu_3(CO_3)_2(OH)_2$ Chrysocolla $Cu[SiO_3]$ ÈnH_2O Malachite $Cu_2CO_3(OH)_2$ Cuprite Cu_2O	0.2%–4%	Selective flotation of oxide and sulfide Cu minerals	Xanthates, alkyl hydroxamates, fatty acids, fuel oil†	Na_2CO_3, lime	Na_2SiO_3 Na_2S Polyphosphates Na_2CO_3	Na_2S NaSH	6–9.5
	Bornite Cu_5FeS_4 Chalcopyrite $CuFeS_2$ Chalcocite Cu_2S Covellite CuS	0.2%–1.5%	Selective flotation of Cu sulfide minerals	Aliphatic and aromatic dithiophosphates, thionocarbamates, xanthate esters, alkyl sulfides, dithiophosphinates, hydrocarbon oil†	lime	Na_2SiO_3, synthetic polymeric modifiers, sulfoxy compounds	—	8–12
	Enargite Cu_3AsS_4 Tennantite $Cu_{12}As_4S_{13}$ Tetrahedrite $Cu_{12}Sb_4S_{13}$							
Mo (molybdenum)	Molybdenite MoS_2	0.005%–0.1%	Primary Mo	Xanthate esters, hydrocarbon oil†	—	Sodium silicate	—	7–9
		0.1%–0.4%	Cu-Mo ores	See Cu section	—	—	—	—

(continues)

Table 3 Reagent selection guidelines for the flotation of common base metal sulfide minerals* (continued)

Valuable	Major Minerals	Common Ore Concentration	Ore Types	Collector†	Modifier‡	Depressant or Dispersant‡§	Activator‡	pH Range
Ni (nickel)	Millerite NiS Niccolite NiAs Pentlandite $(Fe,Ni)_9S_8$, Mackinawite $(Fe,Ni)_{1+x}S$ Violarite $Fe^{2+}Ni_2^{3+}S_4$	0.2%–2%	Ni, Ni-Cu Massive sulfide ores	Xanthates, alkyl dithiophosphates, dithiocarbamates, dithiophosphinates	H_2SO_4, lime	Na_2SiO_3, NaCN, synthetic polymeric modifiers, DETA	$CuSO_4$	5–10
			Serpentine-hosted Ni ores		H_2SO_4, Na_2CO_3	CMC, dextrin guar gum	$CuSO_4$	
Pb (lead)	Cerussite $PbCO_3$ Anglesite $PbSO_4$ Plumbojarosite $Pb_{0.5}Fe^{3+}{}_3(SO_4)_2(OH)_6$	0.5%–2%	Sulfidization-flotation or direct flotation	Condensate amines, xanthates, mercaptobenzothiazoles	Na_2CO_3	—	Na_2S	7–9
	Galena PbS	0.2%–3%	Selective flotation from Cu/Pb or bulk	Dithiophosphinates, xanthates, aryl dithiophosphates	Lime, Na_2CO_3	NaCN, $Zn(CN)_2$, $ZnSO_4$, MBS, SO_2, dextrin, nigrosine	NaCN	6–9.5
Pt/Pd/Rd (platinum group metals, or PGMs)	Braggite (Pt,Pd,Ni)S Sperrylite $PtAs_2$ Cooperite PtS Metallics/alloys such as: Platiniridium (PtIr) Ferroplatinum (FePt)	0.5–5 ppm	Selective flotation of PGM mineral values disseminated in Ni and Cu minerals	Xanthates, mercaptobenzothiazoles, dithiocarbamates, thionocarbamates, dithiophosphates	—	CMC, dextrin, guar gum, synthetic polymers	$CuSO_4$	7–9
Zn (zinc)	Hemimorphite $Zn_4Si_2O_7(OH)_2{*}H_2O$ Hydrozincite $Zn_5(CO_3)_2(OH)_6$ Smithsonite $ZnCO_3$ Willemite $ZnSiO_4$	0.2%–5%	Bulk flotation of oxide and sulfide zinc minerals	Condensate amines, ether amines, alkyl hydroxamates, xanthates, hydrocarbon oil†	NaOH	Na_2SiO_3	Na_2S NaSH	7–9
	Sphalerite ZnS Marmatite (Zn,Fe)S	1%–15%	Activation and flotation of Zn in polymetallic ores	Xanthates, dithiophosphates, dithiophosphinates, thionocarbamates	Lime	—	$CuSO_4$	9–12
Others (e.g., Bi, Sb, and Hg)	Bismuthinite Bi_2S_3 Aikinite PbCuBiS3	0.2%–0.5%	Selective flotation in polymetallic ores	Xanthates, mercaptobenzothiazoles, alkyl dithiophosphates	Lime	Na_2CO_3 NaCN Activated carbon	$Pb(NO_3)_2$	8–12
	Stibnite Sb_2S_3	0.5%–1%	—	Xanthates, hydrocarbon oil†	Na_2CO_3	Excess NaCN, Na_2SiO_3	$Pb(NO_3)_2$ $CuSO_4$	6.5–8
	Cinnabar HgS	0.2%–5%	—	Xanthates, alkyl and aryl dithiophosphates	—	Na_2SiO_3	—	7–10

*Notes:
- This table is based on industry-wide averages and is therefore meant to be a starting-point recommendation and not an end-means conclusion. A reagent scheme cannot be determined for a given ore type without proper laboratory and plant testing.
- Frothers are also subject to these parameters but are not listed here because all essentially work. Short-chain aliphatic alcohols include methyl isobutyl carbinol (MIBC) as the most widely used frother for sulfide flotation; others are cresylic acid and alkoxy-substituted paraffins (e.g., triethoxybutane); hydrocarbons may be needed to stabilize the froth, but caution is needed because they could act as a nonselective collector. Long-chain polyglycols include both unsubstituted and alkyl monoethers.

†Collectors of hydrocarbon and related oils can be used and may be specified above as fuel oil, kerosene, and so forth.

‡pH modifiers, depressants/dispersants, and activators (type and dosage) depend on value and gangue mineralogy, ore type, flow-sheet and operating philosophy, equipment, particle size, and process water chemistry.

§CMC = carboxymethyl cellulose; DETA = diethylene triamine; MBS = sodium metabisulfite.

Table 4 Reagent selection guidelines for flotation of common nonsulfide minerals*

Valuable	Minerals	Weight % in Ore	Processing Strategy	Collector†	Frother‡	Extender	Modifier	Depressant§	Activator	pH Range
Aluminum	Bauxite $Al_2O_3ÈH_2O$ Gibbsite [Al(OH)$_3$], Diaspore [α-AlO(OH)] Boehmite [γ-AlO(OH)]	30–70	Direct selective flotation of bauxite from silica (current practice); reverse flotation of gangue (silica, etc.) is an option	Fatty acid (primarily), primary amine for reverse flotation	Alcohol and/ or glycol as needed	Hydrocarbon oil	Na_2CO_3	Na_2SiO_3, Na_2CO_3	—	8–9
Barite	Barite $BaSO_4$	0.2%–0.5% Ba	Flotation of barite from calcite, fluorite, and other silicates	DASS, alkyl sulfate, PS, fatty acid	Alcohol and/ or glycol as needed	—	NaOH, Na_2CO_3, Na_2SiO_3, citric acid	$AlCl_3$, $FeCl_3$, F, $K_2Cr_2O_7$, quebracho	$BaCl_2$ Pb salts	8–10
Coal	Coal	15%–40%	Flotation of coal from ash	Hydrocarbon oil and frothers (mixture of aliphatic alcohol, ethers, and esters)	See Collector column	—	—	—	—	Natural
Feldspar	Feldspar [Na,K] (Al Si_3O_8), $CaAl_2Si_2O_8$	7%–8% Na_2O or K_2O	Removal of mica, silica, and Fe oxides from feldspar	Alkyl diamine, alkyl ether diamine for mica removal, fatty acid for Fe removal, amine again for feldspar removal	Alcohol and glycol as needed	Hydrocarbon oil	—	Na_2SiF_6, Na_2SiO_3, H_2SO_4, $Al_2(SO_4)_3$	HF	1.5–3
Fluorspar	Fluorite CaF_2	2%–10% CaF_2	Flotation of fluorite from silica	Fatty acids, PS	Alcohol and glycol as needed	PAE, modified sodium carboxylate, hydrocarbon oil	Na_2CO_3, Na_2SiO_3, quebracho starch	$K_2Cr_2O_7$, $BaCl_2$, Na_2SiF_6, citric acid, Al^{3+} salts, Na_2SiO_3	Hot pulp	7–9
Garnet	Garnet minerals (Mg,Fe,Mn,Ca)$_3$ (Al,Fe,Cr)$_2$ [SiO_4]$_3$	Variable	Flotation of garnet from silicates	AEM, alkyl amine, fatty acids, sulfonated fatty acids, PS	—	Hydrocarbon oil	H_2SO_4	Excess acid	—	3–4
Glass and foundry sands	Quartz SiO_2	1%–3%	Removal of Fe and Ti impurities from sand	AEM, alkyl ether diamine, quaternary alkyl amine, fatty acid, PS, DASS	Alcohol and glycol as needed	Hydrocarbon oil	NaOH	H_2SO_4, HF, NaF Na_2SiO_3	$CaCl_2$	8.5–9.5
Heavy mineral sands (Ti, Zr)	Anatase TiO_2 Brookite TiO_2 Ilmenite $FeTiO_3$ Rutile TiO_2 Titanite $CaTiSiO_5$ Zircon $ZrSiO_4$	2%–5% THM (total heavy minerals)	Flotation of heavy minerals from silica, optional feldspar removal	PAE, fatty acids, AS, fatty amine to remove feldspar	Alcohol and glycol as needed	Hydrocarbon oil	H_2SO_4, Na_2SiO_3, Na_2Co_3	Caustic starch, dextrin, synthetic polymeric modifiers	$CuSO_4$	8–9
Iron	Hematite Fe_2O_3 Magnetite Fe_3O_4	32%–50% Fe	Reverse flotation of silica from hematite	Fatty and etheramines, fatty acids, PS	Alcohol or glycol as needed	Hydrocarbon oil	H_2SO_4	Caustic starch, tannic acid	Desliming (optional)	3–6

(continues)

Table 4 Reagent selection guidelines for flotation of common nonsulfide minerals* (continued)

Valuable	Minerals	Weight % in Ore	Processing Strategy	Collector†	Frother‡	Extender	Modifier	Depressant§	Activator	pH Range
Kaolin	Kaolinite $Al_4Si_4O_{10}(OH)_8$	2%–3% TiO_2	Flotation of TiO_2 and Fe_2O_3 from kaolin	Alkyl hydroxamates	—	—	—	—	—	2–2.5
Kyanite	Kyanite Al_2SiO_5	Variable	Flotation of sulfide impurities followed by flotation of kyanite	Sulfides removed using dithiocarbamate, fatty acids, and PS; DASS for floating kyanite	Alcohol or glycol as required	—	H_2SO_4	—	—	8 then 2.5–3
Limestone	Calcite Aragonite $CaCO_3$ Dolomite $CaMg[CO_3]_2$	1%–9% Silica	Flotation of limestone from silica	Fatty acid or amine, PAE	Alcohol and/ or glycol as needed	—	Na_2CO_3,	Excess Na_2SiO_3, caustic starch, NPPE, quebracho $K_2Cr_2O_7$	—	Natural
Lithium	Lepidolite $KLi_2Al(Al,Si)_3O_{10}(F,OH)_2$	Li_2O_5- 5%–8%	Flotation of Li minerals from other silicates	Fatty acids, alkyl amines	Alcohol or glycol as needed	Hydrocarbon oil	H_2SO_4	Lactic acid HF, starch, dextrine, Na_2SiO_3	—	7–11
	Spodumene $LiAl[Si_2O_6]$								$Pb(NO_3)_2$	
Manganese	Pyrolusite MnO_2	8%–15% Mn	Flotation of Mn minerals away from silicates	Fatty acid, PS, DASS	Alcohol or glycol as needed	Hydrocarbon oil	Na_2CO_3, NaOH, Na_2SiO_3	Phosphates quebracho, excess Na_2SiO_3	Mn^{2+}can be used	4–5 and 8–10
	Rhodochrosite $MnCO_3$									8–10
Mica	Muscovite $KAl_2(AlSi_3O_{10})(OH,F)_2$	5%–15%	Flotation of mica away from silicates	PAE, fatty acids, AS, fatty amines	Alcohol as needed	Hydrocarbon oil	H_2SO_4 polyphosphates	HF, starch, Na_2SiO_3	—	2.5–4
Niobium	Columbite $FeNb_2O_6$	0.5%–7%	Reverse flotation of carbonate/silicates followed by direct flotation of Nb	Fatty acids for gangue, alkyl amines, imidazoline for Nb	Alcohol or glycol as needed	—	Na_2SiO_3	Caustic starch	H_2SiF_6, oxalic acid	8 for reverse, 3–5 for Nb
	Pyrochlore $(Na, Ca)_2(Nb, Ti, Ta)_2O_6(OH,F,O)$									
Phosphate	Apatite $Ca_5(PO_4)_3(F,OH)$ Collophanite $Ca_5(PO_4, CO_3)_3F$ Phosphorite $Ca_5(PO_4)_3(F,OH,CO_3)$	5%–23% P_2O_5	Double flotation of phosphate then purification by silica removal	Fatty acids, PS, DASS, PAE modified sodium carboxylate, alkyl ether diamine	Glycols	Hydrocarbon oil	NaOH, Na_2CO_3, NH_3, caustic starch, synthetic polymeric modifiers	NaF, H_2SiF_6, H_3PO_4, Na_2SiO_3, excess NaOH	Pb salts	9.0–10.0 then 6–7

(continues)

Table 4 Reagent selection guidelines for flotation of common nonsulfide minerals* (continued)

Valuable	Minerals	Weight % in Ore	Processing Strategy	Collector†	Frother‡	Extender	Modifier	Depressant§	Activator	pH Range
Potash	Sylvite KCl	25% K_2O, 40% KCl	Flotation of sylvite from halite (NaCl)	Alkyl amines	Alkyl polyglycols as needed	Hydrocarbon oil	—	Dextrin, guar, starch, CMC, synthetic polymeric modifiers	—	Natural
Rare earth elements (REE)	Bastnaesite [REE(CO_3)]F Monazite (REE)PO_4 Xenotime (REE)PO_4	1%–5% REE	Flotation of rare earths from silicates	Fatty acids, alkyl hydroxamates	Alcohol and glycol as needed	—	Na_2S Na_2CO_3, Na_2SiO_3	Strong acids	Hot pulp	5–8, 9.5–10 with hydroxamates
Talc	Talc $Mg_3Si_4O_{10}(OH)_2$	40%–55%	—	—	Alcohol as needed	—	—	—	—	7–9
Tantalum	Tantalite $MnTa_2O_6$	0.5%–1%	Removal of sulfides then flotation of Ta and Sn	Xanthate to float sulfides, DASS, PAE, AS for Ta	Alcohol and glycols as needed	Hydrocarbon oil	—	Caustic starch Na_2SiO_3	H_2SiF_6	5
Ti, Zr (heavy mineral sands)	Anatase TiO_2 Brookite TiO_2 Ilmenite $FeTiO_3$ Rutile TiO_2 Titanite $CaTiSiO_5$ Zircon $ZrSiO_4$	2%–5% THM (total heavy minerals)	Flotation of heavy minerals from silica, optional feldspar removal	PAE, fatty acids, AS, fatty amine to remove feldspar	Alcohol and glycol as needed	Hydrocarbon oil	H_2SO_4, Na_2SiO_3, Na_2Co_3	Caustic starch, dextrin, synthetic polymeric modifiers	$CuSO_4$	8–9
Tin	Cassiterite SnO_2	0.5%–1%	Removal of sulfides then flotation of Sn	Xanthate to float sulfides, PS, DASS styrene phosphonic acid to float Sn	Alcohol and glycols as needed	Hydrocarbon oil	H_2SO_4	NaF, $BaCl_2$, lime, tannic acid, Na_2SiO_3	Desliming	3–5
Tungsten	Huebnerite $MnWO_4$ Scheelite $CaWO_4$ Wolframite (Fe,Mn)WO_4	0.1%–5%	Flotation of W away from gangue silicates	Fatty acids, modified sodium carboxylate	Alcohol or glycol as needed	—	Na_2SiO_3, Na_2CO_3, NaOH	Quebracho, citric acid	—	9–10.5
Uranium	Uraninite U_3O_8 Carnotite $K_2(UO_2)_2(VO_4)_23H_2O$	0.1%–1%	Flotation of U minerals from silicates	Fatty acid	Alcohol and glycol as needed	—	Na_2CO_3	Na_2SiO_3	Pb salts	8–9
Vanadium	Vanadinite $Pb_5(VO_4)_3Cl$	0.5%–2%	Flotation of calcite then silica	Fatty acid	Alcohol or glycol as needed	—	Na_2CO_3	—	Na_2S, $CuSO_4$, Pb salts	8–9
Wollastonite	Wollastonite $Ca_3[Si_3O_9]$	20%–50%	Flotation of impurities away from wollastonite	Alkyl amines, AS	Alcohol or glycol as needed	—	H_2SO_4	Tannic acid	Pb salts to activate pyroxene	9 then 3–5

*This table is based on industry-wide averages and is therefore meant to be a starting-point recommendation and not an end-means conclusion. A reagent scheme cannot be determined for a given ore type without proper laboratory and plant testing, as detailed in Chapter 7.5, "Flotation Chemicals and Chemistry."

†AEM = alkyl ether amine; AS = alkyl sulfate; DASS = dialkyl sulfosuccinamate; PAE = phosphoric acid esters; PS = petroleum sulfonate

‡Frothers: Although specified in this table, the footnote regarding frothers in Table 3 generally applies here too.

§ CMC = carboxymethyl cellulose; NPPE = nitrophenyl pentyl ether

CHAPTER 1.2

Analytical Testing

Meg Dietrich LeVier

Sampling and analytical testing are the basis of decision making in exploration, mineral evaluations, and/or environmental dispositions. Even before the 16th century, comprehensive schemes of assaying ores were known, using procedures that do not differ materially from some of those employed today. Most conventional methods of chemical analyses used today have evolved using sophisticated analytical instrumentation to detect and estimate quantities of elements in ores and minerals, particularly at low concentrations, to develop a detailed understanding of the ore characteristics and the corresponding response in mining and processing operations. This allows financial decisions to be made regarding treatments to optimize each ore type for extraction, increase production, or reduce costs.

The primary objectives of analytical testing are generally for one of the following reasons:

- For evaluating a mining property
- To develop a detailed understanding of the physical properties of the ores examined
- For plant quality control (QC)
- To provide comprehensive recommendations for process performance improvements
- For metallurgical accounting and inventory requirements
- For environmental considerations

ANALYTICAL TESTING STRATEGY

Selecting appropriate analytical tests can be a confusing process. Balancing cost and risk with quality fit for purpose is the goal of every test program. Costs for analysis can range from a few dollars per sample for "ballpark" exploration methods to hundreds of dollars per element when high accuracy and precision are required for contract settlements. The purpose and expected quality requirements need to be communicated to the laboratory when a program is initiated so that a suitable quality assurance and QC program can be included. Some examples of different assay programs and the associated QC procedures are as follows:

- **Exploration geochemistry.** This analysis is intended for exploration program rock samples with low metal contents. The frequency of inserted QC materials is low and the assay methods have low upper-measurement limits.
- **Ore-grade assaying.** This assay level is used when exploration geochemistry has identified higher-grade mineralized rock or core samples. The data from these analyses are often used for resource/reserve estimation. The frequency of inserted QC materials is more frequent, generally about 10%, and methods are designed to measure a wider range of elements.
- **Process control.** This level of analytical testing is used for metallurgical support and/or accounting, or the data are used to monitor a metallurgical process. The frequency of inserted QC materials is about 10%–20%. Methods are designed to analyze samples that contain up to 100% of the target element in some cases. High precision and accuracy is necessary and fewer samples are assayed at a time, but a large number of analytes per sample are measured.
- **Commercial settlement.** This grade of analysis is used when extremely high precision and accuracy data are required for high dollar value of concentrate and bullion shipments for commercial transactions and contract settlements, and to settle disputes between buyers and sellers of traded commodities (umpire analysis). The splitting limits (the accepted spread of analytical results agreed upon by concentrate or bullion seller and buyer) are determined by negotiation. The frequency of inserted QC materials is at least 100% and multiple replicates are analyzed.

Communication with a laboratory is vitally important for a successful analytical project. Once the expected quality of work is established, the detail must follow. Analytical laboratory managers need as much information as possible to prepare quotations to meet budget costs and project schedules, as well as for minimization of mistakes and unreal expectations. A laboratory visit or audit is prudent and one must be prepared to discuss numerous questions.

Meg Dietrich LeVier, Consulting Analytical Chemist, K. Marc LeVier & Associates Inc., Highlands Ranch, Colorado, USA

The following questions can be used as a checklist of analytical requirements to be communicated to the laboratory manager:

- What sample preparation protocols should be used?
- How many samples/day/week?
- What are the sample types? (Rocks, effluents, soils, concentrates, metals?)
- What are the approximate concentration ranges for each sample type and expected reporting limits?
- What are the accuracy and precision requirements? This may determine the analytical procedure required.
- What is the required turnaround time?
- For methods requiring sample dissolution, what method of dissolution will be employed?
- How many and what elements need to be determined per sample type?
- Are there specific QC reporting requirements?
- Are the analytical requirements likely to change on a regular basis?
- Are there regulatory or legal issues to take into consideration (e.g., environmental permit requirement or ISO/IEC 17025 [2005] specific protocol requirements)?
- Will there be liquid samples needing preservation, refrigeration, or that have short hold times?
- What sample volume is typically available?
- How are the data being used?
- What will the budget allow?
- May a subcontracted commercial laboratory be used?
- What is the reporting format? Spreadsheet or are data to be imported?
- Should completed samples be returned, stored, or disposed after 90 days?
- How will samples be transported to the laboratory?

SAMPLING AND SAMPLE PREPARATION FOR ANALYTICAL TESTING

The importance of sampling and sample preparation cannot be overemphasized. Typically, the focus of laboratory results is on the analytical method, and sampling is pretty much taken for granted. However, sampling is perhaps the major source of error in the measurement process and, potentially, is an overwhelming source of error for any metallurgical project. The selection of the actual sample preparation procedures depends on the type and size of the sample, the mineralogy, and the analytical and budgetary constraints. Discussion with the laboratory manager is required to determine the most suitable options to achieve the best outcomes in terms of precision, accuracy, and meaningful analytical data.

Common sense suggests that the larger the mass of the sample, the closer it should resemble the composition of the material it came from. However, preparing large masses by grinding entire lots of material is not practical. Understanding the Gy sampling theory, often presented in a short course by Francis Pitard (1993), will provide guidance for what may seem like an overwhelming amount of work. The simple task of taking a small amount of material out of a laboratory sample bag could possibly be the largest source of error in the whole measurement process, leading to meaningless results and incorrect and costly decisions.

Laboratories recognize that the quality of all analyses is dependent on the quality of sample preparation. Whether a blasthole is being collected, copper concentrate is shoveled from a stockpile, or a scoop is taken from a sample bag, analytical results from these samplings have financial significance. Some are more impactful than others. Fire assaying of gold is typically performed on 30 g of material and may represent thousands of tons of material. How can one know that the sampling is an accurate representation of that material? Biased sampling protocols can have negative consequences such as poor ore/waste selection decisions, deficient profit/loss decisions, inadequate reconciliations, or unbalanced metal accounting. To have confidence in the integrity of a sample and the sampling procedure, one must be sure all samplings and subsequent preparation are entirely random. Preparing a bulk sample, such as a 1,000 kg of HQ drill core, might include sawing the core into quarters, crushing, subdividing, pulverizing, subdividing again, and taking a split for assay. To understand how much mass should be used, one must refer to a *sampling nomogram*, which is a graphical representation that summarizes the relationship between the fragment size, the mass of the sample relative to the mass of the lot, and the relative variance that can be anticipated. Details of creating sampling nomograms and proper sampling procedures are found in Pitard (1993) and other chapters of this handbook.

During mineral sample preparation, geological or process material is broken down into a fine, dry pulp, ideally <75 µm particle size, that can be subsampled to provide a representative sample of the original material. Well-blended small particles are the key to ensuring that the target elements are effectively released from the rock for decomposition and further analysis. Quality protocols must be followed during all stages of sample preparation, including proper handling, safety, and sample tracking. Utilizing a laboratory information management system, or LIMS, for tracking allows each sample to be given a unique identification number. This maintains organization and allows samples to be tracked throughout all stages of analysis and enables easy monitoring of progress.

A few tips on the basic steps for preparing samples for optimal analytical testing, minimization of contamination, and longevity of grinding equipment are discussed in the following paragraphs.

Safe sample preparation procedures must be followed. A very small silica dust particle could easily be aspirated, which then becomes a health hazard that could lead to serious lung disease. An exposure control plan must be in place before work begins. Additionally, fibrous or asbestiform samples and samples of naturally occurring radioactive material require special handling during sample preparation and may result in higher preparation costs. Material must be clearly identified with the hazard on the sample submission and sample packaging. An isolated preparation and storage area is required. Additional personal protective equipment and awareness training must be given to employees (Horak et al. 2016).

Sample drying of all free moisture in mineral samples is important to ensure that particles do not adhere to the preparation equipment and that data are represented on a dry basis. Drying time and temperature depend on the type of sample with respect to mass, moisture, and matrix but is typically dried at 105°C to remove free moisture without compromising the sample. Clays containing crystalline water require much higher temperatures to dissociate the water. Removing this type of water is not usually necessary for analytical test samples. Samples containing volatile analytes, particularly mercury, can be lost during drying and are best dried at 60°C for a longer period. A sample is considered dry when it no longer has a weight change over a given drying period, generally 1 hour or more.

Crushing of rocks and core samples is required when particle sizes are too large for pulverizing equipment. The most common primary crushing tool in mineralogical laboratories is the jaw crusher. Jaw crushers handle rock material up to about 15 cm and can reduce most material down to about 85% passing 2 mm (10 mesh). Jaw crushers have an adjustable crushing diameter. The rocks are loaded, crushed to a target size of 2 mm, and fall to the bottom of the crusher, and, consequently, the crusher can be effectively cleaned between samples. Care must be taken to capture all dust fines. If the material is still damp or contains clay, samples can coat the jaws and the target particle size is not achieved. This can also cause cross contamination if the equipment is not cleaned between samples. A secondary roll crusher may be needed if the jaw crusher fails to crush to 85% passing 2 mm. Roll crushers are often difficult to clean, which can result in serious cross contamination. The contamination risk of using a roll crusher must be weighed against the benefits. If a roll crusher is being used, one must ensure that the crushing procedure specifies a cleaning step between samples to prevent cross contamination.

Splitting (subdividing) is a method of reducing sample volume by dividing the sample into representative subsamples. It is critical that methods used to reduce the ore to those few grams be as accurate as possible, and equipment is available that will assist in doing this. For instance, cone and quartering is the "old" method for splitting a large sample. Cone and quartering induces a margin of error of 19.2% (Allen and Khan 1970), which is not the best way to obtain accurate assays. To ensure that a proper representative sample is obtained, careful consideration is given when choosing the size and type of the splitter and its contact with the sample to split the material without bias. The rotary sample splitter has a relatively low margin of error of 0.125% (Table 1) and is capable of splitting large bulk samples of 6 mm ore. The use of a rotary splitter for large samples can eliminate significant error in obtaining a representative sample. A precision rotary splitter has a margin of error much lower than that of a typical Jones riffle splitter. Utilizing the most appropriate and accurate equipment will make the difference in achieving an unbiased sample. If a sample splitting method with a high margin of error must be used, then the overall error can be reduced by repeating the mixing and subsampling procedure several times.

Milling is altering the material to a smaller particle size by grinding in a container typically made of hardened steel to create a fine homogeneous powder. A particle size of 85% passing 50 μm is ideal, although 85% passing 74 μm (200 mesh) is acceptable in most cases. This allows for a representative subsample to be taken for analysis and is a requirement for chemical analysis. One should not be able to feel any grit when rubbing ground material between two pieces of paper. Contamination from pulverizing bowls, such as Fe in mild steel bowls, may be a concern if Fe is a target analyte. For high-precision whole-rock analysis, tungsten carbide pulverizing is recommended. Very hard material may need to be pulverized in tungsten carbide grinding pots; however, this is not appropriate if tungsten, carbon, or cobalt are target elements. An appropriate grinding bowl needs to be selected for material hardness and potential contaminants (Table 2). Careful

Table 1 Standard deviations of samples produced from a 60%/40% mixture of fine and coarse sand

Sampling Method	Standard Deviation of Samples, %
Cone and quarter	6.81
Grab sampling	5.14
Chute-type sample splitter	1.01
Rotary riffle	0.125
Random variation for a theoretically perfect sampler	0.076

Adapted from Allen and Khan 1970

Table 2 Properties of grinding container materials

Material	Major Elements	Minor Elements	Hardness	Resistance to Abrasion	Durability	Comparative Efficiency
Hardened steel	Fe	Cr, Si, Mn, C	Mohs: 5.5–6 Rockwell: C 60–65	Moderate	High	High
Stainless steel	Fe, Cr	Ni, Mn, S, Si*	Mohs: 5–5.5 Rockwell: C 55–60	Moderate	High	High
Cr-free steel	Fe	C, Mn, Si, Mo	Mohs: 5–5.5 Rockwell: C 55–60	Moderate	High	High
Tungsten carbide	W, C, Co	Ta, Ti, Nb	Mohs: 8.5+ Knoop: 1,400–1, 800	High	Long-wearing, subject to breakage	Very high
Alumina ceramic	Al	Si, Ca, Mg	Mohs: 9 Rockwell: R45N 74–79 Knoop: 1,160	Very high	Long-wearing, brittle	Moderate
Agate	Si	Al, Na, Fe, K, Ca, Mg*	Mohs: 6-7 Knoop: 550–800	Extremely high	Very long-wearing	Moderate
Zirconia	Zr	Y, Mg, Hf	Mohs: 8.5 Rockwell: R45N 74–79 Knoop: 1,160	Extremely high	Very long-wearing	Moderate
Silicon nitride	Si	Y, Al, Fe, Ca	Mohs: 8.5+ Knoop: 1,600	Extremely high	Very long-wearing	Moderate
Plastic	C	—	Mohs: 1.5	Low	Low, disposable	Low for grinding; high for blending

Source: SPEX SamplePrep 2016.
*All reported <0.02%

protocols must be followed when grinding. Overloading the grinding bowl is a common sample preparation error resulting in poor grind. Larger grinding pots are available in cases where larger samples need pulverizing, such as in the case of material with heterogeneity problems like particulate Au. Also, contamination may occur when particulate Au or other metallic smears inside the grinding bowl and must be cleaned out between samples using sand, sometimes multiple times. The laboratory seldom needs more than 150 g to complete several tests.

INSTRUMENTATION FOR ELEMENTAL ANALYSIS

To keep up with advancements in material processing, analytical methods require continuous improvement to be faster, more rugged, provide more information, and be more cost-effective. Qualitative wet chemical methods used during the 19th and early 20th centuries were replaced with more efficient quantitative techniques. The invention of the transistor, microprocessor, and other computer technologies brought electronic instrumentation into the laboratories. Classic wet chemical methods for elemental analysis, considered old-fashioned science, have mostly been replaced by instrumentation in analytical laboratories but are still used in certain applications.

There is considerable overlap of analytical methods, making selection of the appropriate instrument or tool confusing. In fact, all of the techniques may be able to perform a particular analysis at an acceptable level of accuracy and precision, depending on the requirements of a metallurgical project.

Samples may arrive to the laboratory in liquid or solid form. Most instrumental techniques are designed for either solids or liquids with some capable of measuring in either form with modifications. To establish a rational basis for decision making, one must understand the relative strengths and weaknesses of each of the techniques as they may apply to the requirements of a project. The complex technological details of each of the various techniques are not covered in this chapter, but it is helpful to have a general understanding of basic principles. The project requirements must first be determined and discussed with the laboratory manager. The laboratory manager will select methods based on requirements for turnaround time, degree of accuracy, budget, and so forth.

Instrumentation Commonly Used for Solid Material

Spark Optical Emission Spectroscopy

Spark optical emission spectroscopy (arc spark OES or spark OES) is typically used where metallic samples are produced, such as foundries. Sample material is vaporized with a testing probe by an arc spark discharge. The atoms and ions contained in the atomic vapor are excited into emission of radiation. The radiation emitted is passed to the spectrometer (arc spark OES) optics via an optical fiber, where it is dispersed into its spectral components. From the range of wavelengths emitted by each element, the most suitable line for the application is measured by means of a charge-coupled device. The radiation intensity, which is proportional to the concentration of the element in the sample, is compared to calibration curves. This is usually used for measuring concentrations of metals\alloys and impurity analysis in metals such as

- Steel, cast-iron, and high-alloyed steels;
- Nonferrous metals and their alloys;
- Al: wrought alloys, casting alloys, etc.;
- Cu: bronze, brass, cupronickel, etc.;
- Mg, Zn alloys, solders;
- Nitrogen in steel;
- Phosphorus in aluminum; and
- Ultra-low carbon analysis for elements such as Se, La, Te, and so on.

X-Ray Fluorescence

XRF is used on both solid and liquid material. It is fast and known for high precision. It is a useful tool for the elemental analysis of ores, concentrates, tails, metals, glass, and ceramics. When XRF is configured with purge gas, liquids may also be analyzed. X-rays form part of the electromagnetic spectrum. They are on the high-energy side of ultraviolet and are expressed in terms of their energy in kiloelectron volts (keV) or wavelength in nanometers (nm). XRF can be considered in a three-step process occurring at the atomic level:

1. An incoming X-ray knocks out an electron from one of the orbitals surrounding the nucleus within an atom of the material.
2. A hole is produced in the orbital, resulting in a high-energy and unstable configuration for the atom.
3. To restore equilibrium, an electron from a higher-energy, outer orbital falls into the hole. Since this is a lower-energy position, the excess energy is emitted in the form of a fluorescent X-ray.

The energy difference between the expelled and replacement electrons is characteristic of the element atom in which the fluorescence process is occurring; thus, the energy of the emitted fluorescent X-ray is directly linked to a specific element being analyzed.

The elements and concentrations that XRF analyzers can determine depend on the material being tested and the configuration of the instrument used. The two main XRF methodologies are energy-dispersive X-ray fluorescence (EDXRF) and wavelength-dispersive X-ray fluorescence (WDXRF). Each method has its own advantages and disadvantages. The range of detectable elements varies according to instrument configuration and setup, but typically EDXRF covers all elements from sodium (Na) to uranium (U), while WDXRF can extend this down to beryllium (Be). WDXRF systems can routinely provide much better working resolutions (between 5 and 20 eV), depending on their setup, whereas EDXRF systems typically provide less resolution (ranging from 150 to 300 eV or more), depending on the type of detector used. Limits of detection depend on the specific element and the sample matrix, but as a general rule, heavier elements will have better detection limits.

Comprehensive software algorithms using fundamental parameters, characteristics based on theoretical X-ray beam intensity, detector solid angle, matrix effects (element–element interactions), band overlap, and spectral backgrounds are used to calculate element concentrations based on the observed peak intensities. This provides a very fast, standardless quantification method, which will work well for various matrices. However, matrix-matched calibration standards are best for accuracy.

Advantages of XRF over other techniques are that it is a fast analytical tool for elemental composition. XRF can measure concentrations ranging from parts per million to 100% and can be used on solids, liquids, and powders. It can also be nondestructive to the sample, and the technique has high precision.

Whole rock analysis is the determination of major elements of a rock sample. This will total approximately 100% in non-mineralized samples with elements calculated as oxides—Al_2O_3 (aluminum oxide), CaO (calcium oxide), Cr_2O_3 (chromium oxide), K_2O (potassium oxide), MgO (magnesium oxide), MnO (manganous oxide), Na_2O (sodium oxide), P_2O_5 (phophorous pentoxide), Fe_2O_3 (iron oxide), SiO_2 (silicon dioxide), and TiO_2 (titanium dioxide)—and loss on ignition (LOI), which includes moisture and carbonaceous material, to sum to 100%. Where the analytes of concern in many industrial products (i.e., iron ore, cement, ilmenites, rutile, limestone) are the major elements, XRF is the first choice for instrumentation.

Disadvantages of EDXRF and WDXRF are that they suffer from matrix and inter-element interference due to signal enhancement or suppression. Also, the standard and sample must be matrix matched with respect to both the matrix and the particle size. Finding calibration standards with an exact match of sample type is difficult and time-consuming; consequently, sample preparation to match standards on hand becomes a simpler plan.

In general, the energy of the emitted X-ray for a particular element is independent of the chemistry of the material. For example, a calcium peak obtained from EDXRF for $CaCO_3$ (calcium carbonate), CaO, and $CaCl_2$ (calcium chloride) will be in exactly the same spectral position for all three materials. However, the wavelength shifts a bit for Cu when it occurs as CuS (covellite) compared to $CuSO_4$ (copper sulfate), and will require sample preparation by fusion for accurate total Cu measurement.

Sample preparation for semiquantitative analysis is minimal with direct analysis of loose powder. However, preparation for quantitative analysis is more rigorous by either of two methods. Both require the particle size to be <50 μm to minimize grain-size effects. Adding a binder and pressing into a pellet with a hydraulic press is one method. This technique is useful if very low detection is necessary and mineralogical effects are not expected to be a problem. Another preparation method involves mixing the finely ground sample with a flux of lithium tetraborate, metaborate, or a combination of the two, and a small amount of wetting agent in a platinum crucible. The crucible is fused at about 1,000°C and the melt is poured into a platinum mold, forming an amorphous glass disk. (Graphite crucibles may be substituted for platinum if needed for only a few fusions.) Fusion eliminates the physical and mineralogical interferences and allows the analyst to make corrections on only the inter-element effects. Alternatively, the melt may be poured into a solution of weak hydrochloric or nitric acid for analysis on other instrumentation. The level of skill required for routine analysis is low. However, skill level for accurate results of frequently changing ores can be high. The instrument cost for EDXRF is moderate (US$35,000–US$85,000), while the cost of a WDXRF can be quite high (US$150,000–US$275,000).

Other spectroscopic methods use the same principle as XRF but use incident radiation. These include laser-induced breakdown spectroscopy, electron microprobe analyzer, particle-induced X-ray emission, X-ray photoelectron spectroscopy, and auger. Although these techniques are extremely useful for mineral characterization, they are found exclusively in research or university laboratories and are reserved for specific applications. Details of these instruments are outside the scope of this chapter.

Instrumental Neutron Activation Analysis

Neutron activation analysis (NAA) is an analytical technique that measures gamma rays emitted from a sample that was irradiated by neutrons. The rate the gamma rays are emitted from an element in a sample is directly proportional to the concentration of that element. Multiple samples are placed in polyethylene vessels and simultaneously irradiated. Following irradiation, a high-resolution germanium detector coupled to a multichannel analyzing system measures sample decay. Sample data are compared to standard data.

Advantages of NAA over other analytical techniques are numerous and include ease of sample preparation, high precision, simultaneous measurement of multiple elements, high reproducibility, excellent interlaboratory comparability to other methods, large dynamic range from parts per billion to high percentages, and relatively low cost of analysis at approximately US$23 per sample.

The major disadvantage is the requirement to have access to a licensed nuclear reactor.

Combustion Infrared Analyzers

Elemental *combustion infrared analyzers* is a generic term for sulfur and carbon analysis. These types of analyzers are often referred to as Leco analysis although other manufacturers (e.g., Eltra, Horiba) produce similar instruments. Analytes of interest in metallurgical products may include light elements that are difficult to analyze in other ways and include C, S, O, H, and N. Final and intermediate metal and alloy products have tight limits on S, O, C, H, and N because each may be detrimental to the properties of the metal.

Total carbon and sulfur and phases of carbon and sulfur are of particular interest in characterizing ores and metallurgical products for process development, process control, and for estimating acid rock drainage. Carbon and sulfur are usually measured simultaneously on one of two types of carbon/sulfur (C/S) combustion infrared analyzers, resistance furnace or induction furnace. A common error is made when measuring geological and metallurgical process material on a combustion analyzer designed for metallic samples. Sulfur results are often erroneously low if samples contain crystalline water or hydroxide, such as clays and hydrated process material when measured on a combustion instrument with an induction furnace. Also, most induction styles were designed for very low concentrations of carbon and sulfur in metallic samples and are incapable of accurately measuring sulfur concentrations above 10%. Sulfur in geological or metallurgical process samples are best determined using a combustion analyzer with a resistance furnace and a dual concentration detector to accommodate a wide range of sulfur concentrations.

Combustion analyzers are simple to operate. In the case of a C/S instrument, a sample is weighed into a ceramic crucible and an accelerator (such as iron or tungsten oxide) is added to the sample and placed in the furnace. It is melted in a stream of pure oxygen in a high-temperature atmosphere, causing all sulfur forms to oxidize to sulfur dioxide (SO_2) and carbon forms to carbon dioxide (CO_2). The combustion gases pass through a halogen trap and moisture absorber for purification. The SO_2 and CO_2 are detected in infrared cells. Using Leco combustion analysis for estimating carbon and sulfur phases is detailed later in this document.

Instrumentation Used for Liquids

Metallurgical solid samples (e.g., pulps, sludges, concentrates, tailings) requiring elemental analysis are usually digested in acid or fused with salts and dissolved into a solution to be measured using one of the following liquid spectrometric techniques.

Flame Atomic Absorption Spectroscopy

Atomic absorption spectroscopy (AAS), commonly called atomic absorption (AA), is an invention that has since been labeled as one of the most significant achievements in chemical analysis in the last century. Measurement of metals went from hours to minutes. AAS remains a workhorse in most laboratories with little training required to operate the equipment.

A solution is first digested and aspirated into a flame and a ground-state atom absorbs energy in the form of light of a specific wavelength and is elevated to an excited state. The amount of light energy absorbed at this wavelength will increase as the number of atoms of the selected element in the light path increases. The relationship between the amount of light absorbed and the concentration of analyte present in known standards is used to determine unknown sample concentrations by measuring the amount of light they absorb. This is a relatively low-cost and simple instrument to operate but is designed to measure one element at a time. Some newer models are programmed to measure other elements sequentially to increase productivity.

Accessories are used to expand the capabilities of AAS. Hydride generation and flow injection atomic spectroscopy are accessories for AAS to allow hydride generation of hydride-forming elements (As, Bi, Sb, Se, Sn, Te) using sodium borohydride to convert these elements into hydride gases. This technique works well for low detection in water samples. Other techniques must be used when significant amounts of base metals are present, such as those in sulfide concentrates.

AAS for mercury poses special analytical challenges because mercury may be partially or completely volatilized by high-temperature drying or acid digestions. It also amalgamates with gold resulting in low instrument response using the cold vapor method. Two AAS methods are used for Hg analysis:

1. **Cold vapor atomic absorption** is an AAS atomization method limited to only the determination of mercury, due to it being the only metallic element to have a large enough vapor pressure at ambient temperature. A sample is oxidized with nitric and sulfuric acid to Hg^{+2} and then reduced with stannous chloride to Hg^0. An inert gas carries the mercury gas through the reaction mixture. The concentration is determined by measuring the absorbance of this gas at 253.65 nm. Detection limits for this technique are in the parts-per-billion range, making it an excellent trace-level mercury detection AAS method. However, in the presence of Au, Hg is amalgamated and the signal is suppressed by an amount equivalent to the Au concentration, and a low reading results.
2. **Direct thermal combustion analysis** is the preferred method for higher concentrations of mercury typically found in gold processing plants because of gold–mercury amalgamation. A solid or liquid sample is weighed into a quartz or metal boat and dried. It is then thermally decomposed in an oxygen-rich furnace. Mercury and other combustibles are released from the sample and carried to a catalyst section of the furnace, where nitrogen and sulfur oxides, as well as halogens and other interfering compounds, are eliminated. Mercury is selectively trapped, in a separate furnace, through gold amalgamation. The amalgamation furnace is heated and mercury is rapidly released. Mercury vapor flows via a carrier gas through an optical path of the spectrophotometer, where it is quantitatively measured by AAS at 253.65 nm.

Graphite Furnace Atomic Absorption Spectroscopy

Graphite furnace atomic absorption spectroscopy is used when much lower detection limits than AAS can produce are required. A small volume (microliters) of digested sample is pipetted into a small graphite tube (the furnace). Then a series of three electrical heating steps are usually applied to the sample contained in the graphite furnace. The first of these is the drying stage, followed by the ashing stage where the temperature of the furnace is increased in the range of 400° to 800°C. The final stage is atomization where the temperature of the furnace is very rapidly raised to temperatures often as high as 2,700°C, effectively volatilizing the remaining metallic components on the furnace wall. Many of the volatilized metals will come off the walls in the atomic form, creating an atomic vapor. The absorbance measurement that occurs during this stage, using a hollow cathode lamp shone through the furnace, creates transmittance measurement that is common to all forms of spectrophotometry. An integrated signal of absorbance versus time is directly proportional to the actual mass of the element. This technique is relatively slow and most labs have abandoned it for other instrumentation.

Microwave Plasma–Atomic Emission Spectrometry

MP-AES is a relatively new technology, and, although it is not an acceptable technique where a laboratory must use a regulated environmental method for a particular analysis, it offers some advantages over other techniques. Remote minesite laboratories often have difficulty receiving instrument gases, such as acetylene or argon. MP-AES uses nitrogen from a nitrogen generator as a plasma gas. The nitrogen MP is considerably hotter than the acetylene flame used in AAS, reaching temperatures nearing 5,000 K. At these temperatures, atomic emission is very strong for most elements, leading to improved detection and linear dynamic ranges over flame AAS for most elements. Operating costs can be significantly reduced compared to flame AAS by using a nitrogen generator. Using inert nitrogen is a safer choice over flammable acetylene. MP-AES can be a productive tool compared to flame AAS because it uses a scanning monochromator and a solid-state detector for multi-element analysis instead of individual lamps as used in AAS. A disadvantage of MP-AES is that it is limited in the amount of total dissolved solids (TDS) it can tolerate in a solution.

Inductively Coupled Plasma–Optical Emission Spectrometry

ICP-OES has become another laboratory staple. A solution in which elements are dissolved is pumped into a nebulizer, mixed with gas to produce a fine mist, and injected into an argon plasma that has been induced by means of high-frequency waves. The plasma electron temperature can range between ~6,000 K and ~10,000 K (~6 eV and ~100 eV; Shun'ko et al. 2014), comparable to the sun's surface. Charged elements become thermally excited and each element emits radiation in the form of light at a characteristic wavelength.

The intensity of the emission has a proportional relationship to the concentration of that element. In more modern units, the light is focused through a series of optics and separated energies fall upon an array of detectors, allowing for simultaneous measurement of approximately 1 to 60 elements per minute. The orientation of the torch may be aligned in a radial position or end-on axial position. Axial position is used to achieve 3–10 times lower detection of an analyte. Radial alignment is useful for major elements of interest.

Inductively Coupled Plasma–Mass Spectrometry

ICP-MS follows a sample introduction system through a high-temperature argon plasma similar to ICP-OES. A pair of cones, which allow only a fraction of the sample to enter the chamber, reduces the number of excited ions. The positively charged ion stream is focused through an electrostatic lens, which also has a positive charge, and collimates the ion beam, focusing it through a quadrupole where the ions are filtered by their mass-to-charge ratio. This ability to filter ions on their mass-to-charge ratio allows ICP-MS to take advantage of isotopic information, since different isotopes of the same element have different masses. A detector receives an ion signal proportional to the concentration. As ions hit the surface of the detector and are converted to electrons, the active film coating is consumed and detectors must be replaced if bombarded with high concentrations of ions. Care must be taken to protect the costly detector by diluting samples or choosing a less abundant isotope.

Interferences may be caused by chlorides from hydrochloric acid used in sample digestion or plasma gas combining with ions and forming masses that overlap with analytes of interest. High-resolution power is required to separate interferences from masses close to the analytes of interest. A skilled ICP-MS operator is required to understand how to overcome such interferences. High-resolution or magnetic sector mass spectrometers have become more common in ICP-MS, allowing the user to eliminate or reduce the effect of interferences due to mass overlap. Reaction or collision gases remove interfering ions through ion–neutral reactions (Tanner et al. 2002). Ammonia, methane, oxygen, and hydrogen have been used as reaction gases to eliminate interferences.

Analytical Interferences and Detection Limits

Interferences—matrix effects in particular—are the changes in the sample composition (non-analyte constituents) of the test material and are inherent in most chemical analyses. Often the bulk of the analyst's time is directed toward eliminating matrix effects, either by separation of the analyte from the matrix, chemical treatment measures, or by calibration procedures to minimize the effects. Steps taken to minimize interferences can compromise the expected detection limits of a particular method or instrumental response. If budget constraints require choosing to purchase only one instrument, ICP-OES is usually chosen over ICP-MS and XRF because the systematic errors are much easier to eliminate and the random measurement error can be as low as 0.5% (Gaines 2012). Addressing specific interferences for each instrument or analytical method is out of the scope of this chapter. Thompson (2005) describes strategies for minimizing interferences and calculating uncertainties.

Published detection limits for analytical instrumentation are useful when comparing relative responses between analytical instruments but seldom represent actual reporting limits. A relative comparison of detection ranges for instruments measuring elements in solution are shown in Figure 1. However, matrix effects and other factors contribute to the overall uncertainty and method reporting limits. Most acid solutions of ores and metallurgical samples contain high amounts of TDS, causing unstable or suppressed instrument response. Accessories are necessary to improve detection, stability, and uncertainty. Specialized nebulizers for high TDS and gas humidifiers aid in analysis.

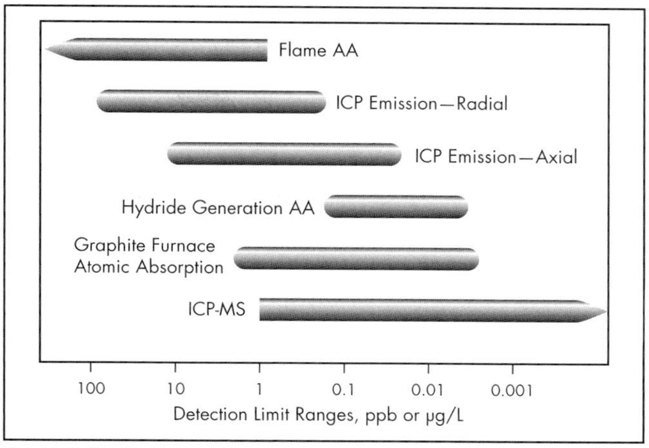

Courtesy of PerkinElmer, Inc.

Figure 1 Typical detection limit ranges for the major atomic spectroscopy techniques

SAMPLE DECOMPOSITION

Before solution analysis is possible, the samples must first undergo dissolution, a process to convert solids to a solution ready for instrumental measurement. Total dissolution is the goal; however, it is not always possible to liberate all elements of interest in a single procedure. Losses occur from volatilization, precipitation, and adsorption onto the vessel walls or incomplete digestion. This may lead to underestimation of elements of interest. Contamination is nearly impossible to avoid, as no single digestion vessel is perfect for all analytes. One must understand the limitations and advantages of each approach.

Acid digestion and salt fusion are the typical approaches, with acid digestion as the preferred method of dissolution of mineralogical matrices into a solution containing metals. Acids are reagents of choice because of low sample contamination with other metals compared to fluxing salts. Larger sample sizes may be used in acid digestions, and final solutions are low in dissolved solids.

The metals below hydrogen in the electromotive series only need to displace hydrogen and will dissolve in nonoxidizing acids with the evolution of hydrogen, but some are too slow to react, such as with hydrochloric acid on lead, cobalt, nickel, cadmium, and chromium.

Oxidizing acids must be used to dissolve the metals above hydrogen in the electromotive series. The most common of the oxidizing acids are nitric acid, hot concentrated perchloric acid, or some mixture that yields free chlorine or free bromine.

In many cases, the sample will not dissolve completely, but the constituent to be determined may readily dissolve out of the insoluble matrix. For example, acid-soluble oxides or sulfides may occur with insoluble silicates. If acid digestion is not sufficient, the residue may be fused by one of the procedures discussed later.

Common Strong Acids for Sample Decomposition

Hydrochloric Acid

Hydrochloric acid, or HCl, is a nonoxidizing, complexing strong acid. Concentrated hydrochloric acid (about 12 M, 37%) is an excellent solvent for many metal oxides as well as those metals below hydrogen in the electromotive series. It is often a better solvent for the oxides than the oxidizing acids. Carbonate and oxide ores of Ba, Ca, Fe, and Mg are directly soluble in HCl. Chlorides are useful as complexing agents to prevent acid-resistant metal oxides from precipitating. Some elements volatilize as chlorides or other halogens when heated to fumes. For example, chlorides of As, Sb, Ge, and B at temperatures >140°C and $AuCl_3$ (gold trichloride) at >265°C will volatilize (Dulski 1996).

Nitric Acid

Concentrated nitric acid, or HNO_3 (about 16 M, 70%), is an *oxidizing* solvent for attacking metals. It will dissolve most common metallic elements except Al, B, Cr, Ga, Hf, In, Nb, Sb, Sn, Ta, Th, Ti, W, and Zr that become passive by forming a film of oxide on the surface of the metal. Nitric acid is more useful in combination with other acids. Some metals, such as metallic Cu, are somewhat insoluble in concentrate nitric acid but readily dissolve in a 1:1 dilution of nitric acid with water.

Perchloric Acid

Hot concentrated perchloric acid, or $HClO_4$ (about 12 M, 72%), is an extremely strong oxidizing agent and solvent. It must be used after nitric acid reaction ceases to prevent any easily oxidizable material from exploding. The hot acid attacks ferrous alloys and stainless steels that are insoluble to the other mineral acids. This acid also dehydrates and rapidly oxidizes organic materials. Dry metal chlorate salts must be wetted with water prior to adding acid for salt dissolution to prevent explosion. The use of perchloric acid requires specially designed fume hoods with water sprays throughout the ductwork to prevent explosive perchloric crystals from forming. A trained technician is required for safe usage.

Hydrofluoric Acid

The primary use for hydrofluoric acid, or HF (about 29 M, 49%), is for decomposition of silicate rocks and minerals *where silica is not to be determined* because the silicon volatilizes from the solution as silicon tetrafluoride. This releases the elements encapsulated within silica grains to be attacked by the other mineral acids. After decomposition is complete, the excess HF is driven off by evaporation with perchloric acid to near dryness. Since HF attacks glass and quartz, polytetrafluoroethylene (PTFE) laboratory ware must be used throughout. **CAUTION:** HF can cause serious painful injury, even death, when exposed to skin. Calcium gluconate ointment should be kept nearby during use and emergency procedures put in place. Only trained personnel should attempt digestions with HF.

Sulfuric Acid

Hot (near boiling point, 340°C) concentrated sulfuric acid, or H_2SO_4 (about 18 M, 96%), is often used as an oxidizing agent; otherwise, it is useful at ambient temperature for partial digestion of oxide minerals. Sulfuric acid is a good desiccant and will dehydrate and decompose most organic materials when concentrated. Sulfuric acid may be used for digestion of ores of Al, Be, Mn, Th, and U.

Mixtures of Acids for Optimal Sample Decomposition

Maximum sample decomposition of minerals is rarely achieved using one type of acid. Blending two or more acids together produces properties, such as oxidizing strength, greater than the sum of the individual acids. The most common of these acid mixtures are discussed in the following sections.

Aqua Regia

The aqua regia digestion method is optimally used with a 3:1 mixture of hydrochloric (HCl) and nitric (HNO_3) acids. The two acids react together and produce synergistic strong oxidizing products: nitrosyl chloride and chlorine gas:

$$3HCl + HNO_3 \rightarrow 2H_2O + NOCl + Cl_2$$

Aqua regia is considered adequate for dissolving most base element sulfates, sulfides, oxides, and carbonates but only provides a "partial" extraction for most rock-forming and refractory elements because it does not attack silicates. For example, aqua regia extraction might give complete extraction of Cd, Cu, Pb, and Zn and also the volatile elements (Sb, As, Bi, Se, and Te) while it is known to provide only partial extraction of Cr, Ni, and Ba. These latter elements can only be efficiently recovered by using HF. However, the aqua regia digestion method may be acceptable for scoping these elements when they are not of primary interest. *Aqua regia* (in Latin) means "royal water" because of its ability to dissolve Au and other precious metals. However, this is not the case when organic "preg robbing" carbon is present and a stronger oxidizing digestion is required (e.g., by sodium chlorate salt addition). Other ratios of nitric to hydrochloric are often used for various digestions.

4-Acid Digestion

4-acid digestion is a blend of nitric + hydrochloric + hydrofluoric + perchloric acids and is more vigorous than aqua regia and provides satisfactory dissolution of most silica matrices. A PTFE digestion vessel must be used. This method is used on a hot plate for a near "total" sample decomposition. Perchloric acid decomposes organic or preg-robbing carbonaceous material. The acid mixture with HF will dissolve silica matrices by the following reaction:

$$4HF + SiO_2 \rightarrow SiF_4 + 2H_2O$$

SiF_4 (silicon tetrafluoride) is volatilized and is not measured. This is a very effective method for trace analysis; however, there can be a loss of other volatile elements (e.g., As, B, Ge, Hg, Pb, Sb), and some refractory minerals (especially oxide minerals) are only partially digested or precipitate quickly as metal oxides (Al, Ba, Cr, Hf, Mo, Mn, Nb, Pb, Sb, Sn, Ta, Ti, W, Zr). Excess HF must be volatilized to prevent precipitation of metal fluorides. The use of perchloric acid requires specially designed fume hoods with water sprays throughout the ductwork to prevent explosive perchloric crystals from forming.

Other reagents to aid in sample dissolution are hydrobromic acid and bromine for sulfides and for volatilizing As, Se, and Sb. Hydrogen peroxide (H_2O_2) may be added as a clean oxidant. Potassium chlorate in nitric acid provides a strongly oxidizing mixture but must be used with care by experienced technicians. Citrate and tartrate aid in complexing certain elements to prevent precipitation.

Sample Decomposition/Digestion Techniques

Four common techniques for sample decomposition are hot plate and hot block open-vessel acid dissolution, leaching, microwave digestion, and fusion.

Open-Vessel Acid Dissolution

Following are the two types of open-vessel acid dissolution.

1. **Hot plate** samples are acid-digested in an open Pyrex beaker. PTFE beakers are used when HF is added. Digestion is on a high-temperature hot plate when silicates or other elements are purposefully volatilized.
2. **Hot block** is used for an acid digestion at a lower controlled temperature when the desire is to prevent element volatilization. This technique is useful for preventing loss of volatile elements such as Sb and Se; however, the temperature is not high enough for a total digestion.

Leaching

Leaching techniques are used to determine the soluble fraction of an element of interest under specific leaching conditions. Results are not intended as a total decomposition. Changing leaching condition variables (temperature, time, leachate strength, agitation speed) usually impacts reproducibility and differs from laboratory to laboratory.

Microwave Digestion

Microwave digestion ovens have solved several problems. Digestion in a microwave oven in a closed container is more powerful than hot-plate heating because of the closed pressurized vessels, and the specialized fume hoods for perchloric acid digestions is eliminated. Strong oxidizing conditions are created particularly when H_2O_2 is added and carbon or other carbonaceous material may be digested, liberating Au. Vessels are sealed and volatile elements are retained. When PTFE vessels are used, HF may be added to dissolve silicates; however, one must be cautious with solubilized samples because HF is still active and dangerous. Contrary to popular belief, addition of boric acid does not render HF into a harmless ionic molecule. Boric acid simply acts to complex HF to prevent precipitation of calcium fluoride (CaF_2) and other fluorides. Sample introduction components (i.e., injectors and spray chambers for AA and ICP) must be HF resistant. The downside to microwave digestion is that the number of vessels and the sample size are limited in these systems.

Fusion

Fusion is a technique to decompose all minerals by melting a finely ground sample with a salt-based flux and is reserved for the most refractory material or for eliminating mineral interferences for XRF instrumental analysis. Many elements of interest in ores are often not completely decomposed after a 4-acid total digestion, particularly samples containing Al, Ba, Be, Cr, Ge, Hf, Mn, Mo, Nb, Pb, Re, Sn, Ta, Ti, W, and Zr. High temperature in conjunction with a flux to lower the melting point decomposes the refractory material. The downsides are that (1) the ratio of flux to sample is often 10:1, diluting a sample beyond detection by XRF or producing a solution high in dissolved solids, which is problematic for sample introduction and precision in ICP-OES and ICP-MS; (2) contamination inherent in fluxes requires meticulous blank correction; (3) the Li and Na in flux are quite damaging to instrument quartz torches; and (4) some elements are volatilized at the high fusion temperature. A skillful analyst is able to avoid many of these issues with third-party accessories, but the analysis cost is at least twice the cost of acid digestion.

A complete decomposition of a sample with a flux is achieved by first grinding the sample to a very fine powder. An appropriate flux is selected and mixed well with the sample in a compatible crucible (Table 3) and covered. The temperature is raised slowly in a furnace or over a flame. The maximum temperature used varies considerably and depends on the flux and the sample. Gentle agitation during fusion improves homogeneity. Fusion is complete when a visual inspection shows a total melt without particles. The mass is allowed to cool somewhat and then dissolved into a liquid using a solvent—depending on the method—of water, dilute hydrochloric acid, or nitric acid. Unattacked material may require a change in flux.

Fluxing and fusing into a glass disk or bead is another option if the sample will be analyzed using XRF or laser ablation ICP-MS. Blends of lithium metaborate or tetraborate are mixed well with a sample, fused at about 1,000°C, and poured into a platinum mold. An automated system can produce several hundred glass disks per day for whole rock analysis. Care must be taken when choosing a crucible since fusion crucibles can be a costly consumable. Fusions must have negligible attack on the crucible and the material from which the crucible is made and should not contain any of the elements for which the sample is being analyzed.

The following should be taken into account when choosing a vessel for fusion:

- Platinum crucibles are excellent with lithium meta/tetraborate fusions but are expensive and easy to destroy if one attempts to clean them in aqua regia. However, they can last through hundreds of fusions with no impact if properly used. They should never be used for a sodium peroxide fusion.
- Zirconium, nickel, and iron crucibles may be used for sodium peroxide fusions where Zr, Ni, and Fe are of no consequence in the analysis and temperatures do not need to be much higher than 600°C. Zr is the best for complete fusion of sulfide minerals since Ni and Fe are usually of interest.
- Graphite crucibles are the least costly but have the shortest life. They may be used with caution and for small samples by heating in an induction furnace with an argon purge gas. The recommended flux is lithium metaborate or lithium tetraborate.

The following should be considered when choosing a flux for fusing:

- Alkaline fluxes are used for attack on acidic materials and include sodium carbonate, potassium carbonate, potassium or sodium hydroxide, borates, and sodium peroxide. Sodium peroxide also has oxidizing properties, making it a popular choice. The basic fluxes may be used individually or as mixtures to lower the fusion melting point. Lithium tetraborate is used for dissolving basic oxides such as alumina. Lithium metaborate is more basic and

Table 3 Common fusion fluxes, temperatures, and preferred crucible type

Flux (melting point)	Fusion Temperature, °C	Type of Crucible	Types of Sample Decomposed
$Na_2S_2O_7$ (403°C) or $K_2S_2O_7$ (419°C)	Up to red heat	Pt, quartz, porcelain	For insoluble oxides and oxide-containing samples, particularly those of Al, Be, Pu, Ta, Ti, Zr, and the rare earths
NaOH (321°C) or KOH (404°C)	450–600	Ni, Ag, glassy carbon	For silicates, oxides, phosphates, and fluorides
Na_2CO_3 (853°C) or K_2CO_3 (903°C)	900–1,000	Ni, Pt for short periods (use lid)	For silicates and silica-containing samples (clays minerals, rocks, glasses), refractory oxides, quartz, and insoluble phosphates and sulfates
Na_2O_2	600	Ni, Ag, Au, Zr; Pt (<500°C)	For sulfides; acid-insoluble alloys Fe, Ni, Cr, Mo, W, and Li; Pt alloys; Cr, Sn, and Zn minerals
H_3BO_3	250	Pt	For analysis of sand, aluminum silicates, titanite, natural aluminum oxide (corundum), and enamels
$Na_2B_4O_7$ (878°C)	1,000–1,200	Pt	For Al_2O_3; ZrO_2, and zirconium ores, minerals of the rare earths, Ti, Nb, Ta, aluminum-containing materials; iron ores and slags
$Li_2B_4O_7$ (920°C) or $LiBO_2$ (845°C)	1,000–1,100	Pt, graphite	For almost anything except metals and sulfides. The tetraborate salt is especially good for basic oxides and some resistant silicates. The metaborate is better suited for dissolving acidic oxides such as silica and TiO_2 and nearly all minerals.
NH_4HF_2 (125°C), NaF (992°C), F (857°C), or KHF_2 (239°C)	900	Pt	For the removal of silicon, the destruction of silicates and rare earth minerals, and the analysis of oxides of Nb, Ta, Ti, and Zr

Adapted from Dean 1995 and Bock 1979

better for dissolving acidic oxides such as silica or titanium dioxide. A blend of both is usually used. Boron oxide (B_2O_3) is used to decompose silicates and oxides when Li, Na, or K will be determined.

- Acid fluxes are the pyrosulfates (of sodium, potassium, and ammonium), bisulfates, potassium hydrogen sulfate, the acid fluorides, and boric acid. They are used for insoluble oxides and oxide-containing samples.
- Oxidizing fluxes contain sodium peroxide, potassium nitrate, sodium nitrate, or potassium chlorate, usually in mixture with alkaline fluxes. They are used for sulfides; acid-insoluble alloys of Fe, Ni, Cr, Mo, W, and Li; Pt alloys; and Cr, Sn, and Zn minerals.
- Neutral fluxes are the most commonly used for destroying silicates and include sodium fluoride, borax, and lithium fluoride.

Table 3 summarizes proper flux and crucibles for different material types.

Sample Decomposition Losses and Contamination

Losses are likely to occur but may be minimized if understood. Acid dissolution may present losses in many ways, such as when complexed ions (e.g., tetrachloroaurate, or $AuCl_4^-$) adhere on glass surfaces. Vigorous shaking of the solution may bring some of the adsorbed species back into solution (Twyman 2005). Plastics also absorb ions, and acids in strengths >10% are recommended for storage. Filter paper can also absorb analytes of interest. Centrifugation is preferable over filtration when the filtrate solution is to be analyzed. Losses from volatilization of halides may be minimized with lower digestion temperatures or the use of closed-vessel microwave.

Silver loss is a common occurrence, usually by precipitation as silver chloride (Equation 1a). The key to consistent Ag analysis is to avoid all chloride or add excess chloride as hydrochloric acid (at least 15%–30% in final solutions) to resolubilize the silver as a very stable silver chloride complex (Equation 1b).

$$Ag^+ + Cl^- \rightarrow AgCl\ (solid) \quad \textbf{(EQ 1a)}$$

$$AgCl\ (solid) + Cl^- \rightarrow Ag(Cl)x\ (aqueous) \quad \textbf{(EQ 1b)}$$

Since avoiding chloride is virtually impossible, excess chloride is always required to stabilize Ag (Hg and Pb behave similarly). Very concentrated silver, however, is best measured by a titration method or by fire assay.

Contamination may occur in a variety of laboratory ware. PTFE beakers are difficult to clean when a "sticky" inorganic species is formed such as silver chloride, which may require an ammonium hydroxide rinse. Other contamination of Al, B, Na, and Si is typically found in borosilicate glass test tubes, which should be replaced with plastic when these are elements being analyzed. Dust in laboratory mining labs is a concern as it is drawn into fume hoods. All reagents should be purchased in a quality suited for the purpose.

FIRE ASSAY FOR GOLD AND PLATINUM GROUP METALS

Fire assay is a quantitative method for accurately determining precious metals of gold and platinum group metals (PGMs). Silver is also fire assayed in higher concentrations. Fire assay principles have been used for hundreds of years with little change except for the integration of AAS finish in the 1960s, which brought higher productivity and improved detection limits for gold determination. An advantage of fire assay over other wet chemical methods is that a larger sample may be used, it is relatively inexpensive, precious metals are quickly separated from gangue minerals, and it is widely applicable to ores, concentrates, and rocks.

Edward Bugbee, a professor at the Massachusetts Institute of Technology (United States) and highly regarded author of *A Textbook of Fire Assaying*, believed that a course in fire assaying is the logical place to introduce the study of metallurgy (Bugbee 1922). Many metallurgical principles are utilized within the fire assay process, such as the thermochemistry of the metals, oxide and sulfide ores, the nature and physical constants of slags, the characteristics of refractories and fuels, the principles of ore sampling, the behavior of metallic alloys on cooling, and the chemical reactions of oxidation and reduction. Training in fire assay fundamentals is difficult to find, although Montana Tech (Butte, Montana, United States) offers a course each summer (Montana Tech 2017).

A typical crucible fire assay involves five steps: fluxing, fusion, cupellation, parting, and finish. The sample is mixed with a flux consisting primarily of lead oxide (litharge), as a collector of precious metals, along with other reagents customized for the characteristics of the material being assayed. The sample is blended well in a crucible with the flux mixture and fused at high temperature to produce two molten layers, a complex borosilicate slag and metallic lead. The gangue material follows the slag while the precious metals collect in the lead. The slag is discarded and the lead "button" is placed in a vessel made of absorbent bone ash or magnesia, called a cupel. During cupellation, the lead metal is oxidized to lead oxide and absorbed into the cupel. A bead containing gold, silver, and possibly some of the PGMs remains on the cupel surface. This bead is then "parted" in acid to remove silver and other impurities. The gold remaining after parting is either finished gravimetrically by weighing on a microbalance or digested in acid and the Au measured using AAS or ICP-OES.

The sample weight used in a fire assay charge is referred to as an "assay ton" or some fraction of an assay ton. An assay ton is exactly 29.167 g. The weight in milligrams of precious metal obtained from an assay ton of ore equals the number of ounces to the ton. With modern electronic balances and automated calculations, the concept of using an exact assay ton is obsolete, however the terminology is still used. Half-assay and quarter-assay ton are 14.583 and 7.292 g, respectively, and are used for more difficult sulfides and complex ores.

The five steps of fire assay are described in detail in the following sections.

Fluxing with Litharge

Accurate fire assay requires characterizing the material for proper fluxing. The material is broadly classified as neutral, acidic, basic, oxidizing, or reducing. In general, nonmetals form acidic oxides with SiO_2 and Al_2O_3 being the most common, while metal oxides are basic. Carbonate metals liberate CO_2 during heating, leaving base metals behind. Some metal oxides, such as zinc oxide (ZnO) and Al_2O_3, are amphoteric with both acid and base properties. Table 4 summarizes oxides in the earth's crust (Bugbee 1940).

Table 4 Oxides in order of abundance in the earth's crust

Slag-Forming Constituents	Acid or Base
Alumina, Al_2O_3	Acid
Calcium oxide, CaO	Base
Cuprous oxide, Cu_2O	Base
Ferrous oxide, FeO	Base
Lead oxide, PbO	Base
Magnesium oxide, MgO	Base
Manganous oxide, MnO	Base
Potassium oxide, K_2O	Base
Silica, SiO_2	Acid
Sodium oxide, Na_2O	Base
Zinc oxide, ZnO	Base

Source: Bugbee 1940

Fluxing acidic, siliceous ores is fairly simple and straightforward with only a decision of how much extra sodium carbonate to add, if any. Fluxing basic ores or pyrite and metallurgical products is more complex. Ideal slags are produced when metals are converted to their bivalent state either by (a) oxidation with PbO, (b) oxidation with heat, or (c) reduction to a lower valence using carbon or metallic iron (Beamish and Van Loon 1977). For example,

a. Oxidation to bivalent oxide
   $FeS_2 + 5PbO \rightarrow FeO + 2SO_2 + 5Pb$
b. Oxidation of a metal by roasting
   $2Cu + O_2 \rightarrow 2CuO$
c. Reduction of a trivalent oxide to a bivalent oxide
   (with carbon) $2Fe_2O_3 + C \rightarrow 4FeO + CO_2$
   (with metallic iron) $Fe_2O_3 + Fe \rightarrow 3FeO$

Samples with small amounts of pyrite may be reduced to divalent pyrrhotite (FeS) using (d) metallic iron, (e) roasting, or (f) niter (potassium nitrate, or KNO_3) addition. Examples (e) and (f) are more effective for high sulfide pyrite concentrates:

d. $FeS_2 + Fe \rightarrow 2FeS$
e. $4FeS_2 + 11O_2 + \text{heat} \rightarrow 2Fe_2O_3 + 8SO_2\uparrow$
f. $4FeS_2 + 10KNO_3 + \text{heat} \rightarrow 4FeO + 5\,K_2SO_4 + 5N_2 + 3SO_2\uparrow$

Roasting sulfides must be carefully controlled to avoid gold losses to volatilization (Strong and Murray-Smith 1974). Ores containing As, Sb, and Bi as sulfides require special treatment where they are partially oxidized. They must be carefully roasted, starting in a cool muffle furnace and raising the temperature slowly to 650°C and maintaining for 2 hours. In the case of As, the sample must be oxidized only to the trivalent state and not all the way to its pentavalent state. If pentavalent arsenic forms, it will combine with iron to form arsenates, or speiss. Speiss compounds can occlude gold with subsequent losses. Speiss arsenates must be eliminated before progressing to the fusion process. To remove pentavalent speiss arsenate, the sample is roasted again under reducing conditions with carbon to reduce pentavalent to trivalent arsenic where it is volatilized. Samples containing chlorides must not be roasted. Substantial gold losses may occur through volatilization when 10% or more of sodium chloride is present (Young 1980).

The ultimate goal of a proper fusion is to produce a lead button of consistent mass containing all the precious metal, none of the non-precious metals, and separates easily from a highly fluid slag. A lead button weight of approximately 28 g is ideal, although in the range of 25–35 g is acceptable. A button too small may not have collected all the precious metal. Too large a lead button could result in losses during the subsequent cupellation step. A larger lead button, while a waste of lead, is preferable to a button too small, since it may be scorified (see the "Fusion" section later in this chapter) to reduce base metals.

One must determine the reducing or oxidizing power of an ore or other metallurgical material and whether it is acidic, basic, or neutral. The reducing power is the amount of lead that 1 g of sample will reduce when fused when excess lead is available. The oxidizing power is the opposite, the amount of lead that 1 g of sample will oxidize during fusion.

Determining the Reducing or Oxidizing Power of Test Material

Visual examination. Visual examination of the ore as received is often a quick and reliable method of approximating the mineral composition of a sample for an experienced assayer. To understand the composition of the material, beginner assayers must also review the ICP elemental analysis, carbon/sulfur results, X-ray diffraction (XRD) report, and discuss the ore composition with a mineralogist or metallurgist.

Vanning. Vanning is a less common way of classifying an ore using a rapid technique to determine the minerals present when Leco sulfur or ICP analysis is not available. Vanning, applied with a few variations between laboratories, consists of taking 0.5–1 g of the ground sample with a spatula and placing it in a porcelain dish. The sample is moistened with a few drops of water and stirred with a glass rod. Two to three drops of 1:1 HCl are added. Any effervescence is generally due to evolution of CO_2. When effervescence stops, more drops of HCl are added, and any additional effervescence is noted. 1:1 HCl addition is continued until effervescence ceases. While a small fume extractor is recommended, an odor of hydrogen sulfide (H_2S) during vanning suggests sulfides are present and is particularly characteristic of sphalerite. Many sulfides, including pyrite, do not give an odor. Sulfides reduce litharge, and therefore, their presence and amount should be known so that the proper flux components can be selected. A few drops of 10% HNO_3 are then added. Brown fumes of nitrogen oxides (NOx) indicate the presence of sulfides. With practice and experience, the amount of carbonate and sulfide in the sample can be estimated rather closely; this amount determines how much of each flux component should be used. These acid tests are often used in conjunction with a Leco total sulfur and carbon, and carbonate calculation and an ICP-OES analysis to establish the presence of carbonate or sulfide minerals.

Preliminary fusion. A preliminary fusion is first undertaken using the following charge mixed in a small crucible: 2 g ore, 10 g sodium carbonate, 46 g lead oxide, 3 g silica, and 1 g $Na_2B_4O_7$ and the resulting lead button is weighed to determine the reducing power. If, for example, the resulting lead button weighs 10 g, then the reducing power is determined:

$$\text{reducing power} = \frac{\text{Pb button weight}}{\text{sample weight}} = \frac{10\text{ g}}{2\text{ g}} = 5$$

Using a half-assay ton (~15 g) as a sample size, one can calculate the amount of niter (KNO_3) required to add to the charge to oxidize the sample in order to achieve a lead button 28–30 g. From Table 5 the oxidizing power of niter is 4.2.

total reducing effect of ore =

sample weight(g) × reducing power = 15 g × 5	= 75 g Pb
lead button target weight	= 28 g Pb
difference, ore equivalent that must be oxidized by niter	= 47 g Pb
1 g of niter oxidizes (Table 5)	= 4.2 g Pb
niter required = 47/4.2	= 11.2 g niter

Flux components. A general flux usually consists of the following four reagents: litharge (PbO), sodium carbonate (Na_2CO_3), borax ($Na_2B_4O_7$), and silica (SiO_2). Other components are added as needed.

Table 5 Approximate reducing power or oxidizing power of some common minerals and reagents

Mineral or Reagent	Reducing Power	Oxidizing Power
Carbon/charcoal, C	18–30	—
Sugar	14.5	—
Cornstarch	11.5–13	
Pyrite, FeS_2	11	—
Flour	10–11	—
Pyrrhotite, FeS	9	—
Argol (potassium tartrate)	8–12	—
Chalcopyrite, $CuFeS_2$	8	—
Sphalerite, ZnS	8	—
Arsenopyrite, FeAsS	7	—
Stibnite, Sb_2S_3	7	—
Chalcocite, Cu_2S	5	—
Cream of tartar	4.5–4.6	—
Metallic iron, Fe	4–6	—
Galena, PbS	3.4	—
Niter, KNO_3	—	4.2
Pyrolusite, MnO_2	—	2.4
Hematite, Fe_2O_3	—	1.3
Magnetite, Fe_3O_4	—	0.9
Magnetite-ilmenite	—	0.4–0.6

Adapted from Hafty et al. 1977; Bugbee 1940

- **Litharge (PbO)** is also a strong basic flux to react with metallic oxides. It acts as an oxidizing and desulfurizing agent. Together with the addition of flour, it is reduced to metallic lead and collector for precious metals. It is, by far, the most expensive consumable component of the fire assay and must be disposed as a hazardous waste. Litharge should always be tested for impurities. Red granular litharge is preferred over yellow finely powdered litharge from a safety and hygiene aspect because dust is less of a problem.
- **Sodium carbonate or soda ash (Na_2CO_3)** is a powerful basic flux that is usually the principal component of fire assay flux. It reacts with silicates to form alkali glassy slag.
- **Borax (anhydrous sodium tetraborate, $Na_2B_4O_7$)** is an acidic flux component that lowers the fusing point of all flux materials. It forms fusible complexes with limestone and magnesite. It is often used as a cover over the fluxed material to minimize losses due to dusting or overflow. It also protects the crucible from flux attack. Too much borax is detrimental in acidic ores, causing difficulty in separating slag from lead phases.
- **Silica (SiO_2)** is the strongest acid component of the flux and combines with metal oxides to form silicates, the primary component of almost all slags. It also provides an acid constituent for basic ores and protects the crucible from the corrosive action of litharge.
- **Calcium fluoride (CaF_2)** increases the fluidity and is used especially for high-grade aluminum-bearing samples, as much as 8 g. Other samples not necessarily containing high aluminum but requiring an excessive amount (4 g or more) of CaF_2 are black sands, magnetite, and calcium phosphate (bone ash).

Table 6 Suggested grams of flux per 15-g sample

Sample Composition	Na_2CO_3	PbO	SiO_2	$Na_2B_4O_7$	CaF_2	Flour	KNO_3
Basic and ultra-basic rocks, including mineralized basic rocks. Also used for silicates where the Fe and Mg are each 5%–10% or more.	30	35	4	35	1	3.2	0
Quartzite, high silicates	20	50	1	3	0	2.8	0
Copper concentrate	20	100	10	5	1	0	7.5
Pyrite concentrate	30	60	12	10	0	0	30

- **Reducing agent (flour, cornmeal, carbon)** is required to reduce PbO to Pb metal evolving CO and CO_2.
- **Niter (potassium nitrate, KNO_3)** is a strong oxidizing agent. It melts at 334°C, but at a higher temperature it expels oxygen, which oxidizes sulfur, carbon, and many of the metals. The reducing power of the ore must be calculated because excessive amounts of niter may cause excess silver loss and boiling-over of the charge.
- **Inquart metal.** Silver, in the form of either silver nitrate or silver prills, is most commonly added in very quantitative amounts. A few milligrams of silver is "inquarted" into the crucible prior to fusion to protect from gold losses, to aid in handling the precious metal bead, and to facilitate parting of silver and gold following cupellation. The ratio of silver to gold should be about 3:1. Other metals (e.g., palladium and gold) are also used as inquarts to improve silver or other PGM recoveries.
- **Other metal collectors.** In addition to lead, other metals may be used, particularly for collection of PGMs. Nickel sulfide is used for PGMs because all six PGMs are collected and the fusion temperature is lower. The disadvantage is that gold recovery is not as good as with lead. Other collector metals include tin, copper, and iron, but these are not as effective as lead for gold. Silver, platinum, and palladium are good collectors but uneconomical.

In-depth flux calculations for all material types are beyond the scope of this chapter and are covered in other sources (Shepard and Dietrich 1940; Lenahan and Murray-Smith 2001; Bugbee 1940; Hafty et al. 1977). Table 6 lists a few common metallurgical types of samples and appropriate ratios of flux components for a half-assay ton.

Fusion

Crucible Fusion

Once the charge is mixed well, the crucible is loaded into a fusion furnace initially set at about 800°C and finishing at 1,100°C for about 1 hour when slag is thoroughly fluid. The lower start temperature prevents rapid CO_2 evolution. The crucible contents are poured into a conical mold or the slag onto a metal table and the lead into a button mold. After cooling, the slag is discarded and the lead button is pounded into a cube shape to simplify moving with tongs and to remove adhering slag particles.

Scorification Fusion

If the lead button is brittle, residual sulfur and base metals may still be present, which will impact cupellation losses. The brittle lead button must be scorified to remove sulfur and base metals. Scorification is an oxidizing fusion carried out in a shallow fire-clay dish called a scorifier. The button is placed in the scorifier with test lead, borax, and sometimes silica. The dish is placed in the furnace under appropriate conditions for scorification to take place. Scorification may also be performed instead of a crucible fusion but is limited to very small sample sizes.

Cupellation

Cupellation is the separation of precious metals from lead. The lead button is placed in a preheated cupel. A cupel is a porous vessel made most commonly of bone ash or magnesia. Under proper temperature conditions, the lead oxidizes to PbO. The molten PbO is absorbed by the cupel. One to two percent of the PbO volatilizes and is captured in a baghouse. Precious metals are not absorbed and a small precious metal bead, also referred to as the doré, remains in the cupel. Cupels should weigh about one-third more than the lead button.

Once cupellation begins and the buttons appear to have melted, the furnace door is cracked open to allow air to flow over the cupels. Very pure lead requires a cupellation temperature of 850°C and impure lead requires higher temperatures of 1,000°C. The rate of cupellation should be about 1 minute per gram of lead. Crystals of litharge resembling feathers on the sides of the cupel are an indication of correct cupellation temperature. The temperature is increased by 100°C toward the end to drive off the last of the lead. Experience is required to recognize brightening and other signals indicating completion of cupellation.

Cupellation losses are inherent in the cupellation process. There is always some loss of gold but not as much as silver. Temperature is the most important factor influencing precious metal losses. Losses are more pronounced when the temperature is too high or when the lead button is too large and cupellation goes on too long. Silver is added to protect against gold losses. Impurities including As, Sb, Se, Bi, Te, Tl, and especially Cu interfere and must be removed by scorification prior to cupellation. Extensive discussion of approaches to minimize losses may be found in fire assay texts (Bugbee 1940; Shepard and Dietrich 1940).

Parting

Parting is the separation of silver from gold alloys by dissolution of silver in hot, dilute nitric acid. Less common are the few special methods that use sulfuric acid. Acids must be free from any form of chlorides, as they tend to precipitate the dissolved silver or even combine with the nitric acid to dissolve gold. (This is not a problem if instrumental finish is used.) Either of these reactions would adversely affect the outcome of the gravimetric finish assay. The most frequent strength of nitric acid used ranges from a ratio of 1:4 to 1:8 parts acid to water. If the parting is incomplete, it is likely due to the incorrect ratio of silver to gold. The ratio of silver to gold must be about 3:1. Too much silver causes the bead to break into pieces and losses are likely, while too little silver will mean parting may be incomplete. In many determinations, a secondary parting using a stronger acid is needed to ensure a complete separation of the silver–gold alloy. The completion of the parting process is indicated when the evolution of nitrogen oxide gas bubbles has ceased. At this time, the presence of platinum may be evident by a brown-colored solution, while the presence of palladium is indicated by a distinctive orange color.

Finish

The finish is usually determined by the grade of precious metals present in the final bead after parting. A large bead will be by a gravimetric finish, while small beads will be digested and follow an instrumental finish for lower detection.

Gravimetric Finish

The gold that remains after parting in dilute nitric acid is black in color or is a large, brassy mass. Annealing the gold in a muffle furnace at 700°C for 5 minutes changes it to the more familiar gold color. This also avoids the possible effect of absorbed gases, preventing added weight other than gold. The final bead is weighed on a microbalance accurate to 0.000001 g (1 µg). Gravimetric finish is suggested for gold concentrations >5 ppm and always for bullion analysis. There is a possibility of other metals (e.g., Ag, Pb, Pt, and other PGMs) interfering with the final weight.

Instrumental Finish

The doré bead is cleaned, pounded flat, and placed in a test tube. One mL of 1:1 dilute HNO_3 is added and test tube is heated in a water bath at 80°C for 15 minutes to dissolve Ag. The gold bead remaining is then dissolved in the same test tube by adding 1 mL of concentrated HCl in a water bath at 80°C. After 15 minutes, 8 mL water is added for a final total volume of 10 mL. Silver chloride solids are separated by centrifuging. The solution is read using AAS or ICP-OES. Measuring metals in solution by AA or ICP-OES only measures the element of interest. Instrumental finish is recommended for gold concentrations <5 ppm or when other metals are suspected to be present. The AA finish has an approximately 10× lower reporting limit than the gravimetric finish.

Screen Fire Assay

Screen fire assay should be performed when apparent coarse particulate gold causes poor reproducibility between same samples or when the concentration of gold is greater than 15 ppm and very accurate results are required. The sample preparation requires significantly more time to complete. One kilogram of sample is screened through a 200-mesh (75-µm) screen. Screen size may vary depending on gold particle size. The weights of each fraction are recorded. The oversize portion contains the coarse gold and is assayed in its entirety. The undersize fraction is rotary split and portions are assayed in duplicate. The assay results are multiplied by the weight fraction of each and added together. A gravimetric finish is typically used.

Gold and other precious metals may be by-products associated with other industrial metal products such as those related to aluminum, bismuth, cadmium, chromium, copper, indium, iron, lead, molybdenum, nickel, palladium, platinum, rhodium, silicon, silver, tantalum, tin, titanium, tungsten, vanadium, and zinc. Appropriate analytical procedures for measurement of precious metals and impurities may be found in standard reference books (Dillon 1955; Furman 1962; (Young 1971).

ELEMENTAL PHASES AND DIAGNOSTIC TESTING BY CHEMICAL DISSOLUTION

Chemical phase analysis determines, by chemical dissolution, a characteristic of an element within a rock or ore. That characteristic might be its oxidation state or a separation of an element associated with a particular mineral. Diagnosing the phase of an element is of interest in hydrometallurgical process optimization and plant design. Sometimes called hydrogeometallurgy, chemical phase analyses are not a replacement for full mineralogical evaluations but offer quick, inexpensive, qualitative results for testing many samples. They are seldom quantitative because of partial solubility of other minerals within the same material. The key to meaningful results is performing these analyses under consistent and reproducible conditions (e.g., leach time, temperature, agitation speed, leach concentration, solid-to-liquid ratio, and particle size).

Gold Diagnostic Testing

Gold by fire assay is the preferred method of analysis for total gold concentration. However, process development and ore control may require other gold leaching tests.

Cyanide-Soluble Gold

Cyanide-soluble gold (AuCN) is the amount of gold soluble in a defined concentration of cyanide. The ratio of cyanide-soluble gold to total gold by fire assay, often referred to as the "cyanide-to-fire ratio," is an important parameter in process development and in ore control. Results from the AuCN procedure is an indicator of non-refractory gold that is amenable to cyanide leaching and is not a predictor of actual deposit Au recovery. For assay results of Au >0.1 ppm, cyanide-soluble gold (AuCN) may be determined by measuring the Au soluble in 30 mL of 0.3% NaCN and 0.3% NaOH solution (or another agreed-upon concentration) mixed for 1 hour at room temperature using 15 g or more of finely ground sample. The result will give an indication of leaching difficulty due to several factors that may include the presence of sulfides, large particulate Au, preg-robbing organic carbon or clay content. Samples containing particulate gold will require averaging of replicates and larger sample sizes and cyanide leachate volumes. Strong cyanide leaching (>3% sodium cyanide solution) is used for testing of material from deposits found to contain free gold >50 µm and some Cu/Au deposits during early stage development. Stronger cyanide (3% NaCN [sodium cyanide] plus 1% NaOH [sodium hydroxide]) is required to appropriately characterize the Au. This test is for characterization of non-refractory free Au and is not a predictor of actual Au extraction.

Preg-Robbing Cyanide

Preg-robbing analysis in cyanide is a diagnostic test for estimating the preg-robbing characteristic for gold (AuPR). It is identical to the AuCN test but also contains a gold spike. The spike concentration of 3.4 ppm (0.10 opt [ounces per ton]) is commonly used. The AuPR index, which is calculated using the AuCN and AuPR assays, typically applies to any deposit containing organic carbon. The cyanide concentration used in the AuPR must match the cyanide concentration in the AuCN test. The degree of preg-robbing is determined by a preg-robbing index (PRI). Preg-robbing may also be expressed as a spike recovery percentage.

Table 7 Selective pretreatment leach stages and the minerals destroyed

Pretreatment Stage	Minerals Likely To Be Destroyed
1. NaCN washes	Precipitated gold
2. NaCN	Gold
3. Na_2CO_3	Gypsum and arsenates
4. HCl	Pyrrhotite, calcite, dolomite, galena, goethite, calcium carbonate
5. $HCl/SnCl_2$	Calcine, hematite, ferrites
6. H_2SO_4	Uraninite, sphalerite, labile copper sulfides, labile base metal sulfides, labile pyrite
7. $FeCl_3$	Sphalerite, galena, labile sulfides, tetrahedrite, sulfide concentrates
8. HNO_3	Pyrite, arsenopyrite, marcasite
9. Oxalic acid washes	Oxide coatings
10. HF	Silicates
11. Acetonitrile elution	Gold adsorbed on carbon, kerogen, coal

Source: Lorenzen 1995

- PRI = 3.4 (spike solution) + AuCN – AuPR
- If PRI = 0, material is non-preg-robbing
- If PRI > 3, material is extremely preg-robbing

Bulk Gold Diagnostic Testing
Bulk Au diagnostic testing is performed when Au mineral associations are unknown and troubleshooting a process requires knowledge of gold deportment. Detailed diagnostic testing is performed on a large sample (0.5–1 kg). The sample is chemically treated to liberate Au associated with various mineral phases (Table 7). All the mentioned pretreatment stages can be varied according to the matrix of the material. Temperature, potential, concentration, treating time, and so on, play a major role in the selection of the desired pretreatment stage. Details are thoroughly discussed in Tumilty and Schmidt (1986) and Lorenzen (1995).

Copper Mineral Phases
Total Cu >0.02% is leached to estimate its behavior under metallurgical processes. Unlike total Cu, partial digestions have a much higher degree of error. Consistency is the key to useful data for estimating Cu mineral variability. Stronger leach solution, aggressive agitation, leach temperature, agitation time, and particle size are all variables that affect results. These mineral-selective leaches are run in sequence on a single sample to semiquantitatively identify the potential recovery by various copper ore processing methods. The actual amount of each mineral dissolved in each leach may vary, depending on the sample mineralogy, grain size, and other physical characteristics. Detailed differences between laboratories must be agreed upon before analysis and meticulously documented. Cu mineral assumptions should be confirmed with scanning electron microscopy (SEM).

Acid-Soluble Copper (CuAS) Test
Ore is typically leached with weak sulfuric acid (100 g/L) for 1 hour at room temperature. CuAS estimates easily leachable oxidized copper minerals such as malachite, azurite, chrysocolla, and portions of cuprite and tenorite as well as "acid-soluble" content of copper in secondary copper sulfides as a result of ferric iron oxidation. If the purpose is to differentiate oxide copper from sulfidic copper in a float process, then the copper oxide (CuOX) test is more appropriate. In this case, ferric iron is chemically reduced to ferrous to prevent oxidation. Addition of citric acid minimizes oxidation of chalcocite from Fe(III).

Copper Oxide (CuOX) Test
Ore is leached with weak sulfuric acid plus a reducing agent (100 g/L H_2SO_4 + 15 g/L sodium sulfite) for 1 hour at room temperature. CuOX is reported since the ferric iron is reduced to ferrous, and the copper from secondary sulfides is not leached. Alternatively, sulfurous acid may be used but is more costly.

Copper Quick Leach Test (CuQLT)
Ore is leached with weak sulfuric acid plus Fe(III) (100 g/L H_2SO_4 + 5 g/L ferric ion as sulfate) for 1 hour at room temperature. The CuQLT estimates easily leachable copper, such as in a dump leach, and includes oxide and some portion of copper secondary sulfides. The CuQLT has utility for bioleaching applications.

Cyanide-Soluble Copper (CuCN) Test
Ore is leached with a solution of 3% NaCN/3% NaOH for 1 hour at room temperature. Cyanide leach will dissolve the oxides (with the exception of chrysocolla, which is only partially digested), secondary sulfides including chalcocite and covellite, and most bornite and enargite. The chalcopyrite content in ores will remain largely undissolved. Cyanide consumption from copper minerals can be estimated.

Sequential Copper Leach (Cu-Seq) Test
This leach is a combination of CuAS, CuCN, and total Cu measured on a single sample aliquot sequentially. This is a special request assay used occasionally to estimate copper minerals present.

Water-Soluble Copper (CuWS) Test
With this method, copper ore is leached in deionized water. This is sometimes used in conjunction with other copper leaches, including EDTA (ethylenediaminetetraacetic acid) as a leach reagent, for characterizing oxidized ore (i.e., stockpiled ore) to understand low recovery or potential environmental impacts.

Sulfur and Carbon Phases
Total carbon and sulfur are first measured on combustion infrared analyzers. A second aliquot of the same sample undergoes a leach, or roast, and the residue is again measured for sulfur and carbon on the combustion infrared analyzers. The leach solution and/or the roast temperature depend on the sulfur mineral, and a knowledge of mineralogy is required to make the best decision. Table 8 summarizes sulfur-phase methods, assumptions, and mineral interferences for each method.

Sulfur Phases
Sulfide sulfur and sulfate sulfur are required on samples for both metallurgical process selection and environmental assessments, such as acid generation potential estimation. Determination of sulfide sulfur can be challenging and a variety of methods are used by different laboratories. It may be measured directly or through subtraction of sulfate sulfur from total sulfur.

Table 8 Sulfide sulfur assay procedure mapping

Assays	Procedure	Assumption of Method	Minerals Determined	Minerals Interference
S(total)	Leco	All sulfur is converted to gas and measured	All sulfur minerals	No interference
Sulfide sulfur	Method A	SS=SCIS	Refer to SCIS	
	Method B	SS=SCIS/HCl	Refer to SCIS/HCl	
	Method C	SS=S(tot)-SAP 550°C	Refer to SAP 550°C	
	Method D	SS=S(tot)-SAP 650°C	Refer to SAP 650°C	
	Method E	SS=SAI	Refer to SAI	
	Method F	SS=SAICONC	Refer to SAICONC	
Sulfate sulfur	Method G	SO_4=S(tot)-SCIS	Refer to SCIS	
	Method H	SO_4=S(tot)-SCIS/HCl	Refer to SCIS/HCl	
	Method I	SO_4=SAP 550°C	Refer to SAP 550°C	
	Method J	SO_4=SAP 650°C	Refer to SAP 650°C	
	Method K	SO_4=S(tot)-SAI	Refer to SAI	
	Method L	SO_4=S(tot)-SAICONC	Refer to SAICONC	
SCIS	Sodium carbonate digestion, filter/Leco sulfide	Sulfate dissolves and sulfides remain.	Sulfur contained in pyrite, pyrrhotite, marcasite, and base metal sulfide minerals measured as sulfide	Barite, alunite, and lead sulfate do not fully dissolve, resulting in high-sulfide results. Orpiment and realgar dissolve, resulting in low-sulfide results.
SCIS/HCl	Sodium carbonate digestion, filter, dissolved arsenic sulfides precipitated with HCl, filter/ Leco sulfide	Sulfate minerals and orpiment and realgar dissolve, sulfides filtered, orpiment and realgar are re-precipitated during acidification and filtered.	Sulfur contained in pyrite, pyrrhotite, orpiment and realgar, and base metal sulfide minerals measured as sulfide	Barite, alunite, and lead sulfate do not fully dissolve, resulting in high-sulfide results.
SAP 550°C	Pyrolysis at 550°C/Leco sulfate	Sulfides are lost, leaving sulfate.	Sulfur contained in pyrite, pyrrhotite, marcasite, and base metals—except copper, lead, and zinc—are volatilized, sulfate remaining measured	Carbonate minerals adsorb sulfur dioxide, resulting in high-sulfate results. Copper minerals are not volatilized, giving high-sulfate results. Jarosite may be partially lost, resulting in low-sulfate results.
SAP 650°C	Pyrolysis at 650°C/Leco sulfate	Sulfides are lost, leaving sulfate.	Sulfur contained in pyrite, pyrrhotite, marcasite, and base metal sulfide minerals are volatilized. Remaining sulfate is measured.	Carbonate minerals and lime adsorb sulfur dioxide, resulting in high-sulfate results. Jarosite may be partially lost, resulting in low-sulfate results.
SAI	HCl digestion, filter/Leco sulfide	Sulfates are dissolved; sulfides remain.	Pyrite, marcasite, and base metal sulfides measured as sulfide in residue	Pyrrhotite dissolves in acid, leading to low-sulfide results. Barite, alunite, and lead sulfate are not fully dissolved, leading to high-sulfide results.
SAICONC	Strong HCl digestion, filter/ Leco	Sulfates are dissolved; sulfides remain.	Pyrite, marcasite, and base metal sulfides	Pyrrhotite dissolves in acid, leading to low-sulfide results. Barite, alunite, and lead sulfate are more fully dissolved, leading to improved sulfide results.
Elemental S <5%	Chloroform/ultraviolet-visible	Elemental sulfur dissolves and is measured in solution.	Elemental sulfur	Elemental S ≥5% will not dissolve in chloroform. Smaller weight can improve results.
Elemental S ≥5%	Carbon disulfide/gravimetric	Elemental sulfur dissolves and is measured in solution residue.	Elemental sulfur	Elemental S <5% poor precision.

HCl = hydrochloric acid
SAI = sulfur-hydrochloric acid–insoluble
SAICONC = sulfur-strong acid–insoluble after digestion (represents sulfide in a concentrate)
SAP = sulfur after pyrolysis
SCIS = sodium carbonate–insoluble sulfur (represents sulfide sulfur in most cases)
SS = sulfide sulfur

If a roast, or pyrolysis, technique is selected for measuring sulfate sulfur, it may be determined by roasting to drive off sulfide sulfur and then measuring the residue, also referred to as sulfur after pyrolysis. The temperature used to roast sulfides must be agreed upon prior to analysis and documented since 550°C is used for estimation of acid generation potential from pyrite while 650°C is used for total sulfide estimation (ASTM E1915 2013). Another approach to determining sulfide sulfur is to hot leach the soluble sulfates with dilute sodium carbonate or dilute HCl to report sodium carbonate–insoluble sulfur or hydrochloric acid–insoluble sulfur. The sulfide residue is washed, dried, and measured on the combustion infrared analyzer. XRD analysis should be completed before selecting an appropriate method since sulfide sulfur can be overstated or understated, depending on which sulfur-bearing minerals are present. Common interfering minerals for selecting a sulfide sulfur method are alunite, barite, jarosite, orpiment, realgar, pyrrhotite, and carbonates. See Table 8 to determine the most appropriate method for determination of sulfide sulfur.

Elemental sulfur or native sulfur is occasionally present naturally in some ores or is a product of a metallurgical process. If elemental sulfur is present in amounts >0.01%, the assumptions in Table 8 are invalid and will require discussion of appropriate sulfide determination with a project metallurgist. Elemental sulfur is considered to be a potentially acid-generating mineral for stockpile and waste rock characterization.

Carbon Phases

Organic and inorganic carbon are of interest in gold process development. Organic carbon, also called acid-insoluble carbon (CAI), is determined by hot leaching a sample with dilute HCl, driving off CO_2, and measuring the residue by combustion infrared analysis. Organic carbon is detrimental in the gold cyanide extraction process and CAI may be used to predict preg-robbing potential of an ore. CAI analysis may also include non-preg-robbing graphitic carbon and is correlated with the preg-robbing index to be certain. Graphitic carbon may be determined by roasting followed by an HCl wash. Graphitic carbon is also separated using thermogravimetric analysis. CAI subtracted from total carbon represents inorganic carbon (carbonate carbon) and is used to estimate the acid neutralization potential in tailings from an environmental perspective.

Mineral Phases of Other Materials by Chemical Dissolution

Use of XRD, SEM, and other mineralogical tools simplify identification of mineral phases. However, when mineral identification instrumentation is unavailable, chemical methods for phase separation are still useful for hydrometallurgy and ore control. Full mineralogical assessment should be performed to validate chemical phase assumptions. A shortened list of lixiviants used for quick phase identification of elemental phases in multiple ore types is shown in Table 9. More detail of each is available in Steger (1976).

CLASSICAL WET CHEMICAL ANALYSES

Prior to about 1960, classical wet chemical analysis was the only means of determining the chemical composition of minerals. Such analyses were limited by the factors previously discussed in terms of sampling size, although techniques were developed for analyzing very small samples. It is important for the extractive metallurgical student to recognize that many hydrometallurgical processes (i.e., dissolution, separation, concentration, purification, and metal plating) involve chemical principles of oxidation, reduction, chelation, hydrolysis, and so on. These steps are also found in traditional analytical chemistry; thus, it is possible to regard hydrometallurgy as applied analytical chemistry with the only difference being the scale of the involved steps. The identification with analytical chemistry suggests that the solutions to the relevant industrial unit operations may be found in the procedures already developed by analytical chemists.

Wet chemical analyses always involve dissolving the mineral into a solution. In order for the dissolution to take place completely, the mineral is usually first ground into a fine powder (to increase surface area) and the appropriate solvent must be used. Wet chemical analysis can be classified into three types: gravimetric, volumetric, and colorimetric.

Gravimetric Analysis

Gravimetric analysis is where the element of interest is precipitated as a compound. The precipitate is then weighed to determine its proportion in the original sample (e.g., $BaSO_4$ [barium sulfate] precipitation for sulfate measurement).

Volumetric Analysis

Volumetric analysis is a quantitative chemical analysis that is used to determine an unknown concentration of an analyte. A titrant is prepared as a standard solution. A known concentration and volume of titrant reacts with a solution of analyte. The volume of titrant reacted is recorded and the concentration of unknown is calculated. Following are common titrations routinely used in an analytical laboratory on metallurgical samples:

- Silver nitrate titration for free cyanide, where Ag forms the $Ag(CN)_2^-$ complex. Another example is silver nitrate titration for free chloride where Ag^+ and Cl^- are precipitated together as silver chloride (AgCl). A silver electrode measures the change in potential in both titrations.
- Short iodide titration is a very accurate method for determining Cu in copper concentrates for settlement between shipper and buyer. Cu is first dissolved in mixed acid. Cu(III) oxidizes potassium iodide (KI) to I_2. Sodium thiosulfate ($Na_2S_2O_3$) is then used to titrate I_2, equaling the moles of Cu present.
- Dimethyl glyoxime precipitates nickel followed by titration with EDTA for settlement of Ni in nickel concentrates.
- Redox titration of Fe in iron ore by acid decomposition, followed by reduction with stannous chloride and then titration with an oxidant such as potassium dichromate.

Colorimetric Analysis

In colorimetric analysis, a reagent is added to the solution that reacts with the element of interest to produce a color change in the solution. The intensity of the color is proportional to the concentration of the element of interest, and thus, when compared to standard solutions in which the concentration is known, the concentration of the element in the unknown solution can be determined (e.g., thiocyanate complexed with ferric iron).

Many classical wet chemical analyses require a skilled analytical chemist, and tests are usually very time-consuming and costly when performed for metal content on settlement material between buyer and seller. These types of testing are usually used only where extremely high precision and accuracy are required.

The high dollar value of concentrate and bullion shipments means that the regular analytical methods used for other stages of mineral resource development are not appropriate and classical wet chemical methods must be used. As an example, the splitting limit for %Cu in a copper concentrate is 0.15%–0.2%. This represents a requirement for less than 1% relative accuracy and precision in the analysis. As a result, the analytical procedures for these types of samples are much more rigorously controlled. To ensure that high accuracy and precision are obtained, the QC protocols are expanded. Four

Table 9 Chemical phase analysis of ores and rocks

Mineral Associated with ...	Treatment	Mineral Separation
Aluminum	Determined by rates of dissolution in KOH (potassium hydroxide)	Bauxite minerals, gibbsite, boehmite, and diaspore, and "available aluminum"
	NaOH + HCl	Gibbsite only
	Boiling (1:1) HCl	Aluminum phosphate minerals such as crandallite and millesite separated from kaolinite
Antimony	Tartaric acid	Antimony oxides separated from sulfides
Beryllium	3% HCl and roasting	Multiple beryllium minerals are separated
Bismuth	5% Thiourea solution in 0.5N H_2SO_4	Bismuth oxide minerals from other bismuth minerals
	0.1M Silver nitrate in 0.5–1.0M nitric acid	Native bismuth from other bismuth sulfide
Calcium	Sucrose solution	Free CaO is extractable from $Ca_3(PO_4)_2$, CaF_2, $CaCO_3$, and $CaSO_4$
	5%–10% acetic acid	Calcite, $CaCO_3$, can be selectively dissolved in the presence of fluorite, CaF_2
Copper	Dilute HCl in ethanol plus stannous chloride	Cuprite, Cu_2O, selectively extracted from ores that contain native copper and tenorite, CuO
	Ferric chloride in 3M HCl	Native Cu from tenorite
	(a) 1% Solution of unithiol in 5% HCl	Azurite, malachite, and cuprite from other Cu minerals
	(b) 2% Sodium sulfite in 3% H_2SO_4 after (a)	Chrysocolla from Cu sulfides
Germanium (in coal ash)	0.1M Ammonia	GeO_2 from other Ge minerals
Indium	Water	$In_2(SO_4)_3$
	3% Bromine solution in methanol	In_2S_3
	3M HCl	In_2O_3
Iron (in reduced ores)	Bromine in alcohol	Metallic Fe
Iron	HCl in an inert atmosphere, then ferrous titrated with dichromate	FeO and Fe_2O_3 in reduced iron ores (after metallic Fe removed)
	1% HCl in H_3PO_4 plus hydrogen peroxide	Magnetite separated from hematite
	Mixture of sulfuric and hydrofluoric acid or ammonium vanadate. Ferrous is titrated with permanganate.	Ferrous separated from ferric iron in rocks and ores
	Bromine in methanol	Pyrrhotite, chalcopyrite, and bornite extracted from pyrite
Lead	25% sodium chloride solution	Anglesite, $PbSO_4$
	15% Ammonium acetate solution in 3% acetic acid	Cerussite, $PbCO_3$
Manganese	6N Ammonium sulfate acidified with H_2SO_4 to pH 2	$MnCO_3$ separated from MnO_2
Molybdenum	20%–50% Ammonia solution	MoO_3 from roasted MoS_2
Nickel-cobalt	Ascorbic acid–hydrogen peroxide	Ni, Co, and Cu sulfide minerals in rocks
Tantalum-niobium	Mixture of HF and HCl at slightly elevated pressure	Pyrochlore and microlite from simpsonite
Uranium	5% Solution of tri-n-octylamine in xylene	U(VI) separated from U(IV)
Zinc	Boiling 5% acetic acid	Oxidized zinc from sphalerite

Adapted from Steger 1976

or six replicates are analyzed with matrix-matched certified reference samples. This ensures that no matrix-caused bias affects the analytical result.

The mineralogical makeup of the reference samples should be as similar to that of the samples as possible. Titer proofs (pure metals titrated along with samples to monitor the concentration of the titrating solution in instances of volumetric control assay work) can also be performed as one determination. It is important that the buyer or seller discusses all requirements with the analytical scientists to ensure that the data generated are adequate for their intended uses.

ANALYTICAL TESTING OF HYDROMETALLURGICAL SOLUTIONS

Controlling hydrometallurgical processes requires analysis of different streams of solvents. Leaching processes may include a variety of leaching reagents. The most commonly used leaching agents are sulfuric acid (Cu, Zn), sodium carbonate (U, V, Mo, W), sodium hydroxide (W), ammonia (Cu, Ni), cyanide and thiosulfate (Au, Ag), sulfite (Sb, Hg), and chlorine and chloride (PGMs and rare earth metals; Bazhko 2009). The elements of interest, the solvent itself, and often the degradation created from the solvent must be measured.

For the most accurate results, solvents containing elements of interest must first be digested to destroy the solvent, usually in aqua regia, before instrumental analysis. Measuring solvents directly by flame-AAS, ICP-OES, ICP-MS, while common practice, often causes erratic results because of solvent decomposition during sample introduction. XRF may sometimes be considered and does not require predigestion. Calibration standards must contain the same solvent and digestion matrix as the samples.

Analytes of interest are difficult to measure in matrices with large concentrations of other elements. Separating metals may be accomplished by precipitation or ligand complexation. Solvent extraction techniques have been developed for analytical purposes to separate elements. Chelating agents added to organic solvents, resins, or filter papers are used to preferentially separate trace elements from major interfering elements. Details of separating metals through chelation may be found in Rydberg et al. (2004).

Measuring ions in solution is necessary for a variety of reasons, including process control, cost estimation, and environmental assessment.

Instrumentation for Analytical Measurement of Hydrometallurgical Solutions

Ion Chromatography

Ion chromatography (IC) is a process used to separate ions into their constituents based on their affinities for column material. A common application is utilizing a conductivity detector to measure the response for Cl^-, F^-, NO_3^-, NO_2^-, PO_4^{3-}, and SO_4^-. Detectors and columns may be customized for other anions or cations.

Discrete Analyzers

A discrete analyzer is an automated chemical analyzer that measures a chemical change in a discreet sample vial instead of a continuous flow. Measuring nitrite, nitrate, or ammonia/ammonium using colorimetric chemistry in process or wastewater is a common application.

Ion-Specific Electrode

A sensor converts the activity of a specific ion in a solution into an electrical potential. The voltage is dependent on the logarithm of the ionic activity, according to the Nernst equation. Common applications are pH, Cl^-, and F^-.

Continuous Flow Analysis

Continuous flow analysis, or similarly, flow injection analysis, is a process to measure ions in solution. A sample is injected into a flowing carrier solution passing rapidly through small-bore tubing. The sample is mixed with a reagent, which reacts to form a color or form a gas, as is the case in weak acid dissociable (WAD) cyanide measurement. The gas passes through a semipermeable membrane where an amperometric Ag electrode detector measures change compared to standard solutions.

Cyanide measurement. Flow injection analysis is used to measure cyanide complexes. Accounting for cyanide losses presents its own set of unique challenges. Reporting cyanide consumption is an important parameter for operational cost estimation of leaching, as well as the cost of detoxification of cyanide in tailings pond and mill water. Cyanide forms both strong and weakly bound complexes with transition and precious metals (Table 10). Measuring cyanide complexes and understanding the terminology for these complexes is necessary for gold process development. Cyanide is measured and reported as free WAD cyanide, or total cyanide, and includes the following:

free cyanide = CN^- and HCN

WAD cyanide = free cyanide + weak and moderately strong metal–cyanide complexes

Table 10 Cyanide metal complexes

Free Cyanide	WAD Cyanide Weak and Moderately Strong Metal–Cyanide Complexes	Strong Metal–Cyanide Complexes
Not Complexed	Log β (stability constant) <30	Log β (stability constant) >30
HCN CN^-	$Ag(CN)_2^-$ $Cu(CN)_2^-$ $Ni(CN)^+$ $Ag(CN)_3^{-2}$ $Cu(CN)_3^{-2}$ $Ni(CN)_4^{-2}$ $Cu(CN)_4^{-2}$ $Ni(CN)_5^{-3}$ $Cd(CN)^+$ $Hg(CN)^+$ $Zn(CN)^+$ $Cd(CN)_3^-$ $Hg(CN)_2^0$ $Zn(CN)_3^-$ $Cd(CN)_4^{-2}$ $Hg(CN)_3^-$ $Zn(CN)_4^{-2}$ $Hg(CN)_4^{-2}$ $Zn(CN)_5^{-3}$	$Au(CN)_2^-$ $Co(CN)_5^{-2}$ $Co(CN)_6^{-3}$ $Fe(CN)_6^{-3}$ $Fe(CN)_6^{-4}$

Table 11 Analytical techniques for cyanide deportment analysis

Analyte	Technique
Free CN^-	Titration with silver nitrate or flow injection analysis
WAD CN^-	Flow injection analysis
Total CN^-	Flow injection analysis after ultraviolet digestion
Anions	Ion chromatography (NO_3^-, NO_2, Cl, F, SO_4, SCN^-, OCN^-, $SeCN^-$)
NH_4	Ion-specific electrode
OxySulfur	Ion chromatography (S_2O_3, S_4O_6, SO_3, S^{2-}, etc.)
Metals	ICP-OES for Ag, As, Ca, Cd, Co, Cu, Fe, Hg, Mo, Na, Ni, Se, Tl, Zn

total cyanide = free cyanide + WAD cyanide + strong metal–cyanide complexes

"Total cyanide" analysis does not include oxidized forms of cyanide such as cyanate, thiocyanate, or selenocyanate. A complete cyanide deportment analysis is accomplished by measuring all cyanide complexes and oxidized cyanide decomposition products. A variety of ways exist to measure each component. The preferred methods are listed in Table 11.

Other environmental analytical testing. Extensive procedures and policies to ensure protection of the environment must be determined. Underground, open pit, and tailings can become unstable or hazardous over time as they weather and deteriorate. When stable un-oxidized material is exposed to the atmosphere, it can become reactive and hydrolyze, generating acid and potentially liberating toxic elements into the environment. Artificial leaching to simulate rainwater action is an example of short-term laboratory tests used to predict long-term models. Analytical testing for water balance and meeting discharge limits is also an important part of a process development plan. Details of acid–base accounting, tailings management, and testing procedures for predicting tailings reactivity are found in other chapters of this handbook.

DATA MANAGEMENT AND QUALITY

Laboratories using a laboratory information management system (LIMS) will be more efficient and organized, resulting in better quality of data. Analysts can dedicate more time to sample analysis and less to menial tasks. A LIMS speeds up the retrieval of information and includes other data such as analytical laboratory costs, data trends, outstanding work, and sample turnaround times. Transcription of data errors can be eliminated. Tracking the location of each sample and each sample movement effectively creates a chain of custody with time, date, and operator identification stamping. Storage location of each sample is known at any given time if additional analyses are added.

With few exceptions, instrumentation must be calibrated with a series of calibration standards traceable to a certified standard, within the expiration window, and their calibrations validated with a second source (different from the source used to prepare calibration standards). The laboratory must establish acceptability criteria for each analyte and determine appropriate wavelengths or mass, linear range, limit of quantification, limit of detection, correlation coefficients, and relative standard deviations, and determine frequency of measuring the continuing calibration verification and certified reference materials.

Limit of Detection

The detection limit of an individual analytical procedure is the lowest amount of analyte in a sample that can be detected but not necessarily quantified as an exact value. One approach is to determine the signal-to-noise ratio by comparing measured signals from samples with known low concentrations of analyte with those of blank samples and by establishing the minimum concentration at which the analyte can be reliably detected. A signal-to-noise ratio of 3:1 is generally considered acceptable for estimating the detection limit. The limit of detection (LOD) must not be confused with method detection limit, a number higher than LOD, and takes into account variables such as the matrix of the actual sample, the instrumentation at a particular laboratory, skill level of staff, and chemicals used.

Limit of Quantification

Lower detection is achieved in a "clean" matrix but quantifying the same analyte in a solution with high TDS is a different story. The limit of quantification (LOQ) of an individual analytical procedure is the lowest amount of analyte in a sample that can be quantitatively determined with suitable precision and accuracy in sample matrices of interest. Determination of the signal-to-noise ratio is performed by comparing measured signals from samples with known low concentrations of analyte with those of blank samples in similar matrix and by establishing the minimum concentration at which the analyte can be reliably quantified. A typical signal-to-noise ratio is 10:1.

Quality Control Checks

After instrument configuration and method development are complete, an ongoing quality program is followed, including insertion of internal QC checks. These are essential for continual assessment of analytical procedures. QC analyses include the use of blanks, internal standards, check standards, matrix spikes, duplicates, replicates, check samples, certified reference materials, initial calibration verification, and continuing calibration verification standards. The frequency of insertion of QC samples depends on how the data are used. Greenfield exploration samples will have less QC insertions than metallurgical accounting data. Participation in control charting and proficiency testing programs are also prudent and a requirement for accreditation.

ISO/IEC 17025 Accreditation

ISO 9001 (2015) is a management tool to evaluate quality systems but does not evaluate the technical competence of a laboratory and does not assure you or your customers that the test, inspection, or calibration data are accurate and reliable. International standard ISO/IEC 17025 (2005) is used for evaluating laboratories throughout the world. This standard specifically addresses factors relevant to a laboratory's ability to produce precise, accurate test and calibration data. It includes several components (e.g., technical competency of staff, validity and appropriateness of the methods, traceability of measurements and calibrations to certified standards, appropriate application of measurement uncertainty, calibration and maintenance of test equipment, sampling and handling of test items, quality assurance of test, inspections, and proficiency testing).

Analytical procedures used to analyze samples for metal accounting or commodities exchange should, where relevant, be accredited to the ISO/IEC 17025 (2005) standard. Accreditation boards have been established in most countries with significant mining and metallurgical operations. Accreditation is seen as a requirement for laboratories performing analyses on payable material such as concentrates that are part of a commercial transaction between a buyer and seller and for other materials that are part of a metal accounting system. To maintain accreditation, a laboratory's quality management system is thoroughly evaluated on a regular basis to ensure continued technical competence and compliance with ISO/IEC 17025 (2005). Accreditation improves the reputation and image of the laboratory. A laboratory may achieve accreditation for just a few methods of analysis. Achieving accreditation in a research laboratory with a large variety of test methods is a business decision and not necessarily a requirement for delivering high-quality data.

COSTS FOR ANALYTICAL TESTING

Building an analytical chemistry laboratory with a proper ventilation system and building code compliance is estimated to cost between US$388–US$430 per gross square foot (Gering 2015). Instrumentation and sample preparation equipment costs to support a *small* operation would add another US$1–US$2 million in capital. Table 12 lists a few pieces of common equipment used in the analytical laboratory with price ranges.

Operating expenses for an on-site laboratory include costs associated with labor, consumables, maintenance, utilities, waste management, and other miscellaneous items. Justification for building an on-site laboratory for a new operation must be weighed against many factors, including but not limited to the estimated mine life, distance to off-site laboratories, turnaround time requirements to meet the mine plan, and availability of skilled workers. Even with the relatively large expense of building and operating an on-site laboratory, the decision may be worth the investment. Babacan (2001) describes experiences at the Çayeli Bakir İşletmeleri A.S.

Table 12 Capital cost of common laboratory equipment

Common Laboratory Equipment	Range of Cost, US$, in thousands
Crusher/rotary splitter combination	60–80
AA	15–30
ICP-OES	75–120
ICP-MS	120–220
Leco C/S analyzer	50–70
Fire assay furnace	10–30
WDXRF	150–250

on-site laboratory in Turkey and demonstrated that a minesite analytical laboratory has a very high rate of return on investment and makes a great contribution to the operation and the community. The author strongly advised that a laboratory be established early in the life of any mining project.

Commercial laboratories, on the other hand, may be the only option in many cases. They publish list pricing for all services. Typical costs may include shipping, sample preparation, test preparation (e.g., digestion or fusion), instrumental measurement, and sample storage or shipping return. Other costs may include quarantine services for soils. Rush charges can be significant and may range from a 300% upcharge for one-day turnaround to a 100% upcharge for a less-than-two-week standard turnaround.

Analytical costs for a short-term metallurgical project can be a large portion of the overall budget. Estimating those project assay costs can be an arduous task and, in general, are directly proportional to the complexity of the ore. Table 13 lists many of the tests and current associated costs per sample required to characterize ore for a complex gold project using a commercial (ISO/IEC 17025 certified) analytical/geochemistry laboratory. Pricing for large amounts of analytical test work is usually negotiated.

SUGGESTED ANALYTICAL TESTING PLAN FOR COMMON COMMODITIES

Gold Ore

For the recommended analytical testing plan for gold ore, see Table 13.

Table 13 Analytical testing plan for a complex gold project with approximate cost per sample

Analytical Test	Description of Test	Approximate Cost per Sample, US$
Sample prep: drill core, rock, and chip samples	Crush to 70% less than 2 mm, riffle split off 1 kg, pulverize split to better than 85% passing 75 μm.	8.60 plus 0.75/kg
Specific gravity	Pycnometer reported as a ratio	12.25
Au—ore grade	Fire assay, 30 g, AA finish	16.60
Pt, Pd—ore grade	Fire assay, 30 g, ICP-MS finish	21.00
Cyanide-soluble Au, Ag, Cu	Cyanide leach, 30 g, AA finish	9.85 plus 4.90/element
Preg-robbing of Au	Au spike cyanide leach, 30 g, AA finish	11.00
Au on carbon	Au on carbon by ashing, aqua regia digestion, and AAS; duplicate analysis	39.25
Au and Ag in a concentrate	Au and Ag by fire assay and gravimetric finish, 30-g sample weight required	85.00
Au and Ag in bullion	Routine (non-umpire) bullion assays by fire assay with gravimetric finish	129.80
Multi-element analysis	33 Elements using 4-acid digestions, ICP-OES finish	20.85
Hg in ore	Acid digestion and ICP-AES	73.60
Paste pH	Paste pH on 10-g sample saturated with water	7.50
Whole rock analysis for use with mineralogy/XRD reporting	14-element package by lithium borate fusion and XRF	30.65
Total C and S	Total carbon and sulfur by Leco furnace	20.55
C (noncarbonate) and S (sulfide)	Organic carbon and sulfide sulfur by HCl (25%) leach of carbonates and sulfates, Leco furnace	27.45
Acid–base accounting for predicting tailings behavior	Nevada Bureau of Land Management's acid–base accounting requirements. Includes the Nevada modified Sobek neutralization potential (NP) method, the siderite-corrected NP method, total sulfur, and inorganic carbon.	216.15
Storage	Storage of samples after testing is complete	0.30/sample/month

Adapted from ALS 2017

*Extra costs for rotary splitting, cleaning between samples with silica sand, or handling extra-large samples.

Copper Ore

Following is the suggested analytical testing plan for copper ore:

- Full 62-element suite using 4-acid digestion and ICP-OES/ICP-MS: Ag, As, Ba, Be, Bi, Ca, Cd, Ce, Co, Cs, Cu, Fe, Ga, Gd, Ge, Li, Mn, Mo, Ni, Pb, Re, Sb, Se, Sn, Sr, Te, Th, Ti, Tl, U, V, W, Zn
- Metaborate fusion for major analytes by XRF: Al_2O_3, CaO, Cr_2O_3, Fe_2O_3, K_2O, MgO, MnO, Na_2O, P_2O_5, SiO_2, SO_3, TiO_2
- Au by fire assay
- Acid-soluble copper
- Ferric sulfate–soluble copper
- Hg by combustion AAS
- F by carbonate fusion and ion-specific electrode
- Cl analysis
- C and S by Leco analysis
- Sulfide sulfur

Iron Ore

In world practice, no minimum standards have been set for iron, silica, alumina, calcium, and magnesium percentages in commercial iron ores, although certain generalizations can be made (Dobbins and Burnet 1982).

Total Fe in iron ore is measured in many ways. Titration with potassium dichromate following reduction is a common method. The iron ore is first digested in HCl. Stannous chloride, a strong reductant, is added slowly dropwise until the dark ferric color changes to a light ferrous yellow, indicating complete reduction of iron. Strong H_2SO_4 and H_3PO_4 (phosphoric acid) are added to prevent air oxidation. Ferrous iron is then titrated with standardized potassium dichromate using barium diphenylamine sulfonate indicator to a purple color. Ferrous Fe (FeO) is measured similarly but without the stannous chloride reduction.

Iron ore samples are also fused with $Na_2B_4O_7$ and the fused beads are measured by XRF. Major analytes of interest are Al_2O_3, CaO, Cr_2O_3, Fe_2O_3, K_2O, MgO, MnO, Na_2O, P_2O_5, SiO_2, TiO_2, LOI, and S as SO_3. Other impurities, Ba, Co, Cu, Mg, Mn, Ni, P, Pb, Ta, V, Zn, and Zr, are acid digested and run by ICP-OES. C and S are determined by high-temperature combustion with infrared detection. Chloride content is determined by a gravimetric method in accordance with internal testing procedures. The recommended limits are

- $Na_2O + K_2O < 0.8\%$,
- $Zn < 0.02\%$,

- $P < 0.04\%$, and
- Cd and $S < 0.01\%$.

Silica and alumina, as well as small amounts of sulfur and phosphorus in iron ore, have significant deleterious effects during processing and in the final properties of products. Mineralogical study using XRD and microstructure determination using SEM and/or FEG-SEM (field emission gun scanning electron microscopy) is prudent.

Coal and Coke

Following is the suggested analytical testing strategy for coal and coke:

- **Proximate and sulfur analysis.** The proximate and sulfur analysis is basic to all coal and coke evaluations. The proximate analysis consists of the following suite of tests: moisture, ash, volatile matter, and fixed carbon (by difference).
- **Ultimate analysis.** Coal is analyzed for the following as part of an evaluation for its use as a fuel: carbon, hydrogen, nitrogen, oxygen (by difference), ash, and sulfur.
- **Ash chemistry.** Major elements of coal or coke ash provide important information for blast furnace and coal utility operations. The following elements are typically tested: SiO_2, Al_2O_3, Fe_2O_3, TiO_2, CaO, MgO, Na_2O, K_2O, P_2O_5, MnO_2, SO_3, BaO, SrO.
- **Trace elements analysis.** Trace elements analysis is critical to evaluating coals for environmental regulations. The following are often requested: As, B, Be, Cr, Hg, Pb, Se, V.

Uranium

Laboratories must be qualified and experienced in handling samples of naturally occurring radioactive material in the jurisdiction with active uranium exploration and mining. The recommended analytical testing plan for uranium is as follows:

- Full 62-element suite using 4-acid digestion and ICP-OES. Table 14 lists the typical specifications for uranium concentrates.
- Pb isotope analysis by aqua regia and ICP-MS.
- U by lithium borate fusion and XRF.
- Other radiometry tests as needed.

Table 14 Typical specifications for uranium concentrates*

Concentrate Composition	Maximum Impurities Permissible
Impurity	Weight percent based on the U_3O_8 content
Na	7.5
H_2O	5.0
SO_4	3.5
K	3.0
Th	2.0
Fe	1.0
Ca	1.0
Si	1.0
CO_3^{2-}	0.50
Mg	0.50
Zr	0.50
PO_4^{2-}	0.35
Cl, Br, I	0.25
V_2O_5	0.23
Rare earths	0.20
P	0.15
Mo	0.15
As	0.10
B	0.10
Extractable organic matter	0.10
Insoluble U	0.10
Ti	0.05
^{226}Ra	740 Bq/g (20,000 pCi/g)
Particle size	6.35 mm (¼ in.)

Source: IAEA 1992, reprinted with permission from IAEA.
*Minimum U_3O_8 content, with a partial size of 65%. Only natural uranium concentrates (nonirradiated material) containing 0.711 wt % of the isotope 235U are acceptable.

Clays, Limestones, and Dolomites

Following is the suggested analytical testing plan for the industrial minerals of clays, limestones, and dolomites:

- Samples are fused with $Na_2B_4O_7$ and the fused beads measured by XRF.
- Major analytes by XRF: Al_2O_3, CaO, Cr_2O_3, Fe_2O_3, K_2O, MgO, MnO, Na_2O, P_2O_5, SiO_2, SO_3, TiO_2.
- LOI at 1,000°C.

Rutile, Zircon, and Ilmenite

Following is the recommended analytical testing strategy for the industrial minerals of rutile, zircon, and ilmenite:

- Samples are fused with $Na_2B_4O_7$ and the fused beads measured by XRF.
- Major analytes by XRF: Al_2O_3, CaO, Cr_2O_3, Fe_2O_3, K_2O, MgO, MnO, Na_2O, Nb_2O_5, P_2O_5, SiO_2, SO_3, TiO_2, V_2O_5, ZrO_2 (add HfO for zircons).
- LOI of 1,000°C.

Nickel Laterites

The following plan is recommended for analytical testing of rare earths:

- Samples are fused with $Na_2B_4O_7$ and the fused beads measured by XRF.
- Major analytes by XRF: Al_2O_3, BaO, CaO, Cr_2O_3, Fe_2O_3, K_2O, MgO, MnO, Na_2O, P_2O_5, SiO_2, SO_3, TiO_2, V_2O_5.
- Major analytes by 4-acid digestion for As, Co, Cu, Ni, Pb, and Zn by ICP-OES.
- LOI at 1,000°C.
- Chloride by silver nitrate titration.
- C and S by Leco analysis.

Phosphate Rock

Fertilizer capabilities for phosphate rock include nutrients such as nitrogen, phosphates, and potassium. The following plan is recommended analytical testing of phosphate rock:

- Acid, insoluble
- Al, Ca, Cd, Fe, K, Mg, Na, P, SiO_2
- Water analysis (moisture content)
- Organic matter
- Carbonate fusion and ion-specific electrode for fluoride
- Forms of sulfur analysis
- ICP trace metals

- Specific gravity
- Total nitrogen analysis
- LOI at 1,000°C
- Total suspended solids
- Viscosity

Rare Earths

Following is the suggested analytical testing plan for rare earths:

- Samples are fused with $Na_2B_4O_7$ (or sodium peroxide) and the fused beads are measured by XRF.
- Major analytes by XRF: Al_2O_3, BaO, CaO, Cr_2O_3, Fe_2O_3, K_2O, MgO, MnO, Na_2O, P_2O_5, SiO_2, SO_3, TiO_2, V_2O_5.
- Minor analytes by XRF: La, Ce, Pr, Nd, Sm, Eu, Gd, Tb, Dy, Ho, Er, Yb, Lu, Th, U, Y. (The elements are recalculated as oxides and reported as such.)
- LOI at 1,000°C.

Bauxite

Lithium borate fusion followed by XRF is the industry-standard analytical method for bauxite analysis. Results are reported on a dry weight (110°C) basis. The suggested analytical testing plan for bauxite is as follows:

- Fused-disc XRF
- Major analytes by XRF: Al_2O_3, Fe_2O_3, SiO_2, TiO_2
- Minor analytes BaO, CaO, Cr_2O_3, K_2O, MgO, MnO, Na_2O, P_2O_5, SO_3, SrO, V_2O_5, Zn, ZrO_2
- Total organic carbon
- Reactive silica
- Available alumina
- Multiscreen sizing to determine the optimum screen size for recovery and subsequent wet beneficiation
- Hot hydroxide leach for soluble Al by ICP-OES
- Sodium oxalate leach for measuring organic carbon contaminants by ion chromatography
- LOI at 1,000°C

For reporting, bauxite = Al_2O_3 + SiO_2 + Fe_2O_3 + TiO_2 + LOI + trace elements.

Lithium

Lithium-hosted minerals, pegamites and jardites, as well as lithium-bearing clays and sediments can co-occur with economic grades of rare earth elements and other rare metals such as cesium and rubidium. Li in sedimentary rocks dissolves easily in aqua regia and may be a simpler approach when other analytes are not of interest. Li is not typically determined by XRF because it is such a light element. The suggested analytical testing plan for lithium is as follows:

- Samples are fused in sodium peroxide and measured for multi-elements on ICP-OES and ICP-MS.
- Major and minor analytes are measured by ICP-OES and/or ICP-MS (depending on concentration): Ag, As, B, Ba, Be, Bi, Ca, Cd, Ce, Co, Cs, Cu, Dy, Er, Eu, Fe, Ga, Gd, Ge, Ho, In, La, Li, Lu, Mn, Mo, Nb, Nd, Ni, Pb, Pr, Rb, Re, Sb, Se, Sm, Sn, Sr, Ta, Tb, Te, Th, Ti, Tl, Tm, U, V, W, Y, Yb, and Zn.

If material has been determined to contain rare earth elements or other economic elements, further analytical testing of major analytes is appropriate. Samples are fused with sodium peroxide, and analytes are measured by ICP-OES: Al_2O_3, As, B, CaO, Co, Cr_2O_3, Cu, Fe_2O_3, K_2O, Li, MgO, MnO, Na_2O, Ni, Pb, S, SiO_2, TiO_2, and Zn.

Lithium Brine Analysis

Lithium brines have their own unique set of analytical difficulties due to viscosity and because Li is such a light element surrounded by much heavier elements. The suggested analysis for lithium brine is as follows:

- Slurries are digested with aqua regia and measured by ICP-OES and ICP-MS.
- ICP-OES analytes include Al, As, B, Ba, Ca, Co, Cr, Cs, Fe, K, Li, Na, Ni, Mg, Mn, Na, P, Pb, S, Sc, Se, Sr, Ti, V, and Zn.
- ICP-MS analytes include Br, Cs, I, and Rb.
- Cl by silver nitrate titration
- F by ion-specific electrode
- PO_4 (phosphate) and NO_3 (nitrate) by colorimetric method
- pH
- Conductivity
- Total dissolved solids
- Sulfate (calculated from S)
- HCO_3 (bicarbonate) and NH_4 (ammonium) by volumetric method

Potash

Potash ores are used primarily in the fertilizer industry but also in various chemical applications in different potassium forms. Most potassium forms are water soluble but major impurities are not. XRF is suggested for total analysis of major analytes. Following is the suggested analytical testing plan for potash:

- Samples are fused with lithium metaborate fused-disc XRF.
- Major analytes by XRF: Al_2O_3, CaO, Fe_2O_3, K_2O, MgO, MnO, Na_2O, P_2O_5, SiO_2, SO_3, SrO, TiO_2
- Additional 4-acid digestion for penalty elements, As, Cd, Ce, Cr, La, Pb, Se, Sm, Th, U, Y, Yb
- Combustion analyzer for Hg
- Carbonate fusion and ion-specific electrode for F
- Chloride by silver nitrate titration
- Color, visual
- Moisture content
- Particle hardness
- pH
- Potassium content (K_2O, KCl)
- Screening
- Forms of sulfur
- Water residues (water-insoluble matter)
- LOI at 1,000°C

Chromite and Manganese Ores

The extreme refractory nature of these materials prevents the use of simpler acid digestions. Following is the suggested analytical testing plan for chromite and manganese ores:

- Samples are fused with lithium metaborate, fused-disc XRF.
- Major analytes by XRF include Al_2O_3, BaO, CaO, Cr_2O_3, Fe_2O_3, K_2O, MgO, MnO, Na_2O, P_2O_5, SiO_2, SO_3, SrO, TiO_2.
- LOI at 1,000°C.

REFERENCES

Allen, T., and Khan, A.A. 1970. Critical evaluation of powder sampling procedures. *Chem. Eng.* 238:108–112.

ALS. 2017. *ALS Geochemistry Schedule of Services and Fees: 2017 USD*. North Vancouver, BC: ALS.

ASTM E1915. 2013. *Standard Test Methods for Analysis of Metal Bearing Ores and Related Materials for Carbon, Sulfur, and Acid-Base Characteristics.* Vol. 3.05. West Conshohocken, PA: ASTM International.

Babacan, H. 2001. A mine site laboratory from exploration to closure: A case study. In *IMCET2001: Proceedings of the 17th International Mining Congress and Exhibition of Turkey.* Ankara: Chamber of Mining Engineers of Turkey.

Bazhko, O. 2009. Application of redox titration techniques for analysis of hydrometallurgical solutions. In *Hydrometallurgy Conference 2009*. Johannesburg: Southern African Institute of Mining and Metallurgy.

Beamish, F.E., and Van Loon, J.C. 1977. *Analysis of Noble Metals-Overview and Selected Methods.* New York: Academic Press.

Bock, R. 1979. *A Handbook of Decomposition Methods in Analytical Chemistry*. New York: John Wiley and Sons.

Bugbee, E.E. 1922. *A Textbook of Fire Assaying*. New York: John Wiley and Sons; London: Chapman and Hall.

Bugbee, E.E. 1940. *A Textbook of Fire Assaying*, 3rd ed. New York: Wiley.

Dean, J. 1995. *Analytical Chemistry Handbook.* New York: McGraw-Hill.

Dillon, V.S. 1955. *Assay Practice on the Witwatersrand.* Johannesburg: Transvaal and Orange Free State Chamber of Mines.

Dobbins, M.S., and Burnet, G. 1982. Production of an iron ore concentrate from the iron-rich fraction of power plant fly ash. *Resour. Conserv.* 9:231–242.

Dulski, T. 1996. *A Manual for the Chemical Analysis of Metals.* ASTM Manual 25. West Conshohocken, PA: ASTM International.

Furman, N.H. 1962. *Standard Methods of Chemical Analysis*, 6th ed. Vol. 1. Princeton, NJ: Van Nostrand.

Gaines, P. 2012. Compare ICP-OES and XRF for determination of metal composition in catalyst powder samples. Christiansburg, VA: Inorganic Ventures. https://www.inorganicventures.com/advice/compare-icp-oes-and-xrf-determination-metal-composition-catalyst-powder-samples. Accessed November 2017.

Gering, J. 2015. Laboratory construction outlook. *Lab. Design.* Aug. 11, 2015. LabDesignNews.com.

Hafty, J., Riley, L.B., and Goss, W.D. 1977. *Manual on Fire Assaying and Determination of the Noble Metals in Geological Materials.* Geological Survey Bulletin 1445. Washington, DC: U.S. Government Printing Office.

Horak, J., Faithfull, J.W., Price, M., and Davidson, P. 2016. Identifying and managing asbestiform minerals in geological collections. *J. Nat. Sci. Collect.* 3:51–61.

IAEA (International Atomic Energy Agency). 1992. *Analytical Techniques in Uranium Exploration and Ore Processing.* Technical Report Series No. 341. Vienna: IAEA.

ISO 9001. 2015. *Quality Management Systems—Requirements.* Geneva: International Organization for Standardization.

ISO/IEC 17025. 2005. *General Requirements for the Competence of Testing and Calibration Laboratories.* Geneva: International Organization for Standardization.

Lenahan, W.C., and Murray-Smith, R. de L. 2001. *Assay and Analytical Practice in the South African Mining Industry.* Monograph Series M6. Johannesburg: Southern African Institute of Mining and Metallurgy.

Lorenzen, L. 1995. Some guidelines to the design of a diagnostic leaching experiment. *Miner. Eng.* 8(3):247–256.

Montana Tech. 2017. Fundamentals and Applications of Fire Assay Short Course. Butte, MT: Montana Tech. http://mtech.edu/academics/mines/metallurgy/fire-assay.

Pitard, F. 1993. *Pierre Gy's Sampling Theory and Sampling Practice: Heterogeneity, Sampling Correctness, and Statistical Process Control*, 2nd ed. Boca Raton, FL: CRC Press.

Rydberg, J., Cox, M., Musikas, C., and Choppin, G.R. 2004. *Solvent Extraction Principles and Practice*, revised and expanded 2nd ed. New York: Marcel Dekker. pp. 549–594.

Shepard, O.C., and Dietrich, W.F. 1940. *Fire Assaying.* New York: McGraw-Hill.

Shun'ko, E.V., Stevenson, D.E., and Belkin, V.S. 2014. Inductively coupling plasma reactor with plasma electron energy controllable in the range from ~6 to ~100 eV. *IEEE Trans. Plasma Sci.* 42(3):774–785.

SPEX SamplePrep. 2016. *Handbook of Sample Preparation and Handling*, 13th ed. Metuchen, NJ: SPEX SamplePrep.

Steger, H.F. 1976. Chemical phase-analysis of ores and rocks: A review of methods. *Talanta* 23(2):81–87.

Strong, B., and Murray-Smith, R. 1974. Determination of gold in copper-bearing sulphide ores and metallurgical flotation products by atomic absorption spectrometry. *Talanta* 21(12):1253–1258.

Tanner, S., Baranov, V., and Bandura, D. 2002. Reaction cells and collision cells for ICP-MS: A tutorial review. *Spectrochim. Acta B* 57:1361–1452.

Thompson, M. 2005. A review of interference effects and their correction in chemical analysis with special reference to uncertainty. *Accred. Qual. Assur.* 10:82–97.

Tumilty, J.A., and Schmidt, C.G. 1986. Deportment of gold in the Witwatersrand system. In *Gold 100: Proceedings of the International Conference on Gold.* Johannesburg: Southern African Institute of Mining and Metallurgy. pp. 541–553.

Twyman, R.M. 2005. Atomic emission spectrometry: Interference and background correction. In *Encyclopedia of Analytical Science*, 2nd ed., Vol 1. Edited by P. Worsfold, A. Townshend, and C. Poole. London: Elsevier Science. pp. 198–203.

Young, R.S. 1971. *Chemical Analysis in Extractive Metallurgy.* London: Charles Griffin.

Young, R.S. 1980. Analysis for gold: A review of methods. *Gold Bull.* 13:9–14.

CHAPTER 1.3

Automated Mineralogy

Mark G. Aylmore

Automated mineralogy is a term that has been used to describe the automation of the analytical process of quantifying minerals, rocks, and production materials. Analytical techniques, such as X-ray powder diffraction and optical and electron microscopy, have been extensively used over the years to describe the mineralogy in geological and metallurgy applications (Petruk 1976; Petruk and Hughson 1977; Cabri 1981; Petruk and Schnarr 1981; Henley 1983; Petruk 2000; Gu, 2003).

With advances in computing power and speed, automated mineralogy techniques have been developed for the minerals industry and used in research institutes for more than two decades. This has primarily been in the areas of X-ray analysis and commercialized in the form of scanning electron microscope (SEM) techniques such as mineral liberation analysis (MLA) and Quantitative Evaluation of Minerals by Scanning Electron Microscopy (QEMSCAN); and more recently, Tescan Integrated Mineral Analyzer (TIMA), Zeiss Mineralogic, INCAMineral (from Oxford Instruments), and Advanced Mineral Identification and Characterization System (AMICS), developed by Yingsheng Technology and now marketed through Bruker Inc. As a result of the improvement in technology, automated mineralogy systems are now regarded as essential by most large companies and used on a routine basis to supply metallurgical operations with regular data (e.g., Barrick Gold, Kennecott Utah Copper Corporation [KUCC], Anglo American, Freeport-McMoRan, Richards Bay Minerals, and Northparkes). There are more than 150 automated SEM systems in operation worldwide (Schouwstra and Smit 2011).

Comprehensive and rapid analysis techniques play an important role in geometallurgy evaluations, such as for scoping; prefeasibility and feasibility studies; process design; and optimization of gold, copper gold, copper molybdenum, nickel, and iron projects (Williams and Richardson 2004; Dobby et al. 2004; Bulled 2007; Bulled et al. 2009; Lotter et al. 2013; Kormos et al. 2013; Montoya et al. 2011; Hatton and Hatfield 2013; Baumgartner et al. 2011, 2013; Hoal et al. 2013; Zhou and Gu 2016).

The application of automated mineralogical techniques can have the following benefits:

- Reduce and circumvent risks associated with the variability of ore bodies during project evaluation, process development, plant design, mine planning, and performance of mineral processing/metallurgical unit operations.
- Provide ongoing ore characterization analysis tools to track and reduce costs in mining, processing, and tailings handling.
- Coupled with other diagnostic sampling techniques used in a plant, automated mineralogy systems can improve the statistical reliability of mineralogical and process measurements for plant surveys (Henley 1983; Lotter 1995, 2011).
- De-bottlenecking of existing or recent design and plant restarts can be executed faster and more cost efficiently with process mineralogical support (Baum 2014).
- Automated central laboratories or smaller mine site laboratories can provide efficient, better quality, and fast laboratory data for exploration and mine geology samples, geometallurgy programs, daily blastholes, and all other production samples (Best et al. 2007).
- Track the deportment of minerals or metals in process streams. Routine analyses to track precious metals associated with other minerals is more efficiently handled.
- Allow semiskilled personnel to operate and free mineralogical staff to perform research into improving and understanding ore and processing issues.

This chapter provides an overview of the main analytical techniques used in automated mineralogy routines that are available to metallurgists. The first two parts cover the main X-ray analysis and digital imaging techniques. The limitations and some applications illustrating these techniques are also described. Semiautomated microprobe techniques are described in the subsequent sections and are included, as they are commonly used to confirm and back up routine automated techniques. High-resolution X-ray computed tomography is briefly discussed at the end of this chapter.

Mark G. Aylmore, Senior Research Fellow, John de Laeter Centre, Faculty of Science & Engineering, Curtin University, Perth, Western Australia, Australia

To utilize the analytical techniques described in this chapter effectively, it is important that the metallurgist first define the type of information required, with consultation of experienced analysts before choosing an analytical method or strategy. All analytical techniques require some validation by other techniques and need to be tailored for the process and operation under consideration. For personnel to operate automated mineralogy techniques requires training and a basic understanding of physics concepts to maximize benefits.

X-RAY ANALYTICAL TECHNIQUES

The two most commonly used X-ray methods are X-ray powder diffraction (XRPD) and X-ray fluorescence (XRF) spectrometry. Although other techniques based on the scatter, emission, and absorption properties of X-radiation can be used in different scenarios, such as surface analysis, they are not discussed here.

X-Ray Powder Diffraction

XRPD is one of the most powerful techniques for qualitative and quantitative analysis of crystalline compounds and is widely used in most analytical laboratories for mineral or phase identification. The information that can be obtained includes

- Types and nature of crystalline phases present,
- Relative abundance of major mineral phases,
- Microstrain and crystallite size, and
- Isomorphic substitution trend mineral phases.

All these attributes can affect the reactivity of minerals in metallurgical processes.

Instrumentation

A typical X-ray powder diffractometer setup with sample changer is shown in Figure 1. Bragg-Brentano focusing geometry is the most widespread geometry used in XRPD methods in both research and industrial laboratories (Klug and Alexander 1974). The geometry of an X-ray diffractometer is such that the sample rotates in the path of the collimated X-ray beam at an angle θ while the X-ray detector is mounted on an arm to collect the diffracted X-rays and rotates at an angle of 2θ. Copper anode is the most common and cheapest target material used in most applications for generating an X-ray source. However, copper anode with a post-sample graphite monochromator or cobalt anodes are frequently used in samples that contain high cobalt, iron, and manganese content, where fluorescence in the X-rays produced from a copper target yield a low peak-to-background ratio in powder diffraction data. Solid-state strip detectors can replace point detectors in diffractometers, which allows a decrease in measurement scanning times from several hours to 10 minutes.

Most of the X-ray diffractometers from the major manufactures have the facilities to handle and collect data for multiple samples. In addition, there are several commercial-size machines developed specifically for industry, such as the Malvern Panalytical CubiX^3 and Bruker D4 Endeavor.

Information Provided from X-Ray Powder Diffraction

Details on the theory and fundamentals of XRPD are well documented, both on websites and in the literature (e.g., Zevin et al. 1995; Bish and Post 1989; Cullity 1978; Klug and Alexander 1974). X-ray diffraction (XRD) is based on constructive and destructive interference of monochromatic X-rays in a sample. Most (95%) solid materials are crystalline. The interaction of the incident X-rays with the sample may produce constructive interference and a diffracted X-ray when conditions satisfy Bragg's law ($n\lambda = 2d \sin \theta$, where n is an integer, λ is the wavelength of incident beam, d is the interplanar spacing, and θ is the scattering angle).

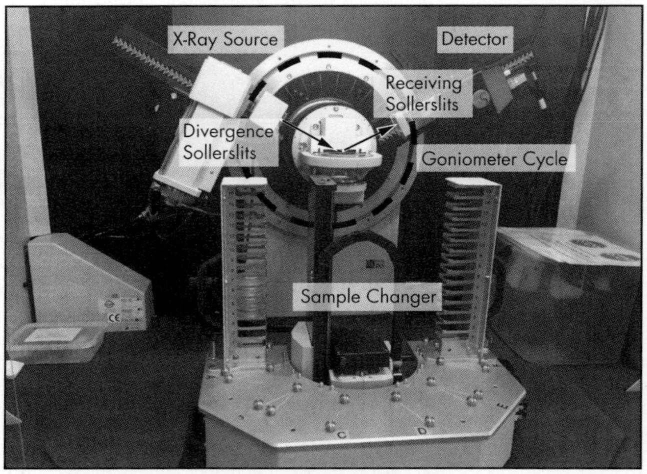

Figure 1 Typical X-ray powder diffractometer with conventional Bragg-Brentano focusing geometry

The position and intensity of peaks in a diffraction pattern are determined by the crystal structure. The unique set of d-spacings derived from a powder diffraction pattern can be used to *fingerprint* the mineral. Using software normally provided by the manufacturers of XRPDs, phases that are present in a sample can be compared with databases such as the Powder Diffraction File (PDF) produced by the International Centre for Diffraction Data (ICDD), which contains the d-spacing, intensity lists, crystal structures, and literature reference information for thousands of crystalline phases. An example of an XRPD pattern for a titanium-rich material with mineral phases identified from the ICDD-PDF database is shown in Figure 2. The minimum detectable limit found by routine qualitative procedures is of the order of ~1%, which restricts ore mineral quantification to concentrates and upgraded feed and tailings samples. Complete analysis requires several hours to complete, although this depends on the experience of the analyst and the complexity of the problem under investigation. For typical powder patterns, data can be collected over a short 2θ range from ~5° to 90° for qualitative analysis of most minerals; full 2θ range is required for full structural refinement applications.

The appropriate data-collection strategy will depend on the nature of the sample, for example, on how well it scatters, peak-broadening effects, and the degree of peak overlap (McCusker et al. 1999). Overall step size should be at least one-fifth of the full width at half-maximum of a diffraction peak, and time per step should approximately compensate for the gradual decline in intensity with 2θ (Madsen and Hill 1990, 1994). The types of information that can be extracted from XRPD are listed in Table 1.

Most industrial processes are interested in changes in the abundance of particular mineral phases and not always concerned with detailed mineral structural properties. However, depending on the analysis method used, it is important that sufficient detail is available for a mineral phase of interest to

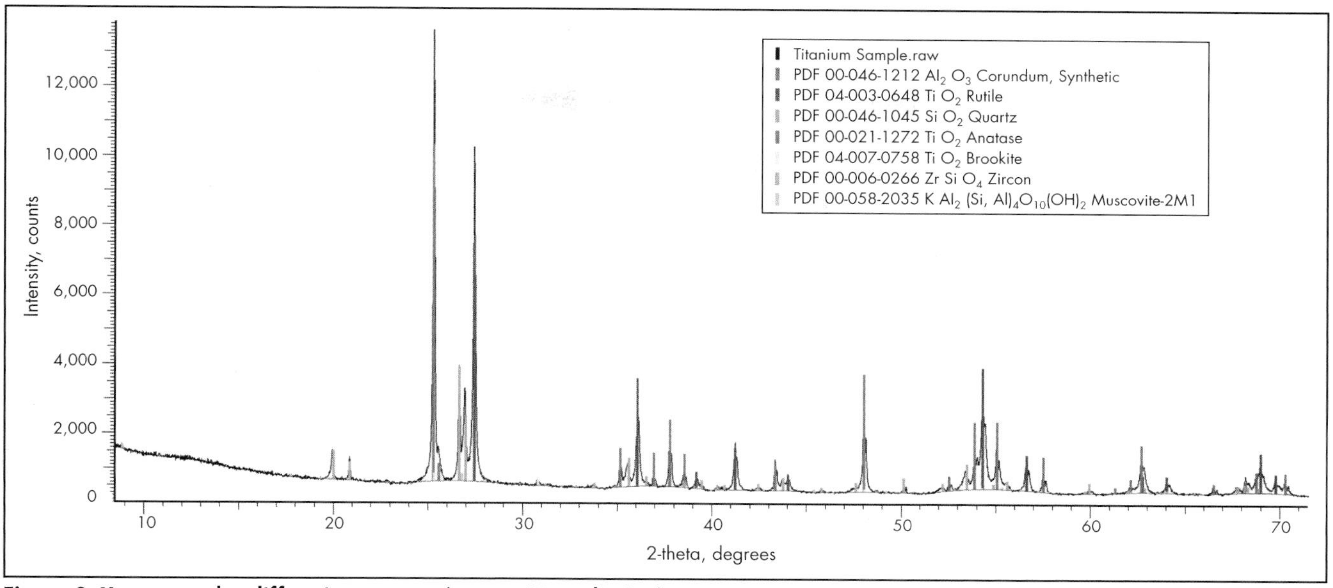

Figure 2 X-ray powder diffraction pattern (copper Ka radiation)

Table 1 X-ray powder diffraction pattern information

Peak Reflection Information					Background Fit
Position		Intensity		Profile, Width and Shape	Sample
Qualitative phase analysis	Lattice parameters	Quantitative phase analysis	Crystal structure	Sample broadening	Diffuse scattering
Identification of individual phase(s)	Composition Macrostrain Space group	Abundance of phases	Atomic positions Temperature factors Site occupation factors	Crystalline size Microstrain	Amorphous fraction

either model or use data from powder diffraction patterns to confidently ascertain the correct information. Furthermore, details on crystallite size, strain, *amorphous* content, and isomorphic substitution of elements within minerals (e.g., Al or Ni in goethite) are important for consideration in many metallurgical applications.

Sample Preparation

How samples are prepared affects the quality of the data collected. The issues and best practices for sample preparation are well documented (Hill and Madsen 2002; Buhrke et al. 1997; McCusker et al. 1999). The main sample issues that affect the powder diffraction data are as follows:

- **Sample not representative.** The sample is not homogenized during sample preparation.
- **Particle size.** Coarse-size particles yield insufficient diffracting particle statistics (i.e., low peak-to-background ratio).
- **Preferred orientation.** Particles lie in a preferred orientation, resulting in enhanced diffraction at the angle (Dollase 1986; Li et al. 1990; O'Connor et al. 1991).
- **Microabsorption effects.** For high X-ray absorbers (e.g., rutile), only a fraction of grains diffract, resulting in underestimation of relative intensity; whereas for low absorbers (e.g., corundum), the beam penetrates more grains, resulting in more diffracting volume and an overestimation of the relative intensity.
- **Extinction.** Reduction in the intensity is caused by rediffraction of a Bragg reflection toward the incident beam.
- **Sample transparency.** For materials with light elements, the X-rays penetrate too deeply into the sample, yielding shifts in peak positions and relative peak intensities.

Fortunately, many of these issues can be reduced by ensuring that samples are ground fine enough to irradiate or dilute these effects. In addition, a sample spinner is used to improve sampling statistics. Although there are methods for modeling and correcting these effects, good sample preparation and correct setup of instrumentation is essential. This reduces the need for corrections.

For small volumes of sample, an agate mortar and pestle can be effective, but micronizing is the most efficient method for generating particle sizes of ~10 µm with a small spread of sizes. For large sample volumes, there are several commercial units for automated sample preparation on the market. Overgrinding can cause merging of particles, solid-state phase transformations (e.g., calcite to aragonite, wurtzite to sphalerite, kaolinite to mullite), loss of crystallinity, peak broadening, or amorphization of phases in samples (Buhrke et al. 1997).

There are many different ways samples can be mounted for XRPD analysis. These include top loading and pressing, flat plate, back loading and pressing, and side drifting. Using back-pressing and side-drifting loading methods reduces the issues associated with preferred orientation compared with top-loading or front-mounted samples. Solid diluents (e.g., gum, glass, gelatin) or binders can be added to reduce

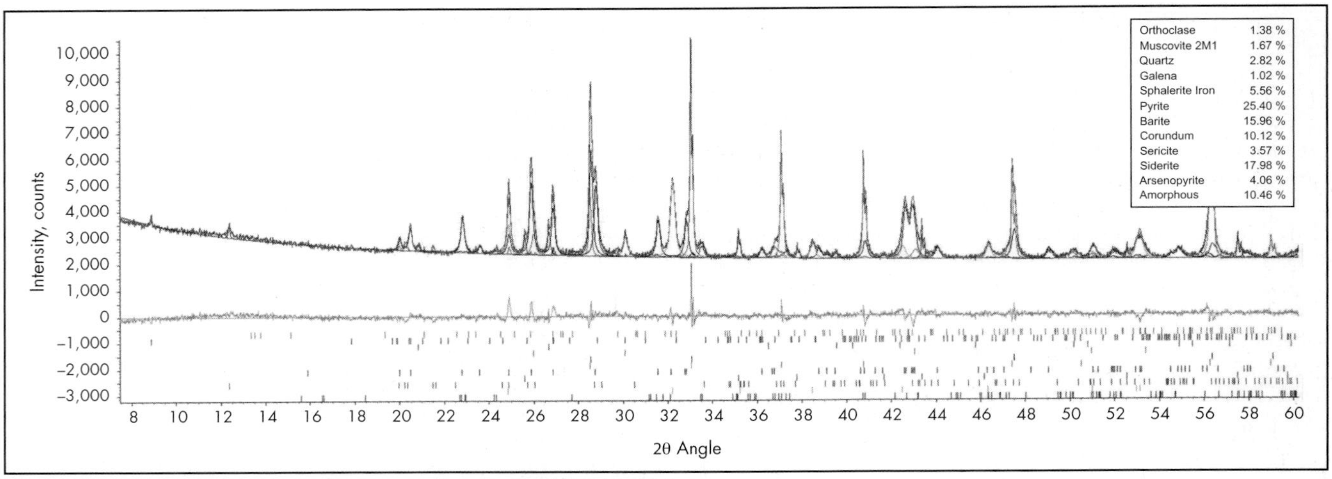

Figure 3 Output patterns for refinement of a zinc and lead sulfide ore

preferred orientation effects, but result in sample contamination, increased transparency issues, amorphous scatter, and/or extra peaks in powder diffraction patterns. For large-scale preparation operations, automated soft-pressing machines (e.g., FLSmidth ASP100) have been designed to prepare mounts for quantification by XRPD and other techniques (e.g., near-infrared spectroscopy, XRF) (Hem et al. 2009).

Some phases have crystal habits (platy or needle like), which promote orientation along specific crystallographic directions (e.g., micas, clay minerals, albite, sanidine, chlorite); therefore, preferred orientation issues cannot be eliminated. In some cases, the preferred orientation is deliberately enhanced for the identification of swelling clay minerals (Brindley 1980). The characterization of clay layers is carried out by XRD on air-dried and glycol-saturated oriented preparations on ceramic holders, sometimes after saturation with different cations (e.g., K, Mg, Li) (Bouchet et al. 1988). The addition of glycolation to the sample is used to identify the nature of the nonswelling clays (e.g., illite or chlorite); discriminate between the swelling clays (di- or trioctahedral high-charge or low-charge smectite, vermiculite); and quantify the nature, degree, and ordering of the mixed layering within phyllosilicate systems (Mosser-Ruck et al. 2005).

Quantitative Analysis

Detailed information on quantitative X-ray diffraction (QXRD) can be found in books such as Zevin et al. (1995) and the chapter by Madsen et al. (2012). Once the presence of phases has been established in a given specimen, the abundance of a phase or many phases present can be determined by use of the intensities of diffraction lines from each phase.

Rietveld and profile-fitting methods. Nowadays, the Rietveld method (Rietveld 1969; Young 1995) is the most employed methodology to achieve quantitative phase analysis (QPA) of crystalline materials (Snyder and Bish, 1989; Madsen et al. 2001; Scarlett et al. 2002; Hill and Howard 1987; Hill 1991; Bish and Post 1993; Hill et al. 1993). It allows an accurate estimation of changes in the mineralogical composition of solids or slurries. Detailed information on the theory of the Rietveld method can be found in Young (1995). The principle of the method is that the intensities calculated from a model of the crystalline structure are fitted to the observed X-ray powder pattern by a nonlinear least-squares refinement. The crystal-structural data (unit cell dimensions, space group, atomic positions, and thermal parameters) for each phase are obtained either within the software or sourced from the Inorganic Crystal Structure Database. To obtain high-quality quantitative results, the X-ray diffractometer is aligned and calibrated with a suitable standard (e.g., lanthanum hexaboride [LaB_6], National Institute of Standards and Technology standard reference material [SRM] 660a).

The refinement of crystal-structural and peak-profile parameters enable physical and chemical details of each particular phase in the mixture to be modified. In addition to phase abundance derived from the relative intensity of phases modeled in the XRPD pattern, the refinement of the lattice parameters of the phase can be used to calculate compositions in solid solution (e.g., Al substitution in goethite) and crystallite sizes determined from peak line broadening. Other crystal structure parameters, such as atomic coordinates, atomic site occupancies, and thermal parameters can be refined or modified in the initial setup and fixed for analysis of samples on a process production scale. The weighted profile R factor is monitored for convergence, often expressed as a percentage, and should be less than 10% for a good fit.

Most software packages provide a graphical representation of the calculated and measured XRD pattern and the difference plot to allow any misfit of data to be easily recognized. Figure 3 is an example of an output for the refinement of a zinc and lead sulfide ore, showing observed, calculated, and difference patterns. The hkl positions for each mineral are also give in order of the minerals listed.

Where all phases in the mixture are known, are crystalline, and have known crystal structures, the relative mineral abundances can be calculated from the Rietveld refined scale factor and a *phase constant* ZMV (Z is the number of formula units in the unit cell, M is the molecular mass of the formula unit, and V is the unit cell volume) for each phase (Hill and Howard 1987).

Where phases have a partial or no known crystal structure, there are methods that can be employed to quantify phases. Such an approach has been described by Scarlett and Madsen (2006) (partial or no known crystalline structure, or PONKCS method) and used in applications such as quantifying cement (Madsen and Scarlett 1999; Scarlett et al. 2001; Taylor and Rui 1992). In these methods ZM and V are empirical values derived by preparing a mixture of the phase of interest with a known, well-characterized standard.

There are a few Rietveld-type software packages available for commercial use. One of the most popular is the TOPAS software, which has two modes of operation. A graphical user interface mode is available where all data entry is handled through parameter windows and is useful for routine operation in profile-fitting and Rietveld applications. The launch mode is designed for crystal structure solutions and for automation or batch operation applications. Data entry is through text files and allows advanced users to write their own functionality.

Approach to modeling and quantifying phases in samples. For analyzing and quantifying the mineralogy of a group of samples by XRPD, the following approach can be used:

- Preliminary setup
  - Identify minerals and grouping of sample XRPD data with similar mineralogy to facilitate defining mineral suites that can be modeled.
  - Ensure that XRPD instrumental parameters are defined to allow for accurate modeling of artifacts in XRPD scans associated with the XRPD instrument (e.g., detector aberrations, tungsten radiation peaks, unfiltered peaks).
  - In some cases, assess synthetic mineral mixtures of like minerals to improve confidence in the ability to model and quantify phases.
- Preparation of mineral templates for modeling XRD scan data
  - Compile structural data for each mineral phase identified in the sample.
  - Refine and fit profile to sample XRPD data.
  - Confirm fit with chemical assay data and/or synthetic mineral mixtures.
- Batch refinement on materials with similar compositions and areas (e.g., lithological zones)
- Correlation and cross checking of chemical data derived from mineral refinements with chemical assay data or other analysis methods such as a thermogravimetric analysis and digital image analysis techniques.

Amorphous content. In many naturally occurring or synthetic systems, poorly ordered phases may be present that cannot be accurately modeled by published structure information in the literature. The presence of poorly ordered phases will have an influence on the reactivity and dissolution of ores, such as dissolution of amorphous silica and polymerization in acid uranium extraction circuits, or may affect the flotation response of some minerals.

Amorphous phases cannot always be detected directly by XRD analysis, because they do not produce visible peaks in the XRD pattern, but only increase the background. There are several methods that can be used for determining amorphous content by XRPD (Madsen et al. 2011). The most common one is the internal method where the sample is *spiked* with a known mass of standard material (e.g., calcium fluoride [CaF_2], corundum). The weight fractions of the crystalline phases present in the sample are estimated using the Rietveld analysis. For a sample containing an amorphous phase, the standard will be overestimated in the analysis. From the overestimation of the standard, the amorphous content of the investigated sample is derived. However, the determination of amorphous content value is sensitive to how well the experimental and calculated data fit, as well as the assumption that all phases present in the sample have been identified and modeled.

Main Sources of Error and Limitations in Quantitative Analyses

Sources of errors have been identified in several round-robin surveys conducted by the International Union of Crystallography Commission on Powder Diffraction to evaluate QPA by powder diffraction (Madsen et al. 2001; Scarlett et al. 2002):

- In some cases a poor understanding of issues in data collection and analysis leading to Rietveld software to be treated as a *black box*
- Incorrect crystal structure data chosen or inappropriate profile models used
- Omission of phase(s) from the analysis and/or errors in identification of phases present
- Failure to refine important parameters such as unit cell and thermal values during modeling of phases
- Refinement of parameters not supported by the data
- Inappropriate use of correction models, such as preferred orientation and microabsorption correction
- Acceptance of physically unrealistic parameters (e.g., over refinement of thermal parameters) or incomplete refinements (high R factor values)

These issues can be overcome through continuing education of users of diffraction methodology and Rietveld-based software. The need for plant metallurgists to consult experienced XRPD analysts before placement of the analysis is emphasized by the surveys.

Problems still arise in the calculation of mineralogical phase abundance in cases where large extinction, preferred orientation, or microabsorption between different phases occur. The correction for these effects can be implemented in some Rietveld analysis programs (e.g., Madsen and Hill 1990; Taylor and Matulis 1991), but with caution. Microabsorption may not be avoided, but awareness of the effect gives an indication of over- and underestimation of phases. To improve precision and accuracy, it is useful to compare the results with alternative analysis methods (i.e., chemical analysis, automated SEM techniques). Precision can be determined through replicates. Calibration methods may also help for multiple samples of similar concentration and composition.

Applications

Rietveld-based QXRD has been extensively used for mineralogical characterization to understand the reaction chemistry and kinetics of mineral leaching and formation (e.g., Scarlett et al. 2008). The formation of new phases as a result of precipitation is often encountered during ore leaching processes, especially under high-temperature conditions, and the formation of these phases can be monitored by QXRD analysis (Whittington et al. 2003; Madsen et al. 2005). Some examples of the applications of Rietveld-based QXRD are illustrated in the following.

Bauxite Materials

The application of the Rietveld method to quantify minerals in bauxites has been reported by various workers (e.g., Aylmore and Walker 1998). A typical Rietveld refinement plot output from the assessment of a suite of bauxite samples from the

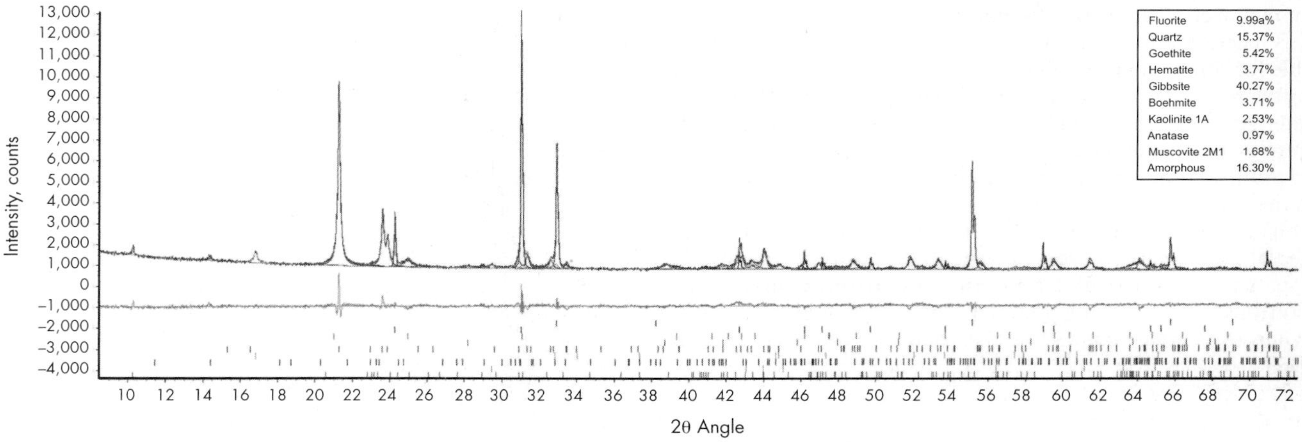

A. Typical output for Darling Range bauxite

Bauxite Sample	Gibbsite	Quartz	Goethite	Hematite	Boehmite	Kaolinite	Muscovite	Anatase	Maghemite	Al Substitution in Hematite	Al Substitution in Goethite
	wt %*									mol % (±0.6 mol %)	
1	47.7 (8)	21.9 (5)	16.1 (6)	6.7 (5)	—	1.0 (3)	4.5 (6)	1.8 (3)	—	8.4	27.7
2	55.1 (8)	21.0 (5)	7.5 (6)	5.1 (5)	5.0 (5)	2.3 (5)	2.3 (5)	1.3 (3)	—	16.5	30.4
3	42.4 (9)	15.5 (5)	22.1 (6)	8.5 (5)	4.8 (6)	—	2.4 (6)	1.6 (3)	—	9.5	28.5
4	46.8 (9)	22.9 (6)	10.2 (6)	6.4 (5)	5.4 (6)	1.5 (4)	2.5 (6)	1.1 (2)	1.6 (2)	10.1	33.3
5	56.2 (8)	23.8 (5)	9.8 (6)	2.6 (6)	3.2 (6)	2.4 (5)	1.5 (6)	1.1 (2)	—	9.9	32.8
6	59.5 (9)	1.9 (3)	19.5 (6)	9.7 (6)	2.6 (6)	2.5 (5)	—	6.6 (3)	—	5.3	9.3
7	69.6 (9)	0.5 (2)	15.7 (6)	6.4 (5)	1.9 (2)	1.5 (4)	—	5.6 (3)	—	8.6	20.6

*Estimated standard deviation of the least significant figure is shown in parentheses. Calcium fluoride (CaF_2) is used as internal standard to calculate amorphous content.

B. Rietveld analysis with calculated Al substitution values for hematite and goethite

Figure 4 (A) Rietveld refinement plot and (B) mineral abundance in bauxites

Darling Range in Western Australia and Rietveld-derived mineral abundance is shown in Figure 4. Observed, calculated, and difference patterns, together with hkl positions for each mineral, are shown in the figure. Refinement of the unit cell parameters enabled the level of Al substitution for Fe in structures to be calculated from the diffraction peak profiles of goethite and hematite. Variations in unit cell dimensions for goethite and hematite associated with Al substitution in the structures are well established for these minerals found in soils and bauxites (Fitzpatrick and Schwertmann 1982; Schulze 1984; Anand et al. 1989). The adjustment of site occupancy values in the structure files for goethite and hematite to reflect the substitution improved the correlation between predicted and calculated values for each mineral.

Poorly ordered phases cannot be accurately modeled by published structure information in the literature. However, if there is sufficient knowledge about the parameters that cause stacking faults, twins, or other extended imperfections within a phase, then it can potentially be modeled and quantified (e.g., kaolinite; Bookin et al. 1989; Neder et al. 1999). However, this type of work requires a specialist with crystallography knowledge to set up a model that can then be used by a metallurgist.

Investigating Reaction Pathways in Pressure Oxidation of Nickel Laterite and Sulfide Materials

Rietveld-based QXRD analysis can be applied to investigate reaction pathways and specifically the oxidation of sulfide minerals (e.g., Madsen and Scarlett 2007; Madsen et al. 2005; Whittington et al. 2003). Some good examples are illustrated by the work of Li et al. (2014) where the hydrothermal conversions of pyrrhotite and pentlandite were demonstrated and quantified using Rietveld-based QXRD analysis during the pressure oxidation of a nickel laterite concentrate. In this work, an intermediate nickel sulfide mineral phase with a cubic structure similar to vaesite and bravoite was identified. In another study, Rietveld-based QXRD analysis showed the co-processing of nickel laterite and sulfide inhibits the formation of basic ferric sulfate normally formed when treating sulfides in pressure oxidation. Quantitative XRD analysis was also applied to monitor the solid formation in the bioprocessing of iron-containing leach liquors and to characterize the mineralogy and crystallinity of the precipitates.

Cement Analysis

The Rietveld method has been successfully employed in industrial applications for QPA in complex hydrated systems such as in the cement industry. The precision of quantifying clinker in laboratory and industrial approaches has been extensively evaluated in the cement industry (Walenta and Füllmann 2004), where precise and reproducible analysis of cement constituents of sulfate phases of gypsum, hemihydrate, and anhydrite, as well as calcite and portlandite are required. XRD studies were used to characterize the hydration reactions of mixed systems containing portland cements and calcium aluminate cements qualitatively. An online XRD instrument for continuously monitoring phase abundances was constructed and installed in an operational portland cement manufacturing plant (Scarlett et al. 2001). XRD data were simultaneously collected using a wide-range (120° 2q), position-sensitive detector to allow rapid collection of the full diffraction pattern. The data were then analyzed using a Rietveld analysis method to obtain a quantitative estimate of each of the phases present with high precision.

Elemental Analysis by X-Ray Fluorescence

XRF spectrometry is a well-established technique that can routinely measure elements from Na through U to complement the mineralogy measured by XRPD and other techniques. XRF is used for both qualitative and quantitative analysis of elements at low concentrations (sub–parts per million) in a wide range of samples to higher concentrations in limited quantities of materials.

A wide range of commercial instruments are available, which include benchtop models, handheld devices, and online process monitors. A schematic diagram of an XRF spectrometer is shown in Figure 5. XRF uses an X-ray tube similar to XRPD, providing an X-ray source, except that end window tubes are used (rather than side window tubes) for higher efficiency. Rhodium (Rh) is used as the standard anode material, as the characteristic energies of this element are simultaneously suitable for exciting both heavy and light elements. Some instruments are also equipped with a vacuum system for improving sensitivity for low-atomic-number elements.

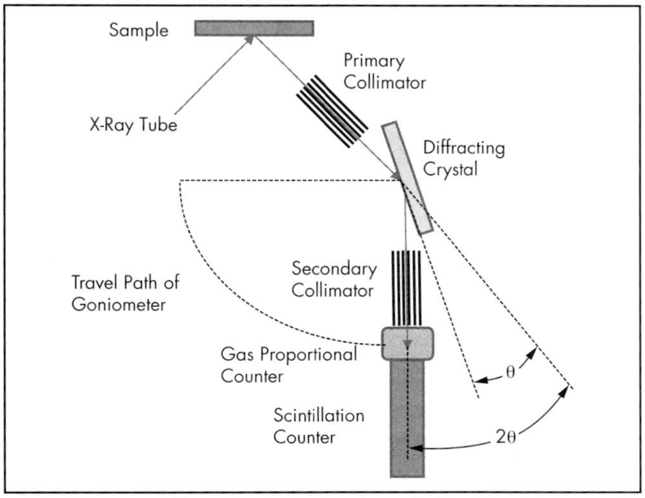

Figure 5 X-ray fluorescence wavelength dispersive spectrometry

Two kinds of instruments are used:

1. **Energy-dispersive X-ray (EDX) detection systems.** With energy ranges from 0.1 to 120 keV, these typically are benchtop size down to handheld models. They are usually employed for dedicated material-specific applications and general monitoring.
2. **Wavelength dispersive X-ray (WDX) systems.** These are widely used for more accurate, fast, and precise process and quality control applications in industry with 0.1–10 nm energy ranges.

EDX spectrometers tend to be more popular than WDX spectrometers because of their lower cost and ability to capture and simultaneously display information for various elements. However, they are more prone to spectral interference and are less accurate than WDX systems. EDX systems are better suited for the analysis of the transition elements and high-atomic-number elements.

Sample Preparation

The method of sample preparation for XRF analysis is dependent on the elements of interest, level of accuracy and precision required, detection limits, and turnaround time for analysis. For high-quality analyses of major and minor elements, fusion of the sample into glass disks with a meta- or tetraborate flux is used (Norrish and Hutton 1969). Sample preparation for the analysis of trace elements is usually accomplished by pressing powder briquettes, where the sample is mixed with a binder and subjected to elevated pressure. This method avoids the unnecessary dilution of the trace elements caused by a fluxing agent. Major element analyses can be carried out by this method, but samples have to be finely milled to reduce the absorption effect caused by coarse particles.

XRF analysis on slurries or loose powders can be done using online measurement instrumentation. However, the levels of precision and accuracy are compromised when compared to those obtained on fusion disk and powder briquette prepared samples.

Data Analysis

XRF analysis is prone to interference effects, but can be successfully addressed and corrected for utilizing computer-based algorithms (Rousseau 2006). The sensitivity of the spectrometer significantly varies, and the background varies by about two orders of magnitude over the wavelength range of the spectrometer. The sensitivity of the X-ray spectrometer decreases abruptly toward the long wavelength limit of the spectrometer mainly because of low fluorescence yields and the increased influence of absorption. Hence, the elements with lower atomic numbers at the long wavelength end of the spectrometer yield poorer detection limits (e.g., F, Na ~0.01% rather than parts per million; C, O the order of 3%–5%).

Reference spectra of thin film standards or pure element material are acquired, and individual elemental spectra are stored. These reference spectra are used in the standard deconvolution and mathematical separation of overlapping peaks of the unknown spectra. Spectral interferences (e.g., elemental peak overlap, escape peak, and sum peak interferences) are analyzed and attenuation corrections made.

DIGITAL IMAGING ANALYSIS TECHNIQUES

Digital imaging analysis involving microscopy techniques can provide the following information:

- Modal mineralogy
- Mineral associations
- Size-by-size liberation analysis
- Digital textural mineral maps
- Bright-field search for precious metal
- Grain size and shape
- Porosity

The advantages of automated digital image analysis techniques for mineralogical investigations over traditional microscopy techniques include fast acquisition time that provides a more statistically representative analysis of a sample as well as the ability to distinguish fine-grained or complex intergrown minerals at the micrometer scale. It also reduces the potential for operator bias and human error.

Optical Microscopy

Before performing any detailed quantitative analysis using automated mineralogy systems with SEM, it is good practice to examine samples with a standard optical microscope at the plant and/or at the analysis laboratory (Pirard 2016). Direct observation of specimens at low to intermediate magnification can provide information about microstructural defects, such as cracks, fissures, cleavages, pores, and grain boundaries, which can be overlooked when observing materials at the micrometer scale. In addition, some minerals are more readily identified by color and habit (e.g., flaky structure), which are not distinguished by backscatter imaging or X-ray analysis used in electron microscopy techniques. Optical microscopy can be carried out by either reflected light microscopy, which is suited for the imaging of most ore minerals, or transmitted light microscopy, which is more suited for the imaging of silicate gangue minerals.

Sample preparation can be similar or a precursor to electron microscopy techniques and does not require carbon coatings. Manual point counting techniques have traditionally been carried out on optical microscopes, but can be automated with camera-mounted systems to allow for digital process applications such as modal analysis. Optical microscopy resolution is not as good as that of SEM systems and does not distinguish minerals with variation in stoichiometries. In general, a larger number of measurements can be readily made by the automated SEM methods than can be attained with optical microscopy.

Electron Microscopy

In addition to the original commercial QEMSCAN and MLA systems now owned by Thermo Fisher Scientific, more recent developments include TIMA, Mineralogic, INCAMineral, and AMICS.

Instrumentation

Automated SEM systems consist of the following:

- The platform is SEM hardware.
- A backscatter electron (BSE) detector is standard with SEM equipment.
- One to four EDX spectroscopy detectors are mounted on the SEM. Most digital imaging analysis systems now use much faster liquid nitrogen–free Si drift detectors.
- User friendly operating and processing software with simplified or ready-made data outputs provide information on mineral speciation, composition, liberation, association, and size distribution.

Details on the theory and fundamentals of electron microscopy can be found on websites and in the literature (e.g., Amelinckx et al. 1997). The electron source in SEM can be a conventional tungsten filament; however, the latest generation of microscopes uses a field emission gun (FEG), which provides a more stable source, longer life span, and smaller spot size for improved BSE image resolution.

Automated SEM systems use a combination of backscatter imaging and X-ray analysis only for mineral identification and properties studies. Secondary electrons that provide images of the surface topography and other interactions (e.g., cathodoluminescence with appropriate detector) are not utilized in this process, although data can be collected simultaneously during analysis.

Backscattered electrons, which are elastically scattered electrons with energies close to the primary electron beam energy generated from tens of nanometers at the surface, are used for imaging compositional variations between phases. Minerals composed of heavier elements (e.g., zircon) backscatter more of the incident electrons of the SEM and appear

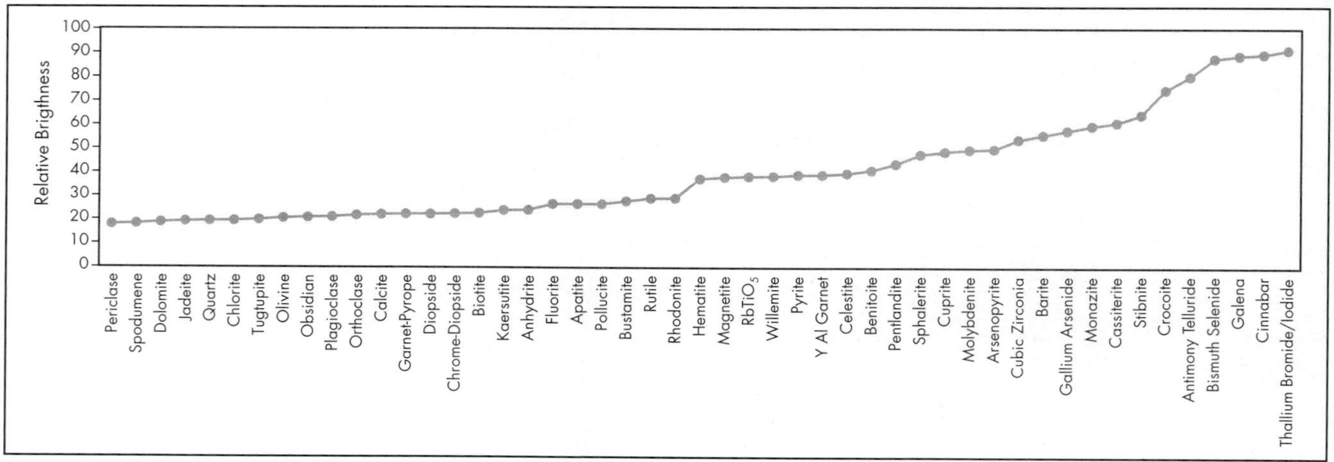

Figure 6 Relative backscatter electron intensity comparison

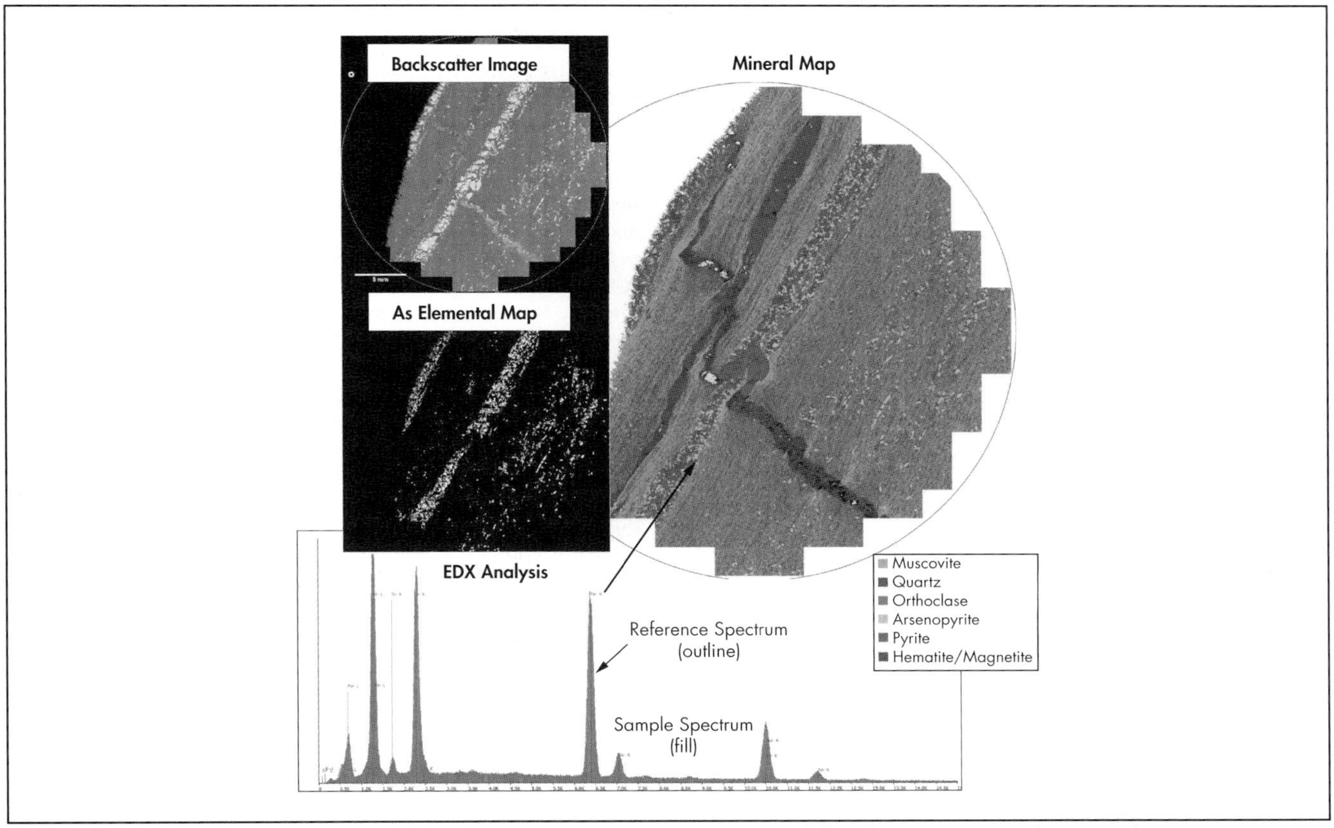

Figure 7 Mineral composition map with backscatter image, arsenic elemental distribution map, and energy-dispersive spectrum

brighter in the BSE image, whereas minerals composed of lighter elements (e.g., quartz) backscatter fewer electrons and appear darker. Figure 6 shows the relative intensities of some common minerals and phases.

The intensity variations (gray level) for different mineral phases in the BSE images provide an effective method of distinguishing the boundaries between mineral grains while the relative intensity values themselves provide a first-order identification of the minerals.

X-rays are produced by inelastic collisions of the incident electrons with electrons in the inner shells of atoms in the sample and are used to obtain the chemical composition of phases. Mineral classification by X-ray analysis is based on matching the entire spectrum of energy peaks collected on an unknown mineral to a library of X-ray spectra of known reference minerals collected using the same instrument parameters. From a combination of BSE contrast and matching EDX spectrum data files, mineral composition maps such as that shown in Figure 7 can be generated for the whole sample mount. The mineral maps can be used to describe the texture and mineral associations in each of the samples. As shown in the figure, the backscatter image depicts the contrast between sulfide minerals (light) and silicates (gray). The As elemental map illustrates the distribution of arsenic within grains in the polished mount and EDX analysis shows the arsenopyrite reference spectrum matching the sample grain spectrum.

The information is largely presented as two-dimensional (2-D) surface information with three-dimensional (3-D) electron beam interactions. Hence, grain boundaries and composition of mineral grains generated from X-ray data include interactions from 1 to 3 µm below the surface, depending on the beam conditions used. The size and shape of the electron beam surface interaction volume depends on the atomic number of the material being examined (a higher atomic number absorbs more electrons) and accelerating voltage being used (higher voltage results in deeper penetration and larger interaction volume).

Sample Selection and Preparation

As with any analytical technique, sample preparation is crucial to obtain reliable information. Samples analyzed can be large hand- or drill-core specimens, rock chips from percussion drilling or crushing circuits, or processed slurried samples from flotation and leach circuits. Some microscopes and software can accommodate polished blocks of different sizes.

Metallurgical plant samples are usually sized using dry and wet screening to eliminate agglomerated fine particles formed from previous filtering and dry procedures. To minimize biases associated with nonrepresentative sampling, riffling and subsampling prior to mounting are important. The selected samples for analysis are usually mounted in epoxy resin of 25- or 30-mm-diameter blocks or normal glass slides. In some cases, square blocks are used to maximize the area available for analysis.

It is important that the resin used for mounting has backscattering electron intensity less than the minerals or phases being measured to avoid misclassification with the background. In cases where carbon components are being measured, such as coal or graphite, mounts can be prepared with the addition of carnauba wax and/or mounted in polyester resin (Liu et al. 2005; O'Brien et al. 2011). Polished sample mounts require a thin layer of carbon (25–50 nm) using a carbon evaporator or carbon coater to prevent charging during the analysis. The carbon-coated polished mounts are then loaded onto the multisample stage and placed in the SEM for analysis.

The main issues that affect the quality of the data that have to be taken into account when preparing polished mounts are as follows:

- The number of coarse particles (>500 μm) that can be mounted in a polished section are low and can lead to issues associated with statistical analysis representative of the size population. Hence, many polished mounts or mounts with a large surface area are analyzed.
- Very fine particles <~15 μm pose practical problems for mounting because of possible bias caused by particle loss during sieving and polishing and tend to agglomerate during the sample preparation. Generally, very fine-particle sizes are not measured and rely on other techniques such as XRPD to measure the mineral composition data or can be inferred to some extent from data at a coarser fraction size.
- Variation occurs in the specific gravity of mineral and phase components within a sample. Samples require careful sample preparation to prevent segregation of heavy particles from lighter gangue occurring within the section yielding a biased distribution. The process of double mounting to get representative samples can be used (Zhou and Gu 2016).
- The measurements of the size and shape of elongated particles can be biased by variable orientations. The use of a low-viscosity resin can allow preferential settling of particles with their largest face parallel to the surface of the mount (Tsikouras et al. 2011).
- For studies of minor to trace heavy minerals in rock or metallurgical samples, the sample should be concentrated to increase the number of heavy mineral grains available prior to mounting and analysis. In particular, the dense nature and low concentration of gold found in ores requires careful sample preparation to monitor.

Sample preparation can be the bottleneck when considerable volumes of samples are required to be processed. Batch productions of polished sectioned mounts are available (e.g., Struers Hexamatic). Automated sample loading and analysis are now available for batch analyses in some SEM systems.

Although surfaces should be polished for best analysis, the geometry of several detectors mounted around the microscope can allow for analysis of unprepared surfaces.

Collection and Analysis of Data

Different methodologies and philosophies have been used and developed over the years for collecting and analyzing data. The general concepts are discussed in the following sections.

Data Collection

In automated SEM systems, the voltage of the electron beam energy typically used is 25 kV, but lower energies can be used to reduce the interaction volume and improve imaging resolution. At 25 kV in a TIMA with FEG, the probe current is 5.78 nA and produces a spot size of 39.5 nm at a working distance of 15 mm. The automated software controls the SEM to collect BSE images and X-ray data by stepping across the polished mount or thin section by a series of fields in an equally spaced regular rectangular mesh.

Measurements are taken in each field based on chosen analysis type and mode of analysis. A typical run time can be anywhere between 2 and 5 hours for analysis, depending on the resolution of the image required, covering ~200 frames, each 1.5 × 1.5 mm, with a typical resolution of 800 × 800 pixels. Sampling statistics are important and therefore the grid spacing used in, for example, point counting should be larger than the largest grain or particle size in the sample.

Mode of Analysis

The method of analysis is dependent on the information required and the methodology supported by the vendor software (Zhou and Gu 2016). All methods generally use a combination of BSE and EDX spectrum data measurements. Overall, a large number of points can be analyzed using a BSE image than by X-rays (e.g., 819,200 pixels vs. 40,000–60,000 X-ray points in 100s). However, the accumulation of X-ray data can be reduced by increasing the number of detectors used on the instrument and the development of methods in combining low-count X-ray data from various analyses (Gottlieb et al. 2015). Spatial resolution is also higher for BSE imaging than X-ray analysis with a resolution of between 0.1 and 0.2 μm compared to between 2 and 5 μm for X-ray analysis. However, difference in particles or grains cannot be distinguished by BSE where BSE intensities of the minerals of interest are similar (e.g., distinguishing different micas or clay minerals). The main measurement methods used in quantitative mineralogy systems are illustrated in Figure 8.

The technique used in a TIMA is illustrated in Figure 9. The image is divided into different segments. Each segment is assigned to a mineral or left unclassified. Adjacent segments located within a single particle are classified as the same mineral and joined to define a grain. For each grain, the total number of pixels within each segment and the mineral name is registered. The number of pixels for a specific mineral is derived as a sum of pixels of each grain that is assigned to that mineral. The relative volume and mass are then derived using the number of pixels.

Two approaches to analyses designed for speeding up the analysis are the bright phase search and the line scan mode. The bright phase search mode uses the backscatter brightness distribution obtained from the backscatter detector on the SEM to filter out particles that contain minerals of interest with high BSE contrast. This method eliminates analyzing grains that are of no interest and thereby reducing the analysis time. This is useful for searching sparse mineral phases such as gold and platinum group metals (PGMs) as described later.

For line-scanning mapping, the electron beam is scanned along in a series of lines across the polished mounts. The distance between adjacent horizontal lines and the distance between the measurement points along each horizontal line are defined. The lines are divided into linear sections divided by measurement points below the threshold to find sections through individual particles. The combination of the BSE level and the spectroscopic data can be used to determine transitions between distinct phases. This method can be used to provide modal analysis, but is useful for mineral grain size analysis also described later.

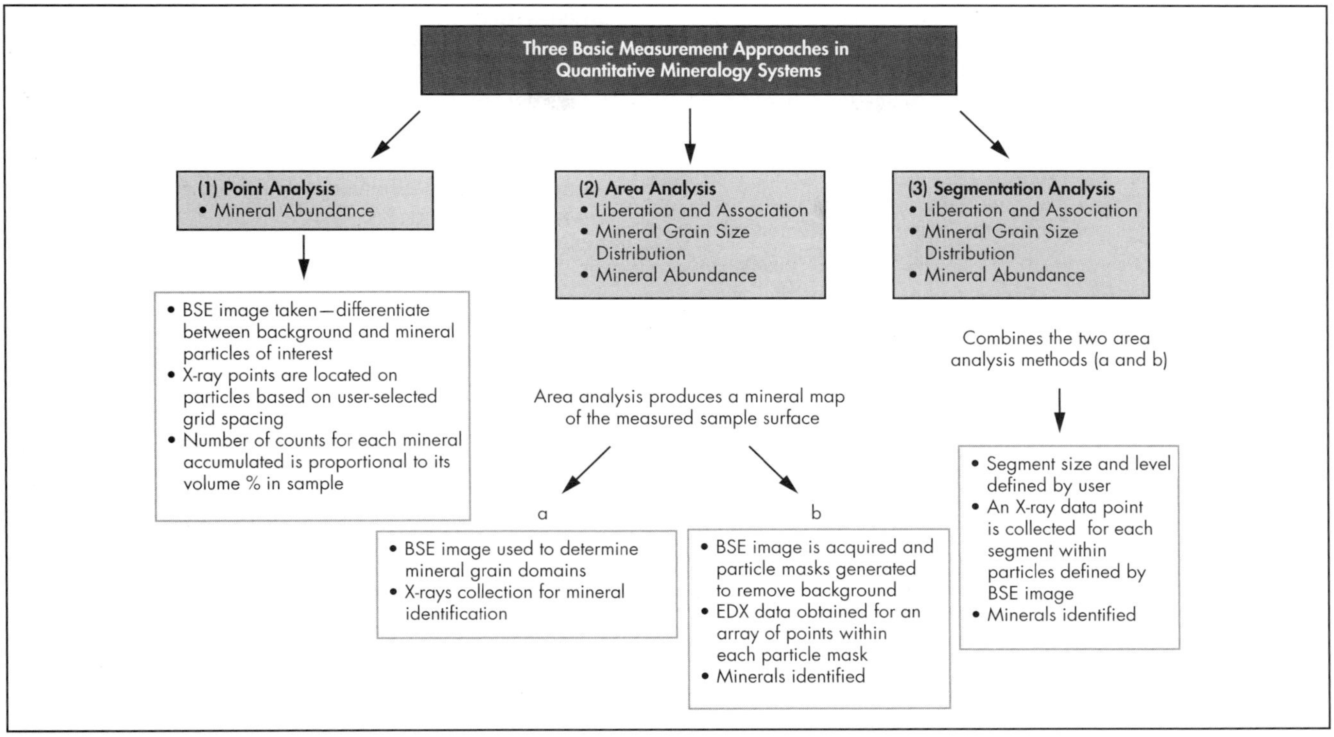

Three Basic Measurement Approaches in Quantitative Mineralogy Systems
(1) Point Analysis
• Mineral Abundance
(2) Area Analysis
• Liberation and Association
• Mineral Grain Size Distribution
• Mineral Abundance
(3) Segmentation Analysis
• Liberation and Association
• Mineral Grain Size Distribution
• Mineral Abundance
• BSE image taken—differentiate between background and mineral particles of interest
• X-ray points are located on particles based on user-selected grid spacing
• Number of counts for each mineral accumulated is proportional to its volume % in sample
Area analysis produces a mineral map of the measured sample surface
a
b
• BSE image used to determine mineral grain domains
• X-rays collection for mineral identification
• BSE image is acquired and particle masks generated to remove background
• EDX data obtained for an array of points within each particle mask
• Minerals identified
Combines the two area analysis methods (a and b)
• Segment size and level defined by user
• An X-ray data point is collected for each segment within particles defined by BSE image
• Minerals identified

Source: Zhou and Gu 2016

Figure 8 Measurement approaches for quantifying mineral phases

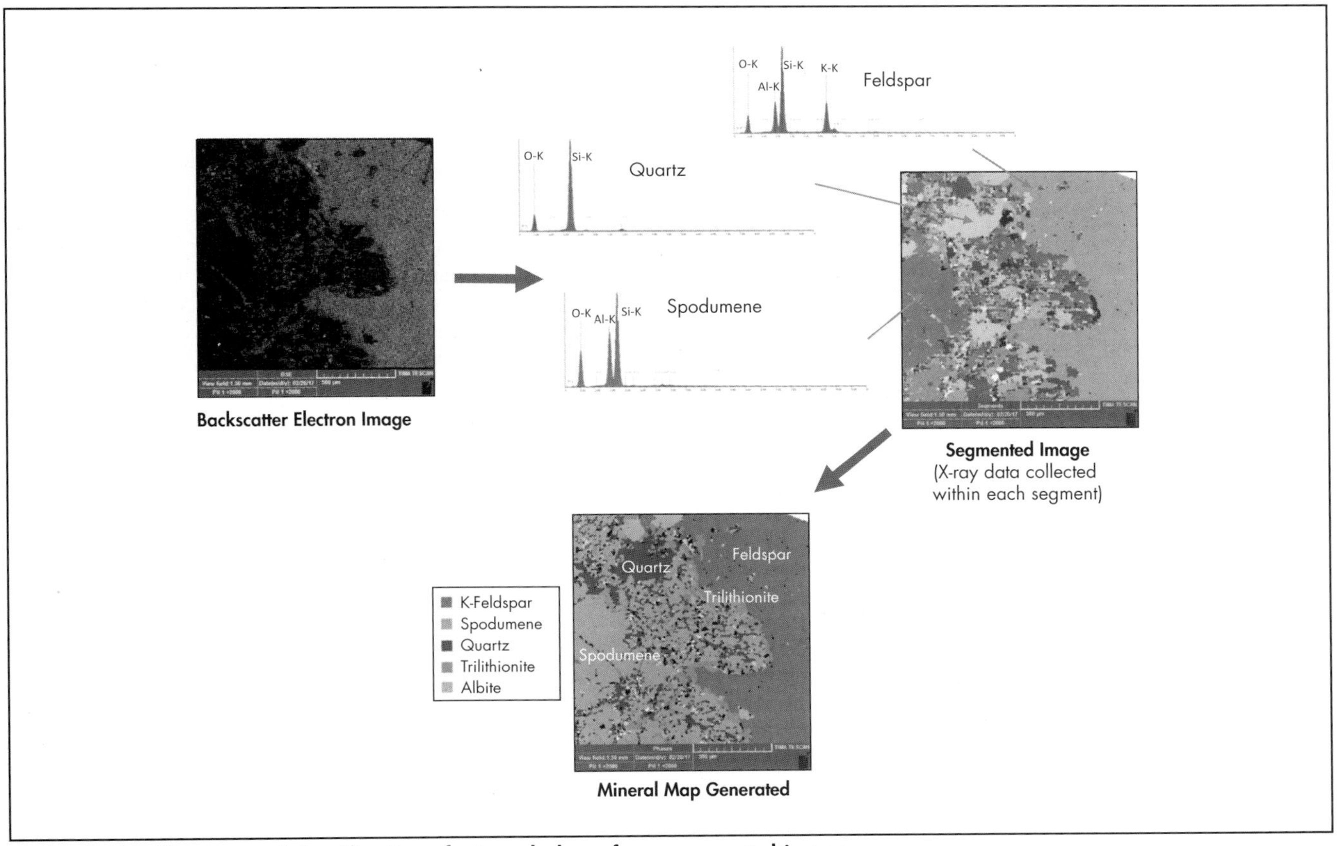

O-K
Al-K
Si-K
K-K
Feldspar
O-K
Si-K
Quartz
O-K
Al-K
Si-K
Spodumene
Backscatter Electron Image
Segmented Image
(X-ray data collected within each segment)
Quartz
Feldspar
Trilithionite
Spodumene
K-Feldspar
Spodumene
Quartz
Trilithionite
Albite
Mineral Map Generated

Figure 9 Identification and classification of mineral phase from segmented image

Classification of Mineral Phases

An important precursor to the measurement of ore characteristics in automated mineralogy analysis is the creation of the mineral library or mineral classification scheme, which represent the minerals expected in the ore or metallurgical sample that is being evaluated (Johnston 2016). Mineral classification by X-ray analysis is based on matching the entire spectrum of energy peaks collected on an unknown mineral to a library of X-ray spectra for known reference minerals collected under the same instrument parameters. The recorded backscatter intensity (gray level) can also be used to help distinguish the unknown mineral. Each mineral record in the library consists of a mineral name, chemical formula, density, X-ray spectra, and backscatter data. Ideally, the data collected for the listed minerals are from standards where the chemical composition has been determined by other analytical techniques (e.g., electron probe microanalyzer [EPMA], laser ablation–inductively coupled plasma mass spectroscopy [LA-ICPMS] described later). For minerals with variable compositions, specification of the elemental composition is more difficult, and some additional measurements may be required to determine an average value of the element in the mineral in the system.

Minerals that have large composition variations can be classified into two or more categories based on their composition ranges. Examples of where this is important in metallurgists' applications include defining sphalerite compositions with variable zinc and iron levels, siderite containing variable zinc levels, freibergite containing variable silver levels, and different types of pyrite grains containing arsenic and associated gold. In cases where fine minerals are intergrown and cannot be distinguished at the resolution of the SEM system, a phase mixture can be defined as a separate phase identity in the mineral library. The correct identification of silicate gangue minerals, such as clay minerals, talc, and pyrophyllite, is always important in leaching systems and froth flotation.

A condensed version of the minerals in the list can be generated by most automated systems. This is particularly useful in flotation applications where minerals can be grouped into nonsulfide gangue minerals, sulfide gangue, and minerals of interest categories. Such applications include bulk sulfide concentrate for gold and separation of copper minerals from pyrite and pyrrhotite and gangue minerals (Johnston 2016).

Type of Information Obtained by Automated SEM Systems

Following is a summary of the type of information that can be obtained from automated SEM systems, which is discussed in detail in the following sections:

- **Modal mineralogy.** The percentage of each mineral in the sample.
- **Elemental deportment.** The distribution of each element across all minerals containing that element.
- **Mineral liberation.** The distribution of the mineral of interest across particle composition classes, based on the bulk composition or surface composition of particles.
- **Mineral locking.** A measure of particle composition distribution based on quantifying the proportion of the mineral of interest.
- **Mineral association.** A measure of the degree to which pairs of minerals are adjacent to one another in the sample.
- **Mineral grain size.** Distribution of sizes of the grain cross-sections.
- **Phase specific surface area.** The surface area per unit volume of a mineral providing a single value, representing the mean mineral grain size distribution.
- **Particle shape.** Particle shape on a mineral-by-mineral basis that can be determined using a variety of conventional shape descriptors.

Importantly, in these analyses, the mineralogical identification and quantification is derived from compositional rather than structural analysis. Comparing quantitative XRPD with automated SEM analysis can therefore be valuable in identifying amorphous phases and correcting bias in either method.

Modal Mineralogy

A modal analysis is a listing of the minerals present and their mineral abundance expressed as weight percent. The volumetric fractions of each mineral are determined by converting the number of pixels of each mineral in a polish sectioned mount into the relative area as a percent. The sum of all relative volumes is calculated for all phases where the relative volume is equal to the relative area for a given phase (i.e., measurement from one-dimensional or 2-D mineral abundance data is equivalent to the true, 3-D value based on the Delesse principle). The relative mass of a phase as a percent is derived using the density value defined for each mineral in the mineral classification scheme. The smallest particle that can be measured is typically about 5 µm in size as the interaction volume of the electron beam is close to this size of the particle being analyzed.

The results can be presented in the form of analysis of different size fractions and whole sample as illustrated in Figure 10 for a crushed micaceous Li pegmatite. A typical table of modal analysis with errors is also shown. The relative error expressed as a percentage is calculated at the 95% confidence level (i.e., 2 standard deviations above and below the measured value; Napier-Munn 2014). Lepidolite represents a solid solution series between polylithionite and trilithionite and is defined as a composite in the mineral classification scheme.

The quality of the modal data depends on the reliability of the mineral recognition and the number of measurements made (Wightman et al. 2016). To increase the confidence of the user in the measurements produced, it is common to convert the mineral assays from an automated system to elemental assays, which are then compared with the independent elemental assays from a chemical laboratory.

Mineral Liberation Analysis and Characteristics

An understanding of the liberation characteristics or particle composition distribution of an ore is important in comminution circuit design and for optimizing beneficiation processes such as flotation, density, and magnetic and electrostatic separations in operating plants. This type of analysis can provide the metallurgist with details on what proportion of the valuable minerals are liberated, the amount of unliberated (composite) particles that require further grinding, and the proportion of the liberated gangue minerals that can be directed to the tails. Equally important is the understanding of how unliberated particles vary in composition and texture (Evans and Morrison 2016). The form of information required is dependent to some degree on which separation process is used to treat the ore and what physical property of the particles is required to be exploited in the process (e.g., surface area

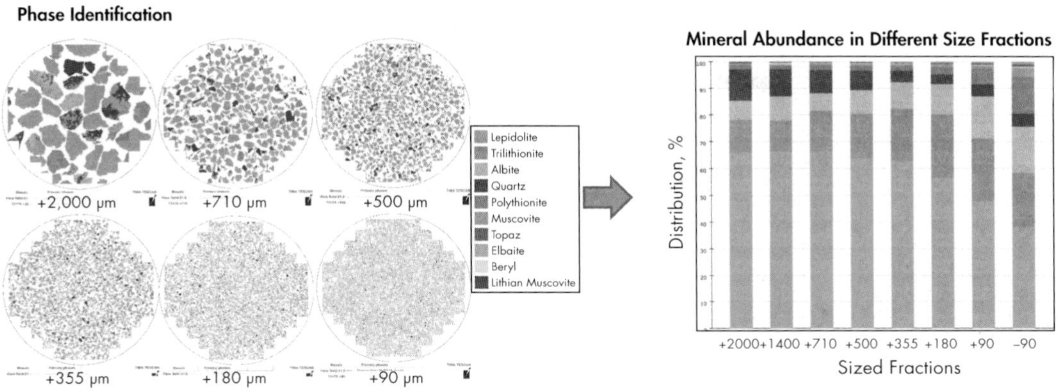

Modal Mineralogy

Mineral	Weight, %	Area, %	Points	Relative Error, %
Lepidolite	58.4	58.2	27,038,849	0.02
Trilithionite	17.0	17.0	7,895,228	0.1
Albite	10.0	11.3	5,267,193	0.1
Quartz	7.3	7.7	3,599,810	0.1
Polylithionite	3.5	4.0	1,846,015	0.1
Muscovite	1.0	1.2	543,919	0.3
Topaz	0.4	0.3	153,638	0.5
Beryl	0.2	0.2	97,903	0.6
Spodumene	0.1	0.1	33,409	1.1
The rest	0.1	0.0	3,684	3.3
Total	99.9	100.0	46,479,648	

Note: Relative error is calculated based on 2 times the standard deviation (i.e., 95% confidence level).

Figure 10 Modal analysis in sized fractions and whole sample for a crushed micaceous Li pegmatite

exposed for leaching and flotation, volume for gravity or magnetic separation techniques). In automated SEM techniques, the liberation of a particle with respect to a given mineral of interest can be computed and expressed in one of the two following forms:

1. **Surface area.** Liberation of a particle is the length fraction on the outer perimeter of the particle covered by a mineral(s) of interest with respect to the whole outer perimeter of the particle expressed as a percentage.
2. **Volume.** Liberation of a particle is the area fraction of a mineral(s) of interest with respect to the total area of the particle normally expressed as a percentage.

Mineral *locking* expresses which phases a mineral of interest is locked within and how much free surface is exposed. The level of locking is a function of the surface area of a particular mineral. Overall, while the liberation characteristic function provides information on the composition of the mineral of interest, the mineral locking function provides information on the texture of each particle.

Mineral association yields information about spatial relation between phases inside particles. This analysis reports the percentage of the mineral of interest that is fully liberated, in binary and in ternary particles. However, reporting the mineral associations as a modal analysis for each liberation class is advised for concentrator plants (Lastra and Paktunc 2016).

To obtain accurate liberation data, samples are sized and polished mounts are made of the different size fractions to be measured separately. This is to reduce the stereological bias that is inevitable when attempting to measure 3-D objects with their 2-D sections (Spencer and Sutherland 2000). The magnitude of the stereological bias is small for particles with complex textures, reaches a maximum for a given ore for binary particles with simple texture, and then reduces to zero for particles that are fully liberated (Gottlieb et al. 2000; Spencer and Sutherland 2000; Evans and Morrison 2016). Fortunately for most systems with a natural range of particle composition distributions, the measured liberation characteristics have been observed to be similar to true liberation values, and only systems with narrow composition ranges require major adjustments (Petruk 2000).

For liberation analysis, it is important to remove agglomerated particles. Agglomerated particles affect the results of mineral liberation assessment, because they are considered as one particle. Samples of dry flotation concentrates commonly have agglomerated particles. Samples can be subjected to a strong attrition treatment process (e.g., ultrasonic bath) to remove the agglomerated material prior to mounting. Alternatively, many automated software systems have particle separation routines that set boundaries between touching particles within agglomerates.

For analysis, the particles are divided into composition or surface exposure bins based on their liberation. Figure 11 illustrates a liberation analysis for sphalerite in a massive sulfide sample expressed as cumulative liberation yield of sphalerite by particle surface area, theoretical grade–recovery curve, and as a graphic representation in terms of degree of liberation and size. In this example, the data is separated into 10 bins at 10% intervals, representing particle surface area liberation classes (≥0 and <10%, ≥10 and <20%,

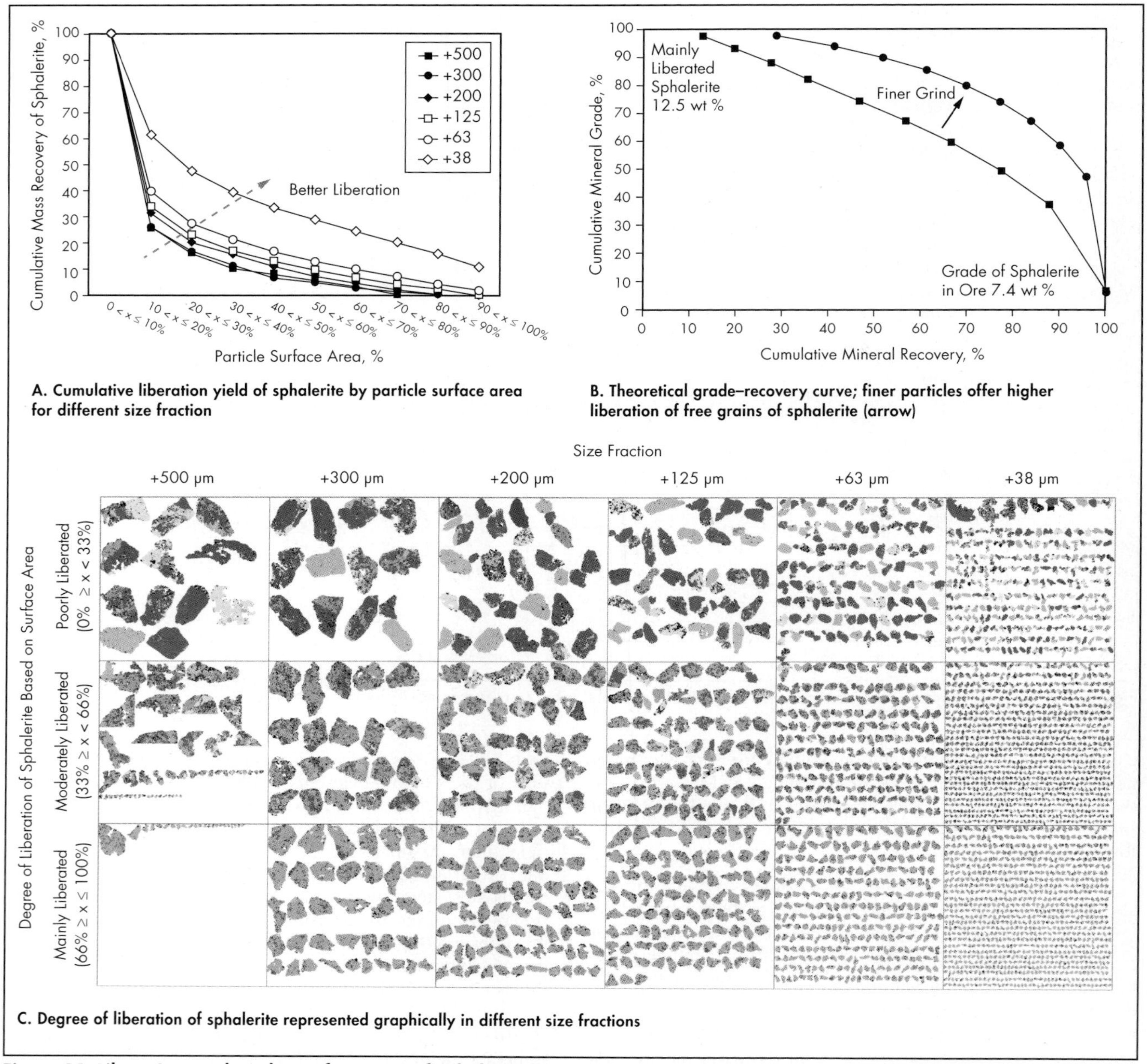

A. Cumulative liberation yield of sphalerite by particle surface area for different size fraction

B. Theoretical grade–recovery curve; finer particles offer higher liberation of free grains of sphalerite (arrow)

C. Degree of liberation of sphalerite represented graphically in different size fractions

Figure 11 Liberation analysis by surface area of sphalerite in a Zn-Pb massive sulfide sample

etc.) for different sized samples, and shows finer grain size promotes exposure of more separate sphalerite grains for beneficiation.

Grade–recovery curves are a common method for representing the performance of industrial flotation operations and batch flotation tests, for example, the liberation of sphalerite in a Zn-Pb massive sulfide sample (Figure 11B). The theoretical grade–recovery curve for an ore is defined as the maximum expected recovery that can be achieved by physical separation of a mineral at a given grade (McIvor and Finch 1991). This is determined by the surface area liberation of the mineral of interest and is directly related to the grind size. A combination of automated mineralogy SEM and flotation test work can be used to define the efficiency of a process and allow for the diagnoses of where potential improvements may be made. The position of a theoretical grade–recovery curve is calculated based on liberation values and the mineral assay data that is sorted by the software in descending order of particle quality. In most automated SEM software, the data can also be screened or filtered to categorize particles or mineral phases and then be visually displayed (Figure 11C).

The disadvantage of producing five or six sized subsamples instead of one sample for liberation studies is that the cost of the study can be expensive. In some circumstances, it is possible to use uncorrected liberation data from unsized samples for comparative studies where samples have been taken from a single process operation (Lastra and Petruk 2014).

Theoretical grade–recovery curves provided by automated mineralogy techniques are generated from 2-D liberation measurements and overestimate the true liberation by a certain amount and therefore only provide a guide. Test work is always required to validate the results.

Mineral Grain Size and Shape Analyses

Automated imaging systems can be used to assist in the measurement of grain size as part of the property describing ore texture. Mineral grain size distribution data are a useful predictor of liberation when developing or modifying existing comminution circuits (McIvor and Finch 1991; Evans and Wightman 2009; Johnston 2016).

Grain sizes should ideally be estimated by measurements on polished mounts of rock samples or blocks of ore that are significantly coarser than the mineral grains of interest (Sutherland 2007). Direct estimate of the grain size can be made by measuring the equivalent circle diameter and maximum diameter of a grain particle. However, this approach is stereologically biased because of the sectioning of grains in the polished mount where the 2-D image of grains represents less than or equal to the true size of the grain. The phase specific surface area (PSSA) measurement function can be used and is not stereologically biased. To collect data for grain size estimation using line scan measurements (line scan mode) is preferred over area image mapping. The parameter provides a single value representing a mean grain size of each mineral in a mount but does not use actual grain measured distribution. Overall, estimation of grain size distributions should be used with caution and in conjunction with the PSSA for each mineral. The results should be collated and cross checked with actual experimental data for validation.

In some software systems, such as MLA, the grain shape can be measured in terms of aspect ratio, shape factor, and angularity (Tsikouras et al. 2011). Particle shape analyses, along with the measurement of porosity, are particularly useful in sedimentary rock studies such as those used in the oil and gas industry.

Statistical and Error Analysis

When comparisons are being made between different size fractions in grinding and flotation circuits, it is useful to quantify the errors in the measurements to determine whether any differences are significant. As the particle size increases, the number of measured particles decreases, which is caused by limitations in the number of particles that fit into one polished block. The reduction in the number of particles measured increases the errors associated with the measurement, and therefore multiple polished blocks may be required to be measured. As many processes are now investigating coarse rejection of gangue as a method to reduce operating costs, the analysis of coarse particles is becoming more prevalent. Based on statistical analysis by various workers, at least 2,000 particles are required to be measured in polished mounts (Evans and Napier-Munn 2013; Mariano and Evans 2015). Alternatively, combining and reducing the particle composition class in the distribution increases the number of particles in each composition class, is also effective, and allows for the reduction in costly analytical time.

Analysis of Low-Grade and Precious Metal Ores

One of the main advantages and applications of automated digital mineralogy techniques is in the measurement of low-grade components in samples, particularly in analyzing precious metals (PGMs and gold) or for investigating indicator minerals (e.g., zircons or rutile) in exploration studies. The benefits of the use of automated mineral mineralogy techniques and some of the challenges in gold deportment studies have been discussed in the literature (Chryssoulis and Cabri 1990; Goodall 2008; Chryssoulis and McMullen 2016).

In gold metallurgy studies, the feed and tailings gold grades are generally very low (<10 ppm), and the low frequency of occurrence of grains containing target minerals means that they are difficult to locate in statistically representative numbers.

Normally, both free gold and sulfide mineral grains associated with gold have a higher density compared with the gangue minerals. This allows for them to be concentrated using techniques such as a shaking Wilfley table or gravity gold recovery test (Laplante et al. 1995) for coarse-particle analysis. For finer particles, a combination of heavy liquid separation (bromoform, 2.889 g/mL; methylene iodide, 3.325 g/mL; water-based Li polytungstate solids, 1–2.95 g/mL) and super-panning (Zhou and Cabri 2004), as well as hydroseparation techniques (Lastra et al. 2005; Cabri et al. 2005, 2008), depending on the type of sample under investigation, can be used to concentrate and improve the efficiency and representation of the analysis. However, care must be taken that gold grains associated with gangue minerals are not discounted from the final gold balance. Coarse free gold minerals are usually best identified by reflected light microscopy as stereomicroscope provides 3-D vision about the flakiness of gold grain (Chryssoulis and McMullen 2016).

In the determination of gold and PGM deportment studies, bright phase search mode is used where each of the fields on the polished mount are fast scanned for precious metals (gold or PGMs) based on the backscatter brightness distribution obtained from the backscatter detector on the SEM. The BSE brightness threshold in the software is set at a level that filters out particles that contains minerals of interest and eliminates analyzing grains that are of no interest. This helps reduce the analysis time. The practical limit of detection for gold grains is 1–2 μm by this method but is dependent on the step size used in the scanning grid. Once a bright phase is found, the whole host particle can be mapped to yield information about the mineral association. Some bright-field scanning modes have been developed to allow more detailed evaluation of grains (e.g., liberation, second grain mapping) such as found in the MLA software (Fandrich et al. 2007). An example of output using the bright phase search in the TIMA is shown in Figure 12, where the grains containing detected gold, sylvanite, and their mineral associations can be listed and documented.

Minerals such as galena and bismuth, however, which also contain high BSE brightness values close to that for gold can cause an issue, which means they also are detected and analyzed along with gold. This is prevalent when analyzing materials such as auriferous massive sulfide Zn-Pb deposits that can have finely disseminated galena grains within pyrite grains. Hence, BSE brightness threshold setting needs to be set to filter out other bright-field minerals.

Although automated mineralogical image analysis can be used to define the mode of gold grain occurrence, associations, and gangue mineralogy, target sulfide minerals must be identified and selected for analysis of submicroscopic gold concentrations by microbeam and surface analysis techniques

Figure 12 Pyrite particles containing gold (light) and sylvanite $(Au,Ag)_2Te_4$ (arrows) grain inclusions

described in the "Semiautomated Microprobe Techniques" section. The combination of the abundance and textural types of gold carriers determined is then used to calculate the cumulative gold concentration in the sample. Combining automated mineralogical techniques with diagnostic chemical assaying has also been effective in applications, such as in analyzing a flotation tailings stream of a Cu–Au operation in North Queensland, Australia, where gold loss was determined to be caused by the presence of coarse free gold (Goodall 2008).

Bright phase search can also be effective for uranium ores for locating uranium minerals, although there, abundance is much higher compared to gold grades.

Main Sources of Error and Limitations in Quantitative Analyses

Although automated quantitative mineralogy estimation provides a powerful tool for analysis, the accuracy and data interpretation depends on many operator, physical, chemical, geological, sampling, and statistical factors. Some sources of error and limitations are as follows:

- The analysis of data requires some knowledge of mineralogy, metallurgical, and geological processes when training the system and interpreting the results, not simply automatic readout of data.
- An understanding of basic physics concepts in accounting for the spatial resolution limitations of electron-induced X-ray emissions (energy, density, and generation volume) is important, particularly as they vary with the composition of phases under examination.
- The quality of the data in the mineral classification scheme (library) used in automated SEM systems for identifying minerals is validated and is clearly documented where the data and information originated. Mineral abundance, assay reconciliation, and mass balancing are reliant on the library data.
- Correct sample selection, subsampling, and processing of the samples ensure that the right sample is analyzed. The sample preparation, knowledge, and skill in creating representative polished sections is also important.
- Appropriate measurement mode, pixel spacing, and measurement time must be selected. The conditions used have to take into account the available time as well as application.
- Estimates for 3-D properties from 2-D measurements (stereology) are limited. The results should always be checked with experimental data to ensure continuity.
- Lack of information in estimating the proportions of low-atomic-number elements from EDX spectra (absorption of low-energy photons by the detector window) can result in error. The data for elements below and including carbon are not or are only partially represented in EDX spectra, which can reduce its effectiveness in classifying some mineral phases (e.g., Li in spodumene, carbon in carbonate minerals).
- The overlapping of some X-ray lines can reduce detection for some elements, such as rare earth elements in monazite or between Pb and S in galena, and ultimately the identification of the mineral phase.
- Minerals should not be identified based on their chemical composition alone (e.g., polymorphs aragonite and calcite, sphalerite and wurtzite).
- Interaction volume can result in a mixed spectrum of minerals for complex samples.
- The practical limitation of grain identification is ~0.5 μm.

Applications

Applications of the use of automated SEM are well documented in the literature for just about every metal commodity. A recent book published on process mineralogy provides many detailed applications where automated mineralogy has been used in industry (Becker et al. 2016). Other applications have included the characterization of Ni-rich and Ni-poor goethite in laterite ores (Andersen et al. 2009) and rare earth element deportment studies (Smythe et al. 2013). A selection of results illustrating the usefulness of automated digital mineralogy is shown in the following subsections.

Liberation Analysis of Sulfide Minerals

The application of automated mineral techniques in optimizing the liberation of sulfide components in concentrators from

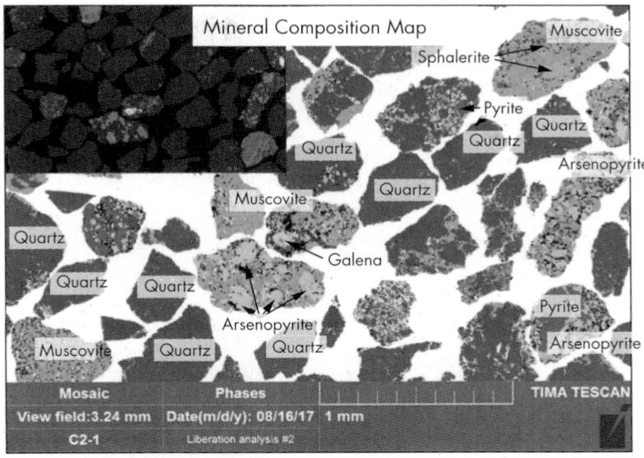

A. Mineral map with embedded backscatter image

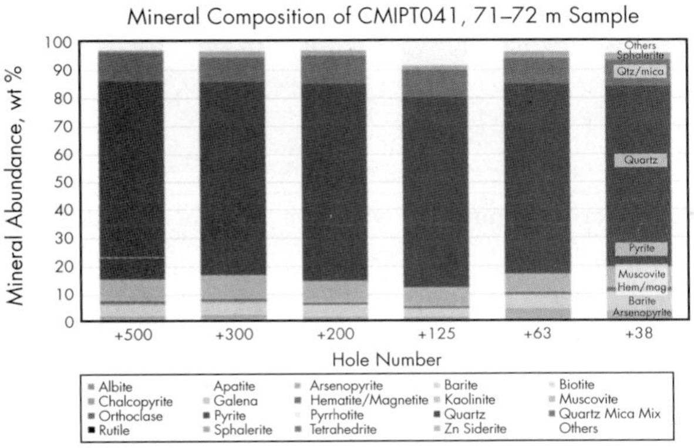

B. Modal mineralogy of different sized fractions

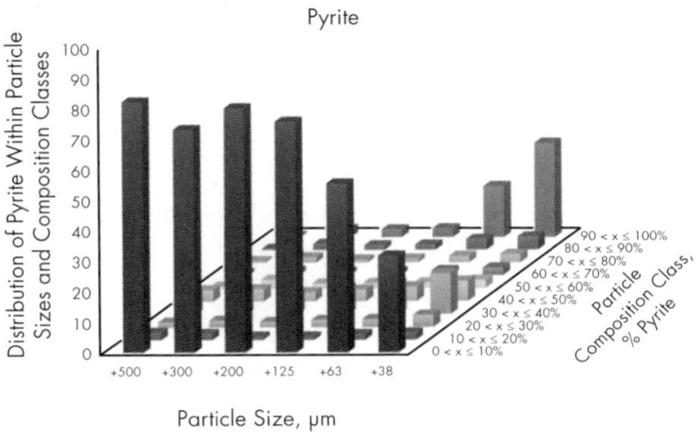

C. Measured particle composition distribution for pyrite

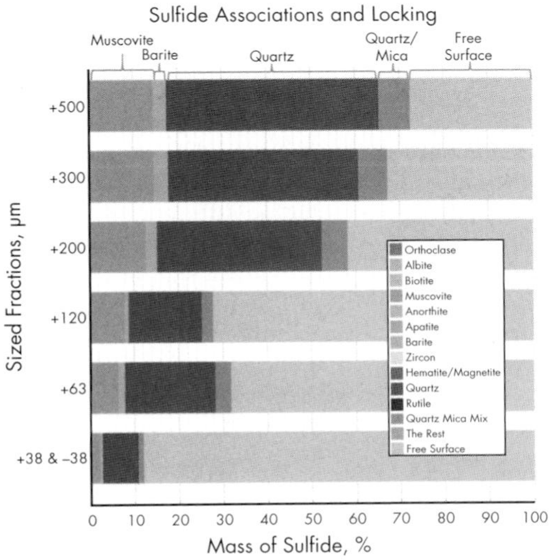

D. Proportions of sulfide minerals free and locked in gangue minerals

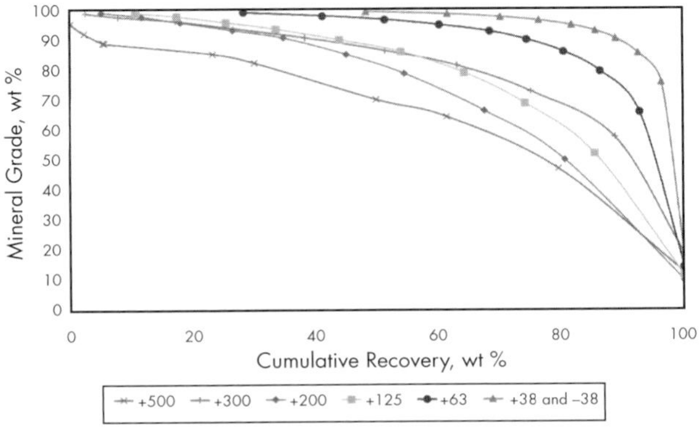

E. Theoretical grade–recovery curve for sulfide minerals

Figure 13 Mineral liberation analyses of a feldspar porphyry rhyolite-rhyodacite sample

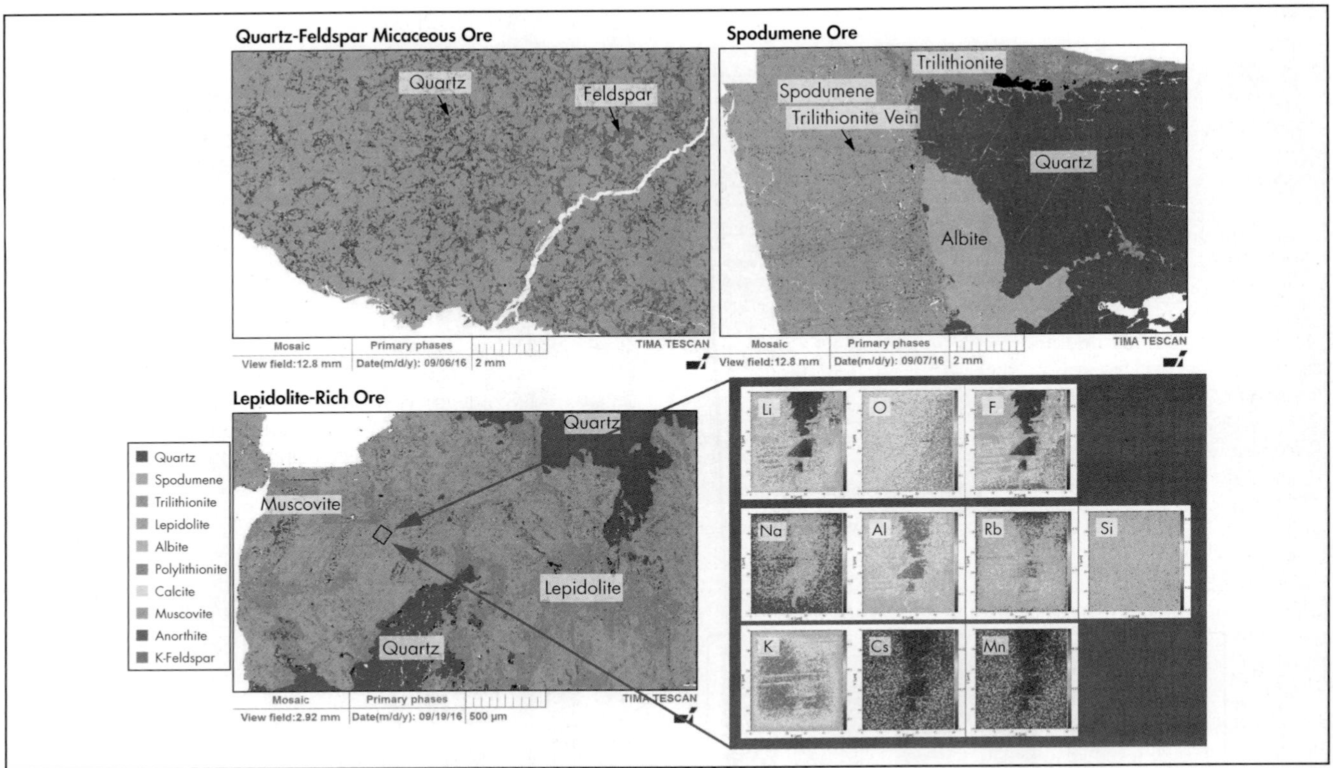

Figure 14 Mineral composition maps for three different textured lithium-bearing pegmatites with time-of-flight secondary ion mass spectrometry analysis

Zn-Pb sulfide ores (e.g., McKay et al. 2007; McKay 2016; Fosu et al. 2015; Johnston 2016) and copper ores (MacDonald et al. 2011) are documented in the literature. Figure 13 illustrates partial results for an assessment of a feldspar porphyry rhyolite-rhyodacite deposit showing the mineral composition map, the modal mineralogy, and the liberation and concentration of sulfide minerals for gold recovery. LA-ICPMS analysis performed on the sulfide grains indicated solid solution gold in pyrite (averaging 4.8 ppm, 0–15 ppm) and arsenopyrite (averaging 86 ppm, 0–550 ppm) grains. A backscatter micrograph together with the mineral map image and modal mineralogy for different sized fractions illustrates most of the sample is made up of silica minerals (Figures 13A and 13B). The particle composition distribution for pyrite results was obtained using the mineral liberation function in the TIMA software. The amount of the major sulfide minerals in each particle composition class, ranging from 0% to 100% in 10% intervals, was calculated for each of the particle size fractions. The measured particle composition distribution for pyrite, the assessment of the amount of sulfide minerals locked in gangue minerals, together with the theoretical grade–recovery curve for sulfide minerals are presented in Figures 13C–13E. Overall liberation analysis generated from the data shows most sulfide minerals are in composite grains with quartz and muscovite, mainly in the coarse-size fractions, and require fine grinding to liberate particles.

This information along with other information on mineral deportment, their size distributions, compositions, and associations with other minerals are being used to assess the economic viability to treat this ore.

Lithium Ores

The demand for lithium has increased significantly in recent years as a result of an increase in demand for lithium-based rechargeable batteries for portable electronic devices and electric passenger cars. Hence, a comprehensive understanding of the deportment of lithium and associated minerals in potential ore bodies is essential to allow the industry to predict the response of ore reserves to metallurgical treatment options. Figure 14 illustrates the complexity of mineral intergrowths in some potential lithium-rich ores using the high-resolution mapping scan mode of TIMA and the integration of techniques such as time-of-flight secondary ion mass spectrometry (ToF-SIMS) to allow for the measurement and correlation of lithium (not measured with standard EDX spectrometry) and other elements associated within different minerals. The mineral composition maps illustrate lepidolite, a lithium-bearing mica, intergrown with quartz and feldspar in a quartz-feldspar-micaceous sample; a spodumene grain with fine veins of lithium-bearing micas (lepidolite, trilithionite) attached to quartz and albite in a spodumene ore; and a lepidolite-rich sample, showing alteration to muscovite. The ToF-SIMS elemental scans illustrate lithium is associated with F, enriched K, Rb, Cs, and Mn in lepidolite; and muscovite is deficient in Li and associated with Na and high Al content. The characterization of different lithium minerals in pegmatite ores and its implication to processing have been published elsewhere (Aylmore et al. 2018a, 2018b).

Precious Metals Deportment Studies

Automated SEM techniques can be used to determine how much gold and silver is free or liberated and what percentage is attached to each of the major minerals such as galena, chalcopyrite, sphalerite, pyrite, and nonsulfides. There are many examples that can be found in the literature (e.g., Goodall et al. 2005; Goodall and Butcher 2012). Figure 15 illustrates results from using bright-field scans and detailed mineral map scans in the high-resolution mapping mode of TIMA to assess precious metal content in six mounts on a massive sulfide flotation feed (Au 1.8 g/t, Ag 140 g/t, 13.6% Fe, 19.4% S). Coarse gold associations and one of the silver-bearing minerals observed, together with their energy spectrum analysis and chemical composition analysis, are illustrated in the figure. The deportment of silver in different minerals and their associations with other minerals is also shown. The results indicated the potential for gravity recoverable gold, whereas silver is present in several different minerals in the ore.

Carbon Analysis

A combination of reflected light microscopy and automated SEM analysis have been used in the determination of mineral characteristics of carbon materials such as coal (O'Brien et al. 2011; Van Alphen 2007; Gottlieb et al. 1991). Coating or mounting samples in carnauba wax is often done when preparing coal samples, as the BSE image provides better discrimination between the mounting medium and the coal particles. In some cases, the chemical signature of the resin can be sufficiently different to carbonaceous material to allow for particles to be differentiated. Figure 16 illustrates mineral compositional maps of in polished cross-sectional mounts of rock and fragmented sample showing graphite particles (dark gray) distinguishable from the resin (light gray).

SEMIAUTOMATED MICROPROBE TECHNIQUES

Automated mineralogical systems require a mineral library with elemental compositions of minerals expected in the ore or residue of interest. Although literature references will provide a basis for these calculations, variations in composition and low-level solid solution values within some mineral species require the use of other analytical electron beam techniques. For accurate mass balance derivations and deportment, assessments require quantitative compositional analyses, which can be obtained by techniques briefly described in the following subsections. Detailed compositional data yielded from microprobe analysis and mass spectroscopy techniques can be entered back into the automated mineralogy software, and mineralogical measurements can be reclassified to update overall modal and deportment data. The suitability of a technique for analysis is dependent on the detection limits and spacial resolution of microbeam techniques used.

Electron Microprobe with Wavelength-Dispersive Analyses

EPMA, using such equipment as a Cameca SX-100 electron microprobe, can be used to acquire precise, quantitative analyses of major elements down to trace with high spatial resolution spot sizes in the order of 100–200 nm in modem FEG-based instruments. Individual grains and textures as small as 2–5 µm can be targeted (Pownceby and MacRae 2016). The high electron beam currents and beam stability of EPMA, coupled with high-resolution WDX spectrometry, allow for detection limits and accuracy to 100 ppm for trace element compositions within many mineral species.

EPMA has the same basic principle of operation as SEM, except EPMA is designed primarily as a quantitative tool. Modem EPMA instruments are usually outfitted with one or more EDX spectrometers as well as an array of several WDX spectrometers mounted to the column for the simultaneous measurement of multiple elements (Pownceby and MacRae 2016). Samples supplied as carbon-coated, polished sectioned mounts prepared for automated mineralogy can be analyzed and allows the same sample to be analyzed using the two techniques.

In comparison with EDX spectrometers, WDX spectrometers have the advantage of superior peak resolution, high peak-to-background ratios, and minimal problems of overlapping X-ray lines, which provides better detection for some elements such as rare earth elements in monazite, or between Pb and S in galena, and trace elements. High count rates (>100,000 counts/s) can be accommodated by WDX spectrometers without loss of peak resolution, which allows them to detect elements at an order of magnitude lower in concentration than can be detected by EDX spectrometers. However, X-ray data can only be obtained for one element at a time, with each element's spectrum acquired sequentially as the wavelength range is scanned, which means acquisition time, particularly in a complex mineral with many elements, can take longer for collection and analysis of data. Recent development of silicon drift detectors, which offer high throughput and improvements in resolution, have led to routine analyses at detection limits in EDX spectrometers approaching that of WDX-based analysis.

The measured X-ray intensities from the detectors are corrected for X-ray absorption, secondary fluorescence, and electron backscattering using one of the matrix correction methods (Pouchou and Pichoir 1985; Armstrong 1988; Goldstein et al. 2003). For quantitative microanalysis, a comparison of X-ray emission from the unknown specimen with that from one or more SRMs of known composition is made. SRMs are readily available for this type of analysis.

Laser Ablation–Inductively Coupled Plasma Mass Spectroscopy

For minerals that require a measurement with a lower detection limit than can be provided by EPMA, LA-ICPMS can be considered (e.g., Cabri et al. 2010; Bowell et al. 2005). Inductively coupled plasma mass spectroscopy (ICPMS) is one of the most important analytical techniques available for trace element analysis, mainly because of its high degree of accuracy and precision. The mineral samples can be thin or thick sections or polished mounts of different dimensions, such as those for automated SEM applications. LA-ICPMS can have sub-parts-per-million sensitivities for elements such as rare earths and one part per million for most elements of the periodic table. However, it has a very large microanalytical volume, which can limit the grain size evaluated and resolution.

LA-ICPMS systems consist of a laser ablation unit and a mass spectrometer. (Klemd and Bratz 2016). The most widely used laser ablation systems are either the solid-state Nd-doped, Y, Al, garnet (Nd:YAG) using a wavelength of 266 nm or 213 nm and the 193-nm excimer gas laser ablation system. The Nd:YAG systems are cheaper and designed to cover a large range of applications such as ablations of metals, ceramics, plastics, glass, coatings, minerals, and environmental matrices. However, the use of shorter wavelength sources improves the absorption of the laser radiation and the resulting

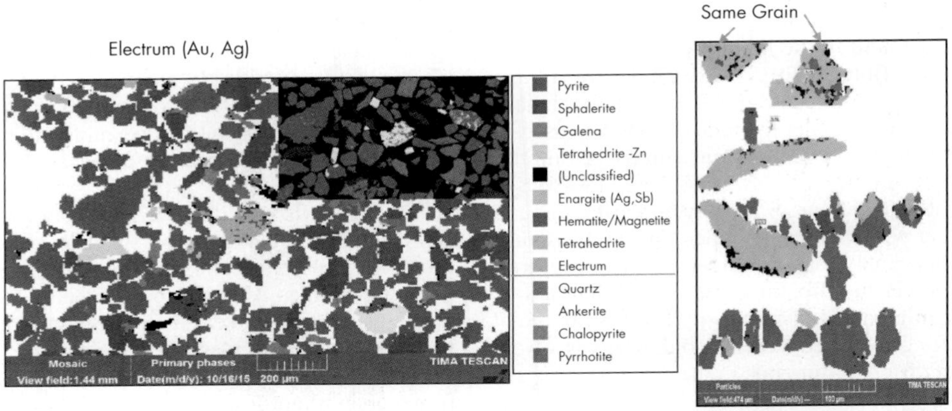

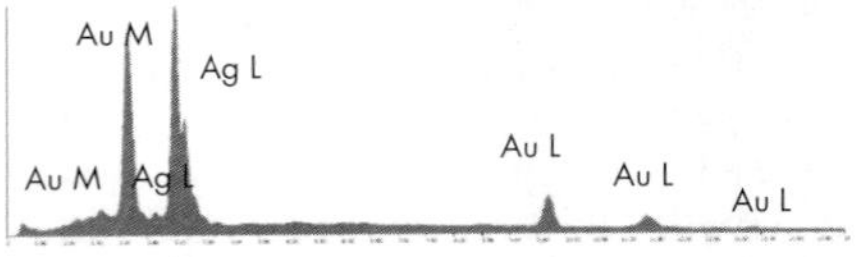

Element	Electrum, wt %	Average Composition, wt %	Range, wt % (8 grains)
Ag	>20	53.8	39.6–63.4
Au	<80	45.4	59.3–36.2
Fe			0–2.5

A. Free electrum grains with chemical analysis

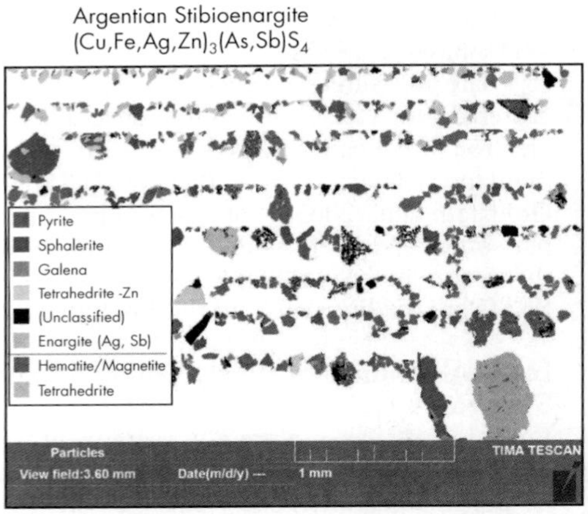

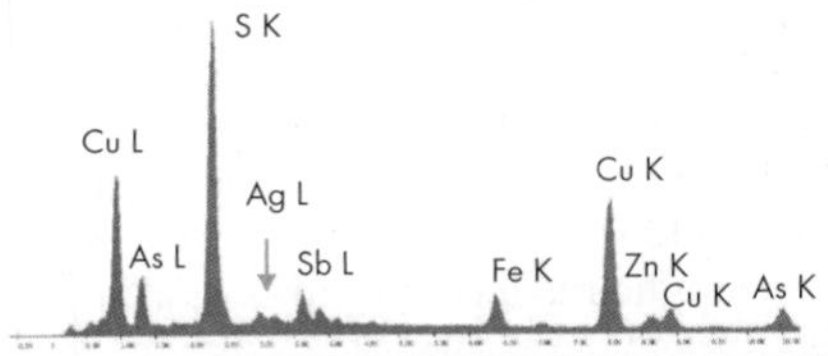

Element	Pure Enargite	Average Composition, wt %	Range, wt % (8 grains)
Cu	48.4	38.1	34.7–39.8
S	32.6	28.6	27.7–29.1
As	19.0	13.2	11.4–13.2
Sb		9.3	7.2–11.1
Zn		4.6	3.8–5.4
Fe		3.7	2.0–4.5
Ag		2.7	1.5–3.7

B. Enargite with energy-dispersive X-ray spectrum and chemical analysis

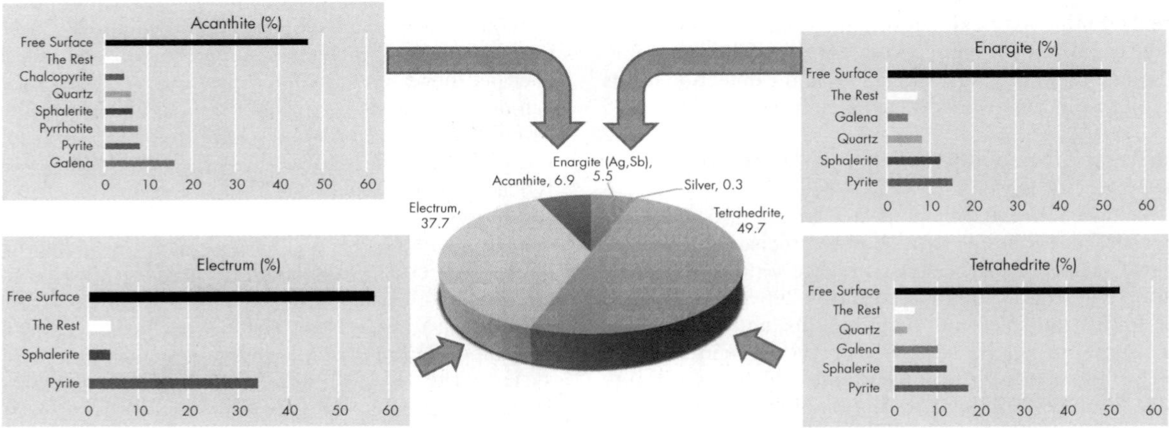

C. Deportment of silver and mineral associations

Figure 15 Gold and silver deportment, mineral analysis, and associations in a massive sulfide flotation concentrate

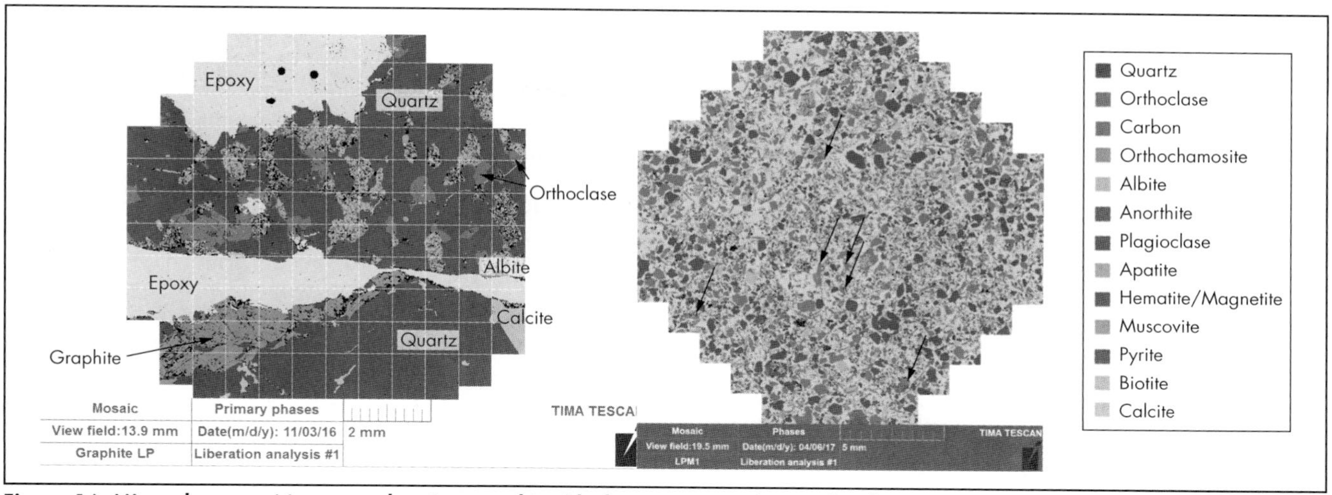

Figure 16 Mineral composition map showing graphite (dark gray; arrows) particles distinguishable from epoxy resin (light gray)

ablation characteristics, especially for colorless minerals and materials.

The laser beam is focused on the sample surface and vaporizes and partially ionizes the material at a spot on a particle. Both spot and detailed mapping analysis can be carried out on different mineral grains or particles. The diameter of the laser beam (1–400 µm), the repetition rate (1–100 Hz), the irradiance (measured in gigawatts per square centimeter), and the fluence (measured in joules per square centimeter) are variables that are adjusted for different applications. The signal intensity obtained for the respective elements is directly proportional to the amount of ablated material transported to the ICPMS. Quadrupole and single collector instruments are normally used for quantitative (moderate-high precision) element analyses. Multicollector magnetic sector field instruments are able to simultaneously determine isotope ratios at a very high precision (Harris 1998; Longerich 2008).

The main issue with LA-ICPMS analysis is the availability of suitable matrix-matched standards, which are essential for the calibration of the elemental analyses and in the destructive nature of the technique.

HIGH-RESOLUTION X-RAY COMPUTED TOMOGRAPHY

The application of high-resolution X-ray computed tomography (XCT) or micro-CT, a nondestructive 3-D imaging and analysis technique for the investigation of internal structures of large objects, has increasingly gained interest as a mineral characterization tool in geoscience and metallurgy applications (Cnudde and Boone 2013; Miller and Lin 2018). XCT consists of an X-ray source, a series of detectors that measure X-ray intensity attenuation along multiple beam paths, and a rotational geometry with respect to the object being imaged. Based on the magnitude of the X-ray attenuation coefficient, differentiation of mineral phases within the sample is possible. A specialized algorithm is then used to reconstruct the distribution of X-ray attenuation in the volume being imaged.

X-ray tomography has been used to quantify particle size, shape, and composition, based on the linear attenuation coefficient of particles as X-rays pass through a sample. Applications to the study of particle fractures resulting from blasting and comminution; liberation characterization to describe expected separation efficiency; grain surface area analysis to explain the flotation of locked particles; and floc size, shape, and water content for polymer-induced flocculation, have been described by Miller and Lin (2018). XCT provides a complementary technique to automated SEM systems in that larger sample volumes can be observed by XCT, and particle size and shape analysis are not affected by stereographic issues, whereas automated SEM systems can provide accurate mineral identification and high-definition images to calibrate mineral identification in XCT images. Recent developments in the use of dual-energy XCT, where two X-ray images at two different X-ray energies are collected simultaneously or sequentially, improves the range of mineral densities that can be measured to aid in differentiating between minerals (Lin et al. 2013; Wang et al. 2014). Recent advances in XCT instruments have included high-speed scanning and image analysis processing to describe the features within multiphase particles within minutes. This now provides opportunities for plant site–based 3-D coarse-particle characterization to be developed (Miller and Lin 2018).

PLANT-BASED OPERATIONS

Process diagnosis, flow-sheet design, and optimization are effectively and efficiently achieved through the use of metallurgical test work combined with modern quantitative mineralogical techniques (Lotter 2011; Gu et al. 2014). Several companies have applied the technology on a more routine basis to supply metallurgical operations with regular data. These have included Rio Tinto (KUCC concentrator), Barrick Gold Corporation, Freeport-McMoRan, Anglo American Platinum, and Glencore operations (formerly Xstrata and Falconbridge). The application of XRPD and XRF techniques have been extensively used over many years, in particular in areas such as the analysis of bauxites, fly ash, iron ore, sulfide mineralogy, and cement manufacturing.

Several case studies have been published that have demonstrated the effectiveness of incorporating automated mineralogy data for process improvements, particularly in optimizing grinding and flotation circuits (Baum et al. 2004; Dai et al. 2008; Rule and Schouwstra 2012; MacDonald et al.

2011; Lotter 2011; Lotter et al. 2010; examples in Becker et al. 2016).

Anglo Platinum was one of the pioneers in using automated SEM techniques to track the mineralogy of processing plants and ore sources (Schouwstra and Smit 2011). A detailed study into the losses of valuable minerals (PGMs and sulfides) using automated SEM techniques showed that an increase in the liberation of the valuable minerals was required to improve recoveries. Following piloting to demonstrate that finer grinding would increase the liberation and recovery of the valuable minerals, large-scale implementation was undertaken in 2009. The implementation yielded a 50% reduction in tailings grade with a corresponding increase in recovery of more than 5% of PGMs.

The most detailed evaluation incorporating all facets of analysis, including representative sampling routines, has been reported by Lotter and coworkers (Lotter 2011; Lotter et al. 2010). For example, the work illustrated for two case studies from Xstrata Nickel's (now part of Glencore) Nickel Rim South mine in Sudbury and its Raglan concentrator in Quebec, which incorporated the use of automated mineralogical techniques. A series of statistical benchmark surveys to describe the flow-sheet behavior under typical mining and milling conditions were conducted over many years (Lotter and Laplante 2007). At plant operations scale, a procedure was developed and validated whereby a representative suite of flow-sheet samples could be taken from an operating concentrator for mass and value balancing, followed by automated SEM (QEMSCAN) measurement across a series of closed size fractions. This work was used to update a concentrator's performance across several years and allowed for identifying opportunities for improving performance in the process.

CONCLUSIONS

Automated mineralogy methods, particularly automated SEM and XRPD systems, are now widely used for ore characterization, process design, and process optimization. These techniques deliver mineralogical information at fast acquisition times, can produce statistically representative analysis of a sample, as well as provide the ability to distinguish fine-grained or complexly intergrown minerals at the scale of micrometers. Several published case studies have demonstrated that large gains can be obtained through grinding and flotation optimization guided by automated mineralogy data. Improvements in both hardware and software will continue that will allow these techniques to be more versatile on mine sites and make them an essential tool for the metallurgist. The key factors that are important in using these automated techniques include

- Obtaining representative samples from the field and/or metallurgy process and suitably mounting them for analysis;
- Generating a mineral classification library and/or a suitable set of standards for analysis, which are representative of the mineral phases present in samples under examination;
- Establishing the limitations of analysis and what other complementary techniques can be used;
- Validating analytical results with other techniques; and
- Providing adequate training to staff.

For accurate mass balance derivations and deportment, assessments require quantitative compositional analyses from techniques such as EPMA and/or LA-ICPMS to validate mineral compositions. Emerging techniques, such as XCT, which allows measurement of materials in situ, will undoubtedly play an important role in the future.

REFERENCES

Amelinckx, S., van Dyck, D., van Landuyt, J., and van Tendeloo, G. 1997. *Electron Microscopy: Principles and Fundamentals*. Weinheim, Germany: VCH Wiley.

Anand, R.R., Gilkes, R.J., and Roach, G.I.D. 1989. Geochemical and mineralogical characteristics of bauxites, Darling Range, Western Australia. *Appl. Geochem.* 6(3):233–248.

Andersen, J.C.O., Rollinson, G.K., Snook, B., Herrington, R., and Fairhurst, R.J. 2009. Use of QEMSCAN for the characterization of Ni-rich and Ni-poor goethite in laterite ores. *Miner. Eng.* 22(13):1119–1129.

Armstrong, J.T. 1988. Quantitative analysis of silicates and oxide minerals: Comparison of Monte Carlo, ZAF and Phi-Rho-Z procedures. In *Microbeam Analysis*. Edited by D.E. Newbury. Berkeley, CA: San Francisco Press. pp. 239–246.

Aylmore, M.G., and Walker, G.S. 1998. Quantification of bauxites by the Rietveld method. *Powder Diffr.* 13(3):136–143.

Aylmore, M.G., Merigot, K., Quadira, Z., Rickard, W., Evans, N., McDonald, B., Catovic, E., and Spitalny, P. 2018a. Applications of advanced analytical and mass spectrometry techniques to the characterisation of micaceous lithium-bearing ores. *Miner. Eng.* 116:182–195.

Aylmore, M.G., Merigot, K., Rickard, W., Evans, N., McDonald, B., Catovic, E., and Spitalny, P. 2018b. Assessment of a spodumene ore by advanced analytical and mass spectrometry techniques to determine its amenability to processing for the extraction of lithium. *Miner. Eng.* 119:137–148.

Baum, W. 2014. Ore characterization, process mineralogy and lab automation a roadmap for future mining. *Miner. Eng.* 60:69–73.

Baum, W., Lotter, N.O., and Whittaker, P.J. 2004. Process mineralogy—a new generation for ore characterization and plant optimization. SME Preprint No. 04–12. Littleton, CO: SME. pp. 1–5.

Baumgartner, R., Dusci, M., Gressier, J., Trueman, A., Poos, S., Brittan, M., and Mayta, P. 2011. Building a geometallurgical model for early-stage project development—A case study from the Canahuire epithermal Au-Cu-Ag deposit, Southern Peru. In *First AusIMM International Geometallurgy Conference: Proceedings*. Edited by S. Dominy. Melbourne, Victoria: Australasian Institute of Mining and Metallurgy.

Baumgartner, R., Dusci, M., Trueman, A., Brittan, M., and Poos, S. 2013. Building a geometallurgical model for the Canahuire epithermal Au-Cu-Ag deposit, Southern Peru—past, present and future. In *Second AusIMM International Geometallurgy Conference: Proceedings*. Edited by S. Dominy. Melbourne, Victoria: Australasian Institute of Mining and Metallurgy.

Becker, M., Wightman, E.M., and Evans, C.L., eds. 2016. *Process Mineralogy*. Indooroopilly: Julius Kruttschnitt Mineral Research Centre, University of Queensland.

Best, E., Baum, W., Gilbert, R., Hohenstein, B., and Balt, A. 2007. The Phelps Dodge central analytical service center: Step change technology implementing robotics systems and lab automation in the copper industry. In *Proceedings of the Sixth International Copper–Cobre Conference*, Vol. 7. Montreal, QC: Canadian Institute of Mining, Metallurgy and Petroleum. pp. 119–126.

Bish, D.L., and Post, J.E., eds. 1989. *Reviews in Mineralogy*. Vol. 20, Modern Powder Diffraction. Edited by P.H. Ribbe. Chantilly, VA: Mineralogical Society of America.

Bish, D.L., and Post, J.E. 1993. Quantitative mineralogical analysis using the Rietveld full-pattern method. *Am. Mineral*. 78:932–940.

Bookin, A.S., Drits, V.A., Plancon, A., and Tchoubar, C. 1989. Stacking faults in kaolin-group minerals in the light of real structural features. *Clays Clay Miner*. 37(4):297–307.

Bouchet A., Proust, D., Meunier, A., and Beaufort, D. 1988. High-charge to low-charge smectites reaction in hydrothermal alteration processes. *Clays Clay Miner*. 23:133–146.

Bowell, R.J. Perkins, W.F., Tahija D., Ackerman, J., and Mansanti, J. 2005. Application of LA-ICPMS to troubleshooting mineral processing problems at the Getchell mine, Nevada. *Miner. Eng.* 18:754–761.

Brindley, G.W. 1980. Quantitative x-ray mineral analysis of clays. In *Crystal Structures of Clay Minerals and Their X-Ray Identification*. Edited by G.W. Brindley and G. Broen. London: Mineralogical Society. pp. 411–438.

Buhrke, V.E., Jenkins, R., and Smith D.K., eds. 1997. *A Practical Guide for the Preparation of Specimens for X-Ray Fluorescence and X-Ray Diffraction Analysis*. New York: Wiley.

Bulled, D., 2007. Grinding circuit design for Adanac Moly Corp using a geometallurgical approach. In *Proceedings of the 39th Annual Canadian Mineral Processors Conference*. Edited by J. Folinsbee. Montreal, QC: Canadian Institute of Mining, Metallurgy and Petroleum.

Bulled, D., Leriche, T., Blake, M., Thompson, J., and Wilkie, T. 2009. Improved production forecasting through geometallurgical modelling at iron ore company of Canada. In *Proceedings of the 41st Annual Canadian Mineral Processors Conference*. Edited by R. Henderson. Montreal, QC: Canadian Institute of Mining, Metallurgy and Petroleum.

Cabri, L.J. 1981. Relationship of mineralogy for the recovery of PGE from ores. In *Platinum-Group Elements: Mineralogy, Geology, Recovery*, special vol. 23. Edited by L.J. Cabri. Montreal, QC: Canadian Institute of Mining, Metallurgy and Petroleum. pp. 233–250.

Cabri, L.J., Beattie, M., Rudashevsky, N.S., and Rudashevsky, V.N. 2005. Process mineralogy of Au, Pd and Pt ores from the Skaergaard Intrusion, Greenland, using new technology. *Miner. Eng.* 18:887–897.

Cabri, L.J., Rudashevsky, N.S., Rudashevsky, V.N., and Gorkovetz, V. 2008. Study of native gold from the Luopensulo deposit (Kostomuksha area, Karelia, Russia) using a combination of electric pulse disaggregation (EPD) and hydroseparation (HS). *Miner. Eng.* 21:463–470.

Cabri, L.J., Choi, Y., Nelson, M., Tubrett, M., and Sylvester, P.J. 2010. Advances in precious metal trace element analyses for deportment using LAM-ICP-MS. In *Proceedings of the 42nd Annual Canadian Mineral Processors Conference*. Edited by D. Fragomeni. Montreal, QC: Canadian Institute of Mining, Metallurgy and Petroleum. pp. 182–196.

Chryssoulis, S.L., and Cabri, L.J. 1990. Significance of gold mineralogical balance in mineral processing. *Trans. Inst. Min. Metall.*, Sect. C. 99:C1–C10.

Chryssoulis, S.L., and McMullen, J. 2016. Mineralogical investigation of gold ores. In *Gold Ore Processing: Project Development and Operations*, 2nd ed. Edited by M.D. Adams. Amsterdam: Elsevier. pp. 57–93.

Cnudde, V., and Boone, M.N. 2013. High-resolution X-ray computed tomography in geosciences: A review of the current technology and applications. *Earth Sci. Rev.* 123:1–17.

Cullity, B.D. 1978. *Elements of X-Ray Diffraction*, 2nd ed. Reading, MA: Addison-Wesley.

Dai, Z., Bos, J-A., Lee, A., and Wells, P. 2008. Mass balance and mineralogical analysis of flotation plant survey samples to improve plant metallurgy. *Miner. Eng.* 21:826–831.

Dobby, G., Bennett, C., Bulled, D., and Kosick, G. 2004. Geometallurgical modelling—the new approach to plant design and production forecasting/planning, and mine/mill optimization. In *Proceedings of the 36th Annual Canadian Mineral Processors Conference*. Montreal, QC: Canadian Institute of Mining, Metallurgy and Petroleum.

Dollase, W.A. 1986. Correction of intensities for preferred orientation in powder diffractometry: Application of the March model. *J. Appl. Crystallogr*. 19:267–272.

Evans, C.L., and Morrison, R.D. 2016. Mineral liberation. In *Process Mineralogy*. Edited by M. Becker, E.M. Wightman, and C.L. Evans. Indooroopilly: Julius Kruttschnitt Mineral Research Centre, University of Queensland. pp. 219–233.

Evans, C.L., and Napier-Munn, T.J. 2013. Estimating error in measurements of mineral grain size distribution. *Miner. Eng.* 52:198.

Evans, C.L., and Wightman, E.M. 2009. Modelling liberation of comminuted particles. SME Preprint No. 09-088. Littleton, CO: SME.

Fandrich, R., Gu, Y., Burrows, D., and Moeller, K. 2007. Modern SEM-based mineral liberation analysis. *Int. J. Miner. Process*. 84:310–320.

Fitzpatrick, R.W., and Schwertmann, U. 1982. Al-substituted goethite, an indicator of pedogenic and other weathering environments in South Africa. *Geoderma*. 27:335–347.

Fosu, S., Skinner, W., and Zanin, M. 2015. Detachment of coarse composite sphalerite particles from bubbles in flotation: Influence of xanthate collector type and concentration. *Miner. Eng.* 71:73–84.

Goldstein, J., Newbury, D., and Joy, D., eds. 2003. *Scanning Electron Microscopy and X-Ray Microanalysis*. New York: Springer. pp. 391–451.

Goodall, W.R. 2008. Characterisation of mineralogy and gold deportment for complex tailings deposits using QEMSCAN. *Miner. Eng.* 21:518–523.

Goodall, W.R., and Butcher, A.R. 2012. The use of QEMSCAN in practical gold deportment studies. *Miner. Process. Extr. Metall.* 121(4):199–204.

Goodall, W.R., Scales, P.J., and Butcher, A.R. 2005. The use of QEMSCAN and diagnostic leaching in the characterisation of visible gold in complex ores. *Miner. Eng.* 18:877–886.

Gottlieb, P., Argon-Olshina, N., and Sutherland, D.N. 1991. The characterisation of mineral matter in coal and fly ash. In *Proceedings of the Engineering Foundation Conference*. Edited by S.A. Benson. New York: American Society of Mechanical Engineers. pp. 135–145.

Gottlieb, P., Wilkie, G., Sutherland, D., Ho-Tun, E., Suthers, S., Perera, K., Jenkins, B., Spencer, S., Butcher, A.R., and Rayner, J. 2000. Using quantitative electron microscopy for process mineralogy applications. *JOM.* 52(4):24–25.

Gottlieb, P., Motl, D., Dosbaba, M., and Kopřiva, A. 2015. The use of the TIMA automated mineral analyzer for the characterization of ore deposits and optimization of process operations. Presented at the Austrian Mineralogical Society MinPet 2015.

Gu, Y. 2003. Automated scanning electron microscope based mineral liberation analysis. *J. Miner. Mater. Charact. Eng.* 2(1):33–41.

Gu, Y., Schouwstra, R.P., and Rule, C. 2014. The value of automated mineralogy. *Miner. Eng.* 58:100–103.

Harris D.C. 1998. *Quantitative Chemical Analysis*. New York: W.H. Freeman.

Hatton, D., and Hatfield, D. 2013. A geometallurgical approach to flotation process design using optimised grade-recovery curves. In *Proceedings of the 45th Annual Canadian Mineral Processors Conference*. Westmount, QC: Canadian Institute of Mining, Metallurgy and Petroleum. pp. 59–70.

Hem, S.R., de V. Louw, J.D., Gateshki, M., Göske, J., Rindbæk Larsen, O., and Strangeways, J. 2009. Sample preparation for quantitative Rietveld analysis, phase identification and XRF in one step: Automated sample preparation by Centaurus. In *Heavy Minerals Conference: "What Next?"* Symposium Series S57. Johannesburg: Southern African Institute of Mining and Metallurgy.

Henley, K.J. 1983. *Ore-Dressing Mineralogy: A Review of Techniques, Applications and Recent Developments*. Special Publication No. 7. Edited by J.P.R. De Villiers and P.A. Cawthorn. Johannesburg: Geological Society of South Africa. pp. 175–200.

Hill, R.J. 1991. Expanded use of the Rietveld method in studies of phase abundance in multiphase mixtures. *Powder Diffr.* 6:74–77.

Hill, R.J., and Howard, C.J. 1987. Quantitative phase analysis from neutron powder diffraction data using the Rietveld method. *J. Appl. Crystallogr*. 20:467–474.

Hill, R.J., and Madsen, I.C. 2002. Sample preparation, instrument selection and data collection. In *Structure Determination from Powder Diffraction Data*. Edited by W. David, K. Shankland, L. McCusker, and C. Baerlocher. New York: Oxford University Press. pp. 98–117.

Hill, R.J., Tsambourakis, G., and Madsen, I.C. 1993. Improved petrological modal analyses from X-ray powder diffraction data by use of the Rietveld method. Part I. Selected igneous, volcanic and metamorphic rocks. *J. Petrol.* 34:867–900.

Hoal, K.O., Woodhead, J., and Smith, K.S. 2013. The importance of mineralogical input into geometallurgy programs. In *Second AusIMM International Geometallurgy Conference: Proceedings*. Edited by S. Dominy. Melbourne, Victoria: Australasian Institute of Mining and Metallurgy.

Johnston, N.W. 2016. Sphalerite and galena liberation levels for the Mount Isa concentrator. In *Process Mineralogy*. Monograph Series in Mining and Mineral Processing, No. 6. Edited by M. Becker, E.M. Wightman, and C.L. Evans. Indooroopilly: Julius Kruttschnitt Mineral Research Centre, University of Queensland. pp. 234–249.

Klemd, R., and Bratz, H. 2016. Laser ablation ICP-MS. In *Process Mineralogy*. Edited by M. Becker, E.M. Wightman, and C.L. Evans. Indooroopilly: Julius Kruttschnitt Mineral Research Centre, University of Queensland. pp. 120–132.

Klug, H.P., and Alexander, L.E. 1974. *X-Ray Diffraction Procedures for Polycrystalline and Amorphous Materials*. New York: Wiley.

Kormos, L., Sliwinski, J., Oliveira, J., and Hill, G. 2013. Geometallurgical characterization and representative metallurgical sampling at Xstrata process support. In *Proceedings of the 45th Annual Canadian Mineral Processors*. Westmount, QC: Canadian Institute of Mining, Metallurgy and Petroleum.

Laplante, A.R., Woodcock, F., and Noaparast, M. 1995. Predicting gravity separation gold recovery. *Mineral. Metall. Process.* (May): 74–79.

Lastra, R., and Paktunc, D. 2016. An estimation of the variability in automated quantitative mineralogy measurements through inter-laboratory testing. *Miner. Eng.* 95:138–145.

Lastra, R., and Petruk, W. 2014. Mineralogical characterization of sieved and un-sieved samples. *J. Mineral. Mater. Charact. Eng.* 2(1):40–48.

Lastra, R., Price, J., Cabri, L.J., Rudashevsky, N.S., Rudashevsky, V.N., and McMahon, G. 2005. Gold characterization of a sample from Malartic East (Quebec) using concentration by hydroseparator. In *Proceedings of the International Symposium on the Treatment of Gold Ores*. Edited by G. Deschenes, D. Houdin, and L. Lorenzen. Montreal, QC: Canadian Institute of Mining, Metallurgy and Petroleum. pp. 17–29.

Li, D.Y., O'Connor, Roach, G.I.D., and Cornell, J.B. 1990. Use of x-ray powder diffraction pattern fitting: Characterising preferred orientation gibbsites. *Powder Diffr.* 5:79–85.

Li, J., McDonald, R.G., Kaksonen, A.H., Morris, C., Rea, S., Usher, K.M., Wylie, J., Hilario, F., and du Plessis, C.A. 2014. Applications of Rietveld-based QXRD analysis in mineral processing. *Powder Diffr.* 29(S1): 589–595.

Lin, C.L., Hsieh, C.H., Tserendagva, T.A., and Miller, J.D. 2013. Dual-energy rapid scan radiography for geometallurgy evaluation and isolation of trace mineral particles. *Miner. Eng.* 40:30–37.

Liu, Y., Gupta, R., Sharma, A., Wall, T., Butcher, A., Miller, G., Gottlieb, P., and French D. 2005. Mineral matter-organic matter association characterisation by QEMSCAN and applications in coal utilisation. *Fuel.* 84(10):1259–1267.

Longerich, H. 2008. Laser-ablation inductively coupled plasma-mass spectrometry (ICPMS). In *Laser-Ablation-ICPMS in the Earth: Current Practices and Outstanding Issues: Short Course*, Vol. 40. Edited by P. Sylvester. Vancouver, BC: Mineralogical Association of Canada.

Lotter, N.O. 1995. Review of evaluation models for the representative sampling of ore. *J. S. Afr. Inst. Min. Metall.* 95(4):149–155.

Lotter, N.O. 2011. Modern process mineralogy: An integrated multi-disciplined approach to flowsheeting. *Miner. Eng.* 24(12):1229–1237.

Lotter, N.O., and Laplante, A.R. 2007. Statistical benchmark surveying of production concentrators. *Miner. Eng.* 20:793–801.

Lotter, N.O., Di Feo, A., Kormos, L.J., Fragomeni, D., and Comeau, G. 2010. Design and measurement of small recovery gains: A case study at Raglan concentrator. *Miner. Eng.* 23:567–577.

Lotter, N.O., Oliveira, J.F., Hannaford, A.L., Amos, S.R., and Broughton, D.W. 2013. Flowsheet development for the hypogene geomet unit in the Ivanplats Kamoa copper project. In *Proceedings of the 45th Annual Canadian Mineral Processors Conference*. Westmount, QC: Canadian Institute of Mining, Metallurgy and Petroleum. pp. 71–85.

MacDonald, M., Adair, B.J.I., Bradshaw, D.J., Dunn, M., and Latti, D. 2011. Learnings from five years of on-site MLA at Kennecott Utah Copper Corporation. In *Proceedings of the 10th International Congress for Applied Mineralogy (ICAM)*. Heidelberg, Germany: Springer. pp. 419–426.

Madsen, I.C., and Hill, R. 1990. QPDA: User-friendly, interactive program for quantitative phase and crystal size/strain analysis of powder diffraction data. *Powder Diffr.* 5(4):195–199.

Madsen, I.C., and Hill, R. 1994. Collection and analysis of powder diffraction data with near-constant counting statistics. *J. Appl. Crystallogr.* 27(3):385–392.

Madsen, I.C., and Scarlett, N.V.Y. 1999. Cement: Quantitative phase analysis of portland cement clinker. In *Industrial Applications of X-ray Diffraction*. Edited by F.H. Chung and D.K. Smith. New York: Marcel Dekker. pp. 415–440.

Madsen, I.C., and Scarlett, N.V.Y. 2007. Quantitative phase analysis. In *Powder Diffraction: Theory and Practice*. Edited by R.E. Dinnebier and S.J.L. Billinge. London: Royal Society of Chemistry.

Madsen, I.C., Scarlett, N.V.Y., Cranswick, L.M.D., and Lwin, T. 2001. Outcomes of the International Union of Crystallography Commission on Powder Diffraction round robin on quantitative phase analysis: Samples 1*a* to 1*h*. *J. Appl. Crystallogr*. 34(Part 4):409–426.

Madsen, I.C., Scarlett, N.V.Y., and Whittington, B.I. 2005. Pressure acid leaching of nickel laterite ores: An in situ diffraction study of the mechanism and rate of reaction. *J. Appl. Crystallogr*. 38:927–933.

Madsen, I.C., Scarlett, N.V.Y., and Kern, A. 2011. Description and survey of methodologies for the determination of amorphous content via X-ray powder diffraction. *Z. Kristallogr.* 226(12):944–955.

Madsen, I.C., Scarlett, N.V.Y., Riley, D.P., and Raven, M.D. 2012. Quantitative phase analysis using the Rietveld method. In *Modern Diffraction Methods*. Edited by E.J. Mittemeijer and U. Welzel. Weinheim, Germany: Wiley-VCH.

Mariano, R.A., and Evans, C.L. 2015. Error analysis in ore particle composition distribution measurements. *Miner. Eng.* 82:36–44.

McCusker, L.B., Von Dreele, R.B., Cox, D.E., Louer, D., and Scardi, P. 1999. Rietveld refinement guidelines. *J. Appl. Crystallogr*. 32:36–50.

McIvor, R.E., and Finch, J.A. 1991. A guide to interfacing of plant grinding and flotation operations. *Miner. Eng.* 4(1):9–23.

McKay, N. 2016. Red Dog concentrator optimisation study 2009. In *Process Mineralogy*. Edited by M. Becker, E.M. Wightman, and C.L. Evans. Indooroopilly: Julius Kruttschnitt Mineral Research Centre, University of Queensland. pp. 250–260.

McKay, N., Wilson, S., and Lacouture, B. 2007. Ore characteristion of the Aqqaluk deposit at Red Dog. In *Proceedings of the 39th Annual Canadian Mineral Processors Conference*. Montreal, QC: Canadian Institute of Mining, Metallurgy and Petroleum. pp 55–74.

Miller, J.D., and Lin, C.L. 2018. X-ray tomography for mineral processing technology—3D particle characterization from mine to mill. *Miner. Metall. Process*. 35(1):1–12.

Montoya, P.A., Keeney, L., Jahoda, R., Hunt, J., Berry, R., Drews, U., Chamberlain, V., and Leichliter, S. 2011. Geometallurgical modelling techniques applicable to prefeasibility projects–La Colosa case study. In *First AusImm International Geometallurgy Conference: Proceedings*. Edited by S. Dominy. Melbourne, Victoria: Australasian Institute of Mining and Metallurgy.

Mosser-Ruck, R., Devineau, K., Charpentier, D., and Cathelineau, M. 2005. Effects of ethylene glycol saturation protocols on XRD patterns: A critical review and discussion. *Clays Clay Miner.* 53(6):631–638.

Napier-Munn, T.J. 2014. *Statistical Methods for Mineral Engineers: How to Design Experiments and Analyse Data*. Indooroopilly: Julius Kruttschnitt Mineral Research Centre, University of Queensland.

Neder, R.B., Burghammer, M., Grasl, T.H., Schulz, J.H., Bram, A., and Fiedler S. 1999. Refinement of the kaolinite structure from single-crystal synchrotron data. *Clays Clay Miner*. 47(4):487–494.

Norrish, K., and Hutton, J.T. 1969. An accurate x-ray spectrographic method for the analysis of a wide range of geological samples. *Geochim. Cosmochim. Acta.* 431:33.

O'Brien, G., Gu, Y., Adair, B., and Firth, B. 2011. The use of optical reflected light and SEM imaging systems to provide quantitative coal characterisation. *Miner. Eng.* 24:1299–1304.

O'Connor, B.H., Li, D.Y., and Sitepu, H. 1991. Strategies for preferred orientation corrections in x-ray powder diffraction using line intensity ratios. *Adv. X-ray Anal*. 34:1–7.

Petruk, W. 1976. The application of quantitative mineralogical analysis of ores to ore dressing. *CIM Bull.* 767:146–153.

Petruk, W. 2000. *Applied Mineralogy in the Mining Industry*. Amsterdam: Elsevier.

Petruk, W., and Hughson, M.R. 1977. Image analysis evaluation of the effect of grinding media on selective flotation of two zinc–lead–copper ores. *CIM Bull.* 787:128–135.

Petruk, W., and Schnarr, J.R. 1981. An evaluation of the recovery of free and unliberated mineral grains, metals and trace elements in the concentrator of Brunswick Mining and Smelting Corporation Limited. *CIM Bull.* 833:132–159.

Pirard, E. 2016. Optical microscopy. In *Process Mineralogy*. Edited by M. Becker, E.M. Wightman, and C.L. Evans. Indooroopilly: Julius Kruttschnitt Mineral Research Centre, University of Queensland. pp. 51–66.

Pouchou J.L., and Pichoir F. 1985. "PAP," procedure for improved quantitative microanalysis. In *Microbeam Analysis*. Edited by J.T. Armstrong. Berkeley, CA: San Francisco Press. pp. 104–106.

Pownceby, M.I., and MacRae, C.M. 2016. Electron probe microanalyser. In *Process Mineralogy*. Edited by M. Becker, E.M. Wightman, and C.L. Evans. Indooroopilly: Julius Kruttschnitt Mineral Research Centre, University of Queensland. pp. 79–96.

Rietveld, H.M. 1969. A profile refinement method for nuclear and magnetic structures. *J. Appl. Crystallogr*. 2(2):65–71.

Rousseau, R.M. 2006 Corrections for matrix effects in X-ray fluorescence analysis—a tutorial. *Spectrochim. Acta, Part B*. 61:759–777.

Rule C., and Schouwstra, R.P. 2012. Process mineralogy delivering significant value at Anglo Platinum concentrator operations. In *Proceedings of the 10th International Congress for Applied Mineralogy (ICAM)*. Edited by M. Broekmans. Berlin: Springer.

Scarlett, N.V.Y., and Madsen, I.C. 2006. Quantification of phases with partial or no known crystal structures. *Powder Diffr.* 21(4):278–284.

Scarlett, N.V.Y., Madsen, I.C., Manias, C., and Retallack, D. 2001. On-line x-ray diffraction for quantitative phase analysis: Application in the portland cement industry. *Powder Diffr.* 16:71–80.

Scarlett, N.V.Y., Madsen, I.C., Cranswick, L.M.D., Lwin, T., Groleau, E., Stephenson, G., Aylmore, M., and Agron-Olshina, N. 2002. Outcomes of the International Union of Crystallography Commission on powder diffraction round robin on quantitative phase analysis: Samples 2, 3, 4, synthetic bauxite, natural granodiorite and pharmaceuticals. *J. Appl. Crystallogr*. 35:383–400.

Scarlett, N.V.Y., Madsen, I.C., and Whittington, B.I. 2008. Time-resolved diffraction studies into the pressure acid leaching of nickel laterite ores: a comparison of laboratory and synchrotron X-ray experiments. *J. Appl. Crystallogr.* 41: 572-583.

Schouwstra, R.P., and Smit, A.J. 2011. Developments in mineralogical techniques—What about mineralogists? *Miner. Eng.* 24(12):1224–1228.

Schulze, D.G. 1984. The influence of aluminum on iron oxides. VIII. Unit cell dimensions of Al-substituted goethites and estimation of Al from them. *Clays Clay Miner.* 32:36–44.

Smythe, D.M., Lombard, A., and Coetzee, L.L. 2013. Rare earth element deportment studies utilising QEMSCAN technology. *Miner. Eng.* 52:52–61.

Snyder, R.L., and Bish, D.L. 1989. Quantitative analysis. In *Reviews in Mineralogy*. Vol. 20, Modern Powder Diffraction. Edited by P.H. Ribbe. Chantilly, VA: Mineralogical Society of America. pp. 101–144.

Spencer, S., and Sutherland, D. 2000. Stereological correction of mineral liberation grade distributions estimated by single sectioning of particles. *Image Anal. Stereology* 19:175–182.

Sutherland, D. 2007. Estimation of mineral grain size using automated mineralogy. *Miner. Eng.* 20:452–460.

Taylor, J.C., and Matulis, C.E. 1991. Absorption contrast effects in the quantitative XRD analysis of powders by full multiphase profile refinement. *J. Appl. Crystallogr.* 24:14–17.

Taylor, J.C., and Rui, Z. 1992. Simultaneous use of observed and calculated standard profiles in quantitative XRD analysis of minerals by the multiphase Rietveld method: The determination of pseudorutile in mineral sands products. *Powder Diffr.* 7:152–161.

Tsikouras, V., Pe-Piper, G., Piper, D.J.W., and Shaffer, M. 2011. Varietal heavy mineral analysis of sediment provenance, Lower Cretaceous Scotian Basin, eastern Canada. *Sediment. Geol.* 237:150–165.

Van Alphen, C. 2007. Automated mineralogical analysis of coal and ash products: Challenges and requirements. *Miner. Eng*. 20:496–505.

Walenta, G., and Füllmann, T. 2004. Advances in quantitative XRD analysis for clinker, cements, and cementitious additions. *Powder Diffr*. 19:40–44.

Wang D., Yuan X., Gu Y., Case T., Feser M., and Schouwstra R.P. 2014. Investigation of microCT (computed tomography) for 3-dimensional mineral characterization: A dual energy approach. In *Third International Symposium on Process Mineralogy (Process Mineralogy '14)*. Cornwall, UK: Minerals Engineering International.

Whittington, B.I., Johnson, J.A., Quan, L.P., McDonald, R.G., and Muir, D.M. 2003. Pressure acid leaching of arid-region nickel laterite ore. Part II. Effect of ore type. *Hydrometallurgy.* 70:47–62.

Wightman, E.M., Evans, C.L., Becker, M., and Gu, Y. 2016. Automated scanning electron microscopy with energy dispersive spectrometry. In *Process Mineralogy*. Edited by M. Becker, E.M. Wightman, and C.L. Evans. Indooroopilly: Julius Kruttschnitt Mineral Research Centre, University of Queensland. pp. 97–107.

Williams, S., and Richardson, J. 2004. Geometallurgical mapping: A new approach that reduces technical risk. In *Proceedings of the 36th Annual Canadian Mineral Processors Conference*. Montreal, QC: Canadian Institute of Mining, Metallurgy and Petroleum.

Young, R.A., ed. 1995. *The Rietveld Method*. Vol. 5, International Union of Crystallography Monographs on Crystallography. New York: Oxford University Press.

Zevin L.S., Kimmel, G., and Mureinik, I. 1995. *Quantitative X-Ray Diffractometry*. New York: Springer.

Zhou, J., and Gu, Y. 2016. Geometallurgical characterization and automated mineralogy of gold ores. In *Gold Ore Processing: Project Development and Operations*, 2nd ed. Edited by M.D. Adams. Amsterdam: Elsevier. pp. 95–111.

Zhou, J.Y., and Cabri, L.J. 2004. Gold process mineralogy: Objectives, techniques and applications. *J. Miner. Met. Mater. Soc.* 56(7):49–52.

CHAPTER 1.4

Ore Liberation Analysis

Dee Bradshaw, Greg Wilkie, Megan Becker, Cathy Evans, and Norman O. Lotter

Effective ore liberation is at the heart of successful mineral processing; without it, no economic separation process is possible. "Too much" liberation through expensive overgrinding can harm performance; "too little" can cause losses, resulting in the "Goldilocks" effect, where "just right" is required. Therefore, understanding why ore liberation is important; what determines "enough"; and how it can be measured, monitored, and optimized is fundamental to successful mineral processing. The purpose of this chapter is to provide a practical guide of ore liberation analysis, with suitable references to more detailed texts available so that typical problems encountered can be addressed. By understanding the role of liberation and therefore being able to characterize and interpret the "current state" of a circuit, it is possible to diagnose weaknesses and potential for improvement. It is also important to know when liberation is not the reason for poor performance in a circuit.

Bradshaw (2014) identified four steps as being critical to understand and address ore liberation, summed up by the four *I*s:

1. *Information* (appropriate measurements and accurate data)
2. *Interpretation* (correct analysis of information)
3. *Implication* (understanding of the context and application of the information and its interpretation)
4. *Implementation* (execution of the changes necessary to address the implications of the measurement to either the existing circuit or the process design)

If any of these steps are missing, the whole process is threatened, and much of the value can be lost. The first three steps relate to the measurements made on the sample obtained and the knowledge that is being generated, generally at considerable cost and requiring specialist expertise and experience. The fourth step relates to the action that is needed to obtain the value of the investment in the costs of the first three steps. If the fourth step of implementation is not actioned, then the knowledge gained in the first through third steps remains an expensive exercise with little potential value or return on investment (Gu et al. 2014). This chapter does not deal with the sampling methodology needed to obtain a representative sample, although it is acknowledged that no amount of measurement or interpretation can compensate for mistakes at this level (Gy 1979).

This chapter is structured to answer five questions:

1. What is liberation, and why is it critical to mineral processing?
2. How is liberation measured and modeled?
3. What is the best way to report liberation information?
4. When is there enough liberation?
5. How are liberation problems diagnosed, and how are improvements proposed?

The first question deals with the importance of liberation and how it is achieved. The second question deals with the various measurements that can be used and each one's limitations or constraints. The most effective reporting on these measurements depends on the questions being asked and the budget available, as addressed by the third question. The fourth question deals with the drivers that determine "enough." Case studies are cited to illustrate the value of diagnosing and addressing inadequate liberation to answer the fifth question.

WHAT LIBERATION IS AND WHY IT IS CRITICAL TO MINERAL PROCESSING

In mineral process engineering, the material being treated is in the form of either a particulate stream or a lot containing valuable minerals and minerals of no value. To separate these two categories into a valuable concentrate and a less valuable reject product, it is first necessary to liberate one from the other. *Mineral liberation* can be defined as the extent to which the particles are made of discrete mineral grains. If a particle is composed of one mineral only, it is regarded as

Dee Bradshaw, Emeritus Professor, Department of Chemical Engineering, University of Cape Town, Cape Town, South Africa
Greg Wilkie, Program Coordinator, CRC Ore, Brisbane, Queensland, Australia
Megan Becker, Associate Professor, Department of Chemical Engineering, University of Cape Town, Cape Town, South Africa
Cathy Evans, Senior Research Fellow, Julius Kruttschnitt Mineral Research Centre, Brisbane, Queensland, Australia
Norman O. Lotter, President, Flowsheets Metallurgical Consulting Inc., Sudbury, Ontario, Canada

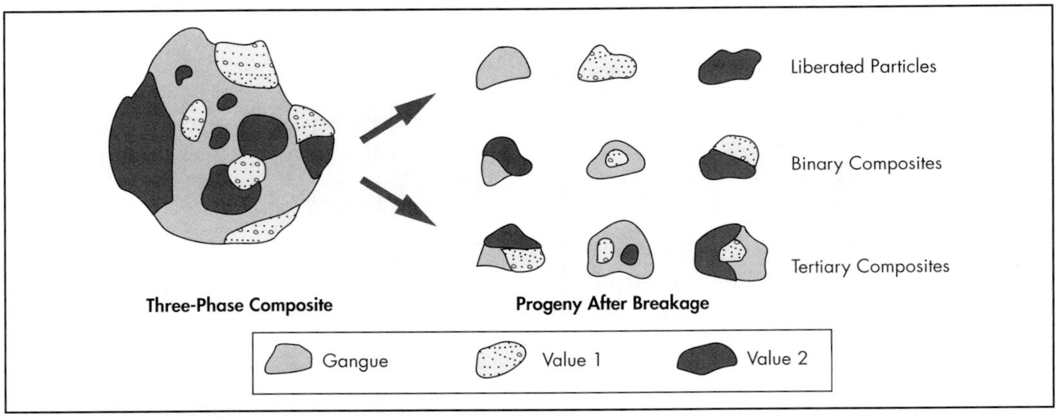

Source: Napier-Munn et al. 2005

Figure 1 Mineral associations in daughter particles from breakage of a complex ore

	Which unliberated particle texture allows the valuable phase to be recovered? (All particles are the same grade.)		
Typical particles of valuable (dark phase) and gangue (light phase) shown in cross section	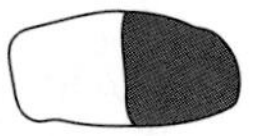	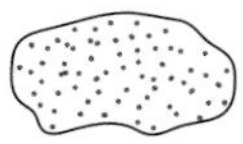	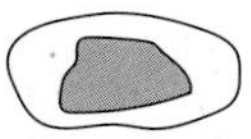
By flotation	✓ Recovered	✗ Not recovered	✗ Not recovered
By density separation	✓ Recovered	✓ Recovered	✓ Recovered
By leaching	✓ Recovered	✗ Not recovered	✗ Not recovered

Source: Evans and Morrison 2016

Figure 2 Schematic showing how ore texture affects liberation and recovery by different processes

fully liberated. Composite particles are made up of mixtures of mineral grains, and a locked particle refers to one having no surface exposure of the selected mineral. Thus, a feed stream to a separation unit contains a distribution of particles of a range of sizes and liberation, and this distribution determines the range of particle properties and the maximum potential for separation success. A stream of well-liberated particles—with the value and gangue minerals in different classes of particles—will separate sharply, whereas a stream of composite particles will not. Figure 1 illustrates the products from rock fragmentation and shows that the daughter fragments of a complex ore can be liberated particles, binary composites, or tertiary composites.

The different processes that the ore is being prepared for require different amounts of liberation to be effective, as shown in Figure 2. Although the amount of valuable mineral in the three examples is similar, the response to the different processes is not because of the difference in texture. Texture, in this context, describes the spatial arrangement, orientation, shape, size, and interrelationships of the constituents in the ore, which include both value and gangue minerals and the mineral aggregates. Textural characterization has been fundamental across the geoscience disciplines, and its application extends throughout the disciplines of mining, mineral processing, and metal extraction, particularly affecting breakage and subsequent liberation (Petruk 2000).

Of particular importance is the surface exposure of the selected or valuable mineral relative to the unwanted gangue mineral. Flotation, for example, is based on the surface properties and requires more of the exposed mineral surface than leaching. There are other factors that can reduce the separation achieved, depending on which separation process is being used (e.g., flotation, gravity leaching). In the case of flotation, these include factors that affect the surface of the particles, such as particle oxidation or surface coatings by fine gangue particles or problems in the equipment operation. Density separation depends on the differences in the particle specific gravity and not the surface properties. Particles can be separated based on their relative movement in response to the force of gravity and one or more other forces (e.g., centrifugal forces, magnetic forces, buoyant forces, or drag forces). This generally takes place in water or in a viscous medium such as heavy media (also known as heavy media separation).

The flotation rate is highly dependent on particle size (Trahar 1981; McIvor and Finch 1991), with ore liberation as an underlying consequence of size. At finer sizes, the particles will tend to be more liberated and will be less liberated at coarser sizes. Figure 3 shows that the recovery is reduced in the finer and coarser sizes. Thus, flotation performance depends on both size and liberation, although they are not the only factors that must be considered.

Figure 4 shows that complete liberation does not guarantee complete recovery. The liberated minerals have to be concentrated in an optimal performance size range by the grinding circuit. Sutherland (1989) identified the interaction between mineral liberation and particle size by demonstrating the different recovery behaviors of chalcopyrite-bearing particles of different sizes with similar liberation characteristics. Figure 4 shows that at the same degree of liberation, coarse particles floated more slowly than intermediate-sized particles but that poorly liberated particles of optimum flotation size could float more rapidly than liberated coarse particles. In some instances, additional collector is added to improve the recovery of coarse liberated particles, but this also leads to the recovery of composite particles, leading to unselective recovery and lower grades.

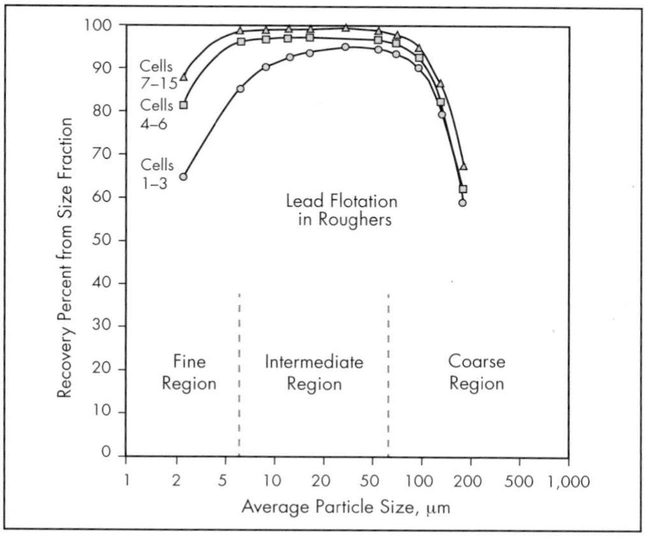

Source: Trahar 1981

Figure 3 Cumulative recovery down the flotation bank at Broken Hill South Ltd. showing the reduced recoveries at fine and coarse sizes

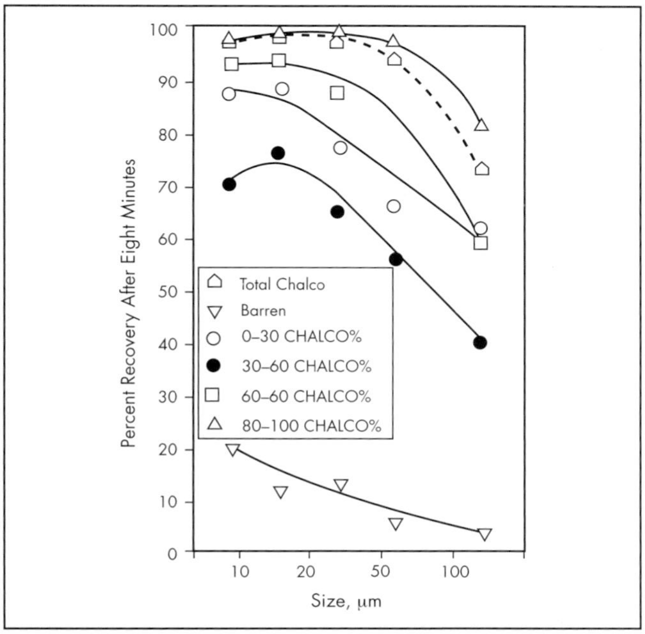

Source: Sutherland 1989

Figure 4 Effects of particle size and liberation on chalcopyrite flotation after 8 minutes

Apart from different size-by-size recoveries by liberation class, different sulfide minerals present different size-by-size recovery curves (Jowett 1980). An extract of these data is shown in Figure 5.

Methods to directly measure the grain size distribution (GSD) of minerals within a polished section using scanning electron microscopy with energy-dispersive X-ray spectrometry (SEM-EDS) techniques have been developed. Miller and Lin (1988) noted that ores with smaller grain sizes—often measured as the equivalent circle diameter—are indicative of a more disseminated mineral texture.

The specified particle size at which the required degree of liberation occurs depends on

- The original ore texture and GSD of the minerals making up the ore,
- The economical balance between the cost of grinding more finely and the value of increased recovery and concentrate quality (grade, impurities), and
- The separation process selected.

The contribution of texture to liberation was first documented by Gaudin (1939). He noted its effect on liberation and discovered that the size of the target mineral grain influenced the way in which it liberated during grinding. Figure 6 shows his four textures, classified as (1) simple texture, (2) coated or rimming, (3) disseminated or emulsion, and (4) stockwork or veining. Gaudin noted that types 1 and 4 are expected to behave in accordance with the relative abundance of the phases present. Type 2, on the other hand, would not be expected to respond as well as the relative abundance of the minerals present would suggest; instead, it would respond in proportion to the makeup of the exposed particle surface area. Particles representing type 3 typically respond to flotation as one would expect the host phase to respond. In this case, the target mineral is at very low quantities. For intergrowth types 2 to 4, downstream liberation through regrinding is also significantly more challenging compared to the simple texture in type 1. Knowledge of the presence of these intergrowth classes is thus important to understand the processing challenges of ores.

As an alternative to estimating the GSD for a mineral, the phase-specific surface area (PSSA) of that mineral has been shown to have very strong correlations to processing

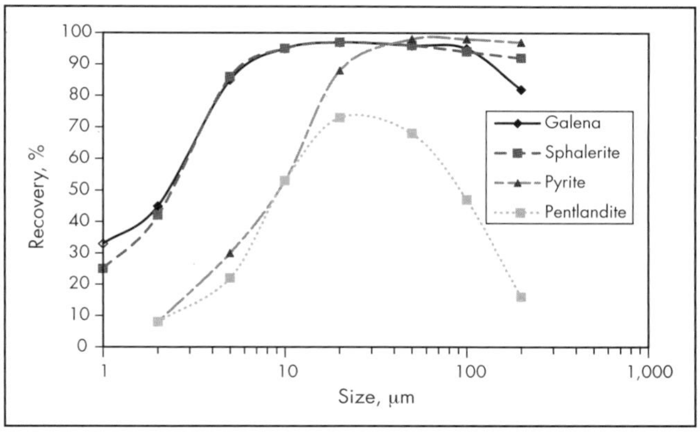

Source: Jowett 1980

Figure 5 Size-by-size recovery curves for different minerals

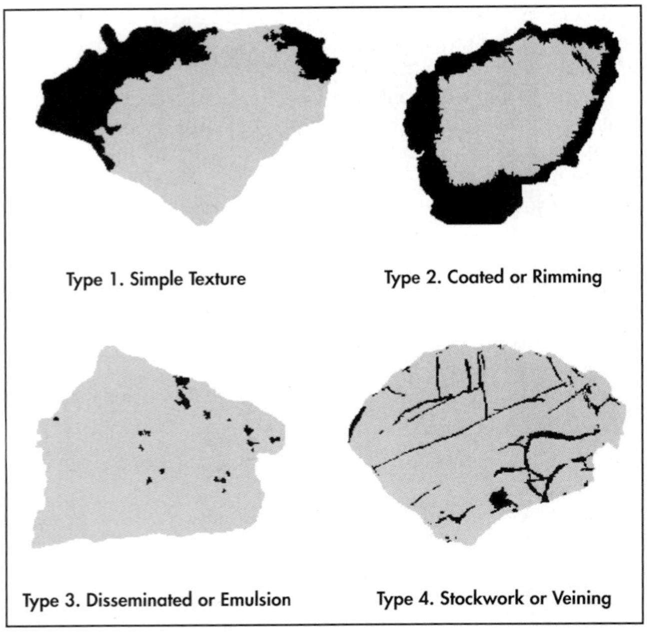

Source: Vos 2017

Figure 6 Examples of SEM-polished section images showing the four intergrowth textures defined by Gaudin (1939) that influence mineral processing

characteristics (Jackson et al. 1988; Sutherland et al. 1991). This quantity, when determined from sectioned images, measures the surface area per unit volume of all mineral grains of a particular phase per total surface area for that phase. A higher PSSA value for a measured population of particle sections indicates finer mineral grain inclusions and hence has a correlation with GSD estimations.

Liberation is achieved through size reduction of the ore. The first size-reduction step is blasting in the mine. In some cases, depending on the ore texture and properties, it is possible to liberate barren gangue material even at large sizes, and this can be treated accordingly. The next step is comminution, which is the most energy-intensive part of the process. It consists of a series of milling and classification units to reduce the feed stream to a specified particle size distribution. This is generally predetermined for a particular ore type to generate the appropriately liberated particles suitable for the specified separation process.

HOW LIBERATION IS MEASURED AND MODELED

Liberation Measurement Techniques

Because mineral liberation depends on the relationship between mineral grain size and particle size, liberation is characterized by the mineral grade per size fraction. The first step is to obtain different size fractions either by physical screening or by using heavy liquid separation or elutriation to obtain particles falling into narrow density classes (Gaudin 1939; Sutherland et al. 1988). There is, however, a known limitation with the latter approach when the density and size of particles vary. Particles of different sizes and density can have the same overall mass and be collected in the same fraction obtained when using a cyclosizer. At finer size fractions where the particle size is equal to or less than the mineral grain size, the particle will be fully liberated. Conversely, at coarser size fractions where the particles contain various mineral grains, the particle will be less liberated. Thus, the first proxy for liberation is the varying chemical assay or grade with particle size. In many cases, characterizing the variation of grade with particle size is enough to characterize where the extent of liberation can become a problem, and more sophisticated, expensive measurement is not necessary.

Formal mineral liberation measurements require observing and characterizing the mineral grain distribution in the particles that make up the stream to be separated. These fall into two main categories:

1. Two-dimensional (2-D) measurements obtained from polished sections of ore particles
2. Three-dimensional (3-D) measurements obtained from whole ore particles by either physical measurement or nonintrusive methods such as 3-D tomography

Within these two categories, the measurement techniques are further subdivided into classifications based on the technology platform being used. Two-dimensional instruments are also known as image analysis instruments and include optical microscopy and SEM-EDS systems. Both techniques can be automated; the former uses light microscopy to identify minerals, whereas the latter uses characteristic X-rays measured from an electron beam to identify minerals.

Three-dimensional techniques include the recently developed computer-aided tomography that has been applied to liberation studies to obtain 3-D images of whole ore particles to overcome the effects of stereological bias, which is described in the next section. The preferred technology platform is the X-ray micro-computed tomography, or μCT (Miller et al. 1990; Morrison and Gu 2016).

Each of the aforementioned techniques has their own individual strengths and limitations; these techniques are discussed in greater detail in the study by Wilkie (2016). Image analysis methods using optical and SEM-EDS platforms are by far the most widely used and most mature of the technologies. Automated image analysis instruments (e.g., Quantitative Evaluation of Minerals by Scanning Electron Microscopy [QEMSCAN], Mineral Liberation Analyzer [MLA], Mineralogic, and TESCAN Integrated Mineral Analyzer [TIMA-X]) are now widely available as commercial products and can be used directly on the site of a mineral processing operation. Their fundamental limitation is that measurements of liberation in polished sections are fundamentally biased (see the following section). There is a substantial body of work in the literature—most notably from King (1983), Barbery (1987), Gay (1996), Leigh et al. (1996), and Latti (2006)—that attempts to transform 2-D measurements into three dimensions. Despite this work, most liberation measurements performed by commercial image analyzers present untransformed 2-D liberation data, and the effect of the bias is often unknown in three dimensions. In practical terms, use of the 2-D data for liberation is useful, and the following typical categories of liberation are successfully used to interpret the flow-sheet and processing implications of the data (Lotter et al. 2002, 2018a):

1. Liberated particles are fragments in which more than 90% of the particle is made up of the mineral of interest.
2. Middling particles are those fragments in which between 30% and 90% of the particle is made up of the mineral of interest.

3. Locked particles are those fragments in which less than 30% of the particle is made up of the mineral of interest.

With regard to 3-D techniques, the physical separation of particles into grade classes using heavy liquid separation is the oldest and simplest technique. This technique works well on simple binary ores where there are two minerals with large differences in density. The technique is less useful on ores where there are a large number of minerals with varying densities present in the particle. X-ray µCT has the ability to directly measure the 3-D liberation of particles, thus overcoming the stereological bias of 2-D image analysis instruments (Miller et al. 2009). The current generation of X-ray µCT instruments is limited, however, in terms of both spatial resolution and phase resolution, so small objects in a particle of complex mineralogy are difficult to resolve. Heavy liquid separation is also more problematic for separations of finer particles.

Stereological Bias in 2-D Measurements of Liberation

There are several methods available for the measurement of liberation; however, the industry standard is based on 2-D image analysis instruments. Therefore, it is important to discuss the effects of stereological bias when using 2-D image analysis liberation data. Figure 7 shows a schematic that demonstrates the stereological bias associated with sectioned measurements.

In this stylized view, a locked particle containing three phases (A, B, and C) is mounted in a 3-D cylinder and sectioned at two points through the cylinder. The two sections on the right of the cylinder show two different particle compositions; the first (through B) shows a completely liberated section of Phase B, whereas the second (through C) shows a particle of Phase C completely rimmed by Phase B. Both of these sections were obtained from the same locked particle, which demonstrates the fundamental stereological bias. A locked particle can be sectioned into a variety of liberated and locked sectional views, whereas a liberated particle can only be sectioned into liberated sectional views. Hence, there is a fundamental bias in overestimating liberation from 2-D measurements.

Work by several authors (Stewart and Jones 1977; Miller and Lin 1988; Spencer and Sutherland 2000; Latti 2006) has shown that the degree of stereological error is variable and depends on the texture of the ore as well as the degree of liberation present. Figure 8 shows a representation of the stereological bias associated with varying degrees of textural complexity as described by Spencer and Sutherland (2000).

Wilkie (2016) analyzed this diagram in the context of using image analysis methods for predicting the liberation of

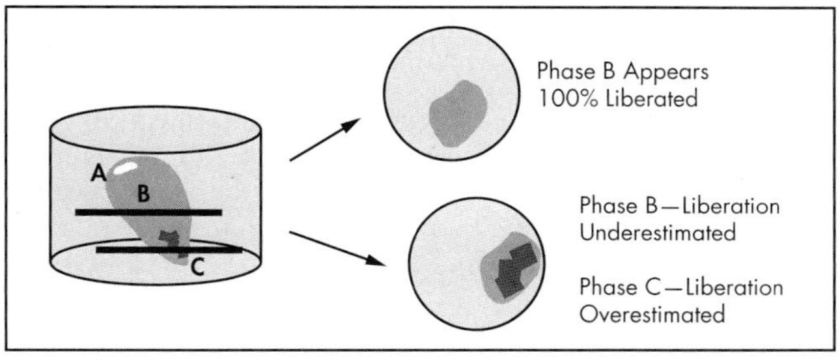

Source: Latti 2006

Figure 7 Stereological bias obtained from sections through a locked particle

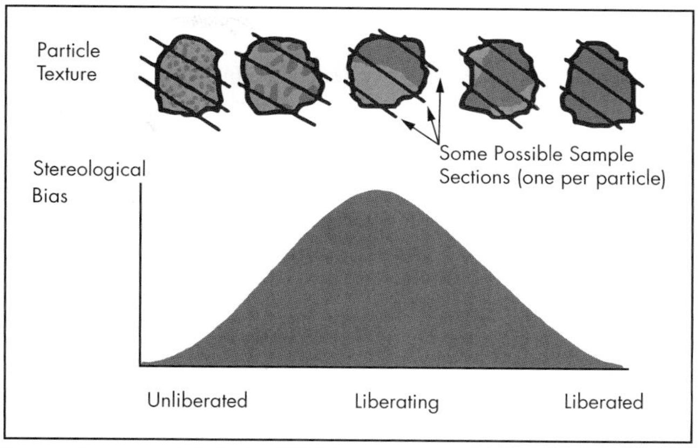

Source: Spencer and Sutherland 2000

Figure 8 Stereological bias associated with textural complexity

gangue for ore-sorting applications. His analysis of the schematic is summarized as follows:

> In essence, the greater the complexity inherent in the texture, the less bias is seen in the untransformed one- and 2-D data. Hence for complex disseminated textures measured at coarse particles, little or no stereological bias is present in the data whilst the presence of simple textures and barren matrix is indicative of stereological error.

The reference to one-dimensional data in this quotation is the use of linear measurements of particle grades performed on image analysis instruments. We can therefore expect to see little or no stereological bias in sample streams that are largely unliberated or highly liberated. The main streams we need to carefully assess are those that contain a large proportion of simple binary particles.

In addition to textural complexity, the stereological bias may also be exacerbated by the modeling procedures used to quantify liberation from image analysis measurements. Latti (2006) concluded that "in multiphase systems evaluated, the stereological bias is small when comparing the overall two- and three-dimensional liberation distributions but significant when liberation distributions are determined from one dimensional data" and that "it was also found that the practice of combining minerals into groups in order to simplify the problem may increase the stereological bias, particularly for particle sizes that are greater than the liberated size of the majority of the minerals in a given dataset." In essence, the practice of modeling multiphase binary particles as simplified binaries between the phase of interest and gangue may result in reallocating particles to the left of the curve in Figure 8 to the middle of the curve, resulting in a high degree of stereological bias. This is due to the modeling method used rather than inherent bias in the raw image analysis method.

In summary, although the debate between researchers on the extent of the liberation bias in 2-D measurements of liberation remains largely unresolved, the work by Latti (2006) and Spencer and Sutherland (2000) indicates that the bias is problematical in a limited set of particle textures, predominantly simple textures in binary particles. To this end, it is common practice in the mineral processing industry to use untransformed 2-D liberation data for modeling liberation around a circuit because the use of stereological transforms is complex and can often exacerbate the bias, as discussed previously. Provided that the untransformed data are viewed in relative terms as opposed to absolute terms (i.e., used to measure changes in liberation as opposed to absolute values of liberation), the use of untransformed liberation data still provides useful insights into the comminution behavior of mineral streams in a unit operation.

Sample Methodologies for Liberation Analysis

The general framework for preparing samples for liberation analysis is common across the various mineral processing operations, including flotation, gravity separation, and leaching. All applications are measured on a size-by-size framework for three main reasons. The first is that most mineral separation techniques perform differently on different particle sizes, as shown by Figures 1 and 2 for flotation. Second, the stereological bias, as shown in Figure 8, is significantly reduced when narrow size fractions are measured. It is typical to use a root-two or factor-two series of screens when classifying the sample into narrow size fractions. The third reason is that the presentation of the sample to a SEM in known size classes allows the application of quality control to disqualify certain particles in view of the SEM because they are either coarser or finer than the stated size limits of that size class. The general framework may be summarized as follows:

1. Representative sampling of the product stream
2. Subsampling of the product stream using rotary rifflers and classification of the stream aliquot into specific size fractions
3. Subsampling of product size fraction and mounting into polished sections
4. Measurement of polished sections by selected image analysis technique (optical, SEM-EDS)
5. Allocation of particles into grade classes (typically 12 grade classes of 10% intervals) ranging from the barren class (0% value) to the fully liberated class (100% value)
6. Analysis, data validation, interpretation, and reporting of changes in liberation across the unit operation

There are, however, several differences in the specific application, primarily due to grade and particle size. To build a statistically relevant grade distribution across 12 grade classes, typically more than 100 particle sections need to be allocated across the 12 grade classes. Lower grades and coarser particles therefore require more polished sections to achieve the particle statistics required for quantitative liberation measurements (Wightman et al. 2016). Table 1 shows a typical experimental design for a liberation analysis around a porphyry copper flotation circuit.

Typical porphyry copper concentrators have a flotation P_{80} feed size of around 150–200 µm, and a factor-two series of size fractions results in six size fractions being analyzed. Because of the grade differences among the feed, concentrate, and tailings stream and the number of value-bearing particles in the various size fractions, concentrate streams require only one polished section to be measured across all size fractions, whereas feed and tailings streams require more polished sections in the coarser size fractions to be measured. This experimental design will ensure that assay reconciliation can be achieved across all size fractions and provide a statistically relevant population of value-bearing particles being measured to build the particle grade distribution.

Table 1 Number of polished sections in each size fraction

Product Stream	Grade, %	Size Fraction, µm					
		−10	−20	−38	−75	−150	+150
Feed	0.5–1	1	1	1	2	2	3
Concentrate	20–40	1	1	1	1	1	1
Tail	0.05–0.1	1	1	2	3	4	6

To understand and characterize the statistical robustness of the liberation measurements, Leigh et al. (1993) proposed estimating the confidence limits of liberation measurements made on ore particles using the "bootstrap" technique. This technique was extended to measurement of mineral grain size by Evans and Napier-Munn (2013). A simple analytical technique also proposed by Leigh et al. (1993) has been applied in a case study described by Lotter et al. (2018b) to estimate the confidence limits of liberation measurements on concentrate and tailings samples of a chalcopyrite ore. In the case study example, approximately 15,000 particle sections were measured for the concentrate sample (grade 15.5% Cu) and 20,000 particle sections were measured for the tailings sample (grade 0.11% Cu). The estimated confidence intervals for the two samples are shown in Figure 9, which clearly shows that although more particle sections were measured in total for the low-grade sample, the confidence limits are much larger for this sample because the low grade of the tailings stream results in fewer particles of interest being present in each particle composition class. To overcome this problem, specialized measurement routines for low-grade samples have been implemented in most automated SEM-EDS-based mineralogy systems to specifically search for and measure only particles that contain the minerals of interest. Using these specialized routines maximizes the number of particles containing the minerals of interest measured in standard measurement times and increases the confidence in the resulting liberation data.

Errors in Liberation Measurements

Whereas correct sampling of particulates provides true sample material with known confidence limits to the mineralogist, the subsequent mineralogical measurements are also subject to measurement errors (Lotter et al. 2018b). Early methods to estimate the errors in the measurement of mineral proportions using optical microscopy were published by Chayes (1944, 1945) and Van der Plas and Tobi (1965). While the development of automated mineralogy systems in recent years has made it easier to collect larger volumes of data, these measurements are still affected because the measured properties of a subsample of particles (typically around 10,000 to 30,000 particles) are used to represent the larger population from which they were sampled. In the early days of these automated mineralogy systems, Leigh et al. (1993), working with QEMSCAN data, developed an analytical method to estimate the confidence intervals for liberation data in cumulative liberation distribution (also known as the cumulative liberation yield) form. They noted that the bootstrap method of Efron (1987) would give a definitive answer, but at that time, the computationally intensive bootstrap technique was not practical to apply. Now, the restrictions imposed by lack of computing power no longer apply and the bootstrap technique has

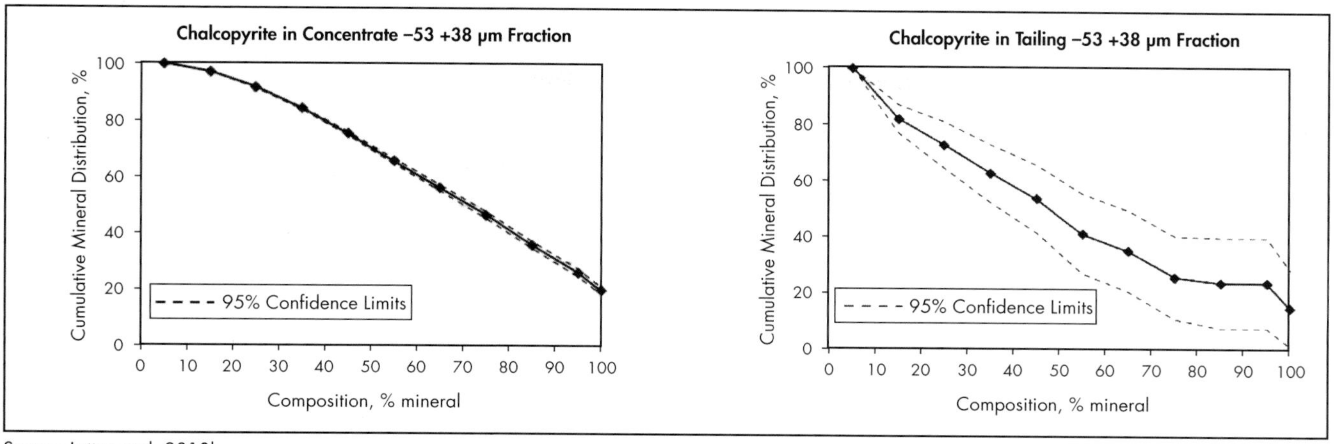

Source: Lotter et al. 2018b

Figure 9 Comparison of the confidence intervals on mineral liberation in concentrate and tailings samples

been successfully used to estimate confidence intervals for mineral GSDs (Evans and Napier-Munn 2013) and for non-cumulative liberation distributions (Mariano and Evans 2015) from MLA data. Typically, in this application, larger data sets are associated with higher levels of confidence in the liberation measurement.

The analytical approach developed by Leigh et al. (1993) uses the following equation to estimate the 95% confidence interval, *CI*, at particle composition *C* as a function of the number of particles in the liberation classes:

$$CI_C \approx 1.96 * \left(1.25 Y_C^2 (1 - Y_C)^2 \left(\frac{1}{N_0} + \frac{1}{N_1}\right)\right)^{0.5} \quad \textbf{(EQ 1)}$$

where

Y_C = value of cumulative liberation distribution at particle composition *C* (as a proportion)

N_0 = number of particles in composition classes with composition less than *C*

N_1 = number of particles in composition classes with composition greater than or equal to *C*

A case study on process streams from a copper concentrator from Lotter et al. (2018b) illustrates the importance of understanding the confidence intervals that apply to liberation data. In this example, the liberation distribution of chalcopyrite has been measured in a concentrate and a tailings stream, and after applying Leigh et al.'s (1993) analytical method to the liberation data for these high-grade and low-grade samples presented in Tables 2 and 3, it is clear that the estimated confidence intervals are much wider for the low-grade material that has fewer observations of chalcopyrite.

When the estimated confidence limits are included in the plots of the cumulative liberation distributions of chalcopyrite in the two streams, as in Figure 9, the difference between the low-grade and high-grade streams is obvious. The level of confidence in the liberation data for the lower-grade sample of tailings with a copper grade of 0.11% is considerably less than the confidence in the higher chalcopyrite content concentrate sample that has a copper grade of 15.5%. The low chalcopyrite content of the tailings stream results in fewer chalcopyrite-bearing particles in each particle composition class (N_1 in the equation of Leigh et al. [1993]) and this is the reason for the wide confidence intervals. The difference in confidence levels, which results from lower numbers of particles in each liberation class in low-grade streams, is the reason why modern automated mineralogy systems include specific measurement procedures to scan for and only measure particles that contain the minerals of interest. This approach maximizes the number of particles containing the minerals of interest that are measured when analyzing low-grade samples and increases the confidence that the end user can have in the resulting liberation distribution data.

The experimental design shown in Table 1 also works well for sulfide nickel operations in which the grade and particle sizes can be similar to those of porphyry copper operations. Lead/zinc flotation operations typically have much higher feed and tailings grades (up to an order of magnitude higher), so the number of polished sections in the coarser fractions can be reduced substantially while still maintaining assay reconciliation and statistically reliable populations of value particles for particle grade distributions.

In the case of leaching operations, the problems associated with particle size are further exacerbated by the coarser feed P_{80} that is typical for leaching. Leach feed particle sizes range from approximately 2 mm to run-of-mine consisting of uncrushed material. Latti and Wilkie (2016) outline an experimental design for a leaching circuit composed of 100% passing 12-mm material. In this study, the size fraction grid consisted of five size fractions represented by the +14 mm, +8 mm, +1 mm, +150 μm, and –150 μm fractions. Sample preparation followed the aforementioned generic workflow with up to 12 polished sections being prepared in each size fraction. Standard 30-mm mounts were prepared for the fractions finer than 4 mm, whereas specially prepared 100-mm square mounts were prepared for the fractions coarser than 4 mm to obtain a reasonable population of particles in the polished section. Figure 10 shows examples of the large-area polished sections from the study by Latti and Wilkie (2016). As with concentrator applications, the polished sections are measured by automated image analysis instruments to obtain the particle grade distribution and model the liberation of the ore at discrete particle sizes.

Modeling of Liberation

The earliest work on liberation modeling was conducted by Gaudin (1939), who conceptualized the liberation process

Table 2 Confidence interval estimation for a concentrate sample (−53 +38 µm, 15.5% Cu)

Mean Grade of Composition Class	Cumulative Distribution of Mineral	Number of Particles in Class	N_0	N_1	95% Confidence Interval	Lower Confidence Interval	Upper Confidence Interval
Barren	100.0	6,185	—	—	—	—	—
5	100.0	2,624	6,185	10,518	0	100.0	100.0
15	97.2	1,484	8,809	7,894	0.09	97.1	97.3
25	91.8	1,222	10,293	6,410	0.26	91.6	92.1
35	84.4	1,006	11,515	5,188	0.48	83.9	84.9
45	75.5	906	12,521	4,182	0.72	74.8	76.3
55	65.8	685	13,427	3,276	0.96	64.8	66.7
65	56.3	580	14,112	2,591	1.15	55.1	57.4
75	46.6	489	14,692	2,011	1.30	45.3	47.9
85	36.0	400	15,181	1,522	1.36	34.7	37.4
95	26.2	235	15,581	1,122	1.31	24.8	27.5
100	19.9	887	15,816	887	1.20	18.7	21.1

Source: Lotter et al. 2018b

Table 3 Confidence interval estimation for a tailings sample (−53 +38 µm, 15.5% Cu)

Mean Grade of Composition Class	Cumulative Distribution of Mineral	Number of Particles in Class	N_0	N_1	95% Confidence Interval	Lower Confidence Interval	Upper Confidence Interval
Barren	100.0	19,908	—	—	—	—	—
5	100.0	62	19,908	102	0	100.0	100.0
15	82.1	12	19,970	40	5.10	77.0	87.2
25	72.9	4	19,982	28	8.19	64.7	81.1
35	62.7	4	19,986	24	10.46	52.3	73.2
45	53.7	6	19,990	20	12.19	41.5	65.9
55	41.2	2	19,996	14	14.19	27.0	55.4
65	35.1	4	19,998	12	14.41	20.7	49.5
75	25.7	2	20,002	8	14.80	10.9	40.5
85	23.7	0	20,004	6	16.16	7.5	39.8
95	23.7	2	20,004	6	16.16	7.5	39.8
100	14.7	4	20,006	4	5.10	1.0	28.4

Source: Lotter et al. 2018b

associated with size reduction of the rock as a cubic aggregate containing both valuable and waste mineral grains. The aggregate is then broken into particles of uniform size by a cubic fracture lattice that is randomly superimposed on the aggregate parallel to the grain lattice. This model was further developed by Wiegel and Li (1967) to produce the Gaudin random liberation model, which is conceptualized in Figure 11.

Both Gaudin (1939) and Wiegel and Li (1967) derived a set of equations to calculate the proportion of liberated values, liberated waste, and locked waste and value minerals, producing the plot shown in Figure 12.

In the case represented by Figure 12, the diagram represents the proportion of liberated values, liberated waste, and locked values and waste for an ore containing 25% values. The diagram is divided into three distinct fields:

1. Coarse particles (grain size–particle size ratio < 1) where locked values and waste dominate the population

Source: Latti and Wilkie 2016

Figure 10 Large-area polished sections for coarse particles

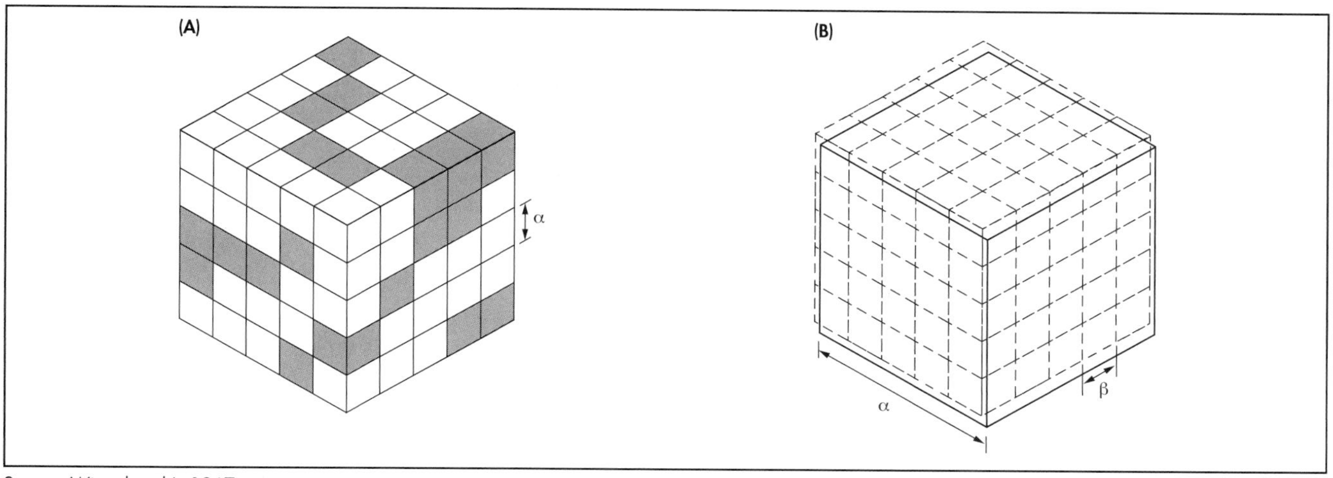

Source: Wiegel and Li 1967

Figure 11 (A) Random arrangement of mineral grains and (B) fracture of a single mineral grain into particles

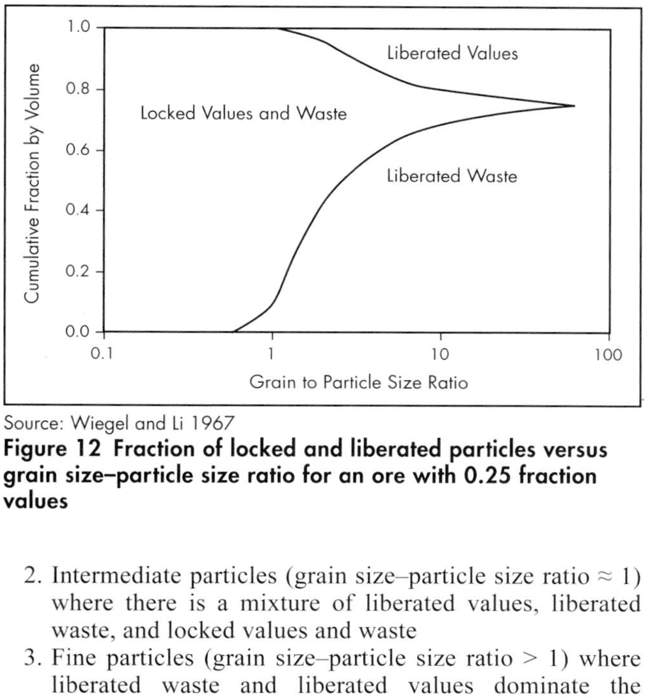

Source: Wiegel and Li 1967

Figure 12 Fraction of locked and liberated particles versus grain size–particle size ratio for an ore with 0.25 fraction values

2. Intermediate particles (grain size–particle size ratio ≈ 1) where there is a mixture of liberated values, liberated waste, and locked values and waste
3. Fine particles (grain size–particle size ratio > 1) where liberated waste and liberated values dominate the population

Other researchers in the field of liberation modeling who have taken various mathematical, theoretical, or combined approaches include King (1982, 1990), Barbery (1984, 1991), Gay (1996, 2004), Evans (2002), Evans and Morrison (2016), Subasinghe and Dunne (2016), and Wilkie (2016). All the methods are aimed at predicting the particle grade distribution within a given particle size class but differ in the mathematical approach that they use to quantify the distribution. Many of these methods rely on image analysis measurements that require stereological correction.

King's (2000) approach, for example, is to model the particle grade distribution with a beta distribution, which is widely used in mathematical statistics. The particle grade distribution is conceived as being made up of three types of particles:

1. Liberated gangue particles, denoted as L0
2. Liberated value particles, denoted as L1
3. Mixtures of gangue and value particles that have a grade distribution given by the probability distribution density function $p(g)$, which is modeled by the beta distribution

By contrast, Barbery's (1984, 1991) method is based on the concepts of geometric probability originally developed by Davy (1984). Barbery's method relies on three fundamental assumptions:

1. Breakage is assumed to be random, uniform, and isotropic.
2. A Poisson mosaic or Boolean texture is assumed.
3. A particle shape model is also assumed.

The value of this liberation modeling is that the proportion of each particle class can be calculated for any given particle size and grain size inherent in the ore. An important aspect to the incorporation of the liberation of particles into process modeling and simulation is the mass balancing of the data. Methods have been developed to mass balance particle liberation data in process streams (Lamberg and Vianna 2007), and a simulation framework has been developed to provide simulations of comminution, liberation, and separation in one integrated system (Evans et al. 2013).

THE BEST WAY TO REPORT LIBERATION

Optical

Traditional optical microscopy methods generally allocate ore-bearing particles into categories represented by liberated particles, binary particles with other minerals, and ternaries that are locked with two or more minerals. This is shown in Table 4, which summarizes the liberation characteristics of chalcopyrite on a size-by-size and overall basis (Henley 1983).

Hence, on an overall basis, 82% of the chalcopyrite is liberated with 6% present as binaries with magnetite, 5% as binaries with pyrite, and 4% present as ternaries. Analysis of the size-by-size distributions shows that between 93% and 100% of the chalcopyrite is liberated in the finer –45-μm size

Table 4 Examples of liberation at the different size fractions

Size Fraction, µm	% Liberated Chalcopyrite	% Chalcopyrite Locked in Binary Particles with ... Magnetite	Pyrite	Nonopaques	Pyrrhotite	% Chalcopyrite Locked in Ternary Particles
+180	45	4	8	22	5	16
−180/+90	50	20	10	10	0	10
−90/+45	88	2	2	0	2	5
−45/+22	93	2	5	0	0	0
−22/+8	95	0	5	0	0	0
−8	100	0	0	0	0	0
Total	(82)	(6)	(5)	(2)	(1)	(4)

Source: Henley 1983

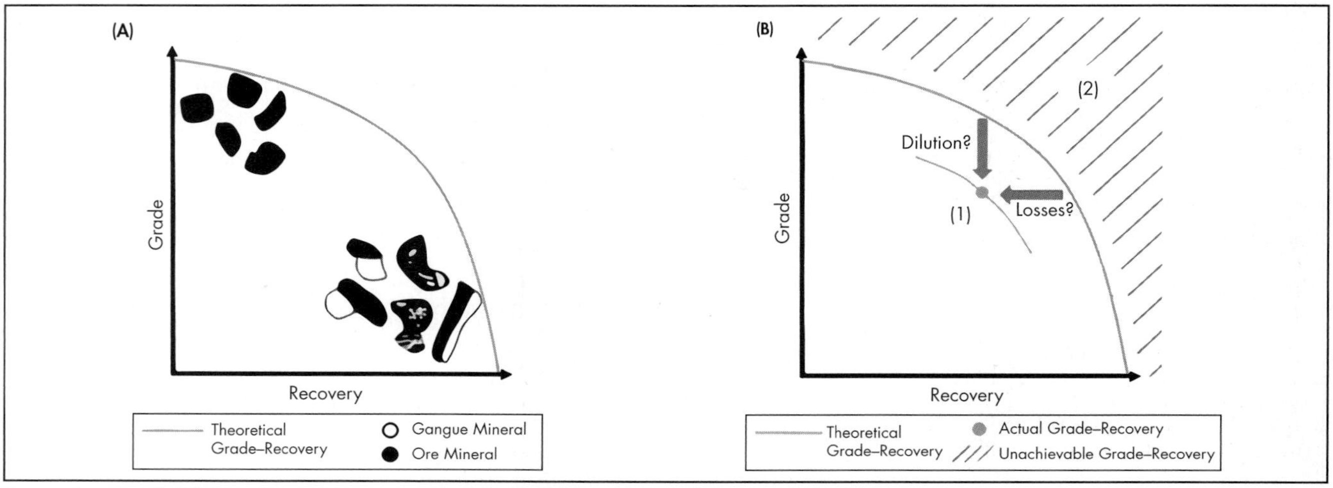

Source: Cropp et al. 2013, reprinted with permission from the Australasian Institute of Mining and Metallurgy

Figure 13 (A) Schematic of the theoretical grade–recovery curve with typical particle images included; and (B) the position of the actual grade recovery relative to the theoretical potential, where (1) operational changes can be made to improve performance and (2) milling and classification changes are needed for improved performance

fractions, with liberation decreasing dramatically to 50% and 45% in the 90- and 180-µm fractions, respectively. These two size fractions show significant proportions of chalcopyrite binaries locked with magnetite, pyrite, and nonopaques along with ternary particles. The overall conclusion is that the chalcopyrite is well liberated in the floatable size ranges but less liberated in the coarser fractions.

Automated SEM-EDS

In addition to the traditional approach of reporting liberation as a proportion of liberated binaries and ternaries, commercial automated SEM-EDS mineralogy instruments have numerous built-in software features to report liberation in several different ways. Furthermore, researchers are developing new and interesting ways to report liberation in a mineral processing context. The following sections illustrate some of these as well as a few newer liberation concepts.

Theoretical Grade–Recovery/Mineral Potential Curve

The mineralogical limit to flotation or theoretical grade recovery is a means of describing or characterizing the recovery of particles based on the ore mineralogy and texture of the ore present in the feed stream. It assumes perfect separation and does not take process kinetics or material recovered by entrainment into account. Therefore, it cannot necessarily be considered a realistic target but does provide the maximum limit. It takes into account the composition of the particles containing the valuable mineral and gives the maximum potential separation of it. The location of the limit is determined by the quality and quantity of composites, as shown in Figure 13.

Figure 13 shows how particle composition defines the theoretical grade–recovery curve. Images of particles are used to show how high recovery of the target mineral will typically also mean recovery of gangue minerals, thus reducing the grade of the concentrate. If there are more composite particles and the feed stream is consequently lower grade, the curve in Figure 13A shifts left (such as in a scavenger stream). Conversely, if it is a stream containing a lot of liberated particles (such as a cleaner stream), the curve shifts to the top right-hand corner.

Figure 13B shows the position of the actual grade–recovery curve relative to the theoretical potential. At point 1, the operational conditions such as reagent addition or circuit modifications can be altered to improve performance, but only to the limit imposed by the curve. If grade recovery above the theoretical curve is required, such as for point 2, then the liberation potential of the feed will need be changed by altering the milling and classification conditions. This curve does not include entrainment or naturally floating gangue, so it will always be an overestimate of what can be achieved practically. The procedure to calculate the theoretical grade–recovery curve is discussed by Johnson (2010).

Cumulative Liberation Yield

The cumulative liberation yield (CLY), as shown in Figure 14, represents the proportion of all of the mineral for a sample contained in particles with a composition greater than some predefined value. For example, CLY90 is the proportion of all of the mineral for a particle contained in particles with a composition greater than 90%. Thus, the CLY cumulates from right to left. An alternate representation of the CLY is to cumulate particles from the lowest quality (least liberated) to highest quality (most liberated), that is, from left to right.

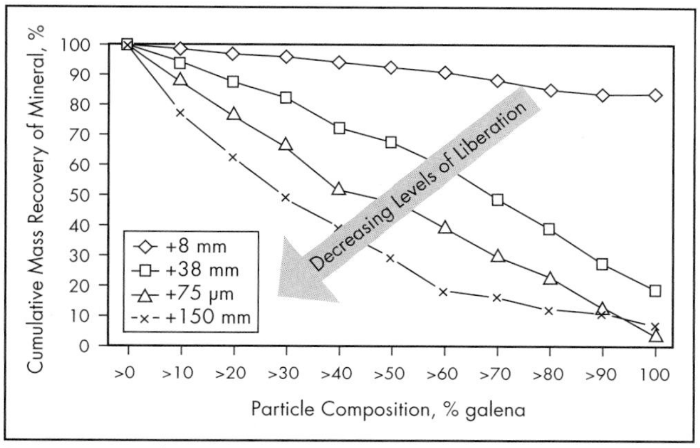

Source: Evans and Morrison 2016

Figure 14 Graphical presentation of particle composition data as a cumulative liberation yield

Liberation and Association

In addition to representing liberation data in discrete categories of particle composition, such as the CLY (Figure 14) or categories of locked–middlings–liberated particles, a user of an automated SEM-EDS instrument is able to interrogate the liberation data to further understand the characteristics of unliberated particles. This could be through visual inspection of the false color particle images provided by the automated SEM-EDS instrument or more quantitative characterization using discrete user-defined categories of interest (Figure 15).

Liberation Spectrum

Wightman and Evans (2013) developed a technique using information obtained from automated SEM-based systems called the "liberation spectrum." This spectrum represents the compositional distribution of a particle population in a form that is useful to mineral processors, particularly to determine grinding targets for separation processes.

Figure 16 shows the graphical relationship of the liberation spectrum by plotting the grade of pyrite containing particles at various particle sizes. It provides insights into the particle size at which liberation begins (point B), above which the linear relationship shows the particle size at which the grains of the mineral of interest remain unbroken (feed grade A). Point C shows the distribution of grain sizes present and whether multiple populations of grain sizes are present in the sample. This type of information can be used to select targets for grind size and includes mineral grade information, which can be used to infer the response to the grind targets.

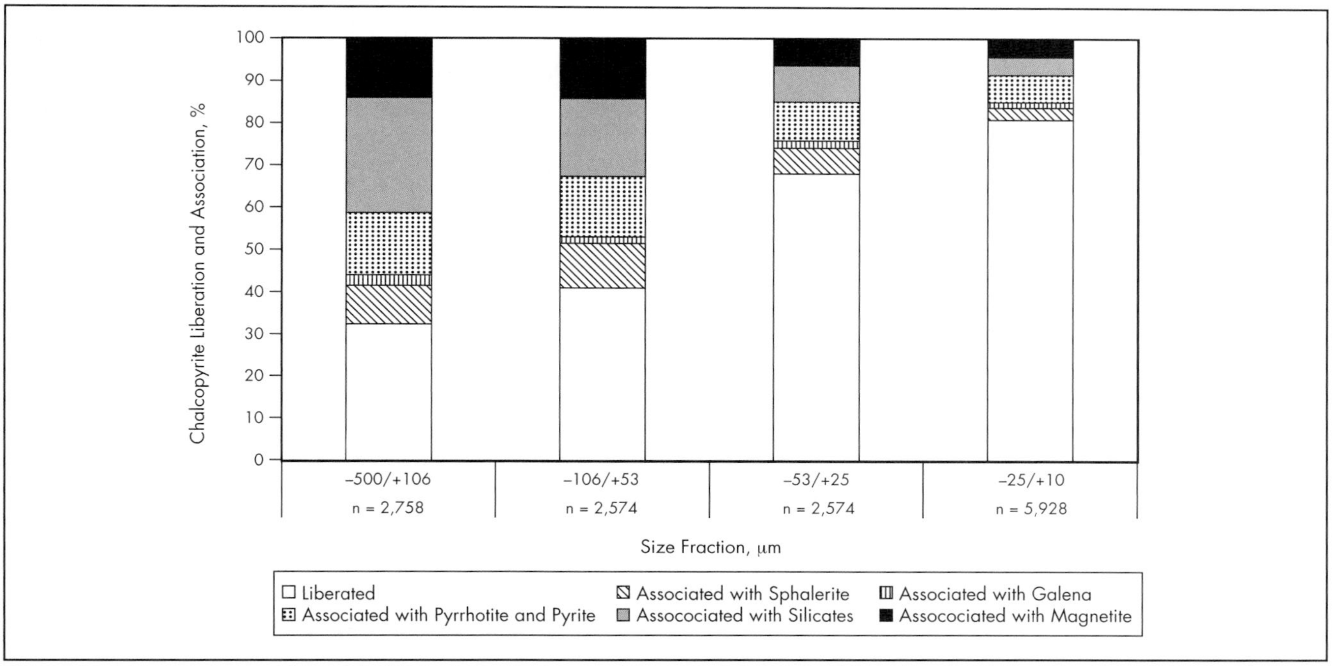

Figure 15 Representation of particle liberation and association on a size-by-size basis (*n* = number of particles measured containing the mineral of interest). The association quantifies the grain boundary characteristics of unliberated valuable minerals with the mineral of interest.

X-Ray Microtomography

The development of the application of X-ray tomography (3-D) to ores and minerals has provided an opportunity to measure mineral liberation (and other characteristics) in three dimensions. These measurements, which were originally developed for medical purposes, have been successful in measuring differences in the density of different minerals nonintrusively (Miller et al. 1990; Evans et al. 2012; Ghorbani et al. 2013; Morrison and Gu 2016). This method has the advantage of removing the effect of stereological bias from measurements of particle liberation. The method also has the advantage of much easier, quicker, and more cost-effective sample preparation in that no polished sections are necessary. The sample size does determine the voxel size obtainable in the output image, and if high resolution is needed, the sample size should be small enough (Morrison and Gu 2016).

The limitations of the measurements also include the lack of elemental determination, and the mineral composition of an ore has to be calibrated using SEM techniques. If the density and resulting X-ray attenuation of minerals are too close, discrimination is difficult, but techniques to address this are being developed. Analytical methods combining the merits of both techniques are also being developed. Large amounts of data are generated, and this requires significant computing power for the reconstruction.

It is anticipated that as software and hardware innovations are developed and computing power continues to increase that this methodology will be developed to give real-time, online liberation measurements of particle streams.

Figure 17 shows that the 2-D measurements overestimate liberation relative to the 3-D measurements, which correlate to the flotation results.

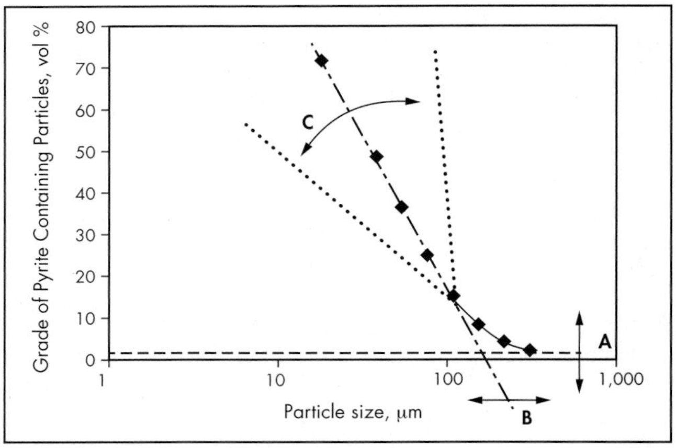

Source: Wightman and Evans 2013

Figure 16 Representation of the liberation spectrum from the automated SEM-EDS measurement, where point A represents the feed grade of the sample, point B represents the size at which liberation begins, and point C indicates the distribution of sizes of the mineral grains

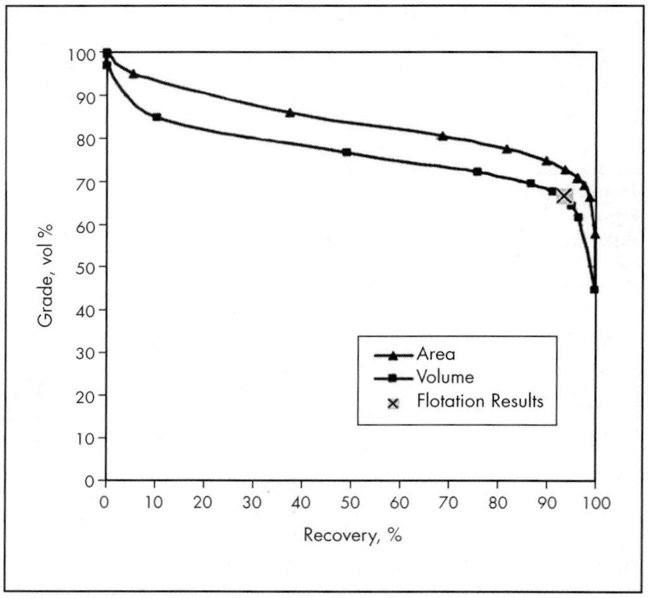

Source: Miller et al. 2009

Figure 17 Difference in theoretical grade–recovery/mineral potential curves for a phosphate ore obtained in two dimensions (area) and three dimensions (volume) as well as the actual flotation results achieved

DETERMINING WHEN THERE IS ENOUGH LIBERATION

Ultimately, the decision on whether there is enough liberation is determined by the mineralogy of the ore and the concentration process being used. Most concentration processes exhibit a size recovery curve, which means that not all particle sizes and particle grades are recovered equally, resulting in a Goldilocks particle size for optimum liberation and recovery. However, there are further technoeconomic (and environmental) considerations that will influence the decisions made regarding the degree of liberation at which a plant will operate, and these need to be taken into account. The following sections describe the need for optimum liberation and particle size for three of the most common extraction processes: flotation, gravity separation, and leaching.

Flotation

The typical influence of mineralogy and texture on flotation performance is summarized in Figure 18; this is accompanied by a schematic showing the size fraction affected.

The losses of valuable mineral and dilution to concentrate in the three major size categories can be commonly attributed to the following:

- Coarse particles can be composites (i.e., contain multiple grains) of value-bearing minerals and gangue minerals. Recovering these particles will increase overall recovery but lower the grade of the concentrate, as the locked, attached gangue will, by necessity, also be recovered. Rejecting these will mean that the grade is not compromised, but the target grains will be lost to tailings, thereby reducing overall recovery. The mineral grains within these particles may be fine grained or have a bimodal GSD of the valuable mineral(s) that may result from either inequigranular grains in one ore or blended material from ores with different size distributions where the target grind size is optimized for the largest of the distribution sizes. Alternatively, they may arise from the processing of a different ore type that requires finer grinding than the current ore's target grind size. These may not be recovered in the flotation circuit because they may

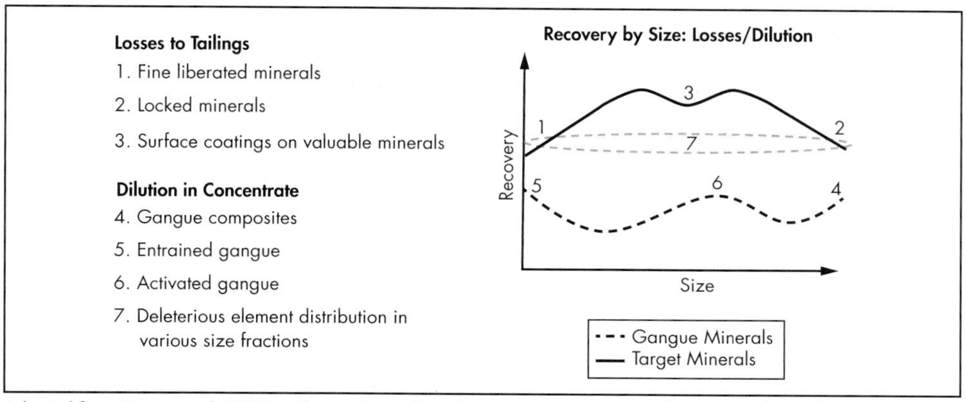

Adapted from Cropp et al. 2013, with permission from the Australasian Institute of Mining and Metallurgy

Figure 18 Summary of common causes for lower-than-anticipated grade or recovery in the flotation concentrate (graph highlights these regions relative to particle size)

be too heavy to float or lack the free surface area to be collected.

- Fine particles largely consist of liberated mineral grains. They have a lower probability of particle–bubble collision and therefore have a reduced chance of being recovered. Some mineral species are more susceptible to this phenomenon than others. For example, chalcopyrite floats reasonably well, even at very fine particle sizes. However, fine pentlandite flotation is challenging, especially when trying to maintain the same selectivity against sulfide gangue as in the more floatable size range. Very fine gangue material may also be recovered by entrainment to the final concentrate.
- In the midrange, particles are typically considered to be fully or well liberated and therefore amenable to separation in a flotation circuit, with the ore minerals being recovered without significant degradation of grade; however, this is not always the case. Surface coatings can reduce the susceptibility of an ore mineral's surface-to-bubble adherence, reducing recovery, whereas concentrate grade may be lowered when the surfaces of gangue minerals become activated (e.g., by copper ions), causing them to float.

From this discussion, it can be concluded that there is an optimum particle size and degree of liberation for flotation recovery. Too little liberation results in poor recoveries in coarse size fractions, whereas too much liberation results in fine liberated particles being lost due to poor flotation recoveries at the fine end of the particle sizes.

Gravity Separation

Like flotation, gravity separation has an optimum particle size for recoveries. Figure 19 shows the typical operating size ranges for a range of gravity separation units.

From Figure 19, it is apparent that some unit operations are more efficient at coarse sizes, such as jigs, heavy media drums, and heavy media cyclones. Others are more efficient at the fine end, as represented by the Falcon, Knelson, and Kelsey jigs. Hence, like flotation, poor liberation needs to be corrected to the particle size at which the gravity concentrators work most efficiently, which is a characteristic of the ore as well as the concentration method being used. Minerals that are amenable to gravity separation tend to have higher specific gravity (e.g., cassiterite, ferroplatinum, native gold, electrum). This effect needs to be considered in the layout of the grinding circuit, lest the classifier in the circuit (e.g., cyclone) inversely classifies the heavies into the circulating load, leading to overgrinding of these minerals and lower gravity recovery. For this reason, it is common to see an open-circuit rod mill performing the duty of primary grinding, followed by a unit gravity recovery process before the ground ore is further ground to product size, in which case gravity gold recoveries approach 90% (Laplante et al. 1990). The operation of a gravity recovery process inside the circulating load of a grinding circuit improves the kinetics of downstream cyanidation (Laplante and Staunton 2004).

Another important point about gravity recovery in the case of gold is that not all of the assayed gold may occur as native gold or electrum, both of which respond to gravity concentration. Detailed studies on the selection and breakage of discrete gold and electrum grains in the presence of silicate gangue showed that the discrete gold and electrum were prone to folding as a breakage mechanism and that this was combined with a smearing effect in which some of the gold entered the silicates as a texture, making it nonamenable to gravity recovery (Banisi et al. 1991). Thus, the achievement of good liberation of gold grains does not necessarily guarantee good recovery thereof. In cases where the total gold assay of an ore is distributed between a gravity-recoverable fraction and a refractory (non–gravity recoverable) fraction, test work has to be performed on a sample of that ore to determine how much gold occurs in either category. This subject was studied in detail by Laplante et al. (1994, 1996) and resulted in a standard procedure called gravity-recoverable gold.

Leaching

Similar to flotation and other separation processes, tank leaching requires valuable mineral particles to have optimal liberation and particle size. Ultimately, the decision on whether there is sufficient liberation is an economic one driven by the amount of extraction that can be achieved at a given particle size (Petersen 2016). This in turn depends on the cost of the comminution versus the value of the metal recovered. In tank leaching where residence times are in the window period of hours, high degrees of liberation and fine particle sizes with a high surface area are generally desired. When tank leaching is used for concentrates, it can be assumed that the valuable

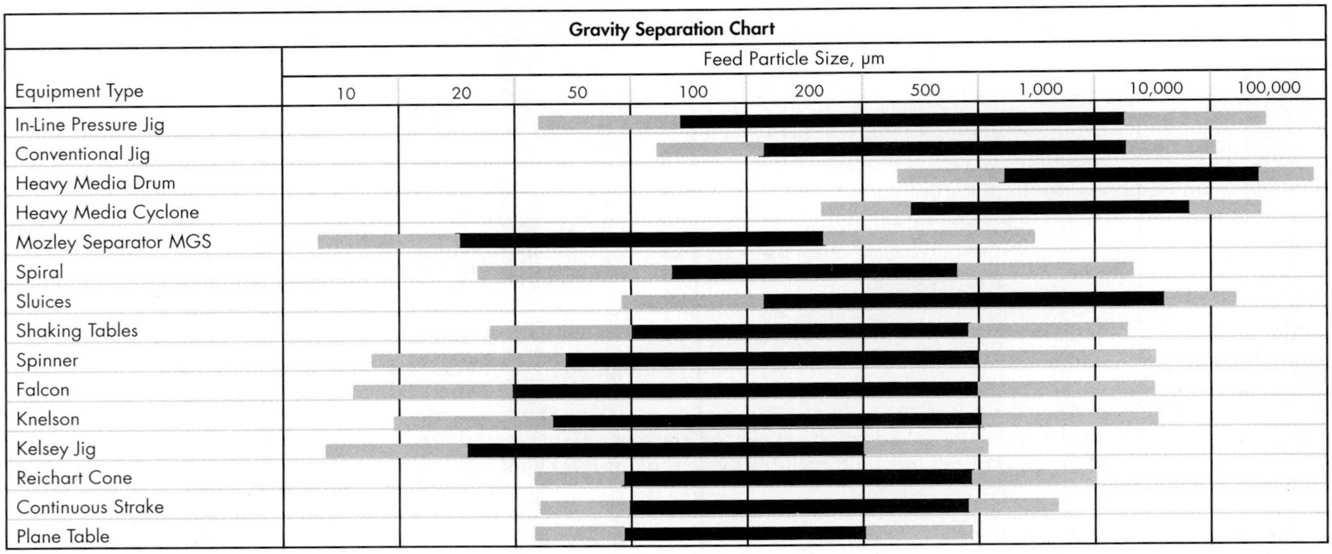

Adapted from Abols and Grady 2006

Figure 19 Operating sizes for a range of gravity separation processes

minerals are already well liberated because they have been recovered in a prior separation step such as flotation. This does not negate the need to measure the valuable mineral liberation itself, however, because the measurements provide accompanying information on the associated minerals (potentially flagging associations to deleterious minerals) or changes in valuable mineral deportment. Over and above is the certainty that the deportment and mineralogy of the valuable elements are amenable to leaching (e.g., refractory gold ores that need chemical pretreatment to liberate the gold from the crystal lattice of the host sulfide minerals) (Zhou et al. 2004; Vaughan 2004).

In the case of heap leaching, high degrees of liberation are not required for extraction, as the phase of interest need only be exposed to the leach liquor. Different processes have different cost–recovery curves, with the best example being the choice of flotation versus heap leaching for porphyry copper ores. In the case of flotation, ore feed is typically milled down to a P_{80} grind size of between 125 and 250 μm to achieve a copper recovery of between 85% and 95%. Heap leaching has a different cost–recovery curve, where ore is typically milled down to 25–75 mm and heap leach recoveries of 60%–80% are achieved. Flotation therefore requires a finer grind to achieve economic recoveries as compared to heap leaching because liberation in flotation is aimed at producing a significant proportion of liberated particles to achieve an economic grade–recovery curve. By contrast, heap leaching only requires enough liberation for the copper minerals to be exposed to the leach liquor. Hence, copper contained in complex particles is still leachable if there is sufficient surface exposure or cracks in the particles to allow the leach liquor to penetrate the internal minerals in the particle.

Unlike flotation and gravity separation, extraction from heap leaching does not suffer from the production of a large quantity of liberated particles in the fine size ranges during crushing. In heap leaching, however, there are good reasons for not having a large quantity of fines in a heap.

DIAGNOSIS OF LIBERATION PROBLEMS AND PROPOSED IMPROVEMENTS

There are many examples of the inclusion of additional milling capacity in a circuit. The particular location of the intervention in an existing circuit and the particle size obtained in most cases have been based on the mineralogy of the ore being processed, the liberation desired, and the point of maximum benefit identified. The following case studies provide examples and details of where liberation problems have been identified and the process improvements that resulted from the liberation studies. These examples are not comprehensive, and the reader should refer to the literature for others, particularly in the monograph "Process Mineralogy" (Becker et al. 2016).

Flotation

New Brunswick Mining and Smelting Corporation Ltd.

The New Brunswick Mining and Smelting Corporation operations near Bathurst, New Brunswick, Canada, were treating the largest zinc ore resource in the world from start-up in 1964 to shutdown in 2014. The ore consisted of 60% sulfide gangue (mostly as pyrite), 20% silicates, and 20% valuable minerals. Four salable concentrates were produced: lead, zinc, copper, and lead/zinc. Differential froth flotation was used to deliver these products from the grinding circuit product. Until 1998, the comminution circuit comprised two crushing stations, three rod mills, and three primary ball mills. Thereafter, these were all replaced by an autogenous grinding (AG) mill with secondary ball mills, the former being converted to semiautogenous grinding (SAG) in 1999 (Cormier and Cooper 2002).

A performance history of zinc grade and recovery at this site is shown in Figure 20 (Orford et al. 2005). Figure 20 shows three distinct domains of endeavor:

1. Back to basics—early 1980s to 1995
2. Point C—1995–2002
3. Six Sigma—after 2002

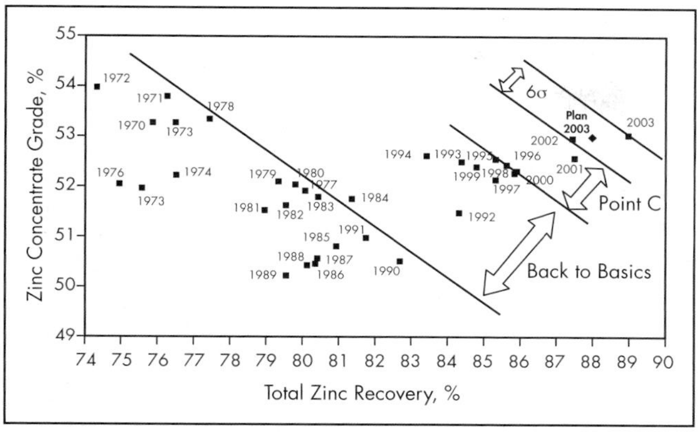

Source: Orford et al. 2005, reprinted with permission from the Canadian Institute of Mining, Metallurgy and Petroleum

Figure 20 Zinc grade and recovery performance history at the Brunswick operations

The improvements since 1990 were accrued from a range of focused efforts, including

- Staged flotation of the primary grinding product,
- Implementation of AG, and
- Streamlining of multiple process lines.

Supporting test work at various scales demonstrated that the baseline plant performance was less than optimum. From 1990 to 1995, improvements focused on optimizing equipment capability and reassessing reagents and control variables. The guideline here was to optimize the existing asset (equipment) before applying for capital to install new machinery. Effort was put into information on liberation, mineralogy, pulp and mineral surface chemistry, flotation kinetics, and entrainment.

From 1996 to 2000 (marked as point C in the figure), capital expenditure projects were approved and implemented at this site to improve the global competitiveness of the operation. The main features of these projects included the AG/SAG mill retrofit, paste backfill, stage grinding, and consolidation of the process into one continuous processing line.

At this stage, one of the tasks in the "staged grinding project" reported by Cormier and Cooper (2002) included the attainment of finer grinding from the existing grinding circuit by means of reducing the slurry density of the cyclone feed. The staged grinding part of the task was to insert the lead rougher float between the primary and secondary grinding stages to remove as much lead as possible from the downstream copper and zinc flotation stages.

Post-2002 saw the Six Sigma phase, which addressed smaller performance opportunities with better tools, such as the Six Sigma measurement system analysis, which, for example, was able to reduce the zinc grade in lead concentrate from 6.0% to 2.5% Zn.

An earlier investigation at the same site had already concluded that liberation was a limiting factor in metals recoveries. The behavior of minerals during flotation of this fine-grained, pyrite-rich, base metal ore was investigated in 1981 by studying the feeds, concentrate, and tailings from every flotation cell (Petruk and Schnarr 1981). The samples were analyzed for Zn, Pb, Cu, Fe, Ag, Sb, Sn, Bi, In, Hg, Mo, and As. Mass balances were computed, and recoveries of all metals from each cell, from each circuit, and from the mill were calculated. The percentages of sphalerite, galena, and chalcopyrite occurring as free grains in each product were determined with an image analyzer—as were size analyses for free mineral grains and middling particles—and were compared with screen analyses. The results showed that the behavior of minerals is defined much better by the recovery of free mineral grains than by the recovery of elements. The recovery of free sphalerite in the zinc circuit was 96%, whereas Zn recovery from the same circuit was 88%, the difference being due to unliberated sphalerite. The recovery of free sphalerite from the mill feed was 88% and Zn recovery was 78%. The recovery of free galena from the mill feed was 88% and Pb 64%. The recovery of free chalcopyrite was 75% and Cu recovery was 45%. Size analyses showed that more sphalerite and chalcopyrite could be liberated by regrinding, but the galena was so fine-grained that regrinding would not necessarily improve liberation. Silver occurs mainly in tetrahedrite and galena; the main loss in the tailings was due to large (+26 μm) and small (–1 μm) free and unliberated tetrahedrite grains.

Kennecott Utah Copper Concentrator

In cases when the ore types in a deposit are treated as blends, particular ore types can cause problems, and if the appropriate operating conditions cannot be identified, it can be better to process that ore type separately. Rio Tinto's Kennecott Utah Copper Concentrator (KUCC) is one of the world's longest-running and largest producing porphyry copper mines also rich in gold, silver, and molybdenum. Because of the quantity of ore being processed at any one time, the feed consists of several blends of its various ore types. In 2006, one of the ore types was identified as problematic, and a project was initiated to identify the causes (Triffett et al. 2008; Bradshaw et al. 2011, 2016).

The recovery from two ore types from KUCC was compared: a monzonite (MZME3) ore representing the ore with a typical copper and molybdenum recovery and a limestone skarn (LSN) ore with poor recoveries (see the reflected-light micrographs of samples in Figure 21). With similar amounts of total copper but very different recovery rates, the nature and proportions of the texture and gangue minerals were examined, including the particle size distribution after milling. Results showed that the LSN had a different particle size distribution, with more materials reporting to both the coarse and fine fractions, possibly due to the wider range of both harder (such as andradite and garnet) and softer minerals than those found in the MZME3 ore. The copper minerals were found to be finer in the LSN ore, allowing the potential for a greater quantity of unliberated copper-bearing grains in the coarse size fractions.

The theoretical grade–recovery curves produced from QEMSCAN mineralogical data shown in Figure 22 highlight the lower liberation of copper minerals in the coarse particles of the LSN ore compared to those of the MZME3 ore. These results are supported by the lower actual recovery to the flotation concentrate in laboratory tests seen in these fractions for the LSN ore shown in Figure 23.

The coarse size fraction in the LSN ore displays a lower grade–recovery curve than the equivalent fraction in the MZME3 ore due to the finer copper minerals staying locked (Bradshaw et al. 2011). Thus, it was also possible at KUCC

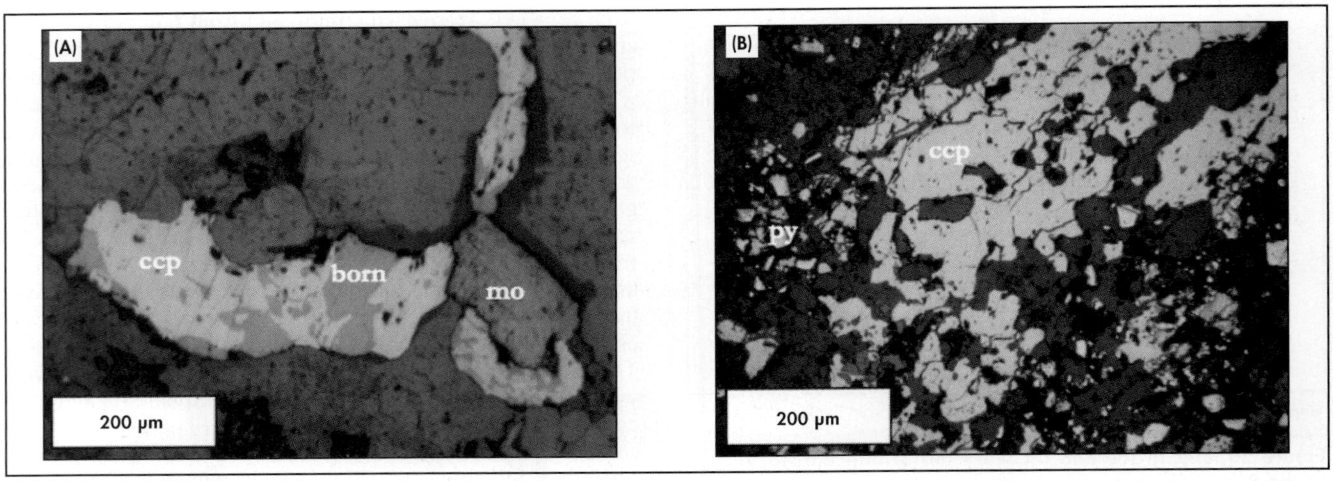

Source: Bradshaw et al. 2016

Figure 21 Reflected-light photomicrographs of (A) MZME3 ore and (B) LSN ore (Ccp = chalcopyrite; born = bornite; mo = molybdenite; py = pyrite)

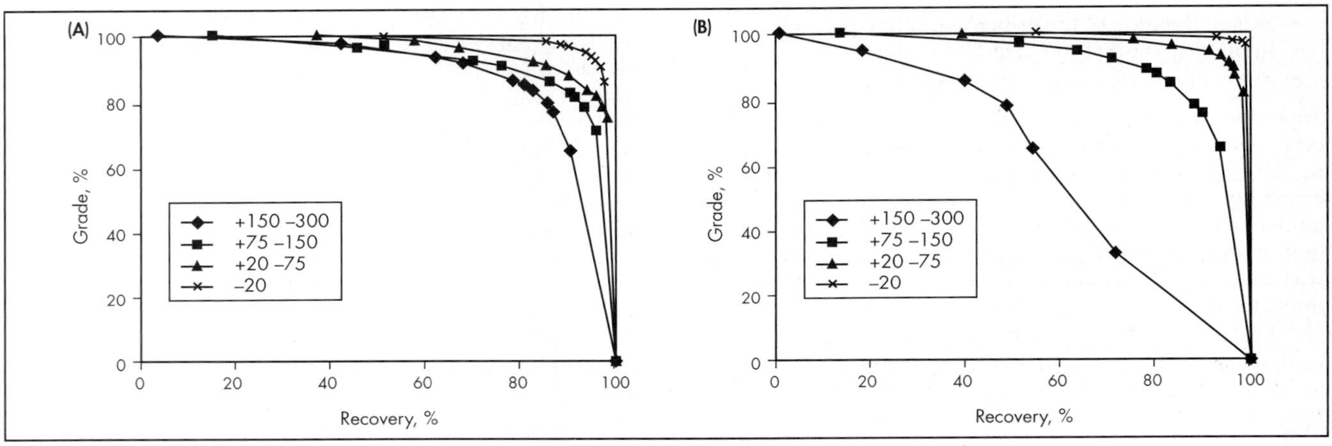

Source: Bradshaw et al. 2016

Figure 22 Theoretical grade–recovery curves based on the mineralogy of (A) the MZME3 ore and (B) the LSN ore

to identify the reasons for the poor performance of one of the LSN ores when it was processed in a blend; it was decided to process this ore type in batches and not to blend it.

Gravity Separation

The most oft-cited argument for installing a gravity circuit to recover gold in a plant is increased metallurgical recovery. The nature of the downstream circuit, as either flotation or cyanidation, greatly affects the impact of gravity recovery on overall metallurgical recovery. When installed inside the circulating load of a grinding circuit, the gravity recovery unit operation bleeds the high circulating load of gold that develops due to its high specific gravity—suffering inverse classification in the hydrocyclone (Laplante et al. 1994). Gold and electrum tend to fold rather than break in the grinding process due to their malleability. In a study on Hemlo ore, it was shown that the selection function of quartz was four times that of gold (Banisi et al. 1991). As a result, when gold particles respond to breakage in a grinding circuit, they tend to only move to the next screen size down in the particle size distribution, whereas quartz tends to break down to much finer size classes. This implies that free gold grains will have a longer residence time in the mill and that liberation measurements may not necessarily imply that this gold is immediately recoverable. Gravity-recoverable gold is limited above a certain lower sieve size below which the gravity recovery units tend to fail; thus, ultrafine liberated gold found by mineralogical measurements should not be added to the total of gravity-recoverable gold.

Leaching

In the case of heap leaching, the phases to be extracted do not require a high degree of liberation, as the aim is to obtain sufficient liberation for the leach liquor to contact the mineral phase of interest. This only requires surface liberation or internal cracks and pathways into the particle to allow liquor to flow into the particle. Hence, heap leach pads typically comminute to a particle size of 12.5–50 mm as compared to 150–200 μm for flotation. The case study described in this section shows a mineralogical and metallurgical comparison of three different ores leached in a column under identical leach conditions. This case study was reported by Latti and

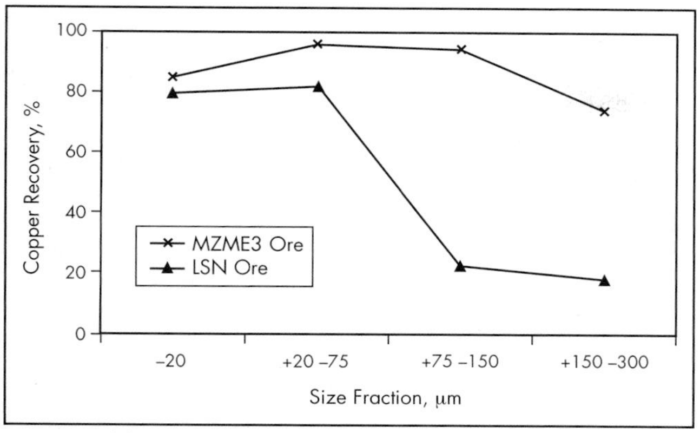

Source: Bradshaw et al. 2016

Figure 23 Copper recovery by size fraction for laboratory flotation tests of the MZME3 and LSN ores

Table 5 Leach conditions for three different ores

Ore	Feed Grade, % Cu	Ore Size, mm	Copper Extraction, % Cu	Acid Consumption, kg/t ore
Baseline	0.22	−12.5	67.7	108
Ore A	0.52	−12.5	29.4	60
Ore B	0.35	−12.5	51.0	232

Source: Latti and Wilkie 2016

Table 6 Mineralogy of the three ore feeds

	Percentage Mineral in Feed		
Mineral	**Baseline Ore**	**Ore A**	**Ore B**
Chalcopyrite	0.66	0.84	0.87
Bornite	0.00	0.20	0.06
Covellite	0.00	0.00	0.00
Chalcocite	0.00	0.16	0.01
Copper-arsenic sulfides	0.00	0.00	0.00
Other copper minerals	0.00	0.02	0.03
Pyrite	2.13	0.16	0.10
Other sulfides	0.13	0.08	0.01
Quartz	82.27	11.88	10.42
Plagioclase	0.03	15.46	33.08
Orthoclase	3.07	16.82	5.45
Altered orthoclase	0.41	27.43	8.92
Muscovite/illite	0.65	1.74	1.30
Biotite/phlogopite	4.21	3.21	12.23
Chlorite	0.86	4.22	14.32
Aluminum silicate	0.67	6.29	0.35
Amphibole	3.77	0.00	1.45
Andradite	0.00	0.00	0.02
Epidote	0.00	0.01	0.09
Calcite/dolomite	0.03	0.17	3.64
Iron oxide	0.43	0.47	2.90
Gypsum	0.00	8.00	0.00
Other	0.66	2.82	4.74

Source: Latti and Wilkie 2016

Wilkie (2016). Table 5 summarizes the leach conditions and extractions obtained from the three different ores. Although all three ores were crushed to the same particle size of 100% passing 12.5 mm, there was a large difference in the extractions obtained from the columns. Surprisingly, ore A had the highest feed grade of 0.52% Cu and the lowest extraction of 29.4%, whereas the baseline ore had the lowest feed grade at 0.22% Cu and the highest extraction of 67.7%. Ore B had a feed grade of 0.35% Cu and an extraction of 51%.

Table 6 compares the mineralogy of the three ore feeds. All three ores are chalcopyrite dominant, but ore A does contain a small amount of bornite and chalcocite. As it is chalcopyrite dominant, it can be concluded that the difference in extractions is not due to the different leaching rates between copper minerals but is rather due to a difference in the texture of the chalcopyrite. There is, however, a large difference in the gangue mineralogy between the two ores. The baseline ore is dominated by quartz with minor orthoclase, biotite/phlogopite and amphiboles, whereas ores A and B have considerably less quartz and significantly more plagioclase, orthoclase, biotite/phlogopite, and aluminum silicates. Ore B has more calcite/dolomite than either the baseline ore or ore A, which is important for acid consumption.

Figure 24 compares the size-by-size copper deportment of the three ore feeds. The baseline ore stands out as being different from both ore A and ore B in that the copper is uniformly distributed across all size fractions, whereas ores A and B have their copper heavily biased in the coarser size fractions. More than 40% of the copper in the baseline ore is in particles that are less than 4 mm. By contrast, only 25% of the copper in ore A is in particles finer than 4 mm, and ore B exhibits 34% in the fractions of less than 4 mm. Hence, although ore A has the highest grade, it also has the lowest proportion of copper in the fine fractions, which corresponds to the lowest extraction.

Conversely, the baseline ore has the lowest grade but the highest proportion of copper in the fine fractions of less than 4 mm, which corresponds to the highest extraction. The copper deportment in ore B is intermediate between the baseline ore and ore A with a corresponding intermediate extraction between the benchmark ore and ore A. It is clear from this result that ores A and B are liberation limited because the higher proportion of copper in the coarser 4-mm particles is unavailable for the leach liquor to contact the chalcopyrite grains. This can only be improved by crushing the ore more finely to send the larger proportion of copper in the coarse 4-mm fractions down to the fractions of less than 4 mm to make it available for leaching.

CONCLUDING REMARKS

This chapter shows that optimum liberation is a fundamental prerequisite to the separation and beneficiation of ores to achieve maximum recoveries across a range of mineral processing technologies. The first section of this chapter answered the question *Why is liberation critical to mineral processing?* The answer is that mineral ores, by their nature, are heterogeneous mixtures of ore minerals and gangue minerals, so to concentrate or extract the valuable mineral or metal, the ore must be broken down to a particle size where the mineral processing technology can efficiently concentrate or extract the valuable mineral or metal from the unwanted gangue minerals.

Because the optimum liberation size of a phase is a function of both the ore and the process technology being used, it

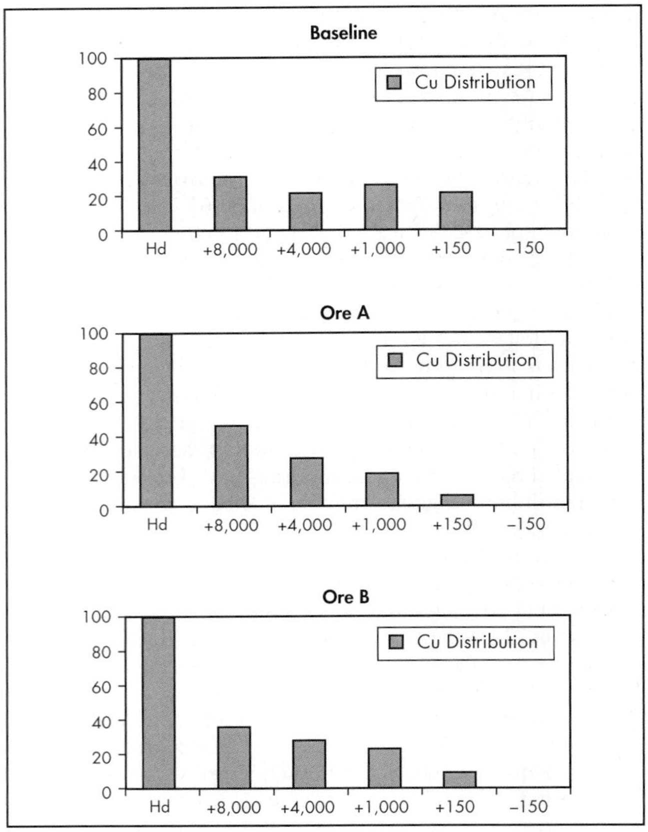

Source: Latti and Wilkie 2016

Figure 24 Size-by-size copper deportment for the three ore feeds

is important to measure the degree of liberation that the ore can produce at the optimum particle size for recovery. The second section of this chapter answered the question *How is liberation measured?* and discussed the two main methods of measurement: 2-D measurements performed on polished sections using manual and automated technologies, and 3-D measurements using either physical methods to produce liberation classes from heavy liquid separation or 3-D visualization of particles using tomography instruments. Each method was explored with respect to its strengths and weaknesses as well as the use of liberation modeling to determine the liberation properties of an ore at a given particle size.

The third section of this chapter answered the question *What is the best way to report liberation?* It provided a description of the traditional analysis using optical microscopy to classify particles in terms of proportions of liberated particles, binary particles with other minerals, and ternaries that are locked with two or more minerals. In addition to this classical approach to reporting liberation, commercially available automated SEM techniques provide additional ways of reporting liberation in the form of theoretical grade–recovery curves, liberation and association, false-color particle images, liberation spectrum, and CLYs. Examples of these forms were also presented.

The fourth section answered the question *When is there enough liberation?* Most mineral processing technologies have an optimum particle size for the efficient recovery of liberated and near-liberated particles, so it is important that the feed stream be prepared to maximize liberation within these particle size constraints. Not enough liberation will result in loss of recovery and grade due to poor exposure of the mineral of interest to the reagents being used to recover the mineral. Too much liberation may result in the liberated particles being too small for the mineral processing technology to efficiently recover fine particles. This section discussed the need for optimum liberation using a range of mineral processing technologies, including flotation, leaching, and gravity separation.

Finally, the fifth section of this chapter provided three case studies of diagnoses of liberation problems and the process improvements required to deliver a more optimum liberation solution.

REFERENCES

Abols, J.A., and Grady, P.M. 2006. Maximizing gravity recovery through the application of multiple gravity devices. Presented at the Gravity Concentration ’06 Conference, Perth, Australia, March 13–14.

Banisi, S., Laplante, A.R., and Marois, J. 1991. The behaviour of gold in the Hemlo Mines Ltd. grinding circuit. Presented at the 23rd Annual Meeting of the Canadian Mineral Processors, Ottawa, ON, Canada, January 22–24.

Barbery, G. 1984. Mineral liberation analysis using stereological methods: A review of concepts and problems. In *Applied Mineralogy*. Edited by W.C. Park, D.M. Hausen, and R.D. Hagni. New York: AIME. pp. 171–190.

Barbery, G. 1987. Random sets and integral geometry in comminution and liberation of minerals. *Miner. Metall. Process*. 4:96–102.

Barbery, G. 1991. *Mineral Liberation: Measurement, Prediction and Use in Mineral Processing*. Quebec City, QC: Les Editions GB.

Becker, M., Wightman, E.M., and Evans, C.L. (eds.). 2016. Process mineralogy. In *JKMRC Monograph Series in Mining and Mineral Processing No 6*. Brisbane, Queensland: Julius Kruttschnitt Mineral Research Centre.

Bradshaw, D.J. 2014. The role of “process mineralogy” in improving the process performance of complex sulphide ores. Presented at the XXVII International Mineral Processing Congress (IMPC 2014), Santiago, Chile, October 20–24.

Bradshaw, D., Triffett, B., and Kashuba, D. 2011. The role of process mineralogy in identifying the cause of the low recovery of chalcopyrite at KUCC. In *Proceedings of the 10th International Congress for Applied Mineralogy (ICAM)*, Trondheim, Norway, August 1–5. Berlin, Germany: Springer. pp. 71–80.

Bradshaw, D., Triffett, B., Latti, D., Wilkie, G., and Adair, B. 2016. Dealing with a problematic ore type at Kennecott Utah Copper Concentrator. In *JKMRC Monograph Series in Mining and Mineral Processing No. 6*. Edited by M. Becker, E. Wightman, and C. Evans. Brisbane, Queensland: Julius Kruttschnitt Mineral Research Centre. pp. 261–272.

Chayes, F. 1944. Petrographic analysis by fragment counting; Part 1, The counting error. *Econ. Geol.* 39(7):484–450.

Chayes, F. 1945. Petrographic analysis by fragment counting; Part 2, Precision of micro-sampling and the combination error of sampling and counting. *Econ. Geol.* 40(8):517–525.

Cormier, J.A., and Cooper, M. 2002. Implementation of stage grinding at the Brunswick concentrator. Presented at the 34th Annual Meeting of the Canadian Mineral Processors, Ottawa, ON, Canada, January 22–24.

Cropp, A., Goodall, W., and Bradshaw, D. 2013. The influence of textural variation and gangue mineralogy on recovery of copper by flotation from porphyry ore—A review. Presented at the Second AusIMM International Geometallurgy Conference (GeoMet 2013), Brisbane, Queensland, Australia, September 30–October 2.

Davy, P. 1984. Probability models for liberation. *J. Appl. Probab.* 21(2):260–269.

Efron, B. 1987. Better bootstrap confidence intervals. *J. Am. Stat. Assoc.* 82:171–185.

Evans, C.L. 2002. *Mineral Liberation Research at the JKMRC—Past, Present and Future.* P9 project report. Brisbane, Queensland: Julius Kruttschnitt Mineral Research Centre.

Evans, C.L., and Morrison, R.D. 2016. Mineral liberation. In *JKMRC Monograph Series in Mining and Mineral Processing No. 6.* Edited by M. Becker, E. Wightman, and C. Evans. Brisbane, Queensland: Julius Kruttschnitt Mineral Research Centre. pp. 133–147.

Evans, C.L., and Napier-Munn, T.J. 2013. Estimating error in measurements of mineral grain size distribution. *Miner. Eng.* 52:198–203.

Evans, C.L., Wightman, E.M., and Yuan, X.M. 2012. Characterising ore micro-texture using X-ray microtomography. Presented at the 44th Annual Meeting of the Canadian Mineral Processors, January 17–19.

Evans, C.L., Andrusiewicz, M.A., Wightman, E.M., Brennan, M., Morrison, R.D., and Manlapig, E.V. 2013. Simulating concentrators from feed to final products using a multi-component methodology. SME Preprint No. 13-094. Englewood, CO: SME.

Gaudin, A.M. 1939. *Principles of Mineral Dressing.* New York: McGraw-Hill.

Gay, S.L. 1996. Liberation modelling using particle sections. Ph.D. dissertation, University of Queensland, Brisbane, Queensland, Australia.

Gay, S.L. 2004. Simple texture-based liberation modelling of ores. *Miner. Eng.* 17(11–12):1209–1216.

Ghorbani, Y., Becker, M., Petersen, J., Mainza, A.N., and Franzidis, J.-P. 2013. Investigation of the effect of mineralogy as rate-limiting factors in large particle leaching. *Miner. Eng.* 52:38–51.

Gu, Y., Schouwstra, R.P., and Rule, C. 2014. The value of automated mineralogy. *Miner. Eng.* 58:100–103.

Gy, P.M. 1979. *Sampling of Particulate Materials—Theory and Practice.* Amsterdam: Elsevier.

Henley, K.J. 1983. Ore dressing mineralogy—A review of techniques, applications and recent developments. Presented at the First International Congress on Applied Mineralogy (ICAM '81), Johannesburg, South Africa.

Jackson, B.R., Gottlieb, P., and Sutherland, D.N. 1988. A method for measuring and comparing the mineral sizes of ores from different origins. Presented at the Third Mill Operators' Conference, Cobar, New South Wales, Australia.

Johnson, N.W. 2010. Existing methods for process analysis. In *Flotation Plant Optimisation: A Metallurgical Guide to Identifying and Solving Problems in Flotation Plants.* Edited by C.J. Greet. Melbourne, Victoria: Australasian Institute of Mining and Metallurgy. pp. 35–64.

Jowett, A. 1980. Formation and disruption of particle–bubble aggregates in flotation. In *Fine Particles Processing, Vol. 1.* Edited by P. Somasundaran. New York: AIME. pp. 720–754.

King, R.P. 1982. The prediction of mineral liberation from mineral texture. Presented at the XIV International Mineral Processing Congress, Toronto, ON, Canada.

King, R.P. 1983. Stereological methods for the prediction and measurement of mineral liberation. Presented at the First International Congress on Applied Mineralogy (ICAM '81), Johannesburg, South Africa.

King, R.P. 1990. Calculation of the liberation spectrum in products produced in continuous milling circuits. Presented at the 7th European Symposium on Comminution, Ljubljana, Slovenia.

King R.P. 2000. *Technical Notes 10: Mineral Liberation.* R.P. King. http://mineraltech.com/MODSIM/ModsimTraining/Module7/TechnicalNotes-10-Liberation.pdf.

Lamberg, P., and Vianna, S. 2007. A technique for tracking multiphase mineral particles in flotation circuits. In *Proceedings of the 7th Meeting of the Southern Hemisphere on the Mineral Technology, Ouro Preto, Brazil.* pp. 195–241.

Laplante, A.R., and Staunton, W.P. 2004. Gravity recovery of gold—An overview of recent developments. Presented at the International Symposium on the Treatment of Gold, Calgary, AB, Canada.

Laplante, A.R., Liu, L., and Cauchon, A. 1990. Gold gravity recovery at the mill of Les Mines Camchib Inc., Chibougamou, Québec. Presented at the 22nd annual meeting of the Canadian Mineral Processors, Ottawa, ON, Canada, January 16–18.

Laplante, A.R., Huang, L., and Vincent, F. 1994. Practical considerations in the operation of gold gravity circuits. Presented at the 26th annual meeting of the Canadian Mineral Processors, Ottawa, ON, Canada, January 18–20.

Laplante, A.R., Vincent, F., and Luinstra, W.F. 1996. A laboratory procedure to determine the amount of gravity recoverable gold—A case study at Hemlo Gold Mines. Presented at the 28th Annual Operator's Conference of the Canadian Mineral Processors, Ottawa, ON, Canada, January 23–25.

Latti, A.D. 2006. The textural effects of multiphase mineral systems in liberation measurement. Ph.D. dissertation, University of Queensland, Brisbane, Queensland, Australia.

Latti, A.D., and Wilkie, G. 2016. Mineralogy and leaching of copper ores. In *JKMRC Monograph Series in Mining and Mineral Processing No. 6.* Edited by M. Becker, E. Wightman, and C. Evans. Brisbane, Queensland: Julius Kruttschnitt Mineral Research Centre. pp. 291–302.

Leigh, G.M., Sutherland, D.N., and Gottlieb, P. 1993. Confidence limits for liberation measurements. *Miner. Eng.* 6(2):155–161.

Leigh, G.M., Lyman, G.J., and Gottlieb, P. 1996. Stereological estimates of liberation from mineral section measurements: A rederivation of Barbery's formulae with extensions. *Powder Technol.* 87:141–152.

Lotter, N.O., Whittaker, P.J., Kormos, L.J., Stickling, J.S., and Wilkie, G.J. 2002. The development of process mineralogy at Falconbridge Limited and application to the Raglan Mill. *CIM Bull.* 95:85–92.

Lotter, N.O., Baum, W., Reeves, S., Arrué, C., and Bradshaw, D.J. 2018a. The business value of best practice process mineralogy. *Miner. Eng.* 116:226–238.

Lotter, N.O., Evans, C.L., and Engström, K. 2018b. Sampling—A key tool in modern process mineralogy. *Miner. Eng.* 116:196–202.

Mariano, R.A., and Evans, C.L. 2015. Error analysis in ore particle composition distribution measurements. *Miner. Eng.* 82:36–44.

McIvor, R.E., and Finch, J.A. 1991. A guide to the interfacing of plant grinding and flotation operations. *Miner. Eng.* 4(1):9–23.

Miller, J.D., and Lin, C.L. 1988. The treatment of polished section data for detailed liberation analysis. *Int. J. Miner. Process.* 22:41–58.

Miller, J.D., Lin, C.L., and Cortes, A.B. 1990. A review of X-ray computed tomography and its application in mineral processing. *Miner. Process. Extr. Metall. Rev.* 7:1–18.

Miller, J.D., Lin, C.L., Hupka, L., and Al-Wakeel, M.I. 2009. Liberation-limited grade/recovery curves from X-ray micro CT analysis of feed material for the evaluation of separation efficiency. *Int. J. Miner. Process.* 93:48–51.

Morrison, R.D., and Gu, Y. 2016. X-ray computed microtomography. In *JKMRC Monograph Series in Mining and Mineral Processing No. 6*. Edited by M. Becker, E. Wightman, and C. Evans. Brisbane, Queensland: Julius Kruttschnitt Mineral Research Centre.

Napier Munn, T.J., Morrell, S., Morrison R.D., and Kojovic, T. 2005. *Mineral Comminution Circuits: Their Operation and Optimisation*. Indooroopilly, Queensland: Julius Kruttschnitt Mineral Research Centre.

Orford, I., Cooper, M., Larsen, C., Deredin, C., Gauthier, J.-G., Fortin, C., Côté, L., Trusiak, A., and Scott, D. 2005. The approach to process improvements at Brunswick Mine. Presented at the 37th annual meeting of the Canadian Mineral Processors, Ottawa, ON, Canada, January 18–20.

Petersen, J. 2016. Heap leaching as a key technology for recovery of values from low-grade ores—A brief overview. *Hydrometallurgy* 165:206–212.

Petruk, W. 2000. *Applied Mineralogy in the Mining Industry*. Amsterdam: Elsevier.

Petruk, W., and Schnarr, J.R. 1981. An evaluation of the recovery of free and unliberated mineral grains, metals and trace elements in the concentrator of Brunswick Mining and Smelting Corp. Ltd. *CIM Bull.* 74(833)132–159.

Spencer, S., and Sutherland, D. 2000. Stereological correction of mineral grade distributions estimated by single section of particles. *Image Anal. Stereol.* 19:175–182.

Stewart, P.S.B., and Jones, M.P. 1977. Determining the amounts and the compositions of composite (middling) particles. Presented at the Twelfth International Mineral Processing Congress, São Paulo, Brazil.

Subasinghe, G.K.N., and Dunne, R. 2016. Predictive model of mineral liberation for geometallurgical applications. Presented at the Third AusIMM International Geometallurgy Conference, Perth, Australia, June 15–16.

Sutherland, D.N. 1989. Batch flotation behaviour of composite particles. *Miner. Eng.* 2(3):351–367.

Sutherland, D., Gottlieb, P., Jackson, R., Wilkie, G., and Stewart, P. 1988. Measurement in section of particles of known composition. *Miner. Eng.* 1(4):317–326.

Sutherland, D.N., Gottlieb, P., Wilkie, G., and Johnson, C.R. 1991. Assessment of ore processing characteristics using automated mineralogy. Presented at the XVII Mineral Processing Congress, Dresden, Germany.

Trahar, W.J. 1981. A rational interpretation of the role of particle size in flotation. *Int. J. Miner. Process.* 8:289–327.

Triffett, B., Veloo, C., Adair, B.J.I., and Bradshaw, D.J. 2008. An investigation of the factors affecting the recovery of molybdenite in the Kennecott Utah copper bulk flotation circuit. *Miner. Eng.* 21:832–840.

Van der Plas, L., and Tobi, A.C. 1965. A chart for judging the reliability of point counting results. *Am. J. Sci.* 263:87–90.

Vaughan, J.P. 2004. The process mineralogy of gold: The classification of ore types. *J. Miner. Met. Mater. Soc.* 56:46–48.

Vos, C.F. 2017. The effect of mineral grain textures at particle surfaces on flotation response. Ph.D. thesis, University of Queensland, Brisbane, Queensland, Australia.

Wiegel, R.L., and Li, K. 1967. A random model for mineral liberation by size reduction. *Trans. SME* 238:179–189.

Wightman, E.M., and Evans, C.L. 2013. Representing and interpreting the liberation spectrum in a processing context. *Miner. Eng.* 61:121–125.

Wightman, E.M., Evans, C.L., Becker, M., and Gu, Y. 2016. Automated scanning electron microscopy with energy dispersive spectrometry. In *JKMRC Monograph Series in Mining and Mineral Processing No. 6*. Edited by M. Becker, E. Wightman, and C. Evans. Brisbane, Queensland: Julius Kruttschnitt Mineral Research Centre. pp. 97–107.

Wilkie, G.J. 2016. Assessing the sorting potential of mineral sulphide ores. Ph.D. dissertation, University of Queensland, Brisbane, Queensland, Australia.

Zhou, J., Jago, B., and Martin, C. 2004. *Establishing the Process Mineralogy of Gold Ores*. SGS Technical Bulletin 2004-03. Geneva, Switzerland: SGS Lakefield Research Limited.

CHAPTER 1.5

Surface Chemical Control in Flotation and Leaching

Roger St.C. Smart, Andrea R. Gerson, and Brian R. Hart

This chapter provides a guide to plant metallurgists seeking to improve recovery and/or selectivity in froth flotation and in recovery from leaching where this is limited by mineralogy and surface chemistry of mineral phases. It aims to provide a strategy for recognition of these limitations, to provide advice on strategic analysis of the limiting mineralogy and/or surface chemistry, and to offer potential changes to operation suggested by this analysis. Some of the information used here is derived from previous compilations of research from many authors and mineral processing laboratories that have provided the basis for the development of this strategy and operational control. In particular, "Innovations in Measurement of Mineral Structure and Surface Chemistry in Flotation: Past, Present and Future" (Smart et al. 2014a) has provided much of the basic information on the analysis techniques. For more detailed references, other compilations (Smart et al. 2003a, 2003b, 2007, 2014b; Hart and Dimov 2011; Chandra and Gerson 2006) contain all the references to the development of the strategy described in this chapter. The focus here is on using these techniques with examples from plant analyses to suggest operational improvement.

WHEN TO USE MINERALOGY AND SURFACE CHEMICAL ANALYSIS: DIAGNOSIS OF LIMITING SPECIES

Froth Flotation

Losses of value in froth flotation have multiple possible causes, as the multivariable froth flotation models illustrate (e.g., JKTech 2011; Ralston et al. 2007). In this context, it is useful to identify some of these complexities that will be relevant to later discussion, including

- The extent of liberation of individual mineral phases by grinding or, conversely, the extent of remaining composite particles;
- The presence of a wide range of particle sizes in the ground sulfide ores;
- Chemical alteration of the surface layers of the sulfide minerals induced by oxidation reactions in the pulp solution;
- Galvanic interactions between different sulfide minerals to produce different reaction products on the mineral surfaces;
- Reaction and dissolution of gangue (e.g., oxide, silicate) minerals releasing ionic species interfering with surface chemical control;
- Interaction between particles in the form of aggregates and flocs;
- The presence of colloidal precipitates arising from dissolution of the sulfide minerals, gangue minerals, and grinding media;
- The mechanism of adsorption of reagents onto specific surface sites; and
- Competitive adsorption between oxidation products, conditioning reagents, and collector reagents.

Liberation and liberation analysis, discussed in detail in Chapter 1.4, "Ore Liberation Analysis," is normally the first consideration for improving flotation performance (Johnson and Munro 2002) by optimizing the size profile by mineral. The techniques of Quantitative Evaluation of Minerals by Scanning Electron Microscopy (QEMSCAN), mineral liberation analysis (MLA), and X-ray diffraction (XRD) are well established in routine use to define both valuable and gangue minerals in an ore to be processed. Some examples can be found in Lastra (2007). Three levels of analysis should be considered: mineral recovery–size relationships for all valuable minerals and any naturally floatable, hydrophobic gangue minerals (e.g., talc); mineral recovery–liberation–size relationships after grinding (liberation analysis); and surface analysis of relevant minerals in the defined size fractions after grinding. The operational basis of the first two levels of analysis are the subject of other chapters in this handbook. The third level is the subject of this chapter concerning analysis of these minerals after grinding and in subsequent flotation stages.

In operational terms, mineralogical interferences and surface chemical control tend to be considered after most other operating variables have been adjusted with fixed reagent doses and further improvement is still required or when a

Roger St.C. Smart, Emeritus Professor & Senior Consultant, Univ. of South Australia & Blue Minerals Consultancy, Mawson Lakes, South Australia, Australia
Andrea R. Gerson, Professor & Managing Director, Blue Minerals Consultancy, Middleton, South Australia, Australia
Brian R. Hart, Adjunct Professor & Senior Research Scientist, Univ. of Western Ontario & Surface Science Western, London, Ontario, Canada

deleterious change has occurred without explanation. After liberation, pulp density, bubble size, gas flow and holdup, agitation, froth control, and other cell parameters have been adjusted with no further improvement, it is the attachment of collector-conditioned mineral particles to bubbles and the stability of this bubble–particle attachment, in both pulp and froth phases, that determines recovery and grade. If this is still unacceptable, the reasons for losses likely lie in the mineralogy and surface chemistry of the conditioned feed. In froth flotation, it is the chemistry of the top few monolayers of different mineral surfaces that determine the recovery and grade in operation. The chemistry of these surface monolayers is determined by reaction of the mineral phases in pulp (e.g., oxidation of sulfides, selective leaching of different minerals) by interferences arising from fine particle or precipitate attachment and by addition of collectors and other reagents, including activators and depressants. The collector is designed to adsorb to the valuable mineral particle surfaces, making them selectively hydrophobic, so that these surfaces prefer to attach to bubbles (displacing water to expose gas–solid interfaces) rather than to water molecules. The surface of all other minerals are assumed or designed (by depressant addition) to be hydrophilic. In reality, however, the surface of each individual mineral particle in the flotation pulp is a complex, distinctly nonuniform array of precipitates (e.g., calcium sulfate or $CaSO_4$, amorphous aluminosilicates), oxidation products from sulfides (i.e., oxyhydroxides, oxy-sulfur species), adsorbed ions, and attached fine and ultrafine particles of other mineral phases, including clays—all hydrophilic surface species—with hydrophobic collector and metal-collector complexes. In feed mineralogy, in addition to particle size distribution and liberation of value minerals normally measured, interference from rapidly soluble minerals, clays, and amorphous phases can seriously compromise recovery and grade. Bubble attachment is therefore largely dependent on the ratio of hydrophobic to hydrophilic surface species on individual mineral surfaces. If this ratio is too low for the value mineral particles, recovery will be low and flotation kinetics too slow. This ratio varies widely between different particles of the same mineral, requiring statistical analysis, and some nonvalue mineral phases will also have adsorbed hydrophobic collector in some form, contributing to gangue recovery and lower grades.

This hydrophobic/hydrophilic ratio may seem an esoteric measure to plant operators, but it has been shown to determine whether particles of both value and gangue report, correctly or incorrectly, to concentrate and tail. To understand and improve poor flotation recovery, it is first necessary to know whether this is being caused by unrecognized interferences in the mineralogy of the feed or poor hydrophobic/hydrophilic conditioning of the value mineral surfaces or inadvertent hydrophobic conditioning of gangue minerals. It is then possible to define the reasons for this before changes to mechanical or chemical conditioning can be usefully made.

A complete strategy has been developed for identification of the reasons for losses in recovery and/or grade due to changes or complications in mineralogy and in mineral surface conditioning, and their subsequent correction. The principle of this strategy involves comparison of surface chemistry of selected mineral phases (value and misreporting gangue) between flotation feed, concentrate, and tail. Ideally, comparison should be made between these samples from the plant in normal operation (baseline) and the same set of samples in poor operation for root-cause analysis. This is not usually possible because baseline surface analysis is not often done. Instead, it is used more in troubleshooting mode when an unexplained loss of recovery or grade has occurred. In this mode, it is first essential to fully define the mineralogy to identify interference from rapidly soluble minerals, clays, and amorphous phases. To understand and improve poor flotation recovery where it is due to surface chemistry, it is the variation of specific hydrophobic and hydrophilic species by particle and as a statistical distribution between different mineral phases across a flotation circuit (e.g., feed, successive concentrates, tails) that one needs to know. This requires surface analysis of a large number of particles with high spatial resolution and chemical speciation. This is a difficult proposition, but surface analysis has come a long way toward this goal. A sequence of measurements using this strategy, with examples of diagnosis and outcomes, is outlined in this chapter.

Leaching Operations

This strategy is equally applicable to recovery of metals by hydrometallurgical leaching, where leach kinetics and selectivity can be cost limiting. Rate-determining surface species and surface layers forming during leaching may limit access of reactants to the surface and of reaction products to the solution. They may also indicate the mechanism of the reaction pathway for optimization of operation. The limiting actions of complexants, oxidation products, fine gangue interference, precipitates as colloidal products, and thick surface layers (amorphous and crystalline) need to be defined before the subprocesses requiring change can be usefully addressed. In leaching operations, the analytical principle involves comparison of surface chemistry of selected minerals taken during the leach progress and, where possible, comparison of good and poor leach operation.

METHODS OF MINERALOGICAL AND SURFACE ANALYSIS

This section outlines the analytical techniques used to diagnose limiting mineralogical and surface chemical factors and the approach the chapter authors used in their application to plant operation.

Sampling from Operating Slurries

Ideally, samples are collected directly from the operating plant using rigorous sampling protocols to ensure that the surface chemistry is as close as possible to that in the flotation or leach circuit at the time of sampling. They may also be collected from lab floats or leach trials using the same protocol where these are being used to test improved processing. Comparison of the feeds, concentrates, and tails or successive leach samples, where possible, between conditions that give varied recovery and grades is a more direct way to recognize the mineralogical or surface chemical factors differentiating in the flotation selectivity or kinetic leach profiles. This requires that the sample is representative of the stream, does not oxidize or react in the sample tube (i.e., is frozen) after sampling, and that these conditions are maintained in delivery to the lab and analytical instruments. In the flotation plant, the feed is normally sampled after reagent addition, the concentrates are sampled by cutting the lip of the cell or in the launder, and the tails are sampled from the final cell using a splitter box in each case. Protocols for these requirements have been developed and verified by correlation with flotation response over more than two decades.

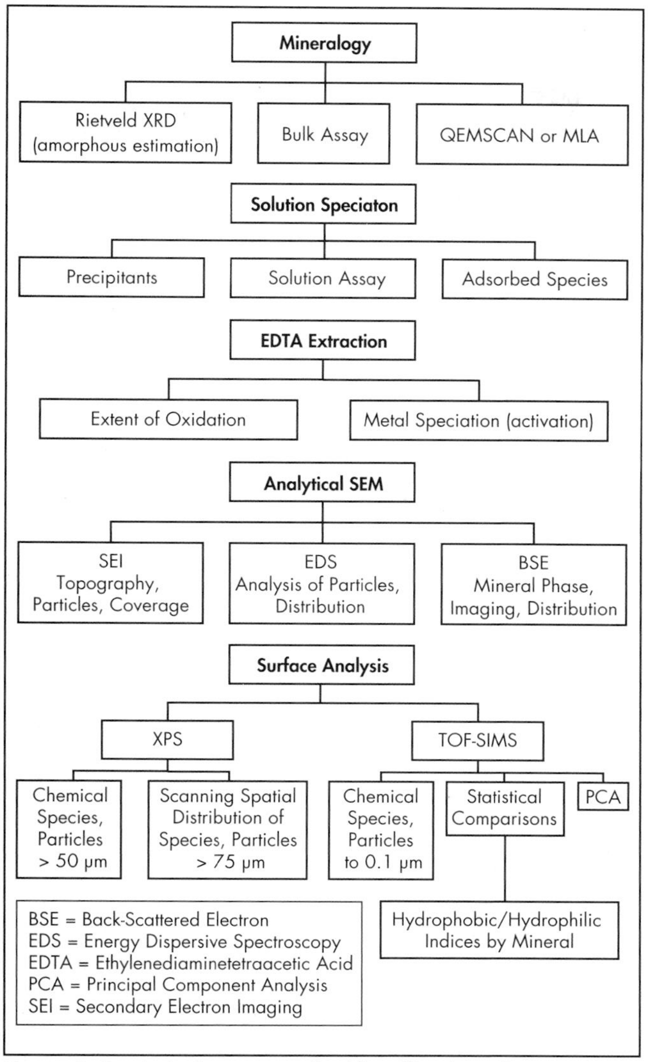

Figure 1 Analytical sequence for identification of mineralogical and surface chemical factors that may be limiting recovery and grade in flotation and leaching operations

The essential steps for sampling are described in Smart et al. (2007). In all cases, representative subsamples should be collected in a specimen tube (10–20 mL polypropylene vials), deoxygenated by bubbling argon (e.g., Ar/H_2 welding gas) or oxygen-free N_2 through the tube for 5 minutes, then closed with polytetrafluoroethylene (plumbers') tape and tightly screw capped. The tube should then be immediately frozen in liquid N_2 (or in a freezer if N_2 is not available on-site). The vial should not be completely filled to the top because it will split when frozen. Each vial should be clearly marked with a permanent marker. The pH and Eh of each stream (feed, concentrate, and tails) must be recorded at the time of sampling and supplied with the frozen samples.

It is essential for the sample to remain frozen until analysis. The sample can be delivered in a cryogenic container or, if delivery will be less than 48 hours, in a polystyrene six-pack cooler box. A cryogenic container licensed for international transport of liquid-nitrogen-frozen specimens is usually supplied by the lab conducting the surface analysis. It can maintain a temperature of less than −150°C for up to 21 days. The cooler box, if used, can be packed with freezer packs or, even better, Carbice (solid carbon dioxide, or CO_2) around the sample tubes and sealed with tape. The frozen sample vials should be transferred to the freight office immediately after freezing for fastest possible transfer to the lab. The key to success with this analysis and diagnosis is engagement from the outset with a professional surface analytical lab working with minerals processing companies. They will provide advice and support around sampling, containers, shipping, interpreting, and acting on results.

At the lab, the sample is thawed and, if necessary, the supernatant solution replaced by decantation with MilliQ water adjusted to the initial plant pH using either sodium hydroxide (NaOH) or hydrochloric acid (HCl). Adjustment of Eh is not normally required after O_2 removal and freezing but may be checked. This procedure is used to avoid free fines and solution species (not adsorbed) drying down onto the mineral surfaces in the vacuum of the microscope or spectrometer. It is necessary only if the ionic concentration of the solution exceeds 10^{-3} M. Samples can then be analyzed by XRD, scanning electron microscopy (SEM), X-ray photoelectron spectroscopy (XPS), or time-of-flight secondary ion mass spectroscopy (TOF-SIMS) as explained in the next section.

Sampling from Laboratory Tests

The same principles apply to sampling from lab flotation where changes of reagents (collectors and collector combinations, activators, depressants, flocculants) or conditioning (e.g., agitation, aeration) are being tested before possible changes to plant operation. The samples of feed (successive) concentrates, and tails should be deoxygenated, frozen, and delivered to the surface analysis lab in this form. This is normally done only for samples from tests that have shown significant improvement in flotation recovery or grade. The surface and other analysis can then define the parameters that have changed in the improved flotation for information to the plant operation.

ANALYTICAL SEQUENCE

The most direct and cost-effective sequence of analysis to determine the factors likely limiting recovery and/or grade is that suggested in Figure 1. The techniques (and acronyms) in the sequence are explained in more detail in this chapter, but some initial comments may be helpful. More-specialized mineralogical analysis used as described here may directly indicate that there are factors unrecognized in standard mine mineralogy and liberation studies (e.g., amorphous minerals, residual thin surface layers) or that there has been a change in the proportion of fine and ultrafine particles, including clays and silicates, likely to interfere with value mineral surfaces.

Solution assays of the pulp after conditioning are often available in routine site analysis (but they need to be complete for all species, as discussed below). They can be used in freely available software to model potential precipitates and solution species (at thermodynamic equilibrium) that may adsorb on value mineral surfaces, particularly where this solution speciation has changed because of run-of-mine (ROM) or processing changes. This information should be used with some caution because the time from comminution through flotation is not as long as required for steady state of some of the predicted species, and some of the modeled reactions may have metastable species. It is valuable, however, in correlation with the surface species (precipitates and adsorbed species) interfering with flotation identified from surface analytical techniques. It may indicate the origin of these interferences in process water and pulp solutions.

The relatively simple ethylenediaminetetraacetic acid (EDTA) chemical extraction can be done in most site labs and can also provide insight to changes in ore oxidation (e.g., oxidized zone, longer-term stockpiling) and the presence in solution of metal ion species, particularly Cu, Pb, and Ag, which are likely to induce inadvertent activation of gangue minerals after collector addition.

The use of SEM in different imaging (i.e., secondary electron, backscattered electron) modes with energy dispersive spectroscopy (EDS) analysis is well known, readily accessible, and low cost. In many cases, it can provide direct observation of surface interference by revealing attachment of hydrophilic particles not available from standard QEMSCAN or MLA imaging.

Finally, the more specialized surface analytical techniques of XPS and TOF-SIMS are now used to specifically define the chemical species (including reaction products, adsorbed species, activators, collectors), their spatial and statistical distributions on the same mineral between feed, concentrate, and tail, or between successive leach samples.

A brief introduction follows for each step in the sequence.

Mineralogy

Mineralogy of the flotation feed is normally assessed as part of liberation studies and whenever the ROM changes significantly. The excellent primary tools for mineral structure and liberation analysis, QEMSCAN and MLA, are well known and widely used (Smart et al. 2007). They are essentially based on compositional identification of minerals. They do not, however, separate crystalline from amorphous phases, identify different mineral structures with the same composition (e.g., sphalerite, wurtzite) or different elemental substitution, and are practically limited to >5 μm particles. The problems with mineralogy in low grade and recovery can be in unrecognized fractions of ultrafine (<200 nm) clays and clay aggregates (<2 μm; e.g., kaolin, smectites, illites) and in amorphous minerals (e.g., silica, talc- and chlorite-like fines) that surface-attach more readily than crystalline forms. Hydrophilic clays and amorphous minerals can be attached to value mineral surfaces with direct influence on the hydrophobic/hydrophilic ratio adversely affecting bubble attachment and flotation. High-clay ores are well known to be problematic, often requiring greater collector dosages. The effects of clay aggregates and mineral surface forms in both the pulp and froth phases affecting rheology, viscosity, and flotation, particularly entrainment, are now well known.

Less well recognized is the presence of relatively high concentrations (5–40 wt %) of amorphous minerals, usually hydrated, in oxidized and high-clay ores (e.g., Smart et al. 2014b). These amorphous mineral particles are normally fine (<10 μm) and may attach to hydrated areas of partially reacted surfaces increasing the hydrophilic load. They are not recognized in XRD mineralogical analysis, and they may be interpreted in QEMSCAN or MLA compositional analysis as crystalline minerals although they will behave very differently in processing (e.g., quartz versus hydrated silica).

Amorphous identification and content can be estimated using a combination of quantitative XRD, bulk assay, and QEMSCAN or MLA. Table 1 from a high-clay ore (Smart et al. 2014b) shows an example where the amorphous material is found to be mainly hydrated silica with additional implications for slime coatings. Amorphous calcium sulfate and chlorite-like phases are also common in other examples.

Table 1 Comparison of mineralogy of a high-clay ore feed between Rietveld XRD and QEMSCAN analyses

Minerals	Rietveld XRD, wt %	QEMSCAN, wt %	QEMSCAN Minus Rietveld XRD, %
Quartz	42	57.7	15.7
Kaolinite	18	11.5	–6.5
Smectite	—	4.5	4.5
Albite	20	15.9	–4.1
K-feldspars	—	0.3	0.3
Muscovite	9	5.4	–3.6
Other silicates	—	0.3	0.3
Carbonates	—	0.3	0.3
Ti(Fe) oxides	—	1.1	1.1
Fe(Mn) oxides	—	0.4	0.4
Pyrite	1	0.4	–0.6
Cu(Fe) sulfides	—	0.3	0.3
Ca sulfates	—	0.0	0.0
Phosphates	—	0.0	0.0
Others	—	1.8	—
Amorphous	11	—	–11.0

Adapted from Smart et al. 2014b

The principle is to use quantitative XRD estimation (e.g., Rietveld analysis using TOPAS software (Coelho Software 2016) with a known (e.g., 15 wt %) added corundum (crystalline Al_2O_3 in the same size fraction). This refinement of other crystalline minerals in Bruker-AXS's TOPAS is then calculated relative to the known and defined 15-wt % corundum addition. Whereas the total of all these minerals and corundum is less than 100 wt %, the additional fraction defines the total amorphous content (not detected in the XRD) in the ore. Finally, these fractions are recalculated to 100 wt %, eliminating the corundum to estimate the wt % amorphous content of the ore (i.e., 11 wt % in Table 1). Other more-detailed methods for determining amorphous content from XRD are reviewed in Madsen et al. (2011). This set of crystalline minerals and amorphous material can then be compared with the compositional analysis of the sample by QEMSCAN analysis or MLA to identify the composition of the XRD amorphous fraction. It is important to emphasize that these two techniquues identify minerals only by their composition although the results are often, or usually, taken to imply crystalline structure. In Table 1, the majority of the amorphous content is in the "quartz" fraction of the QEMSCAN analysis and is likely hydrated silica or silicate content. The QEMSCAN analysis also has lower clay content in kaolinite and muscovite than XRD but assigns some smectite content that would normally be clearly seen in XRD. All of these differences have implications for processing of the ore, particularly high-clay ores (e.g., Cruz et al. 2013, 2015a, 2015b; Zhang and Peng 2015; Farrokhpay et al. 2016; Xu et al. 2014).

Mineralogy can now also be probed for single particles using micro-diffraction with a routine spatial resolution of 10–20 μm when using a laboratory X-ray source and 1–2 μm when using a synchrotron radiation X-ray source.

Micro-diffraction enables specific identification of trace phases that would otherwise not be identifiable in bulk diffraction patterns of complex mineral assemblages. Micro-diffraction does not require complex sample preparation. For analysis of the same sample area, it may be carried out on thin sections or, where this is not necessary, particles may simply be attached to adherent with a noncrystalline tape. Unlike bulk diffraction (with internal standard), micro-diffraction cannot be readily used for mineral quantification because of limitations in the area that may be scanned (i.e., counting statistics) and technical issues in quantitative refinement arising from preferential mineral particulate orientation.

Micro-diffraction is particularly useful where high-value elements with significant compositional and structural variation (e.g., U-containing phases [Gerson 2016] and Ni-containing phases [Fan and Gerson 2011]) are the focus of interest in tailings or leach residue samples as a means for process optimization and improved recovery. Whole areas may be subject to micro-diffraction or entire micro-diffraction maps collected. This enables correlation of variation of mineral phases to flotation or leaching recovery. These different mineralogical forms are often missed by techniques such as QEMSCAN and MLA.

Solution Speciation Modeling

Determination of solution speciation (i.e., predicted relative abundance of specific dissolved and precipitated chemical species) is readily accessible from solution assays but is not widely used in mineral processing control despite routine use in environmental studies of adsorption and examination of oxidation states for toxic elements. Speciation simulation programs (e.g., GEOCHEM-EZ [Shaff et al. 2010], PHREEQC [Toran and Grandstaff 2002], MINTEQ [Allison et al. 1990]) are available free to download. The caution in their use is that solution concentrations of all cation and anions, pH, temperature, and particularly Eh or oxidation–reduction potential (often not measured) are required input. Gas partial pressures and the presence of specific solids phases can be specified in the estimates. The programs come with their own databases of equilibrium constants, solubility products, and redox couples, including some collector and complexant species. These databases can often be supplemented or edited as required.

The value of these simulation programs for minerals processing is to take a "snapshot" of the system, assuming equilibrium, so that undersaturated but potentially adsorbing species and precipitating supersaturated solids may be identified. This can be compared with SEM and surface analyses to extend the correlation of indirect and direct information on hydrophobic and hydrophilic species. The assumption of species and precipitates at equilibrium should be treated with caution in rougher flotation because of the relatively short time from comminution to flotation, with some species being in metastable states (although most dissolved and some precipitating species will equilibrate in this time). With longer flotation (e.g., scavengers) or in recycle, the equilibrium can be approached for these metastable species. Where the assay data are available, this calculation is nevertheless worth doing to direct the interpretation of the surface analysis to these potential species.

An example of the value of this correlation of solution modeling and surface analysis can be found in defining correct procedures for copper (I) activation of sulfide flotation, avoiding precipitation of hydrophilic copper hydroxide ($Cu(OH)_2$) discussed below in more detail (Gerson and Jasieniak 2008; Gerson et al. 1999). In this case, the equilibrium states of both dissolved and precipitated species are found in the time of the conditioning.

EDTA Extraction

This technically simple chemical-based analysis dissolves oxidized metal ions (i.e., Cu^{2+}, Fe^{3+}, Zn^{2+}, Ni^{2+}, Pb^{2+}) as EDTA complexes these ions in situ from surface reaction layers, adsorbed colloids, precipitates, and ions to give solution assays that provide a bulk estimate of the extent of oxidation of all minerals. It does not dissolve these ions from the unoxidized mineral surfaces of sulfides or from crystalline silicates and other bulk minerals. Hence, it provides a direct measure of the extent of oxidation (total extraction) and of specific metal ion availability in the sample (pulp, concentrate, tail, and leach). Comparison with chemical information (i.e., species and composition) from specific minerals in the surface-sensitive analyses of the top 1−10 nm can then provide a more complete picture of the hydrophilic species likely to be interacting with the bubbles and potentially interfering with flotation or of gangue-activating ions.

An example of the laboratory method is to make up a 3-wt % solution of EDTA disodium salt adjusted to pH 7.5 with NaOH. Then 95 mL of this solution is vigorously stirred and purged with N_2 for 5 minutes. A slurry sample (5 mL) collected from the plant is added to the EDTA solution with 5 minutes of continuous N_2 conditioning. The slurry is then filtered through a 0.45-µm Millipore filter. The filtrate is analyzed for metal ions by standard methods. The solids are retained to obtain the sample dry weight and specific surface area, enabling calculation of the mass of metal oxidation species per unit mass of solid. Examples of the methodology and its use in this way can be found in Gerson and Jasieniak (2008).

An example of its use in diagnostic leaching of galena and its oxidation products (Greet and Smart 2002) demonstrated that all oxygen-containing galena oxidation products (i.e., sulfate, hydroxide, oxide, and carbonate, but not polysulfide or sulfur) are rapidly solubilized in EDTA. EDTA does not extract lead from unreacted galena. Continued extraction of lead with increasing conditioning time is due to continued galena oxidation, not to the extraction of lead from galena. Argon gas purging minimizes this oxidation, whereas the use of air or N_2 gas purging does not. Using information gathered from these experiments, an improved EDTA extraction technique was developed and compared with other techniques used in the mining industry.

Analytical Scanning Electron Microscopy

The liberation classes by size and by mineral are well described in the automated QEMSCAN and MLA procedures routinely used in control of comminution for optimization of flotation feed, leach preparation, electrostatic and magnetic separation, and separation by gravity (Smart et al. 2007). These techniques are based on recognition of SEM-imaged particles from pixelated back-scattered electron (BSE) images and stereoscopic EDS spectra-identifying minerals by composition in individual particles or locked composites based on their elemental X-ray counts obtained from EDS detectors (four in QEMSCAN; two in MLA). The MLA discriminates many minerals of interest using only BSE information and combines this with information from EDS only

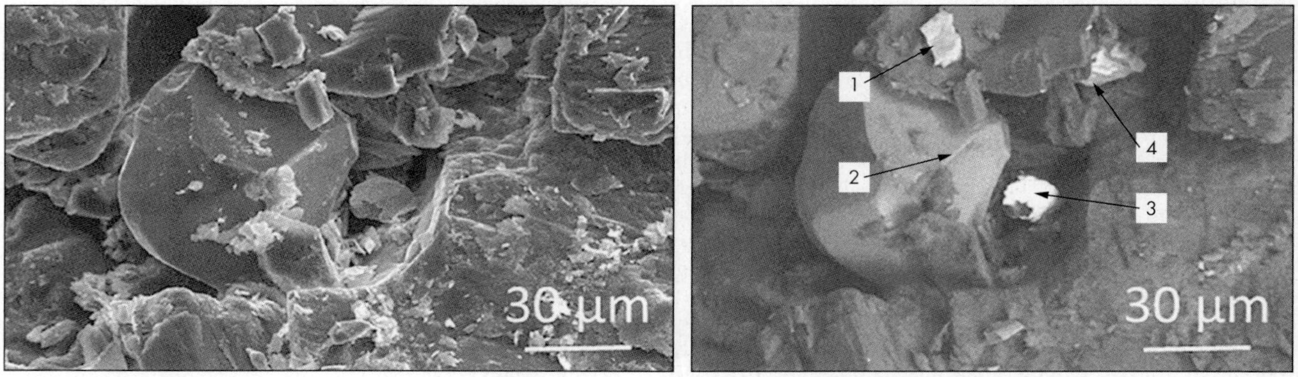

A. Scanning electron image

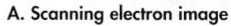

B. BSE image with EDS analysis that identifies 1–chalcopyrite, 2–chromite, 3–Pt-Fe alloy, 4–pentlandite

Figure 2 Micrographs from a UG2 concentrate for platinum group metals recovery

when needed to discriminate certain specific minerals. These analyses are compared with extensive EDS databases for optimized compositional selection. They can provide quantitative estimates of the percentage of each phase in different particle size fractions and the proportion of each phase in locked composite particles.

In some cases, however, these automated systems may not identify specific mineralogical or surface chemical features interfering with processing. As part of the strategy recommended here, it is always worthwhile to analyze the samples using conventional, low-cost SEM/EDS investigation of individual particles and their surfaces. Identification of different phases and their distribution can be made using SEM/EDS in two imaging modes, illustrated in Figure 2. The scanning electron (SE) image (Figure 2A) primarily reveals pores, cracks, grain edges, and other topographical features. The typical BSE contrast (Figure 2B) is due to compositional variations arising from the atomic number (or density) dependence of the yield of BSEs. More-dense minerals appear in light contrast from higher BSE yield, less dense in darker contrast. In many cases, it is possible to distinguish different phases by the difference in BSE contrast, as in MLA, and thus to map the distribution of each phase. The BSE image does not reveal the structure of pores, cracks, and other topographical features that are more easily seen in the SE image. A "locked," or composite, mineral particle comprising two or more phases can be analyzed by use of the sequence SE imaging, BSE imaging, and EDS analysis.

There can be some forms of incomplete liberation that are not detected in MLA or QEMSCAN analyses that may require more-specific SEM/EDS analysis. Residual locked particles below the size limit of liberation analysis can be responsible for incorrect reporting of gangue or value. For example, at the Kanowna Belle gold mine (Western Australia), a bulk sulfide float for Au in pyrite and arsenopyrite from a predominantly sericite gangue with good liberation assessment was producing good gold recovery but too much sericite in rougher concentrate and consequent smelting issues. SEM BSE images of particles from the rougher concentrate (Figure 3) explained this sericite flotation. The large (>20 µm) sericite (mica) particle in the center foreground of Figure 3 (dark in BSE imaging) has residual pyrite particles <3 µm (light in BSE image) attached to the surface. EDS point analysis confirms this differentiation by recording signals from Fe, S, and O in the partially oxidized pyrite (light region) and from K, Si, and O in the region with dark contrast. These pyrite particles would not be detected in the liberation analysis. With collector addition, they provide multiple hydrophobic points of attachment to bubbles and flotation of large sericite gangue. Outcomes from this study resulted in plant changes to reduce pulp density and increase agitation to decrease recovery of this large sericite, an investment that was recovered in 18 months with ongoing benefit.

Importantly, SEM investigations in these SE, BSE, and EDS modes can provide direct information on fine particles and adsorbed precipitates, at sizes down to 0.5 µm, attached to value mineral surfaces, the first-level identification of interfering particulate forms. Comparison between sample streams will then suggest whether they may be discriminating in recovery or grade. Then they can also be compared with mineralogy, solution speciation, and EDTA extractions for a more complete diagnosis. This may immediately suggest problems with circulating loads of fines (e.g., silicates, talc, micas, clays) or precipitated material (e.g., amorphous $CaSO_4$, iron hydroxides) requiring control.

SEM analyses will not generally identify specific chemical species at the molecular level (e.g., metal-collector precipitates, adsorbed ions). EDS analysis depth is of the order of 1–10 µm and, with imaging limitations (effectively 200–500 nm), it cannot examine the outer molecular surface layers. This requires specific surface analysis techniques of XPS or TOF-SIMS.

Surface Analysis

The introduction and evolution of direct surface analytical techniques in their application to mineral surface chemistry has provided much more comprehensive statistical analysis of ore samples and process stream products. The significance of these techniques is that they provide not only a compositional analysis of the outer molecular surface layers but also information on chemical states (e.g., oxidation, bonding), spatial and statistical distribution of adsorbed species between individual particles, and mineral phases in complex mixtures as a function of depth through the surface layers.

This surface analysis is not carried out in on-site analysis because of cost and specialized operation and interpretation. It

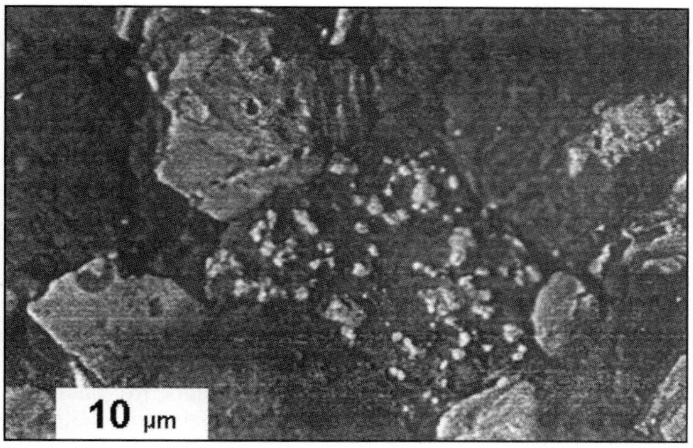

Source: Smart et al. 2014b

Figure 3 BSE image from a rougher concentrate. The large (>20 μm) sericite particle in the center (dark) has pyrite particles <3 μm (light) attached to the surface.

is, however, used in the research and development (R&D) laboratories of some companies, governments (e.g., the Canada Centre for Mineral and Energy Technology), and university minerals processing R&D units. Its main limitations are not in the acquisition of samples or analysis time but in the interpretation of the data. This does require expertise and experience in both the technique and the minerals processing operations to derive the most useful correlations from the large amount of data and implies a dedicated person or group working with company staff. This is normally done as part of the overall application of the strategy working in either project (e.g., industry/government) or in contract mode with the operation.

X-Ray Photoelectron Spectroscopy

In XPS, the sample surface is irradiated with monoenergetic X-rays producing photoelectrons that are analyzed to determine their binding energy. From the binding energy and intensity of a photoelectron peak, the elemental identity, chemical state, and percentage of an element in the top 2–10 nm are determined. These are the first two to five molecular layers in which collector attachment to value minerals may be reduced, obscured, or ineffective in inducing bubble attachment. On gangue minerals, these layers may have inadvertent activated-collector or hydrophobic surfaces. Two-dimensional imaging can now provide species mapping of surface chemistry with a practical spatial resolution of 100 μm, but data collection is relatively slow. In minerals processing, again for practical collection times, it has mostly been used in larger area-average mode (100 μm–1 mm) with limited application to statistical analysis of plant samples. In this mode, it is very useful for defining limitations in chemical processing with collectors, activators, or depressants that are affecting recovery or grade or for defining deposition covering surfaces in flotation or leaching.

XPS has been applied to define many surface oxidation products, including oxidized sulfur species (from sulfide, disulfide, polysulfide, and elemental sulfur to thiosulfate, sulfite, and sulfate) and adsorbed xanthate species on the major base metal sulfide surfaces defining oxidation, collector, activator, and other adsorption mechanisms as well as interfering species, giving more direct insight into the factors affecting recovery and grade. Most of the hydrophobic species, including both reagent and reaction (e.g., polysulfide) products, and hydrophilic species affecting flotation were defined in these studies. The mechanisms and products of in-pulp oxidation and the actions of collectors in partial removal of oxidized species and hydrophobization of the surfaces were studied using XPS combined with other surface analytical methods (Smart et al. 2003a, 2003b, 2007). In particular, the mechanisms of copper sulfate as a flotation activator in sphalerite, in ion exchange with Zn^{2+}, reduction in situ to Cu^+ with collector attachment, and adjacent oxidation of sulfide to polysulfide-like species were revealed. This has defined the correct plant use of copper sulfate ($CuSO_4$) to optimize sphalerite activation without overdosing, producing hydrophilic $Cu(OH)_2$ precipitation. Inadvertent activation by species such as Pb^+ and Ag^+ and actions of depressants, such as cyanide and bisulfite, were also defined in XPS studies. XPS has probably contributed more to understanding of mechanisms of surface chemical actions in flotation than any other surface analytical technique.

In practice now for plant investigations, XPS is normally used in area-averaged mode (100–1,000 particles) to define the ratio of different chemical states between feed, concentrate, and tail or between successive leach samples to define species selectively contributing or interfering with the process. It is differences in the chemical states between the streams that give insight to the selectivity in practice. For instance, sphalerite particles in Cu-activated flotation should be predominantly Cu(I) (complexed with collector) in concentrate, but sphalerite with predominantly Cu(II) as hydrophilic hydroxide is often found in losses to tail (Gerson et al. 1999).

Where grain sizes are larger than 100 μm, XPS may also be used in scanning or imaging mode to define the distribution of specific reagents between mineral phases. An example illustrating the comprehensive information from these techniques is a study of the Inco Bessemer Matte Processing Plant suffering a loss of chalcocite selectivity in flotation separation from Heazlewoodite down the rougher bank (Biesinger et al. 2007). SEM/EDS analyses of the feed, concentrate, and tail samples revealed a decrease in Cu and corresponding increase in Ni recovery through the primary rougher banks with tail samples also containing ≈7 wt % Cu. Suggested mechanisms included inadvertent activation and depression by the dissolution and solution transfer of Cu and Ni ions, lack of collector selectivity, and possible requirement for reaction of the chalcocite surface prior to collector adsorption. The combination of information from SEM/EDS, area-averaged XPS, and XPS imaging clarified the surface chemical mechanisms and control factors in the rougher flotation separation. Chalcocite in the feed to the rougher circuit was unoxidized (i.e., all Cu(I) surface species), whereas the surfaces of the Heazlewoodite appeared to be entirely oxidized to hydroxide species. Chalcocite in the tails from the rougher circuit is more oxidized with evident precipitates corresponding to $Cu(OH)_2$ in both composition and morphology adhering to their surfaces. The primary adsorption of diphenylguanidine (DPG) collector appears to be, correctly, to Cu(I) sites on the unoxidized chalcocite surfaces but is inhibited by the formation or concentration of fine Cu(II) hydroxide precipitates in the later rougher cells. The crystalline morphology of some of the discrete, attached Cu(II) hydroxide precipitates suggested formation in recycle water streams rather than by surface oxidation of the chalcocite through the circuit, although this may not be the only form of Cu(II) hydroxide on chalcocite surfaces in later cells. Heazlewoodite flotation appeared to be associated with the attachment or locking of fine, unoxidized chalcocite rather than any direct adsorption of DPG; Heazlewoodite fines were also found attached or locked to some chalcocite particles in concentrates contributing to loss of grade. Heazlewoodite was not completely liberated. Heazlewoodite particles in tails had relatively high surface concentrations of attached more-oxidized chalcocite fines, contributing to loss of recovery.

Other examples where XPS has been the technique used to identify interferences can be found in the investigation of galena recovery problems at Mount Isa Mines (Grano et al. 1997) and references therein. The chemical nature of carbonaceous surface layers retained on particles after grinding varies across ore bodies from Mount Isa to Hilton and McArthur River (Saxby and Stephens 1973). In Mount Isa ore, hydrophobic graphitic carbon seen in XPS causes excessive, early flotation of pyrite in the lead/zinc circuit (Grano et al. 1990). In the Hilton concentrator (Grano et al. 1997), XPS revealed the surface carbonaceous material to be less in hydrophobic hydrocarbon and more in oxidized hydrocarbon form, causing slower flotation of pyrite into the concentrate. The XPS signals for graphite and hydrocarbon are distinctly different. XPS surface analysis also found that the presence of significant precipitated calcium sulfate (from lime addition for pH control) on the surface of samples taken from the Hilton concentrator reduces the exposure of metal sulfides in the feed to low levels, restricting the flotation rate of galena when using ethyl xanthate collector.

Time-of-Flight Secondary Ion Mass Spectrometry

In the final stage, if no obvious differences have emerged in the earlier analyses or if confirmation of a suggested limitation is required, it may be necessary to examine the molecular surface chemistry in detail. In flotation, it is the variation of the ratio of hydrophobic/hydrophilic species by particle and as a statistical distribution between different mineral phases across a flotation circuit (e.g., feed, successive concentrates, tails) that may be needed. In leaching, the successive leach samples may require statistical variation of changing surface species. This requires surface analysis of numerous particles with high spatial resolution and chemical speciation, which is not practical with XPS or SEM/EDS. Analyses of a few individual larger particles carries the risk that they may not be representative of the surface chemistry that is causing low value recovery or interfering gangue flotation. It is therefore important to analyze a sufficient number of particles (normally more than 20) of the selected mineral to produce 95% confidence intervals that are discriminatory between the species (e.g., adsorbed, precipitates, collectors) being compared among feed, concentrates, and tails. Alternatively, principal component analysis (PCA) of all particles in the field of analysis provides complete statistics and can be selected by mineral phase.

The technique of choice for this analysis is TOF-SIMS. The parallel development and application of TOF-LIMS, combining laser ionization of small samples with time-of-flight mass spectrometry, by Chryssoulis and colleagues (1992, 1995) at the Advanced Materials Testing and Evaluation Laboratory (AMTEL), reviewed in Smart et al. (2007), and applied particularly to gold flotation interferences, is also acknowledged. The focus here is on TOF-SIMS since this is where statistical analysis has been developed in advanced form.

In TOF-SIMS, a focused (90 nm), pulsed primary beam of heavy ions is directed at the mineral surfaces in fixed, rasterized, or pixelated mode. A fixed beam will analyze a chosen area in the practical range from 90 nm to 1 mm in diameter. A rasterized beam produces lateral images of secondary ion products selected from the total mass spectra collected in the scan. In the pixelated mode, a full mass spectrum is recorded at each pixel (up to 256 × 256 pixels) in a chosen area, giving more than 10^7 data points. Images for any ion can then be selected from this stored data, but, importantly, the data can be statistically analyzed for correlation of surface species within and between mineral phases. For analysis, a very low flux of the heavy ions impacts surface layers so that, in the time of routine measurement, fewer than 5 surface atoms in 1,000 are impacted. The secondary ions, both positive and negative, ejected from these impacts are mass analyzed by charge (m/z) with reversed polarities using their time of flight to the detector. The very high mass resolution $m/\Delta m$ in the range 7,000–10,000 is able to easily separate almost all closely similar ion masses (e.g., O_2^- from S^-). Secondary elemental ions and molecular fragment ions detached from the first two molecular layers of the surface provide a very detailed set of positive and negative mass fragments from simple ions (e.g., Na^+, OH^-) through to large molecular ions of specific reagents (e.g., isobutyl xanthate $(CH_3)_2CHOCS_2^-$). This data collection generally takes less than 1 hour.

This technique can be applied to rapidly, statistically analyze detailed differences in surface chemistry of selected mineral phases across a flotation circuit or in successive leach samples. Its unique advantages are in being able to analyze the first few molecular layers on a mineral surface for both chemical species (e.g., oxidation products, reagent adsorption, surface layer formation) and spatial distribution (<0.1 μm). The three main forms of information from this technique are selection of mineral particles and their distribution by imaging for specific mineral-selective elements; statistical analysis of hydrophobic and hydrophilic species by comparison of intensities from selected mineral particles between feed, concentrates, and tails; and full statistical analysis by PCA of all species on particular mineral phases. This hydrophobic/hydrophilic ratio varies widely between different particles of the same mineral, and some non-value mineral phases will also have adsorbed hydrophobic collector in some form, contributing to gangue recovery and lower grades (e.g., Trahar 1976; Stowe et al. 1995; Crawford and Ralston 1988; Piantadosi and Smart 2002; Smart et al. 2003a, 2003b, 2007; Boulton et al. 2003; Malysiak et al. 2004; Shackleton et al. 2007).

TOF-SIMS analysis in these modes to identify surface chemical reasons for losses in recovery or selectivity has now been applied to samples from pilot and operating plants, including

- Mount Isa Mines (Australia),
- BHP Billiton's Cannington operations (Australia),
- Ok Tedi Mining Ltd. (Papua New Guinea),
- Teck Cominco Red Dog mine (Alaska, United States),
- Falconbridge (Strathcona, Canada),
- Anglo Platinum (South Africa),
- Mineração Caraiba (Brazil),
- Inco Matte Concentrator (Sudbury, Ontario, Canada),
- Rio Tinto Kennecott Utah Copper (United States),
- LaRonde, Quebec (Agnico Eagle),
- Kittila, Finland (Agnico Eagle),
- Clarabelle mill (Glencore),
- Matagami mine (Glencore),

- Musselwhite mine (Goldcorp),
- Niobec mine (Magris Resources),
- Highland Valley Copper (Teck Resources),
- Thompson (Vale),
- Goldstrike operations (Barrick Gold),
- Kumtor (Centerra Gold),
- Pueblo Viejo, Dominican Republic (Barrick Gold),
- Cerro-Verde, Peru (Freeport–McMoRan),
- Donlin Creek, Alaska (Barrick Gold), and
- Nechalacho, Northwest Territories, Canada (Avalon Resources).

Some details are discussed in the "Case Studies" section later in this chapter.

In some of these proprietary studies and in plant diagnosis follow-up, it has been possible to change plant processes and conditioning to improve flotation. Examples of plant changes have included the following areas:

- **Mineralogy:** ore blending/mixing/storage to control reactivity (pre-oxidation, dissolution), gangue (sulfide, sulfate, talc), and impurities
- **Grinding/liberation:** grind size for liberation but reduced gangue fines; cyclone cuts (narrow); shear, attritioning to control residual and precipitated surface layers
- **Surface oxidation:** control residence time after grind to reduce oxidation products and inadvertent activation; water recycle quality to remove activators or depressant ions
- **Eh/pH:** measure and control for oxidation, collector adsorption, (e.g., galena <200 mV SHE, or standard hydrogen electrode); dithionite, metabisulphite addition; aeration control
- **Surface cleaning:** collector choice; dosages; time after conditioning; cyclone action; high-intensity conditioning; attritioning
- **Activation:** copper ions: monolayer only; critical activation/oxidation time
- **Inadvertent activation:** control of Cu, Pb, Ag ions; oxidation/dissolution reactions; water (recycle)
- **Recycle water:** assay; pH; Eh; depressant ions (e.g., Ca, Mg, sulfate) and fines; settling time; overflow total dissolved solids

In gangue recovery and grade reduction, as in the Kanowna Belle gold mine case study using SEM/EDS, other examples of residual, locked, hydrophobic surface layers causing gangue flotation were found using TOF-SIMS (Smart et al. 2007). However, in these cases, the locked layer was less than 10 nm thick and could not be imaged using SEM/EDS, illustrating the advantage of the TOF-SIMS technique. These examples relate to recovery of gangue pyroxene and chromite in platinum group metals (PGMs) flotation plants. The Merensky (Bushveld Complex, South Africa) ore is processed by bulk PGMs and sulfide mineral flotation, but normally more than 2 wt % of the gangue minerals—principally orthopyroxene with minor chromite, which constitute more than 60 wt % of the ore—report to the concentrate, contributing to subsequent processing costs. Flotation of this gangue can occur through inadvertent copper-collector complexation, but this mechanism does not account for the true flotation of large pyroxene (20–150 µm) particles. Statistical comparison of pyroxene particles between concentrate and tails revealed no significant difference in Cu and collector (isobutyl xanthate or IBX, diethyl dithiophosphate or DTP) TOF-SIMS signals, but surface exposure of talc-like Mg, Si, and O is favored in the concentrate (Figure 4). A statistical comparison of TOF-SIMS analysis of pyroxene particles from concentrate and tails at two different pH values reveals no significant difference in Cu and collector (IBX, DTP) signals, but surface exposure of Mg, Si, and O is favored in the concentrate (Figure 4). Conversely, statistically significant surface species on pyroxene in the tails include the hydrophilic Na, Al, Ca, OH, and SO_3 (oxidation) ions, as might be expected. Collectorless flotation of pyroxene has confirmed this statistical discrimination. The Mg/Si/O association (normally expected to be hydrophilic as adsorbed species) is explained by areas of hydrophobic talc-like residual layers, probably from partial serpentinization of the pyroxene, which can be imaged in TOF-SIMS. Flotation of pyroxene without collector confirmed this statistical discrimination. These ultra-thin (<10 nm) residual layers result from shearing through the relatively soft talc-like material during autogenous grinding. A similar observation of residual talc-like layers has also been found in flotation of chromite in the UG2 Bushveld ores. Plant control through effective depressant action for talc has been extensively tested.

The chapter authors have also observed thin residual layers of soft chalcocite on pyrite surfaces causing excessive misreporting to copper concentrates and requiring resort to

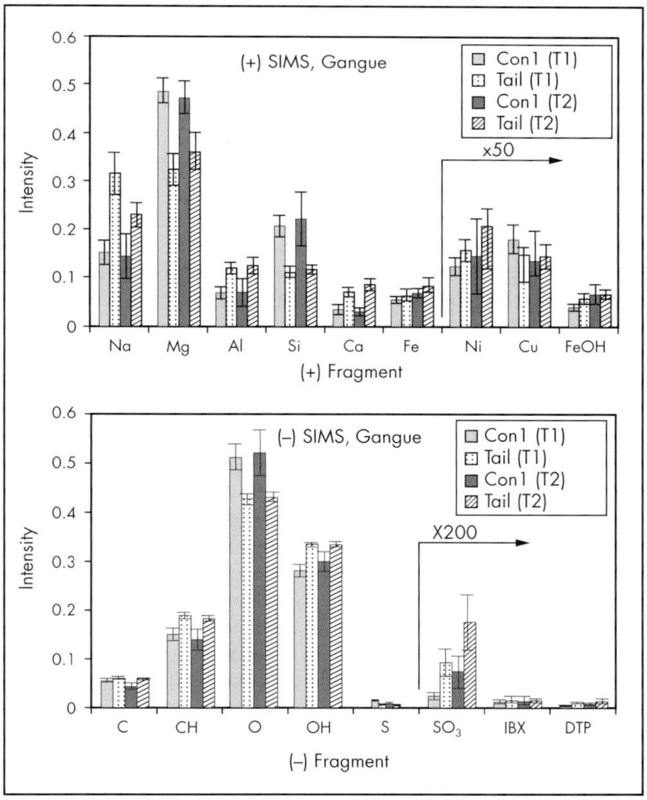

Source: Jasieniak and Smart 2009

Figure 4 Statistical comparison of TOF-SIMS normalized intensities with 95% confidence intervals (error bars) from (+) and (–) mass spectra of pyroxene gangue particles from Merensky Reef ore flotation. Results are shown for flotation concentrates and tailings from two tests (T1, T2) at different pH values.

hydrothermal leaching rather than flotation. This sometimes unrecognized liberation issue, requiring grinding, chemical, flotation, or process changes, can be identified and understood using information from both SEM and TOF-SIMS.

Two other examples illustrate the advantages of the statistical identification of factors discriminating in flotation. A Brunswick Mines plant survey found sphalerite reporting excessively to the copper–lead concentrate. Plant samples were taken in the copper–lead circuit from the rougher feed (RF), rougher bank A concentrate (Con A) and tail (Tail A), and rougher bank F concentrate (Con F) and tail (Tail F). Sphalerite particles were selected using mineral-phase imaging to give reliable statistics. Increased Cu and Pb intensities in the Con A/Tail A pair and Con F relative to sphalerite in the feed (RF) suggested dissolution and adsorption processes in conditioning. TOF-SIMS normalized intensities (Figure 5) displayed as a standard box plot show the range and distribution of both Cu and Pb on the surface of sphalerite grains. The box plot is a standardized way of displaying the distribution of data based on the five-number summary: showing successively minimum, first quartile, median, third quartile, and maximum. The central shaded rectangle spans the first quartile to the third quartile (the interquartile range), which is 1.35σ (standard deviation).

The data suggest that for the Con/Tail pairs in the rougher banks A and F, Cu on the surface of sphalerite grains is discriminating in flotation of the sphalerite in the copper–lead circuit. There is (normally) considerable spread in the surface copper intensities on different particles from plant samples, but the median values with standard deviations discriminate for both sets of Con/Tail samples. Pb, on the other hand, does not show this discrimination in the earlier rougher bank A, but there is large variation in the Pb content on the surface of the sphalerite grains reporting to Tail A. There is discrimination in the down-bank F median values with standard deviations between Con F and Tail F, although intensities are only slightly higher on sphalerite particles from Con F relative to the tail. Cu activation is favored 3–4× over Pb at pH >9 (Sui et al. 1999; Ralston and Healy 1980) so that Pb activation becoming more significant in these later samples suggests decreased Cu concentration in solution. The sphalerite grains that are indeed showing surface Cu in Con F may have been activated much earlier in the stream. A prolonged residence time for sphalerite at pH values ~9 in a Cu-depleted solution would also facilitate the adsorption of $PbOH^+$, enhancing the flotation of grains difficult to recover.

The most recent advance in the statistical analysis of TOF-SIMS data is the use of multivariate PCA, which improves image contrast, mineral phase recognition, and separation of topographic from chemical effects, and it identifies related surface species on particular mineral surfaces. PCA can identify statistical relationships between secondary ions from species that contribute to variation in surface chemistry between minerals and between flotation streams (i.e., feed, concentrates, and tail). This form of analysis can be applied to an entire mapped area of TOF-SIMS data without the need to manually define regions of interest. As each set of correlated components is identified, the correlated data are removed from further analysis to enable further correlations to be defined. Each set of correlations is termed a *principal component* (PC) while the degree of correlation of each species within a given PC is termed the *factor loading* (in both positive and negative sets in the PC). The factor loading is not a measure of amount or intensity of that species but of its correlation with other

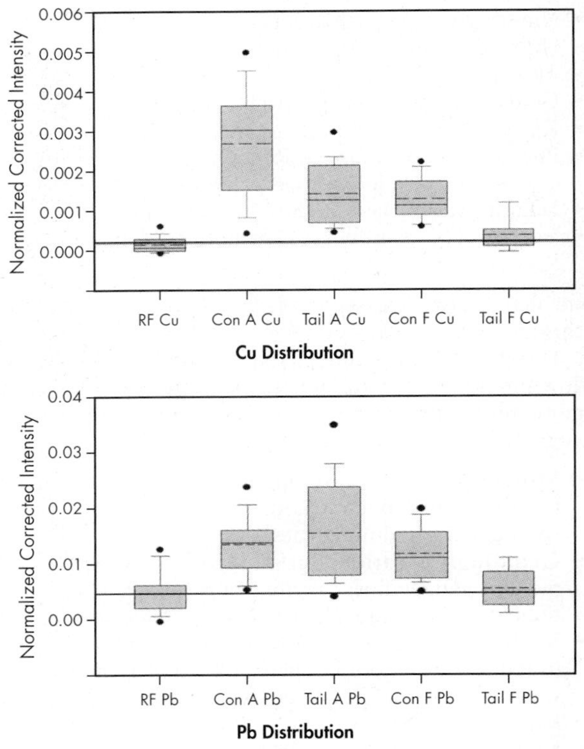

Note: The solid horizontal line in each box identifies median Cu or Pb content of the sample. The solid horizontal line across the entire plot identifies the mean Cu or Pb content in the sample group and is used as a baseline.

Source: Smart et al. 2014a

Figure 5 Vertical box plots for TOF-SIMS analyses of sphalerite surfaces from Con and Tail samples at Brunswick Mines

species in that PC. Many of these PCs immediately identify specific minerals with their associated, correlated surface species. In practice, the first principal component, PC 1, is found to be associated with topographic effects (like a SEM image) and does not hold any chemical information. This surface chemical information appears, after removal of the topography, in PC 2 onward. This analysis can be processed in much shorter time (<1 hour) compared with other techniques.

An example from PCA applied to TOF-SIMS data from a Kennecott Utah Copper (KUC) study to provide chemical diagnosis illustrates the value of the method. This study was also the first to use the full strategy of mineralogy, EDTA extraction, solution speciation modeling, SEM/EDS, and TOF-SIMS. The full set of data, combining mineralogy, liberation, solution speciation, microscopy (SEM), surface analysis, and flotation testing, can be found in Gerson et al. (2012).

The initial flotation stage in the KUC concentrator flotation circuit is the bulk recovery (maximized) of copper and molybdenum sulfides, which are then separated via subsequent flotation steps. The Bingham Canyon porphyry copper deposit, the source of the ore, is geologically complex but can be simplified into limestone skarn (LSN) ore, containing economic concentrations of Cu sulfide minerals, and monzonite (MZ) ore, containing economic concentrations of both Mo and Cu sulfide minerals. It had been proposed, as a result of

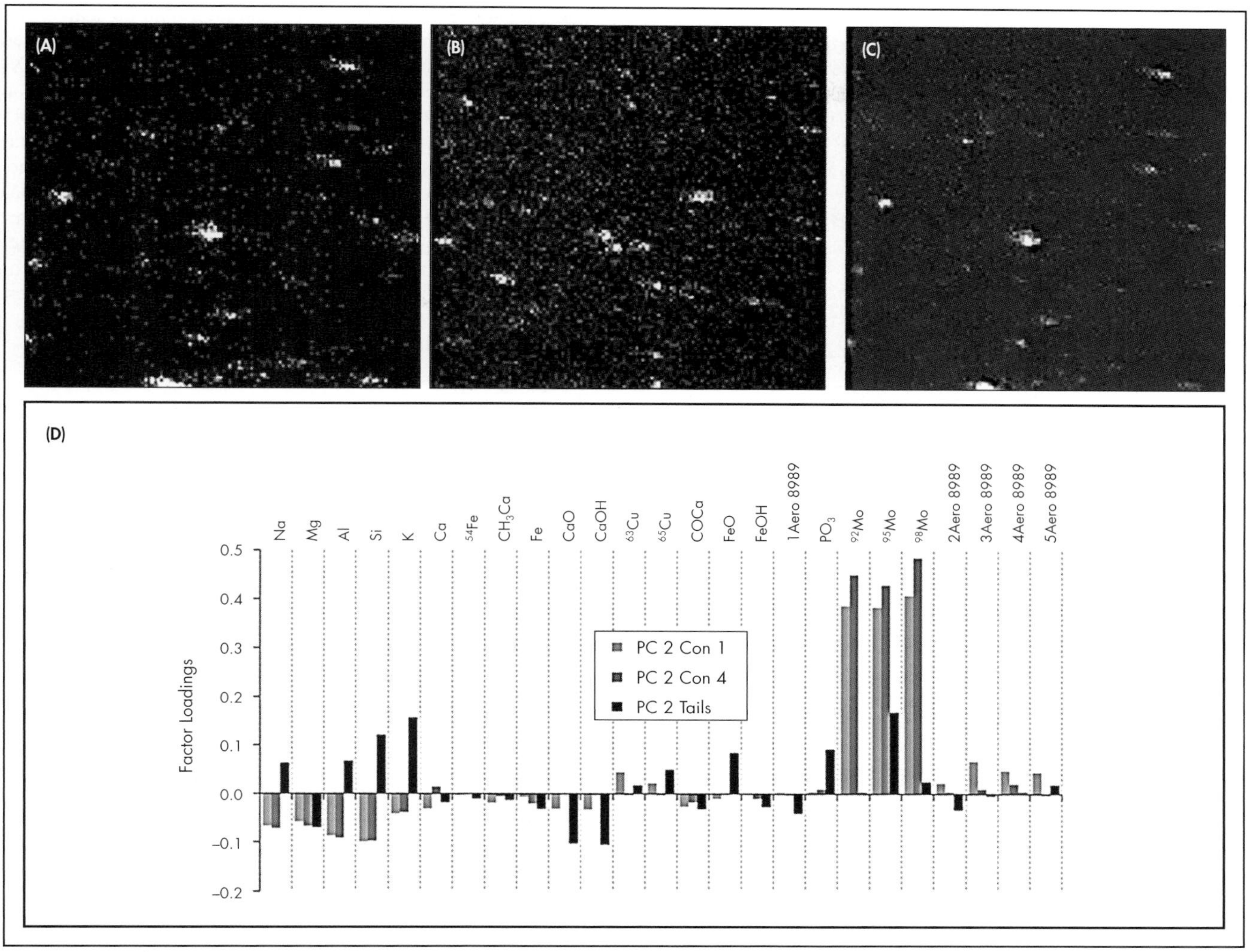

Source: Smart et al. 2014a

Figure 6 500 × 500 μm images of MZ Con 1: (A) TOF-SIMS ^{98}Mo data, (B) TOF-SIMS ^{63}Cu data, and (C) PC 2 showing areas of positive factor loadings as pale. It is clear that there is a high degree of correspondence between PC 2 and ^{98}Mo. (D) PC 2 factor loadings for MZ float Cons 1 and 4 and tails.

plant-based flotation observations, that that blending of these two ore types leads to "poisoning" of the flotation response (Triffett et al. 2008; Triffett and Bradshaw 2008). Laboratory flotation demonstrated that chalcopyrite and bornite recovery was near pro rata in the blend, but molybdenite recovery was substantially adversely affected (Gerson et al. 2012). PCA TOF-SIMS on the laboratory concentrates and tails from MZ, LSN, and blended (70:30, respectively) feeds was used to explain these differences in behavior.

PCA factor loadings from the MZ float experiments found a very strong correlation of all three Mo isotopes in PC 2 from Con 1 specifically identifying molybdenite surfaces (Figure 6D). An indication of the significance of a species identified in a given PC may be obtained by examining maps of the PC as compared to images of the actual mass fragment distribution (e.g., Figures 6A, 6B, and 6C). The lack of significant correlation to other species, except a minor correlation to hydrophobic Cu-collector fragments (dicresyl dithiophosphate), shows very clean molybdenite surfaces. In contrast, significant correlation of Cu isotopes with Fe, suggesting identification of bornite and chalcopyrite, is only obtained in PC 4 where this is correlated to both hydrophobic collector fragments and hydrophilic FeO, FeOH fragments, K, and Mg species. MZ Con 4 still displayed a very strong correlation of the Mo isotopes in PC 2 (Figure 6D). In this later concentrate, no PCs identifying strong factor loadings for Cu and Fe were found, suggesting that the surfaces of the copper-containing minerals were largely obscured by oxidation and/or attached particles. For the MZ tail, no Mo isotopes were found in PC 2 (Figure 6D) and were only weakly selected in PC 4, but no Cu and Fe association was observed. These statistical correlations show that, while the surfaces of molybdenite were relatively clean in Cons 1 and 4, the surfaces of the Cu-containing minerals carried a much higher hydrophilic loading and were substantially obscured.

A similar trend was found for the Cu-containing components of the LSN ore. An increasing surface loading of Ca-containing and other hydrophilic components was observed as demonstrated in Table 2. For the blended ore, PCA TOF-SIMS from copper-containing minerals in concentrates and tails was similar to that for the MZ and LSN ores with a relatively high hydrophilic species surface loading that increased through the flotation process (i.e., from Con 1 to Con 4 to tails). However, the PCA TOF-SIMS analysis for the blended ore after Con 1 did not select Mo in any PC for either Con 4 or the tails, in contrast to the MZ ore analyses, showing either very low concentrations or high surface coverage of the molybdenite.

Table 2 Absolute factor loadings for PC 4 MZ flotation experiment

Species	Con 1	Con 4	Tails
^{63}Cu	0.37	0.17	0.07
Fe	0.20	0.04	0.01
Ca	0.10	0.13	0.28
CH_3Ca	0.01	0.07	0.13
CaO	0.06	0.07	0.21
CaOH	0.01	0.14	0.21

Source: Smart et al. 2014a

These PCA TOF-SIMS analyses show that the surfaces of the copper minerals within both the MZ and LSN ores already carried significant Ca-containing and other hydrophilic components so that, on blending, their hydrophobic/hydrophilic ratio after collector addition did not change significantly and their flotation response remained acceptable. However, the surfaces of the molybdenite component of the MZ ore before blending were largely clean. On blending, partial transfer of the Ca-containing and other hydrophilic components as precipitates, colloidal, and adsorbed ions in the LSN ore took place on the MZ molybdenite particle surfaces, resulting in apparent poisoning of the flotation response of this component. Unless other factors apply (e.g., stockpiling, throughput), there is no flotation benefit from blending these ore types. KUC no longer blends these ores.

Hydrophobic/Hydrophilic Indices

The TOF-SIMS statistical analysis has been applied to the effects of calcium ion depression on a galena flotation (Piantadosi et al. 2000) using collector adsorption of both IBX and DBPhos. This analysis illustrates hydrophobic/hydrophilic ratio differences between minerals in concentrates versus tails. Linear regression and mean analyses of surface species on galena particles in the first concentrate and tail (sets of 26 particles) were used to estimate hydrophobic (e.g., $DBPhos^-/SO_3^-$) and hydrophilic (e.g., Ca/Pb, Al/Pb, PbOH/Pb, SO_3^-/S_2^-) indices. The results correlated closely with the flotation response. Hydrophilic species indices such as calcium ($\approx 2\times$), but also aluminum and metal hydroxides, were statistically greater on tail than on concentrate particle surfaces. This method suggested that it was possible to quantitatively assess conditioning of sulfide surfaces for optimum selectivity although this full ratio determination is not normally required in plant sample determination of interfering factors.

A second statistical study (Piantadosi and Smart 2002) of the effect of iron hydroxide oxidation products and collector, IBX, on galena and pyrite flotation separation at pH 9 confirmed a clear separation of IBX normalized intensities with galena in concentrate $5 \pm 1.5\times$ greater than those in either feed or tail. A hydrophobicity/hydrophilicity index, based on a ratio of signals from the hydrophobic collector (IBX) to ions from hydrophilic oxy-sulfur products (SO_3) and iron hydroxide (FeOH), gave a value for the concentrate of 45 $\pm$ 14 compared with 7 $\pm$ 2 for the tail. A similar differentiation was found using DBPhos collector. These hydrophobic/hydrophilic indices systematically decrease across the flotation sequence between concentrates 1, 2, 3, and tail (Smart et al. 2003b) and can be used to diagnose parts of the circuit in which this ratio is not effective for bubble–particle attachment, therefore requiring changes to conditioning procedures.

EXAMPLES OF INTERFERENCES WITH SELECTIVITY

Common Surface Layers

This section summarizes and provides comments on surface layers and species often found in the chapter authors' analyses of plant samples, which include the following:

- Iron hydroxides in colloidal and floc form (2 nm–5 µm), resulting from oxidation of iron sulfides and precipitation above pH 3. They are generally seen on all mineral surfaces and are not discriminatory for value mineral flotation between concentrate and tail. At high surface loads, often caused by high concentrations in recycle water, they can reduce flotation kinetics of value minerals, causing lower grades in later cell recoveries.
- Adsorbed calcium ions and amorphous hydrated calcium sulfate (not crystalline gypsum) precipitates, both strongly hydrophilic species, often significantly reduce recovery and grade. These species can result from dissolution of soluble calcium minerals, as in the KUC case study, or from lime addition to control pH and inadequate removal in thickeners and recycle water. In some plants, the authors have seen up to 70% coverage of value mineral surfaces by these species, severely reducing availability for attachment of collector. Replacing lime with soda ash for control up to pH 9 is an option for improved recovery, but relative cost factors need to be evaluated and this is not often carried out.
- Generally reduced exposure of value mineral surfaces due to extensive attachment of fine silicate minerals, clays, and micas from comminution of the ore-reducing collector attachment and flotation or leaching kinetics. This is often seen where prior oxidation or reaction/dissolution of the ore has occurred in ROM, storage, or excessive aeration before flotation and the surfaces of the value minerals are already sufficiently hydrophilic that these additional hydrophilic fines readily attach. This may be indicated in EDTA extraction and requires checking back to the point of oxidation increase. This is exacerbated at higher pulp densities and can often be reduced by lowering pulp density and/or increasing agitation as in the Kanowna Belle gold mine case.
- Inadvertent activation of a value mineral in the wrong part of the circuit, particularly sphalerite in lead circuits, or of gangue minerals generally. This is usually caused by dissolution of other minerals providing the activating ions to which collectors will attach, including Cu(II), Pb(II), and Ag(I). This may again be due to ROM, storage, or excessive aeration before flotation. The BHP Billiton case study later in the chapter illustrates this problem. Where excessive gangue mineral flotation occurs with

collector identified on these surfaces, this may be due to the combination of the activating ion dissolution and collector overdosing. Formation of ion-collector complexes in solution, rather than on the value surfaces, can result in indiscriminate attachment to all mineral surfaces, severely reducing grade.
- Other factors discussed previously, including residual layers and hydrophobic mineral attachment (particularly talc) activating surface layers on gangue minerals, can be identified in this strategy.

EXAMPLES OF INCORRECT CONDITIONING

Collector Attachment, Inadequate Coverage, and Comparisons Between Collectors

The attachment of specific collectors to specific minerals can be imaged in TOF-SIMS, and their effectiveness in flotation selectivity can be assessed in statistical comparisons of their signals between concentrate and tail samples. Where flotation kinetics are too slow, causing low selectivity in later cells, the coverage of value mineral surfaces by collector can be imaged to assess whether this is due to inadequate coverage and consequent low hydrophobic/hydrophilic ratios. Early studies by Brinen and co-workers at Cytec (Brinen and Reich 1992; Brinen et al. 1993) and by Chryssoulis and his AMTEL group (Stowe et al. 1995; Chryssoulis et al. 1995) confirmed that this technique can identify parent and fragment ions from collectors and map their distribution on single-mineral grains. Comparisons between the effectiveness of different collectors separately or in mixtures can now be assessed in the statistical mode. An example from a BHP Billiton Cannington plant survey (described in more detail in the "Case Studies" section) illustrates this discrimination. A relatively high recovery of sphalerite to the lead circuit cleaners represented losses for the zinc circuit. The flotation used a combination of ethyl xanthate (EX) and DBPhos collectors. Following a formal plant survey analysis, 26 particles of sphalerite from the lead cleaner concentrate and tails were investigated by imaging for Zn-containing secondary ions. In the statistical analysis, the separation of intensities from specific species at more than the 95% confidence interval indicates that these species discriminate between the two streams. As Figure 7 shows, because of the wide spread of intensities, there is no separation at more than the 95% confidence interval between sphalerite particles from cleaner concentrate and tail for either collector. The mean value for the collector DBPhos molecular ion on concentrate particles is, however, 50% greater than in the tail, but EX mean values are not very different between cleaner concentrate and tail. The EX intensities still indicate that this collector would have contributed to sphalerite being recovered in hydrophobic complexes with Cu, Ag, and Pb into the lead roughers and scavengers before the cleaners (as explained later in the case study). The results suggest that the EX does not cause as much sphalerite loss as DBPhos in the lead cleaner separation. This result led to a reexamination of the collector mixture used in the lead cleaner circuit.

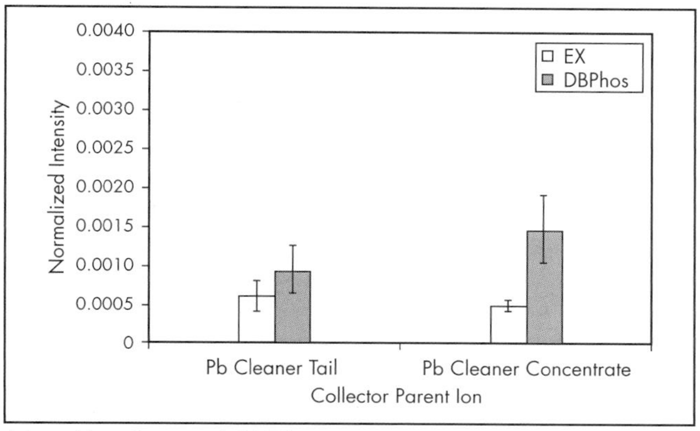

Adapted from Smart et al. 2003b

Figure 7 Statistical comparison of TOF-SIMS normalized intensities for collector ions EX and DBPhos on sphalerite particles from a lead cleaner concentrate and tail (bars = 95% confidence intervals)

Effectiveness of Conditioning Reagents

The detail from surface analysis can suggest changes to conditioning, residence time, and Cu deactivation reagents. For instance, Zn activation control in an industrial Cu flotation circuit treating a complex Cu-Pb-Zn-Au-Ag sulfide ore at the LaRonde Division of Agnico Eagle Mines Limited (Quebec, Canada) was investigated through COREM (Olsen et al. 2012). The action mechanisms of zinc sulfate ($ZnSO_4$), triethylenetetramine (TETA), and sodium bisulfite ($NaHSO_3$) used as depressants to limit ZnS activation were examined. A combined TOF-SIMS/XPS examination of sphalerite surfaces in copper plant samples with Cu^{2+} addition was performed to evaluate the response to the three depressants. Although copper transfer occurred under all test conditions, both the $ZnSO_4$ and TETA test samples indicated that Cu was partially inhibited from attaching to the surface of the sphalerite grains. The operational mechanisms, however, were probably different: the former likely related to the development of oxidative species on sphalerite grains, and the latter to the chelating capacity of TETA (Tukel et al. 2010). Sphalerite surfaces in both the $ZnSO_4$ and $NaHSO_3$ tests reported the highest proportion of species, indicative of oxidation such as $Zn(OH)_2$ and SO_3. It is possible that greater development of sulfoxyl and hydroxide species on the surface of the sphalerite, as identified by TOF-SIMS, may inhibit collector attachment, produce hydrophilic surfaces, and, in combination, result in reduced sphalerite flotation (Chandra and Gerson 2006). In another example, carboxymethyl cellulose (CMC), commonly used as a depressant for talc-like minerals, was studied in the TOF-SIMS imaging mode using CMC fragments by Parolis et al. (2007). These authors showed that CMC coverage of the talc particles was homogeneous after initial adsorption, but subsequent desorption in distilled water partially restored flotability and this change could be related to an inhomogeneous CMC distribution on the talc surface. Additional exposure to Ca^{2+} ions gave homogeneous CMC redistribution and depressant action.

Incorrect Cu Activation—Overdosing

There is considerable misunderstanding of activation mechanisms and their control in flotation, resulting in incorrect conditioning in some plants. A short explanation of the mechanism (Chandra and Gerson 2006) may be helpful to consider in plant control.

Both sphalerite and pyrite can be activated by ions such as Cu^{2+}, Fe^{2+}, Ag^{+}, and Pb^{2+} in solution; however, it is only $CuSO_4$ that is normally added in flotation practice. Sphalerite will not normally adsorb thiol collectors and float without copper activation. Copper addition is also used to assist bulk sulfide and pyrite/pyrrhotite flotation for associated gold or PGM recovery. The other ions resulting from reaction and dissolution are normally considered contaminants, which can cause inadvertent activation and loss of selectivity, particularly where sphalerite flotation and pyrite depression are required. XPS studies have shown that copper activation of sphalerite proceeds with a 1:1 exchange of Cu^{2+} with Zn^{2+} from the first few atomic layers of the sphalerite surface, after which a surface Cu^{+}-S^{-} species is formed. XPS can clearly separate Cu(II) and Cu(I) species to allow assessment of correct conditioning in formation of the Cu(I) species. This surface Cu(I)-S(−I) species formed is hydrophobic and can induce some collectorless flotation, but the addition of collector to form the stable Cu(I)-collector (e.g., xanthate X(−I)) complex induces stronger hydrophobicity. With time after Cu(II) addition, it will continue substituting with Zn and slowly migrate into the bulk, potentially reducing the surface concentration for finite dose. Under conditions of long activation time between copper activator and collector addition, concentrations of Cu(I)-X(−I) formed on the surface may be relatively low, which may affect flotation. Hence, both the total amount of Cu(II) and time after addition before collector addition are important parameters for correct conditioning.

Pyrite activation, however, follows a single fast step involving Cu(II) adsorption to the surface without any exchange with iron in the pyrite lattice. Unlike sphalerite, pyrite adsorbs thiol collectors and floats well even without copper activation because of the activating nature of Fe^{2+} on pyrite surfaces, where Fe-X_2 structures can form along with neutral dixanthogen X_2. This flotation is stronger in low-pH conditions as the surface Fe^{2+} sites oxidize to form iron hydroxide species at higher pH, which retards collector adsorption. This gives improved selectivity for sphalerite/pyrite separation but generally lower kinetics for bulk sulfide and pyrite/sphalerite recovery.

These mechanisms dictate that copper addition should ideally be just sufficient to form the adsorbed outer surface monolayer on the mineral to be floated with collector addition as soon as possible after activation. This will maximize the hydrophobicity of these surfaces. Overdosing copper, sometimes done in PGM flotation to control froth structure, works against the selectivity of these mechanisms. At pH above about 6, hydrophilic colloidal $Cu(OH)_2$ will precipitate onto mineral surfaces and a thick obscuring layer of $Cu(OH)_2$ may form if the conditioning copper concentration is relatively high. $Cu(OH)_2$ precipitates nonselectively and has the potential to cause flotation of unwanted minerals or gangue, through slow subsequent reduction of Cu(II) and reaction with xanthate to form Cu(I)-X(−I) and dixanthogen X_2, both hydrophobic species, particularly if the collector is also overdosed. The $Cu(OH)_2$ layer also obscures the Cu(I)-S(−I) layer on sphalerite, further reducing surface hydrophobicity.

Overdosing collector is equally disadvantageous to selectivity. At excess dosages, hydrophobic dixanthogen forms in solution as nano-colloidal oily droplets and tends to adsorb nonselectively, causing unwanted flotation of non-value minerals, especially gangue. Xanthate may also react with Cu^{2+} in the conditioning solution, causing precipitation of Cu-xanthate in the solution and, again, nonspecific adsorption. This limits the amount of xanthate available for floating the desired mineral, thus increasing reagent consumption. Sequential addition of copper and xanthate is consequently preferred over simultaneous additions of these reagents.

The presence of a high lattice iron concentration in sphalerite, which occurs in some deposits or is a variable in flotation recovery, is an intermediate case. It has caused increased adsorbed copper and subsequently higher xanthate adsorption but lower adsorbed copper and xanthate adsorption in other conditions. The reason for this discrepancy lies in the higher reactivity imparted to sphalerite by high iron content. Sphalerite containing high iron oxidizes much faster than low-iron sphalerite and at high pH is likely to be covered by hydroxides possibly at iron sites, which may be located near steps or defects. Control of surface pre-oxidation (e.g., storage, processing time, and aeration conditions) and pH is therefore critical for optimized flotation of high-iron sphalerite.

An example of the combined use of EDTA extraction, solution speciation, XPS, and TOF-SIMS has also illustrated the importance of pre-oxidation in control of copper activation of pentlandite, and monoclinic and hexagonal pyrrhotite in a PGM recovery circuit (Gerson and Jasieniak 2008). It has been established in a laboratory simulation that the addition of $CuSO_4$ in conjunction with xanthate collector increases the flotation rate of pentlandite as compared to the addition of collector alone. However, it remained unclear whether the addition of $CuSO_4$ has a similar effect within mineral processing circuits and whether this also applied to the alternative forms of pyrrhotite. Activation experiments were carried out at pH 9 using a total solution content of copper ions equivalent to one monolayer. Copper loss from solution was extremely rapid in all cases with less than 10% remaining after 10 seconds. Immediately after collector addition, EDTA extraction, solution analysis, and XPS clearly demonstrated that copper activation on all three minerals was via direct surface adsorption and not an ion exchange process, as is the case for sphalerite, and that most copper on the surfaces is in the form of Cu(I).

Testing increased conditioning time prior to copper activation, and TOF-SIMS analyses of the mineral surfaces subsequent to copper activation demonstrated increased surface oxidation. Regardless of the length of pre-activation oxidation time, only Fe(III) (no Fe(II)-S) was observed by XPS analysis on all three minerals, indicating a highly oxidized surface layer dominated by Fe-O-OH species. The presence of Cu(II) on the mineral surfaces is indicative of the formation of Cu(II)-containing precipitates and indicates non-mineral-specific activation, resulting in increased recovery but lower grade. Mineral-specific copper adsorption, as evidenced by the presence of Cu(I), is required to improve both grade and recovery. Prolonged oxidation prior to copper activation causes a reduction in the Cu(I)-to-Cu(II) ratio on the mineral surfaces as compared to either fresh surfaces or surfaces oxidized for one minute only. This strongly indicates that the copper activation process is less effective on highly oxidized surfaces. However, there is a significant decrease in surface oxidation with time on comparison of samples exposed to copper solution rather than water. This suggests that copper activation stabilizes the surface and reduces the rate of oxidation, but this oxidation still requires control for maximum recovery and grade.

Other examples of combined information from solution modeling, EDTA extraction, and surface analysis (XPS) include the flotation study on Cu and Pb speciation as a function of contrasting Eh and pH conditions during grinding (Peng et al. 2012). In the flotation of copper ores, several processing plants reported that copper recovery is affected by the proportion and reactivity of pyrite in the ore with the effect becoming more intense when the feed particles are finer as a result of regrinding. Flotation tests showed that chalcopyrite flotation rate, recovery, and grade, as well as the pulp oxidation potential decreased with increasing pyrite content although pyrite recovery increased. Surface analysis (XPS, TOF-SIMS, and EDTA) revealed that inadvertent copper activation of pyrite increased with increasing pyrite content, facilitating pyrite recovery (Owusu et al. 2014). Decrease in chalcopyrite recovery was also attributed to increased surface oxidation in association with higher pyrite content or finer grind, both factors reducing recovery and grade.

CASE STUDIES

Base Metal Copper, Lead, and Zinc Flotation

This case study illustrates the sequence of TOF-SIMS analysis normally used for plant samples where the mineralogy, EDTA, solution speciation, and SEM/EDS have not given a clear indication of controlling factors. The operating BHP Billiton Cannington plant in Australia was floating chalcopyrite, galena, and sphalerite in sequence. Sphalerite being recovered to the lead circuit cleaners represented losses for the subsequent zinc circuit. No $CuSO_4$ had been added in the circuit to this point.

Samples were collected from lead cleaner concentrate and tail using the sampling procedure described previously. After initial scans for all species, imaging the spatial distribution of the main surface chemical species is usually the first step in TOF-SIMS diagnosis of surface chemical species controlling recovery and grade. This step is fast and provides the basis for focusing on species that appear to be different between concentrate and tail for the specific mineral phase. TOF-SIMS images of the particles in total ion yield mode as in Figure 8A are similar to those from SEM imaging. Scanning for specific signals (e.g., Pb, Zn, Cu, Fe, Fe/Cu), can then be used to identify the particles of a specific mineral phase (e.g., galena, sphalerite, covellite, pyrite, chalcopyrite) for specific analysis. The region of interest (ROI) facility in the software allows definition of selected particles corresponding to a specific mineral phase with the boundary for analysis set at a fixed position inside the contrast edge. When sufficient particles of the mineral phase have been identified for reliable statistics, a mass spectrum from each particle is recorded and stored. The statistical analysis then determines a mean value for each atomic and molecular species with 95% confidence intervals for each signal.

Sphalerite particles in each sample were identified from the total ion image (Figure 8A) by imaging for Zn as illustrated in Figure 8C. Then other species associated with Zn likely to be discriminating between Con and Tail—in this case Cu, Pb, Ag (identified in these initial scans), and collectors EX and DBPhos—can be extracted from the images, as in Figures 8D, 8B, and 8E, respectively. The spatial distributions in the image analysis show that sphalerite reporting to the cleaner concentrate has adsorbed Cu, Ag, and Pb ions. These ions complex strongly with collectors to produce hydrophobic surface species, inadvertently activating the sphalerite for bubble attachment.

To confirm these discriminating factors causing sphalerite flotation, a statistically valid selection of sphalerite particles from the lead cleaner concentrate and tails was investigated. This statistical comparison is illustrated in Figure 9 where 26 sphalerite particles were selected from plant samples of concentrate and tail (discussed earlier in the "Collector Attachment" section) and images like Figure 8. The TOF-SIMS spectra from each bounded sphalerite particle was collected for the statistical comparisons. The separation of intensities from specific species at a 95% confidence interval indicates that adsorbed Cu, Ag, and Pb positive ions discriminate between sphalerite particles in the cleaner concentrate and those in the tail although these species are present on all sphalerite particles. The negative ions used to discriminate species in the concentrate included C, CH, and collector fragments of DBPhos and EX. These signals all represent hydrophobic species associated with collector adsorption. The hydrophilic surface species that discriminate sphalerite particles into the cleaner tail are positive ions of Mg, Ca, Al, and Si, and negative ions of O, all of which represent aluminosilicate gangue species as fine particles adsorbed to the sphalerite surfaces, decreasing the hydrophobic/hydrophilic ratio on these surfaces. Notice, however, that there are still significant intensities of these species on sphalerite that have reported to the concentrate, confirming the heterogeneity of mineral surfaces in real pulps and the importance of the hydrophobic/hydrophilic ratio in bubble–particle attachment. This "knife-edged" selection of particles reporting to concentrate versus tails is the reality of the flotation process and the reason for surface chemical control.

As noted earlier, the results suggested that the EX does not discriminate as clearly as DBPhos in sphalerite reporting to the lead cleaner concentrate. These combined results caused a reexamination of the collector mixture used in the lead cleaner circuit and of the residence time for conditioning, during which oxidation, dissolution, and adsorption of these activating ions occurred.

Another example of results from samples supplied by the client from rougher and rougher-scavenger flotation is shown in Figure 10. Poor flotation kinetics were exhibited by fine sphalerite (−10 μm) copper-activated down the rougher-scavenger banks. The study aimed to determine whether the poor flotation response by fine sphalerite was due to differences in mineral surface chemistry rather than hydrodynamic collision frequency factors alone. The process characteristics included pH 10.5 adjusted with lime, collector addition of IBX, and copper sulfate activation. Sphalerite particles in the −10 μm size range were selected using the ROI methodology described previously so that the surface chemistry of this mineral phase was examined selectively. The bars in Figure 10 show the median value of each positive and negative ion signal with the 95% confidence intervals indicated by the smaller intervals at the top of each bar. Comparison between the rougher concentrate, scavenger concentrate, and scavenger tail eliminates all species for which confidence intervals overlap as not statistically significantly different (at least to the first level of statistical analysis). It is clear that other signals are statistically different with diverse magnitudes in this comparison. In the selected positive ion species, discrimination into the concentrate streams is indicated for Zn, Cu, and Na with discrimination into the tail for increasing Fe, K, Si, Al, and, particularly, Mg. Low surface concentrations of Ca appear to favor the rougher concentrate but are apparently depressed into the scavenger concentrate and tail. In negative ion SIMS, the concentrates are statistically favored by high exposure of S, CH, and by low surface exposure of F, OH, and O. More detailed examination of the collector IBX signal shows close correlation with the Cu signals. But the oxy-sulfur SO_3 signals are not statistically different between the three streams.

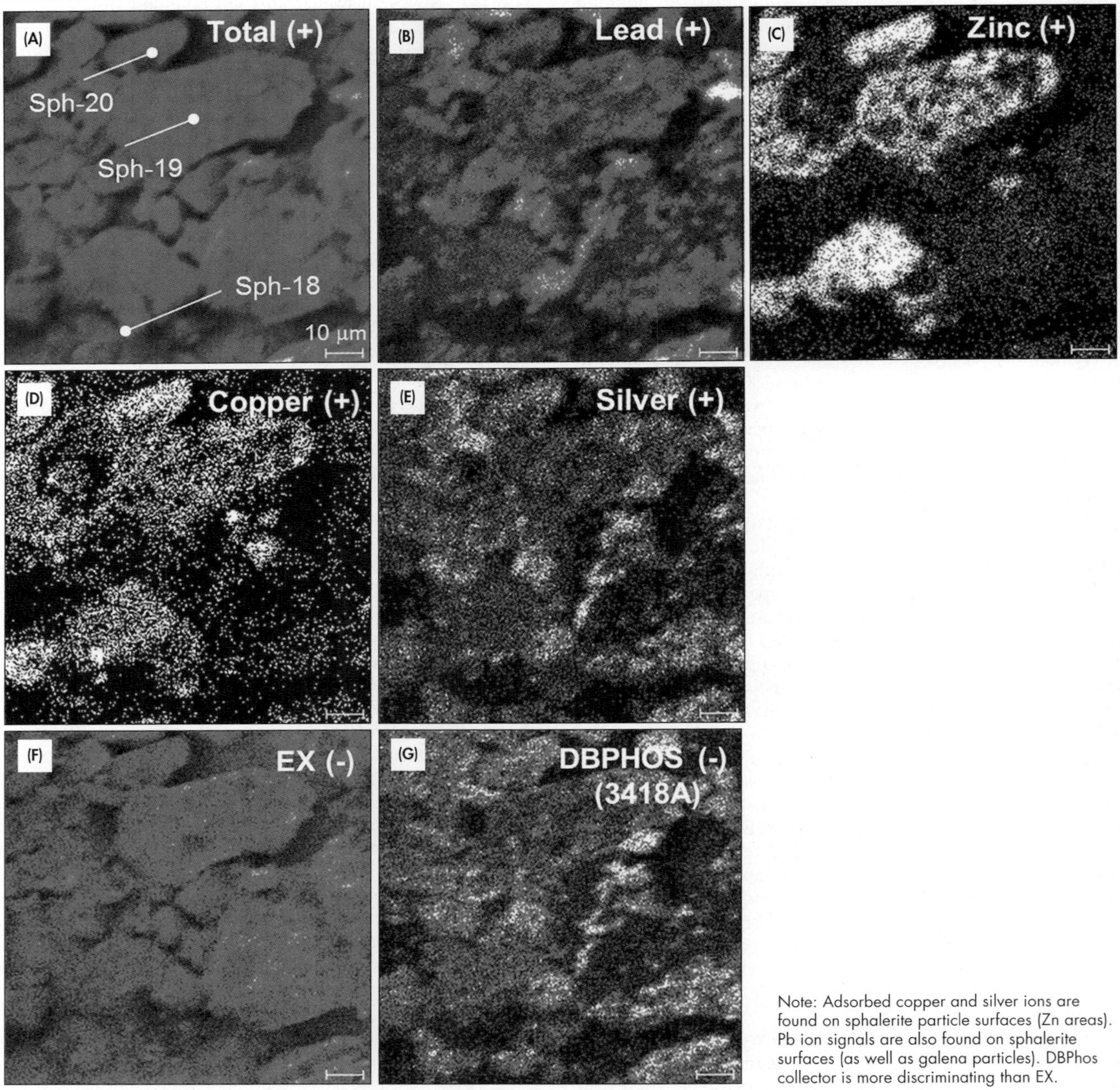

Note: Adsorbed copper and silver ions are found on sphalerite particle surfaces (Zn areas). Pb ion signals are also found on sphalerite surfaces (as well as galena particles). DBPhos collector is more discriminating than EX.

Figure 8 Selected area images from a lead cleaner concentrate for total ion yield, lead, zinc, copper, silver, collector EX, and collector DBPhos

Hence, comparisons show that the surface chemistry of the fine sphalerite is significantly different between concentrates and tails and, for some species, even between the rougher concentrate and scavenger concentrate. The most important difference is the absence of exposure of copper and associated IBX on fine sphalerite in the tail stream, indicating low hydrophobicity of these particles. This difference is exaggerated by the presence of high concentrations of Mg, Ca, Al, OH, and F ions apparently in the form of hydroxides and (alumino)silicates obscuring copper activation. Oxidation to oxy-sulfur species is not a major factor in the depression of fine sphalerite. Calcium ions, in particular, appear to have a depressant role between the rougher and scavenger flotation stages. In the plant, this problem was addressed by improving Ca removal in recycle and staged dosing of Cu down the rougher and scavenger banks.

Non-Sulfide Flotation: Pyrochlore Recovery

In a Niobec mine (Quebec, Canada) plant survey, the composition of the pyrochlore grains reporting to the tails were identified as having a greater concentration of Fe than those reporting to the concentrate. A statistical compositional

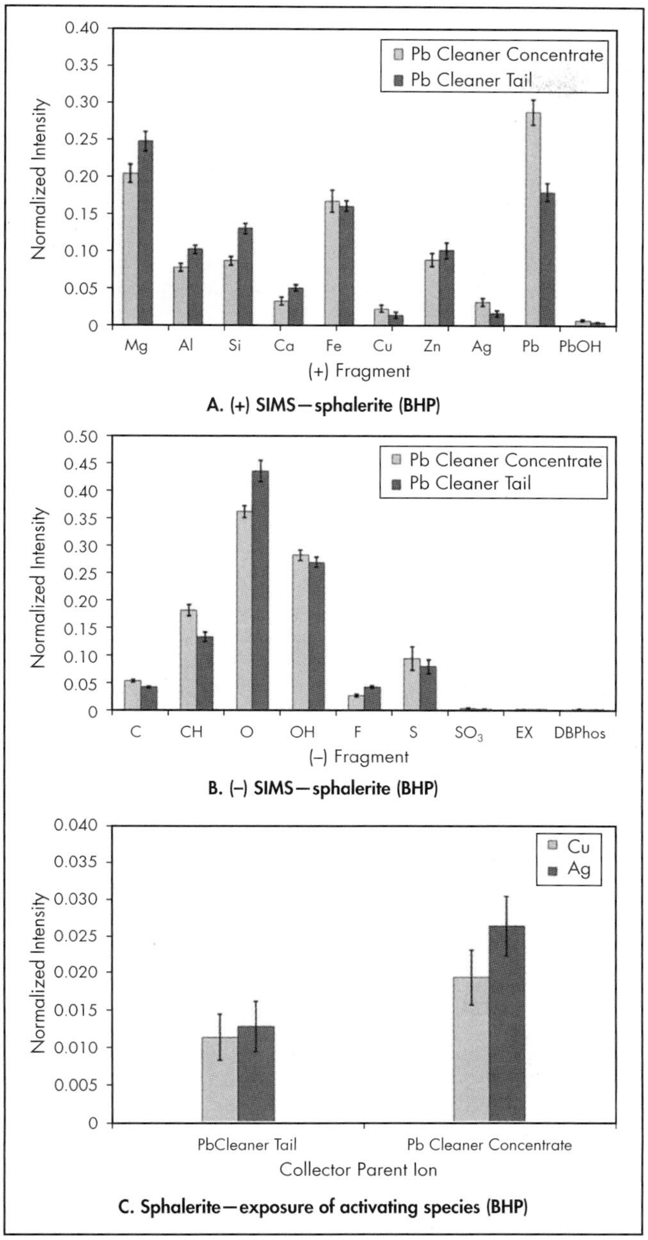

Adapted from Smart et al. 2003b

Figure 9 Statistical comparison of TOF-SIMS normalized intensities for sphalerite particles from a lead cleaner concentrate and tail (bars = 95% confidence intervals) for (A) positive ions, (B) negative ions, and (C) Cu and Ag ions

analyses of +200 grains from the concentrate and tails samples showed that the Fe content was not related to Fe-rich inclusions but rather occurred in the mineral matrix, substituting for Na or Ca. TOF-SIMS surface chemical analyses identified that the grains reporting to the tail (higher matrix Fe content) had a greater content of FeOH (along with other FeOx species; Figure 11A) and significantly less tallow diamine (TDM-1) collector on their surfaces (Figure 11B). The TOF-SIMS surface analyses established a link between mineral Fe content, the degree of surface oxidation, collector loading,

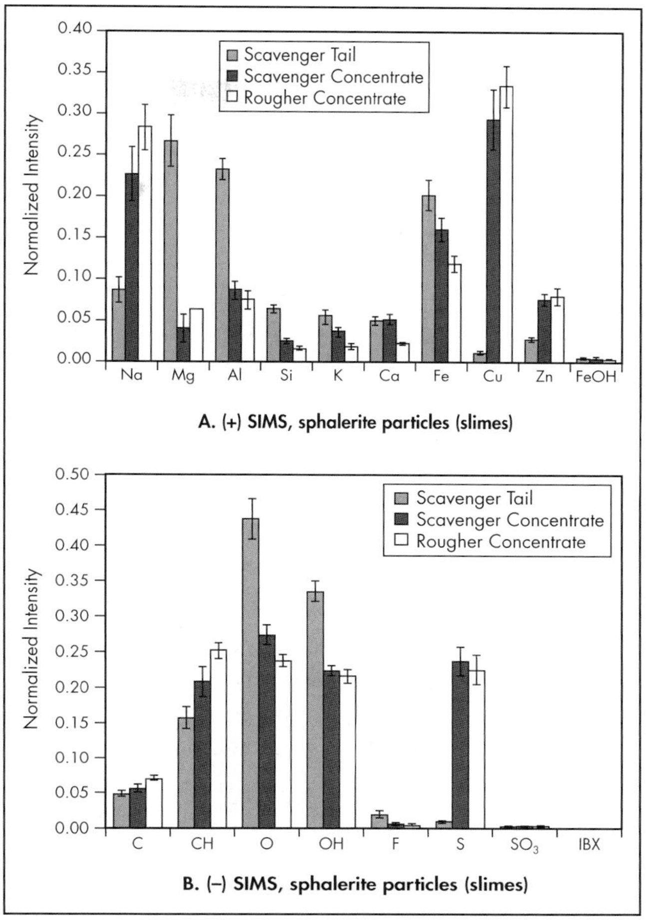

Source: Smart et al. 2003b

Figure 10 Statistical TOF-SIMS spectra for sphalerite particles in the scavenger tail, scavenger concentrate, and rougher concentrate (bars = 95% confidence intervals)

and the observed poor recovery of high Fe-pyrochlore grains (Chehreh Chelgani et al. 2012a, 2012b).

To verify the relationship between pyrochlore matrix Fe content and the identified surface chemical variations, a series of bench conditioning tests were set up to determine the effect of accelerated oxidation on pyrochlore grains of different Fe content. XPS broad-scan data and high-resolution scans for Fe revealed that, under controlled lab conditions, a greater proportion of oxidative Fe species developed on the surface of the pyrochlore grains with higher matrix Fe content (Chehreh Chelgani et al. 2013, 2014; Figure 11C versus 11D). These data support the hypothesis reached from the TOF-SIMS analyses, which suggest that the development of surface oxides favors pyrochlore grains with higher Fe content, ultimately impeding collector adsorption and reducing recovery.

Rare Earth Element Metals and Minerals

The Nechalacho deposit is a world-class resource of rare earth element (REE) metals and minerals in Canada. Concentration of REEs from the host rocks is accomplished by flotation. In this deposit, heavy rare earth elements (HREEs) are present in fergusonite ((Y, HREE)NbO_4) and zircon ($ZrSiO_4$), whereas the light rare earth elements (LREEs) are present in

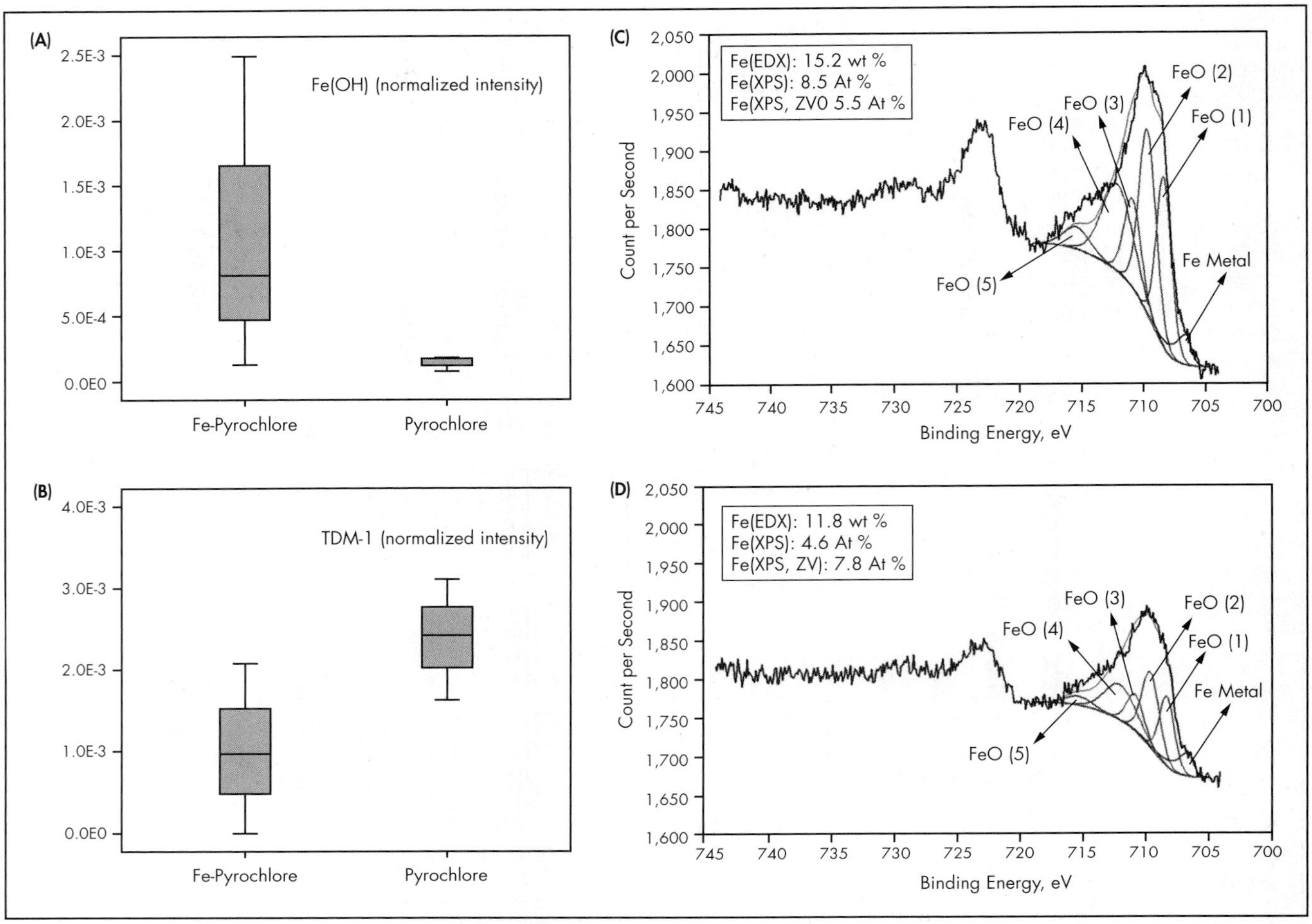

Source: Chehreh Chelgani et al. 2014

Figure 11 (A and B): Vertical box plots of TOF-SIMS normalized intensity for FeOH and collector (TDM-1) on the surface of high Fe pyrochlore (Fe-pyrochlore) and low Fe pyrochlore (pyrochlore) grains from the Niobec flotation plant. (C and D): High-resolution XPS spectra for grains with high Fe content (Fe-pyrochlore, C) versus low Fe content (pyrochlore, D) after conditioning for 30 minutes in O_2-oversaturated water. Included for each grain is the wt % Fe (EDS), the relative atomic proportion (at %) of surface-oxidized FeO components (Fe(XPS)), and Fe metal (ZV-zero valent) proportion of total surface Fe as determined by high-resolution XPS scans.

bastnaesite (Ce, La)CO_3F, synchysite Ca(Y, Ce)$(CO_3)_2$F, allanite (Ce, Ca, Y$)_2$(Al, $Fe^{3+})_3(SiO_4)_3$(OH), and monazite (Ce, La, Nd, Th)PO_4. Niobium and tantalum are hosted in columbite (Mn, Fe^{2+})(Nb, Ta$)_2O_6$ as well as fergusonite (Paul et al. 2009; Cox et al. 2010).

As part of the development work to optimize a REE mineral recovery process flow sheet, several surface chemical tests tied to the flotation recovery process were used to evaluate the effectiveness and mechanism of hydroxamate adsorption (Xia et al. 2015a, 2015b; Jordens et al. 2013). Laboratory flotation tests revealed that the use of hydroxamates in conjunction with lead nitrate resulted in an obvious improvement in recovery of REE minerals reporting to concentrate. The results suggest that lead nitrate could promote rare earth flotation, but the flotation behavior of each rare earth mineral (REM) is not clear.

The REE flotation grade response results from the Xia et al. (2015a) study in the presence and absence of lead nitrate with benzohydroxamic acid (BHD) as collector is shown in Figure 12. The results with the addition of various doses of lead nitrate ($Pb(NO_3)_2$) show that at low doses (500 g/t or 1,000 g/t), LREE recovery improves significantly, whereas only high doses of lead nitrate (up to 2,000 g/t) result in significant recovery improvements for Zr and Nb. The data imply mineralogical control in flotation recovery of various REMs. Jordens et al. (2013) have identified REE niobates and carbonates as showing the highest recovery rates and REE silicates the lowest. The data from Figure 12 partially supports this contention as La, Ce, Nd, and Y (along with the LREEs) are most commonly found in the easily (fast) recovered carbonate phases in the deposit, whereas Nb and Zr (along with the remainder of the HREEs) are found with the more difficult to recover (slower) oxides and silicates.

In the work by Xia et al. (2015a), surface chemical evaluation of the flotation testing products by TOF-SIMS was performed to potentially identify factors promoting stream partitioning in relation to the addition of $Pb(NO_3)_2$. The surface analysis was performed on samples from flotation tests using 2,000 g/t $Pb(NO_3)_2$ along with the BHD collector. The data show that higher normalized intensities for the BHD collector (Figure 13) are present on the surface of the grains reporting to the concentrate. The data also show that there is no difference in BHD collector loading rates between the test with and without $Pb(NO_3)_2$ addition, and furthermore, there is no improvement in collector loading discrimination between concentrate and tails grains from the $Pb(NO_3)_2$ addition test. The value of $Pb(NO_3)_2$ then does not appear to be associated with an increase in collector adsorption. The surface chemical data for Pb, on the other hand, shows a strong intensity discrimination on Con/Tail pair grains from the test with the 2,000-g/t $Pb(NO_3)_2$ addition. The intensity data indicate that there is significantly more PbO and PbOH on the surface of the grains reporting to the concentrate than the tails. This would suggest that lead addition can reverse the surface charge, making it more efficient for collectors to adsorb onto REE surfaces. It is also possible that $PbOH^+$ is potentially acting as a point activator for BHD adsorption.

In flotation tests using upgraded REE samples from the same deposit, Jordens et al. (2013) identified similar recovery characteristics using BHD and lead chloride ($PbCl_2$) rather than $Pb(NO_3)_2$. TOF-SIMS surface chemical analyses of REMs showed elevated levels of both PbO and PbOH on the REM grains from the concentrates relative to the tails and that there is selectivity of the Pb species on the REMs over feldspars and Fe-oxides. This greater proportion of PbO and PbOH on grain surfaces from the concentrate would then indicate that $Pb(NO_3)_2$ is indeed participating in the discrimination process and that it may well be providing some degree of selectivity favoring the surface of REM grains.

The ultimate value of both studies was the identification that variability in REE recovery is linked to the flotation characteristics of the various REMs. The recovery process of each

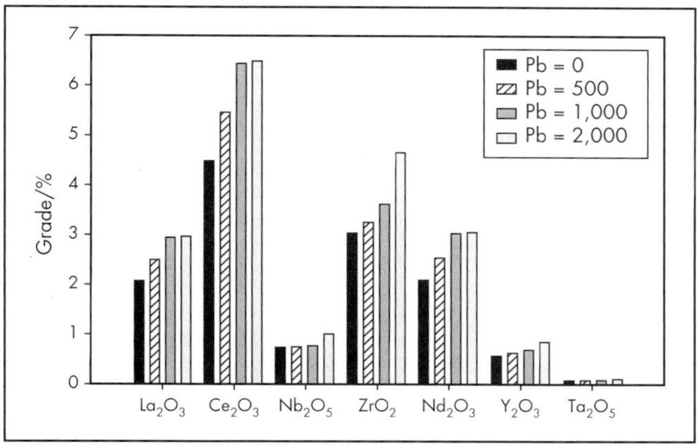

Source: Xia et al. 2015a

Figure 12 Grade of concentrates with different doses (g/t) of Pb as $Pb(NO_3)_2$ using 2,000-g/t BHD as collector

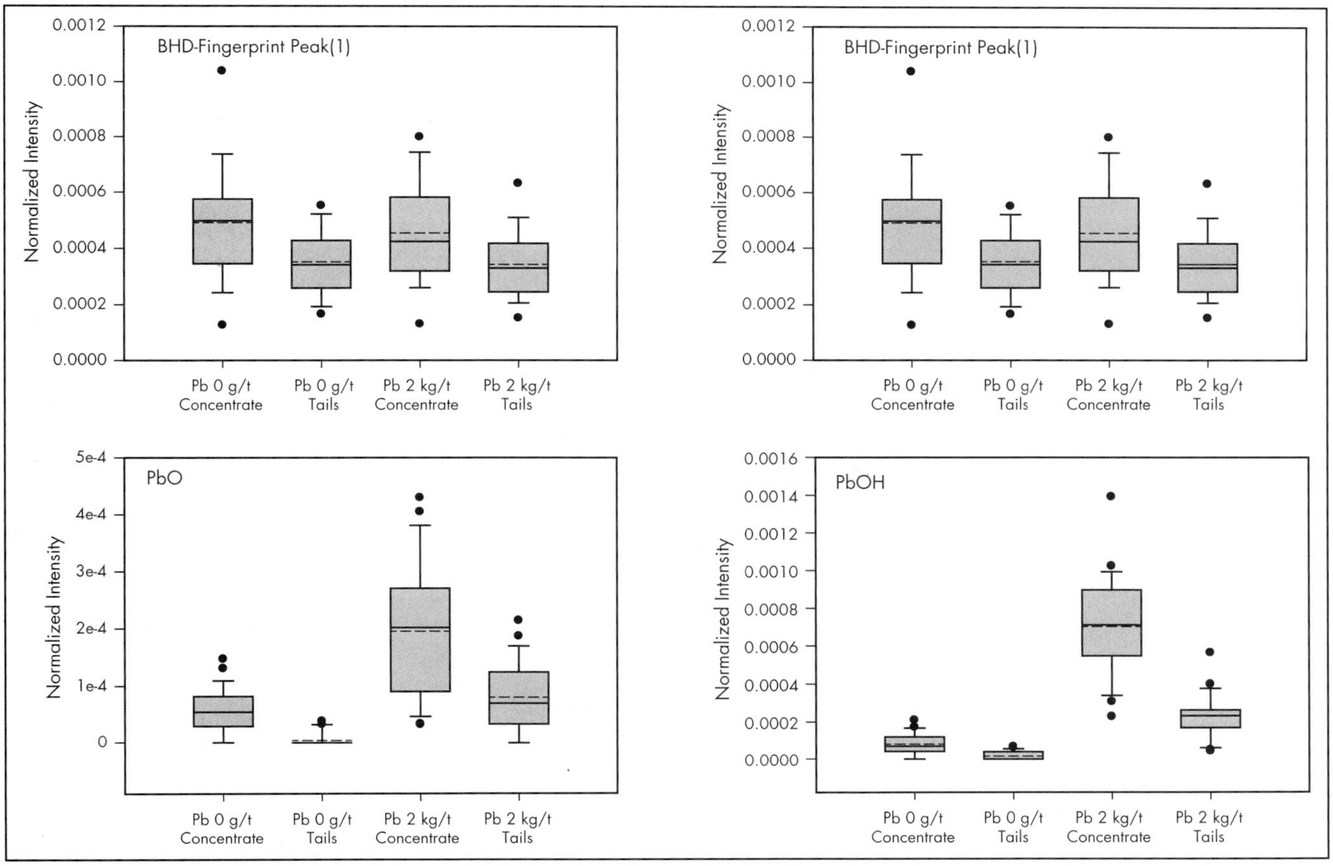

Source: Xia et al. 2015a

Figure 13 Normalized intensity of mass positions indicative of the collectors BHD along with PbO and PbOH on grain surfaces from the control test (no $Pb(NO_3)_2$ addition) and the test with 2,000-g/t $Pb(NO_3)_2$ addition

REE phase is controlled by the interaction of activators and collectors with the mineral surface. The data here indicate that this may well be linked to mineral solubility given that the easily recoverable carbonates are significantly more soluble than the REE silicates. The results of the testing suggest that for the Nechalacho deposit, the most efficient flotation recovery process for the REEs will likely involve a staged addition of both collectors and activators.

Chalcopyrite Leaching

Industrial acid heap and in situ leaching of copper minerals is now used for more than 20% of the world's copper recovery. Copper-containing oxide, hydroxide, sulfate, and silicate minerals acid-leach relatively quickly with Cu(I) oxides (e.g., delafossite and cuprite) being slower. Copper metal and chalcocite-series minerals leach more slowly still with chalcopyrite and bornite being the slowest to leach (Dreier 1999). As chalcopyrite holds approximately 70% of the world's copper reserves, the development of an effective chalcopyrite leach methodology is of considerable industrial interest but has yet to be developed in an economic process.

A considerable number of factors affect leaching rate and cost, including the acid consumption by gangue minerals (e.g., calcite) or, alternatively, acid production through pyrite dissolution. A thorough understanding of bulk mineralogy is therefore vital for calculation of acid balance. This is best approached using more than one technique if the mineralogy is relatively unknown. QEMSCAN and XRD quantitative analyses can be compared directly, but each has its own limitations as described earlier in the "Mineralogy" section. Alternatively, the bulk composition calculated from the mineralogy derived from either of these techniques can be compared to a bulk elemental analysis to check whether there is reasonable agreement.

A further factor that can affect leach rate is galvanic interactions with other minerals. If the rest potential of the value mineral is low as compared to other semiconducting minerals present, then the leach rate will either not be affected or may increase. In this case, the value mineral acts as the anode while the mineral with greater rest potential, often pyrite, acts as the cathode. If the rest potential of the value mineral is greater, leaching may be slowed (Li et al. 2013). The actual physical requirements for galvanic interactions to occur are not yet clear, that is, whether this can happen via solution "contact" or actual physical contact is required. Misdiagnosis of galvanic interaction can occur when leaching gives increased concentrations of Fe^{3+} either directly or via oxidation by O_2 or microbial activity. Importantly, galvanic interactions cease to be a consideration when the solution redox potential becomes greater than the rest potentials of each mineral (e.g., at >660 mV SHE, both pyrite and chalcopyrite will leach independently [Li et al. 2015]).

Solution speciation calculations can be very useful to understanding the behavior of a leach system, but these must be coupled to Eh and pH measurements to evaluate redox pairs—most commonly Fe^{2+}/Fe^{3+} and, in particular, their activity, not overall concentration. It is very common for detailed solution elemental analyses to be done on a regular basis but Eh to never be measured. Solution speciation simulation can be done rapidly and cheaply and is far more accessible than most surface analysis but is seldom carried out. It is not uncommon for an increase in some form of surface speciation or layer formation, most often measured using XPS, to be incorrectly correlated to slow or "passivated" leach kinetics (Crundwell 2013). The most common example of this is the formation of different species in surface layers on chalcopyrite, including polysulfide, elemental sulfur, and iron-containing precipitates (e.g., jarosite [Figure 14]). Although it is not possible to state that surface layer formation is never limiting, many recent studies have suggested otherwise, particularly where leach rates are seen to increase as surface layers are developing and thickening. Particularly if jarosite is present, this would suggest a low solution Fe^{3+} concentration and hence O_2-controlled oxidation, which is considerably slower. In these cases, a focus on solution speciation and gangue mineralogy together with the surface analysis is most valuable.

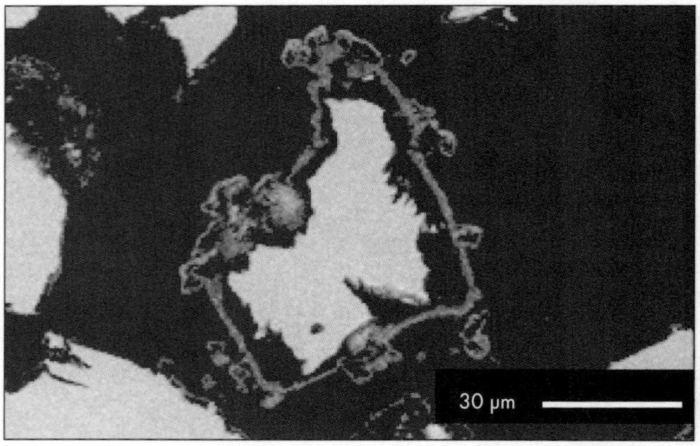

Source: Absolon 2008

A. SEM image of a polished section of chalcopyrite residue from *Sulfolobus metallicus* bioleach after 96 hours (36.5% Cu extraction) showing an unreacted chalcopyrite core surrounded by an S-rich Cu-depleted region containing P and an outer layer rich in Fe containing P and K. Chalcopyrite, sulfur, and spheniscidite were detected by XRD. The leach rate was slow but steady regardless of surface layer buildup.

Source: Li et al. 2016

B. SEM image collected from chalcopyrite leached with added 4 mmol iron at 750 mV and 75°C for 72 hours (approximately 100% leached) showing the formation of extensive surface elemental sulfur. The leach rate of the chalcopyrite was very rapid and showed no signs of inhibition.

Figure 14 Examples of non-passivating surface layers

Nickel Laterite Leaching

In contrast to Cu-ore leaching where the value minerals are almost always readily identifiable Cu-minerals, the Ni within nickel laterites is generally present at small concentrations in solid solution across a range of minerals. This complicates the accurate identification of the location of the nickel for process optimization and, importantly, also means that identification of unleached nickel-containing phases can be very difficult.

Because the nickel is not generally present at high concentrations, it is often not detectable using SEM (or QEMSCAN, detection limit typically 1 wt %) but can be quantitatively analyzed using electron probe microanalysis (EPMA). EPMA is similar to EDS as both rely on the measurement of element-specific X-ray fluorescence. EPMA is generally carried out using a small-incident electron beam size, 1–2 µm, and wavelength dispersive spectroscopy, which provide greater spectral resolution than available with EDS. Traditionally, EPMA has been accomplished as a spot analysis on previously identified grains. EPMA-based elemental mapping remains relatively uncommon but has been conducted on nickel laterites (Santos et al. 2015). This analysis is not yet practical on a regular basis, but where identification of losses is crucial, it is a reasonable approach. It is important to remember that this measurement is based on elemental analysis and is not a direct measure of crystallographic structure (i.e., phase), unlike diffraction techniques. Unfortunately, traditional bulk diffraction techniques are not of great use. Although leach residue minerals that may contain Ni can be identified, the precise location of Ni cannot be determined with confidence.

An alternative analytical approach to identify the nature of unleached nickel-containing phases is to use synchtrotron microprobe analysis. Microanalysis may be carried out using diffraction, X-ray fluorescence (XRF), and X-ray adsorption spectroscopy (XAS). XAS enables identification of the local environment around a specific element out to approximately 5 Å. This is useful to understand the location of the nickel within or on a mineral as identified using micro-XRD. A synchrotron provides highly intense X-ray beams that are many orders of magnitude greater than laboratory X-ray sources, enabling analytical approaches otherwise untenable. These facilities, which are widespread around the world, are available to the industry on a user-pays basis or via an experienced collaborator. Although not an approach to be undertaken lightly, the outcomes can be appreciable. For instance, Gerson and Smart undertook XAS analysis of a nickel laterite leach residue from an industrial pilot plant. Identification of the mineralogy of the nickel in the residue enabled process changes to be made, enabling recovery to increase from 70% to 96%. In Fan and Gerson (2015), synchrotron microprobe analyses (XRF, XRD, and XAS) were used to define the nickel mineralogy in saprolite and limonite feeds and leach residues (Figure 15).

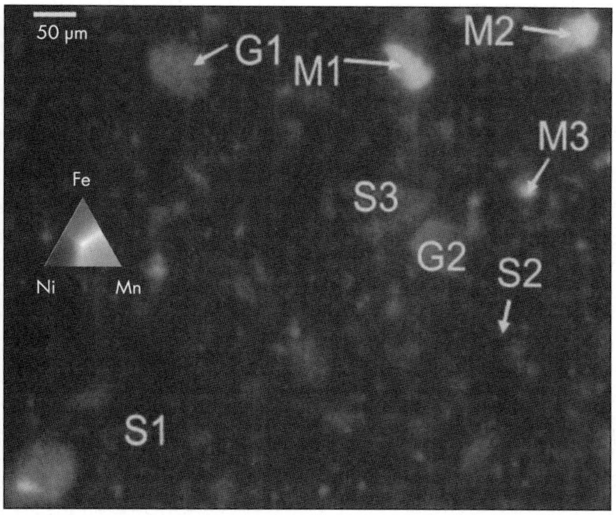

Adapted from Fan and Gerson 2015

A. 600 × 750 µm synchrotron micro-XRF map of a region of a nickel laterite ore showing the distributions of Fe, Ni, and Mn. It is clear that the Ni distribution is highly heterogeneous.

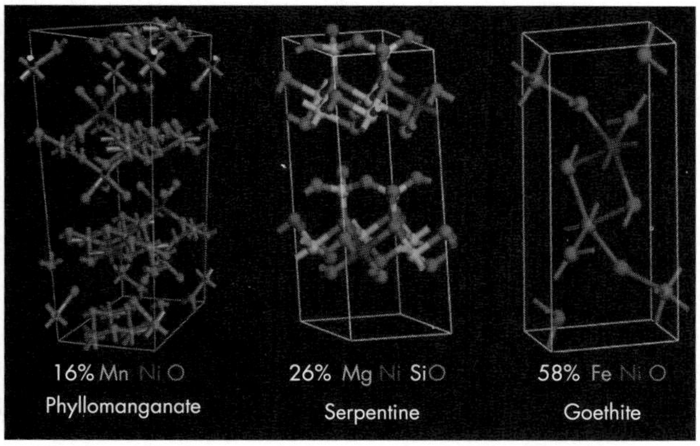

B. Percent location of nickel within the ore. The phase is determined by micro-XRD, and the amount and exact location of the nickel in each crystal structure is derived using micro-XAS analysis.

Figure 15 Location, mineral identification, structure, and distribution of Ni in a nickel laterite ore using synchrotron analysis

Gold Processing: Carbonaceous Matter and Cyanide Leaching

A major obstacle for effective gold recovery during the cyanidation process is the presence of inherent active carbonaceous matter (CM) that has the ability to adsorb, or preg-rob, gold from the cyanide leach solution. Probably the best example is found in the Carlin-type carbonaceous sulfide ores that show varying and significant degrees of preg-robbing activity (Stenebraten et al. 1999; Schmitz et al. 2001; Helm et al. 2009), with excellent reviews on the treatment on preg-robbing ores given by Miller et al. (2005) and Dunne et al. (2013). Many tests are available to determine the preg-robbing capacity of an ore; however, in many instances, the predicted values do not match those of the operating plant. Much of this is due to variability in the nature and presentation of the CM (Hart et al. 2011). In other instances, it may be related to the operation characteristics (pulp chemistry, competing species, etc.). This case study illustrates the combined use of SEM/EDS, TOF-SIMS, and Raman spectroscopy to understand Au losses linked to CM in the ore during the carbon-in-leach (CIL) process. An outline of a procedure where SEM/EDS and Raman spectroscopy, TOF-SIMS, and assays are used to evaluate the preg-robbing capacity of an ore is given in Hart et al. (2011).

The feed to the CIL is a flotation concentrate with upgraded Au values of ≈35 g/t. The CIL residue contained ≈6.5 g/t;

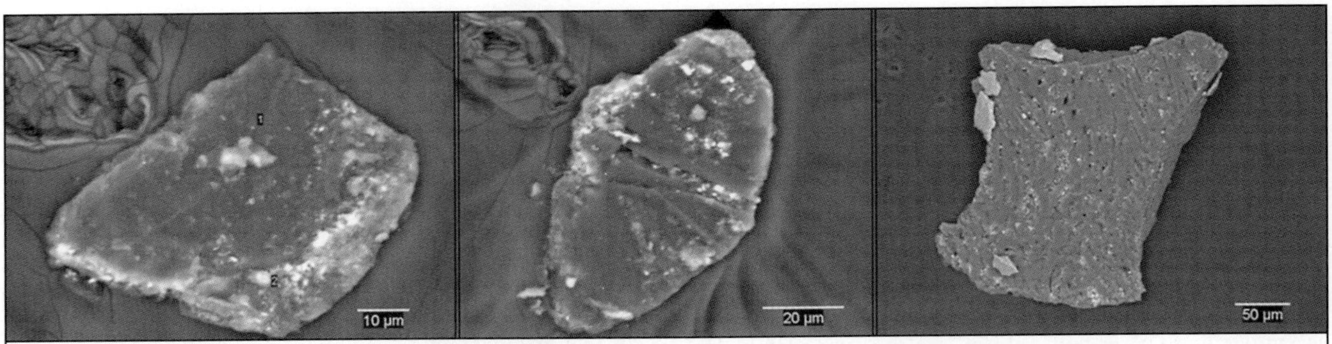

A. Disseminated CM, where the core is CM and the rim on the right side is composed of Si, Al, O, and K with minor Ca and Fe

B. Free CM with small siliceous particles on surface

C. Blocky, coarse, porous CM

Figure 16 BSE images of various forms of CM examined

≈20% of the Au is lost to the tails. The deportment study identified that approximately 10% occurs as locked Au grains and about 20% occurs as invisible Au in sulfides. The remaining lost Au was associated with CM. Detailed SEM/EDS evaluation of the CM in both the CIL feed and leach residue identified both disseminated CM and free CM (Figures 16A and 16B). Also identified were the blocky, porous, coarse grains (Figure 16C). These were only observed in the CIL tails sample.

TOF-SIMS analyses of the CM from the CIL residue (Figure 17) identified gold cyanide Au(CN) on the surface of a significant proportion of the CM grains, verifying that the CM in the sample did indeed preg-rob.

As part of the deportment study, several preg-robbing tests were performed to identify the maximum preg-robbing capacity of the ore. In a geometallurgical context, the measured variability in CM disorder in the ore can be very useful for predicting the preg-robbing response of various blocks to be mined. The degree of disorder of carbon in the CM has been linked to the preg-robbing capacity (Helm et al. 2009)—the less graphitic in nature, the greater the potential to preg-rob.

The CM was analyzed using Raman spectroscopy to determine the internal organization of the carbon (or degree of disorder) and to link this to the degree of preg-robbing. The method of deconvolution of the Raman spectra using the ratios of width (W) to height (H) of the D and G bands is described in Hart et al. 2011 and illustrated for activated carbon in Figure 18A. The rationale for the difference in CM disorder is the significant difference in the saddle height between the D and G bands. The Raman ratio data for the CM from the feed to the CIL process and that from the leach residue of the CIL in this particular case study are given in Figure 18B. Also included on the diagram are the Raman ratios for graphite, activated carbon (from coconut), and the activated carbon that is added to the process for Au recovery. The Raman ratio plot shows two distinct fields: the CIL feed and the CIL residue. The data for the CIL feed samples span the region from graphite to activated carbon, whereas the CIL residue samples show a much greater span and have a significant number of analyses clustered around the Raman ratio for the added activated carbon. The Raman data suggest that many of the carbon grains examined in the CIL residue sample may represent fragments of the carbon added to the process. The SEM/EDS examination of these particular grains shows that their morphology and texture are more representative of activated carbon that supports the interpretation from the Raman analyses.

In the final analyses of the 20% of the Au lost to the tail, slightly more than half of this was due to the inclusion of fragments of added activated carbon; the other half was preg-robbed Au on the inherent "native" carbon within the ore. These fragments of added activated carbon likely result from attrition of the larger chunks added to the CIL process. There is some speculation that the carbon regeneration process may result in a product that is more susceptible to breakage and therefore produces fine particles that are not recovered and are lost to the tails.

RESEARCH AND DEVELOPMENT DIRECTIONS

The current use of these techniques in flotation has been in troubleshooting where plant performance is not meeting targets, has deteriorated unexpectedly, or a change of ore (or stockpile) has occurred. Hence, most analyses have used some of the techniques (e.g., QEMSCAN, SEM, EDTA, TOF-SIMS), but they have not been integrated into the more complete problem-solving strategy (Figure 1). The KUC case study (Gerson et al. 2012; Triffett and Bradshaw 2008; Triffett et al. 2008) provides the model for the most effective use of this technique and information (i.e., where it is correlated to mineralogy, liberation, solution speciation and flotation conditions, and performance with feedback from changes suggested by the results). This is the direction in which development of its application in flotation is proceeding. Practical use of the TOF-SIMS application to process mineralogy has been described in the JKMRC handbook (Smart 2016). Other methods for surface characterization and innovations in research can be found in *Froth Flotation: A Century of Innovation* (Smart et al. 2007) and *Mineral Processing and Extractive Metallurgy: 100 Years of Innovation* (Smart et al. 2014a). Development of the statistical analysis integrated with plant performance is also being pursued within the problem-solving strategy (Gerson and Napier-Munn 2013).

The underutilized areas of development are in application to concentrate and heap leaching operations, where the potential of the full analytical strategy to reveal critical information on the reasons for slow and slowing rates of leaching is considerable, and in extraction processes in gold recovery where interferences can be identified and corrected. It has not yet been used in the statistical analysis mode for these purposes.

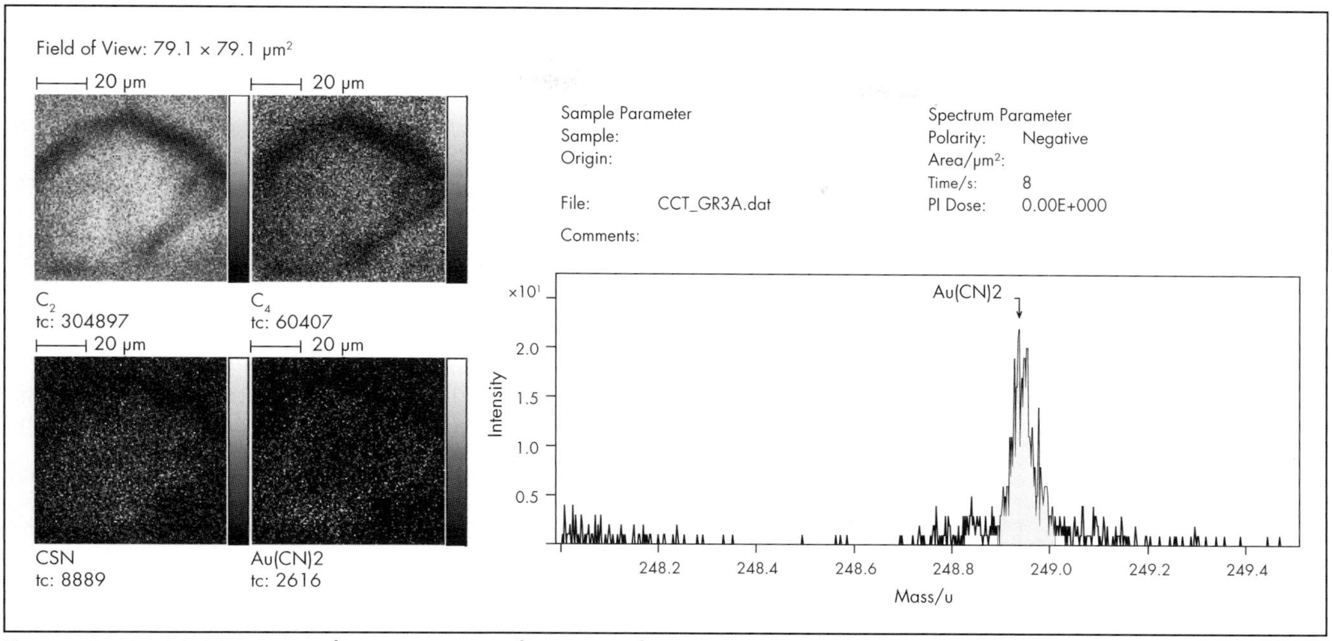

Figure 17 TOF-SIMS images and mass spectra in the region of Au(CN) on CM grain from cyclone classified tailings sample

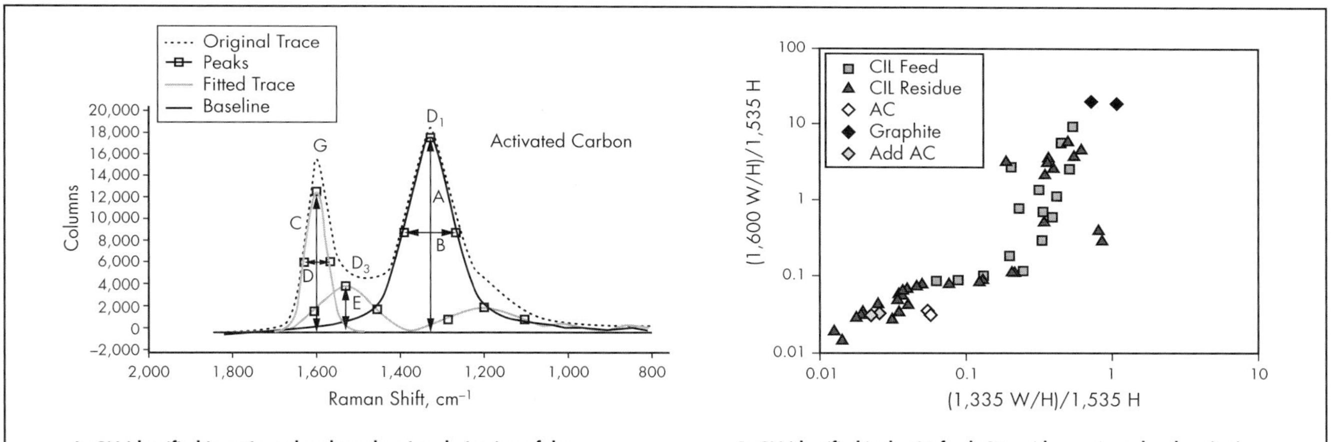

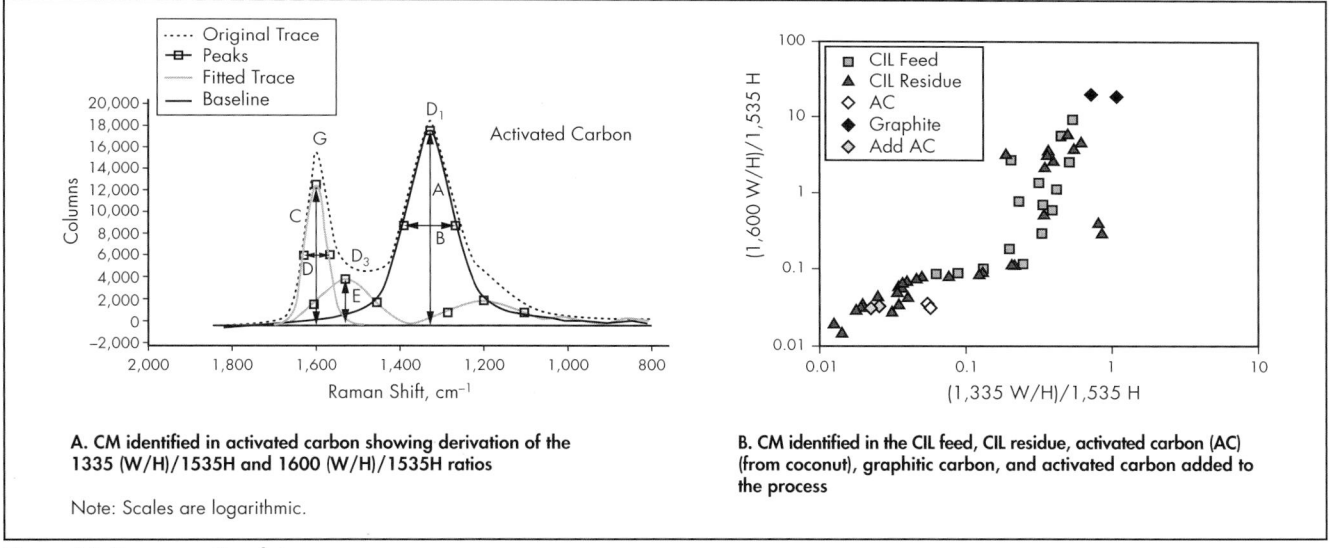

A. CM identified in activated carbon showing derivation of the 1335 (W/H)/1535H and 1600 (W/H)/1535H ratios

Note: Scales are logarithmic.

B. CM identified in the CIL feed, CIL residue, activated carbon (AC) (from coconut), graphitic carbon, and activated carbon added to the process

Figure 18 Raman ratio plots

SUMMARY

Poor flotation recovery or grade and misreporting losses can be caused by poor hydrophobic conditioning of the value mineral surfaces or inadvertent hydrophobic conditioning of gangue mineral surfaces. In leaching operations, interfering solution species, mineral dissolution, and surface layers can limit kinetics and total recovery. To define these limitations, a staged approach can be used to examine mineralogy, solution modeling, oxidation and dissolved metal ion interference, and analytical SEM before it is necessary to examine the surface chemistry in full detail. In addition to MLA and QEMSCAN analysis, losses due to mineralogy (e.g., minor phases, amorphous material, locking at the submicrometer level, and solid solutions) can now be probed using both compositional and diffraction analysis. Finally, however, the hydrophobic/hydrophilic ratio by particle and as a statistical distribution between different mineral phases across a flotation circuit from roughers, scavengers, and cleaners to tail may be needed before changes to mechanical or chemical operation can be effective. Surface analysis using statistical methods in TOF-SIMS can now provide this ultimate information. Direct comparison of intensities and PCA from hydrophobic (principally collectors) and hydrophilic species between feeds, concentrates, and tails can identify surface species discriminating in bubble–particle attachment. These techniques open new opportunities for improvements in recovery and in unrealized value if they are used in plant surveys.

REFERENCES

Absolon, V.J. 2008. A comparison of biological and chemically induced leaching mechanisms of chalcopyrite. PhD thesis, University of South Australia.

Allison, J.D., Brown, D.S., and Novo-Gradac, K.J. 1990. *MINTEQA2/PRODEFA2—A Geochemical Assessment Model for Environmental Systems—Version 3.0 User's Manual.* Athens, GA: Environmental Research Laboratory, Office of Research and Development, U.S. Environmental Protection Agency.

Biesinger, M.C., Polack, R., Hart, B.R., and Kobe, B.A. 2007. Analysis of mineral surface chemistry in flotation separation using imaging XPS. *Miner. Eng.* 20:152–162.

Boulton, A., Fornasiero, D., and Ralston, J. 2003. Characterisation of sphalerite and pyrite flotation samples by XPS and TOF-SIMS. *Int. J. Miner. Process.* 70:205–219.

Brinen, J.S., and Reich, F. 1992. Static SIMS imaging of the diisobutyl dithiophosphinate on galena surfaces. *Surf. Interface Anal.* 18:448–452.

Brinen, J.S., Greenhouse, S., Nagaraj, D.R., and Lee, J. 1993. SIMS and SIMS imaging studies of adsorbed dialkyl dithiophosphinates on PbS crystal surfaces. *Int. J. Miner. Process.* 38:93–109.

Chandra, A.P., and Gerson, A.R. 2006. A review of the fundamental studies of the copper activation mechanisms for selective flotation of the sulfide minerals, sphalerite and pyrite. *Adv. Colloid Interface Sci.* 145:97–110.

Chehreh Chelgani, S., Hart, B., Marois, J., and Ourriban, M. 2012a. Study of pyrochlore surface chemistry effects on collector adsorption by TOF-SIMS. *Miner. Eng.* 39:71–76.

Chehreh Chelgani, S., Hart, B., Marois, J., and Ourriban, M. 2012b. Study of pyrochlore matrix composition effects on froth flotation by SEM-EDX. *Miner. Eng.* 30:62–66.

Chehreh Chelgani, S., Hart, B., and Xia, L. 2013. A TOF-SIMS surface chemical analytical study of rare earth element minerals from micro-flotation tests products. *Miner. Eng.* 45:32–40.

Chehreh Chelgani, S., Hart, B., Biesinger, M., Marois, J., and Ourriban, M. 2014. Pyrochlore surface oxidation in relation to matrix Fe composition: A study by X-ray photoelectron spectroscopy. *Miner. Eng.* 55:165–171.

Chryssoulis, S.L., Reich, F., and Stowe, K.G. 1992. Characterization of mineral surface composition by laser probe microanalysis. *Trans. Inst. Min. Metall. Sect. C* 100:C1–C6.

Chryssoulis, S.L., Weisener, C.G., and Dimov, S. 1995. Detection of mineral collectors by TOF-LIMS. In *Proceedings of Secondary Ion Mass Spectrometry, SIMS X.* Edited by A. Benninghoven, B. Hagenhoff, and H.W. Werner. Chichester: John Wiley and Sons. pp. 899–902.

Coelho Software. 2016. TOPAS-Academic, version 6. Brisbane: Coelho Software.

Cox, J.J., Moreton, Ch., Goode, J.R., and Hains, D.H. 2010. *Technical Report on the Thor Lake Project, Northwest Territories, Canada.* NI 43-101. Toronto, ON: Avalon Rare Metals. pp. 1–350.

Crawford, R., and Ralston, J. 1988. The influence of particle size and contact angle in mineral flotation. *Int. J. Miner. Process.* 23(1-2):1–24.

Crundwell, F.K. 2013. The dissolution and leaching of minerals: Mechanisms, myths and misunderstandings. *Hydrometallurgy* 139:132–148.

Cruz, N., Peng, Y., Farrokhpay, S., and Bradshaw, D. 2013. Interactions of clay minerals in copper–gold flotation: Part 1—Rheological properties of clay mineral suspensions in the presence of flotation reagents. *Miner. Eng.* 50-51:30–37.

Cruz, N., Peng, Y., Wightman, E., and Xu, N. 2015a. The interaction of clay minerals with gypsum and its effects on copper–gold flotation. *Miner. Eng.* 77:121–130.

Cruz, N., Peng, Y., Wightman, E., and Xu, N. 2015b. The interaction of pH modifiers with kaolinite in copper–gold flotation. *Miner. Eng.* 84:27–33.

Dreier, J. 1999. The chemistry of copper heap leaching. http://jedreiergeo.com/copper/article1/Chemistry_of_Copper_Leaching.html.

Dunne, R., Staunton, W.P., and Afewu, K. 2013. A historical review of the treatment of preg-robbing gold ores—What has worked and changed. In *Proceedings of World Gold 2013.* Melbourne, Victoria: Australasian Institute of Mining and Metallurgy. pp. 99–110.

Fan, R., and Gerson, A.R. 2011. Nickel geochemistry of a Philippine laterite examined by bulk and microprobe synchrotron analyses. *Geochim. Cosmochim. Acta* 75:6400–6415.

Fan, R., and Gerson, A.R. 2015. Synchrotron micro-spectroscopic examination of Indonesian nickel laterites. *Am. Mineral.* 100(4):926–934.

Farrokhpay, S., Ndlovu, B., and Bradshaw, D. 2016. Behaviour of swelling clays versus non-swelling clays in flotation. *Miner. Eng.* 96-97:59–66.

Gerson, A.R. 2016. Chapter 14, Synchrotron based process mineralogy techniques. In *Process Mineralogy.* Edited by M. Becker, E.M. Wightman, and C.L. Evans. JKMRC Monograph Series. Indooroopilly, Queensland: Julius Kruttschnitt Mineral Research Centre.

Gerson, A.R., and Jasieniak, M. 2008. The effect of surface oxidation on the Cu activation of pentlandite and pyrrhotite. In *Proceedings of the XXIV International Minerals Processing Congress.* Edited by W.D. Guo, S.C. Yao, W.F. Liang, Z.L. Cheng, and H. Long. Beijing: Science Press. pp. 1054–1063.

Gerson, A., and Napier-Munn, T. 2013. Integrated approaches for the study of real mineral flotation systems. *Minerals* 3(1):1–15.

Gerson, A.R., Lange, A.G., Prince, K.P., and Smart, R.St.C. 1999. The mechanism of copper activation of sphalerite. *Appl. Surface Sci.* 137:207–223.

Gerson, A.R., Smart, R.St.C., Li, J., Kawashima, N., Weedon, D., Triffett, B., and Bradshaw, D. 2012. Diagnosis of the surface chemical influences on flotation performance: Copper sulfides and molybdenite. *Int. J. Miner. Process.* 16:106–109.

Grano, S., Ralston, J., and Smart, R.St.C. 1990. The influence of electrochemical environment on the flotation behaviour of Mt. Isa Mines copper and lead/zinc ores. *Int. J. Miner. Process.* 30:69–97.

Grano, S., Lauder, D.W., Johnson, N.W., and Ralston, J. 1997. An investigation of galena recovery problems in the Hilton concentrator of Mount Isa Mines Limited, Australia. *Miner. Eng.* 10(10):1139–1163.

Greet, C., and Smart, R.St.C. 2002. Diagnostic leaching of galena and its oxidation products with EDTA. *Miner. Eng.* 15:515–522.

Hart, B.R., and Dimov, S.S. 2011. Evolution of ion beam technology towards the needs of the mining and mineral processing industry: A historical Canadian perspective. In *50 Years of COMS.* Edited by J. Kapusta, P. Mackey, and N. Stubina. Westmount, QC: Canadian Institute of Mining, Metallurgy and Petroleum.

Hart, B.R., Dimov, S.S., and Mermillod-Blondin, R. 2011. Procedure for characterization of carbonaceous matter in an ore sample with estimation towards its preg-robbing capacity, World Gold. In *Proceedings of the Conference of Metallurgists: COM 2011.* Montreal, QC: MetSoc. pp. 35–50.

Helm, M., Vaughan, J., Staunton, W.P., and Avraamides, J. 2009. An investigation of carbonaceous component of preg-robbing gold ores. In *World Gold 2009 Conference.* Johannesburg: Southern African Institute of Mining and Metallurgy. pp. 139–144.

Jasieniak, M., and Smart, R.St.C. 2009. Collectorless flotation of pyroxene in Merensky ore: Residual layer identification using statistical TOF-SIMS analysis. *Int. J. Miner. Process.* 92:169–176.

JKTech. 2011. JKSimFloat software. Indooroopilly, Queensland. www.jktech.com.au/jksimfloat.

Johnson, N.W., and Munro, P. 2002. Overview of flotation technology and plant practice for complex sulphide ores. In *Mineral Processing Plant Design, Practice, and Control.* Edited by A.L. Mular, D.N. Halbe, and D.J. Barratt. Littleton, CO: SME. pp. 1097–1123.

Jordens, A., Cheng, Y.P., and Waters, K.E. 2013. A review of the beneficiation of rare earth element bearing minerals. *Miner. Eng.* 41:97–114.

Lastra, R. 2007. Seven practical application cases of liberation analysis. *Int. J. Miner. Process.* 84:337–347.

Li, Y., Kawashima, N., Li, J., Chandra, A.P., and Gerson, A.R. 2013. A review of the structure, and fundamental mechanisms and kinetics of the leaching of chalcopyrite. *Adv. Colloid Interface Sci.* 197-198:1–32.

Li, Y., Qian, G., Li, J., and Gerson, A.R. 2015. Chalcopyrite dissolution at 650 mV and 750 mV in the presence of pyrite. *Metals* 5:1566–1579.

Li, Y., Wei, Z., Qian, G., Li, J., and Gerson, A.R. 2016. Kinetics and mechanisms of chalcopyrite dissolution at controlled redox potential of 750 mV in sulfuric acid solution. *Minerals* 6(3):83.

Madsen, I.C., Scarlett, N.V.Y., and Kern, A. 2011. Description and survey of methodologies for the determination of amorphous content via X-ray diffraction. *Z. Kristallogr.* 226:944–955.

Malysiak, V., Shackleton, N.J., and O'Connor, C.T. 2004. An investigation into the floatability of a pentlandite–pyroxene system. *Int. J. Miner. Process.* 74:251–262.

Miller, J.D., Wan, R.-Y., and Diaz, X. 2005. Preg-robbing gold ores. In *Developments in Mineral Processing.* Vol. 15. Edited by M.D. Adams. Amsterdam: Elsevier. pp. 937–972.

Olsen, C., Makni, S., Hart, B., Laliberty, M., Pratt, A., Blatter, P., and Lanouette, M. 2012. Application of surface chemical analysis to the industrial flotation process of a complex sulphide ore. In *Proceedings of the 26th International Mineral Processing Congress: IMPC 2012.* New Delhi: Indian Institute of Metals.

Owusu, C., Brito e Abreu, S., Skinner, W., Addai-Mensah, J., and Zanin, M. 2014. The influence of pyrite content on the flotation of chalcopyrite/pyrite mixtures. *Miner. Eng.* 55:87–95.

Parolis, L.A.S., van der Merwe, R., van Leerdam, G.C., Prins, F.E., and Smeink, R.G. 2007. The use of TOF-SIMS and microflotation to assess the reversibility of CMC binding onto talc. *Miner. Eng.* 20:970–978.

Paul, J., Stubens, T.C., Arseneau, G., and Maunula, T. 2009. *Avalon Rare Metals Inc. Thor Lake—Lake Zone Mineral Resource Update.* Toronto, ON: Avalon Rare Metals. pp. 1–193.

Peng, Y., Bo, W., and Gerson, A. 2012. The effect of electrochemical potential on the activation of pyrite by copper and lead ions during grinding. *Int. J. Miner. Process.* 102:141–149.

Piantadosi, C., and Smart, R.St.C. 2002. Statistical comparison of hydrophobic and hydrophilic species on galena and pyrite particles in flotation concentrates and tails from TOF-SIMS evidence. *Int. J. Miner. Process.* 64:43–54.

Piantadosi, C., Jasieniak, M., Skinner, W.M., and Smart, R.St.C. 2000. Statistical comparison of surface species in flotation concentrates and tails from TOF-SIMS evidence. *Miner. Eng.* 13:1377–1394.

Ralston, J., and Healy, T.W. 1980. Activation of zinc sulphide with Cu^{II}, Cd^{II} and Pb^{II}: II. Activation in neutral and weakly alkaline media. *Int. J. Miner. Process.* 7(3):203–217.

Ralston, J., Fornasiero, D., and Grano, S. 2007. Pulp and solution chemistry. In *Froth Flotation: A Century of Innovation.* Edited by M.C. Fuerstenau, G. Jameson, and R.-H. Yoon. Littleton, CO: SME. pp. 227–258.

Santos, R.M., Audenaerde, A.V., Chiang, Y.W., Iacobescu, R.I., Knops, P., and Van Gerven, T. 2015. Nickel extraction from olivine: Effect of carbonation pre-treatment. *Metals* 5:1620–1644.

Saxby, J.D., and Stephens, J.F. 1973. Carbonaceous matter in sulphide ores from Mount Isa and McArthur River: Ark investigation using the electron probe and electron microscope. *Miner. Depositor* 8:127–137.

Schmitz, P.A., Duyvesteyn, S., Johnson, W.P., Enloe, L., and McMullen, J. 2001. Ammoniacal thiosulfate and sodium cyanide leaching of preg-robbing goldstrike ore carbonaceous matter. *Hydrometallurgy* 60(1):25–40.

Shackleton, N.J., Malysiak, V., and O'Connor, C.T. 2007. Surface characteristics and flotation behaviour of platinum and palladium arsenides. *Int. J. Miner. Process.* 85(1):25.

Shaff, J.O., Schultz, B.A., Craft, E.J., Clark, R.T., and Kochian, L.V. 2010. GEOCHEM-EZ: A chemical speciation program with greater power and flexibility. *Plant Soil* 330(1-2):207–214.

Smart, R.St.C. 2016. Chapter 12, Time of flight secondary ion mass spectrometry. In *Process Mineralogy.* Edited by M. Becker, E.M. Wightman, and C.L. Evans. JKMRC Monograph Series. Indooroopilly, Queensland. pp. 149–161.

Smart, R.St.C., Amarantidis, J., Skinner, W.M., Prestidge, C.A., LaVanier, L., and Grano, S.G. 2003a. Surface analytical studies of oxidation and collector adsorption in sulfide mineral flotation. In *Solid–Liquid Interfaces.* Topics in Applied Physics series, Vol. 85. Edited by K. Wandelt and S. Thurgate. Berlin: Springer-Verlag. pp. 3–60.

Smart, R.St.C., Jasieniak, M., Piantadosi, C., and Skinner, W.M. 2003b. Diagnostic surface analysis in sulfide flotation. In *Flotation and Flocculation: From Fundamentals to Applications.* Edited by J. Ralston, J.D. Miller, and J. Rubio. Adelaide: Ian Wark Research Institute, University of South Australia. pp. 241–248.

Smart, R.St.C., Skinner, W.M., Gerson, A.R., Mielczarski, J., Chryssoulis, S., Pratt, A.R., Lastra, R., Hope, G.A., Wang, X., Fa, K., and Miller, J.D. 2007. Surface characterization and new tools for research. In *Froth Flotation: A Century of Innovation.* Edited by M.C. Fuerstenau, G. Jameson, and R.-H. Yoon. Littleton, CO: SME. pp. 283–338.

Smart, R.St.C, Gerson, A.R., Hart, B.R., Beattie, D.R., and Young, C. 2014a. Innovations in measurement of mineral structure and surface chemistry in flotation: past, present and future. In *Mineral Processing and Extractive Metallurgy: 100 Years of Innovation.* Edited by C.G. Anderson, R.C. Dunne, and J.L. Uhrie. Englewood, CO: SME. pp. 577–602.

Smart, R.St.C., Xu, N., Fan, R., and Gerson, A.R. 2014b. A strategic approach to flotation losses due to mineralogy and surface chemistry. In *Proceedings of the XXVII International Mineral Processing Congress 2014: IMPC 2014.* Santiago, Chile: IMPC. pp. 1–12.

Stenebraten, J.F., Johnson, W.P., and Brosnahan, D.R. 1999. Characterization of Goldstrike ore carbonaceous material. *Miner. Metall. Process.* 16(3):37–43.

Stowe, K.G., Chryssoulis, S.L., and Kim, J.Y. 1995. Mapping of composition of mineral surfaces by TOF-SIMS. *Miner. Eng.* 8(4-5):421–430.

Sui, C.C., Lee, D., Casuge, A., and Finch, J.A. 1999. Comparison of the activation of sphalerite by copper and lead ions. *Miner. Metall. Process.* 16(2):53–61.

Toran, L., and Grandstaff, D. 2002. PHREEQC and PHREEQCI: Geochemical modeling with an interactive interface. *Ground Water* 40:462–464.

Trahar, W.J. 1976. The selective flotation of galena from sphalerite with special reference to the effects of particle size. *Int. J. Miner. Process.* 3:151–166.

Triffett, B., and Bradshaw, D. 2008. The role of morphology and host rock lithology on the flotation behaviour of molybdenite at Kennecott Utah Copper. In *Ninth International Congress for Applied Mineralogy: ICAM 2008.* Melbourne, Victoria: Australasian Institute of Mining and Metallurgy. pp. 465–473.

Triffett, B., Veloo, C., Adair, B.J.I., and Bradshaw, D. 2008. An investigation of the factors affecting the recovery of molybdenite in the Kennecott Utah Copper bulk flotation circuit. *Miner. Eng.* 21(12-14):832–840.

Tukel, C., Kelebek, S., and Yalcin, E. 2010. Eh-pH stability diagrams for analysis of polyamine interaction with chalcopyrite and deactivation of Cu-activated pyrrhotite. *Can. Metall. Q.* 49:411–418.

Xia, L., Hart, B., and Loshusan, B. 2015a. A TOF-SIMS analysis of the effect of lead nitrite on rare earth flotation. *Miner. Eng.* 70:119–129.

Xia, L., Hart, B., Douglas, K., and Zhong, H. 2015b. Two new structures of hydroxamate collectors and their application to ilmenite and wolframite flotation. In *Proceedings of the 47th Annual Meeting of the Canadian Mineral Processors.* Westmount, QC: Canadian Institute of Mining, Metallurgy and Petroleum. pp. 185–193.

Xu, N., Fan, R., and Smart, R.St.C. 2014. Cryo-SEM investigation of aggregate structures of clay minerals in flotation pulp and froth. In *Proceedings of the XXVII International Mineral Processing Congress 2014: IMPC 2014.* Santiago, Chile: IMPC. pp. 1–11.

Zhang, M., and Peng, Y. 2015. Effect of clay minerals on slurry rheology and the flotation of copper and gold minerals. *Miner. Eng.* 70:8–13.

CHAPTER 1.6

Laboratory Test Work and Equipment

Stefan Norgaard

This chapter provides a broad introduction to metallurgical laboratory test work, and specifically the common types of equipment used in commercial laboratories for various unit processes. The discussion focuses on bench-scale testing, with large piloting equipment excluded.

Metallurgical laboratory test work always features as the basis of any mining and mineral processing operation. Test work typically commences shortly after initial exploration drilling and, thereafter, concurrently proceeds throughout the various phases of the project.

Initial test work, which together with drill-hole interval analysis for ore reserve/resource modeling, consists of simple bench-scale tests to determine overall physical and metallurgical (extraction, beneficiation, and chemical) characteristics of a limited number of sample composites. As a project develops toward construction, metallurgical laboratory testing continues at a higher level of complexity to further clarify physical and metallurgical characteristics over a greater range of samples, optimize processing parameters of differing treatment techniques and processes, and investigate variability across an ore deposit using a constant testing regime. In the final phases of project studies, test work is conducted on larger sample masses (bulk batch testing) to produce various final saleable products and tailings streams for marketing and vendor testing. Finally, pilot testing is often completed using small-scale continuous plant equipment to demonstrate the viability of a selected processing circuit inclusive of all recirculating streams and site water.

Ultimately, the test-work results form the basis of engineering design criteria for the final plant circuit issued for construction. In essence, the more thorough and comprehensive the metallurgical test work conducted in the study phases, the better the probability of success the operation has of achieving design objectives.

Test work often continues during mine site operation on plant samples for ongoing project optimization and investigations, marketing samples, toll treatment umpiring, and to produce data for metallurgical accounting purposes. In addition, brownfields expansion projects can see a repeat of the initial project development test work for newly discovered ore bodies, capacity increases, or different processing requirements to accommodate changing mineralization at mining depth.

METALLURGICAL TEST-WORK LABORATORIES

Laboratories that conduct metallurgical test work can be grouped into four major categories, which are discussed next.

Private Group and Commercial Laboratories

The modern commercial metallurgical laboratory had its genesis in the private laboratory infrastructure set up by the larger mining houses in the latter half of the 20th century. Similar to the university facilities, these laboratories were equipped with an array of equipment to encompass most mineral processing techniques, but with the specific purpose of testing site samples for industrial project development. During the same period, private businesses were also setting up laboratories with metallurgical testing as a sole core business and service stream, offering industry-competitive test work without the heavy cost of sustaining internal infrastructure through a cyclical market. As of 2018, with few exceptions, most of the large mining houses have closed down and/or divested in-house laboratory infrastructure to the independent private sector. In turn, many of those same independent businesses have been acquired by the three main testing and inspection corporations: ALS (formerly Australian Laboratory Services), Bureau Veritas, and SGS (formerly Société Générale de Surveillance), all of which have worldwide footprints with very large laboratory hubs that cover virtually all testing processes. There still remain some notable independent exceptions such as Hazen Research in the United States, and Mintek in South Africa. In most of the world's main mining regions, there are numerous small-scale laboratories providing niche and expert laboratory services.

Private Research Laboratories

These facilities are set up and maintained by the larger equipment vendors and manufacturers (e.g., FLSmidth, Outotec, Metso, Delkor, and many others) and can be very well equipped. However, testing focuses exclusively on the

Stefan Norgaard, Operations Manager, Australian Laboratory Services, Australasian Institute of Mining and Metallurgy, Perth, Western Australia, Australia corporation's core intellectual property and/or processing techniques to match its equipment inventory (thickening, filtering, calcination/roasting, pumping, magnetic separation, etc.). The ultimate aim of all test work is to promote equipment sales, whether it be proving output parameters against equipment designs and models or general sample performance characteristics.

On-Site Mine Laboratories

These laboratories range in size commensurate with the processing operation but fundamentally are geared toward the preparation and analysis of plant control samples. Metallurgical testing equipment is limited to directly match the unit processes of the colocated industrial operation (small crushers, grinding mill and a flotation cell, leaching vessel or magnetic separator, etc.). Increasingly in recent decades, on-site laboratories have been downsized, with most metallurgical testing outsourced to better-equipped commercial laboratories, which precludes having to maintain the cost of a standing staff complement and equipment inventory.

University and Institute Research Laboratories

Situated on a university campus, these laboratories can be equipped with extensive small-scale equipment covering a wide range of comminution, beneficiation, and extraction unit processes. Some of the larger laboratories conduct a niche range of commercial services to augment private and government research grant revenue, but the primary purpose of these laboratories is conceptual research in conjunction with student learning.

The following sections describe laboratory equipment in terms of unit process, type and purpose, listing brands, and suppliers, if available, bearing in mind that some equipment is often custom fabricated. Where appropriate, some commentary is made on laboratory procedures related to the equipment and required sample masses.

CRUSHING AND SCREENING

As with most industrial-scale mineral processing operations, crushing is the first unit process requirement for most laboratory test-work programs. Avoidance of excessive fines production to produce an accurate and measurable target particle size distribution (PSD) is as imperative in the laboratory as it is in industrial practice. As such, closed-circuit screening at a target aperture is usually employed on crushed product at every stage (screening is discussed separately in the "Size Analysis" section). However, nominal crushing to a set crusher gap is sometimes conducted at very coarse sizes where an accurate PSD is not required, or where the sample is destined to be pulverized for assay.

Types of laboratory crushers tend to mirror industrial practice, but at smaller scale. Conventional single-toggle jaw crushers are typically used for initial coarse crushing through to midrange particle sizes. There are numerous manufacturers, makes, and models, of laboratory-scale jaw crushers, including Rocklabs, Essa (now part of FLSmidth), Jacques, and Shanghai Jianshe Luqiao Machinery. Laboratory-scale jaw crushers can range from 310 × 250-mm units that produce a 50–75-mm product, 250 × 150-mm units that produce a 40–15-mm product, and 100 × 75-mm units that can produce <6-mm products.

Finer top particle sizes suitable for treatment by laboratory rod and ball mills to produce material for metallurgical test work require secondary crushing using small-scale cone crushers, such as the Wescone, which can take up to 15-mm feedstock and produce a <2- to 3-mm product. The modern Rocklabs Boyd crusher is a dual-swing flat-faced jaw crusher that can take coarse 90-mm feed and produce a <3-mm product with multiple passes. Small rolls crushers can then take the cone or Boyd crusher product and generate finer distributions <1 mm; however, this is rarely required.

Figure 1 Single-toggle jaw crushers

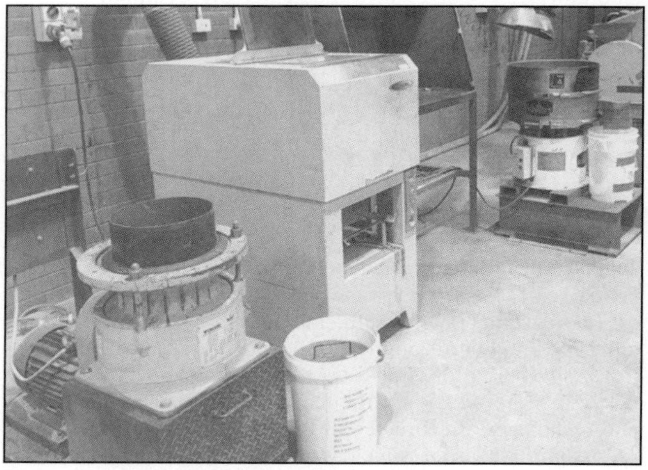

Figure 2 Wescone crusher on left

Figure 1 shows two single-toggle jaw crushers decreasing in size right to left. Figure 2 is the accompanying fines crushing setup with a Wescone crusher on the left, a Boyd crusher located centrally, and the vibrating screen to the right installed for control crushing (i.e., product removal). This sort of setup facilitates the full scope of control crushing preparation from coarse to fine in a relatively efficient batch operation. All of the previously mentioned crushers have a degree of closed-side-setting adjustment to enable sample PSDs to be tailored to target.

Prior to commencing any crushing of test-work samples, process engineers and metallurgists need to carefully consider the particle size requirements for each component of downstream test work. Always starting from coarse through to fine, this determines the type of equipment and crushing methodologies employed.

Selection and collection of coarse-particle specimens must be conducted before crushing. These specimens can be taken directly from drill-core trays and/or from larger run-of-mine (ROM) particles and are used for tests such as uniaxial compressive strength testing, apparent specific gravity (SG), and Bond impact crushing test work. Once test-work composites have been defined and specimens selected, a preliminary phase of coarse crushing is often necessary to facilitate subsamples at various coarse size targets for hardness and comminution testing. When the coarse program is completed, and if required, any comminution reserves recombined with feedstock, then control crushing to mid and fine sizes for grinding can commence. The most common laboratory crusher product sizes for mill feed are <2.0 mm and <3.35 mm.

Crushing of small sample parcels for most bench-scale test-work programs (10–1,000 kg) is usually conducted by manually pouring the material into the crushers from buckets or suitably sized containers. Similar sized crushers as described earlier can be integrated with small conveyor systems to emulate continuous closed-circuit crushing systems for larger parcels between 2 and 100 t (metric tons).

BLENDING AND SAMPLE SPLITTING

Together with control crushing, accurate and precise blending and splitting of crushed samples, and downstream dried ground/processed samples, is a fundamental requirement of quality test work. This cannot be emphasized strongly enough, as any poor blending and splitting practices can compromise the veracity of subsequent test-work results and lead to incorrect conclusions. Best sampling practice should therefore be implemented throughout, and staff should be trained accordingly. There are several standards available from the International Organization for Standardization to assist in this training.

The objective is to either split out a small representative subsample from a larger mass, and/or produce a series of replicate charges for downstream test work. These test-work charges must have a high degree of precision to enable accurate comparative analysis of multiple parallel test-work programs.

Blending and splitting are invariably conducted in the same action. As for crushing, blending and splitting always commences at the largest size and mass and proceeds down to smaller quantities and size requirements. The equipment type and size and by extension, the methodologies used, are dependent on the mass to be blended.

Bulk and Large Samples

For bulk ore samples in the range of 2–6 t, large conveyor-fed rotary drum (usually 200-L capacity) dividers are available. These are usually custom engineered and fabricated. Figures 3 and 4 show 12-drum and 4-drum rotary splitters, respectively.

For larger ore parcels (>5 t), initial blending is often conducted using a skid-steer loader (Bobcat), which effectively is a large-scale cone and quartering equivalent. Large stockpiles are moved and blended three times before being spread out into windrows. The Bobcat then collects perpendicularly across the windrows to place the material into drums. At this point, the material is well blended; however, when a high level of precision is required on a plant feed, additional homogenization of the blended drums can be conducted using rotary splitters and/or large riffle splitters. The drums are presented to the splitters in a set order to ensure that every drum is mixed together.

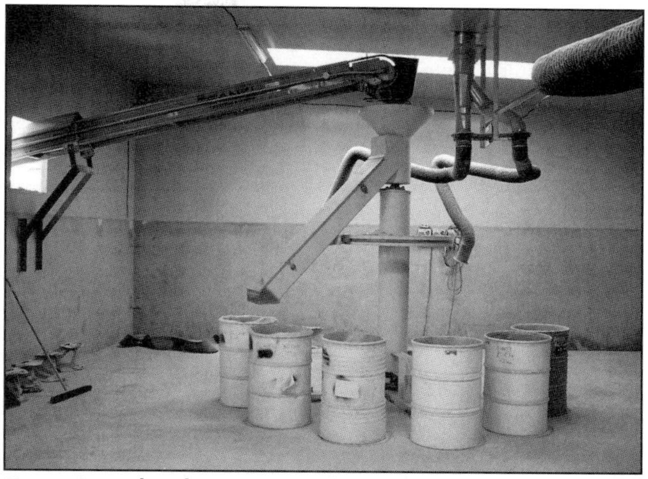

Figure 3 Twelve-drum rotary splitter

Figure 4 Four-drum rotary splitter

Standard Samples

Standard samples measure in the 12- to 300-kg range. Well-equipped laboratories invariably use a rotary sample divider (RSD) to blend and split samples greater than 12 kg. RSDs are specialist laboratory machines whereby sample is poured into a feed hopper equipped with a vibrating feeder. The feeder transfers a slow continuous stream of material from the hopper into a rotating series of circular segmented containers. Each of the segmented containers is thus a representative split of the original mass. Laboratory RSDs are available in various sizes, from small-scale 12-segment units designed to process 12–100 kg, to large 18-segment units capable of handling 300 kg and more. With multiple feed and discharge drums in certain combinations with multiple passes, RSD units can be used to blend and split much larger parcels than the capacity of the feed hopper. The most common brand is FLSmidth-Essa, but there are many others, including Marc Technologies, Rocklabs, Qingdao Yosion Labtech, and Retsch, and local engineering shops can expediently custom fabricate to specification if required. Figures 5 and 6 showcase RSDs.

Blending is done by recombining the split segments back into the feed hopper three times before removing segment sub-splits. Selected segments from larger RSDs can be transferred to smaller RSDs to continue with the blending and splitting process down to very small subsamples from large masses.

Small Samples

Samples of <12 kg are typically blended and split using a standard riffle splitter. Riffle splitters are common items that can be purchased in various sizes from a vendor, with differing riffle vane widths to handle a range of particle sizes. Stainless steel (S/S) is the most common material of construction. Given riffle splitting is not unique to mineral processing, there are numerous suppliers available worldwide, including FLSmidth-Essa, Eriez, Marc Technologies, Westernex, Humboldt, and Sietronics. In each riffle, a diagonal plate forces the flow of material bidirectionally into two collection boxes. Figure 7 illustrates riffle splitters.

Blending is achieved by splitting the sample, recombining the halves, and resplitting three times. Subsampling is conducted by splitting the sample once and setting aside one half, then taking the other half and resplitting into quarters. This process is repeated until the appropriate size sample is achieved.

PULVERIZERS

The primary purpose of pulverizers is to take crushed and ground dry subsamples and very quickly impart high energy into the material to pulverize the sample into a fine powder (often referred to as *pulp*). Pulverized product PSDs are time dependent but are typically between a P_{80} of 70 and 20 µm.

Pulverizing is a batch process and occurs via a set of rings or a puck, oscillating at high speed within a closed bowl. The different bowl arrangements for the common Essa LM2 pulverizer are show in Figure 8. Bowls are manufactured in different sizes to facilitate various sample masses, typically from 50 g to 3 kg. Realistically, 5 g is a minimum requirement for a small bowl pulverizer.

Trace elemental contamination from the bowl, rings, or puck to the sample can be a significant issue in some pulverizing applications, and for this reason, bowls, rings, and pucks are made from a range of construction materials including carbon steel, chrome steel, tungsten carbide, and zirconia.

There are few metallurgical test-work applications where a pulverized product is suitable, because of the excessive fines generation. Pulverizers are mostly employed to prepare subsamples for assay analysis, and for this reason, sample-to-sample contamination is an extremely important consideration. Typically, a barren silica flush is conducted between samples to clean the bowl internals of any remnant of the previous sample. Free coarse gold, however, can sometimes smear onto the bowl internals and cannot effectively be removed with silica flushing. These bowls can, therefore, be reserved for very-high-grade samples.

Mechanical assistance and automation is becoming increasingly prevalent in modern laboratory pulverizers to increase sample processing productivity. The Herzog machines pictured in Figure 9 are a good example of modern automated pulverizers.

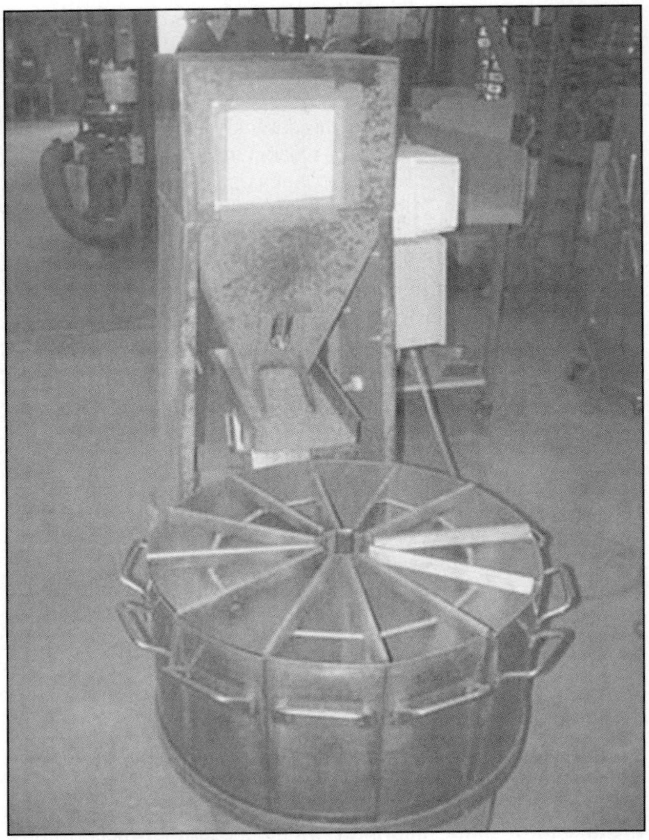

Figure 5 Rotary splitter, front view

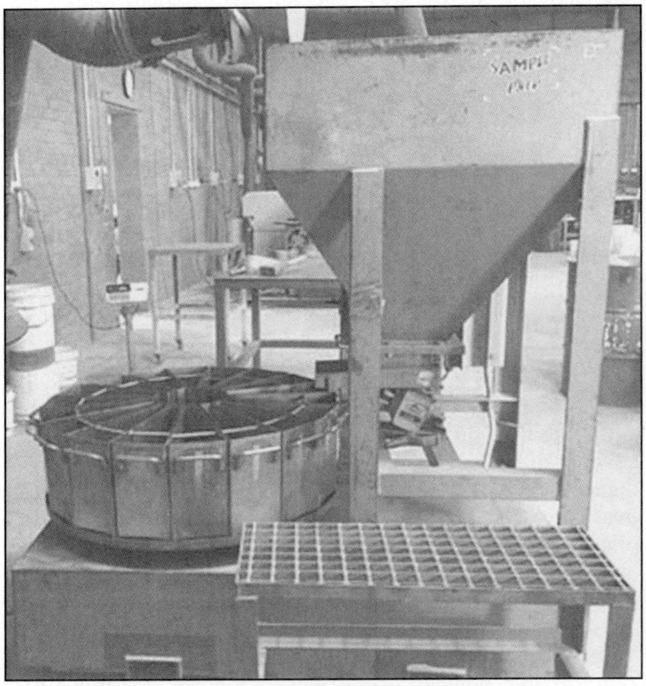

Figure 6 Rotary splitter, side view showing hopper and vibratory feeder

Courtesy of Cooper Technology

Figure 7 Types of riffle splitters

Figure 8 Essa LM2 pulverizer bowls

Figure 9 Herzog automated facility

SIZE ANALYSIS

Size analysis, sieve sizing, or screening is a critical component of metallurgical test work to determine PSD and is typically employed throughout every phase of a laboratory program. Individual chemical assays on discrete particle size fractions also serve to ascertain the metal deportment in the sample. Size analysis can be conducted wet or dry, and the employed methodologies and equipment are dependent on mass and PSD.

Dry Screening

Drying screening is routinely conducted on freshly crushed samples from 200 mm down to 2 mm, and on previously wet screened dried oversize material down to 25 µm. Laboratory screening of very coarse (+125-mm) particles is typically conducted by hand, using custom-cut frames or static screen mesh. Stacks of hand-shaken proprietary fabricated sieves can be used to screen 5- to 1-kg sample lots with particle distributions larger than 31.5 mm, as very little vibratory energy is required to allow passage of the particles through the screen mesh, and in fact, the particles can be oriented and selected by hand.

Although uncommon, orbital screen-shaking machines are available to sieve coarse particles from –125 mm to +10 mm. Notable suppliers are Marc Technologies and its version of the original *AC Cheers* screen design (Figure 10) that shakes a strapped-down stack of robust steel square 450-mm sieves. Endecotts also supplies the Titan 450 sieve shaker that can hold a stack of seven of the larger 450-mm-diameter circular sieves, with the added advantage of also being able to hold the smaller, finer aperture 200- and 300-mm sieves (Endecotts 2018).

Manufacturers of orbital sieve shakers accommodating the smaller mass-produced 200–300-mm sieves in the range of 30 mm to 20 µm are numerous and include W.S. Tyler (Figure 11), Endecotts, Haver and Boecker, Gaofu Sieving, Retsch, and Fritsch. All of these manufacturers supply the accompanying range of sieves over the full range of apertures. Typically, the sieves come in either S/S or brass casings; and in practice, there is very little difference in price or durability of the two materials.

Figure 10 Steel square shaker with screens

Figure 11 Ro-Tap with screens

Material for the sieve mesh itself is S/S wire or for the very coarse apertures, punched plate. Brass wire mesh is often used down to 250 µm but is unsuitable for acidic applications. Woven S/S can now be manufactured down to 20 µm and has a much greater corrosion resistance in acidic oxygenated environments. Woven nylon fabrics can also be used for particles smaller than 45 µm.

For sample masses, coarse screening using +50-mm static screen mesh can treat large sample masses in the hundreds of kilograms, which can mean screening the entirety of the sample mass. However, appropriately sized representative subsample splits are required to process through the mechanical sieve shakers.

Dry screening efficiency is a function of open area and the number of particle strikes presented to the screen aperture. Thus, subsample feed masses to the uppermost sieve in the stack are dependent on the aperture range and, moreover, the size of the bottom sieve aperture.

Typical sample masses for the dry coarse sieving units (–50 +2.0 mm) range between 25 and 5 kg depending on the PSD. Dry screening of finer distributions between top sizes of <10 and <2 mm is invariably done on dried oversize fractions after wet screening to remove fines. For a bottom screen aperture of 300 µm and larger, a 200-mm-diameter screen stack can comfortably receive lots up to 1 kg. Below 300 µm, the lot size needs to be reduced to 200–500 g to prevent overloading and potential blinding of the finer apertures.

Wet Screening

Laboratory wet screening is routinely conducted to assist the passage of particles through the screen mesh, especially at finer apertures smaller than 500 µm. There is also an additional washing action of fines that otherwise adhere to larger particles, or agglomerate with each other, and provide a biased size distribution under dry screening conditions. The medium for wet laboratory screening is almost exclusively the local site tap water, although for water-repellent materials, a wetting agent or dispersant can be added.

Laboratory wet screening of coarser size distributions is always conducted in stages—from coarse to fine—as finer screen meshes are fragile and cannot withstand the impact and abrasion of larger particles. The number of stages is dependent on the range of the size distribution. For coarse top sizes larger than 20 mm, dry screening is usually conducted first with the fines washed and captured from the fractions, either collectively on the stack of screens or from individual fractions.

For samples and fractions of –20 mm +250 µm, wet screening is typically conducted using a single vibrating screen for each stage to achieve sufficient screening efficiency and thereby accuracy. The wet screening stages (apertures) can range in decreasing order from 6.3 mm, 3.0–1.0 mm, 800–500 µm, and 300–250 µm. In all cases, the captured screen undersize can be fed directly to the next screen aperture as a slurry. The amount of wash water applied to the screen typically increases with decreasing aperture size. It may, however, be appropriate to settle, decant, and/or filter each screen undersize prior to the next wet screening stage.

Size fractions or milled slurries smaller than 300 µm are by far the most common samples requiring size analysis in laboratory test programs. These samples can usually be screened directly down to 38 and 25 µm, thus only requiring one stage of wet screening. The exception is magnetic material, such as magnetite, which can form robust agglomerates at fine sizes and usually requires wet screening over every aperture, as described earlier for coarse fractions.

At the final wet screening stage, both the wet screen oversize and undersize are captured, filtered, and dried. A sample of the dried oversize is then dry screened over a nest of sieves as described earlier. Importantly, the dried wet screen undersize must be added to the dry screening pan undersize to achieve the correct size distribution. These two samples must be blended together if an assay subsample is required. It is not

Figure 12 Wet screening station

meaningful to only take the dry sieve nest pan undersize for assay, as the subsieve PSD is quite different.

Wet screening coarse material larger than 10 mm is usually conducted on larger sample masses using continuous 600–1,200-mm Kason- or Russell-style vibrating screens. Feed masses are adjusted accordingly to achieve an appropriate loading on the mesh. Wet screening below 10 mm is conducted on smaller subsamples using Eriez-style wet screening stations for 200-mm Endecott sieves, as shown in Figure 12. Feed masses are similar to dry screening described earlier and decrease commensurately with sieve aperture to prevent overloading of the mesh.

Cleaning and Inspection of Screens

Screens require regular inspection for aperture (mesh) damage and of aperture blinding. Damaged meshes require immediate replacement. Screens with mesh blinding can be cleared by careful brushing using soft wire brushes. This must be conducted on the underside of the screens to avoid forcing oversize particles through the mesh, which can damage or compromise the actual aperture opening size. Ultrasonic cleaners can be employed to dislodge blinded particles. This method minimizes damage to the screen cloth.

Cyclosizer

The Warman Cyclosizer still remains the most common method of subsieve analysis when assays are required on the subsieve fractions. Currently, the manufacturing license for Warman Cyclosizers is held by Marc Technologies in Western Australia, part of ALS Global. A photograph of a typical Cyclosizer is shown in Figure 13.

The practical feed size distribution for a Cyclosizer is <50 μm +10 μm. Feed masses are restricted to 50 g for higher-SG material, such as magnetite, and up to 80 g for lower-SG siliceous material. Any excess of feed mass can overload the top collection chamber of each cone, which deleteriously affects the resultant size distribution. Multiple passes of sample splits can be conducted to increase recovered fractional masses from each cone.

Figure 13 Cyclosizer

An accurate solids true SG determination for each sample is required, and care must be taken to apply the correct elutriation flow rate and correction factors for water temperature. It is also important to be aware of the calculated empirical nature of Cyclosizer results when combining with the sieve size distribution data. Even under the best testing conditions, there must always be a small stepwise variation at the transition point between methodologies. Selecting a final sieve size of 45 μm in place of 38 μm or 25 μm can smooth out the data transition.

Laser Sizing

Sizing using laser diffraction has become a widely accepted alterative to sieve sizing and Cyclosizers in certain circumstances, especially when no chemical analysis is required. Laser diffraction determines PSDs by measuring the angular variation in intensity of light scattered as a laser beam passes through a particulate sample dispersed in an agitated liquid medium. Larger particles scatter light at small angles relative to the laser beam, and small particles scatter light at wide angles. The angular scattering intensity data is then processed with software using the Mie theory to calculate the size of the particles, and particle size is reported as a volume equivalent sphere diameter (Malvern Panalytical 2018).

Theoretically, the technology can measure particles into the millimeter range; currently however, from a practical laboratory perspective, laser sizers are most effective for PSDs below 100 μm. Larger particles can suffer inherent error from the equivalent sphere calculations together with shape correction factors. The machines use very small subsamples of <1 g and so representative subsamples can be questionable for coarse PSDs. Coarser particles are also abrasive on the machine lenses. However, for a large proportion of primary grinds, and especially regrind products where syringe or dip sampling is effective on stirred slurries, laser sizers can provide an accurate and expedient determination of PSD.

Caution needs to be exercised with materials susceptible to agglomeration, such as clays and magnetite, as the beam

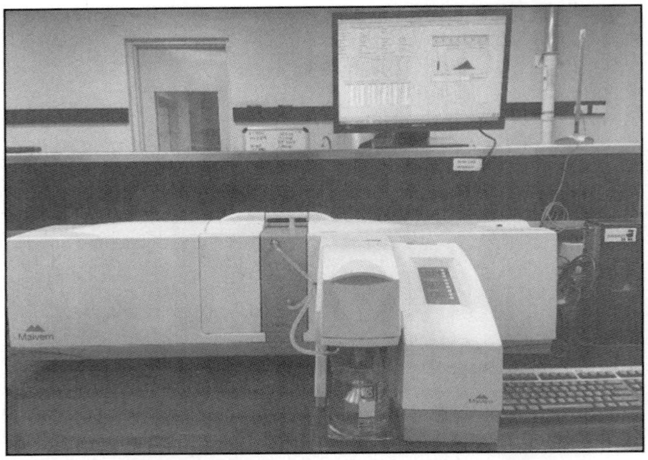

Figure 14 Malvern laser sizer

interprets agglomerates as a single particle and biases the size distribution result. In these cases, dispersants and/or ultrasonics can be employed to break up the agglomerates. A Malvern laser sizer is shown in Figure 14.

COMMINUTION TESTING

Different types of comminution testing are performed in metallurgical testing laboratories. Following are the most common.

JKMRC Drop Weight Test

The full drop weight test is a proprietary procedure developed at the Julius Kruttschnitt Mineral Research Centre (JKMRC) in Queensland, Australia. The test is conducted under license by approved laboratories using the drop weight apparatus.

Achieving the individual particle selections for a drop weight test typically requires between 75 and 90 kg of feed material consisting of either full PQ or HQ drill core or suitably sized ROM ore. The feed material is carefully control-crushed and screened for individual particle selection. Approximately 25 kg of actual sample material is selected in coarse-particle sets in the range of 63 to 13.2 mm (5× fractions of 30 particles) for drop weight impact breakage testing at three energy levels per fraction: 30 particles of –31.5 m +26.5 mm for apparent SG determinations and two 3.0-kg subsamples of –55.0 +38.0 mm for low-energy breakage and abrasion testing via a tumble mill. Where mass availability is limited, all unused material, including broken test products, can be recombined and prepared for further metallurgical testing.

The drop weight test is highly prescribed by the JKMRC proprietary data sheets, and all elements of the test must strictly adhere to the procedures and tolerance parameters. Raw laboratory data must be sent to JKMRC to generate the final test data and report. The test generates the appearance function (e.g., breakage pattern) of the ore under a range of impact and abrasion breakage conditions, which is subsequently reduced to three parameters: A, b (impact breakage), and t_a (abrasion breakage). These ore-specific parameters are used in the JKSimMet mineral processing simulator software and JKSimMet crusher model.

SMC Test

The SMC Test is another proprietary test developed by SMC Testing Proprietary Limited and is conducted under license using the exact same drop weight apparatus just described for the JK drop weight test. The intent of the test is to provide a cost-effective means of obtaining impact breakage parameters from drill core or broken ROM rock samples in situations where limited quantities of material are available.

Achieving the individual particle selections for an SMC Test typically requires approximately 25 kg of feed material consisting of either full or half PQ/HQ drill core. NQ core can also be used, or suitably sized ROM ore. Approximately 5 kg of actual sample material is selected to achieve a set of 120 particles within a single size fraction. The SMC Test can use one of three size fractions depending on the material characteristics.

The current recommended size fraction is the coarsest (–31.5 +26.5 mm range), as these particles provide superior confidence in the database. The next size fraction of –22.4 +19.0 mm is still common because it allows for the full use of quartered core feed samples, and the results are still reliable given the comprehensive data set gathered since the inception of the SMC Test. The finest size fraction of –16.0 +13.2 mm should only be used where the sample characteristics cannot provide the coarser particles. As per the drop weight test, all unused material, including broken test products, can be recombined and prepared for further metallurgical testing.

The SMC Test is also highly prescribed by the proprietary data sheets provided by SMC Testing Pty Ltd., and all elements of the test must adhere strictly to the procedures and tolerance parameters. Raw laboratory data must be sent to SMC Testing Pty Ltd. to generate the final test data and report, and this must include the source information of the sample. Raw data expressly cannot be provided to third parties.

The test generates the drop weight index (DW_i), which is a measure of the strength of the rock when broken under impact conditions and has units of kilowatt-hours per cubic meter (kW·h/m^3), as well as the A and the b parameters. The test also provides an estimated value of the t_a, as well as the M_{ia}, M_{ic}, and M_{ih} comminution parameters. The M_{ia}, M_{ic}, and M_{ih} parameters together with the M_{ib}, which is obtained as part of the Bond ball mill grindability test, can be used to evaluate autogenous grinding mill/ball mill and high-pressure grinding roll/ball mill circuits. The A and b values can also be used directly in JKSimMet (refer to Chapter 2.5, "Modeling and Simulation") for plant design, expansion, and optimization. The SMC Test does not generate the crusher parameters, which must be obtained through a full drop weight test.

Bond Impact Crushing Work Index

The Bond impact crushing work index (CW_i) test (Bond 1946) uses selected specimens in the nominal size range of +75 –51 mm. The specimens are placed on a pedestal with the parallel surfaces (and the smallest dimension) perpendicular to the raised striking hammers, and subjected to progressively higher impact energy levels until fracture.

The selection of the CW_i specimens is particularly important for the test and also in the context of the test-work program. Consideration for specimen selection must be made in the initial stages of the preparation process, usually before any material is crushed. The selected specimens should be *natural* and represent the type of broken material to be presented in the feed to a full-scale crusher. Specimens should be of even aspect ratio and have two near-parallel faces for the hammers to strike. Specimens should not be dried. Typically, between 10 and 20 specimens are selected for each test, and obviously the larger sample set provides better statistical confidence in the results.

Worked samples, such as drill core or material cut with a saw, ideally should be avoided, but practically, this is often not possible when core is the only material available. As such, consideration must be given to the results in that worked specimens tend to give lower breakage energy values. Half core should only be tested as a last resort.

Under the best testing conditions, there is still a high degree of variability in Bond CW_i results, as there are many sources of variability relating to the quality of the contacts between hammers and specimens, the selection of the specimens, and the measurement of the force.

Bond Abrasion Index

Abrasivity is a term that relates to the ability of rock material to cause wear to structural materials after repeated contacts. Wear can be described as the process of damaging, diminishing, eroding, or consuming by long or hard use, attrition, or exposure.

The Bond abrasion index (A_i) test measures flow wear, which is the movement of material across a surface with little or no normal force (Bond 1963; Angove and Dunne 1997). The test relies on measuring the loss of mass from a metal paddle that counter rotates in a drum of rock particles sized to a specific range. The testing machine consists of an impeller rotating at 632 rpm within a contra-rotating drum. The impeller incorporates a Bisalloy 500-Bhn (Brinell hardness number) steel paddle.

The test requires 4 × 400-g charges control crushed and screened to –19 +12.7 mm. For planning purposes, a 5-kg subsample of –20-mm material, rotary or riffle split during the initial stages of sample preparation, suffice. The four lots of 400 g are successively processed for 15 minutes each in the testing machine. At completion of the test, the weight lost by the paddle is measured to 0.10 mg. The A_i is equivalent to the paddle weight loss in grams but is expressed as unitless value.

The test provides consistent results if the correct paddle metal is used and the other test parameters are closely followed. Various laboratories use slight variations in test parameters, such as drum speed, paddle material, and particle size, so caution should be exercised when comparing results from different sources.

Broad guidance for interpreting Bond A_i results is as follows:

- Nonabrasive, <0.01
- Moderately abrasive, 0.01–0.2
- Abrasive, 0.2–0.4
- Highly abrasive, 0.4–0.6
- Extremely abrasive, >0.6

Bond Rod Mill Work Index

Although rod milling is rarely used in modern comminution circuits, the Bond rod mill work index (RW_i) determination still holds significant value in observing the breakage behavior of larger particle sizes comparative with finer breakage mechanisms seen in ball mills (Bond 1961a; Angove and Dunne 1997). As such, RW_i is still a common requirement for a comminution program and is invariably conducted in conjunction with the Bond ball mill work index (BM_i) determination. The two indices are frequently compared as inputs to modern simulation programs that can mathematically model comminution circuits and predict power requirements.

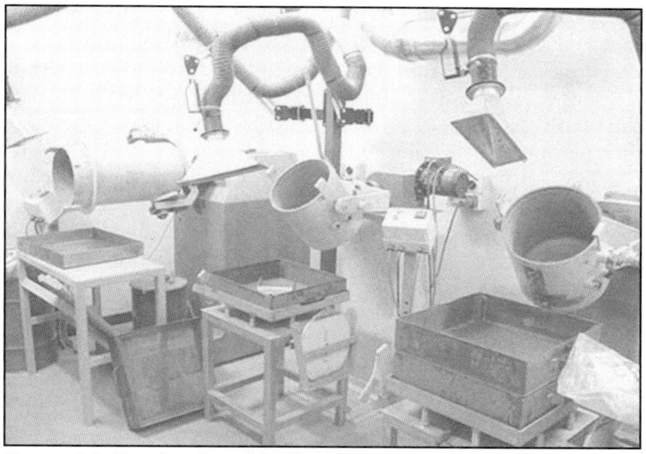

Figure 15 Bond rod and ball mill comminution station

The Bond RW_i requires feed material to be control crushed to <12.5 mm, and so this is also conducted during the initial stages of a sample preparation program before any fines crushing. Typically, a 10–15-kg subsample of coarse crushed material is riffle or rotary split for a Bond RW_i and at least 8–10 cycles are allowed. In situations of limited mass, unused feed charges can be recombined and crushed further for downstream test work. Feed charges are prepared by filling and compacting a 1,250-mL cylinder. The closing screen size for a Bond RW_i is fixed at 1,180 µm.

FLSmidth-Essa is a well-known supplier of Bond mills, and in many cases, the units are fabricated and supplied by local engineering vendors according to the Bond specification. Modern engineering, together with electronics, allow for excellent sample and media handling ergonomics, safety interlocks, and revolution counters. Figure 15 shows a Bond RW_i and BW_i comminution testing station.

Regarding the laboratory Bond RW_i, over many decades, some laboratory regions have replaced the original wave liner mill profile for a smooth shell, which produces slightly higher RW indices in the order of 1.0 to 1.5 kW·h/t. This only has significance if results of one mill type are being compared to that of the other as outliers against the prevailing engineering data sets for the project or region.

Bond Ball Mill Work Index

With ball milling still a fundamental grinding unit process of many mining operations, likewise the Bond BW_i determination is a critical inclusion in most comminution test-work programs (Bond 1961b; Angove and Dunne 1997).

The Bond BW_i test requires material crushed to <3.35 mm, and so representative subsamples can be split at the final stages of sample preparation. For sample mass planning purposes, 10–15 kg comfortably allows for 8–10 test cycles. For low mass availability, 6 kg usually facilitates a minimum of six cycles for material of typical bulk density, however, this is insufficient for high bulk density samples such as magnetite.

Special care should be taken during the control crushing of the feed sample to avoid excessive fines generation. Overcrushed samples with a high fines distribution will bias results against the true BW_i. Zero cycles (whereby product material is screened out of the feed lot without milling) can be conducted on materials with naturally fine PSDs. There is a point, however, where it is practical for other tests, such as the Levin grind test (Levin 1989), to be used for finer feed size distributions.

BW_i differs from RW_i in that the closing screen aperture is routinely changed to suit the likely PSD of the project. The most common closing screen size is 106 μm, but engineers should be aware that this is not standard and that there is little value in assessing the grindability (BW_i, in kilowatt hours per metric ton) of a sample at a product size well away from the projected grind size target.

Rules of thumb for selection of closing screen apertures in the root 2 series relative to the target P_{80} are as follows:

- 1 × root 2 series above P_{80} for samples exhibiting medium hardness/competency
- 2 × root 2 series above P_{80} for very soft samples
- ½ × root 2 series above P_{80} for very hard/competent samples

Common selections are 75 μm (for finer distributions) or 150, 212, 255, 300, and 355 μm for coarser applications.

All aspects of the Bond BW_i test are highly prescribed and standardized, so theoretically, there should not be significant variances between laboratories testing the same samples and ore types. BW_i, however, is susceptible to errors in precision and accuracy if any elements of the test are not properly conducted.

Insofar as the test rig itself, the mill shell integrity, mill speed, and revolution counts, together with maintenance of the media, are elements that need strict control. The mill charge inside a Bond mill is shown in Figure 16. The prescribed mass of the steel ball media charge must be exact for every test, which can be achieved by minor manipulations of the ball distribution. However, balls must be replaced periodically to account for wear.

Ancillary practices, such as accurate sieve sizing and sample splitting, have a high impact on the BW_i result. Having trained technicians able to discern when the test has reached equilibrium is also imperative for meaningful results. Slippage in any of these areas gives decreased precision and accuracy error.

LABORATORY GRINDING MILLS

Grinding holds the same fundamental importance in the laboratory as in commercial operations, with most test-work programs requiring further size reduction following dry crushing and of course, the addition of water to change the material into a liquid-slurry medium. Laboratory dry grinding is rare and is not discussed in this chapter.

Laboratory grinding is specifically focused on achieving a target PSD and much less on specific energies and efficiencies. A very high degree of sample recovery from the mill is also imperative for mass reconciliation of the subsequent test, and for this reason, small-scale laboratory mills do not have the bolt-in liner/lifter arrangement seen in industrial mills, which can trap material. The mill shell and liners are usually integral with each other and mostly metal. Modern mills are increasingly smooth profiles, although lifters can be welded in, or cast inserts can be manufactured to slide into shell sleeves.

Grinding is usually carried out in small lots to match the test-work mass requirements, which often requires multiple grinds per test. Typically, 1-, 2-, and 10-kg sample splits of

Figure 16 Ball media charge in a Bond mill

<2.0 mm or <3.35 mm crushed material are ground in discrete parcels to feed the downstream test work.

Achieving the target PSD is done using two methods: grind establishment or stage grinding. A grind establishment (or grind curve) is conducted when sample mass permits. Representative parcels of crushed ore are ground for various times, with the mill product sized over the appropriate sieve series. The P_{80} of the products at certain apertures can then be plotted with time to determine the grind time required to achieve a target PSD.

Stage grinding is conducted when sample mass is limited. In this procedure, the limited mass of material is ground for short increments, and the mill product for each grind is sized over the desired sieve aperture. The oversize is returned to the mill for another cautiously estimated time increment.

An alternative stage-grinding method is to take a small dip sample (from the agitated mill slurry charge), size it, and return the sample to the mill, then grind further for an additional time increment. However, inherent error from the dip sampling is introduced by this alternative method and it is not recommended.

One common request that process engineers often ask of laboratory metallurgists is to grind a bulk batch of slurry to an agitated tank, and then subsample for subsequent test work. Once again, the inherent subsampling error in this method is large, greatly affects the results of downstream testing, and should be immediately disregarded.

Mill selection is dependent on the sample material characteristics and the target PSD. Typical laboratory mills are rod, ball, and stirred media mills, which are discussed in the following sections.

Rod Mills

Despite being ostensibly obsolete in plant practice, rod milling still accounts for the vast majority of laboratory primary grinding. The reason for this is the natural tendency for tumbling rod media to preferentially grind the coarsest particles and thus prevent overgrinding and excessive fines production. The rod mill allows laboratories to accurately and consistently achieve a target PSD and thereby include grind size as an investigative variable, or alternatively, effectively eliminate grind size as a source of error in the test.

Courtesy of Eriez
Figure 17 Barrel mill and roller

The old-style independent barrel mill on a roller (Figure 17) is still in widespread use in many laboratories. It is a simple system whereby the grinding mill barrel is loaded with media, ore charge, and water on the bench, closed/sealed, and manually transferred on and off the rollers. Mill speed (measured in revolutions per minute) and grind time control are done through the roller mechanism. The manual lifting requirement introduces a weight limitation to the size of the barrels and also the potential for staff injuries. Some roller mill designs can have a central end opening to allow air into the grinding chamber, which can be advantageous to promote pulp-oxidizing conditions when using mild steel (M/S) media.

The modern suspended shaft and yoke design (Figure 18) alleviates most health and safety issues with a three-way pivot-locking pin arrangement that allows the supported mill barrel to be tilted up for loading and media removal and cleaning, down for slurry discharge, and horizontal for milling. Enclosures and cubicles, with associated timers and control interlocks, provide additional safety measures. The independent skid platform also improves general loading and discharge ergonomics over roller mills by allowing the entire grinding procedure to be conducted in the same space.

FLSmidth-Essa is a well-known supplier of laboratory grinding mills and Eriez also supplies a range of roller mills. Increasingly, there are many manufacturers that can be sourced quickly online by searching for "laboratory grinding mills."

The most common mill size is the 10-L barrel (~300 mm long × 210 mm in diameter), which grinds 1-kg charges. The suspended yoke design can also be fabricated in larger sizes such a 40-L barrels, which can grind 10–15-kg charges. Other types of mills are readily available, such as small-scale trunnion bearing mills and girth-driven tilt mills, but these are usually larger-volume units for bulk batch milling or small-scale pilot milling.

Materials of construction for the mill shells and rod media are typically M/S and S/S. The most common arrangement is a combination sealed S/S shell with S/S rods, as this has been proven to correlate well to industrial mills. However, all other combinations are available, that is, M/S shell–M/S rods, M/S shell–S/S rods, and S/S shell–M/S rods. The roller-style mill barrels are also available in ceramic, and ceramic rods can also be sourced.

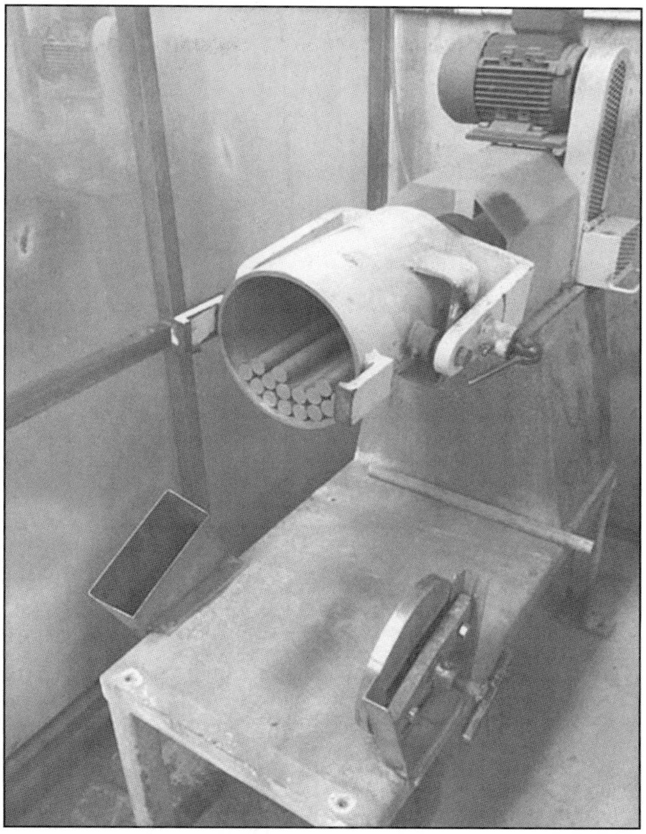

Figure 18 Rod mill showing suspended yoke design

Rod mills comfortably grind charges down to a target particle size of a P_{80} of 38 µm. Finer grinds as low as 20 µm can be achieved with highly extended grind times; however, in these cases, a ball mill or stirred mill are more suitable.

Ball Mills

Laboratory ball mills are most often the exact same rod mill shells or roller barrels simply filled with ball media instead of rods, and most of the preceding points on overall design and materials of construction can be applied. Other media types can be used such as cylpebs, which are slightly tapered cylindrical grinding media with a ratio of length to diameter of unity (Shi 2004).

Graded ball charges do not preferentially grind coarse particles, and as such, ball mills can be expected to produce a greater proportion of fines for the equivalent rod mill grind time, especially for the coarser P_{80} target particle grind sizes. For the same reason, ball mills can be advantageous for finer target P_{80} size distributions and can realistically grind down to 20 µm in practical grind times. Finer particle size targets require a fine grinding (stirred mill) apparatus.

Fine Grinding Mills

General fine grinding duties in the laboratory to achieve target particle P_{80} size distributions of 25–10 µm can typically be facilitated through a stirred media mill. Applications include very fine grinding conducted as a second stage after rod or ball

Figure 19 Laboratory stirred mill

milling and regrinding of intermediate test product streams, such as flotation rougher concentrates.

Few laboratory equipment manufacturers supply small-scale stirred media mills suitable for metallurgical test work; however, the apparatus is simple, and custom fabrication can be readily achieved by local engineering shops. An example of a laboratory stirred media milling setup is pictured in Figure 19. In some cases, a simple overhead drill machine can be modified to suit. Typically, the *pots* are S/S to hold media charges of 1, 2, and 4 kg. The stirrers are usually simple horizontal rods and pins or rings attached to the shaft. Types of media vary from steel, zirconia, alumina, glass, and sand. Media size is generally in the range of 1–5 mm.

Many of the major industrial fine grinding mill suppliers, such as Glencore (IsaMill), Outotec (HIGmill [high-intensity grinding mill]), Metso (stirred media detritor [SMD] and Vertimill), and FLSmidth (VXPmill) do produce laboratory-scale testing rigs that can run batch continuous tests to determine specific energies for various sample streams and ore types. Mass requirements for these mills vary from 20 to 100 kg per test, which obviously precludes these machines for general laboratory duties; however, such machines can be applied in bulk batch grinding and piloting duties. A 20-L IsaMill is pictured in Figure 20.

Figure 20 IsaMill

GRAVITY SEPARATION

Although gravity separation encompasses an array of industrial unit processes in the recovery of heavy minerals and precious metals, from a laboratory bench-scale test-work perspective, an indicative value of gravity recovery (mass yield and grade) suffices for most engineering studies. Larger-scale bulk or continuous testing can be commissioned by equipment vendors on specific unit processes (i.e., spirals and jigs) to firm up design criteria closer to construction.

The two most common laboratory machine types to provide reliable indicative laboratory gravity separation data, however, are shaking tables and centrifugal concentrators, which are described in the following sections.

Shaking Tables

The two major laboratory shaking tabling options are the Wilfley 800 (0.8 m^2; 640 × 1280 mm) and the Gemeni GT60 (1.3 m^2; 894 × 1490 mm). The Wilfley table is shown in Figure 21, and the Gemeni table is shown in Figure 22.

The Wilfley table is available from Holman-Wilfley in the United Kingdom and Motive Traction in Australia. The table has an adjustable table angle and oscillation stroke. Laboratory machines typically have the end launder compartmentalized to capture up to 10 graded products across the width of the table, and the front tailings launder can also be split into two to four compartments. The Wilfley can be operated batch continuous with samples from 1 to 25 kg, or continuously at <75 kg/h. Feed slurry is typically via an agitated header tank at 40%–70% solids with wash water added between 2 and 12 L/min.

The Gemeni table is available through Mineral Technologies in Australia. The table is designed to operate on a flat floor but has an adjustable oscillation stroke. The table is double sided and can be operated with one side only or both. There are three launder outlets per side. Batch samples as little as 1 kg can be treated with one side and ~20 kg with both sides in operation. The Gemeni can be operated continuously at 40 kg/h at 60%–70% solids using an agitated constant header tank with wash water added at ~12 L/min.

Centrifugal Concentrators

Centrifugal concentrators are primarily applied to the gravity recovery of gold and platinum group metals but can also find test-work applications in other heavy metal ores, such as mineral sands, chromite, tin, tantalum, tungsten, iron ore, and cobalt, where there are sufficient differences in the SG of the valuable mineral compared to the gangue mineralization.

The two laboratory centrifugal concentrators of note remain the KC-MD3 Knelson, supplied worldwide by FLSmidth, and the Falcon L40, supplied worldwide by Sepro Mineral Systems.

The KC-MD3 Knelson concentrator is supplied with a 50-mm 4× rib bowl that spins at 60 *G* and collects ~60–80 g of product material per batch run. Typical batch evaluation masses for the Knelson concentrator range from 1 to 20 kg.

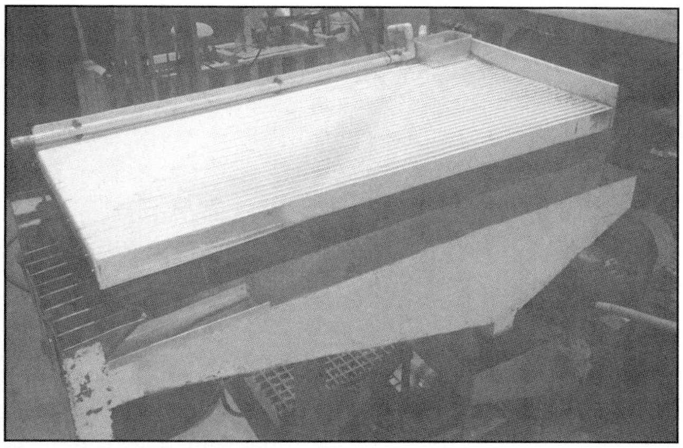

Figure 21 Wilfley shaking table

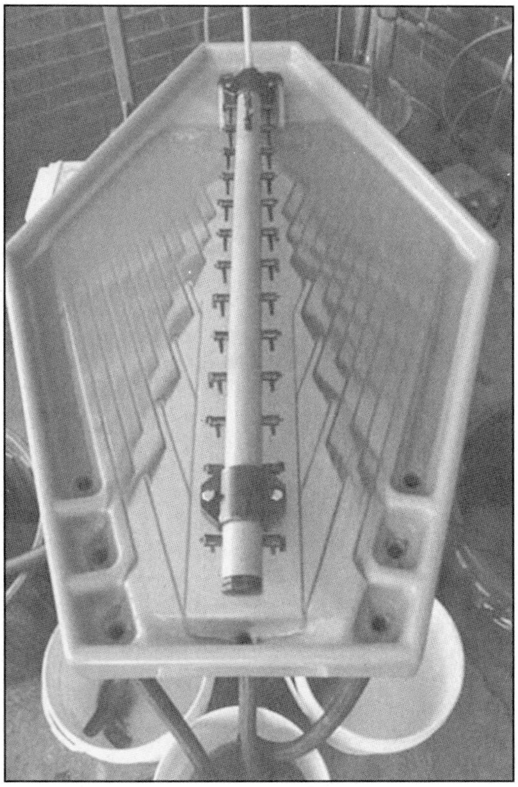

Figure 22 Gemeni shaking table

Figure 23 Knelson laboratory separator

Figure 24 Bowl for Knelson separator

Slurry feed is usually managed from a small header tank to ensure constant feed between 20 and 50 kg/h but can be manually tipped in for small samples. Fluidization water is supplied at up to 1 bar at 3–5 L/min and can be manually adjusted and measured prior to the test or by using a flowmeter. The KC-MD3 is pictured in Figure 23 and a photograph of the bowl in Figure 24. FLSmidth is currently also supplying two slightly larger laboratory units in the KC-MD4.5 and KC-MD7.5, which can process small pilot-scale flow rates (FLSmidth 2017).

The Falcon L40 is supplied with two types of 100-mm bowls that spin at 100–250 *G* using variable speed. The SB (semi-batch) dual-ribbed bowl takes fluidization water at <1.5 bar at 5–9 L/min and holds 80–120 g of product per batch run. The SB bowls have two bowl angels: 14 and 28 degrees. The 28-degree bowl is used for coarser samples. The UF (ultrafine) ribless bowl requires no fluidization water and holds ~200 g of product per batch run.

Typical batch test evaluation masses for the Falcon range from 1 to 50 kg. The slurry feed rate is typically managed from a header tank to ensure constant feed between 50 and 250 kg/h (with the lower feed rate for coarser samples) but can also be manually tipped in for small samples. The Falcon L40 is shown in Figure 25 and a photograph of the two bowl types in Figure 26, with the SB bowl on the left and the UF bowl on the right.

Figure 25 Falcon laboratory separator

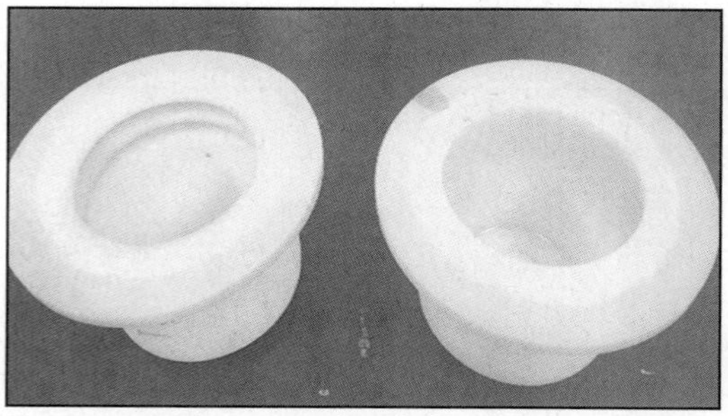
Figure 26 Falcon semi-batch bowl (left) and ultrafine bowl (right)

Gravity assessment via centrifugal concentrators is routinely added in bench-scale test-work programs ahead of small-scale tests, such as flotation and leaching, and sometimes after testing, as required. The bowl concentrates are recovered for downstream testing and/or analysis. For gravity gold concentrates, the two main treatment methods are mercury amalgamation and intensive cyanidation. Because of the tendency to over-recover precious metals on a bench-scale relative to full-scale operation, mercury amalgamation is the preferred method, as it generally only recovers the coarse/free gold, whereas intensive cyanidation also extracts fine-grained or partially liberated gold particles. Thus the gravity recoverable gold (GRG) content can be overestimated.

For gravity testing of larger sample parcels, an assessment of the batch size needs to be made relative to the estimated heavy mineral content versus the bowl size. Laplante and Doucet (1996) have described the industry-accepted method for the determination of GRG, whereby 50–100 kg of sample is passed sequentially through the Knelson concentrator at three successive grind sizes, ranging from a P_{80} of 850 μm to 75 μm. At each stage, the Knelson concentrate is collected and subjected to size by assay gold analysis. The final-stage Knelson tailings is also subjected to size by assay gold analysis, and using these data, an accurate GRG content can be determined.

FLOTATION

Froth flotation remains a widely used industrial beneficiation technique for most base metal mines and an increasing number of other nonsulfide ore bodies, such as iron ores (reverse silica flotation), phosphates, oxides, and rare earth minerals.

In recent decades, the design of industrial flotation cells has advanced dramatically, with manufacturers now producing ever larger and increasingly more efficient cells for the modern beneficiation plant. Traditional small-scale benchtop flotation programs are therefore still required to determine the fundamental flotation performance parameters respective to the ores, such as grind size, residence times, and reagent optimization. In this regard, bench-scale flotation test-work procedures and equipment have remained relatively static, as described in the following sections.

Bench-Scale Flotation Cells

The two stalwart benchtop flotation cells used in the overwhelming majority of test-work programs worldwide are the Metso Denver D12 and the Agitair LA-500. Both machine types can be bought under the original specification and branding; however, equivalent copies are now available from many online suppliers.

Both types of cells share a similar overall configuration (Figures 27 and 28), with a flotation mechanism suspended over a cell (tank), supported by a rear column. The Denver mechanism can be raised and lowered into and out of the cell using a rack-and-pinion gear set in conjunction with a spring-loaded shaft and can be locked at any desired height. The Agitair has a fixed mechanism whereby the shaft/rotor is removed through a flange. Typically, both cell types have installed air/nitrogen injection with the appropriate pressure and volume gauges and manual regulators. The Denver cell, however, can operate fully self-aspirating if required. Other standard features are power actuation and rotor-revolutions-per-minute control. The differences between the two cell types lies in the rotor and stator (diffuser) design together with the associated cell. Figures 29 and 30 showcase each design for comparison.

The Agitair design shows the larger rotor sitting centrally in the stator, which is installed as part of the cell. The rotor supplies the energy to suspend the slurry in the cell as well as shear the air/gas bubbles, while the stator ensures a balanced fluid-flow pattern inside the cell to further diffuse/shear the air bubbles. Agitair cells typically run at rotor speeds of 600–900 rpm and can realistically treat PSDs smaller than 300 μm. Coarser sizes tend to sand around the stator.

The Denver cell has an integrated rotor and diffuser design, which comes in large and small sizes and either open or closed; as can be seen, the cell has no other installed parts. The proportionally smaller rotor in the Denver cell typically runs at higher speeds of 900–1,400 rpm and delivers commensurately higher energy inputs and tip shear. This does draw criticism from some industry sectors; however, the Denver cell can suspend PSDs up to 1 mm, which can be advantageous in flash flotation test-work applications.

Each cell design has its supporters and detractors in different laboratories and regions; however, the arguments are

Figure 27 Agitair LA-500

Figure 28 Denver D12

Figure 29 Agitair-type stator and rotor

Figure 30 Denver-type stator and rotor

Figure 31 Small continuous pilot flotation cells

largely moot if one cell type or another is used consistently throughout a test-work program and, moreover, matches with local engineering scale-up conventions.

Masses that can be treated in these bench-scale machines range from ~250 g to ~4 kg, which is facilitated by a range of cells and rotor sizes. Cell size selection is also related to the target test percent solids, which is typically 25%–40% for rougher tests and <20% for cleaning duties. A range of cell sizes is pictured in Figure 31 for both types of machines and usually includes 0.25, 0.5, 0.75, 1.0, 1.6, 2.2, 4.4, and 6.6 L for the Agitair cell, and 0.5, 0.75, 1.0, 2.0, 4.0, and 8.0 L for the Denver cell.

Materials of construction for the cells are clear Perspex, S/S, and ceramic. Agitair cells are usually Perspex, and Denver cells are typically S/S. Both rotors and cells can be purchased readily online and/or fabricated by local engineering shops.

Larger Batch Cells

At some point, most test-work programs for the larger feasibility studies require a larger mass of final flotation concentrate and also tailings to be produced to facilitate vendor testing (thickening, filtration, geotechnical, viscosity, etc.) and samples for marketing. This can be achieved by conducting multiple 4-kg tests; however, a more expedient and cost-effective method is to use a larger batch flotation cell. In this case, the Agitair cell type comes readily available in 20-L, 40-L, 75-L, 130-L, and 300-L cell sizes. Obviously, flotation residence times need to be scaled up (usually by two to three), and consideration must be given to rotor sizes and energy inputs comparative to the smaller cells. Some comparative optimization tests may be required to ensure similar test performance when using the larger cell size. An example of a 40-L Agitair bulk batch cell is shown in Figure 32.

Other Ancillary Equipment for Flotation Test Work

In the wider context of metallurgical test work, flotation testing cannot be conducted without a range of other ancillary laboratory equipment and infrastructure. This includes items such as a stop watch, a range of syringe sizes and droppers for reagent addition, beakers and stirrers for reagent storage and preparation, a pH and oxidation-reduction potential (ORP) meter and probe setup, hand scrapers and paddles for froth removal, wash bottles to clean the cell and probes between stages, and aluminum trays for capturing the recovered froth.

Figure 32 Agitair bulk flotation cell

Other post-test requirements include Buchner vacuum filtration flasks and funnels, vacuum filters, filter papers, sample tags, and drying ovens. Larger external infrastructure includes a compressed air system, vacuum pump system, nitrogen bottle storage and delivery plumbing, and possible air-extraction systems to remove noxious reagent odors from the laboratory. Figure 33 shows a selection of the benchtop ancillary items mentioned.

MAGNETIC SEPARATION

Laboratory magnetic separation is a frequent requirement for many metallurgical test-work programs, particularly for iron ore, mineral sands, lithium, and vanadium projects. In this regard, laboratory testing has always been particularly well served by the equipment manufacturers of industrial-sized magnetic separators of all forms, and this continues to be the case.

Generation of the magnetic field is via either permanent magnets or electromagnets, and there is a vast array of machine types available to handle either dry or wet applications. Common small-scale laboratory dry and wet units are discussed in the following sections.

Dry Magnetic Separators

Many dry magnetic separators are available with a small benchtop footprint. These are induced roll magnets, permanent rare earth magnets, and electromagnets. Typically, these machines can treat dry material from 10 µm to 5 mm at gauss settings between 0 and 16,000. With this range, the entire spectrum of highly susceptible particles to weakly susceptible

Figure 33 Ancillary equipment for batch flotation test work

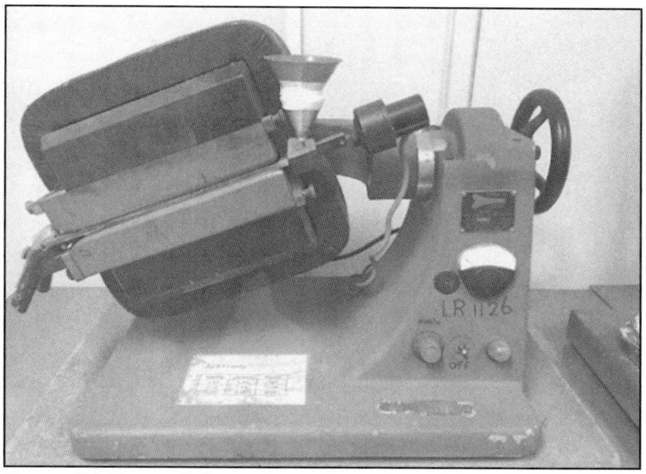

Figure 34 Frantz laboratory separator

paramagnetics can be assessed. Sample masses range from as little as 5 g to 10 kg, although most of the units can effectively be continuously run if product containers are replaced as required.

Laboratory data can be easily scaled for industrial-sized plants. Common brand names and types are

- S.G. Frantz LB-1 magnetic barrier separator (United States),
- Sepor MIH(13) high-intensity induced-roll magnetic separator (United States),
- Mineral Technologies Reading magnetic separators (Australia),
- Mecal magnetic separators (Australia), and
- Eriez 534 rare earth magnetic roll separator (worldwide).

Bench-scale dry magnetic separators are pictured in Figures 34 to 37.

For test-work programs requiring larger masses, Eriez in particular produces a range of laboratory-scale dry low-intensity, medium-intensity, and high-intensity magnetic separators that can process tens to many hundreds of kilograms of sample. These machines have an extremely long lifespan of many decades when maintained correctly, with negligible loss of field intensity. As such, the age of a machine should generally provide no barrier to achieve accurate results when employed by competent laboratory staff.

Wet Magnetic Separators

Wet magnetic separation is usually applied to samples post milling, typically <1 mm down to slimes fractions. Just as for dry magnetic separation, field strength is applied using either rare earth permanent magnets or current-induced electromagnets.

By far, the most common laboratory wet magnetic separation apparatus is the Davis tube, which remains essentially unchanged since its inception in 1921. Pictured in Figure 38, the Davis tube supports an angled oscillating glass tube placed between the poles of two magnets up to 4,000 G. Sample masses of 20 g are utilized for the test at PSDs of <1 mm. Typically, the sample charge is ground or pulverized to a range

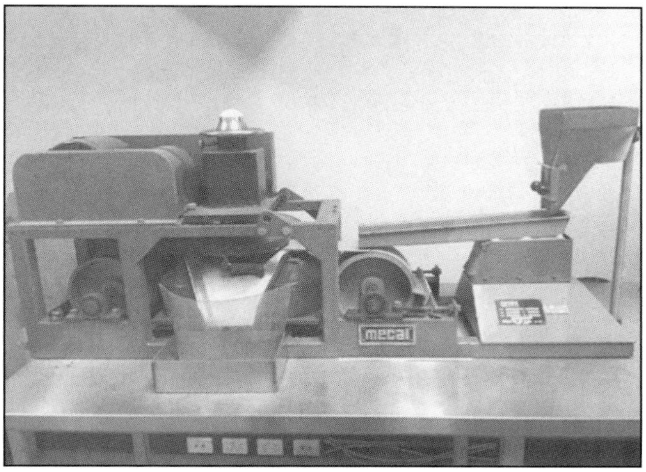

Figure 35 Mecal magnetic separator

Figure 36 Reading magnetic separator

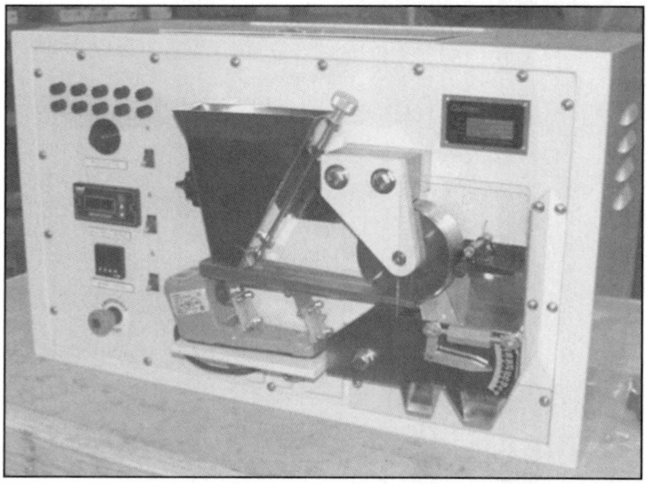

Courtesy of Sepor
Figure 37 Sepor magnetic separator

Figure 38 Davis tube

between 150 and 20 μm. Wash water is applied to the tube and, in conjunction with the rotating oscillation action, facilitates a vigorous washing action to reject weakly magnetic particles. Multiple Davis tubes are often set up together with LM2 pulverizers to enable mass production of Davis tube wash tests, which are sometimes incorporated into both geological and metallurgical modeling.

Proprietary wet separators can be grouped into three broad categories:

1. Low-intensity and medium-intensity magnetic separators (LIMS and MIMS)
2. Small-chamber vertical wet high-intensity magnetic separator (WHIMS)
3. Vertical pulsating high-gradient magnetic separator (VPHGMS)

Laboratory-scale units are available in all types. Notable brands are Eriez, Steinert, Longi, and SLon/Outotec.

A well-known and versatile laboratory LIMS is the Eriez L8 drum electromagnet (Figure 39A), which can characterize 5–10-kg samples at 500–1,100 G in batch continuous mode. These units can also be used in small-scale pilot operations in cascade mode.

For WHIMS applications, the traditional small-chamber vertical-flow WHIMS units are still available from Eriez (Figure 39B); however, these units are not effective in recovering fines (<100 μm) and have largely been superseded by VPHGMS units.

The VPHGMS system involves a vertical ring of rod matrices rotating through a bath of slurry surrounded by a high-gradient electromagnet (up to 1.8 T [tesla]). Magnetic particles are attracted to the rod in the matrix, and the bath is continually pulsed to allow washing and separation of the nonmagnetic particles. The magnetics are then washed from the matrix into a concentrate trough once the ring rotates out of the field. Different-sized matrix rods and spacing configurations are available to treat various particle sizes up to 3 mm. Cooling water, at approximately 4 m^3/h, is required for the electromagnets in the VPHGMS units. To avoid excessive water consumption, a cooling tower or heat exchanger infrastructure is usually needed.

The laboratory SLon 100 VPHGMS is a system using a single matrix in a pulsating bath that requires continuous filling with water to maintain the level and wash the matrix. Sample masses at 230 g per test can be processed. The slightly larger SLon 500 and the equivalent Longi 500 can treat slightly larger ore samples of 30–60 kg. Both the SLon 100 and the Longi 500 are pictured in Figures 40 and 41.

DENSITY MEASUREMENT AND SEPARATION

This section describes the broad methodologies and equipment used for density measurements and heavy liquid separation.

Density Measurement

Density measurement typically falls into three categories: bulk density, apparent SG (or in situ SG), and true SG. Bulk density determinations are simple tests whereby a vessel of known volume is filled with mass of sample and weighed. The result can be uncompacted or compacted (effected by tapping or vibratory energy applied to the vessel). The vessels themselves are typically M/S or S/S pots of various volumes, typically 5 or 10 L. For finer crush sizes, the Bond rod and ball mill pots can be used.

Apparent SG, or in situ SG, determinations use the principle of individual particle pycnometry. This method calculates density based on the mass differential observed between weighing individual particles in air and suspended in water. It can measure density of quite large particles, such as drill-core specimens, and is only limited by the mass capability of the balance and the support screen aperture. Typically, 8–100 mm particles are readily processed. However, to provide accurate sampling statistics, large numbers of particles must be processed (100–2,000).

The equipment required is a balance with underpan weighing capability, underpan weighing cage, a suitable support frame, water tank, temperature measurement capability, and the ability to calculate the density of each particle in real time to allow separation into the various density bands. Wax coating can be utilized around porous or broken samples.

True SG determinations are most commonly conducted using a gas pycnometer, utilizing ideal gases such as helium. These machines determine the difference in gas pressure between a sealed reference chamber and a testing chamber, each of known volume. The reference chamber is first brought

Figure 39 (A) Eriez L8 electromagnet and (B) Eriez WHIMS

Figure 40 SLon separator

Figure 41 Longi separator

to a stabilized pressure, which is recorded. The gas is then released into the testing chamber containing a known mass of sample. The drop in gas pressure from the reference chamber is proportional to the volume of the sample in the testing chamber, and thus the true SG of the sample can be calculated using calibration standards. This density accounts for all voidage space in the sample, including surface pores accessible to the gas, but not internal or *locked* pores.

The machines usually come with different-sized chamber cups to cater for sample sizes ranging from 5 to 180 g, depending on the sample type and PSD. Samples charges are typically crushed and/or ground to a P_{80} of 75 μm to 2 mm. Well-known brands are the Quantachrome Ultrapyc 1000 and the Micromeritics AccuPyc 1330.

Heavy Liquid Separation

Heavy liquid separation (HLS) utilizes liquids of varying densities to separate ores or minerals based on the density of the minerals. A sample is mixed with a heavy liquid with a density intermediate between the mineral densities to be separated and allowed to separate by gravity. Any material with a density lower than that of the liquid floats, while material of higher density sinks. The two fractions are then recovered for further testing or analysis.

The two basic HLS techniques are static separation and centrifugal separation. Static separations are typically performed on coarse material with particle sizes in the range of 0.5 to 30 mm, while centrifugal separations are performed on samples with particle sizes <1 mm, with fines <38 μm usually removed from the sample. There is some overlap, and techniques may be adjusted for specific sample programs.

Some equipment is common to both techniques, while centrifugal separation uses some additional specialized equipment. Essentially all of the equipment can be sourced and purchased through generic laboratory equipment suppliers such as Rowe Scientific, Cole-Parmer, John Morris, and Retche.

Typical equipment inventory includes the following:

- Various sized beakers up to 2 L
- Buchner vacuum filtration flasks and funnels
- Measuring cylinders
- Hydrometers in the range of liquid densities used, for example, 2.0–4.5 kg/dm^3
- Coarse sieves and nylon sieve mesh (38 μm or similar)
- Plastic funnels
- Laboratory centrifuge (Boeco C28A or similar, equipped to use 100 × 40-mm tubes)
- Drying ovens
- Balances (4 kg with increments of two decimal places or better)
- Rotary evaporators (Buchi Rotavapor R-220 or similar)

Although it is possible to separate a sample over the full 0.5–30-mm size range, typically samples are sized into several narrower ranges. This improves the separation efficiency and enables an assessment as to whether mineral liberation is affected by particle size.

Sample mass is limited by the volume of the separation vessels; average sample density; the combined mass of the sample and heavy liquid required to separate it; and the ability to manipulate the combined mass of container, sample, and liquid. A 1.5-kg sample typically requires 1–1.5 L of solution to affect a separation, so assuming a density of 3.5 kg/dm^3, the combined mass may be in excess of 7 kg.

Typically, a sample is placed into a beaker and the appropriate heavy liquid added. The float layer is removed by scooping off the surface with an appropriate sieve or by decanting the float onto a sieve. Residual heavy liquid is then decanted off for recycling. If separation is required at more than one density, then the next-higher-density solution is added to the sinks, and the process repeated. Where it is necessary to change liquids to achieve the required density, the sample needs to be thoroughly washed and dried prior to the next separation.

For samples with particle sizes <1 mm, centrifugal separation is usually used. Sample mass is limited by the capacity of the centrifuge tubes to approximately 200 g. Particle size is limited by the vacuum equipment used to recover floats to ~1.5 mm on the coarse size and by the screen used to collect the samples to 38 μm. Finer sizes can be recovered by filtration with a consequent reduction in throughput. Separations on iron ores sized to 5 μm have been successfully completed and on unsized pulverized gold ores. Where it is necessary to separate larger masses than either process can accommodate, multiple batches of material can be processed and the products combined.

Table 1 Commonly used heavy liquids

Name	Density, kg·dm^{-3}
Lithium polytungstate (LST)	2.920
1,1,2,2-Tetrabromoethane (Muthmann solution [TBE])	2.967
Sodium polytungstate	3.100
Diiodomethane (methylene iodide [MI])	3.325
Thallium formate + thallium malonate (Clerici solution)	4.250

Heavy Liquids

The range of available materials has reduced considerably from previous decades because of the cost and toxicity of heavy liquids. Principally, these liquids are high-density organic compounds or aqueous salt solutions. Commonly used liquids are shown in Table 1.

With the exception of the polytungstates, all of the heavy liquids in Table 1 are considered hazardous and/or highly toxic. Clerici solution has a very high potential for causing harm at low exposures and requires special precautions during manufacture, handling, and use. In most jurisdictions, these poisons are available only to specialized or permitted users under strict regulations.

Solution Recovery

Heavy liquids currently vary in price from ~US$100/L (Muthmann solution) to US$5,600/L (Clerici solution), and together with extreme toxicity, HLS laboratories take steps to recover as much of the liquid as possible after test work has been completed. This reduces cost, minimizes risk or harm to end users of the samples, ensures accurate analysis of the products, and minimizes potential environmental harm.

For organic liquids, the samples are repeatedly washed with the solvent used to dilute the reagent, typically acetone because it is both inexpensive and readily available. All washings are recovered and filtered prior to reprocessing. For inorganic liquids, the samples are repeatedly washed with hot demineralized water. All washings are recovered, the water is removed by evaporation or distillation, and the distilled water is discarded. Discarded liquids must meet local requirements.

DEWATERING

Most slurry dewatering of bench-scale metallurgical test-work products is conducted using either vacuum or pressure filters.

Vacuum and Pressure Filters

Small, readily filterable samples of low percent solids and/or small masses (approximately <2 L and <300 g) can be dewatered quickly and efficiently using the standard Buchner funnel and vacuum flask. The flask is connected to the vacuum hose, and the funnel is pressed into the neck with a rubber stopper seal. A sheet of filter paper is inserted into the funnel and wetted, the vacuum applied, and the slurry sample poured into the funnel. Water and solution are sucked through the filter paper and captured in the flask. Once all solution has passed through, the vacuum is closed, and the filter paper with wet filter cake is removed, usually for oven drying.

Funnels come in various sizes and diameters, usually 100, 200, and 300 mm, and are constructed of either plastic or ceramic. Buchner funnels are typically off-the-shelf items that can be purchased through the main laboratory supplier chains such as John Morris, Cole-Parmer, and Rowe Scientific.

The next size apparatus is the benchtop barrel vacuum filter. These units also employ a standard vacuum flask to capture solution and consist of a base plate and removable barrel. Barrels are usually S/S. Operation is similar to the Buchner funnel with a filter cloth and paper placed on the base plate and held in place by the barrel. The barrel can then be lifted off the filtered sample and the paper and filter cake removed for oven drying or further testing. These benchtop pressure filters are also freely available in laboratory catalogs and also supplied by Marc Technologies and others. Both the Buchner funnel and benchtop barrel vacuum filter are shown in Figure 42.

Pressure filtering is required for larger test-work masses and volumes, and/or less filterable material types. The almost universal filter type for laboratory pressure filtering is the barrel filter, which comes as a benchtop model that can filter up to 6 L (Figure 43) and also as a floor-mounted model (Figures 44 and 45). The benchtop model is similar to the vacuum version except with a sealed removable lid and clamp frame. Compressed air is applied through a regulator valve and gauge through the top lid.

The floor-mounted pressure filter comes in either single free standing or skid mounted in sets of four to six. The central barrel is fixed with a filter cloth and filter paper placed on the base, which is then screwed into place in the frame. The sample is poured into the barrel, and the lid is then also screwed into place to create a seal. Often, a piece of solid bar is used as a lever to achieve firm pressure on the rubber ring seals in the base plate and lid.

The standard 200-mm-diameter × 400-mm-height barrel can effectively filter 10-L slurry volumes and masses between 100 g and 5 kg, depending on the material PSD and filterability. It is recommended to limit filter cake sizes to <3 kg. Manufactures of barrel pressure filters include FLSmidth-Essa, Marc Technologies, Sipor, and many local manufacturers that have generic designs. These units are classed as pressure vessels and need design certification to be compliant with the relevant government standards, together with regular structural integrity testing.

Two critical operating considerations of the barrel pressure filter are (1) thorough cleaning of the barrel around the base plate seal between samples and (2) seal maintenance. Slurry blowouts can occur under pressure as a result of dirty and/or worn seals, which result in loss of sample mass; poor cleanliness can result in cross-sample contamination.

Figure 42 Barrel vacuum filter and Buchner funnel

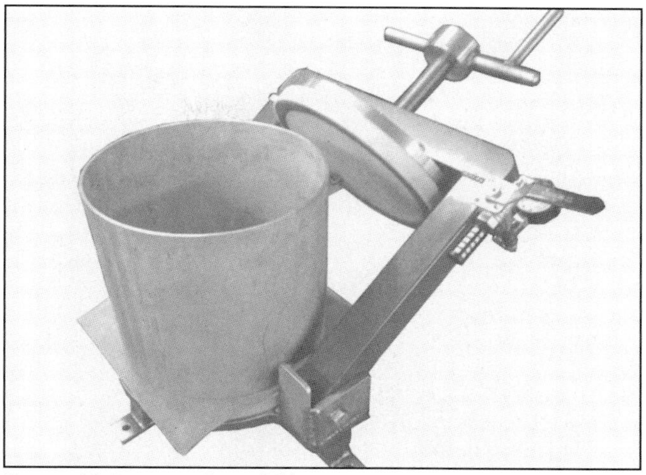
Figure 43 Pressure barrel filter

Bench-Scale Settling, Thickening, and Filtration Test Work

Flocculant selection test work is undertaken on slurry samples with the aim of selecting one or two optimum flocculent types that ideally result in rapid solids settling and high supernatant solution clarity. The test requires 1 kg to evaluate a large array of different flocculant types and dosages. The test is performed in 100-mL measuring cylinders, where the lowest dosage of flocculant (i.e., 10 g/t) is added to the slurry at the conditions to be evaluated, mixed, and the time taken for the solids to settle past free settling recorded. Additional flocculant is then added and the test repeated sequentially up to the highest flocculant dosage (e.g., 100 g/t). A typical flocculant screening testing setup is shown in Figure 46.

The next level of bench-scale testing is a settling rate test based on the Talmage and Fitch method. These tests typically require ~1 kg of solids in slurry form, which can potentially evaluate as many as ten tests, limited only by the percent solids. The test is performed at a selected solids grind size and slurry percent solids in a 1,000-mL measuring cylinder. The required dosage of flocculant is added to the cylinder, mixed, and placed on the bench to allow the solids to begin settling. Using the graduated marks on the side of the cylinder, at specified time intervals, the volume of settling solids is recorded over a period of 24 hours. Notes on the supernatant clarity are recorded during the test. The data of time versus settled volume is then used to determine free and hindered settling rates as well as the calculation of the required conventional thickener area.

Due to the engineering guarantees required for modern construction contracts, settling test work for equipment design has moved in recent decades from the commercial laboratory to the equipment vendors. These vendors, including Outotec, FLSmidth, Delkor, Diemme, and Ishigaki, perform flocculation, thickening, and filtration test work in-house and sign off on performance guarantees on their own test work. This is especially the case for filtration test work, which requires

Figure 44 Floor-mounted barrel filter

Figure 45 Barrel filter station

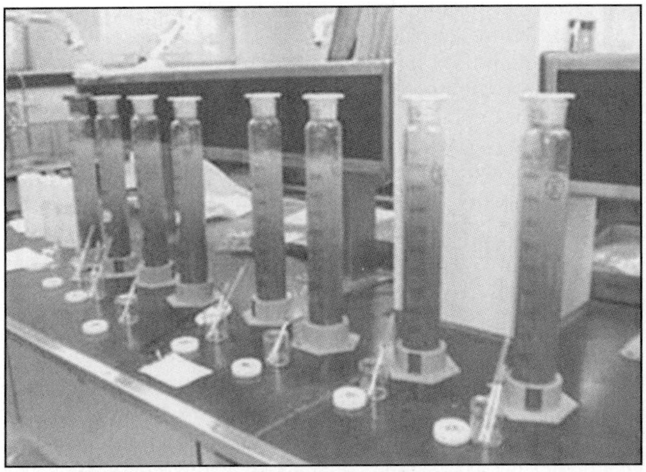

Figure 46 Flocculant screening station with measuring cylinders

complex apparatus and analysis to provide accurate information for equipment sizing.

Samples should be provided to the vendor laboratories at the appropriate grind size and under conditions where testing is required (i.e., leach feed or tailings, flotation concentrate or tailings, etc.). For both thickening and filtration test work, 15 kg of solids should be allowed for testing; however, where solids are not easily generated, that is, low mass pull flotation concentrates, a minimum of 6 kg should be allowed for evaluation. The sample should be provided to the vendor laboratory as a slurry (not filtered or dried prior to testing), and no flocculant/coagulant should be added.

VISCOSITY TESTING

Slurries generated from mineral processing operations and metallurgical test-work equivalents are invariably non-Newtonian fluids. Factors affecting the viscosity of a non-Newtonian fluid include percent solids, particle shape and size, temperature, and reagent additions. As all of these factors are target parameters in metallurgical test work, characterization of slurry viscosity is usually included at some point in most programs, from freshly milled feed to intermediate concentrates to final leach tails and so forth. Viscosity data assist design engineers in identifying potential rheology issues for a wide spectrum of shear rates for typical unit processes such as pumping, classification, agitation, and screening.

There are many different types of viscometers; however, rotational (cup and bob) viscometers are the most common in metallurgical laboratories and are ideally suited for efficient bench-scale testing of pre-prepared slurries taken directly from a range of test types. This category of viscometers is divided into two types. The Searle type has a rotating bob in a static cup of slurry, and the Couette type has the cup rotating the slurry around the stationary bob. Both types create a defined shear rate in the slurry and measure resistance to flow by the torque generated on the bob.

The Searle type is by far the more common, given the design can be more flexible and temperature can be better controlled. Trusted brands that cover all laboratory viscometer needs are Malvern Panalytical Bohlin Visco 88, Brookfield Ametek digital viscometer series, and ThermoFisher Haake Viscotester series. The Bohlin Visco 88 and Brookfield DV1 viscometers are shown in Figures 47 and 48. Currently, most modern viscometers have color digital screens and electronic outputs (not shown in the figures).

Viscosity testing determines the results for viscosity (measured in centipoise or pascal-seconds) and shear stress (measured in pascals) across shear rates ranging from 4.2 to 209.9 s^{-1}.

The standard sample requirement for bench-scale slurry viscosity test work is 1 kg of ore ground to the required PSD and subjected to the relevant test conditions, that is, after cyanidation leaching, flotation testing, thickening, and so forth. It is important to evaluate the slurry as soon as practical to negate the effects of aging (i.e., shear thinning or thickening) and settling in the cup. A 1-kg sample mass is sufficient to evaluate many different conditions and percent solids, testing at an initial high slurry percent solids then diluting to a lower percent solids and testing again, or pH adjustment from lowest to highest adding lime at each testing point.

Some rules of thumb for the interpretation of the test results are as follows: slurry viscosity <100 cps at shear rates of 119.4 s^{-1} is considered acceptable for high-shear-rate environments such as pumping applications; and slurry viscosity <3,500 cps at shear rates of 4.2 s^{-1} is considered acceptable for low- and medium-shear-rate environments such as mixing and screening applications.

During pilot-plant-scale and full-scale operations, the use of a Marsh funnel (Fann Instrument Company 2013) can provide a simple measurement of slurry rheology by observing the time it takes a known volume of slurry to flow from a cone through a short tube. The Marsh funnel should not be used as a substitute for the aforementioned viscometers, as it only provides measurement under one flow condition. The sample requirements for the test are generally ~950 mL of the slurry under the conditions to be evaluated.

LEACHING

Leaching of various descriptions (e.g., water, acid, caustic, ammoniacal, and cyanidation) often constitutes a large component of metallurgical bench-scale test-work programs. Leaching test work is predominately by batch, using a range of different vessels and equipment. Test parameters, especially chemical resistance and temperature, determine the vessel materials of construction, and also whether the vessel operates at atmospheric pressure (i.e., <100°C) or as a pressurized vessel.

The various types of atmospheric leaching and pressure leaching vessels are described in the following sections.

Atmospheric Leaching Using Bottle Rolls and Agitated Vessels

The most expedient and cheapest method of atmospheric leaching is the bottle roll. The *bottles* themselves are, in reality, cheap plastic screw-top jars and barrels of various sizes (500 mL and 1, 2, 4, 6, 8, and 20 L). Glass and ceramic vessels can be used but are rare for minerals testing.

Typically conducted at ambient temperature, milled slurry at the target P_{80} is transferred to the bottle and made up to the desired percent solids before chemical conditioning (i.e., reagent addition and air-gas sparging). The lid is closed, and the prepared bottle is placed on a parallel pair of rollers (one drive roller and one neutral) for a nominated period whereby the rolling action provides the required mixing/agitation of the conditioned slurry to facilitate leaching.

The bottles can stay on the rollers for the entire test duration or, alternatively, be periodically taken off at nominated times for reagent monitoring-maintenance and subsampling for analysis to provide kinetic data. The roller apparatus is usually variable speed and often has an intermittent start–stop capability. Typical speeds for leaching of 2–4.5 L bottles are 30–40 rpm. Suppliers of bottle roller units suitable for metallurgical testing are available from FLSmidth-Essa, Marc Technologies, and Wheaton.

Small-scale agitated tanks, or vats, provide the next level of laboratory leaching capability at ambient or mild temperatures. The vessels are typically custom fabricated and constructed of clear Perspex in various volumes and design geometries, complete with side baffles, and represent miniature versions of industrial-sized leach tanks. S/S is also a viable construction material, as is high-density polyethylene (HDPE), but these are less common. Sample masses are commensurate

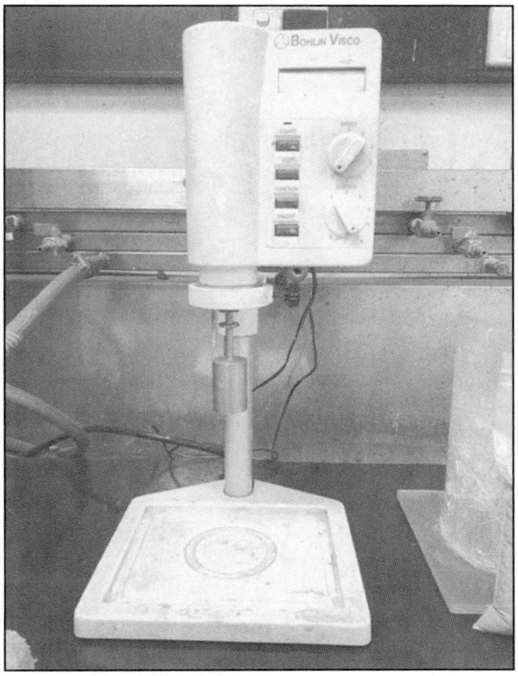

Figure 47 Bohlin viscometer

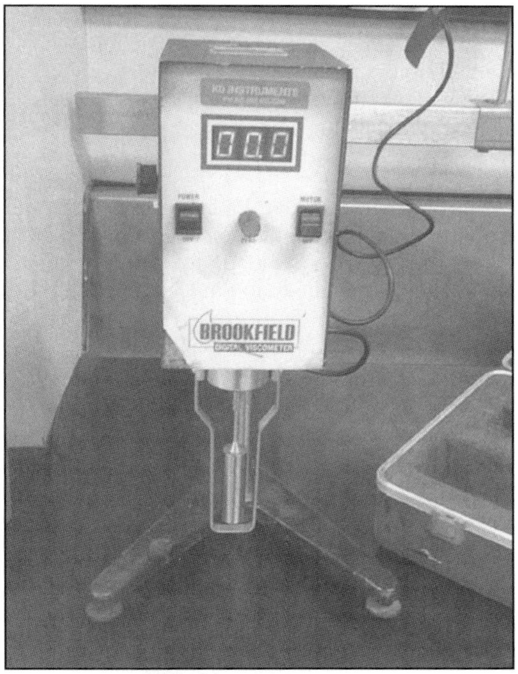

Figure 48 Brookfield viscometer

with the vessel size but typically range between 500 g and 20 kg on the bench. Agitation is provided by an overhead stirrer. The most common laboratory overhead stirrer is the IKA brand, which comes in a range of various powered units. All units are variable speed and some come with increasingly sophisticated digital displays and data outputs. Agitator speed is dependent on the agitator type, slurry PSD, and vessel volume. Typical agitator speeds for metallurgical leaching applications are around 200–300 rpm, sufficient to keep particles in suspension without excessive turbulence. The units allow for interchangeable shafts and impellor designs (i.e., flat blade, Rushton radial flow, axial flow, turbine, propeller, etc.).

Unlike the bottle roll, once the milled slurry has been placed in the vat, slurry conditioning, air-gas sparging, monitoring, and subsampling can take place continually or at any desired interval throughout the test. This is especially advantageous for testing the kinetic response of various test parameters. Heat input can also be facilitated with immersion heaters or external water baths, usually up to 40°C with Perspex tanks, and up to 70°C with S/S tanks or HDPE tanks.

Air or gas addition is supplied via a pressurized ring main with multiple outlets along the bench at each leaching station. Each outlet has an isolation valve followed by a regulated gas rotameter for flow-rate metering. A typical bottle roll testing apparatus is pictured in Figure 49, and Figure 50 shows a typical vat leaching setup.

Higher-temperature leaching applications, especially for acidic or corrosive conditions, are usually conducted in standard laboratory Pyrex glassware. These vessels are fabricated in a multitude of design shapes and volumes, with compatible ported lids, and can be sourced from any of the major laboratory equipment suppliers such as Cole-Parmer, John Morris, and Rowe Scientific. The upper volume sizes for standard glassware are usually limited to 5 L, and as such, sample masses for these setups are also limited to <2 kg. The overall configuration is similar to the preceding vat leaching apparatus, with an overhead stirrer providing agitation. Heat input, however, can be applied with a base hotplate up to 95°C, and outer insulation can be applied. The ported lids can also facilitate other additions such as condensers and scrubbers together with probe/pipette inlets. Figure 51 shows a typical Pyrex high-temperature leaching setup.

Column Leaching

Column leaching is the standard metallurgical testing method to assess heap leaching applications of coarse crushed PSD material, by either gold cyanidation or acid leaching. More detailed information can be found in Chapter 10.3, "Heap and Dump Leaching."

Sample crushed particle sizes range from 6 mm to 75 mm, or greater if investigating dump leach, and usually in 6.3-, 12.7-, 19-, 25-, 51-mm measurements, and so forth. Column heights are typically from 0.5 to 6 m. Preliminary test work is usually undertaken at 2-m height minimum, with confirmatory testing increasing in height to 4–6 m using customized rigs. Large-scale demonstration testing can be conducted in columns as large as 2-m diameter (or square cribs), requiring many metric tons.

Column diameter selection is based on an empirical factor of 6× the maximum particle diameter (i.e., a 50-mm crush size are performed in 300-mm-diameter column). Sample mass requirements are commensurate with the column height and diameter, PSD, and bulk density. Sizes and estimated masses for 2-m laboratory columns are shown in Table 2.

Columns are usually custom designed and fabricated from clear Perspex. Columns higher than 2 m are constructed in flanged 1-m sections to enable easier takedown and material removal at test termination.

Solution irrigation is facilitated with small positive-displacement peristaltic pumps (such as MasterFlex) that can take different-sized heads and tubing. The initial solution reservoir is typically a 20-L plastic bucket, which can be switched out during the test and also monitored and adjusted for chemical concentrations. Irrigation rates typically range from 8 to 20 L/m^2/h.

Cyanide solution strengths for gold leaching applications range from 250 to 1,000 ppm cyanide, but most often <500 ppm. Acid solution strengths for copper and uranium leaching applications range from 5 to 10 g/L free acid. Nickel columns usually require higher strengths at ~50 g/L. Cyanidation column leaching test durations range between approximately 30 and 70 days. Acid leaching column testing durations are much longer, for example, between 6 and 9 months.

Prior to actual leach testing, samples of the material are tested in columns for permeability, ponding, or ratholing. This is done by sealing the bottom of the column, flooding it

Figure 49 Leach using rolling bottle

Figure 50 Agitated vat leaching equipment

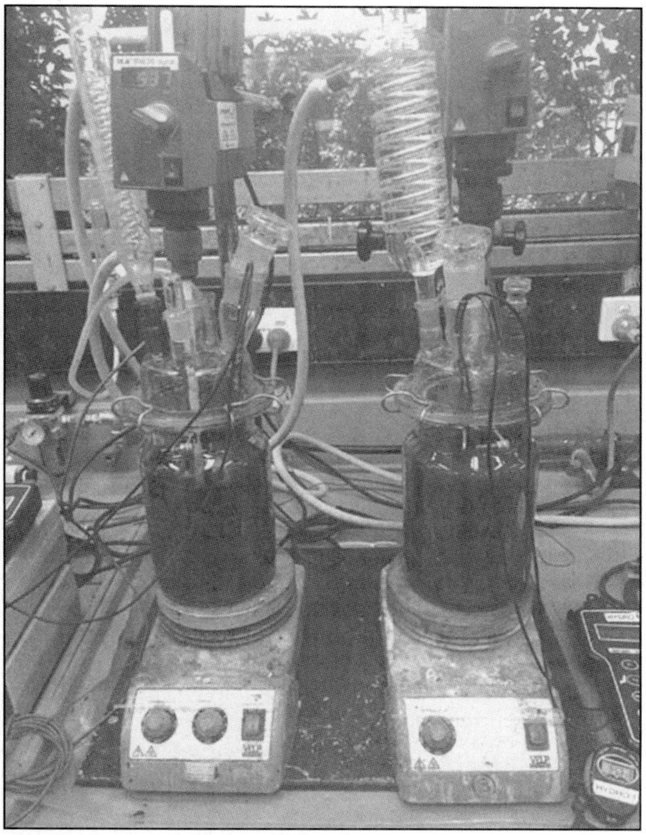

Figure 51 High-temperature leaching setup

Table 2 Laboratory columns

Diameter, mm	Height, m	Mass, kg
80	2	15
100	2	~24
150	2	~50
200	2	~95
300	2	~210
500	2	~580
1,000	2	~2,300

with water, and then examining the flow of water through the material and time of complete drainage to ensure that the total column contents are contacted by the water passage. If any concerns arise from this test, agglomeration of the material may be required.

Pressure Leaching Using Autoclaves

Higher-temperature laboratory leaching applications >100°C, such a high-pressure acid leaching (HPAL) or pressure oxidation (POx), require a pressure vessel or autoclave. Safety and regulatory requirements rise exponentially when dealing with high temperatures, high pressures, and highly corrosive solutions comparative to other bench-scale test work. These vessels require extremely corrosion-resistant materials of construction, such as titanium or tantalum, and precision fabrication to pass certification and, as such, are typically very expensive items to procure. In turn, this means the associated testing is also more expensive than atmospheric leaching; however, in many cases, there is no alternative to achieve acceptable extractions.

The most well-known brand in laboratory autoclaves is provided by Parr Instruments (United States), parts for which can be sourced via the John Morris Group. There are other alternatives; however, considering the safety risks, laboratory metallurgists need to be extremely confident of the construction schedule and certification. A single out-of-specification nut or bolt can result in rapid catastrophic failure under aggressive HPAL testing conditions.

Parr produces two sizes of autoclave (in U.S. gallons) at 3.8 and 19 L. The base vessel is situated in an insulated wraparound heater, and an immersion heater is also fitted to the lid. An internal agitator is connected via a magnetic coupling to an overhead drive mechanism. The lid has multiple ports to facilitate gas injection, nitrogen overpressure, temperature and pressure probes, and sampling bombs.

The 3.8-L autoclave takes sample masses of ~700 g, and the 19-L version can receive up to 3.5 kg. HPAL testing conditions operate at 220–250°C at steam pressures of 2,200–2,800 kPa with nitrogen overpressure up to 4,500 kPa. Maximum acid additions for the smaller unit are ~350 kg/t of sulfuric acid. Test durations range from 90 minutes to 4 hours. POx testing conditions operate at 105–220°C at oxygen overpressures ranging between 200 and 22,000 kPa. Test durations range from 5 minutes for partial oxidation to 3 hours.

The autoclaves also require a temperature and pressure control box, which can be sourced directly from Parr or from John Morris. A typical 3.8-L Parr autoclave skid arrangement is shown in Figure 52.

Leaching Ancillary Equipment

The following is a list of critical ancillary laboratory equipment and infrastructure associated with leaching:

- Grinding and sizing equipment as described in the preceding sections
- Pressure filters, filter cloths, and filter papers
- Drying ovens
- Air compressor and ring main plumbing
- Gas bottle storage and ring main plumbing
- Weigh scales and balances to handle small high-accuracy masses (100 g), midrange masses between 2 and 5 kg, and larger masses up to 30 kg
- ORP and pH meters and probes
- Dissolved oxygen meter and probe
- Burette and Pyrex beakers for titrations
- Pipettes, bulbs, test tubes, and racks
- Equipment for solution subsampling and assay analysis
  - A range of 50–100-mL plastic screw-top phials for solution and slurry subsamples
  - Aluminum trays and plastic buckets for sample collection
  - Aluminum tags, printable stickers, and a range of plastic bag sizes for sample storage

Figure 52 Parr autoclave

ACKNOWLEDGMENTS

The author extends thanks to co-contributors including Russel Philip for the section on heavy liquid separation, Peter Wilkinson for the section on column leaching, Jack Smith for the sections on settling and viscosity, and Hamid Sheriff and Rob Dunne for the invitation and advice in facilitating the chapter. Appreciation also goes to John Angove for his time and efforts in reviewing this chapter and for his valuable edits and commentary.

REFERENCES

Angove, J., and Dunne, R.A. 1997. Review of standard physical ore property determinations. In *World Gold '97 Conference Proceedings*. Melbourne, Victoria: Australasian Institute of Mining and Metallurgy. pp. 139–144.

Bond, F.C. 1946. Bond impact crushing work index (CWi): Crushing test by pressure and impact. *Trans. AIME* 169:58–65.

Bond, F.C. 1961a. Bond rod mill work index (RWi): Crushing and grinding calculations. *Br. Chem. Eng.* 6(6):381–382.

Bond, F.C. 1961b. Bond ball mill work index (BWi): Crushing and grinding calculations. *Br. Chem. Eng.* 6(6):382.

Bond, F.C. 1963. Bond abrasion index (Ai): Metal wear in crushing and grinding. Presented at the 54th Annual Meeting of the Institute of Chemical Engineers, Houston, TX.

Endecotts. 2018. Test sieves. www.endecotts.com/products/sieves/. Accessed August 2018.

Fann Instrument Company. 2013. Marsh funnel viscometer. Drilling fluids testing product information. Houston, TX.

FLSmidth. 2017. Essa Australia. www.flsmidth.com/en-US/About+FLSmidth/Our+History/Our+Product+Brands/Essa/. Accessed August 2018.

Laplante, A.R., and Doucet, R. 1996. A laboratory procedure to determine the amount of gravity recoverable gold. SME Preprint No. 96-174. Littleton, CO: SME.

Levin, J. 1989. Observations on the Bond standard grindability test, and a proposal for a standard grindability test for fine materials. *J. S. Afr. Inst. Min. Metall.* 89(1):13–21.

Malvern Panalytical. 2018. Mastersizer range: World's most widely used particle size analyzers. www.malvernpanalytical.com/en/products/product-range/mastersizer-range/. Accessed August 2018.

Shi, F. 2004. Comparison of grinding media—Cylpebs versus balls. *Miner. Eng.* 17(11-12):1259–1268.

CHAPTER 1.7

Laboratory Automation

Wolfgang Baum and Kevin Ausburn

The mining industry is undergoing comprehensive change regarding automation and related robotics technology in all facets of its activity. Today, and more so in the future, automation use extends from exploration through mine-site drilling, mining, and metallurgical processing to smelting and refining. As part of this seismic shift, all laboratory activities pertaining to the preceding operational steps have been extensively impacted by automation. The authors emphasize that, once this text is published, the high-speed development of the hardware and software used in automation will have already surpassed some of this chapter's statements and descriptions. The reader is therefore advised to use this chapter in consultations with relevant automation-related magazines, equipment providers, and personal benchmarking of robotics technology in the laboratory field. It is of equal importance that the reader and technical staff active in lab automation also track and monitor the developments, practices, and innovations in nonmining laboratories since the pharmaceutical, chemical, cement, steel, and related industries have inherently been pioneers in automated laboratory use.

The scope of this chapter is to characterize the current use of automation and robotics technology in the most important laboratories for mining (i.e., sample preparation, chemical laboratories, mineralogical laboratories, metallurgical laboratories, and special production labs such as cathode, refinery, bullion, etc.). In addition, this chapter covers the use and application potential for automated analysis technology for online/cross-belt installation and downhole purposes, specifically for blastholes. To meet the extreme challenges of future mining (e.g., harder ores, deeper mining, water/power/steel-wear, increased production rates, lack of skilled labor, extending mine life, safety demands, and extended use of robotics technology) and remain competitive, mining companies need to aggressively modernize their lab facilities. In short, automation of laboratories is imperative for future mining (Allen et al. 2007; Baum 2007; Ausburn 2013; Baum 2014a, 2014b). The *ore factory*, a term coined in Australia, is becoming a mandatory strategic development for most miners. The industry's move toward remote control and operations centers that integrate multiple sites and multiple production segments all the way to ship-loading will require more lab automation to maintain the critical and fast data flow from mine geology, ore feeds, process samples, and end products.

BACKGROUND

The nonferrous metals mining industry established its first large-scale and remarkable entry into laboratory automation with the construction of semi- and/or fully automated labs in the Nevada gold mining operations of Newmont Mining in the late 1980s. Much earlier applications occurred in the cement and steel industries, some iron ore operations (Fer et Titane), heavy minerals (Richards Bay, South Africa), and other select industrial mineral applications. Select automation modules and partial lab automation were installed primarily in South Africa, Namibia, and in an aluminum smelter in Mozambique. Lab automation received more attention when breakthrough robotics labs such as Anglo's EBRL and the Phelps Dodge–Freeport Central Analytical Service Center labs were constructed (Best et al. 2007). The start-up of the Freeport AXN lab (an automated X-ray diffraction [XRD] and near-infrared [NIR] mineralogy lab) represents the first 24/7 plant mineralogy support lab in copper mining (Baum 2009). Central or other automation labs followed in Western Australia and elsewhere (e.g., BHP Newman, Rio Tinto Tom Price, Kumba Sishen, FMG Solomon).

Commercial labs have been slower in adapting to lab automation, some of which is related to challenges associated with installation. After the initial use of robotics technology by various labs, considerably more growth can be expected in the lab service industry to handle high sample volumes, deliver better quality, and reduce turnaround times. Container labs with automation modules were also introduced. The innovation leaders in large-scale lab automation clearly were mining operations in South Africa and Australia, followed by North American mining. During the last 15 years, automated fire assay labs were installed in the commercial lab business for gold, platinum, and nickel mining. Although the success of the previously mentioned labs was impressive, the mining

Wolfgang Baum, Managing Director, Ore & Plant Mineralogy LLC, San Diego, California, USA
Kevin Ausburn, Chief Mineralogist, FLSmidth USA Inc., Midvale, Utah, USA

industry as a whole requires considerably more "automation" efforts to update heritage labs and/or replace them with cutting-edge robotics labs. Without this activity, the single most important data-generation segments in mining operations, that is, laboratories, may not be as efficient and competitive as they could be.

The leading laboratory automation equipment and service providers in the mining business are Herzog, IMP, ThyssenKrupp's Industrial Solutions, FLSmidth, and Rocklabs. Various other providers deal with specialty equipment in the fields of comminution, screening, fusion, dissolution, digestion, solution handling, filtration, leaching, flotation, and the entire range of chemical and mineralogical analyses.

FEASIBILITY OF LAB AUTOMATION

In a mining operation, four types of laboratories are generally considered for automation: small mine-site labs, central labs for multiple operations, in-plant modules, and mobile/containerized labs. The overall equipment assessment involves an evaluation of whether the labs require retooling, expansion, or conversion to 24/7 operation and whether they should be fully automated or semiautomated or represent a completely new lab construction effort. The size of the operation is of considerable importance as automation and robotics technology become more economic and efficient with larger sample and determination volumes. A key parameter for the feasibility evaluation is also the type of laboratories required (e.g., sample preparation, chemical, mineralogical, cathode, fire assay).

With the rapidly growing need and use of ore characterization for geometallurgy, any existing and new operation considering an automated laboratory should evaluate the need for more than only chemical analyses. Today, mineralogical analyses are as critical as chemical data. Because of recent advances in modern XRD, Fourier transform near infrared (FT-NIR), NIR, Fourier transform infrared (FTIR), and automated mineral analyzers (to name the most important technologies), it is recommended that any laboratory expansion, upgrade, or conversion to automation should seriously assess the option of adding mineralogical lab capabilities.

Benefits

To determine the viability of a greenfield and/or upgraded automated laboratory, a central laboratory or a containerized lab for one operation, multiple mines, a commercial lab service organization, and/or in other applications such as test or development centers, a thorough evaluation of all business drivers and technical aspects, including turnaround times, requires a two-stage process:

1. **Preengineering Study (PES).** The PES consists of determining all baseline parameters for these labs including (but not limited to) sample types, sample sources, sample volumes, sample delivery during a 24/7 period, type of pulp preparation (client standard operating procedures [SOPs]), analytical data needed, site visit, conceptual layout, turnaround time for all lab clients, sample delivery schedule, site suitability, utility requirements, equipment lists for each lab, redundancy plans, preliminary staff needs, schematic architectural plans, and other factors. Various business cases should be assessed from a preliminary point of view.
2. **Feasibility Study (FS).** The FS details all aspects of the conceptual design, selection of equipment, detailed architectural plans, including construction documents, delivery and installation, confirmed capital cost for building and equipment, consumables and reagent list, hazard and operability study, training program, factory acceptance tests, shakedown tests, start-up support, service, spare parts and maintenance program, laboratory information management system (LIMS) use, safety program, waste handling, quality control (QC) program, operating cost, per-analyte cost, uninterruptible power supply system, and so forth.

Lack of good and reliable PES and FS work may result in a variety of problems, including increased downtime, lower availability, throughput and turnaround below nameplate, start-up delays, undersized equipment, substantial lack of redundancy, variance in sample preparation and related analytical variance, and higher operating costs.

After completion and review of the preceding studies, the project can move into the following phases:

- Approval and purchasing
- Construction
- Training of staff
- Factory acceptance testing
- Start-up and full operational functionality

Overall, the benefits of high-throughput automated labs are evident and have been proven in numerous operations:

- 24/7 operation
- Better safety and hygiene features
- Large sample weights (i.e., more representative samples)
- More reliable sample drying, comminution, and splitting
- Lower operating cost
- Better analytical data
- Faster turnaround

Budget

As each project requires an individual assessment related to budgetary plans, it is strongly recommended that any efforts targeting lab automation undergo a robust PES (which should typically include a 20%–30% cost redundancy) and keep a long-range perspective of cost-efficient capabilities to expand the lab in the near future. Without the PES, any budgetary estimates remain highly questionable.

Based on the preceding information, detailed economic parameters should be calculated, including (but not limited to) commissioning cost, cost per sample and cost per analysis, utility costs, spare parts, staff cost and savings, economic impact of better analytical data, improved turnaround time, safety improvements, return on investment of capital, service contract costs, continuous training costs, and intangible factors.

Expectations

Key parameters for the PES must include a detailed definition and listing of expectations of both management and the lab's clients. These expectations represent a critical design baseline for the PES. Summarized, these expectations encompass (but are not limited to) sample volume, sample types, specific sample preparation procedures, analytical parameters, turnaround for the analyses, QC, lab and equipment redundancy to prevent interruption of production support, and options for new analytical equipment, expansions, and potential rush work.

Sample Types and Volume

The lab's equipment and operational characteristics will be governed by the type of samples—from rock or process materials—to be processed. They may include a large range of materials from one or multiple commodities and/or by-products. Samples may include such material types as exploration samples, mine-site geology samples, concentrator samples, leach operation samples (run-of-mine [ROM] and/or heap leach), solutions, organic materials, solvent extraction and electrowinning of specific samples, metallurgical accounting samples, smelter samples, refinery samples, water, and/or other environmental samples.

During the PES, the sample evaluation should also include giving detailed attention to optimizing the sampling and sample size features, all of which inherently contribute to the largest errors for any laboratory work. For example, the coefficient of variation of fundamental errors in blasthole samples can be large (16%–>25%) and small splits may introduce substantial reproducibility problems. For example, in one case study the authors have worked on, an increase of the blasthole sample weight by 7 kg per sample reduced the variance by 5%.

Industry-wide, one of the problematic areas continues to be the current methods of blasthole sampling. These issues may be solved in the future by introduction of automated blasthole samplers (either from drilling companies, miners, and/or sampling equipment manufacturers). The current blasthole sampling practices (pipe, shovel, auger, grab, etc.) are not ideal and should be optimized. Any optimization will have significant beneficial features for reducing notorious variances or massive misrepresentations in resource assessments (Carrasco et al. 2004, 2005). As indicated by these authors, improper sampling and sample preparation can cause monumental economic losses. Therefore, it is recommended that any laboratory equipment and SOP change be accompanied by an evaluation of sample requirements. This applies to all types of sampling and not just to the blasthole material.

Staffing and Lab Management

Automated or robotics labs require a change in operational thinking, key personnel with a relevant skill set and sound pretraining, sufficient start-up time prior to accepting full production, and continuous staff training. When selecting lab personnel, a small core team of lab-specific experienced chemists, mineralogists, metallurgists, automation engineers, seasoned lab technicians, and QC/LIMS staff is required. A training program should be prepared with the equipment manufacturers, and it is highly advisable to send one or two equipment and robotics/software specialists to the lab for an extended start-up period. It is also beneficial to involve key operators during the factory acceptance tests (FATs). Eliminating proper FATs has, in the authors' experience, never been a good economic and technical decision.

Continuing benchmarking is the responsibility of the senior lab staff (e.g., lab manager, lab superintendent) and should include a good reference library, regular review of pertinent technical publications and attendance at lab-specific conferences, and the important "look across the fence," that is, fact-finding in automated lab environments outside of mining.

Without a doubt, any new or even upgraded lab will require a break-in period during which the operators can fine-tune the overall lab performance and functionality. Because shift operation will be required, equal performance of all shifts will be critical. New and adjusted safety measures will be of the essence for the automation environment, and a continuous effort by a dedicated team should zero in on lab economics and cost optimization. Training, as with any best-practice operation, is imperative for a successful automated lab, and so is a well-functioning LIMS.

Lab Location

Social and neighborhood issues, floodplain and floodway issues, noise, traffic, ground vibrations, air emissions, conflict with communities, and waste handling and disposal should be assessed during the lab selection process. Sample transportation and receiving logistics in the case of a central lab is of paramount importance.

Lab Quality Control

The use of existing lab technology and/or the introduction of new equipment and automated operating procedures mandates a thorough review and, potentially, reassessment of the practices used. The QC procedures are driven by lab type, equipment, material, and analysis. Typically, these procedures should encompass National Institute of Standards and Technology or internal standards, reruns, repeats, a minimum percentage of duplicates, blind samples, determination of repeatability, round-robin work, tracking of differences, assessment of data reporting, housekeeping, safety relations, maintenance, and training. External and internal lab audits or surveys are essential to detect and filter out problematic practices, less-than-optimal sample preparation, or analytical procedures and equipment issues.

Start-up, Optimization, and Avoidance of Problems

From the PES and FS, lab management should pay attention to the architectural layout and equipment functionality. A 24/7 lab operation with robotics technology requires considerably different philosophies and extremely good reliability for equipment, throughput, meeting of product specifications, dust control, and buffer capacity. Frequent engineering staff changes and rotation of liaison engineers from the manufacturers should be avoided under all circumstances.

The design team, including the equipment providers, should pay utmost attention to reducing lab downtime and availability of sample preparation and analysis systems. Consistent checks of abnormalities are mandated, along with routine status reports, weekly start-up team meetings, and the preparation of robust building, equipment, staff, safety, and SOP checklists. Lab crew issues need to be foreseen and require preventive efforts, including (but not limited to) skill sets, attitudes, responsible performance, shift cross-communication, housekeeping, repair checks, use of personal protective equipment, and spare-parts management with foresight, efficiency, and overall facility maintenance.

LAB AUTOMATION MODULES: MOBILE, MINE-SITE, AND CENTRAL

The authors' most important observation in lab design and construction has been that 90% of new labs, central labs, expanded labs, and conversions to automation and/or container labs are conceptualized or built with insufficient redundancy and expansion potential. Clearly, the historic and classical "bucking room" is being replaced by modern sample preparation labs. With the current and future trend to ever-higher throughput operations, *many labs will require substantial expansions, optimizations, or addition of new analytical capabilities*.

As a result of the rather conservative initial design concepts, the capital requirements for follow-up lab additions or capacity increases are unnecessarily high. Consequently, the industry's tendency to prioritize capital expenditures related to tonnage, throughput, overall mining, and plant issues has resulted, in many cases, in a severe neglect of laboratories, even as the laboratory segment (sample preparation, chemical, mineralogical, metallurgical) constitutes the most critical baseline tool for getting the economics right and for optimizing an operation. As such, it is hoped that the mining industry will pay substantially more attention to modern and efficient laboratory facilities in the future as cutting-edge competition caused by lower grades, higher throughput, capital-intensive larger plants, increased operating and sustainability cost, and a shrinking skill and talent base becomes more predominant in the market.

As the mining industry is moving underground, the lab automation business and operational thinking need to be adjusted accordingly. Many underground operations will offer significant opportunities for "subsurface labs," which could have a positive effect on mining and materials handling and reduce overall lab costs.

LAB AUTOMATION FOR SAMPLE PREPARATION

Virtually 90% of today's sample preparation demand in mining operations is amenable to fully automated equipment and procedures. This pertains to the major steps in sample preparation such as drying, crushing, grinding and pulverizing, splitting, dosing, pulp packaging, labeling, bar-code or other tracking technology, and reject material handling including bagging and storage. A large range of operating robotics sample preparation labs around the world include various levels of automated or partially automated sample preparation in commercial labs, specialty labs, central service labs, and container labs. An online search can be conducted to review equipment and assessment studies from key suppliers. Many specialty equipment providers exist, and the reader is advised to evaluate which contact is the most appropriate for his or her application.

In many cases, the flow sheets of sample preparation labs for multiple mining operations have to accommodate the following materials in regard to converting the as-received ore and rock samples into a highly representative assay pulp:

- Miscellaneous geology samples (core, cuttings, bulk, etc.)
- Blasthole samples
- Mill feeds, process streams, concentrates, flotation tailings
- Samples from ROM or heap leach operations
- Smelter samples
- Refinery samples

Mine and plant operators need to assess, in a team effort, how to conceptualize an automated lab to accommodate all sample materials in regard to preparation as pulps and analytical work.

LAB AUTOMATION FOR CHEMICAL ANALYSES

Although most analytical equipment manufacturers have had supplemented systems with automation features and autosamplers for many years, a full integration of standard chemical lab equipment into a line, circuit, or other full robotics configuration is not yet a standard feature in mining. This includes (but is not limited to) reagent dosing, glassware handling, glassware cleaning, solution filtration or centrifuging, and multiple other analytical steps. Established use of robotics lab circuits today encompass the following:

- Large dosing stations
- Fire assay labs
- Solids filtration
- Auto titration
- Microwave digestion
- Automated hot-plate digestion
- Automated cathode analysis
- Auto dilution

For a mine-site chemical lab, operators should first assess the specific sample types and related analytical needs. Thereafter, a process flow diagram needs to be established to conceptualize how far and how deeply automation can be implemented. Here again, redundancy and extra-potential expansion capacity are of utmost importance. For a standard chemical production support lab, this diagram may include bar-coded samples shipped → assay pulp preparation → analysis → data report in LIMS → data transfer to data warehouse → mine customers access their results.

In terms of lab layout for blastholes, geology samples, and other bulk rock/ore samples, a general configuration may be as follows:

- Automated: drying–crushing–splitting–pulverizing–dosing
- Automated elemental analyses and/or selective leach tests plus manual/semiautomated analysis
- Automated/semiautomated concentrate analysis: separate automated preparation followed by analysis
- Automated or semiautomated specialty samples: solvent extraction, cathodes, mineralogy

Quality Assurance and Quality Control Efforts

QA/QC efforts should include a rigorous program that encompasses the following areas:

- A certain percentage of duplicates
- Upper and lower limits on all QCs
- Business rules
- Cross-checking of trend-drift analyses
- Instrument calibration and standards
- Round robin with other labs
- Monthly QA/QC reports for customers
- Annual lab audit by external auditor

LAB AUTOMATION: A KEY TOOL

Over the last 15 years, geometallurgy has become a focal point for better ore control, modeling, forecasting, and optimal process operation. In the early days, geometallurgical work primarily focused on hardness indices in models and simulation. Today, with the operation of multiple automated labs (both chemical and mineralogical), it has become apparent that robust geometallurgical planning and application can only be performed with the help of lab automation. Automated labs enable operations to submit and analyze large volumes of samples, thus improving overall ore-body profiling and providing daily production support data from mine geology to concentrators or heap leach operations.

LAB AUTOMATION FOR MINERALOGICAL ANALYSES

Mineralogy has three distinct analytical techniques that have been automated to varying degrees. The NIR/FT-NIR and/or FTIR analyses have been completely automated by mining companies. This includes all the sample preparation, analyses, and data reporting. This also allows the operations to collect quantitative mineralogy on selected alteration phases on hundreds of samples per day. The data can be transferred directly through LIMS programs into the common software used by the mine planners.

XRD analyses have been partially automated. The automation includes all the necessary sample preparations (crushing, splitting, pulverizing, mounting samples) and the transferring of the prepared sample mount into the XRD unit for analysis. The data (i.e., diffractogram) from the instrument can be sent directly to a server. Typically the data still require manual interpretation by an analyst to produce the quantitative mineralogy. This manual interpretation does become quite routine for an operator but can still take 5–10 minutes per sample to complete, which means that a single analyst can only complete 50–80 samples per day. Currently, no software programs are available that can automate the data interpretation of the variable, complex rocks present in virtually all mining operations.

Microscopy, either by light microscopy or scanning electron microscopy techniques (e.g., automated mineral analyzer units), has also been partially automated. The automation for these analyses is generally focused on selected areas of the sample preparation. Such areas include automated polishing and automated potting of the sample into the epoxy mount. Other necessary steps, such as screening, microsplitting, and carbon-coating the mounts, have not been automated, but it should be possible to automate these areas as well. The main challenge would be to design a way to connect all these areas together to automate the entire process, which would avoid the need for manual intervention to move the samples between each preparation step. Perhaps the main reason these analyses have not been fully automated is because the microscopy analyses are quite labor- and/or time-consuming. As such, these analyses are typically used for daily trend analyses or for determining causes for issues after the ore has been processed. Consequently, the automation of these microscopy techniques has been mainly pursued by high-capacity centralized support labs or by large commercial labs to improve their turnaround time.

REFERENCES

Allen, D., Baber, B., Baum, W., and Eady, S. 2007. Ore characterisation, alteration coding, and plant feed control in copper operations as directed by routine semi-automated mineralogical analyses. In *Proceedings of Cu2007*. Montreal, QC: Canadian Institute of Mining, Metallurgy and Petroleum. Vol. IV (Book 2), pp. 247–258.

Ausburn, K. 2013. Rise of the machines. *Min. Mag.* (December). https://www.researchgate.net/publication/277012698_Rise_of_the_Machines_-_Interview_in_Mining_Magazine. Accessed September 2017.

Baum, W. 2007. Material characterization on the cutting edge. *Process (CSIRO)* (June):12.

Baum, W. 2009. The new automated XRD-NIR mineralogy lab—High capacity tool for short and long-range ore characterization. Presented at SME 2009 Arizona Conference, Tucson, AZ, Dec. 6–7.

Baum, W. 2014a. Ore characterization, process mineralogy and lab automation—A roadmap for future mining. *Miner. Eng.* 60:69–73.

Baum, W. 2014b. Innovations in process mineralogy and laboratory automation. In *Mineral Processing and Extractive Metallurgy: 100 Years of Innovation*. Edited by C.G. Anderson, R.C. Dunne, and J.L Uhrie. Englewood, CO: SME. pp 569–575.

Best, E., Baum, W., Gilbert, R., Hohenstein, B., and Balt, A. 2007. The Phelps Dodge Central Analytical Service Center: Step change technology implementing robotics systems and lab automation in the copper industry. In *Proceedings of Cu2007*. Montreal, QC: Canadian Institute of Mining, Metallurgy and Petroleum.

Carrasco, P., Carrasco, P., and Jara, E. 2004. The economic impact of correct sampling and analysis practices in the copper mining industry. *Chemometr. Intell. Lab. Syst.* 74(1):209–213.

Carrasco, P., Carrasco, P., Campos, M., Tapia, J., and Menichjetti, E. 2005. Heterogeneity and Ingamells' tests of some Chilean porphyry ores. In *Proceedings of the Second World Conference on Sampling and Blending*. Melbourne, Victoria: Australasian Institute of Mining and Metallurgy. pp. 139–150.

CHAPTER 1.8

Sampling Practice and Considerations

S. Komar Kawatra and Howard Haselhuhn

Taggart and Behre (1945) defined sampling as "The operation of removing a part convenient in size for testing, from a whole which is of much greater bulk, in such a way that the proportion and distribution of the quality to be tested (e.g., specific gravity, metal content, recoverability) are the same in both the whole and the part removed (sample)."

Sampling procedures cover the practice of selecting representative quantities of material in the field to evaluate bulk material properties. Examples of the test materials are bulk granular solids, slurries, sludges, grains, solid fuels, process liquids, and process gases. It is necessary to be able to sample bulk materials during shipment and during processing operations to ensure that the bulk material meets specifications.

When sampling a bulk material, it is very important that the sample be representative of the bulk material. In reality, the condition that the sample be completely representative of the bulk material in all aspects except amount is never fulfilled when heterogeneous materials are sampled. For instance, you can never get a perfectly representative sample of water from the Pacific Ocean. You may even ask what good it is or what purpose such a sample would serve. It is important, before collecting a sample, to decide what purpose the sample would be, and then plan the sampling procedure accordingly so that it will be sufficiently representative for one's purposes.

PROPER SAMPLING

Proper sampling techniques ensure that measurements performed on a sample are representative of the material that was sampled. This is extremely important in all industries that process solid, liquid, and gaseous materials. To expand on this topic, the following key terms must be defined:

- **Lot:** A large (bulk) quantity of material from which a sample must be taken. If proper sampling is performed, analysis of the sample will be representative of the lot within a specified degree of certainty (or confidence interval).
- **Homogeneous material:** A material that is either one substance or a perfect mixture of multiple substances with constant thermodynamic properties throughout. By definition, perfectly homogeneous materials do not exist and cannot be created.
- **Heterogeneous material:** A material that has different properties depending on where the material is analyzed. Examples include slurries, immiscible fluids, and even tap water.
- **Component:** A particular substance within a lot of heterogeneous material. For example, silica, pyrite, and gold are typical components of gold ore.
- **Critical component:** The component of the heterogeneous material that is of interest. For example, gold is the critical component in gold ore.
- **Content:** The concentration of a component within a lot of heterogeneous material. For example, a gold ore may have a silica content of 65%.
- **Critical content:** The concentration of the critical component within a lot of heterogeneous material. For example, a gold ore may have a critical content of 0.01% gold.

Example of Poor Sampling: The Bre-X Gold Scandal

The Bre-X gold scandal is regarded as the biggest and most sophisticated scam ever in the history of mining. The scandal began when Canadian-based mining company, Bre-X Minerals Ltd., purchased the mineral rights to land in the jungle of Borneo, Indonesia, in 1993. Bre-X hired a geologist to take core samples of the site and assay the samples for gold. In just a few years' time, Bre-X had discovered the largest gold reserve on the planet and stock soared from roughly US$1 to US$300. At this time, larger entities began to take notice, such as the Indonesian government. The Indonesian government then involved the American mining company Freeport-McMoRan Copper and Gold. To verify the claims made by Bre-X, Freeport sent in a group of their own to take core samples and assay for gold. Freeport reported that they were unable to find any significant gold reserves on-site. Four weeks later, it is believed that the lead geologist for Bre-X "jumped" to his death from a helicopter 800 feet in the air. So how did Bre-X fool everyone into believing it had discovered the largest gold reserve on the planet?

S. Komar Kawatra, Professor & Chair, Department of Chemical Engineering, Michigan Technological University, Houghton, Michigan, USA
Howard Haselhuhn, Applications Engineer—Industrial Minerals, Solvay Technology Solutions, Stamford, Connecticut, USA

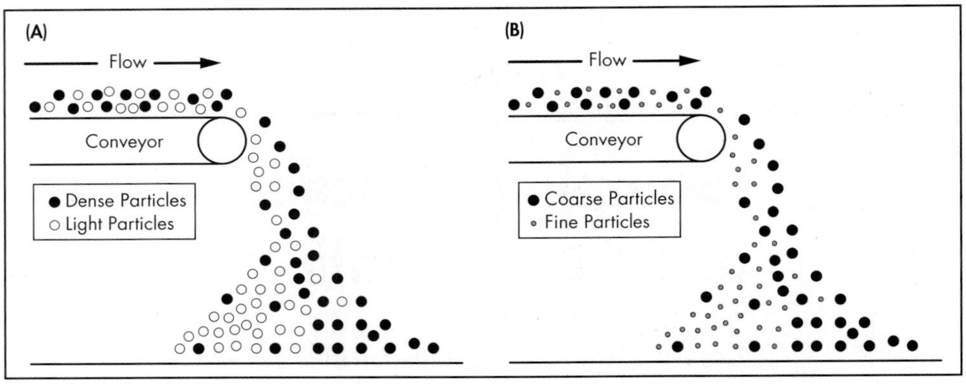

Adapted from Pitard 1993

Figure 1 Segregation of a heterogeneous material (A) of different density and (B) of different particle size when stockpiled from a conveyor belt

Bre-X was able to deceive everyone into believing it had discovered the world's largest gold reserve by conducting a fraudulent sampling campaign. It is believed that before the core samples were to be assayed, the geologist "salted" the samples with gold shavings from his wedding ring or the local river bed. With all the fervor over having found the largest gold reserve in the world, many of the early warning signs of fraud were ignored. For example, an independent contractor was brought on-site to examine the sampling procedure being carried out by the lead geologist. The contractor discovered that Bre-X was crushing and grinding the entire core sample for gold assay (it is typical to save half of the core for outside analysis). By doing this, Bre-X was able to destroy any evidence that may have disputed their claims to finding the world's largest gold reserve. When analyzing sampling results, it is important to always be skeptical of the findings, and remember: If it sounds too good to be true, it probably is.

SCOPE OF THIS TEXT

This chapter is designed to provide an overview of sampling methods without the cumbersome calculations and complicated error analysis of other works. The intended audience for this work is the technician, operator, or student that is interested in the most practical solutions to their sampling issues without complex calculations. For a more advanced and in-depth discussion of sampling and the errors involved, several fantastic and comprehensive books should be referenced:

- *Sampling of Particulate Materials: Theory and Practice* (Gy 1979)
- *Wills' Mineral Processing Technology*, 8th edition (Wills and Finch 2016)
- *Pierre Gy's Sampling Theory and Sampling Practice*, 2nd edition (Pitard 1993)
- *Principles of Mineral Dressing (*Gaudin 1939)
- *Handbook of Mineral Dressing* (Taggart and Behre 1945)
- *An Introduction to Metal Balancing and Reconciliation* (Morrison 2008)

SAMPLING A LARGE LOT

Obtaining a representative sample from a large (>1 metric ton) lot of material can be a very tedious and time-consuming task when done correctly. Many engineers and scientists make the mistake of merely shoveling a bucket of material from a stockpile, splitting it in a laboratory, and assuming that the properties of these splits are representative of the entire lot. Unfortunately, this is rarely the case with a heterogeneous material such as an ore. With a heterogeneous material, particles of varying particle size, density, particle shape, and friction tend to segregate when stockpiled, as indicated in Figure 1 (Pitard 1993).

Because of this segregation upon stockpiling, proper sampling techniques must be used to obtain a representative sample from a large lot of material. Several techniques can be used to split a large lot of material, some being better than others. Two common techniques are the "grab sample" and "coning and quartering."

Grab Sample

A grab sample is a convenient procedure for sampling large lots of material and can be conveniently done using shovels or even front-end loaders for very large samples. A grab sample involves removing equal-sized increments from a well-mixed lot, creating numerous smaller samples until the lot is eliminated, as shown in Figure 2. The sample for laboratory testing is then randomly selected from the increments. This procedure should be performed on a clean surface that will not contaminate the lot. Before incrementing the lot, it is advisable to use

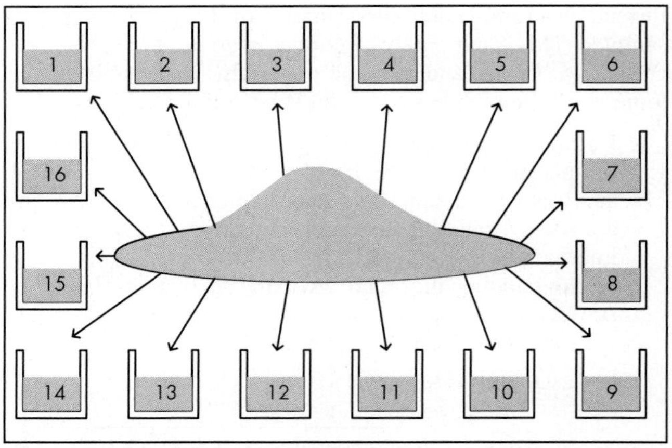

Figure 2 Using the grab sample technique to separate a lot into 16 representative increments

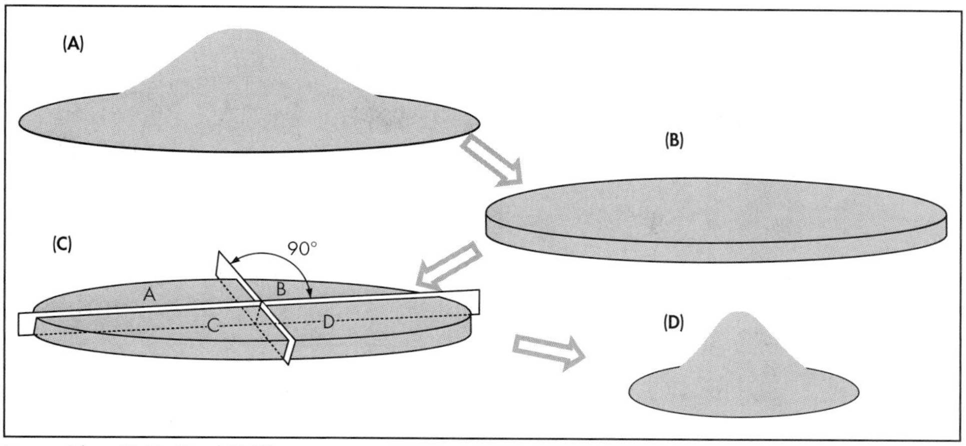

Source: Khan 1968; Pitard 1993

Figure 3 Coning and quartering a lot of material to acquire a sample that is one half the size of the entire lot (A+C or B+D)

an appropriately sized shovel (hand shovel, front-end loader, etc.) to mix the lot thoroughly by scooping from the sides of the lot and dumping the scoops in the center of the lot to create revolutionary homogeneity. Increments are removed from the lot and placed in a specified number of containers until the lot has been completely split. A container is selected at random and used as a sample. Grab sampling is a good way to obtain a representative sample from a large lot of material; however, this method is very time-consuming. Properly using the grab sample technique has been shown to give a standard deviation of ~5.14% between samples (Khan 1968).

Coning and Quartering

This method is well-suited for large lots of material and can be conveniently done using shovels or front-end loaders for very large samples. First, the material is mixed and shoveled into a uniform conical pile, as shown in Figure 3A. The pile must be made so that the natural segregation in the cone is radially symmetrical. The cone is then spread from the center to form a flattened disk of material, as shown in Figure 3B. This disk is then divided into quarters at exactly 90° using perpendicular boards, as shown in Figure 3C. One pair of opposite quarters (either A and C or B and D) is removed, and the other pair is used as the sample, as shown in Figure 3D. The choice between the two should be random (flip a coin). If the sample is too large, it can be coned and quartered again until the desired sample size is obtained. In addition to error generated from segregation, this procedure is prone to human error with a standard deviation of 6.81% between samples (Khan 1968; Pitard 1993).

SAMPLING A PROCESS FLOW

Heterogeneous process flows, including conveyor belts, slurry lines, liquid streams, and gas streams, tend to segregate based on the properties of the process flow. Sampling from a segregated process flow can present large errors if proper procedures are not followed. In general, the sample must be taken from a free-falling process stream, be it liquid, slurry, or solid, using a traversing or rotating cut. The cut must span the entire stream and minimize loss of material from dust generation or liquid splashing. Numerous cuts (increments) must also be taken to ensure that the overall sample is representative.

Rules of Cutting Flowing Liquid or Solid Material

Several methods for sampling material from a conveyor belt are available, and the method selected depends upon the accuracy desired, labor available, and the cost. However, the following basic principles should always be observed:

1. The cutter must cut the entire stream.
2. The speed of the cutter must be constant and lower than a specified maximum.
3. The cutter must have an opening width that is at least three times larger than the largest particle to be sampled, and it should be wide enough to prevent bridging (at least 1 cm for dry solids, and at least 0.5 cm for slurry streams).
4. The cutter opening must have parallel edges.
5. The flowing stream to be sampled must be in free fall.
6. The sample must pass quickly through the sampler to avoid blockage.

Sampling Flowing Solid Material (Conveyor Belt)

Linear (Traversing) Cut

A linear (traversing) cut sampler (or high cross-shear sampler) is the best way to cut a falling stream from a conveyor belt. The cutter moves across the falling stream in a straight-line path, as shown in Figure 4. Usually the path is perpendicular to the direction of the flow. The sample cutter travels linearly across the entire falling stream at a constant velocity. The primary sample is diverted from the main product flow onto a conveyor or into a collection bin. The cutter must travel across the entire falling stream because particle segregation occurs on the belt. If adjusted properly, this is the most accurate type of sampler because it follows all six of the principles given previously, but it is also the most expensive and hardest to maintain. These samplers typically use multiple identical cutters attached to a moving chain to take equal-sized cuts at a constant velocity.

The cutter must also be designed properly to ensure that each particle in the slurry has the same chance of being taken into the sample. The cutter opening width (or aperture) should be at least 10 mm or 3 times the size of the largest particle. Also, the speed the cutter traverses the falling stream must remain constant and below the maximum speed, given in the following equation:

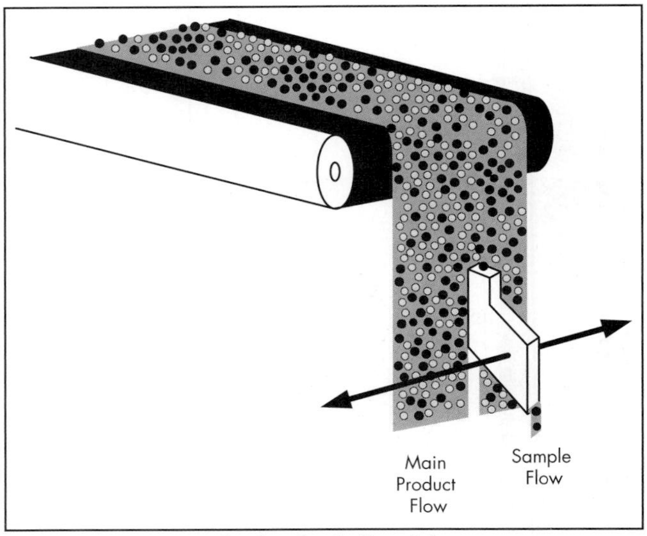

Figure 4 Linear sampler for dry bulk solids

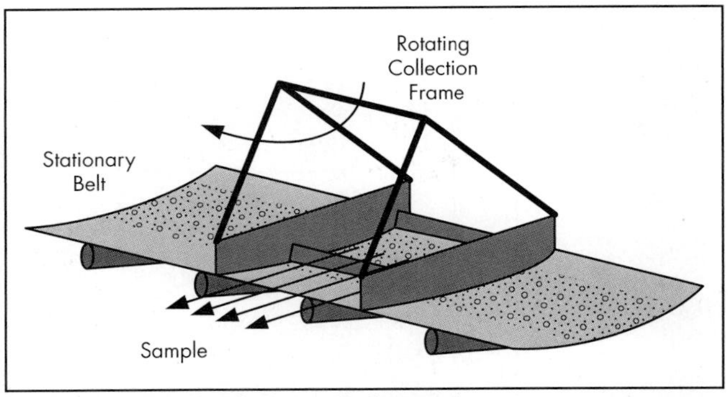

Figure 5 Example of a cross-belt sampler operating on a stationary conveyor belt

$$V_C = 0.3\left(1 + \frac{w}{w_0}\right)$$

where

V_C = maximum cutter speed, m/s
w = cutter opening width (or aperture), mm
w_0 = minimum cutter opening width (or minimum aperture) as defined previously, mm

Cross-Belt Sampler

Collecting a sample from on top of a conveyor belt is possible using a cross-belt sampler, as shown in Figure 5. Representative sampling requires that the end plates are made of a rigid material that exactly conforms to the curvature of the belt. The type of cross-belt sampler shown requires that the belt be stopped while the sample increment is being taken. This limitation can cause issues in continuous processes but can work well when synchronized with belt filters that move by indexing incrementally. After the belt is stopped, a cross section of the material on the belt is scraped off the belt into a sample container. This type of sampler provides excellent representability of samples and is typically used to ensure that other automated samplers do not introduce bias.

Example of a Poor Solid Sampler: Stationary Cut

The stationary cut method, as depicted in Figure 6, represents a common but poor sampling method for sampling off of a conveyor belt. The cutting device is stationary and does not traverse across the entire material flow. Without the linear traversing motion, the segregation of particles on the belt causes the sample to not be representative of the entire process flow. This method ignores the heterogeneity and segregation of the solids being conveyed on the conveyor belt.

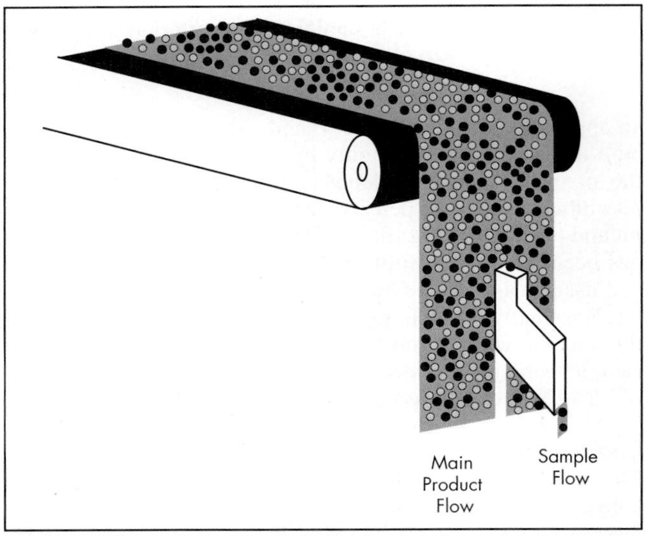

Figure 6 Example of a poor sampling technique: the stationary cut

Sampling Flowing Liquids and Slurries

Sampling liquids and slurries presents inherent problems due to splashing and segregation of solids and liquids within process lines. Solid particles in slurries tend to settle in process lines unless the fluid flow is sufficiently turbulent. Nonhomogeneous liquids can also segregate or be segregated based on the miscibility of the liquid mixtures or insufficient mixing at reagent addition points. To sample these fluids, the rules outlined previously still apply. Most importantly, the fluid must be in free fall during sample collection to reduce any gravitational segregation. This poses issues when dealing with pressurized systems and requires specialized sampling devices, such as a Vezin sampler.

Rotating Cut (Vezin Sampler)

A Vezin sampler consists of a sealed enclosure through which process liquids (or solids) flow vertically. A sample cutter rotates at a constant angular velocity to collect the sample increments and discharge them through a sample port, as shown in Figure 7. The key to correct sampling with a Vezin sampler is both the geometry of the cutter and the distance (u) between the top of the cutter and the liquid inlet. The cutter must be larger than the falling stream and be angled appropriately so that there is no bias in the part of the stream where the sample is being taken; see the top view in Figure 7. The distance, u, must be at least 3 times the size of the largest particle, or 0.5 cm for liquid streams. The sample cutter spins at a constant angular velocity and cuts increments from the process flow. The geometry of the cutter is important to obtain an unbiased cut; note the angular geometry of the cutter. If

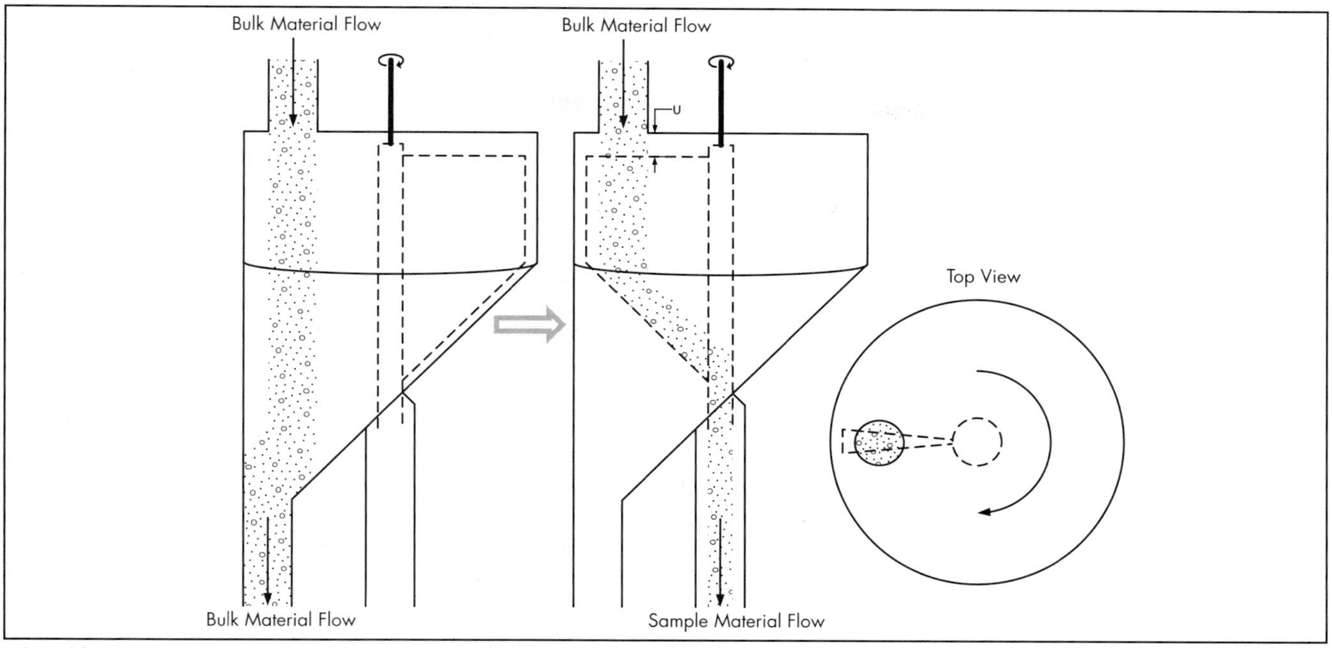

Adapted from Wills 1992

Figure 7 Vezin sampler. The top view illustrates the geometry of the cutter.

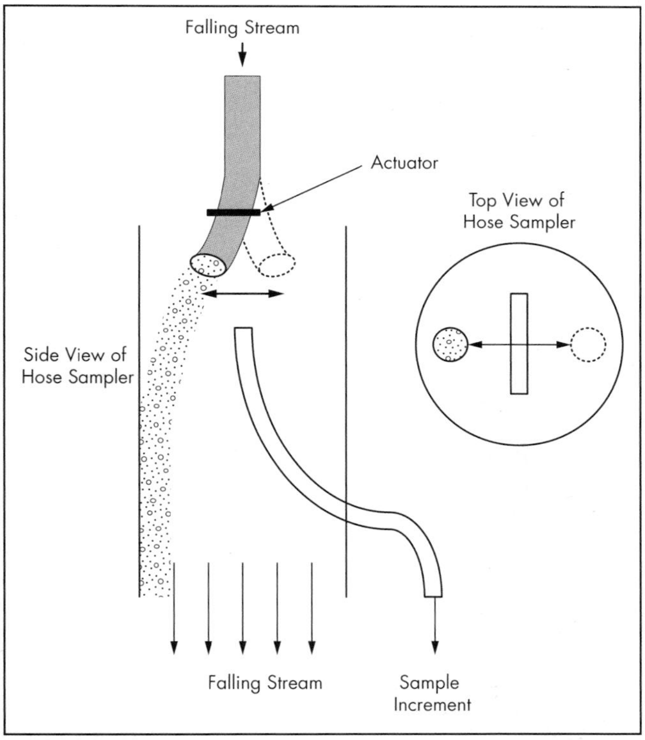

Figure 8 Hose sampler

using a Vezin sampler for sampling solids, this distance should be minimized to prevent dust buildup (Pitard 1993). Further design considerations can be found in Pitard's book, *Pierre Gy's Sampling Theory and Sampling Practice: Heterogeneity, Sampling Correctness, and Statistical Process Control*.

Hose Sampler

A hose sampler is a common method of taking representative samples with enhanced control over the time span between increments (Figure 8). A falling stream travels through a flexible hose. The trajectory of the falling material is changed using an actuator that moves the hose back and forth across the vessel at a constant velocity over the sample cutter. Each time the hose moves past the sample cutter, an increment is collected. Sampling by using this method offers greater flexibility than a Vezin sampler because the time interval between increments as well as the velocity of the stream crossing the cutter can be altered. With a Vezin sampler, only the angular velocity of the cutter can be altered.

Examples of Poor Liquid and Slurry Samplers

Many common types of sampling devices are available that attempt to collect a representative sample; however, they do not satisfy all the conditions outlined previously for process sampling. Two examples of in-stream sampling devices, an in-stream probe and a swing-gate sampler, can be seen in Figure 9A and 9B, respectively. The in-stream probe does not cut the entire stream and hence ignores radial heterogeneity. The swing-gate sampler opens on one side of the pipe and only captures a portion of the stream. This type of valve ignores any revolutionary heterogeneity. Both of these types of samplers work for a homogeneous or well-mixed process flow, if such a flow exists practically or statistically.

Sampling Process Gases

Taking a representative sample of a gas or liquid is a difficult task. This task is made even more difficult if the gas or liquid is flowing through a pipe. It is often difficult to obtain a representative cross-stream sample due to segregation of material caused by laminar flow. Laminar flow often causes the heavy fluid to settle out to the bottom of the pipe while the light fluid floats atop the heavy fluid (Figure 10).

There is currently no correct way to sample a gas or liquid flowing in a pipe, only recommendations to minimize bias. To minimize bias caused by segregation, it is important that the fluid be well mixed before sampling. Static mixers, as seen in Figure 10, can be used to mix gases and liquids. For a stream flowing horizontally, the sample probe should come out of the side of the pipe for liquids and out of the top for gases. The location of a sampling probe should be far enough downstream of a static mixer that the velocity profile of the fluid is well established.

When sampling flowing fluids, it is important to remember:

- Increasing the stream velocity does not ensure a well-mixed stream.
- A strainer is not sufficient for mixing; use a static mixer.

Sampling from Vessels

Sampling from vessels can be problematic because of the amount of material present and how the material may segregate over time. First, consider a large amount of *solids* being stored in a large vessel, as shown in Figure 11A. It is extremely time-consuming to cone and quarter a large quantity of material and then properly split it to achieve a representative sample. Therefore, tools have been designed that allow for easy sampling of bulk solids. One of the more common tools for bulk solids sampling is called a thief probe (Figure 11B). A thief probe is chiefly used for obtaining vertical core samples from bulk solids. The knob at the top is turned to open and close the sample windows along the length of the shaft. The pointed tip at the end is used to penetrate through the material. To use a thief probe, one must make sure the sample windows are completely closed before penetrating the material. Once the probe has been pushed through the material, the windows are opened to collect the sample. A thief probes work well for loosely packed solids.

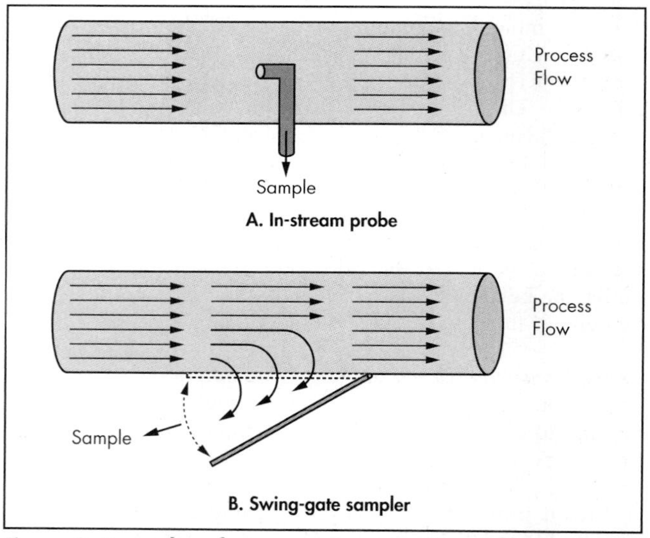

Figure 9 Examples of poor process sampling devices

Before using a thief probe, it is important to be aware of some of its limitations:

- The windows may not seal properly, causing sample leakage into the ports before the probe is fully inserted.
- Material may be too large to fall into the sample ports.
- If the solids are tightly packed, it will be difficult to push the probe through the material.

Now, consider obtaining a representative sample from a large amount of *liquid* being stored in a large vessel. As liquids are stored for long periods of time, the immiscible liquids tend to segregate from one another. Next, consider a mixture of oil and water. The water will eventually settle out on the bottom of the vessel and the oil will float atop the water, as shown in Figure 12A.

Obtaining a representative sample in this case is difficult because the immiscible nature of the liquids creates two distinct layers. If a sample were poured off the top, if would only contain oil; if a sample were taken from the bottom, it would only contain water. Therefore, a tool that utilizes the same principles as a thief probe is commonly used to obtain representative liquid core samples. The tool is called a coliwasa (composite liquid waste sampler), as shown in Figure 12B.

To operate a coliwasa, the instrument must be inserted to the bottom of the tank slowly with the bottom seal open. The coliwasa should be inserted as deep as possible to obtain a representative sample. Once the fluid from the tank has filled the coliwasa, the handle must be pulled to seal the bottom of the instrument. The coliwasa can then be slowly removed from the tank. The coliwasa must be sealed while pulling it from the tank or it will leak fluid back into the tank and the sample will not be representative.

Modern Sampling Devices

Slurries and liquids pose many issues to proper sampling, such as how does one split a continuously flowing stream down to a continuously flowing sample stream for online analysis? Recent developments in sampling technology have

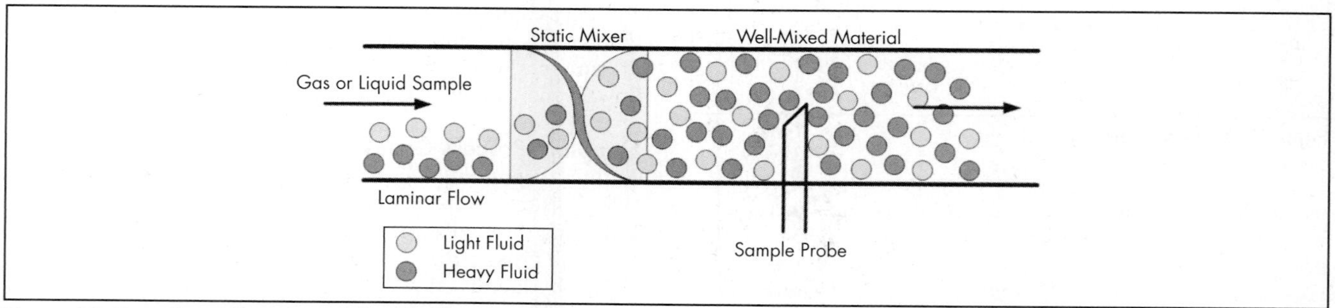

Figure 10 Sampling of gases and liquids from a pipe

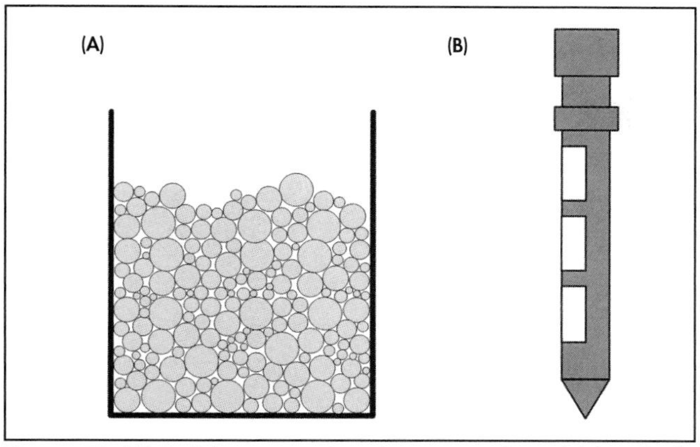

Figure 11 (A) Solids storage in a large vessel, and (B) a thief probe used for bulk solids sampling

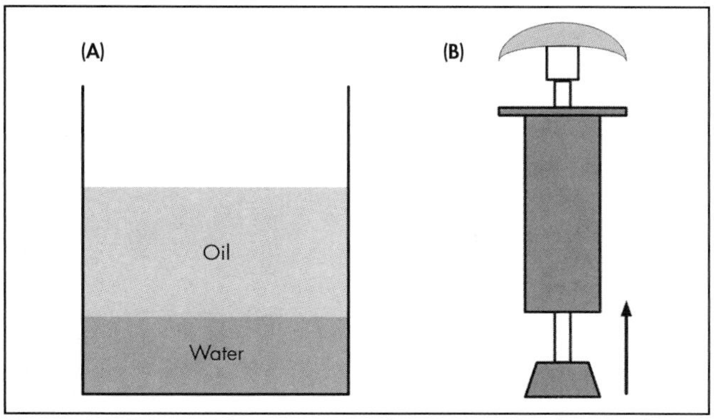

Figure 12 (A) Liquid storage in a large vessel, wherein the immiscible phases have separated over time. (B) The coliwasa is used to sample large quantities of liquid.

addressed some of these issues. One such sampling device is the Outokumpu slurry sampler depicted in Figure 13. It is used to sample high flow-rate slurries. This sampler reduces the flow rate of the slurry by taking multiple stages of cuts from the incoming process flow. Each stage has a lower incoming slurry flow rate, and the final stage is cut using a traditional linear traversing sample cutter (Alfthan 2003).

This type of sampler has two stages of stationary cutters. Although there are many stationary cutters per stage, it still violates the rules of sampling process flows. To remedy this issue, the first and second cut could be taken using a linear traversing cutter.

SPLITTING SAMPLES FOR LABORATORY ANALYSIS

Laboratory Grab Sample

A laboratory grab sample, otherwise known as the "four-corners method," is the simplest, quickest, and most flexible method, as it can be carried out on small quantities using spatulas, or on large quantities using shovels, and can divide the material into the number of samples desired. The material is first homogenized by thorough mixing on a rolling mat, as shown in Figure 14. The mat should be a smooth, flexible sheet that the sample will not stick to, such as glazed paper, hard vulcanized rubber, or smooth vinyl. The material is then divided into samples by randomly grabbing small amounts from the homogenized pile on the cloth, not unlike the method shown in Figure 2. This method uses the least equipment, but also is the most prone to human biases and has a higher variance between samples than other methods.

Laboratory Coning and Quartering Methods

Coning and quartering can be performed on small amounts of material in the laboratory as well. The method is the same as discussed earlier except with smaller equipment. This is typically carried out on a smooth, clean laboratory surface such as a lab bench or lab floor.

Riffle Splitters

Riffle splitters consist of a series of chutes that run in alternating directions, so that when material is poured into the top of the splitter, it flows through the chutes and is randomly divided into two equal-sized fractions. An example of a riffle splitter is shown in Figure 15. One of the fractions can then be split again, and the procedure can be repeated until a sample of the desired size is obtained. To work properly, these splitters must be fed using a special pan that is the same width as the top of the chutes, otherwise the amount of material entering the two end chutes will be different and the sample will not be representative. Also, if a material is repeatedly split into smaller fractions using a riffle, the errors from each stage of splitting will be added together, resulting in increasing variance between samples.

Rotary Riffle Splitters

The rotary, or spinning, riffle is the best method to use for dividing material into representative samples. It produces the lowest variance between samples of all sampling methods discussed thus far. It can also produce many samples in a single operation.

The material to be sampled is fed from a feed hopper to a feeder (usually a vibratory feeder, although screw feeders and small conveyors work as well). The feeder drops the material at a uniform rate into a series of bins (sample containers) on a rotating table, as shown in Figure 16. The turntable speed is set so that each sample container will pass under the end of the feeder numerous times before the feed hopper is emptied, but slowly enough that the bin edges do not strike the falling particles hard enough to bounce them into a different container or throw them out of the machine entirely (generally about 10 to 25 rpm). The turntable rotates at a constant angular velocity, providing representative increments.

Comparison of Laboratory Sampling Methods

A comparison of the relative standard deviations of samples made by the previously mentioned methods is given in Table 1. It can be clearly seen that rotary riffling is the best method of sample division and approaches the standard deviation that would be expected from an ideal sample divider where division of material into samples is perfectly random. Both coning and quartering and grab sampling perform relatively poorly, indicating that they should only be used when there are no other practical methods that will work with a given material. Riffle splitters give intermediate performance, indicating that they are suitable for routine, noncritical work.

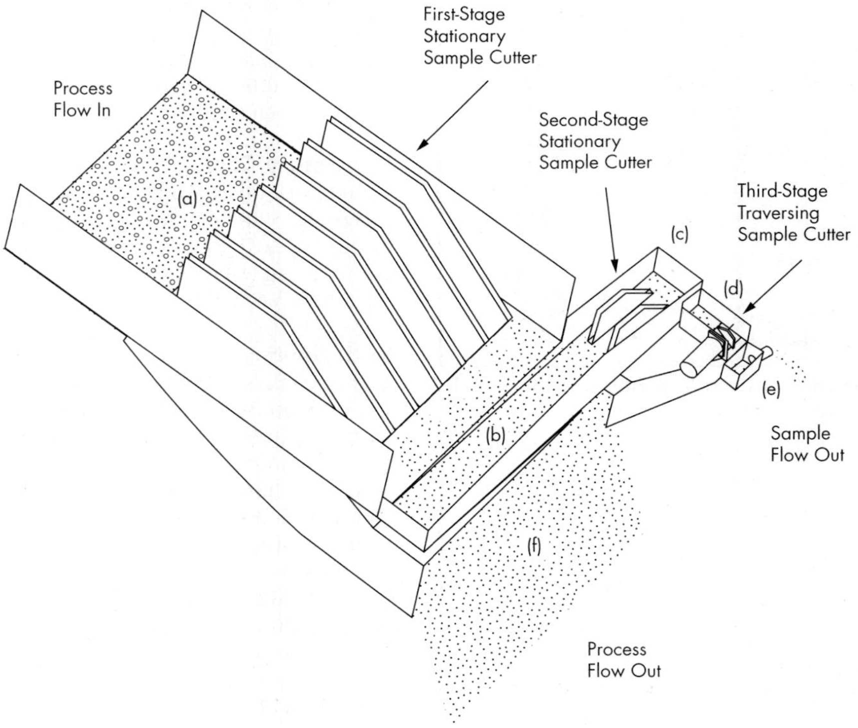

(a) The feed slurry enters the sampler and is cut by several stationary cutters while the slurry is in free fall.
(b) The cuts are mixed in a trough and redirected toward a second stage of cutters. Again, the cuts are taken while the slurry is in free fall.
(c) The cuts are mixed in another trough.
(d) The cuts are redirected to a third stage where the slurry flow is cut in free fall using a traversing cutter.
(e) The resulting cut is the sample.
(f) The slurry that did not get cut in three stages of cutters is mixed in the outlet trough and sent to downstream processes.

Adapted from Alfthan 2003
Figure 13 Modern sampler

TYPICAL SOURCES OF ERROR IN SAMPLING

Poor sampling occurs because of poor technique selection, poor equipment maintenance, carelessness, haste, and lack of knowledge. When developing a sampling strategy, it is important to consider the following:

- Is the appropriate sampling equipment being used?
- Where could possible errors in sample handling occur?
  - Contamination
  - Loss
  - Unintentional mistakes

When sampling, always keep in mind the principle of correct sampling: Every part of the lot has an equal chance of being in the sample and the integrity of the sample must be preserved during and after sampling.

Appropriate Sampling Equipment

Choosing the appropriate equipment for sampling is important in obtaining a representative sample. Many times, equipment for sampling solids may not be suitable to sampling liquids and gases. For example, the rotary riffle splitter is great for

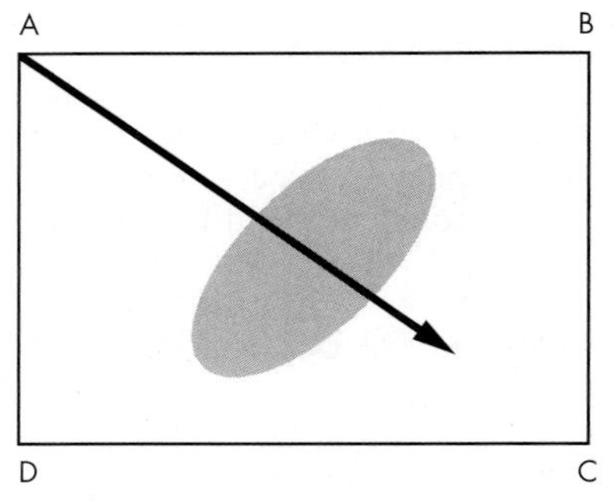

Note: Mix by first drawing corner A so that the sample rolls toward corner C. Then draw corner B to corner D, corner C to corner A, and corner D to corner B. This should be repeated at least 40 times to ensure proper mixing.

Figure 14 Four-corners method of mixing a sample for grab sampling

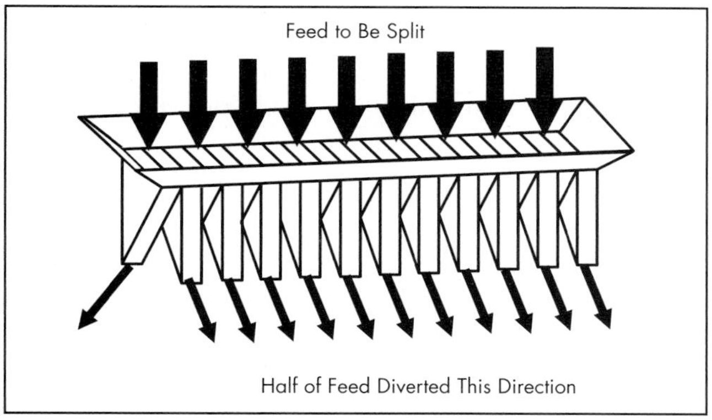

Figure 15 Riffle splitter, wherein the material flows through alternating-direction chutes and is split in half

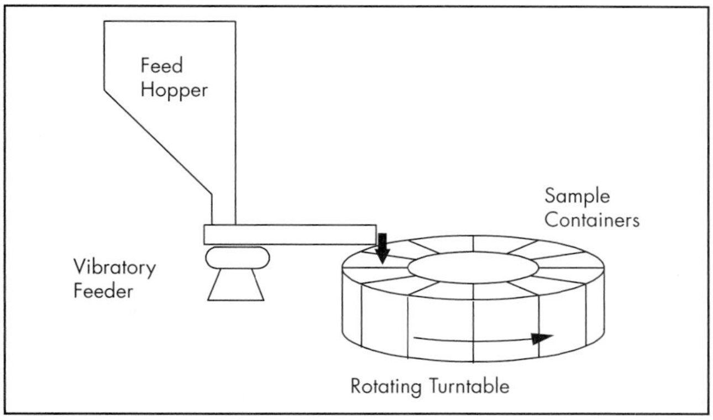

Figure 16 Rotary riffle splitter

Table 1 Standard deviations of samples produced from a 60%/40% mixture of fine and coarse sand

Sampling Method	Standard Deviation of Samples, %
Cone and quarter	6.81
Grab sampling	5.14
Riffle splitter	1.01
Rotary riffle splitter	0.125
Random variation for a theoretically perfect sampler	0.076

Source: Khan 1968; Allen and Khan 1970

splitting and sampling solids, but using a rotary riffle splitter for a liquid would be a poor choice of sampling equipment. Materials spills and splashing may occur, leading to errors in sample handling.

Sample Contamination

Sample contamination occurs when unwanted material is added to the sample after sampling but before chemical or physical analysis. The simplest source of contamination occurs because the sampling equipment has not been cleaned. If equipment is not cleaned, it is possible that material from a completely different product may contaminate the current sample. For example, because of static electricity, fine material often adheres to the sampling tool. If not cleaned properly after sampling, these fines may end up in the next sample. Sampling lines can become contaminated when they are not properly purged of old material. After using a sampling line, it must be purged the length of the line to avoid buildup of old material.

Chemical reactions between the sample material and sample container can also lead to unwanted atoms or molecules contaminating the sample. For example, if a sample is to be analyzed for sodium, it should not be stored in a glass container. When choosing a sample container, it is important to choose an inert environment for storage.

Sample Loss

Sample loss occurs when some of the sample mass, or a percentage of a specific component, is not retained after the sample is taken. Spills and splashes during or after sampling are the most common source of material loss during sampling. Material is also often lost during the crushing and grinding stages. Precious metals tend to smear onto the surfaces of crushing and grinding equipment, causing some material to remain behind and not become part of the sample. Care needs to be taken when transferring material from one storage container to another. When transferring material to another storage container, fines tend to be left behind. It is important that the number of transfers between containers be minimized to reduce the amount of fines lost.

Sample loss will also occur with materials that are susceptible to chemical reaction. For example, consider that the moisture content of a material is of interest. If the material is not stored in a humidity-controlled environment, moisture may be lost to the surrounding atmosphere, causing a loss in material.

Unintentional Mistakes

Unintentional mistakes are innocent errors that can severely compromise analytical results. Mislabeling of a sample is a common unintended mistake. For example, labeling a sample for ambient storage when it needs to be refrigerated may compromise the analytical results. Many times, unintentional mistakes can be avoided by simply paying more attention to what one is doing.

CALCULATIONS FOR DEVELOPING SAMPLING STRATEGIES

The best and most impractical way of determining the characteristics of a lot is to analyze the entire lot as a sample. However, this is not feasible in most situations. A smaller sample is typically analyzed and is assumed to be representative of the entire lot. So far, the various methods of taking these smaller samples have been discussed, but the amount of sample required to be considered representative has not yet been examined. This section details the calculations required to determine the amount of sample necessary to be statistically representative of the entire lot.

Gy's Method for Calculating Required Sample Size

Gy's method is a general-purpose calculation to determine the minimum size of sample needed to ensure that it will be representative of the whole lot, within specified statistical limits. Before using this method, approximate estimates of the following will be needed:

Table 2 The number of standard deviations away from the mean that can be expected at a given confidence interval

Confidence Interval	h
0.90	1.645
0.91	1.705
0.92	1.750
0.93	1.812
0.94	1.881
0.95	1.960
0.96	2.054
0.97	2.170
0.98	2.326
0.99	2.576
0.999	3.291
0.9999	3.890
0.99999	4.417
0.999999	4.892

- The content of the species of interest in the lot (critical content)
- The general shape of the particles
- The densities of the various species and phases present
- The particle size distribution
- The degree of liberation and the grain size

Basic Equations for Gy's Method

The basic calculation to determine the required sample weight, M, from a lot of total weight, W, is shown in Equations 1 and 2. This equation provides the mass of a sample required to be representative of the entire lot within a specific confidence interval, s.

$$s^2 = \left(\frac{1}{M} - \frac{1}{W}\right)Cd^3_{max} \quad \textbf{(EQ 1)}$$

or

$$\frac{WM}{W - M} = \frac{Cd^3_{max}}{s^2} \quad \textbf{(EQ 2)}$$

where

s = value of the standard deviation that is needed for desired confidence interval
M = minimum sample weight needed, g
W = weight of entire lot being sampled, g
C = mineralogical factor for the material being sampled, g/cm^3
d_{max} = size of the largest particle in the lot (or top size), cm

When the weight of the lot being sampled, W, is much larger than M, this equation can be simplified to

$$M = \frac{Cd^3_{max}}{s^2} \quad \textbf{(EQ 3)}$$

Determining the Statistical Parameter: s

To calculate the desired value for s:

1. First select the *desired certainty range* (e.g., you might want a copper assay to be accurate to within ±0.1% copper).
2. Divide this value by the *mean probable assay value* (so for a 5% copper sample, you get 0.1/5 = 0.02).
3. Select the number of standard deviations, h, that will be needed to give the desired confidence interval (see Table 2).

$$s = \frac{\text{desired certainty range}}{\text{mean probable assay value}} \cdot \frac{1}{h} \quad \textbf{(EQ 4)}$$

Determining the Mineralogical Factor: C

The mineralogical factor for an ore body is a constant that depends on the physical characteristics of the ore particles. It can only be calculated if extensive information about the ore has been determined. In most circumstances, the variables used to determine the mineralogical factor are assumed from general observations and previous experience working with the particular ore. The mineralogical factor, C, can be calculated as follows:

$$C = f \cdot g \cdot l \cdot m \quad \textbf{(EQ 5)}$$

where

f = shape factor
g = size distribution factor
l = liberation factor
m = composition factor

Shape factor. The shape factor, f, corrects the mineralogical factor for particles that are not perfect cubes (f = 1). For most natural minerals, the shape factor is taken to be equal to 0.5. Flaky minerals, such as mica, have a shape factor of about 0.1. Materials that contain soft solids, such as gold, have a shape factor of about 0.2. Needle-like materials, such as asbestos, have a shape factor ranging from 1 to 10 (Pitard 1993).

Size distribution factor. The size distribution factor, g, is determined by estimating the 95% passing size, d_{95}, and the 5% passing size, d_5, and using the following guidelines:

- If $d_{95}/d_5 > 4$ (broad size distribution), then g = 0.25.
- If $2 < d_{95}/d_5 < 4$ (moderate size distribution), then g = 0.5.
- If $1 < d_{95}/d_5 < 2$ (narrow size distribution), then g = 0.75.
- If $d_{95}/d_5 = 1$ (mono-sized particles), then g = 1.

When size distribution information is not available, g should be estimated conservatively as a narrow size distribution having a value of 0.75.

Liberation factor. The liberation factor, l, is a measure of the degree of dispersion of the valuable material through the bulk, and of the homogeneity of the material. The liberation factor varies from 0 to 1, with 0 representing a perfectly homogeneous material and 1 representing a perfectly heterogeneous material. It is calculated from the following expression:

$$l = \sqrt{\frac{d_l}{d_{max}}} \quad \textbf{(EQ 6)}$$

where

d_l = liberation size, cm
d_{max} = size of the largest particle in the lot (or top size), cm

Generalizations can be made when insufficient data are provided to calculate an exact liberation factor. These generalizations are shown in Table 3.

Table 3 Generalizations for liberation factor when sufficient information is not available to calculate it

Degree of Heterogeneity of the Material	Approximate Liberation Factor
Very heterogeneous	0.8
Heterogeneous	0.4
Average	0.2
Homogeneous	0.1
Very homogeneous	0.05

Adapted from Pitard 1993

Composition factor. The composition factor, m, is calculated from the following formula:

$$m = \left(\frac{1-a}{a}\right)((1-a)r + at) \qquad \textbf{(EQ 7)}$$

where

a = fractional average assay of the critical component
r = specific gravity of the critical component, g/cm^3
t = specific gravity of the bulk material, g/cm^3

The fractional average assay of the critical component, a, is the estimated critical content of the bulk material in decimal format (not percent).

Example 1: Calculation Using Gy's Method

In this example, a nickel ore has a nickel sulfide (NiS) critical component containing 3.6% Ni (5.567% NiS) by weight. The top size of the ore particles is 1.2 cm, and the NiS grain size is 0.004 cm. The desired sampling accuracy is ±0.01% Ni (0.0155% NiS), with a 99% confidence interval (2.576 standard deviations). The specific gravity of the NiS is 5.3, the specific gravity of the bulk ore is 3.1, and it has a "broad" size distribution.

First, determine the shape factor, f. Nickel sulfide is a common natural mineral:

$$f = 0.5 \text{ (regular shape)}$$

Next, determine the size distribution factor, g. The problem statement claimed a broad size distribution:

$$g = 0.25 \text{ (broad size distribution)}$$

Determine the liberation factor, l. The liberation size and top size are given as 0.004 cm and 1.2 cm, respectively:

$$l = \sqrt{\frac{d_l}{d_{max}}} = \sqrt{\frac{0.004}{1.2}} = 0.05774$$

The critical content is given as 5.567% NiS:

$$a = 0.05567$$

The specific gravity of the critical component (NiS) and the bulk ore are given as 5.3 and 3.1, respectively:

$$r = 5.3 \text{ g/cm}^3$$

$$t = 3.1 \text{ g/cm}^3$$

With this information, the composition factor can be calculated as follows:

$$m = \left(\frac{1-a}{a}\right)((1-a)r + at)$$
$$= \left(\frac{1-0.05567}{0.05567}\right)((1-0.05567)5.3 + 0.05567 \cdot 3.1)$$
$$= 87.826 \text{ g/cm}^3$$

All of the variables in the mineralogical factor, C, have now been determined. C can now be calculated as follows:

$$C = fglm = 0.5 \cdot 0.25 \cdot 0.05774 \cdot 87.826 = 0.634 \text{ g/cm}^3$$

The statistical parameter, s, can be calculated using the desired sampling accuracy, 0.0155% NiS, the critical content, 5.567% NiS, and the number of standard deviations correlating with the 99% confidence interval, 2.576:

$$s = \frac{0.0155\%}{5.567\%}\frac{1}{2.576} = 0.001079$$

All required information has now been calculated to predict a sample size that would be considered representative of the lot at a 99% confidence interval:

$$M = \frac{Cd_{max}^3}{s^2} = \frac{0.634(\text{g/cm}^3)(1.2 \text{ cm}^3)}{0.001079^2}$$
$$= 940{,}572 \text{ g, or } 0.9 \text{ metric tons}$$

Gaudin's Method for Calculating the Required Sample Size

Gaudin's method is a derivation of Gy's method using assumptions typically found in ores with the critical component being a precious metal (gold, platinum, diamonds, etc.). This method is limited to ores where the critical content is a small fraction (a few percent or less) of the total volume (Gaudin 1967).

Basic Equations for Gaudin's Method

The basic calculation to determine the required sample weight, S, from a lot is shown in Equation 8. This equation provides the mass of a sample required to be representative of the entire lot within a specific confidence interval, y:

$$S = \frac{n}{n'} \qquad \textbf{(EQ 8)}$$

where

S = sample weight required for representativity, g
n = number of particles that are required for a representative sample
n′ = number of particles in 1 g of material in the lot, 1/g

Determining the Number of Particles Required: n

To determine the number of particles required for a representative sample, the approximate volumetric critical content, x, and the allowable volumetric error, y, must be known:

$$n = 0.45\frac{x}{y^2} \qquad \textbf{(EQ 9)}$$

where

x = approximate volumetric critical content of the lot, decimal fraction
y = allowable volumetric error, decimal fraction

Often, the critical content and allowable error is given as a weight percent instead of a volume fraction. These can be converted to volume fractions given the density of the critical component and the density of the bulk material in the lot.

$$x = \frac{\text{critical content, \%/wt}}{\text{critical component density}} \cdot \frac{\text{bulk material density}}{100\%}$$

$$y = \frac{\text{allowable error, \%/wt}}{\text{critical component density}} \cdot \frac{\text{bulk material density}}{100\%}$$

Determining the Number of Particles per Gram: n'

To make a conservative estimate of the number of particles per gram of material, n′, it is necessary to know the density as well as the top size of the bulk lot:

$$n' = \frac{6}{\rho_s \cdot d_{max}^3} \qquad \textbf{(EQ 10)}$$

where ρ_s is the density of the bulk material, in g/cm^3.

Example 2: Calculation Using Gaudin's Method

In this example, a nickel ore has a nickel sulfide (NiS) critical component containing 3.6% Ni (5.567% NiS) by weight. The top size of the ore particles is 1.2 cm, and the NiS grain size is 0.004 cm. The desired sampling accuracy is ±0.01% Ni (0.0155% NiS), with a 99% confidence interval (2.576 standard deviations). The specific gravity of the NiS is 5.3, the specific gravity of the bulk ore is 3.1, and it has a "broad" size distribution.

First, the critical content and allowable error must be converted to volumetric fractions from weight percent:

$$x = \frac{\left(\frac{5.567}{5.3}\right)}{\left(\frac{100}{3.1}\right)} = 0.0326\%$$

$$y = \frac{\left(\frac{0.0155}{5.3}\right)}{\left(\frac{100}{3.1}\right)} = 0.000091\%$$

Next, the number of particles required to obtain a representative sample can be calculated:

$$n = 0.45\frac{x}{y^2} = 0.45\frac{0.0326}{0.000091^2} = 1{,}783{,}000 \text{ particles}$$

Using the provided lot density and top size, the number of particles per gram can be estimated:

$$n' = \frac{6}{\rho_s \cdot d_{max}^3} = \frac{6}{3.1 \cdot 1.2^3} = 1.12007 \text{ particles/g}$$

Finally, the sample size required for representativity can be calculated:

$$S = \frac{n}{n'} = \frac{1783000}{1.12007} = 1{,}591{,}617 \text{ g, or 1.6 metric tons}$$

This calculation was performed on a nickel sulfide ore, not a precious metal ore. The sample size predicted by Gaudin's method is much larger than necessary because of the assumption made when defining this method.

Comparison of Gy's and Gaudin's Methods

Gy's method is a general-purpose sampling equation. Gaudin's method is specifically intended for use in calculating sample sizes for precious metals, and it makes a number of simplifying assumptions based on the fact that precious metals make up a minute fraction of the mass of the ore (typically only a few grams per metric ton).

Gaudin's equation can be put into the same terms as Gy's equation by using the same symbols for equivalent values: S = M. The value of y (a volume fraction) can be converted into s (a weight fraction) by multiplying it by a constant, k, that depends on the relative densities of the ore and the valuable minerals, and so, y = s/k. If these substitutions are made, then Gaudin's equation becomes

$$M = 0.075 \cdot x \cdot \rho_s \cdot k^2 \frac{d_{max}^3}{s^2} \qquad \textbf{(EQ 11)}$$

If we now take $(0.075 \cdot x \cdot \rho_s \cdot k^2) = C$, it can be seen that Gaudin's equation reduces to the same form as Gy's equation. Gaudin's equation is therefore simply a special case of Gy's equation.

Hassiali's Method for Development of Incremental Sampling Strategy

This method is intended for designing a mechanical sampling system that will account for normal variations in the characteristics of the process stream being sampled. A sample cutter is used to collect a series of test increments over a particular time interval, and then each increment is assayed separately. The variations in the assays between increments are then used to calculate the number of increments that the sample cutter needs to collect over that span of time to give the desired confidence interval in the results. Correct increment sampling should produce a normal distribution curve where individual increments are plotted against increment frequency.

Basic Equations for Hassiali's Method

To calculate the number of increments required from a process flow, n, to reach a desired confidence interval of the measured critical content, the mean, $\bar{x}$, and standard deviation, σ_s, between experimental measurements of the critical content of individual increments is necessary (Taggart and Behre 1945). The basic calculation for Hassiali's method is as follows:

$$n = \left(\frac{h \cdot \sigma_s}{\bar{x} \cdot z}\right)^2 \qquad \textbf{(EQ 12)}$$

or

$$n = \left(\frac{\sigma_s}{\bar{x} \cdot s}\right)^2$$

where

- n = number of increments required for desired confidence interval
- h = number of standard deviations corresponding to desired confidence interval
- σ_s = experimental standard deviation of the measured critical content
- $\bar{x}$ = experimental mean of the measured critical content
- z = allowable sampling error
- s = value of the standard deviation that is needed for desired confidence internal as described by Equation 4

The number of standard deviations corresponding to the desired confidence interval, h, can be found in Table 2. The allowable sampling error, z, is expressed as a fraction of the critical content. The value of z is equivalent to the first fraction in Equation 4:

$$z = \frac{\text{desired certainty range}}{\text{mean probable assay value}}$$

Example 3: Calculation Using Hassiali's Method

For this example, the following 11 test increment assays were obtained from increments collected at 30-minute intervals (total elapsed time of 5.5 hours). Values are weight % Ni.

3.61	3.69	3.58	3.65	3.56	3.63	3.61	3.66	3.57	3.59	3.60

Calculate the number of increments that must be collected over this same time interval so that the probability of the assay of the final composite sample is accurate to within 1% of the true assay at a 99% confidence interval.

First, calculate the mean and standard deviation of the increment assays:

$$\bar{x} = \frac{\Sigma x_i}{N} = 3.614$$

$$\sigma_s = \sqrt{\frac{\Sigma(x_i - \bar{x})^2}{N-1}} = 0.0403$$

where

x_i = assay value of each increment
N = total number of increments

Next, refer to Table 2 for h at a 99% confidence interval:

h = 2.576 for 99% confidence interval

The value of z is given as 0.01:

z = 0.01 for 1% accuracy

Finally, all of the information has been determined to calculate the number of increments required for a 99% confidence interval and ±1% accuracy:

$$n = \left(\frac{h \cdot \sigma_s}{\bar{x} \cdot z}\right)^2 = \left(\frac{2.576 \cdot 0.0403}{3.614 \cdot 0.01}\right)^2 = 8.26$$

Round up to 9 increments because one cannot take fractional increments.

TERMS AND SYMBOLS USED IN THIS CHAPTER

Term	Definition
ρ_s	Density of the bulk material (Gaudin's method)
σ_s	Experimental standard deviation of the measured critical content (Hassiali's method)
a	Fractional average assay of the critical component (Gy's method)
C	Mineralogical factor for the material being sampled (Gy's method)
Component	A particular substance within a lot of heterogeneous material. For example, silica, pyrite, and gold are typical components of gold ore.
Content	The concentration of a component within a lot of heterogeneous material. For example, a gold ore may have a silica content of 65%.
Critical component	The component of the heterogeneous material that is of interest. For example, gold is the critical component in gold ore.
Critical content	The concentration of the critical component within a lot of heterogeneous material. For example, a gold ore may have a critical content of 0.01% gold.
d_5	5% passing size of the material being sampled (Gy's method)
d_{95}	95% passing size of the material being sampled (Gy's method)
d_l	Liberation size of the material being sampled (Gy's method)
d_{max}	Size of the largest particle in the lot (or top size) (Gy's method)
f	Shape factor for the material being sampled (Gy's method)
g	Size distribution factor for the material being sampled (Gy's method)
h	Number of standard deviations corresponding to the desired confidence interval (Gy's method and Hassiali's method)
Heterogeneous material	A material that has different properties depending on where the material is analyzed. Examples include slurries, immiscible fluids, and even tap water.
Homogeneous material	A material that is either one substance or a perfect mixture of multiple substances with constant thermodynamic properties throughout. By definition, perfectly homogeneous materials do not exist and cannot be created.
l	Liberation factor for the material being sampled (Gy's method)
Lot	A lot is defined as a large (bulk) quantity of material from which a sample must be taken. If proper sampling is performed, analysis of the sample will be representative of the lot within a specified degree of certainty (or confidence interval).
M	Minimum sample weight needed for a representative sample of a lot (Gy's method)
m	Composition factor for the material being sampled (Gy's method)
N	Total number of increments (for calculating standard deviation)
n	Number of particles that are required for a representative sample (Gaudin's method) or number of increments required for desired confidence interval (Hassiali's method)
n′	Number of particles in 1 g of material in the lot (Gaudin's method)
r	Specific gravity of the critical component (Gy's method)
s	Value of the standard deviation that is needed for the desired confidence interval (Gy's method and Hassiali's method)
S	Minimum sample weight needed for a representative sample of a lot (Gaudin's method)

Term	Definition
t	Specific gravity of the bulk material (Gy's method)
W	Weight of the entire lot being sampled (Gy's method)
x	Approximate volumetric critical content of the lot, decimal fraction (Gaudin's method)
$\bar{x}$	Experimental mean of the measured critical content (Hassiali's method)
x_i	Assay value of each increment (for calculating mean)
y	Allowable volumetric error, decimal fraction (Gaudin's method)
z	Allowable sampling error (Hassiali's method)

REFERENCES

Alfthan, C.V. 2003. *Method and Apparatus for Taking Slurry Samples*. G01N 1/20 ed. Finland: Outokumpu.

Allen, T., and Khan, A.A. 1970. Critical evaluation of powder sampling techniques. *Chem. Eng.* 238:108–112.

Gaudin, A.M. 1939. *Principles of Mineral Dressing*. New York: McGraw-Hill.

Gaudin, A.M. 1967. *Principles of Mineral Dressing*. New York: McGraw-Hill.

Gy, P.M. 1979. *Sampling of Particulate Materials: Theory and Practice*. Amsterdam: Elsevier.

Khan, A. 1968. Critical evaluation of powder sampling procedures. Master's thesis, University of Bradford, West Yorkshire, United Kingdom.

Morrison, R.D. 2008. *An Introduction to Metal Balancing and Reconciliation*. Indooroopilly, Queensland: Julius Kruttschnitt Mineral Research Centre.

Pitard, F.F. 1993. *Pierre Gy's Sampling Theory and Sampling Practice: Heterogeneity, Sampling Correctness, and Statistical Process Control*, 2nd ed. Boca Raton, FL: CRC Press.

Taggart, A.F., and Behre, H.A. 1945. *Handbook of Mineral Dressing: Ores and Industrial Minerals*. New York: John Wiley and Sons.

Wills, B.A. 1992. *Wills' Mineral Processing Technology*, 5th ed. New York: Pergamon Press.

Wills, B.A., and Finch, J.A. 2016. Wills' *Mineral Processing Technology*, 8th ed. Oxford: Elsevier/Butterworth-Heinemann.

CHAPTER 1.9

Viscosity and Rheology

Michael G. Nelson

Operations in mineral processing and hydrometallurgy frequently include the handling of fluids—gases, liquids, and slurries. A fluid is a substance that undergoes continuous deformation, or strain, when subjected to shear stress; the relation of shear stress and shear strain in a fluid is characterized by the fluid's viscosity and rheology.

Cowper (1992) provides an excellent description of the importance of fluid viscosity in mineral processing and hydrometallurgical operations. Cowper notes the following effects:

- Grinding is most efficient at viscosities of 100–300 mPa·s, and coal grinding mill throughputs fall off rapidly when the slurry yield stress exceeds 10–20 Pa. Typical gold ore mills operate with slurry yield stress of 5–7 Pa, while cement plant ball mills run at 20–40 Pa.
- Classification is much less efficient at high viscosity, which implies high solids content.
- Flow through screen openings is generally laminar. For a Newtonian fluid at a given head, screen throughput is approximately inversely proportional to the viscosity of the feed. For non-Newtonian fluids, throughput depends mainly on the slurry yield stress, and the dependence is nonlinear.
- In hydrocyclones, when the slurry yield stress is too low, the cut point decreases and undersized particles may be unnecessarily returned to the mill, with a consequent reduction in throughput and increase in circulating load. On the other hand, if the yield stress is too high, oversized particles may pass through the cyclone for processing, with values unliberated.
- Within practical constraints, leach tanks should be operated at the highest possible slurry density, increasing retention time for a given solids throughput. However, high pulp densities can hinder carbon-in-pulp leaching and adsorption kinetics, so a balanced approach is appropriate.
- The pumping of thickener underflows—concentrates and tailings—is very sensitive to slurry rheology. A small change in solids concentration can cause a significant change in slurry yield stress. In normal operation, the underflow pipeline operates at a constant flow and a relatively constant head in turbulent flow. If the solids concentration in the slurry increases, the flow may transition to the laminar region, and the system head will increase, and if the system head exceeds the pump capacity, flow will stop. Similarly, on the suction side of the pump, an increase in solids concentration may reduce the available suction head of the pump. Because most centrifugal slurry pumps do not provide sufficient net positive suction head to prevent cavitation, the thickener outlet may become plugged with a high-density, high-viscosity slurry, effectively shutting down the thickener.

VISCOSITY AND RHEOLOGY

Figure 1 shows the behavior of a fluid experiencing a simple shear stress, which results in a simple shear flow. The fluid is between two large parallel plates of area A, separated by a small distance H. The bottom plate is fixed and a force F acts on the upper plate, causing it to move at a velocity V. The fluid will continue to deform as long as the force is applied, unlike a solid, which will experience a finite deformation and then fail. The shear stress in this illustration is $\tau = F/A$. Under a no-slip condition, the fluid at the lower plate will have a velocity of zero, and the fluid at the upper plate will move with the upper plate at velocity V. This will result in a velocity gradient in the fluid between the plates, given by $v_y = Vy/H$. The velocity gradient, dv/dy, is defined as the shear rate and has units of s^{-1}.

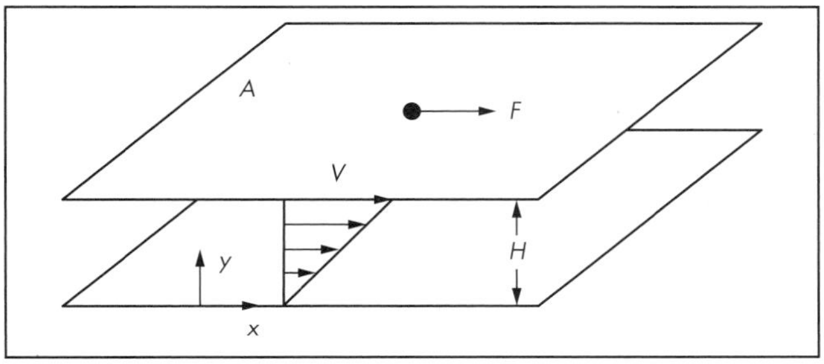

Source: Tilton 2007

Figure 1 Movement of a fluid under simple shear stress

Michael G. Nelson, Professor & Chair, Mining Engineering Department, University of Utah, Salt Lake City, Utah, USA

Dynamic viscosity is simply defined as the ratio of shear stress to shear rate, thus

$$\mu = \tau/(dv/dt) \quad \textbf{(EQ 1)}$$

In the metric system, viscosity has units of kilograms per meter-second (kg/m·s), Newton-seconds per square meter (N·s/m^2), or pascal-seconds (Pa·s). It is also expressed in poise (P) or centipoise (cP), where 1 Pa·s = 10 P or 1,000 cP. These are the standard units used for viscosity. The viscosity defined by Equation 1 is also known as the absolute or shear viscosity. The kinematic viscosity (υ) is the ratio of viscosity (μ) to density (ρ) given by

$$\upsilon = \mu/\rho \quad \textbf{(EQ 2)}$$

and has units of m^2/s or stokes (St), where 1 St = 1 cm^2/s, and 1 cSt (centistoke) = 1 mm^2/s.

Experience shows that fluid movement is often more complex than that just described. *Rheology* is the study of the relationship between fluid deformation and stress. Most solids exhibit some degree of elasticity, in which they show reverse deformation when stress is removed. Fluids that do not exhibit any solid-like, elastic behavior are called *purely viscous fluids*, while those that exhibit viscous and elastic behavior are classified as *viscoelastic*.

The behavior of purely viscous fluids may be either time independent or time dependent. In the first case, the shear stress depends only on the instantaneous shear rate, while in the second case the shear stress also depends of the rate of deformation in the past, as the result of changes in structure or orientation during shear deformation.

Fluid rheology is usually characterized by a rheogram for the fluid. A rheogram is a plot of the shear stress versus the shear rate for a fluid in simple shear flow. Figure 2 shows rheograms for six types of fluids, as classified by rheology:

1. The rheogram for a Newtonian fluid is linear and intersects the origin of the graph. Newtonian fluids are the most common and include gases and liquids of low molecular weight such as water, solutions of salts in water, gasoline, kerosene, and light oils.
2. Pseudoplastics are shear-thinning materials whose apparent viscosity drops as flow or shear rate increases. Some substances exhibit a yield stress above which the apparent viscosity drops, so that a unit increase of driving force results in more and more flow. Pseudoplastic materials include catsup, paper pulp, and printer's ink.
3. Dilatant materials show shear thickening. Their apparent viscosity increases as the flow increases, and more and more stress is required to obtain the same increase in flow. A mixture of cornstarch and water is dilatant; other dilatant materials include quicksand, peanut butter, and many candy compounds.
4. Plastic solids are true plastics in the sense that they normally behave like solids, but when the shear stress reaches their yield point, they behave as viscous fluids and start to "cold flow." Most plastics, chewing gum, tar, and some oils exhibit this behavior.
5. Thixotropic materials are usually pseudoplastic, shear-thinning substances, but they exhibit hysteresis. For example, a thixotropic material will require less power when re-agitated than was required during the first agitation. Thixotropic substances include asphalt, lard, silica gel, most paints, glues, and fruit juice concentrates.

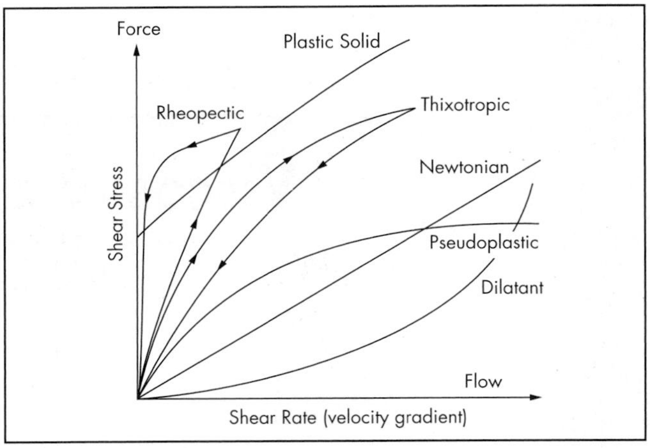

Source: Kim and Liptak 2003a

Figure 2 Rheograms for various types of fluids

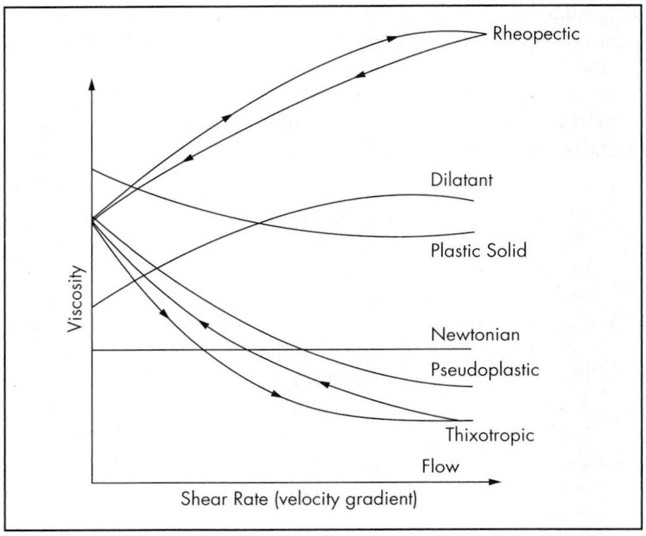

Source: Kim and Liptak 2003a

Figure 3 Variation of viscosity with shear rate for various types of fluids

6. Rheopectic substances also display hysteresis, but instead of a shear thinning, they thicken with increasing shear. Their viscosity appears to increase, and some will "set" after a period of agitation. Gypsum in water is an example of this behavior.

Non-Newtonian fluids are those in which the viscosity varies with the shear rate. For these fluids, the viscosity is usually referred to as the *apparent viscosity*, to emphasize their distinction from Newtonian fluids. Figure 3 shows how viscosity varies with shear rate for the six types of fluids previously described.

LABORATORY VISCOMETERS AND RHEOMETERS

All instruments for measuring viscosity and rheology in the laboratory are based on one of two principles—measuring the movement or flow of the fluid in a system with a standard configuration, or measuring the movement of a standard object or mechanism in the fluid. Some laboratory instruments are designed solely for use in the lab and some in the lab and the plant.

Bubble-Time Viscometer

In the bubble-time viscometer, a liquid flows down in the annular zone between the wall of a sealed tube and the perimeter of a rising air bubble. The kinematic viscosity of the fluid is determined by comparing the rising velocity of the bubble in the subject fluid with the bubble velocity in a tube of the same size filled with a fluid of known viscosity. This method is easy to use, and the equipment requires no calibration. While it is usually used in a laboratory, it can be used in an operational setting if a sample of the test fluid can be drawn. It is suitable only for Newtonian fluids.

Capillary Viscometers

The manual capillary viscometer measures the time required for the sample fluid to flow through a capillary tube. The Ostwald viscometer, shown in Figure 4, is commonly used. A fixed volume of the sample fluid is introduced to the lower receiving vessel, and the vessel is placed in a constant-temperature bath. After the sample temperature has stabilized (usually about 5 minutes), the sample is sucked up into the efflux vessel until the fluid level is above the upper index line. Suction is released so the fluid flows down through the capillary, and the time of flow between the upper and lower index lines is recorded. The flow time is multiplied by the viscometer's calibration constant to obtain the kinematic viscosity. The method is limited in application to Newtonian fluids. Because the measurement is conducted under atmospheric conditions and given that it takes several minutes, it should not be used with volatile or atmospherically unstable fluids. The sample fluid must also be free of solids. The method is used in both laboratory and industrial applications. Automatic capillary viscometers are available and can be used for online process control when coupled with an automatic sampling system.

Capillary-Extrusion Viscometers

Capillary-extrusion viscometers were designed to overcome the limitations of capillary-tube viscometers, which can measure only low-viscosity, Newtonian fluids. These instruments use a controlled pneumatic or hydraulic pressure system to force the sample fluid through an orifice. The kinematic viscosity is determined from the measured flow rate through the orifice and the driving pressure. These instruments are used only in the laboratory, and care must be taken to correct for errors that result from nonuniformity in the sample shear rate, entrance and exit effects at the orifice, fluid compressibility, loss of pressure resulting from flow in the sample chamber, and uneven temperatures produced by shear-induced heating.

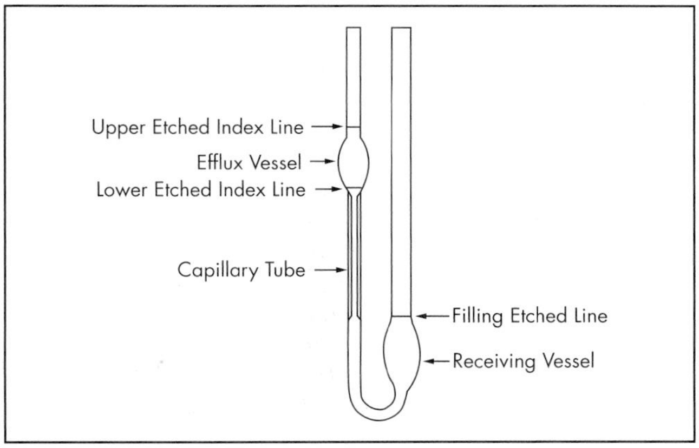

Source: Kim and Liptak 2003b

Figure 4 Ostwald manual capillary viscometer

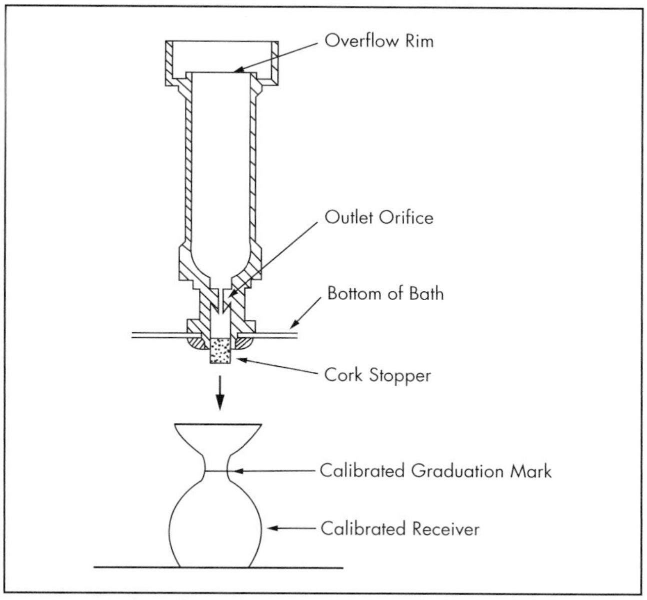

Source: Kim and Liptak 2003b

Figure 5 Saybolt viscometer

Efflux-Cup Viscometers

Efflux-cup viscometers are widely used in the field for measuring the viscosity of oils, syrups, paints and varnishes, and bitumen emulsions. The governing principles and operating procedures are similar to those for capillary-tube viscometers. The user measures the time required for a known volume of the sample fluid to flow through a fixed orifice at the bottom of a cup and into a receiving vessel. The efflux time is converted to kinematic viscosity using a conversion chart or formula. There are many types of efflux-cup viscometers. The most common is the Saybolt viscometer, shown in Figure 5. It is commonly used for testing petroleum products. Viscosity measured with a Saybolt viscometer is expressed directly in Saybolt universal seconds, or SUS—the time required for 60 cm^3 of sample to flow through an orifice with a diameter of 0.176 cm and a length of 1.225 cm. Saybolt universal seconds, t, can be converted to kinematic viscosity, υ, using one of the following equations:

$$\upsilon = 0.226\,t - 195/t \text{ cSt, when } t < 100 \text{ s} \quad \textbf{(EQ 3a)}$$

$$\upsilon = 0.220\,t - 135/t \text{ cSt, when } t > 100 \text{ s} \quad \textbf{(EQ 3b)}$$

Ford-cup and Zahn-cup viscometers are used in measuring low-viscosity liquids, primarily paints, varnishes, and other finishes. Automated Zahn-cup viscometers are used for online measurements in industrial applications.

Falling-Ball Viscometers

The falling-ball viscometer uses a precision-bored glass tube, usually about 200 mm long and 16 mm inside diameter, fitted with a capillary plug and an assortment of calibrated glass or steel balls. The viscometer is enclosed in a water jacket, so temperature can be held constant during tests. Measurement is made by timing the fall of the ball through an accurately calibrated distance, and absolute viscosity is calculated as a function of the fall time, the specific gravities of the ball and the fluid at the measuring temperature, and a ball constant, based on Stokes' law. In the range of 0.01 to 0.6 Pa·s, experienced personnel can determine absolute viscosity with an accuracy of ±0.1%. Measurements of this accuracy also require a viscometer that is clean and well maintained, a stopwatch with an accuracy of 0.02 seconds, and careful control of temperature during the measurements.

Falling-needle viscometers are similar to falling-ball devices but use a needle to provide a more stable falling motion and minimize wall effects. Automated viscometers of both types are available.

Rotational Viscometers

Rotational viscometers operate on the principle that the torque necessary to overcome the viscous resistance to the movement induced by the rotation of a spindle is directly proportional to the viscosity of the fluid. The entire sample is subjected to a uniform or near-uniform shear rate, and viscosity is determined directly by measuring the corresponding shear stress. A properly equipped rotational viscometer can measure fluid viscosities ranging from 10^{-5} to 10^{7} Pa·s at a range of shear rates from 10^{-4} to 10^{4} s^{-1}. Rotational viscometers can perform continuous measurements under varying conditions on the same sample and can also operate at a given set of boundary conditions for an extended time period.

Figure 6 illustrates a rotational viscometer. All rotational viscometers have a mechanism for driving a spindle at a constant speed and a torque-measuring device. Shear rate is determined as a function of spindle diameter and speed of rotation. There are several design variations. Some involve the rotation of a disc or spindle in the fluid; others rotate the container around the fluid and measure the torque required to restrain rotation of the suspended spindle.

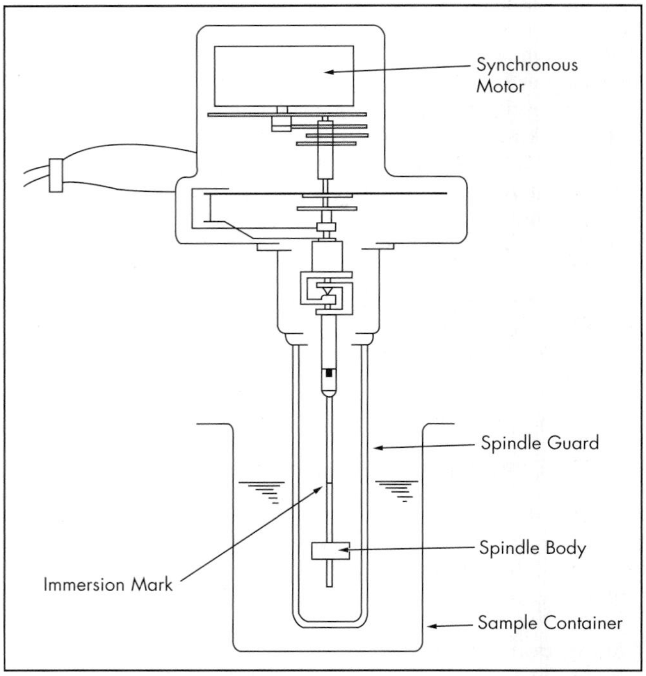

Source: Kim and Liptak 2003b
Figure 6 Simple rotational viscometer

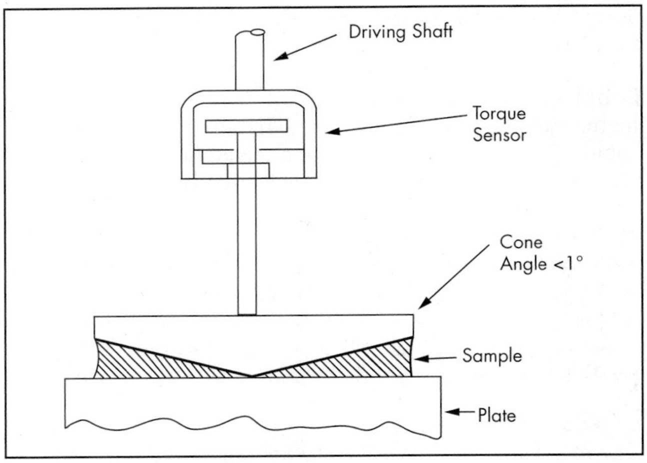

Source: Kim and Liptak 2003b
Figure 7 Cone-and-plate viscometer

Spindle-type rotational viscometers work well with Newtonian fluids, but errors can arise when they are used with non-Newtonian materials, because the property that is being measured—viscosity—is affected by the shear applied to it. Further inaccuracies can arise from the end effects around the disc and from heat generated at high shear rates. Alternative designs can overcome these errors. One approach is to use two coaxial cylinders; another is the cone-and-plate viscometer, shown in Figure 7. At a constant angular velocity, the cone-and-plate configuration provides a uniform shear rate and stress throughout the fluid sample, and heating effects are reduced because the sample is contained in a relatively thin layer.

Straight-Pipe Viscometers

Using the following Hagen–Poiseuille equation, the kinematic viscosity μ of a fluid flowing through a pipe section of length L can be calculated as

$$\nu = \Delta P \pi R^4 / 8LQ \qquad \textbf{(EQ 4)}$$

where

ν = kinematic viscosity of the fluid
ΔP = pressure drop over length L
R = radius of the pipe
Q = volumetric flow rate

This relationship provides a useful way to determine the viscosity of a fluid, literally *inline*, using only a straight pipe, a flowmeter, and a differential pressure sensor. It is used in the petroleum, chemical, and food processing industries. Unfortunately, the Hagen–Poiseuille equation applies only to laminar flow of Newtonian fluids. Application to other conditions requires approximations based on constitutive relationships. A method described by Hollinderbäumer and Mez (1998) uses the yield stress and the shear rate to calculate the *consistency* of a non-Newtonian fluid. The consistency, η, is comparable to the dynamic viscosity of a Newtonian fluid and is given by

$$\eta = \tau - \tau_o + \gamma^* \quad \text{(EQ 5)}$$

where

τ = shear stress, Pa
τ_o = yield stress, Pa
γ^* = shear rate, s^{-1}

The method also includes an empirical structure exponent, n, an integer selected to account for the pseudo-plastic behavior of slurries. The resulting equation, shown below, provides a relationship among flow velocity, pressure drop, and consistency, but it is rather unwieldy and has not been widely used:

$$w_m = \frac{1}{R^2} \cdot \frac{1}{\frac{1}{n}+1} \cdot \left(\frac{1}{2} \cdot \frac{1}{\eta} \cdot \frac{\Delta p}{\Delta l}\right)^{\frac{1}{n}} \cdot \begin{bmatrix} R^2 \cdot (R - r_o)^{\frac{1}{n}+1} - 2 \cdot \dfrac{(R-r_o)^{\frac{1}{n}+3}}{\frac{1}{n}+3} \\ -2 \cdot r_o \dfrac{(R-r_o)^{\frac{1}{n}+2}}{\frac{1}{n}+2} \end{bmatrix} \quad \text{(EQ 6)}$$

where

w_m = linear velocity, m/s
R = pipe radius, m
$r_o = 2\tau_o/(\Delta P/L)$
$\Delta p/\Delta l = \Delta P/L$, Pa/m, from Equation 4

INDUSTRIAL VISCOMETERS AND RHEOMETERS

In industrial or process applications, there are two types of viscosity measurement, *behavioral* and *analytical*. A behavioral measurement determines the viscosity of the fluid at the process temperature; an analytical measurement includes the viscosity at one or more reference temperatures. In many applications, a behavioral measurement is sufficient. This is the case in mineral processing—the behavior of the fluids is the primary concern. When the qualities or properties of a fluid must be controlled, an analytical viscosity measurement may be required. For example, when blending lubricants, the viscosities of the components at both 40°C and 100°C are required.

Industrial viscometers are understood to be those that can be installed in a tank or a pipe and operated continuously as process control instruments. Eight types of industrial viscometers are described in detail by Kim et al. (2003): continuous capillary, Coriolis, falling-element, float, oscillating, plastometer, rotational, and vibrational. Continuous capillary instruments are suitable only for Newtonian fluids; plastometers are used solely to characterize the molecular weight distribution of polymers and the plastic behavior (melt flow) of plastic materials. The other six types can be used to measure non-Newtonian fluid viscosities, but each has limitations. Of the standard process viscometers, Coriolis meters are the best suited to use with mineral slurries. Hollinderbäumer and Mez (1998) propose the use of a straight-pipe viscometer for controlling the flow of paste backfill.

Four types of commercially available viscometers are commonly used in measuring viscosity of slurries.

1. Falling-element instruments are not recommended for shear-sensitive fluids, nor for slurries with particles larger than about 0.5 mm or high solids content. They require a sample extraction system for online use and frequent cleaning when used with slurries.
2. Float viscometers will be plugged by particles larger than about 2.5 mm. They also require a sample extraction system for online use, frequent cleaning when used with slurries, and constant flow rate and pressure.
3. Oscillating, rotational, and vibrational viscometers can be installed directly in a pipe or tank and are not susceptible to plugging by large particles. However, in each of these instruments, the sensing element is immersed in the fluid and is therefore subject to wear from flowing fluid, especially if the fluid is a slurry.
4. Coriolis meters were originally designed to measure fluid flow, by sensing the torsional movement of a specially designed and instrumented pipe section. Because these meters have no internal components, problems with plugging and wear are minimized. However, because of the precision design and construction of the meters, they are available only in sizes up to 80 mm inside diameter.

The following discussion is limited to instruments that are suitable for measuring the viscosity of slurries, as that is the most common need in mineral processing. Other process instruments are discussed in detail in Kim et al. (2003). Most of the instruments used for measuring slurry viscosity function by measuring a stream of slurry diverted from the main process line. Some of these instruments can function continuously, while others require periodic washout or flushing.

Coriolis Mass Flowmeters

Coriolis mass flowmeters are widely used in the process industries. They are capable of measuring mass flow, density, temperature, and viscosity. They have no moving parts and can be installed directly in a process flow line. Coriolis meters are provided by several vendors with variable designs. The design used by Endress+Hauser (2018) is discussed here as an example of the measuring principle, which is based on the controlled generation of Coriolis forces. These forces are always present when both translational and rotational movements are superimposed, and are described by the following equation:

$$FC = 2\Delta m(v\omega) \quad \text{(EQ 7)}$$

where

FC = Coriolis force
Δm = moving mass
v = velocity of the moving mass in a rotating or oscillating system
ω = rotational velocity

The amplitude of the Coriolis force depends on the moving mass Δm, its velocity v in the system, and thus on the mass flow. Instead of a constant angular velocity ω, the Endress+Hauser sensor uses induced oscillation of the tube through which the fluid flows. The Coriolis forces produced at the measuring tubes cause a phase shift in the tube oscillations.

Referring to Figure 8, when there is no flow in the pipe, the oscillation measured at points A and B has the same phase, so there is no phase difference (1). Mass flow causes deceleration of the oscillation at the inlet of the tubes (2) and

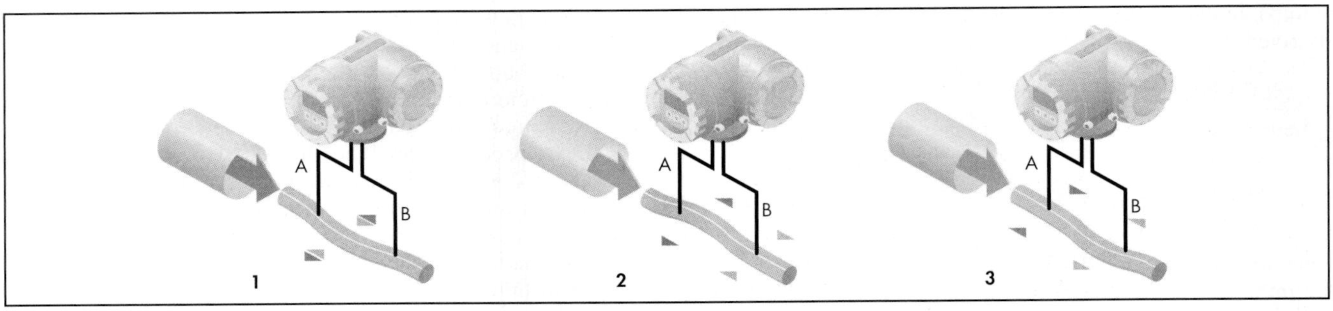

Courtesy of Endress+Hauser

Figure 8 Operation of the Endress+Hauser Proline Promass 80I and 83I mass flowmeters

acceleration at the outlet (3). The phase difference (A-B) increases with increasing mass flow. Electrodynamic sensors register the tube oscillations at the inlet and outlet. The measuring tube is continuously excited at its resonance frequency. A change in the mass and thus the density of the oscillating system (comprising the measuring tube and fluid) results in a corresponding, automatic adjustment in the oscillation frequency. The resonance frequency is thus a function of fluid density. The microprocessor utilizes this relationship to obtain a density signal. The measuring principle operates independently of temperature, pressure, viscosity, conductivity, and flow profile.

The use of Coriolis mass flowmeters for determination of viscosity was described by Kalotay and Van Cleve (1994) and expanded by Van Cleve and Loving (1997). The oscillation of the tube also creates a velocity profile in the cross section of the tube. Viscosity measurement is based on the shape of that profile, on the shear forces, and on the corresponding shear rates developed in the fluid. The shear forces dampen the tube oscillation, so a larger driving force is required to maintain that oscillation, and viscosity can be measured as a function of the excitation current for tube oscillation and torsional movement. Coriolis mass flowmeters are available in standard pipe sizes up to 80 mm. Coriolis meters can be used to measure viscosities up to 500 cSt, depending on the diameter of the meter, with accuracy of ±1% of the span. Recent application of Coriolis meters for viscosity measurement in South Africa is described by Drahm and Matt (2004).

Pressure Vessel Rheometer

The rheometer described by Bakshi et al. (1999) extracts a sample from a process line. An absolute pressure transducer measures the pressure of the air inside the vessel, and a differential pressure transducer measures the pressure difference across the tube. The flow rate is calculated based on the pressure change in the vessel, and the measurements over time are used in the Hagen–Pouiselle equation (Eq. 4) to calculate viscosity. Correlation of measurement made with this instrument showed excellent correlation with measurements made using standard bench-top viscometers. This instrument must be drained after each measurement, but the filling and draining cycle can be automated.

Reeves Online Viscometer

Napier-Munn et al. (1989) describe an online viscometer developed by Reeves (1985). Slurry drawn from the process line is pumped to a steady head box, where tramp oversize is removed, before flowing under gravity to a rotating bobbin. Current drawn by the bobbin motor is a function of drag on the bobbin and rotation rate, which can be related to viscosity using a simple calibration table prepared using fluids of known viscosity.

Slump Plate

Although not really a viscometer, this portable unit gives a quick visual indication of whether a slurry is likely to be pumpable with a centrifugal pump. It consists of a 300-mm^2 by 2-mm-thick stainless, flat plate with concentric rings inscribed on one surface. The central ring is 50 mm in diameter and the other diameters are progressively larger by 20 mm each. A short, thin pipe, 50 mm long with a 50-mm inside diameter and having machined and squared ends, is placed in the middle of the plate and a sample of slurry is poured into it. The pipe is then gently lifted vertically off the plate, allowing the slurry to spread or slump out evenly all around. If the slurry does not spread out to at least the third ring, it is too thick and a centrifugal pump will probably not be able to pump it.

VISCOSITY AND FLUID FLOW

Fluid flow can be laminar or turbulent. In laminar flow, the fluid velocity components vary smoothly with position, and with time if the flow is unsteady and smooth streamlines exist. The fluid flows in parallel layers, with no disruption between the layers. In turbulent flow, there are no smooth streamlines, and the velocity shows chaotic fluctuations in time and space. Figure 9 shows these two types of flow in a pipe cross section.

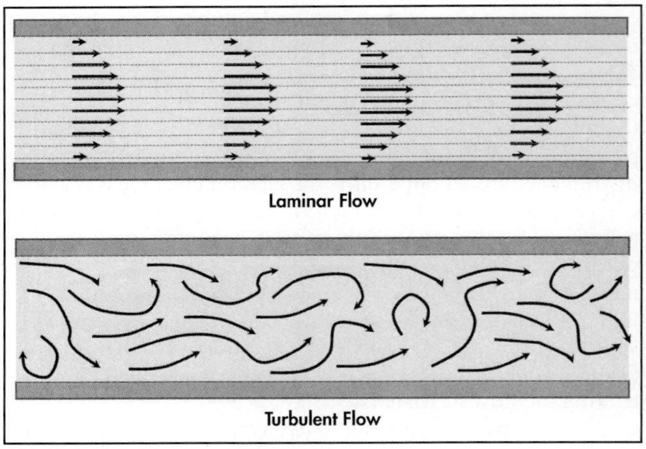

Adapted from Joseasorrentino 2014

Figure 9 Laminar and turbulent flow

In laminar flow, the energy input to the fluid functions to move the fluid in the direction of flow, to overcome friction in the flowing fluid and between the fluid and the wall of the pipe. In turbulent flow, some energy is dissipated in fluid turbulence, as indicated by the small, multi-directional arrows in Figure 9.

The nature of flow in a pipe depends on the pipe diameter, the density and viscosity of the fluid, and the flow velocity. These quantities can be combined into a single, dimensionless number called the Reynolds number, N_{Re}. It is one of several dimensionless numbers used to characterize dynamic similitude in a hydraulic system. Others are the Euler, Froude, Cauchy, and Weber numbers. Each number is the ratio of inertia to a particular force that may affect fluid flow. The Reynolds number, N_{Re}, is the ratio of inertia to viscous force, expressed in fundamental quantities as

$$N_{Re} = Ma/\tau A \quad \textbf{(EQ 8)}$$

where

M = mass
a = acceleration
τ = shearing stress
A = area over which the shearing stress acts

For fluids flowing in a pipe, the Reynolds number is expressed as

$$N_{Re} = DV/\upsilon \quad \textbf{(EQ 9)}$$

where

D = inside pipe diameter
V = fluid velocity
υ = kinematic viscosity

The Reynolds number may also be calculated as

$$N_{Re} = \rho DV/\mu \quad \textbf{(EQ 10)}$$

where

ρ = fluid specific gravity
μ = dynamic viscosity

While the Reynolds number is dimensionless, care must be taken to use consistent units in its computation and use.

The transition from laminar to turbulent flow begins at a Reynolds number of about 2,000. This is called the critical Reynolds number and is important in computing fluid friction losses in pipes. In fact, critical Reynolds numbers may vary substantially with different fluids, but engineering practice assumes that laminar flow occurs below a Reynolds number of 2,000, turbulent flow occurs above 4,000, and a transition zone exists between these values. In the transition zone, both laminar and turbulent flow occur. Examination of the expressions for the Reynolds number shows that the number increases with pipe size, fluid flow velocity, and specific gravity, whereas it decreases with viscosity.

MINERAL PROCESSING APPLICATIONS

It may initially seem desirable to transport all fluids in laminar flow, since in laminar flow no energy is lost to turbulence. However, this is not the case. Almost all the fluids handled in mineral processing and hydrometallurgical processing fall into one of three categories:

1. Liquids and solutions of soluble solids, which are pumped in turbulent flow at low velocities from 0.5–2 m/s. This usually provides an optimal compromise between operating costs, which result from pipe friction losses, and capital costs for pumps and piping.
2. Slurries of liquids and settling solids, which are pumped in turbulent flow at higher velocities, from 2–5 m/s or more. Although higher flow velocity results in increased friction losses (and consequently higher operating costs), it keeps the solids in suspension, preventing pipe blockages.
3. Slurries of liquids and fine, non-settling solids usually form homogeneous slurries, which behave as Bingham plastic slurries, which are described below. These are pumped at intermediate velocities between 1 and 3 m/s, with Reynolds numbers close to 2,000, and exhibit near laminar flow.

The effects of viscosity and rheology on laminar and turbulent flow in pipes are discussed in some detail. This discussion, and the accompanying example, are closely adapted, with permission, from Grzina and Roudnev (2002).

Laminar Flow in Pipes

Recall that the viscous behavior of a Newtonian fluid is linear, as shown in Figure 2. When a shear force is applied at the boundary of a fluid, the fluid begins to move in the direction of the force, developing shear stress between adjacent layers. The dynamic viscosity μ will be the slope of the a graph showing the shear stress τ versus the velocity gradient or shear rate dv/dy, thus

$$\tau = \mu \, dv/dy \quad \textbf{(EQ 11)}$$

where y is the incremental distance in the radial direction.

When a pressure differential P is applied to a fluid over a pipe length L, in a pipe with radius R and diameter D, a frictional shear stress τ_w is generated between the fluid and the pipe at the wall of the pipe, as shown in Figure 10. The axial force causing fluid motion is given by $PD2\pi/4$, and the force resisting this motion (i.e., the wall friction) is given by $\tau_w D\pi L$. Equating these forces and rearranging terms gives the shear stress at the wall:

$$\tau_w = PD/(4L) \quad \textbf{(EQ 12)}$$

All of this happens only in laminar flow. However, in both laminar and turbulent flow, the shear stress τ decreases linearly from τ_w at the pipe wall to zero at the centerline.

The solution of the differential equation (Eq. 11) for laminar flow yields a parabolic velocity distribution with local velocities $v = 0$ at the wall and $v = v_x$ (maximum) on the centerline, as shown in Figure 10. The average velocity of flow V is equal to one-half of v_x.

Pipe friction loss is also given by the Hagen–Poiseuille equation (Eq. 4):

$$P = 32\mu LV/D^2 \quad \textbf{(EQ 13)}$$

Combining this with Equation 11 gives

$$\tau_w = \mu 8V/D \quad \textbf{(EQ 14)}$$

and comparing Equation 14 with Equation 11 shows that the velocity gradient is equal to

$$dv/dy = 8V/D \quad \textbf{(EQ 15)}$$

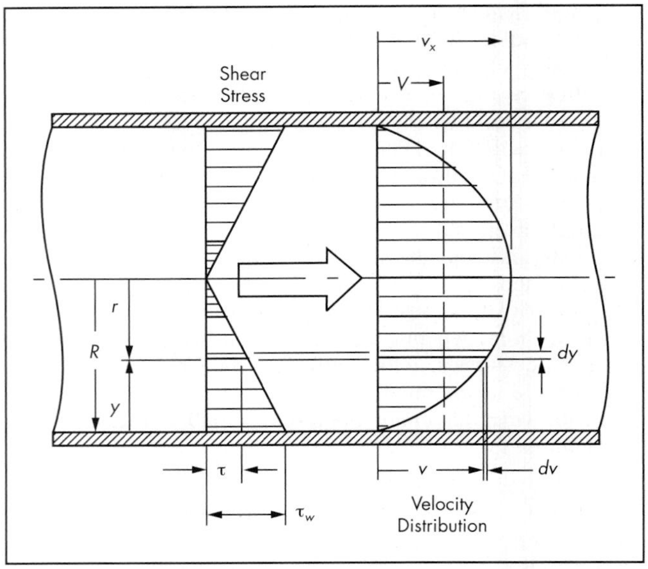

Source: Grzina et al. 2002

Figure 10 Newtonian laminar flow velocity profile in pipe

In laboratory tests, the flow through a pipe is varied, and velocities and corresponding pressure losses are measured. A plot of these measurements, τ_w $(=PD/4L)$ versus $8V/D$, will be linear for a Newtonian fluid, and the slope of that line will provide a good estimate of the dynamic viscosity, μ.

FLOW OF BINGHAM FLUIDS IN PIPES

A Bingham fluid can be described as a Newtonian fluid with an additional parameter, namely a *yield stress*, τ_o. This material behaves like a jelly when stationary and like a fluid when moving. If a shear stress below τ_o is applied to it, it flexes like jelly, and when the stress is removed, it returns to its original shape. However, if the applied shear stress is above τ_o, the material begins to flow. A well-known example of a Bingham fluid is the red mud tailings produced in the Bayer process for recovering alumina from bauxite.

A plot of τ against dv/dy yields a straight line, which intercepts the τ-axis at τ_o (greater than zero), and has a slope η known as the *coefficient of rigidity*, as shown in Figure 11. The equation of this line is

$$\tau = \tau_o + \eta dv/dy \quad \textbf{(EQ 16)}$$

Figure 11 also shows the behavior of two Newtonian fluids, 1 and 2, whose dynamic viscosities are μ_1 and μ_2, respectively. The lines intersect the Bingham fluid line at points 1 and 2, and so it can be said that the Bingham fluid has two *apparent viscosities*, μ_{a1} and μ_{a2}, identical in value to μ_1 and μ_2. Similarly, an infinite number of apparent viscosities can be located on a Bingham fluid line, but only two points are needed to define the slope η.

Shear stress τ of a flowing Bingham fluid in a pipe varies from τ_w at the wall to zero at the centerline, just like in Newtonian fluids. Solution of the differential equation (Eq. 16) yields a parabolic velocity distribution in an annulus between the pipe wall and an inner radius r_o, at which $\tau = \tau_o$, as shown in Figure 12.

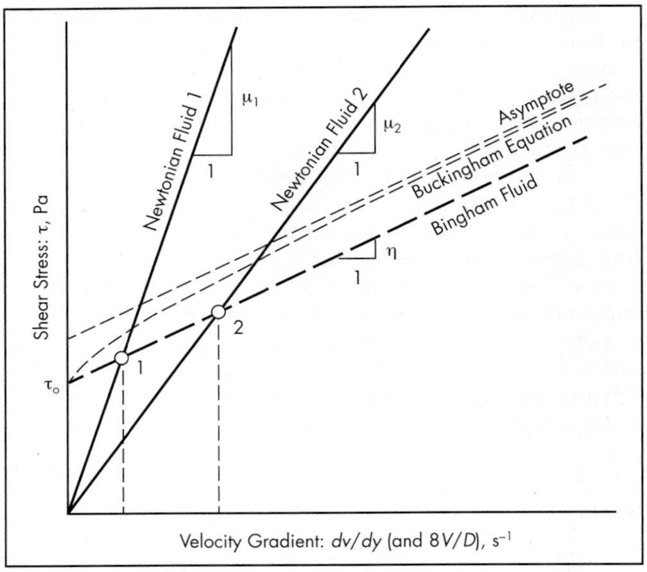

Source: Grzina et al. 2002

Figure 11 Shear stress diagram

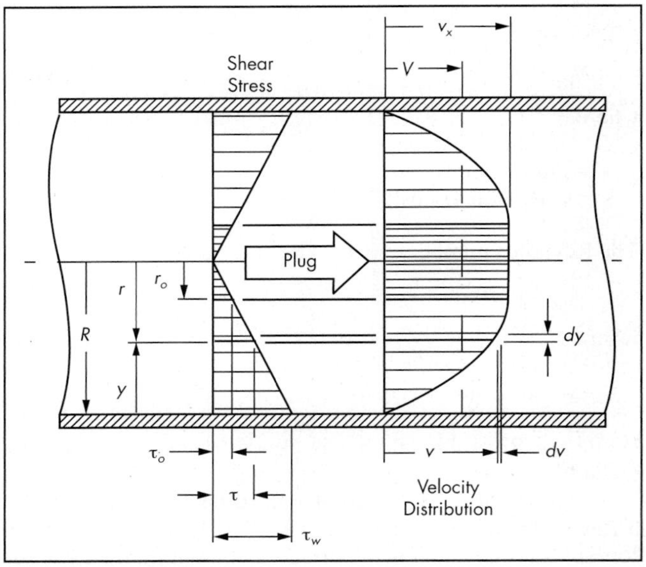

Source: Grzina et al. 2002

Figure 12 Bingham laminar pipe velocity profile

The maximum local velocity v_x is at r_o and the entire central plug also flows with this velocity. The average velocity of flow is given by

$$V = D\tau_w/(8\eta)\ [1 - 4/3(\tau_o/\tau_w) + 1/3(\tau_o/\tau_w)^4] \quad \textbf{(EQ 17)}$$

Equation 17 is known as the Buckingham equation, after its developer. The ratio τ_o/τ_w is always less than one, so $(\tau_o/\tau_w)^4$ is, of course, also smaller. Neglecting this term and rearranging the remaining terms gives

$$\tau_w = 4/3\tau_o + \eta 8V/D \quad \textbf{(EQ 18)}$$

Setting $dv/dy = 8V/D$, Equation 18 produces the line labeled *asymptote* in Figure 11, parallel to the line described by Equation 16, but higher by an amount of $\tau_o/3$.

The Buckingham equation (Eq. 17) produces a true curve of the fluid behavior, shown as a dotted line between the asymptote and the line of Equation 18. However, this equation cannot be used to find τ_o and η because it already contains them. Values of τ_o and η must be found using Equation 16 or 18. Both yield reasonable solutions, and the simpler of the two, Equation 16, is usually selected. The value of dv/dy is set to $8V/D$ in Figure 11, which then becomes a pseudo shear stress graph. Test points usually fall on a straight line and it is then a simple matter to measure its slope η and its vertical intercept τ_o. Based on this procedure, Equation 16 for Bingham fluids is modified to

$$\tau_w = \tau_o + \eta 8V/D \quad \textbf{(EQ 19)}$$

Equation 19 and Figure 11 can be used to derive an expression for the apparent dynamic viscosity:

$$\mu_a = \eta + \tau_o/(8V/D) \quad \textbf{(EQ 20)}$$

This value can be used in the equation for Reynolds number,

$$N_{\text{Re}} = \rho_m DV/\mu_a \quad \textbf{(EQ 21)}$$

where ρ_m is the specific gravity of the fluid mixture. The Reynolds number is used to calculate V_c, the critical average velocity at the end of the laminar flow range. This usually occurs at a Reynolds number of 2,000. After some substitutions and manipulation of terms, the critical velocity can be calculated, thus

$$V_c = X_1 + \sqrt{(X_1^2 + X_2)} \quad \textbf{(EQ 22)}$$

where

$$X_1 = \eta N_{\text{Re}}/(2\rho_m D) \quad \textbf{(EQ 23)}$$

and

$$X_2 = \tau_o N_{\text{Re}}/(8\rho_m) \quad \textbf{(EQ 24)}$$

At velocities lower than V_c, flow is laminar, and total head H_m varies relatively little with changes of flow rate Q. Pumping costs are therefore approximately directly proportional to flow rate. At velocities higher than V_c, flow is turbulent and head and pumping costs increase proportionally to the square of the flow rate. The most economical pumping velocity is therefore the one nearest to V_c.

Example: Pumping a Bingham Slurry

Slurries of water and finely ground limestone are used in the production of cement. These slurries are usually Bingham plastic fluids. To determine the lowest-cost flow rates for various pipe diameters and slurry densities, a pilot plant was built comprising a slurry pump, a holding tank, flow control and measuring equipment, instrumentation, two 100-m-long pipes of 0.150- and 0.200-m diameter, return pipes, and accessories. The static head was equal to zero.

The solids were limestone (88%) and clay (12%) with an overall density of $\rho_m = 2{,}650$ kg/m^3. The mass screen analysis is shown in Table 1 and the particle size distribution is plotted in Figure 13. The d_{50} particle size was 20 µm.

Slurries for testing were mixed and prepared on-site to C_w concentrations of 50%–65%, in 5% increments. Flow rates, pressure losses, slurry temperatures, and C_w were recorded. Hundreds of test points were collected and plotted and curves of constant C_w were fitted through all the points. In this example, only the slurry with a C_w of 65% (a density $\rho_m = 1{,}680$ kg/m^3) is described. A few data points from the tests are listed in Tables 2 and 3. These points were selected to include the transition from laminar to turbulent flow.

Figure 14 is a plot of the (Q, H_m) test points, in which the shallow-angled portions represent laminar flow and the

Table 1 Mass screen analysis

Screen Opening Size, µm	Cumulative % Passing
51	78
88	89
149	96
297	99
841	100

Adapted from Grzina et al. 2002

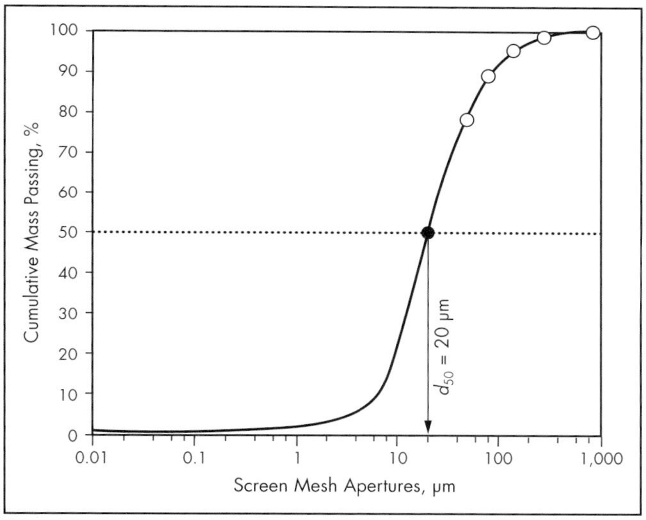

Source: Grzina et al. 2002

Figure 13 Size distribution of solids in slurry

Table 2 Tests with 0.150-m-diameter pipe

Point	V, m/s	Q, L/s	H_m, m slurry	$8V/D$, s^{-1}	τ_w, Pa
1	0.67	12	3.37	35.7	20.8
2	1.06	19	3.50	56.5	21.6
3	1.86	33	3.76	99.2	23.2
4	2.38	42	5.85	126.9	36.1

Adapted from Grzina et al. 2002

Table 3 Tests with 0.200-m-diameter pipe

Point	V, m/s	Q, L/s	H_m, m slurry	$8V/d$, s^{-1}	τ_w, Pa
1	0.68	21	2.49	27.2	20.5
2	1.28	40	2.60	51.2	21.4
3	1.82	57	2.70	72.8	22.2
4	2.23	70	3.86	89.2	31.8

Adapted from Grzina et al. 2002

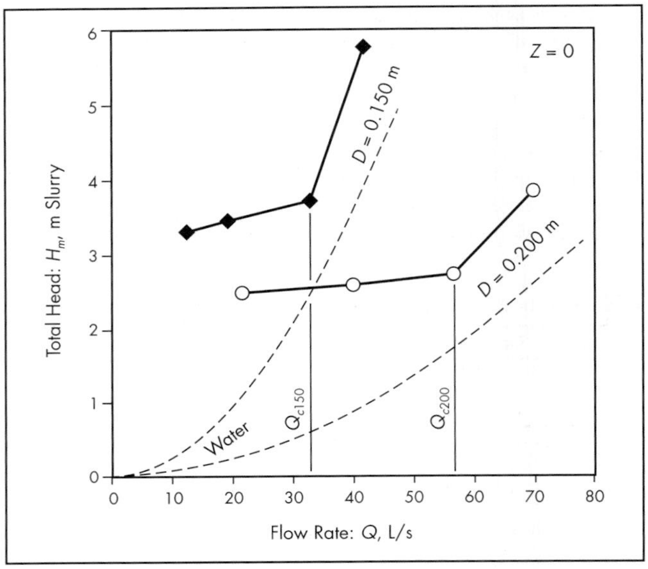

Source: Grzina et al. 2002
Figure 14 Pipe friction losses

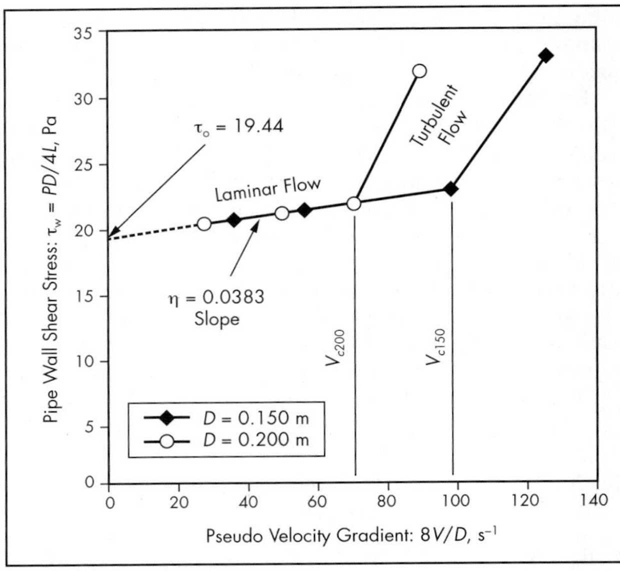

Source: Grzina et al. 2002
Figure 15 Rheology graph

steeper portions, turbulent flow. Water friction curves for the same pipes are also shown for comparison, as determined from the Moody diagram shown in Figure 2 of Chapter 5.3, "Pumps." Tables 2 and 3 also show derived values $8V/D$ and τ_w [$=PD/(4L) = g\rho_m H_m D/(4L)$]. These points are plotted in Figure 15 to show the slurry rheology in this case.

Laminar Flow

The coefficient of rigidity η may be determined from the test values by calculating the slope of the laminar-flow portion of the rheology graph shown in Figure 15. The endpoints of the laminar-flow line are point 1 of the graph for 200-m pipe (27.2, 20.2) and point 3 of the graph for the 0.150-m pipe (99.2, 23.2). The coefficient of rigidity is then given by

$$\eta = (23.2 - 20.5)/(99.2 - 27.2) = 0.0375 \text{ Pa·s} \qquad \textbf{(EQ 24)}$$

The yield stress τ_o is calculated with Equation 18, using any point on the laminar flow line. For example, using point 2 of the graph for the 0.200-m pipe (51.2, 21.4):

$$\tau_o = \tau_w - \eta 8V/D = 21.40 - (0.0375)(51.2)$$
$$= 19.48 \text{ Pa} \qquad \textbf{(EQ 25)}$$

The critical pipeline velocities V_c for the various pipe diameters can now be determined. As mentioned, critical velocity for a slurry of this type usually occurs at a Reynolds number of about 2,000. The observed rheologies shown in Figure 15 show the critical velocity at point 3 for both pipe diameters. The Reynolds numbers calculated for these two points using Equation 21 are 2,004 and 1,961 for the 0.150-m pipe and 0.200-m pipe, respectively—reasonably close to the rule-of-thumb value 2,000.

Critical velocities may be calculated from Equations 22–24 for four pipe sizes, using a Reynolds number of 2,000. Those values are listed in Table 4, with corresponding critical volumetric flow rates Q_c.

Table 4 Critical flow rates

D, m	V_c, m/s	Q_c, L/s
0.100	1.94	15.2
0.150	1.86	32.9
0.200	1.82	57.2

Adapted from Grzina et al. 2002

REFERENCES

Bakshi, A.K., Eisele, T.C., and Kawatra, S.K. 1999. On-line rheometer for mineral slurries. SME Preprint No. 99-82. Littleton, CO: SME.

Cowper, N.T. 1992. A newly developed and commercially proven on-line viscometer offers significant benefits in mineral processing. SME Preprint No. 92-139. Littleton, CO: SME.

Drahm, W., and Matt, C. 2004. Coriolis mass flowmeter with direct in-line viscosity measurement. *S. Afr. Flow Measure. Control* (July).

Endress+Hauser. 2018. Proline Promass 80I, 83I Coriolis flowmeters. https://www.us.endress.com/en/Field-instruments-overview/Flow-measurement-product-overview/Product-Coriolis-flowmeter-Proline-Promass-83I. Accessed June 2018.

Grzina, A., and Roudnev, A. 2002. *Slurry Pumping Manual*, 1st ed. Warman International Ltd. www.pumpfundamentals.com/slurry/WeirSlurryPumpingHandbook.pdf.

Hollinderbäumer, E.W., and Mez, W. 1998. Viscosity controlled production of high-concentration backfill pastes. In *Proceedings of AusIMM '98—The Mining Cycle*. Melbourne, Victoria: Australasian Institute of Mining and Metallurgy.

Joseasorrentino. 2014. Transition from laminar flow to turbulent flow [Wikimedia]. https://commons.wikimedia.org/wiki/File:Transicion_laminar_a_turbulento.png.

Kalotay, P.Z., and Van Cleve, C.B. 1994. Viscometer for sanitary applications. U.S. Patent 5,359,881.

Kim, C.H., and Liptak, B.G. 2003a. Viscometers—Application and selection. In *Instrument Engineers' Handbook: Process Measurement and Analysis*, 4th ed. Vol. I. Edited by B.G. Liptak. Boca Raton, FL: CRC Press.

Kim, C.H., and Liptak, B.G. 2003b. Viscometers—Laboratory. In *Instrument Engineers' Handbook: Process Measurement and Analysis*, 4th ed. Vol. I. Edited by B.G. Liptak. Boca Raton, FL: CRC Press.

Kim, C.H., Liptak, B.G, and Jamison, J.E. 2003. Viscometers—Industrial. In *Instrument Engineers' Handbook: Process Measurement and Analysis*, 4th ed. Vol. I. Edited by B.G. Liptak. Boca Raton, FL: CRC Press.

Napier-Munn, T.J., Reeves, T.J., and Hansen, J.O. 1989. The monitoring of medium rheology in dense medium cyclone plants. *Proc. Aus. Inst. Min. Metall.* 294(3):85–93.

Reeves, T. J. 1985. On-line viscometer for mineral slurries. *Trans. Inst. Min. Metall.* 94:C201–C208.

Tilton, J.N. 2007. Fluid and particle dynamics. In *Perry's Chemical Engineers' Handbook*, 8th ed. Edited by D.W. Green and R.H. Perry. New York: McGraw-Hill.

Van Cleve, C.B., and Loving, R.A. 1997. Coriolis viscometer using parallel connected Coriolis mass flowmeters. U.S. Patent 5,661,232.

Van Wazer, J.R., Lyons, J.W., Kim, K.Y., and Colwell, R.E. 1963. *Viscosity and Flow Measurement*. New York: Interscience.

CHAPTER 1.10

Geometallurgy

Dean M. David

TERMINOLOGY

Geometallurgical concepts are as old as the Bronze Age, but the term *geometallurgy* is one of the newest mining words. According to Jackson et al. (2011), the first mention of the term *geometallurgy* in literature was in 1968 by McQuiston and Bechaud. The oldest paper in the OneMine database to use the term geometallurgy (Martens et al.) is dated 1998. In the late 1990s, geometallurgy was adopted for general usage in the mining industry. Like most language additions, the term was used and understood within some localized mining communities, such as the Chilean copper industry, before it was adopted as a globally understood term.

Apparently, even in its earliest context, the term embraces geology, mining, and metallurgy (not simply geology and metallurgy as the name implies). The *geo* part of the term refers to the ore in the ground and the way that ore is removed from its natural surroundings so it can be delivered to the process plant. The *metallurgy* part refers to all the processes that are applied to the ore to generate a revenue stream for the organization.

Geometallurgy has been loosely applied as a general term to describe geological measurements that relate to metallurgical outcomes. In its simplest (and least successful) form, these measurements are ones that have been routinely performed by geologists and have simply been used to describe properties of geological ore types. This is a low-cost and hopeful geometallurgical approach.

Similarly, it is insufficient to have a process engineer dictate a new set of measurements to geologists and new planning methods to miners. Geometallurgy requires teamwork between the disciplines and a shared understanding of all aspects of the project. Rigorous implementation of geometallurgy in a project requires a detailed understanding of the project cost, risk factors, and revenue drivers and how the mining and processing influence these drivers. The following is a useful definition of *geometallurgy*: "Geometallurgy is a scientific discipline in which geological data, mining data, and processing data are co-analysed to generate useful information and knowledge to optimize resource profitability" (David 2014).

The important points in the definition are that all disciplines are involved and that the outcomes should be real in monetary terms, not simply nice-to-know information. When developed and implemented in an appropriate way, geometallurgy unlocks value in a resource that is kept hidden by the silo mentality that often exists in most modern operations. The knowledge silos of geology, mining, and metallurgy become interactive and inter-responsive, revealing limitations, realities, and opportunities, some of which have lain dormant for years or decades. For those who have successfully implemented geometallurgical principles at sites, geometallurgy is much more than something to add on to the business to help improve profitability—it is the heart of the business.

ARTISANAL MINING

A modern mining operation employs hundreds and often thousands of people, and they all carry out individual important functions. The result is that a resource that nature has provided is converted into a steady revenue (and sometimes profit) stream. The revenue stream supports all the employees, their families, the managing company, the shareholders of the company, the economy of the region in which it is located, and the various levels of government with responsibility for the locality. Mining projects are usually big and expensive, and it is almost impossible for individuals to grasp the complexity, let alone optimize it. It is much easier to describe a mining operation at the other end of the operations scale by examining the artisanal (often illegal) miner, working alone or in a small group, who can be thought of as the embodiment of geometallurgy.

It is useful to walk in the shoes of an artisanal miner to appreciate the breadth of realities that must be managed just to secure an ounce of gold. Imagine you have a small operation centered on a vein of gold-bearing sulfides that have supported you and your family for three years. However, the vein has split in two directions, and you need to make a choice. Going one way looks like the right direction for the highest grade, but the ground is bad. Your own son could become one of the next forgotten victims of the goldfield if you chase this vein. Going the other way, the ground looks good, but the grade is lower and the ore is harder. You also know that what you can see today does not hold any guarantees for tomorrow.

Dean M. David, Technical Director–Process, Mining and Minerals Australia Division, Wood, Perth, Western Australia, Australia

You decide to get some help from the bad-ground specialist on the field (at the cost of a few ounces) while chasing the safe vein with your team in the short term.

Although you are poor, mining is in your blood, and you have been working this goldfield for a decade. You are rich in knowledge and hoping that life's lottery will favor you and your family one day soon. From knowing the vein to follow, the pain of seemingly endless time at the face, the dust and danger of hand drilling, the small-scale blasting you do every couple of days, hauling the ore out in a wheelbarrow, breaking the big lumps with a sledgehammer followed by hours of crushing in a mortar and pestle (that is big enough to hold 5 kg at a time)—only then is the ore ready for you to get the gold out. You have taught your daughter to manage the cyanide, zinc, and mercury, which is something you did for six years on another field before you were strong enough to work the ore. Sometimes the cyanide followed by zinc (or iron scrap) provides the majority of the gold, but sometimes it is the mercury when the gold is coarse.

You also know the financial pressure from the richest miner on the field—the one that supplies you with the explosives, cyanide, mercury, and even the new wheelbarrow trays you need every three months. You also know where the best and worst prices are to be found for the gold you extract. As an illegal miner, you can only sell on the black market at the prices they offer. All the time, there is the threat of prosecution or being shut down by the government, and it is necessary to keep on the good side of the local authorities.

To keep the gold revenue flowing, there is no real alternative for the artisanal miner who is fulfilling the roles of exploration geologist, production geologist, mining engineer, crushing and grinding supervisor, gold extraction supervisor, health and safety officer, process supervisor, purchasing officer, marketing manager, and government liaison officer. In short, the artisanal miner must be a geometallurgist (and more), and some of those who are less experienced and are working the ore in the team will also be learning essential geometallurgical skills on a daily basis.

GEOMETALLURGY IN EXPLORATION

The discovery of ore bodies is the job of exploration geologists, and it is an unfortunate reality that most exploration efforts end in failure. Only a few exploration targets progress to be considered as mining projects, and even fewer make the full journey to become producing mines.

The prospects that progress to the point of economic assessment have (by definition) many of the attributes that are essential for success. These include the most fundamental, such as deposit size, grade, location, and the ability to mine it economically. However, it is also necessary to be able to process the ore using proven methods, in an economic fashion, in an environmentally acceptable way and produce something a customer will want to buy. A prospect that advances to the point where diamond core drilling has been approved has already passed many geological hurdles but may have passed no metallurgical hurdles. Basic application of geometallurgy late in exploration is about placing metallurgical hurdles in front of the project at an early stage, and it is also about making sure that all ongoing geological data collection is relevant to potential metallurgical outcomes.

A potential consequence of early geometallurgical intervention is that exploration on a prospect falters, or even fails, because of a metallurgical hurdle. Typical metallurgical problems include low recovery values and the inability to generate a marketable product using conventional processing means. The hurdles are there to modify exploration and economic assessment behavior at the earliest possible time, because that is when the least money has been spent. The behavior change on that prospect could be as final as stopping the program and using exploration resources on a more attractive prospect. However, it could also be in the form of a change in the way drilling is targeted or modifications to the standard assay suite for the project core samples.

Nickel sulfide mineralization is particularly prone to the problem of lower than expected recovery. An exploration core that appears to have excellent grade, in terms of total nickel, may only have half of the nickel in a form that is recoverable by the separation process of choice—flotation. The unrecoverable nickel is bound up at very low grades within some of the silicate minerals. Early metallurgical intervention quickly identifies this relatively common problem and results in the introduction of appropriate assaying techniques that distinguish sulfide nickel from silicate nickel. The prospect grade assessment basis is then changed to sulfide nickel (rather than total nickel) and a realistic reassessment of economic prospects for the mineralization is performed. Not only are decisions being made on the appropriate basis, any future routine drilling on the prospect will be assessed correctly.

Some copper sulfide mineralization is unable to generate a saleable product. One example is where a deposit shows good copper grades but has unusually high arsenic levels. The majority of the copper and arsenic report to flotation concentrates in early testing, and all efforts to reject arsenic also reject large quantities of copper. The problem is almost certainly that a large amount of the copper is contained in the mineral enargite with the formula Cu_3AsS_4. Copper and arsenic are chemically bound together in a mineral that floats as readily as chalcopyrite or bornite. Enargite is rich in copper (almost 50% Cu), but it is also 20% arsenic. It may be impossible to generate a copper concentrate with less than 5% arsenic, which is a contamination level that prevents the concentrate from being shipped to many countries, including China, and a level that is unacceptable to copper smelters. A metallurgical solution to a problem such as this would be expensive on-site concentrate processing using chemical (rather than physical) separation to make a saleable product. Immediately, the problem of enargite is recognized, and the economic and environmental assessment criteria for the prospect change significantly.

These two examples are somewhat uncommon, but both have been encountered on many occasions in the author's own experience. A more common occurrence is early or late identification of the refractory nature of a gold deposit. Early identification through appropriate metallurgical intervention allows geologists to set up a routine assay suite suited to potential processing options. Late identification occurs when scoping or prefeasibility study (PFS) testing identifies the refractory problem, but the only assays performed on the entire drill library are gold and (maybe) silver. For a refractory gold ore, the most important assay is sulfur. If core or drill chips have oxidized in storage, they will not be suitable for determining sulfur content retrospectively. (To assess oxidation, photos of the core when drilled can be compared with the core currently in the trays.) The only solution is to drill new core. If the core has not oxidized, then all mineralized core intervals need to be resampled and assayed for sulfur, which is achievable but also expensive and time consuming.

Other metallurgical problems best identified at the earliest possible time include extremely fine-grained mineralization (below 10–20 µm) that is difficult to liberate; high levels of manganese in sphalerite, which could make an unsaleable zinc concentrate; or extreme comminution properties.

The ideal process engineer to assist an exploration team is someone experienced in sampling and testing, comminution and physical separations, and with a comprehensive knowledge of flow sheet development techniques for the commodity in question. In ores where it is obvious that the flow sheet will be complex (such as nickel laterites or copper leaching), expertise in both the physical (ore handling, comminution) and the chemical (hydrometallurgy) is needed at an early stage, and this usually requires two individual experts.

With the increasingly proscriptive public reporting requirements, such as National Instrument NI-43-101 in Canada and the Joint Ore Reserves Committee (JORC 2012) in Australia, an early metallurgical intervention can also avoid the release of false information that may need to be retracted in a later release and may even leave the owner and report signatories open to legal proceedings.

It is preferable that problems with projects be identified (and acted on) as early as possible. Once the author asked an experienced metallurgist what he would do with the difficult ore body being worked on. The learned response was “find another ore body.” Time, and a lot of misspent company money, proved him right.

GEOMETALLURGY IN PROJECT DEVELOPMENT

Geometallurgy is best applied during project development by understanding its ultimate purpose and source of value in the project. In the description that follows, the time progression of the project development is reversed so that the end point is understood first and the steps taken to get to that point are developed in that context. The steps are outlined in Figure 1.

Operational Phase

In a geometallurgically functioning operating environment, the mine site will seamlessly speak the same language: from the geologist carrying out grade control activities, to the short-term mine planners devising this week’s blending tactics, to the process superintendent needing to get an extra 50 t/h

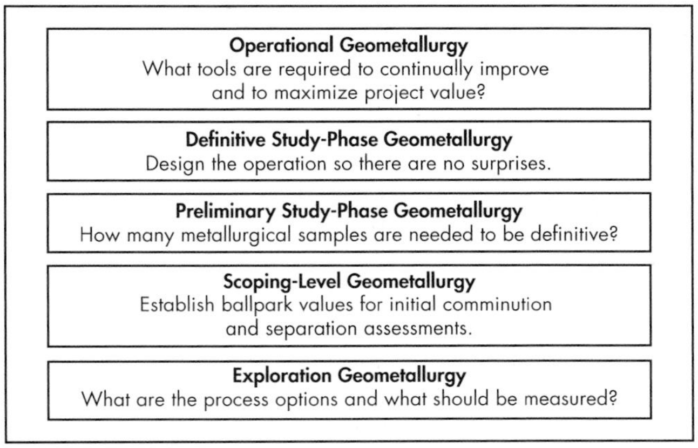

Figure 1 Geometallurgy steps through a project’s life cycle

(metric tons per hour) from the plant at the end of the month. For this to be a reality, the geologists need to have ore type definitions that feed through to the mine planning process, and the mine planning outputs must indicate key factors, such as likely throughput rates, recoveries, and product qualities for the mix of ores. Finally, the process engineers must have a reasonable level of faith that the process information in the mine plan is reliable.

This sort of environment may exist on many sites worldwide, but it is certainly not normal practice. More typically, the process engineer, wanting more throughput, would ask the geologist if there was any soft ore about. The geologist would consider soft ore to be a particular geological type that has been globally assigned a single work index and would accordingly identify where such ore was in the pit. When the mining engineer delivers it to the plant, the throughput mysteriously drops by 50 t/h because the ore is hard. The *geological* ore type was not a *geometallurgical* ore type and, consequently, the mine planning information about ore from that particular location was unreliable.

The latter case is encountered on the majority of sites and is often characterized by poor relationships between the disciplines. The geometallurgically driven site is rare with one of the best published examples being the work done at Batu Hijau in Indonesia (Wirfiyata and McCaffery 2011). Other published examples of sites that are working geometallurgically are Cerro Corona in Peru (Baumgartner et al. 2016), Quebrada Blanca in Chile (Chait and Schiller 2016), and Cripple Creek and Victor gold mine in Colorado, United States (Leichliter and Larson 2013).

To underpin a successful operational phase, the following geometallurgical tools need to be available:

1. There need to be established and reliable relationships between easily obtainable measures (from tests conducted on the ore) and the metallurgical performance of the plant. This can be as simple as a recovery relationship based on head grade or more complex, such as a concentrate grade prediction based on mineralogical mixture. In the comminution area, this could be semiautogenous grinding (SAG) throughput predicted using a simple measure of ore competence or ball mill performance based on a grinding work index. All such relationships are recorded, along with their proofs, in a single document (typically a spreadsheet) termed the *mine and process agreement document* and owned by the site process engineers and mining engineers.
2. Measurements on the ore to develop the preceding relationships must have been performed in the past (during exploration and infill drilling) and must be continually conducted in a program of testing upcoming ore (using samples from grade control drilling) in time to be useful for production planning.
3. A mechanism is in place to enter this metallurgical information into the geological database, including assignment of the ore to a particular geometallurgical domain.
4. The geological model recognizes geometallurgical domains, and these are likely to be different than geological domains.
5. The geometallurgical domains and the metallurgical properties are carried electronically from the geological database through to the mine planning procedures, providing properties for ore blocks.

6. The geometallurgical domains, metallurgical properties, and metallurgical predictions are a standard component of mine planning reports.
7. An ongoing program of geometallurgical verification, reconciliation, and improvement is in place.

Putting such a system in place after commencement of operations is expensive and is not trivial. First, dedicated personnel who intimately understand the operation, but who are removed from day-to-day operational responsibility, need to analyze historical data to show that relationships exist between easily measured parameters and plant performance. Second, a clear economic case needs to be developed for implementing an extensive ore measurement system and linking this to mine planning. Third, the geometallurgical system needs to be institutionalized as standard practice at the site (parallel to existing grade control), rather than being an occasional measurement and adjustment system. Not surprisingly, the sites where implementation after start-up has been successful have required major changes from the status quo, accompanied by site-wide cultural changes. The preferred pathway to achieving an effective operational geometallurgical mindset is through developing a geometallurgical focus in the early project phases, commencing with the first geologists, miners, and process engineers to work on the resource at about the time where the discovery is considered to have potential. From that point onward, a model, such as the one shown in Figure 2, can be used to keep the project geometallurgically on track.

Recently, it has been shown (in confidential studies by the author) that geometallurgical prediction based only on data collected during the definitive design phase can provide useful, but limited, operational guidance. A metallurgical data set of about 200 samples measured for multiple comminution properties has provided useful plant throughput predictions. These predictions are well correlated (±10%) when the prediction time frame ranges from two to four weeks. On a daily or shift basis, the scatter in the actual-versus-predicted throughput graph is large (equivalent to the full span of throughput rates experienced by the plant), and these predictions should be ignored.

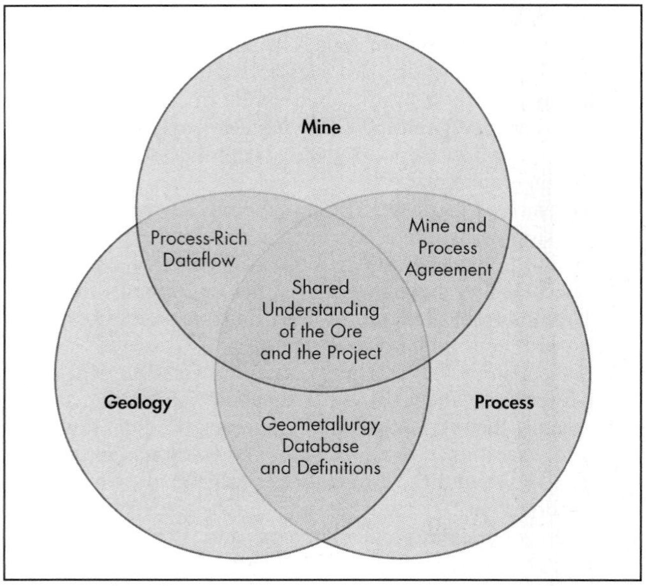

Source: David 2017

Figure 2 Simple operational geometallurgy model

If operational metallurgical predictions are required for periods shorter than two weeks, then there are two choices. The first is to test many (thousands) spatially distributed samples during the design period, similar to the program instituted at Olympic Dam in South Australia (Liebezeit et al. 2016; Liebezeit 2011; Turner et al. 2013). The second is to test hundreds of samples so that medium-term planning is geometallurgically informed and then institute an ongoing geometallurgical testing and verification program.

Definitive Phase

The preliminary phases must set the scene for the definitive phase. The definitive project phase is the last phase before a project moves into implementation and is also called the *feasibility phase* (JORC 2012). Typically, a single option for taking the project to implementation is designed and costed out to a level of accuracy suitable for company boards to approve implementation, for banks to make decisions on lending, and for equity partners to evaluate the investment potential. By referring to this stage as *definitive*, there is much more than a suggestion that the project, as presented, will achieve the stated outcomes. From a process perspective, definitive means that the operation will be implemented with the projected capital expenditure level; it will achieve the nameplate annual throughput; and it will achieve the design metallurgical outcomes at the estimated operating cost.

Process designs that fail by a margin of more than about 5% to achieve stated physical performance have damaging business outcomes and often fail financially (unless the metal price gods smile at the ideal moment).

The key question arising from this discussion is "How much metallurgical test work is necessary to make a design definitive?" As with all things metallurgical (and indeed, all things connected to the development of mineral projects), the answer is "It depends!" The factors it depends on include resource size, number of geometallurgical domains, variability of properties within the geometallurgical domains, flow sheet complexity, and product-quality requirements.

Many owners (especially those in a hurry) present a case for conducting a feasibility study (FS) to an engineering group without conducting the preliminary work (i.e., a PFS), and while an active exploration campaign is ongoing on the resource, they wish to evaluate definitively. This situation is a massive contradiction, and it fails the tests that will be imposed in attempting to meet reporting codes such as NI 43-101 or JORC (2012). Invariably, the exploration work will expand the resource during the term of the FS, and the owner will demand the FS now be based on the new larger resource. However, the inputs to an FS need to be frozen at the 10%–15% elapsed time mark if it is to be completed in the agreed schedule. Of course, it is possible to allow for a full FS rework at the 50% effort mark, but the risk is high that much of the work done up to that point will be wasted if there is a relatively simple change to the ore delivery schedule in terms of annual ore tonnage, feed grade, ore type, or key ore properties such as hardness. It is much better for the owner to be improving mineralization status during an FS with infill drilling than expanding the resource and changing the FS fundamentals. In the worst-case scenario, the owner discovers a new nearby deposit during the FS and switches to it as the only ore source on the assumption that it has similar properties to the one the FS is based on. This is an actual case example in which all geometallurgically based risk reduction was lost, the ore was not the same, and the project quickly failed.

The main message is that once a decision is taken to go to the FS stage, then expansive exploration activity should cease and work should shift to infill drilling with a target of finalizing mine planning and process selection at the 10%–15% mark in the FS. The FS should be allowed to run its course and either be rejected as uneconomic or accepted based on the physical limits imposed on the mine and the test sample selection before FS commencement. If exploration is ongoing and identifies a better prospect during the FS then there are two paths: either cancel the FS and return to the PFS on the new prospect or continue the FS on the existing prospect and commence a separate PFS on the new discovery.

Fundamental to the concept of being definitive in design is understanding the inherent variability in the most important (and most variable) process input—the ore. Given the importance of this topic, it is dealt with in some detail in the following sections.

Ore Variability

A common statement heard from less-enlightened geologists, miners, and process engineers is that ore type X is all the same and we have tested it. Further investigation usually reveals that ore type X is dominated by one lithology and, in that limited sense, it is all the same. A few tests may have been conducted on this lithology and the answers came out close to each other. Consequently, the average property measurement has been assigned to that particular geological ore type. This is not an adequate basis for assigning a design point to an ore type, and not only because the average is an inappropriate design point. For an ore type to be metallurgically confirmed as being "all the same," an adequate number of tests need to be conducted on spatially disparate examples of the one lithology to determine a reliable standard deviation for the property in question. If there are no metallurgical design implications arising from the plant being fed ore at the low end of the property spectrum (average minus two standard deviations) or the high end of the spectrum (average plus two standard deviations), only then can the ore be considered all the same with respect to this property. A common example of this is specific gravity (SG), which has a very low relative standard deviation in many (but certainly not all) ore bodies. The band of values encompassed by two standard deviations either side of the mean accounts for 95% of the expected values (for a statistically normally distributed property). Many ore properties, especially within a lithology, are either normally distributed, log normally distributed, or close enough to these distribution types. Provided enough samples are tested, it is possible to determine the standard deviation of the measure in question.

Therefore, the starting point for having a definitive design basis is understanding ore-body variability. Unfortunately for the impatient, this is a two-step process. The first step is estimating the inherent degree of variability by testing an adequate number of samples. The second step is testing adequate samples to confidently define the variability envelope for design purposes. These steps sound almost the same but the difference between them is crucial to this discussion.

Coefficient of Variance Concept

The principles of understanding variability in respect to comminution properties have been outlined by Morrell (2011), and this section will follow Morrell's conventions. First, the concept of coefficient of variance (COV) must be understood. This is sometimes called the *relative standard deviation*. COV is simply the standard deviation of a set of values divided by the mean of the same set of values and expressed as a percentage. For example, a data set with a mean of 25 and a standard deviation (SD) of 5 has a COV of 20%. For such a distribution, the band containing 95% of the measurements (±2 SDs from the mean) spans from a low value of 15 to a high value of 35. This means the maximum expected value (35) is much more than twice the expected minimum (15). If this is a critical design parameter (such as a grinding work index), then this high level of variability must somehow be handled by the process design or mitigated by an ore blending strategy.

A more benign case might be a siliceous ore with a mean SG of 2.5 and an SD of 0.05, giving a COV of 2%. The 95% band of SG values will be from 2.4 to 2.6. Most plant designs that are required to treat such an ore could be safely based on the average 2.5 SG without allowances for variability.

Morrell has examined the distribution of COV for the ore competence measurement (drop weight index [DWI] in kilowatt hours per cubic meter) of 650 ore bodies within his database of SAG mill competency (SMC) test results. The relationship between the impact energy imparted to a particle and the degree of breakage achieved in that particle is used to calculate the DWI. The results are useful in designing crushing and grinding circuits, and the test has been designed to be applied in geometallurgical programs. The graph showing the distribution of variability outcomes across the 650 ore bodies is reproduced in Figure 3.

Only 12% of the measured ore bodies have COV values of 10% or less, and half of the ore bodies tested have COV values greater than 26%. Competence is one basis for estimation of a SAG or autogenous grinding mill throughput capacity because, simplistically, the milling power requirement per metric ton is directly proportional to DWI.

At the median COV of 26%, the spread of DWI values expected is 100% of the average value (52%, or 2 × SDs below the average and 52% above). For example, an ore body with an average DWI of 5.0 kW·h/m^3 and an SD of 1.3 kW·h/m^3 (COV = 26%) will have ores with competence values ranging from 2.4 to 7.6 kW·h/m^3. The upper DWI value is more than

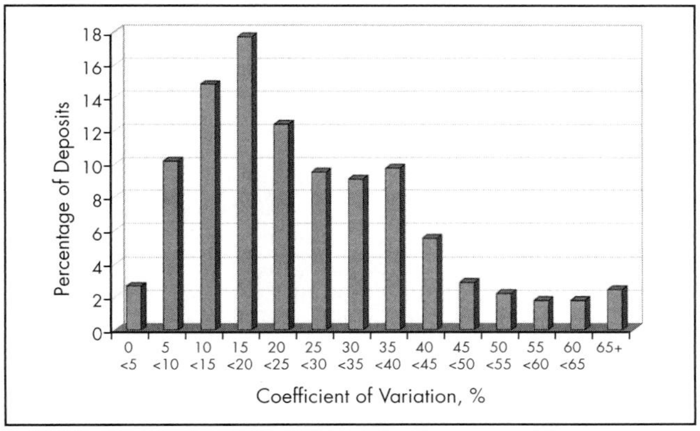

Adapted from Morrell 2011

Figure 3 Distribution of ore-body COV values for competence measurements

three times the lower value, and this is a range for which it is difficult to design a milling circuit without allowing for some blending. Note, however, that many ore bodies have well above average COV values and present even greater design challenges.

Coefficient of Variance Estimation

Asking how much information is needed to estimate the COV is the next obvious question. The answer is somewhere between 7 and 20 samples. The more samples, the better the estimate of COV will be. However, as COV should be estimated in the PFS stage, it will be covered in the "Preliminary Phases" section.

Assuming the COV of the measurement is approximately known (because a good PFS test program was conducted), it is then necessary to test enough samples to adequately define the distribution. The larger the COV, the more samples needed to achieve an acceptable level of confidence that the distribution average, and its limits, are understood. Of course, it is possible to design with a lower level of confidence, but this demands either acceptance of risk that the project will not perform or the inclusion of educated levels of overdesign to compensate.

Confidence Level in the Results

To assess the raw test data for variability, it is sensible to determine the average together with the SD and then calculate the COV. This is a simple matter in a spreadsheet software application, such as Microsoft Excel, which provides standard functions for these purposes. It is statistically correct to say that the more samples of ore that are tested, the more meaningful the statistics will be for the property being measured. An alternative expression for meaningful is *confidence*.

Mathematically (also available as an Excel function), the confidence level to which the mean or average is known can be calculated for any data set. A confidence calculation returns an absolute number (z), which is then used to define a range within which the true average should lie (measured average ± z). This range is the *confidence interval* (CI), and its very existence attests to the fact that any average from test work is only ever an estimate of an ore property.

One of the inputs to the calculation is the alpha value (α), and this is directly related to the confidence level to which you need to know the average. An α of 0.05, for example, is equivalent to a confidence level of 95%. At 95% confidence level, the answer you are looking for can be relied on to lie within the CI 19 times out of 20. At 99% confidence level, the certainty is 99 times out of 100 that the answer will be within the CI. To achieve this higher level of certainty (99% confidence), the interval calculation results in a much wider range within which the average value could reside.

How does the confidence level concept relate to the real world? When samples are chosen for metallurgical testing, they are usually selected from available core. If there is 1,000 kg of core available to sample, and each sample needs to be 25 kg, then (simplistically) there are 40 possible samples available for testing. Money has only been allocated for 10 tests, so only 10 samples are chosen, somewhat at random. Tests are performed and an average property value is calculated. However, if a different set of 10 samples was randomly chosen and tested, then the average value would have been different, but the method of arriving at the average has not changed. Calculating the CI for the mean using the 10 chosen tested samples provides the designer with a measure of what could have happened if a different random set of 10 samples had been chosen from the same batch of core for testing. For example, if the calculated CI ranges for the Bond ball mill work index (BWI) ranges from 8–12 kW·h/t, then the average value is not very well known at all, and the design basis (whatever it is) will carry a lot of risk. Had a different set of 10 samples been chosen for testing, then the average might not have been 10 kW·h/t but could easily have been as low as 8 kW·h/t or as high as 12 kW·h/t. However, if the CI ranges from 9.5 to 10.5 kW·h/t then the average of 10 kW·h/t is well known. It is highly likely that a similar average would have been determined for any other set of 10 test samples, so the design basis risk is low.

Apart from the confidence level being sought (α of 90%, 95%, etc.), the other two properties used in calculating the CI are the number of samples tested (n) and the SD of the data set (mathematically, CI = f [α, n, SD]). Once sufficient samples have been measured to arrive at a credible SD, there is little change to the SD value as more samples are tested. Therefore, the only method of improving confidence in the mean (making the CI narrower) is to test more samples.

The concept of confidence must be understood by process designers because it is fundamental to knowing if sufficient tests have been performed to size equipment. The confidence a designer has in a design point translates directly to the confidence the designer has in the equipment selected.

Being Definitive

Being definitive in design means that the resulting plant will achieve the intended performance with the assumed inputs, and within a stated tolerance of accuracy. It is sensible, then, to adopt a conservative design position once the data has been analyzed. Conservatism is normally achieved by selecting a value that is some known distance above the average but less than the maximum measured value. A typical approach is to use the 80th percentile value, but an alternative valid approach would be to use the upper confidence limit in combination with a second factor (e.g., the SD). In general, design factors and operating margins are also added to the calculated process requirement. One clear warning that insufficient samples have been tested is when the upper confidence level (UCL) of the mean is greater than the 80th percentile value in the data set. This is a red flag, and it implies that if another set of samples is chosen and randomly tested, then the average of that set could actually be greater than the so-called safe design point, being the 80th percentile value. The obvious solution is to randomly select enough additional test samples to at least double the data-set size and ensure that the new UCL is significantly less than the 80th percentile value.

An example data set is shown in Figure 4 with the CI for the mean shown as crosses on the 50th percentile rank line. The 80th percentile point of the data set (10.6 kW·h/t) is circled and is clearly close to (i.e., in conflict with) the largest potential value for the mean of the data (10.4 kW·h/t).

An alternative approach for selecting the design point from the data set in Figure 4 is to adopt the 80th percentile of the fitted normal distribution (the dashed line) rather than the 80th percentile data point. The normal distribution line places the 80th percentile value (10.9 kW·h/t) more than twice as far above the measured mean (10 kW·h/t) as the UCL (10.4 kW·h/t), and this would be considered a safe design selection. Testing more samples is also an option, but it is not warranted in this particular case.

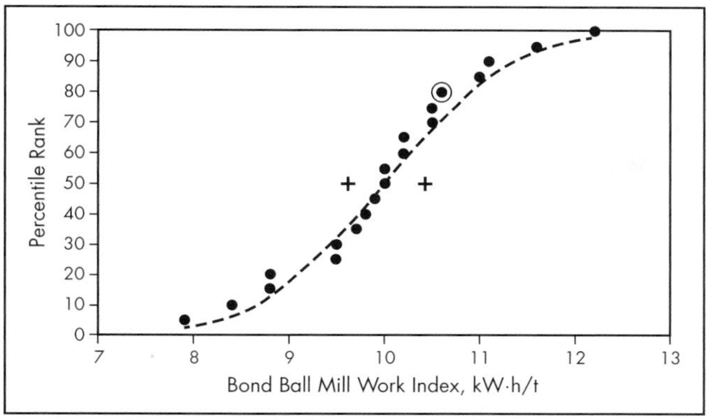

Source: David 2015, reprinted with permission from the Australasian Institute of Mining and Metallurgy

Figure 4 Close proximity of average value and 80th percentile value

The ultimate message from this statistical discussion is that enough samples need to be measured to provide confidence that the average value is known and, at the same time, provide confidence that the design position is demonstrably conservative.

Measuring Variability

Given the correct tools, it is a relatively simple matter to statistically analyze a set of numbers. However, it is essential that the numbers being analyzed are the right numbers. In designing a plant, it is essential that ore variability is understood because this drives maximum equipment capacities and minimum flow requirements. All operating metallurgists will know that their plant will become throughput limited when a particular ore property (e.g., grindability, competence, or grade) approaches an extreme. They also know that this type of limitation can last from a few minutes to months. Therefore, what is a sensible time frame for understanding ore variability? Typical starting points are the operating day or the operating shift, because this is the time frame over which operational control is exercised.

An immediate implication of this typical starting point is that mine plans prepared on a monthly or annual basis do not contain variability information immediately relevant to design calculations. They effectively contain grouped average data that significantly understates variability. By extension, a more inappropriate basis for design is the use of life-of-mine ore-body average properties.

By definition then, the analysis of variability can be considered successful if the method used to estimate variability generates answers that approximate daily or shift fluctuations. To achieve such an answer, it is necessary to examine the sampling basis, the mining method, and the available short-term data.

The short-term data available for analysis refers to information available in equivalent daily (or smaller) parcels. The two data sets representing the shortest time frames on an operating mine site are usually the geological drill database and the mining block model. The drill database typically contains short interval (1–5 m) chemical analysis data together with longer interval geotechnical, lithological, mineralogical, and alteration data. A single 1-m interval will typically weigh 5–10 kg and will provide a representative analysis of only a fraction of a day's production, generally measured in minutes. A single metallurgical variability sample may represent 5–10 m of core, weigh up to 60 kg, and represent up to 30 minutes of production. Although the blocks in a block model do not represent physical samples, they are mathematically generated from the drill database (e.g., using kriging). Given sufficient raw data, block properties are probably the most useful data set available for design purposes. Typically, the ore represented by a mine block will represent between a few hours and a day of production. In a geometallurgical program, variability testing is performed on core samples with the aim of embedding useful metallurgical information into the ore blocks.

Metallurgical samples are usually selected from the geological core set and, if the ore-body geometry permits, represent single geometallurgical domains. To get the required mass for comminution and separation testing, it is common to take contiguous samples over lengths of 10–50 m, depending on the core diameter and how much of the core the geologists are able to release for testing. A single metallurgical sample can intercept anywhere between one and five mine plan ore blocks depending on the length of core selected. The variability of assays of metallurgical samples is typically between the variability present in the drill sample assays and the variability present in the block model.

To illustrate how variability varies with time frame, an analysis has been conducted on assays for a copper and magnetite deposit where drill data is available as is about six years of mine block model data. The mine block data set is arranged in a typical mining sequence and has been subjected to mine planning grade control practices in forming that arrangement. This allows the block data set to be subgrouped into shift, day, week, month, and annual data sets, all of which can be analyzed for variability. All intervals under 0.1% Cu (the cutoff grade) have been excluded from the drill data set and the block model data set.

The average, 80th percentile values, and the upper and lower 95th percentile bands were determined for each data set. A comparison is made of the variability levels within the data sets in Figure 5.

Variability within the "Year" set is very low and far less than the variability expected on a shift basis. This shows the

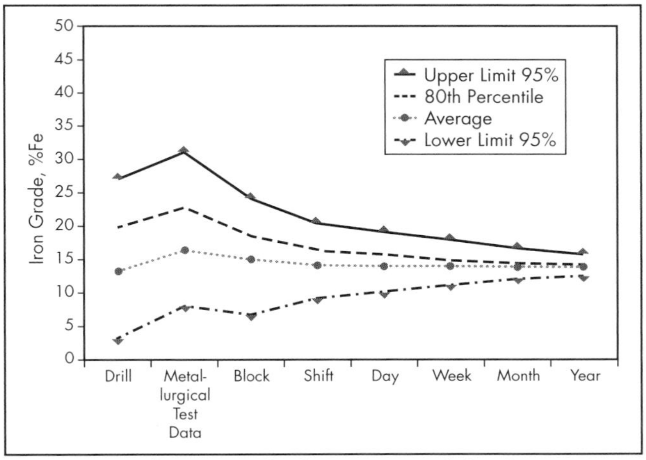

Figure 5 Single iron ore-body assay variability for various data sets

folly of using annual plan data to select a design point for any property. On an annual basis, the 80th percentile and the average are almost the same number (14% Fe). At the preferred design time frame (the shift data set), the 80th percentile value is much higher at 16% Fe while the average is effectively unchanged (as it should be) at 14% Fe. Figure 5 also demonstrates the overestimation, compared to shift data, of variability inherent in drill assay data (1-m interval basis) and in the metallurgical test data set. The average value for the metallurgical test data set is offset from the general average value because of the relatively small number of samples.

Out of 365 available days per year, the feed is expected to be within the 95% band in Figure 5 for 347 days. This means that there are nine days per year where the feed assay is expected to be greater than the upper limit (>20% Fe) and nine days where it is expected to be below the lower limit (<9% Fe). On a three-shift-per-day basis, this translates to 27 shifts of very high grades and 27 shifts of very low grades a year, about one of either instance per week. If the upper and lower feed grade expectations are set using yearly data, then for about 50% of the time, the feed % Fe will be higher or lower than these annual extreme limits suggest. Unless realistic expectations of shift operational variability are established by process engineers during design and operation, regular conflict can be expected between the process engineers, geologists, and mine planners on almost a weekly basis.

There is a valid alternative to accepting the shift variability outcomes as shown in the Figure 5 analysis. The alternative is to rerun the mine planning model with tighter limits on grade control. The disparity between the monthly variability and the shift variability will remain, but the "Upper Limit 95%" shift value may be reduced from 1% to 0.9% Cu or even 0.8% Cu, and this would lead to a design capacity reduction in the concentrate handling sections of the plant. Tighter grade control results in reduced plant capital expenditure but has a cost in terms of mining capital expenditure and operating cost (such as increased rehandling of ore and the possible use of smaller equipment). For each project, the design team needs to balance grade control cost, capital reductions, and plant operational flexibility. For demonstration purposes, this analysis has been limited to the iron feed grade. Similar consideration is necessary for each critical design input including, for example, comminution properties; impurity assays; acid consumption values for leach design; and concentrate mass yields for flotation, gravity, or magnetic separation circuits.

Importance of the Definitive Phase

It must be remembered that the definitive phase study (FS) design outcomes have the inherent potential to become an implemented project, often with associated process performance warranties for the designer and vendors. If the critical variability issues have not been measured adequately, then the implemented outcome will suffer financially. In many of these cases, the project may never achieve nameplate design, and some of these projects will fail to perform to the point where closure or major capital expenditure are the only options.

Although not wise, it is possible to complete a study with inadequate variability measurements, but such a study is far from definitive. A designer in this position must incorporate a prudent design contingency (i.e., select larger equipment) and argue the reason for it statistically, using experience and through appropriate benchmarking. A low level of confidence in a key design value has a real cost as higher capital expenditure.

It is only possible to achieve optimal definitive stage outcomes in a timely fashion if the preceding phases have been performed adequately. This includes thinking about how many individual test samples may be required for the definitive stage test program. This, in turn, influences the type of drilling conducted, the diameter of core being extracted, and even the way the drill samples are stored after extraction.

Evaluating the Adequacy of Outcomes

When evaluating definitive phase (alternatively bankable, or FS phase) outcomes, the key word is *definitive*. Are there undefined elements, such as the true variability of critical measures, which are missing from the evaluation and are not reflected in the design outcomes? All raw test data pertaining to these critical measures should be sighted, and both the adequacy of the raw data set and the logic behind the selection of subsequent design points must be assessed.

The following are key questions:

- Have enough samples been tested to establish a reliable value for the variability within the definitive ore body (as contained in the definitive pit shell or underground workings) and within major ore types?
- Do the sets of ore property data confirm that the various ore types have geometallurgical meaning?
- Do all predictive relationships (linking geology measures and process outcomes such as throughput or recovery) have strong enough correlations to be useful?
- Has the ore on which the plant will most likely be commissioned been evaluated as a set of separate samples?
- Has the ore that will be treated within the period in which the capital is repaid been evaluated as a set of identifiable samples?
- Are there any key process design aspects not supported by adequate test work?
- If a choice has been made to overdesign, rather than conduct definitive testing, this must be documented and be reflected in the equipment selections.

It is recommended that all samples to be tested are identified, collected, and in process before any definitive phase engineering commences. Testing at the same time as definitive engineering is proceeding is virtually guaranteed to negatively impact the definitive study schedule or the study quality. This often means that a definitive testing subphase should exist between the preliminary and definitive study phases.

Preliminary Phases

The preliminary project phases must set the scene for the definitive phase and, eventually, operations. Quantification of variability in the earlier phases allows informed decision-making in relation to the ultimate numbers of samples to be tested.

Variability Estimates for Each Critical Measure

A typical aim of a PFS phase is to evaluate many potential project development alternatives and select a best case to take forward to the definitive stage. A typical costing accuracy target at PFS is ±25%. To ensure that a believable comparison is achieved between options, it is necessary to have a relatively good understanding of the variability of the critical ore properties so that comparable design points can be selected for each option.

In some PFS comparisons, the key design decisions for the options being compared rely on totally different measures. An example of this for gold processing is comparing a flotation-based circuit with a whole ore leaching-based circuit. The flotation-based circuit needs to be understood with respect to concentrate grade, mineralogy, and recovery by flotation. The leaching circuit needs to be understood with respect to leach recovery, reagent consumption, and leach residence times (among others). When comparing these two options, it is tempting to take shortcuts, such as using the average values (rather than design values) when selecting equipment. This approach is likely to be misleading if the key parameters in each option have significantly different levels of inherent variability.

It is recommended that the prefeasibility test work aims to establish the variability levels in all critical design inputs (e.g., those that leverage more than 5% of the final capital or operating cost) by major ore type. In this way, the results provide a level of accuracy appropriate for prefeasibility and also guidance as to how much more testing is needed before a definitive study can commence.

It is normally recommended that general flow sheet development (such as establishing the flow sheet steps and the grind sizes for each stage) be conducted on composite samples. Once a reliable flow sheet has been developed, it is possible to test variability in a sensible way. For variability estimation, the recommended number of tests (as indicated previously) is between 10 and 20. For cost control, and to limit consumption of valuable core, select 10 variability samples and test each sample for every critical measure. Continuing the example of the gold ore where direct leaching is to be compared with pyrite flotation, the program for each major ore type might look something like the following:

- 10× SMC Tests for SAG sizing, general comminution circuit sizing, and comminution power consumption estimation
- 10× BWI tests for ball mill sizing and power estimation
- 10× Bond abrasion index tests to measure ore abrasivity and estimate consumptions of grinding media and mill liners
- A sensibly weighted composite of the 10 samples to develop a flotation flow sheet and develop leach conditions, including establishing separate grind sizes for optimal liberation
- 10× variability flotation tests to estimate all major flotation design parameters
- 10× variability leaches to estimate all major leach design parameters

When these results are available, it will be possible to compare the two circuit types at the required level of costing accuracy, and it will be possible to estimate how many more tests of each type are required before design can be elevated to the definitive level. The number of additional tests will vary from measure to measure. For example, it is possible that 40 additional SMC Tests may be necessary although only 10 more BWI tests are required to achieve acceptable accuracy and confidence.

Sampling needs for the definitive stage must be identified early in the PFS program so that any additional drilling or sample extraction can be conducted without delaying the project schedule. It is also essential that the sample selection locations be matched to likely mining areas. Sample location selection should be left until a mine plan is available that approximates the plan that will form the basis of the FS.

Essential Tests

A precursor to the preliminary test program is identifying what tests must be performed. The list of tests needed is a direct consequence of the various flow sheets and ore types being considered. Each flow sheet that is to be evaluated needs to be identified, and the types of equipment or processes being considered need to be listed.

The test type has geological implications because many tests require particle sizes that can only be delivered by core that is PQ size or larger. In the special case of pilot testing of fully autogenous milling, it is essential that either broken ore up to 200-mm topsize or large-diameter diamond core (PQ 83 mm and larger) is available.

If it is necessary to compare two technologies in the PFS (e.g., semiautogenous milling compared with crushing and ball milling), then tests appropriate to each technology must be performed for the comparison to be fair. It is not adequate to use Bond measurements for the ball mill case, for example, and then infer SAG performance from Bond measurements to provide the basis for the SAG design. In some cases, using the Bond inference works, but if the ore is high competence (DWI >8 or A × b <35), the Bond methods will typically underestimate SAG power and overall power for the case, giving a false comparison with high risk inherent in the SAG option alone.

Best Sample Size

Once the tests are known, then sample masses for conducting each of the tests can be obtained from the relevant laboratories. The overall sample mass requirement to conduct all tests, and prepare composites, is then used to determine how long the single sample contiguous core interval must be. The length will depend on the core diameter and how much is available for metallurgical testing (whole, half, or quarter core are the usual options). In virtually all instances, reverse circulation chips are unsuitable for metallurgical testing, especially comminution testing.

If the drill run intersections in the ore zone are shorter than the required length, then multiple holes at the same location (where the mineralization is shallow) or multiple wedge holes from the one parent hole (for deeper drilling) may be needed to provide suitable samples comparable to contiguous core. In all instances, the selection of metallurgical test samples must be a collaborative effort between the geologists and process engineers.

Optimum Testing Schedule

In the ideal world, PFS metallurgical test work should be completed before any PFS engineering design commences. This is relatively easy to arrange for standard comminution tests but much more difficult for concentrator flow sheet development work. If test work is ongoing while a plant section is being designed (using scoping design parameters), then there is a high risk that the results will change the flow sheet, the mass balance, and then the equipment selections. If any level of meaningful mechanical, civil, or electrical design or cost estimation work has been carried out based on the old parameters, it will all be wasted and have to be redone with the new inputs. When the test work cannot be completed in the suggested sequence and time period, a risk review must be conducted, the test work prioritized, and the PFS schedule rearranged to match the design timing to the availability of results. An alternative approach is to add design contingency, but it must be remembered that this will increase costs and reduce estimation accuracy.

Conducting PFS level testing then skipping the PFS rigor to commission a fast-track FS is a common mistake made by mining companies, both junior and multinational. The usual result (based on direct experience of many examples of this approach) is that the information set is far from adequate for conducting an FS, and additional un-costed testing and planning occurs during the fast-track study. At the end of the exercise, the FS is usually inadequate because it is out of date (e.g., more and better ore has been discovered), and the study schedule has blown out by months. These months would have been better spent conducting the PFS in an orderly fashion before deciding on the testing necessary for FS. The reality of a fast-track FS is that it takes about as long as a PFS followed by an FS, but for the entire time, an engineering team is consuming the owner's resources much faster because they have set in place FS staffing levels rather than the much lower PFS staffing levels. Worse, the delays shift the schedule for the preparation of the FS report to the extent that the promised expert resources (e.g., in estimating and financial evaluation) have moved on to another project, and lesser (but available) so-called expert resources are allocated to do this work. Where possible, one should strongly advise against conducting a fast-track FS and propose a professional PFS.

Evaluating the Adequacy of Outcomes

When evaluating PFS outcomes in preparation for a media release (e.g., acting as a Qualified Person under National Instrument NI 43-101, a Chartered Professional under JORC [2012], or in a similar manner under other jurisdictions), many key questions need to be answered, including the following:

- Has the test work been conducted on a mix of true variability samples and composites as appropriate?
- Can all the major design inputs and decisions be linked to both the average value and an estimate of its variability?
- Have the correct tests been performed to adequately assess the various processing options under consideration?
- Has a plan been prepared for additional testing ahead of the definitive phase?

Scoping Level

At scoping level, it is highly likely that no preceding metallurgical work exists, unless an old ore body is being reevaluated. The aim of a scoping study is to determine if this particular prospect is better than alternative prospects or investments vying for funding from one entity. As such, a sensible flow sheet of some sort needs to be evaluated at an accuracy of ±30% or greater. To select a sensible flow sheet for the major ore type(s), it is necessary to know how hard the ore is, according to the various comminution measures, and if a typical separation technique (such as flotation for copper sulfide mineralization) is effective. It is not necessary to optimize the flow sheet or compare alternatives, provided the selected flow sheet has a high probability of being successful. For example, if the evaluation is performed using a crush and ball milling circuit, and the investment is definitely attractive, this conclusion would not change (within scoping accuracy) if the same project were to be evaluated with a SAG or high-pressure grinding roll circuit. Where the fundamentals of recovery change (such as comparing a copper heap leach with a tank leach), it is legitimate to conduct both evaluations at scoping accuracy. Similarly, if a high-quality scoping evaluation shows the investment potential is marginal, then shuffling the equipment or efficiency deck chairs is not going to change this conclusion within scoping accuracy. Additional sampling and testing may show the scoping information was poor, and this is probably the best pathway to shifting the investment potential status. At the geological level, the best way to change the investment potential is to identify more ore. This leaves a marginal prospect in the situation where either more money is spent on exploring or testing, or it is shelved until conditions are more appropriate.

It is legitimate to work with composites at the scoping stage because there is little point in understanding ore variability until an ore body has been defined. In the author's experience, available scoping samples for a greenfield exploration prospect will have little relevance during PFS and FS evaluations because the resource knowledge and size expand exponentially (for a good prospect) in this time period.

Using Geology for Guidance and Sample Selection

If a prospect has reached scoping level, the geology should be well advanced. Geological domaining is the main source of guidance for metallurgical sample selection, and there is often a shortage of good core to test. First, the selection should be based on the major lithologies and five examples of each lithology should go into any test composite that constitutes more than 5% of the attractive mineralization. If a particular lithology is dominant (50% or more of the mineralization), then subdivision should be sought. Subdivision can be by alteration type, grain size, depth, and so forth. The aim is to ensure that no one scoping composite represents more than about 35% of the mineralized material.

These samples should be tested as if the most logical and common flow sheet for the ore type was to be implemented. If there is more than one obvious alternative, the test regime should be expanded to include the additional possibilities. In case the logical processing pathways become unattractive as a result of the initial test outcomes, there should be composite material in reserve.

When the test results are received, there should be initial indications if knowledge of the lithology (or another geological domaining basis) provides any guidance for metallurgical response. If lithology has no influence on metallurgical response, then another basis for understanding the mineralization geometallurgy will be needed. There will definitely be strong guidance as to how well the material responds to conventional processing, and this will provide a basis for estimating comminution and separation requirements to a scoping accuracy. If the separation test work was unsuccessful, there may also be grounds for selecting a more complex and expensive processing method with some confirmatory further test work.

Evaluating an Outcome

A scoping study should quickly arrive at a conclusion, indicating if the prospect is worth spending additional money on. A negative conclusion at scoping level is valuable as it allows the owner's scarce resources to be redirected to more favorable prospects. A strong positive conclusion is also valuable as it has identified real potential. With either of these definite scoping outcomes, the main point to evaluate is whether there is a strong plan for future work.

In the case of a strong negative outcome, the plan should relate to the conditions that may turn the prospect into a positive and how this should be addressed. This will typically be linked to a change in the metal price regime or future drilling success on the same prospect. The recommendations should include an ongoing geometallurgical plan and should be on a similar scoping level basis. There is no point planning for PFS level testing unless a positive scoping outcome is achieved.

In the case of a strong positive outcome, the evaluation should ensure that the correct preliminary test work has been conducted and the results have been correctly interpreted. The evaluation should also determine if a strong geometallurgical framework has been recommended for progressing the prospect to PFS level.

An unconvincing positive or small negative is problematic as, within the accuracy of the scoping study, the outcome could statistically be either a strong positive or a strong negative with deeper investigation. An evaluation of this type of inconclusive study should focus on methods of improving the evaluation accuracy with the minimal amount of test work or via consideration of alternative processing options. Again, an evaluation of geometallurgical issues, such as the representativeness of the samples that have been tested, requires close scrutiny. Often the best approach with such a prospect is to explore more and see if the resource itself improves before repeating scoping sampling, testing, and evaluation.

CONCLUSIONS

The last thing a project needs on commissioning is to encounter ore that was unanticipated or, at the least, not adequately designed for. Many projects have failed, or suffered a near-fatal setback, as a result of being unable to process the initial ore. Many other projects have faltered because the plant was designed for the average case, or the operational variability was considerably underestimated because variability for design was sourced from an annual mine plan. Still more projects operate in a state of semi-blindness because the mine planning outcomes do not provide the process plant operators with meaningful predictions of throughput, product grade, or other essential economic factors.

Actions to avoid these sorts of mistakes must begin as early as possible in the project study process and, if at all possible, in the exploration phase. It is the duty of qualified persons across the geological mining and metallurgical disciplines to understand what constitutes a strong geometallurgical project basis and what does not. The sooner that issues such as potential flow sheets, variability, and geometallurgical ore type definition are addressed within a study framework, the greater are the chances that the common operational problems will be avoided or that the owner will shelve the project at an appropriately early stage. It must be remembered that metallurgical test work is relatively inexpensive compared to drilling, and test work is a trivial cost compared to fixing an operating plant that fails to match the ore it is being fed.

Another important consideration is that the big-picture project is not all about the capital cost. Provided capital can be sourced, the real big-picture project is minimizing the time between accessing that capital and subsequently paying it back. Anything that can justifiably de-risk this payback period (and geometallurgy is at the top of the de-risking list) should be seen as an essential component of the study phases. The only asset a minerals project has is the mineralization, and all sensible measures should be taken to understand how value is generated by the asset so that it can (possibly) become a resource, then a reserve, and finally an operating project.

Conducting an effective geometallurgical evaluation of a project, especially at PFS and definitive levels, requires the correct metallurgical test samples and the performance of the correct metallurgical tests. This requires both appropriate time allowances in the schedule and the ability to access the appropriate geological sample types. Realistic allowances in the budget and schedule at, and between, the various study phases are a sign that such issues have been adequately considered in the project plan. Small test work budgets, inadequate time allowances, and, probably the worst of all options, the fast-track project that skips essential stages such as the PFS, are all geometallurgical red flags.

REFERENCES

Baumgartner, R., Gomez, P., and Escobar, G. 2016. Comprehensive mineralogical characterisation at the Cerro Corona Cu-Au porphyry mine—the fundamental key for geometallurgical applications. In *Proceedings of the Third AusIMM International Geometallurgy Conference (GeoMet) 2016*. Melbourne, Victoria: Australasian Institute of Mining and Metallurgy. pp. 221–230.

Chait, E., and Schiller, R. 2016. Adding copper recovery and acid consumption variables to the geological model of Quebrada Blanca. In *Proceedings of the Third AusIMM International Geometallurgy Conference (GeoMet) 2016*. Melbourne, Victoria: Australasian Institute of Mining and Metallurgy. pp. 257–266.

David, D. 2014. Geometallurgical guidelines for miners, geologists and process engineers: Discovery to design. In *Mineral Resources and Ore Reserve Estimation: The AusIMM Guide to Good Practice*, 2nd ed., Monograph 30. Melbourne, Victoria: Australasian Institute of Mining and Metallurgy. pp. 443–450.

David, D. 2015. Measuring and taking notice of orebody variability, an essential ingredient for reliable plant design. In *Metallurgical Plant Design and Operating Strategies—World's Best Practice (MetPlant) 2015*. Melbourne, Victoria: Australasian Institute of Mining and Metallurgy. pp. 10–26

David, D. 2017. An Introduction to Geometallurgy [presentation]. One-day course in geometallurgy.

Jackson, J., McFarlane, A.J., and Olson Hoal, K. 2011. Geometallurgy—Back to the future: Scoping and communicating geomet programs. In *Proceedings of the First AusIMM International Geometallurgy Conference (GeoMet)*. Melbourne, Victoria: Australasian Institute of Mining and Metallurgy. pp. 125–131.

JORC (Joint Ore Reserves Committee). 2012. *Australasian Code for Reporting of Exploration Results, Mineral Resources and Ore Reserves*. Melbourne, Victoria: Australasian Institute of Mining and Metallurgy, Australian Institute of Geoscientists and Minerals Council of Australia. Clauses 37–40.

Leichliter, S., and Larson, D. 2013. Geometallurgy for two recovery process operations at Cripple Creek and Victor gold mine. *Min. Eng.* 65(1):29.

Liebezeit, V. 2011. Geometallurgy data management—a significant consideration. In *Proceedings of the First AusIMM International Geometallurgy Conference (GeoMet)*. Melbourne, Victoria: Australasian Institute of Mining and Metallurgy. pp. 237–245.

Liebezeit, V., Ehrig, K., Robertson, A., Grant, D., Smith, M., and Bruyn, H. 2016. Embedding geometallurgy into mine planning practices—Practical examples at Olympic Dam. In *Proceedings of the Third AusIMM International Geometallurgy Conference (GeoMet) 2016*. Melbourne, Victoria: Australasian Institute of Mining and Metallurgy. pp. 135–144.

Martins, M.A.S., Medeiros, H., and dos Santos, B.Y.B. 1998. Geometallurgical studies for the Mineração Rio Do Norte Bauxites, Plateau Saracá III, Brazil. In *Latin American Perspectives: Exploration, Mining, and Processing*. Edited by O.A. Bascur. Littleton, CO: SME. pp. 155–162.

Morrell, S. 2011. Mapping orebody hardness variability for AG/SAG/crushing and HPGR circuits. In *International Autogenous Grinding, Semiautogenous Grinding and High Pressure Grinding Roll Technology 2011*. Edited by K. Major, B.C. Flintoff, B. Klein, and K. McLeod. Westmount, QC: Canadian Institute of Mining, Metallurgy and Petroleum.

McQuiston, F.W., and Bechaud, L.J. Jr. 1968. Metallurgical sampling and testing. In *Surface Mining*. Edited by E.P. Pfleider. Littleton, CO: SME-AIME.

National Instrument NI 43-101. 2011. *Standards of Disclosure for Mineral Projects*. Toronto, ON: Ontario Securities Commission.

Turner, N.L., Kuhar, L.L., Francis, N., Liebezeit, V., Ehrig, K., and Robinson, D.J. 2013. Development and testing of a small-scale uranium leach for geometallurgical applications. In *Proceedings of the Second AusIMM International Geometallurgy Conference (GeoMet) 2013*. Melbourne, Victoria: Australasian Institute of Mining and Metallurgy. pp. 345–350.

Wirfiyata, F., and McCaffery, K. 2011. Applied geometallurgical characterisation for life of mine throughput prediction at Batu Hijau. In *International Autogenous Grinding, Semiautogenous Grinding and High Pressure Grinding Roll Technology 2011*. Edited by K. Major, B.C. Flintoff, B. Klein, and K. McLeod. Westmount, QC: Canadian Institute of Mining, Metallurgy and Petroleum.

PART 2

Management and Reporting

CHAPTER 2.1

Health and Safety

Stacy Kramer and Ed McGowan

THE PROTECTION ENVELOPE

Everyone in the mining and mineral processing business is lured to the same means of making ends meet, with the expectation that they will do so without harm. Consequently, to meet this goal, a substantial commitment and continuous improvement will always be expected. These are fundamental requirements of a *mature safety culture.* All levels of management need to be participate, and employee empowerment is critical to ensure success. Writings in this handbook or in company policy manuals are just that, words to provoke a better way. Regulations are provided for everyone's benefit as well, with each standard providing a specific level of protection. Yet everything businesses commit to is contingent on people of authority implementing best practices and ensuring that those practices prevail. Competence is essential regarding accident prevention, and implementation of a successful safety and health program is not likely without it. People in authority are the first tier of protection. It may sound like a coach cheering on his or her team, but it holds true nonetheless: "If not us, then who? If not now, then when?"

What gives at-risk workers the greatest advantage? Knowledge does. Savvy workers are far less likely to be injured or suffer ill consequences from potential exposures. The premise is simple, yet in practice it requires a robust effort by many to ensure worker safety and positive outcomes every day. In the world of mining and mineral processing, the risks are real and ever present. History has proven that. Without empowerment of every employee, the outcome of dangerous work is left to chance. Statistically, if those in authority allow it, something bad will happen. However, it is not about statistics; it is about people. People can control the outcome of each aspect of their work by applying knowledge, effort, and engineering principles that have proven records of success. There is a safe way to do every job; all accidents can be prevented. The process of accident prevention involves workers, under the assumption they are knowledgeable and completely enabled to make the right choices. A tier of empowered, savvy workers is fundamental to the development of the safety envelope. On this same note, not all workers are cut out for this.

From concept to reality, a safety and health program requires a lot of hard work. Nothing good comes easy. The "protection envelope" involves numerous layers of involvement. At-risk workers are best protected when the work culture recognizes hazards and implements appropriate controls. Communicating the sequence of work and maintaining consistent checks from start to finish is as essential as employee competency. Safety processes should keep workers out of harm's way even when something goes wrong. There is no spin on words here; the protection of workers is a conscious thought. Competent people know what it takes to not only do things right, but to do the right things. Ethos is core to people in the first place. When it comes to employee safety and health, everyone faces this challenge together. This chapter discusses elements that are essential for a successful health and safety program as the Society for Mining, Metallurgy & Exploration (SME) strives to communicate and expand the protection envelope.

HEALTH AND SAFETY CULTURE

Modern health and safety management systems establish building blocks for a variety of operation types. In all cases there is a leadership expectation. Without top-level involvement, the success of the program will be limited. Likewise, there must be an evaluation process, an audit, which ensures that each aspect of the system is functioning as planned. One could argue that the audit is most essential because the scorecard that leaders are challenged to fulfill would also be the upper limit of their overall success. Therefore, knowing the safety culture becomes an integral part of the overall health and safety audit. What are the expectations, who is responsible for what, and what processes are in place to provoke the desired outcomes? A business can deliver a healthy environment and accident-free workplace every day when its commitment, resources, and finances are in alignment. The culture model helps organizations answer these questions and prioritize their tasks.

Most people in the industry have seen graphics that chart safety programs as either successful or pathetic. Such models

Stacy Kramer, Vice President Global H&S, Freeport-McMoRan Inc., Phoenix, Arizona, USA
Ed McGowan, Safety Manager, Kinross Gold Mining (Bald Mountain) Inc., Elko, Nevada, USA

represent a snapshot in time, but in reality, the system represents only the current team and their commitment to health and safety. An organization may own the strategy, but it is people who make up the results. It would be unwise not to recognize that the culture is subject to frequent change for this reason. People always have something tugging on them. Anything that influences their behavior will impact their decision making. Consequently, a work culture that places people first provides the best health and safety environment. When employee protection is incorporated throughout the system audit, the higher levels of achievement will always be targeted. Members are asked to make a specific note that certified safety and health programs are in the center of the cultural curve. Mature safety cultures recognize that a certified system is a key building block, but not the ultimate goal, for achieving safety and health success. Members of SME and company officials can score themselves using the same criteria. There will not be a test, nor will they have to complete assignments, but they should strive to stay to the right of the curve shown in Figure 1. This figure typifies a conventional health and safety approach for defining culture.

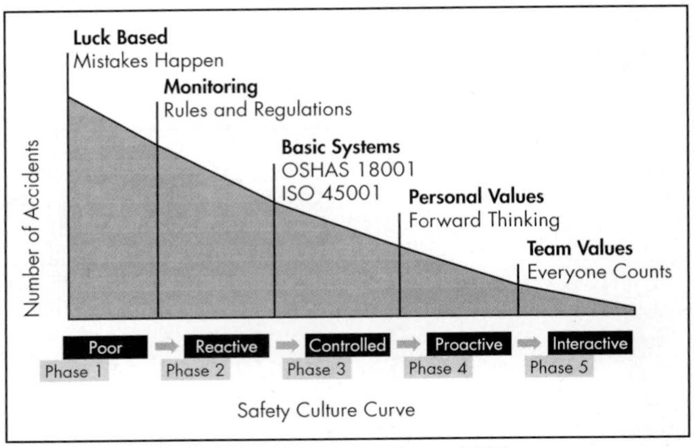

Adapted from FLSmidth Safety Management System 2016

Figure 1 Health and safety culture curve

LEADERSHIP EXPECTATIONS AND THE EMPLOYEE ENGAGEMENT PROCESSES

The first question in the system audit might be: Are all levels of management engaged in safety and health processes? And the second question might be: Are leadership requirements measured in "activities, outcomes, or both"? The questions assume that lack of leadership will be negative and active engagement is necessary for success. Active involvement might include a safety component in every staff meeting, specialized training for key individuals, participation in safety committee functions, or routine participation in plant safety meetings or inspections. Ultimately, at-risk workers need to see support, not just hear of it. When worker confidence in management is high, the motivation of the entire team is elevated. Work is therapeutic when pride is the driving force. Motivation is spurred by recognition of achievement and the opportunity to do interesting work. Recognizing this and being highly visible as a leader is of extreme value. For these reasons, the engagement process should include both individual and team opportunities. Both lagging and leading indicators, including the use of near-miss reporting, can and should be used to determine leadership activities that focus on the prevention of injuries. The audit structure should tally activities that are in alignment with this.

A leader also needs to be cognizant of hazards and critical risks. Some cultures have defined serious risk as a new paradigm where not all safety rules have the same weight. The theory suggests that mobile equipment, working at heights, confined space entry, electrical hazards, certain chemical exposures, lockout/tagout, and rigging/material handling type injuries are less forgiving. Fatal events are more likely to occur in these areas. In turn, critical safe work procedures, the enforcement thereof, and training (including practice) must be used accordingly. The leader must be keenly aware of this because, in the desire to be highly visible, one could unknowingly endorse poor practices within the paradigm. Sam Walton of Walmart got it right in his strategy of management by walking around. He simply shopped at his own stores to determine how things were. The strategy of safety by walking around is of equal value. Are leaders practicing what they say they are? Leadership safety walks, for the purpose of employee engagement, are one of the best tools for ensuring positive outcomes. Leaders should make sure they are well versed on critical safe-work procedures before walking around. They must be respectful of the culture.

In the spirit of taking advantage of every detail, leaders should make it clear they are in the field. They should be in uniform. How leaders dress not only is expected to comply with personal protective equipment (PPE) requirements but also adds to the element of respect. Worker apparel should be distinct, making it clear at a glance who makes up the team. The same is expected of a visitor or vendor; workers should have no doubt who is approaching. Those who are new to a job require a safety briefing. This places the burden on each newcomer to ensure that they check in with the person in charge before they proceed. Doing so is respectful and wise; not doing so is a mistake.

MEASURING PERFORMANCE

Most major mining companies recognize safety performance not only as essential but also as a key performance indicator (KPI). It would be unreasonable to expect a top-notch organization to have poor safety performance. Likewise, it would be unreasonable to expect top-notch safety performance when other KPIs are lagging. Sound safety results and overall efficiencies are one in the same. All safety results are driven by people; thus, accurately measuring performance is essential. An organization's performance indicators ensure the correct focus for financial components, training initiatives, and resources. Watching the trends is important, yet the lessons learned from any event are equally important. Conducting a root-cause analysis digs deeper into the cause of any event. The standard tools to measure safety performance are provided within this chapter to include insurance rating in the form of workers' compensation experience modification factors.

Naturally, incident rates tell us one story. A "zero injury" rate is something all organizations desire. How they have done in the past helps them predict the future. The Occupational Safety and Health Administration (OSHA) and Mine Safety and Health Administration (MSHA) provide ample information regarding reporting criteria and safety performance (OSHA 2001; MSHA 1986). Incident rates can be used for general purposes and are commonly used as a starting point to measure performance. The reporting criteria is the same for all operations in the United States since such reporting is regulatory driven and typically overseen by a safety professional. A recordable event is the same for all. Major mining companies typically use total rates on a worldwide scale (all-incident rate) since this provides a more accurate measurement of overall performance. Total injury frequency rates use a broader numerator than lost-time incident frequency rates (LTIFRs) and assume any incident has the same potential of loss and is therefore more telling of the success or failure of the accident prevention program. There can also be discrepancies with lost-time events based on an operation's overall capabilities to provide alternate work options for injured workers. Internationally, LTIFRs are still used because a lost-time event is typically defined the same way in any country. Nonetheless, showing incident rates over time (usually annually) is more telling of performance trends. A trend that consistently shows a "zero point something" incident rate is an indicator that the operation is a top-notch performer.

Conventional incident rates are based on 200,000 equivalent hours worked. This is derived from exposure of 100 workers who worked 40 hours per week, 50 weeks per year. It compares the safety performance of 100 workers per year rather than a percent of injury. The formula used by both OSHA and MSHA is as follows:

$$\text{incident rate} = \text{number of injuries} \times 200{,}000/\text{hour worked}$$

This rate is a good barometer to use if an operation has 100 workers or more. If an operation recorded three events in a year and employed 100 workers, the incidence rate would be 3.0, meaning roughly 3% of the workforce had a recordable event that year (even though one person could have been injured more than once). Had this team worked more than 200,000 hours in the year, the incident rate would be slightly less, even though 3% of the workforce had been injured. If this same performance prevailed for a worker's entire career, one could extract from the same numbers that it would be likely for that worker and all others to be injured within a 35-year time frame (with the odds being worse for those performing the riskiest of tasks). For this reason, it is extremely important to sustain at or near-zero incident rates year after year after year.

Performance can be measured using injuries per hours worked, with no association with the 100-worker equivalent, but this is less common. Also, some international operations use 1 million hours rather than the 100-person equivalent-hours work formula. The trend line would mimic that of conventional rates, but the rate would calculate to be five times higher. Both methods mean the same when used as trend indicators and for overall goal-setting. Having an all-incident rate goal of less than 1.0 conventional would be the same as having a goal of less than 5.0 per million hours worked. The latter would scare most major mining companies if they were unaware of the alternate calculation that is sometimes used globally.

Although not applicable to all countries, a common tool to evaluate performance in the United States is the workers' compensation experience modification rating (EMR). It is the insurance rating assigned to an operation or company based on claim costs, frequency of injury, and historical performance as it relates to that company's industry type. A rating of 1.0 is average, below 1.0 is favorable, and above 1.0 is undesirable. In general, the rating is based on a company's performance over the past three years exclusive of current claims that have yet come to maturity. In short, it reveals what it would cost to insure the operation based on criteria used by mandatory no-fault insurance programs. Insurers must know the comparative risk to insure the operation. A low rating indicates that costs to insure are lower, medical costs are lower, and, most important, fewer injuries are occurring in this group. Although saving money is not often cited as the motivation behind accident prevention, as a KPI, less money spent on claims means lower incident and severity rates and lower insurance premiums. The reverse is true of high EMRs: the cost to do business is more, more injuries are occurring, and the operation is viewed as one with higher-than-normal risk.

Figure 2 depicts safety performance using EMRs. Each bar measures three years of performance; therefore, the chart is accurate. As noted previously, 1.0 represents the norm for the related industry. A lower limit of the rating system is identified in the chart as well. This varies by state but is typically set at 0.57 (lower limit = 57% of premium).

By using safety as a KPI, one can quickly assess a company's overall efficiency. The operation depicted in Figure 2 is very efficient and continuously improving, although not incident free. Having an injury-free workplace is both personally and financially wise. For these reasons, an operation's EMR is commonly the first issue addressed on a prequalification questionnaire. Bidding processes commonly use EMR performance comparisons. Mine operators want to know an operation's safety KPI in advance of any alliance or work.

An organization's leaders expect to have safety and health targets associated with both performance and activities. Watching trends is important, yet one must know the limitations of incident rates. Regarding safety and health, the

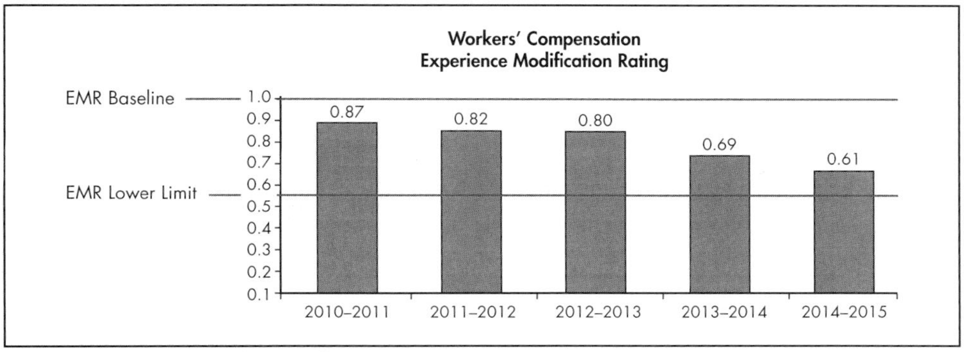

Figure 2 Measuring safety performance using EMR

system simply wants actions that enhance the culture of "zero injuries." The open-door policy, as commonly expressed in business, is part of this as well, as it is likely in use by the organization. It not only gives everyone access to leadership, it also suggests the leader should get out of the office and into the operation. And as the leader does so, he or she should do so with caution, be well read, know the risks, and seek out the persons in charge. The power of perception reminds us that *how* one proceeds is as important as proceeding.

SYSTEMS TO GUIDE HEALTH AND SAFETY MANAGEMENT SYSTEMS

A health and safety management system (HSMS) is a process an employer puts into place to minimize workers' risk of injury and illness. Developing the system is accomplished by identifying, assessing, and controlling risks to workers in all workplace operations (Figure 3). The scope and complexity of an HSMS will vary according to the type of workplace and the nature of its operations. To be effective, the following components should be in place:

- **Health and safety policy.** Develop a statement of the organization's commitment to health and safety. Use this policy as a framework for planning and action.
- **Health and safety hazards and risks.** Identify health and safety hazards, assessment of risks, and the implementation of necessary control measures.
- **Legal and other requirements.** Identify and ensure access to relevant laws and regulations (and other requirements to which the organization adheres).
- **Objectives and targets.** Establish goals for the organization, in line with its policy, health and safety hazards and risks, legal mandates of the organization, and the view of interested parties.
- **Health and safety programs.** Plan actions to achieve objectives and targets.
- **Structure and responsibility.** Establish roles and responsibilities and provide necessary resources.
- **Training awareness and competence.** Ensure that employees are trained and capable of carrying out their health and safety responsibilities.
- **Consultation and communication.** Establish processes for internal and external communications on health and safety management issues.
- **Health and safety documentation.** Maintain information on the HSMS and related documents.
- **Document control.** Ensure effective management of procedures and other system documents.
- **Operational control.** Identify, plan, and manage the operation and activities in line with the policy, objectives, and targets.
- **Emergency preparedness and response.** Identify potential emergencies and develop procedures for preventing and responding to them.
- **Monitoring and measurement.** Monitor key activities and track performance.
- **Accidents, incidents, nonconformance, and corrective and preventive action.** Identify and correct problems and prevent recurrences.
- **Records and record management.** Keep adequate records of health and safety performance.
- **Audit.** Periodically verify that the HSMS is operating as intended.
- **Management review.** Periodically review the HSMS with an eye to continual improvement.

CONTRACTOR REQUIREMENTS AND PREREQUISITES

Contractors are a significant aspect of doing business in the mining industry. Whether the contractors are working side by side with employees in the operation, as part of small projects to improve a site, or as part of a major green- or brownfield construction project, a company has an obligation to ensure that the contract companies and their employees are working within its safety and health programs. Companies that have the best safety performance treat the contractors as they do their own employees by setting clear expectations and then following through to ensure these expectations are met. Contractors are typically responsible for establishing, implementing, and maintaining their safety programs to meet the goals and objectives as stated by the hiring company. This responsibility also applies to subcontracted work that the general contractor may require. In general, most companies have an established process for bidding work and choosing contractors. The following section provides recommended processes for lining up contractors, commencing with the initial bid.

Safety in the Bidding Process

Often, a company addresses safety after the contractor is already on-site. However, it is much more effective to involve contractors as early as possible in the bidding and selection process. Ensuring that the contract company has information on what is required prior to starting will ensure that it has trained employees and the proper equipment when the bid is awarded. Doing so makes the overall project more productive and saves time on making adjustments later. The expected commitment to safety should be clear from the start. Most projects perform some sort of bid walk to review the project scope of work. This typically involves the area where work will be performed, and the site health and safety management group should be involved. This is the time to set expectations and review specific safety processes required. There should be no surprises. After a bid is received, a contractor must make a series of verifications. Some of the prequalifications are regulatory driven, and others are legally driven. Most mining companies have their own customized safety prequalification package, but others use subscription safety verification programs such as BROWZ, ISNetworld, and ComplyWorks, to name a few.

Safety Plan

Prior to beginning work at the site, the contractor should be required to prepare and submit for review a site/task/project-specific safety plan that reflects the contractor's intentions for full and complete compliance for the scope of contracted work. This plan should be reviewed by the project manager and health and safety manager or designated staff to ensure that it meets the minimum standard set by the company. To coincide with this, a project safety meeting should be used to engage all parties. This process will confirm that expectations are clear and health and safety requirements will be met by the contractor's plan. Any site-specific details not addressed

1. Obtain Management Consent

The first step in building the health and safety management system (HSMS) is gaining top management's commitment to supporting the HSMS. Management must understand the benefits of HSMS and what it will take to put an HSMS in place. **Management commitment** and vision must be clear and communicated across the organization.

2. Choose a Champion

It is critical to choose a site **project champion**. The champion should have the necessary authority, an understanding of the organization, and project management skills. The champion should be a "systems thinker" and must have the time to commit to the HSMS building process.

3. Prepare a Budget and Schedule

The project champion should prepare a preliminary **budget and schedule** for developing the HSMS. Costs will likely include staff and employee time, training, some consulting assistance, materials, and possibly some equipment (computer—applications). The schedule should consider the various tasks described below, among others.

4. Build Project Team

A **team** with representation from key management functions and production or service areas can identify and assess issues, opportunities, and existing processes. Consider including contractors, suppliers, and other external parties to be part of the project team where appropriate. This team will need to meet frequently, especially in the early stages of the project.

5. Involve Employees

Employees are a great source of knowledge on health, safety, and environmental issues related to their areas as well as on the effectiveness of current health and safety processes, tools, and procedures. They can help the project team in drafting procedures. **Employee ownership** of the HSMS will be greatly enhanced by meaningful employee involvement in the HSMS development process.

6. Conduct Preliminary Review

The next step is to conduct a **preliminary review** of your current health and safety programs and system and compare these against the criteria for your HSMS (such as OHSAS 18001, ISO 45001). Evaluate your facility's structure and its procedures, policies, hazards, risks, training programs, and other factors. Determine which elements of your current system are in good shape and which need additional work. You may want to use consultants to guide you in this step.

7. Modify Plan

The **project plan** might need to be modified based on the results of the preliminary review. The modified plan should describe in detail the key actions needed, who will be responsible, what resources are needed, and when the work will be completed.

8. Prepare Procedures and Documents

At this point, you are ready to **develop procedures and other system documents**. In some cases, this might involve modifying existing procedures or developing new procedures. Get help from employees and the cross-functional team.

9. Plan for Change

In building your HSMS, make sure that the system is sufficiently **flexible**. While you will likely need to modify your HSMS over time, try to avoid making your HSMS so rigid that you must change it frequently to reflect the realities of your operation.

10. Train Employees

Once procedures and other documents have been prepared, you are ready to implement the HSMS. As a first step, **train** your employees on the HSMS, especially with regard to the risk management process, new procedures, and any new responsibilities.

11. Assess Performance

After the HSMS is up and running, be sure to **assess system performance**. This will be accomplished through periodic HSMS audits, and ongoing monitoring and measurement. Assessment of HSMS performance provides the opportunity to improve the system and your health and safety performance over time.

Figure 3 Step-by-step process for building a health and safety management system

by the plan can be implemented or modified at this time. Amendments or changes to the plan should be reviewed and approved by the company before being implemented. Ongoing meetings of a similar nature will be required of the project as the work progresses.

Health and Safety Professional Requirements

To ensure proper health, safety, and environment (HSE) oversight, setting a minimum requirement for a dedicated on-site safety professional is a necessary component of the plan. This professional is in addition to the competency expected of the person in charge of the work. Most companies require a minimum of one full-time, qualified safety professional whenever the contractor's workforce meets or exceeds 50 employees. Additional full-time, competent safety professionals are recommended for each additional 250 employees. As a guide, all contractor health and safety professionals, through education, training, and experience, should be capable of the following:

- Identifying existing or potential risks, including unsafe acts and behaviors
- Identifying and implementing controls to mitigate the risks of tasks
- Identifying working conditions that are unsafe, hazardous, or dangerous to the safety and health of employees and the environment
- Identifying nonconformance with health and safety requirements, including at-risk behavior
- Authorizing prompt action to maintain a healthy and safe work environment

Some companies might require the site safety professional to be degreed or certified in the field of HSE, but in all cases this individual is to be a person of authority. The scope of work will determine the level of expertise and number of professionals needed. When developing the contracts and setting expectations for safe work, a company typically requires that the contract company commits to additional formal processes as follows:

- Complying with national, regional, and local health and safety laws and regulations, the company contractor safety guidelines, and any requirement imposed by the local site where the work will be conducted.
- Providing all contract personnel with PPE for the work for which they are responsible, including safety glasses, hard hats, protective footwear, fall protection, hearing protection, respiratory protection, high-visibility apparel, and/or other safety equipment as may be required.
- Maintaining the highest standards of housekeeping. Workplaces must be kept organized with all debris, waste materials, and so on, cleared as work progresses. All wastes shall be properly disposed of according to the site-specific policies.
- Participating fully in the management of risk by implementing controls in the workplace. Defining and controlling risk is a key element of an HSMS.
- Verifying that all contract employees have received project safety orientation as well as other training that is required specific to the job function being performed (i.e., lockout/tagout/tryout, confined space entry, working at heights, excavation, chemical handling and use, etc.). The contractor must determine that workers understand and are knowledgeable.
- Providing a disciplinary action policy, including exclusion from the site if necessary, for individuals who violate health and safety procedures or drug and alcohol policies, or otherwise work in a careless or unsafe manner.
- Providing the first response for emergencies (first aid, emergency, fire, etc.) while activating site response for supplementary action, treatment, and support.
- Keeping all registers, records, and reports up to date and properly completed, stored in a safe place on-site, and maintained for review by legal or regulatory agencies.
- Immediately removing from the project site any contractor's manager, supervisor, owner, or other person in charge that requires, condones, asks, or allows employees to work in or around unsafe acts or conditions.
- Attending regular site safety meetings as set by the company.

Employee Training

Each contractor is required to provide regular and continuing health and safety training for all employees and to monitor subcontractor training programs. Training should include a site safety orientation as well as task-specific training as required by regulatory agencies or identified by the hiring company. All training should be documented and a process implemented allowing a quick verification of training received by any individual. Maintaining verification of training will ensure regulatory compliance as well.

Auditing Field Performance

It has been stated that what gets measured gets done. After expectations are set, one must verify that those expectations are being met. An audit schedule for the project should be established as part of the scope of a work and safety plan. Scheduling should be jointly organized by the project manager and contractor site manager and performed on a monthly or quarterly basis depending on the project. Area supervisors and safety professionals (contractor and company) should accompany the project and site managers in their respective areas. Audit results should be documented, and corrective actions identified and tracked to completion.

Equipment Inspection and Operation

Any equipment brought to the site that requires inspections (daily, monthly, annually, etc.) should be accompanied by that documentation and made available for review upon request. Equipment added or changed after the project has commenced should be identified by the contractor and be subject to the same requirements. Requirements for pre-shift inspection and safe operation should be made known to the contractor as part of the bidding process so that equipment can be made available that meets company requirements.

Safety Meetings with Contractors

Communication is a key element of success in any safety plan. Having multiple contractors on-site can make it difficult to communicate effectively. Holding a monthly meeting, at a minimum, with all contractors on-site is a good way to keep those communication lines open. Pertinent health and safety issues should be discussed frequently. Reviewing new procedures or policies, talking about any concerns or incidents that have occurred, and reviewing existing procedures are several suggested agenda items for this meeting. As the sequence of work changes, workers need to be kept up to date. In so doing, appropriate safeguards can match the nature of the work.

CONTINUOUS RISK ASSESSMENT AND FIELD-LEVEL EVALUATIONS

Many struggle with the concept of zero injuries being achievable. Are all injuries really preventable? Is it possible to eliminate minor incidents like cuts and bruises? Whether one agrees that zero is attainable, everyone can agree that serious injuries and fatalities can be prevented and should rise to the top when a company is prioritizing its efforts. It is hard to focus on everything and do it well. This is why many companies are shifting to a focus on prevention of serious injuries and fatalities. This does not mean other incidents are acceptable or ignored. It means that time and resources are going to be spent on the high-risk tasks that have the potential to permanently affect employees' lives and the lives of their family members.

Figure 4 is a version of the Heinrich pyramid. It depicts the theory presented by H.W. Heinrich in the 1930s that is still used today. According to the theory, for every serious incident there are many less serious incidents that present the opportunity to address issues and prevent a fatality or other life-altering event.

In looking at incident trends, the Heinrich theory applies only to the area outlined in the box. By focusing on near misses and their causes, one can track the less serious incidents and make changes to help reduce the severity rate as well as the number of total reportable incidents. Lessons learned from minor and near-miss events are invaluable; companies do not need to wait for something bad to happen to realize that improvements are necessary.

The narrow triangle in the center of the Heinrich pyramid accurately represents serious injuries and fatalities. The theory suggests that the higher potential of severity events usually results in severe consequences. There may not have been an opportunity for a fatality to be a near miss first; thus, in most cases, companies lack the ability to prevent the situation. For example, if a haul truck goes over a berm, typically a serious injury or fatality will result. If an electrician makes contact with 4,160 V in a processing plant, the resultant injuries will be awful. Therefore, modern safety systems look at fatal incident prevention differently. Companies are now expected

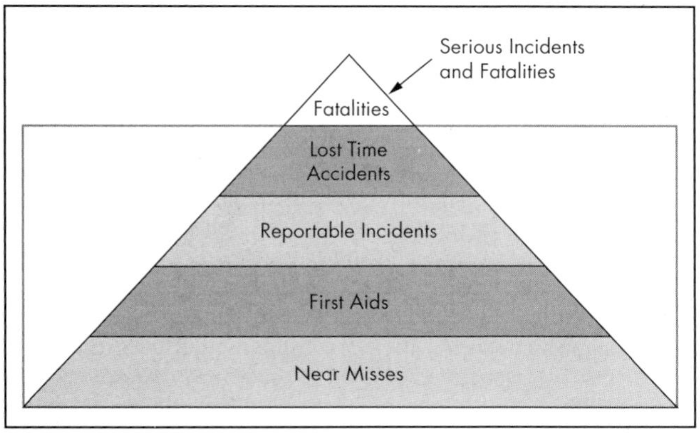

Adapted from the National Mining Association

Figure 4 Heinrich theory and the serious injury paradigm

to have both critical injury prevention programs as well as standard health and safety programs. The new paradigm was well stated in International Council on Mining and Metals conference presentations by both Goldcorp and Newmont mining companies (Krause 2012; Newmont Mining 2012). This recent shift in focus of major mining companies initially drew the concern that fatal injury prevention would allow the total incident rate to trend upward. Interestingly, the major mining companies have learned that total reportable incidents and serious injury rates have improved. Fatal injury prevention should always be a focus, but programs that incorporate both safe work procedures and critical safe work procedures see favorable downward trends. The application of work procedures alone is not enough. Oversight of implementation is required at the field level.

HAZARD IDENTIFICATION AND FIELD-LEVEL RISK ASSESSMENT

To prevent serious incidents and fatalities, one must first understand the high-risk tasks. Then processes must be put in place to eliminate or mitigate the risk. A sound risk management process should contain the following elements:

- **Hazard identification.** Identify hazards that exist, define their characteristics, and assign severity of probability that harm will occur.
- **Risk assessment.** Identify tangible and intangible risks and impacts of a certain task, function, or work environment.
- **Critical control identification and implementation.** Determine appropriate and feasible risk controls using the hierarchy of controls (elimination, substitution, engineering, administrative, PPE).
- **Communication of the risks and controls.** Engage employees and field supervisors in hazard identification and control implementation. Ensure that all employees understand the controls and the expectation that controls are used.
- **Verification and auditing of controls.** Develop a plan for field verification of controls to ensure effectiveness and audit to ensure that controls are being used appropriately. Use accountability as appropriate where not in use.

Leaders have a responsibility to evaluate tasks and determine the level of risk for these tasks by identifying the hazards and developing controls to eliminate or mitigate the risk. A risk matrix, such as the example in Figure 5, can be used to assist with the determination of risk level. Risk level is determined by multiplying the likelihood of occurrence by the consequence. Pure risk should be determined without consideration for controls. Mitigated risk is then determined with consideration of controls. Likelihood and consequence categories are further defined based on company criteria. High numbers in the matrix reflect the need to implement controls.

CRITICAL CONTROL IDENTIFICATION AND IMPLEMENTATION

A critical control is a safeguard that is crucial to preventing an event or mitigating the consequences of an event. The absence or failure of a critical control would significantly increase the risk despite the existence of other controls. In addition, a control that prevents more than one unwanted event or mitigates more than one consequence is normally classified as critical. A control is an act, object (engineered), or system (combination

5 × 5 Risk Matrix

Likelihood					
Frequent	5	10	15	20	25
Probable	4	8	12	16	20
Occasional	3	6	9	12	15
Remote	2	4	6	8	10
Improbable	1	2	3	4	5
Consequence	Negligible	Low	Moderate	Significant	Catastrophic

Risk Rating: High / Medium / Low

Courtesy of Freeport-McMoRan

Figure 5 Risk matrix

of an act and an object) intended to prevent or mitigate an unwanted event.

When implementing a new control, consider the following questions:

- Is it an action, object, or system?
- Does it prevent or mitigate an unwanted event?
- Is performance specified, observable, measurable, and auditable?

If the answer to any of these questions is no, then it is not a control.

When determining critical controls, consider these questions:

- Is the control crucial to preventing the event or minimizing the consequences of the event?
- Is it the only control, or is it backed up by another control in the event the first fails?
- Would the control's absence or failure significantly increase the risk despite the existence of the other controls?
- Does the control address multiple consequences of the unwanted event? (In other words, if it appears in several places on the risk assessment, or on several risk assessments, this may indicate that it is critical.)

Although each of these risk-reduction strategies can be effective, the hierarchy of controls should be used (elimination, substitution, engineering, administrative, PPE) to find controls that rely less on human behavior and thus reduce the potential for failure. As shown in Figure 6, as one moves up the hierarchy, there is reduced reliance on behavior and therefore more reliability of the controls. Continuous improvement efforts should be focused on reducing the dependency on human behavior for the control of significant risks.

Communication of Risks and Controls

After these hazards, risks, and controls are identified, steps must be taken to ensure that the employees doing the work understand them. Each job should start with a "job briefing" in which the sequences of work and safety controls are discussed. Employees should confirm that they understand the hazards, risks, and controls. Each person is expected to check that controls are in place prior to starting work. This can be accomplished through audits, prejob planning, task monitoring, and reevaluating the task periodically. If controls are not in place, the expectation should be to stop the work and reassess before continuing.

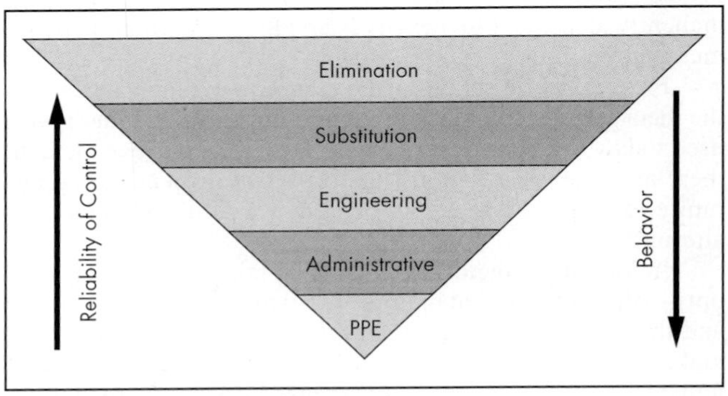

Adapted from the National Safety Council

Figure 6 Hierarchy of controls

Verification and Field Auditing of Controls

Leaders must make oversight of high-risk tasks a daily priority. If several jobs are occurring, those with the highest risks should be visited the most frequently. In some cases, there may be a need to assign dedicated resources to monitor the task, be it supervision or safety personnel. Critical risk audits should be conducted on a regular basis based on the risk level and the level of reliability of controls. Tasks with lower risks can be audited less frequently. Task with higher reliability controls, such as engineered controls, can be audited less frequently as well. Tasks that rely more heavily on PPE controls and employee actions (administrative and PPE) should have frequent oversight. For tasks where a critical control is missing or inadequate, work should not continue until it is corrected. Action plans should be established and followed up by management following the company's tracking process.

SYSTEM EVALUATIONS AND AUDITING

An HSE audit is very personal. Individuals' responsibilities, actions, and accountabilities are fundamental to safety success; thus, all associated processes need to be scored. In so doing, each person's commitments, or lack thereof, can be determined. Both internal and/or external audits can be used, and in some cases, cross-departmental oversight can be implemented. The latter process allows work groups to compare their efforts to those of their colleagues. Using the premise that "the best way to learn is to teach" suggests that workers are most likely to improve themselves if they are asked to evaluate others doing similar tasks. It is difficult for a person to insist someone else improve in one area if he or she would not be willing to do the same. Regardless of the psychology of an audit, the expectation is relatively simple. Each element of a health and safety system needs to have a corresponding checklist. There is no limit how deep this should go. Questions to consider include: Are all levels of competency intact? Do the necessary resources match up? Are regulatory requirements at the forefront? Does a culture that promotes "zero harm" exist? A comprehensive audit will capture the answers. Because the work environment is in constant change, the same audit must encourage both sustainability and continuous improvement. Organizations know the risks, the value of their workers, and the economic well-being or products and operations they bring to the world. The purpose is clear, and self-evaluation is a must.

REFERENCES

ISO 45001. 2018. *Occupational Health and Safety Management Systems*. Geneva: International Organization for Standardization.

Krause, T. 2012. Eliminating fatalities. In *Proceedings of the International Council on Mining and Metals*. London: International Council on Mining and Metals. p. 10.

MSHA (Mine Safety and Health Administration). 1986. *Report on 30 CFR Part 50*. PC-704. Arlington, VA: U.S Department of Labor, MSHA.

Newmont Mining. 2012. Vital behaviors. In *Proceedings of the International Council on Mining and Metals*. London: ICMM. pp. 2, 5.

OHSAS 18001. 2002. *Occupational Health and Safety Management*. London: British Standards Institution.

OSHA (Occupational Safety and Health Administration). 2001. 29 CFR 1904.7. *Recording and Reporting Occupational Injuries and Illnesses, General Recording Criteria*. Washington, DC: OSHA.

CHAPTER 2.2

Community and Social Issues

Jose A. Botin

The index of the *SME Mineral Processing Handbook* (Weiss 1985) does not contain the words *sustainability* or *social responsibility*. More than three decades have passed since its publication, and today these words, and the underlying concerns about mining's environmental and social challenges, have taken on heightened importance in industry and business, especially in the minerals industry.

This chapter seeks to address community and social issues in the context of the interaction between the minerals industry and nearby communities and social stakeholders. It begins by reviewing the ideas behind social engagement and sustainability in a broad sense, and then goes into their implications for the minerals industry and discusses aspects specific to mining companies and communities. It also reviews the leading best practices standards in mining and extractive metallurgy sectors and their role in social acceptance. Finally, the community impacts and social risks resulting from mining and metallurgical activities through their life cycle are analyzed in the context of sustainable development.

SOCIAL ENGAGEMENT FOR SUSTAINABILITY IN THE MINERALS INDUSTRY

In business, the term *social engagement* refers to the interaction of companies with their employees, communities, and other stakeholders to address economic and social issues to achieve business objectives and stakeholder expectations. Social engagement is an important issue in every industrial sector, but in the minerals industry, where the social and political environments are fixed by the location of mineral resources, social engagement becomes critical.

In the context of the minerals industry, the broader concepts of sustainability, sustainable mining, responsible mining, and corporate social responsibility (CSR) are used to describe the social and environmental contributions and consequences of business activity. These concepts are tridimensional and refer to integrating environmental, economic, and social aspects of mining business activity. A notable reference to this concept was made in 2002 by the United Nations Environment Programme in a publication titled *Berlin II: Guidelines for Mining and Sustainable Development* (UN 2002). *Berlin II* stated:

> If sustainable development is defined as the integration of social, economic and environmental considerations, then a mining project that is developed, operated and closed in an environmentally and socially acceptable manner could be seen as contributing to sustainable development. Critical to this goal is ensuring that benefits of the project are employed to develop the region in a way that will survive long after the mine is closed.

Later in this section, the stakeholders of a mining company will be identified, and the different concepts related to social engagement in the minerals industry will be analyzed, including CSR, sustainability, the social license to operate (SLO), and the modern concepts of value sharing and the business case on sustainability.

Stakeholders

The key strategic issue for social engagement is relationships with the stakeholders. In his landmark reference on stakeholders, Edward Freeman (1984) defined *stakeholder* as "a party which affects, or can be affected negatively or positively by the activities of the company." In the context of CSR, a proactive interaction between companies and stakeholders is always important, but in the minerals industry, an efficient and trustful interaction between a company and social stakeholders becomes of greatest strategic interest. This is due to the greater importance and magnitude of the social impacts (positive and negative) of the mining activity and the controversial legacy of mining throughout history.

There is ample literature on community participation and the impacts of mining activity on the community, but, with a few exceptions, literature prior to 2002 focuses on CSR concepts and issues rather than stakeholder engagement and "social license," which is gaining social acceptance. For early references to the subject, see Cragg (1998), Cragg and Greenbaum (2002), and Azinger (1998).

Probably the first systemic approach to a stakeholder-based corporate strategy in mining is presented in the publication *Breaking New Ground: The Report of the Mining,*

Jose A. Botin, Retired Professor, Universidad Politecnica de Madrid, Madrid, Spain

Table 1 Typical minerals industry stakeholders

Business-Related Stakeholders
Stockholders
Employees
Financial and lending institutions
Market partners (suppliers, clients, contractors, etc.)
Social Stakeholders
Directly affected parties (hosting communities, landowners, etc.)
Indirectly affected parties (craftsmanship, farming, cultural, private business, etc.)
Social activists/nongovernmental organizations (environment, human rights, etc.)
Artisanal miners
Media (press and leaders of opinion)
Political groups
Labor unions
Minorities and other historically marginalized groups
Government-Related Stakeholders
National, regional, and local authorities
Judicial authorities
Government agencies

Adapted from Ovejero Zappino 2009

Minerals, and Sustainable Development Project (IIED 2002). Later in 2007, the International Finance Corporation published *Stakeholder Engagement: A Good Practice Handbook for Doing Business in Emerging Markets* (IFC 2007). Also worth noting is the *Stakeholder Research Toolkit* published by the International Council on Mining and Metals (ICMM 2007), which provides tools to quantify the reputation of a mining company among its stakeholders.

No hard rules exist to identify critical stakeholders, but the World Bank (1996) suggests that a good way to start the identification process is by asking the following questions:

- Who are the "voiceless" for whom special efforts may have to be made?
- Who are the representatives of those likely to be affected?
- Who is likely to mobilize for or against what is intended?
- Who can make what is intended more effective through their participation or less effective by their non-participation or outright opposition?
- Who can contribute financial and technical resources?
- Whose behavior has to change for the effort to succeed?

A complete analysis of the interactions between mining companies and stakeholders is presented in the book *Sustainable Management of Mining Operations* (Botin 2009). Table 1 lists typical mining company stakeholders (Ovejero Zappino 2009). The first group of stakeholders listed in the table is the business-related stakeholders who are part of the business structure, and therefore, their interest is economic. In this group, stakeholders of special significance are the employees. They represent the interests of both the company and the community, and therefore, their position is critical for social engagement. Companies tend to view employees as a resource necessary to achieve economic objectives; however, community perspective is based on value systems, and in many cases, the values of a mining company are attested and measured through the actions of its employees. It follows that the needs of the community are only understood to the degree that employees understand and can communicate these needs to the company in such a way that management can take action. As such, the ability of a company to achieve and retain its social license largely rests on the actions, knowledge, and perceptions of its employees.

Just as important to social engagement are contractors and suppliers whom, to a large degree, are managed by local people. Besides, they are a leading force and a major source of information during feasibility and construction stages of mining projects. A local focus is therefore important in the selection and management of contractors and suppliers.

Engagement with local industry managers and entrepreneurs is also important to social engagement since, depending on their perception or their role as potential business partners with the company, they are likely to mobilize for or against the company objectives. Furthermore, contact with local industry leaders and entrepreneurs, seeking for shared professional and business interests, can provide important management and cost advantages.

The second group in Table 1 is the social stakeholders. This group is comprised of social parties whose interests and concerns are varied, difficult to assess, and potentially conflicting with other parties and with the mining company. Regarding the engagement with the hosting communities, the focus should be placed on reconciling corporate and community views of the advantages and disadvantages of mining operations.

Landowners are important stakeholders of a mining project. Land acquisition is often challenging, since it implies negotiation with several types of landowners (e.g., private, corporate, municipal, communal) and sometimes reaching separate deals with land leases (e.g., timber, fishing/hunting, tourism). In countries where mining regulations are based on Roman law, mineral resources are state owned, and therefore, the holder of a mining concession may claim the right of expropriation when the exploitation of mineral resource is declared of "public utility" by the government. However, expropriation is an act of force that may generate social conflict and project delays. Therefore, every reasonable effort must be made to reach negotiated agreements with landowners to buy or rent the affected properties.

Interacting with social activists and nongovernmental organizations (NGOs) may be an important part of a mining business strategy. NGOs are geared toward a cause rather than a profit. The cause of an NGO may combine geography (e.g., local, national, or international) and a subject (social development, ecology, human rights, etc.), and its positions on mining can range from pro-engagement to antimining. Some of these groups maintain a close scrutiny of the mining company and government regulators, which is often constructive and beneficial but sometimes obstructive and unfair.

Dealing with the media involves providing information on corporate issues, projects and operations, environmental performance, safety, and so on, and responding (silence is not an option) to arising issues. Ideally, a personal relationship with media management and journalists should be maintained. Mining organizations should establish different levels for media interaction: (1) media specializing in business and economy; (2) local or provincial information sections in the general media, used by the general public; (3) editors, managers, and owners of the primary media (press, TV, radio, etc.); and (4) regional and local media present in the municipalities where the project or mine is sited.

Corporate Social Responsibility

A generally accepted definition of CSR comes from the International Organization for Standardization (ISO 26000):

> Social responsibility [is the] responsibility of an organization for the impacts of its decisions and activities on society and the environment that, through transparent and ethical behavior, contributes to sustainable development, including the health and the welfare of society, takes into account the expectations of stakeholders, is in compliance with applicable law and consistent with international norms of behavior, and is integrated throughout the organization and practiced in its relationships.

The concept of CSR, considered as a business's concern for society, can be traced back to the late 1800s during the Industrial Revolution, when industrial corporations developed welfare systems to provide for hospitals, bathhouses, and other facilities for employees, focusing on preventing labor problems and improving performance. However, the notion of CSR as an integrated, socially focused corporate strategy is a product of the 20th century, from the early 1950s to the present.

The late 1950s and 1960s saw a growing interest in the concept of CSR. One of the leading writers in this period was Keith Davis. Davis was the first to debate the "business case of CSR" when he asserted that some socially responsible business decisions can be justified by their potential to bring long-term economic gain to the company, a debate that is very alive today (Davis 1960, 1967). Another key contributor to the concept of CSR in the 1960s was Clarence Walton, who stated that the essential ingredient of CSR actions is voluntarism, as opposed to coercion, a concept that is at the core of CSR today (Walton 1967).

The 1970s were marked by the introduction of the "stakeholder approach" to CSR, a key aspect in today's definition of CSR. Along this line, Steiner (1971), in his book *Business and Society*, discusses specific spheres in which CSR might be applied and the criteria for determining the social responsibilities of business. Opposed to the stakeholder approach, Friedman, in a landmark article, stated that in a free society, "there is one and only one Corporate Social Responsibility of business—to use its resources and engage in activities designed to increase its profits so long as it stays within the rules of the game" (Friedman 1970). Obviously, Friedman did not think it was inappropriate for corporations to do all they could to be socially responsible, but he was arguing that applying a shareholder's profit-maximizing strategy allied with an appropriate framework of law would be more likely to produce social good than exhortations that firms be socially responsible.

In a different line of thinking, Davis (1973) maintained that "CSR refers to the firm's consideration of, and response to, issues beyond the narrow economic, technical, and legal requirements of the firm," a definition that, in line with Walton's (1967), implies that voluntarism is an essential ingredient of CSR actions. Another significant contribution to previous definitions of CSR was a publication by the Committee for Economic Development (CED 1971), which related CSR to a "social contract between business and society," a concept that may be considered a precursor of SLO, a core aspect of today's approach to CSR.

In the 1980s and thereafter, the understanding of CSR as the voluntary acceptance by corporations of a certain obligation to stakeholders beyond that required by law had become commonplace. A key development in that decade was the book *Strategic Management: A Stakeholder Approach* (Freeman 1984). In the book, Freeman stated that business can be understood as a system to create value for stakeholders, considered as those who affect or are affected by the business. Another significant feature in this period was the increasing evidence of the relationship between corporate social performance and financial performance (Griffin and Mahon 1997), anticipating today's central problem of the "business case on sustainability." A major book cataloging best CSR practices was written by Kotler and Lee (2005). The authors set out to characterize a new way of doing business that combines the success and the creation of value with a respectful and proactive attitude toward stakeholders.

In the minerals industry, CSR takes the form of voluntary, nonregulated actions by mining companies intended to gain the acceptance of the nearby communities; that is, to gain SLO. The voluntary actions are carried out through the implementation of environmental and social best practices and standards. Therefore, the concept of CSR in mining integrates the concepts of "sustainable mining" and "responsible mining."

Typical legal and voluntary frameworks for mining are presented in Figure 1 (Botin 2009). The voluntary framework for sustainability refers to the company's voluntary adherence to international sustainable development frameworks and standards. Some of these standards are listed in the figure and described in detail later in this chapter.

The legal framework refers to the body of international, national, regional, and local regulations that any mining project must comply with to receive the necessary permits. The process to achieve all necessary permits is referred to as permitting and is a complex management process, sometimes taking several years to complete. The permitting requirements vary widely for different countries, regions, and local governments. Table 2 list some of the most frequently required permits.

Business Case for Sustainability

Although CSR standards are voluntary, market pressures and stakeholders' demands are gradually leading to a situation where noncompliance with international CSR frameworks may bear an increasing social and economic risk. On the other hand, there is ample evidence of the link between environmental and social performance and financial performance (Salzmann et al. 2005). However, the benefits of sustainability are sometimes subjective and difficult to quantify.

As an example, Figure 2 shows a business case matrix proposed by the IFC (2002) that presents the business benefits and risks from social and environmental improvements based on 240 cases from Africa, Asia, Latin America, and Central and Eastern Europe, some of them relating to the minerals industry.

In the context of the mining, the business case for sustainability has also been studied by McKinsey and Company (Bonini and Gorner 2012). The study concluded that companies from the extractive sectors (mining, oil and gas) expect to add significant value from their sustainability activities,

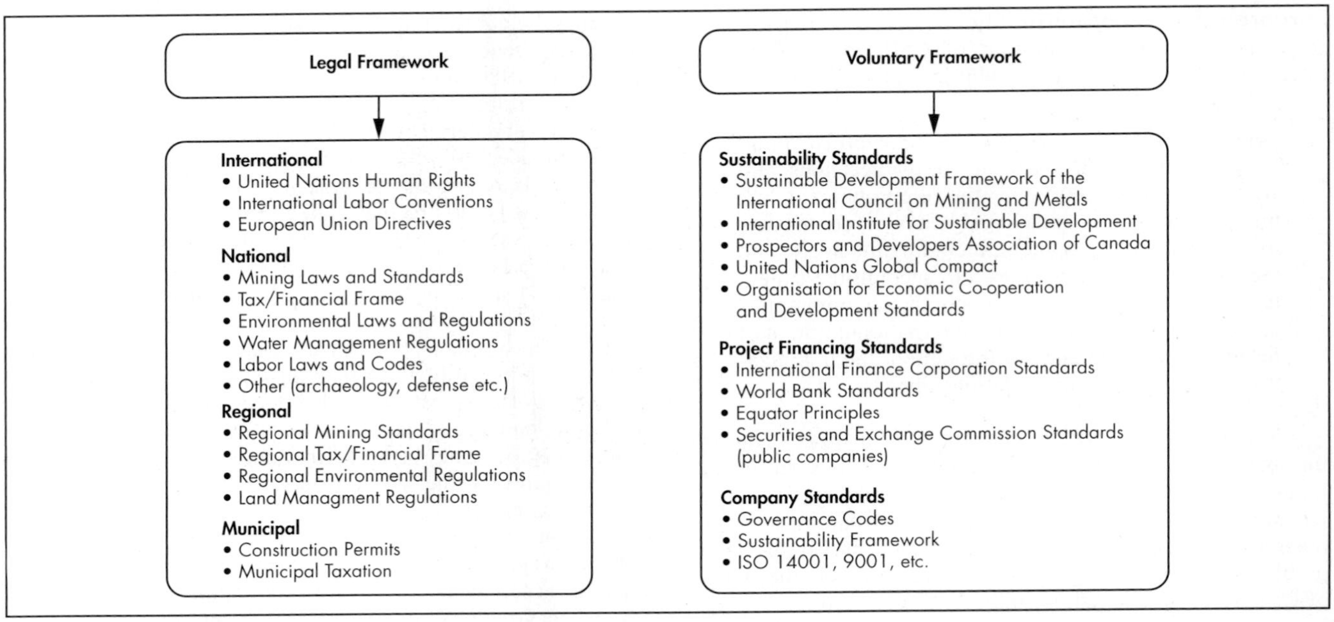

Source: Botin 2009

Figure 1 Typical legal and voluntary frameworks for mining projects

especially from those related to resources development, environmental operations, and regulatory management. Also, intangible assets, such as reputation, were identified to account for a substantial and growing portion of the value of companies in the minerals industry.

Social License to Operate

In general, a SLO refers to a business gaining social acceptance. The term is often attributed to Jim Cooney (1997), a Placer Dome executive, who used it at the Roundtable on Mining: The Next 25 Years, a meeting organized by the World Bank in 1989. Today, the term *social license to operate* has been widely adopted by the minerals sector and civil society in general. However, what constitutes a social license and the underlying processes to obtain one are less well understood, and limited research has been conducted on what factors contribute toward and/or undermine acceptance of mining developments by hosting communities.

Gaining a SLO is about "going beyond legal compliance." In the context of mining, Parsons et al. (2014) find that the SLO may be understood as a set of "best practices" aiming to legitimize mining activity in the minds of local community. SLO may involve listening to, and engaging with, community members, and perhaps allowing them to participate in some company decision-making processes. Furthermore, Moffat and Zhang's (2014) investigation on the critical elements for social license concluded that the key factors in securing and holding a social license are mutual trust and high-quality engagement of mining companies with communities, alongside mitigation of operational impacts.

Any management approach to the SLO must be based on the engagement of social stakeholders in an appropriate process to identify and balance business objectives and the expectations of the community and local stakeholders. The challenge of this process (Botin 2010), illustrated in Figure 3, lies in the need to reconcile corporate and community views of the advantages and disadvantages of mining operations. In fact, a company's view mainly focuses on long-term potential benefits derived from public reputation, positive risk profile, and the economic benefits resulting from improved operational efficiency, trouble-free permitting, and privileged access to debt financing, ore resources, and human capital. On the other hand, a community's perception of environmental and social impacts of mining are generally negative, and the benefits derived from job creation and the contribution of the company to fund community projects are generally undervalued.

Table 2 Main project areas requiring permits

Environment	Mining	Water
Environmental (environmental impact assessment)	Exploration permit	Water concession
Effluent discharges	Mining concession/ project	Water management
Air emissions	Health and safety plan	Streams diversions
Solid residues disposal	Emergency plan	Water reservoirs, ponds
Noise	Explosives	Effluent recycling
Hazardous materials	Yearly work plans	
Tree-cutting permits		
Infrastructures	**Others**	**Municipal Permits**
Power lines and substations	Public domains	Town zoning plans
Ancillary installations	Streams, roads, public paths, etc.	Construction permits
Mining camps/towns	Archaeological permits	Activity permits
Access roads	Construction permits	Opening permits
Railways	Construction materials	
Pipelines		
Ports		
Telecommunications		

Adapted from Ovejero Zappino 2009

A practical approach to this challenging process is the Seven Questions for Sustainability, a methodological framework for community engagement developed by the International Institute for Sustainable Development (IISD

	Corporate Governance	Stakeholder Engagement	Environment Performance	Community Development	Human Resources Management
Production/Sales	Probable	Probable	Proven	Proven	Probable
Costs/Productivity	Probable	Probable	Proven	Probable	Proven
Access to Capital	Probable	Probable	Probable	Probable	Probable
Social Risk/Social License	Probable	Proven	Probable	Probable	Possible
Access to Human Capital	Possible	Probable	Possible	Probable	Proven
Reputation	Probable	Probable	Proven	Probable	Probable

Legend: Proven, Probable, Possible

Adapted from Van Heel et al. 2001 and IFC 2002

Figure 2 Business case matrix

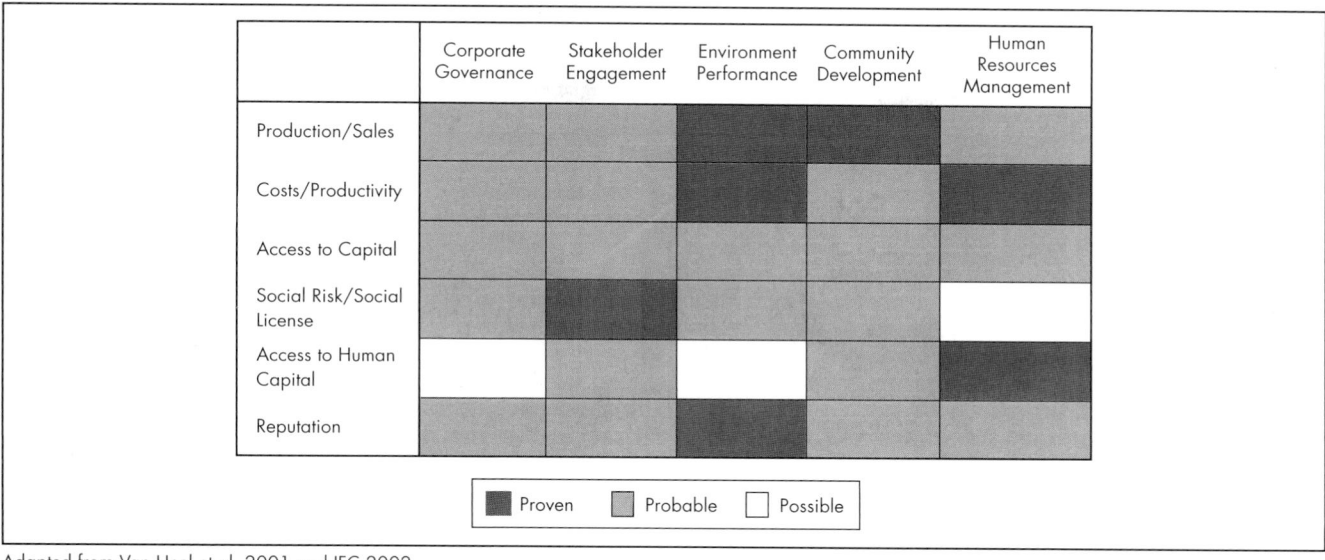

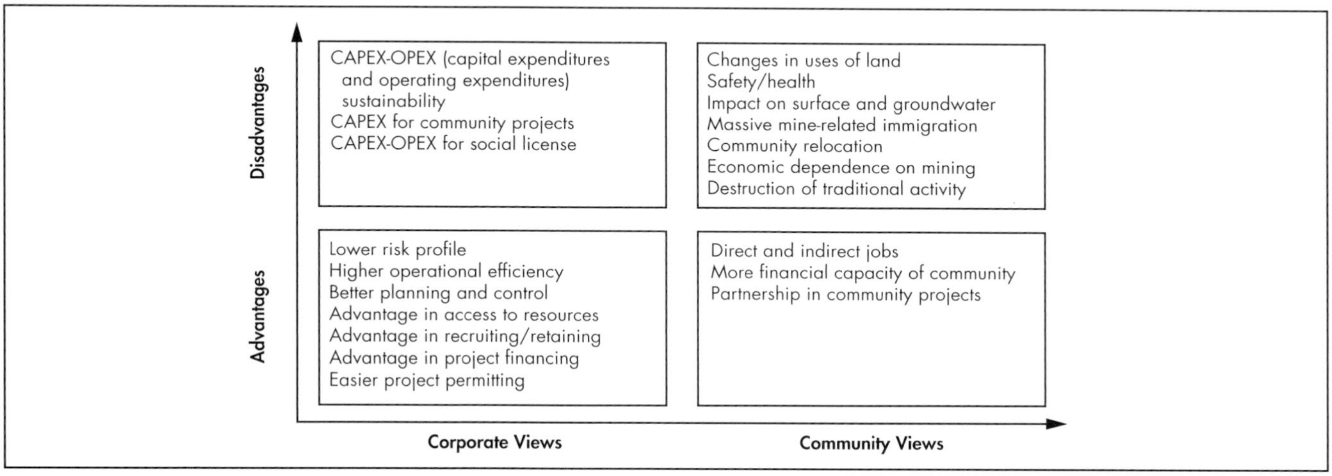

Source: Botin 2010

Figure 3 Corporate and community views of CSR and the social license

2002; Hodge 2004) as part of the Mining, Minerals and Sustainable Development Project (MMSD; IISD 2002). This framework was successfully applied to the experience of the Tahltan First Nation (British Columbia, Canada) with mining, where several gold mines have operated in the last 60 years (Tahltan First Nation 2003).

The Seven Questions framework aims to determine whether the mining project or operation generates a net positive contribution to sustainability over the longer term. Each broad question serves as a starting point for a management process, including project-specific objectives, indicators and metrics, dispute resolution mechanisms, reporting, and verification. The Seven Questions include the four major items that are commonly referred to in all sustainable development considerations—community, environment, economics, and governance—and other topics that are significant in a mining context. The following paragraphs describe some of the aspects associated with each of the questions/themes.

1. **Engagement: Are engagement processes in place and working effectively?** The engagement process includes the identification of stakeholders and active participation in discussions with them, leading to better implementation of local expectations in the project design, development, operations, and closure. Engagement is not a "one-way street" public relations exercise; rather it is the willingness to listen and respond with actions based on expectations. It can result in some loss of the power of independent decision making, but the outcomes will be better appreciated and accepted. Besides, stakeholder engagement is never a static process. Ongoing communications, reporting, verification, and other activities are necessary.
2. **People: Will people's well-being be maintained or improved?** This includes all the people who come into contact with the project, whether directly or indirectly. Workers' health and safety is regulated in most

jurisdictions; however, it is important to go beyond regulatory compliance and be proactive rather than reactive to all issues affecting the well-being of nearby communities, such as wind-blown dust from a tailings impoundment or potential impacts on the availability or the quality of drinking water. The Millennium Ecosystem Assessment (2005) lists the following components of human well-being: security, basic material for good life, health, good social relations, and freedom of choice and actions. It further identifies the role of ecosystem services as supporting (e.g., nutrient cycling), provisioning (e.g., food), regulating (e.g., climate), and cultural (e.g., recreation). These aspects are part of all human interaction with the earth, including mining activities.

3. **Environment: Is integrity of the environment assured over the longer term?** Extensive regulatory regimes have been developed to protect the environment affected by mining operations. However, going beyond the regulatory requirements by implementing concurrent reclamation and other activities meant to enhance the well-being of the environment should be "on the radar" of mining operations. Sometimes this includes a better understanding of the local ecosystem by funding research.
4. **Economy: Is the economic viability of the project or operation assured, and will the economy of the community and beyond be better off as a result?** All mining projects must make money; this is essential in the positive contribution of the mine to local and regional sustainability. The economic successes of a mine are shared through employment, local taxes, and so forth. Ongoing operational reviews to improve efficiencies at a mine are a big part of this theme. Many mining companies are also regular contributors to local activities such as schools, libraries, and universities. Contributions to sustainable causes must be a consideration; multigenerational thinking is an essential part of sustainable development.
5. **Traditional and nonmarket activities: Are traditional and nonmarket activities in the community and surrounding area accounted for in a way that is acceptable to the local people?** Understanding the traditions and customs of local indigenous people must take high priority at all mines. Having access to certain plants or places at a mine site may be essential to local indigenous people. There are also many other nonmarket issues associated with mines, such as volunteerism associated with local fire departments, emergency medical activities, churches, or other social activities such as search and rescue. Mining operations are usually very aware of the need to participate in such activities.
6. **Institutional arrangements and governance: Are rules, incentives, programs, and capacities in place to address project or operational consequences?** Mining companies operate in well-regulated environments, and regulations for land ownership, corporate financial management and reporting, taxes, labor relations, safety, and so forth are typically well developed in most jurisdictions. Regulatory compliance is the first order of business for all companies. However, many companies now have corporate policies and requirements for sustainable development, environment, health and safety, and many other aspects. Many of these policies go well beyond the governmental and regulatory frameworks.
7. **Synthesis and continuous learning: Does a full synthesis show that the net result will be positive or negative in the long term, and will there be periodic reassessments?** Ongoing review and continuous improvement is part of the culture of most mining operations. Regular synthesis of all aspects of the sustainability policies and applications at a mine site should be done to review whether the contribution to people and the environment will be net positive in the long term. This theme also includes the ongoing review of project and activity alternatives to improve operational efficiencies and overall project performance with respect to the six other themes.

Concept of Shared Value

Porter and Kramer (2011) studied the apparent contradiction of "shared value" and concluded that, in most cases, companies remained stuck in a wrong (philanthropic) approach to CSR, in which societal issues are not part of core business strategy. Porter and Kramer make a point that shared value is not about "sharing" the value already created by firms, but about expanding the total pool of economic and social value by reconceiving products and markets, redefining productivity in the value chain, and building supportive industry clusters at the company's locations. Each of these is part of the virtuous circle of shared value; improving value in one area gives rise to opportunities in the others. In this line of thinking, Porter and Kramer state that the concept of creating shared value supersedes CSR in guiding the investments of companies in their communities. Although CSR focuses mostly on reputation and has only a limited connection to the business, making it hard to justify and maintain over the long run, creating shared value is integral to a company's profitability and competitive position.

In recent years, several international platforms have been created to promote the shared value concepts. The Shared Value Initiative is a nonprofit community funded in 2012 with the sponsorship of several mineral resource companies. In the report *Extracting with Purpose* (Hidalgo et al. 2014), the authors present several case studies describing specific programs incorporating the application of the shared value concepts in mineral resources companies and geographies. However, they conclude that today, the creation of shared value is deterred by (1) inadequate organizational structures and behaviors; (2) incomplete measurement of cost and benefit; (3) low levels of motivation; and (4) the role of governments in incorporating shared value principles into concession agreements, becoming operating partners for shared value strategies, incentivizing shared value investments, and so on.

Also in 2012, Engineers Without Borders Canada created Mining Shared Value, a framework to promote increased engagement between community providers and the international mining sector (EWB Canada 2012). The framework ensures the relationships between local communities and global corporations are reciprocal and mutually beneficial, meaning both parties share in the value generated from mining.

BEST PRACTICES FOR SUSTAINABLE DEVELOPMENT

Referring to sustainable development in a mining context, the concept of best practices for sustainable development is particularly difficult to grasp. In fact, it refers to a management practice aiming to legitimize mining activity in the minds of local community and stakeholders. In this view, the general

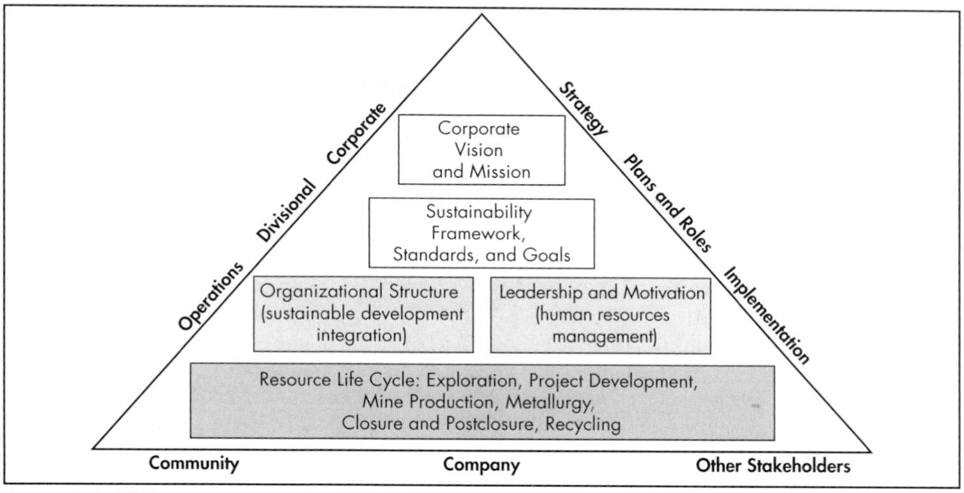

Source: Botin 2009

Figure 4 An integrated model for sustainable management of mineral resource companies

concept of best practice fits in the scope of SLO, where the desired output is stakeholders' acceptance rather than sustainable development.

In this line of thinking, a best practice for sustainable development must aim for the effective integration of sustainability into a company's strategy and structure from the chief executive officer (CEO) down to the lowest operational levels. In addition, integration should be efficient and therefore must lead to improved profitability and added value to shareholders.

Integrating Sustainability into Corporate Strategy

In the author's vision, integrating sustainability into corporate strategy refers to the management process of achieving top-down implementation of sustainable development principles and values at all levels in the organization. The author proposed the model in Figure 4 (Botin 2009), where sustainability is vertically integrated at three organizational levels (corporate, divisional, and operations) and three functional levels (strategy, planning, and implementation). Sustainability goals are integrated throughout the resource life cycle (exploration, project development, production, etc.). In addition, integration requires an organizational structure with adequate integration mechanisms and a business culture where sustainability is a high professional and business value.

This integration process, often difficult and challenging, requires an organizational structure with integration mechanisms, management roles, plans, and systems ensuring proper communication, coordination, and control. Integration roles are individual positions or ad hoc committees with accountability for the integration of sustainable development values and objectives. The integration plans and systems are the policies, standards, and management tools that are required to carry out sustainable management at the operations level.

As an example, a typical organizational structure for a large minerals corporation where several integration mechanisms are present is shown in Figure 5 (Botin 2009). This model would also be valid for medium and small companies, but in those cases, the divisional level would not exist.

At corporate level, three roles are key to sustainable management:

- The CEO role, developing a business vision, formulating challenging goals, and reinforcing motivation toward the values of ethics and sustainability.
- The sustainable development committee, assisting the board in overseeing the sustainable development framework and standards, and evaluating performance, public reporting, and external auditing.
- The vice president of sustainability, a key leader who typically is on the CEO's staff and is accountable for setting strategy, establishing goals, and integrating sustainability down and across the organization. Among other tasks, the vice president of sustainability assists the human resources department in developing sustainability values in employees.

At the divisional level, divisional managers' roles are related to the efficient integration of sustainability standards, the implementation of management systems for planning and control of sustainable development goals, and the reporting system. At operational levels, the site manager of sustainability reports to the operations general managers with overall accountability for the implementation of sustainability procedures at the operational level.

The sustainable development standards are a set of best practices, policies, and planning and control systems at all management levels. In most cases, sustainable development standards are designed in alignment with international frameworks (ICMM, Global Compact, etc.).

The sustainable development public reporting systems are often designed in alignment with international reporting standards. The most widely used minerals industry reporting standard is the Global Reporting Initiative. GRI, in cooperation with the ICMM, has published a public reporting guide (G3) specifically designed for the minerals industry (ICMM 2010).

External auditing and independent assurance systems are the last steps in integrating sustainable development in corporate strategy. Only publicly listed companies are legally required to issue an independently assured public report. However, many non-listed mining companies seek independent assurance considering the positive effect on reputation and credibility.

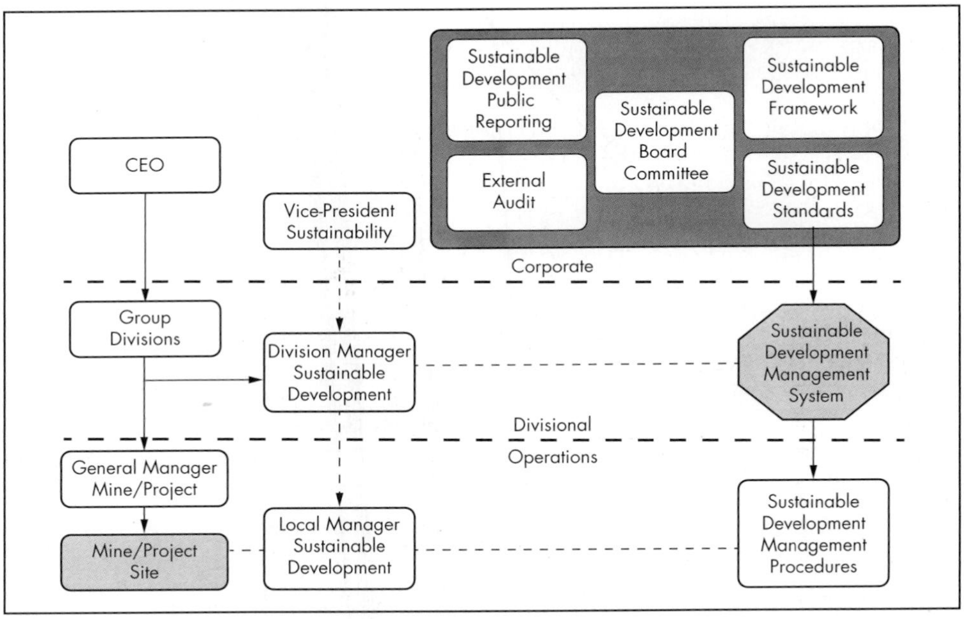

Source: Botin 2009

Figure 5 Process of integrating sustainability in the company structure

Regarding business culture, the integration of sustainability requires of business standards and codes in which sustainable development values are regarded as key to the professional success of company employees, so that sustainability performance may rely on personal commitment rather than compliance with policy and regulations. The integration of sustainability as a business value is critical since it allows for maximum decentralization of decision making in sustainability issues. In this context, a business culture based on personal conviction and commitment is critical to corporate reputation and how the company is perceived by local communities.

Best Practice Initiatives for Sustainable Mining

The *Cambridge Dictionary* (2018) defines *best practice* as "a working method or set of working methods that is officially accepted as being the best to use in a particular business or industry, usually described formally and in detail." Best practice may refer to a technique, a methodology, or a process (operating, management, manufacturing, etc.) and, in a broad sense, to a specific way of accomplishing a task. The only requirement is that it must be based on repeatable procedures that have proven to be more effective at delivering a particular outcome than any other. The desired outcome of a best practice may be effectiveness, efficiency, quality, low risk, or other outcome.

In line with *Berlin II: Guidelines for Mining and Sustainable Development* (UN 2002), a best practice for sustainable mining may be defined as "the management approach which efficiently integrates economic, environmental and social issues into operations, aiming to create long-term benefits to stakeholders, including shareholders, and to secure the support, cooperation, and trust of the local community in which the company operates."

Broad, Site-Based, Multi-Participant Initiatives

In 2002, Anglo American plc (United Kingdom), BHP Billiton (Australia), Rio Tinto (United Kingdom), and other global mining companies created the Global Mining Initiative to launch the MMSD project, an independent review of the key issues to understanding the relationship between sustainable development and mining. The MMSD's report led to the creation of International Institute for Environment and Development, which started workshops and expert-group meetings with more than 150 commissioned studies to produce the *Breaking New Ground* report (IIED 2002). The GMI and MMSD gave rise to the creation in 2001 of the ICMM, an international organization dedicated to a safe, fair, and sustainable mining industry (ICMM n.d.). Today, ICMM is composed of 23 mining and metals companies and more than 30 regional and commodities associations. In 2003, ICMM developed the 10 Principles of Sustainable Development, a framework for best practices on sustainable mining (ICMM 2015). Over the years, a series of position statements were developed to accompany and strengthen the 10 principles.

Between 2001 and 2003, the World Bank commissioned the Extractive Industries Review (World Bank 2004), a comprehensive assessment of its activities and its role in the extractive industries. The review concluded that the World Bank does have a role to play in the extractive sectors, but only if its activities promote sustainable development and poverty alleviation. The detailed findings of the review are available at the World Bank website (www.worldbank.org), and the World Bank's response can be viewed at the IFC website (www.ifc.org).

The Global Compact (GC), initiated in 2000, is a voluntary network of businesses and other organizations committing to adhere to appropriate corporate behavior in the areas of human rights, labor practices, the environment, and corruption reduction and prevention (UN 2004).

The Equator Principles (EP) are a voluntary set of guidelines for banks and other financial institutions to use in managing environmental and social issues that arise in lending to investment projects (Equator Principles Association 2013). The principles were developed in 2003 and revised in 2006

(EP-II) and 2013 (EP-III). Currently, 91 financial institutions in 37 countries have officially adopted the EP, covering more than 70% of international project finance debt in emerging markets.

The Intergovernmental Forum on Mining, Minerals, Metals and Sustainable Development (IGF) was created in 2002 as a voluntary initiative for national governments with an interest in advancing the priorities identified in the Johannesburg Plan of Implementation and—more recently—the United Nation's Sustainable Development Goals and Agenda 2030 (UN 2015). Today, the IGF supports 60 nations committed to leveraging mining for sustainable development (IGF 2017). The IISD has served as secretariat for the IGF since October 2015. The IGF's flagship work is the Mining Policy Framework, a best-practice framework that sets out objectives and includes guidance reports for governments on key sustainable development issues. The last IGF guidance report for governments, *Managing Artisanal and Small-Scale Mining*, was presented in early 2017.

Assurance, Auditing, and Reporting Initiatives

Besides the broad sustainability frameworks (ICMM, GC, etc.), many mining companies are joining broad-based assurance, auditing, and reporting systems, offering independent, third-party auditing and assurance to enable companies to obtain certification. Some examples follow.

The GRI was founded in 1998 as voluntary public reporting standard for sustainability reporting. Although GRI provides multisector reporting standards, it has published a *Mining and Metals Sector Supplement* (G3) in cooperation with the ICMM (2010). Today the GRI is probably the most widely used standard worldwide.

The AccountAbility AA1000 assurance standards were created in 2008 (AccountAbility 2008). AA1000 is a standard for assessing, attesting to, and strengthening the credibility and quality of organizations' sustainability reporting and their underlying processes, systems, and competencies. AccountAbility has also launched *The SIGMA Guidelines*, consisting of a set of guiding principles that help organizations understand sustainability and their contribution to it, and a management framework that integrates sustainability issues into core processes and mainstream decision making (AccountAbility 1999).

The International Standard for Integrated Sustainability offers a single certificate to verify a commitment to conserving natural resources and respecting social, environmental, and occupational safety issues (ISIS 2018).

Several international institutions for standardization have developed voluntary standards for auditable third-party verification on sustainability and social responsibility. The best-known standards are ISO 26000:2010 (social responsibility), ISO 14001:2015 (environmental management), SA8000:2014 (social accountability), and OHSAS 18001:2007 (occupational health and safety). Other well-known, issue-focused initiatives include the United Nations Voluntary Principles on Security and Human Rights and initiatives of the Alliance for Responsible Mining, the Public–Private Alliance for Responsible Minerals Trade, and the World Gold Council.

Issue-Focused Initiatives

In addition to the broad initiatives, there are many initiatives for best practice standards that focus on a single issue (operating stage, a management process, a technology, etc.). The following refer to those directly or indirectly related to the minerals industry.

In 2009, the Prospectors and Developers Association of Canada (PDAC 2018) launched E3 Plus, an online information resource focusing on minerals exploration. E3 Plus is designed to help companies improve their social, environmental, and health and safety performance in their exploration projects. More recently, PDAC has launched toolkits for environmental stewardship, health and safety, and social responsibility.

The Extractive Industries Transparency Initiative promotes full disclosure and independent verification of company payments and government revenues from oil, gas, and mining. Launched in 2003, the initiative aims to improve private and public governance in resource-rich nations by making it more difficult for companies and governments to undertake activities that the public at large finds inappropriate (EITI 2017).

The Responsive Minerals Initiative and its flagship, the Conflict-Free Smelter Program (CFSP), were founded in 2008 to help companies make informed choices about conflict minerals in their supply chains (RMI 2018). The CFSP uses an independent third-party audit of smelter/refinery management systems and sourcing practices to validate compliance with CFSP protocols and current global standards. The audit employs a risk-based approach to validate smelters' company-level management processes for responsible mineral procurement. Companies can then use this information to inform their sourcing choices.

The Responsible Jewellery Council is a not-for-profit, standards setting and certification organization. It has more than 1,000 member companies that span the jewelry supply chain from mine to retail. The council was founded to focus on supply chain assurance for minerals in jewelry, to develop a voluntary system of standards and site-based assurance. The RJC's Chain-of-Custody Certification for precious metals supports these initiatives and can be used as a tool to deliver broader member and stakeholder benefits (RJC 2018).

The International Cyanide Management Code, launched in 2002, is a voluntary initiative for the gold and silver mining industries and the producers and transporters of the cyanide used in gold and silver mining (ICMC 2018). The code focuses exclusively on the safe management of cyanide that is produced, transported, and used for the recovery of gold and silver, and on mill tailings and leach solutions.

The ISEAL Codes of Good Practice (ISEAL Alliance 2018) cover all steps in the standards and certification process, including standard-setting, impact evaluation, and assurance (certification and accreditation). Through membership in ISEAL, standards systems show a commitment to supporting a unified movement of sustainability standards.

The European Union, through the Institute for Prospective Technological Studies (IPTS 2018), has developed several best available techniques (BATs). The BATs are good practice standards of industrial processes with a focus on air pollution and compliance with the Industrial Emissions Directive, major environmental legislation that regulates approximately 50,000 installations in the European Union dealing with a wide range of industrial and agricultural activities.

The European Commission's guidance on the practical arrangements for the exchange of information under the Industrial Emissions Directive (2010/75/EU), including the collection of data, the drawing up of BAT reference documents, and their quality assurance as required by Article 13(3)(c) and (d) of the directive, can be found in the *Official Journal of the European Union*. BATs on mineral processing of ferrous metals, iron and steel production, nonferrous metals and cement, and lime and magnesium oxide, among others, are available online.

The International Atomic Energy Agency (IAEA) was created in 1957 under U.S. President Dwight Eisenhower's "Atoms for Peace" address to the United Nations on December 8, 1953. IAEA has developed many best practice guides and standards with focus on safety engineering and management. For detailed information on the resources offered, visit the IAEA website at www.iaea.org/. Also see the *Guidebook on Good Practice in the Management of Uranium Mining and Mill Operations and the Preparation for Their Closure* (IAEA 1998).

Commodity-Focused Initiatives

A group of international multi-sectorial voluntary initiatives focus on best-practice standards and certification of sustainability of the mine-to-retail value chain of specific commodities.

In the early 2000s, downstream jewelry companies joined with civil society organizations to promote the idea of responsible sourcing and certification for diamonds and then gold, known as the Kimberley Process. Today, the Kimberley Process unites administrations, civil societies, and industry in 80 countries in reducing the flow of conflict diamonds ("rough diamonds used to finance wars against governments") around the world (Kimberley Process 2018).

The Aluminium Stewardship Initiative (ASI), created in 2009, is a global, multi-stakeholder, nonprofit standards setting and certification organization (ASI 2018). ASI standards are applicable to all stages of aluminum production and transformation, specifically bauxite mining, alumina refining, primary aluminum production, semi-fabrication (rolling, extrusion, forging, and foundry), material conversion, and refining and remelting of recycled scrap, as well as material stewardship criteria relevant to downstream users of aluminum.

The Swiss Better Gold Association (SBGA), created in 2013, is a not-for-profit association created by Swiss players of the gold supply chain, from refiners to retailers (SBGA 2018). SBGA's aim is to create a simple market-driven mechanism that enables formalized gold mining entities to adopt more socially inclusive and better environmental practices.

The RJC is a not-for-profit standards setting and certification organization, with more than 1,000 member companies that span the jewelry supply chain from mine to retail (RJC 2018). RJC members commit to and are independently audited against the *Responsible Jewellery Council Code of Practices*, an international standard on responsible business practices for diamonds, gold, and platinum group metals (RJC 2013). The *Code of Practices* addresses human rights, labor rights, environmental impact, mining practices, product disclosure, and other important topics in the jewelry supply chain.

The Green Lead initiative, created in 2004, is a commodity-focused stewardship program aimed at contributing to broader and better sustainable development outcomes for the lead industry through management of the lead product life cycle. Organizations that adhere to these procedures will be Green Lead certified, as will the lead products they produce or handle.

Table 3 Community Development Toolkit structure

Section	Key Points
Introduction	Background, objectives, and target audience for the toolkit.
Mining and community development	Definition of community development, key principles for sustainable community development phases of the mining and metals project cycle, and stakeholder roles and responsibilities.
Community development tools • Relationships tools • Planning tools • Assessment tools • Management tools • Monitoring and evaluation tools	Twenty practical tools for community development supported by step-by-step guidance to assist in using them.
Glossary and references	Glossary and a list of referenced sources.

Source: ICMM 2012

Sustainable Development Toolkits and Handbooks

A toolkit may be defined as a structured set of tools (procedures, guidelines, criteria, etc.) prepared for a specific purpose to ensure a desired or required result or prevent oversights. A handbook is a collection of reference works, instructions, case studies, and other material that is intended to provide ready reference on a subject or area of knowledge. Unlike toolkits, handbooks are not oriented to any specific objective, other than providing ready reference.

In the context of mining, these types of documents are provided by many sources, but the three leading sources are (1) broad, multi-participant initiatives focusing on the minerals industry, mainly those referred to earlier in this chapter; and (2) government sources from leading mining countries, especially the United States, Canada, and Australia.

The intent of this chapter is not to provide a comprehensive reference to all available toolkits and handbooks on the subject but to give an example of each type of document from two well-established, easy-access sources: (1) ICMM and (2) handbooks on leading practices on sustainable mining from the Australian Government (n.d.[a]).

Community Development Toolkit. Developed by the ICMM, this toolkit consists of a set of 20 tools aimed at a constructive engagement between mining companies and communities (ICMM 2012). The objective is to build capacity and improve opportunities for the sustainable development of communities around mining and metals operations.

The toolkit is structured in five groups depending on the scope and objective (see Table 3):

1. Relationship tools (tools 1–5) include guidelines and detailed methodology for the characterization of all stakeholders, consultation, and community engagement.
2. Planning tools (tools 6–10) are intended for the design, resource evaluation, and implementation of development programs and policies.
3. Assessment tools (tools 11–13) contain comprehensive information and methodologies for the identification, characterization, and evaluation of impacts (positive and negative) generated by the mining activity on the nearby communities. Also, they provide guidance on mitigation methods and potential opportunities for partnership with stakeholders.

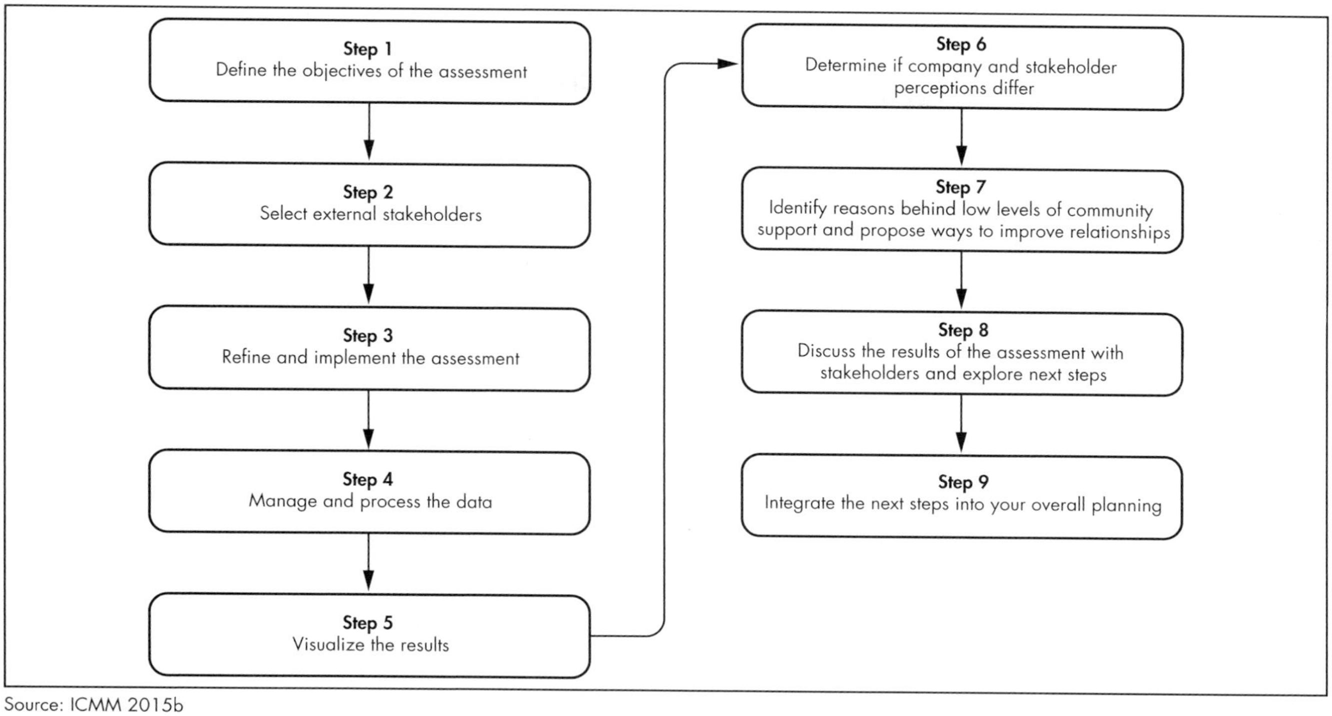

Source: ICMM 2015b

Figure 6 Understanding Company–Community Relations Toolkit

4. Management tools (tools 14–18) are designed for the management of development programs.
5. Monitoring and evaluation tools (tools 19 and 20) are included for measuring progress toward program goals.

Understanding Company–Community Relations Toolkit. This ICMM toolkit was developed in 2015 to address a fundamental drawback in most company–community engagement processes in the past: the lack of a positive and resilient relationships with host communities (ICMM 2015b). The toolkit focuses on three key objectives:

1. To help companies understand the factors that influence community support and measure the level of community support for a project or operation.
2. To provide a tool to visualize the levels of community support that different stakeholders or stakeholder groups have for a project or operation.
3. To offer practical guidance on how this tool can be used to monitor and strengthen community support and, ultimately, community–company relationships.

The overall flow of the activities in this toolkit is shown in Figure 6. The process is stepwise and comprises two sections. The first section (steps 1–5) measures the current level of community support. The second section (steps 6–9) identifies and evaluates the factors influencing the current level of community support, and how it can be improved. The overall objective of the process is understanding the community support mechanisms and obtaining the information needed to plan for a continuous improvement process based on the plan-do-check-act cycle.

Handbooks on leading practice sustainable development in mining. The Leading Practice Sustainable Development Program (LPSDP), launched in 2006, is a program for the mining industry that promotes sustainable development and industry self-regulation. The program is administered by the Department of Industry, Innovation and Science of the Australian Government (n.d.[a]). LPSDP has developed 17 handbooks to address the key issues affecting sustainable development (Australian Government n.d.[b]). These handbooks are directed to mining and mineral processing operators, communities, and regulators and contain many methodologies, data, and case studies to assist all sectors of the mining industry in implementing good practice standards within and beyond the requirements set by legislation.

IMPACT ON COMMUNITIES

Breaking New Ground defines a mining community as a community whose population is affected by nearby mining operations (IIED 2002). Regarding community impacts, one should consider three different categories of mining communities:

1. Occupational communities, which are generally villages whose population relies extensively on local mining for all or most of their income from mining.
2. Residential communities, preexisting or built as a result of mining, where households or families live within the geographical area affected by mining impacts (economic, social, or environmental), in close proximity to or far away from the mine.
3. Indigenous communities, a special category for households or families with an ancient and cultural attachment to the land where mining occurs or has an impact.

This classification is relevant with regard to impact mitigation priorities. In occupational communities, the leading social consideration is that people must have the means to survive and prosper, either in the same place or elsewhere, when mining ceases. In residential communities, however, the leading

Table 4 Environmental and social concerns during the production stage

Aspect	Ore Extraction and Processing Concerns	Comments
Land disturbance	Involves impact on landscape and groundwater	Larger in open pit than underground mining
Waste rock disposal	Involves trucking, runoff, and leachate management; dust and aesthetic considerations	Larger in open pit than underground mining
Tailings	Involves handling and storage of large volumes of mill tailings	Long-term risk control of stability and effluents
Acid drainage	May be associated with both mine and waste rock areas	Long-term risk control
Reclamation	Major concerns for mine, dumps, and tailings storage	Long-term risk control
Noise/vibrations/dust	Main concerns are for haul-truck traffic, waste dumps, and blasting	Potential impacts on community well-being
Water effluents	Ammonium from blasting, sediment loading, metal dissolution, low pH	Potential impacts on community well-being

Adapted from Environment Canada 2009

consideration is minimizing the environmental footprint of mining. Indigenous people present a special case of community. They have specific social needs and very distinct systems of decision making, social and political institutions, and systems of wealth generation and distribution, often closely associated with land and natural resources.

These categories are not mutually exclusive. Indigenous communities may work in a mine and therefore be occupational communities too, whereas faraway communities may be outside the geographical area affected by mining impacts but host fly-in, fly-out migrant labor.

Community Impacts Through the Mining Life Cycle

Mining's impacts on community are complex and are variable throughout the different stages of the mining life cycle. During exploration, social and environmental impacts are limited and have a positive effect on the local economy. During project evaluation and feasibility stages, a significant amount of fieldwork and studies are performed, including the social and environmental baseline studies and project planning and engineering. In this stage, company–community interaction and information sharing becomes intense and potentially conflicting. The construction stage is comparatively brief but may have great short-term impacts with long-term implications. It may bring a boom in jobs but can also cause considerable physical and social upheaval, opening up remote areas through the development of infrastructure and stimulating migration to the area. The production stage has the longest-term positive impacts (e.g., income and infrastructure) but also negative and often unintended repercussions. The environmental impacts in this stage are mainly associated with day-to-day operations. Shocks and vibrations as a result of blasting in connection with mining can lead to noise, dust, and collapse of structures in surrounding inhabited areas. Unless fully controlled, mining operations also may be a source of social impacts and potential conflict (MINEO Consortium 2000). The impact of the closure phase depends largely on the degree of forward planning and the available means to sustain benefits, such as institutional capacity and financial resources.

During the exploration stage, activities are relatively nonintrusive and have limited, short-term impacts on the community (Government of Canada 2017). Access is seldom intensive and supports a few people for short periods of time. In most cases, environmental concerns are associated with noise from airborne surveys (impacts on wildlife), land geophysics, and earthmoving for diamond drilling. Exploration activities may require access roads, site preparation, transportation and storage of fuels, operating materials and core samples, the establishment of campsites for exploration crews, and other temporary infrastructure.

The construction stage can have potentially important environmental implications. The activities for site preparation (clearing, stripping, and grading) and infrastructure construction (plants and plant services) may have significant impacts on air quality, water quality, aquatic ecosystems, soil quality, and terrestrial ecosystems. For the establishment of mine workings (mine construction phase), the main concerns are the management of waste rock and mine water, and the generation of dust, noise, and vibration from drilling and blasting activities.

During the production stage, the primary concerns for social impacts are the disposal of waste rock and mill tailings and the management of mine and mill effluents (Table 4). In addition, ore extraction can generate impacts on community from dust, noise, and vibration, which are mainly the result of drilling and blasting, and ore and waste haulage. Potential impacts are also associated with the risk of spills and accidents, which could result from tailings dam failures and the release of chemicals used in ore processing.

Mine reclamation may overlap with operations. Regarding open pit mining, large areas of land may be disturbed through ore extraction and other mining activities. Disturbed areas are susceptible to erosion caused by both wind and water, which can lead to dust generation and water quality problems. Therefore, landscape rehabilitation, including reshaping and restructuring of the landscape and erosion control measures, may occur during the operations stages. These early reclamation works may also include the restoring of stockpiled soils in preparation for revegetation.

The mine closure stage should mainly focus on safety and the reclamation of land, water, and ecosystems to ensure that the mine site is self-sustaining and to prevent or minimize long-term impacts (Table 5). Land reclamation aims to restore waste dumps and tailings dams to achieve long-term physical and chemical stability, and also to rehabilitate disturbed areas to their natural or other acceptable use. The safety objectives include the capping of the shaft, raises, declines, and adits to ensure public and wildlife safety.

Community Perception of Sustainability

The community perception of a mining and metallurgical project depends on the type of community and how the community generally perceives mining projects. Community perception can change through time because of poor community

Table 5 Main components to be addressed in the mine closure plan

Components	Aspects to Be Addressed
Mining	Sealing of shafts, tunnels, and raises to prevent unauthorized access
	Slope stability, security access and fencing, wildlife entrapment
	Groundwater and rainwater, drainage into and from pit, ice plugs
Processing	Removal of infrastructure and disposal of scrap and building material
	Cleanup of workshops, fuel and reagent, site reprofiling and revegetation
Waste dumps and tailings dams	Effluents from seepage of surface and groundwater, surface water, and rainfall discharge
	Dump and tailings dam stability, long-term physical and chemical stability and changes
	Visual impact, dust generation, with special consideration to EM radiating ore
	Access and security, wildlife entrapment
Water management facilities	Restoration or removal of water reservoirs, settling ponds, culverts, pipelines, and spillways no longer needed; maintenance of permanent water facilities.
	Surface drainage of the site and discharge of drainage waters
Other infrastructure and facilities	Disposal or removal from site of hazardous wastes
	Removal of sewage treatment plant and disposal/stability of treatment sludges
	Removal of roads, power and water supply
	Access security and prevention of illegal dumping and groundwater contamination

Adapted from Environment Canada 2009

engagement, unintended events such as accidents and social conflict, or changes external to the mining operation. In the following sections, community perceptions are analyzed from the three perspectives of sustainable development (economic, environmental, and social).

Economic Perspective

From an economic perspective, communities can expect to receive compensation and substantial flows of revenue from mining, which can transform the economic and social basis of the communities and contribute to sustainable development at the community level. In addition, mining often provides local communities with new jobs and better salaries. In some regions, mining provides the bulk of job opportunities. As an example, the Grasberg copper and gold mine in West Papua, Indonesia, which employs 14,000 people, generated 75,000 indirect jobs as a result of Freeport-McMoRan's mining activities (Freeport-McMoRan 2017). Elsewhere, with the exception of the construction phase, many mines no longer generate significant numbers of local jobs.

Another important source of economic benefits to communities is the input services provided to mining operations. Companies are increasingly required to assist local business development, outsource services, and give preference to local businesses. However, increased demand may cause the prices of goods and services to rise locally. Moreover, the concentration of economic activity centered around the mine often increases the community's dependence on the mining operation, making it vulnerable to downsizing or other changes and exacerbating the power imbalance. On the other hand, since the company may also depend on the community for employees and services, a well-organized community can potentially make numerous demands on the company.

Environmental Perspective

The environmental problems related to mining and extractive metallurgy have been known since ancient times. In *De Re Metallica*, first published in 1556, German mineralogist Georgius Agricola, who traveled extensively in Saxony and its neighborhoods to visit mines and smelters, referred to mining and smelting of ores generating poisonous fumes at the workplace (Agricola 1950).

In this subsection, the main environmental problems of mining from the community's perspective are summarized. For a more detailed description of environmental impacts of mining projects, see the *Guidebook for Evaluating Mining Project EIAs* (ELAW 2010).

Problems related to water management and impacts on aquatic ecosystems. The effective management of mining water and effluents constitutes a primary environmental problem and community concern at most mines, especially in metal and coal mining and metallurgy. The main water management concerns include (1) segregating clean and contaminated water flows and recycling process water to minimize water usage and effluent generation and (2) addressing surface and groundwater contamination from seepage from mine rock and tailings containment structures. In this context, community concerns relate to water needs and the following environmental impacts:

- Most metallic ores and waste rocks contain sulfide minerals which, in the presence of water and oxygen, induce "acidic drainage," an oxidation process that generates metal-laden effluents of low pH. Acidic drainage results in the release of metals in effluents and, if not carefully managed, can significantly affect aquatic ecosystems and long-term liability for effluent treatment by mine owners/operators (Earthworks n.d.). Potentially, defective mine water and inadequate effluent management may also affect wildlife, such as plants, because the release of airborne particulate matter and seepage of groundwater or surface water may transport metals to areas adjacent to mine sites.
- Acidic drainage collected from waste dump mines is commonly treated with lime. A by-product of this treatment is sludge, a residue that may contain a wide range of metals. The volume of sludge produced may be large and is generally disposed of on-site or sent to smelters for recycling. There are uncertainties about the long-term chemical stability of many sludges and concerns that sludge could become an additional source of the release of metals.
- Flotation and other ore separation processes require high pH (alkaline), thus requiring the addition of lime and flotation reagents to ore pulps. As a result, alkaline effluents are produced and, if the water balance is positive, a well-controlled pH adjustment of effluents is necessary prior to discharge.
- Cyanide is used in the recovery of gold by heap leaching and other gold recovery processes. As a result, wastewater from processes using cyanide may contain cyanide and cyanide compounds. Cyanide is also used as a reagent

in flotation separation circuits. Thus, cyanide compounds may also occur from the recycling of process water from tailings ponds in some base metal flotation mills.

- Ammonia from ammonium nitrate and fuel oil (ANFO) and other blasting agents is soluble in water. Ammonium nitrate spilled in preparation for blasting or left over after a blast may reach aquatic ecosystems. In addition, ammonia may occur as a decomposition product from cyanide wastes.
- Wastewater may contain suspended solids ranging from colloidal (non-settleable) to settleable materials. Depending on its composition, settling of these solids may result in the contamination of sediments with metals and other contaminants, which may cause a range of problems in aquatic environments, including impeded oxygen intake by fish and reduced light availability for aquatic plants.
- Thiosalts are sulfur oxide compounds that result from partial oxidation during the milling, grinding, and flotation of some sulfide ores under high-pH conditions. Thiosalts are a concern because they can oxidize in water to form sulfuric acid, which lowers the pH of the receiving water and affects metal mobility, and significantly affects resident aquatic organisms.

Problems related to waste rock and tailings disposal. Tailings disposal has become a serious matter for the mineral industry. During the beneficiation of valuable metals and industrial minerals from their ores, large volumes of waste materials or tailings may be produced and these tailings may be harmful to the environment. It is now necessary to design and build tailings dams to store these waste materials for the modern mining industry. The possible social and environmental impacts related to mine tailings disposal are highlighted in the following list.

- Groundwater seepage is of concern for both waste rock piles and tailings management facilities. Seepage of rainwater through these structures could result in the release of contaminants through a permeable foundation layer or generate instability (IFC/World Bank 2007).
- Air quality impacts associated with the generation of dust are of concern. Dust may result from mining operations (e.g., drilling and blasting, crushing, loading, hauling, and transferring by conveyor). Also, wind-blown particulate matter may be released at open pits, waste rock piles, tailings management facilities, and stockpiles.
- Failure of dams or other containment structures for tailings can lead to severe environmental impacts and significant risks to human health. The risk of failure of these containment facilities (Kemp et al. 2016) is difficult to assess, mainly because they are built in several phases over many years and use waste materials from the mining operations, with limited or nonexistent quality assurance procedures for construction and maintenance. In addition, earthquakes can destabilize these structures, and catastrophic storms have the potential to flood tailings ponds. Regarding the risks related to heavy rains potentially eroding or destroying these structures, Environment Australia states that "It is self-evident that a Sediment and Erosion Control Plan is a fundamental component of a Mine Site Water Management Plan" (Environment Australia 2002).

Table 6 Main environmental concerns in extractive metallurgy

Industry	Concerns
Iron and steelmaking	Gases in coke production, slags, blast furnace, cyanides, electric furnace dust, pickle solution
Aluminum industry	Mercury, red mud, fluorine compounds, toxic organic compounds, cyanides
Sulfide ores: copper, lead, zinc, and nickel	SO_2, mercury, selenium, arsenic
Hydrometallurgical processes: gold, silver, copper, and zinc	Arsine, phosphine, cyanides
Radioactive ores: uranium and thorium	Radon gas (radioactive)
Industrial minerals: coal, phosphate rock, ilmenite, asbestos	Sulfur, ash, trace metals, nitrogen oxides, phosphogypsum, waste acid, toxicity of fibers, tailings

Adapted from Habashi 2012

Problems related to land management. Mining projects are characterized by extensive land requirements for the mine, plants, and the storage of waste rock and tailings, and the additional land development required to host employees and contractors. This large land requirement is of concern for hosting communities because it may affect migration routes, breeding grounds, or nesting areas and affect species that have special cultural significance to local communities. In addition, cattle and wild mammals may be dislocated from mine sites, which could have economic and cultural impacts.

Problems related to extractive metallurgy operations. Extractive metallurgy operations may take place at mine sites as part of mining operations in occupational communities or at custom metallurgical facilities treating concentrate for several mines and located near residential communities. In both cases, the communities near extractive metallurgy operations are concerned about environmental problems related to air emissions and the disposal of solid and fluid waste products (Habashi 2012), as summarized in Table 6.

Regarding air emissions, the smelting of sulfide ores results in the emission of sulfur dioxide gas (SO_2), which, once in the atmosphere, mixes with liquid and solid particles into raindrops and snowflakes that are carried back to the ground. Acidic rain increases the acidity of soils, streams, and lakes, harming the health of vegetation, fish, and wildlife populations. Inhaling SO_2 emissions causes eye and respiratory irritation in people at the plant and in the nearby community. Emissions from smelters also may contain arsenic, cadmium, and mercury in oxide form, which at concentrations above certain limits may be toxic.

A well-documented case history of pollution from extractive metallurgy in Sudbury (Ontario, Canada) is described by Winterhalder (1996). During the 1960s, after decades of smelting in the area, air pollution had generated more than 100,000 ha of barren or semibarren lands and acidified lakes. The situation changed in the late 1960s and 1970s when, under stricter environmental regulations (Buhr 1998), an extensive remediation plan was executed. The plan included the construction of a 380-m super stack and a regrinding program for regenerating 3,435 ha of barren land and planting more than nine million trees.

A lesser environmental risk is that related to the slags. In general, extractive metallurgy slags contain impurities of toxic elements, such as Cu, Pb, Zn, Co, Cr, Ni, As, Cd, and so forth. These slags are recycled through the cement industry or disposed of as landfill. Before disposal, they are chemically stabilized, but environmental liability cannot always be excluded (Barcza et al. 1993).

Treatment of ores by hydrometallurgical methods produces residues and waste solutions. Solid residues filtered off aqueous solutions contain soluble compounds with the potential to contaminate surface waters or groundwater. Consequently, direct disposal into tailing ponds may be hazardous unless they are properly constructed (Reuter et al. 2004). Waste solutions containing toxic reagents must be treated before being discharged in streams (Habashi 2001).

Electrometallurgical processes are used in the aluminum industry and the electrorefining of copper and zinc. The processing of aluminum generates large amounts of gases and dust. Copper electrorefining and zinc electrowinning also cause pollution, but in lesser amounts since these processes have undergone intensive improvement.

Social Perspective

It is difficult to separate the economic impacts of mining operations from the social impacts. Mining helps a community become prosperous, and may tackle social ills such as malnutrition, illiteracy, and poor health. On the other hand, mining activities may cause economic hardship by polluting rivers and damaging fish stocks, for instance, or by appropriating grazing land and forestry resources.

Although mining generates a positive economic impact on the community, it may exacerbate existing social problems or create new ones, such as social inequality resulting from changes in the traditional income distribution and social status. For example, mining districts in Peru have larger income per capita and lower poverty rates than other districts, but consumption inequality within mining districts is higher than in comparable nonproducing districts (Loayza and Rigolini 2016). Also, a social audit of the Grasberg mine showed that the worsening inequalities in income distribution favor young adults, modifying their position and prestige vis-à-vis their elders and affecting traditional social structures (McMahon and Remy 2001). If people in a community perceive social inequality, this can result in social tension and even violent conflict within the community or between the community and the mining company or government.

Physical displacement, relocation, and resettlement associated with large-scale mineral development are widely acknowledged as posing enormous social risk (Owen and Kemp 2015). Communities may lose their land, and thus their livelihoods, disrupting community institutions and power relations. Entire communities may be forced to shift to purpose-built settlements, or into areas without adequate resources. They may be left near the mine, where they may bear the brunt of pollution and contamination. Involuntary resettlement can be particularly disastrous for indigenous communities with strong cultural and spiritual ties to the lands, who may find it difficult to survive when these are broken (IIED 2002). A paradigmatic example of the massive relocations near the Chuquicamata (Chile) and Butte (Montana, United States) mines is described by John Hillman (1999).

One of the most significant impacts of mining activity is the migration of people into a mine area, particularly where the mine represents the most important economic activity. For example, at the Grasberg mine, the local population increased from fewer than 1,000 in 1973 to between 100,000 and 110,000 in 1999 (Freeport-McMoRan 2017). Sudden increases in population can also lead to pressures on land, water, and other resources as well as introduce problems of sanitation and waste disposal. In San Ramon in Bolivia, for instance, migration led to an increase in land and housing prices and the saturation of public services, including schools.

The construction of a large mine can produce significant infrastructure improvements. Most mining operations of any size are served by airstrips, roads, water supplies, sanitation systems, and electricity. With some planning and a willingness to consult with the community, these can bring lasting benefits at little or no added cost. And the development of infrastructure may facilitate other forms of economic activity, such as tourism.

Health services typically increase markedly with the advent of mine development as companies develop facilities for employees and their families, but these may not necessarily translate into overall improvements in community health if the facilities are not made available to the broader community or if the introduction of new diseases and health risks associated with the mine are taken into account. Relatively isolated communities, including indigenous peoples, may be particularly vulnerable to diseases brought by miners, such as influenza, malaria, and HIV/AIDS (human immunodeficiency virus/acquired immunodeficiency syndrome).

A key issue is sustaining health services and benefits in the community after mine closure, which might depend on the approach taken during the life of the mine. Training local health paraprofessionals, for example, might provide higher benefits in the long term than importing contract doctors. Also, some of the detrimental health effects of mining on communities (silicosis, asbestosis, etc.) may surface years after mining has ceased.

Mining companies are often involved in the provision of educational infrastructure or educational opportunities through scholarships. These can come in the form of corporate support or through trust funds or foundations, such as the Rio Tinto Aboriginal Foundation in Australia. However, because of a highly competitive market and the need to improve efficiency, there is a tendency to move away from providing services such as housing, schools, and health care for mine workers and their families, except in remote regions.

The social benefits of minerals development must be seen in the context of the many problems associated with mining-induced social change. Large mine development may be accompanied by the widespread availability and consumption of alcohol, an increase in gambling, the introduction of or increase in prostitution, and a widely perceived breakdown of law and order. Male-dominated mining camps and communities, such as those found in South Africa, often attract prostitutes and may lead to high levels of sexually transmitted diseases.

Mining operations are often perceived as widening gender disparities within communities. Women benefit from mining from better living standards and services such as water and electricity in occupational communities. There are also benefits from improved nutrition and access to medical services. However, women tend to bear a disproportionate share of the social costs and receive an inadequate share of the benefits. In occupational communities, women are more often spouses of mine employees, and are therefore passive recipients of benefits. There are few job opportunities for women in mining communities, and this lack of job opportunities is aggravated by other limiting factors, including the relative isolation of many mine sites, the absence of local markets to support other economic activities, a lack of higher education institutions, and so forth. Increasing female employment at mine sites would not only bring direct benefits from increasing their incomes but would also help to mitigate many of the social ills, such as alcoholism and prostitution, found in some occupational communities. Clearly, strategies need to be developed for integrating women into the sector.

Conflict in and around mining operations usually stems from poor governance. It is also more likely to take place where the distribution of mineral revenues and benefits are nonexistent or perceived to be unjust, or where the community opposes and actively resists any mining activity on their land. Companies or even central governments may have little understanding of the customs and traditions of those living in and around the mines, and they may therefore be insensitive in their dealings with local communities, potentially fueling further conflict. It has been suggested that in many cases of conflict involving local communities and mining interests, radical environmental NGOs (often headquartered in a foreign country) have been involved whose primary aim is to contribute to tension in the community through misinformation and fearmongering.

In some cases, human rights abuses by police or security forces acting in the interests of the company may occur. Several complaints recently brought to the Community Aid Abroad Mining Ombudsman concerned Australian companies operating in various developing countries that had been removing people, sometimes violently, from their land or homes (Oxfam Community Aid Abroad 2004). In some instances, their houses, mining equipment, or other assets have been destroyed.

Mining activities can cause considerable disruption to local cultures, especially when the operations occur, as is increasingly the case in areas occupied by indigenous people who have had little contact with the outside world. Although some of the "Western" values imported by the mining company and its workers may be admirable or suitable, this is by no means always the case. Cultural clashes may occur, with deep-reaching, destabilizing effects on traditional ways of life.

Some local cultural traditions and practices decline, or their significance alters, which may be particularly lamented by older members of communities. At some locations, companies may deliberately intervene and try to support cultural institutions or events. A related cultural issue is that of geographic boundaries between groups. Borders that may have been fluid may become more precise and fixed as they become critical to obtaining benefits from a development. For example, a group's traditional rights to hunt in an area may not be recognized in the distribution of benefits from a mine if there are groups with a more complete set of rights (such as residence) to the area.

SUMMARY

Community and social issues are of paramount importance for sustainable mining, as acknowledged by the United Nations in *Berlin II: Guidelines for Mining and Sustainable Development* (UN 2002). In this report, sustainable mining is defined as "developing mining projects in a socially acceptable manner and ensuring that its benefits are employed in a way that will survive long after the mine is closed."

Three mining-specific factors explain why community and social issues are key for a mining company:

1. When an ore body is discovered, the geographical, social, and political environment become locked for the mine life.
2. Mining interaction with its social environment is strong, and sometimes communities rely extensively on mining for all or most of their income.
3. Mining activity generates employment and other social benefits for the community, but mining of an ore body is not sustainable in time and therefore neither are its socioeconomic benefits unless invested in sustainable social fabric.

Companies interact with communities through CSR actions; these are voluntary contributions beyond compliance with laws and regulations. Although CSR practices are voluntary, market pressures and stakeholders' demands are gradually leading to a situation where noncompliance with international CSR frameworks may bear an increasing social and economic risk. In fact, there is a growing evidence of the link between environmental and social performance and company financial performance, which eventually may lead to the "business case of sustainability," where the potential benefits in terms of higher operational efficiency, lower risk profile, and better reputation would balance the costs of CSR. In this context, many international institutions and multiparticipant initiatives (ICMM, GC, and others) have developed sustainable development frameworks, standards, and toolkits for voluntary adherence and assistance, as well as provided independent third-party auditing and assurance.

REFERENCES

2010/75/EU. Directive 2010/75/EU of the European Parliament and of the Council on industrial emissions (integrated pollution prevention and control). In *Official Journal of the European Union*. https://www.ecolex.org/details/legislation/directive-201075eu-of-the-european-parliament-and-of-the-council-on-industrial-emissions-integrated-pollution-prevention-and-control-lex-faoc109066/. Accessed July 2018.

AccountAbility. 1999. *The SIGMA Guideline: Putting Sustainable Development into Practice—A Guide for Organisations*. www.projectsigma.co.uk/Guidelines/SigmaGuidelines.pdf. Accessed June 2018.

AccountAbility. 2008. *AA1000 AccountAbility Principles Standard 2008*. London: AccountAbility UK.

Agricola, G. 1950. *De Re Metallica*. Translated by H. Hoover and L.H. Hoover. New York: Courier.

ASI (Aluminium Stewardship Initiative). 2018. ASI performance standard. https://aluminium-stewardship.org/asi-standards/asi-performance-standard/

Accessed January 2018.

Australian Government Department of Industry, Innovation and Science. n.d.(a) Leading Practice Sustainable Development Program for the Mining Industry. https://industry.gov.au/resource/Programs/LPSD/Pages/default.aspx. Accessed March 2018.

Australian Government Department of Industry, Innovation and Science. n.d.(b) Resources: handbooks. https://industry.gov.au/resource/Programs/LPSD/Pages/LPSDhandbooks.aspx. Accessed May 2018.

Azinger, K.L. 1998. Methodology for developing a stakeholder-based external affairs strategy. *CIM Bull.* 91(1019):87–93.

Barcza, N.A., Hutton, C.J., Freeman, M.J., and Shaw, F. 1993. The treatment of metallurgical wastes using the Enviroplas process. In *Extraction and Processing for the Treatment and Minimization of Wastes*. Edited by J. Hager, B. Hansen, W. Imrie, J. Pusatori, and V. Ramachadran. Warrendale, PA: The Minerals, Metals & Materials Society. pp. 941–962.

Bonini, S., and Gorner, S. 2012. *The Business Case for Sustainability: Putting It into Practice*. New York: McKinsey and Company.

Botin, J.A. 2009. Chapters 1, 5, and 8. In *Sustainable Management of Mining Operations*. Edited by J.A. Botin. Littleton, CO: SME.

Botin, J.A. 2010. Integrating sustainability down to the operational levels of a mining company. *Dyna* 77(161):43–49.

Buhr, N. 1998. Environmental performance, legislation and annual report disclosure: The case of acid rain and Falconbridge. *Account. Audit Account. J.* 11(2):163–190.

Cambridge Dictionary. 2018. Best practice. In *Cambridge Dictionary.* New York: Cambridge University Press. https://dictionary.cambridge.org/us/dictionary/english/best-practice. Accessed April 2018.

CED (Committee for Economic Development). 1971. *Social Responsibilities of Business Corporations*. New York: CED.

Cooney, J. 1997. Presentation given at Roundtable on Mining: The Next 25 Years. Washington, DC, March 19–20.

Cragg, A.W. 1998. Sustainable development and mining: Opportunity or threat to the industry? *CIM Bull.* 91(1023)45–50.

Cragg, W., and Greenbaum, A. 2002. Reasoning about responsibilities: Mining company managers on what stakeholders are owed. *J. Bus. Ethics* 39(3):319–335.

Davis, K. 1960. Can business afford to ignore social responsibilities? *Calif. Manage. Rev.* 2(Spring)70–76.

Davis, K. 1967. Understanding the social responsibility puzzle: What does the businessman owe to society? *Bus. Horiz.* 10(Winter):45–50.

Davis, K. 1973. The case for and against business assumption of social responsibilities. *Acad. Manage J.* 16: 312–22.

Earthworks. n.d. Earthworks Fact Sheet: Hardrock Mining and Acid Mine Drainage. https://earthworks.org/publications/fact_sheet_acid_mine_drainage/. Accessed January 2018.

EITI (Extractive Industries Transparency Initiative). 2017. Home page. www.eiti.org. Accessed September 2017.

ELAW (Environmental Law Alliance Worldwide). 2010. *Guidebook for Evaluating Mining Project EIAs*. https://www.elaw.org/mining-eia-guidebook. Accessed March 2018.

Environment Australia. 2002. *Overview of Best Practice Environmental Management in Mining*. Canberra, Australian Capital Territory: Environment Australia.

Environment Canada. 2009. *Environmental Code of Practice for Metal Mines*. Ottawa, ON: Environment Canada.

Equator Principles Association. 2013. *The Equator Principles III*. www.equator-principles.com/. Accessed March 2018.

EWB (Engineers Without Borders) Canada. 2012. Mining shared value. https://www.ewb.ca/en/venture/mining-shared-value/. Accessed July 2018.

Freeman, R.E. 1984. *Strategic Management: A Stakeholder Approach*. Boston: Pitman.

Freeport-McMoRan. 2017. Workforce. https://www.fcx.com/sustainability/workforce. Accessed October 2017.

Friedman, M. 1970. The social responsibility of business is to make profit. *NY Times Mag*. September 13.

Government of Canada. 2017. Environmental concerns through the mine life cycle. In *Environmental Code of Practice for Metal Mines*. https://www.ec.gc.ca/lcpe-cepa/default.asp?lang=En&n=CBE3CD59-1&offset=5&toc=show. Accessed October 2017.

Griffin, J.J., and Mahon, J.F. 1997. The corporate social performance and corporate financial performance debate: Twenty-five years of incomparable research. *Bus. Soc.* 36(1):5–31.

Habashi, F. 2001. Clean technology in the metallurgical industry. In *Proceedings II Mining Meeting of South America/VII Mining Meeting of Tarapaca*. Edited by J.P. Ibáñez, E. Patiño, and X. Velose. Iquique, Chile: Arturo Prat University. pp. 15–28.

Habashi, F. 2012. Rev. del Instituto de Investigación (RIIGEO). *FIGMMG-UNMSM* 15(29):49–60.

Hidalgo, C., Peterson, K., Sith, D., and Foley, H. 2014. *Extracting with Purpose: Creating Shared Value in the Oil and Gas and Mining Sectors' Companies and Communities*. https://www.sharedvalue.org/sites/default/files/resource-files/Extracting%20with%20Purpose_FINAL_Exec%20Summ_Single%20Pages_0.pdf. Accessed March 2013.

Hillman, J. 1999. Tracing the veins of copper, culture, and community from Butte to Chuquicamata. *Can. Rev. Sociol. Anthropol.* 36(3):449.

Hodge, R.A. 2004. Mining's seven questions to sustainability: From mitigating impacts to encouraging contributions. *Episodes* 27(3):177–184.

IAEA (International Atomic Energy Agency). 1998. *Guidebook on Good Practice in the Management of Uranium Mining and Mill Operations and the Preparation for Their Closure.* IAEA-TECDOC-1059. Vienna: IAEA.

ICMC (International Cyanide Management Code). 2018. The Cyanide Code. https://www.cyanidecode.org/about-cyanide-code/cyanide-code. Accessed January 2018.

ICMM (International Council on Mining and Metals). n.d. Home page. https://www.icmm.com/. Accessed March 2018.

ICMM (International Council on Mining and Metals). 2007. *Stakeholder Research Toolkit*. London: ICMM. https://www.icmm.com/en-gb/publications/mining-and-communities/stakeholder-research-toolkit. Accessed March 2018.

ICMM (International Council on Mining and Metals). 2010. *Sustainability Reporting Guidelines and Mining and Metals Sector Supplement.* https://www.icmm.com/ru/publications/commitments/gri-mining-and-metals-sector-supplement. Accessed June 2018.

ICMM (International Council on Mining and Metals). 2012. *Community Development Toolkit.* London: ICMM. www.icmm.com/website/publications/pdfs/social-and-economic-development/4080.pdf. Accessed March 2018.

ICMM (International Council on Mining and Metals). 2015a. ICMM ten principles. https://www.icmm.com/ru/about-us/member-commitments/icmm-10-principles/the-principles. Accessed June 2018.

ICMM (International Council on Mining and Metals). 2015b. *Understanding Company–Community Relations Toolkit*. London: ICMM. www.icmm.com/website/publications/pdfs/social-and-economic-development/9670.pdf.

IFC (International Finance Corporation). 2002. *Developing Value: The Business Case for Sustainability in Emerging Markets*. Washington, DC: IFC.

IFC (International Finance Corporation). 2007. *Stakeholder Engagement: A Good Practice Handbook for Doing Business in Emerging Markets*. Washington, DC: IFC.

IFC/World Bank. 2007. *Environmental, Health and Safety Guidelines for Mining*. www.ifc.org/wps/wcm/connect/topics_ext_content/ifc_external_corporate_site/sustainability-at-ifc/policies-standards/ehs-guidelines. Accessed March 2013.

IGF (Intergovernmental Forum on Mining, Minerals, Metals and Sustainable Development). 2017. *IGF Guidance for Governments: Managing Artisanal and Small-Scale Mining*. Ottawa, ON: IGF.

IIED (International Institute for Environment and Development). 2002. *Breaking New Ground: The Report of the Mining, Minerals, and Sustainable Development Project*. London: Earthscan.

IISD (International Institute for Sustainable Development). 2002. *Seven Questions to Sustainability: How to Assess the Contribution of Mining and Minerals Activities*. www.iisd.org/sites/default/files/publications/mmsd_sevenquestions.pdf. Accessed March 2017.

IPTS (Institute for Prospective Technological Studies). 2018. Best available techniques. http://ec.europa.eu/geninfo/query/index.do?queryText=2010%2F75%2FEU+on+industrial+emissions&summary=summary&more_options_source=global&more_options_date=*&more_options_date_from=&more_options_date_to=&more_options_language=en&more_options_f_formats=*&swlang=en. Accessed January 2018.

ISEAL Alliance. 2018. *ISEAL Codes of Good Practice*. https://www.isealalliance.org/. Accessed January 2018.

ISIS (International Standard for Integrated Sustainability). 2018. Certifying sustainable business practices. www.isis-standard.com. Accessed January 2018.

ISO 9001. 2015. *Quality Management Systems—Requirements*. Geneva: International Organization for Standardization.

ISO 14001. 2015. *Environmental Management*. Geneva: International Organization for Standardization.

ISO 26000. 2010. *Guidance on Social Responsibility*. Geneva: International Organization for Standardization.

Kemp, D., Worden, S., and Owen, J.R. 2016. Differentiated social risk: Rebound dynamics and sustainability performance in mining. *Resources Policy* 50:19–26.

Kimberley Process. 2018. Kimberley process. https://www.kimberleyprocess.com. Accessed January 2018.

Kotler, P., and Lee, N. 2005. *Corporate Social Responsibility: Doing the Most Good for Your Company and Your Cause*. Hoboken, NJ: John Wiley and Sons.

Loayza, N., and Rigolini, J. 2016. The local impact of mining on poverty and inequality: Evidence from the commodity boom in Peru. *World Dev.* 84:219–234.

McMahon, G., and Remy, F. (eds.). 2001. *Large Mines and the Community: Socioeconomic and Environmental Effects in Latin America, Canada, and Spain*. Ottawa, ON: World Bank/IDRC.

Millennium Ecosystem Assessment. 2005. *Ecosystems and Human Well-Being: Wetlands and Water*. Vol. 5. Washington, DC: World Resources Institute.

MINEO Consortium. 2000. Review of potential environmental and social impact of mining. www2.brgm.fr/mineo/UserNeed/IMPACTS.pdf. Accessed January 2018.

Moffat, K., and Zhang, A. 2014. The paths to social license to operate: An integrative model explaining community acceptance of mining. *Res. Policy* 39:61–70.

OHSAS 18001:2007. *Occupational Health and Safety Management Certification*. Dublin, Ireland: Certification Europe.

Ovejero Zappino, G. 2009. Project management and stakeholders. In *Sustainable Management of Mining Operations*. Edited by J. Botin. Littleton, CO: SME.

Owen, J.R., and Kemp, D. 2015. Mining-induced displacement and resettlement: A critical appraisal. *J. Cleaner Prod.* 87:478–488.

Oxfam Community Aid Abroad. 2004. *Mining Ombudsman Case Report: Tolukuma Gold Mine*. Fitzroy, Victoria, Australia: Oxfam. http://actnowpng.org/sites/default/files/Oxfam%20Mining%20Ombudsman%20Report.pdf. Accessed July 2018.

Parsons, R., Lacey, J., and Moffat, K. 2014. Maintaining legitimacy of a contested practice: How the minerals industry understands its "social licence to operate." *Res. Policy* 41:83–90.

PDAC (Prospectors and Developers Association of Canada). 2018. Responsible Exploration: E3 Plus. www.pdac.ca/priorities/responsible-exploration/e3-plus. Accessed April 2018.

Porter, M.E., and Kramer, M.R. 2011. The big idea: Creating shared value. *Harvard Bus. Rev.* 89(1):2.

Reuter, M., Xiao, Y., and Boin, U. 2004. Recycling and environmental issues of metallurgical slags and salt fluxes. In *VII International Conference on Molten Slags Fluxes and Salts*. Johannesburg, South Africa: Southern African Institute of Mining and Metallurgy. pp. 349–356.

RJC (Responsible Jewellery Council). 2013. *Responsible Jewellery Council Code of Practices*. London, RJC. https://www.responsiblejewellery.com/files/RJC_Code_of_Practices_2013_eng.pdf. Accessed April 2018.

RJC (Responsible Jewellery Council). 2018. Chain of custody certification. https://www.responsiblejewellery.com/chain-of-custody-certification/. Accessed January 2018.

RMI (Responsive Minerals Initiative). 2018. Responsible minerals assurance process. www.responsiblemineralsinitiative.org/responsible-minerals-assurance-process/. Accessed January 2018.

SA8000:2014. *SA8000 Standard*. New York: Social Accountability International.

Salzmann, O., Ionesco-Somers, A., and Steger, U. 2005. The business case for corporate sustainability: Literature review and research options. *Eur. Manage. J.* 23(1):27–36.

SBGA (Swiss Better Gold Association). 2018. About SBGA. www.swissbettergold.ch/en/about. Accessed January 2018.

Steiner, G.A. 1971. *Business and Society*. New York: Random House

Tahltan First Nation. 2003. *Out of Respect: The Tahltan, Mining, and the Seven Questions to Sustainability. Report of the Tahltan Mining Symposium, April 4–6, 2003, Dease Lake, British Columbia*. Winnipeg, MB: International Institute for Sustainable Development.

UN (United Nations). 2002. *Berlin II: Guidelines for Mining and Sustainable Development*. New York: UN.

UN (United Nations). 2004. *The UN Global Compact: The 10 Principles*. https://www.unglobalcompact.org/. Accessed March 2018.

UN (United Nations). 2015. *Transforming Our World: The 2030 Agenda for Sustainable Development*. A/RES/70/1, October 21. New York: UN.

Van Heel, O.D., Elkington, J., and Fennel, S. 2001. *Buried Treasure: Uncovering the Business Case for Corporate Sustainability*. London: SustainAbility and the United Nations Environment Programme.

Walton, C.C. 1967. *Corporate Social Responsibilities*. Belmont, CA: Wadsworth Publishing.

Weiss, N.L., ed. 1985. *SME Mineral Processing Handbook*. Littleton, CO: SME-AIME.

Winterhalder, K. 1996. Environmental degradation and rehabilitation of the landscape around Sudbury, a major mining and smelting area. *Environ. Rev.* 4(3):185–224.

World Bank. 1996. *The World Bank Participation Sourcebook*. Washington, DC: World Bank.

World Bank. 2004. *Striking a Better Balance: The World Bank Group and Extractive Industries—The Final Report of the Extractive Industries Review*. Washington, DC: World Bank.

CHAPTER 2.3

Mill Reports

Deepak Malhotra

The primary objective of mill reports is to provide plant, senior, and corporate management with an ongoing status of plant operations in respect to historical performance, predicted performance, and a road map for future improvements. The type of reports prepared by mill staff may include daily, monthly, and yearly reports metallurgical test work reports; capital project reports; and general cost reports.

Generally, comprehensive daily mill data are entered into a database or spreadsheets and then reviewed by metallurgists for the individual unit operations. Data from this database are summarized for management, namely, tons treated, head grades, recovery, and metal production. These data are compiled and used to prepare monthly and annual mill reports by metallurgical staff. There are no standard formats established by the mining industry. The data compiled for mill reports are influenced by the size of the operation, desire of corporate management, and the complexity of the operation. A monthly report prepared for a 400-t/d (metric tons per day) operation may be significantly different from one prepared for 40,000 t/d.

The advancement in instrumentation and information technology has enhanced the compilation of data. Prior to recent developments in technology, data were compiled by hand in log books and then transferred manually into metallurgical reporting software. This has now changed with real-time data often being transferred directly from process control software into an operation's metallurgical database.

The type of data gathered in a site database on a daily basis is given in Table 1. This information is reviewed daily to keep track of the performance of the mill. Very often, mill staff use computers to generate graphs showing relationships between different process parameters.

The general format for monthly or annual mill reports is given in Table 2. The major topics are briefly discussed in the following sections.

SAFETY

Safety is generally the first item discussed in the mill report. The number of accidents that occurred and injuries recorded during the reporting period are provided in this section. A brief description of each incident is discussed, and the control measures that were identified or implemented to prevent similar occurrences in the future are presented. The management will also report the number of workplace inspections, safety observations, hazard identifications noted, and any training and refresher courses provided.

Table 1 Partial list of daily data collected by mill personnel

Feed rate	Power consumption
Percent moisture	Reagent mixing and inventories
Pulp densities	Labor reports
Particle size monitoring	Abrasion, sickness, vacations
pH	Reagent consumption
Assays	Grinding media used
Concentrate shipped	Metallurgical logs
Concentrate on hand	Safety report
Production calculations	Water and tailing management report
Operating hours	Processing cost breakdown
Reasons for lost time	

Table 2 Outline of a typical monthly or annual mill report

1. Safety
2. Mill production summary
3. Processing plant
   a. Crushing
   b. Grinding and classification
   c. Concentrate handling/packaging or gold production
   d. Tailings pond
   e. Laboratory
   f. Metallurgical balance
4. Maintenance
   a. Summary
   b. Preventive maintenance
   c. Workforce
5. Capital projects
6. Cost summary totals
   a. Cost and production—actual vs. budgets
   b. Major cost variances

Deepak Malhotra, President, Resource Development Inc., Wheat Ridge, Colorado, USA

Most of the operations will paraphrase the information. However, some operations may report the information in a table with a typical format, such as illustrated in Table 3. A detailed discussion about the reasons for the compilation of the information is provided in Chapter 2.1, "Health and Safety."

MILL PRODUCTION SUMMARY

This section usually provides production data for the month. It could be a few sentences stating the tons of concentrate produced and/or the ounces of precious metals produced during the month or detailed tables, as shown in Tables 4 and 5, with in-depth discussion providing the highlights of issues encountered and the reasons for the variance between budgeted and actual production. Table 5 is typical of a three-stage crushing ball mill circuit for a flotation plant. Many of the comminution circuits today are primary crush followed by an SABC (i.e., semiautogenous, ball mill, and pebble crusher) configuration.

PROCESSING PLANT

This section provides a more detailed discussion of production achieved during the month as well as any reasons for variance of targets for the month.

The discussion will generally follow the plant flow sheet, starting with crusher circuit performance and leading into the grinding circuit. Depending on the plant flow sheet, the performance of other unit operations (flotation, thickening, filtration, leach, solvent extraction electrowinning, etc.) will also be discussed in detail.

Very often the tailings pond is under the environmental group. However, the mill report should discuss the status of the tailings and detoxified leach residue, including the mitigation efforts to ensure that tailings are benign.

The report may also include a table summarizing reagent consumption on a total and per-ton basis. Actual consumptions will be compared against budget figures with the reasons for any variance being discussed. The reasons for variance could be due to less or more tonnage processed; harder ore, which resulted in higher media consumption; or higher feed grade, which requires more flotation reagents.

Laboratory activities are generally directed at working on projects to improve plant metallurgy and reduce costs (e.g., to evaluate cheaper alternative reagents). A summary of these activities should highlight the results achieved during the reporting period. Additionally, the laboratory personnel undertake plant sampling, testing reagents, or pilot new equipment. These activities are also reviewed in the mill report.

Metallurgical staff prepare a daily metallurgical balance. These data are accumulated and the monthly and year-to-date metallurgical balances are calculated and reported in this section. When the flotation plant produces more than one concentrate, as in polymetallic processing plants, the methodology used for metallurgical balance is also discussed in the report. A detailed discussion on the metallurgical balances is provided in Chapter 2.4, "Metal Accounting." However, the monthly report will highlight the outcome of the monthly metallurgical accounting with commentary if there are issues and what is being done to rectify them.

MAINTENANCE

The maintenance section presents a brief summary of both the major work completed and major downtime recorded during the reporting period. Some operations break down their activities and time spent for the different areas and/or types, namely, unit operations or mechanical, electrical, and miscellaneous activities. Typical formats are shown in Tables 6 and 7.

The preventive maintenance scheduled for the following month is also reported in the monthly report. Very often, the tables will report the availability and utilization of the equipment. Availability is the number of hours the equipment can be operated, whereas the utilization is the number of hours the equipment was operated divided by the number of hours the equipment was available for operation.

Table 3 Typical data in a safety report table

Item	Month	Year to Date	Last Year
Active number of employees			
Total worker-hours			
Lost-time accident (LTA) Medical accident (MED) First-aid accident			
Incidence rate LTA Incidence rate MED Incidence rate all accidents			
Days without LTA			
Off-the-job accidents Off-the-job days lost			

Table 4 Typical monthly mill production table for gold operation

	Month			Year to Date		
Process Parameters	Actual	Budget	Variance	Actual	Budget	Variance
Crusher throughput, dmt (dry metric tons)						
Crusher throughput rate, t/h						
Crusher operator time, %						
Mill throughput, dmt						
Mill throughput rate, t/h						
Crusher throughput time, %						
Gold head grade, g/t Au						
Gold recovery, %						
Combined tail grade, g/t Au						
Gold poured, oz						

Table 5 Typical monthly mill production table for three-stage crushing ball mill comminution followed by flotation plant

Process Parameters	Month			Year to Date		
	Actual	Budget	Variance	Actual	Budget	Variance
Primary crushing plant						
Cars dumped, no.						
Operating time, h						
Crushed, dmt						
Secondary crushing plant						
Operating time, h						
No. 1: standard						
No. 1: short-head						
No. 2: short-head						
Product						
Crushed (weightometer), wet t						
Moisture, %						
Crushed, dmt						
Grade, % metal						
Grinding						
Ball mill						
Feed, wet t						
Moisture, %						
Operating time, h						
Feed, dmt/h						
Flotation						
Grind P_{80}, mesh						
Feed grade, % metal						
Rougher concentrate						
Recovery, wt %						
Recovery, % metal						
Final concentrate						
Recovery, wt %						
Recovery, % metal						
Grade, % metal						
Concentrate stockpiled, t						
Concentrate shipped, t						

Table 6 Breakdown of worker-hours by category for maintenance

Category	%	Actual	Year to Date
Actual worker-hours worked			
Budget worker-hours			
Hours in backlog			
Work requests received (mechanical)			
Work requests received (electrical)			
Work requests received (miscellaneous)			
Work requests completed (mechanical)			
Work requests completed (electrical)			
Work requests completed (miscellaneous)			
Total work requests completed			
Worker-hours for planned repairs (ppm running)			
Worker-hours for breakdowns			
Worker-hours for modifications			
Worker-hours for corrective action			
Worker-hours for fabrication			
Worker-hours for crane service			
Worker-hours for shutdowns			
Worker-hours for safety			
Operator damage			
Miscellaneous			
Mechanical availability			
Electrical availability			

Table 7 Breakdown of maintenance worker-hours by unit operation

Unit Operation	%	Hours	Year to Date
Primary crushing			
Course ore reclaim			
Secondary crushing			
Tertiary crushing			
Conveyors			
Dust suppression			
Agglomeration			
Leach pad spray system			
Wet plant and gold room			
Auxiliary power generation			
Potable water			
Process water			
Light vehicles			
Maintenance workshop			
Laboratory			
Administration			
Exploration			
Miscellaneous			

Table 8 Operating cost summary

Unit Operation	Month			Year to Date			Comments
	Actual	Budget	Variance	Actual	Budget	Variance	
Crushing							
Grinding							
Flotation							
Thickening							
Filtration							
Tailings							
Technical services							
Supervision and overhead							
Total							

Table 9 Summary of milling costs

Account	This Month		Year to Date	
	Total	Cost, $/t Milled	Total	Cost, $/t Milled
Operation				
Labor				
Supplies				
Power				
Other				
Total operations				
Maintenance				
Labor				
Supplies				
Power				
Other				
Total maintenance				
Total direct milling costs				
Indirect costs				
General expense				
Net vacation allowances				
Net holiday allowances				
Pension accruals				
Taxes				
Insurance				
Salaries				
Miscellaneous maintenance				
Receiving and buying				
Shipping and selling				
Safety and welfare				
Total indirect costs				
Total operating costs				
Cost of ore milled				

CAPITAL PROJECTS

If the plant has ongoing modifications to improve the operation, the mill report will provide a summary of the status of these capital projects. The summary will include the work completed and ongoing costs.

COST SUMMARY TOTALS

The cost summary section presents the cost breakdown by area for the month and year to date. A brief description of the reasons for the variance between the actual and budget costs is also generally given. Typical tables for the cost summary are given in Tables 8 and 9. Many operations will report their costs as shown in Table 9.

MISCELLANEOUS REPORTS

Metallurgical staff will also prepare reports on development test work directed at enhancing plant performance and/or reducing operating costs.

Management may also require mill staff to prepare reports on capital cost projects, detailing the modifications in the plant, the reasons for doing so, and the projected benefits that are anticipated.

CHAPTER 2.4

Metal Accounting

Robert D. Morrison

Metal accounting is still treated as a "poor cousin" at many sites even though the situation has improved in recent years. However, a dependable metal accounting system should provide a reliable basis for estimating production, losses, and important key performance indicators (KPIs). Metal accounting also provides a key tool in the search for "low-hanging fruit" and is probably the only way to confirm with some reasonable degree of confidence that they might have been harvested.

A five-year, industry-funded, collaborative project (AMIRA International Project 754 [P754], Metal Accounting and Reconciliation) was undertaken from 2004 to 2009 to address the challenges of metal accounting. The project produced a draft code of practice (AMIRA International 2007), trained seven graduate students, and culminated with publication of a textbook (Morrison 2008). The project carried out detailed metal accounting studies at sponsor-nominated sites.

These studies disproved some common myths and identified some serious issues. Metal accounting problems are often attributed to poor sampling practice. In a few cases, sampling was an issue. However, the industry is well supported by sampling consultants and International Organization for Standardization (ISO)-certified laboratories. Hence, at many sites, sampling and assaying are well executed and no longer offer unresolved issues. Nevertheless, it is still worth comparing the sampling methods in use for metal accounting at a particular site with those recommended in Chapter 1.8, "Sampling Practice and Considerations," and rectifying any shortcomings.

Just how much material has been processed and produced is, of course, critical to metal accounting. The belt weightometer is usually regarded as the "gold" standard. Modern belt weighers are sophisticated, robust, and generally excellent measurement devices when correctly installed and operated. In practice, they are rarely installed properly, which seriously compromises their potential performance. Standards of operation also vary widely. However, the biggest challenge for metal accounting is that it is intrinsically a collision between technical and financial cultures.

AMIRA P754 developed a systematic approach to aligning these cultures so that the estimated balances would be statistically identical. Hence, only one set of numbers need be considered for reporting, for generation of KPIs, and, more importantly, for decision support.

In short, the process of metal accounting should transform measured data into performance information that can be used for reliable decision making. This chapter works through the key features of the process and provides some practical examples.

THE BASICS

The underlying purpose of metal accounting is to "process" a set of measurements and historical data into a set of self-consistent numbers that can be used as a sound basis for making decisions. Hence, a more succinct description of *metal accounting* is to "turn data into information."

It is also necessary to select the "best" streams for sampling and mass measurement. The obvious streams are the feed, concentrates, and tailings for the total plant. However, measuring each stage of a complex plant may provide useful information about its performance and aid in diagnosis of problems. Mass measurements and sampling points are also needed wherever a transfer of custody occurs. Masses and metal contents measured in more than one way may seem excessive but are vital to the development of cumulative sums, or "cusums," which allow biases to be detected and tracked over time.

Metal accounting information is typically used as the basis for a series of period-based reports. These reports usually provide estimates of input to the process and where those inputs report to, as well as changes in accumulation (i.e., stockpiles, bins, or tanks) between the beginning and the end of the reporting period. These estimates should be the "best available" in terms of the reliability of the measured data within the period and information related to previous operating periods.

As there will usually be many more measurements than are required to generate a single balance, it is tempting to create different balances for different purposes. This means

Robert D. Morrison, Former Chief Technologist, Julius Kruttschnitt Mineral Research Centre, University of Queensland, Brisbane, Queensland, Australia that many "best estimates" can be generated. They need not be self-consistent, and decision making will be influenced by opinion about which balances might be best.

Metal accounting also covers the transfer process between different parts of an organization (mine to concentrator) and to other organizations (concentrator to smelter.) The relationships with external organizations are usually defined by comprehensive sales contracts. Within organizations, these contracts are often informal. However, the potential for internal strife can usually be reduced with at least a semiformal "contract" describing the points at which custody is transferred and how measurements and calculations are to be carried out.

As with all contractual processes, especially those that involve commercial transfers, some definitions acceptable to both parties are essential. Hence, the major challenges for metal accounting are measurement (and quality of measurement) as well as management of custody transfers between various parties.

Given that technical and financial cultures are different, metal accounting also involves managing those relationships in a structured and cooperative manner, as they offer substantial opportunities for confusion and conflict.

Until recently, there have not been any broadly applicable definitions or a robust, published methodology available. An important driver for development of the code was to improve corporate governance. Developing a common language was also important. Hence, the code developed an extensive glossary with detailed definitions of key terms and concepts.

THE DRAFT CODE OF PRACTICE

For many years, there has been some degree of acceptance that as much as 15% of the metal in a body of ore might fail to appear at the concentrator. In an era where measurements were few (at both mine and concentrator) and mechanized samplers even fewer, this might have been acceptable. In the current era, hugely better mass measurement and fully automated samplers are available, as well as better assaying and essentially unlimited computational capabilities. Hence, there are no valid excuses for using 19th-century approaches in the 21st century. In a few cases, application of the code might have detected fraud more quickly, but in almost all cases it will likely identify problems and opportunities as they develop. Furthermore, a sound metal accounting system linked to records of which ore types were being processed in each period offers a basis for a sound production plan.

The development of the draft code is well documented (Morrison and Gaylard 2008). Hence, only a brief summary of the objectives and principles is covered in this chapter. The code itself is freely available from AMIRA International (AMIRA 2007).

The draft code's glossary also includes definitions that are used in the remainder of this chapter. Formal definitions drawn from the code are included in this chapter.

The code defines *metal accounting* and *reconciliation* as follows (AMIRA 2007). These definitions and all quotations from the code are provided by kind permission of AMIRA International:

> **Metal Accounting.** The system whereby selected process data (pertaining to metals of economic interest) is collected from various sources including mass measurement and analysis and transformed into a coherent report format that is delivered in a timely fashion in order to meet specified reporting requirements.
>
> **Reconciliation.** A metallurgical balance which relates production of saleable and reject or waste materials from a process back to its source as ore or other feed material. It should be provided with defined and stated errors as for any other metallurgical balance.

The code is based on the following 10 principles (AMIRA 2007):

1. The metal accounting system must be based on accurate measurements of mass and metal content. It must be based on a full Check in-Check out system using the Best Practices as defined in the Code, to produce an on-going metal/commodity balance for the operation. The system must be integrated with management information systems, providing a one-way transfer of information to these systems as required.
2. The system must be consistent and transparent and the source of all input data to the system must be clear and understood by all users of the system. The design and specification of the system must incorporate the outcomes of a risk assessment of all aspects of the metal accounting process.
3. The accounting procedures must be well documented and user friendly for easy application by plant personnel, to avoid the system becoming dependent on one person, and must incorporate clear controls and audit trails. Calculation procedures must be in line with the requirements set out in the Code and consistent at all times with clear rules for handling the data.
4. The system must be subject to regular internal and external audits and reviews as specified in the relevant sections of the Code to ensure compliance with all aspects of the defined procedures. These reviews must include assessments of the associated risks and recommendations for their mitigation, when the agreed risk is exceeded.
5. Accounting results must be made available timeously, to meet operational reporting needs, including the provision of information for other management information systems, and to facilitate corrective action or investigation. A detailed report must be issued on each investigation, together with management's response to rectify the problem. When completed, the plan and resulting action must be signed-off by the Competent Person.
6. Where provisional data has to be used to meet reporting deadlines, such as at month ends when analytical turn-around times could prevent the prompt issuing of the monthly report, clear procedures and levels of authorisation for the subsequent replacement of the provisional data with actual data must be defined. Where rogue

data is detected, such as incorrect data transfer or identified malfunction of equipment, the procedures to be followed, together with the levels of authorisation must be in place.

7. The system must generate sufficient data to allow for data verification, the handling of metal/commodity transfers, the reconciliation of metal/commodity balances, and the measurement of accuracies and error detection, which should not show any consistent bias. Measurement and computational procedures must be free of a defined critical level of bias.
8. Target accuracies for the mass measurements and the sampling and analyses must be identified for each input and output stream used for accounting purposes. The actual accuracies for metal recoveries, based on the actual accuracies, as determined by statistical analysis, of the raw data, achieved over a company's reporting period must be stated in the report to the Company's Audit Committee. Should these show a bias that the Company considers material to its results, the fact must be reported to shareholders.
9. In-process inventory figures must be verified by physical stock-takes at prescribed intervals, at least annually, and procedures and authority levels for stock adjustments and the treatment of unaccounted losses or gains must be clearly defined.
10. The metal accounting system must ensure that every effort is made to identify any bias that may occur, as rapidly as possible, and eliminate or reduce to an acceptable level the source of bias from all measurement, sampling and analytical procedures, when the source is identified.

As the project to develop the draft code neared its conclusion, it became clear that the code would provide a strong guide to what a metal accounting system should achieve but not a great deal of guidance about how to actually achieve the desired outcomes. Therefore, the project team and some invited experts produced a textbook (Morrison 2008) that provided practical guidance on how to account for a wide range of processes.

The underlying process of metal accounting is custody transfer. The remainder of this chapter considers custody transfer, followed by measurement (mass and sampling), and approaches to calculation. How to report the outcomes of the system and to detect longer-term errors are considered.

CUSTODY TRANSFER (CHECK IN–CHECK OUT)

A custody transfer is an ancient process where a seller and a buyer agree on what is to be bought, how it is to be measured, and what the price will be based on. The formal definition from the code is as follows (AMIRA 2007):

> **Check In-Check Out System.** The system whereby all streams into and out of the Process or Plant, for which the balance is being performed, are measured, sampled and analysed.

In essence, a formal custody transfer is very much like a sales contract for a concentrate or a contract for toll treatment of a parcel of ore. Applying an agreed set of definitions and a check in–check out (CI/CO) process to the transfers will greatly reduce variations due to "opinions" of both buyer and seller. As noted earlier, it is certainly worth considering at least an informal "contract" within organizations and it should still be written down.

Occasionally, the transfer process will break down. In these cases, an exception report is required (AMIRA 2007):

> **Exception Report.** A report generated, in terms of this Code, whereby each non-compliance with the requirements of this Code is motivated and approved.

This is similar to a yellow or red tag for a process plant malfunction and is likewise a problem to be diagnosed and fixed as quickly as possible.

In the flow sheet shown in Figure 1, the CI/CO points are fairly clear. The transfer from mine to concentrator happens at the mass measurement (and sampling) point between the stockpile and the mill. Within the concentrator, the CI/CO points between grinding and conventional flotation, and conventional flotation and tailings management, are also obvious. However, flash flotation is a part of the milling circuit with a major impact on overall flotation performance. It will

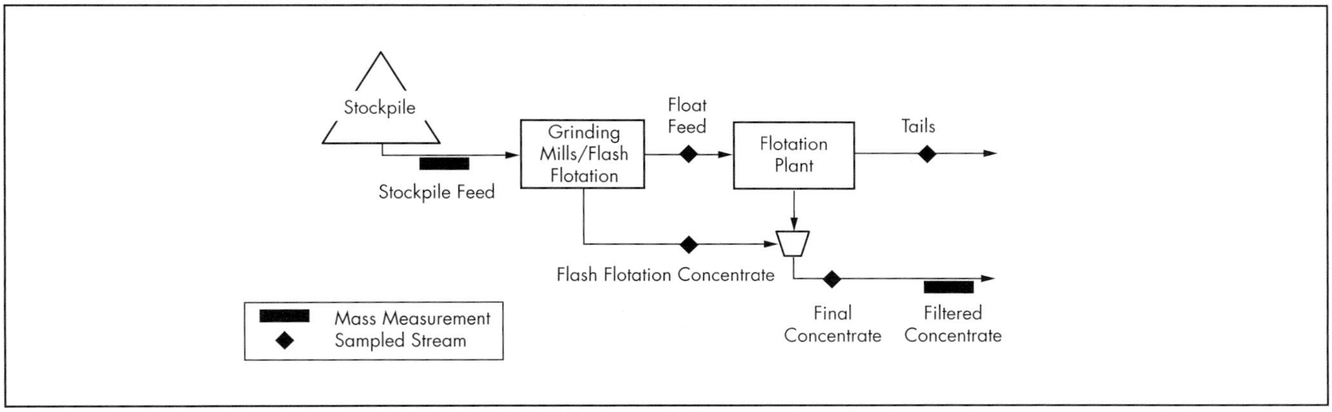

Source: Jansen et al. 2007, reprinted with permission from the Australasian Institute of Mining and Metallurgy

Figure 1 Overview flow sheet and measurement points at the Northparkes copper–gold concentrator

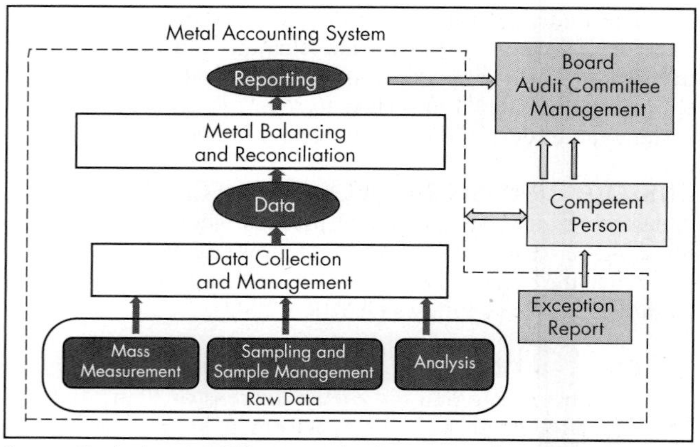

Source: Gaylard et al. 2009, reprinted with permission from the Southern African Institute of Mining and Metallurgy

Figure 2 Structure and data flows within a typical metal accounting system

need some careful consideration for effective custody transfer boundary. One possible approach is provided as an example.

MEASUREMENT

The data-gathering component of a metal accounting system is mostly concerned with measurement. An accounting system must have at least one (and preferably several) point for mass measurement. A sampling and assaying procedure will usually be associated with each measurement point and always with each transfer of custody, as shown in Figure 2.

Measurement Quality

All measurement and sampling processes are associated with an error distribution. (See Chapter 1.8, "Sampling Practice and Considerations," for further guidance about sampling.) Given that metal accounting is very much concerned with how well measurements are defined, it is worth briefly considering the types of errors that contribute to the quality of a measurement and to the credibility of calculated outcomes based on those measurements.

Three types of error are of particular interest: accuracy, precision, and bias. The code definition for *accuracy* is rather long-winded, but it is also self-explanatory (AMIRA 2007):

> **Accuracy.** A measurement is accurate if it, or the average of a number of measurements, is close to the true value. In metallurgical operations this true value is unknown.
>
> In addition, there is often misunderstanding between the terms accuracy and precision, which is the measure of the spread of a number of measurements around their mean value.
>
> For these reasons, it has been decided to adopt the following definition for accuracy, which incorporates the concept of precision, and is based on that given in ISO 5725-1:1994.
>
> A measurement that is accurate is one that is free of bias and has a dispersion (standard deviation) that is lower than a defined dispersion or indeed a probability density of a particular nature. The level of dispersion or the nature of the probability density is defined with the purpose of separating measurements that are entirely fit for a particular purpose or use and those that are not.

The definition of *precision* is also self-explanatory. Many engineers use *reproducibility* interchangeably with *precision*. Because metal accounting is usually concerned with absolutes in terms of payment for product in a custody transfer, one must be most concerned about accuracy in the short term and avoidance of bias in the longer term.

Precision of a measurement depends on the closeness of the outcomes of a repeated measurement or test procedure. Hence, it depends only on the distribution of random errors and not on any relationship to a "true" value. It is usually expressed as the standard deviation of the test results. That is, by a measure of imprecision. The code defines *precision* as follows (AMIRA 2007):

> **Precision.** Precise measurements have a dispersion about their mean value, which is lower than a defined dispersion or indeed a probability density of a particular nature. The level of dispersion or the nature of the probability density is defined with the purpose of separating measurements that are entirely fit for a particular purpose or use and those that are not. A precise measurement may not be accurate; its mean may differ from the true value of the measured quantity by an arbitrary amount. Standards often quantify the term precision as a value corresponding to the magnitude of a 95% confidence interval around a result.

If the n measurements of the quantity to be estimated are normally distributed, "the interval is $\pm\, ts$, where t is the two-sided Student-t value at 95% confidence and $n - 1$ degrees of freedom, and s is the estimate of the standard deviation" of the estimated quantity.

Metal accounting is also concerned with long-term quality of data. Hence, bias—or the detection and correction of it—is also critical. The formal definition is perhaps not as clear as the others but may become clear after the reading the discussion in the "Detecting Bias" section later in this chapter. The code definition for *bias* follows (AMIRA 2007):

> **Bias.** Bias is the difference between the mean result of one or more measurements and the true value of the quantity being measured. It also can be seen as an ongoing difference between two, or more, measuring systems. (e.g. sender's and receiver's weights, the analysis of the same sample by two laboratories etc.)

Figure 3 summarizes these error types.

Mass Measurement

Mass measurements can be continuous (flow rates on conveyors or in pipes), discrete (by mine ore truck or concentrate rail wagon), or more or less static (stockpile or tank surveys). For continuous measurements, the integral of the measurement over an accounting period is the key piece of data. However, while the integral of dry mass is the most important, most systems measure wet mass.

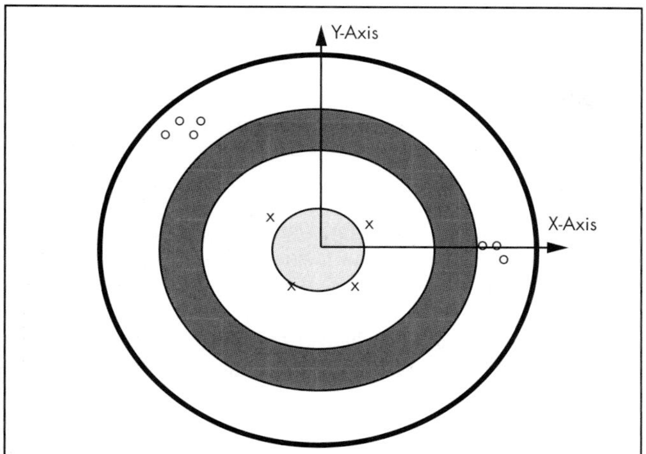

Note: The x's are accurate but are not very precise. The small circles are less accurate and more precise but exhibit bias in X alone in the right-hand set and in both X and Y in the other set.

Source: Gaylord et al. 2009, reprinted with permission from the Southern African Institute of Mining and Metallurgy

Figure 3 Accuracy, precision, and bias

Belt weightometers (or scales) are usually considered to be the "gold" standard. These devices are certainly capable of high-quality measurements if correctly installed, calibrated, and maintained. One of the more startling outcomes of the investigations carried out during the development of the code was that these requirements are very rarely fully satisfied.

Hence, a good first step toward a robust metal accounting system is to check the manufacturer's guide for recommended installation requirements for the weigh scale. The code contains five pages of these requirements. Dominant areas of noncompliance include using a weighing point that is not on a horizontal section of the belt and failing to keep the weigh frame clean. The latter is important because the weigher will indicate that more ore has been treated than has actually passed over the frame. A quick check at the start of each shift for rocks that have "fallen" onto the weigh frame is good practice.

In an ideal situation, a belt conveyor should be able to be diverted to a truck that can then be weighed on a weigh bridge that has been "certified for trade" by the relevant authority. This is not possible for many primary ore feed conveyors. A more workable alternative is to weigh several truckloads and then load them onto the belt. A roller chain for calibration should be standard, as should regular, manual belt speed checks—given the wide range of ways in which belt speed sensors can malfunction. The roller chain may be described as "certified," which means it has a well-defined mass per unit length of chain. This does not mean that the measurement the weigher produces is "certified for commercial use," as considered further below.

Static weights can be attached to the weigh frame to check it for zero and ranging errors. This will check if the load cell is drifting, but it will not help with variable belt tension or belt speed errors.

Belt scale manufacturers specify measurement quality as an "error" that is a percentage of full scale. They are often reluctant to state just how many standard deviations the claimed error represents. Assuming that the stated error represents one standard deviation is a sensibly conservative point to start at, given the myriad possible sources of error.

Consequently, about two-thirds of the time, the belt integral should be expected for, say, a shift to be within plus or minus the stated error range. Obviously, a well-loaded belt will be better measured (i.e., be subject to a lower relative standard deviation, as the specified error is absolute) than a lightly loaded one. It should be made clear that the belt loading should be as steady as possible with only slow variation. This is achievable with multiple feeders drawing from a stockpile that is not close to empty. With direct (and always intermittent) loading from a truck or front-end loader, the belt load per meter will vary widely and reduce the reliability of the mass measurement. Almost all belt weighers produce an increment. This means that the weigher integrator will advance by one unit. An increment results from the product of belt movement and weigh-frame load. Some also produce an analog signal for an instantaneous product of weigh frame load and belt speed. Integrating that signal in the control system—smoothed or not—is not recommended because the weigher integrator itself should be the most reliable. If the weigh frame analog signal is available, rapid fluctuation will usually indicate problems with belt loading or the hardware. If the signal is reaching its maximum, the belt load is certainly high but not actually measured.

So-called automated calibration, where the belt is empty or loaded with a roller chain and is run for several revolutions, should provide an indication of just how much the zero point and slope have been adjusted. Otherwise, it becomes very difficult to track variations and biases in the integrated mass records.

International standards do provide further guidance, but ore-feed weigh scales are rarely certified for commercial use. *Certification* is also defined in the code (AMIRA 2007):

> **Certification (Test).** A calibration of a scale or flow meter in terms of accepted standards defined by the relevant government authority to permit its use for trade purposes and custody transfer.

A comprehensive review of weightometer application—and mass measurement in general—is provided by Wortley (2009). Wortley also contributed to the code and to the textbook. By far, the most comprehensive reference for mass measurement is by Colijn (1983), but it is out of print and difficult to obtain.

Moisture Measurement

As the estimated metal content of a stream is based on a dry assay, measuring or estimating the moisture content of the ore feed stream is important.

Before semiautogenous milling became dominant for primary grinding, plants typically employed three stages of crushing that preceded milling; this was fairly simple because a representative sample of the final crushed product was usually a practical size for lab processing. With the almost universal dominance of semiautogenous milling, the problem has become much more challenging. An accurate measurement would require periodic large samples, which would then be crushed and subsampled. Unless the mineral of interest is of low grade but very high value, this strategy will be too expensive to install and maintain. In practice, most of the moisture is usually contained in the –6 mm size fraction. The percentage of fines can be evaluated as a proxy for moisture when multiplied by a reasonably constant factor, as almost all of the moisture is contained in the fines fraction of the stream.

Because of the challenge, a common approach (which is almost standard practice) is to combine the metal masses in tails and concentrates to estimate the feed grade to the semiautogenous grinding (SAG) mill. Hence, the downstream mass measurements become critical. This challenge is discussed further in the case study of Northparkes, reported in detail by Jansen et al. (2007) and summarized later in this chapter.

Possibilities exist for automation of moisture measurement. Measures based on capacitance and response to microwaves are commercially available. However, these meters also need to be kept calibrated, and some common minerals—magnetite and many sulfides—interfere with the measurements to some degree.

Flow Measurement

The flows of most interest in mineral processing are usually slurries. Therefore, using flowmeters that have no internal protrusions is good practice. For good-quality solids flow measurement, a combination of a magnetic flowmeter and a nuclear density gauge constitutes the traditional approach. Various acoustics-based approaches are also available, and their technologies are both improving and offering an ever-wider choice.

For almost all types of flowmeters, several pipe diameters of straight pipe should be provided for approach and retreat. Instrument manufacturers typically recommend 10 or more diameters. This is often impractical. However, it is critical that no mass holdup or any degree of partial filling is occurring within the meter. Hence, upward vertical flow is preferred—never on a gentle downward slope. For downward slopes, a weir with level measurement should be used.

If a slurry meter is installed vertically, it is also essential that there are no coarse particles in the flow that can settle at a rate approaching the flow velocity and generate misleading density measurements and consequent overestimates of flow rate. In this situation, acoustic flowmeters that depend on the Doppler effect will be biased low as they measure an average particle velocity. Hence, for accurate, integrated flow measurement, flow rate and density should be kept within specified ranges.

The ideal way to calibrate a flowmeter is by timed flow into a large tank. For slurries, the stream can be sampled to measure density. In practice, a portable ultrasonic flowmeter can often provide an adequate check.

Nuclear density gauges are calibrated assuming an average particle density. If the mineral mix varies much, the solid flow estimates can be biased in either direction. A nuclear gauge installation should allow for calibration against a pipe full of water and have a set of high-density plates that can be introduced into the beam path for a multipoint calibration.

As the pipe diameter increases with wear, flow rate and density measures will read low. This is important because the integrated error will become a bias, and flow at a particular velocity depends on diameter squared. Therefore, flow and density must be periodically calibrated for pipe wear.

For a more detailed discussion, see Section 3.2.2 of the draft code (AMIRA 2007).

Shipping Measurements

For bulk transport, there are very good train wagon weighers that can even be certified while the wagons are moving slowly through a tipping or dumping station. Truck weighers can also provide good-quality results but can only be certified for static operation.

The major transport method for many concentrates and bulk commodities is by ship. Strangely, the method of estimating net cargo mass is by surveying how far the ship sinks into the water while unloaded and loaded. Corrections are used for water temperature and salinity, but the method is inherently limited, with a typical standard deviation of about 5% relative. Its great advantage is that it is difficult to interfere with the measurement (see Section 3.4.6 of the code for more detail [AMIRA 2007].)

For bulk commodities, the purchaser often divides the measurement and payment process into two steps. A proportion of the payment (a provisional payment) is made based on the measured load and sampled assays going into a ship or onto a train. A final payment is then made on the as-received mass loadout and assays. This is a complex CI/CO process, which will be covered by a detailed sales contract. The contract will also detail how disputes are to be resolved—for example, by using another laboratory to assay duplicate samples.

Some additional factors need to be considered when sampling containers. After the material of interest is placed into the container, it becomes very difficult to take a representative sample. A continuous feed stream to a container by pipe or conveyor provides a nearly ideal situation for crosscut sampling and precise weighing. Section 4 of the code (AMIRA 2007) and Chapter 1.8, "Sampling Practice and Considerations," provide more detail on this topic. The most difficult containers to sample are stockpiles and bins. Measuring mass and composition in and out of the stockpile is the favored strategy. Frequent zero points should also be part of the system. If there are two concentrate storage tanks, they should operate alternately, not in parallel.

A large stockpile can often be managed as two smaller ones, more than doubling the number of zero points. Frequent zero points help to detect biases that may be creeping into the system if any of the calibrations suffer from drift.

ASSAYING

The assays of most interest for metal accounting are usually for water (i.e., % solids or % moisture) and for paid metals and (penalty) contaminants, as these will determine the price paid for the concentrate. For moisture, particularly in arid climates, not keeping samples sealed from dry air is often a source of bias.

As for all accounting samples, they should be cut for equal increments of mass flowing past the sampling point. The next preferred strategy is random sampling with a well-controlled average cutting frequency. Given that the processing of large samples often leads to errors in sample splitting, multistage samplers should be considered for large flows of slurry. Wills and Napier-Munn (2006) provide a good guide to the design of multistage systems.

Assaying was once an issue, as many errors can be generated in the process. Today, many laboratories are ISO 9000 certified. These labs must have strong procedures and quality assurance (QA) procedures in place. Many will also be using error models to monitor quality.

The standard practice is to process duplicate samples and do a third assay if a trigger discrepancy between the first two occurs. Section 4 of the code elaborates on QA methods (AMIRA 2007).

Error Models

Some examples of error models are given in the case study of Northparkes, reported in detail by Jansen et al. (2007) and summarized later in this chapter.

MASS BALANCING AND RECONCILIATION*

It needs to be emphasized that mass balancing for metal accounting is not the same as mass balancing of, for example, a flotation circuit survey, even though similar mathematical tools may be used. Blindly applying a standard mass balancing package is an effective way of concealing bias. Detecting and avoiding bias should always be a major objective of a balancing and reconciliation system for metal accounting.

Objectives

As previously mentioned, the code recommends a full CI/CO methodology at each transfer of custody. Therefore, any mass balancing technique must be compatible with CI/CO procedures. Perhaps the most valuable aspect of a sound mass balance is that it can be used as a go/no-go test for each CI/CO transfer. This satisfies the first principle in the 10 principles of the draft code of practice (AMIRA 2007). Another way to phrase this principle is as a question: Is there a significant discrepancy, or are the variations within expected measurement error at some agreed level of confidence?

The second objective is to pinpoint any measurement problems or biases without delay. These problems should be rectified as quickly as possible, not allegedly "corrected" by the balancing process.

To set up the mass balancing problem to suit metal accounting and reconciliation, a sound knowledge of error distributions in data measurement is required. All measurement processes—such as instrument readings, sampling and assay measurements, or any other kind—are subject to statistical error. In addition, two useful concepts need to be revisited:

1. A measurement is *accurate* if it is sufficiently close to an accepted standard. A good example is the process of certifying a weighbridge or scale for commercial use, which is a classic example of a custody transfer. A certified mass measurement device usually provides a key input to a CI/CO transfer of value.
2. A measurement is *precise* if it gives very similar results when repeated. If a precise measurement is not close to the result that is accepted as accurate, then there is bias. A small bias is not a problem in a short-term experiment, such as a flotation survey. However, where key results are accumulated over many measurement periods, a small bias will accumulate and cause problems with reconciliation and achieving fair CI/CO transfers.

Figure 3 summarizes these concepts.

One way to state the problem mathematically is to define an adjustment between each measured and balanced value. For n measurements, each of which has a counter value of i, an adjustment delta can be defined as follows:

$$\Delta_i = x_i - \bar{x}_i \qquad \textbf{(EQ 1)}$$

where x_i is each measured value, and $\bar{x}_i$ is an adjusted value of x_i.

Each adjustment can be scaled by its measurement accuracy, estimated by its standard deviation σ_i or some other measure of dispersion. The standard deviation in this case should include all of the measurement errors—including sampling, and sample preparation and analysis, where appropriate.

Now a mathematical criterion can be defined—the weighted sum of squares, WSSQ—which can be minimized to estimate a "best" set of data adjustments:

$$\text{WSSQ} = \sum_i [\Delta_i/\sigma_i]^2 \text{ subject to } C[x_i] = 0 \qquad \textbf{(EQ 2)}$$

where C is a matrix of constraints that must be satisfied to be consistent with the complete flow sheet. The flow sheet in this case includes variations in holdup in bins, stockpiles, or in the process circuit itself.

If the required adjustment delta i is small compared with the measured variable, then intuitively the required adjustment is not a *discrepancy* in CI/CO terms. For normally distributed measurement variation, the required adjustment is expected to be within plus or minus one sigma for about two-thirds of the measurement and within plus or minus two sigma for about 95% of it. Hence, adjustments of more than twice the standard deviation need to be carefully examined.

Bias is easy to include in this model but not so easy to detect:

$$\Delta_i = (x_i' + b_i) - \bar{x}_i \qquad \textbf{(EQ 3)}$$

where b_i is a bias associated with measurement i, and x_i' is the unbiased measurement. If b_i is small compared with the measurement, it is not detectable in a single data set. However, as each mass and metal content increment is accumulated, the expected relative error of each accumulated sum reduces by the inverse of the square root of the number of increments.

Consequently, the bias becomes easier to detect and, in many cases, impossible to ignore. A maximum of one bias at a time can be tested for as part of the least-squares approach. Practical approaches to bias detection are considered later in this chapter.

Solving the Problem

Equations 1 and 2 constitute a standard problem for constrained minimization of the WSSQ, provided reasonably accurate initial estimates are available. Mass balancing provides a reasonably linear problem, and many simplifications are possible. The two-product formula is the most venerable of these, but it provides no information about self-consistency. Standard techniques for error propagation can be used (Cutler and Eksteen 2006). Where multiple components are measured, the standard deviation of the mass split can be estimated.

Two methods are available to attack the general mass balancing problem. The first is to estimate mass splits based on assay differences. Appendix E of the code of practice provides an example (AMIRA 2007). This is the traditional mass balancing approach. Although it can be extended to include process holdups (Morrison and Richardson 1991), it is not generally useful beyond the processing plant.

Most custody transfers occur into or out of separation plants rather than within them. The traditional approach was

* This section is drawn from Gaylard et al. 2009, with permission from the Southern African Institute of Mining and Metallurgy.

very attractive for hand calculation because it is very computationally efficient.

With computational power available from modern computers, the second approach using simulation has much to recommend it. The simulation approach is well suited to stream splitters that provide undefined mass splits because they should not generate assay differences. For the simulation approach, each stream is considered in terms of volumetric flows, as well as of each metal and gangue of interest. Each process block (or node) is modeled as a set of splitters.

If each splitter is modeled in flow-sheet order (with iterations as required for recycle streams), the constraints of Equation 2 become implicit (i.e., they are automatically included). Hence, only the split factors and input flows need to be adjusted. Constraining split factors between 0 and 1 is also recommended. For a detailed description of both techniques, see Chapters 6 and 7 of Morrison (2008).

Reconciliation

The simulation approach is also well suited to reconciliation. The objective of reconciliation of metal accounts is to develop a set of data that is numerically consistent with the CI/CO values at its boundaries (for which the operation paid and/or was paid for) and the measured values within that scope. This can be carried out in two steps:

1. Carry out the balance and check that all values are within the expected ranges of error.
2. If step 1 is satisfied, "fix" the CI/CO values and rerun the balance.

If the measured values are still within the expected ranges, now there is a fully consistent and statistically validated set of data suitable for generation of KPIs and reports. A few values beyond ±2 standard deviations are expected.

Detecting Bias

The traditional approach is to use cumulative sums (or "cusums") of the difference between cumulative metal flows (at any point in the process chain) where they can be estimated by two more-or-less independent methods. Over time, the differences should fluctuate to about zero due to random error. A bias will cause a positive (or negative) accumulation and should be easy to detect.

For the mass balancing approach, it should be clear that any bias will become part of the data adjustments. If there is no bias, the weighted adjustment is expected to have an equal chance of being positive or negative. Even though the bias will be distributed to some degree, the adjustments will tend to be predominantly positive (or negative) if a bias develops. Hence, both cusums and weighted adjustments should be trended for early detection of bias.

APPLYING THE CODE

Processing plants come in many different configurations. However, for the purposes of metal accounting, they can be divided into several types (Table 1). Each type is suited to a particular measurement and accounting strategy. The textbook devotes a chapter to each of these applications (Morrison 2008). For this chapter, a more general case is considered, and then two typical cases are summarized.

Figure 4 shows a generic processing plant flow sheet. It receives feed from four sources (F) via four stockpiles (S). Within the plant there are several holdups (H_1–H_3). The process plant produces rejects (R) for tailings and several products that are stored in inventories (I_1–I_3). During an accounting period, total product sales (P_1–P_3) are drawn from each inventory. For this generic case, there are three valuable components with assays (c1, c2, c3) and two contaminants (c4 and c5).

Check In–Check Out Points

The trucks that deliver feed pass over a certified weightometer. That measurement provides a wet weight F to check in to each stockpile. The wet weight is converted to an estimated dry weight using an average moisture content or a truck factor based on experience. Stocks are reclaimed over a belt weightometer to provide a check-out value for feed stocks and a check-in value for the process plant.

Conveyor reclaim also provides a convenient place for sampling. The samples provide assays c1 to c5 for defined increments of process plant feed. The process block is the most interesting because it produces three products and has three holdups and a single reject product stream.

For each accounting period, the feed-stocks drawdown should be equal to the sum of the rejects and the check-out flows over a conveyor weightometer, with quality control sampling plus the net change in process holdups (H_1–H_3). H_1–H_3 might be positive or negative, while all of the mass and component increments should, of course, be positive.

The products are drawn from the inventories (I_1–I_3) on an as-required basis using a front-end loader with real-time load weighing. The trucks loaded with product exit the site over the same certified weightometer. They will typically be weighed into a customer's plant and then will be sampled as they are unloaded and assayed for quality control.

Hence, there are many CI/CO stages to be considered—for example, feed source out and into feed stocks; feed stocks out into plant and out into products and reject.

Given that all flows in and out per accounting period are measured and assayed (at least at some stage), this balance is completely arithmetical. The total masses and components masses in and out of each stage are simply added, with adjustments for changes in stock and holdups.

How well these arithmetic balances match at each stage is sometimes called "accountability." Statistical variation suggests that it will be unreasonable to expect them to match perfectly every time.

When a stockpile is empty, there is an opportunity to check the accumulated balance and make an adjustment to stock if necessary.

In some industries—typically those with high-value products—an "empty zone" is moved through the process by stopping all feed streams and measuring stocks and holdups as they empty. This is called a bubble method or audit. It will incur some production costs, but it will identify biases in terms of holdups and stocks that contain too much or too little of valuable components.

The separation plant of this generic accounting process can also be combined with the two-product approach, particularly when each product is batched to inventory. This strategy is considered in the two practical examples (case studies) that are presented later in this chapter.

ESTIMATING ACCURACY

A very useful rule called the propagation of variance (or error) allows the accuracy of calculated values to be estimated and the credibility of measurements to be assessed. The rule states

Table 1 Suggested metal accounting strategies for a range of processing plants

Type of Operation	Examples and Recommended Strategies
Type 1 • High-grade/high-value feed • Medium flow rates • High-value product • Significant plant inventory • Weigh and sample all inputs: check in • Weigh and sample all outputs: check out	**Smelter, Metal Refinery** The recommended strategy is check in–check out.
Type 2 • Low-value feed—medium to high flow rates, often of coarse feed • High-value product • Low-mass, high-value plant inventory • Weigh and sample product	**Gold Operation** The recommended strategy is to reconstitute tailings at measured feed rate and production.
Type 3 • Low-value feed—medium to high flow rates—sampled after primary grind • Low-value product—sampled as concentrate shipments • Low plant inventory • Conveyor weigh scale or weigh feeder for feed mass into plant • Accurate sampling of fine tailings, feed analyses, and tailings analyses at measured feed rate and production rates	**Base Metal Concentrator** The recommended strategy is to mass-balance feed analyses, concentrate, and tailings.
Type 4 • High tonnage of feed and products • Minor rejects; low plant inventory • Sometimes direct shipment to customer • Detailed product specifications • International standards for sampling and characterization • Custody transfer mass will often be a draught survey or a train loadout weight	**Coal Operation, Iron Ore** A commodity sales contract will detail prices and penalties as well as acceptable measurement techniques for both producer and buyer. The recommended strategy is to weigh and sample feed and product streams.
Type 5 • High-value, low-tonnage feed • High-value products • Minor reject streams; significant plant inventory • Weigh and sample all inputs: check in • Weigh and sample all outputs: check out	**Precious Metals Refineries** The recommended strategy is check in–check out.
Type 6 • Low-value, high-tonnage feed • Minor rejects, usually based on particle size or density • Minimal plant inventory	**Aggregates** A commodity sales contract will detail prices, usually based on tonnage and size specification. The recommended strategy is to weigh feed and product streams.
Type 7 • Low-grade, high-tonnage feed • Residues remain in situ in some cases • Very high, difficult-to-measure process inventory • Non-steady-state, two-phase, slow reaction kinetics	**Heap Leach** Weigh, sample, and analyze feed and product. Perform accurate solution balance. The recommended strategy is to perform periodic checks of heap solution inventories and heap residue metal contents.
Type 8 • Medium- to low-value, high-tonnage feed • Low process inventory • Requires use of mineralogical analysis • Products may be in bulk, bags, or other containers	**Industrial Minerals** May be treated as a commodity. May be produced as different grades. The recommended strategy is to weigh feed and product streams.

Reprinted by permission of AMIRA International Limited

that, provided the errors are reasonably small (i.e., less than a few percent relative), the variance of a calculated variable is simply the sum of the products of the variance of each input variable and their partial derivative (with respect to each variable of interest) squared.

As most of the inputs are either fixed or linear, these derivatives are typically 0 or 1, and the error estimation process is much simpler than it sounds. Cutler and Eksteen (2006) and Morrison (2008) provide several examples, as do most texts on introductory statistics. Alternatively, a simple Monte Carlo analysis can provide indicative results as long as the error distributions for integrated assays and masses are reasonably realistic.

As a rule of thumb, the larger the assay differences achieved by the separation processes, the better the split factor will be defined. Hence, the two product mass and metal split estimates across a complete plant should be more accurate than any similar calculation for each single stage of separation.

Case Study 1: Northparkes Concentrator Example (adapted from Jansen et al. 2007)

Figure 1 shows the metal accounting flow sheet in use at an Australian copper–gold concentrator. Northparkes was the site nominated by Rio Tinto as sponsors of the code's development. Because of sponsor involvement, statistical errors in the measurement and accounting process could also be measured.

The Northparkes concentrator has two SAG/ball grinding circuits with flash flotation followed by a conventional flotation plant. The concentrator is unusual in that it has two processing lines, one of which is much larger in capacity than the other.

For a more detailed description, see Jansen et al. (2007) and Butcher et al. (2013). Figure 1 shows the key measurement

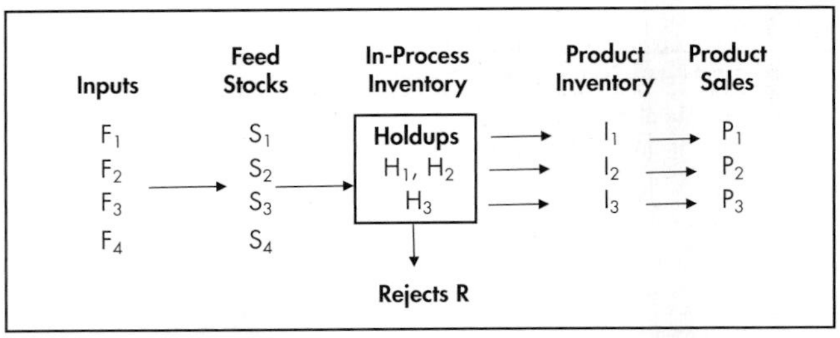

Source: Morrison 2008

Figure 4 Generic metal accounting flow sheet for a typical mineral processing plant

Table 2 Data usage and cut frequencies for on-stream analyzer samples

			Cut Frequency, minutes	
Slurry Stream	Process Control	Metallurgical Accounting	Module 1	Module 2
Flotation plant feed	X	X	30	30
Flash flotation concentrate	X	X	60	20
Final concentrate (combined)	X	X	40	15
Final tailings	X	X	10	40
Rougher tailings	X	—	—	—
Cleaner scavenger feed	X	—	—	—

Source: Jansen et al. 2007, reprinted with permission from the Australasian Institute of Mining and Metallurgy

points. The points at which custody transfer occurs should be fairly obvious. External transfers occur from mine to concentrator at the stockpile feed and of the filtered concentrate to the rail-head silos.

The work reported by Jansen et al. (2007) was carried out under the AMIRA P754 project. The accuracies of almost all of the inputs to metal accounting were investigated. The concentrate weightometers were calibrated with static weights supported by the weigh frame, and conveyor speed was checked by multiple direct measurements of time of travel for 40 m of feed belt.

The weigh-frame calibration resulted in a correction of 3%, but the conveyor speed tests indicated that the speed measurement was both precise and accurate. Some concentrate buildup on the conveyor weigh frame might have contributed to the required adjustment.

Moisture

Northparkes used periodic samples of 3–5 m of SAG mill feed to obtain measurements of feed moisture. This is a labor-intensive process and an average moisture of 2% was the result of this testing. Hence, 2% was used as a factor thereafter. Although this assumption has little effect on the variance of the dry tonnage estimate, it could easily lead to a bias in the totalized feed measurement.

A much finer feed material and regular crossbelt sampling make concentrate conveyors a much more tractable target. Average measurements of 7.5%–8.5% moisture achieved relative standard deviations of less than 1.5%.

Mass Balance

The key mass balance is around the flotation plant. It resembles but is not exactly a two-product balance because the flash flotation cells treating cyclone underflow bring the grinding circuit into the balance. Key streams are sampled by pressure pipe samplers and directed to the onstream analysis system for process control. A subset of samples is used for metallurgical accounting (see Table 2).

Jansen et al. (2007) investigated the sampling variances and assaying variances for samples taken immediately after one another to minimize the impact of process variation. The reproducibility of the sampling process was 4% to 8% (relative standard deviation) for sample mass per increment.

Given that many samples were taken per accounting period of 24 hours, any effect on a composite sample should be small. Because the standard deviations of assays across the entire circuit were measured, Jansen et al. (2007) could develop an error model for copper assays, including sampling. Figure 5 shows that a relative error model with a small offset is appropriate for absolute standard deviations of copper assays. The error model shown in Figure 5 provides an estimate of the expected Cu standard deviation for any stream in the circuit.

Error models can significantly improve the accuracy of the balancing process because they bring quantitative process experience into the balance. Conversely, unrealistic error estimates can lead to inappropriate balances.

CALCULATIONS AND REPORTABLE FIGURES

The true product of any metal accounting system is a series of reports. These reports will vary in the time spans they cover and the level of management they are prepared for. What they have in common is that they should be sufficiently reliable to support sound decisions. Figure 6 shows a section of a typical monthly production report.

The Northparkes system uses a "back to front" strategy because of the difficulty of sampling –200 mm SAG feed. This approach is made more complicated by the flash flotation circuit. Without that circuit, the two-product approach could be applied directly.

Starting with an overall dry tonnage balance for the period:

$$F = C + T$$

$$\text{feed} = \text{concentrate} + \text{tailings}$$

The measured values of feed and concentrate are "wet." Therefore, some factors (designated by Q) are applied based on an arbitrary sampling series to obtain realistic values for sample moisture:

$$F = Fwet - (Fwet*Qfeed)$$
$$C = Cwet - (Cwet*Qcon)$$

For each element (designated by lowercase letters) around the complete module:

$$Ff = Cc + Tt$$

where lowercase letters indicate an assay in that stream. The assays that are important are for copper and gold.

Recalling that the feed assays are not available but they can be directly calculated:

$$f = (Cc + Tt)/F$$

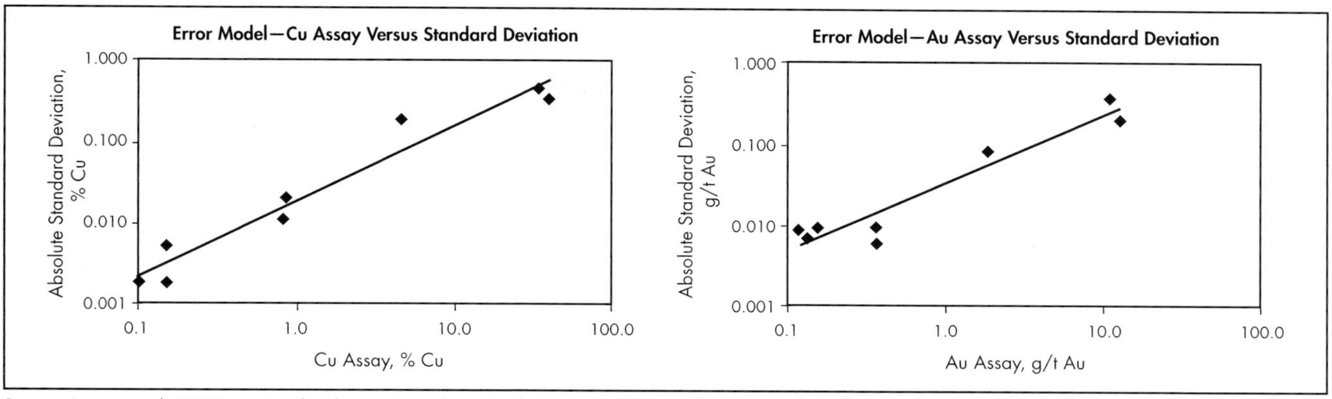

Source: Jansen et al. 2007, reprinted with permission from the Australasian Institute of Mining and Metallurgy

Figure 5 Error models for copper and gold assays, including sampling

Monthly Production Statistics—Ore Processing				
Parameter	**Unit**	**Module 1**	**Module 2**	**Total**
SAG Mill Feed	dmt	129,831	202,631	332,462
Feed Grade	Cu%	0.83	1.08	0.98
	Au g/t	0.52	0.37	0.43
Concentrate Grade	Cu %	32.89	36.17	35.02
	Au g/t	17.26	9.81	12.42
Recovery	Cu%	91.12	91.49	91.37
	Au%	76.76	71.6	74.03

Source: Jansen et al. 2007, reprinted with permission from the Australasian Institute of Mining and Metallurgy

Figure 6 Excerpt from a typical metal accounting monthly report

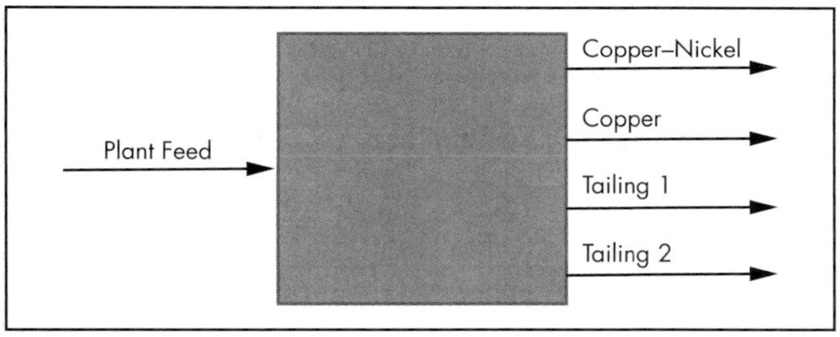

Source: Lachance et al. 2012, reprinted with permission from the Canadian Institute of Mining, Metallurgy and Petroleum

Figure 7 Strathcona flotation plant primary metallurgical accounting flow sheet

Similarly for recovery, R (%), in each element of interest by definition:

$$R = 100^{*}\ (Cc/Ff)$$

However, recovery can also be estimated from the assays:

$$R = 100^{*}[c(f - t)/f(c - t)]$$

The last reportable amount, M, is the metal contained in the concentrate:

$$M = Cc$$

for either copper or gold.

As for the generic balance, the rule for propagation of error can provide some error bounds for calculated variables. Error models can provide very useful input to this calculation.

Perhaps unsurprisingly, this error analysis comes to similar conclusions to one based on three measured products, except that the feed grade is estimated. The feed-grade estimate is most sensitive to errors in the wet concentrate measurement. Because that value is part of the feed calculation, it also dominates recovery errors where the tailings error would usually dominate.

Case Study 2: The Strathcona Mill

Lachance et al. (2012) provide an excellent case study of a two-stage update of the metal accounting system at Xstrata's Strathcona mill from a four-product balance to a code-compliant system. This case study summarizes some key aspects of that paper.

The Strathcona mill is a complex operation treating copper–nickel ores from Xstrata and non-Xstrata sources. Some of these ores contain significant platinum group metals. For metal accounting purposes, the concentrator can be considered as a single processing block with one feed, copper–nickel concentrate, copper concentrate, and two tailings streams (Figure 7). In the original system, three assays (nickel, copper, and sulfur) were used in a so-called four-product formula (Hodouin et al. 2011). Clearly, if only three assays are available, the various flow splits are singly determined. However, as many stages of separation of the minerals (which are summarized by their assays) have occurred within the accounting block, the flow splits may still be reasonably well determined.

As a general rule, the mass split across the entire concentrator is usually better defined than for single stages of separation because the differences between feed, products, and rejects assays are larger compared both in absolute terms and in comparison with measurement errors. However, the lack of redundancy in this example means that there is little chance of error detection within those flow splits. The only mass flow measured was the feed to the mill. The feed stream was sampled for assay and for moisture measurement.

The system upgrade amounted to an audit based on the principles of the draft code (AMIRA 2007). To achieve compliance, it was necessary to replace a spreadsheet-based system with a relational database and a well-controlled mass balancing environment for reconciliation. For this application, the team applied a commercial system (Metallurgical Account, which uses the well-documented BILMAT algorithm). These changes increased integration and provided a much-improved degree of compliance with the code.

The code strongly recommends against spreadsheets, both for data analysis and for data storage, because they provide an environment with extreme ease of editing and one in which it is almost impossible to maintain integrity of either data or calculations (AMIRA 2007). The elemental assay suite was increased in two stages to a total of 13. This number of assays provided a strong degree of redundancy, which should much improve error detection. LaChance et al. (2012) concluded after a trial period that the addition of Fe, MgO, and SiO_2 assays provided the best benefit to balance in terms of accuracy.

Overall, this paper provides a practical strategy for upgrading a "traditional" metal accounting system to one that is strongly compliant with the draft code of practice. However, it is also worth noting that retaining the block separation approach does not expose opportunities for enhanced long-term process understanding. That is, it does not improve opportunities to add value through metal accounting. If that has not already been done, it might provide a third-stage upgrade for the Strathcona operation.

CONCLUSIONS

Metal accounting based on the draft code of practice (AMIRA 2007) provides a structured approach to development, management, and audit of credible systems across a wide range of types of mineral processing. It also offers strategies for detection of short-term errors and long-term bias.

Achieving a single set of well-supported flow rates and assays—with error bounds—should greatly simplify decision making based on performance-related claims and decisions.

The draft code is most strongly focused on performance of the processing plant. However, there will almost always be further benefits in extending this approach to resource to disposal accounting and to balancing around each section within a processing plant.

For those who believe this might entail too much time and effort, it is worth remembering that not losing a kilogram of metal is appreciably more profitable than having to mine and process additional ore to replace it.

ACKNOWLEDGMENTS

This chapter, the draft Code of Practice and Guidelines, and the metal accounting textbook are all outcomes of AMIRA P754. The author acknowledges the contributions from the industry sponsors of P754, the Code Team, and the researchers and students from the universities of Cape Town, Stellenbosch, and Queensland. The AMIRA project manager, Richard Beck, and the leader of the Code Team, Peter Gaylard, deserve special mention for their outstanding contributions.

REFERENCES

AMIRA International. 2007. *P754: Metal Accounting—Code of Practice and Guidelines: Code Release 3.* Lonehill, South Africa: Amira International. www.amirainternational.com.

Butcher, A., Cunningham, R., Edwards, K., Lye, A., Simmons, J., Stegman, C., and Wyllie, A. 2013. Northparkes Mines, Rio Tinto. In *Australasian Mining and Metallurgical Operating Practices*, 3rd ed. AusIMM Monograph 28. Edited by W.J. Rankin. pp. 861–878. Melbourne, Victoria: Australasian Institute of Mining and Metallurgy.

Colijn, H. 1983. *Weighing and Proportioning of Bulk Solids.* Clausthal-Zellerfeld, Germany: Trans Tech Publications.

Cutler, C.J., and Eksteen, J.J. 2006. Variance propagation in toll smelting operations treating multiple concentrate stockpiles. *J. S. Afr. Inst. Min. Metall.* 10:221–227.

Gaylard, P.G., Morrison, R.D., Randolph, N.G., Wortley, C.M.G., and Beck, R.D. 2009. Extending the application of the AMIRA P754 code of practice for metal accounting. In *Base Metals Conference 2009.* Johannesburg: Southern African Institute of Mining and Metallurgy. pp. 15–38.

Hodouin, D., Lachance, L., and Poulin, É. 2011. Suboptimal flowrate estimators and their application to the design of measurement strategies. *J. Process Control* 21(2):235–245.

ISO 5725-1:1994. *Accuracy (Trueness and Precision) of Measurement Methods and Results—Part 1: General Principles and Definitions.* Geneva: International Organization for Standardization.

Jansen, W.M., Morrison, R.D., and Dunn, R. 2007. Metallurgical accounting in the Northparkes concentrator—A case study. In *Ninth Mill Operators' Conference, Fremantle, Western Australia.* Melbourne, Victoria: Australasian Institute of Mining and Metallurgy.

Lachance, L., Leroux, D., Gariépy, S., Halsall-Whitney, H., and Tuzun, A. 2012. Implementing best practices of metal accounting at the Strathcona mill. In *Proceedings 2012, 44th Annual Meeting of the Canadian Mineral Processors.* Westmont, QC: Canadian Institute of Mining, Metallurgy and Petroleum.

Morrison, R.D. 2008. *An Introduction to Metal Balancing and Reconciliation.* Edited by R.D. Morrison. JKMRC Monograph Series 4. Indooroopilly, Queensland: Julius Kruttschnitt Mineral Research Centre.

Morrison, R.D., and Gaylard, P.G. 2008. Applying the AMIRA P754 code of practice for metal accounting. In *Proceedings of MetPlant 2008: Metallurgical Plant Design and Operating Strategies.* Melbourne, Victoria: Australasian Institute of Mining and Metallurgy.

Morrison, R.D., and Richardson, J.M. 1991. JKMBal—The mass balancing system. In *Proceedings of the Second Canadian Conference on Computer Applications in the Mineral Industry (CAMI '91).* Vol. 1. Vancouver, BC: University of British Columbia and Canadian Institute of Mining, Metallurgy and Petroleum. pp. 275–286.

Wills, B.A., and Napier-Munn, T. 2006. *Wills' Mineral Processing Technology*, 7th ed. Amsterdam: Elsevier.

Wortley, C.M.G. 2009. Mass measurement for metal accounting—Principles and practice. In *Proceedings of the Fourth World Conference on Sampling and Blending.* Johannesburg: Southern African Institute of Mining and Metallurgy. pp. 121–128.

CHAPTER 2.5

Modeling and Simulation

Khosrow Nikkhah, David Wiseman, Michael J. Dry, and Hsin-Hsiung Huang

In mineral processing and extractive metallurgy, process modeling can be summarized as the numerical analysis of process data, generating the information required for the design of process equipment and the calculation of capital and operating costs for one or more configurations of unit operations for processing ore or concentrate from a given deposit. Process modeling is also a powerful tool for analyzing and optimizing an existing or a conceptual process and evaluating potential improvements.

Taking a mineral deposit from discovery to reality begins with quantification of the amount of ore present and the concentrations of valuable elements. This work includes drilling and analysis of samples, or re-compilation of results from earlier work. The results can be conveniently summarized as a total amount of ore and an elemental analysis, which, in conjunction with relevant prices, leads to an estimate of the potential value of the deposit. When the potential value is deemed to be sufficient, attention turns to the work needed to further quantify the deposit, generate appropriate samples, and evaluate the extraction technology.

This chapter covers process modeling and simulation for mineral processing and extractive metallurgy in hydrometallurgical and pyrometallurgical processes. Process modeling has a broad range of applications in these fields. It is an effective technique for better understanding of processes and their design, and it can be used to maximize the efficiency of the process and to minimize costs. In an operating plant, process modeling allows understanding of the process, thereby enabling better operation of the plant. Applications of process modeling can be categorized by both field of operation (mineral processing, hydrometallurgy, and pyrometallurgy) and the stage of project (conceptual study, feasibility, and detailed design).

Mining and metallurgical operations are commercial ventures, governed by the rules of investment for profit, safety, and environmental compliance. Within this framework, the objective of technology selection, therefore, must be to maximize the commercial benefit to be had from the deposit in question. Preliminary selection of technology for any given deposit is based on knowledge of the technology applied to similar deposits and sometimes also on information about novel technology. Although appropriate technology might be mineral processing, hydrometallurgy, or pyrometallurgy, within these categories are many individual combinations of unit operations. For the purposes of this chapter, the assumption is that the technology falls into one or more of these domains, but the principles illustrated would apply equally to all domains even though the unit operations differ from one domain of extractive metallurgy to another.

Modern computers and commercially available software for process simulation enable the evaluation of existing and envisaged circuits much earlier than was previously feasible, substantially enhancing the efficiency of the work required for the development of new ventures. This power also enables much more detailed study of existing circuits, leading to improved plant operation and uncovering improvements with considerably less risk than would otherwise be possible.

Although sophisticated software is available for process modeling, acquiring and mastering it entails appreciable cost, time, and effort. It is also possible to use spreadsheet calculations for preliminary evaluation of envisaged circuits for processing ore or concentrate.

SIMULATION AND MODELING IN MINERAL PROCESSING

To arrive at the best design and sizing of mineral processing plants and improve efficiency of operations, process modeling can be used to efficiently study the effect of a variety of process concepts, design criteria, and capacities. In addition, process modeling can be used to evaluate process economics for different design alternatives and at the same time show flaws in process design, thus avoiding unnecessary problems during commissioning and plant operation. Having proven the

Khosrow Nikkhah, Senior Process Engineer, AMEC Foster Wheeler, Vancouver, British Columbia, Canada
David Wiseman, Principal, David Wiseman Pty Ltd., Blackwood, South Australia, Australia
Michael J. Dry, Owner, Arithmetek Inc., Peterborough, Ontario, Canada
Hsin-Hsiung Huang, Professor, Department of Metallurgical & Materials Engineering, Montana Tech, Butte, Montana, USA

soundness of the process design, it can be used in raising of capital for the intended development.

In modeling of mineral processing plants, sizing of equipment and optimization of mineral processing circuits are the usual goals. These plants have large capital and operating costs, with electrical power consumption in comminution plants being a significant factor. This necessitates the use of process modeling for sizing of equipment to arrive at increased capacity and improved particle size distribution, while at the same time reducing capital and operating costs. Process modeling in mineral processing uses characterization of feed material, known plant data, and test work to size and specify equipment and optimize flow sheets and operating conditions.

The uses and requirements for modeling and simulation by those in academia and education and by those in industry differ. Although research and training are important in advancing the technology, this chapter focuses mostly on the current "state of the art" of the availability and use of modeling and simulation tools for mineral processing industry practitioners.

Historical Perspective

Herbst et al. (2002) provide a comprehensive overview of the historical background to modeling and simulation in the minerals processing sector of the industry.

The *SME Mineral Processing Handbook* (Weiss 1985) included a simulation and modeling section, specific to comminution circuits, covering much material also published earlier by Lynch (1977), which detailed model equations for grinding mills and hydrocyclones, and suggested that such models could be implemented in programming languages such as Fortran. However, it is unlikely that such use occurred outside of academia and the advanced research departments of larger companies.

Almost coincidental with the publication of the 1985 handbook (Weiss 1985) was the introduction and commercialization of a generation of "general-purpose," "user-friendly" simulation systems specifically aimed at minerals processing applications, which included CANMET-SPOC (Argyropoulos et al. 1984), MODSIM (Mineral Technologies International 2010), JKSimMet (JKTech 2018), and USIM PAC (Caspeo 2018). By providing a prebuilt simulation structure and suitable models, these simulation systems aimed to remove the need for specialized programming skills on the part of those interested in applying the models. After further development, by the time of the 2002 SME Mineral Processing Plant Design, Control and Practice conference in Vancouver, British Columbia, Canada (Herbst et al. 2002), these were reported as the "de facto industry standards."

The period between these two SME publications, from 1985 to 2002, also saw the publication of several important monographs that provided the background for the various data handling, process models, and modeling approaches. These included monographs by Austin et al. (1984), Napier-Munn et al. (1996), Wills (1997), and King (2001). A third contributing factor to changes in approach that occurred during this period was the introduction and subsequent ubiquity of personal computers, which resulted in increased accessibility to the models and the developing simulation frameworks. Coincident with the introduction of personal computing was the advent of general-purpose spreadsheet applications that largely replaced both calculators and mainframe computing and programming languages for mineral processing calculations.

Current Status

In the intervening years since 2002, the modeling and simulation of mineral processing landscape has seen additional changes. Apart from specialized academic and research applications, with the introduction of commercial packages, user programming of complex simulation calculations has largely ceased. For simpler and ad hoc calculations, spreadsheets (typically Microsoft Excel) and general-purpose engineering calculation packages such as MATLAB (MathWorks 2018) and Mathcad (PTC 2018) have become the norm.

In the commercial sphere, there have also been changes to the four reported "de facto industry standards":

1. The CANMET-SPOC applications appear to be no longer in general use, although the mass balancing component of that system now forms part of the commercial data reconciliation and material balancing package Algosys Bilmat and Algosys Metallurgical Accountant from Triple Point (2018a, 2018b).
2. A free version of the MODSIM software was released on a CD with King's 2001 monograph. Current support is through Minerals Technologies Inc., a spin-off from the Utah Comminution Center at the University of Utah.
3. USIM PAC development has kept pace with these changes to the underlying system environment and continues to be commercially available through Caspeo in France, offering around 150 process models for a wide range of minerals.
4. JKSimMet remains in active development and provides support with a focus on simulation of comminution and classification circuits. Its latest version is commercially available and includes a flotation-specific analysis and simulation package in JKSimFloat and a two-dimensional (2-D) size by assay mass balancer in JKMultiBal. The JKMetAccount simulation-based approach to metallurgical accounting was spun off to Mincom/ABB and is now part of the ABB Ventyx suite (JKTech 2018; ABB 2018).

Further developments in the commercial realm include application in the minerals processing area of general-purpose simulation systems originally developed for use in other domains, including the following:

- Aspen Plus (Aspentech 2018), originally chemical/hydrocarbon process focused
- IDEAS (Andritz 2018), originally for pulp and paper and then developed for mining and metals
- METSIM (Proware 2018), originally for pyrometallurgical processes
- SysCAD (2018), originally for dynamic control system validation and operator training
- HSC SIM (Lishchuk 2016), graphical and spreadsheet type process unit models. This package can be used to model many types of processes in metallurgy and mineralogy.
- IGS (Integrated Geometallurgical Simulation) (Lishchuk 2016). This tool minimizes risk using a geometallurgical model of the ore body by integrating geometallurgical data with process simulation. It combines these data with mining, operational, and environmental inputs that affect the process simulation results, capital, and operating costs. The equipment for mineral processing and the associated capital and operating costs can then be optimized based on the geometallurgical inputs. Utilization of

the geometallurgical models of the ore body are becoming the standard for development of both steady-state and dynamic models of mineral processing plants.

With the use of spreadsheets becoming more commonplace, several other spreadsheet-based applications have also emerged. These include the following:

- Moly-Cop Tools (Moly-Cop 2018). Now in version 3.0, this package consists of a set of Excel spreadsheets covering individual process models and calculations for most of the common comminution circuit unit processes as well as their combination in flow sheets. It is available free of charge.
- Limn—The Flowsheet Processor (Maelgwyn Mineral Services Ltd. 2009). Originally developed as a general-purpose simulation system that could model almost any process, this Excel-hosted package has found a niche in the simulation of physical separation processes, thus finding widespread use in the coal, diamond, iron ore, and mineral sands industries. It is under active development and support. Also available are several products aimed more at the aggregates industry but that may find mineral processing applications, particularly in the comminution area. These include Metso's Bruno (Metso 2008) packages and the AggFlow package from BedRock Software Inc. (AggFlow 2018).

Aspects of Commercialization

Most simulation work in the industry is performed using commercial packages. Although low-cost or no-cost "open-source" software has received some backing, particularly in the academic sphere, without a commercial arrangement, support and updating of applications in the fast-moving computer software and hardware environment is unlikely.

Such commercial arrangement can take the form of free or relatively low-cost software supported as part of a corporate identity (e.g., Moly-Cop Tools, Metso Bruno, MinOOcad [Herbst and Pate 2018], and ProSim [2018]), by the commercial arms of research groups, or spun off into new companies (e.g., MODSIM, JKSimMet, USIM PAC), or by small groups of focused individuals producing software targeted at a particular market niche (e.g., METSIM, Limn [Maelgwyn Mineral Services Ltd. 2009]).

With commercialization comes the need for competitive advantage, and this is evidenced most in the area of process models. Unlike much of chemical engineering and perhaps also some areas of pyrometallurgy and hydrometallurgy, where general theoretical principles can be more readily applied, or extractive test work and pilot-plant results be used to confirm chemical species, equations, and degree of conversion, the detailed models of processing units available to mineral processors are typically more empirically based.

Although many papers have been written over the years describing modeling approaches and model development, the most successful application of models has come from those that have been implemented in a suitable simulation framework.

Where a model development group possesses such a simulation framework, there is no incentive for model developers or "owners" within that group to make those models available to other parties for use in other frameworks. This represents a considerable impediment to entry into the mineral processing simulation area by software originally developed for other domains, for example, Aspen Plus, IDEAS, and SysCAD.

This also means that users, rather than purchasing multiple simulation systems to gain access to specific specialized models, often need to use consultants and specialists (including those within the simulation system supplier organization) to carry out the actual simulation studies, using the appropriate software. Given the specialized nature of some of the models and the reliance on experience and accumulated databases of parameters, and lack of cost and time-effective in-house expertise, this may be entirely appropriate.

Mineral Processing Simulation Regimes

In the minerals processing area, there are three main simulation regimes of interest: steady-state, dynamic, and general multiphysics numerical modeling techniques (e.g., computational fluid dynamics–discrete element method).

Historically, steady-state simulation has been the most common area of application, and although this continues to be the case, the rapid rise of computer power has seen increasing use of dynamic simulation and multiphysics approaches.

Steady-State Simulation

A simulation at steady state produces a representation of the plant or subprocess in a condition that is unlikely to arise in practice, that is, the condition where the sum of inputs equals the sum of outputs both for the overall plant and for each individual unit process.

Despite this, steady-state simulation remains extremely useful and represents the most widely used form of simulation in industry. The complexity and degree of accuracy required of steady-state models can be tailored to match the outcomes desired. The Pareto principle applies, in that if little data are available, simple models can still be used to get a useful basic understanding of a circuit. If more-detailed or more-accurate results are required, more (often considerably more) effort can be directed to fine-tune and validate more comprehensive unit models.

With appropriate models, steady-state simulation finds application in the following, for example:

- Detailed equipment assessment for design
  - Crusher selection: cone versus HPGR (high-pressure grinding roll)
  - Screen aperture selection
  - Ball and AG/SAG (autogenous grinding/semiautogenous grinding) mill sizing
  - Hydrocyclone component selection
- Production of process balances for design
  - Multiple scenarios and what-if analysis for equipment envelope definition
  - Evaluation of throughput options for operating envelope definition
  - Examination of effect of multiple feed types on design
- Evaluation of circuit operations for alternative equipment selection
  - Spirals versus teeter bed separation versus flotation
  - Jig versus dense medium separation versus ore sorting
  - Wet versus dry processing
- Assessment of circuit operations for alternative flow configuration
  - Rearrangement of existing grinding capacity

  - Variation of size cut points to examine effect on downstream density separation
- Evaluation of proposed circuit modifications
  - Effect of additional regrinding capacity on flotation
  - Effect of upgrading water recovery devices on plant water balance
- Optimization of equipment settings
  - Selection of relative density cut points for heavy medium circuits
  - Selection of screen and hydrocyclone apertures to match changes in ore type
- Benchmarking of operations: Comparison of current operation with predefined "best practice" for current plant, other similar plants within the company, or similar global operations
- Geometallurgy and future ore test work
  - Estimation of future plant performance for mine and plant design
  - Evaluation of design modifications for existing plants
- Mine planning and production scheduling
  - Mine design
  - Production schedule optimization
  - Treatability studies
  - Blending strategy evaluation
- Online product estimation for grade control
  - Mine planning
  - Plant feed scheduling
  - Plant product estimation
- Online in-plant parameter estimation to calculate
  - Milling circuit product size
  - Optimum separation circuit cut points for current feed conditions

Dynamic Simulation

Dynamic simulation aims to produce a system that replicates the dynamic behavior of a process unit or of an overall circuit. Similar models to those applied in steady-state simulations may be used where the unit process residence time is less than the time increment of interest (e.g., hydrocyclones within a grinding circuit). However, internal dynamics need to be addressed where residence times are significant (e.g., within a SAG mill).

Whereas in steady-state simulations, material transfer between unit processes (via pumps, conveyors, pipes, etc.), is ignored, in dynamic simulation these interprocess aspects are of similar importance to the units themselves. Setting up a dynamic simulation thus requires significantly more effort than is required for steady-state simulation, and this has contributed to somewhat restricted use in industry.

In the mineral processing domain, examples of the use of dynamic simulation include the following:

- Ore flow modeling for design and analysis of
  - Conveyor systems
  - Stockpiles and stockyards
  - Feed and product bins
  - Train loadouts
- Examination of blending strategies for
  - Mine planning
  - Plant feed stockpiles
  - Bulk commodity (e.g., coal, iron ore) rail and port facilities
  - Conveyor systems
- Dynamic plant models to assist in plant control implementation
  - Test bed for control system programming and commissioning and determining stop–start logic and safety interlocks
  - Preliminary loop tuning
  - Plant controllability testing and validation
- Modeling of circuits for control, for example,
  - Grinding circuit mill load control
  - Flotation plant circulating load control
  - Slurry pumping
- Modeling of circuits for operator training, for example,
  - Plant familiarization, *ab initio* training
  - Start-up/shutdown training
  - Unusual operating modes training
  - Safe operations training

Multiphysics Numerical Modeling Techniques

Multiphysics numerical modeling techniques encompass such techniques as computational fluid dynamics (CFD), smooth particle hydrodynamics, discrete element method, and discrete particle method techniques, as well as their combinations.

Because of its computation, intensive nature, and the time taken for solution, the multiphysics approach is not one that currently lends itself to direct application in minerals processing plant operations.

Although commercial packages (e.g., ANSYS Fluent [ANSYS 2018] and EDEM [EDEM 2018]) and open-source solutions (LIGGGHTS, CFDEM [CFDEM Project 2018], and OpenFOAM [OpenCFD Ltd. 2017]) exist, the specific difficulties presented in modeling of many processes in the mineral processing regime (particulate slurries at high percent solids, particles in dense media, etc.) mean that much manual configuration/modification is required. However, consultants and specialist service providers do offer services in several areas applicable to minerals processing, particularly in the bulk materials handling area, including the following:

- Stacker/reclaimer design
- Ore flow and bin design and validation
- Conveyor and transfer point design and validation

On the process hardware side, these techniques have also found use within equipment design and manufacturing companies to assist in design optimization in areas such as the following:

- Crusher liner profile optimization
- Grinding mill pulp lifter design
- Flotation agitator design
- Slurry pump design
- Studies of erosion and wear

Mineral Processing Modeling Implementation

Where circuit or plant-wide simulation implementation is required, two main approaches to steady-state simulation have been used: the equation-solving method and the sequential modular method.

Steady-State Regime

The equation-solving method involves writing the equations for the relationships between all simulation components across the circuit (e.g., in a flotation circuit, the relationships between the individual masses of copper and gangue minerals and the mass of water at each point in the circuit—both mass balance and separation equations) as well as the constraint and specification equations. These equations are then solved using standard mathematical techniques for the solution of sets of nonlinear equations. Although solution can be relatively efficient, the number of equations for solution of even a modest circuit can be daunting to set up, and the approach does not lend itself readily to incorporation into a prepackaged simulation system.

The sequential modular approach involves linking individual unit models by data structures representing process streams and having those unit models determine the composition of output streams based on the input streams and the unit model parameters. As each unit model is calculated in sequence, output streams from some unit models become inputs to the further downstream process models. As recycle material subsequently alters the feed streams to processes back upstream, the process unit calculations are repeated for successive iterations until some suitable completion criterion is achieved. (The approach is analogous to starting up a real plant and waiting until steady-state operation is achieved.) Although generally slower in solution than the equation-solving approach, advances in computing power have made speed of solution less of an issue.

Largely because of ease of implementation and the way prebuilt process unit models can be robustly implemented in modular form, the sequential modular approach is by far the most commonly applied.

Of the simulation packages mentioned, JKSimMet/JKSimFloat, Limn—The Flowsheet Processor, and USIM PAC primarily target steady-state simulation. JKSimMet has the strongest set of predictive comminution models, whereas JKSimFloat focuses on flotation. Moly-Cop Tools, Bruno, and AggFlow also have a comminution focus.

Limn—The Flowsheet Processor and USIM PAC are broader in application, covering general minerals processing plant operations, but generally with simpler, more basic public domain models, often sourced from the published literature. Although not specifically targeted at minerals processing applications, IDEAS, METSIM, and SysCAD also have steady-state capability and have been used in this area, again with more general public domain models.

The output requirements from steady-state simulation are relatively modest (e.g., tables of flows and component assays), and most commercial packages have both built-in data displays and export facilities to a spreadsheet to allow further analysis as required.

Dynamic Domain

The two steady-state approaches can be extended into the dynamic domain as well, with the sequential modular methodology being the most widely adopted.

The additional computing load required by dynamic simulation has been somewhat addressed by the exponential rise in computing power indicated by Moore's law, which states that overall processing power for computers will double every two years. The trade-off between solution time and fidelity of results (which generally translates to the magnitude of the time increment between successive simulation "slices") remains an issue, however. The use of computing hardware other than the personal computer (e.g., online "software-as-a-service" and cloud-based systems such as Microsoft's Azure) or personal computer–hosted numerical calculation enhancements (e.g., graphics processing units or GPU-accelerated computing) has provided some benefits in this area.

IDEAS and SysCAD focus strongly on dynamic simulation. METSIM and Aspen Plus also have dynamic simulation capability.

The display requirements for dynamic simulation are much more extensive than those required for steady-state work and include trend charting and operator interfaces. In some cases, this is handled by a pseudo-SCADA (supervisory control and data acquisition) system built into the simulator. Configuration of these subsystems contributes to the higher setup levels required for dynamic simulation. In operator training systems, the dynamic simulator may even be connected to a version of the actual plant human–machine interface hardware and software. Facilities for data export to spreadsheets are also commonly available.

Multiphysics Approaches

The numerically intensive multiphysics approaches depend on the availability of suitable computing platforms, with solutions taking hours to days rather than the seconds to minutes typically observed in steady-state and dynamic simulations. Similar approaches to cloud-based and accelerated computing options that apply to dynamic simulation also find use in the multiphysics area. Where personal computer use for these applications is driven by cost or availability, because end use tends to not be time limited, the calculations can be run on dedicated machines over extended time periods.

The results of multiphysics simulations lend themselves well to false color images and three-dimensional (3-D) graphics, and users can make good use of the advances in these fields.

Process Models

This section applies largely to models used only in the steady-state and dynamic simulation categories.

Stream Data Structures

Whereas the calculations that describe the process (the process unit models) determine the overall simulation results, the process stream data structures define the components of the system over which those calculations must be applied.

Mineral processing modeling and simulation is dominated by concerns about tracking particles, and although some of the multiphysics approaches do address individual particles, within the constraints of current hardware and software, the number of particles present precludes such approaches in general steady-state and dynamic modeling. The work-around is to combine particles into appropriate groupings or classes and to use the average properties of each class as the property for all particles within such a class.

The classes used usually match the physical groupings of interest within the processes, which are themselves often driven by the analysis techniques available. If one considers, for example, particle size, although there are techniques that report actual particle size, sizings are generally reported as the fractional (or percentage) mass of material falling between two sieve sizes in a series or range of such sieves. For calculation purposes, all particles within the size fractions are assumed to have the geometric mean size of that sieve interval. In cases where the raw analysis data are sparse, or intervals are irregular, interpolation is often performed to regularize the data.

Particle size is important in almost all mineral processing unit operations and in commercial simulation systems that focus on comminution circuits; it forms the main property of interest. In these cases, the stream data structure becomes a single vector, tracking the amount of material in each size fraction, plus additional individual components such as water, which may be included as required. Such stream data structures are said to be one dimensional, or 1-D.

Where an ore has multiple phases (e.g., hard and soft, kimberlite and clay, heavy mineral and gangue, coal and rock), even in comminution circuit simulation it is sometimes necessary to extend the 1-D approach to accommodate the difference in properties and hence the comminution response. In that case, the data structure becomes 2-D, with each size fraction further separated according to the phase properties of interest.

Moving beyond the comminution sections of a mineral processing plant, other properties of the process stream also become important, and the use of a multidimensioned stream structure approach is more common. For example, in coal, iron ore, mineral sands, and diamond plants, where separations based on relative density are important, the stream structure is extended to include a density distribution for particles in each size interval.

Although most simulation systems treat size as a distributed property describing a continuum of sizes, the second dimension tends to be more diverse. Thus, although it may be possible to model a mineral sands circuit using relative density as a continuous distributed property within each size fraction, it may also be possible to model it using the size distributions of the individual minerals present.

Higher dimensions are also possible. For example, in a flotation circuit, the data may be expressed as particles within a size range, of a particular mineral type, each having a range of floatability indexes, thus making a 3-D stream structure. Similarly, in a diamond processing circuit, the distributed properties of size and relative density may be tracked, and within each of those categories, a diamond size distribution may be tracked.

Simplification, to minimize the calculation load and/or data handling requirements, can sometimes be applied. In modeling a coal plant, it is often assumed that a single "washability" applies across the plant (i.e., if a particle is found anywhere within the circuit, if it has a certain size and a certain relative density, it will be assumed to have the same assay). This simplifying assumption means that a 3-D problem can been handled as a 2-D problem.

Liberation

Because different plant sections tend to be analyzed in different ways, it is common practice to simulate the different plant sections separately also, changing streams data structures to suit.

Crossing the boundary from one stream data structure to the other often requires more information to be added (e.g., going from a 1-D comminution circuit to a 2-D flotation circuit structure) or information to be compressed (e.g., from a 2-D size by density structure to the 1-D comminution structure required by a re-crush circuit model in a diamond processing plant). The major issue in such cases is determining how properties in one of the dimensions vary with changes in another dimension. For example, a breaker in a coal circuit and regrind mills in flotation circuits both result in a movement of particles from coarser sizes to finer sizes. Because of this, the resulting progeny particles will have different secondary or second-dimension properties from their parent particle.

Describing this liberation of material and allocating the progeny particles to the correct secondary property class remains one of the difficulties for simulation of mineral processing plants. Although improvements in analysis methods and advances in the theory have been made, their application in commercial simulation systems remains scarce. No standard approach to the problem of liberation has emerged, and apart from some basic heuristic and rudimentary empirical methods, applied particularly where comminution equipment is embedded within separation circuits, the problem remains largely unsolved, contributing to the tendency to model the separate plant sections individually.

Unit Process Models

Unit process models include models for comminution, classification, and separation as well as those for mineral recovery and based on mass recovery approach.

Comminution. Comminution remains the costliest component of many minerals processing plant flow sheets; consequently, considerable effort has been expended in model development in this area. The most common comminution simulation approach in both steady-state and dynamic simulation is the size-based "matrix model" population balance approach (Austin et al. 1984; Napier-Munn et al. 1996; Wills 1997).

Jaw and cone crushing (after Whiten). The Whiten (1972) crusher model was one of the earlier matrix comminution models developed at the Julius Kruttschnitt Mineral Research Centre (JKMRC) in Queensland, Australia, and remains in widespread use. It has been published in sufficient detail by Lynch (1977) to be effectively in the public domain and has been included in most of the comminution-focused simulation systems (e.g., JKSimMet, Limn). The basic model was extended by Andersen to form the Whiten–Andersen crusher model (Andersen 1988). This model uses a series of regression equations to relate ore and machine parameters to the underlying model parameters. Although it is present in several simulation systems in this form, in the absence of sufficient data to populate the regression equations, it is most often used in the simpler form originally published by Whiten (1972). Although the default breakage characteristics published for use with this model are generally sufficient, the model has the ability to make use of ore specific breakage data from ore testing via the JKMRC drop weight test. Estimation of the machine characteristics, however, requires a suitable database of "typical" parameters, or backfitting to measured data from operating crushers. Where backfitting is used, the model is not predictive for the effect of changes in machine geometry and is then only used to estimate the effects of changes in feed material type and size distribution.

Other jaw and cone crushing. In a greenfield situation, where no other data are available, the Whiten model approach has limitations. In such a case, models based on a manufacturer's "grading curves" can be used. Although these models are built into some packages (e.g., Bruno, AggFlow), they can be implemented in most simulation systems by simply entering tables of data extracted from the available manufacturer's literature. In some cases, multiple curves are available for different feed rate or crusher settings and these can be incorporated into such tables. Care does need to be taken using this approach, as it applies a predetermined distribution to the product stream, taking no account of feed size distribution, ore type, and so on.

Impact crushing (vertical shaft impactor and hammer mill). Both the Whiten crusher model and the manufacturer's grading curve approaches can be applied to impact mills.

Rolls crushing, toothed roll and sizers. Both the Whiten crusher model and the manufacturer's grading curve approaches can again be applied.

Rolls crushing (HPGR). The Whiten crusher model has been extended by Daniel and Morrell (2004) to handle modeling of HPGR crushing. In this case, the device is divided into three sections, pre-, bed, and edge crushing, and an individual basic Whiten model applied in each section. For simulation systems that implement the Whiten model but not the specific HPGR version, with reference to the Daniel and Morrell (2004) paper, the HPGR model can be implemented using three instances of the basic crusher model, with the precrusher in series with the bed and edge crushers operating in parallel.

AG/SAG. The general-purpose Whiten perfect mixing mill model (Whiten 1974) was incorporated into an energy-based, ore-specific AG/SAG model by Leung (1987). Further extended with power modeling calculations due to Valery and Morrell (1995) and implemented in the JKSimMet package, this model is believed to be the most widely used for design and optimization of SAG circuits. Although the model has been published in sufficient detail to allow implementation by others, it does not appear to be available in any of the other referenced commercial packages. Where other packages include calculations for AG/SAG mills, they appear to follow a more general approach such as that outlined by King (2001). In the absence of a suitable database of operations on which to base parameter selection, both the JKMRC model and the more general versions require backfitting of parameters from operating data. This has limited the application of the models for design. Coupled with the requirement for a deep understanding of the model limitations and experience in the design and operation of the physical equipment, most simulation work in this area remains the province of consultants and specialists.

Rod mill grinding. With rod mill use declining, development of new models in this area has been limited. The original Calcott–Lynch matrix model (Valery and Morrell 1995) is available in commercial packages such as JKSimMet and Limn and in the public domain (Lynch 1977; Napier-Munn et al. 1996) but does require backfitting to obtain useful parameters. For greenfield design situations, simple tables of estimated product size distribution will often suffice, since these units rarely operate in closed circuit.

Ball mill grinding. The Whiten perfect mixing model (Whiten 1974) also forms the basis of the most commonly used ball mill model. It is available in the public domain (Napier-Munn et al. 1996) and is widely used in commercial simulation systems. It has the facility for ore-type specific breakage information (appearance function) to be used but does require back calculation of combined breakage/discharge rate parameters from survey data. Although variations in mill size are accounted for in the model form, the fitted, combined breakage/discharge rate parameters can be further scaled to allow for variation in mill load, mill speed, and ball size. The more general approaches (King 2001) are also available in the public domain and have been implemented in commercial systems and in spreadsheets by some practitioners.

Stirred grinding (tower mill and IsaMill). The Whiten ball mill model (Whiten 1972) has been successfully used to model these devices.

New developments. A major issue with the use of the matrix comminution models for both design and optimization is the requirement to backfit breakage parameters to plant survey data to tune the model to particular applications. Daniel and Morrell (2004) have described an energy-based approach to comminution modeling, which, in conjunction with an extensive database from which the specific equation parameters have been derived, shows promise as a design strategy for comminution machine selection. The published approach and equations are in the public domain; however, the individual specific parameters are proprietary. When combined with a simulation system (the current implementation is used in conjunction with Limn—The Flowsheet Processor), the system may be used for overall comminution circuit design. Commercial use of the package CM-DOCC (Tian and Morrell 2015) is in its infancy but is available as a service through CITIC-SMCC Process Technology (2018).

Classification and separation. Although comminution is a high-cost component of plant flow sheets, the ultimate aim of a mineral processing plant is to upgrade the valuable components of the feed ore. Modeling of the separation stages required to achieve this has historically received less input compared with comminution, although more recently, considerable effort has been expended on modeling the flotation process.

Three main methods are typically applied to the modeling of separation devices:

1. Efficiency or partition curve approach,
2. Mineral recovery/mass recovery approach, and
3. Kinetic approach.

Partition curve approach. The partition curve approach reflects the reality that in any separation process, a particle must report to only one of the product streams. Where the measured property of interest driving the separation (e.g., particle size, density, color, X-ray fluorescence) is discretized into suitable groups or classes, the term *partition number* (also *partition coefficient*, *recovery factor*, *separation number*, *split factor*) is used to describe the fraction of material in that class that reports to the product stream of interest.

In simulation systems where the stream structure for input and output (feed and product) streams is defined in terms of those property classes, a similarly structured table in a unit model can be filled with the partition numbers for each class or grouping. Where a unit process has more than two products, the model can be set up as multiple partition tables, arranged in series or parallel as required to describe the process.

This partition table forms the most basic model of separation. However, with a single parameter for each property class, such models are not particularly useful.

To reduce the overall number of parameters, it is usual to attempt to link the partition numbers in some logical manner related to the physical separation process. Where the property classes represent subranges within a distribution (size ranges, density intervals), the partition numbers can be related by a sigmoidal or S-shaped curve. At one extreme of the property distribution, most particles in that property class report to one product; at the other extreme, most particles report to the alternate product; and at some intermediate property value, there will be equal numbers of particles reporting to both products.

The literature contains many examples of equations to describe partition curves (King 2001). The application of the basic partition equation forms of these "second-level" models has proven successful for obtaining reasonably detailed balances useful for benchmarking and mine-planning type studies. They find application in the physical separation processes, such as separations by size and density, where the few model parameters describing the basic curve shape do not tend to vary with ore/material properties and are reasonably well understood. The basic configuration of the Limn package (Maelgwyn Mineral Services Ltd. 2009) makes extensive use of models at this level in modeling, for example, coal, diamond, iron ore, and mineral sands circuits. The models are also relatively easily implemented in stand-alone spreadsheets.

After a partition curve has been defined (or a set of partition curves defined in a 2-D simulation framework), the parameters of that curve can themselves be related to machine operating characteristics. These higher-level relationships often comprise empirical equations whose parameters require backfitting from test data.

The two most commonly encountered examples of this "multilevel model" approach are size separation on screens, such as outlined in the Vibrating Screen Manufacturers Association handbook (VSMA 2014), or in hydrocyclones (Nageswararao 1990) and in comminution circuits and density or gravity-based separations, such as those operating in spirals or dense medium cyclones (Luttrell et al. 2005).

Mineral recovery–mass recovery approach. Apart from the partition curve approach, there are many other ways to represent separation data (Drzymala 2007). The approach of Holland-Batt (1989) has been implemented in at least one commercial simulation system (Fourie 2007) and appears useful for modeling gravity separators, particularly those found in mineral sands circuits. Based on the relationship between cumulative mineral recovery versus cumulative overall mass recovery, it provides a description of the process operating "separation window" and uses the concepts of separation efficiency and proportionality of efficiency to allow scaling for changes in feed grade. It is, however, an empirical approach and does require test work to derive the requisite relationships.

Kinetic approach. Although partition curve and mineral/mass recovery techniques are applicable to all separation processes, it is the layers built upon those base techniques that provide the framework for a more deterministic understanding of the underlying process.

Particularly in the modeling of flotation unit processes, an understanding of the kinetics of the process is important, and most of the commercially available flotation models are based on process kinetics, usually assuming a first-order rate equation. Different rate constants may be applied by size, mineral, floatability class, or liberation class. At least one proprietary simulation package, SUPASIM (Hay 2005), has been reported as successful in applying simple laboratory test work determined rate constants together with experiential heuristics in process unit and circuit models to build a better understanding of the issues in running the circuits.

Much research work has focused on measurement of many cell and bank operational parameters and in relating those parameters to the rate constants, resulting in significantly increased model complexity. The research effort from the JKMRC (JKTech 2018) and associated Australian Mineral Industries Research Association (AMIRA) (Schwarz et al. 2006) collaborative projects in this area is embodied in the commercially available JKSimFloat package (Schwarz and Richardson 2013). Calibration of the models remains a nontrivial exercise, and simulations of this nature are typically performed with a combination of experienced plant personnel and expert consultant assistance.

Ancillary processes (dewatering, water recovery, medium recovery devices, etc.). Although detailed theoretical phenomenological models exist for many of these devices, the implementation and incorporation in a circuit simulation of such models is not straightforward. They are typically modeled in only a most basic form in most commercial simulation packages and their role even ignored in certain instances. This is entirely appropriate where the goal of the simulation is, for example, to determine the likely plant recoveries for various ore blends in a geometallurgical or mine planning study.

Where inclusion of such unit processes is important (e.g., determining a water balance for a process design, or evaluating medium losses in a dense medium circuit), simple recovery-type models are often used.

Material handling (conveyors, pumps, bins, stockpiles, blending/ore handling). Generally, models of these process items are not required in steady-state simulations. Where needed to complete a flow sheet, a simple mixer is usually sufficient.

The most common dynamic simulations using these unit processes are longtime slice ore-flow-type simulations. The models are usually spatially related with internal data storage of property information at a spatial resolution to match the required output information.

Example of Steady-State Simulation

With the availability of commercial simulation packages, the task of the process engineer becomes one requiring an understanding of the simulation process, data requirements, and potential shortcomings, rather than one requiring the ability to program in a computer language or to have a deep understanding of the mathematical approaches behind the provided models. As an example, the following section describes the use of a typical steady-state simulation package to examine the potential for inclusion of a flash flotation unit within a grinding circuit. It is based on work described by Yan et al. (2005). Although this study details the modeling work, the example described here focuses on the simulation task and the user inputs and analysis required.

Project Planning

Prior to beginning a simulation exercise, whether it is performed personally, by an in-house specialist group, or by an external consultant, several basic steps are required.

Use expected outcomes to select the simulation environment. With any simulation project, the first question is: "Why am I doing this?" The answer to that question will determine the appropriate level of complexity required and generally point to appropriate tools to apply. For example, if the aim of the project is simply to produce a solids and water balance about a circuit, a simulation environment that requires size distribution data, grinding rates, and complex flotation models is probably not necessary and quite simple models may be suitable. Alternatively, if estimation of maximum mill throughput is one of the desired outcomes, more-detailed mill models are probably required.

In the example project covered here, the aim is to see whether flash flotation units will benefit an existing grinding and flotation circuit. Because particle size is expected to be important, the models must be able to take that into account. The existing plant produces a bulk sulfide concentrate, so for being an initial study, one probably does not need to track individual minerals but will, at a minimum, need to distinguish between sulfide and gangue species. The simulation system therefore must be capable of handling the size distributions of the two species present in each stream.

Choose flow sheet extent and detail. Focusing on just those processes that affect the required or expected outcomes can considerably simplify the tasks of collecting data and setting up the simulation. In a steady-state simulation, bins, conveyors, pumps, sumps, and pipes do not affect the outcome; therefore, omitting them will simplify the simulation with no loss of fidelity.

Similarly, if the focus is on the grinding and flotation sections, the feed to the grinding circuit can perhaps be taken as a constant, removing the need to simulate the crushing plant. Any considerations of dewatering units or other downstream processing may also be ignored.

In some cases, it is also possible to ignore certain aspects of the process streams. For example, in a mine planning or plant design study to determine the effect of ore block variability on a dense medium circuit and the consequent effect that will have on product quality and production rate over the life of the mine, it may be possible to simulate the ore solids only. Where necessary, percent solids, medium density, ore to medium ratio, and similar metrics can be used as model parameters to determine the requirements for these stream components in a particular process unit model, but it may not be necessary to include them in the overall circuit balance. Simplifying the stream description by removing any requirement to balance the accompanying water and dense medium components of the stream will simplify the overall simulation calculations and eliminate the need to include models for the water and medium recovery aspects of the circuit.

In the example flow sheet shown in Figure 1, the aim is to perform a qualitative assessment of the addition of flash flotation cells within the grinding circuit, and the following definitions, choices, and simplifications have been made:

- The simulated circuit ore feed is the feed to the fine grinding (ball mill) circuit.
- The circuit products will be the combined flotation concentrate and tailings.
- Simulation of conveyors, pumps, and sumps is not required.
- The stream ore description will include size distributions for two subspecies.
- Simulation of both solids and water stream components is required.
- Water addition points will be required.

Draw the flow sheet. Having made initial decisions regarding flow-sheet extent and detail, it is useful to begin to draw the flow sheet at this point because that task will direct attention to the important aspects of the simulation study.

A graphical flow sheet or process flow diagram forms the primary interface between the user and the simulation system. Most modern commercial simulation systems provide a means of defining the flow sheet in graphical form. This usually takes the form of a basic drawing package with a library of shapes representing items of process equipment and some facility for joining these shapes using lines to represent process flows.

The drawn flow sheet will reflect the decisions made in the project planning stage, ideally being as simple as possible, while still addressing the reason for conducting the study.

For the example task, the simplified flow sheet required to allow assessment of the effect of adding a flash flotation unit in the grinding circuit is shown in Figure 1.

Select suitable unit process models. Having drawn the flow sheet, the process units that will have to be modeled have been identified. Of course, model choice is driven by the availability of specific models in the selected simulation package. Compromises may need to be made. In the minerals processing arena, most models in regular actual use are empirical in nature and require some form of fitting to plant data, or tuning, before they can be applied. Availability of such data will also drive model choice.

Appropriate models. Most commercial simulation packages have a set of available process models. Some systems explicitly associate a model with the flow-sheet picture chosen to represent the process unit. Other systems associate a class of models with the picture, offering a limited choice, whereas some systems are completely general, relying on the user to make appropriate choices.

Where a suitable model does not exist, most systems have mechanisms to allow user-developed models to be incorporated, either as a standard part of the package or as an add-on "model developer's kit."

Where suitably detailed models do not appear to be present in any available simulation package, the user will need to choose between the additional costs required to develop such models (perhaps by some external consulting or research group) versus the loss in verisimilitude associated with choosing a more general model of lower accuracy.

The difficulty of implementing a simulation model from published papers should not be underestimated. Among other issues, the paper may have too narrow a focus for the intended application of the m3odel, may be reporting work in progress, or, for intellectual property reasons, may publish the broad framework of a model but omit detailed values for parameters that are only obtainable from a proprietary database. Similarly, models detailed in academic theses may require significant additional work to make them suitable for use in solving a real-life problem.

Appropriate levels of detail. As with the flow sheet itself, the models for the individual process units should be chosen to be as simple as possible while meeting the overall project goals. For example, although models exist that will allow optimization of cyclone physical size and orifice dimensions, they may require more data for tuning than are readily available, or be overly complex when compared with the result sought. In such cases, it may be more sensible to use simpler separation curve–type models with fewer parameters that can be estimated from a basic understanding of the process.

Models for the Example Flow Sheet

The flow sheet shows that models are required for the following:

- **Ball mill.** It is assumed that particle size will be an important property in determining how effective the flash flotation units will be and that removal of sulfide material in those units may affect the grinding process.

  The ball mill model will therefore need to respond appropriately to changes in feed size distribution. It

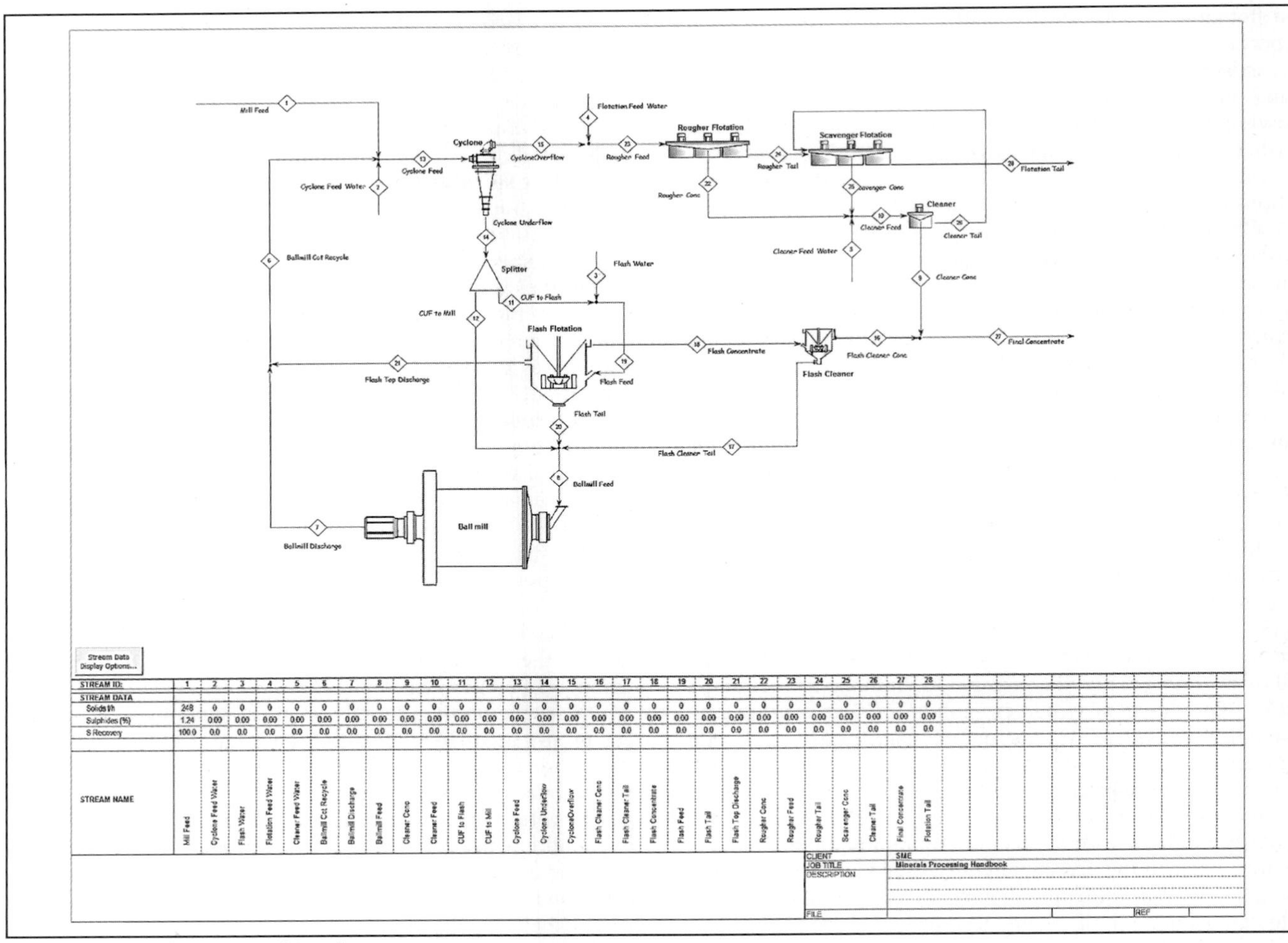

Figure 1 Example process flow sheet

will need to handle at least two ore subspecies with different grinding behavior.

The Whiten perfectly mixed ball mill model (Whiten 1972), using separate appearance functions determined from ore testing of each ore subspecies, was chosen for the example.

- **Cyclone.** If percent solids values are maintained in a similar range, the cyclone could be expected to behave in a similar manner under all regimes to be simulated. Consequently, a relatively simple "efficiency curve" model should be adequate.
- **Splitter.** The splitter model for steady-state simulation is trivial, requiring just a single split value to determine product stream makeup.
- **Primary and cleaner for flash flotation**. A size-by-size, two-parameter single-rate equation, recovery-based flash flotation model with the addition of solids and liquid bypass streams as in plant practice, was described by Mackinnon et al. (2003) and used for the flash flotation units in the example flow sheet given here.
- **Rougher, scavenger, and cleaner for conventional flotation.** A similar model to that used for flash flotation (without the bypass streams) was also used for the conventional flotation units.

Data collection. The choice of model will generally indicate the data required to tune or adapt the model to the circuit of interest. Good data are expensive and in any simulation study will invariably be in short supply.

Existing equipment. If the flow sheet or part of the flow sheet encompasses an existing plant, there may already be data available that can be used. Alternatively, fresh test work may be commissioned to obtain data explicitly directed at tuning the models for the proposed simulation task. Generally, each model has its own data requirements for tuning, and a series of specific targeted surveys about individual process units to meet these requirements may be a better use of resources than a complete circuit survey. If the test work is focused on one or more individual unit processes, achieving overall circuit steady-state conditions is less of a consideration.

New equipment. Where the process units in the flow sheet do not form part of an existing plant, either because the flow sheet represents part of a greenfield study or because the individual units are proposed new additions to an existing circuit, laboratory or pilot-plant test work may be required. Some experience with the model to be used will be invaluable in planning this test work.

Industry standard operational data. In some cases, it may be possible to use "industry standard" parameters as an alternative to generating them for a study. This is particularly the case for a separation process relying on physical separations, for example, size and density-based separations in coal or iron ore upgrading plant operations. In these cases, the equipment is expected to operate within a range of efficiencies and the parameters that represent these efficiencies and can, as a useful starting point, be considered as independent of plant operation, feed type, or equipment manufacturer and installation.

Potential for progressive improvement in model accuracy. As always, the decision about the level of data required should be driven by the cost/value trade-off for the level of accuracy required given the answers sought.

There can also be a benefit in performing initial simulation runs using more general models or more general model parameters to get a "feel" for how the available data fits the existing circuit, as well as a better understanding of the areas in which the data are deficient.

Data Collection for the Models in the Example Flow Sheet

For the example, the following data were available or obtained:

- **Ball mill.** From the work leading to the Mackinnon et al. paper (2003), a set of ball mill feed and product data is available to allow estimation of the breakage rate parameters for the Whiten ball mill model.

  Ideally, test work (drop weight breakage testing) to determine the ore-specific appearance functions for the sulfide and gangue components should be performed. In this case, it was not available and published "typical" values were used.
- **Cyclone.** The survey results used for the ball mill also included sufficient data to model the cyclone.
- **Splitter.** The splitter model split is a user-controlled value and no data are required.
- **Primary and cleaner flash flotation.** The Mackinnon et al. (2003) paper had described a model suitable for use for the flash cells.
- **Rougher, scavenger, and cleaner conventional flotation.** The work by Yan et al. (2005) describes the laboratory flotation test work that was required to enable application of the two-parameter, single-rate flotation model used for the flash units to also be applied to the conventional cells. These data were available for use in the example shown here.

Although the data available for the example are limited, it is of an appropriate level to address the reason for carrying out the simulation: a qualitative assessment of the addition of flash flotation cells within the grinding circuit.

Fit or tune models to match available process data. The empirical nature of most minerals processing unit models means that fitting those models to data is usually a necessary step.

Many simulation packages incorporate parameter estimation/model fitting routines (usually nonlinear, least squares based) within their framework. Other packages rely on external tools such as Microsoft Excel's Solver, Solver extensions from Frontline Systems Inc., or similar software.

There are advantages in having the parameter estimation routines available as part of the simulation package, since the actual models to be used are available in that environment without transcription to other tools. Also, where fitting of multiple units in a circuit is required, it is often advisable to fit them simultaneously to allow the effects of circulating loads to be included in the parameter estimation.

Multiple data sets. When using simple empirical models, where relationships between model parameters and operating variables are required (e.g., screening efficiency versus feed rate), it becomes necessary to fit multiple sets of data. Typically, each data set is fitted to the basic model, and regression equations are then determined between the fitted parameters and the measured operating variables. Again, there are trade-offs to be made between cost of test work and analysis against expected improved versatility of the model.

Model Fitting for the Example Flow Sheet

Because data were limited to the results of a single plant survey, it was not possible to derive higher-level relationships for the ball mill model in the example case.

The Whiten efficiency curve (Whiten 1972), however, can be extended to include an estimation of the effect of relative density of ore components using the relationship reported by Napier Munn et al. (1996), thus minimizing the number of parameters that need to be fitted to account for the two ore components being considered. This is shown in Figure 2.

Keeping in mind the reason for performing the study, the researchers felt that, provided percent solids and circulating load values could be kept similar between the measured data set and the simulation runs, this would not impinge on the broader desired outcomes and there was no incentive to collect additional data to model fit more complex models for mill or cyclone.

Because the laboratory flotation test work had been performed at four different aeration rates, it was possible to incorporate these data into an interpolative aeration rate model calculation layer built over the simple two-parameter, single-rate flotation model.

Determine input conditions. After the flow sheet has been drawn, and models have been selected (or developed) and suitably tuned by extracting parameters from data sourced from operating plants or laboratory test work, the remaining task is to input the process stream data to the simulation.

Ore feed streams. Depending on the simulation task to be undertaken, the input or feed streams to the simulation may be a single typical/representative set of component values or a set of data extracted from a table of such component values via an index, thus allowing the simulation of the circuit with multiple feed compositions.

A single scenario or data set for the feed stream is most typically used when the process itself is under scrutiny. Multiple scenarios or data sets for the feed are used, for example, to examine the treatment of the total ore body in a chosen plant configuration.

Ancillary feed streams. Additional input streams may be required where component makeup is required at various places in the circuit. Water makeup is the most obvious, but there may also be a requirement to include addition points

Model for Unit: Cyclone

Model Summary	Feed	Product 1	Product 2	Audit Check
Solids t/h	953.881	246.314	707.566	OK
Water t/h	655.36	455.79	199.57	OK
% Solids	59.275	35.082	78.00	-
Pulp SG	0.948	0.451	1.54	-
Volume m3/h	1,006.256	546.707	459.55	OK
Sulphides (%)	1.473	0.741	1.727	-
Gangue (%)	98.527	99.259	98.273	-
Sulphides (tph)	14.046	1.826	12.22	OK
Gangue (tph)	939.834	244.489	695.346	

Model Parameters	
Water Split	
Required % Solids in Underflow	78.00
Water split to Product 1	0.695
Efficiency Curve Parameters - to Product 1	
Nominal d50c	160.00
d50c Scale Factor - K2	1.166
Alpha - sharpness of cut	1.736
rf - bypass fraction	0.395
Feed SG Average	2.718

		Species	
		Sulphides	Gangue
		5.00	2.70
SG specific d50c		59.719	162.023
Size (µm)			
+6700	7,978.10	0.000	0.000
+4750	5,641.37	0.000	0.000
+3350	3,989.05	0.000	0.000
+2360	2,811.76	0.000	0.000
+1700	2,003.00	0.000	0.000
+1180	1,416.33	0.000	0.000
+850	1,001.50	0.000	0.000
+600	714.14	0.000	0.001
+425	504.98	0.000	0.012
+300	357.07	0.000	0.057
+212	252.19	0.002	0.152
+150	178.33	0.016	0.271
+106	126.10	0.066	0.375
+75	89.16	0.166	0.451
+53	63.05	0.285	0.502
+38	44.88	0.384	0.535
+25	30.82	0.462	0.558
-25	20.00	0.518	0.576

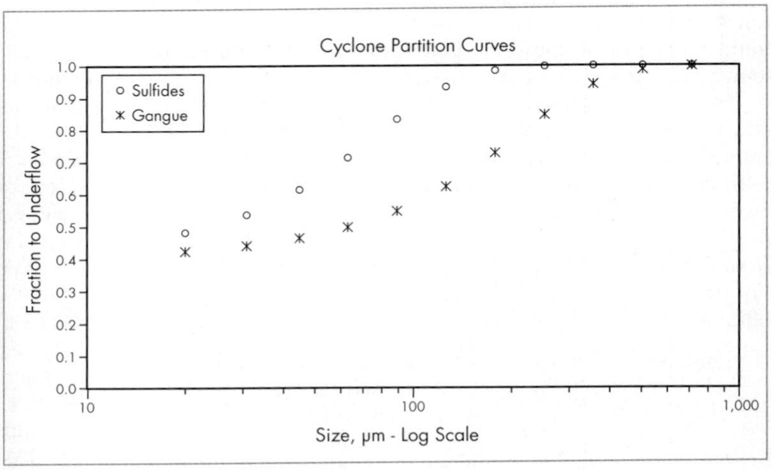

Figure 2 Fitted hydrocyclone model (Whiten efficiency curve equation for multiple minerals)

for reagents, medium solids makeup in a dense medium plant or carbon in a gold plant. Such requirements will depend on the degree of detail required for the simulation task to be undertaken.

Input Streams for the Example Flow Sheet

Because the example flow sheet simulation is to be used to assess large-scale circuit changes and given that the data available are limited, only a single "typical mill feed" ore stream was used. To maintain stream densities, water addition points were also included. The chosen simple models do not allow or require reagent additions or other such components.

Setup constraints. In any simulation, there are typically conditions that must be achieved or operational regimes that must be adhered to if the simulation results are to be realistic. In initial simulation runs, this may be achieved manually by observing the results and adjusting the inputs (and possibly unit parameters) to drive the simulation into the domain of interest.

After the simulation system is running in approximately the correct manner, however, it is an advantage to have available some form of constraint control built into the simulation system, which can ensure that the required regime is maintained. There are two main approaches to achieving this:

1. **Direct calculation of required values.** It is sometimes possible to build direct calculations into the models such that the required operating points become parameters of the model. For example, it would be possible to specify the percent solids of both the concentrate and tailings in a flotation model and to use this to determine the feed water additions required to achieve the water balance deriving from those percent solids constraints. However, if such a facility is not available within the model, the calculations must be performed in some other manner, perhaps via an ad hoc calculation facility built into the simulation system.
2. **Generic constraint controllers.** An alternative approach that has proven successful and useful is to provide constraint control via a separate model that is solved as part of the simulation sequence. A simple implementation of this is a controller acting in much the same manner as a conventional plant controller. The user specifies the dependent variable to be monitored, the independent variable to adjust, and the required value or set point for the constrained value. Based on the difference between the current and required constraint value, the simulation system adjusts the independent variable as part of the solution, such that the constraint calculation converges along with overall simulation model convergence.

Constraints for the Example Flow Sheet

In the example flow sheet, a generic constraint controller facility was used to control water additions to the ball mill and cyclone, as well as allow control of final concentrate grade if required.

Refine models and simulation conditions. During the preceding initial steps, the simulation model generally will have been run many times to identify areas that needed adjusting and to begin to understand the response of the overall simulation system to the available inputs.

If deficiencies are identified, or it is obvious that the system will not produce suitable answers to the questions posed of it, the deficiencies must be corrected and the models further refined, which may require additional test work, model fitting, and so on. This should be regarded as normal practice; development of a useful simulation model tends to be an iterative process.

Investigate operational scenarios. After the base case simulation runs have been performed and all stakeholders in the exercise are comfortable with the results, the true power of the simulation study will become apparent. Where "back of the envelope" studies can give an indication of results from a limited range of conditions, a properly structured simulation study can be used to examine the full envelope of options quickly, consistently, and with a high degree of confidence that mistakes have not been made in manual calculations or transcriptions, and that errors have not been made during cut-and-paste operations typically required in extending spreadsheet or similar software calculations to handle multiple inputs to the model.

Although all simulation systems will allow the user to enter a new set of equipment parameters or input conditions, rerun the simulation, and store the results, many systems also have built-in arrangements to automate the process of examining simulation scenarios. A table of inputs and conditions may be specified, simulation runs automatically performed over all the sets of conditions, and results logged for later analysis.

Example Scenarios

For the example shown here, the scenarios manager tool available in the simulation system used was set up to examine a range of cyclone underflow split to flash rougher cell values from 0%–70%. Figure 3 shows a summary of scenarios for the example flow sheet and its models.

Assess results. The state of the art of modeling of unit processes in the minerals processing arena remains largely empirically based. Thus, the results of any simulation study will only be a good as the data used in the modeling process. Models are usually limited in the range of applicability, and where processes themselves are ore-type dependent, the associated models are often restricted to one or, at best, a limited number of different sets of feed conditions.

With these factors in mind, there is a need to critically examine the results of the simulation runs by asking such questions as: Are they reasonable? Do they involve large extrapolations from the base data? Do they answer the original question?

Example Circuit Results

Figures 4 and 5 show the results from the multiple runs performed at varying cyclone underflow split to the flash flotation. As the split to the flash circuit increased, a slight increase in grade is observed, accompanied by a sizeable increase in sulfur recovery. Increases in size and recovery are unusual in normal flotation circuit operation, so it appears that the flash flotation unit has moved the simulated plant to a new operating point.

If the example was for a new plant design, given the reduction in sulfur recovery in conventional flotation indicated by the simulation, appropriate sizing of flash and conventional flotation sections for the recovery originally targeted could result in capital savings in the overall design.

Apart from the grade aspects, the simulation can also give an indication of the changes in size distributions within the circuit as well as the changes in flows.

Iterate the aforementioned steps as necessary. If the simulation predicts an answer to the question that is sufficiently attractive to justify additional work, the next step is to examine the models to determine where that work should take place. Most commonly, this will take the form of collecting more data—or more accurate data—to improve the veracity of the models. It may also take the form of additional laboratory or even pilot-scale test work to verify the simulation results. Further simulation can then be performed using this additional data and the process repeated until the original question is answered to the satisfaction of all concerned.

Example Circuit

The increase in recovery predicted would appear to be sufficient to encourage further work.

The flotation models particularly have been derived from limited test work and contain significant simplifying

Run the Scenarios

	Variables			*Results*									
Cell Address	0.25	1.00	22.677	6.0057	63.4142	22.677	79.0998	1.9254	36.3976	42.6676	1.4856	44.8408	40.5582
					Final Conc			Flash Conc			Flash Cleaner Conc		
Cell Description	Split to Flash	Concentrate Grade Controller Enabled	Concentrate Grade Setpoint	Cleaner Air Rate	Solids Flow t/h	%S Grade	%S Recovery	Solids Flow t/h	%S Grade	%S Recovery	Solids Flow t/h	%S Grade	%S Recovery
	0.70	0	-	6.00	61.38	22.53	85.15	3.61	28.38	62.44	2.44	39.21	58.18
	0.625	0	-	6.00	61.30	22.68	84.50	3.37	29.37	60.30	2.31	39.99	56.35
	0.60	0	-	6.00	61.28	22.72	84.26	3.29	29.72	59.52	2.27	40.26	55.67
	0.50	0	-	6.00	61.15	22.90	83.19	2.94	31.27	56.03	2.09	41.41	52.62
	0.40	0	-	6.00	60.98	23.04	81.84	2.57	33.06	51.71	1.88	42.67	48.79
	0.30	0	-	6.00	60.78	23.12	80.06	2.16	35.18	46.16	1.63	44.07	43.76
	0.20	0	-	6.00	60.51	23.09	77.55	1.67	37.75	38.46	1.32	45.66	36.66
	0.10	0	-	6.00	60.16	22.79	73.47	1.05	40.98	26.28	0.87	47.51	25.18
	0.00	0	-	6.00	59.70	21.54	64.59	0.00	0.00	0.00	0.00	0.00	0.00

Figure 3 Scenarios run for the example flow sheet and models

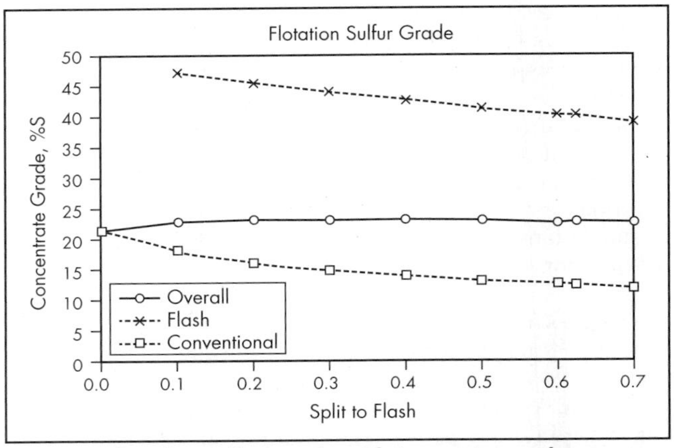

Figure 4 Flotation sulfur grade with increasing cyclone underflow split to flash flotation

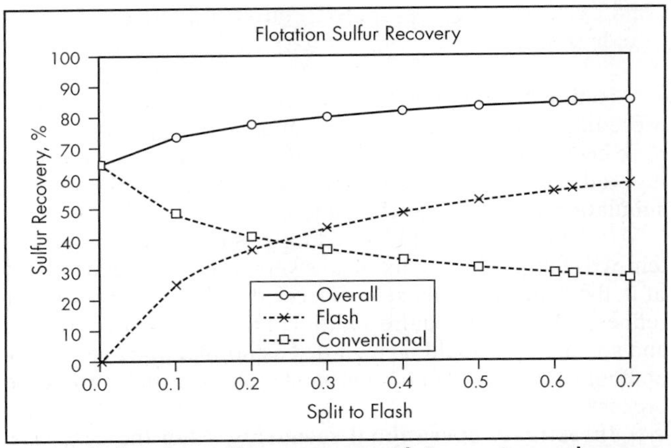

Figure 5 Flotation sulfur recovery with increasing cyclone underflow split to flash flotation

assumptions, namely, that the flotation rates in the conventional cells do not change as the split to the flash circuit is increased. This is clearly unlikely and would require additional test work at varying feed grades to the conventional flotation circuit to produce a more accurate model.

The models of the grinding circuit were also based on a single survey, and additional test work, particularly with ore of reduced feed sulfur if possible, would help to pin down the effect of removing sulfides from the grinding circuit circulating load.

Of course, if the results are of such magnitude and the models are conservative, the simulation could provide justification to proceed to installation of a pilot flash unit in the cyclone underflow that would provide further information. The simulation could then be repeated and used to justify (or discourage) the idea of proceeding to a full-scale installation.

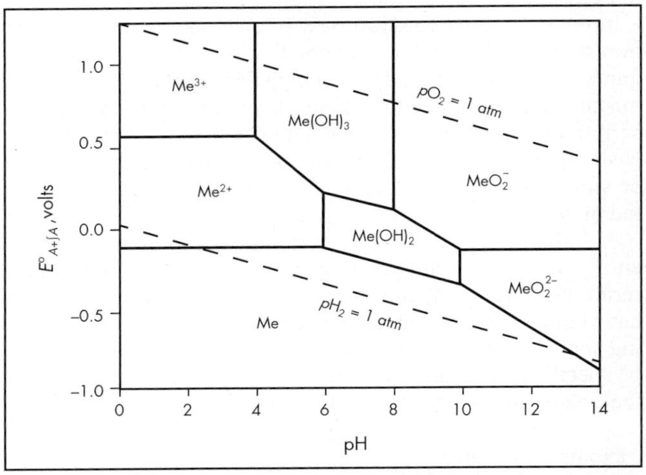

Source: Gilchrist 1989

Figure 6 Eh–pH diagram for the system metal Me–water as a function of pH and reduction potential

COMPUTER MODELING IN EXTRACTIVE METALLURGY

Process modeling in extractive metallurgy can start with determining chemical speciation based on known process conditions and test work followed by detailed chemical and physical models of unit operations to provide a more accurate representation of the extractive metallurgy in the process plant.

Modeling of Chemical Speciation in Extractive Metallurgy

Initial use of simulation packages for modeling of extractive metallurgy involving hydrometallurgical and pyrometallurgical processes is beneficial in clarifying the chemical speciation associated with the process concept. Although test work and information from operating plants are used to develop the conceptual chemistry as well as heat and mass balance, in addition or alternatively at initial stages of design, process modeling can be used to perform chemical and equilibrium calculations for aqueous and solid systems. Still, pilot plants and test work are important in providing information for rates of chemical reaction and non-ideality of solutions.

Establishing the process chemistry for hydrometallurgical reactions is fundamental in understanding the process and its conditions. For this reason, computer modeling of aqueous thermodynamics is used for understanding system chemistry and determining preliminary information on the behavior of aqueous/solid systems in the form of Eh–pH diagrams. In hydrometallurgy, a Pourbaix diagram, also known as an Eh–pH diagram, outlines stable (equilibrium) phases of an aqueous electrochemical system. Predominant ion boundaries are represented by lines. An Eh–pH diagram is like a standard phase diagram and does not contain any information on reaction rate or kinetic effects.

Figure 6 shows a characteristic Eh–pH diagram for a typical metal Me for which the standard half-cell potentials are represented as follows:

$$Me^{n+} + ne^{-} = Me \quad \textbf{(EQ 1)}$$

$$Me^{m+} + (m - n)e^{-} = Me^{n+} \quad \textbf{(EQ 2)}$$

$$Me\,(OH)_n + nH^{+} + ne^{-} = Me + nH_2O \quad \textbf{(EQ 3)}$$

The third potential is a function of pH, assuming unit activity for Me and Me $(OH)_n$. For a given metal, these potentials may be plotted in an $E°_A$ versus pH diagram. This is represented in Figure 6 for the arbitrary metal Me.

The lines in Eh–pH diagrams show the equilibrium conditions, where the activities are equal, for the species on each

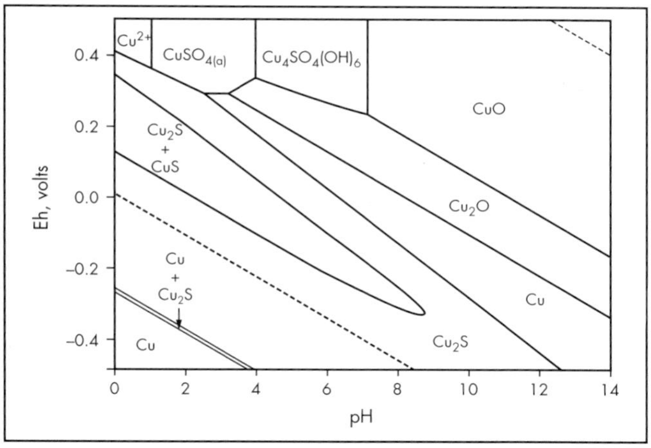

Figure 7 Eh–pH of Cu-S showing coexistent of two solids in one area

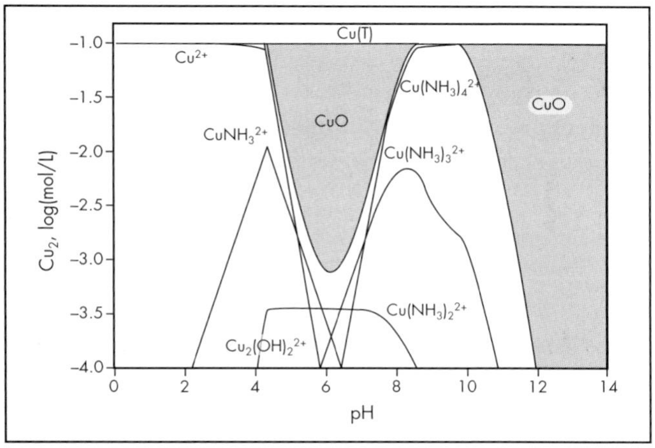

Source: Baes and Mesmer 1976

Figure 8 Distribution pH of Cu(II) with NH_3 generated by Stabcal program

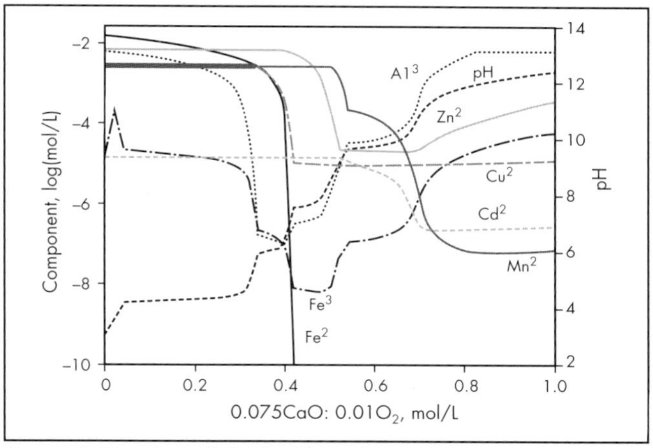

Figure 9 Titration of the Berkley pit water

side of that line. On either side of the line, one form of the species is predominant.

STABCAL (2015) (stability calculation for aqueous and nonaqueous systems) is an integrated Windows program that performs comprehensive stability calculations for complex systems for hydrometallurgical, pyrometallurgical, and environmental engineering; geochemistry; and chemistry for academic research and industrial applications. The program is broken down into two separated groups: aqueous and nonaqueous. Each group function independently has its own computer code and databases taken from the literature.

STABCAL can produce Eh–pH diagrams showing the coexistence of several solids in one area. For instance, the Eh–pH diagram in Figure 7 shows that both covellite (CuS) and chalcocite (Cu_2S) are stable in a large area of the Eh–pH diagram of Cu-S-H_2O system.

Another important diagram for aqueous system is the distribution versus pH (Baes and Mesmer 1976). Besides showing the hydrolysis of metal ions on the diagram, STABCAL reveals the formation of complexes from metal with ligands such as Cu(II) and NH_3. This is shown in Figure 8.

STABCAL is also capable of simulation of systems where chemicals are added to an aqueous solution. For instance, by addition of CaO and O_2 stepwise to the Berkeley pit water (Butte, Montana, United States), as depicted in Figure 9, the program will show how the metal hydroxides precipitate, and while Fe(II) is being oxidized, Eh and pH change.

Another program with similar capability in simulation of speciation in extraction of metals and mineral processes is HSC Chemistry by Outotec (Lishchuk 2016). It can show this information by drawing simple Eh–pH diagrams for systems as basic as an element and water, as well as more complicated diagrams with several elements.

Fundamentals of Heat and Mass Balance in Extractive Metallurgy

In extractive metallurgy, steady-state computer modeling and simulation is usually limited to producing a heat and material balance based on known process design criteria. Process design criteria for well-defined operations include process conditions such as temperature and pressure, feedstock and fuel flow rates and compositions, physical properties including thermochemical properties for each input and output material stream, and the extent of chemical reactions based on the known chemical stoichiometry for the process. Knowing the extent of chemical reactions for a commercial or pilot-plant operation means that reaction kinetics are well known or have already been determined from test work or similar operating plants. The heat and mass balance is then determined based on the following criteria:

- Mass flow rates

$$\text{rate of mass in} = \text{rate of mass out} \quad \textbf{(EQ 4)}$$

- Chemical reactions and their extent and phase changes for each process unit
- Stoichiometry (for chemical-reacting systems in addition to the law of conservation of mass)

The non-steady-state form for the rate of an individual component, j, is as follows:

$$\frac{dmj}{dt} = \sum_{i=1}^{n\,\text{in}} \frac{dmj}{dt},\ \text{in},\ i - \sum_{i=1}^{n\,\text{out}} \frac{dmj}{dt},\ \text{out},\ i + rp,\ j - rc,\ j \quad \textbf{(EQ 5)}$$

where

m = unit mass
j = representative component in all input and output streams
rp, j and rc, j = its rates of production and consumption by chemical reactions

Similarly, for steady states, the component mass balance equation is reduced to

$$\sum_{i=1}^{n\,\text{in}} \frac{dmj}{dt}, \text{ in, } i + rp, j = \sum_{i=1}^{n\,\text{out}} \frac{dmj}{dt}, \text{ out, } i + rc, j \qquad \textbf{(EQ 6)}$$

In extractive metallurgy where there is chemical or physical change, system energy can be gained or lost from the following sources:

- Sensible heat for each input and output material
- Latent heat because of phase change
- Heats of reaction for all chemical reactions
- Heat inputs and outputs to and from the system by a heat source or cooling

Energy balance at a steady state is defined as total heat input minus heat output. Sources of heat input considered are heat of reaction at 298 K, heat of fusion and evaporation (latent heat), and sensible heat in reactants. Heat balance is an important exercise, and it is used to determine the theoretical temperature reached during the extractive metallurgy process step. Knowledge of heat balance and temperature reached is useful in determining the materials of construction in hydrometallurgy as well as refractory material and overall roaster or furnace design in pyrometallurgy.

The non-steady-state form of energy balance is given as

$$\frac{d(mh)}{dt} = \left(\frac{dm}{dt}\right)_{\text{in}} \cdot h_{\text{in}} - \left(\frac{dm}{dt}\right)_{\text{out}} \cdot h_{\text{out}} + rh\,p = rh\,c \qquad \textbf{(EQ 7)}$$

where

m = unit mass
h = associated enthalpy
$rh\,p$ and $rh\,c$ = rates of heat production and consumption because of chemical reactions, heat input to the system, or cooling

For steady-state balance, the energy balance reduces to

$$\left(\frac{dm}{dt}\right)_{\text{in}} \cdot h_{\text{in}} + rh\,p = \left(\frac{dm}{dt}\right)_{\text{out}} + rh\,c \qquad \textbf{(EQ 8)}$$

Understanding the energy balance as part of extractive metallurgy process design is important. For example, in pyrometallurgy, concentrates are often roasted with coal to convert metal sulfide into oxide. Heat generated by roasting and combustion raises the temperature of the products of coal combustion and roasting. In this case, one goal of pyrometallurgical process simulation is to calculate the temperature under adiabatic conditions, that is, no heat input or loss. This calculated temperature can then be adjusted by incorporating heat losses. The steady-state process simulation would then consider an adiabatic process of roasting with the following heat sources: heat of combustion, heat of reaction, and the sensible heat in reactants. The products consist of solids and gases.

The rate of change of enthalpy of a substance with temperature at fixed pressure is called the heat capacity at constant pressure, Cp. Given the function $Cp\ (T)$, the change in enthalpy of a substance as its temperature is raised:

$$\sum_{j=1}^{n} (H_T - H_{298}) = \sum_{j=1}^{n} (H_T \frac{dmj}{dt}, \quad Cpj \int_{298}^{T} Cp\ (T)\ dT \qquad \textbf{(EQ 9)}$$

where T is the adiabatic temperature attained in a pyrometallurgical process.

Thus, for adiabatic roasting and combustion, Equation 9 can be used together with the liberated or consumed heats of reaction and combustion. If the material balance for roasting is known, as well as temperatures of fuel and air, heat capacity (Cp), and heat of reaction, the temperature of the process can be calculated a part of a steady-state process model. The final step in this case is accounting for the input and output components, and determining heat and mass balance by calculating the difference between input and output streams.

In hydrometallurgy, water is a major system component, and the latent heat of evaporation of water is an important variable in the heat balance. In processes operating at temperatures greater than the atmospheric boiling point of water, the pressure is elevated and the design of equipment for such processes is strongly dependent on the vapor pressure of water at the required temperature.

Application of Process Modeling in Hydrometallurgy

In its simplest form, hydrometallurgy can be described as a technique within the field of extractive metallurgy involving the use of aqueous chemistry for the recovery of metals from ores, concentrates, solutions, and recycled or residual materials. In hydrometallurgy, although sophisticated software is available for process modeling, acquiring and mastering it entails appreciable cost, time, and effort. It is also possible to use spreadsheet calculations for preliminary evaluation of envisaged circuits for processing ore or concentrate.

The modeling software METSIM (Proware 2018), used extensively in extractive metallurgy, comes with models that include chemical reactions and hydrometallurgical unit operations to simulate leaching, autoclaves, solid–liquid separation, precipitation tanks, and numerous other pieces of equipment such as tanks, pumps, bins, stockpiles, dryers, kilns, furnaces, boilers, and steam and gas handling equipment. In METSIM, an interface to FactSage is also available to access Fact's extensive databases.

METSIM is also used for development of simulation models for heap leaching that include a mine block and plan that take into consideration the geometry of heap, its leach cycles, and metal recovery in conjunction with gold carbon-in-circuit and copper solvent extraction and electrowinning operations. The results of such models would be dynamic mass balances over extensive periods of heap leaching operation.

Marsden and Botz (2017) have given a complete review of current approaches to metal production forecasting in heap leach modeling. In their review, heap leaching models are divided into six categories based on type and functionality. The first three categories are Excel based, with the most basic having no scale-up or time delay functionality and the other two as having empirical scale-up and time delays and in-heap inventory-adjusted production estimation. The fourth category consists of off-the-shelf heap leach modeling software capable of some of the aforementioned capabilities.

Off-the-shelf heap modeling includes METSIM and SysCAD described previously. At this level, geometry of the heap is represented in the model. The heap is represented as many "blocks" or "reactors" that contain solids, liquids, and gases. These models can depict the chemical reactions based on test work such as leach test curves. Drainage factors are also used to calculate moisture contents within each block of the heap.

The last two categories, known as "phenomenological" modeling, are based on use of CFD, thermodynamics, and kinetics of the reactions involved and with the most advanced version having 3-D heap geometry capability.

Another software package that can be used during all stages of a hydrometallurgical plant for numerical analysis and process evaluation is IDEAS (Andritz 2018). This software package includes unit operation objects in libraries that enable users to effectively simulate a nickel laterite plant. IDEAS libraries feature a flexible and easily customized database that contains the material properties for components commonly used in the mineral industry. IDEAS has powerful steady-state capabilities. It performs steady-state mass and energy balances; tracks components, compounds, and element flow and concentration; and handles particle size distributions. IDEAS has been used to model complex plants that include high-pressure acid leaching, heat recovery circuits, neutralization, countercurrent decantation (CCD), autoclaves, precipitation, filtration, separation, solvent extraction, and electrowinning.

Steady-state models created in IDEAS can be easily converted to a dynamic environment to include detailed dynamic specifications and process control logic. By using IDEAS for control logic verification, plants can reduce costly design errors that could otherwise delay start-up. In fact, studies have shown that using simulation to help with start-up can correct a great percentage of control logic problems before field implementation.

The dynamic simulator can also be used for operator training. Working just like a flight simulator, IDEAS allows trainees to gain realistic, hands-on experience without inflicting harm on themselves, the environment, or the plant.

Example of Process Modeling in Hydrometallurgy

This section uses an example to best illustrate the hands-on use of process modeling software and spreadsheet-based calculations in hydrometallurgical process development. The example chosen for this exercise is the processing of laterite. This example is generic and simplified, but the principles illustrated can be applied to actual projects. The exact numbers used are less important than the principles illustrated.

Laterite processing. This example is based on general process knowledge and published information, including the results of laboratory leaching tests (Dutwa Project 2018). The key finding of the experimental work was that this particular laterite ore responds well to leaching at low temperature and reasonably low additions of sulfuric acid, leading to the speculation that lower-cost atmospheric leaching might be appropriate in this instance, rather than the more established high-pressure acid leaching (HPAL) route.

Mineralogy. In spreadsheet calculations and in process modeling using specialized software, the elemental analysis of the feed ore or concentrate must first be translated into a suite of valuable and gangue minerals, such that the mineral suite accounts for the total mass of the feed and correctly reflects the available data. Modeling software such as METSIM and Aspen Plus can be used as a better alternative to spreadsheets to model a variety of processes in extractive metallurgy. In this example, feed assay and acid leach data are available. Selecting the minerals in this suite would initially be guided by the geology of the deposit, and later refined via mineralogical analysis and other experimental data. Knowledge about how the various minerals respond to different chemical environments can then be used to compile one or more conceptual processes for extracting the valuable elements. While in spreadsheets, the user must compile the thermochemical properties of the elements and components that are used in the heat and mass balance; the commercial modeling software provide the advantage of having an associated compilation thermochemical database.

Table 1 contains the available information on the composition of the laterite ore used in this example. In laboratory leaching tests this laterite was found to consume 210 kg H_2SO_4/t (metric ton) of dry laterite ore and to release 92% of the nickel into solution during the leach.

Typically, nickel occurs in laterites as nickel oxide incorporated in goethite-type minerals and in magnesium silicate. The iron is usually present as goethite (FeOOH) and hematite (Fe_2O_3), the relative proportions depending on the degree of weathering of the ore. Magnesium is generally present as magnesium silicate minerals. Acid leaching generally decomposes the magnesium silicate minerals and the goethite but not the hematite, so the proportions of goethite and hematite influence the overall consumption of acid. The silicate in the magnesium silicate minerals is converted to silica in the solid residue. Table 2 shows the acid leach chemistry in simplified form. Table 3 shows a simple representation of the laterite as

Table 1 Laterite assay, mass percentage

Laterite Assay	Mass Percentage
Ni grade	1.1
Co grade	0.034
Cu grade	0.007
Fe grade	8.5
Mg grade	3.5

Source: Dry 2013

Table 2 Acid leach chemistry

$$NiO + H_2SO_4 \rightarrow NiSO_4 + H_2O$$

$$CoO + H_2SO_4 \rightarrow CoSO_4 + H_2O$$

$$CuO + H_2SO_4 \rightarrow CuSO_4 + H_2O$$

$$MgSiO_3 + H_2SO_4 \rightarrow MgSO_4 + H_2O + SiO_2$$

$$2FeOOH + 3H_2SO_4 \rightarrow Fe_2\,(SO_4)_3 + 4H_2O$$

Source: Dry 2013

Table 3 Laterite representation, mass percentage

Laterite Representation	Mass Percentage
NiO	1.40
CoO	0.04
CuO	0.01
MgO	5.80
FeOOH	3.22
Fe_2O_3	9.26
SiO_2	80.18

Source: Dry 2013

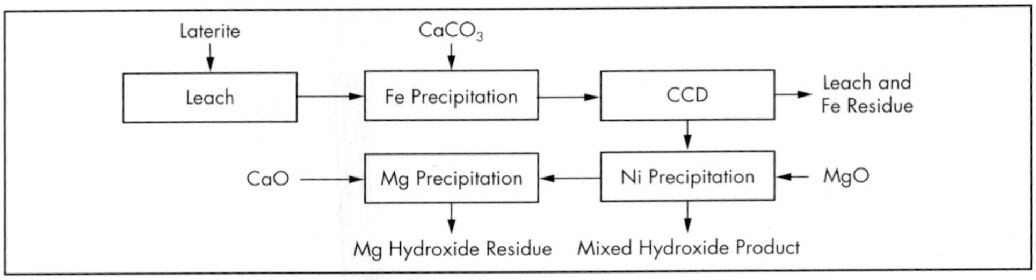

Figure 10 Block flow diagram of the laterite circuit

a mixture of oxides of the valuable metals, goethite, hematite, magnesium oxide, and silica.

The knowledge that goethite is dissolved and hematite is not enables the calculation of the amount of goethite required to give the observed overall consumption of sulfuric acid. This is a simple acid balance.

In molar units, 1,000 kg of dry ore contains the following:

- 0.187 kmol of NiO
- 0.006 kmol of CoO
- 0.001 kmol of CuO
- 1.440 kmol of MgO
- 1.522 kmol of iron, split between FeOOH and Fe_2O_3

The total consumption of H_2SO_4 is 210 kg, or 2.141 kmol per 1,000 kg of ore. The dissolution of NiO, CoO, CuO, and MgO accounts for 1.634 kmol of the acid, leaving the balance for the dissolution of FeOOH, making the amount of FeOOH dissolved 0.338 kmol per 1,000 kg of dry ore. Assuming the percentage dissolution of FeOOH to be the same as those measured for nickel, cobalt, and copper (92%) and that the dissolution of MgO is also 92% gives the mineral assemblage listed in Table 3. Hematite accounts for the difference between the total iron and the iron present as goethite. Silica is simply the difference between 100% and the sum of the other components. This mineral assemblage correctly reflects the available measured data because the feed sample was not assayed for silica, hence the use of silica as the difference between the total mass and the mass of the other minerals. In more detailed work, the silica assay would also be available and the mineral assemblage would very probably be somewhat more complex, but this example still illustrates the principle.

Modeling calculations for laterite processing. Formulating a conceptual process includes choosing the form of the products and by-products. For example, a laterite deposit could be processed to an intermediate hydroxide product containing the nickel and cobalt, or all the way to metallic products. Selling prices need to be assigned to each product and by-product so the potential revenue from the deposit can be estimated.

The next step would be to define one or more potentially suitable process routes and evaluate the potential cost of processing the ore for each route selected. When an envisaged process is hydrometallurgical, that generally requires laboratory leaching tests to measure the extractions of valuable metals and the consumption of reagents such as acid. It is reasonable (but not essential) to calculate the consumption of, say, acid from the mineral suite and knowledge of the chemistry of the various minerals before commissioning laboratory tests. It is, however, always prudent to obtain experimental confirmation of the leach at an early stage.

In this example, the circuit entails leaching with sulfuric acid, neutralization of the resulting slurry with limestone, solid–liquid separation, precipitation of a mixed nickel–cobalt hydroxide with magnesium oxide, and precipitation of magnesium and sulfate from the barren solution using lime. Figure 10 is a block diagram illustrating a basic circuit.

Two variations of the basic circuit are covered in this example: leaching in agitated tanks under atmospheric pressure (ATL) and leaching at high pressure and high temperature (HPAL) in autoclaves. For the ATL case, the leach chemistry is the same as that shown in Table 2.

Leaching laterite at elevated temperature and pressure (HPAL) is done when the ore is more limonitic in nature, meaning that most of the nickel is contained in goethite-like iron minerals and those minerals make up the bulk of the ore. For such laterite ores under atmospheric leaching conditions, dissolving essentially all the iron minerals to release the nickel into solution and converting the iron to dissolved ferric sulfate would require large amounts of acid and then large amounts of limestone to neutralize the leach solution and precipitate out the iron. At the elevated temperatures generally used in HPAL (typically 240°–270°C), most of the dissolved ferric iron hydrolyzes and is precipitated as hematite, releasing the acid consumed in its dissolution and thereby greatly reducing the overall acid requirement. Leaching laterite under HPAL conditions also greatly accelerates the leach reactions, relative to leaching at or below the atmospheric boiling point of the solution.

At the elevated temperatures of HPAL, the second dissociation of sulfuric acid essentially does not occur. The top half of Table 4 gives stoichiometry illustrating the HPAL leach chemistry. On cooling after the high temperature leach reaction, the bisulfate salts revert to sulfuric acid and sulfate salts, as illustrated by the stoichiometry shown in the bottom half of Table 4. Although the overall chemistry is the same as for the atmospheric leach, the behavior of sulfuric acid under HPAL conditions is such that more than half of the sulfuric acid added must be neutralized after the leach.

In saprolitic laterites, which contain higher levels of magnesium, the acid requirement in HPAL is high even though the dissolved iron is re-precipitated as hematite, because the magnesium dissolves but does not re-precipitate.

The extent of precipitation of ferric iron in HPAL is influenced by the amount of residual sulfuric acid remaining in the solution. Figure 11 shows published data relevant to the hydrolysis of ferric iron under HPAL conditions (Papangelakis et al. 1996). In these measurements, the slurry contained 22% solids and the temperature was 250°C. This can be used to calculate the acid consumption that leaching the mineralogy in Table 3 under HPAL conditions would require for the same dissolution of nickel as in the heap leach option, assuming the same

Table 4 HPAL stoichiometry

$NiO + 2H_2SO_4 \rightarrow Ni^{2+} + 2HSO_4^- + H_2O$
$CoO + 2H_2SO_4 \rightarrow Co^{2+} + 2HSO_4^- + H_2O$
$CuO + 2H_2SO_4 \rightarrow Cu^{2+} + 2HSO_4^- + H_2O$
$MgO + 2H_2SO_4 \rightarrow Mg^{2+} + 2HSO_4^- + H_2O$
$2FeOOH + 6H_2SO_4 \rightarrow 2Fe^{3+} + 6\ HSO_4^- + 6H_2O$
$2Fe^{3+} + 6\ HSO_4^- + 3H_2O \rightarrow Fe_2O_3 + 6H_2SO_4$
$Ni^{2+} + 2HSO_4^- \rightarrow NiSO_4 + H_2SO_4$
$Co^{2+} + 2HSO_4^- \rightarrow CoSO_4 + H_2SO_4$
$Cu^{2+} + 2HSO_4^- \rightarrow CuSO_4 + H_2SO_4$
$Mg^{2+} + 2HSO_4^- \rightarrow MgSO_4 + H_2SO_4$
$2Fe^{3+} + 6\ HSO_4^- + 6H_2O \rightarrow Fe_2(SO_4)_3 + 3H_2SO_4$

Source: Dry 2013

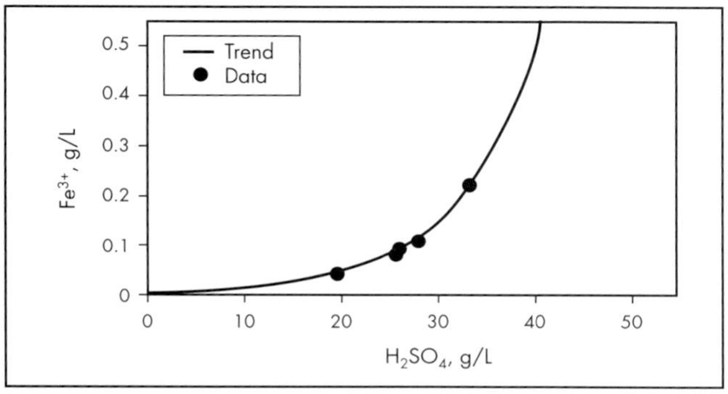

Source: Marsden and Botz 2017

Figure 11 Ferric iron, hydrolysis data

conversion for the first five reactions in Table 4 and adjusting the conversion of the sixth reaction at the same final acid concentration as for the atmospheric leach (5 g/L) in a leached slurry containing 25% solids, such that the dissolved ferric iron lies on the trend line shown in Figure 11. This calculation gives an acid requirement of 323 kg/t of ore, compared to 210 kg/t measured in the leach tests at atmospheric pressure. For generating the acid from elemental sulfur, the sulfur requirements would be 69 kg/t for the atmospheric leach and 106 kg/t for the HPAL case.

The pregnant liquor from the leach is treated with limestone ($CaCO_3$) to neutralize residual acid and to precipitate the dissolved iron. Table 5 shows the overall chemistry associated with the iron precipitation step. The amount of limestone consumed is fixed by the amount of iron dissolved in the leach and the level of free acid in the solution after leaching.

The nickel, cobalt, and copper are precipitated from the iron-free pregnant liquor as a bulk hydroxide, using magnesium oxide. Table 6 shows the overall chemistry assumed for this step. The precipitate would be washed, dried, and sold. Its nickel content would be about 50% by mass. The consumption of magnesium oxide is fixed by the amounts of nickel, cobalt, and copper precipitated and the amount of residual MgO in the precipitate.

Calculating the residual magnesium oxide in the mixed hydroxide precipitate (MHP) is straightforward, as follows:

Table 5 Iron precipitation

$H_2SO_4 + CaCO_3 \rightarrow CaSO_4 + H_2O + CO_2$
$Fe_2(SO_4)_3 + 3CaCO_3 + H_2O \rightarrow 3CaSO_4 + 2FeOOH + 3CO_2$

Source: Dry 2013

Table 6 Base metal precipitation

$NiSO_4 + MgO + H_2O \rightarrow Ni(OH)_2 + MgSO_4$
$CoSO_4 + MgO + H_2O \rightarrow Co(OH)_2 + MgSO_4$
$CuSO_4 + MgO + H_2O \rightarrow Cu(OH)_2 + MgS_O4$

Source: Dry 2013

Table 7 Magnesium precipitation

$MgSO_4 + Ca(OH)_2 \rightarrow Mg(OH)_2 + CaSO_4 + 2H_2O$

Source: Dry 2013

- Ni dissolved from the laterite: 1.01 kg/t ore
- Assumed Ni in MHP: 50 mass %
- MHP produced: 0.202 kg/t ore
- $Ni(OH)_2$ in MHP: 0.160 kg/t ore
- $Co(OH)_2$ in MHP: 0.005 kg/t ore
- $Cu(OH)_2$ in MHP: 0.001 kg/t ore
- Therefore, MgO in MHP: 0.037 kg/t ore

From this calculation, the residual MgO in the MHP is small compared to the amount required to precipitate the base metals.

The barren solution remaining after precipitation of the base metals contains magnesium and sulfate that are precipitated with slaked lime, according to the chemistry shown in Table 7. This is necessary to avoid a buildup of magnesium sulfate in the circuit as the process water is recycled.

The amount of slaked lime required can be calculated from the amount of magnesium, which can be calculated from the amount of magnesium leached and the amounts of nickel, cobalt, and copper precipitated.

Lime (CaO) is produced by heating limestone ($CaCO_3$) to between 900°C and 1,000°C, driving off the carbon dioxide and leaving reactive calcium oxide. A conservative estimate of energy consumption for this is 6 mJ/kg CaO produced. One source of energy is coal. A typical heating value for coal is 30 mJ/kg, so a rough approximation of the amount of coal needed to produce 1 t of lime from limestone would be 200 kg.

Table 8 lists the amounts of sulfur, limestone, magnesia, and coal calculated for a circuit processing the laterite in this example, for ATL and HPAL technology, in kilograms per metric ton of the dry feed laterite.

For a venture to be commercially viable, the costs associated with any process need to be sufficiently less than the revenue generated. Commodity prices are not static; therefore, a single current price for any given commodity will probably not be reasonable in a study that would entail starting a commercial operation, say, 5–10 years into the future and operating it for the next two or more decades. One way around this problem is to use price ranges instead of single numbers for the different commodities. Figure 12 shows historical commodity prices (Kelly and Matos 2014) relevant to this exercise in constant 2013 U.S. dollars, over the past century for nickel, cobalt, sulfur, lime, and magnesia and over the past half century for coal (U.S. Energy Information Administration 2018). The dashed lines on these charts are the long-term arithmetic average price, plus or minus one standard deviation. Although the various commodity prices have moved outside the "channels" depicted by the dashed lines in these graphs, they have always returned and each one has spent the bulk of the time covered within a standard deviation of the long-term average. In the absence of more detailed pricing for the relevant commodities, it is not unreasonable to assume the range indicated by the dashed lines for the relevant commodity.

Table 8 Reagent consumption from spreadsheet calculations

	Consumption, kg/t	
Reagent	**ATL**	**HPAL**
S	69	106
$CaCO_3$	214	172
MgO	7	7
Coal	17	17

Source: Dry 2013

Another cost item that needs to be included in these calculations is the cost of mining the laterite and delivering it to the processing facility. Laterite mines are usually opencast, and a mining cost model reference from the public domain (InfoMine 2018) gives a mining cost of US$7.32/t of ore mined. For the laterite in this example, that translates to US$0.33/lb of nickel in the hydroxide product.

Table 9 lists the variable costs from the spreadsheet model calculations and the commodity price ranges shown in Figure 12, along with the calculated revenue, assuming that

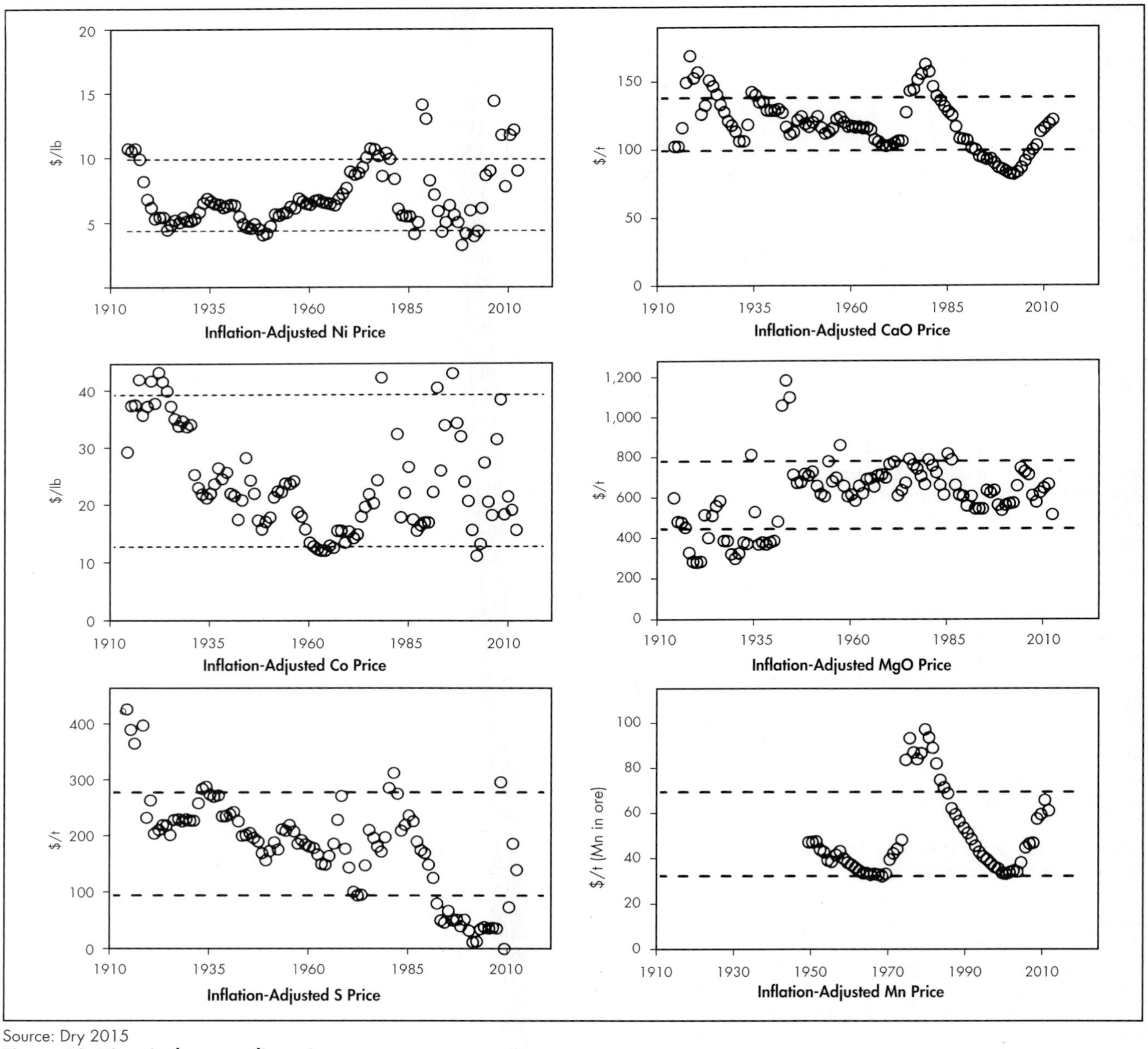

Source: Dry 2015

Figure 12 Historical commodity prices

Table 9 Variable costs from spreadsheet calculations

	Consumption Ore, kg/t		Reagent Price, $/t		Variable Cost, $/lb Ni			
					ATL		HPAL	
Reagent	ATL	HPAL	Low	High	Low	High	Low	High
S	69	106	96	278	0.30	0.86	0.46	1.32
$CaCO_3$	214	172	100	138	0.96	1.33	0.77	1.07
MgO	7	7	449	782	0.14	0.25	0.14	0.25
Coal	17	17	33	70	0.03	0.05	0.03	0.05
Mining cost ($7.32/t ore)					0.33	0.33	0.33	0.33
Total calculated variable cost					1.75	2.82	1.72	3.02
Revenue for Ni hydroxide (75% of Ni price)					3.33	7.48	3.33	7.48
Revenue for Co hydroxide (75% of Co price)					0.30	0.91	0.30	0.91
Revenue minus variable costs					1.88	5.57	1.90	5.37

Source: Dry 2013

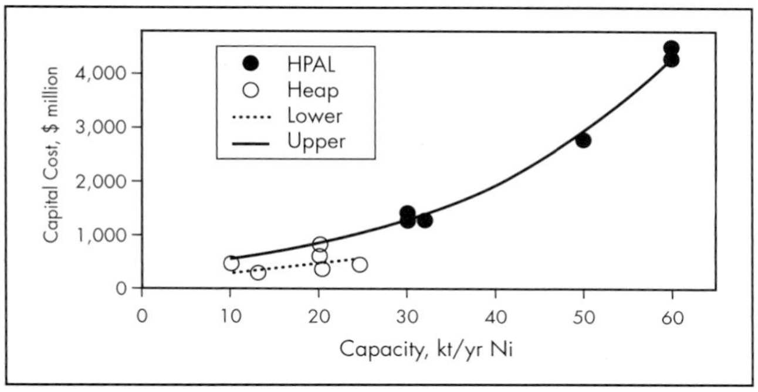

Source: Wedderburn 2010

Figure 13 Capital costs for laterite projects

the intermediate hydroxide precipitate can be sold for 75% of the price ranges shown for nickel and cobalt.

The calculated revenues exceed the calculated variable costs for both process routes examined. Had the variable costs approached or exceeded the revenue at this stage of the analysis, there would be no incentive to proceed further, apart from a careful check of the assumptions made and the calculations done. Since the calculated revenue does appreciably exceed the calculated variable costs, further evaluation is justified. The differences between the ATL and HPAL routes are not large enough, at this level of analysis, to differentiate between the two, so further examination of both is merited.

The costs for any venture are the variable operating cost, fixed operating cost, and capital cost. The fixed operating cost comes from salaries, spares, inventory, and so on. This number has to be estimated at this stage. For this exercise, the initial guess is 100 people at an average employment cost of US$50,000/yr, which translates to about $0.08/lb of nickel for 30,000 t of nickel per year.

The capital cost is a number that comes from engineering design work that has not even been contemplated at this stage of the exercise. However, an initial guess can be formulated using numbers from other similar projects. The capital costs shown in Figure 13 are for projects processing laterite to a mixed hydroxide intermediate via both HPAL and heap leaching.

Three of the data points for heap leaching are very close to the regression line through the HPAL data points (upper), whereas the other three lie well below this trend (lower). Assuming that the ATL capital cost can be approximated by the trend represented by the lower data points, extrapolating the lower regression line through these three points gives a capital cost of about US$600 million for 30,000 t/yr of nickel in the hydroxide product. Assuming HPAL for the leach step, interpolating the upper regression line to the same capacity gives a capital cost of about US$1,300 million.

Having evolved these very preliminary estimates for the costs and revenues associated with the laterite project chosen for this exercise, simple cash-flow analysis can be used to help decide whether the project is worth more substantial evaluation. Table 10 gives the results of this analysis for the case of ATL technology. The assumptions used are a two-year construction period with half of the capital expenditure in each of those years, followed by 20 years of production, with the throughput at 50% of design in year 3 and at 100% of design thereafter. Corporate tax was assumed to be 20% of the gross margin after recoupment of all the capital expenditure. For this calculation, the higher reagent costs and the lower metal prices were used.

The cash-flow analysis predicts an after-tax internal rate of return (IRR) of about 26% in the ATL case, for the average variable cost and revenue values.

Table 11 shows the cash-flow analysis for the HPAL case. For the average variable cost and revenue, the after-tax IRR comes to about 12%. The first conclusion to be drawn from these very preliminary calculations is that, for processing this laterite, the ATL case appears to be the stronger one, economically.

This process analysis does not require calculations that cannot be done quite easily on a spreadsheet, along with some rather persistent searching of open-domain information. The results can either greatly enhance confidence moving forward or rationally discourage further work if the numbers do not indicate a potentially viable project.

Whether these results lead to further effort would be a business decision based on information relevant to the project in question. These preliminary results will be taken as sufficient inducement to proceed to the next level of evaluation, which does get more sophisticated, computationally. From the preliminary numbers outlined in Table 11, it would seem logical to drop the HPAL option at this stage, but for completeness of the exercise, the HPAL option is also examined in more depth.

Table 10 Cash-flow analysis for ATL case, US$ million

Year	1	2	3	4	5	6	7	8	9	10–22
Capital costs	300	300	—	—	—	—	—	—	—	—
Fixed costs	—	—	17	17	17	17	17	17	17	17
Variable costs	—	—	75	151	151	151	151	151	151	151
Revenue	—	—	198	397	397	397	397	397	397	397
Margin	–300	–300	106	229	229	229	229	229	229	229
Corporate tax	0	0	0	0	0	38	46	46	46	46
Profit	–300	–300	106	229	229	190	183	183	183	183

Internal rate of return: 26%

Source: Dry 2013

Table 11 Cash-flow analysis for HPAL case, US$ million

Year	1	2	3	4	5	6	7	8	9	10–22
Capital costs	650	650	—	—	—	—	—	—	—	—
Fixed costs	—	—	17	17	17	17	17	17	17	17
Variable costs	—	—	78	157	157	157	157	157	157	157
Revenue	—	—	198	397	397	397	397	397	397	397
Margin	–650	–650	103	223	223	223	223	223	223	223
Corporate tax	0	0	0	0	0	0	0	0	28	45
Profit	–650	–650	103	223	223	223	223	223	195	178

Internal rate of return: 12%

Source: Dry 2013

Process modeling for laterite processing. One approach could be to first proceed with spreadsheet modeling of the process. When the results of the preliminary evaluation using spreadsheet calculations indicate a sufficiently strong business case for the project, the next stage can be pursued, which is to refine the calculated cost and revenue numbers so that the financial modeling can in turn be strengthened. At this stage, further experimental work becomes more than appropriate, as does a more detailed evaluation of the envisaged process. Process modeling can be used to generate and study a mass–energy balance of the envisaged circuit, optimally while more detailed experimental work is being done. This gives two major benefits: (1) a substantially enhanced understanding of the envisaged circuit enables the experimental work to be done much more cost effectively than would otherwise be possible; and (2) when experimental results become available, they can be incorporated into the mass–energy balance, which then becomes a very good basis for subsequent detailed engineering design and costing work, should the project reach that stage.

In this example, two detailed process models were constructed using commercially available process simulation software known as Aspen Plus, and the mass–energy balances generated were used to refine the analysis of the project. Figure 14 is a diagram of the ATL circuit modeled and Figure 15 is a diagram of the HPAL circuit. The process chemistry is the same as before. Incoming laterite is crushed and slurried with recycled and makeup water, the slurry is milled and the milled slurry is thickened, with the thickener overflow returning to the slurry makeup step and the underflow being pumped to the leach.

The process models include a standard sulfur burning acid plant. Incoming elemental sulfur is stockpiled as a solid, then reclaimed and melted in a sulfur melting pit. The molten sulfur is pumped to a sulfur burner and burned in air. The energy released is captured as high-pressure steam. The gas from the burner passes through three catalyst beds in which the sulfur dioxide is oxidized to sulfur trioxide. The oxidation is exothermic and the hot gas generated passes through heat exchangers, raising more high-pressure steam, between the catalyst beds. The gas from the third catalyst bed passes through an absorber in which the sulfur trioxide is captured into circulating strong acid and water, making 98% H_2SO_4, some of which is used to leach the milled laterite. The gas from the absorber passes through more heat exchangers and a fourth catalyst bed, then through another heat exchanger to a second absorber in which the sulfur trioxide is captured into the balance of the 98% sulfuric acid from the first absorber. The resulting strong acid returns to the first absorber and the residual gas is vented to a stack.

The leach is done at atmospheric pressure and 95°C in the ATL case, and at 250°C and elevated pressure in the HPAL case. In both situations, the leach is heated by injection of live steam from the acid plant, but in addition the HPAL case recycles steam flashed off after the leach to preheat the incoming slurry.

The residence time for atmospheric leaching was assumed to be 20 hours. For the HPAL case, the leach residence time was taken to be 75 minutes.

The leached slurry is pumped to a neutralization stage, where limestone is added to neutralize the residual free acid and precipitate the dissolved ferric iron. The incoming

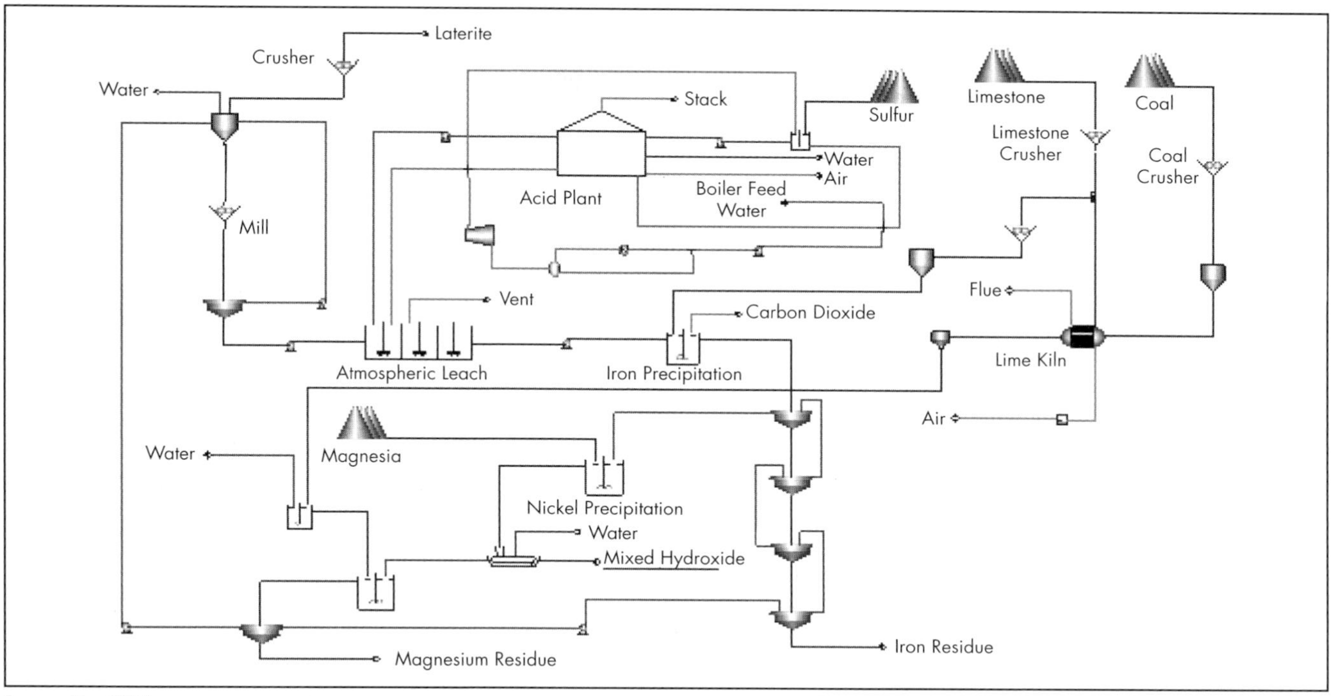

Source: Dry 2013

Figure 14 Process model (ATL case)

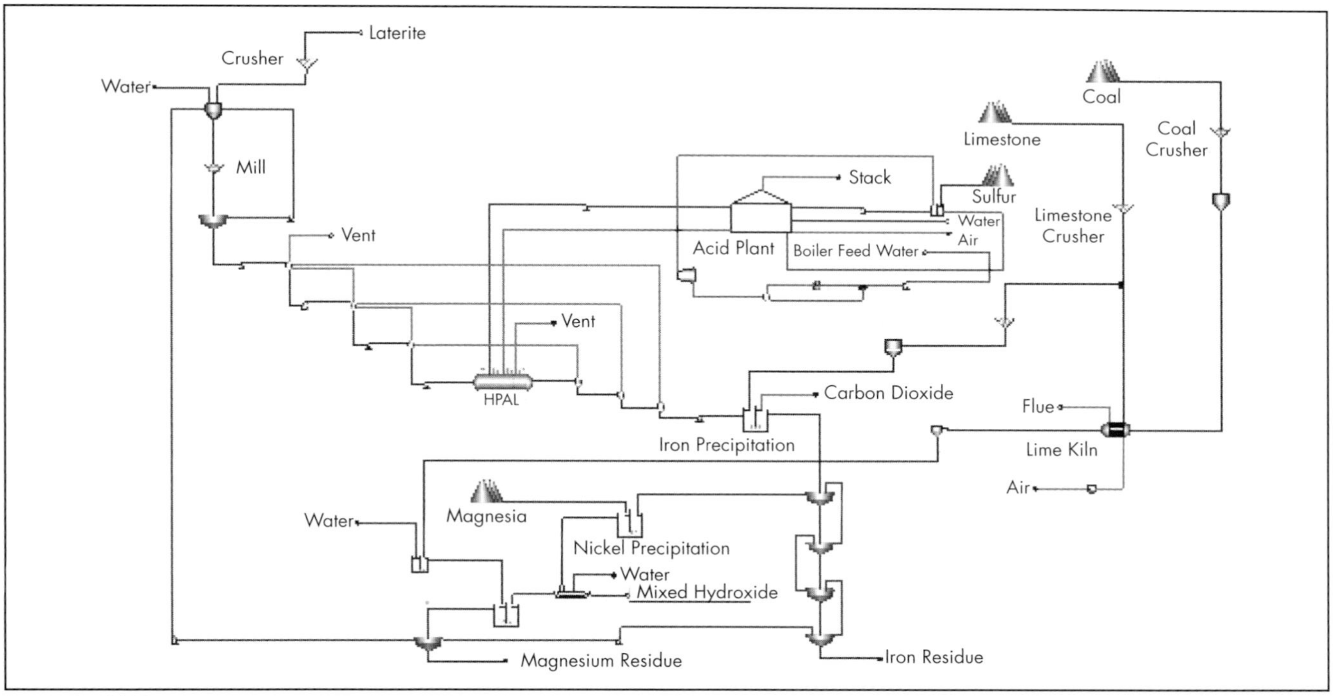

Source: Dry 2013

Figure 15 Process model (HPAL case)

limestone is crushed, and a part of the crushed limestone is used in the neutralization and iron precipitation step.

The neutralized slurry passes through a CCD train in which the solids are washed with recycled water. The washed solids leave the circuit. The supernatant from the CCD train is contacted with magnesium oxide to precipitate nickel, cobalt, and copper as a mixed hydroxide that is filtered out and washed with water. The washed hydroxide leaves the circuit as the desired intermediate product.

Table 12 Reagent comparison, in kilograms per ton of laterite

	Atmospheric Leach		HPAL	
Reagent	**Spreadsheet**	**Process Model**	**Spreadsheet**	**Process Model**
S	69	69	106	106
$CaCO_3$	214	227	323	348
MgO	7	8	7	8
Coal	17	22	17	22

Source: Dry 2013

The filtrate from the base metal precipitation step is contacted with slaked lime to precipitate the magnesium. The resulting slurry of magnesium hydroxide and calcium sulfate is thickened, the thickener underflow leaves the circuit, and the supernatant is recycled as wash to the CCD train and to the slurry step ahead of the leach. The slaked lime is made by calcining the crushed limestone not used in the iron precipitation step. A kiln was assumed for this duty, fired by coal and air. In this model, the ash from the coal accompanies the lime to the slaker.

The first comparison of interest between the results of the spreadsheet calculations and the balances from the process models is between the reagent consumption numbers, as shown in Table 12. The numbers for limestone are the sum of the amounts of calcium carbonate used directly in the iron precipitation step and the amounts needed to make the calcium oxide used in the magnesium removal step. Except for sulfur, the process models predict higher reagent consumption numbers than the spreadsheet calculation. The differences are because of the solution volumes in the case of $CaCO_3$ and MgO, and the heating of solids and air in the case of coal. These were ignored in the spreadsheet calculations.

Using the reagent consumption numbers from the process models in the cash-flow analyses done before reduces the IRR numbers by only 1%, which is not a significant difference at this level of evaluation. The relative differences in acid and limestone consumption predicted by the process models are in line with those from the purely stoichiometric calculations.

Cost modeling for laterite processing. Process engineering, equipment design, and cost estimation have traditionally been undertaken by different groups of specialists not automatically available to the people developing mining projects. Particularly in the case of junior mining companies, access to cost estimation has not been feasible until a substantial body of data has been generated, at which stage the investors allow the project team to commission engineering companies to do the necessary process engineering, equipment design, and cost estimation. The desired and anticipated outcome of this work is that the project in question is economically viable, but until the work is done, that outcome is uncertain, however much the project proponents believe in it. There is a risk that the time, effort, and money spent up to that stage of any given project will not result in a commercially viable operation.

A second consequence of the traditional separation between the process and estimation disciplines is that when there are alternative process options, comparison of options can be based more on opinion than on dispassionate evaluation, simply because dispassionate evaluation requires capital cost estimation, traditionally unavailable for reasons of cost and time. This is why approximately 80% of the capital cost of the process plant for any given project is sometimes determined by choices made in the conceptual design phase of the project, before any capital costs are known. In the modern world of global competition, investment decisions bringing projects to fruition faster, at less cost, and with minimum risk and uncertainty become ever more difficult.

It would be better to know sooner rather than later if a project is not sufficiently robust, economically, to justify its continued existence. It would also be better to know sooner rather than later that the most appropriate process option has been selected. Although perfect prescience is not achievable, it is quite possible to evaluate the capital and operating costs of a project, and hence the potential economic viability of that project, significantly earlier than has previously been practical. Just as process engineers have developed their tools, so have cost estimators. Equipment sizing and costing algorithms have been developed that generate equipment sizing and cost numbers from minimal process information. These algorithms have been embedded into modern computer software that can be used to automate the interface between process design, equipment sizing, and cost estimation. This enables the project team to iteratively evaluate project costs and compare options rapidly and cost effectively.

Preliminary equipment sizing. The following discussion illustrates the use of commercially available equipment sizing and costing software to estimate project costs. In the example used in this chapter, the two cases are similar except for the leach section. The cost estimation software automatically sizes the various pumps in the circuit, assuming standard design codes and carbon steel as the material of construction, but the material of construction can also be specified: stainless steel or rubber-lined carbon steel, for example. The various vessels in the circuit are automatically sized, assuming a specified residence time and default or specified material of construction.

The software used for this exercise is known as Aspen Process Economic Analyzer (APEA). This software can be activated directly from the Aspen Plus process simulation software; therefore, the entire mass–energy balance is electronically transferred from the process model to the corresponding process cost model. The various unit operations in the process model are computationally mapped to materials of construction and process equipment that is automatically sized, based on either default parameters or parameters that are specified for any item of equipment. The sizing calculations give the amounts of material and labor required to fabricate and install the process equipment, as well as the materials and quantities required for foundations, supporting structures and process piping, instrumentation, electrical wiring, and so on. In the exercise presented in this chapter, the software was set to give costs for a project in Africa. The equipment sizing was set up as follows.

Ore preparation. In the ore preparation section, incoming ore is crushed, slurried in water, and milled. The milled slurry is thickened, and the thickened slurry proceeds to leaching. The surplus water from the milled slurry is recycled.

The minimum process information required for the crusher by the sizing/costing algorithm is the power input, which can be approximated from the work index of the ore, the crushed product size, and the throughput. In this example, the work index has not been measured, so a number from the published literature must be used instead. One published number for an Indonesian laterite is 9.7 kW·h/t, and the ore throughput for this exercise is 360 t/h. Crushing the ore to

3-mm particles requires a crushing power of 638 kW (or 1.8 kW·h/t of ore). The crusher was assumed to be fed from a belt conveyor 100 m long and 1 m wide.

The information required for costing the mill is the internal diameter and the length. Outokumpu has published useful information on preliminary mill sizing, specifically a correlation linking the mill power to the mill type and dimensions. Assuming the same throughput and work index as for the crusher, and that the ore is milled from 3–100 mm, gives a milling power requirement of 2,854 kW (7.9 kW·h/t of ore). Using that and the standard factors tabulated in the Outokumpu pamphlet gives a mill diameter of 3.7 m and a length of 17 m.

The slurry tank between the crusher and the mill was assumed to be a carbon-steel agitated tank with a residence time of one hour, to allow the crusher to be run intermittently if necessary.

The mill thickener was assumed to be a standard thickener made from carbon steel. The thickening rate of the milled slurry not having been measured yet, 0.1 t/h/m^2 was assumed. The same number is used for the CCD train after leaching, which was inferred from published data from a pilot plant (detailed in the section on CCD later in this chapter).

Acid plant. The sulfur melting pit was assumed to be an agitated tank holding 2 hours of feed to the sulfur burner and heated by a steam coil using steam from the acid plant. The material of construction was set to stainless steel. The cost of the sulfur burner was approximated by the cost of a process furnace, the material of construction being stainless steel. The converters were sized from a standard design, assuming the same superficial gas velocity entering the first catalyst bed, which gave the vessel size and the volume of the catalyst beds. The catalyst bulk density and price were taken from a catalog of sulfuric acid catalysts. The various pumps and heat exchangers were sized automatically by the software. The material of construction was stainless steel throughout the acid plant.

Atmospheric leach. In the circuit using atmospheric leaching, the underflow from the mill thickener is pumped into the leach train, where it is leached for 20 hours. The reactors were specified as 10 agitated tanks, each having a residence time of 2 hours, with the tanks and agitators made of rubber-lined steel. At a height-to-diameter ratio of 2, that makes the diameter of each tank 9.5 m. The material of the pump after the leach train was specified as stainless steel.

HPAL leach. In the circuit using HPAL for the leach section, the milled and thickened ore slurry is passed through three direct-contact heat exchange towers in which it is progressively heated by condensing steam flashed from the hot leached slurry. These heat exchange towers were assumed to have a residence time of 10 minutes each, made of titanium-clad steel and lined with acid-resistant bricks to protect the vessel against erosion.

The preheated ore passes to the leaching step. The autoclave design was based on data from the Ambatovy (Madagascar) pilot plant. Anaconda's Murrin project had four autoclaves, each 35 m long and just less than 5 m in internal diameter, made from titanium-clad steel. The residence time in HPAL given in the work from the Ambatovy pilot plant is 75 minutes and the slurry into the preheating step contained 30% solids. That solids content and residence time translate to a required active volume of 1,500 m^3 for the 360 t/h throughput of the exercise in this chapter. Assuming the active volume to be 60% of the total internal volume of the autoclave, that requires four autoclaves similar to the ones used by Murrin. For costing purposes, each autoclave was assumed to be a multicompartment agitated horizontal tank reactor with the wall thickness designed for a working pressure of 40 bar, made from titanium-clad steel (8 mm titanium). The agitators were assumed to be six per autoclave, fitted with mechanical seals and made from titanium.

The hot pressure-leached slurry is flashed to atmospheric pressure in three let-down vessels. The residence time was left as the default, and the material of construction was set to titanium-clad steel lined with acid-resistant bricks.

Iron precipitation. The leached (and depressurized, in the case of the HPAL option) slurry is neutralized with solid limestone. The process equipment was assumed to be two agitated tanks in series. At a residence time of 20 minutes/tank and a height-to-diameter ratio of 2, the required tank diameter is 4.3 m. The material of construction for the tank and the agitators was set to rubber-lined steel.

The incoming limestone was assumed to be crushed to 3 mm and then milled to 100 μm. A typical work index for limestone is 103 kW·h/t and the limestone throughput is 82 t/h for the atmospheric leach case, giving a required crusher power of 153 kW and a mill of 3.7 m diameter and 4.1 m long.

Countercurrent decantation. The solids are separated from the neutralized leach solution and washed in a four-stage CCD train. The Ambatovy nickel project pilot-plant data were used to calculate a thickening rate, as follows:

- The working volume of the pilot autoclave was 30 L, the feed slurry contained 30% solids, and the residence time in the autoclave was 75 minutes. That gives a solids feed rate of 7.2 kg/h.
- The thickeners in the CCD train were 30 cm in diameter. Assuming the solids rates entering the autoclave and leaving the first neutralization step to be approximately the same means that each thickener settled about 7.2 kg/h of solids, and because the cross-sectional area of each thickener was 0.071 m^2, the solids settling rate was about 0.1 t/h/m^2.
- In the exercise presented here, the solids flows through the CCD train are 375 t/h for the atmospheric leach and 351 kg/h for the HPAL case. At the assumed settling rate of 0.1 t/h/m^2, that gives thickener diameters of 69 m for the atmospheric leach case and 67 m for the HPAL case.

The required parameters for thickener costing are the diameter and the total volume of the thickener. Assuming a sidewall depth of 3 m and a cone angle of 10° allows calculation of the total volume in the thickener. Optional specifications selected were that the bridge is from the center to the edge of the thickener, the drive and rake are heavy duty with overload alarm, a flocculation system is included, and the whole thickener is made from rubber-lined steel. Each thickener includes an underflow pump made from stainless steel.

Nickel precipitation. The supernatant from the CCD train is contacted with powdered magnesium oxide to precipitate the nickel, cobalt, and copper in two agitated tanks in series. The residence time in each tank was set to 20 minutes. The material of construction was left at the default, carbon steel. At a height-to-diameter ratio of 2, the required tank diameter is 10.5 m. The MgO was assumed to be purchased, stored in a conical bottomed hopper, and metered into the precipitation tanks via a screw feeder.

The resulting slurry is filtered. A rotary drum filter was specified for costing purposes. Since no filtration test results were available, the filtration rate was set to a default representing slow filtration. The material of construction was specified as epoxy-lined steel.

Magnesium precipitation. The filtrate from the nickel precipitation step is contacted with slaked lime to precipitate the dissolved magnesium. For this, crushed limestone is calcined in a coal-fired kiln, and the burnt lime is held in a surge vessel before being slaked with water in an agitated tank for 10 minutes. The slaked slurry overflows into a second non-agitated tank where it is diluted with more water, and grit is settled out and removed by a screw conveyor. The diameter of the two tanks is 2.5 m, and the height-to-diameter ratio is 2. The dry lime is fed into the first tank from one of two hoppers by a rotary feeder (24 hours of storage capacity, CaO bulk density of 0.5 t/m^3, and volume of 777 m^3 per hopper). The material of construction for this equipment was left as carbon steel throughout.

The filtrate and slaked lime are contacted in two agitated tanks in series (carbon steel, residence time 30 minutes each, height-to-diameter ratio 2, tank diameter 10.8 m).

In the absence of measured information, the thickening rate of the limed slurry was assumed to be the same as the rate in the CCD train, that is, 0.1 t/h/m^2. That gives a thickener diameter of 40 m for that thickener in both variations of the process. The optional parameters for this thickener were set to the same as the thickeners in the CCD train, except that in this thickener the material of construction was left as carbon steel.

Infrastructure. For ranking alternatives, the infrastructural cost components can be ignored because these items would be common to the competing alternatives. In this exercise, the infrastructural items were included because the capital cost for a new project does not consist only of the cost of the process equipment. The software also has cost models for various items of infrastructure.

The process model for the ATL case predicts 155 m^3/h solids and 615 m^3/h water in the leach residue plus the magnesium precipitate (45% solids by mass, combined). Assuming the final consolidated tailings to contain 80% solids, the volume would be 379 m^3/h. Taking a rise rate of 4 m/yr gives a tailings dam area of 793,000 m^2, with a perimeter of 3,567 m.

The process model for the HPAL case predicts 364 m^3/h of consolidated tailings. At the assumed rise rate of 4 m/yr, that gives a tailings dam area of just over 764,000 m^2 and a perimeter of 3,497 m.

A tailings slurry pump station and a pipeline 2,000 m long and 0.4 m in diameter were assumed for both cases. The pipe material was assumed to be fiberglass-reinforced polymer (more to illustrate that there is a variety of material in the cost modeling database than for any real reason).

For both cases, the tailings dam was assumed to require a starter dyke 5 m high. The base of the tailings dam was assumed to require brush clearing and excavation to a depth of 0.5 m. The bottom of the tailings dam was assumed to be a clay liner between two geosynthetic membranes, plus a grid of piping to assist drainage. Each year, the dyke would need to be raised by at least 4 m. The cost of the starter dyke (US$0.7 million) was thus added to the fixed operating cost to cover the annual raising of the dyke around the tailings dam.

The circulating process water was assumed to circulate through a fenced pond holding 48 hours of the process water circulating to the CCD stage as wash and to the feed preparation section to slurry the incoming crushed laterite.

Makeup fresh water was assumed to come from 25 km away via a pipeline and pumping station. Cooling water, 4,767 t/h in the atmospheric leach case and 1,197 t/h in the HPAL case, mostly for condensing the residual steam from the acid plant after electricity generation, was assumed to circulate through an evaporative cooling tower.

The project was assumed to require a heavy-duty rail link 80 km long and a paved road 20 km long. The required buildings were assumed to be a plant office and a laboratory, each 200 m^2. A parking lot (250 m^2) and a security fence (3 m high, 2,000 m long) were assumed. A power line (11 kV, 20 km) and a standby generator (1,500 kVA) were also assumed.

Cost estimates. Tables 13 and 14 show the cost estimates generated from the two mass–energy balances and the equipment sizing described previously, using the cost modeling software.

These costs exclude the capital cost for the mine, which would be the same in both cases. The direct field cost is the cost of the process equipment and its installation, including items like an electrical substation that the software adds automatically. The equipment costs are the material and labor costs for manufacturing and installing the various items of process equipment. Process piping, instrumentation, electrical wiring, foundations, and structural steel are from the default piping allocations, electrical allowances, foundation standards, and so on, in the software. These are based on data from numerous projects. When more detailed engineering work is done, these defaults are replaced by the results of the relevant work. The indirect field costs cover home-office costs, field supervision, start-up, and commissioning. The non-field costs are for freight, taxes and permits, basic and detailed engineering, procurement, overheads, and contract fees. These are all default numbers at this stage, but defaults based on extensive real data. The contingency of 15% was set to allow for the process being new.

Interestingly, although the extrapolations from published capital cost information predict that the HPAL case would cost about twice the cost of the ATL case, the numbers in the exercise presented here indicate that the total capital expenditure for the HPAL case will be only about one-third higher than the capital cost of the ATL case. A caveat here is that the cost database does not have cost information on the specialized slurry pumps needed in the pressure leach section, and these pumps were approximated by using cost information on less specialized pumps. This probably means that the capital cost for the HPAL case is underestimated.

The total installed equipment costs calculated are US$259 million for the ATL case and US$367 million for the HPAL case, or about 90% of the total direct field cost. The remaining 10% of the direct field cost covers items like electrical substations and other hardware added by the software.

Figure 16 shows how the installed equipment costs are distributed between the various parts of the process. The leach section makes up a much greater proportion of the total in the HPAL case, with the estimated cost of the HPAL autoclaves and ancillaries being about 10 times the estimated cost of the ATL reactors. The acid plant is also bigger and therefore more expensive in the HPAL case because of the higher overall acid requirement arising from the nature of sulfuric acid at the HPAL leach temperature, that is, having only one proton per H_2SO_4 instead of two at the lower temperature of the ATL.

Table 13 Capital cost estimate for the ATL case, US$

Account	Worker-Hours	Wage Rate, $/h	Labor Cost, $	Material Cost, $	Total Cost, $
Equipment	23,257	29.88	694,875	96,680,284	97,375,159
Piping	284,527	29.62	8,428,760	34,890,531	43,319,291
Civil	1,939,524	24.22	46,982,556	85,233,942	132,216,498
Steel	7,397	27.79	205,591	1,157,698	1,363,289
Instruments	21,634	30.24	654,107	6,181,606	6,835,712
Electrical	24,446	29.17	713,103	3,271,639	3,894,742
Insulation	70,116	22.57	1,582,808	1,639,548	3,222,355
Paint	35,490	22.30	791,586	653,635	1,445,222
Direct field costs	**2,406,390**		**60,053,386**	**229,708,882**	**289,762,268**
Indirect field costs	**271,091**				**86,243,305**
Total field costs	**2,677,481**				**376,005,573**
Freight					27,565,100
Taxes and permits					9,188,401
Engineering and head office	234,866				25,088,702
Other project costs					27,395,372
Contingency					83,729,360
Non-field costs	**234,866**				**172,886,935**
Project total costs					**548,892,508**

Source: Dry 2013

Table 14 Capital cost estimate for the HPAL case, US$

Account	Worker-Hours	Wage Rate, $/H	Labor Cost, $	Material Cost, $	Total Cost, $
Equipment	39,595	29.93	1,185,033	168,614,804	169,799,837
Piping	595,940	28.40	16,927,161	67,948,917	84,876,078
Civil	1,820,265	23.95	43,601,902	85,879,760	129,481,662
Steel	9,743	27.77	270,587	1,533,526	1,804,113
Instruments	25,527	30.24	771,942	11,297,602	12,069,544
Electrical	31,101	29.20	908,251	5,040,725	5,948,977
Insulation	114,134	22.57	2,576,385	2,652,741	5,229,126
Paint	37,033	22.31	826,080	671,824	1,497,904
Direct field costs	**2,673,337**		**67,067,342**	**343,639,900**	**410,707,243**
Indirect field costs	**317,312**				**96,176,705**
Total field costs	**2,990,649**				**506,883,948**
Freight					41,237,000
Taxes and permits					13,745,601
Engineering and head office	250,055				26,650,802
Other project costs					35,664,822
Contingency					112,352,808
Non-field costs	**250,055**				**229,651,033**
Project total costs					**736,534,981**

Source: Dry 2013

The algorithms inside the cost modeling software predict the fixed operating costs listed in Table 15. Inserting the capital and operating cost estimates generated by the cost modeling software into the cash-flow analyses done before gives the results shown in Dry (2013).

Results are shown in Table 16 for the ATL case and in Table 17 for the HPAL case. In this case the capital cost is the total capital cost calculated for the process plant plus the capital cost for the mine (US$14 million added to the cost of the process plant).

Even though the capital cost of the HPAL option decreased from the US$1,300 million assumed initially to the US$750 million calculated from the process and cost modeling exercise, the other changes to the reagent consumption numbers and fixed operating costs emanating from the modeling keep the HPAL case a distant second to the ATL case at best. The point is that the calculations based purely on the overall stoichiometry and interpolations from the literature gave results that appear to be reliable, certainly in light of the results from the process and cost modeling work. A reasonable generalization would be that if preliminary calculations

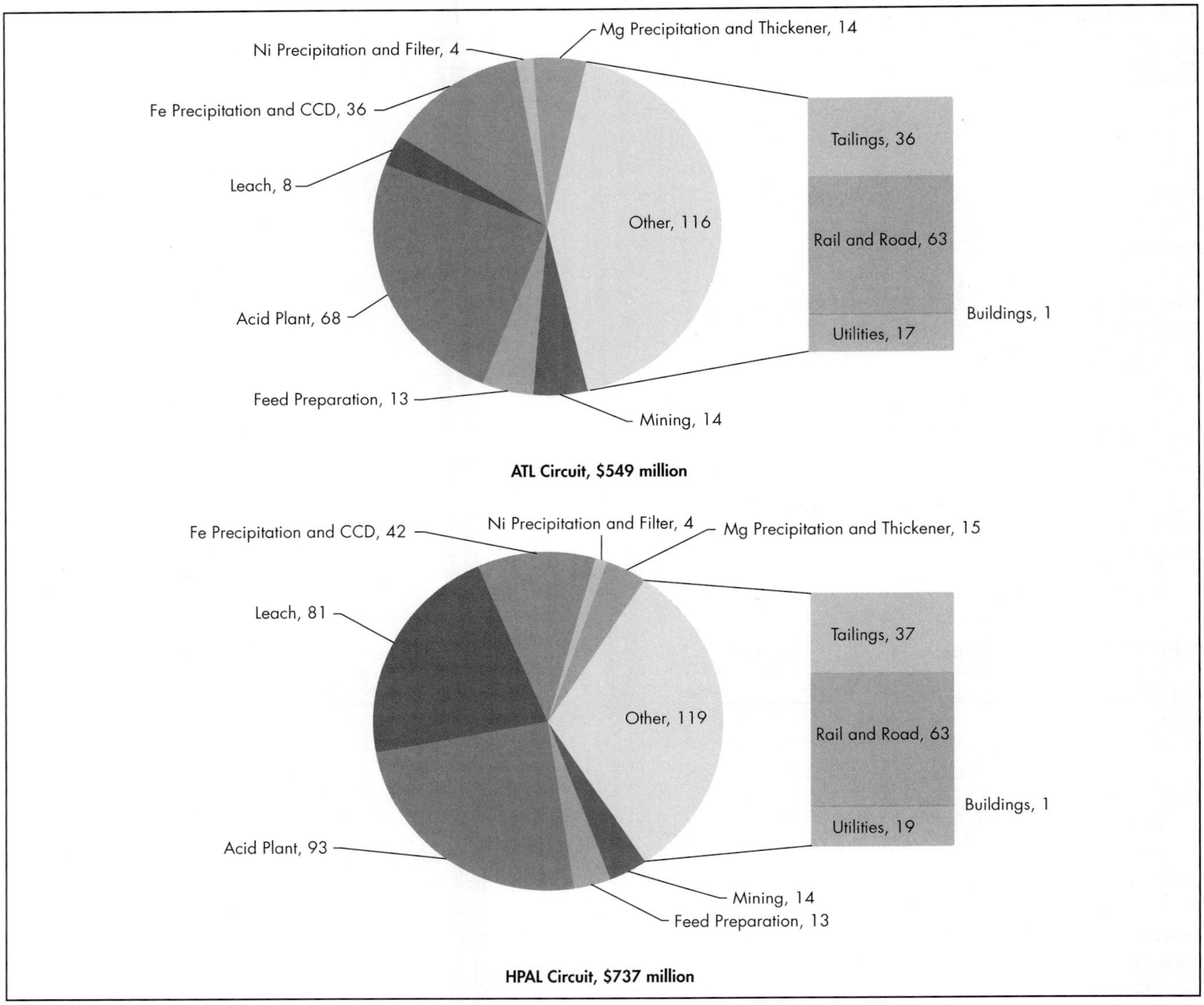

Source: Dry 2013

Figure 16 Breakdown of the installed equipment costs

indicate that the project is economically weak, it most probably is economically weak.

Design changes. After the process models and the costing models have been set up, it is quick and easy to evaluate different scenarios. For example, if the settling rates in the various thickeners are actually half or double the value assumed (i.e., 0.05 or 0.2 instead of 0.1 t/h/m^2) the thickener diameters would change accordingly, leading to a corresponding change in the estimated capital cost. Evaluating the impact of the thickening rate on the projected economics of the project is a simple matter of changing a few numbers in the thickener sizing calculations and the input to the cost model, rerunning the cost model, and saving the output to another spreadsheet.

To illustrate this, Figure 17 shows the impact of changing the settling rate on the installed equipment cost for the ATL case. The settling rate in the base case is 0.1 t/h/m^2; in the slow settling variation, it is 0.05 t/h/m^2; and in the fast settling variation, it is 0.2 t/h/m^2. The projected total capital cost changes from US$549 million in the base case to US$638 million for 0.05 t/h/m^2 instead of the assumed 0.1 t/h/m^2, and to US$504 million if the settling rate is 0.2 t/h/m^2. The cost of the equipment for the iron precipitation and CCD section

Table 15 Fixed operating costs, US$ million/yr

Cost Item	ATL	HPAL
25 operators/shift (S20/h, 4 shifts, 2,000 h/shift)	3.5	4.0
2 supervisors/shift (S35/h, 4 shifts, 2,000 h/shift)	0.6	0.6
Maintenance cost	3.9	5.8
Utility cost	9.0	16.1
Operating charges, plant overhead, general and administrative	9.9	9.9
Tailings dam dike	0.7	0.7
Total operating cost	26.0	3.7

Source: Dry 2013

Table 16 Cash-flow analysis, ATL case, US$ million

Year	1	2	3	4	5	6	7–22
Revenue	—	—	267	535	535	535	535
Costs	281	281	150	273	273	273	273
Margin	–281	–281	117	262	262	262	262
Tax	0	0	0	0	0	0	0
Profit	–281	–281	117	262	262	246	210
Before-tax internal rate of return			34%				
After-tax internal rate of return			32%				
Capital cost estimate, $ million			563				
Net present value at 10% discount rate, $ million			808				

Source: Dry 2013

Table 17 Cash-flow analysis, HPAL case, US$ million

Year	1	2	3	4 to 8	9	10	11–22
Revenue	—	—	266	533	533	533	533
Costs	375	375	212	387	387	387	387
Margin	–375	–375	54	146	146	146	146
Tax	0	0	0	0	7	29	29
Profit	–375	–375	54	146	139	117	117
Before-tax internal rate of return			–5%				
After-tax internal rate of return			–5%				
Capital cost estimate, $ million			750				
Net present value at 10% discount rate, $ million			–452				

Source: Dry 2013

changes from US$35 million in the base case to US$61 million in the slow-settling variation and to US$23 million in the fast-settling variation. Settling rates in thickening steps are not always measured in preliminary experimental work, but given their potential impact on the capital costs, they should be. The point of this illustration is not that the settling rate is or is not any particular value; the point is that the settling rates in the various thickeners have an appreciable impact on the overall capital cost. Therefore, settling rates should be measured at as early a stage of the experimental work as can be arranged, even knowing that later work may well generate improved or more accurate settling rates. In this illustration, doubling the settling rate has a substantially smaller impact on the calculated capital cost than halving this number, which means that early-stage experimental work would be more focused on finding out whether the various slurries settle "normally."

If early work uncovers especially slow settling, appropriate work can be planned to improve the settling rate, confirm that it is slow enough to make the circuit of which it is a part unviable, or to find a better solid–liquid separation technique for the project in question. If reasonably fast settling rates are found early on, there would be little rational incentive for extensive experimental work aimed at further improving the settling rate, as the impact of further improvements on the capital cost of that project would be small.

Capacity changes. A question that commonly arises when processes are under evaluation is what effect the throughput has on the economics. If the changes in throughput are not excessively large, the fixed operating cost is, at least to a first approximation, unaffected and the variable cost is directly proportional to the throughput. That leaves the capital cost, which is traditionally scaled using the "0.6 rule"; that is, the baseline capital cost is multiplied by the ratio of the new throughput to the baseline, to the power 0.6. The cost modeling software used for the exercise presented in this chapter contains a feature that enables the effect of throughput on the capital cost to be evaluated more rigorously. As the capital cost is based on equipment sizing results that come from a numerically rigorous mass–energy balance, and that balance can be scaled linearly, the cost modeling software can easily recalculate the capital cost for throughputs other than the baseline by resizing the various items of process equipment, support structures, foundations, and so on, and recalculating the capital cost from the various quantities of material and labor so recalculated. If any item of process equipment, when resized, falls outside the range covered by the capital cost database, the software warns the user and then the user would specify duplicate items or make whatever other design change is required.

Figure 18 shows the results obtained by using the 0.6 rule and using the cost modeling software to calculate the capital cost of the ATL case, from half to double the baseline capacity of 30 kt/yr of nickel in the mixed hydroxide product. In this exercise, the mine cost was left unchanged.

The 0.6 rule gave rather good results for this exercise, compared to the more rigorous calculation from the cost modeling software (APEA; Aspentech 2018). The 0.6 rule does seem to underestimate slightly for scaling down, but the difference is unlikely to be significant at the preliminary stage of evaluation assumed for the exercise presented here.

By way of calibration of the process and cost modeling illustrated in this example, a press release in 2009 reports a

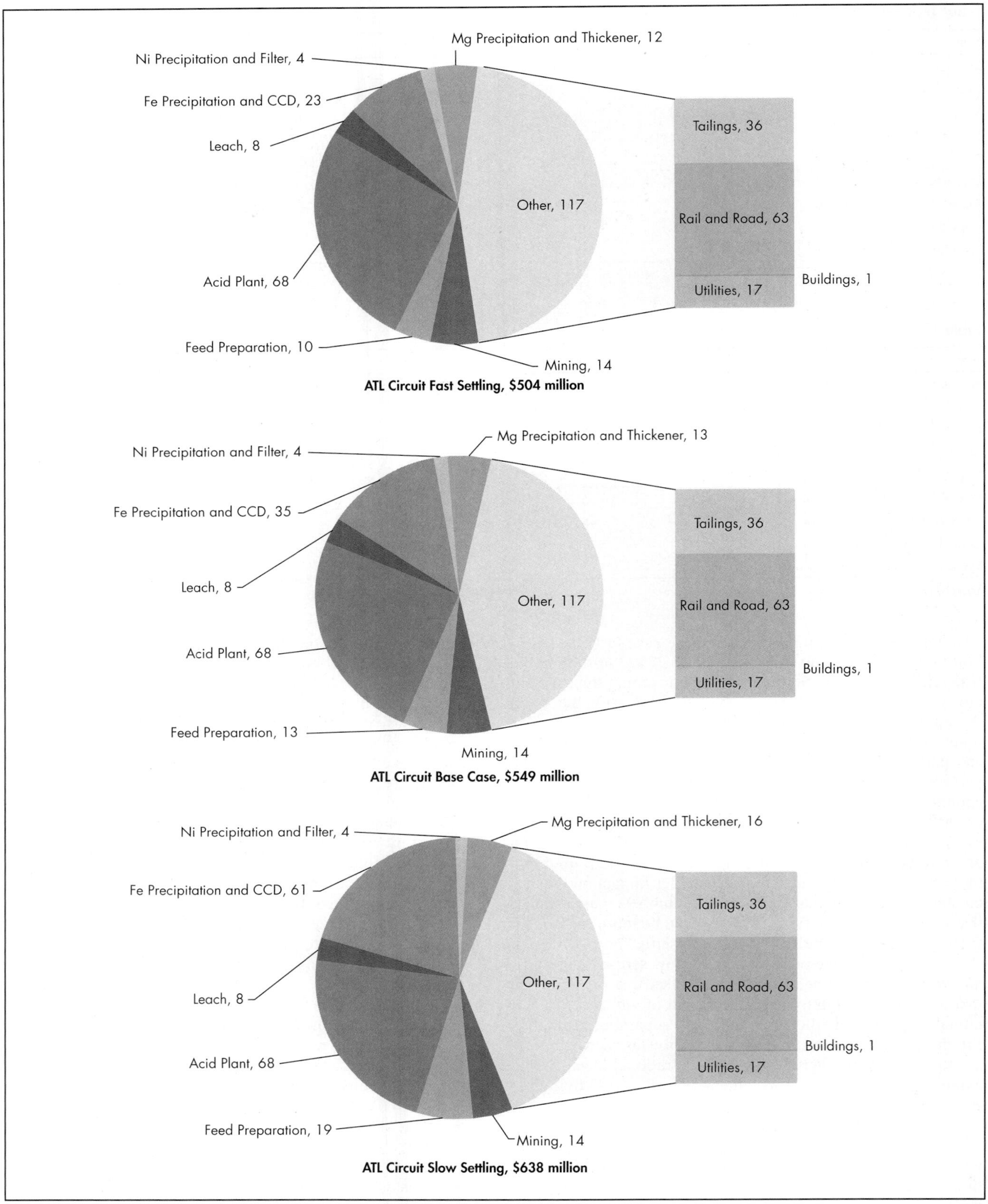

Source: Dry 2013

Figure 17 Effect of settling rate, ATL case

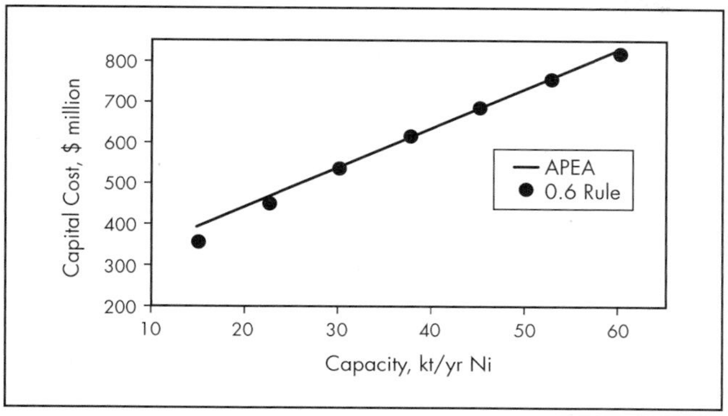

Source: Dry 2013

Figure 18 Effect of capacity on capital cost, ATL case

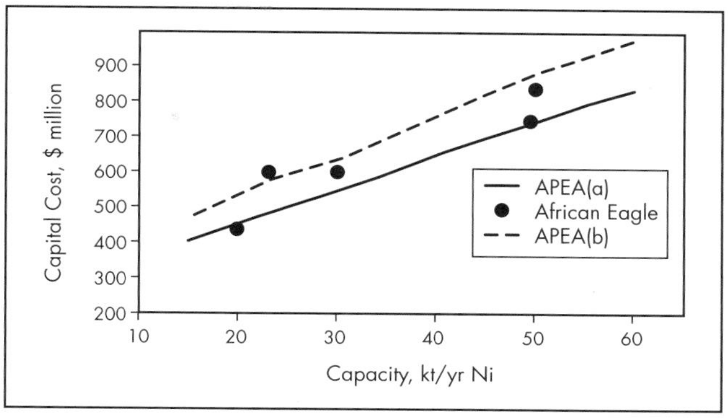

Source: Dry 2013

Figure 19 Calculated and published capital costs

capital cost estimate of US$435 million for Black Down Resources' Dutwa nickel project (Tanzania) (Dutwa Project 2018), at an annual throughput of 2 million t/yr of ore. Another press release from African Eagle, released in December 2010, gives a capital cost estimate of US$600 million for a production rate of 23 kt/yr of nickel. A third release gives capital costs of US$600 million and US$840 million for facilities producing 30 kt/yr and 50 kt/yr, respectively, of nickel in a mixed hydroxide precipitate.

In Figure 19, the symbols plot the capital costs published for the Dutwa nickel project (Dutwa Project 2018), and the lines are the capital costs calculated using the cost modeling software over the capacity range shown. The solid line is the capital cost calculated using the assumption that the settling rate in the various thickeners is 0.1 t/h/m^2, and the dashed line is the capital cost calculated assuming that the settling rate is 0.05 t/h/m^2.

The settling rate of 0.1 t/h/m^2 was inferred from HPAL pilot-plant data. In the HPAL, much of the iron is precipitated in the autoclave as hematite, which tends to have better settling properties than the goethite that is precipitated under the conditions found in the ATL circuit. Therefore, it would not be implausible to assume that the settling rate in the ATL circuit could be lower than the rate that would be expected in the HPAL circuit. If the settling rate is actually 0.05 t/h/m^2, the capital costs calculated in the exercise presented in this chapter pass through the numbers published by African Eagle, which, having been generated by process engineering companies, are most probably based on much more engineering design work than was used in the cost modeling exercise presented in this chapter.

Even without adjusting the settling rate, the capital cost numbers calculated using the capital cost modeling software are within 20% of the corresponding published numbers, except for one published number that is out of line with the others and about 26% off the corresponding number from the cost modeling software. That is well within the uncertainty generally associated with initial capital cost estimates.

The objective of the exercise presented in this example is not to show whether the Dutwa nickel project has business muscle. This project was chosen for mimicking because there is enough information published about it to enable sensible comparison between the calculated capital costs and corresponding published numbers that were generated more traditionally.

Conclusions in Hydrometallurgical Process Modeling

The hydrometallurgical modeling example presented assumes that the quantity of ore is sufficient to sustain a viable rate of production over a useful project life. In the case of a new deposit, the calculation of reserves would entail fieldwork and consume substantial amounts of effort, time, and money. Doing calculations such as in the examples presented here, based on very preliminary experimental work, yields valuable information relevant to business decisions on whether to undertake the expense of proving up any particular deposit. If the process is not sufficiently strong economically, the size of the deposit is irrelevant and money spent proving the deposit is money wasted.

An important consideration in the example presented is what happens in the leach. The extraction of metals from ores necessarily involves minerals that are complex substances. The first experimental work necessary in developing any hydrometallurgical circuit is leaching tests. After information on the leaching behavior of the feed is available, either from appropriate experimental work or as knowledge from a suitable similar process, the modeling calculations become meaningful.

New projects can be sensibly evaluated early in the development cycle, as soon as enough is known about the ore to enable preliminary estimates of its overall composition and its leaching behavior. Preliminary calculations based on the overall stoichiometry and information in the open literature can inexpensively generate useful evaluations of whatever process options are under consideration. When the preliminary calculations and process modeling indicate a process option to be potentially viable, process and cost modeling can be used to predict rational capital and operating costs from limited hard data, and these numbers can be used to confidently rank process options and justify work or the termination of work on any given project much earlier in the development cycle than would be possible without these tools.

No skilled person would knowingly spend time, effort, and money on a doomed project, but all too often the realization that the project is doomed comes after the expenditure of considerable amounts of time, effort, and money. Since process modeling and cost estimation are inevitable steps in the development of a project, it would be wise to not commit to extensive experimental and engineering design work before doing the process and cost modeling that will be needed anyway to evaluate its economics. It is far better to use preliminary experimental results for setting up the process and cost models and then use the models to evaluate the project and, assuming that the models predict sufficiently strong economics, use the models to guide the experimental work, feed the data back into the models, and so on.

In the case of viable projects, judicious use of process modeling can make the whole project development cycle much more cost- and time-effective, especially for dispassionately ranking process options and determining the best program of experimental work. Identifying and testing the parameters that most strongly affect the process economics is also possible using these tools, again leading to better results and less wasted effort.

Applications of Process Modeling in Pyrometallurgy

Process modeling and simulation of pyrometallurgical processes require an in-depth understanding of high-temperature treatment of materials. Typical extractive operations in this category are calcining, roasting, smelting, and refining. In each case, physical and chemical transformations are brought about in materials at elevated temperatures.

Pyrometallurgy is extensively used in iron and steel production, refining precious metals, and extraction of base metals. About 90% of world copper output is produced pyrometallurgically from sulfide-based minerals, and in many cases from high-arsenic feedstock. Although some pyrometallurgical processes are relatively straightforward, others involve extremely complex reactions between multiple phases (gases, solids, and liquid slag and metal) that can lead to the release of large quantities of acidic, metallic, and other toxic compounds. In addition, the development and study of pyrometallurgical operations require very expensive piloting and laboratory test work. These factors, together with the need to evaluate pyrometallurgical processes thoroughly as part of their development, increase the importance and value of computer modeling and simulation. This section outlines modeling and simulation of pyrometallurgy and covers its application to the study of extraction of metals and the processing of other materials.

Calcining takes place in absence of air or oxygen to bring about thermal decomposition, a change of phase, or removal of a volatile fraction. The calcining process normally takes place at temperatures above the thermal decomposition temperature but below the melting point of the product materials. The *decomposition temperature* is usually defined as the temperature at which the standard Gibbs free energy for a reaction is equal to zero. For example, in limestone calcination (a decomposition process), the chemical reaction is

$$CaCO_3 \rightarrow CaO + CO_2(g) \quad \textbf{(EQ 10)}$$

The standard Gibbs free energy of limestone calcination reaction is approximated as $\Delta G°r = 177{,}100 - 158\ T$ (J/mol) (Mackinnon et al. 2003). The standard free energy of reaction is zero in this case when the temperature, T, is equal to 1,121 K, or 848°C, the temperature for decomposition of calcium carbonate.

Roasting involves more complex gas–solid reactions with carbonates and sulfides at elevated temperatures, with the aim of purifying the metal and eliminating unwanted carbon and sulfur, leaving behind a metal oxide that can be directly reduced. Roasting can be initiated by burning a fuel with the ore. This raises the temperature of the ore to the point where its sulfur content becomes the fuel. Roasting then continues autogenously, with no need for an external fuel source.

Before roasting, the metal concentrate is mixed with other materials to facilitate the process. Typical roasting thermal gas–solid reactions include oxidation, reduction, chlorination, sulfation, and pyrohydrolysis (Vahed 2004). In roasting, the ore or concentrate is treated with hot air. The sulfide is converted to an oxide, and sulfur is released in gaseous form. For example, roasting of chalcocite and sphalerite ores (Cu_2S and ZnS) can be represented as

$$2Cu_2S + 3O_2 \rightarrow 2Cu_2O + 2SO_2 \quad \textbf{(EQ 11)}$$

$$2ZnS + 3O_2 \rightarrow 2ZnO + 2SO_2 \quad \textbf{(EQ 12)}$$

The gaseous product, sulfur dioxide (SO_2), is often used to produce sulfuric acid. Many sulfide minerals contain other components, such as arsenic, that are released into the environment if there were no abatement measures.

Smelting is a form of pyrometallurgy for extraction of metal from its ore. It includes production of silver, iron, copper, and other metals. Smelting uses heat and a chemical reducing agent to decompose the ore, driving off unwanted elements as gases or slag, and leaving just the metal behind. The reducing agent is commonly a source of carbon, such as coke. The carbon or carbon monoxide produced during this process removes oxygen from the ore, leaving behind elemental metal. The carbon is usually oxidized in stages, producing first carbon monoxide and then carbon dioxide. As most ores are impure, it is often necessary to use flux, such as limestone, to remove the accompanying rock gangue as slag. Plants for the electrolytic reduction of aluminum are also generally referred to as smelters.

Refining is the step that follows smelting, with the purpose of purifying the metal. For example, the discharge stream from the copper converter contains a significant amount of dissolved oxygen and copper oxide. The oxygen is removed by adding a hydrocarbon as a reducing agent:

$$4CuO + CH_4 \rightarrow 4Cu + CO_2 + 2H_2O \quad \textbf{(EQ 13)}$$

Pyrometallurgy Modeling Regimes

Types of pyrometallurgical process simulation can be categorized as follows:

1. Steady-state simulation based on known process reactions and equilibrium
2. Steady-state simulation with predictive modeling where chemical reactions and pathways are not clearly understood
3. Dynamic simulation based on known pyrometallurgical reactions and kinetics, to determine effects of process variables with time
4. Dynamic simulation where nature of chemical reactions and kinetics are not clearly known and must be predicted from thermochemical properties of the reactants

Steady-state modeling in pyrometallurgy. The simplest approach to modeling pyrometallurgical processes is when the process is clearly defined and the chemical reaction pathways and kinetics are well known. When Gibbs free energy is minimized, it means that a system has reached equilibrium at constant pressure and temperature. Therefore, it is a convenient indicator of the spontaneity of processes to proceed. For example, roasting of base metals such as zinc, iron, lead, and copper to their respective oxides is most effectively carried out in the temperature range 600°–700°C, because

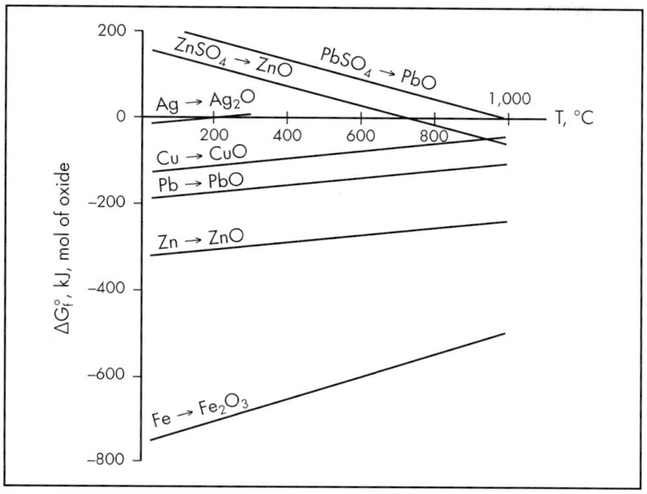

Source: Nicol et al. 1987

Figure 20 Free energy of formation of metal oxides from elements and by decomposition of sulfates: ρSO_2 = 0.001 atm

the values of free energy of formation of metal oxides are negative at this range, enabling metal oxides to form. Hence, the temperature range for the process is predetermined and no predictive simulation is needed to identify the outcome. Figure 20 shows values of free energy of formation for metal oxides associated with refinery gold materials as a function of temperature for partial pressure of sulfur dioxide gas, ρSO_2 = 0.001 atm (Nicol et al. 1987).

In similar fashion to hydrometallurgy, understanding of chemistry in pyrometallurgy is the first step, and the similar use of simulation packages could be beneficial in clarifying the chemical equilibrium associated with the process concept. Furthermore, in pyrometallurgy, performing test work and obtaining information from operating plants that are used to develop the conceptual process is costlier and challenging. Based on the same reasoning as in hydrometallurgy, process modeling can be used to perform chemical and equilibrium calculations for the aforementioned systems at elevated temperatures.

In similar fashion as use of Pourbaix diagrams (Gilchrist 1989) in hydrometallurgy, the standard Gibbs free energy for formation of a product in pyrometallurgy can be used to determine the chemical equilibrium for the reactions involved. The Ellingham diagram as shown in Figure 21 (Ellingham 1944) is a plot of the standard free energy of metal oxidation reactions as a function of temperature, and it determines the equilibrium composition of a pyrometallurgical process. These plots are useful in understanding the conditions under which extraction of metals is possible by pyrometallurgical process reactions of a gaseous phase (the oxidizing gas) with a pure solid or liquid phase (the metal and oxidized compounds). The Ellingham diagram also allows calculation of the equilibrium composition of the system and is useful in other ways: it can be an indicator of process conditions for reduction of a metallic oxide to pure metal, as well as partial pressure of oxygen that is in equilibrium with a metal at a set temperature. Also, the position of the line for a given reaction on the Ellingham diagram shows the stability of the metal oxide as a function of temperature. Reactions closer to the top of the diagram are for the most inert metals (e.g., gold and platinum), and their oxides are unstable and easily reduced. Toward the bottom of the diagram, metals become progressively more reactive and their oxides become harder to reduce.

Fundamentals of heat and mass balances in extractive metallurgy, described earlier in this chapter, also apply to pyrometallurgy. Many examples of steady-state modeling in pyrometallurgy include determination of energy balance based on known process design criteria, such as fraction of heat input lost to the surroundings. The following example is a heat and mass balance analysis for zinc roasting, based on previously determined conversion rates that determined 90% of total iron charged converted to $ZnO \cdot Fe_2O_3$. The goal of the steady-state calculation here could be to verify the furnace bed temperature at, for example, 10% heat lost to the environment. Table 18 shows the basis of this exercise.

Using quantities of each component for input and output streams, the associated sensible heat, and the heat of reaction, total heat liberated is calculated. This heat, minus the heat lost to the environment, is assumed to be available to raise the products' temperature in the furnace, and the temperature attained by the products is calculated. This simple steady-state modeling can be performed by any software program that can tally the chemical and physical changes and produce results with a steady-state heat and mass balance. For example, simple Microsoft Excel spreadsheets can be developed for a specific flow sheet such as a copper roaster. A review of kinetics of copper roasting processes for removal of impurities (Haque et al. 2012) also shows that when adequate kinetic data are available, impurity conversion levels obtained from test work can be used directly in steady-state models. In a 1997 study by Fan, conversion levels of arsenic as a function of time and roasting temperature were determined for arsenic-bearing copper concentrate in a fixed bed. Converted fractions of arsenic as a function of roasting time are shown in Figure 22 (Fan 1997). Similar test work can form the basis of arsenic conversions in roasting and afterburner sections of a pyrometallurgical plant. In addition to spreadsheets, simulation software programs such as METSIM, SysCAD, USIM PAC, and IDEAS are extensively used to provide a more user-friendly and understandable interface that provides access to previously identified component thermochemical and physical properties.

Here, as part of the steady-state model of the pyrometallurgical process, the intention can be to achieve an arsenic reduction target by partial roasting, and pilot tests can be conducted using high-arsenic concentrate to determine arsenic reductions in the roaster and the afterburner. The expected arsenic content in treated concentrate for commercial roasting is modeled based on this result. The heat and mass balance for the roaster package is based on typical design criteria for the roaster balance produced in a spreadsheet, as shown in Table 19.

Although its use is not prevalent, METSIM software can be used in pyrometallurgical process modeling in place of simple steady-state spreadsheets. Its advantage is that its array of unit operations includes several types of dryers, furnaces, and fluid bed reactors that can be modeled (Haque et al. 2012). The user provides details of these unit operations including number of stages, dimensions, area of heat transfer, bed area, feed solid characteristics, and flow rate. Heat input and losses can be provided to simulate heat transfer operation. Heat input or loss can be specified as a fraction of the local heat input to the unit operation. Heat generation as the result of

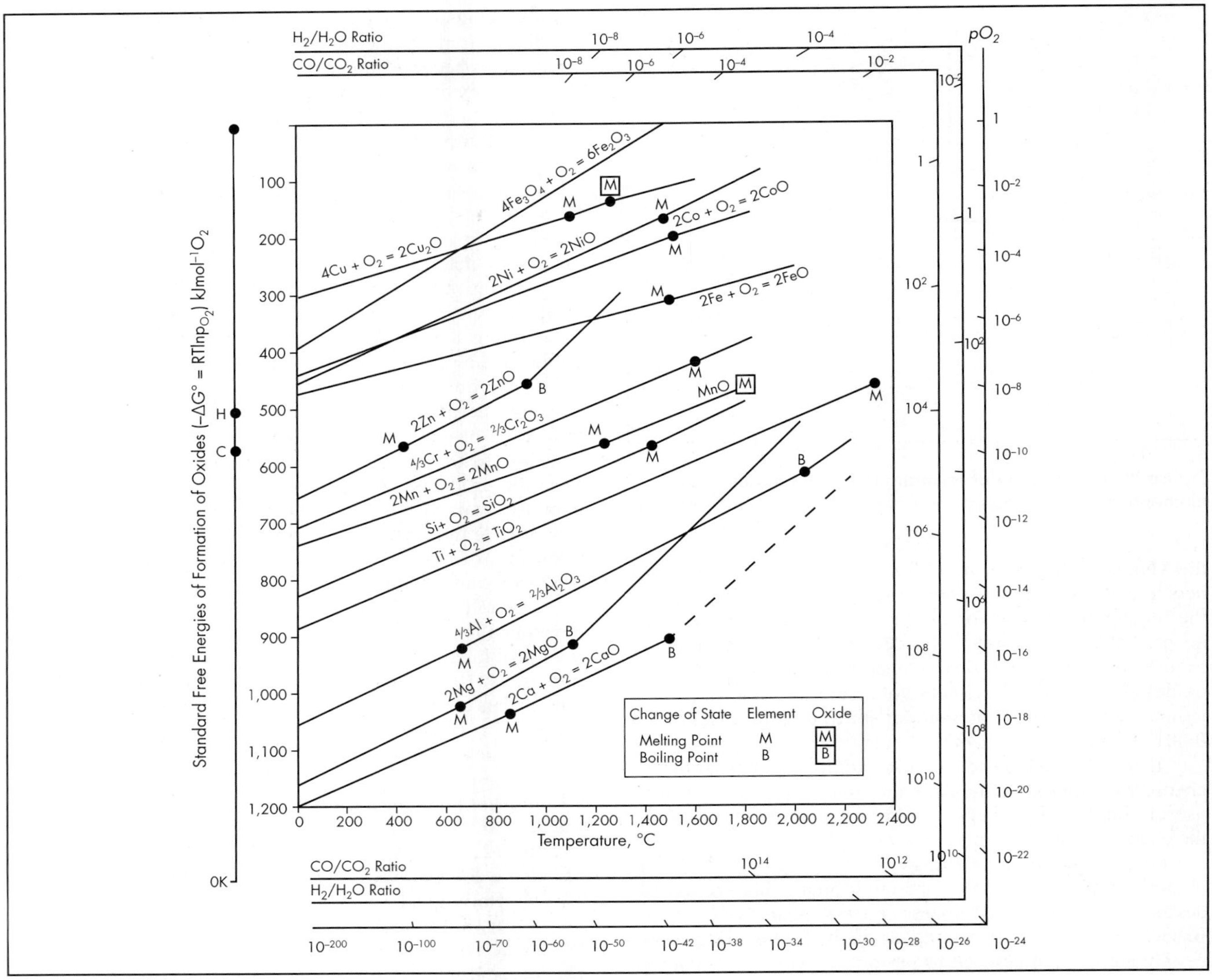

Source: Ellingham 1944

Figure 21 Ellingham diagram

Table 18 Typical zinc concentrate roasting model design criteria

	Zinc Concentrate Roasting Design Criteria				
Feed components	ZnS	FeS	PbS	SiO_2	H_2O
Weight %	75	18	3	3	1
Roaster reactions	$ZnS = 1½O_2 = ZnO + SO_2$ (99%); (−ΔH° = 105,950 kcal/kg mol ZnO) $2FeS + 3½O_2 = Fe_2O_3 + 2SO_2$; (−ΔH° = 292,600 kcal/kg mol Fe_2O_3) $PbS + 2O_2 = PbSO_4$; (−ΔH° = 197,000 kcal/kg mol $PbSO_4$) $ZnO + Fe_2O_3 = ZnO{\cdot}Fe_2O_3$; (−ΔH° = 4,750 kcal/kg mol $ZnO{\cdot}Fe_2O_3$) (Test work shows that 90% of total iron is tied up with ZnO.)				
Heat losses	10% of heat input is lost to the surroundings.				
Feed rate	Material basis is 1,000 kg Zn concentrate feed				

Source: Dry 2013

pyrometallurgical rections is determined by METSIM, based on assumed conversion rates and thermochemical properties. This level of modeling can also be produced using IDEAS modeling software with capabilities over a range of industries. It can start with a simple steady-state model and end with predictive and dynamic modeling of the pyrometallurgical operation (Nikkhah and Anderson 2001).

Predictive steady-state modeling for pyrometallurgical operations. Before including enough detail about kinetics of pyrometallurgical reactions to observe the dynamic effect

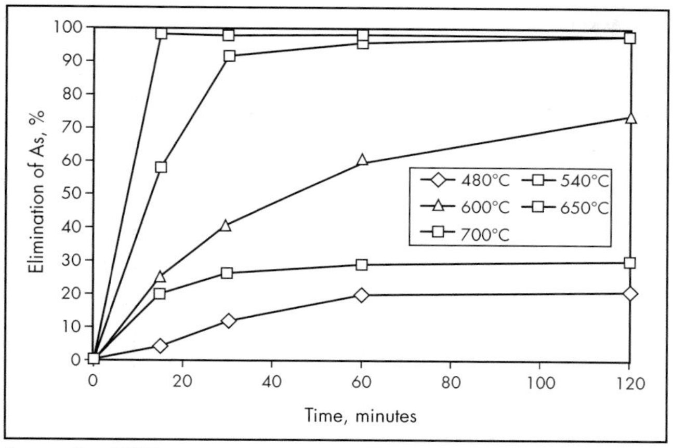

Source: Fan 1997

Figure 22 Arsenic removal as a function of time and temperature during roasting of arsenic-bearing copper concentrate in a fixed bed

Table 19 Typical process design criteria for steady-state pyrometallurgical computer modeling (arsenic removal from copper concentrate)

Design Criteria	Roaster	Afterburner
Roasting temperature, °C	670	250
Air to concentrate mass flow rate ratio	38.7/100	26/100
Sulfur conversion, %	95	95
AsS oxidization, %	5	5
Arsenic fraction volatilized, %	90	100
Sulfur fraction volatilized, %	20	100

Source: Dry 2013

of changes in process parameters such as feed type and flow rate, computer modeling can be used to predict the steady-state operation of a pyrometallurgical operation. Mintek's Pyrosim (Jones 1987) is an early example. Pyrosim software relied on the complete input of process conditions such as temperature, pressure, and heat losses, as well as knowledge of the process to be operated under isothermal or adiabatic conditions. Appropriate thermochemical attributes, such as enthalpy and entropy of formation and the heat capacity of minerals, were used to calculate the enthalpy of each component. Total enthalpy of each stream was then determined as the sum of the enthalpies of all components in each stream.

Pyrometallurgical operations take place at highly elevated temperatures. At these temperatures, reaching equilibrium is not prevented by kinetic limitations, and a general method for predicting composition at the process temperature is required. This is performed by minimizing the Gibbs free energy of the system, which is the fundamental description of chemical equilibrium, based on the fact that at a specified temperature and pressure, the most stable products are those associated with the lowest free energy. The Gibbs free energy of the system is defined as follows:

$$G = \sum_{i}\sum_{p} n_i^p \mu_i^p \quad \textbf{(EQ 14)}$$

where

$n_i{}^p$ = number of component i moles in phase p
$\mu_i{}^p$ = chemical potential of species i in phase p

Chemical potential is given by

$$\mu_i^p = \mu_i^0 + RT\ln a_i^p \quad \textbf{(EQ 15)}$$

where

$\mu_i{}^0$ = reference chemical potential for species i
$a_i{}^p$ = activity of species i in phase p

Assuming that $\mu_i 0 = 0$ for all elements in their standard states

$$\mu_i^0 = \Delta G_{fi}^{\circ} \quad \textbf{(EQ 16)}$$

where ΔG_{fi}° is the Gibbs free energy of formation of species i at the process temperature.

Dynamic modeling for pyrometallurgical operations. Using the aforementioned principle of minimizing Gibbs free energy, IDEAS was used to develop a dynamic model of a Kivcet lead smelter (McGarry and Watson 1998) to assist in staging the ABB control system and in operator training. The Kivcet model began with existing IDEAS furnace and boiler modules, modified to suit the process requirement. Adding Gibbs free energy was a fundamental improvement to the furnace software module, along with the solution of more than 50 stoichiometric, equilibrium, and phase equilibria equations that accurately characterized the process. Results of the dynamic model were shown to be within 3% of values predicted by pilot-plant research and test work. This gave the furnace operators confidence in the proposed changes to the process control strategy. The dynamic model was successfully used as a testing platform for modifications to the smelter process control system.

IDEAS process simulation capabilities have been used for the entire pyrometallurgical plant: from initial design concept to start-up and process control verification and training of operators (Rajoria et al. 2014). IDEAS can extend from steady-state models that provide heat and mass balance to dynamic models that simulate material and energy flows over time (Nikkhah and Anderson 2001).

The goal of a process simulation model is to accurately represent real equipment operations. The better it does this, the higher its fidelity is said to be. IDEAS is a high-fidelity software, with accurate descriptions of material, chemical, and energy changes in pyrometallurgical processes.

The graphical nature of the IDEAS platform allows the user to construct a dynamic model by retrieving icon-based objects from various libraries and assembling them on a drawing-like worksheet. The user can develop a model by creating an intelligent process and instrumentation diagram of the process. Individual equipment characteristics are provided by filling in the process information for each object. Every time an object (unit operation) is retrieved from a library and placed onto a worksheet, a computer simulation of that unit is created. Each instance of the same object is identical in terms of the code (programming) executed when the model is used.

The objects can be connected in numerous plug-and-play combinations to create complex models. After the desired flow sheet is created, the necessary information is entered into input sheets (dialog boxes) and the simulation is started. Flows in different sections of the model at different temperatures and pressures throughout the model are solved and compiled in the output dialog boxes. These outputs are also used in the model's virtual plant and transmitters, controllers, and plotters. The IDEAS steady-state program structure can evolve

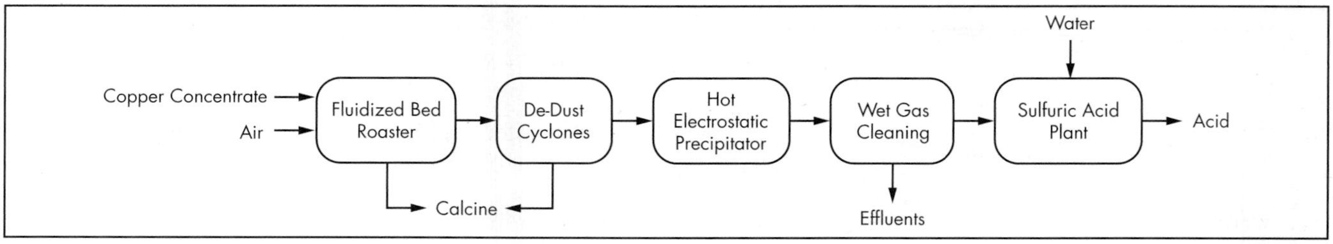

Source: Rajoria et al. 2014

Figure 23 Simple block diagram of a high-arsenic copper concentrate roasting plant

into a dynamic simulation that is highly accurate and can be used by the process engineering design team.

IDEAS' pyrometallurgy unit operations include a catalytic converter, drying tower, feed bin, packed-bed tower, post-combustion (afterburner), quench tower, radial flow scrubber, roaster, and the thermo analyzer ChemApp. ChemApp is a stand-alone thermochemistry programming software that has an extensive set of subroutines based on the thermodynamic phase equilibrium calculation module of FactSage (Bale et al. 2002). FactSage is used on a Microsoft Windows platform and as a dynamic link library. FactSage consists of a series of information, database, and calculation modules that enable utilization of databases for pure substances and mixtures of compounds. It can be used to perform a wide variety of thermochemical calculations and to generate tables, graphs, and figures of interest for pyrometallurgical processes. It allows the calculation of complex, multicomponent, multiphase chemical equilibria and their associated energy balances (Bale et al. 2002).

For pyrometallurgical process modeling, IDEAS software generally takes one of two approaches. When pyrometallurgical reactions are clear and well-defined, the standard IDEAS reactor object (either steady-state or dynamic) is used to calculate heat and mass balance based on first principles as described earlier in this chapter. Where pyrometallurgical reaction paths are unclear, ChemApp is used offline to calculate reaction paths and kinetics, which are then input to the IDEAS model. The latter approach was used to predict fluidized bed roasting of copper concentrates with high arsenic content (Rajoria et al. 2014).

Case Study: Application of Dynamic Modeling for Roasting of High-Arsenic Copper Concentrates

A dynamic process model was used to predict the fluidized bed roasting of copper concentrates with high arsenic content, to show its benefits and to outline the steps required (Rajoria et al. 2014). Figure 23 shows a simple outline of this process. The model was used to eliminate arsenic from copper concentrate before smelting, and it included a fluidized bed roaster for conversion of copper sulfide concentrate to copper oxide and SO_2, along with the volatilization of arsenic and other impurities. The desulfurized product from the roaster was cooled and used as smelter feedstock to recover the metal. Off-gas from the roaster, containing sulfur dioxide, dust, and volatilized impurities, was de-dusted in cyclones, cooled, and then further de-dusted in a hot-gas electrostatic precipitator (ESP). Wet gas cleaning was used to humidify the de-dusted gas and scrub it in a radial-flow venturi. Arsenic was mostly captured in the scrubbing liquor, which was sent to effluent treatment. Residual impurities in the gas were removed in wet ESPs. SO_2 was converted to SO_3, and the SO_3 was absorbed in acid to produce sulfuric acid (H_2SO_4). Dust from the roaster hot-gas cleaning was cooled and stored for further processing.

The steps in constructing an IDEAS computer simulation model in this case were as follows:

1. Pyrometallurgical process mechanism determination
2. Thermodynamic analysis
3. Process model development

The roaster feed consisted of enargite (Cu_3AsS_4), pyrite (FeS_2), and covelite (CuS). The chemistry of decomposition of this type of roaster feed has been outlined in detail (Fan 1997; Rajoria et al. 2014) and is summarized in Table 20 (Jones 1987).

Based on earlier work by Devia et al. (2012), roasting reactions for any component, *X*, can be described as the product of a temperature-dependent kinetic constant, *k*, and a temperature-independent concentrate component concentration function in the following form:

$$\frac{dx}{dt} = -kf(X) \quad \textbf{(EQ 17)}$$

This approach essentially implies that the roasting reactions described in Table 20 follow first-order kinetics (Devia et al. 2012). The temperature dependency of *k* is expressed by the well-known Arrhenius equation (Arrhenius 1889):

$$k = Ae^{-Ea/RT} \quad \textbf{(EQ 18)}$$

where

A = Arrhenius constant
Ea = activation energy
R = gas constant
T = absolute temperature

In addition to the heats of reaction for the roasting reactions obtained from ChemApp (ChemApp 2018; Bale et al. 2002), test results obtained by Devia et al. (2012) were necessary to determine the kinetic parameters needed in the process model. Rates of arsenic removal by roasting (Fan 1997) indicated that maximum arsenic volatilization takes place at 700°C.

Mass and energy balances were performed based on the roasting mechanism derived from thermodynamic data of species. The reaction rate data, obtained from pilot-plant studies (Devia et al. 2012), were used together with the reaction mechanisms to model the fluidized bed reactor using IDEAS, as shown in Figure 24. Roasting reactions shown in Table 20, and their kinetic parameters taken from pilot-plant studies at the University of Concepcion in Chile (Fan 1997), were used to model the roaster. The process model includes the

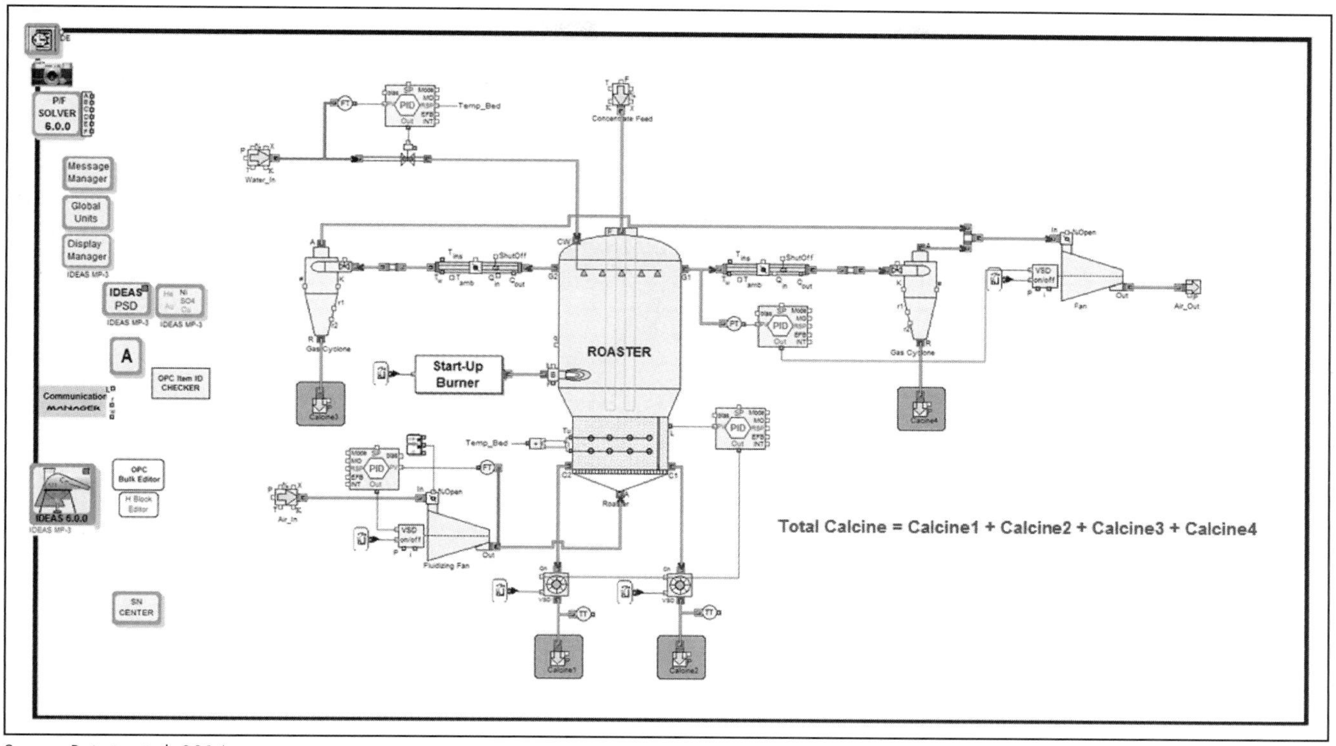

Source: Rajoria et al. 2014

Figure 24 Outline of the IDEAS process model of a fluidized bed roaster

roaster and the auxiliary units for collection of formed calcine. Concentrate is fed from top of the roaster. A start-up burner preheats the initial sand bed of the roaster. Bed temperature is controlled by water sprinklers. Discharged gases are then passed through cyclones to collect the entrained calcine. Pressure inside the roaster is controlled by a fan placed after the gas cyclones. Calcine is collected with the rotary feeders placed at the bottom of the roaster. The speed of these feeders is controlled by the level of the fluidized bed. This dynamic model was then used to investigate the case study outlined in this section.

Based on the pilot-plant study, the composition of the roaster concentrate feed, conditions and feed rates were used in the simulation model as outlined in Table 21 (Rajoria et al. 2014).

Typically, in a dynamic simulation of a pyrometallurgical process, effects of parameters such as feed rate, airflow rate, cooling water rate, and feed composition are of interest. In this case study, behavior of a copper concentrate roaster was simulated with the intention of investigating these parameters (Rajoria et al. 2014). Process conditions shown in Table 21 represent the steady-state operation of the fluidized bed roaster. However, during cold start-up, the roaster goes through several stages that IDEAS can model. Typically, the roaster burner is turned on to heat the fluidized bed with no concentrate and airflow. After the bed temperature reaches 450°C, airflow starts and is increased incrementally as concentrate feed rate and temperature also increase. Ultimately, as concentrate feed and airflow reach their steady-state levels, cooling water is used to maintain thermal equilibrium in the roaster. Figure 25 shows the dynamic computer simulation of the fluidized bed start-up and its transition to steady-state

Table 20 Major copper concentrate reactions

Roaster Reactions	Heat of Reaction, kJ/kg
$2Cu_3AsS_4(s) = 3Cu_2S(s) + As_2S_3(g) + S_2(g)$	–132.42
$2FeS_2(s) = 2FeS(s) + S_2(g)$	1,118.32
$4CuS(s) = 2Cu_2S(s) + S_{2(g)}$	455.96
$10Cu_2S(s) + 4FeS(s) + S_2(g) = 4\ Cu_5FeS_4(s)$	–257.75
$0.6FeS(s) + O_2 = 0.2Fe_3O_4(s) + 0.6SO_2(g)$	–6,457.35
$S_2(g) + 2O_2 = 2SO_2(g)$	–1,1255.48
$2As_2S_3(g) = 9O_{2(g)} = As_4O_8(g) + 6SO_2$	–6,160.16

Source: Fan 1997; Rajoria et al. 2014; Jones 1987

Table 21 Composition of roaster feed and simulation model process conditions

Mineral	Weight %
Cu_2S	30
Cu_3AsS_4	22
Cu_5FeS_4	4.8
$CuFeS_2$	7.3
CuS	3.9
FeS_2	26
SiO_2	6
Process Conditions	**Value**
Concentrate feed rate, t/h	75
Fluidized airflow rate, Nm^3/h	40,000
Fluidized air temperature, °C	250
Fluidized bed temperature, °C	700

Source: Rajoria et al. 2014

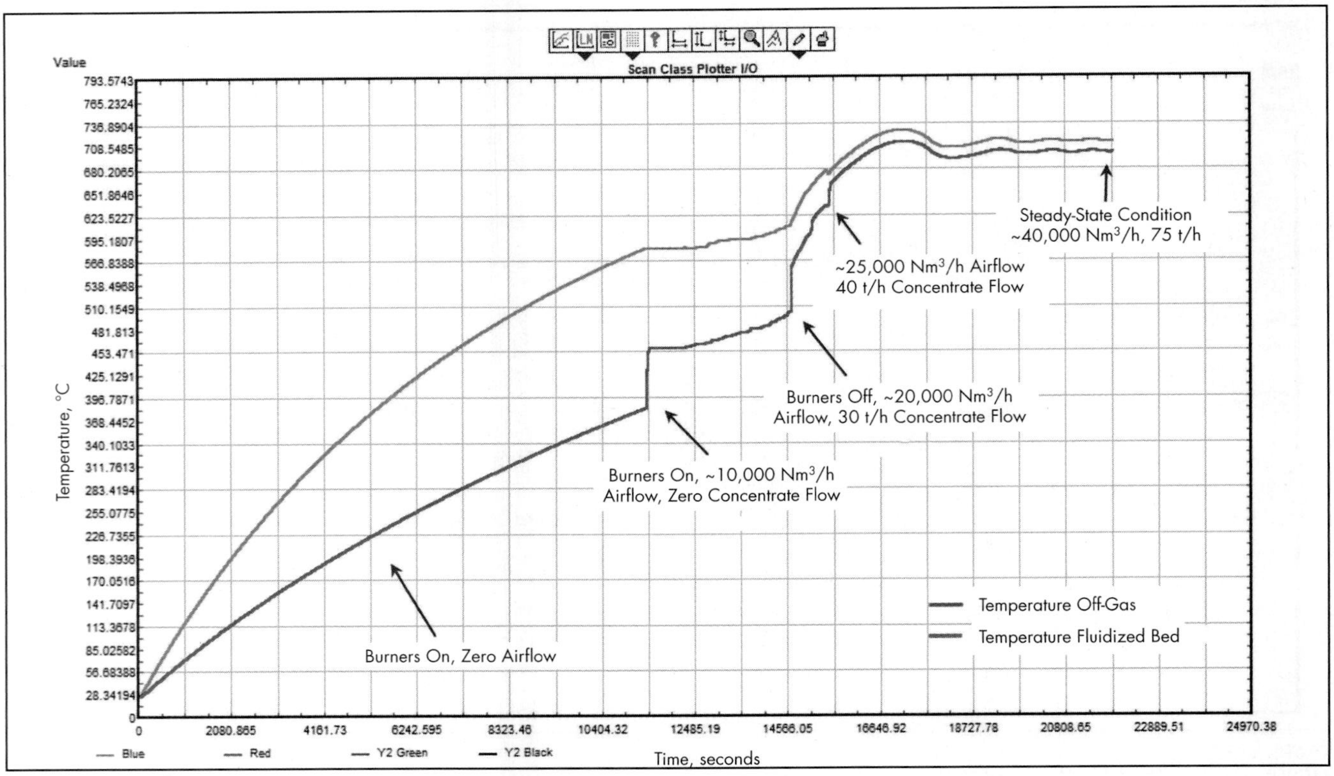

Source: Rajoria et al. 2014

Figure 25 Dynamic process model of copper concentrate roaster start-up and transition to steady state

behavior during a 6-hour period. The model also verified a 25°C gap between fluidized-bed and off-gas temperatures.

As part of this case study, several process conditions were modeled to understand the effect of variation of feed flow with constant airflow. The concentrate feed flow rate was varied between 55 and 90 t/h, while keeping airflow constant at 40,000 Nm^3/h (Rajoria et al. 2014). The results are shown in Figure 26.

At the maximum concentrate flow rate of 90 t/h, the computer simulation verified that the temperature of the roaster starts to decrease. This is in line with actual operation of the roaster under these conditions; as during plant operation, insufficient oxygen to oxidize the concentrate sulfur content limits heat generation. Plant operators can use IDEAS to understand the effect of concentrate feed and airflow rates on the roaster temperature. Computer simulation showed that increasing airflow or decreasing the concentrate feed rate increased the roaster temperature.

Another behavior that can be investigated using IDEAS is the effect of variations in the fluidized airflow rate at constant concentrate feed rate. This is essential for the plant operator to understand the effect of air-flow rate on temperature profile and arsenic volatilization at a constant rate of concentrate feed. Figures 27 and 28 show these effects (Rajoria et al. 2014).

As the flow rate of fluidized air was increased to 55,000 Nm^3/h, the temperature of the fluidized bed increased to around 900°C (Figure 27). As more oxygen was passed through the bed, more sulfur was oxidized. When the flow of air was decreased to 35,000 Nm^3/h, the temperature of the bed started declining. With the change in the fluidized airflow rate, arsenic volatilization (weight % of arsenic going to off-gas) was also analyzed for all three cases and plotted as shown in Figure 28. The model provides a realistic depiction of arsenic volatilization rate with airflow rate as a function of time and temperature. Note that as the flow of air is reduced, off-gas arsenic content increases and, as a result, less arsenic is trapped in the calcine formed.

The model of the roaster operation was also used to study the effect of the flow of cooling water rate set-point on the roaster temperature profile. This is shown in Figure 29.

Figure 30 shows the effect of concentrate feed composition on temperature profile of the fluidized bed. The main change in feed composition was the arsenic content. A similar graph was plotted for the three cases with different feed compositions, and the same procedure was followed to cold-start the roaster as mentioned previously, to maintain a steady-state condition at 700°C with the concentrate flow of 75 t/h, with fluidizing air maintained at 40,000 Nm^3/h.

As shown in Figure 30, because of more heat generated from decomposition of enargite (Cu_3AsS_4) than that associated with decomposition of pyrite (FeS_2), the higher arsenic content of the concentrate results in more heat being generated.

Benefits of Process Modeling in Pyrometallurgy

Steady-state modeling, leading to a material and energy balance of a process, is useful in determining flow-sheet alternatives and the flow rates of process input and output streams, as well as heating and cooling requirements. Dynamic process models have many additional uses, which include the following:

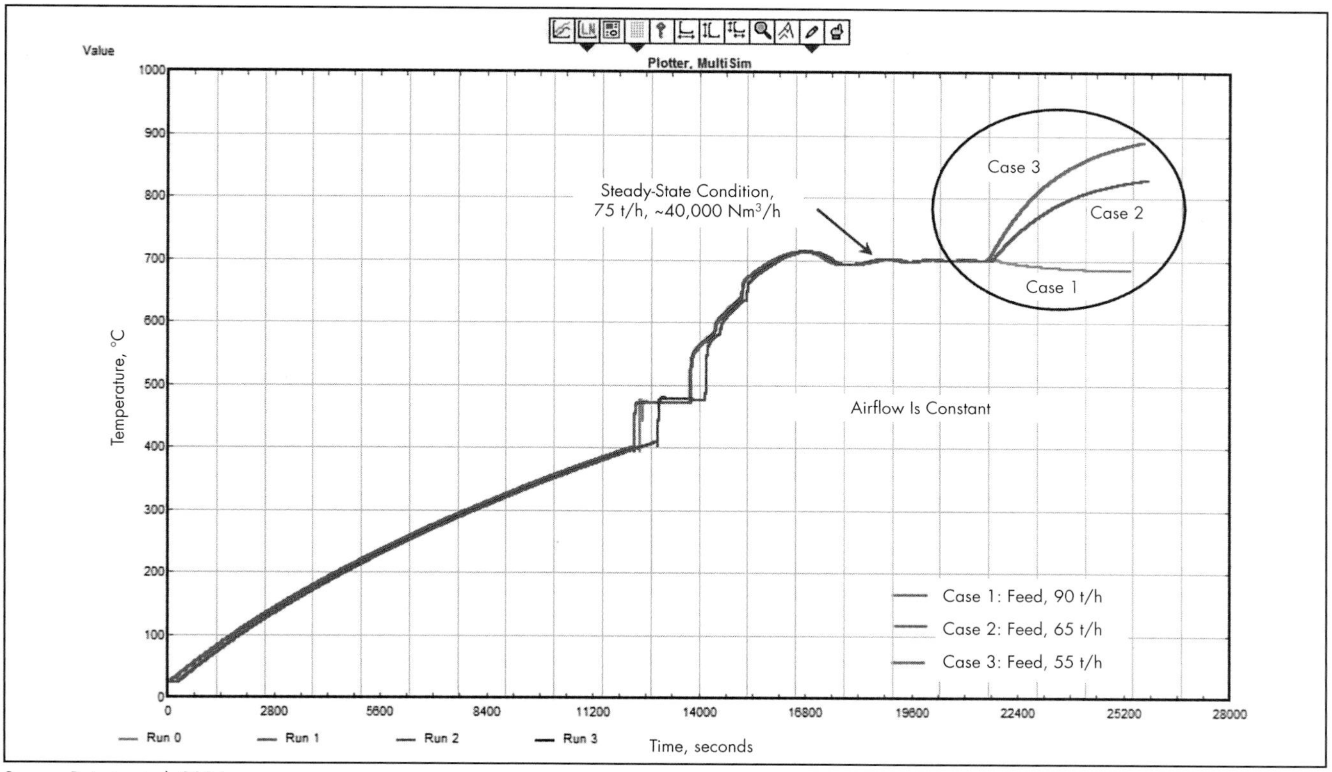

Source: Rajoria et al. 2014

Figure 26 Effect of concentrate flow rate on roaster temperature

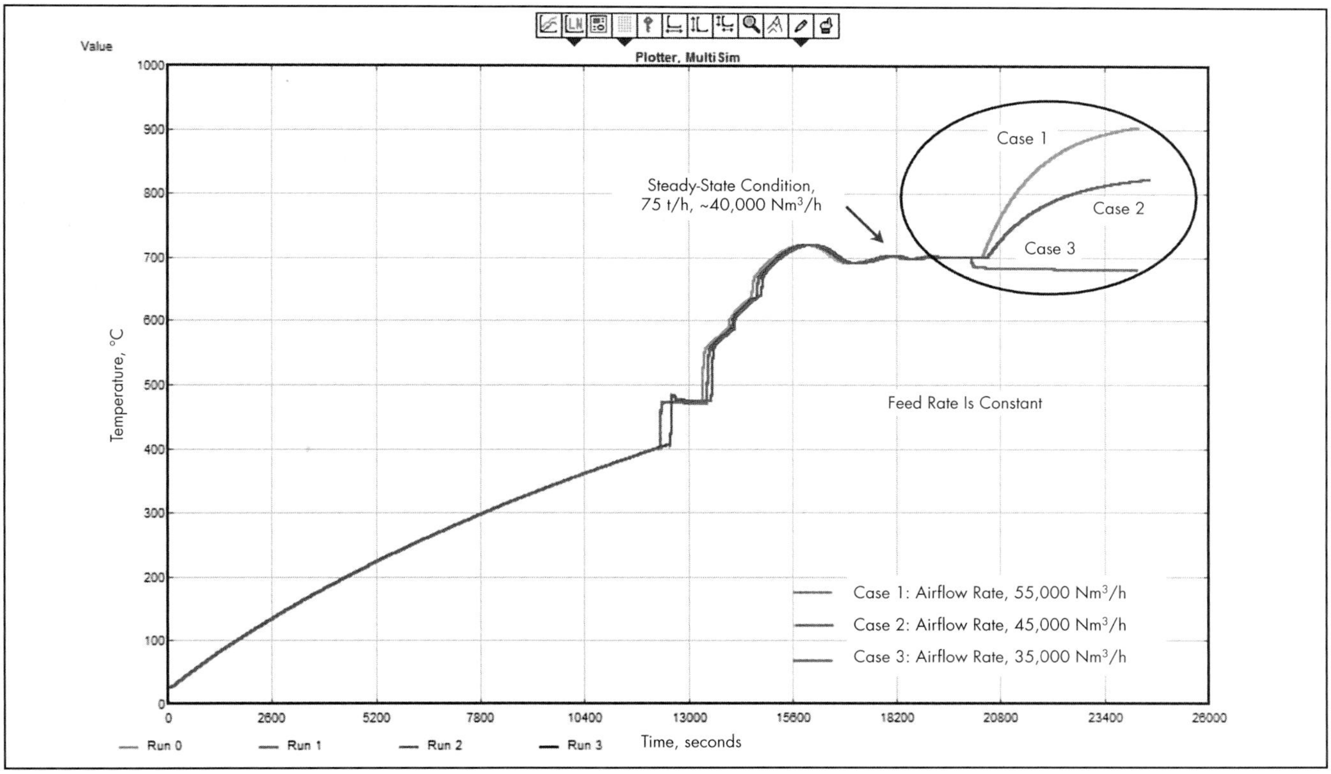

Source: Rajoria et al. 2014

Figure 27 Effect of roaster airflow rate on process temperature at constant concentrate feed rate

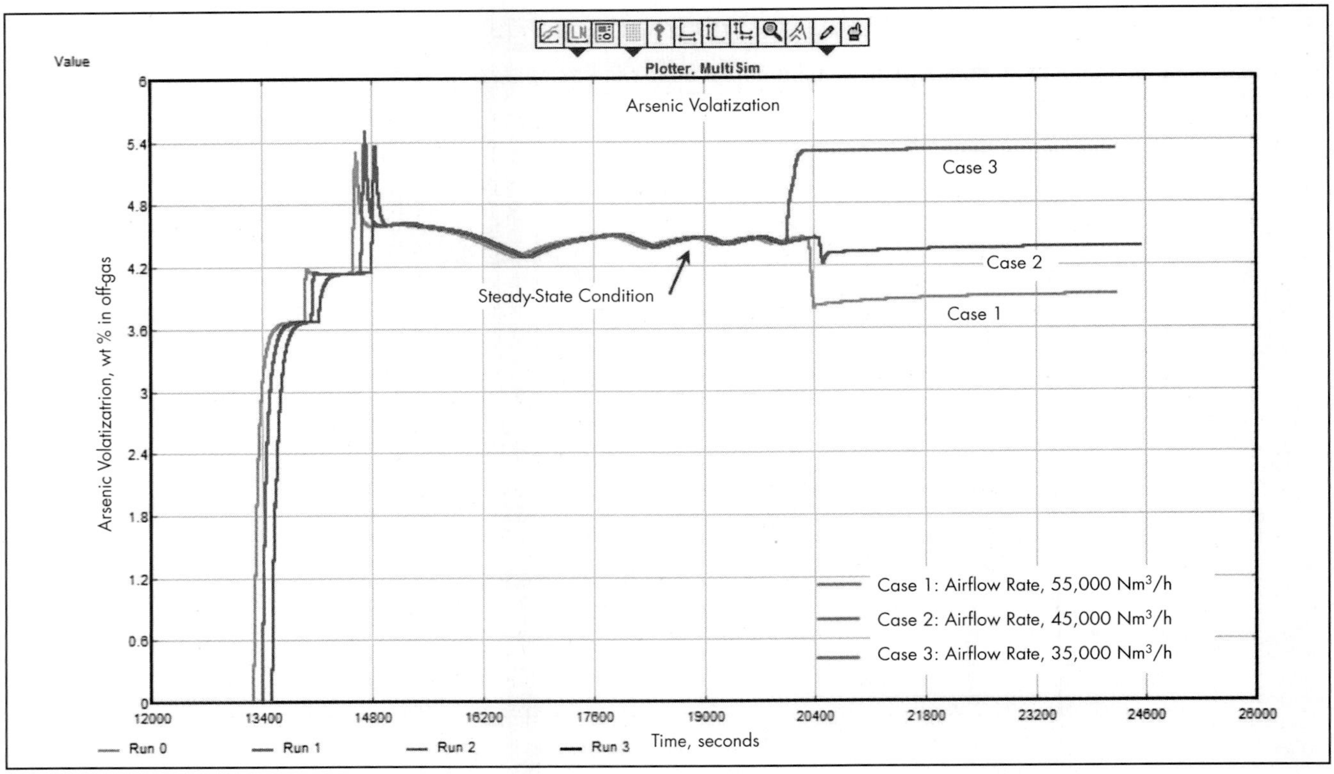

Source: Rajoria et al. 2014

Figure 28 Effect of roaster airflow rate on arsenic volatilization at constant concentrate feed rate

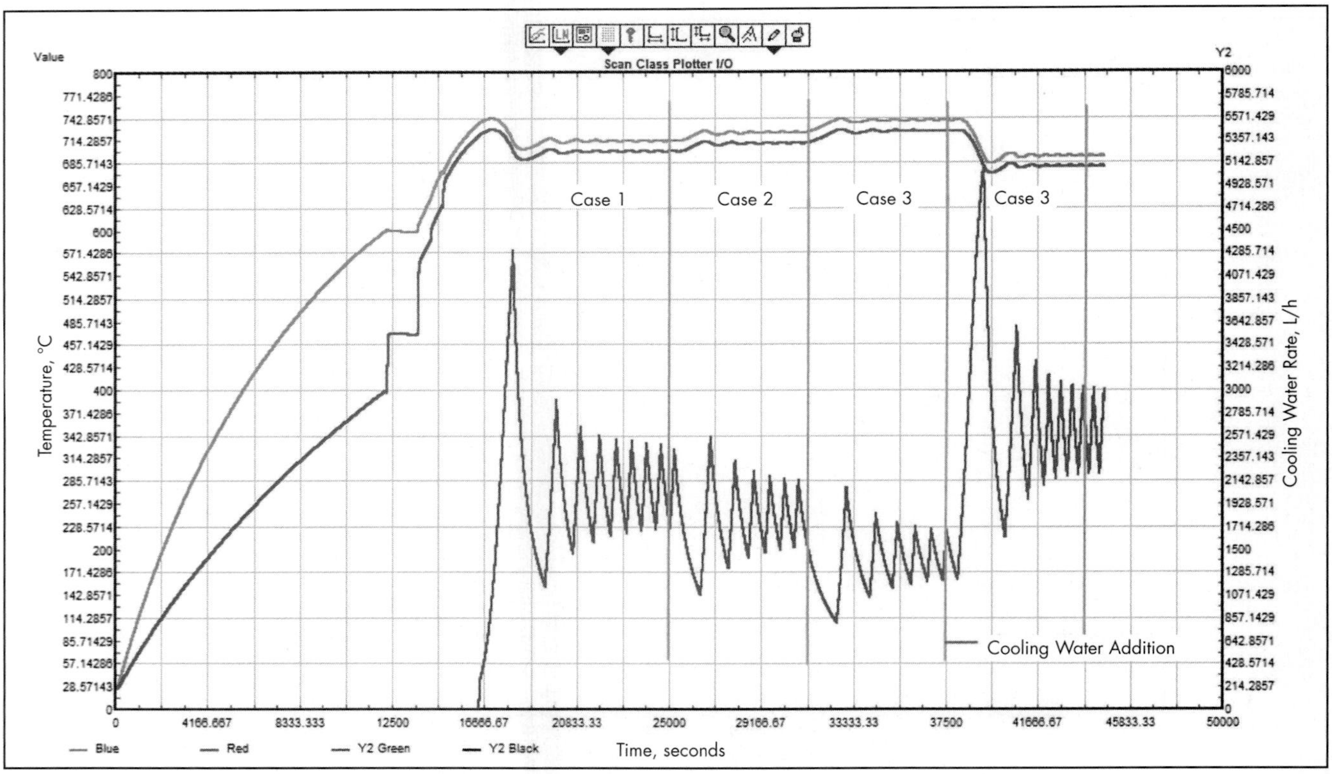

Source: Rajoria et al. 2014

Figure 29 Effect of roaster temperature set-point variation on the cooling water rate

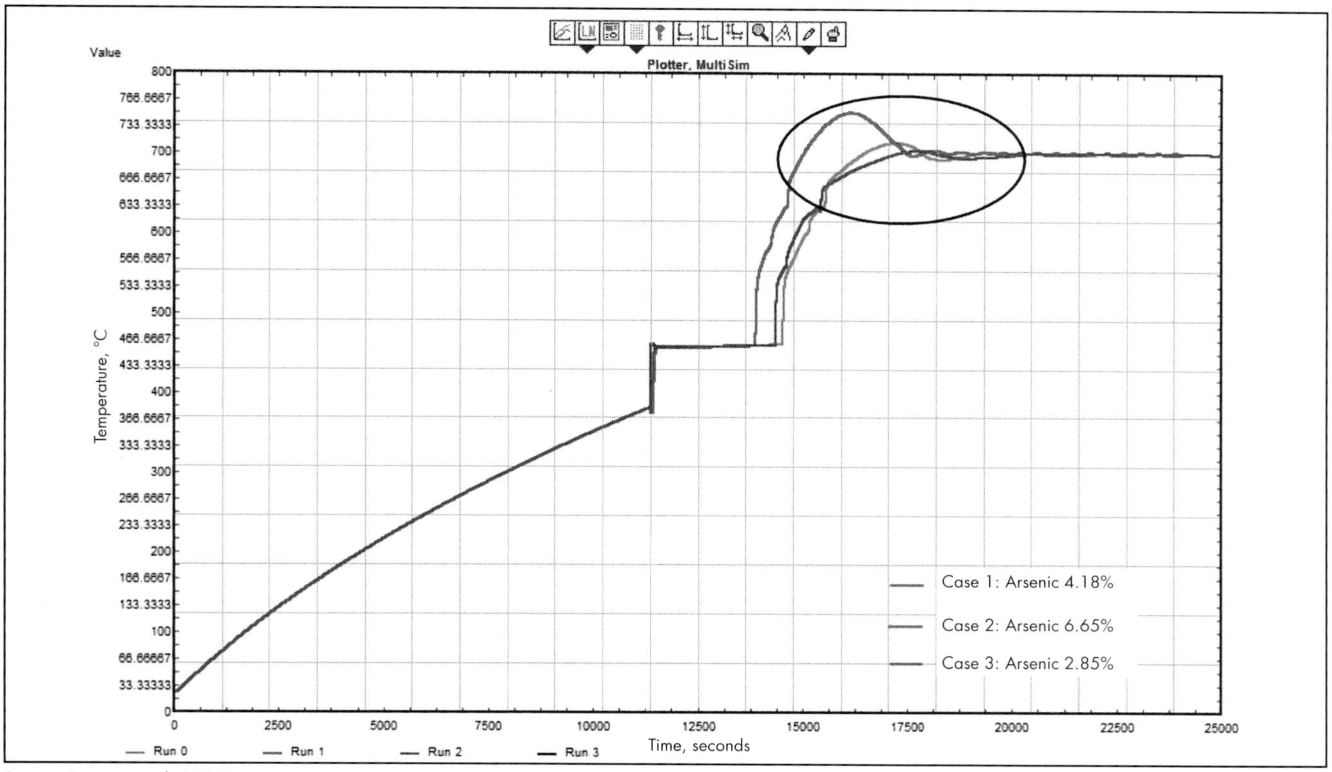

Source: Rajoria et al. 2014

Figure 30 Effect of concentrate arsenic content on roaster bed temperature

- **Validating operating procedures.** As part of dynamic modeling of a pyrometallurgical process, a virtual plant is created that can validate the operating process for unsteady-state conditions in real time. These include start-up, shutdown, and emergency situations such as loss of power or cooling water. In the case study of a fluidized bed roaster (Figure 24), the operating procedure for start-up was validated with the use of IDEAS.
- **Investigating what-if scenarios.** The simulation can test a pyrometallurgical process under different conditions and investigate their effect on the process. In the case study, several scenarios were tested (change in concentrate feed flow rate, airflow rate, cooling water rate, and concentrate feed composition), and their effects were shown on the temperature profile of the fluidized bed and arsenic volatilization.
- **Verifying process control logic.** The model can seamlessly interact with the control system—distributed control system and/or programmable logic controller—and be used for logic and control loop verification prior to the plant start-up.
- **Training operations personnel.** The dynamic model can be linked to the actual plant control system and used for operator training. Operators can use real control screens (human–machine interface) to be trained in operating the virtual plant, gaining an understanding of the process and equipment dynamics and interactions.

REFERENCES

ABB. 2018. ABB Enterprise Software. http://new.abb.com/enterprise-software. Accessed February 2018.

AggFlow. 2018. AggFlow Design. https://aggflow.com/. Accessed February 2018.

Andersen, J.S. 1988. Development of a Cone Crusher Model. M. Eng. Sc. thesis, University of Queensland, Julius Kruttschnitt Mineral Research Centre, Brisbane, Australia.

Andritz. 2018. IDEAS software simulation downloads. www.andritz.com/products-and-services/pf-detail.htm?productid=18154. Accessed February 2018.

ANSYS. 2018. ANSYS Fluent. www.ansys.com/Products/Fluids/ANSYS-Fluent. Accessed February 2018.

Argyropoulos, S.A., Closset, B., and Valiveti, R. 1984. Documented Fortran programs for ball-mill modelling. *CIM Bull.* 870(77).

Arrhenius, S.A. 1889. Über die Dissociationswärme und den Einflusß der Temperatur auf den Dissociationsgrad der Elektrolyte. *Z. Physik. Chem.* 4:96–116.

Aspentech. 2018. AspenPlus. www.aspentech.com/products/engineering/aspen-plus/. Accessed February 2018.

Austin, L.G., Klimpel, R.P., and Luckie, P.T. 1984. *Process Engineering of Size Reduction: Ball Milling*. New York: SME-AIME.

Baes, C.F., and Mesmer, R.E. 1976. *The Hydrolysis of Cations*. New York: Wiley-Interscience.

Bale, C.W., Chartrand, E.P., Degterov, S.A., Eriksson, G., Hack, K., Ben Mahfoud, R., Melançon, J., Pelton, A.D., and Petersen, S. 2002. FactSage thermochemical software and databases, computer coupling of phase diagrams and thermochemistry (Calphad). *Int. Res. J. Calculation Phase Diagrams* 26(2):189–228.

Caspeo. 2018. USIM PAC. www.caspeo.net/USIMPAC. Accessed February 2018.

CFDEM Project. 2018. LIGGHTS open source discrete element method particle simulation code. www.cfdem.com/ liggghts-open-source-discrete-element-method-particle -simulation-code. Accessed February 2018.

ChemApp. 2018. Home page. www.gtt-technologies.de/ chemapp. Accessed February 2018.

CITIC SMCC Process Technology. 2018. Home page. www .citic-smcc.com. Accessed February 2018.

Daniel, M.J., and Morrell, S. 2004. HPGR model verification and scale-up. *Miner. Eng.* 17(May):1149–1161.

Devia, M., Wilkomirsky, I., and Parra, R. 2012. Roasting kinetics of high arsenic copper concentrates: A review. *Miner. Metall. Process.* 29(2):121–128.

Dry, M. 2013. Early evaluation of metal extraction projects. Presented at ALTA 2013, Perth, Australia, May 25–June 1.

Dry, M. 2015. Technical and cost comparison of laterite treatment processes, Part 3. Presented at ALTA 2015, Perth, Australia, May 23–30.

Drzymala, J. 2007. *Mineral Processing Foundations of Theory and Practice of Minerallurgy* [sic]. Poland: Wroclaw University of Technology.

Dutwa Project. 2018. Home page. www.blackdownresources .com/dutwa-overview.asp. Accessed February 2018.

EDEM. 2018. EDEM Software. www.edemsimulation.com/ software. Accessed February 2018.

Ellingham, H.J.T. 1944. Reducibility of oxides and sulfides in metallurgical processes. *J. Soc. Chem. Ind.* 63(1944):125–133.

Fan, Y. 1997. Cinética y mecanismos de vaporización de sulfuros de arsénico desde concentrados de cobre. Tesis de magíster en ciencias de la ingeniería, Mención Metalurgia Extractiva, Universidad de Concepción, Concepción, Chile.

Fourie, P. 2007. Modelling of separation circuits using numerical analysis. In *Proceedings of the 6th International Heavy Minerals Conference "Back to Basics."* Johannesburg: Southern African Institute of Mining and Metallurgy. pp. 1–6.

Gilchrist, J.D. 1989. *Extraction Metallurgy*, 3rd ed. Oxford: Pergamon Press.

Haque, N., Bruckard, W., and Cuevas, J. 2012. A techno-economic comparison of pyrometallurgical and hydrometallurgical options for treating high-arsenic copper concentrates. Presented at the XXVI International Mineral Processing Congress (IMPC), New Delhi, India, September 24–28.

Hay, M.P. 2005. Using the SUPASIM® Flotation Model to Diagnose and Understand Flotation Behaviour from Laboratory through to Plant. In *Proceedings of 37th Annual Meeting of the Canadian Mineral Processors*. Montreal, QC: Canadian Institute of Mining, Metallurgy and Petroleum. pp. 463–479.

Herbst, J.A., and Pate, W.T. 2018. Dynamic simulation of size reduction operations from mine-to-mill. Kealakekua, HI: J.A. Herbst and Associates.

Herbst, J., Rajamani, R.K., Mular, A.L., and Flintoff, B. 2002. Mineral processing plant/circuit simulators: An overview. In *Mineral Processing Plant Design, Practice, and Control*. Edited by A.L. Mular, D.N. Halbe, and D.J. Barratt. Littleton, CO: SME. pp. 383–403.

Holland-Batt, A.B. 1989. Spiral separation: Theory and simulation. *Trans. Inst. Min. Metall.* 98(Jan.-Apr.):C46–C60.

InfoMine. 2018. Mining cost models. http://costs.infomine .com/costdatacenter/miningcostmodel.aspx. Accessed February 2018.

JKTech. 2018. Software. http://jktech.com.au/software. Accessed February 2018.

Jones, R. 1987. Computer simulation of pyrometallurgical processes, APCOM 87. In *Proceedings of the Twentieth International Symposium on Application of Computers and Mathematics in Mineral Industries*. Vol. 2: Metallurgy. Johannesburg: Southern African Institute of Mining and Metallurgy. pp. 265–279.

Kelly, T.D., and Matos, G.R. 2014. Historical statistics for mineral and material commodities in the United States (2016 version). US Geological Survey Data Series 140. http://minerals.usgs.gov/minerals/pubs/historical -statistics. Accessed February 2018.

King, R.P. 2001. *Modeling and Simulation of Mineral Processing Systems*. Oxford: Butterworth-Heinemann.

Leung, K. 1987. An energy based ore specific model for autogenous and semi-autogenous grinding mills. PhD thesis, University of Queensland, Australia.

Lishchuk, V. 2016. Geometallurgical programs: Critical evaluation of applied methods and techniques. Licentiate of engineering thesis in mineral processing, Luleå University of Technology, Sweden.

Luttrell, G.H., Barbee, C.J., Bethell, P.J., and Wood, C.J. 2005. *Dense Medium Cyclone Optimisation*. Final Technical Report to the U.S. Department of Energy, DOE Award Number DE-FC26-01NT41061.

Lynch, A.J. 1977. *Mineral Crushing and Grinding Circuits: Their Simulation, Optimization, Design and Control.* New York: Elsevier Scientific Publishing.

Mackinnon, S., Yan, D., and Dunne, R. 2003. The interaction of flash flotation with closed circuit grinding. *Miner. Eng.* 16 (11):1149–1160.

Maelgwyn Mineral Services Ltd. 2009. Limn: The Flowsheet Processor. www.maelgwyn.com/technology/limn-the -flowsheet-processor/. Accessed February 2018.

Marsden, J.O., and Botz, M.M. 2017. Heap leach modeling: A review of approaches to metal production forecasting. *Miner. Metall. Process.* 34(2):53–64.

MathWorks. 2018. MATLAB. www.mathworks.com/ products/matlab.html?s_tid=hp_products_matlab. Accessed February 2018.

McGarry, M., and Watson, M. 1998. Dynamic simulation: A maturing technology generating real benefits. https://www.andritz.com/resource/blob/14790/d17a6a c5824afe0c442c065bb93a47c6/aa-automation -simulation-maturing-technology-benefits-data.pdf. Accessed February 2018.

Metso. 2008. Bruno: Metso's simulating tool. www.metso .com/miningandconstruction/mm_segments.nsf/ WebWID/WTB-081113-2256F-FD63A/$File/Bruno _presentation.pdf. Accessed February 2018.

Mineral Technologies International. 2010. MODSIM: Modular simulator for mineral processing plants. www.mineraltech.com/MODSIM/. Accessed February 2018.

Moly-Cop. 2018. Moly-Cop Tools 3.0 Software. www.molycop.com/supporting-customers/molycop-tools. Accessed February 2018.

Nageswararao, K. 1990. Reduced efficiency curves of industrial hydrocyclones: An analysis for plant practice. *Miner. Eng.* 12(5):517–544.

Napier-Munn, T.J., Morell, S., Morrison, R.D., and Kojovic, T. 1996. *Mineral Comminution Circuits: Their Operation and Optimisation*. Indooroopilly, Queensland: Julius Kruttschnitt Mineral Research Centre.

Nicol, M.J., Fleming, C.A., and Paul, R.L. 1987. The chemistry of extraction of gold. In *Extractive Metallurgy of Gold*. Edited by G.G. Stanley. Monograph Series M7. Johannesburg: Southern African Institute of Mining and Metallurgy. pp. 894–899.

Nikkhah, K., and Anderson, C. 2001. Role of simulation software in design and operation of metallurgical plants: A case study. SME Preprint No. 01-176. Littleton, CO: SME.

OpenCFD Ltd. 2017. About OpemFOAM. www.openfoam.com/. Accessed February 2018.

Papangelakis, V.G., Georgiou, D., and Rubisov, D.H. 1996. Control of iron during the sulfuric acid pressure leaching of limonite. In *Proceedings of the Second International Symposium on Iron Control in Hydrometallurgy*. Edited by J.E. Dutrizac and G.B. Harris. Montreal. QC: Canadian Institute of Mining, Metallurgy and Petroleum.

ProSim. 2018. Software and Services in Process Simulation. www.prosim.net. Accessed February 2018.

Proware. 2018. METSIM: The premier steady-state and dynamic process simulator. www.metsim.com/. Accessed February 2018.

PTC. 2018. PTC Mathcad. www.ptc.com/engineering-math-software/mathcad. Accessed February 2018.

Rajoria, A., Lucena, L., Das, S., Szatkowski, M., and Wilkomirsky, I. 2014. Dynamic simulation of processing high arsenic copper concentrates in a fluidized bed roaster. SME Preprint No. 14-071. Englewood, CO: SME.

Schwarz, S., Alexander, D., Whiten, W.J., Franzidis, J.P., and Harris, M. 2006. JKSimFloat V6: Improving flotation circuit performance and understanding. Presented at the XXIII International Mineral Processing Congress, Istanbul, Turkey.

Schwarz, S., and Richardson, J.M. 2013. Modeling and simulation of mineral processing circuits using JKSimMet and JKSimFloat. Presented at the SME Annual Meeting, Denver, CO.

STABCAL. 2015. Stability Calculation for Aqueous and Nonaqueous System. Version 2015. Butte, MT: Montana Tech.

SysCAD. 2018. Plant simulation software. www.syscad.net/. Accessed February 2018.

Tian, J., and Morrell, S. 2015. Operation and process optimization of Sino Iron's AG milling circuits. Presented at SAG Conference 2015, Vancouver, British Columbia, Canada.

Triple Point. 2018a. Algosys Bilmat: Accurate and timely data reconciliation. www.tpt.com/bilmat. Accessed February 2018.

Triple Point. 2018b. Algosys Metallurgical Accountant: Powerful metal accounting tool. www.tpt.com/metallurgical-accounting. Accessed February 2018.

U.S. Energy Information Administration. 2018. Coal. www.eia.gov/coal/. Accessed February 2018.

Vahed, A. 2004. Development of a fluid bed pyrohydrolysis process for Inco's Goro Nickel Project. Presented at the TMS Annual Meeting, Charlotte, NC.

Valery, W., Jr., and Morrell, S. 1995. The development of a dynamic model for autogenous and semi-autogenous grinding. *Miner. Eng.* 8(11):1285–1297.

VSMA (Vibrating Screen Manufacturers Association). 2014. *VSMA Vibrating Screen Handbook*. Milwaukee, WI: Association of Equipment Manufacturers.

Wedderburn, B. 2010. Nickel heap leaching study. Presented at the ISNG Environmental and Economics Session, April 27. www.insg.org/presents/Mr_Wedderburn_Apr10.pdf.

Weiss, N.L., ed. 1985. *SME Mineral Processing Handbook*. Littleton, CO: SME-AIME

Whiten, W.J. 1972. Simulation and model building for mineral processing. PhD dissertation, University of Queensland, Australia.

Whiten, W.J. 1974. A matrix theory of comminution machines. *Chem. Eng. Sci.* 29:589–599.

Wills, B.A. 1997. *Mineral Processing Technology*, 6th ed. Oxford: Butterworth-Heinemann.

Yan, D., Wiseman, D., and Dunne, R. 2005. Predicting the performance of a flotation circuit that incorporates flash flotation. Presented at the Centenary of Flotation Symposium, Brisbane, Queensland, Australia.

CHAPTER 2.6

Process Control and Operational Intelligence

Osvaldo Bascur

This chapter presents a summary of the state of the art for mineral–metallurgical process control and "plant operational intelligence." Operational intelligence is a way of providing an augmented view of real-time data rather than using traditional fixed plant information management reports. The data are transformed into insights for further analysis using business intelligence tools.

Basic process control is an integral part of most mineral and metal processing plants. It has helped many operations reduce costs and increase productivity and performance. Developments in process control in the mineral–metallurgical processing industry over the past decade have been greatly influenced by new measurements, advances in computer software and communications, and internet/cloud technologies. Another extremely important factor has been the selection of strategies to link control actions to process measurement to form an overall control and plant management system. These three components—measurements, process control strategies, and computer hardware—form a triangle of integrating elements, as shown in Figure 1. The knowledge workers who are the operators and process engineers are at the center of these elements. These individuals operate the process units. They are learning about the process and designing and maintaining the process control strategies. These control strategies are integrated with an operational management support system, combining all forms of information in a mineral processing plant. A continuous improvement and innovation culture is required for enabling operational excellence. In the following paragraphs, a control triangle is used to describe the process control and operational management technologies in the digital age. This is called "Industries 4.0" and refers to the digital revolution that is currently taking place.

The decades between 1990 and 2010 saw the parallel development of three important aspects of modern mineral processing control: (1) measurement devices designed especially for the particulate systems encountered in these industrial plants, (2) mathematical models constructed for the analysis of these systems, and (3) control strategies developed using the capabilities of digital computers. In recent years, the development of intelligent sensors (advanced control systems based on object technologies and communications networks) have transformed the way plants are managed.

Plants require many sensors to measure the process and equipment variables to monitor conditions and to process sensor data using a variety of algorithms to assist in the stabilization of operations and the optimization of resources to minimize operating costs for an environmentally safe, profitable operation. The "Process Measurements" section provides an overview of this topic. The lower right side of the triangle in Figure 1 shows the control hardware and computer systems that provide the environment for delegating the transformation of the measurements into real-time controls for manipulating the final control elements for all process units. Shown at the top of the triangle are the control and performance management strategies that process the sensor data to manipulate the final control elements in a process plant. The "Process

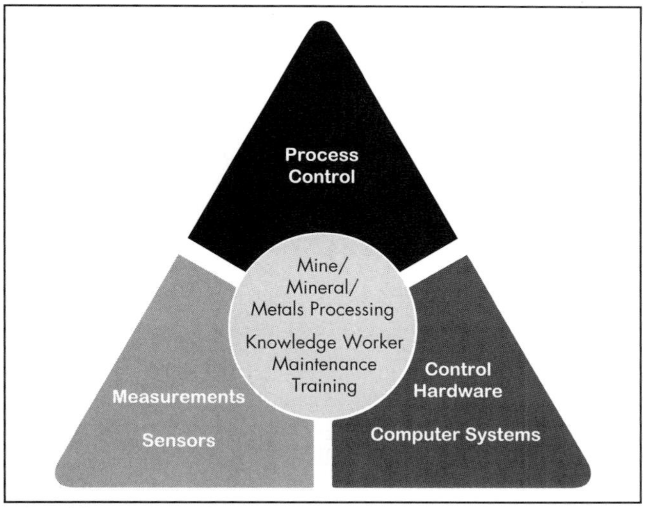

Adapted from Herbst and Bascur 1984

Figure 1 Control triangle—basic components and interactions in automatic control

Osvaldo Bascur, Principal, Academia–Industry Innovation, OSIsoft LLC, Houston, Texas, USA

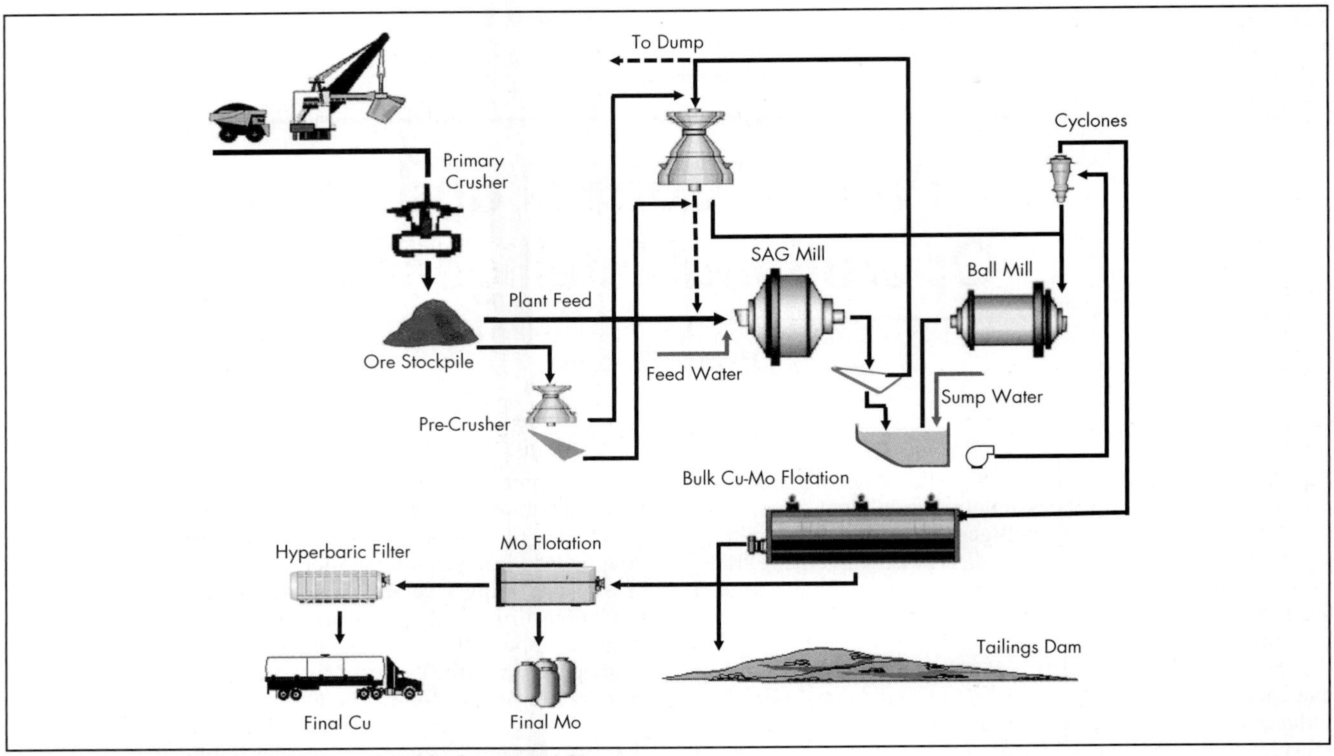

Figure 2 Integrated mineral processing block process diagram

Control Strategies" section provides an overview of commonly used process control techniques. At the center of the triangle are the people who design, operate, and maintain the systems necessary to keep the overall process of the triangle in working order.

Operators are usually part of the control loop because they must oversee the plant to achieve the desired target set daily by management. Process engineers are the support personnel who use the process data to develop strategies to manage constraints; process advisors also support plant operations. Real-time data are used to assist with condition-based maintenance to prevent costly equipment downtime or to prevent process excursions due to failure to detect these abnormal situations in real time. Condition-based equipment maintenance, process excursions, and fault detection support functions belong to the plant data infrastructure, which provides data for performance monitoring, statistical process control, overall performance management, predictive analytics, evolutionary optimization, metallurgical mass balances, and energy management. These systems include ore blasting to the final tailings ponds and route deliveries to the filter plants and port.

Figure 2 shows a typical mineral processing plant. It starts with the mine trucks feeding the ore to a primary crusher, followed by comminution circuits to liberate the metal-bearing mineral in the ore for further separation in a flotation circuit to capture the metal-bearing species and discard the waste in to a tailings dump prior to the thickening of the waste to reprocess as much water as possible. The flotation circuit uses the mineral characteristics to float the desired species and to hinder the waste material that needs to be separated from the rock. The major process variables are recovery and the grade of the metals, depending on the type of ore being processed that is sent from a mine area or several mines.

These plants have traditionally used automation in the form of regulatory controls to stabilize process operation by rejecting the main disturbances.

Table 1 shows the variables that are used as controlled variables and those that are used as manipulated variables to achieve the desired business objectives in the regulation and stabilization of crushing, grinding, flotation, leaching, thickening, filtering, slurry transport, drying, smelting, and converting and refining the minerals and final metal products. The process control and operational management technologies are very unique. However, the basic principles are applicable to all industrial process systems. First, the control objective needs to be defined and then the behavior of the process units needs to be observed (and modeled). The best pair of manipulated variables needs to be coupled to the desired control variables while avoiding process and equipment constraints for an environmentally safe, profitable operation.

Today's technologies are enabling mineral and metallurgical processors to develop competency centers where they can manage their mine and mineral processing plants remotely to coordinate mining operations with downstream operations. The objective is to find the optimal cut of grade at the mine and the optimal cut size at comminution to achieve the best recovery grade combination to maximize profits. The typical process control objective is to maximize the metal yield while lowering operating production and mine costs. Because grades at mines have been lowered, large volumes of ore are processed to achieve economical production targets. As such, it has become more expensive to transport the ore and dispose

Table 1 Controlled and manipulated variables for crushing, grinding, flotation, thickening, slurry transport, and filtration

Process Units	Controlled Variables	Manipulated Variables
Crushing	Product fineness	Feed rate
	Circulating load	Closed-side setting
	Power draw	Screen area
	Bin level	Screen speed
	Crusher level	
Grinding	Product size distribution	Feed rate
	Feed size	Sump level
	Sump level	Pumping rate
	Circulating load	Solids in feed
	Holdup of solids	Mill speed
Flotation	Recovery	Aeration
	Grade	Agitator speed of rotation
	Circulating loads	Pulp levels
	Froth levels	Reagents
	Froth speed	Frother
	Percent solids	Collector
		Modifiers
		Depressants
		Water spray
Thickening	Slime level	Flocculant addition
	Rake torque	Pumping rate
	Underflow % solids	Rake position
	Underflow viscosity	Water dilution
	Bed level	Viscosity modifier
Slurry transport	Slurry flow	Water dilution
	Slurry density	Pump speed
	Slurry viscosity	Upstream pressure
	Suction and discharge pressure	Tank level
	Slurry pH, oxygen	Slurry inhibitor
	Slurry–water interface	Viscosity modifier
Filtration	Cake humidity	Feed rate
	Cake thickness	Pressure
		Water removal
		Time

Adapted from Herbst and Bascur 1984

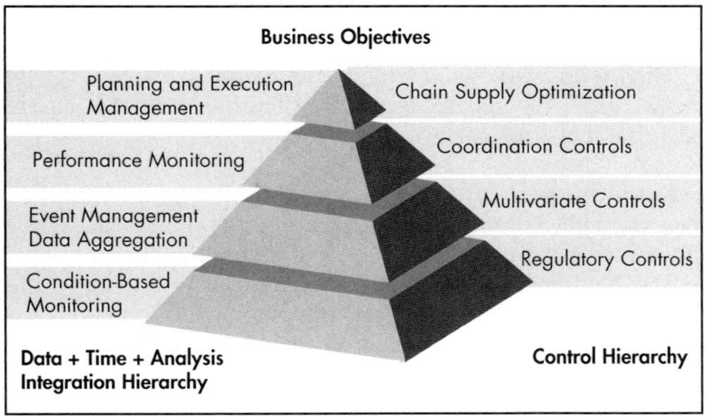

Source: Bascur 2016

Figure 3 Plant data and control hierarchies for process control and operational intelligence

of the waste. At mills, energy costs have increased to liberate the metal-bearing particles. Large volumes of water are also required by the wet grinding circuits. Final water recovery is achieved at tailing impoundment facilities at large mineral processing complexes. The center of excellence can be located nearby to integrate the mine and the mill, or it can be based remotely to enable a virtual, cloud-based environment. Integration of the operational data and events transforms the organization into a continuous improvement and innovation culture. An operational excellence program is an ongoing process of improvement to achieve environmentally safe, profitable operations.

The overall objective of process control and management systems is to stabilize operations to be able to maximize the economic profit of the operations while adhering to equipment, production, quality, environmental, and safety constraints.

The "Plant Operational Intelligence Management" section briefly describes the disruptive technology in process industries. In the past, process controls have lacked maintenance and training. Distributed control system (DCS) hardware is expensive, so this hardware is not updated very often. As such, many vendors provide alternative cost-effective solutions on other, more modern platforms. However, the software on which all these tools are built requires constant upgrades to perform well and deliver value. The rapid rate of change in computer systems and software applications requires modern maintenance practices. DCSs and programmable logic controllers (PLCs) can provide robust data to external systems that have professional data and analytical capabilities. Being able to access or share data with external services allows plant employees to augment their acquisition of knowledge and allows extended support of remote plant operations.

The next section, "Industrial Internet of Things: Disruption in Automation," presents new ways of using real-time data and remote connectivity for collaboration using the cloud.

The "Conclusions and Future Implications" section summarizes process control and data infrastructure in the mining, mineral processing, and extractive metallurgy industries. It provides an overview of the benefits and methodology for the assessment of advanced process controls and newer predictive analytics.

PLANT DATA AND CONTROL HIERARCHY

Figure 3 shows a data and control hierarchy diagram with two sides. The right side of the pyramid shows the traditional real-time control levels fed by the sensors and data collected from the field. At the lowest level is the instrumentation level, which consists of devices for acquiring data from sensors, field displays, and hardware safety interlocks for ensuring emergency shutdowns. The "Process Measurements" section of this chapter lists the basic concepts to explain the many factors involved in implementing the right measurements in this very harsh environment for the mineral and metals industry.

The instrumentation level reports the data to the regulatory control level, which is implemented using control hardware such as DCSs and PLCs. This level provides integration of the real-time data for the regulatory control level. It is one of the most important levels because it has to be extremely robust for industrial process continuity and operational safety. (This is equivalent to the brain stem and cerebellum in the brain.)

Regulatory Controls

Regulatory controls maintain the process variables at their prescribed set point, stabilizing the variations due to local disturbances occurring at a time scale of seconds to minutes. The disturbances can be caused by many factors, including weather conditions, changes in the raw material characteristics, ore hardness changes, particle size distributions, or mineralogical compositions' start-up and shutdown at the other sections of the chain supply of the plant. In addition, this level allows the operator to take control of the plant in an emergency, and it can be controlled manually. The DCS and PLCs require software tools to configure the control strategies and require a human operator interface to monitor the behavior of the plant under control. The level sends the streaming data collected from the plant to a dedicated industrial historian (like the black box of an airplane). The stream data are used for enhanced equipment condition-based maintenance and for control tuning and improvements.

The "Process Control Strategies" section provides additional details about ways to capture the process dynamics and the configuration of the possible control strategies. The proportional–integral–derivative (PID) controller is perhaps the most commonly used process control algorithm to implement local single-input, single-output (SISO) control loops. The DCS is used to implement multiloop control strategies such as feedforward, cascade, ratio, and constraint controls. DCSs have augmented their capabilities so that the next level is also part of a modern DCS.

Multivariate Controls

The next level is called multivariable process control or model-based process control. This level has evolved in recent years due to advances in hardware and computing capabilities. In a processing plant, the problems are typically multivariable with many control interactions due to the nonlinearities of the process, process equipment constraints, and unknown process disturbances. Because of the possible interaction among the variables, all the control movements must be coordinated. Control actions are taken to accommodate longer duration disturbances at a time scale of approximately minutes. This is also due to the slow processing time of the online process sensors or the instream process analyzers. Mineral and metal operation process dynamics are important, and a dynamic process model is necessary. The output of this level is used by the DCS as a set point to drive the final control actuators that operate the plant. The final control elements are the valves, variable-speed pumps, and weight feeders, among others. This level is metaphorically equivalent to the hypothalamus, which maintains brain chemistry and controls appetite, thirst, body temperature, and libido. This last statement stresses the importance of having a hierarchy in process control design. It is highly recommended that priorities are clarified when implementing and maintaining overall control strategies.

Coordination Controls

The plant coordination controls are implemented to balance the overall constraints to find the optimal steady-state operating conditions of the plant based on the current production requirements and factors, such as raw materials, energy and consumable costs, and production demand. These process controls require the left-side level to analyze operations to identify the current process and equipment constraints.

Chain Supply Optimization

The planning and scheduling activities are usually determined by the integration of the plant's industrial data infrastructure with the enterprise business systems, which contain the production plans and utility costs. For example, in mineral processing plants, the mine production, mill, tailing, and port production areas are integrated to optimize the overall energy and water use based on the type of ore being mined (drilling, explosives, and blasting strategies). The lowering of the ore grades requires a much more detailed analysis for classifying the operating strategies and lineups of the plant for the best and most profitable alternative. Mill data are now being used to optimize mine-blasting operations for minimum operating costs while maximizing throughput.

Condition-Based Equipment/Production Monitoring

This section describes alternative ways of reusing data, generating events to create actionable insights based on operational intelligence. It is critical to be vigilant about maintaining equipment to achieve high productivity and availability levels to remain competitive in the mining and mineral processing industry. Maintenance activities were not carried out in the past because the technology was not available. Reusing data from the industrial data infrastructure is the most cost-effective way of achieving high productivity levels. Today, operations and maintenance teams have to work with the same data to keep the instrumentation, process controls, and equipment online 24/7 for optimal production levels. Using the process sensors and capabilities of a modern industrial data infrastructure, maintenance personnel can implement condition-based maintenance strategies to monitor and predict the performance of moving equipment in real time. Now there is powerful data analytics to filter data to detect problems prior to a failure.

Because the process and equipment data are always available, employees can use the data to generate notifications to initiate a programmed shutdown as opposed to dealing with an unscheduled shutdown.

When regulatory controls and multivariable control loops underperform, this can cause process variability that adversely affects profitability. Having the stream data available in an easily accessible format for advanced analysis simplifies the continuous improvement process required to support the controls at all levels for all process units in the plant.

The increasing use of real-time data is elevating the role of plant personnel to that of knowledge workers who are *expected* to make real-time decisions to improve business performance. The high variability of ore qualities, rapid changes in energy costs, reduced availability of process water (or, in some cases, too much water in equatorial regions) require adaptive real-time performance strategies.

Event Management Data Aggregation

The left side second level supports the validation and classification of the data to develop actionable information. This actionable information is the key ingredient for implementing a continuous improvement and innovation program in process industries. This is called an operation intelligence program. It allows the sensor data and the operational events to enable online analytics to evaluate the production and operational costs on a shift-by-shift basis. The transformation of data and operational events into actionable information using these operational events to obtain production, energy, water, reagents, and other consumption variables at a high level of detail is the new currency in business. These operational events are the new transactions to optimize production and reduce operating costs. Transactional systems are now integrated with this new data to proactively improve the overall production performance of mineral and metal processing plants (Bascur et al. 2017).

Performance Monitoring

The left side of the pyramid represents the support center, which provides actionable information to all functions of the plant to improve productivity. The operational metrics can be calculated by using proper data classification and aggregation at a high level of detail. This new strategy enables faster communication and collaboration within the functional teams at the plant and within the global business via the cloud. At this level, the business and time contexts are added to the process measurement. This level is used to generate tangible benefits for improvements. The data and events are analyzed to determine the best course of action for the plant.

Planning and Execution Management

The integration of operational analysis and business intelligence is the next level. At this level, plans are constantly being adapted by defining the best targets to optimize all process areas, beginning with the mine, process plants, and tailings management. Botin (2009) describes the latest strategies to integrate sustainability into a mining organization. Data management and automation are what allow these strategies to be successfully implemented.

PROCESS CONTROL FUNDAMENTALS

Many excellent books have been written about process control and there are many different approaches, but it is best to follow the advice that best suits the needs and goals of one's specific company.

In his textbook, Stephanopoulos (1984) states: "All the requirements (safety, environmental constraints, product specifications, operational constraints and plant economics) listed above dictate the need for continuous monitoring of the operation of a chemical plant (process), and external intervention (control) to guarantee the satisfaction of the operational objectives."

This is accomplished through a rational arrangement of equipment (e.g., measurement devices, valves, controllers, computers) and human intervention (e.g., plant designers, plant operators), which together constitute the control system. In fact, one's working definition of process control is shaped by one's specific interests. The point is that to be successful in the long term, one must recognize that a systems or holistic approach will be essential (i.e., the broader definition of *process control*).

Why Do We Need Controls?

A critical problem in the process of ore extraction is the variability of the different elements that constitute the ore; these unmeasured variables are called *disturbances* in the process control vocabulary. Each section of the mine has different ore types, hardnesses, mineralogical compositions, and geological characteristics; in addition, there are disturbances due to weather changes, people, equipment deterioration, power oscillations, and water qualities. Marlin (2014) outlines seven major categories of control objectives, which are discussed in the following sections.

Safety and Health

The safety of the people in the plant and in the surrounding community is of paramount importance. Plants are designed to operate safely at expected temperatures and pressures; however, improper operation can lead to equipment failure and the release of potentially hazardous materials.

Environmental Protection

Protection of the environment is critically important. This objective is mostly a process design issue—that is, the process must have the capacity to convert potentially toxic components to benign material. For example, in smelters, the gases are cleaned and processed in sulfuric acid plants, where the sulfur dioxide (SO_2) generated in the smelting of sulfides is converted into sulfuric acid (H_2SO_4). Many sensors are used to monitor the gas cloud of smelters. Tailing ponds are always very well designed to prevent issues with the re-treatment of the water and effluents.

Equipment Protection

Much of the equipment in a plant is expensive and difficult to replace without costly delays. Therefore, operating conditions must be maintained within constraints to prevent damage. The types of control strategies for equipment protection are similar to those for personnel protection, that is, controls to maintain conditions near desired values and emergency control to stop operations safely when the process reaches boundary values. Many sensors are used to robustly maintain the equipment to prevent exceeding the mechanical strength or the chemical limits of the material. This is very important for the steel used in smelters or in hydrometallurgical plants. Pumps, compressors, and blowers are typically subjected to high temperatures and corrosive gases. Installing the proper sensors in addition to process control sensors to protect the equipment will prevent costly downtimes and loss of production. All these support strategies have become an important part of implementing environmentally safe, sustainable process control strategies.

Smooth Operation

A mineral or metallurgical plant includes a complex network of interacting processes; thus, the smooth operation of a process is desirable because it results in few disturbances to all integrated units. A typical example is the comminution circuits. These are highly unstable due to the large variation of the ore qualities (soft and hard ore); these variations are directly sensed by the rougher flotation banks. Having a stable grinding operation is necessary to have an optimum flotation performance. Most of the disturbances in flotation circuits come from the grinding circuits.

Product Quality

The final products and quality specification in a metallurgical complex are determined by customer requirements. Having the right humidity for the concentrate and the right mineral composition for the downstream processes such as smelting or pellets used in blast furnaces is extremely important. It is extremely expensive to remove high silica content in a blast furnace if the pellet quality has reached the target level. The same is true for the quality of the concentrates—they must have the right amount of sulfur and the right metal grade. The specifications may be expressed in terms of composition (e.g., percentage of each component, chalcopyrite, pyrite), physical properties (e.g., density, humidity), or a combination of these factors. Process control contributes to efficient plant operations by maintaining the operating conditions required for excellent product quality. Improving product quality control is a major economic factor in the application of digital computers and advanced control algorithms. This is where process models are used to estimate the quality variables in the process due to the inadequacy of the quality sensors to measure the particle size distributions, mineralogical compositions, or chemistry of the process streams. Laboratory samples can be correlated with process variables using machine learning to accurately determine the quality conditions in a plant. The augmentation of knowledge provided by the operational intelligence data infrastructure is changing how mineral and metallurgical plants are operated.

Economics

Naturally, the goal of the plant is to return a profit. There are several strategies based on sustainable long-term profits. Process control improves plant performance by reducing the variation of key variables. When variations are reduced, the desired value of the controlled variable can be adjusted to move closer to constraints, enabling an increase in profits. Furthermore, it should be as economical as possible in its use of raw materials, energy, capital, and human labor. Thus, operating conditions must be controlled at optimum levels of minimum operating costs, maximum yields, and minimum metal losses.

Monitoring, Fault Prevention, and Diagnosis

Complex mineral and metallurgical complexes require excellent automation processes. Plant control and computing systems generally provide monitoring features for everyone in the organization to properly manage the instrumentation calibration, process control identification and tuning of the control algorithms, stream samplers for metallurgical stream laboratory data, equipment sensor maintenance, environmental remote monitoring, and so on. Because employees cannot monitor all variables simultaneously, the control system includes alarms that can draw attention to a problem. Now there are new ways of alerting operations and maintenance personnel to problems using plant computers and competency centers. Operational intelligence and machine learning algorithms are now used to predict faults or quality. Many opportunities exist to reduce minor losses that previously may not have been prevented in remote plants. New ways of using data and events are discussed later in the chapter.

All seven of the aforementioned requirements should be addressed simultaneously; failure to do so could lead to unprofitable or, worse, dangerous operations. Today, ever-increasing environmental and safety regulations require personnel to be more knowledgeable to avoid penalties due to process incidents in mineral and metallurgical plants.

Automatic controls are devices that assist operators with performing routine tasks faster and better than they could do manually without assistance. Mills run for 24 hours a day, and it is impossible for a human worker to remain alert at all times. Automation using the feedback principle is not new; there are many examples throughout history of automatic control principles being used for many applications. Process control is a subdiscipline of automatic control that involves the selection and tailoring of methods for the efficient operation of chemical and metallurgical processes.

Proper application of process control can improve the safety and profitability of a process while maintaining consistently high-quality products at a production target set by the economics and planning department. The automation of selected functions has relieved plant personnel of tedious routine tasks, allowing them more free time to analyze data, supervise, and learn from the process. The operational data and metadata available in process information systems help to improve processes. The data can be used to analyze the current operating conditions for ore types and economic conditions. As such, there are many different approaches to running a mill these days, especially with the high volatility of metal prices and energy and water costs.

The variability of ore qualities and lower grades requires plants to process larger volumes of ore; at the same time, plants must also implement environmentally safe process controls.

Process Control Loops

A typical control loop can be described in terms of input and output process variables. These variables are measured using online sensors, if available, or inferred from online calculations.

The output variables are associated with the control objective. The input refers to a variable that causes an output. The input is the tonnage flow rate to the mill, and the output is the % circulating load of the ball mill circuit, for example. The input could also be the water addition to the sump, and the output of the controlled variable could be the % solids of the hydrocyclone overflow.

The input causes the output, and this relationship cannot be inverted. The causal relationship inherent in the physical process forces one to select the input as the manipulated variable and the output as the measured variable. Selected examples in mineral and extractive metallurgical processing are presented later to illustrate the process control pairing of controlled variables and manipulated variables. The final pairing is the responsibility of the mineral processing engineer who understands the process units and has a clear objective for the design strategy. A basic list of the possible available variables for selected process units is also given later in this chapter.

For example, for a grinding circuit control, the objective could be to maximize the product throughput while maintaining the cut particle size on target for the best recovery and grade possible. The engineer selects sensors that measure important variables rapidly and with sufficient accuracy. The engineer must provide the manipulated variables that can be adjusted by the control calculation.

The engineer must decide which variable should be manipulated for each controlled variable. He or she should determine the most logical structure, or pairing, that requires a causal relationship between the final control element of the manipulated variable (usually a valve, pump, or belt conveyor) and the controlled variable. Many other issues must be considered in a complex mineral processing plant, such as favorable dynamic responses and reducing the interactions among controllers.

The development of mathematical models for the analysis of dynamics and the control of mineral processing systems is extremely important. Digital process simulators for crushing,

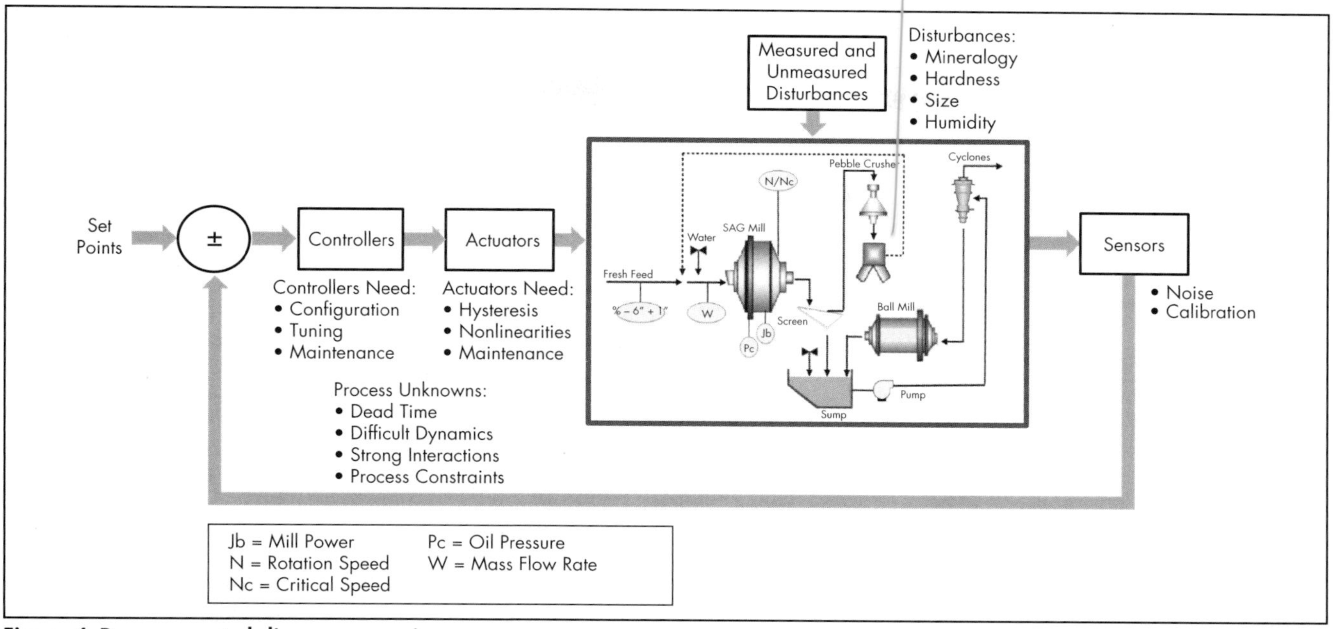

Figure 4 Process control diagram overview

grinding, classification, flotation, dewatering, and agglomeration models assist with defining the control algorithms. Dynamic process simulators can help with process control system training and selection of alternative control strategies. The design of the operator interface is also vital; this is where workers can visualize and respond to problems within the plant. Figure 4 shows a simple process control diagram displaying the controller and the selected variables.

Based on the control algorithm, the manipulated variables change the output of the controlled variables to adjust for process disturbances. Every process control strategy has a clear objective and a profit objective function to evaluate the performance of the setting and methods used to analyze the manipulated variables.

The set point is the target that is determined for the controlled variable. Figure 4 also shows a feedback control loop. The process outputs are measured by sensors that have noise and require good calibration. These controlled variables are compared with the set point to come up with a way, using the controller, to manipulate a variable that will maintain the desired target. The manipulated variable connects to an actuator that drives the pump or conveyor to the desired target—say, for example, the correct flow of water into the sump pump to achieve the right change in the hydrocyclone overflow particle size.

Control

The *control* maintains the desired condition in a physical system by adjusting a selected variable in the process. A control strategy can be defined in several ways. The most common one is a feedback controller.

Feedback Control

The *feedback controller* uses an output of the process to calculate the value of the manipulated variable to produce the desired effect on the desired output by setting a set point.

Feedforward Control

A *feedforward controller* does not use an output of the process—it uses the measurement of an input disturbance to the plant. This measurement provides an early warning that the controlled variable will be disturbed sometime in the future. With this warning, the feedforward controller has the opportunity to adjust the manipulated variable before the controlled variable deviates from its set point. A good process model is required to produce this advanced calculation for assessing the calculation. For example, knowing the hardness variation of the ore can drive the feed to the mill. If it is difficult to measure the ore hardness, the controller can reduce the feed flow rate to prevent overloading the mill, for example. This type of controller provides the value for the set point of the controlled variable.

The operating conditions of the process are measured—that is, all control systems use sensors to measure the physical variables that are to be manipulated near the desired values (operating targets). Each process has a control calculation or algorithm that uses the measured and desired values to determine a correction to the process operation. The control calculation can be very simple or sophisticated, depending on the stability of the given process circuit. The results of the calculation are implemented by adjusting a valve or motor in the process for the manipulated variable to compensate for the disturbances in the process inputs (measured or unmeasured).

Process Control Documentation

Good control engineering always requires a thorough understanding of the process. A good process flow diagram showing the major sensors provides a solid start when designing a process control system (Bhattacharyya et al. 2012; Turton et al. 2012). A process flow control diagram (PFCD) is different from a traditional piping and instrumentation diagram (P&ID), which is a great visual for detailed process documentation. A P&ID is a detailed graphic used in the process industry that shows the piping and vessels in the process flows along with the instrumentation and control devices. These drawings are used for many purposes, including designing plants, purchasing equipment, and reviewing operating and safety procedures. Having a good PFCD of the plant is necessary to design the process control strategy well. In addition, it serves as a good way for the operator to understand the behavior of the process and have a clear picture of the overall progression. Therefore, many people use PFCDs, and to avoid misunderstandings, standards have been developed by the International Society of Automation (ISA) for all countries worldwide (ANSI/ISA-S5.4-1991; ANSI/ISA-S5.5-1985).

It is important to use the best possible process control diagrams to document the process and its controls not only to document the strategies but also to teach users about the cause and effect of the manipulated and controlled variables in the overall circuit. This is also valid for performing a mass balance of the process units or the combined process systems in a plant. It is good practice to be able to close a mass balance with the process control data available. These diagrams can be used for process diagnostics and troubleshooting. They might also be used for developing process control graphic diagrams to build a human graphical interface for the operators.

Referring to Figure 4, all process equipment—piping, vessels, valves, and so forth—is drawn in solid lines. Sensors are designated by a circle connected to the point in the process where they are located. The first letter in the PFCD indicates the type of variable measured. Some of the most common designations are as follows: A = analyzer, F = flow rate, L = level, P = pressure, W = mass flow rate, J = power measurement, T = temperature, C = control, and I = indicator.

If the signal is used in a real-time calculation, it is also shown in a circle in the PFCD. The second letter in the symbol indicates the type of calculation. Two possibilities can be considered: C for feedback control and Y for another calculation. A noncontrol calculation might be used to measure the flow and temperature around a furnace to calculate its work level, that is, $Q = \mathrm{r}C_P F(T_{\mathrm{in}} - T_{\mathrm{out}})$.

The first reason for control is to maintain the temperature at its desired value when disturbances occur. The second reason is to be able to navigate to the desired variable (set point) without overshooting. The desired values are based on a thorough analysis of the plant operations and objectives. The selection of sensors and the proper manipulated variables and their physical control elements (valves, pumps, motors) with the right controlled variables is very important for good process control performance.

Today, virtually all large and medium-sized mineral processing plants around the world practice some form of automatic control. The interactions between the basic elements of the control triangle can be summarized as follows: The state of the art involves the use of several stabilizing control loops involving a mixture of control strategies. These control strategies include a variety of classical control strategies, advanced control strategies using models, and expert controls based on best practices. By capturing and analyzing the process variables, employees can maintain the instrumentation and process control loop performance to achieve a high level of productivity and simplify the maintenance of the control loops.

The vast majority of control schemes can be classified as classical control strategies. The feedback control law involving proportional and integral actions is used to compute changes in the controller output (manipulated variable) in response to measured deviations from a set point for a specific controlled variable. For such strategies, each manipulated variable is linked to a controlled variable on a one-to-one basis with the formation of a SISO control system. The computation capabilities of the control systems allow several process measurements to derive estimates as controlled variables. These soft sensors are derived from early work on advanced model-based control strategies. However, they require additional information for successful implementation. The operator requires training to understand this additional information embedded in the control loops. These strategies are usually called advanced controlled strategies. Many algorithms are available in modern DCSs to account for many of the time domain industrial aspects of a plant.

Control handling is delivered using a human–machine interface. The engineers building, tuning, and maintaining the control strategies design these tactics. These engineers need to design the visualization of the process controls for operator interaction. This is something that can be overlooked. Bascur (1991a) presents some guidelines that were used to successfully implement process controls in several plants. Li et al. (2011) also discusses the new role of the operator and his or her training requirements. The design of the control room also plays a very important role in the delivery of the controls and the management of the operations (Lundmark 2008).

PROCESS MEASUREMENTS

Table 2 summarizes the mineral processing instrumentation available along with the principle of measurement and the types of circuits in which it is most often used. Many of the sensors used in the mineral industries take advantage of microprocessors, digital imaging, and sophisticated signal processing required for particulate system measurements.

Table 2 also lists the many types of measurement devices used in mining and metal processing. Flow, pressure, temperature, load, vision, sound, particle size, and chemical online analyzers of all types and shapes are available. The newest measurement devices use optics in video cameras to transform the images into process variables by measuring particle sizes at the mine and the mill. Other novel vision measurement devices have been used in semiautogenous grinding (SAG) mills to analyze particle sizes, mill liner conditions, and flotation circuits.

Recently, additional vision sensors have entered the market (Coker 2015). By combining the application of standard imaging technology with new liner wear monitoring hardware, the Kaltech Sentinel reveals live information about a running mill. The Kaltech Sentinel is a system that incorporates two main components: the MillWatch camera unit and the electronic bolt (eBolt). By providing visualization along with the process, the data can be analyzed to optimize the grind cut and the overall grinding circuit availability.

Soft Sensors

Software sensors (or soft sensors) are another important process monitoring tool. Soft sensors essentially use other easy-to-measure process variables to estimate the value of an important property of a piece of equipment or process unit that is otherwise difficult, costly, or time-consuming to measure. Predictive analytics capabilities to obtain empirical models using a hierarchical data model are shown later in Figure 13 (Bascur 2016). After data classification and assessment of the process unit operating conditions, the process variables can be

Table 2 Mineral–metallurgical processing instrumentation

Measurement	Devices	Type of Circuit
Aeration, oxygen, gases	Magnetic rotameter Orifice plate Turbine Delta pressure	Flotation, hydrometallurgy, pyrometallurgy
Belt conveyor speed	Measurement of revolutions per minute	All operations
Bin level	Sonic sensor Capacitance probe Laser Radar	All operations
Crusher power, mill power, flotation power, process unit power	Watt meter Torque meter Motor current	All rotating process equipment
Feed rate dry solids	Weightometer Two-dimensional profile	All operations
Flotation cell froth level	Digital vision Capacitance probe Conductivity probe	Flotation
Flotation cell level	Bubble tube Float Float and ultrasonic Resistance tape Conductivity probe	Flotation
Mill load, converter load, process unit load	Load cells Watt meter Torque meter Sound meter Vibration sensors Bearing temperature Oil pressure Vision sensors	Grinding
Mineral species composition	X-ray fluorescence analyzer Neutron activation analyzer	Flotation, hydrometallurgy
Particle size	Ultrasonic particle size analyzer Light-scattering size analyzer Digital vision (dry solids) (two-dimensional, three-dimensional)	All operations
pH	Electrode Conductivity	All operations
Pressure	Load cells Many technologies	All operations
Pulp density	Gamma nuclear gauge U tube/load cell Differential pressure cell	All operations
Pulp level	Capacitance probe Sonic sensor Conductivity probe Delta pressure Radar Bubbling methods	Flotation
Sludge level	Float Ultrasonic sensor Light attenuation	All operations
Slurry flow rate	Magnetic flowmeters Ultrasonic flowmeters Sonar flowmeters	All operations
Torque	Amperage Load cell Torsion bar	All operations
Viscosity	Shear stress Flow	Thickening
Water (liquid) flow rate	Orifice plate and other delta P devices Turbine meter Magnetic flowmeters Sonar flowmeters	All operations

Adapted from Herbst and Bascur 1984

used to predict other variables using a deterministic first principle model or a semi-empirical model. The use of a first-order filter is recommended to smooth the data to remove noise. It is also possible to use a Kalman filter to determine the optimal estimation if a dynamic process model is available. An extended Kalman filter is capable of providing an estimate of the process kinetics. As such, the flotation kinetics or comminution kinetics can be obtained (Bascur and Herbst 1985b, 1986; Herbst and Harris 2007).

The most common procedure is to run experiments in the plant and measure the critical variable while recording all process variables in the history log. Then, operators typically run a procedure to obtain the process variables at the same time as the sample measurement and run a multilinear regression using Excel data analysis tools or other software tools such as Python, R, or Microsoft Azure Machine Learning Studio.

The objective of soft sensors is to make this information immediately available to operators and to advanced process control systems. Soft sensors combine critical process variables to infer quality variables such as particle size, flotation cell air holdup, % metal content in a process stream, inventory, and many other operational variables required to optimize a process plant. These estimates can be easily obtained by using the current analyzers to develop these regression models.

Detection and diagnosis of process faults and critical conditions are essential for efficient operations. Faults can typically be determined from the measured data by identifying a given or normal process operating pattern. If the composite measurements or indicators fall outside the original or normal operating pattern, then a fault must be present in the process. Using time-derived variables or using statistical process control or multivariate statistical control can become a valuable tool to perform fault diagnosis and to prevent problems.

Sampling Systems

Mining, mineral, and metallurgical plants and ports require special sampling systems. Automated slurry sampling to analysis is typically used when systems cannot deliver the accuracy required to manage the plant. Samples are automatically taken from the slurry streams, then automatically filter pressed, dried, pulverized, and analyzed, with results quickly returned to the plant for plan control. Typical systems process four feed lines, four concentrator lines, and four tailing streams, giving the plant metallurgists a picture of the process every few minutes. All the data are captured by a plant historian at the original resolution in real time. The metallurgical laboratory manual data are time-stamped for correlation process data when the sample is taken. A data infrastructure system can generate quality empirical models to produce lab estimates from process data. At port laboratories, the entire process, from sampling to analysis and cargo certification, is fully automated. In the case of iron ore processing, for example, primary cuts of up to 1 t are taken. Each cut is split within the sampling system to produce a sample for the automated laboratory. A portion representing each primary cut is transported automatically to the robotic laboratory via a conveyor system. These aliquots are composited and split to produce sublots. The particle sizes and moisture levels are determined for each sublot. Depending on the quality requirement, sublot chemical analysis is carried out; sometimes, however, only a final composite is done. Chemical analysis as well as particle size and moisture analysis is performed automatically, with data then transmitted to the plant control system. There are variations of sampling, such as truck and train integrated sampling and bulk bag sampling.

There are dry product samplers for primary sampling, such as heavy-duty rotating samplers, continuous reverse-discharge samplers, continuous forward-discharge samplers, and front-discharge cross-belt samplers, among others. There are also secondary samplers such as Vezin samplers, rotating plate dividers, and rotating sample divider multi-output.

Slurry sampling and analysis systems with in-line automated analytical and particle size measurement techniques are also used. In-line analysis provides very fast feedback for plant control. There are, however, applications in which offline sampling and analysis are required. Some examples include metal accounting or cases in which the analyses cannot be measured by online techniques or in which online techniques do not achieve the required detection limits. Applications include phosphate ore and precious metals.

Typical slurry samplers include primary samplers such as gravity samplers, linear cross-out samplers, rotary cross-out samplers, and pipe offtake samplers. Secondary samplers include gravity samplers, launder (cross-cut) samplers, primary slurry samplers, two-in-one samplers, pipe offtake samplers, and timed and continuous Vezin samplers.

All mining metallurgical plants require analyses of plant in-process streams to monitor plant performance. Typically, automated cross-stream cutters are used to provide shift or daily composites that are then sent to an analytical laboratory for analysis. The turnaround time for plant personnel to receive the analytical data is typically 24 to 48 hours after the laboratory receives the samples.

Thus, the data report the past performance of the plant and cannot be used as a real-time tool to control plant performance. This is where new machine learning tools come in handy to model the ore types with the process and laboratory data to determine quality estimates that are modeled and then used as predictors.

Keeping a plant running at peak performance is mainly dependent on the skill and experience of plant operators. They should notice changes, such as the absence of bubbles in a flotation circuit, and then immediately apply corrective measures. Advances in video analysis are also being made; operators can then use video analysis to manage the plant rather than physically monitor and analyze occurrences.

Specialized Online Analyzers

Online analyzers are available for slurry streams but typically require a great deal of maintenance, are not very accurate due to calibration difficulties, and have high detection limits due to the high dilution factor in the slurry streams. IMP Automation provides near-online analysis for metallurgical plants for fire assay (precious metals). X-ray fluorescence spectrometry using fusion discs for major elements and pressed pellets for minor and trace elements can also be used (see Braden et al. [2002] for more information).

However, these analytical methods provide data for individual elements. If a process in a metallurgical plant involves the separation of phases containing the same element(s), these analytical methods produce less useful data. In addition, online and in-line analyzers can be used for particulate systems.

Belt product analyzers use advanced neutron-gamma technology to provide elemental composition measurements that are both rapid and accurate. This technology uses the penetrating power of neutron radiation to interrogate a large volume of material flowing on the conveyor belt. When neutrons interact with the material, gamma radiation is emitted promptly with energy signatures that are characteristic of elements present in the material.

Online X-ray diffraction provides information on the mineralogy of the product. In-line rapid mineralogical analysis can significantly improve process control for an improved recovery and grade. These analyzers can be integrated with sampling and sample preparation equipment, presenting the analyzer with a dry representative product for analysis. Robotics technology is used to automate this process in the laboratory. On-belt moisture measurement allows for the accurate measurement of moisture in real time from belt conveyors.

There are many video camera–based particle size analyzers, such as the split online analyzer, which are mounted on trucks and on belt conveyors in comminution systems. These provide reduction ratio factors based on the operating conditions and enable improvements from the blast to the final concentrate products by integrating all data points using a real-time industrial data infrastructure.

For pyrometallurgical integrated sampling, IMP invented a special system to automatically sample an Ausmelt converter with a hearth depth of 17 m.

Instrumentation and Devices Network

Some examples of control networks are Profibus, DeviceNet, Modbus Plus, Ethernet/IP, Fieldbus, and remote input/output systems. Ethernet networks and security gateways are typically used for information networks. For more information, refer to the studies by Stuffco and Sunna (2002), Medower and Cook (2002), and Lukas (1986).

PROCESS CONTROL STRATEGIES

This section discusses the types of controls that are needed to optimize the operations of a process plant. There is a control hierarchy in the implementation of these controllers. Regulatory controls govern these controllers, just as the brain controls the involuntary responses of the body (located in the brain stem and the cerebellum). As the human body needs to control its chemistry for dealing with emotions, thus enters the limbic system. These can be called the body's advanced controls, which are based on multivariate controls due to the integrations of the many variables that need to be considered. The more sophisticated plant-based controls need to coordinate several process units—from the raw materials, to the products that provide targets, to all the process units to optimize the plant. This is similar to the function of the thalamus, which relays sensory impulses from receptors throughout the body to the brain and controls motor skills, language skills, vision, and emotions. This is called the performance monitoring management system, as shown on the left side of the triangle in Figure 3.

Process Dynamics

One of the most important aspects in process control is understanding process unit dynamics to identify the best way to reduce the variations of the output or controlled variables to achieve a certain process business objective. Process response modeling and analysis can help us obtain this required knowledge.

The process reaction curve is probably the most widely used method for identifying dynamic models. It enables us to identify the reaction times between the controlled and manipulated variables. The analysis provides a good foundation for coupling the regulatory basic controls and methods to tune a PID feedback controller. From a steady-state operating mode, a step change is introduced with the manipulated variable, and the response data of the process-controlled variables are collected until the process again reaches a steady state. Figure 5 shows the process reaction curve. $MV\Delta$ is the magnitude of the input change of the manipulated variable, $PV\Delta$ is the magnitude of the output or controlled variable, and S is the maximum slope of the output versus the time plot. This slope represents the rate of change or velocity of the response. The sign of this slope can be positive or negative or both. In addition, the velocity of the response can be fast or slow compared to other output variables in the process. The values of the real-time plot can be related to the model parameters according to the following relationships for a first order with a dead-time process model. A typical model of the form $y(t) = KpMV\Delta(1 - \exp[-(t - \theta)/\tau)]$ is used to calculate the parameters Kp, τ, and θ. The variable t is the time while performing this response test.

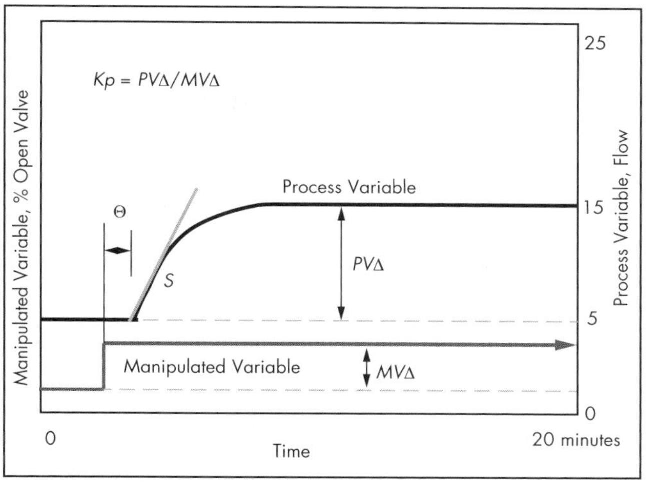

Figure 5 Dynamic process reaction curve for modeling and analysis purposes

- $Kp = PV\Delta/MV\Delta$: This constant is called the gain of the reaction curve.
- $\tau = PV\Delta$/slope of the reaction curve: This time constant is called the time constant of the reaction curve.
- θ = intercept of the maximum slope: This time constant is called the delay for reacting to the input.

The process reaction curve is a way to obtain basic knowledge about the controlled and manipulated variables. The typical process to obtain a process control model is

- Perform step testing,
- Perform time domain modeling using curve fitting, and
- Perform pseudorandom pulse tests and other frequency domain modeling.

The following section presents the results of this type of process testing based on dynamic simulators and industrial experience from tests in these types of processes.

Process Knowledge Table for Mineral Processing Operations

Mineral processing unit operations have many nonlinearities and interactions that complicate their behavior. This complex behavior can be visualized in a process matrix table for pairing the possible combinations using traditional processing control loops and adding decoupling actions to compensate for the behavior. Table 3 shows the typical process output as control variables measured subject to an increase in the manipulated variable. It shows the direction of a step response from the manipulated variables—increase (+) and decrease (–)—and the speed of the response (slow or fast). The speed of the response is the combination of the controlled variable and the manipulated variable. A + and – means that there is an inverse response. This means that the variable will start going up very fast and then will decrease, which is called nonlinear behavior. For example, in a closed-loop grinding circuit, when water is added to the sump pump, the % solids in the cyclone overflow will shoot up immediately but will come down once the overall circuit settles down. The time is usually about the overall residence time of the circuit (volume/feed rate)—15 minutes or so. This table can be validated using the process history of the variables from a history log to be calculated using a spreadsheet with mathematical analysis multivariable statistical tools.

The process response matrices for crushing, ball mill grinding, flotation, and thickening provide a qualitative guide to strategy selection. All other factors being equal, it is desirable to perform manipulations that will result in the largest possible change in the process output variable to be controlled and the quickest rate while minimizing undesirable changes in the other variables. Without models, this must be translated into quantitative terms for each application with the possibility that, under some circumstances, incorrect control actions may be taken. It is recommended that a clear uncontrolled or open-loop mode is used while performing the step responses to perform a test.

Table 3 shows a process control knowledge table that recommends combinations of controlled and manipulated variables to comply with process control objectives. It presents cause-and-effect and speed-of-response scenarios of these pairings that can be obtained from plant process data or with process dynamic simulators. In addition to these pairings of variables, one must take into account additional observed variables that are used in the diagnosis of process constraints of the ever-changing operating conditions in a mill due to ore variability disturbances, the age of the equipment, maintenance issues, and operator strategies. The concept of such a matrix is important and can provide insight with respect to the pairing of the variables available in the field. It is a good way to track the control performance of the process units once the computer automatically evaluates the parameters.

Table 3 was derived using mineral processing dynamic models and empirical data collected from plant trials (Herbst and Bascur 1984; Bascur and Herbst 1985a, 1986). Having the process control matrix available simplifies the implementation of both traditional regulatory PID controllers and multivariable process controllers.

Regulatory Controls

Regulatory controls are the foundations of process control and must perform well for supervisory control to succeed. In a mineral processing plant, this class of control typically includes flow, level, power, composition, loops, and so on. These are implemented with proportional integral (PI) controllers that are available in DCS operating software. In some instances, cascade or ratio control structures are used to achieve operational objectives. Because PI control has been well studied and because most instrument technicians and process/control engineers are exposed to the theory and application, it would be logical to conclude that regulatory control is not usually a significant problem.

Continuous feedback control offers the potential for improved plant operation by maintaining selected variables close to their desired values. The PID control algorithm has been successfully used in process industries since the 1940s and remains the most commonly used algorithm today. This algorithm is used for single-loop systems, which have one controlled and one manipulated variable. Usually, many single-loop systems are implemented simultaneously in a process, and the performance of each control system can be affected by interaction with other loops. As such, one needs to rely on other technologies to deal with these nonlinearities, constraints, and interactions.

The Proportional Integral and Derivative Control Algorithm

This algorithm is based on measuring the error between the controlled variable, CV(t), and the set point, SP(t). Based on this difference, we can calculate the manipulated variable, MV(t), setting to correct for the error (Marlin 2014). The error, E(t), is defined in the following equation:

$$\mathrm{E}(t) = [\mathrm{SP}(t) - \mathrm{CV}(t)] \qquad \textbf{(EQ 1)}$$

Proportional Mode

The first mode to take the control action (i.e., the adjustment to the manipulated variable) is proportional to the error signal because as the error increases, the adjustment to the manipulated variable should increase. As such, MV(t) = KC E(t) + IP. This is a very simple way to calculate the moves of a valve, for example, to control the flow rate of the liquid in a pipe to maintain the flow at a desired value. KC is the proportional controller gain, which will need to be estimated to tailor the controller to the desired level. The controller gain has engineering units of (manipulated variable)/(controlled variable). IP is a constant or bias that is used during the initialization of the algorithm.

Integral Mode

Because the proportional mode does not eliminate the error (because it uses the error to move the output), the next mode should be persistent in adjusting the manipulated variable until the magnitude of the error is reduced to zero. These results are achieved by the following integral mode:

$$\mathrm{MV}(t) = \frac{\mathrm{KC}}{\mathrm{TI}} \int \mathrm{E}(t)\mathrm{d}t + \mathrm{IT} \qquad \textbf{(EQ 2)}$$

The new adjustment parameter is termed the integral time, TI, which has units of time. IT is an initialization constant. The manipulated variable increases linearly with the slope of E(t) KC/TI. This behavior is different from that of the proportional mode, in which the value is constant over time for a constant error.

Derivative Mode

If the error is zero, both the proportional and integral modes give zero adjustment to the manipulated variable. This is a proper result if the controlled variable does not change; however, when the disturbance just begins to affect the controlled

Table 3 Process knowledge table for pairing control and manipulated variables

Manipulated Variables	**Controlled Variables**				
Crushing	**Product Finesse**	**Circulating Load**	**Power Draw**	**Bin Level**	**Crusher Level**
Feed rate	(±) Fast	(0) Slow	(–) Slow	(+) Slow	(+) Fast
Closed-side setting	(±) Fast	(–) Slow	(–) Fast	(–) Slow	(–) Fast
Screen area	(±) Slow	(–) Slow	(–) Slow	(–) Slow	(–) Slow
Grinding	**Product Finesse**	**Circulating Load**	**Sump Level**	**% Solids in Mills**	
Sump water addition rate	(+) Fast	(+) Fast	(0) Fast	(±) Slow	
Fresh feed solids rate	(–) Slow	(+) Slow	(+) Slow	(+) Slow	
Cyclone feed-pumping rate	(±) Fast	(+) Fast	(–) Fast	(+) Fast	
Feed water addition rate	(±) Slow	(+) Slow	(+) Slow	(–) Fast	
Number of hydrocyclones	(±) Fast	(+) Fast	(–) Fast	(–) Slow	
Mill speed	(±) Fast	(±) Fast	(–) Slow	(–) Slow	
Semiautogenous Grinding	**Mill Load**	**Power Draw**	**Product Size**		
Feed rate	(+) Slow	(±) Slow	(0) Slow		
Water addition rate	(±) Fast	(±) Fast	(±) Slow		
Mill speed	(±) Slow	(±) Fast	(±) Slow		
Feed size coarse ratio	(0) Slow	(±) Fast	(±) Slow		
Flotation	**Grade**	**Recovery**	**Froth Depth**	**Air Holdup**	
Aeration rate	(–) Fast	(+) Fast	(+) Fast	(+) Fast	
Agitation rate	(0+) Slow	(±) Slow	(+) Slow	(–) Slow	
Pulp level	(–) Slow	(+) Fast	(–) Fast	(–) Fast	
Froth addition rate	(–) Fast	(+) Fast	(+) Fast	(–) Fast	
Collector addition rate	(+) Slow	(0) Slow	(+) Slow		
Depressant	(+) Slow	(–) Slow	(–) Slow		
Tailings opening gate	(+) Slow	(–) Slow	(+) Slow		
Thickener	**Sludge Level**	**Settling Rate**	**Torque**	**Underflow Viscosity**	
Pump speed	(–) Fast		(–) Fast	(±) Slow	
Flocculant addition rate	(–) Fast	(±) Slow	(+) Slow	(–) Fast	
Rake position	(+) Fast		(–) Fast	(0+) Fast	
Feed rate	(+) Fast	(0+) Fast	(0+) Fast	(0+) Fast	
Rake speed	(0–) Slow		(+) Fast	(+) Fast	

Adapted from Herbst and Bascur 1984

Key:

(+) For an increase in the manipulated variable, there is an increase in the controlled variable.

(–) For an increase in the manipulated variable, there is a decrease in the controlled variable.

(±) The controlled variable has an inverse response. It will start one way and then change to another direction. This is a nonlinear response, which is very common in grinding circuits or systems with a large recycle stream.

Fast means that the response is quick, and *slow* means that it will take a few minutes.

variable, the error and integral error are nearly zero, but a substantial change in the manipulated variable would seem appropriate because the rate of change of the controlled variable is large. This situation is addressed by the following derivative mode:

$$MV(t) = KC\,TD\frac{dE(t)}{dt} + ID \qquad \textbf{(EQ 3)}$$

The final adjustable parameter is the derivative time, TD, which has units of time, and the mode gain has an initialization constant for the derivative action, ID. The derivative mode provides rapid correction based on the rate of change of the controlled variable and can cause an undesirable high-frequency variation in the manipulated variable. Therefore, the PID control algorithm can be set as follows:

$$MV(t) = KC\left[\frac{E(t)+1}{TI}\int E(t)dt + TD\frac{dE(t)}{dt}\right] + I \qquad \textbf{(EQ 4)}$$

This is the ISA form, also known as the ideal form.

Control tuning must be performed using the same algorithm that is applied in the control system. The implementation should be done in a DCS or by using PLCs. This configuration will present several options depending on the type of hardware. The implementation of the control algorithm will require presenting the information to the operator, and the settings available will depend on the overall control strategy. There are many good strategies to tune the desired PID or PI controllers.

It is very important to understand the manipulated variable actuator. This actuator is usually mechanical in nature, so it will have what is known as hysteresis and backlash. This means that for a given value, it might operate physically differently than normal. A process identification of the actuator is recommended, and an adjustment should be made in the implementation using the control block of the DCS or PLC.

The dynamic behavior of both the controlled and manipulated variables must be understood to evaluate the performance of a feedback control system.

Regulatory strategies maintain key local operational variables associated with the different process units in their set point values, compensating for the effects of high-frequency disturbances. In addition, they ensure safe start-up and shutdown, minimizing the risk of damage to people and equipment by keeping the variables within the known operational and safety limits. The regulatory control strategies act locally, and they are usually designed around PID controllers and set standardized control functions such as selectors, multipliers, time lead or lag functions, and so on. The proper maintenance of these strategies is fundamental for sustaining any advanced control strategy. These types of controllers include cascade controls in which a process flow rate controls the temperature of a process furnace. Here, rather than acting directly on a valve, the control design prefers to control the flow, which will control the opening and closing of the valve to regulate the temperature of the interior of the furnace. Usually, the process controls for overflow density and/or particle size controls in grinding circuits are driven by a cascade controller for the water flow controllers, which direct the opening and closing of the valves. It is easier to control the flow than the position of the valve (or that of a pump or belt conveyor). It is simpler to use a cascade controller to control the flow and let the underlying controller do the positioning of the manipulated variables to open or close the flow with whichever actuator is in use.

Regulatory controls require the tuning of the algorithm parameters. These are usually obtained from the process dynamics response using some suggestions provided by Ruel (2010a, 2010b). Many tuning strategies are available by performing an online Google search. Rice (2015) also wrote a PID tuning guide.

Multivariable Process Control Strategies

The optimal operation of a mineral processing plant requires acting simultaneously over different control variables and taking into account operational constraints. Thus, the objective of advanced control strategies is to keep the process operating at near-optimum conditions expressed by a proper objective function defined in a time horizon. The strategy calculates the set points of the controllers associated with the regulatory strategies to optimize this objective function (Flintoff 1995, 2002; Flintoff et al. 2014; Sbarbaro and del Villar 2010; Ferrarini and Veber 2009).

A discrete mathematical model is used to predict the process responses and calculate over the time horizon the future control actions based on the available measurements and predictions of the disturbances. According to model predictive control (MPC), only the first action is actually implemented, and the optimization process is repeated at the next sampling time. In these strategies, the dynamic process model is very important because it is used explicitly in the calculation of the control variables. Good mathematical models are necessary for the successful implementation of a model-based control strategy. The parameters of simple models can be obtained by performing a dynamic test during the process or by using open-loop historical data. As in regulatory strategies, it is very important to establish proper maintenance procedures to update the models and check the tuning of the different parameters associated with the control strategy. MPC models are obtained based on the process matrices shown in Table 3. There is a relationship between the controlled variables $y(t)$ and the manipulated variables $u(t)$.

Effective MPC depends on a model that is representative of the process dynamics. After several months, the process tends to differ enough from the model that performance deteriorates. Determining a good model that matches the current process is a challenge that can be easily overcome by using current data analytics capabilities.

Holistic and Expert System Process Controls

If the process models are difficult or impossible to obtain, then the control strategy can rely on the knowledge of an operator or an expert in the field to calculate the control variables. This knowledge is expressed in terms of rules describing the experiences of the operator. These heuristics are written in statements with an if–then form. A set of rules constitutes the knowledge base required to control a given system. Because the expert knowledge is expressed in linguistic terms, a suitable framework to describe the ambiguity of the natural language is provided by fuzzy logic. Even though these strategies can deal with complex control problems, they cannot be easily extended to multivariable systems with many inputs and outputs. In addition, because they do not use detailed information about the evolution of the process and/or disturbances, their performance is no better than the one obtained by model-based strategies.

The decision to use a model-based approach or a rule-based approach depends on models. If the process and disturbances can be modeled, the controller will be developed to handle constraints, combat disturbances, and react accordingly to expected performance. If the process cannot be modeled, can the operator be modeled? If so, a rule-based approach can be used and the controller will mimic the best operator.

Instead of using process models, these types of controls use process rules based on heuristics and fuzzy logic controls. In rule-based approaches, an expert operator manipulates set points on control loops. A fuzzy control system is a control system based on fuzzy logic; decisions are made using analog inputs that represent a value ranging from 0 (false) to 1 (true). The logic deals with partially true and partially false values. In many projects, the result will be a combination of model- and rule-based systems.

Time-Derived Variables

Time-derived variables are used in operator mimic controls and in fault diagnosis procedures. Object-oriented expert systems are discussed later. Using an industrial data infrastructure that provides the analytics and the data-derived process algorithms allows for a smoother implementation of dynamic process analytics, and it is easier to maintain the analytics in the long term. This also allows for the classification of data to implement model-based inferential controls. Table 4 lists several time-derived variables and observed variables that are used to develop process specifications and detect abnormal situations (Bascur 1990a, 1999). Many algorithms are available to filter the process signal to eliminate noise and to infer additional information from the raw data (Otnes and Enochson 1978; Bascur 1988).

Example for MPC Strategy Implementation of SAG and Ball Mills

Figure 6 shows a typical SAG and ball mill control strategy (Henriquez et al. 2012).

Table 4 Time-derived variables used in digital control systems and real-time information systems

Measured Variables	Snapshot	Moving Average (time)	Rate of Change (time)	1-Hour Average	1-Hour Standard Deviation
Temperature	High	High	High rate	Medium	Maximum
Flow 1	Normal	—	—	—	—
Level	—	—	—	—	—
Flow 2	Normal	—	—	—	—
Flow 3	Normal	—	—	—	—
Press 1	—	High	High rate	High	Maximum
Press 2	—	High	—	High	Maximum
Additional Information					
Weather: Rainy/sunny/cloudy			Unit age: Old/middle/new/unknown		
Unit location: Geographical coordinates			Number of cells in series: Few/many/very many		

Source: Bascur 1990b

Process Control Objectives

The objectives of process control are to maximize throughput, decrease variability of the product size and % solids to flotation, improve overall equipment availability, and reduce specific power consumption. The implementation of the control strategies consists of three steps:

1. The first step focuses on the standardization of the process measurements and process knowledge through characterization of the disturbances and the manipulated and controlled variables.
2. The second step focuses on the reduction in process variability by properly selecting the manipulated variables and providing extensive operator support to test and tune the system.
3. The third step consists of process optimization by collecting stable metallurgical results, managing process flow, and maximizing throughput while optimizing particle size (P80).

The main characteristics of process control are as follows:

- Particle size control is based on the main hydrocyclone pressure. Solids concentration control maintains the pump box to prevent overflowing or pump cavitation.
- Fine hydrocyclone pressure control is obtained by manipulating the pump speed; coarse pressure control is regulated by the available number of hydrocyclones. Thus, depending on the incoming feed flow, more or fewer open hydrocyclones will be required to maintain the same working pressure.
- Solids concentration control (% solids) is obtained by manipulating dilution water to the pump box.
- It has intelligent hydrocyclone rotation control and monitors the running hours for each hydrocyclone (for maintenance purposes). The system also tries to keep at least one hydrocyclone available.
- The operator sets the desired working pressure and density. The proper selection of both parameters and stable control (provided by the MPC) will result in an optimal working P80.
- In the case of two running hydrocyclone batteries, the number of open hydrocyclones in each battery is equalized (to balance throughput).
- The control strategy uses a combination of MPC and expert logic for adapting to the changing conditions of the feed flow and the available hydrocyclones.

In the case of the P80 variability from the data feed of the rougher flotation line, Figure 7 shows the improvement results for circuit 8 with the use of the MPC systems for two consecutive periods (69 days each). Despite the increase in the mean value from 197 µm to 217 µm, the variance decreased by 54%. The plant tonnage operating range is 145,000 to 200,000 t/d, the hydrocyclone pressure is more than 68.9 kPa, the hydrocyclone feed % solids is more than 50%, and the sump box level is more than 70%.

In conclusion, the MPC system reduces variability in pressure and % solids in the hydrocyclone batteries. The reduced variability allows for reduced variability in the P80 to flotation and thus leads to a better metallurgical performance. The reduced P80 variability enables the reduction of pulp transportation flow variability, therefore improving the dewatering process. Reducing the variability is the main control objective of the MPC implementation. This MPC enables the operation to approach the operating constraints to optimize the process.

Additional comminution circuit process control strategies are discussed by Bascur (1990b, 1991b) for traditional grinding mills and by Edwards et al. (2002), Karageorgos et al. (2001, 2006), Fuenzalida and Olivares (2012), Baas et al. (2014), and Ruel (2012) for SAG and ball mill grinding circuits.

Example of a Copper Flotation Rougher and Scavenger MPC Implementation

The flotation process is a complex physicochemical process (Fuerstenau 1999). Table 5 lists the many operational variables that influence the flotation of minerals. The detailed dynamic process shows that it is a zero-order process that is controlled by the interface between the froth and the pulp (Bascur 2005). A zero-order process is unstable by nature. A flotation model is a dynamic process where the froth interface is always in a dynamic state, receiving bubbles with attached particles and draining them back to the pulp hydrophilic particles. A flux model defines the hydraulics of a flotation system. As such, maintaining a proper interface and froth height is crucial for the cleaning action in a flotation cell. The attachment and detachment of particles in the pulp zone are attributed to the physicochemical characteristics of the minerals and the turbulence in this zone to achieve bubble generation and the bubble–particle attachment (Bascur 2010, 2012).

The most important control variables for the flotation process are categorized for process control and optimization as disturbances, measured variables, manipulated (physical

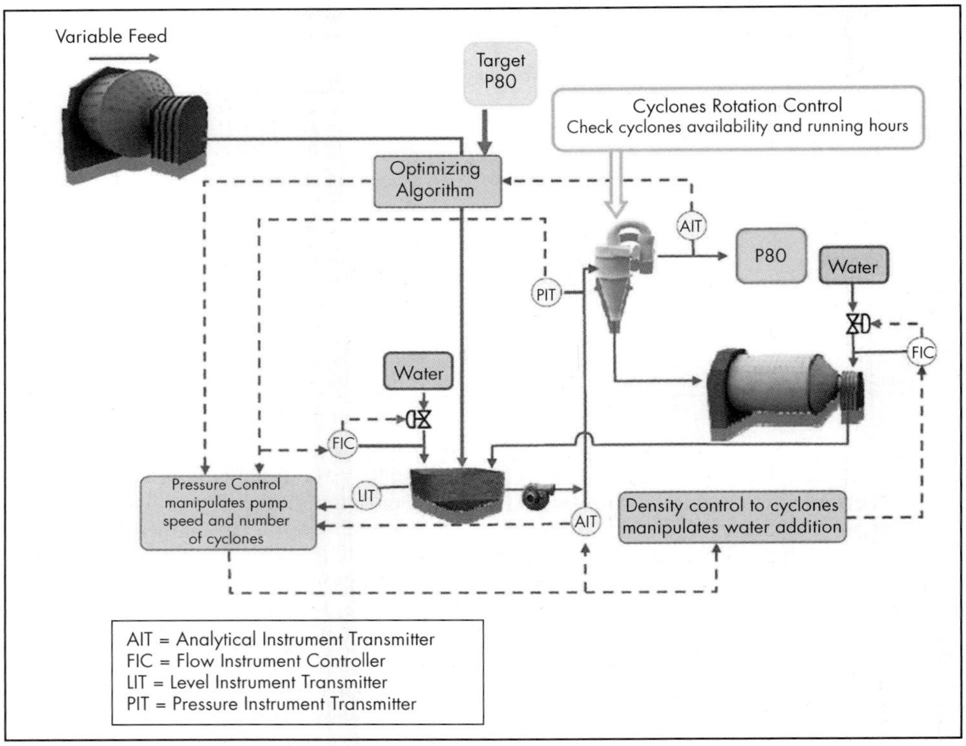

Source: Henriquez et al. 2012

Figure 6 Example of SAG/ball mill multivariable process controller at Minera Los Pelambres in Chile

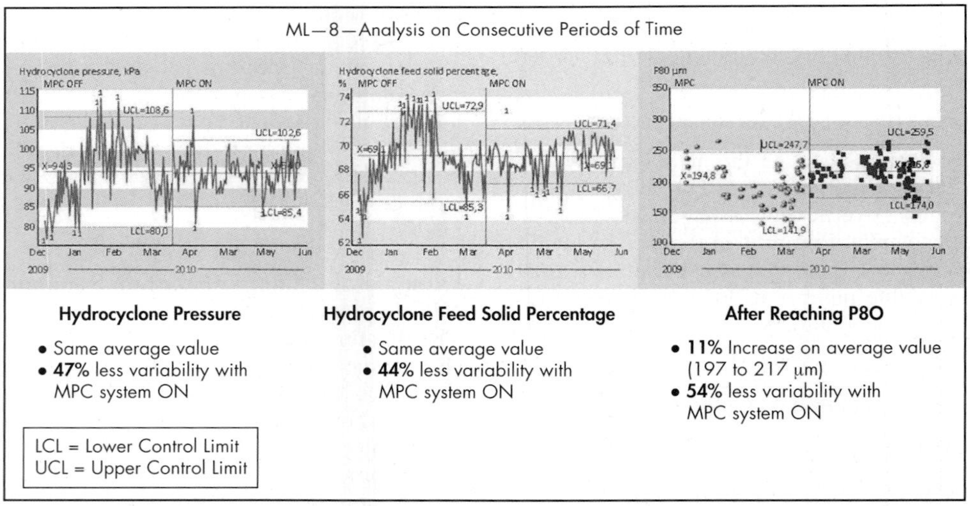

Source: Henriquez et al. 2012

Figure 7 Ball mill hydrocyclone pressure, % solids, and product size P80 results before and after MPC implementation

and chemical) variables, and controlled variables, as shown in Table 5.

The flotation process is highly multivariable, which means that there are strong couplings between the various control variables and the important stages of the process. The choices depend on the particular design and application of the flotation system and the materials to be separated.

Controlling an entire flotation circuit correctly is still one of the most difficult process control problems in a processing plant. Figure 8 shows a typical flotation circuit, including the rougher section, the scavenger circuit, and the concentrate cleaner circuit.

It is difficult to control a flotation circuit because the majority of the measurements that are required for control rely on various analyzers and instruments that require regular

Table 5 Variables of importance for a flotation process

Disturbance Variables	Measured Variables
Mineralogical composition	Metal analysis
Degree of surface oxidation	Volumetric flow rate
Fluctuating feed size distribution	Pulp densities
Surface modifications	Size distribution
Fluctuating feed rate	Pulp level
Fluctuating feed composition	Froth level
Pulp viscosity and density	Eh and pH
Water characteristics	Collector addition
	Frother addition
	Power intensity
Manipulated Variables	
Physical	**Chemical**
Airflow rate	Frother addition rate
Pulp level	Collector addition rate
Impeller speed	Modifier addition rate (activation, pH, etc.)
Conditioning time	Reagent addition points
Stream diversion	Electrochemical potential
Froth sprinkling rate	
Coarse/fine split	
Controlled Variables	
Recovery	Feed % solids
Grade	Pulp level
Tonnage throughput	Feed size distribution
Circulating loads	

maintenance and calibration checks. Sometimes the regular maintenance checks or required calibration checks are forgotten or delayed in a maintenance schedule, resulting in these measurements becoming unreliable. When they become consistently unreliable, the control system cannot operate correctly, and the control room operators then tend to run in manual mode.

Even though correct flotation control is one of the most difficult control problems in a processing plant, if each required flotation control function is broken down into manageable "bite size" control strategies, it is possible over time to develop a robust flotation process control system that highlights the required critical measurements as each part of the flotation control strategy is implemented. By implementing such a control system over time and breaking the system down into various sections, the value to the business of each of these parts can be demonstrated and thus justified. With this, the cost of the aforementioned required maintenance and calibration checks can then be justified to management and scheduled as part of a routine preventive maintenance plan.

The control functions that are required to develop a good robust flotation control strategy can be divided into three parts:

1. Interface-level stabilization control for each flotation section,
2. Rougher and scavenger mass pull with grade/recovery control as limits, and
3. Cleaner and recleaner grade/recovery control.

The first two control functions are relatively easy to design and commission; however, designing and implementing the third function of cleaner and recleaner grade/recovery control is more complicated. Baas et al. (2007) describe good, robust grade/recovery control as the "holy grail" of flotation control. Figure 9 shows a schematic of how the first two levels of flotation control are broken down into manageable bite-size functions.

Grade/recovery control is so complex because the recovery of some minerals, especially those that are finely liberated, can be affected by factors such as surface and pulp chemistry changes. To date, there have been very few methods of successfully measuring the required pulp chemistry parameters. (Some companies can successfully measure the pulp chemistry; however, these companies are in the minority.)

Even when some of these chemistry parameters are successfully measured online, attempting to control these has proven to be difficult because the effect of changing the pulp chemistry does not lead to consistent changes to the grade/recovery in the cleaner and recleaner circuit.

In recent years, new instruments and analyzers have been developed that can help accurately measure some of the parameters in a flotation cell; previous methods were either unavailable or only somewhat reliable. These parameters include the actual particle size of the fine fractions, the vertical and horizontal bubble velocity in the froth zone, the froth depth and froth density, the pulp depth and pulp density, and the chemistry of the pulp zone. As these new instruments are introduced and tested, flotation control will likely become a less complex and difficult process. Processing plants take the regular maintenance of these new analyzers and instruments quite seriously.

There have also been advances in froth video analysis in the control of flotation circuits. An additional sensor can provide more information in order for the plant to manage the cleaning action of the froth and to stabilize the flotation mass balance of the flotation banks. Kewe et al. (2014) incorporated froth visualization analysis to improve flotation circuits using a combination of froth visualization and a professional control system.

For additional information, refer to the studies by Johansson et al. (1999), Herbst and Pate (1999), Herbst and Harris (2007), and Baas et al. (2014).

Example of Thickener Multivariable Process Control

Developing thickener control is similar to the Cinderella story (Karageorgos et al. 2009; Weidenbach and Lombardi 2012). To date, the majority of the process control applications on a mineral processing site have predominantly focused on SAG mill control, general grinding circuit control, and various aspects of flotation control. Unfortunately, control of thickeners and countercurrent decantation (CCD) circuits is only performed well by a small number of companies worldwide. Thickeners are zero-order processes, which are unstable by nature. The generation of an interface is modeled using a flux curve (Concha 2014). Bascur and Herbst (1986) implemented an industrial variation of an online model to obtain additional information for the thickener inventory, the interface level, and the torque of the rake. This online model was very useful in identifying the many variables that affect the thickening operation. The most important factors are the large variations of the mineral specific gravity and the amount of clay in the feed, which affect the overall behavior of the thickener underflow density and rheology. However, in this model, the addition of flocculant was not capable of aggregating the particles to control their settling rate. In addition, changes to the feed well design were necessary. An inventory prediction was used to modify the scheduling of the plant, which was constrained

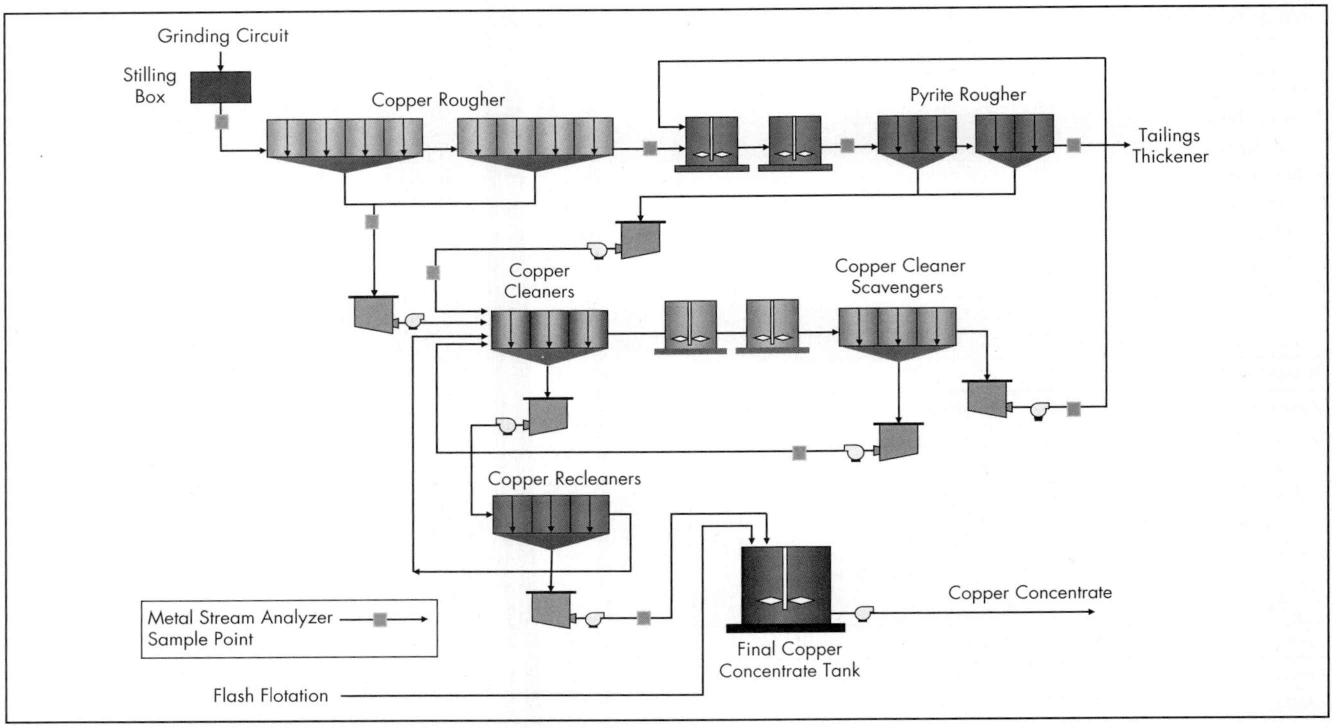

Figure 8 Typical copper flotation circuit

by the capacity of the pipeline due to the large variations in the underflow product.

Those who have the opportunity to begin the process of controlling a thickener or a CCD circuit will go through several stages. At first, the initial perception may be that controlling a thickener is easy, but this perception soon ends with the realization that obtaining good thickener control is more difficult than SAG mill or grinding circuit control and even some parts of flotation control. Once one begins to develop and implement thickener control, the process may be considered to be refreshing and exciting. Because there has been very little published in textbooks or studies on thickener control, it can be exciting to work with innovative technology. New concepts have to be defined, developed, and implemented to improve good thickener control and eventually create exceptional thickener control.

As with controlling any unit process, the first step is to define the outcome that is required and how this will benefit the processing plant. Improving the control on a thickener is initially performed to improve the amount of process water that can be recovered; another important benefit is the ability to treat a higher throughput than what was previously achievable without good control. A reduction in flocculant of approximately 20% to 30% can also be achieved, which is important, as this is a consumable.

The objective of thickener control is to improve the amount of water that can be recovered; this then translates to improving the underflow density. Therefore, it would follow that underflow density control should form a major part of the thickener control strategy; however, a paradigm shift must be realized.

As a general rule, it is not possible to directly control the density of any given thickener, and any thickener control strategy that mainly relies on thickener underflow density will inevitably fail or operate poorly.

To design and implement a robust control strategy for a thickener, three main objectives must be achieved:

1. Correct inventory control (it is important to measure the inventory correctly in a thickener)
2. Particle settling control
3. Thickener protection control

The underflow controller monitors the mud level (the true compacted bed level), the rake torque, the bed pressure, and the underflow density. It calculates the optimum underflow flow rate (manipulated variable) to achieve the best underflow density (controlled variable subject to the other constraint variables) while safely maintaining the thickener.

A key sensor was developed to measure the density profile of a thickener called SmartDiver. The SmartDiver is an instrument patented and made by Precious Light and Air. The SmartDiver uses an ultrasonic probe to measure the clarity of the liquid. The probe is lowered and retracted by a cable drum into the thickener approximately every 5 minutes, depending on the rake position (Weidenbach and Lombardi 2012).

The flocculant controller examines the settling band (controlled variable) within the thickener, the thickener overflow clarity, and the rake torque. It calculates the optimum flocculent rate (manipulated variable) to achieve the required particle solids settling rate (controlled variable subject to the other constraint variables) just above the compacted bed.

The three aforementioned thickener control strategy requirements do not address underflow density control (Hartog et al. 2014; Weidenbach and Lombardi 2012). Figure 10 shows a process graphic for the implementation of a multivariable thickener control strategy using the Manta Cube system. This

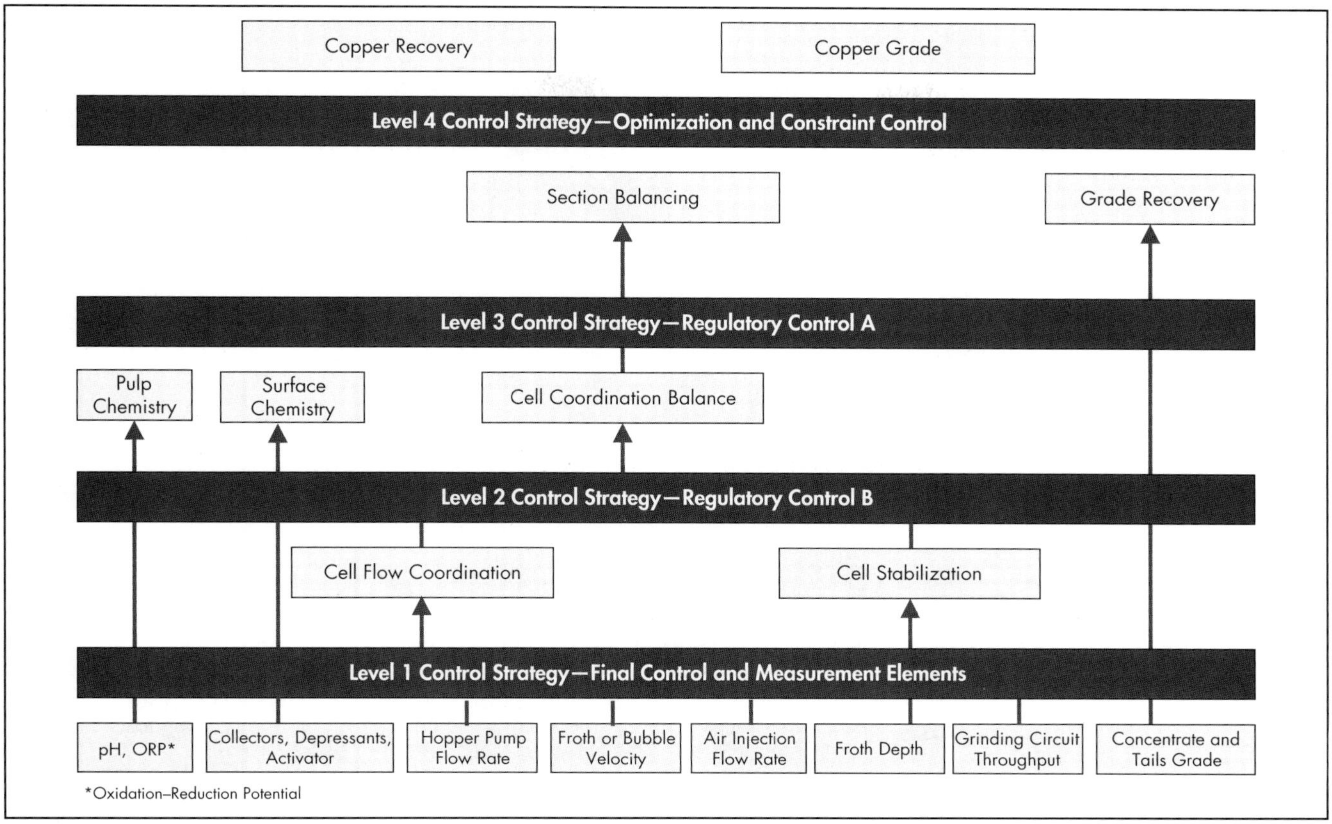

Source: Baas et al. 2007, reprinted with permission from the Australasian Institute of Mining and Metallurgy

Figure 9 Flotation hierarchy of control for interface-level stabilization and mass pull with grade/recovery control for the rougher and scavenger circuit

multivariable control strategy has improved the stability of the underflow density by 70%, reduced the risk of bogging the tailing thickener, reduced the reliance on the water supply, and improved water recovery rates.

Once one has realized the paradigm shift, it becomes clear that improving the underflow density simply depends on correctly controlling the thickener. That is, once the inventory is controlled correctly, the particle settling is controlled correctly, and the thickener protection control is implemented, the thickener will operate in a more stable and reliable manner. The suggested control hierarchy allows the particles to settle and compact correctly, which all translates to improved underflow density control. The thickener can then safely process a higher throughput than would be possible without the aforementioned control strategy in place.

Additional thickener controls are described by Schoenbrunn et al. (2002). The authors use an expert system strategy and present a good overview of the process instrumentation and the rules to consider to meet the control objectives.

Example of a Gold Cyanide Leach Multivariable Controller

Figure 11 shows a graphical interface that illustrates the implementation of a multivariable control strategy for gold cyanide leaching as reported by Hartog et al. (2014). The leach circuit consists of two leach tanks (TA-111 and TA-112) and seven absorption tanks. The slurry is fed to the leach feed thickener, and after leaching the slurry flows to the tailings thickener. The cyanide leach is supplied to the leach circuit by two variable-speed pumps in a standby configuration (manipulated variable). The barren eluate from the elution circuit is returned as a batch to the first leach tank.

Originally, the cyanide in the solution was measured in the first leach tank by a Cyantist cyanide analyzer. The cyanide concentration was originally manually controlled by the control room operator. The operator would manually adjust the cyanide flow rate into Tank TA-111 depending on the cyanide concentration in Tank TA-111.

In 2012, a process control strategy configured an automatic cyanide concentration sample-and-hold controller that automatically determined the required cyanide solution flow rate depending on the cyanide concentration in Tank TA-111. This flow controller adjusted the speed of the standby cyanide pump to maintain the desired cyanide flow rate set point. The control system was configured so that the cyanide flow rate was stopped when the measured cyanide concentration reached a certain limit.

The cyanide set point for the first leach tank would be set with the aim of achieving a desired cyanide profile across the adsorption tanks. When the measured cyanide concentration showed large-amplitude cycles, the sample-and-hold controller were retuned to achieve better control.

Improved process control in the cyanide leach circuit involved both equipment and control strategy changes. With the Cyantist cyanide analyzer and the associated sampling system at the end of its maintainable life, the analyzer was

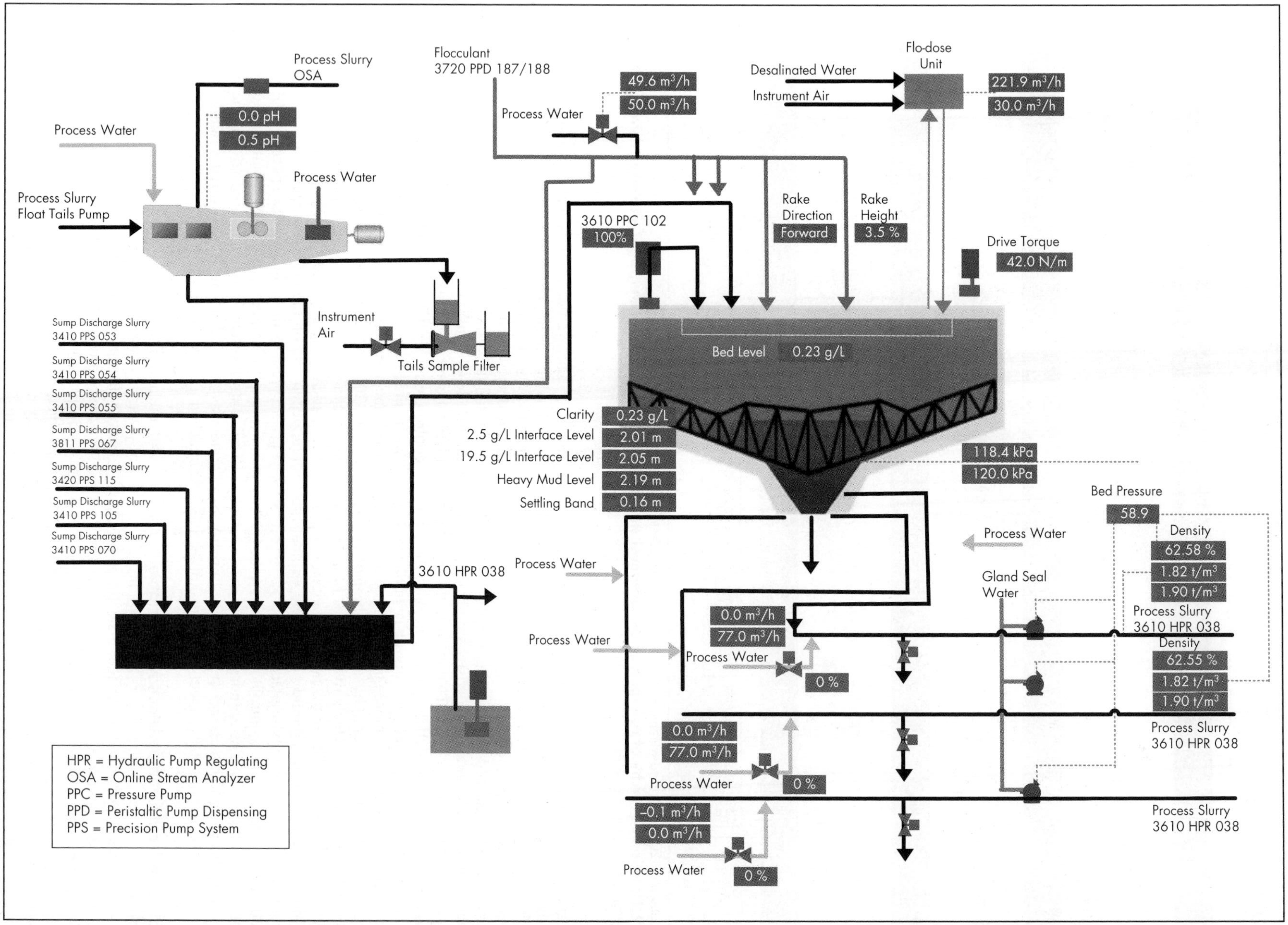

Source: Weidenbach and Lombardi 2012, reprinted with permission from the Australasian Institute of Mining and Metallurgy

Figure 10 Thickener multivariable control strategy human interface

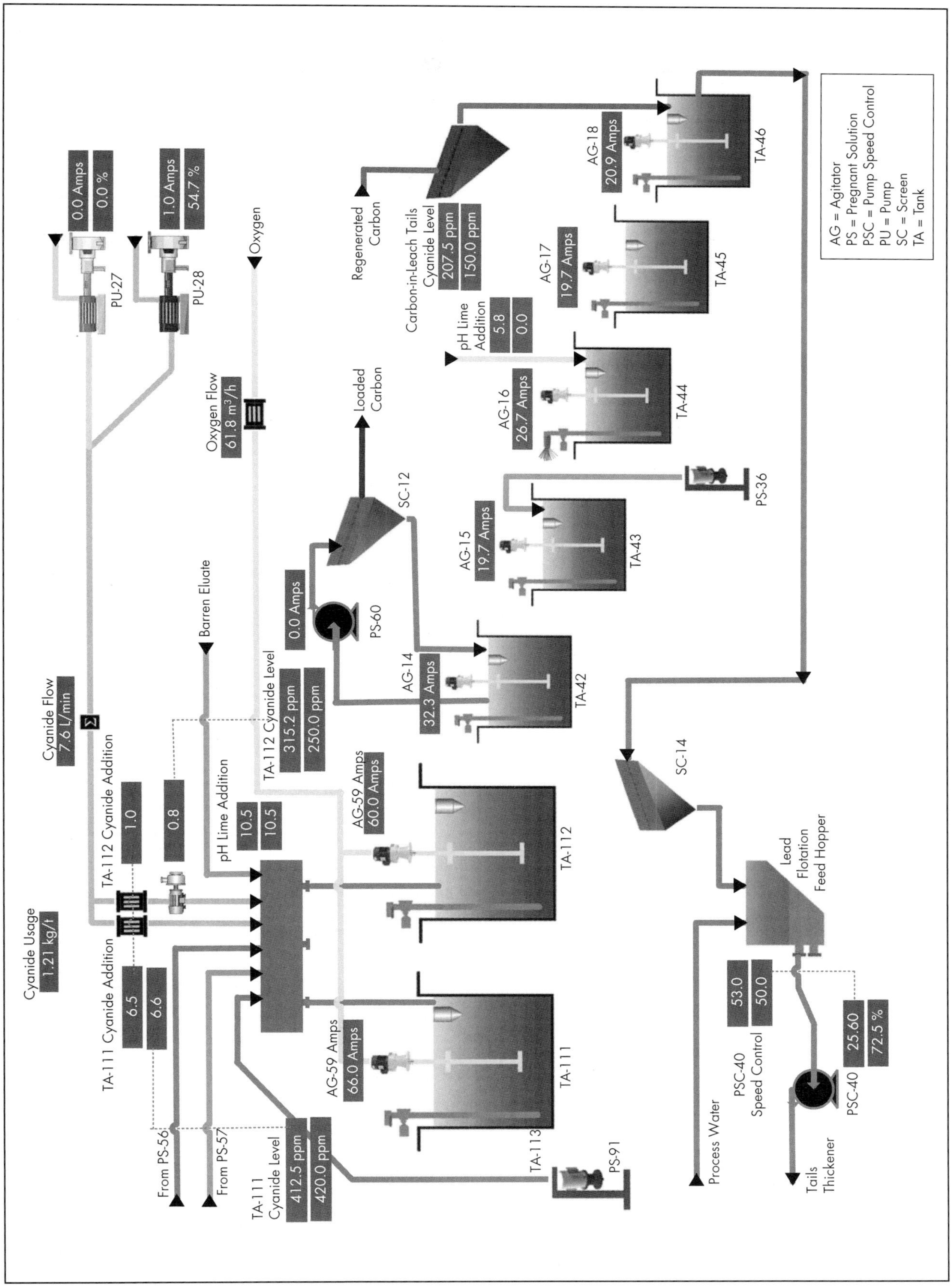

Source: Hartog et al. 2014, reprinted with permission from the Australasian Institute of Mining and Metallurgy

Figure 11 Gold cyanide leach operation multivariable control diagram

replaced by a Cynoprobe cyanide analyzer (sensor and controlled variable). This new cyanide analyzer was configured to draw samples from the first leach tank, the second leach tank, and the final adsorption tank. To expand the flexibility of the cyanide dosing, new cyanide dosage points were added to the second leach tank and the fourth adsorption tank that consisted of a flowmeter and flow control valve.

A cyanide leach multivariable controller called Manta Cube was configured to determine the cyanide solution flow set point for the first and second leach tanks and the cyanide set point for the second leach tank based on the cyanide measurements from the cyanide analyzer. The multivariable control strategy for the cyanide leach may be configured for cyanide dosing to the first leach tank only, split cyanide dosing to the first and second leach tanks, or control of the cyanide level in the leach circuit tail. The multivariable controller leach system for cyanide addition was commissioned in conjunction with a new free cyanide analyzer. The cyanide multivariable controller was installed to ensure that maximum gold recovery was obtained in the leach circuit while cyanide use was minimized.

Following the commissioning period, a significant reduction in the variation of the free cyanide measurement was observed. As a result of this decreased variation, the free cyanide set point could be reduced, as there was confidence that the system would maintain free cyanide levels at or above those required for effective leaching. The free cyanide set point was reduced from 600 ppm in December 2012 to current levels of 450 ppm. Following the commissioning of this system, a clear reduction in cyanide consumption of 0.18 kg/t (>10%) was observed. This was due to both the reduced free cyanide set point and the reduced process variable variation, resulting in less cyanide waste.

The net present value for the reduction in cyanide consumption is calculated using a 5% discount rate for the current 5-year mine life based on the data and assumptions. The net present value is calculated to be approximately $1.9 million for the $150,000 capital project, with a payback period of only 106 days.

The work by Bascur et al. (2008) and Steyn et al. (2018) provides additional details about hydrometallurgical process plants.

Pyrometallurgy: Implementation Example of an Integrated Furnace Electric Arc Control for Ferronickel

The efficient operation of a shielded-arc smelting furnace requires good control of the power and feed delivered to the furnace to maintain the arc cover and the slag bath temperature.

The electric power input to the furnace has two components: arc power and bath power. Figure 12 from Voermann et al. (2004) shows the arc power labeled as Pa. This is power delivered to the furnace by the long, powerful arc between the electrode tip and the surface of the slag bath. The slag bath has an average height (ys), as shown in Figure 12. The charge material covers the electrode tip, and the arc power heats the charge. The bath power is labeled Pb. This is power delivered to the slag batch by the resistance or Joule heating caused by the electrode current flowing through the slag level.

The furnace power controller is part of the integrated control system that regulates the power delivery to the furnace. The furnace power controller controls not only the total power delivered to the furnace but also the power balance between electrodes and the ratio of the arc to bath power (Pa/Pb). The ratio of the arc to bath power is a critical control parameter: Too little slag bath power will make the slag colder and more viscous, making it more difficult to cleanly separate the matte or metal and also difficult to tap. Too much slag bath power will increase heat transfer from the furnace walls due to increased slag temperature and more vigorous stirring. This increased heat flux results in increased refractory wear and excessive furnace losses, according to Janzen et al. (2004).

The furnace feed material—or calcine—is supplied from a rotary kiln and stored above the furnace in the feed bins. The hot calcine in the feed bins is metered into the furnace through feed ports in the roof of the furnace, as shown in Figure 12.

The heat in the furnace crucible is shown in Figure 12 as the bottom heat losses (Qbottom) and sidewall losses from the metal and slag levels (Qmetalsidewall and Qsidewall, respectively). The tapped slag (msarc + msbank at Tsbulk) and the tapped metal (mmarc + mmbank at Tm) also lose heat. The slag and metal accumulation in the furnace are represented by the slag height (ys) and the metal height (ym), respectively, in Figure 12.

The furnace incorporates a significant number of instruments to provide a great deal of information about the process. Some of the instruments on the furnace are

- Power consumption meters,
- Electrode slip meters,
- Load cells on the feed bins,
- Water temperature thermocouple at the entrance and exit of each water-cooled element,
- Water flowmeters at the entrance and exit of each water flow circuit,
- Two level refractory thermocouples,
- Furnace pressure monitor,
- Furnace temperature thermocouples,
- Airflow meters, and
- Air thermocouples.

For maximum production and efficiency, coordinated control software architecture is needed.

The integrated furnace control system for the furnace includes five main modules or controllers:

1. **Power controller:** The power controller provides operating points tailored to the measured slag batch resistance.
2. **Feed controller:** Feeding is based on the actual amount of material charged (using bins on load cells) and the actual power delivered to the furnace.
3. **Online heat balance module:** This continuously monitors surface temperatures, water flow, and temperature increases to calculate energy losses and adjust the power and feed balance accordingly.
4. **Slipping control:** A slip is recommended for each electrode based on actual conditions (power, current, limit switches).
5. **Supervisory controller:** The supervisory controller coordinates the work of other controllers.

These modules are described in detail in the study by Janzen et al. (2004).

The furnace power controller controls the total power delivered to the furnace to

- Control slag bath power to maintain target metal and slag temperatures and acceptable sidewall heat flux,

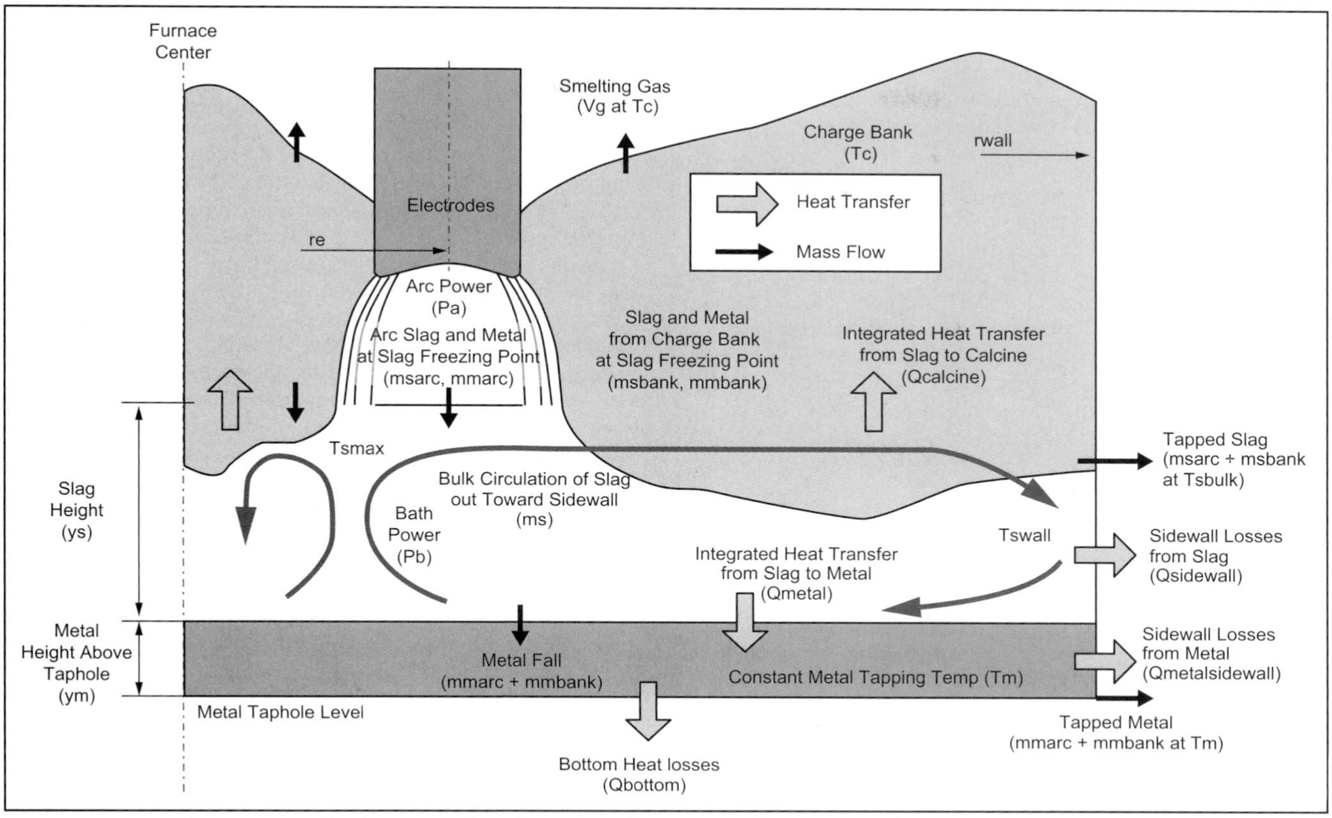

Source: Voermann et al. 2004, reprinted with permission from the Southern African Institute of Mining and Metallurgy

Figure 12 Ferronickel furnace diagram

- Provide stable long-arc operation to enable high furnace power production without excessive sidewall heat flux, and
- Reduce the occurrence and duration of loss of arc.

The power controller is used to control the power delivered to the furnace, including balancing the power delivered by the three electrodes and controlling the ratio of arc to bath power (Pa/Pb).

The feed control system distributes the charge to maintain a covered-arc operation, which is vital to a stable, reliable, and efficient operation.

The online heat balance (OHB) module tracks the energy losses from the furnace and estimates the metal accumulation in the furnace. The OHB module also collects sensor data to establish and track the furnace net specific energy use in terms of kilowatt-hours per ton (kW·h/t). Heat from air infiltration and electrode consumption are also accounted for.

The furnace supervisory control module provides overall control of the furnace and governs the interaction between the various modules of the furnace control system. The furnace supervisory controller integrates and coordinates the feed controller, the furnace power controller, and the online furnace heat balance. The feed rate and net specific energy requirement (kW·h/t) are used to determine the power set point.

The automatic electrode positioning and feeding used on the furnace provide stable operation at the design power of 75 MW. The integrated control system provides the operators with both control and process control information for optimizing the overall metal production. It enables the operator to increase the power with minimum excursions and upsets.

Another advanced control module is used in plants where there are several furnaces on the same high-voltage bus. This module balances the total power draw of a group of furnaces. For example, if an electrode on one of the furnaces strikes a limit switch and as a result unbalances the furnace power draw, the individual electrode set points on the other furnaces on the grid are temporarily altered to balance the total draw of all the furnaces (Voermann et al. 2004).

In a smelting furnace, the ratio of arc to slag bath power is one of the key parameters. If the power delivered to the slag bath is too high, the slag bath temperature and stirring will increase, resulting in increased refractory wear. If it is too low, the metal and possibly the slag will be too cold and difficult to tap. Hatch has developed a system to provide an online estimate of slag bath resistance; this system has been installed on several smelting furnaces.

For nonferrous metals, additional information can be found in the study by Boulet et al. (1997). Tan and Vix (2005) also report advances in controlling the magnetite formation in copper converting.

Maintenance of the Instrumentation and Process Control Systems

Mineral processing and extractive metallurgy process control and instrumentation present unique challenges in the maintenance of process measurements and control devices. The typical harsh environment and difficult process conditions require special consideration of the ruggedness, reliability, and maintainability of these devices. Most advanced control strategies fail due to lack of maintenance, being in the wrong location, and incorrect calibration of the sensors and instruments. Sienkiewicz (2002) provides some guidelines:

- The controls should be regarded as another part of the process and must be used and maintained.
- Annual maintenance and support should be established in the initial plans and likely reinforced to obtain the highest return on the capital and intellectual investment.
- A maintenance program should be designed and implemented that will allow for preventive maintenance of the systems, instruments, controls, and operator interface. A program for change management should be put in place to keep up with expansions and modifications.
- Treat sensors and field devices as critical plant assets if good control usability and performance are to be achieved. Planned maintenance causes the production availability of particular equipment to decrease to approximately 90%–95%. Unplanned downtime may reduce equipment availability by another 5% or more. Low availability and unplanned downtime translate into lost production opportunities that can represent a significant portion of revenues and profits. In fact, unplanned downtime is one of the largest factors that erode production performance. Combining the maintenance of the instrumentation and process controls into a condition-based maintenance program is recommended.

The following section provides additional information on enhancing the traditional maintenance and tuning of the process control algorithm by instituting regular maintenance and performance monitoring of all instruments, actuators, and control algorithms. A change management policy and updates of these vital procedures should go hand in hand with the Occupational Safety and Health Administration (OSHA) 29 CFR 1910 (OSHA 2017a) and *Process Safety Management for Petroleum Refineries* guidelines (OSHA 2017b).

PLANT OPERATIONAL INTELLIGENCE MANAGEMENT

Today, operational efficiencies (such as improved equipment availability and use, increased tonnages, and reduced dilution) are necessary to increase production and lower operating costs in mining, mineral, and metallurgical processing plants. In addition, continuous improvements, innovation, and understanding where opportunities exist are the key to increasing profits. This is where measuring, managing, and maximizing mineral/metallurgical performance is enabled by using an industrial data infrastructure. Transforming the sensor data into performance metrics becomes a reality. To do so, a data hierarchy strategy like the one shown in Figure 13 is needed. Real-time streaming data are transformed into information by using online analytics, generating operational events to aggregate the production and consumable data into improvement workflows generated by the automation of current operational knowledge. The classification of the data enables us to aggregate the data at the desired level of detail to determine where opportunities for improvement are. As such, collaboration among production, economics and planning, maintenance, and all the safety and environmental support become active and not passive as they were in the past.

Figure 13 shows the various levels that can be included in the implementation of process controls and plant management (Bascur 1988, 2016; Bascur and Kennedy 1999). Figure 14 shows a typical process control network with the process information and enterprise network integration. On the left side are data collectors within the systems. The security of the process control network is protected by a demilitarized zone. In the plant information network, the servers that manage the streaming data, the relational information, and the context data model feed the local data visualization and analysis tools to local users. This level feeds the enterprise global server information for integration with the enterprise resource planning (ERP) or a connection to the cloud for external services collaboration.

All the data are captured by sensors through the diverse control systems. These are specialized data acquisitions and

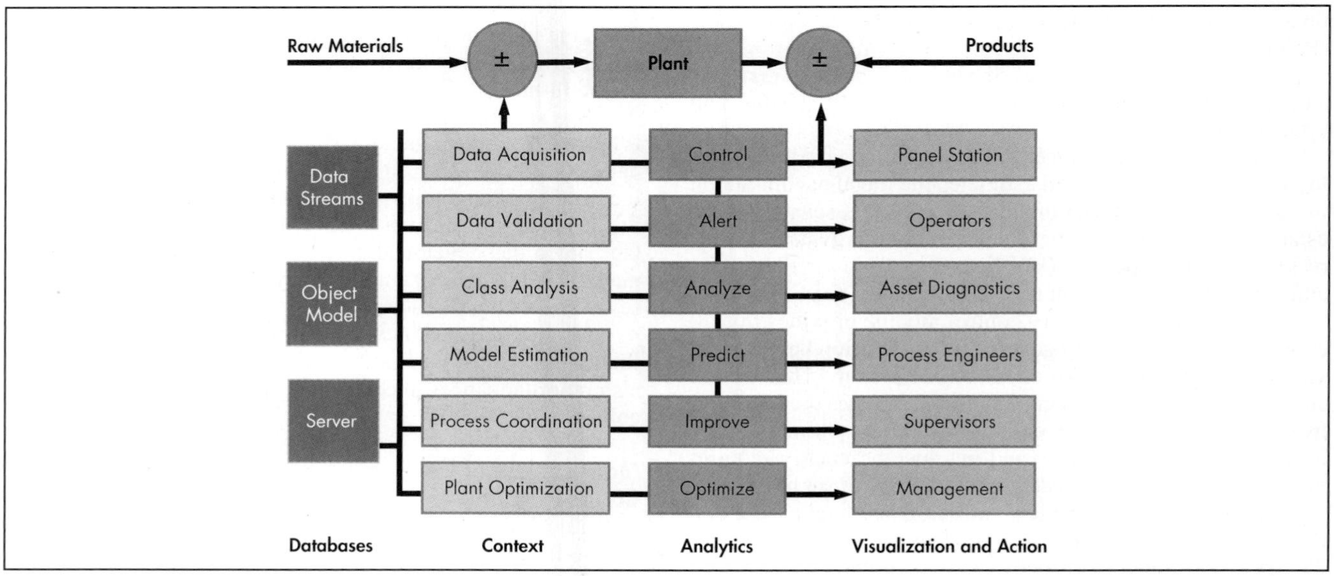

Source: Bascur 2016

Figure 13 Data hierarchy infrastructure for transforming data into actionable information

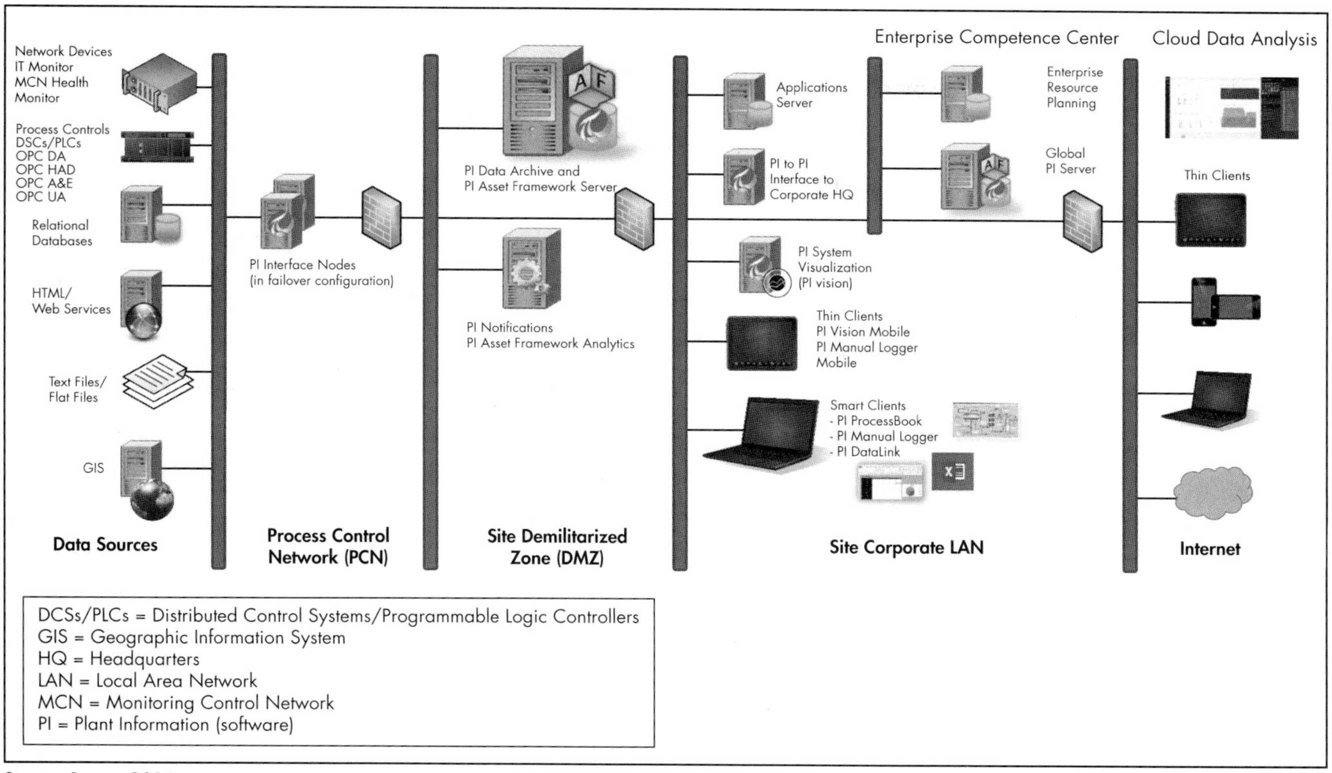

Source: Bascur 2016

Figure 14 Typical process controls and data management infrastructure

control devices that reside near the process equipment or in mobile equipment. These control systems are integrated with PI data servers. The first level is the traditional process control level. This is where the process equipment is regulated for stabilization, start-up, and shutdown of the units. Data validation is required to check the validity of the sensor signal. This is required when advanced process controls are implemented. Data are classified to coordinate the integrated processes for overall plant optimization. Data modeling is conducted when an operating mode is known to perform performance calculations, such as equipment condition alerts, process efficiencies, energy management, and estimation of inferred process variables based on line process models or real-time simulations.

Process coordination is used to integrate the chain supply to extend the analysis from all processes (such as in a mine to a mill to a port, for example). Mine-to-mill business analysis and mine-to-mill mass balancing are discussed later in the chapter.

Once the process chain is well tuned, chain supply optimization becomes a reality. Through energy and mass balancing, it is now possible to identify opportunities to move closer to operating constraints. Operational (real-time) data can be used to explore new opportunities while using business analysis, visualization, and analysis of data and events.

Table 6 summarizes the three types of enhanced initiatives for real-time process improvements and production and operations management using operational data (Fountaine 2014). The table provides real-time operations information, process improvements based on abnormal evaluations, and traditional production and operation management reports.

For daily and real-time operations, basic users will benefit from all the visualization tools that have been discussed so far. These users include operators, craftsmen, and supervisors. The visualization objectives are to achieve daily targets, resolve immediate issues, maintain scheduled plans, and follow environmental regulations. However, users can also benefit from using a context data model of the plant. Users can receive real-time alerts and notifications, and different functions can collaborate automatically.

The process stability and improvement team usually consists of process engineers (local), production superintendents, and center of excellence experts (either regional or global). These are whom might be considered expert and advanced users. Their objectives are to detect process or equipment excursions and develop analytics to configure online performance equations and notifications. These users also maintain process stability by implementing condition-based equipment and process monitoring and diagnostics; maintain process control performance, as described previously; and maintain and plan for production and quality improvements. This team should be directly involved with the economics and planning team to improve scheduling activities and to update the optimization parameters of the forecasting models. This will involve periodic support to the operations and management teams. Continuous improvement and innovation are required to achieve a successful operational excellence program (Bascur 2016).

The production and operations management team is composed of local managers, regional and global managers, and business leadership. This team uses the system on a daily and

Table 6 Real-time analysis and reporting strategies for operational intelligence

	Strategy		
Usage	**Real-Time Operations Actionable Information**	**Process Improvements Analysis**	**Production and Operations Management Reporting**
Frequency	Real time to daily	Daily to annually	Daily to monthly
Type of tool	Visualization tools	Analysis tools	Reporting tools
Audience	Operators Craftsmen Supervisors	Process engineers Production superintendents Center of excellence experts (regional and global)	Local managers Regional and global managers Business leadership
Objectives	Achieve daily targets Resolve immediate issues Maintain scheduled plan Maintain safe operations Achieve environmental regulations compliance	Detect excursions, root cause analysis Maintain process stability Reduce operating costs Improve productivity Improve quality	Understand grade performance Adjust expectations Establish plans Calculate forecasts Adapt for business long term

Source: Bascur 2016

weekly basis to assess the overall performance of the plant and collaborate with external support teams. These users also

- Define the grade performance based on the raw materials and customers' quality and delivery requirements;
- Define the production plan and operations targets;
- Review improvement initiatives for the operational intelligence team;
- Interact and collaborate strategically on a daily and weekly basis rather than monthly, as has been done in the past; and
- Work with the operational intelligence team to improve their forecast and projections based on the current state of the plant's operating conditions and local environmental and safety regulations.

A competency center usually exists where mine and mill activities are coordinated and where integration with the business's ERP transactional systems is provided.

Figure 14 shows the integration of process controls with a data management infrastructure. The left side of the diagram shows the types of data providers using interfaces and connectors to capture data in the PI system. The asset framework organizes the data, analytics, and event generation to transform data into operational insights. Once the data are transformed into operational insights, they are visualized and analyzed by client-based tools such as PI Vision or Microsoft Power BI. The data are made available through PI integrators to ERP systems such as SAP, geographical information systems such as ESRI, and cloud systems such as Microsoft Azure.

The most common way to organize contextual data is to use the ISA standard guidelines provided by the S95 Enterprise to Device model of the American National Standards Institute (ANSI/ISA-95 and ANSI/ISA-99), as shown in Table 7. Using common process industry practices, as presented by Turton et al. (2012), is also recommended; these are basic methods for process analysis and the optimization of processing plants from a chemical process viewpoint.

Collaborative Operational Performance Management

Process engineers assigned to a unit area are responsible to track and improve key metrics using process performance monitoring. Having real-time data and events available simplifies efficiency calculations and yields calculations for constant evaluation and improvement. For example, in a mineral processing plant, it is important to maximize grade recovery and to find the best cut size for grinding to lower energy and water costs. It is in this area where process engineers can implement statistical process control, fishbone analysis, and multivariate statistical monitoring and diagnosis using principal component technologies, which require data and advanced product quality estimators for advanced control implementation. With today's open systems and technologies, process engineers can also connect to process simulators to provide estimates based on process data to optimize process conditions and improve control strategies.

Improvements in thickening process control and availability and dewatering processes can be enhanced by analyzing geometallurgical data and using real-time operating control strategies to optimize water recovery and tailings disposal. Tailing impoundment remote operating conditions are now monitored carefully to manage mine and mill environmental performance, risks, and compliance to governmental and community regulations. In the mining industry, pumping tailings, which consist of ground rock and process effluents, present several challenges to operating personnel. Online real-time monitoring of pumps is now feasible using condition-based monitoring strategies. These strategies combine real-time process variables such as flow rates, pump motor amperage, running time, pressure, temperature, % solids, pump vibration, oil ferocity, and other measurements to reduce leaks, increase operational availability, decrease maintenance costs, and diminish safety incidents by preventing downtime.

The field of multivariate statistical process control enabled by industrial data infrastructures has been growing in recent years. Principal component analysis methods enable the reduction of a set of variables to the minimum level required to discover unusual process events that may occur during regular operations. A good example of this is the implementation of caster breakout by ArcelorMittal Dofasco at each of its casters (Bascur and Kennedy 2002). Applications in the mineral industry have been reported by Hodouin et al. (1993), Garrigues et al. (2000), Bascur et al. (2006), and Romero et al. (2006).

Rigorous model-based technology is also becoming popular due to its ability to access real-time data through dynamic simulators. These simulators require validated and classified data for the estimator to be robust. State estimators are a great way to detect equipment or process changes (Bascur 1999).

Table 7 ANSI/ISA-95 Enterprise to Device decomposition model

Physical Model	Abstraction	Typical Model
Enterprise	A collection of sites	A process block diagram of a plant, including inventory, shipping, and receivables and a geographical model integrated with a geographic information system and enterprise resource planning
Site	A set of plants within a site, including inventory, shipping, and receivables	A plant layout map, integration with geographical information systems
Plant	A set of processing units within a plant area, including inventory, shipping, and receivables	Process block diagram
Plant area	A set of units	Process flow diagram
Unit	A unit with all sensors defined	Process flow control diagram
Equipment	Equipment with all sensors included	Process equipment diagram
Device	Detailed sensor locations	Piping and instrumentation diagram

Source: Bascur 2016

These techniques are now evolving into machine learning algorithms.

Collecting and reusing process equipment data from DCS and PLC systems and laboratory information systems is usually done within an industrial data infrastructure (data, analytics, event framing, visualization, and collaboration tools).

Operational excellence is one of the most valuable enterprise programs to boost morale. Employees feel empowered knowing that they are contributing to the company's overall mission. Implementing a culture of continuous improvement and innovation is necessary to add value to process controls and data management.

Planning and Production Operational Intelligence

The intention of online performance monitoring, as shown in Figure 15, is to detect and predict causes that could lead to a problem. Maintaining a plant and keeping the process flow on target according to the business plan is key for an operational manager. A reduction in process flow is an issue that requires investigation. Using new ways of transforming data into actionable information is the new transaction of the future, and submitting and analyzing any problems requires proper action. The proper alert mechanisms are email, process graphics, and visualization tools. These events can be monetized by estimating losses, having the necessary time intervals available, and integrating the data to the desired level of detail. The alerts can trigger collaboration and further analysis, as shown in Figure 15. Operations must be strict to ensure that control loops operate at peak performance—they are essential to operating the plant safely and efficiently.

Figure 15 is a conceptual diagram that shows the process of detecting abnormal operating modes for continuous improvement and innovations. The schematic shows a process workflow that inputs the targets from the daily plant schedule and the process inputs into a process unit analytics template that classifies the operating modes of all the process units in an industrial plant. The variance is the difference between the expected and actual results. The expected results are generally specified in the operational budget or in the current production schedule. This variance can be used as an exemption operating strategy when implementing an operational intelligence program. As such, the classification of operating times can be used to aggregate the information for all the process units. The schematic presents two questions: (1) Are we on target? and (2) Are we satisfied? (Bascur and Kennedy 1996; Bascur et al. 2016). In essence, it is an innovative strategy that automates the theory of constraints for a digital plant (Goldratt 1984).

The data for the first question can be monitored in real time by defining the interval of time that the process units are running on target, are in trouble, are idle, are down, or are in maintenance. These operational events can then be used by the continuous improvement team to aggregate the data to look for opportunities. The large amount of data calls for the use of modern tools such as Microsoft Power BI to visualize the information and to assess operational losses and gains (Bascur 2016).

Figure 16 shows a typical smelter process block diagram (Bascur and Kennedy 2004b). The smelter data are modeled using process unit templates to standardize the nomenclature for the information to be accessed by the process analytics and visualization tools. Figure 17 shows a diagram of the data transformation to detect operating events and provide further analysis.

Figure 18 shows the results of the data aggregated by the operational events for all the units in the smelter process diagram presented in Figure 16. For this time interval, all the units exhibit many events in "Trouble" mode. This means that the daily production set for these units was not satisfactory, resulting in many minor losses, as presented in this unique set of tools. Presenting the information using this innovative tool enables users to ask questions about the data and view graphics that present the results for evaluation and further analysis. Power BI downloads the event data, and Cortana assists in providing the best visualization of the information.

Using these tools allows users to access the dashboards via the cloud either onsite or remotely. These new tools provide a collaborative environment to enhance improvements in the operating plant. As such, a user can upload a report to the cloud and send an email from his or her tablet or smartphone to others to request assistance with analyzing the data.

Equipment Condition-Based Monitoring

Condition-based monitoring is a very beneficial practice when real-time data from the process and equipment become available. Equipment performance monitoring provides accurate performance information on equipment such as boilers; compressors; furnaces; heat exchangers; and general rotating equipment such as pumps, conveyors, turbines, and so on. This monitoring focuses on increasing throughput, determining when to stop the equipment before it breaks, and reducing

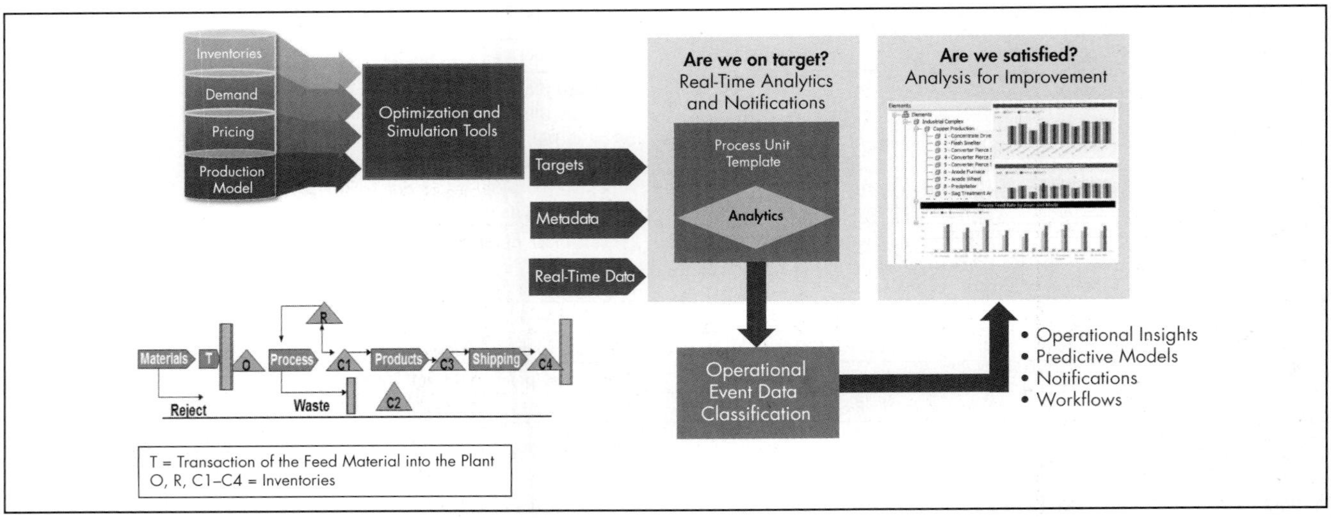

Source: Bascur 2016

Figure 15 Real-time operational intelligence strategy

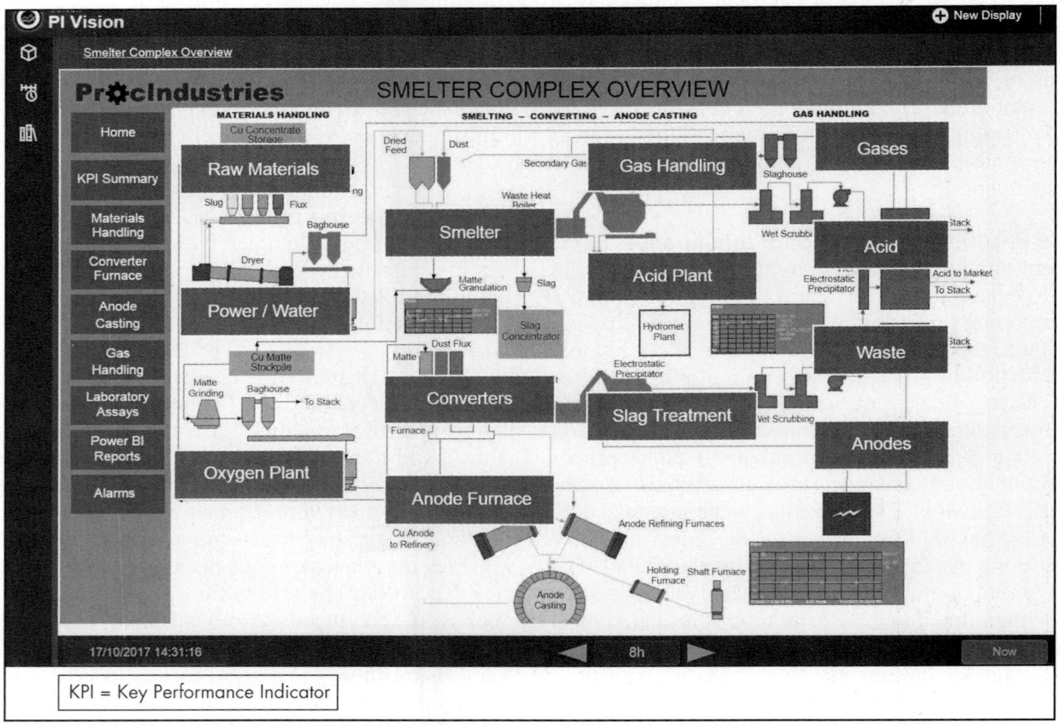

Adapted from Bascur and Kennedy 2004b

Figure 16 Smelter process model context for data modeling for performance management

unscheduled equipment downtime. Thus, increasing plant availability and consequently reducing the overall specific consumption of utilities and production losses by analyzing real-time data enable operators to detect the first indications of an increase in the motor amperage standard deviation or to analyze the lube chemistry of the rotating equipment. This condition-based monitoring strategy directly supports the elimination of waste, which is consistent with an operational excellence program (Pierce 2015). Mojtabai (2009) provides an overall description of mine planning and production management. The author discusses the latest concepts in equipment reliability, functionality, and the maintenance life cycle in mining and mineral processing operations.

Figure 19 shows equipment conditions over time. Over time, the performance of the equipment can be captured in real time using sensors to monitor the overall quality of the equipment (Pierce 2015). The *P* in Figure 19 stands for potential failure. Potential failure event detection is the moment when

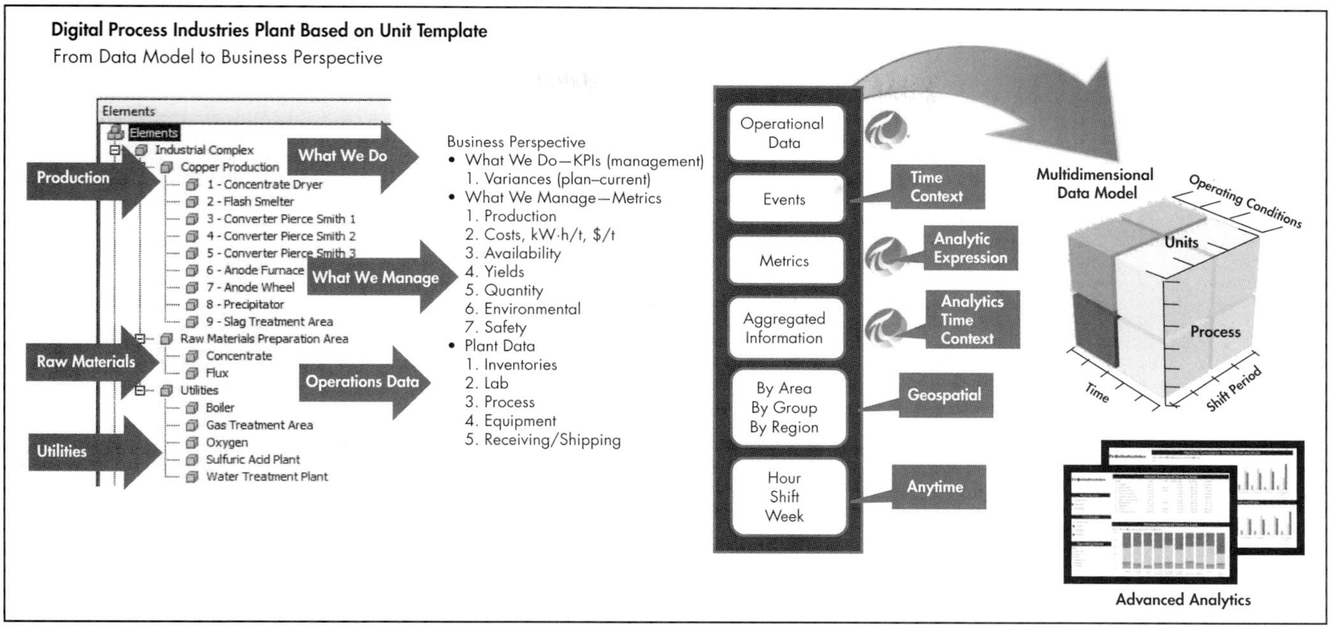

Source: Bascur 2016

Figure 17 Digitizing a smelter plant for performance monitoring in an operational excellence program

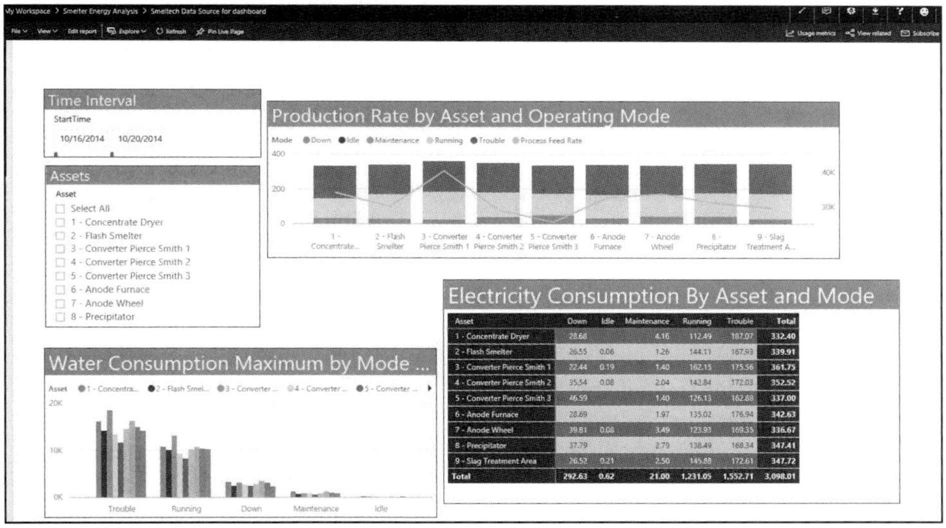

Figure 18 Microsoft Power BI with Cortana presenting the smelter electricity consumption and operating states for all process units

the first sign of deterioration is seen and reported. There are additional measures that can help assess the condition of the equipment over time.

The “run to failure” portion of the graphic in Figure 19 represents the risk of asset failure, assuming that the test method indicates that a failure is imminent. The response opportunity time varies based on the assets, the type of test, and the frequency of the testing method. Quite often, there is little time to respond, depending on plant conditions, time of the notification, and so on. Most commonly, alarms for critical items are sent to operations to ensure that equipment can be moved to a safe location prior to catastrophic failure (represented by F). At this point, a work order is generated and work begins. Pierce (2015) describes several alternatives to this process.

Control Loop Performance Monitoring

Another strategy enabled by an enterprise data infrastructure program is control loop monitoring. This strategy focuses on process control loop monitoring, instrumentation calibration, and process stabilization and optimization. The goal is to monitor the key metrics for the unit and unify the process control loops associated with the unit optimization strategy. Control loops require continuous monitoring and tuning to maintain their performance and usability. Degradation of the sensors, final control elements, and process changes in the equipment

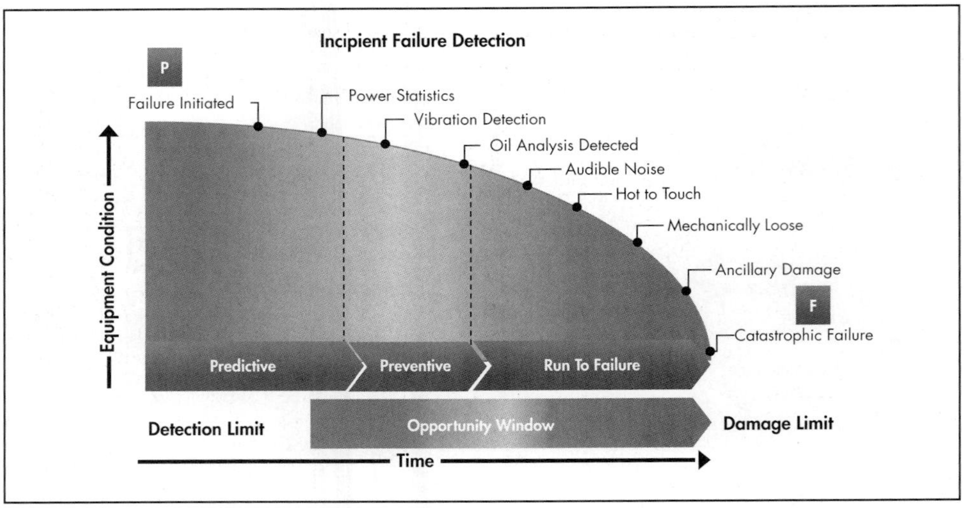

Courtesy of Michael Santucci

Figure 19 Condition-based monitoring: Proactive detection of equipment issues

must always be taken into account. Maintaining all the control loops is a challenge for all process plants. Having the capability to track the controlled and manipulated variables against process disturbances allows for the improvement of product quality and the overall management of the process on a day-to-day basis.

Control loop performance monitoring not only helps identify faulty control valves but also indicates that properly functioning valves may have been targeted for maintenance through a preventive maintenance program. The costs resulting from not performing maintenance can be significant (Bascur and Soudek 2014; Ruel 2010a, 2010b; Van Scholkwyk 2012).

Several control performance goals should be taken into consideration. These are discussed in the following sections.

Controlled Variable Performance

The standard deviation, variability, percentage of time inside limits, percentage of time not saturated, noise level, and oscillation index should be monitored.

Model Error

The variability and integral moving average should be monitored.

Control Loop Performance

The percentage of time in the highest mode, percentage of time in service, percentage of time in normal range, oscillation index, and maximum error should be monitored.

Manipulated Variable Behavior

The percentage of time not saturated, error due to manipulated device stiction, valve travel, and valve reversals should be monitored.

Sensor monitoring is intended to validate data and detect failures to reduce the risk of damaging equipment and to improve overall production performance and availability. The primary benefits of intelligent field devices are improved process control and optimization, as well as the ability to better manage the health and life cycle of the devices. Rigorous equipment monitoring focuses on increasing throughput, determining when maintenance is needed, and increasing plant availability.

The ideal approach is to use real-time data and equipment information to classify the current operating data to provide alerts and notifications to prevent process downtime.

Mine-to-Mill Big Data Analysis Example

One of the most valuable tools today is the availability to provide context to real-time data for big data analysis using self-service tools such as Microsoft Power BI and machine learning tools.

Consumables such as power and water are monitored in real time, as well as the average, maximum, and minimum targets for each operational mode by shift, month, or any interval of time. The average, maximum, and minimum targets are calculated using an algorithm from the data infrastructure. These are called time-derived variables for operational modes, as given in Table 4. The variability of consumption is therefore monitored and provides the engineer with information to investigate abnormal consumption and identify the root cause of this occurrence. The status and availability of the assets in the grinding circuit are also monitored to provide vital information on events—whether an asset is running properly, idling, in maintenance, down, or having trouble, as well as how often it has been in these states and for how long. Monitoring process events enables operators to aggregate the data by ore type, shift operator, and process modes to evaluate the copper grade, recovery energy, and water consumption based on the ore type (e.g., Bascur and Soudek 2014; Bascur et al. 2017).

Figure 20 shows a process data model for a typical mineral processing concentrator. A process unit template is used to organize and perform online calculations. These calculations generate event frames to aggregate the data and events into real-time insights.

A cause-and-effect diagram, also called a fishbone diagram, is used to organize the operational data for business analysis. The process data are aggregated by the events (people, quality, and equipment operating modes) for analysis and visualization.

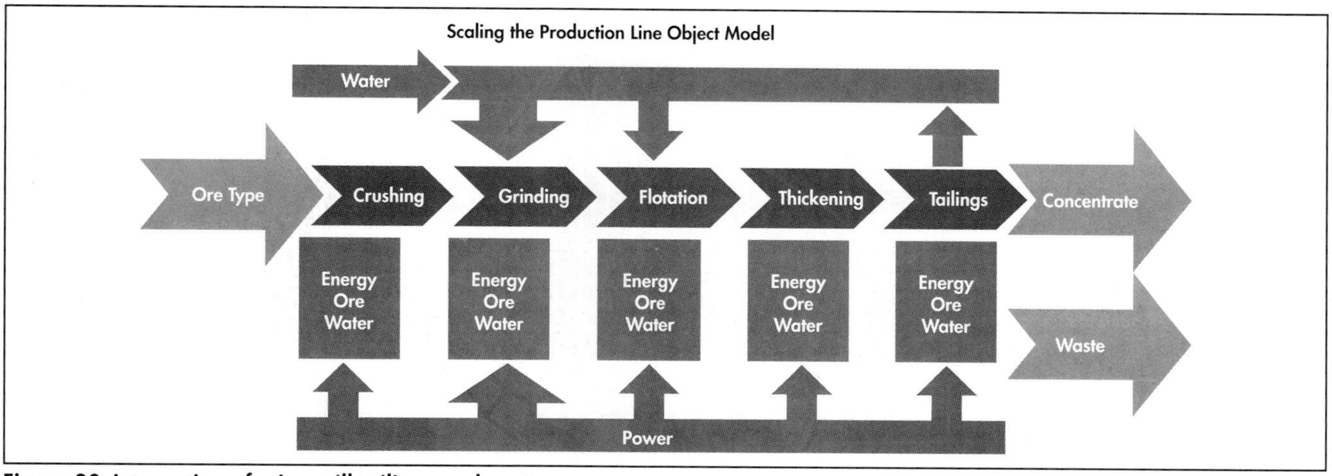

Figure 20 Integration of mine mill tailings and port

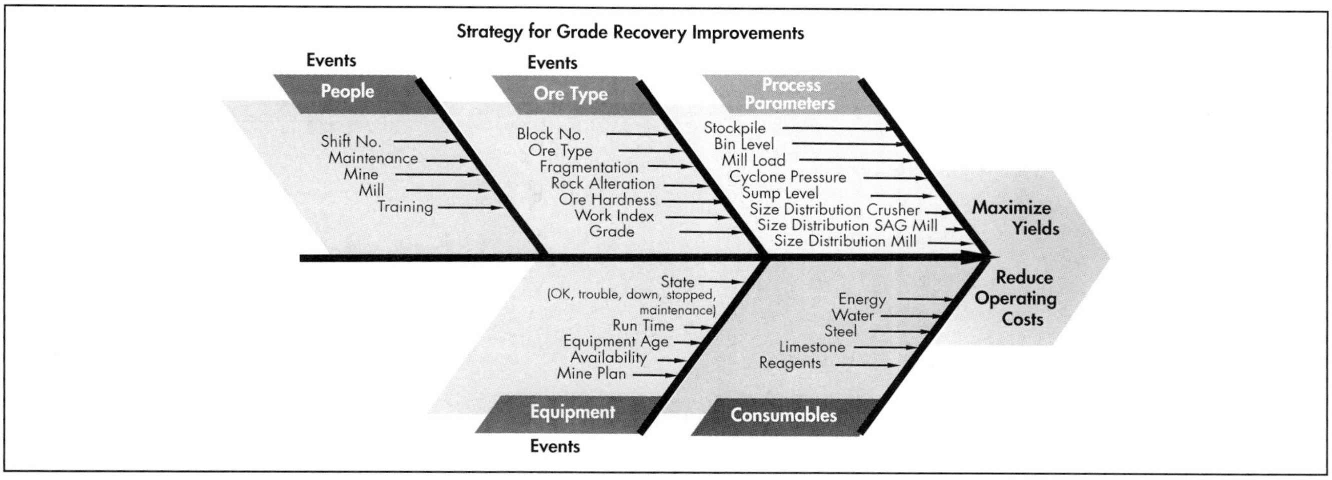

Source: Bascur 2016

Figure 21 Cause-and-effect fishbone (Ishikawa) analysis design structure

Figure 21 shows an example of a fishbone structure for concentrator analysis. It contains five key categories: people, ore type, process parameters, equipment, and consumables. Each of these branches contains subcategories of variables that can be organized depending on the type of cause-and-effect analysis to be done.

In this example, the assets are organized by production line with the monitoring of consumables. Real-time statistics are used to calculate the minimum, maximum, and standard deviation of each consumable by asset, by production line, and by mill. Users can modify the time range they are interested in from, say, 1 hour to a shift or a day. The statistics are recalculated in real time based on this time range. This way, operations and the plant floor are using the same assets as the business systems, but these departments have different views of the same data for their own purposes and needs (Bascur et al. 2017). Figure 22 shows the handling of the process data, process events, equipment status, process areas in consideration, material type (ore/blend), and process time. This figure shows the analysis of the results shown using Power BI in Figures 18 and 23.

Figure 23 shows a Microsoft Power BI dashboard, where the selected event is sliced into data segments. These filters are used to modify the table based on the events chosen by the user. These can be based on many selected criteria, such as block number, process area, ore type, and shift team for all the variables selected, such as total production, reagent consumption, water consumption, and electricity consumption for the process lines in the mill. The variables are time interval, ore type, ore type block number, number of production lines (bin, mill, flotation), and production line availability (running, in trouble, down, stopped, in maintenance) (Bascur et al. 2012). Kanchibotla (2014) and Bennett et al. (2014) discuss advances in integration with blasting and mine-to-mill implementations.

Predictive Analytics and Machine Learning

Previous sections of this chapter discussed strategies to perform gross operational data classification, as shown in Figure 13, as a prerequisite to perform online modeling of the process plant. Predictive analytics is an important subfield of data analytics. Predictive analytics is the art of building and using models that make predictions based on patterns

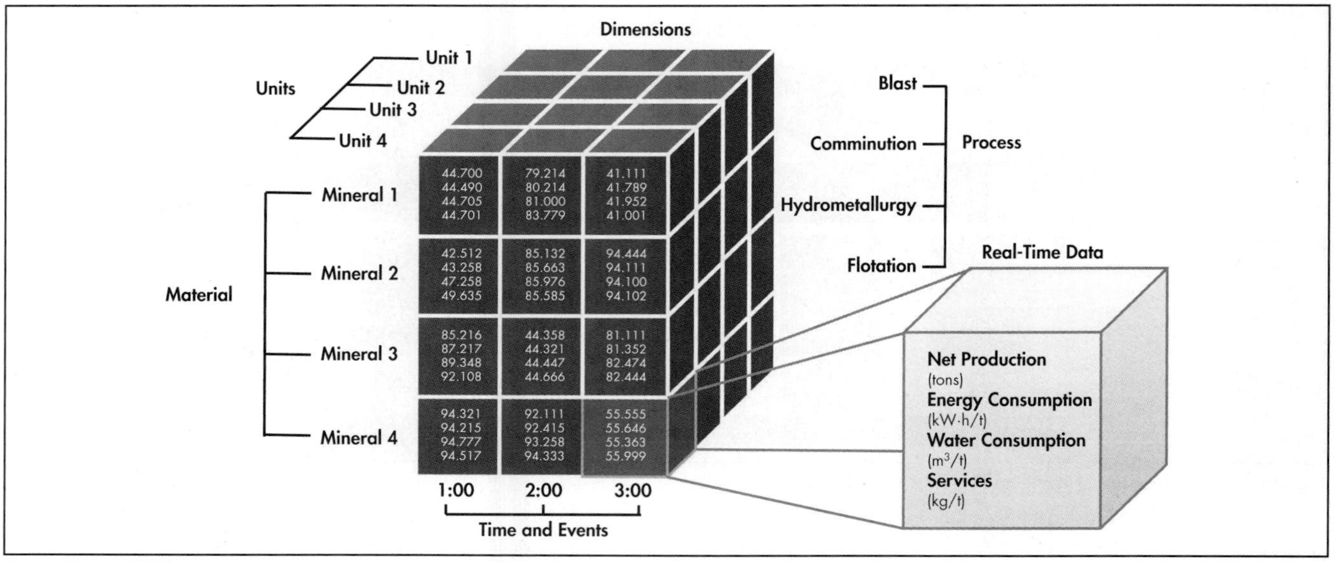

Source: Bascur 2016

Figure 22 Multidimensional cube of real-time data with context and data services

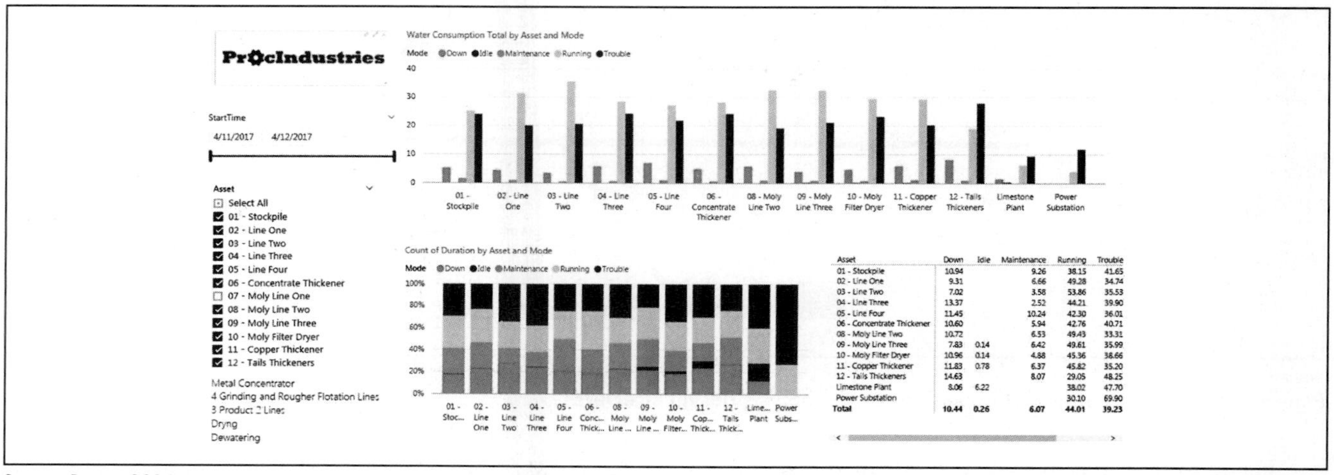

Source: Bascur 2016

Figure 23 Example of a multidimensional analysis report to perform dynamic operation analysis using Microsoft Power BI

extracted from historical data. Some examples are particle size distribution; Gaudin module estimation; particle size; flotation air holdup in the pulp; metal recovery; grade predictions based on process conditions and feed grade material; equipment conditions based on their current speed, power, and vibration to prevent downtime; and quality data versus process variables. Identification of the proper manipulated variables to achieve the right size distribution or the best strategy to minimize metal losses in the tails can predict emissions based on process parameters (Steyn et al. 2018).

An empirical model makes predictions to advise operators. The model is obtained by training an algorithm to make predictions based on a set of historical examples. This is called field machine learning based on the evolutions of the algorithms to classify and fit the data to empirical models. Machine learning is defined as an automated process that can extract patterns from data. An example of this is called supervised machine learning. In this process, a model of the relationships among a set of descriptive features and a target feature based on a set of historical data examples is created. The generated data from an algorithm model are used to make predictions (Kelleher et al. 2015; Raschka 2015). The operational data subset is obtained from the gross operational mode classification, as shown in Figure 15. Defining an operating mode for each process unit can help provide a good set of historical data to develop process models for plant optimization (Steyn et al. 2018; Plourde et al. 2017).

Figure 24 shows an integrated grinding flotation circuit using an online predictive model for recommendations and/or supervisory controls. A yield-based model can be derived from the fishbone analysis shown in Figure 21. The airflow rate to the flotation cell is a measured variable and can be controlled. The air holdup is a soft sensor calculation based on a flotation model (Bascur 2012). Having the right air holdup

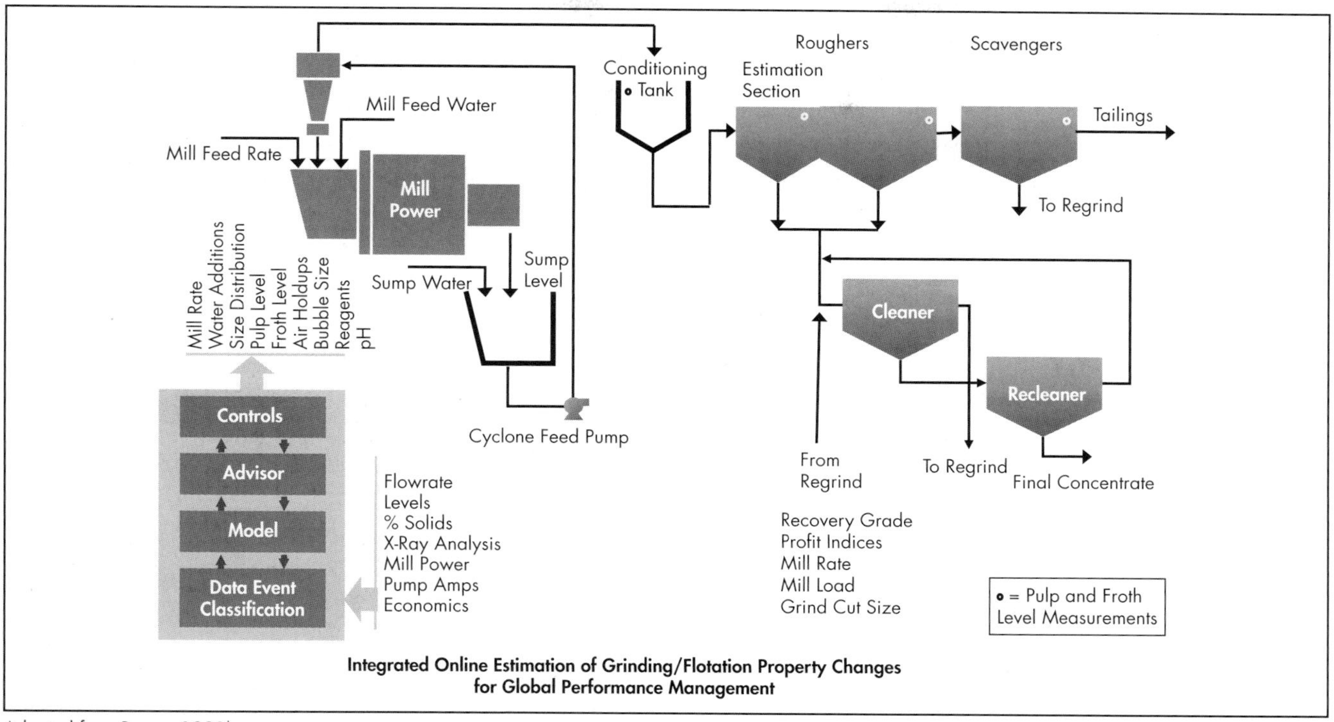

Adapted from Bascur 1991b

Figure 24 Online model predicting quality variables for integrated grinding/flotation optimization

profile in a flotation bank improves the overall recovery of metal into the concentrate. It is a critical variable that can be used with these new tools.

The fishbone analysis shows the effect of the recovery grades and losses based on the operating parameters; the equipment events; the operating shifts; the material grades; and the amount of energy, water, and reagents used to achieve the recovery and grades. Classification of the operating data enables operators to build this predictive model using regression analysis and other models provided by the new tools that are available.

Once a training data set is obtained, a search for the best algorithm to fit the data can begin. To do so, one must have a good understanding of the business and the problem to be solved, a good understanding of the data, and proper preparation of the data. The key variables shown in the proposed fishbone cause-and-effect diagram are used in designing the predictive analytic model. The data hierarchy model shown in Figure 13 can then be used.

There are many algorithms to choose from in modern machine learning tools. Least-squares multiple regression, neural networks, and random forest trees are the most traditional models used. These models are found in Microsoft Azure Machine Learning Studio, or it might be useful to learn how to use Python and R. Python and R are two of the most commonly used programming languages for predictive analytics. They are not especially difficult to learn, but learning these programs may be more difficult for someone who has not been a process control engineer or process engineer. One can also use the Analysis ToolPak available in Excel to develop regression models.

Mass Balances and Data Reconciliation

There are several requirements for developing a methodology to implement a data reconciliation system. First, the algorithm that is able to balance and reconcile the plant data must be robust and must perform correctly against any process topology or configuration. In the past, several mathematical and statistical tools have been developed to solve this reconciliation problem. However, a mathematical algorithm without an information infrastructure is of little value. Second, the right infrastructure to connect the object-oriented model to the real-time process data must be in place. A database that allows storage and manipulation of elements is necessary. The system has to adapt to changes in the process topology because, for example, a meter going out of service is enough to change the mass balance of a process network. Additionally, this object-oriented database should be able to communicate with the real-time information system. Third, the real-time system acts as a repository of process data and inventory data and as a metallurgical laboratory of both the raw and reconciled data. The results are distributed to all staff, including the operators and the plant manager.

The typical problems with process data in industrial plants are

- An overwhelming amount of data,
- Low confidence in the available data,
- Lack of consistency—the data do not make sense,
- The data violate known constraints (mass and energy balances), and
- Poor data quality creates a decision-making "fog" at all levels of an organization and results in a financial penalty (fine).

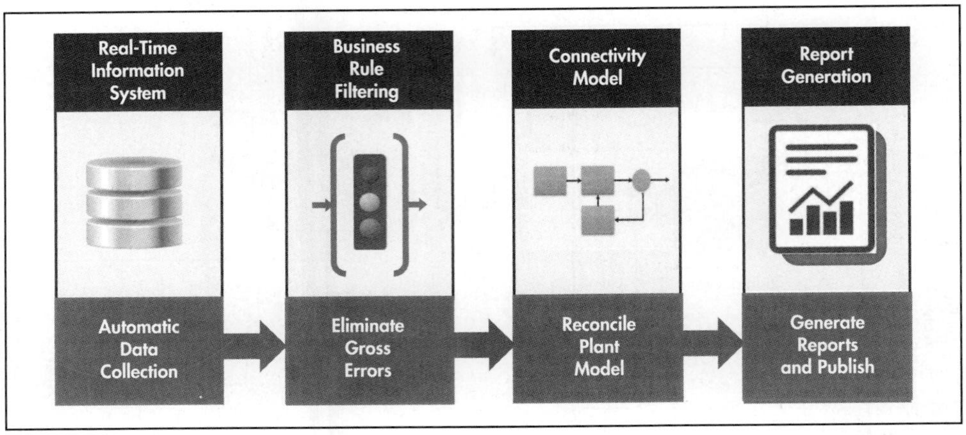

Source: Bascur and Linares 2005

Figure 25 Daily procedure for data reconciliation

These problems have been addressed in the past using traditional methods to capture the required information. However, the main problem with these traditional methods is that human errors can occur during manual data entry, and operational events, which are typical in metallurgical complexes, may be ignored.

The instream analyzer measures the metal compositions for the period of the balance analysis. The mass balancing algorithm runs daily, producing the operational reconciled reports. Figure 25 describes the daily procedure to reconcile process data. The unbalanced data are collected from the real-time information system. Once the data are in the system, a set of analysis rules is executed to detect gross errors that can negatively affect results. These gross errors are eliminated before the final data reconciliation is run. The results provide a unified balance for the whole complex. The assay errors are identified and reported.

A concentrator plant mass balance model includes the operational data, metal assays, metal inventories and receipts of mineral from the mine, stockpiles, bins, thickeners, crushers, grinding, flotation circuits, and the overall flow and composition balance of the chemical components. Once the mass flow, stream, and inventory compositions are available from the system, many calculations and reports can be performed. In the process of solving the network problem, the measurements are validated and gross errors are detected prior to providing a solution (Bascur and Kennedy 2002, 2004a).

Once the data are reconciled, they can be sent to the business information system, where they can be distributed to all users, from operators to engineers and managers. The infrastructure of the data reconciliation system has to adapt to any changes to the process flow or the measuring system because both the process topology and the data are not static.

The reconciled data can then be used to improve yield performance, balancing the optimal recovery and grades while minimizing metal losses in the tails. The data can be reused to improve process planning and to determine the optimal set points for steady-state optimization of the plant (Bascur and Linares 2005; Bascur and Soudek 2009a, 2009b). Hodouin (2011) also described additional uses for this type of methodology.

INDUSTRIAL INTERNET OF THINGS: DISRUPTION IN AUTOMATION

Industrial companies have pursued horizontal and vertical connectivity within their operations for some time now in their ongoing efforts to improve performance and achieve operational excellence. Most existing sensor and actuator points in an industrial automation system are there to support process or production control, safety, and regulatory compliance. However, in the past, adding sensors to support condition-based maintenance or other noncontrol uses was done infrequently due in part to the costs of adding the sensors and associated software systems to existing hierarchical control systems. However, new approaches, including technologies such as less expensive strap-on sensors, Wi-Fi connectivity, predictive analytics, and cloud computing can make condition-based maintenance and other "connected world" applications practical.

Large mining and metal companies rely on many vendors and partners. There is an increasing degree of outsourcing and integration between an operating company and its partners. There is also constant pressure to improve performance and reduce costs. Minerals are becoming more complex. All of these challenges require companies to innovate and simplify access to process and equipment information. Data transfer is a manual process; it is very time-consuming and has very little auditability and security.

In many mining operations, ore is crushed and wet milled to liberate the valuable mineral. This slurry is concentrated by flotation and then filtered to form a dry mineral concentrate that is shipped to refineries to produce metallic products. The type of filtration equipment required depends on the particle size, mineralogy, and shipping requirements. As with all mining operations, the required equipment is robust and designed to be reliable even under the toughest operating conditions.

The availability of a central data infrastructure database allows operational information to be shared with external collaborators. As shown in Figure 26, PI Cloud Connect can be used to share and expand the use of the operational data. Cloud-based strategies simplify access to real-time information through simple registration to PI Cloud Services running on Microsoft Azure. A plant asset framework data model is selectively shared using PI Cloud Connect between the

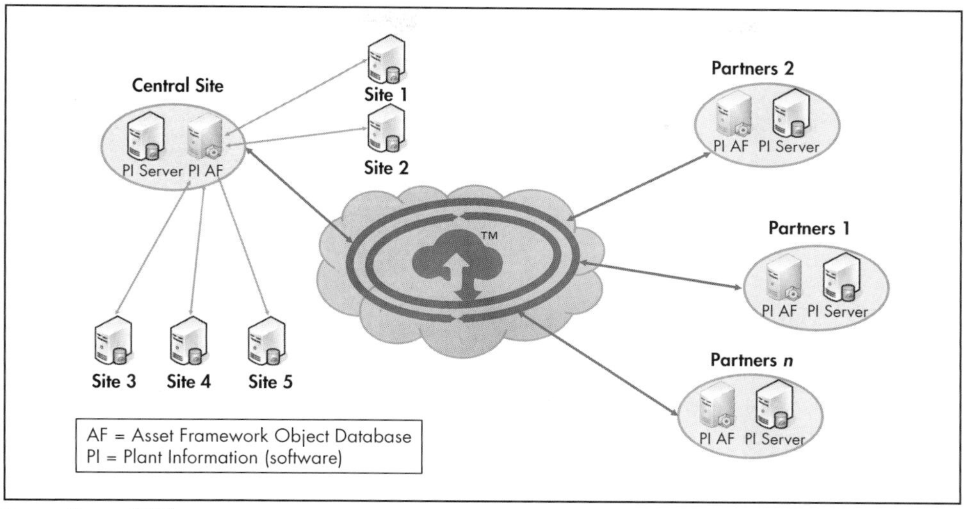

Source: Bascur 2016

Figure 26 PI Cloud Connect computer architecture example connecting an enterprise to its service partners

enterprise and an equipment vendor. The data published by the mineral processing company pertain to a set of equipment used for material separation such as filters. As such, the operational data of the filters are shared by the customer and the service provider in real time. This new capability provides a secure, auditable, and low-maintenance exchange of real-time processes and equipment. It drastically reduces the time spent accessing data and enhances the use of data for equipment vendors. As such, vendors can make recommendations about the optimal use of energy. Many exciting new possibilities are currently being explored by service providers using the data at the original resolution. This allows for the monitoring of sophisticated equipment by experts, integration with utilities, automatic restocking, and integration with joint ventures (Bascur et al. 2016).

Plant Operational Intelligence Strategy Conclusions

Despite the advances in automatic data collection and archiving, business decision makers face the problem of exploiting information that is relevant for plant operations and the sustainability of the business enterprise.

The plant operational intelligence strategy adds operational context to the data. The data can be transformed into operational insights for further analysis using machine learning and business intelligence tools. This novel strategy improves collaboration and provides real-time feedback on decisions made at all levels of the organization. This strategy allows the operating and support team to proactively detect shifts in process performance; the team can use statistical tools and process knowledge to identify shifts in performance. It can provide structure and a means of determining the root cause of the shifts and, if needed, propose corrective action. The competence team is in charge of identifying process improvement opportunities and providing sufficient information and root-cause diagnostics to identify areas of improvement and identify new performance indicators.

The efficient use of water, energy, and resources is critical; the best approach is to have the infrastructure in place to be able to conduct small focused projects, collaborate among different teams in the organization, and understand that this is a continuous improvement process.

CONCLUSIONS AND FUTURE IMPLICATIONS

Many advances in sensors, control systems, and information systems are now converging into what is being called the "internet of things." It is always a challenge to assess the situation in operating plants moving forward, but it is clear that an operational excellence program that takes into account the sustainability of the company is mandatory. Process analysis, control, and optimization are part of the ecosystem.

For new plants, it will be a challenge to select the best practices in this field to be able to exceed the traditional state of the art. This chapter has discussed the state the art of what is available and most prevalent today.

For existing plants, plant optimization will generate the best return on investment because if equipment or software is not added, optimization consists of using what one has on hand. A step-by-step process is proposed in the study by Ruel (2014b).

The first phase is to remediate instrumentation problems, including the ones pertaining to valves, variable-speed drives, analyzers, transmitters, and so on. When equipment is in order, loops are put back in automatic mode. The key performance indicator (KPI) for this phase is the proportion of loops operating in the highest mode. If a loop oscillates, standard default values can be used for this type of loop.

The second phase consists of validating the control strategies, properly configuring the systems (control system, programs and parameters, human–machine interface, alarms), and then tuning the loops and control strategies. The KPI for this phase is the number of loops in service (right mode, nonsaturated, nonoscillating, low variability). Soft sensors should also be validated.

The third phase consists of analyzing the process and expected performances to optimize the control loops and control strategies. When mixing products, for instance, each flow loop should move at the same speed to ensure the proper recipe when increasing the total flow. The KPI for this phase is variability (per loop, per unit) and economic weight.

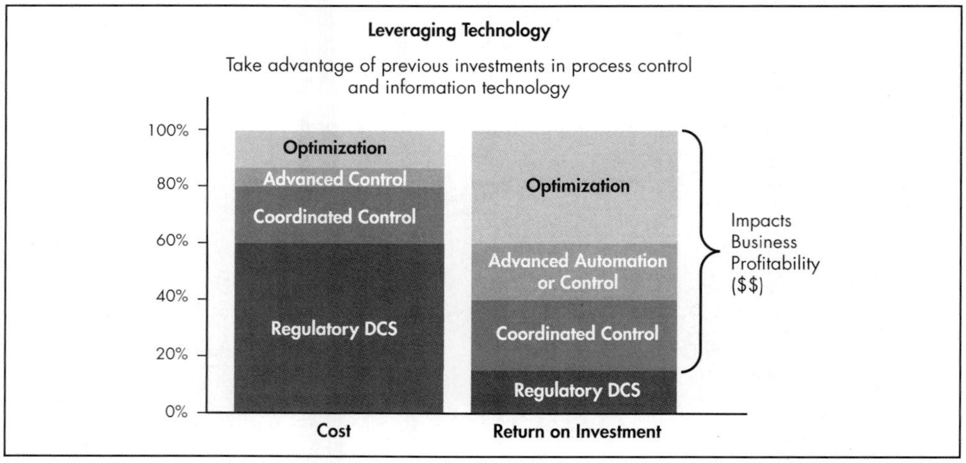

Source: Ruel 2014b

Figure 27 Process control investments and return on investment based on knowledge strategies

The fourth phase is necessary when performance is not sufficient. In this phase, advanced control should be added: advanced regulatory control, MPC, or fuzzy logic control. A decision tree should be presented to determine the right approach. The KPI is the same as that of the third phase, but a percentage of time use should also be added.

Once all steps have been completed, control performance monitoring software should be used to sustain the results and pinpoint equipment or loops that are not performing as stated. Also, after an audit on the alarm system, alarm management and rationalization should be performed if needed.

To calculate benefits, the following need to be analyzed: the past and benchmark production, recovery, chemical product consumption, water consumption, energy usage, and other factors. KPIs (e.g., energy/t, $/t, %/t) need to be estimated.

Once an area is optimized, the KPIs need to be reviewed and the return on investment needs to be calculated. In most cases, the optimization projects will be funded within a week.

Figure 27 shows that most plants have invested in automation hardware, which is the most costly item, as discussed by Ruel (2014b). It is clear that process control hardware is a large part of the investment in new plants. However, the plants are living things—they evolve, and the process must adapt to new raw materials. As such, plants must change and adopt new technologies to operate efficiently based on current economic and local environmental regulations. Additional information can be found in the studies by Ruel (2010b, 2014a, 2014b).

Boufard (2015) discussed the benefits of process control systems in mineral processing grinding circuits, concluding that the benefits are undisputable: a 1%–16% gain in throughput, up to a 1% gain in recovery, fewer operator interventions, and a payback in less than 6 months (Bouche et al. 2005). Gupta (2016) described the new set of skills required to manage process control and advanced operational techniques and stated that "technology has both complicated process control and made it easier."

Bauer and Craig (2008) provided an economic assessment of advanced process controls. The framework provides a method to evaluate the advanced process control program for benefits estimation. The main profit contributors from their survey are as follows:

- Throughput increase (67%–70%)
- Process stability improvements (45%–67%)
- Energy consumption reduction (45%–65%)
- Increased yield mass of more valuable product (40%–60%)
- Quality giveaway reduction (30%–40%)
- Downtime reduction (15%–22%)
- Better use of raw materials (15%–19%)
- Reprocessing cost reduction (11%–12%)
- Response increase (11%–14%)
- Safety increase (9%–11%)
- Operating labor reduction (5%–15%)
- Other (7%–15%)

The data generated by the evolution of sensors, process control, vision, sound, and other fields will continue to grow. This big data opportunity is a fact. It requires a reengineering of the current situation. The integration of real-time data and operational data with business systems will converge. Today, new large file systems can handle the large data requirements. In addition, new machine learning algorithms are becoming available to develop online predictive models to improve the early detection of problems within plants.

Mill availability remains one the most significant indicators of efficient mill operation and a KPI for concentrate managers. Maintaining the optimum charge and composition in a SAG mill remains a fundamental challenge for efficient mill processes.

Continuous improvements of liner life prediction and reline planning are now a great augmented alternative for improving availability and optimizing the grinding behavior for optimal metal recovery.

Tailings are a necessary by-product of mining activities. An operational excellence program is a holistic view that requires data analysis to conserve energy, conserve water, and manage geowaste materials such as tailings and waste rock.

Wireless and cellular technologies can transmit operational data from supervisory control and data acquisition systems to operational site computers for additional analysis to classify the data to detect problems. Historical, operation mode–based data can be used to develop models for the proactive generation of alerts using a real-time industrial data infrastructure.

ACKNOWLEDGMENTS

The author sincerely thanks the following people for taking the time to read the draft of this manuscript and provide feedback to improve the content of the chapter: John Karageorgos, Mantra Controls, Australia; Michel Ruel, BBA Inc., Canada; Daniel Sbarbaro, University of Concepción, Chile; Gina Laviste, OSIsoft Asia; and many other dear colleagues who have contributed to the author's work over the years.

REFERENCES

ANSI/ISA-95. *Enterprise to Control System Integration*. New York: American National Standards Institute.

ANSI/ISA-99. *Control System Integration*. New York: American National Standards Institute.

ANSI/ISA-S5.4-1991. *Instrument Loop Diagrams*. Research Triangle Park, NC: International Society of Automation.

ANSI/ISA-S5.5-1985. *Graphics Symbols for Process Diagrams*. Research Triangle Park, NC: International Society of Automation.

Baas, D., Hille, S., and Karageorgos, J. 2007. Improved flotation process control at Newcrest's Telfer operation. Presented at the Ninth AusIMM Mill Operators' Conference, Fremantle, Australia.

Baas, D., Bennett, D., and Walker, P. 2014. Developing process control standards for optimal plant performance at PanAust Limited. Presented at the 12th AusIMM Mill Operators' Conference, Townsville, Australia.

Bascur, O.A. 1988. A control data framework with distributed intelligence. In *Advances in Instrumentation: Proceedings of the ISA/88 International Conference and Exhibit*. Research Triangle Park, NC: Instrument Society of America.

Bascur, O.A. 1990a. Profit based grinding controls. *Miner. Metallurg. Process.* 7:9–10.

Bascur, O.A. 1990b. Expert process operator advisor. In *Control 90: Mineral and Metallurgical Processing*. Edited by R. Rajamani and J.A. Herbst. Littleton, CO: SME. pp. 67–77.

Bascur, O.A. 1991a. Human factors and aspects in process control use. In *Plant Operators' Forum*. Edited by D.N. Halbe. Littleton, CO: SME.

Bascur, O.A. 1991b. Integrated grinding/flotation controls and management. Presented at the Copper 91/Cobre 91 International Symposium, Ottawa, Ontario, Canada.

Bascur, O.A. 1999. The industrial desktop—Real time business and process analysis to increase the productivity in industrial plants. Presented at the Second International Conference on Intelligent Processing and Manufacturing of Materials, Vancouver, British Columbia, Canada.

Bascur, O.A. 2005. Example of a dynamic flotation framework. Presented at the Centenary of Flotation Symposium, Brisbane, Australia.

Bascur, O.A. 2010. A flotation model framework for dynamic performance monitoring. In *Proceedings of the 7th International Mineral Processing Seminar*. Santiago, Chile: Gecamin.

Bascur, O.A. 2012. Dynamic model-based flotation performance monitoring. In *Separation Technologies for Minerals, Coal, and Earth Resources*. Edited by C.A. Young and G.H. Luttrell. Englewood, CO: SME. pp. 709–718.

Bascur, O.A. 2016. A journey toward a digital transformation in the process industries. https://pisquare.osisoft.com/docs/DOC-2935-journeydigitaltransformation-draft-c1-4-binder1pdf. Accessed December 2017.

Bascur, O.A., and Herbst, J.A. 1985a. Dynamic simulators for training personnel in the control of grinding/flotation systems. *IFAC P. Ser.* 18(6):315–324.

Bascur, O.A., and Herbst, J.A. 1985b. On the development of a model-based control strategy for copper-ore flotation. In *Flotation of Sulphide Minerals*. Edited by K.S.E. Forssberg. Amsterdam: Elsevier. pp. 409–431.

Bascur, O.A., and Herbst, J.A. 1986. Improved thickener performance through the use of an extended Kalman filter. In *Design and Installation of Concentration and Dewatering Circuits*. Edited by A.L. Mular and M.A. Anderson. Littleton, CO: SME. pp. 835–845.

Bascur, O.A., and Kennedy, J.P. 1996. Measuring, managing and maximizing refinery performance. *Hydrocarb. Process.* 75(1):111–116.

Bascur, O.A., and Kennedy, J.P. 1999. Real time business and process analysis to increase productivity in the process industries. Presented at the 1999 ISA Conference, Houston.

Bascur, O.A., and Kennedy, J.P. 2002. Reducing maintenance costs using process and equipment event management. In *Mineral Processing Plant Design, Practice, and Control*. Edited by A.L. Mular, D.N. Halbe, and D.J. Barratt. Littleton, CO: SME. pp. 507–527.

Bascur, O.A., and Kennedy, J.P. 2004a. Are you really using your information to increase the effectiveness of assets and people? In *Plant Operators' Forum*. Edited by E.C. Dowling and J.O. Marsden. Littleton, CO: SME. pp. 47–62.

Bascur, O.A., and Kennedy, J.P. 2004b. Improving metallurgical performance in pyrometallurgical processes. *JOM* 56(12):22–27.

Bascur, O.A., and Linares, R. 2005. Grade recovery optimization using data unification and real time gross error detection. *Miner. Eng.* 19(6-8):696–702.

Bascur, O.A., and Soudek, A. 2009a. Real-time integration of mining and metallurgical information for efficient use of energy and water. SME Preprint No. 09-103. Littleton, CO: SME.

Bascur, O.A., and Soudek, A. 2009b. Real time information management infrastructure—Collaboration at mine-mill for asset optimization. In *Recent Advances in Mineral Processing Plant Design*. Edited by D. Malhotra, P. Taylor, E. Spiller, and M. LeVier. Littleton, CO: SME. pp. 490–498.

Bascur, O.A., and Soudek, A. 2014. Implementation strategies for energy effectiveness and sustainability—Example of Anglo American Platinum. Presented at the 12th AusIMM Mill Operators' Conference, Townsville, Australia, September 1–3.

Bascur, O.A., Linares, R., and Yacher, L. 2006. Improving mine mill performance in large metallurgical complexes. Presented at the XXIII International Mineral Processing Congress, Istanbul, Turkey, September 3–8.

Bascur, O.A., Linares, R., and Riveros, E. 2008. Hydrometallurgical operational management strategies. Presented at the Fifth International Copper Hydrometallurgy Workshop, Antofagasta, Chile, May 13–15.

Bascur, O.A., Hertler, C., and Benavides, N. 2012. Specific energy and water reductions in mine to mill operations. In *Proceedings of the International Mineral Processing Congress*. New Delhi, India: Indian Institute of Metals.

Bascur, O.A., Halhead, M., Garrigues, L., and Jarvis, M. 2016. Mineral processing plant asset and energy optimization: The calming cloud over operations. In *Proceedings of the XXVIII International Mineral Processing Congress*. Westmount, QC: Canadian Institute of Mining, Metallurgy and Petroleum.

Bascur, O.A., Plourde, M., Paquet, S., Morissette, S., and Gervais, D. 2017. A journey towards mine to port operational intelligence. Presented at the 2017 SME Annual Conference and Exposition, Denver, CO, February 19–22.

Bauer, M., and Craig, I.K. 2008. Economic assessment of advanced process control—A survey and framework. *J. Process Control* 18(1):2–18.

Bennett, D., Tordoir, A., Walker, P., La Rosa, D., Valery, W., and Duffy, K. 2014. Throughput forecasting and optimization at the Phu Kham Copper-Gold Operation. In *Proceedings of the 12th AusIMM Mill Operators' Conference*, Townsville, Australia, September 1–3. Melbourne, Victoria: Australasian Institute of Mining and Metallurgy.

Bhattacharyya, D., Shaeiwitz, J.A., Turton, R., and Whiting, W.B. 2012. Diagrams for understanding chemical processes. In *Analysis, Synthesis and Design of Chemical Processes*, 4th ed. Edited by R.A. Turton, R.C. Bailie, W.B. Whiting, J.A. Shaeiwitz, and D. Bhattacharyya. Indianapolis: Pearson Education.

Botin, J.A. 2009. Integrating sustainability into the organization. In *Sustainable Management of Mining Operations*. Edited by J.A. Botin. Littleton, CO: SME. pp. 71–132.

Bouche, C., Brandt, C., Broussaud, A., and Drunick, V.W. 2005. Advanced control of gold ore grinding plants in South Africa. *Miner. Eng.* 18:866–876.

Boufard, S.C. 2015. Benefits of process control systems in mineral processing grinding circuits. *Miner. Eng.* 79:139–142.

Boulet, B., Vaculik, V., and Wind, G. 1997. Control of nonferrous electric furnaces. *IEEE Can. Rev.* (Summer):1–13.

Braden, T.F., Kongas, M., and Saloheimo, K. 2002. On-line composition analysis of mineral slurries. In *Mineral Processing Plant Design, Practice, and Control*. Edited by A.L. Mular, D.N. Halbe, and D.J. Barratt. Littleton, CO: SME. pp. 2020–2021.

Coker, R. 2015. A clearer picture. *Min. Mag.* (November):38–42.

Concha, F.A. 2014. *Solid–Liquid Separation in the Mining Industry*. Cham, Switzerland: Springer.

Edwards, R., Vien, A., and Perry, R. 2002. Strategies for the instrumentation and control of grinding circuits. In *Mineral Processing Plant Design, Practice, and Control*. Edited by A.L. Mular, D.N. Halbe, and D.J. Barratt. Littleton, CO: SME. pp. 2130–2151.

Ferrarini, L., and Veber, C. 2009. *Modeling, Control, Simulation, and Diagnosis of Complex Industrial and Energy Systems*. Research Triangle Park, NC: International Society of Automation.

Flintoff, B. 1995. Control of mineral processing systems. In *Proceedings of the XIX International Mineral Processing Congress*. Littleton, CO: SME.

Flintoff, B. 2002. Introduction to process control. In *Mineral Processing Plant Design, Practice, and Control*. Edited by A.L. Mular, D.N. Halbe, and D.J. Barratt. Littleton, CO: SME. pp. 2051–2065.

Flintoff, B., Guyot, O., McKay, J., and Vien, A. 2014. Innovations in comminution and control. In *Mineral Processing and Extractive Metallurgy: 100 Years of Innovation*. Edited by C.G. Anderson, R.C. Dunne, and J.L. Uhrie. Englewood, CO: SME.

Fountaine, L. 2014. Leveraging data as an enabler for operational excellence. Presented at the OSIsoft Regional Seminar, São Paulo, Brazil.

Fuenzalida, R., and Olivares, J. 2012. Experiences and projections on SAG mill multivariable predictive control. SME Preprint No. 12-123. Englewood, CO: SME.

Fuerstenau, D.W. 1999. The flotation century. In *Advances in Flotation Technology*. Edited by B.K. Parekh and J.D. Miller. Littleton, CO: SME. pp. 3–21.

Garrigues, L., Kettaneh, N., Wold, S., and Bascur, O.A. 2000. Multivariate process analysis and optimization in mineral processing. In *Control 2000*. Edited by J.A. Herbst. Littleton, CO: SME. pp. 41–50.

Goldratt, E.M. 1984. *The Goal: A Process of Ongoing Improvement*. Great Barrington, MA: North River Press.

Gupta, M.S. 2016. Rethinking the role of process control. www.arcweb.com/blog/rethinking-role-process-control. Accessed December 2017.

Hartog, J., Beehan, V., Karageorgos, J., and Beeson, D. 2014. Optimization of the Peak Gold Mine's processing plant through advanced process control. Presented at the 12th AusIMM Mill Operators' Conference, Townsville, Australia.

Henriquez, F., Silva, D., and Jiménez, C. 2012. Improving ball mills operations by applying MPC and advanced control systems at Minera Los Pelambres. Presented at the XXVI International Mineral Processing Congress 2012, New Delhi, India.

Herbst, J.A., and Bascur, O.A. 1984. Mineral processing control: Realities and dreams. In *Control '84: Mineral Metallurgical Processing*. Edited by J.A. Herbst, D.B. George, and K.V.S. Sastry. Littleton, CO: SME. pp. 197–215.

Herbst, J.A., and Harris, J. 2007. Modeling and simulation of industrial flotation processes. In *Froth Flotation: A Century of Innovation*. Edited by M.C. Fuerstenau, G. Jameson, and R.-H. Yoon. Littleton, CO: SME. pp. 757–777.

Herbst, J.A., and Pate, W. 1999. Developments in flotation control: One part evolution, two parts revolution. In *Advances in Flotation Technology*. Edited by B.K. Parekh and J.D. Miller. Littleton, CO: SME. pp. 399–412.

Hodouin, D. 2011. Methods for automatic control, observation, and optimization in mineral processing plants. *J. Process Control* 21(2):211–225.

Hodouin, D., MacGregor, J., Hou, M., and Franklin, M. 1993. Multivariate statistical analysis of mineral processing data. *CIM Bull.* 23–34.

Janzen, J., Gerritsen, T., Voermann, N., Veloza, E.R., and Delgado, R.C. 2004. Integrated furnace controls: Implementation on a covered-arc furnace at Cerro Matoso. In *Proceedings of the Tenth Ferroalloys Congress*, Cape Town, South Africa, February 1–4. Marshalltown, South Africa: Southern African Institute of Mining and Metallurgy.

Johansson, B., Bergmark, B., Guyot, O., Bouche, C., and Broussaud, A. 1999. Model-based control of Aitik bulk flotation. *Miner. Metall. Process.* 16(2):41–45.

Kanchibotla, S. 2014. Mine mill value chain optimization: Role of blasting. In *Mineral Processing and Extractive Metallurgy: 100 Years of Innovation*. Edited by C.G. Anderson, R.C. Dunne, and J.L. Uhrie. Englewood, CO: SME.

Karageorgos, J., Skrypniuk, J., Valery, W., and Ovens, G. 2001. SAG milling at the Fimiston Plant (KCGM). Presented at the SAG 2001 Conference, Vancouver, British Columbia, Canada.

Karageorgos, J., Genovese, P., and Baas, D. 2006. Current trends in SAG and AG mill operability and control. Presented at the SAG 2006 Conference, Vancouver, British Columbia, Canada.

Karageorgos, J., Davies, S., Broers, E., and Goh, J. 2009. Current trends in countercurrent decantation and thickener circuit operability and control. Presented at the 10th AusIMM Mill Operators' Conference, Adelaide, Australia.

Kelleher, J.D., Namee, B.M., and D'Arcy, A. 2015. *Fundamentals of Machine Learning for Predictive Analytics*. Cambridge, MA: MIT Press.

Kewe, T., Moffatt, N., and Schaffer, M. 2014. Porgera flotation circuit upgrade and expert system installation. Presented at the 12th AusIMM Mill Operators' Conference, Townsville, Australia.

Li, X., McKee, D.J., Horberry, T., and Powell, M.S. 2011. The control room operator: The forgotten element in mineral process control. *Miner. Eng.* 24(8):894–902.

Lukas, M.P. 1986. *Distributed Control Systems: Their Evaluation and Design*. New York: Van Nostrand Reinhold.

Lundmark, P. 2008. Collaboration with the operator in focus. Presented at Automining 2008, Santiago, Chile.

Marlin, T.E. 2014. *Process Control: Designing Processes and Control Systems for Dynamic Performance*, 2nd ed. New York: McGraw-Hill.

Medower, R.A., and Cook, R.E. 2002. The selection of control hardware for mineral processing. In *Mineral Processing Plant Design, Practice, and Control*. Edited by A.L. Mular, D.N. Halbe, and D.J. Barratt. Littleton, CO: SME. pp. 2077–2103.

Mojtabai, N. 2009. Mine planning and production management. In *Sustainable Management of Mining Operations*. Edited by J.A. Botin. Littleton, CO: SME. pp. 261–356.

OSHA (Occupational Safety and Health Administration). 2017a. 29 CFR 1910. *OSHA General Industry Regulations*. Washington, DC: OSHA.

OSHA (Occupational Safety and Health Administration). 2017b. *Process Safety Management for Petroleum Refineries*. OSHA 3918-08. Washington, DC: OSHA. https://www.osha.gov/Publications/OSHA3918.pdf.

Otnes, R.K., and Enochson, L. 1978. *Applied Time Series Analysis*. New York: John Wiley and Sons.

Pierce, K. 2015. *A Guidebook to Implementing Condition-Based Maintenance (CBM) Using Real-Time Data*. San Leandro, CA: OSIsoft.

Plourde, M., Bascur, O.A., Paquet, S., and Gervais, D. 2017. Digital innovation in modern engineering and operational excellence. Presented at the 2017 SME Annual Conference and Expo, Denver, CO, February 19–22.

Raschka, S. 2015. *Python Machine Learning*. Birmingham, UK: Packt Publishing.

Rice, R.C. 2015. *PID Tuning Guide*. Bethlehem, PA: NovaTech.

Romero, F., Yacher, L., and Bascur, O.A. 2006. Extended semiautogenous milling: Smooth operations and extended availability at C.M. Doña Ines de Collahuasi SCM, Chile. In *Advances in Comminution*. Edited by S.K. Kawatra. Littleton, CO: SME. pp. 181–190.

Ruel, M. 2010a. Control system performance assessment—Best practices. Presented at the Second International Congress of Automation in the Mining Industry (Automining 2010), Santiago, Chile.

Ruel, M. 2010b. Closed loop tuning vs. open loop tuning: Tuning all your loops while the process is running is now possible. Presented at the 2010 TAPPI PAPTAC International Chemical Recovery Conference, Williamsburg, VA, March 29–April 1.

Ruel, M. 2012. Fuzzy logic control: A successful example. Presented at ISA 2012 Automation Week, San Diego, CA.

Ruel, M. 2014a. Advanced control decision tree. In *Proceedings of the 46th Annual Canadian Mineral Processors Operators Conference*. Westmount, QC: Canadian Institute of Mining, Metallurgy and Petroleum.

Ruel, M. 2014b. Successful methodology to select advanced control approach. Presented at the Process Control and Safety Symposium of the International Society of Automation, Houston, TX, October 6–9.

Sbarbaro, D., and del Villar, R. 2010. *Advanced Control and Supervision of Mineral Processing Plants*. London: Springer-Verlag.

Schoenbrunn, F., Hales, L., and Bedell, D. 2002. Strategies for instrumentation and control of thickeners and other solid–liquid separation circuits. In *Mineral Processing Plant Design, Practice, and Control*. Edited by A.L. Mular, D.N. Halbe, and D.J. Barratt. Littleton, CO: SME. pp. 2164–2173.

Sienkiewicz, J.R. 2002. Basic field instrumentation and control system maintenance. In *Mineral Processing Plant Design, Practice, and Control*. Edited by A.L. Mular, D.N. Halbe, and D.J. Barratt. Littleton, CO: SME. pp. 2104–2113.

Stephanopoulos, G. 1984. *Chemical Process Control: An Introduction to Theory and Practice*. Englewood Cliffs, NJ: Prentice Hall.

Steyn, J., Bascur, O.A., and Gorain, B. 2018. Metallurgy analytics: Transforming plant data into actionable insights. SME Preprint No. 18-029. Englewood, CO: SME.

Stuffco, T., and Sunna, K. 2002. Well balanced control systems. In *Mineral Processing Plant Design, Practice, and Control*. Edited by A.L. Mular, D.N. Halbe, and D.J. Barratt. Littleton, CO: SME. pp. 2066–2076.

Tan, P., and Vix, P. 2005. Control of magnetite formation during slag-making in copper smelters. In *Converter and Fire Refining Practices*. Edited by A. Roos, T. Warner, and K. Scholey. Warrendale, PA: The Minerals, Metals & Materials Society. pp. 247–258.

Turton, R., Bailie, R.C., Whiting, W.B., Shaeiwitz, J.A., and Bhattacharyya, D. 2012. *Analysis, Synthesis, and Design of Chemical Processes*, 4th ed. Upper Saddle River, NJ: Prentice Hall.

Van Scholkwyk, T. 2012. PI applications in Anglo American Platinum: Control loop monitoring and reporting. Presented at the OSIsoft Regional Seminar, Johannesburg, South Africa.

Voermann, N., Gerritsen, T., Candy, I., Stober, F., and Matyas, A. 2004. Developments in furnace technology for ferro-nickel production. Presented at the 10th International Ferroalloys Congress, Cape Town, South Africa.

Weidenbach, M., and Lombardi, J. 2012. Tailing thickener operability and control improvements at Prominent Hill. Presented at the 11th AusIMM Operators' Conference, Hobart, Tasmania.

CHAPTER 2.7

Project Management

Henry Schnell

Projects come in many shapes and sizes. Recent data show that more than 50% of projects have cost overruns and more than 40% have scheduling problems. The reasons behind these problems vary, but in most cases it comes down to poor project definition and preparation. Environmental concerns and the evolution of the associated actions and remediation requirements have also affected project management in the industry. Increasing social licensing issues also further complicate projects. These issues are evident worldwide, are difficult to estimate, and can greatly upset scheduling efforts.

Financing of a new project, particularly for smaller developing companies, depends on the availability of funds. Investors and major financial markets are demanding greater transparency, and many have special reporting needs, such as National Instrument 43-101 (2011) in Canada and the Australian Joint Ore Reserves Committee standards (JORC 2012). Although the definitions of various required studies are somewhat flexible, there are accountability issues. Investors are becoming more astute, all pointing toward the need for improved project planning, implementation, and a multidiscipline approach.

This chapter is intended to emphasize major concepts but is not a detailed look at all project aspects. Project management has developed into a professional art, requiring many special skills to successfully implement a project. We tend to mostly talk about new developments when discussing projects, which is the case in mining and will be the case in this chapter. However, a project can be a modification; a change of methodology; a change of philosophy; an operation suspension; a remediation; or another, less evident change. Project principles and organization are required for all types of these changes.

PROJECT CYCLE

Mining, by its nature, is a temporary activity, with a beginning, an operation, and an end; there may even be repetition of the same cycle as prices change or exploration and development provide added resources. All phases of the mining cycle require some level of planning and financing, and there is a need for a holistic, complete-cycle approach. The cycle is long, with major investments required over a long period of time before the cash flow becomes positive, as illustrated in Figure 1.

Exploration

The first group on the ground is comprised of exploration prospectors and geologists. They have the huge task of discovery and resource delineation. At the preliminary exploration stage, large areas are often evaluated by airborne or ground-based mapping or sampling surveys of the earth's surface done by prospectors and geologists. From maps and existing data, specific areas are singled out for more detailed studies. If valuable mineral potential is indicated, a "claim" is staked by way of an online, computer-based application system.

The second stage of exploration involves more detailed surveys, including mapping, sampling, and diamond drilling (often at great depths) to determine the size and shape of the mineral deposit. In addition, data collection for environmental studies begins at this stage. Some preliminary samples should also be taken for a preliminary metallurgical evaluation.

There is a tendency to become enamored early when mineral indicators are encountered. This excitement must be tempered as costs increase with further exploration and the reality of a resource base to justify further expenditures must be established.

A "mineral resource" is a concentration or occurrence of solid material of economic interest in or on the earth's crust in such form, grade (or quality), and quantity that there are reasonable prospects for eventual economic extraction. The location, quantity, grade (or quality), continuity, and other geological characteristics of a mineral resource are known, estimated, or interpreted from specific geological evidence and knowledge, including sampling. Mineral resources are subdivided, in order of increasing geological confidence, into inferred, indicated, and measured categories (JORC 2012).

For a resource to become a reserve, all aspects of a project must be addressed, including various modifying factors. The typical breakdown of inferred, indicated, and measured resources required for various study levels (conceptual study, prefeasibility study, and definitive feasibility study) is illustrated in Figure 2 and Table 1.

Henry Schnell, Retired, Formerly with AREVA (Paris, France), Eagle Bay, British Columbia, Canada

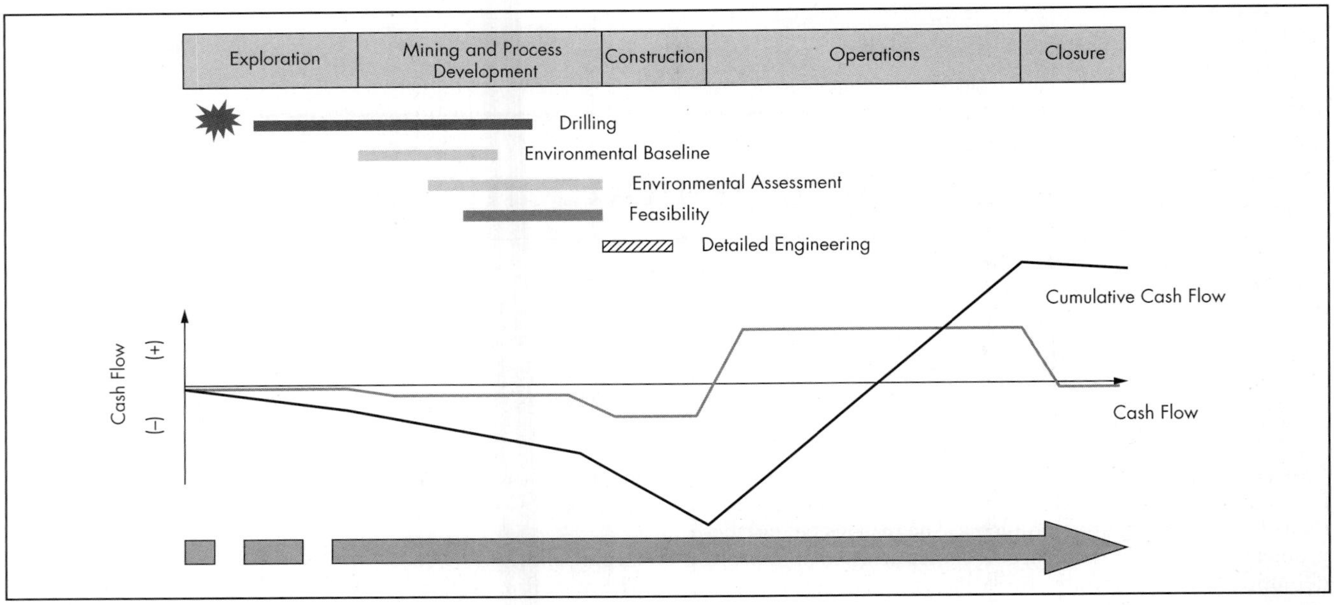

Source: Schnell 2013

Figure 1 Typical mining cycle

Mining Development

Mining is initially investigated based on preliminary ore-body knowledge and geological engineering data. Mine design can only be optimized once geology is sufficiently understood at the deposit and more local scales. Optimization is achieved through an iterative and collaborative process that takes into account the vast array of geological, mine engineering, scale of operation, cost, and processing factors. Ultimately, successful mine design and production scheduling deliver ore that is suited to the flow-sheet and plant requirements.

Process Development

Metallurgical data are initially collected from core or chip sampling through laboratory test work with input from mineralogical and geological logging. Flow-sheet optimization occurs after several stages of test work have taken place and is achieved through an iterative and collaborative process based on geology, feed grade, physical properties, throughput, and costs and is closely linked to the mine plan. Finalization of the flow-sheet and plant design requires a demonstrated environmentally acceptable method for tailings disposal and consideration for future closure and remediation.

As a project moves from conceptual study to definitive feasibility study evaluation, there is a growing degree of interaction among geology, mining, and processing, and these relationships become more complex. Common examples of such relationships include the deposit structural framework (faults, shear zones, rock mass fabric) that may influence mining; the alteration type and mineralogical associations, including contaminants that may influence processing; and the ore hardness, which may influence both mining and processing. Assessing the sensitivity of the ore reserves to the cutoff grade is a further example. This is an area that requires a good understanding of the relationship between the grade distribution and a range of mining and processing parameters (e.g., the selective mining unit size, head grade, and metal recovery). Poor

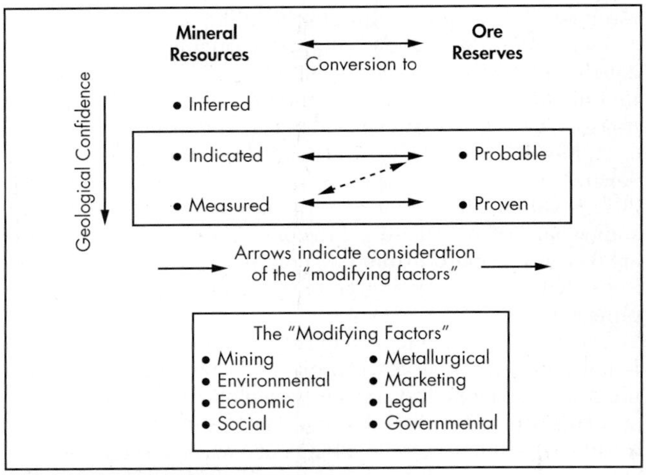

Adapted from JORC 2012

Figure 2 Moving a resource to a reserve

handling of the geology–mining–processing interface usually leads to serious issues such as loss of value, missed opportunities, and project failure.

There is a tendency to focus on mineralogy, with a geological emphasis, but flow-sheet development and processing is most influenced by the gangue ore matrices. All testing also needs to look at other possible sub- or by-product options. Modern process development is more and more considering not only sustainability but also development of "comprehensive" extraction technology.

Construction

The construction phase will be variable, often with some preliminary construction or phased construction proceeding during the study phase. Part of this is due to needs for exploration

Table 1 Typical resource/reserve breakdown for study levels

Study	Inferred, %	Indicated, %	Measured, %	Typical Breakdown, Inferred/ Indicated/Measured, %
Conceptual study	70–100	0–30	—	80/20/0
Prefeasibility study	30–40	30–60	0–20	40/50/10
Definitive feasibility study	10–30	40–60	20–60	20/30/50

Table 2 Historical study nomenclature

Author	Date	Level I	Level II	Level III	Level IV
Taylor	1977	Preliminary	Intermediate	Feasibility	—
Hustrulid and Kuchta	1995	Conceptual	Preliminary	Feasibility	—
Barnes (reported from Australia)	1997	Resource calculation	Preliminary evaluation	Feasibility	Basic engineering
		Preliminary	Indicative	Definitive	
		Scoping	Preliminary	Detailed	
		Order of magnitude	Prefeasibility	Bankable	
Mackenzie	2006	Scoping	Prefeasibility	Definitive	—
Bullock	2011	Preliminary feasibility	Intermediate feasibility	Final feasibility	Basic engineering
Most common usage	2000–present	Scoping	Prefeasibility	Feasibility	Basic engineering

Adapted from Bullock 2011

or start of the mine pre-stripping development. Early in the project, there is a need to have a conceptual, holistic vision to avoid future demolition, prevent interference with mine development, and avoid environmental and social issues.

The construction phase on many high-investment or high-risk projects is phased to minimize capital investment, reduce risk, and allow for early production or long start-up cycles. These decisions need to be based on well-defined economic models, taking into account future production interferences and considering expansion options and flexibility as part of the facility design, reducing future investment.

Operations

Mine production involves the extraction of ore, separation of minerals, disposal of waste, and shipment of ore minerals. Additional exploration may lead to the discovery of additional mineralization that leads to expansion of the operation during the mine life. These expansions must involve the full cycle of studies, evaluations, and permitting processes that are required for a new mine development.

Closure and Reclamation

Mine closure is the last phase of the mining cycle. Shutdown and decommissioning involve the removal of equipment, the dismantling of facilities, and the safe closure of all mine workings. Reclamation, which in fact occurs at all stages of the mine life cycle, involves earthwork and site restoration, including revegetation of waste rock disposal areas. The final stage is monitoring, which includes environmental testing and structural assessments that commonly continue long after the mine is closed.

Mining is a temporary land-use activity. The goal of a reclamation plan is for areas affected by mining activity to host self-sustaining ecosystems that provide a healthy environment for fish, wildlife, and humans.

STUDY DEFINITIONS

The various stages of the project cycle are not absolute, and part of a cycle may be repeated several times before the project advances. It is important that the whole project team, the owners, and the investors have a common understanding of the language and expectations of each part of the project cycle. Each stage of the study needs to have an end, including a formal report, with a review process and approval for advancing to the next stage.

Terminology

In general, there is no international agreement on the terminology for each stage of a feasibility study, and there is no agreed-upon standard for quality or accuracy. Various publications do provide a set of standards that are becoming more widely used. Although it is convenient to refer to scoping studies, prefeasibility studies, and final feasibility studies, in reality the study process is iterative and increasingly detailed (McCarthy 2015).

There is a large variety of titles used for the various studies. Historically, the various studies were just categorized as Levels I to IV, with Level III signifying feasibility and Level IV representing basic engineering. Most common today are the terms *scoping*, *prefeasibility*, and *feasibility*. Various teams have used some poetic license to name studies, generally because there may be a gap in applying all required criteria (e.g., while waiting for environmental permits, social interaction, or even final resource figures). Table 2 lists the nomenclature that has been used by various authors.

Scoping Studies

A scoping study is carried out very early in the exploration phase as a basis for acquiring exploration areas or making a commitment for exploration funding. At this stage, the investment risk may be relatively small, but it is obviously undesirable to spend money on something that has no chance of coming to fruition.

The major risk at this stage is for a viable mining project to be relinquished due to an inadequate assessment. For this reason, it is essential that experienced people be involved in the scoping study. The intended estimation accuracy is usually 30%–35%, although some companies accept ±50%.

It is acceptable for scoping studies to be based on very limited information or speculative assumptions in the absence of hard data. The study is directed at the potential of the property rather than a conservative view based on limited information.

A sensitivity analysis, however, should present the likely range of possible outcomes so that decision making, including investment decisions that may follow a public release of the study results, is not biased to the optimistic end of the range. It is important in this phase to identify all the potential risks and determine what trade-offs are required to minimize these risks in the next phase.

Prefeasibility Studies

Prefeasibility studies are an essential part of mining projects as a basis for committing to a major exploration program following a successful preliminary geo-metallurgical program. These studies are used to attract an investor to the project or partner or as a basis for a major underwriting to raise the required risk capital. A prefeasibility study may also be prepared in full or in part by potential purchasers as part of the due diligence process. It is also necessary to provide a business case for proceeding to a final feasibility study.

The prefeasibility study must be prepared with great care by experienced people, and its conclusions should be heavily qualified wherever necessary. Assumptions and trade-off studies should be realistic rather than optimistic, because it is very difficult to bring management and markets back to reality in the event that the final feasibility study is significantly less favorable.

It is also very important to include *all* potential mining, processing, environmental, social, and location options with associated trade-off studies. At this stage, many options can be more easily and cost-effectively evaluated at minimal cost, whereas in the final feasibility study it is much more expensive. Failure to review a large variety of process options generally leads to "second thoughts" during the subsequent feasibility study or even during engineering, causing project delays and cost overruns. A single preferred go-forward process alternative is the objective of the prefeasibility study and is supported by a solid business case and risk analysis.

Final Feasibility Studies

The final feasibility study is usually based on the most attractive alternative for the project as previously determined in the prefeasibility study. Moving forward into a feasibility study with more than one option increases costs significantly. The aim of the study is to remove all significant uncertainties and to present the relevant information with backup material in a concise and accessible way. The final feasibility study has several key objectives, including

- To demonstrate with reasonable confidence that the project can be constructed and operated in a technically sound and economically viable manner;
- To provide a basis for detailed design and construction;
- To provide a Class 2 capital estimate, operating cost, and project schedule for the go-forward design case; and
- To enable the raising of funds for the project from banks or other sources.

The term *bankable* is sometimes used in connection with final feasibility studies. This just means that the study achieves a quality and standard that would be acceptable for submission to bankers. Whether a bank will actually lend against the project is another question, depending on many matters that are outside the control of the feasibility study team.

Whether the project design has been optimized in the feasibility study will depend on the time and budget allowed. Often a suboptimal—but acceptable—design is used as the basis of the feasibility study, with further optimization undertaken (or not) once the project has been approved. Outstanding issues and options at the end of the feasibility study will lead to cost uncertainty and (most likely) cost overruns.

The feasibility study is only one step along the design path. Much more work must be undertaken during the detailed engineering phase that follows project approval. The engineering work usually overlaps and is ongoing at some level through to project completion, commissioning, and early production.

STUDY COST AND QUALITY

Projects are costly—cost reduction is ever present, and studies are particularly heavily scrutinized because they occur early in a project and do not add to production. However, the cost of a study, its content, and its level of detail directly affect the final project, control costs, and help limit cost overruns.

Study Variables

A final feasibility study will cost approximately 3%–5% of the final project costs. Some mining companies will exceed this investment to reduce risk. In addition, innovative solutions or technology will increase risk and costs at all levels of studies. These costs exclude the exploration costs required to achieve a suitable reserve to support a project. Interestingly, companies may announce the start of feasibility study work, suggesting that the study will be positive—but undertaking a study does not guarantee positive results. Table 3 lists the typical variables that are used to determine study accuracy at various levels.

Costs

The earlier the team is assembled, the more opportunities there are for the team to challenge and affect the overall project budget. It is the combination of all disciplines early on that creates the balance needed to create a successful project. This signifies that early in the project, seemingly small decisions made by a small project group can significantly affect the final project cost. As the project progresses, the concepts are better defined, with costs increasing and making changes less possible.

It is normal to continue to question project costs and estimates. Experienced cost experts should estimate project costs, keeping in mind the estimate accuracy in relationship with the amount of work undertaken at each study stage. Project cost estimate reviews, known as "tollgate reviews," should be undertaken at each stage in an open, multidiscipline environment to vet the technical and economic concepts.

"Value engineering" is often used to achieve an optimum balance among function, performance, quality, safety, and

Table 3 Typical study variables

Factor	Conceptual	Prefeasibility	Feasibility
Objectives	To generate a range of viable options	To examine all options and select one option for feasibility	To maximize "value" and make a decision on mine development
Database	Limited, mostly assumed	Better, mostly assessed	Large to very large, calculated
Reserves	None	Probable	Proven
Accuracy, %	±30 to ±50	±20 to ±25	±10 to ±15
Typical duration, months	2–9	9–18	12–24
Typical cost, US$	50,000–200,000	0.5 million–1 million	1 million–5 million
Typical staffing, no.	2–6	5–20	15–50

Adapted from Bullock 2011

cost. The proper balance results in the maximum value for the project. This should really be part of the philosophy within each discipline as the project progresses. Be aware that the compromises made in such value engineering exercises will result in hampering or even preventing project goals from being achieved. In recent years, management at many companies has tried to help the project team better understand the business drivers. Decisions are then based on the key performance indicators for the project and any other key risk areas for the business.

Study Content

Most studies have similar content, adjusted in certain cases to emphasize project priorities. It is recommended that the main study volume be kept to a reasonable length—say, a thickness of 2 to 3 in. for a feasibility study to make it readable, with the details contained within a series of extensive appendices. A typical table of contents is as follows:

- Volume 1: Summary Report
- Volume 2: Geology and Ore Reserves
- Volume 3: Mine Planning
- Volume 4: Mine Plant
- Volume 5: Design Criteria, Mineral Processing, and Metallurgy
- Volume 6: Consultants' Reports
- Volume 7: Side Studies
- Volume 8: Environmental and Socioeconomic Factors
- Volume 9: Quotations and Proposals
- Volume 10: Project Planning and Execution
- Volume 11: Cost Estimates
- Appendices of all background information

ENVIRONMENTAL IMPACT STUDY

One part within the feasibility study is the "Environmental and Socioeconomic Factors" section. These studies need to start very early in the project cycle, even with the first prospector on the ground. The environmental impact study (EIS) is often done by a consultant who specializes in this type of work. It is becoming more and more important to establish the go-forward process option for the EIS. Changing direction later has the potential to delay the permit issue and the possible start of the permit schedule. A typical table of contents for the EIS includes the following but may require other specific areas in consultation with state and local authorities and stakeholders:

- Executive Summary
- Statement of Project Objectives
- Identification and Description of Project Alternatives
- Rationale for Selection of Preferred Alternatives
- Detailed Project Description of Preferred Alternatives
- Description of Existing Environment
- Description and Evaluation of Predicted Impacts
- Social and Economic Impacts to Local Communities
- Identification of (and Commitment to) Mitigation and Enhancement Measures and Appropriate Post-EIS Studies
- Impacts to Historic and Cultural Sites, Particularly Sites of Significant Importance to Indigenous Peoples
- Documentation of Public Participation Program
- Cost Analysis of Alternatives

PROJECT ORGANIZATION

Projects require a dedicated team assigned to various tasks. The team will change, expand, and add support groups as a project evolves. Oftentimes, key roles are not specifically identified or their function is not clear. It is critical that the right team members have lead roles during project evolution.

Organization

There is no single organizational approach to projects. Each project is organized to accomplish the work effectively and efficiently. Several factors influence the organizational approach to execute a project. The complexity profile of a project, the culture of the parent organization, the preferences of the project manager, the knowledge and skills of the team, and whether the project management office is in-house or outsourced are some factors that influence the project's organization. Most projects have similar functions that are important to successfully manage the project. These include

- The sponsor,
- The project manager,
- Project controls,
- Procurement,
- Technical management,
- Quality control,
- Commissioning planning, and
- Administration.

Most of these functions will continue throughout the various phases of a project. The primary leadership during a project will vary, with technical needs taking precedence early in a project until handover to the project engineering/construction/execution team and finally to the operator. Continuity is important, but the leadership should be appropriate at each stage. For example, operator input is vital, but a technical group leads during the conceptual study, prefeasibility study, and feasibility study stages. A suggested leadership model is shown in Figure 3.

Consultant Interface and Contracting

Engineering contracting, even when projects are carried out in-house, requires some form of contract between the owner and the engineer, consultant, and/or contractor. Service agreements range from fixed price to cost-plus arrangements. The former is generally preferred by purchasing departments, as this puts all the risk on the contractor but requires a high level of detail in engineering prior to award. A cost-plus contract requires careful control by the owner's team but allows for maximum project flexibility and less up-front completion of detailed engineering. Table 4 shows the available contract options for engineering that are also applicable to other types of services.

Test Work

Exploration requirements are addressed early in the project, and environmental studies are part of specialist programs for the EIS. Metallurgical and process studies are often minimized and shortened because they tend to affect the study schedule by being carried out after the geological results. As mentioned previously, it is at the prefeasibility stage that various options need to be reviewed, and in most cases this requires a wide range of metallurgical testing. It is important that samples be

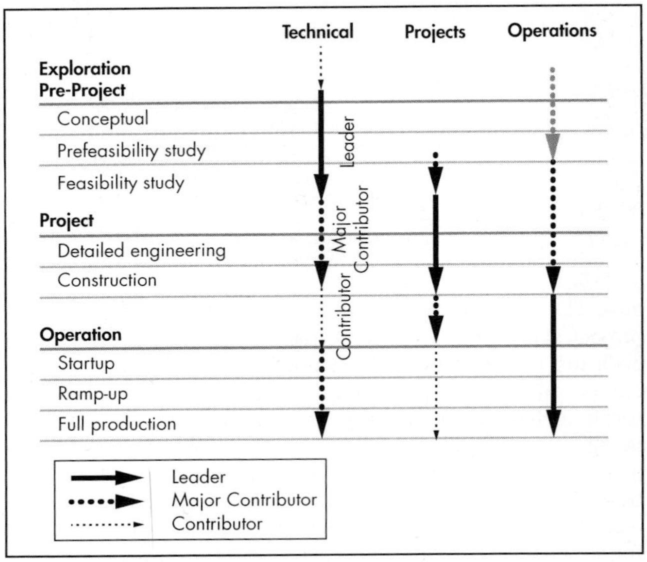

Source: Schnell 2013

Figure 3 Project leadership roles

Table 4 Characteristics of engineering contract types

Type	Advantages	Disadvantages	Owner Influence	Comments
• Fixed price or lump-sum engineering	• Known capital cost • Maximum third-party liability	• Must have detailed scope • Change order dependent • Profit and risk costs added to project	• Minimal	• Good for established and well-known facility
• Engineering and procurement (EP) lump sum	• Initial cost estimated in proposal • Requires smaller owner team	• Needs good scope • Interface to construction difficult • Change order dependent • Construction errors are common	• Little influence by project team or its organization • Owner monitoring from outside only	• Profit built into lump sum • Engineering company strives to reduce its costs (to increase profit) by using vendor engineering, reduced details, and increased "field fit" construction • Difficult for modifications to existing plants • Generally, construction depends on a single general contractor
• Engineering, procurement and construction management (EPCM) lump sum	• Same as EP, except construction now integrated with engineering • Control only with EPCM costs	• Same as EP	• Same as EP	• Construction contracting is more flexible
• EPCM cost plus	• Maximum integration among owner, engineering, and construction • Availability of all systems and standards of engineering company • Available experts from within other parts of engineering company	• Requires owner control of EPCM scope and budget • Requires experienced owner team for project control • Target budget not available at beginning of engineering	• Owner control of project team • Maximum influence	• Profit is built into the hourly rates • Possible to integrate owner team into engineering team • This can be considered similar to in-house but under a third-party umbrella

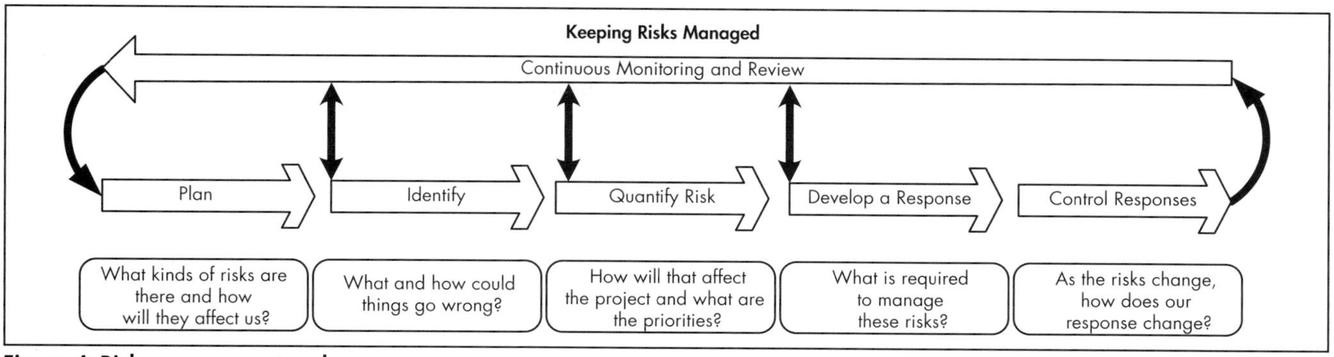

Figure 4 Risk management cycle

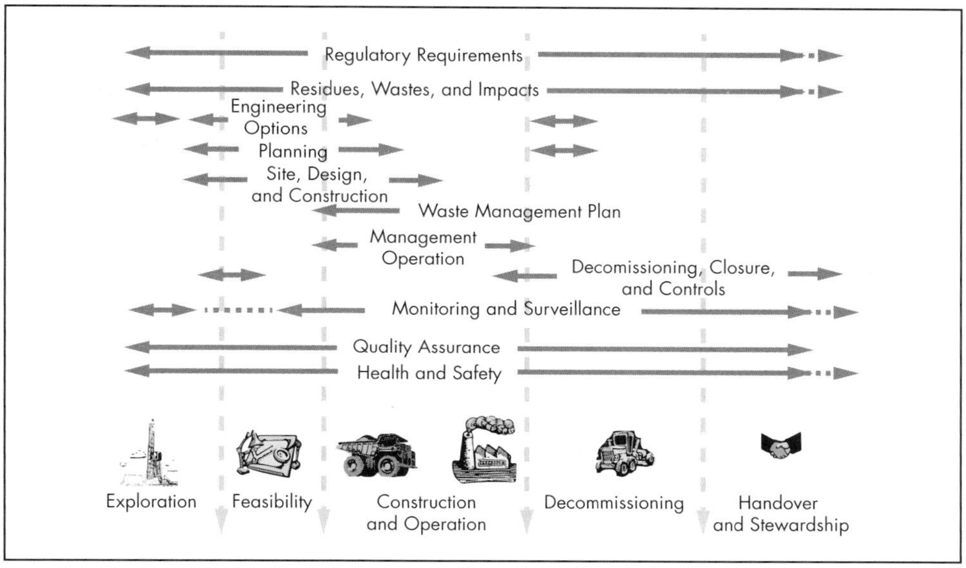

Source: Schnell 2010

Figure 5 Project development challenges

collected early, allowing for early mineralogical work aimed at metallurgy. Metallurgical testing covers many aspects, including the following:

- Grindability and work index
- Flotation and concentrate/tailings settling testing
- Acid and alkaline leaching
- Agglomeration and percolation
- Agitation leaching—batch and pilot
- Pressure leaching
- Heap leaching (bench and engineering)
- Recovery from solution using solvent extraction and/or ion exchange
- Precipitation/reduction
- Solution purification
- Resin in pulp—batch and pilot
- Resin in solution—batch and pilot
- Resin–ion exchange—batch and pilot

Most test work generally uses composite-type sampling, but variability testing from different parts of a deposit and different ore types is critical to avoid any future surprises.

Change Management

Changes during a project cycle are inevitable and require control. The completeness of prior trade-off studies and engineering will help reduce the magnitude and effect of project changes. These changes need to be effectively managed on both the technical side and the people side. A focus on the technical side ensures that the change is developed, designed, and delivered effectively. The discipline of project management provides the structure, processes, and tools to make this happen. A focus on the people side ensures that the change is embraced, adopted, and used by the employees who must do their jobs differently as a result of the change. The discipline of change management provides the structure, processes, and tools to make this happen.

Everyone working on a project needs to ensure that changes are communicated and subsequently controlled so that the project does not derail. Every project needs a formal change management system that evaluates the effect of a change on all areas—cost, schedule, safety, environmental permits, team morale, and so on. Project success depends on a fully developed project scope, and the investors' and stakeholders' needs must be met within the approved business case.

Risk

Mining has many risks that are continually evolving. Productivity, capital appropriation, and obtaining a social license to operate are now the top three risks faced by mining and metals companies globally (Elliott 2015). Currently, the industry faces low commodity prices, making productivity the most important issue. This will again change as the price cycle improves, but obtaining a social license to operate and environmental issues will continue to be long-term priorities. Assessment and control of risks are part of a successful project with a need to identify the risks, develop and quantify a plan, and develop response scenarios followed by control of the project. This risk cycle must be continually reviewed and modified to keep the project on track. A typical risk cycle is shown in Figure 4.

CONCLUSION

Mining projects and project management have many aspects that are not covered in this short chapter, but there are many capable consultants and professionals available to carry out the details required for each stage of a project. Mining projects are complicated and need to be systematically developed to improve the current trend of cost overruns, scheduling issues, and overall failure to deliver the business case. There are many parallel factors in a project, as shown in Figure 5. A project from initial discovery to initial production varies with the commodity, but it can take many years for a deposit to come into successful production.

REFERENCES

Bullock, R.L. 2011. Accuracy of feasibility study evaluations would improve accountability. *Min. Eng.* 63(4):78–85.

Elliott, M. 2015. *Business Risks Facing Mining and Metals*. London: Ernst and Young.

JORC (Joint Ore Reserves Committee). 2012. *The Australasian Code for Reporting of Exploration Results, Mineral Resources and Ore Reserves (The JORC Code).* Gosford, New South Wales: JORC.

McCarthy, B. 2015. Obstacles and opportunities for effective life-of-mine (closure). *AusIMM Bull.* (February):48–51.

National Instrument 43-101. 2011. *Standards of Disclosure for Mineral Projects*. Toronto, ON: Ontario Securities Commission.

Schnell, H. 2010. Supporting sustainable uranium mining in less prepared areas. In *UMREG, Techno-Economic Feasibility Studies and Testing* (on-line study guide). Project 2010-PUI-NE-03-NFCMS. Vienna, Austria: International Atomic Energy Agency.

Schnell, H. 2013. Les relations entre Etats et entreprises minières. IAEA Workshop (PowerPoint presentation). Morocco, North Africa, November.

CHAPTER 2.8

Research and Development

Nick Hazen and Joanna Robertson

Human prosperity has advanced throughout history largely by the implementation of new technologies made possible by new ideas and new knowledge. Research is a systematic search for new ideas and new knowledge and is often classified as either fundamental or applied. Fundamental research, although it may yield game-changing concepts, is generally thought of as driven purely by a quest for knowledge or understanding rather than practical application. The driver of applied research, however, is to develop solutions to specific industrial problems or challenges. Development generally concerns itself with producing information for implementation, that is, scale-up, of a technology from bench scale to an economically feasible production operation. The term *research and development* (R&D) is used here to mean applied research and development, or industrial research, and includes what is often called process development.

Business managers invest in research because they believe this effort will ultimately provide a return on the investment. The amount invested in research as a percentage of sales, often called the R&D intensity, varies considerably from one industry to another. The R&D intensity of the minerals industry is significantly lower than that of many other industries (Hirschey et al. 2012). This limited investment implies a relative lack of confidence in the value of research in the mineral industry compared with other industries. Much has been said, however, about failures of major mining investments over the last two decades (Kuestermeyer 2016). A study by T.P. McNulty (1998) indicated that some of the failures stemmed, at least in part, from insufficient quantity and/or quality of the R&D work.

Research is one of the services used by corporate management to achieve corporate goals. It is critical to recognize, therefore, that corporate management must not only set meaningful and realistic research goals, but that these goals must be transmitted and understood by operating management and those individuals performing the research.

The value of research, in terms of future profit, is usually related more to the quality or relevance of the research rather than to the monetary size of the investment. Poorly directed or inexpertly executed research can rapidly waste resources and corporate patience. Therefore, it is critical that the function of R&D is clearly understood by management, and that high-caliber thought, judgment, and expertise be brought to bear on the correct problems to deliver a profitable result. This chapter discusses the function, the guidance, and the execution of R&D in the minerals industry.

THE FUNCTION OF R&D IN THE MINERALS INDUSTRY

Mining and minerals companies apply technology to convert minerals in the ground into salable products. When a conversion technology is successfully demonstrated at commercial scale with a specific feedstock, it can often be used for similar mineral occurrences at other locations. Typically, at least slight modifications are required to fit the technology to a particular prospective deposit. Sometimes major modifications to the technology are needed, and in some cases a totally new and innovative technology must be developed. Therefore, the extent of experimental and process engineering work required covers a wide range, from simply accessing the results achieved by applying the technology to samples from a new prospective deposit up to and including developing advanced processes. Supplying the technical knowledge required over this entire range is one of the functions of R&D in the minerals industry.

It also is worth noting that much more than technical knowledge is required to fit technologies to deposits. Evolving environmental requirements (including greenhouse gas emissions), market demands, sustainability considerations, and new social or governmental constraints must also be considered and may be overriding. Variations in the costs of operating supplies often have a great impact on production costs and profitability. Such criteria and constraints should be considered during the R&D phase of a project so that constraints are addressed and projected operating costs are optimized. Uppermost in determining the technical fit of a process and modifying the process to fit a deposit is a clear understanding of the mineralogy and its variability in the deposit.

Nick Hazen, President, Hazen Research Inc., Golden, Colorado, USA
Joanna Robertson, Director of Metal Recovery, Freeport-McMoRan Technology Center, Tucson, Arizona, USA

Table 1 Comparison of R&D spending by industry

Industry	R&D Spending as a Percentage of Sales		
	2008–2010	2000–2002	1976–2010
Metal mining	0.0	0.2	0.8
Coal mining	0.0	0.0	0.2
Oil and gas extraction	0.2	0.4	0.6
Mining nonmetallic minerals	2.4	0.4	1.2
Primary metals	0.8	0.6	0.7
Industrial machinery and computer equipment	3.7	4.7	4.3
Electronic equipment	10.1	7.4	5.3
Transportation equipment	4.0	3.7	3.2

Source: Hirschey et al. 2012

Mineral industry R&D serves to adapt operating plants to foreseen or unforeseen changes. Mineralogy changes are common as new ore zones are encountered. Operating plants may need to alter their processes to meet safety requirements or environmental criteria, or to achieve cost and production targets. R&D teams may also be called upon to resolve issues with efficiency or bottlenecking. Generally, changes at operating plants only occur out of necessity, such as in response to a threat. There is always risk in implementing a new idea; therefore, it is "almost axiomatic that no new idea will be put into practice until the promised results ... are seen as essential to the health of the enterprise" (Simons 1983).

Still another function of R&D in the mineral industry is to apply new technical advances or innovations that are discovered or developed, often in unrelated fields such as the oil and gas industries. Examples of these advances are the application of computers, the global positioning system (GPS), robotics, autonomous vehicles, drones, X-ray tomography, automated quantitative scanning electron microscopy and energy dispersive X-ray spectra, and others. It seems probable that the Internet of Things will be another outside advance for the mining industry over the next decade or so (Lasiker 2017). Often, advances outside the industry are brought to the mineral industry by technology vendors seeking market expansion.

Although most R&D is directed at innovation, the preponderance of advancements deal with improving, modifying, and sustaining current technologies (incremental improvements). Disruptive innovations that cause industry-wide changes in technology (froth flotation, solvent extraction, heap leaching, autoclaving, flash furnaces, submerged lance smelting, open pit mining, semiautogenous grinding mills, etc.) are usually harder to achieve as part of a systematic research effort. Although there are examples to the contrary from other industries, including Bell Labs from the 1920s to the 1980s (Gertner 2013), and present-day Apple and Google innovations, disruptive innovations in the minerals industry primarily arise from a need to solve specific problems related to specific deposits or have originated outside the minerals industry, such as computer modeling or GPS, and are usually unrelated to the size of the research budget.

INNOVATION AND R&D BUDGETS

Innovation is vital to the minerals industry, both to realize near-term profitability and to maintain long-term asset value. Resources such as ore reserves and processing plants represent long-term investments; hence, the value of these assets must be maintained for decades to ensure continued profitability. Innovation can enhance the value of such assets by increasing throughput or lowering processing costs, and, potentially, increasing ore or mineral reserves. However, the decision to employ innovative technologies can add to project risk, particularly when such technologies depend on complex principles that may be familiar to R&D personnel but are not fully understood by those who are charged with operating and maintaining the plant and its processes.

As ore grades decrease in operating mines and financing for new operations becomes more difficult to obtain, it is vital that companies innovate mindfully to minimize risk and to sustain the value of mines and processing plants throughout the life of the resource. Innovation can improve the health and safety of employees, allow companies to comply with new regulations, and maintain operations in the face of water and energy shortages. Thus, R&D not only plays a vital role in the profitability of a mining project, it often makes the difference between success and failure.

Although innovation is vital, the minerals industry spends much less on R&D than other industries. Additionally, the amount of spending as a percentage of sales has decreased more for mining-related industries than for other industries over the last three decades, as illustrated in Table 1.

The data show a disparity between R&D spending for mining and that for other manufacturing industries whose drivers are linked to product life cycle and end use. Manufactured products become obsolete and designs are continually renewed to provide consumers with new, state-of-the-art products. New models of vehicles and cellular phones, for example, are issued every year or so. Some of this continual design renewal is required to ensure that new safety and operational features are incorporated, whereas other, more cosmetic, design changes are driven by customer preferences. In contrast with manufactured products, basic metals and minerals do not become obsolete because they are the building blocks of other products. The basic forms of minerals and metals provided to the market do not change from year to year. Although copper concentrate sold in 2017 is essentially identical to that sold in 1980, items such as vehicles and computers have undergone vast transformations. Hence, R&D spending for product manufacture is consistently higher than for minerals and metals production.

Another reason that yearly R&D spending for the mining industry is lower than that spent on manufactured products is related to the capital intensity and useful service life of minerals processing facilities. Mining is capital intensive, and the capital expenditure per annual value output is high.

For example, billions of dollars are spent on new copper milling facilities, but the investment must be good for many years given that the facility is designed to last for the economically foreseeable life of the ore reserve. Capital costs have risen dramatically in recent years because of two main factors: (1) larger and more complex equipment to effectively accommodate declining ore grades and (2) new infrastructure required for mines in very remote locations (Mills 2012). In the year 2000, the average capital investment needed for a new copper mine was between US$4,000 and $5,000 per metric ton of copper produced. By 2012 this cost had risen to more than US$10,000 per metric ton (Mills 2012).

The capital portion of recent mining projects ranges from US$3.5 billion for the Esperanza Sur (Chile) copper/gold expansion (Mills 2012) to US$5.95 billion for the Cobre Panama copper/gold project (*Canadian Mining Journal* Editor 2015). The large capital expenditures required to bring new metal mining projects to market demand that processing and extraction technologies are robust and well understood. The inclusion of newer technologies or processing schemes into new operations may present an unacceptable risk. A good example is the high-pressure acid leach approach for extracting nickel and cobalt from lateritic ores; much effort and investment has been put into making the process work. There have been several commercial plant failures, but the process has succeeded in certain circumstances. Process development faces not only the challenge of developing new and cost-effective methods of mineral extraction and recovery, but also the challenge of reducing operational risk to an acceptable level. In the face of these challenges, many companies have all but abandoned R&D efforts and have chosen to use tried-and-true methods of mineral extraction for new projects. R&D dollars are often spent more on the optimization of existing processes rather than the development of new ones.

Given these challenges, it is not surprising that in the metals mining industry, R&D spending as a percentage of sales has decreased drastically in recent times. In the four years between 1976 and 1980, R&D spending was 1.6% of sales, whereas during the four-year period from 2006 to 2010, R&D spending had diminished to less than 0.1% of sales (Hirschey et al. 2012). Commodity price slumps occur periodically; hence, the decrease in R&D funding previously noted was at least partly caused by the 2008 financial pullback during which metals prices fell, making many large projects uneconomic. As a result of this downturn, many companies cut exploration funding. Without the driving forces of new projects and new exploration targets, the R&D focus narrowed and funding for pilot plants and large development projects became rare.

This recent narrowing of R&D focus led to a shift in allocation of resources to favor near-term, low-risk projects and the optimization of existing operations. Solving near-term problems and addressing pressing needs is the major consumer of R&D dollars. Such work is often related to problems arising within a specific deposit negatively affecting process recovery. An unforeseen change in ore hardness or mineralogy, for example, can cause bottlenecks in crushing and grinding circuits and deficits in recovery. Such issues, if not foreseen and planned for well in advance, may represent production emergencies. This is especially true when production shortfalls are large enough to lower critical cash flow. Companies may augment in-house research with consultants and outside laboratory work to obtain solutions in as short a time frame as possible.

Projects whose time frame is between two and five years include optimization of technologies for brownfield projects or laboratory experimental work to provide updated criteria for expansion projects. Such R&D may be deemed economically necessary to provide older plants with more up-to-date technologies, such as advanced instrumentation and process control. Processing plants and facilities typically have long lives that often exceed those projected at start-up. It is not uncommon for processing facilities built in the 1940s and 1950s to still be in operation. To keep such facilities current with industry standards, R&D activities may focus on such issues as process and equipment constraints, safety improvements, energy efficiency, water conservation, or compliance with new regulations. R&D for an existing, or brownfield, operation brings specific challenges and constraints not present when work is focused on an exploration target or greenfield opportunity. The degrees of freedom in design may be limited by existing equipment and space constraints, the availability of power and water, as well as by new regulations that were not in force when the plant was constructed, thus requiring R&D to find a solution.

A few large, greenfield projects may clear initial economic hurdles so that R&D spending is justified. This R&D is directed at developing optimized flow sheets for exploration prospects and design criteria for projects that are between 5 and 20 years away from commissioning. Such R&D activity often takes place during the long permitting cycle required for new mining projects. Because of the extended time required before project commissioning, researchers must consider the use of breakthrough technologies that could change the way a company does business. Technologies such as driverless haul trucks, for example, may not be in wide use during the early R&D stage of the project, but may be an inevitable industry trend. Thus, the use of or transition to such technologies must be considered in the initial design phase. Technology is changing rapidly, driven by such forces as robotics and automation as well as by the development of materials with improved properties that allow the construction of much larger-scale equipment than has been previously employed. Inclusion of the newest technologies in long-term design plans affects both capital and operating cost and thus may influence whether a particular project meets the required financial hurdles to receive funding.

DRIVING FORCES FOR R&D

Mining and minerals companies have many reasons for undertaking R&D efforts. First, they must create value out of minerals in the ground. As ore grades decline or ores become more mineralogically complex, innovative technologies may be required. For example, a copper ore containing significant levels of arsenic or mercury may be unsuitable for copper smelting and would require a different processing approach, necessitating flow-sheet development through an R&D program. Companies may also undertake R&D efforts to keep from going out of business or to expand revenues. For example, ore deposits often change throughout the life of a mining operation and a technology suitable for processing the original ore may be unsuitable for processing ores having different mineralogies. In some cases, trace metals such as molybdenum or rhenium become economically valuable and by-product extraction becomes economically attractive. Often, profitability is a driver for an R&D program, requiring development of technologies that allow for increased extraction or decreased processing costs.

A key goal of the minerals industry is to convert mineral deposits to economic ore reserves. A mineral deposit may be reclassified as an ore reserve, and vice versa, for several reasons including changes in metals prices and projected processing costs, updated information on ore grades, energy and water costs, or regulatory compliance costs. R&D activities are often focused on decreasing processing unit costs by increasing productivity and optimizing resource utilization. For example, despite the decreasing copper ore grades in the United States, the volume of copper generated has increased while the production cost has dropped via implementation of new technologies. In 1970, mine output was 1.56 million t of contained copper. By 1995, mine output had risen to 1.89 million t. In 1970, copper production costs were US$1.04 per pound, and by 1995 production costs dropped to US$0.61 per pound (Tilton and Landsberg 1997). Although it may be assumed that the quality of the deposits had increased, the opposite is true. On average, the head grade for copper mines declined from 0.73% in 1971 to 0.60% in 1993 (Tilton and Landsberg 1997). The implementation of new technologies such as solvent extraction and electrowinning (SX-EW), which were developed through R&D efforts in the 1960s and 1970s, increased operational productivity and thus led to increases in output and decreases in processing costs.

Compliance with changing laws and regulations is yet another driving force for R&D spending. For example, enactment of the U.S. Clean Air Act in the 1970s required copper smelters to reduce sulfur dioxide emissions (Ober 2002). This placed an economic burden on the U.S. copper smelting industry and led to the eventual closure of several smelters. Those smelters that remained in operation modified equipment and operational practices to more effectively capture sulfur dioxide–laden gas and added acid plants to convert the gas to sulfuric acid (Ober 2002). The sulfuric acid produced by the copper smelters was not a value-added product, as there was little to no market for it. Some mines began using the acid for leaching copper from oxide ores that had been stockpiled because they were unsuitable for beneficiation by froth flotation. Such leaching operations were, at the time, of more use for disposing of excess acid than for producing copper. The copper that was solubilized by the acid leaching was recovered by cementation onto iron scrap metal, which was then fed to a copper smelting operation (White 1999). The market for the low-cost acid improved when heap leaching and SX-EW were developed in the 1960s and were found to produce high-quality cathode suitable for feeding directly to rod mills. Thus, compliance with environmental regulations altered the economics of a related processing operation and helped to spread the heap leach SX-EW process in the 1970s. R&D efforts played a key role in bringing about these changes, not only in the design and development of acid plant technology and the advantageous use of the resulting low-cost acid, but in the development of the SX-EW technology employed to extract copper from heap leach solutions. The process development for copper extraction via SX-EW progressed from laboratory work at Hazen Research in May 1965 to the successful start-up of the Ranchers Bluebird (Arizona, United States) plant in March of 1968. The spread of SX-EW was ensured by the timely development of an especially effective copper extraction reagent, LIX 63, by General Mills, which was piloted side-by-side with another reagent at Ranchers Bluebird (Hazen 2014).

Finally, R&D projects are often undertaken to solve current operating problems. Such work is often the reason that large companies maintain in-house R&D departments staffed with professionals who are very familiar with company-owned plants and facilities. Such groups, at least in theory, can solve problems more quickly than outside consultants who must take time to become familiar with the operation before attempting to solve the problem. A common example of an operational issue requiring quick research is an unforeseen mineralogy change that negatively affects metal recovery and revenue. Often, the first indicator of a problem is the unexplained inability to hit production targets. Research efforts will quickly attempt to identify the problem while development teams will seek to find strategies to mitigate the issue and restore production. For example, problems with a copper heap leach operation may be due to the occurrence of copper in slow-leaching clay minerals. Researchers will work to identify the clays while development teams will conduct experimental work to determine whether changing production parameters, such as acid strategy or leach cycle time, will alleviate the problem.

PROJECT SELECTION

Research is effective only if it is focused on problems whose solutions will accomplish the corporate objectives. These objectives, and therefore the research goals, must be defined and clarified by top management and transmitted to the researchers. Clear definition of what is to be accomplished is necessary and can be a strong motivating force for a research team. Dictating how to accomplish the goals, however, can have the opposite effect. Projects selected to receive R&D effort and funding can be divided into three broad categories.

1. High-priority, short-term projects that minimize risk to company existence or profits, such as encountering new ore mineralogy, safety, or production bottlenecks.
2. Growth projects such as new geological acquisitions, new processes, or the production of new products. These are often medium-term two- to five-year projects aimed at reducing the risk of introducing changes.
3. Projects that can increase profits of present operations, such as new technology. These projects typically have a longer-term horizon of 5 to 20 years and are often seeking a completely new, sometimes disruptively new, solution to a problem.

High-Priority, Short-Term Projects

High-priority short-term projects usually require immediate attention to resolve a production or safety challenge. Short-term R&D projects are often illustrative of the saying that "necessity is the mother of invention," in that a need is determined and a solution is developed to meet that need. Although solutions developed very quickly may not provide the most elegant outcomes and may suffer from unrelated drawbacks, they either solve the problem that led to their development or provide some breathing room until a longer-term solution can be developed. An example of this scenario is the unexpected release of toxic gas in the workplace that requires R&D to find the source and mitigate the problem. Another example is the unexpected lowering of underflow density in a countercurrent decantation washing circuit that results in high soluble losses in a hydrometallurgical plant.

Medium-Term Projects

Many medium-term projects focus on applying an existing or modified technology to a new deposit. These projects are conducted to reduce the risk inherent in introducing a change in technology or bringing a new ore body into production. Technology changes should be evaluated, experimentally or otherwise, for metallurgical performance, operability, robustness, and safety before being employed in operations. Examples of medium-term projects include using different equipment, such as a filter, or ideas currently used in other fields, such as communications or control strategies. Such projects generally begin with awareness of a change or improvement that could be made to a plant or an operation. Sources of ideas may include presentations from vendors or concepts gleaned from benchmarking, either within a given industry or in a field of seemingly unrelated endeavor. For example, heap leaching operations almost exclusively used sprinkler-type equipment to introduce leach solutions to stockpiles. Now, many operations use drip emitter technology originally developed for the agricultural industry. This technology, which reduces solution losses to evaporation, is used in locations where saving water is important. The technology was investigated to show that using drip emitters instead of sprinklers would not adversely affect copper recovery or operational costs (Uhrie 1998). When preliminary demonstrations were complete and risks were sufficiently addressed, using drip emitters became common. During periods of strong investment in mining, much of the work of the R&D group may be developing the experimental evidence to support the various stages leading to feasibility studies for expansion projects. These are typically medium-term projects with strong corporate drive.

Long-Term Projects

Long-term projects seek to address problems for which a solution is not known. For example, a project may be undertaken to develop an economic process for a specific exploration target or to develop novel technology for an existing potential resource. Technology development for exploration targets is complicated by the interrelationship of specific process economics and project viability. For example, a targeted mineral deposit may have the advantages of high grade and an accessible location but may contain environmentally or metallurgically challenging trace elements. The flow sheet that is developed to process the ore and the desired plant capacity will have estimated capital and operating costs. These costs will determine whether the project is economic. Thus, an iterative approach is necessary to prevent the complete development of a processing flow sheet that is too complex or expensive to employ. At times, it may be helpful to know project cost constraints at the beginning of R&D efforts. For example, the R&D team may be asked to keep processing costs below a cost-per-weight value of the primary metal to be processed. Such constraints serve to direct research toward robust and simple processes that can be started up quickly and provide expeditious payback.

Long-term R&D projects may also be undertaken to develop innovative technology for existing resources. Such projects may be complicated by having no specific deadlines because the solution being sought is not known. One example is the heap leaching of low-grade chalcopyrite ores. High-grade chalcopyrite is most economically processed by froth flotation and smelting, but when copper grades are below the economic cutoff for a specific mining operation, the ore is often treated by heap leaching. Leach recoveries from chalcopyrite ores are low; usually, only about 10% of the copper is recovered, even when leach cycles are extended. An economic solution is needed because as copper mines worldwide extend their operations into deeper primary sulfide deposits, increasing amounts of below-cutoff-grade chalcopyrite will be available for heap leaching. R&D efforts for such a problem may focus on basic research as well as analyzing solutions proposed by vendors. Research efforts often cast a wide net in search of potential solutions and may suffer from a lack of focus if research objectives are not carefully laid out.

Abandonment of Projects

As important as deciding which R&D projects to undertake is the decision regarding which projects not to begin at all. Several factors may be taken into account when making the decision to abandon an idea before the R&D stage. It may be that the preliminary economics are unfavorable or the technical hurdles are deemed insurmountable, or both. Often, the process that would be required to obtain a valued product is simply too complicated and the technical risk of operating such a process would be too high. At the beginning of a project, before much work has been done, some of these decisions are simply judgment calls. Information from batch laboratory work, preliminary flow sheets and mass balances, and economic estimates can help R&D personnel discard unwieldy processes without missing out on opportunities.

Deselecting projects should be a regular continuing task of the selecting group, which should include top corporate management. Without this regular pruning, projects tend to take on a life of their own, particularly if perpetuated by a corporate champion. Projects originally thought to be driving toward accomplishing corporate goals may be outdated or have veered off course, or goals may have changed.

MANAGING AND CONDUCTING R&D

The most important factor in providing high-quality useful results on time is ensuring that the technical team is working on the proper problems (Hazen 1965). The critical function of top management in setting the goals for R&D cannot be overemphasized. Clear goals that are clearly transmitted are the only measure against which the productivity of R&D can be measured. The goals cannot be so broad that they are meaningless ("create something that makes money") or so strictly defined that valuable ideas are ignored. Guidance of R&D is, in this way, a rather delicate balance.

Much has been written about how to provide a creative atmosphere for industrial research. The kind of creativity that is required for technically productive research is 99% problem solving. Motivation to solve technical problems is the key ingredient required. Research workers want to feel that their efforts are meaningful to their employers, they want the inner satisfaction of accomplishment, they want to be recognized by their peers, and they want money (Hazen 1965). These desires are much the same for many occupations. The key skill required of research management is common sense. The atmosphere will happen by itself if the challenges are real and the goals are clear.

Historical Development

Although there is a tendency to take for granted that technologies begin in the laboratory and progress through piloting and demonstration before implementation, it was once common for projects to simply begin with an engineered demonstration phase. One example of this approach is the use of carbon for gold recovery. The first industrial carbon-in-pulp plant, San Andres de Copan in Honduras, had a capacity of 250 t/d. In 1950, the Getchell mine in Nevada (United States) operated a 500-t/d carbon-in-pulp plant (Marsden and House 2006). Another example of rapid advancement to a demonstration phase is uranium solvent extraction. After laboratory work conducted at Oak Ridge National Laboratory in Tennessee (United States) and brief verification in a makeshift laboratory, many thousands of metric tons of ore were treated at the Kerr-McGee plant in Shiprock, New Mexico (United States), beginning in July 1955 in a batch demonstration plant built from mothballed tanks and equipment (Hazen and Henrickson 1957). In addition to providing important process learnings, this experimental procedure added significantly to plant uranium production while a continuous circuit was constructed.

As the capital requirements for plants increased over the years, minerals companies needed a way to reduce the risk that a large operation would fail. Because of this situation, closed-cycle laboratory work and the extensive piloting of processes became much more common, with some R&D efforts costing tens to hundreds of millions of dollars. For example, during the development of its successful copper pressure leaching technology, in addition to piloting, Phelps Dodge spent US$40 million on a demonstration plant in Bagdad (Arizona, United States). These development dollars were well spent. The operational lessons learned from this demonstration plant led to the smooth start-up and effective implementation of copper concentrate leach technology in a full-scale plant in Morenci, Arizona (Marsden et al. 2003). The implementation of concentrate leach technology has provided operational synergies and significant cost savings for every metric ton processed by elimination of shipping, smelting, and refining charges. There are other examples in which reducing the risk of failure can mean abandoning an unworkable technology during the development phase.

Process Development

A major R&D activity in the mineral industries is process development. Process development consists of all the activities required to specify how a mineral deposit can be converted into a salable product. It is undertaken to reduce the risk of technical or economic failure when a technology is scaled up to commercial operation. The literature offers good discussions of the sequence of tasks required to conduct successful process development in the mineral industries (O'Callaghan 2001; Smith 2005).

Process development is a critical part of project development. Project development proceeds through a path from securing legal rights to a mineral property through technology review, desktop studies, scoping studies, securing environmental and social permissions, prefeasibility studies, definitive engineering, construction, and commissioning. Generally, more and more experimental evidence is required as the steps progress from scoping studies to definitive engineering. Process development consists of providing evidence-based information for designing the process and establishing the design criteria.

The evidence is typically collected initially from thorough mineralogical characterization and scoping level batch laboratory work. Then, more extensive laboratory work, including continuous benchtop operations for parts of the process, variability studies for all major ore zones, continuous piloting, and, if needed, demonstration scale work are conducted. To keep the work on track, timely mineralogical work must be performed on materials derived from experiments to correctly interpret the results, and the flow sheet and mass balance must be updated frequently as new experimental results become available. Colocation of the laboratory, analytical, mineralogical, process engineering, and owner–representative personnel during critical periods in development make keeping the project on track much easier to accomplish. And setting up a pilot plant adjacent to an existing operating plant provides many synergies, including access to feedstock, energy, steam, and facilities for handling waste streams.

Each ore body is unique, and each offers a unique challenge to develop an effective metallurgical route that brings success to the venture. The variety of process choices for mineral deposits is so wide that it is impossible to accurately list all the steps necessary to reject weaker solutions and develop the best course. The following is a list of matters to consider when faced with a specific process development challenge. Not all items fit every process.

1. Begin at the end. Identify the final product of the process in question. Is it a metal, an oxide or other salt, or an intermediate product to be converted by others either internally or externally? What purity is required? What impurities will be penalized and at what concentration level? Knowing what you are bringing to market establishes the first process constraint and begins to define the process.
2. Thoroughly characterize the mineralogy of each ore-body zone. In what phases and how do the values occur? In what phases do the impurities occur? What are the gangue components, alteration products, and lithology? What does the mineralogy predict about liberation size? Some of these questions can now be answered quantitatively using Quantitative Evaluation of Minerals by Scanning Electron Microscopy (QEMSCAN), mineral liberation analysis (MLA), or Tescan integrated minerals analyzer (TIMA).
3. Relevant literature should be reviewed and summarized so that relevant information may be taken into account when designing the process and planning the experimental program. Some have said that "three months in the laboratory can save you two weeks in the library." Literature searches should include journal articles, conference proceedings, and patent searches.
4. Draw a simplified flow sheet, prepare an abbreviated mass balance, and estimate order-of-magnitude capital and operating costs. Start with the simplest operable process possible and target the first salable product for which there is a reliable market for the quantity you plan to make. If more than one process appears to fit, draw flow sheets, mass balances, and economic estimates for all. Do not plan for multiple products at first. If you believe, based on initial mineralogical characterization, that a mineral processing concentrate can be made, assume you can make one in your preliminary economic estimates. If costs are unacceptable, determine what needs to change

to bring the costs in line, and let that guide the next step in development. Remember, there are ore deposits that are not profitable at prevailing prices. Deselecting a process or project can be the most economic option.

5. Use the mineralogical information to guide physical beneficiation experiments. Make sure you have adequate expertise brought to bear on the potential for physical beneficiation. Exhaust the options for physical separation before accepting the result that you must treat the whole ore. A difficult physical separation is almost always cheaper than a chemical separation. It is possible that making and selling a concentrate is the best economic answer.
6. Address the key reaction and type of process reactor. The key step may be a leach or a furnace or a roast-leach. This step needs to be verified experimentally, the variables explored, and the chemistry validated even if it is thought that the chemistry is well understood. The response of samples from different parts of the ore body should be determined and results understood. Resulting laboratory products need to be examined mineralogically to establish a complete understanding of the results.
7. When the main reaction is understood, move through several layers of process steps in the laboratory using the onion analogy (Smith 2005). The next steps are as follows:
   a. Separation and recycle systems
   b. Energy requirements and heat recovery systems
   c. Heating and cooling utilities
   d. Water and effluent treatment systems
8. Periodically throughout the work, update in increasing detail the flow sheet, mass balance, and economics in response to process findings, and modify the metallurgical work program in response to flow-sheet and economic findings. The program can also change because of newly imposed environmental or social requirements or even because the form of the product changes as a result of market considerations.
9. Call in outside expertise early and often to keep the project from going astray. If the technical team is heading down a wrong path, getting the right practical experience and technical expertise in at the right time can save months of effort.
10. For a greenfield project or a significant brownfield project, the effect of the project costs on mineral reserves must be calculated. When the process costs have been estimated, mine models such as MineSight may be used to estimate the metric tons of ore for any deposit that may be economically mined and processed. If the process is too expensive, the economic tons will be too few to bear the capital cost of the plant and create a viable profit.
11. Throughout the stages of the project, it is vital that clear communication between all technical teams be a priority. In his paper about the Moa Bay project in Cuba, Stu Simons (1973) describes the level of communication necessary to keep a project running smoothly and to obtain the key objective of a smooth start-up. He recommends having a key person who was involved in the laboratory work involved in the pilot plant and having laboratory and pilot plant personnel involved with engineering, and so on. As the project team grows, it never loses the original focus of the researchers who designed the process.
12. The process design should be as simple as possible. Complex flow sheets lead to high capital cost projects, are difficult and unwieldy to commission, and may not be economic to operate. If possible, a plant should be started with the minimum number of unit operations it takes to make a salable product. Key characteristics of financially successful processing plants are that they are operable and start up smoothly. Additional stages may be added to the commercial plant later to make more refined products or by-products if the economics justify the addition of the equipment and labor.
13. The line between research and development is not well defined. Here we define *research* as the work conducted to arrive at a process design, design criteria, a mass and energy balance, capital and operating cost estimates, and a profitability analysis, all based on laboratory work done at bench or continuous bench scale. When this stage has been completed, a decision must be made as to the next steps, called "development."
    a. Development is characterized by scale-up activities, such as larger-scale experimental work, to provide confidence for full-scale design. If the process is relatively simple and if it can be implemented at a reasonable cost, the decision may be made to build a demonstration plant, or an expandable plant. This option has the advantage of creating an opportunity to learn about the process while producing a salable product.
    b. If, however, the process is more complex and the risks are higher, a pilot plant may be constructed and operated. Data generated from the pilot plant may be used in feasibility studies. If desired, a demonstration plant can be constructed and operated to minimize the risk of scaling up directly from a pilot plant to a full-scale plant.

CONCLUSION

The purpose of R&D in the minerals industry is to improve existing processes or create new ones. Such improvements may make mining operations more profitable, safer, or more environmentally sustainable (or all of these). Because of the nature of the commodities produced by the mineral industry, R&D dollars do not represent as large a proportion of sales as for manufactured products, but, wisely spent, these funds can provide many benefits. Given the high capital cost of operating plants, the risk of using new technology is large. T.P. McNulty has written a series of papers (1989, 2002, 2004) providing data on degrees of ramp-up success and failure for several metallurgical processing plants. He attributed generalized reasons for the successes and failures. Among other valuable insights, he points out the high cost of poorly executed or insufficient R&D work. Thus, despite the costs of R&D and the time needed to conduct the work correctly, the costs of neglecting it are much higher.

In addition to several well-publicized recent start-up failures, however, there are also several successful new mining ventures and new technology implementations. Although there are many risks inherent in creating a new venture or technology, the technical risk can be largely mitigated by proper execution of R&D. The list of information required from R&D to mitigate technical risks is long and detailed. To provide that information accurately, R&D must be governed by addressing the right problems, developing a thorough understanding of all materials to be processed and of each process step, and demonstrating the operability of each process step as part of the whole. The work is challenging, but for those so inclined, the joy of seeing a new operation start up smoothly makes it all worthwhile.

REFERENCES

Canadian Mining Journal Editor. 2015. COPPER: First Quantum cuts costs at Cobre Panama development. *Can. Min. J.* October 6. www.canadianminingjournal.com/news/copper-first-quantum-cuts-costs-at-cobre-panama-development/. Accessed March 2016.

Gertner, J. 2013. *The Idea Factory: Bell Labs and the Great Age of American Innovation.* New York: Penguin Books.

Hazen, N. 2014. Agile process development and early use of solvent extraction in the mining industry. In *Hydrometallurgy 2014*. Vol. 1. Westmount, QC: Canadian Institute of Mining, Metallurgy and Petroleum. pp. 5–16.

Hazen, W.C. 1965. Research for the mining industry. Presented at the 1965 Mining Show, American Mining Congress, Las Vegas, NV, October 11–14.

Hazen, W.C., and Henrickson, A.V. 1957. Solvent extraction of uranium at Shiprock, N.M. *Min. Eng.* 9:994–996.

Hirschey, M., Skiba, H., and Wintoki, M.B. 2012. The size, concentration, and evolution of corporate R&D spending in U.S. firms from 1976 to 2010: Evidence and implications. *J. Corp. Finance* 18(2012):496–518.

Kuestermeyer, A. 2016. Capital cost overruns in the mining industry—Mistakes, misjudgements, or ? Presented at Colorado School of Mines, February 4.

Lasiker, R. 2017. Modernizing the mining industry with the Internet of Things. In *Internet of Things and Data Analytics Handbook*, 1st ed. Edited by H. Geng. Hoboken, NJ: John Wiley and Sons.

Marsden, J.O., and House, C.I. 2006. *The Chemistry of Gold Extraction*. Littleton, CO: SME. p. 10.

Marsden, J.O., Brewer, R.E., and Hazen, N. 2003. Copper concentrate leaching developments by Phelps Dodge Corporation. In *Hydrometallurgy 2003*. Vol. 2. Edited by C.A. Young. Warrendale, PA: The Minerals, Metals & Materials Society. pp. 1428–1446.

McNulty, T.P. 1989. Research and development. *Mater. Soc.* 13(2):189–191.

McNulty, T.P. 1998. Developing innovative technology. *Min. Eng.* 50(10): 50–55.

McNulty, T.P. 2002. Overview of metallurgical testing procedures and flowsheet development. In *Mineral Processing Plant Design, Practice, and Control.* Edited by A.L. Mular, D.N. Halbe, and D.J. Barratt, Littleton, CO: SME. pp. 119–122..

McNulty, T.P. 2004. Minimization of delays in plant start-ups. Presented at the SME Plant Operators' Forum, February 25.

Mills, R. 2012. Global copper production under stress. *AheadoftheHerd.com*. http://aheadoftheherd.com/Newsletter/2012/Global-Copper-Production-Under-Stress.htm. Accessed March 2016.

Ober, J.A. 2002. *Materials Flow of Sulfur*. Open File Report 02-298. Reston, VA: U.S. Geological Survey. p. 33. https://pubs.usgs.gov/of/2002/of02-298/of02-298.pdf.

O'Callaghan, J. 2001. Minimising risk—The role of testwork in engineering design and project development. Presented at ALTA 2001 Nickel/Cobalt-7, Perth, Western Australia, Australia, May 15–18.

Simons, C.S. 1973. The interface between the research laboratory and a profitable metals processing plant. In *International Symposium on Hydrometallurgy*. Edited by D.J. Evans and R.S. Shoemaker. New York: AIME. pp. 753–769.

Simons, C.S. 1983. Necessity as the mother of invention: Development of the extractive technology of oxide nickel ores. Presented at SME-AIME Fall Meeting and Exhibit, Salt Lake City, UT, October.

Smith, R. 2005. *Chemical Process Design and Integration*. Hoboken, NJ: John Wiley and Sons.

Tilton, J.E., and Landsberg, H.H. 1997. Innovation, productivity growth, and the survival of the U.S. copper industry. Discussion Paper 97–41. Washington, DC: Resources for the Future. pp. 4, 15.

Uhrie, J.L. 1998. Use of electrical resistivity geophysics in the stockpile comparison of raffinate application techniques: Drip emitters versus wobblers. Presented at the Fourth Annual Randol Copper Hydromet Roundtable 98, Vancouver, BC.

White, H. 1999. Morenci: Making the most of a world class resource. In *Copper Leaching, Solvent Extraction and Electrowinning Technology*. Edited by G.V. Jergensen. Littleton, CO: SME. pp. 229–238.

PART 3

Comminution

CHAPTER 3.1

Mine-to-Mill Optimization

Walter Valery, Kristy-Ann Duffy, and Alex Jankovic

A mining operation is essentially a series of interconnected processes, with the performance of each stage affecting the subsequent ones. Optimizing each stage in isolation can result in suboptimal performance of the overall operation. For example, isolated changes in drill-and-blast can have detrimental impacts on downstream crushing and grinding processes, which subsequently affect the downstream separation process. Therefore, it is important to analyze and optimize each process stage in the context of the whole operation.

Blasting is the first stage of comminution in most mining operations. It is a very cost-effective and energy-efficient preparation step for the subsequent crushing and grinding stages, which will then provide much finer size reduction to liberate valuable minerals for separation. An example of energy consumption and cost breakdown by comminution stage for a relatively hard ore is shown in Table 1. This illustrates the increasing cost and energy consumption of comminution stages.

Table 1 Example of energy consumption and cost breakdown for relatively hard ore

Stage	Specific Energy, kW·h/t	Cost, US$/t
Drill-and-blast	0.1–0.25	0.1–0.25
Crushing	0.5–8	0.5–1
Grinding	10–35	2–5

Note: Data from Hatch database of industrial data.

Significant improvement of overall mine and plant performance can be achieved by optimizing blast fragmentation. This optimization requires utilization of appropriate levels of blasting energy according to the different types of ore but must also be tailored to the specific downstream processes, circuit flow sheet, types and sizes of comminution and classification equipment, installed power, and final product specifications.

There are many cases where the additional costs of increasing and better distributing blasting energy are more than compensated for by the improved throughput and recovery and reduction of energy consumption in downstream processing stages. For example, at the AngloGold Ashanti Iduapriem gold mine in Ghana, mine-to-mill optimization delivered finer run-of-mine (ROM) fragmentation that increased throughput by 20%–30% while reducing the 80% passing feed size (F_{80}) to leaching from 155 to 132 µm. The finer leach feed size increased gold recovery by about 0.5% (Renner et al. 2006). At the Gold Fields Cerro Corona operation, mine-to-mill optimization increased throughput by almost 15% for a particular hard ore type and 6% across all ore types while the semiautogenous grinding (SAG) specific energy was reduced by more than 9% (Diaz et al. 2015). The major changes recommended for the Cerro Corona project and the predicted impact on throughput and final grind size (P_{80}) for the ore type investigated are shown in Figure 1.

The potential for optimized blasting fragmentation to improve downstream processing performance is well recognized within the industry. Unfortunately, this has resulted in oversimplifying the concept to the idea that increasing explosive consumption in the mine results in optimization of the entire value chain (mining and processing). This oversimplification is often referred to as mine-to-mill. However, for true holistic optimization, blast intensity is not always increased but rather adjusted to best suit the different types of ores and also the circuit configuration, equipment, and installed comminution power downstream. It is a fully integrated effort with many aspects of optimization in the mine, comminution, and separation processes.

Various mine-to-mill initiatives have been implemented over the past 20 years with varying degrees of success. The long list of well-documented successes includes the examples provided in Table 2 and many more.

However, there are also anecdotal references of failures from some operations on how "mine-to-mill did not work for us." These failures can often be attributed to several common reasons, such as the following:

- Necessary structure and integrated methodology were lacking.
- Powder factor (kilograms of explosive per ton of ore) increased indiscriminately with no real optimization

Walter Valery, Global Director, Consulting and Technology, Mining and Minerals Processing, Hatch, Brisbane, Queensland, Australia
Kristy-Ann Duffy, Process Consultant, Consulting and Technology, Mining and Minerals Processing, Hatch, Brisbane, Queensland, Australia
Alex Jankovic, Principal Consultant, Consulting and Technology, Mining and Minerals Processing, Hatch, Brisbane, Queensland, Australia

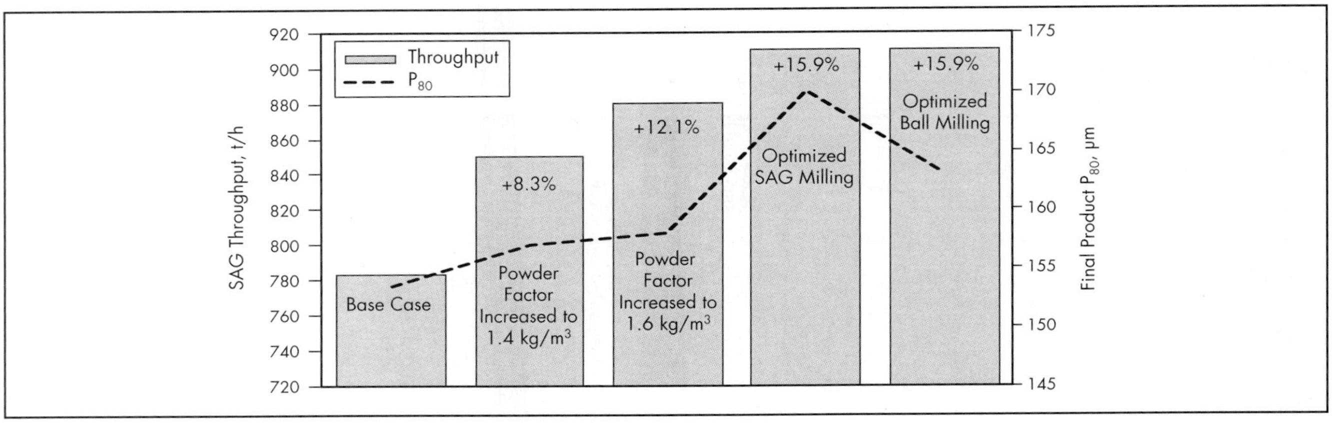

Source: Diaz et al. 2015

Figure 1 Mine-to-mill improvements at Cerro Corona

carried through the blasting and downstream processes to realize the full benefits.

- Results were not properly measured, followed through, well documented, or supported by management.
- Costs and performance were measured in isolation in the mine versus the mill instead of the overall mine and mill costs and benefits.

To be successful, mine-to-mill projects require a structured methodology supported by extensive auditing, surveys, and data analysis. Training and incorporation into site procedures, practices, and key performance indicators (KPIs) or metrics are also essential to ensure that benefits are quantified and maintained in the long term.

The objective of mine-to-mill optimization is to develop and implement integrated mining and processing strategies tailored to the operation to minimize the overall cost per ton treated and maximize company profit in a sustainable manner. The specific targets and objectives will vary for each project depending on prevailing site and economic conditions. For example, in times of high commodity prices, the main objective is often to increase production, whereas in harder economic times, the key driver is usually to reduce operating costs.

A comprehensive mine-to-mill approach recognizes that every ore body and mining operation is different, and to get the best results it is important to integrate and optimize each stage in the context of the whole operation. Understanding the ore body, and the characteristics of the ore within, allow the process to be tailored to suit the ore properties and key business drivers of the operation. Thus, the benefits from mine-to-mill will vary depending on the particular targets and objectives but can include many of the following:

- Maximization of system throughput (mine and mill), including increased excavation and loading efficiencies.
- Minimization of the overall operating cost with minimal adverse impacts, such as dilution or ore loss, structural damage, or environmental nuisance.
- Improvement of overall process stabilization (by reducing variability), resulting in superior processing performance in terms of production (throughput and recovery) and product quality (grade).
- Development of accurate throughput forecast models, because of the understanding of the impact of the ore types on blasting and processing. These models can be incorporated into the block model for mine planning and to establish optimum blending strategies for the operations. This should enable an improved revenue cutoff relationship to be determined to truly optimize blending and ore delivery strategies for maximizing revenue.
- Reduction in energy consumption and greenhouse gas emissions. At the Antamina copper–zinc mine, the fine feed from the high-energy blasts reduced the specific energy consumption by 25% (Rybinski et al. 2011).

These improvements can usually be achieved with minimal or no capital expenditure.

METHODOLOGY OVERVIEW

The performance in mining and mineral processing activities is governed by in situ ore properties. Therefore, a mine-to-mill optimization should start with ore characterization; then ore domains should be defined based on blasting, comminution, and metallurgical properties, and the spatial distribution of these domains mapped across the ore body. Detailed data from blasting and processing operations should be collected and analyzed for defined ore types, and used along with historical operating and benchmarking data. These data can be used to develop site-specific predictive models for each operation (blasting, comminution, separation). Together, these models indicate how the whole process will respond to different ore types and operating conditions in the mine and the plant. Simulations combined with extensive operational experience may be used to identify problems, process bottlenecks and opportunities for improvement, and then inform strategies to optimize the entire process for different ore types. The blast design should be optimized to generate optimal and consistent ROM fragmentation for all ore types, and downstream processes can be adjusted and optimized accordingly.

The models also allow prediction of throughput and recovery performance for each ore domain. By linking these with the block model and mine plan, a geometallurgical model can be created. This model can be used for forecasting, planning, and longer-term optimization purposes.

The following project structure has been successfully applied to deliver mine-to-mill outcomes:

- **Scoping:** Historical data and information are collected to identify problems, bottlenecks, and opportunities for improvement in the mine and processing plant.

Table 2 Examples of successful implementation of mine-to-mill optimization

Commodity	Operation	Country	Benefits	Reference
Coal	Hunter Valley	Australia	Decrease in fines production	—
Copper	Phu Kham	Laos	8% Production increase (simulated)	Bennett et al. 2014
Copper/gold	PT Newmont Nusa Tenggara Batu Hijau	Indonesia	10%–15% Production increase	Burger et al. 2006
Copper/gold	Cadia	Australia	15% Production increase	Hart et al. 2001
Copper/moly	Los Bronces	Chile	10%–15% Production increase (simulated)	Powell et al. 2006
Copper/zinc	Antamina	Peru	45% Production increase (+10% more in 2010); 23% less SAG specific energy	Rybinski et al. 2011; Samuel et al. 2012; Valery and Rybinski 2012
Gold	Iduapriem	Ghana	21%–32% Production increase; 0.5–1% recovery increase	Renner et al. 2006
Gold	Porgera	Papua New Guinea	15% Production increase	Lam et al. 2001
Gold	Boddington	Australia	Steady improvement in performance over time	Hart et al. 2011
Gold	Yanacocha	Peru	Reduction of ROM fragmentation and consumption of power and steel in SAG mill	Burger et al. 2011
Gold	Rio Paracatu Mineração	Brazil	Operating strategy (mine and plant) defined by ore type for circuit with new SAG mill	Tondo et al. 2006
Gold	Kalgoorlie Consolidated Gold Mines	Australia	8%–12% Production increase	Valery et al. 2001
Gold/copper	Cerro Corona	Peru	15% Production increase (hard ore); 6% production increase (overall)	Diaz et al. 2015
Iron ore	Marandoo	Australia	Increase in lump production	Valery et al. 2001

- **Ore characterization:** Measurements of rock structure, strength, and breakage characteristics are used to define how the ore will perform during blasting, crushing, and grinding. Mineralogical properties are also required to define the recovery and separation performance.
- **Benchmarking, audits, and surveys:** Audits and surveys are conducted for key processes (drill-and-blast, crushing, grinding, flotation/leaching, gravity concentration, etc.). The data collected are analyzed and can be compared with data from other mines and processing plants, allowing benchmarking against past performance and/or other operations.
- **Development of site-specific process models and simulations:** Data from benchmarking, audits, surveys, and ore characterization can be used to develop site-specific regression or predictive/mechanistic simulation models for key processes (drill-and-blast, comminution, separation). These models can be integrated to represent the main operations of the entire process. Simulations may be conducted using these models to evaluate different operating parameters for each ore type (domains) to determine alternative designs, modifications, and operating strategies to improve throughput and overall efficiency of the operation. The use of mathematical modeling and simulations is a rapid and cost-effective way to select potential optimization strategies and achieve results as opposed to expensive trial and error in the blast and plant.
- **Validation and implementation:** A detailed plan should be developed to implement the optimization strategies that have merit based on mine and plant constraints and a cost–benefit analysis. KPIs should be measured to quantify improvements and fine-tune the operating strategies.
- **Sustaining the benefits:** Recommended changes should be incorporated into managerial and site operating procedures. Training of operators and engineers is necessary to ensure that the benefits are maintained over the longer term.

ORE CHARACTERIZATION

Common ore characterization tests and measurements used in mine-to-mill optimization are shown in Table 3.

Rock structure is a measure of the natural fractures and discontinuities in the rock. It is determined by the in situ joints and fractures and can be estimated with rock quality designation (RQD), fracture frequency, and joint mapping. The energy and gases from the blasting tend to detach the rock along these natural fractures, so the rock structure generally controls the coarse end of blast fragmentation size distribution.

Rock strength is a measure of the hardness of the rock matrix and can be measured with laboratory tests such as point load index (PLI), JK drop weight test parameters (A, b, ta), and SMC drop weight index. Unconfined compressive strength is a common measure of strength and can be estimated from PLI values to reduce laboratory testing costs. Rock strength (hardness) affects the generation of fines in blasting. Rock strength is sometimes related to rock structure, but not always.

Measurement of crushability and grindability with laboratory tests such as the Bond crushability work index, equipment supplier proprietary tests, SAG power index, Bond rod mill work index, and Bond ball mill work index is important for comminution circuit evaluation and modeling.

Improved plant throughput can often be achieved by manipulating ROM fragmentation through optimization of blasting to reduce top size and increase fines. Using data that are typically available in geotechnical block models, such as RQD (for structure) and PLI (for strength), is a cost-effective and fast method to characterize ore for breakage. Examples of where this approach has been applied include the PanAust Phu Kham operation in Laos (Bennett et al. 2014) and Batu Hijau in Indonesia (Burger et al. 2006).

The range of ore properties are mapped out and ore domains defined in the example in Figure 2. The domains cover the complete range of rock properties that are present. Comminution practices (blasting, crushing, and grinding) can then be optimized for each of the ore domains.

BLAST OPTIMIZATION

Blasting is the first stage of comminution in most mining operations and should not be seen solely as a means of sufficiently reducing rock size for load-and-haul activities. The ROM size distribution has a large impact on downstream crushing, grinding and separation process efficiency, and consequently overall mine-site profitability.

Applying the appropriate intensity and better distributing the blast energy according to ore type produces a ROM size distribution that has a controlled top size and the required amount of fines. The blast intensity is not always increased; it is adjusted to best suit the different types of ores, providing a consistent and optimized fragmentation to the downstream processes.

The in situ ore characteristics, drill-and-blast pattern, and properties of the explosive govern the fragmentation size distribution and energy efficiency of the blast. Detailed auditing of the blasting and processing practices may be used to develop site-specific predictive models to predict blast fragmentation resulting from different ore types and blast designs. These models can be used to assess the optimum blast conditions required for a particular ore type.

To assess, model, and improve blasting for downstream processes, it is necessary to measure the resulting fragmentation. Image analysis of resulting muck piles, sieving of belt cut samples, and online size measurement tools can be used to estimate ROM fragmentation, crusher product, and mill feed size distributions.

An example of predicted blast fragmentation resulting from simulations of different blast designs for the Gold Fields Cerro Corona operation is shown in Figure 3. The powder factor (PF) indicates the amount of explosive used per cubic meter of blasted rock, and it can be increased by using a tighter blasting pattern (shorter burden and spacing; i.e., less volume of rock per hole) or more explosive per hole. Other changes to blast design that can impact fragmentation include explosive type, blast-hole diameter and arrangement (staggered or not), burden-to-spacing ratio, stemming material type, stemming length, bench height, initiation timing, wet or dry holes, and so on.

Blast fragmentation modeling is used to determine the optimum blast design for each ore domain considering the rock structure and strength and blasting parameters as

Table 3 Ore characterization measurements

Rock Structure and Strength	Crushability and Grindability	Flotability/Separability
Fragmentation Modeling	**Comminution Modeling**	**Flotation/Separation Modeling**
• Rock quality designation	• Bond crushability work index	• Detailed assaying of head samples
• Fracture frequency	• JK drop weight test parameters (A, b, ta)	• Size-by-size assays and mineralogy
• Joint and plane mapping with Sirovision system	• SMC drop weight index	• Liberation analysis (optical microscopy, mineral liberation)
• Unconfined compressive strength	• Bond rod mill work index	• Laboratory flotation tests
• Point load index	• Bond ball mill work index	• Flotation kinetics and locked-cycle tests
• Young's modulus	• Bond impact work test	• Any other specific separation tests according to the flow sheet (e.g., gravity, magnetic, leach)
	• Grind establishment tests: SAG power index, SAGDesign test, etc.	

Adapted from Valery et al. 2013

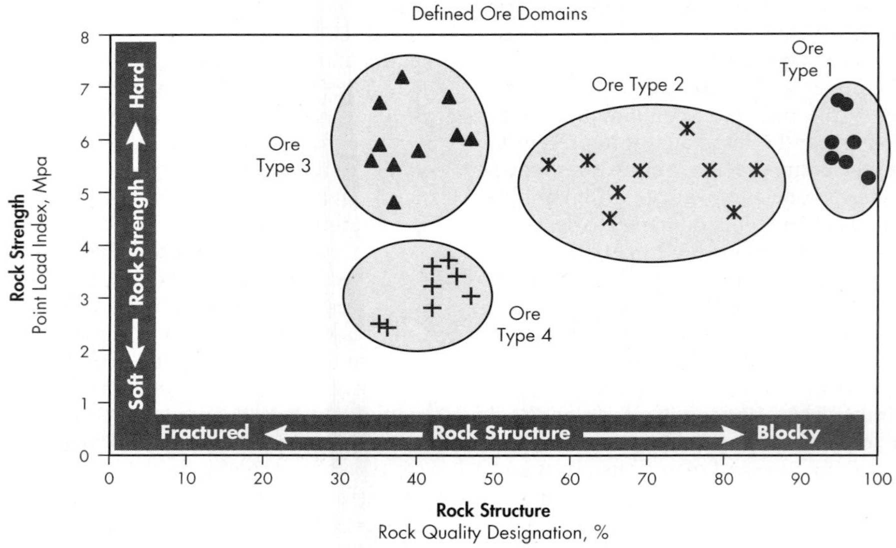

Figure 2 Example of ore domain mapping based on strength and structure

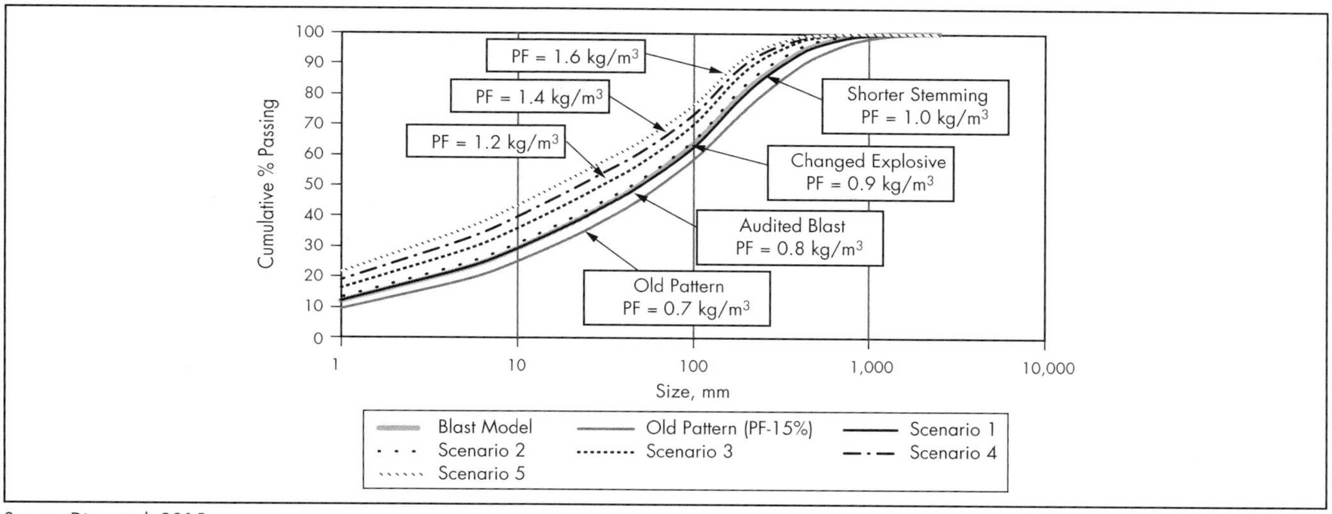

Source: Diaz et al. 2015

Figure 3 Example of blast simulation results for the Cerro Corona operation

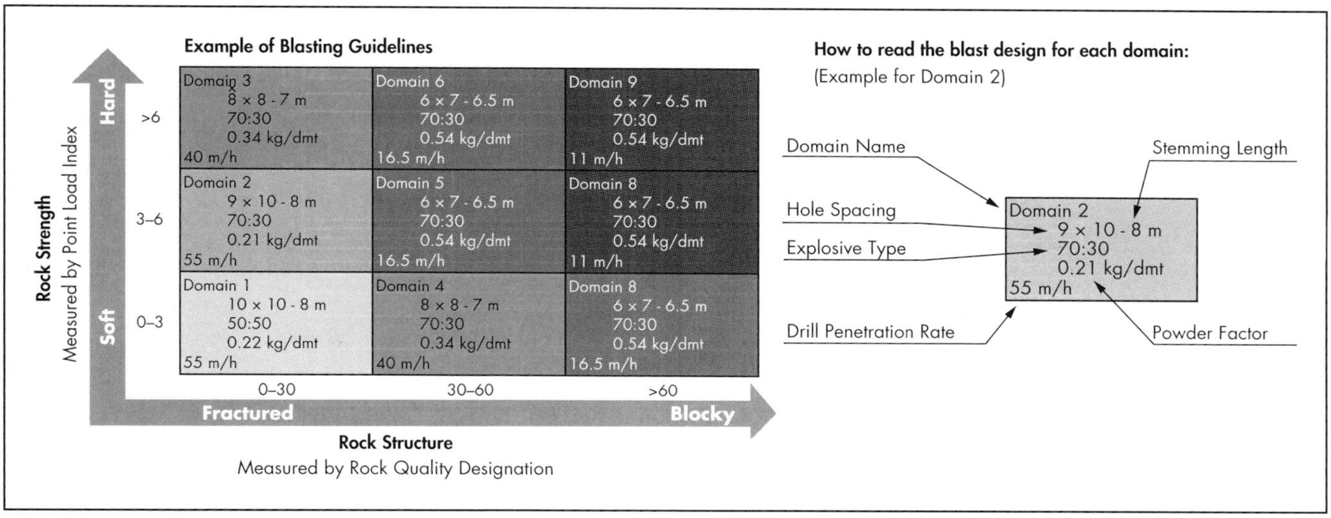

Adapted from McCaffery et al. 2006

Figure 4 Example of blasting guidelines for the Newmont Batu Hijau operation

mentioned earlier. Blasting guidelines are specified, providing an optimized blast design for each ore domain (see example in Figure 4). Blasting according to these guidelines provides a more consistent and optimized feed size distribution to the downstream processes, increasing throughput, process stability, and efficiency. Following the guidelines also avoids excessive blasting in softer ore domains, thus reducing energy consumption and costs, and preventing the excessive production of ultrafine material that can be detrimental to some downstream processes (e.g., in heap leach operations or flotation).

To realize the full benefits of improved ROM fragmentation from changing blasting practices, optimization must be carried through the downstream processes. This may be achieved by conducting comprehensive plant surveys while the material from an audited blast is treated in the processing plant. An ore tracking system, such as radio-frequency identification (RFID) tags, can be used to track material from the audited blast into the plant. Alternatively, it may be possible to use the mine dispatch system to identify when audited material is treated in the plant or to arrange campaign material from the audited blast through the plant by direct dumping of material into primary crushing. However, it is important to also consider residence times and mixing in intermediate stockpiles such as coarse ore stockpiles.

COMMINUTION CIRCUIT OPTIMIZATION

The ROM size distribution has a significant impact on the throughput and performance of downstream processes. Finer ROM fragmentation from blasting optimization may allow the primary crusher gap to be reduced, as the finer ROM enables this change to be made without compromising crusher throughput or exceeding installed power.

The ideal ROM size distribution, which will result in maximum throughput and performance, will depend on the breakage characteristics of the ore, equipment type, and circuit arrangement as well as operating conditions. The optimum feed size requirements for autogenous grinding mill, SAG mill, or multistage crushing circuits are very different.

For SAG mill circuits, higher throughput may be achieved when the SAG mill feed has

- As fine a top size as possible,
- The smallest possible amount of 25–75 mm intermediate size material, and
- The maximum amount of –10 mm fines.

This is shown schematically in Figure 5, and an example of the impact of feed size on SAG mill throughput (for the Antamina operation) is shown in Figure 6.

A change in comminution circuit operating strategy can often result in a reduction in power usage or an increase in circuit capacity. Comprehensive plant surveys, historical operational data, and benchmarking can be used to develop site-specific models for the comminution processes. The site-specific models allow simulation of different operating conditions, alternative circuit configurations, expansion options, and so on. This facilitates evaluation of many scenarios, avoiding trial-and-error experimentation in the processing plant, which is risky and expensive because of lost production.

Trend and variability analysis of historical operating data, power calculations, and benchmarking with similar operations can also be used to determine the bottlenecks and constraints in the current circuit and identify opportunities for improvement. For example, consideration should be given to the following:

- Availability and utilization
- Power draw, utilization of installed power
- Gap measurement and feed conditions (choke feeding) for crushers
- Pressure, roll speed, gap for high-pressure grinding rolls (HPGRs)
- Media size, ball load, charge and lifter/liner design for tumbling mills
- Trommel aperture and design
- Pebble crushing
- Media size in stirred mills
- Transfer size between various comminution equipment
- Recirculating loads
- Screen aperture, open area and loading
- Cyclone size, internals and feed density, and so on

However, the constraints and limitations are not always associated with major pieces of equipment but can also include ancillary or material transport issues such as pumping and conveying limitations, and water or power constraints.

Comminution flow sheets using technologies such as HPGRs and stirred mills are becoming more common because of growing challenges with regard to supply and cost of grinding media, energy, and water. The holistic mine-to-mill approach is equally relevant for different technologies and flow-sheet arrangements, and requires detailed understanding of the ore characteristics and processes. While mine-to-mill optimization has been more widely applied (and benefits documented) for operations with SAG mills, there are several examples with HPGR circuits. The Newmont Boddington gold mine (with HPGR circuit) in Australia is a good example of the significant impact a focused mine-to-mill program can have on multistage crushing plant efficiency and operating costs (Hart et al. 2011).

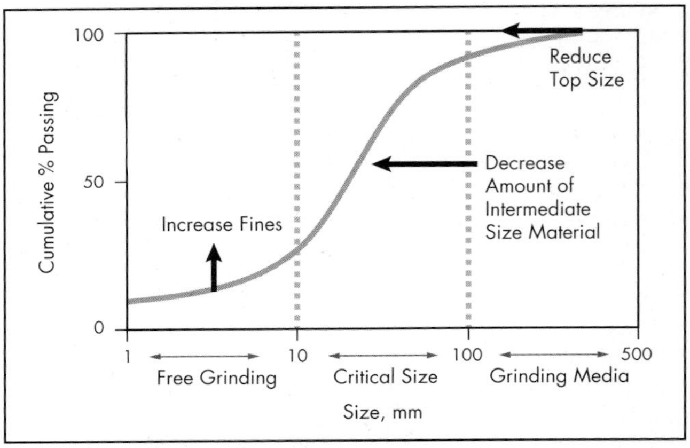

Adapted from Valery et al. 2015

Figure 5 Ideal SAG mill feed size distribution

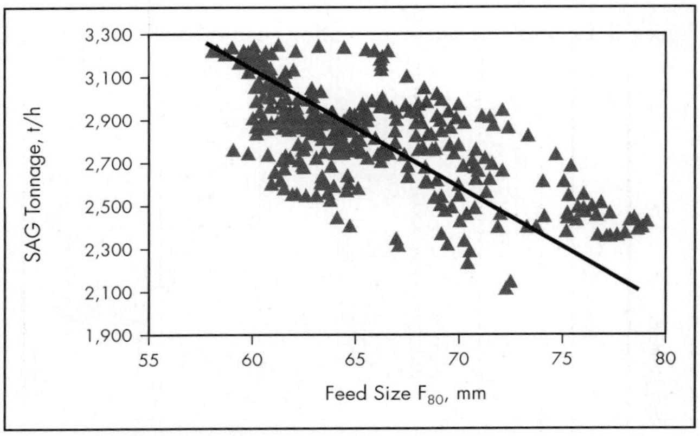

Source: Rybinski et al. 2011

Figure 6 Example of the impact of feed size on SAG throughput for the Antamina operation

Comminution is the most energy-intensive part of mineral production (typically accounting for more than half of the total energy consumption in the processing plant), and energy consumption increases as particle size decreases. Therefore, particle size should only be reduced as much as required for effective separation. The relationship between final grind size and separation performance should be understood to find the optimum trade-off between grind size (throughput and energy consumption) and recovery.

FLOTATION CIRCUIT OPTIMIZATION

In many mine-to-mill projects, for brownfield (existing) operations, the grinding circuit capacity is limited. Without significant capital investment for additional equipment, increasing throughput can result in coarsening of the grinding circuit product size, even with optimization of the circuit. Depending on the ore characteristics and liberation, this can have a detrimental effect on metal recovery in the flotation circuit. (Although in some cases the reduction of ultrafine material can have a positive effect on flotation recovery.)

Comminution and flotation circuits have traditionally been surveyed, analyzed, and optimized separately. This often results in less-than-optimal outcomes because of the complex interactions between the two processes. Flotation recovery is a strong function of the valuable mineral particle size distribution in the flotation feed. The highest recoveries are achieved for intermediate-sized particles, with lower recovery of ultrafines because of poor flotation kinetics, and lower recovery for the coarser particles because of poorer liberation and settling. Thus, to optimize flotation recovery and grade, it is usually best to grind to a fine size. This requirement needs to be balanced by the cost of grinding finer and the revenue that results from an increase in grinding throughput, which often results in a coarsening of the flotation feed. The flotation circuit can be adjusted and optimized for coarser feed if necessary, often making greater use of regrinding capacity where available.

There is a cost related to grinding to a fine size. Power usage increases as the grind size decreases, and the tonnage able to be processed by a grinding circuit can be limited by a grind size target. The overall objective should be to optimize profitability of the entire operation. It is therefore important to understand the relationship between grind size and the flotation grade–recovery performance. One possible technique, described by Bazin et al. (1994), uses simple relationships, calibrated with data routinely measured by an operation, to predict flotation recovery as a function of grind size.

This enables optimization of the comminution and flotation operations with respect to the overall operation by finding the optimum trade-off between grind size (throughput) and flotation recovery. In some cases, the effect of increased throughput outweighs the lost recovery to provide an overall increase in metal production and profitability. A good example is documented by Runge et al. (2014) for the Barrick Osborne copper–gold operation in Australia. An economic trade-off demonstrated that while higher flotation recovery could be achieved with lower throughput and finer grind size, the operation generated more revenue per hour by operating at higher throughput despite the lower flotation recovery because of higher metal production per hour and lower costs. See Figure 7.

Besides the integration between comminution and flotation circuits, in a mine-to-mill optimization study, the flotation circuit itself can be evaluated and optimized. This involves studying the flotation circuit to identify opportunities for improving circuit metallurgical performance (grades and recoveries). The scale of the flotation optimization exercise can range from a simple Bazin-type analysis to determine optimum grind size to a comprehensive circuit evaluation including surveys, cell characterization measurements, batch flotation testing, liberation analysis, mass balancing, model development, and simulations. Flotation parameters such as bubble size, bubble load, superficial gas rate, gas holdup, bubble surface area flux, carrying capacity, lip loading, residence time, froth stability, froth transportation, and froth depth can be measured or estimated to identify possible avenues for improving cell and bank performance. This information, along with detailed surveys, historical operating data, and flotation kinetics can be used in modeling and simulation to evaluate and optimize flotation circuits. These analysis techniques are used to

- Determine characteristics of the valuable mineral not recovered (e.g., size and degree of liberation) and the reason they are not being recovered (e.g., insufficient circuit residence time, poor selectivity, insufficient grinding, poor liberation, insufficient reagent, oxidation surface coatings);
- Ascertain optimum circuit configuration;
- Determine optimum feed and regrind particle size;
- Establish strategies for optimizing cell operation (e.g., froth management, air profiling); and
- Quantify downstream flotation circuit benefits associated with changes to comminution circuits (e.g., tonnage rate, grind size).

ENSURING EFFECTIVE CHANGE MANAGEMENT TO SUSTAIN BENEFITS

The mining industry typically experiences a high turnover rate of employees, particularly at many remote operations. As a result, the benefits from mine-to-mill optimization (or any other improvement initiative) can be short-lived if the changes are not effectively embedded into systems, site operating practices, and procedures. Even in locations where turnover is not high, the changes may not be sustained if the operating staff does not understand the reasons for the changes or where measures of personnel performance are not aligned with the mine-to-mill objectives.

Therefore, to sustain the benefits in the longer term, it is important to debrief site operating personnel and management on the outcomes of mine-to-mill optimization, and the reasons for changes and benefits. The results need to be properly measured, documented, and supported by management. Furthermore, the changes need to be incorporated into managerial and site operating systems, procedures, and ongoing training of operators and engineers. In many cases

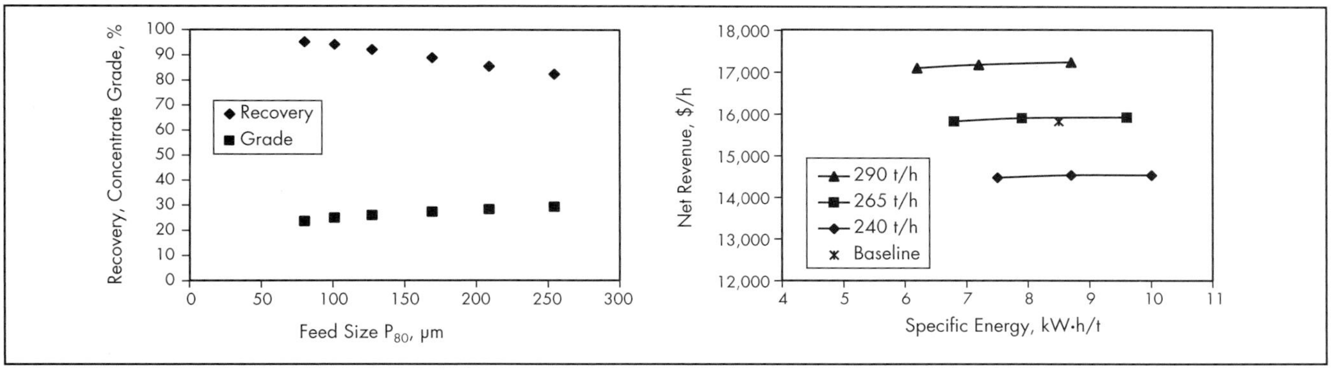

Source: Runge et al. 2014, reprinted with permission from the Australasian Institute of Mining and Metallurgy

Source: Crosbie et al. 2009, reprinted with permission from the Australasian Institute of Mining and Metallurgy

Figure 7 Example of economic trade-off between throughput/grind size and flotation recovery

where mine-to-mill has been particularly successful, regular meetings have been established involving both mining and processing personnel. These mine-to-mill meetings ensure an ongoing integrated effort to maximize the profitability of the overall operation.

Integration of mine and processing databases, including block models, mine dispatch systems, mine plans, and processing plant distributed control systems (DCSs), will certainly facilitate effective implementation and management of mine-to-mill optimization strategies. These systems provide the quantification of key operational and business drivers to allow informed decisions that maximize the profitability of the overall operation. Common centralized mine and processing plant operation and control centers will also play a major role in ensuring effective communication and aligned focus on common goals.

It is also important that KPIs or other measures of performance are adjusted to reflect the mine-to-mill strategies. Metrics and KPIs help to focus people and measure success; however, if inappropriately used, they can detract from broader operational goals. For example, if the performance of the mining department is measured only on cost and tons mined, better performance could be achieved by decreasing blasting intensity, but this would result in coarser fragmentation and lower throughput in downstream processing. To prevent this, KPIs and rewards or bonus payments for mining and processing plant personnel should be aligned and focus on profitability of the entire operation. As well stated by Eliyahu Goldratt (1990), "Tell me how you measure me and I will tell you how I behave."

MINE-TO-MILL OPTIMIZATION CASE STUDY: ANTAMINA

Numerous case studies demonstrate successful implementation and sustained benefits from mine-to-mill optimization for a wide variety of operations around the world. One such case study, Antamina, is discussed briefly here and in more detail by Rybinski et al. (2011), Samuel et al. (2012), and Valery and Rybinski (2012).

Compañía Minera Antamina is a polymetallic mining complex situated in the central Peruvian Andes around 4,300 m above sea level. Antamina produces copper and zinc concentrates as primary products, and represents one of the main complexes of the Peruvian mining industry.

The advent of harder ores presented a problem for the operation, significantly reducing plant throughput. A comprehensive review was conducted of existing operations at the mine and in the comminution circuits. Ore sources were characterized into ore domains based on rock structure and strength. Blast audits were conducted, and the blasted material was tracked from the mine through the process using RFID tags. Process surveys were conducted while treating the ore from audited blasts, allowing the performance in comminution circuits to be linked to the ore characteristics and blast fragmentation measured during the audited blasts. Historical data and current operating practices were also reviewed.

The collected data were used to develop mathematical models of the drilling-and-blasting, crushing, and milling operations. The resulting site-specific models were linked and simulations were conducted to identify optimum operating strategies for the overall process. Conditions such as different blast designs, primary crusher setting, SAG mill grate design, SAG mill ball charge level, ball mill operating conditions, changes in classification with hydrocyclones, and use and operation of recycle/pebble crushers were all considered and evaluated in the simulations. The proposed strategies were discussed and analyzed in detail with site technical staff.

The optimal and most cost-effective integrated operating strategy for the entire process (mine and plant) was then implemented, significantly increasing throughput. After the initial success, further investigation of higher-intensity blasts produced even finer ROM fragmentation and subsequently finer feed to the SAG mill. Additional optimization of the crusher, SAG, and ball mill circuits was conducted to achieve maximum throughput and best overall performance. The additional blasting energy was more than compensated for by the large energy savings in the crushing and grinding processes, thus minimizing the specific energy consumption of the entire operation. The sustained throughput benefits and energy savings achieved at Antamina are demonstrated in Figure 8.

At the beginning of the project, the operation achieved a throughput of 2,600 t/h with a specific energy consumption of 14 kW·h/t. At the end of the project, throughput of 4,500 t/h was achieved and maintained with a specific energy of 10.5 kW·h/t. This represents a significant increase in resource efficiency by reducing the total environmental impact of metal concentrate production with an energy savings of 25% for Antamina.

GEOMETALLURGICAL MODELING: OPTIMIZING MINE-TO-MILL OVER LIFE-OF-MINE

Mine-to-mill optimization should not only be considered in the short term. To remain viable and competitive in the face of increased challenges in extracting valuable minerals, the mining industry requires methodologies and tools to mitigate risks and to ensure the highest operational profitability throughout the life-of-mine (LOM). To maintain high profitability in the long term, it is crucial to be able to accurately predict the future performance of the process as well as the interaction with mining practices. This requires an accurate understanding of how each stage of the operation behaves for different ore characteristics as well as a detailed knowledge of the relevant ore characteristics and their distribution throughout the ore body.

The concept of geometallurgy integrates the disciplines of geology, mining, and metallurgy with the aim of developing proactive operating strategies as a function of the ore variability. Geometallurgical modeling requires a detailed understanding of the relevant ore characteristics, and models of how these will affect the performance of the blasting, crushing, grinding, and separation stages in terms of throughput, recovery, and product grade. This technique provides a link between the performance of the process, or efficiency of the extraction, and the physical properties of the ore. It facilitates throughput forecasting and allows strategic planning and optimization for different ore types to maximize the profitability of the operation in the long term.

The mine's existing block models can be used as the framework for developing geometallurgical models. The process starts with the mine-to-mill methodology described earlier in this chapter. The site-specific predictive models generated for blasting, comminution, and separation allow prediction of throughput and recovery performance for each ore domain and, when combined with the mine plan, enable a production forecast for the LOM. This can be used for long-term planning and to identify potential bottlenecks, process

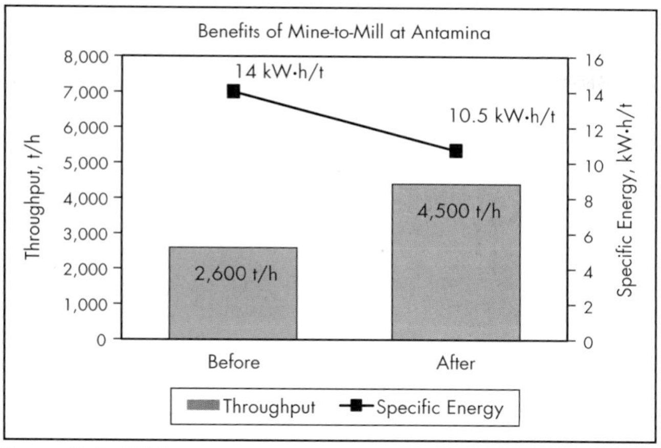

Source: Duffy 2012

Figure 8 Increased throughput and energy savings achieved at Antamina

constraints, and opportunities for improvement. For example, this was well executed at the PanAust Phu Kham operation as discussed by Bennett et al. (2014). Geometallurgical modeling and production forecasting determined that the harder ore later in the mine life would prevent production targets from being met for several years. This facilitated projects and planning to account for the harder ore and process constraints.

FUTURE TRENDS OF MINE-TO-MILL OPTIMIZATION

As high-grade and easily accessible mineral deposits are exhausted, new mineral deposits are expected to be low grade, more complex, and difficult to extract. In addition, the mining industry is facing growing challenges associated with the cost and supply of energy, limited water resources, and more stringent legislative requirements. Therefore, the search for more economical and sustainable technologies and practices is becoming increasingly important.

Many strategies for improving sustainability of mining operations are not new but rather involve novel application of existing technologies and tailored solutions based on understanding of the process and ore. Existing technologies from other industries can be used, or technologies can be applied in new ways to maximize efficiency. Also, understanding the impact of each process step on preceding and subsequent stages allows optimization of the entire process.

Mining operations need to be more efficient to remain economically viable but also to meet environmental targets. Improving the resource efficiency increases the economic return of the project and can make the difference between a project being viable or not.

Duffy et al. (2015) proposed that a more resource- and eco-efficient mining process of the future may incorporate the following methods.

- High-intensity selective blasting (HISB) improves blast fragmentation and reduces the energy requirement of downstream comminution.
- In-pit crushing and conveying (IPCC) is a more efficient method of transporting ore and waste than conventional truck-and-shovel operation, and can eliminate the use of diesel.
- Elevated high-angle conveying (HAC) facilitates removal of material from deep pits via the shortest route for IPCC solutions.
- Preconcentration using screening and/or bulk ore sorting could be implemented to discard barren material, consequently reducing haulage and downstream processing requirements per ton of product.
- Alternative energy-efficient and dry technologies such as HPGRs, vertical roller mills (VRMs), and air classifiers may be included in new comminution circuits.
- Partial or full replacement of hydrocyclones with fine screens may be used to improve the efficiency of existing wet grinding circuits.
- Coarse particle flotation can significantly reduce the energy consumed in the previous grinding stages.
- Filtration and dry stacking of tailings can be implemented to reduce water consumption, with a much higher recovery of water than can be achieved from a traditional tailings dam, and reduce reclamation and closure costs.

Each of these concepts is described in more detail in the following sections.

High-Intensity Selective Blasting

HISB involves further increasing and better distributing the energy of blasting, and using advanced initiation systems to provide a more selective blast. The result is improved fragmentation, reducing the energy requirement of downstream comminution. Blasting energy may exceed double the current levels. It may require significant changes in current drill-and-blast practices and possibly adaptation of drilling equipment to allow several blast holes to be drilled simultaneously with a single rig. The result is ROM fragmentation with a controlled top size and more fines, as illustrated in Figure 9. The ROM top size could be decreased to less than 400 mm compared to traditional ROM top sizes of about 1–1.5 m. The improved fragmentation significantly decreases the energy requirements in downstream comminution circuits. Benefits include maximization of system throughput (mine and mill), increased excavation and loading efficiencies, better overall process stabilization, and minimization of the overall operating cost. HISB also improves the viability of IPCC because of the reduced top size.

In-Pit Crushing and Conveying

IPCC is the use of fully mobile, semimobile, or fixed in-pit crushers with a conveyor system to remove material from the pit. IPCC can eliminate the use of diesel and is more efficient than conventional truck-and-shovel operation. The economics of belt conveying are particularly attractive in large-volume operations, since conventional truck haulage costs increase dramatically with pit enlargement.

The main advantages of IPCC are low operating costs, reduced supply requirements, high capacity, reduction in greenhouse gas emissions, and improved safety. Operating costs are typically 20%–60% lower than a truck-and-shovel system, with a fully mobile IPCC system the most frugal (Foley 2012). The benefits depend on the site conditions; however, studies have indicated operating cost reductions of between US$0.18/t and US$0.25/t compared with conventional truck-and-shovel operations (Cooper 2008). In addition, studies by Norgate and Haque (2010) and Koehler (2010) have indicated potential reductions in greenhouse gas emissions of 100,000–150,000 t CO_2/yr.

However, IPCC requires careful evaluation of geological, technical, and economic factors specific to the operation. IPCC needs to be factored into the mine design, which is generally more complex and less flexible than truck-and-shovel operations, and has longer delivery and development time. HAC is an effective system to implement IPCC in operations with steep pit walls, removing the material via the shortest route. The capital cost is higher for IPCC but is typically paid back within four or five years.

Preconcentration

The aim of preconcentration is to discard barren material early in the process, decreasing the amount of material that needs to be treated in downstream processes. This significantly reduces the energy and water consumption and operating cost per ton of product. Resource efficiency is improved by extending resource utilization (upgrading below cutoff grade material) and/or increasing production rates per ton of ore treated.

For some ores, preconcentration can be achieved by screening, as valuable minerals are often softer or more friable than gangue and therefore are concentrated in fine size fractions after initial breakage stages. There is evidence of this for several ores, including massive sulfide, base metal, gold, and some platinum ores (Bamber 2008; Bowman and Bearman 2014; Van der Berg and Cooke 2004). Screening has reasonably high capacity, low costs, and low technical risk.

Where the grade-by-size relationship does not allow effective separation, it may be possible to achieve an upgrade using a sensor to measure a difference between the valuable and waste material. A variety of sensors are available that measure different ore properties, the most common being photometric, electromagnetic, radiometric, and X-ray. The current technology measures and separates individual particles and thus is throughput limited. To be practical in preconcentration applications for large-scale mining operations, sensor-based sorting must be applied to bulk quantities of ore (e.g., on a truck tray or batches of ore on a fully loaded conveyor belt). Most of the existing sensor technologies are currently not suitable for bulk ore sorting because they are either too slow or insufficiently penetrating. However, it may be possible to adapt existing sensors or develop new sensors suitable for bulk quantities of ore. Bulk ore sorting may incorporate more than one type of sensor. It also needs a control system to make an accept-or-reject decision and a diversion system to separate the "batches" of valuable material from the waste. Ore sorting benefits from the natural heterogeneity of deposits and should be implemented as early in the process as possible, at the mine, immediately after loading, or integrated with the IPCC (where the variability is greatest) to maximize the benefits.

Energy-Efficient Comminution Circuits and Technologies

Comminution is the most energy-intensive part of mineral processing operations, consuming more than half of the energy in the mineral processing plant.

New comminution circuits may include alternative technologies to tumbling mills such as HPGRs, VRMs, and stirred mills. These are recognized to be more efficient than conventional tumbling mills. HPGRs and VRMs have the added advantage of being dry processes, thus reducing water losses. They also use no grinding media, which reduces operating costs and embodied energy, and may in some cases improve flotation performance by reducing galvanic interactions on particle surfaces. Both HPGRs and VRMs have shown energy and other process benefits in pilot-scale testing of hard ores. Novel flow-sheet arrangements such as HPGRs followed by stirred mills may significantly reduce the energy consumed in comminution (Valery and Jankovic 2002).

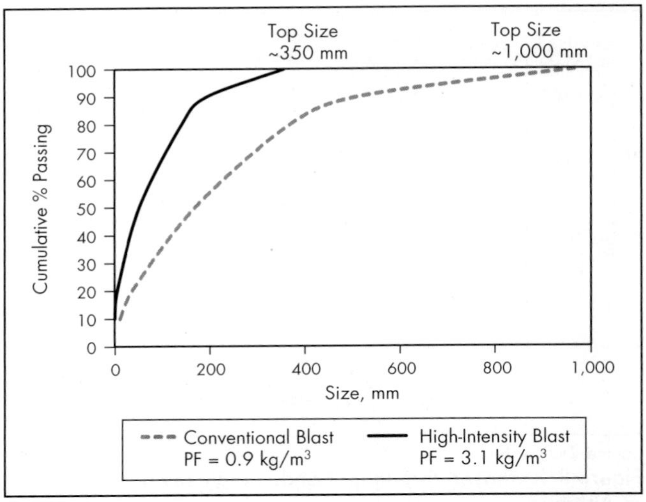

Source: Duffy et al. 2015, reprinted with permission from the Australasian Institute of Mining and Metallurgy

Figure 9 Effect of high-intensity blasting on ROM fragmentation

In existing operations, a typical grinding circuit arrangement uses a ball mill(s) in closed circuit with hydrocyclones. Full or partial replacement of hydrocyclones with fine screens may improve both process and energy efficiencies. A circuit closed with screens is expected to have 15%–20% higher capacity at significantly lower circulating load because of higher classification efficiency (Jankovic and Valery 2012). Two-stage classification with cyclones followed by screens may be able to achieve higher efficiency with less screening capacity. A conservative economic evaluation suggests that replacing cyclones with fine screens could be justified at energy costs of around $US160/MW·h or higher for some operations. Benefits are even greater when the valuable mineral is higher density than the gangue; for example, capacity may be increased by as much as 50% for magnetite ores (Nunna et al. 2014). However, further development is necessary to tailor fine screening to the very large throughputs encountered in most grinding circuits.

Coarse Particle Flotation

Grinding energy increases exponentially as particle size decreases. Therefore, if flotation of coarse particles could be enhanced, the energy consumed in the preceding grinding stages could be significantly reduced by delivering a coarser product to flotation. Typically, flotation recovery decreases for particles coarser than about 0.1 mm. However, if flotation could be performed effectively at 0.3 mm rather than 0.1 mm, the potential energy savings in grinding would be around 30%–50%. A coarse rougher flotation stage could reject barren material prior to regrinding and subsequent cleaning flotation of a much smaller amount of material to achieve the required product quality.

Coarse particle flotation may be enhanced by using alternative technology, such as the fluidized bed froth flotation cell.

This technology uses a teeter bed to maximize coarse particle attachment and shallow froth depths to reduce fallback.

Alternatively, coarse particle flotation in conventional flotation cells may be improved through modifications of cell design and operating philosophy. This may include split conditioning of reagents, manipulating the turbulence regime, optimizing bubble size, increasing froth stability, and minimizing froth residence times (Tabosa et al. 2013).

Water Recovery Optimization

The mining industry typically consumes 0.6–1.0 m^3 of water for each ton of ore processed by flotation (Brown 2003; Norgate and Lovel 2006; Wiertz 2009). Most water is lost from tailings dams in evaporation, entrapment, and seepage. Plants typically recycle water; however, site water balances have revealed that as much as 70% of the water discharged into a conventional tailings dam can be lost.

Dry stacking of filtered tailings is becoming increasingly common, especially in locations with limited water supply or high seismic activity (where risk of dam failure is high). Significantly higher water recovery can be achieved compared to a traditional tailings dam. Additionally, reclamation and closure costs are significantly reduced because of smaller footprint, easier construction, and the possibility to rehabilitate progressively. Currently, there are only a few applications of dry stacking of tailings because of the high cost associated with filtering. However, it is expected that within 10 years, legislation will prohibit the deposition of wet tailings in some countries.

The dry tailings system at the Gindalbie Metals Karara iron ore mine, installed in November 2012, is the first of its type in Australia. Filtered tailings are conveyed and stacked rather than pumped to the tailings storage facility. The system has a capacity of 2,850 t/h and a total conveying distance of 3.3 km. Water consumption is reduced by about one third, providing significant environmental benefits in the arid project location (Read 2012). However, operational issues with the tailings dewatering system have limited the plant's ability to produce at nameplate capacity for sustained periods. Partial reengineering is required to achieve design moisture content in the final tailings product, thereby allowing the tailings filtration circuit to operate at design throughput (Read 2012). This emphasizes one of the main challenges in implementing tailings filtration and dry stacking, which is ensuring correct selection and sizing of technology based on the nature of the tailings.

CONCLUSION

Mine-to-mill optimization has proven to be extremely successful when implemented using a structured methodology and supported by extensive data collection and analysis. Significant increases in throughput and reduction in energy consumption and costs have been achieved for numerous operations worldwide and across various commodities. Some notable recent examples include the Gold Fields Cerro Corona operation, Compañía Minera Antamina operation, and PanAust Phu Kham operation.

The mine-to-mill benefits are sustained in the long term when the outcomes are incorporated into site practices, procedures, and training. Integration of mine and processing databases, including block models, mine dispatch systems, mine plans, and processing plant DCSs, also facilitates effective implementation and management of mine-to-mill optimization strategies.

The production, efficiency, and cost benefits of mine-to-mill optimization will become more critical in the future as the mining industry faces growing challenges with decreasing feed grades, increasing cost pressures, issues with energy and water supply, and stricter legislation. Mining operations will need to operate more efficiently to meet environmental targets and to remain economically viable.

Mine-to-mill optimization is equally applicable for greenfield projects as it is for existing operations. This provides an opportunity to make a fresh start and develop something smarter and more efficient. Novel application of existing or adapted technologies and tailored solutions based on understanding of the process interactions and ore types provides an opportunity to develop solutions that are more efficient and profitable for the overall operation.

REFERENCES

Bamber, A.S. 2008. Integrated mining, pre-concentration and waste disposal systems for the increased sustainability of hard rock metal mining. Ph.D. thesis, University of British Columbia, Vancouver, BC.

Bazin, C., Grant, R., Cooper, M., and Tessier, R. 1994. Prediction of metallurgical performance as a function of fineness of grind. In *Proceedings of the 26th AGM of Canadian Mineral Processors*. Montreal, QC: Canadian Institute of Mining, Metallurgy and Petroleum.

Bennett, D., Tordoir, A., Walker, P., La Rosa, D., Valery, W., and Duffy, K. 2014. Throughput forecasting and optimisation at the Phu Kham Copper–Gold Operation. In *Proceedings of the 12th AusIMM Mill Operators' Conference 2014*. Melbourne, Victoria: Australasian Institute of Mining and Metallurgy.

Bowman, D.J., and Bearman, R.A. 2014. Coarse waste rejection through size based separation. *Miner. Eng.* 62:102–110.

Brown, E.T. 2003. Water for a sustainable minerals industry—A review. In *Proceedings of Water in Mining*. Melbourne, Victoria: Australasian Institute of Mining and Metallurgy. pp. 3–12.

Burger, B., McCaffery, K., Jankovic, A., Valery, W., and McGaffin, I. 2006. Batu Hijau model for throughput forecast, mining and milling optimisation and expansion studies. In *Advances in Comminution*. Edited by S.K. Kawatra. Littleton, CO: SME.

Burger, B., Vargas. L., Arevalo, H., Vicuna. S., Seidel, J., Valery, W., Jankovic, A., Valle, R., and Nozawa, E. 2011. Yanacocha Gold single stage SAG mill design, operation and optimisation. In *SAG 2011 Conference Proceedings*, Vancouver, BC.

Cooper, A. 2008. *Considering the Use of In-Pit Crushing and Conveying (IPCC) as an Alternative to Trucks*. Internal report. Perth: Snowden.

Crosbie, R., Runge, K., Brent, C., Korte, M., and Gibbons, T. 2009. An integrated optimisation study of the Barrick Osborne concentrator: Part B–Flotation. In *Proceedings of the 10th Mill Operators' Conference 2009*. Melbourne, Victoria: Australasian Institute of Mining and Metallurgy. pp. 189–197.

Diaz, R., Mamani, H., Valery, W., Jankovic, A., Valle, R., and Duffy, K. 2015. Diagnosis of process health, its treatment and improvement to maximise plant throughput at Goldfields Cerro Corona. In *SAG 2015 Conference Proceedings*, Vancouver, BC.

Duffy, K. 2012. Unlocking energy efficiency in the mining process. Efficiency Opportunities (EEO) National Workshop. September 6. Perth: Department of Resources, Energy and Tourism.

Duffy, K., Valery, W., Jankovic, A., and Holtham, P. 2015. Resource efficient mining processes of tomorrow. Presented at the Third International Future Mining Conference, Sydney, Australia, November 4–6.

Foley, M. 2012. In-pit crushing: Wave of the future? *Aust. J. Min.* (May/June):46–53.

Goldratt, E.M. 1990. *The Haystack Syndrome: Sifting Information Out of the Data Ocean*. New York: North River Press.

Hart, S., Valery Jr., W., Clements, B., Reed, M., Song, M., and Dunne, R. 2001. Optimisation of the Cadia Hill SAG mill circuit. In *SAG 2001 International Conference on Autogenous and Semiautogenous Grinding Technology*. Vancouver, BC: University of British Columbia, Department of Mining Engineering.

Hart, S., Rees, T., Tavani, S., Valery, W., and Jankovic, A. 2011. Process integration and optimisation of the Boddington HPGR circuit. In *SAG 2011 Conference Proceedings*, Vancouver, BC.

Jankovic, A., and Valery, W. 2012. The impact of classification on the energy efficiency of grinding circuits—The hidden opportunity. Presented at 11th Mill Operators Conference, Hobart, Tasmania, October 23–31.

Koehler, F. 2010. In-pit crushing looms the way into Australia. Mining Magazine Congress.

Lam, M., Jankovic, A., Valery, W., and Kanchibotla, S. 2001. Increasing SAG mill circuit throughput at Porgera gold mine by optimising blast fragmentation. In *SAG 2001 Conference Proceedings*, Vancouver, BC.

McCaffery, K., Mahon, J., Arif, J., and Burger, B. 2006. Batu Hijau—Controlled mine blasting and blending to optimise process production at Batu Hijau. In *SAG 2006 Conference Proceedings*, Vancouver, BC.

Norgate, T.E., and Haque, N. 2010. Energy and greenhouse gas impacts of mining and mineral processing operations. *J. Cleaner Prod.* 18(3):266–274.

Norgate, T.E., and Lovel, R.R. 2006. Sustainable water use in minerals and metal production. In *Proceedings of Water in Mining*. Melbourne, Victoria: Australasian Institute of Mining and Metallurgy. pp. 331–340.

Nunna, V., Suthers, S., Chinthapudi, E., Raynlyn, T., Wills, T., Tsedenbaljie, T., and Douglas, J. 2014. *Effect of classification efficiency on Ball Mill capacity for magnetite ores—Phase 1*. CSIRO and Metso PTI internal report.

Powell, M., Condori, P., Smit, I., and Valery Jr., W. 2006. The value of rigorous surveys—The Los Bronces experience. In *SAG 2006 Conference Proceedings*. Vol. 1. Vancouver, BC. pp. 233–248.

Read, N. 2012. Karara project awards key tailings management contract to BIS Industries. Securities Exchange announcement and media release. Perth: Gindalbie Metals.

Renner, D., La Rosa, D., DeKlerk, W., Valery, W., Sampson, P., Bonney Noi, S., and Jankovic, A. 2006. Anglogold Ashanti Iduapriem mining and milling process integration and optimization. In *SAG 2006 Conference Proceedings*, Vol. 1. Vancouver, BC. pp. 249–264.

Runge, K., Tabosa, E., and Holtham, P. 2014. Integrated optimisation of grinding and flotation circuits. In *Proceedings of the 12th AusIMM Mill Operators' Conference 2014*. Melbourne, Victoria: Australasian Institute of Mining and Metallurgy.

Rybinski, E., Ghersi, J., Davila, F., Linares J., Valery, W., Jankovic, A., Valle, R., and Dikmen S. 2011. Optimisation and continuous improvement of the Antamina comminution circuit. In *SAG 2011 Conference Proceedings*, Vancouver, BC.

Samuel, M., Valery, W., and Rybinski, E. 2012. Mine-to-mill mastery: Antamina boosts throughput for hard ores with support from Metso PTI. *CIM Mag.* 7(7):44–46.

Tabosa, E., Runge, K., Duffy, K., and Morgan, D. 2013. Strategies for increasing coarse particle flotation in conventional flotation cells. In *Proceedings of the Sixth International Flotation Conference (Flotation '13)*. Red Hook, NY: Curran.

Tondo, L.A., Valery Jr., W., Peroni, R., La Rosa, D., Silva, A., Jankovic, A., and Colacioppo, J. 2006. Kinross' Rio Paracatu Mineração (RPM) mining and milling optimisation of the existing and new SAG mill circuit. In *SAG 2006 International Conference on Autogenous and Semiautogenous Grinding Technology*. Vol. 2, pp. 301–313. Vancouver, BC: University of British Columbia, Department of Mining Engineering.

Valery, W., and Jankovic, A. 2002. The future of comminution. In *Proceedings of the 34th IOC on Mining and Metallurgy*. Bor: University of Belgrade, Technical Faculty Bor. pp. 287–298.

Valery, W., and Rybinski, E. 2012. Optimisation process at Antamina boosts production and energy efficiency: Helping a large copper/zinc mine meet the economic challenge of processing harder ore types. *Eng. Min. J.* 213(9):116–118, 120.

Valery, W., Morrell, S., Kovovic, T., Kanchibotla, S., and Thornton, D. 2001. Modelling and simulation techniques applied for optimisation of mine to mill operations and case studies. Presented at VI Southern Hemisphere Conference on Minerals Technology, Rio de Janeiro, Brazil, May 27–30.

Valery, W., Lynch-Watson, S., Valle, R., La Rosa, D., and Duffy, D. 2013. GeoMetso™: A site-specific methodology to optimise production and efficiency over the life-of-mine. In *Proceedings of the 10th International Mineral Processing Conference—PROCEMIN*. Chile: Gecamin.

Valery, W., Benzer, H., King, C., and Schumacher, G. 2015. Comminution circuits for ores, cement and coal. In *Comminution Handbook*. Edited by A.J. Lynch. Melbourne, Victoria: Australasian Institute of Mining and Metallurgy. pp. 167–189.

Van der Berg, G., and Cooke, R. 2004. Hydraulic hoisting technology for platinum mines. In *Proceedings of the International Platinum Conference "Platinum Adding Value."* Johannesburg: Southern African Institute of Mining and Metallurgy.

Wiertz, J.V. 2009. When best water use efficiency is not enough, what can the mining industry do? In *Proceedings of Water in Mining*. Melbourne, Victoria: Australasian Institute of Mining and Metallurgy. pp. 13–18.

CHAPTER 3.2

Rock Blasting

Sarma S. Kanchibotla

PURPOSE OF BLASTING IN THE MINERAL INDUSTRY

In most metal mining operations, the in situ ore is separated from the waste rock and is subjected to a series of breakage and separation processes to convert it into a valuable product. Profitability in this industry depends on how efficiently the in situ rock is converted into the final product. Because the breakage and separation processes that take place at the mine and mill are interdependent, traditionally, mining and milling processes are managed as separate cost centers and optimized with little understanding of the impact of one over the other.

Blasting is the first step in that breakage and separation process, but traditionally, the main purpose of drilling and blasting in most metal mines is to break the in situ rock mass into a muck pile of an appropriate size distribution and looseness so that the excavation equipment can load and haul blasted rock efficiently. The fragmentation and muck pile looseness are expected to be achieved without causing uncontrolled blast movement, which poses a safety risk to people and equipment, without causing damage to surrounding rock mass (or walls), and without causing environmental nuisances such as ground vibrations, airblast/noise, and dust.

During the past two decades, several researchers have demonstrated that all the processes in the mine-to-mill production value chain are interdependent and also understand that the impact of blasting outcomes on the overall process efficiency is critical to improve overall profitability.

The purpose of this chapter is to discuss the importance of the drill-and-blast outcomes in comminution and separation processes in the mine-to-mill value chain and then use it as a lever to improve overall profitability rather than a series of simple stand-alone processes. This chapter includes discussion of the following areas:

- Explosive rock interaction during blasting and an explanation about what happens in a blast—how rock fragments and moves in a blast
- Key blasting outcomes and their impact on downstream process efficiency
- Different modeling approaches used to predict blast fragmentation, ore loss, and dilution
- Measurement of blast outcomes
- The traditional blast optimization approach and its drawbacks compared to a holistic mine-to-mill optimization approach
- Case studies to demonstrate the potential value of the mine-to-mill optimization approach
- Challenges and limitations of this approach and further potential of next-generation mine-to-concentrator ideas and technologies

EXPLOSIVE ROCK INTERACTION—WHAT HAPPENS IN A BLAST

The interaction of the surrounding rock mass with the high-pressure and high-temperature gases released by the detonation of an explosive is a very complex process (Fourney 1993). In a blast, the in situ rock mass fragments and moves as a function of the detonation properties of the explosive, the physical-mechanical and structural properties of the surrounding rock mass, and the confinement provided to the explosion gases. The particle size distribution (PSD) or fragmentation in a blast muck pile is formed by

- The new fractures created by the detonating explosive charge,
- Extension of the in situ fractures in combination with the new fractures, and
- Liberation of in situ blocks and the newly formed blocks (Figure 1).

Detonation is the beginning of the rock breakage process in blasting by explosives. An explosive confined within a blasthole is detonated by a pentolite booster that was in turn initiated by a detonator, as shown in the Figure 2. The detonator can be initiated at any prescribed time and the accuracy of its timing depends on whether is it is an electronic detonator or conventional shock tube pyrotechnic detonator. Once initiated, the explosive ingredients within the blasthole are rapidly converted into gaseous products at very high pressures and temperatures. The chemical reaction or detonation front travels along the explosive column generally at a velocity of

Sarma S. Kanchibotla, Professor, JKMRC, Sustainable Minerals Institute, University of Queensland, Brisbane, Queensland, Australia

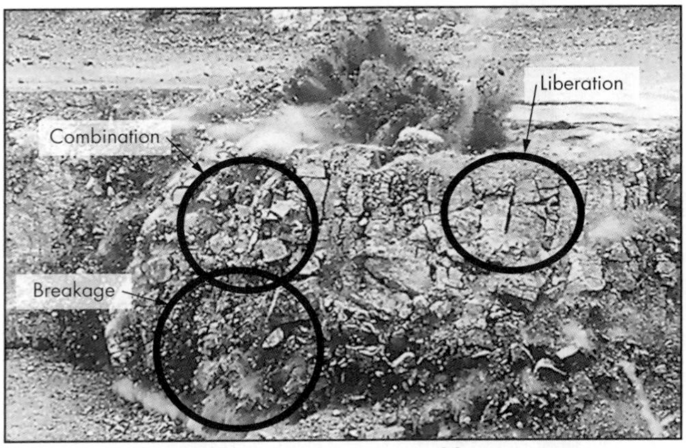

Figure 1 Fragmentation mechanisms in blasting

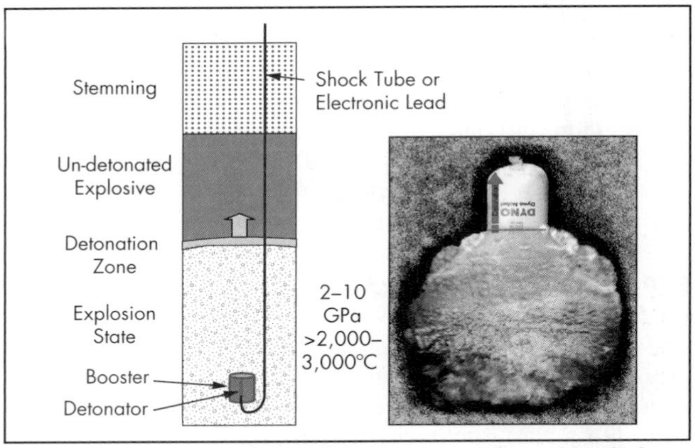

Figure 2 Detonation phase of explosives within a blasthole

Table 1 Properties of commercial and military explosives

Explosive Type	Density, g/cc	Velocity of Detonation, m/s	Explosion Pressure, GPa
ANFO*	0.8–0.85	3,500–4,500	2–4
Heavy ANFO	1.1–1.35	4,500–5,500	6–10
Emulsions	1.1–1.25	5,000–6,000	7–11
Military explosives	1.3–1.4	7,000–7,500	16–20

*ANFO = ammonium nitrate and fuel oil

3,000-6,000 m/s, which is defined as the velocity of detonation (VOD). The pressure of explosion gases within a blasthole for commercial explosives and military explosives such as PETN (pentaerythritol tetranitrate) are given in Table 1.

The explosion pressure of commercial explosives is around 2–10 GPa, whereas the dynamic compressive strength of most rocks will be less than 300 MPa. Therefore, the pressure of explosion gases within the blasthole is much higher (at least 1,000 times) than the dynamic compressive strength of any rock mass.

The sudden impact of the high-pressure explosion gases on the borehole wall transmits a compressive stress wave into the surrounding rock (Figure 3). If the intensity of the stress from this stress wave is more than the dynamic compressive

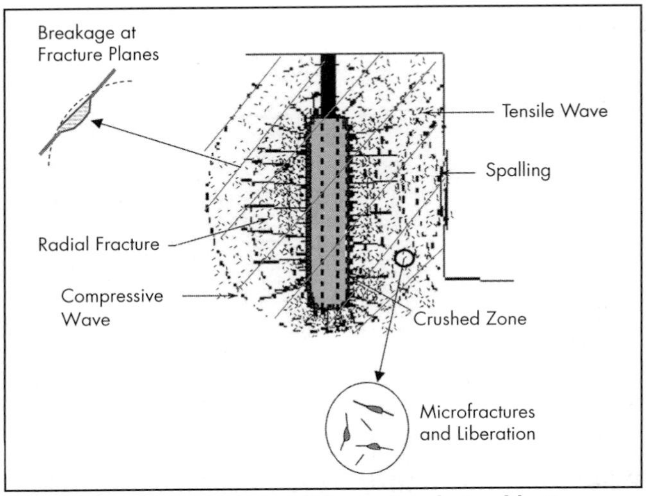

Figure 3 Stress wave–induced fracturing during blasting

strength of the rock, it pulverizes the rock matrix. PSD within this zone is expected to be less than the grain size of the rock matrix.

In ore blasts, the extent of the crush zone can have a significant effect on the efficiency of the grinding circuit because most of the particles within the crush zone require very little additional grinding. The extent of this crush zone depends on the magnitude of blasthole pressure, dynamic compressive strength, and elastic moduli of the surrounding rock mass. In a given rock, high-VOD explosives such as emulsions produce a larger crush zone compared to low-VOD explosives (ANFO [ammonium nitrate and fuel oil] or diluted ANFO). Similarly, a given explosive produces a smaller crush zone in harder rocks compared to a softer rock.

Beyond the crushing zone, the shock wave compresses the material at the wave front and induces a tangential tensile stress known as hoop stress. If the intensity of the hoop stress is more than the dynamic tensile strength of the rock, tensile radial fractures will develop. Both the compressive and tensile components of the wave decay with increasing distance from the blasthole. When the compressive wave hits a free face, nearly all the energy is reflected back as a tensile wave and develops tensile fractures known as spalling.

Microcracks Generation

Since the stress wave is transient, the major radial fractures will not have sufficient time to propagate and hence will branch out by forming microcracks. These microcracks may not propagate very far, but their total surface area in the rock mass may be 10–100 times larger than the total surface of the primary fragments in a blast (Revnivstev 1988). Generally, more microfractures will be formed by an increase in magnitude and the rate of energy release. Therefore, high-energy blasts with high-VOD explosives are expected to produce more intense microfracturing.

Fragmentation and Breakage due to Gas Penetration

During and after the stress wave propagation, the high-pressure gases confined by the blasthole walls and the stemming penetrate into the fractures (in situ fractures and new fractures developed by the stress wave) and expand them further. Some researchers believe that the fracture network throughout the rock mass is completed before the gas pressure phase (Hino 1956; Duvall and Atchison 1957). Other researchers believe that a significant portion of the fracturing process takes place during the gas pressure expansion phase (Kutter and Fairhurst 1971; Brinkmann 1987, 1990).

Impact of Discontinuities on Rock Breakage

Discontinuities and fracture planes in the rock mass play a significant role in the fragmentation mechanism. When the compressive stress wave from the blasthole encounters a discontinuity or a mineralized vein, some portion of the energy is reflected as a tensile wave, creating new fractures along the fracture plane. The propagation of radial fractures formed by the stress wave will be arrested when they encounter a discontinuity, but the energy will be dissipated by forming microfractures along the discontinuity. Subsequent penetration by explosion gases also expands the preexisting fractures, thus liberating the particles bounded by these discontinuities and also by liberating the material within the discontinuities. If the discontinuities are mineralized, then breakage along the discontinuities will have a significant effect on the liberation characteristics of the ore in downstream processes.

Blast Movement

Blast movement is the last stage of the rock-blasting process. In flight, collisions and landing produce minimal rock breakage compared to the shock phase and gas penetration. High-speed photography studies of several large-scale blasts indicated that in many blasting instances the explosion gas remains in the rock mass even after considerable burden movement (Chiapetta et al. 1983; Cameron 1992). The magnitude and direction of rock movement at any location in the blast depend on the confinement conditions, energy intensity, rock mass properties, blast geometry, and timing design. Blast movement can have a significant effect on ore loss and dilution. Mischaracterization of the grade boundaries, both prior to as well as a result of blasting, can lead to significant financial losses because of ore loss and dilution. Ore dilution occurs when waste material is miscategorized as ore and sent for processing diluting the run-of-mine (ROM) head grade and recovery. Ore loss takes place when valuable mineral is miscategorized as waste and sent to the waste dumps, thus never being processed and recovered.

Burden movement also creates a path of least resistance toward the free face; hence, the explosion gas tends to move toward the free face rather than penetrating into the fractures behind the blasthole. This is called relief and is a very important factor for control of blast-induced damage. Proper confinement of explosion gases in the rock mass is necessary to produce the desired fragmentation and displacement. Adequate confinement requires the stemming material to remain in place until the explosion gases release their full potential energy, and this can be achieved by using good stemming material (e.g., crushed gravel instead of drilling cuttings) and by providing sufficient stemming length. Inadequate and poor stemming will result in loss of energy into the atmosphere, reducing the effective heave energy of the explosive energy. It will also result in excessive airblast and noise. On the other hand, if the stemming column is too long, the explosive will be poorly distributed, resulting in the formation of boulders and poor fragmentation from the collar region.

BLAST OUTCOMES AND THEIR IMPACT ON DOWNSTREAM PROCESSES

Traditionally in metal mining, the key objectives of the blasting process are to

- Eliminate any safety hazards such as misfires and fly rock,
- Achieve sufficient fragmentation for efficient digging and primary crushing,
- Achieve an appropriate muck-pile shape and looseness suitable to the loading equipment,
- Not cause any excessive damage beyond the blast boundaries,
- Not exceed statutory ground vibration and airblast limits, and
- Minimize unit drill-and-blast cost per metric ton of rock.

The effects of blasting outcomes on grinding and separation were not considered and limited to the primary crusher. This approach works if profitability of the operation depends only on moving the rock from place A to place B and requires very little downstream processing (e.g., a quarry). However, in most operations, the blasted rock undergoes several other processes before it is sold as the final product; therefore, the traditional approach may not necessarily lead to optimum overall process efficiency. In addition to safety, the other key blasting outcomes that will impact the mine-to-mill value chain are

- Fragmentation,
  - Run-of-mine particle size distribution,
  - Quality of blast fragments (ore softening due to microcracks),
- Blast damage, and
- Blast movement (ore loss and dilution).

Run-of-Mine Particle Size Distribution

Run-of-mine particle size distribution (ROMPSD) has a direct influence on the performance of mining production equipment, crushers and grinding mills, and separation, as shown in Figure 4. The definition of oversize depends on the downstream process conditions and equipment size. Large equipment can tolerate bigger oversize compared to small equipment.

Muck piles with finer fragmentation not only improve the dig rates of excavators but they also reduce maintenance costs due to less stress on loading equipment. The fill factor of trucks is influenced by the size distribution of the muck, as Michaud and Blanchet (1995) have shown that the payload of a truck loaded with finely fragmented muck is greater than a truck loaded with coarsely fragmented muck. Primary crusher throughput and power consumption are strongly influenced by the ROM oversize. Eloranta (1995) and Elliot et al. (1999) demonstrated that reducing the oversize in the ROM ore can increase crusher efficiency.

Semiautogenous grinding (SAG) and autogenous grinding (AG) mills operate most efficiently when they are fed with a bimodal feed size distribution. To run SAG and AG mills efficiently, ore blasts should aim to maximize fines (–20 mm) generation and minimize the critical size (20–100 mm) and maintain the media (+100 mm) around 10%–15% (Figure 5A). The amount of rock media has a more pronounced effect in AG mills than SAG mills. AG mills require around 10%–15% of rock media for efficient grinding, whereas in SAG mills, some of this rock media can be compensated by adding steel media. In highly fractured ores, maintaining rock media may

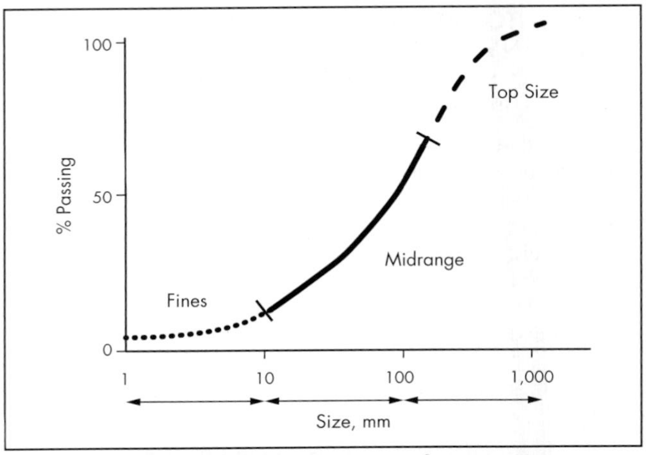

Figure 4 ROMPSDs and their impact on downstream processes

be difficult because it is governed by the in situ size distribution. Hence, blasting strategy in these ores should aim to increase fines and retain in situ grinding media as much as possible (Figure 5B).

In hard and massive ores, generally there is enough rock media; therefore, by adjusting blast patterns and primary crusher settings, it is possible to reduce the rock media to less than 15% (Figure 5C). Primary crushers are mostly used to reduce the top size from the ROM ore. Therefore, it is advantageous to generate as much fines as possible in blasting to maximize mill throughput. Control of critical size during blasting may be difficult, but the negative impact of critical size on mill efficiency can be minimized by optimizing the grates (pebble ports) and crushing the rejected pebbles using conventional crushers.

In some operations, excess fines may not help downstream processes. For example, fine coal is difficult to handle, suffers low yield, carries excessive moisture, and often attracts a lower sales price. Iron ore fines are sold at a reduced price or must be pelletized to regain value. Quarry products are sold on the basis of size and shape and are particularly prone to a reduction in value from over-fragmentation. In certain heap leaching operations, excessive fines can affect acid percolation and impact leach recovery (Scott et al. 1998). In some operations, it is important to understand the material handling properties of fines and their impact on all downstream processes, otherwise excess fines may result in bogging of crushers and storage bins.

Blast Damage: Impact on Separation

Separation of waste and ore starts at the mine, where strip ratio (waste/ore) determines the amount of waste that needs to be removed to expose the ore. Any breakage that takes place beyond the blast boundary limits is considered as damage. Poorly designed and implemented controlled blasts can cause damage near the crest and leave excessive toe near the floor. A combination of crest damage and excess toe will result in the loss of catch berm, which in turn can lead to sterilization of ore and pose a safety risk to both machinery and human life (Figure 6). In addition, the actual pit slope will become flatter, thus increasing the amount of waste.

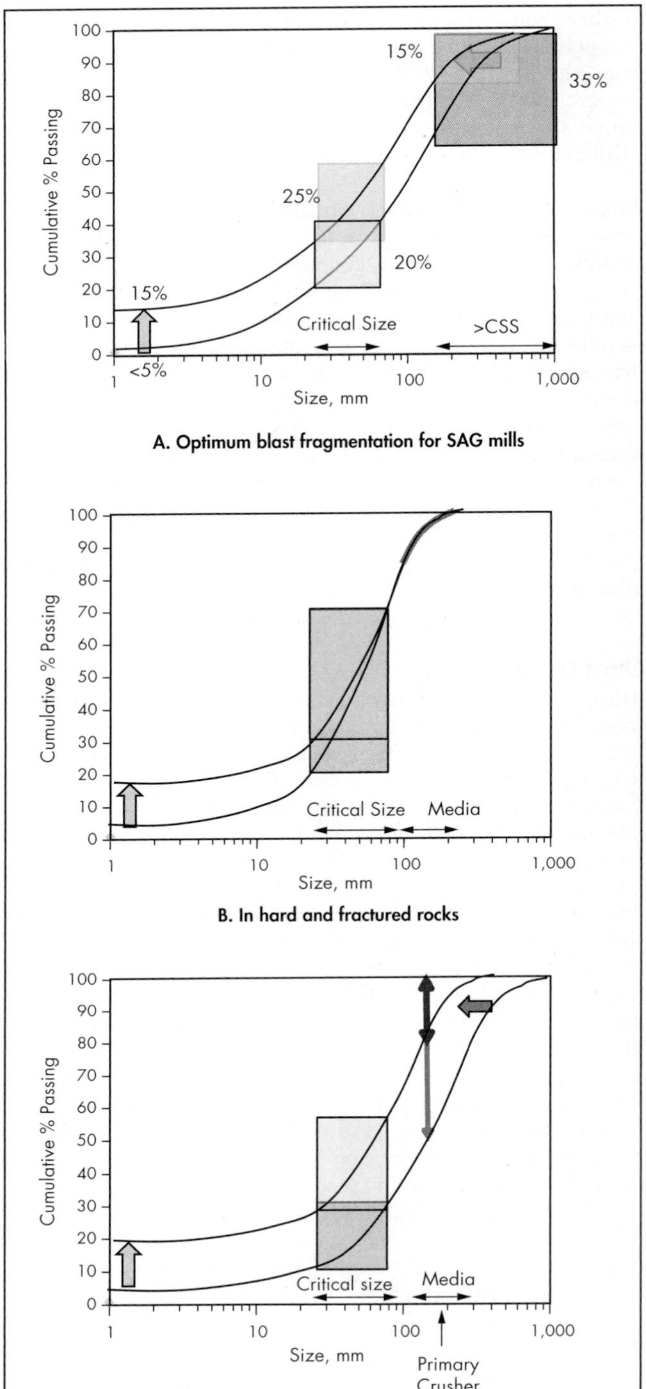

Figure 5 Blasting strategies to control feed size for SAG mills

General thinking in the mining industry is that high-energy blasts produce greater damage than low-energy blasts. This is why most mines implement controlled blasts adjacent to the final walls and interim walls to minimize the blast damage to these walls. These controlled blasts are fired after the

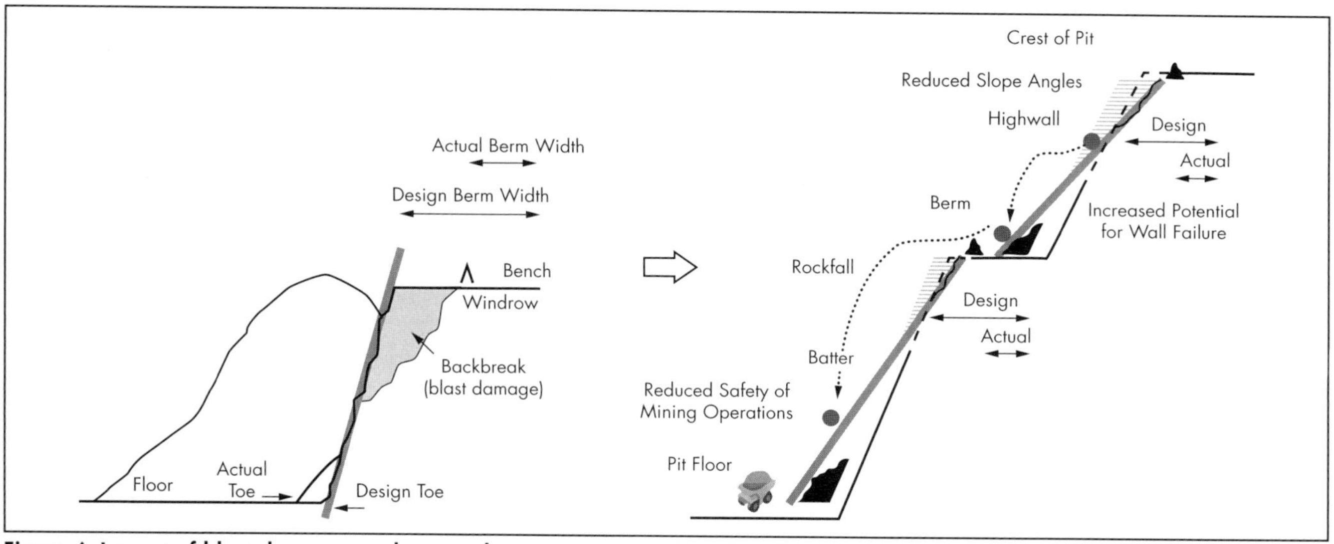

Figure 6 Impact of blast damage on slope angle

Source: Gaunt et al. 2015, reprinted with permission from the Australasian Institute of Mining and Metallurgy

Figure 7 Active split trials and results

production blasts and are generally designed with low energies and good free-face conditions to provide decent relief. It is true that sometimes the fragmentation and blast-movement outcomes from these controlled blasts may not be optimum for downstream comminution and separation processes.

In some cases, production blasts are fired along with the controlled blasts for scheduling and production purposes. In those cases, blast damage can extend to the final walls from the production blasts. However, with proper understanding of the damage mechanisms and with appropriate blast designs, it is possible to minimize blast damage even in high-energy blasts. Gaunt et al. (2015) demonstrated that implementing an active presplit between the high-energy production blast and trim blast can prevent the activation of joints into the wall, and reduce the damage behind the high-energy blasts (Figure 7). Another option is to change the blast sequence such that a carefully designed trim shot is fired before the high-energy production blast, creating a broken rock filter between the production shot and the final wall. This broken rock absorbs the energy from the high-energy production blast and thus limits blast damage to the wall (Scott et al. 1999).

Blast Movement: Impact on Dilution and Ore Loss

In many mines, ore and waste are often blasted together, and the rock mass within the blast volume is fractured and displaced relative to one another. Blast movement can be detrimental to the accurate delineation of post-blast grade boundaries if the movement is not consistent and properly understood (Tordoir 2009). This issue is demonstrated in Figure 8, where the post-blast grade boundaries are assumed to be the same as pre-blast boundaries, and so no account is made for movement.

The consequence of movement can be blast-induced ore loss and dilution. *Ore loss* takes place when valuable mineral is sent to waste dumps, and *ore dilution* occurs when waste rock is mischaracterized as ore and sent to the mill for further processing. Unintended mixing of different ores can also have other negative effects on the quality of the final product. For example, if a high-arsenic ore is mixed with the clean ore, the final concentrate quality will be affected. Similarly, if oxidized clay ores are mixed with fresh sulfide ores, this can result in lower recovery.

Generally where dilution is a major concern, production blasts are designed with low powder factors and choked conditions to reduce any lateral movement during blasting (Crone et al. 1973; Little and Van Rooyen 1988; Davis et al. 1989). Such an approach may reduce blast-induced movements, but it can have significant impact on fragmentation and muck-pile looseness, thus affecting the efficiency of downstream operations. More importantly, this simplistic approach does not address the fundamental cause of ore loss and dilution: Is it blast movement or a lack of understanding of blast movement that causes ore loss and dilution?

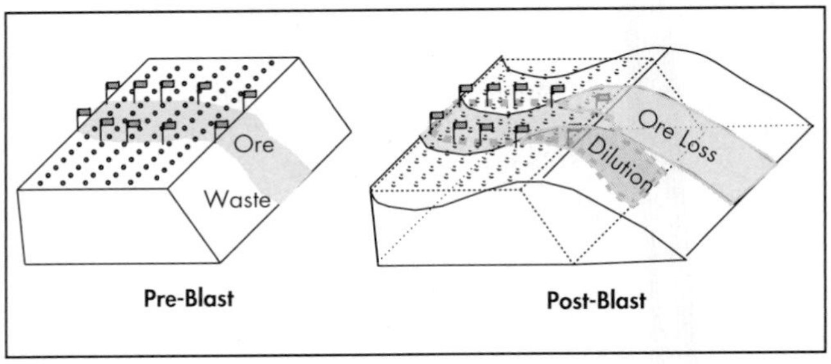

Source: Kanchibotla 2014

Figure 8 Blast-induced ore loss and dilution

Blast-movement studies have shown that the economic impact of blast-induced ore loss and dilution on a mining operation's profit is significant. Taylor et al. (1996) monitored ore control grade versus crusher discharge grades for Coeur Mining's Rochester mine in Nevada (United States) from 1992 to 1994 and estimated that if the mine had accounted for blast-induced movement during 1994 alone, the mine could have produced an additional 200,000 oz of silver and 4,560 oz of gold and would have added some US$2.7 million in annual revenue (assuming a gold price of US$380/oz and silver price of US$4.80/oz). Blast-movement results at an open pit gold mine operated by Placer Dome Inc. indicated that the economic impact of blast-induced ore loss was greater than US$1.5 million and dilution of about US$0.5 million (Thornton et al. 2005). Engmann et al. (2013) from Ahafo gold mine in Ghana estimated that the cost of ore loss and dilution due to unaccounted blast movement of ore polygons was around 33% and approximately US$160,000 for three production blasts. Adjusting ore blocks for blast movement is an effective way to reduce blast-induced ore loss and dilution.

In addition to blast movement, several other factors will affect overall dilution. Some of those factors are shape of ore body, accuracy of ore body modeling, excavator size and shape, production sequence, and so on. Although ore loss and dilution have a significant impact on the overall profitability of a mining operation, very few mines have a real understanding of their current dilution level and the cause.

Impact of Microcracks and Preferential Breakage on Comminution and Separation

The laboratory experiments conducted by several researchers suggest that the microcracks developed during blasting action can reduce the impact breakage resistance of post-blast fragments (Revnivstev 1988; Chi et al. 1996; Nielsen and Kristiansen 1995; Kojovic and Wedmair 1995; Michaux 2005; Kim 2010; Parra 2013; Parra et al. 2016). In some mineral deposits, the valuable minerals are deposited along cracks and fissures, and preferential breakage along these fissures can liberate mineral-rich material. Because of its relatively low to moderate strength, a material deposited within fractures tends to become fragmented into small size fragments (fines). Bulk sampling and screening tests of ROM ore from the Bougainville mine in Papua New Guinea showed better upgrade potential in ores where mineralization was mainly on fracture planes compared to highly disseminated ores (Burns and Grimes 1986). Laboratory blast trials at the Julius Kruttschnitt Mineral Research Centre (JKMRC) showed the beneficial effect of blast-induced preferential breakage and upgrading of the fines fraction from two ores (JKMRC 2002).

Generally, marginal ores and mineralized wastes are blasted and removed or stockpiled without further treatment. However, by exploiting this selective breakage mechanism, it may be possible to further process the high-grade fine fractions. Considering that the amount of low-grade or marginal ore is frequently larger than the amount of ore above the cutoff grade, the proposed approach offers an avenue for substantial increase of a mine's economically recoverable and treatable mineral reserves. Increased reserves and successful implementation of the proposed technology will ultimately result in increased profitability and increased value for mining companies. In recent years, the Cooperative Research Centre for Optimising Resource Extraction in Australia had developed and implemented Grade Engineering technology at several large open pit operations by deploying a range of coarse waste rejection solutions (Walters 2016).

BLASTING MODELS—HOW TO PREDICT BLAST OUTCOMES

A variety of modeling approaches have been used to describe the rock-blasting process and to predict blasting outcomes. They range from highly computationally intensive numerical models to reasonably simple empirical models. Over the years, several numerical models were developed to estimate either breakage and fragmentation or movement. The hybrid stress blasting model (HSBM) is one of the recent numerical models developed in collaboration between the University of Queensland, Itasca Consulting Group, and the Cambridge, Imperial, and Leeds Universities (Sellers et al. 2012). HSBM is a rock-breakage engine to model detonation, wave propagation, rock fragmentation, and muck-pile formation. It also incorporates an ideal and a nonideal detonation performance and predicts fragmentation, displacement, and damage. The main limitation of most numerical models is that the amount of computational power and time required to simulate production blasts is quite large. That is why empirical and semi-empirical models are more popular to predict blast outcomes; some of them are discussed in the following sections.

Blast Fragmentation Models

Kuz–Ram Model

One of the most popular blast fragmentation models is the Kuz–Ram model. It is an empirical model that uses the equation developed by Kuznetsov (1973) to estimate the mean fragment size:

$$x_{50} = A \cdot K^{-0.8} \cdot Q^{0.167} \cdot \left(\frac{115}{E}\right)^{0.633} \quad \textbf{(EQ 1)}$$

where

x_{50} = 50% passing fragment size, cm
A = rock factor:
7 for medium-strength rocks
10 for hard and fractured rocks
13 for very hard and massive rocks
K = powder factor, kg/m^3
Q = mass of main explosive per blasthole, kg
E = relative weight strength of explosive (ANFO = 100)

Cunningham (1983) combined Kuznetsov's model with the Rosin–Rammler equation to calculate the entire size distribution (Equation 2) and developed an empirical equation to estimate the uniformity index "n" using Equation 3.

$$P = 1 - e^{-0.693\left(\frac{x}{x_c}\right)^n} \quad \text{(EQ 2)}$$

where

P = cumulative weight passing size (x)
x = diameter of fragment
x_c = characteristic size (mean size)
n = uniformity index (relates to the slope of the distribution)

$$n = \left(2.2 - 14\frac{B}{D}\right)\left[\frac{1 + \frac{S}{B}}{2}\right]^{0.5}\left(1 - \frac{W}{B}\right)\left(\frac{L}{H}\right) \quad \text{(EQ 3)}$$

where

B = burden, m
D = charge diameter, mm
S = spacing, m
W = standard deviation of drilling accuracy, m
L = charge length, m
H = bench height, m

Lilly (1986) modified the description of the rock factor (A), which is derived from properties of the rock mass:

$$A = 0.06\ (RMD + JF + RDI + HF) \quad \text{(EQ 4)}$$

where

A = rock factor
RMD = rock mass description
JF = joint factor
RDI = rock density influence
HF = hardness factor

Research at the JKMRC and elsewhere has demonstrated that the Kuz–Ram model underestimates the contribution of fines in the ROM size distribution (Kojovic et al. 1998; Comeau 1996; Kanchibotla et al. 1999). The model estimates were reasonable for quarry blasting applications where the impact of fines is not as important as in metal mining operations where the ore needs to undergo crushing and grinding processes.

JKMRC Blast Fragmentation Models

JKMRC researchers have investigated many alternative approaches to improve the fines prediction of fragmentation models (JKMRC 1998). One of the approaches is the crush zone approach developed by Kanchibotla et al. (1999), and the other is a two-component model (TCM) developed by Djordjevic (1999). Both models assume that fragmentation in a blast takes place in two different mechanisms—(1) creation of new fractures and (2) liberation of existing fractures—and therefore estimate the coarse end and fines end of PSD separately. Both models estimate the coarse end of the size distribution similar to the Kuz–Ram with slight modifications to the rock factor estimate.

The main difference between the JKMRC models and Kuz–Ram model is how they estimate the fines end of size distribution. The TCM uses experimentally derived parameters from small-scale blast chamber tests to estimate the fines region of the model. One of the main drawbacks of the TCM is the requirement to conduct a series of small-scale laboratory blasting tests to derive the crushed zone parameters, and the crush zone model has been developed to overcome that limitation.

Crush Zone Model

The crush zone model (CZM) assumes that fines in a blast are generated predominantly by the crushing of rock around the blasthole due to compressive and shear failure. The coarse fragments are generated primarily by tensile failure, by extending the in situ fractures and by liberating in situ block. The CZM uses a rock calibration factor (RCF) instead of a constant 0.06 to estimate the rock factor.

$$A = RCF \times (RMD + JF + RDI + HF) \quad \text{(EQ 5)}$$

The CZM also calculates the RMD and JF differently from the Kuz–Ram model; a detailed description of their calculations is given in the AMIRA International report (JKMRC 1998). The extent of the crushed zone is determined by calculating the point at which the radial stress around the blasthole exceeds the dynamic compressive strength of the rock. The stress around the blasthole is estimated by assuming that explosion gases apply pressure on the blasthole walls quasi-statically by using the following equation (Jaeger and Cook 1979):

$$\sigma_x = p_b\left(\frac{r}{x}\right)^2 \quad \text{(EQ 6)}$$

where

σ_x = radial stress at a distance, x, from the center of the blasthole
p_b = borehole pressure
r = radius of the blasthole

The borehole pressure for a fully coupled explosive is estimated by the following equation:

$$p_b = K\ \rho_e\ VOD^2 \quad \text{(EQ 7)}$$

where

K = constant dependent on explosive density (between 0.25 and 0.125)
ρ_e = density of the explosive
VOD = velocity of detonation of the explosive

For decoupled charges, the explosion gases expand adiabatically and hence the blasthole pressure will decrease. The rock matrix within the crushed zone is assumed to be totally pulverized and the particle size within this zone is understood to be less than the grain size. The default value is assumed as 1 mm. By using the crushing zone radius, the volume and the percentage of fines within the crushed zone is estimated using Equation 7:

$$\% -(\text{grain size}) = 100 \times V_{crush}/V_{br} \quad \text{(EQ 8)}$$

where

V_{crush} = volume of crushed zone
V_{br} = volume of blasted rock per blasthole

The uniformity index and the coarse end of the size distribution are estimated by using Kuz–Ram Equations 1 through 3. The fine uniformity index and fine end of the distribution is then calculated by substituting the % (grain size or –1 mm value) and characteristic % (X_c%) using the following equation:

$$n_{fine} = \frac{LN\left[LN - \frac{(1 - \text{fines}\ \%)}{-X_c\%}\right]}{LN\left(\frac{1}{X_c}\right)} \quad \text{(EQ 9)}$$

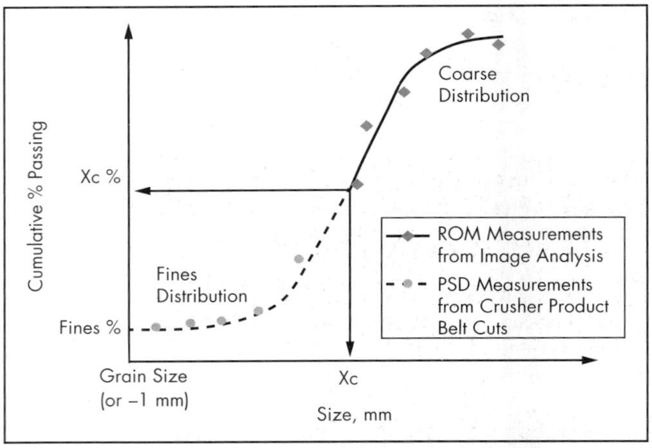

Figure 9 Fines and coarse size distribution in ROM

The X_c% is where the coarse and fine curves join, and initially this was chosen as the mean size or X_{50} (Figure 9). However, subsequent trials indicated better results with a characteristic size of X_{90} with very soft rocks (uniaxial compressive strength [UCS] < 10 MPa). It is likely that for intermediate-strength rocks, the point where the two distributions are joined will vary between the X_{50} and X_{90} values. It is therefore necessary to calibrate the CZM with the measured PSDs. The coarse end of the size distribution is calibrated with the ROM measurements obtained from the muck-pile images or at the back of trucks. Primary crushers are not expected to produce many fines (–25 mm); hence, most of these fines in the primary crusher product or SAG feed are assumed to have come from ROM ore. That is why the fine end fragmentation is calibrated to match the size distributions from the primary crusher product or SAG feed belt cuts. Once the model is calibrated for a given ore, then it is useful to run several what-if simulations to predict the entire ROMPSD.

An example comparison between the JKMRC CZM and Kuz–Ram model is provided at the end of this chapter.

KCO Model

The KCO model (Ouchterlony 2005) is an extended version of Kuz–Ram model and it essentially replaces the Rosin–Rammler function used in Kuz–Ram with a new function, the Swebrec function, as shown in the following equations:

$$P(x) = \frac{1}{\left\{1 + \left[\frac{\ln\left(\frac{x_{max}}{x}\right)}{\ln\left(\frac{x_{max}}{x_{50}}\right)}\right]^b\right\}} \quad \textbf{(EQ 10)}$$

$$b = \left[2 \cdot \ln 2 \cdot \ln\left(\frac{x_{max}}{x_{50}}\right)\right] \cdot n \quad \textbf{(EQ 11)}$$

where

$P(x)$ = percent of material passing size (x)
x_{max} = maximum fragment size (top in situ block size, spacing, or burden)
x = fragment size
x_{50} = mean (50%) passing size (same as Kuz–Ram)
b = fragmentation curve undulation parameter
n = uniformity index (relates to the slope of the distribution)

Limitations of Empirical Blast Fragmentation Models

The following list describes the limitations of empirical blast fragmentation models:

- They require field calibration before they are used for any design purposes.
- They are single-hole models and the rock mass around the blasthole within a bench is considered one material. The impact of any variation in the rock properties across the blasthole is not considered.
- They do not take any blasting mechanisms explicitly. Only the CZM uses an analytical equation to estimate the crushed zone and fines around the blasthole.
- The input variables (e.g., rock mass properties and blast design parameters) used in these models has been assumed to be constant and uniformly distributed within a blast, whereas in reality, these parameters are not uniformly distributed.
- The impact of delay timing between the blasthole is not considered.

Blast Movement Models

Various modeling approaches have been developed to estimate blast movement; however, they have had very limited use in the context of production grade control. Yang et al. (1989) developed a two-dimensional (2-D) kinematic model to predict final muck-pile shape for bench blasts with free faces. Preece (1994) utilized advances in computing power to develop a 2-D model that represented the rock mass as discs. The explosive charge in the blasthole acts upon these discs and their trajectory, and final position defines the final muck pile. Tucker et al. 2006 used the particle flow code, a discrete element model to predict blast movement and muck-pile shape. Use of blast movement models in the production environment is limited because of their need for significant computational power and complexity. Production-grade control typically requires the use of simple and effective techniques that can be implemented on a day-to-day basis without impacting the time constraints imposed by the production environment. An alternative to computationally intensive numerical modeling is a relatively simple movement template approach (Rogers et al. 2012). This approach consists of translating grade boundaries based on average movement vectors measured from previous displacements of blasts with similar design properties (blast type, rock type, initiation type; Figure 10).

W.H. Bryan Mining and Geology Research Centre (BRC) of the University of Queensland has been involved in the development of a three-dimensional (3-D) blast movement model that uses an external "physics engine" (Tordoir 2009). In this model, the blast volume is discretized into equidimensional cubes and are assigned initial conditions based on the energy distribution and detonation times of surrounding explosive decks. Initial velocity vectors and first movement time are calculated for each block. These blocks have only a notional mass, and changing block size does not have a logical impact of the model behavior. The cubes are then moved using a physics engine, which handles all inter-cube collisions and landing logic. The explosive energy in each block is converted to kinetic energy and an energy loss factor. An energy loss factor scales the explosive energy in the blast to match the proportion of work it does in moving the rock. Each block can be assigned with specific attributes such as grade, mineralogy, or hardness, and their post-blast location can be tracked.

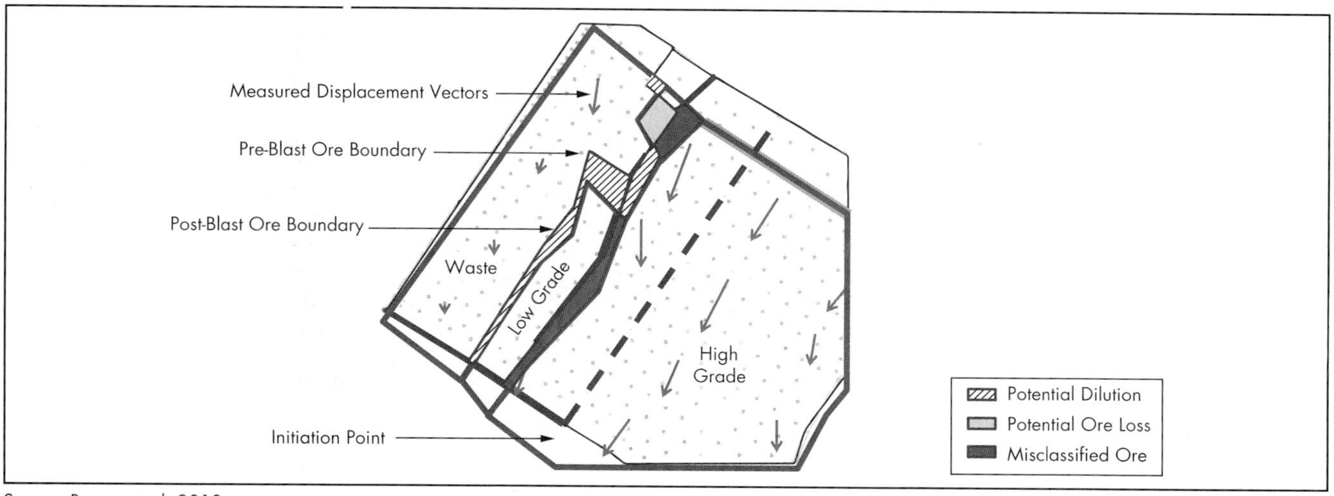

Source: Rogers et al. 2012

Figure 10 Template to adjust ore boundaries for blast movement

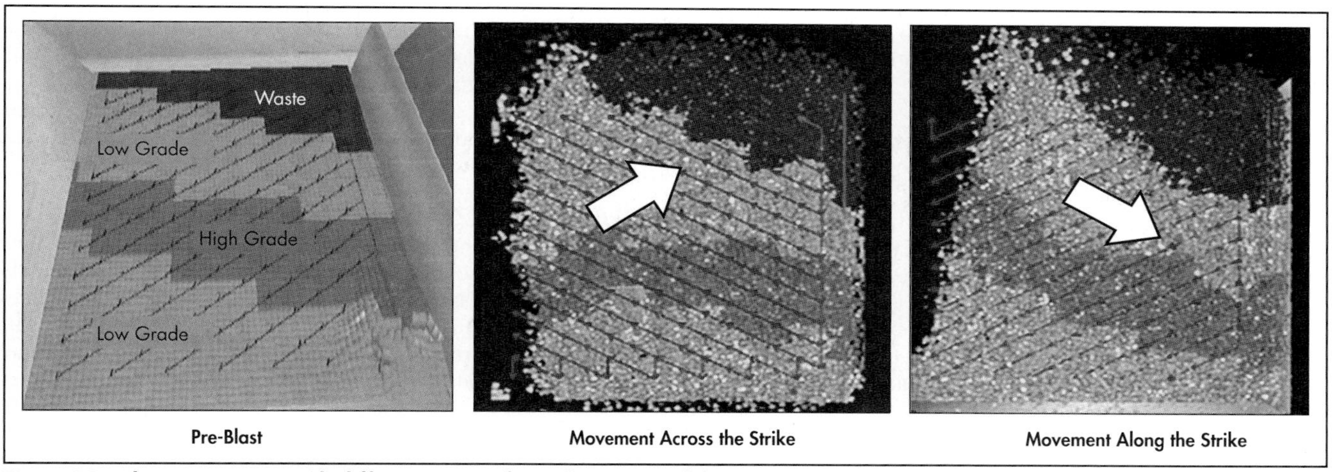

Figure 11 Blast movement with different timing from the BRC model

Figure 11 is an example of ore block movements simulated using the BRC model with different delay configurations. The simulation results clearly show that promoting rock movement parallel to the strike of ore body results in less disruption to the boundaries compared to movement perpendicular the strike of the ore body, which is consistent with field observations.

BLAST MEASUREMENT—HOW TO MEASURE BLAST OUTCOMES

To quantify the impact of blast outcomes on downstream processes and then to sustain the benefits, it is very important to measure the blast outcomes (PSD, blast movements, dilution, ore loss, damage, etc.) in an accurate and timely manner.

Fragmentation Measurement Techniques

Application of digital image processing (DIP) techniques to measure and/or assess fragmentation has become popular in the last few decades. The two primary types of DIP systems are desktop and online. The desktop systems are those that involve the collection of images from a blasted muck pile. The particles in these images are delineated and their sizes are measured using image analysis software. In desktop systems, there can be manual intervention to delineate the particles. In online systems, images are captured automatically at regular intervals and PSD is measured automatically. The online systems can be installed at the crusher feed, moving conveyors, shovel bucket, and so forth, where real-time fragmentation analysis is desired.

One of the main drawbacks of current image analysis systems is their inability to accurately measure fines because fine particles are not always present on the surface of the material on a truckload or conveyor belt because of segregation. Figure 12 shows the plan and cross section of particles on a belt conveyor where it can be clearly seen that most of the fines are at the bottom of the pile. Since the images are generally taken on the top, the fine particles are not seen on the image, and hence not accounted for or measured. Secondly, even if the fine particles were visible on the surface, the individual fragments are too small to be delineated properly because of resolution limitations of the imaging systems.

Figure 12 Hidden fines on a conveyor belt

Noy 2006 suggested using 3-D imaging to improve the accuracy of particle delineation. Thurley (2009) developed a system using lasers and demonstrated that a laser system is more accurate than 2-D image analysis systems. However, even these systems have the same limitation in assessing hidden fines.

Blast Movement Measurement Techniques

Drill-and-blast engineers have been using several methods to measure rock movement during blasting. These methods can be categorized into direct and indirect techniques. Indirect techniques such as high-speed video analysis and mobile radar systems measure rock mass movement (displacement, velocity, and shape) of muck-pile surfaces during blasting. The movement data from the muck-pile surface is used to estimate or infer the subsurface movement inside the exposed surface.

The need to measure rock movements beyond the exposed surface led to the development of several direct methods, such as passive markers (PVC pipes, colored stemming) and interactive remote transmitters (e.g., radioactive, magnetic, and electronic). The simplest direct method consists of marking out or delineating the areas of interest on the bench's horizontal surface with paint, chalk, or flagging tape (Davis et al. 1989; Morely and McBride 1995). Excessive heave, cratering, or stemming ejection could damage these markers and make them untraceable, and they only measure surface movement. The second method consists of installing passive visual markers such as polyethylene pipe, chains, wooden stakes, or colored stemming into specially drilled holes or into the stemming regions of the blastholes such that they protrude from the surface (Davis et al. 1989; Morely and McBride 1995; Scott et al. 1996; Zhang 1994). This method requires surveying of pre- and post-blast locations to estimate the rock movement and thus can disrupt production.

The next generation of markers are those that involve remote detection methods. They include radioactive markers and magnetic markers (Gaidukov 1970; Harris et al. 2001; Gilbride 1995). Usage of radioactive markers is restricted because of health and safety concerns, whereas magnetic markers have limited detection range and poor recovery.

The latest measurement systems involve Blast Movement Monitors (BMMs) developed at the JKMRC at the University of Queensland. The BMM is an electronic target that can be programmed at different frequencies and placed in different areas of the blast at variable depths. After the blast, it is necessary to walk over the muck pile with a radio frequency detector and global positioning system (GPS) to establish post-blast locations of the BMMs (Thornton et al. 2005).

The post-blast depth of the monitor is calculated from its signal strength, and its location is measured using conventional surveying techniques of a GPS. Using this information and pre-blast coordinate data, 3-D movement vectors can be calculated prior to muck-pile excavation (Figure 13).

BLAST OPTIMIZATION

Although blasting is the first step in the comminution and separation process of any mining operation, traditionally the blasting process is grouped with the mining processes and optimized to minimize the total mining costs. In this approach, blast results are considered good when they ensure good digging and loading operations while maintaining safety and environmental standards. That is why graphs (shown in Figure 14) are often used to estimate the optimum blasting effort (McKenzie 1965; Dinis da Gama 1990; Eloranta 1995).

The aforementioned approach may be appropriate if the sole purpose of the operation is to move the in situ rock mass from one place to another. In this traditional approach, the mine and concentrator are managed as separate cost center "silos" and each cost center is optimized independently, with little understanding of the real effect on downstream processes and overall efficiency. However, in the case of most metal mining operations, the in situ ore is separated from the waste rock and is subjected to a series of breakage and separation processes to convert it into a valuable product. Profitability in the mineral industry is the difference between the price of the final product when sold and the costs to produce it and can be estimated as

profit = revenue – total operating cost – fixed cost

where

revenue = (grade × recovery × price) / (1 + dilution)

total operating cost = cost of drilling, blasting, loading, hauling, crushing, grinding, and liberation

fixed cost = capital and overhead cost per annum / total production per annum

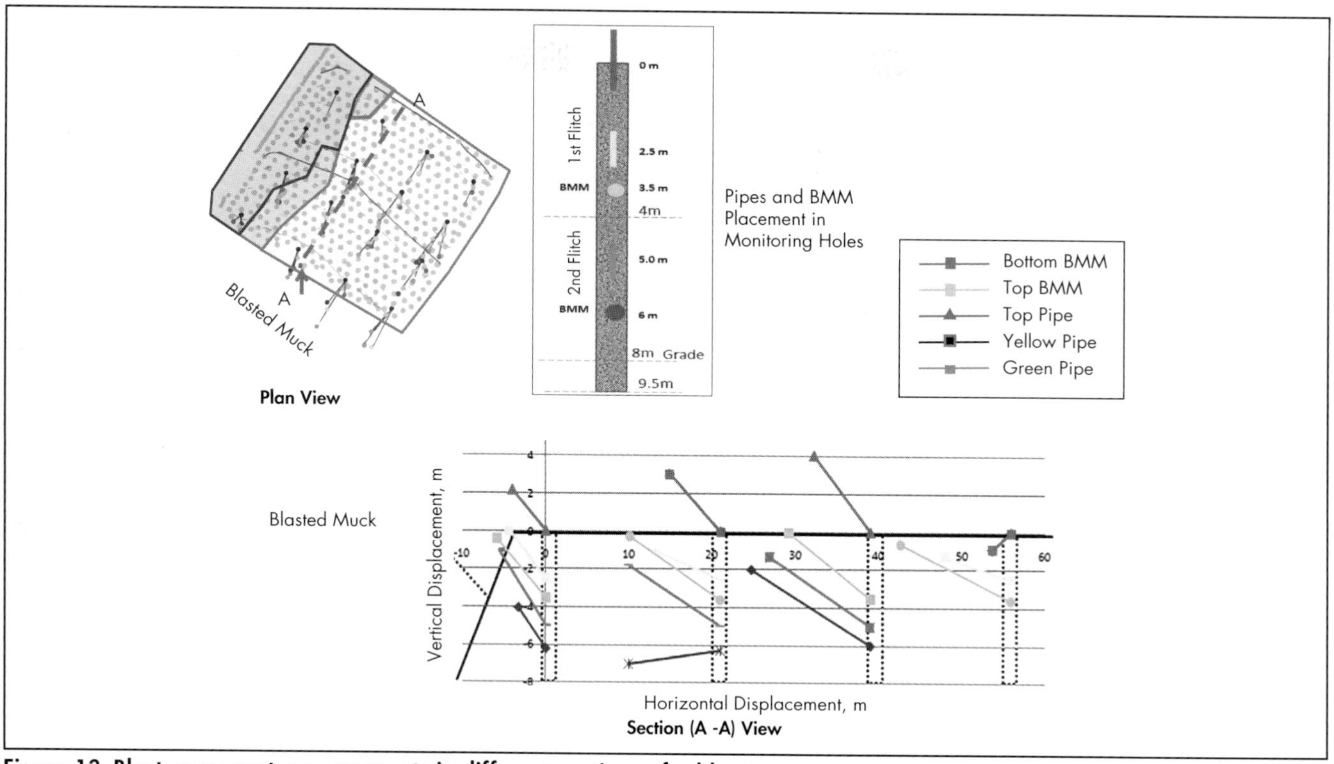

Figure 13 Blast movement measurements in different portions of a blast

In most cases, fixed costs are a significant portion of overall costs, and the increase in throughput without additional capital will reduce the fixed cost per ton of final product. Since blasting cost is an operational cost, any increase in final product throughput should be weighed against the decrease in unit fixed costs. In cases where there is demand for the final product, it is important to consider this aspect. Finally, in mine-to-mill blast optimization, the optimum blasting effort is where overall profit per ton of ore is maximized. In this approach, the aim is to leverage blasting outcomes to maximize the overall profits of operations rather than reduce the unit mining cost or unit operating cost.

Figure 15 shows approximate unit costs of key processes within the mine-to-mill value chain for a large open pit base metal operation treating hard ore. In this operation, a 100% increase in drilling-and-blasting cost can be compensated by a 4% to 5% reduction in downstream grinding costs. Since drill-and-blast is the first step in comminution and separation, its impact will be felt in all the downstream processes.

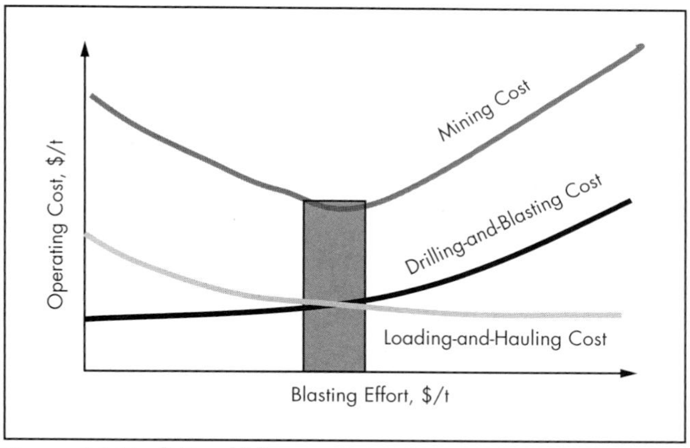

Figure 14 Traditional blast optimization

CASE STUDIES OF MINE-TO-MILL BLAST OPTIMIZATION PROJECTS

The theory of using blasting outcomes to improve the downstream processes is not new and has been implemented at several operations. However, most of the earlier case studies were limited to a large extent to improving the productivity of load and haul or the primary crushing process. During the AMIRA/JKMRC mine-to-mill research project P483A, *Optimization of Mine Fragmentation for Downstream Processing* (JKMRC 1998), the concept has been extended to understand the impact of blasting outcomes on grinding operations. The positive effect of finer feed size distribution on mill throughput, especially on SAG and AG mills, has been demonstrated at several mining operations around the world: Highland Valley Copper in Canada; Minera Alumbrera in Argentina; Cadia Hill, Kalgoorlie Consolidated Gold Mines, and Ernest Henry in Australia; Batu Hijau in Indonesia; Porgera gold mine in Papua New Guinea; and Candelaria in Chile. Most of these operations include conventional SAG-mill/ball-mill circuits and to a large extent were SAG-mill constrained. In all these operations, the blast designs were modified with higher energy, better energy distribution, and better confinement to produce finer fragmentation. Finer blast fragmentation coupled with tighter primary crusher operation produced much finer feed size distribution to the SAG mill, resulting in a throughput increase of between 5% and 25%. In most of the cases,

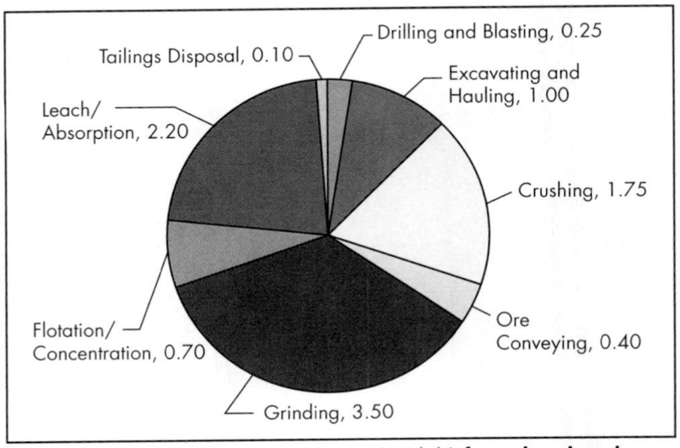

Figure 15 Approximate unit costs (US$/t) for a hard rock open-pit base metal operation

the ball mills have enough power to take care of additional throughput. In operations where there is a limitation of ball-mill power, either the throughput was decreased or the grind size was coarsened, whichever resulted in maximum metal production. A list of operations where finer ROMPSD from blast optimization improved grinding circuit performance is given in Table 2.

Case Study 1—Impact of Blast-Induced Dilution and Ore Loss (taken from Engmann et al. 2013)

Newmont's Ahafo operations in Ghana realized the impact of blasting outcomes, mainly the fragmentation and dilution, on the overall profitability of their operations and therefore implemented a blast optimization program to improve fragmentation and minimize blast-induced dilution. The blast optimization project was implemented in two phases. The first phase resulted in much finer fragmentation and allowed operation of the primary crusher with a much tighter gap. These changes resulted in much finer feed to the SAG mill, resulting in increased throughput in primary ores by up to 30% (Mwansa et al. 2010; Dance et al. 2011). In the second phase, a comprehensive blast movement monitoring program, as shown in Figure 16, was implemented to minimize blast-induced ore loss and dilution. A detailed description of this study is given by Rogers et al. (2012) and Engmann et al. (2013). Based on the understanding of blast movement dynamics from the study, alternative strategies were developed and implemented to reduce blast-induced ore loss and dilution. These strategies included

- Implementation of blast designs to promote consistent movement along the strike of the ore body and minimize inconsistent movement from edge effects, uneven free faces, and cratering, especially along the ore/waste boundaries; and
- Adjusting the post-blast ore boundaries to account for expected blast movement and to ensure that excavation follows the adjusted polygons. The blast movement vectors were estimated based on blast movement measurements.

The alternative strategies developed from this study have been incorporated in standard site operating procedures. Reconciliation data show improved agreement between the mine and mill grade and reduction in diluted tons since the implementation of the blast movement study in April 2011 (Engmann et al. 2013; Figure 16).

In addition to blast movement adjustments, several other improvements, such as increased blasthole sampling and pit supervision, have already been implemented. The geologists at Ahafo mine consider the biggest gains to have been made from post-blast movement adjustments. Post-blast polygon adjustments have enabled ore to be recovered correctly. This project has also improved focus on heave mining and excavation control to ensure that grade blocks are captured correctly in dispatch and the correct sampling techniques are being followed.

Case Study 2—Potential Impact of Microcracks on Mill Throughput (taken from Kanchibotla et al. 2015)

Paddington mill in Australia was expected to treat a significant portion of fresh ore from the Navajo Chief open pit, which was harder compared to the then blend of altered ore. This was expected to have a negative effect on mill performance; hence, a blast optimization project was implemented to improve the feed size distribution from fresh ores. The optimization involved advance blasting practices with changes to the blast energy, energy distribution, type of energy, and, more importantly, priming and delay timing. The advance blast designs resulted in finer fragmentation and a 36% increase in mill throughput. The different factors and their relative contributions in increased throughput are shown in Figure 17.

The results highlighted that the most significant contributor to the increase in throughput between the baseline and validation surveys was the change in ore hardness, followed by the increased blasting intensity (finer ROMPSD) and some improved circuit parameters. The drill penetration rates and the point load tests data indicated that the pre-blast hardness of ore (as measured by the drill penetration rates and point load tests) from the advanced blast designs was harder than the baseline blasts (Figure 18). However, the comminution properties measured as from Bond tests and JKMRC drop weight tests showed that the ore from the advanced blast designs was softer than that from the baseline blasts (Figure 19). The JKMRC drop weight test measures the impact breakage of ores by two parameters, A and b, and $A \times b$ is considered the impact hardness of ore (Napier-Munn et al. 1996).

The study results suggest that the high-intensity advanced blasts with multiple primers and fast timing may have created more microcracks and apparently softened the ore. This phenomenon was observed in several laboratory blasting experiments and few controlled blasting trials (Revnivstev 1988; Chi et al. 1996; Nielsen and Kristiansen 1995; Kojovic and Wedmair 1995; Michaux 2005; Kim 2010; Parra 2013; Parra et al. 2016). This hypothesis needed to be further investigated; consequently, research is being conducted at the JKMRC in this area. However, if this hypothesis is proven to be correct and the advanced blasts can consistently soften the mill feed, it can have a significant impact not only on grinding circuit performance but also on recovery.

CHALLENGES IN IMPLEMENTATION

Even though the benefits of the mine-to-mill blast optimization methodology are obvious, there are many challenges to sustain the benefits in the long term. One of the key challenges is to convince one part of the value chain (e.g., blasting and mining) to increase their costs to realize the benefits at other parts (e.g., grinding) of the value chain. It requires monitoring

Table 2 Select mine-to-mill blast optimization case studies

Mine	Mine-to-Mill Program Description	Process	Reference
Highland Valley Copper, Canada	Blast optimization included pattern changes and stemming practices.	Demonstrated an increase in SAG mill throughput by controlling feed size.	Djordjevic 1999; Dance et al. 2011
Minera Alumbrera, Argentina	Blast designs were matched to ore types, and the primary crusher closed side was reduced.	Mill throughput increased by at least 13%. Increased blasting costs would have been offset by a 5% increase.	Valery 2000
Placer Dome's Porgera gold mine, Papua New Guinea	Blast designs were modified to produce a finer feed size distribution to the SAG mill.	Finer run-of-mine particle size distribution (ROMPSD) from the modified blast increased SAG mill throughput by up to 25%.	Grundstrom et al. 2001; Lam et al. 2001
Barrick/Newmont's Kalgoorlie Consolidated Gold Mines operation, Australia	Blast optimization and primary crusher operations were improved. Powder factor increased 29%.	Mill throughput increased by 13%.	Kanchibotla et al. 1998; Nelson et al. 1996
Cadia Hill, Australia	A finer feed size distribution was produced through modified blasting practices. SAG mill and pebble crushing operating parameters were modified.	Demonstrated an improvement in SAG mill throughput by up to 10% from finer feed size.	Kanchibotla et al. 1999; Hart et al. 2001
Antamina, Peru	Blast designs were modified in two interventions in harder ores to produce finer feed to the mill. The primary crusher and mill operating practices were modified to take advantage of finer feed.	Finer ROMPSD improved the primary crusher, and SAG mill throughput increased by more than 25%.	Rybinski et al. 2011; Kanchibotla and Valery 2010
Newmont's Ahafo gold mine, Ghana	Blasting practices and primary crushing operations were modified to produce finer mill feed.	SAG mill specific energy was reduced by 25% and the throughput in primary ores was increased by up to 30%.	Dance et al. 2011
Newmont's Ahafo gold mine, Ghana	Blast movement was monitored and a system was implemented to adjust ore blocks for blast movement.	Blast-induced dilution was reduced.	Engmann et al. 2013
PanAust's Ban Houayxai mine, Laos	Blasting practices in fresh ores resulted in lower mill throughputs. A blast optimization project was implemented that required changes in blast geometry and energy.	Changes in blasting practices resulted in finer mill feed and in mill throughputs by more than 40% than the design for fresh ores.	Symonds et al. 2015
Paddington gold operations, Australia	Blasting practices with advanced priming and timing designs were modified for harder ores to produce finer feed. This was coupled with changes in digging practices to reduce dilution and in the mill to take advantage of finer feed.	SAG mill throughput was increased by 35%, and energy consumption was reduced by 27%.	Kanchibotla et al. 2015

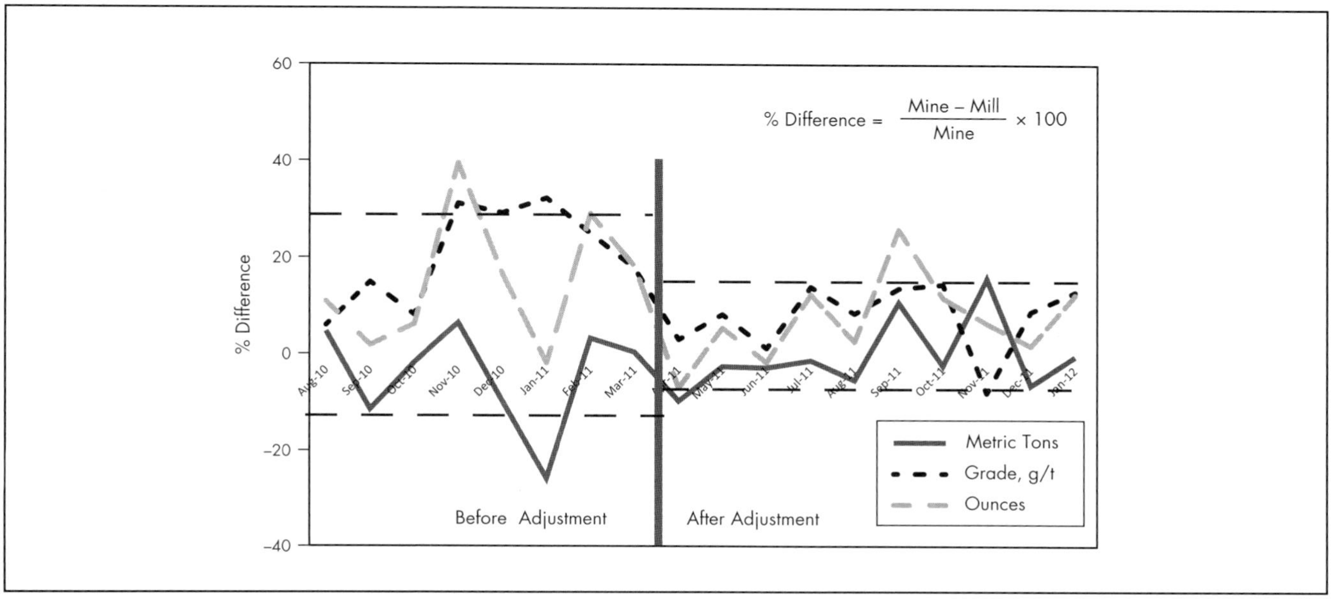

Source: Engmann et al. 2013

Figure 16 Grade reconciliation between mine and mill

of key performance indicators (KPIs) at each stage and effective communication between the key processes.

Another important challenge is to understand the characteristics of the ore body, its variations, and its response to key processes and then implement processes to improve the overall performance of the mine-to-mill value chain. In most operations, rock mass is characterized and tested by geologists, drill-and-blast engineers, and metallurgists separately for different purposes with very little communication between different groups. In a majority of these cases, available information is either late or inappropriate to use for operational decision making. For example, many mines collect blasthole chips and assay them to estimate the ore-grade boundaries, but the ore-grade boundary estimates from these data are not made available to drill-and-blast engineers before the blast. Therefore, reliable online techniques to measure the changes in the geology are very important for successful implementation of this approach. New technologies such as big data analytics, measure while drilling (MWD) techniques, and other geophysical testing methods can assist in this area. Further research is needed to understand the relationships between the MWD parameters and rock mass characteristics to develop process performance indices such as blastability, grindability, and separability using these data.

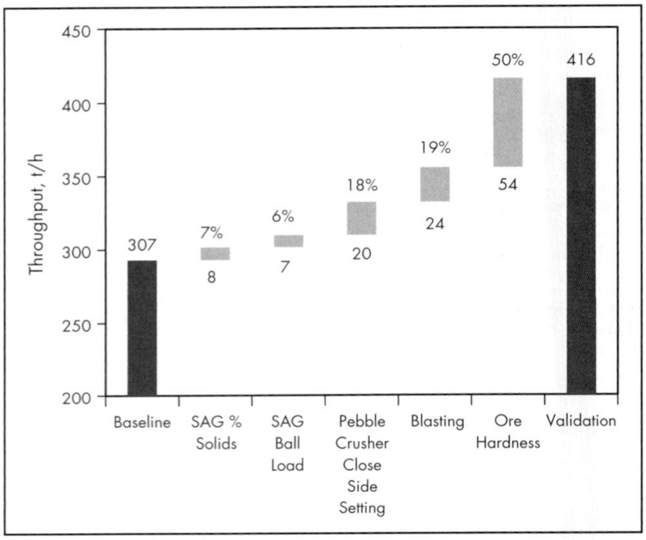

Source: Kanchibotla et al. 2015

Figure 17 Relative contribution of changes to throughput increase (estimates from simulation)

A good understanding of interactions between the key processes and the ability to leverage them to optimize the overall performance is another important element. Reliable models of key processes within the value chain would help not only to understand these interactions but also to control and optimize them. Current models represent only one or two processes within the value chain, and most of the current models represent only one rock/ore. These models cannot handle the variability even within each rock/ore characteristic. Stochastic approaches are needed to understand the impact of natural variability of rock mass on the process performance. Further work is also needed to integrate the models of all key processes in one platform so that the operations can simulate the key mine-to-mill processes and understand the interaction between them.

Establishment of appropriate KPIs and reliable systems to monitor these KPIs accurately and report them on a daily basis is another important element. For example, reliable measurement of ROMPSD, especially fines, is important to quantify the impact of fragmentation, but the current image analysis systems have limitations in this regard. It is therefore important to develop reliable measurement systems that can measure the relevant KPIs in a timely and accurate manner so that the operations can use them to not only detect the changes but to take appropriate actions to mitigate the negative effects. Data from these online measurements can be used to calibrate the process models and then use them for day-to-day process control as well as optimization purposes. A schematic view of such an approach is shown in Figure 20.

It is important that the new methodology, designs, and operating procedures are embedded in the day-to-day standard operating procedures and all key site personnel are well trained to implement them. The measurement systems and/or models need to be embedded in the operational work flow rather than stand-alone separate systems so that the operators are familiar with them and can use them routinely for diagnostic as well as optimization purposes.

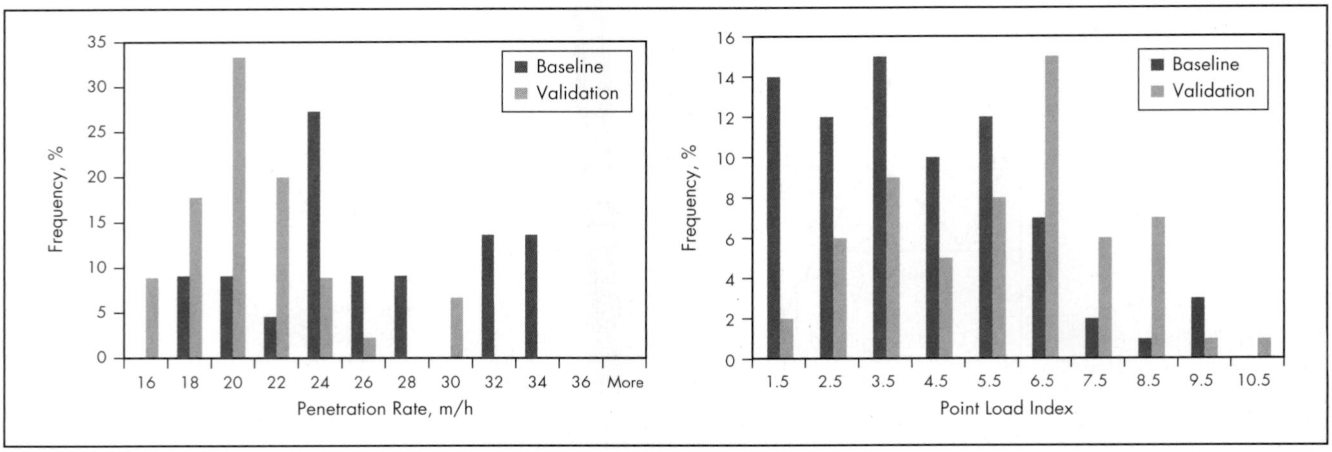

Source: Kanchibotla et al. 2015

Figure 18 Drill penetration rates and point load strength of rock from standard and advanced blasts

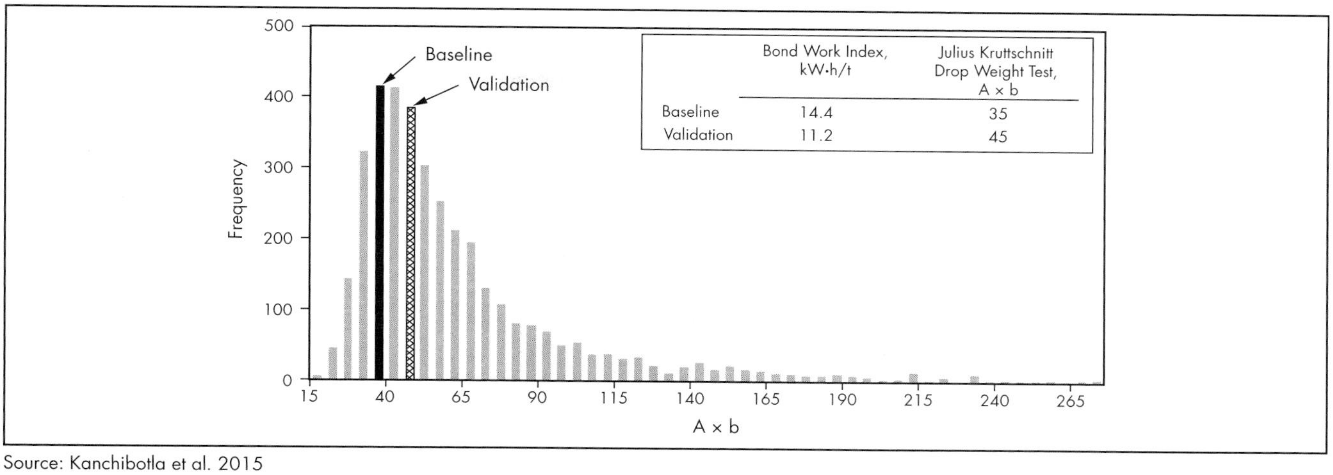

Source: Kanchibotla et al. 2015

Figure 19 Comminution properties of ore from standard and advanced blasts

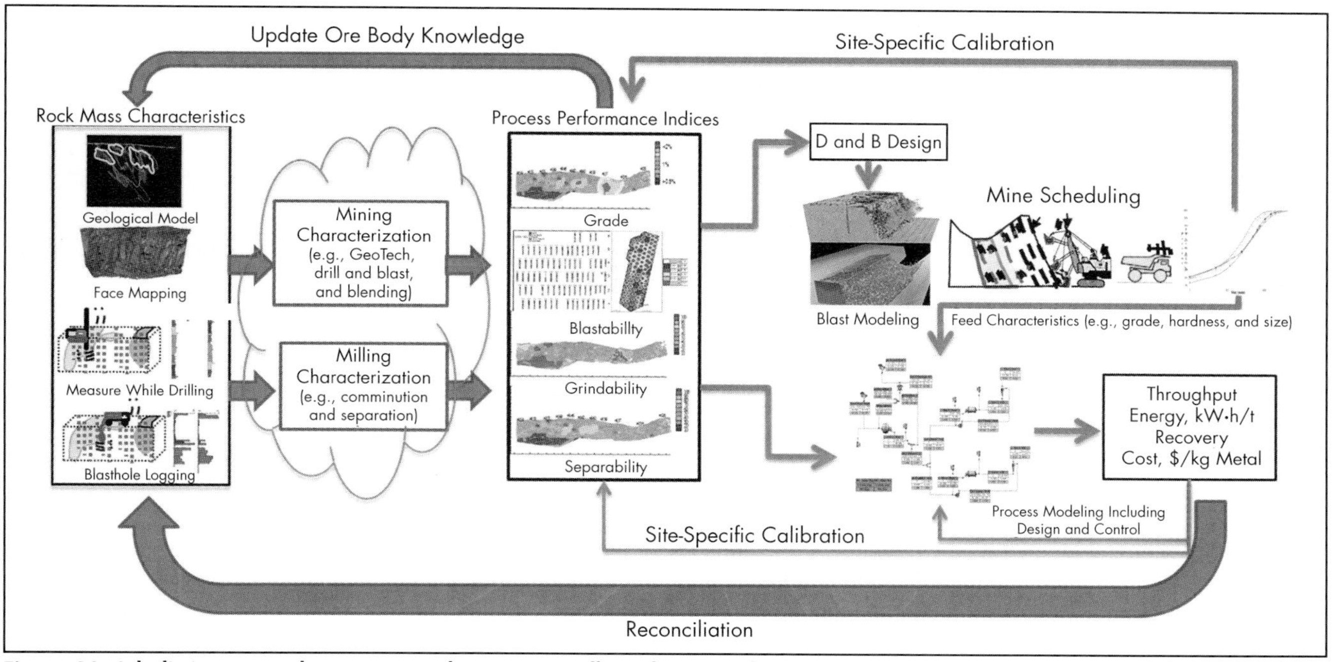

Figure 20 A holistic approach to optimize the mine-to-mill production value chain

CONCLUSIONS

The traditional approach of optimizing mining and milling operations separately can fail to recognize the potential benefits that can otherwise be achieved by optimizing the entire process. All the processes in the mine-to-mill production value chain are interdependent, and the results of the upstream mining processes can have a significant impact on the efficiency of downstream processes. A new holistic approach is needed wherein mining activities are designed to supply the mill with ore characteristics to increase the overall energy efficiency and productivity of the total production value chain rather than an individual unit process without adversely affecting safety, slope stability, and environmental risks. Blasting plays an important role in this approach, because it is the first step in comminution and separation processes within the mine-to-mill value chain.

Several case studies demonstrate the value of mining and concentrator personnel working together to reduce the overall cost of an operation rather than reducing their individual operation costs. The benefits derived from designing blasts to improve the efficiency of crushing and grinding stages far outweigh the additional blasting costs. Ore softening and grade increase in finer fractions with advanced blasting practices can further increase the leverage of blasting on the production value chain. Possible negative side effects, such as ore loss, dilution, and blast damage, can be controlled with improved understanding of the blasting process based on systematic monitoring. Effective communication between the key process areas and reliable tools to measure the KPIs in a timely manner are essential for successful implementation of this approach. A good understanding of ore-body characteristics and the ability to model their impacts on the performance of key processes in the value chain is important to sustain the benefits.

COMPARISON OF FRAGMENTATION ESTIMATES FROM THE KUZ–RAM AND JKMRC CRUSHED ZONE MODELS (adapted from Kanchibotla et al. 1999)

A comparison of fragmentation estimates between the Kuz–Ram model and JKMRC CZM is shown using the data from a blast trial conducted at Cadia Hill, a large copper–gold open pit mine in Australia. The rock mass properties, blast design parameters, and explosive properties for this blast are given in Table 3. ROM size distribution was estimated from 10 muck-pile images and 18 images taken from the back of trucks using Split image analysis software (Split Engineering 2017). Primary crusher product size distribution was obtained by taking a belt cut.

Calculations of the Kuz–Ram Model

The following calculations pertain to the Kuz–Ram model:

- Rock factor (A) is estimated as 10, because the rock is considered to be hard and fractured.
- Powder factor (K) = explosive weight / (burden × spacing × bench height) = 593 / (6 × 7 × 15) = 0.94 kg/m³
- Explosive charge weight (Q) = 593 kg
- Energy factor or relative weight strength (RWS) = 80
- Uniformity index (n) according to Equation 3 = 1.52
- Particle size distribution:
  - Mean fragment size (X_{50}) or 50% passing according to Equation 1 = 0.38 m
  - Cumulative % passing 1 mm or 0.001 m consistent with Equation 2 = 0%
  - Cumulative % passing 10 mm or 0.01 m according to Equation 2 = 0.3%
  - Cumulative % passing 100 mm or 0.1 m consistent with Equation 2 = 8.5%
  - Cumulative % passing 1,000 mm or 1 m according to Equation 2 = 95%

Calculations of the JKMRC Crushed Zone Model

The following calculations pertain to the JKMRC CZM:

- Rock factor (A) is similar to Lilly's blastability index shown in Equation 5 with some modifications:
  - Rock mass description (RMD) is estimated as function of in situ block size (I_{50}). In Figure 21, with an I_{50} of 1.4 m, RMD is estimated to be 50.
  - The joint factor (JF) is based on the assumption that rock is easier to blast if it has many fractures or no fractures in between two holes and is estimated as a function of the number of joints between blastholes, as shown in Figure 22. In this example, with an I_{50} of 1.4 m, the number of joints between two holes is estimated as five and the JF as 33.
  - RDI (rock density influence) = 25*(rock density – 2) = 25 × (2.6 – 2) = 15
  - HF (hardness factor) = Young's modulus (YM) if YM < 50 GPa
    = YM/3 if YM > 50 GPa
    = UCS (uniaxial compressive strength) /5 if 50 GPa > YM < 50 GPa
    = In this case, HF is 25.4 (UCS/4) because YM is 62 GPa
- Rock factor A according to Equation 5 = 0.6 × (50 + 33 + 15 + 25.4) = 7.4
- Powder factor and energy factor are the same as the Kuz–Ram model

Table 3 Rock properties, blast design parameters, and explosive properties for trial blast no. 760043

Rock Mass Properties	
Rock type	Monzonite
Density	2.60 t/m³
Young's modulus	62.00 GPa
Uniaxial compressive strength	127.00 MPa
Rock quality designation	100
Mean in situ block size (I_{50})	1.4 m
Blast Design Parameters	
Burden	6 m
Spacing	7 m
Hole depth	16.5 m
Hole diameter	229 mm
Explosive weight	593 kg
Bench height	15 m
Stemming length	4.50 m
Subdrill	1.50 m
Explosive Properties	
Explosive	Titan 2,070 g
Density	1.2 g/cc
Velocity of detonation	4,950 m/s
Relative weight strength (ANFO* = 100)	80

*ANFO = ammonium nitrate and fuel oil

- Mean fragment size (X_{50}) or 50% passing using Equation 1 = 0.28 m
- This model uses two uniformity indices.
  - Uniformity index for the coarse end is the same as the Kuz–Ram and in this example it is 1.52.
  - Uniformity index for the fines end is estimated using Equations 6 through 9.
    - Borehole pressure P_b = 0.25 × 1,200 × 4,000² = 7.3 GPa using Equation 7
    - Crushed zone radius according to Equation 6 = 0.87 m
    - Fines uniformity index according to Equations 8 and 9 with a grain size of 1 mm = 0.48
- Particle size distribution:
  - Mean fragment size (X_{50}) or 50% passing according to Equation 1 = 0.28 m
  - Cumulative % passing 1 mm (based on volume of crushed zone) = 4.5%
  - Cumulative % passing 10 mm or 0.01 m = 13%
  - Cumulative % passing 100 mm or 0.1 m = 34%
  - Cumulative % passing 1,000 mm or 1 m = 99%

A comparison of ROM predictions from the two models (Kuz–Ram and JKMRC CZM) with the fragmentation measurements from Split Engineering (2017) and crusher product belt cuts is shown in Figure 23.

The results showed good comparison between the Kuz–Ram estimate and Split measurements without fines correction, whereas the JKMRC CZM showed good comparison with crusher product belt cuts and Split measurements with fines correction. It is very well accepted that most image analysis systems underestimate fines.

The Kuz–Ram model estimates very little or no fines (–25 mm), whereas the crusher product belt-cut measurements

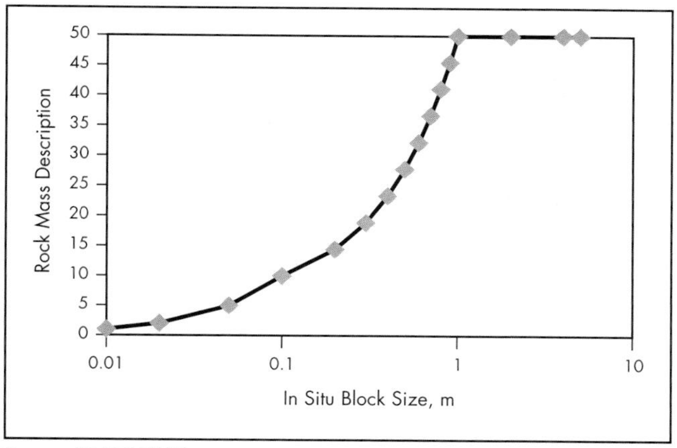

Source: JKMRC 1998

Figure 21 Relationship between RMD and in situ block size

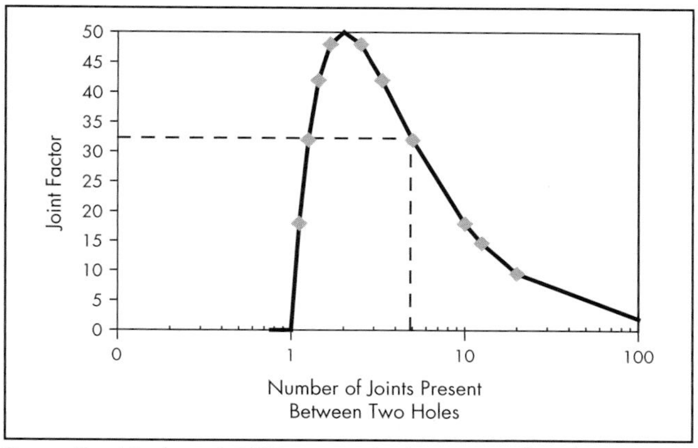

Source: Kanchibotla et al. 1999, reprinted with permission from the Australasian Institute of Mining and Metallurgy

Figure 22 Relationship between JF and number of joints between holes

clearly show around 20% of –25 mm particles. It is very well known that primary crushers do not generate much fines (–25 mm) and most of the fines in the primary crusher product would be coming from the blast. Therefore, it can be concluded that the conventional Kuz–Ram blast fragmentation model underestimates fines (–25 mm) compared to the JKMRC CZM.

REFERENCES

Brinkmann, J.R. 1987. Separating shock wave and gas expansion breakage mechanisms. In *Second International Symposium on Rock Fragmentation by Blasting*. Bethel, CT: Society for Experimental Mechanics.

Brinkmann, J.R. 1990. An experimental study of the shock and gas penetration in blasting. In *Third International Symposium on Rock Fragmentation by Blasting–Fragblast 90.* Melbourne, Victoria: Australasian Institute of Mining and Metallurgy.

Burns, R., and Grimes, A. 1986. The application of pre-concentration by screening at B.C.L. Presented at AusIMM Mineral Development Symposium, Madang, Papua New Guinea.

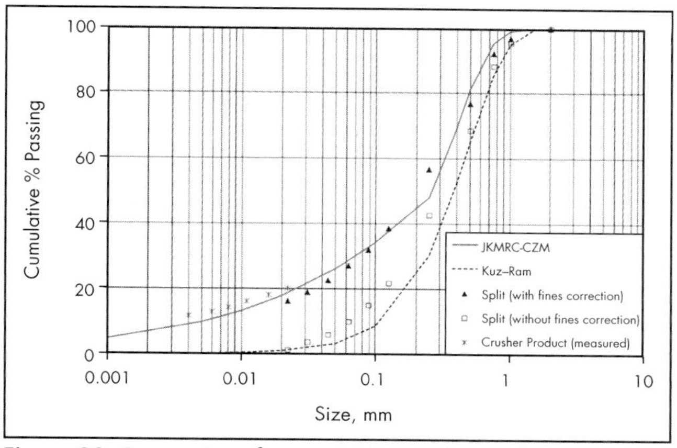

Figure 23 Comparison fragmentation estimates from the models and measurements

Cameron, A.R. 1992. Development of techniques for evaluating the performance of bulk commercial explosives. Ph.D. thesis, University of Queensland, Australia.

Chi, G., Fuerstenau, M.C., Bradt, R.C., and Ghosh, A. 1996. Improved comminution efficiency through controlled blasting during mining. *Int. J. Miner. Process.* 47:93–101.

Chiapetta, R.F., Bauer, A., Dailey, P.J., and Burchell, S.L. 1983. The use of high-speed motion picture photography in blast evaluation and design. In *Proceedings of the Ninth Conference on Explosives and Blasting Technique*. Edited by C.J. Konya. Montville, OH: Society of Explosives Engineers. pp. 258–308.

Comeau, W. 1996. Explosive energy partitioning and fragment size measurement—Importance of correct evaluation of fines in blasted rock. In *Proceedings of the Fragblast-5 Workshop on Measurement of Blast Fragmentation*. Brookfield, VT: A.A. Balkema. pp. 237–240.

Crone, J.G.D., Hammond, J.R., and Ward, T.A. 1973. Quality control at Hamersley Iron. Presented at AusIMM Annual Conference, Western Australia.

Cunningham, C.V.B. 1983. The Kuz–Ram model for prediction of fragmentation from blasting. In *Proceedings of the First International Symposium on Rock Fragmentation by Blasting*. Lulea, Sweden: Lulea University of Technology. pp. 439–454.

Dance, A., Mwansa, S., Valery, W., Amonoo, G., and Bisiaux, B. 2011. Improvement in SAG mill throughput from finer feed size at the Newmont Ahafo operation. Presented at SAG 2011, Vancouver, BC. September 25–28.

Davis, B.M., Trimble, J., and McClure, D. 1989. Grade control and ore selection practices at the Colosseum gold mine. *Min. Eng.* pp. 827–830.

Dinis da Gama, C. 1990. Reduction of cost and environmental impacts in quarry rock blasting. In *Proceedings of the Third International Symposium on Rock Fragmentation and Blasting*. Melbourne, Victoria: Australasian Institute of Mining and Metallurgy.

Djordjevic, N. 1999. Two-component of blast fragmentation. In *Proceedings of the Sixth International Symposium of Rock Fragmentation by Blasting*. Johannesburg: Southern African Institute of Mining and Metallurgy. pp. 213–219.

Duvall, W.I., and Atchison, T.C. 1957. *Rock Breakage by Explosives*. Report of Investigation 5356. Washington, DC: U.S. Bureau of Mines.

Elliot, R., Either, R., and Levaque, J. 1999. Lafarge Exshaw finer fragmentation study. In *Proceedings of the 25th ISEE Conference, Nashville*. pp. 333–354. Cleveland, OH: International Society of Explosives Engineers.

Eloranta, J. 1995. Selection of powder factor in large diameter blastholes. In *EXPLO 95: Exploring the Role of Rock Breakage in Mining and Quarrying*. Melbourne, Victoria: Australasian Institute of Mining and Metallurgy. pp. 25–28.

Engmann, E., Ako, S., Bisiaux, B., Rogers, W., and Kanchibotla, S. 2013. Measurement and modelling of blast movement to reduce ore loss and dilution at Ahafo gold mine in Ghana. In *Proceedings of the 2nd UMaT Biennial International Mining and Mineral Conference*. Tarkwa, Ghana: University of Mines and Technology.

Fourney, W.L. 1993. Mechanisms of rock fragmentation by blasting. In *Comprehensive Rock Engineering*. Edited by J.A. Hudson. Oxford: Pergamon Press. pp. 39–69.

Gaidukov, É.É. 1970. Characteristics, structures, and textures of caved rock. *Soviet Min. Sci.* (5):536–539.

Gaunt, J., Symonds, D., McNamara, G., Adiyansyah, B., Kennelly, L., Sellers, E., and Kanchibotla, S.S. 2015. Optimisation of drill and blast for mill throughput improvement at Ban Houayxai mine. In *Proceedings of the 11th International Symposium on Rock Fragmentation by Blasting*. Melbourne, Victoria: Australasian Institute of Mining and Metallurgy.

Gilbride, L.J. 1995. Blast-induced rock movement modelling for bench blasting in Nevada open-pit mines. Master's thesis, University of Nevada, Reno.

Grundstrom, C., Kanchibotla, S.S., Jankovich, A., and Thornton, D. 2001. Blast fragmentation for maximising the SAG mill throughput at Porgera gold mine. In *Proceedings of the 28th ISEE Conference, Orlando*. Cleveland, OH: International Society of Explosives Engineers.

Harris, G.W., Mousset-Jones, P., and Daemen, J. 2001. Blast movement measurement to control dilution in surface mines. *CIM Bull.* 94(1047):52–55.

Hart, S., Valery, W., Clements, B., Reed, M., Song, M., and Dunne, R. 2001. Optimisation of the Cadia Hill SAG mill circuit. Presented at SAG 2001, Vancouver, BC. September 25–28.

Hino, K. 1956. Fragmentation of rock through blasting, and shock wave theory of blasting. In *Quarterly of the Colorado School of Mines*. Golden, CO: Colorado School of Mines Press.

Jaeger, J.C., and Cook, N.G.W. 1979. *Fundamentals of Rock Mechanics*. London: Chapman and Hall.

JKMRC (Julius Kruttschnitt Mineral Research Centre). 1998. *AMIRA Final Report on P 483—Optimization of Mine Fragmentation for Downstream Processing*. Submitted to Australian Mineral Industries Research Association Limited.

JKMRC (Julius Kruttschnitt Mineral Research Centre). 2002. *Final Report on P 483A—Optimization of Mine Fragmentation for Downstream Processing*. Submitted to Australian Mineral Industries Research Association Limited.

Kanchibotla, S.S. 2014. Mine to mill value chain optimization—Role of blasting. In *Mineral Processing and Extractive Metallurgy: 100 Years of Innovation*. Edited by C.G. Anderson, R.C. Dunne, and J.L. Uhrie. Englewood, CO: SME. pp. 51–64.

Kanchibotla, S., and Valery, W. 2010. Mine to mill process integration and optimisation—Benefits and challenges. In *Proceedings of the 36th ISEE Conference, Orlando*. Cleveland, OH: International Society of Explosives Engineers.

Kanchibotla, S.S., Morrell, S., Valery W., and O'Loughlin, P. 1998. Exploring the effect of blast design on SAG mill throughput at KCGM. In *Proceedings of the Mine to Mill Conference*. Melbourne, Victoria: Australasian Institute of Mining and Metallurgy.

Kanchibotla, S.S., Valery, W., and Morrell, S. 1999. Modelling fines in blast fragmentation and its impact on crushing and grinding. In *EXPLO 99*. Melbourne, Victoria: Australasian Institute of Mining and Metallurgy.

Kanchibotla, S.S., Vizcarra, T.G., Musunuri, S.A.R., Tello, S., Hayes, A., and Moylan, T. 2015. Mine to mill optimisation at Paddington gold operations. Presented at SAG 2015, Vancouver, BC.

Kim, S. 2010. An experimental investigation on the effect of blasting on impact breakage of rocks. Master's thesis, Department of Mining Engineering, Queen's University, Kingston, ON, Canada.

Kojovic, T., and Wedmair, R. 1995. Prediction of fragmentation and the lump/fines ratio from drill core samples at Yandicoogina. Julius Kruttschnitt Mineral Research Centre internal report.

Kojovic, T., Kanchibotla, S.S., Poetschka, N., and Chapman, J. 1998. The effect of blast design on the lump:fines ratio at Marandoo iron ore operations. In *Proceedings of the Mine to Mill Conference*. Melbourne, Victoria: Australasian Institute of Mining and Metallurgy.

Kutter, H.K., and Fairhurst, C. 1971. On the fracture process in blasting. *Int. J. Rock Mech. Min. Sci.* 8(3):181.

Kuznetsov, V.M. 1973. The mean diameter of fragments formed by blasting rock. *Soviet Min. Sci.* 9(2):144–148.

Lam, M., Jankovic, A., Valery, W., Gannon, S., and Kanchibotla, S. 2001. Increasing SAG mill circuit throughput at Porgera gold mine by optimising blast fragmentation. Presented at SAG 2001, Vancouver, BC.

Lilly, P. 1986. An empirical method of assessing rockmass blastability. In *Large Open Pit Mine Conference, 89-92*. Melbourne, Victoria: Australasian Institute of Mining and Metallurgy.

Little, T.N., and VanRooyen, F. 1988. The current state of the art of grade control blasting in the Eastern Goldfields. In *Explosives in Mining Workshop*. Melbourne, Victoria: Australasian Institute of Mining and Metallurgy.

McKenzie, A.S. 1965. Cost of explosives—Do you evaluate it properly? In *Proceedings of the American Mining Congress, Las Vegas, Nevada*.

Michaud, P.R., and Blanchet, J.Y. 1995. Establishing a quantitative relation between post blast fragmentation and mine productivity: A case study. In *Proceedings of Fragblast-5*. Brookfield, VT: A.A. Balkema. pp. 389–396.

Michaux, S.P. 2005. Analysis of fines generation in blasting. Ph.D. thesis, University of Queensland, Brisbane, Australia.

Morely, C., and McBride, N. 1995. Keeping geologists, production personnel and contractors happy—An integrated approach to blasting at Boddington gold mine, WA. In *EXPLO 95*. Melbourne, Victoria: Australasian Institute of Mining and Metallurgy. pp. 29–34.

Mwansa, S., Dance, A., Annandale, D., Kok, D., and Bisiaux, B. 2010. Integration and optimisation of blasting, crushing and grinding at the Newmont Ahafo Operation. www.ceecthefuture.org/wp-content/uploads/2013/02/Ahafo-Mine-to-Mill-Optimisation.pdf?dl=1. Accessed Aug. 18, 2017.

Napier-Munn, T.J., Morrell, S., Morrison, R.D., and Kojovic, T. 1996. *Mineral Comminution Circuits: Their Operation and Optimisation.* JKMRC Monograph Series in Mining and Mineral Processing 2. Edited by T.J. Napier-Munn. Indooroopilly, Queensland: Julius Kruttschnitt Mineral Research Centre.

Nelson, M., Valery Jr., W., and Morrell, S. 1996. Performance characteristics and optimisation of the Fimiston (KCGM) SAG mill circuit. In *International Conference on Autogenous and Semiautogenous Grinding Technology.* Vol. 1. Vancouver: University of British Columbia. pp. 233–248.

Nielsen, K., and Kristiansen, J. 1995. Blasting and grinding—An integrated comminution system. In *EXPLO 95*. Melbourne, Victoria: Australasian Institute of Mining and Metallurgy. pp. 427–436.

Noy, M.J. 2006. The latest in on-line fragmentation measurement—Stereo imaging over a conveyor. In *Proceedings of the 8th International Symposium on Rock Fragmentation by Blasting*. Santiago, Chile: ASIEX. pp. 61–66.

Ouchterlony, F. 2005. The Swebrec function: Linking fragmentation by blasting and crushing. *Trans. IMM Sect. A-Min. Technol.* A29–A44.

Parra, H. 2013. Blast induced fragment conditioning and its effect on impact breakage and leaching performance. PhD thesis, The University of Queensland, St. Lucia.

Parra, H., Onederra, I., and Michaux, S. 2016. Effect of blast-induced fragment conditioning on impact breakage strength. *Min. Technol.* 123(2):78–89.

Preece, D. 1994. A numerical study of bench blast row delay timing and its influence on percent cast. In *Computer Methods and Advances in Geomechanics: Proceedings of the Eighth International Conference.* New York: A.A. Balkema.

Revnivstev, V.I. 1988. We really need revolution in comminution. In *Proceedings of the 16th International Mineral Processing Congress.* New York: Elsevier.

Rogers, W., Kanchibotla, S., Tordoir, A., Ako, S., Engmann, E., and Bisiaux, B. 2012. Solutions to reduce blast-induced ore loss and dilution at the Ahafo open pit gold mine in Ghana. *Trans. SME* Vol. 332.

Rybinski, E., Ghersi, J., Davila, F., Linares, J., Valery, W., Jankovic, A., Valle, R., and Dikmen, S. 2011. Optimisation and continuous improvement of Antamina comminution circuit. Presented at SAG 2011, Vancouver, BC. September 25–28.

Scott, A., Cocker, A., Djordjevic, N., Higgins, M., La Rosa, D., Sarma, K.S., and Wedmaier, R. 1996. *Open Pit Blast Design: Analysis and Optimisation.* JKMRC monograph. Indooroopilly, Queensland: Julius Kruttschnitt Mineral Research Centre.

Scott, A., David, D., Alvarez, O., and Veloso, L. 1998. Managing fines generation in the blasting and crushing operations at Cerro-Colorado mine. In *Proceedings of the Mine to Mill Conference.* Melbourne, Victoria: Australasian Institute of Mining and Metallurgy.

Scott, A., Kanchibotla, S.S., and Morrell, S. 1999. Blasting for mine to mill optimisation. In *EXPLO 99*. Melbourne, Victoria: Australasian Institute of Mining and Metallurgy.

Sellers, E., Furtney, J., Onederra, I., and Chitombo, G. 2012. Improved understanding of explosive–rock interactions using the hybrid stress blasting model. Presented at the Southern Hemisphere International Rock Mechanics Symposium, May 15–17, Sun City, South Africa.

Split Engineering. 2017. Digital image analysis software. Tucson, AZ; Johannesburg: Split Engineering.

Symonds, D., McNamara, G., Adiyansyah, B., Gaunt, J., Kennelly, L., Sellers, E., and Kanchibotla, S.S 2015. Optimisation of drill and blast for mill throughput improvement at Ban Houayxai Mine. In *Proceedings of the 11th International Symposium on Rock Fragmentation by Blasting.* Melbourne, Victoria: Australasian Institute of Mining and Metallurgy.

Taylor, S.L., Gilbride, L.J., Daemen, J.J.K., and Mousset-Jones, P. 1996. The impact of blast induced movement on grade dilution in Nevada's precious metal mines. In *Proceedings of the Fifth International Symposium on Rock Fragmentation by Blasting—Fragblast-5*. Edited by B. Mohanty. Brookfield, VT: A.A. Balkema. pp. 407–413.

Thornton, D., Sprott, D., and Brunton, I. 2005. Measuring blast movement to reduce ore loss and dilution. In *Proceedings of the Thirty-First Annual Conference on Explosives and Blasting Technique.* Orlando: International Society of Explosives Engineers. pp. 189–200.

Thurley, M.J. 2009. Fragmentation size measurement using 3D surface imaging. In *Proceedings of the 9th International Symposium on Rock Fragmentation by Blasting.* Granada, Spain: CRC Press. pp. 229-237.

Tordoir, A.E. 2009. A study of blast induced rock mass displacement through physical measurements and rigid body dynamics simulations. PhD thesis, University of Queensland, Brisbane, Australia.

Tucker, M., Kanchibotla, S., and Ruest, M. 2006. Modelling coal loss in open-pit blasting using PFC2d. In *Proceedings of the 8th International Symposium on Rock Fragmentation by Blasting—Fragblast-8.* Santiago, Chile: Editec.

Valery, W., Jr. 2000. *Mine to Mill Optimisation Study at Minera Alumbrera.* Julius Kruttschnitt Mineral Research Centre internal report.

Walters, S. 2016. *Driving Productivity by Increasing Feed Quality Through the Application of Grade Engineering Technologies.* Kenmore, Queensland: CRC ORE.

Yang, R.L., Kavetsky, A., and McKenzie C.K. 1989. A two dimensional kinematic model for predicting muckpile shape in bench blasting. *Int. J. Min. Geolog. Eng.* 7:209–226.

Zhang, S. 1994. Rock movement due to blasting and its impact on ore grade control in Nevada open pit gold mines. Master's thesis, University of Nevada, Reno.

CHAPTER 3.3

Jaw and Impact Crushers

Richard (Ted) Bearman

JAW CRUSHERS

Jaw crushers have been the mainstay of the mining and quarrying industry for more than a century and, to a great extent, this situation continues. Apart from some advances (in bearing types, manganese wear liners, frame construction, and adjustment systems), the basic frame shape and principles of operation have not changed greatly since the machines were first introduced in the mid-19th century.

As with many types of equipment, there has been a rationalization of the number of manufacturers and the types of machines available over the last 20 years. Weiss (1985) gives a history and overview of a range of crushers, which is important context, but because of the aforementioned rationalization, this chapter only considers the more common single- and double-toggle jaw crusher types, as shown in Figures 1 and 2.

Of these two common types of jaw crusher, there has been a steady decline in the market for the more specialized, high-strength double-toggle crushers compared to the single-toggle crushers. This change in preference can be attributed to the increased strength and resilience of the newer single-toggle crushers and popularity of these machines for transportable and mobile plants.

Jaw Crusher Types

The application of energy dictates how a crusher performs, and in a jaw crusher, the energy applied is heavily dependent on the motion imparted to the swing jaw. For example, the double-toggle crusher uses a pendulum-type motion and hence, the variables in the energy application are related to the amount of eccentricity (stroke), nip angle between the jaws, and speed of the swing jaw. In single-toggle crushers, the motion of the swing jaw is controlled by additional factors including the angle of the toggle plate and the length of the toggle plate, which leads to a more complex motion.

Double-toggle crushers (also known as the Blake type) have always held a special place in mining, where they are regarded as the only machines to be used in the very hard and highly abrasive rock applications. This reputation is well founded and based on the design shown in Figure 3.

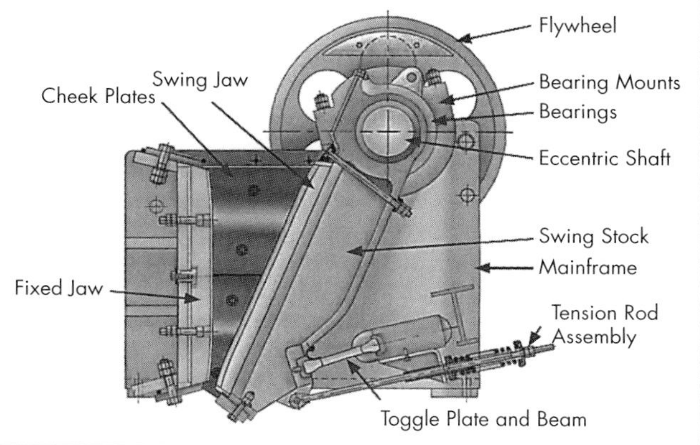

Courtesy of Telsmith

Figure 1 Single-toggle crusher

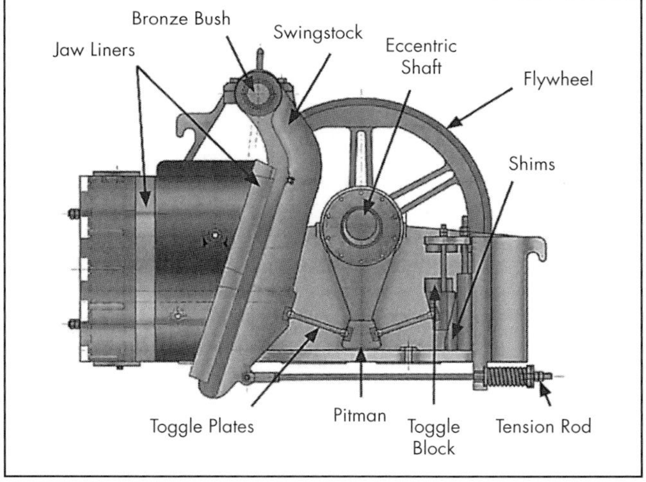

Courtesy of Telsmith

Figure 2 Double-toggle crusher

Richard (Ted) Bearman, Director, Bear Rock Solutions Pty Ltd., Attadale, Western Australia, Australia

Jaw Motion and Actuation

Typical Motion of Swing Jaw

Courtesy of Metso

Figure 3 Double-toggle jaw crusher motion

It is important to note that in a double-toggle crusher, the swing jaw is pivoted around a stationary jaw shaft, and the motion of the swing jaw is imparted via a separate shaft, the pitman, and a double-toggle plate arrangement. These components provide the strength assigned to double-toggle machines, as they generate the aforementioned pendulum-like crushing action.

The action is undoubtedly appropriate for hard feed material, but the crushing action does restrict performance in other ways. The swinging, or pendulum-type motion of the swing jaw, has a major upward component to the motion, and this essentially hinders material flow through the crushing chamber, making the throughput of double-toggle crushers slightly less than that of an equivalently sized single-toggle crusher. The other main restriction is that at the feed opening of a double-toggle crusher, the jaw motion is negligible. At the pivot point, there is zero movement; however, movement increases down the length of the crushing chamber. The lack of motion (at the feed opening) does not encourage material to feed into the crusher, and in the case of larger particles, there is an increased likelihood of bridging.

The lack of motion at the feed opening also leads to the rule of thumb that the maximum particle size for a double-toggle jaw crusher should be no more than 60%–70% of the minimum feed opening dimension. In contrast, the single-toggle crusher can tolerate a maximum feed size of 80%–85% of the minimum feed opening dimension. In an effort to improve the motion of the double-toggle crusher at the feed opening, Kue-Ken developed a machine where the swing jaw pivot point was raised upward and forward. Although this *overslung* design improved the motion, it was to the detriment of the feed opening configuration, and the ingress of feed into the chamber was restricted.

When considering a double-toggle crusher, there are certain features that set it apart from the single-toggle version. As noted earlier, the double-toggle machine is used in tough applications, and this is reflected in the mass of the machine. As an example, the ThyssenKrupp double-toggle 1,500 × 2,000 mm (60 × 79 in.) machine has a total mass of 285,000 kg (628,140 lb). By way of comparison, an equivalent single-toggle machine would weigh approximately 137,000–150,000 kg (302,033–330,693 lb). Simply, this mass difference means that in terms of assembly, maintenance, and installation, the requirements for a double-toggle crusher are significantly different and must be considered.

Single-toggle jaw crushers use an overhead eccentric shaft arrangement to impart motion to the swing jaw, as seen in Figure 4. Unlike the double-toggle machine, a single shaft provides both the pivot point and the source of eccentricity.

Traditionally, single-toggle machines were regarded as only suitable for soft through moderately hard rock types that display low to medium abrasion feed characteristics. Once into the hard and very hard feed types, the double-toggle became the preferred machine.

For process performance, the single-toggle crusher offers benefits in throughput and improved mobility in the feed area. These advantages come from the position of the eccentric shaft, which is at or just above the feed opening.

The position of the eccentric action provides maximum motion at the feed opening, and the subsequent induced motion has an elliptical form with the main axis of the ellipse having a downward inclination in the direction of the material flow. The orientation of the ellipse does change depending on the position in the crushing chamber because of the geometric relationship induced by the combination of the eccentric motion and the constraint of the toggle plate.

These ellipses provide throughput, but at the expense of crushing angle, and this is always one of the reasons why double-toggle machines have been preferred for hard feed types. Improvements in machine design now provide single-toggle crushers with a design that can handle extreme forces and at the same time have optimized the crushing motion ellipses to give a combination of throughput and angle that allow them to be highly effective in hard rock applications.

This elliptical motion, however, has a downside in that single-toggle machines generally experience two to four times the rate of wear of double-toggle machines in the same crushing conditions. Such a difference can be attributed to the scuffing action of feed material against the jaw plates brought about by the elliptical motion of the single-toggle jaw. In hard and highly abrasive feed types, liner wear rates, related costs, and downtime can become a major factor in the selection of single- versus double-toggle jaws. Applying strength and abrasivity limits to single- and double-toggle jaw crushers is difficult, as the different crusher models and manufacturers provide a variety of guidelines. In broad terms, if the strength

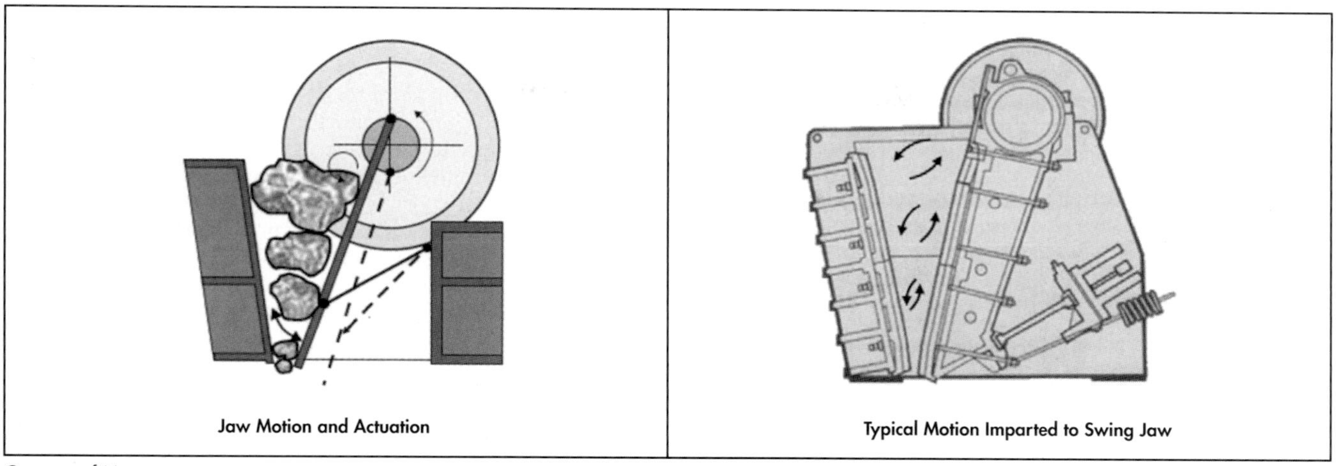

Courtesy of Metso
Figure 4 Single-toggle jaw crusher motion

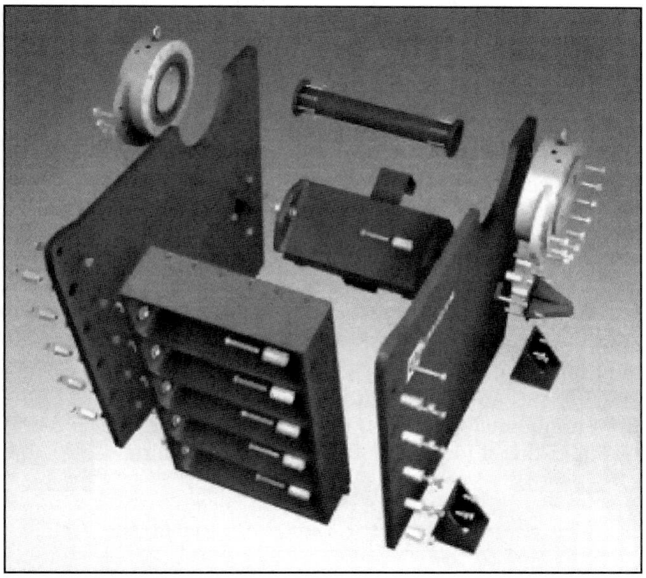

Courtesy of FLSmidth
Figure 5 Typical bolted assembly of jaw crusher mainframe

of the feed material is regularly and consistently at or above 400 MPa compressive strength, then a double-toggle crusher is likely to be more appropriate. In terms of abrasivity, feed material with a Bond abrasion index of 0.6 or greater would cause accelerated wear in a single-toggle jaw crusher, and a double-toggle option should be considered. As with all equipment selection tasks, there are a series of trade-offs required, and therefore, all other performance considerations should be taken into account, not just strength and abrasivity.

Single-toggle machines are considerably lighter than double-toggle machines with similar feed openings. The outcome is that single-toggle machines are more suited to installation on steel structures and for use on mobile plants.

Construction and Design

Currently, both gyratory and single-toggle crushers are now routinely used to crush feed types formerly reserved for double-toggle jaw crushers. This has been driven by changes in the structural design of single-toggle machines and an improved understanding of the forces present through widespread application of finite element analysis (FEA) and validation from strain gauging. FEA is a standard design tool for any piece of mechanical equipment and has been for many years. As such, many manufacturers make reference to the use of FEA in improving design, machine mass, and strength.

Alongside this modification, a manufacturing change has taken place, with a move away from castings for the mainframes and walls of many jaw crushers. Traditionally, castings were the more trusted method for construction of crushers, but the improvements in welding processes (automated and nonautomated), better heat treatment, more consistent steel plate quality, and nondestructive testing have all combined to make fabricated frames the preferred and lower cost method of construction.

In the following paragraphs, the major mechanical and design components of jaw crushers are examined, namely:

- Mainframe
- Pitman
- Fixed and swing jaws
- Flywheels
- Main shaft
- Toggle plates
- Setting adjustment
- Overload protection
- Crusher dimensions

Bolted frames are now the standard form of construction of mainframes in jaw crushers (Figure 5), although some manufacturers still use cast steel end frames, bolted and bossed into fabricated side frames. Use of FEA techniques has allowed stiffness to be designed into the frame using web plates, ribs, and fasteners, rather than relying on heavy, one-piece cast frames. Such an approach is standard across the major manufacturers (Sandvik 2012). Welded joints are also still widely employed in the fabrication of frames, and given modern quality control and nondestructive testing employed during manufacture, there is now a much reduced likelihood of the welds introducing stress raisers and possible weak points in the frames.

Fixed jaw arrangement has varied over time and, depending on the manufacturer, the angle of the fixed jaw has varied between vertical and 80–85 degrees to the vertical plane. The angle of the fixed jaw can assist in providing improved nip angles and also for presenting rocks in relation to the type of motion ellipse that is applied.

In both types, the longer the design of the crushing chamber, the lower the included angle between fixed and swing jaw for a given feed opening and closed side setting (CSS). In very *short* or squat jaw crusher chambers, there is a risk of particle slippage upward during the compression stroke, particularly if the discharge setting used is too tight.

Replaceable crushing liners used on both the fixed and swing jaws are very similar in both single- and double-toggle crushers. In some instances, the liner plates are designed to be reversible, or they can be rotated from top to bottom in the chamber, but plates need to be designed specifically to be moved and swapped in this manner. Because of the impact environment in a jaw crusher chamber, the liner material of choice is manganese steel, which work hardens at the liner surface because of deformation caused by the rock-on-liner interaction. The main design features for the liners is the number and profile of vertical corrugations used on both fixed and swing jaws. Metso (2011) provides a comprehensive description of the various profiles available for various crushing duties. Factors that drive the selection of nonstandard profiles include the following:

- Rock strength
- Feed abrasivity
- Shape of feed
- Requirement to reduce the slabbiness of the product
- Coefficient of fraction of the feed (i.e., slippery feed)
- Fines content of the feed

An example of some of the types of liner available is given in Figure 6.

In some applications, liner designs with a curved, convex surface are used, mostly on the swing jaw. Such curved liners are usually used to restrict compaction occurring in the chamber, particularly where there are larger quantities of fines in the feed.

The *pitman* arm is integral to the design of jaw crushers, but currently the term is mostly reserved for double-toggle crushers. In single-toggle jaw crushers, the word is less widely used and has in some instances been replaced by terms such as *jawstock, jaw holder*, or *swing jaw holder*.

In essence, the pitman is the component that transfers the eccentric motion from the eccentric shaft to the toggle plate(s) and then into the motion of the swing jaw. In the double-toggle crusher, it is a major independent component, but in the single toggle, it is essentially the mass of the swing jaw behind the swing jaw liner plates.

In an effort to improve the process performance of jaw crushers, most manufacturers now target the use of large eccentric throws in combination with appropriate speeds and nip angles suited to the feed material and size reduction being targeted.

Flywheels are integral to the design of jaw crushers, as the embedded energy in the flywheels is required to help overcome the strength of the rock and to smooth the power draw. On each machine, two flywheels are installed, with one acting as a sheave wheel, which is driven by a series of V belts, and the other acting as a balancing unit and also providing additional inertia. The key elements of flywheels are the bearings and the balance of the flywheels. The latter element is of particular importance when the crushers are used on mobile plants. Some modern designs of single-toggle machines are so well balanced that they can be mounted on rubber isolation mounts.

Toggle plates that give their name to the crushers under discussion are integral to the operation of the crushers.

In the double-toggle crusher, a rear toggle plate forms a pivot connecting the stationary toggle block to the pitman, and the front toggle plate links the pitman to the rear of the swing jaw. Toggles are designed to be robust, but in the case of the front toggle plate, it is also designed to act like a mechanical fuse and therefore to fail if uncrushable material enters the

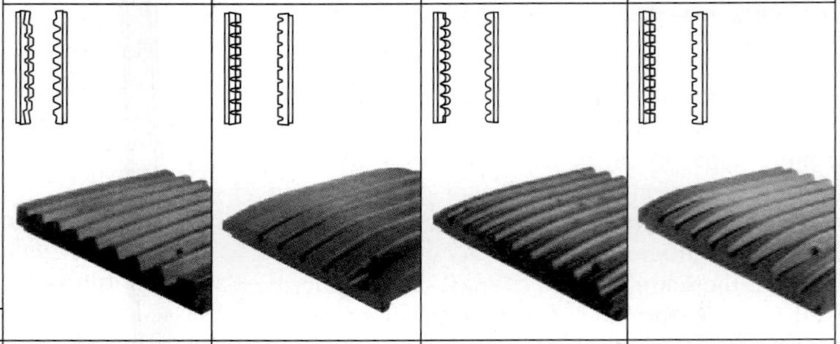

Feed Material Types:	Standard	Quarry	Superteeth	Special + Quarry
Hard Rock, UCS > 160 MPa	•	• • •	• •	• •
Soft Rock, UCS < 160 MPa	• •	• • •	• •	•
Gravel	• •	•	• • •	•
Slabby Soft Rock, UCS < 160 MPa	•			
Slippery Rocks	•	•	• •	•

Note: UCS = unconfined compressive strength. More bullets represent the better choice.

Courtesy of Metso

Figure 6 Example of some types of jaw crusher liner available

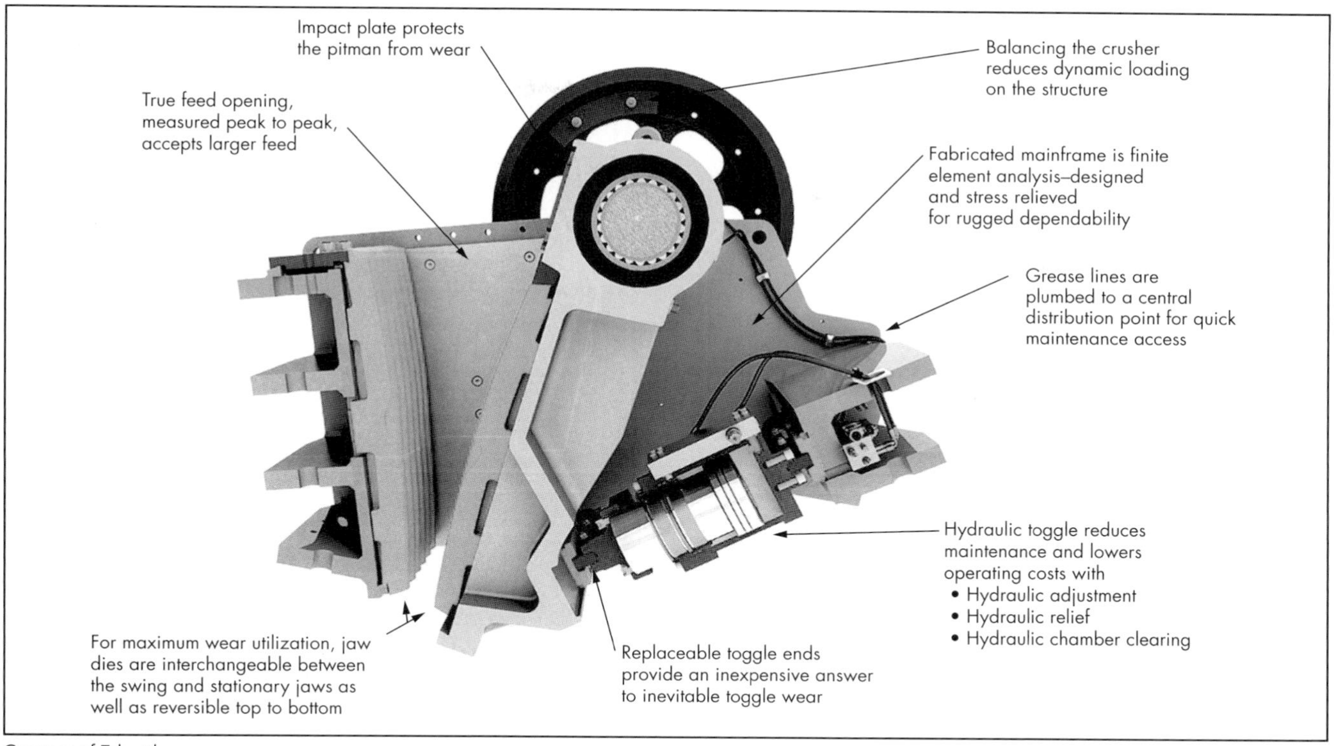

Courtesy of Telsmith

Figure 7 Hydraulic toggle arrangement

crusher. Good maintenance and lubrication of the toggle seats where the plate pivots against the toggle block seat, pitman seats, and swing jaw seat is essential. It is also required that the tension rods are properly maintained to prevent impact between the ends of the toggle plates and the pivot points. Some manufacturers employ *dry* rocking end toggles and seats that are designed for long service life and lubrication-free operation.

In single-toggle crushers, the toggle plate is free to pivot against a fixed toggle block at the rear of the crusher and also against the back of the swing jaw assembly. As with the double-toggle machine, the single-toggle plate must be free to pivot, and the correct tension needs to be applied to prevent any slack in the system. The tension rods used in single-toggle crushers are designed to be of a similar orientation to the angle of the toggle plate.

Setting adjustment in jaw crushers relies on moving the toggle plate, or rear toggle plate in double-toggle machines, which in turn moves the position of the swing jaw. Traditionally, plate metal shims were inserted to move the toggle block forward to reduce the CSS, or plates were removed to increase the CSS. This manual adjustment method is still commonly employed, but numerous attempts have been made to replace either the shimming technique or in some cases the entire metal toggle plate system with hydraulic solutions.

A common replacement for shim-based CSS adjustment is the use of counterposed wedges that can be driven into position to reduce the CSS or, alternatively, withdrawn to open the CSS. Such wedge assemblies can be either linear screw actuated or hydraulically actuated. In the Metso C series single-toggle jaw crushers, the wedge style is standard and the hydraulic version is available as an option. There are obvious advantages of having a wedge system, which include faster adjustment, reduced manual handling, and in the case of the hydraulic version, ability to remotely change the CSS from a control center.

Advancing beyond the wedge-based solution, several companies have replaced the metal toggle plate entirely with hydraulic cylinders (Figure 7). Some early attempts to deploy such a solution failed, but increasingly, such devices are finding favor. In addition to the improvements related to CSS adjustment, the hydraulic versions also offer a more effective solution to the ingress of uncrushables. Numerous patents have been lodged in the area of hydraulic adjustment and relief, including Haven et al. (2002) and Burhoff et al. (2015).

To enhance the hydraulic toggle system, the usual tension rods have also been upgraded. Figure 8 shows the toggle tensioning system deployed by Telsmith. The system uses spring assemblies to connect each end of the hydraulic toggle to the toggle seats so that there is no requirement to adjust the springs when changing the discharge setting.

In some machines, crusher overload protection has been built into the flywheels. One such device uses a torque limiting system to secure the flywheel to the shaft. In the case of uncrushable material, the device allows the flywheel to disengage and spin harmlessly on the shaft.

In some machines, the metal toggle plate is still designed to fail to protect the crusher, but in doing so, the plates must be replaced before production can resume.

If a hydraulic toggle unit is used, the rams retract to open the crusher once a pressure overload is detected; once the uncrushable material has passed, the hydraulic system resets the crusher setting back to normal. Using this approach, production can be resumed much quicker, and therefore operational interruption is greatly reduced.

One of the consequences of the difference in design between single- and double-toggle jaw crushers is the dimensions and masses of the two variants. Tables 1 and 2 show dimensions and masses for typical single- and double-toggle crushers.

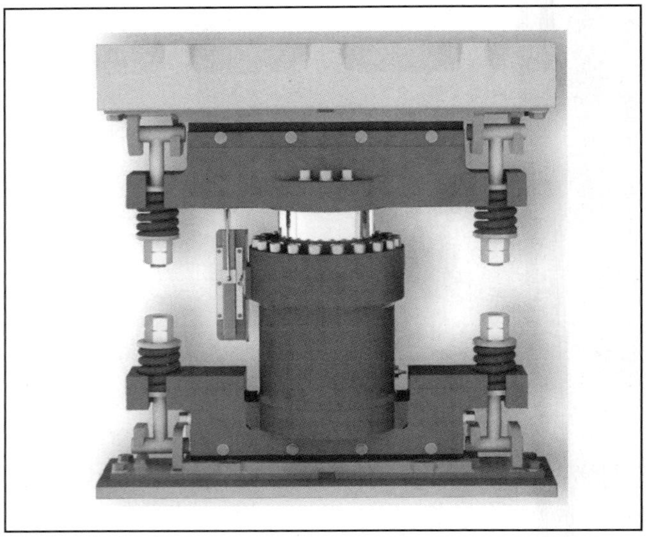

Courtesy of Telsmith

Figure 8 Toggle tensioning system

Operation

Jaw crushers are most commonly used for primary crushing duties, with features such as ease of maintenance, mass, capital cost, throughput, and durability driving the selection. Jaw crushers can be seen in some limited secondary duties; however, in most instances, such applications are based on availability of equipment and not on required process performance.

Jaw crushers are similar to other compression crushers in that in operation, they oscillate from an open side setting to a CSS; however, the reciprocating action of the jaw crusher and the steep chambers are not ideal features for maximizing throughput, or for controlling flow and hence the degree of size reduction (reduction ratios in jaw crushers are normally limited to 4:1).

Jaw crushers are reasonably accepting of fines in the feed, particularly at large CSS values. In practice, a majority of jaw crusher installations are set up to remove fines prior to the crusher, mostly using a vibrating grizzly feeder or heavy-duty screen. The action of the grizzly, or screen, ahead of the crusher also performs the useful function of aligning rocks prior to the feed opening of the jaw crusher, which can reduce blockages in the feed area. The decision to bypass fines is mostly taken to increase the overall capacity of the primary crushing station or to remove troublesome, wet, and sticky fines prior to the crushing chamber.

Table 1 Metso single-toggle dimensions and masses

		C80	C100	C96	C106	C116	C3054	C110	C125	C140	C145	C160	C200
A	mm	800	1,000	930	1,060	1,150	1,380	1,100	1,250	1,400	1,400	1,600	2,000
	in.	32	40	37	42	45	54	44	50	56	56	63	79
B	mm	510	760	580	700	800	760	850	950	1,070	1,100	1,200	1,500
	in.	21	30	23	28	32	30	34	38	43	44	48	60
C	mm	1,526	2,420	1,755	2,030	2,400	2,640	2,385	2,800	3,010	3,110	3,700	4,040
	in.	61	96	70	80	95	104	94	111	119	123	146	160
D	mm	2,577	3,670	2,880	3,320	3,600	3,540	3,770	4,100	4,400	4,600	5,900	6,700
	in.	102	145	114	131	144	140	149	162	174	182	233	264
E	mm	1,990	2,890	1,610	2,075	2,675	2,470	2,890	3,440	3,950	4,100	4,580	4,950
	in.	79	114	64	82	105	98	114	136	156	162	181	195
F	mm	1,750	2,490	1,460	2,005	2,730	2,470	2,750	2,980	3,140	3,410	3,750	4,465
	in.	69	99	58	79	107	98	109	118	124	135	148	176
G	mm	1,200	1,700	755	1,135	1,790	1,080	1,940	2,100	2,260	2,430	2,650	2,800
	in.	48	67	30	45	71	43	77	83	89	96	105	111
H	mm	2,100	2,965	2,500	2,630	2,885	2,950	2,820	3,470	3,755	3,855	4,280	4,870
	in.	83	117	99	104	114	117	112	137	148	152	169	192
I	mm	625	775	465	700	1,255	690	580	980	1,050	1,050	1,300	1,400
	in.	25	31	19	28	50	28	23	39	42	42	52	56
Basic crusher weight*	kg	7,670	20,060	9,759	14,350	18,600	25,900	25,800	37,970	47,120	54,540	71,330	121,510
	lb	16,900	44,240	21,520	31,650	40,920	57,100	56,880	83,730	103,900	120,260	157,280	267,930
Fully equipped crusher weight†	kg	9,520	23,300	11,870	17,050	21,500	30,300	29,500	43,910	54,010	63,190	83,300	137,160
	lb	21,000	51,390	26,170	37,590	47,300	66,800	65,050	96,830	119,100	139,330	183,680	302,440

Courtesy of Metso

Note: Certified general arrangement, foundation, and service space requirement drawings are available from Metso.

*Crusher without options.

†Crusher, hydraulic setting adjustment, flywheel guards, integral motor support, feed chute, automatic grease lubrication system, and typical electric motor.

Table 2 Double-toggle crusher dimensions and masses

Crusher Type	Largest Component	Heaviest Component	Dimensions of Largest Component, m × m × m (ft × ft × ft)	Weight Heaviest Component, kg (lb)	Maximum Weight of Crushing Jaw, kg/unit (lb/unit)
DB 6-4,2	Housing	Housing	2.60 × 1.67 × 1.08 (8.53 × 5.48 × 3.54)	4,590 (10,116)	425 (937)
DB 8-5,7	Housing	Housing	3.00 × 2.00 × 1.30 (9.84 × 6.56 × 4.27)	8,360 (18,425)	700 (1,543)
DB 10-8	Housing	Housing	3.90 × 2.65 × 1.57 (12.80 × 8.69 × 5.15)	17,600 (38,790)	1,600 (3,527)
DB 12,5-9	Housing	Housing	4.05 × 2.80 × 1.93 (13.29 × 9.19 × 6.33)	24,200 (53,337)	2,080 (4,584)
DB 15-12	Housing	Housing	5.60 × 3.90 × 2.40 (18.37 × 12.80 × 7.87)	59,070 (130,190)	4,290 (9,455)
DB 18-14	Side wall	Swing jaw	6.65 × 3.10 × 1.00 (21.82 × 10.17 × 3.28)	23,950 (52,786)	6,260 (13,797)
DB 21-16	Swing jaw	Swing jaw	5.40 × 3.10 × 1.50 (17.72 × 10.20 × 4.92)	28,710 (63,277)	8,030 (17,698)
DB 25-18	Swing jaw	Swing jaw	5.80 × 3.55 × 1.80 (19.03 × 11.65 × 5.91)	39,380 (86,794)	10,780 (23,759)

Crusher Type	A, mm (in.)	B, mm (in.)	C, mm (in.)	D, mm (in.)	E, mm (in.)	F, mm (in.)	G, mm (in.)	H, mm (in.)	J, mm (in.)	K, mm (in.)	L, mm (in.)	M, mm (in.)	N, mm (in.)	O, mm (in.)	P, mm (in.)	RØ, mm (in.)
DB 6-4,2	3,500 (138)	1,420 (55)	1,060 (42)	150 (6)	345 (14)	425 (17)	950 (37)	595 (23)	200 (8)	615 (24)	880 (35)	1,570 (62)	995 (39)	220 (9)	1,080 (43)	1,200 (48)
DB 8-5,7	4,200 (165)	1,780 (70)	1,350 (53)	150 (6)	350 (14)	570 (22)	1,040 (41)	780 (31)	200 (8)	700 (28)	1,010 (40)	2,020 (80)	1,280 (50)	320 (13)	1,300 (51)	1,650 (65)
DB 10-8	5,400 (213)	2,400 (95)	1,850 (73)	150 (6)	510 (20)	800 (32)	1,450 (57)	960 (38)	200 (8)	950 (37)	1,250 (49)	2,180 (86)	1,540 (61)	320 (13)	1,570 (62)	1,820 (72)
DB 12,5-9	5,700 (224)	2,510 (99)	2,000 (79)	200 (8)	480 (19)	900 (35)	1,635 (64)	880 (35)	250 (10)	1,050 (41)	1,380 (54)	2,730 (107)	1,930 (76)	400 (16)	1,930 (76)	2,100 (83)
DB 15-12	7,400 (291)	3,475 (137)	2,780 (109)	300 (12)	750 (30)	1,200 (48)	2,250 (88)	1,210 (48)	660 (26)	1,700 (67)	2,090 (82)	3,220 (127)	2,340 (92)	400 (16)	2,400 (95)	2,500 (98)
DB 18-14	8,700 (343)	4,200 (165)	3,200 (126)	250 (10)	865 (34)	1,400 (55)	2,750 (108)	1,280 (50)	135 (5)	2,020 (80)	2,490 (98)	3,830 (151)	2,830 (111)	500 (20)	2,960 (117)	3,000 (118)
DB 21-16	9,300 (366)	4,200 (165)	3,450 (136)	420 (17)	890 (35)	1,600 (63)	2,960 (117)	1,390 (55)	950 (37)	2,300 (91)	2,800 (110)	4,320 (170)	3,320 (131)	500 (20)	3,360 (132)	3,000 (118)
DB 25-18	10,250 (404)	4,860 (191)	3,900 (154)	420 (17)	1,080 (43)	1,800 (71)	3,460 (136)	1,500 (59)	1,100 (43)	2,250 (88)	3,060 (121)	4,960 (195)	3,860 (152)	600 (24)	3,820 (150)	3,500 (138)

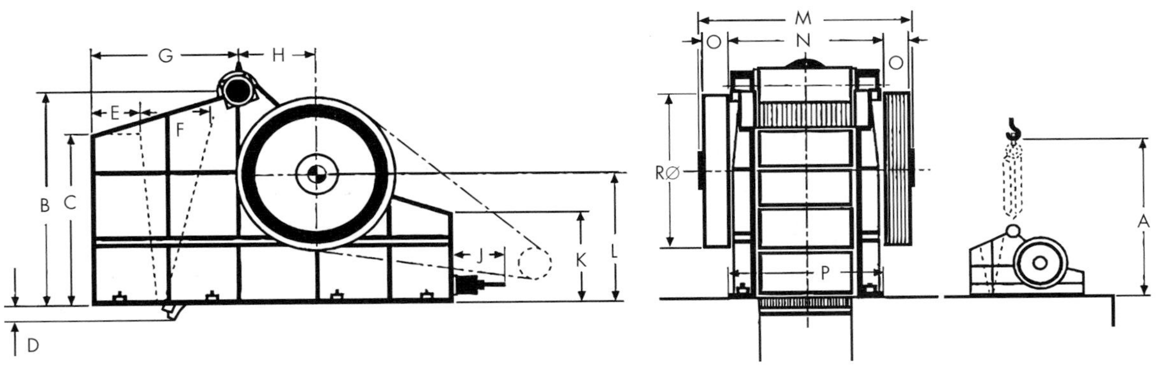

Courtesy of ThyssenKrupp Industrial Solutions

One of the more common feed approaches for jaw crushers is to have a run-of-mine (ROM) hopper receiving direct tips from haul trucks, with an apron feeder pulling material from the hopper onto a vibrating grizzly screen or feeder. As part of such an arrangement, it is also common to include a hydraulic rock breaker with sufficient reach to access oversize on the feed equipment and in the feed opening of the jaw crusher.

In jaw crushers, the size designation is usually quoted as *gape × width*, for example, 914 × 1,219 mm (36 × 48 in.), or 914-mm (36-in.) gape between the fixed and swing jaw plates at the top of the chamber by 1,219-mm (48-in.) wide, where the width of the opening is the dimension of the opening between the cheek plates. Some manufacturers reverse the designation, but regardless of the manner that the model is stated, the higher number is the width of the feed opening between the cheek plates.

In practice, the width determines the throughput of the crusher at any given CSS, and the gape controls the maximum feed size. Currently, jaw crusher sizes range up to 1,500 × 2,000 mm (60 × 80 in.) for single-toggle jaw crushers and up to 1,676 × 2,133 mm (66 × 84 in.) for double-toggle crushers. The installed power for both the largest single-toggle and double-toggle machines is approximately 400 kW (536 hp).

Throughputs for jaw crushers are quoted as the amount of material passing through the jaws. In reality, the throughput of a jaw crusher should always take into consideration the total throughput for the whole station, that is, the amount of material through the jaws and the amount of fines bypassing via the scalping grizzly or screen. The additional throughput from the scalping process often makes jaw crusher throughput and unit cost more comparable to gyratory operating costs.

Throughput rates for single- and double-toggle jaw crushers are shown in Table 3 and Figure 9. Product size distribution data for single- and double-toggle crushers is provided in Figures 10 and 11.

Application

In the vast majority of fixed plant and mobile plant applications up to 1,000–1,200 t/h (1,102–1,322 stph), jaw crushers are still the preferred primary crushing tool. In the larger open pit environment, there is a continuing debate about jaw versus gyratory crusher. The main differences between the two are in the processing rate achievable and the complexity of the feed arrangements.

The feed arrangement for a jaw crusher was discussed earlier: ROM hopper, apron feeder, grizzly screen or feeder, associated chute work, and a rock breaker. In comparison, the minimal arrangement for a gyratory crusher consists of direct tip into the crusher from haul trucks. In cases where jaw crushers can still meet the throughput requirements, the relative complexity of the installation should not be the only consideration. To gain a fair comparison of jaw versus gyratory application, the following should be considered:

- Capital cost
- Operating cost
- Maintenance
- Liner life and change-out
- Machine dimensions and mass
- Physical location

Capital cost difference varies considerably according to the type of installation, geographic location, topography, and the required crusher duty. As such, there is no single answer to which machine is best, but rather it must be determined on a case-by-case basis. Factors that influence the operating cost of a crusher station need to be considered, such as ancillary and feeding equipment, hopper and chute liner wear costs, associated labor, and power.

Jaw crushers are relatively simple and robust machines to maintain, with less complexity than gyratory crushers. As such, the amount of maintenance required and the duration of major maintenance activities is often much shorter for jaw crusher installations.

Probably one of the main examples is the difference in crusher liner change-outs. On gyratory crushers, a full change of the mantle and concave plates can take up to three days, whereas changing the fixed and swing jaw plates is an activity that can be comfortably completed inside one shift. In most jaw crushers, there is also the option of rotating and/or switching the jaw plates to give extended life. In gyratory crushers, this is not an option, but in some instances, the shells (including concave plates) can be turned through 180 degrees. This option only really applies if one-half of the crushing chamber is being worn at an accelerated rate compared to the other side.

A major difference between jaw and gyratory crushers are the dimensions of the crushers and the related masses. As already stated, double-toggle jaw crushers are much heavier than single-toggle jaw crushers, and for an equivalent feed opening gyratory, the mass is close to that of a double-toggle crusher.

In most instances, in an open pit environment, the physical mass differences can be handled by the installation of gantry cranes and other specialized lifting equipment, but a bigger factor in the discussion usually relates to the height of the machines.

For underground mines, height is a significant factor, as it drives the dimensions of the chamber and the need for large excavations (Nixon and Weston 2005). In many instances, the excavations are required for the entire mine life, and this presents geotechnical and cost issues, particularly in areas where the in situ stress state is poor. Minimizing the primary crusher chamber and/or cavern dimension often drives the decision to select a jaw crusher, but where high throughput is required, this causes a mismatch of machine and capability. Various mine sites have taken different views of such issues, ranging from multiple jaw crusher installations to single large (1,524–2,260 mm [60–89 in.]) size gyratory crushers. Once again, the choice is highly site dependent.

Mobile crushing plants have exploded in popularity since the early 1990s, with single-toggle jaw crushers providing the mainstay of these plants' primary crushing duties. For a vast majority of primary mobile applications, single-toggle jaw crushers have no real competition; however, in certain instances, smaller gyratory crushers can be used, but because of height considerations, these can hinder mobility.

The jaw crushers used on mobile plants, which can be track-, wheel-, or skid-mounted, are the normal units used in static applications. In some instances, modifications are made to lower the height of the crushers, but this is the only significant change to the crusher itself. Regarding the mobile plant itself, a range of innovations have been developed to deal with problems of locating and operating equipment within a constrained space. Latimer (2012) reports on the largest mobile jaw application using a Metso C200 jaw crusher.

Table 3 Estimated throughputs for Metso C series single-toggle jaw crushers

Capacities and Technical Specifications		C110	C125	C140	C145	C160	C200
Feed opening width, mm (in.)		1,000 (44)	1,250 (49)	1,400 (55)	1,400 (55)	1,600 (63)	2,000 (79)
Feed opening depth, mm (in.)		850 (34)	950 (37)	1,070 (42)	1,100 (43)	1,200 (47)	1,500 (59)
Power kW (hp)		160 (200)	160 (200)	200 (250)	200 (300)	250 (350)	400 (500)
Speed, rpm		230	220	220	220	220	200
Product size, mm (in.)	**Closed side setting, mm (in.)**	**t/h (stph)**	**t/h (stph)**	**t/h (stph)**	**t/h (stph)**	**t/h (stph)**	**t/h (stph)**
0–60 (0–2⅜)	40 (1⅝)						
0–75 (0–3)	50 (2)						
0–90 (0–3½)	60 (2⅜)						
0–105 (0–4⅛)	70 (2¾)	160–220 (175–240)					
0–120 (0–4¾)	80 (3⅛)	175–245 (195–270)					
0–135 (0–5⅜)	90 (3½)	190–275 (215–300)					
0–150 (0–6)	100 (4)	215–295 (235–325)	245–335 (270–370)				
0–185 (0–7)	125 (5)	260–360 (285–395)	295–405 (325–445)	325–445 (355–490)	335–465 (370–510)		
0–225 (0–9)	150 (6)	310–430 (340–470)	345–475 (380–525)	380–530 (420–580)	395–545 (435–600)	430–610 (475–670)	
0–260 (0–10)	175 (7)	350–490 (390–540)	395–545 (435–600)	435–605 (480–665)	455–625 (500–690)	495–695 (545–765)	630–890 (695–980)
0–300 (0–12)	200 (8)	405–555 (445–610)	445–615 (490–675)	495–685 (545–750)	510–710 (565–780)	560–790 (615–870)	710–1,000 (780–1,100)
0–340 (0–13)	225 (9)		495–685 (545–750)	550–760 (605–835)	570–790 (630–870)	625–880 (685–965)	785–1,105 (860–1,215)
0–375 (0–15)	250 (10)		545–755 (600–830)	610–840 (670–925)	630–870 (695–960)	685–965 (755–1,060)	865–1,215 (950–1,340)
0–410 (0–16)	275 (11)				690–950 (760–1,045)	745–1,055 (820–1,160)	940–1,320 (1,030–1,455)
0–450 (0–18)	300 (12)					815–1,145 (895–1,260)	1,015–1,435 (1,120–1,575)

Courtesy of Metso

In low-abrasion, medium-strength applications, such as limestone and low-strength ores, horizontal shaft impactors (HSIs) and twin-shaft sizers present other options for both fixed and mobile primary crushing. These two machine types are relatively lightweight, compact, and have high throughput capacities for their overall size.

For overall primary crusher selection, the reader is directed to Utley (2002) and Chapter 3.6, "Crusher Selection and Performance Optimization," where selection charts are given for various types of crushers and applications. For context, see Tables 4–5 in which recent installations by the major suppliers are provided.

IMPACT CRUSHERS

The family of impact crushers all rely on applying a high-speed impact to a falling stream of material and accelerating falling material to a high velocity, therefore causing either particle-on-particle or particle-on-metal breakage. The breakage of the particles is unconstrained and this imparts specific performance characteristics to this type of machine.

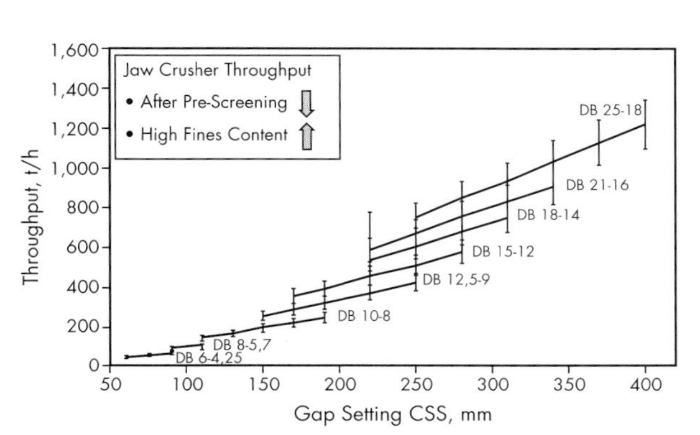

Courtesy of ThyssenKrupp Industrial Solutions

Figure 9 Estimated throughputs for ThyssenKrupp double-toggle jaw crushers

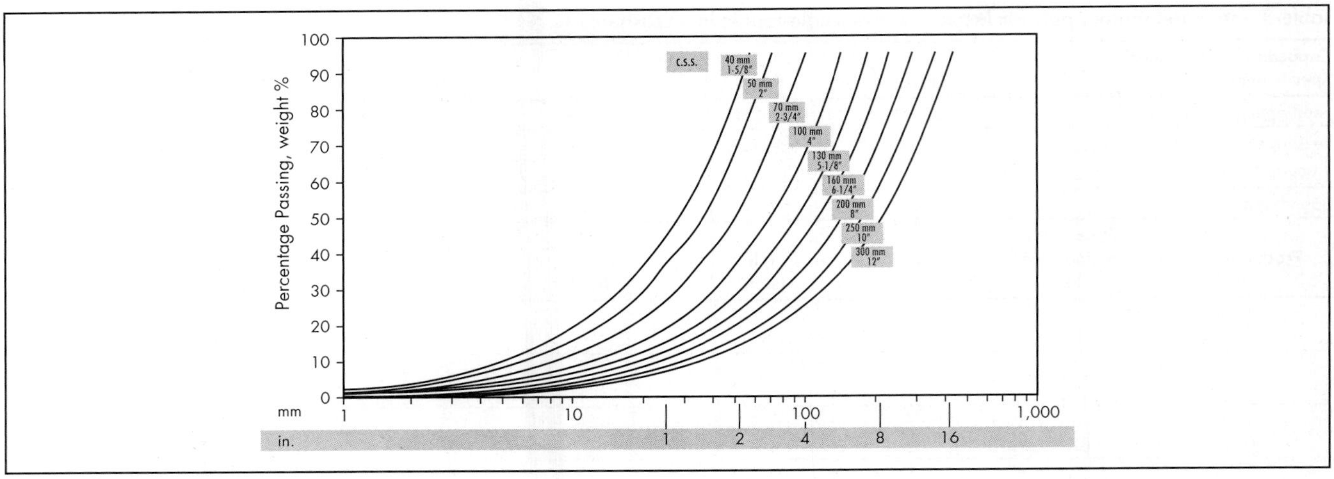

Courtesy of Metso

Figure 10 Typical product size distribution from single-toggle jaw crushers

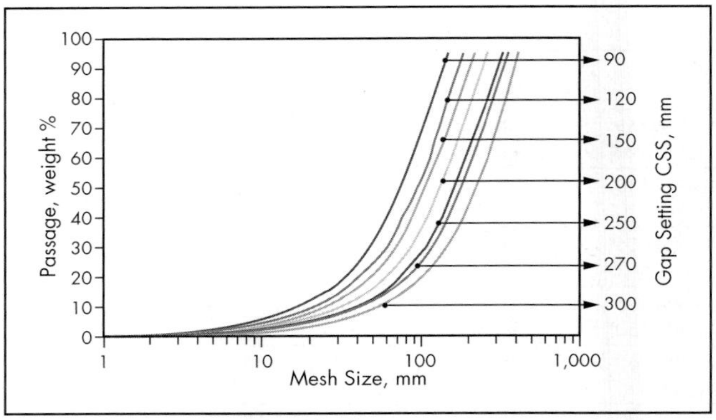

Courtesy of ThyssenKrupp Industrial Solutions

Figure 11 Typical product size distribution from double-toggle jaw crushers

The *SME Mineral Processing Handbook* (Weiss 1985) included a major section on hammer mills, which generally used free-swinging hammers to break by impact and then crush between the path of the hammers and a discharge grate. Such machines have mostly been replaced by fine crushing impact crushers or high-pressure grinding roll machines. Because of the decline in hammer mill use, the reader is directed to Eacret and Klein (1985) for a comprehensive review of the hammer mill, and its variants and applications. The current focus for impact crushers is confined to HSIs and vertical shaft impactors (VSIs).

Principles and Terminology

The HSI has a long history and has been widely applied in a variety of applications including primary, secondary, and tertiary duties. Feed size varies with the stage of application, but in primary duties, the top size of feed can reach 1,400–1,500 mm (55–60 in.) at throughput rates of up to 3,000 t/h (3,307 stph). In finer duties, feed size can be as low as 30–40 mm (1.2–1.6 in.) at throughput rates of 70–100 t/h (77–110 stph). The machines are best suited to treating soft, friable material of low to moderate abrasivity.

In the HSI, hammers rotate around a central horizontal shaft and engage the falling feed stream. The issue of wear has always been the limiting condition for impact crushers, as the high velocity of the rotor and therefore the significant speed differential between the rotor and the feed stream leads to accelerated wear. In HSI crushers, as the hammer surfaces and fixed impact surfaces become worn, the crushing performance of the machine gradually declines.

Guidelines have been given over the years in relation to the maximum abrasivity of the feed, mostly stated in terms of silica content (<8%) or standard abrasion index measures such as the Bond abrasion index. Regardless of the type of abrasion test used, impact crushers mostly do not find economic application above an abrasion level of *low-medium* abrasivity. If applications of impact crushers are examined around the world, it becomes obvious that they are occasionally used in certain applications with very high wear rates caused by abrasive feed. In such applications, the high wear and operating costs are considered acceptable, as the impact crusher offers some other advantage(s) to the user, which are not available with alternative styles of crusher. Examples include the following:

- Ability to better shape the product
- Ability of impactors to handle damp, somewhat sticky feed
- Excellent size reduction without the generation of fines and ultra-fines
- Overall simplicity, which can be a factor where availability of service and maintenance facilities are limited

The base design of the HSI is simply a fabricated rotor with replaceable hammers fitted to a heavy shaft that runs in external roller bearings. Figure 12 shows a typical HSI cross section and the general flow of material through an HSI.

The rotor assembly is mounted within a heavy split casing with a belt and pulley drive system that spins a rotor at the desired speed. The replaceable hammers are also known as *blow bars* or *impact bars*.

The main differences in the design of HSI machines are usually found in the following:

- Steepness of feed entry
- Type of fixing used to secure the hammer
- Number of hammers

Table 4 ThyssenKrupp large single-toggle (ST) and double-toggle (DT) jaw crusher installations since 2010

Type	Feed Opening, width × gape, mm (in.)	Number of Units	Client	Country	Feed Material	Year
ST	1,100 × 800 (45 × 33)	1	Silverstone Crushers and Minerals Private Ltd., Haryana	India	Granite	2017
ST	900 × 700 (37 × 29)	1	Oasis Dale Aggregate Products, Kerala	India	Granite	2016
ST	1,200 × 1,000 (49 × 41)	2	Yamama Saudi Cement Co., Riyadh	Saudi Arabia	Iron ore	2016
ST	1,100 × 800 (45 × 33)	1	Granulats Du Cameroun, Yaoundé	Cameroon	Granite	2015
ST	2,000 × 1,500 (82 × 61)	1	McInnis Cement Co., Quebec	Canada	Limestone	2014
ST	1,600 × 1,200 (66 × 49)	1	China Nonferrous Metal Industry	China	Iron ore	2013
ST	1,400 × 1,100 (57 × 45)	1	China Nonferrous Metal Industry	China	Iron ore	2013
ST	2,000 × 1,500 (82 × 61)	3	Vale—Serra Sul, Serra Sul	Brazil	Iron ore	2012
ST	2,000 × 1,500 (82 × 61)	1	CODELCO (Corporación Nacional del Cobre de Chile)—División El Teniente, Rancagua	Chile	Copper ore	2011
ST	1,600 × 1,200 (66 × 49)	2	Angang Group, Anshan	China	Iron ore	2011
DT	1,500 × 1,200 (61 × 49)	1	Comspain S.A., Youssoufia	Morocco	Phosphate	2011
ST	1,400 × 1,100 (57 × 45)	1	China CAMC Engineering Co., Ltd. (CAMCE), Gol-E-Gohar Iron Ore Co.	Iran	Iron ore	2010
ST	1,600 × 1,200 (66 × 49)	1	China Nonferrous Metal Mining Co., Ltd. (NFC), Jalalabad	Iran	Iron ore	2010

Courtesy of ThyssenKrupp Industrial Solutions

Table 5 FLSmidth large single-toggle jaw crusher installations since 2004

Model	Size, mm (in.)	Crushers Supplied	Client	Location	Commodity	Feed Arrangements	Throughput per t/h (stph)	Installed Motor Power, kW (hp)	Year
TST1200	1,219 × 914 (48 × 36)	1	Uranium Corporation of India Limited (UCIL)	India	Uranium ore	Run-of-mine (ROM) grizzly feeder	200 (220)	150 (201)	2004
TST1550	1,549 × 1,270 (61 × 50)	2	Konkola copper mine	Zambia	Copper ore	ROM grizzly feeder	435 (480)	260 (349)	2007
TST1200	1,219 × 914 (48 × 36)	1	Bozymchak	Kazakhstan	Gold ore	ROM grizzly feeder	190 (210)	150 (201)	2009
TST1400	1,397 × 1,168 (55 × 46)	1	Administración Nacional de Combustibles, Alcoholes y Portland (ANCAP)	Uruguay	Limestone	ROM grizzly feeder	226 (250)	225 (302)	2011
TST1200	1,219 × 914 (48 × 36)	1	Uranium Corporation of India (UCIL)	India	Uranium ore	ROM grizzly feeder	200 (220)	150 (201)	2012
TST1550	1,549 × 1,270 (61 × 50)	1	Akoga	East Guinea	Limestone	ROM grizzly feeder	500 (550)	260 (349)	2013
TST1400	1,397 × 1,168 (55 × 46)	1	Edo Limestone	Nigeria	Limestone	ROM grizzly feeder	435 (480)	225 (302)	2013
TST1900	1,905 × 1,600 (75 × 63)	2	D.G. Khan	Pakistan	Limestone	ROM grizzly feeder	699 (770)	375 (503)	2014

Courtesy of FLSmidth

- Ability to reverse the direction of the rotor
- Presence of more than one rotor
- Presence of a grinding path
- Presence of an adjustable breaker or breakers (used with some fixed grinding path machines)
- Rotational speed control and adjustment through electric or hydraulic variable-speed drives

Model designation with HSI machines varies between manufacturers, but in many instances, nomenclature such as 13/16/4 is used, which means that this machine has a 1,300 × 1,600 mm (51 × 63 in.) feed opening with four blow bars. Additional numbers and letters are added to indicate the presence of a grinding path (i.e., an area where additional breaker bars are positioned to provide finer crushing and improved product size control). Examples of HSI machines, with and without grinding paths, are given in Figure 13.

In the most extreme version of the grinding path concept, HSI machines start to mimic the functionality of a hammer mill. Figure 14 shows an HSI designed for fine crushing; this machine is also reversible, so the rotor can be operated in both directions to get maximum usage from the wear of the stationary blow bars. This design essentially has a continuous grinding path, with the chamber ending with the blow bars and liners in close proximity to give the finest product size possible.

The VSI has a much shorter history. The first machine to market was the Barmac VSI, which started to make a market impact in the late 1970s. The machine was developed in New Zealand by Bryan Bartley and Jim McDonald (McCarthy and Campbell-Hunt 2000), with the original version sold as the Rotopactor.

In the original design, feed was introduced into the top of the machine and the feed then entered through the center of a

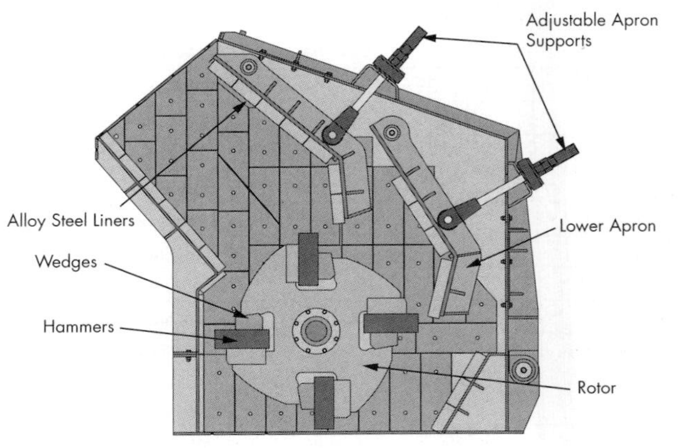

A. Cross section of a typical horizontal shaft impactor

B. Flow of material through an NP model horizontal shaft impactor

Courtesy of Telsmith

Courtesy of Metso

Figure 12 Arrangement and material flow in horizontal shaft impactors

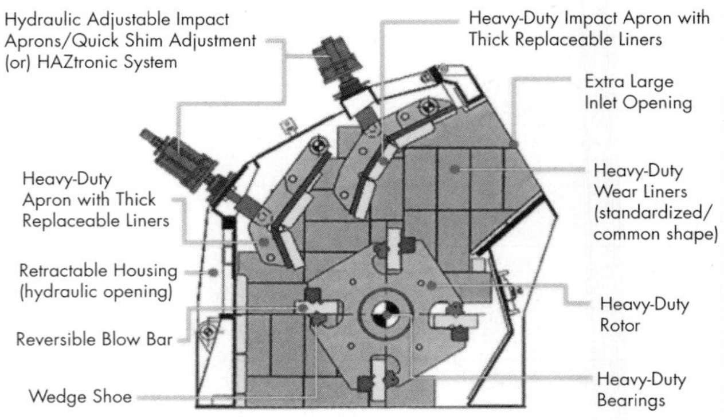

A. Hazemag APS series secondary impactor without third crushing stage

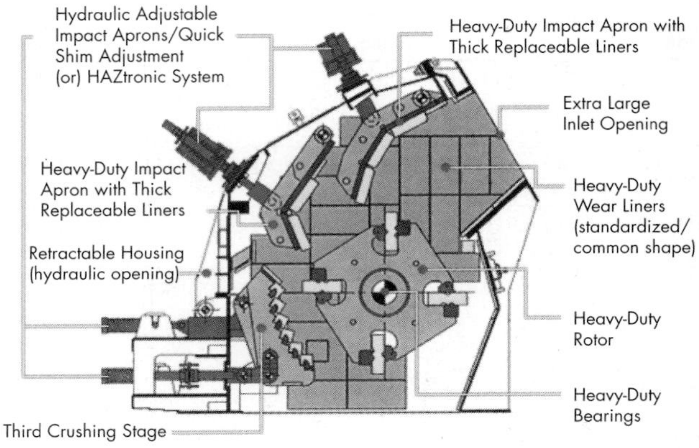

B. Hazemag APS series secondary impactor with third crushing stage, also known as a grinding path

Courtesy of Hazemag USA Inc.

Figure 13 Horizontal shaft impactors with varying configurations of two secondary machines

rotor spinning about a vertical shaft. The rocks were accelerated outward from the rotor and hit beds of material formed on ledges around the chamber. This *rock-on-rock* or autogenous principle (Hamer 1990) was the main target, as this reduced wear on the shell of the machine while allowing particles to break freely. These machines were found to be proficient in improving the shape of misshapen aggregate products, and the particles generated were found to be free of the internal damage often seen as a result of breakage in compression crushers. The size reduction capability of these early VSIs was limited, however, and the energy consumed to achieve that reduction was relatively high. In addition, as the top size of the feed was pushed higher, the effectiveness of the original design declined in that the majority of the larger material would survive the crushing action.

In an effort to improve the crushing action, the Duopactor was developed. In this design, larger rocks are forced to fall in an annular curtain; rocks in the annular curtain are in turn impacted by the high-velocity material exiting the spinning rotor. The action provides the opportunity for the larger rocks to be randomly broken without applying any additional energy to the machine. Shortly after this development, the concept of cascade feed was introduced, whereby rather than channeling coarse feed just to the outside, it is introduced both through the rotor and also around the periphery.

Since the early days, the Barmac machine has been through several ownership changes and arrangements where machines were manufactured under license by distributors in a range of countries. Currently the Barmac VSI technology is supplied through Metso, following their takeover of Svedala. In essence, the B series Barmac VSI operates very much in the same way as the Duopactor, and cascade feed can be used to fine-tune the performance of the machine. The general arrangement of the current Barmac machine is shown in Figure 15. As

Courtesy of SBM Mineral Processing GmbH
Figure 14 Example of a reversible, fine crushing horizontal shaft impactor

can be seen, the Barmac shows a pivoting upper section for ease of access and maintenance and an overall design that has been optimized for safety, simplicity, and speed.

The Barmac VSI is now marketed as a machine for tertiary or quaternary crushing duties, and this extends into the region of manufactured sand for the quarrying business. Crushing in the quarrying sector often needs to consider the shape, or cubicity, of the particles. This is also true of the size ranges found in manufactured sand. The Barmac VSI has found a major niche in this area with its ability to generate improved cubicity because of the rock-on-rock breakage principle. It has had a major impact in the manufactured sand sector (Gonçalves et al. 2007) because of its ability to generate fines but control particle shape.

In the VSI market, there have been many variants and versions developed, but the only other type considered here, in addition to the Barmac, is the type characterized by the Canica VSI, patented in Canada in 1992 by the now defunct Canica Crushers Inc. In the Canica design, a rotor spins around a vertical shaft, with the rotor being either of an open or enclosed design. In the open design, feed drops into the center of the spinning rotor; in the enclosed design, the top plate has a central feed entry point and the enclosed rotor ensures that material is directed outward to provide optimal impact on a

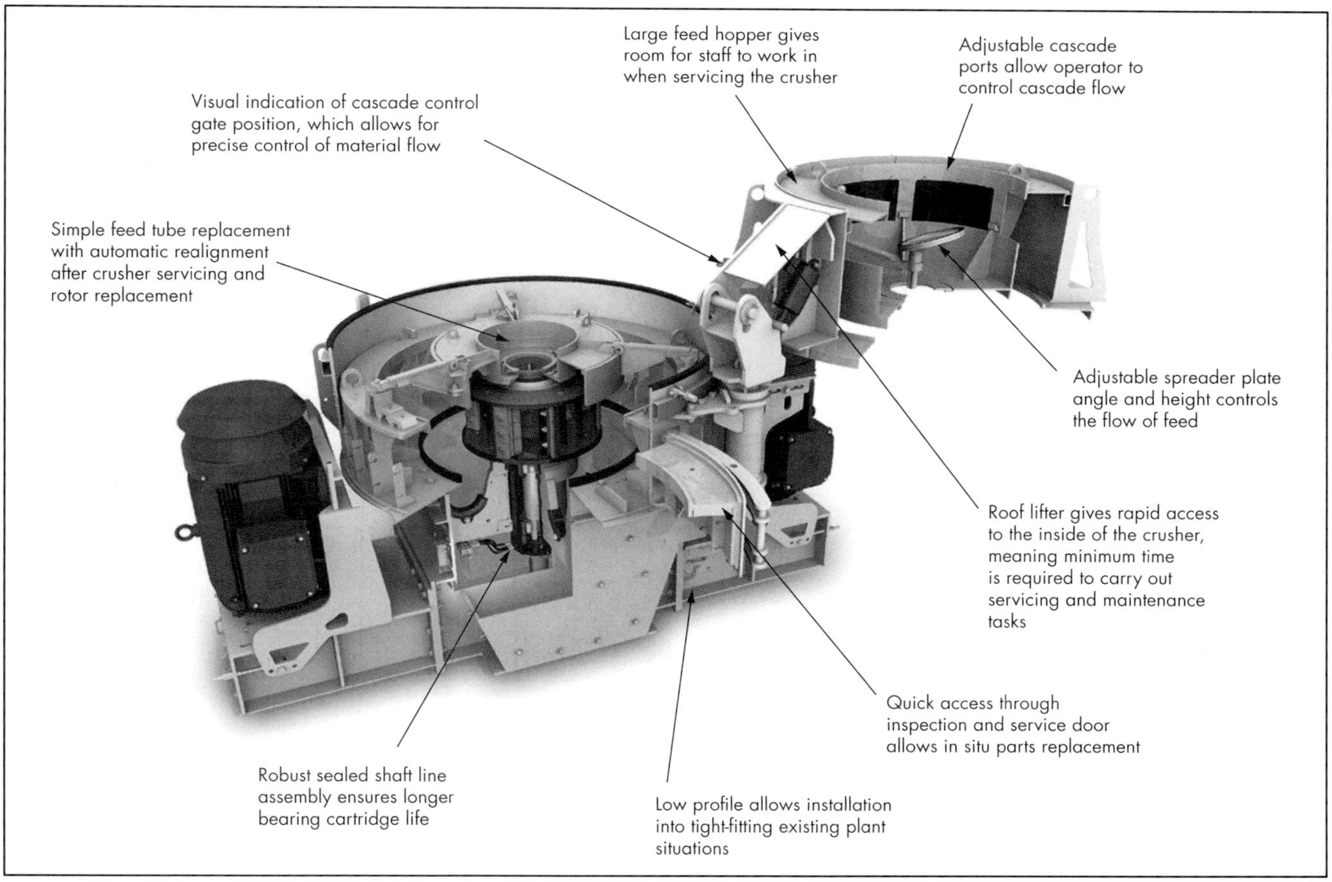

Courtesy of Metso
Figure 15 Current Metso Barmac B series vertical shaft impactor

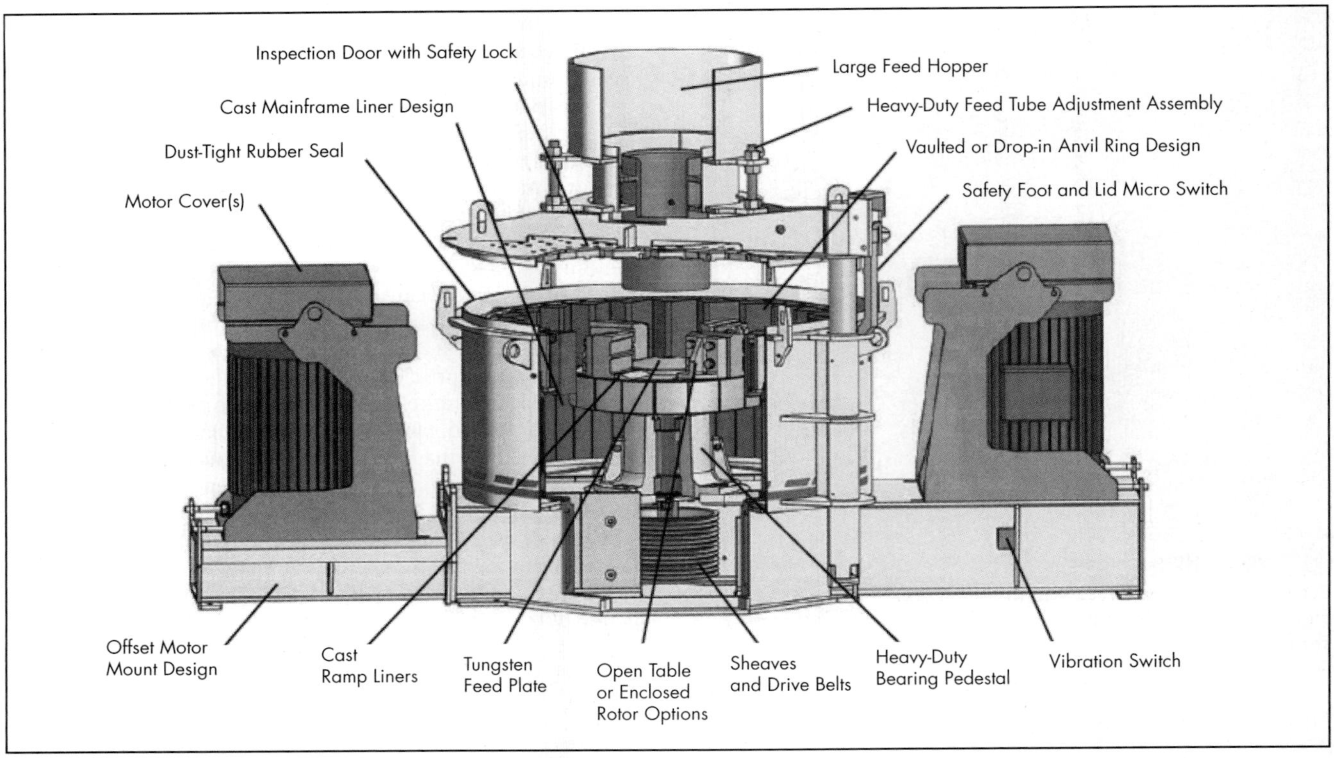

Courtesy of Terex Corporation

Figure 16 General arrangement of a Canica vertical shaft impactor

series of anvils surrounding the rotor or against rock material retained on ledges.

A general arrangement of a Canica-style VSI is shown in Figure 16. As in the Barmac design, there is a commitment to provide the operator with safe and easy access to key maintenance areas.

The notable difference between the autogenous Barmac-style VSI and the stone-on-steel Canica style is that the stone-on-steel machines employ an arrangement of breaking surfaces (anvils) around the inside of the circular casing. The surfaces are arranged to maximize the impact energies and therefore the size reduction achieved.

As a general rule, the stone-on-steel type VSIs achieve much higher particle reduction and at a much lower overall energy input than the autogenous Barmac-style VSIs. These gains, however, have to be traded against higher wear costs when compared with autogenous-style VSIs. As protection against wear in the Canica VSI, ledges can be used to retain rock particles.

When fitted with an open rotor, the stone-on-steel type VSIs can be successfully operated in secondary crushing roles, accepting particles up to 305 mm (12 in.) in maximum size in some applications.

Like the autogenous machines, stone-on-steel VSIs produce good particle shape, even when underfed.

As mentioned, early variants of VSI crushers, both of the Barmac and Canica types, used stone on steel, which led to high wear rates. The developments flagged previously, where crushing can now be provided via semiautogenous or autogenous means, has lifted the ability to crush abrasive feed. In general, stone-on-steel VSI action is still not suited to highly abrasive feed, and although opinions vary, apparently a limit of 8%–15% of abrasive content is commonly accepted. These abrasivity limits do not apply to VSI machines where particle-on-particle crushing predominates, as is demonstrated in the aforementioned prevalence of VSI application for manufactured sand. It is also known that VSI machines generate product with properties that differ from compression crushers (Bengtsson and Evertsson 2006). The free breakage environment tends to lead to intergranular breakage, which reduces microscale damage and generates more competent particles (Briggs and Bearman 1996). At the same time, the product shape generated tends to be more cubic and the attrition products from the shaping process contribute significantly to the increased fines seen in VSI products.

Construction and Design

The construction of HSI crushers is relatively straightforward and consists of an outer housing, stationary breaker plates (or aprons), and a rotor that carries the hammers or impact bars. The outer housing is essentially a steel box that contains the crushing area and provides structural support. The main issue for the housing is to resist flow and high-impact abrasive contacts from the flow of feed material and particles from the breakage process. To deal with the wear issue, most housings are lined with heavy-duty wear plates in areas away from the direct breakage zone. As these zones are generally not the highest wear areas and the wear is mainly flow-based, these can be made of relatively brittle material. In such a duty, materials such as white cast iron can be deployed. Also attached to the housing are the breaker plates, which are situated directly opposite the spinning rotor. These stationary plates are positioned and designed to be the anvil against which rocks can be further broken following impact by the spinning rotor and hammers or impact bars. To withstand this duty, the breaker plates are mostly made of 13%–18% manganese steel, and they are usually designed to be turned to allow extension of the overall wear life. Depending on the manufacturer, the size and arrangement of the breaker bars can vary. Some suppliers prefer to use large single castings, which are often designated as *aprons*, whereas others have designs where multiple smaller plates are used. In considering the right solution for a particular situation, safety, masses, lifting, and maneuverability should all be assessed.

One of the main design features of the HSI relates to the use of a grinding path. Coarse impact crushers tend to simply have breaker bars high in the chamber. Once breakage has occurred in this area, the particles pass through the crusher without any further interactions. If a grinding path is used, secondary breakage occurs toward the exit point of the crushing chamber. The length of the grinding path dictates the amount of fine size reduction delivered. Claims are also made that the length of the path improves the cubicity of the final product. In Figure 17, the final arrangement of liners is shown: In (A), the rotor and the last three blow bars are in close proximity, and in (B), extra length is added to the grinding path by having five liners prior to the discharge point.

In HSI machines, adjustment for wear and therefore control of product size is achieved via adjustment of the static breaker bars or aprons. In the preceding figure, the breaker bars can be seen with a hydraulic device; in some designs, a spring-loaded adjustment mechanism is located on the outside of the crusher body. Using such a system, the position of the breaker bars can be adjusted via the hydraulic, or spring, assemblies. The system relies on each section of breaker bars being pivoted at a point and the adjustment system being connected to a point toward the center of the breaker bars.

In the VSI design of impact crusher, both the Barmac- and Canica-style machines employ the same general approach of dropping feed onto a spinning rotor that then accelerates the material outward for crushing. Crushing energy in VSI machines is imparted through the kinetic energy generated by the acceleration of the particles by the spinning rotor (Nikolov 2002):

$$\text{crushing energy} = \tfrac{1}{2} mv^2$$

where

m = particle mass
v = particle velocity, m/s

The Barmac exclusively uses the enclosed rotor design where feed enters through an orifice in the top of the rotor and then exits through the body of the rotor. Metso uses what it terms the *deep rotor technology* design, which offers advantages in throughput and also the ability to allow buildup inside the rotor to reduce wear at this critical point in the crusher design. As with all types of impact crusher, wear is the constant enemy, and therefore manufacturers put a tremendous effort into minimizing wear through design and materials selection. The Barmac rotor as shown in Figure 18, shows the overall design of the deep rotor.

The rotor tip and associated holder assembly can be configured in many ways depending on the model (size) of machine, the required duty, and the type of feed. Metso provides a detailed summary of their proprietary rotor tips and holders, which is reproduced in Figure 19. The rotor tips are color coded, available in a range of tungsten carbide grades, and can be fitted to a range of hangers. The range of tips, materials, and hangers allows an appropriate selection to be made for a specific crushing duty.

A feature of the Barmac design is the cascade feed arrangement, whereby a feed stream is also directed down the periphery of the crushing chamber to amplify the rock-on-rock crushing action of the machine. In terms of efficiency, it is generally considered that the proportion of feed used in the cascade should not exceed 60% of the feed to the rotor, except where recommended by the manufacturer. The cascade is particularly effective, as the falling stream is impacted by material exiting the rotor at speeds in the order of 80 m/s (262 ft/s). The rock-on-rock area therefore offers a range of

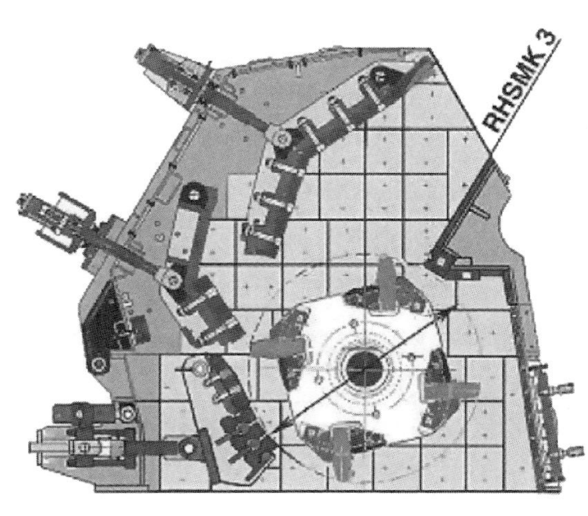

A. Three-row grinding path

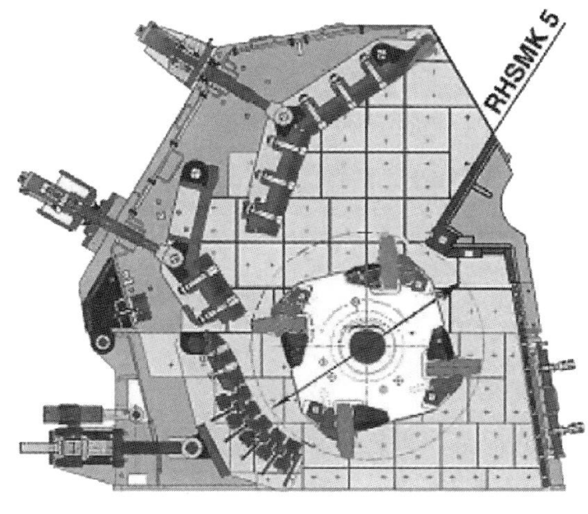

B. Five-row grinding path

Courtesy of SBM Mineral Processing GmbH

Figure 17 Different types of grinding path

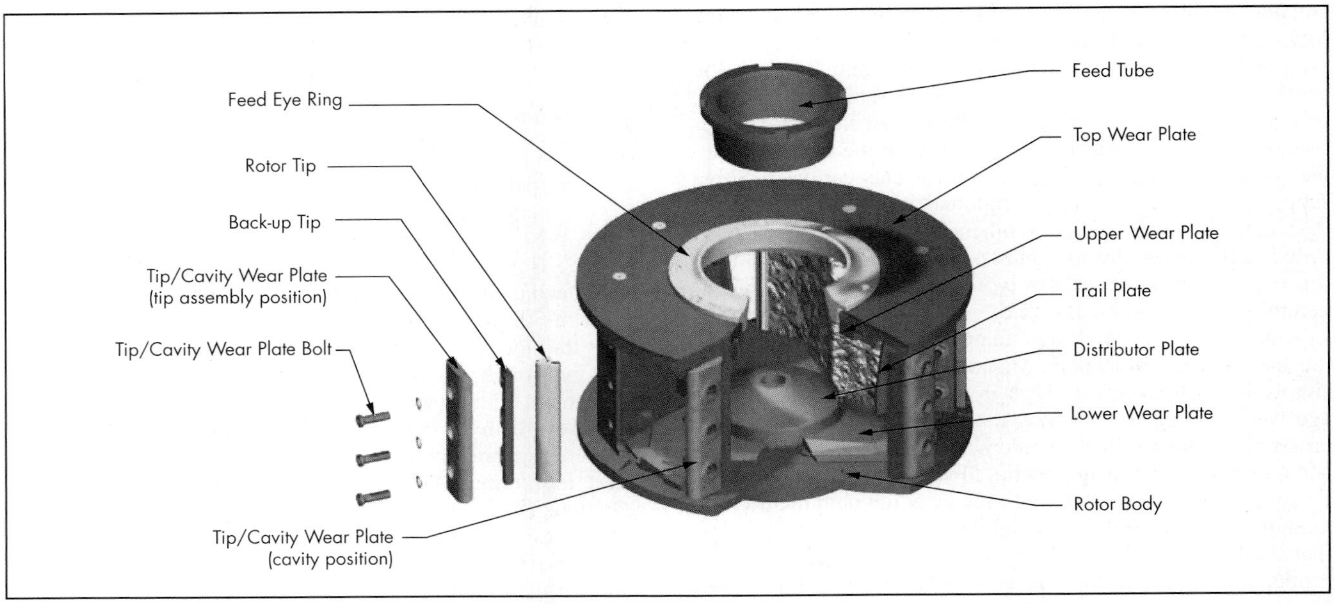

Courtesy of Metso

Figure 18 Exploded section of Metso Barmac rotor assembly

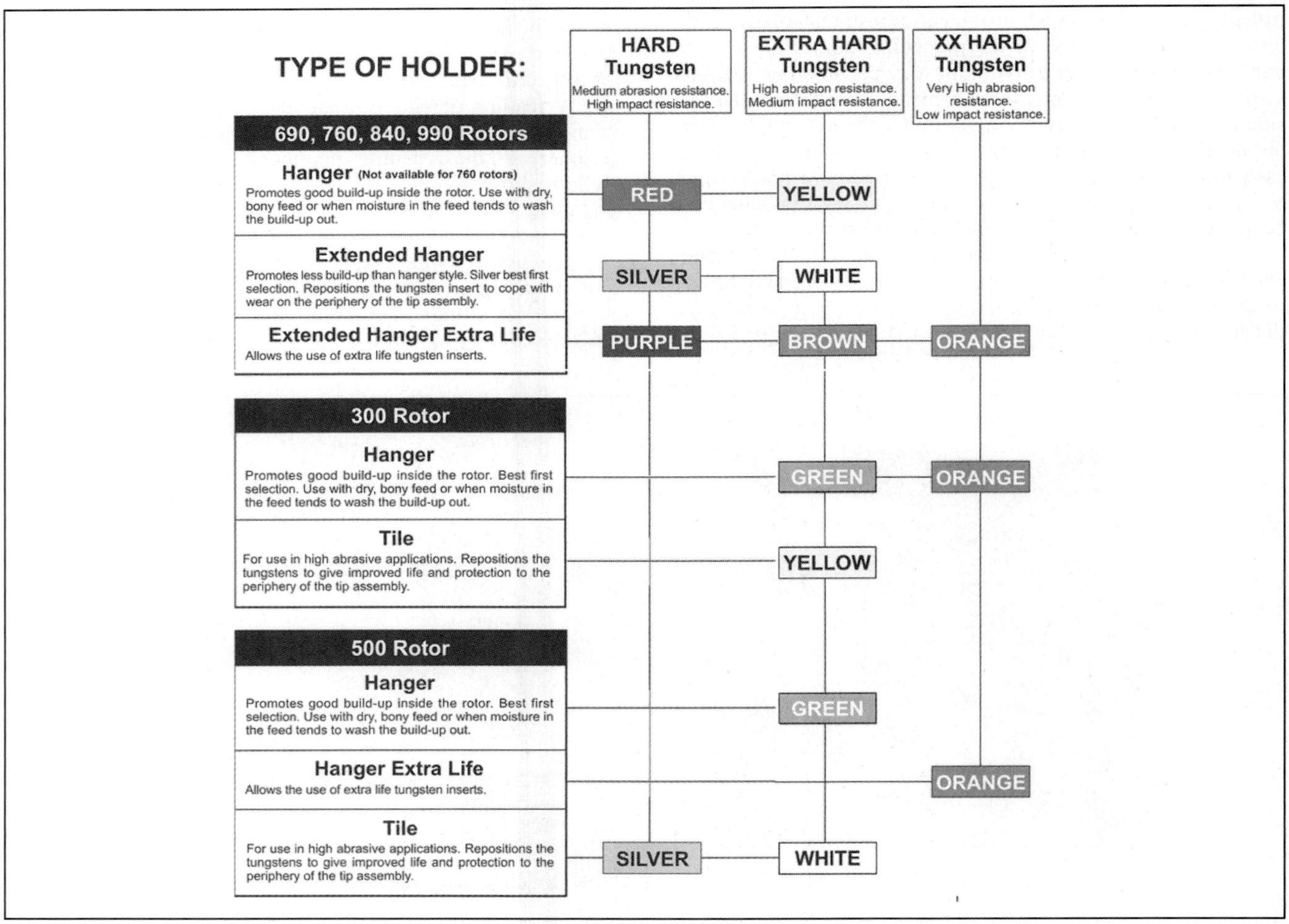

Courtesy of Metso

Figure 19 Rotor tips and holders available for the Metso Barmac

benefits including higher throughput, improved size reduction, and the ability to improve the product cubicity for quarrying applications. The action of rock-on-rock crushing is shown in Figure 20.

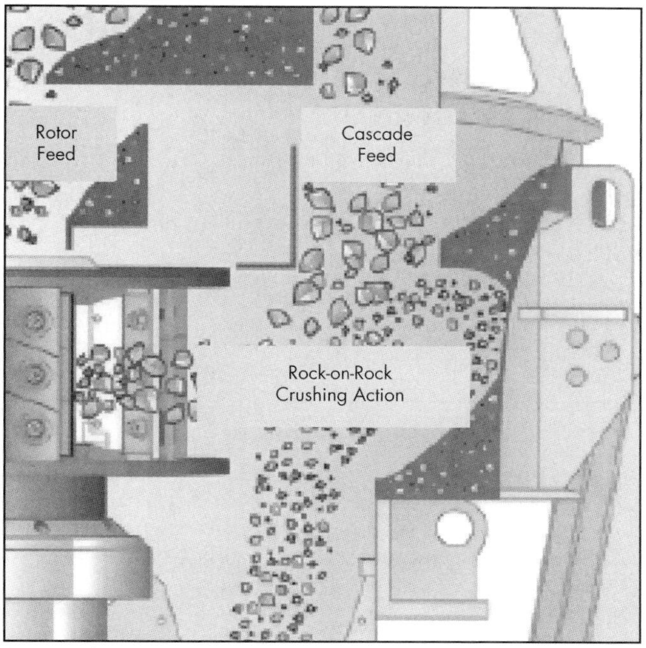

Courtesy of Metso

Figure 20 Rock-on-rock crushing action using the cascade feed arrangement in a Barmac vertical shaft impactor

The crushing action in the rock-on-rock area of the chamber is a combination of impact, abrasion, and attrition. The density of particles in this area is critical to the effectiveness of the breakage, as higher densities promote extra particle-on-particle crushing through the increase in turbulence and residence time (Cunha et al. 2013) in the zone.

The Canica VSI design relies on feed entering the rotor in a central area, but the rotors used can be either open or closed. In addition, the Canica does not have a cascade feed feature and therefore relies on different chamber arrangements to generate either rock-on-anvil or rock-on-rock breakage. Figure 21 shows three typical arrangements for a Canica machine. The shoe-and-anvil arrangement with the open rotor is designed to take much larger feed, with the largest version stated to accept feed with a longest dimension of 305 mm (12 in.). The rock-on-anvil design is more focused on smaller feed size and greater size reduction, and the rock-on-rock design uses a rock pocket (or shelf) arrangement so material accelerated from the rotor impacts a bed of material at the wall.

Application and Operation

As mentioned, HSI machines can be deployed in a wide variety of roles from primary through to fine crushing, depending on the arrangement of the machine. The selection of a specific HSI machine follows the same principles as with any other crusher, which is feed size, reduction ratio, product size, and ability to handle feed characteristics. As an illustration of the main features of the various stages of impactor, a selection from the SBM range is given in Figure 22.

In terms of throughput, the machine variables controlling the capacity are

- Size of feed opening,
- Angle of feed entry,

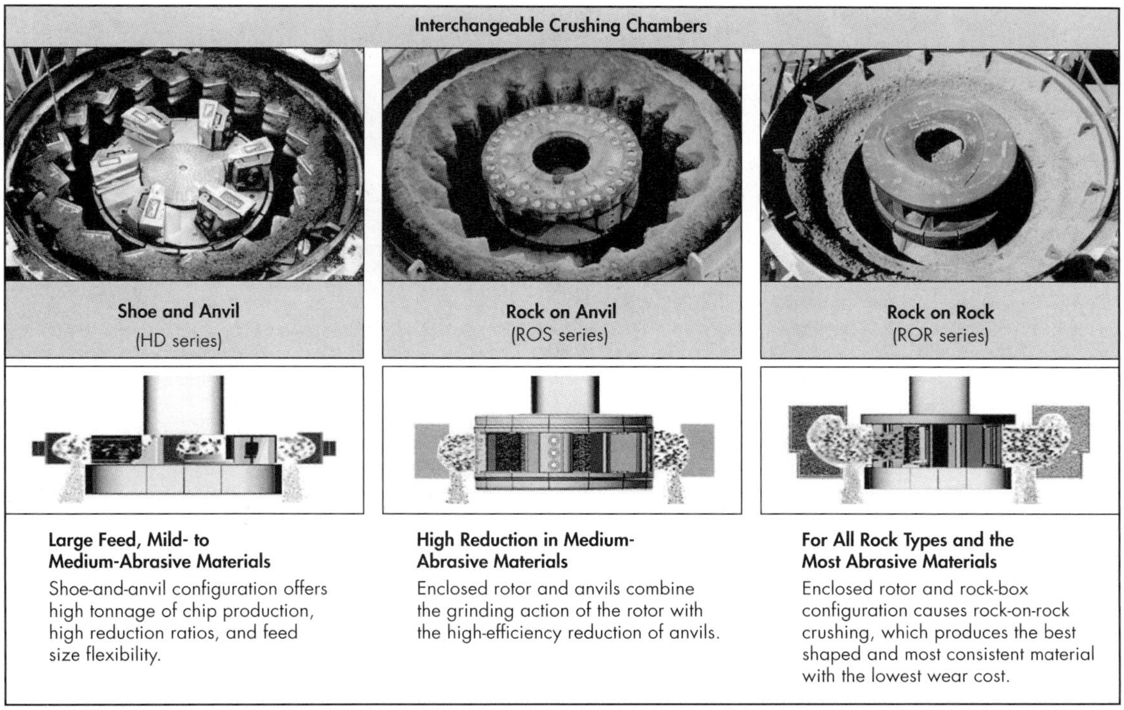

Courtesy of Terex Corporation

Figure 21 Various rotor and anvil configurations for a Canica-style vertical shaft impactor

- Entry velocity of feed,
- Rotor diameter,
- Rotor speed,
- Number of blow bars, and
- Presence of a grinding path.

In terms of the product size distribution, the main factors controlling output are

- Crushing gap, including presence of grinding path;
- Rotor circumferential speed; and
- Number of blow bars.

In the world of HSI crushers, manufacturers offer a great deal of customization to provide a machine suited to the specific duty required. As such, it is difficult to give a full summary of the options. Many manufacturers exist, including Hazemag, Metso, SBM, BJD, and Terex Corporation. It is not within the scope of this chapter to provide manufacturer performance estimates for all suppliers, so examples have been chosen to illustrate the range of applications.

Table 6 shows data relating to large primary HSI specifications, whereas Table 7 and Figure 23 show throughputs and product size distributions for secondary impact crushers. For completeness, Table 8 and Figure 24 shows data relating to tertiary crushing using an SBM HSI machine.

In terms of the VSI machines, the Barmac has come to be well accepted with feed sizes mostly in the –50 mm region, with a strong focus on manufactured sand generation and the ability to provide cubical product shape. This provides a point of difference to the Canica-style VSI crushers, which in the rock-on-rock configuration have similar capabilities but are also positioned to crush much coarser feed.

The application data for the Barmac B series VSI is summarized in Table 9. Metso also provides general guidance on how feed properties and operating parameters influence the key output parameters. This general guidance is shown in Figure 25.

Canica process performance guidelines are provided in Table 10, where the maximum figures are based on the *heavy-duty* version with the open rotor and anvils. Performance

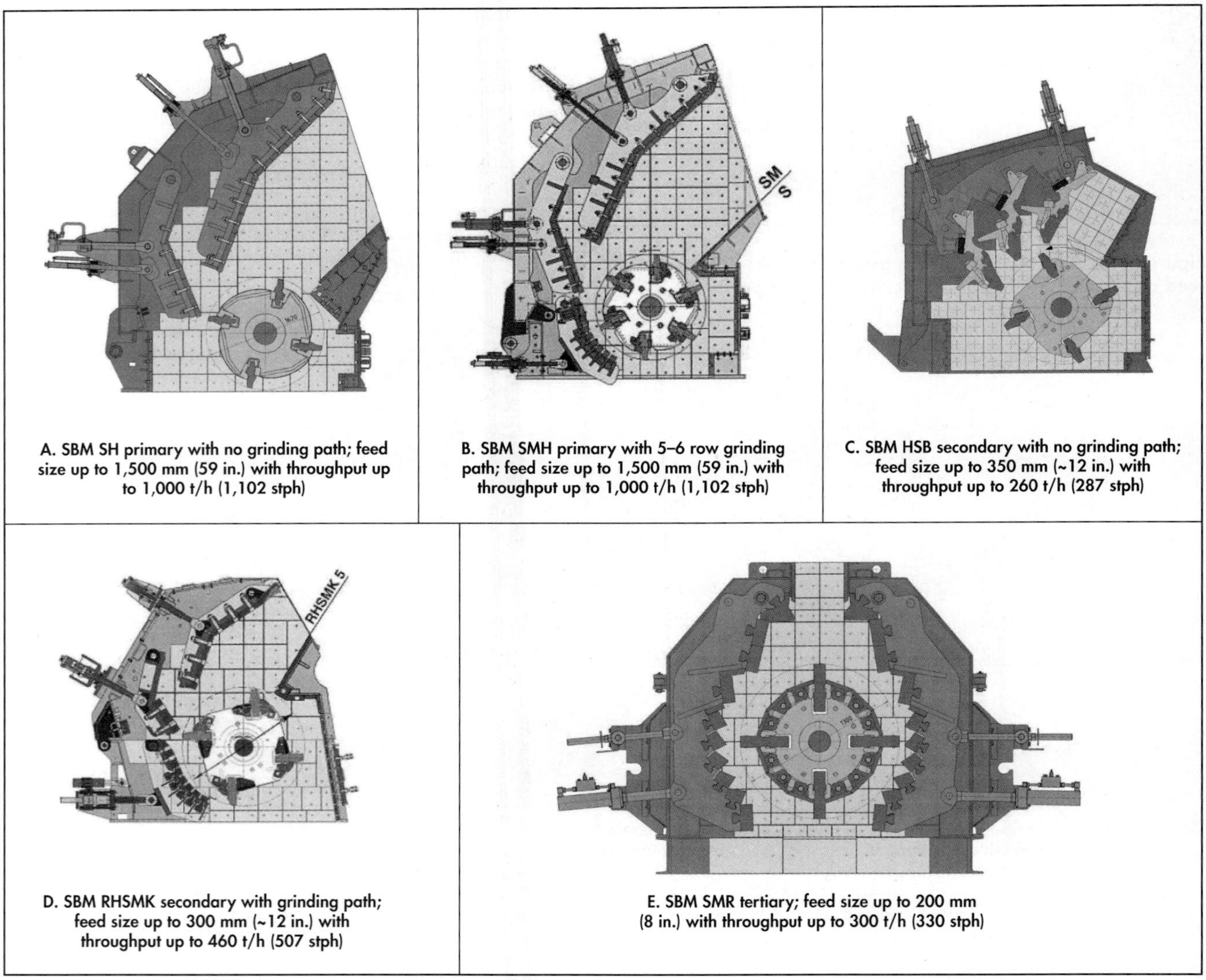

A. SBM SH primary with no grinding path; feed size up to 1,500 mm (59 in.) with throughput up to 1,000 t/h (1,102 stph)

B. SBM SMH primary with 5–6 row grinding path; feed size up to 1,500 mm (59 in.) with throughput up to 1,000 t/h (1,102 stph)

C. SBM HSB secondary with no grinding path; feed size up to 350 mm (~12 in.) with throughput up to 260 t/h (287 stph)

D. SBM RHSMK secondary with grinding path; feed size up to 300 mm (~12 in.) with throughput up to 460 t/h (507 stph)

E. SBM SMR tertiary; feed size up to 200 mm (8 in.) with throughput up to 300 t/h (330 stph)

Courtesy of SBM Mineral Processing GmbH

Figure 22 SBM horizontal shaft impactors for varying stages of size reduction

Table 6 Large primary horizontal shaft impactors*

Model†	Capacity, t/h (stph)	Power Requirements, kW (hp)	Inlet Size, h × w, mm (in.)	Maximum Feed Size, mm (in.)	Rotor Size, d × w, mm (in.)	Weight, kg (lb)
APP-1513	220 (250)	160 (250)	900 × 1,360 (35 × 54)	760 (–30)	1,500 × 1,340 (60 × 53)	21,700 (47,800)
APP-1615	350 (400)	250 (350)	1,290 × 1,520 (51 × 60)	1,000 (–40)	1,600 × 1,500 (63 × 59)	39,500 (86,900)
APP-1622	600 (700)	450 (600)	1,290 × 2,270 (51 × 89)	1,000 (–40)	1,600 × 2,250 (63 × 88)	58,500 (128,700)
APP-1822	800 (900)	750 (1,000)	1,600 × 2,270 (63 × 89)	1,200 (–48)	1,800 × 2,250 (70 × 88)	69,000 (151,800)
APP-2022	1,000 (1,100)	900 (1,200)	1,830 × 2,270 (72 × 89)	1,500 (–60)	2,000 × 2,250 (79 × 88)	88,000 (193,600)
APP-2025	1,090 (1,200)	1,125 (1,500)	1,830 × 2,520 (72 × 99)	1,500 (–60)	2,000 × 2,500 (79 × 98)	108,150 (237,900)
APP-2030	1,500 (1,650)	1,350 (1,800)	1,290 × 3,020 (51 × 118)	1,500 (–60)	2,000 × 3,020 (79 × 118)	110,000 (242,000)
APP-2522	1,360 (1,500)	1,200 (1,600)	2,125 × 2,270 (84 × 89)	1,500 (–60)	2,500 × 2,250 (98 × 88)	135,000 (297,000)
APP-2525	1,800 (2,000)	1,350 (1,800)	2,125 × 2,520 (84 ×99)	1,500 (–60)	2,500 × 2,500 (98 × 98)	151,000 (332,000)
APP-2530	2,300 (2,500)	1,800 (2,500)	2,125 × 3,020 (84 × 118)	1,500 (–60)	2,500 × 3,020 (98 × 118)	180,000 (396,000)

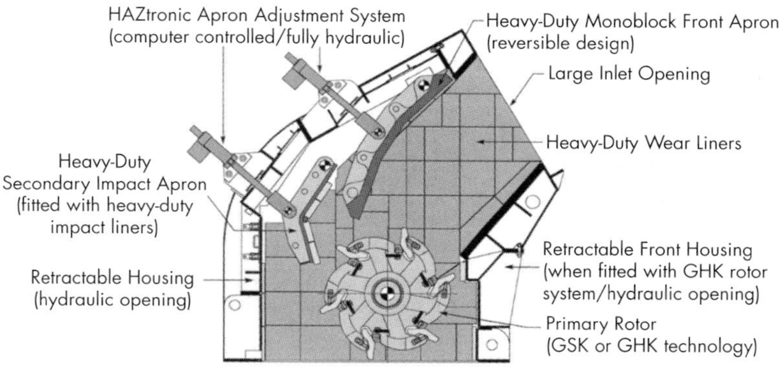

Courtesy of Hazemag USA Inc.
*Performance details relate to medium-hard limestone.
†The naming convention used for crusher sizes and duties is "APP-rotor diameter, inlet width," e.g., APP-1513 means a crusher with a 1,500-mm (60-in.) rotor diameter and a 1,360-mm (54-in.) inlet width.

varies for other types of rotor, and thus guidance needs to be obtained from the supplier.

It is important to note that throughput and feed size effects with impact crushers are highly material specific. In compression crushers—jaw, gyratory, and cone—the progression through the crushing chambers is controlled by the geometry of the chamber and the mechanical features that control the movement of material (eccentric throw, pivot point, and speed). In impact crushers, the throughput limit is primarily controlled by the speed, feed size, and the number of blow bars. Because of these parameters, there is a certain volumetric space between each row of blow bars as the shaft rotates. The space is fixed; thus the other factor is the ability of material to enter the space. Such geometric and flow considerations can lead to situations whereby running with two blow-bar rows is more beneficial than using a higher number.

Another feature of impact crushers is the sensitivity to feed size. In impact machines, the size reduction is dependent on the ability to strike and shatter particles. If the feed size is too fine, it becomes difficult to impart sufficient energy to the smaller particles, and therefore the size reduction is much less effective. Because of the different crushing configuration between the Barmac- and Canica-style VSI machines, it is important to note that the Barmac feed size is usually limited to 50 mm (2 in.) (see Table 9), whereas the Canica style can accommodate feed up to 305 mm (12 in.). Obviously, this variation in feed size is dependent on the duty required; therefore when selecting a type of VSI, a direct comparison of duty and performance is required.

Because of the variability inherent in the crushing process in an impact crusher, it is essential that the manufacturers are consulted, and in many cases, test work is required to confirm the most suitable machine for a given duty.

Regarding the operation of impact crushers, the two styles—HSI and VSI—have differing feed arrangement requirements. In the HSI, it is critical to ensure that the feed is

Table 7 Typical secondary HSI throughput and product size guidance*†

NP Model		Top Feed Size, 400 mm (16 in.) End Product, 60 mm (2½ in.) t/h (stph)	Top Feed Size, 400 mm (16 in.) End Product, 40 mm (1½ in.) t/h (stph)	Top Feed Size, 200 mm (8 in.) End Product, 40 mm (1½ in.) t/h (stph)	Top Feed Size, 200 mm (8 in.) End Product 20 mm (¾ in.) t/h (stph)
Secondary range	NP1110	190 (210)	150 (170)	210 (230)	130 (140)
	NP1213	250 (280)	200 (220)	280 (310)	180 (200)
	NP1315	315 (350)	250 (280)	350 (390)	225 (250)
	NP1520	500 (560)	400 (450)	560 (630)	360 (400)

Courtesy of Metso

*NP secondary crusher of the type shown in Figure 12B.

†Represents capacity through crusher based on instantaneous product sample. Impact crusher capacity charts are developed for use as an application tool to properly use the NP crusher's capabilities. The capacity figures apply to material weighing 1,600 kg/m³ (100 lb/ft³). The crusher is one component of the total circuit. As such, its performance is also dependent on the proper selections and operation of feeders, conveyors, screens supporting structure, electric motors, drive components, and surge bins.

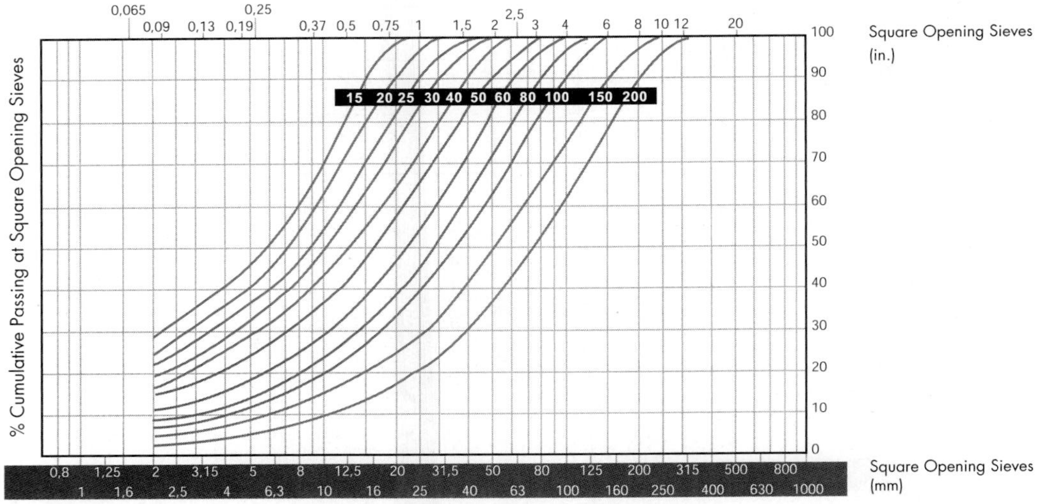

Note: The gradation and capacities shown are dependent on feed gradation, material density, and crushability.

Courtesy of Metso

Figure 23 Typical secondary horizontal shaft impactor product distribution curves

uniformly distributed across the width of the feed opening and that there is no segregation of fines or coarse material to one particular area. Any bias in the feed leads to accelerated wear in localized areas and this leads to a loss in product size control and also the need to prematurely remove the wear components. For the most efficient operation, the feed material should have sufficient velocity to fall deep into the crushing chamber, so that the blow bars can fully engage the feed. Entry velocity is simply a function of the fall height under gravity.

As with cone crushers, it is also useful to remove fines from the feed. In HSI and VSI machines, the fines do not have a detrimental impact in terms of mechanical operation, but the presence of fines simply occupies space, therefore reducing the effective throughput of material requiring crushing. One other potential impact of fines is that with the high velocities, the fines can adhere to surfaces within the chamber, which again may compromise throughput. In terms of the product from the HSI machines, particles can exit the crushing chamber at high velocity, and therefore consideration must be given to wear protection and, if appropriate, the use of rock boxes.

For VSI machines where the feed enters into the center of the spinning rotor, feed distribution is less of an issue. A greater concern is that the top size of the feed is controlled. If it is not, then this can lead to wear issues, and there may be a mismatch between the larger particles and the pathways within the rotor. In the Barmac-style VSI where the cascade feed arrangement is used, it is best to ensure that the feed size distribution is uniformly spread around the periphery of the machine. This should be achieved using feed control via a suitable feeder.

Installed motor power varies widely in HSI machines. The motor selection is usually based on the mechanics of being able to accelerate the rotor to the desired speeds and the added energy required to cause breakage of the rocks. To determine the required breakage energy, most suppliers use the Bond abrasion index approach. In VSI machines, the installed power is more uniform and this relates to the analogy

Table 8 Typical tertiary horizontal shaft impactor data*

Machine Type	Feeding Size Up to, mm (in.)	For Rotor Circumferential Speed, m/s (ft/s)	Throughput, t/h (stph)	Driving Power, kW	Examples of Achieved End Product Over K80 at Delivered Particle Size	Machine Weight, kg (lb)
10/05/2 SMR	80 (3¼)	50–75 (164–246)	80 (73)	75–160	v = 71 m/s; river gravel; K80:0/2.8 mm for feed 8/16 mm	7,900 (17,412)
10/05/4 SMR	150 (6)	30–60 (98–197)	130 (118)	75–160	v = 44 m/s; river gravel; K80:0/14 mm for feed 32/80 mm	8,100 (17,852)
10/10/2 SMR	80 (3¼)	50–75 (164–246)	150 (136)	110–250	v = 50 m/s; river gravel; K80:0/5.6 mm for feed 8/16 mm	12,000 (26,448)
10/10/4 SMR	150 (6)	30–60 (98–197)	200 (181)	110–250	v = 37 m/s; river gravel; K80:0/11 mm for feed 32/150 mm	12,400 (27,330)
13/7/4 SMR	200 (7⅞)	30–60	200 (181)	90–200	—	~14,600 (32,178)
13/10/4 SMR	200 (7⅞)	30–60	250 (226)	110–250	—	~22,300 (49,149)
13/13/4 SMR	200 (7⅞)	30–60	300 (272)	132–315	—	~28,900 (63,696)

Courtesy of SBM Mineral Processing GmbH
*SMR tertiary crusher of the type shown in Figure 22E.

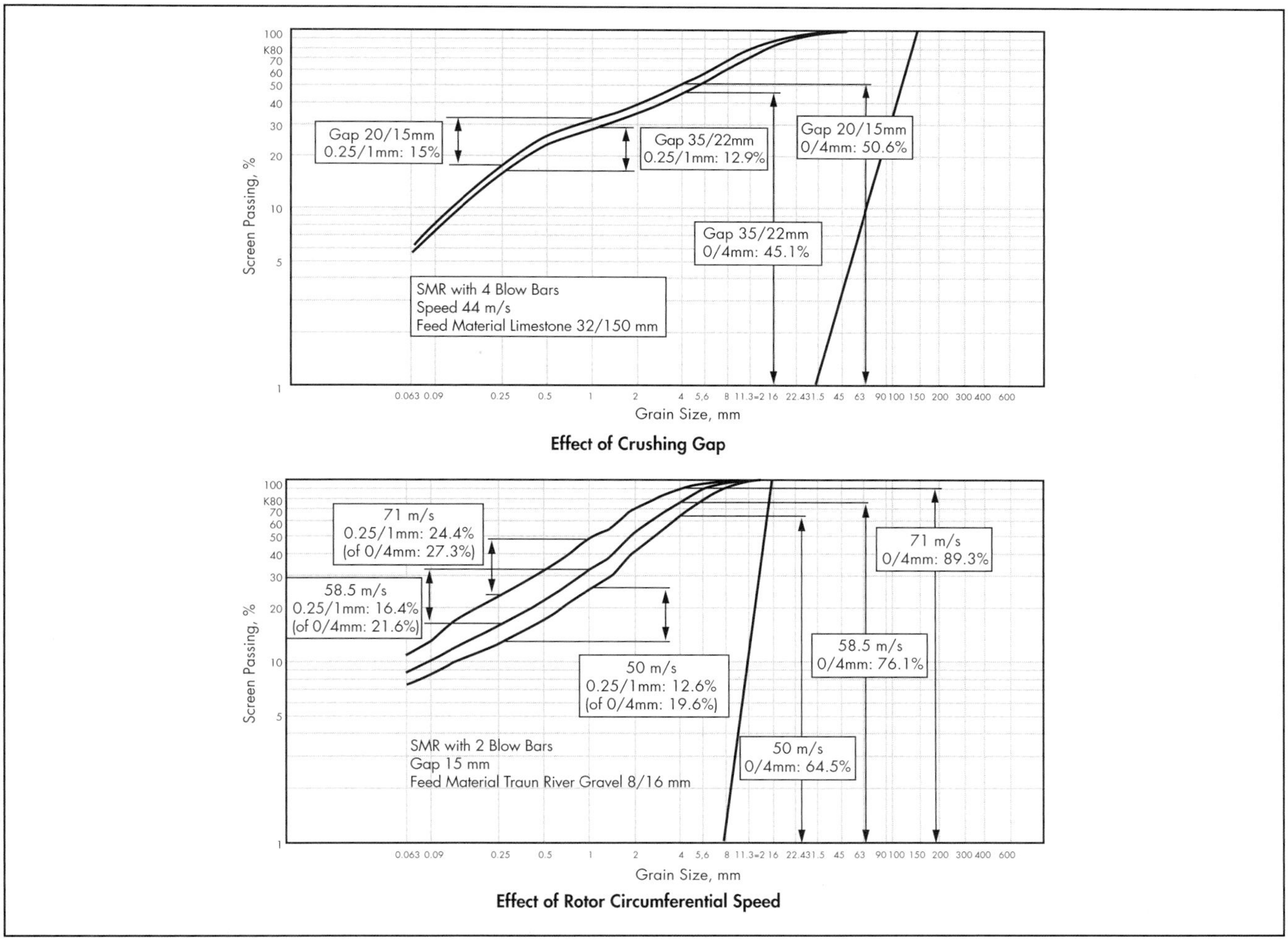

Courtesy of SBM Mineral Processing GmbH

Figure 24 Effect of the crushing gap and circumferential rotor speed on size reduction

Table 9 General operating range for Metso Barmac B series vertical shaft impactor

Barmac Model	Maximum Feed Size, Square Mesh, mm (in.)	Speed Range, m/s (rpm)	Power Range, kW (hp)	Throughput, t/h (stph)
B5100SE	30 (1¼)	45–75 (2,000–3,400)	37–55 (50–70)	15–60 (14–54)
B6150SE	37 (1½)	45–75 (1,500–2,500)	75–150 (100–200)	60–200 (54–181)
B7150SE	45 (1¾)	45–75 (1,250–2,100)	185–220 (250–300) Single drive 260–300 (350–400) Dual drive	110–420 (100–381)
B9100SE	50 (2)	840 Rotor: 45–65 (1,250–1,800) 990 Rotor: 45–75 (1,250–1,700)	370–600 (500–800) Dual drive	180–700 (163–635)

Courtesy of Metso

	Feed Material Characteristics		Operating Parameters		
	Easy Crushability	High Abrasiveness	Rotor Speed	Cascade Increase	Choke Feed
Capacity	→	→	↓	↑	↑
Cubicity of Product	↑	→	↑	↓	↑
Specific Power Consumption	↓	↑	↑	↓	↓
Wear Costs	↓	↑	↑	→	↓

Courtesy of Metso

Figure 25 Guide to the impact of feed on key performance parameters and Barmac operation

Table 10 Canica vertical shaft impactor general performance data

Description*	Canica Model 1200	1400	2000SD	2000DD	2050	100	2300	105	2350	2500	3000
Motor drive	Single	Single	Single	Dual	Dual	Dual	Dual	Dual	Dual	Dual	Dual
Maximum feed size (longest dimension) HD configuration, mm (in.)	33 (1.5)	51 (2)	102 (4)	102 (4)	102 (4)	127 (5)	127 (5)	152 (6)	203 (8)	254 (10)	305 (12)
Maximum throughput capacity HD configuration, t/h (stph)	64 (70)	113 (125)	227 (250)	317 (350)	363 (400)	363 (400)	454 (500)	454 (500)	544 (600)	726 (800)	907 (1,000)
Power requirement for maximum throughput, kW (hp)	37–112 (50–150)	75–186 (100–250)	149–298 (200–400)	298–522 (400–700)	298–522 (400–700)	298–522 (400–700)	298–522 (400–700)	373–594 (500–800)	373–594 (500–800)	447–746 (600–1,000)	522–895 (700–1,200)
Internal configurations available†	HD, HDS	HD, HDS, ROS, ROR	HD, HDS, ROS high-speed, ROR high-speed, ROS HD, ROR HD	HD, HDS, ROS high-speed, ROR high-speed, ROS HD, ROR HD	HD, HDS, ROS high-speed, ROR high-speed, ROS HD, ROR HD	HD, HDS, ROS high-speed, ROR high-speed, ROS HD, ROR HD	HD, HDS, ROS high-speed, ROR high-speed, ROS HD, ROR HD	HD, HDS, ROS high-speed, ROR high-speed, ROS HD, ROR HD	HD, HDS	HD, HDS	HD, HDS

Courtesy of Terex Corporation

*Important: The maximum feed size, throughput, and power requirement are dependent on the internal configuration used. The internal configuration depends on the actual feed material, discharge requirements, and the material abrasive and strength properties.

†HD = heavy-duty (open table/anvils); HDS = heavy-duty sand (open table/anvils); ROR = rock on rock (enclosed rotor/rockshelf); ROS = rock on steel (enclosed rotor/anvils).

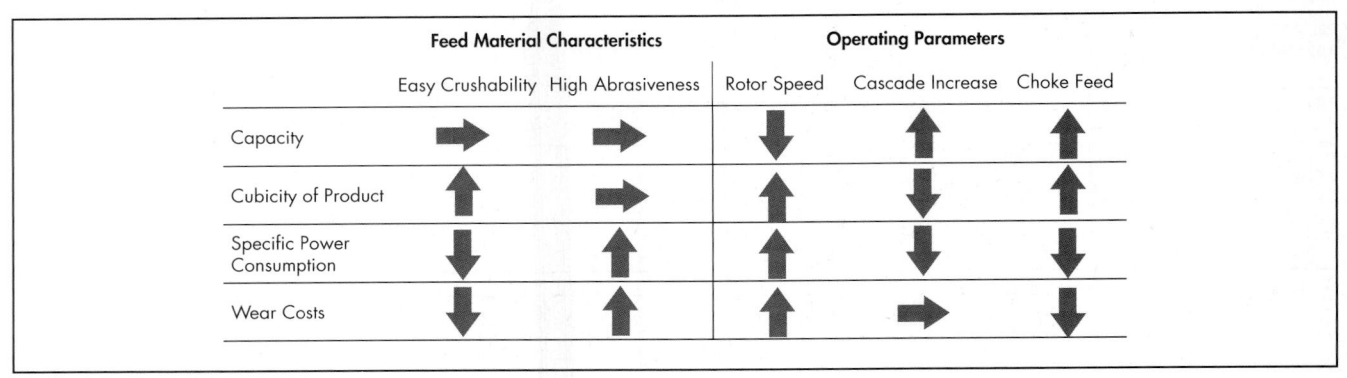

between a VSI and a *rock pump*. Essentially, the VSI has an *impeller*, which is the rotor. The throughput and the required power are more related to allowing material to pass and be accelerated by the rotor. In the case of both the Barmac and the Canica versions, it is common to have dual motor drives in the larger variants.

Regarding limits in VSI performance, the action of the rotor and the impact breakage mechanism employed mean that the main limiting factor is purely volumetric flow, providing sufficient power is available to the rotor to keep the charge moving and crushing. The volumetric constraint is purely a flow issue and not a *packing* issue, as seen in compression crushers.

Because of the nature of the crushing action in a VSI, the rotor simply has to impart sufficient energy to the particles to cause them to fracture either on impact with other particles or the walls or anvils. Motor power and rotor speed are therefore critical to impart the required energy and therefore generate the product size (Sinnott and Cleary 2015). However, they do not represent a limit in the same way as would be seen in a compression crusher, where input power is applied in a much more direct manner. A comparison of the energy considerations in cone and VSI crushers is specified by Lindqvist (2008).

Given that the VSI is analogous to a rock pump, measures of rock strength are still indicative of the ability to generate size reduction and therefore net product, but the effect of strength on gross throughput is much less than is seen in compression crushers. VSI machines essentially employ unconstrained, or free, breakage, either from impacting particles on a steel surface or through interparticle impacts. These mechanisms provide many of the benefits seen in VSI machines, including that all particles are exposed to breakage (unlike compression crushers), fines generation exceeds that seen in compression crushers, and the ability to treat abrasive feed via the particle-on-particle system. As always, the same features that provide benefits also constrain performance, with the main constraint seen in VSI crushers being the trade-off between feed size, reduction ratio (at the P80 level), feed abrasivity, and wear. In this regard, VSIs are a good example of the importance of understanding the fundamental basis underlying the action of any crusher. Such an understanding ensures that the equipment is applied in the appropriate duty and delivers the best overall benefit to the wider circuit.

ACKNOWLEDGMENTS

The author thanks Peter Pluis of Palminco Pty Ltd. for his input and for sharing his experience in relation to the design, application, and maintenance of jaw crushers.

REFERENCES

Bengtsson, M., and Evertsson, C.M. 2006. Measuring characteristics of aggregate material from vertical shaft impact crushers. *Miner. Eng.* 19(15):1479–1486.

Briggs, C.A., and Bearman, R.A. 1996. An investigation of rock breakage and damage in comminution equipment. *Miner. Eng.* 9(5):489–497.

Burhoff, K., Sjoberg, P., and Wood A. 2015. Hydraulic system for controlling a jaw crusher. U.S. Patent 20150107232 A1.

Cunha, E.R., Carvalho, R.M., and Tavares L.M. 2013. Simulation of solids flow and energy transfer in a vertical shaft impact crusher using DEM. *Miner. Eng.* 43-44:85–90.

Eacret, R.L., and Klein, E.F. 1985. Hammer mills and impactors. In *SME Mineral Processing Handbook*. Vol. 1. Edited by N.L. Weiss. Littleton, CO: SME-AIME. pp. 3B-70–83.

Gonçalves, J., Tavares, L.M., Toledo Filho, R.D., Fairbairn, E.M.R., Cunha, E.R., 2007. Comparison of natural and manufactured fine aggregates in cement mortars. *Cem. Concr. Res.* 37:924–932.

Hamer, M.D. 1990. The Barmac autogenous crushing mill—the new development in autogenous crushing and milling. In *Proceedings of the AusIMM Annual Conference: The Mineral Industry in New Zealand*. Melbourne, Victoria: Australasian Institute of Mining and Metallurgy. pp. 207–215.

Haven, M.B., Quella, P., and Jaworski, B.P. 2002. Jaw crusher toggle beam hydraulic relief and clearing. U.S. Patent 6375105 B1.

Latimer, C. 2012. Metso supplies world's largest mobile crusher. In *Australian Mining*. June 29. www.australianmining.com.au/news/metso-supplies-worlds-largest-mobile-crusher/.

Lindqvist, M. 2008. Energy considerations in compressive and impact crushing of rock. *Miner. Eng.* 21:631–641.

McCarthy, K., and Campbell-Hunt, C. 2000. *Competitive Advantage New Zealand (CANZ): Svedala Barmac: Case History*. Wellington, New Zealand: School of Business and Public Management, Victoria University. www.victoria.ac.nz/som/research/research-projects/competitive-advantage/documents/svedalaf.pdf.

Metso. 2011. *Crushing and Screening Handbook*, 5th ed. Helsinki, Finland: Metso.

Nikolov, S. 2002. A performance model for impact crushers. *Miner. Eng.* 15(10):715–721.

Nixon, J., and Weston, A. 2005. Excavation of the lift 2 crusher hydroset chamber and orepass in seismically active conditions at Northparkes mines. In *Proceedings of the 9th Underground Operators Conference*. Melbourne, Victoria: Australasian Institute of Mining and Metallurgy. pp. 193–202.

Sandvik. 2012. Sandvik jaw crushers. Brochure B5-125:1Eng. p. 3.

Sinnott, M.D., and Cleary, P.W. 2015. Simulation of particle flows and breakage in crushers using DEM: Part 2—Impact crushers. *Miner. Eng.* 74:163–177.

Utley, R.W. 2002. Selection and sizing of primary crushers. In *Mineral Processing Plant Design, Practice, and Control*. Vol. 1. Edited by A.L. Mular, D.N. Halbe, and D.J. Barrett. Littleton, CO: SME. pp. 599–603.

Weiss, N.L. 1985. Jaw crushers. In *SME Mineral Processing Handbook*. Vol. 1. Edited by N.L. Weiss. Littleton, CO: SME-AIME. p. 3B-2.

CHAPTER 3.4

Gyratory and Cone Crushers

Richard (Ted) Bearman

GYRATORY CRUSHERS

The gyratory crusher was originally patented by Philetus W. Gates in 1881, with the first crusher installed at the Buffalo Cement Company. During this period of mining and quarrying, the performance requirements for the crushers were modest, as there was little mechanization and production rates were low in comparison to current times. From this early point in the development of gyratory crushers, a range of crusher types evolved. Westerfeld (1985) discussed some of the gyratory types, including the long-shaft suspended and fixed-spindle variants, and noted that both of these variants were falling from favor, with the development and deployment of the hydraulically supported short-shaft gyratory crusher. In the intervening years since Westerfeld's work, this move has now been fully completed. All major manufacturers now use the hydraulic shaft-supported machine, and it is rare to find any of the previous variants.

In all upcoming descriptions and discussions, the only type of gyratory crusher that is considered is the hydraulic shaft-supported machine, and hence all mentions of the *gyratory crusher* refer to this type. For reference, a typical hydraulic shaft-supported machine is shown in Figure 1.

Gyratory crushers are commonly used as primary machines, where the feed material displays moderate to strong breakage characteristics and the throughput rate is in excess of that from an equivalent jaw crusher installation. The gyratory crusher uses a compressive crushing action, with the material progressing through the chamber under gravity. As with other types of cone crusher, the gyratory crusher has an inner mantle that gyrates eccentrically within a fixed outer crushing surface known as the *bowl* or *concave*. The eccentric action generates a crushing position that moves from a fully open position (open side setting [OSS]) through to a minimum distance between the mantle and concave (closed side setting [CSS]). Figure 2 shows two views of the discharge setting.

Because of the rotation of the eccentric driving the mainshaft in a gyrating motion, the position of the closed side and open side between mantle and concave constantly moves around the crushing chamber. Typical primary gyratory crushers have an eccentric speed of 150–200 gyrations per minute. Figure 3 provides a schematic diagram of the gyrating mechanism used in gyratory crushers.

In most gyratory crusher installations, there is no attempt to remove fine material ahead of the crusher; rather the fines pass through the crusher along with the oversize feed. Fines are typically considered as a particle size of less than 50% of the OSS.

The gyratory crusher is regarded as an all-purpose machine and as such is used without restrictions across most feed types. Feed material that is not generally suited to treatment in a gyratory crusher is mostly wet and sticky feed, with, or without clay and in certain instances, very fine feed can compromise the operation of such machines.

Given that a majority of gyratory crushers act as primary machines, the feed size is typically run-of-mine (ROM). Gyratory crusher notation is usually of the form "feed opening–mantle diameter," hence a common size is a 60–89, that is, a 60-in. (1,524-mm) feed opening and an 89-in. (2,260-mm) mantle diameter. This designation is shown graphically in Figure 4.

Essentially, the feed opening dictates the largest rock that can be handled, and the mantle diameter is indicative of the maximum throughput. In terms of feed size, the general rule is that the maximum feed size should not exceed 60%–80% of the maximum feed opening (i.e., in the preceding figure, this represents 914–1,219 mm [36–48 in.]), with the feed measured in the second dimension. As with all compression crushers, the product size is dictated by the OSS and CSS, whereby the maximum product size is equal to or less than the OSS, and the CSS influences the degree of size reduction. The recommendation for the maximum size of feed relative to the feed opening varies, and suppliers should be consulted for advice specific to a given application.

An issue to consider with primary gyratory crushers is that, in essence, their role is to reduce the feed to a manageable size and let fine material pass through as quickly as possible. Because the feed can contain large amounts of fines, the product size distribution from a typical machine is more

Richard (Ted) Bearman, Director, Bear Rock Solutions Pty Ltd., Attadale, Western Australia, Australia

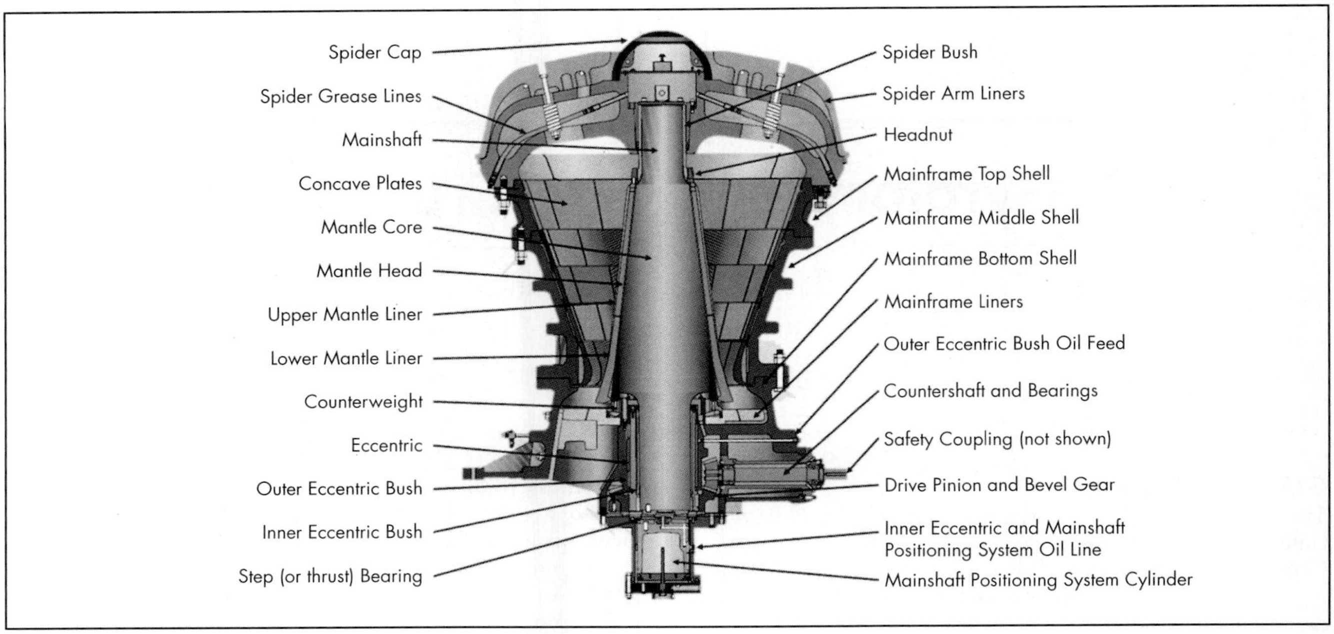

Courtesy of Metso

Figure 1 Sectional view of a typical hydraulic shaft-supported primary gyratory crusher

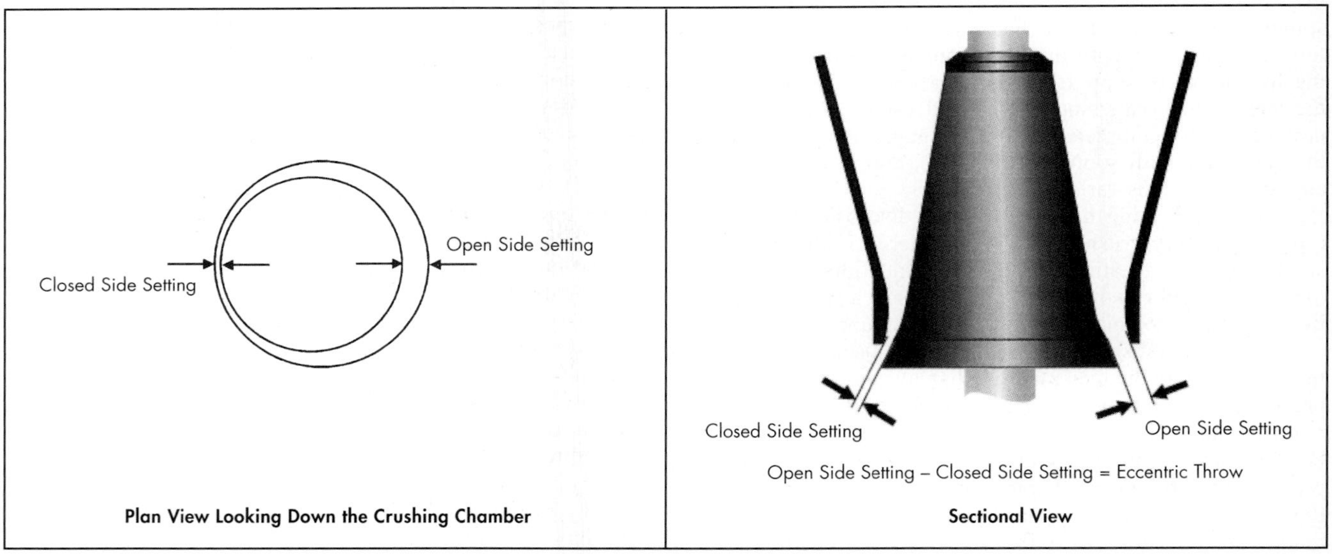

Courtesy of Metso

Figure 2 Discharge setting, comprised of open side setting and closed side setting

controlled by the feed particle size distribution than the OSS or CSS—this makes primary gyratory crushers distinctly different from cone crushers.

The feed particle size distribution is also the major contributor to the throughput. Book values provided by manufacturers all need to be corrected to account for *free throughput*, that is, material passing through and adding to the throughput, but too small to require crushing. All manufacturers have proprietary equations to generate this correction.

The corrections can be large, with many large gyratory crushers deployed on fine feed reporting throughput values equal to twice the book values.

The large throughputs also explain why power consumption values for primary gyratory crushers can appear to be very low. It is not unusual to see values of 0.1–0.2 kW·h/t (0.09–0.18 kW·h/st), but these values are a result of high throughput and relatively low power draw.

Until the 2000s, traditionally installed motor power in primary gyratory crushers was approximately 200–400 kW (268–536 hp). It is now reported that 1,000 kW (1,341 hp) is installed on large crushers of the 60–89 and larger sizes.

As mentioned earlier, gyratory crushers are designed to accept a given feed size, and in most cases, this is fixed for the duration of the life of the machine. In a recent development,

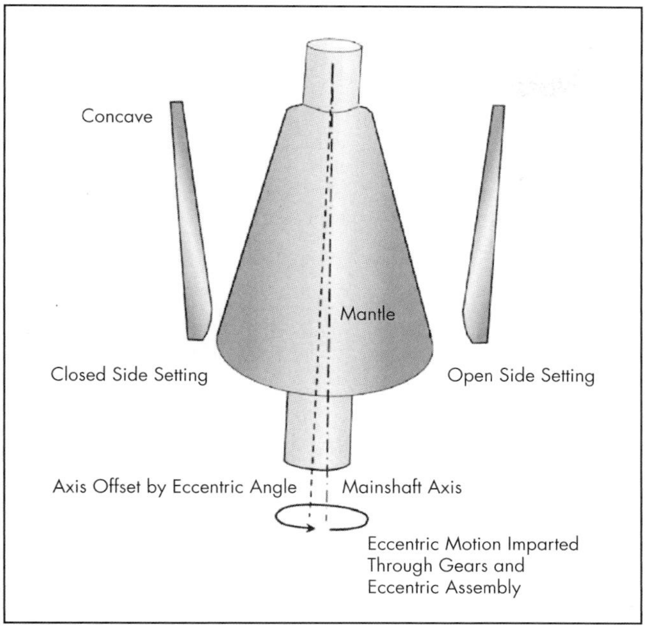

Figure 3 Gyrating action of mantle in a primary gyratory crusher

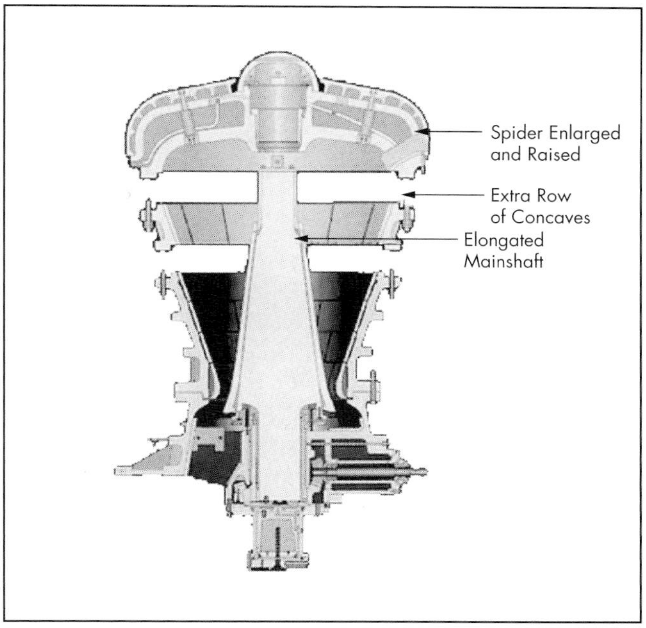

Courtesy of Metso
Figure 5 Metso super spider concept

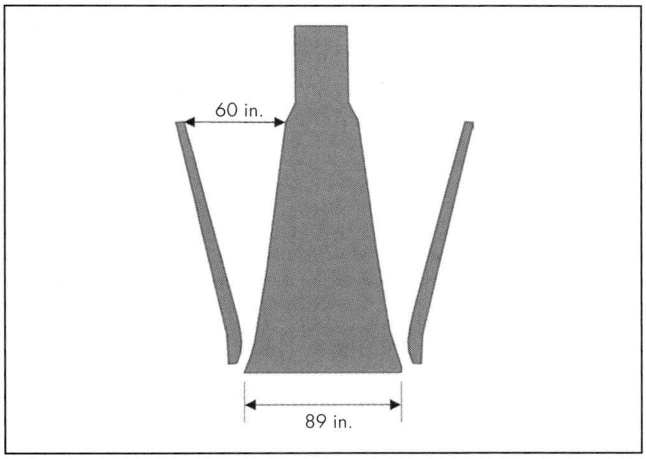

Figure 4 Size designation for primary gyratory crushers

Metso developed a concept known as the super spider (Figure 5).

A super spider S70–89 MK-II primary gyratory crusher is an S60–89 MK-II fitted with an arched spider and also an additional upper top shell (with extra concave now) and new mainshaft. The combination of an arched spider and an upper top shell increases the feed opening to 1,828 mm (70 in.) compared with 1,524 mm (60 in.) for the S60–89.

An existing S60–89 can be upgraded to an S70–89 by installing a new spider, upper top shell, mainshaft, concave set, mantle, arm guards, and rim liners. The crusher top shell, bottom shell, eccentric, and drive components are not changed. Alternatively, the S70–89 is offered as a new machine.

The height of the S70–89 to the top of the rim liners is 756 mm (30 in.) higher than an S60–89. Because of this and in the case of an upgrade, the increase in crusher height may necessitate modifications to the dump hopper structure. Consideration also needs to be given to the design of supporting structure, as Metso reports a marginal increase in the dynamic loads caused by the crusher's center of mass being raised.

In addition to the increased feed size, Metso also suggests that the super spider configuration can increase throughput by 12%–15% compared with an S60–89.

Principles and Terminology

The gyratory style of crusher relies on the same principles as the cone-type crusher, that is, an eccentric movement of the crushing head (mantle) induced by the rotation of an eccentric assembly around a central axis. The mantle liner that provides the moving crushing surface is usually cast in either a single piece or, more commonly with larger machines, in two pieces. Because of the size of the casting, the thickness and diameter, and the required wear resistance, the material favored is an austenitic manganese steel. This type of material exhibits the phenomenon of work-hardening. In the as-cast and heat-treated condition, it will have a hardness of 300 (Brinell hardness [BH]), whereas once crushing takes place and the liner is worked, the surface hardness can increase to 550–650 BH. Depending on the crushing duty, the manganese content can be varied from approximately 12% to 24% and still retain the work-hardening characteristic. The increasing percentage of manganese potentially offers an extended wear life, but it does come at the cost of ductility; that is, the higher manganese castings will be more susceptible to cracking from the ingress of tramp metal. Some manufacturers also add elements such as nickel and chromium to manganese steel crushing parts to improve initial hardness, ductility, and wear characteristics. Typically, an 18% manganese crusher liner may also contain up to 3% nickel and a small percentage of chromium.

The outer fixed crushing surface is made up of a series of circumferential rows, with each row containing multiple concave plates or segments. Depending on the size and design

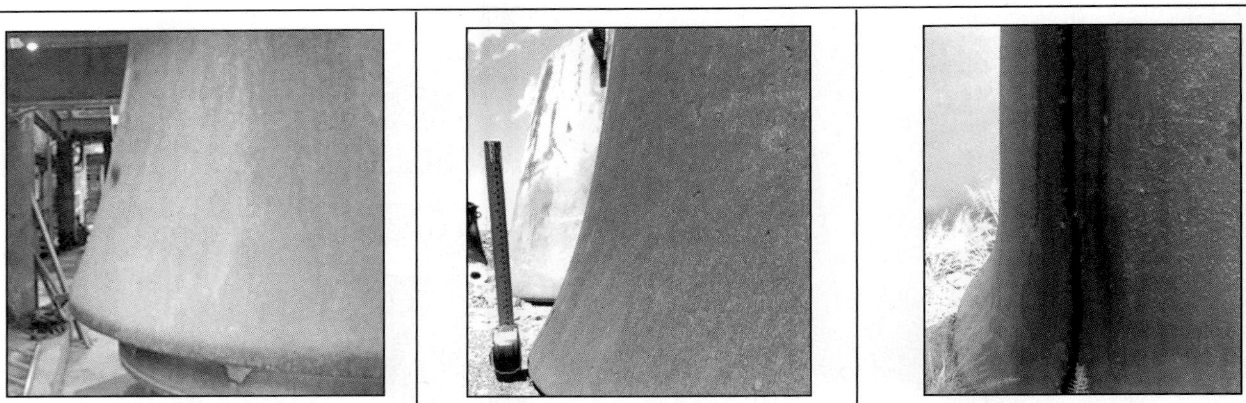

A. Worn mantle showing a normal profile with no localized wear

B. Worn mantle showing wear development toward a ski-ramp or top-hat profile

C. Worn mantle with wear showing a typical top-hat profile

Figure 6 Examples of mantle profiles

of the crusher, the plates are supplied in three-, five-, six-, or seven-row versions. As multiple rows are used, this provides the opportunity to use different materials for the various rows of concave plates. In applications where sliding wear is predominant and the presence of uncrushable items is unlikely, white cast-iron plates can be used in all but the upper row. White cast iron is highly resistant to sliding wear and can provide a longer wear life, but it is brittle and will therefore have a higher chance of cracking, which may lead to loss of the segment and lost production time for the repair. Where gouging wear is more common and uncrushables may be seen, the risk associated with white cast iron may be judged as unacceptable and as such, it can be replaced with materials such as through-hardened martensitic steel alloy. A martensitic steel will have slightly reduced surface hardness but a greater resistance to cracking. Harder materials are mostly deployed in the middle and lower rows, but toward the top of the crushing chamber, where impact can become the predominant concern, a more impact-resistant material is preferred. To address this point, the top one or two rows often revert to a manganese steel composition that can resist the impact and offer work-hardening.

Chamber design and liner profiles in primary gyratory crushers need to be appropriate for the required duty. Many performance issues (both process and mechanical) can be attributed to inappropriate and/or mismatched concave and mantle profiles. Typical problems seen include

- Inability to set a required discharge setting,
- Localized and excessive wear on lower area of the mantle,
- Abnormally high power draw,
- Excessive hydraulic pressure, and
- Low throughput.

Of these problems, one that is commonly seen is the excessive wear just before the discharge zone on the mantle liner. This phenomenon has several colloquial names, including *ski-ramp*, *top-hat*, or *duckbill* wear. An example of such a wear profile is provided in Figures 6B and 6C.

These terms describe a situation whereby the mantle develops a dish of wear just prior to the discharge point, but a lip of material remains at the very bottom of the mantle. Such a wear pattern can have serious effects on the crusher, as the flow of material through the chamber is hampered (low throughput). There is also the tendency for excessive power and pressure, as crushing can occur practically up against the base of the final row of concaves.

The development of such a wear profile is usually associated with either of the following situations:

- A mismatch of the mantle and a concave profile only allows the required discharge setting to be achieved higher on the mantle (compared to the normal position near the discharge area).
- There is a lack of mantle adjustment for wear so that wear is localized. This is often associated with the type of mismatch just highlighted.

One of the complexities of primary gyratory crusher chamber design occurs because optimizing the wear life of the total chamber—concaves and mantles—is required. As the wear of the mantles is more rapid than for the static concaves, the ratio of mantle-to-concave-liner changes can vary from 2:1 to 5:1. As an example, a size 60–110 gyratory crusher may have mantle diameters of

- 2,718 mm (107 in.) (standard),
- 2,769 mm (109 in.) (oversize), or
- 2,819 mm (111 in.) (double oversize).

The design and designation of the series of mantles will vary according to the concave set used, the type of wear, and the required discharge setting, so the sizes just mentioned are simply an example. The main point to consider when seeking to optimize the sequence and size of replacement mantles is to ensure that the required discharge setting can be achieved when the new mantle is installed. At the same time, the position of the mainshaft should be a minimum (i.e., at a position in the range of 0%–10% of operational travel of the mainshaft) when the mantle is new, as this will allow the maximum wear to be obtained for the mantle. In stating that the mainshaft should be low within the operational travel range, under no circumstances should the zero point of the operational travel be lowered into the safety zone between the operational zero set point and the absolute mechanical zero point. Such an action compromises the ability to use the hydraulic system to relieve an overload and provide protection to the machine.

The use of laser, or optical chamber profiling, for gyratory crusher liners has recently grown, and Elias et al. (2015) describe a chamber optimization exercise that was undertaken using a laser scanning technique. The level of detail available from laser scanning systems and the speed of the systems offer a practical option to determine wear over time and invaluable input for the customization of liner profiles.

Another point often discussed is the difference between choking and nonchoking chambers. The essential difference is that nonchoking chambers have a relatively straight concave profile from top to bottom, while choking chambers have a curved concave profile. Nonchoking chambers promote throughput but provide less control of product size and are prone to loss of CSS control because of localized wear. Choke chambers tend to generate reduced throughput but can operate at lower discharge settings, give better control of product size, and usually offer longer wear life. In this author's opinion, the differentiation between these styles of chamber design should be replaced by a discussion based on the required duty of the crusher. Specialized chamber design needs to address feed size distribution, material properties, flow characteristics, and mantle and concave change-out strategy. Given the complexity of this, the choke versus nonchoke discussion is too simplistic, and the specifics of liner designs are therefore best handled by specialists in the area who can provide a design customized to the required crushing duty.

As with all compression-based crushers, the distribution of energy within the crushing chamber is the critical element of the design. In gyratory crushers, the force is transmitted from an electric motor through a bevel gear set into the gear mounted on the outside of the eccentric assembly. The eccentric assembly is then free to rotate within the mainframe via a bronze bush (outer eccentric bush). The eccentric assembly has a concentric outside face, but the inner bore is eccentrically located or offset to the main centerline. The inner eccentric bore houses a bronze bush (inner bush) that allows the mainshaft to freely rotate within the body of the eccentric.

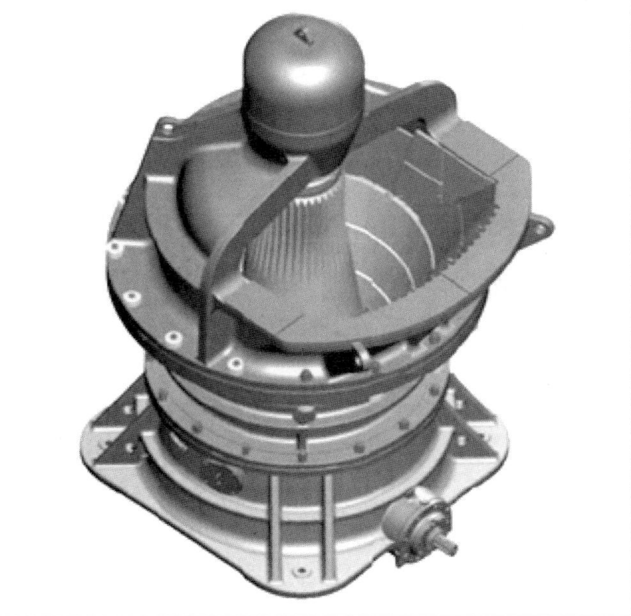

Courtesy of ThyssenKrupp Industrial Solutions

Figure 7 Jaw gyratory crusher showing enlarged, single-sided feed opening

The motion of the eccentric therefore causes the mainshaft to gyrate and, with its integral mantle surface, cycle from the OSS to the CSS in relation to the fixed concave liners.

When running empty, the mantle may be seen to rotate at up to approximately 6 rpm. With the introduction of feed into the chamber, the rocks act like epicyclic gears and cause the rotation of the mantle to reverse and the rotation to slow to 1–2 rpm. This rotation in either direction is simply the result of frictional forces and transfer of momentum between the eccentric and the mainshaft. It should be noted that this rotation is not the crushing action of the machine. Crushing is caused by the gyrating motion of the mainshaft, which creates a constant cycling between OSS and CSS. This cycling applies force to rocks traveling down through the chamber under gravity. As the gap between the mantle and concave reduces the chamber, the degree of size reduction increases.

In the operating crusher, the large rocks are captured and broken high in the chamber with the progeny particles falling lower into the chamber to again be crushed, or exit if the size is sufficiently small. Typically, feed particles entering the top of the chamber may undergo anywhere between 5 and 10 compressive strokes before exiting the chamber.

Use of primary gyratory crushers in underground mines has increased with the popularity of block cave–type operations (Hernandez et al. 2014). In cave operations, the ROM top size to the crusher can be very large and as such, gyratories offer advantages equivalent to jaw crushers. One specific derivative of the gyratory crusher has found particular favor in block cave operations, and this is the jaw gyratory crusher (Figure 7). Essentially, the jaw gyratory crusher is a normal gyratory machine with the top shell modified and the spider enclosed fully on one side to enable only one-sided feeding.

The feed opening is deliberately flared to take larger rocks, and the mainshaft is lengthened to allow capture of the rocks very high in the chamber. Currently, it appears that only ThyssenKrupp Industrial Solutions manufactures and supplies this hybrid type of gyratory crusher.

Construction and Design

The construction and design of primary gyratory crushers has evolved based on the requirement to meet the needs for higher tonnages, greater size reduction (i.e., use of high installed power), and easier maintenance. The main components of the machine include the following (for further information, see Figure 1):

- Mainframe
- Bottom, middle, and top shells
- Spider arm
- Mainshaft
- Eccentric
- Drive train, including the countershaft and gear sets
- Hydraulic shaft support (also known as the mainshaft positioning system)

The mainframe and the outer shells remain as cast steel items, but increasingly, various forms of split shells are being developed, where the shells can be split both horizontally and vertically. Traditionally, the bottom, middle, and top shells have been cast as single items and then bolted together. With the increased need to locate crushers into remote locations and underground spaces, suppliers have designed shells with special flanges that allow each shell to be split vertically. Split spider arms are also available. Figure 8 shows split-shell arrangements from one of the main manufacturers.

The main changes in the mainshaft and eccentric assembly have been in the increased specification of the materials to absorb higher loads and greater consideration of lubrication and control of heat in the system. Essentially, the arrangement of the eccentric assembly remains similar across all manufacturers, with a typical arrangement seen in Figure 9.

Because of the greater motor power and higher forces, the countershaft and the gear sets have received more attention. The countershaft material specification has been increased, and in several models, torque couplings have been deployed to give protection to the crusher under overload conditions. In tandem with this change has been the increasing move to spiral bevel gears. Such gear sets allow greater efficiency and contact area to transmit the high forces. A typical modern direct-drive countershaft arrangement is shown in Figure 10.

Courtesy of FLSmidth

Figure 8 Gyratory crusher split-shell arrangement

A development aimed at improving maintenance access is the Top Service crusher from FLSmidth. In this design, the manufacturer has introduced the ability to maintain all major components through the top of the crusher. Such an innovation is particularly useful for maintenance to the major components lower in the machine, that is, eccentric assembly and hydraulic shaft adjustment components. Previously, the lower components were only accessed from underneath the crusher, which involved clearing the discharge vaults under the machine, working in confined spaces, and the need for a specialized maintenance trolley for removal and refitting of the hydraulic shaft adjustment assembly.

Two important considerations in the operation of gyratory crushers are the lubrication and hydraulic systems. For modern gyratory crushers, there is a trend to install higher motor power. Such input power allows the crusher to work harder and therefore generate more heat, particularly in the eccentric bush region. Because of the duty cycle of crushers, the eccentric bushes still rely on the use of leaded bronze as the base material. One of the key features of leaded bronze is the ability to *bleed* lead if bushes are subjected to localized heating from a high-force event, or alternatively, from frictional heating. To provide lubrication and to mitigate high-temperature events, the flow of lubrication, specification of the oil, and pressure of delivery must all be matched to the system. A critical part of the lubrication system is the monitoring of the oil temperatures in both the feed and return lines, with manufacturers having limits on the differential temperature. If the feed–return differential exceeds the recommended set points, then process control systems are programmed to

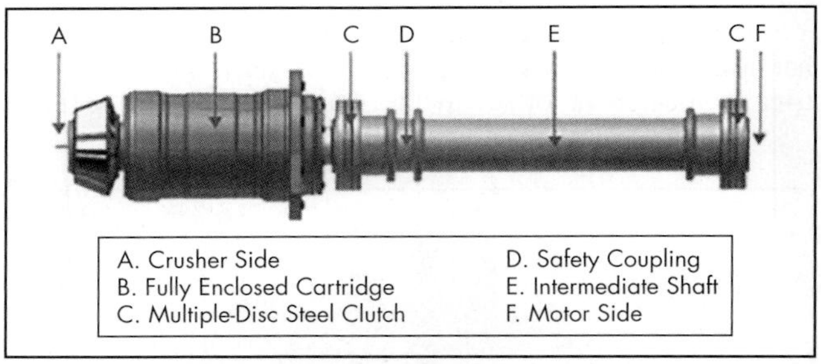

Courtesy of ThyssenKrupp Industrial Solutions

Figure 10 Countershaft showing main components

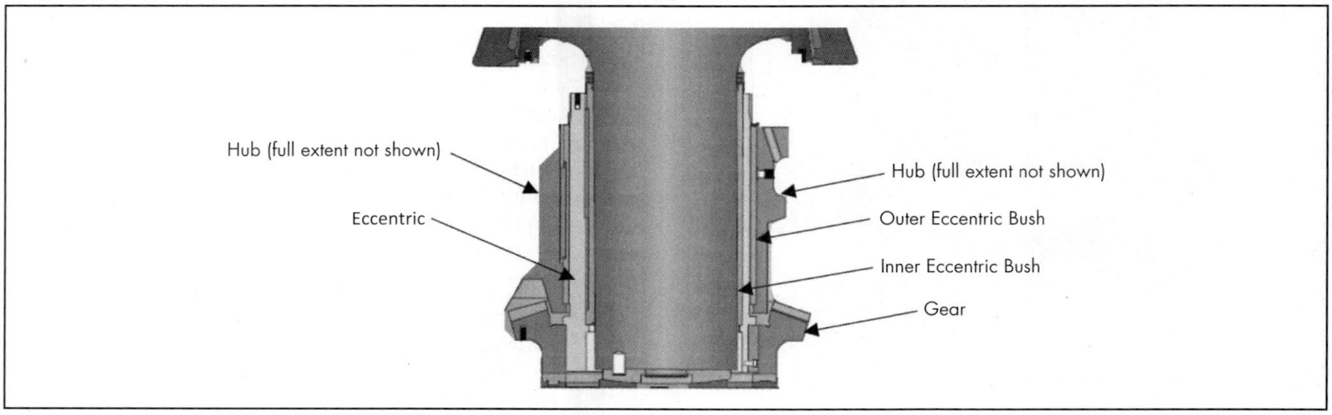

Courtesy of Metso

Figure 9 Eccentric assembly (associated components excluded for clarity)

provide alarms to the operator. To ensure that the temperature differential stays within the acceptable bounds, crushers are fitted with heating and cooling systems, which must be matched to both the duty and environmental and climatic conditions of the operating site.

It is also important to consider the cleanliness of the oil and ensure that any debris or contamination is removed. Gyratory crushers usually have two stages of filtration, with a coarse *trash* screen usually located under an inspection cover in the top of the storage tank and a fine 20-µm cartridge-type filter in the feed line.

Several generations of gyratory crusher have incorporated hydraulic support of the mainshaft to allow

- Easier setting adjustment (and ability to adjust for liner wear),
- Ability to keep pressure on step bearing components in the event of the mainshaft *jumping* upward,
- Overload protection, and
- Chamber clearance following an unplanned stoppage.

The basic design of the hydraulic circuit used by most manufacturers consists of a fluid storage tank, a motor-driven gear pump, control and relief valves, and a cylinder assembly (with or without a separate accumulator).

Under normal operating conditions, the hydraulic circuit is simply used to raise or lower the position of the mainshaft, with the position monitored through a setting measurement device that is integrated into the cylinder below the mainshaft. The measurement device can be either a potentiometer or a linear variable differential transformer. This raising and lowering function can also be used in the event of a crusher stall where the crusher stops with a chamber full of feed.

Under certain crushing conditions, there is a tendency for the mainshaft to jump upward, and if not addressed, this will release load off the step bearing components located under the mainshaft. When the mainshaft drops back, damage can occur to the step bearing plates. To avoid this situation, the balance cylinder in the circuit forces oil into the system to ensure that load is maintained on the step bearing components. The balance cylinder incorporates a section with nitrogen under pressure, and it is this that controls the stiffness of the system and the ability to respond to the rapid upward movement of the mainshaft. Depending on the manufacturer, the cylinder may have an integral nitrogen system, or this can be separate through the use of an accumulator. An example of the function of the Metso system is shown in Figure 11.

Another protection function of the hydraulic system is to detect an overload situation in the crushing chamber. In such an instance, a relief valve in the circuit is actuated and the hydraulic fluid is dumped back to the storage tank.

Outside of the functions just described, there is a strong trend to start using this shaft adjustment as an active crusher control. One example of this approach is the Gyramatic system from ThyssenKrupp Industrial Solutions. In this system, power, pressure, and mainshaft position are continually monitored. Logic can be applied to incrementally open the crusher if power exceeds certain values and then return the crusher to the previous position after the peak power has passed.

Application and Operation

The application of primary gyratory crushers should be based on the information available from manufacturers with the main information relating to dimensions, masses, feed size, feed material characteristics, throughput, product size distribution, and required crushing power. All manufacturers provide this information in a variety of forms, but it is also important to seek examples of similar applications and installations from around the world.

As a guide to the type of information available, manufacturer data is reproduced in the following tables. Table 1 shows typical dimensions and weights of significant components and assemblies. Table 2 shows the typical throughput for various models of crusher based on the feed size opening and mantle diameter. Manufacturers will state the density of the feed material used in the calculation of throughput. For all applications, the actual density of the feed at the site should be compared to that used by the manufacturers and a correction applied.

In Table 1, the KB 63–130 gyratory crusher is listed. This machine is notable, as it is currently the largest gyratory crusher available (Carter 2016). Carter states that "despite a roughly 14% larger mantle diameter of 130 in. (3.3 m), it weighs less than its predecessor, the KB 63–114, at 490 t [metric tons]. Its rated throughput of up to 14,000 t/h is more than 30% higher than that of the preceding model." The other main information supplied by manufacturers is guidance on the potential product size distribution, as seen in Figure 12.

For general information regarding feed, manufacturers also often provide guidelines on the fines content of the feed, moisture content, clay content, and the degree of stickiness. *Fines* can be defined as the percentage of feed passing the CSS, or alternatively, the percentage passing CSS/2, and certain manufacturers will additionally state a limit for the fines content, beyond which the mechanical performance of the crusher may be compromised. The other major consideration with fines content is the direct relationship with throughput. All gyratory crusher throughput tables give a base throughput with defined feed properties. Should the fines content of the feed increase, then there will be a proportional increase in throughput. Most of the calculations for the throughput increase are proprietary and are not publicly stated. Kawasaki (now Earth Technica) states in the technical information for the KG series primary gyratory crushers that the following correction factor should be applied to a base throughput:

$$\text{correction factor} = \alpha_1 \cdot \alpha_2 \cdot \alpha_3 \cdot (\gamma/1.6)$$

where

α_1 = compressive strength coefficient
α_2 = feed size coefficient
α_3 = crushability, presence and frequency of fracture lines
γ = bulk density of the feed

According to the α coefficients, fine material less than the CSS can have a multiplying impact on capacity ranging from 1.0 to 3.0. This is significant compared to the two other coefficients that provide a range of 0.6 to 1.5 for the impact of compressive strength and a range of 0.9 to 2.0 for the crushability coefficient.

Moisture content, clay content, and the degree of stickiness are usually covered by a generic statement that feed material should be free-flowing. There are no specific tests to determine the limits for these factors in relation to crushers, and therefore input from manufacturers is required to assess any constraints that may be caused by elevated levels.

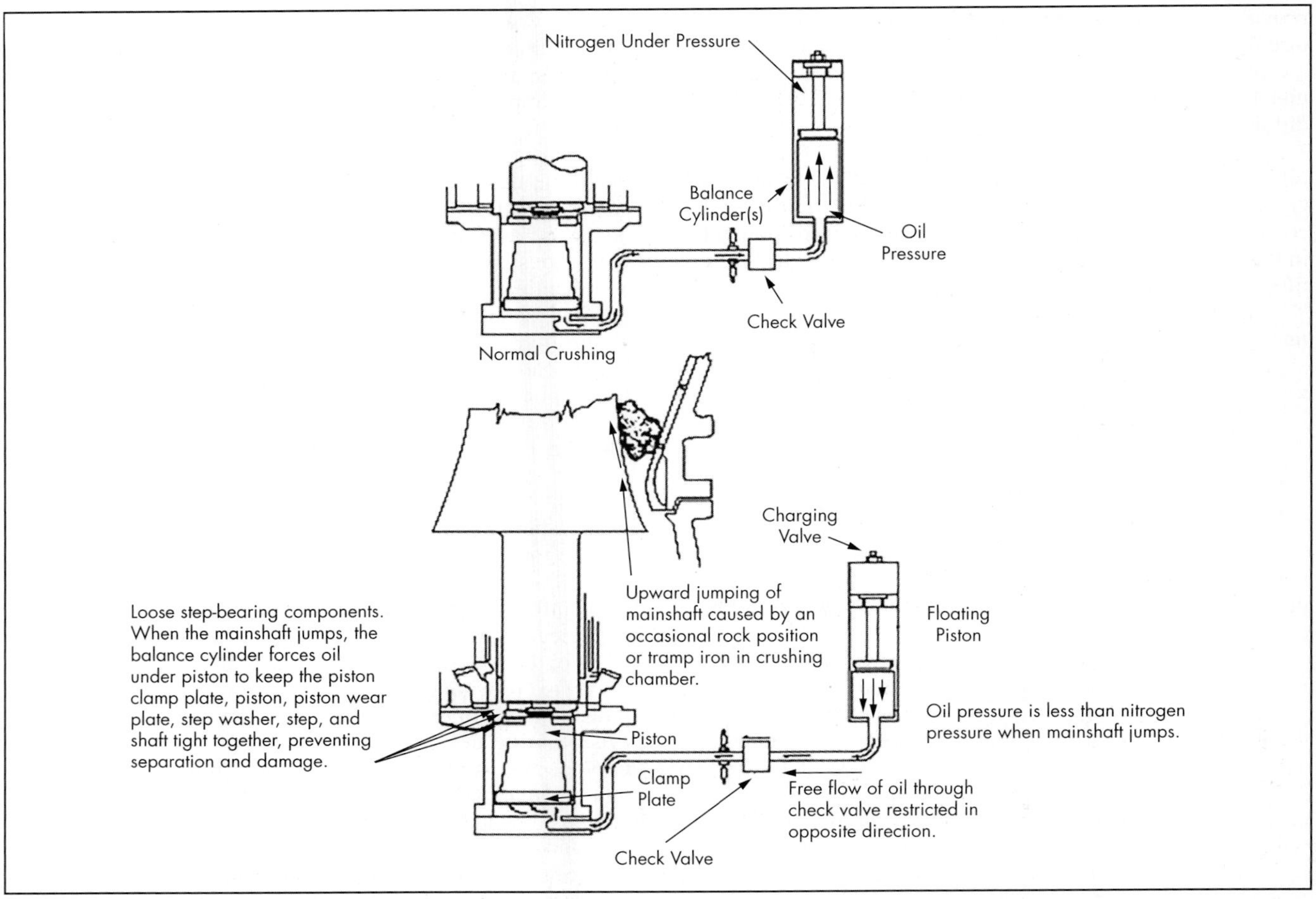

Courtesy of Metso

Figure 11 Balance cylinder operation

Table 1 Typical masses for primary gyratory crushers

								Weights*			
Gyratory Crusher Type	**Feed Opening, mm (in.)**	**Mantle Diameter (oversized), mm/in.**	**Speed of Eccentric Brushing, rpm**	**Maximum Motor Power, kW (hp)**	**Total Weight of Crusher, kg (lb)**	**Spider,† kg (lb)**	**Heaviest Shell, kg (lb)**	**Bottom Shell,‡ kg (lb)**	**Mainshaft Assembly,§ kg (lb)**	**Eccentric Bushing with Bottom, kg (lb)**	**Hydraulic Cylinder, kg (lb)**
KB 54–67	1,370 (54)	1,700/67 (1,750/69)	137	450 (604)	180,000 (396,832)	29,000 (63,934)	30,500 (67,241)	37,100 (81,791)	32,500 (71,650)	7,100 (15,653)	8,100 (17,857)
KB 54–75	1,370 (54)	1,900/75 (1,965/77)	137	650 (872)	215,000 (473,993)	32,000 (70,548)	33,000 (72,753)	43,500 (95,901)	43,000 (94,799)	8,500 (18,739)	9,500 (20,944)
KB 63–75	1,600 (63)	2,030/80 (2,100/83)	137	650 (872)	270,000 (595,247)	37,800 (83,335)	66,000 (145,505)	43,500 (95,901)	50,000 (110,231)	9,500 (20,944)	9,500 (20,944)
KB 63–89	1,600 (63)	2,260/89 (2,370/93)	130	1,000 (1,341)	332,000 (731,934)	46,500 (102,515)	71,000 (156,528)	53,000 (116,845)	69,500 (153,221)	13,000 (28,660)	14,000 (30,865)
KB 63–114**	1,600 (63)	2,900/114 (2,985/118)	127	1,200 (1,609)	530,000 (1,168,449)	65,500 (144,403)	73,000 (160,937)	124,000 (273,373)	124,000 (273,373)	16,000 (35,274)	19,000 (41,888)
KB 63–130	1,600 (63)	3,300/130 (3,400/134)	125	1,500 (2012)	495,000 (1,091,287)	66,000 (145,505)	97,500 (214,950)	85,500 (188,495)	130,000 (286,601)	12,100 (26,676)	13,000 (28,660)

Courtesy of ThyssenKrupp Industrial Solutions

* Average component weights including internals; safety margins for selecting lifting gear not included.

† Including wear cap.

‡ Excluding hydraulic cylinder.

§ With *oversized* mantle diameter, including crushing elements.

** This model of crusher is available on request.

Table 2 Typical throughputs for primary gyratory crushers at various open side settings

Gyratory Crusher Type, mm (in.)	Throughput, t/h* at OSS,† mm‡																	
	130		150		170		185		200		215		240		270		300	
	Min.	Max.	Min.	Max.	Min.	Max.	Min.	Max.	Min.	Max.	Min.	Max	Min.	Max.	Min.	Max.	Min.	Max.
KB 1,372–1,702 (54–67)	1,200	2,300	1,500	2,900	1,800	3,700	2,000	4,300										
KB 1,372–1,905 (54–75)	1,300	2,400	1,500	3,200	1,900	4,200	2,100	4,800	2,000	5,300								
KB 1,600–1,905 (63–75			1,700	3,300	2,000	4,500	2,200	5,100	2,400	5,600	2,600	6,000						
KB 1,600–2,261 (63–89)			2,300	4,500	2,500	5,100	2,700	5,900	3,100	6,700	3,300	7,500	3,500	8,500				
KB 1,600–2,896 (63–114)§					3,000	5,500	3,400	6,400	3,800	7,300	4,000	8,000	4,700	10,300				
KB 1,600–3,302 (63–130)					3,700	7,900	4,300	9,200	4,700	10,100	5,100	10,900	5,400	12,000	5,800	13,200	6,200	14,400

Courtesy of ThyssenKrupp Industrial Solutions

* To convert from metric tons per hour (t/h) to short tons per hour (stph), multiply by 1.1.

† Min. = typical feed material with minimum eccentricity and F80 (680 mm [27 in.]). Max. = typical feed material with maximum eccentricity and F80 (230 mm [9 in.]). Bulk density = 1,600 kg/m³ (100 lb/ft³).

‡ To convert from millimeters (mm) to inches (in.), divide by 25.4.

§This model of crusher is available on request.

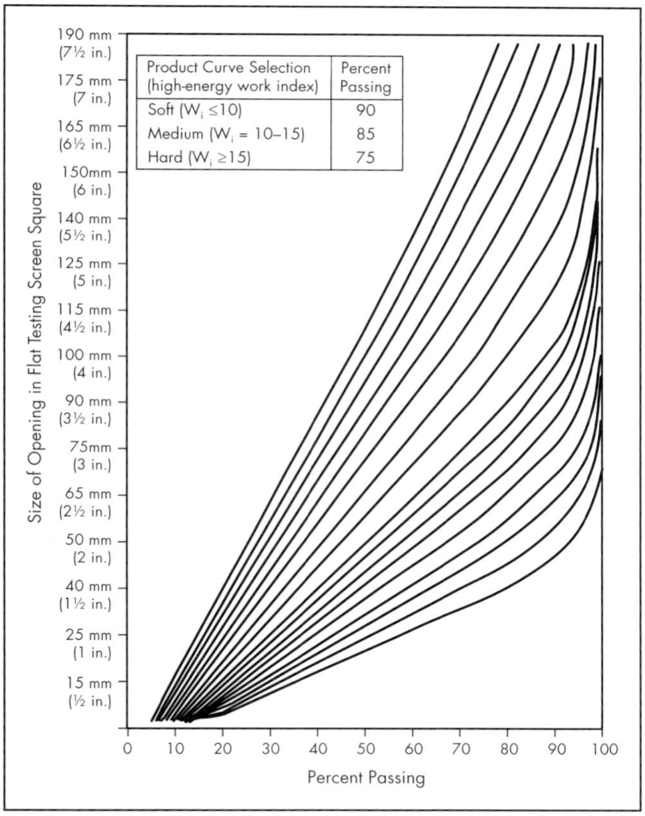

Courtesy of Metso

Figure 12 Product size gradation curves for Metso Superior MK-II primary gyratory crushers

Reference lists are available from suppliers, and these can assist the identification of similar duties throughout the world. Typical lists are shown in Tables 3–5.

Once the ability to achieve the required duty is defined, the major consideration is to examine the installation required for the application in question. Because of site-specific issues, installations are not necessarily the same between sites. The major considerations for installation are

- Feed material and performance estimation,
- Mining method (e.g., open pit blasting vs. sublevel block cave),
- Feed arrangement,
- Discharge arrangement,
- Topography,
- Geographical location, and
- Requirement for relocation.

Ore testing and calculations for gyratory crusher installed power requirements are quite limited and still, to a large extent, rely on the Bond crusher work index equation. Every manufacturer has introduced modifications to the basic equation, which are based on their experience database. The crusher work index uses what is usually referred to as the *impact strength* as the *strength characterization parameter*. This parameter is primarily obtained using derivatives of the twin hammer device. In the use of the twin hammer test, caution is advised in the interpretation of the test data. The spread of values from the test can be significant, and the repeatability of the test should be carefully examined prior to the use of the value in the equation for installed power.

Over the last 10 years, some alternative approaches have been developed to provide improved predictive capability for primary gyratory crushers. Details of the mechanistic and numerical approaches are provided in Chapter 3.6, "Crusher Selection and Performance Optimization."

During the past two decades, finite element modeling (FEM) has grown in use to determine stresses in the mechanical components of crushers. These simulations allow identification of stress raisers in designs and points of weakness. Redevelopment of designs based on FEM is now common and has led to the ability to deploy designs that have required levels of strength, without the need to rely purely on the mass of steel in the design. Such FEM approaches have been particularly useful in the design of split-shell machines so that shell stiffness is maintained while still allowing the machines to safely apply high forces.

For primary gyratory crushers, the main feed arrangement for open pits is by direct tipping from mine haul trucks. Most mine haul trucks are of the back-tipping arrangement, although side-tippers are sometimes encountered, and for these some extra safety considerations will need to be examined.

Table 3 ThyssenKrupp large* primary gyratory crusher models and clients†

Crusher Model	Quantity	Client (Project)	Country	Material	Year
BK 137–170	1	CBMI Construction Co., Ltd. (Azura-EDO)	Nigeria	Limestone	2016
BK 160–190	2	Palabora Mining Company (Lift II)	South Africa	Copper ore	2016
KB 160–290	1	Codelco‡ (Andina)	Chile	Copper ore	2015
KB 160–230	1	OJSC Apatit (Koschinsky)	Russia	Overburden	2015
KB 160–230	1	Mineral Technologies Group (Darkhan)	Mongolia	Iron ore	2014
KB 160–290	1	Southern Peru Copper Corporation (Cuajone)	Peru	Copper ore	2014
KB 160–190	1	Baogang Group (Baiyunebo iron mine)	China	Overburden	2013
KB 160–230	2	Kansanshi Mining (Kansanshi mine)	Zambia	Copper ore	2013
KB 160–190	1	Shougang Hierro Peru S.A.A. (Marcona)	Peru	Iron ore	2013
KB 160–230	3	Kalumbila Minerals Limited (Sentinel)	Zambia	Copper ore	2011
KB 160–230	5	PT Freeport Indonesia (Irian Jaya)	Indonesia	Copper ore	2011
KB 160–230	1	Codelco (Mina Ministro Hales)	Chile	Copper ore	2011

Courtesy of ThyssenKrupp Industrial Solutions
* Large feed openings measure ≥1.6 m (63 in.).
† Installation figures since 2007.
‡ Codelco = Corporación Nacional del Cobre de Chile.

Table 4 FLSmidth large* primary gyratory crusher models and clients†

Crusher Model‡	Crushers Supplied	Site	Country	Commodity	Feed Arrangement and Feed Size	Throughput per Machine§ and Product Top Size, t/h (stph)	Installed Motor Power, kW (hp)	Year
16 × 29 TS	1	Aktogay	Kazakhstan	Copper ore	Run-of-mine (ROM) direct feed	4,250 (4,680)	750 (1,005)	2012
16 × 29 TS	1	Bozshakol	Kazakhstan	Copper ore	ROM direct feed	4,250 (4,680)	750 (1,005)	2012
14 × 21 TS	1	Mission Asarco	United States	Copper ore	ROM direct feed	2,950 (3,249)	450 (603)	2012
16 × 30 TSU	1	Constancia	Peru	Copper/moly ore	ROM direct feed	6,000 (6,608)	1,000 (1,340)	2012
14 × 21 TSU	1	Sukari gold mine	Egypt	Gold ore	ROM direct feed	2,000 (2,203)	600 (804)	2012
16 × 29 TS	5	FLSmidth SA	North Africa	Copper ore	ROM direct feed	4,300 (4,736)	750 (1,005)	2013
11 × 18 TSU	1	Kansuki	DRC	Copper ore	ROM direct feed	700 (771)	450 (603)	2013
14 × 21 TSU	1	Rainy River	Canada	Gold ore	ROM direct feed	1,350 (1,487)	600 (804)	2014
16 × 24 TSU	1	Pena Colorada	Mexico	Magnetite	ROM direct feed	2,850 (3,139)	750 (1,005)	2015
14 × 21 TSU	1	Tasiast	Mauritania	Gold ore	ROM direct feed	1,850 (2,037)	600 (804)	2016

Courtesy of FLSmidth
* Large feed openings measure ≥1.6 m (63 in.).
† Installation figures since 2012.
‡ FLSmidth model designation is of the format 16 × 29 (i.e., 1,600 × 2,900 mm). TS = Top Service; TSU = Top Service ultra-dutya
§ Throughput may not be the maximum possible for the machine as other system-related constraints may apply.

Two-sided tipping is also common for large gyratory crushers, whereby two trucks can tip into the crusher from opposite sides (Figure 13).

Regardless of the tipping configuration, it is normal practice to tip in line with the direction of the spider arm. This practice allows the spider arm to act as a feed splitter to improve the distribution of material to the crushing chamber and to minimize particle size segregation or bias.

In open pits, direct tipping is the most common feed arrangement, although in some instances a ROM bin is used with material drawn from the bin into the crusher using a heavy-duty feeder. There are a range of features that should be considered regarding discharge of material from the crusher.

The primary function of the discharge pocket of a primary gyratory crusher is to physically decouple the crusher from the downstream processes. By the nature of the equipment, the crusher is a discontinuous process with a throughput rate that is largely determined by the characteristics of the feed material. Fine material (with a lot of the feed being smaller than the crusher setting) will flood through quickly, while highly competent large material will be held back as it progresses down the crushing chamber. As most primary gyratory crushers are direct fed by trucks, the discharge pocket allows these variations (which can occur between trucks) to be smoothed out. If sufficiently large, the discharge pocket can also absorb short stoppages in the downstream process without forcing a crusher stoppage. This aspect becomes more important the harder the material is to crush because repeated crusher stoppages and subsequent loaded starts can lead to crusher damage. The size of pocket required will depend on the truck size feeding the crusher. Some crusher manufacturers recommend two times the truck size as the live capacity in the discharge pocket; however, this is not a hard and fast rule. The size of the pocket will be a trade-off between capital cost (of steelwork and the reinforced ROM pad wall) and the flexibility allowed to operations. Accurate level sensing in the pocket is critical for making best use of the pocket volume.

Table 5 Metso large* primary gyratory crusher models installed since 2010

Crusher Model	Quantity	Client	Country	Material	Year
60–89	2	BHP Billiton	Australia	Iron ore	2014
60–110E	1	PJSC Polyus	Russia	Gold ore	2013
60–89	2	Minera México S.A. de C.V.	Mexico	Copper ore	2013
60–110E	1	Terrane Metals	Canada	Copper ore	2012
60–89	1	Russian Copper Company (Miheevskye)	Russia	Copper ore	2012
60–89	1	Consolidated Thompson (Bloom Lake)	Canada	Iron ore	2012
60–89	1	Crushing Services International	Australia	Iron ore	2012
60–89	1	Yichun Luming Mining Co., Ltd.	China	Molybdenum	2012
60–110E	1	TISCO	China	Iron ore	2011
60–89	2	BHP Billiton	Australia	Iron ore	2011
62–75	1	Sagrex	Belgium	Porphyry	2011
60–110	1	Associated Manganese Mines (Beeshoek)	South Africa	Iron ore	2010
60–89	1	Companhia Vale do Rio Doce/Salobo Metals	Brazil	Copper ore	2010
60–89	1	SCT Belgium	Belgium	Limestone	2010
60–89	1	Dexing	China	Copper ore	2010
60–89	1	Similco Mines Ltd.	Canada	Copper ore	2010

Courtesy of Metso
* Large feed openings measure ≥1,524 mm (≥60 in.).

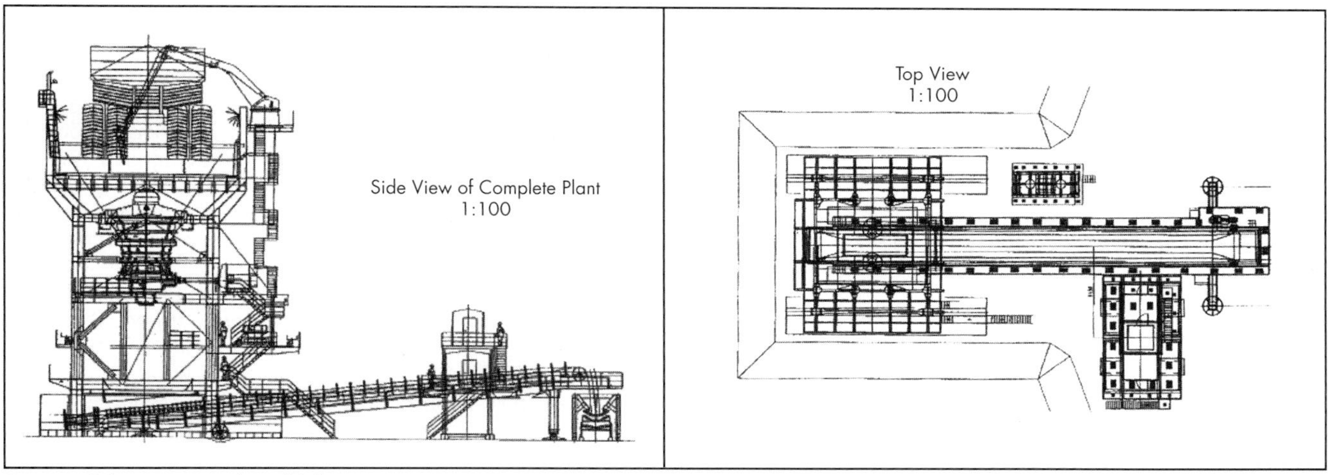

Courtesy of ThyssenKrupp Industrial Solutions
Figure 13 Direct-tip feed arrangement with a dual-tip configuration

The design of the pocket should include wall liners capable of withstanding rolling impact of material discharging from the crusher itself. Although these liners should be very long lasting, maintenance access to change them is still required. The greatest maintenance load in the pocket is the discharge slot, which should be designed as self-relieving (where the width of the slot becomes wider and higher off the receiving feeder as it moves toward the front of the bin). Maintenance access to the bin should be through large barn door–style access, which will be required anyway for crushers that need bottom access for maintenance of the hydraulic shaft adjustment assembly. Large doors allow easier movement of people, tooling, and parts into the pocket. Lifting equipment is also required to enable the heavy lip liners on the discharge slot to be replaced.

Feed control (whether automatic or manual) should be based on limits associated with the levels in the crusher chamber and the discharge pocket or vault. The standard arrangement is to have red and green traffic management lights to indicate to the truck if tipping can be undertaken. In general, the feed level in the crushing chamber should be sufficient to protect the liners from direct impact from the rocks discharging from the truck.

In underground applications, trucks designed especially for underground use may tip into the crusher, or alternatively load-haul-dump vehicles may be used. When compared to open pits, using vehicles with relatively small payloads sometimes requires installation of bin and feeder arrangements to ensure choke feed to the crusher.

Discharge arrangements for gyratory crushers usually consist of a discharge pocket or vault with a heavy-duty belt or apron feeder taking the material from the pocket onto a conveyor for further transport. The *instantaneous* throughput of a large primary gyratory can fluctuate or surge widely depending on feed type and size distribution. The use of the discharge pocket and feeder allows designers and operators to control the rate of discharge to downstream conveying systems, enabling a more efficient, lower-powered conveyor and drive system to be used.

Topography will mainly impact the earthworks requirements for the crushing installation, and in some cases these can be significant, particularly given the height of a large primary crusher and associated equipment.

Geographical location has major impacts on supporting equipment. One of the main factors is the ambient temperature. All crushers have recommendations regarding the maximum absolute temperature of the lubricating oil and the delta between the inflow and outflow. Such cooling and heating systems are integral to the crusher to ensure that lubrication remains effective.

Relocation of crushers and the likely frequency is a matter that must be considered early in the design study. Large primary gyratory crushers are tall and heavy, and relocation is not a trivial exercise. If the mine plan dictates that the primary crusher will need to be moved, then the design of the crusher station will need to incorporate this requirement, and supporting equipment such as low-profile transport crawlers will be needed (Boyd and Utley 2002). The factors that need to be considered in terms of relocation are as follows:

- Required degree of mobility
- Dump and feed arrangements
- Discharge arrangement
- Site topography
- Access and maintainability
- Decommissioning or recommissioning
  - Safety
  - Complexity
  - Time and duration

In particular, the required degree of mobility should be carefully examined, as mobility comes at a cost. For example, is the deposit amenable to strategic planning and scheduling to accommodate and effectively utilize crusher mobility? The following additional elements should also be considered:

- Deposit variability and number of sources or working faces
- Geographical spread of sources and/or working faces
- Speed of advance along strike
- Trade-off between haul distance and haul climb against crusher location
- Deposit topography
- Crusher mobility and the overall materials handling system
- Truck size, fleet number, haul profiles, and economic haul range

CONE CRUSHERS

Of all crushers throughout the world, the cone crusher is, by far, the most common device for post-primary crushing duties. Essentially, the cone crusher employs the same crushing action as the gyratory crusher, but the cone crusher head has a much shallower angle and the eccentric speed is higher. Cone crushers are designed to reduce the size of feed in a number of steps and therefore generate a well-defined product size distribution. Such definition of the product is essential, because in many cases the product size can be the primary determining step in meeting the specification of a final product, as in quarrying, or alternatively, it can be a crucial intermediate, or preparatory step, in the mining process.

The role of crushing plants has changed over the years. Prior to the 1980s, large multistage crushing plants were used to generate suitable feed sizes to rod or ball mill circuits. Major examples of such plants include Bougainville Copper Limited, PT Freeport North-South, and Rössing uranium mine. However, with the introduction of semiautogenous milling, the need for this type of crushing plant fell away. Interestingly, however, the last 10 years has seen renewed interest in cone crushing plants, with this resurgence caused by the introduction of high-pressure grinding roll technology and the movement to examine more energy-efficient comminution circuits that include crushers in semiautogenous grinding (SAG) mill pre-crush and pebble-crushing roles. New plants built with some of these revised views include the plants at Newmont Boddington (Hart et al. 2011), Karara Mining, First Quantum Minerals Sentinel, and FCX Cerro Verde (Koski et al. 2011).

The design evolution has led to two dominant cone crusher designs emerging:

1. Bowl adjustment
   - Fixed-mainshaft, base-supported machines with CSS adjustment and overload protection via movement of the upper frame (containing the bowl or concave liners)
   - Mainshaft mounted within a long, tapering eccentric bush, providing eccentricity to an integral crushing head, with CSS adjustment and overload protection via movement of the upper frame (containing the bowl or concave liners, e.g., Symons cone crusher)
2. Head adjustment
   - Top supported (spider bush) machines, with CSS adjustment via hydraulic raising and lowering of the mainshaft (with integral cone head and mantle liner)

The fixed-mainshaft, base-supported design now predominates in the bowl-adjusted machines. It is seen in a variety of forms, although the most common are the MP and HP series cones from Metso (Figure 14), with others available from FLSmidth (Raptor) and Terex Corporation (Rollercone and MVP). The other type of bowl-adjusted machine using a mainshaft mounted in an eccentric bush has a more direct lineage to the Symons crusher. The Symons crushers (Figure 15) were the mainstay of crushing plants for the majority of the 20th century, and there are still an enormous number of these units in mines and quarries around the world. The Symons style of machine uses bowl adjustment and has a base-supported mainshaft that moves within an eccentric assembly. CSS adjustment is via a screw-threaded bowl assembly in an adjustment ring supported on the upper frame. The overload protection mechanism is generally spring based, although in the later machines, hydraulic cylinders were introduced. Such machines are still produced; for a detailed commentary on the Symons crusher, the reader is directed to Flavel et al. (1985). In some instances, developments of the base concept seen in the Symons machines have spawned modern variants, and one of the main manufacturers of this style is Astec with their range of Telsmith SBS (Figure 16) and SBX crushers. With

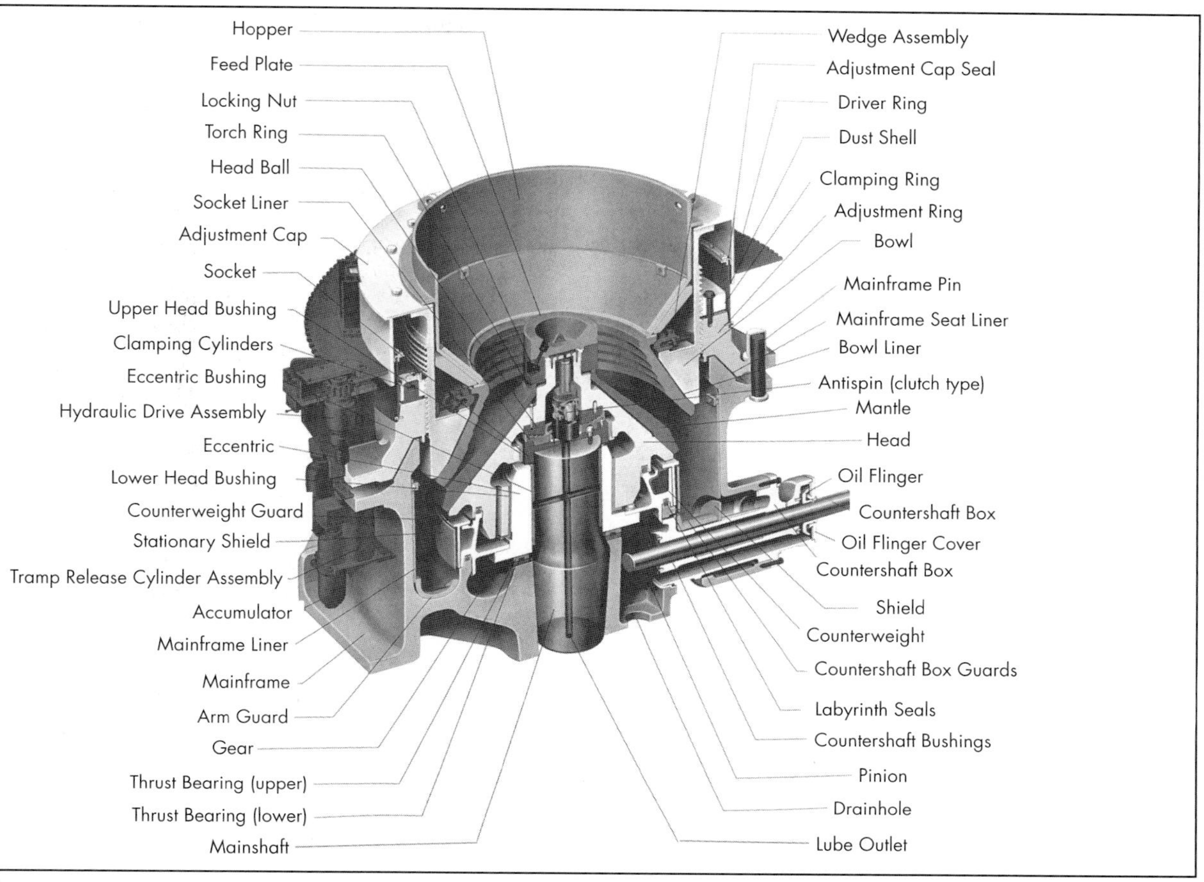

Courtesy of Metso

Figure 14 Metso MP series bowl-adjusted cone crusher

bowl-adjusted machines of both types, the terms *standard* and *shorthead*, are still used, although they were originally used in relation to the Symons style of machines. A standard arrangement is a crusher head and bowl assembly that is designed predominantly for coarse crushing, where the feed size is large and the product size will undergo further crushing or treatment. The term *shorthead* is reserved for machines that have a shorter crushing chamber and are designed to accept smaller top size and target fine crushing.

Head-adjusted machines, or top-supported machines, were first pioneered by Allis-Chalmers under the Hydrocone brand and are still widely encountered. Because of various mergers and acquisitions, the Hydrocone-style cone crusher is now manufactured by Sandvik, although in modern versions, the brand name has been replaced by new designations such as CS and CH. Head-adjusted machines are also available from other manufacturers including Metso (GP cones) and Earth Technica (Cybas cone). An illustration of a Sandvik CH head-adjusted crusher is shown in Figure 17.

Outside of the cone crusher types just mentioned, the only other current models of cone crusher that differ fundamentally are the latest version of the Kubria crusher from ThyssenKrupp and the MX range from Metso. Both machines are new entrants to the market with limited information available. The Kubria has appeared in a couple of forms with the older version being similar in principle to the head-adjusted cone crusher of a hydrocone type. The newer version is also head-adjusted, but there is no supporting spider arm for the top of the mainshaft. In this regard, it is essentially a hybrid machine, and it is claimed to offer the advantages of hydraulic setting adjustment via the mainshaft and head, but with high eccentric throw and without the need for a spider spanning the top of the machine. A description of the new HKB 2300 is provided by a 2014 *Mining Magazine* article ("A Fresh Approach to Cone Crushing"), although further details and the success of the design are not currently known.

The Metso MX cone crusher (Figure 18) is a further attempt to blend the best features of the bowl-adjusted and hydraulic shaft adjustment-style cone crushers. In the ThyssenKrupp HKB 2300, the manufacturer relies entirely on a hydraulic shaft adjustment style for all setting adjustment and overload protection. Metso's MX crusher uses a design where the CSS adjustment is still controlled by a screw thread bowl adjustment, but a hydraulically supported mainshaft is used both as an overload device and to temporarily remove load, so that CSS adjustment can be performed without the need to stop the feed to the crusher. Using this patented "Multi-Action" concept, which allows automated control of CSS adjustment and wear compensation, the manufacturer claims that productivity of the machine is increased and liner life is extended (Metso 2017).

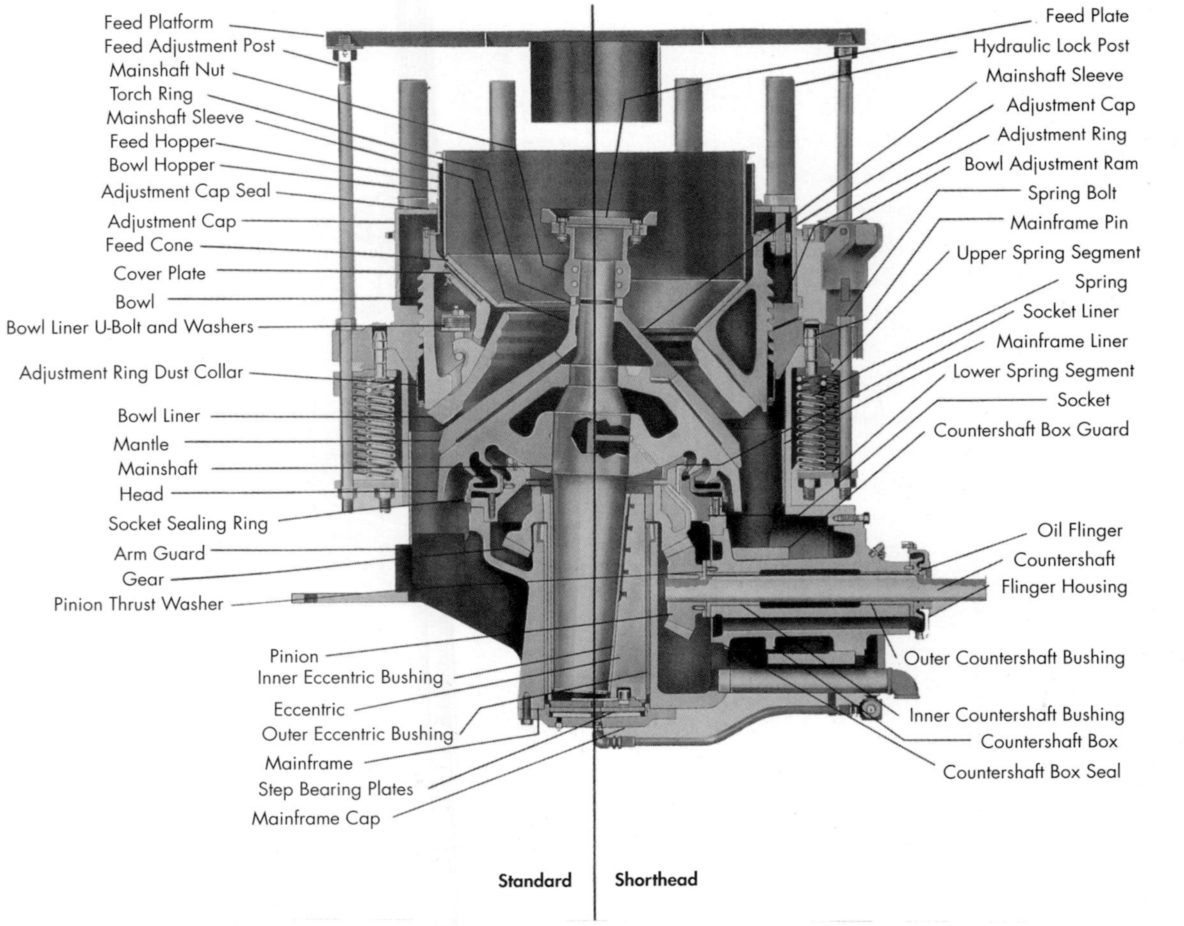

Courtesy of Metso

Figure 15 Cross section of standard and shorthead Symons cone crushers

Principles and Terminology

When considering cone crushers, it is crucial for operators and engineers to understand that the energy distribution within the crushing chamber of the cone crusher is of paramount importance. There are many ways of categorizing the parameters of importance in cone crushers, and one approach (Bearman et al. 2011) is to consider what controls the energy application. The following are the main topics:

- Machine design variables (MDVs)
- Feed material variables (FMVs)
- Machine operating variables (MOVs)
- Machine limits
  - Volumetric
  - Power
  - Force
- Flow-sheet design and interaction with other equipment

MDVs are those parameters that are part of the inherent design of the crusher and will vary between designs and manufacturers. In most cases, the parameters can either not be changed or can only be changed with the input of the manufacturer. In summary, the MDVs are as follows:

- **Pivot point.** In head-adjusted cones, the pivot point of the cone head is within the spider bush area, whereas in bowl-adjusted cones, it is located in space within the geometry of the head or, in modern versions, above the cone head.
- **Eccentric throw and motion.** This is accomplished through an eccentric assembly, which is also sometimes referred to as a *wedge plate* in Rollercone crushers. The assembly sits on the vertical centerline of the crusher, but the assembly has a bore that is eccentric rather than concentric. This eccentricity provides the gyration of the cone head and the movement from OSS to CSS.
- **Eccentric speed.** Most crushers can be run at speeds within a 15%–20% range above and below the design figure set by the manufacturers.
- **Chamber and liner design (including feed opening) parameters** are as follows:
  - Designs vary depending on size reduction duty.
  - Modifications can dramatically influence energy distribution in the chamber.
  - There is variation in the nip angle between the mantle and bowl (or concave) down the length of the chamber.

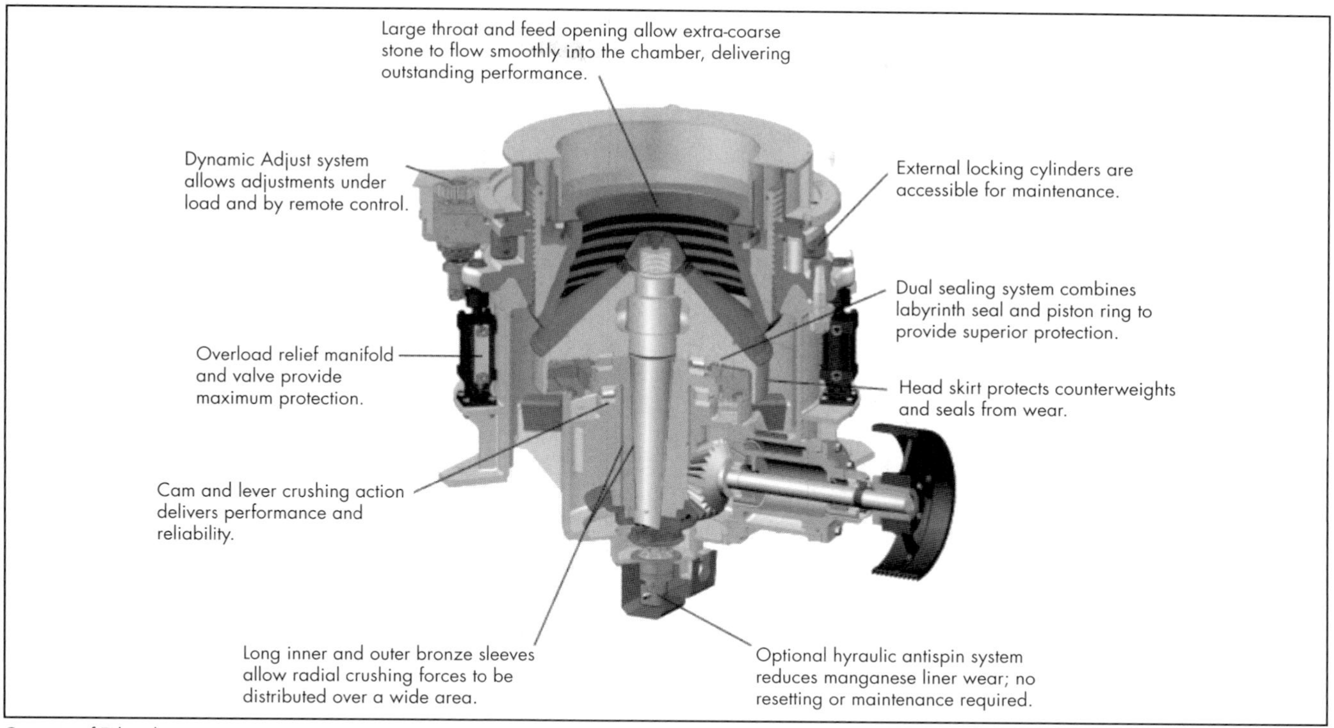
Large throat and feed opening allow extra-coarse stone to flow smoothly into the chamber, delivering outstanding performance.
Dynamic Adjust system allows adjustments under load and by remote control.
External locking cylinders are accessible for maintenance.
Dual sealing system combines labyrinth seal and piston ring to provide superior protection.
Overload relief manifold and valve provide maximum protection.
Head skirt protects counterweights and seals from wear.
Cam and lever crushing action delivers performance and reliability.
Long inner and outer bronze sleeves allow radial crushing forces to be distributed over a wide area.
Optional hyraulic antispin system reduces manganese liner wear; no resetting or maintenance required.

Courtesy of Telsmith

Figure 16 Telsmith SBS-style bowl-adjusted cone crusher

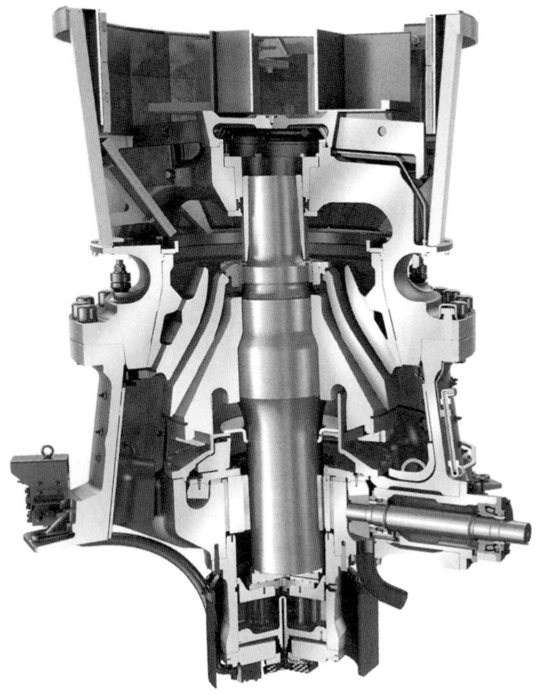

Courtesy of Sandvik

Figure 17 Sandvik CH head-adjusted cone crusher

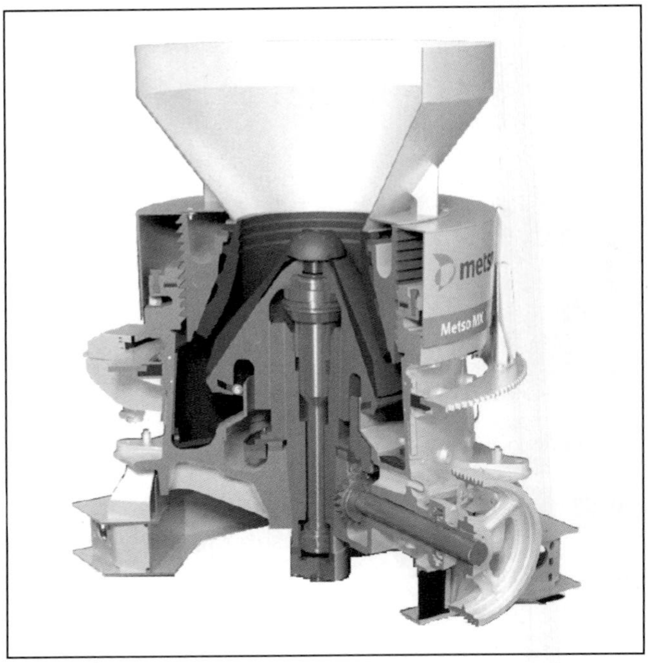

Courtesy of Metso
Figure 18 Metso MX cone crusher

MOVs are those variables over which the operator has some control and these fall into two areas, as follows:

1. **CSS.** Minimum distance between the mantle and bowl (concave) liners varies.
2. **Feed presentation varies:**
   - Feed rate and control
   - Feed positioning and distribution
   - Segregation of particle size into the crusher

FMVs are the properties of the feed that will impact the performance of the crusher. Some FMVs are simply characteristic of the material found at the mine or quarry site, but others such as the feed size distribution or the maximum feed size are somewhat within the control of the site and the flow sheet:

- **Feed material strength and abrasivity**
  - Higher strength requires greater power input to achieve size reduction.
  - Highly abrasive feed will increase liner wear rate and can lead to the development of poor wear profiles and the generation of high crushing forces.

See Chapter 3.6, "Crusher Selection and Performance Optimization," for further information.

- **Bulk density**
  - High bulk density can restrict the amount of crushing possible and can lead to a volumetric limit.
  - Throughput is directly affected.
- **Feed size distribution.** Directly linked to bulk density, the length of size distribution and percentage of fines can dramatically influence performance and crushing force requirements.
- **Maximum feed size.** To meet nip angle requirements, the feed opening and liner design vary.
- **Feed moisture content.** For most materials, moisture content of <4% has no material impact, but higher moisture levels can exacerbate packing and thus crushing force requirements. The handling characteristics of the feed must also be considered. There is no test for materials handling specific to crushers, but if bulk handling tests show a propensity for material to resist flow, then this is a strong sign that crushing problems may be expected. Feed containing clay, either naturally occurring or from alteration of minerals in the parent rock, will cause flow impediments in the chamber and this will greatly increase the chance of packing. Where such feed exists, removal of the material prior to the crusher should be targeted, or in situations where this is not possible, alternative types of crusher should be considered.

Cone crushers have been subject to a significant amount of development over the last 30 years and much of this development has been around the MDVs, increasing the throughput and size reduction and developing systems to protect the crusher from overload conditions.

In the preceding discussion about MOVs, only feed presentation and CSS are given as parameters. This is correct, but recently, manufacturers and researchers have examined the use of variable-speed drives (VSDs) so that the eccentric speed of cone crushers can be varied in real time. The ability to apply VSD technology is now possible because of the significant reduction in the cost of VSD systems, particularly for the larger drive sizes. Hulthén and Evertsson (2008) examined optimization of an aggregate crushing plant, with the VSD system applied to the tertiary Metso HP300 cone crusher. The application of speed control was credited with extending the liner life in the tertiary crusher by 27% and also increasing the operational throughput of the crushing plant by 4.2%. A pilot study examining the issue was presented by Jacobson and Lamminmaki (2013) where one of the main conclusions stated that the speed adjustment is best used to "account for changing ore/feed conditions, or to use the speed as a dynamic input parameter into a control system."

Regardless of the type of crusher used, cone crusher applications can be summarized:

- **In-line** has secondary, tertiary, and quaternary stages giving increasingly fine crush.
- **Recycle** (or pebble) uses the machine to recrush oversize from SAG or autogenous grinding mills.
- **Other** are often not distinctly different from the preceding applications, but they are named to show a distinct purpose. An example would be SAG pre-crush, where a cone crusher is used to reduce the feed size, or a portion of the feed size distribution, prior to SAG treatment.

Secondary crushing is used to treat material that has been primary crushed and therefore usually involves a feed top size ranging from 150 to 350 mm (6–14 in.) with CSS values at approximately 25–60 mm (1–2½ in.). Despite varying opinions, the removal of fine material from the secondary crusher feed is still the main recommendation. The definition of *fines* in this case is usually based on the CSS of the crusher; that is, any fine material less than the CSS should be removed from the feed via scalping screens. Removing the fines from the feed ensures that there is sufficient space inside the crushing chamber to allow crushing and the associated increase in the bulk density of the material within the chamber. Under these circumstances, secondary crushers are usually power limited. If fines are not removed, the bulk density in the crushing chamber is already high before crushing commences. As crushing of the larger particles takes place, the fines generated add to the already high bulk density, and the crushing chamber reaches a point whereby it becomes packed. *Packing* is a term indicating that the volumetric capacity of the crusher has been exceeded and the input power cannot cause further breakage. At this point, the force is transmitted through the packed chamber and into the overload system of the crusher. In fixed-shaft, base-supported machines, the upper frame will relieve, either hydraulically or via springs. The relief system will allow the material to pass and then try to return to the original CSS to continue crushing. In top-supported machines with hydraulically supported mainshafts, the hydraulics will relieve and cause the mainshaft and head to drop before trying to return to the operating CSS. Packing is undesirable and frequent events will eventually cause damage to certain mechanical areas of the crusher. Packing can be described in various terms, with small-scale movement without relief being termed *flutter* and major relief events often described as *tramping* or *ring bounce*. These terms are references to the fact that relief is often caused by the ingress of *tramp* metal or other uncrushable material, and in Symons style machines, this was seen as the upper frame and adjustment ring bouncing.

Operationally, secondary crushers with fines in the feed suffer as they often reach both power and packing limits simultaneously, which prevents the machines from performing to the required levels.

Tertiary crushing applications typically take a feed top size of 30–80 mm (1³⁄₁₆–3⅛ in.) and operate at CSS values of approximately 10–20 mm (½–¾ in.), depending on the feed material and required output. Quaternary crushing is usually the most extreme duty for cone crushers, and this will often involve feed top sizes as small as 6 mm (¼ in.) and up to 20 mm (¾ in.). In this stage of crushing, the aim is often to produce the finest product possible, and hence CSS values can be as low as 3–4 mm (approximately ⅛ in.), but more typically 6–12 mm (¼–½ in.). As with secondary crushers, fines should be removed; for tertiary and quaternary duties, the removal of fines is even more critical, because in most cases, the feed bulk density is already high. The result of the fine, high-density feed is that these stages are more often controlled by packing and volumetric limits. Wear of the liners can often exaggerate the effect of high bulk densities, and in many cases, the need for a liner change is a trade-off between liner life and machine operational stability.

Recycle or pebble crushing has emerged as one of the main uses of cone crushers, and in this duty, the idea is to crush pebbles from SAG mills to improve the efficiency of the grinding. Pebble ports on SAG mills usually have apertures measuring approximately 40–100 mm (1½–4 in.), and hence the top size of the pebble crusher feed is dictated by the apertures. The lower feed size is mostly controlled by the aperture of the screen receiving the pebbles from the mill. It is common to have a lower feed size of 10–12 mm (approximately ½ in.) The deployment of pebble crushers is reliant on the plant design, but the product from the crushers can either be reintroduced into the mill feed or alternatively directed back to the pebble screen at the mill discharge.

Pebble crusher duties are often considered to be a reasonably tough duty because the pebbles themselves tend to represent the harder material. In addition, there is always the potential for worn mill balls to enter the pebble crusher. Various metal detection systems are available to prevent ingress of mill balls with most being linked to a diversion system to dump the part of the feed containing the mill balls prior to the feed chute.

Crushers can also be used for other duties, with one being to treat material prior to semiautogenous milling. This trend has fluctuated in popularity, but under certain circumstances, it can prove to be highly beneficial. Although the duty is referred to as *SAG pre-crush*, it is essentially a secondary crushing duty. Reducing the mill feed size can generate higher throughput, although careful consideration needs to be given to ball charge, mill speed, slurry pooling, and grind size.

Construction and Design

The two styles of cone crusher discussed in this section have a range of common design features, as would be expected. Both styles rely on applying an eccentric motion to generate the usual open side and closed side discharge settings, that is, OSS and CSS, respectively.

The major components of cone crushers can be summarized as follows:

- Upper frame, including adjustment ring and crushing bowl liner
- Crushing head, including mantle liner
- Eccentric assembly
- Mainframe
- Countershaft and gear set
- Overload protection

In the bowl-adjusted machines, the upper frame is a catchall description for the section containing the bowl assembly (including bowl liner or concave), setting adjustment ring (including clamp ring, clamp cylinders, and adjustment mechanism), and feed hopper assembly. The main components of the upper frame are shown in Figures 14 and 15.

The function of the total assembly is to secure and lock the upper stationary crushing surface against the forces generated. The other main functions are to provide adjustment of the CSS via a screw thread and to offer overload protection in the case of uncrushable material entering the chamber. There are not too many major variations from this design, with the only significant exception being in some smaller cone crushers whereby the adjustment and overload is based on the use of hydraulics and a frictional wedge to lock the upper frame in place. The Terex Corporation Automax is one cone crusher that has this type of arrangement.

Within the upper frame, the bowl liner is securely mounted. The liner is the stationary crushing member, and it is usually the liner that controls the maximum feed size that can be presented to the crushing chamber. A full discussion of crushing chambers is provided later in this section.

In the head-adjusted machines, the mainshaft is supported in a top bearing within a spider arrangement, not dissimilar in appearance to a primary gyratory crusher spider. The mainshaft is free to turn and slide within the bearing, and so, unlike the base-supported machine, the upper section of the machine only needs to provide a pivot point for the crushing action, via the location of the mainshaft.

The crushing head in bowl-adjusted machines usually has a head angle of approximately 35–45 degrees. A mantle liner is located on the head via a combination of a head nut and in some cases backing material poured into the cavity between the liner and the head. The mantle liner casting has a machined surface around the lower circumference of the liner, which seats onto the head. In current versions, the head itself is free to turn on the eccentric assembly that, in turn, is free to rotate around the stationary mainshaft. The rotation is facilitated via a spherical seat on the top of the shaft and through the use of bronze bushes or roller bearings near the top and bottom of the head. Head-adjusted machines tend to have a steeper head angle and the head is integrally fitted to the mainshaft. Unlike the bowl-adjusted machine, an eccentric motion is applied to the lower portion of the mainshaft below the crushing head. Mantle liners for these machines tend to use a machined finish so that they are secured to the crushing head via a threaded ring and backing material is not commonly used.

The eccentric assemblies that provide the motion and the total throw vary in design between the two types of cone crusher being discussed. In the bowl-adjustable version, there are two mainshaft arrangements. In the fixed-mainshaft variant, the mainshaft is stationary, being shrunk into the mainframe. With the mainshaft stationary, the eccentric is designed to rotate around the mainshaft and within the crushing head. In most machines, the eccentric is separated from the mainshaft by a bronze bush and then the eccentric is separated from the crushing head by upper and lower bronze bushes. This general arrangement is widespread, and the only major modification is that in some machines, the bushes are replaced by roller bearings. Smaller cone crushers from ElJay, Pegson, and JCI all use some form of the roller bearing design. Because of market rationalization, the Rollercone, Autocone, and Automax now sit within the Terex Corporation as the Rollercone MVP and the TC, respectively. KPI-JCI are now within the Astec organization, and they manufacture the LS and the Kodiak cone crushers, again using a roller bearing design. Figure 19 shows the various methods of imparting eccentric motion to the cone head. The other version for bowl-adjusted crushers is where the mainshaft sits inside a long eccentric bronze bush, freely mounted inside the rotating body of a concentric mass, with the mass driven by bevel gears from a countershaft. This is essentially the principle employed in the Telsmith cone crusher (Figure 16), and it was the basis of the original Symons cone crushers.

Head-adjusted machines use an eccentric assembly similar to gyratory crushers in that the eccentric is located within the mainframe at the base of the crusher. The eccentric is allowed to rotate within the mainframe through the use of a bronze bush. The mainshaft is located within the eccentric and is free to rotate courtesy of another bronze bush. With the eccentric being rotated by the gear driven from the countershaft, the inertia of the system causes the mainshaft to rotate and move from one extreme of the eccentric position to the other, which is essentially moving from a minimum to a maximum throw position.

The mainshaft in a bowl-adjusted stationary mainshaft machine is secured by a temperature-induced fit into the body of the mainframe and is a relatively short, squat shaft. The dimensions of the shaft are deliberately short and robust, as it is only supported by the mainframe and therefore has to resist crushing forces. The Symons style bowl-adjusted crusher has a lighter mainshaft, integral with the crushing head. The mainshaft is supported for a majority of its length in a long eccentric bush; again the mainshaft is relatively short. The mainshaft in a head-adjusted machine is a different proposition and is deliberately much longer, as it is supported within the spider bearing at the top and located within the rotating eccentric at the base. This difference in mainshaft function translates into a difference in the pivot point of the cone head. The eccentric offset is also canted from the vertical, and thus both types of machine will have a point of zero motion, that is, a pivot point. In the head-adjusted machine, the pivot point is a physical position usually at a point on the mainshaft within the spider bearing. In the bowl-adjusted crusher, the pivot point is a location in space at a distance above the center of the crushing head.

Eccentric assemblies are driven by a bevel gear set, with the gear ring fitted to the outside of the eccentric itself and the drive gear mounted on the end of the countershaft. There is a significant speed reduction between the countershaft speed and the speed of the eccentric. Examples include countershaft speeds of approximately 800 rpm for larger cone crushers, which translate into eccentric speeds of approximately 250–350 rpm.

Overload protection is used in cone crushers to protect the machines during the ingress of uncrushables or where the packing density in the crushing chamber becomes too high and the mass in the chamber effectively becomes uncrushable. In Symons style crushers, the protection was implemented using an array of springs around the circumference of the crusher. Bowl-adjusted machines now predominantly use a hydraulic version of the spring system, whereby hydraulic cylinders and rams allow the increased pressure to relieve against a set of (or a single) hydraulic accumulators, thus permitting the bowl and adjustment ring assembly to lift and pivot away from the upper frame, allowing uncrushable material to pass. A hydraulic system for a bowl-adjusted cone crusher is shown in Figure 20.

In top-supported cone crushers, the mainshaft sits on bearing plates located on top of the hydraulic system. When crushing, the mainshaft and integral crushing head operate in a raised position, but if a significant pressure increase is detected because of uncrushable material or packing, the hydraulic system will relieve and allow the mainshaft and head to drop.

Key to the process performance of cone crushers is the design of the crushing cavity that is formed between the bowl and mantle liners. All manufacturers have a range of liner configurations that allow varying feed size and minimum CSS values. Typically, the chamber that will accommodate the largest feed size is designated as *extra coarse*, and then the liner range will move down in feed size through multiple variants to *fine* or *extra fine*. Liner design is crucial to the correct application for a crusher and also for the optimization of crusher performance. Figure 21 shows a typical range of crusher cavities for a cone crusher.

The key design features of a liner set include

- Ability to handle maximum feed size,
- Application of staged size reduction down through the cavity,
- Ability to achieve the required CSS,
- Ability to pull material into the chamber and effectively capture and break material,
- A smooth progression of nip angle throughout the chamber, and
- Ability to maintain an effective chamber profile even as wear takes place.

The crushing chamber dictates the change in sectional volume throughout the cavity, and therefore the design of the profiles defines the choke point. The choke point is the minimum cross-sectional area in the chamber from a plan view perspective. The concept of *choke* is very important in cone crushers, because once the volume of material in the chamber exceeds the minimum sectional volume, the material will build up above the choke point. In this scenario, the choke

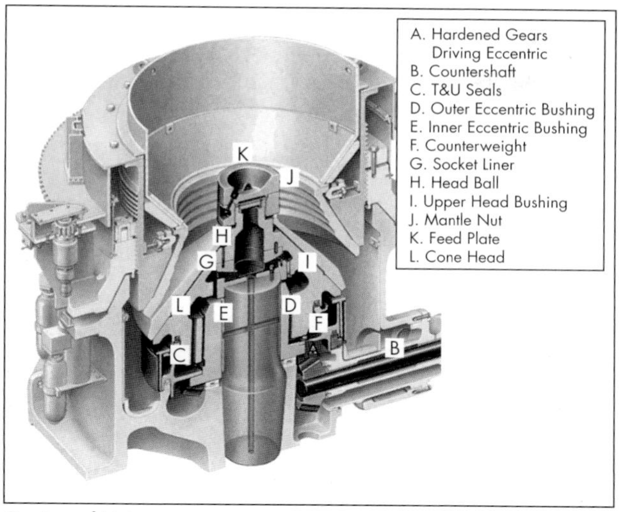

Courtesy of Metso

Eccentric motion imparted by eccentric assembly. Hardened gears driving eccentric (A) and eccentric bushings (D and E) rotating about the mainshaft in an MP series cone crusher.

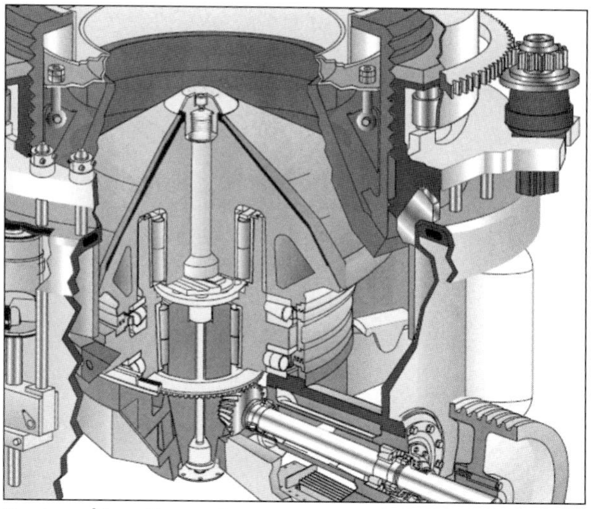

Courtesy of Terex Corporation

Eccentric motion imparted by roller bearings located under the crushing head in a Rollercone MVP

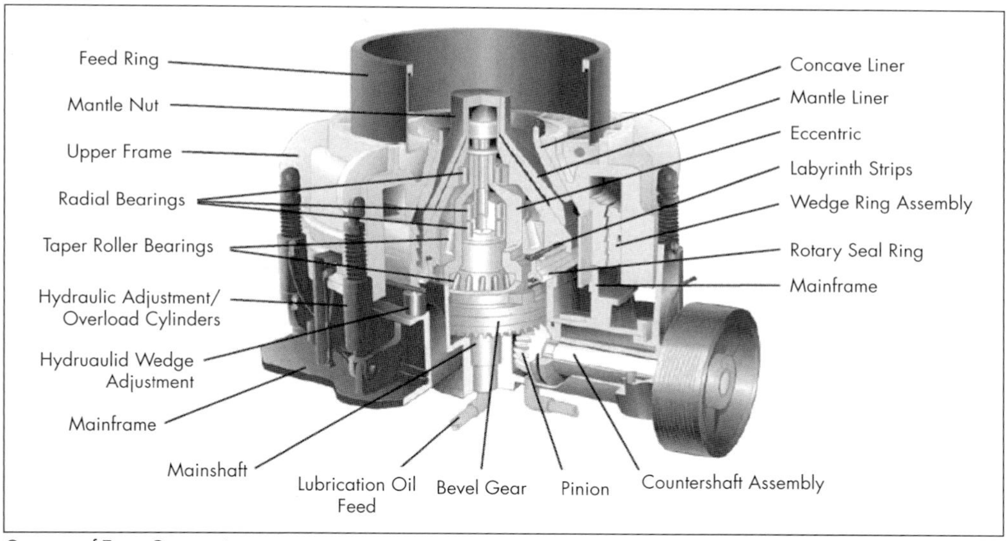

Courtesy of Terex Corporation

Eccentric with roller bearings between mainshaft and eccentric and between the eccentric and the crushing head in a TC cone crusher

Figure 19 Examples of bush and roller bearing based eccentric assemblies

point then controls the flow of material and ensures an orderly progression of material through the chamber, thus ensuring that material passing through the chamber is exposed to the appropriate number of breakage events. Alternatively, if a cone crusher is being fed at rates lower than the rate dictated by the choke point, then the rocks are free to rush through the chamber and undergo a less-than-optimal number of breakage events. It is well documented that *choke feeding* maximizes throughput and size reduction. It is important to note that the choke point is not at the discharge setting, but it is at a point dictated by the liner configuration.

The matter of choke versus nonchoke feed is often raised. In addition to the points mentioned, it is also known that choke feeding stabilizes the crusher operation in terms of the consistency of product size and throughput but also, critically, the forces transmitted from the chamber to the structure of the crusher. The stability referred to in the choke-versus-nonchoke discussion is also seen in cases of feed segregation where the particle size distribution varies at points around the chamber because of feed presentation issues. In Figure 22, Evertsson (as reported in Bearman et al. 2011) shows high-frequency

Hydraulic tramp release is fully automated to pass uncrushable material without disruption of production. Hydraulic accumulators ensure an instantaneous response of the hydraulics to a tramp steel condition.

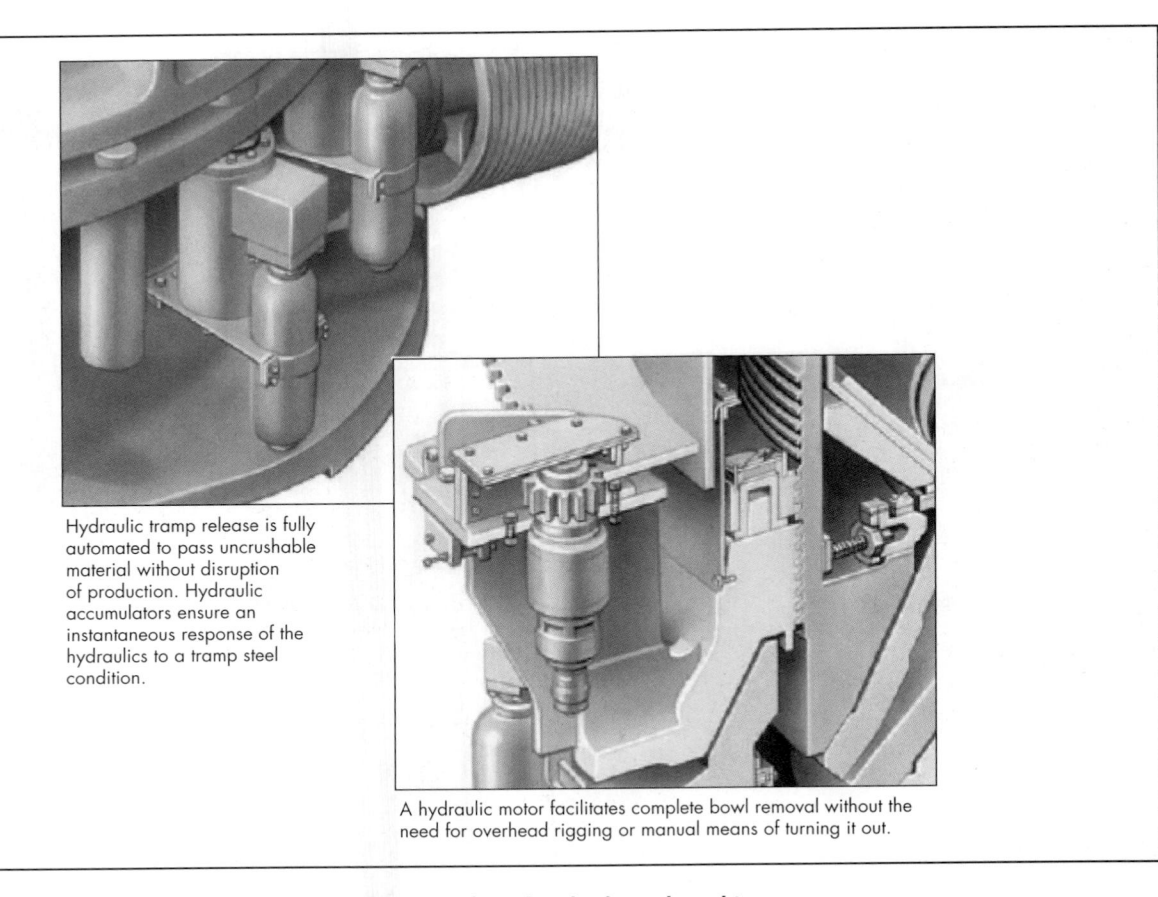

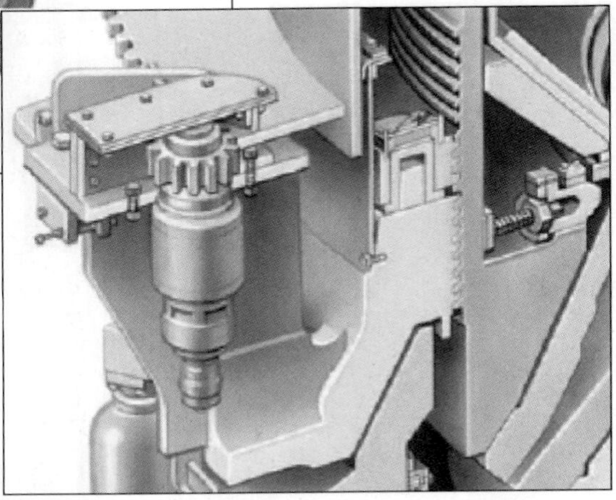

A hydraulic motor facilitates complete bowl removal without the need for overhead rigging or manual means of turning it out.

Courtesy of Metso

Figure 20 Hydraulic overload system and setting adjustment for a bowl-adjusted machine

plots of mainshaft supporting pressure for a segregated and nonsegregated feed.

Figure 22A shows a highly erratic pressure signal with extreme peaks, whereas in Figure 22B the signal is much steadier with lower peaks. The data shown in the figure is highly analogous to the stability issues between choke and nonchoke conditions. Because these forces are transmitted into the crusher structure, the type of signal seen in Figure 22A should be avoided by improving feed presentation and control of feed level, that is, choke.

Because liner profiles largely dictate the energy distribution in the crushing chamber, the issue of wear is not just a maintenance issue, but it significantly impacts processing performance and the force generated in the chamber. Some of the issues seen in association with poor wear profiles include the following (Figure 23):

- Inability to achieve required CSS—power or pressure limits met
- Poor product size distribution and increased recirculating load
- Reduced throughput, both gross and net
- High crushing forces induced with potential impacts on mechanical components (e.g., bushings)
- Early close-off of feed opening, where feed can no longer enter the chamber effectively
- Reduced mean time between failures for components and reduced life of major components

Figure 23 shows significant change in the distribution of the pressure on the mantle. From a gradient of pressure seen with the new liner, there is an obvious move in Figure 23B to a narrow zone of crushing, and this is responsible for high and erratic forces, reduced throughput, and decreased crushing efficiency.

A major move in recent years is the development and deployment of large-diameter, high-powered cone crushers. The Symons style cone crushers were commonly manufactured with heads measuring 2,133 mm (7 ft) in diameter, and there were even machines manufactured as special versions with heads measuring 3,048 mm (10 ft) in diameter. The MP1000 crusher slightly extended this head diameter rating, but over the last two years, both Metso and FLSmidth have deployed even larger units. Metso released the MP2500 cone crusher (*E&MJ News* 2014) with the model designation referring to a 1,864-kW (2,500-hp) installed motor (see Figure 24 where a comparison to the MP1000 is given), and FLSmidth released the XL2000 (Rose et al. 2015). To illustrate the size and capability of these large machines, the MP2500 has a head diameter of 3.29 m (129 in.) and the complete crusher weighs 465 t, or 513 st. In terms of process performance, the coarsest liner configuration has a feed opening of 560 mm (22 in.), and the throughput of a single MP2500 under similar operating conditions is equivalent to that which can be obtained from two MP1250 crushers. The machine is based on the successful MP range but has a range of features including

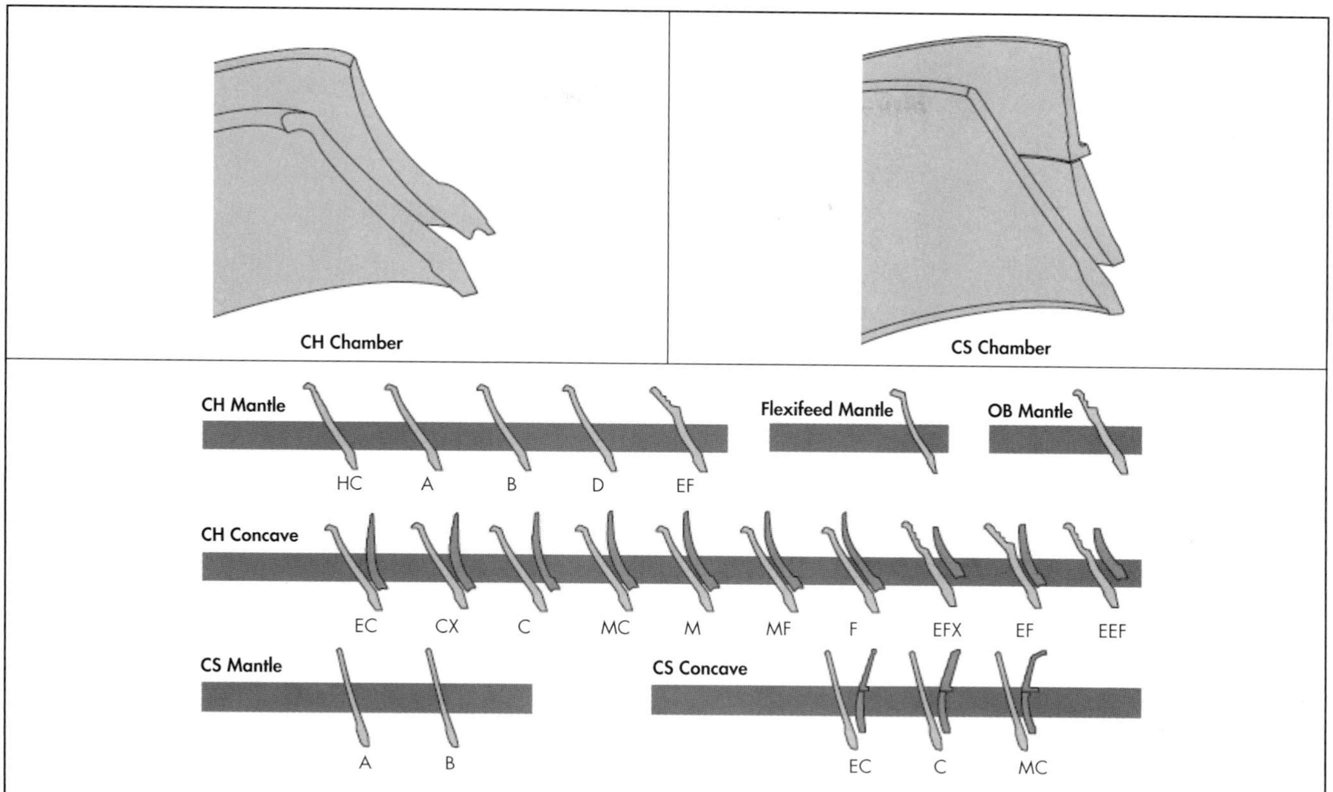

Note: Mantles and concaves are matched to provide the most appropriate chamber for the required crushing duty. The designation (naming convention) of mantles and concaves can change, and therefore the supplier must always be contacted to determine the current designs and the optimum combination of mantle and concave. In this figure, CH mantles and concaves apply to the CH range of crushers, and the CS mantles and concaves apply to the CS range of crushers. CH concaves in association with the relevant mantle give chambers ranging from extra coarse (EC) to extra-extra fine (EEF). CS chambers range from extra coarse (EC) to medium coarse (MC).

Courtesy of Sandvik

Figure 21 A range of cone crusher cavity configurations

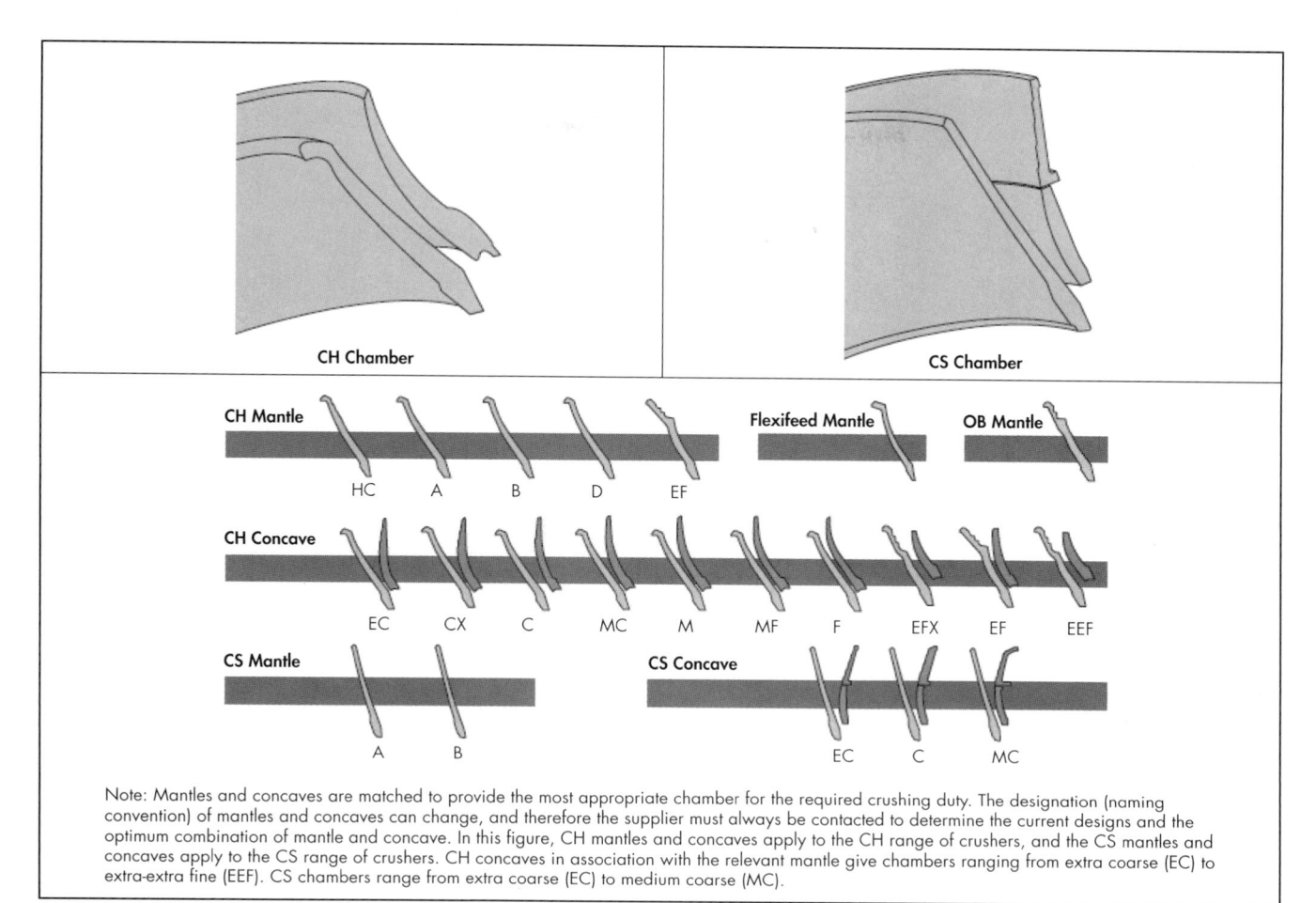

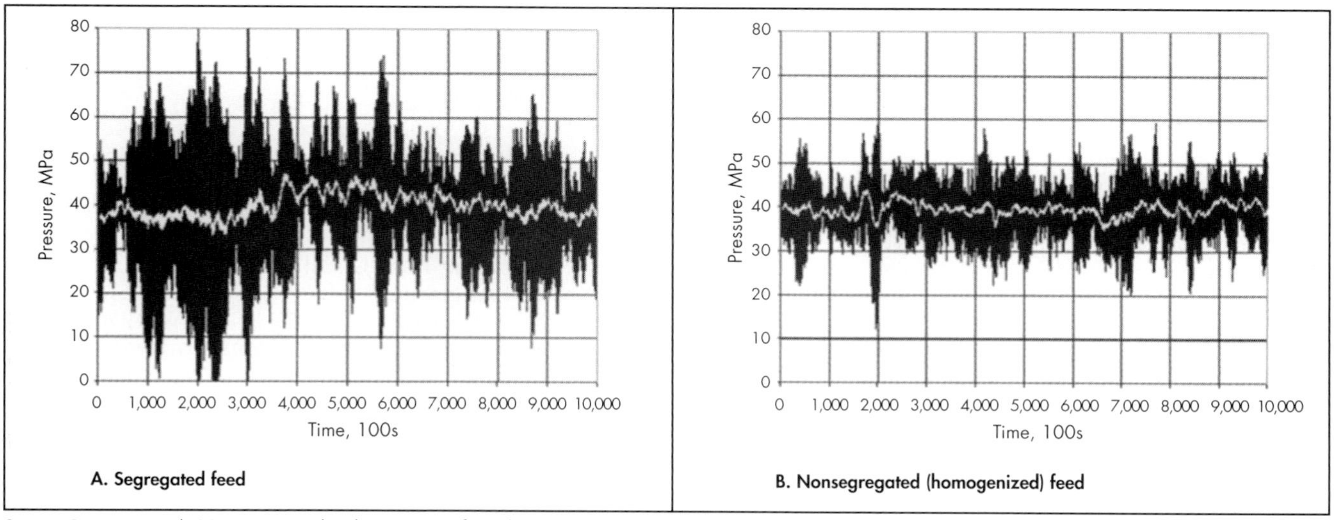

Source: Bearman et al. 2011, reprinted with permission from the Australasian Institute of Mining and Metallurgy

Figure 22 Crusher stability issues caused by feed presentation

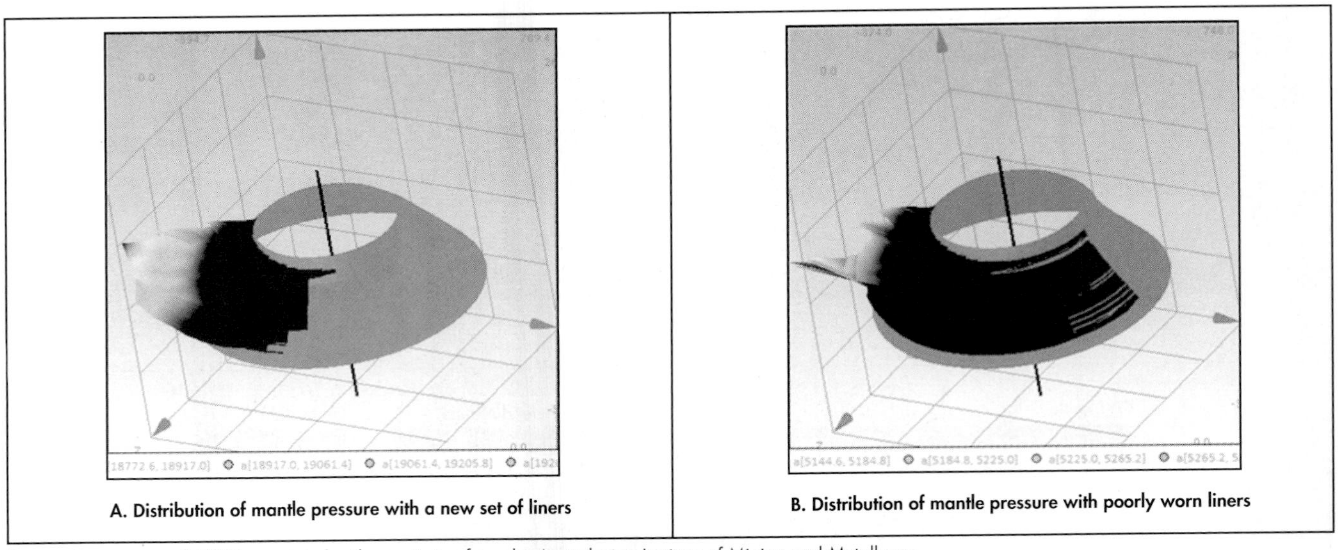

A. Distribution of mantle pressure with a new set of liners

B. Distribution of mantle pressure with poorly worn liners

Source: Bearman et al. 2011, reprinted with permission from the Australasian Institute of Mining and Metallurgy

Figure 23 Changes to mantle pressure in new and worn liners

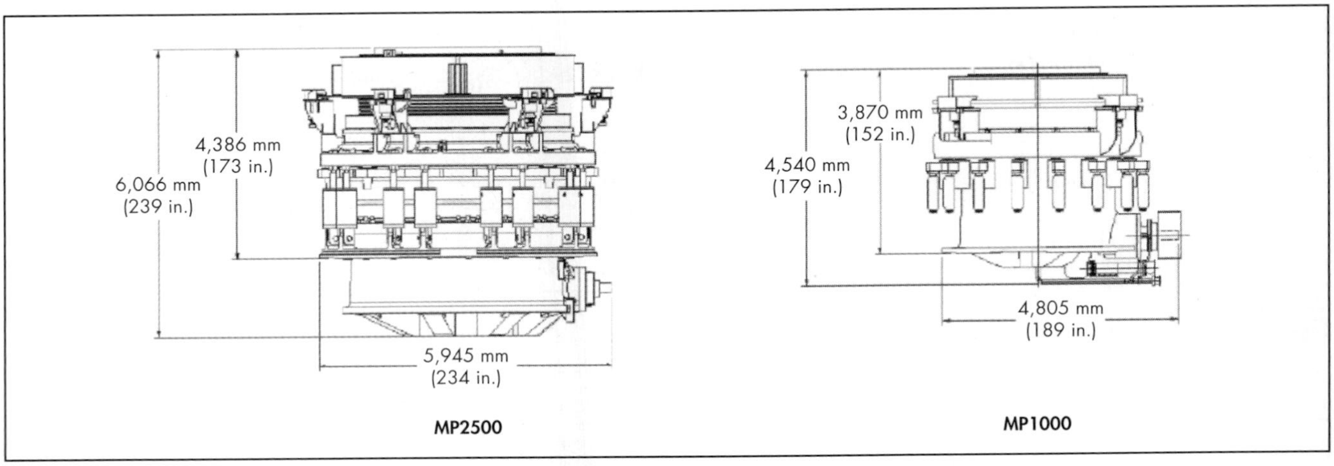

Courtesy of Metso

Figure 24 Metso MP2500 and MP1000 dimensional comparison

- Sleeved eccentric,
- Jackbolt locking nuts,
- Double-acting tramp cylinders,
- Patented MP1250 style counterweight design,
- Patented head-spin reduction (without clutch-style anti-spin device),
- Seven-arm mainframe design, and
- Sloped countershaft arm design to reduce material buildup.

For comparison purposes, the XL2000 is the other cone crusher available at the larger sizes. Figure 25 shows a cross section through an XL2000 that operates with a 1,864-kW (2,500-hp) motor.

The main claims for the XL2000 are

- Crushing force equals 1.9 million kg (4.2 million lb) (XL1100: 1.0 million kg [2.2 million lb]),
- Significant eccentric throw and high pivot-point crushing action,
- Integral countershaft assembly,
- Enclosed counterweight assembly with replaceable liners,
- Noncontacting T and U seal arrangement,
- Spiral bevel gearing,
- Double-acting tramp release and clearing cylinders,
- Easy access to critical load-carrying bearings,
- 2,800 mm (110 in.) head diameter (XL1100: 2,280 mm [90 in.] head diameter), and
- Mainframe inspection ports.

Machines such as the MP2500 and XL2000 and other recently launched competitor machines represent a major commitment to expanding the scope of crushing and deploying machines with the ability to handle large feed sizes and apply significant levels of energy.

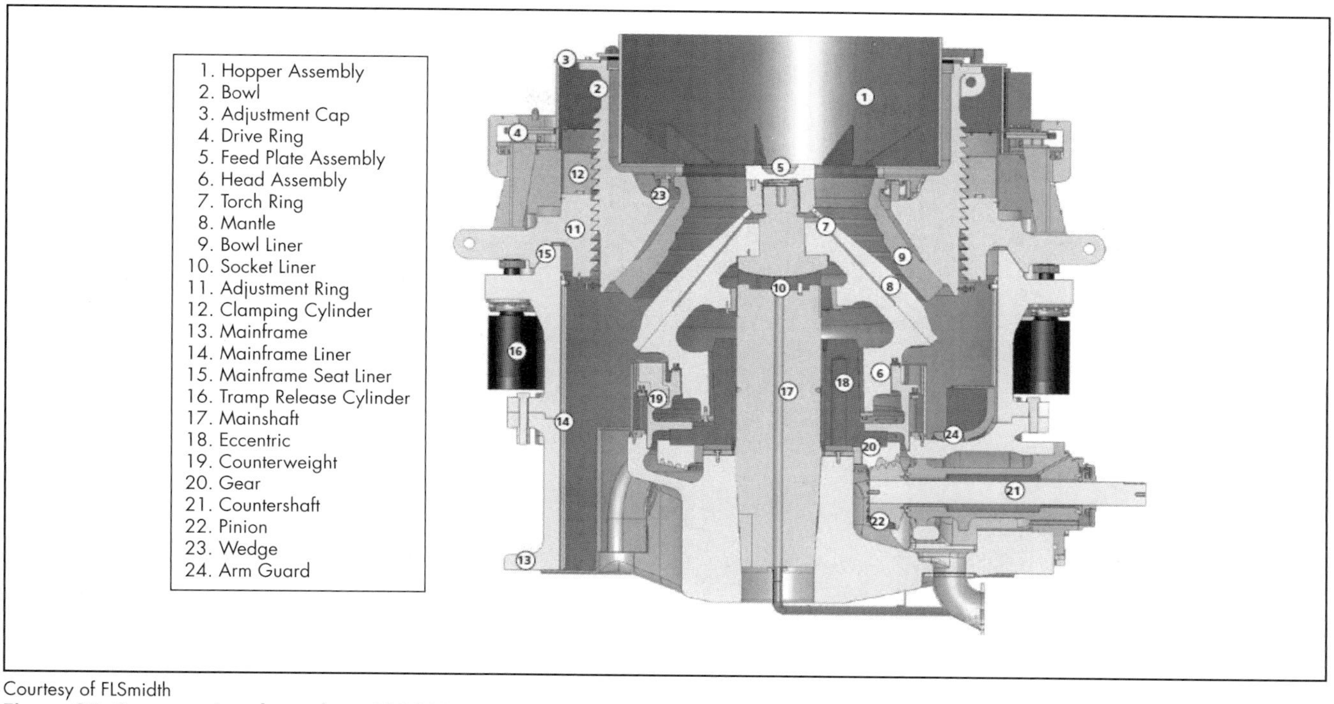

Courtesy of FLSmidth

Figure 25 Cross section through an XL2000

Application and Operations

Cone crushers are selected for a specific duty on the basis of

- Feed size,
- Product size, and
- Required throughput at a given product size.

Cone crushers are usually described in terms of a head diameter or an installed motor power. Examples of a head diameter designation include the Telsmith 38 SBS, which has a crushing head measuring 965 mm (38 in.) in diameter, whereas the Metso MP range of crushers uses the motor power designation, that is, MP1250 has a 932-kW (1,250-hp) motor installed.

For any model of cone crusher, a range of crushing chambers are available, with these chambers being defined by the types of liners used in the bowl (bowl liners or concaves) and on the crushing head (mantle liners). Chambers formed by using different combinations of bowl and mantle liners provide variation in

- Feed opening (maximum feed size),
- Length of crushing chamber, and
- Minimum CSS achievable.

For each chamber, manufacturers provide tables that describe the throughput that can be achieved at varying CSS values. These tables usually state the maximum feed opening and the minimum recommended CSS for the chamber under consideration. Table 6 shows examples of throughput and product size data from Metso relating to the MP range of crushers, and Table 7 gives throughput data for the head-adjusted Sandvik CH crusher. Figure 26 gives product size data.

Table 6 Guidelines for throughput and product size for MP Series cone crusher

Crusher Capacities, t/h*					
	Closed Side Setting, mm†				
Model	**50**	**38**	**25**	**19**	**13**
MP800	1,460–1,935	1,100–1,285	735–980	580–690	495–585
MP1000	1,830–2,420	1,375–1,750	915–1,210	720–900	615–730
MP1250	2,290–3,025	1,720–2,190	1,145–1,510	900–1,125	770–915
Product Gradations					
	Closed Side Setting, mm				
Sieve, mm†	**50**	**38**	**25**	**19**	**13**
100	100				
75	92–98	100			
50	67–81	86–94	100		
38	54–64	68–78	92–98	100	
25	38–54	48–54	65–80	94–98	100
19	30–35	37–42	51–62	82–90	96–99
16	25–29	31–35	43–53	73–82	92–97
13	22–25	26–29	35–44	63–73	83–93
10	18–21	22–24	28–34	52–61	70–91
6	13–14	15–16	19–23	36–44	50–57

Courtesy of Metso

*To convert metric tons per hour (t/h) to shorts ton per hour (stph), multiply by 1.1.

†To convert millimeters (mm) to inches (in.), divide by 25.4.

Table 7 Guidelines for throughput and product size for Sandvik cone crushers*

Crusher	Maximum Motor Size, kW (hp)	Crushing Chamber†	Maximum Feed Size, mm‡	Nominal Capacity in t/h§ with Crusher Running at Closed Side Setting, mm												
				10	13	16	19	22	25	32	38	44	51	57	64	70
CH660	315 (422)	EC	275			177	190–338	203–436	216–464	246–547	272–605	298–662	328–511			
		CX	245			174–194	187–374	200–488	212–519	242–592	268–654	293–521	323–359			
		CX	215			171–190	184–367	196–480	209–510	238–582	263–643	288–512	317–353			
		MC	175			162–253	174–426	186–455	198–484	226–552	249–499	273–364				
		MC	135			197–295	211–440	226–470	240–500	274–502	302–403					
		MF	115		192	207–369	222–396	237–423	252–450	287–451	318–363					
		F	85		195–304	210–328	225–352	241–376	256–400	292–401	323					
		EF	65		211–293	227–316	244–298	261–290								
CH870	500 (671)	EC	300					448–588	477–849	544–968	601–1,070	658–1,172	725–1,291	782–1,393	849–1,512	906–1,331
		CX	240				406	433–636	461–893	525–1,018	581–1,125	636–1,232	700–1,357	756–1,464	820–1,461	876–1,286
		MC	195				380–440	406–723	432–837	492–954	544–1,055	596–1,155	657–1,272	708–1,373	769–1,370	821–1,206
		M	155				400–563	428–786	455–836	519–953	573–1,054	628–1,154	692–1,271	746–1,372	810–1,248	865–1,098
		MF	100			379–424	407–716	434–765	462–814	527–928	852–942	638–789	702			
		F	90		357–395	385–656	414–704	442–752	470–800	535–912	592–857	649–718				
		EF	80	280–405	304–517	328–558	352–598	376–639	400–680	455–775	503–128	551–669				
CH890	750 (1,006)	EC	370						394–459	449–1,309	496–1,446	543–1,584	598–1,745	646–1,883	701–2,043	748–2,181
		CX	330					397	422–774	482–1,404	532–1,552	583–1,700	642–1,873	693–2,020	752–2,193	803–2,140
		MC	300				342–513	365–852	389–1,232	443–1,404	490–1,552	536–1,700	591–1,873	637–2,020	692–2,005	803–1,739
		M	230			267–312	287–670	307–951	326–1,106	372–1,261	411–1,394	450–1,526	496–1,681	535–1,814	580–1,800	720–1,564
		MF	160		204	220–514	237–690	253–921	269–980	306–1,117	339–1,235	371–1,352	409–1,490	441–1,607	598–1,396	638–1,170
		F	120		248–289	268–669	287–838	307–895	326–952	372–1,085	411–1,165	450–1,051	496–827	535–625		
CH895	750 (1,006)	EFX	100		212–423	228–666	245–715	262–763	278–812	317–926	351–994	384–896	423–705	457–533		
		EF	85	185–246	201–585	216–631	232–678	248–724	264–770	301–878	333–970	364–1,063	401–1,170	433–1,010	470–862	502–669
		EEF	75	178–475	193–564	209–608	224–653	239–697	254–742	290–846	321–855	351–761	387–580	417		

Courtesy of Sandvik

*See Figure 21 for details of crusher cavity configurations.

†Guide to Standard Sandvik crushing chambers is available: EEF = extra fine; EF = extra fine; EFX = extra fine extra; F = fine; MF = medium fine; M = medium; MC = medium coarse; C = coarse; CX = coarse extra; EC = extra coarse.

‡ To convert millimeters (mm) to inches (in.), divide by 25.4.

§ To convert metric tons per hour (t/h) to short tons per hour (stph), multiply by 1.1.

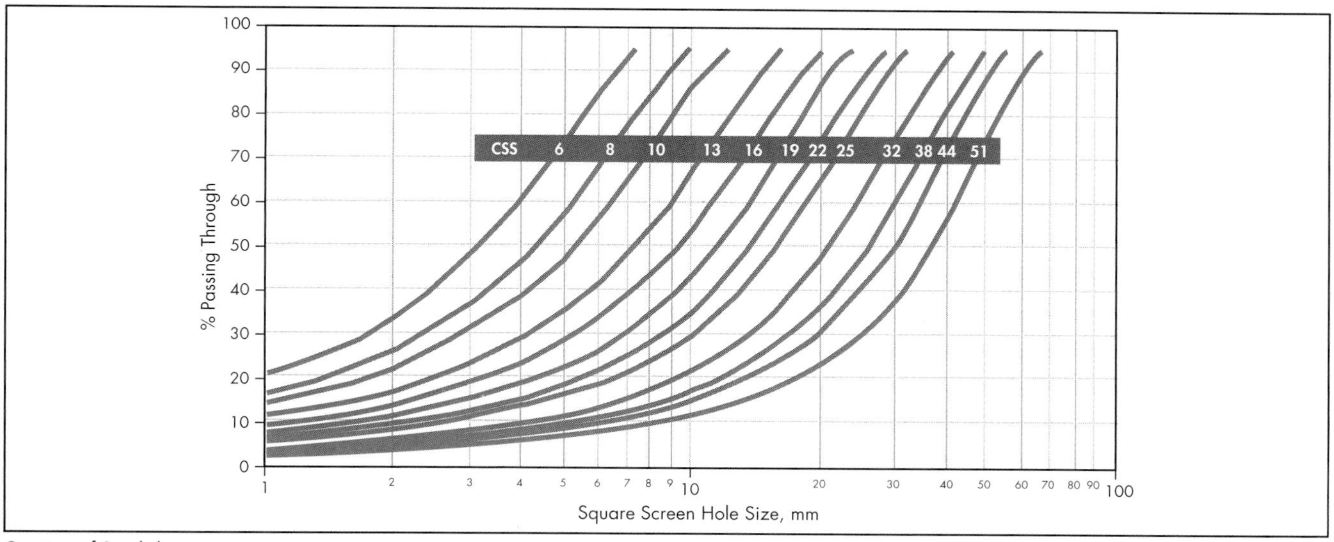

Courtesy of Sandvik

Figure 26 Guidelines for product size distribution for Sandvik cone crushers

ACKNOWLEDGMENTS

The author thanks Peter Pluis of Palminco Pty Ltd. for his input and for sharing his experience in relation to the design, application, and maintenance of gyratory crushers. The author gratefully acknowledges the input of Peter Erlynne of LakePlains Pty Ltd. for his impartial advice and input.

REFERENCES

Bearman, R.A., Munro, S., and Evertsson, C.M. 2011. Crushers—An essential part of energy efficient comminution circuits. In *MetPlant 2011: Metallurgical Plant Design and Operating Strategies*. Melbourne, Victoria: Australasian Institute of Mining and Metallurgy.

Boyd, K., and Utley, R.W. 2002. In-pit crushing design and layout considerations. In *Mineral Processing Plant Design, Practice, and Control*. Vol. 1. Edited by A.L. Mular, D.N. Halbe, and D.J. Barrett. Littleton, CO: SME. pp. 606–620.

Carter, R.A. 2016. New crusher models enhance process design flexibility. *Eng. Min. J.* (May 11). www.e-mj.com/features/6162-new-crusher-models-enhance-process-design-flexibility.html#.WTo8XuvfqUk.

Elias, I., Chaffer S., and Moya, E. 2015. Improvements in service life and cost reduction for gyratory primary crushers through mantle and concaves optimisation—A case study. In *Proceedings of Iron Ore 2015*. Melbourne, Victoria: Australasian Institute of Mining and Metallurgy. pp. 115–127.

E&MJ News. 2014. First Quantum selects Metso crusher for Sentinel mine in Zambia. *Eng. Min. J.* (November 6). www.e-mj.com/news/leading-deveopments/4685-first-quantam-selects-metso-crusher-for-sentinel-mine-in-zambia.html#.Wk4oHd-nG70.

Flavel, M.D., Jergensen, G.V., and Motz, J.C. 1985. Cone crushers. In *SME Mineral Processing Handbook*. Vol. 1. Edited by N.L. Weiss. Littleton, CO: SME-AIME. p. 3B-2.

Hart, S., Jankovic, A., Rees, T., Tavani, S., and Valery, W. 2011. Process integration and optimisation of the Boddington HPGR circuit. In *Proceedings of the Fifth International Conference on Semi-Autogenous and High Pressure Grinding Technology*. Westmount, QC: Canadian Institute of Mining, Metallurgy and Petroleum.

Hernandez, E., Kusumohastuti, W., and Agustian I. 2014. Excavating the first underground gyratory crusher at the Deep Mill Level Zone mine, PT Freeport Indonesia. In *Proceedings of the 12th AusIMM Underground Operators' Conference*. Melbourne, Victoria: Australasian Institute of Mining and Metallurgy. pp. 43–49.

Hulthén, E., and Evertsson, C.M. 2008. On-line optimization of crushing stage using speed regulation on cone crushers. In *Proceedings of XXIV International Mineral Processing Congress*, Vol. 2. Edited by W. Dianzuo, S.C. Yao, W.F. Liang, Z.L. Cheng, and H. Long. Beijing: Science Press. pp. 2396–2402.

Jacobson, D., and Lamminmaki, N. 2013. Pilot study on the influence of the eccentric speed on cone crusher production and operation. SME Preprint No. 13-048. Englewood, CO: SME.

Koski, S., Vanderbeek, J., and Enriquez, J. 2011. Cerro Verde concentrator? Four years operating HPGR. In *Proceedings of the Fifth International Conference on Semi-Autogenous and High Pressure Grinding Technology*. Westmount, QC: Canadian Institute of Mining, Metallurgy and Petroleum.

Metso. 2017. Innovation drives uptime, higher wear liner life. *Quarry* (June):34.

Mining Magazine. 2014. A fresh approach to cone crushing. 2014. *Min. Mag.* (July/August):36–38.

Rose, D., Meadows, D.G., and Westendorf, M. 2015. Increasing SAG mill capacity at the Copper Mountain mine through the addition of a pre-crushing circuit. Presented at the SME Annual Conference, Denver, CO, February 15–18.

Westerfeld, S.C. 1985. Gyratory crushers. In *SME Mineral Processing Handbook*. Vol. 1. Edited by N.L. Weiss. Littleton, CO: SME-AIME. pp. 3B-28.

CHAPTER 3.5

Sizers and Roll Crushers

David J. Bowman

Sizers and roll crushers represent an old family of crushing machines, with roll crushers often described as some of the earliest crushing machines. Roll crushers, as known today, have been in use since the 1920s (Cohen and Porter 1986). Sizers, or low-speed mineral sizers, are a newer introduction, with the Mining Machinery Developments (MMD) company having designed the first modern sizer in 1978. The first sizers were developed for underground crushing of coal in the United Kingdom, where low physical profile and high throughput were required. The terminology is important: *Sizers* are so named because the machine creates a series of sized voids for undersize rock to pass through before presenting oversize to the teeth to be broken in tension to minimize energy used (Flynn 1988). This comminution action specifically *sizes* the rocks, rather than simply crushing them.

The first sizers were installed to combat problems of blockages in more traditional equipment, which is an application that continues today. The other features of both sizers and roll crushers that make them attractive to designers are the efficient application of crushing power (through low speed and high torque) and the smaller footprint compared to an equivalent gyratory or jaw crusher installation. The supporting structure can also be lighter than for a gyratory crusher, as the crushing forces have no eccentric component and are contained within the machine frame. Another feature of sizers that can differentiate them from other coarse comminution devices is that the action tends to produce less fines than a comparative compression-based crusher.

Both sizers and roll crushers consist of two counter-rotating crushing members that grab or *nip* the rocks between the rolls (in the case of inward turning) or between the rolls and fixed combs (in the case of outward turning). The key components and their terminology are illustrated in Figure 1.

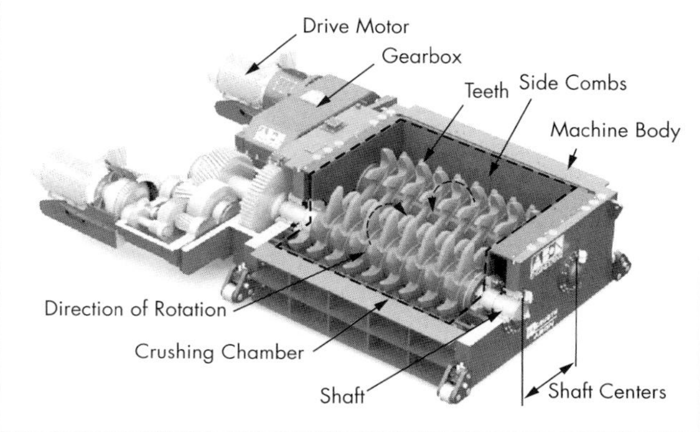

Courtesy of FLSmidth ABON

Figure 1 Key terminology for sizers

Roll crushers are available as a single-roll unit, with the material crushed between the rotating roll and a fixed plate; however, they are most commonly manufactured in a double-roll configuration. In the case of sizers (Figure 2), the breakage is between opposing teeth, with the large, aggressive tooth profile and low rotation speed working to nip the rock. Undersize material falls through the gaps between the rolls and teeth. With roll crushers (Figure 3) breakage is between two sets of much smaller teeth (or smooth drums), with the large roll diameter and high roll speed acting to nip and drag rocks into the crushing chamber. Once material has entered the crushing chamber, the crushing action is more of a traditional compressive crush than in a sizer.

In addition to the primary breakage mechanism between the rolls, some sizers rely on secondary breakage with the use of a breaker bar, as shown in Figure 4. The application of the breaker bar allows fewer teeth to be used in the crushing chamber, creating larger gaps between teeth, thus giving greater opportunity to nip larger rocks, and enabling a greater reduction ratio. The breaker bar can also be adjusted upward toward the rolls as teeth wear, giving the ability to control product top size. The downside to the application of the breaker bar is that it introduces an interruption into the flow, creating a further wear area caused by constant washing by fine material flowing past. In applications with wet and sticky material, the breaker bar can collect this material and force the teeth to be dragged through the fine material, accelerating tooth wear. No equivalent to the breaker bar is employed in roll crushers.

David J. Bowman, Principal Engineer, Bear Rock Solutions Pty Ltd., Attadale, Western Australia, Australia

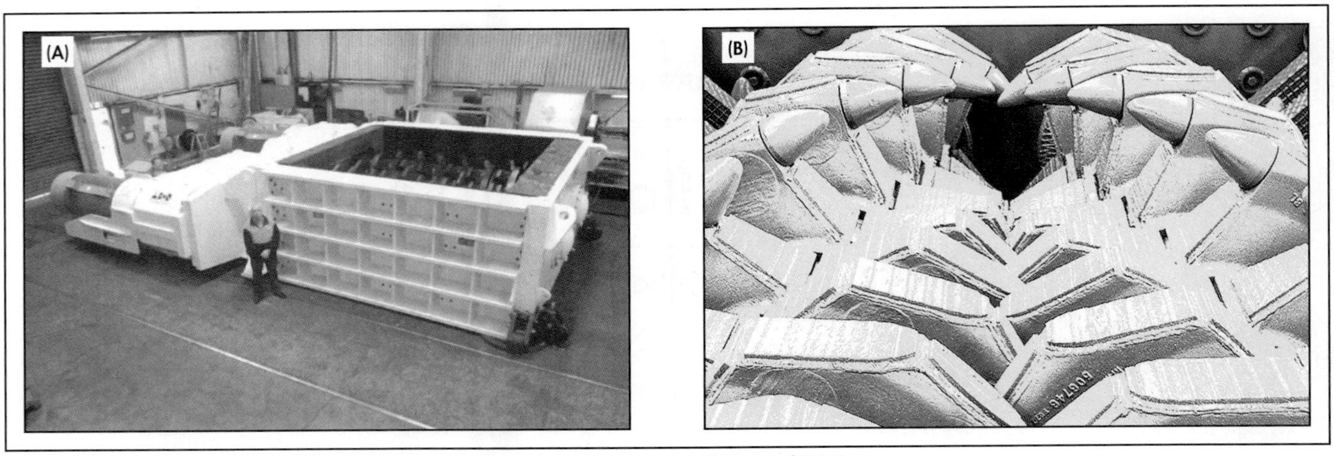

Courtesy of FLSmidth ABON

Courtesy of MMD

Figure 2 (A) Primary sizer and (B) primary sizer teeth

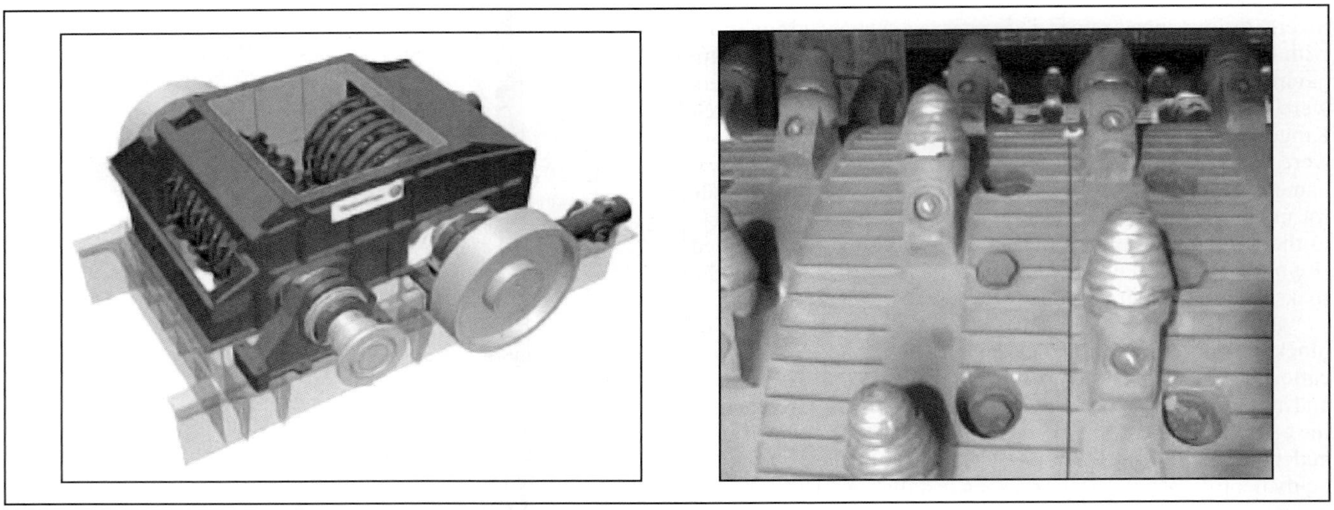

Courtesy of ThyssenKrupp Industrial Solutions

Figure 3 Double-roll crusher showing tooth profile

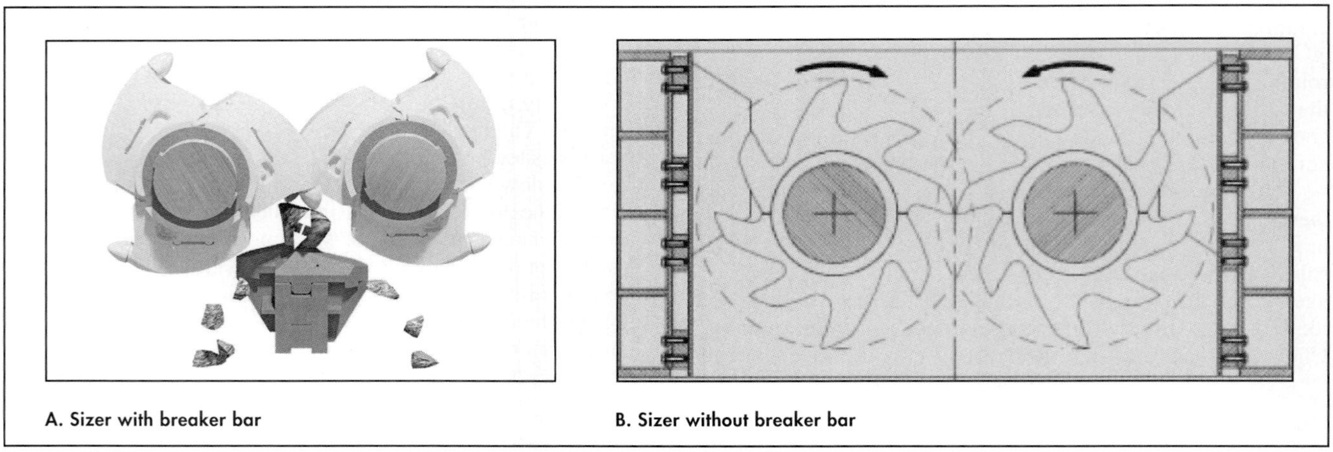

Courtesy of MMD

Courtesy of FLSmidth ABON

Figure 4 Sizer arrangements

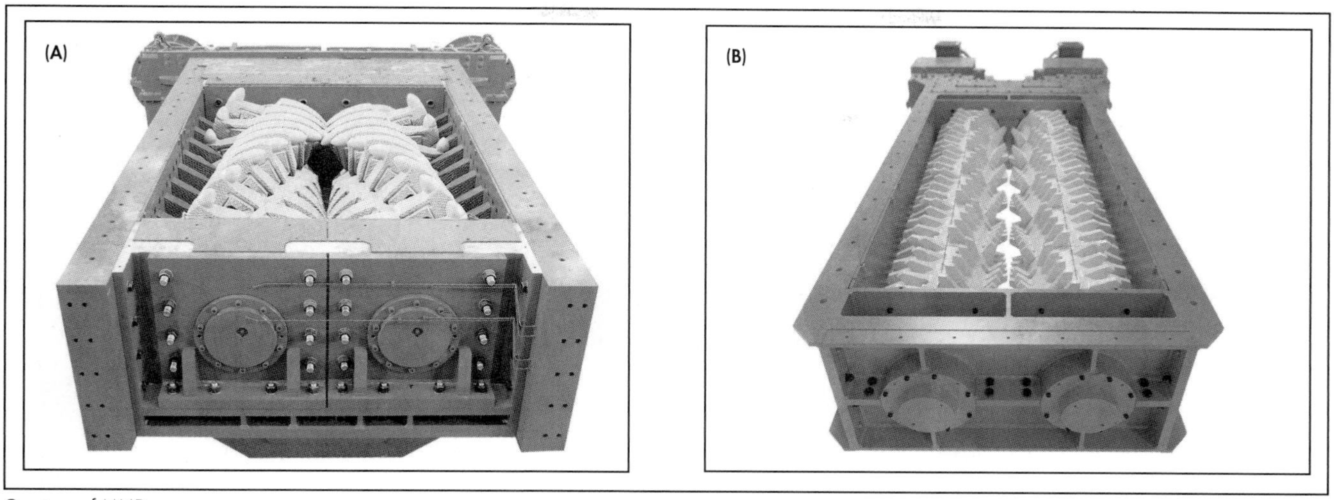

Courtesy of MMD

Figure 5 (A) Primary and (B) secondary/tertiary sizers

Courtesy of ThyssenKrupp Industrial Solutions

Figure 6 Double-roll crusher showing shaft adjustment

Sizers and roll crushers are used at all stages of comminution, from primary to quaternary duties. Machines specified for the different duties have different tooth profiles and shaft centers, in line with the feed and required product size distribution. Sizers are generally specified in terms of the gap between the shaft centers, with the largest primary sizers employing 1,500-mm (59-in.) shaft centers and the smaller sizers using 200-mm (~8-in.) centers (Figure 5). Roll crushers are specified in terms of the diameter and width of the rolls. The smallest machines usually measure about 400 (diameter) × 750 mm (width) (16 × 30 in.). The largest roll crusher currently on the market measures 2,400 × 3,600 mm (94 × 142 in.).

The shafts in sizers are traditionally fixed in place, giving no option for the machine to have the rolls adjusted inward as the teeth wear (to maintain product top size) or to relieve outward to allow tramp metal such as ground-engaging tools to pass. Although sizers have safety devices such as underspeed sensors to protect the machine in this event, tramp metal in the chamber is difficult to remove because it requires either the chamber to be manually dug out or the sizer retracted to the maintenance position with a full chamber (where possible). Both of these activities can create greater hazards to personnel than allowing the tramp metal to pass and collecting from a receiving conveyor. As shown in Figure 6, roll crushers do allow this horizontal adjustment, reducing the potential damage from ground-engaging tools and giving the operation greater control over product top size. The shaft adjustment can be mechanical, including a spring for passing tramp material, or more commonly, a hydraulic arrangement is used.

There are some machines currently on the market that incorporate the adjustable shaft centers on a low-speed sizer, such as the Sandvik hybrid crusher (Figure 7).

Feed to primary sizers or roll crushers is traditionally an *all-in* feed, as with primary gyratories, with the full run-of-mine (ROM) feed distribution going to the machine. The material that is smaller than the desired product size will fall through the machine, although it is important to note that the volume taken up by this fine material will contribute to the volumetric capacity. It is possible to have a primary sizer or roll crusher fully volumetrically utilized, yet drawing very little power when crushing fine feed. Material too large to fall through the gaps between the teeth and rolls will be grabbed by the teeth and broken. There are two mechanisms of breakage at work in the crushing chamber, exploiting compressive and tensile failure. Compressive failure occurs when material is forced into the gap between a tooth and the opposite roll, a space that will reduce in size as the rolls rotate. Tensile failure occurs when rocks are grabbed between opposing teeth, creating sufficient tensile force to cause fracture. Subsequent crushing of the progeny particles can be in either mode, but more commonly compressive as the particles become smaller. The smaller teeth and larger-diameter rolls of a roll crusher (relative to a sizer) lead to the crushing action being primarily driven by the compressive force between the rolls. The teeth in this instance are more for grabbing material than crushing or sizing it.

There is another variant of the roll crusher concept, which is the eccentric roll crusher (ERC). The ERC25-25 has a 2,500-mm (98-in.) feed opening width and a 2,500-mm (98-in.) diameter roll (Szczelina et al. 2017). In the design, the roll is mounted horizontally and moves with an eccentric motion that is consistent across the full width of the roll. Material is fed horizontally to an integral screen arrangement,

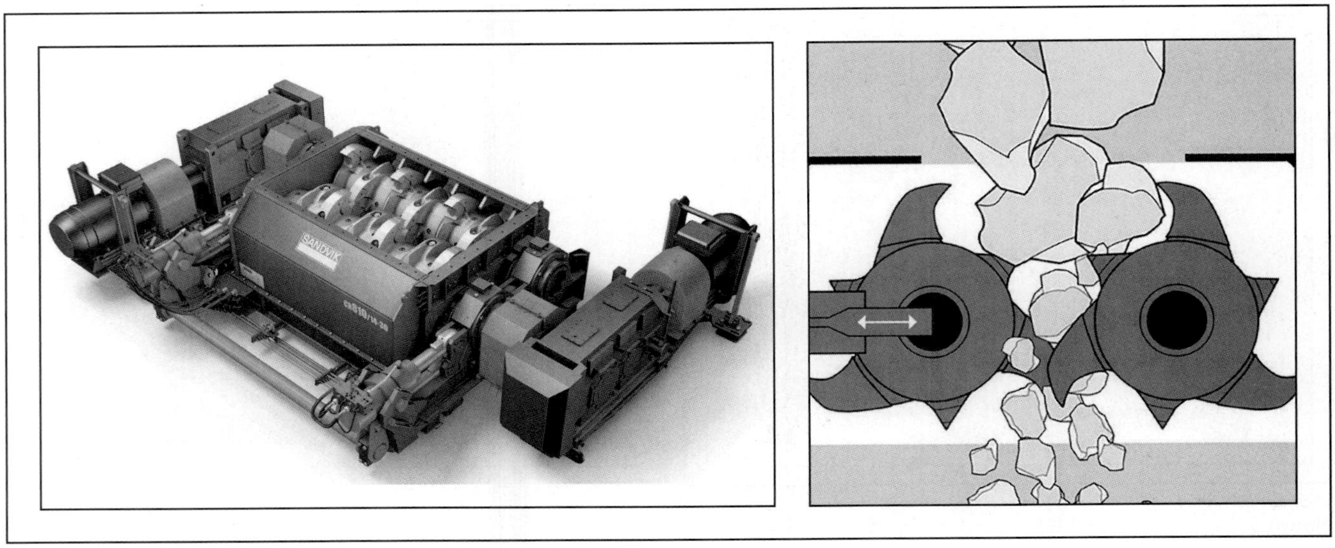

Courtesy of Sandvik

Figure 7 Sandvik hybrid crusher

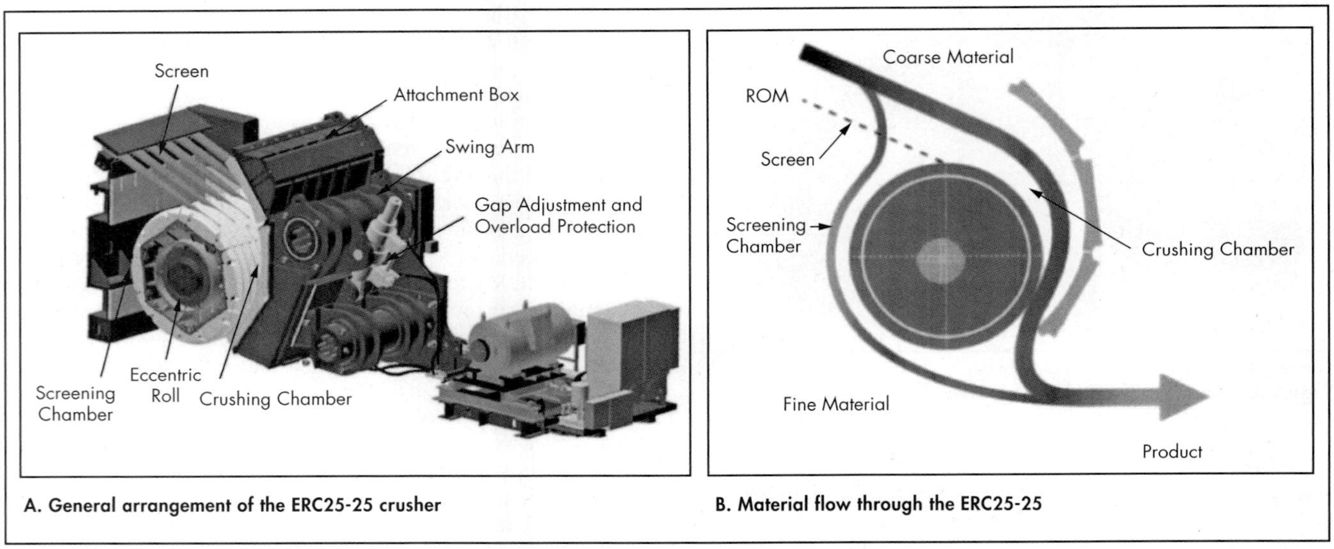

Courtesy of ThyssenKrupp Industrial Solutions

Figure 8 Eccentric roll crusher

where undersize is removed down the front of the crusher, therefore bypassing the crushing chamber. Oversize from the screen falls into the gap between the roll and a stationary curved crushing surface. The flow and general arrangement of the crusher are shown in Figure 8. The machine has a range of features, including a low-height requirement, high throughput (up to 3,000 t/h), integrated gap adjustment and overload protection, low residual forces, and a reduction in the generation of elongated product. As the machine is new to market at the time of writing, it is not possible to gauge the potential for the machine, but it is claimed that it specifically offers advantages for deployment in underground caverns and in mobile applications.

The power required for sizers and roll crushers is generally less than that for other forms of crushers, such as gyratory machines. Primary sizers are typically fitted with twin 350–400-kW (470–540-hp) drives, one for each shaft for a total installed power of 700–800 kW (940–1,070 hp). The most powerful sizers are fitted with 2 × 750-kW (1,005-hp) drives. As a comparison, some 60–89 primary gyratory crushers are now fitted with 1,000-kW (1,340-hp) drives.

PRINCIPLES AND TERMINOLOGY

Both sizers and roll crushers rely on the principle of grabbing rock using rotating elements mounted on a horizontal shaft. For sizers, these shafts generally rotate slowly (<1 m/s [1.3 ft/s] tip speed) to minimize the amount of material movement (boiling) on top of the shafts. Roll crushers operate at higher shaft speeds (typically 130–300 rpm), although in tougher applications this can be as low as 80 rpm (~3 m/s

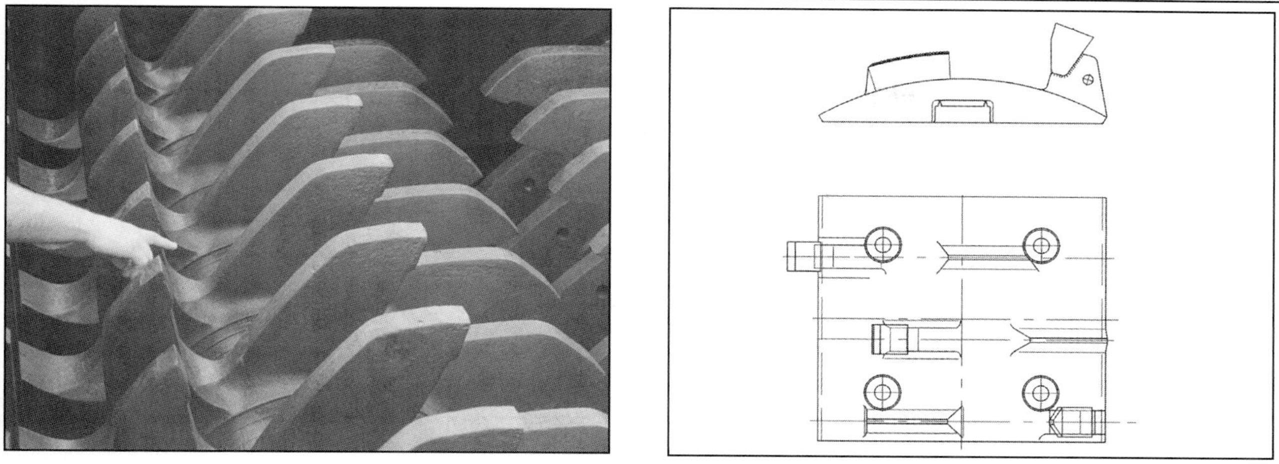

A. Single-piece tooth segments for sizers

B. Tooth segments for roll crushers

Courtesy of FLSmidth ABON

Courtesy of ThyssenKrupp Industrial Solutions

Figure 9 Tooth segments for sizers and roll crushers

[~10 ft/s] at the tips) with the circumferential speed being used to grab rock in the chamber and to match the velocity of the falling material, thus reducing wear. Roll crushers operate with inward-turning shafts, that is, reducing distance between the surfaces of the rolls, moving down the machine to create the compressive crushing force. Sizers are generally inward rotating for primary and secondary crushing, often moving to outward turning for tertiary and quaternary applications. The outward-turning arrangement subjects the machine to higher wear, but crushing between the teeth and fixed side combs gives a fine product size that is difficult to achieve with inward-turning shafts.

The rock-engaging wear materials used in sizers and roll crushers fall into two broad groups: single-piece segments (Figure 9) and replaceable picks (Figure 3). The roll crusher plates can have the teeth replaced if tooth wear has far outstripped the rate of baseplate wear. These picks can be welded in place or mechanically fastened, with mechanical fastening claimed to reduce downtime from tens of hours to a few hours (Good and Zimmerman 2014). One-piece sizer tooth segments are either cast or flame cut, and are often hard faced. The inherent risk of hard facing is that once the weld layers have been worn away, the underlying parent metal suffers accelerated wear because of heat-induced softening from the hard-facing process. The hard-faced layer can either be reapplied in situ or after removal of the shafts for maintenance. Side combs and breaker bar components are also subject to hard facing and suitable hardening treatment to extend operational life.

The transfer of energy in sizers and roll crushers is through a simple drive train (Figure 10). Machines are fitted with a single drive for soft rock applications, such as coal, or multiple drives for harder rock applications. Most hard rock primary sizers are fitted with a single motor for each shaft, driving through a fluid coupling (for machine protection) and reduction gearbox to the main shafts. Some manufacturers also employ a synchronizing gear set to ensure that the two shafts turn with the teeth meeting at the optimum angle for rock nip and ingestion.

This drive arrangement has been optimized for use on mobile machines underground by Joy Global. The Joy Global DRC1200 was designed specifically to achieve the ability to move the sizer through an underground mine on a tracked chassis to enable semimobile operation (Good and Zimmerman 2014).

Sizers for further reduction are frequently designed in a quad-drive arrangement, as this works better with the dimensions of the machine itself. These sizers are generally flatter and longer than a primary machine (driven by the roll diameter), so quad drive allows the drive motors and gearboxes to be smaller and limits the torque loading on the main shafts.

Roll crushers and the hybrid crusher are generally driven through a set of belts following the gear reducer (see Figure 7). The belt drive provides further speed reduction, effective flywheel energy storage in the large pulley, and mechanically isolates the other drive components from impact forces seen by the rolls.

Feed rate to sizers and roll crushers is generally controlled with an apron feeder, separated from the truck tipping forces by a ROM ore bin. The shafts and bearings in some machines are not designed to accept the impact forces imparted by rock rolling directly down from a truck body, only feed from the apron feeder, the crushing forces, and those applied through the use of hydraulic rockbreakers.

Roll crushers are fed with the *ribbon* of feed material falling directly into the crushing chamber, with the apron feeder shafts in the same orientation as the rolls. Sizers can be fed either parallel to the machine shafts, as shown in Figure 11, or perpendicular to the shafts with the machine turned through 90°. Both of these alternatives have positive points, depending on the machine style used. Most scrolling machines are fed perpendicular to the apron feed discharge to allow the scrolling motion to remove uncrushed material from the ore stream. Nonscrolling machines and roll crushers are parallel fed, with the sizer shafts arranged parallel with the apron feeder discharge axis. In general, machines fed this way use a larger *snag tooth* (as seen in Figure 4B).

The sizer discharges to either a receiving conveyor belt or another sizer (known as a *stacked arrangement*). Feed rate is dictated and controlled by many factors including sizer power draw and load on the receiving conveyor.

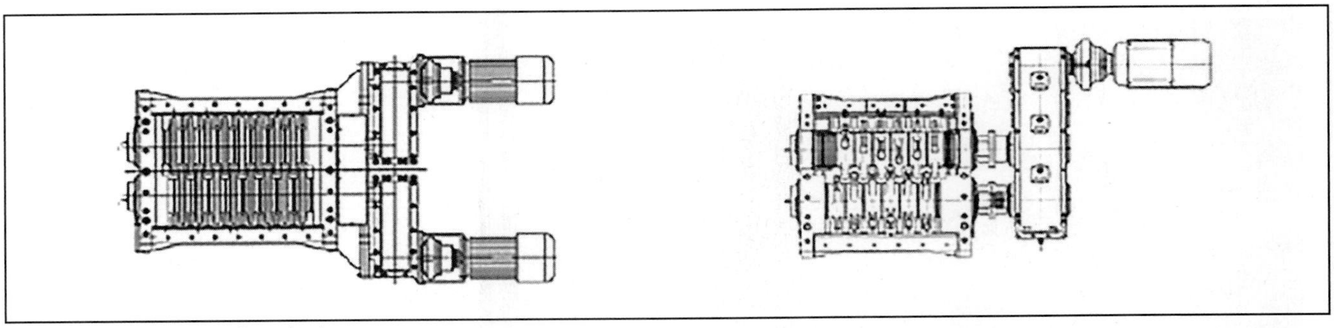

Courtesy of ThyssenKrupp Industrial Solutions

Figure 10 Dual- and single-drive arrangements

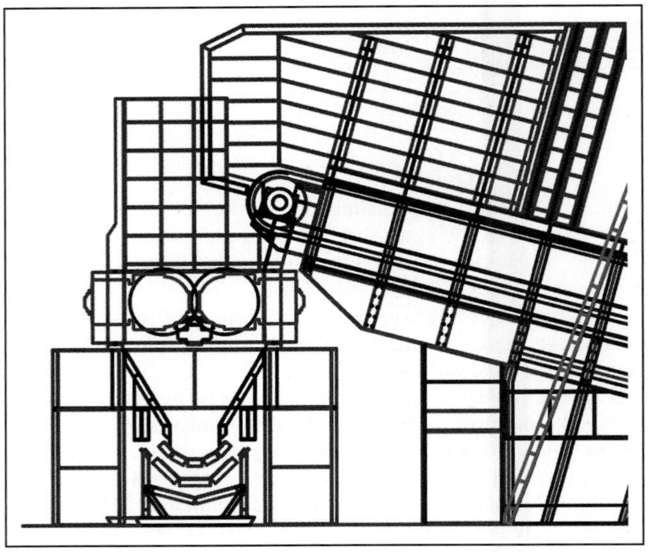

Courtesy of MMD

Figure 11 Parallel sizer feed arrangement

CONSTRUCTION AND DESIGN

As with most mineral processing equipment, sizers and roll crushers have been increasing in load and power as the scope of application grows. The structure of the machines is simple, with a twin-roll secondary sizer shown in Figure 12. The overall construction principles for roll crushers are very similar. The main frames are fabricated from heavy welded plate, not cast elements such as used in gyratory crusher mainframes. Frames are designed with removable end plates that allow the removal of the shafts for refurbishment, reducing the duration of production interruption. The machines are mounted on either fixed or retractable wheels, which are jacked down, thus lifting the machine, for removal along steel rails from under the feed point. This makes machine maintenance safer (less risk of falling material, with clear space overhead for crane lifts) and quicker.

Replacement of teeth, being the main wear component on both sizers and roll crushers, can be done either in the machine or with the shaft removed to a maintenance stand. Access to sizer teeth for in situ work is from the top, with roll crushers generally accessible via maintenance access doors on the sides of the machine (Figure 13).

The teeth in a sizer or roll crusher act to engage or nip the rock, apply force to the point of breakage, and also screen the passing material. Tooth design must work within these design parameters while also ensuring that product top size is controlled as the teeth wear, with an acceptable wear life. Generally, teeth are fashioned in a pick arrangement with an aggressive point to engage the rock (see Figure 9A). The width and profile of the tooth will provide the effective open area for screening, so these areas must be built to handle both flow wear and wear under high normal force. Manufacturers offer various tooth materials on machines to provide the best balance between wear rate and tooth strength.

The shaft removal approach offers the advantages of reducing risk of injury to maintenance personnel by taking the activity to a level surface as well as reducing the time the machine is out of productive service. Many sizers and roll crushers are now sold with rotatable spare shafts as part of the package; worn shafts are taken offsite for repair and refurbishment in a workshop environment.

All sizer manufacturers have different methods to handle clay, excessively hard rock, or other uncrushable material. The scrolling action of some machines removes unnipped material from the main flow area, with other machines being able to stop and reverse to clear blockages. Some, such as the FLSmidth ABON secondary sizer, are fitted with hydraulically actuated side panels that open while the machine reverses to eject uncrushable material.

APPLICATION AND OPERATION

As with any installation, information should be sought from manufacturers regarding the feed size, product size, required power, and throughput. All manufacturers have this information available for stated feed properties, along with reference lists of global installations. A guide to the type of information generally available is shown in Table 1.

As a general rule, the specification of sizers and roll crushers is dictated by the feed size and strength. In a typical primary application, ROM feed will be presented with an F_{80} in the order of 1,000–1,500 mm (39–59 in.). Such feed forces the designer to specify a machine with sufficiently large shaft centers to guarantee that this size material will be nipped and broken. This size of machine may well exceed the necessary throughput requirements of the process plant because of the large volumetric capacity. For example, a large (1,300-mm [51-in.] centers) primary sizer operating in moderate rock will comfortably process up to 5,000 t/h, regardless of the plant requirement. This gives further justification for feeding these

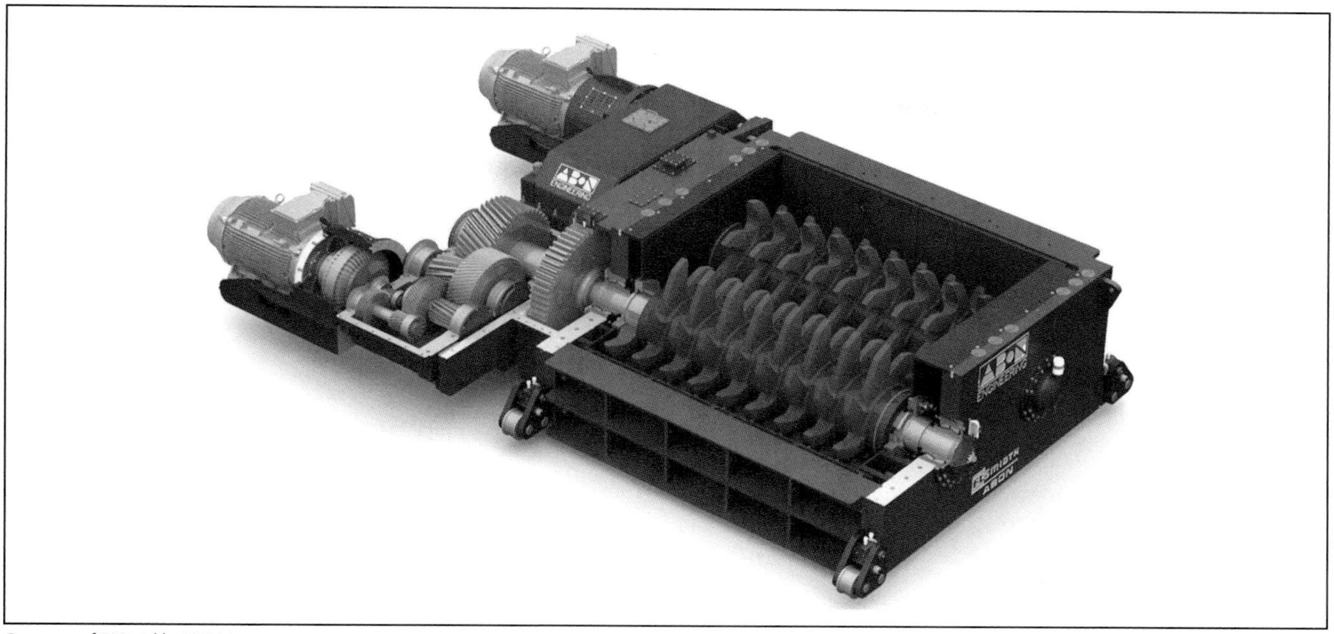

Courtesy of FLSmidth ABON

Figure 12 Cutaway view of a twin-drive secondary sizer

Courtesy of ThyssenKrupp Industrial Solutions

Figure 13 Maintenance access to primary double-roll crusher

machines through a controlled means, such as an apron feeder, to avoid flooding downstream equipment.

The limitation to the application of sizers and roll crushers is usually the strength of the feed rock and/or the abrasivity. The strength and abrasivity must be considered for each area of the material to be mined, not as an ore-body average. There are many examples of crushing plants (of all types) around the world that have failed to perform on plant start-up because of the highly competent cap rock that can be a feature of certain deposit types. Although the average rock strength may be within the necessary limits, small parcels of extreme material can stop production through the inability to nip and break the rock, or through machine damage and accelerated wear.

The measure of rock strength and abrasivity is discussed at length in Chapter 3.6, "Crusher Selection and Performance Optimization." Typically sizers and roll crushers are used in material with an uniaxial compressive strength below approximately 180 MPa, or fracture toughness below ~2 $MN/m^{1.5}$.

Because the machines are fed with an all-in whole of feed stream (including all fines), both sizers and roll crushers are highly vulnerable to wear. They are susceptible to both flow and gouging abrasion, depending on the application. In soft rock applications, wear on the machines is governed by flow abrasion of material running between the teeth and rolls. This can present itself with the back and side surfaces of the teeth wearing away, leaving a slender tooth of almost original length. In harder rock applications, gouging abrasion becomes an issue, with the large force on the teeth acting to wear away the points, leaving a much shorter tooth with almost the original width and depth. As a general rule, feed to sizers and roll crushers should have a Bond abrasion index (i.e., flow measure) less than 0.4 and a CERCHAR (Laboratoire du Center d'Études et Recherches des Charbonnages de France) abrasivity index (i.e., gouging measure) no greater than 3.0, or alternatively, a gouging abrasion index of less than 15. The duty and likely type of wear should be considered in all applications and an economic trade-off applied.

Sizers and roll crushers are fed from a ROM bin (single or dual tip) via an apron feeder. The exception to this is underground, where the ROM bin is replaced by a feed hopper to accept discharge from load-haul-dump machines. There is no requirement for the installation to include a large bin beneath the machine (as with a gyratory) as the surge decoupling is taken up by the ROM bin, and the production of crushed rock is at a controlled rate.

Table 1 Typical sizer manufacturer information

Sizer Shaft Centers, mm (in.)	Maximum Feed Size,* mm (in.)	Throughput Rate, t/h	Total Installed Power, kW (hp)	Machine Mass† (approx.), t
500 (20)	480–650 (18–26)	Up to 1,500	55–220 (75–300)	5
660 (26)	650–850 (26–33)	Up to 2,500	160–500 (215–670)	20
800 (31)	750–1,050 (30–41)	Up to 3,500	200–630 (270–845)	32
1,000 (39)	950–1,300 (37–51)	Up to 5,000	250–700 (335–940)	42
1,250 (50)	1,200–1,600 (47–63)	Up to 7,500	450–800 (600–1,070)	80
1,500 (59)	1,400–2,000 (55–79)	Up to 10,000	750–1,200 (1,000–1,600)	125

Courtesy of ThyssenKrupp Industrial Solutions
* Maximum feed size depends on tooth arrangement.
† Mass of machine without drive.

Courtesy of MMD
Figure 14 Fixed, semimobile, and mobile primary sizer stations

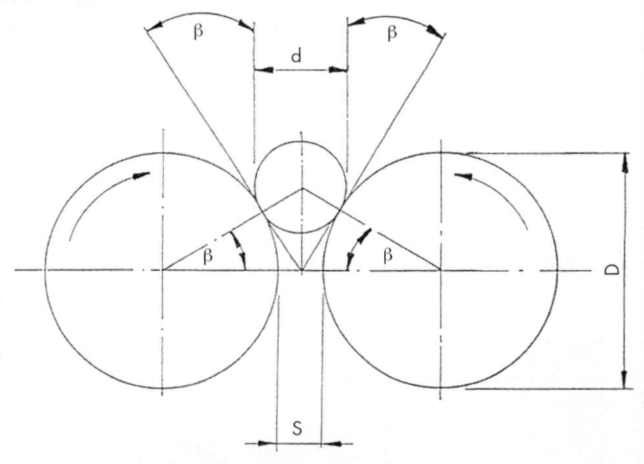

Courtesy of ThyssenKrupp Industrial Solutions
Figure 15 Calculation of maximum feed size in double-roll crushers

A key consideration with the application of sizers and roll crushers is the ability of the machines to be used in fixed, mobile, or semimobile applications. The machines are small, light, and simple when compared with other crushing equipment. There is no requirement for stand-alone hydraulic and oil lubrication systems, further increasing the mobility.

This small size of machine, coupled with the containment of all crushing forces within the machine frame, make sizers and roll crushers ideal for mobile applications. The smaller mass and reduced head height allow machine stability, even when track mounted. The use of a ROM bin and inclined apron feeder to feed the machine also allows a lower ROM pad wall, reducing the cost of installation for fixed or semimobile installations. This reduction in site capital costs for machine installation can make a strong economic case for relocatable units, with the corresponding reduction in both the capital and operating costs of supporting long truck hauls. The trade-off is the additional cost of conveying infrastructure, which is largely driven by site layout and topography. Typical sizer installations are shown in Figure 14.

Mobile applications have been successfully deployed in coal and oil sands, fed directly by shovel, truck tip, or as a dozer trap (Morrison and Lourel 2009). Further considerations for the degree of mobility for in-pit and mobile crushing are contained in Chapter 3.4, "Gyratory and Cone Crushers," and in Boyd and Utley (2002).

Process performance of sizers and roll crushers are not well documented, compared to many other types of crushers, with the general level of detail usually limited to the type of data seen in Table 1. In terms of throughput and maximum feed size, some estimates are possible.

For smooth-surfaced roll crushers, Figure 15 shows the geometry for the calculation of maximum feed size:

$$d = 2\left(\left(\frac{\left(\frac{D}{2}+\frac{S}{2}\right)}{\cos\beta}\right)-\frac{D}{2}\right)$$

where

- d = maximum particle diameter, m
- D = roll diameter, m
- S = gap between rolls, m
- β = half of total included nip angle, degrees

In reality, the calculated feed size is quite conservative, and in the case of toothed rolls, the maximum feed particle size can be significantly larger.

For the theoretical calculation of throughput for a roll crusher, the following equation can be used:

$$Q = 3{,}600\ WDvfs\gamma$$

where

- Q = throughput, t/h
- W = roll width, m
- D = roll diameter, m
- v = peripheral speed, m/s
- f = void factor
- s = gap between rolls, m
- γ = bulk density, t/m^3

In terms of product size distribution from roll crushers and sizers, a majority of data is experiential, and as such, the manufacturers should be contacted for further information.

ACKNOWLEDGMENTS

The author acknowledges the input of Richard (Ted) Bearman (Bear Rock Solutions), Peter Pluis (Palminco), Evan Douglas (Transmin), Mark McVey and Joel Brockett (MMD), and Steve Bond and Phil Mulcahy (FLSmidth ABON).

REFERENCES

Boyd, K., and Utley, R.W. 2002. In-pit crushing design and layout considerations. In *Mineral Processing Plant Design, Practice, and Control*. Vol. 1. Edited by A.L. Mular, D.N. Halbe, and D.J. Barrett. Littleton, CO: SME. pp. 606–620.

Cohen, S.M., and Porter, E.S. 1986. Hydraulic roll crusher. Presented at the AIME/SME Fall Meeting Technical Session, St. Louis, MO, September 8.

Flynn, B. 1988. The concept of primary rock sizing. *Quarry Manage.* 15(7):29–31.

Good, T., and Zimmerman, J. 2014. DRC1200 hard rock sizer: a twin roll sizer designed for mobility. SME Preprint No. 14-021. Englewood, CO: SME.

Morrison, D., and Lourel, I. 2009. In-pit crushing and conveying. Presented at the Iron Ore Conference, Perth, Western Australia, July 19–27.

Szczelina, P., Drescher, F., and Silbermann, F. 2017. Compact, robust and high performance—New ERC25-25 eccentric roll crusher revolutionizes primary crushing in underground mines. *AT Miner. Process.* 58(May):62–72.

CHAPTER 3.6

Crusher Selection and Performance Optimization

Richard (Ted) Bearman

CRUSHER SELECTION

The selection of equipment and flow-sheet design considerations for crushing plants cover a wide range of topics. Boyd (2002) gives a comprehensive summary of the factors involved in the design of crushing plants. The principal design parameters are

- Production requirements,
- Capital cost,
- Ore characteristics,
- Safety and environment,
- Project location,
- Life of mine and expansion plans,
- Operational considerations,
- Maintenance requirements, and
- Climatic conditions.

Many of these parameters directly relate to the selection, sizing, and application of crushing equipment. This chapter focuses on these specific factors. A combination of production requirements, ore characteristics, operational considerations, and maintenance requirements are considered. Related to crusher selection, the main issues are

- Stage of size reduction,
- Circuit arrangement,
- Feed characteristics (feed particle size, strength, abrasivity, and materials handling properties),
- Feed presentation,
- Throughput (gross and net), and
- Size reduction and required product size.

Tables 1 and 2 provide a ranking of the crusher types suitable for various stages of size reduction against the ability to treat material with certain characteristics. The number of dots denotes the suitability of different crusher types in relation to feed and performance characteristics. The first table is based on Utley (2002), but with modifications to the crushers considered and updates to account for changes in performance capabilities. Five dots denotes a high level of suitability, whereas one dot denotes that application can be problematic. Within these rankings, the operational and maintenance performance is integral; that is, a poor ability to deal with highly abrasive feed is intrinsically linked to a high maintenance demand.

Priority should be given for the stage of reduction and the suitability to various feed characteristics when selecting crusher types. Providing the crusher type selected is inherently suited to the application, it is a matter of addressing the other factors that will influence how the machine will perform in the role.

Given that crushers can represent either a minor or major component of an overall comminution circuit, it is difficult to characterize a *typical* configuration. Having said this, one of the most often-mentioned characteristics relates to the issue of *open* or *closed* circuit operation. In the open-circuit configuration, the crusher operates as a single-pass machine, and feed is simply transformed into product, which then goes to another processing stage. In the closed-circuit situation, product from the crusher in question is screened, and oversize material is recirculated back to the crusher, while undersize from the screen is sent for further treatment. In general terms, primary and secondary crushers are most often deployed in open circuit, with secondary crusher feed pre-screened to remove material smaller than the closed side setting (CSS), whereas tertiary and quaternary machines are used in closed circuit. The decision to implement open or closed circuits within an overall flow sheet becomes a trade-off between throughput, degree of size reduction required (or possible), cost, and the need for control of the product size for subsequent equipment and processes. A typical three-stage crushing flow sheet is provided in Figure 1, where primary and secondary crushing are open circuit and the tertiary stage is closed circuit.

Generally, the design considerations for feed presentation are related to distribution, rate, and control. There will always be specific issues related to certain types of crusher, but in general, feed presentation should meet the following criteria:

- Maximum feed size should not be greater than 60%–80% of the maximum feed opening (exact number is related to crusher type).
- Feed should be distributed evenly either across the feed opening or where the opening is an annulus (gyratory, cone, and vertical shaft impactor [VSI] crushers) around

Richard (Ted) Bearman, Director, Bear Rock Solutions Pty Ltd., Attadale, Western Australia, Australia

Table 1 Suitability for primary crushing based on feed and performance characteristics

	Gyratory	Jaw-Gyratory	Double-Toggle Jaw	Single-Toggle Jaw	Sizer	Roll Crusher	Horizontal Shaft Impactor	Hammer Mill
Feed size[1]	•••••	•••••	•••••	•••••	•••••	••••	•••••	•••••
Reduction ratio	•••	••••	•••	•••	••	••	•••••[2]	•••••[2]
Throughput	•••••	•••	••[3]	•••[3]	•••••	••••	••••	••••
Feed strength	••••	••••	•••••	••••	••	••	••	••
Abrasive feed[4]	•••	•••	••••	•••	•	••	••	••
Sticky feed	••	••	•	•••	•••••	••	•[5]	•[5]

Adapted from Utley 2002

1. Large feed size in combination with high-strength, high-elastic modulus feed can present problems with nip and ingress to the crushing chambers, as can rocks with low coefficients of friction (i.e., graphite or talc ores). Such issues particularly apply to compression-type machines.
2. Good reduction ratios can be achieved, but only with feed that has a crushing resistance no greater than moderate. For vertical shaft impactor machines, the reduction ratio at the P80 scale is moderate; however, the reduction ratio in the fines area is much greater.
3. Jaw throughput needs to be considered as a total-station value, where a grizzly is incorporated to bypass fines to avoid the crushing chamber. As the percentage of fines varies according to the feed source and type, the overall throughput of the station will be highly sensitive to the quantity of bypass.
4. Abrasivity covers both the flow and gouging wear mechanisms.
5. In some instances, specialized variants are available that will improve the ability to deal with high clay and/or sticky feed.

Table 2 Suitability for secondary, tertiary, and quaternary crushing based on feed and performance characteristics

	Cone	Sizer	Rolls	Vertical Shaft Impactor	Horizontal Shaft Impactor	Hammer Mill
Predominant duty[1]	S/T/Q	S	S/T/Q	T/Q	S/T/Q	T/Q
Feed size[2]	••••	•••••	•••	•••	•••	•••
Reduction ratio	•••••	••	••	••••[3]	•••••[3]	•••••[3]
Throughput	•••••	•••••	••••	••	•••••	•••••
Feed strength	•••••	••	••	•••	••	••
Abrasive feed	••••	•	••	••••[4]	••	••
Sticky feed	••	•••••	••	•	•	•

Adapted from Utley 2002

1. S = secondary; T = tertiary; Q = quarternary.
2. Large feed size in combination with high-strength, high-elastic modulus feed can present problems with nip and ingress to the crushing chambers, as can rocks with low coefficients of friction (i.e., graphite or talc ores). Such issues particularly apply to compression-type machines.
3. Good reduction ratios can be achieved, but only with feed that has a crushing resistance no greater than *moderate*. For vertical shaft impactor (VSI) machines, reduction ratio at the P80 scale is moderate; however, the reduction ratio in the fines area is much greater.
4. VSI machines can treat highly abrasive feed, providing the internal configuration is appropriate to the operational conditions.

the opening. The evenness of the distribution needs to apply to both the rate of delivery and also the composition of the feed in terms of size distribution.

- The feed size distribution supplied to crushers differs between machine types. The sensitivity of crushers to the length of size distribution also varies significantly. The inclusion of full-length size distributions (also called *all-in* feed), where fines are present, are known to compact in compression-type crushers, particularly at finer discharge settings (Svensson and Steer 1990). In cone crushers, such compaction is commonly seen in tertiary and quaternary applications but less seldom in secondary duties, except where high reduction ratios are being targeted. There has been a significant trend in recent years to use an all-in feed to secondary cone crushers. This specific issue is addressed in Chapter 3.4, "Gyratory and Cone Crushers." Noncompression machines, or machines where voidage is maintained, including impactors, hammer mills, and sizers, do not cause compaction, and hence the length of the feed size distribution is less of a concern. It can become a concern, however, in cases where the fines are moist or sticky, and this adversely impacts the flow characteristics. In such instances, the presence of fines can cause blockages and build-ups, both within the machine and in peripheral areas.
- Feed rate needs to be controllable, and the ability to control feed depends on the type of feed arrangement, the equipment used, and the flow characteristics of the feed. All crushers should be operated with consistency of feed and rate. In some machines, such consistency is absolutely crucial (e.g., choke feed in cone crushers), whereas in others, it is merely recommended (e.g., sizers, horizontal shaft impactors). The presence of a dedicated feeder is the best option for full control, and Carson and Holmes (2002) provide a review of bins, hopper outlets, and feeders, with additional commentary on certain feeder types provided in Carson and Petro (1998).

Given the importance of feed arrangements, the following paragraphs give an overview of feeder types and applications. The main feeder types encountered related to crushing are vibratory pan, vibrating grizzly, belt, reciprocating plate, or apron. In terms of selection for crusher feed, the main points to be considered when selecting a feeder are

- Robustness (including resistance to impact),
- Particle size,
- Capacity,
- Integral fines removal,
- Ability to handle abrasive material,
- Ability to handle wet and/or sticky material,

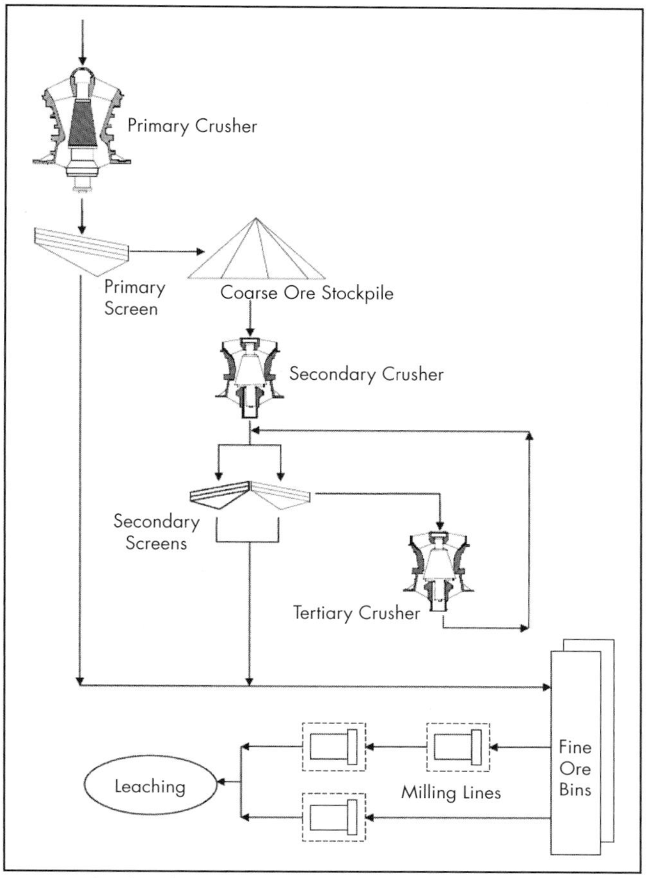

Source: Huband et al. 2006

Figure 1 Example of a typical crushing flow sheet with open- and closed-circuit sections

- Physical location (e.g., primary crusher dump pocket, tunnel, stockpile),
- Spillage,
- Dust control,
- Ease of cleaning, and
- Ease of control.

In broad terms, the following statements can be made regarding feeders for crusher installations (Carson and Holmes 2002).

- Vibratory pan feeder
  - Mainly found in duties requiring less than 500 t/h (metric tons per hour)
  - Capable of handling feed up to 300 mm
  - Tolerates a reasonable level of impact
  - Accommodates abrasive feed
  - Accommodates moist material, but may be subject to packing and build-up with wet and/or sticky material
  - Can be enclosed to control dust
  - Well suited to control
- Vibrating grizzly feeder
  - Integral fines removal
  - Highly robust (can be used ahead of primary crushers)
  - Handles run-of-mine (ROM) feed material
  - Tolerates impact
  - Handles abrasive feed
  - Accommodates wet and/or sticky material
- Belt feeder
  - Feed particle size limited to 150 mm
  - Can achieve high rates (3,000 t/h)
  - Not generally applicable where impact is present, particularly if feed is abrasive
  - Not suited to handling wet and/or sticky feed
  - Well suited to control
- Reciprocating plate feeder
  - Highly robust (can be used ahead of primary crushers)
  - Handles ROM feed material
  - Tolerates impact
  - Handles abrasive feed
  - Better suited to free-flowing, nonsticky feed
  - Lags in material flow (feeder action can make control more difficult)
- Apron feeder
  - Highly robust (often used ahead of primary crushers)
  - Handles ROM feed material
  - Good tolerance of impact
  - Handles abrasive feed
  - Better suited to free-flowing, nonsticky feed
  - Requires a *dribble* system to capture and remove fines spillage
  - Action amenable to control

As discussed earlier, there is a need to consistently provide appropriate feed to any crushing unit to provide peak operational and process performance. In terms of throughput—and providing the crusher operates within the acceptable envelope for the integrity of the machine—the main considerations are scoping the base machine for catch-up capacity and ensuring that any throughput sufficiently allows for recirculating load in a closed circuit environment.

Crushers deal with granular material and bulk transport, and therefore availability and utilization are not as high as subsequent grinding equipment. Typically, crushing plants may have mechanical availability of 90% and an equipment utilization of 80%, which gives an overall operation utilization of 72%, compared to milling circuits with values of 92%. In addition, many crushing circuits are decoupled from subsequent processing stages via stockpiles or bins. In this regard, crushers selected for a specific duty need to have a throughput rating that accounts for the lower available operating hours and also gives the ability to run at higher catch-up rates. In designing circuits, individual crushers and ancillary equipment should therefore have a design factor of 1.25 to 1.5 times the nominal rate.

In relation to closed-circuit operation, the balance between the crusher product size distribution, the screen cut size, and the rate of net final product required is critical. Crushers will typically generate 50%–90% passing a nominal discharge setting, and therefore the screen cut size must be set to a value where the tonnage of recycle to the crusher does not accumulate to the point that the circuit is overwhelmed. To ensure that sufficient capacity is available, it is advisable to undertake a circuit simulation of the proposed crusher and screening arrangement using an appropriate software package. Such programs use a convergence algorithm, and if the circuit fails to converge, this indicates that the settings of the crusher and screen cannot be balanced.

All crushers will have a limit for size reduction, usually based on a ratio of feed size (F80) to product size (P80).

Reduction ratio varies by crusher type and the stage in the overall comminution circuit. To calculate the equipment required for the overall reduction needed in a crushing plant, the total reduction ratio should be defined and then compared to that achievable by using combinations of crushers. Table 3 gives typical reduction ratios for various types of crushers.

As an example, if a plant needs to reduce material from a primary crusher feed size of 600 mm down to a final product size of 25 mm, this represents a 24:1 overall reduction. If the application is in medium-hard rock and a gyratory, cone combination is to be used, then by using the values from Table 3, the number of size reduction stages can be calculated. For such feed, average ratios from Table 3 should be used for gyratory and cone crushers, so therefore a primary gyratory in combination with a secondary cone will deliver an overall reduction of (5.5 × 5):1, that is, 27.5:1. Should the feed be harder, and the minimum ratios are therefore used, then it would only be possible to achieve (3 × 3):1; that is, 9:1, and as such, a tertiary stage would be required.

Table 3 Typical reduction ratios achievable by crushers in varying duties*

	Primary	Secondary	Tertiary	Quaternary
Jaw	3–7:1	—	—	—
Gyratory	3–8:1	—	—	—
Cone	—	3–7:1	3–5:1	2–3:1
Horizontal shaft impactor	4–8:1	4–6:1	3–4:1	2–4:1
Vertical shaft impactor	—	3–5:1	2–5:1	2–4:1
Roll crusher	3–5:1	3–5:1	2–4:1	2–3:1
Sizer	3–5:1	3–4:1	2–4:2	2–3:2

*Reduction ratios based on F80:P80 ratio.
Note: Higher reduction ratio values only apply to low-strength, free-crushing feed.

PERFORMANCE OPTIMIZATION

The optimization of crushers and flow sheets containing crushers needs to be targeted to allow the machinery and the circuit to

- Generate the required size distribution, either for final use, or for subsequent processing;
- Provide the required net throughput;
- Apply an optimal level of power to undertake the duties;
- Operate at a level of maintenance costs (direct and indirect) that is a minimum for the required duty;
- Achieve availability and utilization rates that are required to meet the overall production requirements for the operation; and
- Meet operational cost targets.

All crushers are subject to the preceding optimization goals, but depending on the arrangement and use of equipment, certain trade-offs may be required. Overall process optimization for crushers and crushing circuits relies on

- Understanding the circuit mass balance and the performance envelopes of the equipment involved—not only crushers, but classification and materials handling equipment, that is, screens, feeders, conveyor belts, transfer chutes, bins (capacity and residence time), and other external factors (climatic, environmental, etc.);
- Feed strength and abrasivity;
- Feed behavior, such as the inclusion of adhesive clays;
- Feed size distribution (including impact from the performance of screens);
- Feed rate;
- Feed distribution (including segregation);
- Discharge setting;
- Measurement of performance, in terms of crusher liner profile, power consumption, product size distribution, crushing pressure, and recirculating load;
- How the circuit is impacted by classification; and
- Optimization of maintenance activities and schedules for maintenance.

To optimize any circuit involving crushers, the basic requirement is to understand the circuit. During the design phase of mineral processing flow sheets, it is usual practice to assemble the *process design criteria*, which state key parameters that will need to be addressed in the design and the aim of the flow sheet in terms of the performance requirement, that is, throughput, product size distribution, and so forth.

Because of the inherent variability of natural geological materials, the performance of the actual flow sheet will vary from the original intent. If operational practice is overlaid, then the actual performance of the plant can vary considerably from the original plan.

The first step in the optimization of a crushing plant should therefore refer back to the original design inputs and assumptions. This will help with understanding the basis of the original decisions that led to the subsequent design. This original intent should be compared to the actual site parameters, including feed size distribution, specific gravity, bulk density, moisture, strength, throughput, and equipment performance predictions. It is also critical to include the asset management aspects of the actual operation, which would include mechanical availability, utilization of equipment, maintenance practices, and change-out targets. Encompassing both—process and assets—the cost base being achieved must also be compared to those originally estimated.

A key part of the investigation of the design versus actual performance is to establish a full mass balance around the circuit, including total throughput, recirculating loads, size distributions, bulk densities, moisture content, and power consumption. Once established, the settings and performance of the individual equipment units must be assessed, with the assessment also including process control at both the local equipment level and the circuit level. Such an operational survey is at the heart of understanding circuit performance and optimization. For detailed consideration of circuit sampling for granular material and mass balancing, the reader is directed to Napier-Munn (1996). The existence of a well-developed circuit model is the cornerstone of optimization, and the use of this approach allows bottlenecks, inefficiencies, and mismatches to be identified and articulated. Once into the rectification phase of an optimization project, the same model can be used as a virtual test bed for improvements and upgrades.

Key Variables and Measurement Techniques

For a total crushing plant survey, samples for size distribution (and other physical parameters) and throughput at all the key points (such as screens and crushers) in the circuit need to be obtained. The exact method for obtaining a representative survey will be site specific and dependent on the flow-sheet configuration and the ability to safely and effectively access the key points. As a general guideline, a survey of crushing plants should be scheduled so that it corresponds to a period of known feedstock and, if possible, to provide consistency, the feed should come from metallurgically surveyed stockpiles. The feedstock should be sufficient to allow the plant to run consistently for 2–4 hours, during which time all routine data logging should be active. If possible, additional instrumentation should be brought in for the purposes of measuring information not routinely available. Among such additional measurements are conveyor belt speed, crusher countershaft speed, screen motion, and so forth. Once the operation is considered stable, a *crash-stop* needs to be implemented to enable sampling. In certain instances, this may need to be a staged arrangement to prevent blockages or machine stalls. With granular material in crushing circuits, the most effective sampling is to take a known length of material from the relevant conveyor belts. The sample length must be sufficient to provide a representative sample of material based on the top size of the particles present.

Such manual samples are critical to the process, but the entire data set will need to comprise both manual and online data sources.

Throughput is the easiest crusher parameter to measure, and a range of weightometer systems are available in the marketplace. In most instances, the choice comes down to cost, required accuracy, and physical position. Single-idler weightometers are one of the simplest options commonly used in plant process control, but longer multiple-idler systems can achieve higher levels of accuracy and are less susceptible to localized effects. Whichever system is used, it is crucial that it is maintained, calibrated, and also kept clear of excessive spillage. In crushing circuits, it is often the net crusher production that is of interest: In a closed-circuit application with a crusher and screen, it is the screen undersize that is of primary importance. Although this is the operational imperative, sufficient weightometers should be installed in the circuit to allow assessment of the oversize recirculation back to the crushers, as this can be indicative of crusher issues and/or crusher–screen mismatches. As with any equipment, initial calibration and regular verification are essential. Without this, drift in readings can provide a distorted view.

Regarding full product size distribution for operational sites, the options are either to use a sample cutter to physically take a sample from a falling stream of material or by the use of a belt-plow (although this is generally unpopular to the possibility of damage to the conveyor belt). Falling stream samplers come in a range of designs, and providing they take a representation sample, the full size distribution can be manually assessed. Increasingly, optical or automated sizing systems can be used to analyze the sample taken. The other option that is now commonplace is to use either an optical or laser-based online system to assess product size distribution without the need to take a physical sample. Systems such as WipFrag, VisioRock, and Split-Online all rely on taking photographs or video of material on conveyor belts (although they can also be applied to stockpiles or material in truck bodies). Such systems need to be calibrated for the feed material; as in all cases, the systems see the top layer of material and cannot penetrate underneath to measure the rest of the material. The ability to only see the surface of the belt charge means that a correction needs to be applied to predict the finer portion of the size distribution curve. Once calibrated and under consistent feed conditions, such systems can provide a quick and accurate assessment of size distribution.

Power consumption in crushers is simply obtained from meters logging power draw, although in many instances, the output is quoted in amperes. Obviously there are equations to relate power to amperes, but care must be applied because line voltages can differ depending on the power supply and type of motor. Power factors can also be quite variable depending on the load, and any power factor correction that is employed at the site must be understood and incorporated into the assessment.

Hydraulic pressure can be a useful parameter to measure in crushers, such as the primary gyratory and head-adjusted, top-supported cone crushers. In these machines, the hydraulic pressure measured in the system is directly related to the crushing force in the chamber. In such crushers, a pressure transducer in the hydraulic line provides a direct readout into the control system, which can be recorded and logged.

In terms of the internals of the crushing chamber, most forms of crusher have a discharge setting of some description, whereby a physical distance can be set. This setting is critical to the performance of crushers and as such must be recorded. Changes in the setting caused by adjustments for wear or for operational reasons must be logged. Regarding wear, the condition of the liners will contribute directly to the energy applied in the crushing chamber. As such, the wear should always be recorded using a metric suitable for the particular application, with this being throughput or operating hours with reference to a datum. Failure to understand the wear condition of a crusher can lead to incorrect conclusions in relation to crusher and circuit performance.

Online, real-time measurement of crusher discharge settings, or CSS, are routinely gathered, and the method varies depending on the type of crusher. Primary gyratory crushers use the vertical position of the mainshaft to estimate a discharge setting, and a similar approach is used in head-adjusted cone crushers. To obtain the setting and to account for wear, a zero point is required, which is obtained through measuring a metal-to-metal contact (without the crusher running). In bowl-adjusted cone crushers, a zero point is also required, but the CSS is then estimated via the position of the adjustment ring; that is, each tooth will equate to a change at the CSS. Therefore, knowing the number of teeth used for adjustment can provide an indication of the CSS. More sophisticated techniques have been attempted, including laser distance measurement and sensors embedded into the crusher liners, but these have been stymied by operational considerations such as requirements for line of sight and operational durability.

The role of liner profile and liner condition (the worn state) cannot be overstated. Crusher performance and therefore that of the total circuit can drift over time, and in many instances, this drift is directly related to changes in crusher performance from wear on the liners. As crusher performance degrades, control of the crusher product size distribution is lost. This can lead to excess undersize material in the product, or excessive recirculating load, reducing overall throughput. Every site will be different, and there is an inevitable trade-off between liner change-out (including associated costs and downtime) and degradation in crusher process performance. Such a trade-off needs to be carefully considered, and depending on the economics of the operation, there are many instances where liners are removed *early* (in terms of maintenance life) to maintain process performance at acceptable levels.

The role of separation, or classification, can be underestimated in the analysis of crushing circuits. The most common classification devices for crushing circuits are vibrating screens of various types. Screens dictate what material can pass to the next stage of treatment and how much must be retreated to meet the requirements. As such, screens play a crucial part in the overall circuit performance, and additionally, poor screen performance can adversely impact the process and mechanical performance of crushers.

An in-depth analysis of screens is outside the remit of this chapter (for more information, see Chapter 4.1, "Screens." The main factor in crushing circuit optimization is the efficiency of separation. Poor separation can lead to many issues, including misplacement of material, high recirculating loads, and adverse feed size distribution to crushers. The reasons for poor screening efficiency are mainly

- Ineffective screen motion,
- Feed rate,
- Feed size distribution and percentage of near-cut size material,
- Sticky material causing blinding of the screen media,
- Bed thickness,
- Selected cut size,
- Aperture shape, and
- Stiffness of screening media.

One of the most common negative effects seen in terms of crushing circuit performance relates to the situation whereby a screen is set to recirculate oversize back to a crusher. A worst-case scenario is as follows:

- The crusher product sent to the screen is not efficiently separated (because of the factors stated previously).
- Oversize from the screen is sent back to the crusher with a high content of undersize.
- Undersize is not crushed effectively and again returns to the screen as *near-sized* material.
- Near-sized material close to the aperture size further hinders separation.
- Screen efficiency drops further, and more undersize is sent back to the crusher.

This vicious circle contains both inefficient screens and consequently inefficient crushers. In tandem, the impact on the net throughput of the circuit can be damaging. Outside of the throughput issue, cone crushers also tend to suffer if the fines percentage in the feed is high. The main effect seen is an increased incidence of *packing* in the chamber, and this can also lead to mechanical failures.

Feed Properties

Feed properties have a direct and quantifiable impact on crusher performance. The main factors are

- Strength,
- Feed size distribution,
- Abrasivity and abrasive mineral content,
- Moisture content,
- Clay content, and
- Density.

Crushing Strength

Crushing strength can be measured in many ways. There are other chapters in this handbook that review the Bond work index and the JK drop weight and semiautogenous mill comminution tests.

In addition to these tests, a further range of parameters can be quoted in relation to crushing performance. The following list gives the most commonly encountered *strength* values seen in relation to crushing, with all of these tests coming from the rock mechanics discipline:

- Uniaxial compressive strength
- Brazilian tensile strength
- Young's modulus
- Poisson's ratio
- Point-load strength
- Fracture toughness
- Schmidt hammer
- Sonic (P-wave and S-wave [longitudinal and transverse]) velocities

Most of the tests listed above are index tests; in other words, they are values to allow the user to place a rock in comparison to others. Index tests are useful and can provide an idea as to the relative crushability of rock; equally, some care must be taken in this approach. As context, previous work has examined relationships with crushing and also general correlations between geotechnical parameters. On the basis of these previous studies, a generalized statement is summarized in Table 4.

The following are the most important points to consider when examining index values:

- The source of the material and the potential for the sample to contain pre-weakening cracking sustained in the sample preparation
- The mode of failure induced during the test
- The *standard* used for tests and the adherence to the standard conditions
- The variability of the test data
- Consideration of strength anisotropy because of bedding, banding, or lineation within the rock type

For crushing, the most appropriate tests are those that show the purest form of tensile failure. As tests deviate from pure tensile failure, the results start to reflect the test geometry and the conditions rather than an inherent property of the rock. In this regard, fracture toughness and indirect tensile strength (ITS) are the most tensile-based tests and, therefore, those with the strongest relationship to crusher performance.

Fracture toughness comes from the fracture mechanics domain and is designed to determine the critical stress intensity factor, that is, the stress required to cause catastrophic crack extension. There are several tests designed to deliver a pure tensile condition and to allow measurement of the critical stress intensity (K), that is, fracture toughness. For crushing, it is sufficient to test for the mode I fracture toughness, which relates to tensile failure. Other fracture toughness modes can also be tested, but these introduce mechanisms such as shear and are less appropriate for crushing.

The International Society for Rock Mechanics (ISRM) has introduced standards for a variety of fracture toughness techniques, including the chevron bend (Figure 2), short rod, and semicircular bend (Figure 3).

Details of the test methods can be found in Ulusay and Hudson (2007) and Ulusay (2014). All tests are designed to provide repeatable and consistent results, and in most instances, the choice of test will depend on the availability of the test and/or available sample size. As guidance, the

Table 4 General relationship between crushing resistance and a range of tests

Crushing Resistance	Fracture Toughness, $MN/m^{1.5}$	Brazilian Tensile Strength, MPa	Uniaxial Compressive Strength, MPa	Point-Load Is50, MPa	P-Wave Velocity, m/s	Crushing Work Index,* kW·h/t	Drop Weight, A × B
Minimal (1)	<0.25	<0.5	<5	<0.25	<500	<2	>127
Very low (2)	0.26–0.75	0.5–2.5	5–10	0.25–0.5	500–1,000	2–5	68–127
Low (3)	0.76–1.2	2.6–5	11–50	0.6–2	1,001–2,500	6–12	57–67
Moderate (4)	1.3–1.8	6–15	51–150	3–7	2,501–4,000	13–18	44–56
High (5)	1.9–2.6	16–23	151–250	8–12	4,001–5,000	19–25	39–43
Very high (6)	2.7–4.0	24–35	251–350	13–20	5,001–6,500	26–35	30–38
Extreme (7)	>4.0	>35	>350	>20	>6,500	>35	<30

Courtesy of Bear Rock Solutions Pty Ltd.

*The Schmidt hammer test is not included because of the range limit. Crushing Work Index values are only included for completeness. Crushing Work Index is known to generate highly variable results, and as such, values from these tests should not be relied on. Details and a critical review of the test can be found elsewhere in this handbook.

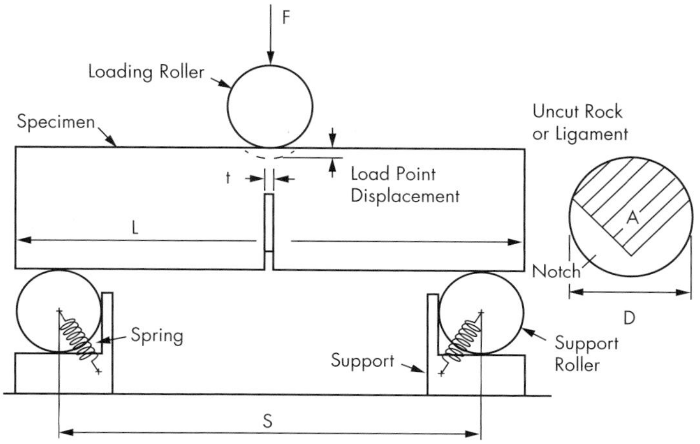

Source: Ulusay and Hudson 2007

Figure 2 Chevron bend fracture toughness

chevron bend uses sample lengths of four times the core diameter, whereas the semicircular bend uses 30-mm-thick discs cut from cores with 76-mm diameters. Fracture toughness has been cited by several authors over the last 20 years as having a close relationship to rock breakage behavior. Bearman et al. (1991) and Briggs et al. (1997) reported links to crushing and comminution, with the 1997 study showing a strong link to energy mapping approaches such as the JK drop weight test. As such, its application is beyond that of a mere index, and it should be considered as a parameter than can be used for quantifying crushing strength and machine performance.

ITS, also known as the Brazilian tensile strength test (Figure 4), was introduced because of the difficulty in undertaking a direct tensile test on rock materials. In the ITS method, there is no attempt to physically pull the sample apart; rather a disc of material is compressed to establish a stress regime that leads to a tensile failure being induced.

The ITS requires little sample preparation and limited core (sample discs need to be a minimum of 54-mm diameter but only require a disc thickness of $\geq 0.5 \times$ diameter), so in this regard, the test has minimal requirements. The test is well established with high repeatability and typically low variability in results. Another advantage is that should enough sample be available, ITS samples can be tested in a variety of directions to determine the extent of any anisotropy in the samples. As a measure of *strength*, providing that the minimum requirements of the standards are followed, ITS shows less variability than the uniaxial compressive strength (UCS) and point-load strength tests, and it also displays improved repeatability. In addition, the ability to obtain many more ITS test discs from a given core length, compared to UCS, provides improved representivity. Given the various benefits of ITS, the use of this test should be preferred over UCS for geotechnical indexing tests for crushing.

UCS and point-load testing (PLT) are often discussed together, as the point-load test was originally designed to be a rapid technique for determining UCS. UCS is a well-known test that relies on the compression of a cylinder of rock core, with the key dimensions being the diameter of the core and the length-to-diameter ratio.

The core is compressed axially (Figure 5), and the force at failure is recorded. The peak force divided by the cross-sectional area of the core gives the UCS for the sample. During the test, it is also possible to measure the Young's modulus of the rock by recording the displacement achieved during the test. Young's modulus is then calculated as the gradient of the stress-versus-strain graph from the test. There are a range of defined calculations for various forms of Young's

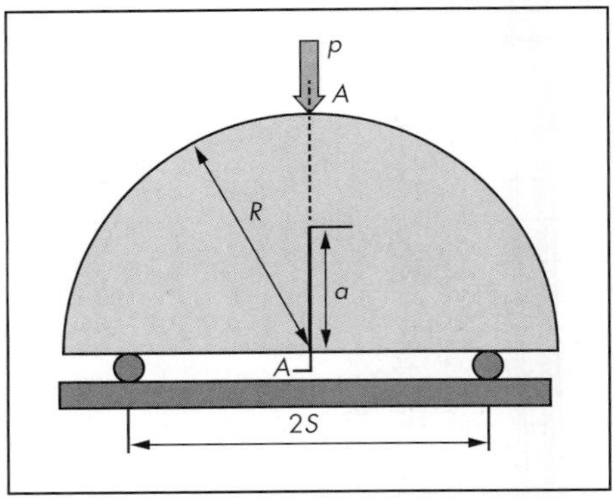

Source: Ulusay 2014

Schematic

Courtesy of Bear Rock Solutions Pty Ltd.

Test Piece

Figure 3 Semicircular bend fracture toughness test

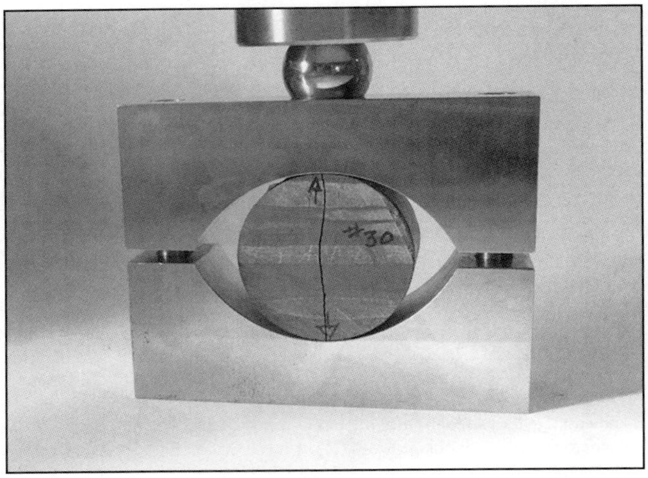

Courtesy of Bear Rock Solutions Pty Ltd.

Figure 4 Brazilian tensile strength test

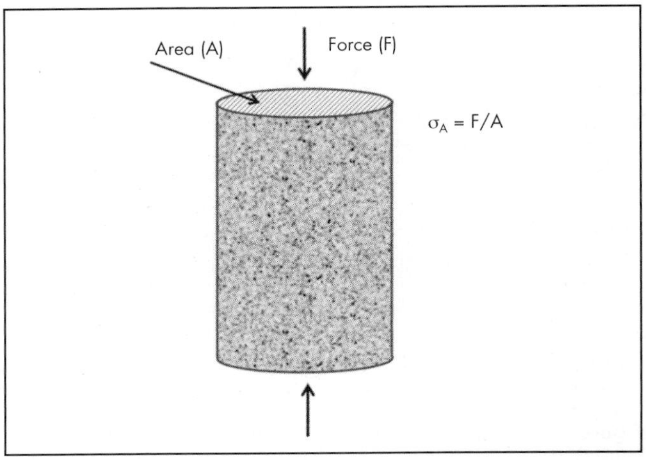

Adapted from Ulusay and Hudson 2007

Figure 5 Geometry for uniaxial compressive strength test

modulus, and a discussion of these can be found in the ISRM publication by Ulusay and Hudson (2007). In an addition to the standard test, strain gauges can be attached to the rock sample to measure axial and lateral strain. Using this data, the Poisson ratio for the rock can be determined. Although sometimes quoted, the use of Poisson's ratio does not appear to be strongly correlated with crushing performance.

One of the drawbacks of the UCS test is the variability seen in the results, which is through samples fracturing under different modes of failure. The tests consume a large amount of core (length-to-diameter ratios of 2.5:1 as required by the ISRM and 2:1 to 2.5:1 as required by ASTM D7012-14e1), and if recommendations for load application rate and the end preparation of the cores are not closely followed, variability can become extreme. The consequence is that useful core is consumed, without any quality data being gained.

PLT was developed as a rapid and portable means for obtaining an estimate of UCS. The ability to undertake PLT in the field is a valuable feature, and it can provide information on general trends. As mentioned previously, variability can be an issue with UCS tests, and this is also observed in PLT. Depending on the test piece geometry, the variability can be higher than in UCS tests. PLT relies on the use of two diametrically opposed platens truncated by 5-mm-diameter points (Figure 6).

The rock sample is placed between the pointed platens, and a load is usually applied with a small portable pump, often of a manual, hand-operated type. The force at failure is used in a standard equation, and the force is then corrected to an equivalent for a 50-mm-diameter core. One advantage of PLT is that the test pieces can be of differing geometries, that is, diametric core, axial core, or irregular lumps. Although this is an advantage for the ability to gain a PLT value, the issue of variability again comes to the fore, with irregular geometry introducing extra variance. Another key parameter to consider regarding variability of results is the maintenance of the pointed platens. With use, the spherical truncations will wear, and this wear impacts the profile of the tips. As the spherical truncation is lost because of wear, the results from PLT will deviate from those expected.

In summary, both UCS and PLT are useful indexing tests when used to examine trends, but because of issues around variability, some care must be exercised in the use of the results.

All the preceding tests are destructive; that is, the core material is fragmented at the end of the tests and is therefore

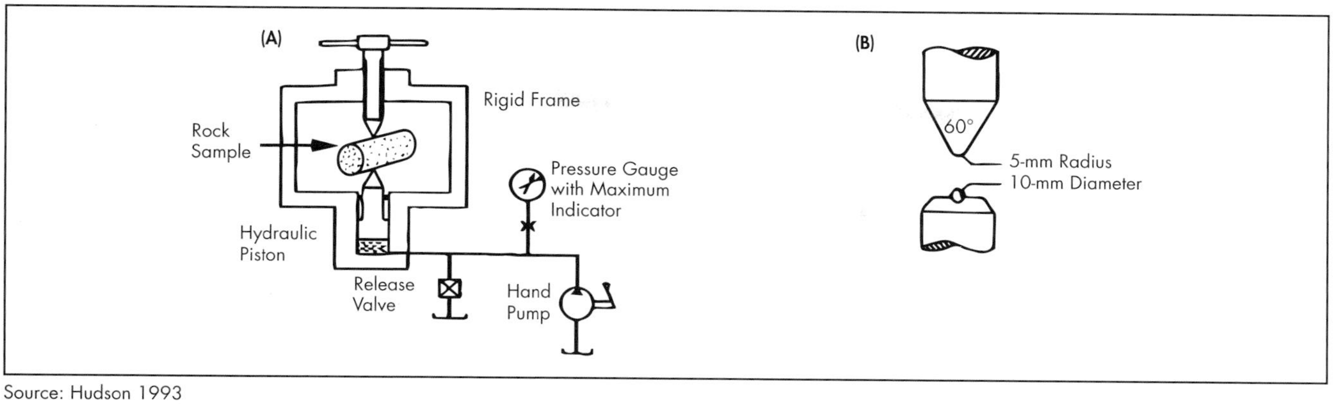

Source: Hudson 1993

Figure 6 Point-load strength test

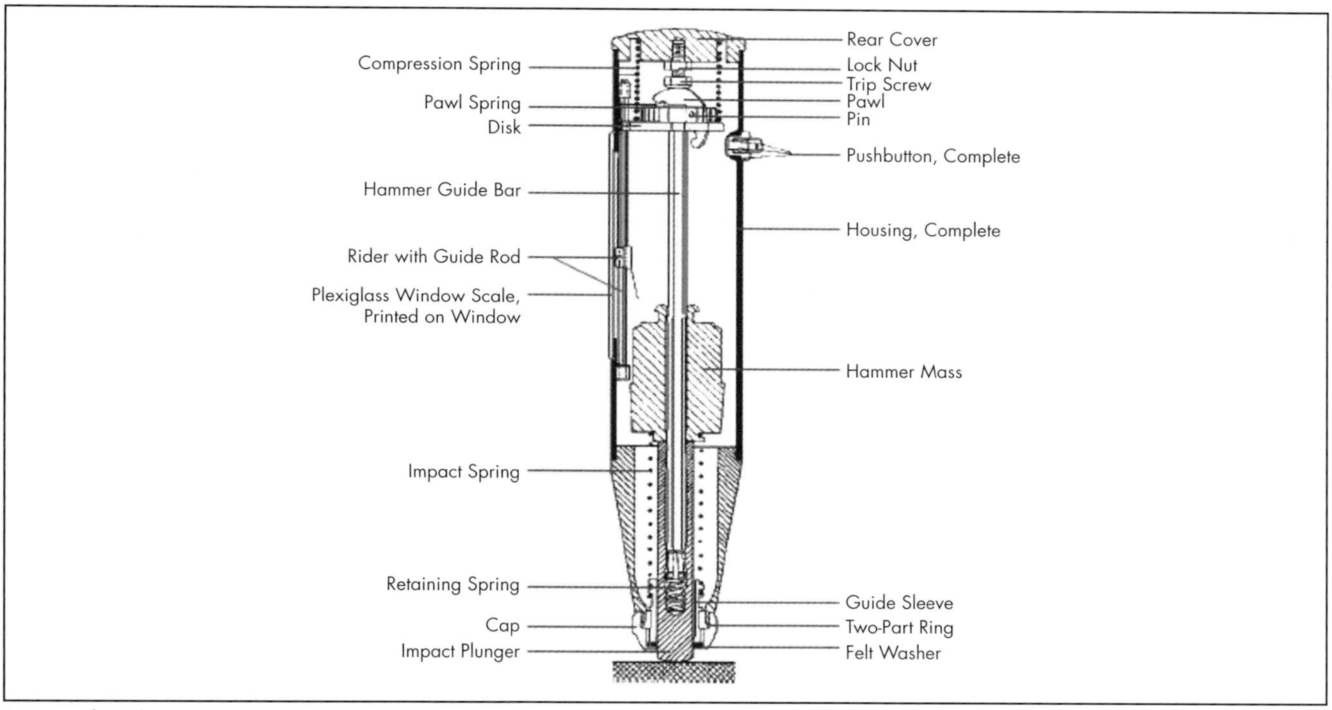

Courtesy of Gardco

Figure 7 Schmidt hammer

not available for any subsequent testing. The geotechnical tests listed at the beginning of this section show two tests that are nondestructive in nature. One is the Schmidt rebound hammer and the other is the sonic velocity.

The Schmidt hammer test (Figure 7) was developed to provide a quick assessment of the strength of concrete blocks. The test has since been deployed as a measure of rock strength and has found acceptance as the quickest and easiest strength assessment technique. The test is covered by the ISRM standard (Ulusay 2014), and the following is a summary of the key points.

The test is based on a handheld unit that has a spring-loaded piston mechanism that is released when the plunger is pressed against the rock surface. The impact of the piston onto the plunger transfers energy to the rock. The rebound value is measured as the amount of energy recovered, which in turn relies on the hardness of the rock. The Schmidt hammer is available in two versions: the L-type and the N-type. The L-type has a lower impact energy and improved sensitivity in the lower ranges, making it better suited to weaker rock types, whereas the N-type has a higher energy and better deals with irregular surfaces. Various workers have provided equations to relate the rebound number to UCS. As a rapid and portable indexing tool, the Schmidt hammer is useful for identifying trends. The main factors to consider are the calibration of the hammer, the flatness of the surface, significant strength differences between the grains or crystals and the matrix, weathering and moisture content plus the presence of anisotropy, and the confinement of the rock under test. Any spring-loaded device will eventually require maintenance to address loss of elasticity from extended periods of use, and hence any unit used must be regularly checked. In addition, the point on the rock surface where the unit is applied must be flat, otherwise there may be some localized crushing or slippage. The final point relates to the stability of the sample under test. If used on a large stable rock, then this is not a concern, but where the unit is applied to a block of rock or a core, then it is essential that the sample is firmly secured to prevent movement and hence an invalid reading. As a general rule, the range of application for the Schmidt hammer is mostly accepted to be in the 20–150 MPa UCS range. Outside of this range, sensitivity or data scatter issues may be encountered.

The other main, nondestructive test available is the sonic velocity test. In this test, the aim is to measure the P- (primary, longitudinal) and S- (secondary, traverse) wave velocities along the length of the sample. The measurement is achieved by measuring the transit time of the waves and dividing by the length of the sample. ISRM (Ulusay 2014) provides full details of the test procedures required. In a laboratory setting, the sample is mostly intact core, but blocks and discs can be used, providing the dimensions meet the requirements of the standard. Using the data generated and the density of the sample, it is also possible to calculate elastic properties, including dynamic Young's modulus and Poisson's ratio. P-wave velocity is most often used to provide a measure of the rock integrity and competence. In general terms, there is a trend between P-wave velocity and strength tests.

Table 5 Typical Bond abrasion index values

Sample Type	Bond Abrasion Index	Classification
Bauxite	0.0005–0.02	Nonabrasive
Pisolitic iron ore	0.005–0.03	Nonabrasive
Dolomite	0.01–0.05	Nonabrasive
Magnetite	0.1–0.3	Slightly abrasive–abrasive
Marra Mamba iron ore	0.2–0.3	Slightly abrasive–abrasive
Basalt	0.2–0.4	Slightly abrasive–abrasive
Diabase	0.2–0.4	Slightly abrasive–abrasive
Gabbro	0.4	Slightly abrasive–abrasive
Amphibolite	0.2–0.45	Slightly abrasive–highly abrasive
Copper ore	0.3 –0.45	Abrasive–highly abrasive
Andesite	0.4–0.5	Highly abrasive
Graywacke	0.3–0.6	Highly abrasive
Lamproite	0.35–0.6	Highly abrasive
Gneiss	0.4–0.6	Highly abrasive
Granite	0.45–0.65	Highly abrasive–extremely abrasive
Hornfels	0.4–0.7	Highly abrasive–extremely abrasive
High-grade hematite	0.8–0.8	Abrasive–extremely abrasive
Diorite	0.4–0.8	Highly abrasive–extremely abrasive
Quartzite	0.7–0.9	Highly abrasive–extremely abrasive

Ore Abrasivity

The abrasivity of the feed material is the ability of the feed to wear the components of the crusher. Many types of wear test exist, and as they relate to milling, there are other mentions in this handbook. For crushers, the ones of interest are those that examine loss of metal because of a certain form of wear. The most widely known test is the Bond abrasion index, where a prescribed charge of feed of a specified fraction is placed in a rotating drum. A paddle within the drum rotates through the mass. The mass of metal lost from the paddle after a given time is then stated as the abrasion index. This test was introduced in the late 1940s and is still in widespread use. This test predominantly examines flow abrasion, that is, low normal force wear caused by differential speed and abrasion. Table 5 gives a summary of Bond abrasion data.

In crushers, such wear occurs, but not at the point where the rocks are nipped and broken. At these points, the wear is enhanced because of the high normal force gouging abrasion. Tests for this type of wear are less well known, and they generally lack an inherent link to actual wear rates in crushers; that is, results are indicative of the magnitude of the wear rates, not explicit to a direct metal loss. The two tests of interest in this area are the gouging abrasion and the Laboratoire du Center d' Études et Recherches des Charbonnages de France (CERCHAR) tests. The gouging abrasion test is a recent development (Figure 8) and relies on sweeping a test piece of the metallic test material across the face of a prepared rock sample.

The metallic test piece is mounted on a sturdy pendulum device, usually a modified Charpy or Izod test machine, and the rock sample is located at the base of the pendulum arc. The pendulum and metallic test piece therefore swings from a given height and gouges a path along the surface of the rock sample. The index is calculated from the dimension of the wear-flat surface on the end of the metallic test piece. In this regard, the test replicates a high normal force plowing or gouging wear regime. As a guide, Table 6 provides measured values for the gouging abrasion index.

The CERCHAR test is a more established test and has been standardized by ISRM (Ulusay 2014). Figure 9 shows the equipment.

In this test, a weighted metallic needle is moved across the flat surface of a prepared rock sample. The CERCHAR abrasion index is calculated from the diameter of the wear-flat surface of the needle. In this instance, the weighted needle does not exert a high force onto the rock surface, so in essence, the wear regime is a low normal force, abrasion-type test. There is little or no gouging involved. Table 7 provides a guide to the classification of wear from the CERCHAR test.

Feed Conditions

Crushers are particularly susceptible to feed conditions, and these play a major part in the optimization of crusher performance. The main factors relating to feed include

- Feed size distribution,
- Feed rate,
- Distribution of feed around the chamber,
- Sized-based segregation of feed, and
- Presence of clay and/or moisture.

The impact of the preceding factors will vary depending on the type of crusher. One often-cited issue is that of choke feeding in cone crushers. For more details on this and related issues, see Chapter 3.4, "Gyratory and Cone Crushers."

Process control of crushers needs to take into account many of the factors examined in the previous section. Because of the lack of crusher variables that can be rapidly adjusted, most control has focused on keeping the crusher running at

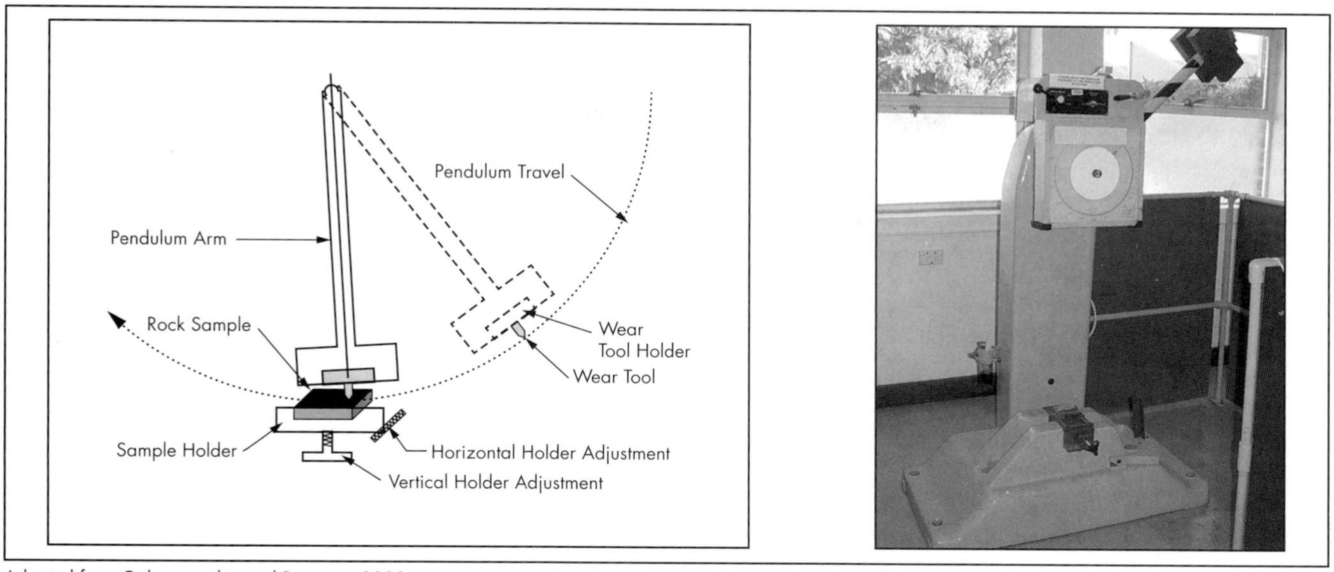

Adapted from Golovanevskiy and Bearman 2008

Photo © Richard Bearman

Figure 8 Gouging abrasion index equipment

Table 6 Typical gouging abrasion index results

Sample Type	Gouging Abrasion Index	Classification
Basalt (weathered)	0.50–1.00	Nonabrasive
Bauxite	2.90–6.90	Nonabrasive–slightly abrasive
Pisolitic iron ore	4.50–9.00	Nonabrasive–moderately abrasive
Basalt	8.30–18.80	Moderately abrasive–highly abrasive
Breccia	16.90–18.00	Highly abrasive
Copper porphyry	16.00–18.00	Highly abrasive
Banded iron formation	16.10–19.80	Highly abrasive
Quartzite	18.20–22.20	Extremely abrasive

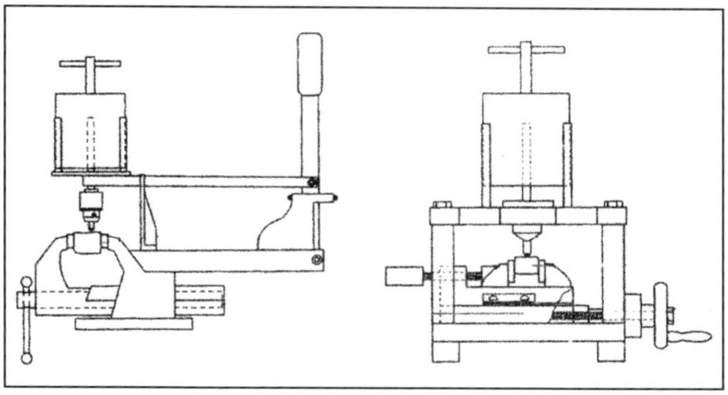

Source: Ulusay 2014

Figure 9 CERCHAR test equipment

Table 7 Typical CERCHAR results

CERCHAR Abrasion Index	Classification
0.1–0.4	Extremely low
0.5–0.9	Very low
1.0–1.9	Low
2.0–2.9	Medium
3.0–3.9	High
4.0–4.9	Very high
>5.0	Extremely high

maximum power to deliver full size reduction, or to keep the crusher at a given discharge setting to generate a specified size.

All manufacturers provide process control systems that can target power or CSS, and within these targets, there is also a compensation for liner wear. Often tied to these systems is instrumentation to detect the level of feed both in the feed bin and in the crushing chamber and feedback loops to control the speed of the crusher feeder.

As part of the process control system for most crushers, power, hydraulic pressure (where applicable), motor temperature, lubricating oil flow rate, lubricating oil in and out temperature, countershaft speed, rotor speed, discharge setting, mainshaft position (where applicable), and feed level and rate are all monitored. To control the crusher, the only control variables readily available are discharge setting, rotor speed (impact crushers), and feed rate. Having such a limited set of control variables constrains the speed and flexibility of control, particularly when it must be remembered that in the case of cone crushers, the discharge setting is usually altered by stopping feed to the crusher.

As mentioned, most crusher suppliers provide control systems to varying levels of sophistication. One of the systems with the longest pedigree is the automated setting regulation (ASR) package for cone crushers provided by Sandvik, with the original system dating back to the Svedala company. In the current configuration, the system is now known as ASRi, where "i" denotes the move to what Sandvik terms an *intelligent* version. In essence, the ASRi system has three operating modes:

1. Auto-CSS—the ASRi system aims to maintain the desired CSS.

2. Multi-CSS—two different product curves can be combined to give a new desired product.
3. Auto-load—the ASRi system regulates the setting so that the crusher operates at a desired load level.

In the Sandvik head-adjusted cone crushers, the system monitors a range of operational parameters including power, hydraulic shaft adjustment pressure, and CSS. Depending on the mode selected, the system will control CSS (auto-CSS and multi-CSS) or it will target a power limit (auto-load). The system is designed to have a user-friendly front-end display and to communicate with wider plant control systems.

The display seen in Figure 10 shows one of the main trends in control interfaces in process plants, with the move to a clean, visual display with a *dashboard* approach, whereby only the important information is shown, which is graphically represented in a form that can be quickly interpreted by the operator.

It has long been known that the eccentric speed in cone crushers has the potential to be used for process control, but because of the cost and durability of older variable-speed drives, there was a reluctance to study this aspect further. In recent years, work at Chalmers University and Metso has examined the use of modern variable-speed drives and concluded that the use of speed can help extend liner life and the generation of targeted product sizes.

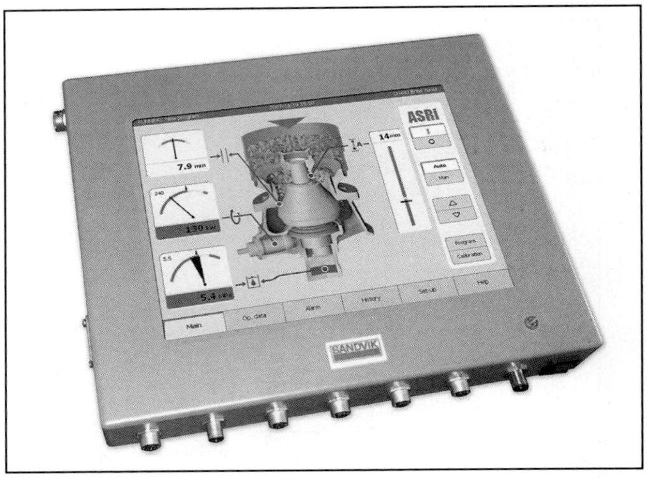

Courtesy of Sandvik

Figure 10 Interface for the ASRi control system

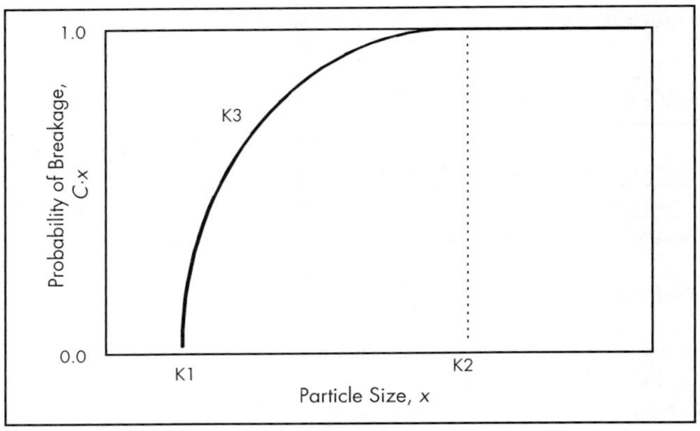

Source: Whiten 1984

Figure 11 Classification function for the Whiten crusher model

Modeling and Simulation

Modeling and simulation of crushers and crushing circuits can be undertaken at a variety of levels:

- Basic uses hard-wired or look-up tables to estimate throughput and size reduction.
- Intermediate is classification-function based and uses mechanistic models.
- Numerical is based on discrete element modeling (DEM) to model the action of individual particles.

At the basic level, there is a range of software options available that provide good representations of crushing and screening circuits and will ensure that the flow within the circuits converge and give a mass balance. These packages tend to use either manufacturer data or simple models to provide the user with estimates of unit and circuit performance.

Intermediate modeling uses sophisticated flow-sheet simulation packages, such as JKSimMet and SysCAD, and can use various types of crusher models:

- Classification function models based on fitted or empirical relationships for crusher parameters
- Mechanistic crusher models based on direct mechanical crusher design parameters

Both of the preceding approaches use detailed experimental descriptions of breakage, that is, appearance function, to describe the breakage in the crusher.

One such example of the classification function approach is the well-documented Whiten crusher model that uses classification parameters K1, K2, and K3 to describe the workings of the crusher and the JK drop weight breakage function to describe the response of the rock. The model framework can be applied to a wide range of crusher types. The K1 and K2 parameters are not inherent crusher parameters, but rather they can be fitted via relationships with operational factors such as F80 or throughput. Models such as the Whiten model are well regarded and often used where performance data for the crushers are available. If a particle is smaller than a certain size (K1), it will fall through the crusher and become product. If it is larger than a certain size (K2), it cannot fall through the crusher but will fall through until it is caught. A typical classification curve is shown in Figure 11.

In the Whiten model, a power curve with exponent K3 is used to join the two defined points. For many types of crusher, $2 < K3 < 2.5$ is a good description.

In the case of a cone crusher, K1 represents the CSS, and K2 the open side setting (OSS). This assignment of K1 and K2 to CSS and OSS is, however, not definitive, and in many instances, the K parameters need to be fitted.

To model the breakage process, the Whiten model uses a set of curves (cubic splines) to describe the size distribution produced by breakage events of increasing size reduction or energy input. This defines the breakage function for the feed material.

The classification and breakage functions are combined to allow the model to classify the output of the breakage function and to ensure that the mass balance is correct. The function of the model is shown in Figure 12.

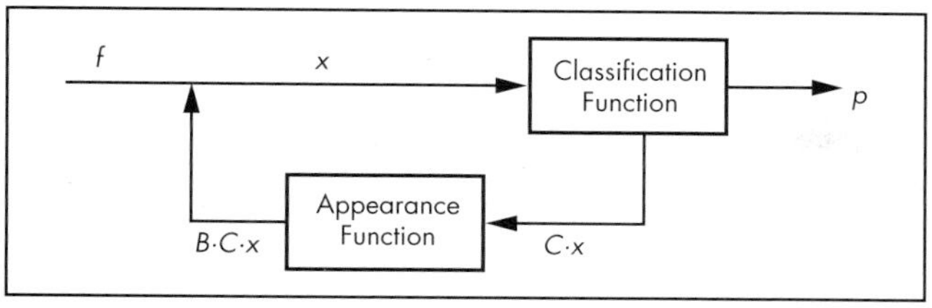

Source: Whiten 1984

Figure 12 Schematic of Whiten crusher model

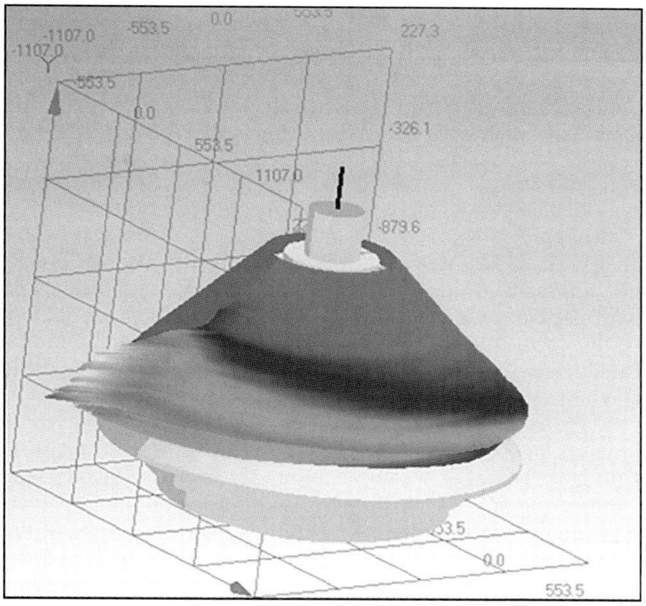

Courtesy of Bear Rock Solutions Pty Ltd. and Met Dynamics Pty Ltd.

Figure 13 Three-dimensional plot of power consumed in cone crushing

Mass balance equations may be written about each node as

$$x = f + B \cdot C \cdot x \qquad x = p + C \cdot x$$

where

x, f, and p = vectors representing the amount in each size fraction in the crusher feed and product, respectively

B = breakage distribution function, a lower triangular matrix giving the relative distribution of each size fraction after breakage

C = classification function, a diagonal matrix describing the proportion of particles in each size interval entering the crushing zone

Combining the mass balance equations gives

$$p = (I - C) \cdot (I - B \cdot C) - 1 \cdot f$$

Therefore, from B and C, it is possible to calculate the product size distribution (p) for any feed size distribution (f).

Mechanistic models take the ideas from the intermediate approach and increase the functionality by removing the need for fitting of parameters. In crushing, mechanistic models have been developed for a variety of machines including secondary sizers, VSIs, primary gyratory crushers, and cone crushers, with the latter two attracting the most attention. In the mechanistic approach, the geometry and mechanics of the crushers are used to generate the energy application within the crusher, and then, this is coupled to a representation of breakage. In some cases, the breakage functions can be the same as those used in the intermediate approach. The advantages offered by the mechanistic approach are as follows:

- They can run as stand-alone models or within a flowsheet simulation.
- Operational data is not required to set up the model.
- They incorporate actual components and dimensions specific to the crusher.
- They track size classes of particles through the crushers and develop energy profiles that can be used for prediction of power, crushing force, liner wear, size reduction, and throughput.
- They can be used to model wear of liners and therefore can be used to track changes in performance with wear.

Mechanistic crusher models are well covered in work from Bearman et al. (2011) for a kinematic crusher model for primary gyratory and cone crushers, Evertsson (2000), Hulthén (2010) for cone crushers, Nikolov (2002) for impact crushers, and Heng et al. (2003) for sizers.

A feature of all mechanistic models is that they couple the crusher chamber geometry with equations of particle motion and a population balance breakage model so that they are not crusher specific.

Using the preceding approach, the models are able to predict a range of critical parameters, such as the following for cone crushers:

- Product size distribution
- Power consumption
- Energy distribution in the chamber
- Hydraulic pressure
- Onset of packing and power overload

The models generate a variety of key process and mechanical diagnostic parameters (Figure 13) for use in the analysis. In combination, these provide a unique view of the crusher application.

Unlike the basic and intermediate approaches detailed previously, numerically based techniques are not currently suited to use in a circuit simulation setting because of the computational requirements. However, the application of numerical techniques to crushers has increased dramatically because of the rapid increase in computing power. In these models, each particle is represented by either an individual element or a group of elements within the computing code. All collisions, interactions, and movements are based on physics models, and the crusher itself is defined via engineering drawings. Various DEM codes have been used to examine crushers of different types, including gyratory, cone, sizers, double-roll (Figure 14), and impact crushers (Figure 15).

The main drawbacks have been computational power and speed and also the ability to reach small particle sizes. A thorough review of DEM in comminution is provided by

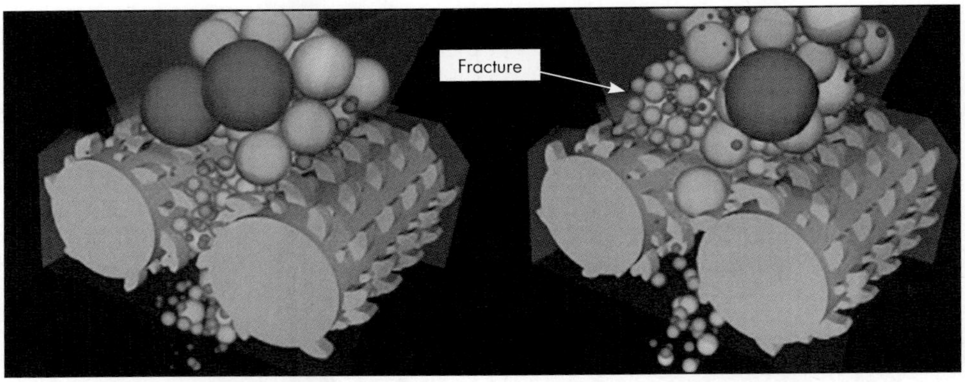

Source: Cleary and Sinnott 2015
Figure 14 Double-roll crushers

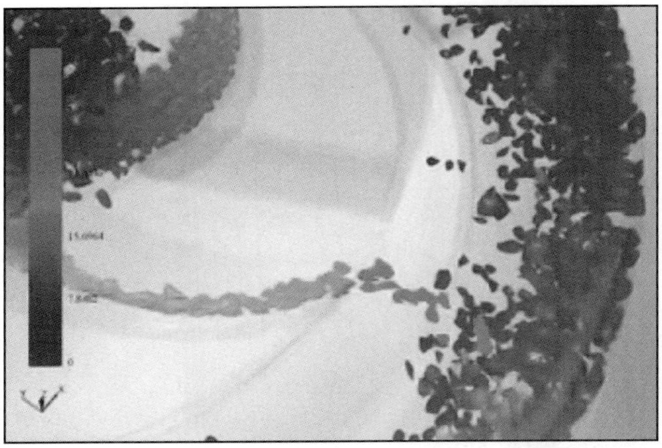

Source: Chalmers Rock Processing Systems 2015
Figure 15 DEM simulation of a VSI crusher (plan view)

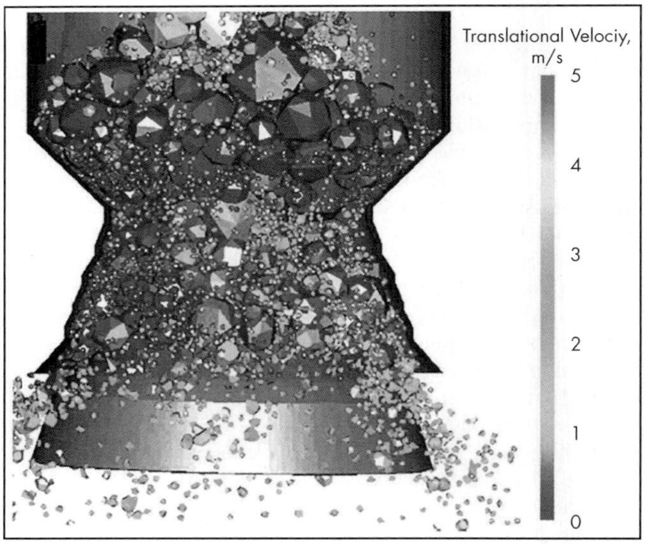

Source: Popatov et al. 2007
Figure 16 Fast breakage model of an HP100 cone crusher

Weerasekara et al. (2013), and in this, the methods applied and the steps taken to overcome issues are well described.

In an effort to deal with some of the main issues associated with DEM, Metso developed its proprietary fast breakage approach (Potapov et al. 2007), which combines DEM with population balance modeling. This allows the modeling of finer size fractions, without the need to specifically model each grain as a discrete particle. An example of the output is shown in Figure 16.

The power of a system such as DEM, or a derivative, is obvious, as it deals with all aspects of the crusher simulation. Currently, however, the application is restricted to stand-alone engineering studies. With increases in computational power, this situation will undoubtedly change over the course of the coming years, and this will allow wider application and the eventual use within a full flow-sheet simulation environment.

REFERENCES

ASTM D7012. 2014. *Standard Test Methods for Compressive Strength and Elastic Moduli of Intact Rock Core Specimens Under Varying States of Stress and Temperatures*. West Conshohocken, PA: ASTM International.

Bearman, R.A., Barley, R.W., and Hitchcock, A. 1991. The prediction of power consumption and product size in cone crushing. *Miner. Eng.* 4(12):1243–1256.

Bearman, R.A., Munro, S., and Evertsson, C.M. 2011. Crushers: An essential part of energy efficient comminution circuits. In *MetPlant 2011: Metallurgical Plant Design and Operating Strategies*. Melbourne, Victoria: Australasian Institute of Mining and Metallurgy. pp. 66–85.

Boyd, K. 2002. Crushing plant design and layout considerations. In *Mineral Processing Plant Design, Practice, and Control*. Vol. 2. Edited by A. L. Mular, D.N. Halbe, and D.J. Barrett. Littleton, CO: SME. pp. 669–697.

Briggs, C.A., Bearman, R.A., and Kojovic, T. 1997. The application of rock mechanics parameters to the prediction of comminution behaviour. *Miner. Eng.* 10(3):255–264.

Carson, J.W., and Holmes, T., 2002. The selection and sizing of bins, hopper outlets and feeders. In *Mineral Processing Plant Design, Practice, and Control.* Vol. 2. Edited by A.L. Mular, D.N. Halbe, and D.J. Barrett. Littleton, CO: SME. pp. 1478–1490.

Carson, J.W., and Petro, G. 1998. How to design efficient and reliable feeders for bulk solids. Westford, MA: Jenike and Johanson. www.web-tech.com.au/wp-content/uploads/Design-efficient-feeders.pdf.

Chalmers Rock Processing Systems. 2015. DEM simulation of theoretical vertical shaft impact crusher (VSI). www.youtube.com/watch?v=LEy854_bh_M.

Cleary, P.W., and Sinnott, M.D. 2015. Simulation of particle flows and breakage in crushers using DEM: Part 1—Compression crushers. *Miner. Eng.* 74(April):178–197.

Evertsson, C.M. 2000. Cone crusher performance. Ph.D. thesis, Chalmers University of Technology, Sweden.

Golovanevskiy, V., and Bearman, R.A. 2008. Gouging abrasion test for rock abrasiveness testing. *Int. J. Miner. Process.* 85(4):111–120.

Heng, R., Cheng, K.D., Tuppurainen, D., Bearman, R.A., and Oswald, S. 2003. Mechanistic modelling of mineral sizers. *Miner. Eng.* 16(9):807–813.

Huband, S., Tuppurainen, D., While, L., Barone, L., Hingston, P., and Bearman, R.A. 2006. Maximising overall value in plant design. *Miner. Eng.* 19(15):1470–1478.

Hudson, J.A. 1993. Rock properties, testing methods and site characterisation. In *Comprehensive Rock Engineering: Principles, Practice and Projects.* Edited by J.A. Hudson, E.T. Brown, C. Fairhurst, and E. Hoek. Oxford, UK: Pergamon Press. pp. 46–51.

Hulthén, E. 2010. Real-time optimization of cone crushers. Ph.D. thesis, Chalmers University of Technology, Sweden.

Napier-Munn, J.J. 1996. *Mineral Comminution Circuits: Their Operation and Optimisation.* Indooroopilly, Queensland: Julius Kruttschnitt Mineral Research Centre, University of Queensland.

Nikolov, S. 2002. A performance model for impact crushers. *Miner. Eng.* 15(10):715–721.

Potapov, A.V., Herbst, J.H., Song, M., and Pate, W.T. 2007. A DEM-PBM fast breakage model for simulation of solid fracture of comminution processes. In *Proceedings of the Minerals Engineering International (MEI) Discrete Element Methods (DEM) 07*, Brisbane, Queensland.

Svensson, A., and Steer, J.F. 1990. New cone crusher technology and developments in comminution circuits. *Miner. Eng.* 3(1-2):83–103.

Ulusay, R. 2014. *The ISRM Suggested Methods for Rock Characterization, Testing and Monitoring: 2007–2014.* Cham, Switzerland: Springer-Verlag.

Ulusay, R., and Hudson, J.A. 2007. *The Complete ISRM Suggested Methods for Rock Characterization, Testing and Monitoring: 1974–2006.* Ankara, Turkey: International Society of Rock Mechanics.

Utley, R.W. 2002. Selection and sizing of primary crushers. In *Mineral Processing Plant Design, Practice, and Control.* Vol. 1. Edited by A.L. Mular, D.N. Halbe, and D.J. Barrett. Littleton, CO: SME. pp. 584–603.

Weerasekara, N.S., Powell, M.S., Cleary, P.W., Tavares, L.M., Evertsson, C.M., Morrison, R.D., Quist, J., and Carvalho, R.M. 2013. The contribution of DEM to the science of comminution. *Powder Technol.* 248:3–24.

Whiten, W.J. 1984. Models and control techniques for crushing plants, control 84. *Min. Metall. Process.* (February):217–225.

High-Pressure Grinding Roll Technology

Michael Daniel, Bern Klein, and Chengtie Wang

The high-pressure grinding roll (HPGR) technology concept originates from earlier roller press technology that has been widely used in the briquetting of fine materials into lumps. This is shown in the roller presses that were manufactured by Köppern as early as 1918. Existing designs of roller presses have since been modified to employ the concept of so-called interparticle crushing under continuous operating conditions; subjecting a bed of particles to a pressure greater than 50 MPa for comminution originated in the 1979 international patent of Klaus Schönert in Germany (Schönert 1982, 1988, 1991).

Schönert's general remarks in his fundamental study was that in any comminution process, the particles are broken by contact forces, which deform the particle and cause a stress field. As the stress level meets the criteria either of yielding or fracturing, then the particle will be deformed inelastically or broken, respectively. The number of contact forces depends on the mode of stressing being either in single-particle mode or multiple particles (Schönert 1988). Schönert states that comminution devices, such as crushers and grinding mills, all stress the material by compression and shear. Both single-particle and bed-particle stressing experiments were conducted as part of his fundamental research, and Schönert concluded that interparticle bed breakage has a lower efficiency than single-particle stressing. Schönert states that the efficiency may drop by as much as a factor of two to three depending on the conditions relating to the number of contact forces (Schönert 1991).

The manufacturers Weir (KHD), ThyssenKrupp (Polysius), and Köppern obtained the rights to Schönert's patent and further developed the idea based on their technology of briquetting machines. This eventually resulted in the modern-day version of the HPGR. Since the introduction of the first commercial HPGR units in 1985, HPGR developments have been mainly driven by the cement industry and the associated benefits that have been gained in reducing comminution energy requirements. The European cement industry widely uses HPGR, where it is accepted as a mature technology. The diamond industry has adopted the technology mainly because of better liberation of particulate high-value diamonds (Knecht 2000). HPGR installations were first introduced to diamond operations in the late 1980s, the iron ore industry in the mid-1990s, and in hard rock applications in the mid-2000s. The Cerro Verde in Peru was the first large-scale hard rock copper mine to install HPGRs in 2006. Since then, the number of installations in each type of operation has sharply increased (Figure 1; Table 1).

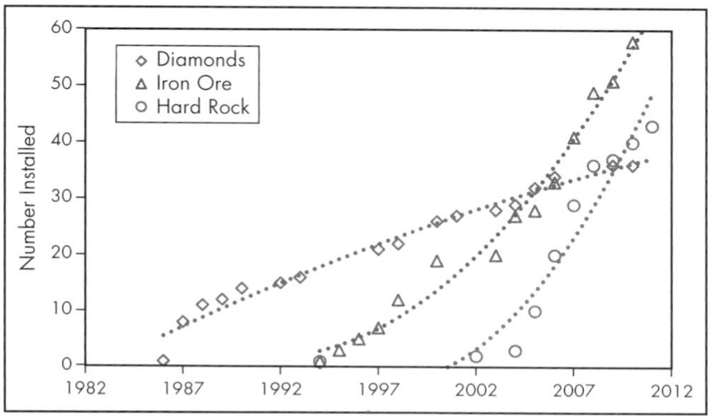

Source: Burchardt 2012

Figure 1 HPGR growth in diamond, iron ore, and hard rock mining

The main potential benefit is reported to be related energy savings resulting in lower overall operating costs. This benefit increases with ore hardness and higher power costs. Another observed benefit is that HPGR-circuit throughputs are relatively insensitive to variations in ore hardness when compared to semiautogenous grinding (SAG) circuits. Additional reported benefits include microfracture formation that lowers the ball mill work index and enhances liberation by promoting intergranular breakage to improve downstream flotation and leaching.

There are number of limitations and perceived risks with the HPGR circuits. Not all ores are amenable to the HPGR

Michael Daniel, Consulting Metallurgist & Director, CMD Consulting Pty Ltd., Brisbane, Queensland, Australia
Bern Klein, Professor, Norman B. Keevil Institute of Mining Engineering, University of British Columbia, Vancouver, British Columbia, Canada
Chengtie Wang, PhD Candidate, University of British Columbia, Vancouver, British Columbia, Canada

Table 1 HPGR installations for precious and base metal operations

Project	Company	Location	Number of HPGRs	Throughput, kt/d	Ore Type	Year	Reference
Toquepala	Southern Peru Copper Corporation	Peru	2	~60	Copper porphyry	2017	—
Cerro Verde 2	Freeport-McMoRan	Peru	8	~240	Copper porphyry	2016	Vanderbeek and Gunson 2015
Sierra Gorda	KGHM/Sumitomo	Chile	4	~110	Copper-molybdenum	2014	López 2011
Morenci	Freeport-McMoRan	USA	1	>63.5	Copper porphyry	2014	Herman et al. 2015; Knorr et al. 2015; Mular et al. 2015
Tropicana	AngloGold Ashanti	Australia	1	~15	Gold	2013	Gardula et al. 2015; Kock et al. 2015
Cuajone	Southern Peru Copper Corporation	Peru	1	~90 (quaternary)	Copper porphyry	2013	—
Salobo	Vale	Brazil	2	~33	Copper, gold	2012	—
Cadia Hill	Newcrest	Australia	1	~55 (HPGR-semiautogenous grinding hybrid)	Copper, gold	2012	Engelhardt et al. 2015
Peñasquito	Goldcorp	Mexico	1	~+100 (pebble crusher circuit)	Polymetallic	2010	—
Boddington	Newmont	Australia	4	~100	Gold, copper	2009	Hart et al. 2011; Tavani et al. 2015
Mogalakwena	Anglo Platinum	South Africa	1	~25	Platinum	2008	Rule et al. 2008, 2015
Grasberg	Freeport-McMoRan	Indonesia	2	~70 (quaternary)	Copper, gold	2007	—
Cerro Verde 1	Freeport-McMoRan	Peru	4	~108	Copper porphyry	2006	Koski et al. 2011; Vanderbeek et al. 2006

because of the elastic properties of the rock, resulting in energy absorption without fracture. Ores with significant clay content do not respond well, although approaches have been developed to address this issue (Rosario and Drozdiak 2015). Material-handling systems are relatively complex, comprising conveyors and storage bins to feed closed-circuit screens at each stage of comminution. In wet climates and cold climates, variations in moisture levels and potential freezing can cause additional operational issues.

Manufacturers of HPGRs and design engineers have published many papers at various international conferences that provide more information on the application of the technology. The literature generally describes the potential benefits of the HPGR in energy savings and/or high throughput or potential benefits to downstream processes (Battersby et al. 1993; Knecht 1994; Kellerwessel and Oberheuser 1995; Gunter et al. 1996; Dunne et al. 1996; Patzelt et al. 1997, 2000, 2001; Stephenson 1997; Rosario 2010; Rosario et al. 2011; Knorr et al. 2013, 2015; Costello and Brown 2015; Sonmez et al. 2016).

HPGR DESCRIPTION

Conventional roll crushers have long been used in the mining industry, but have usually been restricted to clay materials, or very hot materials such as iron sinter (Otte 1988). The conventional roll crushers were largely replaced by gyratory, jaw, and cone crushers, whereas the HPGR is a specialized grinding unit that has the potential of replacing autogenous grinding (AG) mills, SAG mills, and rod mills within comminution circuits.

Comminution within the crushing zone of a conventional roll crusher relies on particles being in contact with the surfaces of both rollers with a set operating gap. Particles finer than the roll operating gap experience minimal or no breakage. In an HPGR, the crushing of particles is accomplished in

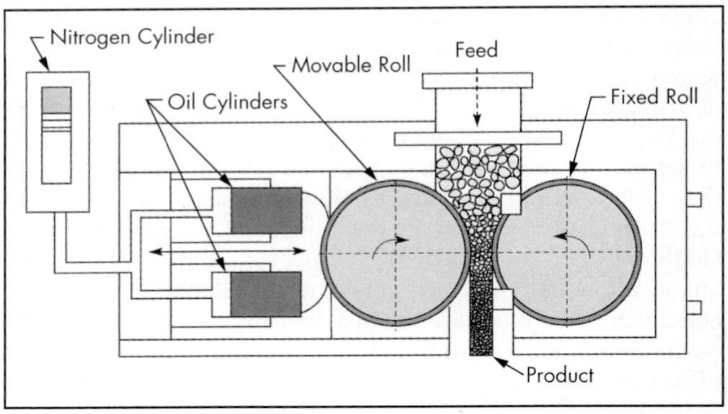

Source: Napier-Munn et al. 1996
Figure 2 HPGR cross section

a bed of material at a well-defined pressure level. As opposed to the process in a crusher, the load onto the particles in an HPGR can be controlled. The resulting thickness of the grinding bed chiefly depends on the roll diameter; the roll surface; and the ore properties, such as the moisture content and the particle size distribution (PSD) of ore feed; and, to a lesser degree, on the press force.

The major physical difference between these machines is the use of a hydraulic system for constant high pressure on the HPGR versus a rigid spring on the roll crusher. Conventional roll crushers traditionally operate at higher roll speeds compared to an HPGR.

The basic HPGR machine concept is very simple. The material is force-fed into the unit by creating a head of material over the machine, as seen in Figure 2 (Napier-Munn et al. 1996). Two counter-rotating rolls allow compression breakage to be used in a continuous rather than batch operation.

Feed System

Crushed ore is fed to the HPGR by a chute usually mounted directly above the gap between the rolls. The *choke* gravity feeding of the HPGR creates a bed of particles to be formed between the rolls. The constrained bed of particles or material mass acts as a wedge in the rolls causing the rolls to resist rotation. The high energy input of the motors provides sufficient torque to overcome the reluctance of the material mass to pass through the rolls. As the packed bed moves through the rolls, the bed pressure increases, and the particles are comminuted by interparticle crushing under very high pressure (Otte 1988). The method of material feeding has an important influence on steady and vibration-free HPGR operation. Depending on the specific material to be processed, the feeding device requires specific geometric designs equipped with a regulating gate and given the most appropriate wear-protection lining. Coarse ores pose very different requirements to all these design aspects of a feed device than iron ore pellet feed.

Rolls

One of the rollers in the HPGR rotates on a fixed axis while the other moves linearly in response to the press force applied to the roll. The movable roller is forced up against the material in the gap between the rollers by a hydraulic oil cylinder system. This oil pressure acts through four or two cylinders (depending on the manufacturer) and transmits the grinding force over the cross section of the diameter of the rolls where the bed has formed.

The roll designs currently consist of three different types and may be used along with a series of different roll surfaces, depending on the application. The roll designs may be a solid roll/shaft arrangement or multiple roll segments or tires secured through a tempered shaft. The length-to-diameter ratio of the roll varies between 0.4 and 0.7 depending on the application and the specific throughput requirements. Special applications bring the ratio closer to 1.0 when a high throughput is required.

Hydraulic Pressure System

The hydraulic pressure system consists of four hydraulic cylinders, two on each side, and are linked via an oil-filled system to the nitrogen cylinder or accumulator. The compression of the nitrogen gas in the nitrogen accumulator acts as the spring of the hydraulic system whose hardness may be adjusted by a preset nitrogen pressure. It is a combination of the initial nitrogen and hydraulic oil pressures that effectively determines the nonlinear stiffness response of the *spring*. Generally, a low preset nitrogen pressure means a *hard* spring, whereas a higher initial preset nitrogen pressure means a *softer* spring. A mechanical stop between the bearing blocks prevents the rolls from touching during operation and may be used to preset an initial gap. The initial gap setting is a preliminary controllable process variable that may be used to alter process conditions or be set according to the detection limit of tramp metal detectors mounted on upstream material-handling equipment. The initial gap essentially defines the starting point from which press forces from the hydraulic system are applied to the feed material. This is significant as variations in the manipulative process variables such as the initial gap, initial nitrogen accumulator pressure, and oil pressures, are important mechanisms for operational control of the energy input to the system (Daniel 2002).

Wear Protection

One of the most important subjects in HPGR processing is the wear of the roll surface (Daniel 2002; Dunne 2006). Most applications in the cement industry have hard-faced smooth rolls, frequently with a hard-facing welded rim pattern (Figures 3A and 3B) on the surface for a better surface grip. The Hexadur wear system (Figure 3D) was recently introduced and is available from Köppern (Schumacher and Theisen 1997, 1998).

In the case of minerals, which typically cause significantly higher wear, such a surface necessitates high maintenance efforts and often requires frequent rewelding of these roll surfaces. For this reason, a patented studded roll surface system was developed in the late 1990s. These rolls provide a longer wear life because of the greater wear resistance of the studs and the embedding of an autogenous wear layer of crushed rock/ore between the studs. This autogenous wear layer is formed by the ore packing itself between the studs on the rolls. The autogenous layer also prevents larger rocks from a direct impingement on the roll surface and provides a shield from the abrasive movement of material parallel to the roll surface. The wear thus becomes principally that of the hard metal tungsten carbide stud inserts, which are much more resistant (Figure 3C).

The introduction of these wear-resistant surfaces has produced added benefits to the process, in that the throughput of units using such designs increases (Lim et al. 1997; Weller and Lim 1999). This result is caused by increased working gaps, which are experienced when using the studded rolls. The increased working gaps are because of the rolls being capable of drawing in more material into the compression zone under similar compression forces. The larger working gap has produced thicker flakes and a corresponding increase in throughput. Even rough surfaces such as the welded-on hard facing (chevron and profiled designs) leads to thicker flakes (Lim and Tondo 1996; Lim and Weller 1997). The diverse stud lining designs and patents have distinguished the various HPGR

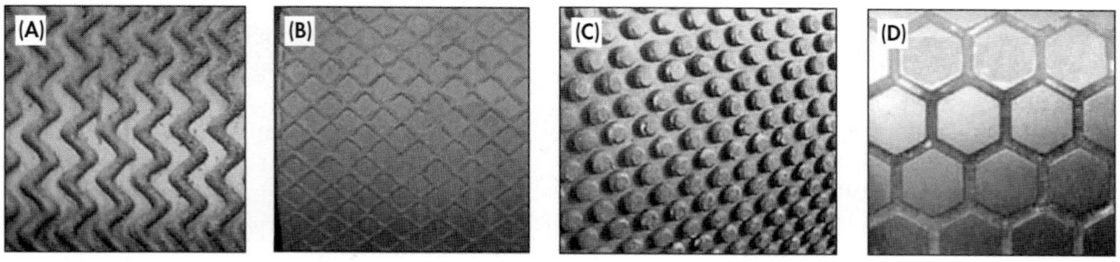

Source: Daniel and Morrell 2004

Figure 3 Various types of roll surfaces

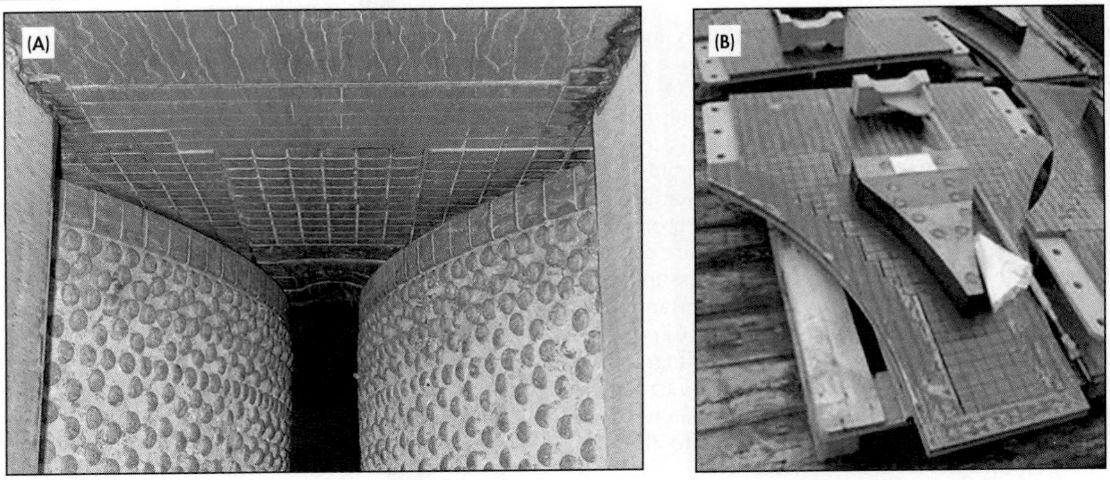

Source: Koski et al. 2011

Figure 4 (A) Cheek plates between rolls and (B) split-design cheek plate

suppliers since 2000; however, since the patent expired in 2012, the development of new designs and new manufacturers on the international minerals market have occurred.

Typical service life of a set of rolls depends on the application and the size of the rolls. The service life ranges from 4,000 to 36,000 hours (Weir Minerals 2010). In hard rock gold and copper applications, the service life is up to 10,000 hours. In comparison, the typical service life of a set of SAG mill liners and lifters is up to 6 months or about 4,000 hours. As a result, HPGR mechanical availability is higher than 96% as compared to SAG circuits at about 92%. The following list offers some examples of stud life with the indicative roll service life for selected minerals treated by HPGR:

- Iron ore (pellet feed), 14,000–36,000 hours
- Iron ore (coarse), 6,000–17,000 hours
- Copper/gold ore (coarse), 4,000–10,000 hours
- Kimberlite rock (coarse), 4,000–12,000 hours
- Phosphate ore (coarse), 6,000–12,000 hours

The wear of the roll surfaces of an HPGR with a specific feed material can be a significant influencing factor on project economics, thus a reliable wear estimation test is critical.

Cheek Plates

Cheek plates are mounted on both sides of the rolls and are vertically and horizontally adjustable (Figure 4A). They ensure that only a minimum of unpressed feed will flow past. The plates are mounted in a manner that enables them to give way on skewing of the movable roll. In such a case, the cheek plate is pressed back to its original position by a preloaded spring assembly after the parallel position of the movable roll relative to the fixed roll has been restored. Usually, the cheek plate is of split design to be capable of separately changing the bottom part of the cheek plate, which is subject to maximum wear (Figure 4B). This is a further contribution to lower the cost of wear components. The cheek plates also provide lateral pressure to the bed of material on the rolls. For HPGRs, the lateral force distribution across the length is not constrained at the ends, resulting in reduced comminution at the edge of the roll and referred to as the *edge effect*. To address this phenomenon, HPGRs are often operated (1) in closed circuit, subjecting the total HPGR discharge to a screening process; or (2) with *edge recirculation*, in which the edge material is either recirculated straight back to the HPGR feed bin or, in some cases, sent to a closing screen. To maintain the horizontal pressure and reduce the edge effect, Metso has installed a flange to one of the rolls of their HRC system.

Tramp Detection and Removal

Tramp metal entering the HPGR circuit of a size greater than the operating gap has the potential to break studs. If left unchecked, a localized zone, where only a few studs are broken, could potentially grow into a large zone of stud washout, leading to premature failure of the tire. To protect against stud damage, two-stage metal protection is recommended to be incorporated into the design. Besides conventional tramp metal removal by self-cleaning magnets, an additional high-sensitivity metal detector coupled to a bypass system is recommended to be installed on the feed conveyor to an HPGR.

OPERATING PARAMETERS

Feed Moisture

The feed of an HPGR should preferably contain some moisture. This assists in generating a competent autogenous wear surface. Generally, HPGR technology facilitates the processing of relatively wet ores, in some cases containing moisture of up to 10% (e.g., iron ore concentrate). This may be of a significant benefit in those applications where moist material is to be comminuted. In a conventional grinding scheme, the ore would require either drying before milling or wet grinding. Drying is obviously a costly process stage, and in some instances, wet grinding may require a significant effort in subsequent sedimentation and filtering of the ground ore. In these cases, HPGR processing can provide a feasible alternative.

The specific throughput drops as moisture content increases, as shown in Figure 5. In some circumstances, the nature of the chemistry of the ore reacts with the water, where it acts as a binding agent. In these ores, the application of the HPGR may not be suitable. Also, high moisture can increase slippage between the roll surface and the packed bed, resulting

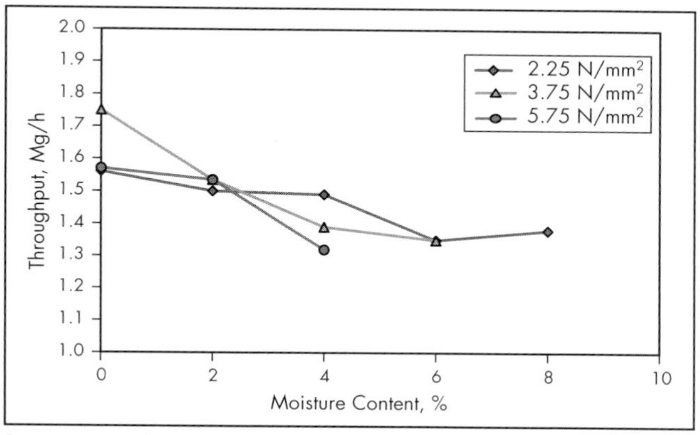

Source: Saramak and Kleiv 2013
Figure 5 Moisture content effect on specific throughput

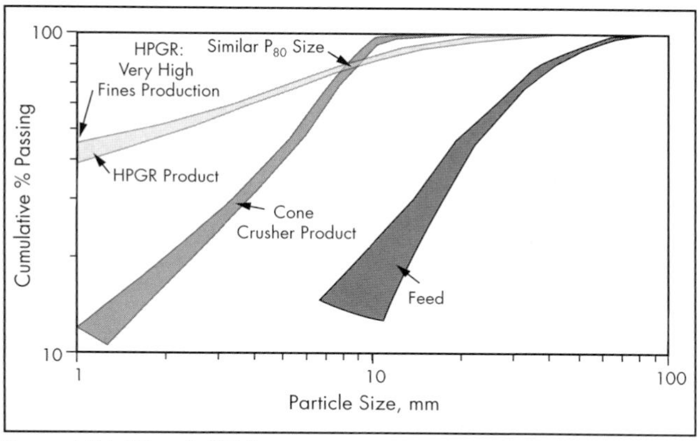

Source: Weir Minerals 2010
Figure 6 HPGR versus conventional cone-crushing particle size distribution

in high wear rates. Some moisture will also help to suppress dust during HPGR operation. Flake competency as a result of ore moisture is an important consideration when selecting an HPGR for comminution duty. For hard rock applications, the preferred moisture levels are typically between 2% and 5%.

Feed Particle Size

The feed particle size is an important parameter that influences an HPGR (Campbell et al. 2001). For ideal interparticle compression grinding, the feed particle sizes should be smaller than the operating gap between the rolls. In actual applications, the feed size distribution can contain particles of up to about 70% greater than the operating gap. For smooth rolls, too large a feed top size could decrease nipping efficiency and accelerate roll wear. For studded rolls, oversized rocks increase tangential forces via direct contact with the roll surface, which can cause stud breakage.

An HPGR produces a PSD that is wider, with more fines, than a tertiary crusher (e.g., cone crusher), as shown in Figure 6. The reason is that the compressive force not only acts on the coarse end of the PSD but throughout the particle bed on both coarse and fine particles, including the fine particles derived from the initially coarser fractions.

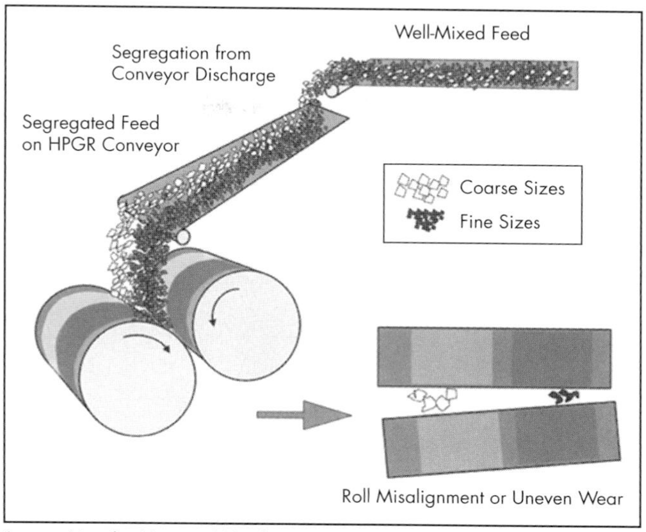

Source: Van der Meer and Maphosa 2012
Figure 7 Rolls skewing caused by uneven feed presentation

Feed Presentation

Feed presentation is very important in industrial applications. If the feed is not presented homogeneously to the rolls, then rolls skewing occurs. *Skewing* is a condition where the axes of the rolls do not stay parallel because of uneven feed distribution for a prolonged period. Although a certain degree of skewing is part of the HPGR operating principle and even desirable, ensuring efficient particle bed comminution caused by an even pressure distribution along the roll width, excessive rolls skewing results in poor performance and ultimately uneven wear of the rolls surfaces. HPGRs are typically equipped with appropriate gap monitoring devices and skew control logics to keep skewing within acceptable limits. Figure 7 shows the concept of rolls skewing.

Roll Speed

Roll speed is an important process control variable (Lim et al. 1997). The power draw and throughput in most cases are directly proportional to the roll speed, which in fact has a limited effect on specific energy consumption and product particle size. However, a choke feed condition can be well maintained if controlled by variable-speed drives; this helps minimize stud breakage and improve HPGR availability. The roll speed at the circumference of the rolls is normally equivalent to the roll diameter in meters per second. This translates to the constant angular velocity between 18 and 23 rpm. Table 2 presents ranges of HPGR roll speeds at some HPGR operations. Variable-speed drives are offered for most HPGR units.

Operating Gap

A bed of particles is normally formed between the rolls where all particles of all sizes are selected for comminution and produce what is commonly known as *flake*. Ore properties generally define the extent of breakage, and HPGRs can often be described as units that facilitate a comminution process where the rock is able to break on itself. Flake product is produced within a small area bounded by the geometry between the rolls defined by a compression zone and is the reason for the claim

Table 2 HPGR roll speed range

Project	Company	Diameter, m	Maximum Roll Speed, m/s	Maximum Roll Speed, rpm
Cerro Verde	Freeport-McMoRan	2.4	2.8	22.3
Morenci	Freeport-McMoRan	3.0	3.3	21.0
Boddington Gold	Newmont	2.4	2.7	21.3
Grasberg	Freeport-McMoRan	2.0	2.6	24.8

that the HPGR working gap is largely a function of the roll diameter regardless of the feed size.

Another way of looking at the compression zone based on the theoretical definition is to view the compression zone as an area within the crushing volume that is not affected by particles that are larger than the working gap. This occurs when the particle size of the feed is less than the working gap, which suggests that it is the characteristics of the bulk material or internal friction between particles within the crushing zone that define the gap. The gap is a function of the amount of material that is drawn into the gap by the rolls and is generally a function of the rolls diameter, the rolls surface, and material properties rather than the press force or PSD.

A wide range of ores appear to generate product PSDs that are self-similar (have a similar PSD when scaled by a critical percentage passing size, usually the P_{80} or P_{50}). This is the result of the product size distribution at a given press force being linked to the achieved gap opening between the rolls. This gap opening depends on the characteristics of a given ore type in producing a material bed that stands up to the applied pressure. For example, for most of the base metal ores tested, the gap opening at a given roll size and an average specific press force is about 2.5% of the roll diameter (Daniel 2002).

The maximum particle size in the product is determined mostly by the achieved gap opening between the rolls or the maximum feed size. With 100% of the product passing the gap opening size, the slope of the PSD would be determined by the breakage characteristics of the ore under high-pressure conditions. A finer size distribution would require a higher press force (except when processing iron ore concentrates) or, in some cases, a higher moisture content.

The amount of material in the gap, or compression zone, can be manipulated to a limited degree to result in optimum operating conditions, but generally, it is a characteristic of the process ore, roll diameter, and surface characteristics. During processing, the particle bed is compressed to a density of greater than 70% solids by volume. The material is usually agglomerated into a cake (flake) that may have to be de-agglomerated before passing on to subsequent processes (Schönert 1988). The hydraulic system in an HPGR allows slight gap variations during the high-compression operation.

Throughput

Several terms are specific to HPGR technology. These terms relate to the performance of the HPGR and were largely formulated to assist in the process of scale-up from laboratory or pilot tests to full-scale machines (Daniel 2002; Daniel and Morrell 2004; Daniel and Bailey 2009; Morley 2008, 2010). The following terms relate to throughput:

- *Unit capacity* (measured in metric tons per hour) is the product of specific throughput and diameter, width, and peripheral speed of the rolls.
- *Specific gap* (percent of roll diameter) is determined by test work and used in scale-up. Typical values are 2.0%–2.5% for full-fines feed and about 1.5%–2.0% for truncated feeds.
- *Specific throughput*, also referred to as *m-dot* (measured in tons-second per cubic meter per hour), is the capacity of a machine with a roll diameter of 1 m, a width of 1 m, and a peripheral speed of 1 m/s. Such a unit will have an operating gap of 25–30 mm. The specific throughput is scalable and is used to predict the throughput capacity of industrial-sized units. It is also known as *specific capacity*. Typical values for studded rolls are in the range of 210 to 260 t·s/m^{3}h for an ore specific gravity (SG) of 2.7. (Values are proportional to SG.) For truncated feeds, that is, feeds with fines removed, then values decrease.

Throughput is a function of ore properties as well as machine parameters. Ore properties that affect throughput are ore bulk density, product (flake) density, and moisture content; and operating parameters that affect throughput are press force, operating gap, roll width, and roll speed. The flake density is normally 85% of the virgin ore density. It is believed that at the time of maximum compression within the rolls, the ore and air between the voids are compressed to effectively convert the compressed ore density to 99% of the ore density. After the compression release, the ore expands somewhat to the levels described as 80%–85% of the virgin ore density.

Pressure and Specific Energy

Parameters that relate to pressure and energy usage are defined as follows:

- *Specific press force* (measured in Newtons per square millimeter) is the applied force divided by the projected area of the roll (width × diameter). The applied force does, of course, increase with machine size, but when normalized in this way, it becomes independent of size. Typical practical operating values are 1–4.5 N/mm^2 for studded roll surfaces and up to 6 N/mm^2 for Hexadur.
- *Net specific energy* (measured in kilowatt hours per metric ton) is the net power draw per unit of throughput. Also independent of machine size and an almost linear function of specific press force for a given application, typical operating values are 1–3 kW·h/t.
- *Specific power* (measured in kilowatts-second per cubic meter) is the product of specific energy and specific throughput; it is used to estimate the shaft power required for a given machine.
- *Shaft power* (measured in kilowatts) is the product of specific power and the diameter, width, and peripheral speed of the rolls.

The shaft power and unit capacity predicted by these formulas can be compared with the measured values of the operating unit to evaluate the accuracy of scale-up in design.

Flakes and De-Agglomeration

Flake is the term used for the compacted products from an HPGR. The feed material is compressed and comminuted

Source: Hilden and Powell 2008

Source: Klymowsky et al. 2008

Figure 8 Compacted flake with stud impressions

within the compressed bed zone between the two rotating rolls. Figure 8 shows examples of the compacted ore with impressions from the studs clearly visible. The thickness of the flake is an important measure of the operating rolls gap, which in turn with the roll speed, determines the throughput capacity of the HPGR. Most of the products of the HPGR are generally discharged in the form of a flake of compacted material; they are, however, of a very fragile consistency.

De-agglomeration with a very low energy input may be required in cases where sizing or classification of the products is involved, for example, in scalping of the finished product sizes ahead of a subsequent grinding stage. Given a certain moisture content and relatively low stickiness of the material, the produced flakes usually break down at transfer points of conveyor belts or during wet screening (Van Drunick and Smit 2006). However, in circuits with dry size classification, a separate de-agglomeration stage may be needed; for example, a cascade tower with a series of impact plates may be required (Van der Meer and Gruendken 2010).

Edge and Product Recycle

Edge and product recycle options use recycling of the HPGR *edge fraction* or partial recycle of the full product stream to achieve additional comminution performance in the HPGR stage and so reduce the inefficiency penalty in the milling stage. The edge recycle option eliminates the need for a screening plant downstream from the HPGR, with an achieved product fineness somewhere between an open-circuit and closed-circuit product. This option offers higher overall efficiencies as it targets the least efficiently comminuted material. However, it requires a splitter arrangement below each HPGR unit to separate the edge fraction, and this is potentially a more complex design with greater maintenance demand than the partial product recycle alternative. The edge recycle option is particularly advantageous for heap leach applications, by recycling the edge material with a high amount of coarse particles with no or very few microfractures.

Dust

Like any dry crushing process, HPGR operation can generate significant dust because of the relatively large quantity of fines in the HPGR product. Appropriate dust collection and suppression systems need to be included in the design of an HPGR circuit, especially around the screening and storage of the HPGR product (Vanderbeek et al. 2006).

INDUSTRIAL HPGR EQUIPMENT

HPGR units are manufactured to suit many varied process streams and ore types and offer a wide range of throughput capacities and power ratings. There are currently six recognized manufacturers of HPGR machines:

1. Maschinenfabrik Köppern, or Köppern
2. ThyssenKrupp, previously known as Polysius, a subsidiary company of ThyssenKrupp
3. Weir Minerals, previously KHD Humboldt Wedag
4. Metso Minerals
5. Citic Heavy Industries Company, or Citic HIC, predominantly for the Chinese cement and minerals market
6. FLSmidth, previously Alpine, for the cement industry

Although the main components of the HPGR units are similar, each manufacturer has design features specific to their units.

Köppern

The Köppern HPGR has both rollers supported on self-aligning spherical roller bearings. A hydraulic system provides for independent control of the hydraulic pressure at the drive and non-drive ends, thus enabling control of roller skew. In addition to controlling the comminution process, the control system has automatic control for the roller skew and motor torque, as well as operation diagnostics. Additional features include automatic lubrication and an adjustable feed supply system. The Köppern HPGR incorporates a multi-hinged frame that allows quick exchange of the rollers from the fixed-roller side of the machine, as shown in Figure 9. Table 3 provides information on the standard equipment sizes of Köppern machines.

Köppern has range of wear-protection systems for different applications. In each case, the two-part design of grinding roller consists of a wear-protected tire shrink-fitted onto a roller core. This system allows roller cores to be reused after the end of the service life of the tires. For the minerals industry (Figure 10), the Köppern wear-protection systems are

Courtesy of Köppern
Figure 9 Multi-hinged frame HPGR

Table 3 Köppern HPGR machines*

Model	Working Surface		Maximum Specific Grinding Force, kN/m²	Typical Installed Power, kW	Capacity, t/h	Heaviest Maintenance Lift (roller assembly), t	HPGR Mass, t
	Diameter, mm	Length, mm					
072-11.0	1,100	750–970	3,500–4,500	2 × 300	200–300	11	60
084-12.4	1,240	870–1,120	3,500–4,500	2 × 450	300–400	16	84
092-13.3	1,330	930–1,190	3,500–4,500	2 × 550	400–500	20	103
500-14.5	1,450	1,010–1,300	3,500–4,500	2 × 725	500–650	26	128
560-15.9	1,590	1,120–1,450	3,500–4,500	2 × 1,000	650–900	34	177
630-17.6	1,760	1,230–1,580	3,500–4,500	2 × 1,300	900–1,200	45	219
710-19.2	1,920	1,350–1,730	3,500–4,500	2 × 1,400	1,200–1,500	59	303
750-20.5	2,050	1,440–1,850	3,500–4,500	2 × 1,600	1,450–2,000	72	350
900-22.8	2,280	1,600–2,060	3,500–4,500	2 × 2,100	2,000–2,500	101	458
950-24.0	2,400	1,700–2,180	3,500–4,500	2 × 2,550	2,400–3,000	116	542
1000-25.4	2,540	1,800–2,310	3,500–4,500	2 × 3,000	2,800–3,600	139	630

Source: Köppern 2018
*Based on typical ore of ~1.7 t/m³ bulk density.

- Studded wear protection using conventional studded wear protection with tungsten carbide plates for armoring the tire edge; and
- Hybridur, developed for ore-grinding applications in the mining industry, using a conventional studded lining with a hot-isostatic pressed edge.

For the cement industry, the wear systems are

- Welded solution for low-wear environments, and
- Hexadur, a wear-resistant surface with a high content of hard phases.

ThyssenKrupp Polycom

The ThyssenKrupp Polycom HPGR consists of two counter-rotating rolls, horizontally mounted and supported in a frame, and is driven by two independent drives (Figure 11). The grinding force applied to the grinding zone is controlled by a hydropneumatic spring system, acting on the floating roll as described earlier, and the roll speed is typically adjustable. The main components of the machine are the rolls, bearings, bearing blocks, frame, pressure beams, hydraulic system, nitrogen accumulators, drive system, lubrication and cooling systems, and adjustable feed chute with feed gates. Information on the range of industrial production machines is shown in Table 4.

The rolls are protected by wear-resistant liners, and the ore is contained within the rolls by stationary cheek plates. The types of liners recommended for hard abrasive ores are shrink-fitted solid tires equipped with tungsten carbide studs and edge protection. ThyssenKrupp conducts the Atwal test to assess the wear properties of the ore to support its design.

Courtesy of Köppern
Figure 10 Wear-protection systems

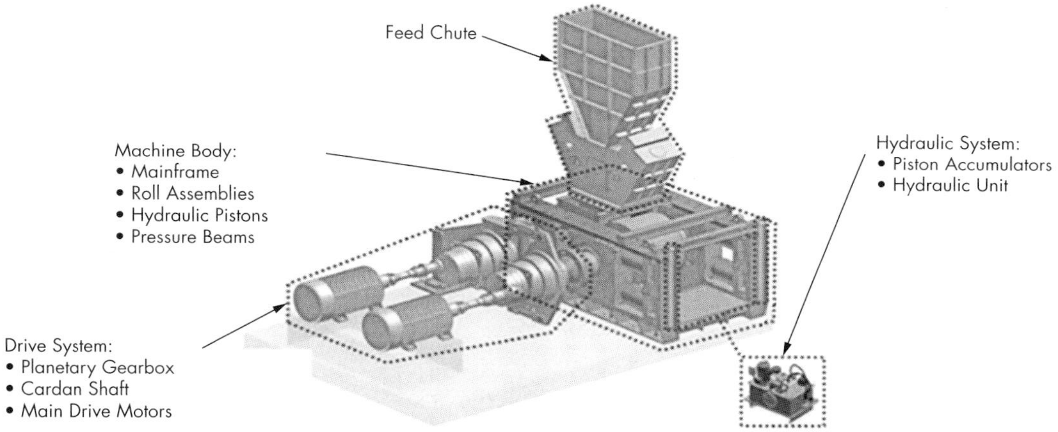

Courtesy of ThyssenKrupp
Figure 11 Polycom HPGR main components

Table 4 ThyssenKrupp Polycom production machines

		Working Surface			
Size	Model	Diameter, mm	Length, mm	Grinding Force, kN	Motor Power, kW
1	9/7	950	650	2,700	2 × 220
2	11/8	1,100	800	3,400	2 × 450
3	14/8	1,400	800	4,300	2 × 500
4	14/10-17/12	1,400–1,700	950–1,200	7,000	2 × 800
5	17/12-20/10	1,700–2,000	1,000–1,200	8,600	2 × 1,600
6	19/15-20/15	1,850–2,000	1,500	11,000	2 × 1,850
7	20/15-20/17	2,000	1,500–1,650	13,500	2 × 2,500
8	22/15-24/17	2,200–2,400	1,550–1,650	17,000	2 × 2,800
9	26/18	2,600	1,750	20,000	2 × 3,400
10	30/20	3,000	2,000	25,000	2 × 5,000

Courtesy of ThyssenKrupp

The specific design of the liners, cheek plates, and other wear-protection methods is selected based on the application. The Polycom HPGR can be equipped with forged, hard-faced roll bodies; chill-cast alloy tires made of bainite; or roll bodies made of compound casting with hard metal studs. For abrasive materials, forged roll bodies (*tires*) with tungsten carbide studs have been developed. This concept relies on the effect of an autogenous wear-protection layer that builds itself up between the studs, protecting the tire from excessive wear. The studded tire is not only able to withstand very abrasive ores but is also suitable for applications with difficult, moist, and sticky feed materials. The design of the hard metal studs (geometry, hardness, and metallurgical composition) and their arrangement on the roll surface (Figure 12) are selected to suit the operating conditions and feed materials and is based on experience collected in the large spectrum of industrial applications. Worn tires can be replaced and potentially refurbished.

Courtesy of ThyssenKrupp

Figure 12 Studs and configuration

Weir Minerals Enduron

Depending on the HPGR type, two different bearing systems can be installed, either multi-row cylindrical roller bearings or self-aligning roller bearings. Smaller HPGRs are equipped with conventional heavy self-aligning roller bearings. The Weir Minerals Enduron HPGR (Figure 13) has a four-row cylindrical roller bearing system for the main bearings. This design reduces the outside diameter of the bearing construction compared to spherical roller bearings. The design allows for smaller tire diameters while compensating for capacity by increasing the tire width. It also allows for oil lubrication, as the bearing is always aligned with the shaft, and proper shaft sealing is possible. The main bearings only deal with grinding load; separate bearings for axial load are applied. Oil-lubricated bearings cool the bearings directly at the rolling elements and flush out possible penetrated pollution.

Another important feature is the combination of polished chromium plates and polytetrafluoroethylene (PTFE)-faced sliding plates that can be replaced when worn without disassembly of the rollers and do not require additional maintenance. They are mounted between the bearing housings and the top and bottom frame, as well as to the external guide mechanism of the bearings for axial fixing of the press rolls (Maxton et al. 2006).

A gearbox torque support system was developed to balance the radial load and therefore reduce load on the gearbox main bearings. The design also allows for fast and safe sideward roller change (<24 hours) without the need to disassemble components above or below the machine.

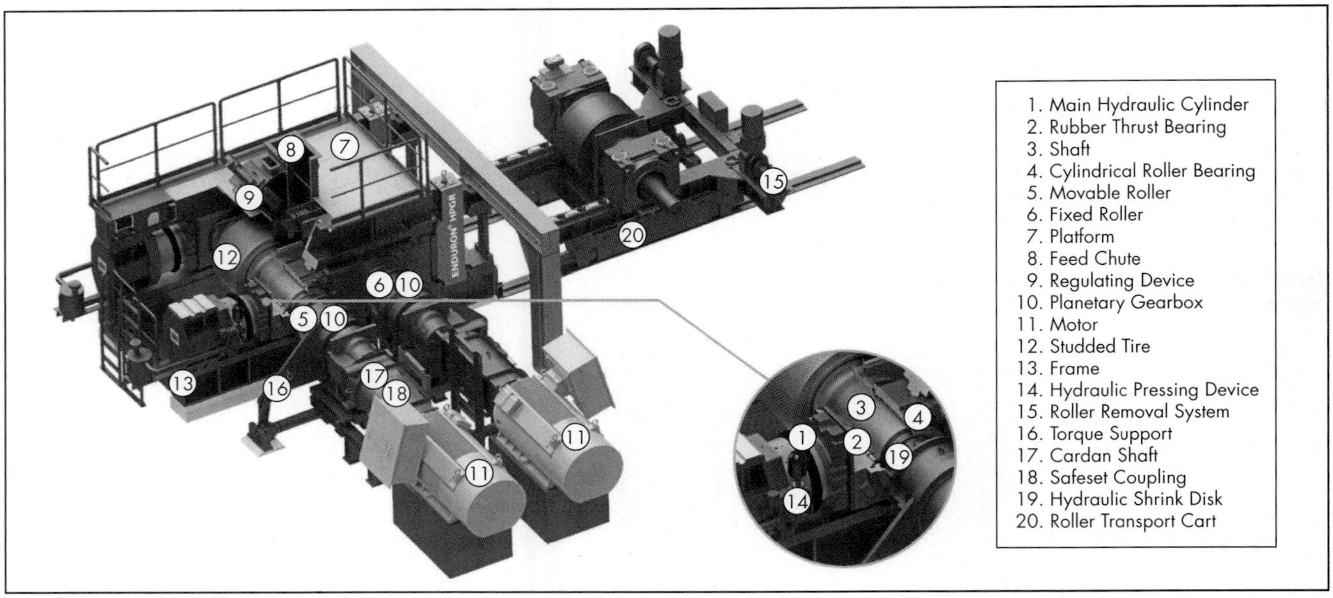

Courtesy of Weir

Figure 13 Enduron HPGRs

Table 5 Weir Enduron production machines

Type	Press Force, kN	Tire Size (D/W), mm	Specific Pressure, N/mm²	Maximum Top Feed Size, mm	Maximum Shaft Torque, Nm	Maximum Power, kW	Throughput, t/h
Enduron RPM2–RPM 32	2,000–32,000	800/250 to 3,200/2,400	Up to 6	Up to 95	Up to 6,000,000	Up to 14,000	Up to 7,200

Courtesy of Weir

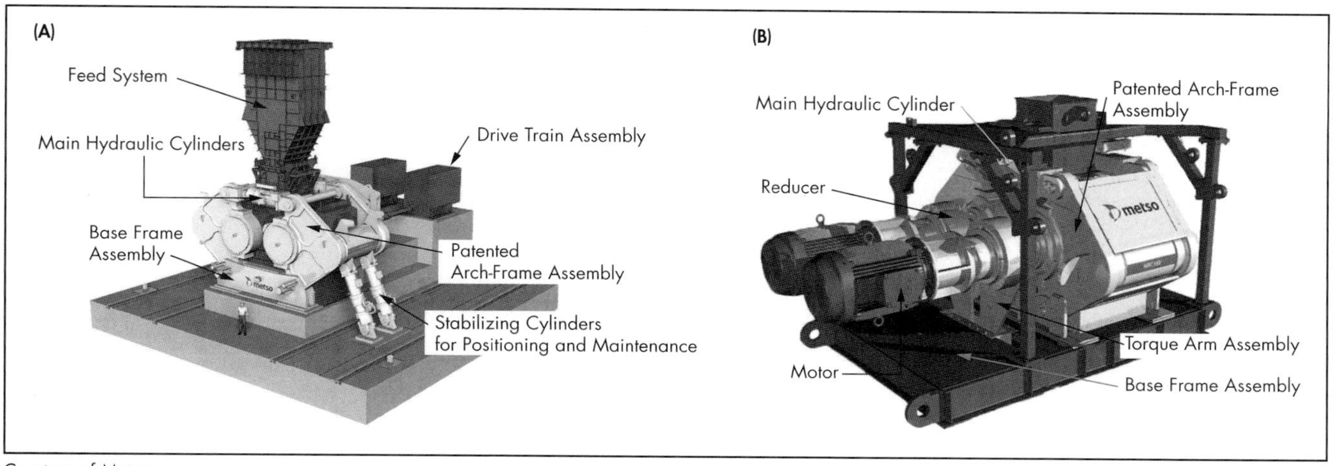

Courtesy of Metso
Figure 14 Main components for (A) HRC 3000 and (B) HRC 800

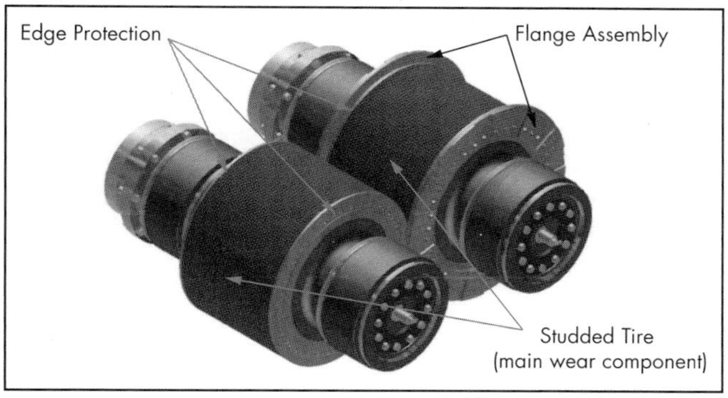

Courtesy of Metso
Figure 15 HRC 3000 flanged and non-flanged roll assembly

The Weir HPGR design allows a large tire-width-to-diameter ratio, creating a higher percentage of center material that generates a finer product and reduces the circulating load. Lateral wall plates that follow the tire edge further reduce the edge effect, as the edge carryover is minimized. Controlled skewing provides an effective pressure distribution across the length of the material bed between the rollers, even with segregated feed. The tire size reduces machine weight by reducing bearing housings and frame size. The roller life is maximized by optimizing the stud length, diameter, and distribution pattern. Smaller stud diameter provides higher local pressure peaks on the material bed. A self-clamping lateral stud prevents falling out or early failure caused by large particle impact. Information on the industrial Weir Enduron HPGRs is given in Table 5.

Metso Minerals HRC

The initial concept of the HRC HPGR began with the arch frame, which mechanically absorbs unbalanced loads to eliminate downtime caused by skewing. With the HRC, the two sides of the Arch-frame are mounted into the base frame with pins. Hydraulic cylinders at the top of the frame apply the crushing force. The cylinders only need to apply roughly half of the required force at the rolls because of the mechanical advantage of the pivoting Arch-frame. This idea was based on how a nutcracker uses a mechanical lever to multiply the crushing force. Figure 14 shows HRC HPGRs and their main components.

The HRC uses a flanged roll design where one roll has a set of flanges, which are bolted onto the side of the roll, as shown in Figure 15. The flanges are designed to reduce edge effect. In contrast to stationary cheek plate arrangements, the flanges are bolted onto the roll and move in the direction and speed of the ore and therefore pull material into the crushing zone. Figure 16 shows a comparison of a traditional cheek plate arrangement versus the Metso HRC HPGR arrangement with flanges (Knorr et al. 2013, 2015).

The flanges are reported to help to pull material into the crushing zone. This component, as well as the cheek plates above the nip angle and the feed chute liners are wear parts. Typically, HRC rolls are fitted with tungsten carbide–studded rolls for most hard rock applications. If the application has a low wear rate, smooth manganese rolls can be used.

The HRC uses spherical roller bearings on the shaft and roll assembly, as seen in Figure 17. The drive train is designed with a planetary reducer that mounts directly to the shaft assembly. The reducer is connected to the squirrel cage induction motors via a cardan shaft. For most HRC models,

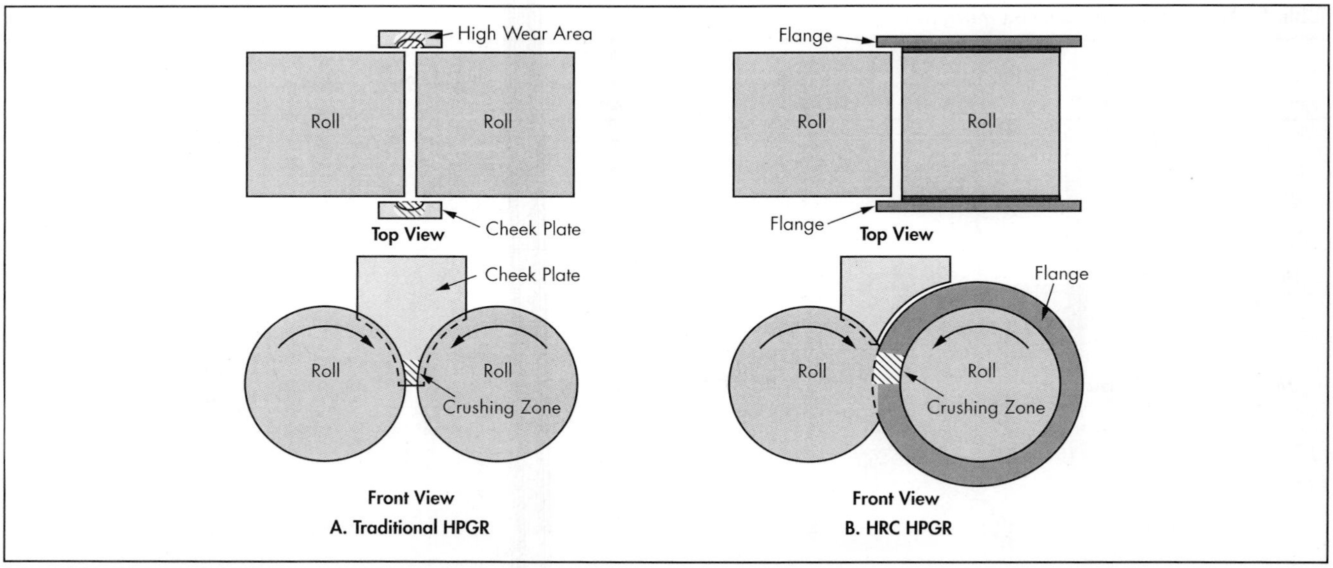

Adapted from Knorr et al. 2013

Figure 16 (A) Traditional HPGR with cheek plates and (B) Metso HRC with flanges

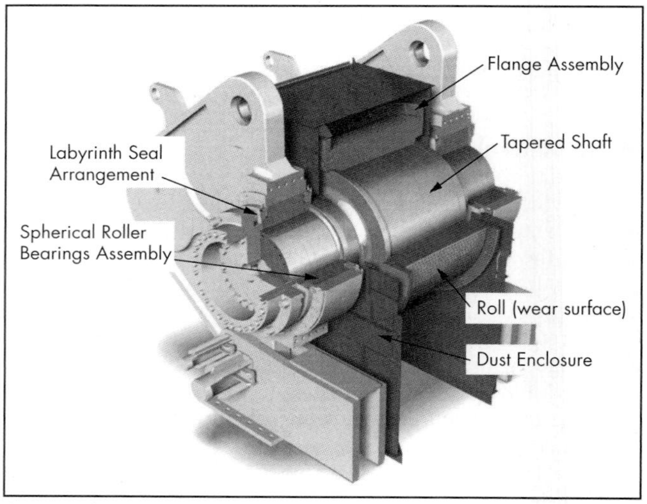

Courtesy of Metso

Figure 17 Metso HRC

the machine equipment with variable-speed drives can operate between 30% and 110% nominal speed (Table 6).

HPGR TESTING AND MODELING

Each of the manufacturers have pilot testing facilities to determine design parameters for the HPGR. Pilot programs are focused on determining specific energy requirements, throughput, and product PSD for an ore and to assess the effect of operating parameters such as ore moisture, specific press force, and roll speed.

The number and types of tests depend on the level of the study, ranging from conceptual to feasibility. Conceptual level studies may involve four to five pilot tests, requiring about 2 t (metric tons) of sample. For a full feasibility study, tests include closed-circuit testing and require up to 20 t of sample plus variability testing that requires another 2 t of each ore type. Results support trade-off studies, circuit modeling for design, and optimization and geometallurgical assessments. For scale-up, pilot test units range in diameter from 700 to 1,000 mm.

Testing of material at smaller test machines (benchtop machines) reduces the sample size but requires additional factors to scale up to industrial-size units. Most HPGR manufacturers do not offer laboratory-scale units for testing, so that vendor specific scale-up methodologies are not well accepted by all HPGR manufacturers. Some advances have been made with the development of piston–press tests that can predict the energy–size reduction relationship and product PSD for HPGRs.

Each manufacturer conducts wear tests to support tire life estimates. Mineralogy is a major driver on the abrasion of HPGR rolls and has a higher impact on HPGR wear rate than feed size, operating pressure, or moisture. Conventional abrasion (Bond abrasion index) and hardness tests do not correctly reflect the conditions and wear mechanics of an HPGR.

Laboratory and Pilot-Plant Machines

Köppern

Pilot-scale testing is available at the University of British Columbia in Vancouver, Canada; ALS Metallurgical in Perth, Australia; and Köppern Technical Centre in Freiberg, Germany. The Freiberg facility is equipped with two HPGR units for comminution tests. One of the laboratory units is a smaller machine particularly suited for wear testing. For full wear assessment, 1–3 t of material is required. An additional apparatus that requires ~100 kg of test material is available for quick wear-rate estimates. A single test run requires a 200-L drum of material. The extent of test-work program varies, and typically 3–12 test runs are required. The maximum particle size for pilot testing is 32 mm. Each of the machines can be equipped with either Hexadur, studded, or welded roller surfaces. A full spectrum of process parameters can be tested.

Table 6 Metso HRC production machines

Description	HRC 800	HRC 1000	HRC 1450	HRC 1700	HRC 2000	HRC 2400	HRC 2600	HRC 3000
Roll diameter, mm	800	1,000	1,450	1,700	2,000	2,400	2,600	3,000
Roll width, mm	500	625	900	1,000	1,650	1,650	1,750	2,000
Maximum grinding force, kN	1,800	2,812	5,873	7,650	14,850	17,820	20,475	27,000
Maximum installed power, kW	2 × 150	2 × 260	2 × 650	2 × 900	2 × 2,100	2 × 3,000	2 × 3,700	2 × 5,700
Typical capacity, t/h*	121–169	211–295	530–742	723–1,012	1,650–2,310	2,376–3,326	2,958–4,141	4,500–6,300
Maximum particle size, mm†	20	25	36	43	50	60	65	75

Source: Metso Minerals 2016

*Varies per application; values based on nominal equipment speed and a specific throughput of 250–350 t·s/m^3/h; values are for the machine and do not include any circulating load.

†Varies per application.

Table 7 ThyssenKrupp Polycom HPGR test units

Test	Description	Target	Requirements
Labwal (laboratory-scale HPGR)	For scoping studies in cases of limited availability of samples	Sizing of Polycom	20–30 kg sample material per test Maximum feed size 12 mm
Magro/Pilotwal (pre-industrial-scale test)	For the more accurate sizing of the industrial Polycom HPGR equipment	More accurate sizing of Polycom	1–5 t sample material Maximum feed size 32 mm/<20 mm
Magro/Pilotwal (pilot-plant test)	For test work with a high amount of available test material on site	Highly accurate sizing of Polycom on industrial level	>50 t sample material per day Maximum feed size 32 mm/<20 mm
Magro (dry grinding test)	For the accurate sizing of the industrial Polycom HPGR/air classifier dry grinding circuit	More accurate sizing of Polycom and air classifier	>1 t sample material Maximum feed size 32 mm
Atwal (wear test)	To determine the wear properties of the ore and to estimate the wear life of the Polycom HPGR	Estimate of roll life time	~100 kg sample material Maximum feed size ~3 mm

Courtesy of ThyssenKrupp

ThyssenKrupp

A variety of test machines are available for testing the characteristics of a given feed material. The available test units are designed to produce specific data and test results to investigate various operational aspects. An overview of existing ThyssenKrupp Polycom HPGR test units is provided in Table 7.

To predict tire wear, direct and indirect methods can be applied. Direct methods use a small-scale version of an HPGR, an Atwal machine. A certain amount of ore is processed and the corresponding loss of weight or loss of radius of the wear liner is measured. From the data obtained, a prediction of the wear life of the industrial unit is made. The test procedure typically accommodates different pressure levels, varying moisture content, and PSD of the ore feed. An indirect method to estimate wear is by abrasion testing. A probe is subjected to the ore sample under a defined mechanical load. The index determined is used to scale up the wear life of the rolls in an industrial application based on industrial experience.

Weir

HPGR test facilities are located in Germany, Chile, United States, Canada, and Australia. All facilities are equipped with the same pilot unit with a tire size of 800 mm in diameter. Testing determines basic design parameters and assesses the effect of operation parameters as described earlier. Sample quantities for the different test procedures vary between 70 kg and 1.5 t of feed material. The measured parameters need to be scaled up to industrial-size machines because the ratio between the feed particle size and tire diameter, length, and speeds under test conditions are not the same as for industrial-scale machines. This scale-up is done with Weir simulation software based on material breakage models that have been validated with industrial machine performance data. In addition, Weir also tests the abrasiveness of the ore using an ASTM standardized test (grinding wheel), which is the basis for the tire lifetime warranty.

Metso Minerals

Laboratory testing is conducted with an HRC 300 test unit with 300- by 150-mm rolls. Test programs include open-circuit, locked-cycle, and continuous closed-circuit pilot tests. For an open-circuit test, Metso requires around 300 kg of sample per test series. The amount of sample for the other tests varies depending on the application and testing plan. To determine which roll type to install and to estimate an expected roll life, wear testing is recommended.

Metso also employs the packed bed test (PBT), which is an instrumented piston and die test. This test provides information regarding the breakage characteristics of the ore and is useful when there is a limited amount of material available or many samples are to be tested to assess variability. Typically, about 3 kg of sample is required per PBT series. The information collected from these laboratory tests is then used to estimate the energy-based selection and breakage function parameters for the selection and scale-up of an HRC HPGR.

Laboratory-Scale HPGR Scout Test

The JKTech laboratory-scale HPGR tests are referred to as *scouting tests* and can be conducted with small amounts of drill core. Scouting tests are also a means of testing the ore's amenability to compressive forces and are complementary to

Figure 18 ThyssenKrupp laboratory-scale HPGR testing, rolls and product

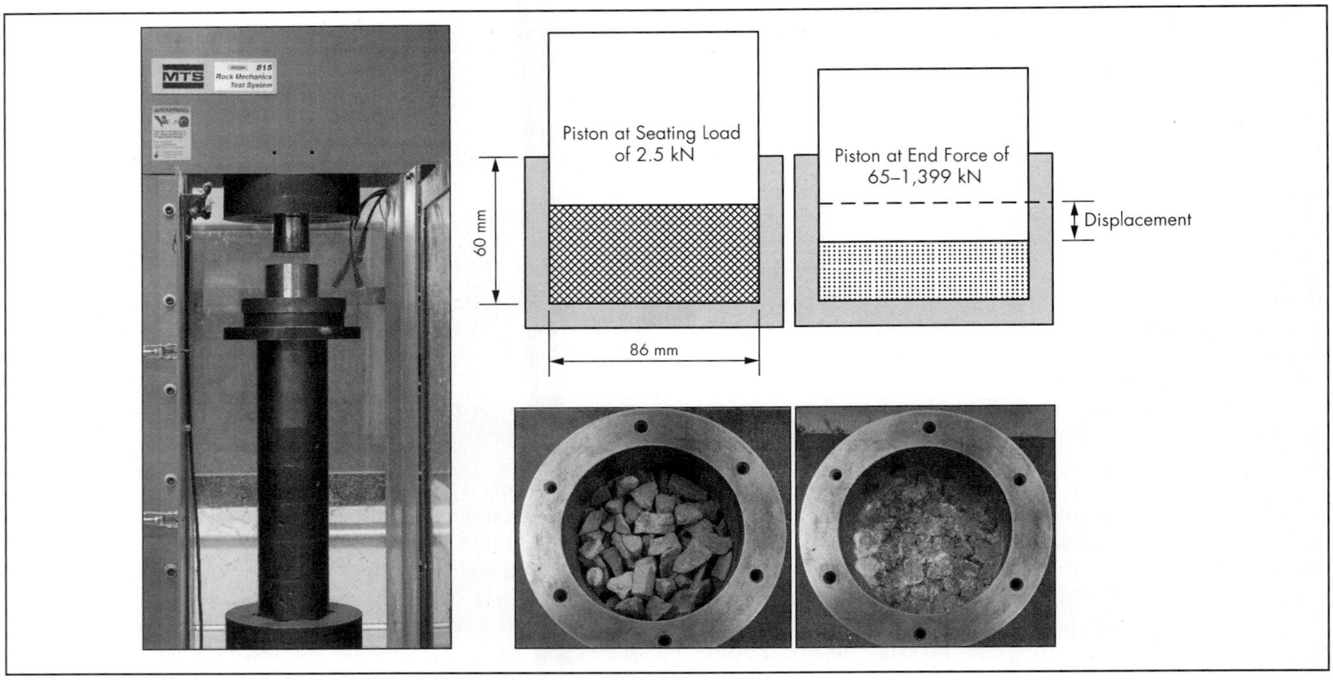

Figure 19 Hydraulic press and the piston–press test

the standard JK drop weight test and the SMC Test that generate HPGR ore properties independently of an HPGR test. Two different HPGR tests are performed using a laboratory-scale ThyssenKrupp machine (Figure 18):

1. **HPGR energy response test.** The test work determines the best comminution-to-energy response and requires a minimum of six tests. The test is carried out to achieve an effective characterization of an ore under different HPGR operating pressures.
2. **HPGR ore variability response test.** Ore variability tests are carried out on as many ore types as can be provided to test the relative response of the ore to the same HPGR process settings.

The recommended sample size for each HPGR *pass* is between 20 and 30 kg. This is the minimum amount considered necessary to obtain an assessment under steady-state processing conditions (Daniel and Bailey 2009).

Piston–Die Testing to Size HPGR

For early-stage studies and new mining projects, collection of large quantities of representative samples is challenging and costly. In addition, mining companies have a preference toward third-party independent testing.

The development of HPGR technology originated from fundamental studies conducted by Schönert (1988) who used piston–press tests to investigate interparticle breakage and measure the breakage rate and breakage distribution functions.

Table 8 Piston–press testing methodologies

Methodology	Sample Size	Pilot Test	Accuracy, %	Application
Direct calibration: involves a limited number of pilot-scale HPGR tests and piston–press tests to calibrate models to predict specific energy and particle size reduction relationship	1 t for HPGR pilot and 5–10 kg for piston–press test per ore type	Yes	±10	Scoping study Prefeasibility study Feasibility study Design
Database calibration: developed from a database of HPGR pilot test work and paired piston–press tests	5–10 kg per ore type	No	±25	Scoping study
Simulation test: similar to the JK drop weight tests and involves piston–press tests on narrow size fractions to calibrate a modified breakage index model	10 kg per ore type	No/Yes	±25	Scoping study Prefeasibility study Feasibility study Design Operation

Source: Davaanyam 2015

Various researchers have investigated the possibility of conducting small-scale laboratory piston–press tests to predict the sizing information for the HPGR (Bulled and Husain 2008; Hawkins 2007; Kalala et al. 2011; Davaanyam et al. 2013).

With support from the Canadian Mineral Industry Research Organization, the University of British Columbia developed a piston–test apparatus and procedures that were correlated to a pilot- and full-scale HPGR operating information (Davaanyam et al. 2013, 2015; McClintock et al. 2018). Figure 19 presents a hydraulic press unit and the developed piston–die test apparatus. For each test, a force-versus-displacement curve is generated such that the integration of the area under the curve represents the energy input. Studies showed that piston–press and HPGR products PSDs are self-similar allowing prediction of HPGR PSD from piston–press results. The information obtained is directly used in models to predict the HPGR energy–size reduction relationship and predict PSDs for an industrial-scale HPGR. The tests allow the evaluation of variables such as pressing force, moisture content, and mineralogical variability.

Three piston–press test methodologies and models have been developed: direct calibration, database calibration, and simulation. Each method is briefly described in Table 8 along with the sample requirements, testing accuracy, and respective applications.

Currently, results from more than 200 pilot-scale HPGR tests on about 30 ores have contributed to the piston–press test database and models. The procedures are in the public domain and can be performed by metallurgical laboratories, and the models can be used by design engineers and mining operations.

HPGR Modeling

Modeling of the HPGR has focused on the development of three submodels covering throughput, power draw, and product PSD (Fuerstenau et al. 1991; Weller and Lim 1999; Klymowsky and Lui 1997a, 1997b). The models have been developed with varied degrees of success, but one overriding conclusion is that the process is strongly dependent on the specific ore characteristics of the material being treated. Performance predictions have therefore had to rely on experiments using the particular ore in question to justify and size new units within a comminution circuit (Otte 1988).

Throughput Modeling

A plug flow hypothesis is the basis of the HPGR throughput model, as represented by Equation 1 (Schönert and Lubjuhn 1990). Throughput is calculated from various measurements made during the process (roll speed, roll width, and working gap) and of the material in the working gap (flake density).

$$Q_{\text{calc}} = 3{,}600 U L \rho_g x_g \qquad \textbf{(EQ 1)}$$

where

Q_{calc} = calculated throughput, t/h
U = roll peripheral speed, m/s
L = roll length, m
ρ_g = flake density of the material within the working gap, t/m^3
x_g = measured roll working gap, m

Power Draw Modeling

HPGR power draw is a direct function of the press force applied on the roller. The applied force is split into tangential and radial components, where only the tangential force creates the torque that drives the rollers. Assuming equal force is applied to both rollers, the power draw of an HPGR can be estimated by Equation 2 (Schönert 1988), which is determined by roll speed, applied forces, and force reaction angle. The force reaction angle in the equation can be estimated as half the compression angle (Klymowsky et al. 2002).

$$P = \Omega T = 2Fu\sin\beta = 2Fu\sin\left(\frac{\alpha}{2}\right) \qquad \textbf{(EQ 2)}$$

where

P = motor power, kW
Ω = roll radial velocity, 1/s
T = roll torque, m
F = grinding force, kN
u = roll speed, m/s
β = force reaction angle, degrees
α = compression angle, degrees

The specific comminution energy consumption is then expressed as the ratio between the power draw (in kilowatts) and the throughput (in metric tons per hour).

Comminution Specific Energy

At the prefeasibility stage, laboratory- and pilot-scale HPGR tests are often not possible because of ore sample constraints. To rectify these limitations, an HPGR ore hardness parameter has been developed (Morrell 2006, 2009) for use with the following power-based equation (Morrell 2004):

$$W_h = K_3 M_{ih} 4\left(x_2^{f(x_2)} - x_1^{f(x_1)}\right) \qquad \textbf{(EQ 3)}$$

Table 9 HPGR modeling and simulation tools

Model	Testing Requirement	Working Environment	Model Inputs	Model Outputs
Morrell et al. 1997a	Laboratory-scale or pilot-scale HPGR test	JKSimMet	1 Critical gap 1 Split factor 3 Breakage indexes 2 Power efficient factors 9 Additional model parameters for each of the breakage zones	Prediction of • Throughput • Power draw • Product size distribution
Torres and Casali 2009	Pilot- or full-scale test	Molycop tools	Measured throughput Measured power draw Measured size distribution	Compression angle Breakage parameters Selection parameters
Davaanyam 2015	Narrow-size piston–press test	Excel	Specific press force • 3 Breakage parameters • 3 Selection parameters	Prediction of • Specific energy • Product size reduction

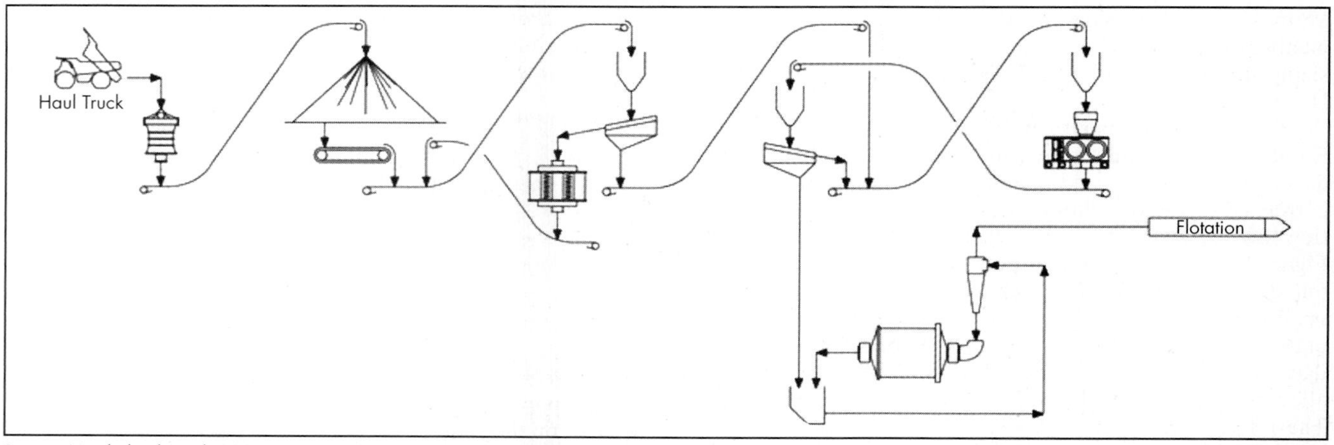

Source: Vanderbeek et al. 2006
Figure 20 Cerro Verde 108-kt/d comminution circuit

where
W_h = specific energy for HPGRs
K_3 = 1.0 for all HPGRs operating in closed circuit with a classifying screen; if the HPGR is in open circuit, K_3 takes the value of 1.19
M_{ih} = HPGR ore work index provided directly by the SMC Test
x_2 = P_{80} of the circuit product, µm
$f(x_j) = -(0.295 + x_j/1{,}000{,}000)$
x_1 = P_{80} of the circuit feed, µm

The equation has been validated using a wide range of data from HPGR laboratory-scale, pilot-scale, and full-scale plant data. The HPGR ore hardness parameter, M_{ih}, is determined from the results of the SMC Test (Morrell 2006) and is generated as part of the standard output when such a test is done. The same test also produces the equivalent hardness parameters for conventional crushers and AG/SAG mills, hence obviating the need for separate (expensive) tests for each type of equipment.

Product Particle Size Distribution Modeling

The most common approach for modeling PSD in the HPGR is through the use of population balance modeling techniques (Fuerstenau et al. 1991; Austin et al. 1993; Klymowsky and Liu 1997a, 1997b; Morrell et al. 1997a, 1997b; Torres and Casali 2009; Davaanyam et al. 2015). The required selection and breakage function can be fitted using either pilot- or full-scale HPGR operation data or can be determined from the well-controlled piston–press testing. Table 9 gives a few examples of such models and tools available for commercial and academic use.

HPGR PROCESS CIRCUITS

The HPGR is applied to a range of materials including cement, diamonds, iron ore, platinum, gold, and base metals (Aydoğan et al. 2006; Liu et al. 2018; Kazerani Nejad and Sam 2017; Rule et al. 2008; Knecht and Patzelt 2004). Flow sheets can be open circuit or closed circuit with edge recycle or size classification that can be wet or dry (Van der Meer and Gruendken 2010). The trend has been toward closed-circuit HPGRs with demonstrated benefits with respect to energy savings and improvements to downstream processes (Altun et al. 2011).

Figure 20 shows the Cerro Verde HPGR process flow sheet, which was the first HPGR installation for a large-scale copper mine. The operation included four HPGRs and was designed for a process rate of 108 kt/d, which was later increased to 120 kt/d. From the primary crusher stockpile, the ore is cone crushed in reversed closed circuit to remove oversize that could damage the HPGR rolls. In addition, a means of metal detection and recovery is required ahead of the HPGR to remove tramp metal that could cause damage to the rolls. A surge bin ahead of the rolls ensures that the HPGR is choke fed. The HPGR product is conveyed to a fine ore

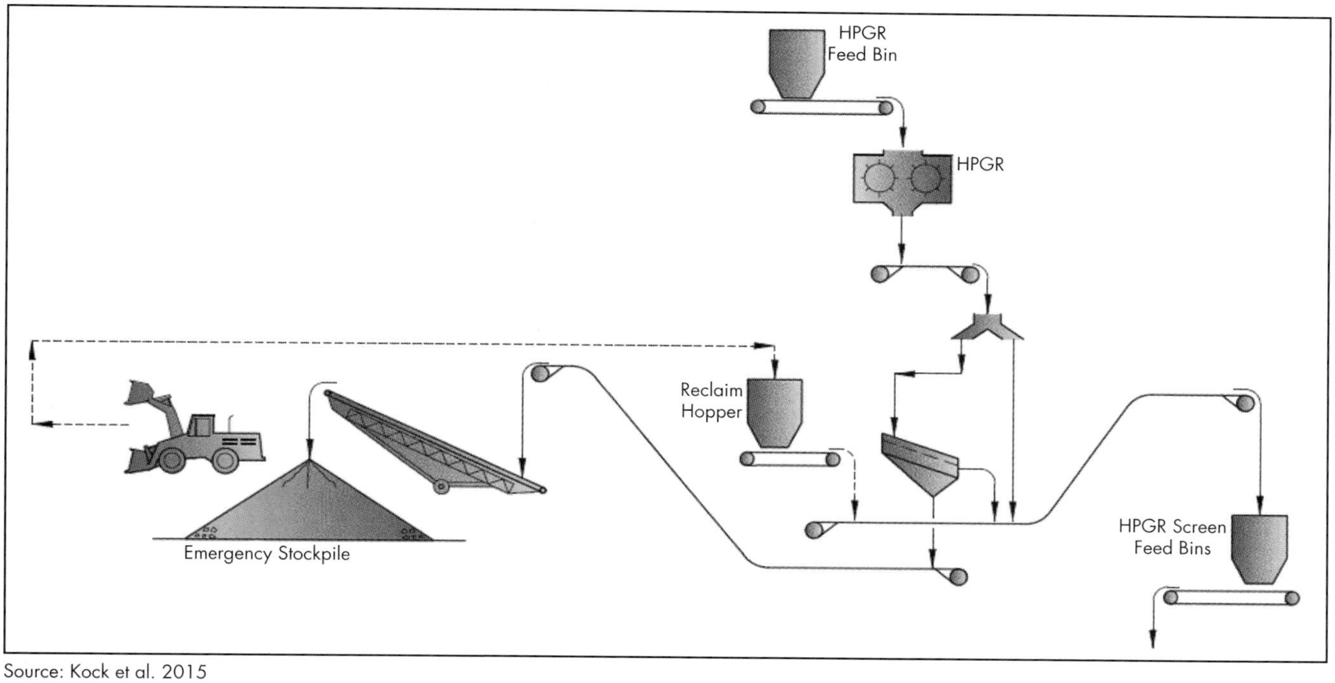

Source: Kock et al. 2015

Figure 21 Tropicana emergency stockpile flow sheet

surge bin ahead of wet screening with oversize returning to the fine ore surge bin and undersize reporting to the ball mill circuit. In 2016, Cerro Verde plant expansion included the addition of eight HPGRs with an added capacity of 240 kt/d. The main modification to the circuit was the addition of a surge bin ahead of the secondary cone crusher (Vanderbeek and Gunson 2015). Surge bins were also included ahead of the cone crushers in the circuits at Boddington in Western Australia (Dunne 2006) and Sierra Gorda in Chile (López 2011).

The Tropicana operations in Western Australia were designed with a single HPGR. To ensure that ball mill feed can be maintained when the HPGR is offline, thus improving grinding circuit availability, an emergency stockpile was added to the circuit (Figure 21). A controlled fraction of HPGR discharge is diverted to a dry screen to produce a stockpile containing –4 mm rock that can be fed to the mill during HPGR-circuit shutdowns. The stockpile has a storage capacity of 48 hours of mill feed.

To simplify the flow sheet with the intent of reducing capital cost, Rosario and Drozdiak (2015; Rosario 2017) proposed an alternative design with open-circuit secondary crushing and having the coarse ore stockpile shifted after the secondary crusher. Despite rigid control of the closed side setting, there is a risk that some coarse particles will cause stud breakage during low choke and starting and stopping of the HPGR. Situating the coarse ore stockpile after the secondary crushing eliminates the need for a large surge bin prior to the grinding circuit.

Rosario and Drozdiak (2015) also propose a variation of the flow sheet that can further simplify the circuit by the elimination of wet screening ahead of the ball mill. This alternative circuit has the HPGR operated with edge recycle (Figure 22). In this option, the ball mill operates without surge capacity, but a higher size reduction is required in secondary crushing to provide an adequate top size to the ball mill. The elimination of wet screening with the adoption of edge recycle can mitigate problems associated with processing a high-moisture HPGR feed, specifically for cold-climate operations, where freezing can occur in bins and ore reclaim systems.

In cold climates, the standard HPGR flow sheet with wet classification significantly increases the risk for operations, where bin hang-ups over HPGR and unstable conveyor operation could occur as a result of mixing the cold fresh feed with wet screen oversize. Without any process modification, this type of closed HPGR circuit could frequently become inoperative during the harsh winter time. To mitigate the negative effect of low temperatures, a wet disintegrator-scrubber could be integrated in the flow sheet ahead of the HPGR.

CAPITAL AND OPERATING COSTS

At the current stage of development, HPGR-based circuit capital costs are generally higher than for the equivalent SAG-based circuit, so that HPGR must offer sufficient operating cost benefits to offset the additional capital cost over an acceptable payback period. It is usually necessary to undertake comprehensive project-specific trade-off studies to quantify this effect (Vanderbeek et al. 2006; Amelunxen and Meadows 2011).

Costello and Brown (2015) compared capital and operating costs of SAG mill, ball mill, and pebble crusher (SABC) and HPGR circuits for seven reference copper porphyry projects and predicted costs for a 140-kt/d case study (Figure 23). The projects ranged in size from 17 kt/d to 240 kt/d, and the boundary limits for comparison were from primary crusher stockpile through to ball mill cyclone overflow. The HPGR/SABC capital cost ratios ranged from 114% to 130% and averaged 121%. Although HPGR machines are often less costly than SAG mills, the higher circuit costs are caused by outlays for crushers, screens, conveyors, bins, feeders, dust control, and tramp metal removal equipment.

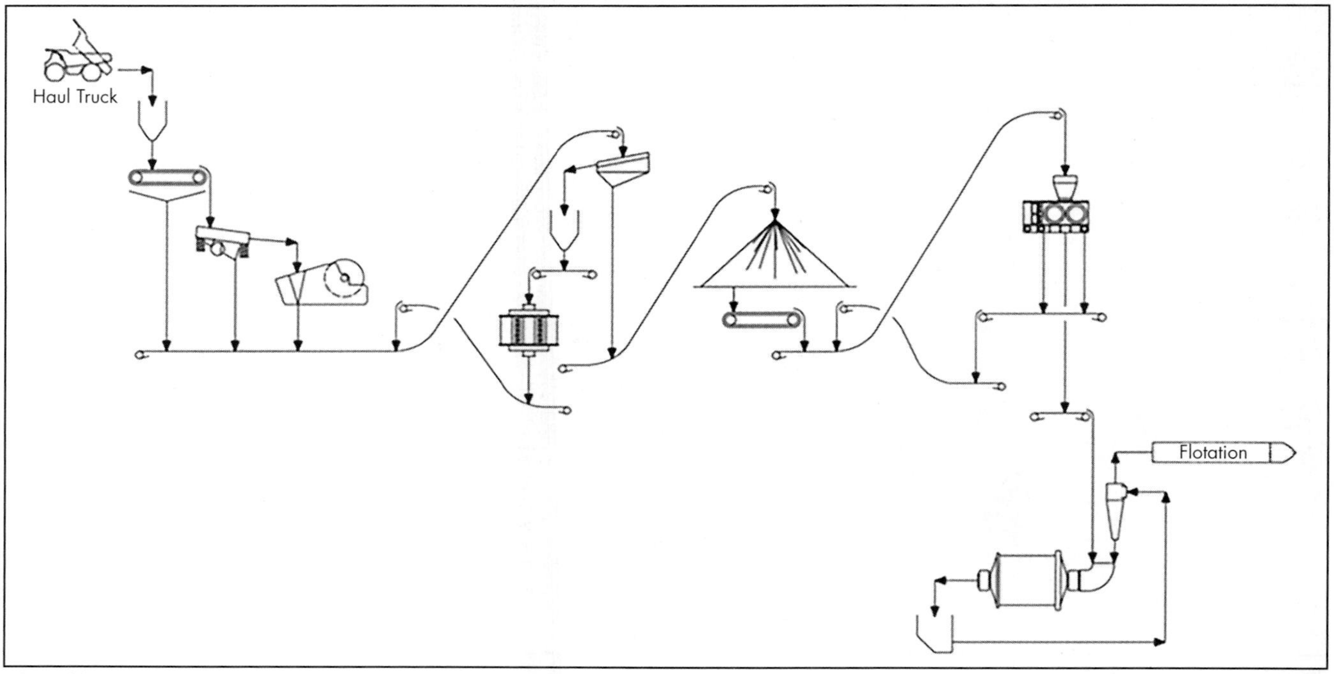

Adapted from Rosario and Drozdiak 2015

Figure 22 Simplified HPGR circuit with HPGR edge recycle

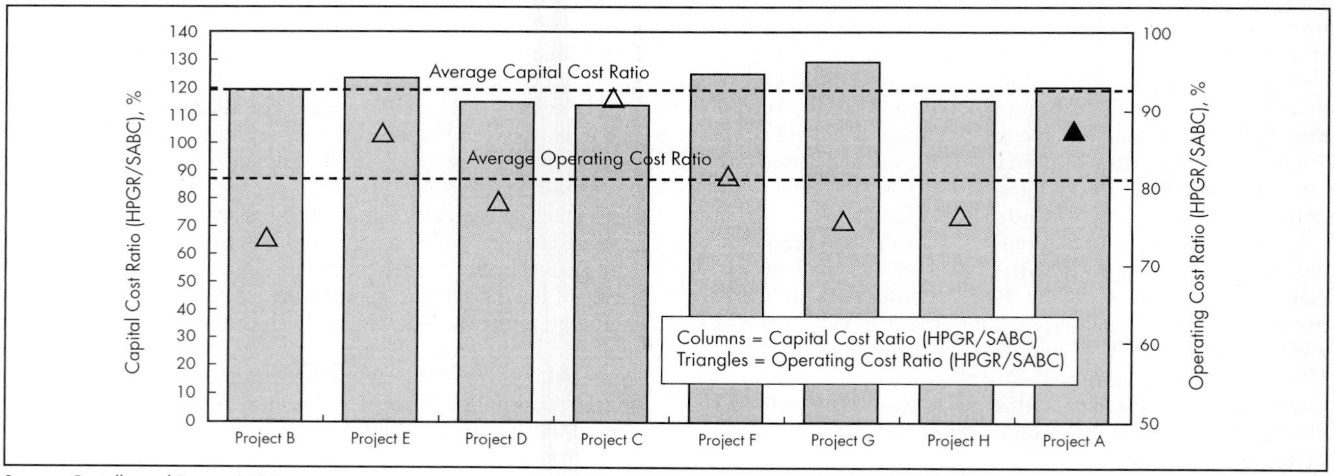

Source: Costello and Brown 2015

Figure 23 HPGR and SABC circuits—capital and operating cost comparison

When comparing operating costs, the HPGR costs are considerably lower with HPGR/SABC operating cost ratios ranging from 74% to 92% and averaging 81% (Figure 23). The main operating cost items for comparison include power, grinding media, crusher and mill/crusher liners, screen panels, maintenance, wear supplies, and labor. Based on a normalized electrical cost of US\$0.11/kW·h and including energy usage of ancillary equipment, energy costs represented the greatest savings in operating costs.

A comparison of capital cost and operating costs are presented in Tables 10 and 11 for 140-kt/d SABC and HPGR circuits (Costello and Brown 2015). Because the primary crusher setup is the same for both, the boundary limits for the cost are from the stockpile feed system to the ball mill circuit cyclone overflow. Although the capital cost is 21% greater for the HPGR circuit at US\$920 million versus US\$762 million, the operating costs are about 15% lower, resulting in a cost per tons savings of US\$0.57/t. Energy costs for the HPGR circuit were US\$1.39/t, accounting for 39.8% of the total operating cost of US\$3.49/t. In comparison, energy costs are 29% greater for the SABC circuit.

As described in the previous section, Rosario (2017) proposed an alternative HPGR circuit to reduce capital costs (Figure 22). Tables 12 and 13 compare capital and operating costs for SABC, conventional HPGR, and alternative HPGR circuits. As the tables show, the capital costs for the alternative

Table 10 SABC and HPGR circuits—capital cost comparison

Cost, 140 kt/d	SABC, US$ millions	HPGR, US$ millions
Direct capital cost	586	709
Indirect capital cost, 30%	176	211
Total	**762**	**920**
Difference		**158**

Source: Costello and Brown 2015

Table 11 SABC and HPGR circuits—operating cost comparison

Cost Item*	SABC, US$/t	HPGR, US$/t
Liners	0.345	0.401
Media	0.999	0.510
Power	1.795	1.389
Repair supply costs	0.522	0.783
Operation and maintenance labor costs	0.410	0.410
Total	**4.070**	**3.493**
Difference		**(0.577)**

Source: Costello and Brown 2015
*For a 140-kt/d plant.

Table 12 SABC, conventional HPGR, and simplified HPGR circuits—capital cost comparison

Case Study, 50 kt/d	Direct Cost, Can$ millions	Indirect Cost, Can$ millions	Total Capital Cost, Can$ millions
SABC	202.1	130.6	332.7
Traditional HPGR	240.2	154.4	394.6
Simplified HPGR	197.6	129.4	327.0

Source: Rosario 2017

Table 13 SABC, conventional HPGR, and simplified HPGR circuits—operating cost comparison

Cost Item*	SABC, Can$/t	Traditional HPGR, Can$/t	Simplified HPGR, Can$/t
Labor	0.67	0.79	0.77
Power	3.65	2.78	2.84
Consumables	1.59	0.94	0.97
Maintenance	0.16	0.18	0.16
Total operating expenditures	**6.07**	**4.69**	**(4.74)**

Source: Rosario 2017
*For a 50-kt/d plant.

circuit design are in line with the SABC costs, while operating cost benefits are maintained.

The operating cost difference is affected by changes in energy costs and variations in ore hardness. Higher energy costs and higher ore competency will increase the operating cost benefits of the HPGR process. HPGR circuits are less sensitive to ore variability than SAG circuits such that for deposits with high variability, the operating costs may favor HPGR operations. Reported increases in tire lives (in some cases greater than 10,000 hours) will lead to greater availability and lower replacement costs, which will further lower operating costs.

CONCLUSION

HPGR technology is an effective comminution device for greenfield projects and brownfield retrofits, to increase plant capacity or to maintain plant capacity when deeper ores become more competent. The HPGR is more energy efficient and reduces overall grinding energy between 15% and 30%. This may vary for specific applications, depending on the target grind size and ore characteristics. HPGR processing has generally shown to be more cost efficient and reduces comminution costs by about 20%. This may vary for specific sites, depending on the unit cost of energy, unit cost of grinding media, and the rate of grinding media wear.

Direct energy, economic, and operating cost efficiencies are traditional criteria employed in the selection of comminution circuits. In addition to these, embodied energy, environmental impact, and emissions criteria are playing an increasing role in decision making for comminution circuits. HPGR has benefits when these criteria are applied for new comminution circuit evaluations.

REFERENCES

Altun, O., Benzer, H., Dundar, H., and Aydoğan, N.A. 2011. Comparison of open and closed circuit HPGR application on dry grinding circuit performance. *Min. Eng.* 24(3-4): 267–275.

Amelunxen, P., and Meadows, D. 2011. Not another HPGR trade-off study! *Miner. Metall. Process.* 28:1–7.

Austin, L.G., Weller, R., and Lim, W.I.L. 1993. Phenomenological modelling of the high pressure grinding rolls. In *18th International Mineral Processing Congress*. Melbourne, Victoria: Australasian Institute of Mining and Metallurgy. pp. 87–96.

Aydoğan, N.A., Ergün, L., and Benzer, H. 2006. High pressure grinding rolls (HPGR) applications in the cement industry. *Miner. Eng.* 19(2):130–139.

Battersby, M.J.G., Kellerwessel, H., and Oberheuser, G. 1993. High pressure particle bed comminution of ores and minerals—A challenge. In *18th International Mineral Processing Congress*. Melbourne, Victoria: Australasian Institute of Mining and Metallurgy. pp. 1403–1407.

Bulled, D., and Husain, K. 2008. The development of a small-scale test to determine work index for high pressure grinding rolls. In *40th Annual Meeting of Canadian Mineral Processors*. Edited by I. Orford. Ottawa, ON: Canadian Mineral Processors. pp. 145–164.

Burchardt, E. 2012. HPGRs in minerals: What do existing operations tell us for the future? In *Fifth Autogenous and Semi-Autogenous Grinding Technology Conference*. Westmount, QC: Canadian Institute of Mining, Metallurgy and Petroleum.

Campbell, J.J., Bearman, R.A., and Thornton, G. 2001. The effect of feed size distribution on high pressure grinding rolls performance. Presented at Comminution 2001, Brisbane, Queensland, March 21–23.

Costello, B., and Brown, J. 2015. A tabletop cost estimate review of several large HPGR projects. In *Sixth International Conference on Semi-Autogenous and High Pressure Grinding Technology*. Edited by B. Klein, K. McLeod, R. Roufail, and F. Wang. Westmount, QC: Canadian Institute of Mining, Metallurgy and Petroleum.

Daniel, M.J. 2002. HPGR model verification and scale up. Master's thesis, University of Queensland, Brisbane.

Daniel, M.J., and Bailey, C. 2009. Uses and benefits of laboratory-scale HPGR tests. In *Procemin 2009: Proceedings of the VI International Mineral Processing Seminar*. Edited by P. Amelunxen, W. Kracht, and R. Kuyvenhoven. Santiago, Chile: Gemacin.

Daniel, M.J., and Morrell, S. 2004. HPGR model verification and scale-up. *Miner. Eng.* 17(11-12):1149–1161.

Davaanyam, Z. 2015. Piston press test procedures for predicting energy-size reduction of high pressure grinding. PhD dissertation, University of British Columbia, Vancouver.

Davaanyam, Z., Klein, B., Nadolski, S., and Kumar, A. 2013. A new bench scale test for determining energy requirement of an HPGR. In *Materials Science and Technology Conference and Exhibition*. Materials Park, OH: Materials Science and Technology.

Davaanyam, Z., Klein, B., and Nadolski, S. 2015. Using piston press tests for determining optimal energy input for an HPGR operation. In *Sixth International Conference on Semi-Autogenous and High Pressure Grinding Technology*. Edited by B. Klein, K. McLeod, R. Roufail, and F. Wang. Westmount, QC: Canadian Institute of Mining, Metallurgy and Petroleum.

Dunne, R. 2006. HPGR: The journey from soft to competent and abrasive. In *Proceedings of the International Conference on Autogenous and Semi-Autogenous Grinding Technology 2006*. Vancouver, BC: Department of Mining Engineering, University of British Columbia. pp. 190–205.

Dunne, R.C., Goulsbra, A., and Dunlop, I. 1996. High pressure grinding rolls and the effect on liberation: Comparative test results. In *Randol Gold Forum '96: Squaw Creek*. Golden, CO: Randol International.

Engelhardt, D., Seppelt, J., Waters, T., Apfelt, A., Lane, G., Yahyaei, M., and Powell, M.S. 2015. The Cadia HPGR-SAG circuit—From design to operation—The commissioning challenge. In *Sixth International Conference on Semi-Autogenous and High Pressure Grinding Technology*. Edited by B. Klein, K. McLeod, R. Roufail, and F. Wang. Westmount, QC: Canadian Institute of Mining, Metallurgy and Petroleum. pp. 1–18.

Fuerstenau, D.W., Shukla, A., and Kapur, P.C. 1991. Energy consumption and product size distributions in choke-fed, high-compression roll mills. *Int. J. Miner. Process.* 32:59–79.

Gardula, A., Das, D., DiTrento, M., and Joubert, S. 2015. First year of operation of HPGR at Tropicana gold mine—Case study. In *Sixth International Conference on Semi-Autogenous and High Pressure Grinding Technology*. Edited by B. Klein, K. McLeod, R. Roufail, and F. Wang. Westmount, QC: Canadian Institute of Mining, Metallurgy and Petroleum.

Gunter, H., Muller, M., and Sonntag, P. 1996. The application of roller presses for high pressure comminution. Presented at the Symposium on Grinding Processes, Toulouse, France, February 14–15.

Hart, S., Parker, B., Rees, T., and Manes, A. 2011. Commissioning and ramp up of the HPGR circuit at Newmont Boddington Gold. In *Fifth Autogenous and Semi-Autogenous Grinding Technology Conference*. Westmount, QC: Canadian Institute of Mining, Metallurgy and Petroleum. pp. 1–23.

Hawkins, R. 2007. A piston and die test to predict laboratory-scale HPGR performance. Master's thesis, University of Queensland, Brisbane.

Herman, V.S., Harbold, K.A., Mular, M.A., and Biggs, L.J. 2015. Building the world's largest HPGR—the HRC 3000 at the Morenci Metcalf concentrator. In *Sixth International Conference on Semi-Autogenous and High Pressure Grinding Technology*. Edited by B. Klein, K. McLeod, R. Roufail, and F. Wang. Westmount, QC: Canadian Institute of Mining, Metallurgy and Petroleum.

Hilden, M., and Powell, M. 2008. *CSRP '08: Delivering Sustainable Solutions to the Minerals and Metals Industries*. Kensington, Western Australia: Cooperative Research Centre for Sustainable Resource Processing.

Kalala, J.T., Dong, H., and Hinde, A.L. 2011. Using piston-die press to predict the breakage behaviour of HPGR. In *Fifth Autogenous and Semi-Autogenous Grinding Technology Conference*. Westmount, QC: Canadian Institute of Mining, Metallurgy and Petroleum. pp. 1–15.

Kazerani Nejad, R., and Sam, A. 2017. Limitation of HPGR application. *Miner. Process. Extr. Metall.* 126(4):224–230.

Kellerwessel, H., and Oberheuser, G. 1995. Scale up of roller presses. In *XIX International Minerals Processing Congress*. Vol. 2. Littleton, CO: SME. pp. 67–70.

Klymowsky, I.B., and Liu, J. 1997a. Modelling of the comminution in a roller press. In *Proceedings of the XX International Mineral Processing Congress: Material Analysis, Plant Design, Operating Practice, Control, and Simulation*. Edited by H. Hoberg. Clausthal-Zellerfield, Germany: GDMB Society of Metallurgists and Miners.

Klymowsky, I.B., and Liu, J. 1997b. Progress in the modelling of comminution in a roller press. *ZKG Int.* 50(9):500–510.

Klymowsky, R., Patzelt, N., Knechtz, J., and Burchardt, E. 2002. Selection and sizing of high pressure grinding rolls. In *Mineral Processing Plant Design, Practice, and Control*. Vol. 1. Edited by A.L. Mular, D.N. Halbe, and D.J. Barratt. Littleton, CO: SME. pp. 636–668.

Klymowsky, R., Patzelt, N., Knecht, J., and Burchardt, E. 2008. Keynote address: HPGR Technology, Present and Future. In *Procemin 2008: Proceedings of the V International Mineral Processing Seminar*. Santiago, Chile: Gecamin.

Knecht, J. 1994. High-pressure grinding rolls—A tool to optimise treatment of refractory and oxide gold ores. In *Fifth Mill Operators' Conference*. Melbourne, Victoria: Australasian Institute of Mining and Metallurgy. pp. 51–59.

Knecht, J. 2000. High pressure grinding—Application and outlook, crushing, grinding and pyroprocessing for minerals. Presented at the Technical Seminar, Brisbane, Queensland, May 25.

Knecht, J., and Patzelt, N. 2004. High pressure grinding rolls—Applications for the platinum industry. In *International Platinum Conference: Platinum Adding Value*. Johannesburg: Southern African Institute of Mining and Metallurgy. pp. 83–88.

Knorr, B., Herman, V., and Whalen, D. 2013. HPR: Taking HPGR efficiency to the next level by reducing edge effect. In *Procemin 2013: Proceedings of the 10th International Mineral Processing Seminar*. Santiago, Chile: Gemacin.

Knorr, B., Elkin, N., Mayfield, N., Mular, M.A., and Whalen, D. 2015. Pilot study of various HPGR circuit arrangements and crusher configurations. In *Sixth International Conference on Semi-Autogenous and High Pressure Grinding Technology.* Edited by B. Klein, K. McLeod, R. Roufail, and F. Wang. Westmount, QC: Canadian Institute of Mining, Metallurgy and Petroleum.

Kock, F., Siddall, L., Lovatt, I.-A., Giddy, M., and DiTrento, M. 2015. Rapid ramp-up of the Tropicana HPGR circuit. In *Sixth International Conference on Semi-Autogenous and High Pressure Grinding Technology.* Edited by B. Klein, K. McLeod, R. Roufail, and F. Wang. Westmount, QC: Canadian Institute of Mining, Metallurgy and Petroleum. pp. 1–20.

Koski, S., Vanderbeek, J., and Enriquez, J. 2011. Cerro Verde concentrator—Four years operating HPGRs. In *Fifth Autogenous and Semi-Autogenous Grinding Technology Conference.* Westmount, QC: Canadian Institute of Mining, Metallurgy and Petroleum. pp. 1–19.

Lim, W.I.L., and Tondo, L. 1996. *Experimental Program and De-Agglomeration Procedure.* Final report of the AMIRA Project P428. Report No. P428/11. Perth: Western Australian School of Mines, Curtin University.

Lim, W.I.L., and Weller, K.R. 1997. Modelling of throughput in the high pressure grinding rolls. In *Proceedings of the XX International Mineral Processing Congress: Material Analysis, Plant Design, Operating Practice, Control, and Simulation.* Edited by H. Hoberg. Clausthal-Zellerfield, Germany: GDMB Society of Metallurgists and Miners. pp. 173–184.

Lim, W.I.L., Campbell, J.J., and Tondo, L.A. 1997. The effect of rolls speed and rolls surface pattern on high pressure grinding rolls performance. *Miner. Eng.* 10(4):401–419.

Liu, L., Tan, Q., Liu, L., and Cao, J. 2018. Comparison of different comminution flowsheets in terms of minerals liberation and separation properties. *Miner. Eng.* 125(12):26–33.

López, L. 2011. *Technical Report for the Sierra Gorda Project, Chile.* Prepared for Quadra FNX Mining Ltd. Publication no. DE-00138. Lakewood, CO: Pincock, Allen, and Holt.

Maxton, D., Van der Meer, F., and Gruendken, F. 2006. KHD Humboldt Wedag—150 years of innovation new developments for the KHD roller press. In *Proceedings of the International Conference on Autogenous and Semi-Autogenous Grinding Technology 2006.* Vancouver, BC: Department of Mining Engineering, University of British Columbia.

McClintock, M., Wang, C., Klein, B., and Kumar, A. 2018. Validation and development of bench scale testing for full scale HPGR. Presented at the XXIX International Mineral Processing Congress, Moscow, Russia.

Metso Minerals. 2016. High pressure grinding rolls: HRC—The evolution of HPGR technology. Brochure No. 2841-01-16-ESBL/SALA. York, PA: Metso Minerals Industries. www.metso.com/globalassets/saleshub/documents---episerver/hrc_brochure_2841_en_lr.pdf.

Morley, C. 2008. HPGR: Frequently asked questions. Presented at Comminution ‘08, Falmouth, Cornwall, United Kingdom, June 17–20.

Morley, C. 2010. HPGR—FAQ. *J. S. Afr. Inst. Min. Metall.* 110(3):107–115.

Morrell, S. 2004. An alternative energy-size relationship to that proposed by Bond for the design and optimisation of grinding circuits. *Int. J. Miner. Process.* 74(1-4):133–141.

Morrell, S. 2006. Rock characterisation for high pressure grinding rolls circuit design. In *Proceedings of the International Conference on Autogenous and Semi-Autogenous Grinding Technology 2006.* Vol 4. Vancouver, BC: Department of Mining Engineering, University of British Columbia. pp. 267–278.

Morrell, S. 2009. Predicting the overall specific energy requirement of crushing, high pressure grinding roll and tumbling mill circuits. *Miner. Eng.* 22(6):544–549.

Morrell, S., Lim, W.I.L., Shi, F., and Tondo, L. 1997a. Modelling of the HPGR crusher. In *Comminution Practices.* Edited by S. Kawatra. Littleton, CO: SME. pp. 117–126.

Morrell, S., Shi, F., and Tondo, L. 1997b. Modelling and scale-up of high pressure grinding rolls. In *Proceedings of the XX International Mineral Processing Congress: Material Analysis, Plant Design, Operating Practice, Control, and Simulation.* Edited by H. Hoberg. Clausthal-Zellerfield, Germany: GDMB Society of Metallurgists and Miners. pp. 129–140.

Mular, M.A., Hoffert, J.R., and Koski, S.M. 2015. Design and operation of the Metcalf concentrator comminution circuit. In *Sixth International Conference on Semi-Autogenous and High Pressure Grinding Technology.* Edited by B. Klein, K. McLeod, R. Roufail, and F. Wang. Westmount, QC: Canadian Institute of Mining, Metallurgy and Petroleum.

Napier-Munn, T.J., Morrell, S., Morrison, R.D., and Kojovic, T. 1996. Mineral comminution circuits—Their operation and optimisation. Julius Kruttschnitt Mineral Research Centre, Monograph. Vol 2. Brisbane: University of Queensland.

Otte, O. 1988. Polycom high pressure grinding principles and industrial application. In *Third Mill Operators' Conference.* Melbourne, Victoria: Australasian Institute of Mining and Metallurgy. pp. 131–136.

Patzelt, N., Knecht, H., and Baum, W., 1997. The metallurgical potential of high-pressure roll grinding. In *Proceedings of the XX International Mineral Processing Congress: Material Analysis, Plant Design, Operating Practice, Control, and Simulation.* Vol. 2. Edited by H. Hoberg. Clausthal-Zellerfield, Germany: GDMB Society of Metallurgists and Miners. pp. 155–164.

Patzelt, N., Knecht, J., Burchardt, E., and Klymowsky, R. 2000. Challenges for high pressure grinding in the new millennium. In *Seventh Mill Operators Conference.* Melbourne, Victoria: Australasian Institute of Mining and Metallurgy. pp. 47–55.

Patzelt, N., Klymowsky, R., Burchardt, E., and Knecht, J. 2001. High pressure grinding rolls in AG/SAG mill circuits: The next step in the evolution of grinding plants for the new millennium. In *SAG Conference Proceedings: Proceedings.* Edited by A.L. Mular. Vancouver, BC: Department of Mining Engineering, University of British Columbia. pp. 107–123.

Rosario, P. 2010. Comminution circuit design and simulation for the development of a novel high pressure grinding roll circuit. PhD dissertation, University of British Columbia, Vancouver.

Rosario, P.P. 2017. Technical and economic assessment of a non-conventional HPGR circuit. *Miner. Eng.* 103:102–111.

Rosario, P.P., and Drozdiak, J.A. 2015. Creative and simpler HPGR circuits may increase their application even in the current restrictive financial environment. In *Sixth International Conference on Semi-Autogenous and High Pressure Grinding Technology.* Edited by B. Klein, K. McLeod, R. Roufail, and F. Wang. Westmount, QC: Canadian Institute of Mining, Metallurgy and Petroleum.

Rosario, P., Hall, R., Grundy, M., and Klein, B. 2011. A novel AG-crusher-HPGR circuit for hard, weathered ores containing clays. In *Fifth Autogenous and Semi-Autogenous Grinding Technology Conference.* Westmount, QC: Canadian Institute of Mining, Metallurgy and Petroleum.

Rule, C.M., Smit, I., Cope, A.J., and Humphries, G.A. 2008. Commissioning of the Polycom 2.2/1.6 5.6 MW HPGR at Anglo Platinum's new Mogalakwena North concentrator. Presented at Comminution '08, Falmouth, Cornwall, United Kingdom, June 17–20.

Saramak, D., and Kleiv, R.A. 2013. The effect of feed moisture on the comminution efficiency of HPGR circuits. *Miner. Eng.* 43-44:105–111.

Schönert, K. 1982. Method of fine and very fine comminution of materials having brittle behavior. U.S. Patent 4,357,287.

Schönert, K. 1988. A first survey of grinding with high-compression roller mills. *Int. J. Miner. Process.* 22:401–412.

Schönert, K. 1991. Advances in comminution fundamental, and impacts on technology. In *XVII International Mineral Processing Congress*. Vol. 1. Freiberg, Germany: Frieberg University of Mining and Technology. pp 1–21.

Schönert, K., and Lubjuhn, U. 1990. Throughput of high compression roller mills with plane and corrugated rollers. In *7th European Symposium [on] Comminution.* Edited by K. Schönert. Ljubljana, Yugoslavia: Fakulteta za Naravoslovje in Tehnologijo-Vtozd Montanistika. pp. 149–158.

Schumacher, M., and Theisen, W. 1997. Hexadur: A novel wear protection of high pressure roller presses for comminution. *ZKG Int.* 50(10):529–539.

Schumacher, M., and Theisen, W. 1998. Hexadur: High wear resistant rollers for high pressure roller presses. *World Cement* 3:35–41.

Sonmez, B., Olivera, R., Jankovic, A., Valery, W., Us, M. 2016. Metso HRC energy efficient comminution technology. In *Proceedings of XVI Balkan Mineral Processing Congress*. Belgrade: Mining Institute, Academy of Engineering Science of Serbia, University of Belgrade. pp. 131–138.

Stephenson, I. 1997. The downstream effects of high pressure grinding rolls processing. PhD thesis, Julius Kruttschnitt Mineral Research Centre, Department of Mining and Metallurgical Engineering, University of Queensland, Brisbane.

Tavani, S., Williams, C., and Hart, S. 2015. Reflections on HPGR circuit operation at Newmont Boddington Gold. In *Sixth International Conference on Semi-Autogenous and High Pressure Grinding Technology*. Edited by B. Klein, K. McLeod, R. Roufail, and F. Wang. Westmount, QC: Canadian Institute of Mining, Metallurgy and Petroleum. pp. 1–21.

Torres, M., and Casali, A. 2009. A novel approach for the modelling of high-pressure grinding rolls. *Miner. Eng.* 22(13):1137–1146.

Van der Meer, F., and Gruendken, A. 2010. Flowsheet considerations for optimal use of high pressure grinding rolls. *Miner. Eng.* 23:663–669.

Van der Meer, F., and Maphosa, W. 2012. High pressure grinding moving ahead in copper, iron and gold processing. *J. S. Afr. Inst. Min. Metall.* 112(July):637–647.

Van Drunick, W., and Smit, I. 2006. Energy efficient comminution: HPGR experience at Anglo research. In *Proceedings of the International Conference on Autogenous and Semi-Autogenous Grinding Technology 2006*. Vol. 4. Edited by M.J. Allan, K. Major, B.C. Flintoff, B. Klein, and A.L. Mular. Vancouver, BC: Department of Mining Engineering, University of British Columbia. pp. 124–139.

Vanderbeek, J., and Gunson, A.J. 2015. Cerro Verde 240,000 TPD concentrator expansion. In *Sixth International Conference on Semi-Autogenous and High Pressure Grinding Technology.* Edited by B. Klein, K. McLeod, R. Roufail, and F. Wang. Westmount, QC: Canadian Institute of Mining, Metallurgy and Petroleum. pp. 1–25.

Vanderbeek, J.L., Linde, T.B., Brack, W.S., and Marsden, J.O. 2006. HPGR implementation at Cerro Verde. In *Proceedings of the International Conference on Autogenous and Semi-Autogenous Grinding Technology 2006*. Vancouver, BC: Department of Mining Engineering, University of British Columbia. pp. 45–61.

Weir Minerals. 2010. *KHD High Pressure Grinding Rolls: First Choice for HPGR Technology and Service*. Product brochure. Glasgow, United Kingdom: Weir Group.

Weller, K.R., and Lim, W.I.L. 1999. Some benefits of using studded surfaces in high pressure grinding rolls. *Miner. Eng.* 12(2):187–203.

CHAPTER 3.8

Grinding Circuit Design

Aidan Giblett and Brian Putland

Other chapters in this handbook reviewed the main grinding technologies used in mineral processing and under which circumstances they will be a more practical flow sheet component than others. For any project, the optimum grinding flow sheet will be subject to many influences and will require detailed consideration of numerous elements, including process plant capital; operating costs; specifics of metal recovery processes; product quality requirements; and geographical, social, and environmental factors. These project-specific considerations can be even more significant than the technical criteria in many cases and have a significant bearing on final grinding technology selection.

This chapter focuses on the technical factors that must be considered during project development to optimize the initial process design, after the preferred grinding circuit flow sheet has been selected. These technical considerations are critical to the successful performance of the project post-commissioning. The authors strongly believe in the value of robustness when it comes to sample selection and ore characterization protocols, and the application of well-validated methods for the sizing and selection of grinding mills. Failure to execute these stages of development with the required level of diligence can have serious ramifications for the profitability of the project. Professionals can readily speculate how many of the circuits that are currently secondary-crushing their semiautogenous grinding (SAG) mill feed could have avoided the added capital expense of that retrofit, and the preceding years of below-nameplate plant production, if they had been more rigorous in their initial ore characterization or in the selection and application of methods to size the original grinding mills.

Fundamentally, the selection of a specific mill for a given grinding application requires several critical inputs:

- Sampling and testing to measure the hardness characteristics of the deposit
- Consolidation of this ore characterization data to establish a design basis for the mill sizing
- Identification of baseline mill operating conditions
- A method to estimate the specific grinding energy requirement for each stage of liberation
- A method to scale the specific grinding energy requirement to a mill input power
- A reliable model to predict the power draw of a mill, and thereby to confirm the selection of an adequate mill size for the task

The methods and protocols of determining the specific energy for a given grinding duty are covered in detail in Chapter 3.12, "Testing and Calculations for Comminution Machines." Implicit in this calculation is an understanding of the input and output particle size distributions of the grinding circuit. In addition, selecting the grinding mills will require estimation of specific operating conditions such as mill rotational speed, ball load, total volumetric filling, sizing screen apertures, recirculating loads, and mechanisms for managing critical size material. Several of these parameters will affect the power draw of the mill; others will influence the specific energy calculation of primary mills depending on the calculation method used. Additional factors that influence mill operability, availability, and grinding efficiency are discussed in this chapter. These include motor selection, liner selection, sizing and classification system design, and grinding media size.

DETERMINING THE DESIGN POINT

The significance of good sampling and testing protocols to determine reasonable grinding circuit design assumptions cannot be understated. The penalties for executing a limited or poorly designed sampling and testing campaign as a basis for grind circuit development can be severe. Production losses and the need to spend significant additional capital on expensive plant upgrades after start-up to achieve target production levels are heavy penalties to pay. In the worst cases, a poorly executed testing program can lead to project failure. The cost of a well-considered, inclusive sampling and testing program will usually be validated by predictable plant performance, achieved in a short time frame after start-up.

The tests that may be applied for grinding circuit evaluations are covered in detail in Chapter 3.12, "Testing and Calculations for Comminution Machines." In respect to which

Aidan Giblett, Senior Technical Advisor—Mineral Processing, Newmont Mining Services, Subiaco, Western Australia, Australia
Brian Putland, Principal Metallurgist, Orway Mineral Consultants, Perth, Western Australia, Australia

samples and how many should be selected for testing, there is no agreed standard. However, as the required accuracy of project design increases, the number of samples tested must also increase. The broad concepts of sampling a deposit in support of grinding circuit design are generally well accepted:

- Major ore domains must be defined in meaningful terms. Typically, especially early in the development of a project, they are defined in geological terms, reflecting characteristics such as host rock types (lithology) and alteration zones. A diligent testing program will challenge the metallurgical validity of the initial geological domains and, as necessary, establish more robust geometallurgical classifications for the deposit. As discussed in Chapter 1.10, "Geometallurgy," a valid geometallurgical domain will exhibit consistent properties throughout the project life. This will, in turn, lead to a consistent and predictable performance response in the process plant when treating either single-domain ore or domain blends. The cost of establishing valid geometallurgical domains is often seen as large, but it is small compared to both the cost of defining ore body geology and the present value of a robust project. Moreover, it is an investment that continually generates project value through reduced design risk, rapid project commissioning and increased confidence in mine planning, and estimating project optimization benefits.
- Sufficient samples within each major domain should be tested to reliably represent the typical domain characteristics as well as the extremes of ore hardness. Ideally, sufficient samples should be tested to allow definition of the full range of variation within each domain at least in spatial terms, and with depth in economic zones of the deposit. Good guidance on the quantity of hardness tests that should be initially conducted is given by Morrell (2011), who specified a minimum of 10 samples where low variability is observed and a minimum of 40 samples when high variability is observed. This guidance is best applied to individual ore domains rather than entire deposits to ensure that sources of variability within the deposit can be isolated and studied in detail. In design studies these initial numbers may be increased as the level of detail required by the project increases from the scoping stage to final bankable feasibility (Lane et al. 2013). It is particularly important that ore variability in the early (payback) years of the project is fully defined, to ensure the financial success of the project.
- Domain hardness modeling must be sufficient to generate a realistic prediction of the hardness progression by period (months, quarters, or years) throughout the project life. Important periods underpinning a commercially successful project are the commissioning period (typically year 1) and the payback period (typically up to year 5). If there are insufficient samples in the set representing the commissioning period, in particular, then urgent supplementary testing is necessary on suitable samples from within the commissioning ore zone. A reliable hardness model can guide designers in the scheduling of milling expansions, allow the evaluation of direct-tip (single handling) and blended-ore (double or even triple handling) plant feeding designs, and may result in a rescheduling of the mine plan to smooth annual production capacity, if the geology and mining allow. Every time the mine plan changes during design, the hardness prediction model needs to be rerun.

Because of the frequently observed variability in ore properties with increasing deposit depth, maintaining the accuracy of geometallurgical models in operating plants requires ongoing sampling and testing of samples, often on a yearly basis. It is not unusual in larger deposits for 100–200 samples to be taken and tested for this purpose per year. The number of samples required is dictated by the variability of the deposit and, contrary to what some literature suggests, is not related to the type of comminution test performed.

In some cases, typically early-stage studies, there may be insufficient information to divide the deposit into well-defined domains that reliably define the physical characteristics of the ore. In these instances, the hardness test results will typically be lumped together and a single design point chosen for provisional grinding equipment sizing and development of the capital estimate. In this case, a conservative design point should be selected to provide confidence that the grinding equipment size will be adequate to accommodate variations in hardness through the deposit. The conservative design point may reflect a database percentile approach, where a higher value than the average, such as the 75th or 85th percentile value of the data set, will be chosen. Where the hardness data set is very small, perhaps consisting of only a handful of results, the design basis is often selected as simply the hardest value in the data set. Such design approaches recognize and address the risk of false representation of the deposit's hardness characteristics that may be provided by small data sets.

As the project progresses, the intent should be to conduct sufficient hardness tests on carefully selected samples chosen to represent the observed variations in geological characteristics, deposit depth, and spatial distribution to provide statistically robust hardness data sets by ore domain. Under such circumstances, the average ore domain characteristics can be combined with tonnage distributions by domain in the mine plan to provide a robust basis for the sizing of grinding mills. When the average ore characteristics of a resource are used as the design basis, it is common to also apply a safety factor to the mill sizing. This factor is normally applied to the estimated grinding power requirement, which provides the design contingency that may otherwise be given using a more conservative percentile value design basis. Safety factors of 10%–20% are typically applied depending on the extent of testing performed and the design point selected. Larger safety factors will be applied when database average ore characteristics are used for the design point, and in early stages of project evaluation when the testing data set is small and less likely to accurately reflect the distribution of ore characteristics through the resource. In deciding to use these factors, the reader must understand that there is an inherent minimum precision of each testing method, limited accuracy of the design methods, and an inability of operating plants to maintain ideal conditions in the process at all times. All these risks and inefficiencies must be understood in selecting a design contingency and operating envelope that will deliver a circuit that will achieve the desired annual production capacity.

DETERMINING THE GRINDING ENERGY REQUIREMENT

The methods available to determine specific energy are discussed at length in Chapter 3.12, "Testing and Calculations for Comminution Machines." The most appropriate method of estimating specific energy can vary depending on the duty at hand. For example, there is no contention that the methods of Bond and Rowland are the optimal approach to size a rod mill. The approaches of Bond, Rowland, and Morrell are suited to sizing ball mills in primary or secondary grinding duties. Several methods, most at least to some degree proprietary, are currently used for the sizing and selection of autogenous grinding/semiautogenous grinding (AG/SAG) mills. In the case of stirred milling and compressive grinding technologies, machine-specific in-house testing procedures and equipment sizing methodologies have been developed that require significant vendor participation in the mill sizing process before the vendor will consider warranting equipment performance.

DETERMINING THE MOTOR POWER REQUIREMENTS

After determining the mill-specific energy requirement at the pinion, or mill shell, as estimated by common power-based calculation methods, the user must convert this figure to a motor sizing that will supply the power demand of the plant at the target production rate. The final motor selection will depend on the design-specific energy requirement (S) in kilowatt-hours per ton, design throughput rate (t) in metric tons per hour, and factors for motor power utilization (PF_1) and drivetrain efficiency (PF_2) by the following equation:

$$\text{motor power, hp} = \frac{1.341 \times S \times t}{PF_1 \times PF_2} \qquad \textbf{(EQ 1)}$$

In operation, variability in the feed plus other plant-based disturbances affect the power draw of tumbling mills, causing it to fluctuate. Consequently, it is not possible to operate mills for any period at their maximum rated output of the motor for fear of overloading the motor when these variations cause high power excursions. The long-term sustainable average power draw of mills is therefore, by necessity, somewhat lower than the motor rated power. The result is that the maximum power that the motor can deliver is rarely fully available for ore processing purposes, and predictions of sustainable throughputs of grinding mills should not use the maximum motor power. For AG/SAG mills, a reasonable assumption is that the sustainable average power draw that will be available for long-term throughput is approximately 90% of the installed motor capacity, although this can approach 0.95 for well-controlled variable-speed mills. This factor can also be as low as 0.85 for ores with highly variable properties. For ball mills, 95% is normally assumed because of the more stable operating environment for these machines. As a result of this stability, the occasional ball mill draws greater than design motor output if the shell size is sufficient and motor winding temperatures are carefully monitored. One issue with ball mills is that the shell dimensions are often undersized, and therefore the mill is incapable of drawing full installed power. This is a waste of applied capital in the design, and sizing should be scrutinized when selecting a ball mill. Typical values for PF_1 are 0.85–0.90 for SAG mills and 0.95 for ball mills.

Drivetrain efficiency (PF_2) reflects the difference between the motor input power and the power delivered to the mill shell or pinion, the latter being the value generated by traditional calculations of specific energy. This efficiency is a function of the type of motor used and the transmission system. Many technical articles describe how system efficiency varies for each of the common drive systems, notably Grandy et al. (2002) and Ploc and Peters (2010). Those authors ascribe similar drivetrain efficiency values, in the context of PF_2 influencing the transfer of motor input power to the pinion shaft, as the efficiency associated with the motor, gearbox, slip system, and ring gear components for the most common drive systems. From this analysis, typical PF_2 values for wound rotor induction motor systems will be 0.93, for low-speed synchronous motor systems will be 0.94–0.96, and for gearless drive systems will be 0.96–0.97.

Ancillary equipment may also be required upstream of the motor for to it operate correctly. For example, a variable-speed single pulse width modulated (PWM) synchronous motor requires an additional transformer and variable-frequency drive, both of which will consume power, approximately 4% total as shown in Table 1. Such ancillaries will not directly influence the size of the motor chosen but must be considered when determining the overall capital cost and total electrical operating cost.

DETERMINING THE MILL DIMENSIONS

After the required motor has been estimated, the dimensions of the mill(s) required to deliver the design power can be estimated using a tumbling mill power model, such as the Morrell model described in Chapter 3.12, "Testing and Calculations for Comminution Machines." Mill vendors also have their own calibrated power draw models. Care must be taken to use practical estimates of mill speed, ball charge levels, and total mill filling to ensure that the design power draw can be achieved under sustainable operating conditions. For example, if the design power draw requires the mill to operate at high speed, high ball charge, and/or high volumetric filling, there is little allowance for normal variations in operating conditions and the difficulty of maintaining precise control over charge levels.

In the case of AG/SAG mills, aspect ratio has a significant influence on the estimated specific energy and is discussed in the following section in more detail. Therefore, the aspect ratio that is used for estimating specific energy must also be used when choosing the mill dimensions to deliver the required power. Conversely, some iteration may be required to ensure that the specific energy values and power draw assumptions used are applicable to the final mill dimensions.

Alternatively, the mill can be estimated with reference to mill installation lists available from mill suppliers, based on the target mill motor size. If a power model is used to determine the mill dimensions, it should also be cross-referenced against the vendor's equipment lists to validate the selected dimensions. A mill with the dimensions and power that a vendor has built previously will be less expensive and probably be delivered more quickly than a mill with dimensions that the vendor has not engineered in the past.

Ultimately, the mill vendor will be charged with supplying a mill that can achieve the design power draw under practical operating conditions. Nonetheless, the project engineer must ensure that there is adequate contingency in the mill design so that the required performance is achieved, that the operating conditions assumed in design remain appropriate, and that the selected mill can deliver the expected performance, as the

Table 1 Mill drive system efficiency

		Component Efficiency, %						
Speed	**Type***	**Transformer**	**Variable-Frequency Drive**	**Motor**	**Gearbox**	**Ring Gear**	**Other**	**Total**
Fixed	Single synchronous	N/A†	N/A	96–97	N/A	98.5	N/A	94.6–95.5
	Single WR	N/A	N/A	95–96	99	98.5	N/A	92.6–93.6
	Dual quad	N/A	N/A	96–97	N/A	98.5	N/A	94.6–95.5
	Dual WR	N/A	97	95–96	99	98.5	97‡	89.9–90.8
Variable	Single PWM synchronous	98.3	97.8	96–97	N/A	98.5	N/A	90.9–91.9
	Single WR	99.7	99.7	95–96	99	98.5	N/A	92.1–93.1
	Single PWM induction	98.3	97.8	96.5–97.5	99	98.5	N/A	90.7–91.4
	Dual PWM synchronous	98.3	97.8	96–97	N/A	98.5	N/A	90.9–91.9
	Dual WR	99.7	99.7	95–96	99	98.5	N/A	92.1–93.1
	Dual PWM induction	98.3	97.8	96.5–97.5	99	98.5	N/A	90.7–91.4
	Gearless	98.8	99	96–97	N/A	N/A	99§, 99.5**	92.5–93.5

Source: Ploc and Peters 2010
*PWM = pulse width modulated; WR = wound rotor induction motor
†N/A = not applicable
‡Slip resistor
§Harmonic filter
**Motor ventilation

owner—not the vendor—will bear the real cost of any loss in available operating power over the life of a project.

MILL DESIGN ASSUMPTIONS

AG/SAG Mill Aspect Ratio

Popular power-based modeling approaches, such as those applied by Morrell (2004; 2006) and Lane et al. (2013), observe that mill aspect ratio has a material impact on AG/SAG mill capacity when operating in primary grinding duty, as illustrated by Figure 1. When discussing mill aspect ratio, note that aspect ratio is quoted in terms of either mill length divided by mill diameter (L:D) or mill diameter divided by mill length (D:L). In this discussion, the diameter-over-length methodology is observed, defining high D:L ratios for high–aspect ratio mills. Figure 1, however, observes the L:D representation, reflecting the visualization produced by Lane et al. (2013).

For the same power draw and feed characteristics, a high–aspect ratio mill will have a lower specific energy, higher throughput, and coarser transfer size than a low–aspect ratio mill when operated in open circuit. This will be reflected in a lower specific energy ratio for the low aspect mill, with reference to Figure 2. For example, a 30 ft × 15 ft effective grinding length (EGL) SAG mill has, by definition, an aspect ratio (D:L) of 2 and a relative specific energy factor of 1.0. Should the mill length be increased to 20 ft to increase the mill power draw, the aspect ratio decreases to 1.33 D:L, and by Figure 2 the specific energy ratio is nominally 1.1 for L:D = 0.75. This reflects a 10% increase in SAG mill specific energy and a commensurate reduction in SAG mill unit capacity. The precise impact of aspect ratio on SAG mill specific energy will be subject to other factors; however, the fact that aspect ratio will influence open-circuit SAG specific energy should not be overlooked.

This effect must be considered when sizing both primary and secondary mills, as these variations in transfer size determined by the primary mill aspect ratio will vary the work load of the secondary grinding stage. Should it be necessary to increase the primary mill power draw at any point of the design process to accommodate harder ore at sustained throughput rates, the designer should attempt to conserve the primary mill aspect ratio wherever possible, or otherwise allow for the

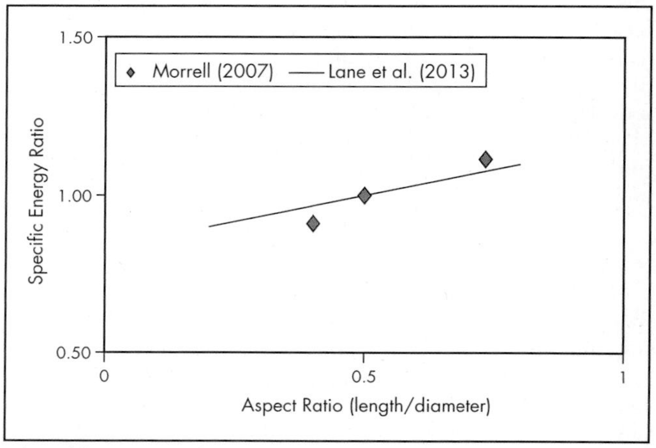

Adapted from Lane et al. 2013
Figure 1 Influence of SAG mill aspect ratio on specific energy

impacts of altering the aspect ratio on primary mill capacity and the transfer size to the secondary grinding stage.

In addition to the estimation of specific energy, modeling of grate and pulp lifter slurry flow capacity should be taken into account when selecting the aspect ratio of the mill, as discussed by Putland et al. (2011) in relation to the design of single-stage SAG mills. Grate open area is particularly important in the case of single-stage SAG mills or very low specific energy open-circuit SAG mills. The diameter of the mill should be selected first based on the required slurry flow with typical grate and pulp lifter performance. The mill length is then selected to achieve the required power draw. The use of too small a mill diameter will result in slurry pooling, which will negatively affect power draw. This may not be a significant problem if considered in the initial selection and design of the mill but can be an issue if the mill is unable to generate design power draw, resulting in reduced capacity. The presence of a slurry pool does not necessarily affect grinding efficiency for single-stage mills, as most of the required energy input is low in intensity. However, inefficiencies and overgrinding can occur when attempting to achieve a coarse

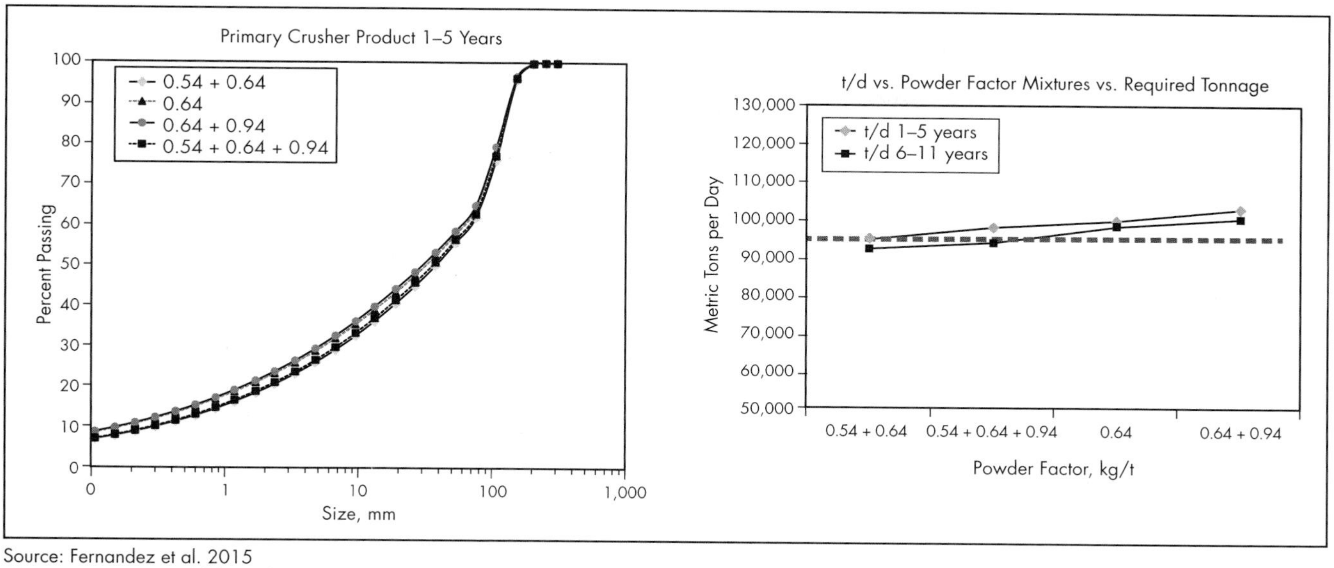

Source: Fernandez et al. 2015

Figure 2 Impact of blast design on run-of-mine size distribution and milling circuit capacity

grind from a competent ore and the slurry pool is providing a cushion, reducing the effectiveness of coarse ore breakage. The issue with accurate assessments of design slurry flow requirements is estimating an appropriate circulating load when not determined by a population model or piloting. In general, very high circulating loads or a coarse grind are generated by a lack of media, while very low circulating loads and overgrinding result from excess coarse rock. These two conditions should occur only if the incorrect single-stage SAG mill configuration has been selected. In the absence of these extreme conditions, circulating load is primarily influenced by the cyclone configuration. The size of cyclone, pressure, and conditions (spigot and vortex selection) that need to be operated to achieve the product size within the target cyclone overflow range determine the circulating load range. Standard vendor cyclone models can be used to estimate optimum cyclone operating conditions and therefore predict the likely circulating load range.

Mill Feed Size Distribution

In practice, mill feed size distribution has a significant impact on AG/SAG mill capacity. Morrell and Valery (2001) provided an example of a SAG mill where feed size variations, in F_{80} terms, of 3–4½ in. were consistent with tonnage rate and SAG specific energy variations of up to 30%. The same authors also illustrated the impact of variations in the fine end of the size distribution on mill capacity, with large variations (50%) in mill throughput achieved for similar F_{80} values, because of a measurable difference in fines content in the SAG mill feed. The amount of material in the SAG mill feed that is finer than the SAG mill discharge screen aperture is an important performance indicator, as this represents material that should readily pass through the SAG mill to the secondary grinding stage, increasing the primary mill capacity.

Such concepts are at the core of mine-to-mill optimization programs and require the design engineer to consider factors that influence mill feed size distribution during the mill design process. It is critical that the design consider the influence of in situ ore parameters, such as fracturing and compressive strength, in addition to drill-and-blast program parameters, such as powder factors, drill-hole diameter, drill-hole spacing, and stemming practices. At the design stage, this requires liaison with the geology and mining engineering groups, and the use of blast simulation software by a suitably skilled practitioner, to ensure these important influences on run-of-mine (ROM) ore distribution can be incorporated in the mill design. The influence of a full ROM particle size distribution data set on grinding circuit performance can only be determined by the application of a population balance simulation process. Such a detailed simulation approach to demonstrate the resultant impact on milling circuit capacity can identify situations where modifications to a standard mill design approach will be required to accommodate deposits and mining strategies that may deliver "nonstandard" ROM feed size distributions. A recent example of such a study was provided by Fernandez et al. (2015), who applied blast simulation software and population balance modeling of the comminution circuit to demonstrate the influence of blast design and geotechnical characteristics of the ore on the ROM size distribution, and subsequently the grinding circuit capacity. Figure 2 reflects the outcomes of this analysis, showing the variation in ROM size distribution and the magnitude of impact on mill capacity. The methods are equally applicable to the optimization of fragmentation in operations and the validation of mill sizing for greenfield projects. Care must be taken to not use an overly optimized design for greenfield projects, as the marriage of optimal mine fragmentation and mill operation (grate and lifter design) may take years to achieve and will impact the ramp-up of a project if used as the basis for design.

Critical Size Management

When the flow-sheet design considers primary autogenous or semiautogenous milling, it will be necessary to estimate how much critical size material will be generated during the primary grinding stage and determine how this material will be managed. In both autogenous and semiautogenous milling, the accumulation of critical size material has been shown to be detrimental to mill capacity as this material is, by definition, slow to grind in the mill. The relative grinding rates of critical size material are effectively illustrated by the typical JKSimMet SAG mill breakage rate curve (Figure 3), which reflects a deterioration of breakage rates in the ½–3 in. size range. While operational considerations can influence the position and relative depth of this trough in the breakage rate distribution, it will exist in some form for all autogenous mills. The most effective method to address the negative impacts of critical size accumulation is to implement pebble crushing in combination with the use of large aperture grates, more commonly termed *pebble ports*. Pebble ports allow ready removal of this critical size material from the AG/SAG mill and more effective breakage by pebble crushing. Pebble crushing is a very effective means to reduce the critical size material to the point where it can be more effectively ground. Ideally, the pebble crusher closed-side setting or gap will be set in concert with the AG/SAG discharge screen panel aperture to produce material that passes the AG/SAG mill discharge screen and directly reports to the secondary grinding stage. Crawford et al. (2009) demonstrated the material impact of effective pebble crushing to produce SAG screen undersize material on a semiautogenous grinding mill, ball mill, and pebble crusher (SABC) circuit throughput at the Telfer gold mine in Australia.

The extent to which an autogenous mill will produce pebbles varies as a function of the mill feed size distribution, the strength of the ore, and the mill operating conditions. To appropriately design the pebble handling system and then determine the size of the pebble crusher, the engineer must estimate the pebble production rates of the mill, normally expressed as a percentage of the fresh mill feed rate. In many cases, the use of generic assumptions based on the pebble production rates of other facilities has been unsuccessful and often leads to an oversized pebble crushing circuit. An oversized pebble crusher is difficult to choke feed without batch processing, which can be detrimental to AG/SAG mill stability. Inability to choke feed makes the crusher prone to rapid and localized liner wear, which in turn reduces the crusher's ability to maintain the target product size gradation. Underestimating the pebble production rate to the extent that the pebble handling system is undersized is similarly undesirable, as the material handling capacity of the pebble crusher feed system can become the milling circuit throughput bottleneck. A good initial estimate of pebble handling system requirements can be generated from rule-of-thumb pebble production estimates, such as 14.5 m^3/h of pebbles per square meter of pebble port open area as provided by Morrell in the JKSimMet training package (Morrell 1999). As with most design efforts, benchmarking of operating plants processing comparable material is invaluable to sanity-check any design inputs.

Ideally, the pebble crushing system will be designed based on an approach that considers the mill feed size distribution, ore hardness and competency parameters, mill grate and pebble port apertures, and mill operating conditions. Pilot mills have been traditionally used to indicate pebble production rates, although they are now less commonly a basis for SAG mill design. Population-based simulation models are the most capable of considering these factors, specifically ball load, mill filling levels, and grate and discharge screen aperture. A simulation approach that considers a range of measured ore characteristics and operating conditions can provide a meaningful estimate of typical levels and ranges of pebble production rates as the basis for crushing system design.

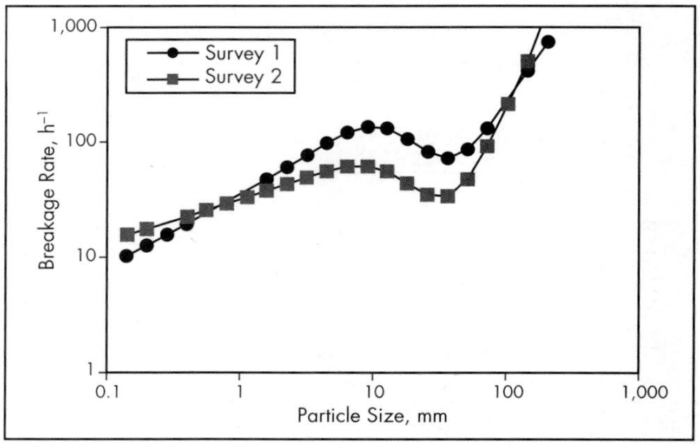

Courtesy of Metso Australia, Ltd.

Figure 3 Example JKSimMet SAG mill breakage rates

Ideally, the pebble crusher will be sized to treat an average or typical pebble production rate, with the screening and conveying system designed with an additional safety factor to allow peak loads to be accommodated. A pebble feed bin and feeder system is desirable to allow the pebble crusher to be choke fed at all times, and an overflow system should be considered to allow pebble production in excess of crusher capacity to be diverted to a static storage pile or directly back to the mill.

Callow and Meadows (2002) provided some practical guidance on pebble crushing system design, with key recommendations including the following:

- Allow for no less than 1 hour of live pebble storage capacity ahead of the crusher. Ideally, a pebble stockpile will be incorporated in the design.
- Favor multiple smaller crushers over a single large crusher for improved operating performance.
- Design for good grinding steel removal ahead of the pebble crushers. Consider combinations of trommel magnets, cross belt magnets, and metal detectors. A minimum of two stages of metal removal prior to a metal detector is recommended.

The pebble crushing system in the Telfer SABC circuit as described by Crawford et al. (2009) provides an example of good but capitally intensive design practices and the significant impact the ability to optimize a pebble crushing circuit can have on SABC circuit capacity. The Telfer circuit includes two-stage screening of the SAG mill discharge using a trommel (⅝-in. aperture) and pebble screen (½-in. aperture), a 3,000-t stockpile ahead of the pebble crushers, pebble crusher feed bins, and two pebble crushers for each grinding line. The consistent crusher feed rate afforded by the stockpile, and the ability to run single or multiple crushers as required were specifically cited by the authors as significant contributors to grinding circuit optimization.

The inclusion of a recycle crusher is detrimental to single-stage milling (Putland et al. 2011), except in certain circumstances. This is because the critical size material that holds up an open-circuit mill causing a fine transfer size is actually the media in the charge required to generate a finished product in single-stage SAG. Recycle crushing in single-stage mills is usually effective only for the treatment of very competent ores where the aim is to produce a coarse product size when the mill is an AG mill.

When used in a single-stage circuit, the pebble crusher is used for balancing load control with grind size. The crusher is brought online if the mill load is building up with the grind size getting too fine, and it is taken out of circuit when the load drops and achieving grind size becomes difficult with a coarser-than-desired value produced. Alternatively, the amount of pebble crushing can be proportionally adjusted with a percentage of the pebbles bypassing the crusher varied by the control system on the same basis. This configuration is best operated with optimization of the pebble crusher closed side set to maximize power draw. If pebbles are being crushed, they should be crushed fine for maximum value. In a single-stage mill with pebble crushing, smaller grate apertures are typically installed when compared to a SAG mill in SABC configuration. Typically, apertures <50 mm are used. This preserves a reasonable portion of coarse rock for use as media. Given the finer grate apertures, finer closed-side sets on pebble crushers and finer trommel or discharge screen apertures can be used.

Mill Speed

The critical speed concept applies to tumbling mills and is loosely described as the speed at which the mill will centrifuge and the charge of the mill will become pinned to the mill shell. At this point, the charge will no longer demonstrate the required tumbling action, and mill performance will significantly deteriorate. The critical speed (*Nc*) for a given mill diameter (*D*, ft) inside liners is determined by the following formula:

$$Nc\text{, rpm} = \frac{76.6}{\sqrt{D}} \quad \textbf{(EQ 2)}$$

Or for a mill diameter (inside liners), in meters:

$$Nc\text{, rpm} = \frac{42.3}{\sqrt{D}} \quad \textbf{(EQ 3)}$$

The concept of peripheral speed (*Np*) of the mill shell is an important factor in the design of rod and pebble mills, as discussed by Crocker (1985). The following equation from Rowland (2002) is used to calculate peripheral speed in units of feet per minute (fpm) or meters per minute (m/min), dependent on what units are used for the mill diameter (*D*, inside liners). *N* is the mill speed in revolutions per minute.

$$Np = \pi DN \quad \textbf{(EQ 4)}$$

High peripheral speeds in rod mills promote rod breakage and increased wear, and in pebble mills will promote faster pebble breakdown. Maximum peripheral speeds recommended by Crocker (1985) were 540 fpm for rod mills and 850 fpm for pebble mills.

AG/SAG mills typically operate in a wide range of critical speeds, dependent on mill operating conditions such as ball charge and total charge filling levels, feed particle size, and hardness. This speed range will generally be between 60% and 80% of critical. Lower speeds are used for softer ores and lower mill filling levels to avoid shell liner damage. Higher speeds are used for higher mill filling levels and more competent ores that require greater energy input to promote breakage. For all grate discharge mills, particularly SAG mills, the impact of mill speed on pulp discharge capacity must be understood. At high speeds, the mill discharge capacity may reduce as flowback or carryover increase. This is most commonly an issue in high-capacity open-circuit SAG mills or in closed-circuit SAG mills where the volumetric throughput of the mill is high.

The speed range of ball mills is much tighter, normally 72%–76% of critical, although higher speeds up to 80% of critical are used in small-diameter mills below 6 ft in diameter. Analysis performed at the Bougainville Copper Limited operation in Papua New Guinea (Plavina and Clark 1986) provided useful insight into the effect of mill speed on the performance of an 18-ft-diameter variable-speed ball mill. A series of trials concluded that variations between 70% and 80% of critical speed had no measurable impact on mill performance. Speeds exceeding 80% were observed to cause coarser grinds. Prior upgrades to increase the speed of the ball mills from 68% to 71% had been observed to increase plant throughput by approximately 3%.

Rod mill speeds are higher in smaller-diameter mills, with mills under 9 ft in diameter operating between 70% and 76% of critical speed; larger mills operate at slower speeds of 64%–70% to minimize liner wear and to prevent tangling of the rods. As noted by Rowland (1985), higher operating speeds increase the elevation of the rod charge and the spread between the rods, resulting in elevated rod wear and a coarsening of the mill discharge.

Pebble mill speeds are normally in the range of 70% to 75% of critical (Rowland 2002).

Ball Mill Volumetric Capacity

High slurry velocities and low slurry retention times in ball mills have been observed to be detrimental to ball mill grinding efficiency by studies performed by authors, including Rowland (1988) and Morrell (2001). Both investigations identified methods for determining critical slurry velocities through a ball mill, above which operational issues will be encountered that include coarse particle short-circuiting, excessive grinding media carryover, and reduced grinding efficiency.

Rowland (1988) provided Equations 5–8 to calculate the slurry residence time and a slurry velocity through the mill, and offered empirical guidance on targets for these values. A target mill residence time not less than 1.2–1.4 minutes and a slurry velocity not exceeding 6 m/min were recommended.

$$Vas = \frac{D^2}{4} \cdot \pi \cdot L(0.4Vp + Vdo - Vp) \quad \textbf{(EQ 5)}$$

where

Vas = mill volume available for slurry
D = mill diameter inside liners
L = mill effective grinding length
Vp = fraction of mill volume occupied by the ball charge
Vdo = level of the discharge opening expressed as a fraction of the mill volume

$$RT = \frac{Vas}{Vs} \quad \textbf{(EQ 6)}$$

where

RT = retention time, in minutes
Vs = slurry volume per minute

Slurry velocity is then calculated with the following equations:

$$Aas = \frac{D^2}{4} \cdot \pi \cdot (0.4Vp + Vdo - Vp) \quad \textbf{(EQ 7)}$$

Table 2 Basic arrangements of mill power transmissions

Motor Type	Speed	Pinion	Motor Speed, rpm	Primary Driver	Secondary Driver	Power Supply
Wound rotor	Fixed	Single or dual	900–1,200	Ring gear	Gearbox	Direct
	Variable	Single or dual	900–1,200	Ring gear	Gearbox	Slip energy recovery
Induction	Fixed	Single or dual	900–1,200	Ring gear	Gearbox	Direct
	Variable	Single or dual	900–1,200	Ring gear	Gearbox	Pulse width modulated
Synchronous	Fixed	Single	180–200	Ring gear	None	Direct
		Dual	180–200	Ring gear	None	Quadramatic
	Variable	Single or dual	180–200	Ring gear	None	Load commutated inverter, pulse width modulated, or cycloconverter
	Variable	None	9–15	Gearless	None	Cycloconverter

Source: Grandy et al. 2002

where *Aas* is the area available for slurry flow.

$$N = \frac{Vs}{Aas} \quad \textbf{(EQ 8)}$$

where N is the slurry velocity in distance per minute.

In a comparable approach, Morrell (2001) calculated a superficial velocity from the slurry flow rate through the cross-sectional area of the slurry pool that exists above the ball charge, extending from the toe of the charge to the discharge trunnion. A critical superficial velocity of 0.32 m/s was defined, above which a ball mill would be expected to demonstrate the operational issues associated with low residence time and high slurry velocities as discussed earlier. The basis for the calculation of Morrell's superficial velocity was not provided; however, his analysis confirmed the influence of pulp velocities on the performance of overflow ball mills.

MILL DRIVE SELECTION

A range of options are available for tumbling mill drives, and the selection is primarily influenced by the amount of mill power required, whether the mill will require fixed- or variable-speed operation, whether bidirectional mill rotation is required, the desired efficiency and mechanical availability of the drive system, and the capital costs associated with the individual options. The continuity of power supply should also be considered and more electrically robust systems preferred where frequent disruptions to power supply are possible. For stirred mills, the mill drive system is a standard component of the mill supply and therefore will only be subject to more subtle refinement during the mill selection process, in close consultation with the mill vendor. Table 2 summarizes the key features of common mill drive systems.

For large tumbling mills, the mill power requirement has a significant influence on the drive selection. Continual advances in the production of pinion drives have increased the capacity of single-pinion drives to 8.5 MW and of dual-pinion drives to 17.0 MW, as reported by Kalra et al. (2013) referencing the operating Wushan (China) copper and gold project ball mills. This change is a material increase over the 15.6-MW twin-pinion drives used on the Boddington (Australia) ball mills (Hart et al. 2011). The increased capacity keeps the dual-pinion drive technologies compatible with all but the very largest ball mills in use today. Coupled with variable-speed capabilities, these developments increase the applicability of gear-driven mills in AG/SAG mill applications. The relative dominance of gearless mill drives in SAG mill applications in recent decades is reflected in the review of Andean SAG mills by Rayo (2014), which reports 21 of 25 mills reviewed using gearless drives, with geared drives in use only in SAG mills with motor sizings of 8 MW or less in that analysis.

Variable-speed capability is an important requirement for AG and SAG mills to enable mill load levels and liner wear to be controlled in response to variations in mill feed size, fragmentation, pebble crusher performance, and ore characteristics. Fixed-speed SAG mill drives ideally should only be specified when ore characteristics are consistent through the deposit, although there are many instances where fixed-speed drives are selected on the basis of cost minimization. Variable-speed drives are rarely justified in the case of rod, ball, or pebble mills, as the same issues with variable mill load levels that can exist with SAG mills are not as significant an influence in environments where the grinding media filling all but completely dictates the volumetric composition of the mill.

Bidirectional capability is often a valued feature for SAG mills, as the ability to change the mill direction and use both faces of the shell lifter as the leading contact surface with the mill charge can allow up to 20% improvement in shell liner life. Continuous rotation in only one direction is, however, required in grate discharge mills that use curved pulp lifter designs. Furthermore, the use of a trommel screen to classify SAG mill discharge often restricts the mill operation to a single direction of rotation by virtue of the internal discharging spiral or scroll, which operates effectively only in one direction.

The optimum motor and drive configuration for a given project is a project-specific decision and must be based on thorough consideration of individual vendor submissions. Balanced consideration must be given to vendor performance records, capital and operating costs, system availability, and process performance requirements such as total power requirement and fixed- versus variable-speed capabilities. As noted by Grandy et al. (2002), this requires input from the process, electrical, and mechanical members of the project team.

INITIAL OPERATING CONDITIONS

Grinding Media Requirements

For rod and ball mills, the optimum ball mill recharge size can be estimated using the equations of Bond (1961). Both models predict the upper media diameter required for effective breakage.

For ball mills, makeup ball diameter (B), in inches, is given by the following equation:

$$B = \sqrt{\frac{F}{K}} \cdot \sqrt[3]{\frac{sg \cdot Wi}{100 \cdot Cs \cdot \sqrt{D}}} \quad \textbf{(EQ 9)}$$

where

F = mill feed size, µm, 80% passing
K = 350 for wet grinding and 335 for dry grinding
sg = specific gravity of ore
Wi = work index, kW·h/st
Cs = mill speed fraction of critical
D = mill diameter inside liners, ft

This empirical ball size estimation model was based on the benchmarking of primary and secondary ball mills, though not ball mills following SAG mills. The model will often predict too fine a ball size when used to predict the optimum ball size for ball mills following SAG mills, as the distribution is typically bimodal, often with particles significantly coarser than the 80% passing size. Neither the SAG transfer size nor cyclone underflow particle size distributions reliably approximate the F value for a primary ball mill or a rod mill product. A good general starting ball makeup blend for ball mills following SAG mills is a mixture of 2-in.- and 3-in.-diameter balls, with the relative distribution adjusted to accommodate variations in SAG screen panel aperture and ore strength and grindability characteristics. In operation, the ball top size should be adjusted downward where possible to maximize ball charge surface area without causing excessive scatting.

For rod mills, makeup rod diameter (B), in inches, is given by the following equation:

$$B = \frac{F^{0.75}}{160} \cdot \sqrt{\frac{Wi \cdot sg}{100 \cdot Cs \cdot D}} \qquad \text{(EQ 10)}$$

Bond (1961) further stated that the rod diameter should be increased by ½ in. when the reduction ratio (F_{80}/P_{80}) is less than 8.

For pebble mills, Crocker (1985) prescribed that grinding media of equal weight to the ball size calculated in Equation 9 is adequate under appropriate grinding conditions. Pebble media must be prepared to allow for this equivalent weight to be achieved after the "rounding up" process, where the particles may lose 15%–35% of their mass while the pebble surface wears to a smooth condition.

Bond (1961) also provided an equation to describe the equilibrium ball size distribution that exists in the mill based on the recharge size (B) determined in Equation 9. Bond speculated that this formula should also hold for rod and pebble mills. The percentage (y) of the total equilibrium charge that passes size (x, in.) is given by the following equation:

$$y = \left(\frac{x}{B}\right)^{3.8} \qquad \text{(EQ 11)}$$

Bond (1958) further provided nominal start-up grinding mill charges for rod and ball mills intending to best reflect the equilibrium ball charge distribution as a function of the optimal recharge size determined by Equations 9 and 10. Bond's start-up, first fill, or graded charge distributions, in terms of percentage weight distribution, are reproduced here for ball mills in Table 3 and rod mills in Table 4.

Another common grinding media size optimization approach uses the following formula developed by Azzaroni (1980) and revised by MolyCop to derive the optimum makeup ball size:

$$dB = \frac{6.06F80^{0.263}\ (ps\ Wi)^{0.4}}{(ND)^{0.25}} \qquad \text{(EQ 12)}$$

where

dB = makeup ball size, mm
F_{80} = 80% passing fresh feed size, µm
ps = ore specific gravity, t/m^3
Wi = Bond ball mill work index, kW·h/t
N = mill speed, rpm
D = mill diameter inside liners, ft

The charge-specific surface or string area (m^2/m^3) produced by this optimal ball size is then replicated as closely as possible

Table 3 First fill charge distribution for ball mills

Makeup Size (*B*), in.	4½ in.	4 in.	3½ in.	3 in.	2½ in.	2 in.	1½ in.
4½	23.0						
4	31.0	23.0					
3½	18.0	34.0	24.0				
3	15.0	21.0	38.0	31.0			
2½	7.0	12.0	20.5	39.0	34.0		
2	3.8	6.5	11.5	19.0	43.0	40.0	
1½	1.7	2.5	4.5	8.0	17.0	45.0	51.0
1	0.5	1.0	1.5	3.0	6.0	15.0	49.0

Source: Bond 1958

Table 4 First fill charge distribution for rod mills

Makeup Size (*B*), in.	5 in.	4½ in.	4 in.	3½ in.	3 in.	2½ in.
5	18					
4½	22	20				
4	19	23	20			
3½	14	20	27	20		
3	11	15	21	33	31	
2½	7	10	15	21	39	34
2	9	12	17	26	30	66

Source: Bond 1958

by using one or more conventional ball sizes in the makeup charge. The same approach can be used to estimate the graded ball size distribution in the mill, which can be replicated in the initial start-up charge.

The empirical method developed by Azzaroni (1980) suffers the same shortcomings as the Bond model when applied to balls mills following SAG mills because of the nature of the benchmarking database. In these SAB/C applications the model predictions should be validated against operating practices for comparable grinding duties. For SAG mills there is no such desktop method to reliably estimate the optimum ball size or top ball size for a given condition, although a suitably validated population balanced based simulation program can provide practical guidance for SAG ball size optimization. In many instances, the selection of the optimum ball size is more arbitrarily based on user experience. That said, other than for very fine feeds, the use of 5-in. balls for SAG mills is common, and the range of ball sizes applied in SAG is narrow, typically between 4½ and 6 in. The use of larger-diameter SAG balls, greater than 5-in. diameter, requires more careful consideration of mill operating conditions and liner design to avoid excessive liner breakage as a result of the nonlinear increase in ball weight and impact energy.

For fine and ultrafine grinding applications, the optimum media size is determined during laboratory- and pilot-scale testing programs, and is specific to the application and target grind size.

The optimum design media filling level can be readily estimated from a good power draw model after the determination of specific grinding energy requirements (kW·h/t) and the design mill capacity. A good example is the Morrell power model described in Chapter 3.12, "Testing and Calculations for Comminution Machines," and available online at www.smctesting.com. For overflow ball mills with high volumetric flow rates and high ball filling levels, the influence of slurry velocity on ball retention must be considered, and in some instances, the use of a ball retaining ring may be required to allow the design ball charge to be maintained. In SAG mills, the effective removal of slurry by the pulp discharge system is required to ensure that the design power draw is achieved at the modeled charge level. In the extreme instance where slurry pooling exists, it is likely that a slightly higher ball charge will be required to deliver the target power draw.

For open-circuit SAG mills, the optimal, nominal, and maximum ball charge levels selected are dependent on the scale of the project, feed size, competency, and design contingency. These parameters become even more important for single-stage mills with the required ball charge and circuit success dictated by the ore characteristics, feed size, and circuit product size (Putland et al. 2011). It is prudent to specify a maximum ball charge level above that ever intended for operation to safeguard the mechanical integrity of the mill. In cases of extreme ore abrasiveness, the mill diameter may be deliberately increased to allow the operating ball load to be reduced or even eliminated to manage the ongoing cost of high grinding media consumption rates.

Consideration should be given at the design stage to the influence of grinding media specifications on metal recovery and final product performance. Several examples of improved flotation response and concentrate quality attributed to the use of high-chrome grinding media have been observed for precious and base metals ores. A recent example was provided by Greet et al. (2012) of the Newcrest Ridgeway Cu/Au concentrator in Australia, where converting the ball mill charge to high-chrome grinding media achieved a 1% increase in copper recovery and a 2%–4% increase in gold recovery by flotation. Reversing the trial and converting the mill back to carbon steel grinding balls resulted in a 1% loss in copper recovery, with an inconclusive influence on gold recovery. The use of other inert grinding media, such as silica sands or ceramic media, in fine and ultrafine grinding applications (Greet 2008; Rule 2011) delivers similar metallurgical process performance and product quality benefits. Where it is essential that final products have minimal iron contamination, the use of pebble milling and primary AG often becomes a preferred technology selection.

Mill Liners

Mill liners are often stated to have two major purposes: (1) to provide an effective wear lining for the body of the mill and (2) to effectively impart energy to the grinding charge to induce effective size reduction. These objectives are noted by Powell et al. (2006) to conflict with each other but also be "optimized simultaneously with suitable liner selection." It is important that the shell liner design achieves the desired charge motion, normally represented by two-dimensional trajectory profiles made popular by discrete element modeling (DEM) techniques, to promote effective size reduction in the mill. The larger the spacing between lifters, the higher the lifter can be above the shell plate and the larger the required relief angle on the lifter. The larger the lifter, the longer the lifter wear life; however, large spacing exposes the shell plates or section to wear. An optimal design must balance production, lifter, and liner wear. These goals often have opposing requirements, and a balance must be found. Feed head liners are designed with the primary objective of prolonging wear life, as is the case for discharge head liners. In the case of grate discharge mills, the discharge head design must also consider the effective removal of material from the mill via grates, pebble ports, pulp lifters, and the discharge cone.

There are many examples of significant issues with the original mill liner design after start-up, specifically packing and liner breakage. Consequently, careful consideration of the mill liner system design prior to mill start-up is well justified. While it is an impractical objective to fully optimize a mill liner profile at the design stage, several factors should be considered to ensure acceptable mill performance on start-up. These include liner material selection and liner profiles, including the appropriate height and spacing of lifters. These are the key considerations to get right with mill liner designs, beyond which further refinement of the liner design is unlikely to deliver significant improvements in grinding efficiency. The objective is to ensure that liner materials are suitably robust for the duty, liner spacing is adequate to avoid packing, and lifter profile is appropriate to mobilize the mill charge but not to overthrow and cause excessive liner impacts.

Liner Materials

A comprehensive review of industry practices by Powell et al. (2006) noted the use of high-carbon chrome moly steel as the most common material used for SAG mill liners, high-chrome irons in rod and ball mills, and rubber liners traditionally used in secondary and tertiary grinding applications. Composite liners, using a rubber base with steel inserts at the wear surface, have been successfully used in many primary

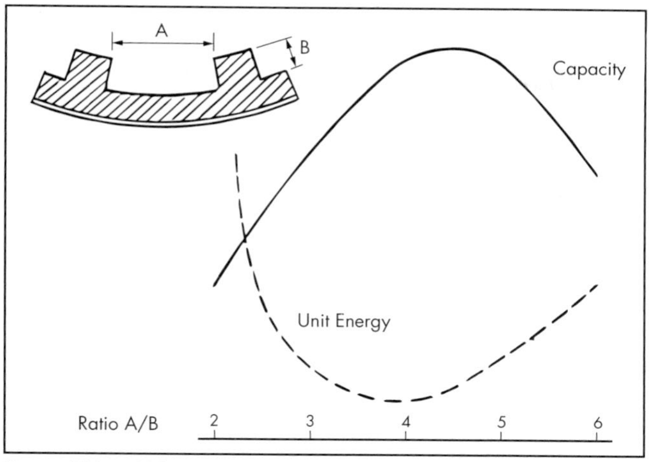

Source: McIvor 1981

Figure 4 Effect of liner spacing-to-height ratio on mill capacity

and secondary milling applications, including large-diameter AG mills (38 ft) and SAG mills (32–34 ft).

Claims of significantly improved grinding efficiency have been made when steel linings are used in preference to rubber in ball mills, although this has not been conclusively demonstrated in published studies. One recent case study (Maclean et al. 2014) of the conversion of the 20-ft Batu Hijau (Indonesia) ball mills from steel to composite linings demonstrated no material difference in grinding efficiency as a function of mill lining in an extended head-to-head trial of composite and steel shell lining systems.

Lifter Spacing

Extensive pilot-scale AG mill test work by Meaders and MacPherson (1964) established the original benchmark lifter spacing ratio, represented as spacing over height, or A:B ratio, as nominally 4:1. This is represented by Figure 4 from McIvor (1981) based on the original pilot work.

The 4:1 spacing ratio was confirmed as an optimum during field trials at Highland Valley Copper (British Columbia, Canada) in the early 1990s reported by Bigg and Raabe (1996) and is implicit in the formula from optimal A:B ranges for rubber-lined mills reported by Moller and Brough (1989), which also considers mill speed. This formula is given in the following equation, where *FCs* is the fraction of critical speed.

$$B = (1 - FCs) \times A \qquad \textbf{(EQ 13)}$$

where A is the spacing between lifters, and B is the lifter height above the liner plate.

AG/SAG Mill Liners

Notwithstanding the supporting information for a spacing-to-height ratio of 4:1 provided by these references, apparently in practice, particularly for large-diameter SAG mills, it is more commonplace to design initially for a lower A:B ratio and accept that the ratio cannot be optimum for the duration of liner life. This is consistent with the recommendations of Powell et al. (2006) to start with a low ratio and finish with a higher value at the time of liner replacement, and consider an ideal variation of A:B ratio of ±1 across the liner life. This is evident in shell liner designs reported for operations such as Yanacocha in Peru (Burger et al. 2011) and Peñasquito in Mexico (Palmer et al. 2011), shown in Figure 5. These designs clearly show an initial A:B ratio of less than 4:1, even when high-low profiles are used to reduce the average height above the liner plate.

DEM is a valuable tool to support initial liner design, particularly to optimize lifter height, face angle, and spacing. This process is particularly important for SAG mills where the potential for shell liner damage because of large-diameter balls overthrowing the toe of the mill charge is more apparent. While increased lifter height is well known to promote longer wear life, when combined with a steep face angle it can result in increased frequency of charge overthrow and liner breakage. The documented experiences at the Freeport Grasberg concentrator in Papua, Indonesia (Coleman and Veloo 1996; Staples et al. 2001) are instructive in this regard, particularly with the use of DEM to optimize the shell liner design to improve the grinding charge trajectory. Ideally, DEM and other instructive simulation techniques, such as smoothed particle hydrodynamics (SPH), and high-fidelity and computational fluid dynamics, should be applied as often as possible in support of good initial liner design in the first instance. They should not be reserved for use only as a tool to correct suboptimal performance after mill commissioning.

Some useful trends in mill liner design for SAG mills are apparent based on recent industry practices and publications presenting optimized liner designs and performance improvements. Quoted improvements are associated with mill throughput, power draw, elimination of packing, reductions in liner breakage, and overall increases in liner service life. The documented experiences at the following operations demonstrate the approaches available for SAG mill liner optimization:

- Grasberg (Coleman and Veloo 1996; Staples et al. 2001)
- Cadia in Australia (Hart et al. 2006)
- Alumbrera in Argentina (Sherman 2001)
- Peñasquito (Palmer et al. 2011)
- Los Pelambres in Chile (Villanueva et al. 2001)
- Collahuasi in Chile (Villouta 2001)
- Candelaria in Chile (Miranda et al. 1996; Kendrick and Marsden 2001)
- Yanacocha (Burger et al. 2011)

As observed by Royston (2007), SAG mill lifter face angles have progressed from steep angles of 5°–15° to a more common range of 22°–35°, large-diameter mills (>24 ft) are tending to favor high-high liner designs over high-low designs, and the number of rows of lifters has on average decreased. Lifter face angles and profile style are relatively easy to change in comparison to lifter spacing, which is dictated by the mill shell drill pattern. Historically, SAG mill linings followed the rule of thumb that the shell should have twice the number of lifter rows as the mill diameter, in feet, often referred to as a 2 × D profile, or comparable variations on that theme. Examples of the 2 × D design are the 72 row lifter sets used in the 36-ft-diameter SAG mills at Batu Hijau (Burger et al. 2006) and Candalaria (Miranda et al. 1996). The Batu Hijau operation is a unique example that reported a different experience than other authors regarding the effects of SAG mill lifter spacing. Trials of reduced lifter rows from the original 72-row design, including 48-row and 36-row designs, were unsuccessful and the operation reverted to the original lifter spacing. In general, however, greater awareness of the occurrence and detrimental impacts of packing on mill performance, compared

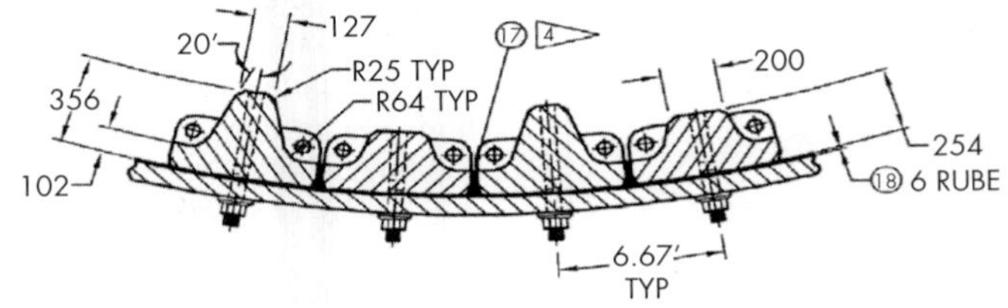

Minera Yanacocha Original Shell Lifter Profile
Source: Burger et al. 2011

Alternative Minera Peñasquito SAG Mill Shell Lifter Profile
Source: Palmer et al. 2011

Figure 5 Recent SAG mill liner profiles

with observations of improved mill performance with wider lifter spacing in practice, have led to a reduction in the number of lifter rows for SAG mills with designs in the range of 4/3D to 5/3D made popular by the experiences of operations such as Cadia (Hart et al. 2006), Alumbrera (Sherman 2001), and Collahuasi (Villouta 2001). This reduction has also been facilitated by the increased use of variable-speed drives on SAG mills. Historically with fixed-speed mills, the presence of packing with a new design was required to prevent ball overthrow and protection of the liners while still maintaining a profile that would maintain reasonable lift at the end of the mill's life. Now with variable-speed drives, the ball trajectory can be optimized throughout the life cycle of the liner. This does, however, impact available SAG mill power draw over the life of the liners. Liner profiles reflecting 1 × D spacing have also been implemented at Yanacocha (Burger et al. 2011) and Los Pelambres (Villanueva et al. 2001). Clearly, the initial drill pattern of the mill shell must consider the optimum lifter spacing, in addition to contingency patterns, should the desired performance not be achieved after commissioning. This has led to the use of approximately 2 × D designs, divisible by 3, so that approximately a 1.33 × D design can be implemented with the same shell drill pattern.

Similar considerations hold for the design of the mill discharge liner system when it comes to the application of advanced modeling techniques in support of discharge liner design, and the planning of the drill pattern for the discharge head to accommodate optimization of the pulp lifter design. A good example of the latter point is provided by Barratt and Sherman (2002) regarding the Inco Clarabelle SAG mill in Canada, which was drilled to accommodate both radial and curved pulp lifter designs. This may prove an important consideration if there is a need to significantly improve mill performance by moving from radial to curved and/or twin-chamber pulp lifter design, as has been achieved at Cadia (Hart et al. 2006) and at Cortez in Colorado, United States (Stieger et al. 2007). Alternatively, the operation may realize up to a 20% increase in shell liner life as observed by Royston (2007) by employing radial pulp lifters and bidirectional rotation of the mill shell. For very high-capacity open-circuit SAG mills, or closed-circuit SAG mills such as in single-stage milling modes, there have been several instances where decreased mill performance was caused by slurry pooling from underdesign of the mill discharge system. Subsequent modeling using techniques such as DEM and SPH has illustrated issues with slurry flow back into the grinding chamber, flow restrictions at the discharge cone, inadequate pulp discharge chamber capacity, and pulp carryover in the discharge vanes. In all cases, such issues can be readily identified at the design stage through detailed simulation and the knowledge applied to optimize the pulp lifter design. Thereby any potential production constraints because of an undersized pulp discharge system can be readily addressed during mill design.

Grate Design

The discharge grate in AG/SAG mills has a significant impact on mill performance, being critical to retaining effective grinding media in the mill charge, while maintaining a working mill load and effectively allowing ground product and critical size material to discharge the mill. The initial grate design can be supported by advanced simulation but will typically be derived from benchmarking mills of comparable size and duty.

Grate apertures will be slots or ports ranging from ⅝ in. to 4 in. Square, slotted, or elliptical designs are common. The amount of the grate surface area that is open for material flow can range from 2% to 14% and is influenced by grate aperture size, ore particle size and competence, and the nature of the pebble management system. SABC-B circuits will tend toward larger apertures and more open area to maximize material flow to the pebble crushers. However, the pebble port size influences the operable closed-side set of the pebble crusher and therefore SAG mill discharge screen aperture. The Los Pelambres operation (Powell and Valery 2006) is an example

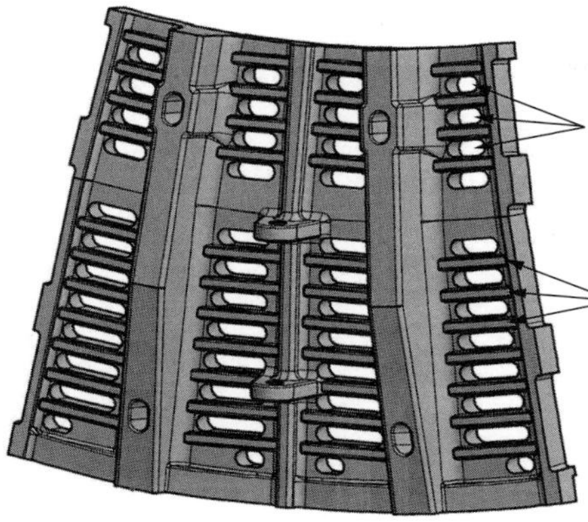

Courtesy of Growth Steel Australia Pty Ltd.
Figure 6 SAG mill grate design

of SAG mills operating with high open area (10%–14%) to maximize pebble discharge rates. The use of very high grate open area can facilitate slurry flowback from the pulp lifter to the grinding chamber of the mill, and the impact of pulp discharge system efficiency should be considered in any grate open-area optimization exercise.

Common issues with discharge grates that need to be managed in operation are peening and plugging of the grates. Peening refers to the closing up of the grate opening because of metal flow from contact with large-diameter grinding steel. Plugging is the blocking of the grate aperture with worn grinding steel, or in some instances with pebbles. Plugging is managed by relieving or tapering the grate apertures so that the aperture is larger on the discharge side of the grate, more readily allowing material to be released through the grate. Rowland described 3.5°–5° relief on either side of the slot as nominal. The use of rubber or composite grates is a common solution to plugging issues with smaller grate apertures because of the greater flex of the rubber while still maintaining strength. Peening can be managed by using protective lifters between grates and using beveled pads surrounding the grate aperture (Figure 6). However, care must be taken not to use too large a protective lifter, as this can result in reduced extraction rates. The beveled pads act as a point to absorb metal impacts, confining the metal flow to the pad rather than on the edge of the slots.

The discharge grate design usually incorporates a lifter as either an integral or a replaceable section. The lifter helps to slow down the wear of the grate face plate section but also has a bearing on the amount of effective grate open area, creating a shadow effect as material flows over the lifter and down onto the grate face as the mill rotates. The grate section thickness and the lifter height must be optimized in unison to ensure both suitable wear life of the grate and effective use of the available grate open area.

Rod Mill Liners

Rowland (2002) reported the popularity of single-wave alloy steel or cast-iron shell liners in rod mills. Figure 7 illustrates a modern rod mill single-wave liner design. Nominal design features include a total number of shell lifters equivalent to double the mill diameter in feet, wave thickness of 2.5–3.5 in. on top of 2.5–3.0-in. liners, and rubber backing to protect the mill shell and heads from washing and corrosion. End liners should be smooth and free of features that may disrupt the rod action and cause tangling. Rubber or wear-resistant cast-iron end liners are prone to damage and are not recommended. Rubber shell liners have been successfully applied in the smaller-diameter rod mills running at slow speeds and reduce the noise generated by the mill. When using rubber liners, close attention must be paid to rod quality and removing broken or thin rods from the charge.

Ball Mill Liners

Wave liners are the most common steel liner design for ball mills, with either single-wave or double-wave designs being popular. Rowland (2002) noted the recommended spacing of 13.1/3.3 × the mill diameter (D) for double-wave liners and $2 \times D + 2$ for single-wave liners when the mill diameter is expressed in feet. Wave liners for ball mills are usually made of alloyed steel or wear- and impact-resistant cast irons. End liners for ball mills conform to the slope of the mill head and can be made of rubber, alloyed cast steel, or wear-resistant cast iron. To prevent racing and excessive wear, end liners for ball mills are furnished with integral radial ribs and/or replaceable lifters. A variety of alloys are suited dependent on the mill duty, with rubber and composite liner designs also suited in many applications including modern, large-diameter ball mills. The use of rubber liners in dry milling applications is not recommended. While rubber liners have been associated with a loss in mill volume and power draw in some primary milling applications, they are well suited for use in ball mills with small-diameter balls. Composite liners have been a viable alternative to steel liners for even the largest-diameter ball mills. Rubber and composite liners are typically designed with separate liner and lifter segments (Figure 8).

Stanley (1987) and Powell et al. (2006) describe the popularity of pocketed grid liners in South African mills, with Stanley observing 26% of South African gold mills using these lining systems at the time of his publication. These liners trap grinding media to act as a self-regenerating sacrificial liner, resulting in a low liner profile and requiring mill operation at high speeds, approximately 90% of critical speed, for effective grinding action. These types of liners are only used in smaller-diameter low-aspect mills. Norrgran (2009) and Ellsworth et al. (2007) describe the application of permanent ceramic magnetic liners in the U.S. iron ore industry, with Norrgran quoting the installation of more than 400 magnetic liner sets worldwide. Magnetic liners are limited by the structural integrity of the liner to application in 10-ft-diameter mills for rubber-encased magnets or 18-ft-diameter mills for steel-hosed magnets. Such systems can offer the advantage of faster and safer installation over conventional steel lining systems, as well as significantly higher liner service life.

Pebble Mill Liners

Crocker (1985) observed that pebble-mill liner design is similar to ball-mill liner design with the primary difference that lifting the lighter specific gravity autogenous media must be accommodated to ensure functional charge motion. Centrifugal force does not hold the light pebble against the shell to the same extent that steel balls are held. Most pebbles are very abrasive, and excessive slip must be prevented to provide reasonable liner life. As in ball mills, the size of media, percent ball load, speed of the mill, and media size are all considered when selecting the proper profile of the shell liner.

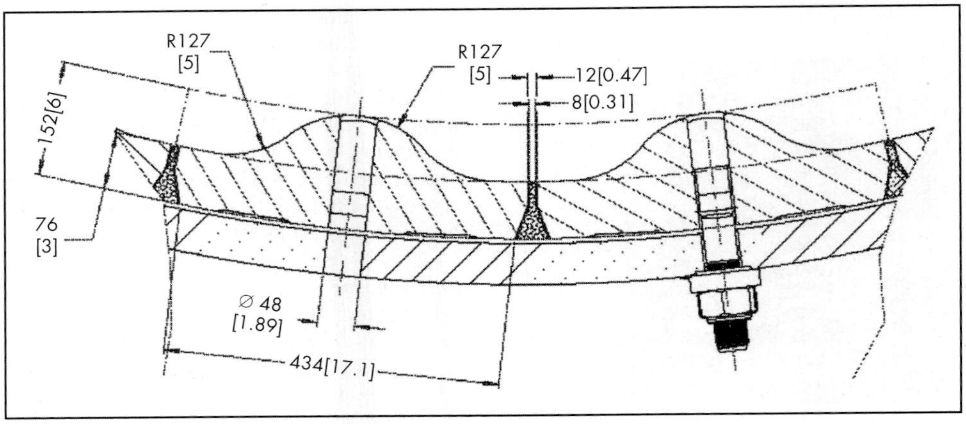

Courtesy of Growth Steel Australia Pty Ltd.

Figure 7 Rod mill single-wave lifter profile

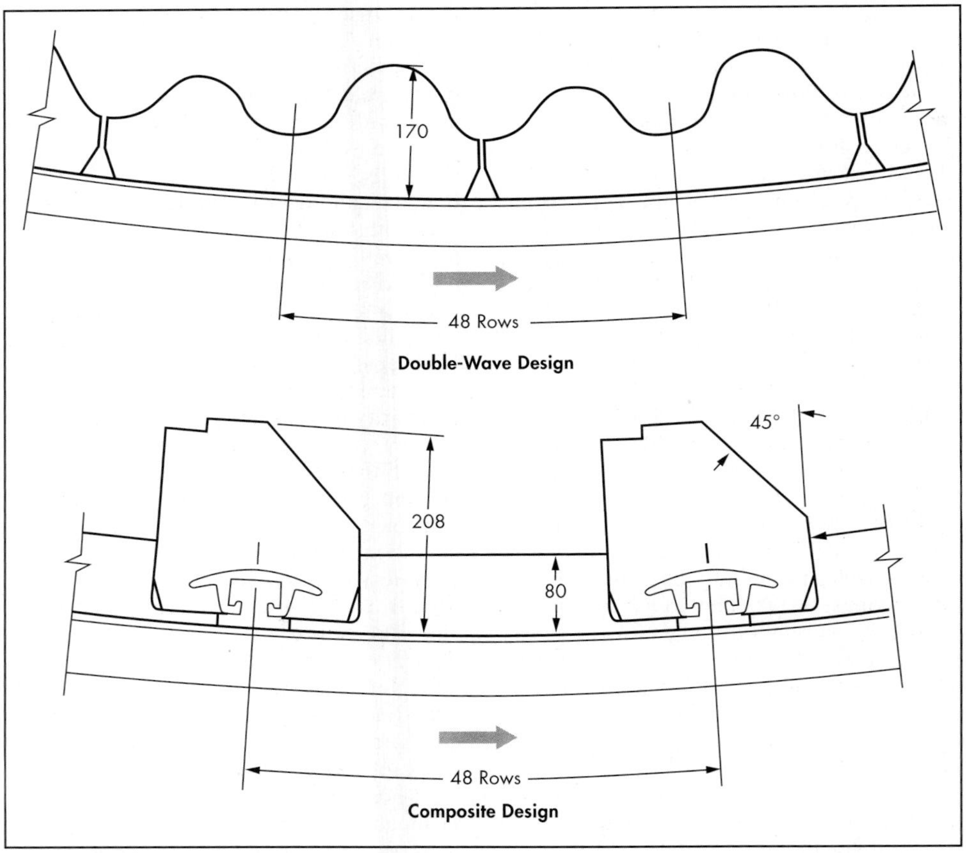

Courtesy of Polycorp Ltd.

Figure 8 Ball mill liner profiles

As pebble mills are not too violent in action, smaller bolts can be used to hold the liners in place. Shell liners may be made of white iron, Ni-Hard, rubber, or chrome moly. The profile of the liner designed to suit the size of media and speed of the mill is the important factor in a successful operation. Feedhead liners are designed with radial and staggered radial lifters. Again, the profile must be higher than would be used with steel balls, usually the same lift as on the shell liners.

CLASSIFICATION

Appropriate classification design is essential for the successful design of a comminution circuit, as poor classification can be a major contributor to grinding circuit inefficiency. Many equipment options exist, including hydrocyclones, screw classifiers, vibrating screens, and sieve bends. However, in most circuits, hydrocyclones are used because of fine product sizes (–250 μm), low capital costs, operability, and their suitability to a wide range of circuit capacities.

Table 5 Hydrocyclone capacity guide

Cyclone Diameter	Capacity, Mt/yr
10 in. or 250 mm	–1.5
15 in. or 400 mm	1.0–5.0
20 in. or 500 mm	2.0–8.0
26 in. or 650 mm	5.0–15.0
33 in. or 800 mm	+10

Screw classifiers were historically used in circuits targeting a coarse cut size but are only applicable to low-throughput circuits and are rarely used in modern designs. Sieve bends are used in some circuits, typically when a coarser cut size is required. A common use of sieve bends is in bauxite grinding where a coarse grind is required and the classifier encounters hot caustic process solutions.

Circuits that may have previously used sieve bends are now often installed with high-frequency vibrating screens. These screens classify more effectively at lower cut sizes with apertures down to 75 μm commonly used. Screens are commonly applied in circuits of low to moderate throughput when the overgrinding of valuable high specific gravity components of the ore is detrimental to downstream processing. Examples include small-tonnage lead–zinc grinding circuits, rare earth, tin, and tantalum mineral comminution.

Recent examples of the successful application of fine screens in small-capacity grinding circuits in duties normally reserved for cyclones are provided by Valine et al. (2012). In these applications, high classification efficiency and low circulating loads are consistently achieved. The authors cite examples where grinding circuit capacity has been increased because of the increased classification efficiency provided by high-efficiency fine screens. Because of the engineering and capital requirements associated with fine-screening technologies, this approach traditionally has been limited to smaller (<300 t/h) grinding circuits, as demonstrated by Valine et al. (2012).

A recent example of the application of screens in preference for hydrocyclones in a coarse-grinding, high-throughput application is the Swakop Uranium Husab mine in Namibia. The operation began commissioning activities in late 2016 and is a 15-Mt/yr nameplate operation treating uranium-bearing Alaskite granite. The comminution flow sheet consists of a primary gyratory crusher, a coarse ore stockpile, and an SABC-A grinding circuit using banana screens for ball mill classification. The design throughput rate is 1,875 t/h and the target P_{80} is 355 μm ahead of leaching with sulfuric acid. Milling is in mild acid conditions, as there is no separation of process waters between comminution and leach. The SAG mill has 10.8-m-diameter inside liners, an EGL of 5.29 m (36 ft × 18 ft shell), and is fitted with twin 6.75-MW variable-speed, geared pinion drives. The SAG mill discharges through a trommel screen with the oversize directed to pebble crushing and crushed pebbles returning to the SAG mill. Trommel undersize is pumped to a distributor box and then to the four (3.66 m × 8.23 m) banana screens, fitted with 0.8-m polyurethane or rubber panels. Screen oversize feeds the ball mill, and the undersize is sent to the leach feed thickener. The ball mill has 6.51-m-diameter inside liners and an EGL of 10.21 m (22 ft × 33.5 ft shell) and is fitted with twin 4.25-MW variable-speed, geared pinion drives. Variable speed was chosen for the ball mill to minimize starting power requirements and is proving a useful control tool when screen classification is employed.

For most grinding circuits, hydrocyclones are used because of their simplicity and suitability for a range of applications. Typically, the size of the hydrocyclone is selected so that it can achieve the target cut size and final grind size at the desired cyclone overflow density. Cyclone diameter and capacity should be selected targeting a total number of operating cyclones between 4 and 12 to ensure process stability under a range of operating conditions. On this basis, there tends to be typical capacity ranges for cyclone diameters. These can obviously overlap if less than 4 cyclones or high overflow densities are targeted (very low circulating loads) or more than 12 cyclones and low cyclone overflow densities (high circulating loads). However, in a broad context, Table 5 indicates a capacity range for a cyclone diameter with lower envelope typical for finer grind sizes and the upper, coarser grind sizes (P_{80} ranging from 200 to 70 mesh).

The importance of the selection of the diameter and number of cyclones operated should not be underestimated, as this defines the slurry density exiting the circuit as cyclone overflow and also the circulating load for a given cut size. Operating a circuit within an optimal circulating load range is important in achieving efficient operation and optimal power draw from the grinding mills. If a circulating load is too low, overgrinding of fines will occur, flattening the shape of the product distribution. This results in higher power consumption for a target P_{80}. If the circulating load is too high, slurry pooling can occur, reducing power draw. This predominantly affects SAG mills and grate discharge ball mills but also affects overflow ball mills as well. Too high a flow through an overflow ball will result in ball ejection, coarse particle short-circuiting, and reduced power draw. Very low residence times and high slurry velocity negatively impact grinding efficiency.

The significance of classification efficiency will vary from project to project based on the nature of valuable minerals and the downstream extraction or separation processes. For example, the extraction of gold by cyanidation in many instances benefits from the preferential recirculation and overgrinding of high-density host sulfides minerals, increasing the liberation and exposure of fine gold inclusions. Gold extraction also does not suffer the reduction in efficiency at fine sizes typically observed for sulfide flotation. By contrast, overgrinding of high specific gravity sulfide minerals ahead of flotation is detrimental to flotation recoveries and more likely to be a factor in the selection of the classification process.

The optimal circulating load for a grinding circuit is not easily defined, although the circuit design and operating conditions should avoid exceeding the volumetric capacity of the tumbling mill. The penalties include operational performance issues associated with high slurry filling levels in grate discharge mills, or high superficial velocities and low residence times in ball mills. The effective management of these conditions during the design process was discussed earlier in this handbook. As a rule of thumb, a coarser grind or a low specific energy consumption grinding duty will be associated with a low circulating load in practice. Conversely, a fine grind or higher specific energy grinding duty will tend to operate at high circulating loads. Coarse grind size or low specific energy grinding duties can often operate effectively at circulating loads as low as 150%, while fine grinding or high specific energy applications can often operate at circulating loads greater than 600%.

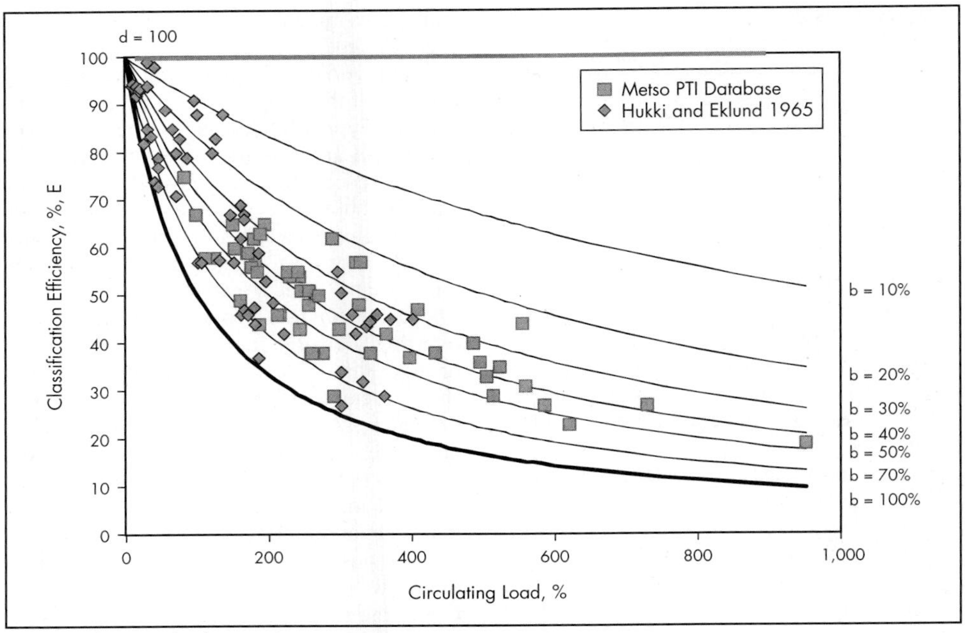

Source: Jankovic and Valery 2012a

Figure 9 Relationship between classification efficiency and recirculating load

Cyclone models are available in the literature and can be used in combination with cyclone vendor capacity charts to establish initial hydrocyclone performance and design criteria. Models such as those developed by Plitt and Nagaswararao (Napier-Munn et al. 1996) are particularly useful for this purpose. Most hydrocyclone vendors have in-house simulation software that can and should be used to support final cyclone selection.

The interdependence of recirculating load and classification efficiency has been well demonstrated by Jankovic and Valery (2012a), as shown in Figure 9. These authors demonstrated that increased recirculating loads are associated with more moderate increasing in grinding circuit capacity than has been proposed by other studies because of the reduced classification efficiency of product size material that is often associated with higher recirculating loads. Figure 9 illustrates the moderate gains in grinding circuit efficiency with increased recirculating load that could be expected from a hydrocyclone system operating with typical efficiency (40% product material recovery to coarse product). As Figure 10 demonstrates, when classification efficiency is higher (15%–20% product size reporting to the coarse product with screening options), the impact of higher recirculating loads on circuit efficiency is much greater.

Jankovic and Valery (2012b) state that up to 15%–25% of the total grinding circuit energy consumption can be attributed to inefficient classification in conventional grinding circuits, highlighting the potential value that can be realized by a well-designed and operated classification system.

SAMPLING

The often-quoted adage that you cannot control what you do not measure is certainly true for grinding circuits, where control of a process that consumes most of the operation's power consumption is a significant ongoing concern. Grinding circuits, particularly tumbling mills, do not require a significant amount of manual sampling for process control and metallurgical accounting purposes, yet there are some streams that need to be accessible for safe and reliable sample collection. The optimization of any grinding circuit requires the periodic sampling of all major process streams, and the coarse particle size distributions and high slurry flow rates encountered around primary and secondary grinding circuits demand close adherence to the principles of cross-stream sampling. These requirements are often poorly addressed in grinding circuit design, leading to invariable compromises in sampling practices and consequent limitation of the plant operator's ability to fully optimize the performance of the circuit.

It is necessary to have some access to the mill discharge stream for wet grinding mills operating in closed circuit, so that the mill operating density can be manually measured and controlled. Access to the final grinding circuit product, often cyclone overflow, is essential for confirming pulp density and achieving grind size control. Where the grinding circuit product is required for metallurgical accounting purposes, the effective sampling of this stream will be carefully considered. In instances where such a sample is not deemed necessary for metallurgical accounting, accessibility for sampling can be considerably diminished. In all grinding circuits, the ability to manually sample the grinding circuit product for performance monitoring is an essential requirement, irrespective of whether automated sampling systems are in place for metallurgical accounting purposes.

For grinding circuit optimization surveys, accessibility to a larger number of sampling points will be necessary. These will generally include the feed and product streams from any individual milling, screening, or classification stage. Data will be mass balanced by size fraction requiring large samples to be

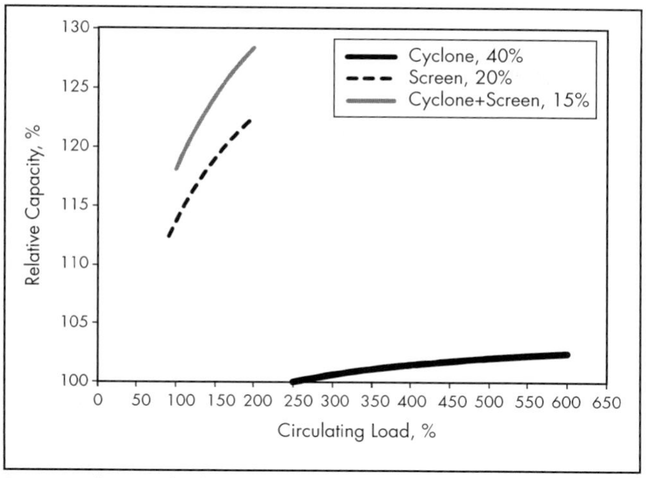

Source: Jankovic and Valery 2012a

Figure 10 Relative circuit capacity as a function of classification mechanism, fines recovery, and recirculating load

collected for statistically representative information on coarse streams. Any conveyors to be sampled require sufficient belt length to be accessible for sampling to allow the required sample mass to be collected. Slurry streams must be accessible for cross-stream sampling of the entire flowing stream. Further fundamental requirements are that any sample point must be accessible in a manner that conforms to acceptable safe work practices, and that sample handling and removal can be readily managed with minimal potential for manual handling issues. In some instances, such as wet screen undersize, a perfect sampling solution can be difficult to achieve in the design stage; however, a minimum requirement should always be to provide some practical means of access to the slurry stream for sampling.

More specific guidance on sampling requirements for grinding circuits is included in Chapter 3.10, "Grinding Circuit Performance Optimization," and these requirements should be reviewed during the design stage of the project.

ACKNOWLEDGMENTS

The authors acknowledge the valued contributions to this chapter from Dean David, Bruce Rattray, Pramod Kumar, and Stephen Morrell.

REFERENCES

Azzaroni, E. 1980. Grinding media size and multiple recharge practice. Presented at the Third Armco-Chile Symposium.

Barratt, D., and Sherman, M. 2002. Selection and sizing of autogenous and semi-autogenous mills. In *Mineral Processing Plant Design, Practice, and Control.* Edited by A.L. Mular, D.N. Halbe, and D.J. Barratt. Littleton, CO: SME. pp. 755–782.

Bigg, A.C.T, and Raabe, H. 1996. Studies of lifter height and spacing: Past and present. In *SAG 1996 Conference Proceedings*. Vancouver, BC: Mining and Mineral Processing Engineering, University of British Columbia.

Bond, F.C. 1958. Grinding ball size selection. *Min. Eng.* 10(5):592–595.

Bond, F.C. 1961. *Crushing and Grinding Calculations*. Milwaukee, WI: Allis-Chalmers Manufacturing.

Burger, B., Hatta, M., McGaffin, I., and Gaffney, P. 2006. Batu Hijau: Seven years of operation and continuous improvement. In *Proceedings of an International Conference on Autogenous and Semiautogenous Grinding Technology*. Vol. 1. Vancouver, BC: Mining and Mineral Processing Engineering, University of British Columbia. pp. 120–132.

Burger, B., Vargas, L., Arevalo, H., Vicuna, S., Seidel, J., Valery, W., Jankovic, A., Valle, R. and Nozawa, E. 2011. Optimization of the single stage SAG mill circuit at the Yanacocha Gold Mill. In *SAG 2011 Conference Proceedings*. Vancouver, BC: Mining and Mineral Processing Engineering, University of British Columbia.

Callow, M.I., and Meadows, D.G. 2002. Grinding plant design and layout considerations. In *Mineral Processing Plant Design, Practice, and Control.* Edited by A.L. Mular, D.N. Halbe, and D.J. Barratt. Littleton, CO: SME. pp. 801-818.

Coleman, R., and Veloo, C. 1996. Concentrator expansion at Freeport Indonesia's Grasberg Operations. *Min. Eng.* (February):25–33.

Crawford, A., Zheng, X., and Manton P. 2009. Incorporation of pebble crusher specific energy measurements for the optimization of SABC grinding circuit throughput at Telfer. Presented at the 10th AusIMM Mill Operators Conference, Adelaide, South Australia, Australia, October 12–14.

Crocker, B.S. 1985. Pebble mills. In *SME Mineral Processing Handbook.* Edited by N.L. Weiss. Littleton, CO: SME-AIME. pp. 3C-94–3C-107.

Ellsworth, A., Hoff, S., Zhou, J., Zhao, M., Jiang, X., and Merwin, R. 2007. Application of metal magnetic liners in the U.S. iron ore industry. SME Preprint No. 07-087. Littleton, CO: SME.

Fernandez, F., Rocha, M., Kemeny, J., BoBo, T., Rodriguez, C., and Fuentealba, R. 2015. Impact of ROM PSD on the crushing and grinding circuit throughput. SME Preprint No. 15-112. Englewood, CO: SME.

Grandy, G.A., Danecki, C.D., and Thomas, P.F. 2002. Selection and evaluation of grinding mill drives. In *Mineral Processing Plant Design, Practice, and Control.* Vol. 1. Edited by A.L. Mular, D.N. Halbe, and D.J. Barratt. Littleton, CO: SME. pp. 819–839.

Greet, C.J. 2008. The significance of grinding environment on the flotation of UG2 ores. Presented at the Third International Platinum Conference: Platinum in Transformation, Sun City, South Africa, October 6–9.

Greet, C.J., Hitchen, C., and Kinal, J. 2012. Conducting high chrome grinding media trials at Newcrest's Ridgeway Concentrator. Presented at the 11th AusIMM Mill Operators Conference, Hobart, Tasmania, October 29–31.

Hart, S., Nordell, L., and Faulkner, C. 2006. Development of a SAG mill shell liner design at Cadia using DEM modelling. In *SAG 2006 Conference Proceedings.* Vancouver, BC: Mining and Mineral Processing Engineering, University of British Columbia.

Hart, S., Parker, B., Rees, T., Manesh, A., and McGaffin, I. 2011. Commissioning and ramp up of the HPGR circuit at Newmont Boddington Gold. Presented at the International Conference on Autogenous Grinding, Semiautogenous Grinding and High Pressure Grinding Roll Technology, Vancouver, BC, September 25–28.

Hukki, R.T., and Eklund, H. 1965. The relationship between sharpness of classification and circulating load in closed grinding circuits. *Trans. SME* (September):265–268.

Jankovic, A., and Valery, W. 2012a. Closed-circuit ball mill—Basics revisited. Presented at the Minerals Engineering International Conference.

Jankovic, A., and Valery, W. 2012b. The impact of classification on the energy efficiency of grinding circuits—The hidden opportunity. Presented at the 11th AusIMM Mill Operators Conference, Hobart, Tasmania, October 29–31.

Kalra, R., Jiangang, J., Druce, I., and Rauscher, M. 2013. Updates on geared vs gearless drive solutions for grinding mills. SME Preprint No. 13-050. Englewood, CO: SME.

Kendrick, M.J., and Marsden, J.O. 2001. Candelaria post expansion evolution of SAG mill liner design and milling performance, 1998 to 2001. In *Proceedings of the International Conference on Autogenous and Semiautogenous Grinding Technology*. Vol. 3. Vancouver, BC: Mining and Mineral Process Engineering, University of British Columbia. pp. 270–287.

Lane, G., Foggiatto, B., and Bueno, M. 2013. Power-based comminution calculations using Ausgrind. In *Proceedings of the 10th International Mineral Processing Conference: Procemin 2013*. Edited by M. Álvarez, A. Doll, W. Krachy, and R. Kuyvenhoven. Santiago, Chile: Gecamin.

Maclean, E., Wirfiyata, F., Khomaeni, G., Jankovic, A., Pasin, G., and Valery, W. 2014. Ball mill Poly-Met liner evaluation at PT Newmont Nusa Tenggara–Batu Hijau Mine. Presented at the 12th AusIMM Mill Operators Conference, Townsville, Queensland.

McIvor, R.E. 1981. The effects of speed and liner configuration on ball mill performance. SME Preprint No. 81-322. Littleton, CO: SME.

Meaders, R.C., and MacPherson, A.R. 1964. Technical design of autogenous mills. *Min. Eng.* (September):81–83.

Miranda, D.G., Bassaure, F.T., and Marsden, J.O. 1996. Evolution of liners for a 36 ft. × 15 ft. (EGL) SAG mill at Candalaria. Presented at the International Conference on Autogenous and Semiautogenous Grinding Technology, Vancouver, BC, Canada.

Moller, T.K., and Brough, R. 1989. Optimizing the performance of a rubber lined mill. *Min. Eng.* (August):849–853.

Morrell, S. 1999. *AG and SAG Mill Circuit Selection and Design by Simulation–A Guide to the JKMRC Model and Its Application*. Indooroopilly, Queensland: JKTech.

Morrell, S. 2001. Large diameter SAG mills need large diameter ball mills–What are the issues? In *SAG 2001 Conference Proceedings*. Vancouver, BC: Mining and Mineral Processing Engineering, University of British Columbia.

Morrell, S. 2004. An alternative size-energy relationship to that proposed by Bond for the design and optimization of grinding circuits. *Int J. Miner. Process.* 74(2004):133–141.

Morrell, S. 2006. Design of AG/SAG circuits using the SMC Test. In *SAG 2006 Conference Proceedings*. Vancouver, BC: Mining and Mineral Processing Engineering, University of British Columbia.

Morrell, S. 2007. The effect of aspect ratio on the grinding efficiency of open and closed circuit AG/SAG mills. In *Proceedings of the Ninth AusIMM Mill Operators Conference*. Melbourne, Victoria: Australasian Institute of Mining and Metallurgy. pp. 121–124.

Morrell, S. 2011. Mapping orebody hardness variability for AG/SAG/crushing and HPGR circuits. Presented at the International Autogenous and Semi Autogenous Grinding Technology Conference, Vancouver, BC, Canada.

Morrell, S., and Valery, W. 2001. Influence of feed size on AG/SAG mill performance. In *SAG 2001 Conference Proceedings*. Vancouver, BC: Mining and Mineral Processing Engineering, University of British Columbia.

Napier-Munn, T.J., Morrell S., Morrison, R.D., and Kojovic, T. 1996. *Mineral Comminution Circuits: Their Operation and Optimization*. Indooroopilly, Queensland: Julius Kruttschnitt Mineral Research Centre.

Norrgran, D. 2009. Magnetic liners increase productivity, reduce energy consumption in iron ore grinding mills. *Min. Eng.* (December):28–30.

Palmer, E., Dixon, S., and Meadows, D. 2011. An update of the SAG milling operation at the Penasquito Mine located in the Zactecas State, Mexico. Presented at the International Conference on Autogenous Grinding, Semiautogenous Grinding and High Pressure Grinding Roll Technology, Vancouver, BC, Canada.

Plavina, P., and Clark, G. 1986. The selection, commissioning and evaluation of a large diameter, variable speed ball mill at Bouganville Copper Limited. Presented at the 13th Congress of the Council of Mining and Metallurgical Institutions, Singapore.

Ploc, M., and Peters, D. 2010. Mining systems and technology–Synchronous electric drives for grinding mills. Presented at the XXV IMPC, Brisbane, Queensland, Australia.

Powell, M., and Valery, W. 2006. Slurry pooling and transport issues in SAG mills. In *SAG 2006 Conference Proceedings*. Vancouver, BC: Universtity of British Columbia..

Powell, M.S., Smit, I., Radziszewski, P., Cleary, P., Rattray, B., Eriksson, K., and Schaeffer, L. 2006. The Selection and design of mill liners. In *Advances in Comminution*. Edited by S.K. Kawatra. Littleton, CO: SME. pp. 331–376.

Putland, B., Kock F., and Siddall, L. 2011. Single stage SAG/AG milling design. Presented at the International Autogenous and Semiautogenous Grinding Technology Conference, Vancouver, BC, Canada.

Rayo, J. 2014. Comparison of semi-autogenous mills operations in Andean countries. Presented at the 12th AusIMM Mill Operators Conference, Townsville, Queensland.

Rowland, C.A. 1985. Rod mills. In *SME Mineral Processing Handbook*. Edited by N.L. Weiss. Littleton, CO: SME-AIME. pp. 3C-44–3C-56.

Rowland, C.A. 1988. Large ball mills–Length and diameter. In *Proceedings of the XVI International Mineral Processing Congress*. Amsterdam: Elsevier Science. pp. 281–292.

Rowland, C.A. 2002. Selection of rod mills, ball mills and regrind mills, In *Mineral Processing Plant Design, Practice, and Control*. Edited by A.L. Mular, D.N. Halbe, and D.J. Barratt. Littleton, CO: SME. pp. 710–754.

Royston, D. 2007. Practical experience in the design and operation of semi-autogenous grinding (SAG) mill liners. Presented at the Ninth AusIMM Mill Operators Conference, Fremantle Western Australia, Australia, March 19–21.

Rule, C.M. 2011. Stirred milling—New comminution technology in the PGM industry. *J. South. Afr. Inst. Min. Metall.* 111:101–107.

Sherman, M. 2001. Optimization of the Alumbrera SAG mills. Presented at the International Conference on Autogenous and Semiautogenous Grinding Technology, Vancouver, BC, Canada.

Stanley, G.G. 1987. Milling and classification. In *The Extractive Metallurgy of Gold in South Africa*. Vol. 1. Johannesburg: Southern African Institute of Mining and Metallurgy. pp. 121–203.

Staples, P., Siewert, H., Stuffco, T., and Mular, M. 2001. SAG concentrator improvements at PT Freeport Indonesia. Presented at the International Conference on Autogenous and Semiautogenous Grinding Technology, Vancouver, BC, Canada.

Stieger, J., Plummer, D., Latchireddi, S., and Rajamani, R. 2007. SAG Mill operation at Cortez: Evolution of liner design from current to future operations. Presented at the 39th Annual Meeting of the Canadian Mineral Processors.

Valine, S.B., Wheeler, J.E., and Albuquerque, L.G. 2012. Applications of fine screens in grinding circuits. In *Proceedings of the 11th AusIMM Mill Operators Conference*, Hobart, Tasmania, October 29–31. pp. 389–394.

Villanueva, F., Ibanez, L., and Barratt, D. 2001. Los Pelambres Concentrator operative experience. Presented at the International Conference on Autogenous and Semiautogenous Grinding Technology, Vancouver, BC, Canada.

Villouta, R.M. 2001. Collahuasi: After two years of operation. Presented at the International Conference on Autogenous and Semiautogenous Grinding Technology, Vancouver, BC, Canada.

CHAPTER 3.9

Grinding Circuit Flow Sheets

Aidan Giblett and Brian Putland

A significant variety of grinding circuit configurations are used in mineral industries, with a similarly wide range of capacity requirements and grinding circuit product specifications. Grinding flow sheets have evolved and continue to evolve as operators and designers strive to improve plant performance efficiencies and reduce new plant capital costs. This chapter covers the major types of flow sheets in use in mineral processing applications today, while also discussing both common and novel variations on the standard concepts. The main types of flow sheets that will be discussed are as follows:

- Semiautogenous circuits
- Autogenous circuits
- Rod/ball mill circuits
- Conventional crush/ball mill and high-pressure grinding roll (HPGR)/ball mill circuits
- Dry grinding circuits
- Fine grinding and ultrafine grinding circuits

For any given metal recovery process, typically more than one style of grinding flow sheet is viable for the project at the design stage. For very high-capacity operations, several recent projects have demonstrated that multiple grinding lines using HPGRs and ball mills in primary grinding duty can be a viable alternative to semiautogenous milling. These projects are Sierra Gorda in Chile (Comi et al. 2015), Cerro Verde in Peru (Koski et al. 2011), and Boddington in Western Australia (Hart et al. 2011). Fully autogenous grinding (AG) equipment for similar high-processing rates has also been used at the Boliden Aitik copper mine in Sweden, as described by Markstrom (2011). The successful adaptation of the HPGR technology, the increased unit capacity of gyratory and cone crushers, the availability of increasingly larger-diameter ball mills, and the expanded applications of stirred milling technologies ensure a wide range of options for the project engineer to consider at the design stage.

When selecting the most suitable grinding flow sheet, several considerations, which can generally be classified as technical or commercial, can influence the final decision. In commercial terms, capital cost and equipment lead times can become significant factors because of impacts on project schedules and critical financial metrics for the project. Technical factors are often the dominant influence on flow-sheet selection and may include the following:

- Equipment unit capacity
- Downstream process requirements, specifically
  - Liberation (grind size)
  - Moisture constraints, for dry grinding processes
  - Pulp chemistry, for wet separation processes
- Material characteristics (abrasiveness, competency, hardness, and material flow properties)
- Safety issues, such as dust generation
- Product specifications (e.g., iron levels)

SAG MILL FLOW SHEETS

Since the 1980s, grinding circuit flow sheets have become dominated by the combination of semiautogenous grinding (SAG) and ball milling technology. This is a direct result of the well-documented decline in average plant feed grades, particularly evident for base metals concentrators (Mudd 2009) and precious metals recovery operations (Mudd 2007). Declining grades necessitate increased plant capacity to achieve sufficient metal production rates for profitable plant operation. Large-diameter SAG mills allow very high single grinding line capacities in comparison to other milling technologies, up to 4,000 t/h (metric tons per hour) when processing low-competency ores. These high capacities per line can generate several advantages in project capital costs, operability, and operating costs.

SEMIAUTOGENOUS, BALL MILL, AND PEBBLE CRUSHER FLOW SHEETS

Over time, the basic semiautogenous/ball mill (SAB) flow sheet has evolved to include pebble crushing (SABC) to increase mill capacity, by either sending crushed pebbles back to the SAG mill (SABC-A configuration [Figure 1]) or sending crushed pebbles forward to the ball mill (SABC-B configuration [Figure 2]), as determined by the relative balance

Aidan Giblett, Senior Technical Advisor—Mineral Processing, Newmont Mining Services, Subiaco, Western Australia, Australia
Brian Putland, Principal Metallurgist, Orway Mineral Consultants, Perth, Western Australia, Australia

of available SAG and ball mill capacity. Early successes in demonstrating the value of including pebble crushing in semi-autogenous milling circuits include the Los Bronces operation in Chile (Vesely and Fernandez 1986) and the Kidston operation in Australia, as described by Bartrum et al. (1988). These authors reported an increase in SAG mill capacity proportional to the tonnage rate fed to the pebble crusher, such that SAG mill capacity increased by 1 t for every metric ton of recycle material crushed. In the case of the Los Bronces operation, this equated to a 10% increase in mill throughput (Suttill 1988).

Grinding circuits that are limited by SAG mill capacity can in some instances be effectively de-bottlenecked by converting to an SABC-B configuration. This requires that some portion of the pebble crusher product be diverted to the ball mill circuit, usually after screening to retain the coarser particles in the SAG mill or pebble crushing circuit. This is an effective, although practically limited, means of increasing the transfer size to the ball mill circuit. Grinding work is consequently transferred downstream from the SAG mill, allowing the SAG mill capacity to be increased. Pilot and industrial testing at the Los Bronces concentrator, reported by Vesely and Fernandez (1986), demonstrated the potential to increase SAG line capacity by 26% under that configuration. After some initial issues with the ball mills handling coarser feed particles as reported by the same authors, the process flow sheet at Los Bronces evolved to an SABC-B configuration (Powell et al. 2006), with pebble crusher product reporting to the ball mill feed. Burger et al. (2006) described the transition of the Batu Hijau (Indonesia) SABC circuit to SABC-B configuration, with pebble crusher product screened and the –10 mm material diverted to the ball mills. Plant trials in this configuration demonstrated an incremental mill capacity increase of 0.5–0.6 t for every metric ton of crushed pebbles bypassed to the ball mill. The SABC-B option is considered where additional ball mill capacity is available or a coarser ball mill circuit product size can be tolerated. Care must be taken to avoid the transfer of excessively coarse particles to the ball mill circuit, because issues with sanding of hoppers and cyclone feed lines, plugged cyclones, high ball mill recirculating loads, charge swelling and reduced power draw, and excessive scat production from the ball mills can result.

Secondary Crush/SABC

The implementation of secondary crushing of the grinding circuit feed has become another common variation on the standard SAG/ball mill flow sheet and is implemented to directly increase the capacity of the SAG mill (Figure 3). This style of flow sheet is a popular post-commissioning retrofit in circuits that are fully using the available SAG mill power but have significant additional ball milling power available, the capacity to readily add more ball milling power, or the ability to tolerate a coarser grinding circuit product size. Early success with this configuration was achieved by operations such as Kidston (MacNevin and Stephenson 1997); Troilus in Quebec, Canada (Sylvestre et al. 2001); and Asarco Ray in Arizona, United States (McGhee et al. 2001); with more recent projects including the Newmont Phoenix operation in Nevada, United States (Castillo and Bissue 2011); Copper Mountain in British Columbia, Canada (Westendorf et al. 2015); Detour

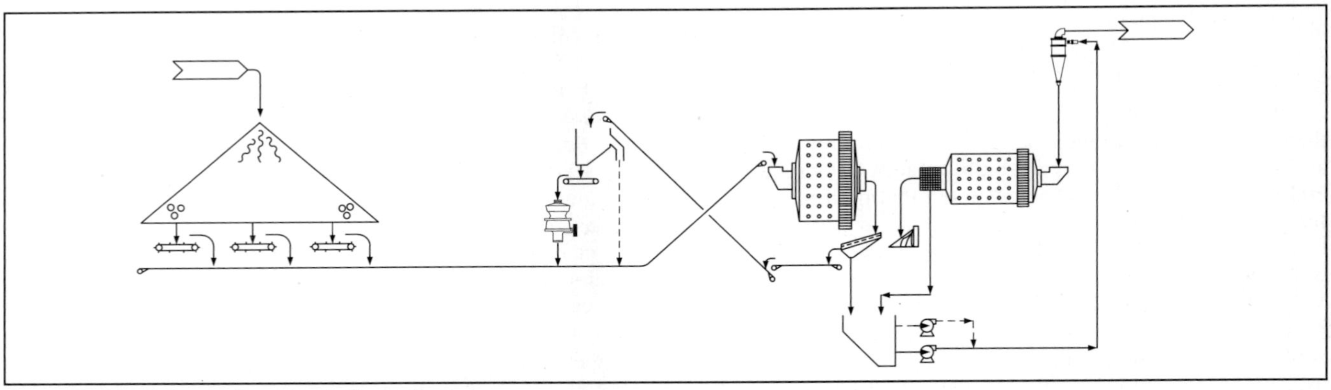

Figure 1 SABC-A grinding circuit flow sheet

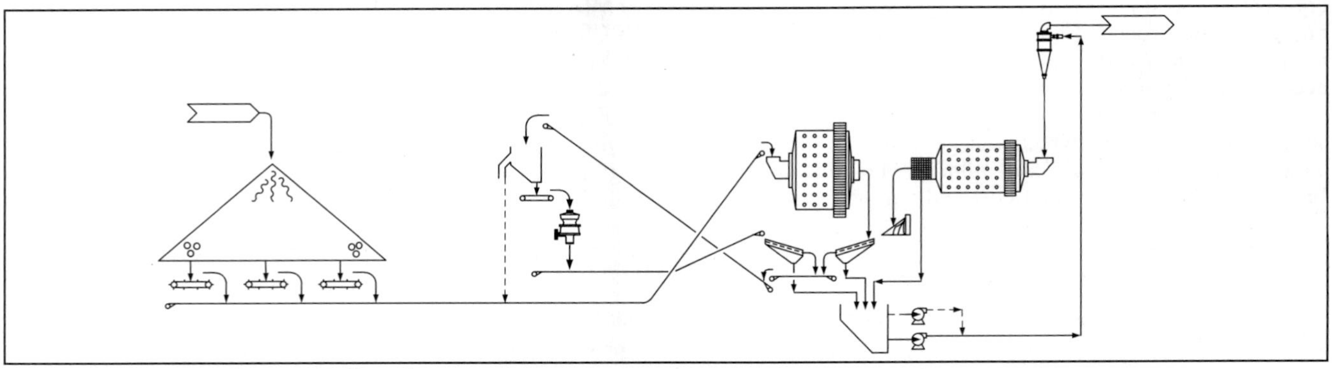

Figure 2 SABC-B grinding circuit flow sheet

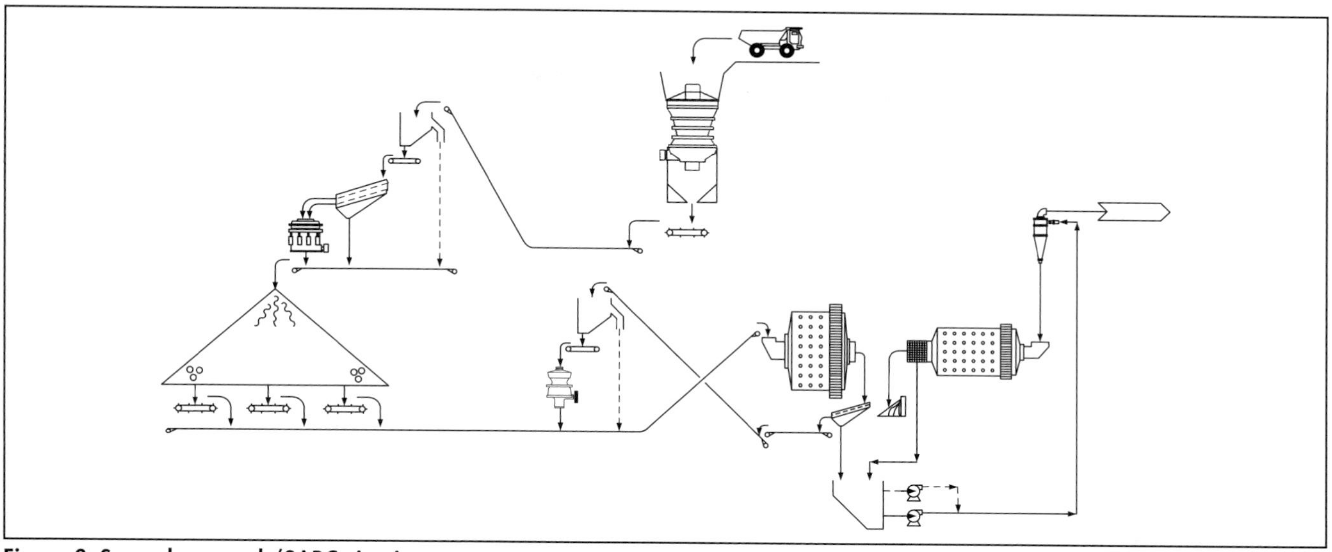

Figure 3 Secondary crush/SABC circuit

Lake in Ontario, Canada (Torrealba-Vargas et al. 2015); and Northparkes in New South Wales, Australia (Sulianto et al. 2016).

Despite early recommendations promoting this comminution flow-sheet configuration for greenfield designs, it is generally well accepted that secondary crush/SAG mill feed flow sheets are best considered for plant expansion or de-bottlenecking exercises where the ability to leverage the underused capital in the secondary grinding mills makes the economics very attractive. A review of several published examples of secondary crush/SAG flow-sheet conversions performed by Siddall and Putland (2007) reflected typical plant capacity increases in the range of 20%–60% for many projects. The extent of the throughput increase is often closely tied to the amount of unused ball mill power that is available or can be readily installed. For greenfield installations, a fit-for-purpose SABC installation will typically have a lower capital cost, and is more amenable to future expansions, than a secondary crush/SABC configuration. Festa et al. (2014) proposed that a secondary crush configuration can become a more practical design option when the design basis exceeds the single-line capacity of the largest available mills or geared drive systems. The Detour Lake project, as described by Torrealba-Vargas et al. (2015), is an example of the successful implementation of this concept for a greenfield installation.

Nelson et al. (1996) provide a thorough account of the operating issues that can occur in secondary crush/SAG mill circuits, based on operating experience at the Kalgoorlie Consolidated Gold Mines (KCGM) Fimiston plant in Western Australia. In this instance, the desire for increased plant throughput rates required a high steel-to-rock ratio in the mill to achieve mill stability. This in turn promotes a coarse transfer size to the ball mill, combined with high metal wear rates and liner breakage in the SAG mill. These issues are particularly evident when the secondary crushers are operated in open circuit, and can be effectively mitigated by closing the secondary crusher circuit with a screen and limiting the mill feed size to 100% passing 1–1½ in. This allows better control of the top size to the SAG mill and allows operation with smaller-diameter grinding media (4–4½ in.) to be considered, in turn reducing the potential for shell liner breakage. SAG mill utilization and operating costs can be further optimized when the mill feed contains a mixture of primary and secondary crushed ore, in what is termed a *partial secondary crush* or *pre-crush circuit.* Examples of partial secondary crushing installations include Porgera (Papua New Guinea), Mount Rawdon (Australia), Geita (Tanzania), and Asarco Ray, as reviewed by Siddall and Putland (2007). Secondary crushing before semiautogenous milling can effectively eliminate the need for pebble crushing because the SAG mill feed contains limited critical size material, and low-operating rock loads in the SAG mill promote effective breakage of critical-size particles.

Single-Stage Autogenous and Semiautogenous Milling

Single-stage autogenous and semiautogenous milling has been widely applied for processing moderately competent to competent ores at small to medium plant capacities in multiple commodities (Figure 4). Applications include St. Ives (Atasoy and Price 2006), Olympic Dam (Alexander and Wigley 2003), Leinster (Fitzmaurice et al. 2000), and Sino Iron in Australia (Tian et al. 2014); and Tarkwa in Ghana (Hothersall et al. 2006). As the grinding is undertaken in a single mill, configuration of the circuit to suit the ore and duty is very important, as discussed by Putland (2011). Configuration options include AG, AG + pebble crusher, SAG, SAG + pebble crusher, high ball charge SAG or run-of-mine ball mill, low ball charge high-speed SAG mill, and partial or full secondary crushed feed. Putland et al. (2011) also noted that the single-stage AG/SAG mill configuration is less suited to grind sensitive ores and high-capacity operations, where two-stage grinding circuits will typically have performance and capital advantages.

There is often speculation regarding the relative suitability of low- and high-aspect-ratio SAG mills for a single-stage grinding duty, although there is no clear demonstration that either style of mill design is any more suitable for the single-stage duty than the other. A notable consideration with single-stage semiautogenous milling duties is the greater exposure to slurry pooling, because of the higher slurry loads associated with the hydrocyclone recirculating load. If not addressed

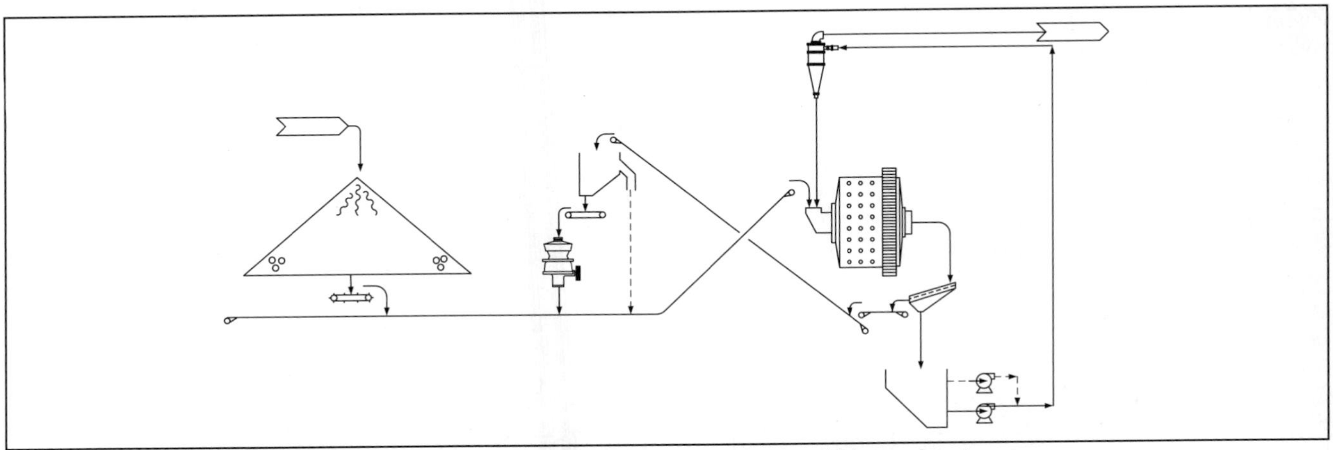

Figure 4 Single-stage autogenous and semiautogenous grinding circuit (with pebble crushing)

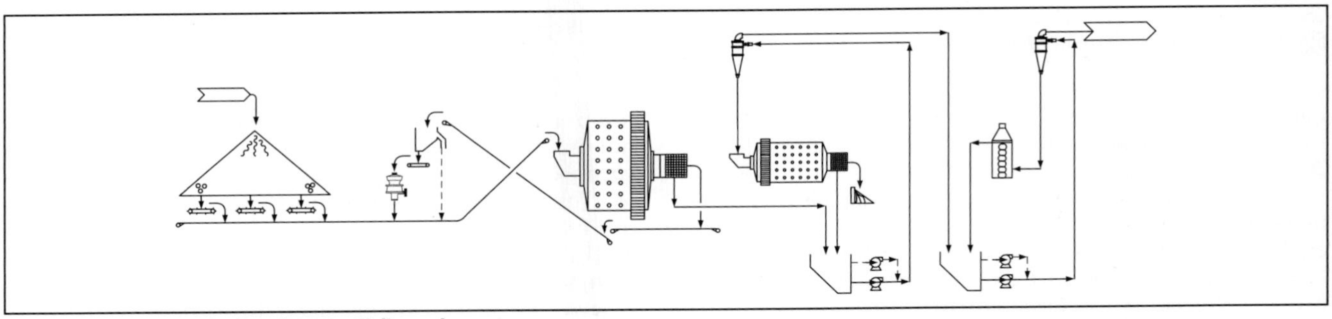

Figure 5 Tertiary grinding/Vertimill flow sheet

when designing the pulp discharge system, resulting impacts can include increased specific energy consumption, reduced grinding efficiency, and high wear rates of grates and pulp lifters. The requirement to have the capacity to move the slurry through the grate and pulp lifters is generally the most important factor in selecting the required mill diameter in single-stage autogenous and semiautogenous milling applications and therefore dictates mill aspect ratio (Putland et al. 2011) as mill length is then selected to achieve the desired power draw.

A major advantage of single-stage autogenous and semiautogenous milling is the low capital cost and flexibility for future expansion (single-stage SAG to SABC), which makes it ideal for starter projects with significant exploration and expansion potential. Another major advantage of the single-stage circuit over the two-stage circuit is the ability to significantly increase throughput by increasing grind size, which is often difficult in two-stage circuits that are constrained by the primary mill (i.e., SAG limited). Alternatively, single-stage mill product size and pulp density are more sensitive to variations in ore properties, and hence are not ideal when downstream metal recovery is sensitive to grind size or viscosity.

MILLING OPTIONS FOR SECONDARY AND TERTIARY GRINDING

Stirred mills have been successfully incorporated into mainstream grinding duties in several base and precious metals flow sheets, after either SAG mills, as shown in Figure 5, or ball mills in the primary grinding duty. An early example of the Vertimill technology in a tertiary grinding duty, as a retrofit to an existing SABC flow sheet, was the Chino operation in New Mexico, United States (Vanderbeek 1997). At Chino, the SABC circuit grind size had coarsened over time to allow increased plant throughput. A tertiary grinding circuit was a practical approach to restore grind fineness and metal recoveries to original target levels while maintaining the higher plant capacity. The incorporation of a VTM-3000 Vertimill into the Ridgeway (Australia) flow sheet in a tertiary grinding duty (Palaniandy et al. 2013) is a more recent example of the same concept.

The Round Mountain operation in Nevada, United States (Henderson et al. 2005), uses a tower mill in secondary grinding duty after a primary SAG mill, fine screening, and a gravity circuit to produce a final grind size of 80% passing 270 mesh. At the McArthur River operation in Australia, described by Anderson et al. (2011), the initial single-stage SAG mill circuit capacity was increased by the addition of a secondhand tower mill processing a bleed of the SAG mill recirculating load. The operation subsequently incorporated two IsaMill M3000s in a similar grinding duty on a SAG mill recirculating load to increase grinding circuit throughput while reducing final grind size.

The economic justification for this style of circuit increases the finer the grind size, with grind sizes below 80% passing 200 mesh viewed as more prospective applications.

An alternative tertiary grinding circuit configuration, consisting of a primary SAG mill and secondary and tertiary ball mills grinding down to 80% passing 500 mesh, is the Sage Mill circuit at the Newmont Twin Creeks operation in Nevada (Yernberg 1996). The 28 ft × 10 ft SAG mill is operated in open circuit, and the 20 ft × 30 ft primary ball mill and secondary ball mill (two 16.5 ft × 29 ft mills) circuits each operate in closed circuit with hydrocyclones. Total grinding circuit specific energy consumption at this grind is nominally 30 kW·h/t and is consistent with the ore hardness properties and the extent of size reduction performed (Giblett and Hart 2016). Specifically, it is an example of fine grinding using ball mills without compromising energy efficiency.

AUTOGENOUS FLOW SHEETS

While the use of autogenous grinding, or autogenous milling, is not as widespread as semiautogenous milling in precious and base metals processing, there are examples of its successful application. The Freeport-McMoRan Bagdad operation in Arizona, United States, is a significant example of the application of an AG, ball milling, pebble crushing (AB/C) grinding flow sheet in base metals concentration, with five grinding lines processing 75,000 stpd of copper–molybdenum ahead of flotation. Because of high pebble loads, circuit capacity is heavily influenced by the utilization of pebble crushers (Clements 1992). The use of 38-ft-diameter AG mills in a copper recovery flow sheet at the Aitik copper concentrator is addressed by Markstrom (2011), where fully autogenous grinding in two AG and pebble mill lines achieves a plant capacity of 33–36 Mt/yr depending on mine production rates. The Palabora copper mine in South Africa (Condori et al. 2011) uses two 32-ft-diameter primary AG mills in an ABC circuit configuration (Figure 6) to process nominally 30,000 t/d of copper ore for flotation, and it is notable for the use of a "permanent bleed-off" to discard pebbles >19 mm in size.

The BHP Cannington operation in Australia (Jankovic et al. 2006) processes silver/lead/zinc ore through a closed-circuit SAG mill, with secondary grinding initially performed by a 1,500-hp Vertimill. In the Cannington case, secondary milling was added to what was initially a single-stage AG mill circuit, to facilitate increased plant throughput at acceptable grind sizes. The primary grinding circuit operates in closed circuit with cyclones, otherwise replicating the flow sheet presented in Figure 5. An expanded secondary grinding flow sheet was later presented by Palaniandy et al. (2014), composed of two VTM-1250-WB Vertimill units.

Typically, the significantly higher mill capacity that results from the conversion of an AG mill to SAG sees semiautogenous milling as the preferred option in the study phase for precious and base metals projects. AG is a rare choice in these instances, where maximum plant capacity per unit capital input is often the critical metric for project success. In the case of the Newcrest Ridgeway concentrator in Australia (Hart et al. 2003), the conversion from ABC to SABC operation resulted in a 25% increase in grinding plant capacity from 4 to 5 Mt/yr, along with an increase in final grind size, reduced pebble recirculating load, and improved primary mill stability.

The use of AG is common in the iron ore industry for grinding magnetite ores because of the high specific gravity of the magnetite, which also generates excellent AG media. In these flow sheets, pebble crushing is not compromised by the presence of steel ball fragments in the mill discharge. The mill product is at a size suitable for primary magnetic separation, which is often employed to reject feed mass before the secondary grinding stage. A recent example of autogenous milling magnetite ores is the massive Sino Iron magnetite operation (Tian et al. 2014). This project has commissioned six grinding lines, each consisting of one 40-ft-diameter × 36 ft effective grinding length AG mill, pebble crushing, and a 26 ft × 44.5 ft ball mill. Powell et al. (2011) describe the use of fully autogenous grinding at the LKAB Kiruna KA3 processing line, with primary autogenous millings and secondary pebble milling to produce a final product of 80% passing 325 mesh. Cleveland-Cliffs Empire mine's Empire IV grinding line (Walqui et al. 2009) applies three lines of 32-ft AG mills with two-stage pebble crushing and dual 15.5 ft × 32.5 ft pebble mills to grind magnetite ore to 90%–95% passing 500 mesh.

ROD MILL/BALL MILL CIRCUITS

In addition to the use of ball mills in secondary and tertiary grinding duties after AG and SAG mills, several other common grinding flow sheets incorporate ball milling: rod mill/ball mill circuits (Figure 7), single-stage ball mill circuits, and two-stage ball mill circuits.

In base and precious metals processing, rod mill/ball mill circuits are often associated with lower-capacity, higher feed-grade operations, and in many cases receive feed exclusively from higher-grade underground mining operations. The use of rod mills in these industries has declined as a result of the widespread use of semiautogenous milling, primarily because of the large difference in unit capacity in favor of SAG mills and the elimination of constraints and costs associated with secondary crushing facilities required to feed rod mill circuits. Rod mills are also useful where the grinding work indices are high and it is necessary to eliminate all coarse particles from

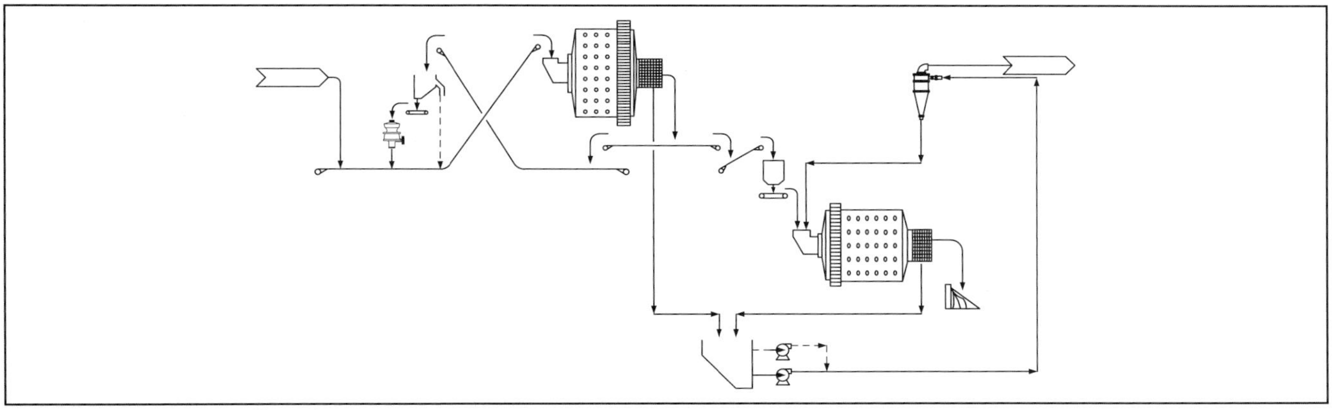

Figure 6 Two-stage AG flow sheet

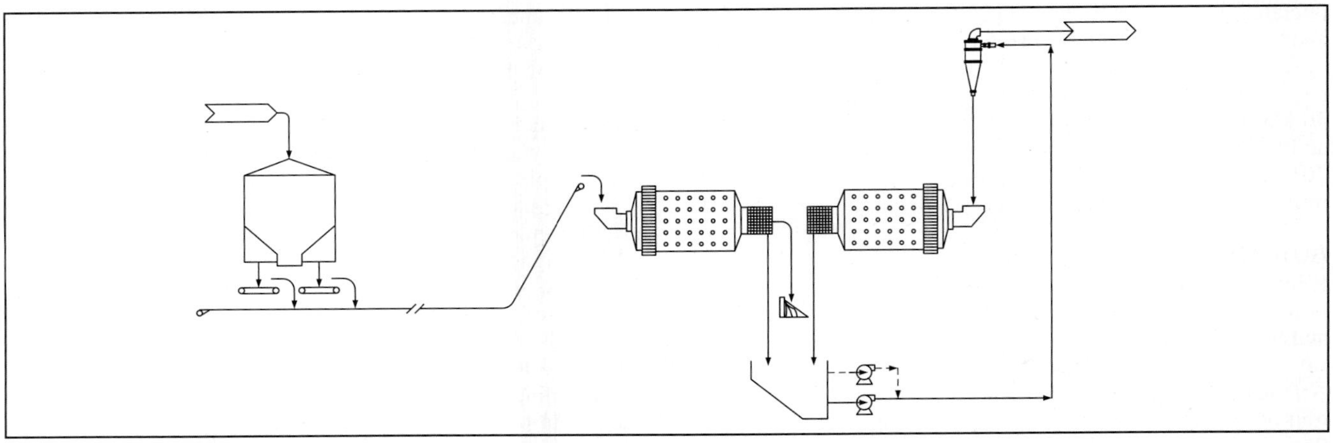

Figure 7 Rod mill/ball mill flow sheet

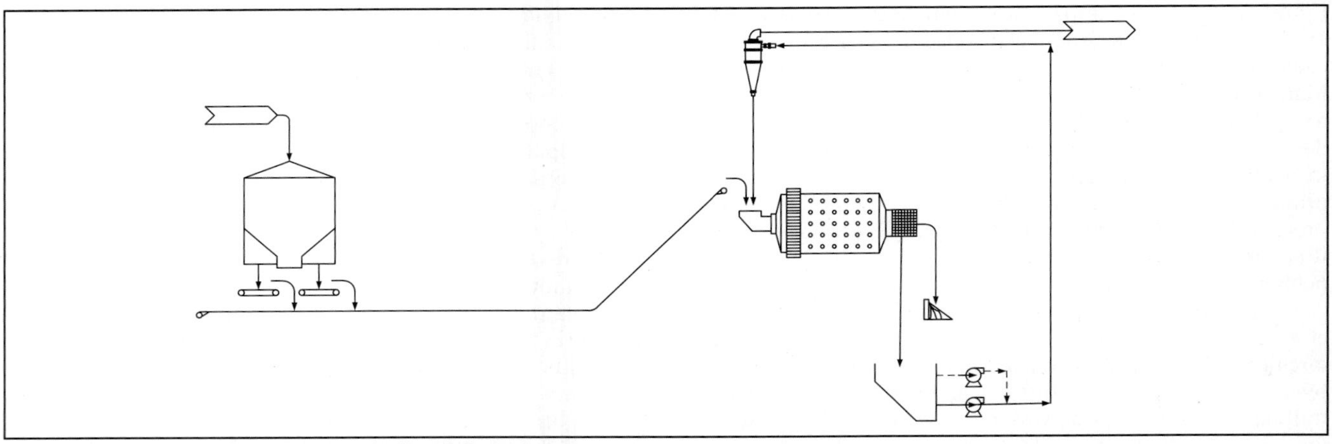

Figure 8 Conventional single-stage ball mill flow sheet

ball mill feed. Eliminating the coarse particles prevents ball mill charge expansion, scatting of oversize from the mill, and associated reductions in grinding efficiency.

Rod mill size is constrained by aspect ratio, rod and mill length, and speed limits required to prevent tangling, bending, and breaking of rods. The largest rod mills are therefore around 15 ft diameter × 21 ft effective grinding length with motor ratings of 3,000 hp or lower, and have significantly lower unit capacity than the average SAG mill. As a function of the mine economics associated with large-capacity, low-grade base and precious metals projects, invariably one or two lines of semiautogenous/ball milling is a more attractive proposition than multiple trains of rod/mill circuits, with the associated crushing and screening plant equipment and plant footprint considerations. Rod tangling is not only a cause of production loss, but it also presents a safety hazard for those tasked with rectifying the issue. However, the conventional rod/ball mill circuit remains a viable option for smaller-capacity operations, particularly those requiring a tight size distribution devoid of excessive fines content. The internal classification performed by the rod mill charge is a recognized advantage over ball mills in reducing the potential for overgrinding, and any recovery increases associated with reduced overgrinding can materially impact project economics, particularly for higher-grade operations.

The use of rod mills is common in the grinding of phosphate ores, where often coarse grinds and low fines content in the flotation feed is advantageous for effective mineral concentration. Van der Linde (1999) and Schmidt (1999) describe the processing of phosphate ores at the Foskor operation in South Africa, where three-stage crushing is followed by 12 8-ft primary rod mills and five 900-hp Vertimill units in secondary grinding duty to produce a flotation feed of 80% passing 370 µm, with <11% passing 400 mesh. The production of apatite concentrate at the Jacupiranga operation in Brazil (Busnardo and De Mineracao 1985) uses four lines of single-stage rod mills to grind 500 t/h of carbonate ore and generate a feed for flotation of 78%–86% passing 50 mesh. Other commodities that often incorporate rod mills in the comminution flow sheet include bauxite, graphite, rare earths, and tin.

PRIMARY BALL MILL CIRCUITS

Single-stage ball mills, conventionally receiving secondary or tertiary crushed feed size of ¼–½ in. (Figure 8), are often used in lower-capacity operations in the 0.5–2.5 Mt/yr range, particularly when processing precious metals. In recent times, the single-stage ball mill flow sheet has been applied on a much larger scale, in the 100,000 t/d range, after tertiary crushing by HPGR technology, at the Boddington gold mine (Hart et al. 2011) and Cerro Verde copper concentrator (Koski et al.

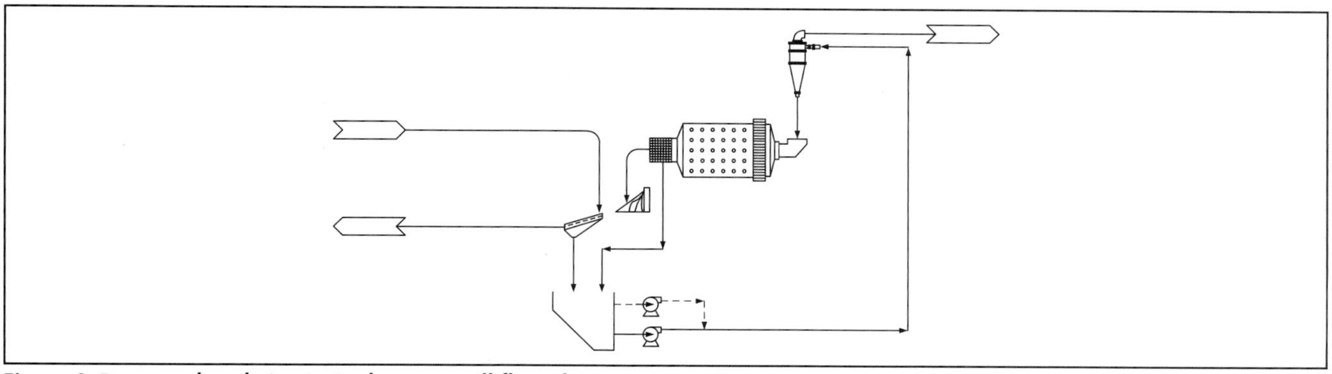

Figure 9 Reverse closed-circuit single-stage mill flow sheet

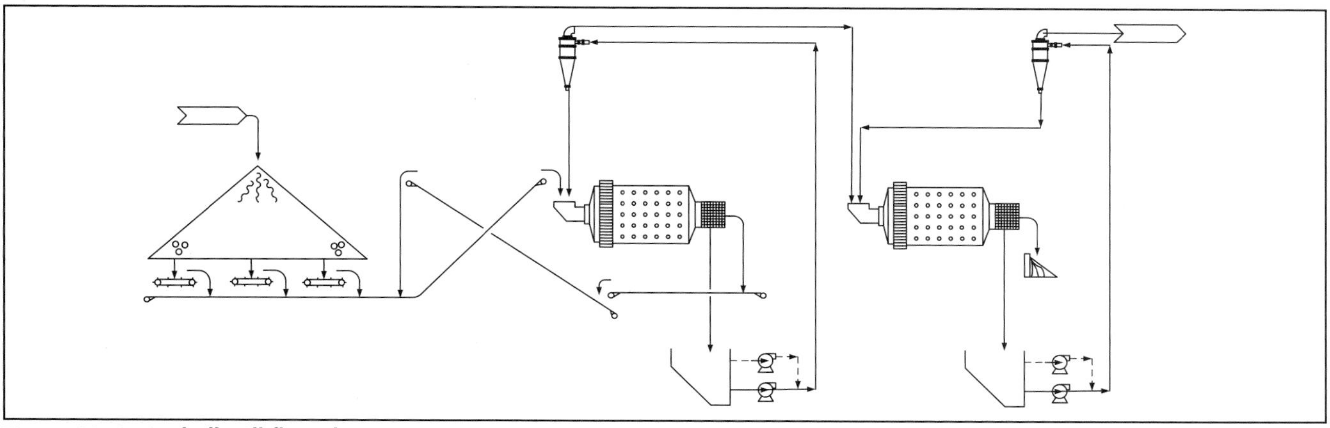

Figure 10 Series ball mill flow sheet

2011). The HPGR/ball mill flow sheet has also been applied at the Sierra Gorda copper–molybdenum project in Chile and the Tropicana gold mine in Australia. These new-generation flow sheets followed from the demonstrated performance of the ThyssenKrupp roll wear protection for hard rock applications at the Newmont Lone Tree operation (LeVier et al. 2004). Furthermore, the availability of very large ball mills and very large-capacity cone crushers in secondary crushing duty are critical requirements for the economic viability of single-stage ball milling in high-capacity applications. Single-stage ball mill applications flowing HPGRs have commonly incorporated a reverse classification configuration (Figure 9) where the HPGR product is wet screened with the screen undersize reporting to the cyclone feed hopper and the ball mill being feed cyclone underflow.

Series ball milling (Figure 10) has been used at several operations, including Sunrise Dam (Nugus et al. 2013), Mount Lyell (Gray and Bisshop 1977), and Marvel Loch (Bird and Briggs 2011) in Australia, with a primary ball operated in closed circuit with cyclones and a secondary ball mill receiving primary cyclone overflow and being, in turn, closed with a secondary cyclone cluster. Series ball milling, with the primary mill receiving secondary or tertiary crushed feed, is a concept more likely to be applied in the fine grinding of gold ores before cyanidation and typically at lower plant capacities (<3 Mt/yr). In the platinum mines of South Africa, it is common to conduct concentration stages between series ball mill grinding stages. The most common circuit of this type is called an MF2, or mill-float-mill-float, circuit as described by Rule (2011).

DRY GRINDING

In specific instances, it is preferable for the downstream recovery process to receive a dry product from the grinding circuit. This is the case in the cement industry where there is extensive use of dry grinding technologies including air-swept SAG mills and ball mills, hammer mills, roller mills, and HPGRs in fine grinding duties, as noted by Burchardt and Brandhoff (2014). Almost all grinding circuits in precious and base metals ore processing flow sheets produce feed for flotation, wet gravity separation, or leaching processes. As a result, there has been limited application of dry grinding technologies in those industries, while the more competent and abrasive nature of these ores also presents challenges for some of the dry grinding technologies. Rowland (1985) stated that "dry grinding requires about 30% more power than wet grinding for comparable size reduction," consistent with the observations of Taggart (1945) and Bond (1961). Rowland (1985) also noted the lower capital cost of equivalent wet grinding circuits, but substantially lower grinding media and liner consumption rates.

Notable exceptions are the use of dry grinding before oxidative roasting of refractory gold ores in Nevada as reported by Tempel (1993) and Thomas et al. (2001), and the use of dry grinding in the processing of nickel laterite ores as practiced at Yabulu in Australia (Stokes 2013). The simplified grinding and classification flow sheet at the Newmont Carlin Trend operation as reported by Tempel (1993) is shown in Figure 11, reflecting the combination of static and dynamic classification of mill products that is common in dry grinding flow sheets.

Burchardt and Brandhoff (2014) observe that the cement industry has over time progressed to the use of vertical roller mills (VRMs) and HPGRs in grinding duties down to product sizes of 450 mesh. Significant improvements in energy efficiency have been observed, and these grinding technologies have significant appeal for application outside cement. The standard VRM flow sheet for finish grinding such as applied to grind phosphate ore at Foskor is shown in Figure 12.

REGRINDING AND ULTRAFINE GRINDING

In the context of this discussion, regrinding is considered to encompass the process where a valuable mineral concentrate, most typically produced by flotation or magnetic separation, is reground to further enhance separation, beneficiation, or metal extraction. Regrinding will typically produce a product size down to 80% passing 25–30 μm (450 mesh). Size reduction by grinding down below 80% passing 15 μm (635 mesh) or finer, to as far as below 80% passing 5 μm in some instances, is defined as ultrafine grinding (UFG) in the context of this discussion.

Vertical stirred mills, specifically the Vertimill or tower mill, have become the dominant technology in conventional regrinding applications. The fine media size that can be used in these mills is a great advantage over overflow ball mills, which are typically limited to a minimum grinding media size around ⅞ in. Vertical mills operating in concentrate regrinding applications have been quoted to operate at 30%–50% lower specific energy consumption than ball mills performing in the same duty. Rule (2011) describes the implementation of the horizontal stirred IsaMill technology to effect size reduction down to 80% passing 53 μm (270 mesh) in a tertiary grinding application in platinum ore grinding flow sheets, demonstrating the suitability of that technology in conventional fine grinding duties. Technologies being applied to this duty also include the VXPmill and HIGmill, both of which are vertical stirred mills.

In UFG duties, defined by Lichter and Davey (2002) as sub-15 μm, several technologies have demonstrated suitability for the task. The UFG technologies will typically receive a concentrated feed stream to minimize grinding of gangue minerals, and grind in open or closed circuit depending on the technology applied. The IsaMill technology is used at the Mount Isa lead–zinc concentrator (Pease et al. 2004) to grind lead and zinc rougher concentrates to 80% passing 12 μm, and zinc cleaner tailings to 7 μm. The circuit flow sheet is shown in Figure 13. Ellis (2003) described the use of IsaMills to grind a gold-bearing pyrite concentrate to 80% passing 11–12 μm at the KCGM operation in Western Australia, before cyanide leaching for gold recovery. Anderson and McDonald (2016) described the recent expansion of the Fimiston UFG circuit, eliminating the roasting of pyrite concentrates at the KCGM operations. Many of the platinum operations in South Africa

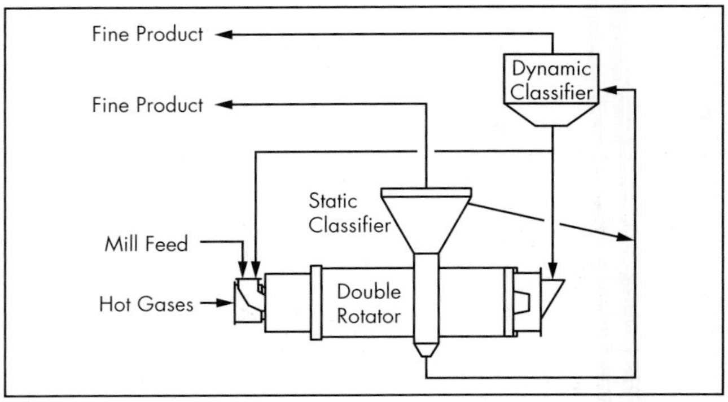

Source: Tempel 1993

Figure 11 Dry grinding schematic for the Newmont Carlin Trend operation

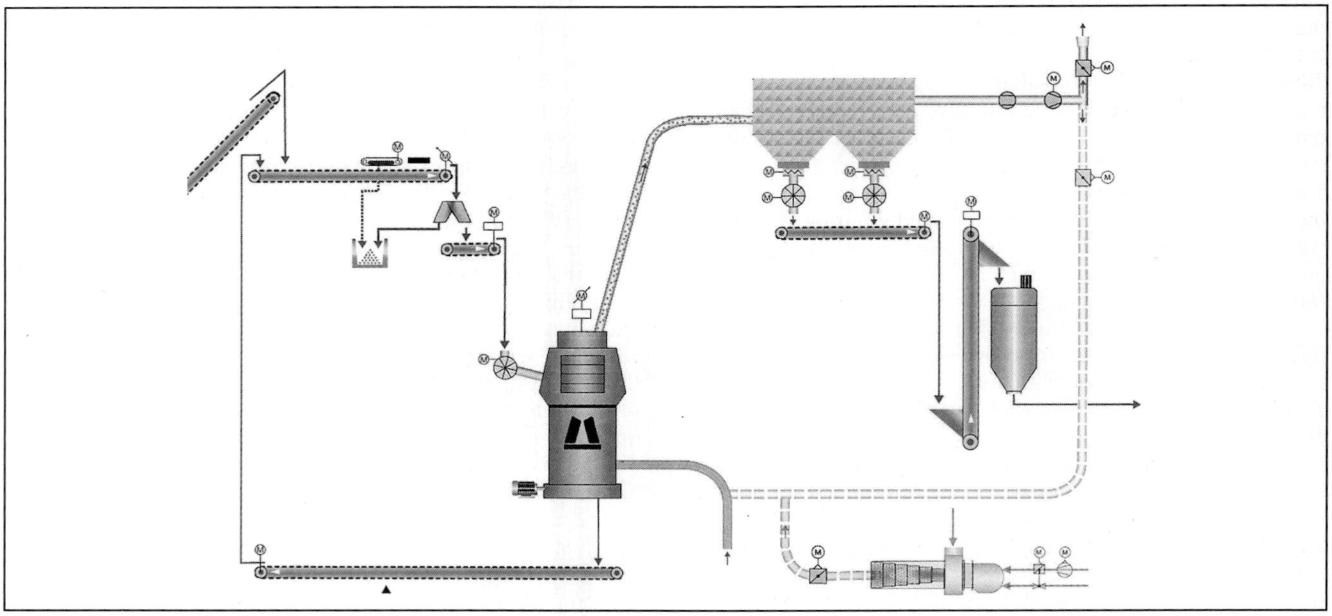

Courtesy of Loesche GmbH

Figure 12 Vertical roller mill circuit flow sheet

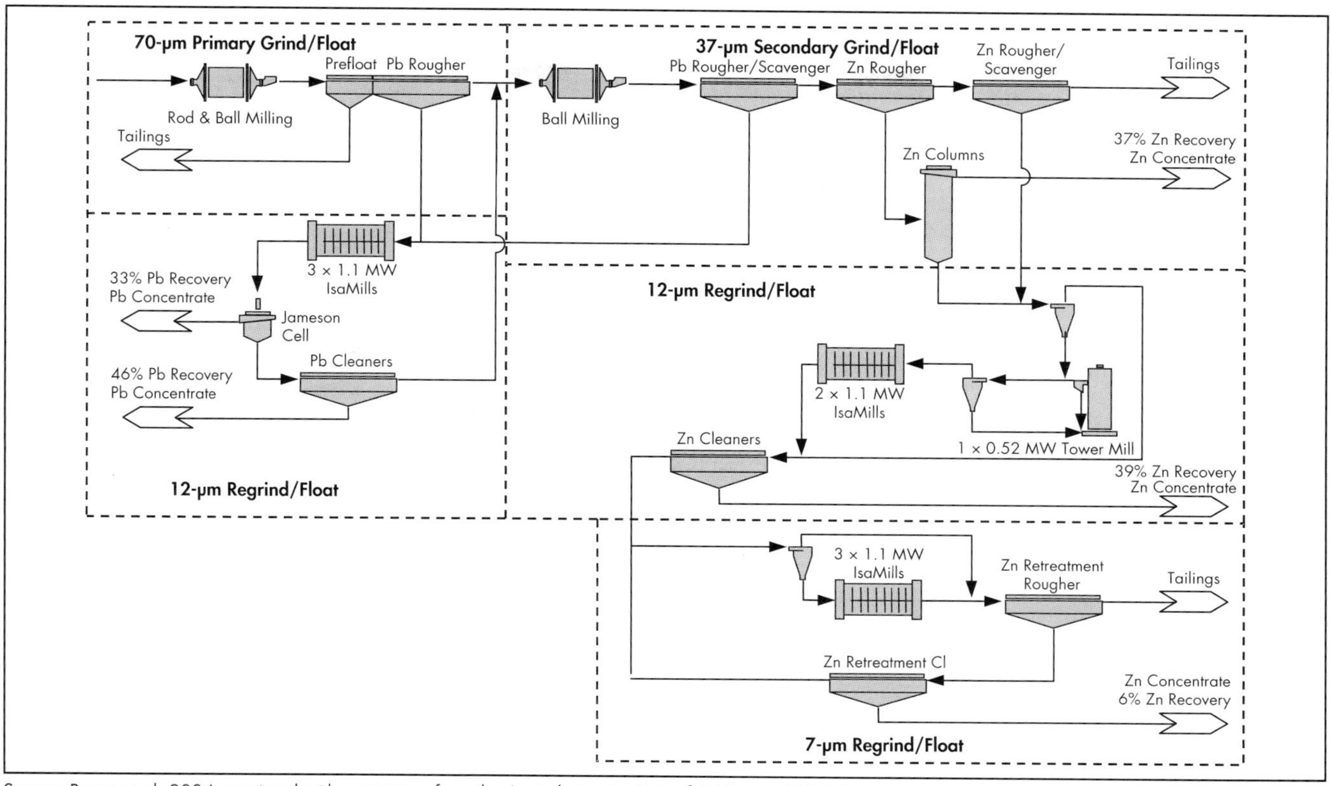

Source: Pease et al. 2004, reprinted with permission from the Australasian Institute of Mining and Metallurgy

Figure 13 Mount Isa lead–zinc concentrator flow sheet

have incorporated IsaMills in their flow sheets in the past decade (Rule 2011).

The Metso Stirred Media Detritor (SMD) is a versatile technology that has found application in both regrinding and UFG duties. The use of SMDs in copper concentrate regrind duties at Copper Mines of Tasmania and Thangala copper mine in Australia has been documented by Davey (2003), and a similar application in the Newcrest Ridgeway flow sheet at the Cadia Hill complex in Australia is described by Hart et al. (2005) and Weidenbach and Cesnik (2007). Davey (2002) details the use of the SMD in both regrind and UFG duty at the Century zinc operation in Australia, with 15 SMDs dedicated to UFG duty producing a P_{80} 6.5-μm product size.

GRINDING FLOW-SHEET SELECTION

The importance of reliably determining ore characteristics and how they vary within a deposit at the design stage can never be overemphasized. Too many operations have been forced into expensive plant upgrades and retrofits because of a failure to fully recognize and account for the fundamental nature of the ore during project development. Even the best flow-sheet design optimization efforts will be undone if design criteria fail to accurately reflect the mined ore, or do not fully consider the impact of measured characteristics. Guidance on how to approach ore characterization, using industry best practices to support grinding circuit design, is given in Chapter 3.8, "Grinding Circuit Design." Guidance on how to characterize an ore body in terms of its comminution property variability is covered in Chapter 1.10, "Geometallurgy."

The resistance of an ore to impact breakage is a particularly crucial parameter when evaluating the optimal comminution flow sheet in all cases. Low-impact breakage resistance will likely result in a shortage of AG media, leading to high steel wear rates in a SAG mill. In these circumstances, higher ball charges coupled with high steel-to-rock charge volume ratios will be required to achieve mill power draw targets. While highly competent ores are certainly harder to break in a SAG mill, those same characteristics make the ore resistant to crushing. When hardness is combined with moderate to high abrasive characteristics, particularly high liner wear rates in crushing applications will result. When the ore is hard in one sense or another, there is rarely a clear-cut equipment selection advantage that will serve as a silver bullet for the designer or, indeed, the operator.

Recognizing this and other conundrums that will influence ultimate flow-sheet selection, the selection matrix in Figure 14 is provided to assist in narrowing down the options for the design engineer, by considering common influences on the suitability of the major grinding technologies for a given project. This flow-sheet selection matrix borrows from the tabular representations published by Lane et al. (2002; 2013), Putland (2006), Putland et al. (2011), and Scinto et al. (2015). The matrix is based on the assumption that grinding will be achieved in no more than two stages, which is intended to evade debate around the use of ball mills or stirred mills in a tertiary grinding duty, or similar flow-sheet refinements that are more a matter of final equipment selection and, in some instances, personal preferences. Figure 14 reflects that certain grinding technologies are more tailored to specific ore

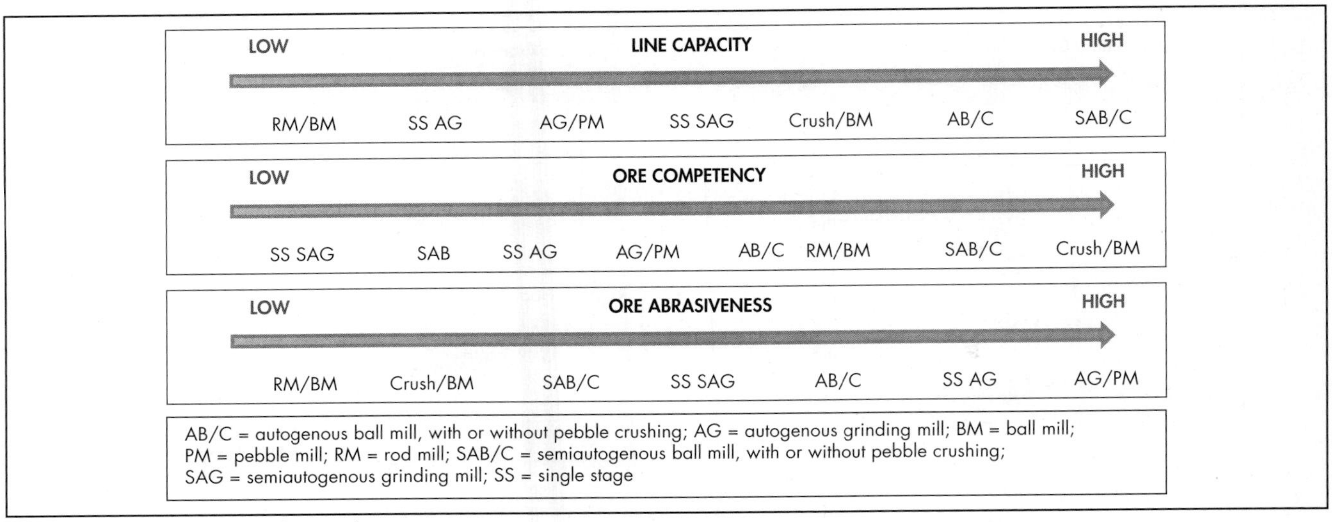

Figure 14 Grinding technology selection matrix

characteristics than others, and that the relative suitability may change based on other project-specific considerations such as life of mine, ore variability, mining method, downstream ore processing requirements, ore density, and clay content.

High specific-gravity ores, such as magnetite, can be more suitable to autogenous milling than others, making that technology a more attractive option in such instances. High-clay-content ores or ores otherwise prone to be problematic in material handling processes have been successfully processed using scrubbers or SAG mills and by minimizing crushing stages before the grinding process. Downstream process requirements can influence grinding technology and flow-sheet selection, as evidenced by the selection of dry grinding technology for the roasting of gold ores in Nevada. If an ore contains hazardous materials (e.g., asbestos or lead), then a wet grinding flow sheet is preferred to dry grinding processes because of the reduced potential for dust generation. Reduced variability in treatment rates and grinding circuit product size can often be achieved by secondary and tertiary crushing before grinding in rod mills or ball mills. Minimizing mineral product contamination with grinding steel can be achieved by the application of AG technologies. Grinding technology selection will be influenced if there is a need for product sizes below the normal operating range of ball mills, and this creates a situation where lower unit capacity grinding technologies such as stirred mills can become viable options. In instances where the minimizing of fines in the grinding circuit product is a critical performance parameter, such as in the flotation of phosphate ores, the use of rod milling technology can be beneficial.

CONCLUSIONS

This chapter has presented several conventional grinding flow sheets and provided references for operations where the many grinding technologies available have been implemented. Some context has been provided in the form of pros and cons of the various flow sheets based on documented operating experience, and guidelines for determining when a particular flow sheet may be applicable have been provided. As was suggested at the beginning of this chapter, there is no single, perfect off-the-shelf flow-sheet solution for a given project; more likely, there will be several flow-sheet variations that can adequately deliver the project's objectives. Similarly, there is no perfect flow-sheet selection guide that can deal with all variables in a desktop analysis, and every project must be considered in detail on its individual merits. Opinions can vary considerably regarding what grinding technologies are better suited to the project and under what conditions. This is largely based on the experience of the engineer, and a significant portion of industry experience and knowledge is unpublished. As a result, there is no substitute for rigorous benchmarking when it comes to final flow-sheet selection and process engineering, ensuring that as much knowledge as possible from others' experience is incorporated in the final design.

The current state of grinding technology is in an interesting position, with established tumbling mill technologies (AG/SAG, rod, and ball mill flow sheets) remaining popular and advanced technologies such as stirred milling and HPGRs establishing a footprint in conventional grinding duties. Worldwide focus on greenhouse gas emissions, increased demand, and rising energy costs support increased focus on grinding circuit energy efficiency during flow-sheet selection. To that end, techniques to define grinding circuit energy consumption using the guidelines described in Chapter 3.12, "Testing and Calculations for Comminution Machines," can be employed. Examples of using these and similar techniques to select more energy-efficient flow sheets have been published by many authors. Notable examples include Parker et al. (2001) and Ballantyne et al. (2016), with both of those assessments supporting the selection of HPGR-based circuits over SAG mills. Other authors have employed similar approaches to promote the use of stirred mills in place of ball mills, and in some publications to promote combining HPGRs and stirred mills as an option for grinding circuits that do not employ any form of tumbling mill technology. How far the industry shifts toward these newer flow-sheet designs will depend on how well the desktop estimates of energy efficiency translate to industrial performance, and to what extent they offset increased capital costs with reduced operating costs.

ACKNOWLEDGMENTS

The authors acknowledge the contributions of Dean David (Wood), Gerhard Sauermann (ThyssenKrupp), and Carsten Gerold (Loesche) to this chapter, and the staff of Orway Minerals Consulting for the drafting of flow-sheet illustrations.

REFERENCES

Alexander, D.J., and Wigley, P. 2003. Flotation circuit analysis at WMC Ltd. Olympic Dam operation. Presented at the Eighth AusIMM Mill Operators Conference, Townsville, Queensland, Australia, July 22–23.

Anderson, G.S., and McDonald, N.W. 2016. IsaMills at Kalgoorlie Consolidated Gold Mines—From the M3000 to the M10000 and replacement of the roasters at Gidji processing plant. Presented at the 13th AusIMM Mill Operators Conference, Perth, Australia, October 10–12.

Anderson, G.S., Smith, D.T., and Strohmayr, S.J. 2011. IsaMill technology in the primary grinding circuit. Presented at the International Autogenous Grinding, Semiautogenous Grinding, and High Pressure Grinding Roll Technology Conference, Vancouver, BC, Canada, September 25–28.

Atasoy, Y., and Price, J. 2006. Commissioning and optimization of a single stage SAG mill grinding circuit at Lefroy gold plant—St. Ives gold mine—Kambalda/Australia. In *SAG 2006 Conference Proceedings*. Vancouver: Mining and Mineral Processing Engineering, University of British Columbia.

Ballantyne, G., Powell, M., Clarke, N., Di Trento, M., Kock, F., and Putland, B. 2016. Introduction to energy curves and AngloGold Ashanti case studies. Presented at the SME 2016 Annual Meeting, Phoenix, AZ, February 22–24.

Bartrum, J., Plyley, W., and Butcher, G. 1988. SABC development at Kidston gold mine. SME Preprint No. 88-150. Littleton, CO: SME.

Bird, A., and Briggs, M. 2011. Recent improvements to the gravity gold circuit at Marvel Loch. In *MetPlant 2011: Metallurgical Plant Design and Operating Strategies*. Melbourne, Victoria: Australasian Institute of Mining and Metallurgy. pp. 115–137.

Bond, F.C. 1961. *Crushing and Grinding Calculations*. Milwaukee, WI: Allis Chalmers Manufacturing.

Burchardt, E., and Brandhoff, W. 2014. Dry grinding: Current and future systems in minerals processing. Presented at Comminution 2014, 9th International Comminution Symposium, Cape Town, South Africa, April 7–10.

Burger, B., Hatta, M., McGaffin, I., and Gaffney, P. 2006. Batu Hijau—Seven years of operation and continuous improvement. In *SAG 2006 Conference Proceedings*. Vancouver: Mining and Mineral Processing Engineering, University of British Columbia.

Busnardo, C.A., and De Mineracao, S.S.A. 1985. Optimization of the grinding circuit of the Jacupiranga carbonatite ore in Jacupiranga, Brazil. SME Preprint No. 85-98. Littleton, CO: SME.

Castillo, G.M., and Bissue, C. 2011. Evaluation of secondary crushing prior to SAG milling at Newmont's Phoenix Operation. Presented at the International Autogenous Grinding, Semiautogenous Grinding, and High Pressure Grinding Roll Technology Conference, Vancouver, BC, Canada, September 25–28.

Clements, B. 1992. Pebble crushing practice at Cyprus Bagdad. *Min. Eng.* (September):1114–1116.

Comi, T., Garcia C., Potulsha, A., and Opazo, J.J. 2015. Sierra Gorda, Chile's new copper–molybdenum operation. *Min. Eng.* (July):23–30.

Condori, P., Fischer, D., Winnett, J., and Makgatho, J. 2011. From open cast to block cave and the effects on the autogenous milling circuit at Palabora copper mine. Presented at the International Autogenous Grinding, Semiautogenous Grinding, and High Pressure Grinding Roll Technology Conference, Vancouver, BC, Canada, September 25–28.

Davey, G. 2002. Ultrafine and fine grinding using the Metso Stirred Media Detritor (SMD). Presented at the 34th Annual Meeting of the Canadian Mineral Processors, Ottawa, ON, Canada, January 22–24.

Davey, G. 2003. Fine copper grinding using the Metso Stirred Media Detritor (SMD). SME Preprint No. 03-086. Littleton, CO: SME.

Ellis, S. 2003. Ultra-fine grinding—A practical alternative to oxidative treatment of refractory gold ores. Presented at the Eighth AusIMM Mill Operators Conference, Townsville, Queensland, Australia, July 22–23.

Festa, A., Putland, B., and Scinto, P. 2014. Shedding light on secondary crushing. SME Preprint No. 14-160. Englewood, CO: SME.

Fitzmaurice, C., Stokes, K., and Wan, E. 2000. Leinster nickel concentrator—Past, present and future. Presented at the Seventh AusIMM Mill Operators Conference, Kalgoorlie, Western Australia, Australia, October 12–14.

Giblett, A., and Hart, S. 2016. Grinding circuit practices at Newmont. Presented at the 13th AusIMM Mill Operators Conference, Perth, Western Australia, Australia, October 10–12.

Gray, L.R., and Bisshop, J.P.W. 1977. Selective grinding to improve copper recovery at Mount Lyell. Presented at the AusIMM Conference, Tasmania, Australia, May.

Hart, S., Griffin, P., Cesnik, F., Gordon, D., and Clements, B. 2003. Design and commissioning of the Ridgeway Concentrator. Presented at the Eighth AusIMM Mill Operators Conference, Townsville, Queensland, Australia, July 22–23.

Hart, S., Green, S., Cesnik, C., and Griffin P. 2005. Design, construction and commissioning of the Ridgeway Concentrator regrind circuit at Newcrest's Cadia Valley operations. Presented at the AusIMM Centenary of Flotation Symposium, Brisbane, Queensland, Australia, June 6–9.

Hart, S., Parker, B., Rees, T., Manesh, A., and McGaffin, I. 2011. Commissioning and ramp up of the HPGR circuit at Newmont Boddington Gold. In *SAG 2011 Conference Proceedings*. Vancouver: Mining and Mineral Processing Engineering, University of British Columbia.

Henderson, R., Warnert, M., Frentress, R., and Donelon, D. 2005. The Round Mountain Gold Project: Operation of a mill and heap leach facility in Nevada. In *Proceedings of the Canadian Mineral Processors—2005*. pp. 125–144.

Hothersall, P., Van Nierkirk, C., Barnard, E., and Nutor, G. 2006. Single stage SAG milling at the Tarkwa Gold Mine. In *SAG 2006 Conference Proceedings*. Vancouver: Mining and Mineral Processing Engineering, University of British Columbia.

Jankovic, A., Valery, W., and Clarke, G. 2006. Design and implementation of an AVC grinding circuit at BHP Billiton Cannington. In *SAG 2006 Conference Proceedings*. Vancouver: Mining and Mineral Processing Engineering, University of British Columbia.

Koski, S., Vanderbeek, J., and Enriquez, J. 2011. Cerro Verde Concentrator—Four years operating HPGRs. In *SAG 2011 Conference Proceedings*. Vancouver: Mining and Mineral Processing Engineering, University of British Columbia.

Lane, G.S., Fleay, J., Reynolds, K., and La Brooy, S. 2002. Selection of comminution circuits for improved efficiency. Presented at the Crushing and Grinding Conference, Kalgoorlie, Western Australia, Australia, October 29–November 1.

Lane, G., Foggiato, B., and Bueno, M. 2013. Power-based comminution calculations using Ausgrind. Presented at Procemin 2013: 10th International Mineral Processing Conference, Santiago, Chile, October 15–18.

LeVier, K.M., Logan, T.C., Patzelt, N., and Klymowsky, I.B. 2004. Latest developments in HPGR technology—Results of a field trial. Presented at MetPlant 2004, Perth, Western Australia, Australia, September 6–7.

Lichter, J.K.H., and Davey, G. 2002. Selection and sizing of ultrafine and stirred grinding mills. In *Mineral Processing Plant Design, Practice, and Control*. Edited by A. Mular, D.N. Halbe, and D.J. Barratt. Littleton, CO: SME. pp. 783–800.

MacNevin, W., and Stephenson, P. 1997. Evolution of the Kidston Gold Mine comminution circuit. Presented at the IIR Optimising Crushing and Grinding Conference, May 1997.

Markstrom, S. 2011. Commissioning and Operation of the AG Mills at the Aitik Expansion Project. In *SAG 2011 Conference Proceedings*. Vancouver: Mining and Mineral Processing Engineering, University of British Columbia.

McGhee, S., Mosher, J., Richardson, M., David, D., and Morrison, R. 2001. SAG feed pre-crushing at ASARCO's Ray Concentrator: Development, implementation and evaluation. In *SAG 2001 Conference Proceedings*. Vancouver: Mining and Mineral Processing Engineering, University of British Columbia.

Mudd, G.M. 2007. Sustainability reporting in the gold mining industry: The need for continual improvement. Presented at the SSEE-07 Conference, Perth, Western Australia, Australia, October 31–November 2.

Mudd, G.M. 2009. Historical trends in base metal mining: Backcasting to understand the sustainability of Mining. Presented at the 48th Annual Conference of Metallurgists, Canadian Metallurgical Society, Sudbury, ON, Canada, August 2009.

Nelson, M., Valery Jr., W., and Morrell, S. 1996. Performance characteristics and optimisation of the Fimiston (KCGM) SAG mill circuit. Presented at the SAG 1996 Conference, Vancouver: Mining and Mineral Process Engineering, University of British Columbia.

Nugus, M., Briggs, M., Tombs, S., Elms, P., and Erickson, M. 2013. Sunrise Dam Gold Mine, Anglogold Ashanti. In *Australian Mining and Metallurgical Operating Practices, The Sir Maurice Mawby Memorial Volume*, 3rd ed. Edited by W.J. Rankin. Melbourne, Victoria: Australasian Institute of Mining and Metallurgy. pp. 985–995.

Palaniandy, S., Powell, M., Hilden, M., Kermanshahi, K., Allen, J., and Mwansa, S. 2013. Vertimill—Development of circuit survey and performance evaluation protocols. Presented at MetPlant 2013, Perth, Western Australia, Australia, July 15–17.

Palaniandy, S., Powell, M., Hilden, M., Allen, J., Kermanshahi, K., Oats, B., and Lollback, M. 2014. Vertimill—Preparing the feed within floatable regime at lower specific energy. Presented at Comminution '14, Cape Town, South Africa, April 7–10.

Parker, B., Rowe, P., Lane, G., and Morrell, S. 2001. The decision to opt for high pressure grinding rolls for the Boddington expansion. In *SAG 2011 Conference Proceedings*. Vancouver: Mining and Mineral Process Engineering, University of British Columbia.

Pease, J.D., Young, M.F., Curry, D., and Johnson, N.W. 2004. Improving fines recovery by grinding finer. Presented at MetPlant 2004, Perth, Western Australia, Australia.

Powell, M., Condori, P., Smit, I., and Valery, W. 2006. The value of rigorous surveys—the Los Bronces experience. In *SAG 2006 Conference Proceedings*. Vancouver: Mining and Mineral Process Engineering, University of British Columbia.

Powell, M., Benzer, H., Dundar, H., Aydogan, N., Adolfsson, G., Partapuoli, A., Wikstrom, P., Fredriksson, A., and Tano, K. 2011. LKAB Autogenous Milling of Magnetite. In *SAG 2011 Conference Proceedings*. Vancouver: Mining and Mineral Process Engineering, University of British Columbia.

Putland, B. 2006. Comminution circuit selection—Key drivers and circuit limitations. In *SAG 2006 Conference Proceedings*. Vancouver: Mining and Mineral Process Engineering, University of British Columbia.

Putland, B., Kock, F., and Siddall, L. 2011. Single stage SAG/AG milling design. In *SAG 2011 Conference Proceedings*. Vancouver: Mining and Mineral Process Engineering, University of British Columbia.

Rowland, C.A. 1985. Ball mills. In *SME Mineral Processing Handbook*. Edited by N.L. Weiss. Littleton, CO: SME-AIME. pp. 3C26–3C44.

Rule, C. 2011. Stirred milling at Anglo American Platinum. In *SAG 2011 Conference Proceedings*. Vancouver: Mining and Mineral Process Engineering, University of British Columbia.

Schmidt, C. 1999. Foskor phosphate beneficiation. In *Beneficiation of Phosphates: Advances in Research and Practice*. Edited by P. Zhang, H.E. El-Shall, and R. Wiegel. Littleton, CO: SME.

Scinto, P., Festa, A., and Putland, B. 2015. OMC power-based comminution calculations for design, modelling and circuit optimisation. Presented at the Canadian Mineral Processors Conference, Ottawa, ON, Canada, January 20–22.

Siddall, B., and Putland, B. 2007. Process design and implementation techniques for secondary crushing to increase milling capacity. SME Preprint No. 07-079. Littleton, CO: SME.

Stokes, P.B. 2013. Palmer nickel and cobalt refinery. In *Australian Mining and Metallurgical Operating Practices, The Sir Maurice Mawby Memorial Volume*, 3rd ed. Edited by W.J. Rankin. Melbourne, Victoria: Australasian Institute of Mining and Metallurgy. pp. 1731–1742.

Sulianto, Y., Trott, T., Morgan, D., Bell, T., Kock, F., and Putland, B. 2016. Optimisation of installed mill power post-installation of secondary crushing circuit at Northparkes Mines. Presented at the 13th AusIMM Mill Operators Conference, Perth, Western Australia, Australia, October 10–12.

Suttill, K. 1988. R. Los Bronces plans major expansion to 30,000 mt/d capacity. *Eng. Min. J.* (May):36–40.

Sylvestre, Y., Abols, J., and Barratt, D. 2001. The benefits of pre-crushing at the Inmet Troilus mine. In *SAG 2001 Conference Proceedings*. Vancouver: Mining and Mineral Process Engineering, University of British Columbia.

Taggart, A.F. 1945. *Handbook of Mineral Dressing*. New York: John Wiley and Sons. pp. 6–15.

Tempel, T. 1993. Newmont Gold Company, refractory ore plant: Selection of a dry grinding process. SME Preprint No. 93-264. Littleton, CO: SME.

Thomas, K.G., Buckingham, L., and Patzelt, N. 2001. Dry grinding at Barrick Goldstrike's roaster facility. In *SAG 2001 Conference Proceedings*. Vancouver: Mining and Mineral Process Engineering, University of British Columbia.

Tian, J., Zhang, C., and Wang, C. 2014. Operations and process optimization of Sino iron ore's autogenous milling circuits: The largest in the world. Presented at the XXVII International Mineral Processing Congress, Santiago, Chile.

Torrealba-Vargas, J., Dupont, J.-F., McMullen, J., Allaire, A., and Welyhorsky, R. 2015. The successful development of the Detour Lake grinding circuit: From testwork to production. In *SAG 2015 Conference Proceedings*. Vancouver: Mining and Mineral Process Engineering, University of British Columbia.

Van der Linde, G.J. 1999. Improved apatite recovery from pyroxenite ore using dry milling. SME Preprint No. 99-74. Littleton, CO: SME.

Vanderbeek, J.L. 1997. Tertiary grinding circuit installation at Chino Mines Company. In *Comminution Practices*. Edited by S.K. Kawatra. Littleton, CO: SME. pp. 241–248.

Vesely, M., and Fernandez, O. 1986. Crusher for critical size material at Los Bronces semiautogenous grinding plant, Chile. *Miner. Metall. Process* (May):104–108.

Walqui, H., Hikade, E., Broeders, F., Whitford, A., Carlson, D., Williams, K., and Sjoholm, C. 2009. Improving Empire's grinding circuit performance using six sigma. SME Preprint No. 09-059. Littleton, CO: SME.

Weidenbach, M., and Cesnik, F. 2007. Recovery improvement project for the Ridgeway Concentrator at Newcrest's Cadia Valley operations. Presented at the World Gold Conference, Cairns, Queensland, Australia, October 22–24.

Westendorf, M., Rose, D., and Meadows, D.G. 2015. Increasing SAG mill capacity at the Copper Mountain mine through the addition of a precrushing circuit. In *SAG 2015 Conference Proceedings*. Vancouver: Mining and Mineral Process Engineering, University of British Columbia.

Yernberg, W.R. 1996. Santa Fe Pacific Gold targets million ounces/year production. *Min. Eng.* (September):39–43.

CHAPTER 3.10

Grinding Circuit Performance Optimization

Brian Putland, Rebecca Sciberras, and Aidan Giblett

Optimization is a critical process for maximizing the profitability of all metallurgical facilities within the constraints of the operation. Comminution is generally a cost-intensive process from both a capital and an operating perspective. It is also the most power-intensive process in the mining cycle and typically consumes more than 50% of total process plant power, ranging from 35% to 80% as indicated by Figure 1, with the distribution varying depending on the nature of the metal recovery process (Connolly and Buckingham 2012). In the U.S. mining industry, 40% of all energy consumption is estimated to be attributed to comminution, inclusive of mining (drilling, blasting, hauling) and processing, based on a study by BCS Incorporated (2007). Grinding power alone can account for more than 20% of direct operating costs in a typical gold plant, inclusive of salaries, wages, fuel, reagents, wear parts, and maintenance (Halbe and Smolik 2002). These percentages can be greater for ore types requiring high power input during comminution. The impact of comminution on project economics is increasing with the observed quantum shift in the minerals industry to lower-grade disseminated ore reserves, with comminution circuits also representing a large component of process plant capital. Consequently, the efficiency of the grinding process has multiple points of impact on project economics and is a key driver for many operations.

The key to successfully optimizing an operation is the definition of the optimization goal, which commonly aims to deliver improved life-of-mine project economics and/or cash flow, often through increased ore treatment rates. The impacts of project economic and cash-flow-based philosophies are not always the same, and the difference is often not well understood by operators focused on short-term process plant performance. Although optimization is project specific, cash flow for the existing asset is typically targeted first, particularly in tough financial times, and life-of-project economics are targeted second, as the latter is often more capital intensive. Optimizing the throughput of the grinding circuit can have economic benefits well beyond the processing plant, requiring optimization of key project features such as cutoff grade in mining, and therefore can significantly alter project economics. Based on an understanding of the goal, optimization of circuit tonnage, grind size, and/or operating cost can occur.

Another key contributor to successful optimization is a holistic view of the comminution process, considering both process fundamentals and operating practicalities. All grinding circuits conform to the same processing fundamentals that influence critical outcomes, such as mill power draw, classification, and grinding efficiencies. At the same time, numerous variables in plant design and operation influence the translation of these fundamentals into practice. Ore viscosity, for example, can create different relationships between pulp density, mill power draw, and grinding efficiency between one grinding circuit and another. Variations in instrument selection, installation, and process control philosophies can create differences in the way that grinding circuits operate and respond to process changes. Consequently, the optimization process requires an in-depth appreciation of both the fundamentals of the grinding process and the specific operating condition of a particular grinding circuit. The optimization team's focus should always be to trace any issues back to their source, which may be associated with the science of the grinding process or constraints imposed by the design and/or operation of the grinding circuit. Focusing on individual fundamentals often does not give the right answer or expected outcome, as the grinding system as a whole must be understood.

Traditionally, optimization has focused on grinding circuit efficiency; however, apart from the odd exception, grinding circuits tend to operate reasonably efficiently. Many significant production improvements are achieved through full use of installed power, improved process control, and optimization of production objectives and operating targets. Configuring the grinding circuit to achieve a target grind size while maximizing the use of installed power and therefore plant throughput is a common objective. Implementing and maintaining control systems that maximize the time that the plant is performing at its optimum is another. Understanding what throughput rate and grind size realizes maximum cash

Brian Putland, Principal Metallurgist, Orway Mineral Consultants, Perth, Western Australia, Australia
Rebecca Sciberras, Lead Metallurgist, Orway Mineral Consultants, Perth, Western Australia, Australia
Aidan Giblett, Senior Technical Advisor—Mineral Processing Newmont Mining Services, Subiaco, Western Australia, Australia

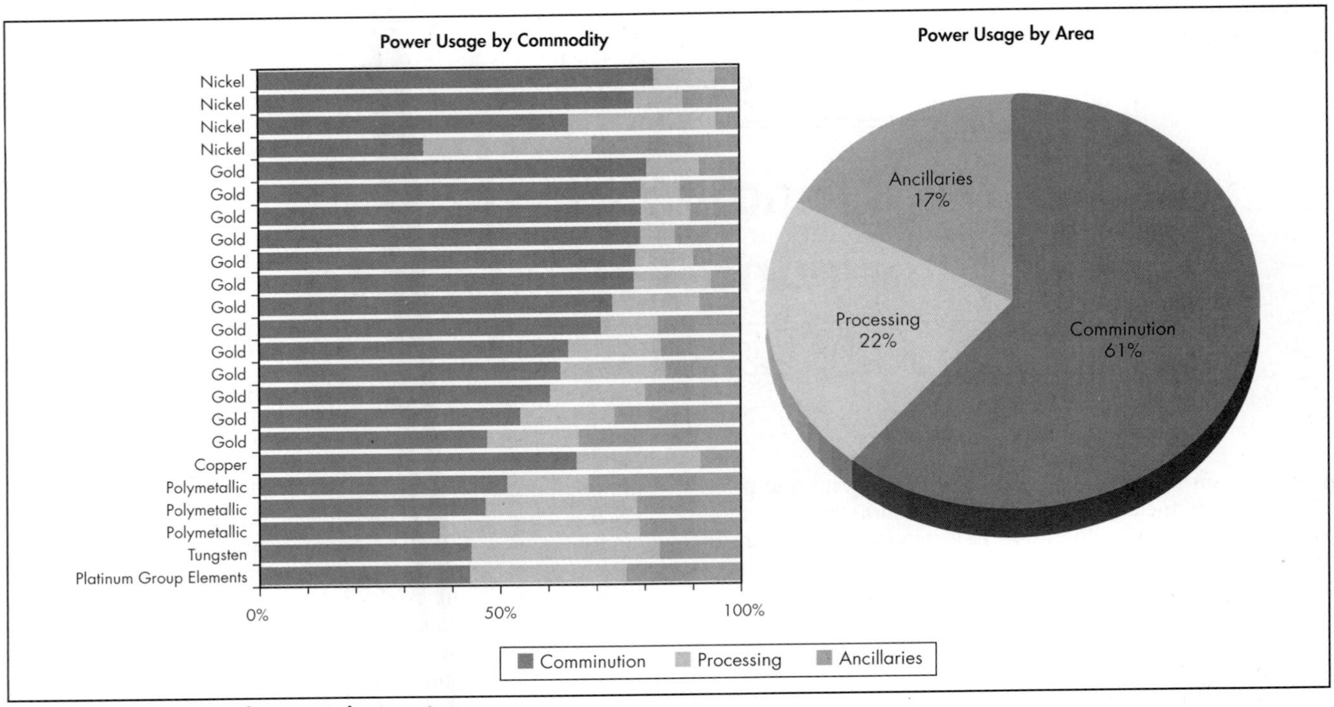

Figure 1 Power usage for several mine sites

flow, or what grind size maximizes net present value (NPV), is essential for a truly optimized operation.

The assessment of what performance targets are optimal must be embedded in continuous improvement programs at an operation. Operational and project teams should continually evaluate production data and process performance to identify bottlenecks, inefficiencies, power use, and economic indicators such as metal production and cash flow. If any such opportunities are identified, then an optimization project should be initiated.

PROJECT PLANNING

For any optimization project to be successful, it is crucial to completely understand what you are setting out to do and why. The possible outcomes of these efforts then need to be critically assessed to determine what impact they may have beyond the immediately obvious. A simple objective may be to increase grinding circuit throughput by 10%. However, if the intention is to achieve this improvement without sacrificing metal recovery, or without a significant capital spend, then these conditions will impact the project and must be recognized early in the project planning phase. Similarly, if operational permits define limitations on plant operating hours or annual plant processing rates to control emissions, then the nature and influence of these constraints must be well understood so that the best outcome can be realized.

In any optimization project, all team members must fully understand the objectives and constraints from the beginning of the project to facilitate effective and expedient execution, and delivery on expectations. The need to backtrack to an earlier stage during the project to negotiate obstacles that could have easily been foreseen is particularly wasteful. This increases the demands on the often-limited project team resources, and negatively impacts project costs and completion schedules. Significant input and support from site management, engagement with key participants in the processing and metallurgy teams, and consultation with other teams outside of the processing department (e.g., mining, maintenance, social, environmental, and finance) are required to ensure that any external influences on project execution can be identified early in the process.

A simple example of grinding circuit optimization that demonstrates the need for stakeholder engagement and consideration of all potential impacts is described as follows. In an operating mine, it was evident that unused ball mill motor power was available. Theoretically, this unused power was enough to deliver a 5% increase in throughput and could be delivered by maximizing the ball charge volume, specifically by increasing the ball charge level to the retaining ring. Consultation with site personnel determined that the simple action of increasing the ball load to the retaining ring would likely be unsuccessful, as it had been tried before and only resulted in the additional balls added being immediately ejected from the mill. Armed with this information, the project team began an investigation to determine the root cause of the low ball mill power draw and had an immediately improved probability of success over previous attempts.

The investigation then focused on what could cause the ejection of the balls from the mill, which was subsequently identified as the increased superficial slurry velocity at the higher ball load. Therefore, there was a mismatch between the desired ball load, the mill volume left for slurry, and the physical slurry flow rate through the mill. The team identified that reducing the circulating load, and therefore the slurry flow through the mill, by approximately 30% was the most practical solution to this problem. However, further investigation into the basis for the current recirculating loads determined that constraints on cyclone configuration were contributing to the high recirculating loads. These constraints needed to be addressed before the reduction in circulating load could be achieved. The most significant of these constraints was the large-diameter cyclone feed line, which was prone to sanding, and low vortex-finder-to-spigot ratio installed on the cyclones to achieve the desired cyclone overflow density to feed the preleach thickener. Prior to increasing the ball load in the mill, the risk of sanding the cyclone feed line had to be eliminated by reducing the cyclone feed line diameter, and the cyclone aperture geometry tuned to optimize classification efficiency. Rather than considering only the immediately obvious step of increasing the ball charge, which in many cases may have been adequate, the required path to optimization in this instance was demonstrated to be somewhat more complex but nonetheless practical and achievable. Effective engagement of stakeholders and diagnosis of the baseline allowed the project team to identify a clear path to deliver successfully on the project objective.

The remainder of this chapter focuses in some detail on the types of parameters and performance in grinding circuits that can be optimized, which are generally aligned with the themes of increasing throughput, grind size optimization, or reductions in operating costs. An optimization project must have a meaningful objective in these terms, and some concept of one or more initiatives to deliver against that objective. This objective must be well considered (practical) and well communicated, which requires effective engagement of the appropriate stakeholders. This engagement should initially seek to solicit input during the idea generation phase based on process knowledge, past experience operating the process, and the equipment in question. This knowledge can provide guidance on immediate obstacles to project success. This engagement applies directly to plant operators, supervisors, and maintenance teams who are intimately invested and engaged with the day-to-day operation of the plant. Successful execution of an optimization project is nearly impossible without their input and buy-in, the latter of which can be ensured by this early engagement. Plant and site management have an important role to play in defining the scope, boundaries, and expectations of the project, and to ensure that the project has an appropriate level of support. Depending on the scope of the project, personnel from other departments may need to be consulted to define potential external impacts on the project, such as ore properties (geology), ore presentation (mining), service availability (power, water), sales contracts (finance), and permit restrictions (water, power, tailings, emissions).

Having progressed to this point, an often-overlooked task is to define the methods by which project success will be measured. What are the key performance indicators (KPIs) and how will they be communicated? Knowing this early can guide the project team to collect the right information, information that may be nonstandard and not available on project completion without adequate forethought. The next task is to develop a clear baseline of current plant performance, through the analysis of existing production data, or the collection of new data as necessary, including in many cases one or more circuit surveys. The analysis of plant operating data, execution of plant surveys, and subsequent use of the information generated form most of the remaining discussion in this chapter. The "Optimization Methods" section later in this chapter discusses the development of options based on the data analysis phase and their assessment, and provides guidance on the ranking of options based on ease of implementation and impact on the project objective.

After these items have been addressed in the planning and scoping stages of the project, the team can progress the optimization effort while leveraging a sound understanding of the process fundamentals.

PROCESS FUNDAMENTALS

In comminution circuits, several fundamental principles underpin circuit design, operation, and optimization. These principles can be condensed into key operating variables and the relationships they have on grinding circuit performance.

The process fundamentals described here specifically influence the throughput capacity of a grinding circuit. They are input variables that dictate either the energy requirements of the ore, the input energy of the machine, or the efficiency of the grinding process itself.

Ore Hardness

Measured using standard testing procedures such as those described in Chapter 3.12, "Testing and Calculations for Comminution Machines," the ore hardness properties dictate the energy requirements for size reduction by conventional crushing and grinding technologies. Relationships between ore hardness and size reduction energy requirements are well demonstrated and form a fundamental physical requirement that defines the grinding process.

Mill Feed Size

The particle size distribution (PSD) entering a grinding stage has a direct impact on the extent of size reduction required and, in combination with the ore hardness and target product size, dictates the grinding energy requirement and the unit capacity of the grinding stage. The feed size is generalized in power-based calculations and general discussion based on the 80% passing size, or F_{80}. However, the nature of the full PSD has a significant impact on grinding energy requirements beyond the 80% passing size. In particular, the quantity of product size or near product size material in the system feed will materially impact grinding energy requirements. As evidenced in many examples for semiautogenous grinding (SAG) mills, drill-and-blast practices in the mine can therefore have a significant effect on grinding circuit capacity as a major influence on the mill feed size distribution.

Grinding Circuit Product Size

The product PSD is also often generalized as the 80% passing size, or P_{80}. This variable has a large impact on stage capacity and grinding energy requirements, evidenced by power-based calculation techniques such as the Bond and Morrell equations. The product size is a commonly manipulated variable when increased grinding circuit capacity is required, depending on the sensitivity of grind–recovery or grind–grade relationships.

Mill Power Draw

The input energy available for grinding has a large impact on the capacity of the grinding stage. This impact is affected by the dimensions of the mill, particularly the mill diameter, but also specific operating conditions such as mill speed, ball filling levels, total mill filling level, slurry level, and pulp viscosity. The objective in operations is usually to manipulate these operating variables to maximize mill power draw, unless some other process constraint necessitates reduced power draw.

Mill Speed

Mill speed is normally expressed in terms of revolutions per minute (rpm) or percentage critical speed. Higher rotational speed will result in higher power draw and increased mill capacity up to a point, but the value of increased power draw must be considered in light of grinding efficiency and charge trajectory. Very high mill speeds will result in a grinding media trajectory that carries the media past the toe of the charge to impact on the unprotected mill liner. This leads to increased grinding media consumption and liner damage. The impact on operating costs and particularly plant downtime far outweigh the benefits of increased mill power draw in these instances. The impact of mill speed on grinding efficiency should also be considered and examples noted where grinding efficiency has been observed to decrease at high speeds (Plavina and Clark 1986).

Mill Filling

Mill filling in this context relates to both grinding media load levels and total mill filling. With the exception of SAG mills, the focus is generally on only one or the other. In rod and ball mills, the grinding media load is typically operated at a maximum level, such that the grinding media level and total level are essentially the same. In autogenous grinding (AG) mills, the ore is the grinding media, so there is no distinction. Grinding media level is directly proportional to mill power draw in these mills, and therefore, the level is managed to deliver the target power draw without overloading the mill.

In SAG mills the relationship is more complex, and the relative levels of steel grinding media and ore are varied to suit the needs of the operation. The makeup of the SAG mill charge can vary widely from 1%–2% to 18% steel by volume and rarely exceeds 35% total volumetric filling. Therefore, the charge composition may be as low at 10% steel and as high as 90% steel depending on the properties of the SAG mill feed. Mills receiving harder and coarser feed tend to have a higher composition of rock in the mill charge, while mills receiving finer and softer feed materials are more dependent on the steel grinding media and tend to have a higher ratio of steel to ore in the mill charge. The appropriate balance of grinding media in SAG mills is critical to provide sufficient media to effect efficient comminution and to allow control over critical size accumulation. In some instances, maximum mill capacity is achieved at less than maximum power draw. This can occur, for example, with secondary crushed SAG mill feeds, where breakage efficiency is maximized at low total filling levels.

Grinding Media Size

Direct relationships exist between grinding media size and the breakage rates of ore particles, although these relationships are influenced by the hardness of the ore particles and the dimensions of the mill and will vary from mill to mill. For each grinding mill, there will be an optimum grinding media charge size distribution that will optimize grinding efficiency. The grinding media size(s) used will be determined by desktop calculation, laboratory testing, simulation, and field testing.

Pulp Viscosity

Pulp or slurry viscosity is typically approximated by slurry percent solids by weight, and influences both mill power draw and the effective transfer of energy from the grinding media to the ore particles. These relationships generally demonstrate parabolic properties, with grinding efficiency and mill power draw increasing to some point with increasing viscosity, after which the pulp becomes too thick for efficient energy transfer. A decrease in power draw and/or mill capacity will be observed beyond this point.

Critical Size Management

The impact of SAG mill charge composition on critical size accumulation has been highlighted previously, with particular reference to the impact on controlling breakage rates and grinding efficiencies. In addition to the SAG mill charge, the configuration of an AG/SAG mill and the operation of any pebble-crushing equipment also significantly affect critical size accumulation. Clearly, the aperture of grate and pebble port inserts influence the holdup of critical size material in the mill itself, while the mill discharge trommel or screen aperture controls the passage of critical size material out of the system. The pebble crusher, specifically its closed-side setting, must operate in combination with the discharge screen/trommel to maximize the removal of critical-size material from the circuit to maximize the capacity of the AG/SAG mill.

Classification System Efficiency

The classification circuit, commonly hydrocyclones but in some instances rake classifiers or screens, controls the flow of final product from the circuit. The efficiency of the classifier used in a comminution circuit affects the perceived grinding efficiency based on the industry standard use of an 80% passing size to describe the feed and product size from a grinding circuit. These systems are operated to deliver the target product-size material, ideally with minimal recirculation of product-size material back to the grinding circuit (i.e., circulating load). Poor or no classification tends to produce a wider product distribution. In these instances, the energy used per ton to grind the ore is increased because material that is already at the target P_{80} size is retained in the system, and energy is used to grind this material unnecessarily to an even finer size. Classification system inefficiency can result in high circulating loads because of a high volumetric load of fines and water in the grinding circuit that can induce a volumetric constraint, directly reducing mill throughput. Metal recovery can also suffer because of an increase in unliberated particles reporting to the feed to metal recovery stages. While high recirculating loads can contribute to increased grinding efficiency under the right conditions, if the higher recirculating load is created by inefficiency in the classification system, it will more likely result in reduced grinding circuit capacity.

Mill Volumetric Capacity

Grinding mills operating at high volumetric capacities can suffer from reduced mill power draw and reduced grinding efficiency because of influences on slurry filling levels and mill residence time. This issue is observed for mills operating in open circuit with high single-line throughput rates and mills operating in closed circuit with high recirculating loads. In SAG mills, the slurry pooling phenomenon has been well documented. It occurs when the slurry flow through the mill exceeds the pumping capacity of the pulp discharge system. Under these circumstances, the slurry filling level inside the mill increases, reducing the power draw and grinding efficiency of the mill. In extreme cases, this is evidenced by a pool of slurry above the toe of the mill charge. In ball mills, high volumetric flow rates through the mill reduce slurry residence time, facilitate short-circuiting of coarse particles, and can cause grinding media to be ejected from the mill.

PERFORMANCE DIAGNOSIS

A recommended approach for optimizing a circuit is to work from a high-level analysis down to more detail. As bottlenecks or possible inefficiencies are identified, they are analyzed in increasing resolution.

At the highest level, analysis of production and operational data will provide a quick and effective way of identifying problems. Reviews of both the power and classification efficiency will provide information specific to each of the unit processes within the circuit and allow a more defined approach to optimization. Finally, surveys can be undertaken to develop a baseline operating point from which final optimization strategies can be simulated.

Operational Data Analysis

The initial high-level analysis can often be undertaken using production data collected in the normal operation of the plant and existing and knowledge and tests on the ore being treated.

Production data typically is not collected at the level of accuracy obtained from surveys, but it nevertheless provides a good picture of the operating circuit because of the quantity of data available. The data also provides trends in circuit operating performance resulting from changes in the feed blend or operating practices that are not captured with a survey, as it is only an isolated snapshot of circuit performance. A good starting point for analysis is daily average data. Most key variables within a circuit are measured at least once a day, be it total power draw for a 24-hour period or the sizing of a composite sample of the product from the grinding circuit. The analysis should recognize that shift or daily data will tend to average out the variability that exists in plant performance. To evaluate true process variability, short interval data (1- to 6-minute averages) are more useful. The inaccuracy in production data is often associated with downtime days with low run time, causing incomplete data sets. Outliers in terms of observed specific energy or operating work index should all be excluded. The focus should be on typical operating performance. The data to be analyzed depends on the circuit configuration and degree of instrumentation. Examples of typical data requirements are listed in Table 1.

Analysis of the data set in terms of average, standard deviation, and maximum and minimum values is useful. Information and relationships to interrogate when reviewing the data include use of installed grinding power, actual grind size versus target, throughput (and specific energy) versus feed size, throughput versus product size, mill weight versus power draw, operating work index versus throughput or grind, and product size versus cyclone pressure or feed density. A general comparison of all operating variables over time to identify trends between measured variables is also useful. The use of cumulative sum (cusum) charts is often instructive to demonstrate the influence of factors external to day-to-day operations (mill relines, crusher relines, and feed blend changes) on operating performance.

Analysis of this data will lead to identification of circuit bottlenecks. Examples include primary mill limitation, grind-size limiting throughput, underuse of installed power, pumping or conveying limitations, and so on.

A list of possible bottlenecks identified in the analysis will provide focus for a higher resolution optimization. For example, if the primary mill is the limitation, modifying the feed size by improving blast fragmentation or crushing can be explored. Maybe pebble extraction can be improved by changes to pebble port and pulp lifter design. If the circuit is ball mill limited, efficiency may be improved by changes to the ball size selection or classification efficiency improvements. It may be possible that the grind size can be coarsened to improve throughput and cash flow.

Table 1 Plant performance data required for analysis

Parameter	Primary Mill	Secondary/ Tertiary Mill	Classification System
Equipment run times	✓	✓	
Crusher power draw	✓		
Ore feed rates	✓		
Ore feed blend	✓		
Pebble recycle rates	✓		
Water additions	✓	✓	✓
Slurry densities	✓	✓	✓
Feed size	✓		
Product size			✓
Mill power draw	✓	✓	
Mill weight/bearing pressure	✓	✓	
Slurry flow rate and density		✓	✓
Number of operating cyclones			✓
Cyclone pressure			✓
Pump speed/current draw			✓

Power Efficiency Analysis

A standard method for quantifying the performance of a grinding circuit is to define the circuit power efficiency. The approach is described in detail by the recent Global Mining Standards and Guidelines Group publication (GMSG 2016). It defines the circuit power efficiency as the energy that is consumed in performing the grinding duty as a percentage of the theoretical energy required for the ore from feed to product.

An example of this approach is the use of power-based modeling techniques to generate a reference specific energy consumption value that is relative to both the hardness of the ore being milled and the extent of size reduction being performed. The observed circuit energy consumption is then compared to this reference value to define the efficiency of the operation. This method involves the following steps:

1. Collection of a representative grinding circuit feed sample. Where the mill feed conveyor has to be stopped for sampling, a mill inspection can also be conducted to assess liner condition and mill charge levels.
2. Measurement of actual grinding circuit energy consumption and throughput rate at the time of sampling, represented as the observed specific energy consumption (kilowatt-hours per ton).
3. Measurement of the grinding circuit product size distribution at the time of sampling, condensed to an 80% passing size value.
4. Determination of the grinding circuit feed size distribution, condensed to an 80% passing size (F_{80}) value.
5. Determination of the competency/hardness of the grinding circuit feed sample using established test protocols (Bond tests, Steve Morrell Comminution tests, SAGDesign, SAG Power Index, etc.).

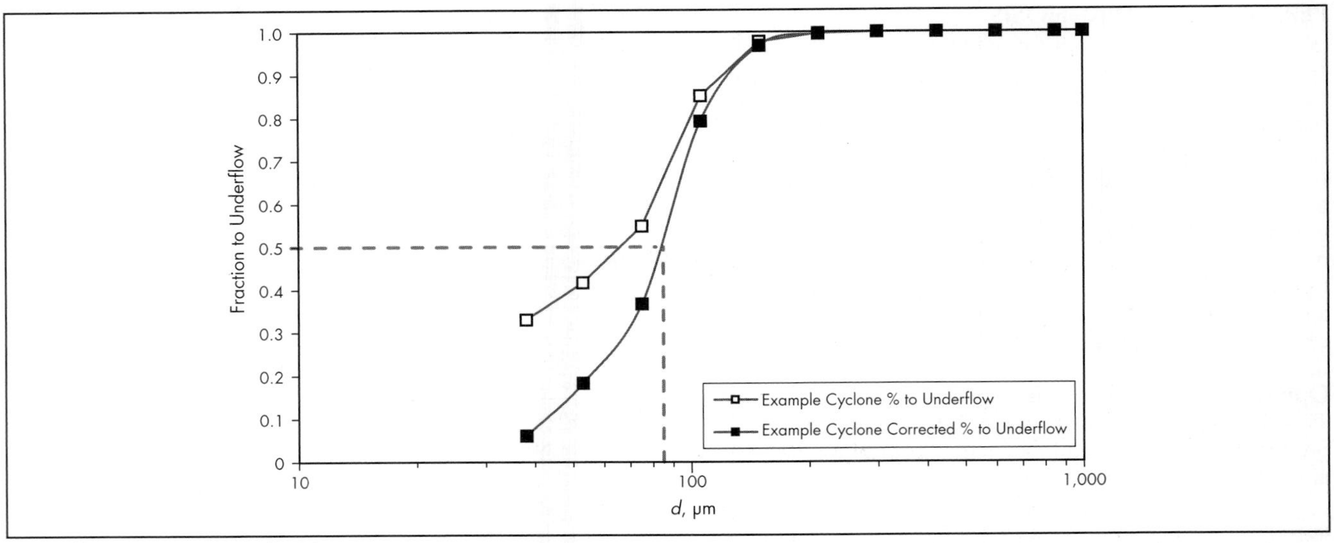

Adapted from Napier Munn et al. 1996
Figure 2 Cyclone partition curves

6. Computation of the theoretical energy requirement based on the actual size reduction (F_{80} size to P_{80} size) and the measured ore properties.
7. Comparison of the actual versus computed energy consumption as a direct measure of grinding circuit efficiency.

In this manner, the overall efficiency of the circuit and the immediate need for optimization can be quantified. Grinding mills that consume more than the specified theoretical energy consumption are perceived to have some level of inefficiency, and conversely, mills consuming less than the theoretical energy consumption are perceived to be well operated. Measurements of this nature are relative comparisons to typical plant performance, and depending on nuances in sampling, testing, and measurement, two circuits of comparable efficiency can score quite differently in this mechanism. Examples exist where grinding circuits consuming well below 100% of the theoretical energy consumption were still able to be optimized, even further reducing the actual energy consumption compared to the theoretical consumption. The precision of the technique is therefore not sufficient to apply it as a basis to identify circuits that do not require optimization, unless it is used to compare relative efficiency under optimized operating conditions. As a result, this technique is best applied to quantify the magnitude of performance improvements between one operating condition and other, and to identify grinding circuits consuming more energy than they should be.

Classification System Performance

Classification systems are often singled out specifically for optimization focus as they immediately control the final production size delivered to downstream metal recovery. The classification system is often the production constraint in grinding circuits as the cyclone feed pump is pushed to a volumetric limit by the drive to maximize grinding circuit throughput.

The cyclone efficiency is typically analyzed using partition and efficiency curves, which require size-by-size analysis of cyclone feed and product streams. A rapid preliminary analysis of efficiency is determined by recirculating load and water split. Survey data are required to conduct the analysis, and this data must be mass balanced prior to analysis.

To study cyclone efficiency, the mass split to underflow must be calculated for each size fraction. Typically, the proportion of material reporting to underflow decreases with decreasing particle size, and the amount of fines reporting to underflow is reflective of the classification efficiency (i.e., a measure of misreporting fines to the coarse fraction). A corrected relationship is also calculated (Napier-Munn et al. 1996) and looks at the cyclone efficiency if the effect of the water split and fines bypass is removed. Both these distributions are plotted as partition curves against the particle size (d) (Figure 2). The corrected partition curve is used to estimate the d_{50} (i.e., the size fraction at which 50% of material in that size fraction will report to the underflow).

The d/d_{50} is calculated for each size fraction and plotted against the corrected percentage to underflow for each size fraction to create what is known as an efficiency curve. The calculated efficiency curve is plotted against a typical cyclone efficiency curve for comparison (Figure 3).

The efficiency of the cyclone is given a numerical value to allow for comparison with other cyclones. This is known as the alpha (α). The alpha for the efficiency curve is calculated by solving Equation 1 and represents the steepness of the slope of the efficiency curve. This is often described as the sharpness of the separation curve, with higher numerical values reflecting increased classifier efficiency. The calculation developed by Whiten and described by Napier-Munn et al. (1996) is performed for each size fraction for the range of α values.

$$Eoa = C\left(\frac{\exp(\alpha) - 1}{\exp\left(\alpha \cdot \frac{d}{d_{50c}}\right) + \exp(\alpha) - 2}\right) \qquad \textbf{(EQ 1)}$$

where

Eoa = actual efficiency to overflow
C = water recovery to overflow
d = particle size of interest

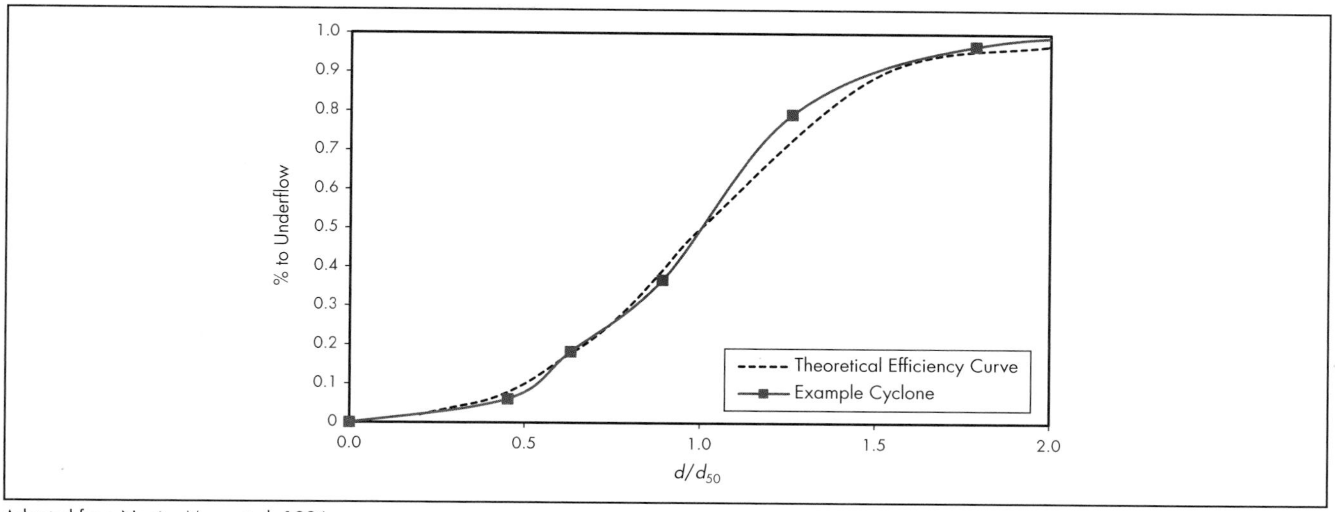

Adapted from Napier Munn et al. 1996

Figure 3 Reduced cyclone efficiency curves

α = efficiency parameter describing sharpness of separation

d_{50c} = size that divides equally between overflow and underflow because of classification forces only

Alternative methods are available for assessing classifier efficiency. The simplest method is the percentage of product-size material reporting to the product stream, quantified by Equation 2. In an efficiently operating cyclone, the proportion of product-size material reporting to the underflow is often in the 40%–50% range, though values from 50%–65% are common.

$$\text{classification efficiency} = \frac{\text{mass of product}}{\text{mass of product size material in the feed}} \quad \textbf{(EQ 2)}$$

The classification system efficiency (CSE) methodology developed by McIvor et al. (1991) also defines the classification efficiency by measuring how well the classifier is removing product-size material from the circuit. It is defined as the proportion of material coarser than the target product size (expressed as the 80% passing size, P_{80}) that is contained in the mill solids and uses plant survey data as the basis for calculation. It is calculated by taking the average percentage of material coarser than the P_{80} of the mill feed and mill product. The CSE is used as an input into the functional performance calculation, a relationship between the classification efficiency and the overall grinding circuit efficiency. Calculating the CSE (McIvor et al. 1991) provides a benchmark on how well the cyclones are removing product-size material from the circuit and can be used in conjunction with the mill grinding efficiency to determine the effect of classification performance on grinding circuit efficiency.

Aside from classification efficiency, most cyclone vendors offer guidance on cyclone capacity that is based on a series of individual cyclone capacity curves. In these types of models, consideration to the cyclone diameter, spigot and vortex finder size, circulating load, pressure, and the number of operating cyclones is considered. These types of models are useful to assess if the cyclones are configured correctly to achieve their purpose. They can also be used as part of the mass balancing process to cross-check the volume flow and operating pressure during the survey.

Physical modeling of the cyclones will also ensure that the cyclones are the appropriate size for the duty. This can also be reassessed during simulation, where changes to the throughput or product size may alter the cyclone feed characteristics (i.e., flow rate, density, etc.).

Grinding Circuit Surveys and Simulation

Grinding circuit surveys and simulation models are commonly used optimization tools but can also be applied for baseline performance diagnosis. In this instance, the objective is to qualify performance against some established benchmark, allowing an assessment of circuit efficiency and the identification of initial optimization initiatives. This method was discussed earlier in this chapter and is a usual starting point for the analysis of grinding circuit survey data. This approach requires a straightforward sampling program focused only on grinding circuit feed and final product. A more in-depth understanding of the circuit can be achieved by conducting a more thorough sampling program, which is typically used as the basis for an integrated, population balance based simulation model of the grinding circuit. It requires the sampling and analysis of all internal process streams. These models allow the simulation of the impact of a range of different operating conditions on the performance of the circuit.

Several population balance style models or discrete element models (steady state or dynamic) can be used to model particle breakage in a grinding circuit. Many of these require significant modeling knowledge by the practitioner, a very high level of data input, and often programming of models. Others have reasonable built-in models but may not be set up to fit model parameters based on collected data. The software most commonly used in recent decades for optimization of comminution circuits is JKSimMet, a steady-state simulator designed specifically for the optimization of comminution circuits. The software includes proprietary models that are reasonably simple to use and provide an adequate level of complexity for the standard optimization exercise. It is most commonly used as a simulation tool, whereby changes to unit variables can be made. The effect of these changes on the circuit performance can then be assessed in conjunction with more traditional modeling methods.

Analysis of production data provides a good deal of high-level information for optimizing the circuit, including throughput, product size, and power use. This process typically identifies the key issues around a circuit that require immediate attention and also indicates whether additional, longer-term strategies will be needed to meet site-specific production targets.

Surveys, on the other hand, provide detailed information, allowing each unit to be analyzed independently from the whole circuit performance. Surveys are an important tool in the comminution circuit optimization process. They are intended to be reflective of a "base case" condition for the circuit and provide a snapshot representation of the circuit under typical operating conditions. This provides detailed data for circuit analysis and forms the basis for modeling and simulation.

To get the most benefit from a comminution circuit survey, it must be planned, executed, and documented with significant attention to detail. The consequences of a poor-quality survey are far-reaching. Poor data potentially can render the survey useless. More important, however, is that if the data do not reflect typical operating conditions, they can lead to misinterpretation of circuit performance resulting in misguided optimization strategies. Likewise, if a model is built using suspect data, the simulations used to predict outcomes of the optimization may be incorrect or overly optimistic.

Surveying comminution circuits is a costly and time-intensive process. It is therefore vital that the process be conducted in accordance with strict procedures to ensure that time is not wasted collecting data that are either misleading or cannot be used. Some guidance on the conduct of successful comminution circuit surveys is provided in the following section.

GRINDING CIRCUIT SURVEY PROCEDURES

Grinding circuit surveys can be an invaluable source of very detailed information about plant performance, which can in turn be used to tune simulation models that can then be used to predict future performance based on any number of input variables. Grinding circuit surveys represent a considerable investment in time and money, so it is essential that they are well planned and designed to generate the best quality information possible. In many cases, you only get one chance to get it right, so spending the necessary time planning the successful execution of the survey should be given due consideration.

Survey Planning

Planning the survey requires development of a project schedule that spans several months. This schedule starts well before survey execution for planning and circuit stabilization and up to and including the issuing of final reports. Items that require long lead times, such as sampling equipment, should be identified, designed, and ordered well in advance of the survey so that they are on hand for practice runs prior to the day of the survey. The sample points must be identified early and inspected to ensure that the sample can be collected safely using the sampler design selected. Other equipment, such as sample drums, buckets, and any lifting aids or machinery required to transport samples, should also be identified and sourced as necessary.

Planning must identify the need for specific work permits for confined space entries and isolations, and allow for the training and availability of competent persons to execute the associated tasks. Survey personnel should be provided with adequate training in the correct execution of the sampling task in advance of the survey. The capacity of the site or contract laboratory to process the samples as required should be confirmed, as should the availability of the necessary laboratory equipment (e.g., drying ovens, scales, screens).

Plans should be made to inspect and calibrate critical instruments such as weightometers, flowmeters, and densitometers prior to the survey. This will ensure the functionality of equipment that must be used or operated during the survey (valves, auto-samplers, blanks) and confirm the basis of any critical information such as power or mill speed readings. Vendor manuals for major equipment such as crushers, mills, screens, and cyclones, and drawings for components such as lifters, liners, grates, and screen panels are required to provide key information into process models and should also be sourced as early as possible.

Pre-survey planning must consider ore preparation schedules to ensure that the ore to be processed during the survey is consistent for the duration of the survey and the pre-survey stabilization period. It is also important to confirm that the ore to be processed meets the expectations for the survey and is representative of future mill feed.

Survey Sampling

Representative sampling of slurry and solids streams is critical to the validity of any grinding circuit survey and can be readily achieved with due consideration to sampling equipment design and sampling procedures. When selecting or designing an appropriate sampler, three general principles should be applied: (1) the width of the opening of the sampler must be at least three times the width of the largest particle in the stream, (2) the length of the opening must be greater than the cross section of the stream to be sampled, and (3) the volume of the container must be sufficient to hold the full sample without any overflow, occurring. Often for high flow rate or inaccessible slurry streams, such as ball mill discharge or SAG screen undersize, some innovative designs and sampling techniques are necessary to collect a suitable sample.

There is a significant amount of information in the public domain regarding appropriate sampling of slurry samples. The generally accepted technique for obtaining a representative slurry sample in grinding circuits requires the full cross section of the stream to be sampled; it is inadequate to randomly dip a sampler into a stream, as there is bound to be segregation of the solids within the stream and bias as a result. The sampler must be swept through the stream at a constant speed, and the sampling speed should be less than the velocity of the stream being sampled. The sampler must also not overflow, as there is a rapid settling of particles, resulting in a bias of smaller particles and water in any overflow. Settled coarse particles should be tapped or scraped out of the sampler but should never be washed out. Generally, five or six samples are recommended over a one-hour survey period, although the cut frequency must be varied to generate larger samples (25–30 kg of dry solids) for coarser streams such as mill discharge, screen oversize, or cyclone underflow.

Cyclone feed slurry sampling is often critical to allow a valid circuit mass balance to be generated and requires consideration and planning prior to the survey. The preferred method to collect this sample requires the cyclone overflow outlet of a spare cyclone to be blanked off for the period of the survey. When the survey is complete, prior to the mill crash stop, the feed valve to this cyclone is opened and the sample collected over a short period from the underflow outlet of that cyclone. This method provides a representative slurry sample and minimizes any impact on circuit stability during the survey period. Alternatively, for grinding circuits with smaller packs of larger cyclones, the implications of diverting an entire cyclone to underflow or the pressure changes with opening and closing cyclones are much more significant. In this instance, a spare cyclone outlet can be modified with reducers to connect to a 50-mm hose for taking of the sample, and may allow the hose to be elevated to the appropriate height that equalizes the head with the cyclone pressure. The flow from the hose should be directed to the cyclone underflow launder. The diameter of the sample line should not be less than 50 mm, nor the valve feeding this line restricted, to prevent obstructed and potentially biased slurry flow.

For SAG mill, ball mill, or rod mill feed conveyors, feed samples are generally collected using belt cut methods on isolated conveyor belts following the completion of the slurry sampling component of the survey. Depending on belt loadings, several meters of material must be removed from the belt to form the feed sample. For coarser streams including primary crushed products, a two-stage sampling method is required to minimize the sample mass collected to a practical level without sacrificing the quality of the sample. This method involves taking the larger material only off a longer length of belt and a "remaining fines" fraction off a shorter length of belt. For example, a conveyor containing primary crushed ore might have 1–2 m of the belt totally removed to constitute the "fines" portion, with an additional 5–10 m of belt length sampled to remove coarse (+75 mm) material. In this way, a suitable number of particles of each size class will be collected for accurate representation of the PSD of the stream.

Sample Analysis

For comminution circuit surveys, samples are analyzed for percent solids by weight and PSD.

PSD is generally determined using a combination of wet and dry screening. Laser sizing will be necessary for samples collected from fine and ultrafine grinding circuits, which requires carefully considered sampling and subsampling practices to ensure that sample representivity is maintained.

Sieve series selection must accommodate the largest particle size (D_{100}) in the mill feed and should extend to at least two screen sizes smaller than the target P_{80} of the cyclone overflow. Ideally, a square root of two sieve series is recommended and the same sieve series should be used for all samples.

Survey Execution

In the hour leading up to the survey, production data must be reviewed to ensure that the plant is operating in a stable manner. This data includes stable feed rate, mill load, power draw, cyclone operation, and water additions. During the survey, these same parameters must be monitored to ensure that the circuit remains in a stable condition. If it does not, the survey should be abandoned and rescheduled.

Information describing grinding circuit performance for the hours leading up to the survey and until the mill stop post-survey must be collected at a high frequency rate on the day of the survey. Typical plant production data are collected, which include fresh mill feed rate, run time, power draw, scats recycle rate, cyclone feed flow rate and percent solids, cyclone pressure, number of operating cyclones, water additions, and online grind size measurement data, if available. These same data should also be compiled daily for at least one month before the time of the surveys with any feed changes (i.e., ore properties or feed size) noted.

At the end of the survey, the primary mill should be crash stopped and secondary ball mills ground out. This will enable a sample of the primary mill feed to be collected from the mill feed conveyor belt and an inspection of the primary mill charge, including slurry and charge levels, to be performed. Mill diameter, length, liner, and grate condition should also be measured. The ball mill ball charge should be measured, and following the completion of belt sampling, the SAG mill should be ground out and a subsequent measurement of the operating ball charge performed. Cyclone component diameters, crusher gaps, and grate and screen apertures should be physically measured to record the operating condition at the time of the survey. The correct execution of these steps is essential to the accurate assessment of circuit efficiency and the development of useful circuit simulation models. Any measurements and equipment inspections should be supported by digital images, with a standard reference present for image scaling, wherever possible.

After the completion of the survey, the operating data should be reviewed to confirm that they demonstrate the desired level of stability and reflect the operating conditions that were being targeted for the survey. Simple visual trend analysis is sufficient for this analysis and can allow comparison to recent operating performance if it was intended that the survey reflect normal operating conditions. After the survey data have been confirmed fit for purpose, the survey data can be mass balanced and used to develop calibrated equipment models for simulation. The mass balance outputs should be carefully inspected for consistency and practicality prior to progressing to the model fitting and simulation stages of analysis.

Good references for a more detailed discussion of comminution circuit survey protocols include Napier-Munn et al. (1996), Mosher and Alexander (2002), and GMSG (2016).

OPTIMIZATION METHODS

Optimization of a circuit can occur in several phases. The first is generally the type of circuit optimization that is performed in normal day-to-day operations and can be as simple as identifying an obvious inconsistency in the operating parameters and taking steps to correct it (e.g., a sudden step change in a trend, such as a sudden change in SAG mill power draw). This type of circuit optimization is generally addressed by process control systems. It is, however, also complemented by well-trained operators who understand how circuit changes affect the plant performance as a whole and tends to be focused on operational trends and scenario analysis. After a scenario is identified, a series of generic steps can be followed to reach a satisfactory result. Such an approach is readily captured in a decision matrix, an approach that is targeted specifically at operators and staff who will be on the front line of circuit operation. An example is provided in Table 2.

The second type of circuit optimization is much more detailed, requiring significantly greater understanding of the technical aspects of circuit operation. It is generally related to

Table 2 Example plant optimization philosophy

Variable	Condition	Qualifier	Response
Cyclone pressure	< Target	Pump amps > minimum	⇧ Pump speed
	> Target	Pump amps > minimum	⇩ Pump speed
Hopper level	> Target		⇩ Sump water addition
	< Target		⇧ Sump water addition
Cyclone overflow density	> Target	Pump amps < maximum	Open a cyclone
		Hopper level > minimum	
		Number of cyclones < maximum	
	< Target	Pump amps > minimum	Close a cyclone
		Hopper level < maximum	
		Number of cyclones > minimum	
Ball mill power	< Target		⇧ Feedwater
	> Target		⇩ Feedwater
Grind size	> Target (coarse)		⇩ Cyclone feed density
	< Target (fine)		⇧ Cyclone feed density

at least one of four key targets: throughput, grind size, circuit efficiency, or operating costs. In each instance, the desire for optimization can often be attributed to common causes, one of which is looming changes in ore properties, often as a new ore source comes into production. Failure to achieve nominated production targets is also a common starting point for optimization programs. Alternatively, it may be considered that current plant performance, while in line with prescribed performance targets, still demonstrates potential for improvement. Commodity cycles, driving lower metal prices or increasing costs of production, place pressure on project economics and can provide strong incentive for performance improvements.

In all cases, the optimization process is the same. Production data should be analyzed and circuit surveys undertaken. Using this information, problems or opportunities can be identified and changes required to achieve the desired outcome can be predicted.

Optimization is most successful when it is a continual process where there is a healthy ongoing relationship between the operations team, corporate team, and consultants. Optimization methodology can be tailored to suit individual project requirements; however, a methodology successfully applied to many projects is shown in Figure 4 and includes the following:

- Production data are analyzed and compared with the theoretical performance for the current equipment and configuration.
- The control philosophy is reviewed, taking into account any constraints identified in the production data analysis.
- Based on the analysis and site observations, recommendations are made regarding changes in operating parameters and/or control philosophy, if applicable. The optimization strategy should be set at this point, but to quantify the likely benefit of certain changes, a detailed circuit survey is generally recommended.
- Online, low-cost, or short execution time changes are trialed first, and if successful are permanently implemented. As a minimum, these changes should at least improve circuit stability.
- After stable operation is achieved, a grinding circuit survey should be undertaken treating ore types expected to be treated in the medium to long term. The survey conditions and performance are benchmarked against the historical operating data to ensure that the survey is within the range of normal operation. The survey data can then be mass balanced and model fitted.
- Simulations can be run to quantify the effect of optimization recommendations on circuit performance. The results of the modeling can then be used to justify further adjustments and/or capital projects if required.
- Training of site personnel is a useful consequence of the optimization process. This training is often undertaken informally as part of the process or as formal training based on the needs of frontline managers (metallurgists and supervisors) and operators as identified during the process. A knowledgeable team is required to implement rapid and lasting improvements.
- Ongoing review of the circuit performance should be conducted with KPIs or set points to maintain optimal performance or obtain ongoing improvement.

Unit Process Optimization

Table 3 summarizes the operating variables that can affect the performance of grinding mills. It provides a high-level guide on parameters to manipulate when optimizing the individual units within the comminution circuit, considering the key process fundamentals presented at the beginning of this chapter. The following section provides more details on common examples of unit operational issues that may be identified in circuit analysis and the simulations that may be implemented to improve performance.

SAG Mills

High SAG mill specific energy and low throughput are often attributable to difficulties in maintaining the target mill load weight. These may stem from several possible changes to the mill feed, such as PSD or feed hardness. Alternatively, they may result from mill operating conditions such as liner design, operating speed, and targeted load level.

Feed size distribution. The feed may be coarser than ideal for the required breakage rate given the SAG mill size and target throughput. Large lumps that are too big to be broken by the largest ball size in the SAG mill and are above the grate aperture size can build up within the mill, drawing excess power and creating a fine transfer size because of lower energy breakage events. Higher ball loads, larger ball diameters, and increased pebble port apertures should be investigated. Often, SAG mill ball charges greater than 16% cause load instability when treating competent primary crushed feed, because of the relationship between the power draw and mill load and the

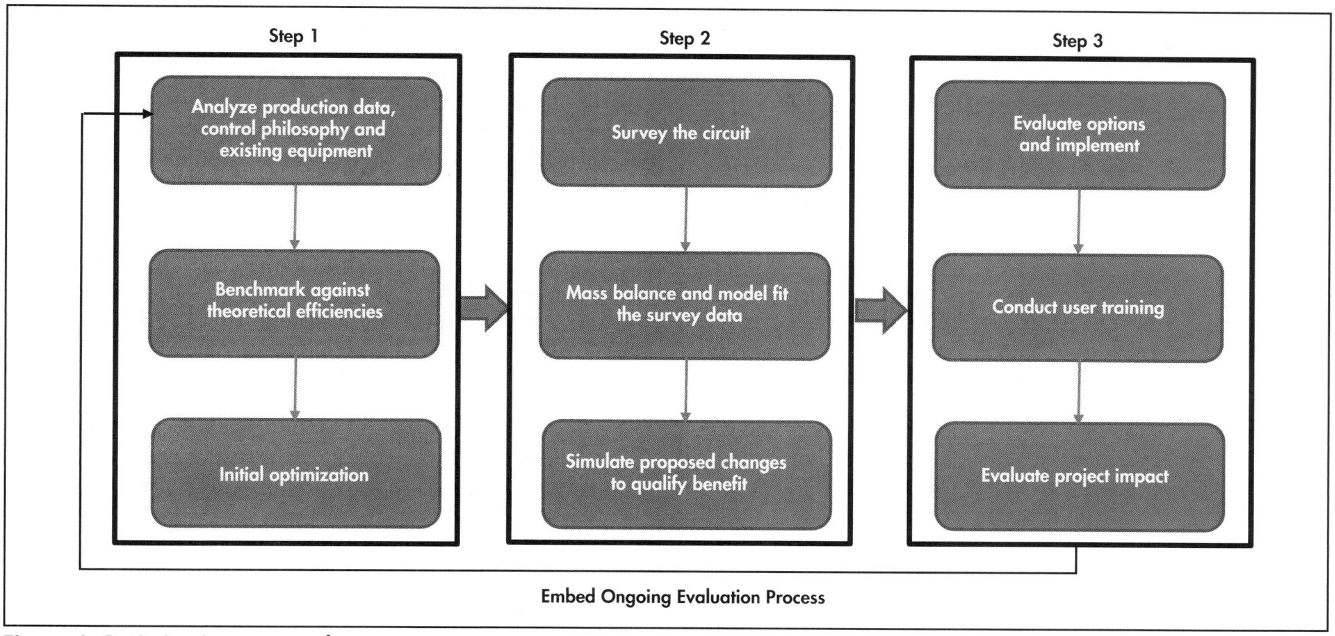

Figure 4 Optimization approach

Table 3 Optimization table: What to look for

Parameter	SAG Mill	AG Mill	Rod Mill	Ball Mill	Pebble Mill	Regrind/ Ultrafine Grinding
Feed size and fines content	✓	✓	✓	✓	✓	✓
Hardness	✓	✓	✓	✓	✓	✓
Milling density	✓	✓	✓	✓	✓	✓
Mill speed	✓	✓				
Media size	✓		✓	✓		✓
Media load	✓		✓	✓		✓
Mill weight/filling	✓	✓			✓	
Volumetric flow rate	✓	✓		✓		✓
Grate aperture	✓	✓				
Recycle crusher operation	✓	✓				
Classifier efficiency	✓	✓	✓	✓	✓	✓
Tramp steel removal	✓		✓	✓		✓
Pulp lifter design	✓					
Liner design	✓					

effect on rock ball breakage rates. Another option is optimization of the primary crusher product to minimize the feed topsize to the SAG mill. Increased blast fragmentation through tighter hole spacing, larger blast-hole diameter, higher density explosive, or other means to increase the powder factor has been successful in many instances. This approach is discussed in detail in Chapter 3.1, "Mine-to-Mill Optimization." Installing pebble crushers, converting to SABC-B (SAG mill, ball mill, and pebble crusher) configuration, or implementing secondary crushing of the SAG mill feed are viable options, although more capital intensive and generally considered as expansion options more than optimization strategies. These strategies work well for open-circuit SAG mills; however, they may not be as effective for AG mill circuits and single-stage SAG mills, which can often require a minimum amount of coarse rock to operate efficiently. Excessively fine feeds can result in reduced mill throughput or coarse product size in those instances.

Feed competency. The feed may be more competent than expected from the design criteria or current mining schedule. An expectation of the SAG mill throughput may be unrealistic should the ore be harder than that considered in the original design. Load issues may be encountered at throughputs higher than appropriate for the ore type being processed. This may require an adjustment of expectations in the realistic circuit capacity and require production forecasting to be reviewed in light of the harder ore properties. Short-term strategies to counter more competent ore may include changes to the feed PSD with increased blasting fragmentation, ore blending, or crushing circuit modifications. In the longer term, extra capital may be considered to add extra crushing or grinding power to ensure that production targets are met.

Ball trajectory and load level. The operating load level may be above the toe of impact of the ball trajectory, resulting in a reduction of high-energy impacts within the mill. This encourages the development of critical-size material and a fine transfer size. Lower loads at the toe of impact have the opposite effect, which is difficult to simulate in most models. The JKSimMet model is advancing to include the effect of load level (Hilden et al. 2015). However, to really understand the location of the toe of the charge and the optimal impact zone, multiple surveys at different load levels are required and can be assessed in conjunction with trajectory models. Mill operation should be assessed at multiple load levels before a survey is undertaken to determine the optimal load level. The survey to be used for modeling and simulation should then be taken at this level so that it does not require further optimization during simulation.

If the SAG mill is variable speed, then the speed of the mill can be used to increase or decrease breakage to a certain extent. If the SAG mill is limiting throughput, the SAG mill speed can be increased. Increasing the speed should increase both power draw and breakage, which should in turn reduce the load and allow for the throughput to be increased. The response will vary depending on the characteristics of the ore being processed and the operating conditions of the SAG mill at the time. Note that many variable-speed-drive mills have variable power up to synchronous motor speed and are then constant power with further speed increases. In this case, it should be evaluated which provides better breakage, generating power draw with ball charge or speed. If the SAG mill load is low, with the throughput constrained by other unit processes, the SAG mill speed can be reduced. This variable can be used in simulations to balance the grinding duty between stages of comminution.

Low SAG mill specific energy and slow SAG mill speeds and/or load levels are normally associated with the SAG mill being restricted by external unit processes. Low SAG mill power draw or load level in a two-stage circuit is often coupled with a restriction in the ball mill, preventing the problem from simply being rectified by increasing the throughput. The optimization solution is to distribute the grinding duty evenly between the two stages of grinding. In this case, reversal of the steps taken for the SAG limited circuit scenario detailed earlier in this chapter should be undertaken. The use of a smaller SAG mill ball top-size, lower ball charge levels, bypass of the recycle crusher, coarser feed, and a reduction in the mill speed may all be beneficial. In addition to the ball mill, there may also be other "nongrinding" downstream processes that may be contributing to the circuit constraint. These processes may include trash screens, interstage screens, downstream residence time, and so on, and these may require review and optimization before the milling circuit.

There are many reasons a SAG mill may not draw the power expected for a given ball charge and load level. The two most common contributors to this situation are packing between the lifters and the development of a slurry pool. Although these issues should be identified as part of the survey process (i.e., power modeling and inspection of liners during crash stop), simulating changes to alleviate the problems is not possible in standard modeling software like JKSimMet where the fitted breakage rates or constants are influenced by these conditions. The best that can be done when simulating is to try to compensate with mill speed. A review of the lifer, grate, and pulp lifter design is ultimately required. Improved designs can eliminate packing and minimize slurry pooling. If the SAG mill is operated in single-stage configuration, the recirculating load may be reduced using the cyclones configuration to alleviate the slurry pool and maximize power draw.

Secondary Crushing

Secondary crushing of the SAG mill feed, when implemented correctly, significantly reduces SAG mill and total circuit specific energy. Very fine feed, good process control, and/or the presence of some coarse rock in the feed (partial secondary crushing) is required to implement this strategy successfully.

In scenarios such as these, benefits to throughput and load stability will be achieved by reducing the feed top-size down to a size that can be easily accommodated by the SAG mill using a reasonable ball top-size (full secondary crushing). Alternatively, efficiency gains will be achieved by reducing the intermediate size fractions in a feed where competent critical size is problematic (partial secondary crushing). Crushing all the ore to an intermediate size between true primary crushed feed and secondary crushed can increase throughput, but the predominately critical-sized material generated can make the circuit extremely unstable.

The extent of the grinding throughput and/or energy-efficiency gains are largely governed by the ore characteristics as well as the selection and proper implementation of either partial or total crushing. Feeding 100% secondary crushed ore to a SAG mill requires the mill to essentially be a grate discharge ball mill. Efficient operation without a recycle crusher can only occur if the feed is very fine, less than 80% passing 38 mm (1½ in.). Full secondary crushing of the feed can only be considered if the SAG mill has been designed to operate at the increased installed power and with the structural integrity required to operate at a high ball charge.

A partial secondary crushing circuit can be simulated if the SAG mill cannot be operated at high ball charge levels or when only a moderate increase in capacity is required.

Secondary crushing only a portion of the mill feed has many benefits, including the following:

- **Control of the energy demand in the SAG mill.** This impacts transfer size and therefore ball mill specific energy demand and throughput, providing optimal power use when properly deployed.
- **Sufficient coarse rock for media in the SAG mill.** This allows larger rock loads to be operated in the SAG mill, optimizing power and increasing circuit stability. The higher load level protects mill liners, reducing maintenance and ball consumption. This is important for SAG mills designed with a maximum ball charge of less than 15%.
- **Generation of a bimodal feed consisting of coarse rock and fine secondary crushed material.** This increases stability of the circuit by minimizing critical sized rock in the feed.

In this case, the amount of secondary crushed ore in the feed blend can be tailored to match the required efficiency and power split to match the installed milling power.

Ball Mills

Generally, the ball mill power draw is less likely to deviate from the theoretical compared to a SAG mill. The ball mill power draw is substantially more stable than a SAG mill as there are smaller variances in loading because all the media is in the form of steel balls. Given that load control is typically not an issue in a ball mill, the capacity is related to feed size, feed characteristics, and grinding efficiency.

Power draw lower than maximum. Low power draw is usually attributed to lower than maximum operating ball charge. Modeling of the ball mill will determine the maximum ball charge from both a power and a volumetric perspective. Excessively high flows through a ball mill may also result in reduced power draw.

Excessive scatting. Excessive scatting in the ball mill will be present if the feed size is too coarse for the ball top-size or if the superficial velocity through the mill is too high (very high circulating loads or low specific energy input).

High circulating load. This is predominately controlled by the cyclone performance and setup. High circulating loads can reduce the retention time to a point where grinding efficiency is impacted or the classifier performance is low.

Low circulating load. If too low, the retention time in the mill will be very high and will result in overgrinding per pass, increasing the fines content in the circuit product and reducing grinding efficiency for a target P_{80}.

Coarse product size. A coarser than target product size can be symptomatic of several issues. If the ball mill is operating at optimal conditions, then the ore may be harder than design, the feed coarse, or the throughput too high. In this case, a more realistic target needs to be set or upstream comminution needs to do more work, reducing the feed size to the ball mill. Alternatively, the grind size may be coarse because the mill is not operating efficiently. Causes of inefficient ball milling can be inappropriate liner design, media selection, or milling slurry density. The effect of media size and slurry density can be assessed in simulations.

Mill Ball Charge and Maximizing Power Draw

Full use of available power in a balanced circuit maximizes throughput, provided that it does not cause a downstream constraint such as a ball mill limiting condition. This should be the first step in any optimization. The capacity to increase charge level and power draw can be identified by comparing the measured motor power against the installed power. Small increments in increased ball charge can be made to ensure that the rated motor power is not exceeded and no adverse reactions occur within the circuit. Following successful power increase(s), the throughput can usually be increased proportionally in the SAG mill. In a SAG–ball mill circuit, the ball mill power should only be increased appropriately to ensure that a power excess is not created and a finer than required grind size generated. If ball mill power is still available, ways to balance the required grinding power should be assessed.

Grinding Media Size Optimization

Chapter 3.8, "Grinding Circuit Design," discusses in some detail the estimation of optimum grinding media sizes using the techniques proposed by Bond (1958, 1961) and Crocker (1985). A similar desktop calculation methodology is also provided by Azzaroni (1981) and reproduced by Dunn (1989). These desktop methods are a useful starting point, although often there is further opportunity to improve grinding efficiencies by revising the grinding media sizes in use. In many cases, using multiple media recharge sizes will give improved performance compared to a single recharge size. Laboratory testing with ability to scale to full-scale performance (Herbst and Rajamani 1982) is a technique to identify optimal media size blends for ball mills, and it has been used for optimization by several operations. Villanueva et al. (2011) describe a similar program using a 30.5 cm × 30.5 cm (12 in. × 12 in.) laboratory ball mill to identify the optimum ball size distribution at PT Freeport Indonesia, realizing a 5%–10% throughput improvement at that operation.

Changes in grinding media diameter will also affect media wear rates based on influences on media weight and impact energies as well as total charge surface area and abrasive interactions. When attempting to increase grinding efficiencies through media size optimization, the practitioner must remain aware of side effects that can erode the intended benefits for the operation.

Balancing Circuit Power Split

Like trucks in the pit, effective use of capital is a practical optimization philosophy. Use of capital is achieved in a grinding circuit by balancing the required energy between grinding stages while maximizing power draw. First, the circuit must be analyzed to determine the amount of available power in each stage and which stage is currently rate limiting. If the primary mill is limiting throughput, then it will be necessary to examine options that transfer the power required from the primary to secondary stages. Common options for this include reducing the feed size by crushing finer, increasing grate open area and aperture size, or adding or optimizing a recycle crusher. Increasing the trommel/screen aperture from, for example, 9.5 mm to 15.9 mm (3/8 to 5/8 in.) is a relatively quick and low-cost option, although the benefit is typically minor. Adding a recycle crusher will only be of benefit if the material first breaks down to below grate aperture. If a recycle crusher already exists, it may be possible to increase the amount of material crushed by increasing the grate/port size from, for example, 51 mm to 76 mm (2 in. to 3 in.) and increasing pulp lifter efficiency. Typically, a standard radial pulp lifter design can be replaced by a curved pulp lifter design. Even if a recycle crusher is not installed, increasing the grate aperture and removing the steel with a magnet can increase the average ball diameter within the mill.

If the secondary grinding stage is rate limiting, it may be necessary to divert a portion of the cyclone underflow to the primary mill feed to gain additional grinding. The amount to be diverted will vary depending on the available power in the primary mill. Alternatively, the pebble crusher can be bypassed or smaller balls or grate apertures can be used in the SAG mill.

Classification

In a comminution circuit, the product size is dictated by the grinding power available and the feed characteristics and is influenced by classifier efficiency.

As stated in the previous section, analysis of cyclone efficiency is usually done with survey data to define the classification system efficiency. The data are analyzed to determine the proportion of the product size material that is reporting to the underflow, if there is any size fraction in the distribution that seems to report preferentially one way or the other (often referred to as "short-circuiting"), and also to obtain a mathematical value for the efficiency, alpha (α).

Typically, a modern-style cyclone is operating efficiently if its α value lies between 2 and 3; however, cyclones can often demonstrate α values of less than 1. Screens and many other devices have much higher efficiencies, with α values closer to 10. Screening options, however, are adopted less often because of the low capacity per unit and higher capital cost. This is the case where industry trends are moving toward larger-capacity circuits and minimization of capital in the face of decreasing head grades.

The efficiency of the cyclones depends on several variables that can be classified as physical parameters or operating parameters. Physical parameters relate to the configuration of the cyclone and include the diameter of the cyclone and the inlet, vortex finder, and spigot inserts. Operating parameters are tuned in the plant and include cyclone pressure drop, feed, and product percent solids. Table 4 summarizes each cyclone variable and a generic description of how it affects separation. Some of these variables cannot be changed easily, such as the physical parameters of the cyclone. Therefore, the cyclone should be modeled using a population-based modeling package before any changes are made. This type of modeling

Table 4 Cyclone variables and responses

Variable	Change in Variable	Effect on Cyclone
Feed inlet diameter	Larger diameter	Higher flow rate
	Smaller diameter	Lower flow rate
Vortex finder diameter	Larger diameter	Higher flow rate
		Increased water split to overflow
	Smaller diameter	Lower flow rate
		Decreased water split to overflow
Spigot diameter	Larger diameter	Increased flow/water split to underflow
		Increased circulating load
	Smaller diameter	Decreased flow/water split to underflow
		Decreased circulating load
Vortex: Spigot ratio	High	Low circulating load
	Low	High circulating load
Cyclone overflow density	High	Low circulating load
	Low	High circulating load
Feed density	Higher	Coarser product size
	Lower	Finer product size
Feed pressure	Higher	Finer product size
	Lower	Coarser product size

enables you to assess how each of the variables interact and affect the overall separation without affecting production or circuit stability. Changes that alter separation efficiency can be difficult to simulate because the efficiency in most models is a fitted parameter.

When simulating changes to unit-specific variables, the operational limits of the piece of equipment should be considered. Each cyclone has a range of recommended inlets, vortex finders, and spigot sizes. It is best to obtain a list of these from the manufacturer before embarking on any simulations.

The flow limitations of the mill (grate or overflow) should be considered when making changes to the cyclones to ensure that slurry pooling or excessively high superficial velocities through the mill are minimized.

With non-unit specific cyclone variables such as the vortex finder to spigot ratio, feed density, and overflow density, typical operating ranges as well as other circuit constraints should be considered.

Generally, the vortex finder to spigot ratio ranges from 1.4 to 2.0. A high ratio and high overflow density will give the lowest circulating load. A low vortex finder to spigot ratio and low overflow density will give the highest circulating load. Circulating load is not a direct function of ore properties but is predominately controlled by the setup of the cyclones.

Higher feed density usually results in decreased efficiency because interparticle interaction is greater within the unit. The feed density also affects the product size, with a higher density related to a coarser product. Typical operating density ranges for cyclones in mineral processing circuits are between 45% and 65% solids w/w.

Table 5 Cyclone efficiency based on water recovery to cyclone underflow

Water Recovery to Underflow Stream, %	Cyclone Efficiency
>50	Very poor
40%–50	Poor
30%–40	Reasonable
20%–30	Good
10%–20	Subject to underflow density and roping
<10	Available only with an underflow valve for dewatering, producing a product for conveying or stockpiling

Source: Napier-Munn et al. 1996

The overflow density targeted should be as low possible without breaching downstream constraints (e.g., less than the optimal flotation feed density) and without causing issues related with high circulating load (lower overflow densities require more cyclones online and therefore higher circulating loads). The installation of a thickener following grinding decouples the downstream processing from grinding, allowing more scope for optimization of classifier and grinding efficiency. The achievable overflow density depends on the cyclone size selected and target grind sizes. Densities typically range from 25% to 50% solids w/w. Lower overflow densities are typically used for finer product sizes with ores that are recovery sensitive to grind and have a thickener following grinding. Higher overflow densities are typically associated with coarser product sizes and ores that are relatively recovery insensitive to grind with no thickener following grinding.

The operating cyclone pressure and spigot size should be set up so that the cyclone operates with a high cyclone underflow density, typically a few percent solids below roping conditions. The actual density depends on the specific gravity of the ore and slurry viscosity. Historical texts and experience suggest that this limitation occurs at approximately 60% solids by volume. Running the underflow density as high as possible without roping minimizes the water going to underflow and therefore the entrained fines returning to the mill to be overground, maximizing efficiency. Napier-Munn et al. (1996) provided a guide for assessing cyclone efficiency based on the extent of water recovery to the underflow stream, which is reproduced here as Table 5. This efficiency guide is provided on the basis that the water split in conjunction with the volume flow to the cyclone will have a large impact on both the circulating load in the circuit and the product density.

In scenarios where high specific energy, high reduction ratios, and a fine product size are targeted, high circulating loads will also be observed. Conversely, a low specific energy, low reduction ratio, or coarse grind size grinding duty will result in low circulating loads. In the first case, the feed rate to the mill will be low and specific energy high. The opposite applies to the second case, when the feed rate will be high and the specific energy low. Consider these two scenarios in the context of the same-sized mill and the capacity restriction being related to volumetric flow constraints. The volumetric flow through the mill will be comparable in both cases when optimized; however, as a percentage, the circulating load will be much lower when the mill throughput is higher.

Optimizing for Circuit Throughput

Aside from the product size, throughput is a major driver for the operation of the comminution circuit. Often, a circuit throughput will be limited by either the amount of power available to perform the grinding or downstream processing constraints. In these cases, additional capital in the form of more grinding power and/or downstream equipment will be required to lift these constraints. If this is not the case and the circuit has the capacity to accommodate higher throughput, identifying how to best achieve this is key.

Options that may provide a basis for optimization of throughput include altering the feed size by way of changes to the existing crushing circuit parameters, modifying blast fragmentation, blending, and secondary crushing. Relaxing the targeted product size can also be considered if the throughput gain outweighs the downstream recovery loss in terms of metal production and cash flow.

Optimizing for Product Size

For most mineral processing operations, the main objective of the grinding circuit is to achieve liberation of the valuable minerals from the gangue host rocks. Nevertheless, most downstream mineral separation equipment also operates more effectively at well-controlled PSDs, and this can also influence the product size targeted from the grinding circuit.

The relationship between the grade, recovery, and product size is ore specific; therefore, each mining operation will have its own target product size that has been derived from metallurgical testing and/or operating experience. Some ore types are more sensitive to product size than others, and the target product size may change for each ore type within one deposit. These adjustments may require circuit modifications as feed changes are presented to the plant.

The grind size selected during a feasibility study to maximize project NPV may not be the same in the constructed plant and certainly may not be the product size that generates the highest cash flow. After the grinding circuit becomes a fixed asset, increasing throughput at the expense of grind-size typically improves cash flow. Other factors affecting the grind size that should be considered are commodity price and operating costs. An economic model should be developed for the project and the target grinding circuit product size reviewed on a regular basis after the throughput–grind size–recovery relationship for the circuit is understood.

The optimal grind size in a plant can also often be coarser than that determined in laboratory work to achieve an equivalent metal recovery. This difference is often because of the finer PSD realized in the plant for higher specific gravity minerals and metals compared to the bulk ore constituents that normally reflect the gangue of barren components of the ore. The influence of gravity on classification by hydrocyclones in a production plant, an effect that is absent in the laboratory, is the cause. As a result, a finer grind is often required in the laboratory to make sure that sufficient liberation of the valuable minerals is obtained to ensure that the desired metal recovery efficiency is achieved.

Several variables within the grinding circuit dictate the product size achieved. Changes to mill operating parameters, such as milling density, ball size and mill speed, and so on, can have an effect. In most cases, however, there are two main reasons a target product size is not achieved: (1) insufficient power to achieve the target product size at the operating throughput and (2) poor classification. Therefore, these should be the initial focus of the optimization.

Optimizing for Overall Circuit Efficiency

Optimizing the energy efficiency is generally quite complex. The *efficiency* of the grinding circuit can be defined as the amount of energy that grinding an ore from a feed size to final product size physically consumes, compared with a theoretical calculation of the energy required to perform the same amount of work. The difference between the actual and theoretical energy requirement defines the potential for optimization. However, most of these calculations have less than ±15% accuracy for standard circuits and even less for more complicated configurations with multiple products. Furthermore, the laboratory tests used as input to the theoretical models may have an accuracy of ±10% in comparing different commercial laboratories (Kaya 2001).

If the specific energy differs significantly (+10%) from that of the theoretical calculation based on the ore properties (Bond work index and Autogenous Media Competency Test, JK drop weight, SAGDesign, etc.), the circuit may not be operating in an optimal manner. This assessment becomes more complex when nonstandard configurations and technologies are used, such as two-stage pebble crushing, stirred mill, or tertiary grinding flow sheets. In these more complicated cases, the assessment of efficiency of the individual stages of comminution is essential. However, in all cases the high-level, overall circuit power-based analysis provides a highly useful baseline efficiency analysis and reference point that can be used to quantify the impact of future optimization initiatives.

If there is a significant differential between the actual and theoretical energy requirement, then it is likely that the feed rate and product size are not balanced for the circuit power available, or there is something inherently wrong with the way that the circuit is operated. This could mean exceptionally high circulating loads, imbalance in the power split between two grinding stages, incorrect milling densities, inefficient cyclones, and so on.

Process Modeling and Simulation

Process simulation is a powerful tool that can be used to estimate the impact of process changes on machine performance, and to describe how the performance of individual grinding and classification units is interrelated. For example, a change in vortex finder size has an immediate impact on grind size and recirculating load. It may further influence changes in cyclone operating pressure and cyclone feed density, affecting circuit water additions and particle size, slurry density, and flow through the ball mill. Assessing the overall impact of the change on grinding circuit performance requires the application of advanced process models, typically generalized as population balance models. These models describe in detail, on a size-by-size basis, the breakage, transportation, and classification of ore particles through the grinding circuit. In so doing, they facilitate a detailed understanding of grinding circuit performance.

Process modeling can take simpler forms; examples include the power-based modeling techniques developed by individuals such as F.C. Bond, D.J. Barratt, S. Morrell, and J. Starkey. As much simpler empirical approaches compared to population balance modeling, they provide less detailed outputs but are useful for high-level efficiency analysis and throughput estimation. In some instances, this is an appropriate level of analysis.

Where a more detailed understanding of grinding circuit performance and the impact of process changes is required, population balance modeling is the preferred approach. This requires diligent surveying of the grinding circuit under controlled conditions, measurement of ore hardness properties and stream size distributions, collation of operating data, mass balancing of flows by particle size, process model fitting, and simulation. It is a lengthy and involved process; however, the effort is essential if a full and accurate understanding of the grinding circuit performance drivers is required.

After a model of the comminution circuit is built, simulations are performed by changing operating parameters within set limits. Simulations should be prioritized with unit processes optimized first. The mills need to be simulated at their optimal ball charge and load, mill speed, milling density, ball size, and so on. Changes to cyclone feed density, vortex finder, and spigot diameter can be made at this point.

The model should not be simulated with inputs that are too far away from the surveyed conditions when conducting the optimization. Most of the modern modeling techniques described earlier in this chapter were developed using regression models based on a database of operating data. They are most suited to being simulated within a tested calibration range where the confidence is high. This method also applies to population models. In these cases, the vendor of the package generally provides guidelines on the simulation constraints, and it is important to take heed of these limitations to ensure that the simulation results are valid. Models are predictive tools and are not perfect representations of reality, and should not be assumed to be valid outside of their range of calibration. Different models also have different sensitivities, and understanding each model's limitations is crucial when evaluating the results of simulations.

Simulations of the entire circuit may include changes to the ore type and/or feed size (crushing finer, partial secondary crushing, full secondary crushing, etc.). Additional grinding units or a recycle crusher may be added. With these changes, the units may require adjustment to maintain good performance. These simulations focus on the three core optimization targets of throughput, grind size, and circuit efficiency.

Simulation Techniques

After it is mass balanced, the survey data must be modeled so optimization simulations can be conducted. Knowledge of the process detailed in the previous section forms the basis of the modeling process from which a list of optimization strategies can be identified for simulation.

Mill Unit Process Modeling

The power draw in the grinding mills should be modeled and compared with the theoretical calculations. The mill dimensions, mill speed, ball charge, and total mill load (in the case of SAG mills) measured during the survey can be used to calculate the theoretical power draw of the mills. This result is then compared with the measured value for power draw during the survey.

In SAG mills, trajectory modeling will provide a qualitative insight into whether the mill speed is appropriate for the liner design and load conditions. Changes to the liner profile can be simulated if the existing design is problematic. This information is also useful when combined with the measured load level. As discussed previously, the relationship of the toe of the charge in the mill to the impact zone of ball trajectories affects the breakage rates in the mill. Ball trajectories that fall into the toe of the charge promote impact breakage, while an impact point that falls within the charge promotes higher levels of abrasion and lower levels of impact breakage. When optimizing a semiautogenous/ball mill or SABC circuit for maximum throughput, load conditions must be targeted that promote impact breakage so that the toe of the charge is being struck by the balls, resulting in a coarser transfer size (SAG mill limited scenarios). Conversely, when lower throughput and a finer product are the objective (ball mill limited scenarios), load conditions that promote abrasion breakage (where balls fall above the impact zone) are better suited. This principle also applies to balancing breakage rates in a single-stage SAG mill.

In the case of grate discharge mills, the volumetric flow and pebble extraction rate must be assessed to determine whether the grate is adequate in design. Without sufficient grate flow, some reduction in power draw can occur as the restriction on slurry flow causes the development of a slurry pool. Furthermore, an increased mill-specific power may be observed if the extraction of pebbles is below that required, resulting in the buildup of critical size rock in the mill load. A slurry pool can also cushion the toe of the charge from ball impacts, reducing impact breakage.

In ball mills, calculation of the retention time and superficial velocity will provide information about the volumetric flow through the mill, which is of interest if circulating loads are problematic. In optimization, the key is to keep these calculated values within boundaries defined as acceptable for efficient option. The calculation of these parameters and the boundaries for best practice for design are discussed in Chapter 3.8, "Grinding Circuit Design."

Classification Modeling

Screens within the comminution circuit are generally assessed for their loading and separation efficiency. Typically, screens and trommels are not the focus of primary optimization strategies unless they have significant efficiency issues or a significant change in product size is required. Screen performance, however, must be reviewed during the simulation process, particularly if increased throughput is the main driver for the optimization. As a guide, an efficient screen will operate with 80% load at a separation efficiency of approximately 90%–95%.

Modeling of the cyclone performance will include analysis of the circulating load, the water split to underflow, the partition curve, and calculation of the separation efficiency factor, α. This type of modeling indicates if there is short-circuiting within the cyclones, sending product-size material back to the grinding mills. This process often provides insight as to where the problems associated with circulating load and variable product size lie. Typically, a modern-style cyclone is operating efficiently if its α value lies between 2 and 3.

The partition curve is used to assess the proportion of material in each size fraction reporting to the underflow. In an efficiently operating cyclone, the proportion of product-size material reporting to the underflow is often in the 40%–50% range, although values from 50% to 65% are common. In some instances, the partition curve can kick up at fine particle sizes. This is known as the fishhook effect and relates to the fines content that follows the water split. Bumps in the partition curves can also be seen at times and can be related to the concentration of higher specific-gravity minerals in some size fractions.

In some instances, the partition curve can kick up at fine particle sizes. This is known as the fishhook effect and relates to the fines content that follows the water split. Bumps in the partition curves can also be seen at times and can be related to the concentration of higher specific gravity minerals in some size fractions.

PROCESS CONTROL

Process control is often an area where significant gains in performance can be achieved with minimal effort. In general, circuits tend to be either under- or overcontrolled. Undercontrolled circuits are operated manually and offer the advantages of increased surveillance and measured adjustment. Overcontrolled circuits can constrain throughput via inefficient or counterproductive control.

The primary role of process control should be stable production, and the secondary goal, optimized performance. Most modern control systems can efficiently stabilize the operation of a comminution circuit; however, most cannot optimize performance. The key is applying an effective control philosophy to the installed system.

A good control philosophy is based on an understanding of the interrelationship of unit processes. Too many systems individually control unit processes and do not control the system well as whole, resulting in a system that is out of equilibrium and therefore unstable. Frequent causes include suboptimal load levels in a SAG mill and constrained cyclone conditions.

SAG Mill Load and Throughput

The SAG mill load level must be selected to balance grinding requirements between impact breakage and abrasion. The load level stabilizes and optimizes operation in the case of a single-stage SAG mill and balances the energy requirement between stages of grinding in a two-stage circuit. High load levels increase abrasion while lower loads associated with the impact toe of the charge increases impact breakage. Because of this phenomenon, the interaction between mill throughput and mill load is nonlinear and simple PID (proportional–integral–derivative) loops struggle to undertake this function on most mills. Expert systems, gain scheduling, or fuzzy logic are typically used for the best results. These types of systems generally target constant load but not constant throughput. Throughput is never constant because of continuous changes in feed characteristics: primarily ore hardness and PSD. Changes in stockpile fill levels can result in significant changes in the feed PSD. To improve the resulting instability, feeder contributions on a multiple feeder system are often manipulated.

Cyclones

Cyclones must be operated so that they cut at the appropriate product size to suit the throughput, ore characteristics, and grinding power available. Alternately, the throughput or grinding power must be adjusted to meet the cut point of the cyclone. A mismatch in these variables results in circuit instability (variable circulating load). Changes in circulating load are influenced by changes in feed characteristics (hardness, PSD, feed rate). The cyclone components (diameter, vortex, spigots, etc.) and number of cyclones are selected so that an appropriate overflow and underflow density are produced at a specific cyclone feed density and pressure that achieves the target product size. Cyclone overflow density is a constraint set by downstream processing and is typically only an issue if high slurry densities are required in the absence of a thickener following classification. The underflow density should be as high as possible without causing roping of the spigot, minimizing the water to underflow and therefore fines content, and maximizing grinding efficiency.

The cut point of a cyclone is influenced by both the flow (pressure effect) and slurry density (hindered settling effect). As a result, two methods or philosophies exist for operating cyclones, given that one component must be an adjusted variable and the other a set point of operation. The two methods are (1) constant feed density, variable pressure, and (2) constant pressure, variable feed density.

Constant Feed Density

In constant feed density control, the water addition to the discharge hopper is varied to maintain a constant feed density to the cyclones. The pump speed is varied to maintain a loosely constant level in the discharge hopper.

Theoretically, this configuration should maintain the most constant product. However, if the cut point of the cyclone does not match the grinding being undertaken, an increase in the circulating load occurs, which increases the flow to the cyclones and thus the cyclone pressure. This decreases the cut point of the cyclone, which causes the circulating load to increase even more. Therefore, the operator or control system must keep the operation of the circuit within a tight range to avoid total loss of control. Bringing cyclones into and out of circuit is, at best, a short-term solution. Only with a decrease in throughput or increase in grinding power can the circuit be stabilized. If this situation is not quickly brought under control, coarse material builds up in the circulating load and is eventually discharged via the overflow.

A decrease in the circulating load decreases the flow to the cyclones, and thus the cyclone pressure decreases. This increases the cut point of the cyclone, which causes the circulating load to drop further. This configuration does tend to minimize overgrinding resulting from a decreased grinding efficiency affected by the poor classification in this scenario.

Constant Pressure

In constant pressure and variable feed density configuration, the pump speed is set to maintain a set cyclone pressure. Water addition to the hopper is varied to maintain the hopper level set point. As the circulating load changes, the feed density to the cyclone is altered, changing the cut point of the cyclone. Under such an operating system, the product size varies; however, the grinding power is always fully used. As the circulating load increases, the cyclone feed density increases and thus the product size increases. If the circulating load decreases, the cyclone feed density decreases, thus lowering the product size. In this way, the product varies but is self-compensating and prevents rapid changes in circulating load resulting from a mismatch in the product size and the cyclone cut point. The circuit will reach an equilibrium product for a given feed (ore characteristics, PSD, etc.) and feed rate.

The aim of a circuit targeting a constant product size is constant feed density and cyclone pressure. Constant pressure is the authors' preferred regulatory control philosophy because it best achieves the primary object of the control system, stability. Cyclones do not grind material and therefore cannot be used to effectively dictate the product from the circuit. Throughput, and in the case of a SAG mill circuit, SAG mill operating conditions, should be varied to maintain the P_{80} within the target range. In this way, the target cyclone feed density is also controlled, although not by water addition.

Cyclone switching, or the number of cyclones operated, should be adjusted to maintain the cyclone overflow within a target density range. Cyclones should not be taken in and out of circuit based on changes in circulating load. Cyclones should be taken in and out of circuit based on the calculated or measured cyclone overflow density, with a low set point triggering a cyclone to be closed and a high set point triggering a cyclone to be opened.

If a tight control of the overflow density is required, the control philosophy becomes more complicated regardless of the philosophy employed, with the pressure or cyclone feed density target changing as the pressure or densities change with the water input controlled to the overflow density. In this case, cyclone efficiency is sacrificed to achieve constant overflow density. To achieve good cyclone efficiency at a tight feed density to downstream processing, a thickener must be included in the circuit post-grinding or the cyclone overflow density set above that required for downstream processing so that post-classification dilution can be varied to maintain a constant slurry density.

Pebble Crusher Control

A pebble crushing circuit can sometimes be difficult to optimize if the circuit is not well engineered. For a circuit to be optimized, it should have a feed surge bin and feeder or be slightly undersized for the duty with excess pebbles overflowing the crusher feed canister or chute. Either one allows the pebble crusher to choke feed. Choke-feeding the crusher is key to minimizing recirculation of near discharge screen aperture material.

If the crusher has a bin and feeder, the crusher should be batch fed so that it operates under choke-fed conditions. Oversizing of the pebble crusher can cause operational issues. If the crusher is oversized, the pebble crusher frequently cycles on and off, sending pulses of material through the circuit. In this instance, compensatory feed control should be considered to minimize the disturbance. The crusher gap should be operated as fine as possible within the limitations of the crusher, typically no more than a reduction ratio of 4:1 (gap control). Alternatively, the crusher may be operated to maximize power draw (minimization of gap to optimize power draw). This second method is often employed when hydroset crushers are installed, as this is easily programmed into the control system.

If the circuit is SAG mill limited, the pebble port open area should be optimized to maximize pebble extraction. Often it is better to extract too many pebbles and have some pebbles bypass the crusher, so that it is constantly choke fed, rather than to underfeed the unit so it is not choke fed or so the crusher is cycling on and off. Curved pulp lifters are often used to optimize pebble extraction, frequently achieving 50% higher pebble extraction compared to radial pulp lifters.

Preventing steel grinding media from entering a pebble crusher can hamper optimization. Belt magnets should be well located to maximize ball removal. Metal detection should be located close to the bypass chute, as spherical balls can slowly roll back on the conveyor as they are being transported. The longer the distance between the detector and the chute, the larger the possible change in a ball position following detection.

REFERENCES

Azzaroni, E. 1981. Mill grinding media size and multiple recharge practice. Presented at the Second Asian Symposium on Grinding, Manila, Philippines.

BCS Incorporated. 2007. Mining Industry Energy Bandwidth Study. www1.eere.energy.gov/manufacturing/resources/mining/pdfs/mining_bandwidth.pdf. Accessed June 2013.

Bond, F.C. 1958. Grinding ball size selection. *Min. Eng.* 10(5):592–595.

Bond, F.C. 1961. *Crushing and Grinding Calculations.* Milwaukee, WI: Allis-Chalmers Manufacturing.

Connolly, J., and Buckingham, L. 2012. Barrick's grinding efficiency improvements through targeting performance. Presented at the Procemin 9th International Mineral Processing Conference, Santiago, Chile.

Crocker, B.S. 1985. Pebble mills. In *SME Mineral Processing Handbook*, Edited by N.L. Weiss. Littleton, CO: SME-AIME. pp. 3C-94–3C-107.

Dunn, D.J. 1989. Design of grinding balls. In *Innovations in Gold and Silver Metallurgy*. Golden, CO: Randol International.

GMSG (Global Mining Standards and Guidelines Group). 2016. Methods to Survey and Sample Grinding Circuits for Determining Energy Efficiency. Ormstown, QC: GMSG.

Halbe, D., and Smolik, T. 2002. Process operating costs with applications in mine planning and risk analysis. In *Mineral Processing Plant Design, Practice, and Control.* Edited by A.L. Mular, D.N. Halbe, and D.J. Barratt. Littleton, CO: SME. pp. 326–345.

Herbst, J.A., and Rajamani, K. 1982. Developing a simulator for ball mill scale-up: A case study. In *Design and Installations of Comminution Circuits.* Edited by A.L. Mular and G.V. Jergensen. Littleton, CO: SME-AIME. pp. 325–342.

Hilden, M.M., Powell, M.S., and Bailey, C. 2015. The new JK variable rates AG/SAG mill model. Presented at the SAG Conference, Vancouver, BC, Canada.

Kaya, E. 2001. Evaluation of bond grindability tests. Presented at the SAG Conference, Vancouver, BC, Canada.

McIvor, R.E., Blythe, P., Finch, J.A., Lavallee, M.L., and Wood, K.R. 1991. Functional performance characteristics of ball milling. SME Preprint No. 90-27. Littleton, CO: SME.

Mosher, J., and Alexander, D. 2002. Sampling high throughput grinding and flotation circuits. In *Mineral Processing Plant Design, Practice, and Control.* Edited by A.L. Mular, D.N. Halbe, and D.J. Barratt. Littleton, CO: SME. pp. 63–73.

Napier-Munn, T.J., Morrell, S., Morrison, R.D., and Kojovic, T. 1996. *Mineral Comminution Circuits: Their Operation and Optimization.* Brisbane, Queensland: Julius Kruttschnitt Mineral Research Centre.

Plavina, P., and Clark, G. 1986. The selection, commissioning and evaluation of a large diameter, variable speed ball mill at Bougainville Copper Limited. Presented at the 13th Congress of the Council of Mining and Metallurgical Institutions, Singapore.

Villanueva, A., Banini, G., Hollow, J., Butar-Butar, R., and Mosher, J. 2011. Effects of HPGR introduction on grinding performance at PT Freeport Indonesia's concentrator. Presented at the Semiautogenous Grinding and High Pressure Grinding Rolls Conference, Paper No. 172.

CHAPTER 3.11

Grinding Technologies

Aidan Giblett

Major grinding technologies in use in mineral processing today can be broadly categorized into tumbling mills, stirred mills, and compressive grinding mills. The term *mill* is ubiquitous and defines an extremely wide variety of mechanical devices that grind material to an equally wide range of product sizes. Perhaps the most widely recognized technology is the tumbling mill, where the action of a rotating steel drum promotes the grinding of mineral ores. Common types of tumbling mills include autogenous grinding (AG) and semiautogenous grinding (SAG) mills, rod mills, ball mills, drum scrubbers, and pebble mills. In tumbling mills, the primary distinction between mill types is the type and size of media used to promote comminution, and to a lesser extent the size and dimensional aspect ratio of the mills themselves.

Stirred mills employ a stationary drum and rotating shaft fitted with a helical screw, pins, or discs to agitate the mill contents and are commonly used to grind particles to fine sizes in concentrate and tailings regrind or ultrafine grinding (UFG) duty. They may be vertically or horizontally configured, reflecting either upward or horizontal flow of material through the grinding chamber. In addition to the orientation of the mill and variations in stirring mechanisms, stirred mills can also differ in energy intensity. Grinding media is commonly screened silica sand, solid steel cylinders, steel balls, or ceramic spheres.

Compression mills employ grinding wheels or rolls to grind particles under compression without the use of grinding media. Vertical roller mills (VRMs) and high-pressure grinding rolls (HPGRs) are technologies extensively used in industrial mineral processes to generate final comminution circuit product material. These are dry grinding technologies in contrast to stirred and tumbling mills, which are more commonly associated with wet slurry systems, although dry applications of those technologies in specific industries are not uncommon.

The range of application and degree of size reduction of the major grinding technologies applied in mineral ore processing are illustrated in Figure 1, progressing from the autogenous/semiautogenous milling of large particles up to 500 mm (20 in.) down to the sub-15-µm range (UFG).

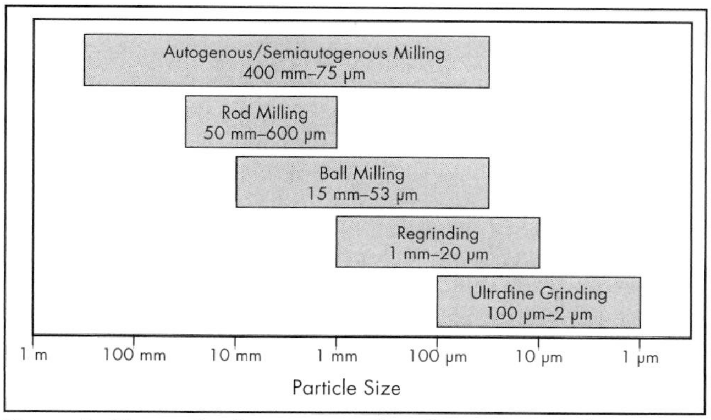

Figure 1 Size reduction ranges of mainstream grinding technologies used in mineral processing

A summary of mill installation data provided by major global grinding mill suppliers (Metso, FLSmidth, ThyssenKrupp, CITIC HIC, and Outotec) is shown in Figure 2, reflecting the applications of the major conventional tumbling mill technologies. Technology selection is influenced by the suitability for the grinding duty, ore characteristics, and the range of mill sizes available. The higher-capacity unit technologies, AG, SAG, and ball mills, are well suited to processing low-grade ore at high volumes. These technologies have the versatility to operate as single-stage mills producing final product material and are all suited to use as primary grinding mills receiving suitably prepared feed material. Ball mills have the additional flexibility of being suited to secondary, tertiary, and regrind milling duties. Because of unit capacity limitations, the use of rod mills has declined as a percentage of total mill sales terms since the 1990s, although they remain a viable technology selection in specific applications. Sales volume data indicate a similar number of rod mill sales in the 1970s and in the 2000s, defying the often-quoted status of the rod mill as an expired technology. A more pronounced decline in the number of pebble mill installations has been observed since the 1980s, although

Aidan Giblett, Senior Technical Advisor—Mineral Processing, Newmont Mining Services, Subiaco, Western Australia, Australia

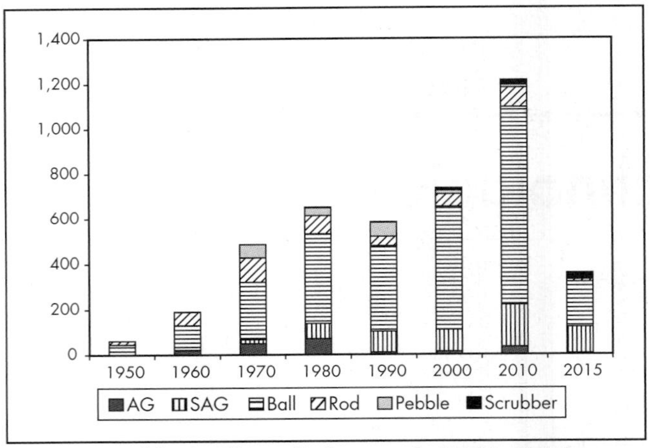

Figure 2 Tumbling mill installation trends (1950–2015)

pebble mills are still used in the grinding of ores in several industries. There are far fewer scrubber installations than the other tumbling devices; however, scrubbers are a frequently used grinding technology in the processing of diamond-bearing ores and iron ores, among others. While the general dominance of SAG and ball mills in the tumbling mill market in recent decades is clear, the other tumbling mill technologies remain relevant in mineral processing.

Stirred milling technologies include the tower mill, Vertimill, IsaMill, pin mill, stirred media detritor (SMD), VXPmill, and HIGmill. These technologies are primarily used in mineral processing for the regrinding and fine grinding of concentrates; however, a growing number of stirred mills are employed in secondary and tertiary grinding duties. The adaptation of stirred milling technology, particularly in very fine grinding applications (sub-15–20 μm), has increased significantly since the mid-2000s based on several successful installations, increasing mineral resource complexity and the availability of larger-capacity mills.

Compressive grinding technologies, most notably VRMs and HPGRs, are extensively used in the cement industry for the dry grinding of limestone, cement clinker, and slags. These technologies are not widely used in mineral processing applications to generate final ground product material, although they have significant appeal based on well-documented energy efficiencies over ball milling. Dry grinding is not immediately compatible with the wet downstream recovery processes typical of the extraction of precious and base metals but may be viable where minimal energy is required for the drying of mineral ores. The significant reductions in energy consumption that these technologies have demonstrated in the cement industry warrant consideration for more widespread application, subject to suitable wear performance. Wear performance in abrasive minerals applications has been demonstrated for HPGRs in tertiary and quaternary crushing duties, although the VRM technology has not been tested to the same extent in this industry and has found more limited application to date.

AUTOGENOUS GRINDING, SEMIAUTOGENOUS GRINDING, AND PEBBLE MILLING

This section adopts conventional definitions for autogenous and pebble milling, which in concept are primarily distinguished by the size of the media in use and the feed size processed. Crow and Lipphardt (1985) defined *autogenous grinding* as the action of a material grinding on itself, as occurs when pieces of ore of different sizes are rotated together in a tumbling mill, and observed that this action is the same in AG mills as in pebble mills. The distinction is in the media size used. AG mills are then further defined as mills that receive primary crushed or run-of-mine (ROM) ore. Mills operating in this duty, which have some part of the mill charge composed of steel grinding media, are called SAG mills. *Pebble milling* is defined as the use of tumbling mills that use rock particles of a specified size class as grinding media. In pebble milling, the grinding media may come from the sizing of crushed or milled ore particles, or from an external source, such as prepared flint or gravels.

Autogenous and Semiautogenous Milling

Autogenous and semiautogenous milling are performed in tumbling mills and in predominantly wet grinding environments. Slurry densities in wet grinding typically range from 60% to 84% solids and can vary considerably from one operation to another based on the rheological characteristics of the ore. The mills are of grate discharge design, meaning that the slurry flows through the mill and is classified by discharge grates with slotted apertures ranging from 25 mm (1 in.) up to approximately 90 mm (3.5 in). Large aperture slots greater than 38–50 mm (1½–2 in.) are known as pebble ports and are intended to allow the ejection of critical size material from the mill. Material passing the grates enters the discharge chamber where pulp lifters convey the material to a central discharge cone and out of the mill. Another less common variation is to have the mill end open after the grate, such that the discharging slurry is immediately free to fall into a sump or be further classified by a screen.

AG and SAG mills (Figures 3 and 4) are most commonly operated as the primary stage of grinding before ball milling, although single-stage semiautogenous milling and multistage AG mill/pebble mill circuits are common flow-sheet variations. Feed material may be ROM but is typically primary crushed to 80% passing 100–150 mm (4–6 in.) The AG or SAG mill will discharge through a trommel screen and/or a vibrating screen to protect the secondary grinding circuit from excessively coarse particles and large mill balls. The screening stage allows oversize material discharging the primary mill to be recirculated, commonly via a pebble crushing circuit, to reduce the oversize to a suitable feed size for the secondary grinding stage. In single-stage AG/SAG applications, the mill will operate in closed circuit with hydrocyclones or fine screens to regulate the final product size.

Mill diameters up to 9.8 m (42 ft) have been constructed, although mill diameter varies significantly by application as a function of design grinding circuit capacity. Mill aspect ratio—the ratio of the mill diameter (*D*) to the length (*L*), expressed as *D*/*L* (Napier-Munn et al. 1996)—varies from high to low. High-aspect mills have greater diameter than length and may have aspect ratios of 3:1 (*D*/*L*); low-aspect mills may have a *D*/*L* ratio up to 1:2½. Generally, a high-aspect mill is preferred where high mill unit throughput is required; a low-aspect mill is preferred where a fine grind size is required, particularly if the mill is not operating in closed circuit with hydrocyclones. When the mill is closed with hydrocyclones, the aspect ratio is of no clear significance, although low-aspect mills are often favored in such applications.

Courtesy of ThyssenKrupp
Figure 3 Prominent Hill SAG mill in Australia

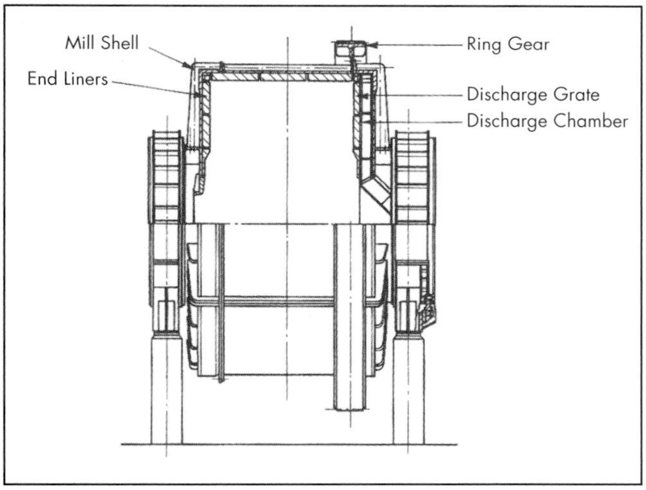

Source: Mular and Jergensen 1982
Figure 4 AG/SAG mill components

When autogenously milling primary crushed or ROM ore in the absence of steel grinding media, the ore must contain some material of sufficient size and strength to effectively perform as autogenous media. Ideally, the feed material will comprise material that has strength at larger sizes but also grinds easily at finer sizes. These characteristics avoid the accumulation of material in the mill that is too small to effectively act as autogenous media, yet not fine enough to be classified as product from the mill. Excessive accumulation of such material is detrimental to grinding performance. It will result in a loss of mill capacity and is subsequently known as critical size material. Critical size material is likely to be in the 15–75 mm (⅝–3 in.) range, coarser than the discharge screen aperture and exhibiting low grinding rates in the mill. Where critical size accumulation is predicted or observed to be an issue, the use of larger grates known as pebble ports and the implementation of pebble crushing are demonstrated means to increase mill capacity. An advantage of autogenous milling is the elimination of costs associated with steel grinding media consumption, and any contamination of final products or reduced recovery because of deteriorated mineral surface condition that results from the presence of grinding media corrosion products or tramp iron. Alternatively, AG mills will have lower power draw, and lower mill capacity than an equivalent mill operating in SAG mode. Autogenous milling will produce more fines than semiautogenous grinding, which can be detrimental to downstream recovery depending on the separation process applied. Ores of higher specific gravity, such as iron ores, are known to be well suited to autogenous milling.

As a result of the potential impact of critical size buildup in AG mills, and the increased mill capacity that comes with the introduction of steel grinding media, the practice of semiautogenous grinding has become the more common application of the semiautogenous milling concept. Grinding steel balls of 100–125 mm (4–5 in.) diameter, steel charge levels ranging from 4% to 15% by volume, and total operating mill charge levels of 25%–35% by volume are traditional characteristics of semiautogenous grinding. Over time the use of steel balls up to 150 mm (6 in.) has become more popular, particularly in copper porphyry operations. The use of up to 18%–20% steel charge by volume has also become necessary to maintain high mill power draws when processing less competent ores, and highly fragmented or secondary crushed SAG mill feed. In these instances, the feed contains an insufficient number of particles of adequate size and strength to serve as autogenous media, and subsequently the mill will often operate at lower total mill filling levels in the 20%–25% by volume range. SAG mills have the added cost of grinding media consumption in comparison to AG mills; however, higher mill capacities are achieved for comparable mill sizes. SAG mills

Courtesy of ThyssenKrupp Courtesy of Allis-Chalmers

Figure 5 (A) Cerro Verde ball mills in Peru and (B) overflow ball mill cutaway

are more prone to liner breakage because of the weight of steel grinding media used, and rubber mill linings are not considered a viable option for the larger-diameter (>34 ft [>10.4 m]) SAG mills, although composite shell liners have been successfully used on small- to medium-diameter mills.

Pebble Milling

As previously described, pebble milling implies the use of screened rock particles as grinding media. These particles may be externally sourced, produced by screening of the primary crushed ore, or generated in the primary milling stage. Where the pebbles are generated from the crushed ore, the process is often termed *secondary autogenous grinding*, and this is largely the focus of this section.

In this secondary form of autogenous milling, grinding is usually performed wet and in closed circuit. Pebble mills typically are used in secondary and tertiary grinding duties after AG or rod mills. The broad range of application of pebble milling is defined by Crocker (1985) as receiving feed material up to 9.5 mm (⅜ in.) and grinding down to as fine as 10 µm. Pebble mills receiving crushed material up to 9–10 mm in size are termed *primary pebble mills* or *lump mills* and would be charged with pebble media ranging from 125 to 250 mm (5–10 in.). Where the pebble mill is a secondary mill receiving rod mill product passing 2–2.5 mm (8–10 mesh), the pebbles are taken from primary crusher product in the –100 mm (4 in.) to +25 mm (1 in.) size range. Pebble size is specific to the grinding duty to be performed, with smaller pebbles required for finer grinding, and the optimum pebble size is defined by Crocker (1985) as the equivalent weight of a steel ball that would be applied to achieve the same grinding duty in a ball mill. Pebble mills operate at high media loads (35%–40%) to maximize mill power draw and are exclusively grate discharge mills to effectively retain the grinding media in the mill at those high load levels. In some instances, combinations of pebble and steel ball grinding media have been applied.

BALL MILLS

Ball mills are the most widely used form of tumbling mill where the ore is ground with a charge of cast or forged steel balls as the grinding media (Figure 5). Ball sizes from 25 to 100 mm (1–4 in.) are used, varying as a function of feed size, ore grindability, and target product size. Ball mills may receive secondary or tertiary crushed feed ranging from 80% passing (F_{80}) 6 to 15 mm (¼–⅝ in.), or classified product from a primary grinding stage producing typically finer than 6 mm (¼ in.) feed F_{80}.

While suited to wet or dry grinding, ball mills are most commonly used for the wet grinding of mineral ores and routinely generate the final product for separation or extraction at product sizes under 80% passing (P_{80}) 300 µm (50 mesh). Dry grinding using ball mills historically has been standard practice in the cement industry, although more recently VRMs have become the dominant dry grinding technology in that industry. Wet grinding overflow discharge ball mills are usually operated at 74%–76% of critical speed, 60%–80% solids, with ball loads between 25% and 35% by volume, although in smaller-diameter, grate discharge or shell supported mills, the ball load may approach 40% by volume.

Ball mills may be of grate or overflow discharge design; however, the overflow ball mill is the most common. The aspect ratio of ball mills is highly variable with no fixed and firm rules governing the optimum aspect ratio for a given application. However, Rowland (2002) did observe that the potential for grinding media charge segregation increased with decreasing aspect ratio (*D/L*) when using grinding balls larger than 64 mm (2.5 in.) in diameter.

MULTICOMPARTMENT MILLS

Multicompartment mills employ two or more grinding chambers to allow grinding stages to be combined in a single mill. Ground material can pass directly from one chamber to the next through a grate arrangement, or material can be removed for classification in between grinding stages. Multicompartment mills can have high length-to-diameter aspect ratios up to 5:1. Combination rod/ball mills (Figure 6) are used in many applications including bauxite grinding at Australian Alumina operations at Gove (Gouma and Bhasin 1993) and Gladstone (Singh and Sisley 2013; Skiba 1993) and grinding cement raw materials as described by Rowland (1985).

Courtesy of Allis-Chalmers
Figure 6 Combination rod and pebble mill

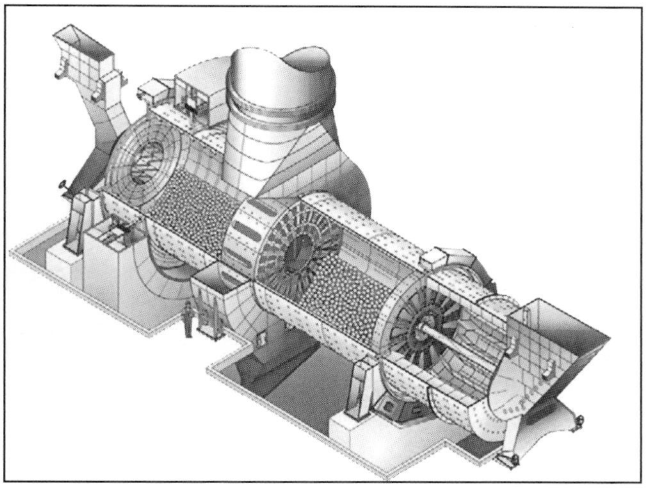

Source: Thomas et al. 2001
Figure 7 Double-rotator ball mill

The double-rotator ball mill is a specialized multicompartment ball mill that was developed for the cement industry, and it has been successfully used for many years to dry-grind refractory gold ores of the Carlin Trend in Northern Nevada, United States. The selection and use of this technology for gold ore grinding has been discussed in detail by Tempel (1993) for the Newmont Carlin operation, and by Thomas et al. (2001) for the Barrick Goldstrike operation in Nevada. In both instances, secondary crushed ore is ground to final product size to provide feed material for oxidative roasting and subsequent hydrometallurgical extraction of gold. Figure 7 shows a cutaway drawing of a double-rotator mill. ThyssenKrupp mill installation data reflects almost 50 double-rotator ball mills in use, primarily in iron ore, gold, and base metals applications.

The double-rotator ball mill consists of four main sections: drying chamber, primary grinding compartment, central discharge chamber, and secondary grinding compartment. The incoming ore is dried by hot air, generated by a gas burner, in the drying chamber prior to migrating to the primary

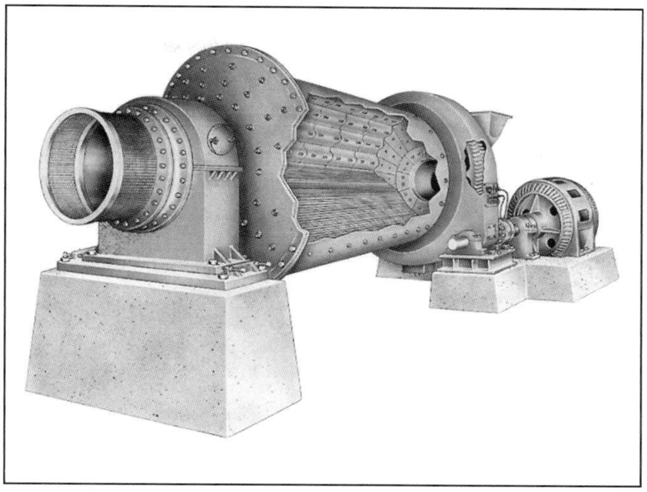

Courtesy of Metso Minerals Industries
Figure 8 Rod mill cutaway

grinding chamber. Ground ore from both the primary and secondary grinding chamber enters the central discharge chamber through diaphragms. The grinding circuit is operated under negative pressure, and fine particles generated in the mill are air swept through the discharge chamber and report to the static classifier. Finished size material from the static classifier, representing approximately 30% of the feed, is collected in a baghouse cluster and transferred into fine ore storage silos. Near-size particles in the air-swept stream are captured in the static classifier and rerouted to the bucket elevators. Coarse particles are discharged from the mill outlet into air slides that feed dual-bucket elevators, which transport the coarse mill products into the feed inlet of a dynamic classifier. Most of the oversize product from the dynamic classifier is returned to the fine grinding chamber of the mill with a small portion being returned to the drying chamber. Finished size product enters a second four-baghouse cluster and is subsequently transferred into the fine ore storage silos.

ROD MILLS

Prior to the widespread application of semiautogenous milling, rod mills (Figure 8) were the favored technology for primary wet grinding applications. While still in use in many operations, dimensional constraints on rod mills typically limit their suitability in new projects to low single-line milling capacity applications or applications where downstream mineral recovery is substantially affected by fines produced in grinding.

As the name implies, rod mills use steel grinding rods as grinding media. Rod mills typically serve in primary grinding duty in wet grinding applications but have also been used as the final stage of grinding. Rod mills have been reported to receive feed as coarse as 50 mm (2 in.), although for optimum performance a feed size of 19 mm (¾ in.) is more often targeted. Grinding rod diameters range from 38 to 115 mm (1½ to 4½ in.), with rod mill operating charge levels of 35%–40% by volume being typical. The upper limit on rod length of 6.1 m (20 ft) provides the upper limit on maximum rod mill length to approximately 6.5 m (21½ ft). Rowland (2002) defined this as the length that rods will stay straight in the mill and will break into smaller pieces that will discharge from the mill when worn. Rod mills are unique in that they need to be stopped to charge the mill, and a rod storage table is required adjacent to the discharge end of the mill. Rowland further specified an optimum aspect ratio of 1:1.4–1:1.6 (*D/L*) to minimize potential for dangerous rod tangling, which then reflects the practical maximum rod mill size of 4.6 m × 6.5 m (15 ft × 21½ ft). Rod mills are commonly overflow discharge mills but can also be configured as end peripheral discharge or center peripheral discharge (Figure 9).

Rod mills have a higher charge packing density than ball mills, which results in higher energy intensity and more effective breakage of the coarsest particles in the mill feed. Rod mills will therefore have an efficiency advantage over ball mills when there are excessive oversize particles in the feed and may be selected where top size control to the next process is essential.

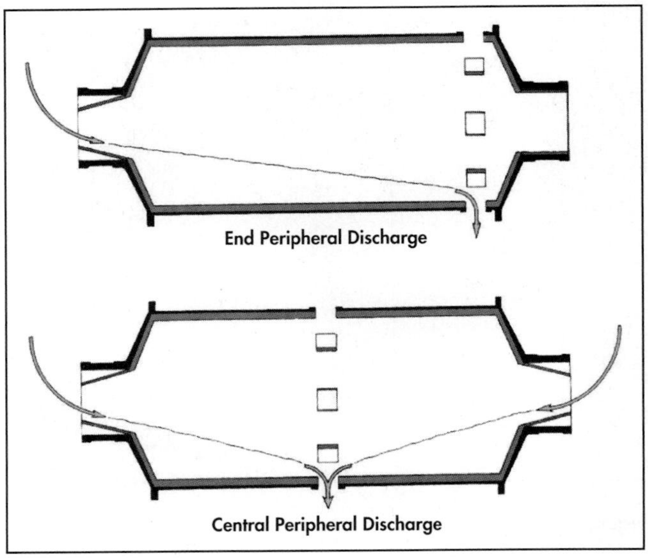

Courtesy of Metso Minerals Industries
Figure 9 Rod mill configuration

Courtesy of ThyssenKrupp
Figure 10 Scrubber in a diamond application

SCRUBBERS

Scrubbers are a specialized low-aspect-ratio mill used in applications where the physical separation of a soft component in the ore, often clays, is required before downstream processing (Figure 10). Scrubbers are associated with processing of bauxite, phosphate, iron, laterite nickel, and diamond-bearing ores. Mill diameter-to-length aspect ratio is typically 1:1.5–1:2.5, and mill diameters are commonly 5 m (16.4 ft) or less. A requirement for effective scrubbing is that the feed contain particles large enough and competent enough to act as media. Mill speeds and installed motor power are lower than for similar sized ball mills, and scrubbing is normally performed in the absence of steel grinding media. Scrubber specific energy consumption is low (typically 0.1–0.5 kW·h/t), as the duty is effectively washing rather than grinding; however, high energies can be required when processing tenacious clays or moderately competent feed material (saprock). Scrubbers are configured with either a simple overflow weir or a combination of weir and internal discharge screen. The weir retains a volume of slurry to ensure washing occurs, and the internal screen retains rocks to act as scrubbing media. The scrubber discharge is classified by fixed trommel or external screens.

STIRRED MILLS

Stirred milling technology has demonstrated increased energy efficiency over ball milling in fine and UFG duties, with quoted energy reductions of 30%–50% for a comparable task. This has given rise to the increasing application of stirred milling technology in mineral processing flow sheets. Stirred mills are horizontal or vertically aligned and use pins, discs, or a spiral screw fixed on a main shaft to transfer energy to the mill charge. Steel balls, steel cylinders, silica sand, and ceramic beads are common forms of grinding media. Lichter and Davey (2002) defined two classifications of stirred mills based on the impeller speed. Low-speed mills stir the media, without fluidizing it. These mills use coarser, high-density grinding media such as steel balls and cylpebs and are better suited to the fine grinding of coarser feeds. Higher-speed mills fluidize the grinding media and use lower-density sand or ceramic media.

Specific energy consumptions for stirred mills 30%–50% lower than ball mills have been demonstrated in fine grinding (<50 µm or 270 mesh) and UFG to <15–20 µm (635 mesh). Fine grinding systems achieve superior grinding efficiencies compared to ball milling because the media size can be well matched to the grinding duty. Fine and UFG mills are constructed in a way that media cannot easily leave the mill with the discharge slurry flow. This allows media as fine as 1 mm to be used in some mills when appropriate. Conversely, overflow ball mills allow particles in the ideal fine grinding media size range (1–15 mm) to leave the mill with the slurry flow and are only suited to media diameters of 20 mm or larger.

Depending on the technology, a stirred mill can accept a feed size up to several millimeters, and as a result, the technology has been applied in secondary and tertiary grinding duties in addition to the more conventional concentrate regrind and UFG applications in which they dominate. The wide range of stirred mills and mill motor sizes is summarized in Table 1.

Tower Mills

As reported by Jankovic (2005), the original tower mill was released in 1953 and has been produced under several owners since that time. The most widely recognizable tower mill brand is the Metso Vertimill (Figure 11), which came into production in 1991. Tower mills are composed of a vertical cylinder with a relatively slow speed screw media agitator, as represented in Figure 11. Because of their vertical arrangement, footprint reductions can be substantial compared to those for horizontal grinding technologies.

Table 1 Stirred mill sizes for mineral processing applications

Installed Motor Capacity, hp	Vertimill	SMD	IsaMill	VXPmill	HIGmill
0–499	VTM-15 VTM-150 VTM-20 VTM-200 VTM-40 VTM-250 VTM-60 VTM-300 VTM-75 VTM-400 VTM-125	SMD-0.75 SMD-7.5 SMD-18.5 SMD-90 SMD-185 SMD-355	M100 M500	VXP250 VXP500	HIG132 HIG300
500–999	VTM-500 VTM-650 VTM-800		M1000	VXP1000 VXP2000 VXP2500	HIG500 HIG700
1,000–1,999	VTM-1000 VTM-1250 VTM-1500	SMD-1100	M3000	VXP5000	HIG900 HIG1100
2,000–2,999			M5000		HIG1600
3,000–3,999	VTM-3000				HIG2300
4,000–4,999	VTM-4500		M10000		HIG3000 HIG3500
5,000–9,999			M15000 M50000		HIG4000 HIG5000

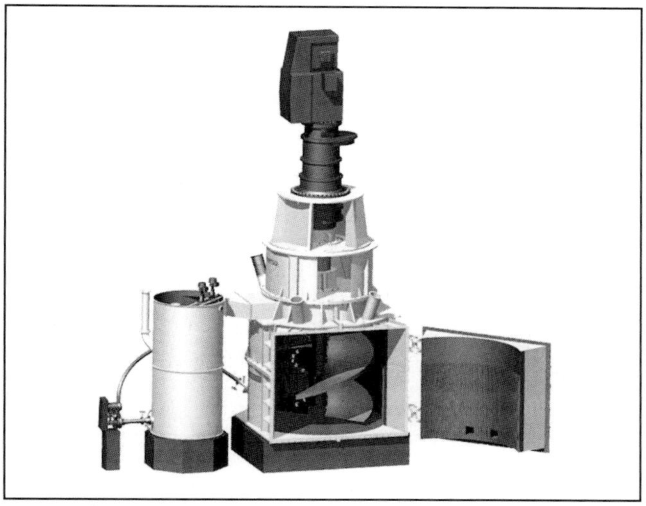

Courtesy of Metso Minerals Industries

Figure 11 Metso Vertimill and feed tank

Courtesy of Metso Minerals Industries

Figure 12 Metso stirred media detritor

Tower mills are commonly used in flotation concentrate regrind applications and are a frequent fixture in base metals flotation circuits. Concentrate regrind duties have a typical feed size of around 100–300 µm (140–50 mesh) and products of 100–15 µm (140 to <635 mesh.) The finest grinds in commercial Vertimill applications achieve approximately 80% passing 12 µm, with finer grinds having been obtained in pilot installations. Energy savings relative to conventional ball mills in regrinding applications can exceed 35%. Media savings relative to ball mills can be 30%–50%, resulting from lower energy consumption and predominance of the attrition grinding mechanism.

The technology has also been extensively used in lime slaking applications, as the grinding mechanism increases the available surface area for reaction and the large mill volume provides sufficient retention time to output a very fine, highly reactive hydrated lime product.

With the advent of 4,000-hp and larger units, Vertimills have also been successfully used in high-capacity secondary and tertiary applications with feed sizes as large 6 mm (¼ in.). In this duty the mills typically use 6–40 mm (¼–1¾ in.) diameter steel balls as grinding media or cast steel cylinders such as cylpebs.

Stirred Media Detritor

The Metso SMD, shown in Figure 12, uses a vertical shaft equipped with pins to provide agitation of the mill charge. The grinding media is ceramic spheres or alluvial silica sand in the range of 1 mm (18 mesh) to 6 mm (¼ in.) in diameter. The agitator speed is high enough to fluidize the media but operates at lower energy intensity when compared to other fluidized media mills to produce lower wear rates at comparable grinding efficiency. Screens are fitted to the upper circumference of the mill body to retain the mill media but otherwise

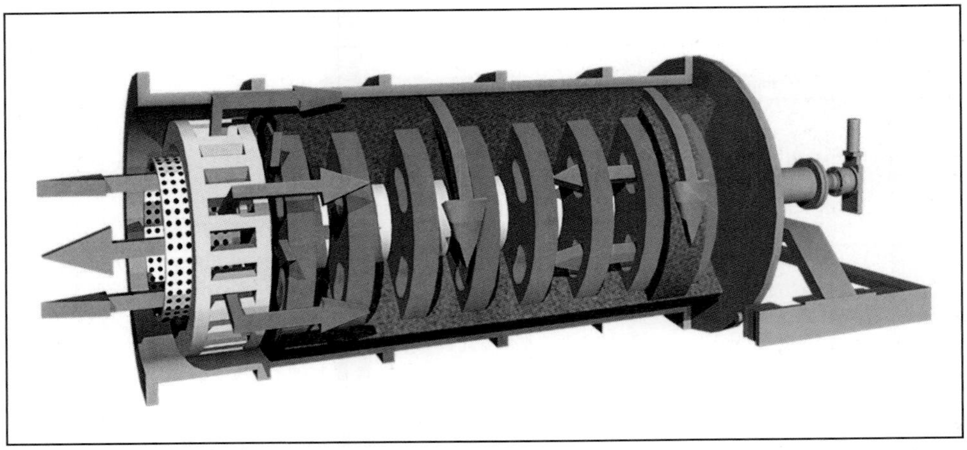

Courtesy of Glencore Technology
Figure 13 IsaMill cutaway schematic

perform no classification of the product material. Feed size is in the range of 100 μm (140 mesh) to <15 μm (635 mesh), and products down to 80% passing 2–6 μm can be produced. The Metso SMD has been applied in lead, zinc, copper, and platinum group mineral flow sheets.

IsaMill

The IsaMill is a horizontal stirred media grinding mill that uses small ceramic grinding media for common regrind and UFG applications (Figure 13). Ceramic media size can range from 1.5 to 6.5 mm (18 mesh to ¼ in.) and accept feeds of up to 80% passing 350 μm in some regrind duties and discharge products as fine as 80% passing 5 μm for UFG duties.

The basic principle of the IsaMill is that it fluidizes a charge of small ceramic grinding media. Media and coarse particles are centrifuged to the outside of each disc where backflow from the product separator keeps them in the mill until the ore is ground enough by attrition and compressive forces to pass down the middle of the mill and discharge. Because of this process, external classification is not always required, but cyclones or thickeners may be required ahead of the mill to achieve a suitable operating density.

The IsaMill comes in a variety of sizes, with the mill model denoted by the chamber volume in liters. For example, the M10000 has an internal volume of 10,000 L or 2,640 gal. IsaMills are installed in several regrind and UFG duties in copper, lead/zinc, platinum, gold, molybdenum, nickel, and iron ore duties, and have been used in mainstream grinding duties at McArthur River mine in Queensland, Australia.

HIGmill

The HIGmill comprises a vertical mill body and shaft fitted with grinding discs (Figure 14). The grinding chamber is filled up to 70% with grinding beads. Rotating discs stir the charge and grinding takes place between beads by attrition. The number of discs is application specific with as many as 30 discs used. Feed slurry is typically processed by a scalping cyclone for the removal of product size material. The scalping cyclone underflow is then pumped into the base of the mill with the slurry flowing upward, through the consecutive grinding stages. Final product discharges at open atmosphere at the top of the machine. In typical cases, the process is single pass and no external classification of the mill discharge slurry is required.

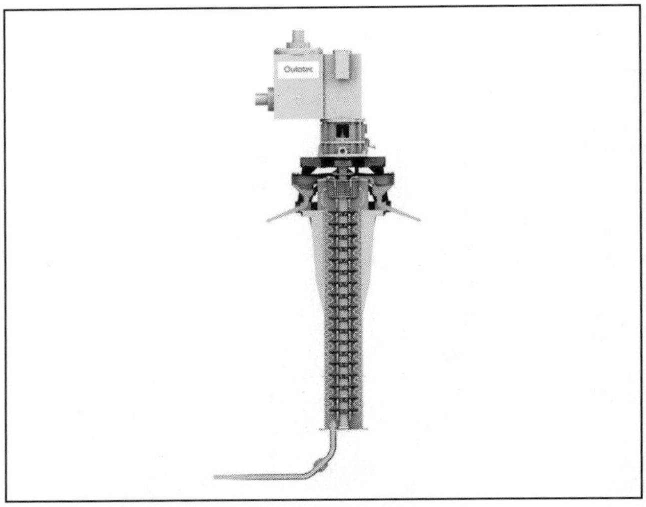

Courtesy of Outotec
Figure 14 Outotec HIGmill

The HIGmill was initially developed for the calcium carbonate industry and has more than 200 installations to date in the industrial minerals area. The technology has been adopted for mineral processing applications by Outotec, which has provided mills for base metal and platinum concentrate regrinding duties.

VXPmill

The FLSmidth VXPmill was originally developed in South Africa for the manufacturing industry and marketed as the Deswik mill before finding application in the gold industry for fine grinding applications (Figure 15). The VXPmill is a vertically oriented stirred mill that is open to the atmosphere. It is designed with a modular impeller that allows the energy distribution within the mill to be varied by changing the disc configuration. The general process flow for the VXPmill is for the slurry to enter at the bottom of the mill, traveling upward through the grinding chamber where the grinding discs agitate the media to induce attrition breakage. The slurry then overflows through a media retention screen at the top of the mill. The mill is also equipped with a variable-speed drive, which allows the mill to be operated over a wide speed range.

The VXPmill uses ceramic grinding media that is selected based on the properties of the mill feed and discharge. The mills are often used in flotation concentrate regrind and precious metals tailings retreatment where the feed size is typically <200 µm. In these applications, a media size of 2–4 mm (5–10 mesh) is commonly used to produce product sizes ranging from 50 µm (270 mesh) to as low as 10 µm (1,250 mesh).

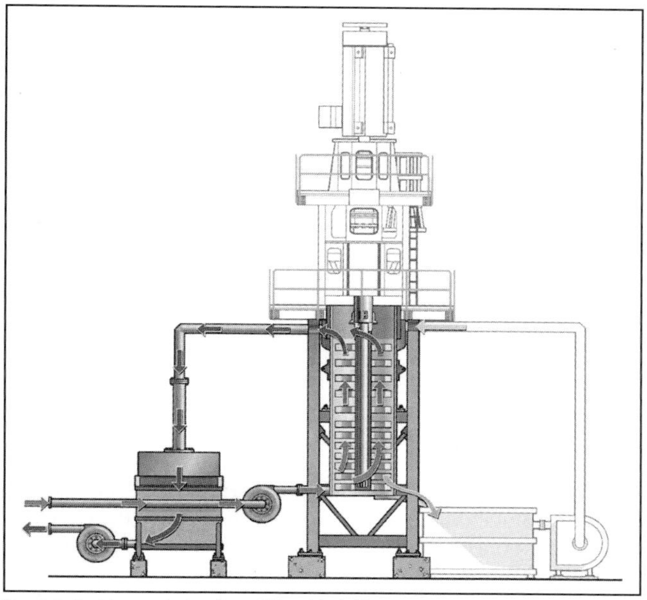

Reprinted with permission from FLSmidth A/S
Figure 15 FLSmidth VXPmill

COMPRESSIVE GRINDING TECHNOLOGIES

Compressive grinding machines are extensively used in the grinding of cement raw materials, cement clinker, slags, coal, coke, electrode materials, paint pigment, clay, intermediate pyrometallurgical products, and industrial minerals. Considerable energy savings have been demonstrated by comparison to dry ball mills operating in these applications, and the high reduction ratios achieved can allow multiple comminution stages to be replaced by a single unit. The abrasive characteristics of the materials ground are low, and demonstrating suitable service life of major wear components in hard-rock grinding applications remains an issue limiting adaptation in mineral processing flow sheets. The successful development of wear lining strategies for HPGRs in tertiary and quaternary hard-rock crushing duties provides reason for optimism that a suitable wear solution can be found for the finish grinding of similar ores. The energy efficiency benefits of dry grinding mineral ores ahead of wet separation or leaching processes can be eroded by the energy requirements of hot gas generation and the power associated with material transport. Therefore, dry grinding technologies of this nature are best suited to ores with low moisture contents and limited abrasiveness, and where subsequent processing stages require a dry product.

Vertical Roller Mills

Grinding occurs by compressive force applied to a grinding bed between fixed grinding rollers, with adjustable shear forces and a rotating horizontal grinding table. Depending on grinding duty and unit capacity, the VRM may employ two, three, four, five, or six grinding rollers. Figure 16 depicts a three-roller Loesche VRM. Ground material flows outward,

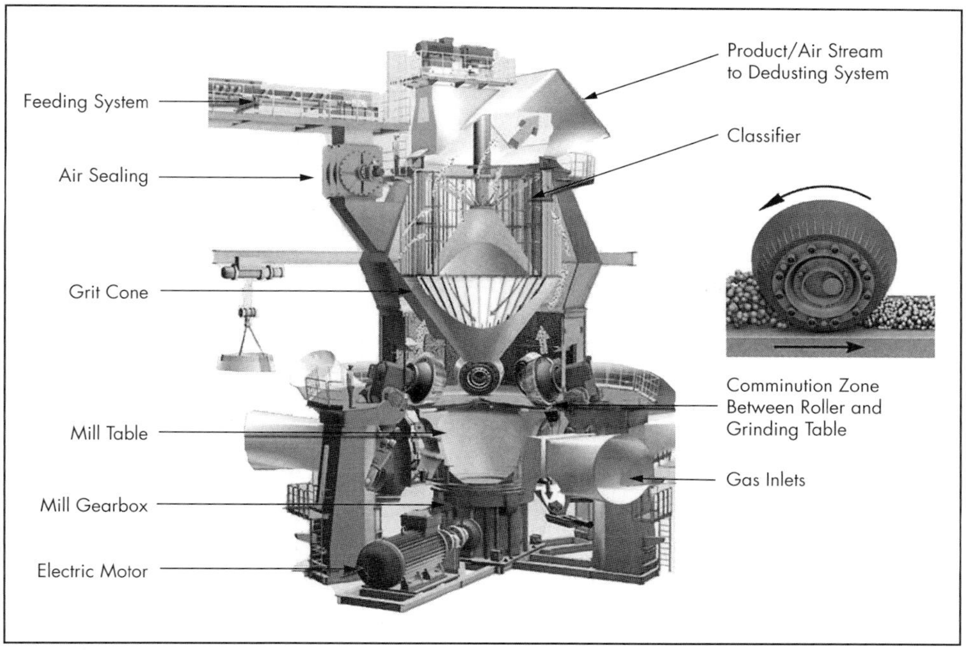

Courtesy of Loesche GmbH
Figure 16 Loesche vertical roller mill

over the edge of the grinding table where the rising gas flow carries finer material to the high-efficiency classification system while allowing coarser, heavier particles to be returned to the grinding table. Very coarse particles leave the mill as reject against the gas flow and are recirculated mechanically by conveyors or bucket elevators. The use of hot gas (recycled or direct heated) can be employed to promote drying of the particles within the grinding chamber. The material ground to specification passes the classifier and is conveyed from the mill with the gas stream to the dedusting process via a baghouse or cyclone baghouse combinations. Classifier coarse fraction or grits are directed back to the grinding table by an internal grit return cone. The whole grinding-classifying circuit is maintained under negative pressure produced by a fan at the end of the process that draws air through the mill and the dedusting process. The ability to combine comminution stages and classification stages results in a significantly reduced plant footprint over conventional crushing and tumbling mill circuits.

Modern VRMs are characterized by a very high reduction ratio. The larger mills can be fed with –150 mm (6 in.) material, and medium sized mills can be fed with –90 mm (3½ in.) material. Grinding down to final product sizes of 80% passing 20 μm (635 mesh) or even finer is possible, although most installations generate product sizes around 80% passing 90 μm (170 mesh). Product size requirements in practice have ranged significantly from 80% passing 600 μm (30 mesh) to 30 μm (450 mesh). A wide variety of machine sizes is available for a given application, with installed power up to 17,400 hp and unit capacities greater than 2,000 t/h.

The Foskor operation in the Phalaborwa region of the Republic of South Africa is a notable application of the VRM technology in finish grinding duty ahead of phosphate flotation. A Loesche LM 50.4 VRM was installed in place of conventional wet rod and ball milling technology in 1999 (Creamer 1999) to dry grind pyroxene ore from a secondary crusher P_{80} of 54 mm (~2 in.) to 80% passing 425 μm (40 mesh). The VRM processes on average more than 500 stph of hard pyroxene ore, of Bond ball mill work index 30 kW·h/t, and operates in parallel with a rod/ball mill circuit receiving tertiary crushed feed at 80% passing 16 mm (⅝ in.).

HPGR Finish and Semifinish Grinding

Burchardt (2014) described the use of HPGRs in the cement industry to grind slag, clinker, and limestone to product sizes of 80% passing 30 μm (450 mesh) to 90 μm (170 mesh), at energy consumptions 30%–50% lower than ball mill grinding systems. This fine grind is reportedly achievable in minerals applications. In case a coarser grind size is required in minerals applications, the "coarse end" of feasible finish grinding products may in fact be as coarse as 80% passing 250–300 μm (nominally 50 mesh). Semifinish grinding can be used to generate an intermediate product sizing less than 1 mm (18 mesh) for grinding in a ball mill or stirred mill. In both finish and semifinish applications, the feed size to the HPGR is around 80% passing 38 mm (1½ in.). Factors that need special considerations in these grinding circuits are feed moisture, material abrasiveness, and throughput capacities. Feed moisture content dictates the need for external dryers and impacts the energy consumption and capital cost of the installation.

A typical finish grinding flow sheet is shown in Figure 17. The key components of the flow sheet are the HPGR, two-stage air classifier (separators), system fan, deduction cyclones, and

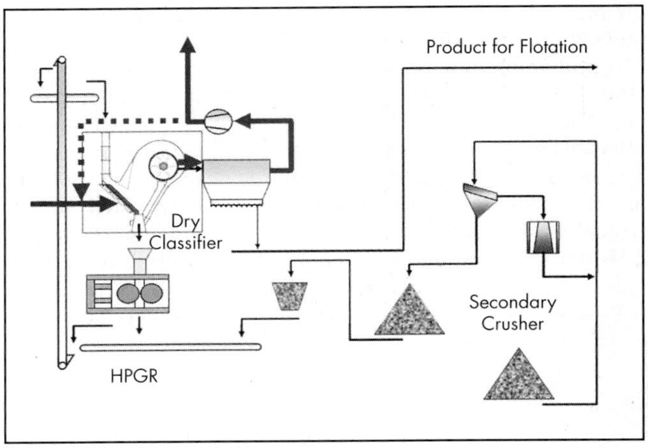

Courtesy of ThyssenKrupp

Figure 17 Finish/semifinish HPGR grinding flow sheet

bucket elevators. New feed containing typically 2%–3% and up to 10% moisture is dried by hot gases in the air classification system, which consists of a static and a dynamic separation stage. The static stage generates an intermediate product that is then classified in the dynamic stage. Rejects from both stages are fed to the HPGR. Finished product is removed via dedusting cyclones or alternatively in a baghouse system. Airflow through the entire system is achieved by the system fan. Bucket elevators, or conveyor belts for high-capacity applications, convey the new feed and HPGR product material to the classification stage. Burchardt (2014) cites the example of the Newmont Minahasa (Indonesia), Newmont Carlin, and Barrick Goldstrike dry grinding applications as evidence that the air classification systems can stand up to the abrasive characteristics of gold ores.

ACKNOWLEDGMENTS

This chapter has leveraged the valuable past contributions to SME publications of several authors, specifically Chester Rowland, Bunting Crocker, Jens Lichter, and Graham Davey, and their works are detailed in the references section at the end of this chapter. Several vendor representatives have provided data, schematics, and technology descriptions that have materially contributed to this compilation, and their efforts are also acknowledged with gratitude. They are Michael Larson (Glencore XT), Gerhard Sauermann and Egbert Burchardt (ThyssenKrupp), Rajiv Kalra and Mahendra Patel (CITIC HIC), Eddie Jamieson (Outotec), Adam Moore and Frank Tozlu (Metso), David Rahal and Jim Cicero (FLSmidth), and Carsten Gerold and Frank Dardemann (Loesche). Dean David of Wood is also acknowledged for his valued input during the review.

REFERENCES

Burchardt, E. 2014. Dry finish grinding with HPGRs: The next step ahead in mineral comminution? Presented at the XXVII International Mineral Processing Congress.

Creamer, M. 1999. Foskor Extension 8. *Min. Weekly*. October 29–November 4, pp. 8–20.

Crocker, B.S. 1985. Pebble mills. In *SME Mineral Processing Handbook*. Edited by N.L. Weiss. Littleton, CO: SME-AIME. pp. 3C94–3C107.

Crow, W.L., and Lipphardt, L.E. 1985. Autogenous mills. In *SME Mineral Processing Handbook*. Edited by N.L. Weiss. Littleton, CO: SME-AIME. pp. 3C57–3C93.

Gouma, J., and Bhasin, A. 1993. Alumina production by Nabalco Pty Limited at Gove, NT. In *Australian Mining and Metallurgy: The Sir Maurice Mawby Memorial Volume*, Vol. 1, 2nd ed. Edited by J.T. Woodcock and K. Hamilton. Melbourne, Victoria: Australasian Institute of Mining and Metallurgy. pp. 763–766.

Jankovic, A. 2005. A review of regrinding and fine grinding technology—The facts and myths. Presented at the IIR Crushing and Grinding Conference, Perth, Australia, March 22–25.

Lichter, J.K.H., and Davey, G. 2002. Selection and sizing of ultrafine and stirred grinding mills. In *Mineral Processing Plant Design, Practice, and Control*. Edited by A.L. Mular, D.N. Halbe, and D.J. Barratt. Littleton, CO: SME. pp. 783–800.

Mular, A.V., and Jergensen, G.V. 1982. *Design and Installation of Comminution Circuits*. Littleton, CO: SME-AIME.

Napier-Munn, T.J., Morrell, S., Morrison, R.D., and Kojovic, T. 1996. *Mineral Comminution Circuits: Their Operation and Optimisation*. Indooroopilly, Queensland: Julius Kruttschnitt Mineral Research Centre.

Rowland, C.A. 1985. Rod mills. In *SME Mineral Processing Handbook*. Edited by N.L. Weiss. Littleton, CO: SME-AIME. pp. 3C44–3C56.

Rowland, C.A. 2002. Selection of rod mills, ball mills and regrind mills. In *Mineral Processing Plant Design, Practice, and Control*. Edited by A.L. Mular, D.N. Halbe, and D.J. Barratt. Littleton, CO: SME. pp. 710–754.

Singh, G., and Sisley, M. 2013. Queensland alumina refinery. In *Australian Mining and Metallurgical Operating Practices: The Sir Maurice Mawby Memorial Volume*, 3rd ed. Edited by W.J. Rankin. Melbourne, Victoria: Australasian Institute of Mining and Metallurgy. pp. 299–306.

Skiba, G.R. 1993. Alumina production by Queensland Alumina Ltd at Gladstone, QLD. In *Australian Mining and Metallurgical Operating Practices: The Sir Maurice Mawby Memorial Volume*, Vol. 1, 2nd ed. Edited by J.T. Woodcock and K. Hamilton. Melbourne, Victoria: Australasian Institute of Mining and Metallurgy, pp. 766–768.

Tempel, T.A. 1993. Selection of a dry grinding process, Newmont Gold Company, refractory ore plant. SME Preprint No. 93-264. Littleton, CO: SME.

Thomas, K.G.L., Buckingham, L., and Palzelt, N. 2001. Dry grinding at Barrick Goldstrike's roaster facility. In *SAG 2001 Conference Proceedings*. Vol. 1. Vancouver, BC: Mining and Mineral Processing Engineering, University of British Columbia. pp. 174–188.

CHAPTER 3.12

Testing and Calculations for Comminution Machines

Stephen Morrell

This chapter deals with the testing of rocks to determine their hardness/strength and the subsequent use of the test results in calculations that, in design, lead to the choice of an appropriately sized comminution machine. The equipment and flowsheet selection processes are discussed in the relevant chapters of this handbook. The usual calculation route to choose the comminution machine incorporates the following steps:

1. Rock samples are laboratory tested to obtain hardness parameter values.
2. An equation (or equations) that uses the hardness parameter values is applied to estimate the full-scale machine specific energy (kilowatt-hours per metric ton).
3. Given the required throughput capacity (metric tons per hour) of the comminution machine, its required power draw is estimated by multiplying the specific energy by the throughput capacity. Contingencies (safety factors) are then applied to this power draw.
4. A comminution machine is chosen that can deliver the required power. In the case of milling, this step requires the application of equations that relate dimensions and operating conditions to power draw (power model) so that the correct size of mill is chosen.

It follows from the preceding steps that a laboratory rock hardness test is of little value without an associated equation(s) that enables the prediction of the comminution machine specific energy. Also, it is paramount that the resultant predictions are accurate. For this to be established, extensive benchmarking is required in which predictions are compared with actual (observed) values from a wide range of operating full-scale plants.

In the following sections, several laboratory rock hardness testing methods will be described together with their associated specific energy equations. An overview of commonly used hardness test methods is provided in Table 1. The final section of this chapter will address tumbling mill power draw modeling.

Stirred mills are not included in this chapter, although are reviewed in Chapter 3.11, "Grinding Technologies." Their testing procedures and associated scale-up techniques and equipment sizing are specific to vendors and their particular mill design.

BOND

Fred C. Bond was the first person to develop laboratory tests and associated equations that could be applied to practical design situations. Interestingly, Bond never published any data to support the validity of his approach, though Blaskett (1969), Rowland (1973), and Moore (1982) did. Bond did conduct experiments on a larger scale to collect data with which he could test and tune at least some of his equations. However, the following quote from his 1961 Allis-Chalmers publication indicates that perhaps not all of his equations were rigorously derived from data: "Many (equations) are empirical, with numerous constants being derived from experience."

Bond's approach covers conventional crushers, and rod and ball mills. However, in his day, autogenous grinding and semiautogenous grinding (AG/SAG) mills and high-pressure grinding rolls (HPGRs) were not common or not yet invented, and hence his approach does not explicitly encompass these machines. Proprietary techniques, however, have been developed that have adapted his equations for use with AG/SAG mills, and some of these are described in the following sections.

Bond Laboratory Tests

Bond developed three laboratory rock characterization test procedures, each one associated with a particular comminution machine (i.e., crushers, rod mills, and ball mills). All the tests require relatively small amounts of material and hence are readily suited for use with drill core samples, as proposed in Table 1. However, ideally the crushing test needs whole PQ core. This tends to be less common (and more expensive) than NQ and HQ cores, which are suitable for use in Bond's rod and ball mill tests.

Crushing Test

For crushing applications, Bond developed an apparatus comprising two opposing 30-lb hammers that came together through the action of two counterrotating wheels (Bond 1946,

Stephen Morrell, President, SMC Testing Pty Ltd., Brisbane, Queensland, Australia

Table 1 Laboratory tests for comminution circuit design

Test	Use	References	Sample Quantity, kg	Material Size	Minimum Core Size
Bond impact crushing test	Conventional crushers	Bond 1946, 1961	40–50	2–3 in.	PQ
	SAG mills	Barratt and Allan 1986			
Bond rod mill work index	Rod mills / ball mills	Bond and Maxon 1943; Bond 1961; Rowland 1982	10–15	100% – 12.7 mm	NQ
	SAG mills	Barratt and Allan 1986			
Bond ball mill work index	Ball mills	Bond and Maxon 1943; Bond 1961; Rowland 1982	15	100% – 3.35 mm	NQ
	SAG mills	Barratt and Allan 1986			
Bond abrasion index	Wear prediction for crushers and tumbling mills	Bond 1963; Giblett and Seidel 2011	2	12–19 mm	NQ
SMC Test	Conventional crushers, AG/SAG mills, HPGRs	Morrell 2004a–c, 2009	20	19–31 mm	NQ
JK drop weight test	SAG mills, conventional crushers	Napier-Munn et al. 1996	75	13.2–63 mm	PQ
SPI Test	SAG mills	Starkey et al. 1994	10	80% – 12.7 mm	NQ
SAGDesign test	SAG mills	Starkey and Larbi 2012	15	80% – 19 mm	NQ

Source: Giblett and Morrell 2016

1961). The hammers were made of steel with a Brinnell hardness of 230. Each hammer was 2 in. in cross section and 28 in. long. The hammers were mounted on the wheels so their centers at rest were 16 in. below the axis of rotation of the wheels. The wheels were 22-in. front bicycle wheels, reinforced with steel bands around each rim. The left-hand wheel was initially rotated clockwise and, through a connecting device, the right-hand wheel was synchronized and rotated counterclockwise by a similar amount. As the hammers were attached to the wheels, the hammers were lifted up by this rotation, and when the wheels were released, the hammers would collide with one another at the 6 o'clock position. By measuring how high the hammers were raised in relation to their final rest position, the potential energy could be estimated. A rock specimen was mounted on a plinth at the point of collision of the hammers and hence was broken with an amount of energy equivalent to the estimated potential energy. Figure 1, taken from a modern machine that is faithful to Bond's original design, shows where the hammers collide.

Bond specified that only particles in the size range of –3 in. + 2 in. should be tested in this machine. Particles outside of this range should not be used. Bond (1946) also stated that slabby or acicular pieces should not be used. He recommended that a minimum of 10 suitable rocks should be broken. Mounting of the rocks was to follow this procedure: "If the longest dimension is designated as A, the longest dimension perpendicular to A as B and the longest dimension perpendicular to both A and B as C; the specimen is placed in the holder in such a position that the hammers strike on both sides of dimension C." For breaking each specimen, Bond (1946) quoted the following procedure: "The first piece is tested with a low-energy blow and the height of fall is gradually increased until the specimen breaks into two or more pieces of approximately equal size. Each succeeding piece is first tested with an energy slightly under that required to break the preceding piece and the height of fall is increased so that the specimen is broken after two or three blows." The aim of the test is therefore to try to measure the energy needed to just break the rock. The so-called crushing work index (W_{iC}) is then calculated using Equation 1, based on the average result from the 10 rocks.

Courtesy of Amdel Mineral Laboratories

Figure 1 Region where breakage takes place in Bond's crushing work index machine

$$W_{iC} = 2.85C/sg \qquad \textbf{(EQ 1)}$$

where
W_{iC} = crushing work index, kW·h/t
$C = E/D$
E = breakage energy, ft-lb
D = rock thickness, in.
sg = specific gravity

Recently, some laboratories have started to use a machine with a design different from that developed by Bond (Figure 2). Experience has shown that machine design influences the outcome of breakage testing devices, and some data have suggested that this machine may not give the same results as Bond's original design (Bailey et al. 2009). Caution should

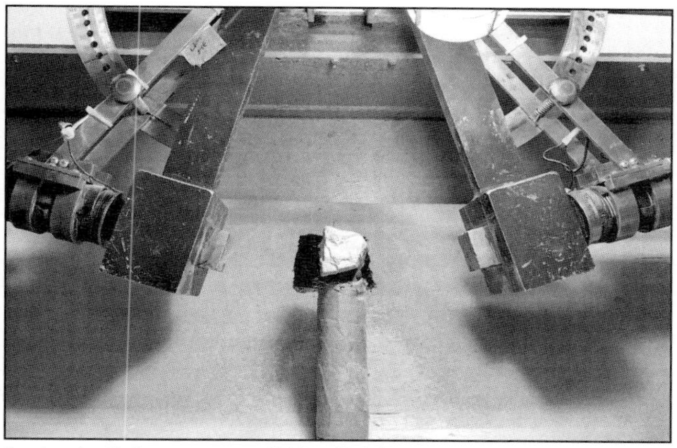

Reprinted with the permission of FLSmidth A/S
Figure 2 Modern version of Bond's crushing work index machine

Courtesy of BICO Braun International
Figure 4 Bond ball mill

Courtesy of JKTech Pty Ltd.
Figure 3 Bond rod mill

therefore be exercised if tests are conducted on machines that are not entirely faithful to Bond's original design.

Rod Milling Test

For rod milling, Bond (1961) developed a dry locked-cycle test that uses a 12-in.-diameter, 24-in.-long batch mill with wave liners and running at 46 rpm (see Figure 3). The mill is mounted on a rocker, which allows for the mill to be tilted up by 5° to the horizontal and down by 5°. The mill is charged with a specified quantity of rods of given sizes. The rock being characterized is stage-crushed to 100% passing 12.7 mm and ground, dry, in closed circuit with a screen of aperture P_1 such that the recycle load is 100%. Depending on the size of the starting material, stage crushing may need to involve an initial relatively coarse crush of the sample using a crusher setting of about 25 mm. The crusher is then set to 12–15 mm and the sample is passed through, with the product screened on a 12.7-mm sieve. The screen oversize is then passed again through the crusher, screened again at 12.7 mm, and so on, until the entire sample has passed through the 12.7-mm sieve. The net grams of final product produced when the recycle load is 100% is measured and inserted into Equation 2 to obtain the rod mill work index, W_{iR}. By inserting this value into Equation 4, Bond claimed that the specific energy of a wet 8-ft overflow rod mill in open circuit could be obtained.

$$W_{iR} = \frac{68}{P_1^{0.23}(\text{Grp})^{0.625}10\left(\frac{1}{\sqrt{P}} - \frac{1}{\sqrt{F}}\right)} \quad \textbf{(EQ 2)}$$

where

W_{iR} = bond laboratory rod work index, kW·h/t
P_1 = closing screen size, μm
Grp = net grams of screen undersize per mill revolution
P = 80% passing size of the product, μm
F = 80% passing size of the feed, μm

Ball Milling Test

For ball milling, Bond (1961) developed a dry locked-cycle test that uses a 12-in.-diameter, 12-in.-long batch mill with rounded corners and smooth liners and running at 70 rpm (see Figure 4). The mill is charged with a specified quantity of balls of given sizes. The rock being characterized is stage-crushed to 100% passing 3.35 mm and ground, dry, in closed circuit with a screen of aperture P_1 such that the recycle load is 250%. Depending on the size of the starting material, stage crushing may need to involve an initial relatively coarse crush of the sample using a crusher setting of about 25 mm. The crusher is then set to 3–4 mm and the sample is passed through, with the product screened on a 3.36-mm sieve. The screen oversize is then passed again through the crusher, screened again at 3.36 mm, and so on, until the entire sample has passed through the 3.36-mm sieve. The net grams of final product produced when the recycle load is 250% is measured and inserted into Equation 3 to obtain the ball mill work index. By inserting this value into Equation 4, Bond claimed that the specific energy of a wet 8-ft overflow ball mill in closed circuit could be obtained. Bond recommended that the screen aperture (P_1)

chosen to close the test with should be such that it gives a final product P_{80} similar to that being targeted in the full-scale plant. Typically, the final product P_{80} is of the order of $1/\sqrt{2}$ of the closing screen aperture.

$$W_{iB} = \frac{49}{P_1^{0.23} (Gbp)^{0.82} 10\left(\frac{1}{\sqrt{P}} - \frac{1}{\sqrt{F}}\right)} \quad \textbf{(EQ 3)}$$

where W_{iB} is the Bond laboratory ball work index, in kW·h/t.

General Equations

Bond's general equation to predict the specific energy requirement for crushers, rod mills, and ball mills to reduce a feed with a specified F_{80} to a product with a specified P_{80} is as follows (Bond 1952):

$$W = 10W_i (P^{-0.5} - F^{-0.5}) \quad \textbf{(EQ 4)}$$

where

W = specific motor output energy, kW·h/t

W_i = work index, kW·h/t, as determined from the relevant laboratory test

Bond (1952) stated that Equation 4 "… conforms with the motor output power to an average overflow ball mill of 8 ft interior diameter grinding wet in closed circuit…."

Having developed Equation 4, Bond proceeded to modify it with a range of efficiency factors (EFs) that attempted to address the differences in feed and operating conditions of different circuit designs and ore types. These EFs applied to rod and ball mill circuits only. Using Rowland and Kjos's (1978) description, the EFs are as follows:

EF1—Dry grinding (rod and ball mills):

EF1 = 1.3 for dry grinding

EF2—Open-circuit milling (ball mills only):

EF2 = 1.2 for open-circuit grinding

EF3—Mill diameter (rod and ball mills):

$$EF3 = (8/D)^{0.2} \quad \textbf{(EQ 5)}$$

where D is the mill diameter, in ft, inside liners.

The diameter factor (see Equation 5) is particularly interesting (and much debated). Bond introduced it because of what he stated was an improvement in mill energy efficiency as the diameter increased above 8 ft. However, his argument as to the reason for this improvement is tenuous, to say the least, as it is based on his assertion in his 1961 paper that mill power draw increases on the basis of $Diameter^{2.4}$ and throughput capacity increases on the basis of $Diameter^{2.6}$. He concluded that the difference in these diameter exponents of 2.6 and 2.4 was 0.2, and this meant that the kilowatt-hours per ton required to grind decreases on the basis of $Diameter^{0.2}$. Interestingly, in 1962 Bond changed his mill power draw exponent from 2.4 to 2.3 but did not change his diameter efficient factor to 0.3, as would be expected from his arguments.

Rowland (1972) modified the application of EF3 and stated that it should be used in mills up to 12.5 ft and then kept at the 12.5-ft value for all larger mills (i.e., EF3 = 0.914 for mills larger than 12.5-ft diameter). The capacity problems experienced by Bougainville Copper (Steane and Hinckfuss 1979) brought into review the efficiency of larger mills, and at the time, some placed blame on the size of the mill, as it was one of the biggest diameter mills of its day (18 ft). This led to the belief by some that mills with larger diameters were in fact less efficient than smaller ones, in direct contrast to Bond's assertion. However, by 1988, ball mills had reached 21 ft with no apparent diameter-related problems (Forsund et al. 1988). Most likely, there is no change in the energy efficiency of mills as they increase in diameter, as Morrell (2001) demonstrated using data from a range of mill sizes. This finding is confirmed by the data in Figure 5, which shows the results of predicting ball mill specific energy using the Bond ball mill work index and Bond's equation (with no correction factors). The figure shows the error of the predicted specific energy as compared to the observed values from 45 operating ball mills with diameters in the range of 12 to 26 ft. If diameter did influence the specific energy efficiency, then there should be a significant trend in the error with respect to mill diameter, hence suggesting that a diameter efficiency factor is warranted. This is not apparent in the data, and therefore the hypothesis that mill diameter influences energy efficiency is not supported, at least within the range of 12 to 26 ft.

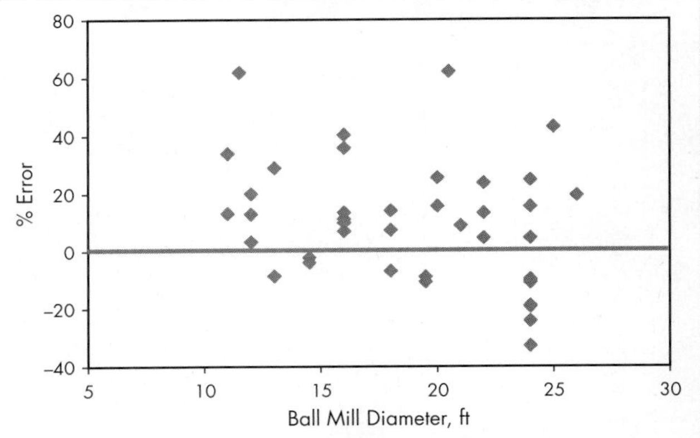

Figure 5 Observed error in ball mill specific energy using Bond's equation versus mill diameter

EF4—Oversize feed (rod and ball mills):

$$EF4 = (R_r + (W_i/1.1 - 7) \times ((F_{80} - F_o)/F_o))/R_r \quad \textbf{(EQ 6)}$$

where

R_r = reduction ratio F_{80}/P_{80}

$F_o = 16{,}000 \times (14.3/W_i)^{0.5}$ for rod milling circuits **(EQ 7)**

$= 4{,}000 \times (14.3/W_i)^{0.5}$ for ball milling circuits **(EQ 8)**

W_i = bond laboratory work index, in kW·h/t; use rod mill work index for rod milling circuits and ball mill work index for ball mill circuits

applied only where $F_o < F_{80}$.

EF5—Fineness of grind (ball mills only):

$$EF5 = (P_{80} + 10.3)/(1.145 \times P_{80}) \quad \textbf{(EQ 9)}$$

applied when grinding finer than a P_{80} of 75 µm.

EF6—High or low reduction ratio (rod milling only):

$$EF6 = 1 + (R_r - R_{ro})^2/150 \quad \textbf{(EQ 10)}$$

where

R_r = reduction ratio F_{80}/P_{80}

$R_{ro} = 8 + 5L/D$ **(EQ 11)**

L = rod length

D = mill diameter

applied only when R_r falls outside of the range $R_{ro}\pm2$.

EF7—Low reduction ratio (ball milling only):

$$EF7 = (2 \times (R_r - 1.35) + 0.26)/(2 \times (R_r - 1.35)) \quad \textbf{(EQ 12)}$$

where R_r is the reduction ratio F_{80}/P_{80}—applied only when R_r is less than 6.

EF8—Rod mill feed size distribution (rod mills only):

EF8 = 1.4 in rod-mill-only circuits fed with feed from an open crushing circuit
= 1.2 in rod-mill-only circuits fed with feed from a closed crushing circuit
= 1.2 in rod-mill-only circuits fed with feed from an open crushing circuit
= 1.0 in rod-mill-only circuits fed with feed from a closed crushing circuit

Equation Accuracy

Rod and Ball Mills

Unfortunately, there is little published data on the accuracy of Bond's original equations or the modifications to them that Rowland (1972) subsequently introduced. Blaskett (1969) provided some relevant data for relatively small ball mills in the diameter range of 5 to 10.5 ft, and Rowland (1973) also published some data on rod mills in the range of 9 to 12 ft and ball mills in the range of 9 to 19.5 ft. These data are shown in Figure 6. The Blaskett ball mill results indicate one data set in particular that has problems. This data set comes from a primary ball mill grinding from 11.6 mm to a relatively coarse 630 µm. The resultant EF4 (coarse grinding factor) was an extremely large 2.35, and hence the predicted power requirement was also relatively large and did not match the relatively small energy requirement observed in the plant. If the mill had been grinding to a much finer 150 µm, for example, the associated EF4 factor would be only 1.3, which seems counterintuitive, as it suggests that the mill would become more energy efficient if it were to grind finer. It is possible, therefore, that there is a problem with the EF4 factor where the grind is relatively coarse.

If Rowland's data only in Figure 6 are considered, the standard deviation of the differences between observed and predicted specific energies is 9.3% with a mean difference of 1.8%. If Blaskett's data are added (and ignoring the one problematic data set), this standard deviation increases to 25.6% with an overall mean of differences of 16.5%.

In the case of the rod mill data, the standard deviation of the differences between observed and predicted specific energies is 12.6% with an overall mean of differences of 21.8%.

Crushing

In the case of crushing, Moore (1982) published several data sets that show the approach has very poor accuracy (see Figure 7) and confirms Flavel and Rimmer's (1981) similar conclusions from a study they conducted while working for Allis-Chalmers (where Bond worked when he originally developed the test). Flavel and Rimmer claimed that the reason for the inaccuracy was that Bond originally developed his

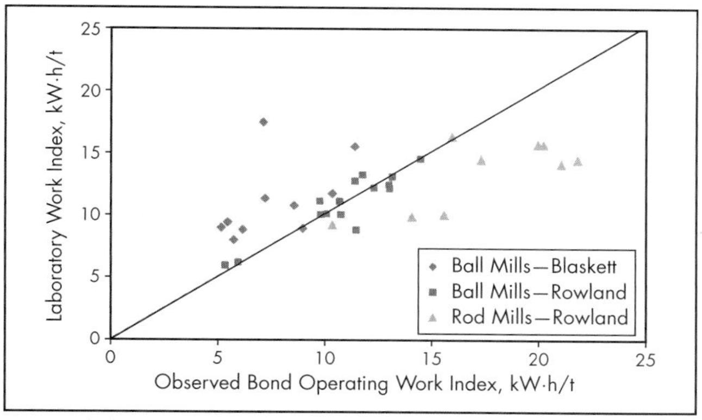

Source: Blaskett 1969; Rowland 1973

Figure 6 Bond laboratory rod and ball mill work index versus observed operating work index

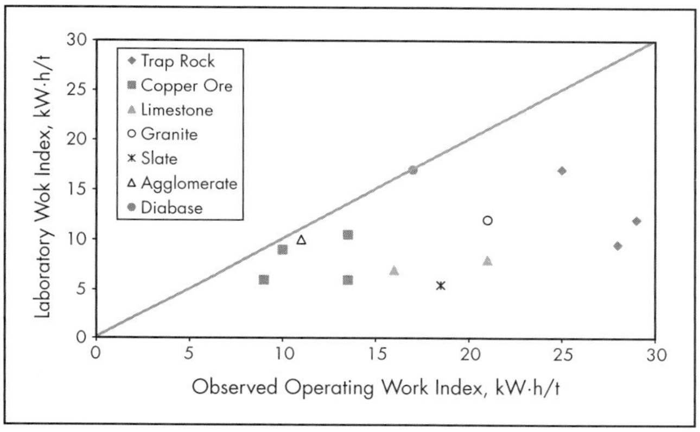

Source: Moore 1982

Figure 7 Bond laboratory crushing work index versus observed operating crushing work index

crushing work index test for primary crushers in which the applied specific energy is low (~0.1 kW·h/t), but it was not suitable for secondary and tertiary crushers where the applied specific energy was much higher (>0.25 kW·h/t). A contributing cause is the inherent very poor reproducibility of the test as reported by Angove and Dunne (1997) in a study they conducted in which three different samples were sent to several laboratories for Bond crushing work index testing. Results indicated a huge variation in work index values, with maximum values consistently almost double the minimum values. This variation is suspected to be caused by both the lack of standardization in equipment used by different laboratories and the operator sensitivity inherent in the use of the machine.

Analysis of the data in Figure 7 indicates that the standard deviation of the differences between observed and predicted values is 22.9% with an overall mean of differences of 39.9%.

Equipment and Procedural Considerations with Bond Tests

Several factors should be addressed when conducting Bond tests to ensure that the most accurate and repeatable results are obtained. These factors can be divided into two principal categories: (1) those related to the equipment and (2) those related to how the tests are conducted.

Equipment factors relate to the design/operation of the equipment. Hence the equipment being used should conform to Bond's specification. A check of the following should help ensure that the most important specifications have been met:

- Crushing tester generally conforms to the design in Figure 1.
- Crushing hammers are the correct mass and shape, and made of the specified material (hardness).
- Ball/rod mill dimensions are correct.
- Rotational speeds are correct.
- Rods are the correct sizes and weight/number per size class.
- Balls are the correct sizes and weight/number per size class.
- Ball mill has curved edges and an access door that closes flush with the inside of the mill.
- Rod mill has liners, and these are the correct thickness and shape.
- Ball mill has no liners.
- Rod mill has a rocker system to allow for the mill to be tilted by 5° up and down.

In terms of how the rod and ball mill tests should be conducted, Mosher and Tague (2001) provide particularly good guidance. Based on their advice and other experience, checking the following should ensure good operational procedures:

- Crushing work index rock specimens must be in the size range of 2 to 3 in.
- Ball mill closing screen aperture (P_1) has been chosen such that it gives a final product P_{80} similar to that being targeted in the full-scale plant. Typically, the final product P_{80} is of the order of $1/\sqrt{2}$ of the closing screen aperture.
- Feed samples have been generated by riffling from the master sample that has been correctly stage-crushed. Depending on the size of the starting material, this may need to involve an initial relatively coarse crush of the sample using a crusher setting of about 25 mm. The crusher is then set to just over the top size required for the test and the sample is passed through, with the product being screened on the required sieve. The screen oversize is then passed again through the crusher, screened again on the sieve, and so on, until the entire sample has passed through the sieve.
- Sizing of feed and products has been done on a full set of fourth root of two sieves to ensure accurate measurement of the F_{80} and P_{80}.
- The final three cycles have shown a reversal in the Grp/Gbp trend and are within 3% of one another.

Worked Examples

The following worked examples have been taken from Rowland and Kjos (1978) to illustrate the application of Bond's approach. The following Bond test data are provided:

- Rod work index = 14.52 kW·h/t
- Ball work index = 12.87 kW·h/t at a closing screen size of 250 μm

Two circuits are considered, a rod mill–ball mill circuit and a single-stage ball mill circuit fed from a crushing circuit, both with a final product size P_{80} of 175 μm. The examples consider the determination of the milling circuit specific energy.

Rod Mill–Ball Mill Circuit

Rowland and Kjos (1978) specified that the rod mill would be fed from a closed-circuit crushing plant producing a P_{80} of 18 mm and further specified that the rod mill would grind to a P_{80} of 1,200 μm. Considering the rod mill circuit:

$$W = 10 \times 14.52 \times (1{,}200^{-0.5} - 18{,}000^{-0.5}) \text{ kW·h/t (uncorrected)} = 3.11 \text{ kW·h/t}$$

EF1, EF2, EF5, EF7, and EF8 do not immediately apply. EF3 and EF6 need to be determined after a mill size has been selected (see later), leaving EF4 to be determined at this stage.

$$R_r = 18{,}000/1{,}200 = 15$$

$$F_o = 16{,}000 \times (14.3/14.52)^{0.5} = 15{,}878$$

As $F_o < 18{,}000$, then EF4 is applied:

$$EF4 = \left(15 + \left(\frac{14.52}{1.1}\right) - 7\right) \times \frac{\left(\frac{18{,}000 - 15{,}878}{15{,}878}\right)}{15} = 1.06$$

Rowland and Kjos select 11.35-ft-diameter mills inside liners, and hence for EF3:

$$EF3 = (8/11.35)^{0.2} = 0.93$$

They then select 17-ft-long mills using 16.5-ft rods. Hence, evaluating EF6:

$$R_{ro} = 8 + 5 \times 16.5/11.35 = 15.3$$

As this value is within the limits $R_r \pm 2$ (15 ± 2), EF6 does not apply. The final (corrected) rod mill specific energy is given by

$$\text{specific energy} = W \times EF3 \times EF4 = 3.11 \times 0.93 \times 1.06 \text{ kW·h/t} = 3.07 \text{ kW·h/t}$$

For the ball mill circuit: The ball mill circuit feed is the product of the rod mill, and hence the ball mill F_{80} is 1,200 μm. The target product P_{80} is 175 μm and hence,

$$W = 10 \times 12.87 \times (175^{-0.5} - 1{,}200^{-0.5}) \text{ kW·h/t (uncorrected)} = 6.01 \text{ kW·h/t}$$

Efficiency factors EF1, EF2, EF5, EF6, EF7, and EF8 do not apply.

Rowland and Kjos chose a mill with a diameter greater than 12.5 ft, so EF3 applies but is fixed at 0.914. The application of EF4 has to be evaluated by reference to F_o. Hence,

$$F_o = 4{,}000 \times (14.3/12.87)^{0.5} = 4{,}216 \text{ μm}$$

As $F_o > 1{,}200$, then EF4 is not applied. The final (corrected) ball mill specific energy is then given by

$$\text{specific energy} = W \times EF3 = 6.01 \times 0.914 \text{ kW·h/t} = 5.5 \text{ kW·h/t}$$

Single-Stage Ball Mill Circuit

Rowland and Kjos (1978) specified for this worked example that the ball mill would be fed by a closed-circuit crusher producing a product with a P_{80} of 9.4 mm, and hence the ball mill circuit F_{80} would be 9,400 µm. Rowland (1972) asserts that in cases where the ball mill circuit receives coarser feed, some of which is larger than 3.35 mm (top size in the Bond ball mill work index test), for this fraction the rod mill work index is more relevant. In such cases, he therefore divides the ball mill specific energy calculation into two steps. The first step uses the rod mill work index from the ball mill circuit F_{80} down to 2,100 µm, while the second step applies the ball mill work index from 2,100 µm down to the final grind. Hence,

Step 1

$$W = 10 \times 14.52 \times (2{,}100^{-0.5} - 9{,}400^{-0.5})\ \text{kW·h/t (uncorrected)} = 1.67\ \text{kW·h/t}$$

Step 2

$$W = 10 \times 12.87 \times (175^{-0.5} - 2{,}100^{-0.5})\ \text{kW·h/t (uncorrected)} = 6.92\ \text{kW·h/t}$$

This totals 8.59 kW·h/t. The efficiency factors EF1, EF2, EF5, EF6, EF7, and EF8 do not apply.

Rowland and Kjos chose a mill with a diameter greater than 12.5 ft, so EF3 applies but is fixed at 0.914. The application of EF4 must be evaluated by reference to F_o. Hence

$$F_o = 4{,}000 \times (14.3/14.52)^{0.5}$$
(note that the rod mill work index is used in this case)
$$= 3{,}970\ \mu m$$

As $F_o < 9{,}400$, then EF4 is applied.

$$R_r = 9{,}400/175 = 53.7$$

$$EF4 = (53.7 + (12.87/1.1 - 7) \times ((9{,}400 - 3{,}970)/3{,}970))/53.7 = 1.12$$

The final (corrected) ball mill specific energy is then given by

$$\text{specific energy} = W \times EF3 \times EF4 = 8.59 \times 0.914 \times 1.12\ \text{kW·h/t} = 8.79\ \text{kW·h/t}$$

BOND TESTS FOR AG/SAG CIRCUITS

Bond did not specifically develop tests for predicting the specific energy of AG/SAG circuits. However, over the years, some have attempted to adapt his techniques to overcome this deficiency. One such approach was developed by Barratt and Allan (1986) in which they used a combination of the standard Bond crushing, rod mill, and ball mill work index tests.

Equations

The equation Barratt and Allan (1986) developed for AG/SAG mill circuits is as follows:

$$\begin{aligned} E_{SAG} = 1.25\cdot[&(10\cdot W_{iC}\,(P_C^{-0.5} - F_C^{-0.5})) \\ &+ (10\cdot W_{iR}\,(P_R^{-0.5} - F_R^{-0.5})\cdot K_R) \\ &+ (10\cdot W_{iB}\,(110^{-0.5} - F_B^{-0.5})\cdot K_B)] \\ &- (10\cdot W_{iB}\,(110^{-0.5} - T_{SAG}^{-0.5})\cdot K_B) \end{aligned} \quad \textbf{(EQ 13)}$$

The equation for the associated ball mill circuit is

$$E_{BM} = 10\cdot W_{iB}\,(P_B^{-0.5} - T_{SAG}^{-0.5})\,K_B \quad \textbf{(EQ 14)}$$

where

E_{SAG} = specific energy of the SAG mill circuit
E_{BM} = specific energy of the ball mill circuit
W_{iC}, W_{iR}, W_{iB} = bond crushing, rod, and ball work indices, respectively
P_C, P_R, P_B = 80% passing size of the product of the stage associated with crushing, rod milling, and ball milling, respectively
F_C, F_R, F_B = 80% passing size of the feed of the stage associated with crushing, rod milling, and ball milling, respectively
K_R = composite of the rod mill (EF) factors (Rowland 1982), excluding the diameter factor
K_B = composite of the ball mill (EF) factors (Rowland 1982), excluding the diameter factor
T_{SAG} = transfer size (80% passing) between the AG/SAG and ball mill circuits

All P and F values as well as the T_{SAG} must be specified/estimated for Equations 13 and 14 to be applied. The value used for T_{SAG} was recognized by Barratt and Allan (1986) as critical to the prediction of the SAG mill circuit specific energy (and by inference also the ball mill circuit specific energy).

Validation

While the method has been used in support of grinding circuit design and optimization on several projects, no definitive published validation data on the method is available yet.

MORRELL

Morrell, like Bond, also developed ore characterization tests (Morrell 2004a) as well as associated equations (Morrell 2004a, 2009, 2010). These enabled the application of the test results in design situations using both power-based equations and simulation modeling. In the course of development, Morrell collected large quantities of operational plant data with which to ensure the validity and accuracy of his approach. Unlike Bond's approach, which does not explicitly cover AG/SAG and HPGR circuits, Morrell's approach does.

Morrell Laboratory Tests

Morrell developed one new laboratory test, the SMC Test, which covers the size reduction in crushers, HPGRs, rod mills, and AG/SAG mills, and adapted Bond's ball mill laboratory work index test to cover ball mills.

SMC Test

The SMC Test was developed by Morrell to provide a range of comminution parameters from the breakage of relatively small amounts of small-diameter drill core. SMC is an acronym for Steve Morrell comminution (not SAG mill comminution as is incorrectly stated in some published literature). SMC Testing Pty Ltd. currently owns the SMC Test.

Normally, 15–20 kg of drill core (or rocks) are required by the laboratory to conduct the test (though under some conditions, much less can be used), from which 1–5 kg is extracted for the actual test depending on the size fraction chosen to do the test with. This material can be added back to the original sample and reused to conduct a standard Bond ball work index test.

The test uses the JK drop weight testing device to break a suite of 100 closely sized particles at five different energy levels. The particle size used for the test is chosen from one of the following three fractions: –31.5 + 26.5 mm, –22.4 + 19 mm, or –16 + 13.2 mm. The choice of which size fraction to use depends on the source material; for example, if only quartered (slivered) BQ (38 mm) core is available, then the smallest size fraction would be chosen, while if larger diameter core were available, then the coarser size fractions could be used. In general, where sample size/quantity permits, the largest size fraction should be used.

The particles for the test can be obtained using two different feed preparation routes. If a relatively small quantity of drill core sample is available (i.e., much less than the usual 15–20 kg required), then the particles can be produced by cutting the drill core into wedges (Figure 8). If sample quantity is not a problem, then the particles are produced by "light" stage-crushing of the core and selecting the required sizes via sieving the product at the end of each stage (Figure 9). Light stage-crushing involves crushing the drill core in stages, with each successive stage having a smaller gap setting and the gap settings chosen so that the reduction ratios (F_{80}/P_{80}) of each stage are relatively small. This ensures that the maximum amount of correct sized particles is produced and at the same time minimizes the amount of fines produced. Research has shown that there is no difference in results from using either diamond-cut or crushed material (see www.smctesting.com/about/technical-information).

The raw data from the SMC Test is processed, and from the results a range of comminution parameters are generated. These parameters fall into two groups. The first group contains parameters that are used in power-based equations for predicting the specific energy of comminution machines. This group includes the comminution indices DW_i, M_{ia}, M_{ih}, and M_{ic}. They cover the following circuits:

1. AG and SAG mills
2. Rod mills
3. Crushers
4. HPGRs

The second group contains parameters that are used for simulation modeling purposes, notably the AG/SAG and crusher models used in the comminution simulator JKSimMet. This group includes the JK rock breakage parameters A, b, and t_a as well as the JK crusher model's t10-Ecs matrix.

When the SMC Test is conducted using relatively small particle sizes (i.e., 14 mm or 20 mm), it may be necessary to confirm the size-adjustment factor that the SMC Test data processing algorithm uses to estimate the A and b parameters for these particle sizes. When 28-mm particles have been used, a size-adjustment factor is not required. The size-adjustment necessary when 14-mm or 20-mm particles are used is typically referred to as *calibration*. The need for such a factor arises from the change in rock hardness that is observed in all rocks as the particle size changes. This change is reflected in the change in the magnitude of the A*b parameter with size that can be observed from the data obtained in a drop weight test. The gradient of this A*b variation with size is determined from a drop weight test. It is used to "calibrate" the SMC Test A and b estimation algorithm, which then estimates the magnitude of the appropriate size-adjustment factor. The average particle size tested in a drop weight test is ~28 mm, and

Courtesy of SMC Testing Pty Ltd.

Figure 8 Particles selected for SMC testing from cutting drill core

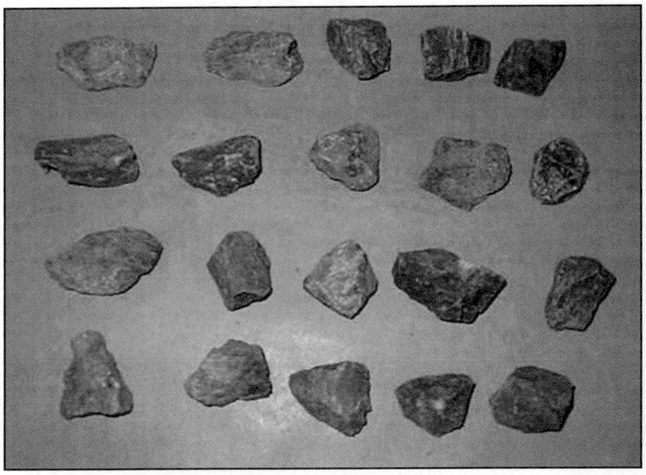

Courtesy of SMC Testing Pty Ltd.

Figure 9 Particles selected for SMC testing from crushed rock

hence, the drop weight test A and b values relate on average to breakage of this particle size. The purpose of the SMC Test size-adjustment algorithm is, through the use the A*b-size gradient, to determine the amount of adjustment to the raw SMC Test results such that they reflect breakage of 28-mm particles (i.e., breakage of the average particle size of the drop weight test). As mentioned earlier, if the SMC Test has been conducted using 28-mm particles in the first instance, little to no size adjustment is necessary (i.e., the size-adjustment factor tends to unity), as this is the same size as the average particle tested in a drop weight test. Note that calibration only applies to the estimation of A and b values and is not used in the generation of the power-based parameters DW_i, M_{ia}, M_{ic}, and M_{ih}.

If a drop weight test is not done in conjunction with a SMC Test, then calibration is conducted using factors obtained from SMC Testing's database of over 500 SMC Test–drop weight test pairs. These data show that most ores have a similar calibration factor and that A and b parameters generated

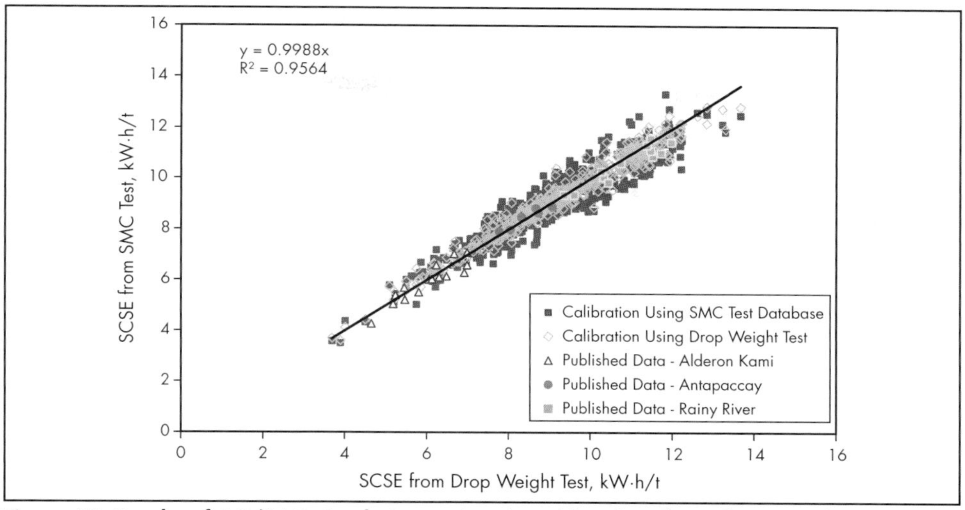

Figure 10 Results of AG/SAG simulations using A and b values from drop weight tests versus those from SMC Tests

from SMC Tests using the database calibration factors result in simulation results similar to those obtained using full drop weight tests or results from processing SMC Tests using calibration factors obtained from associated drop weight tests (see Figure 10). Figure 10 also includes some results based on data from several projects in the public domain to further illustrate this point. The data in this figure relate to the standard circuit specific energy (SCSE) parameter (Matei et al. 2015). The SCSE was adopted by JKTech in 2015 to represent the value of A*b in units of kilowatt-hours per ton that would result from the use of A*b in the simulation of a defined (standard) SAG mill circuit. This approach was adopted for several reasons, including to remove the nonlinear relationship that exists between A*b and the resistance of a rock to breakage by impact. This last factor is particularly important when comparing the A*b values of different ore samples and gives rise to the somewhat counterintuitive phenomenon that the difference in hardness between two samples with A*b values of, for example, 25 and 29 (15% difference) may be statistically different from a specific energy perspective, yet in the case of two samples with A*b values of 250 and 350 (40% difference) they may not be significantly different. The only way to tell is through simulating the influence that the different A and b values have on the AG/SAG mill specific energy.

Statistical analysis of the data in Figure 10 shows that the standard deviation of the differences between the drop weight test SCSEs and the SMC Test SCSEs is 3.8% when calibration using drop weight test results is used and 6.2% when SMC Test database calibration is used. In both cases, the overall mean of differences is 0.0%. The recent extensive Round Robin program conducted by JKTech (Matei et al. 2015) showed that the repeatability of the drop weight test SCSE was 3.8% (one standard deviation). Comparing this result with the statistics of the data in Figure 10 suggests that when drop weight test calibration is employed, the SCSEs from the SMC Test are statistically indistinguishable from those of a drop weight test. If, instead, calibration is conducted using the database, the degree of uncertainty in the SCSE, as measured by the standard deviation, increases only by 2.4 percentage points.

Ball Mill Laboratory Test

Morrell's approach for ball milling takes the raw data from a standard Bond ball mill work index test and uses it to generate a ball mill parameter, M_{ib}, which is compatible with the SMC Test parameters, M_{ia}, M_{ic}, and M_{ih}. The equation to obtain the M_{ib} is as follows (Morrell 2009):

$$M_{ib} = \frac{18.18}{P_1^{0.295}(Gbp)\left(P_{80}^{f(p_{80})} - F_{80}^{f(f_{80})}\right)} \quad \textbf{(EQ 15)}$$

where M_{ib} is the fine ore work index, in kW·h/t. Note that the test should be carried out with a closing screen size that gives a final product P_{80} similar to that intended for the full-scale circuit.

General Equations

From a study of many operating circuits, Morrell (2004b) developed the following general size-specific energy equation:

$$W_i = M_i 4\left(x_2^{f(x_2)} - x_1^{f(x_1)}\right) \quad \textbf{(EQ 16)}$$

where

W_i = specific comminution energy, kW·h/t
M_i = work index related to the breakage property of an ore, in kW·h/t, and the type of equipment used (e.g., tumbling mill, crusher, or HPGR)
x_2 = 80% passing size for the product, µm
x_1 = 80% passing size for the feed, µm

$$f(x_j) = -(0.295 + x_j/1{,}000{,}000);\quad j = 1,2 \text{ (Morrell 2009)} \quad \textbf{(EQ 17)}$$

For tumbling mills, the specific comminution energy (W_i) relates to the power at the pinion or, for gearless drives, the power at shell. For HPGRs it is the energy inputted to the rolls, while for conventional crushers W_i relates to the specific energy as determined using the motor input power, less the no-load power.

Equation 16 is used in conjunction with the following three equipment categories:

1. Tumbling mills (e.g., AG, SAG, rod, and ball mills). Size reduction specific energy is predicted using two indices: M_{ia} and M_{ib}.
2. Conventional reciprocating crushers (e.g., jaw, gyratory, and cone), which use one index: M_{ic}.
3. HPGRs, which use one index: M_{ih}.

For tumbling mills, the two indices relate to coarse and fine ore particle breakage properties. *Coarse* in this case is defined as spanning the size range from a P_{80} of 750 µm up to the P_{80} of the product of the last stage of crushing or HPGR size reduction prior to grinding. *Fine* covers the size range from a P_{80} of 750 µm down to P_{80} sizes typically reached by conventional ball milling (i.e., about 45 µm).

The work index covering grinding in tumbling mills of coarse sizes is labeled M_{ia}. The work index covering grinding of fine particles is labeled M_{ib} (Morrell 2009). M_{ia} values are provided as a standard output from an SMC Test, while M_{ib} values can be determined using the data generated by a conventional Bond ball mill work index test. Note that the M_{ib} is *not* the Bond ball work index. M_{ic} and M_{ih} values are also provided as a standard output from an SMC Test.

The total specific energy (W_T) to reduce in size primary crusher product to final product is given by

$$W_T = W_a + W_b + W_c + W_h + W_s \quad \textbf{(EQ 18)}$$

where

W_a = specific energy to grind coarser particles in tumbling mills
W_b = specific energy to grind finer particles in tumbling mills
W_c = specific energy for conventional crushing
W_h = specific energy for HPGRs
W_s = specific energy correction for size distribution (crushing–ball mill circuits only)

Clearly, only the W values associated with the relevant equipment in the circuit being studied are included in Equation 18.

Tumbling Mills

For coarse particle grinding in tumbling mills, Equation 16 is written as

$$W_a = K_1 M_{ia} 4\left(x_2^{f(x_2)} - x_1^{f(x_1)}\right) \quad \textbf{(EQ 19)}$$

where

K_1 = 1.0 for all circuits that do not contain a recycle pebble crusher and 0.95 where circuits do have a pebble crusher
M_{ia} = coarse ore work index and is provided directly by SMC Test
x_2 = 750 µm
x_1 = P_{80}, of the product of the last stage of crushing before grinding, µm

For fine particle grinding, Equation 16 is written as

$$W_b = M_{ib} 4\left(x_3^{f(x_3)} - x_2^{f(x_2)}\right) \quad \textbf{(EQ 20)}$$

where

M_{ib} = fine ore work index
x_3 = P_{80} of final grind, µm
x_2 = 750 µm

By combining Equations 19 and 20, the total specific energy of the tumbling mill circuit is predicted. In AG/SAG–ball mill circuits, if the specific energies of the AG/SAG mill and the ball mill are required separately, then an additional equation is required that predicts the AG/SAG mill specific energy. The ball mill specific energy is then found from the difference between this value and the total milling circuit specific energy.

The AG/SAG mill specific energy equation in general form is as follows (Morrell 2011a) and contains all of the variables that are known to influence AG/SAG mill specific energy:

$$S = K \cdot F_{80}^{a} \cdot DW_i^{b} \left(1 + c(1 - e^{-dJ})\right)^{-1} \phi^{e} \cdot f(A_r) \cdot g(x) \quad \textbf{(EQ 21)}$$

where

S = specific energy at the pinion
K = an empirical function whose value is dependent upon whether a pebble crusher is in-circuit and whether the crushed pebbles are returned to the SAG mill or sent directly to the ball mill
F_{80} = 80% passing size of the feed
DW_i = drop weight index (from the SMC Test)
a, b, c, d, e, f, g = empirical constants
J = volume of balls, %
ϕ = mill speed, % of critical
$f(A_r)$ = function of mill aspect ratio
g(x) = function of trommel aperture

The empirical constants in Equation 21 were fitted to operating data from 70 different concentrators covering over 110 different ore types. Details of these constants are proprietary and are currently owned by CITIC SMCC Process Technology.

Conventional Crushers

Equation 16 for conventional crushers is written as

$$W_c = S_c K_2 M_{ic} 4\left(x_2^{f(x_2)} - x_1^{f(x_1)}\right) \quad \textbf{(EQ 22)}$$

where

S_c = coarse ore hardness parameter, which is used in primary and secondary crushing situations. It is defined by Equation 23 with K_s set to 55.
K_2 = 1.0 for all crushers operating in closed circuit with a classifying screen. If the crusher is in open circuit (e.g., pebble crusher in an AG/SAG circuit), K_2 takes the value of 1.19.
M_{ic} = crushing ore work index and is provided directly by SMC Test
x_2 = P_{80} of the circuit product, µm
x_1 = P_{80} of the circuit feed, µm

The coarse ore hardness parameter (S) makes allowance for the decrease in ore hardness that becomes significant in relatively coarse crushing applications such as primary and secondary cone/gyratory circuits. In tertiary and pebble crushing circuits, it is normally not necessary and takes the value of unity. In full-scale HPGR circuits where feed sizes tend to be higher than used in laboratory and pilot-scale machines, the parameter has also been found to improve predictive accuracy. The parameter is defined by Equation 23 with K_s set to 35.

$$S = K_s (x_1 \cdot x_2)^{-0.2} \quad \textbf{(EQ 23)}$$

where
K_s = machine-specific constant that takes the value of 55 for conventional crushers and 35 in the case of HPGRs
x_1 = P_{80} of the circuit feed, µm
x_2 = P_{80} of the circuit product, µm

Note that the procedure to determine when the parameter S should be applied is to first use Equation 23 to determine the value of S, and if it is less than 1, then it should be applied. If it is greater than 1, it should be set at 1.0.

High-Pressure Grinding Roll

Equation 16 for an HPGR's crushers is written as

$$W_h = S_h K_3 M_{ih} 4\left(x_2^{f(x_2)} - x_1^{f(x_1)}\right) \quad \textbf{(EQ 24)}$$

where
S_h = coarse ore hardness parameter as defined by Equation 23 and with K_s set to 35, K_3 = 1.0 for all HPGRs operating in closed circuit with a classifying screen. If the HPGR is in open circuit, K_3 takes the value of 1.19.
M_{ih} = HPGR ore work index and is provided directly by SMC Test
x_2 = P_{80} of the circuit product, µm
x_1 = P_{80} of the circuit feed, µm

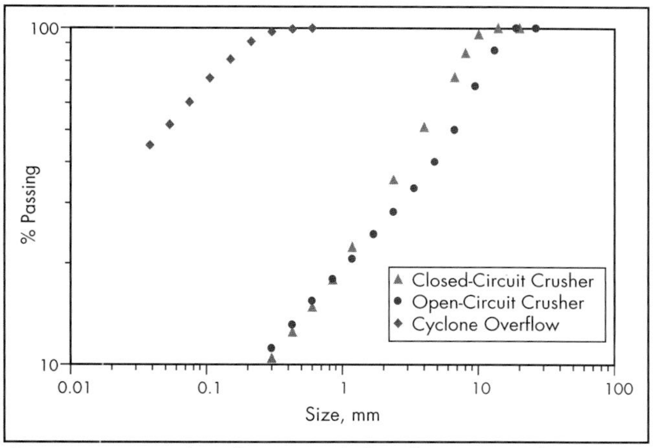

Figure 11 Examples of open and closed-circuit crushing distributions compared with a typical ball mill cyclone overflow distribution

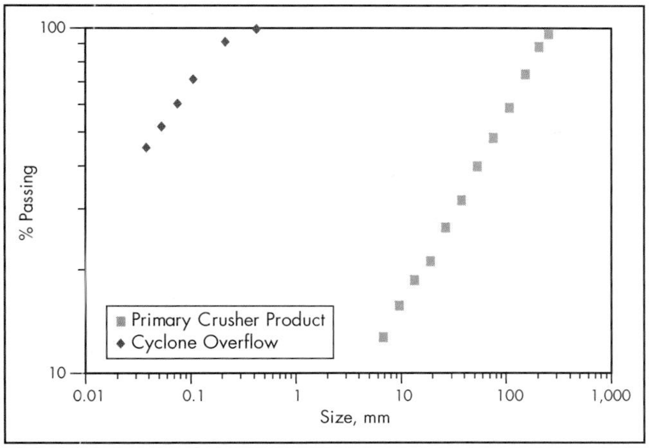

Figure 12 Example of a typical primary crusher (open circuit) product distribution compared with a typical ball mill cyclone overflow distribution

Specific Energy Correction for Size Distribution (W_s)

Implicit in the approach described in this section is that the feed and product size distributions are parallel and linear in log-log space. The same is true of Bond's approach (Bond 1960) and arises from the fact that the distributions are represented by a single point (the 80% passing size). Where they are not parallel, allowances (corrections) need to be made, a point also recognized by Bond (1961). Such corrections are most likely to be necessary (or are large enough to be warranted) when evaluating circuits in which closed-circuit secondary/tertiary crushing is followed by ball milling. This is because such crushing circuits tend to produce a product size distribution that is relatively steep when compared to the ball mill circuit cyclone overflow. This situation is illustrated in Figure 11, which shows measured distributions from an open and closed crusher circuit as well as a ball mill cyclone overflow. The closed-circuit crusher distribution is relatively steep compared with the open-circuit crusher distribution and ball mill cyclone overflow. Also, the open-circuit distribution more closely follows the gradient of the cyclone overflow. If a ball mill circuit were to be fed two distributions, each with same P_{80} but with the open- and closed-circuit gradients in Figure 11, the closed-circuit distribution would require more energy to grind to the final P_{80}. How much more energy is required is difficult to determine. However, it has been assumed that the additional specific energy for ball milling is the same as the difference in specific energy between open and closed crushing to reach the nominated ball mill feed size. This assumes that a crusher would provide this energy. However, in this situation the ball mill has to supply this energy and it has a different (higher) work index than the crusher (i.e., the ball mill is less energy efficient than a crusher and has to input more energy to do the same amount of size reduction). Hence from Equation 22, to crush to the ball mill circuit feed size (x_2) in open circuit requires specific energy equivalent to

$$W_c = 1.19 * M_{ic} 4\left(x_2^{f(x_2)} - x_1^{f(x_1)}\right) \quad \textbf{(EQ 25)}$$

For closed-circuit crushing, the specific energy is

$$W_c = 1 * M_{ic} 4\left(x_2^{f(x_2)} - x_1^{f(x_1)}\right) \quad \textbf{(EQ 26)}$$

The difference between the two (Equations 25 and 26) has to be provided by the milling circuit with an allowance for the fact that the ball mill, with its lower energy efficiency, has to provide it and not the crusher. This is what is referred to in Equation 18 as W_s and for the previous example is therefore represented by

$$W_s = 0.19 * M_{ia} 4\left(x_2^{f(x_2)} - x_1^{f(x_1)}\right) \quad \textbf{(EQ 27)}$$

In Equation 27, M_{ic} has been replaced with M_{ia}, the coarse particle tumbling mill grinding work index.

In AG/SAG-based circuits, the need for W_s appears to be unnecessary, as primary crusher products often have a very similar gradient to typical ball mill cyclone overflows (see Figure 12). A similar situation appears to apply with HPGR product size distributions, as illustrated in Figure 13.

Weakening of HPGR Products

Laboratory experiments have been reported by various researchers in which the Bond ball work index of HPGR products is less than that of the feed (Stephenson 1997; Oestreicher and Spollen 2006; Shi et al. 2006). The amount of this reduction appears to vary with both material type and the pressing force used. Observed reductions in the Bond ball work index have typically been in the range of 0% to 10%. In the approach described in this chapter, no explicit allowance has been made for such weakening. However, if HPGR products are available that can be used to conduct Bond ball work index tests, then M_{ib} values obtained from such tests can be used in Equation 20. Note that these tests must be performed using a feed size distribution that is similar to that produced from preparing material that has not been previously treated in an HPGR (i.e., HPGR feed). The reason for this is that HPGR products tend to be relatively fine compared to conventionally crushed material. If the HPGR products are therefore used as is in a Bond test, their relatively fine distribution introduces a bias in the results. Alternatively, the M_{ib} values from Bond ball mill work index tests on HPGR feed material can be reduced by an amount that the user thinks is appropriate. Until more data become available from full-scale HPGR/ball mill circuits, it is suggested that, in the absence of Bond ball mill work index data on HPGR products, the M_{ib} results from HPGR feed material are reduced by no more than 5% to allow for the effects of microcracking.

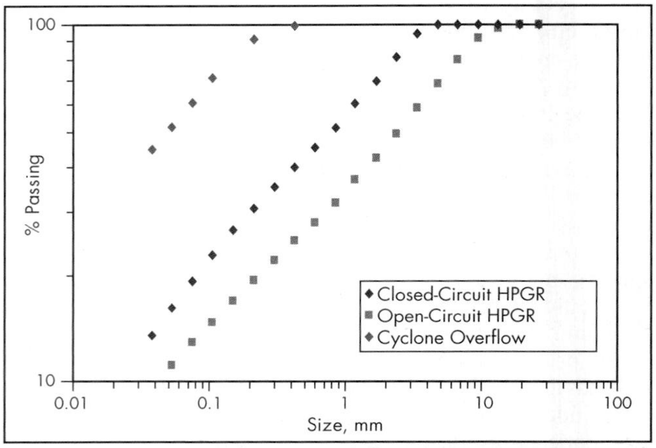

Figure 13 Examples of open- and closed-circuit HPGR distributions compared with a typical ball mill cyclone overflow distribution

Equation Accuracy

Total Specific Energy

In terms of predicting total comminution circuit specific energy, benchmarking has been conducted using data from 72 different concentrators covering more than 110 different ore types. Nine different types of circuits are covered. The results are shown in Figure 14 (Morrell 2011b). Examination of the statistics associated with the data in the figure indicates that the standard deviation of the differences between the observed and predicted values is 6.5% with an overall mean of differences of 0.2%.

AG/SAG Mills

For AG/SAG mills, benchmarking has been conducted using data from 64 different concentrators covering more than 100 different ore types (Morrell 2011a); results are shown in Figure 15. Ball mill data are shown in Figure 16. Examination of the statistics associated with the data in these figures indicates that the standard deviation of the differences between the observed and predicted values is 8.6% and 8.9%, respectively. Associated overall means of the differences are 0.6% and 0.4%, respectively.

Conventional Crushers

The crushing specific energy equations have been benchmarked using 12 different crushing circuits (25 data sets), including primary, secondary, tertiary, and pebble crushers in AG/SAG circuits (Morrell 2010). Observed versus predicted specific energies are given in Figure 17. The observed specific energies were calculated from the crusher throughput, and the net power draw of the crusher as defined by

$$\text{net power} = \text{motor input power} - \text{no-load power} \quad \textbf{(EQ 28)}$$

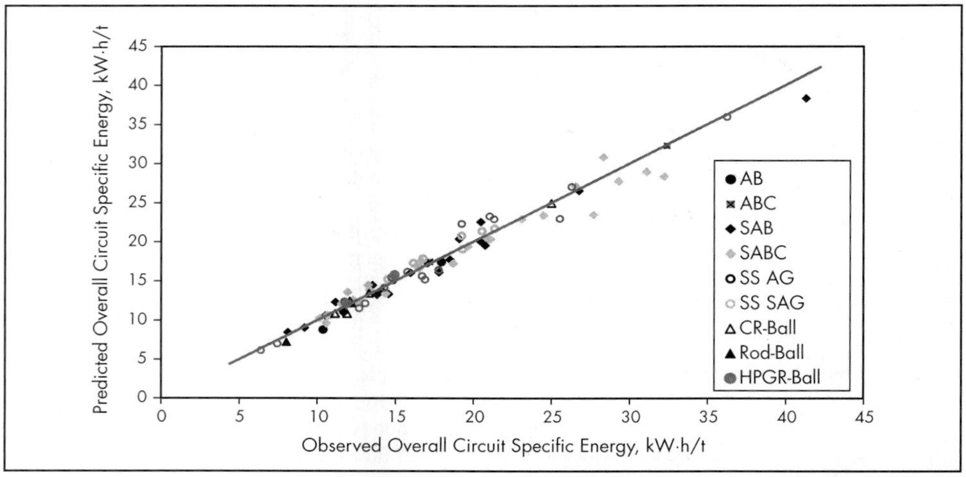

Source: Morrell 2011b

Figure 14 Predicted versus observed overall circuit specific energy

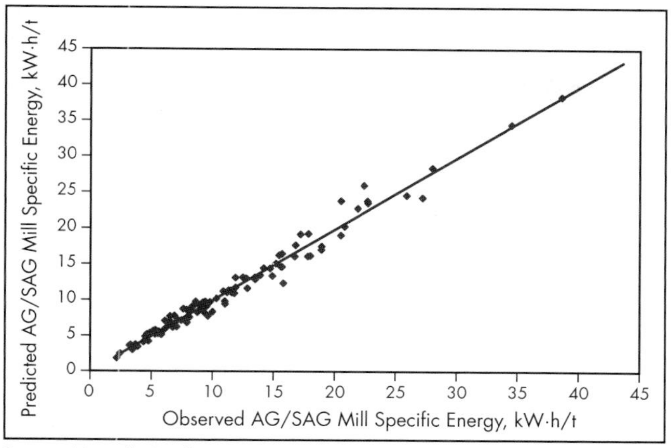

Source: Morrell 2011b

Figure 15 Predicted versus observed AG/SAG mill specific energy

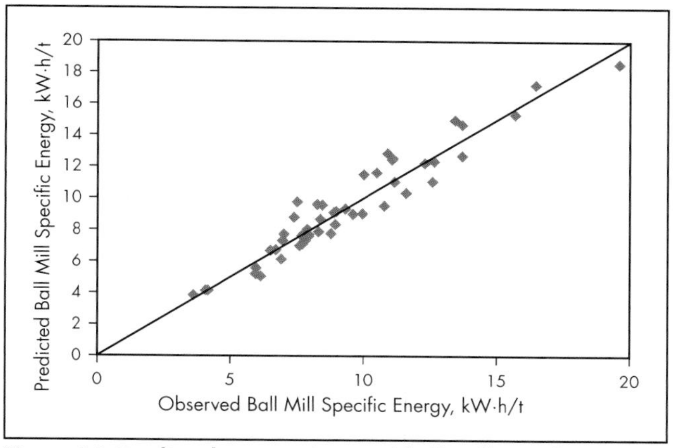

Figure 16 Predicted versus observed ball mill specific energy

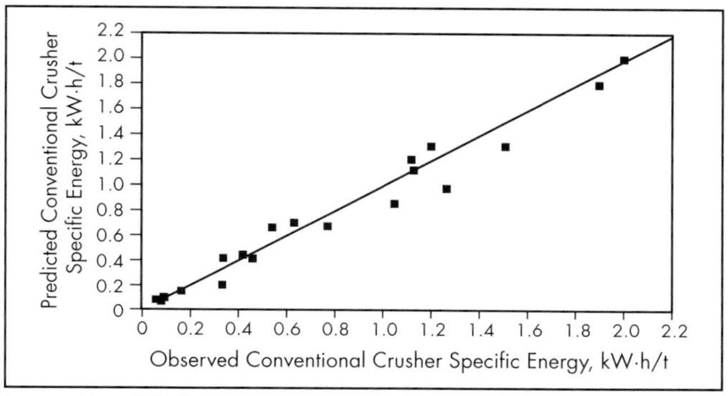

Source: Morrell 2010

Figure 17 Predicted versus observed conventional crusher specific energy

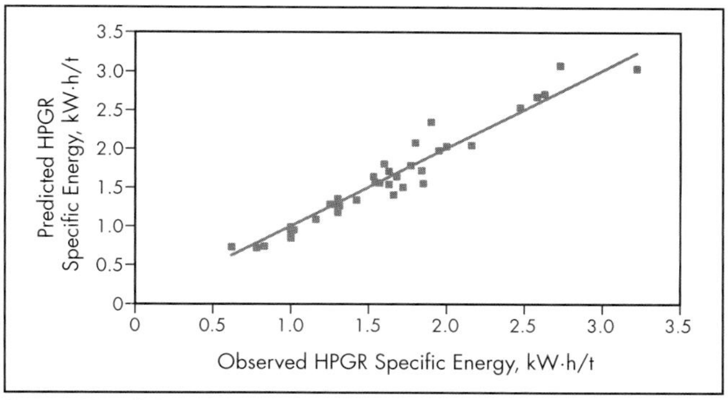

Source: Morrell 2010

Figure 18 Predicted versus observed HPGR specific energy

No-load power tends to be relatively high in conventional crushers, and hence net power is significantly lower than the motor input power. From examination of the 25 crusher data sets, the motor input power was found to be, on average, 20% higher than the net power.

Examination of the statistics associated with the data in Figure 17 indicates that the standard deviation of the differences between the observed and predicted values is 18.1% with an overall mean of differences of 1.0%.

High-Pressure Grinding Rolls

The crushing specific energy equations have been benchmarked using 19 different circuits (36 data sets) including laboratory-, pilot-, and industrial-scale equipment (Morrell 2010). Observed versus predicted specific energies are given in Figure 18. The data relate to HPGRs operating with specific grinding forces typically in the range of 2.5 to 3.5 N/mm^2. The observed specific energies relate to power delivered by the roll drive shafts. Motor input power for full-scale machines is expected to be 8%–10% higher.

Examination of the statistics associated with the data in Figure 18 indicates that the standard deviation of the differences between the observed and predicted values is 8.5% with an overall mean of differences of 1.6%.

At the time of writing, more than 42,000 SMC Tests have been conducted covering more than 1,500 deposits.

Worked Examples

An SMC Test and Bond ball work index test were carried out on a representative ore sample. The following results were obtained from the laboratory.

SMC Test:

- M_{ia} = 19.4 kW·h/t
- M_{ic} = 7.2 kW·h/t
- M_{ih} = 13.9 kW·h/t

Bond ball work index test:

- P_1 = 150 µm
- F_{80} = 2,250 µm
- P_{80} = 106 µm
- Gbp = 1.49 net grams of screen undersize per mill revolution
- W_{iB} = 15.15 kW·h/t
- M_{ib} = 18.8 kW·h/t (from Equation 15)

Three circuits are to be evaluated:

1. SABC
2. HPGR/ball mill
3. Conventional crushing/ball mill

The overall objective of the worked examples is to estimate the specific grinding energy to reduce a primary crusher product that has a specified P_{80} of 100 mm to a final product P_{80} that has been specified as 106 µm.

SABC Circuit

Coarse particle tumbling mill specific energy:

$$W_a = 0.95 \times 19.4 \times 4 \times \left(750^{-\left(\frac{0.295+750}{1,000,000}\right)} - 100,000^{-\left(\frac{0.295+100,000}{1,000,000}\right)}\right) = 9.6\ \text{kW·h/t}$$

Fine particle tumbling mill specific energy:

$$W_b = 18.8 \times 4 \times \left(106^{-\left(\frac{0.295+106}{1,000,000}\right)} - 750^{-\left(\frac{0.295+750}{1,000,000}\right)}\right) = 8.4\ \text{kW·h/t}$$

Pebble crusher specific energy: In this circuit, it is assumed that the pebble crusher feed P_{80} is 52.5 mm. As a rule of thumb, this value can be estimated by assuming that it is 0.75 of the nominal pebble port aperture (in this case, the pebble port aperture is assumed to be 70 mm). The pebble crusher is set to give a product P_{80} of 12 mm. The pebble crusher feed rate is expected to be 25% of new feed tons per hour.

$$W_c = 1.19 \times 7.2 \times 4 \times \left(12,000^{-\left(\frac{0.295+12,000}{1,000,000}\right)} - 52,500^{-\left(\frac{0.295+52,500}{1,000,000}\right)}\right)$$

= 1.12 kW·h/t when expressed in terms of the crusher feed rate
= 1.12 × 0.25 kW·h/t when expressed in terms of the SABC circuit new feed rate
= 0.3 kW·h/t of SAG mill circuit new feed

Total net comminution specific energy is therefore the sum of W_a, W_b, and W_c:

$$W_T = 9.6 + 8.4 + 0.3\ \text{kW·h/t} = 18.3\ \text{kW·h/t}$$

HPGR/Ball Milling Circuit

In this circuit, primary crusher product is assumed to be reduced to an HPGR circuit feed P_{80} of 35 mm by closed-circuit secondary crushing. The HPGR is also in closed circuit and is assumed to reduce the 35-mm feed to a circuit product P_{80} of 4 mm. This is then fed to a closed-circuit ball mill, which takes the grind down to a P_{80} of 106 µm.

Secondary crushing specific energy:

$$W_c = 1 \times 55 \times (35,000 \times 100,000)^{-0.2} \times 7.2 \times 4 \times \left(35,000^{-\left(\frac{0.295+35,000}{1,000,000}\right)} - 100,000^{-\left(\frac{0.295+100,000}{1,000,000}\right)}\right) = 0.4\ \text{kW·h/t}$$

HPGR specific energy:

$$W_h = 1 \times 35 \times (4,000 \times 35,000)^{-0.2} \times 13.9 \times 4 \times \left(4,000^{-\left(\frac{0.295+4,000}{1,000,000}\right)} - 35,000^{-\left(\frac{0.295+35,000}{1,000,000}\right)}\right) = 2.4\ \text{kW·h/t}$$

Coarse particle tumbling mill specific energy:

$$W_a = 1 \times 19.4 \times 4 \times \left(750^{-\left(\frac{0.295+750}{1,000,000}\right)} - 4,000^{-\left(\frac{0.295+100,000}{1,000,000}\right)}\right) = 4.5\ \text{kW·h/t}$$

Fine particle tumbling mill specific energy:

$$W_b = 18.8 \times 4 \times \left(106^{-\left(\frac{0.295+106}{1,000,000}\right)} - 750^{-\left(\frac{0.295+750}{1,000,000}\right)}\right) = 8.4\ \text{kW·h/t}$$

Total net comminution specific energy is therefore

$$W_T = 4.5 + 8.4 + 0.4 + 2.4\ \text{kW·h/t} = 15.7\ \text{kW·h/t}$$

Conventional Crushing/Ball Milling Circuit

In this circuit, primary crusher product is assumed to be reduced in size to a P_{80} of 6.5 mm via a secondary/tertiary crushing circuit (closed). This is then fed to a closed-circuit ball mill, which grinds to a P_{80} of 106 µm.

Secondary/tertiary crushing specific energy:

$$W_c = 1 \times 7.2 \times 4 \times \left(6,500^{-\left(\frac{0.295+6,500}{1,000,000}\right)} - 100,000^{-\left(\frac{0.295+100,000}{1,000,000}\right)}\right) = 1.7\ \text{kW·h/t}$$

Coarse particle tumbling mill specific energy:

$$W_a = 1 \times 19.4 \times 4 \times \left(750^{-\left(\frac{0.295+750}{1,000,000}\right)} - 6,500^{-\left(\frac{0.295+6,500}{1,000,000}\right)}\right) = 5.5\ \text{kW·h/t}$$

Fine particle tumbling mill specific energy:

$$W_b = 18.8 \times 4 \times \left(106^{-\left(\frac{0.295+106}{1,000,000}\right)} - 750^{-\left(\frac{0.295+750}{1,000,000}\right)}\right) = 8.4\ \text{kW·h/t}$$

Size distribution correction:

$$W_s = 0.19 \times 19.4 \times 4 \times \left(6,500^{-\left(\frac{0.295+6,500}{1,000,000}\right)} - 100,000^{-\left(\frac{0.295+100,000}{1,000,000}\right)}\right) = 0.9\ \text{kW·h/t}$$

Total net comminution specific energy is therefore

$$W_T = 5.5 + 8.4 + 1.7 + 0.9\ \text{kW·h/t} = 16.5\ \text{kW·h/t}$$

SAG POWER INDEX

The SAG power index (SPI) was originally developed in the 1990s (Starkey et al. 1994) and is currently owned by SGS S.A. The test uses a 12 in. × 4 in. (D × L) batch laboratory mill loaded with 15% by volume of 1-in. balls. A picture of the mill is shown in Figure 19. The mill is loaded with 2 kg of –19 mm (P_{80} = 12.7 mm) of dry sample and run until it is ground to 80% passing 1.7 mm. The time in minutes taken to reach this grind size is designated the SPI value. An additional laboratory crushing test must also be performed, which requires 10 kg of material and generates a crushing parameter (Ci), the details of which are proprietary.

Source: Amelunxen et al. 2016
Figure 19 SPI mill

Equations

The original equation in which the SPI was used is as follows (Starkey and Dobby 1996):

$$\text{SAG kW·h/t} = (2.2 + 0.1\ \text{SPI}) / T_{80}^{0.33} \qquad \textbf{(EQ 29)}$$

where T_{80} is the 80% passing size of the SAG circuit product (so-called transfer size).

Equation 29 was subsequently modified to the form shown in Equation 30 (Dobby et al. 2001).

$$\text{SAG kW·h/t} = K\ (\text{SPI} \times T_{80}^{-0.5})^n\ f_{sag} \qquad \textbf{(EQ 30)}$$

where

K, n = proprietary constants
f_{sag} = feed size and pebble crusher function

The f_{sag} function is proprietary. According to Dobby et al. (2001), it is a submodel that incorporates the effects of feed size and pebble crusher circulating load (PCCL) and takes the value of unity when the circuit does not have a pebble crusher and is fed with a nominal 6-in. size distribution (F_{80}). Under these conditions, Equation 30 predicts the specific energy of a so-called "reference" or "standard" circuit (Amelunxen 2003). According to Amelunxen (2003), apart from the effects of feed size and PCCL, it also incorporates "some or all of the effects of.... differences in ball charges (or fully autogenous grinding), extremely fine grinding, low aspect-ratio mills, and open-circuit grinding. Grinding circuit audits performed on industrial-sized circuits are required for calibrating the submodel for the target circuit. There are sufficient data in the MinnovEX database to model fine feed or pebble crushing conditions without necessarily collecting plant data; however, when other conditions (such as fine grinding, low-aspect ratio mills, or open circuit SAG mills) are investigated it might be wise to first perform some calibration work before attempting to estimate the value of f_{sag}."

The f_{sag} submodel includes at least the feed size and PCCL (Dobby et al. 2001). In greenfield/brownfield design situations, these have to be predicted and hence require further submodels. Consequently, to predict the feed size distribution, Dobby et al. developed the following approach. They assumed that the feed size distribution followed a Rosin–Rammler function, and by predicting the F_{80} and F_{50} from the following equations, the entire distribution could be generated:

$$F_{80} = Ci^f SPI^g CSS^h \qquad \textbf{(EQ 31)}$$

$$F_{50} = Ci^i SPI^j CSS^k \qquad \textbf{(EQ 32)}$$

where Ci is the crushing parameter from laboratory crushing a sample of circuit feed, and CSS is the closed side setting of the primary crusher. The feed size distribution was then divided into three parts (designated "streams") with percentages θ_1, θ_2, θ_3. A fourth stream (θ_4) was also used and is associated with the pebble crusher circuit.

The PCCL was then estimated using the following equation:

$$PCCL = a(\theta_2 + b\theta_3/SPI^c)^d\ SPI^e \qquad \textbf{(EQ 33)}$$

Equation 30 also incorporates a transfer size (T_{80}), and in greenfield/brownfield situations this also needs to be predicted. A T_{80} submodel is therefore also required and was configured using the following equations:

$$T_{80} = T_{80}(A)\ \theta_1 + T_{80}(B)\ (PCCL \times \theta_4) + T_{80}(C)\ (\theta_2 - PCCL \times \theta_4) + T_{80}(D)\ \theta_3 \qquad \textbf{(EQ 34)}$$

$$T_{80}(A) = a_1\ D_1\ SPI^{b1}\ SF_A \qquad \textbf{(EQ 35)}$$

$$T_{80}(B) = a_2\ D_2 SPI^{b2}\ SF_B \qquad \textbf{(EQ 36)}$$

$$T_{80}(C) = a_3\ SPI^{b3} SF_C \qquad \textbf{(EQ 37)}$$

$$T_{80}(D) = a_4\ P_{64}^{b4}\ SF_D \qquad \textbf{(EQ 38)}$$

where

a, b, c, d, e, f, g, h, i, j, k, a_1, a_2, a_3, a_4, b_1, b_2, b_3, b_4, SF_A, SF_B, SF_C, SF_D = fitted empirical factors
D_1 and D_2 = 80% of θ_1,θ_2, respectively
SF = factors describing the effect of ball load

Including the K and n factors in Equation 30, this gives a minimum requirement of 25 empirical factors that need to be determined before Equation 30 can be used for greenfield/brownfield prediction purposes.

Validation

Kosick and Bennett (1999) reported the fitting of the empirical factors K and n in Equation 30 to data from 13 different plants that had been collected to date (see Figure 20). According to Dobby et al. (2001), further data were added to give a total of 26 plants that were used to fit the f_{sag} and T_{80} submodels in Equation 30.

From data in Starkey and Dobby (1996) and Kosick and Bennett (1999), Amelunxen et al. (2014) were able to estimate the K and n values in Equation 30 and obtained values of 5.9 and 0.55, respectively. They then used these parameters to compare the observed AG/SAG mill specific energies from 58 data sets from 14 different plants with the predicted values of the standard circuit using Equation 30. The resultant observed and predicted specific energies are reproduced in Figure 21. Additional data from Starkey and Dobby (1996) are also included, bringing the total number of data sets to 63. The obvious considerable scatter in Figure 21 results from the differences between the circuits from which the data were drawn and the standard circuit. To correct for these differences the appropriate f_{sag} values are required. Amelunxen et al. (2014) provide guidance as to what these values should be for SAG circuits with pebble crushing (f_{sag} = 0.85), AG circuits with pebble crushing (f_{sag} = 0.62), and finer feed (i.e., F_{80} <75 mm) (f_{sag} = 0.9). When these values are applied to the standard circuit predictions, Figure 22 results. Note that the "unknown" data set is not included as the feed size or circuit configuration of these circuits is not known and hence appropriate f_{sag} values cannot be assigned to them. Scatter is reduced significantly, though there is a problem with the Holt–McDermott data, a problem that Starkey and Dobby (1996) also noted and stated needed further investigation.

Examination of the statistics associated with the data in Figure 22 indicates that the standard deviation of the differences between the observed and predicted values is 18.9% with a mean of the differences of 0.3%. If the Holt–McDermott data are not included, the results are 17.3% with a mean difference of 2.3%. However, the predicted data in Figure 22 were generated from equations that used the observed T_{80}, not modeled ones, which would be the case in greenfield/brownfield situations and would increase the scatter in Figure 22 and hence the degree of uncertainty in the predictions. As Amelunxen

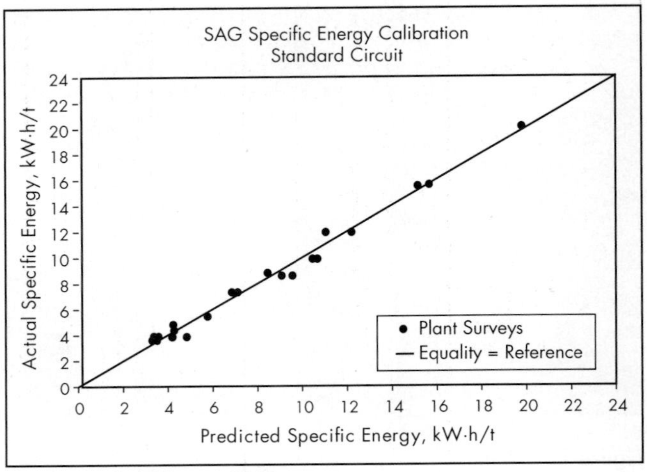

Source: Kosick and Bennett 1999

Figure 20 Calibration scatterplot

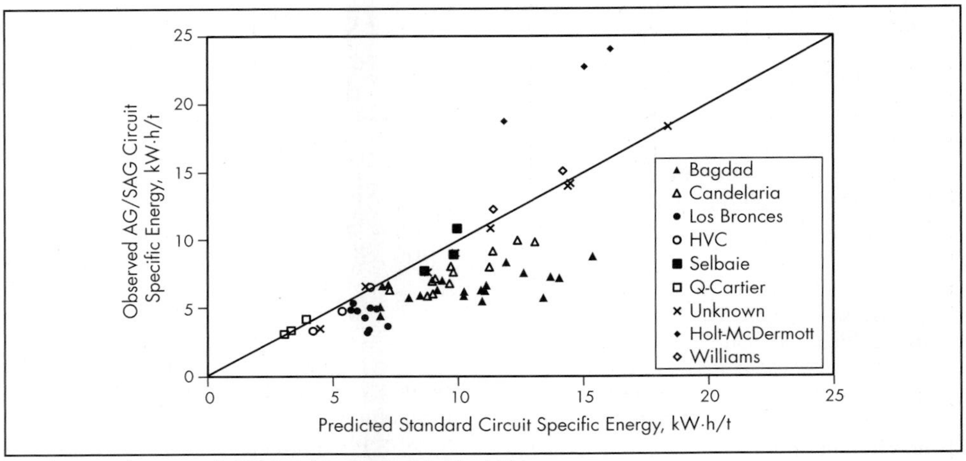

Data from Amelunxen et al. 2014 and Starkey and Dobby 1996

Figure 21 Observed AG/SAG circuit specific energy versus predicted standard circuit specific energy using the SPI

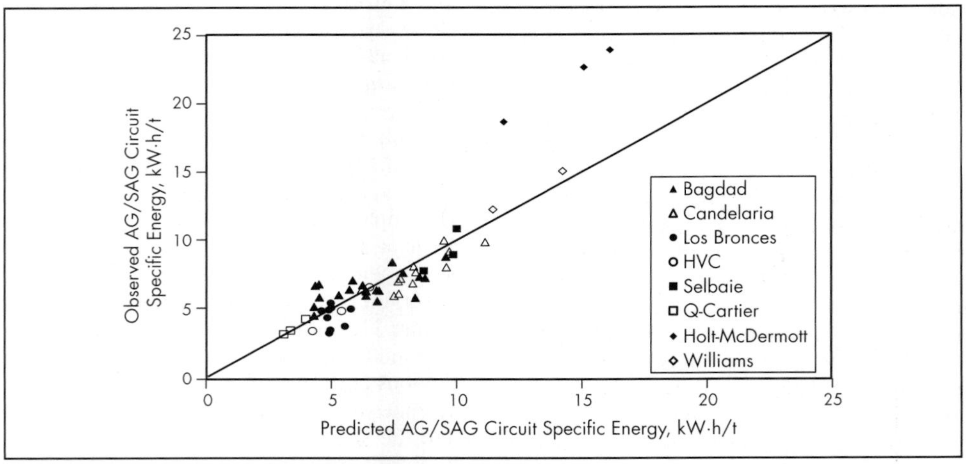

Data from Amelunxen et al. 2014 and Starkey and Dobby 1996

Figure 22 Observed versus predicted AG/SAG circuit specific energy using the SPI

et al. (2014) state, "One of the limitations of the (SPI) scale-up methodology ... is that the transfer size must be known in order to estimate the specific energy required for a given SAG mill circuit," a limitation that applies to all methodologies requiring knowledge of what the transfer will be.

At the time of writing, the website of SGS, which owns the test, states that to date the organization has performed more than 25,000 SPI Tests.

SAGDesign

The SAGDesign test was originally developed by Starkey et al. (2006) and is currently owned by Starkey and Associates Inc. The test is effectively a derivative of the SPI Test and, according to Starkey et al. (2006), during its development, "Initial work was done using SPI Test results for comparative examples." The test uses a batch mill with dimensions (D × L) 488 × 163 mm. The mill is charged with 11% by volume of a mixture of 1.5-in. and 2-in. steel balls. A minimum of 15 kg of ore is required, crushed to a P_{80} of 19 mm. Sufficient ore is added to the mill to obtain a filling of 26% by volume. The test is conducted dry and the mill is run until the charge is ground to 80% passing 1.7 mm.

Equations

The number of mill revolutions required until the charge is ground to 80% passing 1.7 mm is put into the following equation:

$$\text{SAG kW·h/t} = \text{revolutions} \times \frac{(16{,}000 + g)}{447.3g} \qquad \textbf{(EQ 39)}$$

where

SAG kW·h/t = specific energy at pinion of a SAG mill grinding from an F_{80} of 152 mm to a P_{80} of 1.7 mm

g = mass of ore added to the mill

The ground product of the mill is used as feed for a Bond ball work index test. As the size distribution is not the same as that for a conventional Bond ball work index test (should be stage-crushed to minus 3.35 mm), the resultant ball mill work index is not the same as that from a conventional Bond test (Starkey el al. 2006).

To predict the total specific energy of a single-stage SAG mill circuit or a SAG-ball mill circuit with an F_{80} of 152 mm, the kilowatt-hours per ton from Equation 39 is added to the specific energy to grind from 1.7 mm to the final grind of the circuit, which is calculated using Bond's equation following all of Bond's original recommendations for the application of his EFs. How predictions of total energy are adjusted to account for SAG circuits with F_{80} values other than 152 mm or for the inclusion/exclusion of pebble crushers is unclear.

According to Starkey and Larbi (2012), the SAG circuit specific energy is determined by first choosing a T_{80} value. In cases where this differs from 1.7 mm, the SAG kW·h/t in Equation 39 is modified by determining "values [which] are added or deducted from this basic number by calculation, using the BWI [W_{iB}] to calculate the adjustment."

For the ball mill circuit, Starkey and Larbi (2012) state that "the balance of the required power is provided by the ball mill. Here, correction factors are applied when needed."

Validation

At the time of writing, according to the SAGDesign website, more than 900 tests have been performed and benchmarking has been done using six circuits (Starkey and Larbi 2012). Four are single-stage SAG mills, of which one is a pilot mill and two are SAG–ball circuits. Observed versus predicted specific energies from these are shown in Figure 23. Benchmarking data are insufficient to provide meaningful accuracy statistics.

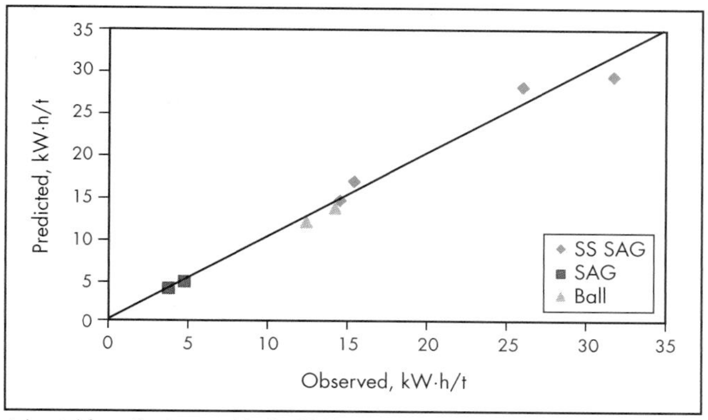

Adapted from Starkey and Larbi 2012

Figure 23 Predicted versus observed specific energy using SAGDesign approach

MACPHERSON AUTOGENOUS GRINDABILITY TEST

The MacPherson autogenous grindability test was originally developed by MacPherson in the 1970s (MacPherson and Turner 1978) and is currently owned by SGS S.A. It is a continuous test performed in an air-swept 18-in. semiautogenous mill, with an 8% ball charge and closed with a 1.2-mm screen. The test requires sufficient feed ore for the mill to reach steady state with a 28% total charge volume. This state can normally be achieved with less than 100 kg, but typically a 175-kg sample is required to allow for soft and/or dense ores. At test completion, all the products are submitted for particle size analysis, and the mill charge is dumped and observed. The charge is submitted to a particle size analysis, and size-by-size specific gravity (sg) determinations. This allows the evaluation of any coarse material build-up, or if any heavier component is present in the mill. The products from the mill at the end of the test are collected and weighed and sized.

Equations

The feed size and product size are determined, and from the measured power draw of the mill and feed rate, a Bond operating work index is determined. This result is called the autogenous work index (AWI). According to Mosher and Bigg (2001), a correction is required when harder ores are treated due to "the shortfall of high energy events in the ... mill." According to MacPherson and Turner (1978), in design situations the AWI is referenced against existing autogenous mills whose ores have also been tested using this approach and hence "the requirements of a full scale autogenous plant to treat the new ore can readily be established." No details are published as to exactly how this is done.

Validation

No published sources supply any benchmarking data that show how accurate the test is in predicting full-scale mill performance. However, according to SGS, about 750 tests have been performed on about 275 deposits.

ADVANCED MEDIA COMPETENCY TEST

The advanced media competency test (AMCT) was developed by Orway Mineral Consultants (OMC) and is a development of the Allis-Chalmers autogenous rock media competency test. The test uses a 6 ft × 1 ft batch mill and is charged with a total of 50 rocks divided equally into five size fractions in the range of 100 to 165 mm. The mill is rotated for 500 revolutions at a speed of 75% of critical, and the charge is dumped and size analyzed. The surviving rocks are sieved into five size fractions in the range of 19 to 100 mm and each size fraction is submitted to breakage in a Bond crushing work index machine. Feed samples are also subjected to the full range of Bond tests plus unconfined compressive strength (UCS) tests. Typically, 300 kg of PQ (85 mm) core is required; 180 kg is used for the tumbling test and the remainder is used for the associated Bond and UCS tests.

Although OMC originally developed the AMCT, it subsequently discontinued its use in mill selection and design due to the test's requirement for large sample masses of large-diameter drill cores (Scinto et al. 2015). OMC's approach now relies on SMC Test and drop weight test results.

OMC POWER-BASED APPROACH

OMC uses the results from the Bond ball work index, the SMC Test, and the drop weight test in a power-based methodology described by Scinto et al. (2015).

Equations

OMC's approach is similar to that developed by Morrell (2011a) in which one equation is used to estimate the total circuit specific energy and a second equation is used to predict the AG/SAG circuit specific energy. The ball mill circuit specific energy is the difference between the two values (Scinto et al. 2015).

The total energy equation developed by OMC is as follows:

$$E_{TOT} = 10 \cdot BWi \cdot [(75^{-0.5} - 150{,}000^{-0.5}) \times f_{sag} - (F_{80}{}^{-0.5} - 150{,}000^{-0.5}) - (75^{-0.5} - P_{80}{}^{-0.5})] \quad \textbf{(EQ 40)}$$

where

E_{TOT} = total grinding specific energy, kW·h/t
BWi = Bond ball mill work index, kW·h/t
f_{sag} = efficiency factor
F_{80} = feed size, 80% passing, µm
P_{80} = product size, 80% passing, µm

The details of how f_{sag} is determined are proprietary but are related to the parameters A,b and t_{10}, which are derived from SMC Test / drop weight test results.

The AG/SAG mill circuit specific energy uses the equation developed by Morrell (2011a) with the constants being fitted to OMC's database:

$$E_{SAG} = a(A \times b)^{b} \cdot F_{80}{}^{c} \cdot (1 + d(1 - e^{-gB}))^{-1} \cdot Sp^{h} \cdot f(Ar) \cdot f(K) \quad \textbf{(EQ 41)}$$

where

E_{SAG} = specific energy at the pinion, kW·h/t
a, b, c, d, e, g, h = empirical constants
A×b = appearance function

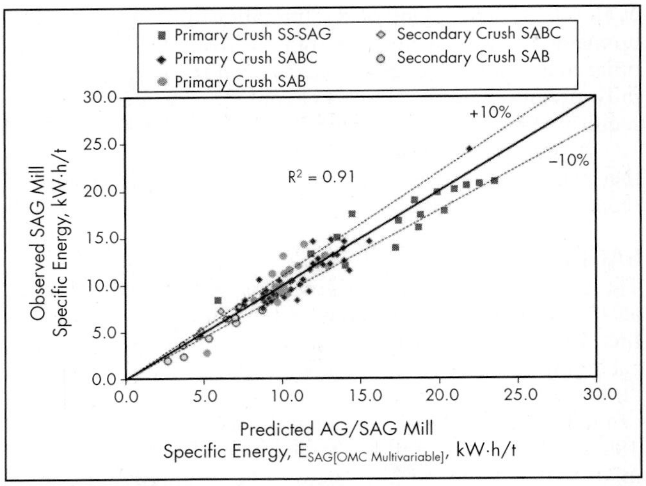

Source: Scinto et al. 2015

Figure 24 Observed versus predicted AG/SAG mill specific energy using OMC approach

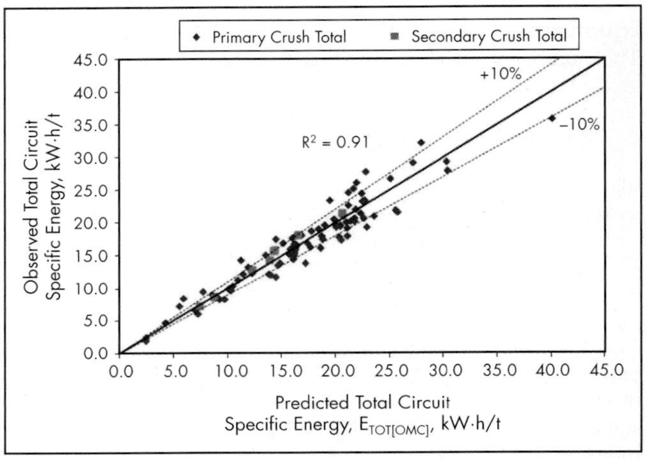

Source: Scinto et al. 2015

Figure 25 Observed versus predicted total circuit specific energy using OMC approach

F_{80} = 80% passing size of the feed
B = volume of balls, %
Sp = mill speed, % critical
f(Ar) = function of mill aspect ratio
f(K) = function for pebble crusher in circuit

Validation

OMC has benchmarked its approach using more than 100 sets of data from operating plants. The results from this benchmarking are given in Figures 24 and 25. According to Scinto et al. (2015), the average absolute error of the differences between the observed and predicted AG/SAG mill specific energies is 8.7%, and for total specific energy the associated value is 8.2%. These values are approximately equivalent to standard deviations of 11.2% and 10.4%, respectively, assuming that the mean overall difference between observed and predicted values is approximately zero.

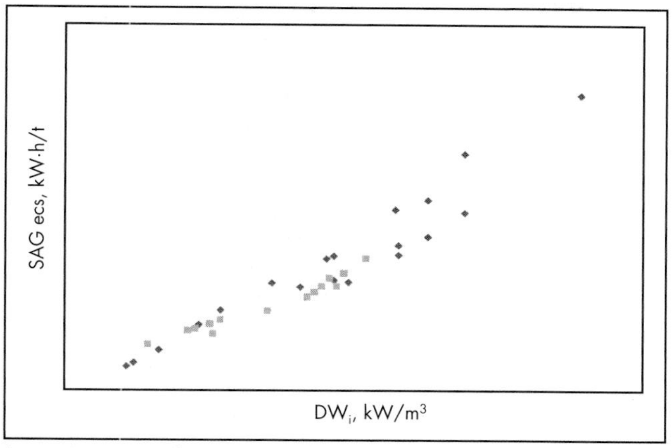

Source: Lane 2010, by permission of Gecamin, www.gecaminpublications.com

Figure 26 Correlation between the SMC Test parameter, DW$_i$, and SAG circuit specific energy

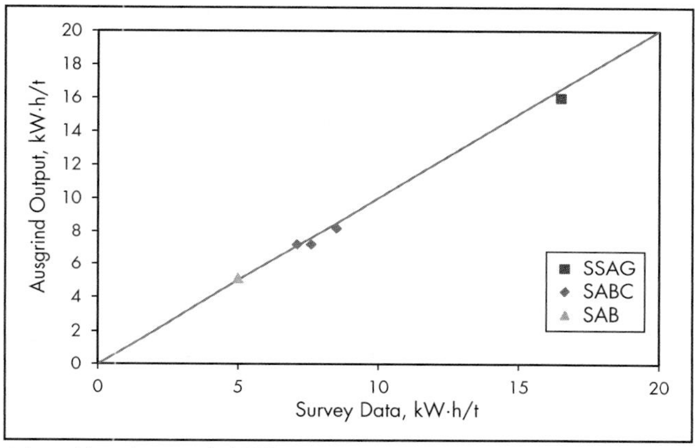

Adapted from Lane et al. 2013

Figure 27 Observed versus predicted SAG mill circuit specific energy using Ausgrind

AUSENCO POWER-BASED APPROACH

Ausenco's approach, called Ausgrind, uses the results from Bond crushing, rod, and ball work indices in conjunction with SMC Tests and is described by Lane et al. (2013).

Equations

As with OMC, Ausenco's general approach follows that developed by Morrell in which one equation is used to estimate the total circuit specific energy and a second equation is used to predict the AG/SAG circuit specific energy, the ball mill specific energy being the difference between the two. Also in line with OMC, Ausenco uses Bond equations combined with an f_{sag} correction to estimate total specific energy, the f_{sag} correction being a function of the SMC Test parameter, DW$_i$ (Lane et al. 2013). Ausenco's total specific energy equation is as follows:

$$\begin{aligned}\text{total } E_{cs} &= (\text{Bond } E_{cs} \text{ to } 150\ \mu\text{m}) \\ &\times (f_{sag} - F_{80} \text{ effect}) \\ &\pm (\text{Bond } E_{cs} \text{ to final } P_{80})\end{aligned} \quad \textbf{(EQ 42)}$$

where E_{cs} is specific energy. In the case of Equation 42, the Bond E_{cs} value is determined following Bond's approach for a crushing/rod/ball mill circuit, though without any of the EFs applied.

For the AG/SAG circuit, Ausenco initially estimates a so-called base case circuit specific energy using a correlation with the SMC Test parameter, DW$_i$, that was developed using operating plant data (Figure 26). Adjustment factors (described graphically in Lane et al. 2013) are then applied to take account of various design and operating conditions such as aspect ratio, ball load, feed size, and pebble crusher performance:

$$\text{SAG } E_{cs} = (\text{base case } E_{cs}) * \text{adjustment factors} \quad \textbf{(EQ 43)}$$

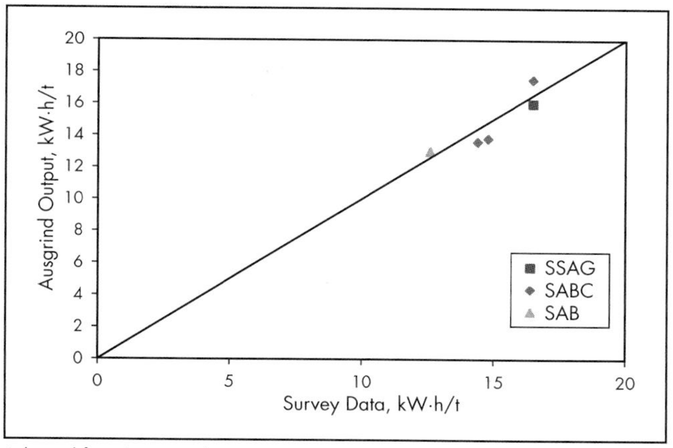

Adapted from Lane et al. 2013

Figure 28 Observed versus predicted total circuit specific energy using Ausgrind

Validation

Ausenco has published benchmarking data from three different circuits (single-stage SAG, SABC, and SAB) to illustrate the accuracy of its technique as shown in Figures 27 and 28. Benchmarking data to date is insufficient to provide meaningful accuracy statistics.

JK DROP WEIGHT TEST

The JK drop weight test uses the JK drop weight tester, which was originally developed in 1992 (Napier-Munn et al. 1996) and was precipitated by the need for a machine that could break relatively large rocks, was simple to use, was easy to maintain, and was relatively precise (i.e., had good repeatability). The device (shown in Figure 29) comprises an impact head with a hardened steel face, which can be raised to a range of heights up to ~1 m. The mass of the impact head can also be varied. The impact head is raised and a single rock is placed on a hardened steel anvil directly under the impact head. The impact head is then released and falls under the action of gravity and impacts and breaks the target particle. Through a combination of different impact head masses and heights, a very wide range of energies can be generated with which to break rocks.

The drop weight test itself typically requires about 75 kg of sample and involves breaking rocks from five different size fractions: –63+53 mm, –45+37.5 mm, –31.5+26.5 mm, –22.4+19 mm, –16+13.2 mm. If the test is conducted on drill core, it normally requires PQ (85 mm) core to provide sufficient material for the largest size fraction. Each rock size

Courtesy of JKTech Pty Ltd.

Figure 29 (A) Drop weight test in operation and (B) close-up of impact head and anvil

fraction is broken with a range of three input energies and the resultant broken particles are sized. In total, 15 sets of data are generated (5 size fractions × 3 energy levels).

The purpose of the test is to generate relationships between the energy used to break rocks and the size distribution of progeny rocks. These data are then used in simulation models of AG/SAG mills and crushers using the comminution simulator JKSimMet.

Given the relatively large sample requirement and the relatively large rock sizes also required, the test is not suitable for use with small-diameter drill core. In cases where only small-diameter drill core is available and JKSimMet modeling parameters are required, the SMC Test can be used.

Equations

Energy is related to size distribution via a data reduction process that involves the size distribution parameter, t_{10}. The t_{10} is determined from the size distribution of broken products and is defined as the percentage passing a sieve aperture 1/10 of the original particle size that was broken. The t_{10} is then determined from each of the 15 sets of size distribution data that the drop weight test generates. The relationship between the resultant t_{10} values and the associated input specific energies (E_{cs}) is fitted using the following equation (Leung 1987):

$$t_{10} = A\,(1 - e^{-b \cdot E_{cs}}) \qquad \textbf{(EQ 44)}$$

where

A and b = ore-specific parameters

E_{cs} = specific comminution energy, kW·h/t

As the impact energy is varied, the resultant t_{10} also varies, with higher impact energies producing higher values of t_{10}, which are reflected in products with finer size distributions (Figure 30).

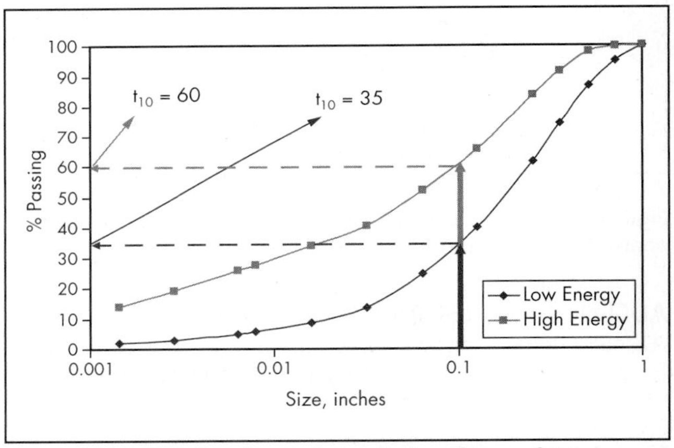

Figure 30 Estimating the t_{10} from the breakage of 1-in. particles

The t_{10} parameter and its associated use in modeling for reproducing size distributions was originally developed by Narayanan (Narayanan and Whiten 1988). Figure 30 illustrates how the t_{10} is estimated from breaking 1-in. rocks at two different energy levels. Figure 31 shows how, for a given rock size, the t_{10} varies with input energy following the shape of curve described by Equation 44. Hence, when modeling comminution machines, if the energies applied in breaking rocks can be estimated, Equation 44 can be used to predict the associated t_{10} values. As simulation models are required to predict product size distributions, this still leaves a link from the t_{10} to a size distribution to be made. This link is made using the relationships given in Figure 32. Just as the t_{10} is the percentage of broken particles that pass a sieve that is 1/10 of the original

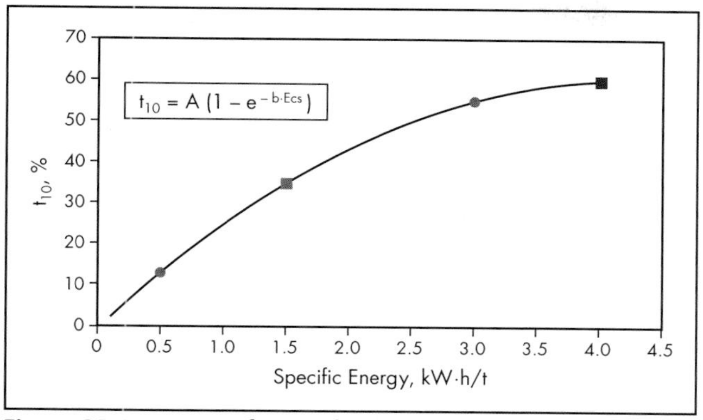

Figure 31 Variation of t_{10} with input energy for breakage of rocks in a fixed narrow size class

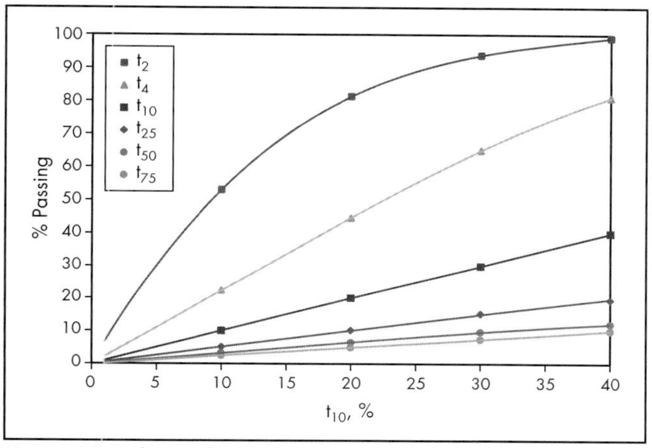

Figure 32 Relationship between t_{10} and the t_n "family" of curves

particle's size, the t_2 is the percentage of broken particles that pass a sieve that is ½ of the original particle's size, the t_{75} is the percentage of broken particles that pass a sieve that is 1/75 of the original particle's size, and so on. With reference to Figure 32, it is therefore possible to determine not only a t_{10} from a particular product size distribution but a t_2, t_4, and so on—in fact, a whole family of t_n values can be determined. Narayanan found that if the t_{10} values from breaking rocks at different energies are plotted against the associated t_2, t_4, ... t_n values, the graph in Figure 32 is obtained. Narayanan found that regardless of the rock being broken, the trends shown in this figure remained largely the same; that is, the graph is a universal "map" of breakage distributions. Hence, if the t_{10} is known (e.g., from Equation 44), then the associated t_n values can be read off Figure 32. These t_n values are different points on the product size distribution curve that can thus be fully reproduced. Figure 32 and its companion Equation 44 have become central to the JK comminution modeling approach.

Whereas the drop weight test is used for simulation modeling purposes, the parameter A*b is often used in a qualitative way to indicate rock hardness. High values of A*b indicate a rock that is easy to break, while low values indicate the opposite so they work in a different manner to other rock hardness indicators. The relationship of A*b to hardness is also not linear. As a result, in terms of simulation, the difference in the predicted specific energy of a SAG mill treating an ore with an A*b of 30 and an A*b of 40 would be about 15%. However, there is only a 5% difference in the predicted specific energy when comparing ores with A*b values of 300 and 400. To help users of the drop weight test to better appreciate the differences between A*b values of different ores, JKTech recently started also reporting the SCSE (Matei et al. 2015).

The SCSE is the specific energy that would be predicted using JKSimMet of a SAG mill treating ore with the given A*b. The SAG mill circuit conditions are kept constant; that is, the SCSE always relates to the same "standard" circuit. Circuit configurations, mill designs, feed sizes, ball loads, and so on that depart from this standard will result in values that are not the same as the SCSE, and in these situations a simulation must be conducted to obtain the correct associated value. The "standard" SAG mill circuit is defined as follows:

- SAG mill:
  - Inside shell diameter to length (effective grinding length) ratio of 2:1 with 15° cone angles
  - Ball charge of 15%, 125 mm in diameter
  - Total charge of 25%
  - Grate open area of 7%
  - Apertures in the grate are 100% pebble ports with a nominal aperture of 56 mm
- Trommel: Cut size of 12 mm
- Pebble crusher: Closed-side setting of 10 mm
- Feed size distribution: F_{80} from the t_a relationship given in Equation 31 (Morrell and Morrison 1996).

$$F_{80} = 71.3 - 28.4 \ln(t_a) \quad \textbf{(EQ 45)}$$

One use of the SCSE is in evaluating the repeatability of the drop weight test. This evaluation is done on a regular basis by JKTech, which sends samples to all of the laboratories that have drop weight testers and reports on the associated statistics (Matei et al. 2015). From the latest results, the standard deviation of the relative differences between laboratories was 3.8% for the full drop weight test and 4.9% for the SMC Test.

Validation

As mentioned in the previous section, the drop weight test parameters are for use in simulation modeling and therefore validation must be in associated with a particular simulation model. The so-called Julius Kruttschnitt Mineral Research Centre (JKMRC) variable rates model is arguably the most commonly used AG/SAG mill simulation model. During its development, a total of 18 AG/SAG mills were used for benchmarking purposes (Morrell and Morrison 1996). Benchmarking was done by running the model under the same feed conditions as the full-scale mill and comparing the power draw of the actual mill with that predicted by the model. The results are shown in Figure 33 with error bars representing the 95% confidence interval. The predicted product sizes were also compared with those measured on the plant and are given in Figure 34.

Analysis of the data in Figure 25 indicates that the standard deviation of the differences between observed and predicted values is 6.4% with a mean of the differences of 3.5%.

More recently, Morrell (2004a) developed a more advanced AG/SAG model than the variable rates one and it was benchmarked using 21 different mills. The model predictions were compared with plant operating data by running the simulation model until it reached the same operating load

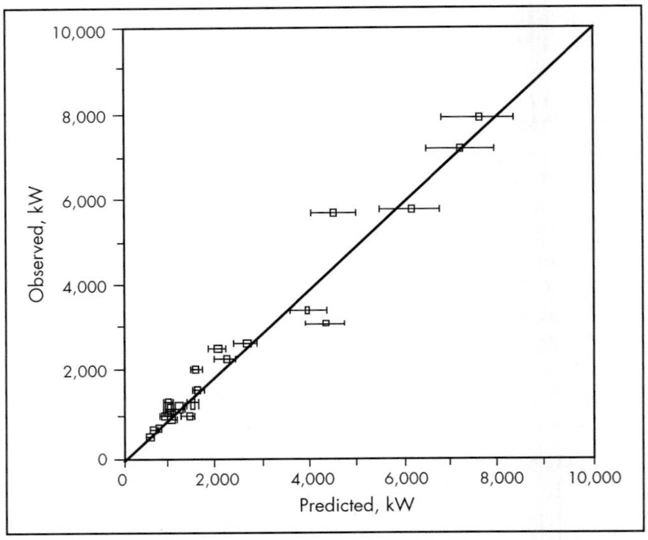

Source: Morrell and Morrison 1996

Figure 33 Observed versus predicted AG/SAG mill power draw using variable rates model

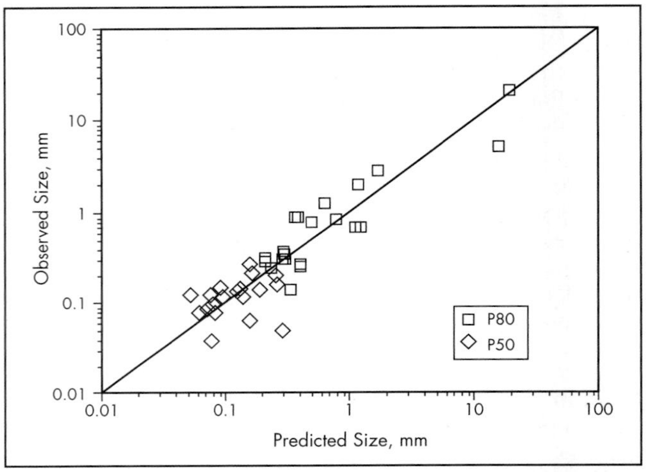

Source: Morrell and Morrison 1996

Figure 34 Observed versus predicted AG/SAG mill product size using variable rates

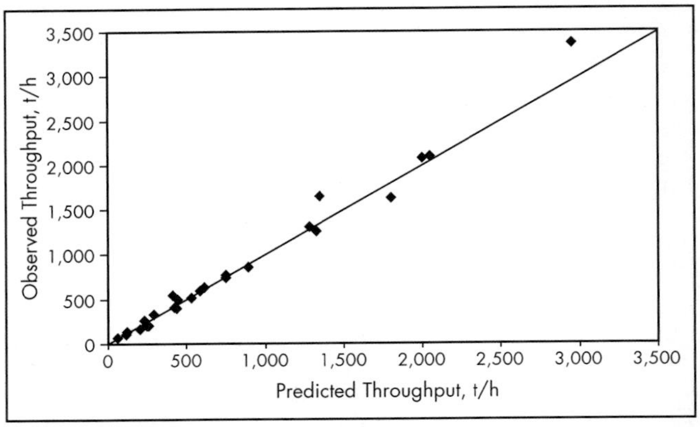

Source: Morrell 2004a

Figure 35 Observed versus predicted AG/SAG mill throughput using Morrell simulation model

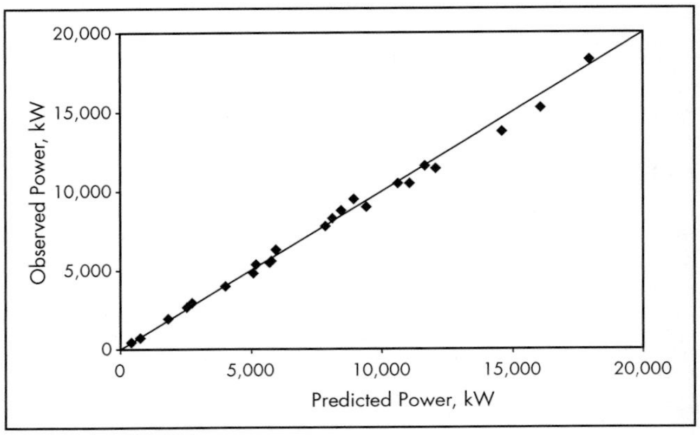

Source: Morrell 2004a

Figure 36 Observed versus predicted AG/SAG mill power draw using Morrell simulation model

that was observed on the plant. Comparisons were then made between the simulated throughput and power draw and the observed values. The results are shown in Figures 35 and 36, respectively. Comparisons of the predicted product size and the observed values were also made (Figures 37 and 38).

Analysis of the data in Figure 35 indicates that the standard deviation of differences between observed and predicted values is 11.0% with an overall mean of differences of 0.6%. For Figure 36, the associated standard deviation and mean difference are 3.8% and 0.0%, respectively.

TEST PROLIFERATION

At first sight it may appear that the proliferation of comminution characterization tests implies little consensus as to the true underlying breakage mechanisms at work in size reduction equipment. This may well be the case from a theoretical viewpoint, but from a pragmatic viewpoint there is a compelling argument that regardless of what the nature of the test is, if it can be demonstrated that it accurately predicts comminution equipment/circuit energy requirements, then it is a suitable test for the mining industry to use. Rayo (2014) illustrated this in his study of South American comminution circuits and the range of characterization techniques used in the mill selection methodology. Although this argument may help address the issue of the lack of consensus, it does not answer the question "Why are there so many tests?" To address this question, it is helpful to take a historic perspective and review when and why the different tests were developed.

The earliest tests were those developed by Bond from work begun in the 1930s and specifically targeted the most popular equipment used at that time (i.e., crushers, rod mills, and ball mills). The tests were designed to generate a single characterization value for each type of equipment (the so-called work index). In turn, the work index was used to predict the specific energy on the basis of given feed and product size distributions, the latter being described by single points (80% passing size). This methodology was termed the *power-based approach*. At this time, engineers relied on hand calculations assisted by slide rules and hand-drawn graphical techniques. Therefore, characterization tests and associated equations, of necessity, had to be simple.

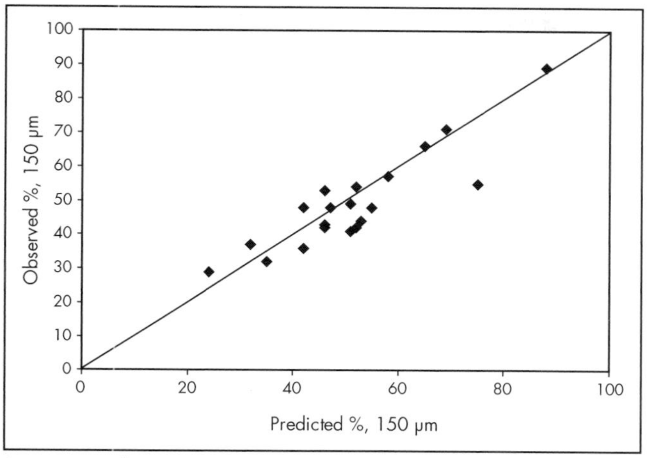

Source: Morrell 2004a

Figure 37 Observed versus predicted AG/SAG mill product size using Morrell simulation model (−150 µm)

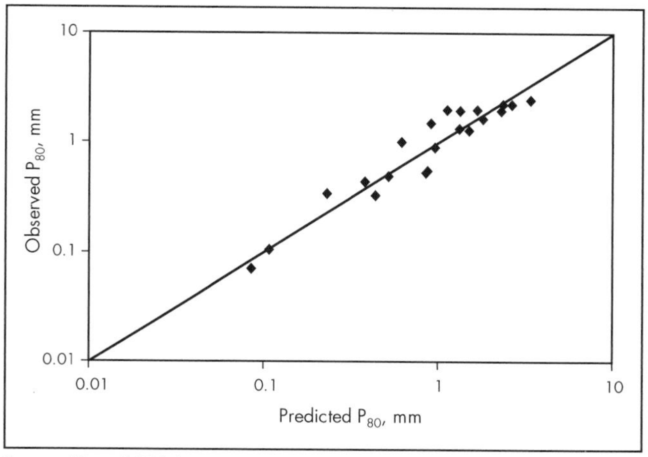

Source: Morrell 2004a

Figure 38 Observed versus predicted product size AG/SAG mill product size using Morrell simulation model (P_{80})

The rapid development of computers, particularly personal computers, in the late 1970s and the simultaneous increase in popularity of AG and SAG mills caused significant changes in minerals processing. Access to relatively powerful and inexpensive computers gave rise to the development of simulation models that attempted to mimic most aspects of the performance of comminution circuits, including power draw and predictions of mass flows and size distributions of all streams throughout the circuit—a quantum leap compared to what the power-based approach offered. Most of the simulation models used the so-called population balance equation, which in turn required ore-specific data relating to entire breakage size distributions and their relationship to breakage energy, not the single-point breakage descriptors such as the Bond tests provided. New characterization techniques were therefore required, and this drove the development of the JKMRC twin-pendulum in the early 1980s and subsequently the JK drop weight tester, which superseded it in the early 1990s. These devices were originally designed solely to provide breakage data for simulation models, which were integrated in the well-known JKSimMet comminution circuit simulator. About 75 kg of PQ core is required for the drop weight test, and therefore one of its drawbacks is that it is not suitable for small-diameter core.

The rise in popularity of AG/SAG mills initially presented designers with problems concerning equipment size selection, as at that time there were no AG/SAG breakage characterization tests similar to what was available for rod and ball mills. Initially, therefore, equipment was sized predominantly from scale-up of specific energy results from pilot tests. Pilot testing is very expensive and requires considerable amounts of ore sample (ideally, at least 400 t per ore type). Consequently, the MacPherson and autogenous rock media competency tests were developed in the late 1970s and early 1980s. They can be thought of as "mini-pilot" tests and require significantly less material than pilot tests (175 kg). Although these tests significantly reduced the amount of material required, they still had limited value for use with small-diameter drill core. At about the same time, practitioners such as Barratt and Allan (1986) attempted to combine the results from Bond's laboratory tests to predict AG/SAG circuit specific energy.

As time went on, the continued and very rapid increase in computer power gave rise to ever-sophisticated and detailed three-dimensional descriptions of ore deposits for mine planning purposes, into which process descriptions increasingly found their way—so-called geometallurgical models. Such models require significant amounts of ore characterization data so that hardness variability throughout the deposit can be adequately described. In most cases, the source of material on which hardness testing can be performed is small-diameter drill core. Given that by the early 1990s AG/SAG mills had begun to dominate circuit design, geometallurgical models increasingly required characterization tests that targeted such mills and were suitable for use with small-diameter core. To fill this need, the SPI Test was developed in the early 1990s. As with the Bond rod and ball mill tests, it uses relatively small amounts of material and grinds it in a tumbling mill to generate a single hardness parameter for use in power-based equations.

In the early 2000s and driven by the same forces, the SMC Test was developed. Unlike the SPI and Bond tests, which produce parameters that can only be used in power-based equations, the SMC Test was developed to produce parameters that can be used for both simulation modeling and power-based equations. To ensure compatibility with what had become the most popular and successful comminution simulator in the industry (JKSimMet), the SMC Test was developed for use with the JK drop weight tester.

By the mid-to-late 2000s, a sufficient number of AG/SAG mill circuits were in operation that databases being collated by some organizations such as OMC, Ausenco, and SMCC (Morrell) had become large enough to enable the development of robust semi-empirical models that reduced the variations in these databases to simple functions of hardness, equipment design, and process conditions. To varying degrees, each of these approaches uses a combination of SMC Test/drop weight test and Bond test data to characterize ore hardness.

At about the same time, an alternative but in many ways similar approach to the SPI also became available with the launch of SAGDesign. Like the SPI and Bond rod and ball mill tests, this test incorporates a tumbling mill and a size-distributed feed in contrast to the single-particle breakage mechanisms employed by the drop weight, SMC, and Bond crushing work index tests. Industry was therefore provided with alternative approaches to predicting specific energy requirements—a prerequisite for the vital independent review stage in project development.

Although from an academic viewpoint it may seem desirable to relate different breakage mechanisms occurring in industrial comminution machines to those employed in laboratory characterization tests, in practice the validity of the test is directly related to the degree to which it correlates with industrial performance and the size and variability of the database used in the correlation, irrespective of the nature of the test. The appropriate criterion from an industry perspective, therefore, is whether the test can be successfully used to predict industrial-scale performance, and this can only be demonstrated by benchmarking using a large and varied database (e.g., Figure 14).

POWER DRAW MODELS

The simplest way to model the power draw of tumbling mills is to consider the power associated with a solid rotating body, in which case the following equation applies:

$$\text{power (P)} = 2\pi N\tau \qquad \textbf{(EQ 46)}$$

where

N = rotational rate, in revolutions per unit time
τ = torque

Torque (τ) is defined as the product of force (F) and distance (s) when the force is applied to assist or retard the rotation of a body. The distance (s) is measured at right angles to the direction of force and joins the line of force to the axis of rotation of the body. An example of a simple system where a force is applied to retard rotation is shown in Figure 39. The torque in this case is given by

$$\text{torque } (\tau) = F \cdot s \qquad \textbf{(EQ 47)}$$

where s is the distance between center of rotation and centroid of gravity of the charge (often referred to as the torque or lever arm distance).

Equation 47 describes the relationship between torque and power in a simple system. This relationship can be extended to tumbling mills, usually by assuming that the charge takes the shape shown in Figure 40. This shape results from the rotation of the mill shell, which "holds the charge up" in approximately the position shown. By considering the charge as a solid mass with a weight (W) and remembering that weight is a force (mass × gravitational acceleration), the torque that it applies against the rotation of the mill is given by

$$\text{torque } (\tau) = W \cdot s \qquad \textbf{(EQ 48)}$$

Hence the power draw is given by

$$\text{power} = 2\pi \cdot W \cdot s \cdot N \qquad \textbf{(EQ 49)}$$

where

W = charge weight
s = lever arm distance
W·s = torque
N = rotational rate (speed)

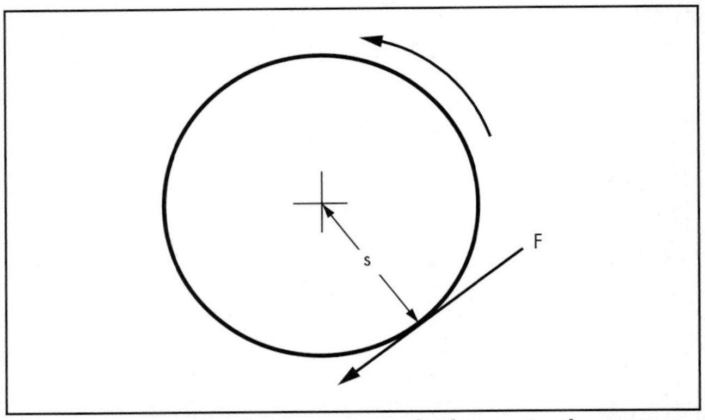

Figure 39 Example of torque as applied to a simple system

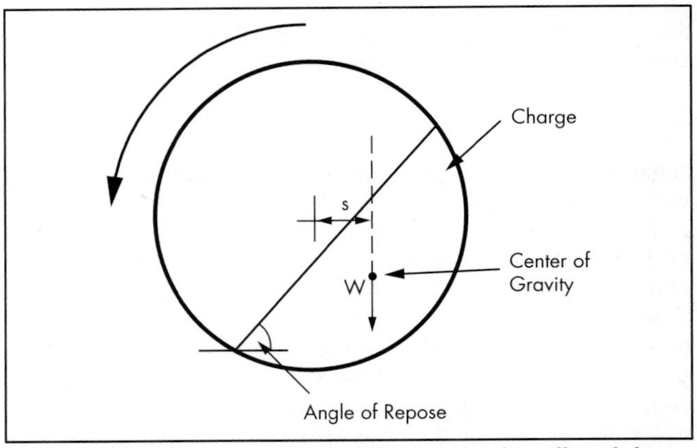

Figure 40 Schematic of grinding charge inside mill and the torque it applies

Equation 49 can be rewritten in a more general form as follows:

$$\text{power} = K\, D^n\, L\, \rho\, f(\phi)\, g(J) \qquad \textbf{(EQ 50)}$$

where

K = constant for a particular design/use of mill and is different for different designs/uses (e.g., overflow ball mill, grate discharge ball mill, rod mill, autogenous mill, or semiautogenous mill)
D = mill diameter
n = the diameter exponent that in theory should be 2.5
L = mill length
ρ = charge density
$f(\phi)$ = function of mill speed (ϕ) where speed is represented as the fraction of critical
$g(J)$ = function of mill charge volume (J) where J is expressed as the fraction of mill volume occupied by the charge

Many published models in the literature can be derived using Equation 50 (Bond 1962; Hogg and Fuerstenau 1972; Harris et al. 1985; Austin 1990). Rose and Evans (1956a, 1956b) and Morrell (1993), however, are some of the few who took a different route.

Bond Model

The original form of Bond's (1961) grinding mill power equation is as follows:

$$kWb = 2.8D^{0.4}(3.2 - 3Vp)\, C_s\, (1 - 0.1/2^{(9-10C_s)}) \quad \textbf{(EQ 51)}$$

where

kWb = mill input kW (power at pinion) per ton of grinding balls in overflow wet grinding mills
D = interior mill diameter, ft
V_p = fraction of mill volume occupied by balls
C_s = fraction of critical speed

Equation 51 was first published in January 1961. However, in April 1962 it was revised, presumably on the basis of further operational data. Equation 51 was therefore changed to

$$kWb = 3.1D^{0.3}(3.2 - 3V_p)\, C_s\, (1 - 0.1/2^{(9-10C_s)}) \quad \textbf{(EQ 52)}$$

For grate discharge mills, kWb was multiplied by

$$1 + \frac{0.4 - V_{pd}}{2.5}$$

where V_{pd} is 0.029 for wet-grinding grate and low-level discharge mills. The previous term, when evaluated using V_{pd} = 0.029, gives a value of 1.1484. Converting Equation 52 into metric units and using a notation consistent with the other equations gives

$$\text{power (kW)} = 12.262\rho_b \phi L D^{2.3} \times J_B(1 - 0.937J_B)\left(1 - \frac{0.1}{2^{9-10\phi}}\right) \quad \textbf{(EQ 53)}$$

Bond's equation was meant for ball mills and therefore in Equation 53 the bulk ball density (ρ_b) is used (i.e., the contribution of the slurry fraction is not explicitly incorporated). To extend the application of the equation to SAG and AG mills, ρ_b was replaced with the bulk density of the ball and/or rock charge (ρ_c) (Morrell 1993). As with Bond's ball mill power equation, the slurry fraction was not included. The charge density was therefore defined as

$$\rho_c = (1 - E)\frac{(J_B\rho_b + J_o\rho_o)}{J_t} \quad \textbf{(EQ 54)}$$

where

E = grinding media voidage = 0.4
J_B = fractional ball filling
ρ_b = steel specific gravity = 7.8
J_o = fractional ore filling (excluding slurried ore)
ρ_o = ore specific gravity
J_t = total fractional filling = $J_B + J_o$

For ball mills, Equation 54 reduces to $\rho_c = 0.6 \times 7.8$, which is the bulk density of the balls as per Bond's original equation. Bond's grate discharge correction was also applied to SAG and AG mills.

Validation

The mill power draw database that was published by Morrell (1993) and is reproduced in Tables 2–4 was used to evaluate the Bond power draw model. Results are shown graphically in Figures 41 and 42. Statistical analysis of results shows that for ball mills, the standard deviation of the differences between observed and predicted values is 4.7% with the overall mean of differences being 2.3%. For AG/SAG mills, the associated figures are 14.7% and 13.1%, respectively. The results indicate that for ball mills the Bond model is very good, but it is quite poor for AG/SAG mills. This is perhaps not surprising, as Bond developed his model with only ball mills in mind.

Harris et al. Model

Harris et al. (1985) adopted a torque-arm approach, envisioning the mill contents as per Figure 43. Their approach led to the following equation:

$$P = \frac{\pi\rho g N L D^3 \sin^3\theta \sin\alpha}{6} \quad \textbf{(EQ 55)}$$

where

P = mill power, kW
ρ = bulk density of the charge, t/m^3
g = gravitational constant, m/s^2
N = mill speed, rps
L = mill length, m
D = mill diameter, m
θ = half-angle subtended by the charge (see Figure 43)
α = angle of repose

This equation is essentially the same as that obtained by Hogg and Fuerstenau (1972), who used a potential energy approach in their work and envisioned the contents of the mill as per Figure 44.

The term $\sin^3\theta$ in Equation 55 was approximated by Harris et al. (1985) as follows:

$$\sin^3\theta = 4L_f(1 - L_f) \quad \textbf{(EQ 56)}$$

This leads to the equation:

$$P = 1.333\pi D^3 L\rho_c N J_t(1 - J_t)g \sin\alpha \quad \textbf{(EQ 57)}$$

where

P = mill power, kW
D = mill diameter, m
L = mill length, m
ρ_c = bulk density of the charge, t/m^3
N = mill speed, rps
J_t = fraction of mill volume occupied by grinding media, measured at rest

The bulk density was calculated as follows:

$$\rho_c = \frac{(J_b\rho_b + J_o\rho_o)}{J_t}(1 - E) + E\rho_s \quad \textbf{(EQ 58)}$$

where

J_b = fractional ball filling
ρ_b = steel specific gravity = 7.85
J_o = fractional rock filling: $J_o = J_t - J_b$
ρ_o = ore specific gravity
E = grinding media voidage = 0.4
ρ_s = slurry specific gravity

To determine $\sin\alpha$, Harris et al. (1985) fitted Equation 57 to a range of manufacturer's data and obtained the following values:

Table 2 Ball mill power draw data

Discharge Mechanism	Diameter Inside Liners, m	Length (belly), m	Length (C/line), m	Length/Diameter Ratio	Mill Speed, fraction of critical	Mill Speed, rpm	Ball Filling, %	Total Filling, %	Ore Specific Gravity	Gross Power, kW
Overflow	4.41	6.10	6.10	1.38	0.74	14.86	35	35	4.10	1,900.00
Overflow	2.30	4.20	4.20	1.83	0.82	22.87	36	36	2.70	299.00
Overflow	2.65	3.40	3.40	1.28	0.77	20.08	36	36	2.70	334.00
Overflow	2.52	3.66	3.66	1.45	0.67	17.98	35	35	2.70	265.00
Grate	1.73	2.44	2.44	1.41	0.68	22.03	35	35	2.70	97.00
Overflow	3.48	4.62	4.62	1.33	0.71	16.10	39	39	2.70	834.00
Overflow	3.54	4.88	4.88	1.38	0.76	17.20	42	42	2.70	1,029.00
Overflow	4.12	5.49	5.49	1.33	0.75	15.57	45	45	2.70	1,600.00
Overflow	4.38	7.45	7.45	1.70	0.75	15.16	30	30	2.70	2,026.00
Overflow	5.29	7.32	7.32	1.38	0.70	12.87	40	40	3.20	3,828.00
Overflow	4.80	6.10	6.10	1.27	0.69	13.32	40	40	3.00	2,498.00
Overflow	3.05	4.27	4.27	1.40	0.70	16.95	40	40	4.50	580.00
Overflow	2.60	3.70	3.70	1.42	0.69	18.10	40	40	4.50	347.00
Overflow	3.05	4.27	4.27	1.40	0.73	17.68	45	45	3.90	600.00
Overflow	3.50	4.42	4.42	1.26	0.74	16.73	35	35	2.75	820.00
Overflow	4.87	8.84	8.84	1.82	0.72	13.80	27	27	2.60	2,900.00
Overflow	4.87	8.84	8.84	1.82	0.75	14.37	30	30	2.60	3,225.00
Overflow	4.87	8.80	8.80	1.81	0.75	14.37	31	31	2.60	3,104.00
Overflow	5.33	8.54	8.54	1.60	0.72	13.23	34	34	2.60	4,100.00
Overflow	3.04	3.05	3.05	1.00	0.82	19.77	45	45	3.50	475.00
Overflow	2.29	2.74	2.74	1.20	0.83	23.11	44	44	3.50	235.00
Grate	1.70	2.70	2.70	1.59	0.81	26.27	40	40	2.70	103.00
Overflow	3.55	4.87	4.87	1.37	0.72	16.16	40	40	2.80	970.00
Overflow	3.50	4.75	4.75	1.36	0.75	16.95	42	42	2.80	921.00
Overflow	0.85	1.52	1.52	1.79	0.71	32.57	40	40	2.90	10.00
Overflow	0.85	1.52	1.52	1.79	0.71	32.57	20	20	2.90	6.80
Overflow	4.75	6.26	6.26	1.32	0.77	14.94	28	28	2.68	2,050.00
Overflow	3.85	5.90	5.90	1.53	0.77	16.60	30	30	2.80	1,300.00
Grate	2.64	3.66	3.66	1.39	0.70	18.22	43	43	2.80	420.00
Overflow	4.12	7.04	7.04	1.71	0.70	14.69	38	38	2.60	1,800.00
Overflow	4.10	5.92	5.92	1.44	0.75	15.67	34	34	3.10	1,525.00
Overflow	4.35	6.56	6.56	1.51	0.70	14.19	40	40	2.72	1,850.00
Overflow	3.48	6.33	6.33	1.82	0.75	17.00	34	34	2.70	1,150.00
Overflow	3.83	4.83	4.88	1.26	0.61	13.29	31	31	2.60	842.00
Overflow	4.68	5.64	5.64	1.21	0.72	14.08	48	48	2.80	2,300.00
Overflow	4.73	7.01	7.01	1.48	0.60	11.76	32	32	2.80	1,840.00
Overflow	5.34	8.69	8.69	1.63	0.73	13.36	28	28	3.20	3,669.00
Overflow	5.34	8.69	8.69	1.63	0.73	13.36	26	26	3.20	3,549.00
Overflow	5.34	8.69	8.69	1.63	0.73	13.36	24	24	3.20	3,385.00
Overflow	5.34	8.69	8.69	1.63	0.73	13.36	23	23	3.20	3,251.00
Overflow	3.87	6.34	6.34	1.64	0.69	14.83	27	27	4.60	1,075.00
Number	41	41	41	41	41	41	41	41	41	41
Mean	3.73	5.58	5.58	1.49	0.73	17.06	35.25	35.32	3.03	1,539.34
Standard deviation	1.23	2.15	2.15	0.21	0.05	4.73	6.79	6.80	0.55	1,223.53
Minimum	0.85	1.52	1.52	1.00	0.60	11.76	20	20	2.60	6.80
Maximum	5.34	8.84	8.84	1.83	0.83	32.57	48	48	4.60	4,100.00

Table 3 SAG mill power draw data

Discharge Mechanism	Diameter Inside Liners, m	Length (belly), m	Length (C/line), m	Length/ Diameter Ratio	Mill Speed, fraction of critical	Mill Speed, rpm	Ball Filling, %	Total Filling, %	Ore Specific Gravity	Gross Power, kW
Grate	w	3.46	3.46	0.45	0.70	10.65	11	11	2.60	1,800.00
Grate	6.50	2.42	3.02	0.37	0.75	12.44	6	21	3.64	1,228.00
Grate	4.35	4.85	4.85	1.11	0.75	15.29	12	29	2.60	1,045.00
Grate	7.05	3.45	3.45	0.49	0.72	11.47	12	33	2.65	2,239.00
Grate	7.05	3.45	3.45	0.49	0.72	11.47	12	12	2.65	1,500.00
Grate	5.30	7.95	7.95	1.50	0.71	13.04	18	30	2.80	3,284.00
Grate	4.05	4.60	4.60	1.14	0.76	15.97	8	26	2.70	688.00
Grate	4.05	4.60	4.60	1.14	0.76	15.97	7	7	2.70	440.00
Grate	4.05	4.60	4.60	1.14	0.76	15.97	6	32	2.70	687.00
Grate	4.05	4.60	4.60	1.14	0.76	15.97	6	34	2.70	706.00
Grate	6.51	2.44	2.44	0.38	0.71	11.77	3	16	4.10	972.00
Grate	1.80	0.59	0.59	0.33	0.75	23.55	6	27	2.74	14.80
Grate	9.59	4.27	5.86	0.45	0.75	10.24	14	14	2.60	5,790.00
Grate	9.59	4.27	5.86	0.45	0.75	10.24	19	31	2.60	7,900.00
Grate	9.59	4.27	5.86	0.45	0.75	10.24	17	30	2.60	7,100.00
Grate	8.39	3.26	5.00	0.39	0.80	11.69	14	18	2.68	4,000.00
Grate	4.12	5.02	5.02	1.22	0.75	15.63	22	22	2.70	1,012.00
Grate	4.12	5.02	5.02	1.22	0.75	15.63	22	33	2.70	1,225.00
Grate	4.16	4.78	4.78	1.15	0.89	18.44	10	38	2.70	1,063.00
Grate	3.90	5.10	5.10	1.31	0.78	16.75	25	34	3.35	1,175.00
Grate	5.08	6.82	6.82	1.34	0.66	12.38	12	31	2.85	2,000.00
Grate	5.05	5.99	5.99	1.19	0.77	14.49	17	21	2.68	2,035.00
Grate	5.82	5.65	5.65	0.97	0.81	14.20	13	33	2.80	2,840.00
Grate	5.80	5.65	5.65	0.97	0.81	14.20	10	27	2.80	2,600.00
Grate	3.85	5.69	5.69	1.48	0.48	10.35	12	12	2.80	424.00
Grate	7.23	3.00	3.00	0.42	0.75	11.80	11	16	2.72	1,920.00
Grate	7.09	2.74	2.74	0.39	0.75	11.91	11	21	3.10	1,900.00
Grate	6.26	2.50	2.50	0.40	0.71	12.00	5	21	2.70	1,200.00
Number	28	28	28	28	28	28	28	28	28	28
Mean	5.79	4.32	4.57	0.84	0.74	13.71	12.16	24.25	2.82	2,099.49
Standard deviation	2.01	1.52	1.55	0.41	0.07	3.03	5.67	8.52	0.34	1,946.55
Minimum	1.80	0.59	0.59	0.33	0.48	10.09	3	7	2.60	14.80
Maximum	9.59	7.95	7.95	1.50	0.89	23.55	25	38	4.10	7,900.00

Mill type	$\sin\alpha$
Autogenous	0.707
Overflow	0.682
Grate	0.809

The mill filling component of Equation 57 was also modified to account for mills with a relatively low filling. Therefore:

$$P = 1.33\pi D^3 L\rho_c NJ_t(1 - J_t)g \sin\alpha; \quad 0.35 < J_t < 0.5 \qquad \text{(EQ 58a)}$$

$$P = 1.33\pi D^3 L\rho_c NJ_t(1.05 - 1.133J_t)g \sin\alpha; \quad 0.2 < J_t < 0.35$$

Validation

The mill power draw database that was published by Morrell (1993) and is reproduced in Tables 2–4 was used to evaluate the Harris et al. power draw model. Results are shown graphically in Figures 45 and 46. Statistical analysis of results shows that for ball mills, the standard deviation of the differences between observed and predicted values is 9.7% with an overall mean of differences of 27.1%. For AG/SAG mills, the associated figures are 11.6% and 11.9%, respectively. The results indicate that for both ball and AG/SAG mills, if the inherent bias were to be corrected, it is a reasonable model.

Table 4 AG mill power draw data

Discharge Mechanism	Diameter Inside Liners, m	Length (belly), m	Length (C/line), m	Length/Diameter Ratio	Mill Speed, fraction of critical	Mill Speed, rpm	Total Filling, %	Ore Specific Gravity	Gross Power, kW
Grate	7.10	2.43	3.47	0.34	0.72	11.43	10	3.57	703.00
Grate	7.10	2.43	3.47	0.34	0.72	11.43	12	4.60	1,009.00
Grate	6.49	2.25	2.48	0.35	0.75	12.45	27	4.00	1,240.00
Grate	6.49	2.25	2.48	0.35	0.75	12.45	19	4.00	960.00
Grate	5.11	5.18	5.18	1.00	0.73	13.63	24	4.20	1,264.00
Grate	1.80	0.59	0.59	0.33	0.75	23.55	25	2.74	12.50
Grate	9.50	4.40	6.40	0.46	0.75	10.29	31	2.90	5,490.00
Number	7	7	7	7	7	7	7	7	7
Mean	6.23	2.79	3.44	0.45	0.74	13.60	21.11	3.72	1,525.50
Standard deviation	2.35	1.53	1.90	0.25	0.01	4.51	7.78	0.69	1,799.05
Minimum	1.80	0.59	0.59	0.33	0.72	10.29	10	2.74	12.50
Maximum	9.50	5.18	6.40	1.00	0.75	23.55	31	4.60	5,490.00

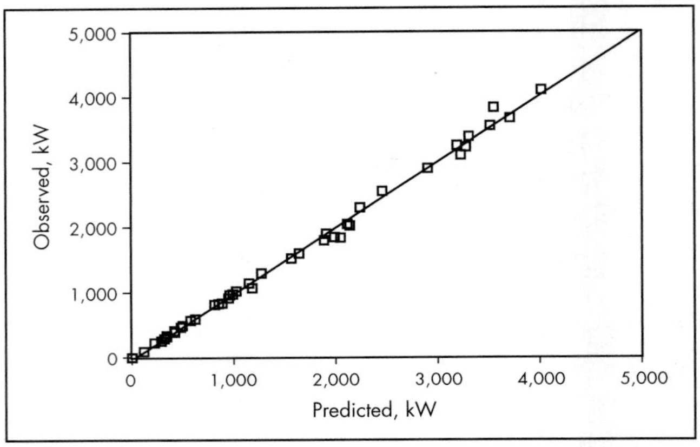

Source: Morrell 1993

Figure 41 Observed versus predicted power draw of ball mills using the Bond model

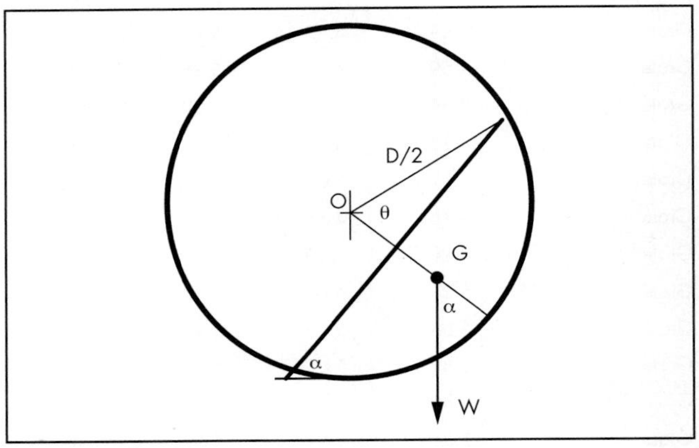

Source: Morrell 1993

Figure 43 Torque-arm treatment of power draw

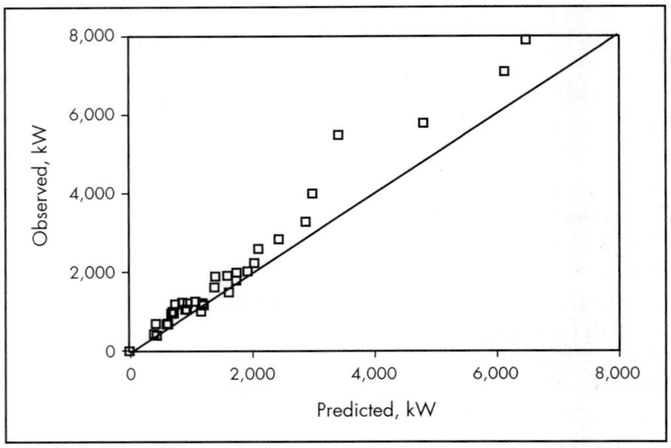

Source: Morrell 1993

Figure 42 Observed versus predicted power draw of AG/SAG mills using the Bond model

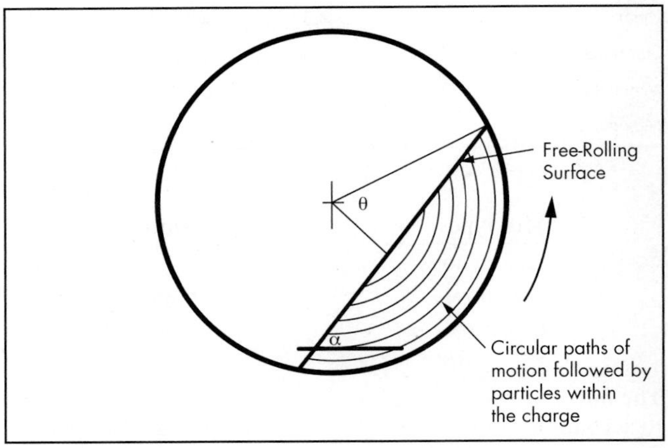

Source: Morrell 1993

Figure 44 Idealized charge motion used by Hogg and Fuerstenau (1972)

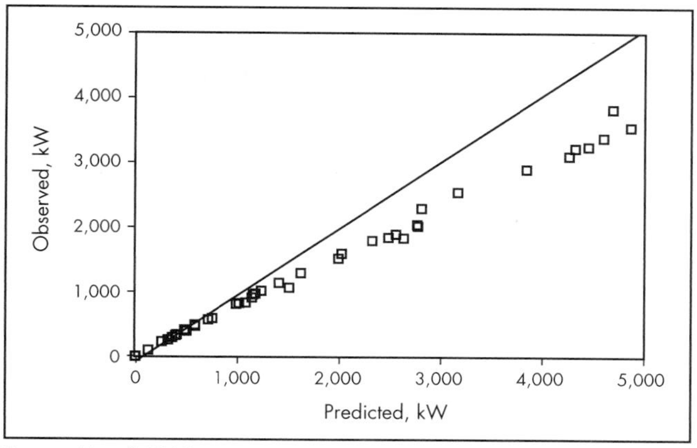

Source: Morrell 1993

Figure 45 Observed versus predicted power draw of ball mills using the Harris et al. (1985) model

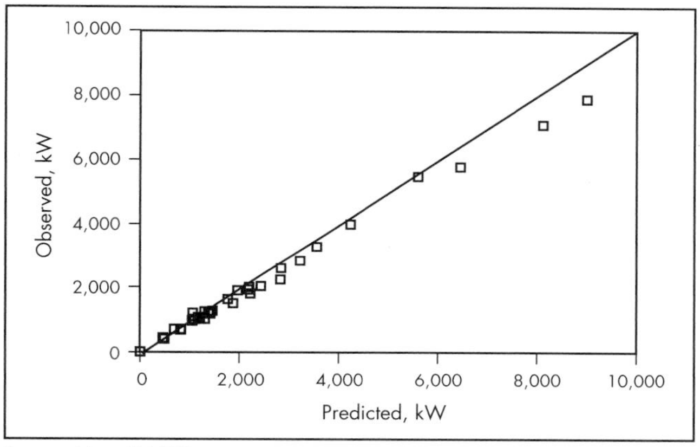

Source: Morrell 1993

Figure 46 Observed versus predicted power draw of AG/SAG mills using the Harris et al. (1985) model

Austin Model

From his literature search, Austin (1990) concluded that for semiautogenous mills there were no generally accepted mill power equations comparable to those of Bond. He therefore used elements of Hogg and Fuerstenau's (1972) and Bond's (1962) equations, plus additional modifications, to provide a model that was claimed to be suitable for SAG mills with both high and low aspect ratios. However, the inherent form of his model appears equally suitable for ball mills. His equation is written as follows:

$$M_p = 10.6D^{2.5}L(1-1.03J_t)[1-E_B)\left(\frac{\rho_s}{w_c}\right)J_t + 0.6J_B\left(\rho_b - \frac{\rho_s}{w_c}\right)](\phi_c)\left(1-\frac{0.1}{2^{9-10\phi_c}}\right)(1+f_3) \quad \textbf{(EQ 59)}$$

where

M_p = net mill power (power at pinion), kW
D = mill internal diameter, m
L = mill length, m
J_t = fractional volume of cylindrical mill filled by total charge
E_B = porosity of total charge = 0.3

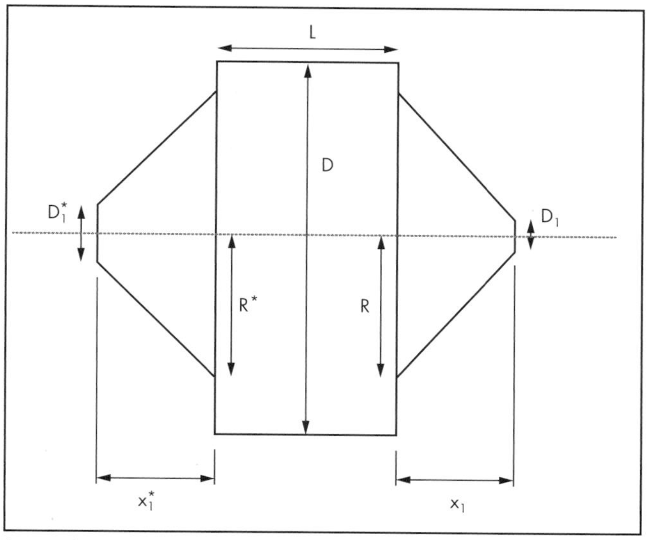

Source: Austin 1990

Figure 47 Schematic of mill with cone ends

ρ_s = mean density of rock, t/m^3
w_c = weight fraction of rock to water and rock in the mill charge ≈ 0.88
J_B = fractional volume of cylindrical mill filled by balls only
ρ_b = density of ball material, t/m^3 = 7.9 t/m^3
ϕ_c = mill rotational speed as a fraction of critical speed
f_3 = conical end correction

Austin developed an equation for predicting the power draw of the conical ends by considering the power draw of the elements shown in Figure 47 and integrating with respect to the filled length of the cone. Rewriting Austin's original correction term (f_3) on the assumption that both cone ends are identical gives

$$f_3 = \frac{2\times 0.046}{J(1-1.03J)}\left\{\frac{\left(\dfrac{x_1/L}{1-D_1/2R}\right)}{\left[\left(\dfrac{1.25R/D}{0.5-J}\right)^{0.1} - \left(\dfrac{0.5-J_t}{1.25R/D}\right)^{4}\right]}\right\} \quad \textbf{(EQ 60)}$$

where

x_1 = length of the cone section, m
D_1 = trunnion diameter, m
R = maximum radius of cone section, m

Validation

The mill power draw database that was published by Morrell (1993) and is reproduced in Tables 2–4 was used to evaluate the Austin power draw model. Results are shown graphically in Figures 48 and 49. Statistical analysis of results shows that for ball mills, the standard deviation of the differences between observed and predicted values is 9.4% with an overall mean of differences of 19.2%. For AG/SAG mills, the associated figures are 13.8% and 9.9%, respectively. The results indicate that as with the Harris et al. model, there is considerable bias, both for ball and AG/SAG mills. For ball mills, Austin's model is on par with Harris et al.'s, but in the case of AG/SAG mills, it is not as good and can be considered relatively poor.

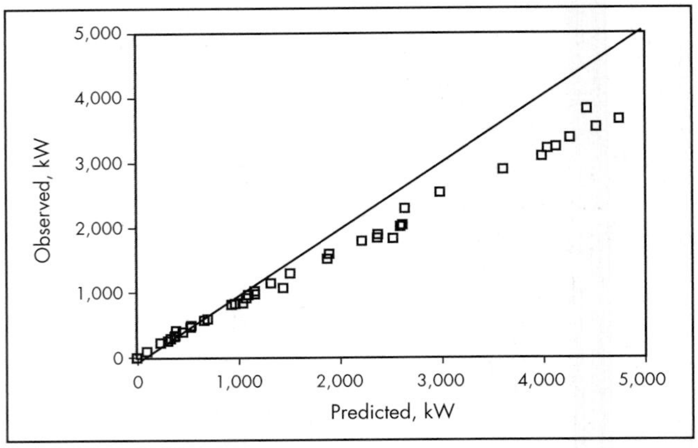

Source: Morrell 1993

Figure 48 Observed versus predicted power draw of ball mills using the Austin model

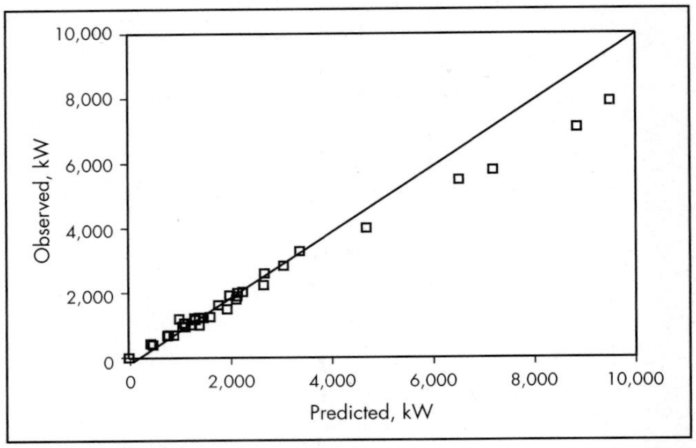

Source: Morrell 1993

Figure 49 Observed versus predicted power draw of AG/SAG mills using the Austin model

Rose and Evans Model

The contributions of the work conducted by Rose and Evans are reported in the first two papers of a trilogy written between 1954 and 1955 and published in 1956 (Rose and Evans 1956a, 1956b; Rose and Blunt 1956).

The work of these two researchers was the first, and so far remains the most thorough, investigation of most of the factors that might reasonably be expected to affect grinding mill power draw. However, the work was conducted using mills of less than 3 in. in diameter.

Dimensional analysis was used to develop the following equation:

$$\frac{P}{D^5N^3\rho} = \phi\Big\{\Big(\frac{h}{D}\Big),\Big(\frac{d}{D}\Big),\Big(\frac{g}{DN^2}\Big),\Big(\frac{b}{D}\Big),\Big(\frac{H}{D^6N^2\rho^2}\Big),\Big(\frac{v}{D^2N}\Big),\Big(\frac{\sigma}{\rho}\Big),\Big(\frac{L}{D}\Big),(J),(f),(e),(v),(u)(n)\Big\} \quad \textbf{(EQ 61)}$$

where

P = power
D = internal diameter of the mill
N = speed of rotation of the mill, rps
ρ = density of ball material
ϕ = denotes some function of each of the dimensionless groups
h = height of lifters
d = diameter of the balls
g = acceleration due to gravity
b = representative particle dimension
H = energy required to bring about unit increase in the specific surface of the powder (specific surface in units of area per unit mass)
v = kinematic viscosity of the mixture of powder and fluid
σ = effective density of the mixture of powder and fluid
L = internal length of the mill
J = volume occupied by the ball charge (including voids), expressed as a fraction of the mill volume
f = coefficient of friction between balls and the mill
e = coefficient restitution between the balls and the mill
v = volume occupied by the powder charge (including voids), expressed as a fraction of the volume of voids in the ball charge
u = volume occupied by fluid expressed as a fraction of the volume of voids in the charge
n = number of lifters

Following analysis of their experimental results, they arrived at the following equation:

$$P = D^5N^3\rho_b\Big(1+0.4\frac{\rho_p}{\rho_b}\Big)\Big(\frac{L}{D}\Big)\times\gamma_1\Big(\frac{N_c}{N}\Big)\times\gamma_2(J)\times\gamma_3\Big(\frac{D}{d}\Big)\times\gamma_4(n)\times\gamma_5\Big(\frac{h}{D}\Big) \quad \textbf{(EQ 62)}$$

where N_c is the critical speed of rotation of the mill. The functions γ_1–γ_5 were fitted to their laboratory mill data and were presented in graphical form. The functions γ_3–γ_5 were found to be equal to unity for most conditions. The function γ_1 was approximated by them as

$$\gamma_1\Big(\frac{N_c}{N}\Big) = 3.13\phi^{-2} \quad \textbf{(EQ 63)}$$

where ϕ is the fraction of critical speed.

The function γ_2 was fitted by Morrell (1993) to Rose and Evans' data to give the following equation:

$$\gamma_2\,(J) = 3.3506J_t + 1.3372J_t^2 - 9.1602J_t^3 \quad \textbf{(EQ 64)}$$

The effect of the discharge type was also addressed by Rose and Evans through the application of an additional function (γ_6), which the product of Equation 62 was multiplied by. Again, the function was presented in graphical form. This function was also fitted by the author as follows:

$$\gamma_6 = 1.7796 - 6.2164J_t + 13.6615J^2 - 8.1923J^3 \quad \textbf{(EQ 65)}$$

Rose and Evans developed their model for ball mills and therefore their charge density (ρ_b) is in terms of the specific gravity of steel. To modify their equation to apply to SAG and AG mills, the ρ_b term was replaced by the mean density of the grinding media (ρ_c) as follows:

$$\rho_c = \frac{J_b\rho_b + J_o\rho_o}{J_t} \quad \textbf{(EQ 66)}$$

where

J_b = volume of mill occupied by ball component of grinding media (including voids)

J_o = volume of mill occupied by ore component of grinding media (including voids)

ρ_o = ore specific gravity

J_t = volume of mill occupied by total grinding charge ($J_t = J_b + J_o$)

For ball mills, where $J_o = 0$ and $J_b = J_t$, Equation 66 reduces to $\rho_c = \rho_b$.

Validation

The mill power draw database that was published by Morrell (1993) and is reproduced in Tables 2–4 was used to evaluate the Rose and Evans power draw model. Results are shown graphically in Figures 50 and 51. Statistical analysis of results shows that for ball mills, the standard deviation of the differences between observed and predicted values is 6.3% with an overall mean of differences of 15.0%. For AG/SAG mills, the associated figures are 9.0% and 27.1%, respectively. The results indicate there is considerable bias of the prediction of AG/SAG mill power draw. However, if this could be corrected, the standard deviation results indicate that it would be better than Austin, Bond, and Harris et al.'s models. For ball mills, the Rose and Evans model also has some bias, and if this could be corrected, it would be on par with Bond's ball mill model.

Morrell Model

Morrell (1993) used an energy balance approach where power was taken to be the rate at which potential and kinetic energy is imparted to the charge. He used a different shape of the charge from that given in Figure 40, opting instead for the one shown in Figure 52.

Morrell argued that this more accurately described the shape of the moving charge in contact with the mill shell (i.e., it excluded material in flight and the stationary material at the

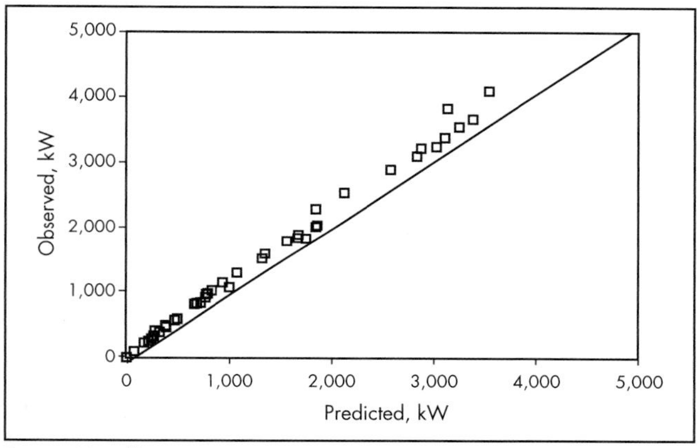

Source: Morrell 1993

Figure 50 Observed versus predicted power draw of ball mills using the Rose and Evans model

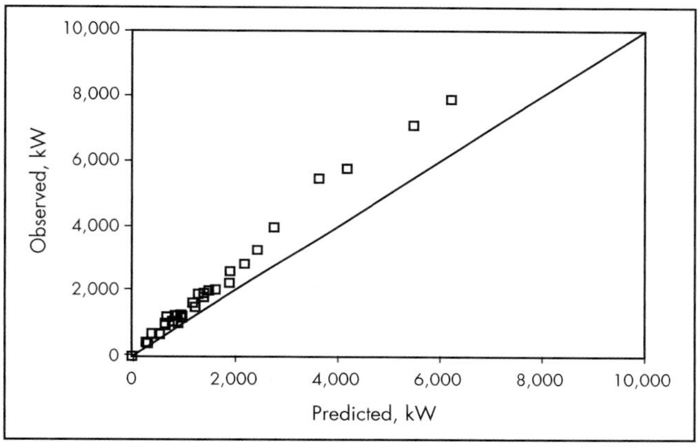

Source: Morrell 1993

Figure 51 Observed versus predicted power draw of AG/SAG mills using the Rose and Evans model

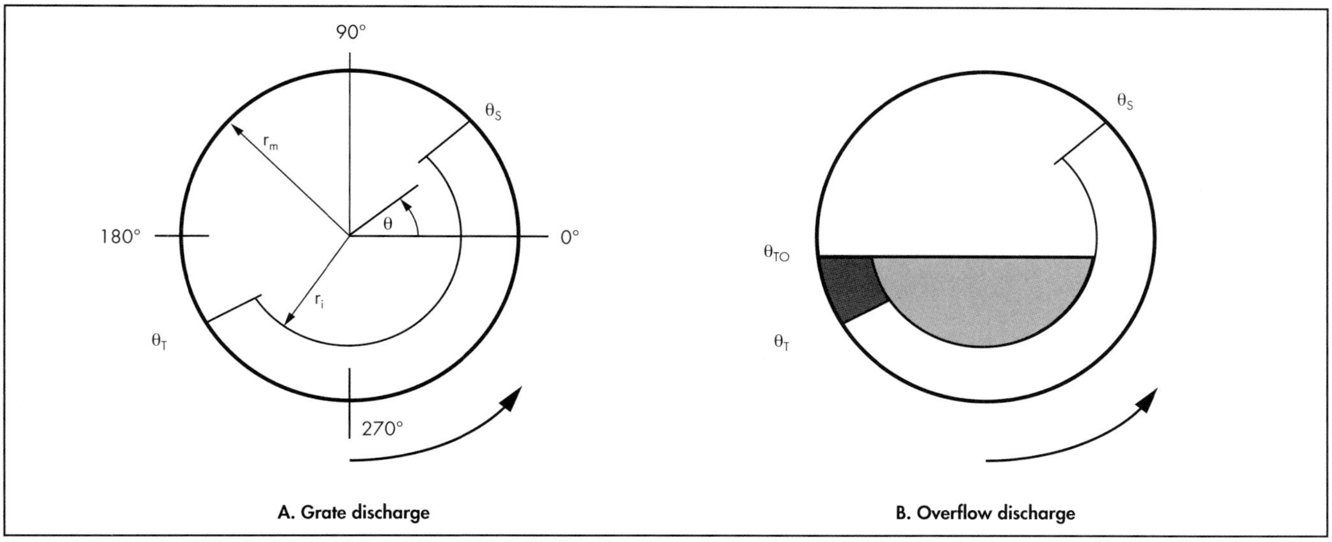

Figure 52 (A) Simplified charge shape for grate mills (no slurry pool); (B) simplified charge shape for overflow mills

center of the charge). These conclusions came from observations of the shape of the charge in a glass laboratory mill (Figure 53).

Equation 67 was therefore developed for the cylindrical section of the mill (a similar equation was also developed for conical ends). From this equation, the power associated with the rock/ball charge is calculated separately from the slurry, the total power being the sum of both.

$$P_{cylinder} = \int_{r_i}^{r_m} \left(V_r L r g \left(\begin{matrix} \rho_c(\sin\theta_S - \sin\theta_T) \\ + \rho_p(\sin\theta_S - \sin\theta_{Tp}) \end{matrix} \right) \right) dr \qquad \text{(EQ 67)}$$

where

$P_{cylinder}$ = power delivered to the charge (net power)
r_m = radius of mill inside liners
r_i = radial position of charge inner surface
V_r = tangential velocity of a particle at radial distance r
L = length of cylindrical section of the mill inside liners
r = radial position
g = gravitational constant
ρ_c = density of rock/ball charge (excluding pulp)
θ_S = angular displacement of shoulder position at the mill shell
θ_T = angular displacement of the grinding media toe position at the mill shell
ρ_p = density of pulp phase
θ_{Tp} = angular displacement of the slurry toe position at the mill shell. For overflow ball mills and AG/SAG mills with slurry pooling, $\theta_{Tp} = \theta_{TO}$ (see Figure 52B).

Equation 67 shows that to execute the model, it is necessary to also have relationships that predict values for parameters such as θ_S, θ_T, θ_{Tp}, θ_{TO}, V_r, ρ, and r_i. All of these relationships are provided by Morrell (1993, 1996a).

Equation 67 describes the power draw by the charge (net power). It does not include electrical losses across the motor as well as the power required to overcome friction in the bearings, and losses in gearboxes/reducers as well as in the gear/pinion coupling, where the mill has a gear-and-pinion drive. Note that this definition of *net power* is not the same as the definition of "power at pinion." The power at pinion cannot normally be measured other than in some pilot mills and is estimated from assumptions about the energy losses of various components in the drive train. Theoretically, it is the power delivered to the pinion shaft in gear-and-pinion drives. It is meant to account for electrical motor and gearbox/reducer energy losses only but does not include the energy losses associated with the pinion gear/ring gear coupling nor bearings.

In practice, the only power draw that is usually measured in full-scale plants is the metered power (i.e., motor input power). Given this fact, and the need to properly validate the model through comparison of measured and predicted values, the power model must also predict motor input power. To do so, the model is configured as follows:

$$\text{motor input power} = \text{no-load power} + \text{net power} \qquad \text{(EQ 68)}$$

Net power is given by Equation 67. A further equation that predicts no-load power is required. The semi-empirical no-load power equation form proposed by Morrell (1996a) is used for this purpose, though it has been modified slightly based on additional data:

$$\text{no-load power (kW)} = K(D^{2.5}\phi(0.667L_d + L))^{0.82} \qquad \text{(EQ 69)}$$

where

K = 2.13 for gear and pinion drives and 1.28 for gearless drives
D = mill diameter
ϕ = fraction of critical speed
L_d = length of cone end
L = length of cylindrical section

The resultant model should be applicable to all tumbling mills.

Validation

The mill power draw database that was published by Morrell (1993) and is reproduced in Tables 2–4 was used to evaluate

Source: Morrell 1993
Figure 53 Charge inside a rotating mill

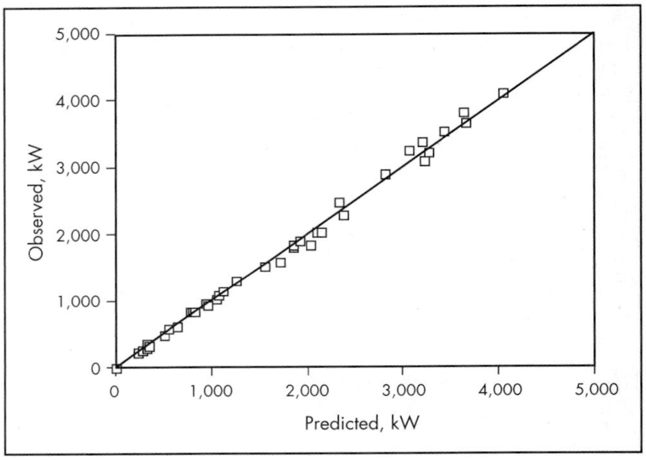

Source: Morrell 1993
Figure 54 Observed versus predicted power draw of ball mills using the Morrell model

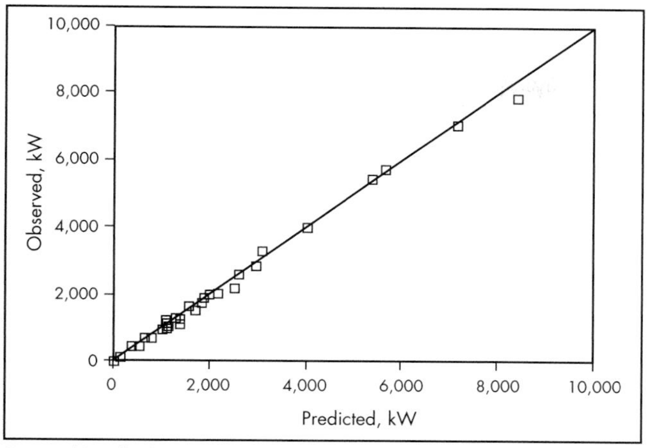

Source: Morrell 1993

Figure 55 Observed versus predicted power draw of AG/SAG mills using the Morrell model

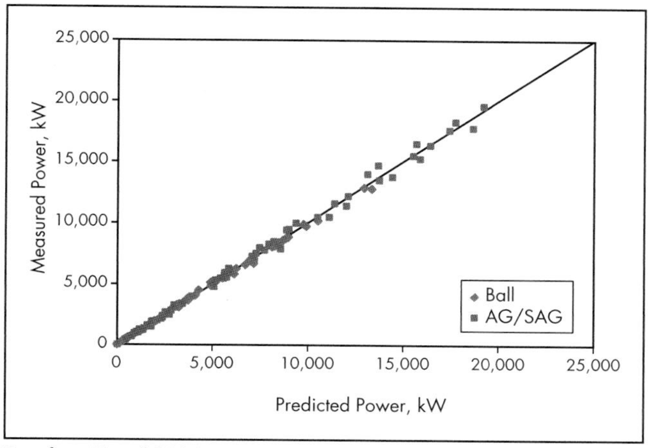

Data from Morrell 2004a

Figure 56 Observed versus predicted power draw of 140 AG, SAG, and ball mills using Morrell power model

the Morrell power draw model. Results are shown graphically in Figures 54 and 55. Statistical analysis of results shows that for ball mills, the standard deviation of the differences between observed and predicted values is 4.4% with an overall mean of differences of 0.3%. For AG/SAG mills, the associated figures are 6.3% and 0.4%, respectively.

Morrell added to his 1993 published database and used it to further benchmark his model (Morrell 2004a), the results of which are illustrated in Figure 56. The data comprise more than 140 ball, AG, and SAG mill data sets. The standard deviation of the differences between ball mill observed and predicted values is 3.1% with an overall mean of difference of 0.0%. For AG/SAG mills, the associated values are 3.8% and 0.0%, respectively.

Power Draw Data

The only comprehensive database of mill power draws and associated mill design and operating conditions that has been published to date is that accumulated by Morrell (1993, 1996b). The details of this database are in Tables 2–4.

In relation to the power draw data that the tables contain, operating plants vary widely in the type and complexity of their instrumentation. As a result, power draw data can be available on-site from a range of devices including kilwatt hour meters, power transducers, and ammeters. Where more than one source of power data was available at a particular site, it was ensured that all sources gave similar readings. If they did not give similar readings, plant electrical staff were asked to investigate and correct the differences. Where this could not be done, the data were not included in the database. If only one source of power measurement was available, efforts were made to ensure that independent checks of the power reading accuracy had been made either prior to the field study or shortly afterward.

The comments in the preceding paragraph are particularly relevant when using some published sources of power draw information, where appropriate checks of the power draw measurements have not been carried out because they may contain significant inaccuracies.

Additional, although limited, data that has been sourced from other published sources can be found in Doll (2013) together with an evaluation of Austin'sand Morrell's model. Daniel et al. (2010) also provide an assessment of the Austin, Hogg and Fuerstenau, and Morrell models.

STEEL WEAR PREDICTION

The most popular laboratory wear test to date is the one developed by Bond (1961). The test machine comprises a drum rotating at 70 rpm into which a sample of ore to be tested is introduced. Also in the drum is a rotating steel impeller comprising a steel paddle with dimensions 3 × 1 × 0.25 in., its center of rotation the same as the drum. The impeller rotates at 632 rpm, and as the drum rotates, the impeller is showered with rock pieces. As a result, the impeller is worn down.

The standard test uses a total of 1.6 kg of rock sample, in the size range –0.75+0.5 in. The 1.6 kg total is divided into four equal batches of 400 g and each batch is placed in the machine for 15 minutes, then removed. The total weight loss (g) of the paddle that results is determined, and this weight loss is designated Ai, the abrasion index.

Equations

Rowland (1982) provides the following equations for applying the abrasion index:

Wet rod mills:

$$\text{rod wear} = 0.159\ (\text{Ai} - 0.020)^{0.2}\ \text{kg/kW·h} \quad \textbf{(EQ 70)}$$

$$\text{liner wear} = 0.0159\ (\text{Ai} - 0.015)^{0.3}\ \text{kg/kW·h} \quad \textbf{(EQ 71)}$$

Wet ball mills:

$$\text{ball wear} = 0.159\ (\text{Ai} - 0.015)^{0.34}\ \text{kg/kW·h} \quad \textbf{(EQ 72)}$$

$$\text{liner wear} = 0.0118\ (\text{Ai} - 0.015)^{0.3}\ \text{kg/kW·h} \quad \textbf{(EQ 73)}$$

For AG/SAG mills, Giblett and Seidel (2011) proposed the following equation from their analysis of a range of measured wear data from Newmont circuits:

$$\text{SAG mill ball wear} = 0.0017(\text{A} \times \text{b}) - 0.0074 \quad \textbf{(EQ 74)}$$

where A and b are the JK comminution modeling parameters that are obtained from SMC Tests/drop weight tests.

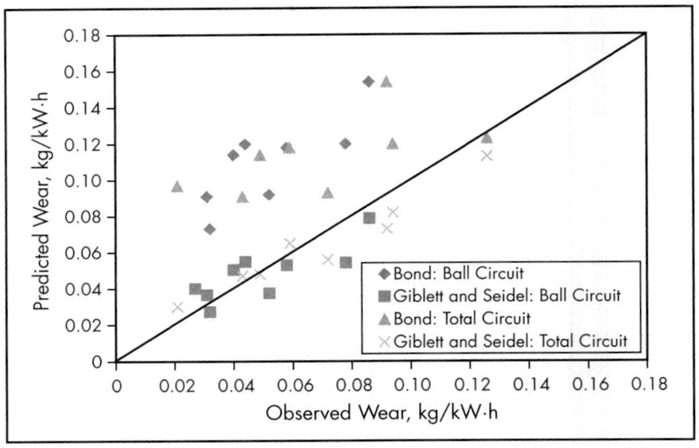

Adapted from Giblett and Seidel 2011

Figure 57 Predicted versus observed ball wear rates

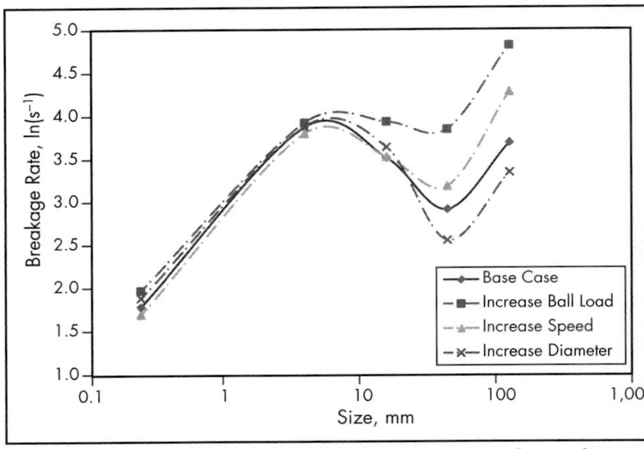

Figure 58 Example of AG/SAG breakage rate dependencies

For ball mills, Giblett and Seidel proposed the following:

$$\text{ball mill ball wear} = 0.0817(\text{Ai})^{0.498} \quad \textbf{(EQ 75)}$$

Validation

The only significant data that have been published are those of Giblett and Seidel (2011). These are plotted in Figure 57 and show that the Giblett and Seidel equations predict the observed wear rates quite well and are far superior to Bond/Rowland's equations.

PROCESS SIMULATION MODELING

The advent and rapid development of high-speed computers provided the opportunity to develop complex models of comminution machines that could execute their calculations very rapidly and could be linked together such that entire circuits could be simulated. These models contrast significantly with the energy–size relationships and are aimed at reproducing the overall response of comminution machines in terms of throughput, power draw, and product size distribution. By their nature, such models are relatively complex and may have many interactions between the various subprocesses that describe the machine in question, and that without computers would be almost impossible to apply in a reasonable time frame. Therefore, in parallel with the development of modern computers, researchers started developing mathematical simulation models of AG/SAG mills, ball mills, crushers, and classifiers (Lynch 1977; Austin et al. 1984; Herbst and Fuerstenau 1973; Whiten 1974). Much of the early modeling work was academically oriented and of little benefit to practicing metallurgists. However, as time went on, user-friendly interfaces were developed that broadened the models' appeal and impact. Such developments gave rise to process simulators such as MODSIM, USIM PAC, METSIM, and JKSimMet.

In all cases, comminution simulation model researchers relied on what is known as the population balance model (or variants of it) as their mathematical framework because it elegantly encapsulated the size reduction process in many comminution machines. The population balance model was originally introduced by Epstein (1947) and can be represented as follows:

$$0 = f_i - p_i + \sum_{j=1}^{i-1} k_j s_j b_{ij} - k_i s_i \quad \textbf{(EQ 76)}$$

where

f_i = t/h of particles of size i in the feed
p_i = t/h of particles of size i in the product
b_{ij} = breakage distribution function
k_i = breakage rate of particles of size i
s_i = mass of particles in the charge of size i

Whiten's (1974) variant of the population balance model essentially uses Equation 76, but to it he added another equation that enabled the introduction of material transportation to be easily incorporated, which greatly aided the development of grate discharge mill models such as those that could be applied to AG/SAG mills. This equation was written:

$$p_i = d_i s_i \quad \textbf{(EQ 77)}$$

where d_i is the discharge rate of particles of size i.

The simplicity of the preceding equations is the source of both their greatest strengths and their greatest weakness. Their greatest strengths are their ease of use and versatility, while their greatest weakness is the lack of any physical description of the subprocesses on which they depend. For them to be used successfully, therefore, a series of supplementary models must be developed and linked to them. Even when such subprocesses are developed, invariably there will remain a range of parameters that must be specified for the models to work. Considering Equations 76 and 77, for example, if a product size distribution and charge must be obtained from simulation of a mill grinding a specified feed, then size-distributed parameters k_i, b_i, and d_i must be specified. In the case of the JKSimMet modeling approach (Napier-Munn et al. 1996), the b_i parameters were derived from the A and b values that are obtained from drop weight or SMC Tests. This left the d_i and k_i, which, in the early years of development, had to be fitted to data from existing mills. Consequently, the use of such models was restricted to process optimization studies, and therefore in design situations they were of limited use. However, even though limited to optimization studies, such models were extremely valuable because they enabled practicing metallurgists to run and assess strategies to improve plant performance. This significantly reduced the time-consuming and potentially costly trial-and-error field experimentation that was the usual methodology until then.

Many models in currently commercially available simulators still suffer from the problem of requiring existing plant data for suitable parameters and therefore still have limited use in design projects. The JKSimMet AG/SAG mill variable rates model (Morrell and Morrison 1996) and Morrell's (2004c) model are two exceptions. Both models used a range of operating plant data to determine the dependency of the model parameters on factors such as mill design and operating conditions. An example of the data that was used to develop these dependencies is shown in Figure 58. It illustrates the highly important breakage rate distribution, which changes its shape depending on the design and operating conditions. By mathematically describing how this shape changes, the resultant equations can be used in design situations to predict all of the required parameters for the simulation to be performed. This gives the design metallurgist a very powerful tool because it enables stream mass and size balances to be generated from the chosen circuit and chosen mill sizes, and allows him or her to run sensitivity analyses to ensure that target throughputs and grind sizes can be achieved under a range of feed and operating conditions. It significantly increases confidence that the chosen circuit and equipment will achieve their specified goals.

No model (regardless of how sophisticated) is an expert system that enables metallurgists with limited knowledge/experience to become "instant experts" by having a simulator at their disposal. Models/simulators are just tools, and it is incumbent on the user to fully understand how they work, what their limitations are, and the operational intricacies of the processes that they simulate. Knowing the model's limitations is particularly important, and this can only be established by benchmarking the model's performance over as wide a range of operating data as possible. According to the published literature, this benchmarking has not often been done, with the exception of the JKMRC variable rates model (Morrell and Morrison 1996) and Morrell model (Morrell 2004c).

REFERENCES

Amelunxen, P. 2003. The application of the SAG power index to ore body hardness characterization for the design and optimization of autogenous grinding circuits. Master of Engineering thesis, McGill University, Canada.

Amelunxen, P., Berrios P., and Rodriguez, E. 2014. The SAG grindability index. *Miner. Eng.* 55:42–51.

Amelunxen, P., De la Cruz, R., Berrios, P., and Panduro, L. 2016. Revisión Tecnica del SAG Grindability Index (SGI and MiniSGI). Presented at 2nd International Seminar on Geometallurgy (GEOMET2016), Gecamin, Lima, Peru.

Angove, J.E., and Dunne, R.C. 1997. A review of standard physical ore property determinations. Presented at the World Gold Conference, Singapore, September 1–3.

Austin, L.G. 1990. A mill power equation for SAG mills. *Miner. Metall. Process.* (February):57–63.

Austin, L.G., Klimpel, R.R., and Luckie, P.T. 1984. *The Process Engineering of Size Reduction: Ball Milling.* New York: AIME.

Bailey, C., Lane, G., Morrell, S., and Staples, P. 2009. What can go wrong in comminution circuit design? Presented at the Tenth Mill Operators' Conference, Adelaide, South Australia, Australia, October 12–14, 2009.

Barratt, D.J., and Allan, M.J. 1986. Testing for autogenous and semiautogenous grinding: A designer's point of view. *Miner. Metall. Process.* 3:65–74.

Blaskett, K.S. 1969. Estimation of the power consumption in grinding mills. In *Proceedings of the Ninth Commonwealth Mining and Metallurgy Congress*, Vol. 3. London: Institution of Mining and Metallurgy. pp. 631–649.

Bond, F.C. 1946. Crushing tests by pressure and impact. *Trans. AIME* 169:58–65.

Bond, F.C. 1952. The third theory of comminution. *Trans. AIME* 193:484–494.

Bond, F.C. 1960. Confirmation of the third theory. *Trans. AIME* 217:139–153.

Bond, F.C. 1961. *Crushing and Grinding Calculations.* Milwaukee, WI: Allis-Chalmers Manufacturing.

Bond, F.C. 1962. *Crushing and Grinding Calculations Additions and Revisions.* Milwaukee, WI: Allis-Chalmers Manufacturing.

Bond, F.C. 1963. Metal wear in crushing and grinding. Presented at the 54th Annual Meeting of the American Institute of Chemical Engineers.

Bond, F.C., and Maxon, W. 1943. *Standard Grindability Tests and Calculations.* New York: AIME.

Daniel, M., Lane, G., and Morrell, S. 2010. Consolidation and validation of several tumbling mill power models. Presented at Procemin 2010, Santiago, Chile.

Dobby, G., Bennett, C., and Kosick, G. 2001. Advances in SAG circuit design and simulation applied to the Mine Block Model. In *Proceedings of the Conference on International Autogenous and Semi-Autogenous Grinding Technology 2001.* Vol. 4. Vancouver, BC: Mining and Mineral Process Engineering, University of British Columbia.

Doll, A. 2013. A comparison of SAG mill models. Presented at Procemin 2013, Santiago, Chile.

Epstein, B. 1947. The material description of certain breakage mechanisms leading to the logarithmic-normal distribution. *J. Franklin Inst.* 244:471–477.

Flavel, M.D., and Rimmer, H.W. 1982. Particle breakage studies in an impact environment. SME Preprint No. 82-124. Littleton, CO: SME.

Forsund, B., Norkyn, I., Sankvik, K.L., and Winther, K. 1988. Sydvarangers 6.5m diameter × 9.65m ball mill. In *Proceedings of the XVI International Mineral Processing Congress.* Edited by K.S.E. Forssberg. New York: Elsevier. pp. 171–183.

Giblett, A., and Morrell, S. 2016. Process development testing for comminution circuit design. *Miner. Metall. Process.* 33(4):172.

Giblett, A., and Seidel, J. 2011. Measuring, predicting and managing grinding media wear. Paper No. 34. In *Proceedings of the Conference on International Autogenous and Semi-Autogenous Grinding Technology 2011.* Vancouver, BC: Mining and Mineral Process Engineering, University of British Columbia.

Harris, C.C., Schnock, E.M., and Arbiter N. 1985. Grinding mill power consumption. *Miner. Process. Tech. Rev.* 1:297–345.

Herbst, L.A., and Fuerstenau, D.W. 1973. Mathematical simulation of dry ball milling using specific power information. *Trans. AIME* 254:343–348.

Hogg, R., and Fuerstenau, D.W. 1972. Power relationships for tumbling mills. *Trans. AIME* 252: 418–423.

Kosick, G.A., and Bennett, C. 1999. The value of ore body power requirement profiles for SAG circuit design. In *Proceedings of the 31*st *Annual Meeting of the Canadian Mineral Processors*, Ottawa, ON: Canadian Mineral Processors.

Lane, G.S. 2010. What is a competent ore and how does this effect comminution circuit design? In *Proceedings of the 7th International Mineral Processing Conference—Procemin 2010*. Santiago, Chile: Gecamin.

Lane, G., Foggiatto, B., and Bueno, M. 2013. Power-based comminution calculations using Ausgrind. Presented at Procemin 2013, Santiago, Chile.

Leung, K. 1987. An energy-based ore specific model for autogenous and semi-autogenous grinding. Ph.D. thesis, Julius Kruttschnitt Mineral Research Centre, University of Queensland, Australia.

Lynch, A.J. 1977. *Mineral Crushing and Grinding Circuits: Their Simulation, Optimisation, Design and Control*. Amsterdam, Netherlands: Elsevier Scientific.

MacPherson, A.R., and Turner, R.R. 1978. Autogenous grinding from test work to purchase of a commercial unit. In *Mineral Processing Plant Design, Practice, and Control*. Edited by A.L. Mular and R.B. Bhappu. New York: AIME. pp. 279–305.

Matei, V., Bailey, C.W., and Morrell, S. 2015. A new way of representing A and B parameters from JK drop-weight and SMC Tests: The "SCSE." Paper No. 16. In *Proceedings of the Conference on International Autogenous and Semi-Autogenous Grinding Technology 2015*. Vancouver, BC: Mining and Mineral Process Engineering, University of British Columbia.

Moore, D.C. 1982. Prediction of crusher power requirements and product size analysis. In *Design and Installation of Comminution Circuits*. Edited by A.L. Mular. New York: AIME. pp. 218–227.

Morrell, S. 1993. The prediction of power draw in wet tumbling mills. Ph.D. thesis, University of Queensland, Australia.

Morrell, S. 1996a. Power draw of wet tumbling mills and its relationship to charge dynamics—Part 1: A continuum approach to mathematical modelling of mill power draw. *Trans. Inst. Min. Metall.* 105:C43–C53.

Morrell, S. 1996b. Power draw of wet tumbling mills and its relationship to charge dynamics—Part 2: An empirical approach to modeling of mill power draw. *Trans. Inst. Min. Metall.* 105:C54–C62.

Morrell, S. 2001. Large diameter SAG mills need large diameter ball mills—What are the issues? In *Proceedings of the Conference on International Autogenous and Semi-Autogenous Grinding Technology 2001*. Vancouver, BC: Mining and Mineral Process Engineering, University of British Columbia.

Morrell, S. 2004a. Predicting the specific energy of autogenous and semi-autogenous mills from small diameter drill core samples. *Miner. Eng.* 17(3):447–451.

Morrell, S. 2004b. An alternative energy-size relationship to that proposed by bond for the design and optimisation of grinding circuits. *Int. J. Miner. Process*. 74:133–141.

Morrell, S. 2004c. A new autogenous and semi-autogenous mill model for scale-up, design, and optimisation. *Miner. Eng.* 17(3):437–445.

Morrell, S. 2009. Predicting the overall specific energy requirement of crushing, high pressure grinding roll and tumbling mill circuits. *Miner. Eng.* 22(6):544–549.

Morrell, S. 2010. Predicting the specific energy required for size reduction of relatively coarse feeds in conventional crushers and high pressure grinding rolls. *Miner. Eng.* 23(2):151–153.

Morrell, S. 2011a. The appropriateness of the transfer size in AG and SAG mill circuit design. Paper No. 153. In *Proceedings of the Conference on International Autogenous and Semi-Autogenous Grinding Technology 2011*. Vancouver, BC: Mining and Mineral Process Engineering, University of British Columbia.

Morrell, S. 2011b. Mapping orebody hardness variability for AG/SAG/crushing and HPGR circuits. Paper No. 154. In *Proceedings of the Conference on International Autogenous and Semi-Autogenous Grinding Technology 2011*. Vancouver, BC: Mining and Mineral Process Engineering, University of British Columbia.

Morrell, S., and Morrison, R. 1996. AG and SAG mill circuit selection and design by simulation. In *Proceedings of the Conference on International Autogenous and Semi-Autogenous Grinding Technology 1996*. Vancouver, BC: Mining and Mineral Process Engineering, University of British Columbia. pp. 769–790.

Mosher, J.B., and Bigg, A.C.T. 2001. SAG mill test methodology for design and optimization. In *Proceedings of the Conference on International Autogenous and Semi-Autogenous Grinding Technology 2001*. Vancouver, BC: Mining and Mineral Process Engineering, University of British Columbia.

Mosher, J.B., and Tague, C.B. 2001. Conduct and precision of bond grindability testing. *Miner. Eng.* 14(10):1187–1197.

Napier-Munn, T.J., Morrell, S., Morrison, R.D., and Kojovic, T. 1996. *Mineral Comminution Circuits: Their Operation and Optimisation*. Indooroopilly, Queensland: Julius Kruttschnitt Mineral Research Centre, University of Queensland, Australia.

Narayanan, S.S., and Whiten, W.J. 1988. Determination of comminution characteristics from single particle breakage tests and its application to ball mill scale-up. *Trans. Inst. Min. Metall.* 97:C115–C124.

Oestreicher, C., and Spollen, C.F. 2006. HPGR vs SAG Mill Selection for the Los Bronces Grinding Circuit expansion. In *Proceedings of the Conference on International Autogenous and Semi-Autogenous Grinding Technology 2006*. Vancouver, BC: Mining and Mineral Process Engineering, University of British Columbia. pp. 110–123.

Rayo, J. 2014. Comparison of semi-autogenous mills operations in Andean countries. In *Proceedings of the 12th AusIMM Mill Operators Conference*. Townsville, Queensland, Australia. pp. 283–295.

Rose, R.E., and Blunt, G.D. 1956. The dynamics of the ball mill, Part III: An investigation into the surging of ball mills. *Proc. Inst. Mech. Eng.* 170(1):793–800.

Rose, H.E., and Evans, D.E. 1956a. The dynamics of the ball mill, part I: Power requirements based on the ball and shell system. *Proc. Inst. Mech. Eng.* 170(1):773–783.

Rose, H.E., and Evans, D.E. 1956b. The dynamics of the ball mill, part II: The influence of the powder charge on power requirements. *Proc. Inst. Mech. Eng.* 170(1):784–792.

Rowland, C.A., Jr. 1972. *Grinding Calculations Related to the Application of Large Rod and Ball Mills*. Allis-Chalmers Publication 22P4704. Milwaukee, WI: Allis-Chalmers Manufacturing.

Rowland, C.A., Jr. 1973. Comparison of work indices calculated from operating data with those from laboratory test data. In *Proceedings of the 10th Commonwealth Mining and Metallurgy Congress*, Ottawa, ON, Canada. pp. 47–61.

Rowland, C.A., Jr. 1982. Selection of rod mills, ball mills, pebble mills and regrind mills. In *Design and Installation of Comminution Circuits*. Edited by A.L. Mular and G.V. Jergensen. New York: AIME. pp. 393–438.

Rowland, C.A., Jr., and Kjos, D.M. 1978 (revised 1980). Rod and ball mills. In *Mineral Processing Plant Design, Practice, and Control*. Edited by A.L. Mular and R.B. Bhappu. New York: AIME. pp. 239–278.

Scinto, P., Festa, A., and Putland, B. 2015. OMC Power-based comminution calculations for design, modelling and circuit optimization. In *Proceedings of the 47th Annual Meeting of the Canadian Mineral Processors*. Ottawa, ON: Canadian Mineral Processors.

Shi, F., Lambert, S., and Daniel, M. 2006. A study of the effects of HPGR treating platinum ores. In *Proceedings of the Conference on International Autogenous and Semi-Autogenous Grinding Technology 2006*. Vancouver, BC: Mining and Mineral Process Engineering, University of British Columbia. pp. 154–171.

Starkey, J., and Dobby, G. 1996. Application of the Minnovex SAG Power Index at five Canadian SAG plants. In *Proceedings of the Conference on International Autogenous and Semi-Autogenous Grinding Technology 1996*. Vancouver, BC: Mining and Mineral Process Engineering, University of British Columbia. pp. 110–123.

Starkey, J., and Larbi, K. 2012. SAGDesign—Using open technology for mill design and performance assessments. Presented at Procemin 2012, Santiago, Chile.

Starkey, J., Dobby, G., and Kosick, G. 1994. A new tool for SAG hardness testing. In *Proceedings of the 26th Canadian Mineral Processors Annual Meeting*. Ottawa, ON: Canadian Mineral Processors.

Starkey, J., Hindstrom, S., and Nadasdy, G. 2006. SAGDesign testing—What it is and why it works. In *Proceedings of the Conference on International Autogenous and Semi-Autogenous Grinding Technology 2006*. Vancouver, BC: Mining and Mineral Process Engineering, University of British Columbia. pp. 240–254.

Steane, R.A., and Hinckfuss, D.A. 1979. Selection and performance of large diameter ball mills at Bougainville Copper Ltd. Papua New Guinea. In *Proceedings of the 11th Commonwealth Minerals and Metallurgy Congress*. Hong Kong. pp. 577–584.

Stephenson, I. 1997. The downstream effects of high pressure grinding rolls processing. Ph.D. thesis, Julius Kruttschnitt Mineral Research Centre, University of Queensland, Australia.

Whiten, W.J. 1974. A matrix theory of comminution machines. *Chem. Eng. Sci.* 29:588–599.

PART 4

Classification and Washing

CHAPTER 4.1

Screens

Brian Flintoff, Ricardo Maerschner Ogawa, and Dusty Jacobson

The classification of particles based on their size is an important step in almost all mineral processing flow sheets. Although the technical focus in plant design, control, and optimization is most often on the comminution (e.g., size control or mineral liberation) and/or separation equipment (e.g., recovery/yield and grade), classification plays a critical role in optimizing process efficiency. This chapter focuses on screens, and in particular on vibrating screens in coarser classification applications.

Peripheral topics such as feeders (Metso 2009, chapter 2), sumps, pumps (Metso 2009, chapter 3.3), slurry pipelines, conveyors (Metso 2009, chapter 3.2), and so on, which are all critically important to the proper operation of screens, are covered elsewhere in this handbook.

To borrow from Matthews (1985), "Screening is defined precisely as a mechanical process which accomplishes a separation of particles on the basis of size and their acceptance or rejection by a screening surface. Particles are presented to the apertures in a screening surface and are rejected if larger than the opening, or accepted and passed through if smaller." The screening process has been a part of mineral processing flow sheets for a very long time, and like many other unit operations, it has seen some interesting mechanical and process developments over the past few years.

Across all the process industries, there are many different kinds of screens and applications. However, as this is a reference source for mineral processing engineers, the discussion is restricted to the most common applications in the mining industry. Figure 1 is a generic graphical representation of the major applications of screening in mining: scalping, size control, and sorting. Table 1 provides more amplification on the type of screen one might find in given applications.

Screen types can generally be categorized as follows:

- **Fixed types:** grizzlies, riffles, sieve bends, and so forth
- **Moving (linear) types:** vibrating, reciprocating, and resonance
- **Moving (rotating) types:** cylindrical trommels and probability

Figures 2, 3, 4, 5, and 6 provide pictures of a grizzly screen, a vibrating screen, a trommel screen for a mill discharge, a Stack Sizer, and a sieve bend, respectively. In general, these are the workhorses of the mining industry, with the vibrating screen standing out as the most widely employed screening unit operation. For this reason, most of this chapter is devoted to the vibrating screen. For the reader with interests in other screening technologies and/or looking for more detail than is presented here, there are many good reference sources, including: basic texts on mineral processing (e.g., Wills and Finch 2016; and numerous SME publications, such as Mular 2003, Bothwell and Mular 2002, Matthews 1985, and Nichols 1982). The latter two sources have companion articles on other aspects of screening as well as associated processes such as bins, feeders, and stockpiles. Original equipment manufacturers (OEMs) also publish handbooks that contain very useful information on the selection and operation of their screen products (e.g., Metso 2009).

VIBRATING SCREEN BASICS

The vibrating screen is ubiquitous in mineral processing plants. One would imagine that with this breadth and history of application, and the screen's apparent simplicity, there would be few remaining unknowns in the screening process. However, the design and operation of screening systems remains a blend of art and science—a statement used by many other authors over the past two decades (e.g., Matthews 1985). It is interesting to note that the screening process is enjoying something of a renaissance today, as new simulation tools offer a closer examination of the physics of the process, providing quantitative insights into mechanisms and their relation to design variables. In addition, new instrumentation offers real-time modulation and measurement of mechanical and process conditions. This chapter deals with vibrating screen basics and applications, and introduces some of the latest tools and techniques in the design, analysis, control, and optimization of vibrating screens.

Brian Flintoff, Consultant, Sigmation Inc., Kelowna, British Columbia, Canada
Ricardo Maerschner Ogawa, Product Manager, Mining Screens, Metso Brasil Indústria e Comércio, Sorocaba, Brazil
Dusty Jacobson, Senior Cushing & Screening Specialist, Metso Minerals Industries Inc., Waukesha, Wisconsin, USA

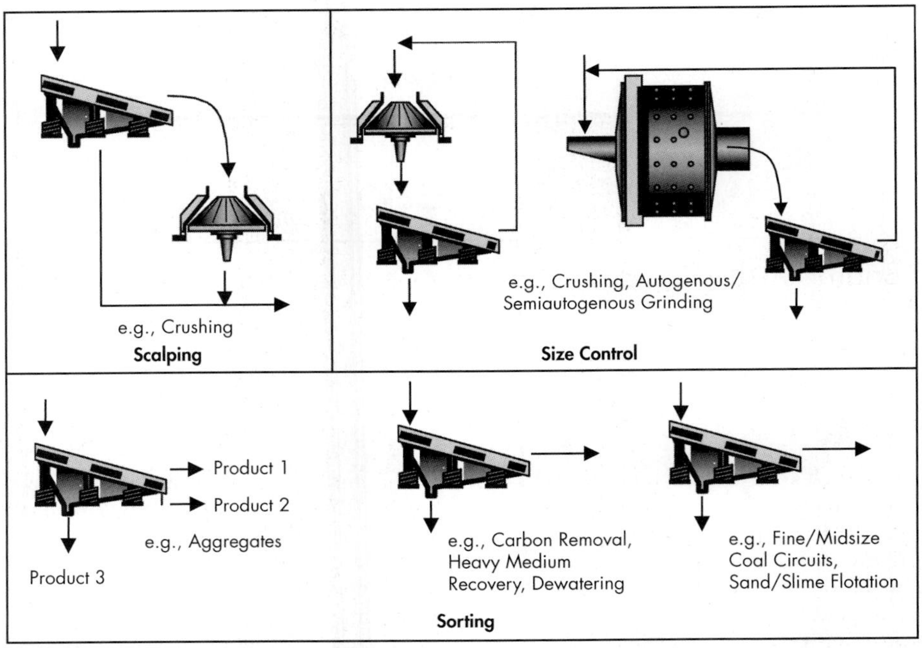

Source: Flintoff and Kuehl 2011

Figure 1 Examples of the application classes for screening in mining

Table 1 Types of screening operations

Operation and Description	Type of Screen Commonly Employed
Scalping: Scalping is strictly the removal of an amount of oversize from a feed that is predominantly fines. Scalping typically consists of the removal of oversize from a feed with a maximum of 5% oversize and a minimum of 50% half size. Practically, it is often used in cases where there is a significant oversize material (e.g., in crushing where there is a desire to remove fines to better utilize the volumetric capacity of the crusher). Coarse scalping is typically smaller than 150 mm and larger than 75 mm.	Coarse (grizzly); fine (same as fine separation); ultrafine (same as ultrafine separation). *Example:* Grizzlies are usually used to scalp oversize material from primary crusher feed or to scalp fines from the feed to crushers or grinding rolls.
Coarse separation: Consists of making a size separation smaller than 75 mm and larger than 5 mm (~4 mesh).	Vibrating screens (banana, incline, or horizontal), and trommel screens. *Examples:* The removal of pebbles from a semiautogenous grinding discharge stream, sizing material for the next crushing stage, and sizing material for leach pile or product pile.
Fine separation: Consists of making a size separation smaller than 5 mm (~4 mesh) and larger than 0.3 mm (48 mesh).	Vibrating screens (banana, incline, or horizontal), which are typically set up with a high speed and low amplitude or stroke; sifter screens; static sieves; and centrifugal screens. *Examples:* Sorting coal into ±0.6 mm fractions for washing, and iron ore processing.
Ultrafine separation: Consists of making a size separation smaller than 0.3 mm (48 mesh).	High-frequency, low-amplitude vibrating screens; sifter screens; static sieves; centrifugal screens. *Example:* Size control on grinding circuit product
Dewatering: Consists of the removal of free water from a solid–water mixture and generally limited to 4.8 mm (4 mesh) and above	Uphill inclined vibrating screens (typically –5°), horizontal vibrating screens, incline vibrating screens (about 10°), and centrifugal screens. *Example:* Wet quarrying operations
Trash removal: Consists of the removal of extraneous foreign matter. This is essentially a form of scalping operation, and the type of screen depends on the size range of processed material.	Vibrating screens (horizontal or incline), sifter screens, static sieves, and centrifugal screens. *Example:* The removal of extraneous organic and other matter from the leach feed in a gold plant.
Other applications: Other applications include desliming (the removal of extremely fine particles from wet material by passing it over a screening surface), difficult-to-screen material; conveying (in some instances transport of a material may be as important as the screen operation), dense media recovery, a combination washing and dewatering operation, and concentration.	Vibrating screens (banana, incline, or horizontal), oscillating screens, and centrifugal screens. *Example:* Dense medium recovery in a coal washing plant, flip-flop screens or a wobbler feeder for conveying, and classifying sticky materials to a crusher.

Adapted from Matthews 1985

Source: Flintoff and Kuehl 2011
Figure 2 (A) Static grizzly (scalping oversize from a primary jaw crusher feed) and (B) vibrating grizzly screen

Source: Flintoff and Kuehl 2011
Figure 3 Vibrating screen with a dust collection system

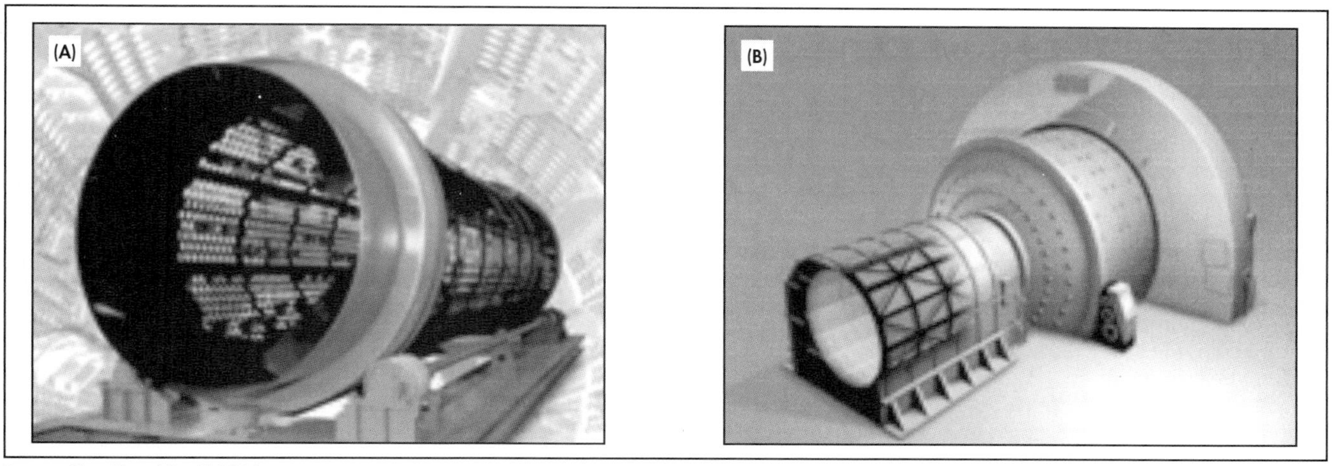

Source: Flintoff and Kuehl 2011
Figure 4 (A) Trommel screen and (B) mill discharge trommel screen

Courtesy of Derrick Corporation
Figure 5 Derrick Stack Sizer

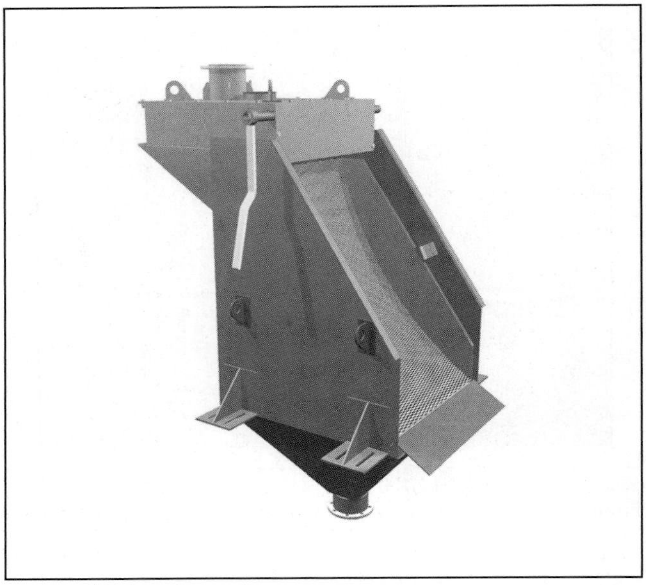

Courtesy of Multotec Process Equipment
Figure 6 Sieve bend

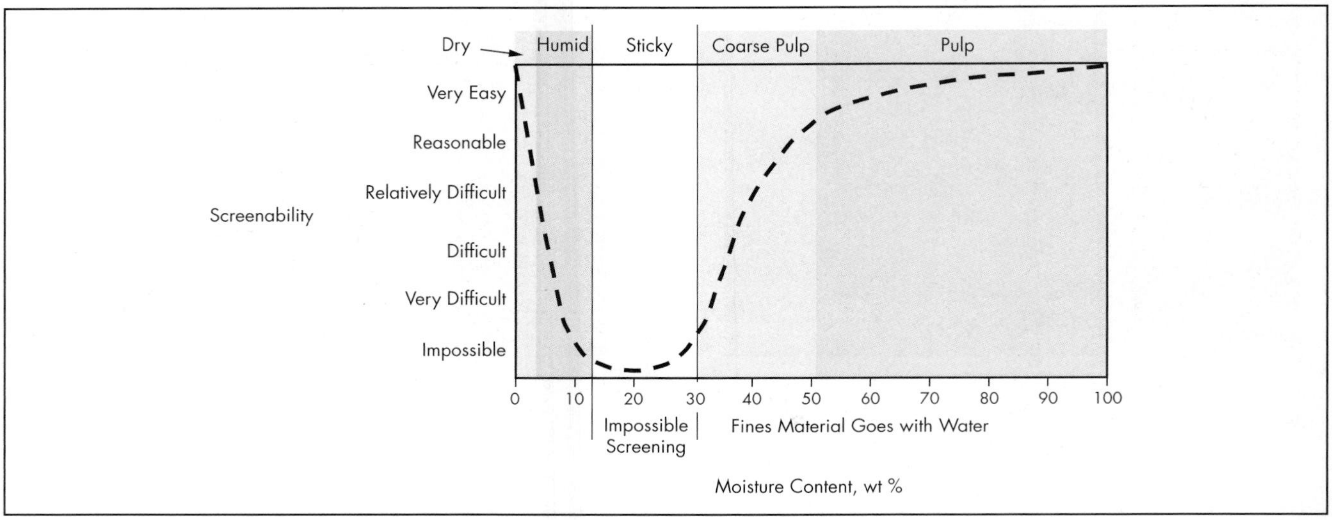

Courtesy of Metso
Figure 7 Screenability as a function of the feed moisture level

Principles of Operation

The vibrating screen is used here as a proxy for all other screens, as the basic operating principles are similar or the same. In the screening process, a feed material comprising particles of varying sizes is presented to the screen media (collectively, the screen deck) in such a way that those particles finer than the screen aperture fall through to the undersize stream, and those that are larger than the screen aperture continue moving along the screen surface to eventually report to the oversize stream.

Screening can be carried out with wet or dry feeds, although coarser particle separation (greater than a 5-mm aperture) is usually performed dry (at the surface moisture of the feed) where possible. Wet screening is not to be confused with dewatering screening, as the process aim remains solids size classification for the former. In dry applications, some combination of vibration ("throwing" the particles off the deck) and gravity is responsible for particle transportation. In wet applications, some form of vibrating mechanisms dictates the movement of material. However, hydrodynamic (drag) and gravitational forces are also at work. Wet screening usually involves either sticky/clay-rich feeds or finer particle sizes where the solid material has been slurried to facilitate transportation and processing (e.g., semiautogenous grinding [SAG] discharge screens). Figure 7 is a graphical illustration of the anticipated degree of difficulty ("screenability") introduced when dealing with moist feeds.

There are many similarities between wet and dry screening, starting with the machines themselves. Figure 8A shows a double-deck, low-head, slightly inclined (~5°) SAG discharge

Courtesy of Metso

Figure 8 Use of water sprays in screening: (A) SAG discharge screen equipped with water sprays and (B) typical water spray deployment on a screen deck

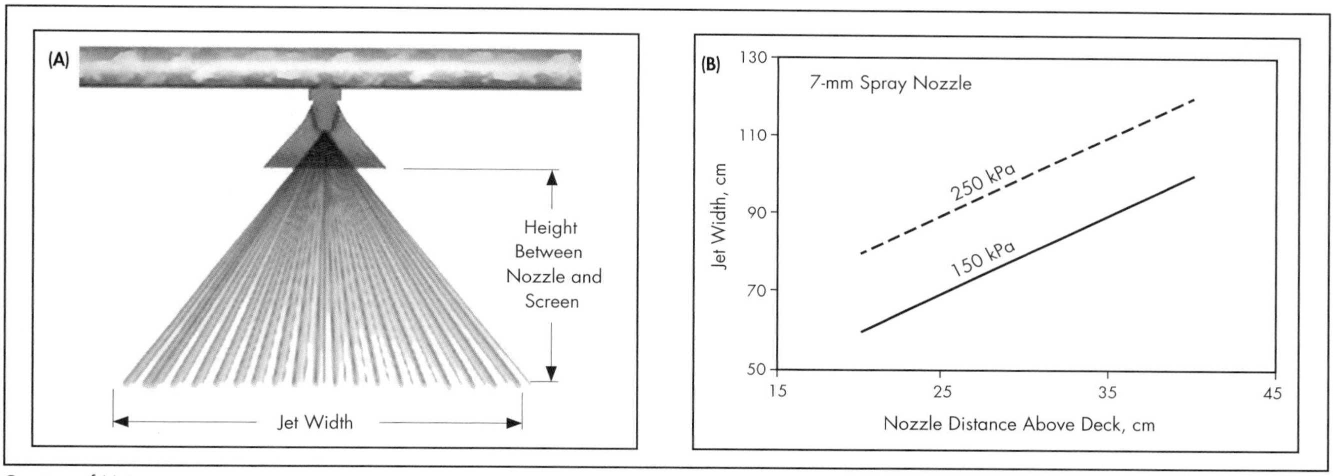

Courtesy of Metso

Figure 9 Water spray geometry: (A) basic dimensional info, and (B) jet width for a particular nozzle opening at different pressures and position above the deck

screen; the additional water supply assembly for the deck sprays is highlighted.

Water can be added to the screen feed in the slurry box feeding the screen, or on the deck through flood boxes or water sprays, the latter being more common. A general rule of thumb is that water added to the feed is more effective than water sprayed on the deck in improving the classification of fines. Spray selection is chosen based on two criteria. One must decide whether the goal is to flush material through the deck or wash off the fines. (In the case of high-pressure grinding roll [HPGR] circuits, water sprays can also be helpful in breaking up the cakes formed in crushing.) For flushing, the nozzles on the spray bar generally operate at lower pressure and higher volumes with the water falling vertically onto the deck. For washing, it is common to operate the nozzles at higher pressure and lower volume, and normally they are oriented to spray against the flow of the solids (e.g., Figure 8B). This is thought to be helpful in "cleaning" fines from the coarser rock surfaces.

The design of the spray water systems is usually in the OEM's purview and depends on the application (e.g., sticky fines, such as magnetite on coal in dense medium circuits) and the water requirements. As Figure 8 indicates, there are often three or more spray bars per deck, and each spray bar is equipped with several nozzles. This is to ensure some overlap in the spray jet from each nozzle and to deliver the required water flow. It is common to try to leave 2–3 m between the last spray bar and the discharge point to allow for dewatering of the oversize materials. From a design perspective, Figure 9 provides typical OEM information relative to the water-jet geometry as a function of the position and operating pressure of the nozzle, here for a 7-mm opening. Nozzle openings typically range from ~4 mm to ~12 mm.

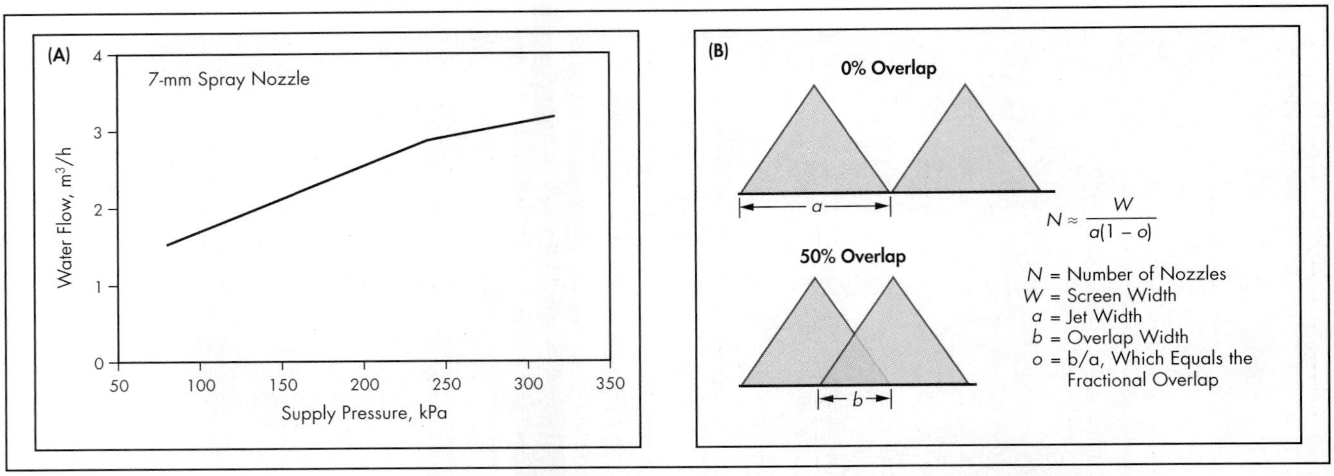

Courtesy of Metso

Figure 10 Water spray flow: (A) water flow for a particular nozzle opening at different pressures and (B) estimating the number of nozzles per spray bar

The companion design information is shown in Figure 10, where the water flow as a function of pressure is illustrated, again for a 7-mm nozzle. The calculation of the number of nozzles relates in part to achieving desired overall water flow to the deck and in part to ensuring efficient operation. For the latter, Figure 10B indicates a couple of options. Since these water sprays use process water, plugging is a possibility and some overlap is generally a good idea. A 50% overlap doubles the number of nozzles and the water flow.

For washing fines (sometimes known as desliming), water is normally sprayed at pressures ranging from 100 to 300 kPa and the volume ranges from 0.7 to 1 m^3/t of solids in the feed. Lower values are possible for relatively clean materials, and higher values are possible for materials with clay or very fine particle content. In the case of treating dry or pit wet feed material, it is common to add 0.3–0.5 m^3 of water per ton of feed to the feed slurry box. For the removal of sticky fines, the spay water addition is usually 0.3–0.5 m^3/t of feed. Again, if the feed is dry or pit wet, it is common to add 0.20.5 m^3 of water to the feed slurry box.

Focusing more on dry screening, Figure 11 is a schematic representation of the results of an experiment with a small horizontal vibrating screen having a square screen aperture of 10 mm. (This screen and test from Hilden [2007] is referenced throughout the text.) The shaded bars depict the relative mass flows, and the annotations show the mass mean size ($\bar{d}$) for each stream. As the feed is introduced onto the deck, it forms a bed that is usually several times the screen aperture in thickness. The bed moves under the influence of gravitation and vibrational motion, and in the process, the particles stratify as the finer materials move quickly through the interstices of the larger particles and find their way to and then through the screen apertures. In this initial zone on the screen, *rapid stratification* (sometimes referred to as segregation) means there is always enough fine material in the layer next to the screen deck that the flow through the screen apertures is more or less constant. This is said to be typical of crowded or saturation screening. However, as the fines are depleted, the bed height is reduced, and at some point the particles begin to act more or less as individual entities. This point is the zone of separated or statistical screening; that is, each time the particle approaches

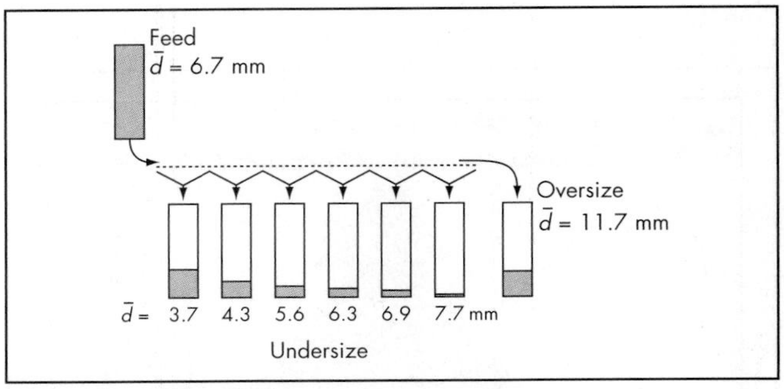

Adapted from Hilden 2007

Figure 11 Relative mass flows and mean particle sizes for a horizontal vibrating screen

the deck, it has a chance (or a statistical trial) to pass through the aperture. (This notion of repetitive trials underpins many semi-empirical screen models.)

Figure 11 shows that the greatest flow of material to the undersize stream occurs closest to the feed end, typical of crowded screening. Stratification ensures that these particles presented to the apertures are small enough that the probability of passage to the undersize stream is ~1.0. The mass flow to the undersize stream decreases along the length of the deck, which is typical of the transition to separated screening, and more typical of the coarser particles that remain in the bed. (These coarser particles are so-called *near-size*, because they are near to the size of the aperture. Formally, near-size is the fraction of the feed material in the size interval 0.75–1.25 h_T, where h_T is defined later in Figure 17.) For these particles, the probability of passage on any one trial is significantly less than 1.0. This probability is a complex function of many parameters (particle size/shape/density, screen deck motion, aperture geometry, etc.), but for illustrative purposes, Taggart (1945) worked out the probability of passage of a spherical particle falling onto a square aperture, and his data are presented in Figure 12. The probability of passage falls very quickly as the size ratio enters the near-size region and approaches 1.

The increase in the mass mean size along the deck in Figure 11 also reflects the fact that the remaining material on

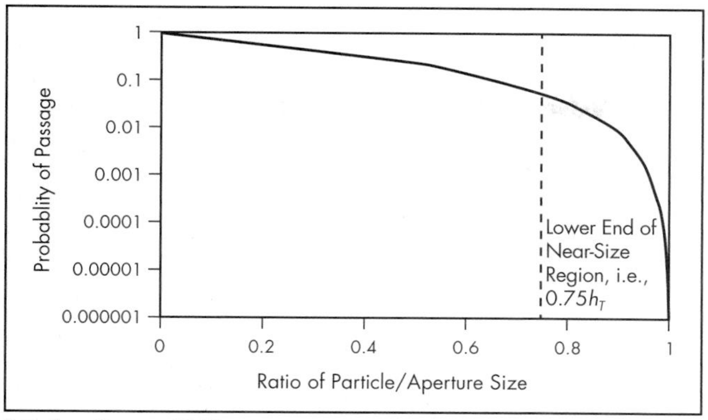

Data from Taggart 1945

Figure 12 Probability of passage as a function of particle size

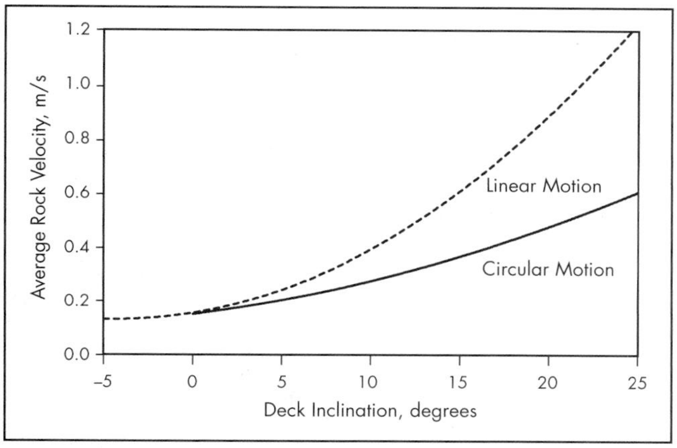

Courtesy of Metso

Figure 13 Rock velocity estimates for an inclined screen as a function of deck motion

the deck becomes coarser as it moves from the feed end to the discharge. This helps to explain why screen length governs efficiency. On the other hand, screen width is related to the amount of material that can be fed to the screen and be effectively separated, and therefore this is the major factor governing capacity.

On this latter point, a rule of thumb for good dry screen operation is for the maximum bed depth at the discharge point to be 3:1 → 4:1 screen apertures in thickness, although operation at 2:1 is quite common. A rule of thumb for wet screening is to design for a maximum bed depth of 4:1–6:1 times the opening. At larger targets, the efficiency of removal of the particles close to the aperture size is reduced, as stratification is more difficult. Such a situation suggests a wider screen, higher transport speeds (increased inclination), or that multiple screen decks probably should be considered, using a larger aperture size on the top deck to scalp out much of the larger oversize and reduce the loading on the lower deck for the final classification. At less than the target bed depths given earlier, these particles will probably bounce excessively, reducing the number of trials and the screen efficiency.

Designers estimate the bed depth using the following equation:

$$B = \frac{(M/\rho_b)}{3.6Wv} \quad \textbf{(EQ 1)}$$

where

B = bed depth, mm
M = solids mass flow, t/h, discharge (or feed) end of the screen (e.g., usually taken as the flow of oversize material in the feed)
ρ_b = bulk density of the solids, t/m^3
W = screen width, m
v = rock velocity across the screen, m/s

The value for the rock velocity, v, usually comes from general correlations developed by the OEMs. Figure 13 shows typical values of rock velocity for inclined screens, as a function of both inclination and the general shape of the orbit. For example, in the case of the screen in Figure 11, the feed rate is 15.7 t/h; the fraction of material coarser than the 10-mm aperture in the feed is 18.7%; the bulk density is 1.62 t/m^3; the screen width is 0.34 m; and, given that the deck is horizontal with linear motion, the estimated rock velocity from Figure 13 is ~0.16 m/s. From this, the calculated bed depth is 9.3 mm, which gives an aperture ratio less than 1 when compared the screen aperture, meaning the screen is underloaded. One would therefore expect some efficiency issues with this screen, and this will be apparent in the discussion on performance assessment later in this chapter.

It is also sometimes useful to install weirs (or dams) on the deck surface, which hold back the flow as an aid to sustaining an active bed and/or facilitating better water removal. These weirs are typically 30–50 mm high and have a maximum spacing of ~1.2 m. Figure 14 is a rather extreme example showing the dams or weirs installed on a screen deck.

Screen design and operation is about matching the screen characteristics to the (range of) feed conditions. Table 2 provides a summary of the important variables for each.

When focusing on design variables, screen area is important because capacity is proportional to width (W) and efficiency is proportional to length (L). It is common for $L \approx 2W$ to $3W$. Because of the mechanical fittings on and around the screen decks, the effective area is often approximated as 90%–95% of the actual (= LW) area. Screens are standard products, so the OEMs dictate the actual sizes. Therefore, it is common to require more than one screen, operating in parallel, to effectively separate a feed.

Open area (OA) is expressed as the ratio of the total area of the apertures over the total active area of the screen deck. Figure 15 illustrates this for a square aperture of 10 mm in a wire mesh screen with a wire diameter of 2 mm, for the screen

Courtesy of Metso

Figure 14 Weirs on a screen deck

Table 2 Design and operating variables for vibrating screens

Design Variables	Operating Variables
Screen area and open area	Particle size, shape, and distribution
Aperture size and shape	Solids feed rate, feed distribution across the deck, and bed depth
Slope of screen deck	Feed moisture content
Speed	Clay content
Magnitude of stroke	
Type of motion	
Feed arrangement	

Source: Bothwell and Mular 2002

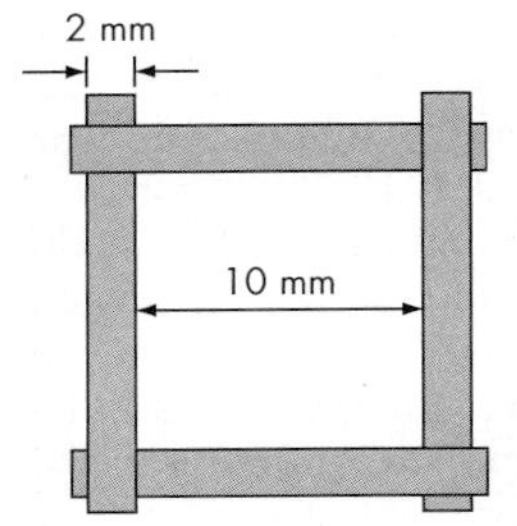

$$OA = \frac{10^2}{12^2} = 69.4\%$$

Source: Flintoff and Kuehl 2011
Figure 15 Computing open area

in Figure 11. From the earlier explanations of screening mechanism, clearly the greater the OA, the greater the capacity of the screen. In practice, OA is limited by the need for sufficient mechanical strength and wear resistance of the deck, and is a function of the media materials of construction (e.g., wire mesh versus perforated plate).

This is a good point to introduce some common, and occasionally confusing, jargon associated with phenomena that act to reduce the OA, and hence capacity. The first is *pegging* (also called clogging or plugging), which occurs when a near-size particle becomes firmly lodged in an aperture, effectively eliminating it from the screening process. The second is *blinding* (also called bridging), which is the phenomena whereby smaller particles aggregate in such a way that they combine to block up the aperture, having the same overall effect as pegging, as shown in Figure 16. Blinding is more of a problem with moist or sticky materials and small aperture sizes.

The aperture size is, by convention, the minimum linear measurement in an opening, for example, the diameter for a round hole, the width (where width ≤ length) for a slot or a square, and so on. (Because screen performance assessments start with sieving samples of process streams, and because the sieve has square apertures, the best correlations occur with square apertures in the screen media.) What is perhaps more important is the so-called throughfall size, h_T, which accounts for the change in effective opening as the screen is inclined. This is illustrated in Figure 17, where the angle of inclination is exaggerated to show the difference between the aperture size, h, and the throughfall size, h_T, for a wire mesh screen surface. (There is a similar effect for rectangular slots.)

Although square apertures may be the most common, there is a multitude of aperture shapes. Some of the more common shapes are illustrated in Figure 18. Square or round shapes generally allow for better control of the separation size, although circular apertures are most often used in coarse screening

Source: Flintoff and Kuehl 2011
Figure 16 Blinding on a screen deck

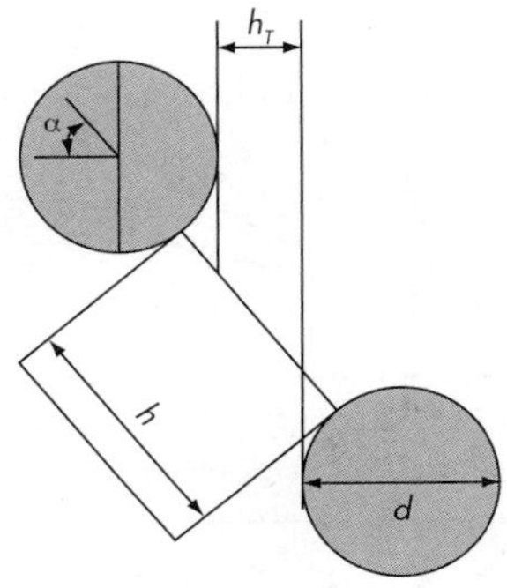

$$h_T = (h + d)\cos(\alpha) - d$$

Source: Flintoff and Kuehl 2011
Figure 17 Relation between the throughfall size (h_T) and the aperture size (h)

applications. Longitudinal slotted apertures allow slabby particles into the undersize stream, so they increase capacity and reduce pegging and blinding. (Aperture shape can have a significant influence on capacity; see, for example, Q2 factor later in Table 6.) Transverse slots are mainly used in dewatering applications. (For steel or polymeric screen media, it is possible to design tapers and other forms of relief into the apertures to minimize pegging.) Finally, the screen deck or panel is designed to withstand the load and maximize wear life, so in the case of a large spread in feed particle sizes and/or some large particles in the feed, it is common to use multiple screen decks to satisfy both mechanical and process requirements.

As observed earlier in Figure 13, inclining a screen deck causes the feed material to move more quickly on the deck, reducing the residence time and increasing capacity and sometimes efficiency. Vibration motion on a horizontal screen deck usually means material velocities of 10–16 m/min, while an inclination of about 20° will increase this to the range of 25–30 m/min. Generally, inclinations range from horizontal to as high as 30°, although ~20° is the usual limit. This led to the evolution of the multislope screen (commonly known as

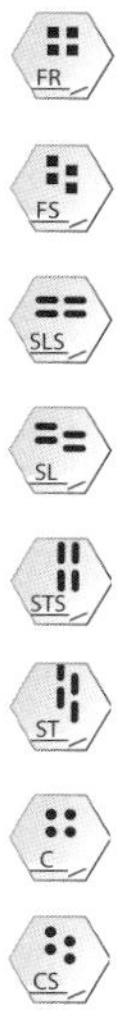

- **FR** - Square apertures - The most common type of aperture. Normally resulting in the most accurate separation.
- **FS** - Square apertures in staggered pattern. Same applications as FR but apertures are arranged to prevent "tracking" of fines.
- **SLS** - Rectangular apertures with the flow typically recommended for screening material with high content of fines or if there are signs of pegging.
- **SL** - Rectangular apertures with the flow typically recommended for screening material with high content of fines or if there are signs of pegging.
- **STS** - Rectangular apertures across the flow. Normally recommended for dewatering applications.
- **ST** - Rectangular apertures across the flow in staggered pattern. Can be used as an alternative in dewatering applications.
- **C** - Round apertures have similar applications to FR openings. Traditionally used in applications such as primary screening.
- **CS** - Round apertures in staggered pattern. Same applications as C but arranged to prevent "tracking" of fines.

Courtesy of Metso

Figure 18 Common screen aperture shapes

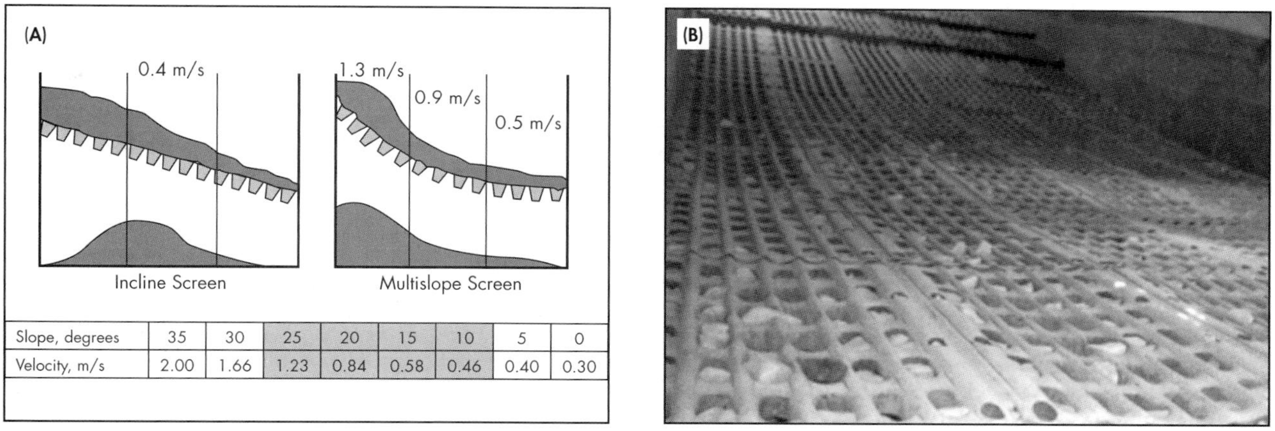

Slope, degrees	35	30	25	20	15	10	5	0
Velocity, m/s	2.00	1.66	1.23	0.84	0.58	0.46	0.40	0.30

Source: Flintoff and Kuehl 2011

Figure 19 Banana screen deck

the banana screen). The banana screen (see Figure 19) has a relatively high slope in the initial screening zone, which takes advantage of the fast kinetics of fine particle removal. The slope then decreases, which slows particle movement and offers an additional g-force to accelerate removal of the larger fine or near-size material. In the final section, the slope is relatively flat, yielding minimal velocity and maximum residence time, for finer near-size particle removal to the undersize stream.

Vibrating screen deck motion (or orbit) is circular, elliptical, or linear, with the vibrating mechanism rotating in the direction of the flow or counter to the flow. In coarser classification operations, strokes are usually in the range of 6–15 mm, and speed ranges from 650 to 950 rpm. Figure 20, for inclined screens treating a material of a specific bulk density, attempts to illustrate the general relationship between stroke and speed. The annotations, for example, ">25.4 mm and <50.8 mm," refer to the top deck aperture range that one would expect with screens operating at or near this combination of stroke and speed. There is overlap—that is, a smaller-stroke, higher-speed machine can also be used in applications up to 38.1 mm. In practice, the OEM will choose the stroke to facilitate the separation at the selected aperture size and then select the speed that will keep the machine within mechanical limitations. However, when the resulting speed is close to the natural frequency, the process is reversed. The main points of Figure 19 are that (1) as stroke increases, speed must decrease; (2) finer cuts require smaller strokes and higher speeds; and (3) aperture changes, especially larger ones (e.g., where a screen is repurposed) should be approached with caution. An OEM's screen offering will generally appear along this curve but not in a continuous fashion. Things like natural frequencies must be avoided.

To elaborate on orbits, circular motion provides for a launch angle of about 70°–80° relative to the screen media

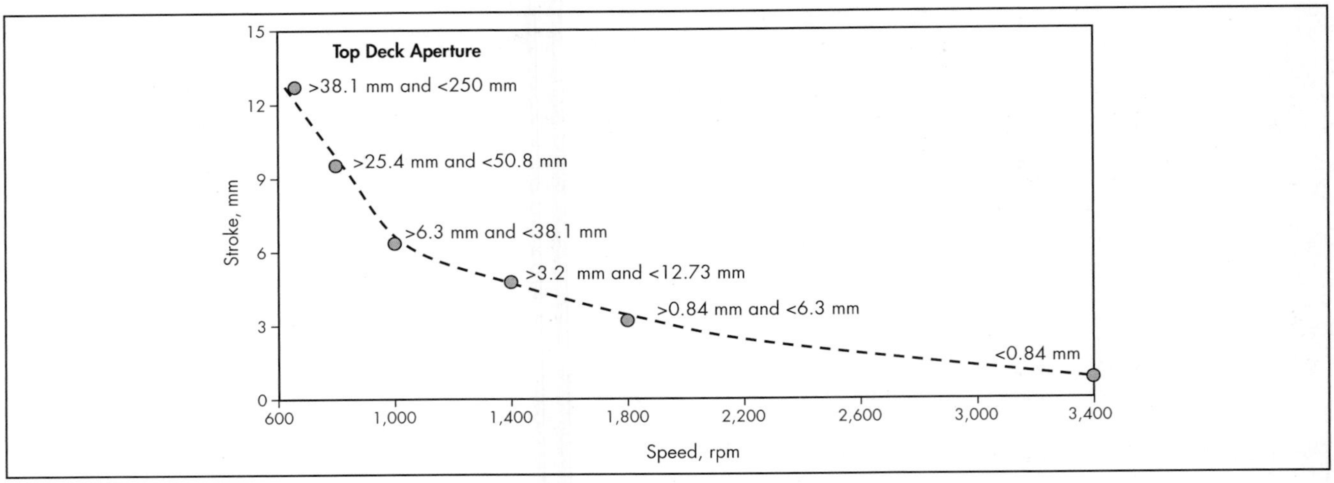

Courtesy of Metso

Figure 20 Example stroke and speed relationship for inclined vibrating screens

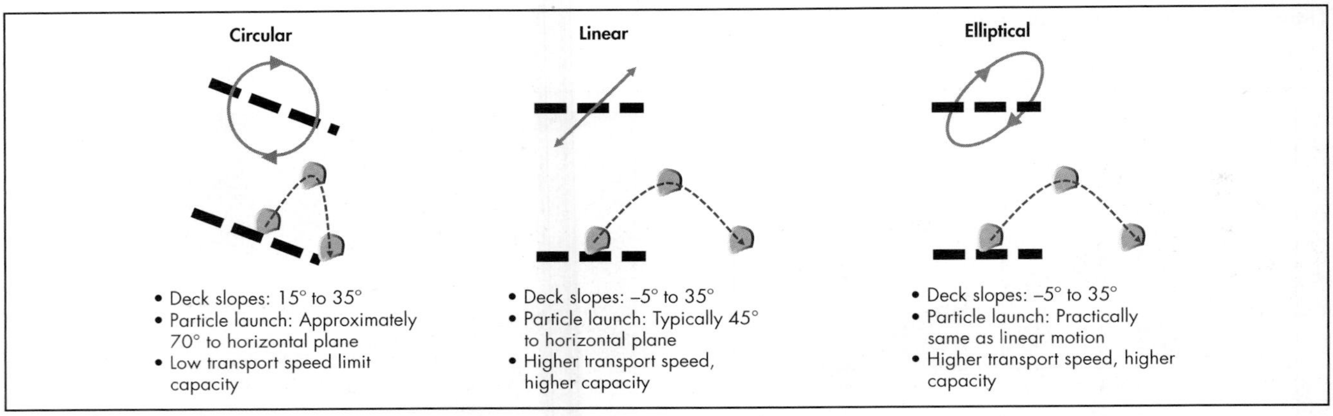

Courtesy of Metso

Figure 21 Particle launch characteristics as a function of deck motion

surface, while linear motion results in a launch angle of 40°–55°. With the high launch angle in circular motion, to achieve a reasonable transport speed (capacity), the deck should have an inclination of 10°–12°. Linear motion screens can function even with a negative deck inclination (e.g., up to –7°). Elliptical motion has the same launching properties as linear motion, and this is summarized in Figure 21.

The advantage of circular motion is that it subjects the particles to forces in all directions, which minimizes the risk of pegging, as shown in Figure 22. Elliptical motion can be thought of as an intermediate to circular and linear motions—it combines the anti-pegging features of circular motion with the ability to work at any screen inclination, as is the case for linear motion.

A commonly referenced screen operating parameter here is the g-force, computed as

$$G = \frac{N^2 S}{1,789,129} \qquad \textbf{(EQ 2)}$$

where

G = g-force (also known as the throw number)
N = speed, rpm
S = stroke, mm

Typical g-forces are in the 3.0–5.0 range, tending to higher values with heavy loads, sticky materials, or where pegging and blinding are issues. The g-force is usually higher in a horizontal screen, as it must provide the means for both stratification and material movement. (For example, with the horizontal test screen in Figure 11, the speed was 981 rpm and the stroke 5 mm, giving $G = 2.69$. As an aside, this would be considered a low value, as horizontal; screens usually run in the range of $4 \leq G \leq 4.7$.) The target acceleration or g-force is usually 3–3.5, perpendicular to the deck. As Figure 23 shows, this requires that a linear screen run at G value of ~5 to achieve 3.5 in the perpendicular plane. OEMs tend to use g-force data mainly for diagnostic purposes, for example, for assessing machine mechanical issues and limitations, understanding pegging and blinding issues, and so on. Table 3 details the conditions prevalent over a broad range of g-forces.

Figure 24 is a simplified graphic of particle motion with different forces and orbits, and Figure 25 shows how deck motion can be induced using weights on the shaft. Of course, there are other methods to induce motion and other more complex orbits.

The stroke angle, measured relative to the length of the deck and in the direction of flow, can affect efficiency and

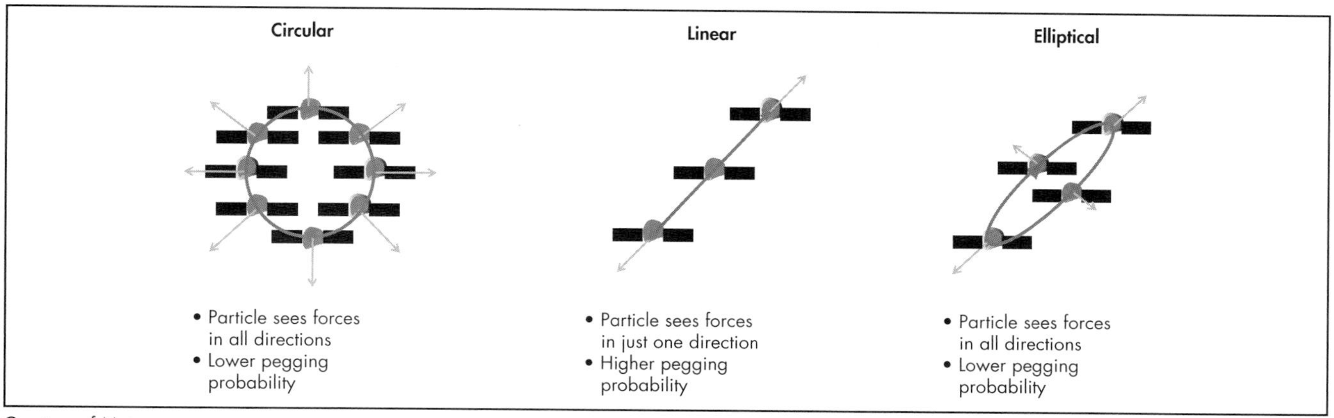

Courtesy of Metso

Figure 22 Forces at work on particles as a function of deck motion

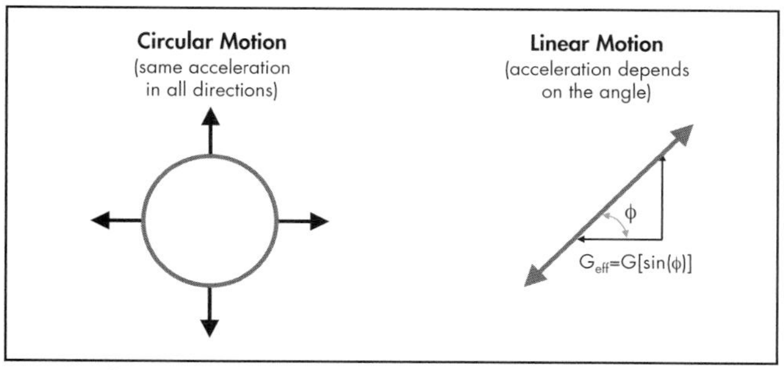

Courtesy of Metso

Figure 23 Resolving the G_{eff} force perpendicular to the screen deck

capacity. Very low (~0°) or high (~180°) angles do little other than accelerate wear. For other angles there is a component of vertical movement (aiding stratification) and a horizontal component aiding (<90°) or retarding (>90°) material flow. (For the screen in Figure 11, the stroke angle was 50°.)

To summarize, the designer for the screening application chooses the orbit motion (circular, elliptical, or linear) based on a compromise between material movement (capacity), which emphasizes linear motion, and efficiency, which emphasizes circular motion. The ellipse is an optimized solution. All give good stratification with the correct stroke angle. For horizontal and banana screens, linear motion is the de facto standard because material movement is critical. For inclined screens, circular or elliptical orbits are more common.

Finally, there is the matter of feeding the screen. Here are three general rules:

1. The feed must be uniformly distributed (at least 75%) across the width of the screen for the unit to achieve rated capacity.
2. The feed rate must be modulated where possible (e.g., using belt feeders) to ensure the screen works efficiently in the face of changing feed conditions.
3. Where possible, it is good to introduce the feed having a motion opposite to the eventual material flow, which helps in distribution and avoids flooding in the initial zone of the screen deck.

Table 3 G-force implications in inclined vibrating screens

G-Force Perpendicular to the Deck	Comment
<1.5	Practically no movement on the deck, so of no use
1.5–2.5	The stratification is limited and relative velocity between the particle and the deck is low. These parameters are suitable for soft screening and where size degradation is problematic and there is a low tolerance to impact.
3.0–3.5	The particle is presented to the screen deck at its maximum relative velocity and offers the best screening condition for inclined screens. The stratification is adequate.
4.0–5.0	The stratification is good, but the relative velocity between the deck and particle is low, offering poor screening efficiency. Bearing life is shortened.
5–6	The stratification is good, but the particle projection exceeds the pitch (particles are in flight too long) and screening opportunities are lost. The relative velocity of screen and particle is low, combining to give poor parameters for efficient inclined screening. The high acceleration forces can act negatively on the screen structure, and bearing life is further shortened.

Source: Moon 2003

This brief synopsis, summarized in Table 4, provides a review of the major design variable of a vibrating screen and indicates usual design ranges where appropriate. Of course, this "one-variable-at-a-time" approach does not always consider the myriad interactions among these variables.

Screen Media

Since screen media is often the single biggest operational expense for a vibrating screen, this chapter would be incomplete without a brief introduction to this topic. The governing objective here is absolute lowest total cost, which implies that availability, cost, wear life, downtime, and so on must all be considered in making the final media selection. OEMs do publish handbooks on this topic, which can be very helpful (e.g., Metso, n.d.).

There are three major materials of construction for screen media: steel (punched plate or wire mesh), rubber, and polyurethane. The polymeric compounds of rubber and polyurethane are usually more expensive but tend to exhibit superior

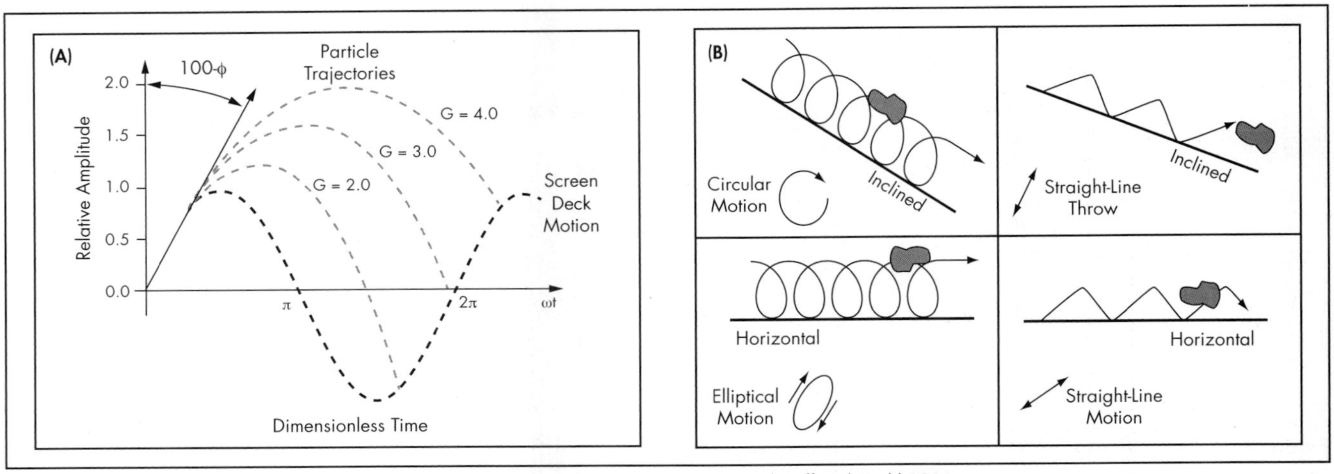

Courtesy of Metso

Source: Flintoff and Kuehl 2011

Figure 24 Particle motion (A) relative to screen deck as a function of g-force and (B) as a function of the deck orbit

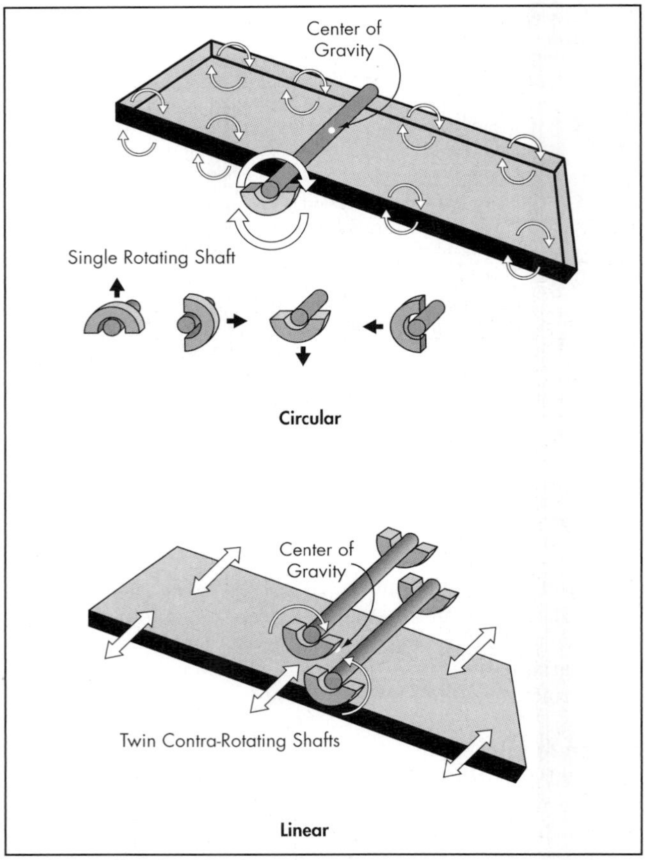

Source: Flintoff and Kuehl 2011

Figure 25 Using weights to induce specific orbits

wear performance in applications that permit alternatives. Service life can be extended by factors of more than four through eight with polymerics but is ore dependent.

Figures 26 through 28 show wire mesh, rubber, and polyurethane media, respectively. Media specialists often break down applications as scalping, standard production (size control and sorting), and washing (sorting).

Table 4 Process matrix for a vibrating screen*

Design Variable	Screen Capacity	Screen Efficiency	Undersize Mean Size	Pegging/ Blinding
Screen area	+	+	0	0
Open area	+	+ or 0	0	0
Aperture size	+	+	+	–
Aperture shape (increasing nonuniformity)	+	+ or 0	+	–
Deck inclination	+	– or 0	–	– or 0
Speed	+	+	0	–
Stroke angle	–	+	0	– or 0
Stroke	+	+	0	–
Orbit (from linear toward circular)	–	+	0	–

Source: Flintoff and Kuehl 2011

*All other factors being equal, an increase in the design parameter will cause an increase (+), decrease (–), or no change (0) in the performance metric.

Wire mesh can have standard weaves, as shown in Figure 26A, or slotted anti-blinding or even custom weaves as shown in Figure 26B. The wires are usually high-carbon or stainless steels.

Typical applications for rubber screen decks (Figure 27) include high-impact scalping type duty or situations where abrasion and/or blinding is a serious concern. The decks are manufactured by punching or molding.

Polyurethane decks (Figure 28) are usually found in standard production applications and in washing or dewatering applications. They are manufactured by casting or injection molding processes.

The polymeric materials lend themselves to modular installation or paneling. Table 5 weighs the advantages and disadvantages of steel versus the polymerics.

As a cautionary note, although there might be a tendency to choose thicker panels to increase wear life, this will reduce capacity and accuracy and will induce higher levels of pegging and blinding. Media suppliers familiar with best practices can be very helpful in these selections, and it is also recommended that the screen OEM is consulted when changing the design of media.

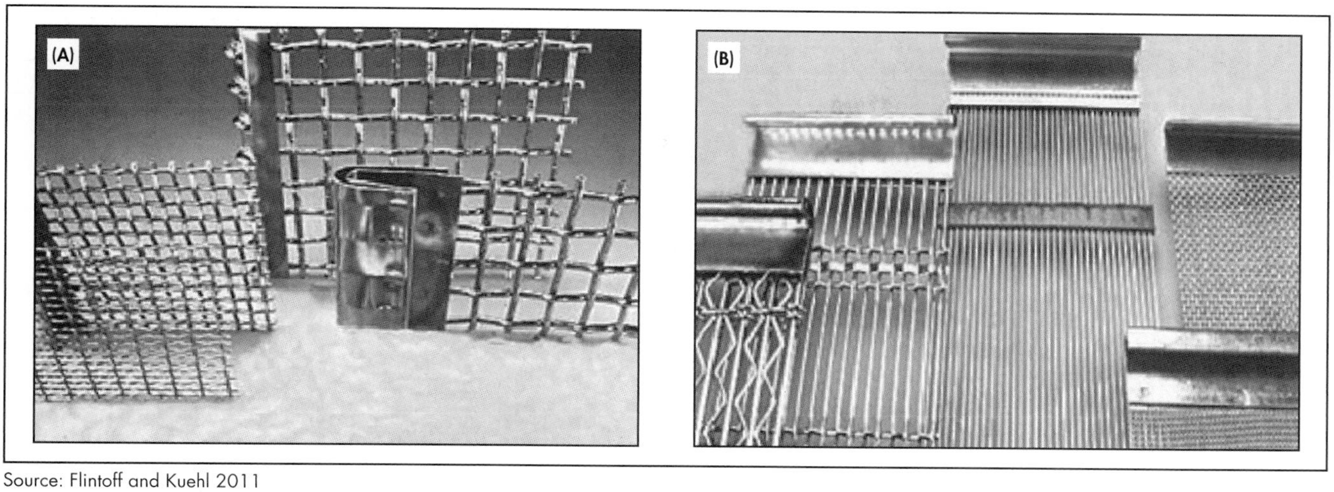

Source: Flintoff and Kuehl 2011
Figure 26 Wire mesh media

Source: Flintoff and Kuehl 2011
Figure 27 Rubber screen decks

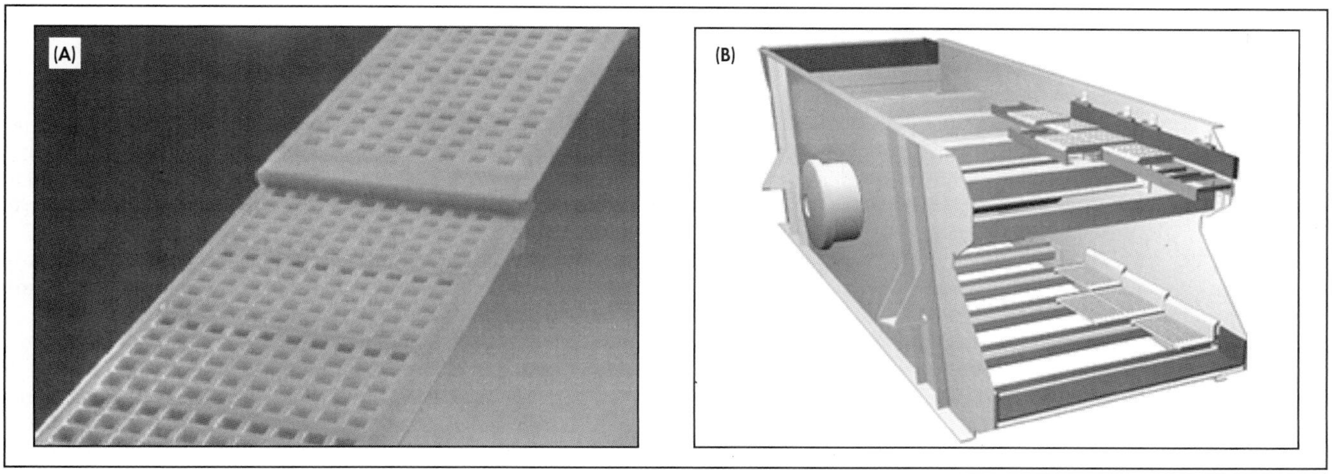

Source: Flintoff and Kuehl 2011
Figure 28 Polyurethane decks

Screen Applications

This section provides more detail on three of the more common applications of screens in mineral processing flow sheets.

Traditional Crushing and Screening Circuit

The more traditional three-stage crushing and screening circuit is illustrated in Figure 29. For open pit operations, the primary crusher would be a gyratory, whereas for underground mining operations, a jaw crusher would most likely be employed. Except for heap leach operations, this type of circuitry has been supplanted by SAG or HPGR circuits. However, there are still quite a few older operations around the world that employ three- or four-stage crushing plants, for example, the iron ore industry in the Commonwealth of Independent States and South American countries. As the figure shows, the traditional three-stage crushing circuit can employ grizzly, scalping, and sizing screens.

Grizzly Screen

In plants where run-of-mine (ROM) ore has a high fines content, vibrating grizzly screens are used to increase the primary crushing capacity by scalping the fines. Grizzly screens are essentially very robust scalping screens. They are most commonly applied with jaw crushers, as gyratory crushers are normally high-capacity machines and can accommodate fines in the ROM feed.

Standard grizzly screens are available in sizes up to 2.4 × 4.8 m and can handle a volumetric capacity up to 1,200 m^3/h (depending on the feed size distribution). (To calculate the mass flow capacity, multiply the volumetric capacity of the screen by the bulk density of the solids.) For higher capacities, custom-made grizzlies can be designed and constructed. When designing a custom-made grizzly, important information includes the feed top size (mechanical robustness considerations), the fines content (OA considerations), and the feeder width (usually an apron feeder for large capacity). The grizzly width must be compatible with the feeder to reduce the drop height of the feeder discharge to the grizzly feedbox, as it can handle large lumps, and the impact provides high stress to the machine body. The use of rubber liners in the feedbox is strongly recommended to absorb the higher impact of the large lumps.

Grizzly machines are usually assembled with grizzly rail bars or with perforated plates, manufactured with abrasion-resistant steel (Figure 30). The grizzly rail bars have a tapered profile and high OA and are assembled in steps with an open end, which minimizes pegging and provides high capacity. Perforated plates have a lower OA, and therefore lower capacity, but are attractive because they reduce the risk of belt damage from large slab-like particles with sharp edges.

In high-capacity mines with finer ROM feed (e.g., Brazilian iron ore mines) that employ larger truck-shovel operations, it is very difficult for the shovel operator to separate the large lumps prior to delivery to the primary gyratory crusher. To avoid grizzly machine mechanical failures, a static grid (Figure 31) is recommended at the truck discharge to remove large lumps (>1,200 mm). Coarse particles remaining on the grid can be broken with a hydraulic rock breaker.

Scalping Screen

Scalping screens are usually intended to remove the fine particles in the crusher feed to avoid packing and the development of extreme forces in the crushing cavity. The aperture size is typically selected to be just smaller than the design crusher closed side setting (CSS), and the usual range is 50 to 120 mm. Because scalping screens are commonly applied in open circuit, efficiencies of 85%–90% are acceptable.

When treating primary crusher discharge, the feed top size to the scalping screens is in the range of 250 to 400 mm, depending on the primary crusher open side setting. To properly handle this top size while ensuring long service life, the use of strong rubber panels on the top deck (Figure 32) is highly recommended. The rubber absorbs the impact of the large lumps, preventing low panel life because of structure bending. In addition, riding bars are used to improve the impact absorption of large lumps. For lower abrasion index ores, perforated plates built with abrasion-resistant steel are also used.

Table 5 Advantages and disadvantages of steel versus polymeric screen decks

Wire	Rubber/Polyurethane
Advantages • Price • Accuracy of cut • Open area is high (on installation)	Advantages • Longer life • Less downtime • Shock resistance • Accuracy • Reduced pegging • Reduced blinding • Open area (in operation) • Noise reduction
Disadvantages • Shorter life • More downtime • Pegging problems • Blinding problems • Reduces open area (in operation) • Noise levels	Disadvantages • Price • Open area (on installation) (e.g., although these screens tend to have lower open areas out of the manufacturing process than wire mesh, they retain open area much better)

Source: Flintoff and Kuehl 2011

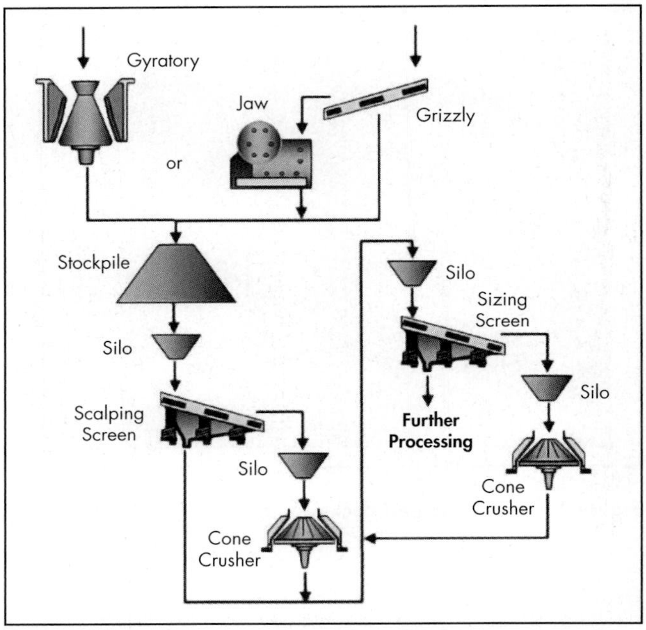

Courtesy of Metso

Figure 29 Traditional crushing and screening circuit

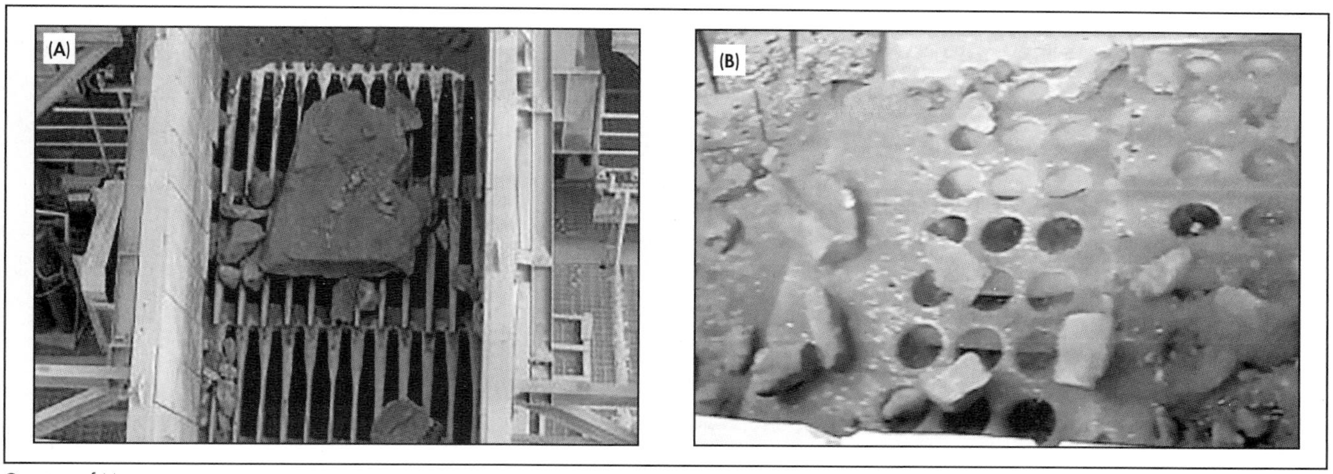

Courtesy of Metso

Figure 30 Grizzly screens with (A) bars and (B) perforated plates

Courtesy of Metso

Figure 31 Static grid at primary crushing dump point

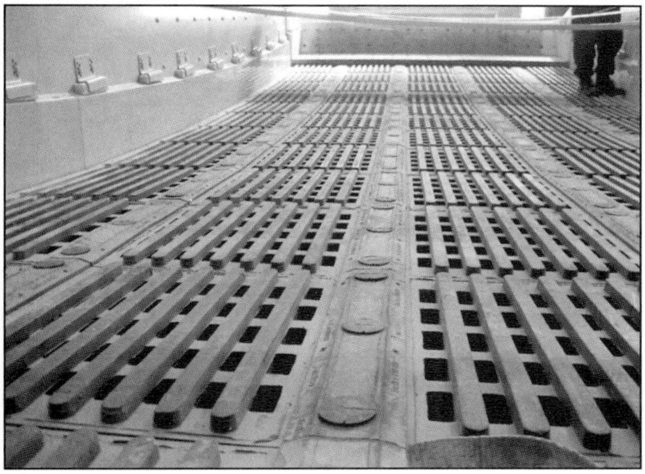

Courtesy of Metso

Figure 32 Rubber panels with riding bars for scalping screening application

The screen types recommended for scalping are the banana (linear motion) and the inclined (circular motion) double-deck type, where the top deck usually performs a relief/protection function. Because the impacts of the large particles are absorbed by the top deck, it is possible to use common modular panels in the lower deck for high abrasion index ores, and wire mesh for low abrasion index ores.

To avoid pegging at large media apertures and to optimize the screening efficiency, the scalping screen stroke must be higher than common sizing screens. The stroke cannot be too large, as that will decrease the screen performance (excessive bounce and lower probability of passage); and it cannot be too small, as that will increase the pegging tendency. A practical way to detect if a scalping screen is operating with an incorrect stroke or speed setting is if the particles pegging the screen deck can be easily removed by hand when the machine stops. If so, the screen has insufficient force to dislodge these particles. Generally, the preferred stroke and speed combination for scalping screens is 6 mm and 820 rpm.

Based on current market trends for large screen selection, and their ability to handle feeds with high fines content, banana screens are the most common scalping machine in new projects. The banana screen shape is the latest screening development, and the screens are designed to maximize the screening capacity as described earlier. The higher the fines content in the feed, the higher the "banana factor" gain in both dry and wet screening. The banana shape starts to exhibit a capacity gain against the more traditional screens when the feed is more than 40% passing the aperture size. As a rule of thumb, banana screens offer 30% higher capacity than a traditional inclined circular motion screen, in about the same footprint.

As larger and larger equipment is required (the economies of scale), the banana screen is a popular choice because it is does not have the same scaling constraints as more traditional screens. Because of roller bearing constraints, inclined circular motion screens are restricted to a 3.0×6.1 m size for scalping. Banana screens can be built with sizes up to 4.8×8.5 m in the double-deck configuration.

Horizontal screens can be applied in scalping, but this makes sense only when there is a height restriction in the building. Because of the horizontal deck, the pegging tendency for

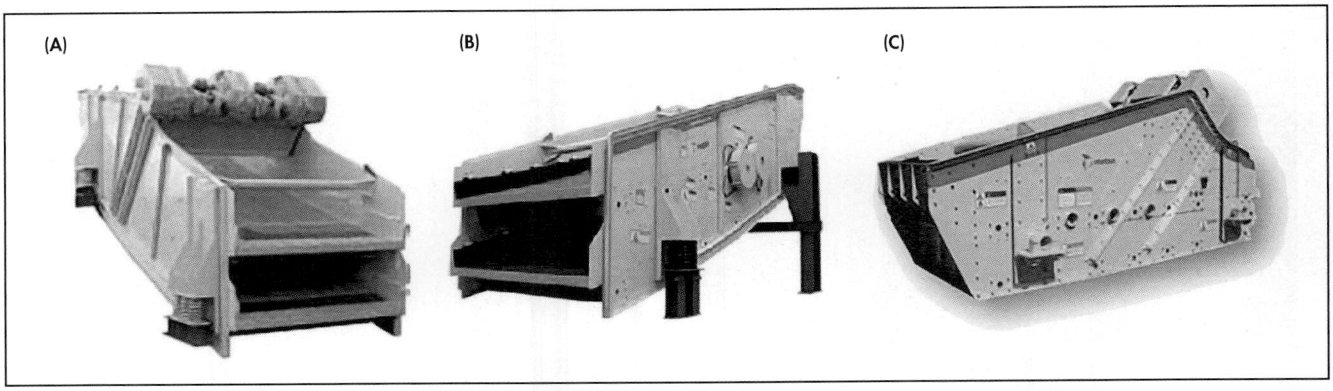

Courtesy of Metso
Figure 33 (A) Banana screen, (B) inclined screen, and (C) horizontal screen

large apertures is high. Good practice is to consider a maximum aperture of 50–75 mm for horizontal screens.

Sizing Screen

Sizing screens, as the name implies, are used to screen out an end product (e.g., aggregate plants) or to deliver a finer product for further processing (e.g., mining plants). The most common application is in closed circuit with a final crushing stage in a wet or dry process, and where the oversize is recirculated to the crusher feed. Efficiencies of 90%–95% are usually required in these applications. In mining applications, the typical apertures for separation are from 3 to 75 mm in dry processes, and from 0.5 to 75 mm in wet processes.

Recommended screens for sizing applications (Figure 33) are the banana (linear motion) and inclined (circular motion) double-deck type. As is the case for scalping screen selection, horizontal screens are recommended only when there is a building height restriction and for fine wet screening (<2-mm separation).

In the past, the "standard" screen size for large mining plants was 2.4 × 6.1 m. However, with the development of engineering tools such as finite element analysis (FEA, which emerged around 1990) and strain gauging and vibration measurements to validate these FEA models, larger screens (e.g., 3.6 m and 4.2 m wide) have been designed and manufactured and perform to expectations. For large greenfield projects, the 3.6 × 7.3 m screen size is becoming the new standard. For some projects, 4.2 × 8.5 m screens are being implemented. To use the full capacity of large screens, it is very important to ensure the correct feed distribution (the feed must cover at least 75% of the width of the deck at the feed end) to avoid significant loss of available screening area. When feeding screens with a width of 3.6 m or 4.2 m from a conveyor belt, it is usually necessary to design a feed chute with rock boxes to distribute the feed properly, as the conveyor generally is much narrower than the screen. The best solution to correctly spread the feed for large screens in dry processes is to use a divergent vibrating feeder (Figure 34). These feeders generally operate with a linear motion and a stroke of ~6 mm. For wet processes, the feed chute must be carefully designed, using the turbulence of the pulp flow to spread the slurry.

In the past, sizing screens were usually sized to reach 90% efficiency, but now engineering companies typically ask for screen designs that offer 95% efficiency. (Screening efficiency will be discussed in more detail later in this chapter.

Courtesy of Metso
Figure 34 Large screen fed by a divergent vibrating feeder

These numbers refer to E_u as defined later by Equation 4.) When doing sizing calculations for a closed circuit, this requirement must be carefully analyzed. For example, to go from 90% to 95% efficiency, the screen may require ~20% more area, which means a larger and more expensive screen. A critical element of this analysis relates to the determination of expected recirculating load (i.e., if the screening efficiency drops from 95% to 90%, the increase in the circulating load must be accommodated by the crusher). Figure 35 illustrates the potential impact of screen efficiency on circulating load.

Although some small mines still use wire mesh because of its lower cost, most modern and larger mining operations use modular panels fabricated from polymeric media (rubber and polyurethane). Panels provide better service life and are easier to handle. In general, due its higher flexibility (impact absorption), rubber media is recommended when the wear is caused by impact or high load, and polyurethane when the wear is caused by abrasion.

When pegging and blinding is a problem, flexibility helps to keep the media clean. Rubber panels can usually be manufactured in 40 and 60 shore A (a hardness measure based on indentation testing). Polyurethane is usually manufactured in 90 shore A, which is hard. It can also be manufactured in lower hardnesses, although this is not common. To mitigate pegging,

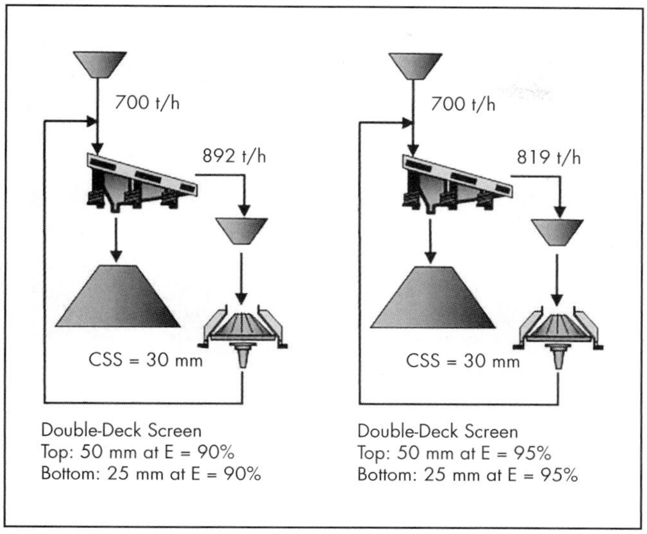

Courtesy of Metso

Figure 35 Closed-circuit load at 90% and 95% efficiency

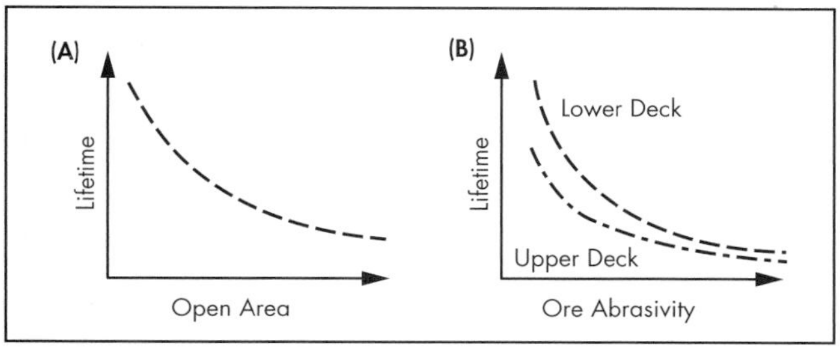

Courtesy of Metso

Figure 36 Relative media lifetime: (A) as a function of open area and (B) as a function of ore abrasivity

polyurethane panels can be manufactured in a zigzag design (e.g., Polydeck Screen Corporation's VR panels). A problem with a zigzag design is that the separation precision is slightly compromised. Some quarry operators choose to use flexible rubber screening media to minimize pegging and blinding in their dry screening operations. In mining sizing applications, separation precision is not as strict a requirement; therefore, both rubber and polyurethane can be employed for an anti-pegging solution.

In terms of the capacity–service life trade-off, the higher the OA of the media, the higher the screening capacity. However, the higher the OA, the shorter the service life, since there is less material to wear and the panel design is weaker (see Figure 36, which also includes ore abrasivity effects). Media suppliers usually have a standard media design that has a good balance of life and OA, and one that emphasizes OA design to maximize capacity. The correct media design is determined by the application. For greenfield projects, where a new screening building must be erected, it is safer to select the screen size and media based on standard panels, but for a revamp of existing plants, where capacity must be maximized, high OA media is probably the best solution. The preferred stroke and speed combination for sizing screens is generally around 5 mm and 900 rpm.

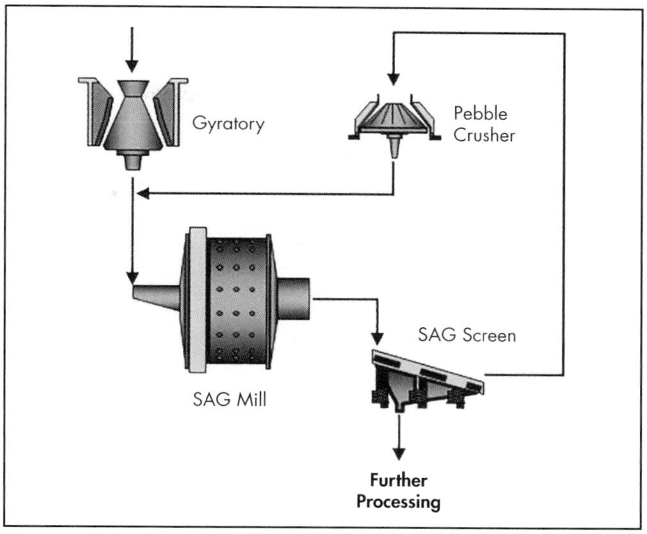

Courtesy of Metso

Figure 37 SAG circuit

SAG Circuit

In a semiautogenous, ball mill, and pebble crusher (or SABC) SAG circuit (see Figure 37), vibrating screens are used to screen the pebbles from the SAG discharge for recirculation to the pebble crusher. The SAG mill may be equipped with a trommel screen, in which case the trommel oversize is the screen feed, and the trommel undersize is combined with the screen undersize.

Because the SAG foundation is expensive and it is usually required to remove the fines adhered to the pebbles, horizontal screen types are often the best compromise for SAG discharge screens (see Figure 38). Horizontal screens offer a low profile and slower pebble travel speeds for more effective washing. The slight modification here is that SAG discharge screens are designed to work at an inclination of 5° to provide a higher capacity (see also Figure 8A).

For high feed solids rates (>1,500 t/h), the screen body must be carefully designed, because the SAG rotation tends to bias the discharge to one side or the other of the screen, creating a torque on the machine body. In addition, the screen body must be designed to accept a high volumetric load dropping from height. Generally, the SAG screen mechanical availability is lower than that of the SAG mill itself, and unlike the pebble crusher, there is no option for bypass. To ensure a high availability for the SAG circuit, the screen should be assembled on a trolley for quick replacement by a spare unit (the "cassette" maintenance concept).

The SAG screen can be used with or without a trommel, although to ensure high screening capacity, the combination of a trommel and a SAG discharge screen is widely employed. The use of a trommel can complicate the screen selection, as it is very difficult to predict the mass balance and particle size distributions for the trommel products. Since vibrating screens have a maximum bed depth (mechanical load and separation performance) to work safely, errors in the predicted mass balance could lead to screen overloads and mechanical damage/failures and/or poor separation performance.

The use of single- or double-deck screens in this application depends on the bed depth. Usually, a standard mining

Courtesy of Metso

Figure 38 SAG discharge screen

screen can handle up to 150 mm of bed depth. However, because of the mechanical design precautions mentioned earlier, a SAG screen can usually handle up to 250–300 mm of bed depth. Only a few OEMs have good (mechanical and process) design experience with SAG screens, and it is best to work only with these firms. Such firms include, but are not limited to, Metso, Schenck Process, and FLSmidth–Ludowici.

Dense Media Circuit

Dense media circuits are common in coal and diamond beneficiation applications. Using dense media (usually magnetite or ferrosilicon), the "fluid" density can be managed to control the separation of lighter (floats) and heavier (sinks) components in an ore. Figure 39 illustrates a typical coal preparation dense medium cyclone circuit (e.g., treating 50 × 0.5-mm particles) with sizing screens to get the right feed size fractions to the correct circuits, and drain-and-rinse screens for media (and some fines) recovery on both dense medium separation products.

Since the media (magnetite and especially ferrosilicon) is expensive, after the separation in dense medium separation cyclones and drums, the media are recovered in drain-and-rinse screens. The first section of the screen is used to recover most of the media and the remaining part to rinse the particles and remove the media adhered to particle surfaces. There are usually separate screens to drain and rinse the media (and ore fines) from the sink and float products, but in some cases, depending on the flow rates, a single screen with a partition wall can be used for both float and sink products, as illustrated in Figure 40.

Horizontal and banana screens (the latter with a slope combination of 25° at the feed end and 0° at the discharge) are widely used for drain-and-rinse applications. These "desliming"-like applications can have apertures as fine as 0.5 mm, so the 0° inclination at the discharge is important to provide good oversize dewatering. Any inclination of this region of the screen deck in the 0.5- to 2-mm aperture range can result in excess water to the oversize stream.

In older coal preparation plants, the combination of a static sieve bend directly feeding a horizontal screen has been successfully applied in drain-and-rinse applications. The sieve bend (with wedge wire) promotes a quicker drainage of the feed pulp, removing all the easy screening fine particles and most of the media pulp. The horizontal screen then washes the remaining particles and, as a result of its slower particle travel velocity, performs the near-size particles screening. The banana screen achieves the same principle, and with the advantage that the entire deck is vibrating, thereby minimizing pegging problems. The sieve bend is static and can experience some pegging issues. In addition, the sieve bend wedge wire can wear to have very sharp edges (a safety matter), and these machines are usually assembled in one big piece, presenting installation and maintenance handling challenges. The banana screen makes it light and easy to handle during maintenance.

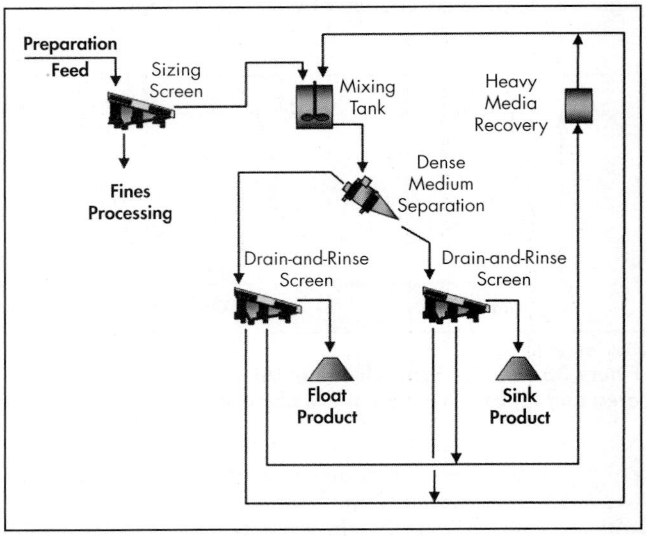

Courtesy of Metso

Figure 39 Dense media circuit

For desliming and drain-and-rinse applications, polyurethane media are usually used because abrasion is the main wear mechanism and rubber media is not generally available with apertures smaller than 2 mm. The preferred stroke and speed combination for desliming and drain-and-rinse screens is the same as the sizing screens: 5 mm and 900 rpm. The OEM list for dense media screening includes, but is not limited to, Metso, Schenck Process, FLSmidth–Ludowici, Joest, Haver and Boecker, Siebtechnik, and Weir–Linatex.

Courtesy of Metso

Figure 40 Drain-and-rinse screen with partition wall

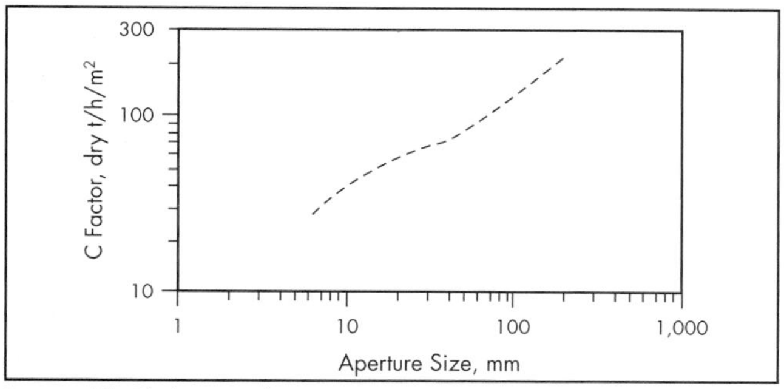

Source: Allis-Chalmers, n.d.

Figure 41 Basic capacity factor (C) as a function of aperture size

Screen Selection

The many opinions and methods of screen selection suggest fertile ground for development and refinement. Three general approaches follow:

- **Class 1:** Empirical methods based on correlations derived from extensive databases of industrial experimental results
- **Class 2:** Lab screening tests with empirical model-based scale-up procedures
- **Class 3:** Fundamental model-based methods (these are relatively new and are described later in more detail)

Class 2 methods have been used sparingly in the past, but recent work (Hilden 2007) suggests there is good scope for improvement and room to combine Class 2 and Class 3 methods very effectively. Despite the potential of using Class 2 and Class 3 methods, most industrial vibrating screens are still selected using Class 1 methods. Because in most cases these screens operate more or less as designed, it is understandable why these methods have survived, with a few improvements, for some 70 years.

Class 1 selection methods employ empirical correlations to deduce screen area. Given the application (e.g., a pebble screen on a SAG mill discharge), the design feed rate of solids, and the typical feed particle size distribution, the design process can begin. Early decisions on numbers of decks, inclination, motion, low-head or high-head screens, and so on, come from the design basis documents and past practices in similar applications. After these parameters are set, the correlations can be used to compute screen area requirements, and the deck with the largest area determines the screen size. The task is then to match one of the standard products to the design. This is often an iterative design process.

Two usual approaches can be used to solve the Class 1 design problem—one based on the total solids feed rate to the screen and the other based on the solids feed rate of undersize material to the screen. Not surprisingly, this is also an area of debate, as the scientific community tends to prefer the latter method, yet the former is very commonly used. The former method is described in an abbreviated way as follows, simply for illustrative purposes. To provide a real example, the data for the screen in Figure 11 is used.

The fundamental equation for determining the screen area required is given in the following equation (Allis-Chalmers n.d.).

$$A = \frac{T}{CMKQ_1Q_2Q_3Q_4Q_5Q_6} \quad \textbf{(EQ 3)}$$

where

A = required screen area
T = mass flow of feed to screen
C = basic capacity factor
M = oversize factor (percent larger than aperture size in feed)
K = undersize factor (percent smaller than ½ aperture size in feed)
Q_1 = bulk density factor (usually ~60% of rock density)
Q_2 = aperture shape factor
Q_3 = particle shape factor
Q_4 = OA correction factor
Q_5 = wet or dry screening factor
Q_6 = surface moisture factor

Figures 41 through 43 provide the basis for estimating the factors C, M, and K. Table 6 provides the means of estimating the Q factors (note that OA and aperture shape come from Figure 15). Table 7 presents the summary data for the calculations, showing both the input data required and the empirical corrections, as deduced from Figures 41–43 and Table 6 (which are adapted from Allis-Chalmers [n.d.]). The required area allows for a 6% loss because of mechanical fittings and so on. Following the rough rule that $L = 3W$, this screen would have the approximate dimensions of 0.34 × 1.03 m. In fact, the screen in Figure 11 is 0.34 × 1.67 m. The width is the same, but the test screen has extra length so one would expect an impact on classification efficiency. This topic is discussed the following subsection.

Although the symbology is slightly different, equations for the relationships shown graphically in Figures 41 through 43 are available in King (2001).

Performance Assessment

Mechanical and process performance can be distinguished, but this subsection only focuses on the latter.

A single parameter can be calculated in several ways that will quantify screen performance. Among these, the most common are E_u, which is defined as the efficiency of removal of undersize material from the oversize stream, and, R_u, which is the efficiency of undersize removal from the feed stream.

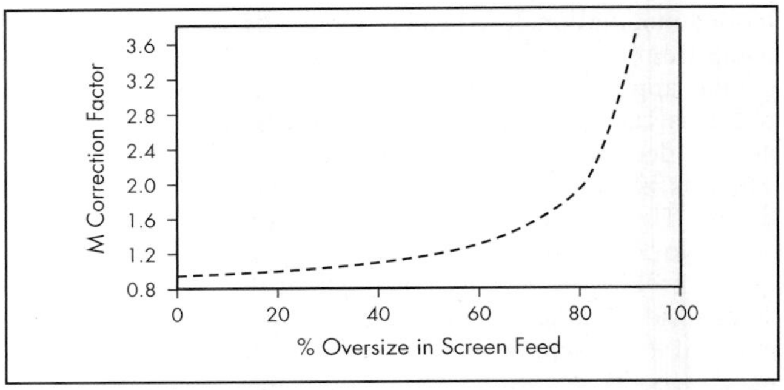

Source: Allis-Chalmers, n.d.

Figure 42 Correction factor (M) as a function of percent oversize

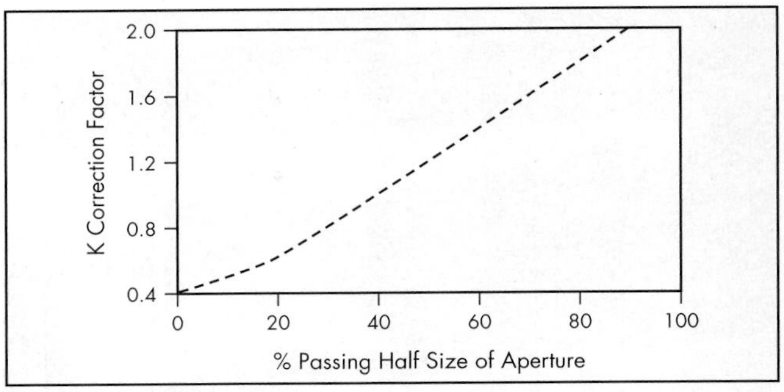

Source: Allis-Chalmers, n.d.

Figure 43 Correction factor (K) as a function of percent half size

Table 6 Q correction factors

Bulk Density Factor	Value
Bulk density, dry t/m^3	Q_1
0.4	0.25
0.8	0.50
1.6	1.00
2.08	1.30
Aperture Shape Factor	
Opening shape	Q_2
Square	1.00
Rounded	0.80
Slotted	
2:1 Slot	1.15
3:1 Slot	1.20
4:1 Slot	1.25
Particle Shape Factor	
Particle shape	Q_3
Cubical	1.00
Slabby	0.90
Open Area Factor	
Open area	Q_4
Calculation	Q_4 = percent open area ÷ 50%
Wet or Dry Screening Factor	
Opening	Q_5
Dry	1.00
Wet	
0.8–3.2 mm	1.25
4.8–6.4 mm	1.40
8–12.7 mm	1.20
14.3–25.4 mm	1.10
Surface Moisture Factor	
Surface moisture	Q_6
≤3%	1.00
3%–6%	0.85
6%–9%	0.70
Wet screen	1.00

Source: Flintoff and Kuehl 2011

Referring to the symbology in Figure 44, Equations 4 and 5 are used to compute numerical values.

$$E_u = \frac{F(1-\hat{f}_u)}{O} = (1-\hat{o}_u) \quad \textbf{(EQ 4)}$$

$$R_u = \frac{U}{F\hat{f}_u} = \frac{(\hat{f}_u - \hat{o}_u)}{\hat{f}_u(1-\hat{o}_u)} \quad \textbf{(EQ 5)}$$

where

F = feed rate
$\hat{f}_u$ = cumulative mass fraction in the feed passing the aperture size
O = oversize mass flow rate
$\hat{o}_u$ = cumulative mass fraction in the oversize stream
U = undersize mass flow rate

To estimate these performance measures, one is required to perform a sampling experiment around the screen while it is running at steady-state conditions. This raises subjects beyond the scope of this chapter (experimental design, sampling theory, sample preparation and analysis, mass balancing, etc.), but these are usually covered in the standard textbooks, such as Wills and Finch (2016). Practical issues, typically associated with sample access as well as the isolation of a screen operating in parallel, tend to make this a challenging experiment. Fortunately, for the screen shown in Figure 11, the balanced data are already available and are summarized in Table 8.

Using these results, E_u = 77.9% and R_u = 93.5%. The data suggest that there is quite a lot of finer (less than the aperture size of 10 mm) near-size material in the oversize stream. In terms of assessing performance, the manufacturers have a curve that relates E_u to the rated capacity of the screen. This is shown in Figure 45, and the rated capacity is the ratio of the actual tonnage (or area) to the calculated tonnage (or area) from Equation 3. The respective area figures for the example are 0.35 and 0.57 m^2, which gives a percentage of rated capacity of 59.9%. This point is plotted on the curve for reference.

This comparison indicates that the test screen in Figure 11 is not performing as well as would be expected at the given operating conditions and further analysis (e.g., low G values,

Table 7 Screen sizing example

Parameter	Value
Feed rate, dry t/h	15.7
Aperture (square), mm	10
Rock density, dry t/m²	2.7
Application	Dry
C Factor, dry t/h/m²	40.12
M Factor	0.972
K Factor	0.88
Q Factors	
Q_1	1.00
Q_2	1.00
Q_3	1.00
Q_4	1.39
Q_5	1.00
Q_6	1.00
Adjusted unit capacity, dry t/h/m²	47.64
Area required, m²	0.34
Size, mm	**Feed Distribution, %**
13.2	5.6
9.5	16.6
8	11.4
6.7	11.1
4.75	24.1
3.35	10.7
2.8	6.5
Pan	14.0
Percent +10 mm	18.7
Percent –5 mm	34.0

Data from Hilden 2007

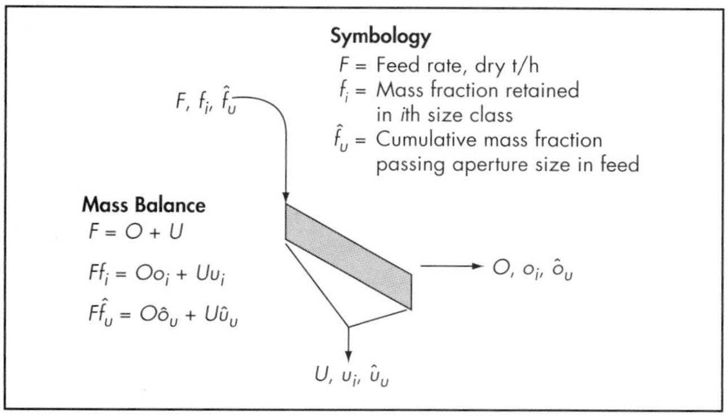

Source: Flintoff and Kuehl 2011

Figure 44 Screen performance parameters

extended length, etc.) of the situation would be required to validate the result and improve performance.

Another very common measure of efficiency is to use the data in Table 8 to compute an experimental partition curve, which can then be modeled to extract performance metrics.

The details related to the estimation and analysis of partition curves are provided in Chapter 4.6, "Partition Curves."

Table 8 Sampled data summary for the horizontal screen in Figure 11

Size	Feed	Oversize	Undersize
13.2	5.6%	20.8%	0.0%
9.5	16.6%	61.8%	0.0%
8	11.4%	13.9%	10.5%
6.7	11.1%	3.5%	13.9%
4.75	24.1%	0.0%	33.0%
3.35	10.7%	0.0%	14.6%
2.8	6.5%	0.0%	8.9%
Pan	14.0%	0.0%	19.1%
Mass fraction	1.00	0.27	0.73
Percent –10 mm	81.3%	22.1%	100.0%
Mass flow, dry t/h	15.7	4.2	11.5

Data from Hilden 2007

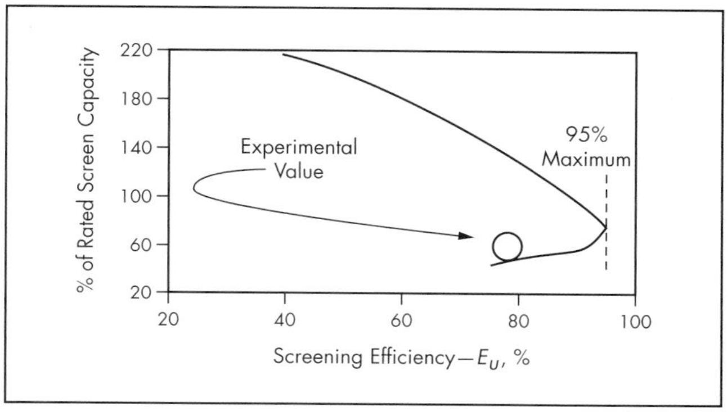

Source: Flintoff and Kuehl 2011

Figure 45 Percentage of rated capacity versus E_u

The authors of this chapter assume that the reader is already familiar with this topic or has read this chapter.

In the case of vibrating screens treating dry materials at relatively large aperture sizes (e.g., >5 mm), it is quite common to use Equation 21 in Chapter 4.6, "Partition Curves," to model the experimental partition curve data; that is, the fines bypass to the coarse product in these cases (Rf) is taken to be zero. Figure 46 presents the results for the screen in Figure 11. The sharpness of separation, α, is very high, which may have something to do with the length of the deck. (Recall that this screen is longer—1.67 m—than would be required for the feed conditions it was seeing, 1.03 m from the earlier screen sizing calculation.)

One would normally expect the cut size to be close to the throughfall aperture size, and clearly the value is a little low here. (This is, in part, because of the way the characteristic size was defined for this study.) The sharpness of separation for screens cutting in this aperture size range is usually around 6, so a value of 9.6 suggests that efficiency is relatively high. Taken together, the implication is that the screen is not processing the finer near-size material very well, which would require further investigation.

The data acquired for the screen in Figure 11 permit the calculation of the cumulative partition curve (here using the experimental data) as a function of the position along the length of the deck. The experimental results are shown in

Figure 47. Given the earlier explanation related to crowded and separated screening, it is no surprise to see that the fine material moves quickly through the screen—collectively, particles do not have to progress far down the deck before all of the very fine material has passed through the apertures (and the corresponding $p_i = 0$). For the fine, near-size material, clearly the process is much slower (as illustrated in Figure 12).

The periodic measurement of screen efficiency is as important to the maintenance of process performance as the periodic measurement of, for example, bearing condition is to the maintenance of mechanical performance. Unfortunately, often the process aspects of screens are ignored. However, screens are robust devices and often have a direct impact on the performance of comminution and separation equipment, for which a "systems thinking" approach must be applied.

Recent Developments

This section provides a brief introduction to some of the developments emerging in the vibrating screen market.

Real-Time Condition Monitoring

Using vibration, thermographic, and occasionally other specialty sensors to automate the more usual manual inspections of predictive maintenance is becoming increasingly commonplace. Where it is done, this is still usually reserved for the typical applications of monitoring bearing health. Manufacturers such as Metso (ScreenWatch) have taken this a step further by using additional wireless accelerometers to compute screen orbits, which can then be compared with expected values from run-in tests and/or finite element modeling to analyze the mechanical and process health of the screen. (This essentially automates the manual throw-card analysis, making the results available in real time, all the time.) Figure 48 is an example of normal orbits and the kinds of trends that develop when there is a problem with the supporting springs. Loose media on the screen deck and other kinds of serious problems have their own orbit "signature."

In the case of the normal orbits (Figure 48A), both are linear, showing the same stroke and angle. In the case of the abnormal condition (Figure 48B), the orbits are now elliptical and the strokes are different.

The frequency spectrum of the accelerometers also provides useful information on the mechanical state of the deck, loading, and other interesting performance parameters. Moreover, when these signals are combined with others (from load cells, vision systems, etc.) in a kind of sensor fusion approach, the analytical detail and reliability can be dramatically increased.

Real-Time Deck Motion Modulation

As seen earlier in the process matrix of Table 4, deck motion (stroke, angle, and orbit) all affect performance, and because screen feeds often change (mass flow and particle size distribution), one school of thought says the ability to modulate these parameters online would support real-time screen performance management. Several manufacturers are exploring this kind of solution on selected screen models. For example, Metso has developed a system for controlling stroke angle on its Ellipti-Flo inclined screens. (Ogawa [2010] has discussed the effect of stroke angle on screen performance.) Figure 49 illustrates an application for screens that are subject to

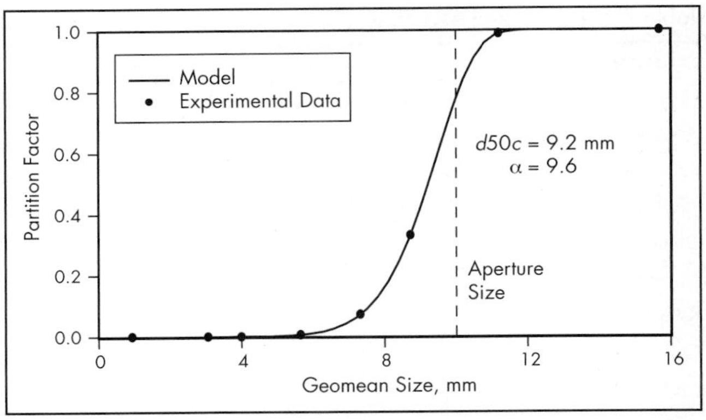

Adapted from Flintoff and Kuehl 2011

Figure 46 Experimental and modeled partition function data

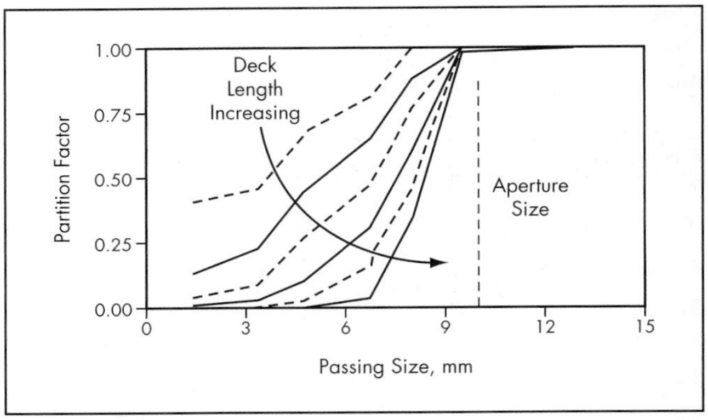

Data from Hilden 2007

Figure 47 Partition factors as function of position on deck

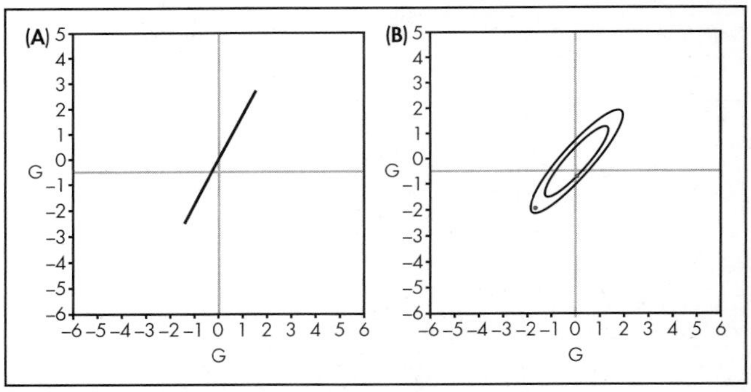

Source: Flintoff and Kuehl 2011

Figure 48 Feed end orbits in (A) normal and (B) abnormal operation

pegging. In this particular case, the feed is interrupted and the angle adjusted to maximize G (see Figure 23) perpendicular to the screen deck. This procedure can also be performed with the feed on, although the angle change is not likely to be as large as shown in this figure. One can also conceive of using this feature to control the rate of movement of material across the deck. For example, if the feed rate drops, the angle can change to increase residence time and avoid screen inefficiencies in the discharge region, and the converse.

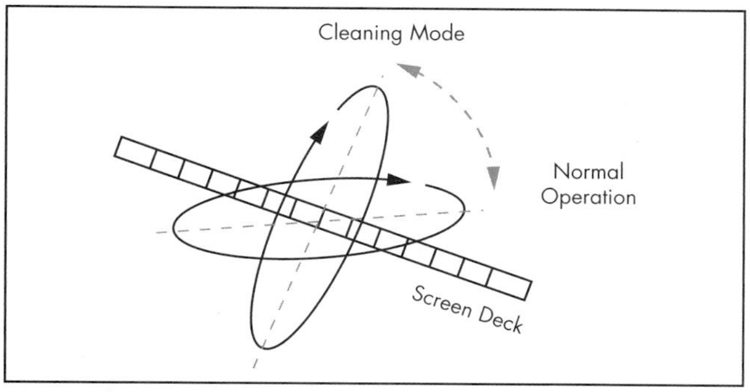

Courtesy of Metso
Figure 49 Controlling stroke angle on a Metso Ellipti-Flo inclined screen

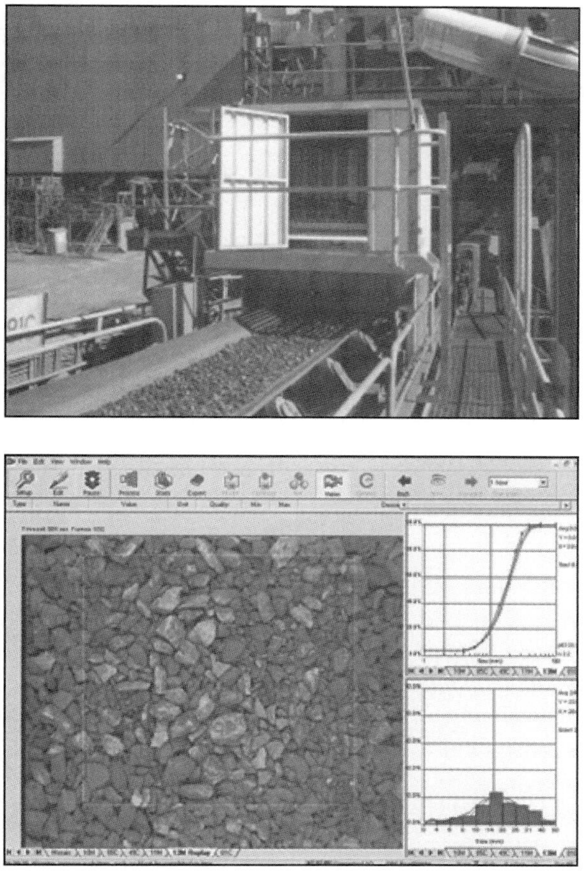

Source: Flintoff and Kuehl 2011
Figure 50 VisioRock installation and typical output display

Instrumentation

One of the exciting areas of potential for performance monitoring is "machine vision," primarily as applied to measuring feed or product size. Typically, dry screening is very dusty and requires hoods covering the entire screening surface for effective dust control. Therefore, applications on the deck itself (oversize velocity, load, etc.) are a challenge. However, in the case where the product streams are discharged onto belts, the standard technology, such as Metso's VisioRock, can be applied. For example, in the production of iron ore, lump size control is very important, and early warning of a problem is essential if, for example, the screen media is broken, thereby producing off-specification material in the undersize stream. One operator is now using machine vision to detect and react to these types of quality assurance events, and Figure 50 shows the VisioRock installation as well as the output from data processing. In this case, images are analyzed at a rate of ~10 per second, which is more than sufficient to monitor everything that passes.

Acoustics also offer interesting potential for analyzing screen performance, but it is very early for research in this area.

Population Balance Simulation Methods

Population balance methods (PBMs) are not new, having been in use for at least two decades, but as the design and operation needs turn increasingly toward simulation to support these activities, higher and higher fidelity is required. These are an element of the Class 2 methods previously mentioned. Although the industry is moving in this direction, older approaches are still in use in most of the simulation packages. One of these is the model by Karra (1979), which is based on a Class 1 selection method, although in this case it is based on the method of sizing relating to the mass flow rate of undersize in the feed stream.

Using Equation 11 in Chapter 4.6, "Partition Curves," Karra (1979) developed a correlation based on the empirical screen sizing factors used in the screen selection approach. The factors have been modified over time, and their estimation is described elsewhere (e.g., King 2001), but the equation for computing $d50c$ is

$$d50c = \frac{G_c h_T}{[(T_u/H)(ABCDEF)]^{0.148}} \quad \textbf{(EQ 6)}$$

where

$d50c$ = cut size
G_c = near-size correction factor
h_T = throughfall aperture (see Figure 17)
T_u = tons of undersize in the feed, dry t/h
H = effective screen area, m^2
A = basic capacity factor
B = oversize factor
C = fine size factor
D = deck location factor
E = wet screen factor
F = bulk density factor

Because several properties for the screen in Figure 11 have been given, Table 9 is a shortened summary of the Karra calculation of $d50c$. In this particular case, the authors elected to use data from the screen in Figure 11 based on a screen length of 1.1 m, closer to what was estimated in the screen selection discussion (1.03 m), and not on the total screen length (1.67 m). This would correspond to the first four undersize "bins" (from the left) in Figure 11. As one can see, the extra screen length does affect efficiency, as there is material in the last two undersize bins. The argument is that the Karra model was developed from screens in near-normal operation, and using the data from a shorter screen complies more closely with that condition.

Karra assumes α is a constant in the partition curve model with a value of ~5.9, and in the authors' experience, numbers

Table 9 Computing Karra's *d50c* for the screen in Figure 11

Symbol	Factor	Units	Result
T_u	Feed	dry t/h	15.67
H	Effective screen area	m^2	0.35
h_T	Throughfall aperture	mm	10.00
G_c	Near-size correction factor	—	0.82
A	Basic capacity factor	t/h/m^2	17.85
B	Oversize factor	—	1.38
C	Fine size factor	—	1.12
D	Deck location factor	—	1.00
E	Wet screen factor	—	1.00
F	Bulk density factor	—	1.01
$d50$	Cut size	mm	7.85

Source: Flintoff and Kuehl 2011

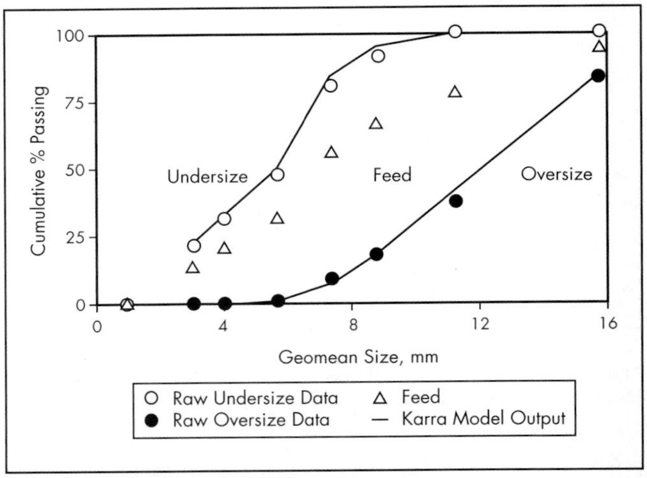

Adapted from Flintoff and Kuehl 2011

Figure 51 Karra's model predictions and actual cumulative size distribution curves for the "shortened" screen in Figure 11

in the region are pretty typical for coarse screening. The *d*50*c* value in Equation 6 is also in decent agreement with the experimental value of 8.2 mm (based here on the shorter screen length data).

Karra's model can be used to predict the expected size distributions for the undersize and oversize products, and these are shown in Figure 51 along with the actual experimental results. The predictions are good. The calculated mass splits of 0.37 (to oversize) and 0.63 (to undersize) compare well with the experimental results of 0.35 and 0.65.

Discrete Element Simulation Methods

In the pursuit of higher fidelity simulation of screen performance, the ultimate tool is probably discrete element modeling (DEM). Very briefly, this approach is based on fundamental models drawn from physics (e.g., Newton's laws of motion). Using a mathematical description of the body and motion of the screen; some typical physical properties; and the feed size distribution, flow rate, and particle shapes, the DEM simulation can be run from only first principles (no correlations, no parameters drawn for experimental data, etc.). Although this sounds simple in concept, solving the problem is onerous, as there are millions of particles, each requiring several differential equations to describe translational and rotational motion in time and space. Because of the interactions of the particles among themselves and with the screen itself, the integration intervals are very short—about one millionth of a second. Simulating several seconds of real-time operation can take hours or days, depending on the problem size, computational capability, and simulation code efficiency. Consequently, the goal is to combine DEM and PBM, the former to work out the real mechanisms and the latter to abstract these and provide speedy calculations.

Because of the vast quantities (gigabytes or even terabytes) of numerical data generated in such simulations, it is common to use visualization techniques to "see the results." However, the numerical data must be used for any calculations. Figure 52 presents a snapshot of visualization, and Figure 53 shows the predicted (lines) and measured (bullets) particle size distributions from a simulation of the banana screen pictured Figure 19B. The agreement in Figure 53 is very good, which provides some insight into the power this tool will bring to screen modeling.

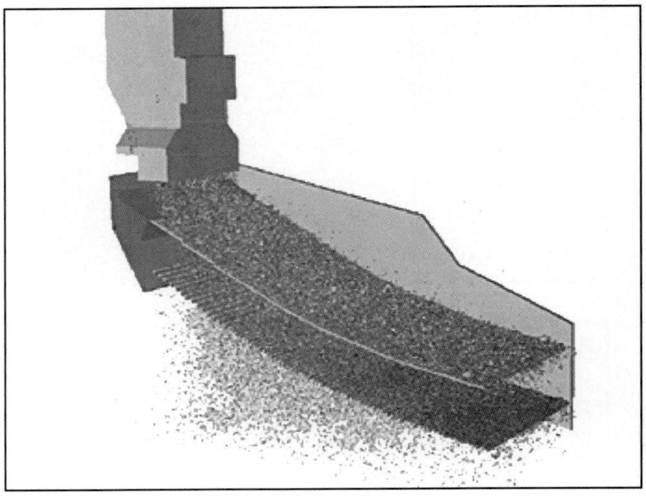

Source: Flintoff and Kuehl 2011

Figure 52 DEM visualization of screen operation

Fine Screening

In general, wet fine particle classification is achieved by mechanical/hydraulic means (e.g., rake or spiral/screw classifiers), hydrocyclones, and fine screens. Although the mechanical devices still have their place, fine screens and especially hydrocyclones dominate fine and ultrafine classification. *Fine screening* is the term generally reserved for fine and ultrafine separations, as defined in Table 1. It is performed dry or wet, although in most mining or quarrying applications, this is a wet process, consequently the emphasis is on wet fine screening.

For the most part, the principles described previously for vibrating screens apply to most screening processes. One notable difference is that fine screening generally involves much higher rotational speeds (e.g., ~3,600 rpm) and therefore much smaller strokes. Another major difference for fine screening is that the selection of the appropriate screen is as

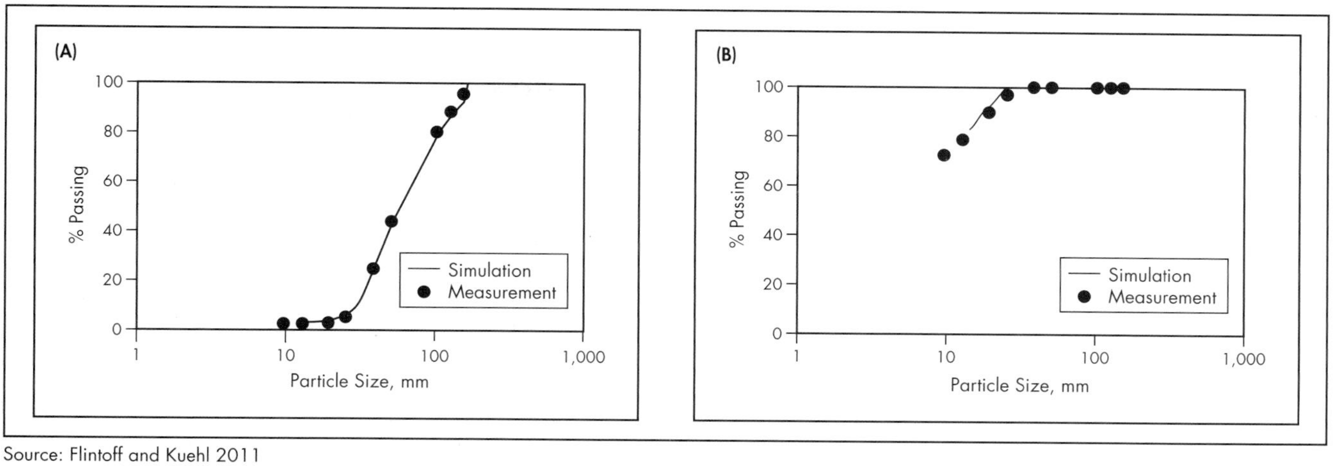

Source: Flintoff and Kuehl 2011

Figure 53 DEM predictions of (A) oversize and (B) undersize particle size distributions

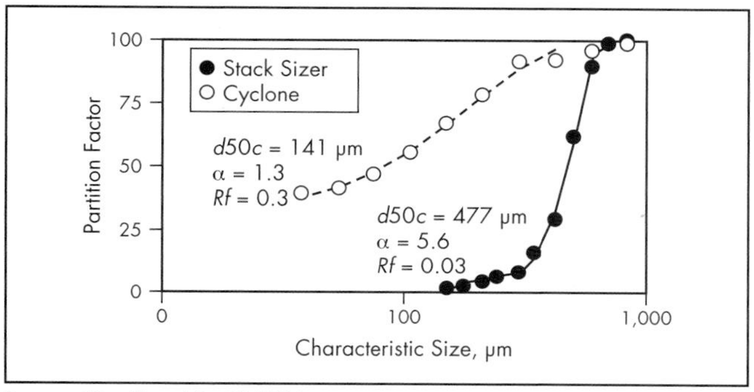

Source: Flintoff and Kuehl 2011

Figure 54 Partition curves for a 10-in. hydrocyclone and a Derrick Stack Sizer

much an art (i.e., past practices in similar applications) as a science (i.e., laboratory testing of specific screens on the expected feed material and supplier databases). This complexity is due in large measure to the critically important aspect of fluid mechanics, as drag forces tend to govern wet fine and ultrafine separations. Suffice it to say that the expertise in fine-screen design generally lies with the major supplier organizations, such as Derrick Corporation, Conn-Weld Industries, Kroosh Technologies, PanSep, Multotec, Wedgewire, and others.

The two most common wet fine screening devices are the sieve bend (i.e., a stationary version of the banana screen, Figure 6) and the more conventional, high-frequency fine screen (i.e., having a more conventional geometry, and represented herein by the Derrick product line, including the Stack Sizer [Figure 5], mulitfeed screen, repulp screen, etc.). Wet fine screening is essentially the only flow-sheet option in several applications, for example, coal preparation desliming, iron ore coarse silica removal. (A more complete review of this topic is available in Valine and Wennen [2002].) There are many other applications where wet screening provides an alternative to hydrocyclones or other mechanical devices, as it offers some significant advantages in classification efficiency (as measured by α or E_u). In some applications, both sieve bends and high-frequency screens are used in conjunction with hydrocyclones to exploit the physical and classification attributes for each type of separator. In other instances, it may also be possible to feed a wet screen by gravity, rather than the pumping systems more commonly associated with hydrocyclone batteries. Nevertheless, in these kinds of trade-off studies, it generally comes down to the footprint advantages of hydrocyclones versus the process efficiency advantages of wet screening. For example, one study (Vince 2007) in coal indicated that purely from a capacity perspective, two to three Derrick Stack Sizers were the equivalent of one 900-mm classifying hydrocyclone. Interestingly, that same study concluded that the Stack Sizer should be considered for coal preparation applications with cut points in the 100- to 350-µm range, while sieve bends could be considered for cut sizes in the 250- to 350-µm range. Of even greater interest was the observation that "similar" equipment made by different suppliers can have quite different performance characteristics—*caveat emptor* (buyer beware).

One area of growing interest for fine screening is to close fine-grinding (ball mill or even stirred mill) circuits (see Wennen et al. 1997). The opportunities to exploit improved efficiency to reduce specific energy and media requirements, and to improve downstream metallurgical efficiency (e.g., by minimizing overgrinding of valuable minerals) are very appealing, both technically and economically. In addition, while hydrocyclones separate based on particle mass (i.e., particle size and density are first-order effects), wet fine screens separate predominantly on size (i.e., density is a higher order effect, mainly related to particle stratification). For example, in iron-ore grinding, the cut size on a screen is the same for the iron minerals and silica, while in a hydrocyclone, the cut size for the iron minerals is about one-third of that for silica, which is to say that a lot of fine liberated iron mineral can be sent back to the grinding mill by a hydrocyclone.

Consider the partition curves shown in Figure 54. Although there are some "apples to oranges" comparisons in terms of the cut sizes, these examples are drawn from the same plant in similar applications (Acquino and Vizcarra 2007); the real interest comes from the comparison of the sharpness of separation (α) and the efficiency (here inversely related to *Rf*, as estimated from the particle size data). The benefits of wet screening are obvious, which makes it a technical option for cut sizes ranging down to 100 µm. For the hydrocyclone

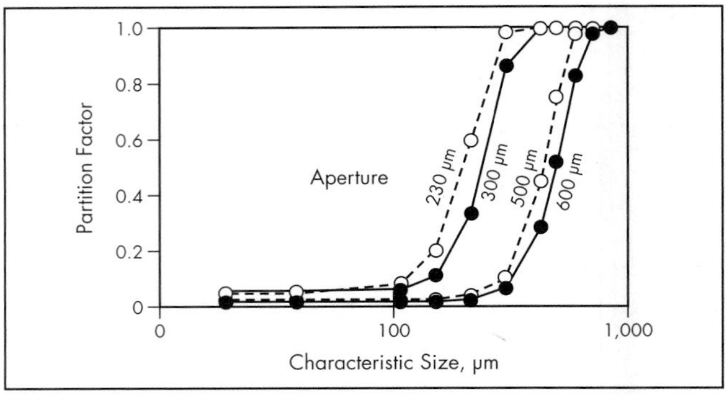

Source: Flintoff and Kuehl 2011

Figure 55 Partition curves for a Derrick Stack Sizer with different screen apertures

partition curve, the inflections and rather poor sharpness of separation are a partial consequence of the existence of liberated heavy and light minerals in the feed (i.e., the particle density effect).

Finally, Figure 55 provides a glimpse of the ability of this fine screen to preserve efficiency over a range of aperture sizes, that is, sharpness of separation is good and bypass is low.

ACKNOWLEDGMENTS

The authors extend their most sincere thanks to Derrick Corporation and its many colleagues at Metso Mining who provided much of the information contained herein, as well as some very helpful feedback.

REFERENCES

Acquino, B., and Vizcarra, J. 2007. Re-engineering the metallurgical processes at Colquirjirca Mine. Socidad Mineral El Brocal S.A.A., Mineria, Peru, October. pp. 30–38.

Allis-Chalmers. n.d. *Theory and Practice of Screen Selection*. Allis-Chalmers Technical Note.

Bothwell, M.A., and Mular, A.L. 2002. Coarse screening. In *Mineral Processing Plant Design, Practice, and Control*. Vol. 1. Edited by A.L. Mular, D.N. Halbe, and D.J. Barratt. Littleton, CO: SME. pp. 894–916.

Flintoff, B., and Kuehl II, R. 2011. Classification by screens and cyclones. In *SME Mining Engineering Handbook*, 3rd ed., Vol. 2. Edited by P. Darling. Englewood, CO: SME.

Hilden, M. 2007. A dimensional analysis approach to the scale-up and modelling of industrial screens. Ph.D. thesis, University of Queensland, Brisbane, Queensland, Australia.

Karra, V.K. 1979. Development of a model for predicting screening performance of a vibrating screen. *CIM Bull.* (April):167–171.

King, R.P. 2001. *Modeling and Simulation of Mineral Processing Systems*. Boston: Butterworth-Heinemann. pp. 81–125.

Matthews, C.W. 1985. Screening. In *SME Mineral Processing Handbook*. Edited by N.L. Weiss. Littleton, CO: SME-AIME. pp. 3E1–3E12, 3E25–3E41.

Metso. 2009. *Crushing and Screening Handbook*, 4th ed. Edited by J. Eloranta. Helsinki, Finland: Metso.

Metso. n.d. *Screening Media Handbook*, Version 1.3. www.metso.com/miningandconstruction/MaTobox7.nsf/DocsByID/EE6685E585F484B2C2257C7E0044BC5F/$File/2977-SM%20Handbook_1-3_150923_FINAL_LR.pdf. Accessed September 2017.

Moon, D. 2003. Understanding screen dynamics for optimum efficiency. Presented at the 34th Conference of the Institute of Quarrying Southern Africa, March.

Mular, A. 2003. Size separation. In *Principles of Mineral Processing*. Edited by M. Fuerstenau and K. Han. Littleton, CO: SME.

Nichols, J. 1982. Selection and sizing of screens. In *Design and Installation of Comminution Circuits*. Edited by A.L. Mular and G.V. Jergensen. Littleton, CO: SME-AIME. pp. 509–522.

Ogawa, R. 2010. Reducing mining environmental impact—Through innovative vibrating screen technology. Presented at the Sustainable Mining Conference, Kalgoorlie, Western Australia, Australia, August 17–19.

Taggart, A.F. 1945. *Handbook of Mineral Dressing*. New York: Wiley and Sons.

Valine, S., and Wennen, J. 2002. Fine screening in mineral processing operations. In *Mineral Processing Plant Design, Practice, and Control*. Edited by A.L. Mular, D.N. Halbe, and D.J. Barratt. Littleton, CO: SME. pp. 917–928.

Vince, A. 2007. *Improved Classification at 100–350 Microns*. Australian Coal Association Research Program (ACARP). www.acarp.com.au/abstracts.aspx?repID=C15051. Accessed December 2017.

Wennen, J., Nordstrom, W., and Murr, D. 1997. National Steel Pellet Company's secondary grinding circuit modifications. In *Comminution Practices*. Edited by S.K. Kawatra. Littleton, CO: SME.

Wills, B.A., and Finch, J.A. 2016. *Wills' Mineral Processing Technology: An Introduction to the Practical Aspects of Ore Treatment and Mineral Recovery*, 8th ed. Waltham, MA: Butterworth-Heinemann.

CHAPTER 4.2

Cyclones

Brian Flintoff and Brian Knorr

Cyclones, which are named for the spiraling motion within the body of the device, are one of a large number of classification devices aimed at separating fine particles in a fluid (water or air; e.g., see Kelly and Spottiswood 1982, Svarovsky 2000, or Aldrich 2015 for additional information on this topic). Unfortunately, the seminal work on this subject (Bradley 1965) is now out of print, although occasionally, copies are available. Cyclone classification is generally reserved for the separation of particles on the basis of size. For that reason, other applications for cyclones, such as thickening and dense-medium separation, are not considered here. Broadly speaking, cyclone classifiers can either treat dry feeds (e.g., the air classifier or air separator) or wet/slurry feeds (e.g., the hydrocyclone). Again, as this is a reference source for mineral processing engineers, the emphasis here is on wet applications, and specifically the hydrocyclone.

Unlike the vibrating screen, which employs a physical means (i.e., the screen media) to effect a separation, the cyclone classifier uses differential settling properties of particles of different mass. The cyclonic action within the body of the device develops g-forces significantly greater than gravity, which accelerate the separation process. In general, cyclones are robust, compact (i.e., having a small footprint), and relatively efficient size separation devices.

Figure 1 shows the most common applications of wet cyclones in mining applications. Air classifiers are also commonly used in these applications, for example, as size control devices in dry grinding (most notably in refractory gold, coal-fired power plant pulverizers, or cement clinker applications) and as sorters, usually in dedusting applications.

Figure 2 shows the geometry and key design components of a hydrocyclone, and it is clear from the shape that it resembles the natural phenomena that give rise to the name. The separator comprises a feed section, cylindroconical section, discharge spigot or apex, vortex finder, and overflow. In some cases, usually for larger diameters, each of the components is supplied as a separate entity, whereas in smaller sizes, they may be cast into a single unit. The capacity of a hydrocyclone is a function of its size and, subject to operational constraints, the rate and solids concentration at which the feed (i.e., slurry) material is supplied. In some cases, hydrocyclones are gravity fed from surge tanks, but the majority of applications utilize a feed sump and pump arrangement. Consequently, it is common to use a group of hydrocyclones operating in parallel to achieve the required process capacity. As Figure 3 shows, the compact design of the units permits the use of a radial header to distribute feed to many hydrocyclones. (This is commonly called a *hydrocyclone cluster* or *battery* or *cyclopac*.) Moreover, as the side view highlights, the use of knife-gate valves to open or close a hydrocyclone is very convenient, both from an online process control point of view as well as

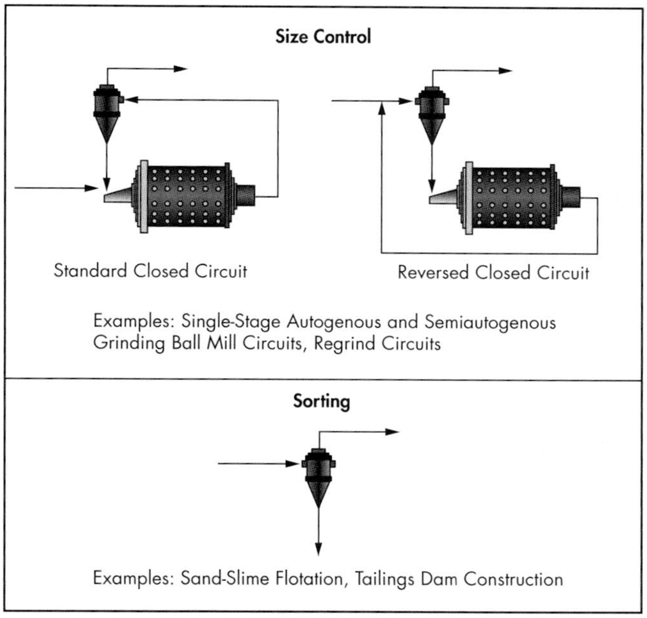

Source: Flintoff and Kuehl 2011

Figure 1 Examples of application classes for hydrocyclones in mining

Brian Flintoff, Consultant, Sigmation Inc., Kelowna, British Columbia, Canada
Brian Knorr, Director, Research & Development, Metso Minerals Industries Inc., York, Pennsylvania, USA

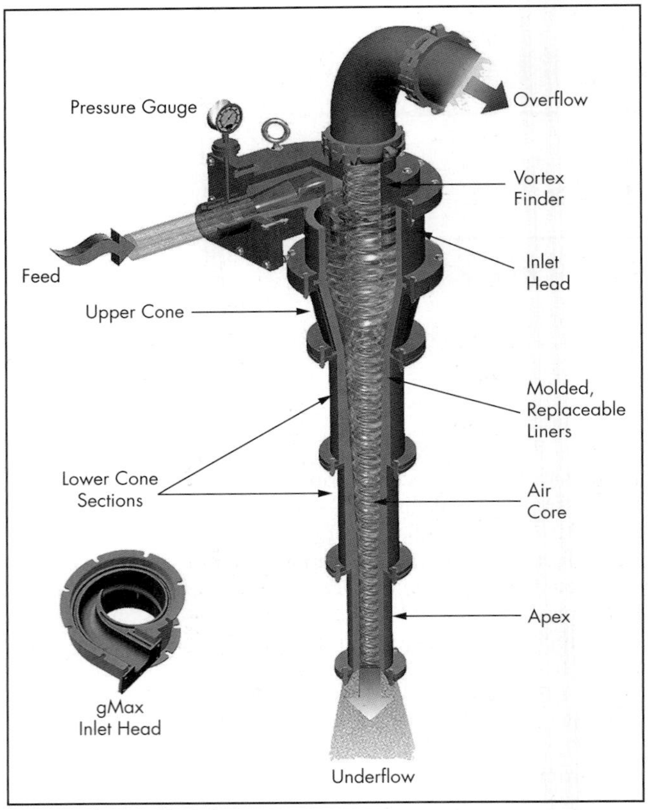

Courtesy of FLSmidth–Krebs

Figure 2 Hydrocyclone design and components

for maintenance purposes. On the latter point, it is typical to design the hydrocyclone cluster with standby hydrocyclone units to allow for removal of a hydrocyclone for maintenance and replacement with a refurbished spare while maintaining operation of the overall hydrocyclone cluster. This is important because the maintenance interval for the hydrocyclone does not typically align with the shutdown schedule of the overall plant.

There are many variants on the more typical hydrocyclone geometry. Figure 4 shows the water-only or automedium cyclone that is frequently used in washing fine (–0.6 mm) coal. (The truncated lower cone section is employed to set up a secondary density separation in a rotating particle bed.) The flat-bottomed hydrocyclone occasionally used in mineral processing applications has a very similar kind of body geometry.

Air separators tend to be designed to treat the whole feed flow, so they are larger units, as shown in Figure 5. These devices can be either static (like the hydrocyclone) or dynamic (include moving blades, etc., to enhance separation).

Figure 5 illustrates a typical dry grinding circuit with the air separator used for classification and a cyclone used for solids recovery. The cutaway view illustrates the internals and the drive for this dynamic air separator.

HYDROCYCLONE BASICS

From about 1950 through 1970, the hydrocyclone evolved into its current role in mineral processing, by which time it had become a key element in the design of most large-scale grinding circuits, offering high capacities in small footprints and arguably enabling the ongoing development of very large grinding circuits. Although they still have their place, mechanical (spiral or screw, rake, etc.) classifiers were effectively rendered obsolete in plant design by the hydrocyclone. Despite the numerous benefits of hydrocyclones, there have always been issues with efficiency. As today's operators look for ways to address energy efficiency, and as competitive technologies emerge, hydrocyclone manufacturers have redoubled their development efforts. As is the case for screens, they are using the very latest simulation tools to help refine existing designs and develop new ones for specific applications, with good success. The following material provides a refresher on hydrocyclone basics and introduces some of the latest tools and techniques for the design, analysis, control, and optimization of hydrocyclones.

Principles of Operation

Figure 6 offers a sketch of a typical hydrocyclone and the main design components. The principal dimensional variable is the diameter of the upper cylindrical section of the hydrocyclone

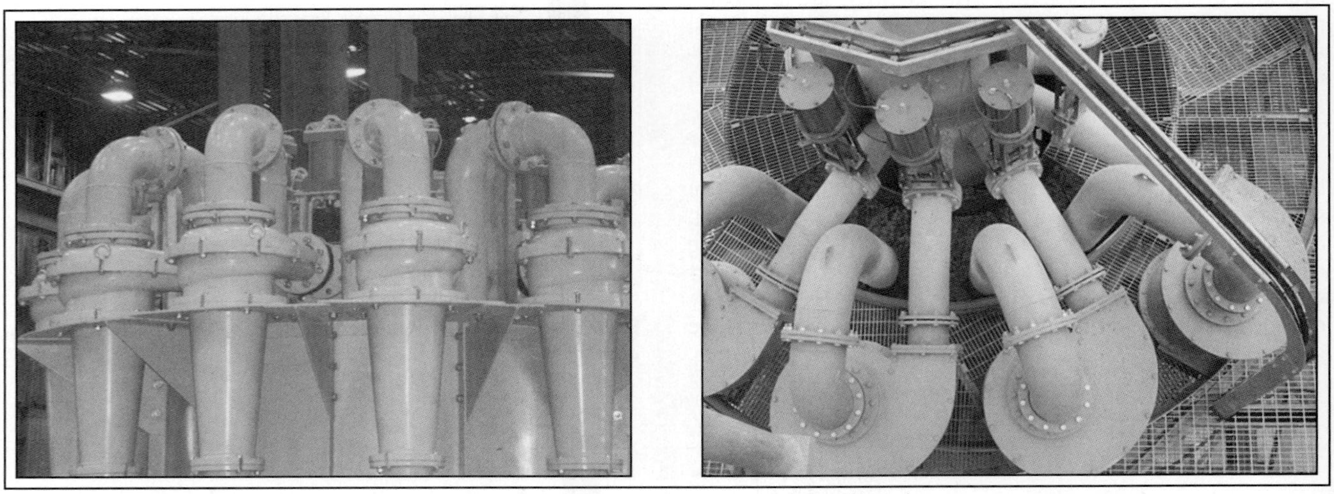

Courtesy of Weir Minerals Courtesy of FLSmidth–Krebs

Figure 3 Side and top views of hydrocyclone clusters

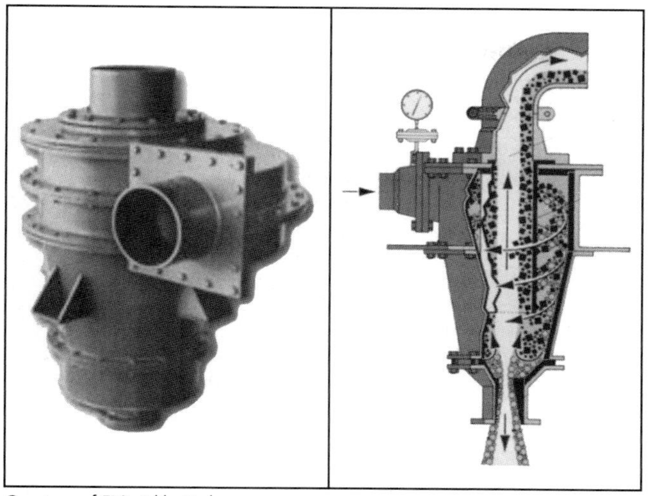

Courtesy of FLSmidth–Krebs

Figure 4 Water-only cyclone

Source: Flintoff and Kuehl 2011

Figure 5 Gyrotor air separator in a typical dry grinding circuit

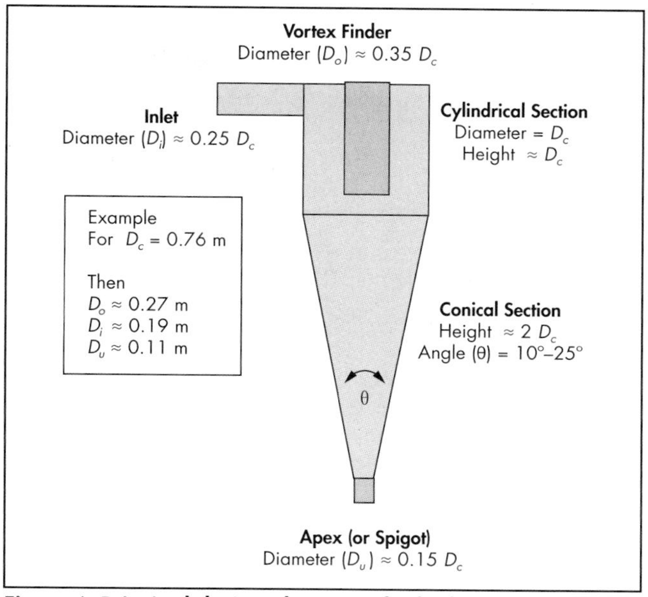

Figure 6 Principal design elements of a hydrocyclone

body, D_c. The inlet (D_i) and discharge (D_o and D_u) dimensions usually lie within some ratio of D_c, for example, $0.2D_c \leq D_o \leq 0.45D_c$. Because the inlet is usually more rectangular, to facilitate a reduction in turbulence on entry (i.e., involute designs; see Olson and Turner 2002 or Aldrich 2015 for examples), D_i is the diameter of a circle with an area equivalent to the actual inlet area. In the past, smaller hydrocyclones have had lower cone angles (~10°), but because of headroom considerations, larger hydrocyclones tend to have larger cone angles (~20°). Depending on the application requirements, some large hydrocyclones have dual cone angles, as this has been found to yield finer cut size ($d50c$) values. Some typical dimensions for a hydrocyclone with D_c = 76 cm are shown in the figure.

The feed enters the hydrocyclone near the top of the cylindrical section, usually through some sort of involute inlet design (see Olson and Turner 2002 or Aldrich 2015 for examples). In the newer hydrocyclone designs, the involute is ramped, which is intended to introduce the slurry into the hydrocyclone with a minimum of turbulence, as this can have a deleterious effect on efficiency and wear. (The top view in Figure 3 clearly shows an involute design.) The rotational motion set up in the body of the hydrocyclone gives rise to the g-forces that accelerate the particle separation process. All of the particles are subjected to the centrifugal forces directing them to the wall of the hydrocyclone, where this material moves downward in a helical fashion to be discharged through the apex, *as underflow*. However, as the particles migrate to the outer wall of the hydrocyclone, they necessarily displace the fluid, which must move inward. The fluid exerts drag forces on the particles and retards their outward motion and, in the case of the finer particles, actually causes them to move inward. This stream moves upward in a helical path to be discharged through the vortex finder, *as overflow*. When the hydrocyclone is operating normally, the centrifugal forces acting on the slurry create low pressure along the central axis, establishing the conditions necessary for development of the characteristic air core.

To look a little more closely at the separation process, consider Stokes' law for laminar settling, given by Equation 1. (Whether it is Stokes' law, Newton's law, or something else—as experimental data would seem to indicate—the trends predicted by Stokes' law are seen in practice.)

$$v_t = \frac{(\rho_s - \rho_l)Gd^2}{18\mu} \quad \textbf{(EQ 1)}$$

where
- v_t = terminal settling velocity
- ρ_s = solids density
- ρ_l = carrier fluid density
- G = g-force
- d = particle diameter
- μ = carrier fluid viscosity

From the preceding explanation, anything that acts to increase the particle settling velocity increases the chances of the particle reporting to the underflow stream. Clearly, the larger the particle, the more likely it is to go to the underflow. Similarly, higher g-forces (meaning higher feed flows or pressure drops with the resulting increased velocities inside the hydrocyclone) will also act to increase the probability that a particle will find its way to the underflow stream. Particles with higher density will also have a higher probability of reaching the underflow stream.

However, if the feed material contains a lot of very fine clay material, thereby increasing the effective viscosity, the particles will be less likely to reach the underflow stream. Similarly, with high-density hydrocyclone feeds, there is a significant concentration of fine particles in the carrier fluid, which increases the fluid density and makes it less likely for particles to reach the underflow stream.

To further complicate things, the size of the vortex finder also has an influence on the separation outcome. Smaller vortex finders will create a finer particle overflow stream, and vice versa. To a lesser extent, the apex size will also have an impact, as a smaller apex tends to limit the underflow capacity. This results in a diversion of some coarser particles toward the upper regions of the hydrocyclone, the finest of which will then be entrained in the overflow stream.

Figure 7 (absent the air core) was developed by Renner and Cohen (1978) and roughly shows the main regions inside a hydrocyclone. Region A is a relatively narrow area in the vicinity of the inlet that is characterized by a size distribution similar to that of the feed. Region B is the finer material with a size distribution very similar to the overflow product. Region C is perhaps the most critical area, as it contains significant quantities of near-size particles, which suggests that they tend to accumulate here until physical capacity constraints force them to move to one product zone or the other. Arguably, this is the region of active classification, and poor operation would be expected if it fails to form properly, possibly because of poor design, excessive feed solids concentrations, or some other reason. Most of the conical section is region D, characterized by a coarser size distribution similar to the underflow.

Rather than produce a process matrix for the hydrocyclone, the authors have chosen to illustrate these concepts with the Plitt model (adapted from Flintoff et al. 1987) given in Equation 2. This is the correlation for the corrected cut size, *d*50*c*. For example, increasing the hydrocyclone, D_c, or vortex finder diameter, D_o, or percent solids in the feed would be expected to lead to a coarser cut size. Similarly, increasing the feed flow, Q, or particle density, ρ_s, would lead to a finer cut size.

$$d50c = \frac{CD_c^{0.46}D_i^{0.6}D_o^{1.21}\mu^{0.5}e^{0.063V}}{D_u^{0.71}h^{0.28}Q^{0.48}\left(\frac{\rho_s - \rho_l}{1.65}\right)^k} \quad \textbf{(EQ 2)}$$

where
- $d50c$ = corrected cut size
- C = calibration constant
- D_c = hydrocyclone diameter
- D_i = inlet diameter
- D_o = vortex finder diameter
- V = volume percent concentration of solids in hydrocyclone feed
- D_u = apex diameter
- h = free vortex height (distance from bottom of vortex finder to top of apex)
- Q = feed volumetric flow rate
- k = settling exponent (theoretically 0.5 for laminar setting and 1 for turbulent setting)

As Equation 11 in Chapter 4.6, "Partition Curves," indicates, *d*50*c* is just one of the parameters in the usual partition curve models, yet it is very strongly correlated with the hydrocyclone overflow particle size as shown in Figure 8. Consequently, one can use *d*50*c* as a proxy for overflow size when contemplating design or operational changes to influence performance.

Before leaving this section, there are a few operating situations worthy of some amplification. The first is the condition known as *roping* (a name based on the appearance of the underflow stream in this regime). Roping occurs when the volumetric flow rate of the underflow stream increases to the point where it causes the air core to collapse. When this happens, the usual helical spray from the apex alters appearance, as shown in Figure 9. Consequently, roping can and eventually will increase the flow of coarser particles to the overflow stream, that is, a coarser cut size. The impact on efficiency (as measured by α in a partition curve) appears to depend on the hydrocyclone feed density, with higher densities having

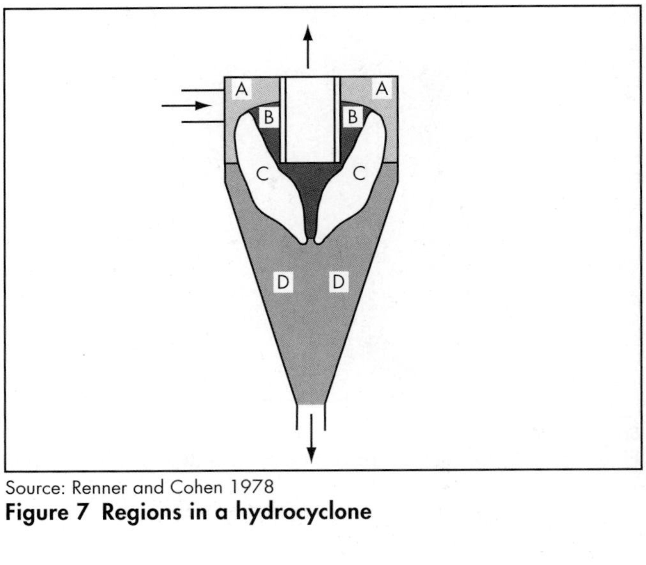

Source: Renner and Cohen 1978
Figure 7 Regions in a hydrocyclone

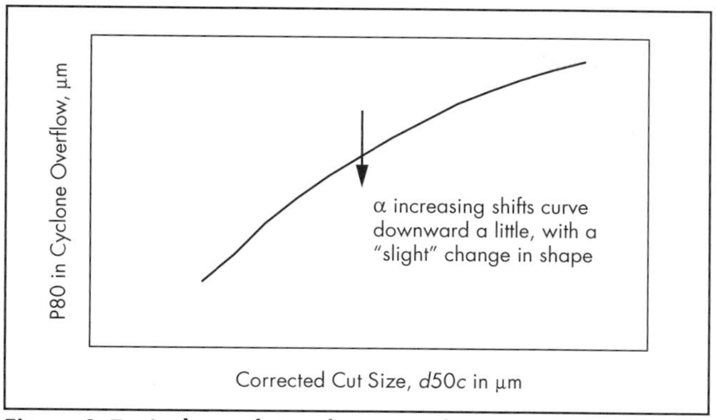

Figure 8 Typical correlation between *d50c* and hydrocyclone overflow particle size (here measured as P80)

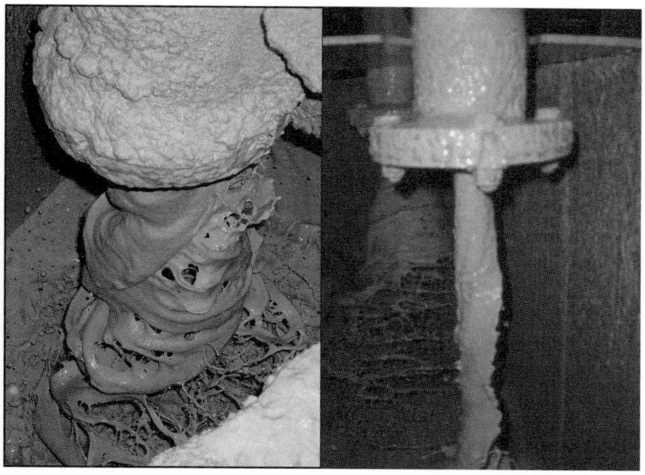

Courtesy of Metso
Figure 9 Spray and rope discharge

a more deleterious effect (see Plitt et al. 1987). Roping does tend to maximize the underflow density, and thus minimizes the fraction of fine material that reports with the carrier liquid to this stream. Figure 10 shows a comparison of partition curves in spray and open discharge operation. In this instance, the sharpness of separation (α) is seen to increase in the roping mode of operation.

The second operating situation is *surging*. With the increased use of variable-speed pumping and the ability to automatically open and close hydrocyclones, this is not the issue it once was. Nevertheless, it can still happen, and with fixed-speed hydrocyclone feed pumps, this usually necessitates a sheave change. Figure 11 shows a before-and-after case where the hydrocyclone overflow particle size was monitored over approximately 24 hours (sampling every 15 minutes), and the impact of the sheave change on the stability of the particle size trajectory is quite obvious.

The last item of interest is hydrocyclone *mounting*. With the development of larger and larger hydrocyclones, mounting them in a vertical orientation puts a significant head on the underflow stream, which can influence both wear and separation performance (e.g., higher *Rf*). However, tilting these hydrocyclones (usually at ~45°) allows operation at lower pressures while still producing a good underflow product, albeit at a coarser cut size. Maintenance savings claims for wear in the lower region of the hydrocyclone are impressive (e.g., 80%; Schmidt and Turner 1993), but it can be more difficult to perform in situ maintenance for this kind of installation. For the interested reader, Asomah and Napier-Munn (1997) have taken a more quantitative look at the impact of mounting angle on hydrocyclone performance.

HYDROCYCLONE SELECTION

The selection of hydrocyclones is normally a task best performed with the supplier, as these firms possess the knowledge, expertise, databases, and tools to ensure that the best results are achieved. In most cases, it is possible to avoid pilot testing; however, there are some instances where this is preferable and even inevitable. One of the more challenging design problems is the closed-circuit grinding application, illustrated in Figure 1 as *size control*. This is challenging because the hydrocyclone feed characteristics very much depend on mill performance, which in turn depends on the hydrocyclone underflow characteristics, which in turn depend on the feed. Experience has given most suppliers a very good understanding of this problem; nevertheless, circuit simulation tools can be useful adjuncts in developing a hydrocyclone design basis. (The use of empirical hydrocyclone models for design decision

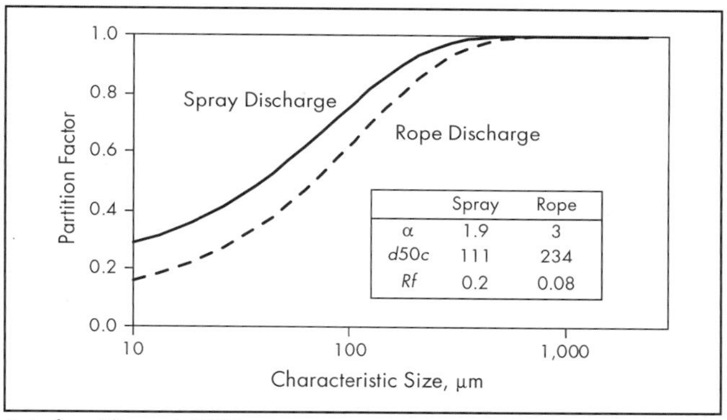

Data from Plitt et al. 1987
Figure 10 Partition curve examples for a hydrocyclone operating in spray and rope discharge modes

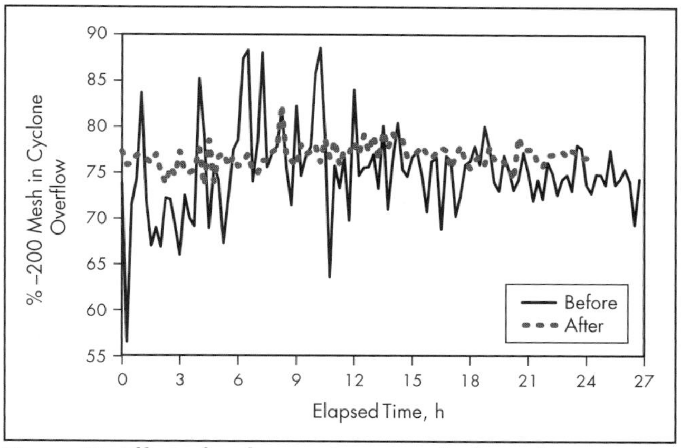

Figure 11 Effect of a sheave change on hydrocyclone performance as measured by hydrocyclone overflow particle size

support is discussed later in the "Hydrocyclone Modeling and Simulation" section.)

The techniques of hydrocyclone selection applicable to older hydrocyclone geometries (for more detail, see Arterburn 1982 and/or Olson and Turner 2002) are briefly reviewed in this section. Although the authors have questioned their current applicability with the leading hydrocyclone original equipment manufacturers (OEMs), the feedback is that these methods are still in use. The newer hydrocyclone geometries, however, require additional correction factors, but these are proprietary to the OEMs. On this basis, the following illustrative example is applicable to older hydrocyclone geometries.

The required information for hydrocyclone selection is summarized in Figure 12. A set of standard tests under very specific design and operating conditions was performed to develop a relationship between what is called the *base corrected cut size* ($d50c_{base}$) and the principal dimensional variable, the hydrocyclone diameter, D_c. The outcome of this work is captured in Figure 13, based on Equation 3a, also given in the alternate form as Equation 3b:

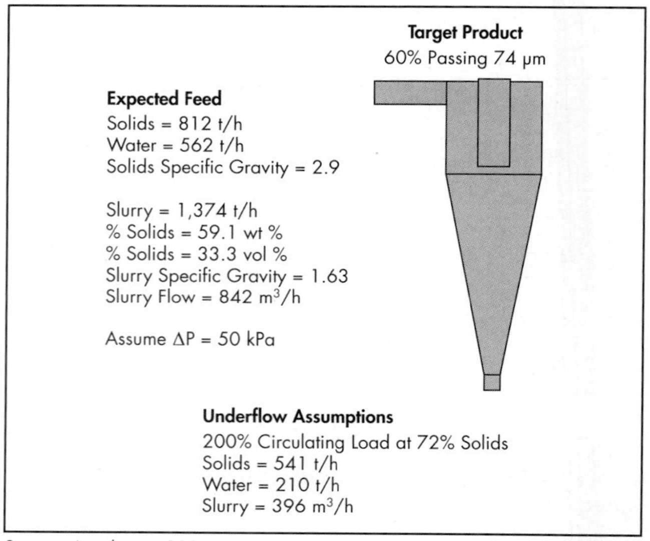

Source: Arterburn 1982

Figure 12 Hydrocyclone design

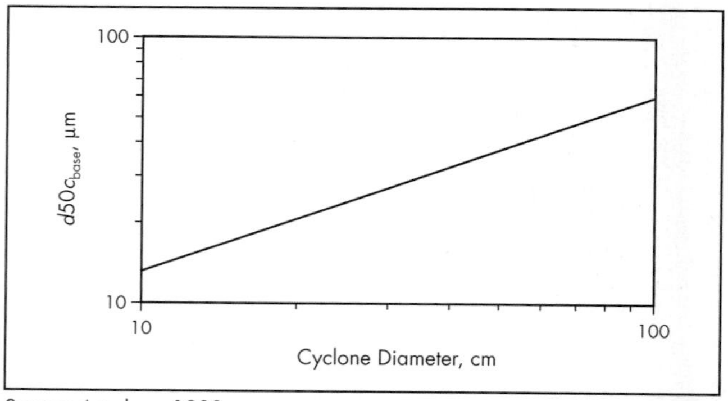

Source: Arterburn 1982

Figure 13 $d50c_{base}$ versus hydrocyclone diameter for standard hydrocyclones

$$d50c_{base} = 2.84 D_c^{0.66} \quad \textbf{(EQ 3a)}$$

$$D_c = 0.206 d50c_{base}^{1.515} \quad \textbf{(EQ 3b)}$$

The hydrocyclone design approach illustrated here relies on many empirical correction factors to scale the performance of the standard hydrocyclone to the expected industrial operation. In this case, the key scaling parameter is the $d50c$, and the key design formula is given by Equation 4a:

$$d50c_{actual} = d50c_{base}(CP1)(CP2)(CP3)(\ldots)(CD1)(CD2)(CD3)(\ldots) \quad \textbf{(EQ 4a)}$$

where

$d50c_{actual}$ = expected corrected cut size in industrial operation

$d50c_{base}$ = corrected cut size for the standard hydrocyclone

CP = process-related correction factors, for example, feed volume (approximate viscosity) correction factor

CD = design-based correction factors, for example, vortex finder diameter

In practice, Equation 4a is rearranged and used in the following form:

$$d50c_{base} = \frac{d50c_{actual}}{(CP1)(CP2)(CP3)(\ldots)(CD1)(CD2)(CD3)(\ldots)} \quad \textbf{(EQ 4b)}$$

With this short introduction, the four basic steps in hydrocyclone design are discussed next.

Step 1. Estimate $d50c_{actual}$

Table 1 provides the relationship between the overflow size specification and $d50c_{actual}$. For the overflow specification in Figure 12, 60% corresponds to a factor of 2.08 that when multiplied by the reference size of 74 µm gives the estimate for $d50c_{actual}$ of 154 µm.

Step 2. Calculate Correction Factors and Estimate $d50c_{base}$

There are three important published process-related correction factors: solids feed volume concentration, operating pressure drop, and solids specific gravity.

Table 1 Empirical correlation between the reference size and $d50c_{actual}$

Required Overflow Size Distribution (% passing of specified µm size)	Multiplier (used to factor the specified size)
98.8	0.54
95	0.73
90	0.91
80	1.25
70	1.67
60	2.08
50	2.78

Source: Arterburn 1982

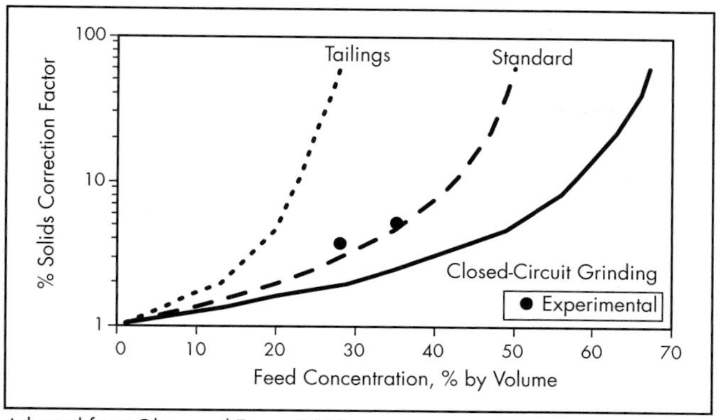

Adapted from Olson and Turner 2002

Figure 14 Correction factors (*CP*1) for hydrocyclone feed solids volumetric concentration

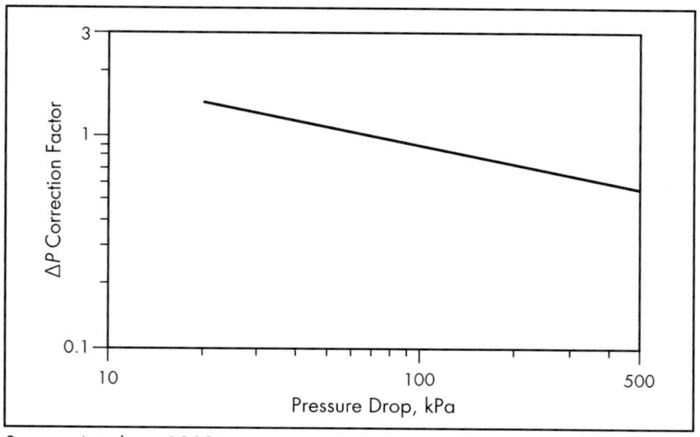

Source: Arterburn 1982

Figure 15 Pressure drop correction factor (*CP*2)

*CP*1 is the correction for the feed volumetric concentration, which is, in effect, a first-order approximation to viscosity. Of course, viscosity also depends on water temperature, clay concentrations, particle shape, and other factors; clays in fine tailings are what gave rise to the tailings curve in Figure 14. In this figure, Olson and Turner (2002) extended Arterburn's method (which was based on the standard curve in Equation 5).

In Chapter 4.6, "Partition Curves," two hydrocyclone sampling experiments are described. With the data from these experiments, it was possible tto back calculate the *CP*1 factor values: These two points appear to fall on the standard curve in Figure 14 and not the closed-circuit curve, as might have been expected. When to use which curve is the purview of the OEM, hence the suggestion to consult it in hydrocyclone design studies.

$$CP1 = \left[\frac{53 - V}{53}\right]^{-1.43} \quad \textbf{(EQ 5)}$$

In this example, we assume the standard curve, and with V = 33.3%, then $CP1 = 4.09$.

*CP*2 is the correction factor for the operating pressure drop, which is usually the difference between the measured pressure in the radial header and atmosphere. Larger hydrocyclones tend to run at lower pressures and vice versa. Figure 15 and Equation 6 show this relationship.

$$CP2 = 3.27\Delta P^{-0.28} \quad \textbf{(EQ 6)}$$

where ΔP is the pressure drop, in kilopascals. In this example, we assume ΔP = 50 kPa, which should give relatively good hydrocyclone performance while minimizing wear. From Equation 6, $CP2 = 1.09$.

*CP*3 is the correction factor for the solids specific gravity and is shown in Figure 16 and Equation 7. For a solids specific gravity of 2.9, the correction factor $CP3 = 0.93$.

$$CP3 = \left[\frac{1.65}{\rho_s - \rho_l}\right]^{0.5} \quad \textbf{(EQ 7)}$$

*CD*1 is the correction factor for the vortex finder diameter. In this example, the default design of $D_o = 0.3D_c$ is used ($CD1 = 1$), and Figure 17 and Equation 8 illustrate the nature of the correction.

$$CD1 = \left[\frac{D_o}{0.3D_c}\right]^{0.6} \quad \textbf{(EQ 8)}$$

Given the correction factors, Equation 4b solves as

$$d50c_{\text{base}} = \frac{154\ \mu\text{m}}{(4.09)(1.09)(0.93)(1)} = 37\ \mu\text{m} \quad \textbf{(EQ 9)}$$

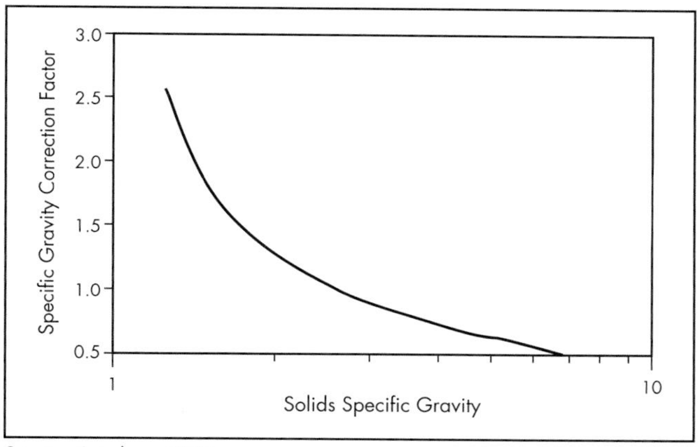

Source: Arterburn 1982

Figure 16 Solids specific gravity correction factor (*CP*3)

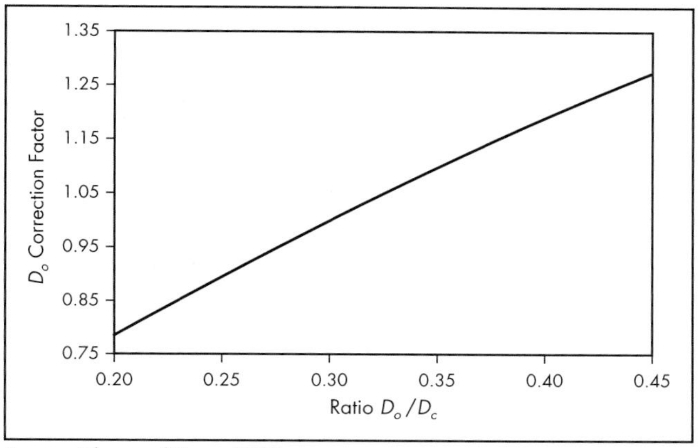

Source: Olson and Turner 2002

Figure 17 Vortex finder correction factor (*CD*1)

Step 3. Select Size and Number of Hydrocyclones

From Equation 4b or Figure 13, the diameter of the hydrocyclone required to give this $d50c_{base}$ is 49.1 cm. Hydrocyclones, like screens, come in certain standard sizes, usually defined by D_c. As shown in Figure 18, the closest size to the *theoretical* value computed earlier is a hydrocyclone with $D_c = 51$ cm.

From Figure 18, for a pressure drop of 50 kPa, the capacity of a single 51-cm hydrocyclone will be approximately 140 m^3/h. From Figure 12, the total feed volumetric flow is 842 m^3/h, so approximately six hydrocyclones will be required in operation. In practice, two or three additional hydrocyclones would be added to accommodate maintenance as well as peaks in the feed flow rate. It is also not uncommon to provide an additional offtake valve in the radial header to be used for sample collection purposes.

Step 4. Select Apex Size

The last step is to determine the apex size, and that comes from Figure 19. For a total underflow flow rate of 396 m^3/h, the flow through each apex would be ~66 m^3/h, corresponding to an apex size of ~9.3 cm. The manufacturer's closest size would be chosen, if possible tending to a smaller size in anticipation of wear.

Of course, once the hydrocyclones are installed and in operation, the reality of the application is known and the optimization process (process variables and design variables such as D_o and D_u) can be modified as required.

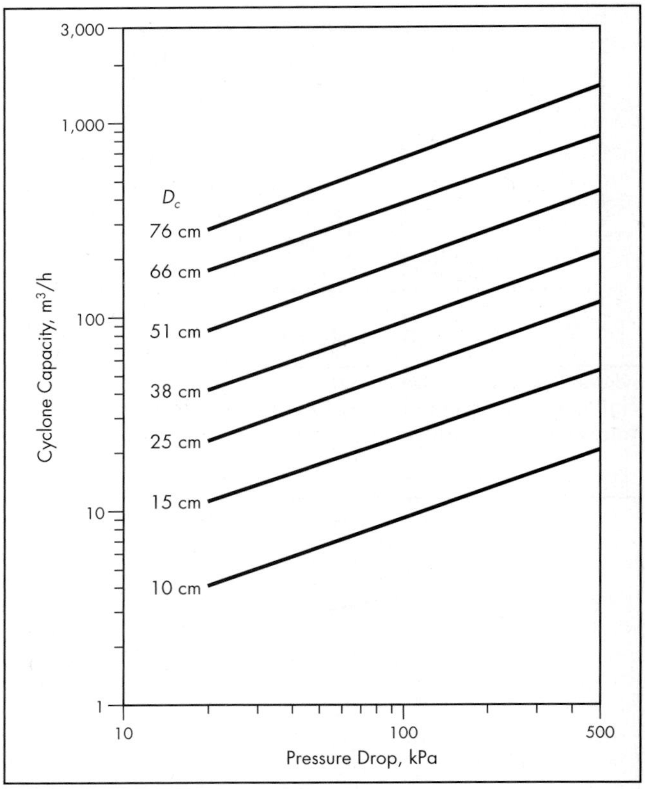

Source: Arterburn 1982

Figure 18 Hydrocyclone capacity correlation

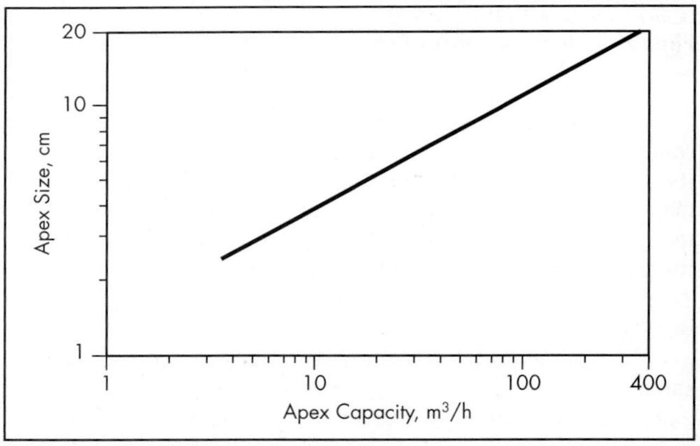

Source: Arterburn 1982

Figure 19 Apex size correlation

HYDROCYCLONE PERFORMANCE ASSESSMENT

Hydrocyclone performance is usually assessed through the partition curve and related efficiency metrics, and this subject is covered in some detail in Chapter 4.6, "Partition Curves."

HYDROCYCLONE MODELING AND SIMULATION

With the advent of mathematical modeling of tumbling mills (also called *population balance modeling*) in the middle to late 1960s, there was a corresponding need to develop hydrocyclone models that would then permit circuit simulations for process analysis, optimization, and design purposes. Research on the mathematical models of hydrocyclones was arguably most prominent from the 1960s to 1980s but continues to this day with both refinements to existing models and development of new models aimed at providing more robust predictions.

Many researchers have contributed to the field of hydrocyclone modeling. A majority of the success has been in the development and application of empirical models. Most notably, the two empirical models in greatest use today are those of Plitt (see Flintoff et al. 1987) and Nageswararao (see Nageswararao et al. 2004). These models have been incorporated into some of the more common mineral-processing simulation packages. The most widely used crushing and grinding process simulation software today employs the Nageswararao model. Additional contributions in empirical modeling have been made by Asomah and Napier-Munn (1997) and Narasimha et al. (2014), but these enhancements have not yet evolved into common usage. The Narasimha et al. work has generally been aimed at extending the Nageswararao model.

In contrast to the history of these empirical models, there has been limited success to date in more theoretical or phenomenological models. This dominance by empirical models is primarily a result of the complex conditions inside an operating hydrocyclone (e.g., hindered settling, viscosity effects). These material-specific complexities introduce significant challenges for theoretical and phenomenological model developers and will be briefly discussed in the next section.

As with all unit operation model development, two separate uses need to be considered for hydrocyclone models: (1) simulation of existing installations for plant optimization and (2) simulation for greenfield plant design. Once calibrated with survey data from an existing plant operation, the empirical hydrocyclone models show strength in providing predictions for changes in operating conditions and/or hydrocyclone fitting sizes (e.g., vortex finder and apex diameter) in that specific installation. Where the current empirical models are less reliable is in greenfield process predictions and equipment selection. For this reason, and as noted previously, readers are recommended to consult with the equipment suppliers having specific expertise in hydrocyclone selection.

Model Development

To provide a comprehensive prediction of performance, the three primary requirements for any hydrocyclone model are as follows:

1. Predicting the corrected partition curve ($d50c$ and α)
2. Predicting the feed water split to the underflow (Rf)
3. Defining the relationship between feed flow rate and operating pressure ($P \propto Q$)

Partition Curve

The partition curve tool is the current standard for the analysis and modeling of physical separation systems. (In this chapter, the partition curve defines the solids split by size to the underflow, whereas in the more recent literature on the Nageswararao model, the partition to the overflow is considered. For consistency purposes, the authors use the former definition.) For the purposes of calculating the solids and water splits around a hydrocyclone, the empirical models are essentially a series of correlations involving design and operating variables, which provide estimates of $d50c$, α, and Rf, as well as P. These first three parameters uniquely define the partition curve for homogeneous (dominated by particles of the same specific gravity) ores.

Feed Flow Rate and Operating Pressure

A relationship must be defined between the feed flow rate and operating pressure. This is instrumental toward understanding the practical capacity of the hydrocyclone as well as determining the effects of operating pressure and/or feed flow rate on hydrocyclone performance.

Nageswararao Model

The standard model nomenclature is summarized in Table 2. Figure 20 provides some explanation for the geometrical variables that characterize the hydrocyclone geometry.

The correlations and auxiliary equations that collectively comprise the Nageswararao model are as follows (JKTech 2014):

- The Euler number, EU, is defined as $Q/\left(D_c^2\sqrt{P/\rho_p}\right)$.
- The dimensionless cut size is defined as $d50c/D_c$.
- The recovery of water to underflow is Rf.
- The volumetric recovery of feed slurry to underflow is R_v.
- Derived from survey data, α is assumed to be a constant.

The following four equations define these dependent variables within the Nageswararao model:

$$\frac{Q}{D_c^2\sqrt{\frac{P}{\rho_p}}} = K_{Q_o}\{D_c^{-0.10}\}\left(\frac{D_o}{D_c}\right)^{0.68}\left(\frac{D_i}{D_c}\right)^{0.45}\left(\frac{L_c}{D_c}\right)^{0.20}\theta^{-0.10} \quad \text{(EQ 10)}$$

$$\frac{d50c}{D_c} = K_{D_o}\{D_c^{-0.65}\}\left(\frac{D_o}{D_c}\right)^{0.52}\left(\frac{D_u}{D_c}\right)^{-0.47}\left(\frac{D_i}{D_c}\right)^{-0.50}\left(\frac{L_c}{D_c}\right)^{0.20}\theta^{0.15}\left(\frac{P}{\rho_p g D_c}\right)^{-0.22}\lambda^{0.93} \quad \text{(EQ 11)}$$

Table 2 Nageswararao hydrocyclone model symbology

Nomenclature	Description
C_v	Volumetric fraction of cyclone feed solids (0–1)
$d50c$	Corrected classification size, µm
D_c	Diameter of the cyclone, cm
d_i	Mean size of the particles in size class i, µm
D_i	Equal area equivalent diameter of the inlet, cm
D_o	Diameter of the vortex finder, cm
D_u	Diameter of the apex (spigot), cm
EU	Euler number
G	Acceleration due to gravity m/s^2
K_{Qo}, K_{Do}, K_{Wo}, K_{Vo}	Material-dependent constants
L_c	Length of the cylindrical section of the cyclone, cm
P	Cyclone feed pressure, kPa
p_i	Partition factor (percent reporting to underflow) for size class i
Q	Cyclone feed volumetric flow rate, L/min
Rf	Mass recovery of water to underflow
R_v	Volumetric recovery of feed slurry to underflow
α	Cyclone efficiency curve shape parameter (sharpness of separation)
β	Cyclone efficiency curve shape parameter (fishhook parameter)
θ	Full cone angle, degrees
λ	Hindered settling factor
ρ_p	Density of the cyclone feed pulp, g/cm^3
ρ_s	Density of the cyclone feed solids, g/cm^3

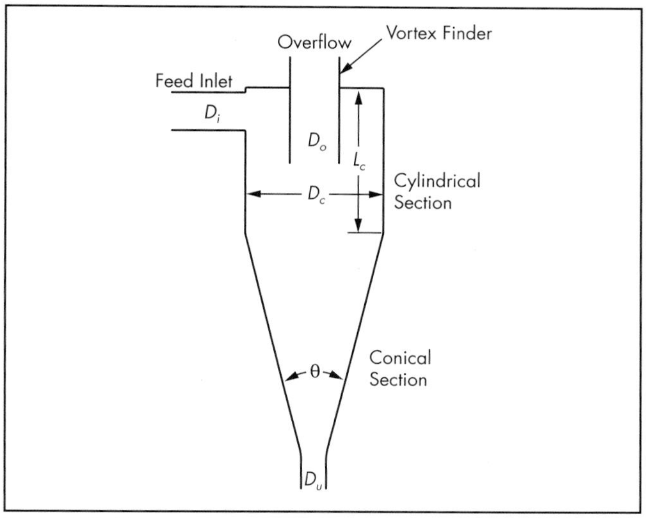

Figure 20 Nageswararao hydrocyclone model geometrical inputs

$$R_f = K_{w_o}\{D_c^{0.00}\}\left(\frac{D_o}{D_c}\right)^{-1.19}\left(\frac{D_u}{D_c}\right)^{2.40}\left(\frac{D_i}{D_c}\right)^{-0.50}\left(\frac{L_c}{D_c}\right)^{0.22}\theta^{-0.24}\left(\frac{P}{\rho_p g D_c}\right)^{-0.53}\lambda^{0.27} \quad \textbf{(EQ 12)}$$

$$R_v = K_{v_o}\{D_c^{0.00}\}\left(\frac{D_o}{D_c}\right)^{-0.94}\left(\frac{D_u}{D_c}\right)^{1.83}\left(\frac{D_i}{D_c}\right)^{-0.25}\left(\frac{L_c}{D_c}\right)^{0.22}\theta^{-0.24}\left(\frac{P}{\rho_p g D_c}\right)^{-0.31} \quad \textbf{(EQ 13)}$$

with the hindered settling factor, λ, defined as

$$\lambda = \frac{10^{1.82C_v}}{8.05 \times (1 - C_v)^2} \quad \textbf{(EQ 14)}$$

There are several points to be made regarding these correlations or model equations. The functional forms of the Nageswararao model equations were not determined via statistical approaches but rather selected based on the assumption that the effects of the independent operating variables followed a monomial power function relationship. In addition, each independent operating variable or factor was selected by the model developers based on phenomenological considerations (see Nageswararao et al. 2004 for amplification). This helps to explain the introduction of terms such as $(P/\rho_p gD_c)$ and λ, aimed at accounting for the centrifugal force field and hindered settling environment, respectively. The Nageswararao model also employs dimensionless variables for most of the hydrocyclone geometrical inputs, dividing each input by the diameter of the hydrocyclone, D_C. The model equations also contain four unique K values, K_{Qo}, K_{Do}, K_{Wo}, and K_{Vo}, or calibration parameters. These are material-dependent constants that allow the model to be tuned to particular feed materials, provided that measured plant data are available. In the absence of measured plant data, the model user is required to source appropriate K values from previous surveys or the literature. Because this has the potential to introduce significant error, the model predictions should be used with caution in any case of sourced K values (Nageswararao et al. 2004). To reiterate, this is one of the reasons that such empirical models are seldom directly used for hydrocyclone selection in greenfield process design.

With the *d*50*c* and *Rf* values from the empirical model along with an α value fit to experimental data, the solids split to the underflow can be computed from the partition curve model given by

$$\hat{p}_i = Rf + (1 - Rf)\left[\frac{e^{\alpha\frac{d_i}{d50c}} - 1}{e^{\alpha\frac{d_i}{d50c}} + e^{\alpha} - 2}\right] \quad \textbf{(EQ 15)}$$

To quickly review, $\hat{p}_i$ is the predicted partition factor for size class i (fraction of feed in size class i reporting to the underflow) and d_i is the corresponding characteristic (*geomean*) size.

The Euler number, defined earlier, provides a means to predict the operating pressure of a hydrocyclone, given the feed flow rate and other relevant geometrical and operating inputs. In a brownfield project (i.e., K_{Qo} known with some confidence), this predicted operating pressure provides the necessary information for selecting the number of required hydrocyclones for a given condition while maintaining a practical operating pressure for hydrocyclone component wear and pumping power requirements. Within the Nageswararao model, the predicted operating pressure also serves as an input for the prediction of the dimensionless cut size, $d50c/D_c$, the recovery of water to underflow, *Rf*, and the volumetric recovery of feed slurry to underflow, R_v.

A weighted average of the direct prediction of *Rf* along with the indirect prediction of *Rf* from R_v allows for the determination of the water balance around the hydrocyclone, meeting the third requirement for a comprehensive prediction of hydrocyclone performance. It was noted by Nageswararao et al. (2004) that this weighted averaging procedure has been applied within the newer Nageswararao model. However, Nageswararao also notes that the direct model equation for *Rf* is a more accurate prediction and could be used independently. The volumetric recovery equation for R_v could then be considered superfluous in application of the model. Nageswararao notes that despite this, the volumetric recovery equation is unlikely to be removed from the newer version of the model as the existing parameter database is based on this described procedure. In the following example, the direct prediction of *Rf* is used in the analysis and R_v is disregarded.

An Example of Empirical Modeling

In Chapter 4.6, "Partition Curves," two data sets are presented for a survey on a ball mill–circuit grinding copper ore: one corresponding to a hard (*hypogene*) feed and the other to a soft (*supergene*) feed. These designations of *hard ore* and *soft ore* are retained here, again for consistency, although they could simply be considered as two independent survey campaigns on the same hydrocyclone cluster.

The general mass balanced test conditions for the hydrocyclone cluster for these two surveys are summarized in Table 3. In addition, the hydrocyclone geometry for both surveys is summarized in Table 4.

For both the hard ore and soft ore, Equation 15 was fit to the experimental partition factor results (from the mass balanced data). The results of this exercise are shown in Figure 21.

To assess the ability of the model to accurately predict hydrocyclone operation, the Nageswararao correlations were calibrated (i.e., K factors estimated) for the conditions of the hard ore survey. The resulting model, with the appropriate operating data, can then be applied to predict the soft ore survey results. A comparison between the actual soft ore survey results and these model-predicted results provides for a measure of the model accuracy, as shown in Table 5. (Note that in this table, the *Rf* value comes from the actual water split to the underflow stream, not from the particle size–based estimates in Figure 21.)

The model predictions for the soft ore survey have errors ranging from 3.8% to 12.6% relative to the observed experimental conditions. This error can be attributed to a range of factors, including model inaccuracy, experimental survey error, and any potential intrinsic changes in the ore characteristics between the two tests. When applying fitted K values from one survey to another operating condition, there is an inherent assumption that any ore-specific variables that could

Table 3 Mass balance in formation summary (hard and soft ore circuit surveys)*

Variable	Hard Ore Survey	Soft Ore Survey
Total feed flow rate, m^3/h	3,917	5,094
Number of hydrocyclones in operation	7	9
Operating pressure, kPa	98	97
Feed flow rate per hydrocyclone, m^3/h	560	566
Feed solids rate per hydrocyclone, t/h	442.0	591.2
Feed solids specific gravity, t/m^3	2.79	2.95
Feed percent solids by mass, %	52.4	61.8
Feed pulp density, t/m^3	1.507	1.690
Feed volume fraction of solids, v/v	0.283	0.354
Overflow solids rate per hydrocyclone, t/h	105.4	144.2
Overflow percent solids by mass, %	29.5	41.1
Underflow solids rate per hydrocyclone, t/h	336.6	446.9
Underflow percent solids by mass, %	69.2	73.7

*See Chapter 4.6, "Partition Curves."

Table 4 Hydrocyclone geometry variables for the surveys

Symbol	Description	Value
D_c	Hydrocyclone diameter, m	0.660
L_c	Cylinder length, m	0.932
D_i	Equivalent inlet diameter, m	0.272
D_o	Vortex finder diameter, m	0.254
D_u	Apex diameter, m	0.178
θ	Cone angle, degrees	22
G	Gravitational acceleration, m/s^2	9.81

influence the hydrocyclone (such as clay content or particle size distribution) are remaining relatively constant. In this particular example, there is certainly room for these kinds of errors, given the markedly different ores.

To illustrate the kind of optimization work that can be undertaken, the following example looks at a change in the hydrocyclone apex diameter—one calculated to reduce the rather large water-based *Rf* value observed in the survey. (Because hydrocyclones are most often used in closed circuits, one must look at the whole system and not just the hydrocyclone. However, for illustrative purposes, the authors look only at the hydrocyclone in this example.) Based on the cursory evaluation, a simulated case can be completed with an adjustment from the existing 178-mm apex diameter to a 152-mm apex diameter. The feed pressure, slurry conditions, and all other hydrocyclone geometry variables were held constant. The results of this evaluation are shown in Table 6 and in Figure 22. It is quite clear that the reduction in apex diameter has a notable change in the predicted cut size and water bypass. These changes have a significant effect on the partition curve, as shown in Figure 22.

Summary

Empirical hydrocyclone models, such as Nageswararao's, are quite commonly used in simulation studies for grinding circuits. They are reasonably accurate when used in process analysis or optimization work, as they can be calibrated to plant survey data. However, the material-related constants (the *K* values) clearly vary over a substantial range, which limits model applicability in process design. That said, most hydrocyclone OEMs use these models as a decision support tool in equipment selection.

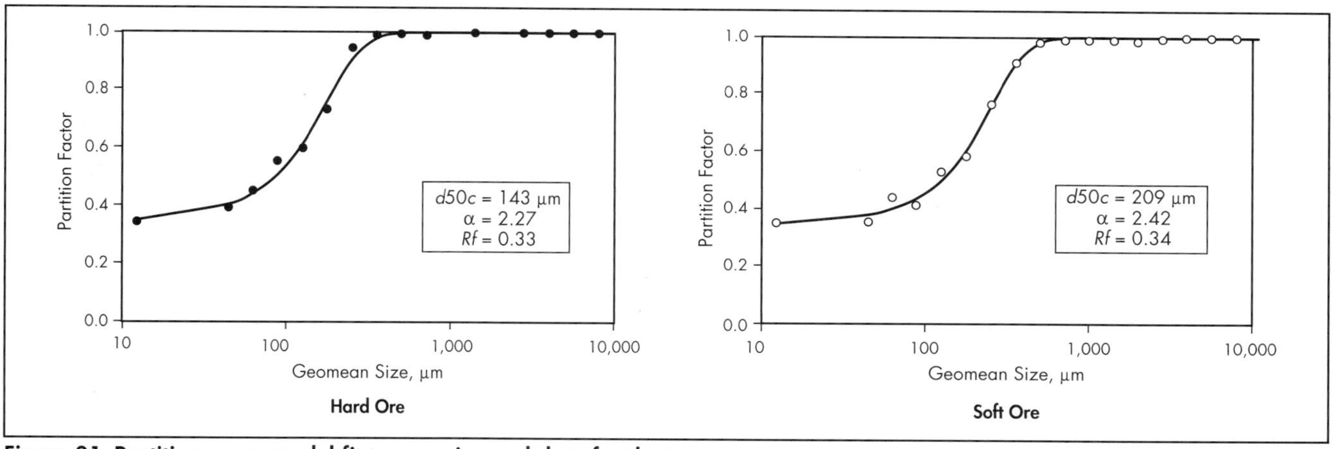

Figure 21 Partition curve model fit to experimental data for the two surveys

Table 5 Calibration of the Nageswararao model to the hard ore data and prediction of the soft ore parameters

	Hard Ore Survey			Soft Ore Survey		
Variable	Experimental	Model Fit	Fitted K Values	Experimental	Model Prediction	Error, %
Feed flow rate, *Q*	560	560	K_{Qo} = 554	566	526	–7.1%
Cut size, *d50c*	143	143	K_{Do} = 0.000114	209	235	12.6%
Actual water bypass, *Rf*	0.37	0.37	K_{Wo} = 12.466	0.44	0.45	3.8%

Table 6 Exploring a change in apex diameter for the hard ore case

Variable	Hard Ore Survey Experimental	Model Fit	Fitted K Values	Simulated Case, 152-mm Apex—Model Prediction
Feed flow rate, *Q*	560	560	K_{Qo} = 554	560
Cut size, *d50c*	143	143	K_{Do} = 0.000114	154
Actual water bypass, *Rf*	0.37	0.37	K_{Wo} = 12.466	0.25

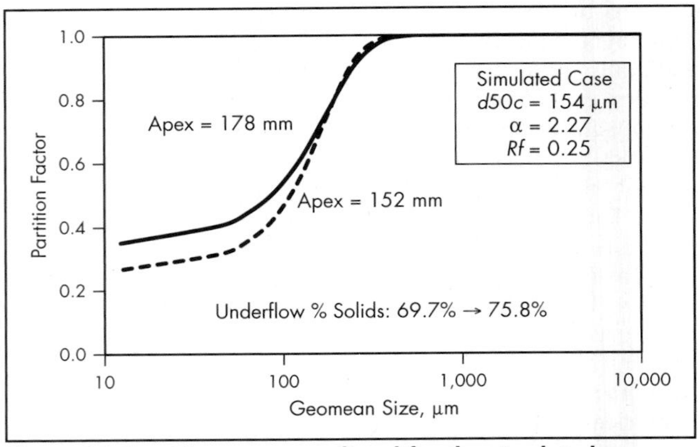

Figure 22 Partition curve predicted for the simulated case, 152-mm apex

PROCESS CONTROL

Equations 10–12 indicate that real-time hydrocyclone performance depends on two operational variables: flow (or pressure) and hydrocyclone feed density (percent solids). To the extent these variables can be modulated, they can be embedded in process control schemes.

Looking just at open-circuit operation, and using the hard ore hydrocyclone data and model, the predicted variations of the hydrocyclone overflow P80 with variations in hydrocyclone feed density and the hydrocyclone feed pressure (or flow) are shown in Figure 23. Keeping the vertical axes scales the same illustrates that the feed percent solids has a larger impact on overflow particle size than does pressure (or flow) over the normal ranges of variation.

In many cases, the feed flow rate to the hydrocyclone varies, and over a broad range. One operating tactic that is frequently employed is to regulate hydrocyclone pressure; that is, to keep it low enough to reduce wear and high enough to preserve stable operation and even influence the overflow particle size distribution. Provided that the hydrocyclones are fit with automatic gate valves, which are integrated into the process control system, this can be automated by opening or closing hydrocyclones. To prevent *chatter*, once a hydrocyclone is opened or closed, it is common to suspend any further changes in hydrocyclone operating status for a suitable length of time (e.g., 15 minutes or more). Figure 24 illustrates the variation of operating pressure and overflow particle size for a given feed flow, where the number of operating hydrocyclones is varied. As expected, pressure is very dependent on the number of hydrocyclones running, but as was seen in the previous figure, hydrocyclone overflow size is not.

Probably the most common application of hydrocyclones in mineral processing applications is in closed-circuit grinding, and a typical ball mill–hydrocyclone subcircuit is illustrated in Figure 25. While the open-circuit results still apply, the closed circuit complicates process control. Choosing the regulatory control structure (i.e., pairing controlled and manipulated variables in the proportional–integral–derivative [PID] controller loops) has been the subject of some debate over the years, but the industry appears to have converged on a best practice.

To amplify, consider the example shown in Figure 26. Here, a relatively small grinding circuit (treating ~325 metric tons per hour [t/h] of fresh feed) was subjected to a small step increase (~7.5%) in the pump box water flow, and the dynamic responses of hydrocyclone feed density, hydrocyclone overflow particle size, and hydrocyclone overflow density were recorded. (The numerical scale for the hydrocyclone overflow density was also missing in the original publication.) The initial effect of the water is to reduce all three variables. However, with a finer cut size, the recirculating load and hydrocyclone underflow percent solids both increase, which acts to increase the hydrocyclone feed density and overflow

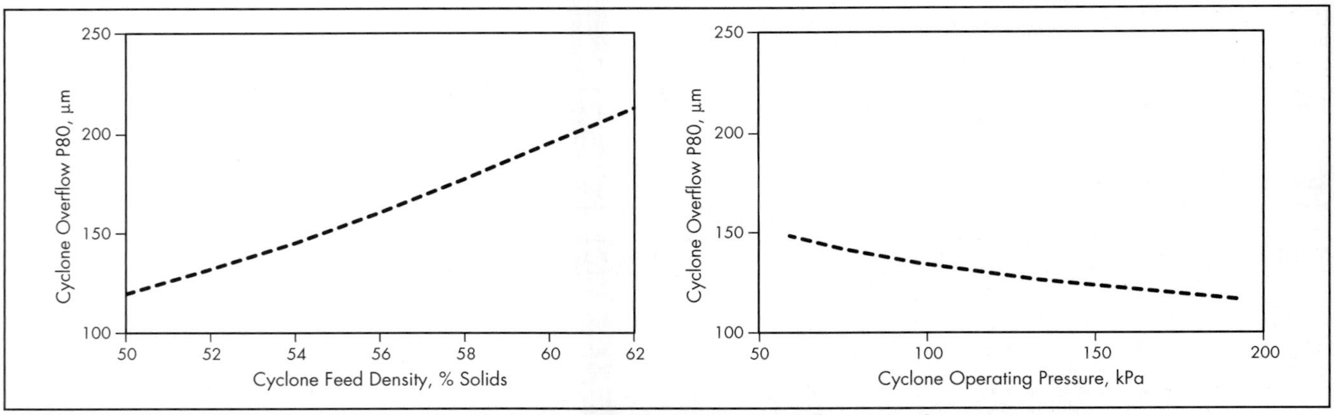

Figure 23 Variation of hydrocyclone overflow size (P80) as a function of feed density and operating pressure

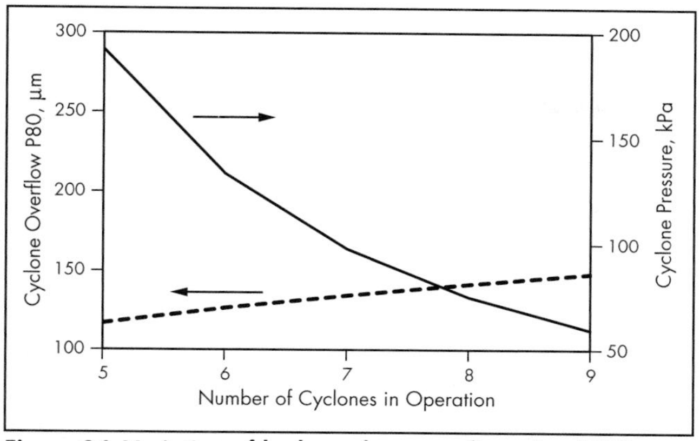

Figure 24 Variation of hydrocyclone overflow size (P80) and operating pressure as a function of the number of hydrocyclones running

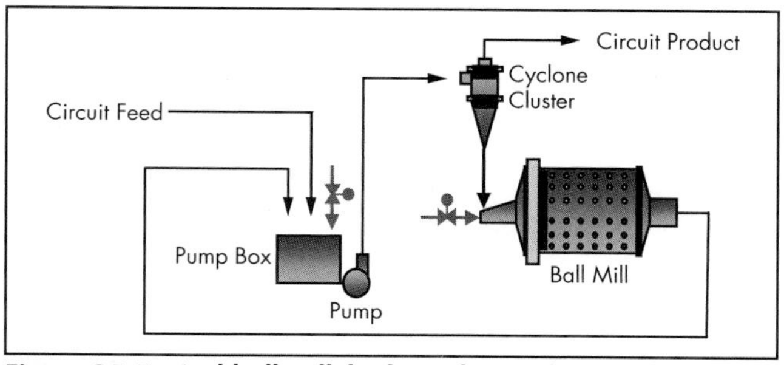

Figure 25 Typical ball mill–hydrocyclone subcircuit flow sheet

particle size. In the end, the variable most affected is the hydrocyclone overflow density, while the other two return almost to the original levels. This example is intended simply to illustrate some of the challenges with hydrocyclone performance management in closed-circuit operation.

Returning to best practice, the typical pairing is illustrated in the process matrix shown in Figure 27. The entries define the response of a controlled variable to an increase in

Process Matrix		Controlled: Cyclone Feed Density	Controlled: Pump Box Level	Controlled: Cyclone Pressure
Manipulated	Pump Box Water	– Fast	+ Fast	None
	Pump Speed	None	– Fast	+ Fast
	No. of Cyclones	None	None	– Fast

Direction: + Increase, – Decrease

Figure 27 Process matrix for regulatory loop pairing for the subcircuit in Figure 25

a manipulated variable in terms of speed and direction. (Here the entries define the response of a controlled variable to an increase in a manipulated variable speed: fast, slow, and none; and the direction: + for increase and – for decrease.) The shaded boxes indicate the preferred pairings, keeping in mind that regulating hydrocyclone feed density may only be practical for rejecting shorter-term (e.g., periods of 10–15 minutes or less) disturbances.

RECENT DEVELOPMENTS

Design Enhancement on Existing Hydrocyclones

During the past 15 years, new and more efficient hydrocyclone designs have been introduced. Design improvements were apparently made in the feed inlet to minimize turbulence when the new feed enters the body of the hydrocyclone, which impacts capacity, wear, and efficiency. Moreover, changes to the hydrocyclone body design were also made. As shown in Figure 28, the potential for efficiency improvements associated with increasing residence time in the critical conical section of the hydrocyclone has been exploited. An extensive redesign campaign yielded optimal combinations of cylinder and cone length, as well as cone design for each standard hydrocyclone diameter.

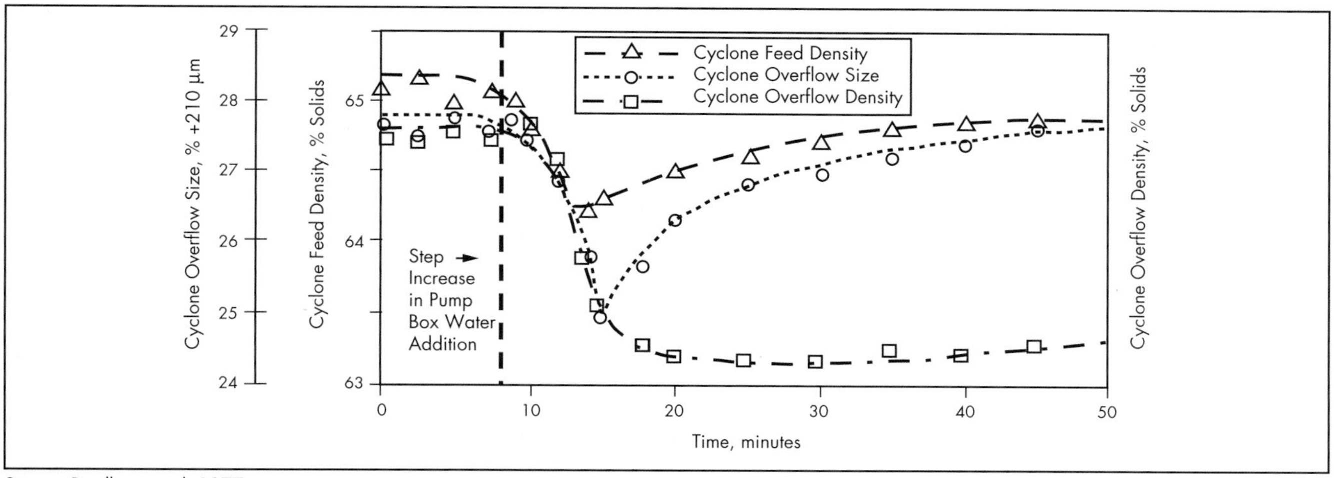

Source: Bradburn et al. 1977

Figure 26 Response of a loaded ball mill–cyclone circuit to a step change of 100 U.S. gallons per minute in pump box water

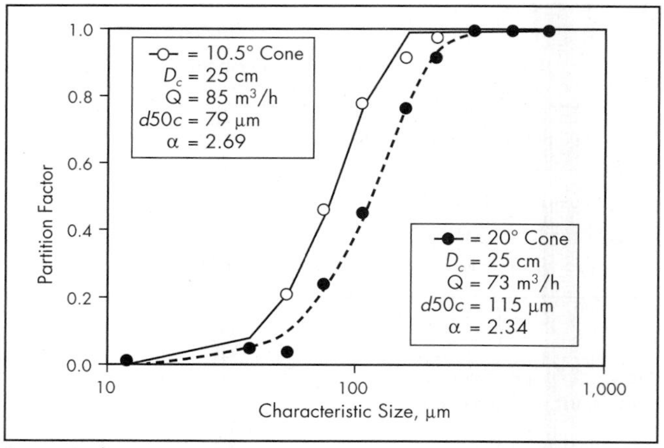

Source: Krebs Engineers 2000

Figure 28 Effect of cone angle on hydrocyclone performance

Mohanty et al. (2002) reported that a conventional hydrocyclone separating a fine coal feed (~45 μm cut target) yielded an efficiency of R_u = 37% (efficiency of fines removal from the feed), while the new design improved this to 68%. Adding an attachment, which offers a means of injecting water into the lower cone section of a hydrocyclone to remove fines, gave a further increase to 70%.

New Hydrocyclone Systems

In addition to the work underway to improve traditional hydrocyclone designs, new designs are also under study by OEMs and researchers alike, although in general, these are application-specific products.

Briefly, one new hydrocyclone design (see Mainza et al. 2004) is intended for use in applications where there are components of distinctly different densities. In conventional hydrocyclones, the heavier minerals concentrate near the hydrocyclone wall, which creates a heavy medium that selectively sends the larger, lighter particles toward the inside of the hydrocyclone where they can be sent to the overflow. In the case of the platinum ores in South Africa, the lighter silica gangue contain the platinum group metals, while the heavier chromite is barren. Larger silica particles that report to the overflow represent potential recovery losses in the flotation process. The new hydrocyclone design uses concentric vortex finders to produce two overflow products. The inner vortex finder collects the fine particles, which are sent directly to flotation. The outer vortex finder collects a coarser middlings overflow, rich in coarse silica, which is then screened to ensure that the coarser material gets to the mill for further grinding, while the finer heavy material joins the flotation feed. Development work continues with this system, which would appear to have other applications in the mineral processing industry.

Another new hydrocyclone design (Castro et al. 2009), which employs two-stage classification as a means to reduce *Rf* and increase α, has been a concept studied and promoted by many academics and a few operators. It does find application in some coal and iron ore flow sheets, but generally, the extra footprint and the capital and operating expenses associated with such an approach have diminished interest among plant designers. One area where this has been extensively applied is in tailings dam construction, where it is critically important to remove as much of the fines as possible from the sand product used to construct the dam. In these cases, two batteries of hydrocyclones (larger primaries followed by smaller secondaries) are common, to ensure that sand percolation rates are sufficiently high to avoid stability problems in the dam. This particular design performs these steps in a single unit.

Figure 29 is an illustration of the two-stage classification unit, and Figure 30 presents partition curves for the primary and secondary units and the overall curve. Note the adjustment system on the primary underflow, which acts conceptually like a variable apex. The effects on *Rf* and hence fine-particle content in the underflow are clear. All of the case studies reported by Castro et al. (2009) are on tailings; however, one can easily envisage other applications.

Instrumentation

Although little has been published thus far, the major hydrocyclone suppliers and some third parties are actively looking at technology value additions that will help improve performance and minimize maintenance costs. For example,

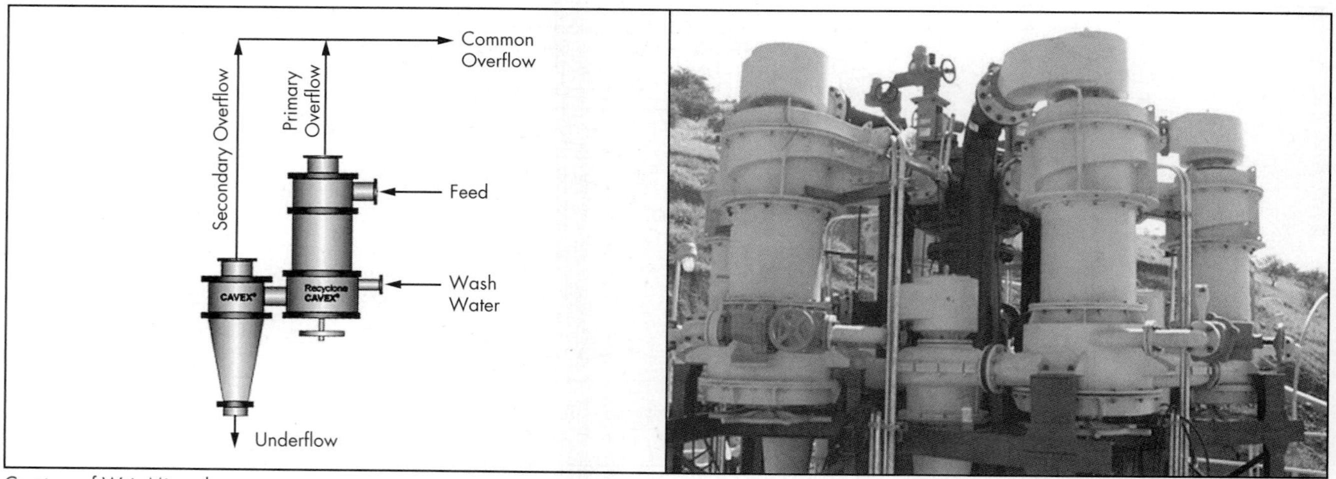

Courtesy of Weir Minerals

Figure 29 Schematic and installation of the two-stage classification unit

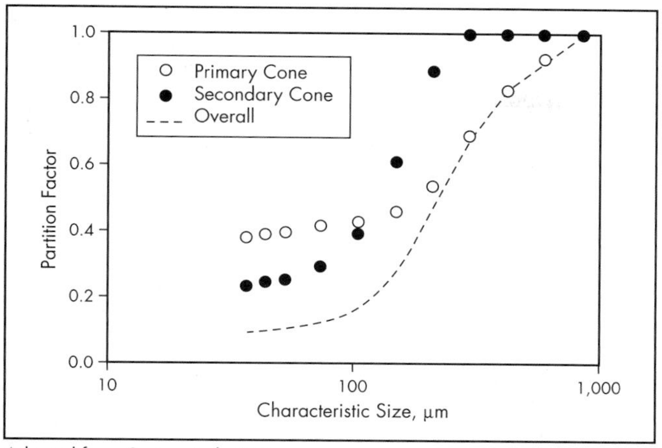

Adapted from Castro et al. 2009

Figure 30 Partition curves for a two-stage classification unit

Table 7 Hydrocyclone lining materials

Rubber	Ceramics
BPC (premium natural rubber)	CR (nitride-bonded silicon carbide)
Polyurethanes	CX (reaction-bonded silicon carbide)
Neoprene	CZ (sintered alpha silicon carbide)
Other synthetics and blends	

Courtesy of FLSmidth–Krebs

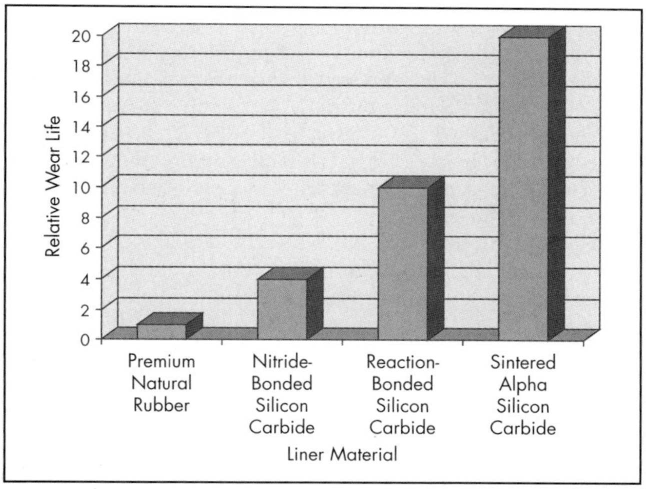

Courtesy of FLSmidth–Krebs

Figure 31 Relative wear life of the more common hydrocyclone lining materials

hydrocyclones are known to vibrate when they transition from normal underflow spray to roping regime, and vibration and/or acoustics (e.g., Wedeles 2007) could be applied to diagnose this event or other events such as apex plugging (e.g., tramp grinding balls in hydrocyclone feed). More recently, an acoustic sensor (Cirulis and Russell 2011) and a vibration application for incident detection (Moore 2016) have been developed. Another area also under study by pump manufacturers is the elusive measurement of the relative wear of hydrocyclone liners. The interested reader should contact suppliers directly on this topic.

Computational Fluid Dynamics

As was seen earlier in the discussion on screens, most major suppliers are rapidly embracing sophisticated simulation tools as way to create and screen virtual machines for the purposes of design and optimization. While this does not preclude the necessary step of field testing prototypes, it does help in the selection of the most appropriate candidate designs, reducing research and design costs and technical risk, and shortening the time to market.

Computational fluid dynamics (CFD) is the simulation technique of choice for fluid systems such as the hydrocyclone. At this stage, however, it is impossible to simulate real slurries, and the standard approach is to simulate the fluid, and then to use these results with some *tracer particles* (within CFD or in the discrete element modeling environment) to gain insight on classification efficiency. This one-way coupling (the fluid influences the particles, but the converse is not true), albeit with dilute suspensions, gives insights into classification efficiency.

Suffice to say, CFD played a major role in the development of new and more efficient hydrocyclone designs and the two-stage classification unit described earlier. For the interested reader, the article by Delgadillo and Rajamani (2007) provides a vision of the benefits of CFD in advancing the hydrocyclone state of the art.

Wear Lining Materials

Wear lining materials is a very important topic for plant operators who need to minimize operating costs and downtime and optimize maintenance cycles. Because hydrocyclones are usually operated in a cluster, the *cassette maintenance model*, which is to close and remove a hydrocyclone for a rebuild and replace that unit with a new or rebuilt one, is considered best practice.

Briefly, hydrocyclones are usually constructed of polyurethane or steel, and less often with other materials, for example, composites. Smaller hydrocyclones, say D_c less than 100–150 mm, are frequently polyurethane, while the larger ones, more common in mining applications, are more typically fabricated from steel or composites. From a wear parts perspective, polyurethane units are usually operated until the end of their service life, and then replaced with a new hydrocyclone.

Conversely, steel or composite hydrocyclones are designed with external housings and lined with wear-resistant materials, such as natural rubber, polyurethane, or ceramics, as summarized in Table 7. The relative wear life of the more common linings is illustrated in Figure 31, and as would be expected, cost is proportional to the service life.

Natural rubber is generally cheaper than polyurethane, and the two typically have similar wear characteristics; consequently, rubber lining is in more common usage. An exception may arise when, for example, the chemical environment (e.g., fuel oils) dictates against natural rubber. Ceramics are the most expensive lining option but offer the best wear resistance.

The challenge is to develop a cost-effective lining solution that essentially ensures that all parts of the lining wear out at about the same time. This is complicated because the wear rates generally vary depending on the location in the hydrocyclone; for example, in order of wear rate, first the apex, then the lower cone, then middle cone, then inlet. To illustrate with an example, one OEM, working closely with a major

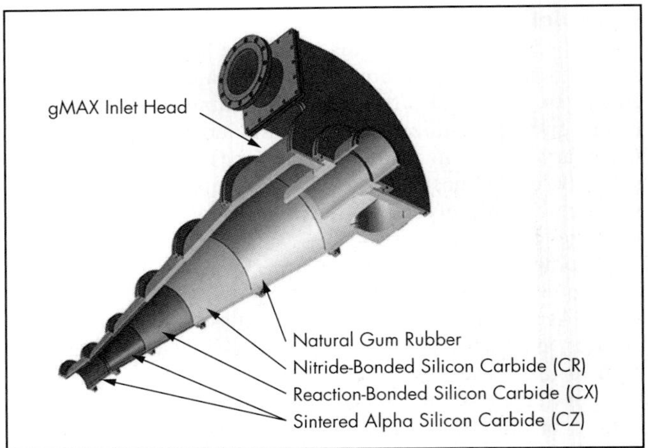

Courtesy of FLSmidth–Krebs

Figure 32 Using multiple lining materials to optimize service life and minimize cost

customer, was able to move from fully rubber-lined hydrocyclones, with an average operating lifetime of 4,000 hours, to the new more efficient hydrocyclones, using a combination of linings, as shown in Figure 32, to achieve average operating lifetimes of 13,000 hours. The upper regions of the hydrocyclone (inlet head, cover plate, and upper conical regions) are lined with rubber, and the lower portions of the hydrocyclone are lined with ceramics, progressing to longer wear-life components as the wear rate increases. Again, the OEM is in a very good position to provide expert application advice on wear liner choices.

ACKNOWLEDGMENTS

The authors extend their most sincere thanks to Weir Minerals, FLSmidth–Krebs, and the many colleagues in Metso Mining who provided much of the information contained herein, as well as some very helpful feedback.

REFERENCES

Aldrich, C. 2015. Hydrocyclones. In *Progress in Filtration and Separation*. Edited by S. Tarleton. London, UK: Elsevier. pp. 1–24.

Arterburn, R. 1982. The sizing and selection of hydrocyclones. In *Design and Installation of Comminution Circuits*. Edited by A. Mular and G.V. Jergensen. Littleton, CO: SME-AIME. pp. 592–607.

Asomah, I., and Napier-Munn, T. 1997. An empirical model of hydrocyclones incorporating angle of cyclone inclination. *Miner. Eng.* 10(3):339–347.

Bradburn, R., Flintoff, B., and Walker, R. 1977. Practical approach to digital control of a grinding circuit at Brenda Mines Ltd. *Trans. SME* 262(7):140–147.

Bradley, D. 1965. *The Hydrocyclone*. New York: Pergamon Press.

Castro, E., Lopez, J., and Switzer, D. 2009. Maximum classification efficiency through double classification using the ReCylone. In *Recent Advances in Mineral Processing Plant Design*. Edited by D. Malhotra, P. Taylor, E. Spiller, and M. LeVier. Littleton, CO: SME. pp. 444–454.

Cirulis, D., and Russell, J. 2011. CiDRA CYCLONEtrac at Kennecottt Utah Copper. Presented at the *Engineering and Mining Journal* (E&MJ) Mineral Processing Conference, Lake Tahoe, NV, October 10–12.

Delgadillo, J., and Rajamani, K. 2007. Exploration of hydrocyclone designs using computational fluid dynamics. *Int. J. Miner. Proc.* 84(1-4):252–261.

Flintoff, B., and Kuehl II, R. 2011. Classification by screens and cyclones. In *SME Mining Engineering Handbook*, Vol. 2. Edited by P. Darling. Englewood, CO: SME.

Flintoff, B., Plitt, L.R., and Turak, A. 1987. Cyclone modeling: A review of present technology. *CIM Bull.* 80(September):39–50.

JKTech. 2014. *JKSimMet User Manual*, Version 6.0. Brisbane, Queensland: JKTech Pty Ltd.

Kelly, E.G., and Spottiswood, D.J. 1982. *Introduction to Mineral Processing*. Hoboken, NJ: John Wiley and Sons.

Krebs Engineers. 2000. *Krebs gMAX Cyclones—For Finer Separations with Larger Diameter Cyclones*. www.krebs.com.

Mainza, A., Powell, M., and Knopjes, B. 2004. A comparison of different cyclones in addressing challenges in the classification of the dual density UG2 platinum ores. In *International Platinum Conference: "Platinum Adding Value."* Johannesburg: Southern African Institute of Mining and Metallurgy. pp. 341–348.

Mohanty, M., Palit, A., and Dube, B. 2002. A comparative evaluation of new fine particle separation technologies. *Miner. Eng.* 15(10):727–736.

Moore, E. 2016. All sorted out: A new cyclone diagnostic system is keeping things moving at Copper Mountain. *CIM Mag.* www.cim.org/en/Publications-and-Technical-Resources/Publications/CIM-Magazine/2016/February/upfront/All-sorted-out.aspx.

Nageswararao, K., Wiseman, D., and Napier-Munn, T. 2004. Two empirical hydrocyclone models revisited. *Miner. Eng.* 17:671–687.

Narasimha, M., Mainza, A., Holtham, P., Powell, M., and Brennan, M. 2014. A semi-mechanistic model of hydrocyclones: Developed from industrial data and inputs from CFD. *Int. J. Miner. Process.* 133:1–12.

Olson, T., and Turner, P. 2002. Hydrocyclone selection for plant design. In *Mineral Processing Plant Design, Practice, and Control*, Vol. 1. Edited by A.L. Mular, D.N. Halbe, and D.J. Barratt. Littleton, CO: SME. pp. 880–893.

Plitt, L.R., Flintoff, B., and Stuffco, T. 1987. Roping in hydrocyclones. In *Proceedings of the 3rd International Conference on Hydrocyclones*. Edited by P.A. Wood. London: BHRA/Elsevier. pp. 21–33.

Renner, V., and Cohen, H. 1978. Measurement and the interpretation of size distributions within a cyclone. *Trans. Inst. Chem. Eng.* 87:C139–C145.

Schmidt, M., and Turner, P. 1993. Flat bottomed or horizontal cyclones: Which is right for your world? *World. Min. Equip.* pp. 151–152.

Svarovsky, L. 2000. *Solids–Liquid Separation*, 4th ed. Boston: Butterworth-Heinemann.

Wedeles, M. 2007. Smart cyclone: A tool for optimizing cyclone operation. Presented at SAG 2007, Vina del Mar, Chile.

CHAPTER 4.3

Fluidized-Bed Classifiers

Michael J. Mankosa, Jaisen N. Kohmuench, and Rick Q. Honaker

The current industry standard for hydraulic classifiers (or separators) is the result of a development process that started more than 40 years ago. For decades, classification was accomplished by mechanical means using rake, screw, or other similar types of classifiers. These machines dominated the industry, particularly for closed-circuit grinding applications, until the onset of classifying hydrocyclones. In parallel, a great deal of work was performed with dense flow hydraulic classifiers, both mechanical and nonmechanical. These devices were an extension of the early cone-type settling classifiers such as sand cones, desliming tanks, and launder classifiers. In these devices, the coarse particles settled into the base of the separator and were extracted via a discharge port. Dense flow separators took this process one step further with the addition of mechanical agitation of the settled solids and/or water injection into the base of the separator to improve efficiency. Examples of these early devices include the Larox cone classifier and the Lavodune classifier (Heiskanen 1993). Unit capacity, size, and weight generally led to the dominance of hydrocyclones for most classification applications, especially in closed-circuit grinding applications. However, several industries, such as fertilizer (phosphate and potash) and aggregate producers, continued to use hydraulic, or fluidized-bed, classifiers because of their very precise sizing relative to the performance of hydrocyclones.

Generally, there are two types of hydraulic classifiers: those operating in the free or partially hindered settling regime and full hindered-bed separators. The latter utilize a type of restricted discharge system in combination with a control system to regulate the exit rate of oversize particles. In both instances, the coarse solids (material larger than the target cut point for the separation) settle against a countercurrent flow of upward rising water injected near the bottom of the unit. Free-settling separators offered an advantage over earlier mechanical devices such as screw and rake classifiers as they did not require a mechanical device to assist with removal of the coarse fraction. Popular units in the 1970s and 1980s included machines such as the Lewis classifier, the Linatex S classifier, and the Krebs Whirlsizer. Although these devices were generally effective, they typically did not offer the capacity of hydrocyclones.

Eventually, better understanding of the benefits of a full hindered-bed device, along with advancements in control systems and discharge mechanisms, led to the proliferation of what is currently recognized as a modern teeter-bed separator. A teeter-bed separator is a device that restricts the discharge of the underflow stream that contains the coarse particles. The holdup of coarse particles, in combination with a rising current of fluidization water across the base of the separator, creates a teetering bed of solids where the settling characteristics of each particle are greatly affected by its proximity to other particles. In this environment, the particles are classified from top to bottom in order of increasing terminal velocity, with the fastest-settling particles migrating toward the bottom of the separation chamber. Most settling in the teeter bed is hindered, but some free settling may take place in the upper portion of the classifier. The high interstitial velocity between the suspended particles in the fluidized bed acts to reject misplaced fine particles from the high-density coarse bed, resulting in a highly efficient separation.

BASIC OPERATING PRINCIPLE

The basic arrangement of a fluidized-bed classifier is shown in Figure 1. These devices work effectively over a wide particle size range from 100 to 1,000 µm. In some extreme applications, such as sedimentary phosphate production, the upper limit may exceed 2,000 µm. These unique applications, however, require extremely high fluidization water rates. As shown in Figure 1, feed is introduced at or near the top of the separator into a free, or partially hindered, settling zone that occurs between the teeter-bed interface and the overflow of the separator. Fluidization, or teeter water, is introduced evenly across the base of the separator and flows upward to the overflow launder. The fluidized bed is formed above the point of introduction of the teeter water and extends upward into the separation chamber. The bed consists of particles with

Michael J. Mankosa, Executive Vice President of Global Technology, Eriez Manufacturing, Erie, Pennsylvania, USA
Jaisen N. Kohmuench, Managing Director, Eriez Manufacturing Pty Ltd., Melbourne, Victoria, Australia
Rick Q. Honaker, Professor, Mining Engineering, University of Kentucky, Lexington, Kentucky, USA

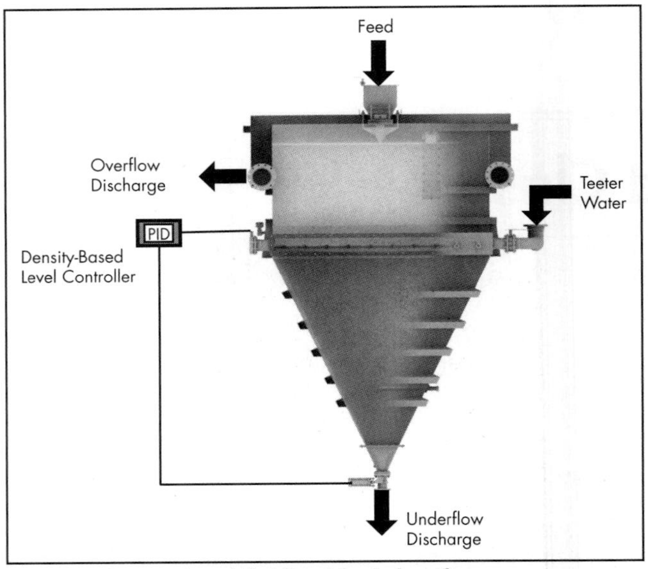

Figure 1 Conventional fluidized-bed classifier

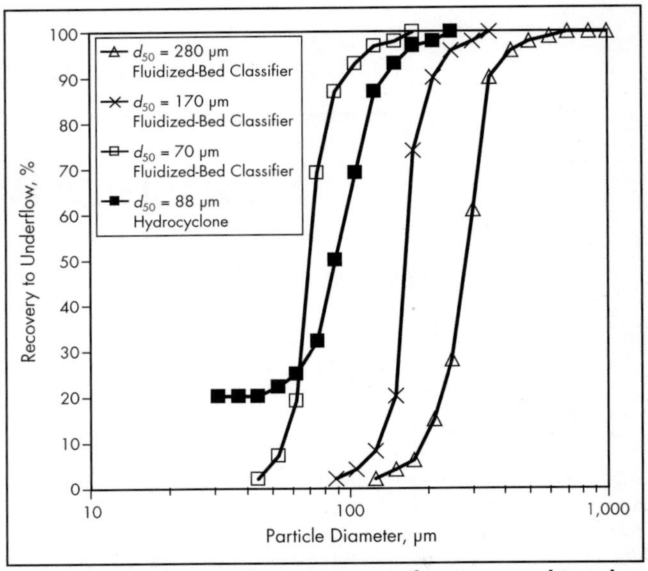

Figure 2 Comparison of separation performances achieved by a fluidized-bed classifier and a classification cyclone

a size gradation resulting from the settling velocity of the particles, with the coarsest material near the bottom and finer material toward the top of the separator. As discussed later in this chapter, the height of the fluidized bed is regulated by a control system that maintains the bed interface at a constant distance from the water injection point. Depending on the method of operation, the fluidized bed may extend to the top of the separator or be maintained at some lower point with a partial fluidized bed in the upper chamber. Fines classification takes place in the upper portion of the separator. As a result of the dense particle concentration in the fluidized bed, high fluid interstitial velocities are created. This feature serves to effectively reject finer particles from the bed. The finer material is subsequently carried to the upper portion of the separator and ultimately transferred to the overflow launder. The unique features of the fluidized bed provide a very effective means of ensuring that fine particles are not displaced to the coarse underflow fraction. As a result, unlike hydrocyclones, fluidized-bed hydraulic classifiers operate with nearly zero fines bypass.

The performance of a fluidized-bed classifier is typically defined by parameters obtained from a partition curve, which represents the recovery to the coarse product stream (typically underflow) as a function of particle size. A particle size corresponding to 50% recovery is known as the cut point (d_{50}), and the slope of the partition curve between the 25% and 75% recovery values is referred to as the imperfection value, as quantified by the following expression:

$$\text{imperfection} = (d_{75}d_{25})/(2d_{50}) \quad \textbf{(EQ 1)}$$

where d_{75}, d_{50}, and d_{25} are defined as the particle size at each respective recovery point. A typical comparison between the classification efficiency of a fluidized-bed classifier and a hydrocyclone is depicted in Figure 2, which shows partition curves for several different cut points. Results for the fluidized-bed classifier are shown for three different applications, with the d_{50} ranging from 70 µm to nearly 300 µm. In each case, the imperfection is approximately 0.1. The hydrocyclone performance curve is less efficient, as indicated by a higher imperfection value (I = 0.27).

Another significant indicator of performance efficiency is the bypass of coarse and ultrafine particles, which can be quantified by the tails of the partition curves. Classifying cyclones are prone to bypass a significant number of ultrafine particles to the coarse product (underflow) stream because of hydraulic entrainment. For the performance comparisons in Figure 2, the hydrocyclone is bypassing 20% of the ultrafine particles entering in the feed stream to the coarse product underflow stream. The fluidized-bed classifier, by comparison, closes out at nearly zero misplacement. The high efficiency results from the deep teetering bed of coarse solids, which is continuously "washed" clean of misplaced fines by the high interstitial velocity of the fluidization water. As such, the selection of the desired classification process should consider the efficiency benefits of a fluidized-bed classifier and the generally higher unit throughput capacity per unit of floor space of a classifying cyclone. In addition, the performance of a fluidized-bed unit is relatively insensitive to changes in feed rate and solids concentration. As such, maintaining a high classification efficiency over the long term favors fluidized-bed units given that these devices provide a constant separation with little or no bypass. Additionally, in cases where a single fluidized-bed classifier can treat a given feed rate, a higher overall sizing efficiency can be achieved compared to hydrocyclones where feed is typically distributed to multiple units. Unequal feed distribution results in a flattening of the partition curve if the operating conditions and geometries of each cyclone are not identical.

THEORY OF OPERATION

Much like hydrocyclones and other classification devices, a great deal of work has been conducted over several decades to develop semi-empirical and fundamental relationships to describe settling characteristics in a hindered environment. Particle settling velocity in a pool of liquid was defined long ago by the work of Stokes and others. In hindered-bed separators, however, particle–particle interactions greatly reduce free-settling rates. The hindered effect is in response to interactions between the upward rising stream lines created by individual settling particles, an increase in the frequency of particle-to-particle collisions, and "near misses" (Littler 1986). The slower settling rate created by the hindered-settling conditions improves classification by reducing fine particle entrainment through the influence of the upward flow of fluidizing medium generated from the volume displacement created by the settling of the high population of coarse particles. According to Littler, hindered settling takes place at a volume concentration greater than approximately 20%. Several expressions have been developed to calculate the hindered-settling velocity of particles (U_t). One of the most commonly accepted expressions was developed by Masliyah (1979) and is represented as

$$U_t = \frac{gd^2(\rho_s - \rho_f)}{18\eta(1 + 0.15\,\mathrm{Re}^{0.687})}F(\phi) \qquad \textbf{(EQ 2)}$$

where

g = gravitational acceleration
d = particle size
ρ_s = density of the solid particles
ρ_f = density of the fluidized suspension
η = apparent viscosity of the fluid
Re = Reynolds number

More recent modeling work by Kohmuench (2000) included the term $F(\phi)$, which corrects for the effects of particle concentration using

$$F(\phi) = (\phi_{max} - \phi)^{\beta} \qquad \textbf{(EQ 3)}$$

where

ϕ = volumetric concentration of solids
ϕ_{max} = maximum volumetric packing
β = dependent on the Reynolds number (Re)

Note that Equation 3 is equivalent to the expression advocated by Richardson and Zaki (1954) when ϕ_{max} = 1. Also, these investigators showed that
for Re < 1:

$$\beta = 4.36/\mathrm{Re}^{-0.03} \qquad \textbf{(EQ 4)}$$

for Re ≥ 1:

$$\beta = 4.4/\mathrm{Re}^{0.1} \qquad \textbf{(EQ 5)}$$

The Reynolds number is calculated using

$$\mathrm{Re} = \frac{d\rho_f\,|U_t|(\phi_{max} - \phi)}{\eta} \qquad \textbf{(EQ 6)}$$

The apparent viscosity (η) in liquid-bed separations can be estimated using a semi-empirical expression suggested by Swanson (1989):

$$t\eta = \eta_w \frac{2\phi_{max} + \phi}{2(\phi_{max} - \phi)} \qquad \textbf{(EQ 7)}$$

where

ϕ_{max} = highest fraction of solids by volume obtainable for a specific material having a given particle size distribution
η_w = viscosity of water or other fluidizing medium

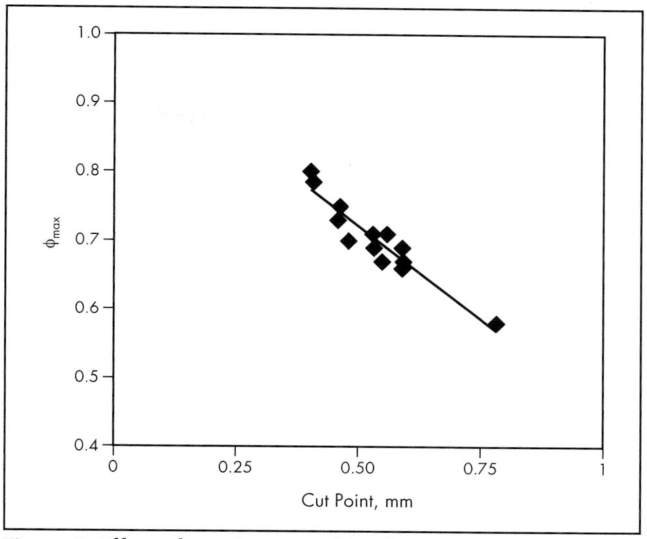

Figure 3 Effect of maximum packing fraction (ϕ_{max}) on the particle size cut point

Empirical methods are normally used to estimate ϕ_{max} (Yu and Standish 1993; Swanson 1989).

Tests conducted with the CrossFlow classifier suggested that changes to the cut point (d_{50}) had a large impact on the maximum particle concentration (ϕ_{max}) of the underflow. This effect should be expected since fine particles tend to fill voids that occur between coarser particles; however, as more fines report to the overflow, these voids remain proportionally empty. To quantify this effect, tests were conducted in which the cut point and maximum packing were determined experimentally. The test data, which are plotted in Figure 3, indicate a linear correlation between ϕ_{max} and d_{50}. A linear fit to these data yielded a coefficient of determination value (R^2) of 0.87.

Using these expressions, the overall hindered settling equation can be derived as follows:

$$U_t = \frac{gd^2(\phi_{max} - \phi)^{\beta}(\rho_s - \rho_f)}{18\eta(1 + 0.15\,\mathrm{Re}^{0.687})} \qquad \textbf{(EQ 8)}$$

With this approach and because of interdependencies between the various equations, an iterative process is required to calculate the hindered settling velocity (U_t) for any particle. This approach is necessary to account for the effect of total particle concentration and other interactions on the settling rate of any single particle.

GENERAL CONFIGURATION AND OPERATION

Although there have been many designs of hydraulic classifiers over the past decades, the current generation of devices can generally be placed into one of three categories: vertical flat-bottom tank, vertical conical-bottom tank, and multicompartment type. Schematic representations of each style are shown in Figures 4A through 4C. The vertical-style separators typically have an aspect ratio greater than 1 and can be designed in a square or round configuration. Generally, past work has shown that the basic shape has no influence on the separation characteristics. Square units have the advantage of higher capacity within the same footprint, whereas round units have a slightly lower overall weight as less reinforcing

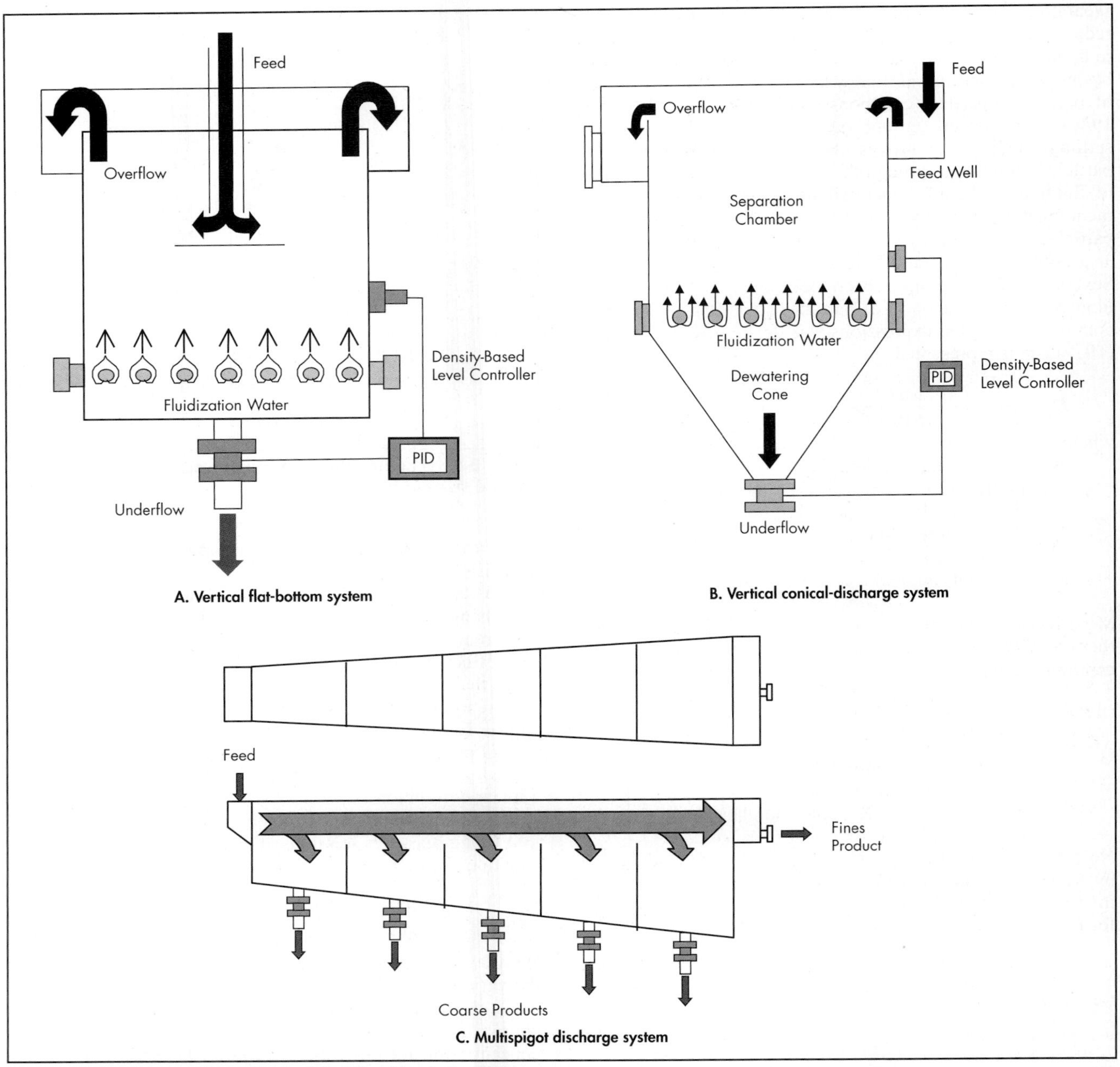

Figure 4 Different configurations of fluidized-bed classifiers

structural steel is required for the sidewalls. The multispigot design, shown in Figure 4C, has a low profile and consists of several distinct compartments along the direction of flow to provide differently sized products from the same device. They are often used for producing various grades of aggregates and foundry sands to maintain specific product criteria.

Various feed introduction designs have been promoted for vertical fluidized-bed classifiers. In general, however, the feed slurry is presented to the device in the center of the separator body and somewhat below the liquid level. A few units use tangential flow introduction to the center-well to break the slurry velocity and provide a more quiescent introduction to the separation chamber. More recently, a side-feed design has been developed that introduces the feed at the surface and to one side of the separation chamber. The intent of this design modification was to remove the bulk feed flow entirely from the separation chamber to reduce disturbances and increase overall separator efficiency by maintaining a constant upward velocity throughout the device (Kohmuench et al. 2002; Luttrell et al. 2006). The multispigot design utilizes a feedbox on one end of the device much like a bank of conventional flotation cells. The feedbox is isolated from the main separation chamber and serves to reduce disturbances and provide even feed distribution.

Feed slurry density typically ranges from 20% to 60% solids. Higher feed concentrations reduce the feed volume

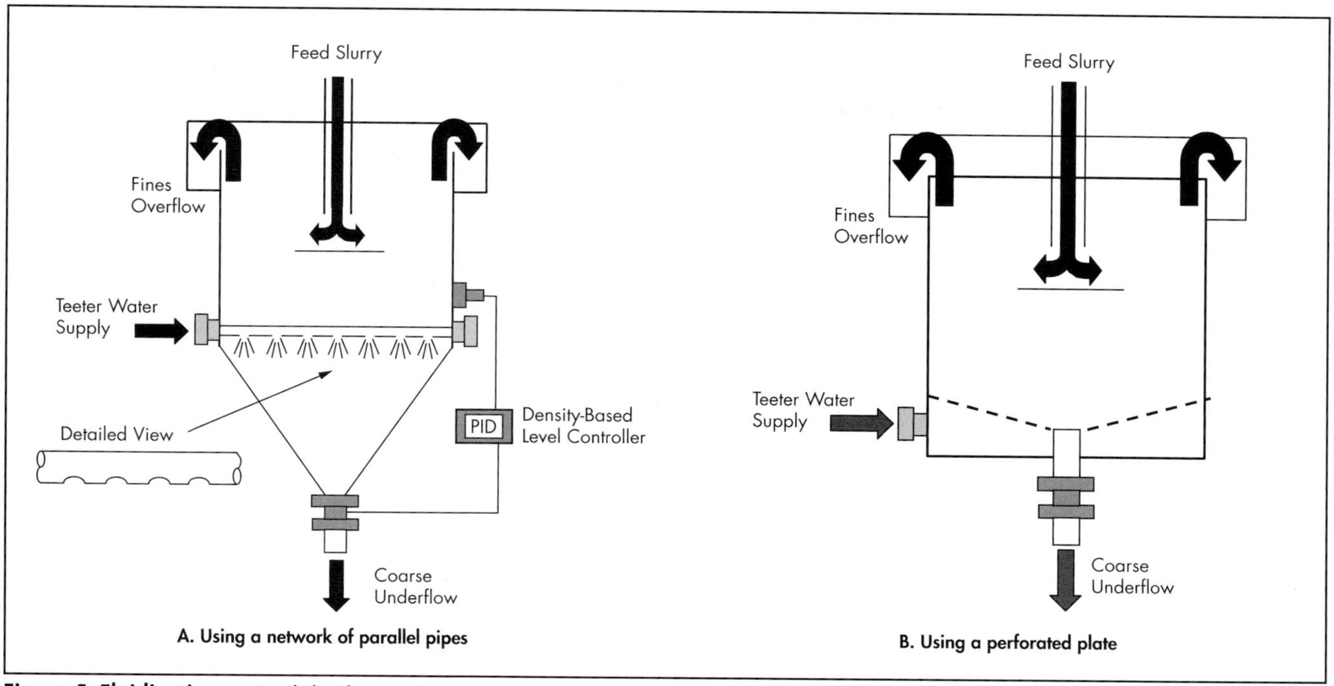

Figure 5 Fluidization water injection systems

flow rate, resulting in less disturbance to the separator. Past work has shown that, in general, the feed percent solids should not exceed the density of the fluidized bed when operating at steady state. At higher solids loading in the feed, viscosity issues can arise that will result in the feed slurry behaving as a slug, penetrating deeply into the fluidized bed, and potentially short-circuiting directly to the underflow product. Additionally, although the solids concentration (% w/w) can be relatively high when treating coarse material, the effect of fines (–0.045 mm) should be considered as their presence can greatly increase the apparent viscosity of the teeter bed and cause a decrease in particle settling rates.

Fluidization/Teeter Water

Two types of teeter water introduction systems have been used for fluidized-bed classifiers. The first, shown in Figure 5A, consists of a network of parallel pipes crossing the base of the separation chamber. The pipes terminate on both ends in water header boxes that are fed from a common supply line. Each pipe contains numerous holes that are sized to provide an even distribution of water across the length of the pipe and, subsequently, across the base of the separator. This type of water supply system offers several advantages, including the ability to remove the pipes for maintenance as well as to replace the pipes with new ones having differently sized injection holes in the event of an application change.

The second commonly used design consists of a simple perforated plate covering the base of the separator. In this design, a chamber is formed between the base of the separator and the perforated plate. Water is injected into this chamber, which acts to distribute the water evenly across the base of the separator. This approach provides a somewhat easier method to ensure equal pressure across all discharge points; however, changeover for maintenance can be more challenging. An example of this design is shown in Figure 5B.

Fluidization water is the primary operating parameter for determining the separation cut point (d_{50}). As a result, the teeter-water rate is a significant process control parameter. On the low end, the fluidization water must be sufficient to maintain the bed in a fluidized state or the device will "channel" and/or "sand out," which is a condition that does not allow for an effective separation. Data show that there is generally a linear relationship between fluidization rate and d_{50}. This relationship is shown for two different mineral applications in Figure 6. One data set shows the effect of fluidization water rate for classification of a 1.0 × 0.106-mm phosphate matrix. In this case, a fluidization rate between 0.5 and 1.5 cm/s provided a range of cut points between 0.25 and 0.65 mm. Results for a 3.0 × 0.85-mm potash application are also shown. In this case, cut points range from 0.5 to 1.2 mm for fluidization rates of 0.75 to 2.5 cm/s.

Given a similar particle size distribution, it is expected that the denser phosphate will require a higher fluidization velocity to affect the same cut point achieved for the less dense potash (2.7 vs. 2.0 SG [specific gravity]). However, in Figure 6, there is only a moderate difference in the fluidization rate for a given cut point. This result is attributed to the influence of other process variables, mainly the feed particle size distribution and the fluidization medium characteristics. In this example, the potash is significantly coarser than the phosphate and classification is carried out using saturated brine solution as the process medium (ρ_f = 1.24 SG).

As discussed previously, mineral density also has an impact on separation performance. Cut-point predictions using fundamental particle-settling models show the effect over a range of mineral density values and fluidization rates in Figure 7. As expected, denser material requires a greater fluid rise velocity to achieve the same separation cut point since, according to Stokes' law, settling velocity increases linearly with solid density.

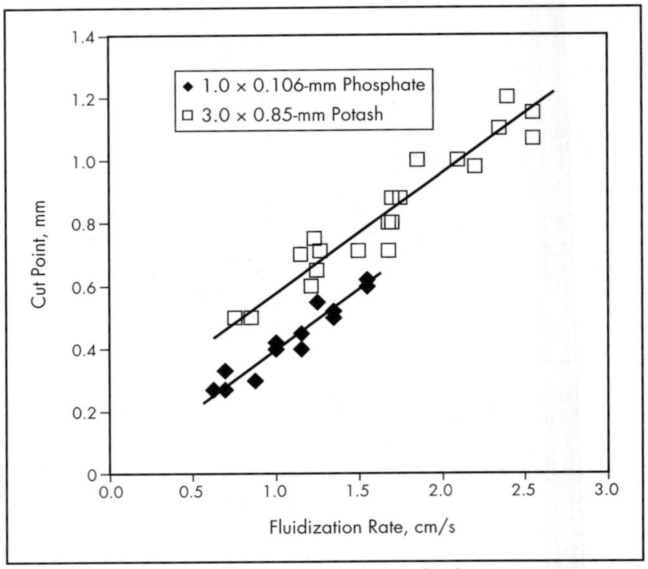

Figure 6 Particle size cut points achieved when treating phosphate and potash over a range of fluidization rates

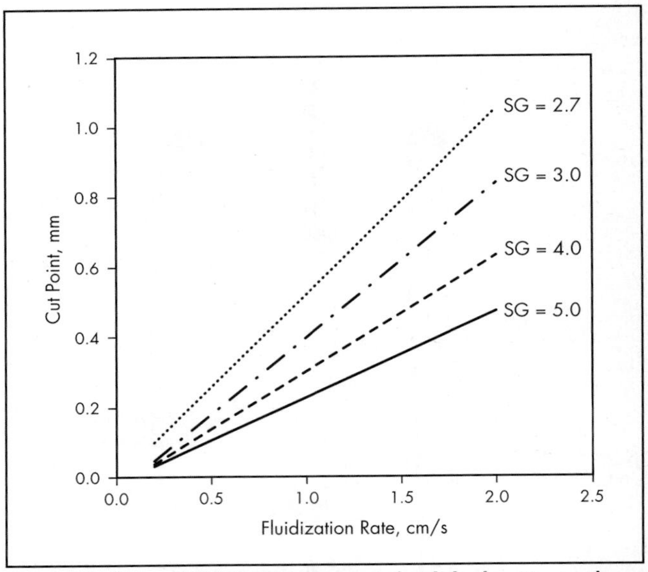

Figure 7 Effect of fluidization rate and solids density on the particle size cut point (d_{50})

Fluidized-Bed (or Teeter-Bed) Level

Unlike the fluidization rate, which expands or contracts the teeter bed, bed-level adjustment is used to modify the proportion of coarse/heavy material retained inside the separator. As such, bed level can be considered a fine-tuning parameter that modifies the height or accumulation of particles within the teeter bed and dictates the particle size that is allowed to be transported into the overflow collection launder. As more coarse material is held up within the separation chamber, the apparent weight of the suspension is increased, which elevates bed pressure. Typical response curves for separation cut point versus bed pressure are provided in Figure 8 for three different fluidization velocities. In each case, as the bed pressure increases, the separation cut point also increases. The absolute value of bed pressure depends on a specific application, separator size, and so on. However, in this case, the total range of bed pressure values shown in Figure 8 represents a physical change in bed height of approximately 30 cm.

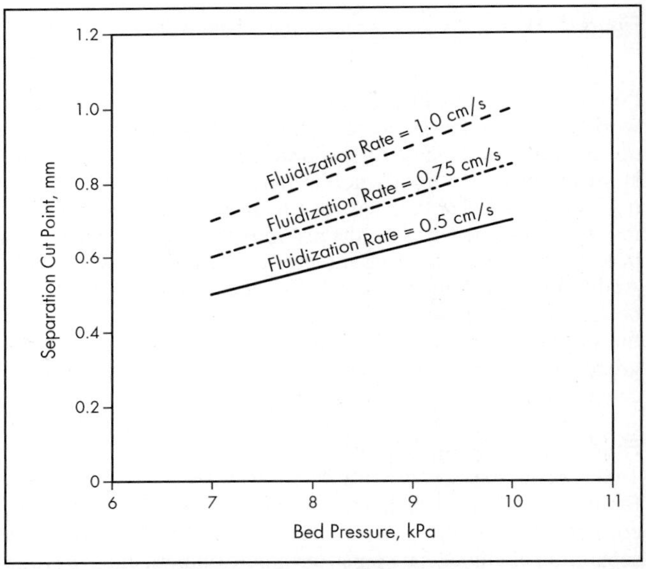

Figure 8 Effect of bed pressure and fluidization water rate on the particle size cut point (d_{50})

Control Systems

Fluidized-bed classifiers are considered operator friendly and simple to control. Traditionally, there are only two primary control variables: fluidization rate and teeter-bed level. Fluidization rate is straightforward and can be either manual or automatic. Manual control simply requires a means to measure and adjust the total volume flow rate of water entering the separator. Modern instrumentation, however, makes automation of this process variable quite simple using a flowmeter, a flow-regulating valve, and a proportional controller as shown in Figure 9. Likewise, level control is accomplished using a pressure sensor in conjunction with a proportional–integral–derivative (PID) loop controller and proportional underflow valve. The pressure sensor measures the weight of material in the separator above the sensing point, and the controller acts to maintain the bed pressure at a constant value by actuating the underflow valve. The mechanism is identical to maintaining water level in a tank or sump. In this case, however, the intent is to maintain the level of suspended solids in a pool of liquid. As such, the response is somewhat slower, requiring different setup parameters for the PID controller as compared to a typical water tank. For instance, in a cone-bottom separator, extracting material too quickly from the separator can cause "rat-holing" in the underflow, which results in short-circuiting of material from the fluidization chamber directly to the underflow, thereby negatively affecting separation efficiency. Recent modifications to the standard control system include linking fluidization water rate to differential pressure measurements as shown in Figure 10. In this instance, fluidization water rate is controlled based on readings from a differential pressure measuring system with the intent of maintaining a constant level of bed expansion (i.e., discrete bed density).

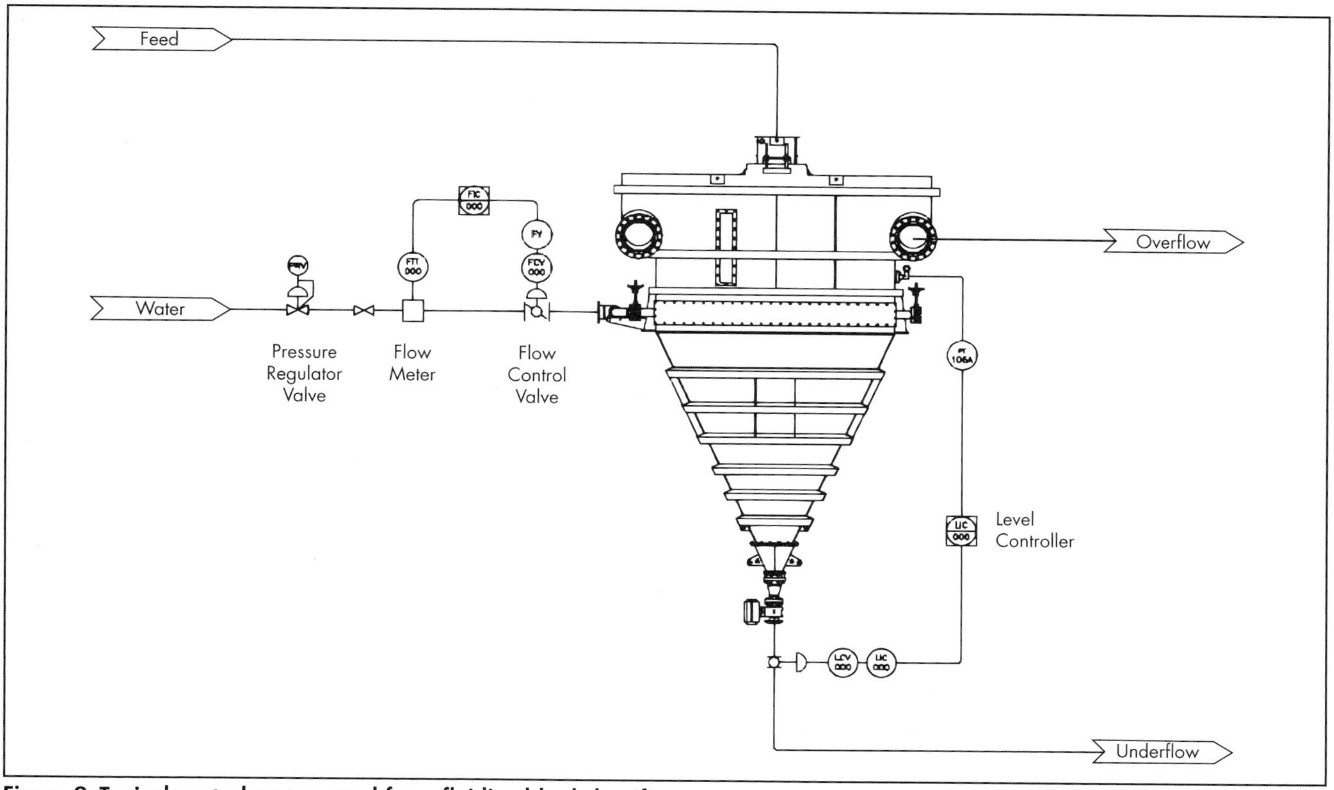

Figure 9 Typical control system used for a fluidized-bed classifier

Additionally, several flat-bottom machine manufacturers incorporate multiple underflow discharge valves. Flat-bottom designs have been developed to reduce overall separator height. In this system, the underflow discharge is drawn directly from the fluidized bed as opposed to the dewatered material in the cone. As such, multiple drawpoints are required to pull evenly from the bottom of the separator. Because of the proximity of the underflow valves to the fluidized bed, a portion of the fluidization medium can be bypassed directly to the underflow stream, which can negatively affect the upward flow rate. In these cases, the fluidization rate can be adjusted in proportion to the underflow valve position to compensate for any bypass. Additionally, each drawpoint is opened sequentially to minimize the total discharge from any one valve.

TYPICAL APPLICATIONS

Particle–particle separation achieved by a system that fluidizes particles using an upward flow of elutriation water was a common industrial operation until it was replaced by more efficient technologies. However, recent advances in control technologies have provided the ability to optimize and maintain a high level of separation performance at a relatively high mass throughput. This section reviews the applications where fluidized-bed classifiers have been installed in industrial processing facilities for treating coal and a variety of ore types.

Classification

Fluidized-bed classifiers are well-proven technologies for particle size classification. These devices offer a high-capacity alternative to hydrocyclones while providing an extremely efficient separation over a wide particle size range from 100 to 1,000 µm. The units are particularly effective at rejecting ultrafine material (Figure 2) while also removing deleterious low-specific-gravity materials such as plastic, wood fiber, and organic contaminants. Additionally, the machines can be gravity fed, have no moving parts, and require no energy input, except for the energy required to pressurize the fluidization water and operate the control system. As such, wear and maintenance are minimal.

Fluidized-bed classifiers are commonly used for sizing applications in the silica sand and fertilizer industries. They are particularly effective for processing sand because of the stringent specifications of the aggregate industry. Complete rejection of fines and a sharp separation curve are a benefit in these applications. Fluidization water rates typically range from 10 to 30 m^3/h per square meter of separator area, depending on the feed particle size distribution and required cut point.

Processing rate is highly dependent on the feed particle size distribution, specific gravity, and required cut point. Fine feeds requiring a low cut point have the lowest capacity, whereas coarse, high-specific-gravity materials may have feed rates over 100 t/h/m^2. Typical feed rates for conventional fluidized-bed classifiers are shown in Figure 11 as a function of cut point for typical sand applications. In this case, the feed rates have been normalized based on the unit cross-sectional area for easy reference and comparison between machine size and shape (round vs. square).

Gravity Concentration

A fluidized-bed classifier can also be used quite effectively as a dense-media concentrator. In fact, early dense-media separators for coal washing, Chance Cones, were basically a simplified

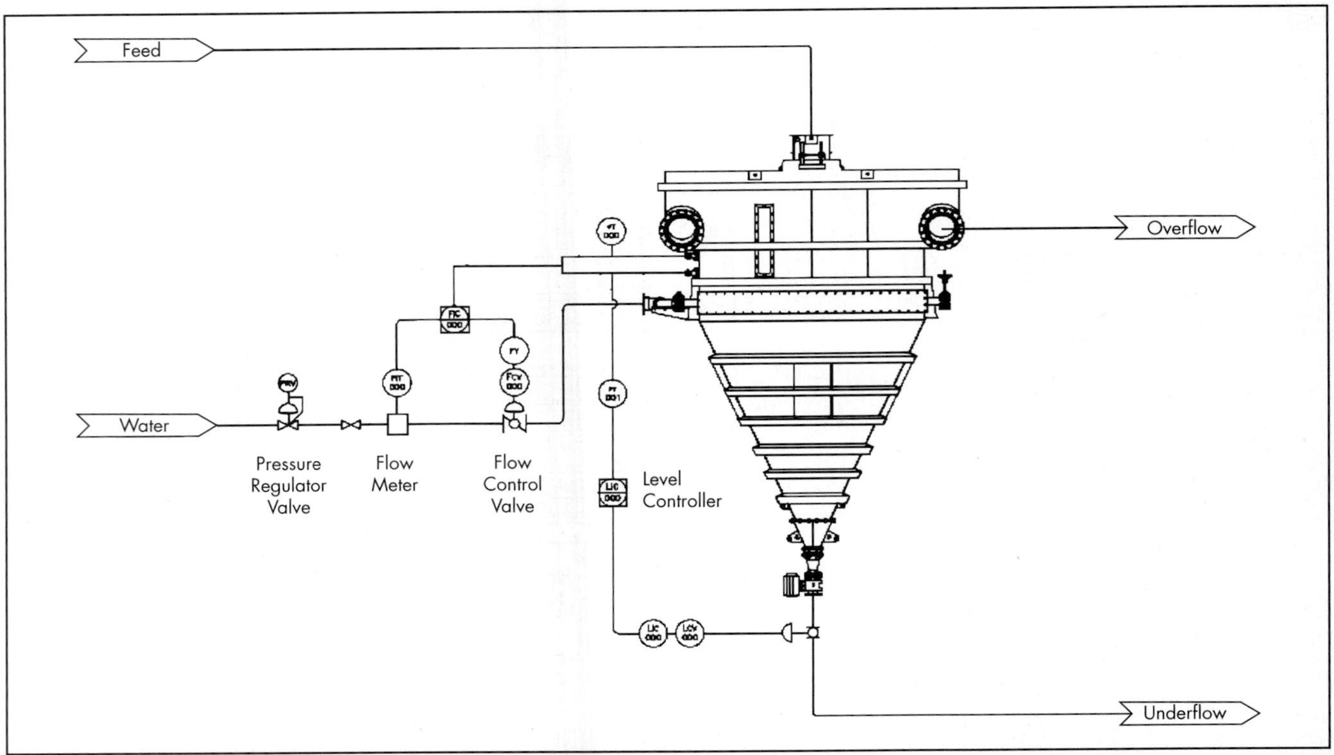

Figure 10 Modification of the standard control system utilizing differential bed pressure to adjust the fluidization rate to control the characteristics of the fluidized bed

version of a fluidized-bed classifier. In this case, a teetering bed of sand was formed in a cone using an upward flow of water, which created a fluidized bed with an effective specific gravity of about 1.6. Coarse run-of-mine coal (nominally 12–100 mm) was added to the device, and the lower-specific-gravity coal/organic fraction would float and report to the overflow along with some of the sand and water. The higher-specific-gravity rock would sink and be rejected using an underflow slide gate system. The sand was subsequently screened from each exiting stream, densified, and returned to the system to maintain the correct specific gravity for separation.

A similar situation occurs in a fluidized-bed classifier when processing a binary mineral system comprised of particles having different specific gravities. Typical applications include coal, iron ore, and heavy mineral sands. As shown in Table 1, there is a substantial difference between the specific gravity of the desired component and the gangue species in each case. In these applications, the higher-specific-gravity component will migrate toward the bottom of the unit and build a fluidized bed. The fluidized bed of high-specific-gravity material effectively creates an "autogenous" heavy media. As such, lower-specific-gravity particles are not able to penetrate the bed and will accumulate above the higher-specific-gravity particles, thereby creating an interface. The lower specific gravity materials will eventually build up and overflow the top of the separator. This approach is quite effective for gravity concentration but is restricted to a top-to-bottom particle size ratio limitation of approximately 5 or 6 to 1. Beyond this limit, the coarsest particles in the low-density material have sufficient mass to penetrate the bed and eventually report to the separator underflow stream, resulting in an efficiency reduction.

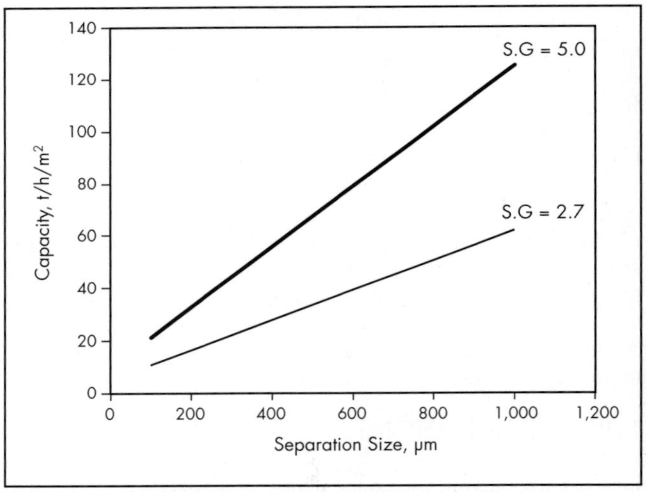

Figure 11 Typical feed mass flux rates for a conventional fluidized-bed classifier over a range of particle size cut points when treating solids with a specific gravity of 2.7 and 5.0

Table 1 Applications of gravity-based separation using fluidized-bed classifiers

Application	Specific Gravity Valuable Component	Specific Gravity Gangue
Coal	<1.6	>2.5
Iron ore	5.0–5.2	2.6
Heavy minerals sands	3.5–4.7	2.6–3.2

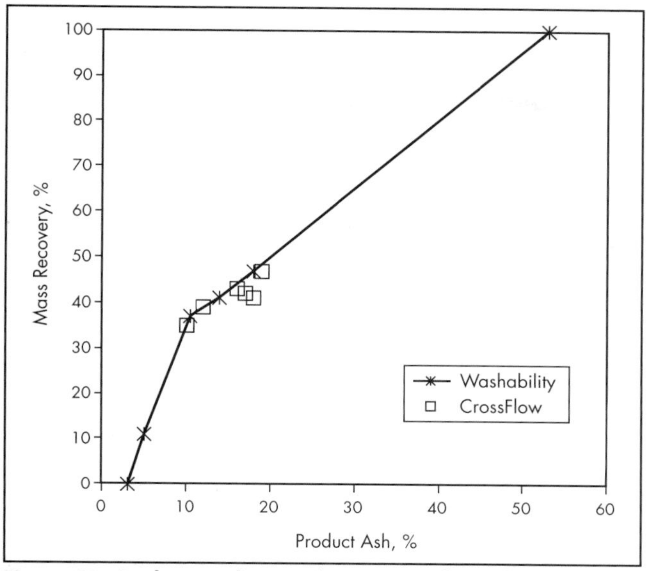

Figure 12 Coal upgrading performance achieved by the CrossFlow fluidized-bed classifier as compared to feed washability characteristics

Table 2 Performance results obtained using a fluidized-bed classifier to treat coal finer than 1 mm and coarser than 0.15 mm

Results	+0.60 mm	0.6 × 0.150 mm	Composite
Ep	0.073	0.105	0.081
Imperfection	0.046	0.059	0.064
SG_{50}	1.569	1.758	1.669

Table 3 Performance results for fluidized-bed classifier for 1.0 × 0.075-mm hematite

Stream Identification	Fe, %	SiO_2, %	Fe Recovery, %	SiO_2 Rejection, %
Fluidized-bed classifier underflow	68.4	0.98	59.1	92.9
Fluidized-bed classifier overflow	51.3	13.9	40.9	7.1
Spiral concentrate	68.6	0.95	89.3	95.4
Circuit product	68.5	0.97	95.7	88.6

Coal

Numerous test programs and commercial installations have been reported over the past decades for coal upgrading applications. Studies reported by others have shown that fluidized-bed classifiers can be quite effective for the recovery of fine coal when treating material in the 1.0 × 0.15-mm size range (Reed et al. 1995; Kohmuench et al. 2003, 2006). Advantages include a relatively low-density cut point, high throughput capacity per unit area, and automatic control.

Results from an eastern Australia coal using a CrossFlow separator are presented in Figure 12. The product yield (weight recovery) versus product ash content is shown in comparison to the washability data. The fluidized-bed classifier was able to produce a low-ash content product at a mass yield approaching the theoretical maximum. In this case, the organic efficiency, which is the ratio of the actual combustible recovery to the theoretical recovery at a given product ash content, approached 99%. The corresponding separation efficiency data are provided in Table 2. The overall specific gravity cut point was 1.67. Separation efficiency was relatively high as indicated by the low Ep and imperfection numbers. Separation efficiency improved with an increase in particle size while the specific gravity cut point was reduced. For the finer particle size fractions (i.e., <0.2 mm), the teeter-water flow rate overcomes the particle settling velocity thereby creating conditions that favor a particle size separation.

Iron Ore

Fluidized-bed classifiers can also be used to upgrade specular hematite. Traditionally, spiral concentrators have been used to upgrade coarse hematite after grinding to a particle size finer than ~1 mm. Although spirals work well in this application, they require a large feed distribution system and multiple stages to achieve the desired product grade and recovery. Work conducted by Venkatraman (1995) showed the advantage of using fluidized-bed classifiers in combination with spirals to simplify circuitry and improve overall metallurgical performance. In this study, a fluidized-bed classifier was used in advance of a spiral to preconcentrate the ore. The objective of the project was to produce a final iron ore concentrate containing less than 1% silicon dioxide (SiO_2) by weight while maximizing overall iron recovery. Because of the relatively large specific gravity difference between hematite and quartz, the process parameters can be set up to maximize quartz rejection with only minimal iron losses. Particle size-by-size analysis of the separated products showed that the bulk of the quartz could be rejected to the separator overflow with only minimal loss of the finest iron. The separator overflow stream containing the fine iron was subsequently reprocessed on a single stage of spirals to recover the lost values. The results from this evaluation (presented in Table 3) show that a concentrate containing less than 1% quartz can be produced with an overall circuit iron recovery greater than 96%.

Heavy Minerals

As noted in Table 1, because of the large specific gravity difference, heavy mineral sands are also amenable to gravity concentration using fluidized-bed classifiers. Opportunities to apply this technology to mineral sand operations were identified by McKnight et al. (1996). In this case, extensive pilot-scale test work was conducted, which led to the installation of multiple full-scale separators for zircon concentration. In this application, the separators were used to preconcentrate material feeding the zircon wet mill, which consists of numerous stages of Reichert cones and spiral concentrators. The primary objective of the wet processing plant was to reject quartz and various non-valuable aluminosilicates. The quartz/zircon separation was quite efficient because of the large density difference between the minerals. Rejection of the various aluminosilicates to achieve an acceptable zircon concentrate grade was more difficult because of the higher specific gravity of these minerals (3.2–4.0). As a result, zircon losses were often too high.

To improve process efficiency, fluidized-bed preconcentration of the zircon wet processing plant feed was evaluated. The findings are presented in Figure 13, which shows the particle size-by-size recovery for the three major components in the feed. Virtually all of the zircon was recovered to the concentrate regardless of the particle size. Only the coarsest particle size fractions (i.e., +0.25 mm) of the silica gangue were recovered, whereas the aluminosilicate recovery was around 50%. Subsequent test work confirmed that rejection of nearly all of the quartz and more than half of the near-gravity aluminosilicates provided an enhanced feedstock that greatly improved the metallurgical performance of the existing spiral circuit.

MANUFACTURERS

Several different commercial units operate based on the same differential settling rate principles, including the Stokes hydrosizer (Mackie et al. 1987), Floatex (Mankosa et al. 1995; Elder et al. 2001), CrossFlow separator, Linatex hydrosizer (Deveau and Young 2005), Allflux separator, and Reflux classifier. The oldest commercial fluidized-bed classifier is the Stokes hydrosizer, which classifies particles fed into a center well located at the top of the unit. On entry, particles begin to settle against an upward flow of fluidization water that is added through a distribution plate at the bottom of the unit. The upward water velocity is set at a rate that equals the settling velocity of a certain class of particles existing in the feed. As a result, the particles in this class become suspended and create a particle bed that cumulates upward until it reaches a controlled height, which is typically a short distance from the bottom of the feed pipe. A pressure tape extends vertically through the Stokes unit to measure the pressure created by the particle bed density and height. When the bed level reaches the control height as indicated by the bed pressure, a controller opens the discharge spigot to allow coarse and/or high-density particles to pass through and report to the underflow discharge pipe. The control is set to allow continuous adjustment to the underflow rate based on the amount of coarse and/or high-density particles reporting in the underflow stream. The fine and/or low-density particles are unable to penetrate the particle bed and are carried by the fluidization water into the overflow stream of the separator.

The other commercial units have similar features but modified designs and different control systems that may enhance classification efficiency, increase capacity, and/or improve the ease of operation. The differences are detailed in the following sections.

CrossFlow Separator

One of the challenges associated with conventional fluidized-bed classifiers is that the feed is injected into the center of the unit at a depth equivalent to about one-third of the total height of the unit. This design feature causes turbulence to occur in a zone in which the large and small particles are settling toward the fluidized particle bed. In addition, the upward velocity of water is increased above the feed injection point because of the additional hydraulic loading from the water in the feed slurry. The increase in velocity results in the possibility of elutriating coarse and/or high-density particles into the overflow stream.

To counter this problem, Mankosa and Luttrell (1999) equipped a fluidized-bed system with a unique feed system that introduces the feed slurry across the top of the separator using a transition box, which is reflected in the commercial name for the technology, the CrossFlow separator (Figures 14A and

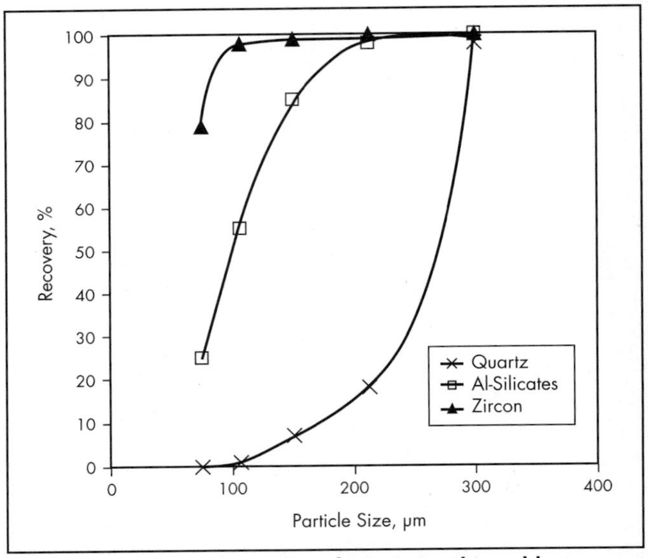

Figure 13 Mineral recovery performance achieved by a fluidized-bed classifier as a function of particle size when treating heavy mineral sands to preconcentrate zircon

14B). The feed transition box increases the area of the feed introduction to the full width of the separator, which reduces velocity and turbulence. The feed flows smoothly over the top of the separator and into the overflow launder. As a result of the feed not being submerged into the teetered-bed unit, changes in feed slurry characteristics do not impact the separation performance. The upward velocity within the separator is constant throughout the vertical plane of the cell.

Linatex

The Linatex unit is a flat-bottom classifier similar to the original Stokes design. This unit also features an adjustable center well to uniformly present the feed slurry into the unit. The fluidization water enters an isolated chamber located at the bottom of the unit and is injected vertically into the fluidized particle bed through a series of nozzles. Multiple discharge values are used on large units to ensure an even drawdown of the underflow solids along the bottom of the unit. The underflow valves are cycled intermittently to minimize short-circuiting directly from the fluidized bed. Like similar units, the underflow valves are actuated based on using a control system to maintain a constant bed density.

Floatex

The Floatex Density Separator, shown in Figure 15, is similar to other devices in that it utilizes a square cross-sectional area with an internal network of teeter pipes to introduce the fluidization water. A conical bottom section is incorporated into the design to assist with dewatering the coarse underflow material prior to discharge. This feature plays an important role in controlling the rate and consistency of the underflow discharge. The high-density zone in the cone prevents material from being drawn directly from the fluidized bed, thereby eliminating the potential for short-circuiting or misplacement to the underflow. A circular feedbox with a tangential feed introduction system is used to introduce slurry into the upper portion of the separation chamber. The tangential design reduces the

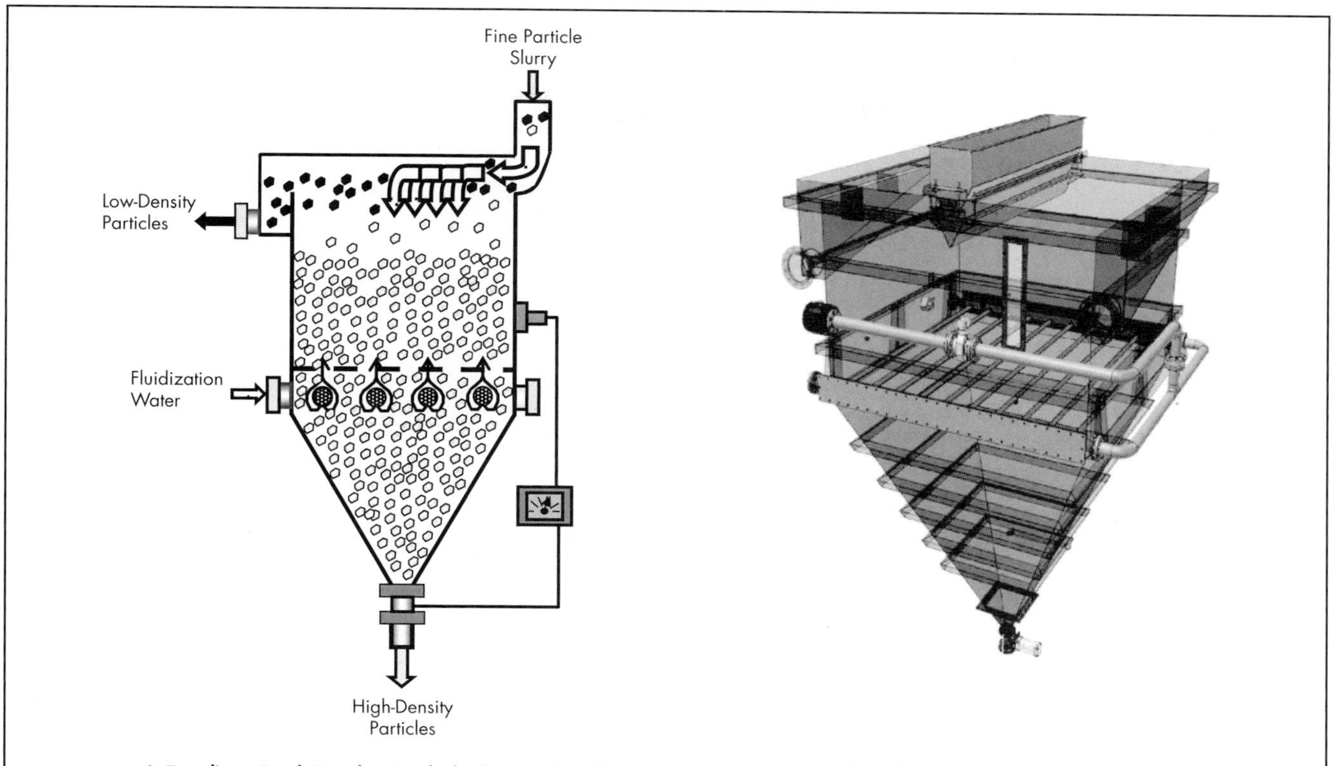

A. Two-dimensional view showing the basic operating principles

B. Three-dimensional view of a commercial unit

Courtesy of Eriez Manufacturing

Figure 14 Schematics of the CrossFlow fluidized-bed classifier

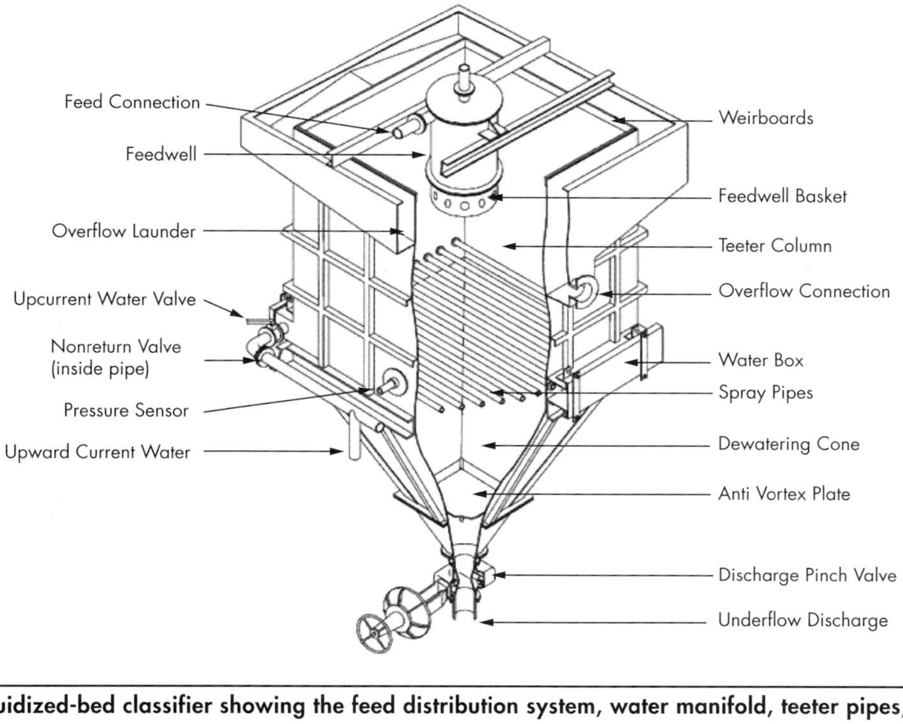

Figure 15 Floatex fluidized-bed classifier showing the feed distribution system, water manifold, teeter pipes, and underflow discharge valve

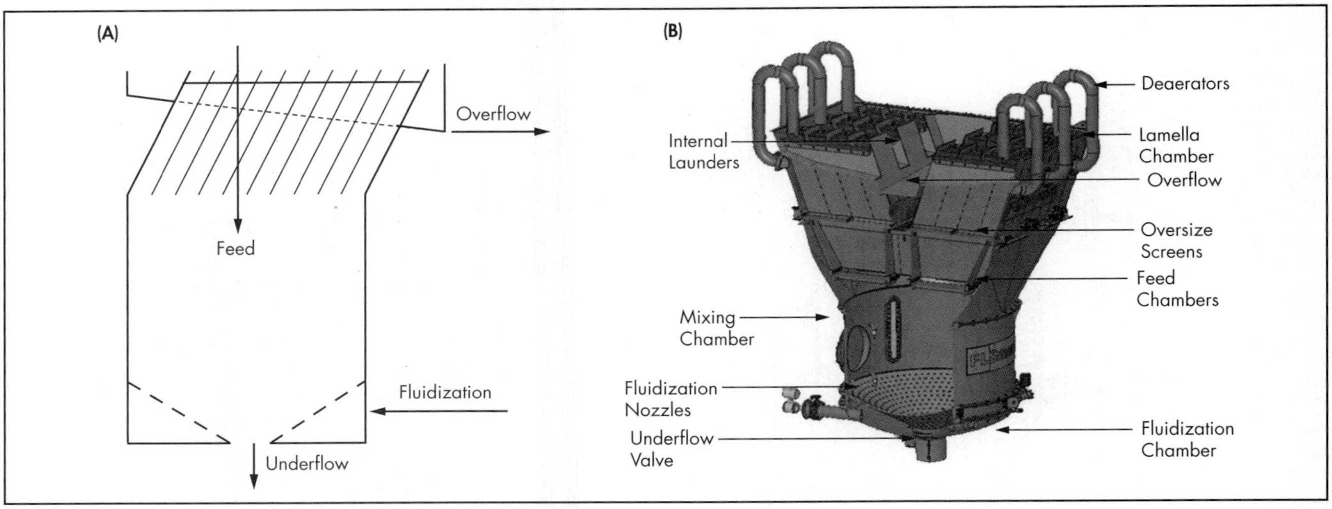

Courtesy of FLSmidth

Figure 16 Reflux classifier schematics showing (A) a basic representation with the fluidization zone and system of inclined channels and (B) a cross-sectional view of a commercial unit

slurry velocity prior to introduction to minimize disturbances to the separation zone. Level is maintained by means of a pressure transducer and automatic underflow control valve.

Reflux

The Reflux classifier is based on the early principle of sedimentation in the presence of inclined planes originally observed by Boycott (1920). More recently, Galvin et al. (2009, 2010) discovered the particle classification benefits of using closely spaced inclined channels and developed a commercial fluidized-bed system known as the Reflux classifier. The classifier consists of a lower fluidization zone and an upper system of parallel inclined channels as shown in Figure 16A. Figure 16B shows a cross-sectional view of a commercial unit. Feed slurry enters in the middle of the machine. Relatively coarse and/or dense particles entering the lamella chamber segregate from the flow and slide downward, eventually returning to the lower fluidized-bed zone. However, fine and/or lower density particles continue to be conveyed upward through the inclined channels and into the overflow launders. This recycling or "reflux" action acts to sharpen separation characteristics, particularly for density-based concentration applications such as fine coal. The cited (Walton et al. 2010) advantage of the Reflux classifier is that the lamella-type plates decrease the unit's sensitivity to wide particle size ranges. Further, the refluxing action of the inclined channels allows the unit to efficiently treat finer particles relative to traditional hindered-bed separators.

Allflux Classifier

The Allflux classifier is a unique unit in which two stages of particle size or density-based separations are achieved in a single unit (Short et al. 2001). As shown in Figure 17, the feed is injected tangentially into the center of a cylindrical tank where the coarse, high-density particles settle to an underflow stream against a rising flow of fluidization water under the influence of a centrifugal field. The low-density particles and the fine, high-density particles overflow a partition and enter an outer separation chamber where the fine, high-density particles and/or coarse, low-density particles are fluidized by a steady stream of water injected through a screen plate and recovered in the underflow stream. The fine, low-density particles overflow into an outer weir. Since the required amount of fluidization water is subject to particle density and size, different screen decks are available to maintain an acceptable pressure drop, ensure a good distribution, and minimize turbulence.

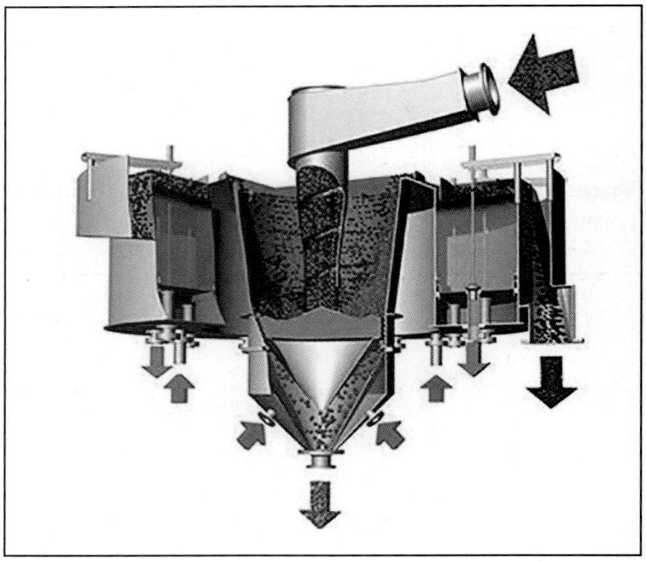

Adapted from Allmineral

Figure 17 Cutaway view of the Allflux classifier showing feed introduction system and multiple separation compartments

REFERENCES

Boycott, A.E. 1920. Sedimentation of blood corpuscles. *Na* 104:532.

Deveau, C., and Young, S.R. 2005. Pushing the limits of gravity separation. SME Preprint No. 05-84. Littleton, CO: SME.

Elder, F., Kow, W., Domenico, J., and Wyatt, D. 2001. Gravity concentration: A better way (or how to produce heavy mineral concentrate and not recirculating loads). In *International Heavy Minerals Conference 2001: Proceedings*. Melbourne, Victoria: Australasian Institute of Mining and Minerals. p. 115.

Galvin, K.P., Walton, K., and Zhou, J. 2009. How to elutriate particles according to their density. *Chem. Eng. Sci.* 64(9):2003–2010.

Galvin, K.P., Walton, K., and Zhou, J. 2010. Application of closely spaced inclined channels in gravity separation of fine particles. *Miner. Eng.* 23:326–338.

Heiskanen, K. 1993. *Particle Classification*. London: Chapman and Hall.

Kohmuench, J.N. 2000. Improving efficiencies in water-based separators using mathematical analysis tools. Doctoral dissertation, Virginia Polytechnic Institute and State University, Blacksburg, VA.

Kohmuench, J.N., Mankosa, M.J., Luttrell, G.H., and Adel, G.T. 2002. A process engineering evaluation of the crossflow separator. *Miner. Metall. Process.* 19(1):43–49.

Kohmuench, J.N., Mankosa, M.J., and Honaker, R.Q. 2003. Advances in teeter-bed technology for coal cleaning applications. In *Advances in Gravity Concentration*. Edited by R.Q. Honaker and W.R. Forrest. Littleton, CO: SME. pp. 115–124.

Kohmuench, J.N., Mankosa, M.J., Bratton, R., and Honaker, R.Q. 2006. Applications of the CrossFlow teeter-bed separator in the U.S. coal industry. *Miner. Metall. Process.* 23(4):187–195.

Littler, A. 1986. Automatic hindered-settling classifier for hydraulic sizing and mineral beneficiation. *Trans. Inst. Min. Metall.* 95:133–138.

Luttrell, G.H., Westerfield, T.C., Kohmuench, J.N., Mankosa, M.J., Mikkola, K.A., and Oswald, G. 2006. Development of high-efficiency hydraulic separators. *Miner. Metall. Process.* 23(1):33–39.

Mackie, R.I., Tucker, P., and Wells, A. 1987. Mathematical model of the Stokes hydrosizer. *Trans. Inst. Min. Metall.* 96:C130–C135.

Mankosa, M.J., and Luttrell, G.H. 1999. Hindered-bed separator device and method. U.S. Patent 6,264,040.

Mankosa, M., Stanley, F., and Honaker, R.Q. 1995. Combining hydraulic classification and spiral separation for improved efficiency in fine coal cleaning. In *High Efficiency Coal Preparation*. Edited by S.K. Kawatra. Littleton, CO: SME. pp. 99–108.

Masliyah, J.H. 1979. Hindered settling in a multi-species particle system. *Chem. Eng. Sci.* 34:1166–1168.

McKnight, K., Stouffer, N., Domenico, J.A., and Mankosa, M.J. 1996. Recovery of zircon and other economic minerals from wet-gravity tailings using the floatex density separator. SME Preprint No. 96-179. Littleton, CO: SME.

Reed, S., Riffey, R., Honaker, R., and Mankosa, M.J. 1995. In-plant testing of the floatex hydrosizer for fine coal cleaning. Presented at Coal Prep '95, Lexington, KY, May 2–4.

Richardson, J., and Zaki, W. 1954. Sedimentation and fluidization: Part I. *Trans. Inst. Chem. Eng.* 32:35–53.

Short, M.A., Snoby, R.J., and Jungmann, A. 2001. Beneficiation of 3mm × 0.15mm fine coal with two-stage hindered settling equipment. SME Preprint No. 01-51. Littleton, CO: SME.

Swanson V.F. 1989. Free and hindered settling. *Miner. Metall. Process.* 6(Nov.):190–196.

Venkatraman, V., Kow, W.S., and Mankosa, M.J. 1995. Industrial application of floatex density separator in processing iron ore and heavy minerals. Presented at the International Conference on Mineral Processing: Recent Advances and Future Trends, Indian Institute of Technology, Kanpur, India, December 11–15.

Walton, K., Zhou, J., and Galvin, K. 2010. Processing of fine particles using closely spaced inclined channels. *Adv. Powder Technol.* 21(4):386–391.

Yu, A.B., and Standish, N. 1993. A study of the packing of particles with a mixture size distribution. *Powder Technol.* 76:113–124.

CHAPTER 4.4

Mechanical Classifiers

Robert C. Dunne

The principal use of mechanical classifiers was in closed-circuit wet grinding; however, this application has been replaced over the past few decades by the hydrocyclone. One mine still using large screw classifiers in the grinding circuit is the Boliden Aitik copper mine in Sweden (Lappi 2015). Other applications still appropriate for mechanical classifiers include desliming, dewatering, and washing operations. Sand–slime separation may be achieved using one or more stages of mechanical classifiers. Typical applications include washing and desliming of concrete aggregate or sand; glass sand; abrasives; oyster shells; phosphate rock; iron, nickel, and chromium ores; alumina; zeolites; solar salt; and precipitates from chemical processing. Again, some of these applications are being performed by hydraulic classifiers (see Chapter 4.3, "Fluidized-Bed Classifiers").

DESIGN AND OPERATING FEATURES

Mechanical classifiers consist of a settling tank with parallel sides and a sloping bottom equipped with a mechanism that continuously agitates the pulp and removes the settled solids. The functions served by mechanical classifiers include the following:

- Allow larger particles to settle in the tank and provide an overflow stream with a minimum of oversized particles.
- Produce an overflow or underflow product of sufficiently high solids content to meet the requirements of the subsequent processing steps.
- Agitate the pulp and allow separation of the entrapped undersize particles so they can report in the overflow product.
- Provide continuous mechanical removal of coarse settled particles.

The underflow product from the mechanical classifier can have a higher solids content and contain less entrapped/entrained undersize particles than the comparable product from the hydrocyclone. Consequently, it is able to operate at higher classification efficiency. The result is a reduced circulating load in closed-circuit grinding applications. The mechanical classifier requires less power and has lower maintenance costs than the hydrocyclone and ancillary feed pump.

The proper selection of mechanical classification equipment requires that the properties of the solids and liquid are adequately defined. The following descriptive information regarding the solids is desired: feed rate, physical composition, density, temperature, size analysis, and desired separation size. For the liquid used, the following data is needed: feed rate, density, viscosity, pH, and temperature.

Settling Area

The pool area required to permit a particle larger than the separation size to settle depends on the density and shape of the particle and on the density and viscosity of the pulp within the classifier pool. These criteria determine the settling rate, which, in turn, determines the settling area required. Determining the settling rate (terminal solids velocity) by the test method of measuring the cylinder batch can be used in calculating the settling area (refer to Chapter 1.6, "Laboratory Test Work and Equipment").

Usually, design information is provided in manufacturer literature. However, for those interested, the required pool area for dilute slurries in open-circuit desliming can be estimated from the following (Heiskanen 1993) equation:

$$A = W \times E = 2Q/V_T$$

where

A = pool area, m^2
W = length of the overflow weir, m
E = weir's mean distance from the feed entry, m
Q = volumetric weir overflow, m^3/s
V_T = terminal settling velocity, m/s

Overflow Weir

The hydraulic head, and therefore the velocity at which the overflow crests the overflow weir, is one of the factors controlling the particle size of classification. Consequently, the length of the overflow weir is designed to provide the overflow velocity that allows the particles larger than the required particle size of classification to settle in the working area of

Robert C. Dunne, Principal, Robert Dunne Consulting, Gooseberry Hill, Western Australia, Australia the pool. Obtaining the necessary weir length for a mechanical classifier of adequate sand-raking capacity may require

- Flaring of the classifier tank toward its overflow end,
- Wraparound overflow weirs, or
- Use of a bowl-rake classifier.

Settling Zones

The pool of a mechanical classifier contains many working zones, as shown in Figure 1. The horizontal overflow zone consists of free-flowing pulp. Essentially, it extends from the feed opening to the overflow weir where it approaches a depth of twice the hydraulic head cresting the weir. In this zone, velocity classification approaching free-settling conditions may be encountered with dilute feed, as in desliming operations. Below the horizontal overflow zone extends the hindered settling zone within which the size of the particles and density of the pulp increase from top to bottom. The less the turbulence and agitation and the lower the pulp density is at the top of this zone, the smaller will be the size of the settled particles. Therefore, the smaller the size of classification desired, the greater must be the depth of the gently agitated settling pool above the turbulence generated by the sand-raking mechanism. The particles settling through the hindered settling zone enter the sand cleaning and removal zone. There, the underflow solids are collected, agitated to remove entrapped water and fine particles, and removed from the pool by the sand-raking mechanism. At the bottom of the settling pool, below the reach of the sand-raking mechanism, is the dead-bed zone. The dead-bed zone serves to protect the tank bottom and has no further part in the classification process. The separation dynamics of solids in mechanical classifiers has been studied by many researchers (Roberts and Fitch 1956; Stewart and Restarick 1967; Reid 1971; Fitch 1973; Schubert and Neesse 1973; Fitch and Roberts 1985).

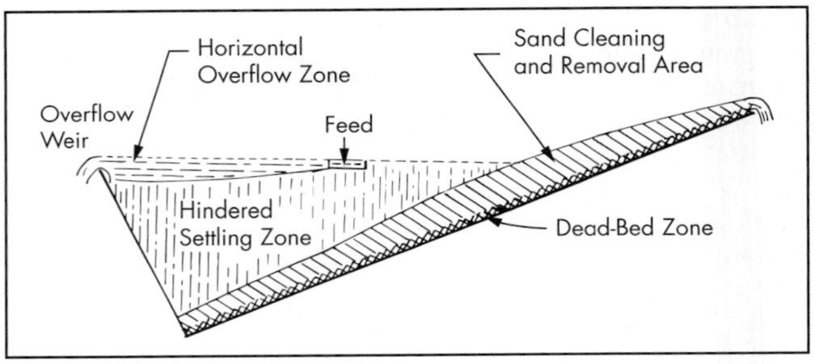

Source: Hitzrot and Meisel 1985

Figure 1 Mechanical classifier zones

Agitation

Agitation generated by the raking mechanism that removes the underflow sands liberates much of the entrapped water and undersize particles, permitting both to rise and report in the overflow product. Therefore, the sand product will contain only a small fraction of entrapped undersize particles, which constitutes a unique benefit from the use of mechanical classifiers. Increasing the speed of the raking mechanism increases agitation and increases the size of the particles reporting in the overflow. Raking speeds are specified by manufacturers of mechanical classifiers for the classification at various particle sizes.

Tank Slope

The speed of the raking mechanism of any given size classifier would have to be increased to maintain its raking capacity at an escalated tank slope. In turn, the increased raking speed compounds agitation and thus boosts the particle size of classification. Consequently, the drainage deck for a sand product of minimum entrapped water and undersize particles requires longer classifiers for classification at finer particle sizes. The slope of the tank bottom has a direct influence on pool area, sand-raking capacity, and sand drainage. It affects the separation size and moisture content of the sand discharge product.

Wash Water

Adequate removal of entrapped water and undersize particles from sand products frequently requires additional wash-water sprays on the drainage deck. However, the sprays wash back much of the solids, which has the net effect of reducing the sand-raking capacity in terms of the final sand product removed.

SPIRAL CLASSIFIER

A spiral classifier consists of an inclined tank in which one or two spirals, mounted parallel to the tank bottom, rotate without contacting the sides or bottom of the tank (Figure 2). The spiral structure provides the necessary pool agitation and

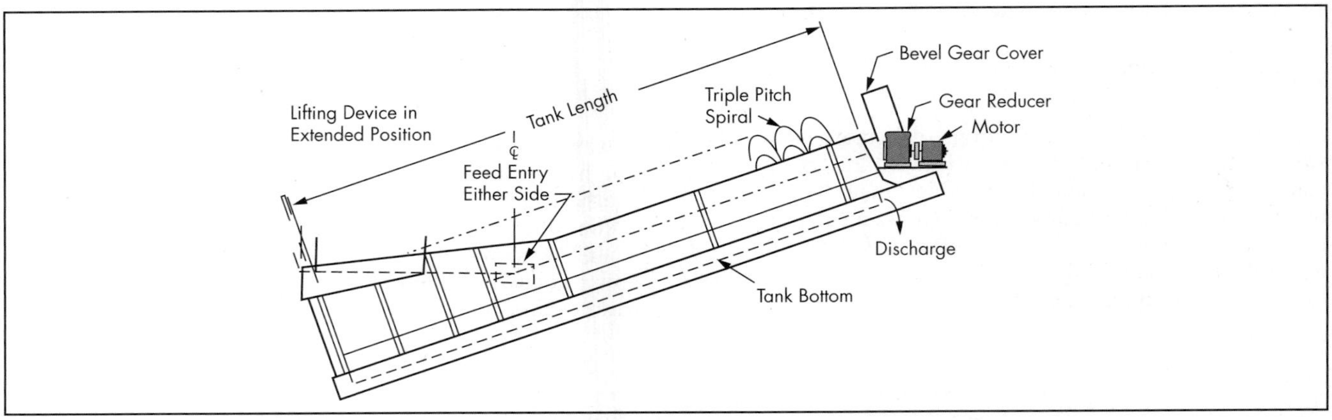

Courtesy of Wemco Division, Envirotech Corporation

Figure 2 Screw classifier

conveys the settled sand to the sand discharge lip. Feed is introduced at the pool level through one or both side walls. Pool level is maintained by adjusting the height of the overflow weir. The classifier overflow discharges from the overflow weir.

Spiral classifiers are built of simplex or duplex design, which defines the number of spiral mechanisms contained in the settling tank (Figure 3). The classifier mechanism is of single, double, or triple helix design, depending on the number of helices mounted on the rotating shaft, which extends through the length of the settling tank. The pitch of the helix varies among manufacturers (Figure 4).

Depending on the desired size of classification, the tank depth or submergence at the overflow weir end generally ranges from 90% to 150% of the diameter of the helix. For velocity classification in open circuit, attaining sufficient weir length and live pool area may require flaring of the weir end of the tank. Flaring of a simplex-classifier tank may increase its width at the overflow weir.

The spiral classifier is made in three different forms defined by the position of the spiral. The high-weir type is usually used in grinding circuits for classification sizes of 150 µm (100 mesh) or coarser. The submerged spiral type has the overflow weir high enough to submerge the lower end of the spiral. This increases the pool area and volume and provides higher capacity for fine separations. The lower end bearing of the spiral is above the pulp level in the low-weir type. This arrangement is typically used for dewatering sand or granular materials and for rough sand–slime separations.

RAKE CLASSIFIER

When the spiral is replaced with a rake, the classifier is called a rake classifier. Rake classifiers are less common than spiral classifiers. The rakes consist of one or more parallel lines of steel plates that hang from a central shaft or shafts. The plates are hinged onto these shafts and have a reciprocating movement. The plates agitate the settling solids and drag the settled particles up the inclined base of the tank. At the end of the stroke, the plates rise sharply and then are lowered back into the tank after an eccentric movement to their original position. When repeating the operation, the settled matter is conveyed up the inclined slope and finally discharged into the sands launder. The overflow stream passes over a weir at the bottom end of the tank and is pumped to the next processing stage. The rake classifier consists of a tank, one or more rakes, and a mechanism for actuating the rakes. A gear case on the tank side near the sand discharge end and appropriate crankshafts and connecting rods move the rakes parallel to the tank bottom. The mechanism, in conjunction with the torque tube, gives the rakes a well-balanced operation at low power consumption. The torque tube is supported at the overflow end on a hydraulic cylinder that lifts the rear end of the rakes when oil is admitted under pressure. Depending on the service and width of the classifier, the tank may be equipped with one, two, or four rake mechanisms. Figure 5 shows a rake classifier.

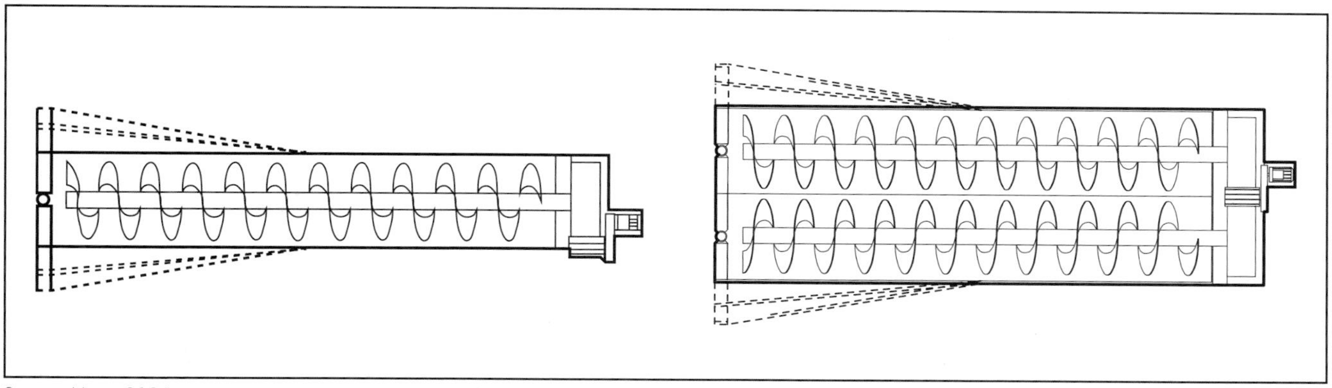

Source: Metso 2006

Figure 3 Single and duplex spiral classifier

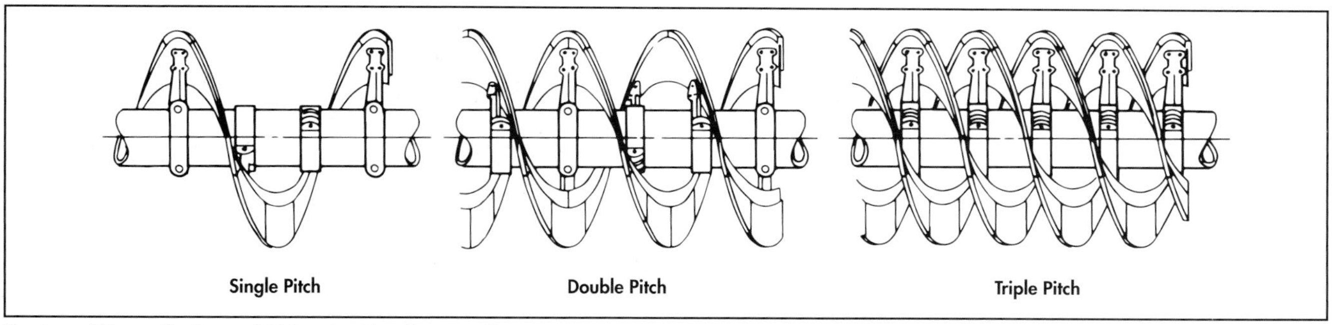

Courtesy of Denver Equipment Division, Joy Manufacturing Company

Figure 4 Different pitch designs

BOWL CLASSIFIER

The bowl classifier consists of a standard unit rake classifier at the pool end in a slightly sloping shallow bowl with a feed well at the center. Bowl diameters range from 3 to 10 m because of the limits on the width of rake classifiers and the limitation on the lower end of the reciprocating rake overhang.

Courtesy FLSmidth
Figure 5 Rake classifier

Courtesy of FLSmidth
Figure 6 Bowl classifier

The heavy particles settle to the bottom of the bowl, which slopes toward the center of the tank. The settled particles are collected by the submerged rakes and guided to the discharge end by a conveyor and dispatched as the underflow fraction. Feed concentration is approximately 65% solids, and overflow and underflow percent solids can vary between 1%–15% and 70%–85%, respectively, depending on the application. The capacities range between 5 and 250 t/h depending on the diameter of the classifier. Figure 6 is a photograph of a bowl classifier.

EQUIPMENT MANUFACTURERS

Mechanical classifiers are produced by many suppliers. The following subsections detail the classifiers available from some of the more well-known manufacturers.

McLanahan Fine-Material Screw Washers

McLanahan (2018) produces fine-material screw washers (FMSWs), which are primarily used to dewater, classify, and wash sand and other fine materials (Figure 7). The selection of an FMSW is based on the type of material, feed size, desired product specification, capacity required, and water volume of a slurry feed. All of these factors must be taken into consideration for proper sizing. FMSW boxes are designed with longer weirs and a larger pool area behind the baffle plate to retain more 75-µm fines. The longer shafts allow for improved dewatering that results in a drier product. They are a low-cost, low-power sand processing solution that ensures retention of the optimal amount of fines. The washers are available in many sizes and configurations to accommodate the feed sand and water capacity. Wear shoes are available in multiple materials, including white iron, hardened metallic plate, and premium urethane to handle different types of feed. Specifications for the different spiral washers are provided in Table 1. The rake speed relationship to percent passing 300 µm in the overflow is provided in Table 2. A screw speed reduction will proportionately reduce rated capacity. For example, a 25% reduction in speed reduces rated capacity by 25%.

Eimco-KCP Spiral Classifiers

Eimco-KCP (2018) manufactures spiral classifiers. Spiral diameters of 0.3–2.25 m are available with single, double, and triple spiral pitches. Adjustable spiral speeds allow for peripheral speeds between 6 and 60 m/min. Replaceable wearing

Courtesy of McLanahan
Figure 7 Fine-material screw classifier

Table 1 Fine-material screw washer specifications

Screw Diameter × Length	Capacity, t/h	Maximum for Retaining Down to, m^3/h			100% Screw, rpm	Power, kW
		150 μm	100 μm	75 μm		
Single-Screw Washer						
508 mm × 7.6 m	31	114	75	42	38	3.7
610 mm × 7.6 m	45	148	97	55	32	7.5
762 mm × 7.6 m	72	179	117	66	26	11
914 mm × 7.6 m	95	207	136	76	21	11
914 mm × 8.2 m	95	207	136	76	21	11
1,118 mm × 10.1 m	158	402	263	148	17	18.5
1,372 mm × 10.7 m	226	460	301	169	14	30
1,676 mm × 11.6 m	362	517	338	190	11	45
1,828 mm × 12.8 m	430	604	395	223	10	56
Double-Screw Washer						
914 mm × 7.6 m	190	312	204	115	21	2 × (11)
914 mm × 8.2 m	190	312	204	115	21	2 × (11)
1,118 mm × 10.1 m	317	597	391	220	17	2 × (18.5)
1,372 mm × 10.7 m	453	704	461	260	14	2 × (30)
1,676 mm × 11.6 m	725	780	511	287	11	2 × (45)
1,828 mm × 12.8 m	861	915	600	337	10	2 × (56)

Adapted from McLanahan 2018

Table 2 Screw speed impact on overflow particle size (relationship to % passing 300 μm)

%	% Speed
0–15	100
16–20	75
21–30	50
31–50	25
51–85	16

Adapted from McLanahan 2018

shoes come in different materials. The lifting device eliminates the necessity to drain the tank during shutdowns. The classifier can be quickly put in operation after shutdown even if the tank is fully sanded. Hydraulic-type lifters are standard on all units of 1.2-m diameter and larger. A handwheel-operated screw-type lifter is standard on units smaller than 1.05 m.

Application

Three classifier models are available (90, 125, and 150). For coarse separations down to 212 μm (65 mesh), series 90 units are used. Series 125 units are used for separations between 300 and 106 μm (48 and 150 mesh); and series 150 units are used for separations of 150 μm (100 mesh) and finer. Eimco-KCP also markets a high-volume classifier for treatment of industrial wastewater, which is designed to settle and dewater relatively coarse particles (~0.5 mm) from high-volume (19 m^3/min), low-percent-solid streams. Flow surges up to 38 m^3/min can be handled. The classifier trough is inclined and flared at the lower end to provide a quiescent settling pool (Figure 8) to prevent mixing of different size particles and carryover of coarse particles to the overflow.

Courtesy of Eimco-KCP

Figure 8 High-volume classifier

SAMM LSX Screw Sand Washing Machine

The SAMM (2017) LSX screw sand washing machine is designed for larger capacities and higher cleaning levels than traditional washing machines. It has three functions—washing, dewatering, and classifying. Consequently, the washer has a relatively small working footprint compared to separate units for comparable duties.

Applications

The LSX screw sand washing machine is used to remove the ultrafine particles in sand. It is widely used for cleaning duties in the following industries: quarry, minerals, building materials, transportation, chemical, and cement mixing stations. Details of the different models available are show in Table 3.

Table 3 Specifications of SAMM LSX models

Model	Spiral Diameter, mm	Tank Length, mm	Number of Spirals	Speed, rpm	Maximum Feed Size, mm	Capacity, t/h	Water Consumption, t/h	Power, kW
LSX920	920	7,585	1	21	≤10	100	9–80	11
2LSX920	920	7,580	2	21	≤10	200	20–160	11 × 2
LSX1120	1,120	9,750	1	17	≤10	175	20–150	18.5
2LSX1120	1,120	9,750	2	17	≤10	350	40–300	18.5 × 2

Adapted from SAMM 2017

Source: Metso 2006

Figure 9 (A) Single- and (B) double-spiral screw classifiers

Metso Spiral Classifier

Spiral classifiers are designed to provide effective pool area and overflow velocity requirements. By combining the appropriate submergence of the spiral with one of the tank designs, a choice of 63 combinations are available in straight, modified, or full flare tank designs from Metso (Jennings Alberts 2018). Polyurethane shoes are standard, and options of iron and Ni-hard are available. The wearing shoes are fastened to the flights with countersunk head bolts. The sealed, submerged bearings comprise two heavy-duty roller bearings mounted in a special housing and protected by spring-loaded grease seals plus a labyrinth seal. The bearing assembly is bolted to the lower end of the shaft. An automatic lubrication system is optional. The spiral assembly is raised and lowered by a simple, positive-action, handwheel-operated screw device for units up to 900 mm in diameter. Larger-diameter units are raised and lowered by an electric screw jack actuator. The adjustable weir overflow system of removable weir bars allows the spiral submergence and pool area to be altered. The classifier is driven by a shaft-mounted gear motor unit or reducer via V-belts and an electric motor. This innovative design allows the reducer bearing to act as the classifier.

Laboratory and pilot-plant spiral classifiers are available in 150-, 225-, 300-, and 400-mm diameter and include many features of the full-sized units. Models of the single- and double-spiral screw classifiers are shown in Figure 9. In addition to the standard-type screw classifier, Metso also supplies purpose-built dewatering and sand screw classifiers.

Spiral Dewaterer

The spiral dewaterer is for the treatment of coarse solids (Metso 2015). The typical application is scale removal in a steel mill. The feed is usually up to 2% solids w/w. The range of units can treat feed rates from 10 to 1,500 m^3/h. An oil skimmer is optional.

Sand Screw

The sand screw is a simpler version of the spiral dewaterer and mainly used for the treatment of natural sand. The objective is to remove the fine fraction, particles smaller than 10–50 µm, in the sand. The sedimentation pool is much smaller compared to that of the spiral dewaterer.

Triveni Spiral Classifier

The Triveni spiral classifier is similar in design features to that of the Metso screw classifier (Triveni, n.d.). Information on sizes and capacities for the Triveni screw classifiers are shown in Table 4.

Xinhai Screw Classifiers

Xinhai Mineral Processing (2017) manufactures two types of spiral classifiers:

1. High weir spiral classifier with capacities from 50 to 1,785 t/d
2. Submerged spiral classifier with capacities from 50 to 1,410 t/d

Both models have automatic spiral-lifting systems. The two models are shown in Figure 10.

Table 4 Triveni screw classifier specifications

Size, mm	Tank Configuration	Overflow Pool Area, m²			Spiral Pitch	Raking Capacity per Spiral Revolution, t/h	Spiral, rpm	Motor, kW
		Model 100	Model 125	Model 150				
610 Simplex	Straight	1.40	—	—	Single	0.5–1.0	6–16	1.1–2.2
	Modified flare	1.50	2.08	2.64	Double	1.0–2.0	—	—
	Full flare	—	2.41	3.19				
760 Simplex	Straight	2.12	—	—	Single	0.8–1.7	5–13	1.1–2.2
	Modified flare	2.09	3.21	—	Double	1.7–3.4	—	—
	Full flare	—	3.72	—				
914 Simplex	Straight	3.47	—	—	Single	1.7–3.5	4–11	1.5–3.7
	Modified flare	3.36	4.53	5.81	Double	3.5–7.0	—	—
	Full flare	—	5.30	7.06				
1,067 Simplex	Straight	4.12	—	—	Single	2.4–4.8	3.5–9	1.5–2.2
	Modified flare	4.61	6.17	—	Double	4.8–9.6	—	1.5–5.6
	Full flare	—	7.25	—				
1,219 Simplex	Straight	5.30	—	—	Single	4.3–8.7	3.2–8	2.2–5.6
	Modified flare	5.96	7.99	10.22	Double	8.7–17.4	—	3.7–7.5
	Full flare	—	9.38	12.45				
1,524 Simplex	Straight	8.23	—	—	Single	8.6–17.3	2.6–6.5	5.6–11.2
	Modified flare	9.25	12.45	15.89	Double	17.3–34.6	—	11.2–14.9
	Full flare	—	14.68	19.42				
1,676 Simplex	Straight	9.94	12.68	15.51	Single	10.2–20.4	2–6	7.5–11.2
	Modified flare	11.19	15.00	19.30	Double	20.4–40.8	—	11.2–14.9
	Full flare	12.63	17.69	23.69				
1,828 Simplex	Straight	11.80	14.96	18.12	Single	13.9–27.8	2–7	7.5–14.9
	Modified flare	13.29	17.74	22.58	Double	27.8–55.6	—	11.2–22.4
	Full flare	—	20.90	27.78				
1,981 Simplex	Straight	13.75	17.47	21.18	Single	15.6–31.5	2–6	7.5–14.9
	Modified flare	15.51	20.81	26.48	Double	31.5–63	—	11.2–22.4
	Full flare	—	24.62	32.61				
1,828 Duplex	Straight	22.58	26.11	29.26	Single	13.9–27.8	2–7	18.6–29.8
	Modified flare	25.64	30.47	35.21	Double	27.8–55.6	—	—
	Full flare	29.08	35.30	42.09				
1,981 Duplex	Straight	26.48	30.29	34.00	Single	15.6–31.5	2–6	22.4–37.3
	Modified flare	30.01	35.40	39.21	Double	31.5–63.0	—	—
	Full flare	34.10	42.09	49.24				

Adapted from Triveni Engineering and Industries 2007

Courtesy of Xinhai

Figure 10 (A) High weir and (B) submerged spiral classifiers

ACKNOWLEDGMENTS

This chapter draws heavily on the theory in the "Mechanical Classifiers" chapter written by H.W. Hitzrot and G.M. Meisel in the 1985 edition of the *SME Mineral Processing Handbook.* The author of this chapter updated the equipment data and information and revised selected material while leaving much of the theory in the chapter as originally written.

REFERENCES

Eimco-KCP 2018. Spiral classifier. http://ekcp.com/our-services/spiral-classifier/. Accessed August 2018.

Fitch, B. 1973. A generalized phenomenological model for classification. In *Particle Technology 1973: Proceedings of the First International Conference on Particle Technology*. Chicago: IIT Research Institute. pp. 92–100.

Fitch, B., and Roberts, E.J. 1985. Classification theory. In *SME Mineral Processing Handbook.* Edited by N.L. Weiss. Littleton, CO: SME-AIME. pp. 3D1–3D10.

Heiskanen, K. 1993. *Classification Handbook.* London: Chapman and Hall.

Hitzrot, H.W., and Meisel, G.M. 1985. Mechanical classifiers. In *SME Mineral Processing Handbook.* Edited by N.L. Weiss. Littleton, CO: SME-AIME.

Jennings Alberts. 2018. Spiral classifiers. www.jenningsalberts.com/Products/176-metso-spiral-classifiers.aspx. Accessed August 2018.

Lappi, S. 2015. A study of cleaner flotation in Aitik. M.S. thesis, University of Oulu, Finland.

McLanahan. 2018. Fine material screw washers. www.mclanahan.com/products/fine-material-screw-washers/. Accessed August 2018.

Metso. 2006. Spiral classifiers. Technical specifications. Helsinki, Finland. www.metso.com.

Metso. 2015. *Spiral Dewaterer.* Brochure No. 1130-09-15-ESBL/Sala. Helsinki, Finland: Metso Corporation.

Reid, K.J. 1971. Derivation of an equation for classifier-reduced performance curves. *Can. Metall. Q.* 10(3):253–254.

Roberts, E.J., and Fitch, E.B. 1956. Predicting size distribution in classifier products. *Trans. AIME* 205 (November):1113–1120.

SAMM. 2017. LSX Sand washing machine. www.trackcorp.co.za/products/lsx-sand-washing-machine.html. Accessed August 2018.

Schubert, H., and Neesse, T. 1973. The role of turbulence in wet classification. In *Proceedings of the 10th International Mineral Processing Congress*. London: Institution of Mining and Metallurgy. pp. 213–239.

Stewart, P.S.B., and Restarick, C.J. 1967. Dynamic flow characteristics of a small spiral classifier. *Trans. Inst. Min. Metall. C* 76:C225–230.

Triveni Engineering and Industries. 2007. Spiral classifiers. www.trivenigroup.com/templates/corporatehome/pdf/triveni-image-10.pdf. Accessed August 2018.

Xinhai Mineral Processing. 2017. Classifying: High weir spiral classifier and submerged spiral classifier. www.xinhaiepc.com/product/classifying. Accessed August 2018.

CHAPTER 4.5

Ore Washing

Richard Williams

To *wash* can be defined as "to remove [a substance] by or as by the action of water" (Dictionary.com 2018). In a mineral processing sense, this commonly refers to the removal of loosely attached particles, often fine clays, from the surface of more competent ore. In this definition, it is distinct from classification, which would be a subsequent processing step that separates the different sized fractions. Despite increasing concerns about the value and availability of water, ore washing remains a common and cost-effective means of beneficiation, as water usage is enhanced by the greater use of water recovery. Advantages that accrue from the scrubbing of ore can range from a reduction of gangue to process further, to improved product quality and higher throughput of downstream equipment. Although generally thought of as removing waste material, in highly weathered deposits, the inverse may apply and the valuable mineral may be contained in the fines. Examples occur in alumina (Husaini and Cahyono 2010), nickel laterites, uranium, and occasionally gold.

DESIGN PRINCIPLES

The overriding principle in ore washing is to provide sufficient energy and time to detach the fine clays but not sufficient to cause comminution of the intrinsic particles to smaller components. Care must also be exercised to provide appropriate energy and flow to maintain the fine particles in suspension and not allow them to agglomerate and grow as *clay balls*. Thus, defining a suitable ore washing process should consider the following:

- Energy input, kW·h/t
- Particle size distribution, mm
- Time
- Throughput

Equipment sizing for ore washing has been largely empirically based on scale-up of laboratory or pilot tests, subsequently benchmarked off operating installations. Miller (2004) created a mathematical model for sizing rotary drum scrubbers that calculated input power and residence times based on physical geometry; however, this still relied on test work to establish the range of acceptable criteria. The high degree of variability in clay mineralogy leads to a preference to do site-specific testing. Standards such as ASTM D4318 (2017), ASTM D4221 (2017), and ASTM D3744 (2003) exist to help define the *plasticity* and/or the *dispersive* characteristic of clay materials and *aggregate durability*, but to date, a measurable correlation between these and energy input has not be published.

TEST WORK

Test work is commonly carried out either in commercial testing laboratories (Figure 1) or by vendors. Equipment used can vary from scaled-down vendor equipment to adapted commercially available substitutes (concrete mixer) to equipment designed for alternative testing (ISO tumble drum [ISO 3271:1995]) to purpose-built pilot plants capable of taking the full size range. A major limitation with all but the last is the inability to process a representative sample size range, particularly for those duties with a feed size of up to 250 mm.

Most small-scale test work is done on a batch basis. This can create some interpretation issues. It is generally acknowledged that batch testing provides an indication of the required solids residence time; however, most equipment is operated in a continuous mode. Subject to the geometry and flow patterns within the equipment, the fluid phase will traverse at a different rate. Given such residence time, distributions of both the liquid and solid phases must be accounted for. Miller (2004) suggested 0.3–0.5 mm as a guideline for the coarse–fine (approximating solids–liquids) split; however, it was also acknowledged that this is application specific. Industry standards use a range of 2.0 to 2.5 to compare the solids residence time to an average. Common design specifications refer only to an average residence time; hence, evaluation of the equipment used for testing is critical for establishing design parameters. Discrete element modeling with computer simulation has been successfully used to approximate both residence time and power draw on a comparative scale when calibrated against known installation.

Richard Williams, Regional Manager Western Australia, McLanahan Corporation, Perth, Western Australia, Australia

Figure 1 Pilot- and laboratory-scale log washer

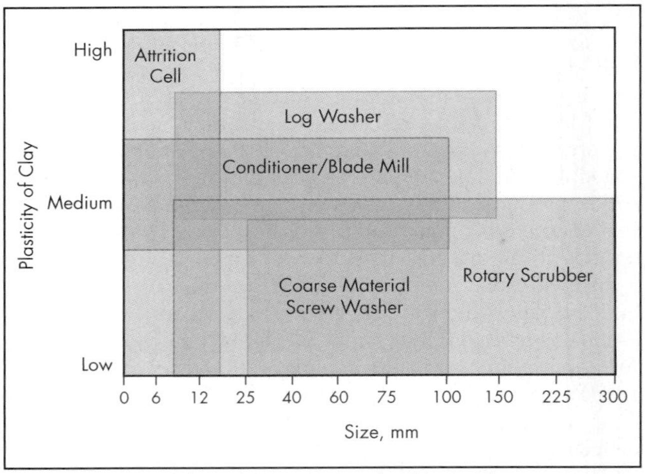

Figure 2 Ore-washing equipment selection guide

KEY EQUIPMENT DESCRIPTIONS

Once the energy requirement, particle size, residence time, and throughput are established, the matrix shown in Figure 2 can provide a guideline for equipment selection.

Attrition Cells

Attrition cells are high-energy devices used to liberate tough plastic deleterious material and remove it from competent material. They are also proven to liberate clays, reducing product turbidity, and to break apart loosely conglomerated clusters in frac sand plants. Manufacturers of high-quality silica sands use them to remove iron staining from the surfaces of sand grains.

These devices are primarily made up of a tank (cell) in which a vertical shaft drives a set of two to three diametrically opposed mixer blades. The cell itself can be round, square, hexagonal, or octagonal, but all are generally baffled to an extent to minimize swirling. Configurations of the blades and whether these are simple axial turbine or propeller in design is largely manufacturer dependent. As the blades rotate, material is forced against itself to cause particle-on-particle collisions, thus abrading the surfaces and liberating the coating material. Because of the intense mixing action, the particle residence time distribution in any one cell can be quite broad, with some particles leaving the cell quickly (bypass) and others remaining for longer. To help ameliorate the bypass, attrition cells are commonly supplied as multiple cells, with two to four cells per machine being most common. Often, alternating cells in series like this will reverse the pitch of the impeller blades to aid direction of the flow toward the outlet at either the top or bottom of the tank. Weirs at the final discharge lip allow the operator to adjust the height of the slurry within the tank and hence increase or decrease residence time.

Attrition cells are typically fed at high solids (60%–80% solids by mass). As such, upstream unit operations will commonly consist of dewatering cyclones or even dewatering screens. These present the material at a high density to achieve maximum particle-on-particle scrubbing action. The feed size is limited to ~6 mm for materials in the range of 2.7 specific gravity. Because the attritioning efficiency is greater at higher solids concentrations, one must be careful not to run the machine into an operating environment that causes plugging. Monitoring of the power input can assist in operating in a more desirable region with simple water addition on demand.

It is also important to understand any changes in the state of the slurry as the attritioning process takes effect. In instances where there is a presence of high clays, or clays that only become broken down and suspended after high shear, the slurry itself may take on an oatmeal-like consistency. The resulting effect is that the competent particles are well lubricated in the mass, and further attritioning is impeded. The solution to this is to use an interstage wash operation between sets of attrition cell units. These wash circuits remove the liberated clays and return the remaining competent particles to downstream attritioning. To help reduce water consumption by the wash stages, a countercurrent approach is often recommended, with the dirtiest discharges being washed with the dirtiest water (Figure 3).

Because of the high-shear environment, it is important to have suitable wear protection (rubber lining and/or hardened steels) to maximize life and minimize time and costs of replacement parts.

Attrition cells are used in the glass, heavy mineral, and frac sand industries to provide intensive scrubbing (Preston and Tatarzyn 2013), with an energy input of 5–10 kW·h/t. To accommodate the high energy, these are small units of 1–10 m^3 and commonly fitted with a direct-drive motor. Residence times may range from 0.5 to 5 minutes.

Coarse and Fine Material Screw Washers

Also known as *spiral classifiers* or *screw classifiers*, these machines are primarily built to wash and classify crushed stone, gravel, and other hard ore minerals, generally ranging from 0.075 to 100 mm. They effectively remove light loamy-type clays, dirt, crusher dust, and coatings that cannot be removed by wet screening alone. They can also be used to take out floating vegetation and soft aggregate, although not all vegetation and waterlogged sticks may be removed completely. Coarse material screw washers are not intended to remove the tough plastic clay that normally requires a more energy-intensive machine.

The sizing and selection of a coarse material screw washer is based on the type of material, desired capacity, and maximum feed size (Figure 4A). The more paddles that are used, the lower the capacity; however, the washing action is increased because paddles do not convey material up the washer box as fast as spiral flights. A coarse material screw washer can be furnished with up to 40 paddles per shaft. When using additional paddles, it is also necessary to lower the slope of the box and increase motor power. This will help convey material to the discharge end.

Subject to the duty, either volumetric and solids capacity limitations may lead to the use of twin-screw units. The larger pool area on the twin-screw unit (Figure 4B) will lower the rise rate, resulting in a finer classification, and the double screw will permit a higher solids loading.

In coarse material screw washers, paddles are used in conjunction with screw flights to provide scrubbing, scouring, and agitation. The turbulent washing action combined with rising current water introduced at the bottom of the box at the feed end results in separation of the lighter fraction from the sound aggregate. The lighter fraction (i.e., clay and floating vegetation) rises to the surface because of the water rising in the box and then overflows the weir located in the back of the box. The desired clean product is then scrubbed and conveyed by the paddles and flights to the discharge end of the box. A rinsing screen is recommended for use after the coarse material screw washer for material classification.

The screw shaft is composed of an extra-heavy steel pipe shaft with inner and outer renewable and reversible abrasion-resistant hard iron paddles bolted to the shaft in the feed area. These paddles are preceded and followed by heavy steel flights equipped with bolt-on abrasion-resistant liners. The screw shaft is flanged at both ends and bolted to stub shafts to facilitate normal maintenance.

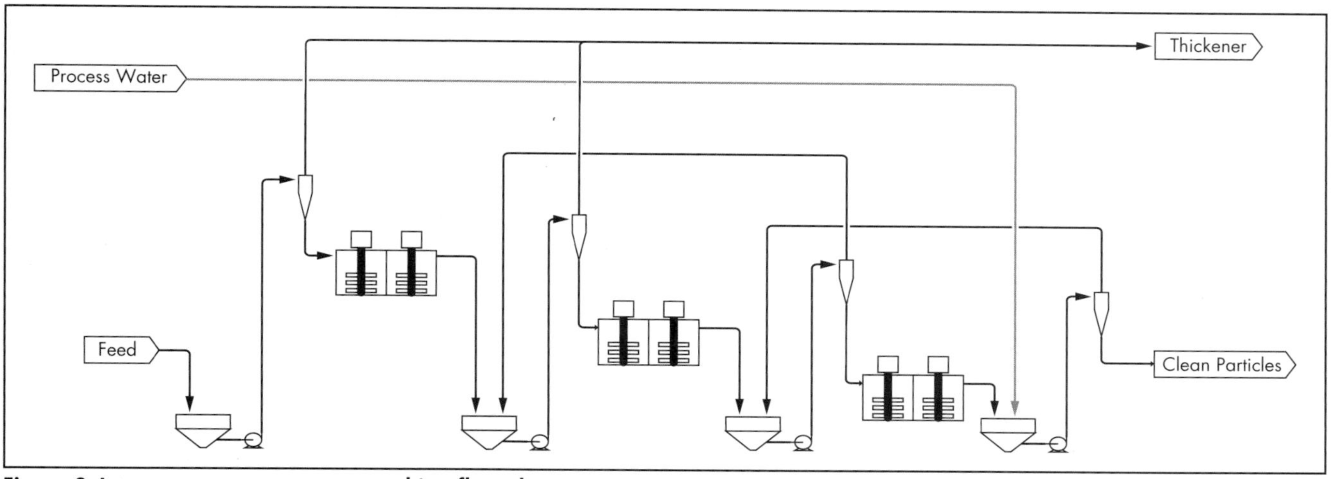

Figure 3 Interstage countercurrent-washing flow sheet

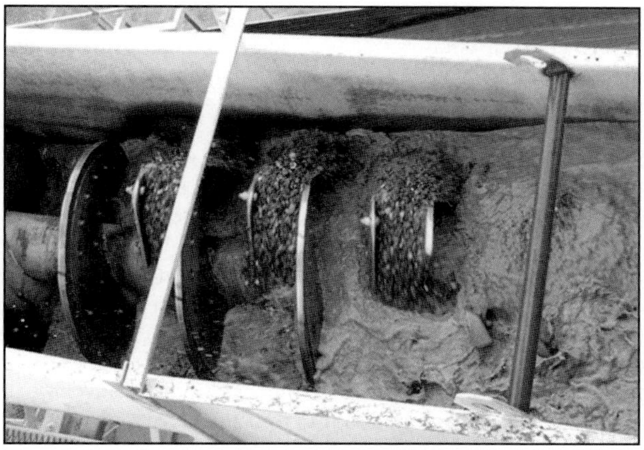

A. Coarse material single-screw washer

B. Fine material twin-screw washer

Figure 4 Screw washers

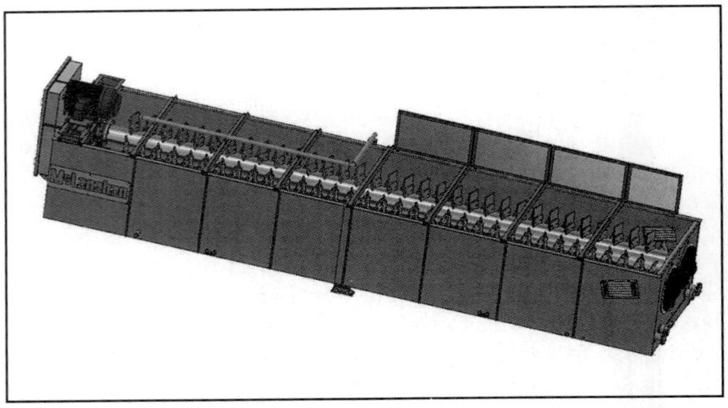

Figure 5 Log washer

Figure 6 Conditioner or blade mill

Log Washers

The twin-shafted log washer is so called as the original design utilized a wooden log as the shaft into which were driven fixed iron paddles. Log washers are used in a variety of material processing applications and are best known for their ability to remove tough, plastic clays from natural and crushed gravel, crushed stone, and ore feeds (Figure 5). Inputs of 2–5 kW·h/t would be typical. One limitation of log washers is that they must have a controlled top size. In general, most units can accept feed material with cubical-shaped particles up to 100 mm. Some larger-diameter units can accept a feed material up to 150 mm. When processing fines, it may be necessary to capture the overflow of the unit to save this fraction for further processing or close off the overflow opening so that material is carried the full length of the box for scouring of the fines.

When selecting a log washer for an application, the amount, type, and percentage of deleterious material (i.e., clay) to be removed should be considered. As the percentage of deleterious material to be removed increases, options exist to

- Vary the length of the wash box,
- Change the angle of the intermeshing paddles to either impart more power or accelerate the conveyance of material,
- Alter the angle of mounting,
- Alter the overflow weir height, or
- Vary water flow rates for rising current and rinsing.

Most log washers come standard with adjustable weirs at the overflow openings that permit the change in water depth, which in turn can affect the quality of the washed product. They are often equipped with a spray bar for rinsing the material prior to discharge.

Critical to the reliability of a log washer is the need for a robust design of the washer box, log shafts, gear box/drive train, and the submerged rear bearings. Depending on the rock or ore being washed, paddle life varies from one to several years between replacements. The submerged-bearing design varies among manufacturers. A separately lubricated seal design provides extended availability. The paddles are often provided with replaceable wear shoes.

Conditioners and Blade Mills

Conditioners, also referred to as blade mills, are designed to help producers begin liberating light, loamy clay or dirt from either coarse rock or sand before further processing. There are many applications where a conditioner can be used. As an example, when washing coarse rock, the conditioner can be placed ahead of a wash screen, resulting in cleaner rock or ore. When washing finer materials, such as sand, the unit can be placed ahead of a screen, sand-classifying tank, or fine material screw washer. Conditioners are able to accept either coarse or fine feed material.

Keep in mind that conditioners are designed to do just that, condition. They are not designed to remove tough, plastic clay or a high percentage of clay. Conditioners are usually selected for use with other processing equipment to improve the downstream washing and separation of material. Because of all the material and water exiting the unit, the discharge material cannot be fed onto a conveyor belt. Generally, the amount of water needed for a conditioner is an additional third of the weight of the material being processed.

Conditioners are very similar in appearance to coarse material screw washers, but function much differently (Figure 6). The major difference is that any material and water that enters the conditioner must exit through the discharge opening located at the bottom of the box opposite the feed end. There is no overflow in this type of unit. Using a combination of paddles and flights arranged in alternating format the entire length of the shaft, they begin to scour, abrade, and break down deleterious material. The shafts can have different configurations, but mainly the idea is to have alternating flights and paddles. Conditioners sit on a slope of 0–5 degrees and have higher capacities than same-size coarse material screw washers. This is caused by the lower machine operating slope. Again, a conditioner is used where the material needs to be worked or wetted before it enters a vibrating screen, fine material screw washer, or other wet processing.

Rotary Scrubbers

Rotary scrubbers are designed primarily for the removal of loam and light clays and soft, soluble waste materials from aggregates, stone, and ore. Scrubbers are not intended to replace log washers in applications where tough plastic clays are contaminating the material to be cleaned. When these tough clays are fed to a scrubber, the clay particles may have a tendency to pelletize through the tumbling action of the material. For example, golf-ball-size pieces of plastic clay may not break down and have been known to increase to the size of a tennis ball.

Scrubbers and coarse material screw washers are often selected to wash out similar contaminants from aggregate and

Figure 7 Chain-driven scrubber

Figure 8 Hydraulic drive scrubber screen

various ores. The major difference is that scrubbers can accept larger feed sizes, whereas coarse material screw washers must have a controlled top size, and scrubbers have much higher throughput capacity. Because large-diameter scrubbers can handle feeds up to 250–300 mm, they are sometimes used as primary washers before any crushing or screening is done.

For long residence times and low energies, a smaller-diameter, longer machine is most effective, whereas for short residence times and high input energies, larger-diameter, shorter machines are more effective.

Capacities are normally determined by analyzing the type of feed material, percent and type of waste material to be removed, and available water; a screen analysis of the feed is also performed. The required retention time in a scrubber will vary for each application because of different types and varying amounts of wastes to be removed. Testing will help establish required guidelines.

Rotary scrubbers may range in diameter from 2.5 to 5.0 m and in length from 2.5 to 18 m, with aspect ratios (L/D) ranging from 1.8 to 3.0. Scrubbers can be sized to process up to 5,000 t/h of ore.

Figure 9 Scrubber internal view with solids pile at dynamic angle of repose

Improved washing can be accomplished by

- Adjusting the water rate to the scrubber,
- Modifying the lifter and liner design,
- Adjusting the discharge weir height (to alter residence time), and
- Altering the rotation speed to vary energy intensity.

The selection of an appropriate drive mechanism depends on the power and variability required. Small units may use a friction, chain, or rubber tire drive system (Figure 7). These styles are limited in power to 300–500 kW and ~3 m in diameter. Larger units may use an end drive stub-mounted hydraulic motor or girth gear with pinion. Power intensity required in most scrubbing duties ranges from 0.2 to 1.0 kW·h/t (Figure 8). Support systems offered include trunnion rollers with steel, rubber, or polyurethane tires or hydrostatic bearing pads similar to ball mills.

Rotary scrubbers work by having process water and solids introduced through the feed chute. As the cylinder rotates, the solid feed fraction is carried up the side of the drum by centrifugal force until its surface reaches the dynamic angle of repose—typically in the range of 35 to 45 degrees. The surface material then cascades and tumbles to the bottom of the solids pile before being carried back under the pile by the rotating drum (Figure 9). Lifter bars on the drum shell ensure that the solids pile cascades and tumbles rather than sliding down the wall. This action helps to break down the softer materials and liberates clays, dirt, loam, and other foreign materials into solution.

For special washing and screening applications, a combination scrubber screen may be used. These units combine the scrubbing characteristics of a rotary scrubber with the screening portion of rotary trommel screens all in one unit (Figure 10). The screen portion can be supplied with either a single- or double-shell design. Scrubber screens can be effective when handling hard-to-wash and hard-to-screen materials, such as oyster shells and phosphate matrix, and offer some benefits in plant layout options.

The maintenance required on rotary drum scrubbers will vary according to the size, hardness, throughput, and

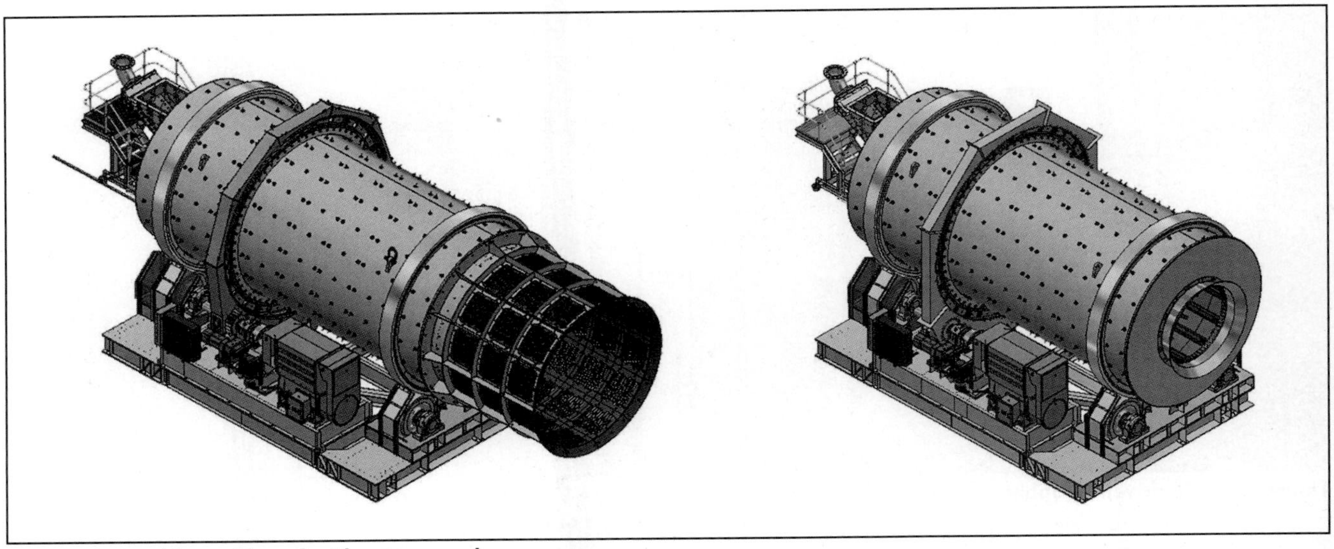

Figure 10 Scrubber with and without trommel screen

abrasivity of material being processed. Replaceable liners on the feed chute, shell, and end wall can be replaced at intervals of 12–36 months.

High-Intensity Washing

Some developments are being considered using high-intensity hydraulic washing drums (Slige 2016). These are suggested to use less water but similar overall power input and are largely adapted from chemical or municipal applications.

ACKNOWLEDGMENTS

Helpful comments and insight have been provided by many of the team from McLanahan Corporation, notably Dave Shellberg, Scott O'Brien, and Ben Freeburn. All figures are courtesy of McLanahan Corporation.

REFERENCES

ASTM D3744. 2003. *Standard Test Method for Aggregate Durability Index*. West Conshohocken, PA: ASTM International.

ASTM D4221. 2017. *Standard Test Method for Dispersive Characteristics of Clay Soil by Double Hydrometer*. West Conshohocken, PA: ASTM International.

ASTM D4318. 2017. *Standard Test Methods for Liquid Limit, Plastic Limit, and Plasticity Index of Soils*. West Conshohocken, PA: ASTM International.

Dictionary.com. 2018. Wash. www.dictionary.com/browse/wash?s=t. Accessed August 2018.

Husaini, H., and Cahyono, S. 2010. Upgrading of Indonesia's bauxite by washing method. In *XXV International Mineral Processing Congress*. Melbourne, Victoria: Australasian Institute of Mining and Metallurgy.

ISO 3271:1995. *Iron Ores—Determination of Tumble Strength*. Geneva: International Organization for Standardization.

Preston, M., and Tatarzyn, J. 2013. Optimising plant efficiency with attrition scrubbers. *Min. Eng.* (October):18–19.

Miller, G.M. 2004. Drum scrubber design and selection. In *Metallurgical Plant Design and Operating Strategies 2004 Proceedings*. Melbourne, Victoria: Australasian Institute of Mining and Metallurgy.

Slige, S. 2016. Iron ore—Improving ore quality. *AusIMM Bull.* (February). www.ausimmbulletin.com/feature/iron-ore-improving-ore-quality/. Accessed July 2018.

CHAPTER 4.6

Partition Curves

Brian Flintoff

The partition curve (also known as a Tromp curve, selectivity curve, efficiency curve, classification curve, or performance curve) is a standard and very useful way to empirically model physical separation systems. This includes separations based on particle size (screens, hydrocyclones, spiral or screw and rake classifiers, air classifiers, deslimers, etc.), particle density (dense medium vessels or cyclones, etc.), particle magnetic properties (magnetic separators), and particle electrostatic properties (electrostatic separators). The partition curve provides an empirically derived description of the separation process within a given system with respect to a specific feed property, such as particle size.

The emergence of multiphysics modeling and simulation tools (e.g., two-way coupled discrete element method and computational fluid dynamics) are leading to more fundamental descriptions based on first principles of the phenomena underpinning the partition curve in these various separators. Nevertheless, the empirical approach will remain as a core tool (1) in process analysis for plant engineers, (2) as an important decision-support tool in process design, and (3) as a key element of most process simulation software.

This chapter provides a relatively in-depth review of the partition curve including details on the data analysis steps. Since the hydrocyclone has a very rich history of study, and because it can exhibit some "unusual" analytical challenges, it will be used as the basis for the descriptive material that follows.

DATA ACQUISITION, SCREENING, AND MASS BALANCES

The construction of a partition curve from plant data requires that a sampling campaign (also known as a survey) be completed. The details around the design of experiments, the identification of good sampling points, the design of proper sample cutters, and the estimation of required sample weights, number of cuts and timing, and so on, are beyond the scope of this discussion. (For more information on these topics, the reader is referred to sources such as Wills and Finch 2016, or, where available, *The SPOC Manual* by Canada Centre for Mineral and Energy Technology [Laguitton 1985]). Suffice to say that, following these principles, two steady-state sampling campaigns were completed, as illustrated in Figure 1.

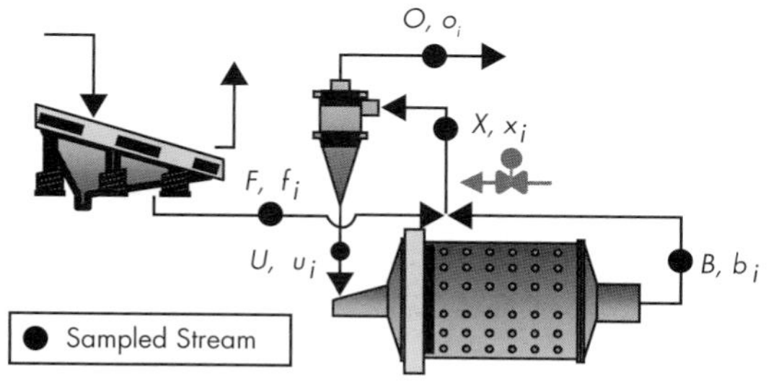

Note: Uppercase is mass flow; lowercase is mass frequency on size.

Figure 1 Sampling experiment in a semiautogenous and ball mill circuit

The grinding circuit in Figure 1 treats a low-grade copper ore with minimal gangue sulfides; that is, it is a relatively homogeneous rock displaying the physical properties of the predominant siliceous gangue. The two campaigns corresponded to plant operation treating both a soft (supergene) ore and a hard (hypogene) ore, and these designations are retained through this chapter. The raw data are presented in Tables 1 and 2.

Notice that a characteristic (geomean, or geometric mean) size has been assigned to the pan fraction, as indicated in boldface in these tables. This is derived by simply assigning a size of $s_{n-1}/3$ to the pan fraction, where s_{n-1} is the finest sieve aperture (often 400 mesh or 37 μm). This approximation, which is explained later, generally gives acceptable results when the partition curve is very well defined by the data, and it expedites the data processing step. However, there is an alternative approach for the pan fraction, which is discussed in the "Challenges with Very Fine Separations" section later in this chapter. In addition, to calculate the geomean size for the coarsest size class, it has been assumed that the upper limiting size (s_0) is 12,500 μm. Since in well-designed experiments the partition factor is always 1.0 in this size range, this assumption or calculation is not very important.

Brian Flintoff, Consultant, Sigmation Inc., Kelowna, British Columbia, Canada

Table 1 Soft ore sampling data

Stream		Screen Undersize (U)	Ball Mill Discharge (B)	Cyclone Feed (X)	Cyclone Overflow (O)	Cyclone Underflow (U)
Mass Flow, t/h		1,298.00				
% Solids		75.0	73.6	61.1	41.3	74.4
Passing Size, µm	Geomean Size, µm	Frequency, % retained	Frequency, % retained	Frequency, % retained	Frequency, % retained	Frequency, % retained
9,525	10,911.6	0.95	2.12	1.71	0.00	6.84
6,700	7,988.6	3.37	2.67	4.56	0.00	5.32
4,750	5,641.4	3.03	2.82	4.96	0.00	4.32
3,350	3,989.0	4.30	3.09	4.95	0.00	4.19
2,360	2,811.8	4.16	3.15	4.53	0.03	4.40
1,700	2,003.0	3.55	2.27	2.59	0.12	3.41
1,180	1,416.3	4.95	3.75	4.20	0.18	5.38
850	1,001.5	4.23	3.96	4.16	0.15	5.52
600	714.1	5.24	6.77	6.58	0.24	8.45
425	505.0	5.65	10.48	9.36	0.70	11.88
300	357.1	5.96	11.81	10.46	4.12	12.06
212	252.2	7.23	9.28	8.83	9.10	8.67
150	178.3	6.24	7.16	6.85	10.45	4.43
105	125.5	6.64	5.91	5.33	10.85	3.73
74	88.1	3.97	3.26	2.62	8.68	1.83
53	62.6	4.00	2.59	2.19	7.12	1.70
38	44.9	2.54	1.91	1.47	5.77	0.92
Pan	**12.3**	23.99	17.00	14.65	42.49	6.95
Checks		100.00	100.00	100.00	100.00	100.00

Table 2 Hard ore sampling data

Stream		Screen Undersize (U)	Ball Mill Discharge (B)	Cyclone Feed (X)	Cyclone Overflow (O)	Cyclone Underflow (U)
Mass Flow, t/h		738.00				
% Solids		59.5	67.0	52.1	29.6	72.1
Passing Size, µm	Geomean Size, µm	Frequency, % retained	Frequency, % retained	Frequency, % retained	Frequency, % retained	Frequency, % retained
9,525	10,911.6	5.20	0.79	1.97	0.00	1.67
6,700	7,988.6	5.91	1.08	2.28	0.00	3.72
4,750	5,641.4	5.78	1.64	2.47	0.00	2.97
3,350	3,989.0	6.74	1.48	2.73	0.00	3.16
2,360	2,811.8	6.80	1.67	2.61	0.00	3.11
1,700	2,003.0	3.11	1.63	2.76	0.00	3.15
1,180	1,416.3	5.18	2.43	3.46	0.00	4.56
850	1,001.5	4.10	2.51	3.14	0.00	4.06
600	714.1	4.88	3.98	4.26	0.19	5.98
425	505.0	5.22	6.39	5.88	0.14	8.52
300	357.1	5.08	10.33	8.58	0.46	12.29
212	252.2	5.13	17.57	13.75	3.24	15.99
150	178.3	5.25	10.81	9.16	11.96	10.61
105	125.5	4.86	9.32	7.97	13.78	6.22
74	88.1	2.72	4.92	4.43	8.68	3.31
53	62.6	2.31	4.43	4.08	7.17	1.86
38	44.9	1.79	2.40	2.31	5.40	1.09
Pan	**12.3**	19.94	16.62	18.16	48.98	7.73
Checks		100.00	100.00	100.00	100.00	100.00

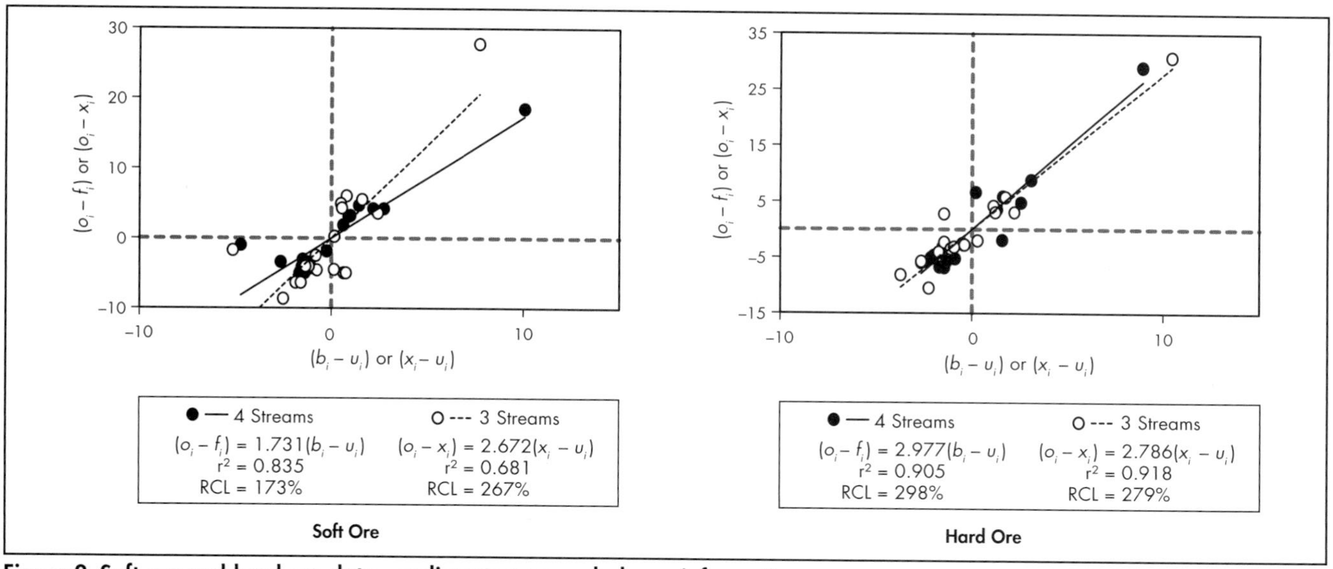

Figure 2 Soft ore and hard ore data—rudimentary mass balance information

Parameter estimation from mass balanced (or massaged) data tends to give better estimates (i.e., minimum variance) than simply using the raw data (Wiegel 1979; Hodouin et al. 1984). Consequently, many operations routinely use the material balance software available in simulation packages (e.g., JKSimMet or USIM PAC) or stand-alone balancing packages (e.g., Algosys Bilmat or Metso MetTools) to process the sample data. In the absence of this kind of software, another less rigorous (and less accurate) method can be used. The authors frequently recommend that this step be conducted in any event, as it gives a good indication of the quality of the data—the so-called data screening step. No mass balance package can "clean" a highly corrupt sample data set—the garbage in, garbage out principle. In such cases, it is usually best to redo the experiment rather than waste a lot of time in data processing and ultimately produce questionable results.

Using the symbology of Figure 1, the basic equation for the experimental partition factor is given in Equation 1. A plot of p_i versus the characteristic size yields the partition curve:

$$p_i = \frac{Uu_i}{Xx_i} = \left[\frac{U}{X}\right]\left[\frac{u_i}{x_i}\right] \quad \textbf{(EQ 1)}$$

where

p_i = partition factor for size class i (the fraction of the feed material in this size class that reports to the underflow or coarse stream)
U = mass flow of solids to the underflow or coarse stream
u_i = mass frequency in size class i in the underflow or coarse stream
X = mass flow of solids in the hydrocyclone feed stream
x_i = mass frequency in size class i in the feed stream

For data screening or to estimate a more rudimentary mass balance (e.g., to deduce the mass split, U/X, in Equation 1), the approach is to develop the basic steady-state mass balance equations for the flow sheet in Figure 1. Since we cannot accommodate "redundant" data, two options are available that relate to the hydrocyclone partition curve. One involves three streams (X, U, and O) and the other involves four streams (F, B, U, and O), and these are treated separately in the following equations.

Three-stream balance:

$$X = U + O \quad \textbf{(EQ 2)}$$

$$Xx_i = Uu_i + Oo_i \quad \textbf{(EQ 3)}$$

therefore,

$$(o_i - x_i) = \frac{U}{O}(x_i - u_i) \quad \textbf{(EQ 4)}$$

Four-stream balance:

$$F + B = U + O \text{ and } F = O \text{ and } B = U \quad \textbf{(EQ 5)}$$

$$Ff_i + Bb_i = Uu_i + Oo_i \quad \textbf{(EQ 6)}$$

therefore,

$$(o_i - f_i) = \frac{U}{O}(b_i - u_i) \quad \textbf{(EQ 7)}$$

where

O = mass flow of solids in the overflow or fine stream
o_i = mass frequency in size class i in the overflow or fine stream
F = mass flow of solids in the screen undersize or fresh feed stream
f_i = mass frequency in size class i in the fresh feed stream
B = mass flow of solids in the ball mill discharge stream
b_i = mass frequency in size class i in the ball mill discharge stream

Plots of Equations 4 and 7 will give a good visualization of the data quality as well as an estimate of the recirculating load (RCL = U/O). Figure 2 shows the results for the data in Tables 1 and 2, and the RCL is reported as both fractional and

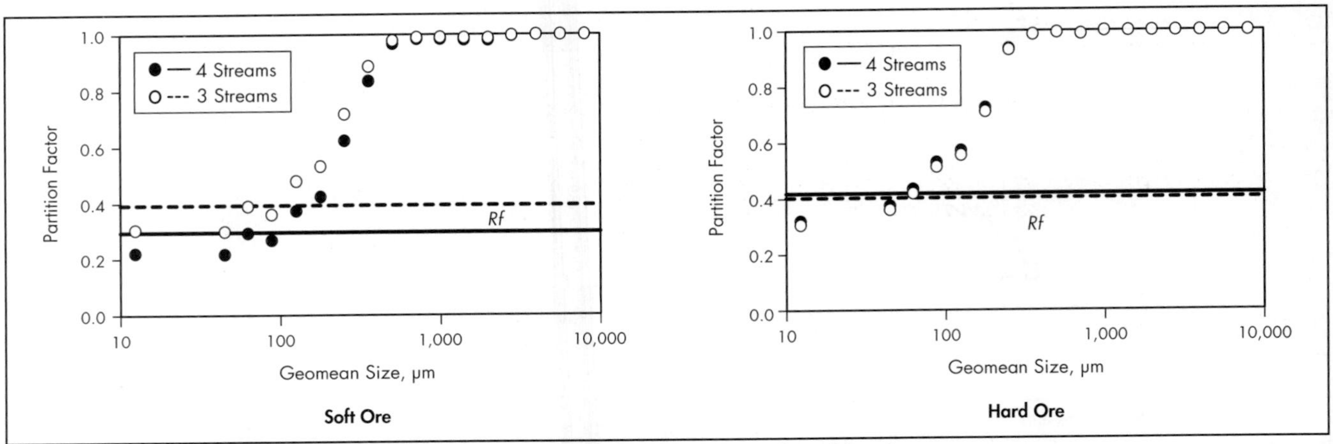

Figure 3 Partition curve results from the rudimentary mass balances for soft and hard ores

percentage values. (Note that for these kinds of plots involving the differences in mass frequency vectors [e.g., $x_i - u_i$], by definition, they must pass through the point [0,0]; hence, the use of a regression that forces the constant to zero.)

It is quite clear from Figure 2 that the hard ore data are of higher quality (better r^2, greater consistency in the recirculating load estimates, visually less data scatter, better agreement of the two methods, etc.). From a data screening point of view, a mass balance will be more difficult on the soft ore and some additional analysis would be required to try to deduce which stream(s) is likely the problem. This stream(s) would then be given less weight in the mass balance solution.

To complete this discussion, the partition factors can be calculated using the (fractional) estimates of the recirculating load value according to Equation 8 and are shown in Figure 3. It is generally good practice to use the underflow and overflow streams for these calculations, as they afford a better definition of the partition curve in the coarse and fine size ranges.

$$p_i = \frac{(\text{RCL})(u_i)}{(\text{RCL})(u_i) + (1)(o_i)} \quad \textbf{(EQ 8)}$$

The parameter *Rf*, shown as the horizontal lines in Figure 3, is discussed later in this chapter, but it is the mass recovery of water from the hydrocyclone feed stream to the underflow stream, or

$$Rf = \frac{(\text{RCL})\left(\frac{100}{u'} - 1\right)}{(\text{RCL})\left(\frac{100}{u'} - 1\right) + (1)\left(\frac{100}{o'} - 1\right)} \quad \textbf{(EQ 9)}$$

where

u' = % solids in the underflow stream
o' = % solids in the overflow stream

For a homogeneous ore in a typical grinding circuit application, the partition factor in the very fine sizes should be quite comparable to *Rf* (i.e., the fine particles behave as part of the fluid). From Figure 3, it is again clear that the hard ore data conform better to this expectation than do the soft ore data.

A detailed discussion on the mechanics of mass balance solutions is beyond the scope of this chapter. (The interested reader could refer to some of the foundational works [e.g., Mular 1979 or Hodouin et al. 1981].) Suffice to say that the adjusted data satisfy all of the mass balance equations (e.g., Equations 2, 3, 5, and 6) and in such a way that the adjustments on all measured data are minimized, in the weighted least-squares sense. Tables 3 and 4 present the mass balance results for the raw data in Tables 1 and 2, respectively. The partition curves for the mass balanced data are shown in Figure 4.

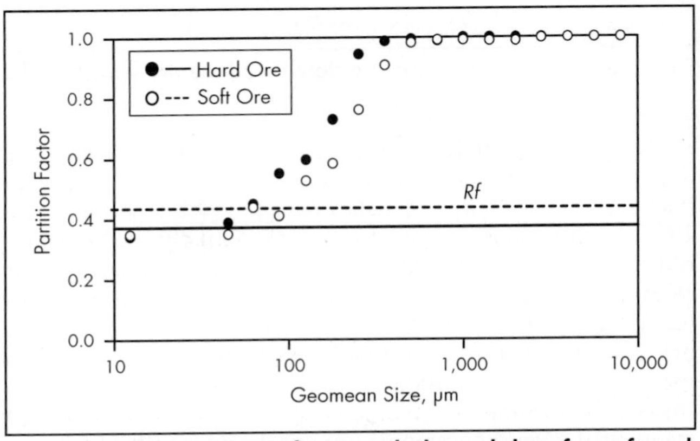

Figure 4 Partition curves for mass balanced data for soft and hard ores

To conclude this section, Figure 5 compares the partition curves from the rudimentary data to those from the mass balanced data. It is clear that for the soft ore (the more problematic data set), significant adjustments in the raw data were made, and even in this case there are still inconsistencies. However, for the hard ore (the better-quality data set), there is generally good agreement between the rudimentary and mass balanced data. This demonstrates the importance of good data screening, followed by mass balancing, to fully understand the quality of the resulting partition curve analysis.

Since the remaining analysis in this chapter utilizes the mass balanced data, henceforth the output from the mass balancing exercise is referred to as simply the *experimental data*.

PARTITION CURVE MODELING AND CURVE FITTING

The preceding partition curve data are interesting in the sense that one can qualitatively assess performance based on factors such as the amount of fine material bypassed to the underflow stream (*Rf*), the sharpness of the separation (e.g., the steepest

Table 3 Mass balance results for soft ore

Stream		Screen Undersize (U)	Ball Mill Discharge (B)	Cyclone Feed (X)	Cyclone Overflow (O)	Cyclone Underflow (U)
Mass Flow, t/h		1,298.00	4,022.39	5,320.39	1,298.00	4,022.39
% Solids		75.0	73.7	61.8	41.1	73.7
Passing Size, μm	**Geomean Size, μm**	**Frequency, % retained**	**Frequency, % retained**	**Frequency, % retained**	**Frequency, % retained**	**Frequency, % retained**
9,525	10,911.6	0.96	2.29	1.97	–	2.60
6,700	7,988.6	3.70	3.29	3.39	–	4.48
4,750	5,641.4	3.23	3.32	3.30	–	4.36
3,350	3,989.0	4.46	3.32	3.60	–	4.76
2,360	2,811.8	4.33	3.41	3.64	0.03	4.80
1,700	2,003.0	3.60	2.31	2.63	0.12	3.43
1,180	1,416.3	5.06	3.91	4.19	0.18	5.48
850	1,001.5	4.32	4.16	4.20	0.15	5.51
600	714.1	5.34	7.13	6.69	0.24	8.77
425	505.0	5.74	11.05	9.75	0.70	12.67
300	357.1	6.05	12.43	10.87	4.09	13.06
212	252.2	7.37	9.68	9.12	8.95	9.17
150	178.3	6.15	6.62	6.50	11.09	5.02
105	125.5	6.50	5.44	5.70	11.05	3.97
74	88.1	3.86	3.00	3.21	7.75	1.74
53	62.6	3.85	2.37	2.73	6.28	1.59
38	44.9	2.42	1.69	1.87	4.96	0.87
Pan	**12.3**	23.07	14.58	16.65	44.40	7.70
Checks		100.00	100.00	100.00	100.00	100.00

Table 4 Mass balance results for hard ore

Stream		Screen Undersize (U)	Ball Mill Discharge (B)	Cyclone Feed (X)	Cyclone Overflow (O)	Cyclone Underflow (U)
Mass Flow, t/h		738.00	2,355.85	3,093.85	738.00	2,355.85
% Solids		59.5	69.2	52.4	29.5	69.2
Passing Size, μm	**Geomean Size, μm**	**Frequency, % retained**	**Frequency, % retained**	**Frequency, % retained**	**Frequency, % retained**	**Frequency, % retained**
9,525	10,911.6	4.45	0.73	1.62	–	2.13
6,700	7,988.6	6.34	1.12	2.37	–	3.11
4,750	5,641.4	5.48	1.55	2.49	–	3.27
3,350	3,989.0	6.51	1.44	2.65	–	3.48
2,360	2,811.8	6.23	1.55	2.67	–	3.50
1,700	2,003.0	3.37	1.85	2.21	–	2.91
1,180	1,416.3	5.44	2.60	3.28	–	4.30
850	1,001.5	4.22	2.64	3.01	–	3.96
600	714.1	4.98	4.14	4.34	0.19	5.64
425	505.0	5.27	6.49	6.20	0.14	8.10
300	357.1	5.11	10.41	9.15	0.46	11.87
212	252.2	5.14	16.78	14.00	3.24	17.37
150	178.3	5.33	11.46	9.99	11.34	9.57
105	125.5	4.87	9.15	8.13	13.74	6.37
74	88.1	2.74	5.05	4.50	8.47	3.26
53	62.6	2.27	3.87	3.49	8.05	2.06
38	44.9	1.78	2.33	2.20	5.64	1.13
Pan	**12.3**	20.46	16.83	17.70	48.73	7.98
Checks		100.00	100.00	100.00	100.00	100.00

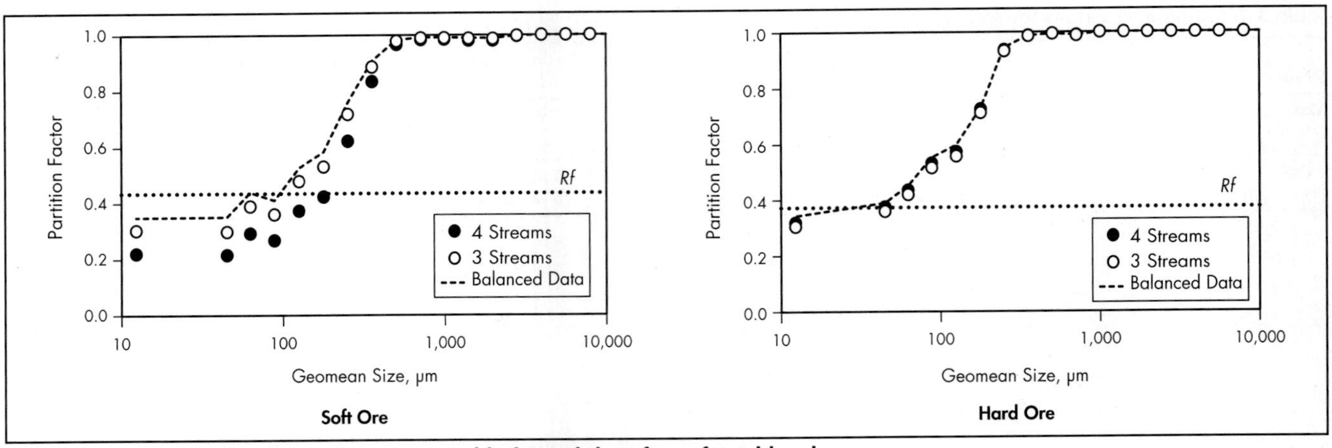

Figure 5 Partition curves for rudimentary and balanced data for soft and hard ores

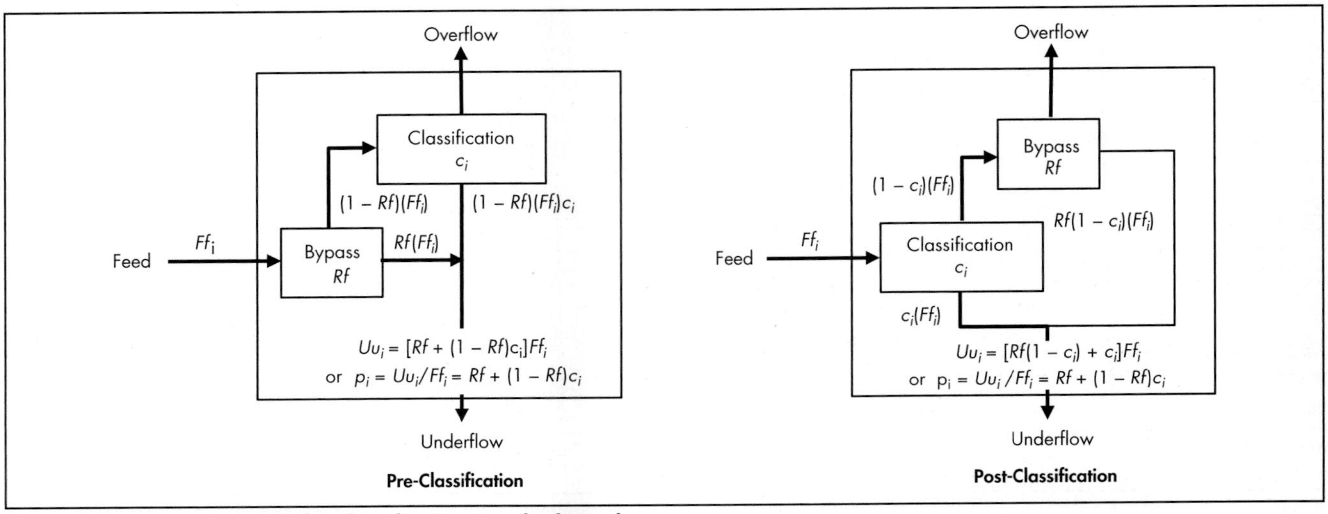

Figure 6 Conceptual models for classification in a hydrocyclone

slope of the partition curve), and effective separation or cut size (e.g., the size at which a particle has a 50% probability of reporting to the underflow stream). Quantitatively, mathematical models are a useful vehicle for data reduction, as the model parameters can provide direct measures of the performance metrics.

Usually there is some conceptual or phenomenological notion(s) that underpins the form of a mathematical model. In the case of the partition curve for a hydrocyclone, the two conceptual models that have been cited are illustrated in Figure 6. Here, a fraction (Rf) of the feed solids (pre-classification) or the overflow solids (post-classification) is bypassed to the underflow stream.

To briefly digress, the preceding conceptual models do not account for another mechanism that is sometimes seen in cyclone performance—the short-circuiting of coarse material across the upper regions of the inlet and into the overflow stream. This mechanism was more common with older cyclone inlet designs and in cases involving vortex finder wear/length issues. The experimental partition curve can give an indication that this may be a performance issue, for example, if the partition factors in the coarser size ranges do not approach the value of 1.0 as quickly as might be expected in this size range (e.g., an apparent offset). Normally this is not an important factor, at least from a partition modeling perspective. However, original equipment manufacturers have noted that new inlet designs effectively eliminate this problem.

Regardless of the way one chooses to think about it, the mass balance calculations in Figure 6 show that the general form of the hydrocyclone partition curve model is

$$\hat{p}_i = Rf + (1 - Rf)c_i \quad \textbf{(EQ 10)}$$

where

$\hat{p}_i$ = predicted partition factor value from the model
Rf = mass fraction of water in the feed stream that reports to the underflow stream
c_i = "corrected" classification partition factor value (i.e., corrected for the bypass, Rf)

As the preceding partition curve figures show, the curve looks sigmoidal in nature, so it is no surprise that quite a number of functional forms have been used to describe c_i. A few of these are summarized in Equations 11 through 16, but the two most common forms are the Rosin–Rammler function used by Plitt (Equation 11—see Plitt 1971; the equation below is shown in a slightly different form than that of Plitt) and the exponential sum function used by Lynch–Rao (Equation 12—see Lynch and Rao 1975). Equations 13 and 14 are variations on the logistic function, which has been used in electrostatic separation (Wei and Realff 2005), Equation 15 is the log normal function, and Equation 16 is an arctangent function that has been used for coal partition curves (De Lange and Venter 1987).

$$c_i = 1 - 0.5^{\left(d_i / d50c\right)^{\alpha}} \quad \textbf{(EQ 11)}$$

$$c_i = \frac{\exp\left(\alpha\left[d_i / d50c\right]\right) - 1}{\exp\left(\alpha\left[d_i / d50c\right]\right) + \exp(\alpha) - 2} \quad \textbf{(EQ 12)}$$

$$c_i = \frac{1}{1 + \left(d_i / d50c\right)^{-\alpha}} \quad \textbf{(EQ 13)}$$

$$c_i = \frac{1}{1 + \exp(-\alpha(d_i - d50c))} \quad \textbf{(EQ 14)}$$

$$c_i = 0.5 + 0.5\left[\mathrm{ERF}\left(\frac{(\ln(d_i) - \ln(d50c))}{\sqrt{2}\,\ln(\alpha)}\right)\right] \quad \text{for } d_i > d50c$$

$$c_i = 0.5 - 0.5\left[\mathrm{ERF}\left(\frac{(\ln(d50c) - \ln(d_i))}{\sqrt{2}\,\ln(\alpha)}\right)\right] \quad \text{for } d_i \le d50c \quad \textbf{(EQ 15)}$$

$$c_i = 0.5\left(\arctan\left[\alpha\left(d_i - d50c\right) + 1/\pi\right]\right) \quad \textbf{(EQ 16)}$$

where

d_i = characteristic size of size class i (defined in the following paragraph)

$d50c$ = corrected cut size parameter (i.e., the size at which $c_i = 0.5$)

α = sharpness-of-separation parameter in most of the preceding cases

ERF = Excel error function

Just as many models are used to describe c_i, there are different ways to define the average (e.g., mass mean) or characteristic size of a size class. The convention is to number the size classes from the coarsest ($i = 1$) to the finest ($i = n$), and Equations 17 through 19 define the three most common characteristic sizes: geomean, arithmetic mean, and bottom size (or top size), respectively. The most commonly encountered is the geomean size, which is what has been used in this chapter.

$$d_i = \sqrt{s_{i-1}s_i} \quad \textbf{(EQ 17)}$$

$$d_i = \frac{(s_{i-1} + s_i)}{2} \quad \textbf{(EQ 18)}$$

$$d_i = s_i \quad \textbf{(EQ 19)}$$

where

s_i = lower size limit for size class i (s_{i-1} is the upper limit)

Common practice is to employ a $\sqrt{2}$ series (i.e., $s_{i-1} = \sqrt{2}\,s_i$) for selecting the sieve sizes, as was the case for the preceding soft and hard ore data. An approximation, such as $d_{pan} \approx s_{n-1}/3$ is used for the pan fraction.

Table 5 Model fits for the hard ore data

Model for c_i	RSS	α	$d50c$	Rf
Equation 11	5.73E-03	1.84	144	0.35
Equation 12	5.50E-03	2.27	143	0.33
Equation 13	1.15E-02	2.78	142	0.37
Equation 14	5.50E-03	0.02	129	0.25
Equation 15	1.08E-02	1.85	141	0.37
Equation 16	1.62E-02	0.02	133	0.27

Fitting the model (Equation 10) to the raw or balanced data is a nonlinear least-squares problem—pick the parameters that minimize RSS below—that can be achieved using tools like Solver in Excel. The general weighted least-squares objective function is given by the following equation:

$$\mathrm{RSS} = \sum_{i=1}^{n} w_i[p_i - \hat{p}_i]^2 \quad \textbf{(EQ 20)}$$

where

RSS = residual sum of squares (unexplained variation in the experimental p_i data)

w_i = weighting factor for size class i (generally an unweighted RSS is used where all $w_i = 1$)

Equation 10 can represent a two-parameter (α, $d50c$) or three-parameter (α, $d50c$, Rf) estimation problem, although if the raw data are truly coherent, the two-parameter model is usually selected and Rf is computed from the mass balance results for water. (In this context, *coherent* means the degree to which the raw data satisfy the mass balance equations. For example, from Figure 2, the hard ore data are more coherent than the soft ore data. Mass balancing renders all the data coherent.) Nevertheless, it is often instructive to fit both forms of the model and then test to see if there is a statistically significant difference in the goodness of fit.

Table 5 presents the fits of the various three-parameter models to the hard ore data. All of the models do a reasonable job of describing the partition curve from the hard ore experimental data. However, only a couple of them provide reasonable agreement with the experimental Rf value and demonstrate the best fits. Specifically, these are the Plitt (Equation 11) or the Lynch–Rao (Equation 12) expressions for c_i, which are then used in Equation 10. For these two equations, the $d50c$ value is very similar, but the α values differ. This difference in the sharpness of separation is consistent for these two models (i.e., the Equation 12 α value is always slightly higher). That is to say, the α's cannot be compared between these two models, so once a functional form for c_i has been chosen, it is best to keep with it. In this chapter, the author uses Equation 11 for c_i, giving Equation 21 as the final partition curve model:

$$\hat{p}_i = Rf + (1 - Rf)\left[1 - 0.5^{\left(d_i / d50c\right)^{\alpha}}\right] \quad \textbf{(EQ 21)}$$

Figure 7 shows the fits of the model in Equation 21 to the experimental partition curve data from the mass balances in Tables 3 and 4. Generally, the goodness of fit is reasonable in both cases, but for the soft ore there is a significant discrepancy in the experimental Rf (from the water balance—0.435) and model Rf (from the size data—0.352) values. As a side note, in the case where the percentage of solids data is not available, or is not used in the mass balance (e.g., obviously

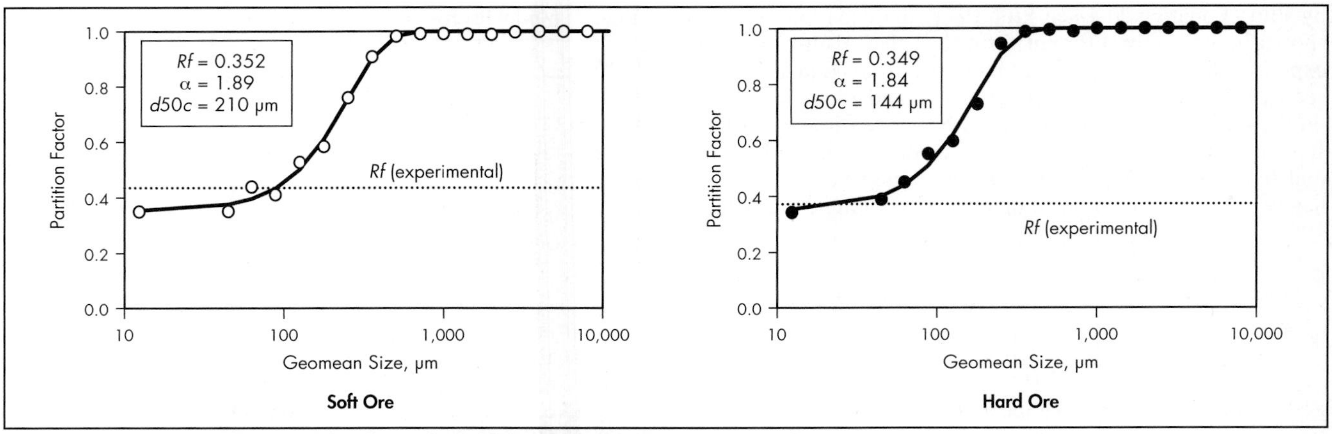

Figure 7 Partition curve model fits for soft and hard ores using experimental data

corrupt measurements), then *Rf* must be estimated from the particle size data.

Turning briefly to the statistical comparison of two- versus three-parameter models, one can apply the Partial F test. In general, the Partial F test compares calculated and tabulated F statistics to deduce whether estimating additional parameters in a model is statistically justified. Equations 22 and 23 show the two *F* statistics:

$$F_{\text{tabulated}} = F(0.05, k_2 - k_1, n - k_2) \quad \textbf{(EQ 22)}$$

$$F_{\text{calculated}} = \left[\frac{\dfrac{\text{RSS}(k_1) - \text{RSS}(k_2)}{(k_2 - k_1)}}{\dfrac{\text{RSS}(k_2)}{(n - k_2)}}\right] \quad \textbf{(EQ 23)}$$

where

k_1 = number of parameters in the most parsimonious model (here, $k_1 = 2$)

k_2 = number of parameters in the "new" model (here, $k_2 = 3$)

n = number of data points to be fit (here, $n = 18$)

In the case of the hard ore, the unweighted residual sum of squares values are RSS(2) = 6.34×10^{-3} and RSS(3) = 5.73×10^{-3}. Applying the preceding equations, $F_{\text{calculated}} = 1.58$ and $F_{\text{tabulated}} = 4.54$, and since the latter is greater than the former, there is no statistical justification for estimating *Rf* from the size data over simply using the experimental value for the hard ore data. That is the expected outcome when the raw data are coherent.

However, for the soft ore, the situation is reversed, with RSS(2) = 1.97×10^{-2} and RSS(3) = 4.79×10^{-3}, giving $F_{\text{calculated}} = 46.8$ and $F_{\text{tabulated}} = 4.54$. In this case, the estimation of the *Rf* parameter leads to a statistically superior fit. Such an outcome requires additional investigative work to determine if the data set can be used or whether the experiment must be repeated. As an example, it is sometimes the case that in sampling a hydrocyclone underflow stream, the cutter overfills and fine material is washed out—that is, the sample is biased coarse. This could be one reason that the partition curve model estimate of *Rf* is lower than the experimental value from the water balance. However, one might also expect the underflow sample % solids to also be biased high in this case, which would likely serve to reduce the experimental *Rf* value. It is a complex forensic problem, and anecdotal information from the persons cutting the samples *may* help rule such factors in or out. Clearly, if the preceding soft ore results are retained, they would have to be discounted relative to the hard ore results.

Table 6 Parameter standard error estimates

Parameter	Value	Standard Error	Coefficient of Variation, %
Rf	0.349	0.018	5.2
α	1.84	0.164	8.9
d50c	144	5.950	4.1
RSS	5.73E-03	—	—

Table 7 Parameter correlation

	Rf	α	*d50c*
Rf	1	0.71	0.77
α	0.71	1	0.61
d50c	0.77	0.61	1

Continuing with statistical considerations, the generalized nonlinear regression packages often provide estimates of parameter standard errors and of covariance/correlation matrices, which can be quite informative. Add-ins are available for Excel (e.g., de Levie 1999), which the chapter author would recommend to the reader. For illustrative purposes, Tables 6 and 7 present the standard errors for the three-parameter model and the parameter correlation matrix, respectively. The approximate 95% confidence limits on the parameters are approximately ±2 times the standard error. For example, the approximate 95% confidence limits on *d50c* are from 132 to 156 µm. The value of these standard errors is just to quantify the effects of data scatter on parameter precision. While it may be an arcane observation, these confidence limits need to be used with caution if one wishes to define, say, a parameter space (i.e., a "rectangular region" defining expected parameter values). As discussed briefly in the text that follows, the significant covariance of these parameters distorts the joint confidence limits.

As is usually the case with nonlinear models, there is a relatively high degree of correlation in the parameters, as

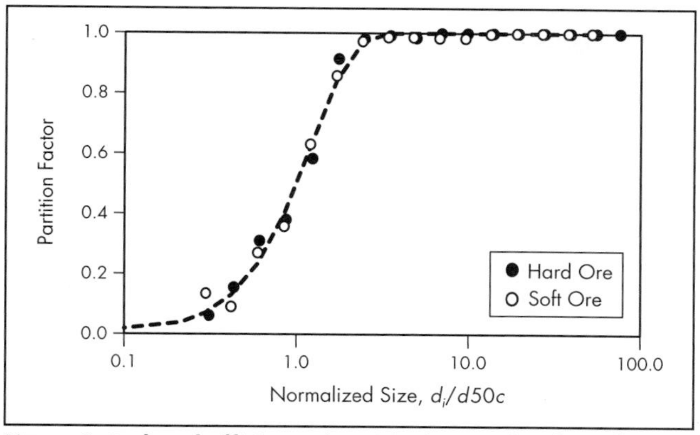

Figure 8 Reduced efficiency (partition) curve for the soft and hard ore balanced data

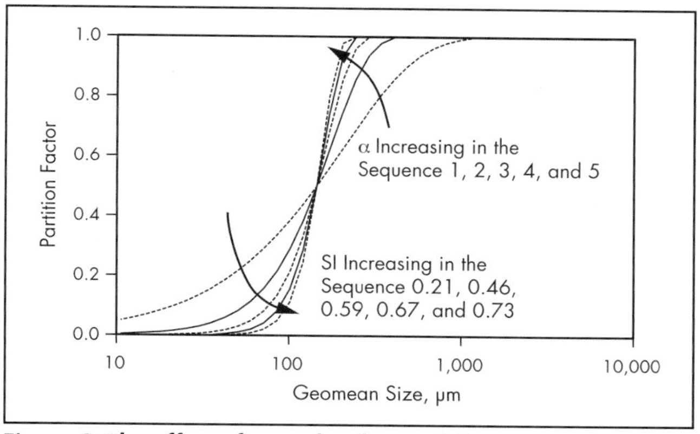

Figure 9 The effect of α on the shape of the corrected partition curve (Equation 11)

changing one will almost certainly require a compensating change in another to maintain a reasonable fit. Consequently, irregularities in one or two data points can have a deleterious effect on all parameter estimates. For example, and with reference to Table 7, if the size data in the very fine range drives an increase in *Rf*, one would observe increases in both α and *d*50*c* to compensate. The point to retain is that these parameters are not independent, and there is a series of combinations of these parameters that will give about the same results when the data are noisy.

To conclude this section, it is good to reflect on the findings of many researchers working on hydrocyclones. Corrected partition curves (c_i) can be normalized with size ($d_i/d50c$ on the x-axis) to form what is referred to as the reduced efficiency curve. (The c_i can be calculated from experimental data by rearranging Equation 10 and using the experimental partition factors, p_i.) For most normal operating regimes, and with geometrically similar hydrocyclones, the reduced efficiency (partition) curve retains the same shape (α is constant, although *d*50*c* and *Rf* can vary significantly). The reduced efficiency curves for the experimental data on both the soft and hard ores are shown in Figure 8.

The common empirical models of the hydrocyclone, as described in the main body of this chapter, have a series of correlations that allow one to estimate *d*50*c* and *Rf*, based on geometrical and operating inputs. These correlations then define the partition curve, which is at the heart of the classification process in this predictive mode.

MEASURES OF EFFICIENCY

In the preceding discussion, the parameter α was defined as the sharpness of separation for the more common corrected partition curve models of Equations 11 and 12. While α represents an important metric for separation efficiency and is explored in a little more detail in this section, it should be observed that *Rf* is also, in a way, a measure of efficiency. Following the thinking of the post-classification model in Figure 6, *Rf* can be seen as a measure of the fines that have been misplaced into the underflow stream. *Rf* is a complex function of many operating and design variables, but values in the range of 0.15 to 0.3 are preferred. Nevertheless, values as high as 0.4–0.5 are sometimes observed in practice. In terms of achieving an efficient separation process, the minimization of *Rf* is equally as important as maximizing α.

For hydrocyclones, Equation 11 α values in the range of 1.5 to 3 are quite common, and as Figure 9 shows, higher values mean sharper separations. As long as the same c_i model is being used, one can compare α's to compare separation efficiencies. Theoretically, perfect separation occurs for $\alpha = \infty$, but in practice, values of 5 or 6 (often seen for coarse vibrating screens) are near perfect.

A few other efficiency metrics were perhaps more common in the past, and they are reproduced for convenience. All are based on a few sizes that are easily extracted from the c_i model used in Equation 21. For these sizes for which $\hat{c} = 0.75$, hereafter termed *d*75*c*, and $\hat{c} = 0.25$, hereafter termed *d*25*c*.

Imperfection, I, is defined as

$$\mathrm{I} = \frac{[d75c - d25c]}{(2)(d50c)} \qquad \textbf{(EQ 24)}$$

Lower values of I imply sharper separations. Typical values for hydrocyclones would lie in the range of $0.25 \le \mathrm{I} \le 0.6$, corresponding to $3.1 \ge \alpha \ge 1.3$. For vibrating screens, these values would be $0.12 \le \mathrm{I} \le 0.2$, corresponding to $6.5 \ge \alpha \ge 3.9$.

The probable error of separation, or ecart probable moyen, Ep, is defined as

$$\mathrm{Ep} = \frac{[d75c - d25c]}{(2)} \qquad \textbf{(EQ 25)}$$

Lower values of Ep imply sharper separations. It is apparent that Ep = (I)(*d*50*c*), which implies that the two will show essentially identical trends as a function of α. In that regard, the dimensionless nature of I may make it a better metric for comparative purposes.

The sharpness index, SI, is defined as

$$\mathrm{SI} = \frac{d25c}{d75c} \qquad \textbf{(EQ 26)}$$

Higher values of SI imply a sharper separation, and perfect separation occurs at SI = 1. Typical values for hydrocyclones would lie in the range of $0.3 \le \mathrm{SI} \le 0.6$, corresponding to $1.3 \le \alpha \le 3.1$. For vibrating screens, these values would be $0.6 \le \mathrm{SI} \le 0.8$, corresponding to $3 \le \alpha \le 6.5$.

For illustrative purposes, the SI values were calculated for the various c_i curves and are also shown in Figure 9. To conclude this section, the three preceding metrics were computed for the hydrocyclones treating the soft and hard ores, as reported in Figure 10. These can be considered typical values.

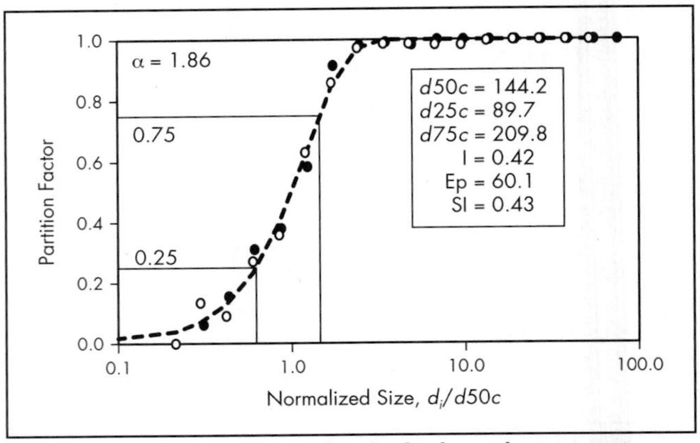

Figure 10 Efficiency metrics for the hydrocyclones treating soft and hard ores

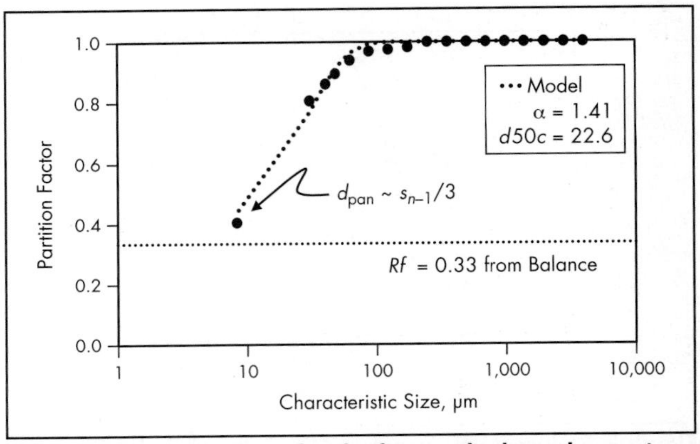

Figure 12 Partition curve for the fine-cut hydrocyclone using the d_{pan} approximation

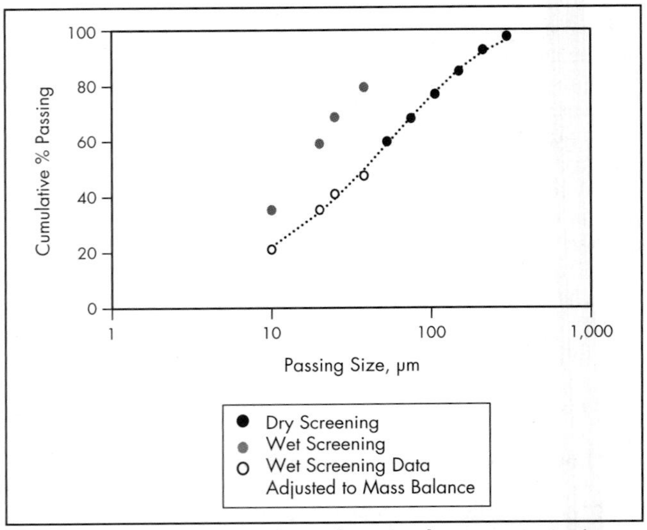

Figure 11 Dry and wet size analysis (with micro-sieving) to determine the full PSD curve

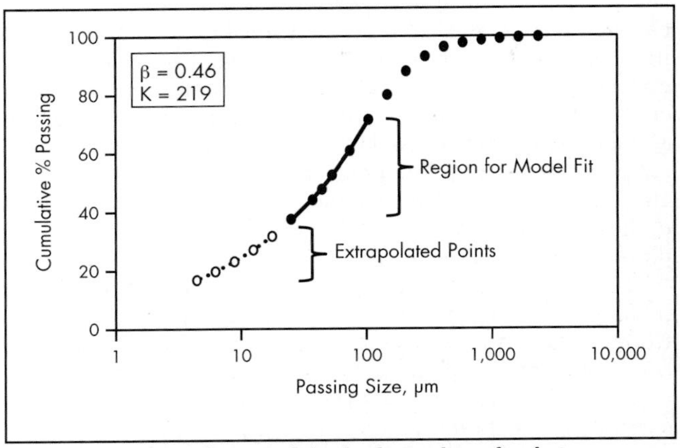

Figure 13 Extrapolation of the hydrocyclone feed size distribution using the Austin and Klimpel method

CHALLENGES WITH VERY FINE SEPARATIONS

In this section, two of the common challenges for very fine separations are addressed. The first is the measurement of particle size itself. Depending on the equipment available, one might be able to use a single device for measuring the particle size distribution, or PSD (e.g., a laser scattering system). However, in some cases one is forced to use sieves down to 400 mesh (37 μm) or 500 mesh (25 μm), combining both dry and wet sieving techniques, and then move to a completely different method such as sedimentation or cyclosizing, or even laser-based techniques. Changing methods means a change in the way the particle diameter is estimated, and that can require some manipulation to get a consistent estimate of the complete PSD. Herbst and Sepulveda (1985) have presented a methodology for making such an adjustment. For the purposes of this discussion, the author recommends that when sieves are employed for the coarser particle sizes, micro-sieving (same separation principles) be employed for the finer sizes whenever possible.

To illustrate, in Figure 11, dry screening was used down to 270 mesh (53 μm), followed by wet screening on the usual sieves for 400 mesh and 500 mesh, and finally wet micro-sieving for the very fine sizes (20 μm and 10 μm). In the figure, the dry and wet screening data are shown separately as solid circles, and the wet data are then adjusted for the sizing mass balance yielding the full PSD. To show the consistency, a Rosin–Rammler model (dotted line in Figure 11) was fit to the full size range, and it is clear that use of dry and wet sieving to very fine sizes essentially requires no additional manipulations.

The second challenge arises when the particle size analysis does not extend to the finer sizes, that is, where $d50c < s_{n-1}$ (the upper limiting size of the pan fraction). This can best be illustrated by example. The data in Table 8 are from a hydrocyclone making quite a fine cut and where the size analysis work was terminated at 25 μm. The geomean size d_1 is calculated by assuming the upper limiting size, $s_0 = \sqrt{2}s_1$.

Figure 12 shows the experimental and modeled partition curve results for the data of Table 8. The model parameters are presented in the figure. It is clear that, with the exception of the point at $d_{pan} = s_{n-1}/3$, the partition curve is only well defined at the coarser sizes. Clearly, the use of this approximation for d_{pan} helps, but the lack of definition of the curve in

Table 8 Mass balance for fine-cut hydrocyclone

Stream	Cyclone Feed (X)	Cyclone Overflow (O)	Cyclone Underflow (U)		
Relative Solids Mass Flow, %	100.00	25.76	74.24		
% Solids	52.41	29.89	70.96		
Passing Size, µm	**Frequency, % retained**	**Frequency, % retained**	**Frequency, % retained**	**Geomean Size, µm**	**Partition Factor**
3,327	0.06	—	0.08	3,956.5	1.00
2,362	0.20	—	0.26	2,803.3	1.00
1,651	0.19	—	0.26	1,974.8	1.00
1,168	0.37	—	0.50	1,388.7	1.00
833	0.50	—	0.67	986.4	1.00
589	0.94	—	1.27	700.5	1.00
417	1.42	—	1.91	495.6	1.00
295	3.16	—	4.26	350.7	1.00
208	5.09	—	6.86	247.7	1.00
147	8.09	0.54	10.71	174.9	0.98
104	8.61	0.82	11.31	123.6	0.98
74	10.55	1.24	13.78	87.7	0.97
53	8.31	1.94	10.52	62.6	0.94
44	4.78	1.94	5.77	48.3	0.90
37	3.59	1.93	4.17	40.3	0.86
25	6.59	4.94	7.16	30.4	0.81
Pan	37.56	86.66	20.51	8.3	0.41
Checks	100.00	100.00	100.00	—	—

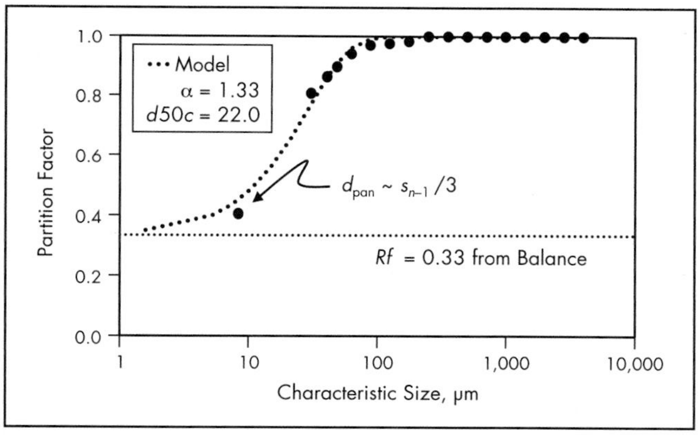

Figure 14 Partition curve for the fine-cut hydrocyclone using the Austin and Kimpel method

this region is a concern in terms of the quality of the parameter estimates.

Austin and Klimpel (1981) developed a clever scheme to analyze classifier data, and this includes the information carried by the pan fraction results. The authors use this method, although only for the pan fraction when $d50_c$ is less than s_{n-1}, as explained below.

Austin and Klimpel noted that the finer size regions of the hydrocyclone feed PSD can usually be well modeled using the Gates–Gaudin–Schuhmann equation, given by Equation 27. By extending the sizes into the very fine regions (say, down to 5 µm or smaller), one can estimate the cumulative fraction passing these finer sizes and, hence, the frequency retained on size data. In this case, a $\sqrt{2}$ series was selected to extend the hydrocyclone feed size distribution from the original s_{n-1} of 25 µm to include the following sizes: 17.7, 12.5, 8.8, 6.3, and 4.4, all in micrometers.

$$y = \left[\frac{s}{K}\right]^{\beta} \quad \text{(EQ 27)}$$

where

y = cumulative fraction passing
s = passing size
K = size modulus
β = distribution modulus

Figure 13 illustrates this process, showing the region over which the model parameters for Equation 27 are estimated and the subsequent results of the extrapolation.

If it is assumed that m new sizes have been added (here, $m = 5$), then the solution for the pan fraction partition factor is given in Equation 28:

$$\hat{p}_{pan} = \sum_{j=n}^{n+m}(\hat{p}_j f_j) \Big/ \sum_{j=n}^{n+m}(f_j) \quad \text{(EQ 28)}$$

Table 9 summarizes the calculations for the balanced data in Table 8, and Figure 14 shows the experimental and modeled partition data. Comparing the parameter values in Figures 12 and 14, it would seem that the approximation of $d_{pan} \sim s_{n-1}/3$ gives reasonable results. (Note that the new pan characteristic size in Table 9 is also estimated as $s_{n+m-1}/3$.)

Based on the Austin and Klimpel method mentioned previously, one can explain the approximation for the

Table 9 Results of applying the Austin and Klimpel technique

Passing Size, μm	Geomean Size, μm	Cumulative % Passing	Frequency, % retained	Extended Model Partition	Experimental Partition	Model Partition
3,327	3,956	99.94	0.06	1.00	1.00	1.00
2,362	2,803	99.75	0.20	1.00	1.00	1.00
1,651	1,975	99.55	0.19	1.00	1.00	1.00
1,168	1,389	99.18	0.37	1.00	1.00	1.00
833	986	98.68	0.50	1.00	1.00	1.00
589	700	97.74	0.94	1.00	1.00	1.00
417	496	96.32	1.42	1.00	1.00	1.00
295	351	93.16	3.16	1.00	1.00	1.00
208	248	88.07	5.09	1.00	1.00	1.00
147	175	79.98	8.09	1.00	0.98	1.00
104	124	71.37	8.61	1.00	0.98	1.00
74	88	60.83	10.55	0.99	0.97	0.99
53	63	52.52	8.31	0.96	0.94	0.96
44	48	47.74	4.78	0.91	0.90	0.91
37	40	44.15	3.59	0.86	0.86	0.86
25	30	37.56	6.59	0.77	0.81	0.77
17.7	21.0	31.69	5.86	0.65	**0.41**	**0.45**
12.5	14.9	27.05	4.64	0.56	Model Fit	
8.8	10.5	23.09	3.96	0.49	*Rf*	0.33
6.3	7.4	19.71	3.38	0.44	*m*	1.33
4.4	5.3	16.82	2.89	0.40	*d50c*	22.0
New pan	1.5		16.82	0.35	RSS	4.95E-03

Table 10 Air classifier partition data

Geomean Size, μm	Experimental Partition Factors	Finch Model Results	Austin et al. Model Results
210.49	1.00	1.00	1.00
148.74	1.00	1.00	1.00
104.88	1.00	1.00	1.00
74.46	1.00	1.00	1.00
52.65	0.96	0.97	0.97
37.23	0.77	0.77	0.77
26.32	0.50	0.50	0.52
18.58	0.41	0.39	0.38
13.14	0.30	0.35	0.33
9.23	0.36	0.36	0.35
6.53	0.41	0.37	0.40
4.69	0.43	0.43	0.44
1.33	0.46	0.46	0.46

Data adapted from Austin et al. 1984

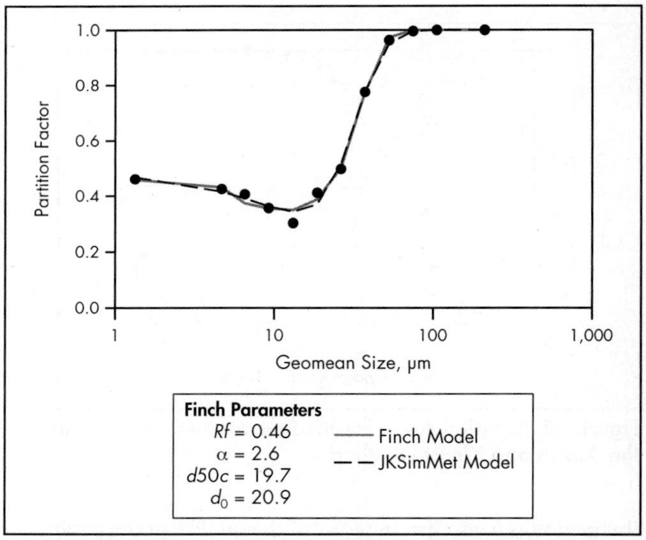

Figure 15 Partition curve data for the air classifier showing the Finch model parameters

characteristic size of the pan fraction, that is, $d_{pan} \sim s_{n-1}/3$, introduced earlier. The frequency form of the Gates–Gaudin–Schuhmann function in Equation 27 is given by Equation 29:

$$\frac{dy}{ds} = y' = \left[\frac{1}{K}\right]^{\beta}\left[\beta s^{\beta-1}\right] \qquad \text{(EQ 29)}$$

The mass mean size for the pan fraction can then be estimated as

$$d_{pan} \approx \left[\int_0^{s_{n-1}} sy'ds \Big/ \int_0^{s_{n-1}} y'ds \right] = \left[\frac{\beta}{1+\beta}\right] s_{n-1} \qquad \text{(EQ 30)}$$

Since the β values in the fine size regions of the hydrocyclone feed PSD often lie in the range of 0.45 to 0.55, assuming an average value of 0.5 yields the approximation $d_{pan} \sim s_{n-1}/3$.

As a rule of thumb, as long as the experimental partition factor for the size class n (p_n) is, say, less than 0.45 or so, using the simple approximation for d_{pan} is acceptable. For values of p_n greater than this level, one should strongly consider applying the Austin and Kimpel method, either as outlined here or the full method described in the reference article.

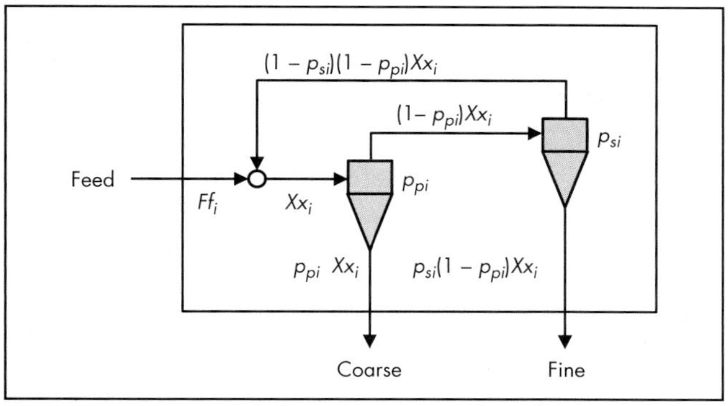

Figure 16 Austin et al. (1984) conceptual model for the air classifier

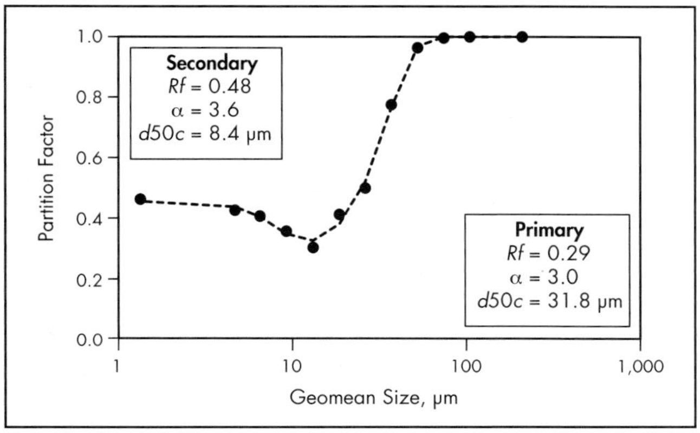

Figure 17 Partition curve data for the air classifier showing the Austin et al. (1984) model parameters

UNUSUAL PARTITION CURVES

The preceding discussion focused more on the usual sigmoidal partition curve; however, in some instances the shape of the experimental partition curve is not strictly sigmoidal. Finch (1983) was one of the first to publish on partition curve inflections in the fines sizes, and over the years many other authors have presented and explained similar findings. One of the best review/research papers was recently published by Eisele et al. (2013), and anyone interested in this topic is strongly encouraged to read this work.

For the purposes of this chapter, where the focus is on the partition curve for a hydrocyclone, there are two phenomena that can alter the shape of the more usual partition curve: (1) fines entrainment in the underflow stream (beyond the effect of Rf) and (2) multicomponent (heterogeneous) ores. Of the two, from an operational perspective, the second item is probably more important.

Fines Entrainment

The "fishhook" phenomenon usually occurs in the finer size ranges, say less than 50 μm and, more often, less than 20 μm. (The term *fishhook* derives from an unexpected upturn in the partition curve in the fine size region.) Explanations for this occurrence can involve flocculation or agglomeration, boundary layer entrainment, wake entrainment, dense medium effects, density effects, settling regimens, or the propagation of experimental errors (see Eisele et al. 2013 for more detail).

The phenomenon is illustrated here by borrowing a data set (Austin et al. 1984) from a sampling experiment around an air classifier, as shown in the first two columns of Table 10.

Finch's approach to describe the partition curve was to reformulate Equation 21 as shown in Equation 31, borrowing from the phenomena of entrainment observed in flotation froths:

$$\hat{p}_i = Rf\left[\frac{d_0 - d_i}{d_0}\right] + \left[1 - 0.5^{\left(d_i/d50c\right)^{\alpha}}\right] \qquad \textbf{(EQ 31)}$$

where d_0 equals the size at which fines entrainment begins.

In this case, the first term is calculated only for the cases where $d_i < d_0$; otherwise, it is taken to be 0. Figure 15 shows the raw data and the Finch model fit, and the fishhook is quite apparent.

Users of JKSimMet will be familiar with the Whiten approach (described in Nageswararao et al. 2004) to account for a fishhook shape in the partition curve data. This four-parameter model is a little more involved, so it is not included here, although the results are shown in Figure 15 for comparative purposes. Nageswararao et al. (2004) have observed that in many cases—including the fishhook parameter in the JKSimMet model—the overall fit was improved, but statistical tests were not applied to assess whether this was simply the effect of additional parameters or whether the partition curve required it.

In their work on air classifiers, Austin et al. (1984) developed a conceptual model based on mechanisms at work in this device, and this is shown in Figure 16. The model that arises from this formulation is given by Equation 32, and the fit is presented in Figure 17.

$$\hat{p}_i = \frac{\hat{p}_{pi}}{[1 - (1 - \hat{p}_{si})(1 - \hat{p}_{pi})]} \qquad \textbf{(EQ 32)}$$

where

$\hat{p}_{pi}$ = primary separation partition curve, same form as Equation 21

$\hat{p}_{si}$ = secondary separation partition curve, same form as Equation 21

Clearly, more complex partition curve shapes call for more complex models. In this case, Finch uses four parameters and Austin et al. (1984) use six parameters. Based on Partial F testing, there is no statistical justification to go with the more complex model of Austin et al.although conceptually it has some merit. For parameters to be useful metrics in assessing performance, parsimony is key (i.e., two or three, and possibly four, parameters are "manageable"). Beyond that, one might just as well use generic functional forms (e.g., high-order polynomials or splines) to connect the points.

Multicomponent Ores

For purposes of this discussion, the author considers multicomponent ores to include significant quantities of minerals with quite different specific gravities. Since hydrocyclone classification involves settling velocities, particle mass is important, and that depends on both size and specific gravity (sg). A particularly good example of this in the mining industry is iron ore processing, where in many cases the ores can be considered to consist of silica (sg = 2.65) and magnetite or hematite (sg = 4.9–5.2).

Eisele et al. (2013) have shown how plateaus or humps are formed in partition curves when synthetic mixtures of

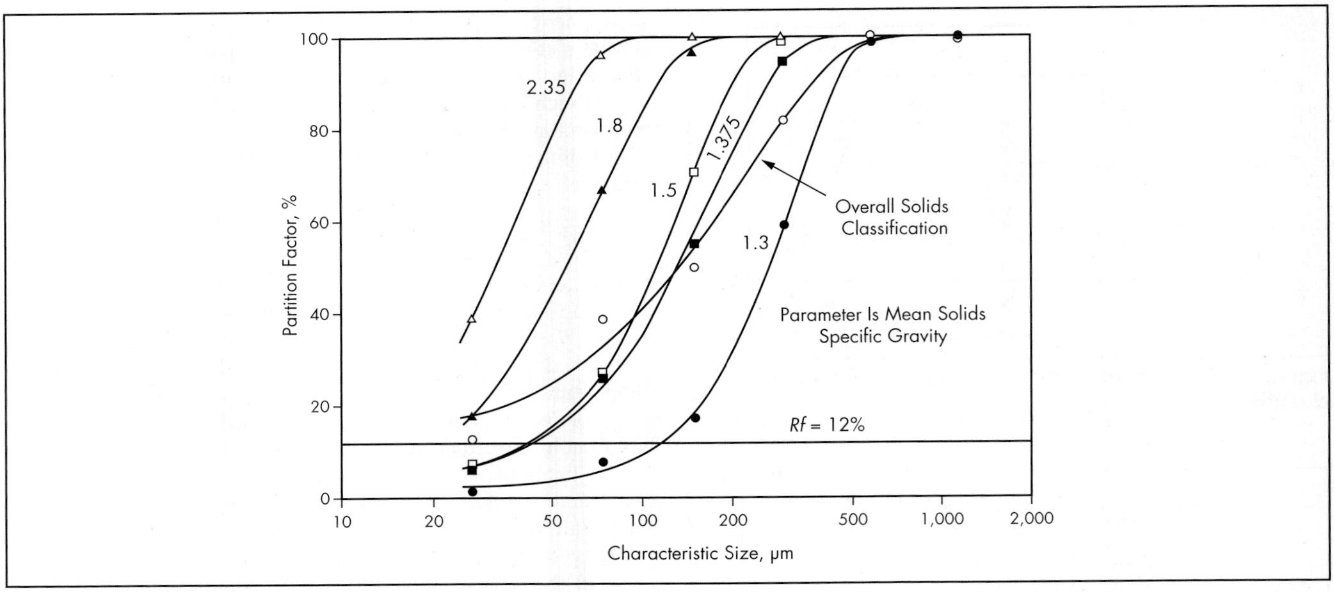

Source: Flintoff et al. 1987

Figure 18 Partition curves for coal slurry in a hydrocyclone

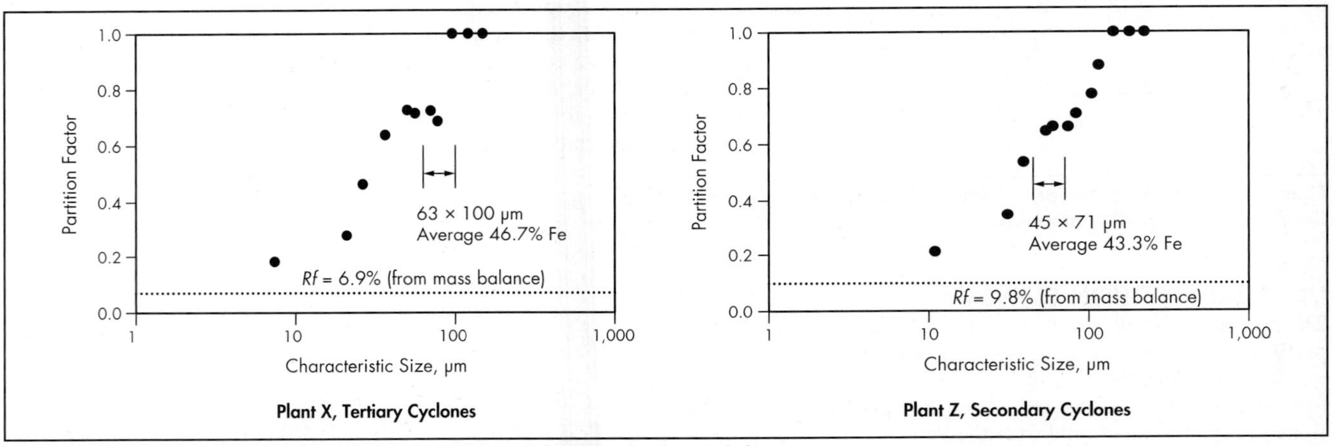

Figure 19 Partition curves for two magnetite plants from experimental data

silica and magnetite are mixed and fed to a hydrocyclone. They have also shown similar partition curve shapes from plant sampling campaigns in a magnetite concentrator.

Not all multicomponent ores give rise to unusual partition curve shapes. A good example of this from the mining industry is a hydrocyclone treating coal. In this case, there is essentially a continuous range of specific gravity components in the solids, and the net effect is that a reasonable overall partition curve shape, albeit with a low α value, is derived. However, if one submits the cyclone products to sink/float analyses to separate the specific gravity classes by size, a clearer picture emerges. Figure 18 illustrates this situation.

It is clear in Figure 18 that the cut size ($d50c$) is dependent on the specific gravity and the functional form follows from the settling theory as shown in Equation 33. Minerals with higher specific gravities have finer cut sizes:

$$d50c \propto \frac{1}{[\rho_s - 1]^k} \quad \textbf{(EQ 33)}$$

where

ρ_s = specific gravity of the solid component

k = exponent 0.5 for Stokesian conditions and 1.0 for Newtonian conditions—experimental values vary from theory (e.g., for the data in Figure 16, $k \approx$ 1.13 [Flintoff et al. 1987])

The behavior of the curves in the very fine sizes may seem a little odd, as they do not appear to asymptotically approach Rf, the lighter fractions showing lower values than the fractional bypass of water. Here, this is likely the result of dense medium separation effects in the neighborhood of the apex, something that is exploited in automedium hydrocyclones for fine coal preparation applications.

Coming back to the two-component iron ore system, Figure 19 shows the results for two different magnetite plants, where the hydrocyclones are producing final comminution circuit product, which, in turn, will get a final cleaning by magnetic separation. The humps or plateaus are very clear in

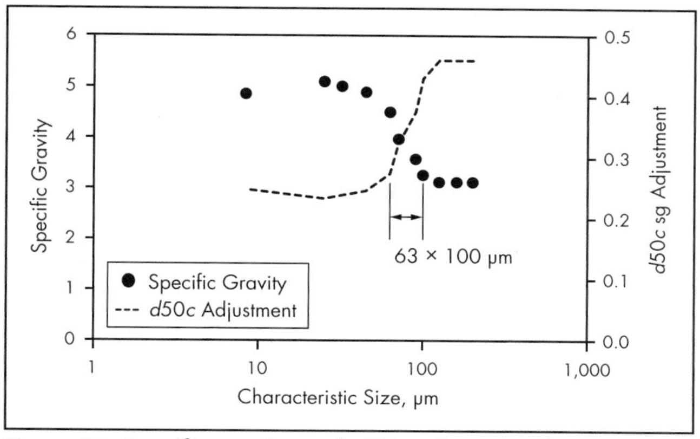

Figure 20 Specific gravity and *d50c* adjustment factors as a function of size for plant X

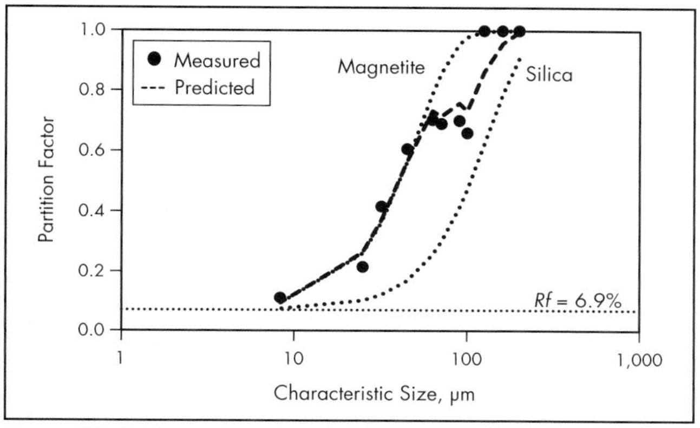

Figure 21 Estimated and actual overall partition curve for plant X

this case, and as Eisele et al. (2013) point out, these are exacerbated by the concentration of fine liberated magnetite in the hydrocyclone underflow stream, which inevitably returns to the hydrocyclone feed after passing through a ball mill.

In both cases, one can see that a significant portion of a middlings size class (~45% Fe) reports to the hydrocyclone overflow, and since magnetic separators are designed to recover all particles that exhibit magnetic properties, this material contaminates the final concentrate and can contribute to a lower grade. This is one of the reasons that fine screens are seeing more use in these circuits (e.g., scavenging the coarse lower-grade material from the hydrocyclone overflow and directing it back to the ball mill to get to premium product grade levels of 67%–68% Fe or higher).

Understanding the reasons for the plateau region in Figure 19 is a little more challenging than making inferences from synthetic mixtures. This is more like the coal problem, although because of prior magnetic separation steps, much of the free silica has been removed, so it is not a natural feed in that sense.

For plant X, there were some liberation data on the ore at the time the sampling campaign was completed, and in all cases, the size fractions were assayed for %Fe. These liberation data, combined with the preceding information, allowed for decomposition of the hydrocyclone feed into three components: free magnetite, locked magnetite, and free silica. In the case of the locked particles, the specific gravity varies with the iron content, which varies as a function of size. Figure 20 shows the estimated variation of particle specific gravity with size. In this case, k in Equation 33 was estimated as 1.03, and the *d50c* adjustment from the equation is also shown on the figure. In the range of the plateau, one can clearly see that while size is increasing (which should increase the partition factor), the density is dropping, which has the effect of translating the actual partition curve to the right and driving the partition factor down. The net result is that during this transient in the average particle density, the overall partition curve will show some sort of inflection, and those displayed in Figure 19 are typical.

Using a model of the form given by Equation 21, with the specific gravity correction of Equation 33, the overall predicted partition curve was computed based on individual component behavior. These results are shown in Figure 21, and given the initial assumptions, they seem quite reasonable. Note that the free magnetite and free silica curves are included for reference.

It may be possible to use simpler modeling approaches to characterize the curve, but for meaningful parameter estimation and understanding of where the inflection forms and its extent, much more information is required. Incorporating this kind of structure into process simulation packages like JKSimMet is in progress (e.g., Narasimha et al. 2012).

CONCLUSIONS

The partition curve is ubiquitous in characterizing classification processes based on the physical attributes of the particles. They serve as a diagnostic tool for assessing the performance of these units through parameter estimation and possible comparison with other parameter data. In the case of some unit operations, like the hydrocyclone and screens, they lie at the heart of more comprehensive empirical models, which are embedded in simulation software, and are useful in process analysis, design, and optimization.

Despite their apparent mathematical simplicity, the work required to get a good estimate of a partition curve is time-consuming, and every effort must be made to comply with the best practices for process surveys. The raw data can be used in that state, but it is better to massage it with mass balance software before getting into the modeling work. In porphyry copper (not massive sulfides) and gold ores, it is common to be able to use two- or three-parameter models. Statistical comparison can be used to select the best model and, in itself, can be another useful diagnostic around data quality. In the case of multicomponent ores (e.g., iron ore, coal, or massive sulfide deposits), specific gravity variation can complicate the partition curve analysis, although work on multicomponent simulation packages should make this easier in the future. The fishhook phenomenon that is visible in some partition curves can usually be handled by adding a single parameter, although a mechanistic interpretation of this can sometimes be difficult.

REFERENCES

Austin, L., and Klimpel, R. 1981. An improved method of analyzing classifier data. *Powder Technol.* 29:277–285.

Austin, L., Klimpel, R., and Luckie, P. 1984. *Process Engineering of Size Reduction: Ball Milling.* Littleton, CO: SME-AIME. pp. 328–329.

De Lange, T., and Venter, P. 1987. The use of simulation Tromp partition curves in developing the flowsheet of plant extensions at Grootegeluk coal mine. In *Proceedings of the 20th International Symposium on the Application of Computers and Mathematics in the Mineral Industries*, Vol. 2. Johannesburg: Southern African Institute of Mining and Metallurgy. pp. 295–311.

de Levie, R. 1999. Estimating parameter precision in nonlinear least squares with Excel's Solver. *J. Chem. Educ.* 76:1594–1598.

Eisele, T., Jeltema, C., Walqui, H., and Kawatra, S. 2013. Coarse and fine fish hook inflections in hydrocyclone efficiency curves. *Miner. Metall. Process.* 30(3):137–144.

Finch, J. 1983. Modelling a fish-hook in hydrocyclone curves. *Powder Technol.* 36:127–129.

Flintoff, B., Plitt, L., and Turak, A. 1987. Cyclone modelling: A review of present technology. *CIM Bull.* 80(905):39–49.

Herbst, J., and Sepulveda, J. 1985. Characterization of test samples—Particle size analysis. *SME Mineral Processing Handbook.* Edited by N.L. Weiss. Littleton, CO: SME-AIME. pp. 30:20–30:43.

Hodouin, D., Kasongo, T., Kouame, E., and Everell, M. 1981. BILMAT: An algorithm for materials balancing mineral processing circuits. *CIM Bull.* 64(833):123–131.

Hodouin, D., Bazin, C., and Trusiak, A. 1984. Reliability of calculation of mineral process efficiencies and rate parameters from balanced data. In *Control '84: International Symposium on Automatic Control in Mineral Processing and Process Metallurgy.* Edited by J. Herbst, K. Sastry, and D. George. Littleton, CO: SME-AIME. pp. 133–144.

Laguitton, D., ed. 1985. *The SPOC Manual.* Ottawa, ON: Energy, Mines and Resources Canada, Canada Centre for Mineral and Energy Technology.

Lynch, A., and Rao, T. 1975. Modelling and scale-up of hydrocyclone classifiers. In *Proceedings of the 11th International Mineral Processing Congress.* Cagliari, Sardinia: Istituto di Arte Mineraria e Preparazione dei Minerali. pp. 245–269.

Mular, A. 1979. Data adjustment procedures for mass balances. In *Computer Methods for the 80s in Mineral Processing.* Edited by A. Weiss. Littleton, CO: SME-AIME. pp. 843–849.

Nageswararao, K., Wiseman, D., and Napier-Munn, T. 2004. Two empirical hydrocyclone models revisited. *Miner. Eng.* 17:671–687.

Narasimha, M., Mainza, A., and Holtham, P. 2012. Multicomponent modeling concept for a hydrocyclone classifier. In *Proceedings of the XXVI International Mineral Processing Congress.* New Delhi: Indian Institute of Metals. pp. 3696–3707.

Plitt, L. 1971. The analysis of solids-solid separations in classifiers. *CIM Bull.* 64:42–47.

Wei, J., and Realff, M. 2005. Design and optimization of drum-type electrostatic separators for plastics recycling. *Ind. Eng. Chem. Res.* 44(10):3503–3509.

Wiegel, R. 1979. The practical benefits of improved metallurgical balance techniques. SME Preprint No. 79-92, Littleton, CO: SME.

Wills, B., and Finch, J. 2016. Chapter 9, Classification. In *Wills' Mineral Processing Technology: An Introduction to the Practical Aspects of Ore Treatment and Mineral Recovery.* Oxford; Waltham, MA: Butterworth-Heinemann.

PART 5

Transport and Storage

CHAPTER 5.1

Solids Storage

John Michael Holden, Michael G. Nelson, Jessica M. Wempen, and Martin Du Plessis

In processing of minerals and other bulk materials, such as grains, salts, wood chips, cornmeal, rice, and so forth, material storage is generally needed at points along the flow stream or at the flow stream's termination. A large storage capacity may be required where feed delivery or product shipment is difficult to schedule; or, depending on the plant's production, the design flow rate is hampered because of unexpected equipment failure(s); or either scheduled or required maintenance must be performed on an as-needed basis. Sufficient storage capacity is generally required to meet the processing plant's requirements for sales, contracts, shipments, whether it is a new processing plant and/or an upgrade to an existing processing plant.

Large-tonnage shipments for overseas export, trade, or with unit trains require both adequate storage and properly sized materials handling equipment for loading and reclaiming material, and appropriate storage facilities to suit the materials being stockpiled or stored.

Within a process plant, storage capacity is required to accommodate differences in production capacity or reliability of the process units and when the various production units need to be taken off-line. The processing plant's designed requirements generally determine the plant's storage surge capacity to sufficiently provide for shutdowns required by emergency or scheduled maintenance. Where plant sections operate on different time schedules, storage capacity adequate for material differential is required. When required storage capacity is extremely large, ground storage stockpiles or multiple storage silos and/or storage domes or barns have proved to be an economic advantage.

The purpose of storage is to provide efficient storage capacity for (1) an adequate reserve of raw materials for the process plant; (2) an adequate raw materials reserve to enable a production rate different from the supply rate; (3) adequate supply reserves where these can be interrupted by shipping delays, weather, strikes, and so on; (4) blending of various grades, sizes, or types of materials; (5) reducing material surges at different process points; (6) collecting and distributing materials at different process points; (7) storing materials ahead of equipment shutdowns for temporary maintenance; (8) adequate buffering between batch and continuous process units; and (9) storing finished processed products/materials to satisfy market demands promptly and economically.

TYPES OF STORAGE

Storage types used in mineral and other bulk materials processing can be divided into open or covered stockpiles, silos, barns, hoppers, railroad cars, fabric tented structures, bins, bunkers, and underground caverns. Bulk materials are stored in ground-level stockpiles that may be exposed, covered, or indoors. They provide economical storage when requirements are large. Stockpiles vary in size and shape, such as conical, elongated, windrow, rectangular, triangular, and trapezoidal, with material placed and reclaimed by various methods and reclaiming equipment suitable for the material's stockpiled and stored characteristics. Bins vary in size from large, for run-of-mine (ROM) ore, to smaller vessels for process additives.

Depending upon process equipment, bins or hoppers vary in design, shape, and size, and are constructed of either masonry, concrete, wood, steel, abrasion-resistant (AR) steel, or synthetic materials. Hoppers are often located under stockpiles and bins to facilitate reclaiming.

The first step in storage facility design is to define storage purpose, capacity required (including surge capacity), and flow rates for stockpiling or storage and for reclaiming materials handling equipment appropriate for the materials to be reclaimed, stockpiled, or stored.

Bulk Materials Characteristics

The physical and chemical properties of the bulk material must be properly classified and described. Attention must be given

John Michael Holden, Materials Handling Engineer, Retired, Littleton, Colorado, USA
Michael G. Nelson, Professor & Chair, Mining Engineering Department, University of Utah, Salt Lake City, Utah, USA
Jessica M. Wempen, Assistant Professor of Mining Engineering, University of Utah, Salt Lake City, Utah, USA
Martin Du Plessis, Lead Mechanical Engineer, Fluor, Perth, Western Australia, Australia

Table 1 Data required for storage design

General Considerations Influencing Choice of Stockpiling and Storage:
1. Bulk material Conveyor Equipment Manufacturers Association (CEMA) Material Classification Code, pursuant to ANSI/CEMA Standard 550-R2009, *Classification and Definitions of Bulk Materials*, according to their individual characteristics (latest edition), and *Belt Conveyors for Bulk Materials* (CEMA 2014)
2. Chemical formulae (if any)
3. Purpose of stockpile or storage
4. Live storage required
5. Permissible dead storage
6. Inflow rate including design flow rate
7. Outflow/reclaim flow rate including design flow rate
8. Location and placement of storage
9. Space requirements
10. History of meteorological conditions/data
11. History of seismic activity
12. Soils report(s)
13. Topography
14. Material inflow velocity/trajectory
Bulk Material Characteristics Influencing Choice of Exposed or Protected Stockpiles and Storage:
1. Bulk density
2. Specific gravity
3. Porosity of particles
4. Surface and inherent moisture content (percentage)
5. Solubility in water
6. Consolidation/compacting tendency
7. Material contamination
8. Environmental contamination
Bulk Material Characteristics Influencing Choice of Equipment for Stockpiling and Storage of Reclaiming Materials:
1. Mohs hardness
2. Particle size distribution (pursuant to screen analysis test-work data)
3. Particle strength
4. Angle of repose
5. Ange of surcharge
6. Angle of drawdown and internal shear angle
7. Abrasiveness and roughness or smoothness
8. Particle shape
9. Percentage of particle fines and lumps
10. Tendency to pack (length of storage)
11. Material temperature
12. Corrosiveness
13. Hygroscopicity
14. Combustibility
15. Toxic fumes and dust
16. Oils or chemicals present? These may affect rubber; ultra-high molecular weight; high-density polyethylene, or HDPE, abrasion-resistant liners; and coating products
17. Floodability, flowability, and tendency to aerate
18. Tendency for interlock or matting
19. Very light and fluffy—material may be wind swept
20. Degradability affecting use or salability

to environmental factors if the materials are exposed. Table 1 presents data required for stockpiling and storage/reclaiming designs.

The CEMA Material Classification Code (CEMA 2014) for a given material provides a simple, standard manner to understand the handling characteristics of that material. Chapter 5.2, "Belt Conveyors," provides a detailed explanation and examples for the use of the CEMA Code. The Material Classification Code for a material is useful in the design of conveying, reclaiming, and materials handling equipment, as well as in the design of stockpiles, storage bins, silos, hoppers, chutes, and other storage vessels.

Characteristics Influencing Exposed or Protected Stockpiles and Storage

Light, fluffy materials that are low in bulk density are readily moved by wind, with resulting losses and nuisance to neighboring properties and plant operators, as well as incoming and outgoing delivery personnel. In such cases, protected storage is desirable. The material's bulking properties are affected by rain, snow, sleet, hail, wind, and climatic conditions. Specific gravity and particle size determine the extent to which wind transports the material. Wind erosion is a factor regardless of specific gravity with particles below 100 mesh such as power plant utility flue boiler dust, coal ash, carbon black, coke breeze (0.635 cm and under), diatomaceous earth, and grains (for ethanol fuel production).

Particles of high porosity pick up and retain moisture. In cold climates, particle freezing may result in degradation and material friability. Freezing of surface moisture binds particles together in a solid mass and makes reclaiming difficult, if not impossible, in cold climates. In such cases, protected storage is necessary. When control of moisture content is important, material should be protected from rain, snow, sleet, hail, and the drying effects of hot sun, ambient air temperature, and wind. In areas subject to blowing dust, protection should be provided to avoid the "drying" effect of dust fines to comply with all applicable environmental and occupational health regulations. Water-soluble materials must be protected. Where historical data show that rainfall occurrence is light, temporary covers may be adequate. The effect of humidity on such material should be considered. Cementing tendency due to long-term storage is closely related to the material's moisture absorption and solubility tendencies and requires similar protection, since it may affect reclaiming and/or discharging from silos, bins, or hoppers. Contamination or the degree to which a material may be diluted or spoiled by dust, with subsequent loss of market value, is a good economic measure for determining desirability of protected storage. Where contamination of the environment or neighboring products by dangerous or controlled material is at risk, protected storage is essential.

Characteristics Influencing Equipment for Stockpiling and Storage of Reclaiming Materials

Mohs particle hardness for a given bulk material provides an indication of the abrasive wear rate that will occur in bins, hoppers, silos, chutes, gates, and other materials handling equipment. Ranges of particle sizes provide an indication of the degree of segregation occurring in transport and drawdown storage. A wide range of particle sizes requires greater materials handling care. Particle strength is important in materials handling methods and materials handling equipment selections.

Material flowability influences stockpile design, storage type, and reclaim system design. The available live storage volume is a function of the material's angle of drawdown, which is generally different than the material's angle of repose. The flowability of materials, which is especially important in bin, silo, and hopper drawdown discharge design, is discussed later in the "Bins" section. Free-flowing materials, such as processed salts and dry sugars, may be reclaimed easily and economically through clamshell, horizontal slide, rolling blade, bifurcated, and adjustable proportional types of discharge gates.

Subsequent conveying systems usually require different types of feeders for flow regulation. For non-free-flowing materials, such as sticky and matting or interlocking materials, equipment that exerts a positive digging or displacing force—such as rotary table discharge plow feeders, hopper/bin air blasters, and adjustable, brute-force vibrators—must be used to dislodge and move material to the recovery point . When the angle of drawdown is steep, power shovels, bucket loaders, and scrapers are used. Stockpile height should be limited as a safety precaution against slides and ratholing. Material abrasiveness is a composite of several different material characteristics, including strength, Mohs hardness, and particle shape. Abrasive wear rate varies directly with the pressure and velocity of particle movement across an inclined surface.

Chutes, hoppers, bins, silos, and slopes, as well as their valley angles, should be designed as closely as possible to the material's angle of slide. Steeper slopes move more material but result in greater wear. AR steel, replaceable AR liners, alloy steels, or coatings may be desirable to reduce wear rates. Material dead beds at impact points reduce wear but may induce another problem in the case of wet, sticky, fibrous, or interlocking materials that show a tendency to build up into an immobile mass. Some loose materials have a tendency to pack and become denser, creating reclamation problems. Such materials usually require a feeder that has a strong positive action. In extreme cases, a digging action such as that provided by a power shovel or bucket wheel excavator may be required. Equipment should not roll or crawl on the materials being loaded.

Material temperature affects the choice of placing equipment. High-temperature materials, such as hot clinker, require special rubber or steel belts. Surface equipment for stockpiles should be chosen with care, and hazards to equipment and operators should be carefully considered. Corrosiveness can be controlled by use of protective coatings, but when abrasion is also present, the coating's effectiveness is reduced. Each passage of abrasive material over a surface exposes a new area of metal to corrosive elements. The combination of corrosion and abrasion requires investigation to keep maintenance costs of equipment and structures at a reasonable level.

Combustibility is a concern only for combustible materials, such as ammonium nitrate, barley, crushed calcium carbide, shelled corn, dry power plant utility flue ash and dust, dry pulverized salt cake, cracked soy cake, and dry beans, as indicated by the Conveyor Equipment Manufacturers Association (CEMA) class code. In addition, systems for sulfide concentrates and fuels should be considered combustible. With such materials, special methods of placing and protection are required, such as instrumentation controls, fire and combustion detection devices, blast gates, blast-relief panels, dust collection, methods of ventilation, and earth grounding.

Spontaneous combustion may occur in stockpiles of coal, sugar, and food powders if they are not properly layered and compacted. When toxic fumes and dust are emitted, remotely operated placing and reclaiming equipment should be used and operator exposure minimized. Flooding, the unstable fluid-like flow of dry solids, results from entrainment of air in material flowing in empty or partially empty hoppers or from collapse of the material arch in the hopper, bins, and silos. Floodable materials may not be suitable for stockpile storage and may require special attention in choice of materials handling equipment.

Material Flow

Successful operation of gravity storage is predicated on uniformity of materials when reclaiming. Flow patterns that occur in stockpiles, bins, silos, and hoppers are funnel, mass, and expanded flows. In funnel flow, material flows through a cylindrical channel extending upward from the discharge outlet, often colloquially called "ratholing." In mass flow, all material flows downward as material is reclaimed successfully. In expanded flow, the size of the flow channel expands upward from the discharge outlet, forming a cone. All three patterns occur in bins, silos, and hoppers, but generally only funnel and expanded flows occur in stockpiles with funnel flow the most commonly observed. Hoppers and bins are usually designed for material mass flow.

Gravity flow of solids occurs by the continuous formation and collapse of domes (arches) over the flow channel. Bulk materials have little or no strength in a loose state, but when under pressure, they consolidate and compact, especially in long-term storage, and gain strength by formation of bonds among particles. If the strength of these bonds is sufficient to support a stable dome or pipe (rathole), a flow obstruction develops. Flow analysis test work and historic flow-pattern data for the same type of material are useful in determining the minimum discharge outlet dimensions for those obstructions to be unstable. Those dimensions have a direct relation to the sizing and selection of bin, hopper, silo discharge gates, and reclaiming materials handling equipment.

Flow properties of bulk materials vary widely. The following characteristics affect flowability of materials from stockpiles, bins, hoppers, chutes, and silos: surface moisture content, atmospheric humidity, temperature, long-term storage, particle size distribution (percentage of fines to lumps), dustiness, stickiness, particle shape, forward velocity of the material as it is being loaded (as this acceleration causes turbulence and particle segregation in the material), and other CEMA class code particle characteristics (such as specific gravity, Mohs hardness abrasiveness, chemical corrosiveness, and hygroscopicity).

Flow analysis is based on mathematical theory and past historical data for the same type of material. This analysis is based on a characterization of the flowability of a solid called the flow function (FF) and the flowability of a solid of hoppers, bins, and silos called the flow factor (ff). These material characteristics can be obtained through laboratory tests and analysis of historical data, and various commercial and academic laboratories offer such services. It is important for the design engineer to supply to the laboratory samples of material that represent the worst cases of the product in question with regard to expected size distribution, screen analysis, particle characteristics, surface area, and inherent moisture content. The material samples should also conform as closely as possible to the anticipated CEMA Material Classification Code designation.

Environmental Factors Affecting Materials Storage

In selecting the type of storage, there must be a matching of storage purpose, overall required storage capacity, live and dead load capacity requirements, and physical and chemical properties of the material with various environmental factors. Table 2 provides recommendations regarding exposed and protected storage.

STOCKPILES

Bulk material stockpiling is defined as storage of bulk material in variously shaped piles on the ground. They may be exposed or protected, with protection ranging from simple

Table 2 Recommendations for exposed (outdoor) and protected (indoor) storage related to material characteristics and environmental factors

Material Property		Environmental Factors														
		Rain and Snow			Humidity			Wind			Dust			Temperature		
Property	Degree	Heavy	Moderate	Little, none	High	Moderate	Low	Frequent and high velocity	Moderate	Infrequent and low velocity	High-dust area	Moderate-dust area	Low-dust area	Dry, arid area and high temperature	Moderate	Extremely low temperature
Bulk density	Light, fluffy	In	In	Out	—	—	—	In	In	In	—		—	—	—	—
	Intermediate	Out	Out	Out	—	—	—	In	Out	Out	—		—	—	—	—
	Heavy, dense	Out	Out	Out	—	—	—	In	Out	Out	—		—	—	—	—
Specific gravity	Low	—	—	—	—	—	—	In	In	Out	—	—	—	—	—	—
	Intermediate	—	—	—	—	—	—	In	In	Out	—	—	—	—	—	—
	High	—	—	—	—	—	—	Out	Out	Out	—	—	—	—	—	—
Porosity	High	In	In	Out	—	—	—	—	—	—	—	—	—	—	—	—
	Medium	In	In	Out	—	—	—	—	—	—	—	—	—	—	—	—
	Low, or none	Out	Out	Out	—	—	—	—	—	—	—	—	—	—	—	—
Moisture content	Control required	In	In	In	In	In	In	In	In	Out	In	In	Out	In	In	In
	Little control	In	In	Out	In	In	Out	In	Out	Out	In	Out	Out	In	Out	In
	None	Out	Out	Out	Out	Out	Out	Out	Out	Out	Out	Out	Out	Out	Out	Out
Soluble in water	Highly	In	In	In	In	In	Out	—	—	—	—	—	—	—	—	—
	Moderately	In	In	In	In	Out	Out	—	—	—	—	—	—	—	—	—
	None	Out	Out	Out	Out	Out	Out	—	—	—	—	—	—	—	—	—
Hygroscopity	High	In	In	In	In	In	In	—	—	—	—	—	—	—	—	—
	Moderate	In	In	Out	In	In	Out	—	—	—	—	—	—	—	—	—
	None	In	Out	Out	In	Out	Out	—	—	—	—	—	—	—	—	—
Cementing tendency	High	In	In	In	In	In	In	—	—	—	—	—	—	—	—	—
	Moderate	In	In	Out	In	In	Out	—	—	—	—	—	—	—	—	—
	None	Out	Out	Out	Out	Out	Out	—	—	—	—	—	—	—	—	—
Contaminable	Highly	—	—	—	—	—	—	—	—	—	In	In	In	—	—	—
	Moderately	—	—	—	—	—	—	—	—	—	In	In	Out	—	—	—
	None or little	—	—	—	—	—	—	—	—	—	Out	Out	Out	—	—	—

covers to enclosed buildings such as domes, A-frame barns, sprung fabric buildings, and prefabricated lightweight sheeted buildings. Stockpile size and location depend upon purpose, process requirements, and climatic conditions. Shape depends on purpose, capacity requirements, material characteristics, available space, methods of materials handling stacking, and reclaiming equipment. All of these factors interact on stockpile system design.

Stockpile Site

The surface selected for a stockpile should be firm, solid, and well drained. A review of the plant site's soils report should be made. The site should be raised, not subject to flooding, and have a good slope for adequate drainage away from the stockpile area. If the site is not well drained, ditches should be provided along the stockpile toe for the full length or circumference of the storage site. If tunnels or culverts are used for reclaim under the stockpile, they should be "daylighted" on both ends with an escape tunnel at one end, and sloped to drain to an open surface or provided with a sump and drainage pump. Before the stockpile is laid down, the area should be compacted to the proper density pursuant to the soils report, and cleared of all debris and organic materials.

The stockpile toe should be accessible for mobile equipment for maintenance purposes and for removing the dead storage and turnover of the larger lumps that normally roll to the toe of the stockpile. Other plant processing or administration buildings, such as the truck shop, warehouse, and operators' change room, should be located a safe distance from the stockpile toe. This distance is determined by the minimum material angle-of-repose boundary; otherwise, an adequate retaining wall(s) should be used to prevent engulfment of various infrastructure.

Stockpile Size

Stockpile volume is based on desired flow rate through the plant, period of retention, and period of plant operation, as determined by the process mass flow and process dynamic modeling. Stockpile size is based on the required storage volume, type of stockpile being used, and the method of stacking and method of reclaiming being used. Plant production rate and market demand should be determined as accurately as possible and a reasonable design safety factor added.

Blending piles are a special case. Their volume must be based on a balance between sources of supply and anticipated rate of demand. Blending normally requires twin stockpiles, so that one stockpile can be blended during stacking while the other stockpile is being reclaimed. In some cases, when the right equipment is available, blending can be done from one stockpile. Reclaim equipment must be designed to minimize segregation, material ratholing, degradation, and production of material fines. In some cases, land area may be adequate but height limitations may restrict stockpile capacity. Maximum height may be limited by the placing and reclaiming equipment; soil conditions; proximity to the loading point (railroad, dock, or stream); height of stacking tubes, if employed; characteristics of the material; and need to ensure a safe operation. Required stockpile height can be decreased by providing additional takeoff points or the choice of materials handling equipment, such as belt conveyor stackers equipped with stacking "slingers," slewing and luffing stackers, windrow stackers, traveling trippers consistent with cross conveyors, belt conveyor traveling belt plows or multiple stationary plows, a series of inclined piggy-back belt conveyors equipped with bifurcated discharge gates, reversible traveling shuttle belt conveyor(s), conveyor head discharge pulleys equipped with a rotating or fixed stinger, and belt conveyor stackers equipped with variable-frequency-drive, wound-rotor motors to adjust the material's velocity and stacking trajectories, each of which will increase available live tonnage. In addition, the delivery system can be altered to permit building additional stockpiles, the overall shape of the stockpile can be changed, or the single pile can be spread over a larger area. In some cases, it may be necessary to consider using mobile equipment to totally reclaim dead storage from the stockpile.

Stockpile Shape

Common stockpile types, based on geometry, are conical, lateral or windrow-parallel, windrow-radial and kidney shaped, ramped, terraced, bunkered, rectangular, triangular, trapezoidal, and elongated.

Conical stockpiles are common for ROM and primary crushed ore storage. *Conical stockpiles* can be defined as a mostly unrestrained stockpile fed at one or more stationary points. A common method of reclaiming a conical stockpile is by gravity from one or more extraction points underneath the stockpile with materials handling equipment.

Live storage represents the capacity available for reclaim by gravity without assistance from auxiliary materials handling reclaim equipment. The reclaiming of the live volume is calculated knowing the drawdown angle. In most cases, the drawdown angle is equal to the angle of internal friction of the specific material, as determined by material sample flow testing and the material's CEMA classification code. For most materials, the angle of drawdown is greater than the material's angle of repose.

For a simple conical stockpile with a single feed and single center extraction point, assuming no ratholing (Figure 1), the total volume can be calculated from the following equation:

$$\text{total volume} = \frac{1}{3}\pi r^3 h \text{ and } \frac{h}{r} = \tan(A)$$

where

r = base radius
h = stockpile height
A = material's angle of repose

The live volume can be calculated from

$$\text{live volume} = \frac{\tan^2(A)}{(\tan(A) + \tan(B))^2}$$

For more complex conical stockpile arrangements where multiple feed points or multiple drawdown reclaim points are present, it is often easier to use a computer model (Figure 2) to determine the total live and dead volumes.

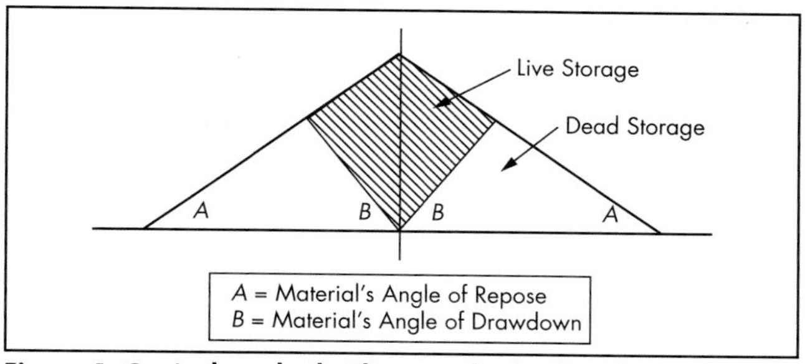

Figure 1 Conical stockpile "live" and "dead" volume

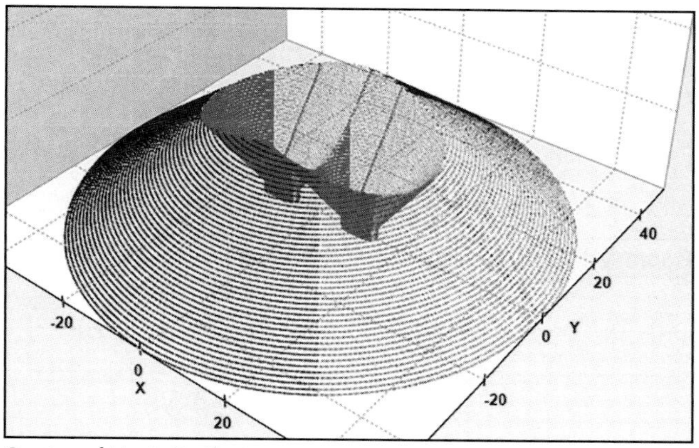

Courtesy of Martin Du Plessis

Figure 2 Computer model of a conical stockpile with two extraction points experiencing ratholing

Windrow stockpiles can be either linear elongated or radial. *Windrow stockpiles* can be defined as a stockpile with one or multiple feed points that either move linearly or radially. Reclaim methods can vary from gravity reclaim from under the stockpile using types of equipment based on the shape and length of the stockpile. Linear elongated stockpiles more than 1.6 km long are not uncommon in port-of-entry and port-of-export facilities.

The volume of a simple linear stockpile can be estimated by

$$\text{volume} = \frac{1}{2}BhL + \frac{1}{12}\pi B^2 h \text{ and } \frac{h}{B} = \frac{1}{2}\tan(A)$$

where

B = base width (see Figure 3)
h = stockpile height
L = stockpile length
A = angle of repose

The cross-sectional profile of elongated windrow stockpiles are seldom simple, which makes calculating total capacity as well as live and dead load capacities nontrivial. These stockpile types may include chamfered toes, multiple peaks, flat-surfaced peaks, compound material angle of repose, or material that maybe slightly compacted and embedded. The exact total capacity volume of these types of stockpiles should be estimated from their respective specific profiles or calculated by direct measurement using photogrammetry or laser scanning.

The volumes of other stockpile shapes depend on their exact geometries. Ramp lengths, step heights, retaining walls, and bunker walls add to the complexity of volume capacity calculations. When material is reclaimed under these type of stockpiles, the determination of live and dead storage is complex for multiple drawdown outlets. Additional design live load factors for reclaim capacity should be considered.

Methods of Placing Material in Stockpiles

Equipment for placing material in stockpiles includes conveyors, buckets, and surface equipment. Conveyors include belt conveyors, screw conveyors, and various types of stackers operating above and discharging material onto the stockpile. Buckets include drag scrapers, draglines, clamshells, bridge cranes, and drag conveyors operating above and depositing or moving material over the stockpile's top surface. The conveyor and bucket-type equipment usually reaches over and is not supported on stockpiled material. Exceptions are mobile and semimobile conveyor and bucket plant systems that can be supported on previously stockpiled material. Surface equipment includes loaders, dozers, and wheeled scrapers (carryalls) operating on and depositing or moving material over the stockpile's top surfaces. Mobile equipment must be supported on stockpiled material. It is often necessary or desirable to employ several different combinations with the more common groupings presented in the following discussion.

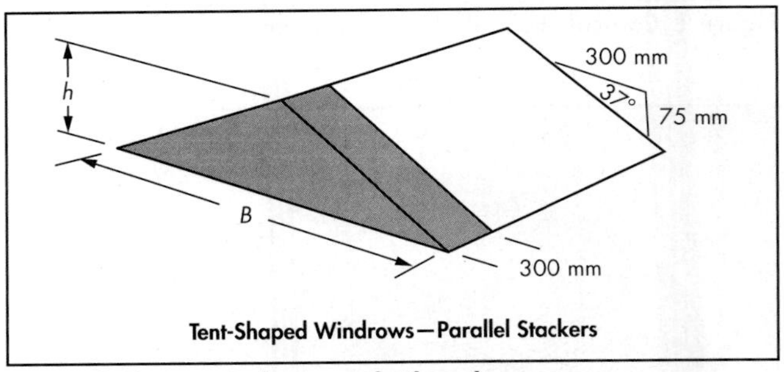

Figure 3 Linear windrow stockpile volume

Conveyors Above Stockpiles

Belt conveyors are the most common method of placing material in stockpiles. Commonly used arrangements of conveyors and auxiliary units are shown in Figure 4. The single stationary conveyor is a satisfactory and simple method for delivering material to a stockpile of moderate size but is often height restricted when supporting the cantilevered section of the feed

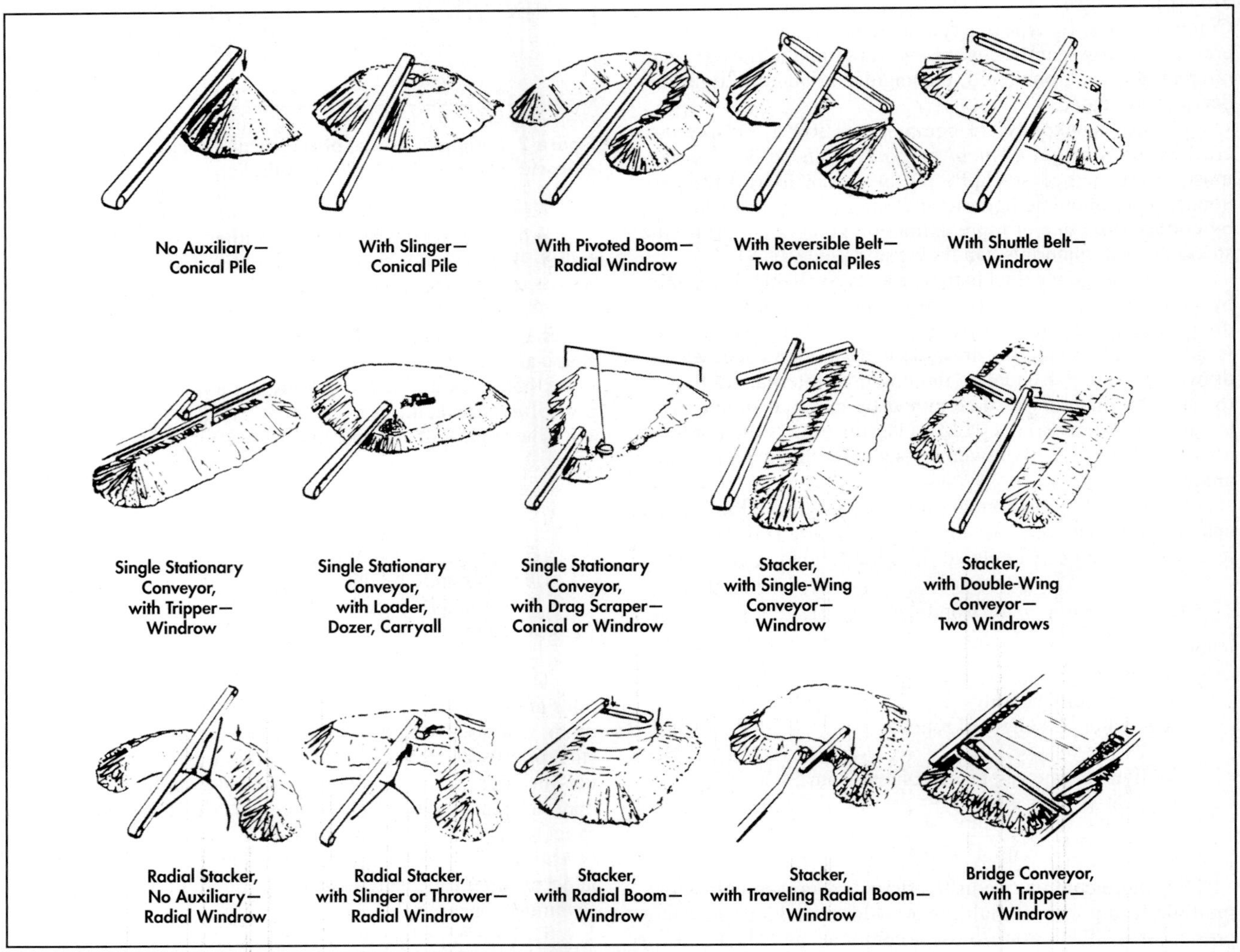

Figure 4 Methods of placing material in stockpiles using conveyors

conveyor. Conical stockpiles up to a height of 50 m have been built. Material segregation may occur with this arrangement.

A gain in storage volume can be obtained by attaching a slinger or thrower to the discharge chute of the stationary conveyor. A slinger is essentially a short, high-speed belt that can increase storage volume significantly. The slinger is not effective when segregation of sizes and densities is a problem, in windy areas, or when fines content is high and material is dry.

A single stationary conveyor with a pivoted boom increases storage volume by adding to the stationary conveyor a horizontal belt supported on a boom truss and arranged to pivot on the centerline of the conveyor discharge. The boom conveyor can discharge material along an arc of about 270°, building a large radial stockpile. The structure required to support a long boom can become elaborate and impose limitations. A single stationary conveyor with reversible belt can be used only if one type of material is being stockpiled. With this arrangement, a short, reversible belt is added under the discharge of the stationary conveyor, doubling storage volume. If more than one type of material is being stockpiled, the reversing cross-belt length should permit formation of two free-standing, nonoverlapping cones.

With a single stationary conveyor that has a shuttle belt arrangement, a reversing shuttle belt is placed on a wheeled frame under the stationary conveyor. The shuttle belt moves back and forth on an elevated track to expand the two cones into a long windrow. The shuttle can also be part of the conveyor, eliminating the transfer point. Volume gained is directly proportional to the length of the shuttle belt. Shuttle belts are often used to handle lumpy, sticky, or abrasive materials that would be troublesome in tripper chutes. They require less headroom than trippers, and their weight is more uniformly distributed along the supporting structure; shuttle belts up to 150 m long have been constructed. A traveling tripper can be installed to the stationary conveyor.

A tripper can be arranged to traverse almost the entire length of the elevated belt and discharge material into a long tent-shaped windrow. If desired, short-wing conveyors can be mounted under the tripper discharge to make a wider windrow. In windy areas, belting should be sheltered from the wind. Trippers may be driven by the conveying belt, powered by electric motors, or moved by a cable system. When using a single stationary conveyor with loader, dozer, or carryall, the conveyor serves as an auxiliary unit placing material within reach of the mobile equipment traveling on stockpiled material. The size and shape of the stockpile depends on the range and maneuverability of the surface units. If degradation of material is undesirable or the weight of equipment causes packing, this arrangement is not satisfactory. A single stationary conveyor with a drag scraper can be used when material properties prevent use of equipment operating on the surface of the stockpile. A conveyor is usually used to place material in a small "cushion" pile in front of the head post of the scraper. The cable-operated scraper moves the material over stored material and places it in piles or windrows.

A stacker usually consists of a cantilevered boom mounted on a wheeled or crawler-tread carriage that travels along a straight path or an arc of up to 270°. A stacker generally receives material from a conveyor–tripper unit. A single-wing stacker has a boom located on one side of its tower structure, usually fixed at right angles to the stacker's line of travel. The boom may be hinged for adjustment of the discharge height, minimizing material fall and reducing degradation, dusting, and wind segregation. A double-wing stacker can discharge to either side of its line of travel. This is accomplished by a single reversible belt conveyor or two separate conveyors. A radial stacker is a single, inclined conveyor pivoted at the tail end, on the loading-point centerline. The conveyor structure is supported near the midpoint by a tower. This support may describe an arc up to 270° in central angle. Units are available with rubber tires, flanged wheels, or crawler tracks. The carriage is usually self-propelled. When additional storage is required, a radial stacker with wing slinger provides an economical solution, sufficing when segregation, degradation, and dusting are not problems. For a stacker with radial-boom side conveyor, the boom is arranged to pivot about a vertical axis for approximately 90° of arc while the stacker moves along its main lines of travel. A hat-topped windrow of greater storage volume can be built. When using a stacker with a traveling radial boom where the boom can pivot through 180° of arc (90° on either side of the line of travel), a stockpile of enormous volume can be produced economically. Maximum movement is provided when the unit is mounted on crawler treads. With wheel-mounted units, the track is usually removed as stacking progresses.

A bridge conveyor is used when material must be placed in a stockpile in shallow, uniform layers to accommodate processing or blending requirements. Feed is delivered by a belt conveyor parallel to the stockpile's length. A moving tripper on the feed conveyor loads material onto the bridge conveyor, supported above the stockpile by a traveling bridge structure. The tripper on the bridge conveyor distributes material across the stockpile width, and the bridge movement distributes material along the stockpile length. The bridge can be equipped with a gantry crane mounted on top of the trusses and equipped with a clamshell to reclaim stockpile material by loading a hopper above the bridge conveyor. By reversing belt direction, material can be removed by the same units that delivered it to storage.

Buckets over Stockpiles

When properties of the material do not permit equipment traveling on the stockpile or when rate of material flow is relatively low, consideration should be given to reaching over the stockpile with drag scrapers, draglines, clamshells, bridge cranes, and drag conveyors (Figure 5).

Drag scrapers are suitable for moving relatively small volumes of material over large areas. The drag scraper consists of a cableoperated bottomless scraper bucket, controlled by a drum hoist, powered by electricity or an internal-combustion engine, and riding on a flow of moving material. When pulled forward, the bucket loads; when pulled backward, it empties. Drag scrapers give good service where material is wet, damp, abrasive, or of high temperature.

Wheel- or crawler-mounted draglines are used to excavate material at or below ground level. The stockpiles built by draglines are usually for short-term storage or surge piles. Pile height is limited to approximately 50% of boom length. This limitation results from the material's angle of repose, allowing for necessary dragline clearance. Clamshells are effective in placing materials into stockpiles because of their reach, dumping height, and dumping accuracy. They can pick up material above or below their level and place it at a higher level than is possible with a dragline or shovel.

Bridge cranes with clamshell buckets are used on bunker-type stockpiles within buildings or partially enclosed

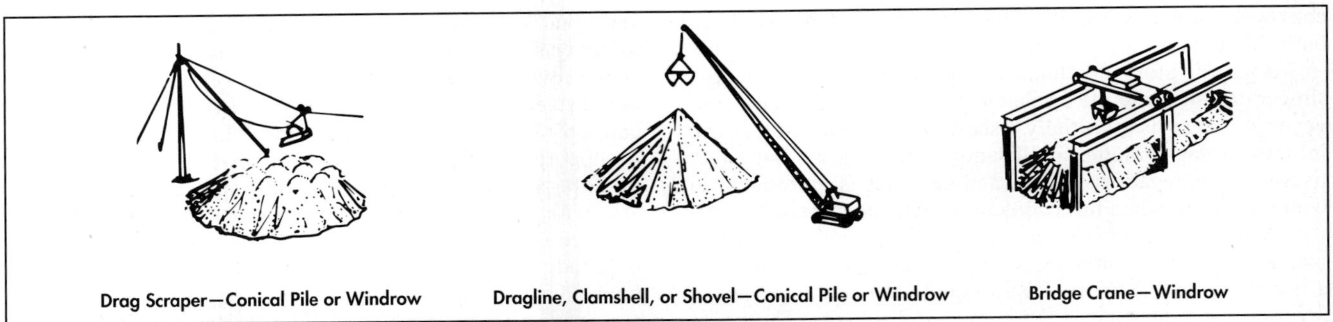

Figure 5 Methods of placing material in stockpiles using buckets over stockpiles

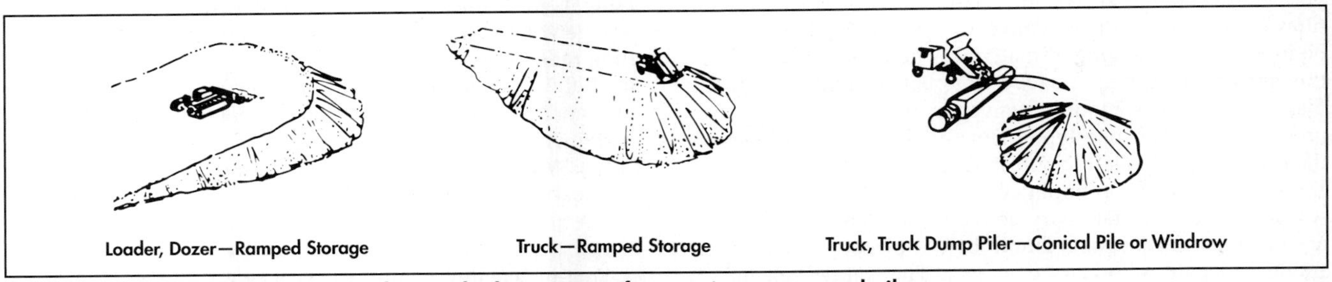

Figure 6 Methods of placing material in stockpiles using surface equipment on stockpiles

structures. The crane runway structure can be readily incorporated into the building's structural frame, making the bridge crane a reasonable choice for indoor bulk materials handling. Drag conveyors are of the bottomless variety in which the stockpile surface provides the bed on which incoming material is moved. Drag conveyors with bottom plates are suitable for many applications described for belt conveyors when handling damp, sticky, wet, or hot materials. Shovels are designed for digging, not stockpiling. Their limited boom and dipper stick length do not allow for development of large stockpiles.

Surface Equipment on Stockpiles

When material characteristics permit use of equipment traveling on the stockpile, loaders, dozers, wheeled scrapers, and trucks can be used (Figure 6 and Table 3). When trucks cannot operate on the stockpile, a portable dump piler can be used. Loaders and dozers are usually restricted to building ramped stockpiles in which material is spread over the ground. Succeeding loads are placed on top of previously placed materials. A ramp is developed gradually from which the loaders and dozers can dump material on existing slopes. The ramp should be kept at a reasonable grade (20%) so that the unit speed is not reduced too much. Rubber-tired units produce a more compact stockpile than crawler units because their wheels exert greater unit pressure. For large stockpiles occupying big areas, a wheeled scraper should be considered. It operates over a large radius and provides required compaction for materials such as coal. In addition, the unit may serve for reclamation.

Trucks are often used for placing materials directly onto small stockpiles, where material quantities are delivered from several sources. Reclamation is usually accomplished by surface equipment such as loaders and dozers. Where trucks cannot be operated on stockpiled material or where increased storage volumes are desired, a portable dump piler can be utilized (Sinden 1967). It can fill storage areas by throwing a steady, compact stream of bulk granular material up to 20 m and piling material up to 12 m high. Unit discharge angle can range from horizontal to vertical. The machine's unique centrifugal thrower action does not require material fall but operates with material fed through a side inlet from a hopper. The standard unit will unload a 5-t truck of 1,200 kg/m^3 material in one minute. Smaller units will discharge 90 t/h into 6-m-high stockpiles. Use of this unit is limited to materials that are not harmed by segregation and degradation and where dusting is not a problem.

Problems Encountered in Placing Material in Stockpiles

Segregation by size and density is a problem when materials have a wide particle size range or a difference in specific gravity. Segregation results from wind, water, and gravity action during use. Wind blowing across a stream of falling material will produce a segregation of fine from coarse particles or light from heavy particles. Fine (or light) particles of small mass will be carried horizontally for some distance while coarse (or heavy) particles of larger mass will fall downward. Depending on particle mass (size and specific gravity), wind-blown material may have considerable horizontal displacement. Rock ladders, telescoping chutes, or lowering wells are effective in countering wind action. Perforated pipe can be used effectively with finer material sizes. Hinged-boom conveyors that lower the discharge pulley to near the top of the pile minimize material freefall distance and reduce dusting. Segregation caused by gravity is the most common and greatest source of difficulty to stockpiling (Table 4).

Figure 7 illustrates theoretically how gravity segregation separates material sizes. Coarse material falling on the cone tip tends to roll down the sloping surface while finer particles tend to stop almost in the position of first impact.

Table 3 Equipment recommendations related to characteristics of stockpiled materials

Properties of Material to Be Stockpiled		Placing			Reclaiming			
Property	**Degree**	**Conveyors over Stockpile**	**Buckets over Stockpile**	**Surface Equipment on Stockpile**	**Conveyors Under Stockpile**	**Buckets over Stockpile**	**Surface Equipment on Stockpile**	**Notes**
Hardness	Very hard	OK	OK*	OK*	OK*	OK	OK*	*Check wear rate
	Moderate	OK	OK	OK	OK	OK	OK	
	Soft	OK	OK	NR†	OK	OK	NR†	†Not recommended
Range of sizes	Very coarse	OK	OK	OK	OK	OK	OK*	*Check maximum size
	Wide range	OK†	OK	OK	OK†	OK	OK†	†Check segregation
	Very fine	OK	OK	OK‡	OK	OK	OK‡	‡Check tire size
Particle strength	Strong	OK	OK	OK	OK	OK	OK	
	Medium	OK	OK	OK	OK	OK	OK	
	Weak–Friable	OK	OK	NR*	OK	OK	NR*	*Not recommended
Degradation of material	Undesirable	OK	OK	NR*	OK	OK	NR*	*Not recommended
	Not a factor	OK	OK	OK	OK	OK	OK	
	Desirable?	OK	OK	OK	OK	OK	OK	
Flowability	High	OK	OK	NR*	OK	OK	NR*	*Check tire size—flotation track
	Moderate	OK	OK	OK	OK	OK	OK	Consider special method
	None, or low	NR†	OK	OK	NR†	OK	OK	†Not recommended
Abrasiveness	High	OK	OK*	OK*	OK*	OK*	OK*	*Check wear rate
	Moderate	OK	OK	OK	OK	OK	OK	Consider abrasion-resistant plate
	None, or low	OK	OK	OK	OK	OK	OK	
Tendency to pack	High	OK	OK	NR*	OK*	OK	NR*	*Rip, scarify
	Moderate	OK	OK	OK	OK	OK	OK	
	None, or low	OK	OK	OK	OK	OK	OK	
Material temperature	Hot: >200°F	OK*	OK	OK*	OK*	OK	OK*	*Use high-temperature rubber,
	Moderate: 200° to 70°F	OK	OK	OK	OK	OK	OK	and check expansion
	Cool: <70°F	OK	OK	OK	OK	OK	OK	
Contaminable	Highly	OK	OK	NR*	OK	OK	NR*	*Not recommended
	Moderately	OK	OK	NR†	OK	OK	NR†	†Not recommended; check limits
	None, or little	OK	OK	OK	OK	OK	OK	
Corrosiveness	High	OK*	OK*	OK*	OK*	OK*	OK*	*Check coatings, etc.
	Moderate	OK	OK	OK	OK*	OK*	OK	
	None, or low	OK	OK	OK	OK	OK	OK	
Combustible	Highly	N/A	N/A	N/A	N/A	N/A	N/A	N/A = No data available
	Moderately	OK	OK	OK†	OK	OK	OK	†Provides compaction
	None, or little	OK	OK	OK	OK	OK	OK	
Toxic fumes and dust	Harmful amounts	OK	OK	NR*	OK	OK	NR*	*Not recommended; check
	Moderate	OK	OK	OK	OK	OK	OK	safety requirements
	None	OK	OK	OK	OK	OK	OK	
Floodability	Highly fluid	N/A	N/A	N/A	N/A	N/A	N/A	N/A = No data available
	Moderately	OK	OK	OK	OK	OK	OK	
	None, or low	OK	OK	OK	OK	OK	OK	

When material is drawn from the pile, it will have a widely varying size distribution; reblending to meet specification grades can be troublesome and costly. Segregation can be avoided by building the pile in shallow layers. For example, when working with a clamshell, the large pile is built up by accumulating many small heaps of material. Since it is not always feasible to prevent segregation, corrective measures are often necessary. Figure 8 is a theoretical model of a conical pile with a single center drawpoint showing probable particle size distribution in the cone's live storage portion. If this pile were drawn down, a sequential tapping of various particle sizes would occur. The first material discharged is primarily fines, followed by medium-sized and then coarse particles. This phenomenon and two corrective procedures have been verified by model tests.

The first procedure (Figure 9) uses four conical piles, each having a single, centrally located drawpoint. By filling these cones and drawing off in a definite sequence, the various particle size zones are reclaimed in approximately the desired proportions. The individual piles are never reclaimed until completely filled, and never filled until completely drawn off. Corrective reblending will not be attained unless this procedure is followed.

The second procedure (Figure 10) uses multiple drawpoints in a single-point loading cone. Illustrated is the theoretical particle size distribution in the cone and the way in which various size fractions can be reclaimed and blended. By regulating discharge rate from the various drawpoints, reblending of particle sizes is accomplished. It is also desirable to have

Table 4 Problems encountered in placing materials in stockpiles

Problem	Preventive Procedure	Corrective Procedure
Segregation, wind	Telescoping chutes Lowering wells, limber pipes Rock ladders Hinged booms, limit freefall of material	Blending operation
Segregation, gravity	Rock ladders Hinged booms Limit height of piles Place material in horizontal layers dropped from low belt; place material in small piles, avoid long slopes on which material can run If segregation must be avoided, do not place material having size ratio over 2:1	Blending operation Multiples drawpoints Sequential loading and reclaiming from several piles
Degradation	Rock ladders, reduce impact Hinged booms, reduce abrading and crushing of particles due to impact Avoid heavy equipment operating on stockpile or use crawler treads Raise scraper buckets above surface of material	
Dusting	Hinged boom, lessen free fall of material and exposure to wind Telescoping chute lowering wells	Use chemical sprays and binders
Windage loss	Place stockpile in building	Use dust-suppressing chemical sprays
Contamination	Cover material Building for the long term Covers or sprays for the short term	
Erosion	Cover material; temporary covers or sprays for the short term	
Abrasion and corrosion	See Table 6.	

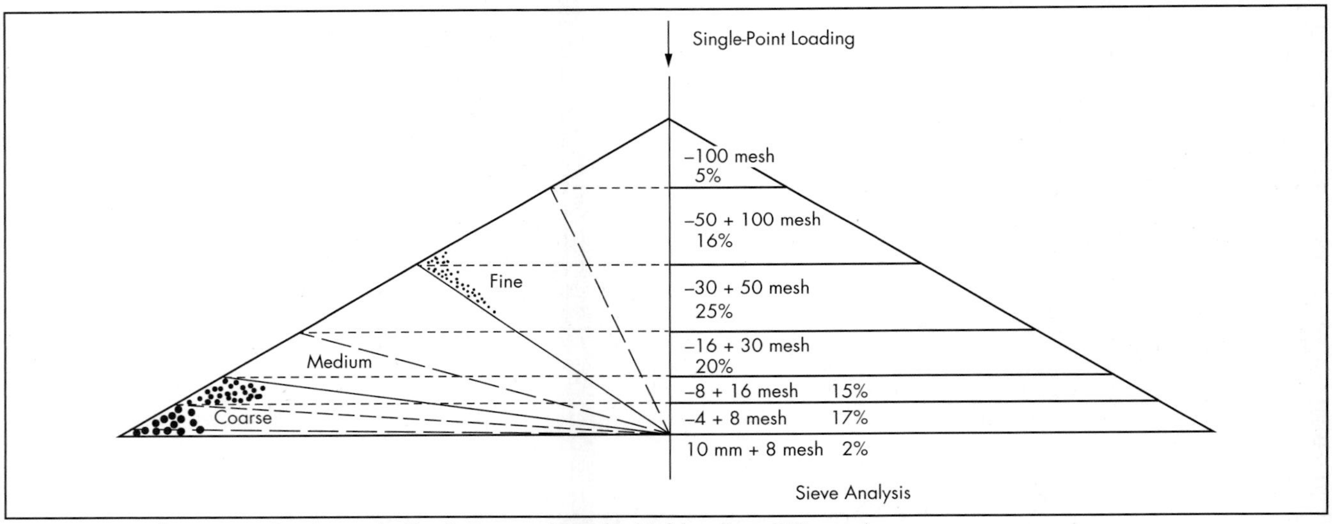

Figure 7 Particle segregation in stockpiles (cross section of initial loading, 31° cone)

the feed conveyor and the reclaim conveyor parallel to each other to improve blending.

Particle size degradation of material occurs wherever there is falling, sliding, or abrading of particles. An increase of fines may be detrimental. Production of excessive fines can make material more susceptible to "dusting" and result in economic losses. However, the greatest problem is the creation of a health hazard or community nuisance. Degradation can be reduced by using "rock ladders" that break the total freefall distance into a series of relatively short drops. Material falls on a dead bed of its own material, slides on the sloping surface, drops over the edge of the ladder "rung," and lands on the next lower ledge. Each time, direction of travel is reversed so that the material starts each subsequent fall with little or no momentum from the preceding drop. Both velocity and individual vertical fall distance are minimized. Degradation resulting from equipment weight can be lessened by proper choice of equipment, such as cushioned crawler treads or flotation tires. If no degradation is permissible, use of equipment moving on the stockpile surface should be avoided and preference given to units suspended above the material.

Dusting and wind losses can be reduced by minimizing the distance through which material must fall. Exposure time to wind action is lessened and material impact is reduced by means of hinged-boom conveyors and "rock ladders." Another device is the telescoping chute, which in extended position reaches from conveyor discharge point to stockpile bed. As stockpile height increases, the chute is gradually retracted, completely sheltering material from all normal wind action. "Lowering wells" can provide wind protection. These wells are constructed of circular pipe sections arranged on a vertical centerline coinciding with that of the discharge chute. At intervals

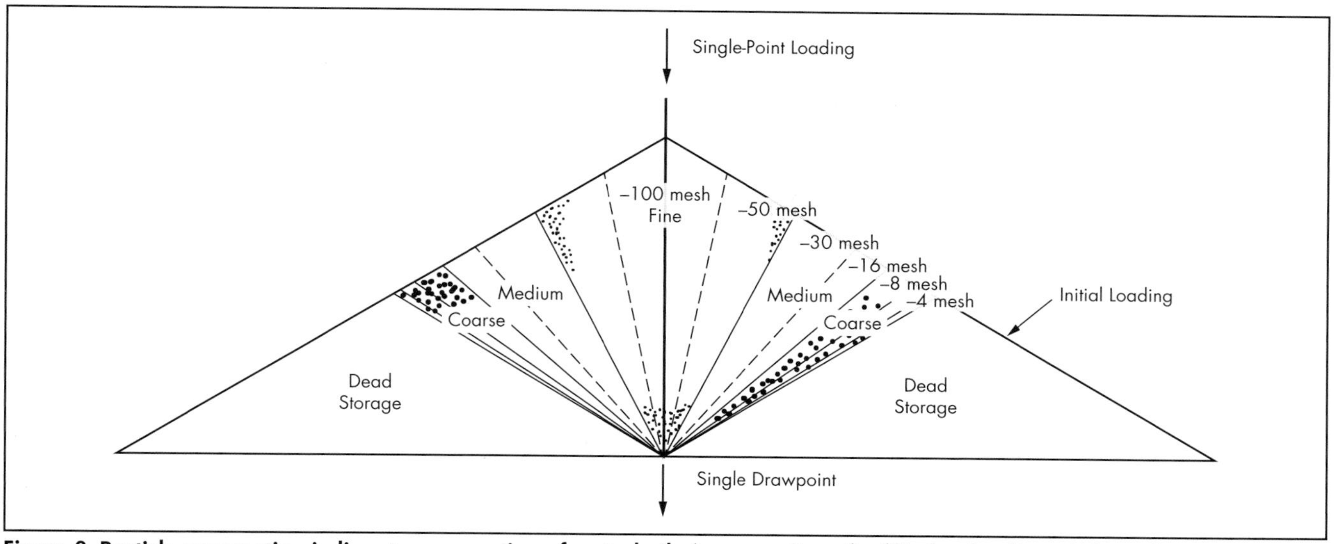

Figure 8 Particle segregation in live storage portion of a stockpile (cross section of refill after loading, 31° cone)

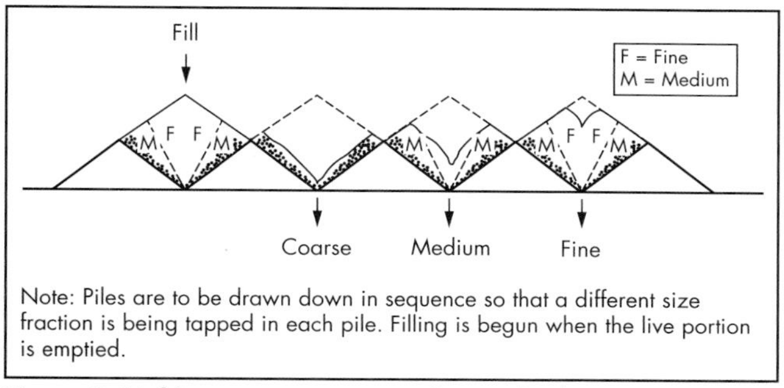

Figure 9 Multi-cone arrangement for stockpiling

along this vertical pipe column, openings are provided through which material can flow when the pipe is filled to the level of that particular opening. Water sprays in the discharge chute of the feed conveyor and over the stockpile surface can significantly reduce dusting during stockpiling. Fine water sprays for settling dust are often made more effective by addition of wetting agents, which reduce water surface tension. Dust suppression only provides temporary relief unless moisture content is maintained by spraying water on the surface. Material in a stockpile can be sprayed with chemicals that bind fine particles together, forming a porous surface crust that is resistant to wind action. Protective covers or buildings do not prevent dust production during materials handling, but they do reduce the wind effect on material already deposited. A permanent weathertight enclosure is the most effective protection from environmental dust, windblown debris, and rain. Where this is not possible, a temporary cover can be considered.

Summary

Equipment selection for placing material in a stockpile is dependent on material characteristics, stockpiling feed capacity, and type of stockpiles. The stockpiling rate provides approximate limits for standard equipment units. While not exact, these values provide a rough basis for eliminating a method on the basis of capacity requirement. Stockpile types and usual protection and materials handling methods are given in Table 5. Common problems experienced with placing material on stockpiles include segregation by size, density, and dust and wind losses. Methods can be applied to limit the impact of these problems.

Reclamation from Stockpiles

Conveyor equipment includes gates, feeders, and conveyors operating under the stockpile and withdrawing material from below the pile. Buckets include drag scrapers, draglines, clamshells, shovels, cranes, rotary reclaimers, bucket-wheel excavators, and drag conveyors operating over the stockpile with material taken from the top of the pile. Conveyor and bucket equipment are not supported by the stockpile. Surface equipment includes loaders, shovels, dozers, and wheeled carryalls, operating on the stockpile with material taken from the top surface of the pile. Surface equipment is supported by the stockpile.

Conveyors Under Stockpiles

Compact, efficient, and economical prefabricated tunnel sections and feeders have encouraged conveyor use under stockpiles. Tunnels generally run below grade for the stockpile length and slope upward to a discharge point above ground level. At intervals along the tunnel, openings in the tunnel roof permit flow of stockpile material into the tunnel and onto the belt conveyor. Material is usually funneled through a hopper. Flow rate is controlled by gates, chutes, or feeders. Tunneling arrangement and related equipment is determined largely by stockpile type and size. For free-flowing materials, a feeder may not be required.

A gate of a radial, clamshell, or slide type in the chute between the drawpoint hopper and transporting conveyor is necessary to shut off material flow. These gates provide a fairly positive shutoff and can be used safely with materials of highly fluid characteristics. With free-flowing material that does not "flood," an adjustable chute can be used. The chute is raised or lowered to vary the slope and thus regulate the flow. Flow is shut off by raising the chute so that the toe of material falls within the chute length.

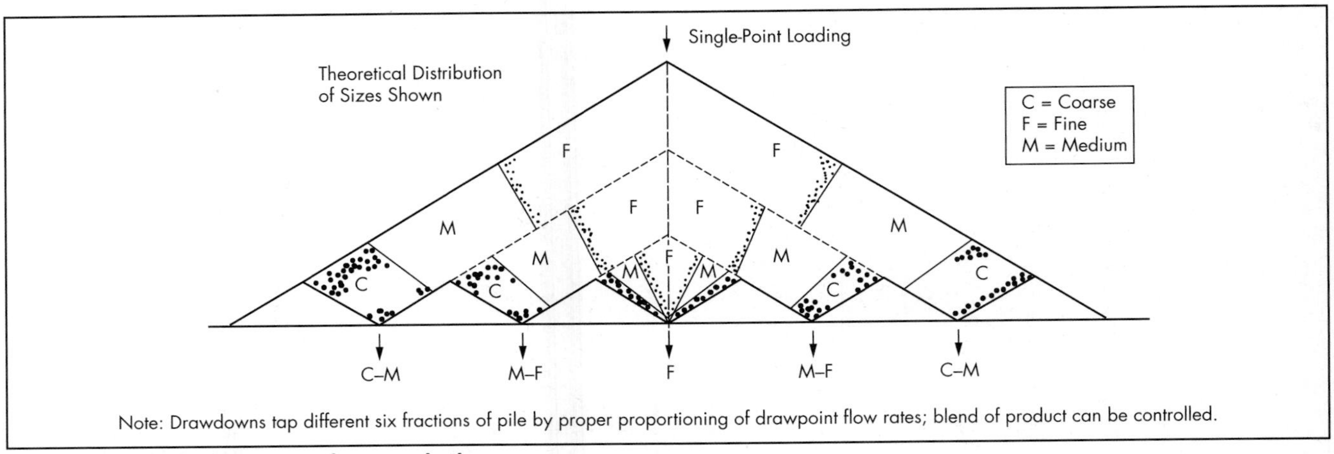

Figure 10 Multiple drawpoints for a stockpile

Table 5 Stockpile types and usual protection and materials handling methods

Type of Stockpile	Methods of Protection	Methods of Placing Material in Stockpile	Methods of Reclaiming Material from Stockpile
Conical piles	Buildings Plastic covers Chemical binders	Conveyors Clamshells Draglines	Conveyors Loaders and dozers Clamshells and draglines Shovels Drag scrapers
Windrows—parallel	Buildings Plastic covers Chemical binders	Conveyors Drag scraper Clamshells Draglines	Conveyors Loaders and dozers Clamshells and draglines Shovels, bucket wheels Drag scrapers, rotary reclaimers
Ramped stockpiles	Compaction Chemical binders Seeding	Trucks Scrapers, tractor-drawn Dozers	Loaders and dozers Clamshells and draglines Shovels Drag scrapers
Cribbed piles	Buildings	Conveyors Trucks Loaders and dozers	Conveyors Loaders and dozers Clamshells and draglines Shovels
Bins—on ground	Buildings Self-enclosed	Conveyors	Conveyors Cranes, with clamshells
Bins, mobile, railway, cars, barges	Self-enclosed	Conveyors Clamshells Loaders	Conveyors (track hoppers) Unloaders, bucket Clamshells Pneumatic conveyors

When control of flow is desired or materials are not free flowing, a feeder is necessary. Vibrating pan feeders, apron feeders, and belt feeders are used depending on the application. Vibrating pan feeders are a cost-effective solution for lower-capacity stockpile withdrawals of small to medium particle size. Higher-capacity stockpile withdrawals are possible for larger-particle-sized material like primary crushed ore using belt and apron feeder equipment. When the material shape and hardness pose a risk of damage to belting when being drawn, apron feeders are preferred. A shutoff mechanism is required to provide safe access to the feeders for maintenance. Sliding gates and spile-bar systems are often used.

Spile bars are large steel rods that can be inserted across the drawpoint hopper to restrict flow through the length of the hopper. Inserting all the spile bars effectively shuts off the flow of larger-particle materials. Partially inserted spile bars are often used to further control drawdown of a stockpile. Insertion and extraction of spile bars can be manual or with hydraulic spile-bar insertion equipment.

Equipment Operating over Stockpiles

Drag scrapers for placing material have been previously described. To adapt a drag scraper to reclaim operations, a sloped loading ramp is added to elevate material. Draglines may be used for reclaiming materials from stockpiles by providing a loading hopper into which the bucket can be emptied. Trucks or cars can be loaded by opening the flow-control gate under the hopper. Small draglines can empty their buckets directly into transport equipment, but this procedure usually results in considerable spillage.

Clamshells are especially suited to stockpile reclamation since they operate in a vertical digging range. Their dumping accuracy is well suited to direct loading into trucks, railroad cars, or elevated hoppers. Power shovels can reclaim material from stockpiles, especially where a strong digging action is required. Material can be placed readily in hoppers, trucks, or railroad cars.

The rotary reclaimer is used successfully to reclaim from both linear and radial windrows. Bucket-wheel boom reclaimers run on tracks parallel to the windrow stockpile over a stationary conveyor onto where the reclaimed material is discharged. The system allows the boom to move in a sweeping and up-and-down movement. By a unique harrow arrangement, this unit minimizes segregation that often accompanies reclaiming operations. Pickup is by rotating buckets which discharge onto a belt conveyor on the boom that discharge onto the stationary conveyor. On linear windrow stockpiles, the bucket-wheel boom reclaimer can be rotated 180° to reclaim a second stockpile that is running parallel with the first. The bucket-wheel boom reclaimer can be combined with a boom stacking conveyor into one unit that can alternate between stacking and reclaiming. Bucket-wheel boom conveyors can also be mounted on a tracked crawler, allowing the reclaiming of irregular-shaped stockpiles. The bucket-wheel stacker/reclaimer, both rail and crawler mounted, was developed extensively in Europe prior to 1960 and has been extensively used in North America since that time. Most American installations are used for coal storage and reclaiming in and around thermal storage plants, although some are located at steel plants for raw materials handling there.

Bucket-wheels are also used in bridge-type reclaiming arrangements. A bridge structure spans across the width of a linear or radial windrow stockpile. One or two bucket-wheels transverse the bridge structure, discharging onto a conveyor running across the bridge. The bridge can move along the stockpile length on tracks while discharging the material onto a stationary conveyor along the stockpile or, in the case of a radial stockpile, into a center chute. Segregation reclaim variations are experienced with single bucket-wheel bridge reclaimers but with a double bucket-wheel bridge reclaimer, the two wheels are arranged in such a manner to smooth out the variation in material due to segregation.

In the Messiter system, a harrow rakes material from the pile face and a drag conveyor moves it laterally from the toe and delivers it to a belt conveyor parallel to the stockpile centerline. In the Dodge system, drag conveyors with adjustable bottoms are used to place stockpile material. In reclaiming, drags without bottom plates are used to rake material from the pile toe into a hopper that loads a belt conveyor. Another system uses the drag to rake material down the windrow slope and into a hopper that loads a belt conveyor running parallel to the pile. The bucket unloader is similar to a bucket elevator without a casing. Such a unit is used principally for unloading from barges, rectangular bins, or other containers. It is well suited for work in close quarters.

Surface Equipment on Stockpiles

The modern front-end loader's operating flexibility and speed provide an efficient unit for reclaiming and cleanup work around stockpile areas. These loaders have bucket capacities up to 40 m^3. The loader can load trucks or railroad cars directly or place material in a discharge hopper. It is sometimes unnecessary to pick up and carry material. The dozer can push material into a "trap" (depressed hopper) that feeds an inclined conveyor for loading transport units. Wheeled scrapers are limited to large stockpiles since the average-sized plant cannot justify their cost. They are used where large tonnages must be handled quickly and inexpensively.

Problems Experienced with Reclaiming

Material characteristics should be checked before deciding on a reclaim method. Some materials pack more under vibration and others fluidize with aeration. Material may often exhibit a tendency toward funnel flow or ratholing over drawpoint openings by flowing only through a cylindrical channel just above the drawpoint. In extreme cases, this could not only reduce the live volume but also make reclaiming inefficient. The drawdown opening size could be maximized and a larger number of closely spaced openings under the stockpile could improve flow. Flow promoters can be used. The friction angle of material to the chute liners is often lower than the internal friction angle of the material. The material will flow more readily against a liner than shear against itself. Internal surfaces can be provided in drawdown hoppers to improve flow.

Some material may be difficult to flow even when great care is taken in engineering the drawdown openings, hoppers, and chutes. For fine nonabrasive materials, liner products like glass and ultra-high-molecular-weight polyethylene show improved friction values for sticky materials. Poke holes provide some relief but are usually inadequate. Air cannons can be installed to shear the material during drawdown in finer materials. Vibrating bin bottoms and vibrating motors mounted on the side of hoppers can improve flow in some cases. When these methods are not a viable option, serious consideration should be given to a different system of reclaiming, preferably the use of buckets or surface equipment.

Packing can be the result of many factors such as adverse moisture content, freezing, cementing action, or pressure from vehicles or overlying material. When material exhibits a tendency to pack, overhead equipment for material placement and equipment with a strong digging action for reclaiming should be used. Drag scrapers, loaders, or shovels working at the pile edges may be satisfactory. When reclaiming with loaders or shovels, the operator must be alert for the formation of "cliffs" that may collapse suddenly, endangering personnel and equipment. In extreme cases, it may be necessary to blast or rip material and to crush the material to regain the required particle size. Plugging of drawpoint openings as a result of interlocking of coarse particles can be corrected by proper sizing of opening dimensions (Figure 11).

Care must be exercised in designing gates for the regulation and flow control of floodable materials. Feeder selections should only be made after consultation with the equipment manufacturer. Abrasive and corrosive properties have little effect on flow characteristics, but they are important in selecting construction materials. Hoppers and chutes for reclaiming or transferring stockpiled materials are usually fabricated from steel plate. Ordinary mild steel is satisfactory when the hardness of the bulk material is 4 or less on the Mohs hardness scale. Mild abrasion occurs when the Mohs hardness is between 5 and 7; severe abrasion is likely for hardnesses between 8 and 10. However, hardness alone is not a valid guide to abrasive wear. Particle shape, weight, and angularity;

sharpness of edges; flow rate; and amount of impact influence the abrasive wear rate.

If severe abrasion is encountered, AR liners can be installed at wear points. Some operators try to anticipate these before starting operation while others install unlined hoppers and chutes, run them for a brief period, and shut down to install liners at points that exhibit the greatest wear. In either case, further abrasion occurs on an expendable liner and the costlier conveying structure is given extended operating life. Table 6 provides a summary of common problems encountered in reclaiming material from stockpiles, including preventive and corrective procedures.

Unprotected Stockpiles

Protection of stockpiles is often omitted when the type of stacking and reclaim equipment and the scale of the stockpile make protection uneconomic. Stockyards where boom stacking conveyors are used for stacking multiple linear-windrow stockpiles and bucket-wheel reclaimers are used to reclaim the material are too large for covering by buildings. Large conical stockpiles are often more than 50 m high, which often makes protecting the stockpile from the elements uneconomical. Water sprays may be used for dust suppression of fine material, with chemicals added to form a crust that is resistant to wind blowing up dust.

Ore Type →

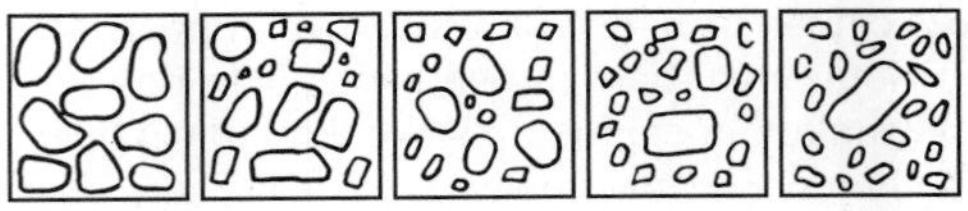

Throat Opening Factor:	3½	3	2½	2	1½
Percent Larger-Size Pieces:	95–100	60–80	40–60	20–30	5–10
Other Conditions:	Screened	Mine Run	Mine Run	—	—

Throat opening selection. To determine the throat opening, (1) select the picture that matches the type of material to be handled, and (2) select the corresponding throat opening factor. Multiply the factor by the largest dimension of the biggest piece of material. This will provide the minimum throat opening to prevent bridging.

Figure 11 Throat opening size

Protection of Stockpiles

Indoor Storage

Building configuration generally is determined by stockpile type and placement, and by the reclaiming equipment used. Buildings are of three general shapes: (1) conical, including domes and semi-domes; (2) tent-shaped, including arches; and (3) box-shaped, rectangular, and flat-roofed.

Conical and radial windrow stockpiles are often stored in dome, semi-dome, or conical buildings where freezing during winter months or high seasonal rainfall may reduce the free flowing of material. Conical stockpiles could also be protected when the stored material poses an environmental or health risk (e.g., because of asbestos in the ore) or where product loss is unacceptable (e.g., with high-grade gold ore fines). Short windrow stockpiles can be protected by tent-shaped or box-shaped buildings.

Table 7 provides information on building type, construction materials, stockpile type, and placing and reclaiming equipment for the various building types. Figures 12 and 13 are examples of indoor storage. Common construction materials are steel, wood, aluminum, and concrete; the choice is usually determined by the corrosive characteristics of stockpiled material. However, protective coatings permit the use of non-corrosion-resistant materials having adequate stiffness and strength. When in doubt, the effect of raw material on structural components should be tested.

Temporary Covers

Plastic, canvas, or other film coverings are often adequate for temporary protection. Nylon-reinforced plastic covers are

Table 6 Problems encountered in reclaiming materials from stockpiles

Problem	Preventive Measures	Corrective Measures
Bridging or arching over drawpoint	Design hopper slopes to be steeper than minimum angle of slide. Make hopper opening as wide as possible. Use multiple openings close to each other.	Use flow-promoting methods. Provide poke holes. Agitate material with vibration, aeration, and pulses.
Ratholing	See bridging above.	See bridging above. Use equipment such as drag scrapers, drag conveyors, or screw conveyors to move the material along the top of pile to the hole.
Packing	Avoid placing weight on the material and compacting it. If excessive moisture is part of the problem, dry the material before stockpiling.	Rip, scarify, or blast material to dislodge it from the pile. In extreme cases, it may be necessary to crush material to regain desired range of sizes.
Plugging of openings	Design the openings to be large enough to handle the largest dimension of maximum particle size. Steepen slopes of chutes and hoppers; use low-friction liners; use coatings to reduce resting and roughening of surface.	Rodding is probably the only effective method.
Flooding	Select gates and feeders that provide positive shutoff of flow.	Install replacement gates and feeders that provide positive shutoff of flow.
Abrasions	Use abrasion-resistant liners, alloy plates, epoxy coatings, fiberglass coatings, or rubber linings. Use dead beds at impact points.	Use abrasion-resistant liners, epoxy or fiberglass coatings, or rubber linings.
Corrosion	Select corrosion-resistant materials. Use protective coatings; avoid the use of dissimilar metals or employ insulators.	Remove products of corrosion by sanding; apply protective coatings to the cleaned surfaces.
Caving materials	Keep shovels and loaders away from toe of wall to avoid damage from fall of material.	Break down "cliffs" with boom or cable before it becomes too high and too steep.

Table 7 Types of buildings used to protect stockpiles from weather

Types of Buildings	Materials of Construction	Types of Stockpiles	Placing Equipment Commonly Used	Reclaiming Equipment Commonly Used
Cones or domes	Frame • Steel • Wood • Aluminum • Plastic Cover • Steel • Wood • Aluminum • Plastic • Protected metal	Conical piles Radial windrows	Conveyors • Belt • Screw • Shuttle • Trippers • Slingers • Pneumatic	Feeders Conveyors Rotary reclaimers Loaders
A-frames or arches	Frame • Steel • Wood Cover • Steel • Wood • Aluminum • Plastic • Protected metal	Windrow (parallel)	Conveyors • Belt • Screw • Shuttle • Trippers • Slingers Drag scrapers	Feeders Conveyors • Belt • Screw • Drag Drag scrapers Loaders Rotary reclaimers
Box or rectangular frames	Frame • Steel • Concrete Cover • Steel • Wood • Aluminum • Plastic • Protected metal • Concrete	Windrows (parallel) Bins on ground, craneways	Conveyors • Belts • Screws • Shuttles • Trippers • Slingers Drag scrapers Cranes, bridge with clamshell bucket	Feeders Conveyors • Belt • Screw • Drag Drag scrapers Loaders Rotary reclaimers Crane, bridge with clamshell bucket

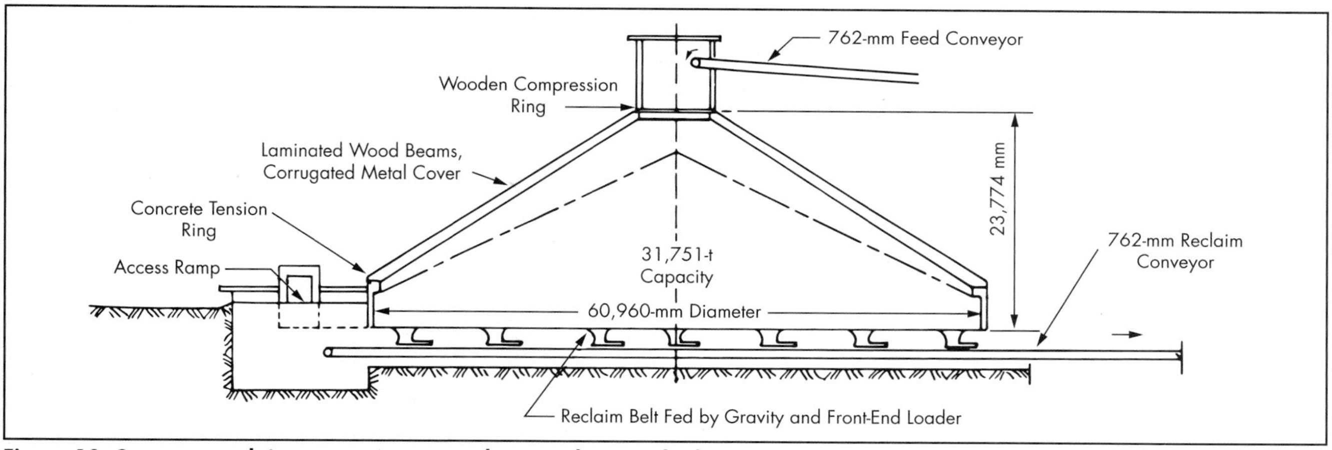

Figure 12 Conveyor reclaim system in a semi-dome, indoor stockpile system

available in different sizes up to 60 × 60 m. Generally, the sheet size is limited by the purchaser's handling capabilities. For cover hold-down, the sheet should be tied at the pile bottom (fastened to a plank border) to prevent stress concentrations. Weights should be placed along the pile top and sides so that wind action does not damage the cover. Sacks filled with stockpiled material can be used as weights to avoid contamination through breaking and spillage of the sacks. Various weights and grades of plastic covers are available. For large stockpiles, three- or four-ply material is recommended.

BINS

The top, vertical portion of a bin is called the *cylinder* and the bottom converging portion, the *hopper*. Purpose and process requirements determine bin capacity and discharge flow rate. Bin type and flow control device depend on storage requirements, material characteristics, discharge rate, anti-segregation and blending requirements, temperature, operating controls, and number of drawpoints. Possible constraints are plant layout, headroom, and foundation restrictions.

Bin Flow Patterns

The three flow patterns in bins are mass, funnel, and expanded (Figure 14). In mass flow bins, all the material is in motion within the bin when the material is drawn from the outlet. The bin hopper must be sufficiently steep, the surface-to-material friction low, and the feeder under the hopper outlet must draw material across the full outlet area. There is material flowing along the wall of the cylinder section and the hopper section of the bin. Since mass flow bins do not channel, they have a first-in, first-out flow sequence with a relatively uniform residence time for all particles and depend only to a small degree on initial position of the particles in the bin. Fine particles or powders flow uniformly and without flooding or flushing from mass flow bins, providing the bin level is maintained, allowing for deaeration, and a shutoff gate is available for initial filling. With mass flow bins, complete discharge of the bin contents is guaranteed.

Materials that degrade and cake have a minimum residence time in mass flow bins. Caking can be prevented and blending achieved by recirculating material through a mass flow bin. Suitably designed mass flow bins also can provide nonsegregating storage. Even though material segregates during placement into a bin, the various fractions remix in the hopper and are withdrawn in the same proportion as charged. Level indicators work reliably in mass flow bins.

In funnel flow bins, material flows toward the outlet through a channel extending upward from the discharge point. The drawdown opening governs the size of a flow channel that develops above the opening. The walls of the flow channel extend slightly, diverge from the opening, and are surrounded by stagnant material. In some extreme cases, the flow channel walls may be vertical. As material discharges, the channel level drops and layers of stagnant material slough from the top of the surrounding mass into the channel. Funnel flow is the result when the bin hopper wall is not steep enough and the surface-to-material friction is too high to achieve mass flow. This pattern leads to a first-in, last-out flow sequence because most of the material that was first deposited at the bin bottom around the channel does not discharge until the bin is emptied.

In all bins, segregation occurs at the impact point of the falling material. In funnel flow bins, there is no remixing of segregated material in the hopper. Funnel flow bins may lead to piping (ratholing) when stagnant material consolidates and they remain stable after the channel empties. Flooding or flushing may result when layers slough from the top of the stagnant mass and hit the channel bottom or as loose material is loaded into a bin and rushes to the outlet. Conventional level indication does not provide a reliable indication of the bin's content or condition.

Funnel flow bins are acceptable only for chemically stable and nonsegregating materials. Coarse, free-flowing material often contains a fine, non-free-flowing fraction. This fine

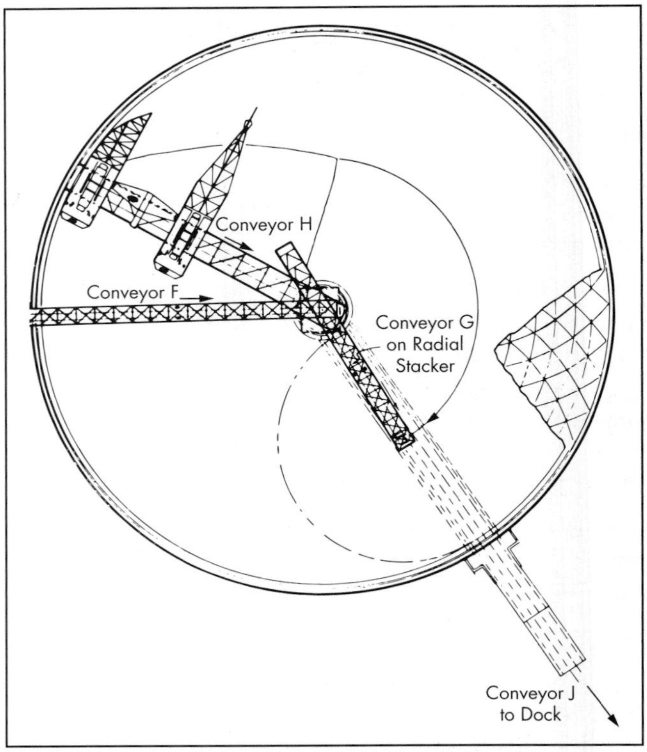

Figure 13 Indoor dome storage with a radial stacker and reclaimer

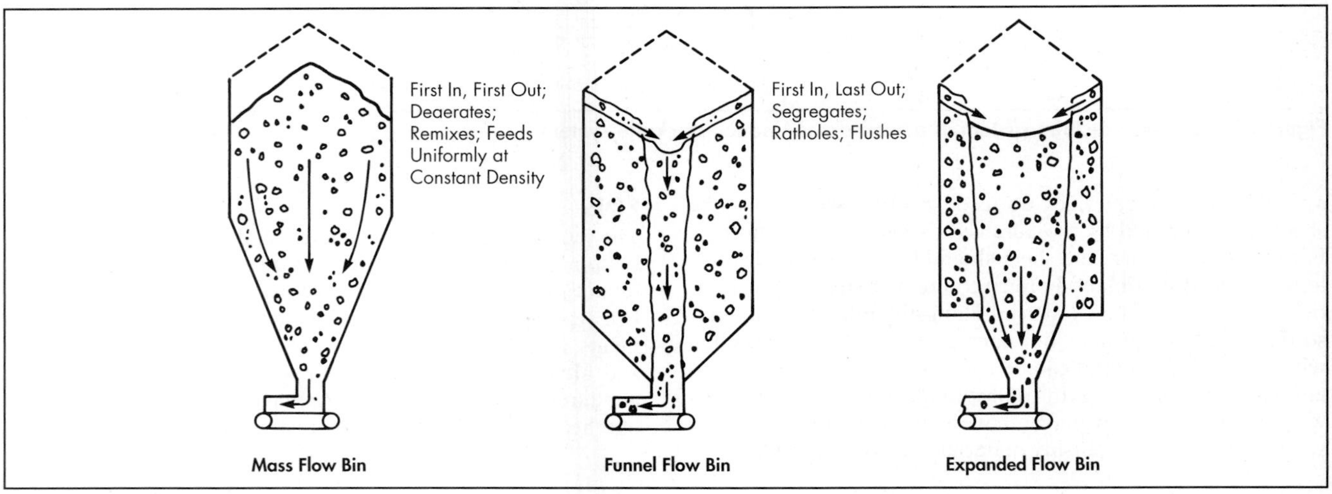

Figure 14 Types of bins

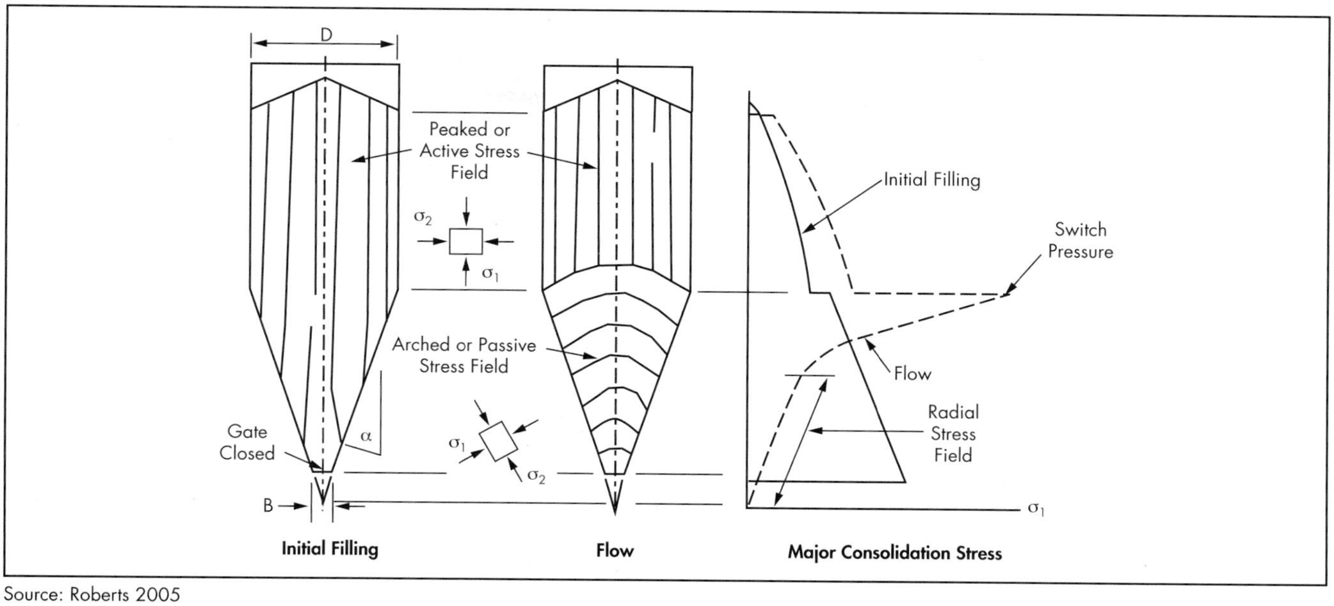

Source: Roberts 2005

Figure 15 Stress fields for initial filling and flow in mass flow bins

fraction may be uniformly distributed throughout the material when placed in the bin, but fines tend to segregate in the bin and percolate into stagnant regions. To prevent loss of live capacity, funnel flow bins for coarse materials should be emptied completely at regular intervals.

Expanded flow combines the wall protection that funnel flow offers with the reliable drawdown of mass flow. The hopper section of the bin is designed for mass flow while the cylinder section for funnel flow. The size of the hopper section opening at the transition should be at least equal to the critical rathole dimension at that height. The mass flow hopper eliminates the possibility of piping (ratholing), ensures deaeration and smooth flow, and reduces segregation. Expanded flow bins are used when storage of large volumes of material is required at acceptable bin heights. This design is also useful in storing large quantities of non-degrading materials. A low-level indicator can be placed within the mass flow hopper. Several mass flow hoppers can be placed under a storage pile or expanded flow bin to reduce dead storage.

Bin Geometry

Bin geometry is determined by volume and flow requirements, taking into consideration maximum pressure limitations. Design for strength of the bin structure needs to take into account the initial and flow stress-field conditions. Resulting stress-field patterns are determined by the bin geometry. When the bin is filled from the empty condition, a peaked or active stress field develops with the major principal stress almost vertically acting on the bottom of the bin and drawdown gate. Once the bin is being drawn down and material flow starts inside the bin, the vertical reaction provided by the gate is removed and the load is transmitted to the hopper walls. The stress field in the hopper changes from a peaked stress field to an arched stress field up to the transition from hopper section to cylinder section. Once a stable stress field is obtained, it is retained even after drawdown is stopped. This means that the load on the gate once the flow has stopped is significantly lower than during initial fill from empty. A major part of the material load in the cylinder section is carried by the upper region of the hopper due to the arched stress field in the hopper. The stresses at the discharge are relatively quite low. Figure 15 presents a graphical representation.

Methods of Placing Material in Bins

Conveyors and buckets are used to place material in bins, as previously described. Material segregation and degradation are minimized during placement in the bin by the same methods used in stockpiling.

Bin Selection

A bin, silo, or bunker generally consists of a vertical cylinder and a sloping converging hopper. Based on the flow pattern that develops, there are three types of bins: mass flow, funnel flow, and expanded flow. The type of bin selected should have the desired capacity, be capable of discharging its contents reliably on demand, and be safely constructed.

Jenike (1964) developed the continuum mechanics approach to hopper design. He defined "mass flow" and showed how the flow properties of bulk solids could be measured as well as how these measurements could be used to design hoppers. Jenike is most strongly associated with the Jenike shear tester. Jenike developed the "rational design method" for bins and hoppers, based on continuum mechanics. As described by Bradley et al. (2011), this method is based on a model of stress distribution in the hopper, calculated using measurements of the flow properties of the material being handled. It predicts the flow pattern that will occur and whether or not flow will be reliable. Bradley et al. note that "... provided the material tested is representative of that loaded into the plant, and any issues of change in moisture content, time in static residence, caking effects etc. are taken into account, it delivers a hopper design that can be guaranteed to work."

Unfortunately, the cost of the testing required for the use of the rational design method can be quite high. Tests using Jenike's shear tester can easily take two days' work by a highly skilled technician to determine the properties of one

Table 8 Important flow properties

Parameter	Measured by	Useful for Calculations of
Cohesive strength	Shear tester	Outlet sizes to prevent arching and ratholing
Frictional properties	Shear tester	Hopper angles for mass flow, internal friction
Sliding at impact points	Chute tester	Minimum angle of chute at impact points
Compressibility	Compressibility tester	Pressure calculations, bind loads, feeder design
Permeability	Permeability tester	Discharge rate calculations, settlement time
Segregation tendency	Segregation tester	Predicting whether segregation will occur
Abrasiveness	Abrasive wear tester	Predicting the life of a liner
Friability	Annular shear tester	Maximum bin size, effect of flow pattern on particle breakage

Source: Carson and Holmes 2002

material. Subsequent designs improved on Jenike's original, and it is now reasonable to test materials in a short time and at a reasonable cost. Several companies offer this service.

A procedure for designing and sizing a storage bin should include the following steps.

Step 1. Define the Storage Requirements

Identify the operating requirements and conditions. Some of the more important ones include the following:

- **Capacity.** This will vary with the plant's operating philosophy and where the bin is to be located (e.g., start of your process, at an intermediate process step, or at the end).
- **Discharge rate.** Consider average and instantaneous rates, minimum and maximum rates, and whether the rate is based on volume or mass.
- **Discharge frequency.** How long will your material remain in the bin without movement?
- **Mixture and material uniformity.** Is particle segregation a concern in terms of its effects on material discharge or, more importantly, downstream processes?
- **Pressure and temperature.** Consider differences between the bin and upstream and downstream equipment.
- **Environmental.** Are there explosion risks, human exposure concerns, and so forth?
- **Construction materials.** Abrasion and corrosion concerns may limit the types of materials you can use to construct your bin.

Step 2. Calculate the Approximate Size of the Bin

Initially, ignore the hopper section. Use the following formula to estimate the approximate height of the cylinder section that is required to store the desired capacity:

$$H = \frac{C}{\gamma_{avg} A}$$

where

H = cylinder height, m
C = bin capacity, m^3
γ_{avg} = material's average bulk density, kg/m^3
A = cross-sectional area of cylinder section, m^2

The actual cylinder height will have to be adjusted to account for volume lost at the top due to the material's angle of repose as well as for the volume of material in the hopper section. In general, the height of a circular or square cylinder should be between about 1.5 and 4 times the cylinder's diameter or width. Values outside this range often result in designs that are uneconomical or have undesirable flow characteristics. It is important to recognize that a bin's storage volume and its active (live, useable) volume are not necessarily the same. With a funnel flow or expanded flow pattern (described later), significant dead (stagnant) volume may need to be taken into account.

Step 3. Determine Your Material's Flow Properties

The flow characteristics of a bulk solid must be known to predict or control how it will behave in a bin or hopper. These characteristics can be measured in a solids flow testing laboratory under conditions that accurately simulate how the solid is handled in the plant. Tests should be conducted on-site if the solid's properties change rapidly with time or if special precautions must be taken.

The most important bulk solids handling properties that are relevant to predicting flow behavior in bins and hoppers are listed in Table 8. Each of these parameters can vary with changes in the following:

- Moisture
- Particle size, shape, hardness, and elasticity
- Temperature
- Time of storage at rest
- Chemical additives
- Pressure
- Wall surface

The appropriateness of these bin design parameters has been proven over many years in thousands of installations handling materials as diverse as fine chemical powders, cereal flakes, plastic granules, and mined ores.

Step 4. Understand the Importance of Flow Patterns

Although it is natural to assume that a bulk solid will flow through storage or conditioning vessels in a first-in/first-out sequence, this is not necessarily the case. Most bins, hoppers, silos, and conditioning vessels move solids in a funnel flow pattern.

With funnel flow, some of the material moves while the rest remain stationary. This first in/last-out sequence is acceptable if the bulk solid is relatively coarse, free flowing, nondegradable, and if segregation is not important. If the bulk material and application meet all four of these criteria, a funnel flow bin is the most economical storage device.

Unfortunately, funnel flow can create serious problems with product quality or process reliability. Arches and ratholes may form, and flow may be erratic. Fluidized powders often have no chance to deaerate. Therefore, they remain fluidized in the flow channel and flood when exiting the bin. Some materials cake, segregate, or spoil. In extreme cases,

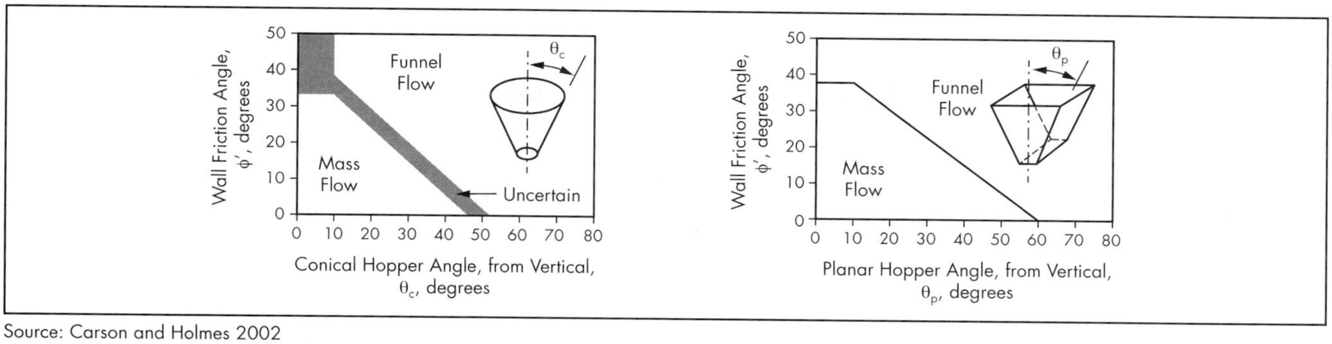

Source: Carson and Holmes 2002

Figure 16 Typical charts for determining mass flow wall angles

unexpected structural loading can result in equipment failure. These problems can be prevented with storage and conditioning vessels specifically designed to move materials in a mass flow pattern. With mass flow, all material moves whenever any is withdrawn. Flow is uniform and reliable; feed density is independent of head solids in the bin; there are no stagnant regions, so material will not cake or spoil and low-level indicators work reliably; sifting segregation of the discharge stream is minimized by a first-in/first-out flow sequence; and residence time is uniform, so fine powders are able to deaerate. Mass flow bins are suitable for cohesive materials, powders, materials that degrade with time, and whenever sifting segregation must be minimized.

A third type of flow pattern is called expanded flow. In this pattern, the lower part of a bin operates with flow along the hopper walls as in mass flow, while the upper part operates in funnel flow. An expanded flow bin combines the best aspects of mass and funnel flow. For example, a mass flow outlet usually requires a smaller feeder than would be the case for funnel flow. This flow pattern is suitable for storage of large quantities of nondegrading solids. It can also be used with multiple outlets to cause a combined flow channel larger than the critical rathole diameter.

Step 5. Follow These Detailed Design Procedures

Step 5A, mass flow. To achieve a mass flow pattern, it is essential that the converging hopper section be sufficiently steep and have low enough friction to cause flow of all the solids without stagnant regions whenever any solids are withdrawn. In addition, the outlet must be large enough to prevent arching and to achieve the required discharge rate.

Typical design charts showing the limits of mass flow for conical- and wedge-shaped hoppers are given in Figure 16. Hopper angle (measured from vertical) is on the abscissa, and wall friction angle is on the ordinate. For example, mass flow will occur in a conical hopper that has an angle of 20° and is constructed from or lined with a wall material that provides a wall friction angle of 23° or less with the stored bulk solid. Making the hopper walls less steep by 4° or more could result in funnel flow. Alternatively, keeping the wall angle at 20° but increasing the wall friction angle to 28° or more would also result in funnel flow.

Step 5B, funnel flow. The key requirements for designing a funnel flow bin are to size the hopper outlet large enough to overcome arching and ratholing, and to make the hopper slope steep enough to be self-cleaning. Minimum dimensions to overcome arching and ratholing require knowledge of the material's cohesive strength and internal friction. The requirement for self-cleaning can usually be met by making the hopper slope 10° to 15° steeper than the wall friction angle.

Step 5C, expanded flow. Consideration must be given to both the mass flow and funnel flow sections. In the lower mass flow section, the procedure outlined above for a mass flow hopper should be followed. In addition, the flow channel must be expanded to a diagonal or diameter equal to or greater than the material's critical rathole diameter, which can be calculated using the procedure in Janike (1964). Here too, the hopper slope in the funnel flow portion should be steep enough for self-cleaning.

Step 6. Consider the Bin's Shape

At first glance, it might appear that a square or rectangular straight-sided section at the top of a bin is preferable to a circular cross section. Such bins are easier to fabricate and have greater cross-sectional area per unit of height. However, these advantages are usually overcome by structural and flow considerations.

A circular cylinder is able to resist internal pressure through hoop tension, whereas flat walls are subjected to bending. Thus, thinner walls and less external reinforcement are required with circular cross sections. In addition, there are no corners in which material can build up. This is particularly important when interfacing with a hopper at the bottom.

Following are several factors to consider when choosing hopper geometry:

- **Sharp versus rounded corners.** Pyramidal hoppers usually cause a funnel flow pattern to develop because of their inward-flowing valleys that are less steep than adjacent sidewalls. Conical, transition, and chisel shapes are more likely to provide mass flow because they have no corners.
- **Headroom.** Typically, a wedge-shaped hopper (e.g., transition or chisel) can be 10° to 12° less steep than a conical hopper and still promote mass flow. This can provide significant savings in hopper height and cost, which is particularly important when retrofitting existing equipment in an area of limited headroom. In addition, a wedge-shaped hopper design is more forgiving than a cone in terms of limiting hopper angles and wall friction.
- **Outlet sizes.** To overcome a cohesive or interlocking arch, a conical hopper must have an outlet diameter that is roughly twice the outlet width of a wedge-shaped hopper (provided the outlet length is at least three times its width). Thus, cones generally require larger feeders.

- **Discharge rates.** Because of the increased cross-sectional area of a slotted outlet, the maximum flow rate is much greater than that of a conical hopper.
- **Capital cost.** Each application must be looked at individually. Although a wedge-shaped hopper requires less headroom or a less expensive liner than a cone, the feeder and gate (if necessary) may be more expensive.
- **Discharge point.** In many applications, it is important to discharge material along the centerline of the bin to interface with downstream equipment. In addition, having a single inlet point and single outlet, both located on the bin's centerline, minimizes flow and structural problems. Generally, conical hoppers are better for these situations, particularly if only a gate is used to stop and start flow.
- **Mating with a standpipe**. If material is being fed into a pressurized environment, a circular standpipe is often preferred to take the pressure drop.

Step 7. Consider Other Important Factors

Some additional considerations include the following:

- **Gate.** A slide gate at the outlet of a bin must generally be used only for maintenance purposes, not to control or modulate the flow rate. Therefore, it should only be operated in a full-open or full-closed position.
- **Feeder.** The feeder's design is as important as that of the bin above it. The feeder must uniformly draw material through the entire cross section of the bin's discharge outlet to be effective.
- **Mating flanges.** The inside dimensions of the lower of two mating flanges must be oversized to prevent any protrusions into the flowing solid. The amount of oversize depends on the accuracy of the construction and erection. Usually 2–3 cm overall is sufficient. If gaskets or seals are used, care must be taken to ensure that these do not protrude into the flow channel. All flanges should be attached to the outside of the hopper, with the hopper wall material being the surface in contact with the flowing solids. This ensures that the flange and gasket do not protrude into the flowing solids.
- **Interior surface finish.** Whenever possible, welding should be done on the outside of the hopper. If interior welding is necessary, all welds on sloping surfaces must be ground flush and power brushed to retain a smooth surface. After welding, all sloping surfaces must be clean and free of weld spatter. The surface finish is most critical in the region of the hopper outlet. Therefore, any blisters in this area from exterior welding must be brushed smooth. Horizontal or diagonal welded connections should preferably be lapped with the upper section on the inside so the resulting ledge does not impede flow. If horizontal butt welds are used, care must be taken to avoid any protrusion into the flowing solid.
- **Liner attachment.** Inside liners, such as stainless sheet or ultra-high-molecular-weight polyethylene, must be placed on sloping surfaces with horizontal or diagonal seams lapped with the upper liner on top in shingle fashion. Vertical seams may be either lapped or butted.
- **Abrasive wear considerations.** In mass flow, a bulk solid flows against the hopper and cylinder walls. Handling an abrasive bulk solid may result in significant abrasive wear of the wall material, including coatings and liners. Therefore, when designing a mass flow hopper, it is important to assess the potential for abrasive wear. Generally, a hopper surface becomes smoother with wear. However, a wall occasionally becomes rougher, which may upset mass flow. The life of a given wall material can be estimated by conducting wear tests.
- **Access doors and poke holes.** In general, poke holes are not recommended in mass flow bin designs because they have a tendency to prevent flow along the walls, thus creating a problem that mass flow bins are intended to solve. Access doors are also a frequent cause of problems. If they are essential, it is better to locate them in the cylinder rather than in the hopper section.
- **Structural design issues.** It is important that the bin be designed to resist the loads applied to it by both the bulk solid and external forces. This is particularly important when designing, or converting, an existing bin to mass flow because unusually high localized loads may develop at the transition between the vertical section and the mass flow hopper.
- **Bulk materials of inferior flowability** (e.g., more cohesive, with larger critical arching and ratholing dimensions than the material upon which the design was based, or more frictional, requiring steeper wall angles) should not be placed in the bin because flow obstructions are then likely to occur. Such obstructions may lead to the development of voids within the bin and impose dynamic loads when material collapses into the voids. Bin failures have occurred under such conditions.
- **Prefabrication drawing review.** Before fabrication of the bin and feeder, an engineer trained in solids flow technology should review all detailed design drawings. This review is necessary to ensure that the design follows the recommendations and that any design details or changes are consistent with reliable bulk solids flow.

Outlet Sizing

In most applications, a feeder is used to control discharge from a bin or hopper. For such applications, the maximum achievable flow rate through the hopper outlet must exceed the maximum expected operating rate of the feeder. This ensures that the feeder will not become starved. This is particularly important when handling fine powders, since their maximum rate of flow through an opening is significantly less than that of coarser particle bulk solids whenever a mass flow pattern is used. In addition, any gates must not interfere with material discharge.

The following step-by-step procedure will assist in properly sizing a hopper outlet.

Step 1. Calculate the Ratio of Outlet Width or Diameter-to-Particle Size

Flow stoppages due to particle interlocking are likely if the diameter of an outlet is less than about six times the particle size. With an elongated outlet, problems are likely if the ratio of outlet width to particle size is less than about 3:1.

Step 2. Determine Whether the Material Is Coarse, Easy Flowing, or Fine

For purposes of flow rate calculations, a bulk solid is often considered coarse if no more than 20% will pass through a 0.7-cm screen. Whether or not a material can be considered easy flowing depends on the cohesiveness of the bulk solid, the dimensions of the container in which it is stored, and whether or not any excess pressures are applied to the material (e.g., the container is vibrated after being filled). If the combination of these factors results in no flow stoppages at the vessel's outlet (due, for example, to arching or ratholing), the aterial can be considered easy flowing in that application.

A fine powder is a bulk solid whose flow behavior is affected by interstitial gas. Common household examples of fine powders include flour and confectioner's (icing) sugar.

Step 3. Determine Maximum Achievable Flow Rates

Step 3A. The bulk material is coarse but not easy flowing. Either the outlet size must be increased or the material's cohesive strength must be decreased to allow the material to flow.

Step 3B. The bulk material is coarse (>500 mesh) and easy flowing. If the ratio of outlet size to particle size is sufficiently large to prevent particle interlocking, then the maximum achievable rate through an orifice of a coarse, easy flowing bulk solid such as plastic pellets is given by the Johanson (1965) equation:

$$\dot{m} = \rho^\circ A \sqrt{\frac{Bg}{2(1+m)\tan\theta}}$$

where

θ = semi-included angle of the hopper
$\dot{m}$ = discharge rate, kg/s
ρ° = bulk density, kg/m^3
g = gravity acceleration, 9.807 m/s^2

Parameters B, A, and m depend on whether the discharge is conical or a symmetrical slot, as shown in Table 9. For a symmetrical slot opening, W (width) is the smaller dimension of the slot and L (length) is the larger.

With a mass flow hopper, the flow channel coincides with the hopper wall; hence, zero is the hopper angle. On the other hand, with a funnel flow hopper, the flow channel forms within stagnant material and, while it is steeper than in mass flow, it is variable. Thus, the maximum flow rate from a funnel flow hopper is generally higher but less predictable than the flow rate from a mass flow hopper having the same outlet size.

A theoretical expression for funnel flow discharge of fine powders is not available. An empirical equation was derived by Beverloo et al. (1961) based on testing a variety of seeds.

Step 3C. The bulk material is a fine powder. Fine powders are often mishandled in funnel flow bins. Fine powders have little or no chance to deaerate in such bins. Instead, they often remain fluidized in the flow channel and flood uncontrollably when exiting the bin. A funnel flow pattern should therefore be avoided when handling fine materials.

Mass flow bins, on the other hand, can provide predictable and controlled rates of discharge of fine powders as well as other bulk solids. Unfortunately, a fine powder's maximum rate of discharge through a mass flow hopper outlet is often several orders of magnitude lower than the limiting rate of a coarse particle material having the same bulk density. This severe flow rate limitation is the result of the upward flow of air through the hopper outlet caused by the slight vacuum condition, which naturally forms in the lower portion of a mass flow hopper as material flows through it.

The limiting rate of material flow through a mass flow hopper outlet can be calculated once the powder's permeability and compressibility have been measured. If the limiting discharge rate is too low, there are several ways to increase it:

- **Increase the outlet size.** Since the limiting rate is approximately proportional to the crosssectional area of the outlet, doubling the diameter of a circular outlet increases the maximum discharge rate by roughly a factor of 4.
- **Decrease the level of material in the bin.** Fine powders do not behave like fluids. Thus, lower heads result in higher discharge rates, although the effect is generally not very pronounced.
- **Provide an air permeation system, as shown in Figure 17.** This has the effect of partially satisfying the vacuum condition that naturally develops. As a result, there is less need for air to be pulled up through the outlet counter to the flow of particles. With such a system, the maximum flow rate can often be increased by a factor of between 2 and 5.

Table 9 Parameters in the Johanson equation

Parameter	Conical Hopper	Symmetrical Slot Hopper
B	D, diameter of outlet	W
A	$\frac{\pi}{4}D^2$	WL
m	1	0

Source: Chase 2015

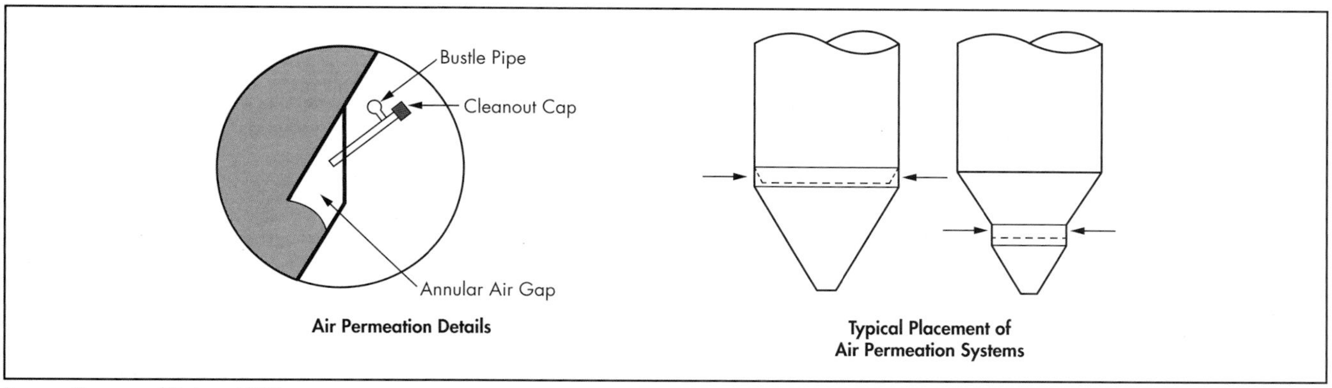

Source: Carson and Holmes 2002

Figure 17 Air permeation system for discharge of fine powders

If none of these will allow the desired discharge rate, fluidization, as discussed in the next step, can be considered.

Step 3D. The bulk material is a fine powder and the required discharge rate is high. Fine particles (<500 μm) tend to flow slower by a factor of 100 to 1,000 than that predicted by the Johanson equation. This is because the effect of air drag on the motion of the particles is much greater for fine particles.

Particle beds need to dilate (increase distance between particles) before the powder can flow. This means air must penetrate into the bed through the bottom surface of the hopper as the powder moves through the constriction formed by the conical walls. For fine particles, the pore diameters in the powder bed are small and there is a significant amount of air drag that resists the powder motion. Carleton (1972) gives an expression for predicting the velocity of fine particles in a moving bed.

If the limiting flow rate from a mass flow hopper is still too low, consideration should be given to fluidizing the fine powder and handling it like a liquid. For this technique to be successful, it is generally necessary that the powder have low cohesion and low permeability. Low cohesion allows the material to fluidize uniformly, so the air does not channel around large lumps. With a low-permeability material, significant pressure gradients can be established and the material takes a long time to deaerate.

The Geldart chart, shown in Figure 18, provides a rough indication of whether a particular material is a good candidate for fluidization. Powders falling within classifications A and B are generally considered good candidates, while category C materials are difficult to fluidize. Category D materials are acceptable for fluidization, but the bed settles quickly and high gas rates are required.

If the bin is small, it may be practical to fluidize the entire contents. With larger bins, this is neither practical nor necessary. However, if only localized regions are fluidized, consideration must be given to the potential for arching and ratholing in non-fluidized regions. In some cases, it is necessary to only fluidize intermittently, while in other cases, continuous fluidization is required during discharge. Whether batch or continuous discharge is required will influence this, but there are other factors to consider as well.

When considering the fluidization option, several operational requirements must be evaluated:

- The bulk density of the discharging material will be lower than if the material were not fluidized. Therefore, a given mass will occupy more volume, which could result in downstream equipment (such as a bulk bag, hopper, or railcar) receiving less than the desired mass even though it is full.
- The material's bulk density will also vary with time, depending on the degree of fluidization of the discharging material. This can present process problems downstream.
- Some materials are hygroscopic, while others are explosive. In such cases, dry air or inert gas may be required for fluidization. Higher energy and gas consumption rates must be taken into account as an additional operating cost.
- The feeder controlling the discharge must be capable of metering fluid-like materials, but this should be avoided if particle segregation is a concern.

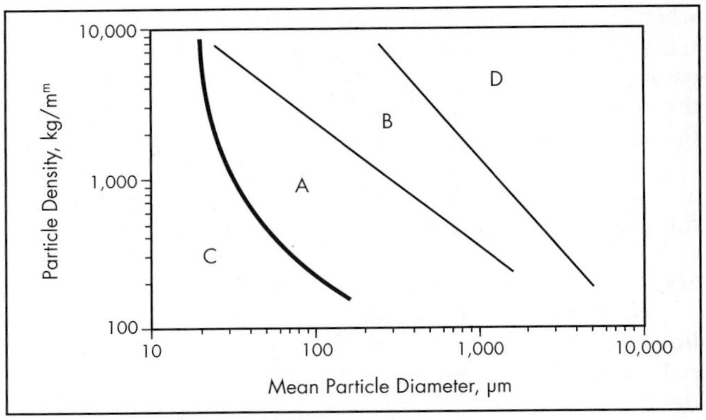

Source: Carson and Holmes 2002

Figure 18 Geldart's fluidization classification

Problems Encountered in Bins

The successful operation of gravity storage bins is predicated on reliable flow of material to the bin outlet. Many problems associated with discharging material from bins are similar to problems encountered in reclaiming material from stockpiles. However, the best solution to these problems is proper bin design.

Many bin problems are related to improper operation of bin and feeder. The principal flow problems are no flow, erratic flow, flooding or flushing, lack of specified storage capacity, degradation, and segregation. No flow occurs when stored material forms a stable dome above the outlet or higher within the bin. Severe vibration, poking, air-lancing, knocking, and even explosives are needed to maintain flow. Erratic flow occurs when momentary domes form within the flow channel and interrupt flow. When the dome collapses, flow resumes and flow rate control is difficult. Dome collapse may result in large impact loads on the bin bottom and causes structural failures. When dry fine materials (powders) flow erratically, they tend to aerate, fluidize, and flush. Material in dead regions is unavailable for processing without an interruption of operations. Furthermore, material may remain in storage for a long time and possibly deteriorate through chemical or physical change.

The flow pattern within the bin is important for minimizing segregation. Materials containing a particle size range will segregate during placement in bins unless special distributing chutes are used. The degree or remixing or the separate fractions as they flow toward the outlet depends on the flow pattern in the bin.

Silo quaking is a problem experienced in mass flow gravity-drawn bin and silos. It is a low-frequency cyclic or pulsating type of flow as a result of changes in density in the material during flow and by changes in the internal friction and boundary wall friction. The pulses are observed as shock waves in the material flowing upward, causing nuisance vibration to structural fatigue failures when the bin's natural frequency is excited by the pulses. Careful design of the bin can prevent or minimize the effect of silo quaking. Scale modeling and discrete element modeling are important tools in the design of bins that would indicate potential silo quaking problems.

ACKNOWLEDGMENTS

This chapter draws heavily from two SME publications. The first is the "Storage" chapter written by C.W. Matthews, W.L. Price, and W.R. Van Slyke in the 1985 edition of the

SME Mineral Processing Handbook. The second is from "The Selection and Sizing of Bins, Hopper Outlets, and Feeders" by J. Carson and T. Holmes in the 2002 volume of *Mineral Processing Plant Design, Practice, and Control*. The authors of this current chapter updated the data and information and revised selected material while retaining much of both chapters as originally written. All figures and tables are from (or adapted from) Matthews et al. (1985), unless otherwise indicated.

REFERENCES

ANSI/CEMA Standard 550-R2009. *Classification and Definitions of Bulk Materials*. Naples, FL: Conveyor Equipment Manufacturers Association.

Beverloo, W.A., Leniger, H.A., and van de Velde, J. 1961. Flow of granular solids through orifices. *Chem. Eng. Sci.* 115:260–269.

Bradley, M.S.A., Berry, R.J., and Farnish, R.J. 2011. Methods for design of hoppers, silos, bins and bunkers for reliable gravity flow, for pharmaceutical, food, mineral and other applications. Presented at BulkSolids India Conference and Exhibition, Mumbai, India, April 6–8. gala.gre.ac.uk/6974/1/WCA091230.pdf. Accessed October 2018.

Carleton, A.J. 1972. The effect of fluid-drag forces on the discharge of free-flowing solids from hoppers. *Powder Technol.* 6(2):91–96.

Carson, J., and Holmes, T. 2002. The selection and sizing of bins, hopper outlets, and feeders. In *Mineral Processing Plant Design, Practice, and Control*. Edited by A.L. Mular, D.N. Halbe, and D.J. Barratt. Littleton, CO: SME. pp. 1478–1490.

CEMA (Conveyor Equipment Manufacturers Association). 2014. *Belt Conveyors for Bulk Materials*. Naples, FL: CEMA.

Chase, G.G. 2015. Hopper design. In *Solids Notes*. Akron, OH: University of Akron.

Jenike, A.W. 1964. *Utah Engineering Experiment Station: Storage and Flow of Solids*. Bulletin No. 123. Salt Lake City, UT: University of Utah.

Johanson, J.A. 1965. Method of calculating rate of discharge from hoppers and bins. *Trans. AIME* (March):69–80.

Matthews, C.W., Price, W.L., and Van Slyke, W.R. 1985. Storage. In *SME Mineral Processing Handbook*. Edited by N.L. Weiss. Littleton, CO: SME-AIME.

Roberts, A.W. 2005. Characterisation for hopper and stockpile design. In *Characterisation of Bulk Solids*. Edited by D. McGlinchey. Boca Raton, FL: CRC Press. pp. 85–131.

Sinden, A.D. 1967. Dump truck piler. U.S. Patent 3,355,038.

CHAPTER 5.2

Belt Conveyors

Michael G. Nelson and Amy J. Richins

Belt conveyors are widely used in transporting bulk materials. They are relatively inexpensive to install and operate, and they are safe, reliable, and versatile. Although there are several types of belt conveyors, the troughed belt conveyor is by far the most commonly used and thus is treated in detail in this chapter. Other types of conveyors are described briefly, with references provided for further information on their design and operation. Detailed information on the design, operation, and maintenance of belt conveyors is found in two readily available publications. The seventh edition of *Belt Conveyors for Bulk Materials*, published by the Conveyor Equipment Manufacturers Association (CEMA), is referred to hereinafter as the *CEMA Handbook* (CEMA 2014). The fourth edition of *Foundations—The Practical Resource for Cleaner, Safer, More Productive Dust and Material Control* is published by Martin Engineering Company (Swinderman et al. 2009) and is referred to hereinafter as *Foundations*.

CONVENTIONAL TROUGHED CONVEYOR

The troughed belt conveyor comprises a continuous belt, supported on rollers called *idlers* and driven by one or more rotating pulleys. Various other components may be added as needed. The system is loaded at one end, usually called the *tail end*, and discharged at the other, the *head end*. The idlers that support the portion of the belt carrying the load and traveling from the tail to the head are called *carrying idlers*. They are almost always configured to form the belt into a trough across its width, increasing the belt's carrying capacity and minimizing spillage. The return idlers, which support the empty belt from the head to the tail, are usually flat. A typical layout for a simple troughed belt conveyor is shown in Figure 1.

Troughed conveyors are very versatile. They can carry a broad variety of materials and have a wide range of capacities over short or long distances. They can be configured to negotiate inclines and declines of 18 degrees or more, and horizontal curves with radii of a few thousand meters. They can be designed to operate in almost any environment. These belts can also operate as components of a materials process, providing blending, sampling, and metered flow as desired.

Belt conveyors have relatively low capital, operating, and maintenance costs, and components are readily available from many sources all over the world. They are capable of throughputs up to 40,000 t/h (metric tons per hour), and conveying speeds up to 10 m/s. Because they are electrically driven, the only potential air emissions are dust and spillage, both of which can be controlled with proper design.

Although each installation is different, the following rules of thumb may be used in assessing the economics of belt conveyors (CEMA 2014):

1. An overland conveyor will have lower operating costs than truck haulage if the haulage distance exceeds 1 km.
2. Beyond the 1-km haulage distance, the ton-per-kilometer cost of belt transport may be as low as one-tenth that of truck haulage.
3. Operating yearly maintenance costs for a belt conveyor may be estimated at 2% of the purchase cost of the equipment (structure and drive) plus 5% of the cost of the belt.
4. Belt replacement is required at approximately 5-year intervals in hard rock applications and up to 15-year intervals in nonabrasive applications.
5. Well-maintained conveyor systems can have an operational availability of 90% or higher.

DESIGN CHARACTERISTICS OF BULK MATERIALS

Material flow may be characterized by many parameters, including internal and effective friction, cohesive and adhesive strength, flowability, boundary friction and adhesion, angle of repose, surcharge angle, particle size distribution, and bulk density. All of these properties are affected by the moisture content of the material and the manner in which it is handled during its transport. In some applications, material property testing by a qualified organization is advisable.

Michael G. Nelson, Professor & Chair, Mining Engineering Department, University of Utah, Salt Lake City, Utah, USA
Amy J. Richins, Research Associate, Mining Engineering Department, University of Utah, Salt Lake City, Utah, USA

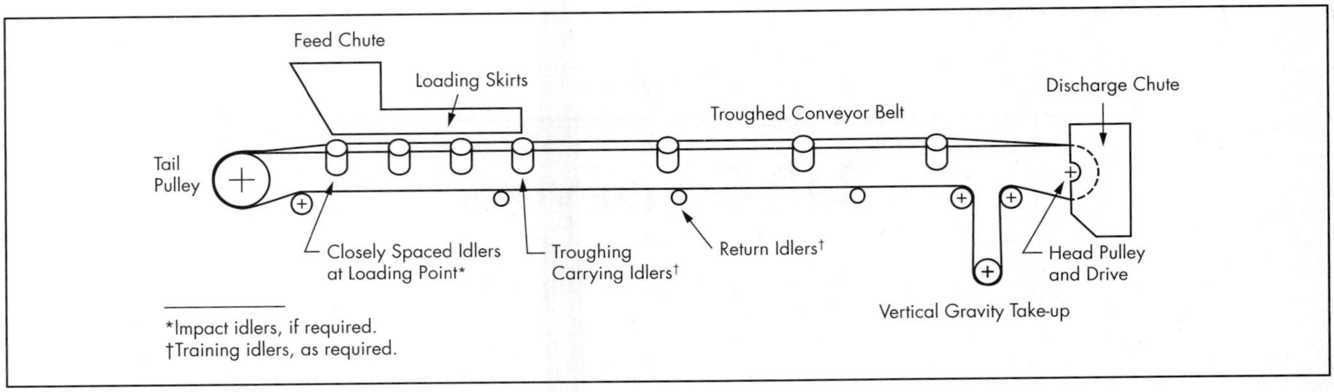

Source: CEMA 2014
Figure 1 Troughed belt conveyor layout

Angles of Repose and Surcharge

The angle of repose ϕ_r, sometimes called the *angle of friction*, is the maximum angle of the stable slope formed when a granular material is poured onto a horizontal surface and forms a conical pile. It is related to the size distribution, density, surface area, particle shape, and coefficient of friction of the material.

The angle of surcharge ϕ_s is the angle to the horizontal formed by material on a moving conveyor. The surcharge angle defines the surcharge area, which is the area of a segment of a circle that is tangent to the surcharge angles at the two points that represent the load width on the belt. The surcharge area is used in the calculation of conveyor capacity, but the value of the surcharge angle is often only an approximation. The surcharge angle used in conveyor design should be compared with that for other conveyors at the same site moving similar material. The angles of surcharge and repose are illustrated in Figure 2.

The carrying capacity in cubic meters per second (m^3/s) for a conveyor system is a function of the cross-sectional area of the material load on the belt, at steady-state conditions, and the linear velocity of the belt. From a dimensional analysis

$$m^2/s \cdot m/s = m^3/s \quad \textbf{(EQ 1)}$$

Figure 2 shows that the cross-sectional area of the material load on the belt is determined by the material surcharge angle, the troughing angle of the belt, and the distance from the material load to the edge of the belt. If the surface of the material on the belt is assumed to approximate a circular arc, this area may be calculated as the sum of two areas, a trapezoid and a circular segment, defined by the points at which the edges of the material load contact the belt. The *CEMA Handbook* (2014) also provides convenient tables showing the cross-sectional areas for belt widths from 500 to 3,000 mm for flat belts and belts troughed at 20, 35, and 45 degrees.

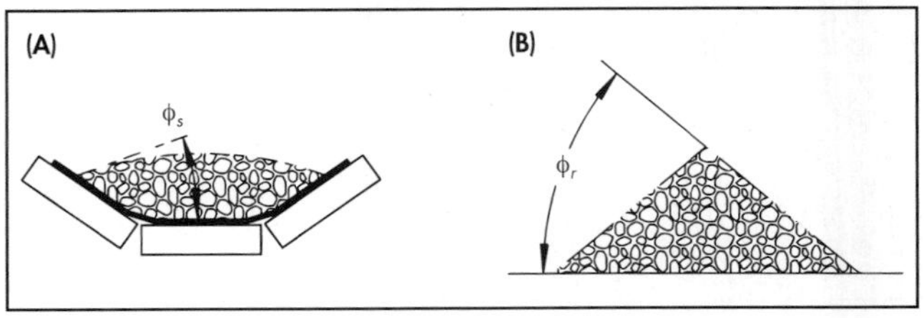

Source: CEMA 2014
Figure 2 Angles of (A) surcharge and (B) repose

Other Properties

Although the angles of repose and surcharge, particle size distribution, and bulk density can be determined with relatively simple and easy-to-use equipment, the determination of other properties requires special testing equipment and standard methods developed and promulgated by such organizations as ASTM International.

The specific gravity of a material is the ratio of the density of that material to the density water at 4°C. It may also be described as ratio of the mass of a substance to the mass of water at its densest for the same given volume. Specific gravity is determined for a single particle of a pure material, without any voids, cracks, or similar defects. The bulk density is defined for powdered or granular material and is the mass per unit volume of many particles of the material divided by the total volume they occupy. The total volume includes particle volume, interparticle void volume, and internal pore volume. Bulk density is not an intrinsic material property. It can change depending on how the material is handled. For example, if a powdered material is poured into a container and then tapped or shaken, the powder particles may move and settle closer together, increasing the bulk density of the powder. This may also happen when a material is on a conveyor, as the material is shaken by the movement of the belt.

CEMA has developed a material classification system, as shown in Table 1 (ANSI/CEMA 550). This system considers particle size, flowability and angle of repose, abrasiveness, and miscellaneous characteristics, assigning a letter or number to characterize each property for a given material. Thus, if a material is granular, average flowing, abrasive, very dusty, and degradable, it would be classified as C36LQ. ANSI/CEMA 550 lists many common materials and their characteristics in their material classification system.

COMPONENTS OF BELT CONVEYOR SYSTEMS

Structure

The structure of a belt conveyor supports all the components associated with the conveyor. The carrying and return run structure, often called the *belt structure*, supports the carrying and return idlers, which in turn support the belt and the material on the belt. Other structures support the head and tail

Table 1 CEMA material classification system

Material Property	Material Characteristics	Code
Size	Very fine—100 mesh and under	A
	Fine—under 3 mm	B
	Granular—under 178 mm	C
	Lumpy—containing lumps under 406 mm	D
	Irregular—stringy, interlocking, mats together	E
Flowability and angle of repose	Very free flowing—angle of repose <19 degrees	1
	Free flowing—angle of repose 20–29 degrees	2
	Average flowing—angle of repose 30–39 degrees	3
	Sluggish—angle of repose >40 degrees	4
Abrasiveness	Nonabrasive	5
	Abrasive	6
	Very abrasive	7
	Very sharp—cuts or gouges belt covers	8
Miscellaneous characteristics (more than one may apply)	Very dusty	L
	Aerates and develops fluid characteristics	M
	Contains explosive dust	N
	Sticky—adheres easily	O
	Contaminable, affecting sale or use	P
	Degradable, affecting sale or use	Q
	Gives off harmful fumes or dust	R
	Highly corrosive	S
	Mildly corrosive	T
	Hygroscopic	U
	Interlocks or mats	V
	Oils or chemicals present—may affect rubber	W
	Packs under pressure	X
	Very light and fluffy—may be windswept	Y
	Elevated temperature	Z

Source: ANSI/CEMA Standard 550 2015

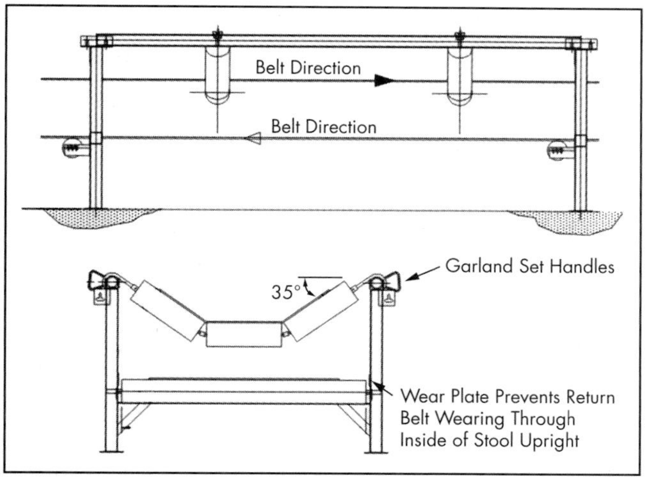

Source: Khodiyar Industrial Corporation 2008
Figure 3 Floor- or ground-mounted basic structure

pulleys, the conveyor drive, and additional components for loading material to and discharging material from the belt.

Figure 3 shows a typical layout for belt structure that sits on the floor or the ground; Figure 4 shows a structure that is supported from above. Such structures are usually made of standard structural steel shapes and must be designed to accommodate the idlers, using CEMA 502 or a comparable standard: Most such standards are based on the idler bolt-hole dimension and the clearance dimensions.

Figure 5 shows a belt structure being installed on the surface in a mine. This is a temporary installation, so the structure is installed and leveled on timbers. Figure 6 shows a structure supported from a mine roof.

The structures for the pulleys and drives are sized to support those components and the loads associated with them. ANSI/CEMA B105.1 and CEMA 501.1 provide the standards for many pulley applications.

Conveyors are installed in a wide variety of conditions and configurations, and conveyor structures vary accordingly. The belt structure often includes a walkway to allow visual inspection and maintenance of the belt. It may also include a cover above the belt, to contain dust and prevent contamination of the load, and a catch plate below the belt, to capture spillage. The structure may also be completely enclosed in a tube.

Overland conveyors may operate at some distance above grade, with the structure supported on towers. In permanent belt installations, all structural elements should be supported on foundations designed to support the dead loads and external loads applied to the system. Internal forces, including the

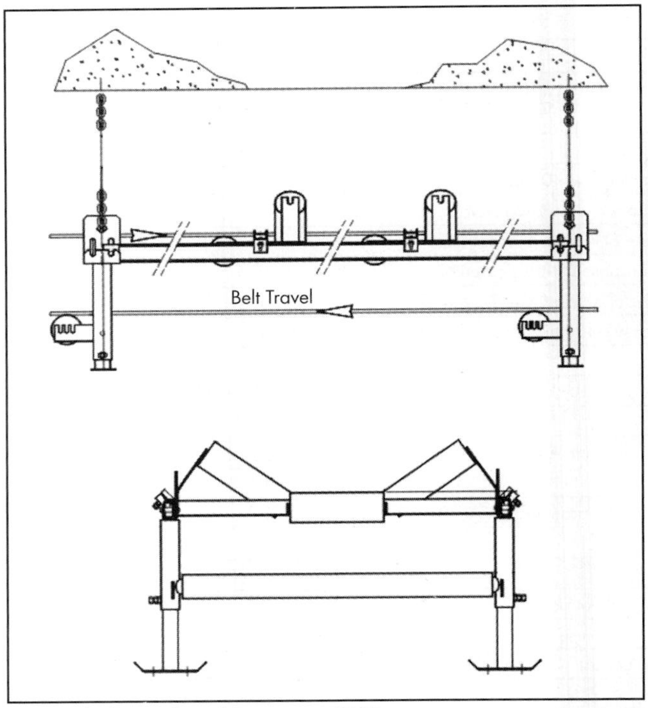

Source: Khodiyar Industrial Corporation 2008

Figure 4 Roof-mounted basic structure schematic

Figure 5 Ground-mounted basic structure

Courtesy of Fenner Dunlop

Figure 6 Roof-mounted structure

belt tension, may be contained within the structure or supported by the foundations, at the discretion of the designer.

Idlers

Idlers are axially mounted cylinders fitted with bearings so they can rotate freely. Belt systems use two types of idlers. Both are shown in Figures 5 and 6. Carrying idlers, on the upper side of the structure, support the belt and the material being conveyed. Return idlers, underneath the carrying idlers, support the empty belt as it returns to the loading point, or tail.

Carrying idlers typically comprise three idler rolls, configured in a shape that will form the belt into a trough. This increases conveying capacity and controls spillage. The troughing angle is the angle to the horizontal formed by the roll on either side of the idlers. Standard troughing angles are 20, 35, and 45 degrees.

Factors in the design of idlers include idler load rating, control of belt sag, and the desired life of the idlers. The idler load rating is determined by an individual idler's share of the weight of the belt and the conveyed material, the belt tension (which will be higher on vertical curves), and the interaction of the belt tension with the vertical misalignment between adjacent idlers. Idler life is determined by the load on the idlers, the idler load rating, and the rotational speed of the idler. Belt sag is determined by the tension in the belt, the idler spacing, the construction of the belt, and the size of the largest lumps of material carried on the belt. Idler design affects the power consumption of the conveyor because energy from the drive is required to overcome the total rolling resistance of all the idlers. Recent developments in idler technology include improved seals and bearings and intrinsic condition monitoring. Motorized idlers are also available, to reduce belt tension and decrease the cost of drives.

Belting

Standard conveyor belts have three components, the carcass, the top cover, and the bottom cover, as shown in Figure 7. The carcass provides the belt its substance and strength; the top and bottom covers respectively protect the carcass. All three components must work together to fulfill the desired function of the belt, providing the required mechanical properties, resistance to wear, abrasion, impact, chemical damage, and extreme temperatures.

Conveyor covers for mining and mineral processing applications are usually made of rubber and rubber-like materials or elastomers, including natural rubber, styrene-butadiene rubber, polybutadiene, acrylonitrile, and various blends. General-purpose covers are used for most heavy industry applications, including mining and mineral processing.

In the United States, elastomers used for conveyor covers are defined by the Association for Rubber Products Manufacturers as either grade I or grade II. Grade I covers are made of natural or synthetic rubber or blends and characterized by high resistance to cutting, gouging, and tearing and very good to excellent resistance to abrasion. These materials are recommended for use with sharp and abrasive materials and for severe impact loading conditions. Grade II covers are made of the same types of materials as those used in grade I.

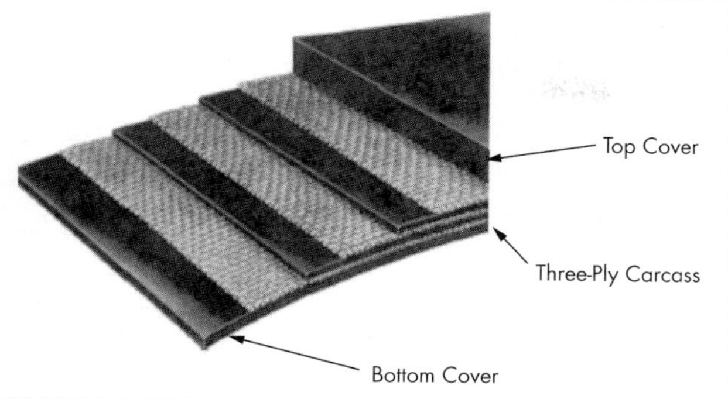

Adapted from CEMA 2014

Figure 7 Cross section of a multi-ply fabric-reinforced belt

They have good to excellent abrasion resistance but may not have the same resistance to cutting and gouging as grade I covers. Conveyor covers may tested in accordance with ASTM D412, to indicate compliance with standard requirements for tensile strength and elongation at break, and in accordance with ISO 4649 Part 8 for compliance with standard volume loss requirements. Specifications for conveyor covers should include elastomer grade, minimum tensile strength, maximum elongation at break, adhesion between adjacent plies, adhesion between cover and ply, and cover thickness.

The belt carcass carries the load and provides reinforcement for resistance to tear and impact and for mechanical fastener retention. Belt carcasses are usually made of woven fabric, with one or more plies as needed for strength. Belts for high-tension use may have a carcass comprising a single layer of steel cable or specially woven aramid fiber.

The belt carcass for a given application is chosen to satisfy the operating requirements. High-strength synthetic fibers are most common, but multi-ply cotton carcass belts are still used in some applications. Belts tension ratings are given in kilonewtons per meter (kN/m) or pounds per inch width, and range from 13 to 220 kN/m for standard belts with rubber covers. Carcasses are available in many configurations, with various fibers and weaving patterns.

The belt design must consider the loading conditions anticipated in application. The belt loading point is often the location of the highest impact and abrasion on the belt, so loading conditions affect selection of the covers and the carcass.

The first factor in assessing the effects of loading conditions is the frequency factor, which is the amount of time required for the belt to complete one traverse of its length. The frequency factor is given by

$$F_f = 2L/V \qquad \textbf{(EQ 2)}$$

where
F_f = frequency factor, min
L = belt length, m
V = belt speed, m/min

When the frequency factor is greater than or equal to 4 minutes, the minimum top cover thickness can be selected based on the loading conditions. As the frequency factor decreases to 0.2 minutes, the top cover thickness should be increased proportionally to twice that use for a frequency of 4 minutes.

Loading conditions that minimize cover wear include the following:

- The direction in which the material feed is traveling is the same as the direction of belt travel.
- The equivalent free fall of material onto the conveyor belt is 1 m or less.
- The loading area of the belt conveyor is horizontal or has a slope of not more than 8 degrees.
- Properly designed chutes and skirtboards are used to form, center, and settle the load on the belt.
- Material temperature is in the range of 0° to 65°C.
- An engineered loading system is used to place the fines on the conveyor belt first, providing a bed for large lumps.

Material transfer and loading onto belts is discussed in a later section of this chapter. When severe loading conditions are anticipated, the belt design may include one or more *breaker* layers. Breaker fabric is usually woven of nylon or polyester, using an open weave to dissipate impact energy and minimize puncturing of the carcass by sharp edges that penetrate the belt cover. Breaker layers are usually between the carcass and the belt cover.

Belt Tension

The motor drive for a conveyor belt provides the power needed to move the belt and the transported material. This power is provided as a torque applied to one or more drive pulleys, which in turn transmit force to the belt by friction. When power is transmitted to the belt, the belt is placed under tension; power is transmitted only when the belt tension entering the drive, T_1, is greater than that when the belt is leaving, T_2. This transmission of power occurs in starting, stopping, and steady-state operation of the belt. The power consumed by the belt, P, at any given time may be calculated as the product of the effective tension, T_e, and the linear velocity of the belt, thus

$$P = T_e \cdot V \qquad \textbf{(EQ 3)}$$

where
$T_e = T_1 - T_2$
V = belt velocity

Correct design of the belt drive must account for all operating conditions. In steady-state operation, power is required only to overcome the friction in the system and to raise the loaded material on an inclined belt. The design must also account for the power required for acceleration on start-up, which will include acceleration of the belt and the load, the drive pulley, all the idler pulleys, and the components of the drive itself—motor, gear drive, and so forth. In steeply declined belts, the drive motor may act as a generator, opposing the acceleration of the material load as it descends, or the energy from the accelerating material may be dissipated into a braking system. The design must also account for the energy required to stop the belt, using brakes. It must also prevent unwanted motion when an inclined or declined belt that is loaded stops unexpectedly using backstops.

Designing the drive for a belt begins with a calculation of the power required to overcome the friction in the system. Power is calculated by estimating the effective tension required in the belt, which requires calculating or estimating the following parameters:

- Tension to overcome frictional resistance of the idlers
- Tension to overcome sliding resistance between the belt and idler rolls
- Tension to overcome resistance of the belt and load to flexure as the loaded belt passes over the idlers
- Tension to overcome frictional resistance of nondriven pulleys, including the tail pulley and snubbing pulleys
- Tension to overcome frictional resistance of accessories such as skirtboards, scrapers, slider beds, and plows
- Effect of low temperature on the system

Power to lift the conveyed material up an incline or power generated by travel of the conveyed material down a decline is added to the power required to overcome friction to provide an estimate of the running power required for the motor. The power required to accelerate the entire system, including the belt, idlers, pulleys, motor, gear drive, and coupling, along with the power required to continuously accelerate the material being loaded, are added to the running power to provide an estimate for the total power required.

A formal method using this approach, now referred to as the *CEMA historical method*, was described in detail in the *CEMA Handbook* for each of its first five editions and is still available from CEMA as a separate publication. The CEMA historical method requires the use of two tables to estimate friction factors, and two-way interpolation in the tables is often necessary. Furthermore, the method requires determination of the rolling resistances of idlers and pulleys, either from vendor literature or standard tables published by CEMA. The calculation is not difficult, but it can be long and tedious, and there are many opportunities for errors. For these reasons, several computer programs exist to make calculation using this method simpler and more reliable. The historical method is still used by some individuals and firms.

The sixth edition of the *CEMA Handbook* (2014) recognized that the historical method did not include five factors that add to the required belt tension—viscoelastic belt deformation, idler misalignment, material liftoff and tramping, and belt bending on the pulleys. The sixth edition recognized three types of conveyor belts, and suggested a method of calculating the power required for each type.

The *basic* conveyor has the following characteristics:

- Single flight of less than 250 m long
- Single, free-flowing load point
- Inclined or horizontal but without curves
- Fabric carcass belt
- Flat or equal-roll troughing idlers
- Single drive
- Unidirectional or reversing up to 150 m/min
- A single gravity or fixed take-up
- Maximum belt tension of 53 kN

Effective tension required in a basic conveyor may be estimated by the following equation:

$$T_e \leq W_m H + 1.72(2W_b + W_m)L \qquad \textbf{(EQ 4)}$$

where

T_e = effective belt tension, kgf
W_m = weight of the material conveyed per unit length of the conveying distance, kg/m
H = height of the incline or decline of the belt, if any, m
W_b = weight of the belt, kg/m
L = conveying distance or length of the conveyor, m

The *standard* conveyor has the following characteristics:

- Single flight of less than 900 m long
- Single or multiple free-flowing load points
- Inclined, declined, or horizontal with or without vertical curves
- Fabric carcass belt
- Flat or equal-roll troughing idlers
- Unidirectional or reversing at any speed
- Single or multiple drives
- Gravity or automatic take-up
- Maximum belt tension of 7,250 kgf

Effective tension required in a standard conveyor may be estimated by the following equation:

$$T_e = LK_t(K_x + K_y W_b + 0.015W_b) + W_m(L K_y + H) + T_p + T_{am} + T_{ac} \qquad \textbf{(EQ 5)}$$

where

L, H, W_b, and W_m are defined as in Equation 4
K_x = idler resistance factor
K_y = belt resistance factor
K_t = temperature correction factor
T_p = tension from belt flexure around pulleys and pulley bearing resistance
T_{am} = tension from the force to accelerate the material as it is fed onto the belt
T_{ac} = tension from accessories

Equation 5 is a formal explication of the CEMA historical method and requires access to the tables for k_x, k_y, and k_t.

The universal conveyor has the following characteristics:

- Any length
- Single or multiple, free-flowing load points
- Inclined, declined, or horizontal flights with horizontal or vertical curves
- Fabric carcass or steel cord belt
- Any belt profile
- Unidirectional or reversing
- Single or multiple drives
- Gravity or automatic take-ups

To calculate effective tension, the belt must be analyzed in sections or flights, starting a new flight at each point where an incline or decline begins or where a drive pulley is located. The following equation shows the calculation for a given flight, n:

$$\Delta T_n = \Delta\Sigma T_{\text{Energy } n} + \Delta\Sigma T_{\text{Main } n} + \Delta\Sigma T_{\text{Point } n} \qquad \textbf{(EQ 6)}$$

where

ΔT_n = total change in belt tension to cause steady belt speed
$\Delta\Sigma T_{\text{Energy } n} = \Delta T_{\text{H}n} + \Delta T_{amn}$
$\Delta\Sigma T_{\text{Main } n} = \Delta T_{ssn} + \Delta T_{isn} + \Delta T_{iWn} + \Delta T_{bin} + \Delta T_{mn} + \Delta T_{sbn} + \Delta T_{sn} + \Delta T_{mzn}$
$\Delta\Sigma T_{\text{Point } n} = \Delta T_{pxn} + \Delta T_{prn} + \Delta T_{bcn}$
ΔT_{Hn} = change in tension to lift or lower the material and belt

ΔT_{amn} = change in tension to continuously accelerate loaded material to belt speed
ΔT_{ssn} = change in tension from the belt sliding on skirtboard seal
ΔT_{isn} = change in tension from idler seal friction
ΔT_{iwn} = change in tension from idler load friction
ΔT_{bin} = change in tension from viscoelastic belt deformation
ΔT_{mn} = change in tension from idler misalignment
ΔT_{sbn} = change in tension from drag in slider beds
ΔT_{sn} = change in tension from sliding of material on skirtboards
ΔT_{mzn} = change in tension material from moving between idlers
ΔT_{pxn} = change in tension from belt bending on the pulley
ΔT_{prn} = change in tension from pulley bearings
ΔT_{bcn} = change in tension from belt cleaners and plows
ΔT_{dpn} = change in tension from discharge plow

The tension along the belt must be continuous, meaning that the tension at the beginning of one flight must equal that of the end of the previous flight. The calculation may require several iterations to achieve continuity. Manual calculation of total tension using this method is possible, but also tedious and time-consuming. It is preferable to do the calculation by computer, and many suppliers of conveyor systems, consultants, and engineering firms have programs designed to do so.

The analysis of belt tension must also include an estimate of the tension required to prevent sag of the belt at any point beyond an acceptable limit. On a fully loaded conveyor, the acceptable sag when running is between 1.5% and 3% of the span between idlers, depending on the troughing angle and the percentage of the load made up of large lumps. Sag on the return side should also be considered. On a level or inclined belt, the minimum tension decreases continually from the head pulley to the tail pulley. If the tension on the belt as it exits the head pulley, T_2, is too low, the belt may sag too much as it approaches the tail pulley.

Belt Specifications

Properties to be considered in specifying belts for a given application include the belt's working strength, number of plies, carcass gauge and weight, carcass fabric type, troughability, and maximum width for empty troughing as a function of the troughing angle and material weight per unit length.

The working strength is often the first factor considered. It is measured in kilonewtons per meter or pounds per inch width. The working strength should be equal to or greater than the maximum tension in the belt divided by the belt width. It is not to be confused with the ultimate tensile strength, which should be greater than the maximum tension experienced by the belt during the most severe operating or start-up conditions.

Conveyor Drives

The drive system includes the motor, the starting mechanism, and the coupling system. In all systems, the motor must be coupled to the drive pulley. The coupling system connects the motor to the drive pulley and must provide the required rotational speed to the drive pulley. It may also provide some isolation between the motor and the moving conveyor to allow controlled acceleration and prevent motor damage under exceptional conditions. Many drive systems also include brakes and backstops. There are many possible combinations of the components in a conveyor drive system. The most common are described here.

Conveyor Motors

Almost all conveyors are driven by electric motors. Criteria for the selection of conveyor drive motors include the following:

- **Required power.** This is based on calculations of belt tension and required acceleration.
- **Operating speed.** An optimum match is provided from the motor to the belt, through the drive system and speed reducer.
- **Voltage.** The plant electrical system must be compatible.
- **Service factor.** The capacity of the motor to operate beyond its rated insulation and winding temperatures for a short period is indicated by the service factor. A service factor of 1.15 is often selected, indicating that the motor can withstand a 15% thermal overload. Higher service factors require coordination with the motor supplier.
- **Torque-speed characteristic.** How the torque is delivered by a motor varies with its rotational speed. Selection under this criterion is primarily made based on the conditions expected when the belt starts. In the United States, motors are rated by the National Electrical Manufacturers Association (NEMA). NEMA designs A–D describe the torque–speed relationships for four types of motors up to 150 kW in size. Designs B and C are most commonly used for conveyors.
- **Operating conditions.** Conditions include operating temperature range; the presence of moisture, washdown water, or precipitation; and the presence of dust or other contaminants.

Coupling Systems

Direct-coupled drive systems include an alternating current (AC) induction motor with a direct coupling and a speed reducer. They are most often used on small belts with moderate speeds and profiles (inclines and declines). Induction motors are widely available and inexpensive, but offer no control over starting torque. These systems require little maintenance, and are capable of absorbing the energy from a descending load, dumping the generated current to other motors in the system. Because the motor experiences high currents during starting and acceleration, the number of starts before failure decreases with increasing motor size.

Direct-coupled drives with reduced-voltage starting systems are the most commonly used for belt conveyors. Control of the terminal voltage to an AC induction motor is used during starting to control the resultant torque. There are many ways to control this voltage, including an autotransformer, a capacitor, or a silicon-controlled rectifier (SCR). The SCR is the most common and most reliable. This system can control starting speed but cannot control speed during operation.

Direct-coupled, variable-frequency control drives provide improved control and increased energy efficiency. Variable-frequency drives (VFDs) control motor speed by electronically modifying the frequency of the voltage applied to the motor. A VFD can start a motor slowly and vary motor speed to optimize energy consumption. VFDs are now relatively small and can be installed in most drive systems, reducing costs and maintenance requirements. VFDs do require additional training for maintenance personnel and the stocking of additional spare parts, but they are now so commonly used that these requirements do not usually result in additional costs. Not all VFDs can function regeneratively. If a belt profile will result in the need for regenerative capacity, the VFD specification must include that capability.

Fluid couplings may be used to provide a gradual transfer of torque from the motor to the belt system. There are two types of fluid couplings—fixed fill and variable fill. Fluid couplings are installed between the motor and the speed reducer and allow the motor to be started at full voltage without putting full torque on the belt system instantaneously, providing soft acceleration. The behavior of a fixed-fill fluid coupling, unlike that of a VFD, cannot be varied. The torque–speed curve of a variable-fill fluid coupling can be varied by changing the fill level in the coupling. Fluid couplings are not normally a good choice for downhill, regenerative conveyors. Fixed-fill fluid couplings are simple and easy to maintain, while variable-fill couplings are much more complex and require careful maintenance. The oil in the fluid coupling gets hot during start-up and operation, and this may limit the number of starts in a given time period.

Speed-reducer mechanisms include V-belts and sheaves, roller chains and sprockets, and many varieties of gear-speed reducers or gear motors. Speed reducers are selected based on cost, power requirements, available space, and user preference. Reducers that use V-belts or roller chains are simple and easy to maintain but usually require more frequent maintenance. All speed reducers use energy, as expressed by their efficiencies. For example, a reducer with V-belts and sheaves has an approximate efficiency of 94%, while that of double-reduction helical gear, shaft-mounted reducer is 97%, and that of a high-ratio worm-gear reducer is 50%.

Installation of belt drive systems requires proper support and alignment of all components—motor, couplings, speed reducer, and drive pulley. An alignment-free drive, containing the couplings and gear-type speed reducer, can reduce installation expenses.

Where local conditions may result in the formation of ice on the conveyor belt, the installation of a creeper drive is advised. The creeper drive uses a small auxiliary motor and associated drive that can be mechanically switched to drive the empty belt at a slow speed.

Brakes and Backstops

Brakes are used to stop or slow down a moving belt. They are required on sloped belts or belts that develop high momentum. After being stopped, almost all inclined or declined belts require the positive holding force of a backstop to keep them from further movement.

Mechanical brakes used on belt conveyors are usually operated electrically. The braking surfaces are engaged by spring action, and disengage by either hydraulic pressure or magnetic action. Such brakes are interlocked with the conveyor motor so that if the motor is shut off after the brake is set, the actuating coil on the brake is also de-energized. Brakes that operate in this way are considered fail-safe. Brakes can be installed on either the high- or low-speed end of the drive train. Low-speed brakes are typically used for downhill conveyors where increased safety and large thermal capacity is required. Eddy-current brakes employ a metallic disk or drum that rotates in a magnetic field. As the drum or disk rotates, eddy currents are generated on its surface and the interactions of these currents with the poles of the magnetic field induce a braking force on the drum. The torque generated by the brake is directly proportional to the field current delivered to the magnet and can be adjusted by a control system. However, eddy-current brakes have no holding action when there is a power failure, so an auxiliary mechanical brake should also be provided.

Braking action can also be provided by the drive motor, in three ways. First, the motor may be *plugged* by reversing the motor current to generate a counter torque on the drive pulley, with the resulting energy being dissipated as heat. When the belt speed reaches zero, the motor must be stopped, or the motor will begin to accelerate the belt in the opposite direction. At zero speed, there is no holding effect. When plugging a motor, energy losses are high, and there is a risk of a belt fire from the heat generated by the reverse motion of the drive pulley against the belt.

In dynamic braking, the motor functions as a generator, with the kinetic energy of the moving belt and load providing the driving force. Use of dynamic braking requires the installation of additional controls, so that when the conveyor is stopped, the AC connection to the motor is disconnected and a direct-current excitation current is connected to the motor primary. Braking force at first increases rapidly as the motor slows down, but then decreases and disappears as the motor comes to a stop. At this point, there is no holding effect.

Finally, when a squirrel-cage motor is used, regenerative braking may be employed. When such a motor is operated above its synchronous speed, power generated by the motor will flow back into the power system. The system must be capable of absorbing this power. Regenerative braking is useful for declined conveyors that drive the motor at its synchronous speed.

Backstops are designed to hold an inclined or declined belt in place when the motor stops, while continuing to allow travel in forward direction. When the force required to lift a load vertically on an incline is greater than the sum of the forces required to move the belt horizontally, a backstop is required. On a long belt, with several inclined or declined sections, each possible condition under which the belt may stop must be considered. For example, if a belt has a short incline followed by a longer decline, it might be thought that a backstop is required only on the declined portion. However, if the belt is stopped after the inclined portion is loaded but before the load progresses to the decline, a backstop may be required on the inclined portion.

High-speed backstops are designed to act quickly. They are mounted on one of the shafts of the gearbox or speed reducer and provide relatively low torque. A high-speed backstop is less expensive than its low-speed counterpart, but because it does not act directly on the drive pulley, a failure in the gearbox or speed reducer will nullify its effect on the motion of the belt. Low-speed backstops are fitted directly to the shaft of the drive pulley shaft. There are several types of backstops, all of which are mechanical so that the absence of electrical power will not affect their proper operation.

Brakes and backstops are usually used in combination. Table 2, adapted from the *CEMA Handbook*, summarizes recommended applications of brake and backstop combinations.

Table 2 Brake and backstop combinations

Type of Conveyor	Backstop	Brake	Forces To Be Controlled
Horizontal	Not required	Required when coasting of the belt and load is not allowable or must be controlled	Decelerating force minus resisting friction forces
Inclined	Required if required lift power is ≥ power to overcome system friction	Not usually required unless preferred over a backstop	In-line load tension minus restricting friction forces
Declined	Not required	Required in every case	Decelerating forces plus incline load tension minus resisting friction

Adapted from CEMA 2014

Pulleys

Pulleys are cylinders that rotate around an axis. In conveyor systems, they provide power input to the belt, change the direction of the belt travel, or modify the tension in the belt as it travels. The carrying and return idlers previously discussed are pulleys, but were considered separately because of their particular function.

Most conveyor pulleys are made of steel, with a continuous, cylindrical surface and two end disks fitted with compression-type hubs. The pulley drum rotates around the shaft, supported by bearings of various designs. The bearings may be protected by the installation of special seals to prevent exposure to conveyed materials, water, and other contaminants. Pulleys used at the tail ends of conveyors are sometimes of the self-cleaning wing type, with open vanes instead of a drum-type surface. These vanes minimize the buildup of material on the pulley face. In the United States, ANSI/CEMA B105.1 and CEMA 501.1 apply to standard drum and wing pulleys, respectively.

Standard-sized pulleys are also available in mine-duty construction and engineering construction. Mine-duty pulleys are designed for heavy use in applications where severe conditions and high installation costs justify a more conservative and costly design. They are often used where conveyors start and stop frequently or where overloads exceed 1.5 times the running tension. The material thicknesses are greater, bearings may be rated for longer wear, and special seals may be included. There are no CEMA standards for mine-duty pulleys; rather, they are designed as specified by the user. Engineered pulleys are custom designed for a particular conveyor and are often used in overland conveyors, high-tonnage mining conveyors, coal power plants, concentrators, and high-capital construction projects.

Conveyor pulleys are often covered with a lagging of rubber, ceramic, urethane, or other material. On drive pulleys, lagging increases the coefficient of friction between the belt and the pulley for more efficient transfer of energy from the drive to the belt. Lagging can also reduce wear and decrease material buildup on the pulley surface. In some cases, lagging material may be selected for specific environmental conditions, for example, to withstand corrosion.

Many types of bearings and seals are available, and their design and selection is best left to experienced individuals. In mines and mills, labyrinth and taconite-duty seals are often used to protect the bearings from dust and water. Diagrams of labyrinth and taconite-duty seals are shown in Figure 8.

As the name implies, labyrinth seals include a complex maze that is usually filled with grease, which blocks passage of contaminants to the bearing or lubricant out of the bearing. Taconite-duty seals were designed for the conditions encountered in the mining and processing of iron ore. They include three separate elements—an inner seal to keep the lubricant in the bearing, an outer seal to keep contaminants out of the bearing, and a chamber between the two seals that can be filled with grease to trap any contaminants that penetrate the outer seal.

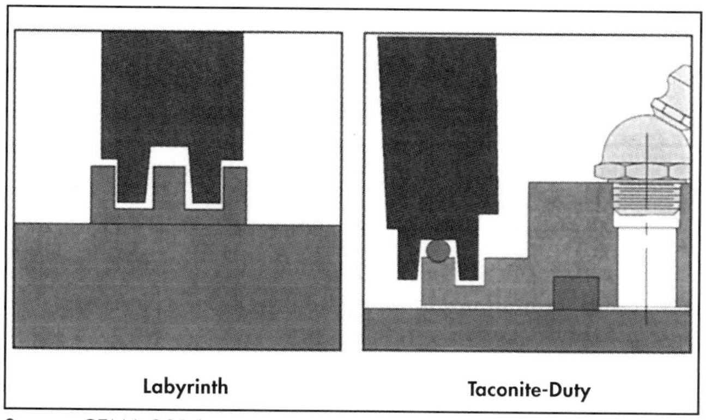

Source: CEMA 2014

Figure 8 Special shaft seals

Material Transfer

In all belt systems, material must be loaded onto and discharged from the belt. In many systems, material must also be transferred from one belt to another. All of these functions are accomplished with transfer systems, at locations called *transfer points*.

All transfer points should be designed with consideration of the following goals:

- Maximize material throughput.
- Contain material adequately.
- Minimize impact, wear, spillage, dusting, and material degradation.
- Match material flow direction and speed across the transfer point.
- Collect and return to flow the belt scrapings.
- Accommodate all anticipated operating conditions.

Transfers are usually made through a chute system, designed to accommodate the material flow and direct it properly, while avoiding plugging and minimizing impact on the receiving belt. Figure 9 shows a standard or conventional transfer chute. The parts of the chute as shown in the figure have the following functions:

- The head chute (A) encloses the head pulley of the feeding conveyor, to ensure that all the discharged material enters the transfer, and to limit release of dust.
- The drop chute (B) transfers the material in free fall.

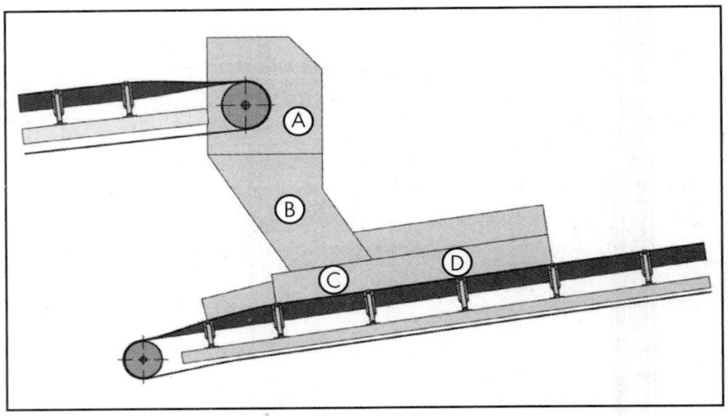

Source: Swinderman et al. 2009
Figure 9 Conventional transfer chute

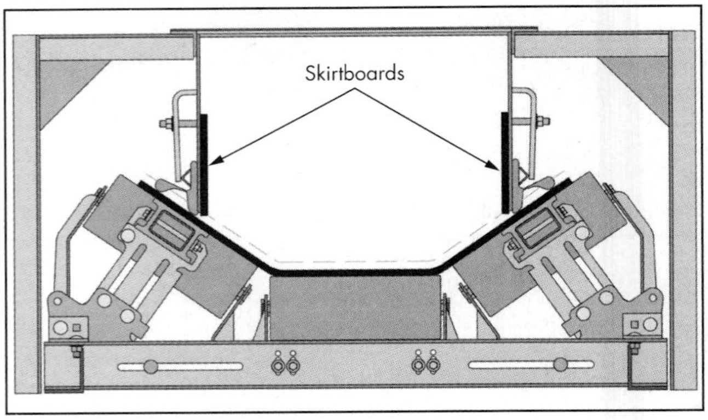

Source: Swinderman et al. 2009
Figure 10 Skirtboards on a conveyor belt

- The loading chute (C) is where the material enters the receiving belt.
- The loading zone (D) is an extension of the loading chute, which may be installed to contain dust as it settles after dropping onto the receiving belt.

The conventional transfer chute design is often modified to achieve the preceding goals. For example, the loading zone may include skirtboards, as shown in Figure 10, and the skirtboards may extend beyond the loading zone portion of the transfer chute. Skirtboards help to center the loaded material on the belt and limit spillage as the material moves away from the transfer point. In transfer points where there is an excessive drop, the loading chute may be modified to decrease the energy of the falling material. At other times, the loading chute may discharge to a curved spoon, which directs the falling material in the direction of the receiving belt; and in other cases, an intermediate belt may be used to accelerate the material as it enters the receiving belt.

It is important to provide extra support for the receiving belt at a transfer point. Rubber-cushioned impact idlers can be installed closely together, but there are drawbacks to their use. First, each idler supports the belt only at the top of each roller, so the belt will deflect under the load and may oscillate or sway. Second, impact idlers are subject to impact damage, especially from unusually large lumps. Third, idlers that are installed closely together are more difficult to service or replace.

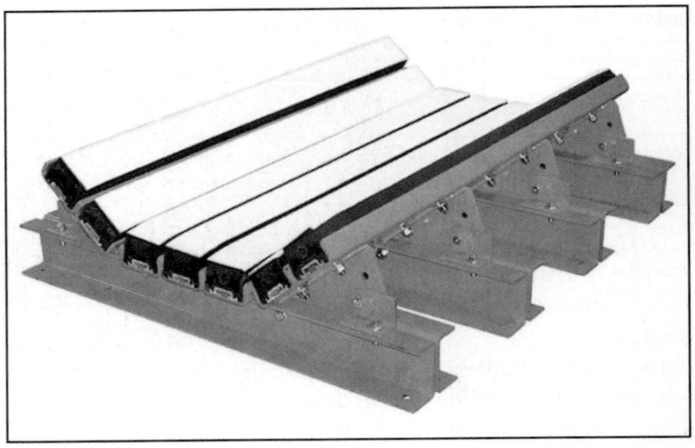

Courtesy of Martin Engineering
Figure 11 Belt-support cradle

Belt-support cradles, also called *impact cradles* or *slider beds*, provide an improved method of belt support at transfer points. Belt-support cradles use low-friction, impact-absorbing bars supported by a steel framework. As shown in Figure 11, the bars are parallel to the direction of belt travel. Because an impact cradle supports the belt continuously, the skirtboards or other devices that contain dust discharge downstream from the transfer point will seal much more effectively. CEMA 575-2013 gives a rating system for impact cradles. It is important the impact cradle is aligned correctly on installation and that the alignment is regularly checked and adjusted.

Open discharge from a conveyor into a bin or onto a stockpile must take into account the discharge trajectory of the flowing material. This trajectory may be calculated using the linear velocity of the material at discharge, the point at which the material leaves the belt, and the effect of gravity on the falling material. The calculation is simple in concept, but the parameters vary considerably with the rotational speed of the pulley, the inclination of the belt, and the material characteristics. The *CEMA Handbook* (2014) gives detailed procedures for calculating the discharge trajectory under various conditions.

Design of transfer points requires careful consideration of all the pertinent data, including material characteristics (which may require laboratory testing of samples), all data for both belts, layout and installation details, method of impact reduction, chute lining materials, and customer preferences. Special care is required when retrofitting equipment at transfer points. Applicable regulations pertaining to noise, dust emission, and spillage must also be considered. The design of the transfer system is often greatly improved by the use of discrete element simulation, as described in the "Developing Technology" section. Most suppliers now offer such simulations in the design of conveyor systems.

Take-Ups

A take-up in a conveyor system is a device that provides the minimum tension on the return side of the belt to prevent slipping of the belt on the drive pulley, compensates for changes in belt length resulting from stretch, and provides extra belt for splicing. All conveyor belts experience stretch, which may be permanent or temporary. Temporary stretch may occur when belt tension changes quickly, as the result of starting, braking, or sudden temperature change. Permanent stretch occurs when the components of the belt elongate or otherwise deform.

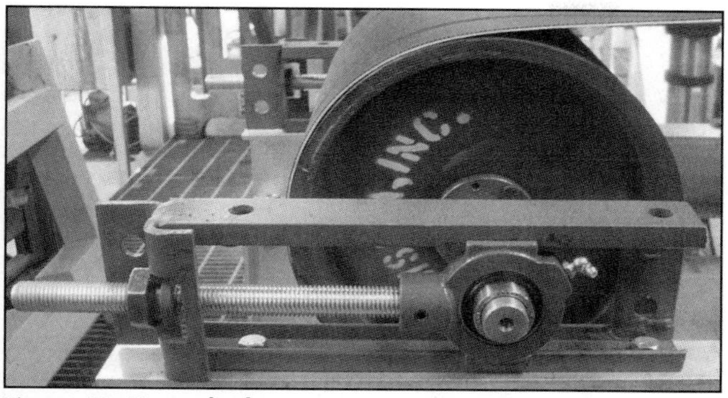

Figure 12 Manual take-up

Where space is constrained, or on short, light-duty conveyors, manual take-ups are used. A manual take-up often consists of two simple, screw-adjusted mechanisms, one on each side of the tail pulley. Use of a manual take-up requires frequent observation of the belt tension and sag to determine when adjustment is needed, and careful adjustment to ensure that the belt tracks properly. When a manual take-up is used, proper guarding must be installed, and adjustments to the take-up must not be made while the belt is running. Figure 12 shows a manual take-up on the tail pulley of a belt designed for laboratory use.

Passive automatic take-ups are actuated by gravity, while active take-ups may be driven electrically, hydraulically, or pneumatically. Some take-ups include spring compression take-ups to accommodate transient loading that occurs during start-up or braking.

Figure 13 shows a vertical gravity take-up. The tension added to the belt by this take-up is equal to half the take-up weight, $W_{tu}/2$. As shown, such a take-up requires three guide pulleys. In some gravity take-ups, the take-up weight may be varied, to account for changing conditions. Because a gravity take-up uses a large, suspended counterweight suspended on a cable, precautions must be taken to allow for failure of the cable or the belt. To prevent damage to the counterweight in a cable failure, an energy-absorbing weight arrestor may be installed.

Design and installation of active, powered take-ups is usually done by specialists. Powered take-ups use stored energy, and often incorporate wire rope, which may be under considerable tension. To ensure proper operation and safety, powered take-ups should be well maintained and have proper guarding installed.

Belt Cleaners

Some of the material carried on a conveyor belt may adhere to the belt after the discharge point. This material, called *carryback*, is usually fine and may be wet or sticky. As the belt travels to its tail end, the carryback is on the underside of the belt, and thus in contact with the return idlers. Carryback results in excessive belt wear, buildup on return idlers, and loss of overall efficiency in the system. It may also lead to misalignment of the belt.

Carryback is removed by belt cleaners. Selection of belt cleaners is usually made by estimating how much material must be removed from the return run of the belt to avoid excessive wear. CEMA 576, *Classification of Application for Bulk Material Conveyor Belt Cleaning*, provides a systematic method for making this estimate, by considering the belt

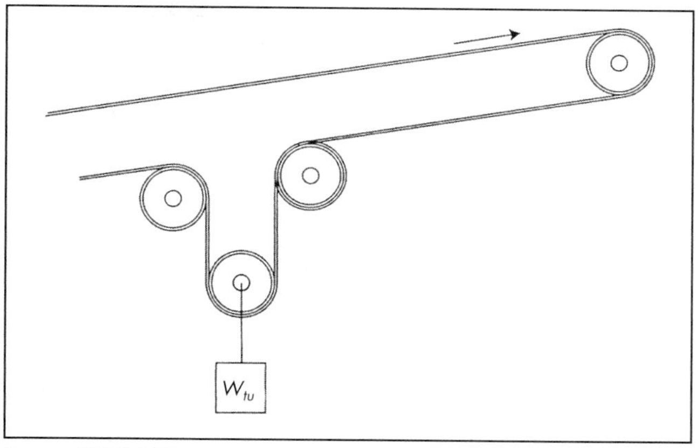

Source: CEMA 2014

Figure 13 Vertical gravity take-up

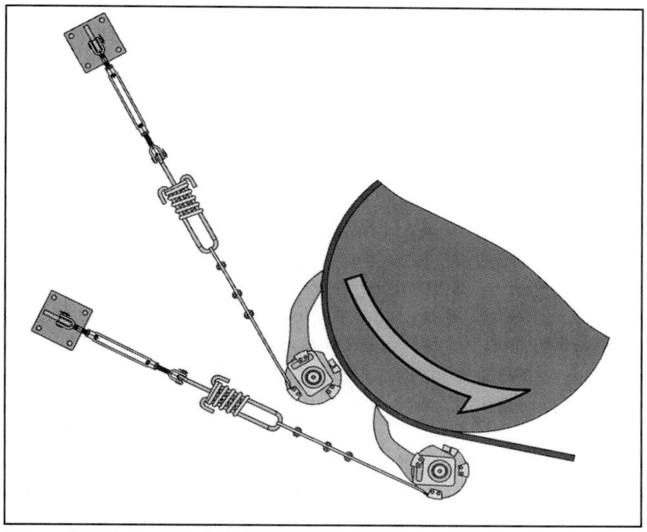

Source: Swinderman et al. 2009

Figure 14 Installation of two primary scrapers

width, belt speed, type of splices used, abrasiveness of the material, moisture content or stickiness of the material, and the location of the scrapers.

On standard, smooth-surfaced conveyor belts, scrapers or blades are usually used for cleaning. Typically two scrapers are installed near the discharge end of the belt. Installation in this location usually makes it possible to capture the material removed by the scraper into the main discharge from the belt.

The first scraper should remove most of the carryback material. It is usually made of rubber or urethane and inclined into the direction of belt travel (a positive rake angle) with moderate tension. The second scraper removes finer, stickier material. It may be made of metal or elastomer and have a positive or zero rake angle. Figure 14 shows how two scrapers would be installed on a belt.

Scrapers are fitted with springs and an adjustment mechanism, so that the tension against the belt can be adjusted. Scrapers and plows must be maintained, adjusted, installed, and removed when the conveyor is not in operation and the system is correctly isolated.

Scrapers may also be used on the return run, to remove material that falls onto the belt from the carrying side. These scrapers are typically a single blade, angled to the width of the belt, or a V-plow, as shown in Figure 15. They are almost always installed at a zero rake angle—perpendicular to the belt surface. Notice the accumulation of material under the cleaner in Figure 15. If a return run cleaner is removing a large amount of material from the belt, the handling of the material may pose a problem. The conveyor should be analyzed to reduce the amount of material that is falling onto the return run.

In some cases, rotary cleaners may be preferred to scrapers. Rotary cleaners can be driven by the motion of the belt or by an electric motor. Spiral cleaners, as shown in Figure 16, provide several continually changing points of contact with the belt surface, resulting in a scrubbing action. They are not affected by mechanical splices and do not affect belt tracking. Brush and finger cleaners are often used on belts that have cleats or some other surface pattern. A motor-driven brush cleaner is shown in Figure 17. A finger cleaner is similar but has flexible fingers—usually made of elastomer—on its surface. Brush and finger cleaners are often used in combination with water sprays

In the selection and installation of belt cleaners, each discharge point should be designed to accept at least one cleaning device. The cleaner should be as close as possible to the discharge point, and the discharge chute should be designed to receive the carryback removed by the cleaner. The discharge area should be designed to allow good access to the chute and the cleaners. If the cleaner is not near the discharge point, special chutes may be needed to direct the cleaner discharge to a point where it can be easily collected and removed.

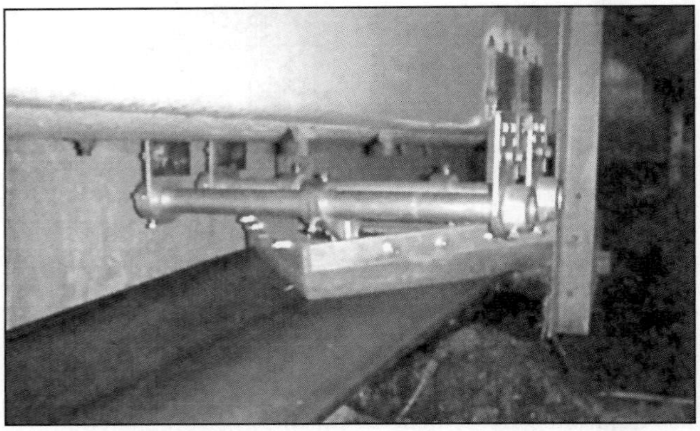

Source: CEMA 2014

Figure 15 V-plow on a conveyor return run

Source: CEMA 2014

Figure 16 Rotary spiral cleaner

Source: Swinderman et al. 2009

Figure 17 Rotary brush cleaner

It is particularly important that installation, maintenance, and adjustment of all belt cleaning devices be done only after the belt is correctly isolated and locked out.

Instruments and Controls

Instruments and controls for conveyor belt systems vary widely in complexity. The simplest systems have controls for routine starting and stopping and for emergency stopping of a single belt. The most complex provide automatic control of a system of several belts, bins, and stockpiles, with starting, stopping, and individual belt speeds controlled automatically, and detection of many operating parameters.

Starting and Stopping

Belts may be started locally or remotely. In simple systems, individual belts are started with locally mounted start and stop switches or push buttons. When two or more belts that operate in sequence are manually controlled, care must be taken to start and stop the belts in the correct order. The last belt in the sequence should be started first, and the first belt in the sequence should be stopped first, with an adequate amount of time allowed between the stopping and starting of succeeding belts. This will prevent overloading of belts on stopping and starting. Many locally controlled belts are also equipped with a *jog* control that allows an individual belt to be briefly energized and then stop. This function may be useful in clearing overloaded belts, in testing the operation of a recently maintained belt, or other situations.

Safe operation requires that every belt have an emergency stop mechanism that can be actuated from anywhere along the length of the belt where personnel may be present. If a belt is accessible from both sides, an emergency stop must be present on both sides. The most common type of emergency stop comprises lengths of cord or cable installed along the edges of the belt, with each end of each cable attached to a switch that cuts the power to the belt drive. When any one of the lengths of cable is pulled, the belt stops. Figure 18 shows such an emergency stop.

Source: Swinderman et al. 2009
Figure 18 Emergency stop, pull-cord type

In complex systems with several conveyor belts, stopping and starting should be automated so the system can be started and stopped with one action, such as an electrical switch or a command through a computer-based control system. The single command will initiate a start or stop sequence, with appropriate timing throughout. The controls for a complex system may be located in a central control room or at an appropriate location in the mine or plant. Although these systems can be started and stopped by the control system, individual belts should still have manual, local start/stop/jog controls to allow troubleshooting of belt functions and clearing of material for maintenance and cleaning.

In systems that handle large amounts of material or operate under other extreme conditions, belt starting and stopping should be controlled so that starts and stops are gradual. This can save energy and avoid overstressing of large or heavily loaded belts. Such starting and stopping may be controlled by gradually actuating motors or brakes in a fixed manner or using more complex systems that apply starting or stopping force based on measurement of system operating parameters, such as current draw in motors, torque on drive pulleys, or belt speeds.

Instrumentation

Instruments are installed in conveyor systems to measure system parameters, for use in control of the system, and for monitoring the conditions of system components. A single instrument often serves more than one of these functions. In the control system, the output from an instrument may be used to control the speed of the belt or the operation of another system component or simply to shut down the system or a system component if an adverse condition is detected.

Individual instruments are regularly used to monitor the following parameters in conveyor belt systems: belt alignment, belt overload, belt slip, take-up overtravel, bin level, chute blockage, belt wear and damage, and splice condition. Instruments also function in important subsystems. Dust suppression systems usually include ambient or atmospheric dust sensors, and fire suppression systems include sensors that measure belt temperature, smoke, combustion gases, and other early indicators of combustion.

Several vendors also provide integrated systems for monitoring the condition of the belt. Such systems are justified when a given belt is very long, difficult to inspect visually, or key to the continuing operation of the plant. Typical belt condition monitoring systems include instruments to monitor belt misalignment, wear, belt tearing or rip, belt speed, and splice condition. They may incorporate high-speed photo or video cameras, with image analysis and pattern detection (Kellis 2000; Nave 2000; Zhu et al. 2011).

Conveyor drives should have instruments and controls typical of large electromechanical systems. This will include monitoring of the electrical power, including current, voltage, and power factor for all systems, and phase imbalance for three-phase systems. Temperature and vibration should be monitored in drives, couplings, and speed reducers.

INSTALLATION

Each conveyor belt is a complex system, with many parts that must work together. Installation must be done carefully by experienced individuals. The *CEMA Handbook* provides detailed installation specifications, based on the following guidelines:

- Conveyor support structures must be parallel, straight, square, and level. If support structures are properly installed, correct alignment of the idlers will follow.
- Pulleys and shafts must be level and square. Alignment should be measured on the shaft, not the pulley.
- Motor and reducer bases can be concrete or steel. Concrete foundations are usually more rigid and secure. Both types of base must be near level and, if not flat, must allow for shimming to level the motor and reducer.
- Motors, speed reducers, and flexible couplings must be installed with proper alignment—angular, parallel, and axial. These alignments should be verified with dial indicators, machine test levels, or other standard means. When shimming of components is required, oily commercial shim stock should be used.
- Idlers must be square, in-line to the conveyor centerline, and parallel to one another. Alignment must be maintained with reference to a squared and leveled terminal pulley. Use of a laser alignment system is recommended, but a tight wire on the conveyor centerline can also be employed, referenced to the squared starting pulley. Typically, a 30-m tight wire is used. After idlers have been installed over a span of about 15 m, the tight wire should be advanced, with a 15-m overlap on the previous position.

Foundations (Swinderman et al. 2009) offers useful guidelines on installing conveyors with proper accessibility. For ease of operation and maintenance, belts and accessories should be easy to see, easy to reach, and easy to replace. Walkways and work spaces around the belt should conform to the minimum distances shown in Figure 19. For belt repair and replacement, an area for the use of vulcanizing equipment on the exposed belt is required. This should be at least 1,000 mm plus the belt width on each side of the conveyor and a distance of at least 3 m along the belt.

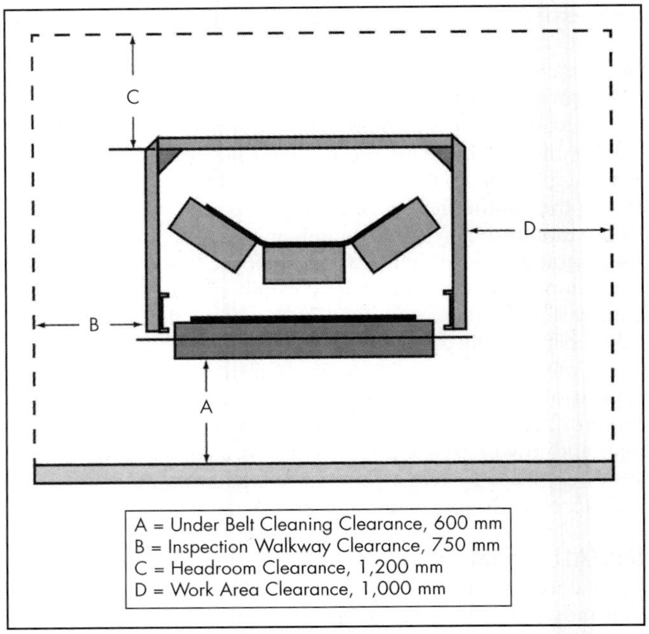

Source: Swinderman et al. 2009

Figure 19 Recommended clearances around a belt conveyor

START-UP

After the support structure, drive and other pulleys, motor and drive system, and idlers are installed, the drive system should be tested for proper rotation. Often the drive motor is bumped—started and then stopped—before it is run for a longer time. During the test of the drive system, experienced individuals should observe the motor and drive system and check the monitoring instruments that are in place, to ensure that no component is vibrating excessively or heating up.

After the belt is installed, it must be aligned or trained. This process is also called *run-in*. Training the belt involves adjusting the idlers and the drive to ensure that the belt does not run to one side or the other. Belt training may require more than one shift and should be done by one individual or crew to ensure that adjustments are made consistently.

The belt is first run empty. It is moved ahead slowly, with careful observation of alignment and other operating parameters. When one or more sections of the belt runs off all along the conveyor, a problem with the belt itself or with a splice is indicated. When the entire belt runs off along a given section of the conveyor, the belt can be centered by moving ahead one side of each of the carrying idlers in that section, and for some distance preceding the region of trouble. The side of the idlers to which the belt runs is moved slightly ahead, but never by more than 6 mm. Belt training usually begins at the head end of the belt and proceeds to the tail pulley, so that the belt can be properly centered on the tail pulley.

If the empty run-in is unsuccessful, the belt may be cambered, or there may be a non-square splice in the belt. Camber occurs when unbalanced warp tensions exist in a conveyor belt, causing the belt to assume a crescent or banana shape when laid flat on a horizontal surface. Slight camber can be instilled into a belt in manufacture, if one of the slitting knives is dull. Camber can also be instilled into a belt during improper storage. Both types of camber can usually be pulled out when the belt is properly tensioned and tracked, but the only correction for a non-square splice is replacement of the splice.

Successful empty run-in is followed by full load run-in. The belt is loaded, and its operation and alignment carefully observed for at least eight hours. If the belt edges remain within the width of the drive pulley face, it is considered properly aligned. If extra-wide pulleys are used, the belt should run within the normal width of the carrying or return idlers.

Again, it is advisable for have one person or crew responsible for training belts. That person should supervise all adjustments. The conveyor should be regularly checked to ensure that it is level. An out-of-level belt may result in misalignment for no other apparent reason. After the belt is run in under full load, the electric power use in the drive should be checked and recorded. These data serve as a reference during future operation, with increased power consumption indicating excessive belt drag because of misalignment or frozen idlers.

When run-in is completed, all belt accessories, including skirtboards, scrapers, shrouds, and transfer point components, should be checked for proper installation and clearance.

OPERATION

A conveyor system that is properly designed, installed, and maintained requires relatively little attention from operators. In a large installation, one or two operators will often suffice for routine operation, inspection, and maintenance. Current systems have extensive monitoring and control systems, incorporating measuring instruments and video-monitoring cameras. Nonetheless, for almost all conveyor systems, the most important operating procedure is the routine inspection and monitoring of the system by the operators.

MAINTENANCE

Because conveyors are often integral to continuing operation, they should be maintained by preventive or predictive maintenance programs, or a combination thereof. Preventive maintenance programs employ regularly scheduled service and rebuilding of major components; predictive maintenance programs rely on inspections and instrument readings to determine when maintenance is required to avoid breakdown.

Correct lubrication of all components is important. This includes idlers and pulleys, the drive system, and moving parts at loading, transfer, and discharge points. A large conveyor will have thousands of idler rolls, all of which have seals and bearings. It is preferable to use sealed bearings, which do not require routine lubrication. When lubrication is required, the manufacturer's recommendations should be thoroughly reviewed. The operating conditions, including operating speed, idler loading, type and size of material being conveyed, and number of operating hours per year should be carefully considered. The operating environment is also important and includes such factors as temperature, dust, material abrasiveness, washdown techniques, and frequency, all of which affect the lubrication and the bearings.

Inspection and monitoring are especially important for the belt itself. The belt is often the most expensive component of the conveyor system. The composition and structure of the belt make it particularly subject to wear and damage, and in modern, high-speed systems, severe damage may occur quickly. Consider a belt with a 300-m conveying distance, and a linear velocity of 3 m/s. If a sharp piece of scrap metal carried in the conveyed material becomes trapped or wedged

Table 3 Regular shift walk-through

Component or System to Be Examined	Notes and Observations
Tail pulley Impact idlers and beds Loading chute Belt Splices Skirtboard seals Belt center loading Carrying idlers Belt tracking and tracking devices Monitoring devices and sensors Guards and covers Safety devices Discharge and dribble chute Head pulley and lagging Belt cleaners Drive Snub bend pulleys Return run idlers Take-up Auxiliary and accessory equipment General housekeeping	Conveyors often give early warning of significant problems in the form of changes in the normal noise patterns of the conveyor, accumulations of fugitive materials, or belt tracking. Every shift an experienced and trained person should inspect the conveyor. The purpose of this inspection is to look for symptoms or early indicators of problems so that they can be dealt with proactively. Each component should be observed, and some minor adjustment or cleaning may be part of the inspection if it can be done safely. Inspection equipment needed is minimal but should always include a flashlight and means of communicating with the operator or control room. A written report, often in the form of a checklist, should be completed and symptoms of major problems addressed as soon as possible. The inspection should take between 30 minutes and 1 hour per shift and is intended as a walk-through only. If problems require urgent attention, they should be dealt with by shutting down the conveyor or immediately scheduling repairs. Perform an inspection of general condition. Test safety features for proper operation. Complete checklist if required.

Source: CEMA 2014

against the belt, the entire belt length of more than 600 m can be damaged or cut in little more than three minutes.

Systematic programs for regular inspection of conveyors are also referred to as *walk-throughs* or *belt surveys*. The *CEMA Handbook* (2014) provides schedules for inspections on per-shift, weekly, monthly, semiannual, and annual bases. Those schedules are shown in Tables 3–9. Tables 3–8 are typical maintenance schedules. Specific procedures should be developed for each site, with reference to information provided by equipment suppliers and experienced personnel in the area.

CONVEYOR SAFETY

Safety incidents associated with conveyors are usually the result of human error, improper maintenance, lack of effective training, or lack of awareness of possible hazards. Padgett (2001) analyzed conveyor-related accidents in non-coal mines in the United States from 1996 to 2000. He identified the following activities related to those accidents:

- Working under or next to poorly guarded equipment
- Using hands or tools to remove material from moving rolls
- Trying to free stalled rolls while the conveyor is moving
- Attempting to remove or install guards on an operating conveyor
- Attempting to remove material at head or tail pulleys while the belt is in operation
- Wearing loose clothing around moving belt conveyors
- Not blocking stalled conveyor belts prior to unplugging, whether flat or inclined, as energy is stored in a stalled conveyor belt
- Reaching behind the guard to pull the V-belt to start the conveyor belt

Padgett's analysis of 459 accidents found that 22 were disabling and 13 were fatal, and that 192 (42%) of the reported injuries, including 10 fatalities, occurred while the injured worker was performing maintenance, lubricating, or checking the conveyor; 179 (39%), including 3 fatalities, occurred while the subject was cleaning and shoveling around belt conveyors.

Kecojevic et al. (2008) analyzed historical fatality data associated with conveyors from 1995 through 2006. Data on belt conveyor–related fatalities from the investigation reports of the U.S. Department of Labor's Mine Safety and Health Administration (MSHA) showed that 49 belt-related fatalities occurred during this period. The authors assigned each accident to one of six risk categories. Fatal incidents that occurred during the belt conveyor cleaning, repair, assembling, or dismantling were classified either as "failure to provide adequate maintenance procedure" or "failure to follow adequate maintenance procedure." The hazard associated with failure of the management to build a crossing facility over the belt conveyor was classified as "failure to provide safe crossing facility," whereas the failure of workers to use such crossing facility or to make shortcuts was classified as "failure to use safe crossing facility." The single fatality caused by roof failure was designated as the hazard "adverse site/geological conditions," and the single fatality caused by the collapse of mechanical structure of belt conveyor was classified as "failure of mechanical components." Table 9 shows the number of fatalities associated with each hazard. The hazards "failure to provide adequate maintenance procedure" and "failure to follow adequate maintenance procedure" contributed to almost 90% of all belt conveyor–related fatalities.

Many countries and organizations issue standards relating to conveyor safety. The following are some of the most widely used:

- ASME B20.1 from the American Society of Mechanical Engineers
- ANSI B11.3 from the American National Standards Institute
- AS 1755-2000 from the Council of Standards Australia
- BS EN 619:2002+A1:2010 from the British Standards Institution
- ISO 7149 from the International Organization for Standardization

Table 4 Weekly belt inspection

Component or System	Activity	Time Required per Conveyor, minutes*
Bend pulleys	Ensure that belt is centered on pulley	10
Carrying idlers	Ensure that all rolls are turning†	10
	Ensure that all idler rolls are free of material buildup†	10
	Ensure that belt touches all three rolls both in loaded and unloaded states†	5
Conveyor belting	Check for belt damage or abuse:	10
	Check for belt cupping	10
	Check for belt camber	10
	Check for impact damage	15
	Check for impingement damage	15
	Check for chemical damage	10
	Check for rips or tears	15
	Check for junction-joint failure	5
	Check for top cover cracking	5
Conveyor drive	Check reducer oil level	10
	Check reducer for oil leaks	5
	Inspect drive coupling	10
	Check oil level in backstop and inspect for leaks	10
	Ensure that all safety guards for drive are in place and in good condition	10
Conveyor structure	Check for rusted, bent, broken, or missing structural parts	20
	Check handrails and toeplates to ensure good condition	20
	Check walkways for material spillage or buildup	10
	Check safety gates to ensure good working order	10
Gravity take-up	Check take-up carriage for free and straight operation† Ensure that belt is centered on pulley* Ensure that all safety guards are in place and in good	10
	Ensure that belt is centered on pulley†	10
	Ensure that all safety guards are in place and in good condition	10
Guards	Check for damage and proper installation	10
Head pulley	Inspect belt cleaners for worn or missing blades	20
	Inspect belt cleaners for cleanliness of frames and blades	20
	Check belt-cleaner tension according to supplier's specification	10
	Ensure that belt is centered on pulley	5
Loading zone	Inspect impact idlers for wear	10
	Inspect impact bars for top cover wear	10
	Inspect seal-support cradles for wear	10
	Inspect and adjust dust seals	10
	Inspect dust suppression nozzles†	10
Return rolls	Ensure that rolls are turning freely	10
	Inspect rolls for material buildup	10
	Inspect mounting brackets for wear from belt tracking problems	10
Safety switches	Inspect cables for correct tension	10
	Ensure that flags are free from material buildup	10
Snub pulley	Ensure that belt is centered on pulley	5
	Inspect pulley for material buildup	5
Splices	Mechanical: Check splice and bolts for wear	10
	Vulcanized: Check splice for separation	10
Tail pulley	Ensure that belt is centered on pulley	10
	Check return plow blade for wear, if installed	5
	Check return plow mounting, if installed	5
	Check return plow tension, if installed	5
Tracking idlers	Check for free pivoting of frame†	5
	Ensure that all rolls are turning†	5
	Check rolls for material buildup	5

Adapted from CEMA 2014
* Some activities may be done simultaneously.
† May require the belt to be running.

Table 5 Monthly belt inspection

Component or System	Activity	Approximate Time Required per Conveyor
Bend pulleys	Check bushings for evidence of movement on shaft	20 minutes
	Check bearing condition and locking collars for tightness	20 minutes
	Check for cracks and wear at face and hub ends	20 minutes
	Check lubrication in shaft bearings	10 minutes
Gravity take-up	Check bushing for evidence of movement on shaft	20 minutes
	Check bearing condition and locking collars for tightness	20 minutes
	Check for cracks and wear at face and hub ends	20 minutes
	Check lubrication in shaft bearings	20 minutes
Head pulley	Check bushing for evidence of movement on shaft	20 minutes
	Check bearing condition and locking collars for tightness	20 minutes
	Inspect pulley lagging for wear and secure to head pulley	20 minutes
	Check for cracks and wear at face and hub ends	20 minutes
	Check lubrication in shaft bearings	20 minutes
Snubbing pulleys	Check bushing for evidence of movement on shaft	20 minutes
	Check bearing condition and locking collars for tightness	20 minutes
	Check for cracks and wear at face and hub ends	20 minutes
	Check lubrication in shaft bearings	20 minutes
Tail pulley	Check bushing for evidence of movement on shaft	20 minutes
	Adjust mechanical take-up for correct belt tension	20 minutes
	Check for cracks and wear at face and hub ends	20 minutes
	Check lubrication in shaft bearings	20 minutes
	Check lubrication in mechanical take-up adjusters	20 minutes
	Check bearing condition and locking collars for tightness	20 minutes
Carrying idlers	Check lubrication of bearings in rolls	20 minutes
Conveyor drive	Check lubrication in backstop bearings	30 minutes
	Check lubrication in shaft bearings	20 minutes
	Inspect drive belts for wear and correct tension	20 minutes
Loading zone	Inspect chutes and chute walls for leaks*	10 minutes
	Inspect entry seals	10 minutes
	Inspect exit seals	10 minutes
	Inspect dust-collection pickups for leaks*	20 minutes
Return rolls	Check lubrication in bearings in rolls	20 minutes
Safety horns and lights	Test to ensure it is working properly prior to conveyor start	20 minutes
Safety switches	Test emergency stop switches	1 hour
Tracking idlers	Check lubrication in rolls and pivot	20 minutes

Adapted from CEMA 2014
* May require the belt to be running.

Table 6 Semiannual belt inspection

Component or System	Activity	Approximate Time Required per Conveyor
Brakes and backstops	Test for proper operation under full load*	60 minutes
Conveyor structure	Check foundations for settling	30 minutes
	Check for corrosion	30 minutes
	Check for structure damage	30 minutes
	Check walkways, handrails, steps, and gates	30 minutes
Loading zone	Inspect wear liners for wear	30 minutes
Safety switches	Test operation for conveyor shutdown*	30 minutes
Warning devices and signs	Test for operation and audible/visual or readable functionally*	30 minutes
	Check traffic control signs for visibility and changes in traffic patterns	30 minutes
Seasonal activities	Winterizing or dry season preparation	8 hours
	Summer or wet season preparation	8 hours

Adapted from CEMA 2014
* May require the belt to be running.

Table 7 Annual belt inspection

Component or System	Activity	Approximate Hours Required per Conveyor
Electrical system	Check for open wiring, damaged conduit, overloads, and system grounds	1
Interlocks	Test to ensure proper interlocking of conveyors*	1
Safety switches	Test conveyor start with flags pulled	1
Documentation	Update maintenance documentation and records	40

Adapted from CEMA 2014
* May require the belt to be running.

Table 8 Routine housekeeping activities

Component or System	Activity	Approximate Time Required per Conveyor
Belt cleaners	Flushing of blades	Daily, 30 minutes
Return plows	Flushing of blades	Weekly, 30 minutes
Magnetic separators	Emptying of bin	Weekly, 1 hour
Central dust collectors	Emptying of bin	As required, 8 hours
Spillage and carryback	Cleaning spillage and carryback	Manually, 1 t/h; Mobile equipment, 4 t/h
Dust	Washdown	Manually, 2 hours per conveyor
Pollution control equipment	Cleaning filters and ductwork	Weekly, 1–4 hours per conveyor
Water treatment systems	Cleaning screens	Weekly, 30 minutes

Adapted from CEMA 2014

Table 9 Conveyor-related fatalities reported by MSHA, 1995–2006

Hazard	Number of Associated Fatalities	Total Fatalities, %
Failure to provide adequate maintenance procedure	28	57
Failure to follow adequate maintenance procedure	16	33
Failure to provide safe crossing facility	2	4
Failure to provide use crossing facility	1	2
Adverse site/geological conditions	1	2
Failure of mechanical components	1	2
Total	49	100

Source: Kecojevic et al. 2008

- *MSHA's Guide to Equipment Guarding* (2000)
- Title 30 in the U.S. Code of Federal Regulations, which covers mineral resources in Parts 1 through 199, especially Parts 56 and 57 (MSHA 1979a, 1979b)
- Title 29 in the U.S. Code of Federal Regulations, regarding the Occupational Health and Safety Administration, especially Parts 1910 and 1926 (OSHA 2004a, 2004b)

CEMA offers general guidelines for conveyor safety. In addition to the detailed requirements given in the preceding standards, the following are useful for workplace discussions and general reflection and are reproduced here in full (CEMA 2014):

- A formal safety training program for operations, maintenance, and supervisory personnel will go a long way toward establishing and maintaining the highest standards of safety in the work place.
- Concurrent with completion of the installation and the trial runs of all belt conveyors and associated equipment, a Safety Checkup is recommended. The checkup should include all mechanical and electrical operating equipment, plus the structures, walkways, ladders, stairs, headroom, and access ways. It is at this time that a detailed physical inspection of the facility and the installed conveyor equipment will often reveal the need for additional guarding, safety devices and warning signs.
- At no time should the conveyors be used to handle material other than that originally specified. Capacity and belt speed design ratings should not be exceeded.
- Only trained personnel should be allowed to operate the conveyor system. They should have complete knowledge of conveyor operation, electrical controls, safety and warning devices, and the capacity and performance limitations of the system.
- The location and operation of all emergency control and safety devices should be made known to all personnel. Surrounding areas should be kept free of obstructions or materials that could impede ready access and a clear view of such safety equipment at all times.
- A program should be established to provide frequent inspections of all equipment. Guards, safe devices, and warning signs should be maintained in their proper positions and in good working order. Only competent and properly trained and authorized persons should adjust or work on safety devices.
- A walking inspection of a belt conveyor system is a good means by which well-trained maintenance personnel can often detect potential problems from any unusual sounds made by such components as idlers, pulleys, shafts, bearings, drives, belts, and belt splices.

- Hands and feet should never come in contact with any operating conveyor component, and no one should be allowed to ride on a moving or operable conveyor. Poking at or prodding material on the belt or any component of a moving conveyor should be prohibited. Contact with, or work on, a conveyor must occur only while the equipment is stopped, following the proper lock out, tag out and test out procedures unless the equipment is specifically design to be safely inspected or serviced and the worker specifically trained and aware of the hazards with the belt in operation.
- No person should be allowed to ride on, step on, or cross over a moving conveyor, nor to walk or climb on conveyor structures, without using the walkways, stairs, ladders, and crossovers or cross unders provided. CEMA R SBP-001: Design and Safe Application of Conveyor Crossovers, which is available for free download at the CEMA website's Safety Page.
- Good housekeeping is a prerequisite for safe conditions. All areas around a conveyor, and particularly those surrounding drives, walkways, safety devices, and control stations, should be kept free of spillage, debris and obstacles, including inactive or unused equipment, components, wiring, and obsolete or non-applicable warning signs or posted instructions. Any conveyor found to be in an unsafe condition for operation or one that does not have all guards and safety devices in excellent condition, should not be used unless adequate supplementary safety devices are installed.
- All persons should be barred by appropriate means from entering an area where falling material may present a hazard. Warning signs and barricades can be used.
- First-class maintenance is a prerequisite for the safest operation of conveyors. Maintenance, including lubrications, should be performed with the conveyor stopped and locked out. Special lubricating equipment, lube extensions, pipes, and the like can be installed so as to permit lubrication of an operating conveyor without any foreseeable hazards.
- Good lighting contributes to a safe working environment.
- During the life of a belt conveyor system, its operational conditions and environment may require changes. There should be a continuing effort to detect and treat promptly any new possible safety hazards associated with these changes. If such a hazard cannot be readily eliminated, warning signs, barricades, or posted instructions should be installed.

Before operation of a conveyor and the associated accessories, safety markings, guards, and warnings must be in place in accordance with governmental regulations and site-specific requirements. The CEMA Safety Committee developed the CEMA *Brochure No. 201: Safety Label Brochure* (2017) to provide advice for the selection and application of safety labels for use on conveyors and related material handling equipment to assist in accident prevention.

Many incidents occur during cleanup operations around conveyors. Most conveyors operate at speeds of 1.5 to more than 5 m/s. Typical reaction time for an average human is 0.75 seconds, so if a person's hand or clothing is caught in a conveyor running at 4 m/s, the belt will move 3 m before a person can react, and the arm or clothing will be pulled into the conveyor. Incidents in which a person is pulled into a conveyor are unfortunately too common. Because of this ongoing hazard, proper conveyor guarding must be used to protect personnel from the injury that may result from contact with conveyor equipment during operation. The regulations and enforcement actions are vary with jurisdiction, so it is important to conform the requirements for guarding specific to each location and type of operation.

Guards and basic safety devices are normally included with major equipment such as drives and as part of the original conveyor construction such as safety shutoff pull-cord switches. Conveyors have other potential hazards associated with the moving belt and conveyor components or accessories that must be accessed by the end user, and may need additional guarding. The most common hazard that must be addressed locally is pinch or nip points, which result from the specific layout and location of the conveyor.

Conveyor guards may be constructed from expanded metal, plastic, or other materials that normally have openings small enough to prevent contact with the hazard being guarded, but allow fines and other materials to pass through and permit inspection of conveyor components during normal operation. The distance around the guard should be maintained so that no one may reasonably reach around the guard into the danger zone. The safe distance around the hazard must also be considered when the guard cannot completely contain the hazard. Although standards vary, a 1-m distance around the hazard is usually protected to prevent reaching around the guard. The guard should be painted to be highly visible, according to the color code for the site. Red is generally reserved for firefighting equipment, e-stops, and similar equipment, so safety orange or safety yellow is frequently used for guards.

Three types of guarding are used to protect personnel from the moving parts of a conveyor. Point guarding isolates the specific moving part that poses the hazard; area or perimeter guarding restricts access to a specific, wider area; location guarding functions when the hazard is a safe distance from normal access points. Each of these is discussed in detail in the applicable standards, in the *CEMA Handbook*, and in *Foundations*.

BELT CONVEYOR SYSTEM CALCULATIONS

Belt Width and Speed

Conveyors are often sized to carry a certain tonnage of a given material, but may be oversized in anticipation of increased production. For this example, consider an application in which an existing conveyor belt is 1,000 mm wide with a trough angle of 35 degrees and a belt speed of 150 m/min. The conveyor has been carrying 2,000 t/h of lead-zinc concentrate. The concentrate has a bulk density (not specific gravity) of 3.5 t/m^3, and its angle of surcharge on the moving belt is 20 degrees. The company wishes to increase concentrate production to 2,500 t/h. Can the existing conveyor carry the additional material?

First, convert the required carrying capacity from t/h (mass/time) to m^3/min (volume/time), thus

$$(2{,}500 \text{ t/h})/(3.5 \text{ t/m}^3 \cdot 60 \text{ min/h}) = 11.9 \text{ m}^3\text{/min} \quad \textbf{(EQ 7)}$$

The required volume-carrying capacity is then divided by the belt speed to give the required cross-sectional area of the belt:

$$(11.9 \text{ m}^3\text{/min})/(150 \text{ m/min}) = 0.079 \text{ m}^2 \quad \textbf{(EQ 8)}$$

The cross-sectional area of the material on a fully loaded existing belt (1,000 m wide with a 35-degree trough and a 20-degree surcharge angle on the load) is 0.091 m^2. This may be calculated from the geometry of the belt and load or determined from tables in a standard reference book, such as the *CEMA Handbook*. This cross section is greater than that required to carry the increased load, indicating that the existing belt is large enough to do the job. In fact, under these conditions, the belt could carry 2,880 t/h.

Idler Selection

Calculate the load on the carrying idlers for a conveyor belt that will operate under the following conditions:

- Material: crushed limestone
- Bulk density: 1.36 t/m^3
- Capacity: 1,200 t/h
- Belt width: 1,200 mm
- Belt speed: 150 m/min
- Belt weight: 15 kg/m
- Belt tension: 5,000 kgf, carrying side; 2,000 kgf, return side
- Idler spacing: 1.2 m
- Misalignment: 6 mm maximum
- Lump size: 150 mm maximum

The total load on each idler is given by Equation 4:

$$L_{it} = [(W_b + (W_m K_1)]S_i + L_{im} \quad \textbf{(EQ 9)}$$

where
L_{it} = total idler load, kg
W_b = weight of the belt, kg/m
W_m = weight of the material load, kg/m
K_1 = lump adjustment factor
S_i = idler spacing, m
L_{im} = idler misalignment load, kg

Idler misalignment load is given by

$$L_{im} = 2 \cdot [(dT)/S_i] \quad \textbf{(EQ 10)}$$

where
d = maximum vertical misalignment between adjacent idlers, m
T = belt tension, kgf
S_i = idler spacing, m

The value of the lump adjustment factor, K_1, is found from Table 10 to be 1.0.

$$W_m = (1{,}000 \text{ kgf/t})[(1{,}200 \text{ t/h})/(60 \text{ min/h})]/(150 \text{ m/min}) = 133 \text{ kgf/m} \quad \textbf{(EQ 11)}$$

The idler misalignment load is

$$L_{im} = 2 \cdot [(\text{dT})/S_i] = 2 \cdot [(0.006 \cdot 5{,}000)/1.2] = 50 \text{ kgf} \quad \textbf{(EQ 12)}$$

Table 10 Lump adjustment factor, K_1

Maximum Lump Size, mm	Material Specific Weight, kg/m³						
	800	1,200	1,600	2,000	2,400	2,800	3,200
100	1.0	1.0	1.0	1.0	1.1	1.1	1.1
150	1.0	1.0	1.0	1.1	1.1	1.1	1.1
200	1.0	1.0	1.1	1.1	1.2	1.2	1.2
250	1.0	1.1	1.1	1.1	1.2	1.2	1.2
300	1.0	1.1	1.1	1.1	1.2	1.2	1.2
350	1.1	1.1	1.1	1.2	1.2	1.3	1.3
400	1.1	1.1	1.2	1.2	1.3	1.3	1.3
450	1.1	1.1	1.2	1.2	1.3	1.3	1.4

Adapted from CEMA 2014

Thus total load per idler is

$$L_{it} = [(15 + (133 \cdot 1)]1.2 + L_{im} = (15 + 133)1.2 + 50 = 227.6 \text{ kgf} \quad \textbf{(EQ 13)}$$

The load on the return idlers is calculated by the same method, but the weight of the material load, W_m, is obviously eliminated, and the return-side tension is used, thus

$$L_{im} = 2 \cdot [(dT)/S_i] = 2 \cdot [(0.006 \cdot 2{,}000)/1.2] = 20 \text{ kgf} \quad \textbf{(EQ 14)}$$

$$L_{it} = W_b S_i + L_{im} = (15)1.2 + 20 = 32.5 \text{ kgf} \quad \textbf{(EQ 15)}$$

These values can be used to select the appropriate idlers for the application from the tables in the *CEMA Handbook* or from manufacturers' literature. The correct selection of idlers is also affected by the operating conditions, the belt speed, and the desired lifetime, or mean time to failure.

Belt Tension and Power

Find the effective tension required in an inclined conveyor and estimate the power required for steady-state operation. The conveyor operates under the following conditions:

- Conveying distance: 200 m
- Vertical lift: 20 m
- Material Weight: 200 kgf/m
- Belt weight: 5 kgf/m
- Belt velocity: 150 m/min or 2.5 m/s

This belt satisfies the criteria for a basic conveyor. Thus Equation 4 is used to calculate the effective tension, T_e. Using 200 m for W_m, 20 m for H, 5 kg/m for W_b, and 200 m for L, T_e can be solved:

$$T_e \leq 200 \cdot 20 + 0.04(2 \cdot 5 + 20)200 = 5{,}680 \text{ kgf}$$

The power required to maintain this tension in steady-state operation is given by Equation 3:

$$P = T_e \cdot V \quad \textbf{(EQ 3)}$$

where
T_e = effective tension
V = belt velocity

Note that 1 W is equivalent to 1 N-m/s or 1 kgf-m/s, so belt velocity (V) must be expressed in meters per second, and the effective tension (T_e) must be multiplied by the acceleration caused by gravity (g) to convert kilogram-force meters per second to watts, thus

$$P = 5{,}680\ \text{kgf} \cdot 9.8\ \text{m/s}^2 \cdot 2.5\ \text{m/s} = 139.2\ \text{kW} \quad \textbf{(EQ 16)}$$

It is important to remember that the calculated power is only that required to raise the conveyed material and overcome the friction in the system. In an actual conveyor, additional power in would be required to accelerate the system when it is started. Furthermore, the nameplate power on the motor would have to be larger than the increased amount, to account for the efficiency of the motor and the drive system, which will always be less than 1.

SPECIAL BELT CONVEYOR SYSTEMS

In addition to the conventional conveyor belt, supported by troughed idlers, there are also special conveyor systems that are useful in particular circumstances. *Foundations* identifies three reasons for considering a special conveyor system. The first is improved environmental control, in which dust, spills, and contamination of the conveyed material must be reduced or eliminated; the second is reduction of labor and capital costs, particularly regarding the footprint of the system and the construction costs; the third is a reduction in the number of transfer points, which is, of course, closely related to the first two considerations. This section offers a brief description of five of the most common special conveyors. All of these systems use flexible belts, but have different methods of supporting and driving the belt. For detailed design and operation of these systems, the reader is referred to the technical literature and to the suppliers of such systems.

Pipe Conveyors

Pipe conveyors are similar to conventional troughed conveyors but have the advantage of enclosing the transported material (Wiedenroth 2010). The components of both systems are similar, but rather than forming a trough, the belt is formed into a pipe using a special transition section after the loading point and opened up through a reverse process before the discharge point, as shown in Figure 20. A photo of a transition section is shown in Figure 21. During transit, the piped belt is supported by assemblies of three or more idlers, as shown in Figure 22.

A pipe conveyor has obvious benefits where dust from the material may a problem or where the material needs to be protected. These conveyor systems can also handle steeper inclines and tighter horizontal curves than conventional belts.

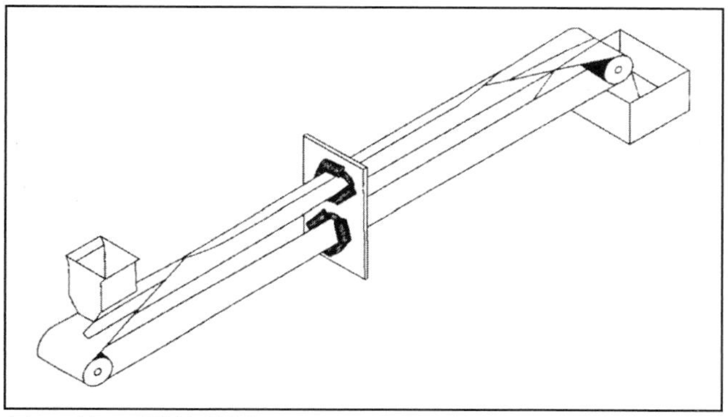

Adapted from CEMA 2014
Figure 20 Operation of a pipe conveyor

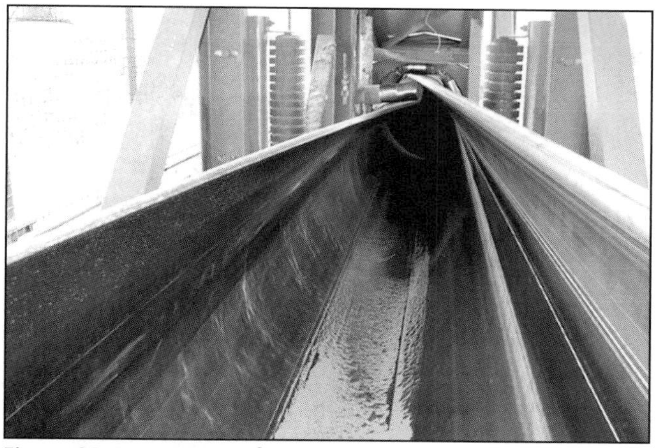

Figure 21 Transition at the entrance to a pipe conveyor

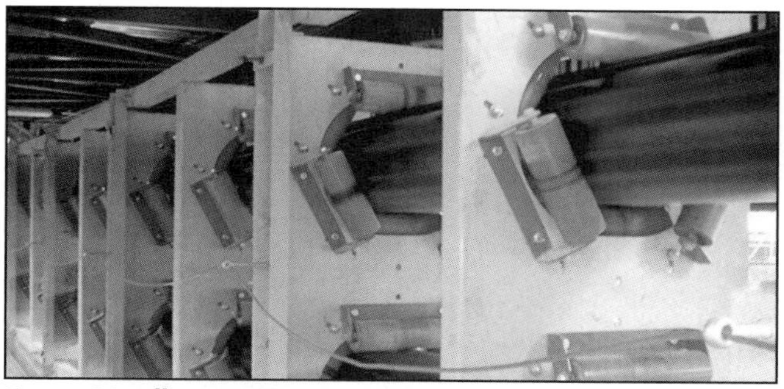

Figure 22 Idlers supporting a pipe conveyor

The disadvantages of this system are the lower tonnages and smaller lump sizes that can be accommodated.

Cable Belts

Cable belt systems use two steel cables to provide the driving force of the system, as shown in Figure 23. The belt has specially molded tracks (sometimes called *shoeforms*) on both sides near its outer edges, so it can be directly supported by the cable, which in turn rides on grooved wheels (Hewett 1957; Anderson and Wilson 1965; Brost and Scanlon 1984). The belt must be more rigid across its width than a conventional belt, so it can support its own weight and the weight of the conveyed material. Figure 24 shows a cable belt with inclined rollers, and Figure 25 is a photo of an operating cable belt.

In long-distance cable belt systems, intermediate drives can be used without the need for transferring material because the driving force is delivered through the cables. Cable belts can also negotiate tighter horizontal curves than conventional systems. A disadvantage is the specialized components that are required.

Air-Supported Conveyors

The air-supported conveyor is constructed with trough-shaped plates that have orifices through which a pressurized airflow provides support for the belt, as shown in Figure 26. The return belt may be supported by air or by conventional return idlers. The belt is covered to contain the escaping airflow, so contamination, spillage, and dust are minimized (Morrell et al. 1984; Hashimoto 1984; Overland Conveyor Company 2017).

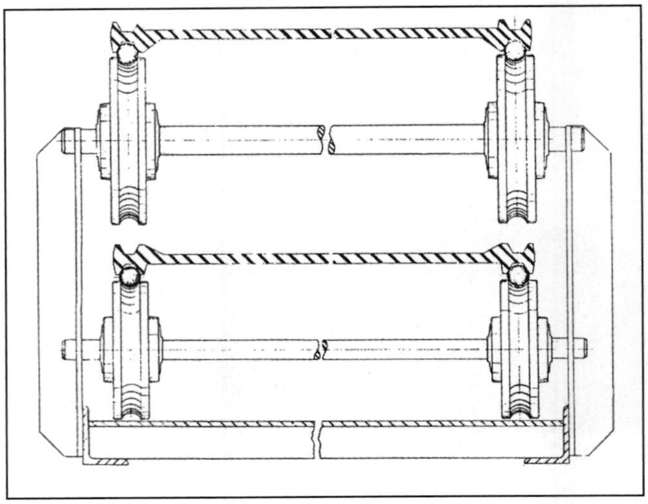

Source: Anderson and Wilson 1965

Figure 23 Cable belt system

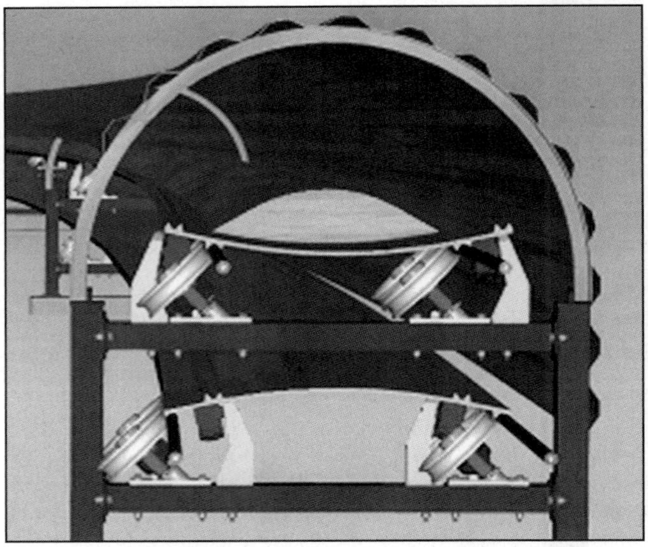

Figure 24 Cable belt system with inclined rollers

Figure 25 Operating cable belt

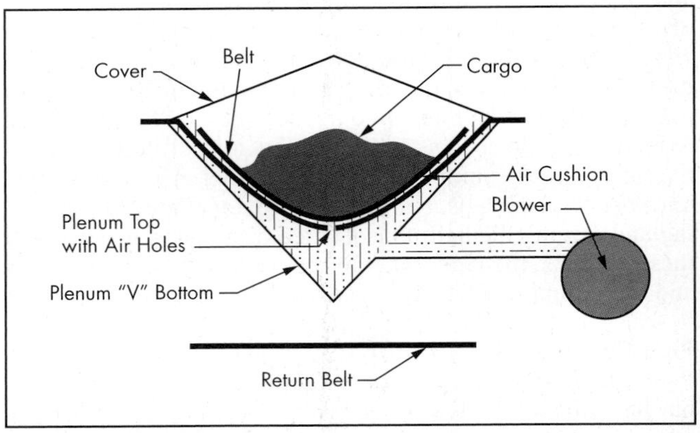

Source: Grisley 2002

Figure 26 Air-supported conveyor

The main advantage of air-supported conveyors is the elimination of carrying idlers. In addition, the belt is uniformly supported and experiences less wear from sag and flexure between idlers. The enclosed design improves dust control and reduces spillages, and there are fewer moving parts. Figure 27 shows an air-supported conveyor with one cover section removed.

The disadvantages of this system are that it requires closer fabrication tolerances and better belt cleaning to ensure smooth, consistent operation. Also, interruption of the air-supported conveyor may be required for belt scales and other accessories. Finally, this type of system cannot accommodate horizontal curves.

RopeCon Conveyors

RopeCon conveyors are manufactured by Doppelmayr Transport Technology (2018). They are similar to conventional conveyors but use flat belts with corrugated side walls, as illustrated in Figure 28. The belt is fixed to axles spaced at regular intervals. Wheels on the axles run on fixed ropes spanning between towers. Towers can be spaced hundreds of meters apart, which allows the conveyor to bridge across valleys or other obstructions. Figure 29 shows a portion of a long overland RopeCon conveyor. The major disadvantage of this system is the specialized components that are required and its higher capital cost.

Steep Angle Conveyor Systems

Steep angle conveyor systems are applicable for inclines exceeding 18 degrees and can operate vertically. Many systems are capable of conveying material up steep inclines (Richmond et al. 1991). One method is to enclose the conveyed material in a pipe conveyor, as described earlier, or in a sandwich belt (Dos Santos 2000), as shown in Figure 30.

Another is to retain the material with molded cleats or chevrons that extend across the width of the belt, as shown in Figure 31, or with cleats and flexible walls as shown in Figure 32. (These flexible walls are similar to those on the RopeCon conveyor.) Some disadvantages of these systems are the specialized components required and the lower tonnages and smaller lump sizes that can be accommodated. Table 11 summarizes the advantages and disadvantages of the special conveyor systems discussed here.

Figure 27 Air-supported conveyor with one cover section removed

Courtesy of Doppelmayr Transport Technology
Figure 28 RopeCon system

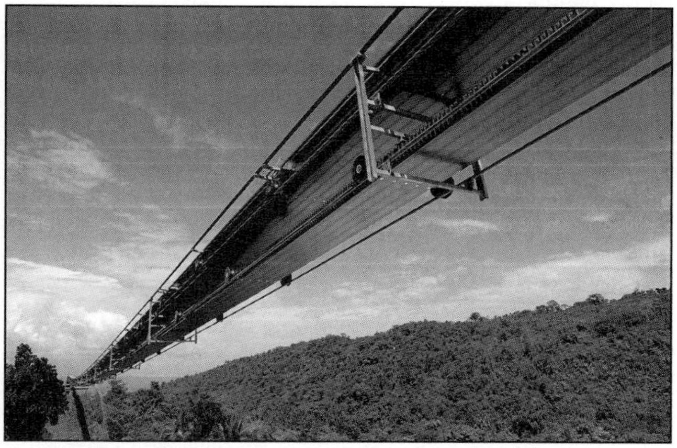

Courtesy of Doppelmayr Transport Technology
Figure 29 Section of a large overland RopeCon system

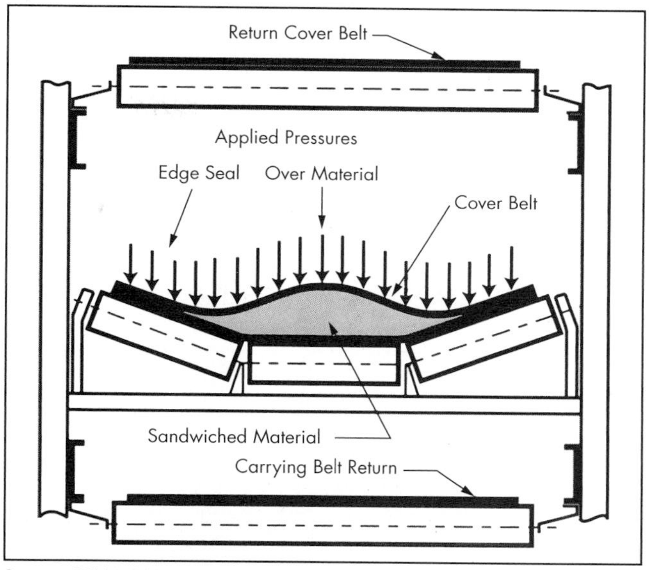

Source: CEMA 2014
Figure 30 Sandwich belt conveyor cross section

Source: CEMA 2014
Figure 31 Molded cleats on a conveyor

DEVELOPING TECHNOLOGY

New conveyor technology for conveyor belts continues to develop in two primary areas. The first is new software for the design of conveyors and conveyor systems; the second is new components for such systems. Examples of each are given here, and references are provided so that the interested reader may easily obtain more information.

Computer Modeling

Belt tension and drive requirement for basic conveyors can be more accurately designed using computer modeling.

Belt Tension and Drive

Belt tension and drive requirements may be estimated for a basic conveyor, with a simple calculation as described earlier. A basic conveyor comprises a single flight, a gravity take-up, a single drive, and a single free-flowing loading point. It is less than 250 m long and may inclined but has neither horizontal nor vertical curves. Its speed is less than 150 m/min, and its maximum tension is 5,400 kgf.

Many modern conveyor systems do not meet the basic conveyor criteria. Although these systems may be designed using manual or spreadsheet calculations, design by a computer program is more accurate and takes into account more of the variables in a complex system.

Such programs can select the radius for vertical curves, to ensure that the belting will not lift out of the idlers on a concave curve or exceed the rating of the belting or idlers on convex curves. The calculation takes into account the mass of the belting; how the belt tensions vary with material loading, especially partially loaded conditions; and how the running forces of the system may vary with temperature and operating conditions. These programs can also calculate the minimum radius for a horizontal curve, balancing the gravitational forces and the weight of the belt and material with the tension forces trying to pull the belt into the curve. To balance these forces, the idlers for the system are angled opposite to the direction of the curve. As in the design of a vertical curve, this calculation must account for varying load conditions and the effects of temperature and operating conditions.

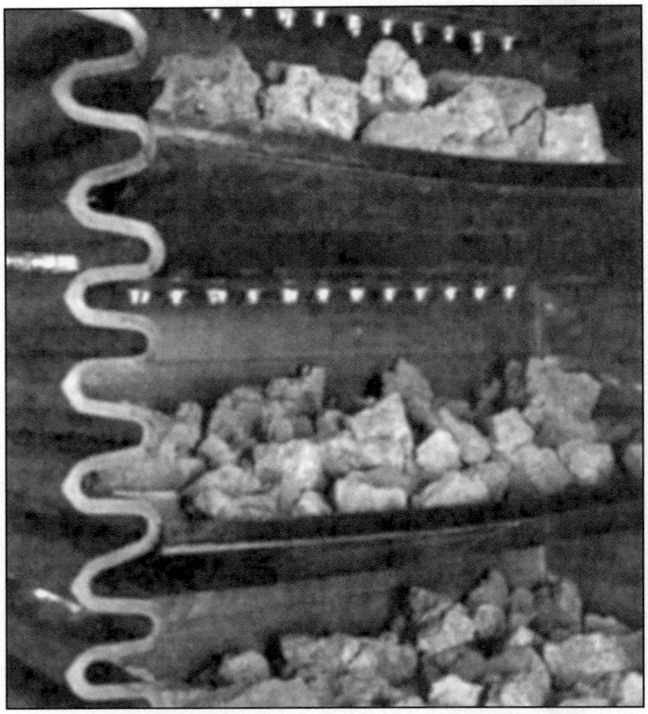

Source: CEMA 2014

Figure 32 Lumps being conveyed vertically on a belt with cleats and sidewalls

In some conditions, dynamic analysis is recommended, simulating the changing tensions and velocities that occur during starting, stopping, and load variations of the system. These calculations are essential in designing long, complicated belt conveyors but are not typically required for standard, straight belt conveyors. For a given conveyor system, these programs can give good estimates for drive torque requirements, test starting control options, identify peak tensions for vertical and horizontal curve calculations, evaluate braking and stopping functions, determine take-up requirements, and identify potential tension problems arising during starting and stopping (Zhang and Steven 2012; Kruse 2002; Lodewijks and Nuttall 2008).

Dynamic analysis may also include design of belt networks, in which several belts receive and deliver material from or to different locations. Such systems are usually analyzed with a network analysis, similar to that used in the modeling of pipeline and electrical distribution systems (Tarshizi et al. 2014; Kumar and Yingling 1994; Yingling et al. 1993).

Some suppliers of belt design software are Conveyor Dynamics, Advanced Conveyor Technologies, Professional Designers and Engineers, and Overland Conveyor Company. Many others may be located by searching online.

Discrete Element Modeling

The discrete element method (DEM) simulates the bulk flow of granular materials and has been widely applied to analysis of the handling and processing of bulk materials. DEM was first used to analyze the flow of materials in chutes and bins, and as computer processing capacity and speed increased, it

Table 11 Special belt conveyor systems key-features comparison chart

Conveyor Type	Tonnage	Lump Size	Parts/ Components	Horizontal Curves	Advantages	Disadvantages
Conventional troughed	High	Large	Readily available	Yes	Up to 18-degree inclines; up to 20,000 t/h	—
Pipe	Low	Small	Readily available	Yes	Enclosed transported material; steep inclines; tight horizontal curves	Small lump size may require more crushing
Cable belt	Low	Small	Specialized	Yes	Some technical advantages for long-distance systems and tight horizontals	—
Air-supported	—	—	Specialized	No	Cleaner and safer because of enclosed design and fewer moving parts	Closer fabrication tolerances and better belt cleaning required
RopeCon	High	Large	Specialized	No	Reduced structural footprint; useful in otherwise impassable terrain	—
Steep angle	Low	Small	Specialized	No	Capable of high inclines, up to vertical	—

was applied to design of conveyor transfer points (Moore et al. 2013; Kruse 2000; Hustrulid and Mustoe 1996; Grima et al. 2013; Fioroni et al. 2007).

DEM analysis can address many problem areas in the design of conveyor transfer points, including uncontrolled flow regions, blockage, off-centered loading, unsteady flow rated, high-impact velocities, particle segregation, spillage, and excessive wear. It can also help optimize the locations of wear surfaces and determine reasonable operational capacities.

Before conducting a DEM analysis, it is important to correctly characterize the material whose movement is being simulated. This characterization should include accurate descriptions of particle shape and size, material friction and cohesion, specific gravity, bulk density, and moisture content. Determination of these properties requires laboratory testing. In some cases, dynamic testing of a material sample is justified.

Recent Developments

In addition to the improvements in the special conveyors described earlier (pipe, cable belts, air-supported, steep angle, and RopeCon), there have been recent innovations in drive technology and idler maintenance.

Gearless Drives

Larger mines and processing plants and the remote locations of many operations have led to the use of longer and higher-capacity conveyor systems wherever possible. In many cases, the entire production process is dependent on the availability, reliability, and cost efficiency of the overland conveyor system. The need for larger-capacity overland conveyor systems places greater demands on the performance characteristics of the conveyor drive train. In almost all cases, electrical drive systems that allow soft starting and stopping are preferred. Such systems protect the conveyor belt from high torque and can extend the life cycle of the conveyor system.

As previously discussed, conveyor drive systems usually comprise an electric motor with a gear reducer and mechanical or fluid coupling. Some of the disadvantages of starting a conveyor with such a drive system include high-tension peaks in the belt at start-up, high starting currents in the network, and higher system and load-sharing losses. All of these conditions can increase maintenance costs and production losses. In particular, gear reducers for longer and higher-capacity conveyors are large and inefficient and require extensive maintenance. The mean time between failures for a gear reducer and associated components is three to four years. The bearings have an operating life of 8–10 years and, when replaced, necessitate a major system overhaul. Finally, the operating life—about 10 years—is often less than half that of the conveyor system.

Gearless drives can reduce or eliminate these problems. The technology for gearless conveyor drives is the same as that used for mine hoist drives. When a bearingless motor is used in the gearless drive system, further advantages accrue. A low-speed, 6-MW gearless and bearingless motor can reduce capital and operating costs by as much as 7% over a 20-year period, compared to a high-speed system with twin gear drives. Gearless drives are more specialized than conventional drives and are offered by only a few vendors. Furthermore, gearless drive installations require additional training for operators and maintenance personnel, and maintenance procedures may require more time.

Gearless drives are offered by several major suppliers of electrical drive systems. Richter (2015) describes gearless drives and provides a detailed case study for the replacement of an existing conveyor drive on a high-capacity mining machine, in which each of the two gearless-drive design options considered gave a weight reduction of about 30%. Richter also provides an economic comparison of four conveyor drive systems, as shown in Figure 33.

Richter (2015) concludes that the major advantages of gearless conveyor drives compared to geared drives are

- Elimination of the main drives gearboxes;
- Higher reliability and availability;
- Lower downtime for maintenance and breakdowns;

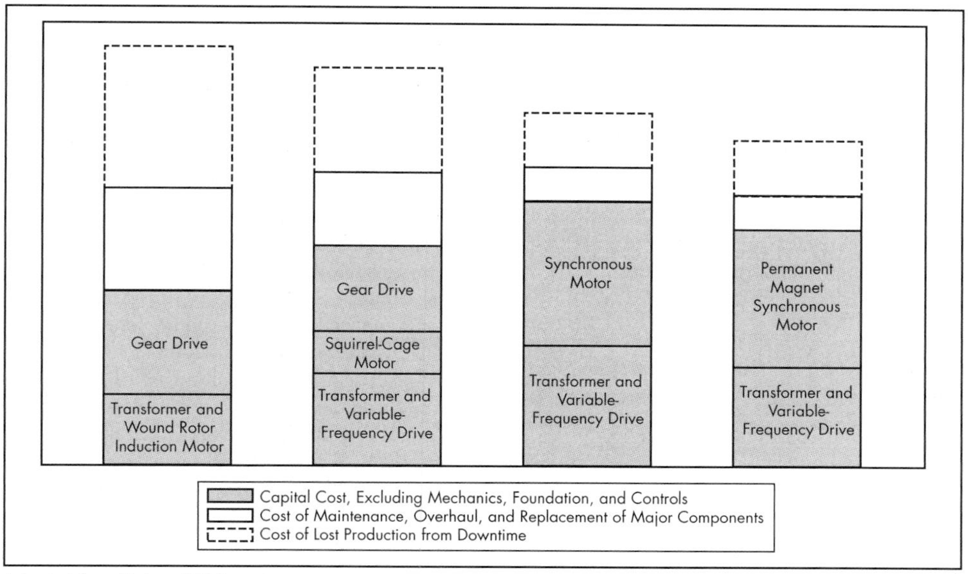

Adapted from Richter 2015

Figure 33 Comparison of costs for four conveyor drive systems

- Higher annual production with the same conveyor;
- Decreased revenue losses;
- Lean drive with lower part count and fewer sensors and spare parts;
- Higher energy efficiency, lower energy cost;
- Elimination of vibration issues caused by parts rotating at high speeds; and
- Much lower noise emission (<75 db).

Richter (2015) also states that gearless drives are especially beneficial in the following types of installations:

- The planned life cycle is longer than about 10 years.
- High availability is required.
- No redundant production lines exist.
- Material buffers such as surge bins and stockpiles are small or nonexistent.
- Maintenance parts are expensive or difficult to procure.
- Maintenance work is hard to perform, for example, at high altitude or extreme temperatures.
- Ambient conditions are harsh.

Idler Maintenance

Maintenance of idlers is an ongoing task on all conveyors. Most conveyors have a set of carrying idlers every 1–2 m. These idlers, of course, rotate rapidly whenever the belt is operating and are thus subject to wear and failure. In addition, replacement of failed idlers normally requires shutting down the belt, following a safe lock out/tagout procedure. Thus idler inspection, maintenance, and replacement are ongoing tasks that can be time-consuming and expensive. Mechanized systems for monitoring or changing idlers are offered by at least two companies. Both systems are relatively new but show great promise.

Scott Technology of New Zealand offers the Robotic Idler Predict (ABHR 2013). It is a six-axis robot arm installed on a track-mounted vehicle that travels alongside a conveyor. The robot arm carries a system that uses thermal, vibration, and acoustic monitoring to detect idler wear. The system can also remove debris from under the conveyor.

Lewis Australia (2017) offers the Spidler, a system mounted on light rail installed on either side of the belt. The Spidler is self-propelled, with a diesel motor that provides power to electrical and hydraulic systems. It travels above the belt to the location of a faulty idler, then lifts a length of the belt off the carrying idlers. A robotic manipulator then removes the faulty idler and installs a new idler in its place.

REFERENCES

ABHR. 2013. Robotic idler changeout technology advances. *Aust. Bulk Handling Rev.* (July/August): 30–31.

Anderson, J.M., and Wilson, H.V. 1965. Downhill cable belt conveyor at Craigmont Mines Limited. *CIM: Can. Inst. Min. Metall. Petrol. Bull.* (April): 435–440.

ANSI B11.3. 2002. *Safety Standards for Machine Tools*. New York: American National Standards Institute.

ANSI/CEMA 550. 2015. *Classification and Definitions of Bulk Materials*. Naples, FL: Conveyor Equipment Manufacturers Association.

ANSI/CEMA B105.1. 2015. *Specifications for Welded Steel Conveyor Pulleys with Compression Type Hubs*. Naples, FL: Conveyor Equipment Manufacturers Association.

AS 1755-2000. *Conveyors: Safety Requirements.* Sydney: Council of Standards Australia.

ASME B20.l. 2009. *Safety Standard for Conveyors and Related Equipment.* New York: American Society of Mechanical Engineers.

ASTM D412. 2016. *Standard Test Methods for Vulcanized Rubber and Thermoplastic Elastomers—Tension.* West Conshohocken, PA: ASTM International.

Brost, F.B., and Scanlon, W.T. 1984. Operating experience with the Eisenhower cable belt overland conveyor. SME Preprint No. 84-335. Littleton, CO: SME.

BS EN 619:2002+A1:2010. 2002. *Continuous Handling Equipment and Systems. Safety and EMC Requirements for Equipment for Mechanical Handling of Unit Loads.* London, UK: British Standards Institution.

CEMA (Conveyor Equipment Manufacturers Association). 2014. *Belt Conveyors for Bulk Materials*, 7th ed. Naples, FL: Conveyor Equipment Manufacturers Association.

CEMA (Conveyor Equipment Manufacturers Association). 2017. *CEMA Safety Label Brochure*, 2nd ed. CEMA Brochure No. 201. Naples, FL: Conveyor Equipment Manufacturers Association.

CEMA 501.1. 2015. *Specification for Welded Steel Wing Pulleys*. Naples, FL: Conveyor Equipment Manufacturers Association.

CEMA 502. 2016. *Bulk Material Belt Conveyor Troughing/Return Idlers: Selection and Dimensions with Metric Conversions.* Naples, FL: Conveyor Equipment Manufacturers Association.

CEMA 575. 2013. *Bulk Material Belt Conveyor Impact Bed/Cradle: Selection and Dimensions*. Naples, FL: Conveyor Equipment Manufacturers Association.

CEMA 576. 2013. *Classification of Applications for Bulk Material Conveyor Belt Cleaning*. Naples, FL: Conveyor Equipment Manufacturers Association.

Doppelmayr Transport Technology. 2018. RopeCon: Powerful and maintenance friendly in difficult terrain. www.doppelmayr-mts.com/en/solutions/ropeconr/. Accessed August 2018.

Dos Santos, J.A. 2000. Theory and design of sandwich belt high angle conveyors according to the expanded conveyor technology. In *Bulk Material Handling by Conveyor Belt III.* Edited by M.A. Alspaugh. Littleton, CO: SME.

Fioroni, M.M., Fúria, J., Franzese, L.A.G., Perfetti, L. de T., Zanin, C.E, Leonardo, D., and da Silva N.L. 2007. Simulation of continuous behavior using discrete tools: Ore conveyor transport. In *Proceedings of the 2007 Winter Simulation Conference*. Edited by S.G. Henderson, B. Biller, M.-H. Hsieh, J. Shortle, J.D. Tew, and R.R. Barton. New York: Institute of Electrical and Electronics Engineers.

Grima, A.P., Curry, D., and Wypych, P.W. 2013. New discrete element modelling calibration technology to facilitate the increased efficiency of ore handling and processing operations. In *Proceedings of Iron Ore 2013*. Melbourne, Victoria: Australasian Institute of Mining and Metallurgy.

Grisley, P. 2002. Air-supported conveying in mines and process plants. In *Bulk Material Handling by Conveyor Belt IV*. Edited by A. Reicks and E.A. Viren. Littleton, CO: SME.

Hashimoto, K. 1984. Introduction and utilization of the "pipe conveyor system." SME Preprint No. 84-334. Littleton, CO: SME.

Hewett, E.F. 1957. The cable belt conveyor and its application. *Proc. Aust. Inst. Min. Metall.* 184:29–48.

Hustrulid, A.I., and Mustoe, G.G.W. 1996. Engineering analysis of transfer points using discrete element analysis. In *Bulk Material Handling by Conveyor Belt*. Edited by M.A. Alspaugh and R.O. Bailey. Littleton, CO: SME.

ISO 4649. 2017. *Rubber, Vulcanized or Thermoplastic: Determination of Abrasion Resistance Using a Rotating Cylindrical Drum Device*. Geneva: International Organization for Standardization.

ISO 7149. 1982. *Continuous Handling Equipment: Safety Code: Special Rule*. Geneva: International Organization for Standardization.

Kecojevic, V., Md-Nor, Z.A., Komljenovic, D., and Groves, D. 2008. Risk assessment for belt conveyor-related fatal incidents in the U.S. mining industry. *Bulk Solids Powder Sci. Technol.* 3(2):63–73.

Kellis, J. 2000. Conveyor belt condition monitoring. *Bulk Material Handling by Conveyor Belt III*. Edited by M.A. Alspaugh. Littleton, CO: SME.

Khodiyar Industrial Corporation. 2008. Some main conveyor structures. http://www.kicdieselengineparts.com/conveyor_structure.htm. Accessed May 2018.

Kruse, D.J. 2000. Applications and advances of the discrete element method in the material handling industry. *Bulk Material Handling by Conveyor Belt III*. Edited by M.A. Alspaugh. Littleton, CO: SME.

Kruse, D.J. 2002. Optimizing conveyor take-up systems using dynamic analysis and the implementation of capstans. In *Bulk Material Handling by Conveyor Belt IV*. Edited by A. Reicks and E.A. Viren. Littleton, CO: SME.

Kumar, A., and Yingling, J.C. 1994. Analytic models for determining capacity and discharge rates for feeders and surge bins in mine conveyor networks. SME Preprint No. 94-270. Littleton, CO: SME.

Lewis Australia. 2017. Robotic idler roller changer. www.lewis.com.au/project/robotic-idle-roller-changer/. Accessed October 2017.

Lodewijks, G., and Nuttall, A.J.G. 2008. Dynamics of long belt conveyors with distributed drives. In *Bulk Material Handling by Conveyor Belt VI*. Edited by M.T. Myers. Littleton, CO: SME.

Moore, B.A., van der Merwe, S., and van Aarde, N. 2013. Implementation of advanced transfer chute designs. In *Proceedings of Iron Ore 2013*. Melbourne, Victoria: Australasian Institute of Mining and Metallurgy.

Morrell, P.D., Weber, J.R., and Jackson, N. 1984. Air-supported belt conveyors for dry phosphate concentrate. SME Preprint No. 84-360. Littleton, CO: SME.

MSHA (Mine Safety and Health Administration). 1979a. *Code of Federal Regulations: Title 30, Mineral Resources, Parts 1–199*. Washington, DC: MSHA.

MSHA (Mine Safety and Health Administration). 1979b. *Code of Federal Regulations: Title 30, Mineral Resources, Parts 56 and 57, safety and health standards—underground metal and nonmetal mines*. Washington, DC: MSHA.

MSHA (Mine Safety and Health Administration). 2000. *MSHA's Guide to Equipment Guarding*, Other Training Material (OT 3, revised). Washington, DC: MSHA.

Nave, M.L. 2000. Condition monitoring of mining belt conveyors. *Bulk Material Handling by Conveyor Belt III*. Edited by M.A. Alspaugh. Littleton, CO: SME.

OSHA (Occupational Safety and Health Administration). 2004a 29 CFR 1910.219. *Mechanical Power-Transmission Apparatus*. Washington, DC: OSHA.

OSHA (Occupational Safety and Health Administration). 2004b. 29 CFR 1926.307. *Mechanical Power-Transmission Apparatus*. Washington, DC: OSHA.

Overland Conveyor Company. 2017. Pipe conveyors. www.overlandconveyor.com/consulting/conveyor-consulting-services/pipe-conveyor.aspx. Accessed November 2017.

Padgett, H.L. 2001. *Powered Haulage Conveyor Belt Injuries in Surface Areas of Metal/Nonmetal Mines, 1996–2000*. Denver, CO: MSHA Office of Injury and Employment Information.

Richmond, A., Atkinson, T., and Scoble, M. 1991. Conveyor systems design and application for surface mining. *CIM: Can. Inst. Min. Metall. Petrol. Bull.* (November-December): 67–71.

Richter, U. 2015. Benefits of high-powered gearless conveyor drives applied to medium-power conveyors. In *Proceedings of the International Conference on Hoisting and Hauling*. Edited by B. Johansen. Englewood, CO: SME. pp. 345–362.

Swinderman, R.T., Marti, A.D., Goldbeck, L.J., Marshall, D., and Strebel, M.G. 2009. *Foundations—The Practical Resource for Cleaner, Safer, More Productive Dust and Material Control*, 4th ed. Neponset, IL: Martin Engineering Company.

Tarshizi, E., Sturgul, J., and Taylor, D. 2014. How simulation and animation model of an aggregate mine assists engineers in the operation of haulage systems. SME Preprint No. 14-208. Englewood, CO: SME.

Wiedenroth, J. 2010. The longest pipe conveyor of the world with double load transport at Cementos Lima in Peru. SME Preprint No. 10-041. Littleton, CO: SME.

Yingling, J.C., Lineberry, G.T., and Kumar, A. 1993. Control of conveyor peak loads by variable feeder discharge rates. *Trans. SME* 292:1918–1922.

Zhang, Y., and Steven, R. 2012. Pipe conveyor and belt: belt construction, low rolling resistance, and dynamic analysis. SME Preprint No. 12-129. Englewood, CO: SME.

Zhu, A.H., Kutsyh, A.Y., Liang, L.Y., and Lu, J.Z. 2011. The latest developments of belt conveyor instrumentation. CIM Maintenance, Engineering and Reliability/Mine Operators Conference (MEMO). Westmount, QC: Canadian Institute of Mining, Metallurgy and Petroleum.

CHAPTER 5.3

Pumps

Michael G. Nelson and Amy J. Richins

A pump is a machine that lifts, pressurizes, and moves liquids or solid–liquid slurries from place to place. The mineral processing industry uses pumps in all phases of wet processing, from grinding to tailings disposal.

TERMINOLOGY AND NOMENCLATURE

Common terms in the pumps industry are described in this section.

- **Measurement of performance.** The amount of useful work that any fluid-transport device performs is the product of the mass rate of fluid flow through the device and the total pressure differential measured immediately before and after the device, usually expressed in the height of a fluid column equivalent under adiabatic conditions. The first of these quantities is referred to as capacity, the second as head.
- **Capacity.** Capacity, Q, is expressed in cubic meters per hour (m^3/h); in U.S. customary units, it is expressed in gallons per minute (gpm). Since all these are volume units, the specific gravity or specific weight of the liquid must be used for conversion to mass rate of flow.
- **Head.** Head, H, is a term for pressure, expressed as the height of a column of liquid. Other expressions for head include the following:
  - **Friction head.** Friction head, H_f, is the pressure required to overcome the resistance to flow in the pipe and fittings.
  - **Net positive suction head.** Net positive suction head, NPSH, is the difference between the liquid pressure at the pump's suction and the vapor pressure of the fluid being pumped. The available net positive suction head, NPSHA, is a measure of how close the fluid at a given point is to flashing, and the inception of cavitation, as described below. The required net positive suction head, NPSHR, is the head value at the inlet of the pump required to keep the fluid from cavitating. In a positive displacement (PD) pump, the NPSHA should be large enough to open the suction valve, overcome the friction losses within the pump liquid end, and overcome the liquid acceleration head.
  - **Total dynamic head.** The total dynamic head, H, of a pump is the total discharge head, h_d, minus the total suction head, h_s.
  - **Total suction head.** Total suction head is the reading h_{gs} of a gauge at the suction flange of a pump, corrected to the pump centerline, plus the barometer reading of atmospheric pressure, atm, and the velocity head, h_{vs}, at the point of gauge attachment:

    $$h_s = h_{gs} + \text{atm} + h_{vs}$$

    If the gauge pressure at the suction flange is less than atmospheric, requiring use of a vacuum gauge, this reading is used for h_{gs} in the preceding equation with a negative sign.
  - **Static suction head.** The static suction head, h_s, is the vertical distance from the free surface of the liquid source to the pump centerline plus the absolute pressure at the liquid surface.
  - **Total discharge head.** The total discharge head, h_d, is the reading, h_{gd}, of a gauge at the discharge flange of a pump (corrected to the pump centerline, or on vertical pumps, to the eye of the suction impeller), plus the barometer reading and the velocity head, h_{vd}, at the point of gauge attachment.
  - **Velocity head.** Velocity head, h_v, is the vertical distance through which a body must fall to acquire the velocity, V.

    $$h_v = V^2/2g$$

    where V is the flow velocity in m/s, and g is the acceleration due to gravity in a pumping system; the velocity head derives from the flow of the fluid at the velocity, V.
- **Speed.** Pump speed, N, is denoted in revolutions per minute, or rpm.
- **Specific speed.** Specific speed, N_S, of a pump is the speed at which a geometrically similar impeller would need to

Michael G. Nelson, Professor & Chair, Mining Engineering Department, University of Utah, Salt Lake City, Utah, USA
Amy J. Richins, Research Associate, Mining Engineering Department, University of Utah, Salt Lake City, Utah, USA

operate to deliver a fixed volume flow rate at a fixed head. The specific speed is calculated using metric units as

$$N_S = (N\sqrt{Q})/(H^{0.75})$$

where N is rotational speed in rpm, Q is capacity in m^3/min, and H is head in m.

Specific speed is usually thought of as a dimensionless number. In fact, specific speeds calculated with this formula are not dimensionless, but the dimensions are not meaningful, so they are not used. It is thus critically important that the units used are reported along with the pump specific speed value. A dimensionless expression for specific speed can be created by multiplying the pump head by the gravitational constant and measuring the shaft rotation in radians, thus:

$$\Omega_S = (N_{rad}\sqrt{Q})/(g\ H^{0.75})$$

where N_{rad} is rotational speed in rad/s, Q is capacity in m^3/s, g is the acceleration due to gravity in m/s^2, and H is head in m.

The dimensionless specific speed, Ω_S, is related to the specific speeds calculated with metric or U.S. units by the following expressions:

$$\Omega_S = N_S\ (\text{U.S.})/2733.016$$

where N_S is calculated with N as rotational speed in rpm, Q as capacity in gpm, and H as head in ft.

Similarly,

$$\Omega_S = N_S\ (\text{metric})/51.64$$

In general, pumps with low specific speed have a low capacity and high head, while pumps with high specific speed have higher capacity and lower head.

- **Velocity.** The linear velocity, V, is found from the ratio of the quantity flowing through a given area in a given time to that area, thus

$$V = Q/A$$

where V is the average flow velocity, Q is the flow quantity, and A is the cross-sectional area of the pipe or channel. For circular pipe, this relationship in metric units is

$$V = 3.54\ Q/d^2$$

where V is the average flow velocity in m/s, Q is the flow quantity in m^3/h, and d is the inside pipe diameter in cm.

- **Viscosity.** The existence of internal friction or the internal resistance to relative motion of the fluid particles is called viscosity. The viscosity of liquids usually decreases with rising temperature. Viscous liquids tend to increase the power required by a pump; to reduce pump efficiency, head, and capacity; and to increase friction in pipelines.

Care must be taken to distinguish between dynamic viscosity and kinematic viscosity. *Dynamic viscosity* is the resistance of a fluid to shearing flows, where adjacent layers move parallel to each other with different speeds. The Greek letter μ is commonly used by engineers and some physicists to denote dynamic viscosity, while η is used by chemists and other physicists. Dynamic viscosity has units of pascal-second, or Pa·s. *Kinematic viscosity*, also called momentum diffusivity, is the ratio of the dynamic viscosity, μ, to the density of the fluid, ρ. It is usually denoted by the Greek letter ν and is measured in units of square meters per second, or m^2/s.

- **Work performed in pumping.** Liquid to flow occurs when work is expended. A pump may raise the liquid to a higher elevation, force it into a vessel at higher pressure, provide the head to overcome pipe friction, or perform any combination of these. Regardless of the service required of a pump, all energy imparted to the liquid in performing this service must be accounted for; consistent units for all quantities must be employed in arriving at the work or power performed.

When analyzing the performance of a pump, it is customary to calculate its power output, which is the product of the total dynamic head and the mass of liquid pumped in a given time. In metric units, power is expressed in kilowatts; horsepower is the conventional unit used in the United States.

In metric units,

$$kW = HQ\gamma/3.670 \times 10^5$$

where kW is the pump power output in kW, H is the total dynamic head in N·m/kg (column of liquid), Q is the capacity or flow in m^3/h, and γ is the liquid density in kg/m^3.

When the total dynamic head H is expressed in pascals, then

$$kW = HQ/3.599 \times 10^6$$

- **Efficiency.** The power input to a pump is greater than the power output because of internal losses resulting from friction, leakage, and so on. The efficiency of a pump is therefore defined as

$$\text{pump efficiency} = (\text{power output})/(\text{power input})$$

In the United States, the power input is often described as the *brake horsepower* (BHP), an anachronistic term that refers to the power output at the shaft of the driving unit, with no accessories attached. A similar term, *brake kW*, is sometimes employed when using the metric system.

The best efficiency point (BEP) is the maximum possible efficiency and is usually shown as an efficiency curve on the head versus capacity chart for a given pump.

HYDRAULICS FOR PUMPING

Properties of Fluids

Design and selection of pumps requires a thorough knowledge of the fluid to be pumped, including specific weight and volume, specific gravity, vapor pressure, and kinematic viscosity. Fluids may be categorized on their shear behavior, as shown in Figure 1. In Newtonian fluids, shear stress versus shear rate is characteristically linear. Thixotropic or pseudoplastic fluids exhibit a shear-thinning characteristic where the shear stress decreases with increasing shear rates. Dilatant fluids exhibit a shear-thickening characteristic where the shear stress increases with increasing shear rates.

Bernoulli's Equation for Energy in Incompressible Fluids

The law of conservation of energy states that energy can neither be created nor destroyed, and that all forms of energy

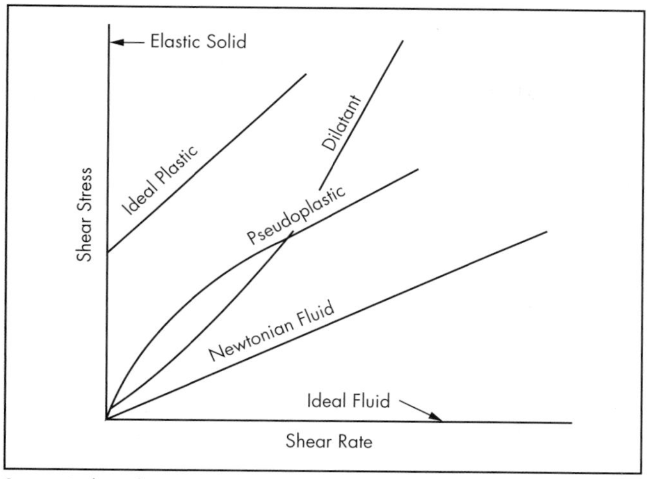

Source: Link et al. 1985

Figure 1 Shear stress–shear rate relationship for various types of viscous behavior

are equivalent. Energy considerations in closed pipe systems are analyzed readily by the following general energy equation, known as Bernoulli's equation:

$$P_1 + V_1^2/2g + Z_1 + H_a - H_f - H_e = P_2 + V_2^2/2g + Z_2$$

The left-hand side of the equation describes the energy at the pump inlet. The first term, P_1, is the pressure head. The second term, $V_1^2/2g$, is the velocity head and accounts for the kinetic energy of the moving fluid. The third term, Z_1, is the elevation head with respect to a defined reference point. The pump head, H_a, is the energy added by the pump. The fifth term is the energy lost to fluid friction, H_f; and the sixth term is the energy extracted, H_e. The terms on the right-hand side of the equation represent similar quantities at the pump discharge. All quantities are expressed in units of length.

Cavitation

Cavitation occurs in a flowing liquid when there is a decrease and subsequent increase in local pressure. The pressure decrease results in localized vaporization and the formation of bubbles. When the pressure recovers, the bubbles implode. Cavitation occurs only if the local pressure declines below the saturated vapor pressure of the liquid and subsequently recovers to above the vapor pressure. If the recovery pressure is not above the vapor pressure, then "flashing" occurs. In pipe systems, cavitation is usually the result of an increase in the kinetic energy through a constricted area, or an increase in the elevation of the pipeline.

The generation, growth, and collapse of cavitation bubbles result in very high-energy densities, and in high local temperatures and pressures at the surfaces of the bubbles. These conditions exist only for a very short time, and the overall liquid medium remains at ambient conditions. Uncontrolled cavitation is damaging and can cause noise, vibration, and a loss of efficiency. Highly localized cavitation can cause erosion and pitting, and can dramatically shorten a pump's lifetime. After a surface is initially affected by cavitation, it tends to erode faster. The cavitation pits increase the turbulence of the fluid flow and create crevices that act as nucleation sites for additional cavitation bubbles. The pits also increase the components' surface area and leave behind residual stresses. This makes the surface more prone to stress corrosion.

Cavitation in pumps may occur in two different forms. *Suction cavitation* occurs when the pump suction is under a low-pressure/high-vacuum condition, and the liquid turns into a vapor at the eye of the pump impeller. When this vapor is carried to the discharge side of the pump, it is no longer under vacuum and is compressed to a liquid state by the discharge pressure. This imploding action can be violent and damaging. It may remove large chunks of very small bits of material, depending on local conditions.

Discharge cavitation occurs when the pump discharge pressure is extremely high, as may be the case in a pump that is running at less than 10% of its BEP. The high discharge pressure causes most of the fluid to circulate inside the pump instead of flowing out the discharge. As the liquid flows around the impeller, it passes through the small clearance between the impeller and the pump housing at very high velocity. This flow velocity generates a vacuum to the housing wall, vaporizing the liquid. A pump that has been operating under these conditions shows premature wear of the impeller vane tips and the pump housing, and the high pressures generated can cause premature failure of the pump's mechanical seal and bearings.

All pumps require well-developed inlet flow to operate efficiently. When poorly developed flow enters a pump impeller, it strikes the vanes and is unable to follow the impeller passage. The liquid separates from the vanes, causing cavitation and vibration resulting from turbulence and poor filling of the impeller. For a well-developed flow pattern, a straight pipe run of about 10 pipe diameters is recommended upstream of the pump inlet flange. Unfortunately, space and equipment layout constraints may not allow this distance. Instead, it is common to use an elbow close-coupled to the pump suction, which creates a poorly developed flow pattern at the pump suction. Cavitation may be avoided by increasing suction pressure, decreasing liquid temperature, throttling back on the discharge valve to decrease flow rate, or venting gases off the pump casing.

Flow in Pipes

When a fluid is flowing in a pipe, the shear stress on the fluid is highest at the walls of the pipe, which is stationary, and is lowest at the center of the pipe. The velocity profile across the pipe cross section is a function of the fluid type.

Liquids flowing in filled pipes at low velocities tend to follow straight lines. This is called laminar flow. As liquid velocities increase, the flow regime eventually becomes turbulent. At intermediate velocities, fluids flow in a transition zone. In the transition zone, the fluid flow is unstable and may oscillate unpredictably between laminar and turbulent flow. In most pumping systems, the flow is either transitional or turbulent.

The nature of flow in a pipe depends on the pipe diameter, the density and viscosity of the fluid, and the flow velocity. These quantities can be combined into a single, dimensionless number called the Reynolds number, N_{Re}, which is defined as

$$N_{Re} = DV/\nu$$

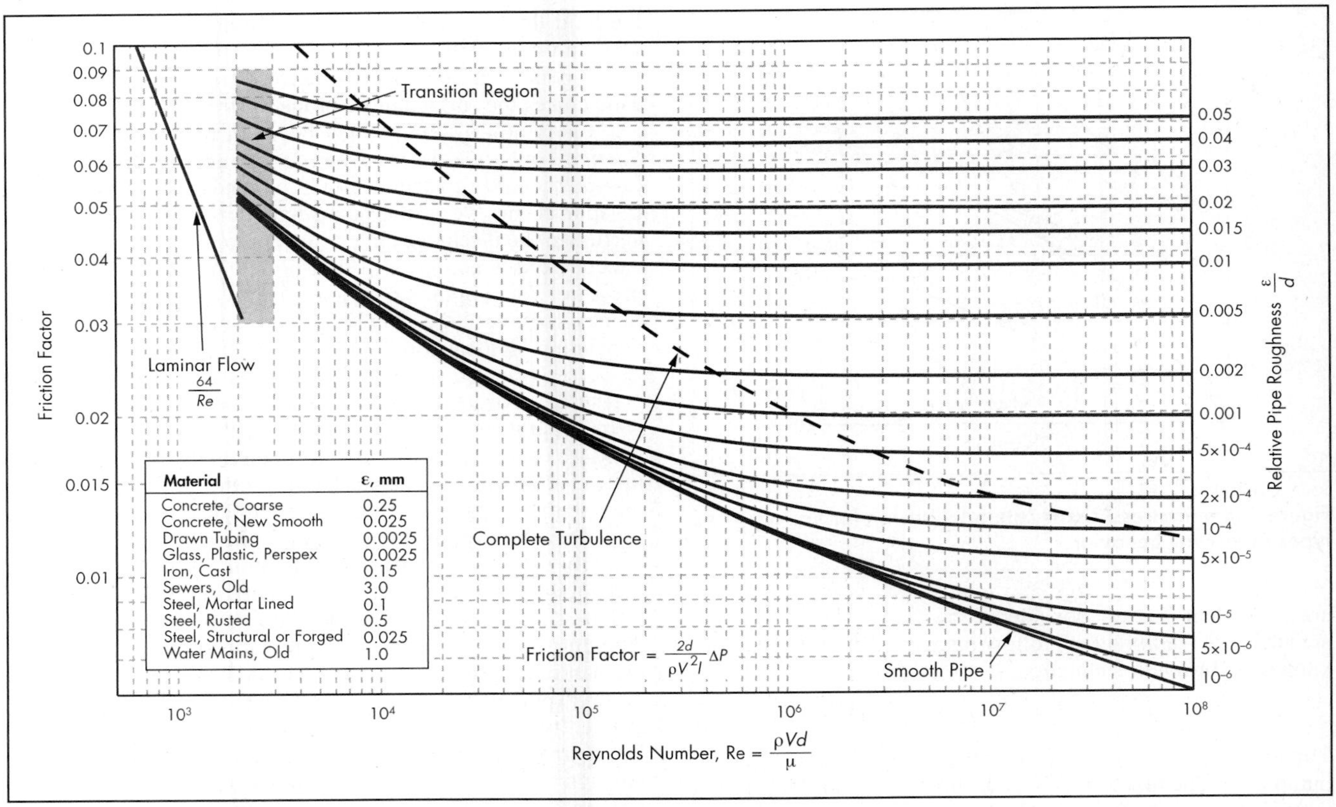

Source: Beck and Collins 2008

Figure 2 Moody diagram

where D is the inside pipe diameter, V is the fluid velocity, and ν is the kinematic viscosity. The Reynolds number may also be calculated as

$$N_{Re} = \rho DV/\mu$$

where ρ is the specific gravity and μ is the kinematic viscosity of the fluid. While the Reynolds number is dimensionless, care must be taken to use consistent units in its computation and use.

The transition from laminar to turbulent flow occurs at a Reynolds number of about 2,000. This is called the critical Reynolds number and is important in computing fluid friction losses in pipes. In fact, critical Reynolds numbers may vary substantially with different fluids, but engineering practice assumes that laminar flow occurs below a Reynolds number of 2,000, turbulent flow occurs above 4,000, and a transition zone exists between these values.

Friction Losses

Liquid flow in pipes is always accompanied by friction losses, which result from internal friction or viscous forces of the liquid, and from the surface roughness of the pipe. These losses are indicated by a reduction in head in the direction of liquid flow. For a given length of pipe, L, this pressure drop can be measured, or predicted with reasonable accuracy using the Darcy–Weisbach equation:

$$H_f = f\,LV^2/2gD$$

where H_f is the pressure drop in m, f is the friction factor, V is the flow velocity in m/s, g is the gravitational constant in m/s, and D is the inside diameter of the pipe in m.

The friction head loss, H_f, is often expressed in meters or feet of loss for a fixed length of pipe. The Darcy friction factor, f, is determined experimentally and varies with the Reynolds number. In laminar flow, the friction head loss primarily results from viscous forces within the liquid. When the Reynolds number is below 2,000, the friction factor may be estimated by

$$f = 64/N_{Re}$$

In turbulent flow, the roughness of the interior pipe wall must be considered in addition to the viscosity of the liquid. The relative roughness, ε/D, is found by dividing the absolute surface roughness of the pipe, ε, by the diameter of the pipe, D. The surface roughness may be thought of as the average height of projections on the pipe wall. With consistent units, the relative roughness is a dimensionless number. The relationship among f, N_{Re}, and ε/D is often plotted on a Moody diagram, as shown in Figure 2, so that values of f may be determined graphically.

Churchill (1977) developed an equation that is too cumbersome for occasional calculation of single values of f, but it is useful when plotting the whole Moody diagram, including laminar, transition, and turbulent flow regions, say, from $N_{Re} = 10^3$ to 10^8 and from $\varepsilon/D = 0$ to 0.01, yielding f-values from $f = 0.01$ to 0.04. The Churchill equation is especially useful in spreadsheets and computer programs:

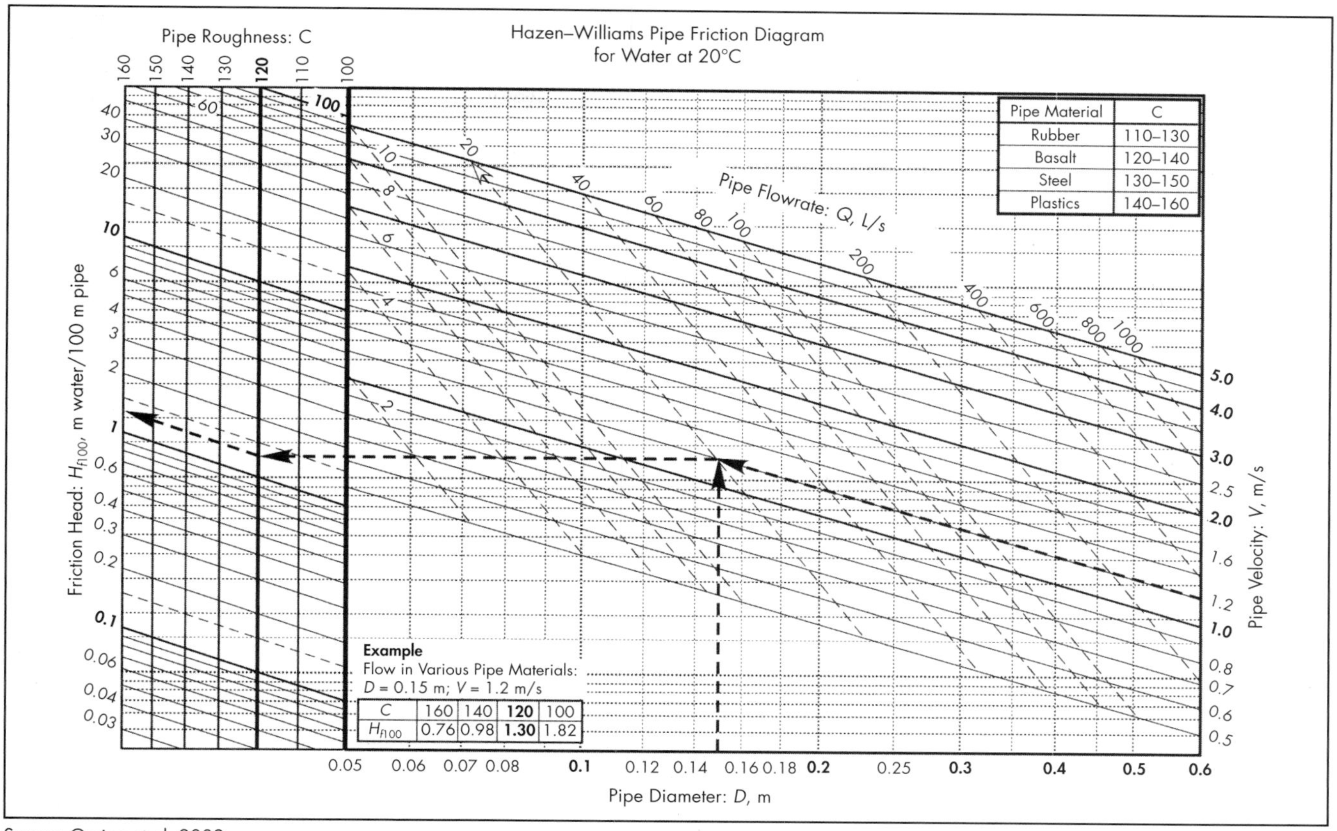

Source: Grzina et al. 2002

Figure 3 Hazen–Williams diagram

$$f = 8\left\{\left(\frac{8}{N_{Re}}\right)^{12} + \left[\left(\frac{37{,}530}{N_{Re}}\right)^{16} + \left\{-2.457 \cdot \ln\left[\left(\frac{7}{N_{Re}}\right)^{0.9} + 0.27\frac{\varepsilon}{D}\right]\right\}^{16}\right]^{-1.5}\right\}^{1/12}$$

To use the Moody diagram, first compute the Reynolds number and the relative roughness for the fluid system. Next, locate the appropriate relative roughness value along the right side of the diagram. Then move to the left, parallel to the group of curved lines. Continue to move left, parallel to the nearest curved line, until the vertical line passing through the appropriate Reynolds number is reached. Then move horizontally to the left side of the diagram and read the friction factor.

Values for the absolute surface roughness for various pipe materials are also included in the Moody diagram, for convenience. These values are for new pipe. Pipe roughness generally increases with age. Changes in pipe wall roughness adversely affect both pipe diameter and friction factor.

The widely used Williams and Hazen (1920) formula provides an alternative to the use of the Moody diagram:

$$V = 0.849CR^{0.63}S^{0.54}$$

where V is the flow velocity in the pipe in m/s, C is the friction factor resulting from roughness only, R is the hydraulic radius (area/perimeter) of the pipe, and S is the friction head loss in m/m. The advantage of the Hazen–Williams equation is that the coefficient C is not a function of the Reynolds number, but the disadvantage is that it is only valid for water. Also, it does not account for the temperature or viscosity of the water. A graphic form of the Hazen–Williams formula is also available and is shown in Figure 3.

Minor head losses are those caused by local disturbances in the liquid system. Disturbances can be caused by entrance and discharge conditions; changes in pipe diameter; pipe fittings, such as valves, elbows, and tees; and minor irregularities in the interior pipe wall. Although these losses are termed *minor losses*, they may become a significant portion of the total head loss in short lines or in in-plant lines where numerous fittings are used. Minor losses are frequently reported as equivalent to a length of straight pipe, or as a constant to be applied to the velocity head. If the latter method is used, the minor head losses, H_{ml}, are given by the following equation:

$$H_{ml} = KV^2/2g$$

where K is an empirically determined constant, and V is the linear flow velocity. Table 1 shows loss coefficients for various pipe fittings.

As liquid flows from a large reservoir into a pipe, turbulence develops at the entrance to the pipe, causing a decrease in pressure head known as an entrance loss. The effect of this turbulence is to restrict the opening size. The degree of restriction depends on the shape of the pipe opening. Figure 4 shows typical entrance configurations and their loss coefficients, K_e.

Table 1 Head loss coefficients for various pipe fittings

Fitting	K	L/D
Globe valve, wide open	10	350
Angle valve, wide open	5	175
Close return bend	2.2	42
T, through side outlet	1.8	67
Short-radius elbow	0.9	32
Medium-radius elbow	0.75	27
Long-radius elbow	0.60	20
45° elbow	0.42	15
Gate valve, open	0.19	7

Source: Scott and Hays 1985

Table 2 Contraction head loss coefficients for various pipe diameter ratios

D_2/D_1	K_c
0.1	0.45
0.2	0.42
0.3	0.39
0.4	0.36
0.5	0.33
0.6	0.28
0.7	0.22
0.8	0.15
0.9	0.06

Source: Scott and Hays 1985

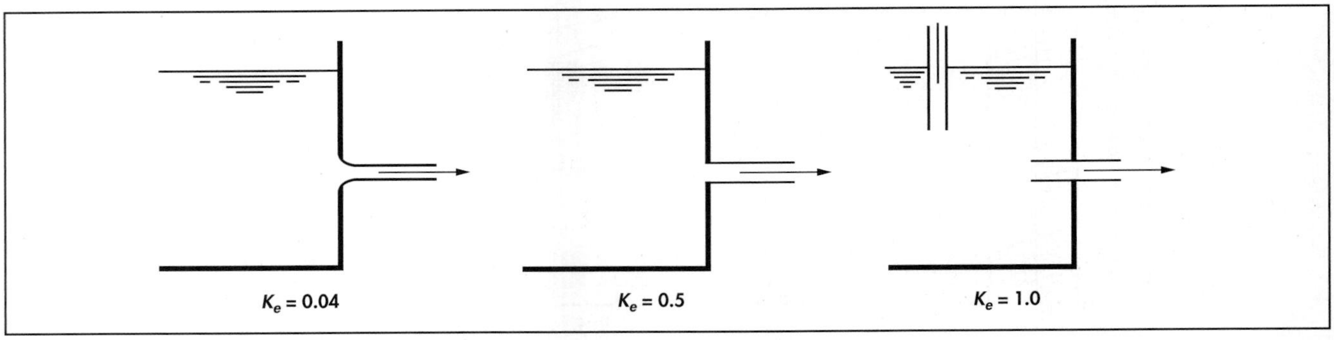

Source: Foust et al. 1964

Figure 4 Entrance losses

As liquids flow from an open pipe into a reservoir, the velocity is dissipated, and the discharge head loss coefficient is usually considered to be 1.0. However, the discharge loss occurs outside the pipe system and thus does not affect the pressure at the pipe end. Sudden contractions in pipe diameters result in contraction head losses similar to those associated with entrance conditions. Table 2 shows loss coefficients for various pipe diameter ratios. To reduce these losses, a conical or smooth-tapered section should be inserted between the two pipe diameters. When the total angle lies between 20° and 40°, the loss is minimized and the coefficient for contraction, K_c, equals about 0.1.

Head losses from a sudden enlargement in pipe diameter, H_{mlc}, are found as follows:

$$H_{mlc} = (V_1 - V_2)^2/2g$$

where V_1 and V_2 are the respective linear flow velocities before and after the enlargement.

The analysis of friction losses in slurry pumping is discussed later in the "Slurry Pumping" section.

Hydraulic Gradient Profile

The total energy or head at any point in a pumping system can be calculated from Bernoulli's equation. If a convenient reference elevation or datum plane is chosen and the total energy is calculated at locations along the system, the calculated values may be plotted to scale to show the energy gradient of the system. Figure 5 shows the variation in total charge energy, H, measured in meters from the suction liquid surface to the discharge liquid surface. The dotted line through the sum of the pressure and elevation heads in Figure 5 is called the hydraulic gradient. The velocity head is the difference between the energy gradient and the hydraulic gradient lines. The total head for the pump is the difference between total energy at the pump discharge, point 2, and the pump suction, point 1.

System-Head Curve

A pumping system may include pipes, valves, fittings, vessels, meters, process equipment, and other components through which flow must pass. When a system is analyzed for pump selection, the flow resistance of the liquid through all these components must be calculated. That resistance increases with flow at a rate approximately equal to the square of the flow through the system. Besides overcoming flow resistance, additional head may be required to raise the liquid from the pump inlet point to a higher level at the discharge. In addition, the atmospheric pressure at the discharge point will usually be different than the pressure at the inlet, and this must be accounted for. Required heads that result from a change in elevation or a change in atmospheric pressure are called *fixed system heads* or *static heads* because they do not vary with rate of flow. Fixed system heads can be positive or negative.

A system-head curve is a plot of total system resistance, variable plus fixed, for various flow rates. It has many uses in centrifugal pump applications. Because centrifugal pumps are rated in units of length, the curve should show head in the same way, and not in pressure units. System-head curves usually show flow in volume per unit time.

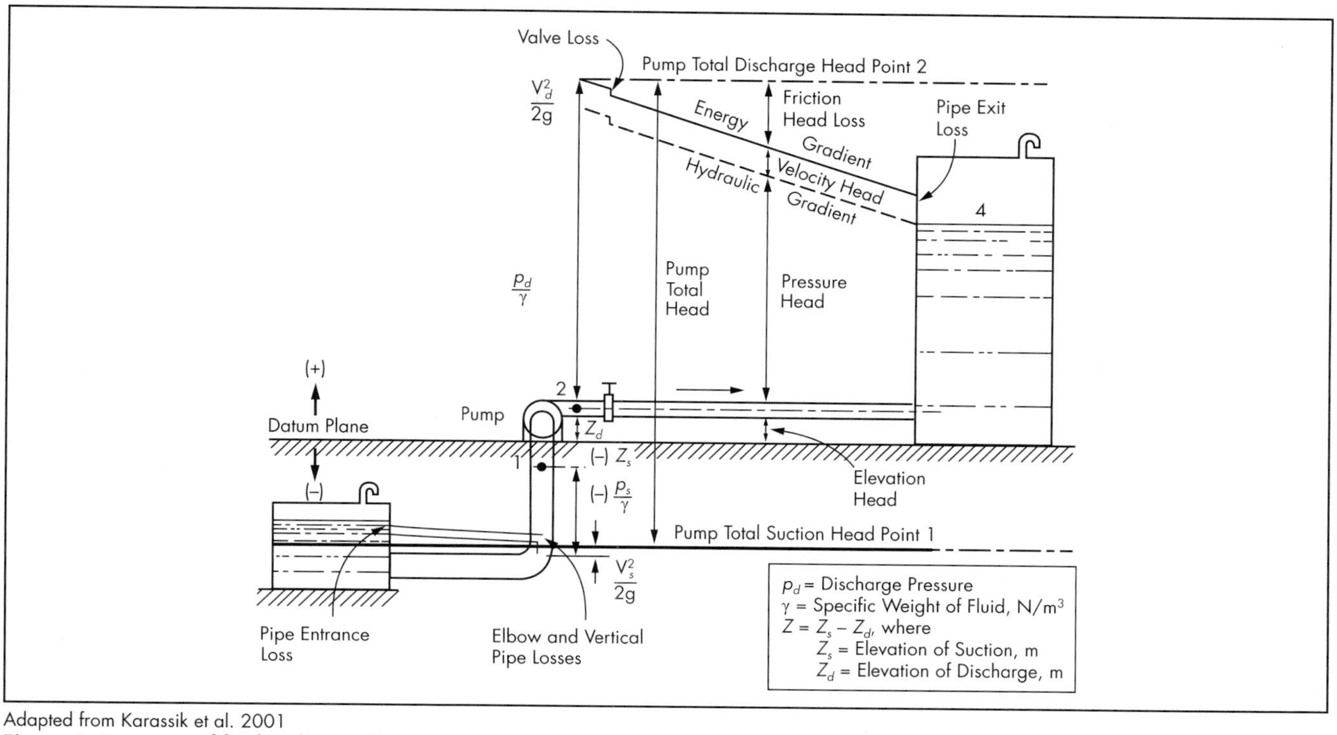

Adapted from Karassik et al. 2001

Figure 5 Energy and hydraulic gradients

When the system head is required for several flows or when the pump flow is to be determined, a system-head curve is constructed using the following procedure:

1. Define the pumping system and its length.
2. Calculate (or measure) the fixed system head, which is the net change in total energy from the beginning to the end of the system due to elevation or pressure head differences. An increase in head in the direction of flow is a positive quantity.
3. Calculate, for several flow rates, the variable system total head loss through all piping, valves, fittings, and equipment in the system.

Figure 6 illustrates construction of a system total-head curve, with a pumping system that starts at point 1 and ends at point 2. In this figure, the system-head curve shows head increasing with increasing flow rate. The fixed system head is the net change in total energy and does not change with flow, because the pressures and liquid levels do not vary with flow. The total head is p_s/γ at point 1 and $p_d/\gamma + Z$ at point 2, where p_s is the suction pressure, γ is the specific weight of fluid in N/m^3, p_d is the discharge pressure, and $Z = Z_s - Z_d$, where Z_s is the elevation of suction in m, and Z_d is the elevation of discharge in m. The pipe, valves, and fittings provide the variable system head. The curve of total system head versus flow is plotted by adding the fixed head and variable heads for several flow rates.

The flow produced by a centrifugal pump varies with the system head, while the flow of a PD pump is independent of the system head. In Figure 6, the head-capacity curve of a centrifugal pump is also shown on the system total-head curve. It shows a decreasing head with increasing flow. The flow of the pump in this system is indicated by the point at which the pump total head and system total head intersect. This intersection should be at or near the pump's best efficiency.

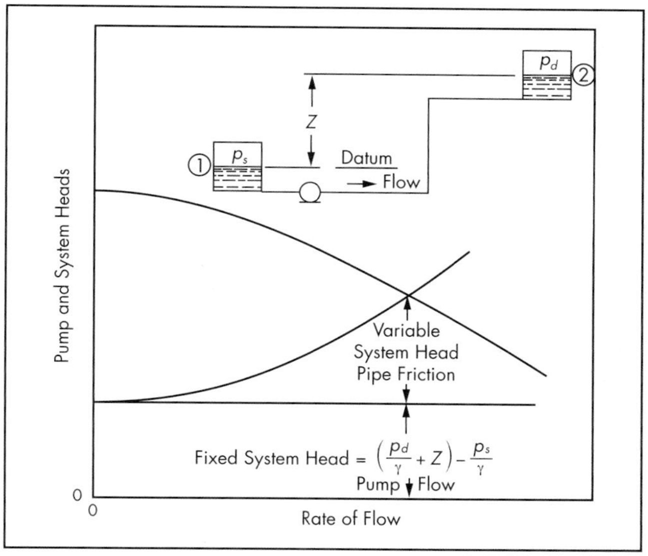

Source: Karassik et al. 2001

Figure 6 Construction of the system total-head curve

TYPES OF PUMPS

The two major pump classes are kinetic and positive displacement. Each of these classes is further subdivided based on operating principle and construction features. The classification system for pumps is shown in Figure 7.

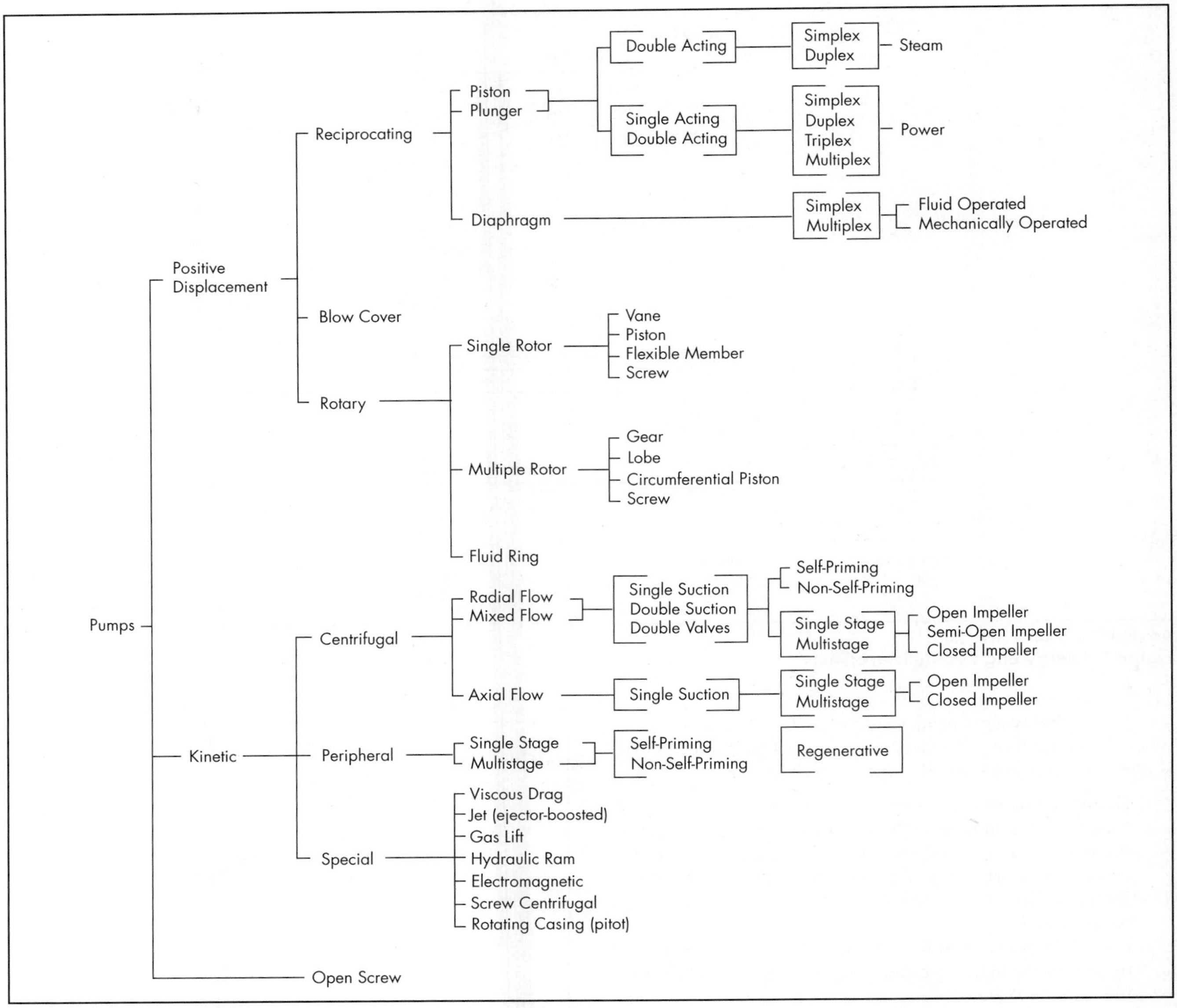

Source: Hydraulic Institute 1975

Figure 7 Classification of pumps

Kinetic (Centrifugal) Pumps

Kinetic pumps impart kinetic energy to the fluid to be pumped and then convert that kinetic energy into an increased pressure, as described by Bernoulli's principle. The most common type of kinetic pump, by far, is the centrifugal pump. As its name implies, this pump converts energy supplied by a prime mover into pressure energy and velocity energy by imparting centrifugal force to the liquid entering the eye of a rotating impeller, as shown in Figure 8. Liquid entering the eye is guided to the outer periphery of the impeller through several vane passages. As the liquid leaves the impeller, it has two components of flow. One component is in the radial direction, traveling through the vane; the other is tangential, resulting from the direction of rotation of the impeller.

The pump casing or volute that surrounds the impeller guides the liquid exiting the impeller to the discharge nozzle. The path between the impeller and casing gradually increases from the closest clearance (called the volute tongue or cutwater) to the discharge nozzle. This is to maintain a relatively constant velocity as more liquid is accumulated from the impeller along the path. Prior to reaching the pump discharge flange, the casing flares out in the diffuser area, as shown in Figure 9. This results in a reduction in velocity with an accompanying increase in pressure energy.

Casing Design

The three basic casing design types used in a centrifugal pump are the true volute, near or semi-volute, and circular volute, as illustrated in Figure 10.

The true volute, Figure 10A, has the highest efficiency because of its tight impeller-to-cutwater clearance. The cutwater is the V-shaped diverter/flow splitter between volute and discharge. This tight clearance increases efficiency by minimizing recirculating flow at the BEP. However, at flows below

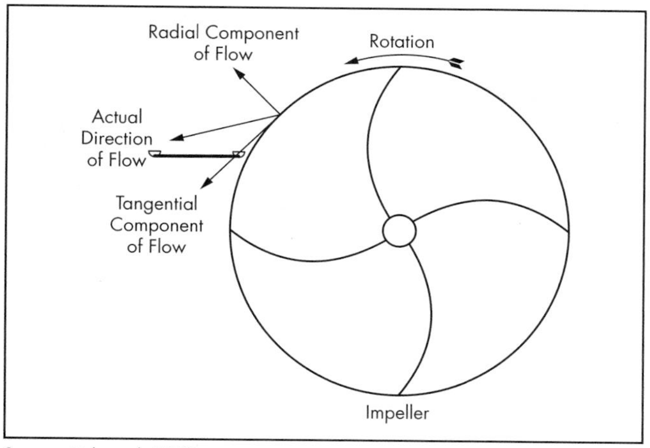

Source: Link et al. 1985

Figure 8 Centrifugal pump force diagram

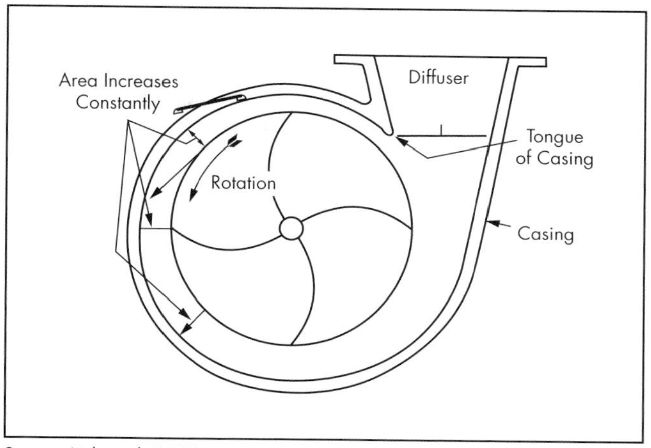

Source: Link et al. 1985

Figure 9 Centrifugal pump terminology

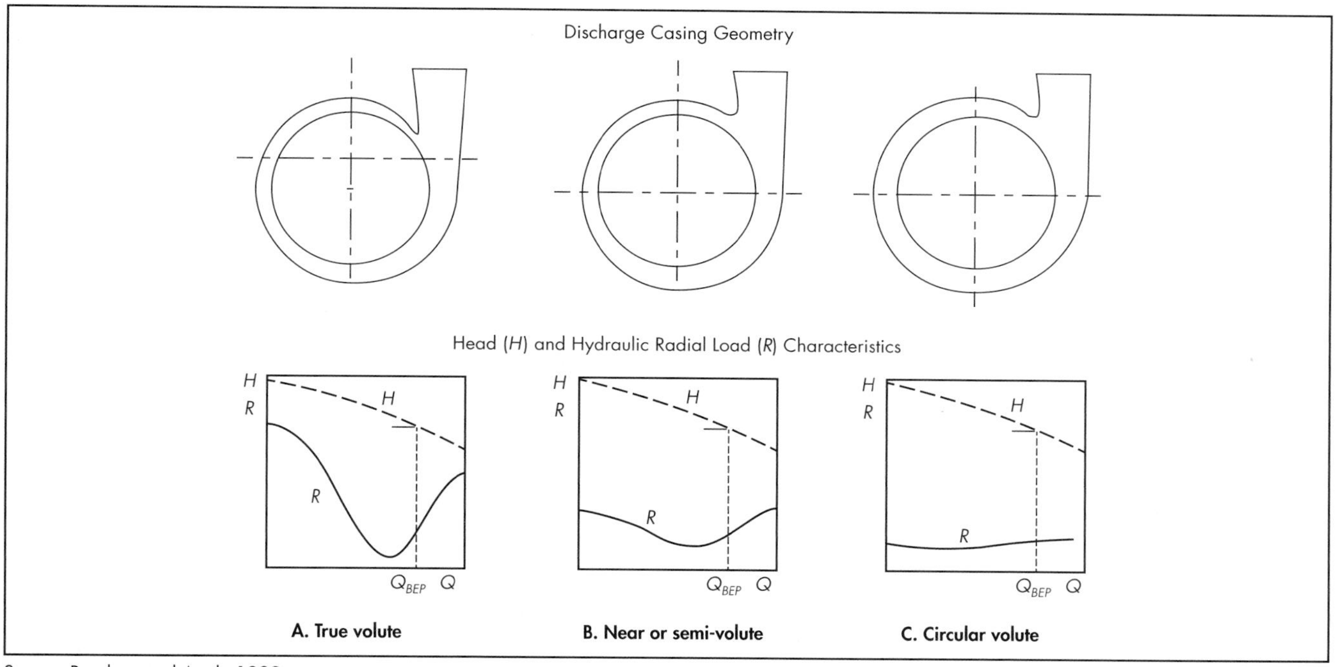

Source: Roudnev and Angle 1999

Figure 10 Slurry pump casing types

the BEP, the increased recirculating flow has a high velocity through the small impeller to the cutwater area. This results in high wear at low flows for this design in the area at and just beyond the cutwater. This design also has the highest hydraulic radial loading at flows other than the BEP.

The near volute, Figure 10B, is similar to the true volute, but the impeller-to-cutwater clearance is increased to reduce velocity at the cutwater area for lower flows. This design results in significantly reduced wear in this area at low flows. This design also has significantly lower hydraulic radial force at conditions away from the BEP. For most slurry applications, this design offers a good balance of wear life and efficiency.

The circular volute, Figure 10C, has uniform area between the impeller and casing all around the volute. This volute design has the lowest hydraulic radial force at conditions off the BEP. This makes it ideal for high-head, low-flow applications.

Normal recommended operating conditions for the three volute types are 80%–120% of BEP flow for the true volute, 60%–110% of BEP flow for the near volute, and 40%–100% of BEP flow for the circular volute. For severe duty, including heavy slurries, the ideal selection would be a pump with an operating point in the middle of one of the previous ranges, to minimize wear on the volute. For light slurries, the preceding ranges can be widened.

Performance

Performance of a centrifugal pump is conveniently characterized in a set of performance curves. These curves consist of plots of total dynamic head *TDH*, total system head *TSH*,

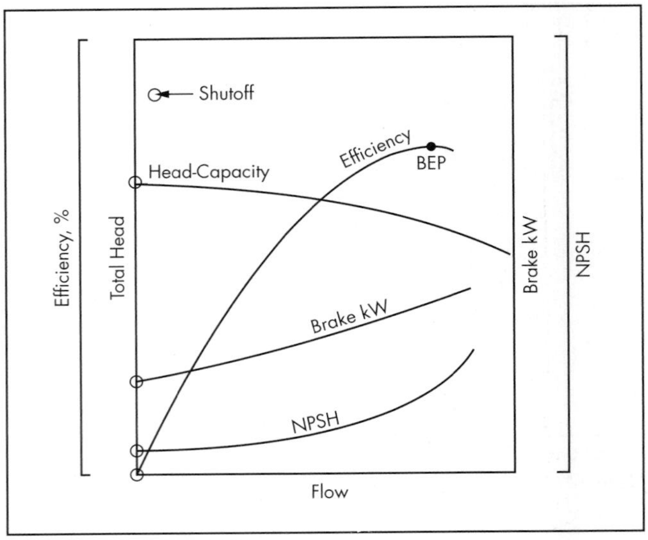

Adapted from Link et al. 1985
Figure 11 Centrifugal pump characteristic curves

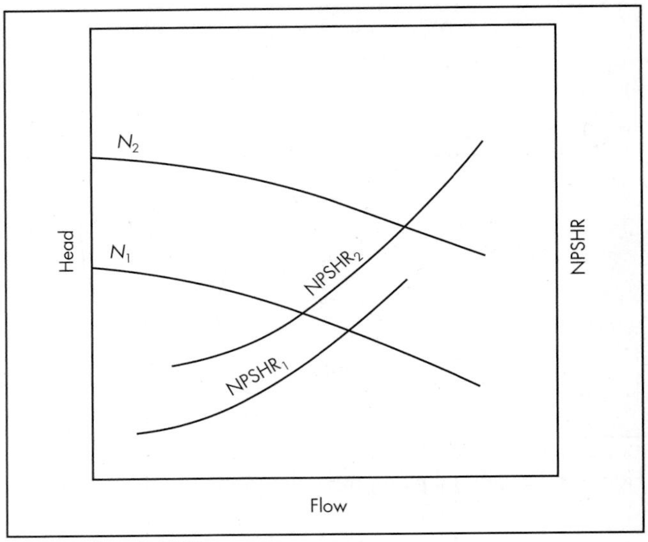

Source: Link et al. 1985
Figure 12 Relationship of *Q*, *N*, and NPSHR

or just head H, along with power and efficiency, all plotted against capacity Q at constant speed N. For a centrifugal pump, the total dynamic head is defined as

$$H = H_d - H_s + (V_d^2 - V_s^2)/2g$$

where H_d is the discharge head as measured at the discharge nozzle and referred to the pump shaft centerline, expressed in meters; H_s is the suction head expressed in meters as measured at the suction nozzle and referred to shaft centerline; and $(V_d^2 - V_s^2)/2g$ is the difference in the suction and discharge velocity heads.

Figure 11 shows a typical set of characteristic curves for one impeller speed, N. Salient points to note are the shutoff head H_{so} or the head at zero flow, and the best efficiency point, *BEP*. Pump efficiency, E, at any given coordinate of H and Q is calculated as

$$E = 9{,}797 QH\gamma / P$$

where H and Q are in units of m and m^3/s, respectively; g is the fluid specific weight; and P is the shaft power in kW.

Efficiency is a measure of the degree of hydraulic and mechanical perfection of the pump. The foregoing ratio is, therefore, that of ideal hydraulic power to brake kW.

In every pump design, the NPSHR must be considered. The NPSHR increases with pumping capacity Q for a given pump speed N and also increases with pump speed at a constant Q, as shown in Figure 12.

The Affinity Laws

Figure 13 shows *H-Q* and *P-Q* curves for two pump speeds, N_1, and N_2. Coordinates of points between the corresponding characteristic curves of two different speeds are related in a definite manner. For example, a point Q_1, H_1 on the *H-Q* curve at speed N_1 has a corresponding point Q_2, H_2 on the *H-Q* curve at speed N_2. The relationships are

$$Q_2 = Q_1 (N_2/N_1)$$

$$H_2 = H_1 (N_2/N_1)^2$$

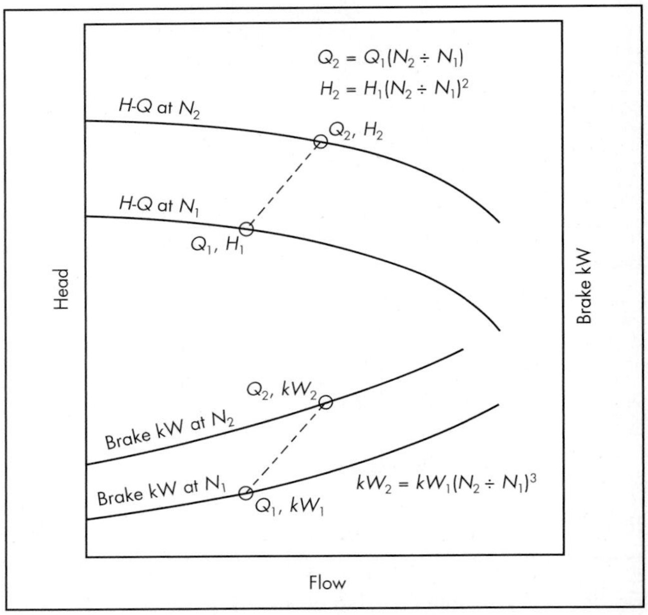

Adapted from Link et al. 1985
Figure 13 Affinity law relationships

$$P_2 = P_1 (N_2/N_1)^3$$

These relationships form the affinity laws. Further, any set of corresponding points occurs at an almost identical pump efficiency.

The affinity laws can be used to generate a family of pump curves such as shown in Figure 14. Accuracy in generating these curves is good if N_1 and N_2 are sufficiently close. However, as the difference between N_1 and N_2, increases, the accuracy of the curve will decrease because of variances in mechanical losses at different speeds.

Pump manufacturers publish a family of curves for each available pump model and standard impeller. These curves

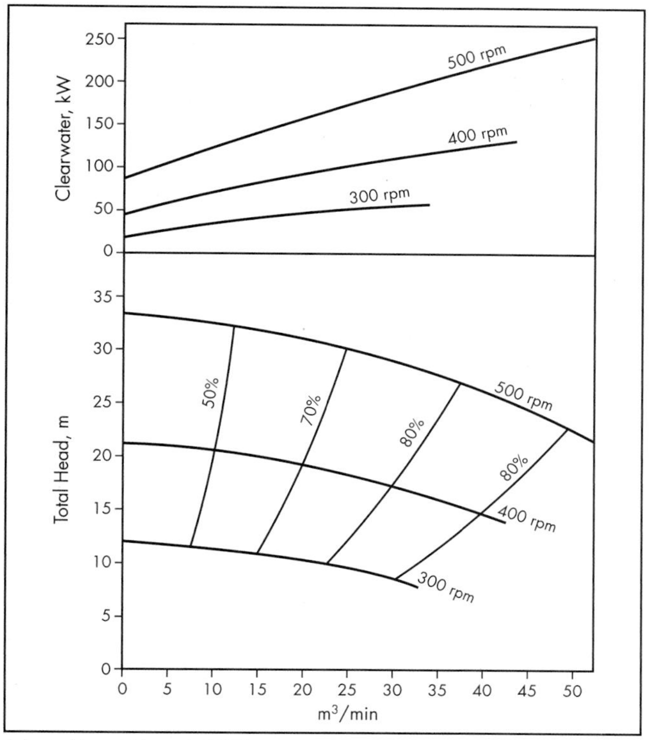

Adapted from Link et al. 1985

Figure 14 A family of pump curves

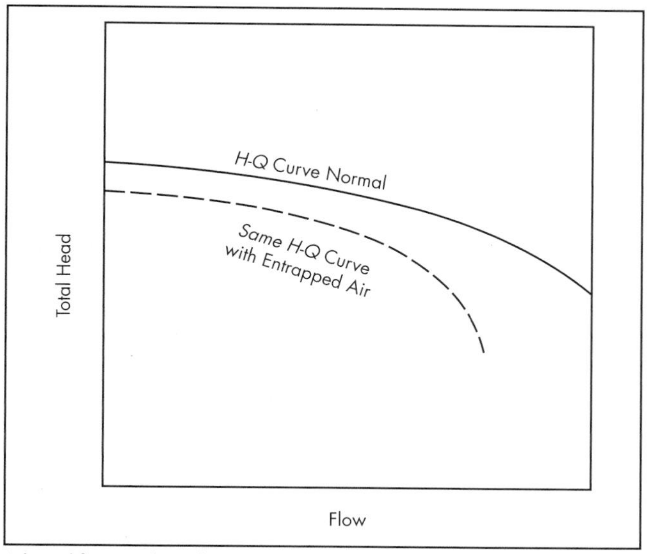

Adapted from Tomb et al. 1985

Figure 15 Effect of entrained air on *H* versus *Q*

are almost always developed by using clear, cold water as the liquid being pumped and are called *clearwater* performance curves. Power is referred to as clearwater power. Because some liquids do not have physical properties identical to those of water, specific gravity and viscosity of the liquid in question must be known before clearwater curves can be used. The corrections to be applied to clearwater performance are considerably more complex if the liquid to be pumped has a viscosity much different from that of water. The pump manufacturer should be consulted in these cases. Pumping of slurries is discussed later.

Pump performance can also be affected by entrained air in the liquid. Because the pump is a centrifugal device, the heavier liquid is thrown away from the impeller eye, while the lighter air will remain behind at the eye until enough air is accumulated to choke off the eye area. This can result in an *H-Q* curve that is actually lower than, and drops off faster than, the desired *H-Q* curve, as shown in Figure 15. Lack of sufficient NPSH will affect pump performance in the same way as entrained air.

Specific Speed

Specific speed, as described previously, is to some extent a characterization of pump type. All pumps of a similar geometric configuration have a fairly constant specific speed. This relationship is shown in Figure 16.

Centrifugal pumps are the most widely used in industry, and impellers for centrifugal pumps have been further classified by various configurations in relation to optimum specific speed for each. Figure 17 shows the classification of centrifugal pump impellers and the associated specific speeds. The four types of impellers are shown in cross section.

Radial-vane pumps develop pressure principally by the action of centrifugal force. In straight-vane radial pumps, the vane surfaces are generated by straight lines parallel to the axis of rotation. Pumps in this class with single-inlet impellers have the lowest specific speeds, with rotational speeds below 4,200 rpm, while pumps with double-suction radial impellers have rotational speeds below 6,000 rpm. In pumps of this class, the liquid normally enters the impeller at the hub and flows radially to the periphery.

In mixed-flow pumps, the head is developed partly by centrifugal force and partly by the lift of the vanes on the liquid. The impeller vanes in these pumps have a double curvature, which imparts both a radial and an axial component to the flow out of the impeller. Mixed-flow pumps that operate at lower specific speeds use Francis vane impellers. This type of pump has a single-inlet impeller with the flow entering axially and discharging in an axial and radial direction. Pumps of this type usually have rotational speeds between 4,200 and 9,000 rpm.

Axial flow or propeller pumps develop most of their head by the propelling or lifting action of the vanes on the liquid. Each has a single-inlet impeller with the flow entering axially and discharging nearly axially. Pumps of this type usually have rotational speeds above 9,000 rpm.

The specific speed equation shows that, for a given head-capacity duty, pumps with higher specific speed will run at higher rotational speeds than those with lower specific speed. Higher specific-speed pumps have smaller-diameter impellers than their lower specific-speed counterparts. Efficiency is generally lower for low-capacity, high-speed pumps and greater for high-capacity, high-speed pumps. Peak efficiencies for ordinary centrifugal pumps range from 90% to 92%.

Specific speed is often the first concept used by a pump designer to estimate the general hydraulic characteristics required for specific markets and applications. Once a pump type is selected, characteristics are refined by the details of volute and impeller geometry—diameter, number of vanes, shape of vanes, and so forth. For example, impellers in radial-flow, end-suction pumps may or may not have a suction shroud

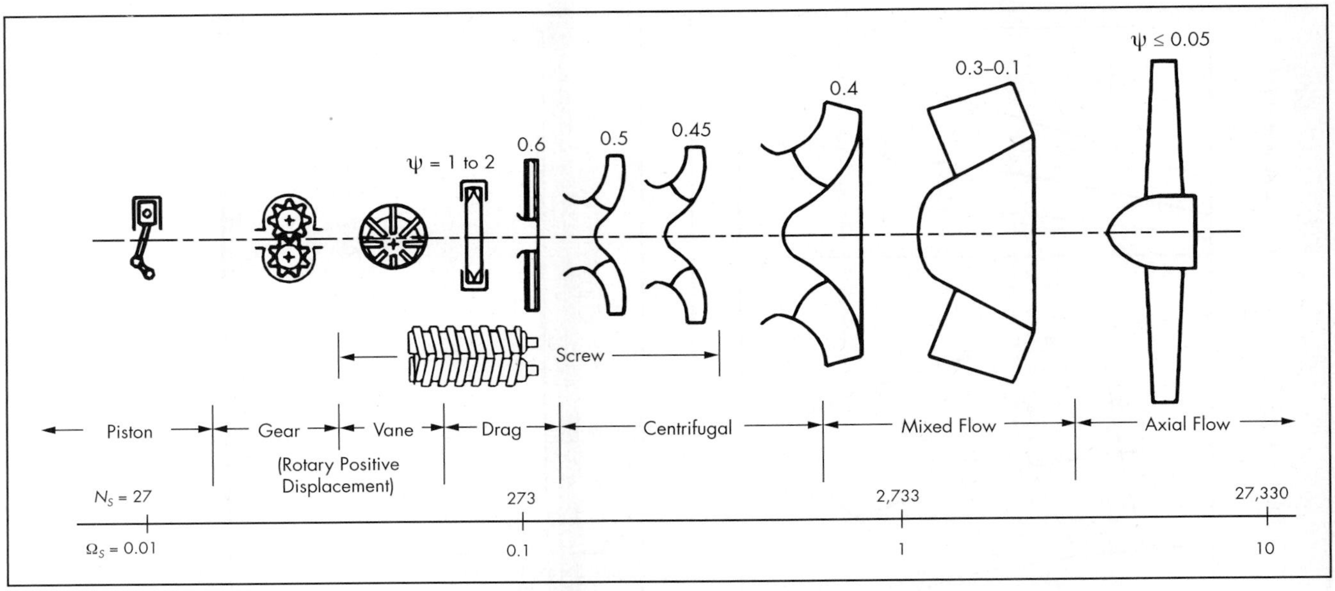

Source: Karassik et al. 2001

Figure 16 Optimum pump geometry as a function of BEP specific speed (unitless) for single-stage rotors

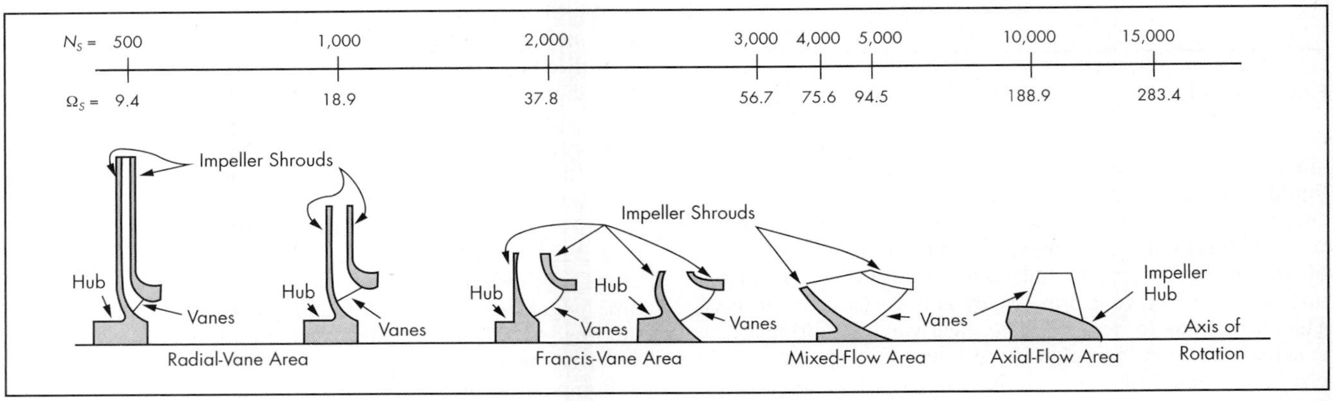

Adapted from Hydraulic Institute 1975

Figure 17 Types of centrifugal pump impellers related to pump specific speed

that covers the outside vanes. An open impeller has such a shroud, while a closed impeller does not. A closed impeller may result in a higher pump efficiency, but an open impeller may be desirable for pumping difficult materials such as froths, to prevent air entrainment at the eye. Both types of impellers are shown in Figure 18.

Figure 19 shows various types of pump characteristics. Head-capacity curves *a*, *b*, *c*, and *d* are called stable curves because only one flow is associated with a given head. In contrast, curve *b* has an unstable characteristic. The power-capacity curves labeled *a*, *b*, and *c* correspond to the head-capacity curves *a*, *b*, and *c*. Curve *a* is called a non-overloading characteristic, since power decreases when the head is either increased or decreased. Curve *b* is called an overloading curve with a reduction in head, and curve *c* is called an overloading curve with an increase in head. Curve *c* is typical of an axial flow pump.

Pumps in Series and Parallel Operation

Centrifugal pumps can be used in parallel or series operation to multiply, respectively, the volume or head capability of a single unit, as shown in Figure 20. In parallel pumping, the discharge lines of two or more pumps are combined into a common discharge line.

Figure 21 shows a combined head-capacity curve for two identical pumps operating at the same speed and installed using identical symmetrical piping to a common discharge point. Curve A is an *H-Q* curve for single pump running at speed N_1. Curve B is the *H-Q* curve for two identical pumps at speed N_1 operating in parallel. The intersection of the system-head curve with curves A and B indicates the increase in flow, $Q_B - Q_A$, possible by going to parallel operation.

Theoretically, each pump in this situation should handle one-half of the pumped capacity. However, imbalances between the two pumps can occur because of minor variations

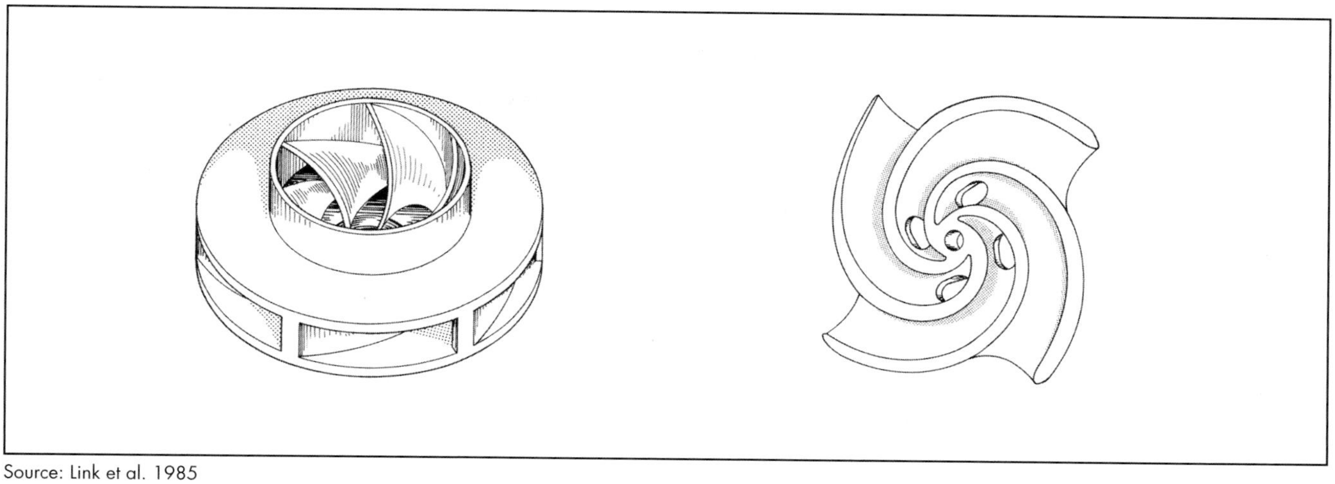

Source: Link et al. 1985

Figure 18 Closed (left) and open (right) impellers

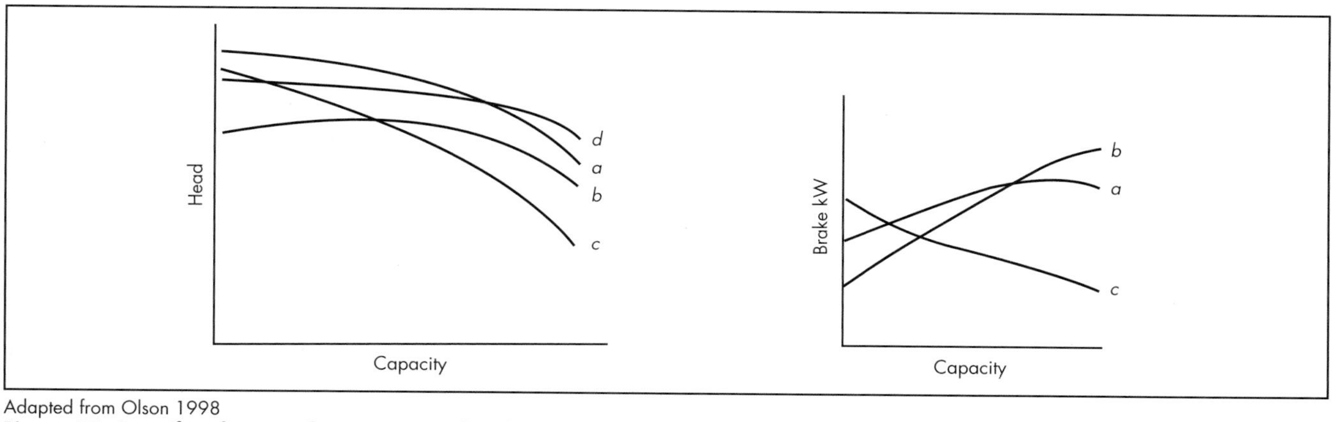

Adapted from Olson 1998

Figure 19 Centrifugal pump characteristic related to head and capacity

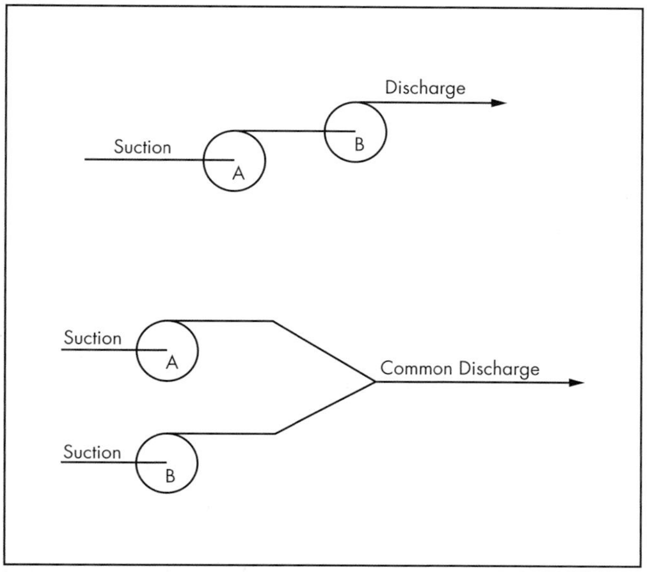

Source: Link et al. 1985

Figure 20 Pumps in series and parallel connections

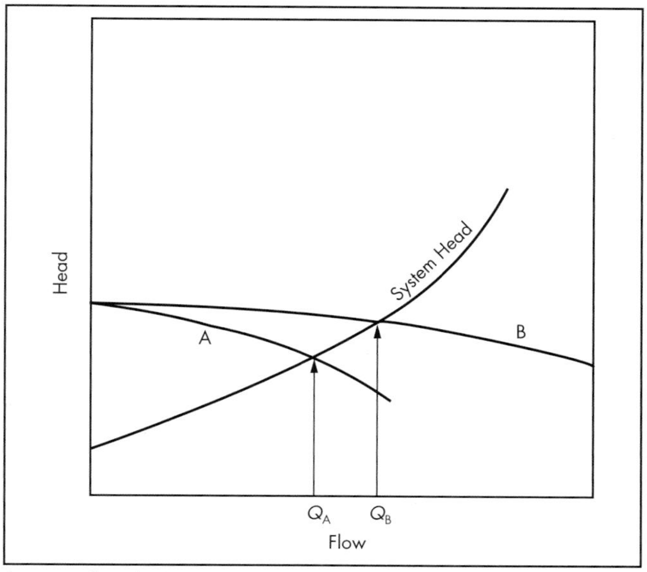

Source: Link et al. 1985

Figure 21 Head versus capacity for centrifugal pumps connected in parallel

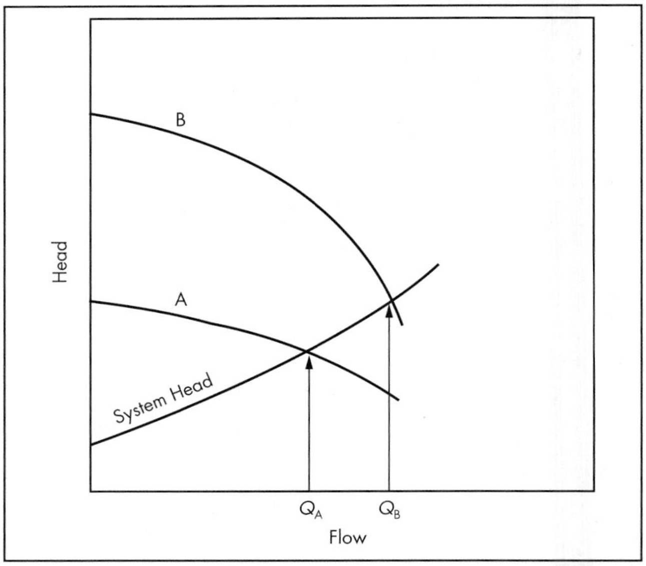

Source: Link et al. 1985

Figure 22 Head versus capacity for centrifugal pumps connected in series

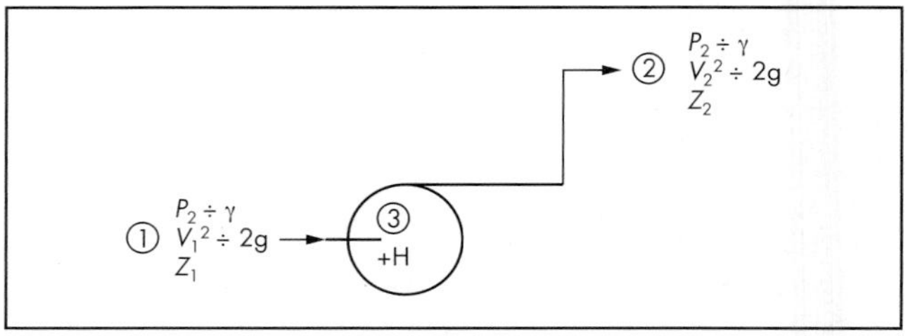

Adapted from Link et al. 1985

Figure 23 Energy in a simple pumping system

in motor speeds, V-belt drive slippage, pump wear, clearance differences, and so on. These imbalances can sometimes be modulated by the use of control valves or variable-speed drives for trimming respective pump speeds.

Parallel operation is best achieved by using pumps with fairly steep head-capacity characteristics. Notably, the flow of one pump will not necessarily be doubled or tripled by the corresponding addition of two or three pumps. Rather, this is a function of the system-head curve with special attention given to the friction head contribution.

In series pumping, the head capability of one pump is increased as long as it is possible to feed the discharge of one pump into the suction of the next without exceeding the pressure rating of the last-stage pump. In multistage pumps, this can be accomplished in a single unit by using one casing with diffuser plates. For example, in boiler feed pumps, thousands of pounds of inlet pressure provide a self-priming ability better than that of other kinetic pumps.

Separate, horizontal centrifugal pumps may be staged such as is done in the transport of mine tailings. Staging is done by connecting pumps, either in series or in parallel.

Figure 22 illustrates the combined head-capacity curve for two identical centrifugal pumps operating at the same speed in series. As for the pumps connected in parallel described

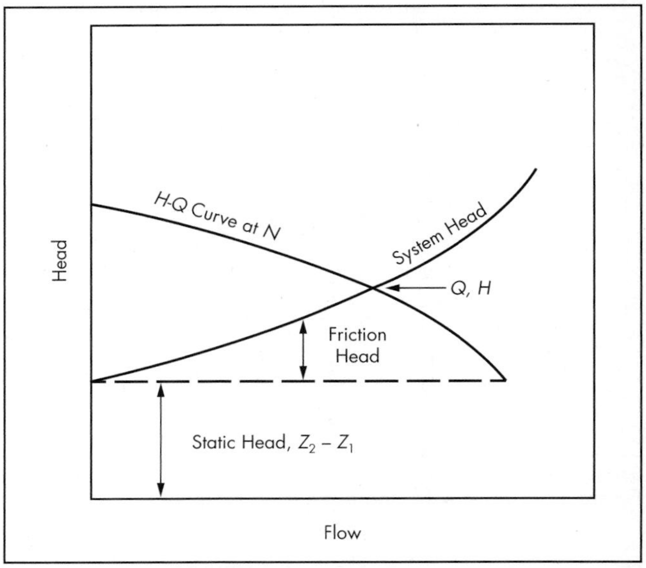

Adapted from Link et al. 1985

Figure 24 Typical system-head curve for a centrifugal pumping system

above, the superposition of the system-head curves shows the flow capability of the pumping system. Curve A is an *H-Q* curve for a single pump running at speed N_1, and Curve B is an *H-Q* curve for two pumps connected in series and operating at speed N_1. The intersections of the system-head curve with the two operating curves indicates the respective flows possible for each configuration. The potential increase in flow quantity by connecting a second identical pump in series is given simply by the difference, $Q_B - Q_A$.

Selection and Operation

Proper pump selection must begin with the nature of the liquid to be pumped. Specialty centrifugal pumps exist for practically every service, such as pumping of corrosive materials, clear liquids or solid suspensions, viscous oils, sludges, high- and low-temperature liquids, and so forth. Many applications can be served by more than one pump type and by a variety of manufacturers.

A system-head curve should be developed as previously described. Consider the simple system in Figure 23. The system-head curve showing the energy requirement in meters of liquid H at each flow Q may look like that shown in Figure 24, depending on the line, valve, and fitting losses that exist in the system.

The system-head curve reflects the static head, $Z_2 - Z_1$, that must be overcome regardless of the flow, in addition to the friction losses in the suction line from point 1 to point 3, and the discharge line from point 3 to point 2 in Figure 23. In Figure 25 the system-head curve is compared to a selected pump curve. The intersection of the appropriate pump curve at point (Q_A, H_A) defines the quantity and head that the pump will provide when operating at rotational speed N_A. From the complete characteristics of the pump at speed N_A, the pump power, efficiency, and NPSHR can also be determined. It is best to have the pump operating point as close as possible to the BEP without going too far to the right of this BEP, as shown in Figure 25. At the BEP, the flow within the pump

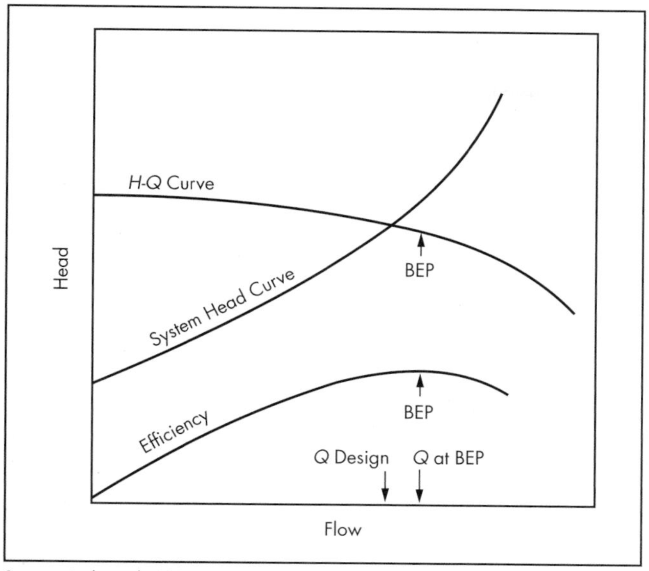

Source: Link et al. 1985

Figure 25 Pump curve showing selection for best efficiency point

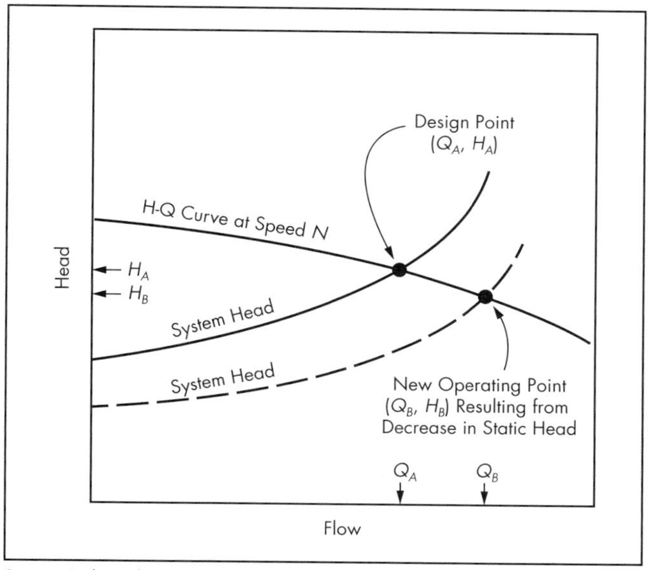

Source: Link et al. 1985

Figure 26 Change in operating curve resulting from change in static head

produces the least turbulence and recirculation is minimal so that efficiency is maximized.

Operation too far to the right of the BEP invites cavitation, and NPSH demands rise rapidly; operation too far to the left of the BEP invites high radial loads and operating instabilities due to the flatter nature of the *H-Q* curve near shutoff. As a rule of thumb, the operating point should be between 40% and 115% of BEP flow.

Variations in the system-head curve should also be considered. System friction may be increased or decreased, for example, throttling a discharge valve. Static head can vary with changes in the suction receiver level or discharge point. Variation in the system-head curve that is imposed will automatically define a new pump operating point, as shown in Figure 26. This is because the pump will operate at the intersection of the system-head curve and pump curve. If such potential variations are disregarded, the pump driver could be underpowered.

Control of Cavitation

Cavitation always increases pump wear. This is especially true for slurry pumps. Severe cavitation can also cause severe damage, leading to failure of the pump. Depending on the severity of the cavitation, high noise and vibration may result, along with a reduced head output, lower flow, or complete loss of prime.

Whether or not cavitation occurs depends on the suction characteristics of the pump and the system in which it is placed. To prevent cavitation, the NPSHA should exceed the NPSHR by 15% to 30%. The NPSHR for a particular pump is determined by the pump design and may be found on the manufacturer's performance curve at the pump duty point in the system. The NPSHR usually increases with flow and pump speed, but an increase can also occur at extremely low flows near pump shutoff.

Values of NPSHA for the system are not simply the height of liquid above the centerline of the pump. Rather, NPSHA is a function of the local atmospheric pressure, the height of the liquid relative to the pump centerline, the friction losses in the suction pipe, and the vapor pressure of the liquid at the pumping temperature. Figure 27 shows how to calculate NPSHA for various suction conditions (Roudnev and Angle 1999).

For calculations under suction lift conditions, as shown in Figure 27A, the atmospheric pressure *pa* should be reduced by dividing it by the specific gravity of the liquid. This is to account for the weight of the liquid in the column and the negative effect this has on the suction pressure at the pump. The application of any positive correction under flooded conditions is not recommended.

Cavitation can be easily prevented during the system design process by ensuring there is adequate suction height, minimal suction friction, and proper pump selection. After system construction, these factors are much more difficult to change.

Installation

The proper selection of a suitably designed centrifugal pump is not sufficient to ensure its success. Equally important is the installation and maintenance of the unit. The pump should be placed as close as possible to the reservoir or suction source as practical. Discharge lines and suction lines should be kept as short and direct as possible to minimize friction losses and hence TDH. Valves and fittings should be kept to the necessary minimum for operation and control. The pump should have ready access for inspection and maintenance. Sufficient headroom should be provided for the installation of a hoist if needed.

Experienced operators report that most pump problems occur on the suction side of the pump. Careful attention should be paid to the suction piping and construction of the sump or reservoir from which the pump is fed. Sump design should provide sufficient volume or retention for the required pump flow with a suitable geometry to prevent the formation of vortices and potential air entrapment in the liquid. Flow

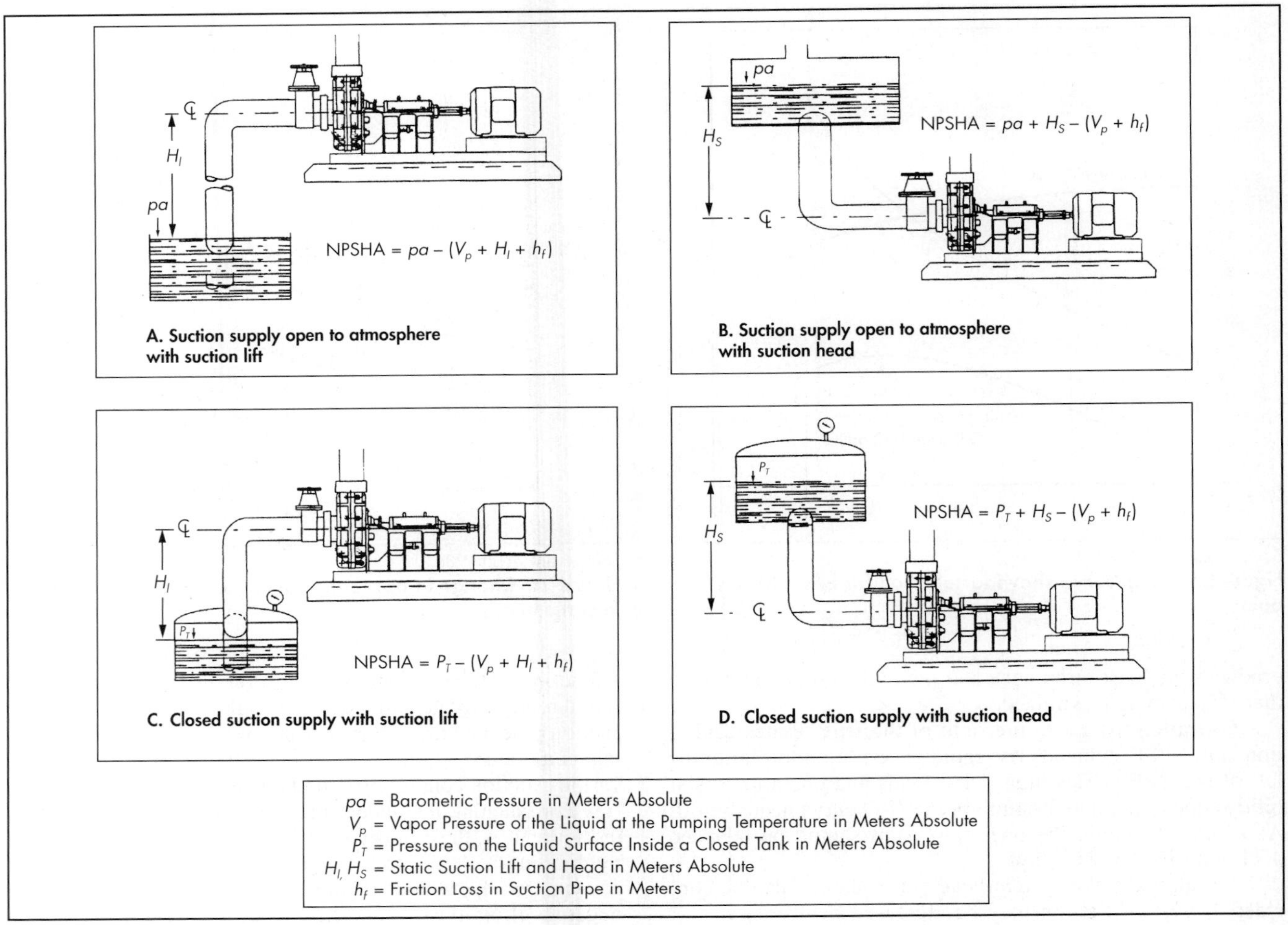

Source: Bootle 2002

Figure 27 Calculation of NPSHA for various pump system configurations

should be directed to the suction piping in evenly distributed streamlines. This applies to both vertical pumps submerged in the sump and horizontal pumps located outside the sump confines. Very often a flared bell-mouthed fitting is employed at the end of a suction pipe protruding into the sump to capture the entering flow at a low average velocity and minimize friction losses.

It is, of course, imperative that sufficient NPSH be available for the pump to operate free from vibration and cavitation. Beyond this, however, there must be sufficient height of liquid in the sump above the suction intake of the pump to preclude vortexing and gulping of air. While each pump manufacturer will have its own recommendation of this minimum height, a safe rule of thumb is to have one meter of height for each meter per second of suction intake velocity.

For flows of 4 m^3/min or less, the sump volume should usually be on the order of one to two times the maximum quantity to be pumped. If the pump is operating in an on–off mode, as in response to a float switch, for example, the sump volume should be sized to allow no more than three to four starts per hour of the pump motor unit. Sump volumes may be minimized using a variable-speed pump drive, or by the action of a discharge control valve that adjusts the system-head curve. Again, sump design and operation should provide sufficient NPSH and liquid-level submergence for all pump conditions.

Generally, inlet flow to the sump should be from the sump wall opposite the pump suction intake. The influent should not jet into the pump inlet or enter the sump in such a manner as to cause a swirling flow. For this reason, rectangular-shaped sumps are usually preferred over cylindrical sumps. Baffles may be used effectively in sumps to break swirls, direct flows to pump intakes as required, and separate intakes when more than one pump is drawing from the sump.

Suction piping, as previously mentioned, should be as short and direct as possible with limited fittings to preserve NPSH characteristics. A surge isolation valve is generally required and should be either totally open or totally closed. It is extremely bad practice to attempt throttling flow in the suction line. A removable suction spool piece, installed with grooved-type couplings, is often a clever idea to permit access to the pump. A drain valve may be installed to remove residual liquid from the piping and associated piping or pump shutdowns. Suction piping may have a rise of at least 3° to 4° from sump to the pump to rid any entrained air from the system. If the piping must be reduced to the suction flange, an eccentric, rather than a concentric, reducer is preferred to

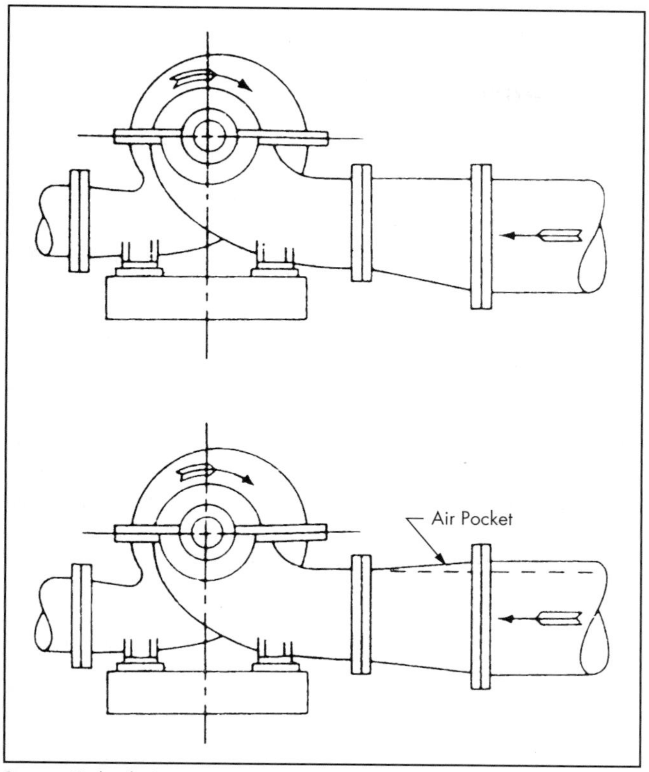

Source: Hydraulic Institute 1975

Figure 28 Correct installation of a reducer in the suction line

prevent trapping air in the inlet line, as shown in Figure 28. Packed-type valves used on the pump suction line should be sufficiently tight to prevent air leakage into the liquid stream.

Discharge piping should usually contain an on–off valve, such as a gate valve, and a check valve placed between the pump and gate valve. The gate valve is used in pump priming, starting (except with radial- and mixed-flow pumps), and shutting down. The check valve is installed to prevent backflow of the liquid through the pump, particularly when a drive power failure occurs. Such backflow can damage the pump and its accessories, especially when a high static head exists.

A pump should never be allowed to operate under the condition of a simultaneously closed suction and discharge. Energy fed by the drive into recirculating liquid within the pump can cause highly dangerous explosive forces.

The pump should be bolted to a foundation that is of adequate strength and design to absorb any normal forces transmitted to the pump by pipe loads, applied motor torque, and counter-produced impeller torque, and so on, to maintain alignment of the pump drive train. Usually the pump and its drive components are mounted on a solid base or pedestal that is anchored to foundation bolts of proper size imbedded in concrete. Piping connected to the pump should be properly supported so that minimum forces and moments are transmitted to the pump flanges.

Axial-Flow Pumps

Axial-flow or propeller pumps are used for conditions of high capacity and low head—15 m or less per stage—and are particularly applicable in closed-loop circulation where the pump becomes a fitting in the line. A typical characteristic curve for an axial-flow pump is shown in Figure 29. Note the discontinuity in the head-capacity curve. Because the power requirement at shutoff may be double that of the design point, the pump discharge valve is often kept open on starting to keep the system head as low as possible. This is opposite to the practice used for radial-flow pumps where the discharge valve is usually shut on starting. The blade angle in an axial-flow pump can often be mechanically changed to alter performance and efficiency to a desired operating condition. An axial-flow pump is very sensitive to suction configurations, and even minor changes in inlet suction conditions can be very detrimental.

Regenerative or Peripheral Pumps

Regenerative pumps, also called peripheral or turbine pumps, depend on a combination of mechanical impulse and centrifugal force to generate heads up to 750 m at flows generally less than 0.5 m^3/min. In these pumps, as shown in Figure 30, the liquid spins within narrow recesses at the periphery of the impeller. This type of pump is excellent for handling vapor-laden liquids and is often used for boiler feed and process applications.

Regenerative turbine pumps typically have good NPSH performance and a steep *H-Q* curve. They are used as small boiler feed pumps, seal water pumps, and chemical process pumps. In a regenerative turbine pump, liquid recirculates

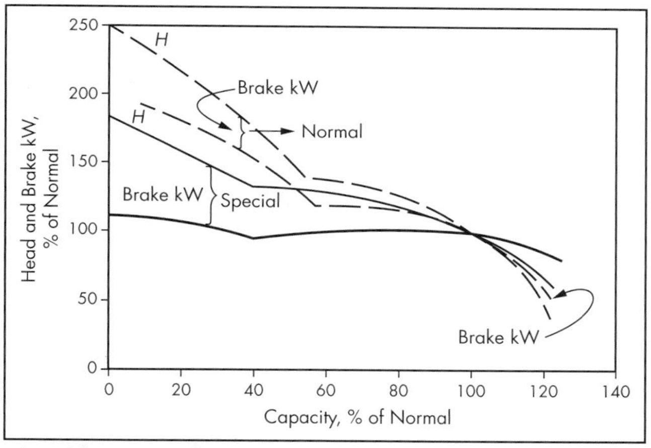

Adapted from Stepanoff 1948

Figure 29 Performance curve for a typical axial-flow pump

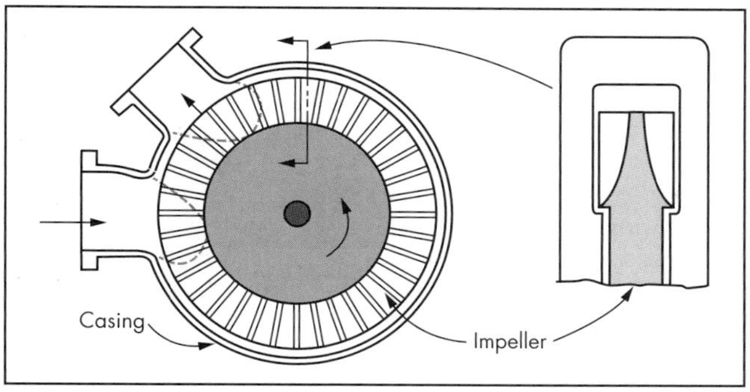

Source: O'Keefe 1972

Figure 30 Regenerative or peripheral pump

between the impeller vanes, so the fluid flows in a helical path as it is carried forward. Consequently, energy is added to the fluid in a regenerative motion by the impeller's vanes as it travels from suction to discharge. This regenerative action has an effect similar to that in a multistage centrifugal pump, wherein the fluid's pressure is the result of energy added in the various stages. In the same way, in a regenerative turbine pump, pressure at its discharge is the result of energy added to the fluid by many impeller vanes.

Special Centrifugal Pumps

Centrifugal pumps can be mounted vertically, with the packing box located well above the highest liquid level. These pumps have the suction immersed in a tank or sump and can occupy less overall floor area than might be needed for a comparable horizontal pump arrangement. Vertical pumps are frequently employed to handle toxic or corrosive process streams as well as for floor wash-down from plant collection pits. A vertical pump is shown in Figure 31.

In a continuous vertical pump, the bearings are located at the top of the shaft. This is the desirable design in most situations. Tapered shafts are used, and hydraulic radial thrust is often balanced using dual suctions and volutes. For extremely long shafts, intermediate and foot bearings must be added. These are difficult to protect when pumping corrosive liquids.

Other specialty pumps include the canned motor pump and the submersible pump. The canned motor pump is a close-coupled pump in which the cavity housing the motor, rotor, and pump volute are interconnected. Consequently, the motor bearings run in the process fluid and seals are eliminated. Flows up to 3 m^3/min and heads from 75 m (one-stage) to 180 m (two-stage) have been achieved. This type of pump has been used for pumping of organic solvents, toxic liquids, and other difficult liquids. An example is shown in Figure 32.

The fully submersible pump is also a closed-coupled, motor-driven unit in which the pump casing and motor are designed for underwater operation. These are excellent pumps for use in flooding conditions such as those encountered in underground mining operations. A main feature of this pump is its slim shape, which facilitates its application in deep boreholes. Flows up to 90 m^3/min can be obtained by multi-staging a series of mixed-flow impellers, and heads in excess of 975 m are possible with the proper choice of impellers.

Positive Displacement Pumps

PD pumps function by displacing a fixed amount of liquid, either by reciprocating or rotary motion. The prime feature of a PD pump is that a definite quantity of liquid will be delivered for each stroke or revolution of the prime mover. Generally, only pump size, design, or suction characteristics will affect the pump capacity. An important consideration for PD pumps is their *volumetric efficiency*, which is defined as the ratio of actual volume displaced to the theoretical or geometric volume of the pump. The difference between these two volumes represents internal leakages within the pump.

Reciprocating pumps add energy to a hydraulic system using a piston or a plunger acting against a confined fluid. In either type, liquid enters the pump cylinder through a check valve that is opened by external pressure acting on the liquid. If the piston carries its own seals or packing with it, the pump is referred to as a piston pump, as shown in Figure 33.

Inflow of the liquid follows the piston movement backward through the cylinder on the inlet stroke. Forward movement of the piston forces the inlet valve to close and the discharge valve to open. The piston must have a close fit with the cylinder walls to minimize slip past the piston seals.

A plunger pump, as shown in Figure 34, uses a close-fitting rod moving through the cylinder past fixed packing. The plunger pump is often used to handle slurries, in which case the plunger may be designed to collapse slightly on the suction stroke. The system provides a small quantity of flush water to protect the seals and minimize wear.

The suction stroke of a reciprocating pump must be designed to prevent cavitation. As the piston moves back on the suction stroke, the valve must open and allow the fluid to pass into the cylinder. Normally the valve head is much smaller than the cylinder, and fluid must pass through the valve at very high speeds. To prevent cavitation within the cylinder or suction lines, piston velocities must be kept as low as possible. Therefore, most reciprocating pumps are operated at 50–100 rpm. Suction lines are also kept relatively large so that velocities will be low and acceleration–deceleration problems minimized. These pumps will operate at relatively large negative suction heads.

The discharge side of a reciprocating pump should be designed to minimize vibrations. As the piston forces the fluid out of the cylinder, local pressures may increase substantially above the downstream line pressure. This results in a series of vibrations that must be damped out by using an air chamber or pressure accumulator. Very high discharge pressures can be attained with reciprocating pumps.

Discharge characteristics of reciprocating pumps are shown in Figure 35. Liquid moves from the discharge of the pump valve to the end of the scroll when the piston stops and begins to reverse. When the pump is a single-acting design, flow stops as the piston reverses. Flow from the discharge of a reciprocating pump can be made more constant using duplex, triplex, or even multiplexed designs. Duplex or double-acting pumps always have a flow moving into the discharge line. In such pumps, the discharge of one cylinder is displaced about one-half stroke from that of the other cylinder, so the total pump flow is the sum of the flow from both cylinders. Nearly pulse-free flow can be achieved by designing for duplex, triplex, or multiplex operation.

Rotary, PD pumps do not rely on valves to control suction or discharge. Rather, they draw liquid at the intake and move it continually to the discharge. This can be done with vanes, reciprocating pistons, gears, a flexible member, lobes, or screws—there are many variations. The pumps shown in Figures 36–38 are typical of those used in mineral processing applications.

The diaphragm pump, shown in Figure 39, is similar to the reciprocating pump. It functions by a back-and-forth motion of a diaphragm made of an elastomer or metal. The diaphragm may be actuated by a piston, as shown in the figure, or by a fluid (usually hydraulic oil or compressed air), which is introduced on one side of the diaphragm, displacing liquid to be pumped from the other side of the diaphragm. This pump requires no packing because the fluid being pumped is isolated inside the diaphragm. It is therefore well suited in pumping difficult materials such as slurries and corrosives. Air-operated diaphragm pumps are often used for slurries.

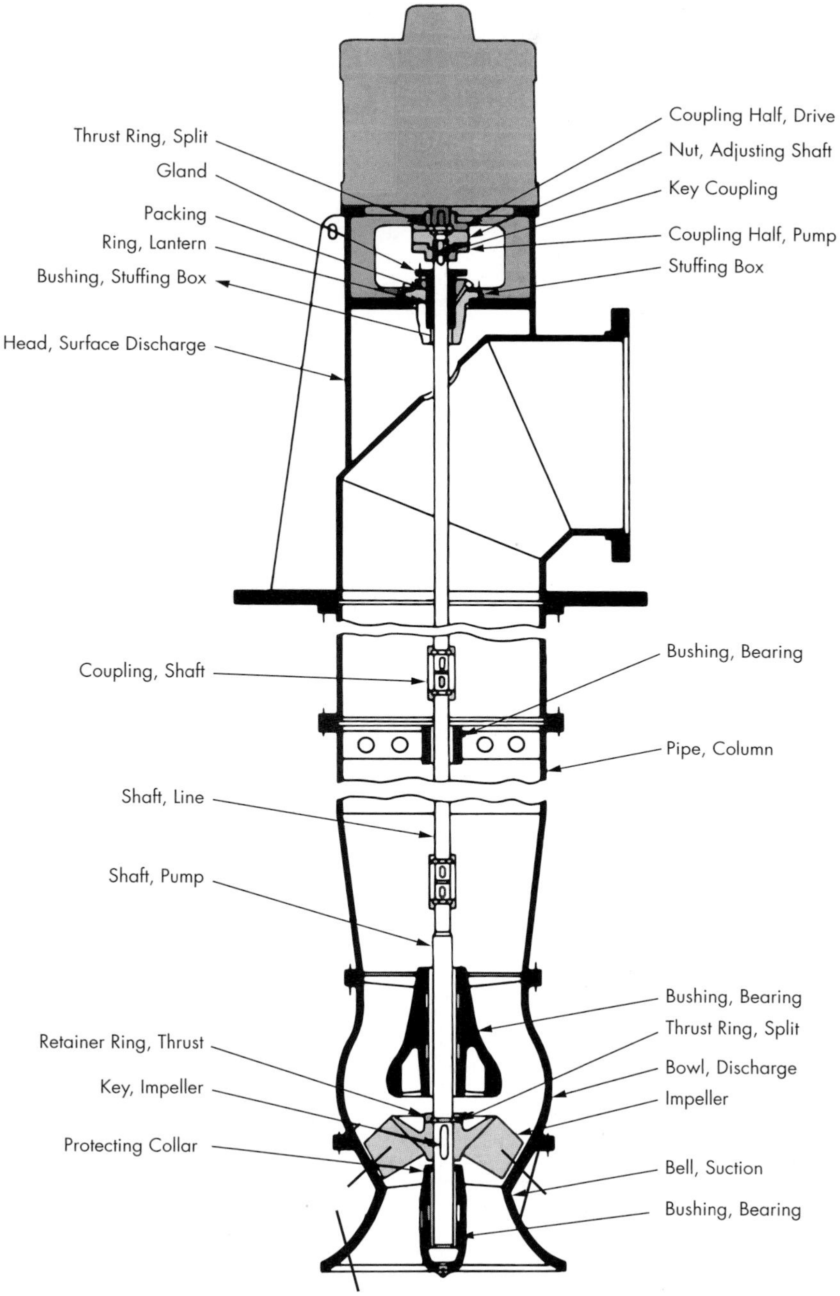

Source: Hydraulic Institute 1975

Figure 31 Vertical centrifugal pump

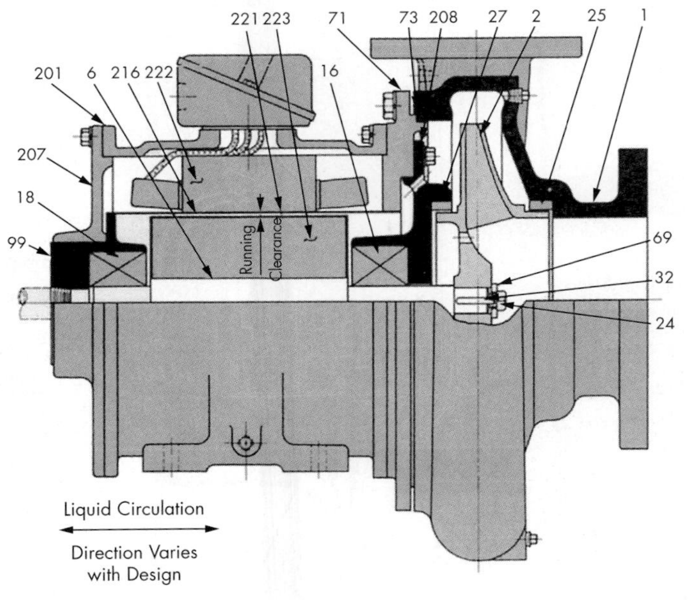

1	Casing	27	Ring, Stuffing Box	207	Cover, Motor End
2	Impeller	32	Key, Impeller	208	Housing Bearing and Wearing Ring
6	Shaft, Pump	69	Lockwasher	216	Can, Stator
16	Bearing, Inboard	71	Adapter	221	Can, Rotor
18	Bearing, Outboard	73	Gasket	222	Assembly, Stator Core
24	Nut, Impeller	99	Housing, Bearing	223	Assembly, Rotor Core
25	Ring, Suction Cover	201	Housing, Stator		

Source: Hydraulic Institute 1975

Figure 32 Canned motor pump

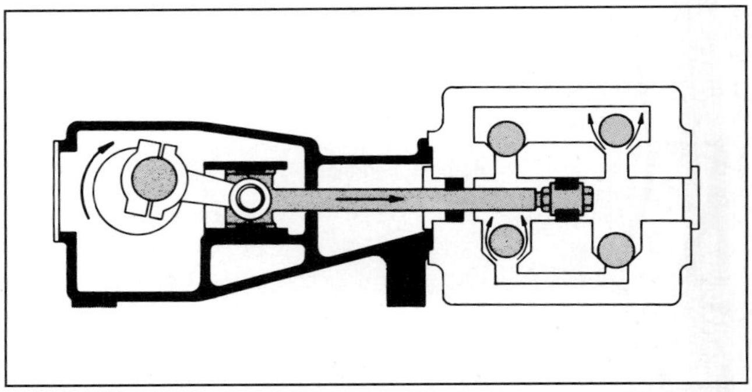

Source: Hydraulic Institute 1975

Figure 33 Horizontal, double-acting piston pump

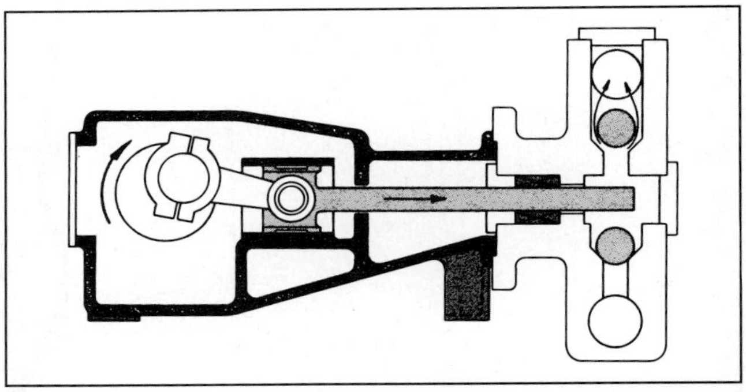

Source: Hydraulic Institute 1975

Figure 34 Horizontal, single-acting plunger pump

Another type of PD pump is the progressing cavity pump, which is often referred to as a "Moyno" pump, after René Moineau, its inventor. Moineau, a pioneer of aviation technology, was designing a compressor for jet engines and realized that this principle could also work as a pumping system. As shown in Figure 40, this pump consists of a rotor turning within a stator. The stator design creates a helical movement where the rotor contacts every surface of the stator. Voids between rotor and stator entrap material that is continually passed from suction to discharge.

CONSIDERATIONS IN PUMP SELECTION

Net Positive Suction Head

As noted earlier, the NPSHR, or net positive suction head required, is very important in the design of a pumping system. If the NPSH available from the system at the impeller eye does not meet the NPSH required by the pump, cavitation will result. The effects of cavitation vary in severity, from pitting of the impeller at the leading edge of the vanes and impeller periphery to large chunking of the impeller. There is

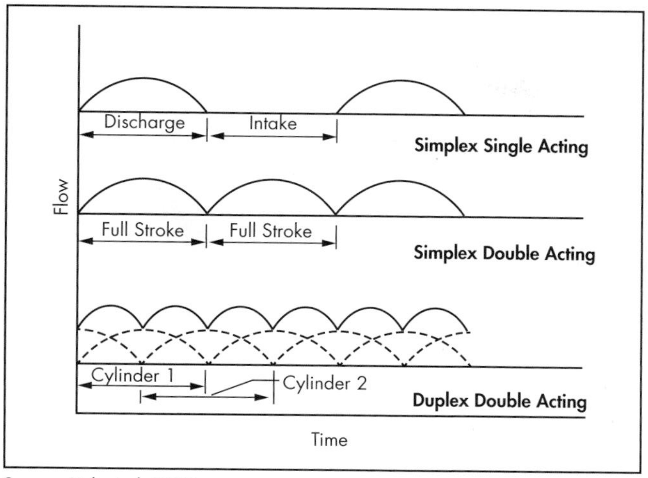

Source: Link et al. 1985

Figure 35 Discharge characteristics of reciprocating pumps

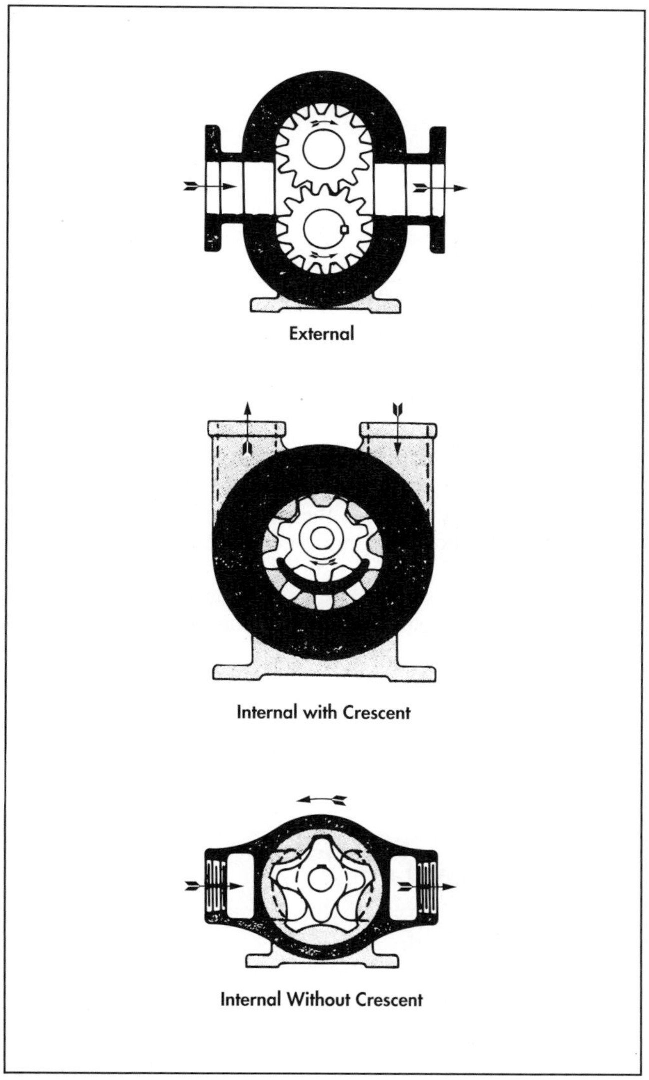

Source: Hydraulic Institute 1975

Figure 36 Gear pumps

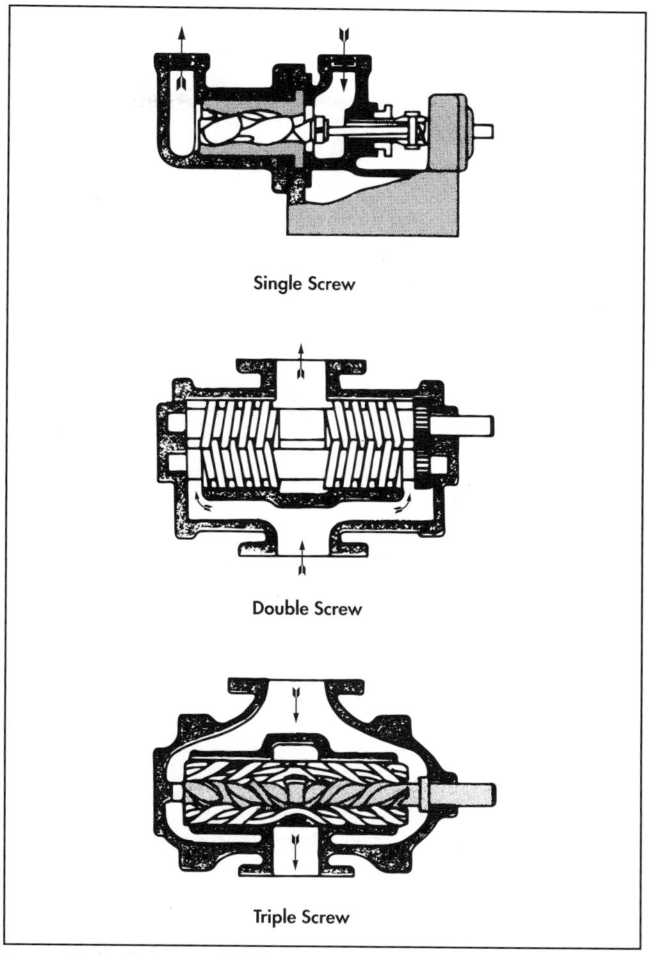

Source: Hydraulic Institute 1975

Figure 37 Screw pumps

generally a loss of pump performance, and a pump operating under a cavitation mode may be relatively quiet or sound as though pebbles are being pumped through it. Severe cavitation can cause vibration severe enough to move the pump off its foundation.

The calculation of NPSHA is a determination of the absolute pressure available at the eye of the impeller over the vapor pressure of the liquid being pumped in terms of meters of head of that liquid:

$$\text{NPSHA} = (H_{\text{atm}} \pm H_s - H_f) - (H_{vp})$$

where H_{atm} is the atmospheric pressure determined by dividing the atmospheric pressure in meters of water by the specific gravity of the liquid being pumped. H_s is the static height in meters of liquid above the center datum line of the pump volute on the suction side of the pump; H_s is positive if the suction head is positive and negative if the pump is lifting liquid to its suction eye. H_f is the friction loss in the suction line to the pump. H_{vp} is the vapor pressure of the liquid pumped in terms of meters of the liquid pumped and is obtained by converting to equivalent meters of water and then dividing by the specific gravity of the liquid being pumped. As a rule, NPSHA should exceed NPSHR by a sufficient margin to account for variations in system head, friction losses, density, temperature, and so forth.

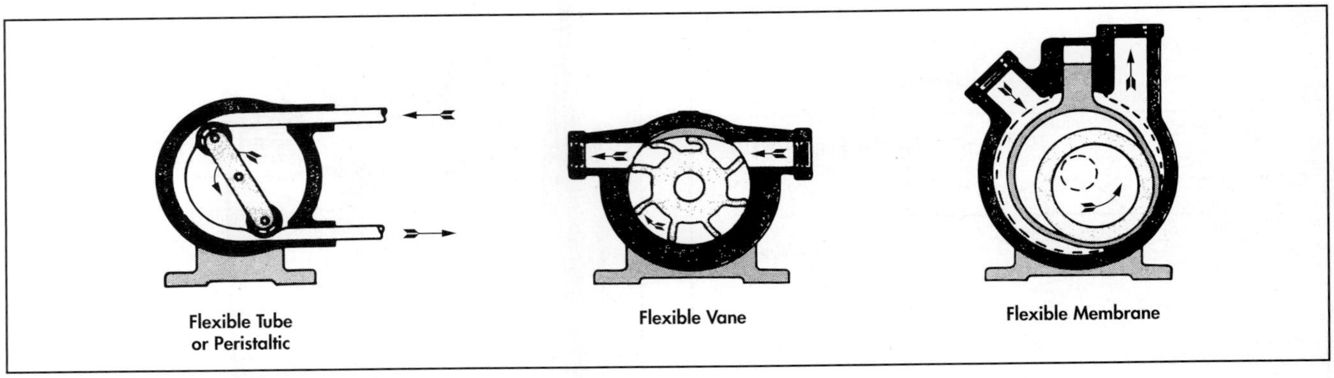

Source: Hydraulic Institute 1975

Figure 38 Flexible member pumps

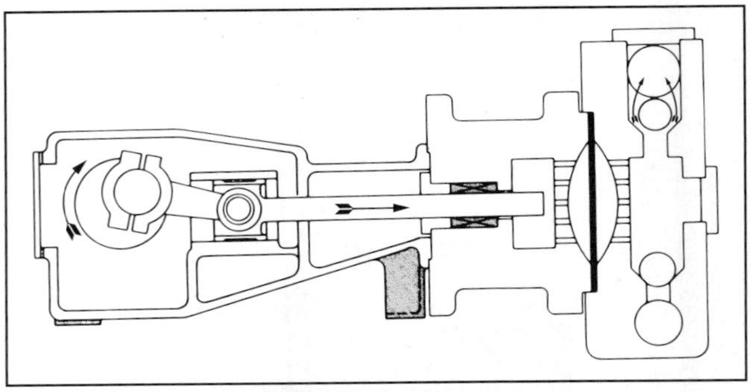

Source: Hydraulic Institute 1975

Figure 39 Horizontal, single-acting, flat diaphragm pump

Mechanical Considerations

Centrifugal Pump Drive Shafts

Mechanical considerations are important in centrifugal pump design. For analysis, consider Figure 41, which shows a pump impeller as an overhung load on a shaft, supported by two sets of bearings. Loads transmitted to the bearings include hydraulic loads and loads imposed by the weight of components attached to the shaft, such as those from the impeller and the drive pulley. Other loads that are applied include belt pull, unbalance loads, and so on.

Hydraulic loads may be divided into thrust loads and radial loads. Thrust loads are directed axially along the shaft and result because of pressure differentials across the shrouds of the impeller. Since the suction eye of the impeller is the low-pressure region, the axial load is directed toward this area. Radial hydraulic loads result from the interaction of the existing liquid from the impeller and the casing. The magnitude and direction of these loads depend on the location of the operating point with respect to the BEP of the head-capacity curve on which the pump is operating, as can be seen in Figures 42 and 43. The farther the operating point is from the BEP, particularly in the direction of shutoff, the greater the radial load.

The thrust factor, K, shown in Figure 42, is found in the empirical equation that describes radial thrust, T_r:

$$T_r = K H D_2 B_2$$

where H is the pump head in m, D_2 is the impeller diameter in cm, and B_2 is the impeller width, including back and front shrouds.

For the pump described in Figures 42 and 43, $D_2 = 27$ cm and $B_2 = 3$ cm. Furthermore, K is given by

$$K = C[1 - (Q/Q_r)^2]$$

where C is a numerical constant that depends on pump geometry and specific speed, Q is the pump operating capacity in m^3/s, and Q_r is the pump rated capacity in m^3/s (Badr and Ahmed 2014).

Pump designers may counteract thrust loads by using auxiliary vanes or expellers on the outside shroud surfaces of the impeller. Another alternative is to use a double-suction impeller, which is fed liquid from both sides. Finally, these

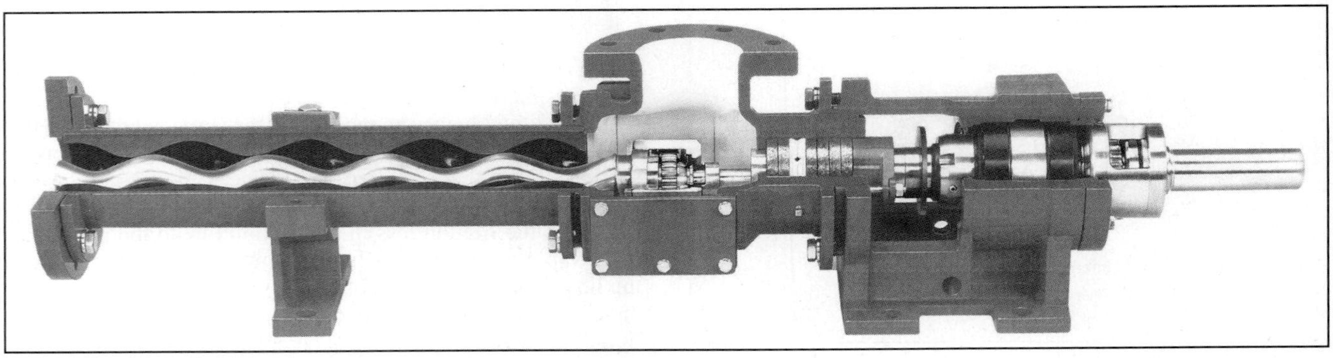

Courtesy of Carl Eric Johnson Inc.

Figure 40 Progressing cavity (Moyno) pump

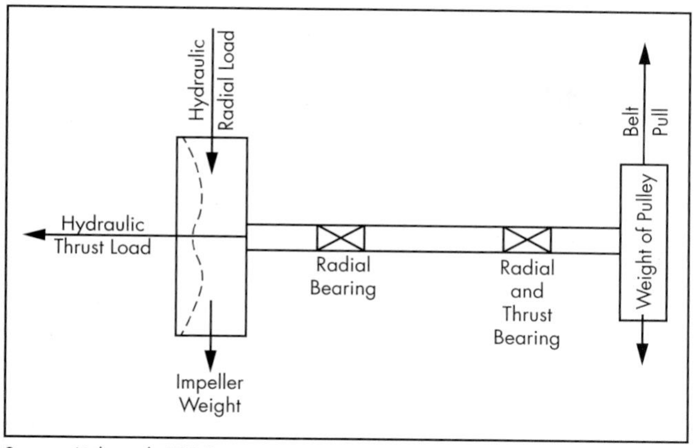

Source: Link et al. 1985
Figure 41 Bearing loads in a pump

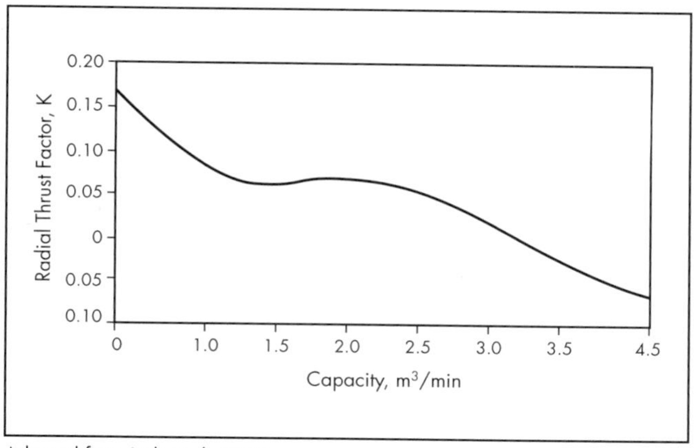

Adapted from Link et al. 1985
Figure 43 Radial thrust factor in a double-volute, 100-mm pump at 1,760 rpm

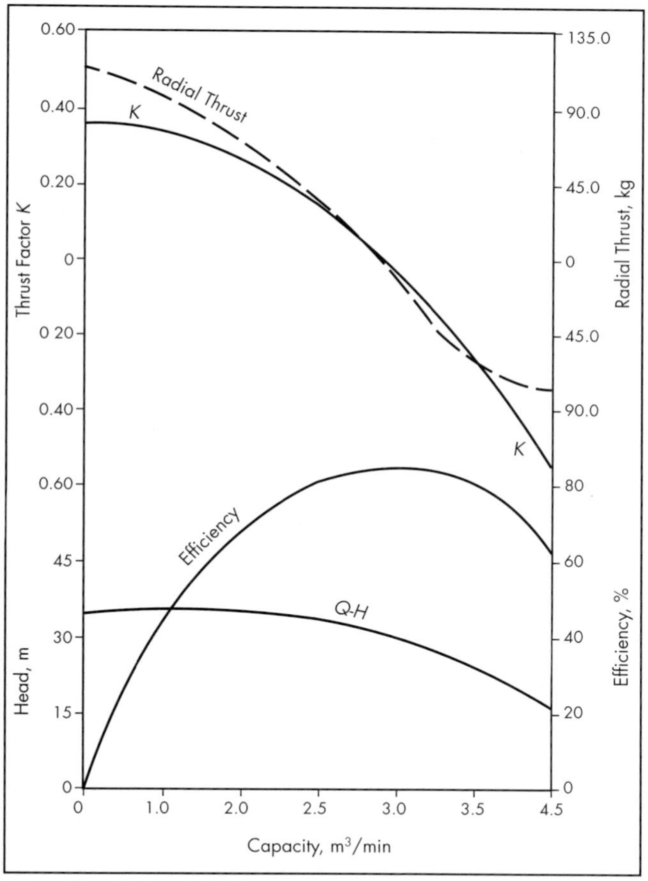

Adapted from Link et al. 1985
Figure 42 Radial thrust in a single-volute, 100-mm pump at 1,760 rpm

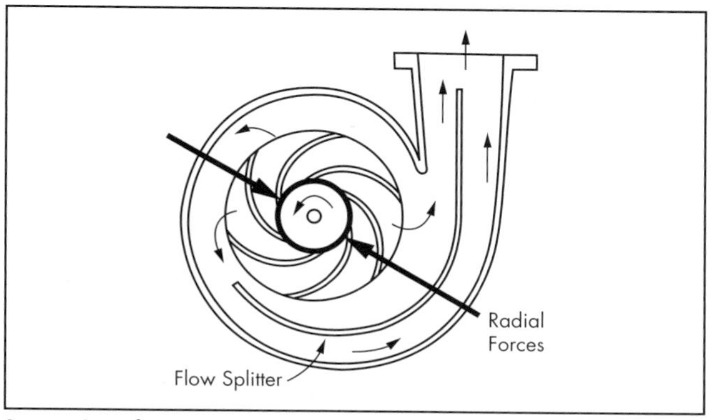

Source: O'Keefe 1972
Figure 44 Double-volute pump casing with flow splitter in discharge passage

forces are sometimes countered by employing a double-volute casing that uses a flow splitter, shown in Figure 44.

All loading imparted to the shaft is absorbed by the bearings. The radial loads are primarily absorbed by the set of bearings closest to the volute on an overhung impeller design, while the bearings nearer the driven end absorb radial and thrust loads. Distances between bearings and between the radial bearing and impeller must be properly established to minimize shaft deflections and stresses.

Lubrication of the bearings, whether by grease or oil, depends on the choice of bearings, running speed range, and service. In some elevated-temperature applications, water-cooled jacketed bearings are used.

Shaft Sealing

Shaft sealing is one of the most challenging areas of pump design and operation. Sealing at the volute or "wet end" of the pump is generally accomplished in a transition area known as the stuffing box, shown in Figure 45. One of the most common ways to seal the shaft is by using packing rings, which can be made from a wide selection of materials such as asbestos, graphite, carbon, polytetrafluoroethylene, and aluminum. The choice of packing depends on factors like pump discharge pressure, corrosiveness, and temperature of the liquid pumped; shaft diameter and speed; and the availability and type of packing lubrication.

Water or grease may be introduced into the packing area to lubricate and cool the packing. In the case of slurry pumps, water is introduced into the end of the stuffing box nearest the volute to prevent solids from entering the packing area. Such injection is normally made through a lantern-ring to ensure

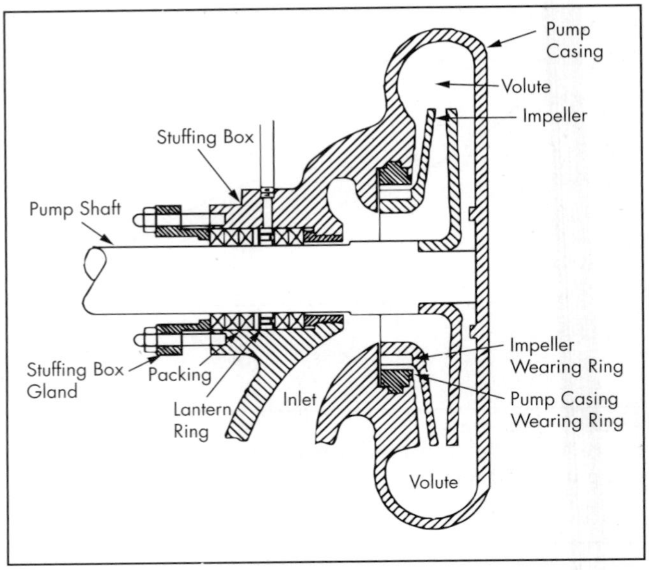

Courtesy of Adamant Valves

Figure 45 Pump sealing arrangement, showing stuffing box and packing

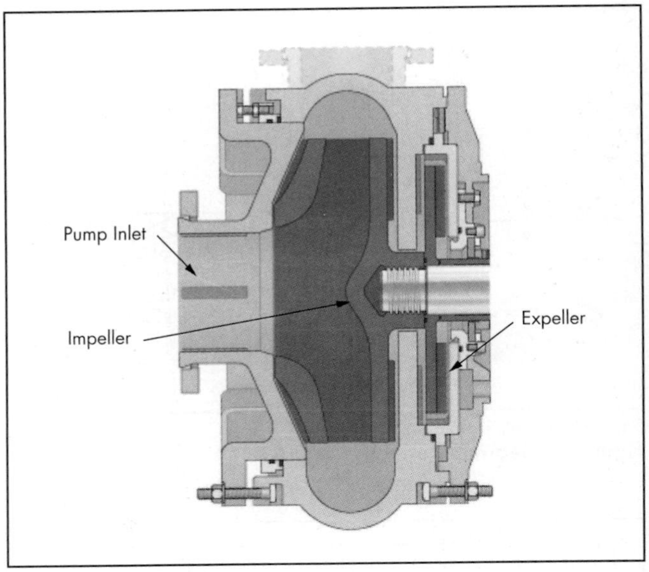

Adapted from Wilfley 2014

Figure 46 Expeller installed in a centrifugal pump

even distribution of water around the shaft. A lantern ring is a perforated hollow ring located near the center of the packing box that receives relatively cool, clean water and distributes the water uniformly around the shaft, to provide lubrication and cooling. The fluid entering the lantern ring can cool the shaft and packing and lubricate the packing. It can also seal the joint between the shaft and packing against leakage of air into the pump when pump suction pressure is below atmospheric pressure. As packing wears, the gland follower is tightened in against the packing until such tightening ceases to be of value.

An alternate method of sealing the shaft is with mechanical seals, which may be installed as a single seal or double seal and may be connected internally or externally of a packing box. Mechanical seals are employed where sealing liquid to the pump must be limited or entirely omitted or where escape of pumped liquid is not permitted. The sealing faces of mechanical seals may be constructed from ceramics, carbon, and carbides as the application warrants. These seals are lubricated and cooled by sealing liquid. Successful operation with a mechanical seal depends on proper alignment, controlled shaft deflections, and a steady supply of clean sealing fluid.

Another method of sealing is the use of expeller vanes on the back shroud of the impeller, opposite the suction eye, as shown in Figure 46. These vanes are generally of a greater diameter than the pumping vane, and their purpose is to move liquid away from the stuffing box area of the impeller. These pumps often have packing that is grease lubricated, and the pumps are designed with a check valve or similar device to permit the entry of air, thus preventing them from forming a vacuum. In some pump designs, the expellers are aided by auxiliary vanes in the stuffing box area. The success of expeller-sealed pumps depends on the operating point on the pump curve and magnitude of the suction pressure at the impeller eye. Higher ranges of pump speed and low suction pressures are favored for successful operation.

Pump Drives

Electric Motors

Electric motors are the most common drives for most process equipment, including pumps. For a given pump application, the motor is usually selected based on the power requirement of the pump. Motors for centrifugal pumps are generally selected based on the normal operating duty point.

The use of variable-frequency drives (VFDs) for AC motors allows pump speeds to be easily varied, in response to process conditions. This capability provides many options for pump operation that were not previously available. The combination of VFD with an AC induction motor provides a drive system that will accommodate considerable variation in operating conditions, requires minimal maintenance, and provides line-to-load efficiency from 90%–97%.

In variable-speed applications, it may be necessary to either derate the motor or provide it with forced-air cooling. The motor cooling fan becomes less effective at lower than rated speed, and the available torque decreases at speeds greater than rated synchronous speed. Motor vendors provide detailed guides on how to select appropriate motors for variable-speed duty.

In centrifugal pumps, the relation between torque and speed is quadratic. Driving these pumps with a variable-speed electric motor generally has no complications. PD pumps may have a significant friction torque at zero speed or may present a constant torque with speed load to the driver. Selection of a motor and VFD for such a load must consider thermal derating (or forced cooling) of the motor as well as the current rating of the VFD.

Internal Combustion Engines

Portable pumps require a nonelectric source of power, as do pumps for critical-service systems such as fire protection backup. Internal combustion engines are the usual choice for

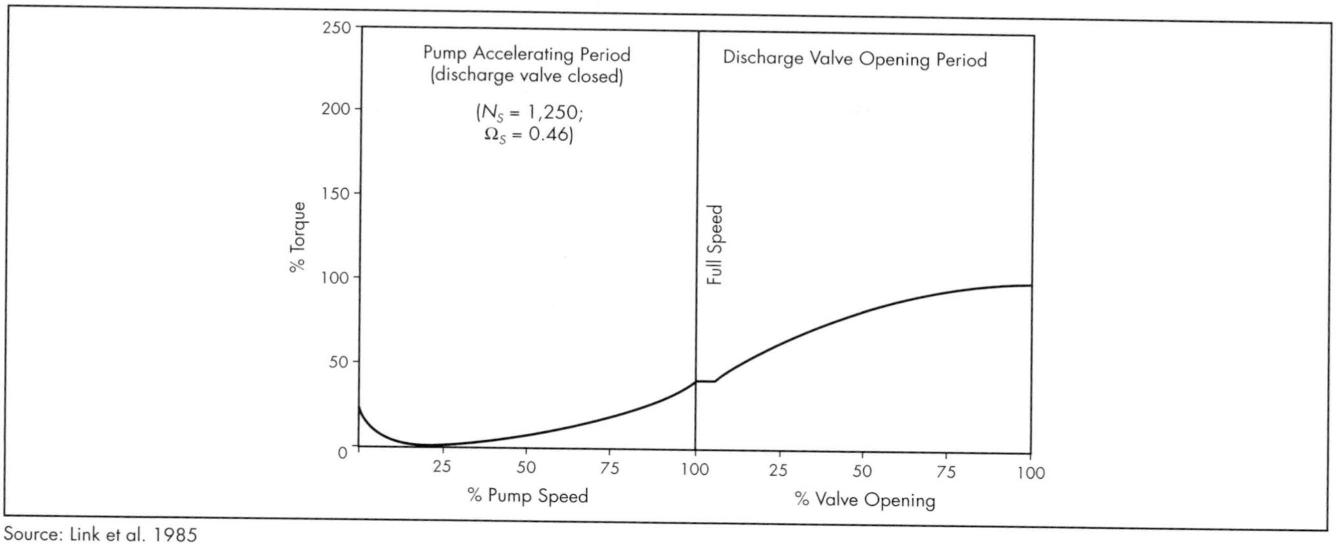

Source: Link et al. 1985

Figure 47 Torque–speed relationship in a start with a closed discharge valve

nonelectric power. Although selecting an electric motor based on power rating and pump speed is straightforward, the torque/speed characteristics of internal combustion engines are significantly different from those of an electric motor. There are also differences in the characteristics of gasoline and diesel engines. The pump speed and torque are used to select the correct internal combustion engine.

Matching Pump and Driver

It is very often useful to make a plot of pump speed versus required torque against a similar curve of speed versus available torque for the motor to determine the time required for the motor to accelerate the pump to full speed. Usually this is not a problem for most plant pumps using today's induction or synchronous motors; however, where there is reduced voltage starting or a relatively prolonged period of operation at low head or "runout" conditions, a decrease in static head analysis of speed–torque is desirable. This analysis also requires an assessment of combined inertia of the pump and driver rotor assemblies, referred to as the motor shaft. The conventional form is to plot percent pump torque on the vertical axis and percent full load speed on the horizontal axis. The pump torque required for any given head capacity is

$$T = (9{,}549 \times P)/N$$

where T is the pump torque in N-m, P is the power in kW, and N is the rotational speed in rpm.

At any speed between zero and full-load speed, FLS, the torque varies as the square of the speed. The curve derived from this is parabolic in shape. For example, at one-half the FLS, the torque will be $(\frac{1}{2})^2$, or one-fourth the torque at FLS.

At zero speed, the theoretical pumping torque required is zero. To overcome the inertia of the rotating element, some torque is required, usually 5%–15% of the full-load torque. After the initial movement when the pump picks up speed, the torque requirement drops to follow the pump's hydraulic characteristic at about 10% FLS. Consider the following scenarios under which a pump may be started, and the speed–torque relationship in each.

Closed Discharge-Valve Start

When the pump is started with the discharge valve fully closed and its speed progresses from zero to FLS, the pump is operating in the shutoff condition at FLS and drawing shutoff power. The torque can be calculated from the power at shutoff. The discharge valve is then opened to its fully open position. The shape of the speed torque curve will vary, depending on the system characteristics and the type of valve. The inertia of the liquid in the pipeline will also have some effect on this portion of the curve. Figure 47 shows this case.

Open Discharge-Valve Start

With an open discharge-valve start, the speed torque curve depends on the head-capacity characteristics of the system into which the pump is delivering. In a system where all the head results from friction, the speed–torque curve is parabolic in shape and rises continuously. At 100% speed, the pump will be operating at the rated point and the torque will therefore be 100%. When there is a combination of static and friction head, the pump must initially generate sufficient head to overcome the static head of the system. This situation is the same as pumping with the discharge valve closed. At the speed where the pump overcomes the static head, the pump then delivers into the system and the torque will then rise to 100% torque at the 100% speed point.

When all the head is static, the pump initially operates as if the discharge valve is closed until all the static head is overcome and then delivers into the system along a constant head line until the operating point is reached.

Long Discharge Line

Another example of a virtually closed discharge-valve system is a long pipeline system where the mass of the liquid becomes so great that the time required to accelerate the liquid in the pipeline is much greater than the time required to bring the motor up to FLS. The pump torque during the starting period will approach that of a pump operating with a closed discharge valve.

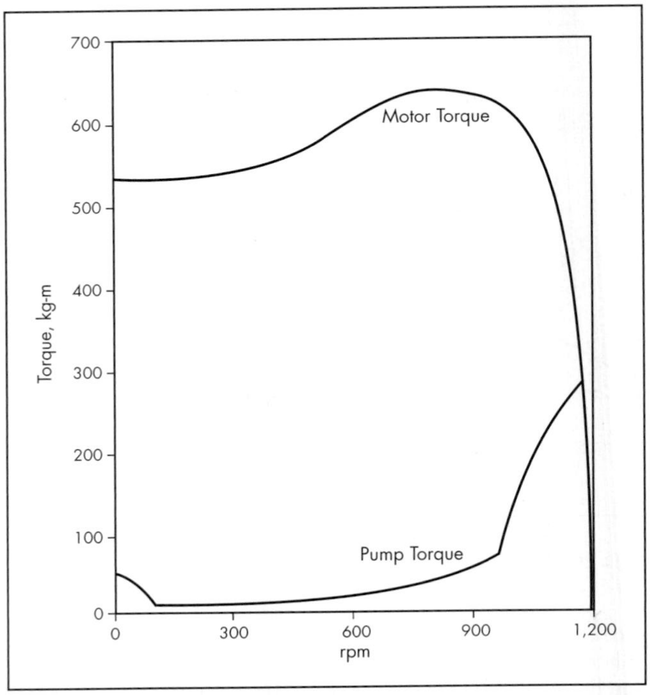

Adapted from Link et al. 1985

Figure 48 Typical squirrel-cage motor curve shown with pump torque curve

In conclusion, if the speed at a given torque of the driver does not drop below that for the pump at the same torque, the pump will reach the operating point. An example of a typical motor speed–torque curve superimposed on a pump speed–torque curve is shown in Figure 48.

Speed-Reducing Systems

Where the pump shaft speed required at the duty point does not match the speed of the driver, it is necessary to provide a means of matching the motor and pump speed. In many cases, the impellers of centrifugal pumps for clear fluids can be trimmed to provide the duty point at an electric motor's operation speed. Trimming is not recommended for the impellers of slurry pumps, which are made of hard materials, so during pump start-up, the speed must be adjusted to achieve the duty point.

V-Belts

The design and construction of material handling pumps require that impeller speed adjustments must sometimes be made after installation, to match pump performance with system demands. At pumps of up to 225 kW, power transmission from the motor to the pump can usually be done efficiently and economically with V-belt drives.

As a rule, the selected ratio of driver speed to pump speed should not be too high. Some manufacturers recommend a maximum ratio of 3:1, while others suggest 4:1. As the ratio increases, the belt wrap angle around the driving pulley decreases, and the power capacity of the belt decreases proportionately. When the driver motor power exceeds 225 kW, V-belt drive selections must be closely evaluated to ensure that overhung loads and belt pull from the belt drive do not exceed pump and motor bearing limitations. At these higher powers, a gear reducer or variable-speed motor may be preferable to a V-belt drive.

Belt sheaves are often mounted to the shaft using a taper lock hub. This allows the sheave to be positioned correctly on the shaft and firmly attached without the need for a press. Sheaves should be dynamically balanced to minimize vibration. Belts can be used in matched sets to increase the power capacity of the drive. Belts should be periodically checked to ensure they remain correctly tensioned.

Maintenance and proper installation and alignment are keys to the successful operation of a V-belt drive. To select a V-belt drive, the following information is needed:

1. Power and speed of motor required (Selected motor power should be 10%–15% greater than the maximum anticipated pump power requirement to overcome losses in the belt drive.)
2. Required speed of driven unit
3. Drive service factor to apply to the motor, usually 1.2

Belt vendors provide catalog procedures for selecting the required belt section, number of belts, driving pulley diameter, power correction factors associated with the drive geometry and candidate belt lengths, and resulting shaft center distances.

Gear Reducers

When power requirements exceed 300 kW, gear reducers are used for power transmission. Gear reducers provide higher drive efficiencies and lower bearing than V-belt drives. The most commonly used gear reducer is a parallel-shaft, single-reduction unit. Gear reducers, motors, and pumps are mounted on a common "ladder-type" sub-base with a fixed high-speed coupling between the motor and gear reducer and a sliding low-speed coupling between the pump and gear reducer. The slide coupling allows for axial impeller adjustment as required by the pump.

When selecting a gear reducer for a specific application, the following factors must be considered:

1. Service factor required, usually 1.25 to 1.5
2. Power required, including service factor
3. Drive ratio (most applications require an exact ratio)
4. Shaft sizes for selection of couplings

Items 1 to 3 are used to pick a reducer that is capable of transmitting the required torque at the stated reduction. Equally important is the thermal power, which is the power (without the service factor) that a gear drive will transmit continually for three hours or more without overheating. Fans or a pump with a cooler may be used to achieve an adequate thermal rating.

Shaft Couplings

Shaft couplings transmit the torque from the driver to the pump. A wide variety of coupling designs is available, all of which are made to allow for misalignment between the driver and pump shafts. A hub is fitted to each shaft, and a flexible element transfers torque from one hub to the other while minimizing radial and axial forces that may arise from the misalignment as the shafts rotate.

Baseplates

Unless the pump incorporates a driver, as in a fully submersible unit, the bare shaft pump, its driver, and any speed-reducer components will require a baseplate to keep these items aligned and transfer the forces between them and the ground or structure on which they are mounted. Generally, bases are of welded steel construction. Jacking bolts are provided on the base frame to allow the bare shaft pump, its driver, and any speed reducer to be positioned relative to each other so that the shafts are aligned within the capabilities of the shaft couplings, or the belt sheaves are parallel to one another and the belt grooves are in the same plane.

Belt-driven pumps must also include a mechanism to move the pulley centers relative to one another to allow the belts to be tensioned correctly. This can be as simple as mounting the motor on threaded rods with adjusting nuts to set the parallelism of the shafts as well as their center distances, or by using proprietary slide rails or spring mounts, available from many vendors.

When a V-belt drive is used, the motor is often mounted to the side of the pump, with the motor shaft and the pump shaft axes in parallel and aligned, for convenient connection. When a gear drive is used, the motor and pump may be side by side, or shaft to shaft, depending on the gear drive configuration. Overhead mounting has been used for motors as large as 260 kW on adjusted tabletop-type bases that straddle the pump. Side mounting is preferred whenever possible, with the motor mounted on an adjustable slide rail base firmly bolted to a concrete foundation.

PUMP SELECTION FOR PLANT APPLICATIONS

Grinding Mill Circuit Pumps

One of the most severe pump duties in the minerals industry is the pumping of the primary mill discharge to cyclones or screens. A centrifugal pump is generally used and is normally required to handle slurries containing 30%–65% solids with top ore particle sizes ranging from 20 mesh to 25- and 50-mm lumps. In addition, the presence of tramp objects, such as lumber and pieces of grinding media, further aggravates pump wear and maintenance problems.

To lessen pump wear problems in this service, tramp protection should be used. This may include ball traps, trammel screens, sump screens, or magnets upstream of the pump suction. In very "dirty" circuits where there is no tramp protection, fully metal-lined pumps with metal impellers will probably give the most satisfactory service. Where tramp protection can be provided and maintained, successful pump service life can be obtained from fully rubber-lined pumps or from pumps with rubber shell liners and metal impellers. Regardless of the pump construction selected, it is imperative for good mill operation and economics that not only must the pump service life be of a reasonably long duration, but it must be reliably repeatable to permit a properly planned maintenance schedule.

Mill discharge pump duties may range from 3 to 300 m^3/min at TSH values from 8 to 40 m and slurry specific gravities to 1.8. In this application, the TSH includes not only static and friction losses, but also the cyclone pressure drop. Since TSH directly relates to pump speed and wear, any lowering of TSH values will result in a significant increase in pump life.

Pumps used for mill discharge should have a relatively large diameter so that a given flow and head can be attained at the lowest possible rotational speed and suction eye speed. Impeller tip speeds for both rubber and metal impellers should be kept as low as possible, preferably under 1,500 m/min. A pump for a mill circuit, however, should not be so oversized that the anticipated operating range is far from the BEP on the operating curve. Operation too far to the left (lower than 50% of BEP) decreases life of pump bearings, as a result of higher radial loading; operation at 75%–80% of BEP is recommended. In addition, wet end port life is decreased because of increased pump recirculation and prerotation of the slurry as it enters the pump suction. The effects of slurry prerotation can be partially offset using straightening vanes in the suction piping or pump suction throat.

Froth Pumps

Pumping of flotation froth can be difficult, especially when conventional horizontal centrifugal pumps are used. Uneven pumping usually results because the centrifugal action of the impeller leads to a separation of air and liquid, with an accumulation of air occurring at the impeller eye. Feed to the impeller eye is thus restricted until the sump level rises sufficiently high enough to compress the entrapped air, which can then be swept away. Where horizontal pumps are used, it is sometimes preferable to use a pump with a semi-open impeller that will be less likely to accumulate air in the pump volute.

Sometimes it is more effective to use vertical-type pumps without stuffing boxes. These pumps permit the ready escape of separated air. There are two versions: the vertical pump with integral tank and the sump pump. The former design consists of a casing placed under the tank that opens upward through a hole in the tank bottom. The shaft passes through the tank and into the hole, allowing clearance for the feed. This pump design requires a controlled inlet flow to the sump. The sump-pump design has its pump casing submerged in the liquid and permits feed to enter from both the top and underside of the casing. The impeller has operating vanes on both sides, and air escapes upward through the surrounding sump enclosure.

A variation of the sump-pump design has operating vanes only on the underside of the impeller, with sealing vanes on the impeller topside. The feed entrance is from the bottom only. The pump casing has two holes through which part of the liquid is sprayed back into the sump. These holes prevent the formation of the liquid ring into which air is trapped. Air release also occurs through an opening surrounding the shaft.

Even though special pumps have been devised to separate and eliminate air from frothy slurries, it is usually necessary to oversize froth pumps, whether horizontal or vertical, to handle the entering and exiting air volumes. This is usually done by applying a froth factor, an integer that is usually in the range of one to four. (Use of the froth factor is described in detail in Chapter 7.1, "Mechanical Flotation.") The actual volume of the slurry pulp is multiplied by the froth factor, and a pump is selected with a volute sufficiently large to handle the increased volume. A higher froth factor indicates that the froth is more highly aerated and, consequently, will likely be more difficult to pump. Application of the froth factor is based on the individual judgment of the process designer. The difficulty in designing froth pumping systems is not only a matter of determining the volume of air entrapped in the froth, but also assessing the ease of separation of air from the slurry. For example, in fatty acid and amine flotation processes, mineral binding to the air bubbles is weaker than that experienced in conventional flotation processes for sulfide ores.

Froth pumping systems should be designed with as low a required TSH as possible to limit the pump's rotational speed. Where horizontal pumps are used, deep, non-cylindrical sumps are recommended. Short, straight suction lines with fully open–closed isolation valves are desirable. Packed-stem valves that can admit air should be avoided. Destruction of froth prior to entering the pump is sometimes attempted using mechanical, chemical, or thermal means.

Thickener Underflow Pumps

Thickener underflow pumps are usually either diaphragm-type PD pumps, where required head and capacity are within the relatively low limits of such pumps, or centrifugal pumps, with their broad capability range. The conventional, gland-water-sealed style of centrifugal slurry pump is most commonly used; dry gland or centrifugally sealed types have the advantage of not re-diluting the underflow slurry, but such pumps are not designed to operate against the high positive static suction heads or high suction lifts usually associated with thickeners.

Pumping systems for thickener underflow applications generally fall into three classes. In the first class, the suction pipe is connected to the center bottom cone. The pump is usually located below the cone and horizontally close to it, in a tunnel or restricted space, but in some cases it may be as far away as ground level outside the thickener. In the second class, the pump is mounted on the thickener bridge some distance above liquid level, the suction pipe extending vertically downward to the bottom, at or near the center. In the third class, the pump is a vertical unit, usually placed in a center well in the thickener. Diaphragm pumps can be used only in the first class. In practice, variations are numerous.

When a pump pulls from the center bottom cone, the distance at which the pump can be located from the thickener's vertical centerline depends on many factors. The bottom cone area is congested, and suction piping arrangements are often complicated, as a result of having to find room for such fittings, valves, flush-out connections, and piping to the pump and spare pump. Since the particle size in the thickened slurry is normally fine, suction pipe velocity can be as low as 1.5 m/s. However, while this low velocity can help keep total suction losses down, there may still be large losses from pumping a high-gravity slurry through relatively small-diameter pipe. Also, under abnormal thickener conditions, the delivery of solids to the pump suction may be at a much higher rate than the design figure, or solids may slough off the accumulation at the bottom, coming down in surges that tend to plug the suction. All these factors make it important that the pump be placed as close as possible to the bottom cone.

Correction of a thickener installation with pump operating difficulties because of pump cavitation can be very difficult and expensive, so the NPSHA of the pumping system should always be checked against the pump's NPSHR as stated by the manufacturer. It is advisable to provide the data necessary for the pump manufacturer to make this check, based on its experience with similar applications. The NPSH check is particularly important when it is proposed to bridge-mount the pump, because with a conventional thickener, the pump centerline will be at least 1.2 m above liquid level. This can result in low NPSHA, even with the absolute minimum length of horizontal pipe following the vertical suction riser on the thickener centerline. Calculations in the design stage often indicate that the pump will not work without being dropped toward the liquid level, and it may even be found that no lift at all can be tolerated. Bridge mounting also requires a reliable means for priming the pump, such as a water-operated exhauster.

When the thickener has a center-column rake-lifting device and bridge-mounted pump, an inverted flexible loop is required in the suction line so it will clear the lifting gear. This introduces an additional factor into designing the pump system for reliable operation, as the loop may rise a few meters above the pump centerline. Here, an analysis similar to the calculation of NPSHA must be made to determine whether the absolute pressure at the high point in the pipe loop is great enough to exceed the vapor pressure of the liquid being pumped, thus preventing a loss of prime through the vaporizing of the liquid.

Placement of a vertical pump, either wet or in a dry well down inside the thickener center column, has obvious advantages over both the bottom-cone and bridge-mounted arrangements. However, vertical pumps available for this type service are of such size that they rarely will fit into the space provided by normal design of such items as the center well and rake drive gearing.

Use of the underflow pump for emergency emptying of the thickener is rarely practical. The normal capacity of an underflow pump is small, compared to the total slurry volume in a full thickener. Furthermore, most thickeners are so deep that pump capacity could approach shutoff, or zero delivery, before the thickener was empty, because of the significant increase in head on the pump.

As with all other slurry pump applications, every item used in thickener underflow pump calculations must finally be expressed in terms of pressure head of the slurry as pumped. In an ideal sump, the whole volume has a uniform specific gravity equal to that entering the pump suction, and it is assumed that this is the case on average actual installation, so the stated level in the sump can be directly used in computing static head. However, a thickener is specifically designed to induce sedimentation, and there can be considerable specific gravity variation between top and bottom with only the slurry near the bottom having a specific gravity as high as that passing through the pump. One method for accommodating this variation is to reduce the elevation of liquid in the thickener when calculating NPSHA. This has the conservative effect of decreasing static suction head in the case of a flooded suction and increasing static suction lift in the case of a negative suction. This correction is used when figuring either the total system head on the pump or the NPSHA of the system, the effect being rather critical on the NPSH figure that is so important in thickener applications.

As a further hedge against abnormal conditions that can occur during thickener operation, good practice calls for a factor of safety of 1.5 m between NPSHA of the system and NPSHR of the pump, versus a conventional specification of 0.7–1 m.

For most mineral processing applications, speed-regulated centrifugal pumps provide adequate control for the underflow pumping system. The high suction head on the underflow pump means that gland seal water is usually needed on a centrifugal pump. To prevent excessive dilution of concentrate slurries that are pumped to filtration stages, variable-speed peristaltic pumps have found increasing acceptance. These pumps can also handle dense slurries. The use of pulsation dampers on the discharge piping is required with this type of pump and is a critical design issue.

The use of gravity discharge for thickener underflows is common in the mineral processing industry. These systems

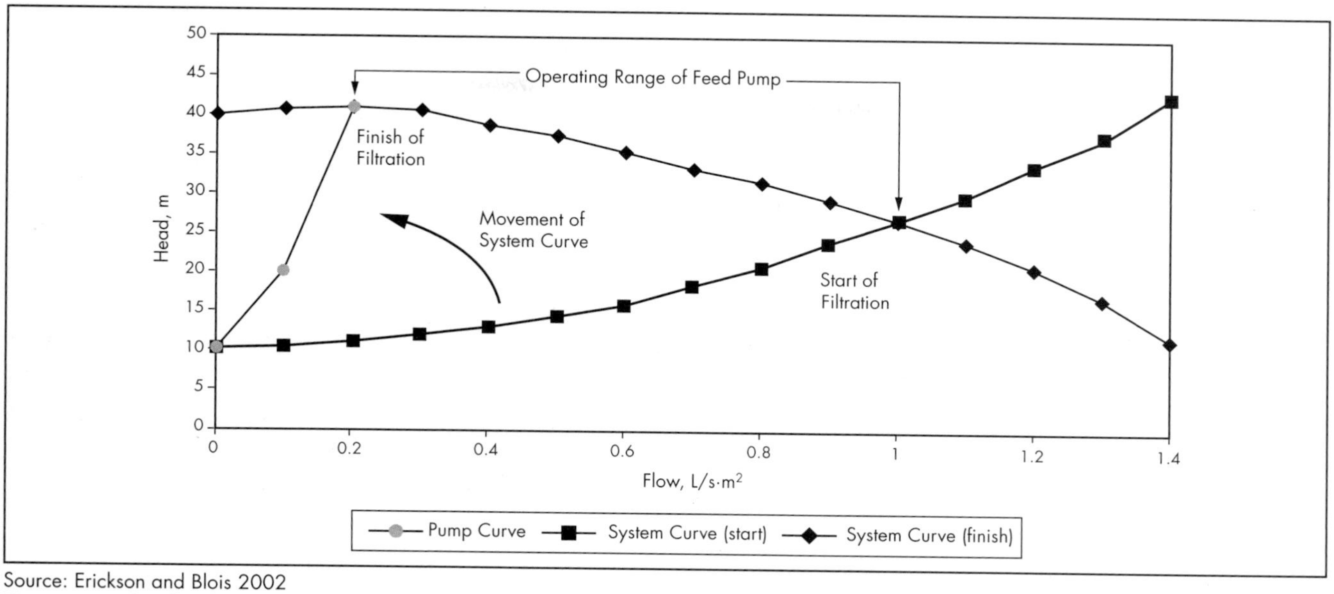

Source: Erickson and Blois 2002

Figure 49 Filter feed pump performance

involve large tonnages and are preferred because larger orifices can be used and plugging from tramp oversize is less likely. This gravity discharge approach can be used where the disposal of the slurry, usually tailings, is downhill from the thickener.

Filter Feed Pumps

A pressure filter is a batch unit, and the filling or filtration stage usually takes only about 20% of the total cycle time. For most mining and metallurgical applications, the filter feed pump runs only during filter feeding, and is started and stopped under control from the filter's control system, although in some cases the feed pump runs continuously, with a recycle at lower pump speed to the feed tank when the filter is not being fed. In systems with no recycle, the instantaneous flow rates during filtration are quite high. Centrifugal slurry pumps with gland seal water are the standard choice. Gland seal arrangements are best determined by the pump manufacturer, but they must account for the start/stop operation of the pump and the wide range of operating conditions. Where full-flow water flushing of the gland is provided, it can cause considerable slurry dilution between filling cycles, and solenoid control of gland water or reduced-flow sealing systems should be considered.

The pump runs against increasing head as the filter cake builds, as shown in Figure 49, and an understanding of the system curve is important for correct pump sizing. Once the chamber is about 50% full, the pressure will rise as the cake resistance increases, causing the operating point to move back up the pump curve. At the end of the filtration stage, the pump will be operating at close to full pressure and the flow will have decreased from 10% to 40% of the initial flow rate. Where the slurry is known to be highly abrasive, the maximum flow rate may have to be lower.

Where pump selection shows that flows greater than the recommended maximum are possible, a variable-speed-drive pump should be used. Pump speed may then be increased as the system head increases. To calculate the flow, the frictional losses for the pipeline and static head must be known. The resultant curve should then be drawn onto the pump curve for the proposed pump. The intersection of the system curve with the pump curve at the proposed speed will then determine the maximum feed flow rate. Where large feed tanks are used, the flow rate should be checked with full and empty tank levels. With flat system curves, there can be a significant change in flow with changes in the tank level.

Where there is significant variation in feed density, the flow with minimum and maximum densities should be checked. Once the maximum flow rate has been determined, the NPSHR should be checked against the NPSHA with the proposed suction pipe and tank design.

Filtrate Pumps

The type of filter determines the approach to the collection of the filtrate. The most important consideration is whether the filtrate flow is continuous or intermittent (batch). The filtrate is usually at atmospheric conditions for pressure filters and can often flow by gravity to the return circuit. However, for rotary drum and horizontal belt filters, the filtrate must be separated from the vacuum system. Design of vacuum pumps and associated equipment is outside the scope of this chapter.

The batch nature of pressure filters means that, like the solids, the filtrate flow will vary from a maximum at the start of slurry feeding to almost zero at the end of the feed cycle. If an air-blow step is included, there is likely to be very little filtrate entrained in the exhaust air stream. Thus the filtrate collection system must be sized for a range of conditions and must recover the fluid from what can be a very high airflow. At the start of the feed cycle, the filtrate can contain a significant quantity of suspended solids, so the filtrate is usually returned to a thickening step for the recovery of the solids. If the thickening stage is omitted, the quantity of the suspended solids must be understood prior to selecting the discharge point of the filtrate.

Process Water Pumps

Process water systems in mineral concentrators are usually closed-circuit recycled streams from which water is added to feed ore at the mill and recovered from the concentrates and tails by thickeners. Volume flow rates can be quite high, and the water is usually stored in lined ponds. This presents a challenging requirement of large flow rates and low available head, sometimes exacerbated by high elevation or high temperature.

A double-volute pump with a horizontally split casing will have a low NPSH requirement because of its large inlet area. However, the sealing rings between the impeller and casing may be prone to wear because of the fine solids in the process water. As an alternative, these pumps can be placed in a pit with suction lines that penetrate the wall rather than going over the top. This improves NPSHA, but does so at the risk of draining the pond should something go wrong.

Fire Pumps

Water is usually used for fire protection. Normal centrifugal pumps for water applications are suitable. Some jurisdictions require that any electrically powered fire pump be backed up by a pump driven by an internal combustion engine, in the case of power failure. Depending on application, foaming agents may be added to the water. This is the case for fighting liquid hydrocarbon fires. Foaming agents can be chemically aggressive in concentrated form. Vendors of fire suppression foam offer expertise in selecting materials of construction and foam-dosing systems for fire pumping systems where foaming agents are required.

Tailings Pond Pumps

Tailings pond pumps return tailings water from the pond to the mill for reuse. In many operations, the pumping load includes groundwater and rainfall in addition to the water that separates from the tailings during its residence in the pond. The rainfall rate at the site and the drainage area involved will determine the required pumping capacity.

As operations progress and tailings accumulate in the pond, the pond may be enlarged, and the location from which water is removed may move. In such cases, barge- or jetty-mounted vertical pumps may be used to allow optimal positioning of the pumps in the tailings pond. A pump barge is shown in Figure 50.

Vertical, cantilevered shaft pumps are often used in tailings ponds. Because these pumps have no submerged bearings, they are usually very reliable and require little attention beyond occasional greasing of the bearings. Where there is significant rainfall, the higher flow rates required may indicate a vertical, double-suction pump. The selection of a double-suction pump may permit the use of a higher-speed unit, but the selection may not be based on hydraulic considerations alone. If a vertical double-suction pump is mounted on a barge, it will work well as long as the water level in the pit is high enough. However, the water level in tailings ponds often varies seasonally or with changes in mill production. If this is the case, a top-inlet, vertical, single-stage pump may be better. A top-inlet pump mounted on a barge can pump the water level down without drawing the mud from the bottom. Of course, the higher capacity of the double-inlet pump will be lost. This analysis may be understood in terms of NPSH requirements.

Courtesy of Technosub.net

Figure 50 Pump barge in a tailings pond

A high-capacity, moderate-head pump may have a specific speed that requires a greater NPSH, which in this case means a greater submergence. When comparing NPSHA and NPSHR, the altitude of the installation and the temperature of the liquid must also be considered. In calculating NPSHA, altitude correction can be conveniently estimated as 0.3 m of head for each 300 m of elevation above sea level. Sometimes the hydraulic conditions will allow the use of either a top-inlet, single-stage pump or a vertical, double-stage pump. In such cases, lower cost and weight of the single-stage pump will make it the preferred option.

When pumps are mounted on a barge or a floating jetty, float design must be carefully considered for safety and convenience of operation. In addition to the weight of the pump and motor, a portion of the discharge line must be supported by the pump float. When the barge or jetty is large enough that workers may stand or walk on it, balance and stability under all conditions must be analyzed by a competent designer, with proper consideration of buoyancy and center of gravity. For example, if several workers gather at one side of the barge, it may overturn. The effects of wind and ice must also be considered.

Finally, seasonal or operational variations in pond levels that result in large variations in pumping rates must be considered in the design of the discharge line. If the friction losses in the discharge pipe are too high, addition of a second identical pump may not double the capacity of the system. In addition, as pipeline velocity increases, transients in the discharge line may damage the pipeline on shutdown. When significant variations in the required pumping rate will at times require much higher pipeline velocities, the preferred alternative is to install a second pump with its own discharge line. Of course, the additional costs must be considered.

PUMPS FOR CHEMICAL SERVICE

Chemical-service pumps are used in several mineral processing applications, including froth flotation, leaching, solvent extraction, and ore treatment in autoclaves and roasters.

Specifications

Pumps used for chemical service are different from those used elsewhere, primarily in the materials from which they are made. General-service pumps are usually made of cast iron, ductile iron, carbon steel, or aluminum- or copper-based alloys. Such pumps can be used for some chemical service, but most chemical pumps are made of stainless steel, Hastelloy, nickel-based alloys, or more exotic metals like titanium and zirconium. Chemical pumps are also available with liners of carbon, glass, porcelain, rubber, lead, and a wide range of engineering polymers, such as thermoplastics, thermosets, epoxies, and fluorocarbons. In every case, these special materials are used solely to eliminate or reduce the destructive effect of the chemical liquid on the pump parts. It therefore follows that the nature of the liquid being pumped determines which materials are used in the pump, and a careful analysis and description of that liquid is thus called for, including the following items:

- **Major and minor constituents.** In many instances, the effects of the minor constituents on corrosion rates will be equal to or greater than those of the major component.
- **Concentration.** Concentrations should be described quantitatively and as precisely as possible. General descriptions such as "concentrated," "dilute," or "trace quantities" should be avoided. Care should be used to clearly state whether concentrations are by weight or by volume. The concentrations of trace materials should be noted, even if they are in parts per million. In some cases, the presence of trace elements can markedly affect corrosion rates.
- **Temperature.** The operating temperature has a distinct effect on corrosion rates, so once again, operating temperature ranges should be described quantitatively, avoiding ambiguous, general terms. Maximum, minimum, and normal operating temperature should be specified. The effects of thermal shock should also be considered when the operating temperature range is large.
- **Acidity and alkalinity.** Acidity or alkalinity should be described in terms of pH. If the solution being pumped is subject to changing pH, during routine or upset conditions, this should also be noted. Some materials that are suitable for handling a given alkaline or acidic solution will not work for a solution whose pH changes during use.
- **Solids in suspension.** Erosion and corrosion of pump components are strongly affected by the solids in suspension and the flow velocity in the system. Solids should be specified quantitatively, again making sure to indicate whether by weight or by volume. See the "Slurry Pumping" section later in this chapter for more information on the effects of solids on pump design.
- **Aeration.** The presence of air in a liquid solution can significantly change the chemistry in a pumping system. For example, a nickel-molybdenum alloy pump is suitable for commercially pure hydrochloric acid, but if the acid becomes slightly oxidizing as the result of entrained air, the alloy will corrode. If the pump is self-priming, such aeration is likely. Furthermore, the presence of air will also affect NPSHR. The maximum amount of air a conventional centrifugal pump can handle is approximately 5% by volume.
- **Transferring or recirculating duty.** In a recirculating pump, the buildup of corrosion product or contaminants may damage the pump.
- **Continuous or intermittent duty.** Continuous or intermittent duty may have a counterintuitive effect on pump life. Intermittent duty can be more destructive than continuous duty if the pump retains some corrosive fluid during downtime, particularly if the presence of air increases corrosion. In such cases, it may be advisable to flush or drain the pump when it is not in service.

Pumps for Widely Used Chemicals

Sulfuric Acid

Sulfuric acid is used in many mineral processing applications, including froth flotation and leaching:

- **Dilution of commercially pure sulfuric acid.** When sulfuric acid is diluted with water, heat is evolved. When this dilution takes place in the pump transferring the acid, special materials are required for the pump components.
- **Sulfuric acid saturated with chlorine.** In a solution containing sulfuric acid and chlorine, the specific weight percentage of sulfuric acid determines whether the chlorine will accelerate corrosion. Concentrated sulfuric acid is hygroscopic and will absorb moisture from the chlorine, so with a sulfuric acid–chlorine solution containing more than 80% sulfuric acid, little corrosion will occur because dry chlorine is essentially noncorrosive. In this case, material selection can be made as if sulfuric acid is the only constituent. If the solution is saturated with chlorine but contains less than ~80% sulfuric acid, the material selection must be based not only on the sulfuric acid but also on the wet chlorine.
- **Sulfuric acid containing sodium chloride.** Addition of sodium chloride to sulfuric acid will result in the formation of hydrochloric acid, and a pump that is handling such a solution must be made of a material that will resist the corrosive action of both acids.
- **Sulfuric acid containing nitric acid, ferric sulfate, or cupric sulfate.** The presence of any of these compounds in sulfuric acid, in quantities of 1% or less, can make a sulfuric acid solution oxidizing, instead of reducing. They can therefore act as a corrosion inhibitor, thus allowing the use of stainless steel. However, the same compounds can act as accelerators for non-chromium-bearing alloys.

Nitric Acid

Nitric acid typically presents fewer problems than sulfuric acid. Nitric acid is strongly oxidizing, allowing the use of stainless steel in many applications. However, this same property limits the application of nonmetallics, and especially plastics. Solutions such as fuming nitric acid, nitric-hydrofluoric acid, and nitric-hydrochloric acid (aqua regia) produce more aggressive corrosion, and careful material selection becomes quite critical.

Hydrochloric Acid

Pumps for both pure and contaminated hydrochloric acid must be carefully designed. Ferric chloride, the most common contaminant, can cause the normally reducing solution to become oxidizing, completely changing the required materials of construction. The presence of only a few parts per million of iron in commercially pure hydrochloric acid can form enough ferric chloride to cause severe corrosion in materials like nickel-molybdenum, nickel-copper, and zirconium. Conversely, the presence of ferric chloride can make titanium completely suitable. Nonmetallic materials are widely used in pumps for hydrochloric acid systems.

Alkaline Solutions
Alkaline solutions such as sodium hydroxide or potassium hydroxide do not result in serious corrosion at temperatures below 93°C.

Organic Acids
Organic acids are much less corrosive than inorganic acids, but their particular properties must be considered carefully when evaluating pump materials.

Salt Solutions
Salt solutions are considered neutral and do not present a serious risk of corrosion. In process streams that are adjusted to a slightly alkaline pH, salt solutions are even less corrosive than when neutral. In process streams that are slightly acidic, the liquid will be considerably more corrosive.

Organic Compounds
Most organic compounds are not as corrosive as inorganic compounds. Although this means that there will be more materials available to choose from when specifying a pump, each application should still be considered carefully. For example, chlorinated organic compounds and compounds that will produce hydrochloric acid in the presence of moisture will require resistant materials. Plastics provide excellent corrosion resistance to inorganic compounds within their temperature limitations, but they show less corrosion resistance to organic compounds.

Organic solvents used in solvent extraction are usually either flammable or combustible liquids. Here, pump construction techniques that minimize the risk of initiation and spread of fire are required. Generally, the American Petroleum Institute standards for pumps API 610 are overly restrictive for atmospheric pressure-settler tank solvent-extraction systems. Pump selection for solvent extraction circuits must consider the shaft sealing arrangement and materials of construction, as most aqueous phases in use are acidic or otherwise reactive.

Autoclave and Roaster Service Pumps
Pumps for service in processes employing autoclaves and roasters must be selected for the particular chemicals in the flowstreams being handled. Such pumps must be capable of pumping high-density slurry at the operating pressure of the systems. Usually, PD pumps of the type manufactured by GEHO, Toyo, and others are used. For example, GEHO series ZPM pumps are used at Newmont's whole-ore pressure oxidation facility at the Twin Creeks operation in Nevada, United States (Eichhorn et al. 2014). These pumps use a design that isolates the piston mechanism for the heated slurry, as shown in Figure 51.

Reagent Pumps
Reagents are usually added to a process based on a ratio or percentage of feed, requiring the flow to be metered. PD pumps have a known volume delivered per revolution or stroke and are thus suitable, in general, for reagent pumping. Helical rotor or lobe pumps are used for shear-sensitive materials such as flocculants. Peristaltic pumps are sometimes used for relatively low flow rates. Piston and diaphragm pumps can be provided with metering valves that allow accurate dosing of reagents and are typically used in water treatment applications.

Many types of chemicals are used as flotation reagents, and the selection of the reagent pump must account for the properties of the reagent it is pumping. For example, the frother MIBC (methylisobutyl carbinol) is considered explosive and requires a non-sparking pump. Furthermore, some frothers contain both an alcohol solvent and a glycol, and the alcohol can cause rubber and other polymers to deteriorate over time. In such cases, a piston-type PD pump would be preferred over a peristaltic pump. Other desirable features in reagent pumps include a totally enclosed, chemically resistant housing and completely enclosed electronics for protection against moisture, corrosive atmospheres, and damage from vibration or mechanical shock. Finally, pumps should have a simple and accurate mechanism for manual adjustment

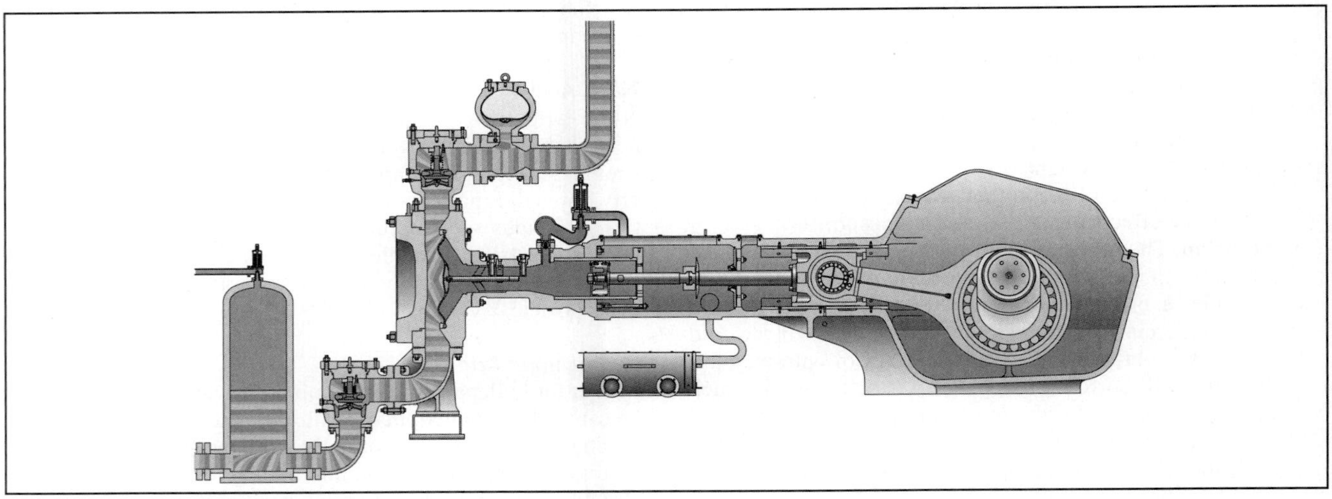

Adapted from Weir Minerals Netherlands b.v. 2009

Figure 51 GEHO ZPM positive displacement piston pump

of the pumping rate, a large turndown ratio. They should be amenable to control through connection to the signal from an automatic control system, usually a 4–20 mA or pulse input.

It is often useful to install a reagent pump in a self-priming configuration. Some manufacturers provide auto-prime valves, which allow automatic degassing for reagents that are prone to gassing in the pumping circuit. The liquid handling portion of the pump should be carefully selected to allow safe and long-term handling of all the reagents whose use is anticipated in the circuit.

SLURRY PUMPING

The design of slurry systems is based on a combination of first principles, rules of thumb, practical operating experience, and common sense. Most in-plant slurry systems can be designed effectively using the present body of practical rules and data. However, design of largescale installations, where slurry is pumped over a long distance, usually require pipeline trials to ensure success. These are almost always justified because of the high costs of incorrect design.

The Nature of Slurry Flow

Slurry flow in pipelines may be classified as homogeneous, pseudo-homogeneous, or heterogeneous. Homogeneous slurries are non-settling slurries that are characterized by a uniform distribution of solid particles throughout the carrier liquid and across the pipeline cross section. Typically, this type of slurry has a high concentration of fine solids (<30 μm) with low specific gravity and exhibits viscous behavior.

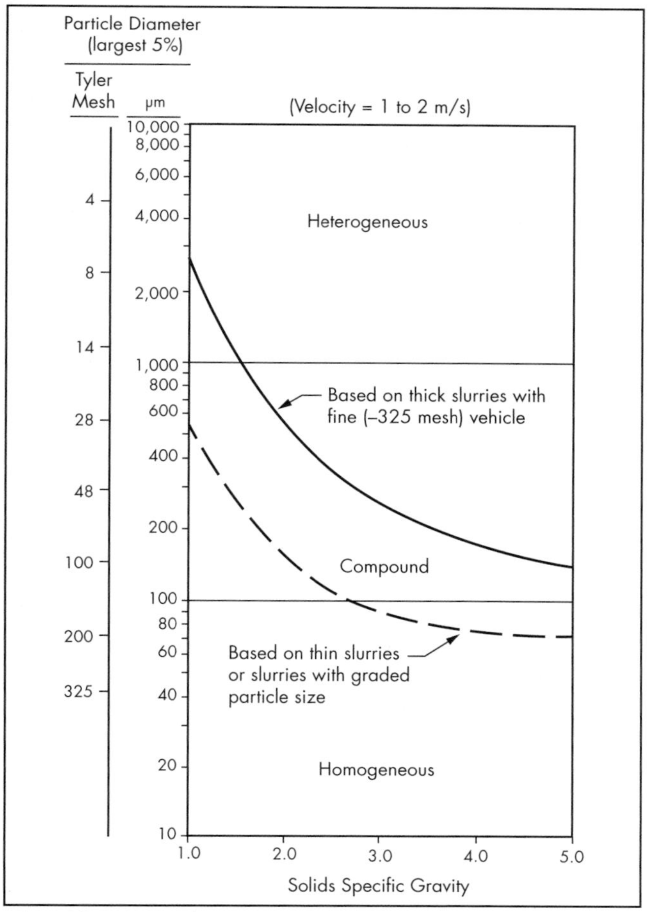

Adapted from Scott and Hays 1985

Figure 52 General guidelines to slurry flow regimes

Heterogeneous slurries are settling slurries characterized by a nonuniform distribution of particles across the pipeline cross section. These slurries usually contain coarse, poorly graded particles that are carried along the pipeline bottom. Pseudo-homogeneous flow of such mixtures may occur, provided the mean flow velocity is sufficiently high to suspend particles uniformly in the pipeline. At a lower velocity, saltation or moving bed flow may occur where the largest particles are alternately dropped out and picked up by the carrier liquid and consequently bounced along the pipe bottom. Figure 52 shows conceptually the relationship among particle diameter, flow velocity V, and solids specific gravity in determining the characteristics of slurry flow. This type of flow has a narrow velocity range and gives the lowest friction losses.

Still lower velocities result in fixed bed flow where all particles are moving in one body in a sliding action along the pipe bottom, which leads to frequent plugging and very high pipe wear, and therefore should be avoided. Usually a practical design in heterogeneous slurry systems is to use a velocity about 0.3 m/s higher than the transition velocity between moving bed and pseudo-homogeneous flow. A combination of homogeneous and heterogeneous flow is also possible where fine particles flow homogeneously and coarser particles flow heterogeneously.

Flow in vertical pipelines approaches homogeneous flow for most solids suspensions, provided that the velocity of the carrier liquids exceeds the terminal settling velocity of the coarsest particle size.

Slurry Pumping System Design and Operation

The design of a slurry transport system begins with an identification of the slurry to be conveyed. Particle size distribution and concentration must be known, a carrier liquid selected, and a minimum practical carrying velocity identified.

Friction losses must be determined as a function of pipe diameter and velocity. Selection of the final pipe diameter will result from an economic balance weighing such factors as initial pipe cost, pipe wear as a function of slurry velocity, solids abrasiveness, slurry corrosiveness, and projected operating costs, which arise mostly from power requirements. The power requirement is a function of system resistance and slurry specific gravity.

System design includes the following: selection of pump type (centrifugal or positive displacement), drives, and drivers; selection of control system and degree of automation; and sizing and layout of sumps and piping. Proper equipment selection and installation are necessary to minimize maintenance costs. Estimation of maintenance costs is often the most difficult task in preparing an operating budget.

Slurry Flow Calculation

In most mineral processing plants, the slurries are mixtures of fine solids and water. Some basic equations used in the analysis of these applications are as follows:

$$S_{SL} = 1 + C_v/100\,(S_S - 1)$$
$$= 1/\{1 - [(C_w/100) \cdot (S_S - 1)/S_S]\}$$

$$C_v = 1/\{1 + S_S\,[(1/(C_w/100) - 1]\}$$

$$Q_{SL} = (M_S/60)/(S_S \cdot C_v)$$

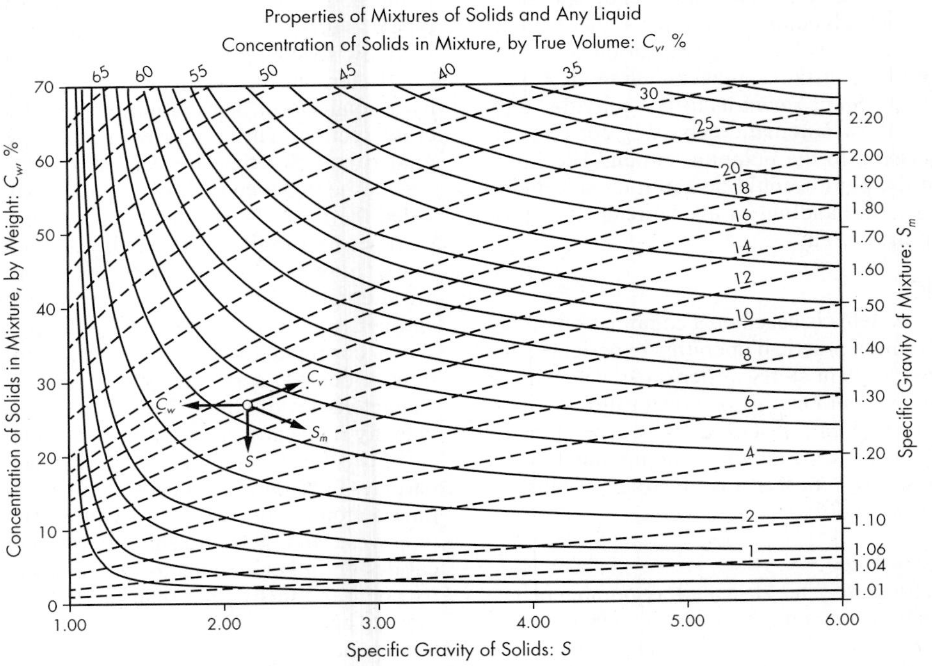

Source: Grzina et al. 2002

Figure 53 Graph for calculating properties of slurries

$$V_{SL} = Q_{SL}/\pi(D/2)^2$$

$$P = (H_{SL}\ S_{SL}\ g)/1{,}000$$

where S_{SL} is the slurry specific gravity, C_v is the percent solids concentration by volume, S_S is the solids specific gravity, C_w is the percent solids concentration by weight, Q_{SL} is the slurry volumetric flow rate in m^3/min, M_S is the solids flow in t/h, V_{SL} is the slurry velocity in m/min, D is the inside diameter of pipeline in m, P is the pressure in kPa, H_{SL} is the head in m, and g is the acceleration due to gravity. These calculations may also be done graphically, using the graph shown in Figure 53.

Design Velocity and Head Loss

In the practical design of a slurry transport system, the designer must choose a velocity range in which to operate. After this determination is made, a pipeline diameter is chosen, the corresponding pressure drop evaluated, and pumping equipment selected. Unfortunately, there are no universal rigorous calculation procedures that can be relied upon to yield velocity and pressure drop information. The designer must rely on previous plant operating data or existing empirical correlations. The latter are, as one may expect, more highly developed for homogeneous slurries than for heterogeneous slurries.

Hydraulics for Homogeneous Slurries

Settling velocity is the rate at which a particle of a given size will settle in a given fluid. It is influenced by particle size, particle shape, particle and fluid specific gravity, and fluid viscosity. Settling velocity must not be confused with deposition velocity. In a pipe with slurry at rest, all solids are settled at the bottom of the pipe and the liquid is at the top. As pumping starts and water velocity increases, the water picks up progressively more solids until, at a certain velocity V_c (at any solids concentration), the last solids at the bottom of the pipe are on the point between moving and staying put. If the flow velocity in a pipe is below V_c, some of the solids will begin to settle in the pipe, and the slurry will become heterogeneous, or settling.

When handling slurries, one must first determine if the slurry is heterogeneous (settling) or homogeneous (nonsettling). If the slurry is heterogeneous, then care must be taken to operate above the settling velocity of the solids, which dictates a maximum pipe diameter for a given required flow. Wilson (1979) produced the nomogram shown in Figure 54 to estimate the deposition velocity V_c as a function of pipe diameter, particle size, and solids density. If the slurry is of the settling type, a determination of the settling velocity must be made to ensure operation above it to prevent solids deposition and possible pipe plugging.

Durand (1952) derived a formula that estimates settling velocity, V_L, as follows:

$$V_L = F_L\sqrt{\frac{2gD(S - S_1)}{S_1}}$$

where F_L is a settling velocity parameter dependent upon particle sizing and solids concentration, g is the acceleration due to gravity (9.81 m/s^2), D is the pipe diameter in m, S is the specific gravity of the solids, and S_1 is the specific gravity of the liquid.

For closely graded particle sizing, F_L is obtained from Figure 55. Closely graded slurries are those in which the ratio of particle sizes does not exceed 2:1 for at least 90% of the weight of the solids in the sample, as described by the expression $d_{80}/d_{20} < 2$ in Figure 55. Figure 55 is conservative for slurries with more coarsely graded solids and significant portions

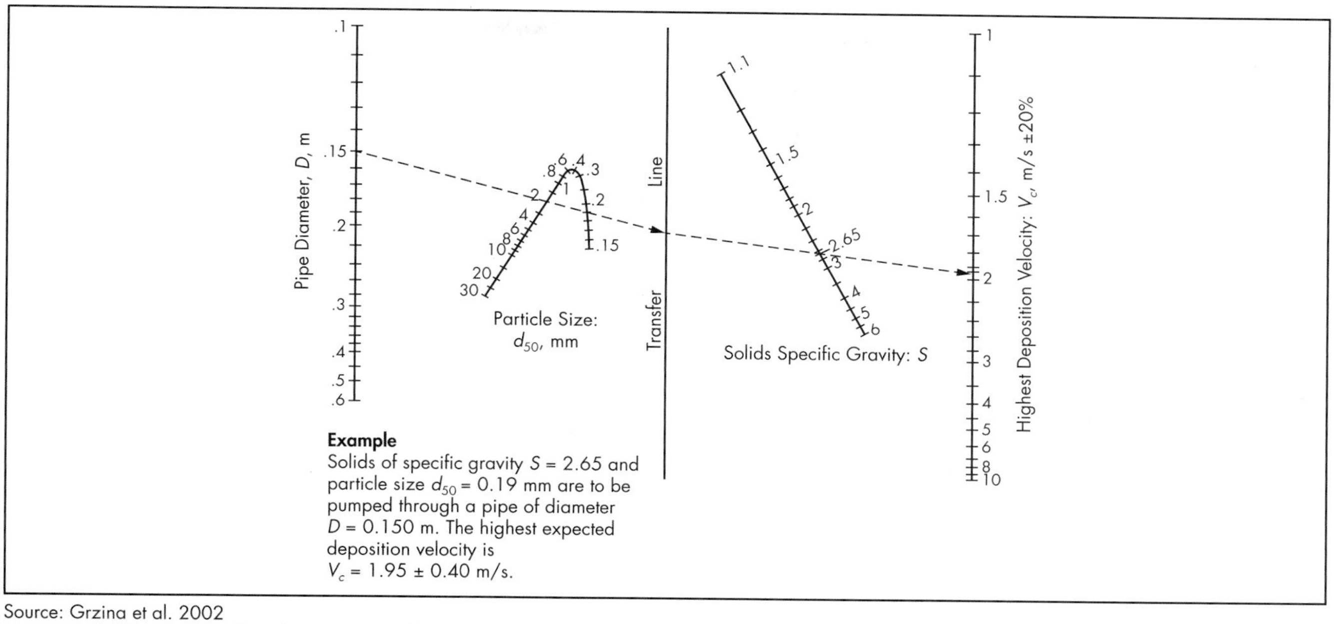

Source: Grzina et al. 2002

Figure 54 Nomogram for determining deposition velocity by Wilson's method

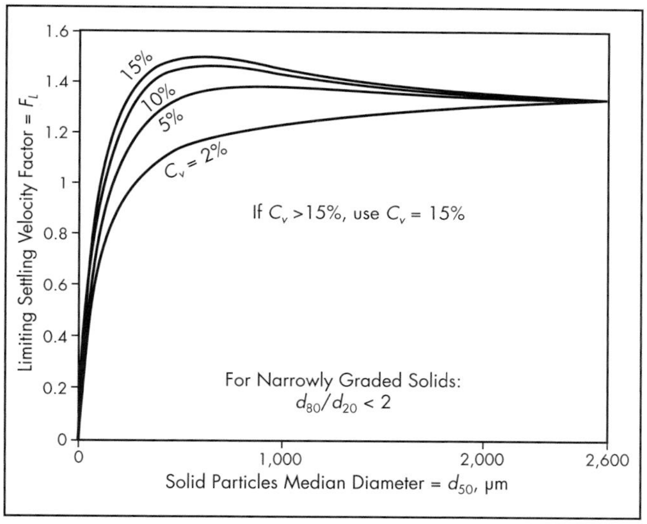

Source: Grzina et al. 2002

Figure 55 Durand's limited settling velocity parameter for closely graded particles

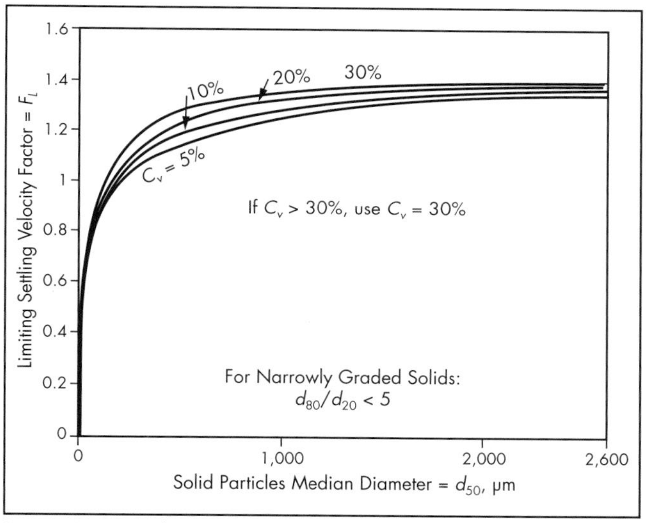

Source: Grzina et al. 2002

Figure 56 Modified Durand's limited settling velocity parameter for widely graded particles

of particles finer than 100 μm. For more widely graded particle sizing, Figure 56 is used to estimate F_L. The value of F_L from Figure 55 or 56 is then used to estimate V_L, thus

$$V_L = F_L\sqrt{2gD(\rho - 1)}$$

where D is the inside pipe diameter in m, ρ is the solids specific gravity, and g is the acceleration due to gravity in m/s^2.

For a heterogeneous slurry, accurately predicting the settling velocity is perhaps the first most critical factor in the design of a slurry pumping system. The settling velocity determines the required pipe diameter for a given desired flow, which is then used to determine the pipe friction. This information is used to make the appropriate pump selection and determine the proper operating speed, the appropriate materials of construction, and the expected power draw.

Performance Derating

The performance curves published by centrifugal slurry pump manufacturers are based on pumping clear water. Many manufacturers suggest that these curves be derated when handling slurries. Arguments for derating are based on considerations that the kinetic energy imparted to the solids by centrifugal action of the impeller cannot be recovered as pressure energy and that slurries exhibit viscous effects that lead to a loss in pump performance. Some pump manufacturers indicate that head and efficiency curves be derated by 5%–10% for all slurry applications, while others propose a more detailed derating procedure based on mean particle size, particle specific gravity, and solids concentration. One such derating procedure is described in the "Heterogeneous Slurry" section.

The operating point for a centrifugal pump is dependent not only upon the pump speed and the system in which it is placed, but also upon the material being pumped. Just as pipe friction can vary with the concentration of solids and the viscosity of the fluid being pumped, pump head output and efficiency can also deviate from the performance expected with water. It is desirable to be able to accurately predict these deviations, or deratings, in order to be able to determine if a centrifugal pump is the appropriate pump choice and to be able to compensate for these deratings by making the appropriate speed changes and providing adequate motor power.

Slurry Classification by Rheology

In addition to classification by deposition velocity, slurries may also be classified by rheological characteristics. A rheological classification is necessary in calculating the appropriate head and efficiency derating so that the appropriate corrections may be applied. Most slurries moved by centrifugal pumps fall into one of the following three categories:

1. **Heterogeneous slurry.** Most solids handling slurries are heterogeneous. This rheological class corresponds directly to the "heterogeneous" class determined by deposition velocity. A heterogeneous slurry is a settling slurry and is typically water based, with most of the solids greater than 100 µm. A relatively low content of solids less than 100 µm results in the carrier fluid, which is the water plus the suspended fines being essentially similar to water.
2. **Viscous Newtonian.** This refers to any slurry or fluid that has no yield stress and a constant viscosity. For this discussion, consider a viscous Newtonian slurry to be one that has a viscosity greater than water but is still free flowing, albeit at a slower rate. Oil is perhaps the most common viscous Newtonian fluid.
3. **Bingham plastic.** In a Bingham slurry, the carrier fluid contains sufficient fines content to provide a yield stress. Bingham plastic fluids are not free flowing unless there is sufficient force to overcome the shear stress. Examples include ketchup, red mud, and concrete.

Figure 57 illustrates the rheological differences among the three slurry types (Wells 1991). For the sake of this discussion, it is assumed that the heterogeneous slurry has a stress-versus-strain plot similar to that of water. For reference, these plots of shear stress versus shear strain are called rheograms. Lab analysis of a small sample of representative slurry for rheological properties and solids composition can provide the most helpful information for the design of pumps and pumping systems.

Heterogeneous Slurry

Following are the principal reasons for derating centrifugal pump performance when handling solids:

- Slip between the water and the solid particles during acceleration and deceleration of the slurry as it passes through the impeller. This results in energy losses.

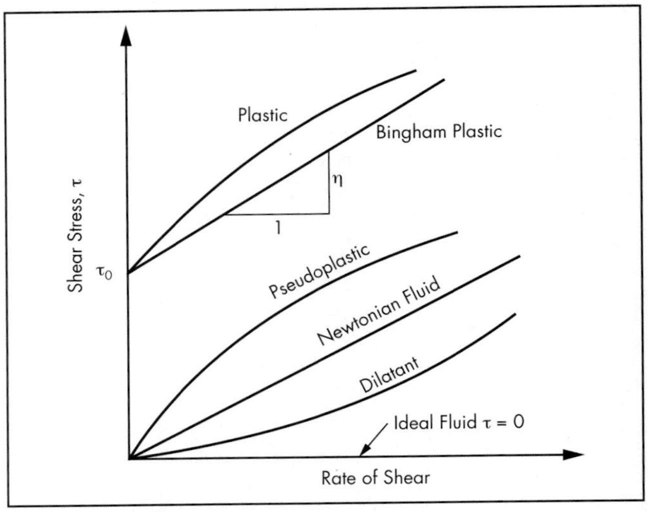

Adapted from Wells 1991

Figure 57 Rheological properties of various fluids

- Increased friction losses. These losses increase with increased solids concentration of the slurry.
- Inability of the suspended particles to store or transmit pressure energy.
- Mechanical friction changes in the gap between the impeller and side walls, which affect energy consumption and thus pump efficiency.

When discussing centrifugal pump deratings for heterogeneous slurries, the terms *head ratio* (HR) and *efficiency ratio* (ER) are used:

$$HR = head_{slurry}/head_{water}$$

$$HE = efficiency_{slurry}/efficiency_{water}$$

Grzina et al. (2002) explain showed that the degree of derating depends upon the solids specific gravity, the percent solids by volume, and the relative particle size (d_{50} particle size/impeller diameter). The results of this work are summarized in the graph in Figure 58. This graph is used by following the dashed lines that connect the circled numbers:

1. Start at the d_{50} particle size for the slurry.
2. Follow the line down to the solids specific gravity, S_s.
3. Follow the line left to the slurry % concentration by volume, C_v.
4. Follow the line down to the ratio of the d_{50} particle size to impeller diameter, d_{50}/D.
5. Then follow the line to the right, again to the C_v, but this time in the fourth quadrant.
6. From point 5, follow the line leading right to point 6 to find the head ratio, HR, and follow the line leading down to point 7 to find the efficiency ratio, ER.

Using the derating example given in Figure 58, consider a slurry that is 30% solids by volume, 2.65 specific gravity dry solids, 1,500-kg/m^3 slurry specific weight, and 350-µm d_{50} average particle size). The derating of a 6/4 pump at 1,200 rpm with the same 365-mm-diameter impeller is shown in the hypothetical system curve and plotted in Figure 59. In this example the intersection of the 1,200 rpm water performance

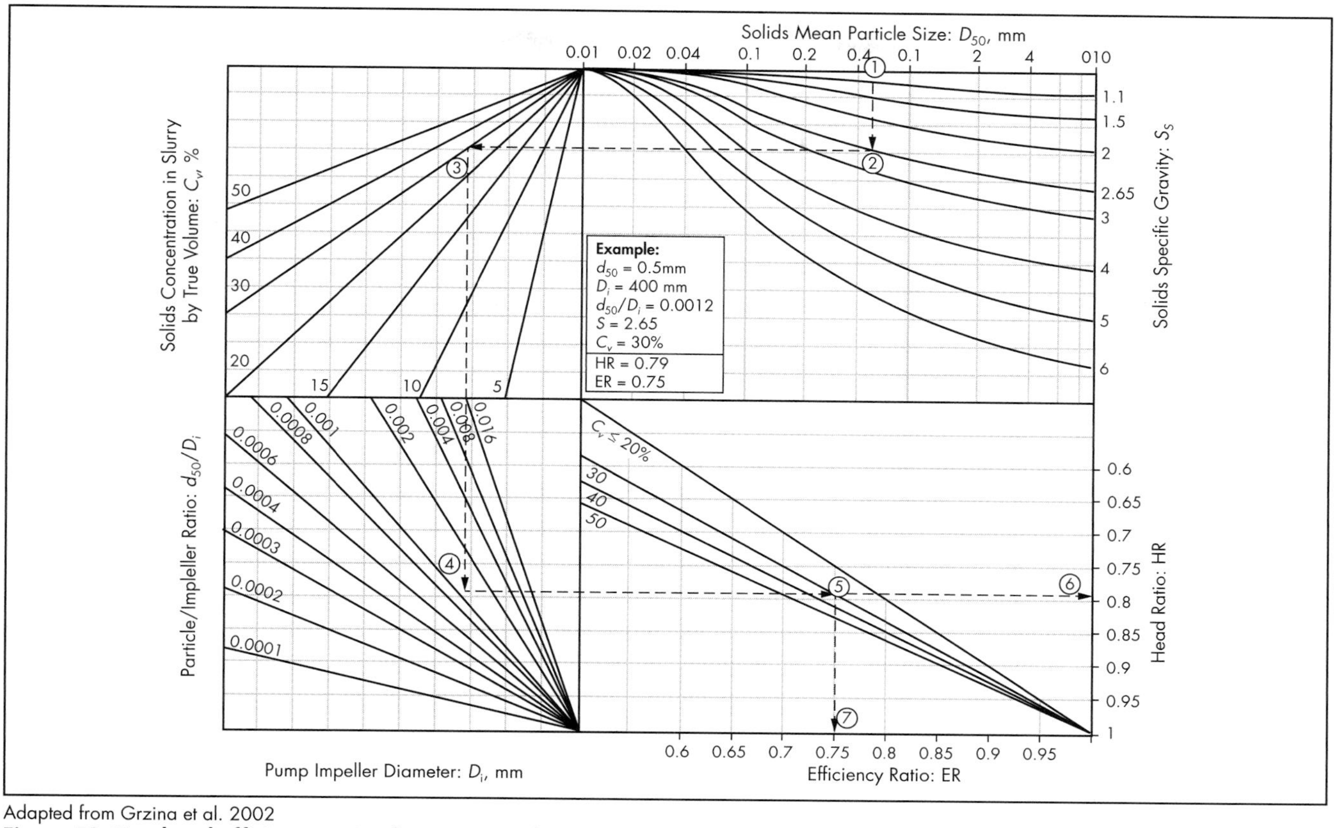

Adapted from Grzina et al. 2002

Figure 58 Head and efficiency ratios for pumping slurries

curve and the system curve occurs at 300 m^3/h, 28 m head, and 70% pump efficiency. With slurry correction (0.84 HR and 0.80 ER), the intersection of the derated pump performance curve and the system curve occurs at only 275 m^3/h at 24.5 m head and the pump's efficiency has been derated to 56%. The derated pump curves (for slurry) are shown in Figure 59. Power draw at a slurry flow of 275 m^3/h is expected to be

$$\text{kW}_{\text{draw}} = [(275\ \text{m}^3/\text{h})(1\ \text{h}/3{,}600\ \text{s})(24.5\ \text{m}) \times (1{,}500\ \text{kg/m}^3)(9.8\ \text{m/s}^2)] / [(56\%/100)(1{,}000\ \text{W/kW})] = 49.1\ \text{kW}$$

For operation at 300 m^3/h in this system, instead of the derated 275 m^3/h referred to previously, the required pump water head would be greater than the 28-m system head at 300 m^3/h by the inverse of the head ratio. This means the pump would have to deliver 28-m head/0.84 head ratio = 33.3 m head on water to achieve a 28-m head on slurry at 300 m^3/h in this system. Using the affinity laws or referring to the manufacturer's characteristic curve indicates this occurs at approximately 1,290 rpm with a water efficiency of ~70%. Derating the water efficiency with a 0.80 ER yields 56% slurry efficiency and the following power draw at 300 m^3/h:

$$\text{kW}_{\text{draw}} = [(300\ \text{m}^3/\text{h})(1\ \text{h}/3{,}600\ \text{s})(24.5\ \text{m}) \times (1{,}500\ \text{kg/m}^3)(9.8\ \text{m/s}^2)] / [(56\%/100)(1{,}000\ \text{W/kW})] = 61.3\ \text{kW}$$

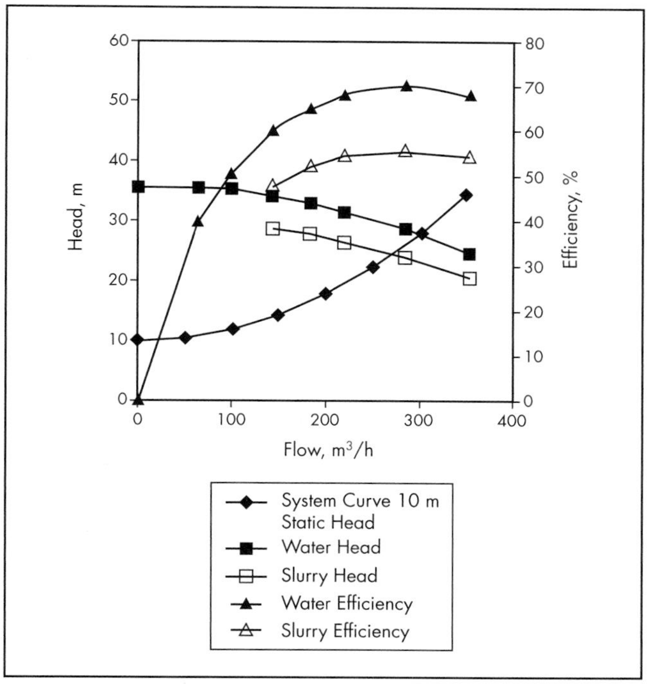

Source: Bootle 2002

Figure 59 Effect of slurry correction on pump performance

Newtonian Viscous Slurry

Purely viscous Newtonian fluids are not very common in slurry pumping. In most instances, the carrier fluid is water with a small percentage of fines in suspension that has a viscosity similar to water. In these instances, when most of the particles are larger than 100 µm, the pump deratings come from the slip and friction caused by the relatively large particles, and the heterogeneous slurry correction discussed previously is recommended.

Recall from Figure 55 that particles smaller than 100 µm with the appropriate specific gravity will be non-settling and remain in suspension. These fines effectively increase the density of the carrier fluid, and if they are there in sufficient quantity, the carrier fluid exhibits a yield stress—it will not be free flowing.

Only when there is no yield stress and when the slurry has a viscosity different from that of water should the traditional viscosity correction charts be used. This indicates the benefit of testing a sample of the slurry for its rheological properties to determine its slurry type.

Bingham Plastic Fluids

A Bingham plastic fluid or slurry is a material that has a yield stress. This means it is not free flowing unless there is sufficient stress to overcome the yield stress. Bingham plastic fluids are often called pastes or paste slurries. As mentioned previously, when non-settling fines are present in a slurry, they effectively increase the density of the carrier fluid. At a certain fines concentration (typically 30%–50% by weight, dependent upon material), the slurry starts to exhibit a yield stress. The addition of flocculants and coagulants can also be used to develop slurry that has a yield stress. Figure 60 is a rheogram of a red mud slurry with Bingham fluid characteristics.

Important parameters for a Bingham plastic fluid are the yield stress, t_0, and the coefficient of rigidity, η, which is also referred to as the plastic viscosity, μ_p. The plastic viscosity is the slope of the shear stress versus shear rate line at high shear rates. The yield stress is the point along the y-axis where this straight line intersects the y-axis. For the example in Figure 60, the yield stress is approximately 23 Pa, and the plastic viscosity is calculated as

$$\mu_p = (38 - 0)\ \text{Pa}/(500 - 0)\ \text{s}^{-1} = 0.03\ \text{Pa·s} = 30\ \text{centipoise}$$

It is important to differentiate between apparent viscosity, μ_a, and plastic viscosity, μ_p. For a Newtonian fluid, there is no yield stress and the viscosity is constant. Thus, its rheological plot is a straight line starting at the origin with the slope equal to the viscosity, as shown in Figure 60. A similar straight line can be drawn from the origin to any point on the rheological plot of a non-Newtonian, Bingham plastic fluid. The slope of this line will be the apparent viscosity for the particular shear rate and shear stress. The apparent viscosity is dependent upon the chosen point. At other shear stresses and rates of shear, the apparent viscosity will differ. For the red mud slurry described in Figure 60, at a shear rate of 200 s^{-1},

$$\mu_a = (29 - 0)\ \text{Pa}/(200 - 0)\ \text{s}^{-1} = 0.145\ \text{Pa·s} = 145\ \text{centipoise}$$

while at a shear rate of 300 s^{-1},

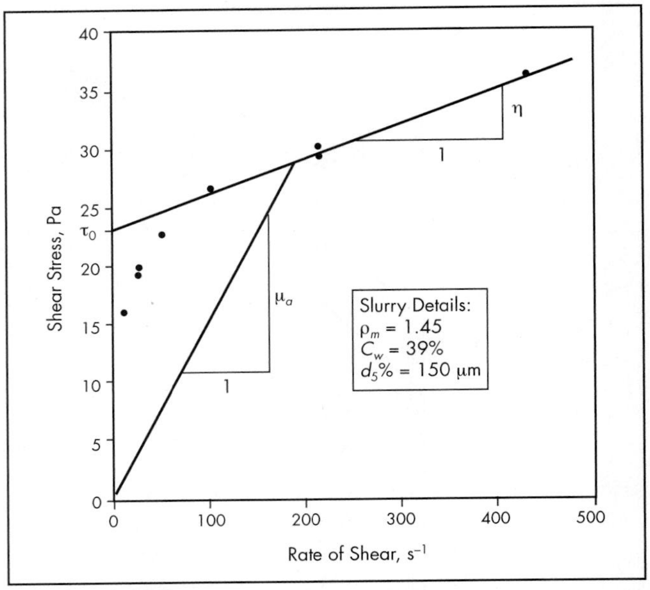

Source: Bootle 2002

Figure 60 Rheology of a red mud slurry with 95% of particles smaller than 150 µm

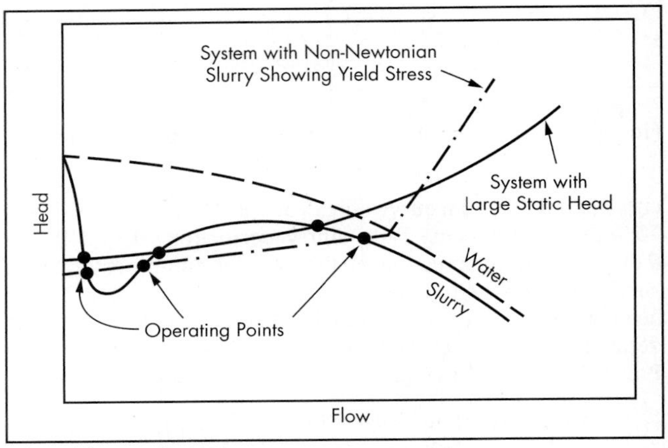

Adapted from Bootle 2002

Figure 61 Effect of high yield stress on pump performance

$$\mu_a = (39 - 0)\ \text{Pa}/(300 - 0)\ \text{s}^{-1} = 0.076\ \text{Pa·s} = 76\ \text{centipoise}$$

Thus the apparent viscosity decreases with shear rate, so this slurry is classified as *shear thinning*.

Walker and Goulas (1984) performed a series of tests with a 76-mm pump using kaolin clay and coal slurries of varying yield stress and plastic viscosity. The tests included three impellers, each at a different specific speed. On high yield stress material (~10–20 Pa), a severe drop in head occurred at low flow, as shown in Figure 61.

The significance of this drop in head is that the pump performance curve crosses over the system curve at three points. For comparison, Figure 62 also shows typical curves for two systems: a typical, turbulent-flow, exponentially rising system curve with high static head, and a high yield stress

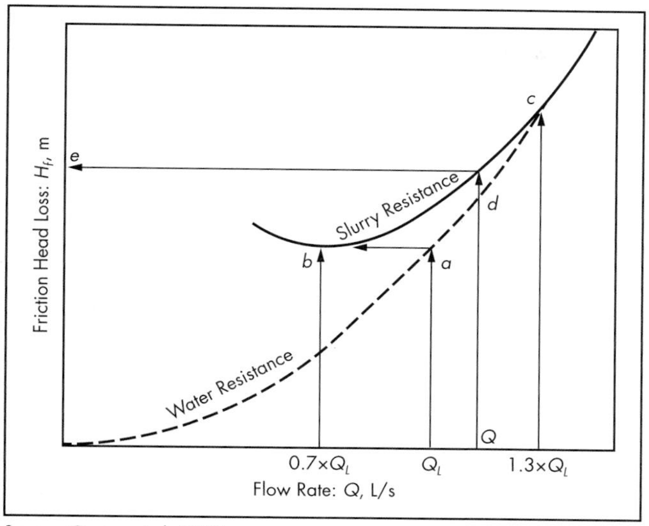

Source: Grzina et al. 2002

Figure 62 Construction of the system resistance curve for a heterogeneous slurry

non-Newtonian system curve with low-flow, laminar, and high-flow, turbulent regions. The intersection of the pump performance curve with the system curves at more than one location indicates that there may be more than one operating point for the slurry pump, so that the pump may swing between high and low flows. This phenomenon has been observed on high yield-stress concrete slurries and thickener underflows (Bootle 2002).

The Hydraulic Institute (2015) provides guidelines for the effects of viscosity on pump performance, but they are limited to specific speeds less than 60 (dimensionless), and liquids exhibiting Newtonian behavior, with kinematic viscosity >1 and <40 cm^2/s.

Performance tests are recommended when a particularly viscous liquid is to be pumped, and this includes most mineral slurries. Bootle (2002) notes that the changes in performance can be approximated using a modified pump Reynolds number, but Rayner (1995) cautions that this approximation is best made by a "manufacturer who can utilize empirical data on a specific pump."

Pipe Friction in Slurry Systems

Several methods can be used for calculating friction losses in slurry pumping systems. In every case, it is difficult to predict friction losses correctly because of the infinite number of combinations of particle sizes that can occur in slurries. The method shown here produces values that are adequate for pipelines up to a few hundred meters long, with static heads less than about 30 m. More rigorous treatment and test work is advised for long pipelines and higher static heads.

It is common in these applications to use V-belt drive systems for the pumps. This is because pump duties often change—and usually increase—after installation. In a V-belt drive system, an increased speed and system throughput are easily obtained by changing the drive sheaves. Of course, power consumption will also increase, so it is wise to select the motors that have 10%–20% more than the initial calculated power requirement.

Calculation of friction losses when pumping liquids has been previously discussed. Friction losses in the pumping of homogeneous slurries of fine non-settling solids are calculated using the method described for Bingham plastic slurries. For determination of friction losses in pumping settling slurries, the limiting settling velocity, V_L, must be determined at the required mass flow, M, of solids. Pumping solids at high concentrations, C_w, requires small flow rates, Q, and small pipe diameters, D; the opposite holds with low concentrations. Usually only one combination will satisfy the required V_L. This value may be found using the Durand method (described here), the Wilson method (using Figure 54), or by test work.

To use the Durand method, proceed as follows:

1. For a fixed solids mass flow M, select a value of C_v and calculate the corresponding Q.
2. Select a standard pipe of inner diameter D.
3. Obtain from the Durand diagram in Figure 55 or 56 the corresponding V_L and multiply it by the pipe's cross section to obtain the limiting flow rate, Q_L.
4. If Q is 10%–15% larger than Q_L, the pipe diameter is correct and the solids will not settle during pumping.
5. If Q is equal to or smaller than Q_L, select the next smaller pipe diameter and return to step 3.
6. If no reasonable pipe diameter satisfies the set requirements, select another value of C_v and return to step 1.

This procedure provides a starting point for use when estimating a system curve, as shown in Figure 62. In this procedure, static head Z is assumed to be zero. If Z is positive (or negative), it must be drawn as a straight line, parallel to the bottom of the graph at a distance Z above (or below) the base. The left point of the water resistance curve then starts from this Z line instead of from zero.

In the method shown in Figure 62, proceed as follows:

1. Calculate the water pipe friction loss, in meters of water at three flow rates, using the Moody diagram in Figure 2.
2. Plot the three points and draw the water resistance curve through them. Next mark the limiting flow rate Q_L estimated previously on the base line, draw a vertical line to the water curve at *a*, and from there draw a horizontal line to the left.
3. Draw two vertical lines up from the base: one from the value $0.7Q_L$ to meet the horizontal line at *b*, and the other from $1.3Q_L$ to the water curve at *c*.
4. Draw a parabola with vertex at *b* and tangential to the water curve at *c*. This is the slurry resistance curve.
5. Finally, mark Q on the base line and draw a vertical line to intersect the slurry resistance line at *d*, and from there a horizontal line to the left axis at *e*, which gives the friction head loss, H_f, in meters of slurry.

This method clearly provides only an estimate of the system resistance curve for pumping slurry, as the two curves are sketched by hand.

Centrifugal Slurry Pumps

The design of a horizontal, centrifugal slurry pump is a balance of design considerations to best meet the requirements of a particular slurry duty. These requirements may include one or more of the following:

- The ability to pump high density, abrasive slurries with adequate wear life

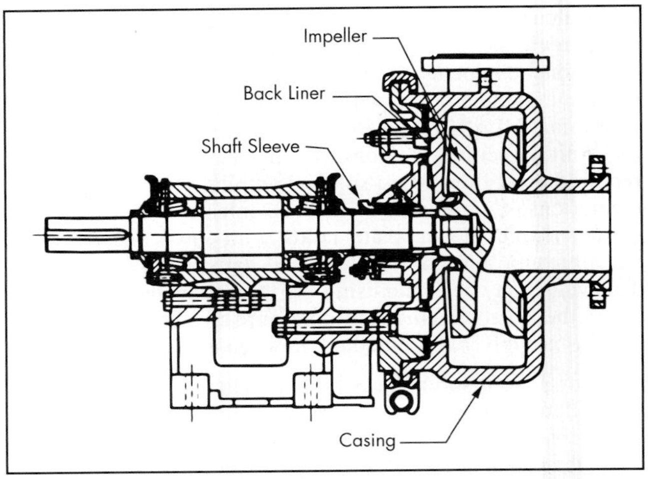

Source: Bootle 2002

Figure 63 Wear components in an unlined centrifugal slurry pump

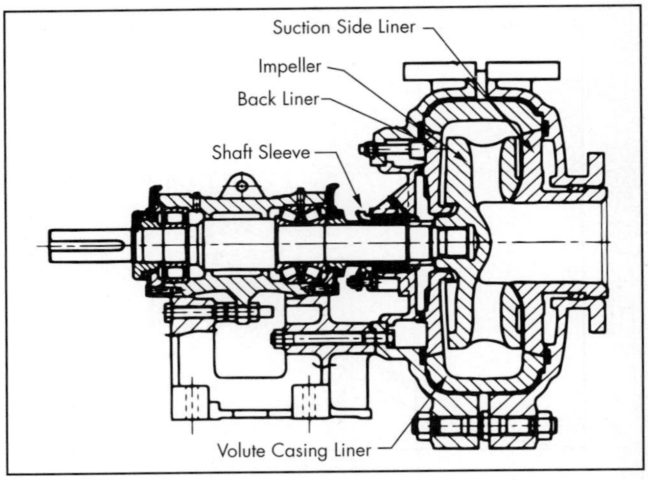

Source: Bootle 2002

Figure 64 Wear components in a fully lined centrifugal slurry pump

- The ability to pass large-diameter solids
- The ability to handle air-entrained or viscous fluids with reliability and minimal performance corrections

When compared with clear liquid pumps, the preceding requirements often result in the slurry pump being larger than its clear-liquid counterpart and sacrificing maximum efficiency in exchange for the ability to achieve the above-mentioned goals.

As they come into contact with abrasive slurry, the inner components of a slurry pump will wear. Wear can be minimized by appropriate pump design, proper material selection, and proper pump application. Figure 63 illustrates a typical unlined, horizontal, centrifugal slurry pump and Figure 64 illustrates a fully lined version of the same pump. The corresponding wear components for each have been appropriately marked. To decrease wear, thick casting sections are provided on the impeller, the casing of the unlined pump, and the wear liners of the fully lined pump. The unlined casing casting thickness is often greater than that of a fully lined casing. This additional thickness is required because the unsupported, unlined casing must safely handle the internal pressure of the pump with an adequate factor of safety to account for wear. The lined pump also allows for the use of a wider variety of materials, such as elastomer liners, which often outperform metal in fine particle and corrosive applications. Therefore, while the unlined pump may offer the lower initial capital cost, the lined version allows for a greater number of material choices, which may have longer wear life and lower replacement spares cost. The lined design is also inherently safer from a pressure containment standpoint. Large clearances are provided within the impeller and casing to allow for the passage of large-diameter solids, while also reducing internal velocities and corresponding wear.

Slurry pump impellers tend to be larger than their clear-liquid counterparts. To minimize speed and maximize wear life of both the impeller and suction side liner, slurry pump impellers are rarely trimmed in diameter to meet the required duty point. For applications below 300 kW, belt drives are the most popular way to achieve the required speed for the duty point. Belt drives are inherently quiet and also allow for relatively easy speed changes if required. For higher-power applications, gearboxes are usually used to meet the desired speed and duty point. On applications where variable flow is required, VFDs are used to provide the necessary continual speed changes. Usually these VFDs operate with other speed reduction devices, such as belt drives and gearboxes, to allow for the use of higher-speed, lower-cost motors.

Bearing Assembly

The bearing assembly of a heavy-duty slurry pump should incorporate a larger-diameter shaft and larger bearings than those found on a clear-liquid pump. Roller bearings are often used to provide further additional load-carrying capacity to handle the higher forces that result from pumping a heavier liquid, the inherent imbalance due to wear, the shock loading from large particles, and the hydraulic imbalance from air entrainment. On belt-drive applications, most of the side load the drive-end bearing results from belt pull, which can be substantial. Most properly sized and properly tensioned V-drives exert approximately 9 kN of belt pull for every 100 kW of motor power.

Figure 63 illustrates a bearing assembly with identical low-angle, angular contact bearings on both the pump end and drive end. The low-angle bearing is well suited to the radial loads from the impeller or belt pull from the drive, if so equipped.

Figure 64 illustrates a higher-capacity bearing assembly, which fits within the same packaging dimensions. This assembly features a duplex angular-contact roller bearing on the pump end and a cylindrical roller bearing on the drive end. The duplex angular-contact bearing has a higher bearing angle than the bearings shown in Figure 63, making it more suitable for higher axial loads, such as those seen with smooth-backed impellers, open-faced impellers, and series pumping applications. In this configuration, the cylindrical roller bearing on the drive end handles none of the axial load but has tremendous radial load capability, making it well suited to belt-drive applications.

When the shaft seal design allows it, shorter shafts with reduced impeller overhangs will result in reduction of bearing loads, shaft stress, and deflection through the seal area. When

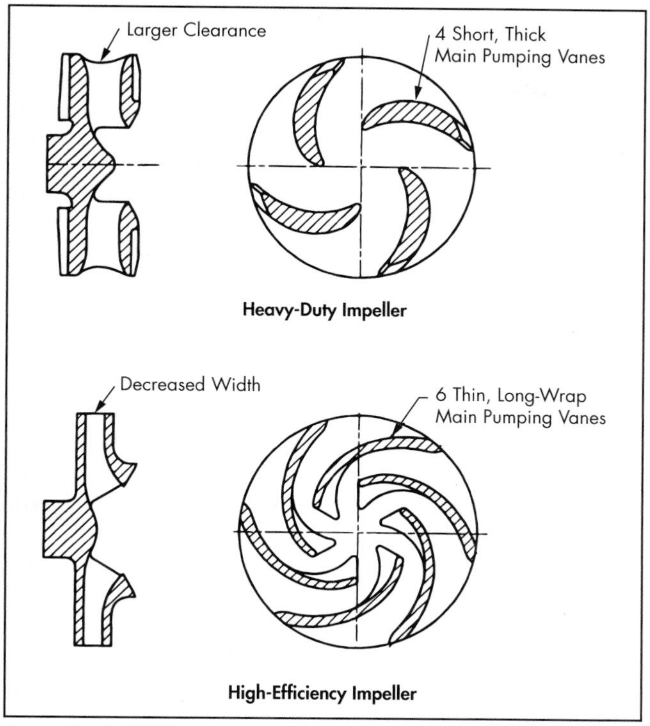

Source: Bootle 2002
Figure 65 Heavy-duty and high-efficiency impellers

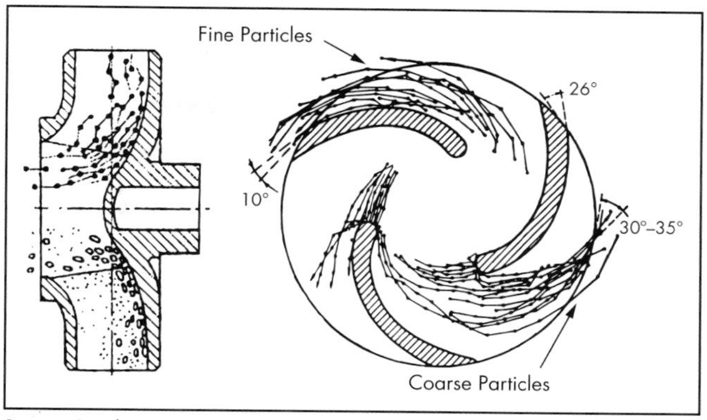

Source: Bootle 2002
Figure 66 Particle trajectories in a centrifugal slurry pump

packing is used as a shaft seal, a hardened or ceramic-coated shaft sleeve is recommended to prevent shaft wear.

Impeller Design

Slurry pumps typically have impellers that are larger than their clear-liquid counterparts. This is to lower the impeller speed required to achieve a given head and to provide more material for wear purposes. For high wear applications, closed impellers are preferred. In coarse particle applications, expelling vanes are recommended on the face of the front shroud. The purpose of these expelling vanes is to prevent large particles from becoming trapped between the impeller and suction side liner and to minimize recirculation. The benefit is a reduction in gouging and recirculation wear, with an acceptable decrease in efficiency.

Expelling vanes are also often used on the back shroud of the impeller in coarse particle applications to prevent the trapping of large particles between the impeller and back liner. In this location they also serve to reduce the forward axial load, improving bearing life by lowering the pressure acting on the back shroud and beneficially reducing the pressure at the hub and packing. This reduces the pressure differential at the shaft seal and reduces the tendency for slurry leakage from the pump. As with expelling vanes on the front shroud, back vanes usually absorb two to three percentage points of efficiency.

To decrease wear and allow for the passage of large-diameter solids, heavy-duty slurry pump impellers feature thicker main pumping vanes and fewer of them. Both of these factors further contribute to a reduction in efficiency when compared with a clear-liquid counterpart. While a clear-liquid impeller usually has five to nine vanes, most slurry pump impellers have two to five, with four and five vane designs being the most common. Two and three vane designs are usually reserved for very large particle passing sizes, as required in dredging applications.

Figure 65 illustrates the difference between a four-vane, heavy-duty slurry design with front and back expelling vanes and a six-vane, high-efficiency slurry design with smooth front and back shrouds. Notice the short "blocky" vanes on the heavy-duty design and the thin, long-length, long-wrap vanes on the high-efficiency design.

In slurries with large particles, the large solids follow a different path than the fluid, because of inertial effects. This is illustrated in Figure 66. This results in gouging wear at any location where the fluid is required to make an abrupt change in direction. For this reason, on large-particle applications, heavy-duty designs with blunt leading edges, wide spacing between shrouds, and thick back shrouds are recommended to reduce the impact of the large particles. Main pumping vanes should be short, with minimal vane overlap to allow the large particles to pass unimpeded. This heavy-duty geometry results in a head-versus-capacity curve that is flatter than the typical clear-liquid pump. The use of fewer main pumping vanes that are also thicker and shorter, along with expelling vanes on the front and back shroud, can reduce slurry pump efficiencies by up to 10% from those of a comparable clear-liquid impeller. These differences are minimized on larger pumps.

In slurries where d_{85} is less than 100 µm, almost all the particles will follow the path shown in Figure 66 for fine particles. In this case, high-efficiency designs that minimize the presence of vortices will improve efficiency and improve impeller life. In such applications, the absence of expelling vanes on the front and back shroud will improve side liner wear life.

Pump Materials

The primary materials used for wear components in centrifugal slurry pumps are hard metals and elastomers. Ceramics are also used, but less often. Hard metals and ceramics resist erosion because of their high hardness values. Elastomers resist erosion by their ability to absorb the energy of the impacting particles, which is due to their resilience and tear resistance. Elastomers generally have better erosion resistance than hard metals. In applications where particle size is smaller than 250 µm, impeller tip speed will be within the limits of the elastomer, and there is no risk of damage from occasional passage of large particles.

Metals

The three basic types of metals used to combat erosion in centrifugal pumps are the martensitic white irons, chromium-molybdenum white irons, and high-chrome irons. ASTM International Standard A532 (2014) provides specifications for these materials, which consist of hard carbides in a supporting ferrous matrix.

The bulk hardness of these materials depends on the carbide and matrix type, and on the volume of the carbides within the matrix. For applications with medium to large slurry particles, the bulk or combined material hardness is most important. For small-particle applications, a fine microstructure with smaller inter-carbide spacing is more important to minimize erosion of the softer matrix. For applications with particles larger than about 100 mm, the fracture toughness of the matrix is most important.

Ni-Hard 1 and Ni-Hard 4, which are martensitic white irons, have high hardness but their erosion resistance is not as good as the other ASTM A532 materials, and their low chromium content provides little corrosion resistance. In heavy-duty slurry applications, martensitic white irons have been largely superseded by chromium molybdenum white irons and high-chrome irons.

The hardness of chromium-molybdenum white irons and high-chrome irons can be increased by proper heat treatment, which can also increase corrosion resistance. The molybdenum white iron 15-3 alloy is suitable for use in slurry pumps where high erosion resistance is required and where impact-loading conditions and corrosion rates are minor to moderate, while the 27% high-chrome iron is suitable for use in slurry pumps where erosion resistance, corrosion resistance, and fracture toughness requirements are moderate to high.

The properties of high-chrome iron can be altered in many ways to suit particular pumping conditions. For example, high-chrome irons with an austenitic matrix and high impact resistance are available for extremely large-particle dredging applications where there are very large particles in the slurry.

Elastomers

Elastomers are typically used in applications with particle diameters <10 mm. They can be broadly broken into two categories: natural rubber and synthetic elastomers. Natural rubber has much greater resilience and tear resistance than the synthetic polymers.

Resilience is a measure of how high a ball of the material will bounce when dropped on a standard surface in relation to the initial drop height. Typically, this value ranges from 65% to 90%, depending on the rubber blend. Resistance to tear initiation for natural rubber is usually 30–110 N/mm, depending on the blend. For natural rubber, tear resistance tends to improve with increasing hardness, while resilience tends to decrease. For particles less than 100 µm, resilience is more important in combating wear, while for particles greater than 500 m, tear resistance is more important. Given that mineral processing slurries have a mixture of particle sizes, the best performing natural rubber will be one with the optimum combination of resilience for fine-particle wear resistance and tear resistance to prevent larger particle damage.

Synthetic elastomers are used in small-particle applications where natural rubber would be subject to chemical attack, causing swelling, hardening, or reversion of the natural rubber. Reversion of natural rubber is the decline in its physical properties, such as modulus and resilience, as a result of weathering or contamination. The most commonly used synthetic elastomers for wear materials and some of their typical applications are as follows:

- **Nitrile:** Generally used in fats, oils, and waxes. Moderate erosion resistance. Limited resistance to acid and alkaline environments.
- **Butyl:** Suitable for hydrochloric acid, phosphoric acid, and sodium hydroxide. Sulfuric acid causes degradation, and chlorinated hydrocarbons cause swelling.
- **Hypalon:** Primarily used in acid conditions with some resistance to vegetable and mineral oils. Not recommended for use in ketones or chlorinated solvents.
- **Neoprene:** Moderate resistance to oils, fats, grease, some hydrocarbons, and some mild oxidizing acids.

Synthetic elastomers also have higher temperature limits than natural rubber. While natural rubber is limited to 75°–85°C, depending on the blend, the above-mentioned synthetic elastomers have temperature limits ranging from 95°C for nitrile to 110°C for Hypalon.

Further comparisons can be made with respect to mechanical properties. The tear resistance for the above-mentioned synthetic elastomers ranges from 30 N/mm for nitrile to 50 N/mm for neoprene. The resilience of neoprene, at 58%, is the highest of the synthetic elastomers. The superior mechanical properties of natural rubber indicate it should be used in preference to synthetic elastomers, unless temperature or chemical resistance are overriding factors.

The high tear strength of polyurethane, 50–100 N/mm, makes it applicable for systems where damage by large particles is likely. Polyurethane is much less susceptible to damage from cutting than natural rubber and other synthetic elastomers. Most polyurethanes cannot be used above 70°C, but special compositions that will function up to 110°C with no degradation are available.

Tip-Speed Limits

Another important factor to consider in the selection of materials is the impeller tip-speed limit. For elastomers, the concern is vibration or fibrillation of the material due to the relative motion of the impeller with respect to the side liners. This vibration can lead to heat generation within the elastomer and a thermal breakdown of the material. For natural rubber, this often results in the elastomer reverting to its natural "gummy" state. Tip speed problems on elastomers are easy to diagnose, as the damage is always the most severe at areas with the highest relative speed. For elastomer impellers, this would be the periphery of the impeller, and for elastomer side liners, this would be the area adjacent to the periphery of the impeller.

The tip speed is the equivalent linear velocity of the impeller tip, and is a function of rotational speed and impeller diameter. The tip-speed limit is determined by the hardness of the elastomer and its ability to dissipate heat. Typically, natural rubber has a tip-speed limit of approximately 27.5 m/s. Highly wear-resistant soft natural rubber can have a tip-speed limit as low as 25 m/s, while proprietary blends with improved thermal conductivity can operate at 30 m/s. For a centrifugal slurry pump, this tip-speed limit is important, as the head generated (meters) is approximately equal to the quantity (0.5 times the square of tip speed in meters per second) divided by the acceleration due to gravity (9.8 m/s^2).

Approximate tip-speed limits and corresponding approximate BEP head limits for various materials are shown in

Table 3 Tip-speed and BEP head limits for various impeller materials

Material	Tip Speed, m/s	Head Limit, m
Highly wear-resistant soft natural rubber	25.0	32
Typical natural rubber	27.5	39
Anti-thermal breakdown rubber	30.0	46
Nitrile	27.0	37
Butyl	30.0	46
Hypalon	30.0	46
Neoprene	27.5	39
Polyurethane	30.0	46
Hard metal (impellers)	38.0	74

Adapted from Bootle 2002

Table 3. The tip-speed limit for hard metal impellers is based on the limited ductility of the material and not on thermal breakdown.

Pump Wear

Wear in a centrifugal pump will include various modes of abrasion and erosion. Abrasion, which is the forcing of hard particles against a wear surface, takes place within a slurry pump only on the shaft sleeve and between the tight tolerance wear-ring section of the impeller and the suction side liner. Erosion is progressive wear loss from the interaction with, or impingement of, the fluid and particles against the wear components. Three primary modes of erosion can take place, each of which usually occurs in a particular location in the pump. Deformation wear results from direct impact to the leading edge of the impeller vanes, the back shroud of the impeller, and the protruding portion of the cutwater within the volute. Random impingement is caused by impacts to the impeller shroud and trailing edge of the main pumping vanes. Low-angle impact wear results from the tangential or near-tangential movement of particles against the volute casing or vane surface.

Of these modes of erosion, deformation wear by direct impact is the most severe and low-angle impact is the least severe. The degree of wear depends on three factors: (1) the kinetic energy of the particle, which is a function of particle mass and velocity; (2) the particle shape—sharp particles have small contact area and produce high local stress, so wear is more severe than with rounded particles; and (3) the slurry concentration—higher percentages of solids result in more impacts for a given flow. Based on these factors, slurries are classified as follows:

- Heavy duty: $C_w > 35\%$, $d_{85} > 400$ µm, s.g. > 2.0, sharp particles
- Medium duty: $20\% < C_w < 50\%$, 150 µm < d_{85} < 400 µm, s.g. > 1.4, angular particles
- Light duty: $C_w < 20\%$, $d_{85} < 150$ µm, s.g., > 1.4, rounded particles

In addition to the general specific speed limits and the material tip-speed limits discussed previously, to maximize wear life, the following general impeller tip-speed limits apply based on the severity of the duty:

- Heavy duty: 25 m/s maximum, 32 m head at BEP
- Medium duty: 32 m/s maximum, 52 m head at BEP
- Light duty: 38 m/s maximum, 74 m head at BEP

Further recommendations for impeller type and flow range are as follows:

- Heavy duty: heavy-duty impeller at 0.60–0.80 BEP
- Medium duty: heavy-duty impeller at 0.70–0.90 BEP
- Light duty: high-efficiency impeller at 0.80–1.1 BEP

Operating Controls

Control by Sump-Level Variation

For lines shorter than about 300 m, flow must be varied to match process conditions while attempting to maintain minimum velocity. Most systems use the sump level to determine flow requirements. Because many centrifugal slurry pumps normally have a relatively flat output curve, with a large flow change for a relatively small head difference, many low-head systems use fixed-speed pumps with deep sumps and allow the fluctuating sump level to regulate flow.

Consider the system shown in Figure 67, delivering design flow Q_1 at a system TDH of H, calculated for a static suction head in the sump of H_{S1}. If inflow to the sump is allowed to vary so that the static suction head can vary up to H_{S2}, or down to H_{S3}, the system-head curve will vary, thus

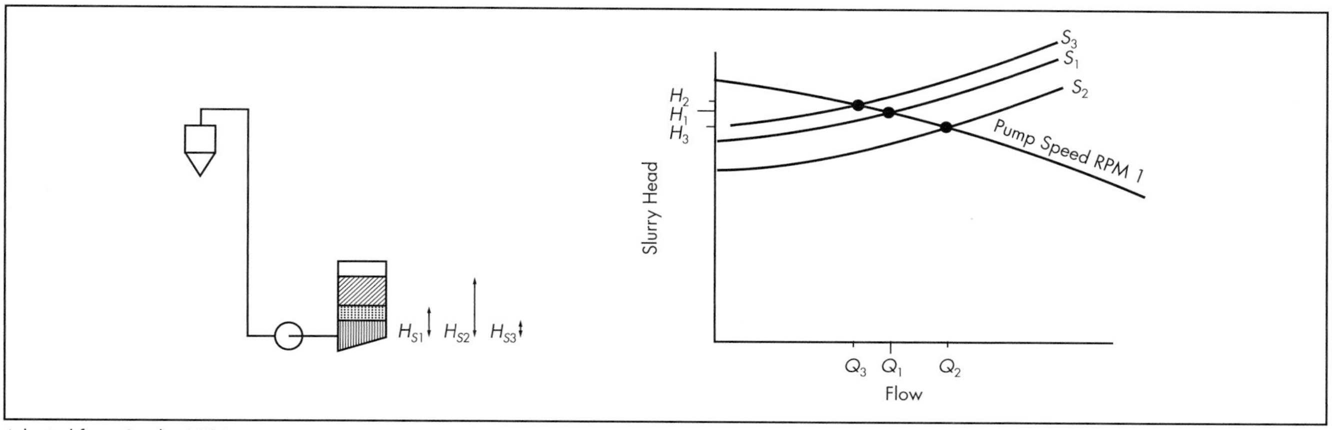

Adapted from Crosby 1985

Figure 67 Effect of sump-level variation on system-head curve

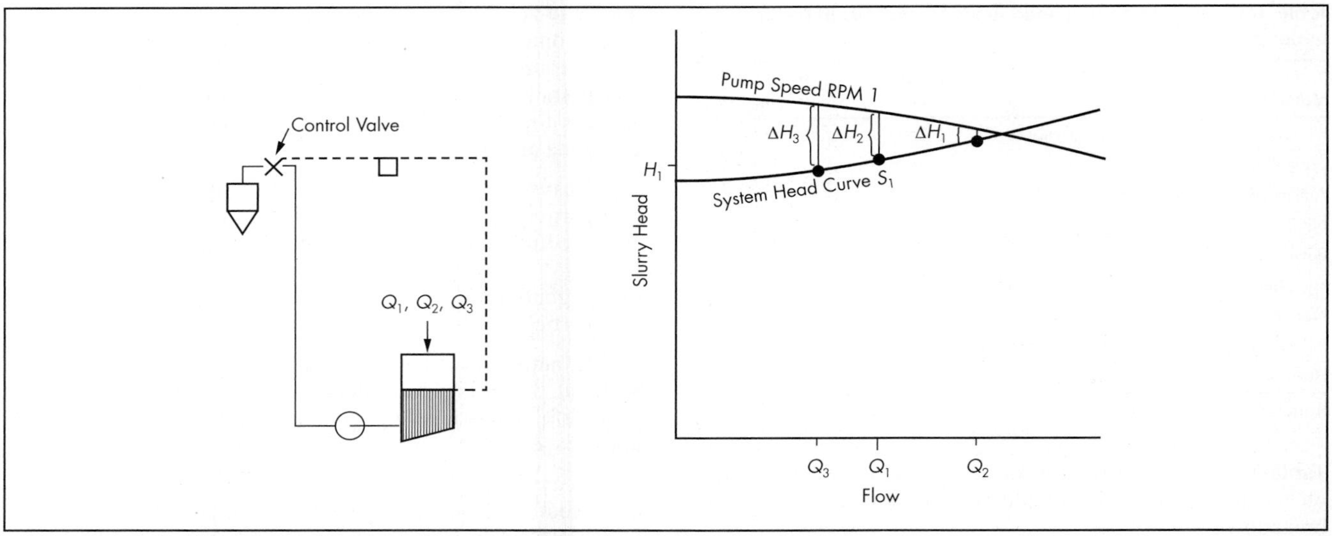

Adapted from Crosby 1985

Figure 68 Use of a control valve to handle varying flow

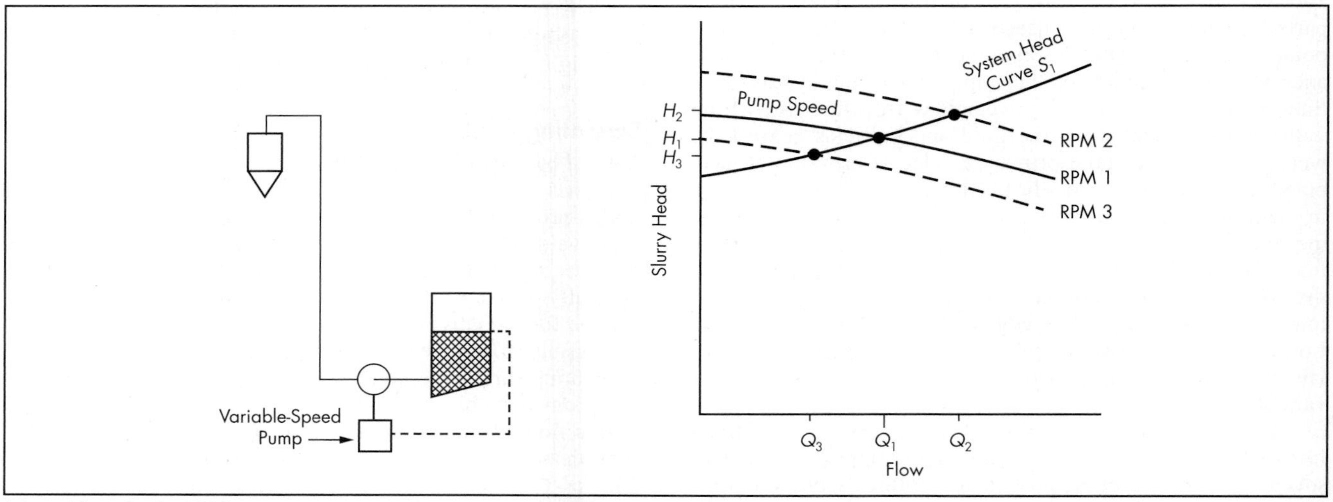

Adapted from Crosby 1985

Figure 69 Use of a variable-speed pump to handle varying flow

yielding corresponding flows Q_2 and Q_3 at new TDH points H_2 and H_3. If the pump is fixed at one speed, it can only operate between Q_2 and Q_3. In the figure, S_1 is the system-head curve at H_{S1}, S_2 at H_{S2}, and S_3 at H_{S3}.

This is an acceptable method of operating, but only if the slope of the system-head curve and pump curve will permit a wide enough variation in flow to accommodate the variation in incoming flow to the sump, flow is not allowed to drop so low as to cause sanding in the line, flow is not allowed to reach a maximum point where there is insufficient NPSH available for pump operation, and the sump static suction head stays high enough to provide sufficient NPSH and prevent vortexing and air entrapment.

Control Valves

Another method for handling varying sump inflow is to install a control valve in the pump discharge line, so that the pump discharge can be throttled to maintain a fairly constant sump level. Now consider the pumping system in Figure 68, again delivering design flow Q_1 at a system head H_1 calculated for a static suction head of H_{S1}. Suppose that sump inflow can vary between Q_2 and Q_3. If the pump curve is that of a fixed-speed unit, a variable resistance, indicated by ΔH_1, ΔH_2, and ΔH_3, is required to operate on the curve. This resistance is provided by a throttling valve in the discharge pipe that operates in response to a sump-level controller. Proper selection and sizing of the control valve is essential for good operation and maintenance.

Variable-Speed-Drive Control

Changing pump speed is another alternative for handling varying flow rates. Figure 69 shows the pumping system, again operating with design flow Q, at a system TDH of H_1 calculated at a static suction head H_S. When incoming flow changes from Q to Q_3, the pump speed can be varied to match the system-head curve requirement. As in the case of the control valve, a sump-level control device or flowmeter is used to generate a signal that slows pump speed or increases pump speed when the sump level rises or falls, respectively. Pump speed can be controlled by a variable-speed coupling, a variable-speed motor, or a variable-frequency drive for the motor.

Water Addition to Sump

Variations in sump inflow can also be smoothed out by the automatic addition of water to the sump. In this case, the pump has a fixed speed designed to handle the expected maximum steady sump inflow. Water is added to maintain the sump level and pumping volume.

Control for Safe Operation

Most centrifugal slurry pump systems require some safety controls to protect pumps, drivers, or pipelines on start-up, during operation, and on shutdown.

Start-Up

When a pump is started, particularly with an empty discharge line, it sees virtually no friction resistance and will instantaneously try to pump as far out on the end of its fixed-speed curve as NPSHA will permit. Operation approaches the curve design point as quickly as static head and friction resistances can be built. Motors must be supplied that will permit an overload for this start-up period. These start-up loads can be reduced by starting with the discharge line full or partially full of slurry carrier liquid, then building pump speed slowly to the design with the use of a variable-speed coupling or motor, and by providing resistance by removable orifices or a throttling discharge valve.

For short, in-plant lines, start-up problems are usually not serious. However, long-distance tailings lines that use centrifugal pumps in series are another matter. These pumps should be started sequentially from the pump in the series train closest to the sump to the last pump that discharges into the pipeline. The time between starts should be sufficient to allow each pump to operate at a point close to its intended design point. Starting intervals that are too rapid invite motor overloads and pump damage due to cavitation. The timing for the start of each successive stage can be controlled by pressure sensors located at the discharge of the preceding stage or alternatively to a single flowmeter connected to the pipeline beyond the discharge of the last-stage pump. When sufficient pressure is reached (depending on start-up slurry density) or the flowmeter records a flow sufficiently on the pump curve, the next stage is started.

During Operation

Probably the most hazardous event that can occur in centrifugal pumping is the blockage of the pump suction or discharge. This leads to heating of the liquid trapped in the volute with subsequent vaporization and explosive pressure buildup. This hazard is more acute in the case of slurry pumping where line blockages can readily occur and the inadvertent closing of either suction or discharge isolation valves can produce dire consequences.

At least two methods of protection are possible. One is to use a current-limiting switch to shut off the pump when a low current corresponding to pump shutoff flow is detected. The other is to provide a rupture disk to relieve pressure in the pump casing. In the case of gland water pumps, this could be placed in the seal water line as close to the pump as possible.

The extent and degree of automatic operation varies considerably, depending on the location and critical nature of pump station operation, as well as the availability of operating personnel. For simple in-plant pumps, seal-water flow alarms and controls may be interlocked into pump start-up and shutdown controls. High sump-level alarms may be used to detect insufficient discharge flow. Low-level sump alarms and detectors may be used to warn of insufficient pump NPSH or submergence, or even to shut down the pump. Automatic pump shutdown may be provided in the case of line rupture, where a considerable decrease in pump resistance results in over-pumping and subsequent motor overloading—similar to what may happen in pump start-up. Large, unattended pumps may incorporate alarms for excess vibration and high bearing temperature.

Shutdown

Emergency or planned shutdowns should be considered carefully to avoid major equipment damage. Most slurry systems are shut down and started up with water. It should be remembered that a centrifugal pump produces head or pressure, while a pipeline system relies on pressure for flow to occur. The pressure drop required to move a pipeline full of slurry is high, and when water only is introduced into the sump, pump discharge pressure will decrease, with an attendant decrease in line flow rate. This may be compensated for by an increase in pump speed if practical. The greater the slurry specific gravity and friction loss, the more closely the changeover from slurry to water must be evaluated, as transient flow instabilities are possible.

For long slurry pipelines, safe draining of the lines must be considered. High spots in the line should be opened to prevent line collapse under vacuum. During emergency shutdowns of such lines (e.g., due to power failures), the accelerating mass of slurry can cause high-pressure shock waves that can burst pipelines and equipment. Sometimes protection can be afforded by installing slow-acting shutoff valves, pulsation dampers, or similar devices. Restarting during reverse flow through pumps can cause equipment damage. During planned shutdown of a multistage pump system, pumps are shut down in reverse sequence to start-up.

POSITIVE DISPLACEMENT PUMPS

The most commonly used types of PD pumps for slurry handling are diaphragm pumps, progressing cavity pumps, and reciprocating piston or plunger pumps, all of which have been described previously. Figure 70 shows an air-operated diaphragm pump used for pumping high-density slurries containing up to 75% solids at a maximum of 90°C. The black spheres marked "A" and "B" in the figure are ball check valves that control flow through the pump during operation.

Progressing cavity pumps are widely used in handling abrasive slurries. Such a pump should be selected to maximize the life of its wear parts. In many cases, a small pump at high speed could handle the pressure and volumes required, but a larger pump operating at a lower speed will last longer and thus be more economical. Progressing cavity pumps have been used for pumping thickener underflow; slurry from dewatering applications; slurries of copper sponge, uranium yellow cake, and molybdenum; pilot-plant slurries; sewage; cement for grouting, at high pressure; and caulking compounds; and pressure spraying of oil on coal.

Reciprocating pumps are used with slurries in which the solid phase has a relatively low susceptibility to degradation during flow, such as with coal slurries. Ordinarily, the piston runs in a renewable metal cylinder or liner made of abrasion-resistant material. Piston rods and plungers are coated to resist wear. Valves and seats on reciprocating slurry pumps are designed to reduce erosion of the valve seat and components. Valve seats are usually replaceable and constructed of an elastomer or plastic material.

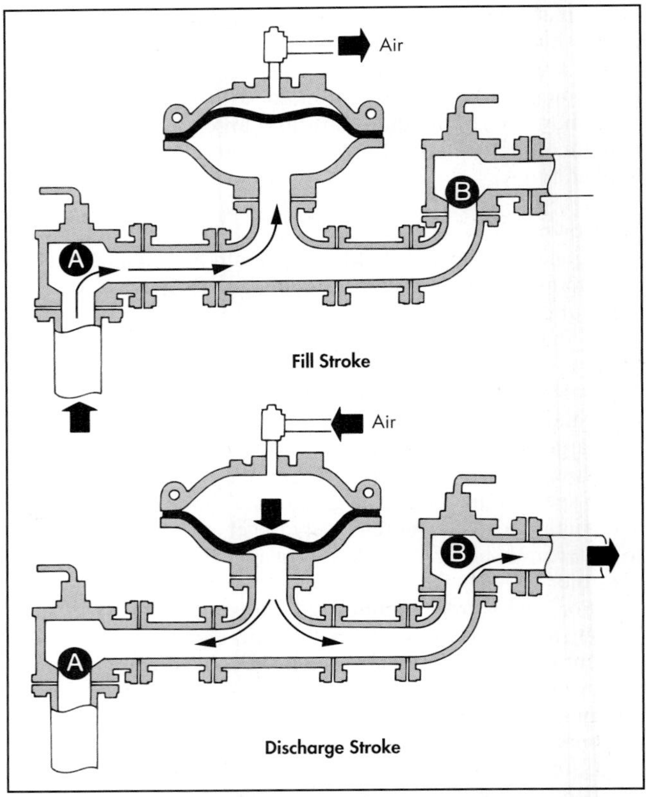

Courtesy of Dorr-Oliver, as cited in Scott and Hays 1985

Figure 70 Air-operated diaphragm pump

Controls for Operation and Safety

Large PD pumps inherently introduce system pressure pulsations as the result of suction and discharge displacements of specific volumes of liquid. In a duplex pump, two pulses are created; and in a double-acting duplex pump, four pulses are created. When multiple pumps are operated in a station, pump speeds must be staggered to keep the pulses out of phase and prevent a sympathetic cycle. Pulsations and vibrations must be accounted for in piping design and the location of instrument sensors, transmitters, and controllers.

PD pumps are constant-volume pumps. A change in volumetric flow rate is affected strictly by an adjustment of pump speed. Speed is normally adjusted using variable-speed motors, or by incorporating fluid or magnetic couplings units.

Sensing devices, gauges, and transmitters should not be located on piping immediately around PD pumps, as they will be subject to vibration failures. Such devices should employ a sealed capillary pressure-sensing system in which a diaphragm separates the slurry from the instrument and an air-free secondary hydraulic fluid transmits slurry variations from the sensor side of the diaphragm.

EXAMPLE CALCULATIONS

This section includes examples of typical calculations made in designing pump systems. Software packages for the design of pump systems include such calculations, of course, but it is still worthwhile for mineral processing engineers to be familiar with underlying concepts and calculations. The notation and units of measure in these examples are not entirely consistent with the preceding text but are used nonetheless to illustrate some of the variety of practice in describing pump systems.

Example 1. Calculating Pump Total Head by Two Methods

A centrifugal pump delivers 227 m^3/h of liquid having specific gravity 0.8 from an open suction tank to a discharge tank,

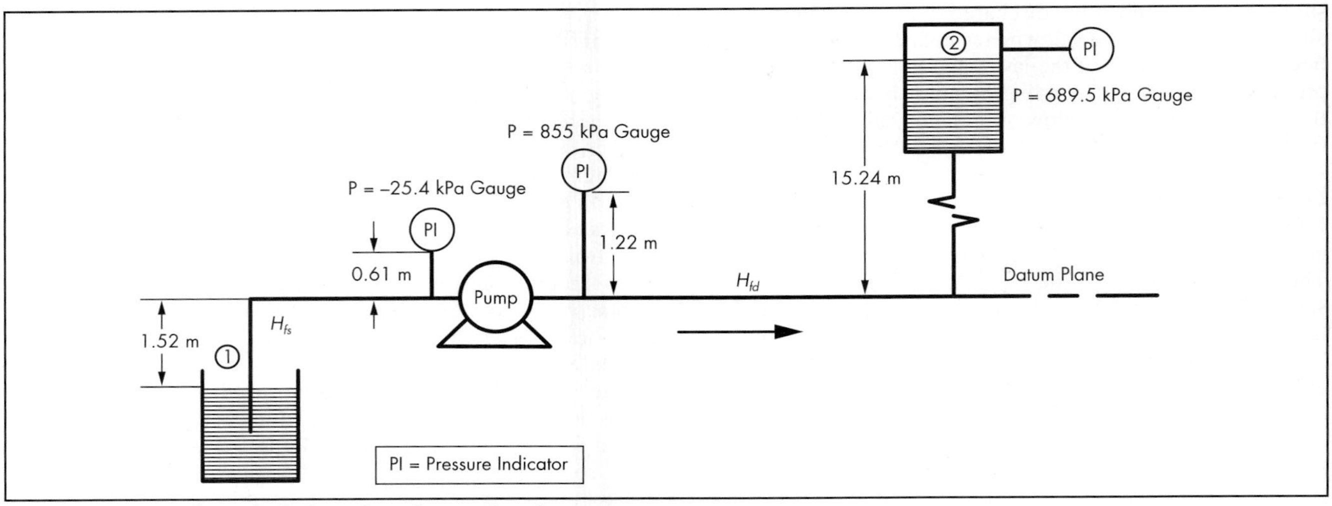

Figure 71 System for calculation of total pump head

through the piping as shown in Figure 71. The suction pipe and discharge pipe have inside diameters of 203 and 152 mm, respectively. The pressure friction heads for the pipe, valves, and fittings are 0.91 m for the suction side, H_{fs}, and 7.62 m for the discharge side H_{fd}.

Figure 71 shows four points at which pressures may be used for calculation of total pump head. The first two points are designated with circled numbers 1 and 2, and are at the suction and discharge tank, respectively. The second two points are at the two pressure gauges installed on either side of the pump.

The pump total head is first calculated using the gauge readings and datum plane shown. The total head, TH, in a pump system is the difference of the discharge head H_d and the suction head H_s, minus the total head loss due to friction ΣH_f, thus

$$TH = (H_d - H_s)$$

which may be expanded as

$$TH = [(V_d^2/2g + (p_d/\rho) + Z_d] - [(V_s^2/2g + (p_s/\rho + Z_s]$$

where

$V_{d,s}/2g$ = velocity head, m
$V_{d,s}$ = flow velocity of the discharge or the suction, m/s
g = acceleration due to gravity, m/s^2
$p_{d,s}/\rho$ = liquid pressure head in the discharge or the suction, m (positive or negative)
$p_{d,s}$ = gauge pressure of the discharge or the suction, N/m^2 (positive or negative)
ρ = specific gravity of liquid
$Z_{d,s}$ = elevation above or below the datum plane of the discharge or the suction, m (positive or negative)

In using this expression to calculate total head, ensure that all pressure are either *absolute* or *gauge* values.

Pipe velocity in m/s is conveniently calculated as

$$V = (m^3/h)(3.54)/(\text{pipe inside diameter, in cm})^2$$

where the required unit conversions are include in the factor 3.54, thus,

$$V_s = 227 \times 3.54/20.32 = 1.95 \text{ m/s, and}$$

$$V_d = 227 \times 3.54/15.22 = 3.48 \text{ m/s}$$

Similarly, liquid pressure head may be calculated as

$$p/\rho = 0.102 \times \text{pressure in kPa}/\rho \text{ m, thus}$$

$$p_d/\rho = (0.102)(855)/0.08 = 109.2 \text{ m, and}$$

$$p_s/\rho = (0.102)(-25.4)/0.08 = -3.2 \text{ m}$$

Finally, from Figure 71,

$$Z_d = 1.22 \text{ m}$$

$$Z_s = 0.61 \text{ m}$$

The total head is then calculated as

$$\begin{aligned} TH &= [(V_d^2/2g + (p_d/\rho) + Z_d] - [(V_s^2/2g + (p_s/\rho) + Z_s] \\ &= (0.62 + 109.01 + 1.22) - (18.65 + 3.24 + 0.61) \\ &= 113.29 \text{ m} \end{aligned}$$

The total head may also be calculated using the pressures at points 1 and 2 and the same datum plane using the following equation:

$$TH = [(V_d^2/2g + (p_d/\rho) + Z_d] - [(V_s^2/2g + (p_s/\rho + Z_s] + \Sigma H_f$$

where ΣH_f is the sum of pressure losses due to friction, m, and all other terms are defined as in the previous calculation.

Note that point 1 is at the surface of the *open* suction tank, so the gauge pressure and the liquid pressure at that point will be zero. Further note that because neither point 1 nor point 2 is in the system that is pressurized by the pump, the suction and discharge flow pressures are not included.

$$\begin{aligned} TH &= [0 + (p_d/\rho) + Z_d] - [0 + 0 - Z_s] + [0.91 \times 7.62] \\ &= [(0.102 \times 689.5)/0.8 + 15.24] - (-1.52) + 8.53 \\ &= 113.20 \text{ m} \end{aligned}$$

The slight discrepancy between the two results (113.29 m and 113.20 m) results from rounding errors.

Example 2. Calculating Head Loss due to Friction

Use the Darcy–Weisbach formula to calculate the frictional head loss for 30 m of 54-cm, smooth concrete pipe. The pipe has an inside diameter of 48 cm. The system operates at 60°C, with water flowing at 2,500 m^3/h.

Recall that the Darcy–Weisbach formula calculates friction loss H_f as

$$H_f = fLV^{2/}2gD$$

where the Darcy friction factor f is determined experimentally or from a chart, the pipe length L and the inside pipe diameter D are in meters, the liquid velocity V is in m/s, and the acceleration due to gravity g is in m/s^2.

The Darcy friction factor may be determined graphically from the Moody diagram (Figure 2) and reproduced in Figure 72. Figure 72 shows that smooth concrete has a roughness ε of 0.025 mm, so that the relative pipe roughness, ε/d (or ε/D) is equal to 0.025 mm/491 mm = 5.09E-05. (Note that ε and D must be in the same units, so that the relative pipe roughness is dimensionless.)

The liquid velocity is found from the volume flow rate and the pipe inside area, thus,

$$V = (2{,}500 \text{ m}^3/\text{h})/[\pi\,(0.48^2)/4] = 13{,}816 \text{ m/h} = 3.84 \text{ m/s}$$

where the denominator in the equation is the inside area of the pipe, in square meters.

It is also necessary to calculate the Reynolds number, N_{Re}, which is defined as

$$N_{Re} = \gamma D v/\mu$$

where γ is the fluid density, D is the inside pipe diameter, v is the fluid velocity, and μ is the dynamic viscosity.

Fluid density and dynamic viscosity of water vary with temperature. In this case, at 60°C, γ = 983.4 kg/m^3 and μ = 4.66E-04 Pa·s. Thus,

$$N_{Re} = \rho D v/\mu = [(998.4)(0.491)(3.83)/4.66\text{E-}04] = 4.03\text{E+}06 \text{ (dimensionless)}$$

The value of the friction factor f is found in Figure 72 by first locating the relative pipe roughness on the right-hand axis of the figure, as shown by the black circle. This defines

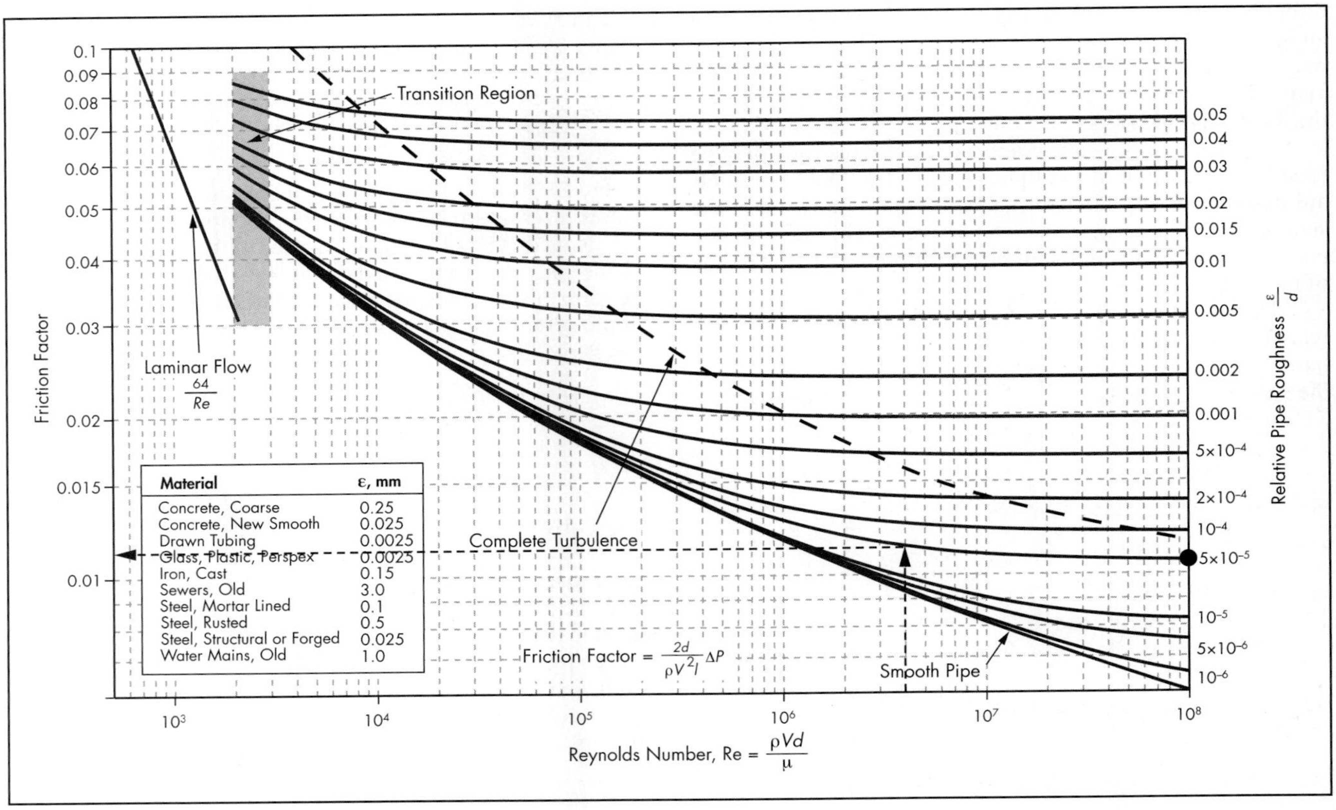

Adapted from Beck and Collins 2008

Figure 72 Moody diagram for Example 2

the curve that will be used. In this example, the relative pipe roughness is almost exactly on the 5E-5 curve, so that curve will be used. If the relative pipe roughness is between two curves, a curve may be lightly sketched on the diagram, maintaining proportional distances from the two bounding curves.

A dashed line is then extended vertically from the point on the horizontal axis corresponding to the calculated Reynolds number to intersect the indicated pipe roughness curve. From that intersection, a dashed line is extended to the left, intersecting the left-hand axis to indicate the value of f for these conditions—in this case, 0.0107.

The friction loss for the 30-m length of pipe may then be calculated from the Darcy–Weisbach equation:

$$H_f = fLV^2/2gD = (0.0107)(30)(3.84^2)/(2)(9.807)(0.491) = 0.482 \text{ m}$$

Example 3. Calculating NPSHA

A tailings pump operates in a mill at sea level. The pump draws from an open sump and moves a fine slurry that has a specific gravity of 1.3. The pump centerline is 2 m above the controlled water level in the sump. Assume a temperature of 15°C and an absolute pressure at the liquid surface of 101.325 kPa (1 atm). The friction head, H_f, in the pump inlet line is 0.8 m.

To calculate the available net positive suction head, NPSHA, at the pump inlet, recall that

$$\text{NPSHA} = (H_{atm} \pm H_s - H_f) - H_{vp}$$

where

H_{atm} = atmospheric pressure
H_s = static height of liquid above or below the pump inlet or the center datum line
H_f = friction head in the pump inlet line
H_{vp} = vapor pressure of the liquid at the operating temperature and atmospheric pressure

NPSHA is expressed in height of an equivalent column of the liquid or slurry being pumped. Two components of the NPSHA, H_s and H_f, are given, but H_{atm} and H_{vp} must be calculated. The atmospheric pressure is converted to meters of water and divided by the specific gravity of the slurry to give meters of slurry, thus,

$$H_{atm} = 1 \text{ atm} \approx 10.34 \text{ m } H_2O \text{ (given) at } 15°C = 10.342/1.3 = 7.96 \text{ m slurry at } 15°C$$

The vapor pressure of water is determined from standard tables at the given temperature, and converted:

$$H_{vp} = 0.0168 \text{ atm} = 0.174 \text{ m } H_2O \text{ at } 15°C = 0.174/1.3 = 0.134 \text{ m slurry at } 15°C$$

Available net positive suction head is then calculated as

$$\text{NPSHA} = 7.96 - 2.0 - 0.134 - 0.8 = 5.03 \text{ m slurry}$$

The values of H_{atm} and H_{vp} will of course vary with climatic conditions, so NPSHA should always exceed NPSHR by an adequate margin.

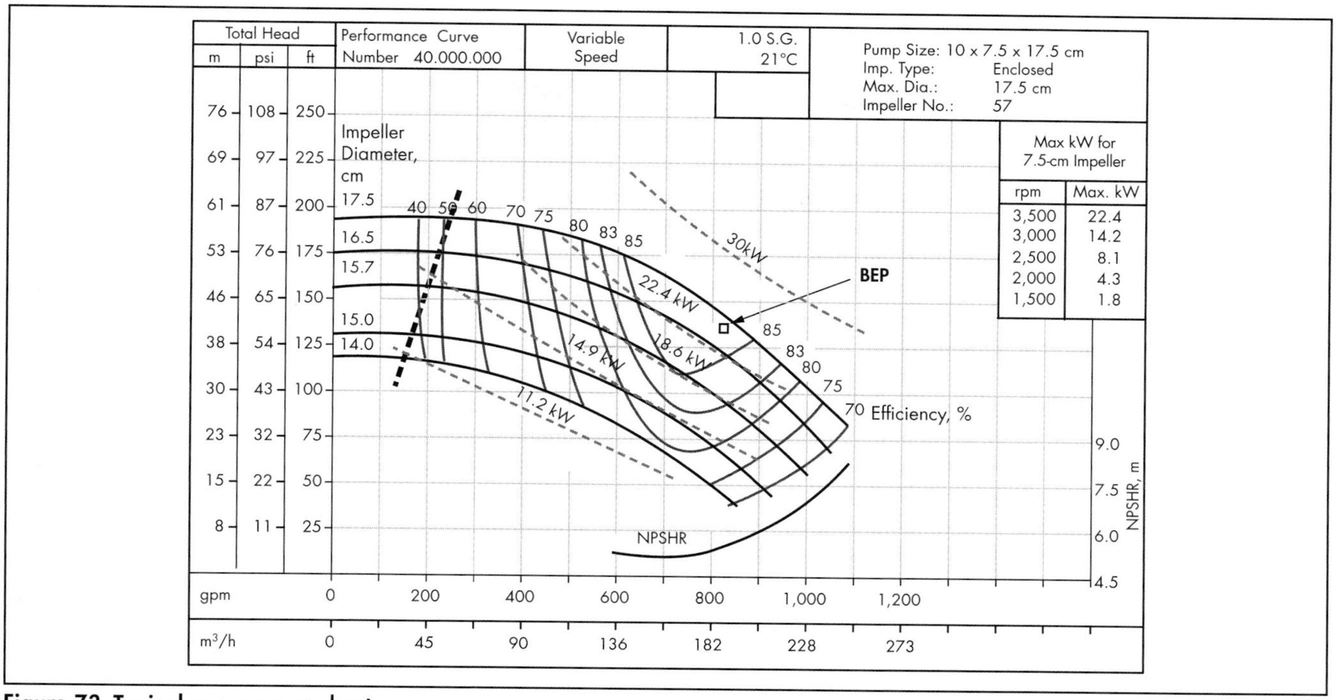

Figure 73 Typical pump curve chart

Example 4. Reading Pump Charts

A pump chart is a graphical representation of the performance characteristics of a pump. A basic curve shows the pump's output pressure or head versus flow rate on an x-y graph. The pump chart usually includes several nearly parallel curves that show the pump's performance for different impeller sizes, different rotational speeds, or some other variable. The chart may also show other operating parameters associated with pump performance, including rotational speed, impeller size, power, efficiency, and NPSHR.

In composite pump charts, all the parameters are shown on the same axes, as in Figure 73. Head versus flow rate is shown for several conditions—usually rotational speed or impeller diameter—by the group of similar heavy, solid curves that trend downward to the right. In Figure 73, these pump performance curves indicate the head versus flow for different impeller diameters 14–17.5 cm. Efficiency is shown by the lighter, solid lines that intersect the performance curves and are labeled at both their ends with numbers from 40 to 85, indicating the percent efficiency along each curve. In many pump curves, the BEP is shown, usually in the trough of the highest efficiency curve. In Figure 73, the BEP is indicated by a small square.

Pump power draw is shown by the dashed, nearly straight curves that are labelled from 11.2 to 30 kW. NPSHR is shown by the single, solid line at the bottom right of the performance curves, and a separate scale is provided for this curve. In Figure 73, the axes for flow and pressure head are labelled in metric and U.S. units. This is common in many pump curves.

A pump chart may also show one or more of the secondary parameters on separate sets of axes below the main graph. In these secondary graphs, the abscissas of the secondary graphs are the same as and aligned with that of the primary graph, indicating flow rate, so that the values of all parameters may be read as functions of a given flow rate.

A pump chart is used by plotting the system-head curve for a pumping system on the pump curve. In doing this, the fixed system head is usually neglected, since it affects both the pump performance curve and the system-head curve equally. The variable system total head loss through all piping, valves, fittings, and equipment in the system is calculated or measured for several flow rates, and plotted directly on the pump curve, as shown in Figure 73 by the heavy, dashed line on the left that extends upward and to the right from the origin. The points at which the pump will operate in the system are shown by the intersections of the system-head curve with the pump performance curves.

Considering the performance curve for the 17.5-cm impeller, it is seen that at zero flow, also known as shutoff or dead-head, the pump will develop about 58 m of head. This is often referred to as the shutoff head or the dead head and is the head the pump would develop operating against a closed valve.

The pressure measured between the pump and the closed valve might exceed this shutoff head value, because a pump adds head to the liquid being pumped. If this pump were operating at shutoff with suction pressure of 6 m, the total head seen at the pump discharge would be 58 + 6 = 64 m.

The location of the BEP relative to the operating condition for the pump is informative. Pumps run best at or near the BEP, and the preferred operating region (POR) is usually considered to be that in which flow is from 70% to 120% of the flow at BEP for most centrifugal pumps. In Figure 73, the BEP is at approximately 178 m^3/h flow and 63 m head, so the preferred operating range is from about 125 to 215 m^3/h. The intersection of the system-head curve with the performance curve for the 17.5-cm impeller shows that for the system in

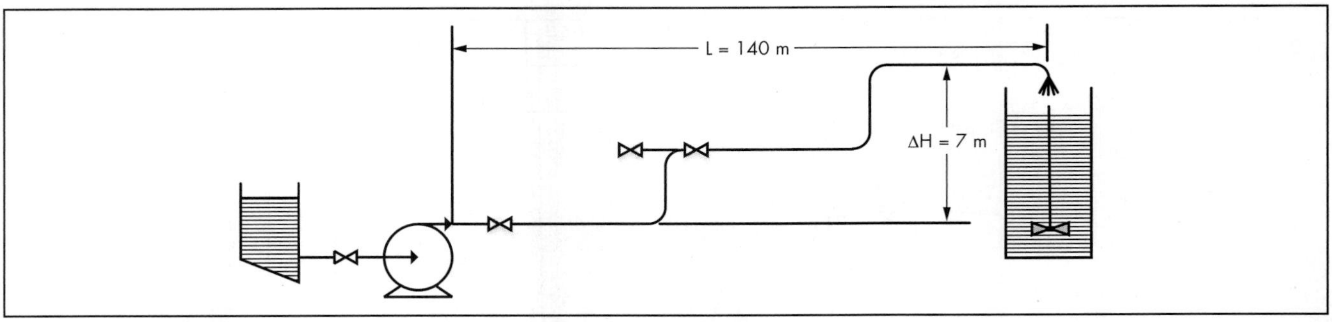

Figure 74 Tailings pumping system

question, the pump will deliver 182 m^3/h, and will thus be operating in the POR.

Many manufacturers also show a continuous stable flow line on their pump charts. The nearly vertical, dotted line that intersects the performance curves on the left side of the chart indicates the conditions at which the manufacturer has determined the pump should not be allowed to operate for any extended time.

Example 5. Calculating Critical Velocity and Pump Head for a Slurry Pumping System

For the tailings system shown in Figure 74 and described in Table 4, determine the critical flow velocity and the pump discharge pressure. The pipe size is specified on a trial basis and may change after the initial evaluation of the system.

Begin by determining the settling velocity for the particles in the slurry. This is the critical flow velocity V_L that must be maintained to keep the slurry in suspension while it is flowing. The settling velocity may be estimated using Durand's method. Assume that the slurry is closely graded, and that the d_{50} size is 270 mesh (53 mm). Referring to Figure 55 earlier in the chapter, extrapolate above the 15% C_v line to estimate the limited settling velocity factor F_L as 1.3.

The critical velocity is then calculated as

$$\begin{aligned} V_L &= F_L\sqrt{2gD(\rho - 1)} \\ &= 1.3\sqrt{2 \times 9.807 \times 0.22(2.7 - 1)} \\ &= 1.22 \text{ m/s} \end{aligned}$$

This estimate indicates that the specified flow velocity, 1.58 m/s, may be too close to the critical velocity. The margin could be increased by using a smaller pipe or a higher-volume flow rate. However, because the flow correction factor was estimated based on an assumed d_{50} particle size in the slurry, it will also be worthwhile to repeat the calculation using a measured d_{50} value. If d_{50} is 40 mm, F_L will be approximately 0.6, and V_L will be 1.04 m/s, in which case the pipe size will be adequate for the specified flow rate.

The calculation of the required pump discharge pressure will be based on an assumption that the specified flow velocity, 1.58 m/s, is adequate. The total discharge pressure includes the pressure to overcome friction and the static head resulting from the increase in elevation from the pump outlet to the pipeline discharge.

To calculate the pressure required to overcome friction losses, H_f, use the the Darcy–Weisbach formula and follow the method used in Example 2. In this case, the Reynolds number is

Table 4 Conditions for a slurry pumping system

Condition	Units
Slurry type	–200 mesh copper tailings
Solids content	60% by weight
Solids specific gravity	2.7
Slurry volume concentration	0.356
Slurry specific gravity	1.61
Slurry density	1,600 kg/m^3
Slurry coefficient of rigidity	0.06 Pa·s
Flow	3.60 m^3/min
Pipe inside diameter, trial basis	0.22 m
Pipe material	Steel, commercial or structural grade
Flow velocity	1.58 m/s
Length (including fittings)	140 m
Inlet elevation	225 m
Discharge elevation	232 m (to atmosphere)

Adapted from Grzina et al. 2002

$$N_{Re} = \rho Dv/\mu = [(1{,}600)(0.22)(1.58)/0.06] = 9.26\text{E+03}$$

Figure 72 shows that commercial steel pipe has a roughness ε of 0.025 mm, so that the relative pipe roughness, ε/d (or ε/D) is equal to 0.025 mm/220 mm = 1.13E–04. The friction factor f is again determined from Figure 72 as 0.036.

$$\begin{aligned} H_f &= fLV^2/2gD = (0.036)(1)(1.58^2)/(2)(9.807)(0.22) \\ &= 0.02 \text{ m } H_2O\text{/m pipe} \end{aligned}$$

The total required discharge pressure is given by

$$H_f = (0.20 \text{ m } H_2O\text{/m})\ (140 \text{ m}) = 2.75 \text{ m } H_2O$$

plus

$$P_{static} = 7.00 \text{ m } H_2O$$

Thus, the total required pump head is

$$P_{total} = 9.75 \text{ m } H_2O$$

However, the pump is moving slurry with a specific gravity of 1.6, so the pump must provide

$$1.6 \times 9.75 = 15.61 \text{ m slurry at } \rho = 1.6$$

Example 6. Ball Mill Discharge Pump and Cyclone Circuit

In the system shown in Figure 75, the discharge from a ball mill reports to a sump, from which it is pumped to a cyclone

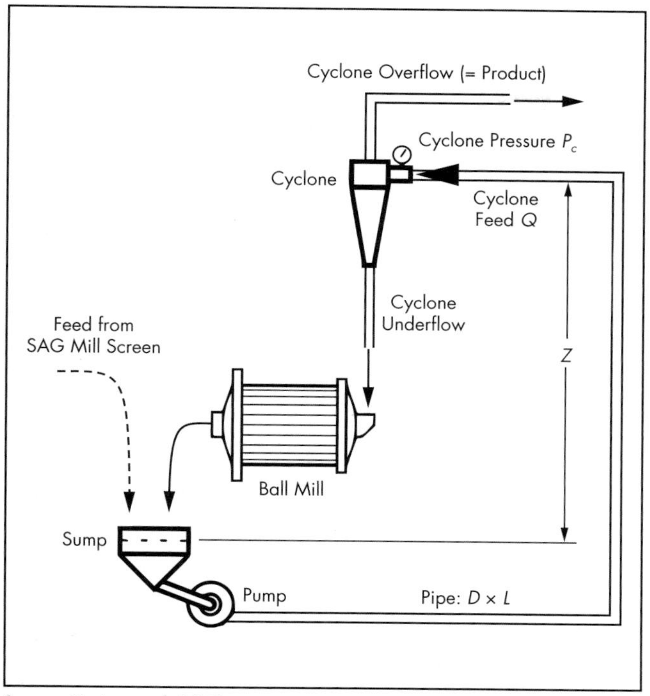

Source: Grzina et al. 2002

Figure 75 Ball mill, pump, and cyclone circuit

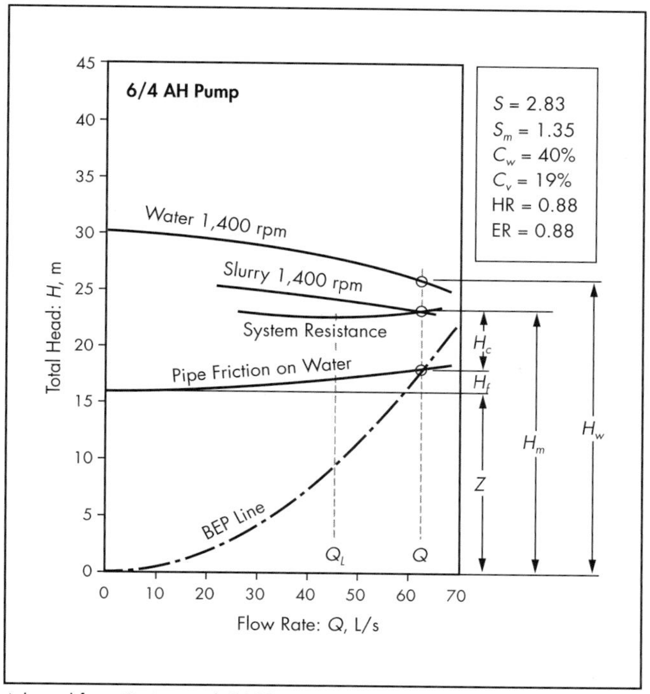

Adapted from Grzina et al. 2002

Figure 76 Mill discharge, pump, and cyclone performance curves

above the mill, for separation at a given particle size. The cyclone overflow, containing the fine particles, moves to the flotation circuit, while the coarser cyclone underflow returns to the mill for additional grinding. The height of the cyclone inlet above the mean water level in the pump sump Z is 16 m, the internal pipe diameter D is 0.150 m, and the total equivalent length of pipe L is 30 m.

The system operates in a concentrator treating copper ore that has a specific gravity of 2.85. The flow of slurry fed to the cyclone Q is 61.7 L/s, the solids concentration in the slurry C_w is 40% (C_v = 19%), and the specific gravity of the slurry is 1.35. The particle distribution of the cyclone feed is d_{20} = 60 µm, d_{50} = 250 µm, and d_{80} = 750 µm. Thus, d_{80}/d_{20} = 12.5.

The cyclone requires an inlet pressure, P_c, of 65 kPa at 61.7 L/s. The pump selected for the duty is a 6/4 AH Warman slurry pump with a hard-metal impeller of outer diameter D_i of 0.365 m. Its performance curves for water and slurry together with the system resistance curves are shown in Figure 76.

A flow of 61.7 L/s in a steel pipe of 0.150 m will have a pipeline velocity of 3.8 m/s. Because d_{80} is more than five times greater than d_{20}. Durand's limiting factor for pipe velocity, F_L, can be estimated from Figure 74 as 1.1 for conditions given. The limiting settling velocity is

$$V_L = F_L\sqrt{2gD(S_S - 1)} = 2.3 \text{ m/s}$$

This is less than the flow velocity, so there will be no settling in the pipe.

To determine the system resistance curve for the pipeline. Using Figure 2, with a relative pipe wall roughness e/D of 2.8×10^{-4}, the Darcy friction factor f is 0.016 and the friction head loss H_f=1.98 m of slurry. The cyclone pressure of 65 kPa and the corresponding cyclone head, H_c, are

$$H_c = P/\gamma g = 65{,}000/(1{,}350 \times 9.81) = 4.91 \text{ m slurry}$$

The units for the density γ are kg/m^3.

The static head Z is 16 m, so the total head required from the pump is

$$H_m = Z + H_f + H_c = 16 + 1.98 + 4.91 = 22.9 \text{ m slurry}$$

The head and efficiency ratios for the pump are estimated from Figure 58. The particle-size to impeller-diameter ratio, d_{50}/D_i, is 0.0007, so the estimated head ratio and efficiency ratio are each 0.88.

Figure 76 shows a portion of the performance curves for the Warman 6/4 AH pump when pumping water. The system resistance and cyclone head when pumping slurry are also shown.

The pump total head when pumping water at the required Q is

$$H_w = H_m/\text{HR} = 22.9/0.88 = 26.0 \text{ m water}$$

Figure 76 shows that the pump speed required to deliver 61.7 L/s of water against this head is 1,400 rpm.

The pump efficiency is 69% when pumping water. When pumping slurry, it is reduced by the efficiency factor to

$$69 \times 0.88, \text{ or } 60.7\%$$

The power input, P_i, required by the pump is given by

$$P_i = QH_mS_{SL}/\eta_m = (61.7)(22.9)(1.35)/(1.02)(60.7) = 30.8 \text{ kW}$$

where η_m is the efficiency when pumping slurry.

A 37-kW motor would give a 20% contingency margin.

REFERENCES

ASTM A532/A532M-10. 2014. *Standard Specification for Abrasion-Resistant Cast Irons.* West Conshohocken, PA: ASTM International.

Badr, H.M., and Ahmed, W.H. 2014. *Pumping Machinery Theory and Practice.* New York: John Wiley and Sons.

Beck, S., and Collins, R. 2008. Moody diagram. https://commons.wikimedia.org/wiki/File:Moody_EN.svg. Accessed July 2018.

Bootle, M.J. 2002. Selection and sizing of slurry pumps. In *Mineral Processing Plant Design, Practice, and Control.* Vol. 2. Edited by A.L. Mular, D.N. Halbe, and D.J. Barratt. Littleton, CO: SME.

Churchill, S.W. 1977. Friction factor equation spans all fluid-flow regimes. *Chem. Eng.* 84(24):91–92.

Crosby, T.C. 1985. Slurry transport: Controls for operation. In *SME Mineral Processing Handbook.* Edited by N.L. Weiss. Littleton, CO: SME-AIME. pp. 10-180–10-182.

Durand, R. 1952. *Hydraulic Transportation of Coal and Solid Material in Pipes.* London: Colloquium of the National Coal Board.

Eichhorn, M., Krumins, T., Zunti, L., and Ruff, F.C. 2014. Capacity enhancement at Newmont Mining Corporation's Twin Creeks whole ore pressure oxidation facility. In *Hydrometallurgy 2014, Volume 1.* Edited by E. Asselin, D.G. Dixon, F.M. Doyle, D.B. Dreisinger, M.I. Jeffrey, and M.S. Moats. Westmount, QC: Canadian Institute of Mining, Metallurgy and Petroleum.

Erickson, M., and Blois, M. 2002. Plant, design, layout, and economic considerations. In *Mineral Processing Plant Design, Practice, and Control.* Vol. 2. Edited by A.L. Mular, D.N. Halbe, and D.J. Barratt. Littleton, CO: SME.

Foust, A.S., Wenzel, L.A., Clump, C.W., Maus, L., and Andersen, L.B. 1964. *Principles of Unit Operations.* New York: John Wiley and Sons.

Grzina, A., Roudnev, A., and Burgess, K.E. 2002. *Slurry Pumping Manual*, 1st ed. Lahti, Finland: Warman International. www.pumpfundamentals.com/slurry/WeirSlurryPumpingHandbook.pdf.

Hydraulic Institute. 1975. *Hydraulic Institute Standards for Centrifugal, Rotary, and Reciprocating Pumps*, 13th ed. Cleveland, OH: Hydraulic Institute.

Hydraulic Institute. 2015. *Rotodynamic Pumps—A Guideline for Effects of Liquid Viscosity on Performance.* ANSI/HI 9.6.7-2015. Parsippany, NJ: Hydraulic Institute.

Karassik, I.J., Messina, J.P., Cooper, P., and Heald, C.C. 2001. *Pump Handbook*, 3rd ed. New York: McGraw-Hill.

Link, J.M., Wasp, E.J., and Horne, C.A. 1985. Hydraulics. In *SME Mineral Processing Handbook.* Edited by N.L. Weiss. Littleton, CO: SME-AIME. pp. 10-142–10-173.

O'Keefe, W. 1972. Pumps. *Power* (June):S1–S24.

Olson, R.M. 1998. Fluid mechanics. In *Mechanical Engineers' Handbook*, 2nd ed. Edited by M. Kutz. New York: John Wiley and Sons. p. 1289.

Rayner, R. 1995. *Pump Users Handbook*, 4th ed. Kidlington, Oxford: Elsevier.

Roudnev, A., and Angle, T. 1999. *Slurry Pump Manual.* Salt Lake City, UT: Envirotech Pumpsystems.

Scott, D.W., and Hays, R.M. 1985. Storage and Transport. In *SME Mineral Processing Handbook.* Edited by N.L. Weiss. Littleton, CO: SME-AIME.

Stepanoff, A.J. 1948. *Centrifugal and Axial Flow Pumps.* New York: John Wiley and Sons.

Tomb, T.F., Mundell, R.L., Taylor, C.D., Custer, J.L., and Martonik, J.F. 1985. Dry bulk material transport: Health and Safety. In *SME Mineral Processing Handbook.* Edited by N.L. Weiss. Littleton, CO: SME-AIME. pp. 10-133–10-136.

Walker, C., and Goulas, A. 1984. Performance characteristics of centrifugal pumps when handling non-Newtonian homogeneous slurries. *Proc. Inst. Mech. Eng.* 98a:(1):41–49.

Weir Minerals Netherlands b.v. 2009. *First Choice for Innovative Positive Displacement Slurry Pumps.* Netherlands: Weir Minerals Netherlands b.v. https://www.global.weir/assets/files/product%20brochures/brochure%20GEHO%20PUMPS-first%20choice-LR-2009.pdf.

Wells, P. 1991. *Pumping NonNewtonian Slurries.* Warman Group Development Technical Bulletin 14. Sydney, Australia: Warman International.

Wifley. 2014. Five things you might not know about the EMW slurry pump. https://www.wilfley.com/blog/five-things-you-might-not-know-about-the-emw-slurry-pump. Accessed July 2018.

Williams, G.S., and Hazen, A. 1920. *Hydraulic Tables: The Elements of Gagings and the Friction of Water Flowing in Pipes, Aqueducts, Sewers, etc., as Determined by the Hazen and Williams Formula and the Flow of Water over Sharp-Edged and Irregular Weirs, and the Quantity Discharged as Determined by Bazin's Formula and Experimental Investigations upon Large Models*, 3rd ed. New York: John Wiley and Sons.

Wilson, K.C. 1979. Deposition-limit nomogram for particles of various densities in pipeline flow. Presented at Sixth Annual Conference on the Hydraulic Transport of Solids in Pipes. University of Kent, Canterbury, UK, September 26–28.

PART 6

Physical Separations

CHAPTER 6.1

Mineral Sorting

Bern Klein and Andrew Bamber

Over the past 40 years, society has put increasing demand on industries to move toward cleaner and more efficient means of production. Industry has generally responded with improved processes and significant increases in efficiency and environmental performance. Over the same period, mining has become a point of focus for governments, nongovernmental organizations (NGOs), and communities as an industry out of tune with this trend. High-grade and easily accessible deposits have become largely depleted, and a large proportion of the remaining deposits are found in either remote locations or at extreme depths or embody other complications such as poor ground conditions, complex structure, adverse mineralogy, or presence of deleterious elements. The grade of new discoveries is declining severely, as is the mined grade at existing operations across most commodities. The general response to this situation is that most companies simply exploit economy of scale to dilute the fixed cost at operations, leading to the prevalence of ever-larger underground and open pit mines (Scoble 1994). Large-scale mining approaches, while improving economics by reducing the contribution of fixed costs and overhead, often incorporate high levels of unit inefficiency, particularly as extraction methods have inherently low levels of selectivity, leading to significant dilution as well as losses. Furthermore, beneficiation processes, particularly grinding, are singularly inefficient; hence, impacts on the efficiency of treating ores with lower grindability, as well as lower value, are particularly acute (Klein et al. 2015). Cost-effectively maximizing the amount of ore extracted while simultaneously reducing the waste content in ore delivered to downstream processes (whether simply transport and crushing or more sophisticated circuits, such as grinding and flotation or leaching) is therefore of key interest for all mining operations and projects struggling with low margins (Scoble et al. 2003; Bamber 2012).

Interest in the preconcentration of ore ahead of transportation and milling, as one way to achieve this, has historically been high. Preconcentration of mined material before it enters the mill reduces waste content and increases the value of ore, reducing overall processing cost (Scoble et al. 2003; Bamber et al. 2005, 2008; Engelbrecht 2012; Robben et al. 2013; Lessard et al. 2014). It may also assist in reducing the capacity required of all downstream processes, including transportation and milling, to deliver equivalent economics (Bamber et al. 2004; Bamber 2008). According to Salter and Wyatt (1991), sorting of ore also helps in providing a more uniform quality feed to the mill, which leads to further improvements in metallurgical recovery. Sensor-based sorting supports efforts to address pressures to transform the mining industry to become safer through mechanization, to improve overall energy efficiency by reducing transport requirements and diverting waste away from comminution, and to become more productive by maximizing resource utilization (Egerton 2004; Klein et al. 2011; Nadolski et al. 2015; Duffy et al. 2015).

There are various benefits to applying sorting in mining, and the best results are observed when it is introduced at very early stages in the mining cycle, potentially as early as the bench or face (Klein et al. 2002; Bamber et al. 2004; Bamber 2008). While several methods of preconcentration are available, including classification by size or density, sorting possibly offers the most potential of all because of its applicability to a wider range of mineralogies and its water- and reagent-free operation benefits.

Electronic sorting of minerals has been commercially available on a small scale since World War II. Early applications were in industrial minerals and more recently, with uranium, gold, diamonds, and massive base metal sulfides. Sensor-based sorting involves the sensing of the quality of granular materials such that the sensor response informs a decision to accept or reject the material. The technology, however, has not been adopted as widely as may be possible for several reasons. These include the traditionally low processing capacity of the technology, resulting in high unit capital and operating costs, limitations on the range of particle sizes that can be treated, as well as pervasive misconceptions about the principles and benefits of sorting among mine operators (Arvidson 2002; Manouchehri 2003; Bergmann 2009; Wotruba 2006; Wotruba and Harbeck 2010). Lack of awareness about what constitutes an opportunity for sorting

Bern Klein, Professor, Norman B. Keevil Institute of Mining Engineering, University of British Columbia, Vancouver, British Columbia, Canada
Andrew Bamber, Principal and Managing Director, Bara Consulting Ltd., London, United Kingdom

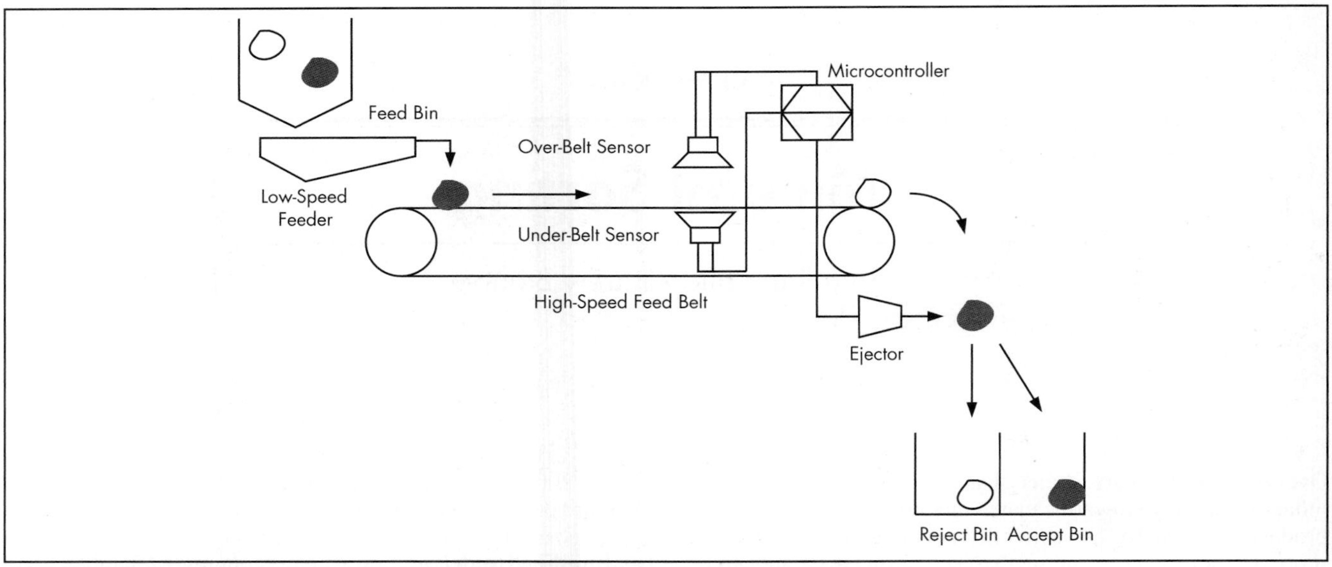

Figure 1 Typical particle sorting system showing sensors and pneumatic diverters

and ignorance of indicators of amenability to sorting has also been a barrier (Klein et al. 2015). However, recent developments in sensor-based sorting, including an increase in the range of applicable sensors, advances in speed, as well as new concepts in bulk- and semi-bulk sorting, have stimulated renewed interest in this unit operation within the industry (Kleiv 2012; Duffy et al. 2015; Bamber et al. 2016). These advances in capacity and range of application suggest that sorting, with its relatively low capital and operating cost, should be considered where good metallurgical response has been indicated, as either a possible final treatment step (in the case of bulk commodities) or a preconcentration step ahead of conventional grinding and flotation or leaching (in the case of base and precious metals).

This chapter seeks to present the current state of sensor-based sorting, the range of sensors, tools for the mineral processor to identify opportunities to sort, approaches to characterize materials for sorting, selection criteria pertaining to the unit operation, circuit designs, capital and operating cost considerations, as well as a range of example applications.

SORTING SYSTEMS

In sensor-based sorting, sensor information is acquired for a material. This information is used in an algorithm to classify the material, for example, according to its value or the concentration of a particular contaminant. Based on the estimated value and comparison of that value to a relevant threshold, the microprocessor informs a mechanical actuator to either accept or reject the material. Sorting can be applied to individual particles (rocks), bulk streams, or batches of material.

Particle Sorting

Sorting has application for a range of deposit types with grades varying from very low to high. For low-throughput high-value operations, particle (rock) sorting can add significant value to the operation by rejecting significant quantities of barren or low-grade waste ahead of more cost- and energy-intensive processes such as grinding (Bamber et al. 2008). In this case, the impact of sorting increases with decreasing grade, because at lower grades, such ore has a greater proportion of waste that can be rejected.

For particle sorting, the material must be prepared through a series of crushing and screening stages to prepare narrowly bounded size fractions, typically 3:1 top-size-to-bottom-size range. Often, washing via sprays or wet screening is needed to clean the particle surfaces for accurate sensing. The sorting systems require feeders that produce a monolayer of spaced particles so that each particle can be sensed and sorted individually.

The most common sorting system involves a belt conveyor, a sensing system to assess the quality of conveyed material, and ejectors consisting of valves delivering compressed air to eject detected waste particles, as shown in Figure 1. Alternative systems sense the rocks while free falling, informing either pneumatic or mechanical deflectors to reject the rock (Figure 2). Sensors used for particle sorting can either detect properties on the surface or assess bulk composition. For some mineralogies, the disposition of surface properties adequately represents the bulk, therefore surface measurement will suffice. This is not true, however, in all cases. Sensing accuracy can be improved by using multiple sensors in different orientations and/or combining responses from more than one type of sensor.

Bulk Sorting

The trend in the mining industry is to mine larger and lower-grade deposits to take advantage of economies of scale. While the throughput of current particle sorters at less than a few hundred metric tons per hour generally prohibits their deployment in such applications, bulk sorting solutions overcome throughput limitations by incorporating sensor systems into the material handling equipment of the mine. While less selective, and delivering generally lower-yield results than rock sorting, bulk sorting has very low cost intensity, and where heterogeneity in the ore is present at a relevant length scale, it can be very effective at delivering increased value. An increasingly wide range of sensors that integrate into conveying equipment, haulage equipment, and more recently into

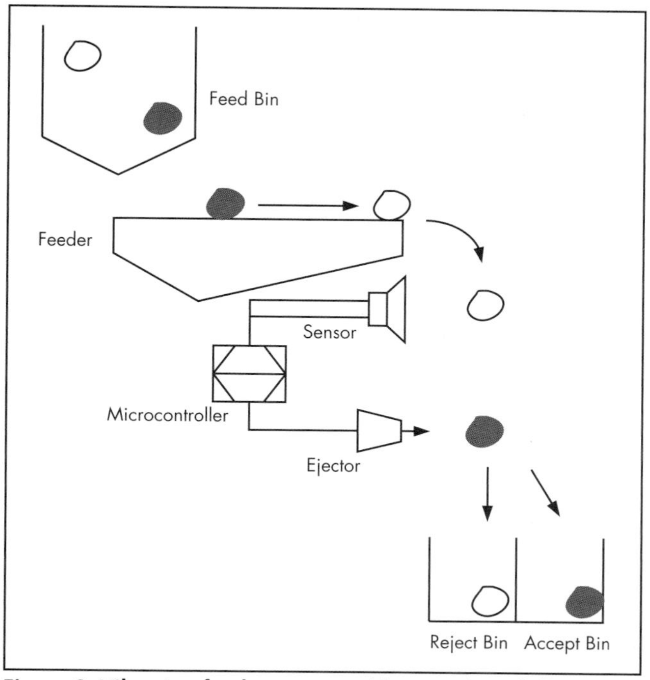

Figure 2 Vibrating feeder system with sensors on falling material and mechanical diversion

Source: Bamber et al. 2016

Figure 3 ShovelSense shovel-based bulk sorting solution at a Cu-Au mine

loading assets (e.g., mining shovels) is now available to support bulk sorting options where these did not previously exist.

Bulk separation has not been widely contemplated as a method, except in specific situations. For example, LKAB (a Swedish mining company) uses laser-induced fluorescence (LIF) for bulk sorting (Kruukka and Briocher 2002). Recent work, however, indicates that a wide potential for the application exists, therefore bulk sorting, whether on shovel (i.e., bulk-batch sorting) or on belt (bulk or semi-bulk sorting) has a key role to play in ore upgrading, particularly at large-scale open pit and underground mines (Murphy et al. 2012; Duffy et al. 2015; Bamber et al. 2016). Real-time bulk sensing systems can be deployed in several ways: Sensing can simply be used to help mines better execute grade control, or it can provide feed-forward process control information to the flotation or leach plant. Integrated with material routing or other diversion systems, material can be physically upgraded, either by rejecting waste or, more importantly, by recovering high-grade ore that would otherwise have been left in situ or sent to the waste dump. Sensors integrated with loading assets, such as shovels, scoops, backhoes, and loaders, sense the material during loading and can support sorting decisions at the bucket or dipper (i.e., 10–50-t scale; Figure 3).

Source: Nield 2002

Figure 4 Overhead scintillometers (left) used for scanning U_3O_8-bearing trucks (right) at Rossing uranium mine in Namibia

Figure 5 Bulk diversion of nickel laterites via X-ray fluorescence analysis to product stockpile (right) and waste stockpile (left)

Sensors integrated with hauling assets, such as trucks, can detect material in transit and support sorting decisions at the truck, for example, on a 300-t scale (Nield 2002; Figure 4). Sensing system deployment for sorting on shovels and trucks can improve the precision, accuracy, and resolution in the routing of material to low-grade stockpiles, leaching heaps, concentrators, or waste dumps.

Sensors installed on belt conveyors with integrated diverters can sense the conveyed material, supporting sorting decisions at a higher resolution than the truck or shovel scales, ranging from metric tons to hundreds of kilograms (Figure 5; Duffy et al. 2015). Probably the best cited example of this type

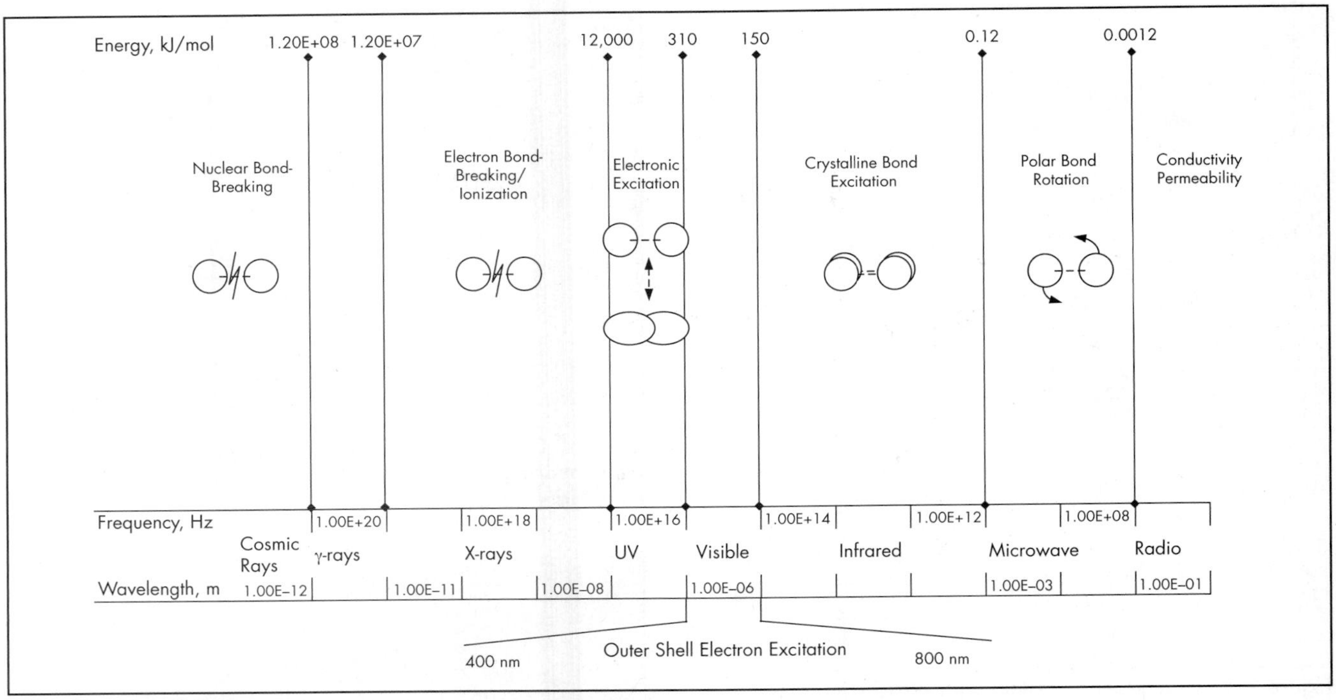

Adapted from Harris 1987

Figure 6 Theoretical potential for analysis using different bands of the electromagnetic spectrum

of sorting is the application of LIF in the bulk sorting of high-phosphorous Fe from low-phosphorous Fe at LKAB iron ore mine in Kiruna, Sweden (Kruukka and Briocher 2002).

MINERAL SENSING TECHNOLOGIES

Most sensors belong to a general class of electromagnetic source–detector combinations, where some form of stimulation using a particular band of the electromagnetic spectrum is used, and the response to that stimulation is recorded by the detector. A summary of the electromagnetic spectrum and its level of interaction with physical materials is shown in Figure 6.

The most common sensor techniques currently used in the industry are optical, conductivity, medium-wave infrared, and X-ray (X-ray transmission [XRT] or X-ray fluorescence [XRF]). Rapidly emerging sensing methods include prompt gamma neutron activation, laser-induced breakdown spectroscopy (LIBS), as well as multispectral reflectance imaging.

A list of sensors that are applicable to ore sorting is presented in Table 1. The sensors can be divided into two broad classes: those that measure properties volumetrically and those that measure surface characteristics only. In general, using a volumetric property to correlate to metal content provides a higher degree of confidence for sorting. For many ores, and particularly in bulk and semi-bulk situations, measuring only surface characteristics can accurately represent the volume characteristic. Sensors also classify according to response type. For example, analytical systems such as XRF or prompt gamma neutron activation analysis (PGNAA) can give a direct measurement of elemental composition, whereas electromagnetic or reflectance-based systems can give only a proxy for material composition. Furthermore, electromagnetic approaches require preexisting conductivity or magnetic susceptibility to be present, whereas other methods, such as PGNAA, have no such prerequisite. There is no general case or rule, however, and it is therefore highly recommended to test the suitability of sensors for each ore type to be sorted (Klein et al. 2002; Fickling 2011; Tong 2012; Altun et al. 2014).

Light-Based Techniques

Light-based techniques are divided into two main categories referred to as photometric and hyperspectral.

Photometric

The most widely known method in mineral sorting is optical sorting. There are many subclasses of photometric sorters, including those that sense visible or near-visible spectrum, minerals that naturally fluoresce under ultraviolet (UV) light, and hyperspectral sensing systems that detect over the visible to shortwave infrared wavelengths.

Visible and near-visible spectrum. Sensors are fundamentally cameras or charge-coupled device (CCD) sensors capturing reflectance phenomena from mineral particles, batches, or flows illuminated by visible wavelengths of light (Figure 7).

Interpretation can be simple, such as aggregate red-green-blue color intensity, or more sophisticated methods can be used, such as image analysis that characterizes texture. Alternate methods of illumination such as UV or near-infrared, coupled with very similar CCD detectors, give capability for characterization and, ultimately, sorting at those wavelengths instead.

Natural fluorescence. Many minerals, such as scheelite and wolframite, are naturally fluorescent under UV lighting conditions and can be sorted, based on responses in the fluorescent spectrum, from non-fluorescing species such as quartz or magnesium aluminosilicates.

Table 1 Sensor types for mineral sorting

Technology	Physical Property	Principle	Surface or Volume	Ore Types	Sorting Applications	Manufacturer*
Radiometric (scintillometer)	Natural gamma radiation level	Radioactivity	Volume	Uranium, Witwatersrand gold ores	Particle and bulk	TOMRA, Rados
Prompt gamma neutron activation analysis (PGNAA), pulsed fast thermal neutron activation (PFTNA)	Elemental composition	Neutron activation/ gamma energy emission	Volume	Iron ore	Bulk	Scantec, ThermoFisher, PANalytical
X-ray transmission (XRT)	Atomic density	Relative absorption of high energy X-rays	Volume	Base/precious metals, coal, diamonds, sulfides, etc.	Particle	Steinert, TOMRA
X-ray fluorescence (XRF) spectroscopy	Elemental composition	Inner shell electron excitation	Surface	Base/precious metals, metal sulfides	Particle and bulk	MineSense, Rados, Steinert, IMA
Laser-induced fluorescence (LIF)	Visible fluorescence under laser stimulation	High-energy photonic emission	Surface	Diamonds, limestone, iron ore, sulfides	Particle and bulk	IMA, AIS Sommer
Microwave-infrared (MW/IR)	Polar bond excitation	Microwave absorption, heat radiation	Volume	Base metals, carbonaceous materials	Particle	Not applicable
Laser-induced breakdown spectroscopy (LIBS)	Elemental composition	Electron excitation/ light emission	Surface	Base metal oxides, sulfides	Particle	Secopta, LDS, LSA
UV/X-ray luminescence (X-Ray/UV-L)	Photonic emission from outer shells	Luminescence through X-ray or ultraviolet (UV) stimulation	Surface	Diamonds	Particle	TOMRA, De Beers
Photometric	Red-green-blue (RGB) color, gray-scale, surface texture, size/shape	Chromatic reflectance/ absorption	Surface	Industrial minerals, gemstones, diamonds, coal, massive sulfides, phosphates	Particle	Steinert, TOMRA
Hyperspectral analysis	Molecular bonds	Reflectance/ absorption	Surface	Hydrated minerals	Particle	Steinert
Electromagnetic (EM)	Conductivity/ magnetic susceptibility	Electromagnetism/ induction	Volume	Base metal sulfides, native metals, massive oxides	Particle and bulk	TOMRA, Steinert, MineSense
Magnetic resonance spectroscopy (MRS)	Magnetic resonance	Resonant frequency of molecules	Volume	Chalcopyrite	Bulk	CSIRO

*AIS Sommer GmbH, Germany; CSIRO (Commonwealth Scientific and Industrial Research Organisation), Canberra, ACT, Australia; De Beers, Johannesburg, South Africa; IMA Engineering Ltd., Helsinki, Finland; LDS (Laser Distance Spectrometry), Petah Tikra, Israel; LSA Laser Analytical Systems and Automation GmbH, Aachen, Germany; MineSense Technologies Ltd., Vancouver, BC, Canada; PANalytical, Almelo, Netherlands; Rados International, London, UK; Scantec, Camden Park, Australia; Secopta, Berlin, Germany; Steinert Global, Walton, KY, USA; ThermoFisher Scientific, Waltham, MA, USA; TOMRA, Shelton, CT, USA.

Hyperspectral Imaging and Reflectance Spectroscopy

Hyperspectral imaging, also known as reflectance spectroscopy in the visible to shortwave infrared regions of light is a rapid, nondestructive remote sensing technique that requires little to no sample preparation. Imaging spectroscopy of geological materials operates on the premise of absorption of incident light in the visible to shortwave infrared region by minerals, whereby diagnostic absorption features are a function of electronic and vibrational processes specific to that mineral's crystal structure and chemistry. Light reflected by mineral targets is collected by a line-scanning imaging spectrometer, producing an *image cube* where each pixel contains a reflectance spectrum (Figure 8).

Spectrometers sample at upward of 400 frames per second with a spectral range from 400 nm to 2,500 nm and ~3 nm spectral resolution. Spectra are processed through proprietary high-speed digital signal processing, pattern recognition, and identification algorithms, providing information about the structural nature of certain detectable minerals, especially clays, carbonates, and OH-bearing minerals. One variation applies industrial microwave treatment of the rock to enhance the infrared response of metal-bearing conducting minerals, which enhances the ability to discriminate from nonconducting minerals (Van Weert et al. 2011).

X-Ray Techniques

Fluorescence

The composition of minerals can be determined by characteristic X-rays or *fluorescence*, generated when atoms are irradiated with low levels of X-ray energy. When X-ray energy strikes a mineral, the atom absorbs X-rays, and electrons are ejected from the inner shells, creating vacancies. These vacancies present an unstable condition for the atom, and the atom returns to its stable condition by transferring electrons

Figure 7 Optical image acquisition system showing lighting (top left), CCD camera with integral ring light (bottom left), and image-processing computer with display (bottom right)

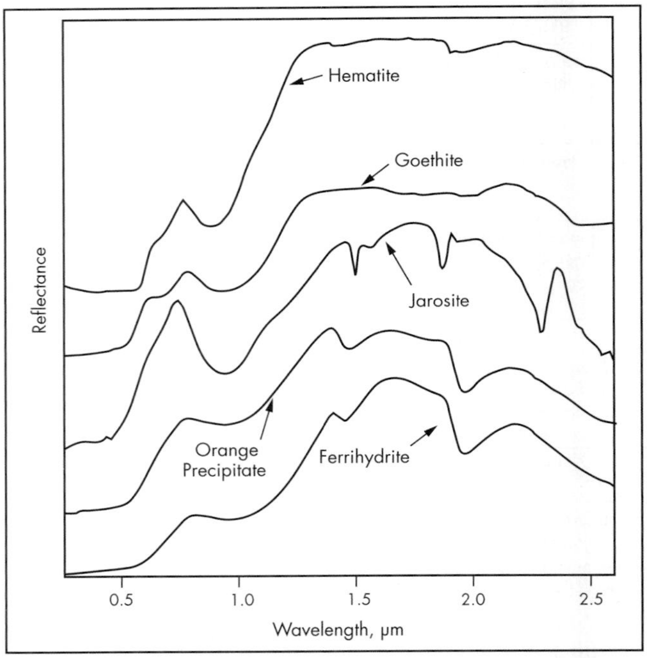

Source: Clark 1999

Figure 8 Example laboratory reflectance spectra of selected iron-bearing geological materials from the visible (~400 nm) to the shortwave infrared (~2,500 nm) light

from the outer shells to the inner shells and, in the process, emits X-rays of characteristic energy related to the difference between the binding energies of the corresponding shells. The X-rays emitted from this process are called *X-ray fluorescence*, or XRF. Emitted X-rays are classified by a spectrometer designed to recognize fluorescence of specific elements in the periodic table. The technology is effective in detecting transition metals and some precious metals (Figure 9).

Mineral samples are irradiated by low-power X-ray energy, and characteristic X-ray backscatter is measured by the spectrometer. Typical deployments require seconds or even minutes for effective detection of mineral composition; in faster approaches, X-ray emissions are sampled more than 60 times per second, with characteristic spectra processed through proprietary high-speed digital signal processing, pattern recognition, and identification algorithms.

Transmission

XRT techniques similarly deploy X-ray sources, but at much higher intensities, which are powerful enough to penetrate the mineral and be captured by an opposing photonic detector (Figure 10).

The method is versatile, and gives an approximation of atomic density based on the degree of X-ray absorption. Principal applications are in coal and fine high-density and low-density minerals. Application is limited, however, as results are compromised in the presence of near-density nonvaluable material (e.g., epidote); applications are also limited by limitations in X-ray penetration to ore minerals typically up to 38-mm particle size (Figure 11).

Laser-Induced Sensors

Fluorescence

Several minerals that do not naturally fluoresce can be made to do so under stimulation from coherent light. Typically, a pulsed laser (e.g., an Nd:YAG [neodymium-doped yttrium-aluminum-garnet] laser) is used as a source to illuminate the sample. Fluorescent elements in the sample respond to bombardment by coherent photons and emit fluorescent backscatter, which can be measured again by CCD (Figure 12). The intensity of fluorescent backscatter is proportional to the concentration of the fluorescing element. The technique has been successfully used in bulk analysis of phosphorus in iron ore and disseminated base metal sulfides.

Breakdown Spectroscopy

Chemical analysis of minerals is also possible by use of laser light. A high-powered, short-duration monochromatic laser beam of approximately 3 GW is used to irradiate the sample from a distance of between 150 mm and 1,000 mm. The high-powered laser pulses ionize the atomic structure of the minerals. Between pulses, however, the atom begins to reorder disrupted electron shells, and characteristic breakdown energies are emitted. Characteristic spectra are passed through a spectrometer specifically designed for the wavelengths of interest, delivered to a CCD for digitization, and analyzed in the embedded computer (Figure 13). Typical deployments of LIBS are for gases and pure alloys, however, application has recently been extended to oxides and sulfides of both base and precious metals. Compact, portable LIBS systems (Fortes and Laserna 2010) and advances in the development of systems to analyze rocks (Senesi 2014) enable application to sorting.

Gamma Techniques

Sensing techniques that use gamma radiation are classified as those that use natural gamma radiation and induced gamma radiation.

Natural Gamma Radiation

Many minerals contain concentrations of standard, naturally occurring radioactive elements K, U, and Th. These include uraninite, pitchblende, and most Witwatersrand gold ores where the gold is associated with uraninite. These elements give off a gamma ray of a unique energy when they decay.

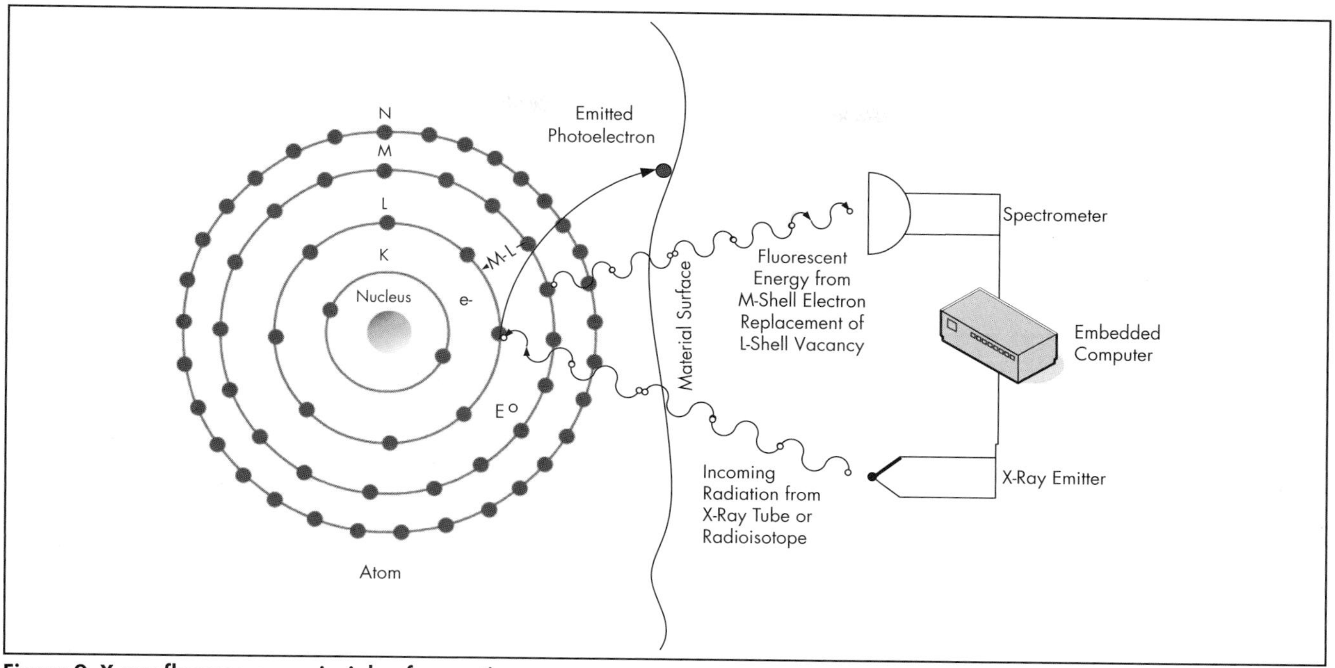

Figure 9 X-ray fluorescence principle of operation

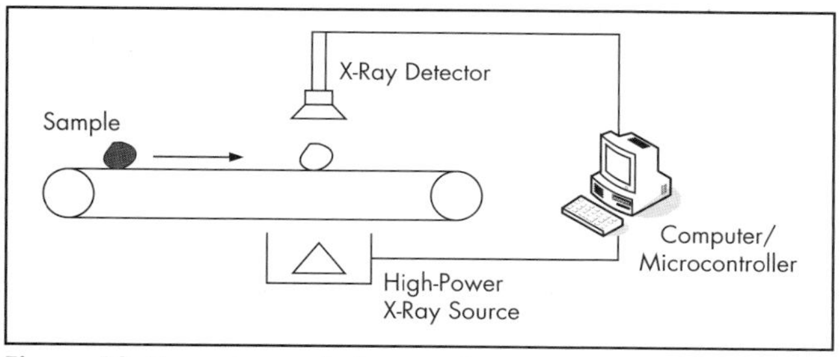

Figure 10 X-ray transmission sensing system

Source: Bamber 2004

Figure 11 False color image for ore minerals in X-ray transmission

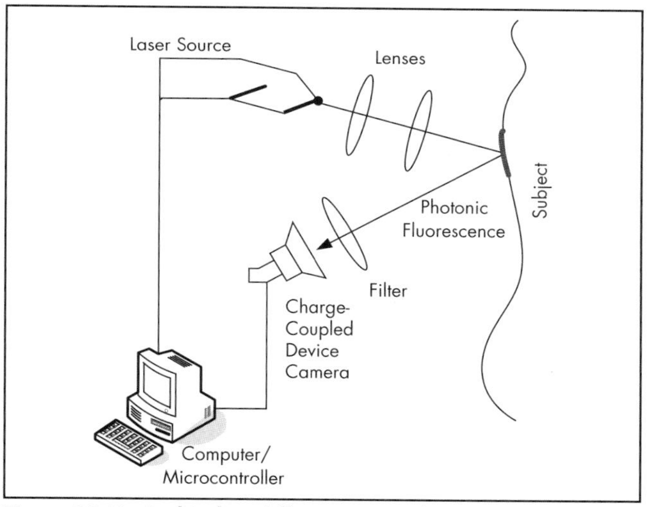

Figure 12 Typical induced fluorescence characterization setup showing source, subject, detector, and analyzer

Natural gamma emissions from these elements are captured in a scintillator crystal, which produces a pulse of visible light whose intensity is proportional to the energy of the gamma radiation. These pulses are in turn detected by a photomultiplier and converted to electrical pulses whose current is proportional to the energy of the gamma ray, and the number of pulses is proportional to the concentration of the emitting element. This is called *pulse-height spectrum analysis* (Figure 14).

Induced Gamma Radiation

PGNAA, pulsed fast thermal neutron activation (PFTNA), and variants are types of analysis that activate neutrons to measure

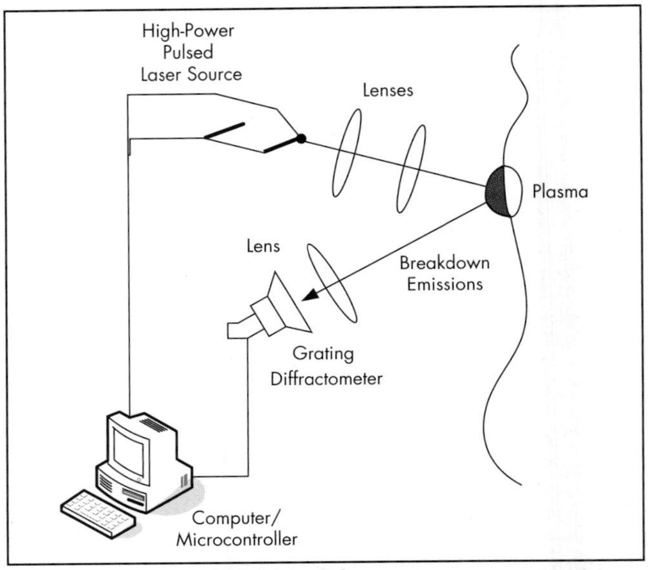

Figure 13 Laser-induced breakdown spectroscopy sensing system

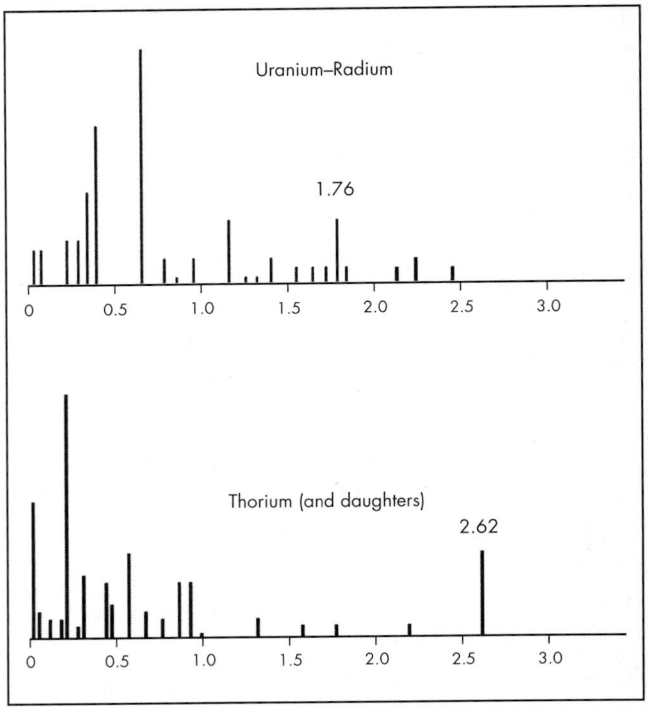

Adapted from SPE 2015

Figure 14 Gamma ray spectral plots for uranium (1.76 MeV) and thorium (2.62 MeV)

elemental composition. The material is bombarded with neutrons that create artificial radioisotopes that decay, emitting gamma rays characteristic of the element from which they are emitted; the intensity of gamma rays at a given wavelength are related to the elemental concentration. For PGNAA, the gamma rays are measured during irradiation (Figure 15). PFTNA is similar but uses high kinetic energy (fast) neutrons to measure gamma rays during irradiation. Both provide bulk chemical composition and require sensor response periods in the order of minutes, making the approaches unpractical for particle sorting but suitable for some bulk sorting applications when positioned over belts.

Electromagnetic Techniques

Many economic minerals of interest are metallic and therefore either conductive, magnetic, or paramagnetic. These properties can be measured using electromagnetic field devices, which is an established principle in mining exploration as well as sorting. In principle, passing a current through a coiled conductor generates an electromagnetic field. In any conductor, inductance (L) is a measurement of the distribution of the magnetic field. The magnitude and phase of the current is determined by the applied voltage and the coil's impedance (Z), which in turn controls the characteristics of the time-varying magnetic field. Impedance is a measure of two factors: resistance (R) to direct current and reactance (X), where j is the imaginary unit, and Ω, ω represent frequency. Reactance is the frequency-dependent opposition to alternating current and is determined by inductance.

$$Z = R + jX\,[\Omega]$$
$$X_L = j\,\omega\,L \quad \textbf{(EQ 1)}$$

Time-varying magnetic fields in turn generate time-varying electric fields. Any object within close proximity to the coil encounters these time-varying fields. Conductive and magnetic minerals that are present distort the field and alter the coil's electrical properties proportional to the amount of material present (Figure 16). Conductive material influences the voltage of the current in the conductor (either positive or negative); magnetic material influences the phase of the current flowing in the conductor, again either positively or negatively. A comparator or other bridge circuit measures these changes; in simple systems, responses above a certain predetermined threshold indicate detection of conductive or magnetic material. In more complex systems, frequency-dependent effects can be monitored, enabling the capability to determine concentration of conductive or magnetic mineral as well.

ORE SORTABILITY

Ore sorting studies are aimed at determining ore sortability and providing design information for sorting systems. There are four components to evaluating the feasibility of sensor-based sorting: ore heterogeneity, sensor response evaluation, sorting analysis, and feasibility as presented in Figure 17. Ore heterogeneity is a fundamental property of mineral deposits and must be assessed independently of the sensor or sorting method to be used. Sensor response evaluation exposes minerals to sensors relevant to the application and confirms the ability of sensors to detect the property of the minerals in question and the quality of correlations between the sensor response and the property (typically ore grade). Sorting analysis considers—in combination—the levels of grade heterogeneity identified and the quality of sensor response to the grade in order to develop theoretical yield–grade–recovery relationships. Feasibility studies would consider the following: the commercial availability of a sorting product that can meet the requirements of the envisaged application, metallurgical performance of the envisaged sorting stage, downstream impacts on throughput and recovery, and capital and operating costs of the envisaged circuit.

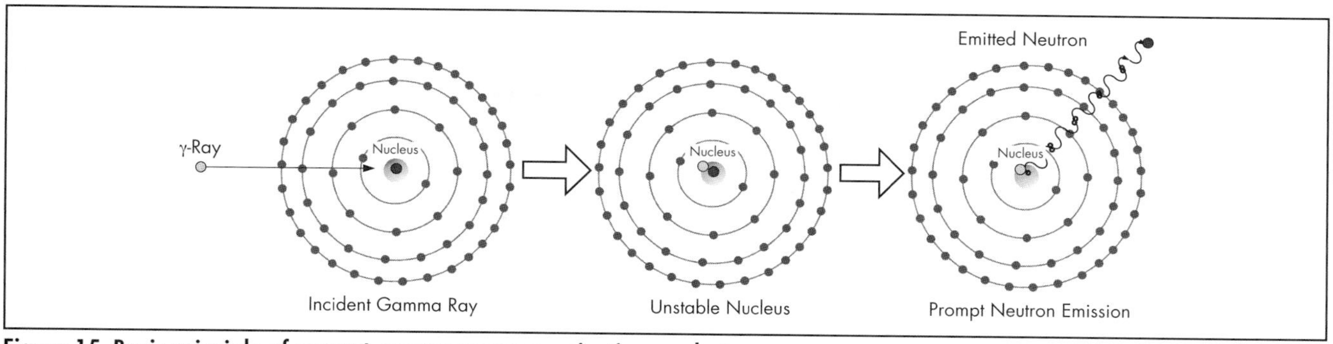

Figure 15 Basic principle of prompt gamma neutron activation analysis

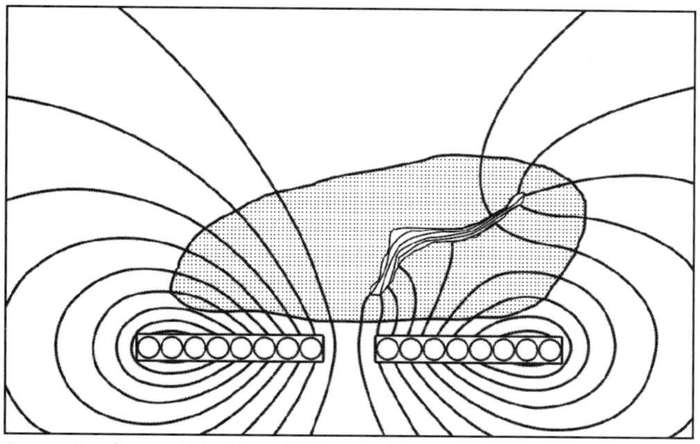

Source: Bamber and Houlahan 2010

Figure 16 Sympathetic fields generated by low concentrations of conductive or magnetic minerals in rock

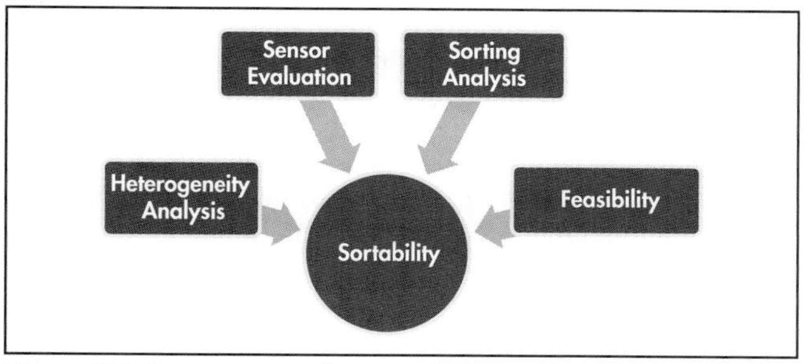

Figure 17 Components of a sortability study

Fundamental Indicators of Sortability in Ores

Ore bodies are complex, highly heterogeneous mixtures of different lithological units with varying mineralogical compositions. A key to successful sorting is knowledge of the lithology, mineralogy, and quantification of levels of heterogeneity as the basis for discrimination.

In mining systems, heterogeneity is generally expressed at the block modeling stage (e.g., 5 × 5-m scale) and is used to distinguish between ore and waste rock, as well as at the beneficiation stage at size ranges where mineral liberation occurs (e.g., 100-mm scale). Sorting is applied at length scales intermediate to these two extremes, and it is therefore important to understand how heterogeneity is expressed between these two size ranges (Figure 18).

Ore Body Lithology and Mineralogy

Much can be learned about heterogeneity and therefore the potential for sorting from knowledge of lithologies within the ore bodies and their mineralogy (Barbanson et al. 2009; Hitch et al. 2015). Preliminary assessment should include a review of reports that describe features of the ore body that would allow sorting (e.g., multiple adjacent lithologies with different metallurgical response). Descriptions of distinctive differences between ore and waste rock lithologies and related mineral associations can provide insights into heterogeneity as well as into physical and chemical characteristics that can be exploited as the basis for sensor discrimination.

Early-stage assessment can come from visual observations of rock or drill core. Providing that samples include a distribution of valuable and non-valuable constituents, a qualitative assessment of sortability is possible and can justify further testing.

When examining the drill core, features that are relevant to sorting are the size of mineralogical features and contacts between valuable and non-valuable constituents (Carrasco et al. 2016). Examination of the drill core, particularly for early-stage projects, reveals the potential for ore sorting (Bamber et al. 2006). Figure 19 shows sharp contacts between ore and waste. It also shows that sections of ore and waste have significant extents, which demonstrates heterogeneity over relatively large sections along the core (5–25 cm). Fragmentation should therefore produce particles that have significant differences in mineralogical and physical properties that are sufficiently heterogeneous to allow sorting.

Simple visual examination of core samples supports early-stage assessments of heterogeneity. For example, within a coarse fraction, such as 5–20 cm, the presence of alternating barren rock and mineralized rock allows conclusion that the rock may be sortable providing a sensor can detect the differences between them. In cases where heterogeneity is not visible, scanning with sensors is required to assess this heterogeneity and the potential for sorting.

Heterogeneity Analysis in Ores

The nature of the distribution of metal grades between individual rocks as well as within and between bulk samples provides an indication of sortability. A wide distribution of grade would suggest that there is potential to beneficially separate at a grade threshold. Conversely, a narrow distribution around a cutoff would indicate that sorting will be difficult. Two additional parameters indicate the potential benefit of sorting. Measures of the constitution heterogeneity (CH) indicate

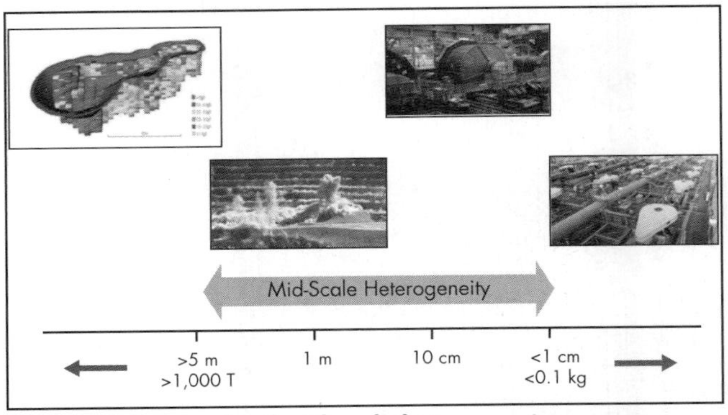

Figure 18 Schematic of mid-scale heterogeneity ranges

Figure 19 Drill-core samples showing sharp contact between lithological rock types and coarse heterogeneity

potential for rock sorting, and measures of the distribution heterogeneity (DH) indicate potential for bulk sorting.

CH characterizes grade variance of rocks such that a high CH implies the potential for rejecting a large portion of barren rock (or alternately recovering a meaningful proportion of valuable rock). DH (or spatial heterogeneity) indicates potential for bulk sorting in a similar mode. The grade and weights of units (rocks or bulk units) allows determination of the potential grade, mass yield, and recovery relationship.

Constitution Heterogeneity

CH was first introduced by Gy (1992) in his theory of sampling in order to define the uncertainty that is inherent in sampling, represented by the relative, dimensionless variance of the heterogeneities associated with each fragment F_i making up a lot of N_F fragments in a sample as shown in the following equation:

$$CH_L = N_F \Sigma_i \frac{(a_i - a_L)^2}{a_L^2} \cdot \frac{M_i^2}{M_L^2} \quad \textbf{(EQ 2)}$$

where CH_L is the constitution heterogeneity; α_i and α_L are the grades of the fragment i and the lot, respectively; and M_i and M_L are the masses of fragment i and the lot.

This definition can be easily extended to the heterogeneity of fragments or particles within a lot of particles within a stockpile, for example, or even fragments within the mineral deposit as a whole. By inputting values for the grade of individual particles in a sample and the average grade of the sample as a whole, CH can be calculated for any sample. The CH parameter has elevated relevance with respect to particle sorting applications. A high CH indicates the potential to reject or separate a greater proportion of the feed (Mazhary and Klein 2015).

The particle heterogeneity is most easily evaluated by measuring the grades of individual particles within a sample, and plotting the measured grades in two possible ways.

Weight Distribution by Grade

First, the frequency distribution within the grade range measured (i.e., the weight percent of the sample falling into specific grade ranges) can be evaluated from the data (Figure 20).

A general indicator of sortability can be evaluated graphically in the form of skewness of the distribution (mode ≠ mean). A normal distribution—0 skewness—indicates low sortability. Left skewness (mode < mean) suggests a high proportion of low-grade material in the sample, and therefore potential for rejection of these fractions. Potential for the recovery of small quantities of high-grade material is also suggested when viewing such a distribution. Right skewness (mode > mean) suggests a high proportion of high-grade material in the sample, and therefore potential for recovery of these fractions.

Of course, simpler indications of heterogeneity are available; for example, taking the quotient of the 90th percentile grade value and the 10th percentile grade value, where for g90/g10 > 20 (where g means grade value), good potential for sorting is indicated. The identification of any sensor-based method that can distinguish between the grades of various particles identified in such an evaluation supports application of sensors to the sorting of these materials.

Grade Distribution by Particle Size

Second, where the individual particles in the sample were first separated into size classes before assaying (i.e., size/assay), the distribution of grade by particle size can be evaluated (Figure 21).

Again, a general indicator of sortability can be evaluated graphically in the form of skewness of the distribution (mode ≠ mean) representing a preferential deportment of grade with respect to particle size. A uniform distribution indicates no preferential deportment. A normal distribution indicates preferential deportment to the middlings size class. Left skewness (mode < mean) suggests higher grades in the fine fractions of the sample, and therefore potential for recovery of this high-value material by size classification. Potential for the rejection of small quantities of low-grade coarse material is also suggested when viewing such a distribution. Conversely, right skewness (mode > mean) suggests higher grade in the coarse fractions of the sample, and therefore potential for the recovery of this high-value material by size classification.

Distribution Heterogeneity

DH is a measure of the heterogeneity of parameters associated between fragments or lots of fragments at different locations in the deposit (Equation 3). The definition can be easily extended to a sample within a set of samples, or one lithological or geometallurgical unit among the set of lithological units of interest (i.e., different ore zones within a deposit). Therefore, the DH parameter has elevated relevance to bulk sorting applications, although it can further inform applications in particle sorting as well.

$$DH = N_g * (\Sigma (a_i - a_L)^2 \times M_i^2) / (a_L^2 \times M_L^2) \quad \text{(EQ 3)}$$

where N_g is the number of groups; α_i and α_L are the grades of group i and lot, respectively; and M_i and M_L are the masses of group i and the lot.

Evaluation of DH is much simpler than for CH. When considering a set of units, such that a lot represents a set of units, Gy (1992) refers to the distribution heterogeneity (DH_L) that depends on three factors: (1) the constitution heterogeneity (CH_L), (2) the spatial distribution of the constituents, and (3) the shape of the lot. Lots can vary in scale, ranging from geological units, to blocks within a block model, to shovel loads, to flows along a conveyor. Initial indications can be found in the ore body model where characteristic *steps* in the grade–tonnage curve of the deposit may suggest significant zones of differential grade (or other characteristics), and therefore potential for classification of these zones by sorting into streams for custom processing (Figure 22).

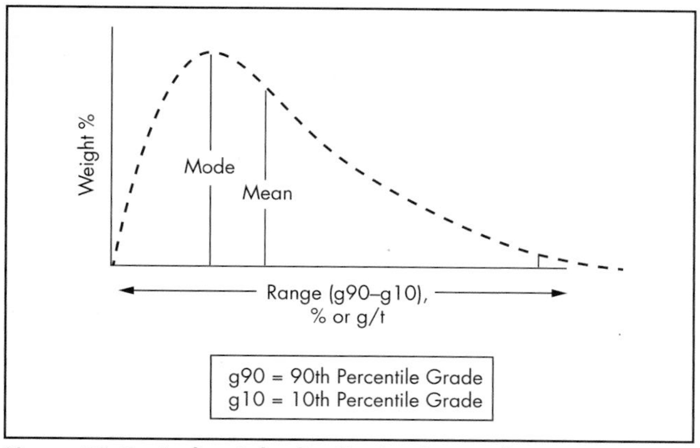

Figure 20 Typical weight distribution by grade

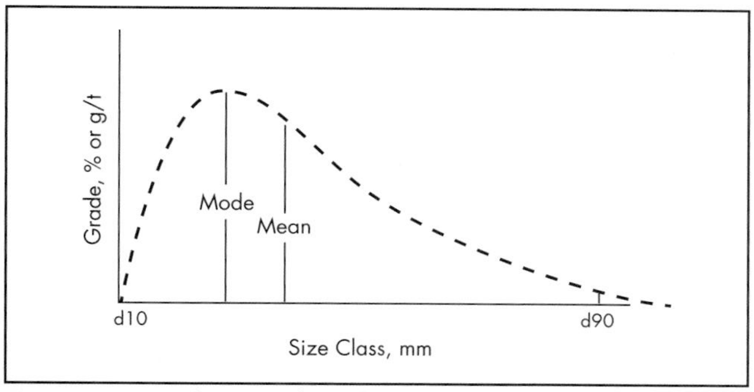

Figure 21 Distribution of grade by particle size

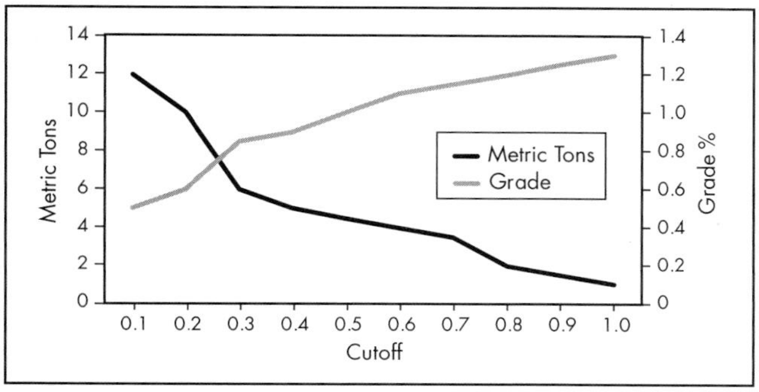

Figure 22 Grade–tonnage curve steps indicating zones of highly differential grade or other characteristics in the deposit

Evidence of DH may be much more elementary. For example, in conventional geological modeling, lithologies of different geological units are usually well defined. These lithologies can and do have widely varying ore and gangue mineralogy, and therefore grade, grindability, and other metallurgical characteristics (thus also making them geometallurgical units). In some cases, heterogeneity between these geometallurgical units can be slight, as in iron ores (Table 2), or distinct, as in highly weathered volcanogenic massive sulfides (VMSs) (Table 3).

Clearly, levels of heterogeneity between the various ore lithologies within the iron ore deposit and between the ore and waste lithologies in the deposit are low; however, separation based on sensor-based characterization may still be possible.

Levels of heterogeneity between the various ore lithologies within the VMS deposit, and between the ore and waste lithologies in particular are high; therefore separation potential based on sensor-based characterization is highly indicated (Bamber et al. 2007).

The identification of any sensor-based method that can distinguish between the various characteristics of such lithological units supports further evaluation of bulk sorting of these materials by the techniques described in the following sections.

Generally, the degree of heterogeneity will increase with the decreasing size of the unit, as shown in Figure 23. However, each ore body will exhibit a different heterogeneity versus unit size relationship, as indicated by the solid and dashed lines. The relationship between DH and unit size range can be estimated from drill-core assay. For example, DH can be calculated from the grades determined at 1-m intervals, 2-m intervals, 3-m intervals, and so on. This approach allows DH assessment from the drill core for early-stage projects.

DH generally decreases as a consequence of the mixing and blending that occurs in material handling (Duffy et al. 2015). Therefore, the greatest heterogeneity and opportunity

Table 2 Low levels of heterogeneity between units in an iron ore deposit

Sample Description	Fe, %	SiO_2, %		Al_2O_3, %	
	Minimum	Minimum	Maximum	Minimum	Maximum
High-grade upper channel iron deposit (CID)	52	0	5.5	0	2.5
Upper CID	52	5.5	11	0	2.5
High-grade detrital	53.5	0	8	0	3
Detrital waste	12	NA*	NA	NA	NA
Blendable upper CID	52	5.5	11	0	2.5
Middle CID	52	6.5	13	0	2
High-grade bedded	52	0	6	0	3

*NA = not applicable.

for bulk sorting is close to the face. CH is affected by mixing, but to a lesser extent.

Test Programs

Three aspects must be considered when developing test programs. First, if amenability test work is preliminary, the tests should be as nondestructive as possible to preserve as much sample as possible for further testing, whether for more scanning or for downstream test work (e.g., grindability or flotation response).

Second, the choice of assay method is critical. Often, if samples are from an operation, the mine will suggest or recommend a specific or custom assay procedure. If working a greenfield project, or if no guidance is available, the assay lab should be consulted beforehand for recommendations as to the appropriate assay procedure for the ore type in question. Inductively coupled plasma (ICP) whole-rock methods are generally suitable for metal sulfides. Specific digestion methods, however, are recommended for Cu porphyries, and multielement X-ray fluorescence (ME-XRF) methods are recommended for oxides such as ferric oxide (Fe_2O_3) or Ni laterites.

Third, the criticality of developing good sensor–assay correlations must be considered. In rare cases, sensor correlations are simple linear functions with high correlation, therefore fewer assays are required to develop a fit. In this case, hand selection of 60–80 specimens from across the grade range (a visual assessment of mineralization may be required) is sufficient to develop the correlation. In most cases, correlations will be more complex requiring cubic functions or multivariate regression, and more assays will be required to develop a good fit. In equally rare cases, no sensor correlation may be apparent from the data, and CH must be judged solely from the underlying assay data. It is, therefore, possible that the entire 500-particle sample set must be assayed. In such cases, the testing is completely destructive, and any rescanning must be done on a separately procured sample.

Sampling and Sample Preparation

Effective sampling is key in assessing the amenability of ores to sorting. A quick referral to Gy (1992) or Pitard (1993) will reveal that representative sampling at the typical particle size distributions and metal grades in question suggests representative sample sizes in the hundreds of kilograms for high-grade fine samples and thousands of kilograms for low-grade coarse samples. As truly representative sampling is generally unfeasible in this domain, the aim of sampling and test work should be to obtain indicative results from as representative a sample as possible. While geometallurgical approaches can be adopted to include many smaller samples from across the deposit to maximize representivity, the individual samples obtained in this mode are by necessity small and therefore may not be optimal in the assessment of sorting parameters.

The fundamental considerations that are applied to sample selection for conventional metallurgical studies are relevant as presented in Chapter 1.8, "Sampling Practice and Considerations." However, considering the constraints of practical sample size and to ensure the capture of as many relevant features of interest as possible, custom, guided sampling is recommended. Sample bias also needs to be avoided. This can arise from three approaches to sampling:

Table 3 High levels of heterogeneity between units in a polymetallic VMS deposit

	Assay			
Unit	Au, g/t	Ag, g/t	Pb, %	Zn, %
Waste	<1	<50	<3	<2
Oxide	6	200	3	2.5
Inner Peñasco breccia	30	500	4	7
Azul breccia	4	300	3.5	7
Sediments	15	500	4	10

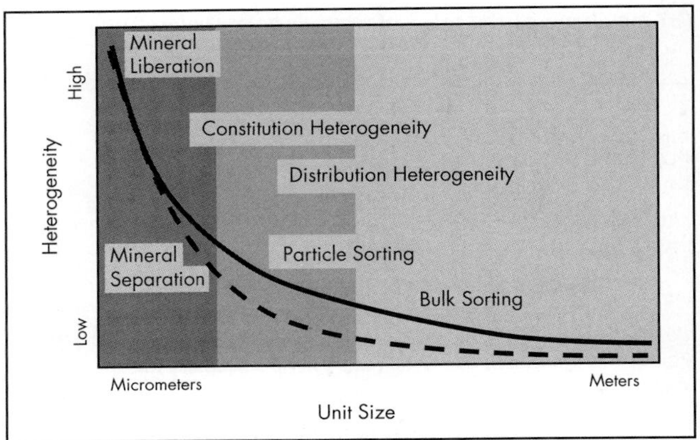

Figure 23 Heterogeneity versus unit size for two different ores represented by solid and dashed lines

1. Preferential exclusion of fines in sampling (fines can be high grade)
2. Preferential exclusion of oversize particles in sampling (oversize particles are difficult to handle)
3. Preferential samplers to prejudge "good" material (bias toward high grade) or "waste" material (bias toward low grade)

Following are three suggested sampling approaches, which are relevant to new (exploration) projects and to surface and underground operations, respectively.

Exploration projects. Where sorting is being evaluated for a new project, and often in underground projects, core samples are the only source of material for testing. In this case, it is not only essential to ensure that samples of every relevant ore type plus waste are sampled but also that a full intersection of ore (i.e., an intersection containing hanging wall, ore zone, and footwall material) is taken. Half-core is generally available and optimal for testing. Quarter-core tends to be too fine at the outset and also tends to break up further during scanning, which is suboptimal. Often, samples provided by geologists or metallurgists for sorting test work are *too good* in that they are too high grade and do not contain relevant quantities of gangue material. Also, diluting *rich* samples with waste will not give meaningful results compared to testing *virgin* samples of similar grade.

Open pit mines. Sampling for sorting test work in open pits is most easily achieved by taking samples directly from the bench or benches in question. Ideally, at least one sample per ore type plus one sample of waste (either overburden or basement material) is recommended. Samples should be full

Source: Bamber 2008

Figure 24 Classification of samples into size fractions for testing

fraction: To be useful, they should contain the largest feasible particle size as well as a representative measure of fines. In open pits, large run-of-mine (ROM) particle size distributions often prohibit the practical collection of samples, and therefore coarse crushing of samples is often necessary. Optimal samples are taken in either 1- or 2-t bulk bags or if absolutely necessary, 210-L drums (approximately 300 kg each). Belt cuts from overland conveyors post in-pit crushing or from the semiautogenous grinding (SAG) mill feed are admissible. Because this type of sample often cannot be traced back to any particular block in the ore-body model, it must be taken under advisement.

Underground mines. Sampling in underground mines is ideally done by sample collection directly from the blasted muck pile in the stopes. Again, one sample per major ore type, plus a sample of waste material as a control, is recommended. Samples can be taken as subsamples of the contents of scoop buckets or manually by hand shovel from the muck pile. Care should be taken to capture both coarse and fine fractions of the sample to ensure maximum representivity and avoid bias as previously described. In the underground context, samples from ore passes and waste passes are also considered valid, as are belt cuts from underground conveyors where the content of the belt can be traced back to a particular heading or section.

Sample Preparation

Samples are ideally weighed and either wet or dry screened into at least three size fractions (+75 mm, –75+25 mm, and –25 mm) although more fractions can be contemplated. Figure 24 shows multiple size fractions: +125 mm, –125+75 mm, –75+53 mm, –53+38 mm, –38+25 mm, –25+13 mm, –13+9 mm, –9+6 mm, –6+3 mm, –3 mm. To deliver clean, fines-free particles for scanning test work, wet screening is preferred; however, dry screening may be used to more closely simulate the condition of particles in a field situation.

Classification into particular size fractions in this way supports assessment of any characteristics of preferential grade deportment by size. Characterization classification of material also supports complementary characterization, such as assessment of the liberation index at the various particle sizes and observation of preferential breakage (i.e., breakage along grain boundaries or contact lines), which might influence presentation of the valuable minerals to the sensor.

Sensor response evaluation. Samples scanning is done to achieve two objectives: (1) to measure the true heterogeneity of the particles in a sample and (2) to generate data for the correlation of sensor responses to the grade of mineralization in the sample. An additional objective that can be achieved in scanning is to screen into an appropriate number of size classes a priori as previously described, and scan each size class separately. To select a method for scanning, Table 1 should be consulted with consideration of the property of interest.

Rocks (or batches of rocks representing a particular lithology) prepared for analysis are then scanned by the selected sensor. Sensor responses are recorded for individual rocks or the batch sample to determine the sensor response distribution with respect to the heterogeneous property. Sensor response distribution is then correlated to the property described by independent assay of the property (i.e., grade). The results indicate the potential for a specific sensor to classify rock according to value. For example, a narrow sensor response distribution close to a threshold grade along with a low CH indicates a low potential for the sensor to classify the material. Conversely, the opposite would indicate a high potential for sensor classification. Similarly, evaluation of sensor response for bulk samples with consideration of the DH parameter would show the potential of the sensor in a bulk sorting application.

Experimental programs are carried out to determine which sensors are capable of sensing ore and waste qualities within each lithological unit in an ore body. Table 1 lists sensors that could be considered as part of an evaluation of sensor-based sorting. Once the sensor responses are mapped against individual rock types, a sorting strategy and system that applies to selected sensors can be investigated and developed further.

Often, scanning using multiple types of sensors simultaneously (e.g., electromagnetic, photometric, and XRF) is recommended, depending on the expected range of properties that may be exploited. In scanning for characterization purposes, all particles in a given size class should be scanned multiple times. At least four scans per particle (one scan per *face*) is recommended, typically turning the particle 90° between each scan. A greater number of scans per particle is possible, even advisable, in the case of coarse particles (e.g., >125 mm).

Fines <25 mm (in the case of ROM samples) and <13.8 mm (in the case of primary crusher underflow) would generally not be scanned, as these are considered too fine, practically, for particle sorting at scale. For low-tonnage and fine-particle applications, evaluation can be done at these particle sizes if desired. When scanning core, each core piece should be scanned as a single sample; multiple scans (four or six) would again be required for core pieces >100-mm long. Core pieces <25-mm long would generally be considered fines and not scanned. If the sample particle size distribution is relatively fine, a practical limit of 500 particles scanned per size class is suggested. Scan data from the various modes of sensor for each size class should be logged together with a sample number and, ideally, the sample mass in a database for later retrieval and analysis (Table 4).

Sorting analysis requires the assessment of the grade and metal distribution in the unit (e.g., by CH and DH metrics previously described) and the development of algorithms that correlate generated sensor responses to variations in grade. Correlations are demonstrated using sensor calibration curves that relate the grades of rocks as determined by whole-rock assay to sensor-predicted responses. Figure 25 shows an example plot

Table 4 Typical data from a sensor amenability test program

Sample No.	Date	Weight	Peak 1	Peak 2	Peak 3	Peak 4	Fe	Cu	Pb	Zn
1	2015-11-24	191	270	20	1,650	20	27.0	2.0	16.5	<1
2	2015-11-25	100	300	30	3,900	20	30.0	3.0	39.0	<1
3	2015-11-26	78	100	40	1,599	20	10.0	4.0	16.0	<1
4	2015-11-27	140	120	10	1,299	20	12.0	1.0	13.0	<1
5	2015-11-28	200	400	30	699	20	40.0	3.0	7.0	<1
6	2015-11-29	156	500	20	679	20	50.0	2.0	6.8	<1
7	2015-11-30	130	270	30	400	20	27.0	3.0	4.0	<1
8	2015-12-01	99	300	40	399	20	30.0	4.0	4.0	<1
9	2015-12-02	238	100	10	700	20	10.0	1.0	7.0	<1
10	2015-12-03	340	120	30	200	20	12.0	3.0	2.0	<1
11	2015-12-04	67	400	20	345	20	40.0	2.0	3.5	<1
12	2015-12-05	180	500	30	899	20	50.0	3.0	9.0	<1
13	2015-12-06	200	300	40	1,237	20	30.0	4.0	12.4	<1

of assayed grades for Pb and Zn against XRF sensor responses for those elements showing good correlation (Tong et al. 2015). It can be concluded from the high correlation that this lead-zinc ore is amenable to sorting by XRF. Correlations may be simple $y = mx + c$ relationships; more complex quadratic or cubic functions; or may require more sophisticated methods, such as multivariate regression analysis to generate a usable correlation. Calibrated sensor responses thus developed can then be used to develop grade, mass yield, and metal recovery relationships discussed in more detail in the following section.

Sortability Analysis

The ability to sort materials is based on the ability to correlate elemental/metal grades to sensor responses with consideration of mineralogical properties of the ore and waste. Most commonly, a single sensor response is correlated to a metal grade. Correlations can be improved by application of several approaches, including

- Combining sensor responses to improve accuracy,
- Analyzing proxies for target metals or minerals (multivariable regression),
- Assessing heterogeneity of gangue phases for rock rejection (e.g., detecting the waste for rejection),
- Applying piece-wise regression to focus models to selected grade ranges, and
- Assessing rock size heterogeneity to determine optimum size ranges for sorting.

Test data from sensor studies are used to evaluate each of the approaches to determine potential metal recoveries, product grades, and mass rejections for a range of thresholds. The results can also be used to test the developed algorithms to indicate separation efficiency by determining the proportion of potentially misplaced rocks or bulk material. Algorithms that can be correlated to mineralogical, physical, or geochemical properties are considered more robust than those that are completely empirical. The assay data are required to provide a calibrating data set against which to correlate the average scan result for each particle to first determine the heterogeneity and second, the degree of correlation between sensor response and the elemental (or mineralogical) composition of the sample.

Heterogeneity and Frequency Distributions

From either the full sample assay data or from calibrated scan data for the sample, a grade distribution can be developed by taking the frequency of occurrence of either grades (in the case of assay data) or magnitudes of sensor response (in the case of scan data) within *bins* of grade or magnitude of response and plotting their relative frequency. Factoring the frequency of occurrence by the mass of particles falling within each bin will give the distribution of sample weight percent in each grade or response interval (Figure 26).

From such results, CH can be calculated using Equation 2. For multiple results across several ore types from the same deposit, DH or spatial heterogeneity (Equation 3) can also be calculated.

Size–Grade Distribution

The size-by-grade distribution can be used to identify the potential for size classification as a means of sorting. The relationship between particle size–grade and metal distribution supports targeting of size classes for sensor-based sorting. Size analysis and assays are shown in Table 5 and are plotted in Figures 27 and 28.

The data clearly suggest that the –75+53-mm and –53+37.5-mm fractions are preferentially carrying the Pb and Zn value. The benefit of recovering by size classification should then be investigated further.

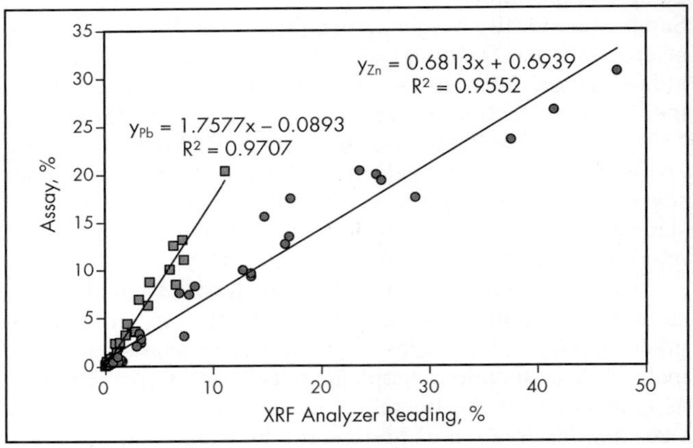

Source: Tong et al. 2015

Figure 25 Correlations between X-ray fluorescence analyzer reading and particle assay

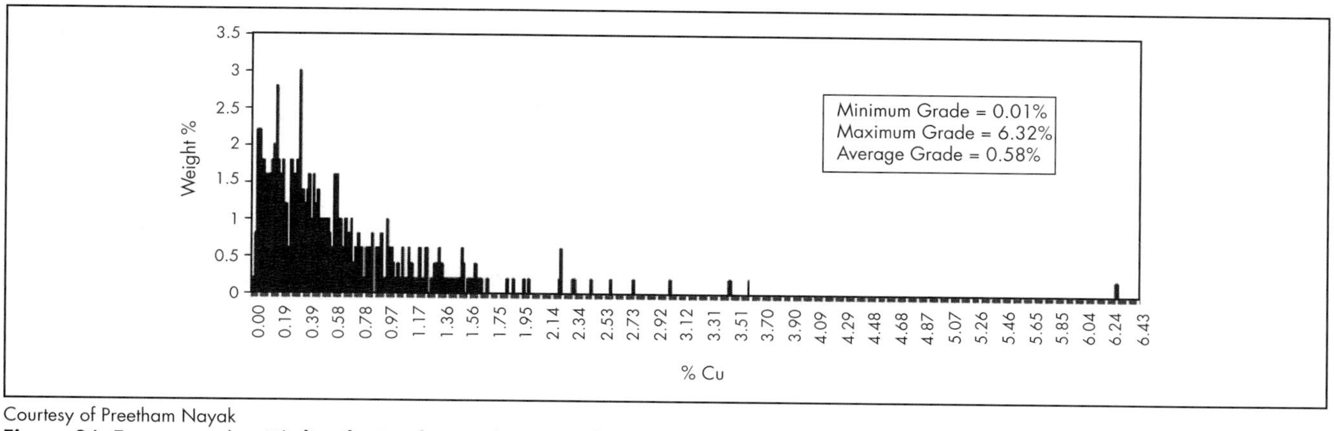

Courtesy of Preetham Nayak

Figure 26 Frequency (wt %) distribution by grade interval in a Cu-porphyry sample

Table 5 Head sample size analysis and assays

Size Fraction		Weight Distribution, %	Cumulative Weight, %		Grade, %		Distribution, %	
mm	in.		Retained	Passing	Pb	Zn	Pb	Zn
+75	+3	24.5	24.5	75.5	2.84	5.58	16.9	14.2
–75+53	–3+2.12	44.9	69.3	30.7	4.66	11.80	50.9	55.1
–53+37.5	–2.12+1.5	14.2	83.6	16.4	5.85	9.97	20.2	14.7
–37.5+26.5	–1.5+1.06	6.3	89.9	10.1	3.65	9.52	5.6	6.3
–26.5+19	–1.06+0.75	2.1	92.0	8.0	1.49	12.80	0.8	2.8
–19+13.2	–0.75+0.5	2.0	94.0	6.0	2.23	7.78	1.1	1.6
–13.2	–0.5	6.0	100.0	0.0	3.10	8.61	4.5	5.3
Total		**100.0**			**4.11**	**9.63**	**100.0**	**100.0**

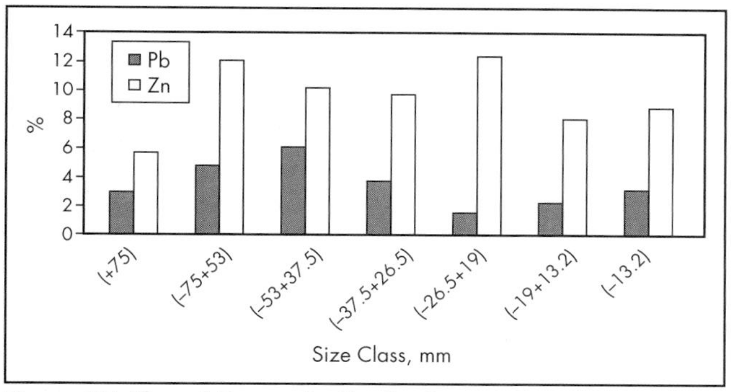

Figure 27 Pb and Zn grade distributions

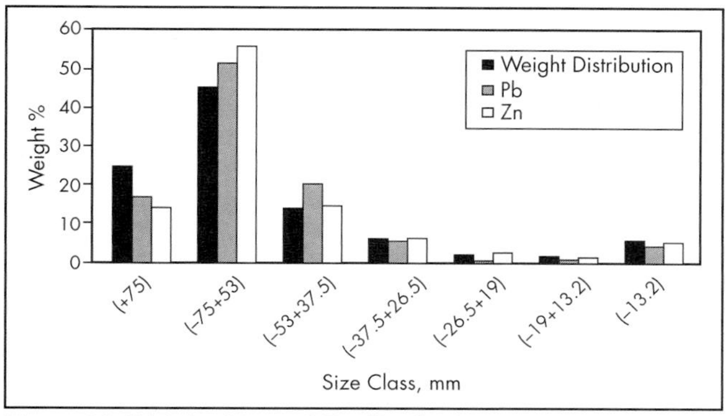

Figure 28 Weight, Pb, and Zn distributions

Grade–Recovery Curves

Theoretical grade–recovery curves for the sorting of the tested sample can be generated from the frequency distribution data. Selective exclusion of the sample mass mathematically by grade, from low to high (or high to low) will generate a theoretical grade–recovery curve for the separation of the sample. Grade–recovery curves can be generated for the ideal case based on the geochemical assay data set, or as derived from calibrated sensor responses.

Relative amenability to sorting can be assessed by examining the yield–grade–recovery characteristics indicated by the curve in question (Figures 29 and 30).

As can be seen, the ideal grade–recovery for the sample with highest CH value is the best; however, even for CH = 4, grade–recovery response may, subject to economic factors, still result in a beneficial outcome in sorting.

Finally, comparison of the ideal grade–recovery relationship as determined by heterogeneity of the sample to the grade–recovery relationship as determined by the sensor method will give an indication of expected performance of the selected sensor in classifying the material according to grade. Figure 31 shows the grade–recovery relationship for Cu, using XRF. It is based on regression models of Cu, Fe, As, and Mo peaks, where Cali-CuFe is calibrated to XRF Cu and Fe, and Cali-CuFeAsMo is calibrated to XRF Cu, Fe, As, and Mo. The figure shows that the regression model based on Cu and Fe alone does not fit the ideal as well as the model based on Cu, Fe, As, and Mo. These elements likely occur in minerals associated with the Cu mineralization.

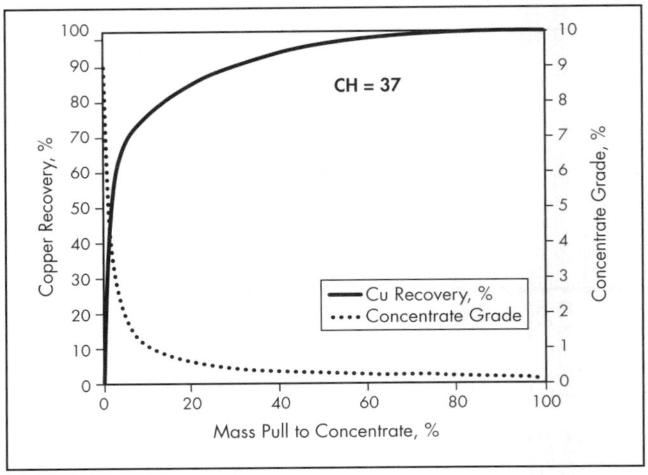

Figure 29 Grade–recovery curve, CH = 37

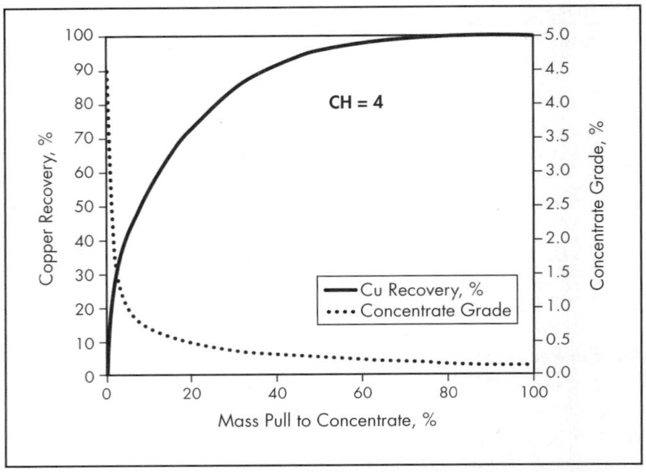

Figure 30 Grade–recovery relationship, CH = 4

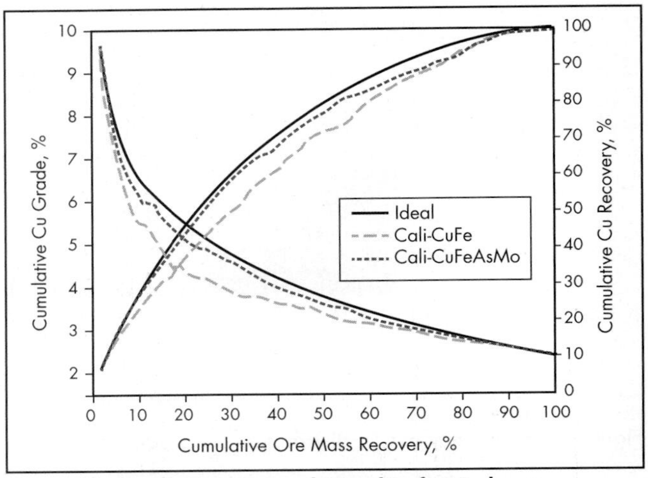

Figure 31 Grade–recovery relationship for Cu by assay and Cu grades estimates

SELECTION DATA FOR SORTING SYSTEMS AND CONSIDERATIONS FOR CIRCUIT DESIGN

Selection of an appropriate sorting system is governed by several factors (Klein et al. 2003, 2010). Mineral type, mine type, and application location are primary factors. Mineral type governs both heterogeneity parameters as well as the type of sensor to be deployed. Mine type largely governs the expected feed particle size distribution and capacity required of the circuit. Application location (e.g., in-pit or underground, pre-mill or within the SAG circuit) similarly governs the expected feed particle size distribution and capacity required of the circuit, as well as placing constraints on the type of sorting system (bulk or batch, semi-bulk or particle) that can be deployed. A final consideration is the optimal or desired yield–grade–recovery of the circuit, where the selectivity of bulk or batch systems, for example, may not allow grade or yield requirements to be met. Alternately, particle sorting systems may not meet capacity (or capacity and cost) requirements. Yield–grade–recovery requirements will also govern the selection and number of stages where delivering a final sorting product in a single stage of sorting, for example, may not be possible, or meeting capacity requirements with a single unit may not be possible, and multiple lines and stages of sorting may need to be considered. Table 6 presents a range of sorting system options and their relevant capacity and selectivity parameters. A range of selection criteria unique to bulk, semi-bulk, and particle belt/chute sorting systems also pertain. For particle sorting, the belt width, belt speed, and particle size distribution range are primary considerations that affect capacity (Table 7). Other parameters that affect capacity are specific gravity and the amount of material ejected.

Example Circuits

Figures 32 through 35 present typical bulk and particle sorting circuit designs. For bulk shovel and truck systems, in-pit or underground locations are typical. For bulk belt systems, locations can be in-pit (crush/convey) or pre-mill. For particle systems, locations are typically pre-mill or possibly applied to treat the pebble stream of the SAG mill circuit. Alternate deployments of both bulk and particle systems can be on the waste dumps of active or inactive mines.

Indicative Particle Sorting Parameters

Particle sorting is generally used for waste rejection ahead of the mill. Mass yield to concentrate can vary from 40% to 80% (20%–60% waste rejection). Recovery to concentrate can vary from 60% for high weight percent rejected, to 99% for low weight percent rejected, or in the case of waste rejection from unusually well-liberated ores. Table 8 presents a range of achievable particle sorting results based on the literature. The reader is reminded that as sortability is more a function of ore characteristics than machine characteristics, results presented here should be viewed not as representative of the particular technology described, but of the ore treated.

Indicative Bulk Sorting Parameters

Bulk sorting is typically applied further upstream than particle sorting, often as a pretreatment step ahead of other sorting stages, or ahead of flotation or leach processes. Selectivity in bulk sorting is inherently lower, therefore is typically applied in either cleaner (removing residual waste from ore material) or scavenger (recovering residual ore from waste material) applications. Mass yield to concentrate typically varies from 85% to 95% (5%–15% waste rejection), although higher rejection rates are possible when entire lithologies must be rejected to waste. It is important to keep in mind that the main value of bulk sorting, particularly shovel-based systems, lies in the ability to recover ore that would otherwise have been left in situ or lost to the waste dump. Recovery rates for ore in waste can be similarly 5%–15%, where ideally, the rate of ore recovery is used to balance the rate of waste rejection to optimize grade and throughput of feed to the mill (or leach). Parameters for bulk sorting often match parameters used in grade control and material routing, as this type of solution is most often deployed in support of those processes at a mine. Two examples of bulk sorting are given in Tables 9 and 10.

Available sorting options are generally traded off against the option of not sorting at all, although several other trade-offs can be considered. In situations where particle sorting is being considered, this option often competes with other coarse gangue-rejection technologies, such as jigging or dense media separation, which can be used in a similar duty with similar effects, or against more selective mining methods where these are possible. A key trade-off is the decision between particle sorting (with higher selectivity and yield–recovery parameters but limited capacity and higher unit capital expenditure [CAPEX] to operational expenditure [OPEX] ratio) and bulk sorting (with lower selectivity and yield–recovery parameters

Table 6 Sorting system selection criteria

Type	Particle Size Distribution Range	Lot Size Range, kg	Capacity, t/h
Fine-particle sorting	4–10 and 10–20 mm	0–0.05	0–40
Coarse-particle sorting	20–80 and 80–300 mm	0.05–6	20–300
Semi-bulk belt sorting	0–300 mm	10–100	100–1,500
Bulk belt sorting	0–600 mm	100–10,000	500–10,000
Bulk sorting shovel	Run-of-mine	10,000–50,000	500–10,000
Bulk sorting truck	Run-of-mine	100,000–300,000	500–10,000

Table 7 Particle sorting system typical selection criteria

Type	Belt Width, mm	Belt Speed, m/s	Particle Size Distribution Range, mm	Capacity, t/h
Ultrafine	600	2.8	0–4	0–10
Fine	600–1,000	2.8–6	4–20	10–60
Standard	1,000–2,000	2.8–6	10–200	20–150
Coarse	2,000–3,000	2.8–6	20–300	40–300

Courtesy of Lütke von Ketelhodt

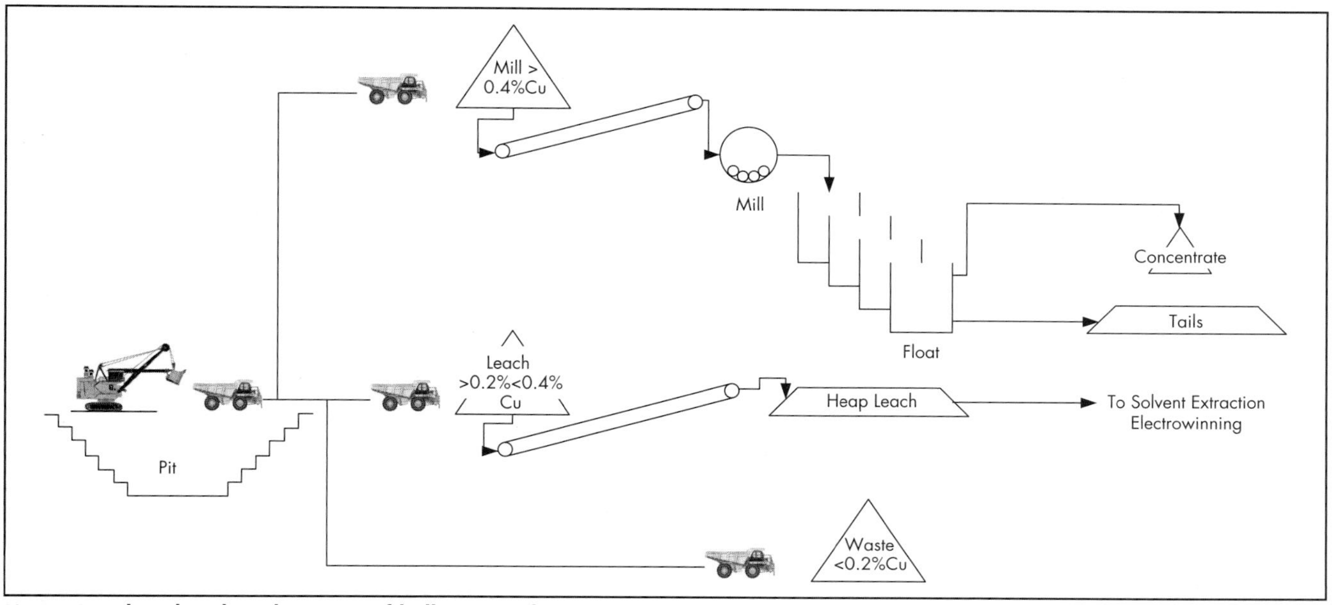

Figure 32 Shovel and truck sorting of bulk commodities (e.g., Cu 100,000 t/d)

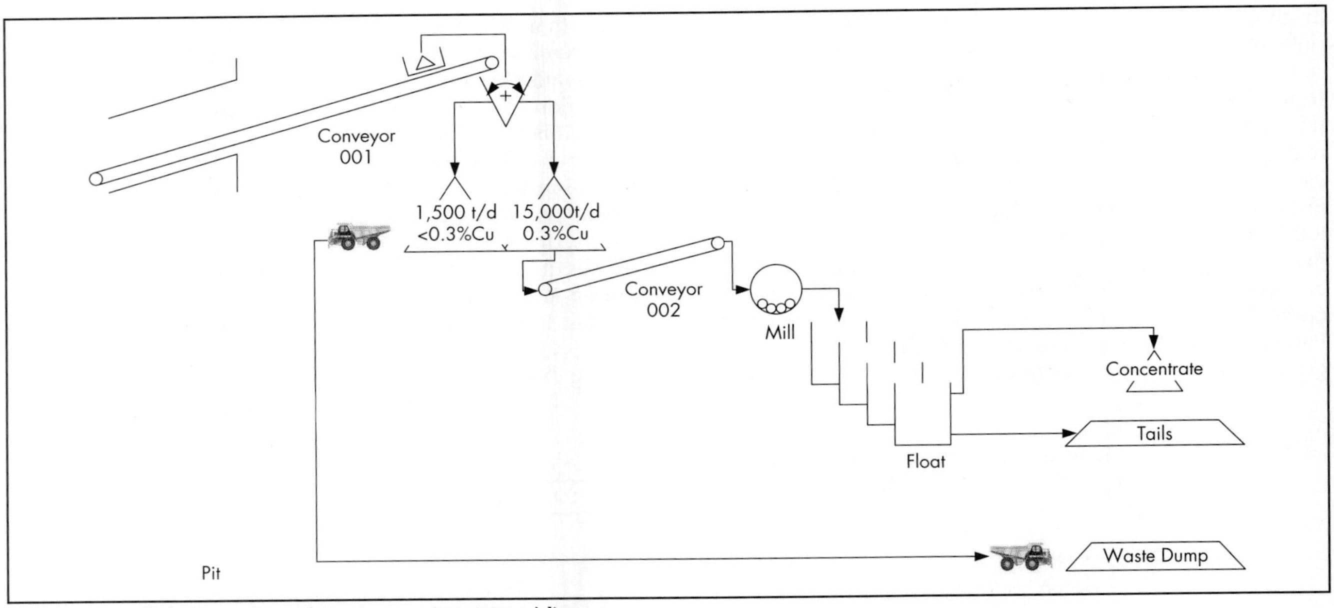

Figure 33 Bulk belt sorting of Cu-Au ore (16,500 t/d)

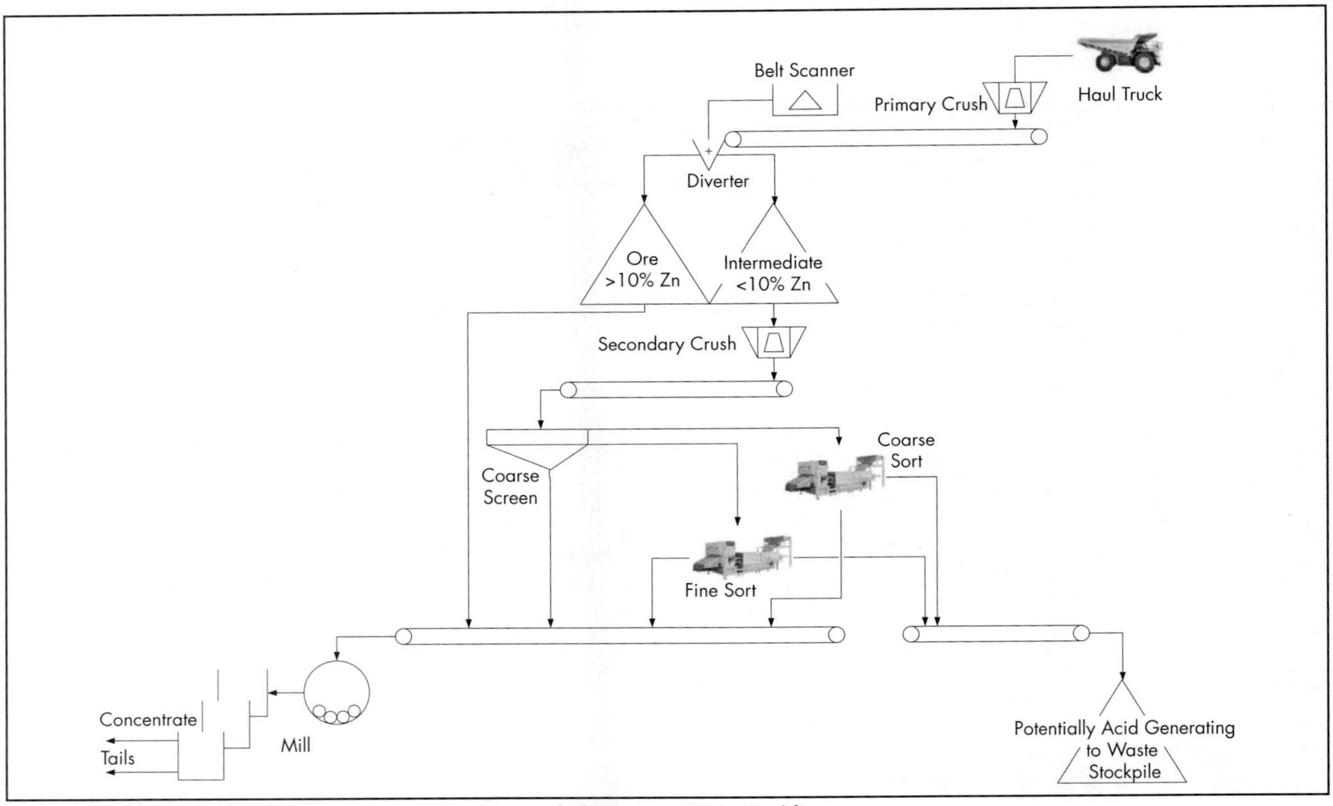

Figure 34 Combination of bulk and particle sorting of Pb-Zn ore (10,000 t/d)

but order of magnitude higher capacity and lower CAPEX to OPEX). In high-capacity (~10,000 t/h) situations, bulk sorting is often the only option. Then, where contemplated, bulk sorting is often traded off against basic capital expansion (pit pushback or plant expansion) options, and commonly against more intensive or sophisticated grade control and material routing strategies. Options are generally constrained by the fundamental degree of heterogeneity in the ore body, the length scale at which this occurs, and the capacity required of the ore recovery or waste rejection solution.

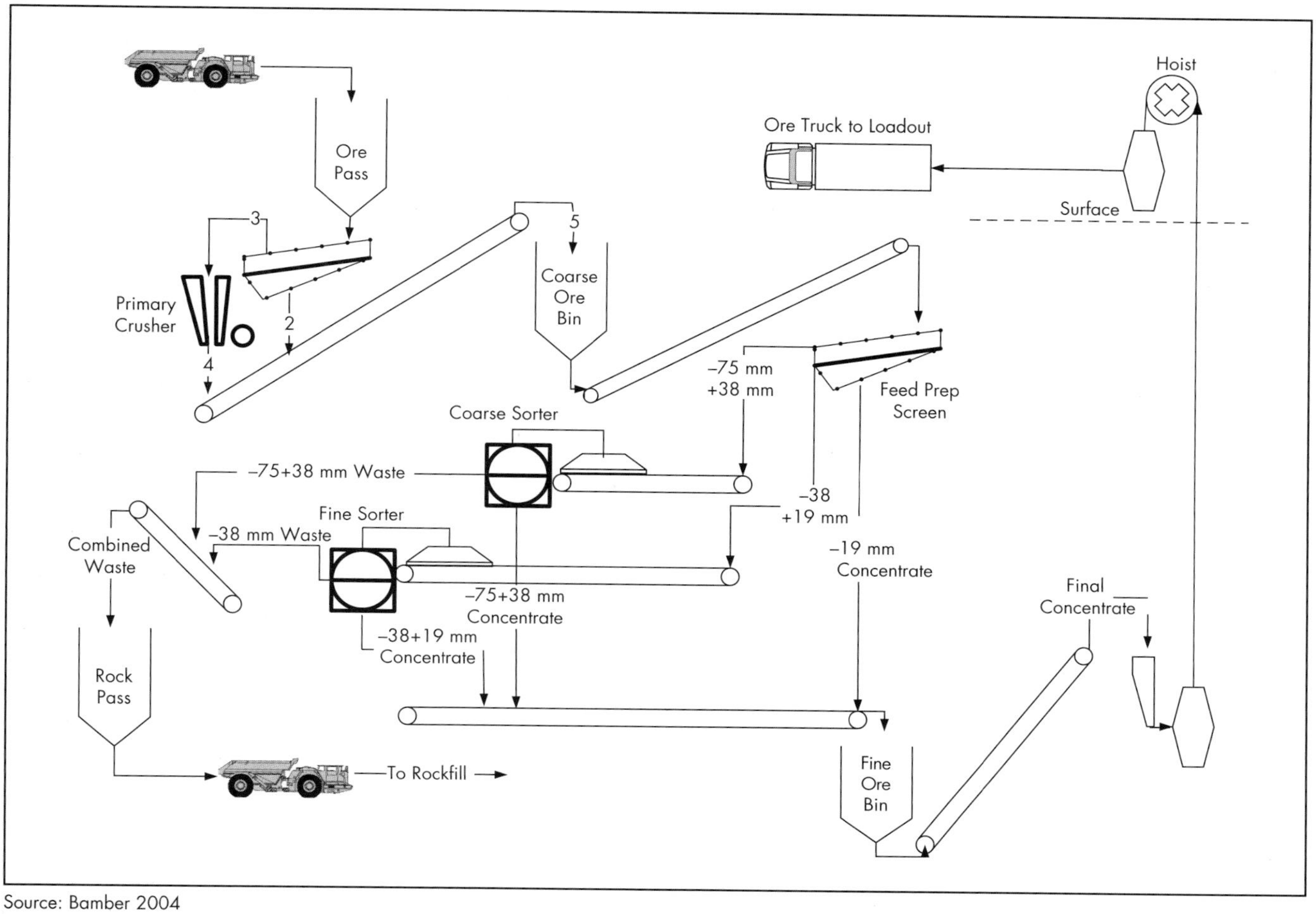

Source: Bamber 2004

Figure 35 Combination design including size classification, coarse and fine-particle sorting for an underground Ni mine (3,000 t/d)

Table 8 Selected published sorting results

Preconcentration Method	Ore Type	Feed Size, mm	Reject, wt %	Metal Recovery	Reference
Model 19 optical	Pb-Zn	150	26.3	94	Collins and Bonney 1995
Model 6 radiometric	U_3O_8	150	39	96.5	
Magnetic	Ni	100	40	96.7	
Radiometric	Au-U_3O_8	50	50	98	Sivamohan and Forrsberg 1991
Conductivity	Ni-Cu	25	80	80	
Conductivity	Ni-Cu	100	60	90	
Conductivity	Ni-Cu	70	27.83	82	
Model 16 optical	Cu	100	32.7	96	
Screening	Cu	100	20	99	
Optical	Au	100	44.8	94.8	Wilkinson 1985
Conductivity sorting	Au	100	54	80.5	
Optical plus conductivity	Au	100	50.2	97.8	
Radiometric	Wits Au	250	44.1	87.9	Kowalcyk 2011
Radiometric	Wits Au	250	29.5	92.6	
Comminution/screening	Cu-Platinum group elements	75	37	99	Bamber 2008
Comminution/screening	Cu porphyry	31.75	54.37	75.75	Burns and Grimes 1986

Adapted from Bamber 2008

Table 9 In-pit ore recovery/waste rejection at a large-scale open pit Cu mine

	Mill	Leach	Waste
Base Case			
Yield, t	95,000	63,000	160,000
Grade, %Cu	0.48	0.35	0.19
Bulk Sorting Products			
Yield, t	23,750	0	25,600
Grade, %Cu	0.22	0	0.55
Combined			
Yield, t	96,850	63,000	158,150
Grade, %Cu	0.56	0.35	0.14

Capital Costs, Operating Costs, and Economic Analysis

In cases where good sorting potential exists, sorting, whether by particle or in bulk, is generally more attractive from a capital cost point of view than competing alternatives. Bulk sorting, whether shovel, truck, or belt, is attractive from an operating cost perspective. Minimal capital investments are required, particularly for instrumenting existing mobile equipment, such as shovels and trucks, making these options attractive from a cost point of view versus alternatives delivering similar value. Nield (2002) reports four truck scanners generating a value of between US$120,000 and $160,000 per month with payback "in months." For belt installations, CAPEX on new conveyors may be required, however, these are again relatively low capacity per cost compared to alternatives and therefore quite feasible. Based on several case studies, typical payback on belt applications, where feasible, varies between three and six months (Kurth 2015; Hilscher 2016).

Selection of particle sorting, however, does incur capacity and cost issues, as applications are inherently limited to low throughputs and where scaling this option to higher throughput requires increasing numbers of machines and therefore circuit complexity, which must be taken into account (Arvidson 2002; McCarthy 2014). For particle sorting installations, pretreatment as well as product handling circuits are required, which can significantly add to the overall capital cost.

Particle sorting also embodies higher operating costs than bulk sorting. This is particularly true in the case of larger models of conventional sorters deploying compressed air pulses as the method of particle diversion. While the sorter itself may be operated at relatively low cost, compressed air generation significantly adds to the capital and operating cost of the circuit. Despite this, particle sorting does find application in lower throughput, higher value situations.

Indicative capacities for typical particle sorting installations, together with indicative capital and operating costs are presented in Table 11. Figure 36 presents some order-of-magnitude operating cost guidelines for particle sorting circuits to further aid in selection.

Barton and Peverett (1980) quote US$1,450,130 initial cost and US$0.28/t operating for a 90-t/h ore sorter model 19 installation at Driefontein gold mine in South Africa. More recently, Rule et al. (2015) reported overall capital costs of US$5 million for a 30-t/h pilot XRF installation, and up to US$100 million capital overall for a 1,000-t/h production facility. Operating costs at 1,000 t/h were estimated at between US$0.20/t and $0.30/t. Nevertheless, even with higher unit capital and unit operating cost requirements, typical reported payback for particle sorting projects, where ore is amenable, vary from one to two years (Hilscher 2016).

Table 10 Ore recovery from Ni-Cu mine waste dump

Distribution/ Grade	Feed, %	Product, %	Tails, %
Weight %	100	33	68
Ni Grade %	0.67	1.57	0.28
Ni Weight %	100	76	24
Fe Grade %	16.55	18.39	9.04
Fe Weight %	100	40	60

Table 11 Indicative throughputs, capital expenditure (CAPEX), and operational expenditure (OPEX) of particle sorters (in US$)

Top size, mm		25	50	100	200
Throughput, t/h		35	60	120	250
CAPEX	Sorter only	$1,312,500	$1,579,000	$1,875,000	$2,025,000
OPEX	Sorter	$0.33	$0.22	$0.10	$0.08
	Ancillaries	$0.50	$0.25	$0.12	$0.09

Courtesy of Matthew Kowalcyk

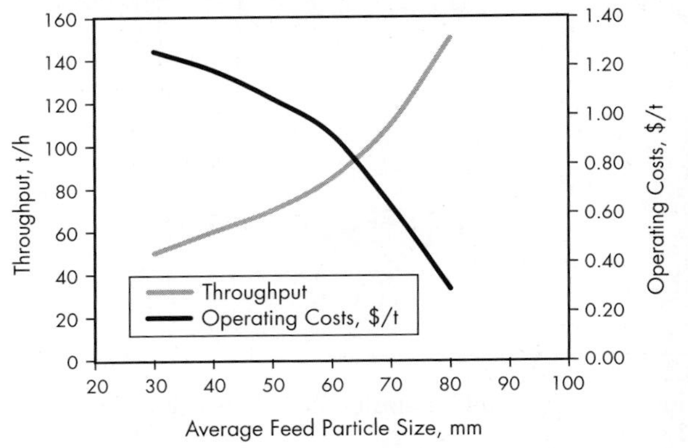

Note: Costs adjusted to 2016 costs using capital expenditure cost index.

Adapted from Robben 2014

Figure 36 Relationship between throughput, particle size, and operating cost

Installation, Operating, and Maintenance Considerations

Shovel- and truck-based systems require little overhead in the way of installation, although Nield (2002) describes a requirement for this type of system to be integrated with the mine's grade control and ore routing systems for maximum effectiveness. Bulk systems, whether shovel, truck, or belt, require little in the way of feed preparation, although bulk belt sorting systems do require additional material handling for reject streams that are created. Particle sorting systems require a higher standard of feed preparation, both in terms of crushing and screening the feed within the particle size ranges previously described (3:1 top size to bottom size) as well as in terms of removing adhering fines, for example, by wet screening for better analysis of the particles to be sorted, particularly with use of surface sensing methods. Strict specifications relating to presentation of the feed to particle sorters (monolayer of particles, controlled velocity, particles preferably not

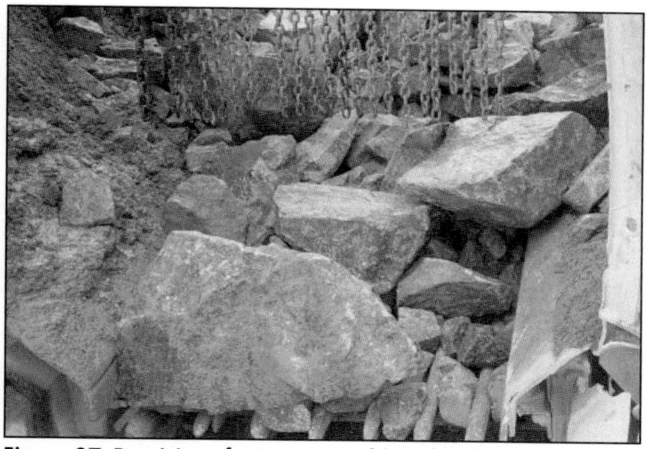

Figure 37 Provisions for impact and/or abrasion resistance in feed and discharge areas are important in primary applications

touching) also need to be observed in circuit design where not catered for by the vendor. Additionally, particle sorting solutions often implicate multiple units in parallel because of their low unit capacity, therefore the feed preparation circuit, the sorting plant itself, as well as the product handling circuit can be complex and require careful design.

In operations, several considerations pertain. Maintaining calibration of the sensors to grade is key, where either step changes in calibration or drift over time can occur; calibration should be regularly monitored and corrected. Changes in the feed material type over time should also be tracked, particularly where feed transitions to a type not previously characterized for sorting; thus recharacterization by the procedures described would then be required to maintain system performance. Barton and Peverett (1980) note fragmentation of the feed on entering the sorter (as this is typically done at high velocity and can be an area of high impact) can affect system performance. Variations in temperature and humidity of the environment, as well as excess moisture content in the feed itself, should also be monitored and, if required, controlled.

Several maintenance considerations should also be noted. Damage to sensors by direct impact of rocks or other environmental factors relating to the aggressive mine environment should be monitored. Sensor and sorter systems should always be designed to *fail-safe*, that is, to default to normal operation on breakdown as well as cause no direct hazard during failure. Barton and Peverett (1980) note the potential wear in feed and discharge chutes, as these are areas of high impact abrasion as well as potential damage to sorter belts by rocks, as these are often not mining-rated belts (Figure 37). Blockages to air valves in air-actuated machines have been noted, as well as wear and actuator failure on mechanically diverted systems.

Recently Reported Sorting Installations

Hilscher (2016) lists several recent sorting examples, their location, application, and technology. Russia, Australia, and South Africa appear as leading practitioners, with Europe and the Americas currently lagging in adoption (Table 12). Figure 38 shows an example of an XRF sorting plant at a South African platinum mine.

Table 12 Recent ore sorting installations and the technology applied

Year	Country	Segment	Application	Type
2000	Russia	Gold	Ore preconcentration	X-ray fluorescence (XRF)
2000	Russia	Gold	Waste rock removal	XRF
2001	Kazakhstan	Manganiferous	Ore preconcentration	XRF
2001	Russia	Gold	Ore preconcentration	XRF
2001	Russia	Lead/zinc	Ore preconcentration	XRF
2002	South Africa	Platinum	Waste rock removal	Color
2003	Russia	Copper/zinc	Ore preconcentration	XRF
2003	Brazil	Aluminum	Not available	Color
2003	Russia	Nickel	Ore preconcentration	XRF
2004	Russia	Nickel	Ore preconcentration	XRF
2005	Australia	Nickel	Waste rock removal	Electromagnetic (EM)
2005	Russia	Copper/zinc	Ore preconcentration	XRF
2006	Kazakhstan	Chromite	Ore preconcentration	XRF
2006	Russia	Copper/zinc	Ore preconcentration	XRF
2006	Russia	Uranium	Ore preconcentration	XRF
2007	Australia	Nickel	Waste rock removal	EM
2007	Austria	Tungsten	Waste rock removal	X-ray transmission (XRT)
2007	Russia	Gold	Ore preconcentration	XRF
2007	Russia	Fluorite	Ore preconcentration	XRF
2010	South Africa	Gold	Waste rock removal	Color
2010	United States	Gold	Waste recovery	EM
2010	Canada	Nickel	Waste rock removal	EM
2011	Australia	Gold	Waste rock removal	XRT/laser
2011	Australia	Gold	Waste rock removal	Color/laser
2011	Australia	Gold	Waste rock removal	Ultraviolet/laser
2011	Russia	Gold	Waste rock removal	XRT
2011	Russia	Magnesite	Ore preconcentration	EM
2011	United States	Copper	Waste rock removal	Not available
2011	Australia	Tungsten	Ore preconcentration	XRT
2012	South Africa	Platinum	Ore preconcentration	XRF
2012	Australia	Tungsten	Ore preconcentration	XRF
2012	South Africa	Manganiferous	Waste recovery	EM
2012	Austria	Tungsten	Waste rock removal	XRT
2013	South Korea	Tungsten	Waste rock removal	XRT
2013	Russia	Gold	Waste recovery	Laser
2013	China	Gold	Waste rock removal	XRT
2014	Russia	Magnesite	Ore preconcentration	XRF
2014	Namibia	Gold	Ore preconcentration	XRF
2014	Australia	Copper	Preconcentration	XRF
2014	Australia	Copper	Ore preconcentration	XRF
2015	United States	Gold	Waste rock removal	XRT

Adapted from Hilscher 2016

Table 12 demonstrates an extensive range of ores where the technology is now proven, including manganese, lead-zinc, chromite, aluminum, and platinum, in addition to traditional applications in nickel, copper, gold, and diamonds. New applications, such as with rare earth elements, lithium,

Source: Rule 2012

Figure 38 X-ray fluorescence (XRF)-based platinum sorting plant

molybdenum, and phosphates, are constantly in development. As can also be seen from the table, the pace of implementation has recently accelerated on the back of improvements in capacity and applicability of the technology, as well as an increased understanding by industry practitioners in how to apply sorting to the benefit of selected operations.

REFERENCES

Altun, N.E., Weatherwax, T., and Klein, B. 2014. Upgrading valuable mineralization and rejecting magnesium silicates by pre-concentration of mafic ores. *Physicochem. Probl. Miner. Process.* 50(1):203–215.

Arvidson, B. 2002. Photometric ore sorting. In *Mineral Processing Plant Design, Practice, and Control*. Vol. 1. Edited by A.L. Mular, D.N. Halbe, and D.J. Barratt. Littleton, CO: SME. pp. 1033–1048.

Bamber, A.S. 2004. An integrated underground mining and processing system for massive sulphide ores. Master's thesis, University of British Columbia, Vancouver, Canada.

Bamber, A.S. 2008. Integrated mining, pre-concentration and waste disposal systems for the increased sustainability of hard rock metal mining. Ph.D. thesis, University of British Columbia, Vancouver.

Bamber, A.S. 2012. The mining industry needs to be shaken *and* stirred. In *Narrow Vein Mining Conference 2012: Proceedings*. Melbourne, Victoria: Australasian Institute of Mining and Metallurgy. pp. 3–14.

Bamber, A.S., and Houlahan, D.J. 2010. Development, testing and applications of an induction-balance sensor for low grade base metal ores. In *Proceedings from Sensorgestützte Sortierung 2010* [in German]. Edited by T. Pretz and H. Wotruba. Clausthal-Zellerfeld, Germany: GDMB.

Bamber, A., Klein, B., Morin, M., and Scoble, M. 2004. Reducing selectivity in narrow-vein mining through the integration of underground pre-concentration. In *Proceedings of the Fourth International Symposium on Narrow Vein Mining Techniques*. Montreal, QC: Canadian Institute of Mining, Metallurgy and Petroleum.

Bamber, A.S., Klein, B., Morin, M., Scoble, M.J. 2005. Integration of pre-concentration underground: Reducing mining costs. In *Proceedings, 14th International Symposium on Mine Planning and Equipment Selection (MPES 2005)*. Edited by R. Singhal, K. Fystas, and C. Chiwetelu. Irvine, CA: Reading Matric.

Bamber, A.S., Klein, B., and Stephenson, M. 2006. A methodology for mineralogical evaluation of underground pre-concentration systems and a discussion of potential process concepts. In *Proceedings of the 23rd International Mineral Processing Congress*. Edited by G. Onal. Istanbul, Turkey: Istanbul Technical University. pp. 253–258.

Bamber, A.S., Klein, B., and Scoble, M.J. 2007. Integrated mining and processing of massive sulphide ores. In *Proceedings, Canadian Mineral Processors 39th Annual Operators Conference*. Ottawa, ON: Natural Resources Canada.

Bamber, A.S., Klein, B., Pakalnis. R., and Scoble, M.J. 2008. Integrated mining, processing and waste disposal systems for reduced energy and operating costs at Xstrata Nickel's Sudbury Operations Institute of Materials. *Min. Technol.* 117(3):142–153.

Bamber, A., How, P., McDevitt, C.A., Munoz-Paniagua, D., Dirks, M., and LeRoss, J. 2016. Development and testing of real-time shovel-based mineral sensing systems for the enhanced recovery of mined material. In *7th Sensor-Based Sorting and Control*. Edited by T. Pretz and H. Wotruba. Aachen, Germany: Shaker Verlag.

Barbanson, L., Sizaret, S., Rozembaum, O., and Rouet, J.L. 2009. Textural characterization using Gy's heterogeneity functions. *Math. Geosci.* 41(8):855–868.

Barton, P.J., and Peverett, N.F. 1980. Automated sorting on a South African gold mine. *J. S. Afr. Inst. Min. Metall.* 80(3):103–111.

Bergmann, C. 2009. Developments in ore sorter technology. Johannesburg: Council of Mineral Technology Developments in Ore Sorting Technologies.

Burns, R., and Grimes, A. 1986. The application of preconcentration by screening at Bougainville Copper Limited. In *Proceedings of the AusIMM Mineral Development Symposium*. Melbourne, Victoria: Australasian Institute of Mining and Metallurgy.

Carrasco, C., Keeney, L., and Walters, S.G. 2016. Development of a novel methodology to characterise preferential grade by size deportment and its operational significance. *Miner. Eng.* 91:100–107.

Clark, R.N. 1999. Chapter 1: Spectroscopy of rocks and minerals, and principles of spectroscopy. In *Manual of Remote Sensing*. Vol. 3, Remote Sensing for the Earth Sciences. Edited by A.N. Rencz. New York: Wiley. pp. 3–58.

Collins, D.N., and Bonney, C.F. 1998. Separation of coarse particles (over 1 mm). *Int. Min. Miner.* 1-2(Apr):104–112.

Duffy, K., Valery, W., Jankovic, A., Holtham, P. 2015. Integrating bulk ore sorting into a mining operation to maximize profitability. In *MetPlant 2015*. Melbourne, Victoria: Australasian Institute of Mining and Metallurgy. pp. 273–287.

Egerton, F.M.G. 2004. Presidential address: The mechanisation of UG2 mining in the Bushveld complex. *J. S. Afr. Inst. Min. Metall.* 104(8):439–455.

Engelbrecht, J. 2012. Potential changes in the physical beneficiation processes that can improve the recovery grade or costs for the platinum group metals. In *5th International Platinum Conference: "A Catalyst for Change."* Symposium Series S72. Johannesburg: Southern African Institute of Mining and Metallurgy. pp. 343–359.

Fickling, R.S. 2011. An introduction to the Rados XRF ore sorter. In *6th Southern African Base Metals Conference.* Symposium Series S66. Johannesburg: Southern African Institute of Mining and Metallurgy.

Fortes, F.J., and Laserna, J.J. 2010. The development of fieldable laser-induced breakdown spectrometer: No limits on the horizon. *Spectrochim. Acta, Part B* 65(12):975–990.

Gy, P.M. 1992. Sampling of heterogeneous and dynamic material systems. In *Theories of Heterogeneity, Sampling and Homogenising.* Vol. 10. Amsterdam: Elsevier.

Harris, D.A. 1987. Chapter 17. In *Spectrophotometry and Spectrofluorimetry: A Practical Approach.* Edited by D.A. Harris and C.L. Bashford. Oxford: IRL Press.

Hilscher, B. 2016. Project revitalization using sensor based ore sorting. In McEwen Mining's Innovation Lunch and Learn Series [YouTube video]. *CIM Mag.* http://magazine.cim.org/en/videos/mcewen-mining-innovation-series/sensor-based-ore-sorting/.

Hitch, M., Bamber, A., and Oka, P. 2015. Pre-sorting of high grade molybdenum ore: A case for enhanced small mine development. *Proc. J. Eng. Appl. Sci.* 2(5):129–135.

Klein, B., Dunbar, W.S., and Scoble, M. 2002. Integrating mining and mineral processing for advanced mining systems. *CIM Bull.* 95(1057):63–68.

Klein B., Hall, R., Scoble M., and Morin, M. 2003. Total systems approach to design for underground mine-mill integration. *CIM Bull.* 97(1067):65–71.

Klein, B., Bamber, A.S., Weatherwax, T., and Altun, N.E. 2010. Comparison of pre-concentration technologies for mine-mill integration. In *Proceedings from Sensorgestützte Sortierung 2010* [in German]. Edited by T. Pretz and H. Wotruba. Clausthal-Zellerfeld, Germany: GDMB.

Klein, B., Bamber, A.S., Altun, N.E., and Scoble, M.J. 2011. Towards tomorrow's 'smart mine'—Embedded sensor telemetry and sensor-based sorting. In *Proceedings from the Second International Future Mining Conference.* Melbourne, Victoria: Australasian Institute of Mining and Metallurgy. pp. 59–68.

Klein, B., Altun, N.E. and Scoble, M.J. 2015. Comminution: Technology, energy efficiency and innovation. In *24th International Mining Congress and Exhibition of Turkey (IMCET 2015).* Edited by M. Kilic. Ankara: Chamber of Mining Engineers of Turkey.

Kleiv, R.A. 2012. Pre-sorting of asymmetric feeds using collective particle ejection. *Physicochem. Prob. Miner. Process.* 48(1):29–38.

Kowalcyk, M. 2011. Sorting opportunities, sensors and applications (Why send waste to the mill?). Presented at the Canadian Institute of Mining, Metallurgy and Petroleum Annual General Meeting.

Kruukka, A., and Briocher, H.F. 2002. Kiruna mineral processing starts underground—Bulk sorting by LIF. *CIM Bull.* 96(1066):79–84.

Kurth, H. 2015. Geoscan elemental analyzer for optimising plant feed quality and process performance. Presented at the 6th International Conference on Semi-Autogenous and High Pressure Grinding Technology (SAG 2015), Vancouver, BC, September 20–23.

Lessard, J., de Bakker, J., and McHugh, L. 2014. Development of ore sorting and its impact on mineral processing economics. *Miner. Eng.* 65:88–97.

Manouchehri, H.-R. 2003. Sorting: possibilities, limitations and future. In *Conference in Minerals Engineering.* Edited by M. Thomaeus and E. Forssberg. Stockholm: Swedish Mineral Processing Research Association. pp. 1–7.

Mazhary, A., and Klein, B. 2015. Heterogeneity of low-grade ores and amenability to sensor-based sorting. *47th Canadian Mineral Processors Annual Meeting.* Canadian Mineral Processors pp. 387–400.

McCarthy, B. 2014. So you have it crushed and on a conveyor—Now what? Optimizing value through ore sorting. Presented at the In-Pit Crushing and Conveying Conference (IPCC), Johannesburg, South Africa, November 16–18.

Murphy, B., van Zyl, J., and Domingo, G. 2012. Underground pre-concentration by ore sorting and coarse gravity separation. In *Narrow Vein Mining Conference.* Edited by S. Dominy. Melbourne, Victoria: Australasian Institute of Mining and Metallurgy.

Nadolski, S., Klein, B., Elmo, D., and Scoble, M. 2015. Cave-to-mill: A mine-to-mill approach for block cave mines. *Min. Technol.* 124(1):47–55.

Nield, T. 2002. Beyond the forked twig. *Geoscience* 12(12).

Pitard, F.F. 1993. *Piere Gy's Sampling Theory and Sampling Practice: Heterogeneity, Sampling Correctness, and Statistical Process Control,* 2nd ed. Boca Raton, FL: CRC Press. pp. 63–87.

Robben, C. 2014. Schriftenreihe zur Aufbereitung und Veredlung [Characteristics of sensor-based sorting technology and implementation in mining]. Aachen, Germany: Shaker Verlag.

Robben, C., Wotruba, H., Robben, M., von Ketelhodt, L., and Kowalzcyk, M. 2013. Potential of sensor-based sorting for the gold mining industry. *CIM J.* 4(3):191–200.

Rule, C. 2012. Pre-concentrating and upgrading of ores. Presented at the Southern African Institute of Mining and Metallurgy (SAIMM) Colloquium on Ore Sorting, Cape Town.

Rule, C., Fouchee, R.J., and Swart, W.C.E. 2015. Run of mine ore upgrading—Proof of concept plant for XRF sorting. Presented at the 6th International Conference on Semi-Autogenous and High Pressure Grinding Technology (SAG 2015), Vancouver, BC, September 20–23.

Salter, J.D., and Wyatt, N.P.G. 1991. Sorting in the minerals industry: Past, present and future. *Miner. Eng.* 4(7):779–796.

Scoble, M. 1994. Competitive at depth: Re-engineering the hard rock mining process. Presented at the 1st North American Rock Mechanics Symposium, Austin, TX.

Scoble, M., Klein, B., Dunbar, W.S. 2003. Mining waste: Transforming mining systems for waste management. *Int. J. Surf. Min. Reclam. Environ.* 17(2):123–135.

Senesi, G.S. 2014. Laser-induced breakdown spectroscopy (LIBS) applied to terrestrial and extraterrestrial analogue geomaterials with emphasis to minerals and rocks. *Earth Sci. Rev.* 139:231–267.

Sivamohan, R., and Forssberg, E. 1991. Electronic sorting and other preconcentration methods. *Miner. Eng.* 4(7-11):797.

SPE (Society of Petroleum Engineers). 2015. Spectral gamma ray logs, fig. 1. http://petrowiki.org/Spectral_gamma_ray_logs.

Tong, Y. 2012. Technical amenability study of laboratory-scale sensor-based ore sorting on a Mississippi Valley type lead-zinc ore. M.S. thesis, University of British Columbia, Vancouver, Canada.

Tong, L., Khoshaba, B., Bamber, A.S., and Klein, B. 2015. Correlation and regression analysis in the x-ray fluorescence sorting of a low grade copper ore. Presented at the 6th International Conference on Semi-Autogenous and High Pressure Grinding Technology (SAG 2015), Vancouver, BC, September 20–23.

Van Weert, G., Kondos, P., and Wang, O. 2011. Microwave heating of sulphide minerals as a function of their size and spatial distribution. *CIM J.* 2(3):117–124.

Wilkinson, S. 1985. Techniques for coarse particle beneficiation of some gold ores and methods of determining potential for sorting any gold operation. In *Proceedings of the Canadian Mineral Processors Annual General Meeting 1985*. pp. 258–284. Canadian Mineral Processors.

Wotruba, H. 2006. Sensor sorting technology—Is the minerals industry missing a chance? In *Proceedings of the 23rd International Mineral Processing Congress* (IMPC). Edited by G. Önal. Istanbul, Turkey: IMPC. pp. 21–30.

Wotruba, H., and Harbeck, H. 2010. Sensor-based sorting. In *Ullmann's Encyclopedia of Industrial Chemistry*, 7th ed. Hoboken, NJ: Wiley.

CHAPTER 6.2

Gravity Concentration

Robert C. Dunne, Rick Q. Honaker, and Aidan Giblett

Gravity separation is second only to hand sorting in being the oldest of methods used to concentrate minerals. Agricola recorded the use of jig screens and sluices for the separation of heavy metals in the 16th century (Hoover and Hoover 1950). The industrial revolution, which occurred from 1760 to 1840, resulted in a demand for large-scale production of minerals and coal, thereby fueling the ingenuity and energy of innovators to develop density-based separators that were used to recover minerals from relatively rich ore deposits. The modern-day separators such as continuously operating jigs and shaking tables were developed, patented, and commercialized in the late 19th century. Inventions that provided enhanced gravity separations in a mechanically applied centrifugal field were patented in the 1890s and served as the basis of recently commercialized technologies.

The early 20th century brought continued development of high-capacity, density-based separators for coarse particle concentration, such as the Baum jig and the Chance cone. Several dry separation technologies were developed and commercialized from 1920 to 1940, mainly for the treatment of coal. As the ore bodies became lower grade and underground mechanized coal mining became more prevalent, development activities shifted somewhat to technologies capable of treating fine particles (1–0.15 mm). Flowing film separators, such as the spiral concentrator and Reichert cone concentrator, were introduced and realized wide commercial use in the middle part of the 20th century. In addition, hydraulic fluidized-bed classifiers used for particle size separations were proven to provide effective density-based separations for a wide variety of fine materials when operated under certain conditions.

The late 20th century was the era of enhanced gravity separator (EGS) development, because of the desire to provide low-cost, highly efficient recovery of ultrafine heavy metals such as gold and tin. Mechanisms commonly employed in conventional technologies such as jigging, flowing film sluices, fluidized beds, and shaking tables were applied in devices that achieve separation in a centrifugal field created by mechanical action. The EGS units have been commercially applied to the recovery of a wide variety of different metallic minerals present in particle sizes as fine as 10 µm. Burt (1984, 1986, 1999) provides a comprehensive historical overview and description of the various density-based separators that have been used by mineral processors since the early 1900s.

HINDERED SETTLING CONCENTRATION

In the free settling of mineral particles in a liquid, the falling particles are at a distance from each other so that no particle is affected by its neighbor. In hindered settling, the concentration of particles is sufficiently high so that each particle is affected by its proximity to other particles in the suspension. Richards and Locke (1940) described the hindered settling phenomenon as the condition "where particles of mixed sizes, shapes and densities in a crowded mass, yet free to move along themselves, are sorted in a rising current of water, the velocity of which is much less than the free-falling velocity of the particles but yet fast enough so the particles are in motion." This is the condition normally encountered in mineral concentration processes.

For free settling of coarse particles (+2 mm), the Newton equation (Symonds 1986; Wilson et al. 2006) applies, whereas for the settling of fine spheres (~106 µm) in water, the Stokes equation applies (Batchelor 1967). For particles whose size lies between ~2 mm and 106 µm, their settling velocity can be determined from experimental data. These data are available in convenient form in the text by Taggart (1951). Alternatively, a Reynolds number coefficient-of-resistance plot may be used to determine the settling rate of such particles (Gaudin 1939).

Particle shape affects the settling rate of both coarse and fine particles. The general effect is to reduce their settling velocities, and the effect is greater for coarse particles and for those settling under hindered settling conditions than for fine particles or free-settling ones. However, nearly all gravity concentration processes (jigs, tables, flowing film concentrators, heavy media separators) and many sizing devices (sizing classifiers, clarifiers, thickeners, hydroseparators) make use of the hindered settling phenomenon.

Robert C. Dunne, Principal, Robert Dunne Consulting, Gooseberry Hill, Western Australia, Australia
Rick Q. Honaker, Professor, Mining Engineering, University of Kentucky, Lexington, Kentucky, USA
Aidan Giblett, Senior Technical Advisor—Mineral Processing, Newmont Mining Services, Subiaco, Western Australia, Australia

Hindered Settling Separators

Jigs

Jigging is an ideal preconcentration process, being relatively inexpensive in construction, operation, and maintenance, and relatively unaffected by feed grade. All jigs utilize a screen (the older-type jigs have a fixed screen while the more modern jigs have a moving screen) and a means to provide pulsations through a particle bed (ragging), which results in particle separation on the basis of density by hindered settling and consolidated trickling (Figure 1). Heavy mineral particles pass through the screen while light particles overflow a weir.

The fundamental principles of all jigs are essentially the same. The basic differences between the various types of jigs are a matter of practical engineering to optimize the operating performance, materials handling, maintenance, and control. The basic design features (Taggart 1945, 1951) of a jig are

- Screen to support the mineral bed,
- Hutch or tank containing the liquid beneath the screen,
- Means of creating a jig stroke or relative motion between the liquid and the bed,
- Method of modulating the jig-stroke waveform,
- Method of regulating the upflow of water,
- Method of supplying feed to the bed, and
- Method of removing products from above the screen and from the hutch.

Ragging is a layer of large, heavy particles at the bottom of the bed and on the jig screen, as shown in Figure 1. The ragging controls the rate at which the heavy fine particles penetrate and percolate through the bed to the hutch. For some ores there is enough coarse heavy mineral to provide this layer. However, with many ores it is necessary to provide an added layer of coarse heavy material, which is called ragging. In general, the ragging particles must be heavy enough to remain at the very bottom of the bed, but light enough for dilation on the upstroke. The particle size must be greater than the screen openings and large enough to provide spaces between the particles to allow concentrate particles to percolate through on the downstroke. Harrison (1962) recommends a ragging particle diameter of four times that of the maximum particle size of the hutch product.

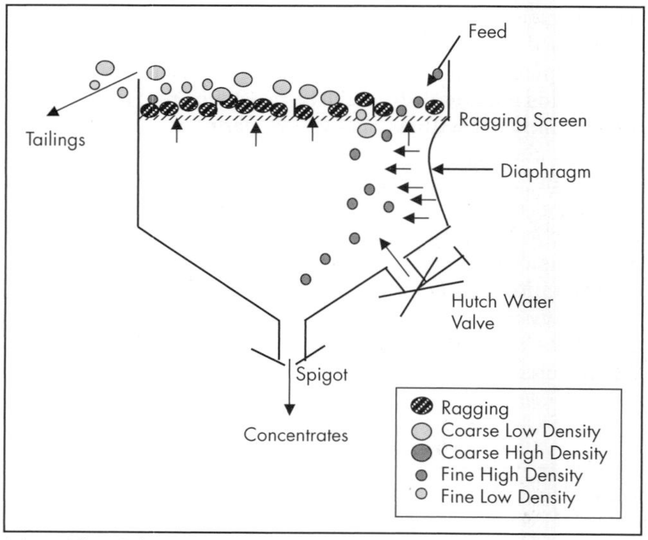

Adapted from Falconer 2003

Figure 1 Generic representation of a fixed-screen jig

For mineral jigs, capacity is heavily dependent on the average density of the feed. This determines the volume of material (dry basis) that is fed to the jig. Jig design also influences capacity, as circular jigs have a higher unit capacity than rectangular jigs.

In any jig application, the size distribution and the density of the particles of the ore will result in a unique situation requiring optimization of each of the jig operating parameters. To concentrate fine heavy particles, the suction phase must be augmented, whereas for coarse heavy particles, the pulsion phase is the most important. These opposed considerations give rise to much of the divergence of opinion about optimizing stroke speed, amplitude, and modulation, and with jig design in general.

The following are some of the major adjustment factors for jig operation:

- **Stroke speed and length influence the water waveform across the screen.** Balancing the stroke speed and length is an important aspect leading to good or poor performance.
- **Water flow.** The jig stroke is modulated by the water upflow. Too much hutch water will result in fine particle values to the tailings. Too little hutch water will dilute the hutch concentrate with non-valuable particles.
- **Ragging.** Choosing the correct type and size of ragging is important to get the desired cut point for separation. Ragging size also limits the size of the coarse material that will pass through the jig screen.
- **Jig slope.** The bed must be sloped from feed to tailings discharge to facilitate transport of solids in the bed. Jig design must allow for adjustment of this slope during installation.
- **Jig area.** The area determines the capacity of the jig.

Pan-American jig. The most common mineral jig in use today is the Pan-American style, although some Yuba and Denver jigs can still be found. The Pan-American (often called just Pan-Am) jig uses a flexible diaphragm in the cone section of the jig, directly below the fixed screen, which is manipulated by a mechanical arm mounted eccentrically to provide an up-and-down movement of the cone and thus a pulsation source to the particle bed (Figure 2). Hutch water is added to reduce the speed of the suction stroke and allow additional time for particle stratification based on density.

The typical Pan-Am jig is normally built as a duplex jig (a pair of balanced jig cells). Each cell consists of an upper hutch (usually rectangular) and a lower hutch (usually conical), and it is joined by an annular diaphragm of flexible rubber to allow up-and down movement of the lower hutch. The standard Pan-Am jig consists of two cells (each 1 m × 1 m) with a common drive for energy savings and the addition of 193 kilos of 4.75-mm steel shot per cell as ragging. It has a stroke length of 19–38 mm and frequency of 20–200 cycles per minute.

Pan-Am duplex jigs are popular in many regions and are operated in Alaska (United States), Yukon (Canada), South America, Africa, and Asia. Much of this popularity is due to their inherent simplicity and relative ease of maintenance.

IHC jig. The IHC circular jig is shown in Figure 3. A trapezoidal-bed diaphragm mechanism provides a sawtooth pulse pattern. The jig can handle a broad particle range from

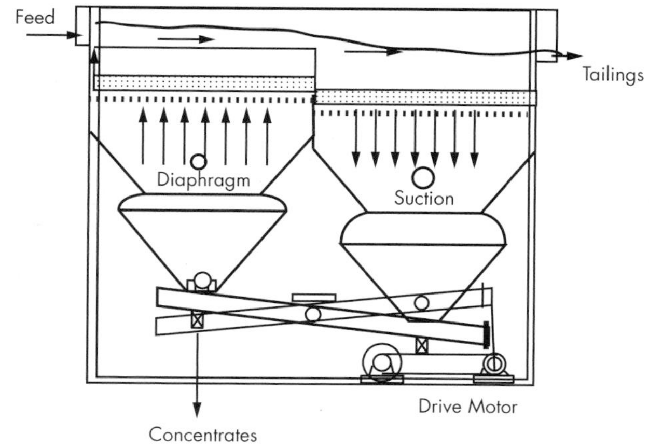

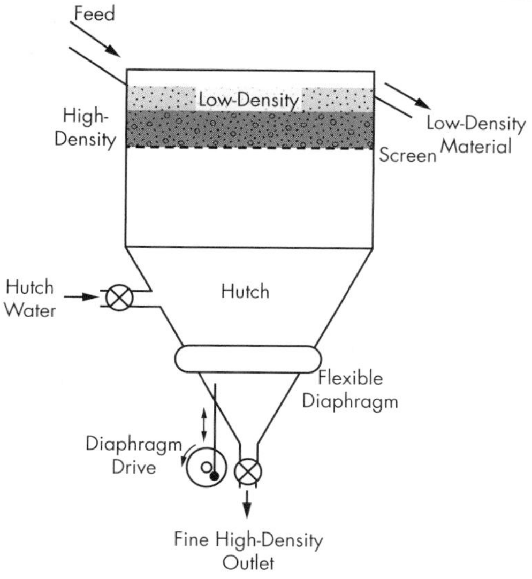

Source: Coggin 1995
Figure 2 Pan-American jig

Source: Honaker et al. 2014

Source: IHC Technical Services, n.d.
Figure 3 IHC circular jig

~50 µm to 12 mm. Capacities of the different sized jigs vary from 260 m^3 to 1,250 m^3 of slurry per hour. Freshwater requirements are ~14 to 336 m^3/h. The power requirements for the jig range from 7.5 kW to 49 kW (IHC Technical Services, n.d.).

The IHC jig employs a sawtooth pulsation pattern that reduces the hutch water requirement. The sawtooth movement consists of a fast upward and a slow downward stroke. The aim of the fast upward stroke of short duration is to prevent the loss of fine heavy material. Each system has its own features; deciding which system to use depends on the number of modules used in an installation. The downward stroke, also called the suction stroke, is much longer. During this phase, the fine particles are drawn into the bed. IHC Merwede has developed two types of jig drives, a mechanical drive and a mechanical-hydraulic drive. Both drives are designed to give reliable performance under the conditions that prevail on mineral dredgers and on mines. Two jig plant options are available:

1. **Standard design (SD) jig series.** The SD jig installation has a modular construction, which makes it a dismountable processing plant and easier to transport. Depending on the required capacity, modules can be combined up to a maximum of twelve. For smaller capacities a mini, micro, and super micro module are available.
2. **Skid frame (SF) jig series.** The SF series of jigs are mounted on skid frames, used for onshore installations. These jig installations are standardized to various capacities. Similar to the SD jig installation, the SF jig is also of modular construction.

Baum jig. The Baum jig, developed in the late 1890s, was a significant development with the replacement of conventional plungers in jigs at that time by compressed air and exhaust through the use of valves (Figure 4). The later improved control of the pulse cycle in larger compartments allowed for more-efficient separation over a wider range of particle sizes at higher throughput capacities (Sanders et al. 2002). The Baum jig became the most popular unit for coal beneficiation in the late 20th century with bed widths of ~2.5 m. The jig pulse cycle is typically sinusoidal, and substantial research has been undertaken to optimize the waveform to improve jig performance (Tanaka et al. 1990; Iijima et al. 1998).

Baum jig capacity is usually 29–59 $t/m^2/h$ of active screen area, with machine capacities ranging from 23 t/h to 635 t/h. The water requirement is usually 3.8–9.5 m^3/min but can be kept at a minimum by operating controls and recirculation. Water is split 30% to the feed, and the remainder to the cells. Because of density and viscosity effects, it is not advisable to operate with recirculated water with densities over 1.04 or if the feed contains more than 15% clay or suspended solids by weight (Leonard and Mitchell 1968; Price and Bertholf 1959). The power requirement is approximately 0.8 kW/m^2 of screen area, although this may be exceeded if the jig is not operated at the correct frequency. Compressed air supply is only 21 kPa.

Operating controls are necessary to control bed density and level, and these are usually achieved by varying the rate of product removal. Level sensors are usually of the balanced-float-compartment type. Some jigs have submerged streamlined floats in the bed to sense density and to automatically control either the air entrance or exit to modify the jig

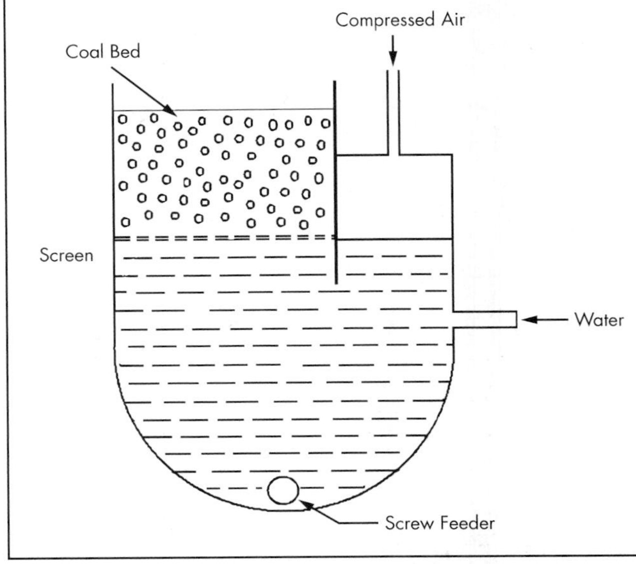

Source: Indian Institute of Technology Kharagpur, n.d.

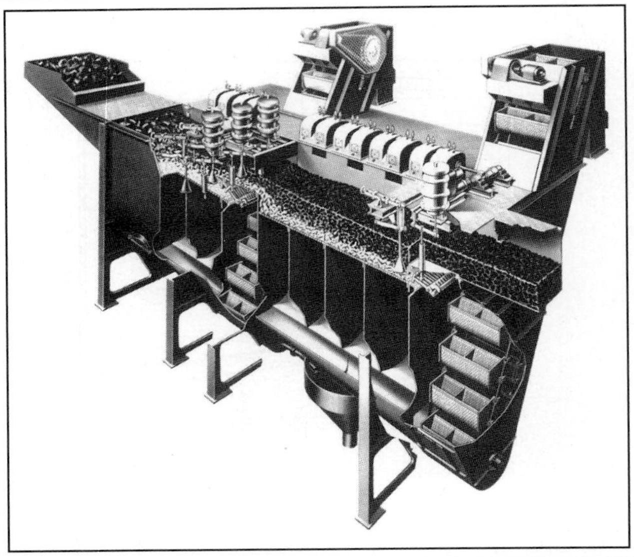

Courtesy of Metso

Figure 4 Baum jig (left) and McNally Baum-type jig (right)

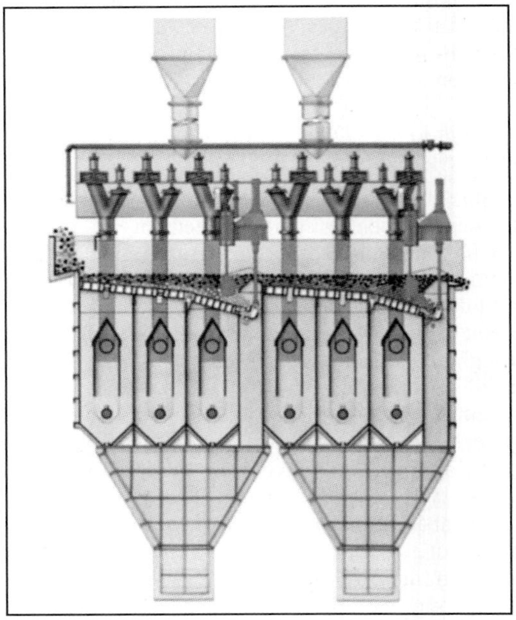

Source: MBE Coal and Minerals Technology GmbH 2011

Figure 5 Batac Jig

stroke. In standard plant practice, at least two jigs are operated in series, the first making a finished clean product and the second making a middling and tailing product. The middling product may be crushed and treated in another jig or by other methods.

Batac jig. The demand to increase throughput that was not achievable in a Baum jig, given the difficulty of generating a uniform bed pulse across a large particle bed, resulted in the development of the Batac jig (Figure 5). The jigging motion in the Batac jig is generated in air chambers located underneath the jigging bed. Low pressure, high-volume air from a blower is intermittently supplied to these air chambers and then discharged by means of an electronically controlled plate valve system. The frequency and profile of the stroke can be easily modified using the jig's programmable logic controller. The stroke is imparted to the water inside the jig as a function of the pressure change generated inside the air chambers. Makeup water is added at the lowest point of every jigging

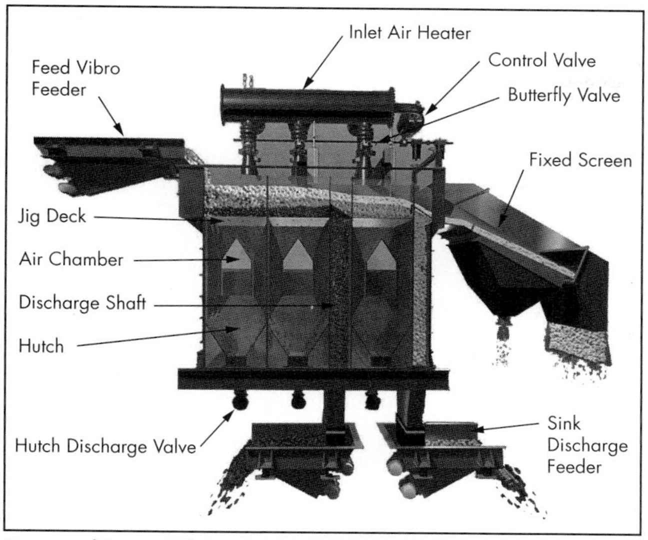

Courtesy of Tenova Delkor

Figure 6 Apic jig

chamber to intensify the upward current and to restrain the downward current. The pulsating motion of the water stratifies the feed material according to its density (Takakuwa and Matsumura 1954). As a result, very efficient separations have been realized in industrial units up to 7 m wide.

The benefits of the Batac jig have been summarized by Sanders et al. (2002). Batac jigs are usually used to treat coal with a size range from 60 to 6 mm. However, the jig is capable of handling run-of-mine coal with a top size up to about 150 mm. Treatment rates range from 68 to 90 t/h/m jig width.

A short Batac jig, which is a jig with a length of only two or three chambers, can be used to de-stone raw coal ahead of a dense medium circuit. The key difference between a Batac jig for de-stoning applications and a Romjig is the size range of the material to be beneficiated.

Apic jig. The Apic jig, shown in Figure 6, is a relatively recent advancement in under-pulsed jigs like the Batac jig. The Apic jig provides enhanced control of the airflow into and out of the chambers and improved gate control for removing low-density particles (Loveday and Jonkers 2002). The Apic gate, a vertical-shaft sink discharge mechanism, added to the advantages of the jig, because it eliminates the vertically operated or conventional discharge mechanism, thereby minimizing the risks of low-density material short-circuiting and in the mixing of bed layers. The other advantage is unhindered smooth discharge of heavy product.

The underbed air pulsation of the water in the jig submits the bed of particles on the screen deck to vertical fluid pulses of extension and compression phases and results in densimetric stratification of the particles. The denser, heavier particles (sinks) settle, and the lighter, less-dense particles (floats) rise to the top of the material bed. The number and widths of the compartments in the jig are selected to achieve the most efficient separation. Where necessary, a ragging bed is installed for the recovery of fine material. The discharge mechanisms of the jig permit a smooth and accurate evacuation of sinks and middlings. The gate discharge is controlled by a float that is continually positioned at the desired density interface within

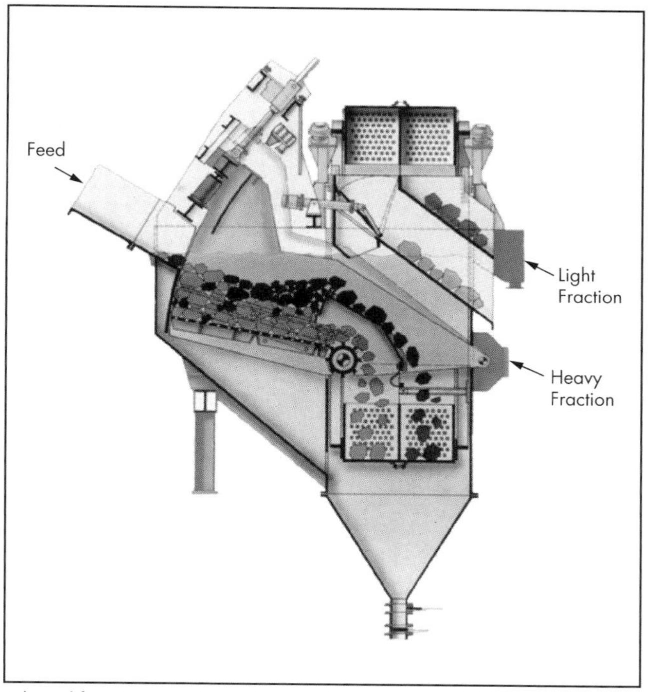

Adapted from MBE Coal and Minerals Technology GmbH 2011

Figure 7 Romjig

the bed. The gate design is chosen to suit the particular feed size distribution and the proportion of sinks.

The pulse characteristics are designed to maximize the separation of products in the jig, and electronic control provides a precise and consistent operation regardless of feed variations. The pulse timings are automatically adjusted for bed-level changes. Each compartment is timed and adjusted independently. The inlet air pressure is automatically regulated to provide the correct pulse for the application. The pulse downstroke can also be regulated to control the suction stroke and thus the settling characteristics.

Romjig. Beneficiation of coal close to the mining face to remove coarse rock from coal was the subject of extensive studies by German researchers in the 1980s. The objective was to reduce the cost and increase production of underground operations (Sanders et al. 2000). The result was the development of the Romjig, shown in Figure 7. The Romjig uses a moving jig screen, the jigging motion provided by a mechanical hydraulic arm. In this way, the feed end of the screen plate moves up and down while the discharge end is pivoted. The lifting cycle is repeated 38–43 times per minute. During the jigging action the material separates, with the heavy particles collecting next to the screen and progressively forming a bed as the material moves along the screen, while the lighter coal moves to the top of the bed. The jig plate is a screen panel with 15-mm slotted openings.

Three factors ensure that the material moves horizontally from the feed end to the discharge end:

1. Slope of the screen deck (pressure from the feed)
2. Linear downstroke
3. Circular upstroke

Table 1 Romjig details

Construction	Units	Process Details	Units
Length	~6.0 m	Feed rate	350 t/h
Width	~6.5 m	Feed size	350 –40 mm
Height	~8.0 m	Stroke amplitude	500 mm
Bed width	~2.0 m	Stroke frequency	38–43/min
Full weight	98 t	Makeup water requirement	10–15 m^3/h
Installed power	110 kW	Specific energy	0.3 kW·h/t
		Cut-point range	1.6–2.1 relative density

Source: Ziaja and Yannoulis 2007

Products are collected in a divided discharge wheel with one compartment for the coal and one for the discard. Particles smaller than the screen panel apertures fall through into the hutch compartment. Material accumulates in the hopper at the bottom and is discharged periodically through a double-gate valve system to join with the –40-mm feed screen underflow material for further treatment or as final product.

To ensure that design and manufacturing remain competitive, the Romjig is available in only one size, with a 2.0-m-wide bed, rated at up to 350 t/h for a nominal 350 –40 mm feed. Details of the construction and operational design are summarized in Table 1. The metallurgical performance of a Romjig is discussed in the paper by Sanders and Ziaja (2003).

Alljig. The Alljig is similar to the Baum jig, with the addition of a control system for discharge, enabling the machine to work autonomously once the control system is adjusted (Figure 8). Since its introduction in the late 1980s, the Alljig has seen many improvements, enabling it to process a growing array of different materials. Electronic sensors are used to automatically monitor and control the precise discharge of heavy particles contained in the feed. Alljig jigging machines create a physically stable and individually adjustable optimal jigging stroke at minimal energy consumption by means of air-pulsed water. These machines provide capacities ranging from 5 t/h to 700 t/h. The process of efficient separation and cleaning of feed material is applicable to particle sizes from 150 mm to <1 mm.

Gekko inline pressure jig. The inline pressure jig (IPJ) is a single-hutch circular jig that is completely enclosed to operate at greater than atmospheric pressures. The IPJ components are shown in Figure 9. An internal screen is pulsed by a hydraulic ram by means of a shaft that extends through the bottom of the cone and is sealed by a diaphragm. The screen frame is sealed to the enclosure by a flexible diaphragm. The IPJ is fed through the center and under pressure to a distributor. The feed is distributed throughout the screen, with the heavy minerals passing through the screen and the light minerals moving to the top and continuously to the tailings outlet. The vessel is completely full of water or slurry. The screen is pulsed by means of the hydraulic ram and the pulse is a sawtooth pattern. Ragging material is on the screen. Hutch water is added at the bottom of the concentrate hutch. The completely submerged feed aids in mineral separation and reduces the hutch water requirement. The quality of hutch water is not an issue, and water of almost any quality, including seawater or water with up to 5% solids, is suitable. The IPJ can recover minerals from 100 µm to 5 mm in size. Because it operates under pressure, it can be incorporated in a cyclone feed line or other pressurized systems. Industrial applications of the jig are provided by Gray (1997).

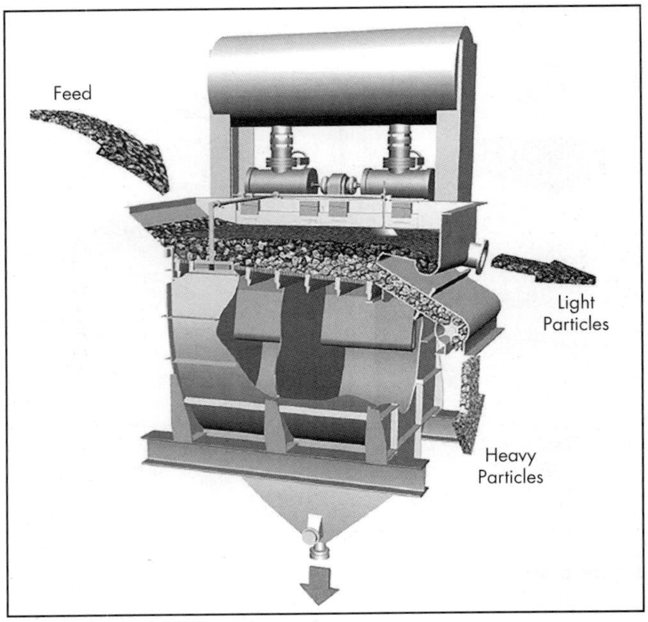

Source: Allmineral 2013
Figure 8 Alljig

The IPJ includes the following operating variables:

- **Feed rate.** The feed rate will ultimately dictate unit performance. Too high a feed rate will impact recovery, though this is mitigated by a higher specific-gravity difference between gangue and valuable mineral.
- **Screen and ragging.** Screen aperture should be slightly greater than the largest particle fed and is generally of wedge-wire design. The ragging material can be natural minerals or steel, lead shot or ceramic beads. Consistent ragging size and shape will enhance unit repeatability and ultimately performance. Synthetic ragging material made from polyurethane or ceramic is available from Gekko.
- **Pulse.** Both stroke length and frequency of stroke, the upstroke and downstroke, can be controlled independently. The upstroke is generally slower than the downstroke, creating a sawtoothed wave profile. This wave profile causes increased acceleration of higher specific gravity particles and thus improves the separation.
- **Hutch water.** Adding water creates a positive flow through the screen bed, aids in concentrate flow, and reduces the suction created during the upstroke. Too low a water flow will produce a low-grade concentrate and cause excess force on the machine.

The IPJ is available in four sizes, with specifications as shown in Table 2.

SLUICES

Sluices can be used to make a rough gravity concentration, provided the valuable mineral is free, not too fine, and possesses a fairly wide size range. Sluice boxes can provide a much higher concentration ratio than most other gravity concentrators. They are also very reliable, inexpensive, and simple to operate, which explains why the sluice box is still the most important placer gold concentrator (Clarkson 1990). The use of sluices has diminished greatly over the last three

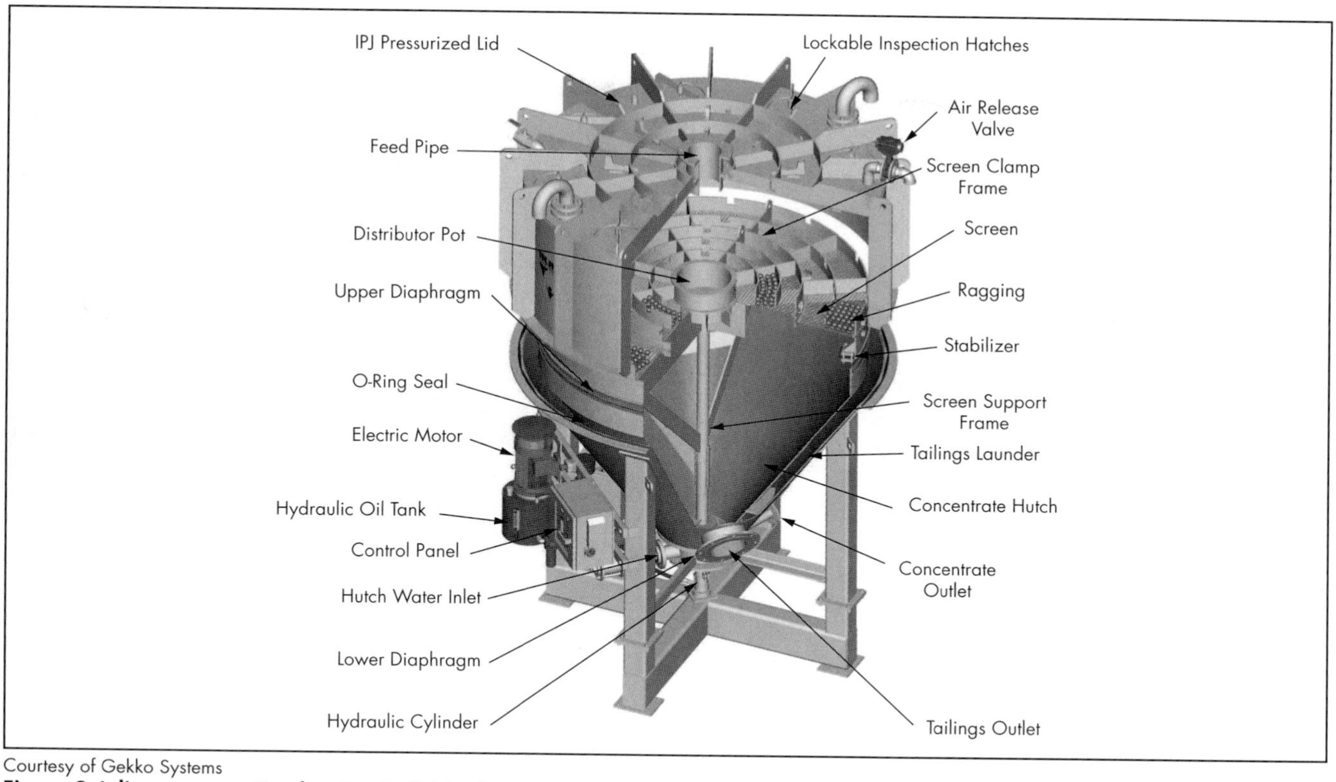

Courtesy of Gekko Systems

Figure 9 Inline pressure jig showing individual components

Table 2 Specifications of inline pressure jigs

	Model Number			
Specification	**600**	**1000**	**1500**	**2400**
Maximum feed rate, t/h	1.5–3	15–25	35–50	80–100
Maximum particle feed size, mm	6	20	20	50
Hutch water flow rate, m^3/h	2.0	10–20	15–35	15–50
Typical yield to sinks/concentrate, t/h	0.6	1–4	5–8	5–40
Overall height, mm	1,300	2,105	2,500	3,200
Footprint diameter, mm	800	1,350	1,800	2,500
Screen diameter, mm	520	1,030	1,330	2,000

Courtesy of Gekko Systems

decades, more so in the tin industry (Min 2006; Hutahaean and Yudok 2013); jigs (Pan-Am and the circular jig) are preferred, especially when using a dredge for mining.

A sluice box is an inclined rectangular flume containing riffles on matting, through which a dilute slurry of water and alluvial gravel flows. Sluice boxes operating under ideal conditions are actually centrifugal concentrators whose riffles overturn ribbons of slurry to form vortices (Figure 10). At the bottom of these vortices, centrifugal and gravitational forces combine to drive placer heavy particles into matting, breaking up clumps of clay and cemented particles. Effective recovery of fine minerals depends upon a fairly loose, active bed of sand between the riffles. At the same time, the rapidly moving coarse gravel above the riffles must not interfere with the access of sand to the riffles or cause too much disturbance. Generally, a smooth flow represents a hard bed and poor recovery.

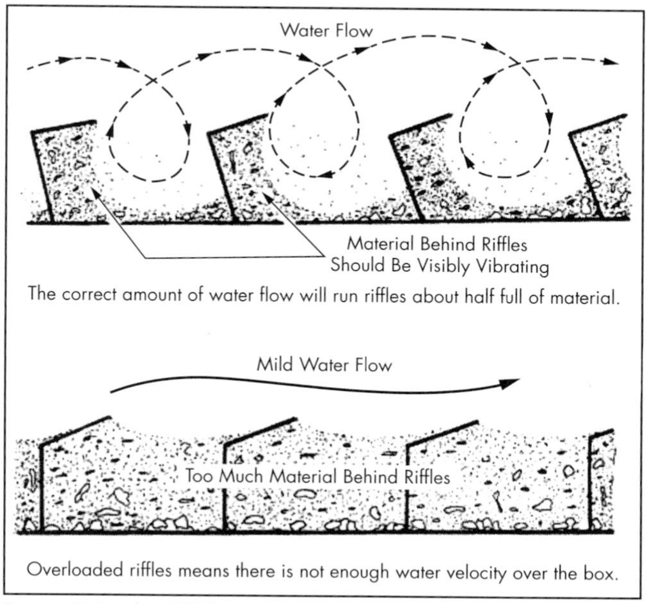

Source: McCracken 2013

Figure 10 Formation of vortices between the riffles of a sluice box

Types of Sluice Boxes

The appropriate selection of material for a sluice box is fundamental to the successful recovery of gold particles of different sizes and shapes.

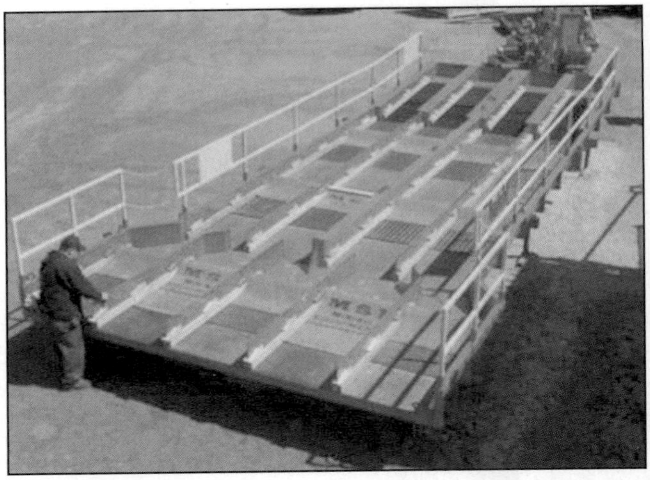

Courtesy of Madden Steel

Figure 11 Large sluice box for gold (~5 × 12 m)

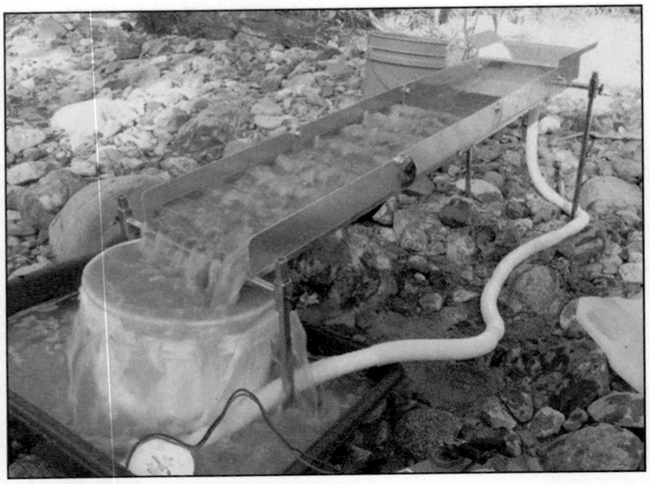

Courtesy of Royal Manufacturing

Figure 12 Small sluice box for miners

- **Standard sluice box.** Sluices come in all sizes depending on throughputs, for single operators to large-scale mining, as shown in Figures 11 and 12. The sluice length can vary from around 1.5 m to more than 30 m (nominally 10 m) and principally depends upon the character of the material to be treated. Coarse and very high-density minerals settle quickly and require only a short length of sluice. The length also depends upon the requirement to break up the alluvial ground. Usual practice is to use sluice boxes between 1.2 m and 1.8 m in width. The slope must be adequate to transport the pebbles along the sluice and also prevent sand packing within the riffles. The slopes can vary between 7° and 14°. High slopes are used where there is a great deal of clay or coarse material. The recommended volumetric feed rate is around 0.75 m^3/h/m width of sluice. Excessive feed rates are one of the major factors contributing to gold losses. Lower feed rates lead to minor improvements in recovery. Water requirements are variable and for low feed rates, a good starting point for gold particles less than 1 mm is 40 L/s/m. For high feed rates, water requirements of 80 L/s/m are adequate for recovering gold particles larger than 1 mm.
- **Oscillating sluice box.** An oscillating sluice box consists of a pair of sluice runs suspended from a frame with cables. A direct current electric motor is mounted between and above the sluice runs and rotates a weighted bent shaft through an angle drive. The motor–drive combination creates a horizontal circular panning motion with a 16-mm-diameter circle oscillating at 130–180 rpm. Oscillating sluice boxes are used for gravels containing a high proportion of high-specific-gravity minerals such as magnetite or a high percentage of clay leading to inter-riffle packing (Clarkson 1990).
- **Triple-run sluice box.** Triple-run sluice boxes rely on the ability of the stationary distributor punch plate to screen fine gravels to the side runs (Figure 13). Most of the water entering the distributor remains above the punch plate to move large rocks. Fine gravels and gold are inevitably trapped in these excessive volumes of turbulent water and are swept off with the boulders at high speed down the center run. At times with high feed rates the distributors become inefficient and gold recovery is reduced by underutilizing the side runs and overloading the center run with large pebbles and fine gravel. Triple-run sluice boxes fabricated with large distributors (>30 m) that contain large holes (+20 mm) in their punch plate, sluice gates to control flows to the side runs, adjustable side run slopes, and manually controlled wash monitors are the most efficient (Clarkson 1990).

Types of Riffles

The appropriate selection of riffle type is fundamental to the successful recovery of gold particles of different sizes and shapes. The attributes of the more well-known riffle types are as follows:

- **Expanded metal.** Coarse expanded-metal riffles are effective at recovering gold particles finer than 1 mm; however, 25-mm angle iron riffles are required to efficiently recover gold coarser than 1 mm. Coarse gold (+1 mm) losses with expanded metal can be very dramatic as the feed particle size increases. Angle iron riffles require higher water flow rates (1.21 m^3/min) and specified gaps (38–64 mm) and inclinations (–15°) for optimum gold recovery. Angle iron riffles of 25 mm do not tend to pack as readily as larger angle iron riffles. Doubled expanded metal and flat bar riffles are not recommended because of their susceptibility to packing and creating excessive turbulence, respectively (Clarkson 1990).
- **Matting.** Unbacked Nomad matting appears to be the best matting in common use because it does not interfere with riffle operation. Matting made from the "fur" of coconuts, artificial turf, and Monsanto matting are not ideal.
- **Hydraulic.** Hydraulic riffles consist of alternating 50-mm flat bar riffles and 25-mm square tubing that is perforated on the bottom of the sluice. Low-pressure water introduced into the square tubing keeps full riffles loose, unlike conventional riffles that rely on the formation of vortices.

FLOWING-FILM CONCENTRATION

The behavior of solid particles in suspension depends, to a large extent, upon the pulp density and the size of the suspended

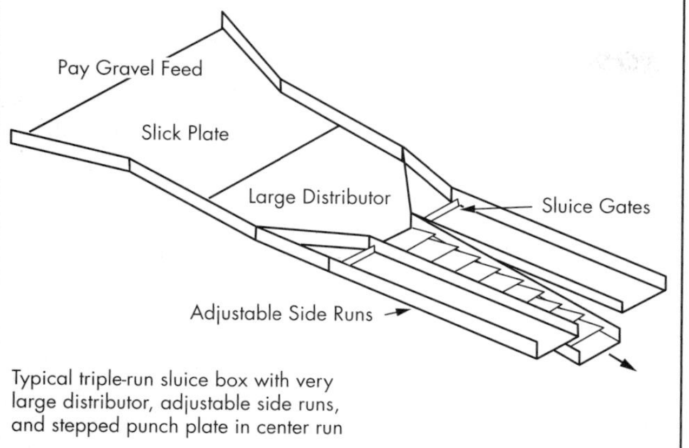

Adapted from Clarkson 1990

Figure 13 Triple-run sluice box

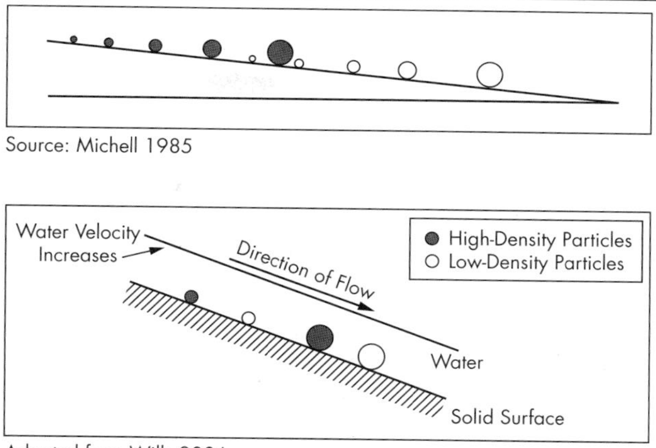

Adapted from Wills 2006

Figure 14 Arrangement of particles of differing size and specific gravity under the influence of a streaming current

particles. In a fairly dilute suspension, such as that normally used when dealing with small particle sizes, the behavior of particles in a flowing film results from two effects. These are the lateral displacement, which is determined by the time taken for each particle to penetrate the flowing film and to reach the solid surface, and the resistance offered by each particle to further displacement after it has reached the solid surface.

The initial penetration through the flowing film depends on the size and density of the particle and the thickness and viscosity of the film. As a result, the smaller particles will migrate further before their movement is retarded relative to larger particles of the same specific gravity. The behavior on reaching the solid surface depends on whether there is a single particle layer or, as is more often the case, a multiple layer or thin bed of material that is sufficiently dilated, which permits the penetration of higher-density particles.

When a thin film of liquid flows over a plane's solid surface, the layer next to the surface remains at rest, but the velocity of the film increases with the distance from the surface and becomes a maximum near, but not quite at, the free surface. Therefore, a particle in suspension in such a film is acted upon by a greater force near the upper part of the film than at the lower part, resulting in an overturning effect. After a particle reaches the separating surface or an accumulated densely packed bed of other particles, the liquid flow causes it to move downstream by rolling, sliding, or a movement involving alternating suspension and deposition (saltation).

In rolling and sliding, which are brought about by a substantially non-eddying stream, the large submerged particles are acted upon to the greatest extent and they move more rapidly than smaller ones, notwithstanding their larger mass. When two particles of the same size but of different specific gravity are considered, the higher-density one moves more slowly because of its greater mass. Consequently, the particles tend to become arranged in the manner shown in Figure 14.

When a thin film of water flows across a flat surface, a fluid shear occurs at the interface between the surface and the water, which creates a fluid velocity gradient. The velocity increases from the surface and through the flowing film. High-density particles within the thin film settle through the high-velocity streamlines and into the low-velocity region at the solid surface, whereas low-density particles are unable to settle from the high-velocity streams and thus report with the majority of the fluid flow.

One of the earliest forms of flowing film separators was the buddle (Davies 1902), followed by the Reichert cone concentrator (Reichert 1965). The unit was developed in response to a need for high-capacity gravity concentration for heavy mineral sand applications (Ferree 1973). Present-day flowing-film gravity separation devices include shaking tables and spirals.

Shaking Tables

The shaking table is a gravity separation device that has been in use for a long time, having been commercialized at the beginning of the 19th century. Little has changed in the design, although multi-deck (up to three levels) tables have led to capacity increases relative to floor area. Shaking tables are normally used only on cleaning stages because of their low capacity. The principle of separation is the motion of particles according to specific gravity and particle size moving in a slurry across an inclined table, which oscillates backward and forward essentially at right angles to the slope. Wash water is provided as a flowing film across the slope of the table. The riffles in the table hold back heavier particles, which are closest to the deck, while lighter particles flow over the riffles. This motion and configuration cause the fine high-specific-gravity particles to migrate closest to the deck and be carried along by the riffles to discharge uppermost from the table, while the low-specific-gravity coarser particles move or remain closer to the surface of the slurry and ride over the riffles, discharging over the lowest edge of the table (Figure 15).

Shaking tables include the following operating variables:

- Angle of deck (a steeper angle means less weight to concentrate).
- Length of stroke (the longer the stroke, the more sideways the motion and hence more weight to concentrate, up to a maximum).
- Frequency of stroke (similar to length; that is, the more frequent the stroke, the more sideways the motion, up to a maximum).

- Splitter positions (the position of the splitters on the concentrate launder will determine the mass to the concentrate).
- Feed rate and density (above a maximum of typically 2 t/h per full-size table and density typically 40% solids, depending on the type and particle size of the feed) separation will be reduced.

Shaking tables have been the most popular of the vibrating separators and remain common today for the recovery of gold, tin, and other heavy metal minerals in the particle size range of 1.65 × 0.074 mm and coal in the 6.7 × 0.15 mm particle size range. Many suppliers provide shaking tables, and of these, Deister and Holman-Wilfley are the best known.

Deister Shaking Table

The deck of a Deister shaking table has a one-piece, ultra-smooth rubber cover, in either black or white, with riffles for sand or slime application (Figure 16). Launders are molded high-density polyethylene with adjustable cutting pans. Five models are available, as shown in Table 3, and technical information is provided in Table 4.

Holman-Wilfley Shaking Table

The Holman-Wilfley shaking table is driven by a head motion system with a 1.5-kW motor and with a nominal stroke adjustment of between 8 mm and 16 mm. The deck is supported on 42 carriers and is vibrated diagonally along the entire length. Good-quality timber with standard 5-mm rubber covers for maximum durability (color white or black options) is used for the decking. Different riffle patterns are available to maximize performance, depending on the application. Table tilt can be easily adjusted using a hand wheel, even while the machine is in operation. All feed/product launders are polyurethane lined.

Three models are available: 2000, 3000, and 8000. The 8000 model is available in single- or double-deck configuration. The Model 8000 is shown in Figure 17 and technical information is provided in Table 5.

Gemeni Shaking Table

The Gemeni shaking table has been specifically designed for the recovery of fine gold to a directly smeltable concentrate (Figure 18). The new direct drive system incorporates a geared motor direct driving a crank connected to the table deck. The crank incorporates a sprung connection system to absorb overrun. The bump stop system has been maintained to provide a fine-tuning mechanism. Table tuning is achieved by adjustment of a single screw (Mineral Technologies, n.d.). Gemeni tables are available in three models, as shown in Table 6.

Spiral Concentrators

The spiral concentrator has been widely used in the minerals industry to achieve efficient fine particle density-based separations since 1943 when the Humphreys spiral was introduced (Humphreys 1975). The separation is a result of a circulating flow, as shown in Figure 19. Light particles are lifted by the circulating flow from the inner part of the trough and carried to the outer area of the trough, whereas high density particles settle onto the trough surface and move toward the inner part of the trough. Internal channels along the vertical axis and splitters located at the end of the trough are used to collect the light and heavy particle streams.

The spiral concentrator incorporates the use of wash water to enhance the separation, and drawpoints are located

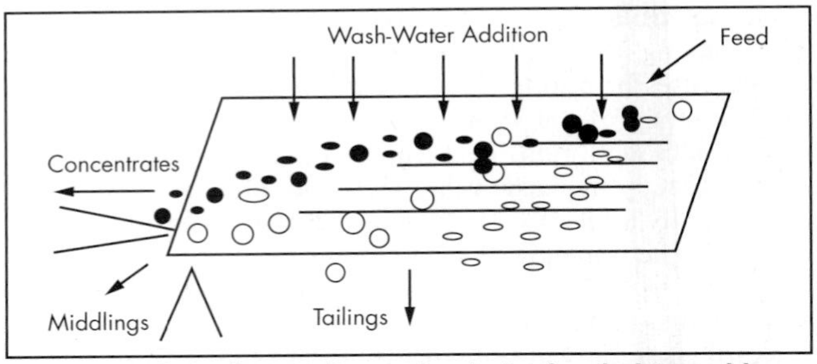

Figure 15 Flow of particles on the deck of a shaking table

Table 3 Deister shaking tables

Model No.	Description
15-S	Laboratory table
14	Pilot-sized table
6	Full-sized table
99	Double deck
999	Triple deck

Courtesy of Deister Concentrator

Courtesy of Deister Concentrator

Figure 16 Deister triple-deck tables and slime deck covering

Table 4 Technical information for Deister shaking tables

Type of Feed	Particle Top Size, mm	Capacity per Deck, t/h			Dressing Water per Deck, m^3/h		
		999/99/6	14	15-S	999/99/6	14	15-S
Coarse sand	2.0	2.3	1.15	0.115	2.3–3.6	1.1–1.8	0.11–0.18
Medium sand	0.30	1.4	0.70	0.070	1.8–2.7	0.9–1.4	0.09–0.14
Find sand	0.15	0.9	0.45	0.045	1.1–2.0	0.7–1.1	0.07–0.11
Slime	0.044	0.4	0.20	0.020	0.7–1.1	0.5–0.9	0.05–0.90

Type of Feed	Feed % Solids	Strokes/min	Stroke, mm	Deck Adjustments, mm/m	
				End Elevation, mm	Side Tilt, mm
Coarse sand	20	260–265	19	20	25
Medium sand	22	265–270	19	15	20
Find sand	23	270–275	13	10	15
Slime	25	275–280	13	5	10

Courtesy of Deister Concentrator

Note: The end elevation is measured parallel to the uppermost riffle, whereas the tilt is measured at a right angle to the riffles.

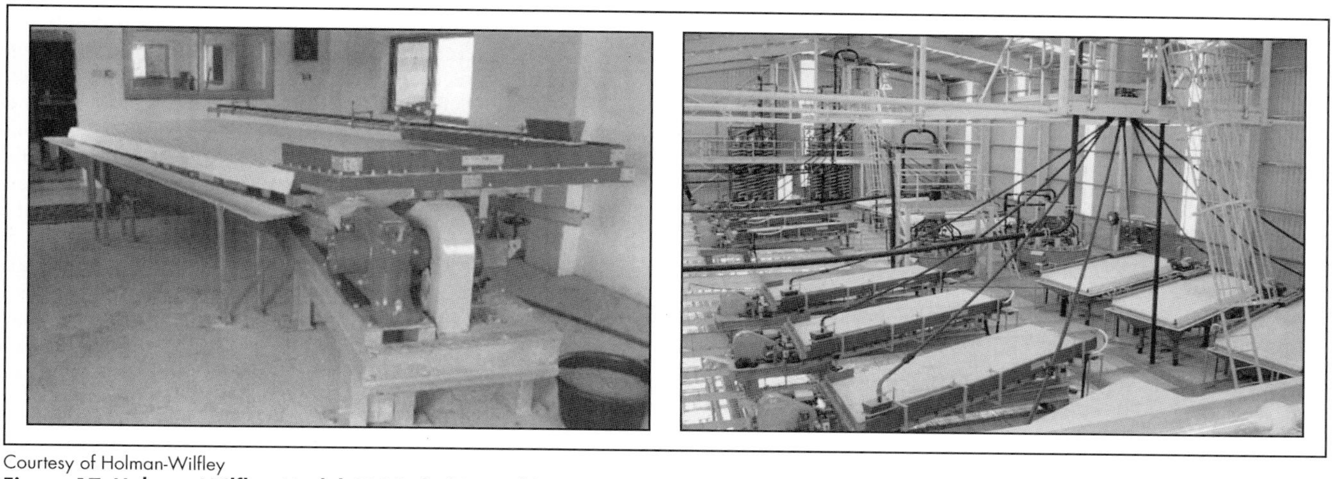

Courtesy of Holman-Wilfley

Figure 17 Holman-Wilfley Model 8000 shaking table

Table 5 Holman-Wilfley shaking table details

Model	Deck Area, m^3	Width, mm	Length, mm	Motor, kW	Water, L/min	Rate, kg/h
2000	2	887	2,445	1.5	5–20	<450
3000	3	1,030	2,515	1.5	10–25	<850
8000	7.5	1,600	4,900	2.2	20–35	<2,500

Courtesy of Holman-Wilfley

at various points along the bottom of the trough to remove the high-density particles. Commercial applications for several decades after its introduction were mainly for easy mineral separations such as enriched chromium placer ores, iron ore, and heavy mineral sands. The application limitations were mainly due to the construction from cast iron or cement that limited the ability to vary the profile and pitch of the trough. Around 1980, the advancement of construction materials led to spirals being constructed of fiberglass and spray coated with polyurethane. The new material construction allowed alterations to the trough geometry to address the needs of more difficult separation applications. The lightweight material construction permitted the entwining of two

Courtesy of Mineral Technologies

Figure 18 Gemeni shaking table

Table 6 Technical information for Gemeni tables

	Model		
Operating Data	**GT60**	**GT250**	**GT100**
Feed rate nominal, kg/h	30	115	450
Fed density recommended, % solids w/w	60–70	60–70	60–70
Feed size nominal top, μm	800–1,000	800–1,000	800–1,000
Nominal wash-water requirements, L/min	12	25	38
Nominal wash-water pressure, kPa	30	30	30

Source: Mineral Technologies, n.d.

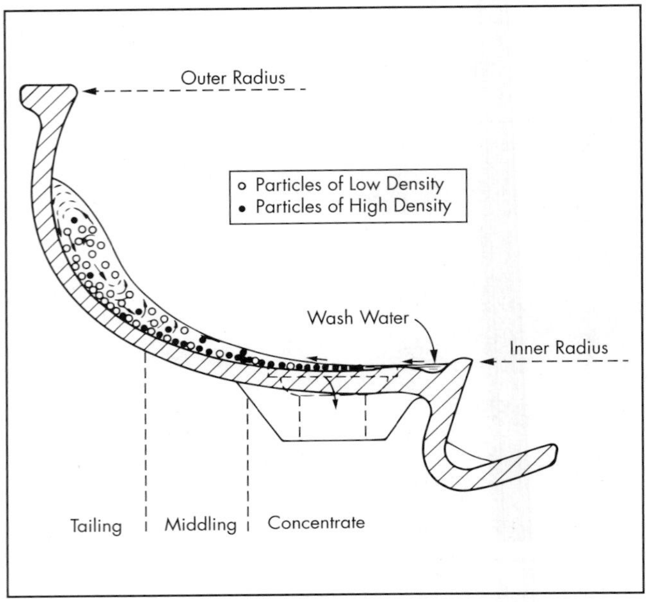

Courtesy of Humphreys Engineering Company

Figure 19 Fluid flow pattern along the cross section of the spiral trough, which provides the density-based particle separation

or three spiral starts around one central vertical stick or axis. For fine coal applications, the trough was designed to allow higher particle retention times by making the trough wider, changing the trough profile, and decreasing the vertical pitch (Richards et al. 1985). Higher throughput capacities per start were realized for mineral applications from the redesign of the trough. A summary of the spiral trough changes is described by Richards and Palmer (1997). Studies conducted by Luttrell et al. (1998) revealed that significant improvements in separation efficiency can be realized when re-treating light particle streams of a spiral concentrator in a second spiral unit in a rougher-cleaner arrangement.

There are many suppliers of spiral gravity separators, such as Mineral Technologies, Multotec, Outotec, and FLSmidth (Krebs spirals).

Mineral Technologies spirals. The Mineral Technologies spirals are fiberglass reinforced and polyurethane lined. The splitters are designed to cater for ease of operation regardless of spiral size and number of splitters. Concentrate diverters improve concentrate grade. Repulping devices incorporated in the spiral further enhance recovery of product. Accurate top or bottom entry feed distribution systems allow for improved recoveries. Replaceable modular cast polyurethane feed and concentrate boxes provide longer spiral service life. Spiral configurations come in single, twin, triple, and quadruple troughs per column to suit capacity requirements. Spiral banks of up to 48 starts are available, to maximize capacity and minimize floor space requirements. High capacity 12 × 3 start spiral modules can process in excess of 150 t/h. Figure 20 shows a single spiral and a bank of spirals.

A multitude of spiral designs are available for particular mineral types and applications in a spiral circuit. A brief description of the different Mineral Technologies designs is as follows (Mineral Technologies 2017a, 2017b):

- MG series—for feed material generally containing up to 25% heavy minerals (up to 40% in some applications).
- HG series—for high-grade feed material generally greater than 25% and as high as 90% heavy minerals.
- VHG model—for feeds with very high levels of heavy minerals.

Courtesy of Mineral Technologies

Courtesy of Iluka Resources Limited

Figure 20 Single spiral and bank of spirals

- FM series—for valuable heavy mineral particles in the range of 30 to 150 μm.
- WW series—utilizes wash-water addition for enhanced grade control in specific applications (e.g., iron ore).
- HC series—super high-capacity spirals specifically designed for more economical and compact plants. The facility to add wash-water is available on some models.
- LC3 coal spiral—a revolutionary new low-cut-point spiral producing a low cut point of 1.4 to 1.5 RD (relative density); and allowing superior product yield at conventional 1.7 to 1.8 RD cut points. The design allows for less buildup than conventional spirals, leading to improved beneficiation of ultrafine material.
- The LD7RC coal spiral—a combined rougher-cleaner spiral featuring seven turns consisting of three rougher turns and four cleaner turns. It offers reduced overall height when compared with standard two-stage systems. This spiral is particularly effective on feeds with high levels of near-gravity material and other difficult coal separations. The unique rougher-cleaner transition stage ensures thorough repulping of the slurry for maximum separation. This system operates within the trough, saving space/height, and dispenses with separate inter-stage components (Figure 20).

Multotec spirals. Six basic models with up to three combinations are available to cater to different mineral applications and circuit design. Units come in three to eight turns with high-, medium-, and low-grade profiles. The spiral trough is sprayed with polyurethane to a thickness of 3 mm ±0.5 mm. Splitters are of solid polyurethane casting. Collected material is channeled into a specific product box outlet.

The following Multotec models are available (Multotec 2016):

- SX4, MX7, and SX7 spirals—The SX4 is a single-stage four-turn spiral used for coal, while the SX7 (seven-turn, for coal) and MX7 are double-stage spirals. Both the primary and secondary stages are contained in one assembly.
- SC20/7 spiral—Used for rougher and scavenger applications.
- SC20 and SC21 spirals—Used for cleaning applications in the mineral sands industry.
- SC20 LG—Twin- and triple-start spirals.
- SC20 HC—High-capacity twin-start spiral with high-flow wash water.
- SC20 VC—Twin- and triple-start spirals.
- SC21—Twin- and triple-start spirals.
- SC21/5—High grade spiral, used for cleaner and recleaner.
- HX3/HX5—Twin- and triple-start spirals.
- HX5 spirals—Shallow in angle and used to treat low-grade ores.
- NHM spirals—For the rougher and scavenger stage. Specifically designed for heavy mineral sands applications.

Outotec spirals. The Outotec urethane fiberglass spirals either have wash water or are wash-waterless. Details of the spirals are provided in Table 7.

Krebs spirals. Krebs spirals are primarily designed to treat pulp streams with a heavy mineral content in the range up to 35% by weight. Each spiral consists of a backfed feed box, single-section helix, and a product box. Finger cutters on the helix divert separated minerals into channel areas. All splitters are located in the product box to divide the stream into concentrate, middlings, and tailings. The spiral can be used in rougher, scavenger, and re-treat duties. The spiral is primarily designed to treat pulp streams with a medium heavy-mineral content up to 15%–80% by weight.

Two models of spirals are available (FLSmidth-Krebs 2014b):

1. LM series—Capacity up to 2 t/h per start with concentrate removal up to 0.7 t/h per start. Feed pulp density up to 45% solids. Models available with three, five, and seven turns in simplex, duplex, and triplex configurations.
2. MM series—Capacity up to 2 t/h per start with concentrate removal up to 0.5 t/h per start. Feed pulp density up to 45% solids. Models available with five and seven turns in simplex, duplex, and triplex configurations.

FLUIDIZED-BED SEPARATION

Fluidized-bed separators (FBSs), also known as teetered-bed or hindered-bed separators, have been used for more than a century in coal and mineral processing plants, primarily for

Table 7 Details for Outotec spirals

Model	Pitch, mm	Starts	Number of Turns	Wash-Water Addition	Concentrate Takeoff	Concentrate Collection	Main Features	Suggested % Solids
CS2000	420	6	Single, double, or triple	No or yes	Splitter at each turn and at discharge splitter box	Inboard collecting trough	Larger diameter, high capacity	20–40
LC3000	410	5	Single, double, or triple	No	Splitter at discharge splitter box	Central channel into discharge splitter box	Low-grade application (wash-waterless)	25–40
LC3700	410	7	Single, double, or triple	No	Splitter at discharge splitter box	Central channel into discharge splitter box	Agitator bumps (optional)	25–40
MC7000	410	7	Single or double	No	Splitter at each turn and at discharge splitter box	Inboard collecting trough	Medium grade and wash-waterless	25–40
HC8000	410	7	Single, double, or triple	No	Splitter at each turn and at discharge splitter box	Inboard collecting trough	High grade and wash-waterless/easy adjustment splitter handle	25–40
H9000W	460	7	Single or double	Yes	Splitter at each turn and at discharge splitter box	Inboard collecting trough	High grade and wash-water/easy adjustment splitter handle	25–40

Adapted from PhySep, n.d.

particle size separations. FBS units can be operated in a manner that provides a very efficient density-based separation by utilizing the finest high-density particles in the feed stream to create an autogenous dense medium.

One of the oldest and perhaps most popular teetered-bed separator is the Stokes unit. The positive features of FBS units include the ability to achieve efficient density-based separations at relatively high mass flow rates (10–20 t/h/m^2), the capability of adjusting online to changes in feed characteristics as well as variations in feed rate, and their overall simplicity of operation. Their inherent disadvantages are the need for tight top-size control in the feed to prevent the recovery of low-density oversize in the underflow stream, the need of a mechanical underflow valve and control system that requires maintenance, the use of clean water in the injection system to prevent plugging, and the existence of components that are susceptible to wear in the bottom of the units.

Many commercial units operate in much the same manner as the Stokes classifier, including the Floatex (Mankosa et al. 1995; Elder et al. 2001), Lewis hydrosizer (Lewis 1990), Linatex hydrosizer (Deveau and Young 2005), Allflux separator, and the Hydrosort (Doerner 1997). The Allflux separator is a unique unit in which two stages of density-based separations using fluidized particle beds occur in a single unit (Short et al. 2001).

A problem associated with conventional FBS units is the turbulence generated by feed injection into the center of the unit at a depth of around one-third of the total height. The turbulence disrupts the settling of large and small high-density particles into the fluidized-bed zone of the FBS, and the upflow velocity of water leads to the loss of high-density particles into the overflow stream. To counter this problem, Mankosa and Luttrell (1999) equipped an FBS system with a unique feed system that gently introduces the feed slurry across the top of the separator using a transition box (i.e., the Crossflow separator).

More recently, Galvin et al. (2009, 2010; Galvin 2012) discovered the benefits of using closely spaced inclined channels, which promote laminar flow and a high shear rate at the plate surface. This leads to enhanced separation efficiency. A new separation mechanism, referred to as the laminar-shear mechanism, develops in the inclined channels. Relatively fine, dense particles segregate from the flow, sliding downward and returning to the lower fluidized-bed zone. However, lower-density particles over a broad range of particle sizes continue to convey upward through the inclined channels. The relatively large, low-density particles experience inertial lift, and hence fail to segregate onto the inclined channels. These relatively coarse particles become exposed to the higher fluid velocities, conveying easily to the overflow. The inclined channels also provide a significant capacity advantage over conventional fluidized beds, especially as the particle size decreases. Moreover, the system is insensitive to low feed-pulp density and can be deployed directly after a classifying screen without the need for thickening cyclones.

In an effort to increase the effective particle size ratio of separation, Mankosa and Luttrell (2002) developed a process referred to as the HydroFloat separator in which air bubbles are injected into the fluidized particle bed. Using this concept, the effective particle size range that can be treated in a modified FBS unit is around 6:1 (Kohmuench et al. 2001; Luttrell et al. 2006). The benefits of the units are discussed by Kohmuench et al. (2007).

WATER-ONLY CYCLONE

The water-only cyclone is a modified classification cyclone that makes specific gravity separations utilizing water as the separating fluid (Figure 21). Since the water-only cyclone does not employ an external heavy medium, the effect of classification forces (forces that separate on the basis of particle size and shape as well as density) must be minimized by the design features of the cyclone if a sharp specific-gravity separation is to be made. To minimize the classification forces, the design of the water-only cyclone, as compared with conventional classification cyclones, has a wider included angle (60°–120°) and a longer vortex finder. Along with these two design elements, other features, such as cyclone diameter and the sizes of the overflow, underflow, and feed orifices, also affect its performance. Extending the vortex finder toward the apex orifice reduces the vertical distance between the end of the vortex finder and the wall of the conical section of the cyclone. This dimension is one of the most critical design variables of the water-only cyclone. The relative shortness of this dimension means that the vertical path travelled by particles in the upward central current of the cyclone is reduced sufficiently so that large low-specific-gravity particles caught in the upward current will not reach their terminal settling velocity and settle out to the wall, but will be captured inside the vortex finder and discharged with the light-specific-gravity fraction.

Operating variables that need to be addressed include the following:

- **Feed solids.** The performance of a water-only cyclone is greatly affected by varying feed concentrations. An increase in feed concentration increases the specific gravity of separation and slightly decreases the efficiency. This factor is responsible for difficulties in the control of water-only cyclones.
- **Feed orifice.** Increasing the diameter of the feed orifice on a water-only cyclone increases the capacity of the device approximately in proportion to the square of the orifice diameter. Increasing the feed orifice diameter decreases the retention time of the feed in the cyclone and thus increases the specific gravity of separation.
- **Overflow orifice.** Increasing the diameter of the overflow orifice has about the same effect as increasing the feed orifice diameter. The capacity is increased, the retention time is decreased, and the specific gravity of separation increases. An increase in the overflow orifice diameter increases the specific gravity of separation more than does a comparable increase in the feed orifice diameter.
- **Underflow orifice.** An increase in the diameter of the underflow orifice increases the capacity of the cyclone for handling refuse or reject material and decreases the specific gravity of separation. Efficiency decreases slightly with an increase in orifice diameter.
- **Orifice ratio.** The performance of a water-only cyclone is affected by the relationship of the diameter of the overflow orifice to that of the underflow orifice. This relationship determines the flow ratio of the cyclone (i.e., the ratio of apex flow to feed flow). Generally, the ratio of the diameter of the overflow orifice to that of the underflow

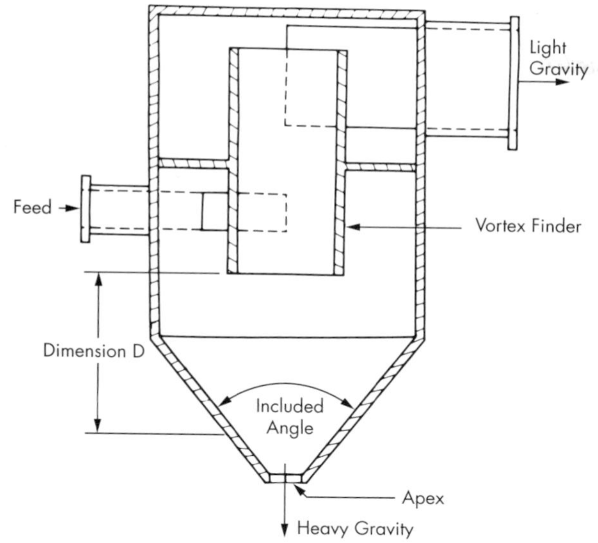

Source: Miller et al. 1985 Source: FLSmidth-Krebs 2014a

Figure 21 Cross-sectional view of a water-only cyclone and production cyclones

Table 8 Water-only cyclones available from Krebs*

Model	Maximum Feed Particle Size, mm	Effective Separation, µm	Dry Feed Range, t/h	Pulp Flow Rate Range, m³/h	Pressure Drop Range, kPa	Maximum Feed % Solids	
						Weight %	Volume %
D10LB-S-218	2	147–104	4–7	43–59	55–103	10	7
D15LB-S-245	3	208–147	11–16	91–132	69–124	12	8
D15LB-S-327	6	208–147	11–23	116–164	69–138	12	8
D20B-S-260	13	295–208	23–41	186–238	83–138	15	11
D20LSB-S-333	13	417–295	32–54	250–341	83–138	15	11
D26-S-224	19	417–295	45–82	338–500	83–152	20	15

Adapted from FLSmidth-Krebs 2014a
*Typical operating parameters.

orifice for water-only cyclones ranges from 1.5:1 to 2:1. The feed orifice is typically slightly smaller in diameter than the overflow orifice.

- **Feed pressure.** Except in those cases where coarse particles, which may cause excessive wear, are being processed, water-only cyclones are generally operated at feed pressures of 100 kPa or greater to ensure high separation efficiencies. An increase in feed pressure increases the separating efficiency and also increases the volume of feed material processed. To extend the separation ability of the cyclone to process the finest sizes, higher feed pressures are required.
- **Cyclone included angle.** This can influence separation efficiency. Some units have an included angle of 60°–95°. Furthermore, modifications of the shape or profile of the conical portion of the cyclone have been developed. Another modification features a spherical bottom for the cyclone.

Table 8 shows cyclones available from Krebs with feed volume and particle sizes that may be processed.

CONCENTRATION CRITERIA FOR GRAVITY SEPARATION

A consistent feed rate and feed density are essential for shaking tables and spiral concentrators, whereas jigs and centrifugal devices are relatively insensitive to feed rate and pulp density. A concentration criterion (CC) provided by Taggart (1945) provides an indication of the practicality of gravity concentration (other than dense medium separation):

$$CC = (D_h - D_f)/(D_l - D_f)$$

where

D_h = density of the heavy particles
D_l = density of the light particles
D_f = density of the fluid (usually water at 1.0)

Gravity concentration will usually be successful if $CC > 2.5$. When $CC < 1.25$, gravity concentration is virtually impossible. When $2.5 > CC > 1.25$, gravity separation is very difficult but may be possible to some extent with narrow feed size classification.

ENHANCED GRAVITY SEPARATION

Enhanced gravity separation involves the application of centrifugal forces to target relatively fine particles. Particles that would normally settle according to Stokes' law experience a centrifugal force with an effective acceleration of G times the normal acceleration, g, due to gravity. This g-force produces a significant increase in the terminal velocity of each particle. Moreover, the settling regime shifts toward the intermediate regime (Vance and Moulton 1965) in which the terminal velocity scales directly with the particle diameter. This means that the dependence of the particle velocity on the particle size decreases, which in turn leads to an enhanced level of separation performance.

The need to utilize centrifugal force to recover granular minerals from ore based on density difference dates back to the late 1800s (Seymour 1893; Ponten 1910; Bradbury 1912). Eccleston (1923) developed the first fully continuous EGS. Several similar types of EGS units were developed and commercially used in Russia and China throughout the late 1900s (Burt 1984). EGS units having capacities from 0.5 to 14 t/h when utilized for the recovery of gold, cassiterite, tungsten, iron, cinnabar, and other heavy minerals.

The developments that led to the EGS units commonly used in the industry today occurred over the last two decades of the 20th century, and detailed reviews of these technologies are provided by Luttrell et al. (1995) and Cole et al. (2012). The technologies vary by their separation mechanism and the magnitude of the applied centrifugal field. Each process is reportedly effective over a particle size range of ~1 mm to 10 μm. For material having a heavy mineral content less than about 1% by weight, semicontinuous units are recommended. Fully continuous units are available for feed streams containing more than 1% heavy minerals. Mass throughput capacities are as high as 1,000 t/h for semicontinuous units and 300 t/h for the continuous units.

Falcon Centrifugal Separators

A review of the historical development of the Falcon is provided by McAlister and Armstrong (1998). Three models are available: a semi-batch (SB) unit, a continuous (C) unit, and a unit to treat ultrafine (UF) material.

The Falcon SB concentrator uses a rotating bowl to generate the high gravitational forces required for mineral separations. Slurry is fed through a fixed center pipe to a distribution plate. The SB bowl has a smooth-walled bottom section and fluidized, riffled upper collection zone. The smooth-walled section allows higher-density particles to stratify and move to the wall, and lighter particles are displaced toward the center of the bowl. The collection zone is vertical and the riffles are fluidized by water injected from the back of the riffles. The fluidization elutriates and cleans the heavy concentrates in the riffles. The term *semi-batch* is used to describe the cycle; the unit is fed for a predetermined time, then the feed and machine are stopped. The concentrate is flushed with water to a concentrate discharge pipe at the bottom center of the unit. The lighter material proceeds over the lip of the collection zone to a tailing launder. The Falcon SB units have a treatment range of 1 to 400 t/h and are typically used in a grinding circuit either on cyclone underflow or cyclone feed. The SB concentrator can operate between 50 and 200 g and typically operates between 90 and 120 g. The g-force operation is determined by metallurgical testing. Clean, high-quality fluidization water is required to operate fluidized-bed-type centrifugal gravity concentrators. The flow is controlled by the process automation system, and water is injected through small holes into the fluidized concentrate bed. The holes through which water is injected in the unit are short, relatively large in diameter, and radially drilled. The material selected is stainless steel that is highly resistant to particle embedment (Sepro Mineral Systems 2018b). Figure 22 shows the features of the Falcon SB unit and operating data for the different SB models is provided in Table 9, including information on the laboratory batch unit, the L40.

The Falcon C concentrator also utilizes a spinning bowl to generate gravitational forces. The bowl is inclined outward and has smooth walls to allow for stratification of higher density particles. At the top of the bowl is a ring of specially designed hoppers followed by pneumatically controlled valves similar to muscle valves. These valves are regulated to produce a constant stream of concentrate. The C machine is able to produce a high concentrate mass. Concentrate masses can vary from 5% to 40% depending on the application. There are four sizes of C concentrators ranging from 1 t/h to 100 t/h. Feed must be screened to 1 mm to prevent blockage of the concentrate valves. These machines are able to operate between 50 and 300 g but typically between 160 g and 200 g. The unit can concentrate particles as fine as 10 μm (Sepro Mineral Systems 2018a). Figure 23 shows a schematic of the Falcon C unit, and Table 10 provides the operating data for different models.

The Falcon UF concentrator is a semicontinuous unit that is focused on the treatment of particle sizes between 37 μm and 3 μm. The UF unit utilizes the same spinning bowl as the SB and C concentrators. The UF concentrator has a smooth, outwardly inclined bowl and retains concentrate through a pneumatically controlled lip at the top of the bowl. The lip slowly expands through the sequence of a cycle. The UF unit operates in semi-batch mode with a shutdown to clean the concentrate. Typical cycle times are approximately 3 minutes. Typical feed for the UF unit is desliming cyclone overflow. The machine operates at up to 600 g and can recover minerals down to 3 μm. The target application is scavenging of slime reject streams (Sepro Mineral Systems 2018c).

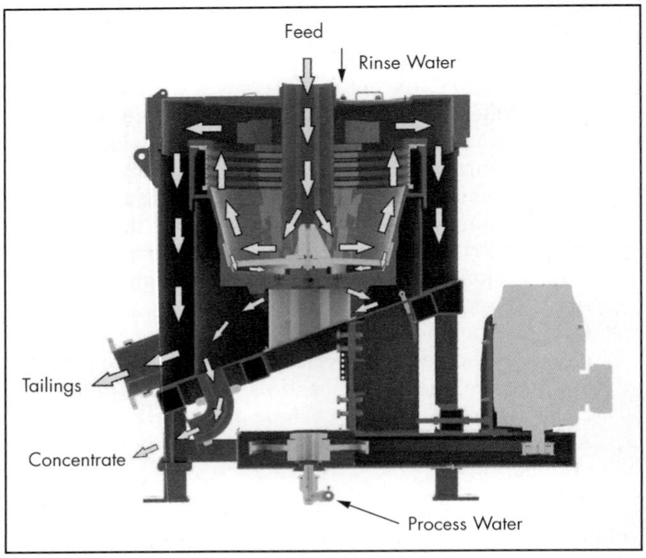

Source: Sepro Mineral Systems 2018b

Figure 22 Falcon semi-batch unit

Table 9 Operating information for the Falcon semi-batch units

Operating Data	Model				
	L40	SB750	SB1350	SB2500	SB5200
Recommended solids capacity, t/h	0–0.25	10–80	50–150	100–250	200–400
Approximate maximum slurry capacity, m^3/h	2.3	100	200	300	450
Concentrating surface area, m^2	0.03	0.46	1.08	2.14	3.37
Upper g-force range	200	200	200	200	200
Lower g-force range	50	50	50	50	50
Machine weight, kg	35	1,250	2,900	4,560	7,720
Motor power, kW (hp)	0.4 (0.5)	7.5 (10)	18 (25)	45 (60)	75 (100)
Process water consumption, m^3/h	0.24–1.2	8–12	12–20	15–28	25–35
Continuous water supply pressure, bar	2–3	2–3	2–3	2–3	2–3
Recommended maximum feed particle size, mm	1.0	2.0	2.0	2.0	2.0

Adapted from Sepro Mineral Systems 2018b

Clean, high-quality fluidization water is required to operate centrifugal gravity concentrators. Water is injected through small holes into the fluidized concentrate bed, and the flow is controlled by a process automation system. The holes through which water is injected in the unit are short, relatively large in diameter, and radially drilled.

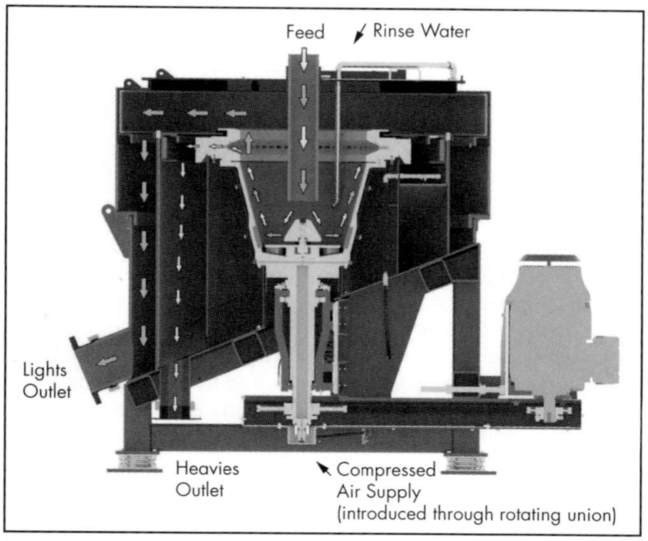

Source: Sepro Mineral Systems 2018a
Figure 23 Falcon continuous concentrator

iCON Concentrator

The iCON concentrator shown in Figure 24 is an innovative, small gravity concentrator. The iCON is designed for artisanal duty. Ore is fed through a fixed pipe in the top of the unit. Fluidization water is introduced through the main shaft to the capture zone in the bowl. The unit utilizes a variable-speed drive capable of 150 g, which allows the capture of very fine gold not recoverable by traditional techniques of artisanal miners. To recover the gold the unit is shut down and gold concentrate flushed through the center bottom to a concentrate outlet pipe. Details of the iCON concentrator are shown in Table 11.

Knelson Centrifugal Concentrator

The concept of the Knelson concentrator was patented in 1986 by B.V. Knelson and has been applied worldwide to the treatment of a variety of minerals (Knelson 1992). Both the semicontinuous and the continuous variable discharge (CVD) machines employ the concepts of fluidized particle bed separation in a mechanically applied centrifugal field.

The Knelson semi-batch concentrator operates by introducing water through a series of fluidization holes located in rings that circle the circumference of a bowl (Figure 25). The bowl, which has a truncated cone shape, is rotated at speeds that provide a centrifugal field from 60 to 200 times gravity. Feed slurry is introduced through a pipe that directs the material toward the bottom center of the machine. The heavy particles settle into the bottom of each slot of the concentrating bowl and are retained until the machine is stopped and the

Table 10 Operating data for the Falcon continuous units

Operating Data	Model			
	C400	C1000	C2000	C4000
Recommended solids capacity, t/h	1–5	5–27	20–60	45–100
Maximum slurry capacity, m^3/h	17	74	210	400
Maximum particle feed size, mm	1	1	1	1
Minimum effective capture size, µm	10	10	10	10
Concentrate % solids	65–72	65–72	65–72	65–72
Maximum feed % solids	40–45	40–45	40–45	40–45

Adapted from Sepro Mineral Systems 2018a

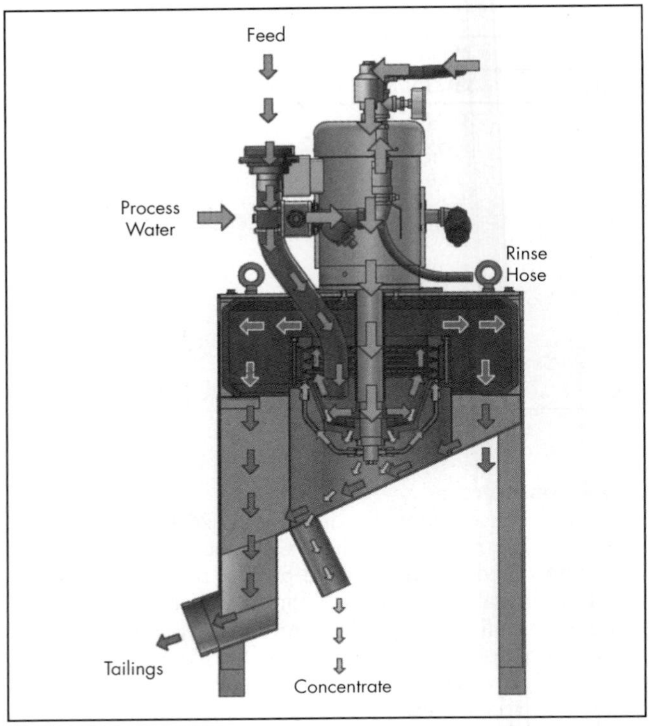

Source: Cole et al. 2012

Figure 24 iCON concentrator

Table 11 Specifications for the iCON concentrator

Operating Data	Units
Solids capacity	2 t/h
Maximum slurry capacity	100 L/min
Concentrating surface area	968 cm^2
G-force range	60–150 g
Machine weight	115 kg
Motor power	1.5 kW
Process water requirement	17 L/min
Water pressure requirement	1.0 bar
Maximum feed particle size	2 mm
Dimensions	0.6 × 0.6 × 1.29 m

Adapted from Cole et al. 2012

bowl is flushed. The duration of the concentrating cycle will vary, depending on the application. In hard rock milling applications, cycle durations typically range from 15 to 90 minutes. In alluvial applications, cycle durations range from 1 to 4 hours.

Four variations of the concentrating cone provide options to adjust the quantity of fluidizing water and so change the quantity of concentrate collected (Table 12). Operating information for the commercially available Knelson semi-batch concentrators is provided in Table 13, while information for the laboratory units is shown in Table 14.

The Knelson CVD concentrator was developed to address the limitation with respect to mass yield (concentrate) of the batch unit. The batch machines are limited to a mass yield of about 0.1%, while the CVD mass yields can be varied from about 0.1% to 50%, depending on the feed characteristics. The operating principles of the CVD concentrator are similar to those of the Knelson semi-batch concentrator but they allow the concentrate to be emitted from the fluidized bed continually. A series of pinch valves, located at the base of the fluidized rings, are kept closed by air pressure. By releasing the air pressure periodically, concentrate can be emitted without an interruption in production. Figure 26 shows a schematic of the CVD-32 model concentrator. The mechanism of separation and recovery is quite similar to the batch machine. The operating variables affecting the quantity of concentrate produced are the pinch valve opening and closing times. Operation information for the CVD unit is provided in Table 15.

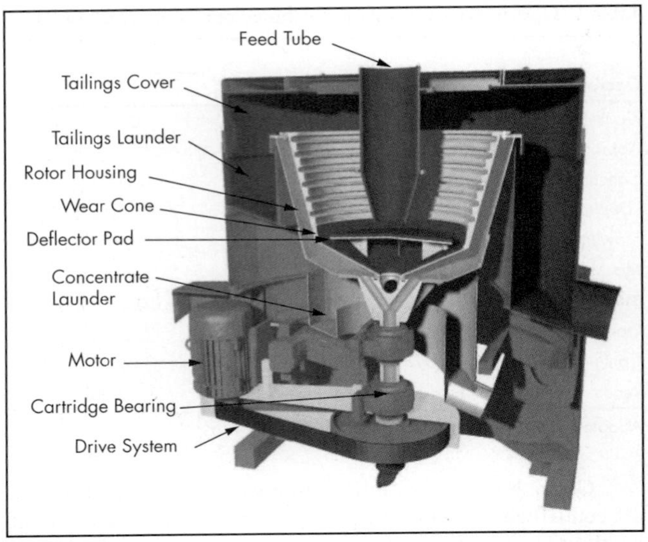

Courtesy of FLSmidth

Figure 25 Knelson semi-batch gravity concentrator

Table 12 Concentrating cone details for the Knelson semi-batch unit

Cone Style	Fluidization Water, m^3/h	Concentrate Weight, kg	G-Force Range, g
G4	30–39	31–45	60
G5	17–24	18–29	60
G6	26–39	31–45	60–120
G7	14–27	34–59	60–200

Adapted from FLSmidth-Knelson 2013

Kelsey Centrifugal Jig

The Kelsey jig consists of a series of hutches that are rotated around a central feed pipe (Figure 27).The unit is capable of generating centrifugal fields up to 100 g. A cylindrical wedge-wire screen is mounted across the top of each hutch to retain ragging material. Feed slurry enters the unit through the central feed pipe and flows outward across the bed of ragging. Hutch water is added under pressure and the water pulsed by means of push arms acting on a flexible diaphragm. The pulsations create oscillations in the bed that differentially accelerate particles based on differences in density. Low-density particles flow across the ragging material and overflow the top of the unit, while high-density particles pass downward

Table 13 Operating information for the Knelson semi-batch industrial concentrators*

Model†	Feed Solids, t/h	Feed Volume, m³/h	Concentrate Solids, kg	Concentrate Volume, L	Fluidizing Water, m³/h	Motor, kW
CD10	8	10	2–3	1.2	3–5	1
CD12	20	27	5–7	1	6–10	1.5–3.8
CD20	80	110	9–11	5	8–11	5.5–7.5
CD30	100	135	23–29	13	17–24	11
XD20	80	110	9–11	5	8–11	5.5–7.5
XD30	150	205	23–29	13	17–24	11–22
XD40	250	340	35–44	19	27–35	30–56
XD48	400	545	34–43	19	41–52	30–75
XD70	1,000	1,360	95–125	81	68–86	150–375
QS30	150	205	23–29	13	17–27	11–22
QS40	250	340	35–44	19	27–35	30–56
QS48	400	545	34–43	19	41–52	30–75

Data from FLSmidth 2014
*Data is for the G5 bowl only.
†CD = Center Discharge; XD = Extended Duty; QS = Quantum Series.

Table 14 Operating data for the batch Knelson laboratory units

Model	Feed Solids, kg/h	Feed Volume, L/min	Concentrate Solids	Concentrate Volume	Fluidizing Water, L/min	Motor, kW
MD7.5	680	95	1,200–1,600 kg	0.7 L	45–68	0.6
MD4.5	275	18	200–300 g	0.18 L	11–19	0.6
MD3	45	8	80–150 g	58 mL	0.7–4.5	0.1

Data from Knelson, n.d.(b)

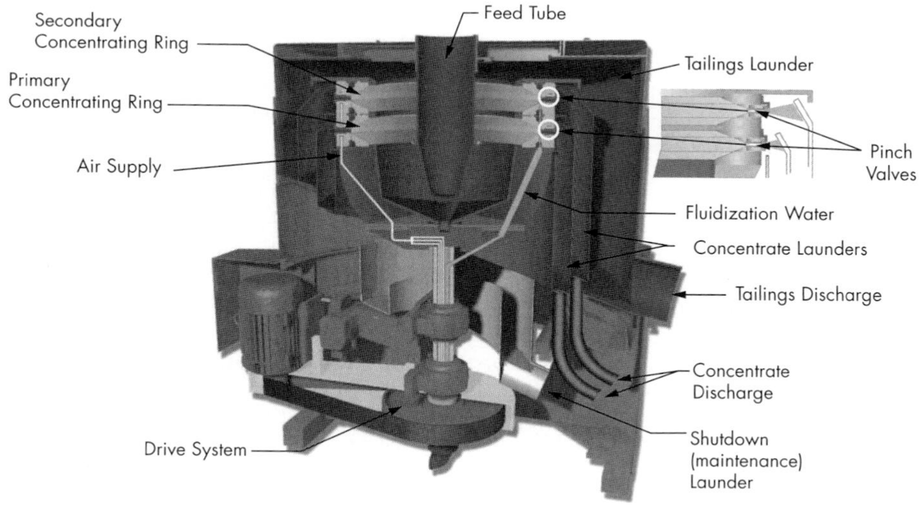

Source: Knelson, n.d.(a)
Figure 26 Knelson CVD separator

through the ragging/screen and are discharged through actuated valves. Key operating parameters include rotational speed, pulse rate, stroke length, specific gravity and particle size of the ragging material, and the hutch water addition rate. In most cases, the unit forms its own ragging material from coarser and heavier feed particles.

Applications for the jig are provided by Richards and Jones (2004) and Cole et al. (2012). The Kelsey jig is available in three models:

- J200—laboratory test unit, with nominal capacity of 15-100 kg/h solids

Table 15 Operating information for the Knelson CVD units

Model	Typical Fluidization Water Requirement, m³/h	Maximum Total Volumetric Throughput, m³/h	Maximum Feed Capacity (solids), t/h	Feed Density (solids weight), %	Maximum Feed Size, mm	Motor Specifications hp	Motor Specifications kW	Concentrate Mass Yield % of Feed Variable	Concentrator Net Weight, kg
CVD64	9–27	635	300	0–50	1.0	100–200	75–150	1–50	18,200
CVD42 1 ring	11–23	250	120	0–50	1.0	40–50	30–38	1–50	7,000
CVD32 2 rings	16–34	170	80	0–50	1.0	40	30	1–50	6,800
CVD20 1 ring	3–9	75	35	0–50	1.0	15	11	1–50	2,500
CVD6 1 ring	1–2	4	2	0–50	1.0	1.5	1	1–50	230

Courtesy of FLSmidth

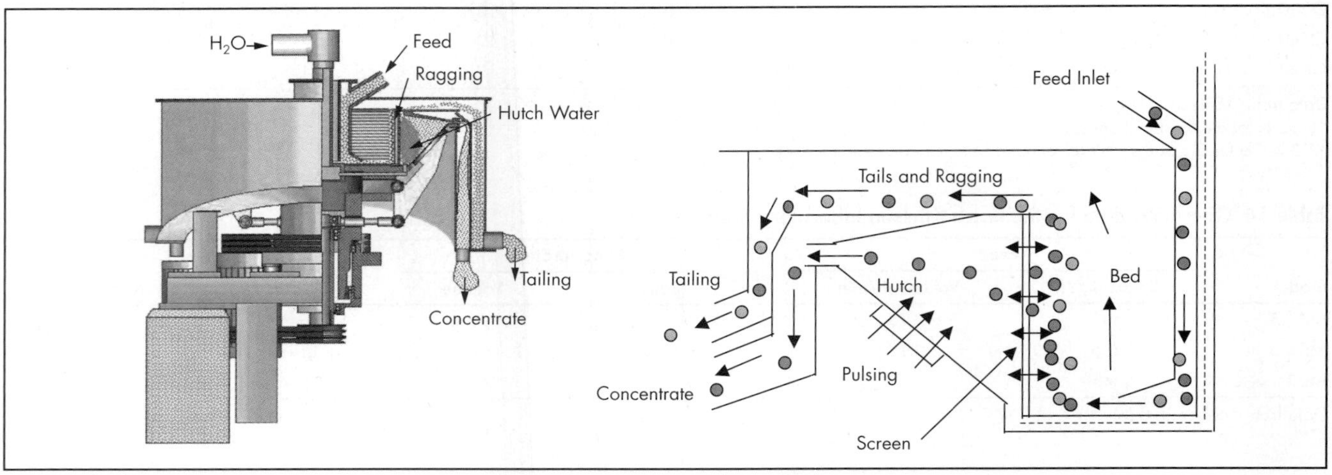

Courtesy of Kelsey Source: Falconer 2003

Figure 27 Kelsey jig

- J1300—smallest commercial unit, with nominal capacity of 2–30 t/h solids
- J1800—largest commercial unit, with nominal capacity of 5–60 t/h solids

Multi-Gravity Separator

Perhaps the most efficient EGS is the multi-gravity separator (MGS), which utilizes table riffling principles (Mozley 1990; Tucker et al. 1992). The unit consists of three main components (i.e., cylindrical rotating drum, internal scraper network, and variable-speed differential drive). Selective separations of fine particles are achieved along the internal surface of the rotating drum using the same basic principles employed by a conventional shaking table. However, replacing the table surface with a rotating drum subjects the particles to many times the normal gravitational pull. This feature allows the MGS to separate much finer particles than would otherwise be possible using conventional flowing-film separators. Figure 28 shows a schematic and industrial units of the MGS. The MGS is suitable for the treatment of fines and ultrafines with a maximum particle size of approximately 150 µm and lower limit of ~5 µm. The maximum treatment rate is around 4 t/h.

The MGS has the following operating variables:

- Drum rotational speed or spin (increased spin enhances the centrifugal g-force imparted to the particles, making it more difficult for the particles to move up the drum, hence resulting in a smaller mass and cleaner concentrate).
- Drum stroke length and frequency (increased length and frequency within limits will tend to increase the forces moving the particles up the drum, resulting in a greater mass of concentrate at a lower grade).
- Drum wash water will increase the washing of the slurry particles as they try to move up the drum, thus producing a cleaner concentrate.
- Drum tilt angle (increased tilt will produce a cleaner concentrate).

Gekko In-Line Spinner

The Gekko inline spinner (ISP) is a batch centrifugal concentrator. Feed is introduced to the base of the ISP spinning bowl. The bowl operates full of slurry and the rotation causes dense particles to flow into the riffles on the bowl surface. The lighter particles travel up the bowl and exit through a tailings launder. The concentrate is flushed through a central concentrate outlet after the unit is taken off-line. Figure 29 shows an ISP cutaway and Table 16 provides the technical specifications for the ISP. The typical application for the spinner is to upgrade low-grade gold gravity concentrates.

DRY GRAVITY SEPARATION

Pneumatic Density-Based Separations

The use of air as a medium to achieve density-based separations was the focus of significant development in the early part of the 20th century. The primary application was in coal

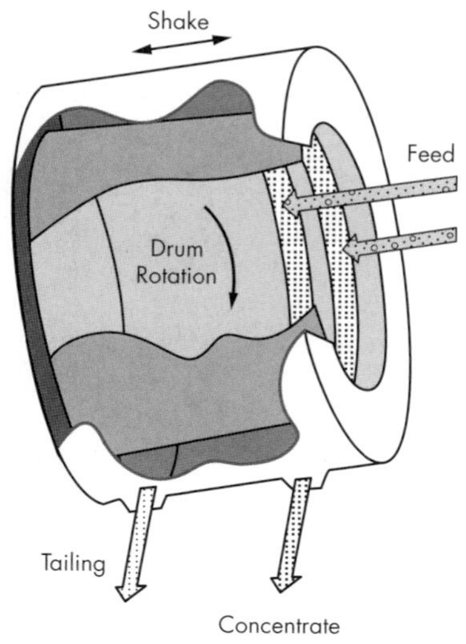

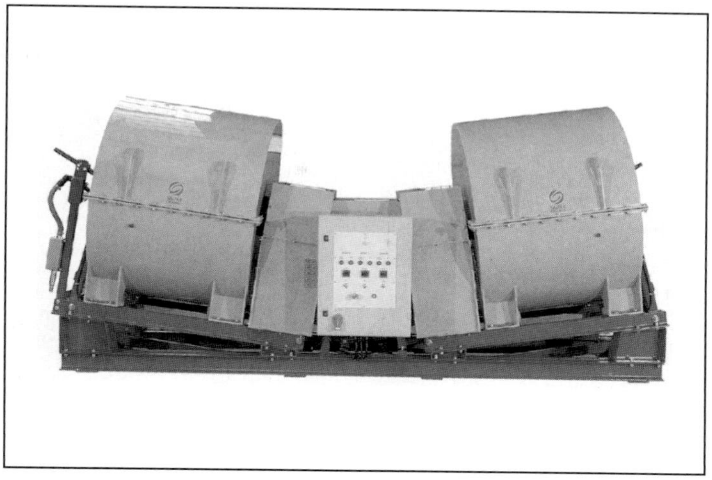

Adapted from Brown 2000

Source: Salter Cyclones 2015

Figure 28 Multi-gravity separator

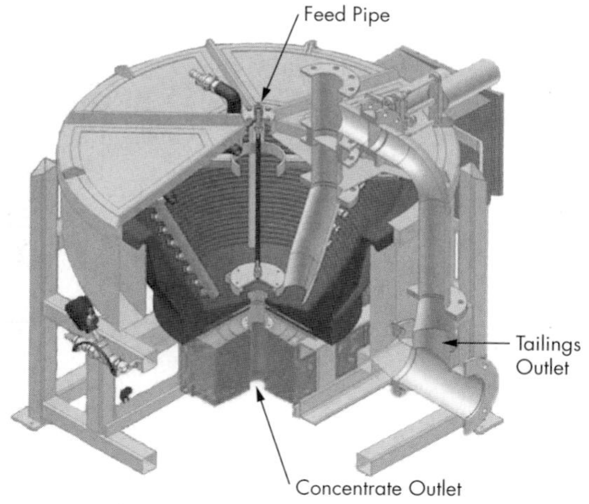

Courtesy of Gekko Systems

Figure 29 Gekko inline spinner

cleaning, with industrial installations mainly in the United States and Europe. The estimated amount of coal cleaned by pneumatic processes in 1939 was 30–40 Mt/yr (million metric tons per year; Gaudin 1939). According to Arnold et al. (1991), the amount of coal processed in the United States through dry cleaning plants reached a peak in 1965 at 25.4 Mt. The largest dry-based cleaning plant processed 1,270 t/h of –19 mm coal using 14 cleaning units.

Pneumatic density-based separators have been widely applied in the food processing industry and the recycling industry for copper recovery from used electric wire and the separation of minerals. Jarman (1942) reported several potential applications for mineral separations, including the recovery of mica from silica sand, gold from sand, fluorspar from quartz, and calcite and garnet from kyanite. In one application, an "air flotation" table was used to concentrate tungsten in a 1 × 0.5-mm particle size fraction from 4% to 64% WO_3 (tungsten trioxide) with only a trace of tungsten in the tailings, which accounted for 85.8% of the feed.

The pneumatic technologies incorporate the same basic mechanisms used in wet separators, including dense medium separations, pulsated air jigging, riffled table concentration, and air-fluidized launders.

Air Jigs

The early development of pneumatic devices using jigging principles resulted in many technologies, including the Plumb and Hooper jigs (Gaudin 1939). These technologies used either pulsating air through a sieve or a moving sieve in a steady stream of air to generate the jigging action. The most commonly used air jig in the United States was the Stump super air-flow jig. It included a vibrating bed, a single fan with a mechanism for producing pulsations, and a multiple-sink product discharge mechanism. Recent modifications to the Stump jig have enhanced the operational characteristics and performance of the unit. The changes include (Kelly and Snoby 2002)

- A star gate to meter homogeneous feed evenly across the entire width of the jigging bed;
- A constant-velocity fan (working air) to loosen the bed of material;
- Installation of external vibrators to enhance particle transport in the jig;
- Addition of a pulsed air fan to optimize stratification by offering independent control of stroke frequency, amplitude, and acceleration; and
- Application of an automatic bed-level control system to achieve a constant separation performance and product quality over large variations in the feed ash content.

The Allair jig shown in Figure 30 has an automatic bed-level sensor that utilizes nuclear density instrumentation. The information from the nuclear density instrument is used to activate a refuse discharge star gate valve to remove refuse at a rate proportional to the amount of impurities entering in the feed. By using only one refuse removal mechanism, the air

Table 16 Gekko inline spinner specifications

Details	Model Mini	02	30
Length, mm	800	1,000	1,650
Width, mm	500	1,000	1,600
Height, mm	710	1,200	1,500
Maximum feed rate, m^3/h	0.25	10	100
Maximum feed rate, t/h	0.05	2	30
Maximum particle size, mm	2	4	6
Typical concentration time, minutes	5–30	5–30	10–30
Typical time off-line for dump, minutes	5	0.5–2.0	1–2
Volume solids per dump, L	<0.5	1–4	8–12
Mass per dump, kg	0.4–0.6	1.5–8.0	12–25
Volume slurry per dump, L	12	45	200
Concentrate discharge	Manual	Automatic	Automatic
Water required per dump, L	5	15	40
Air quality required	Instrument air quality at 600 kPa	Instrument air quality at 600 kPa	Instrument air quality at 600 kPa
Motor speed, rpm	1,440	1,440	1,440
Installed power, kW	0.37	2.2	3
Bowl			
Maximum inside diameter, mm	335	560	882
Depth, mm	100	350	450
Volume, L	6.04	30	160
Material	Polyurethane	Polyurethane	Polyurethane
Revolutions per minute, nominal (40–60 Hz)	100	84–126	88–132
G-force at nominal rpm (top of rotor), g	1.9	2.2–5.0	3.8–8.6

Adapted from Gekko Systems 2010

Courtesy of Allmineral

Figure 30 Allair jig: (A) basic operation and (B) full-scale unit at a coal processing facility

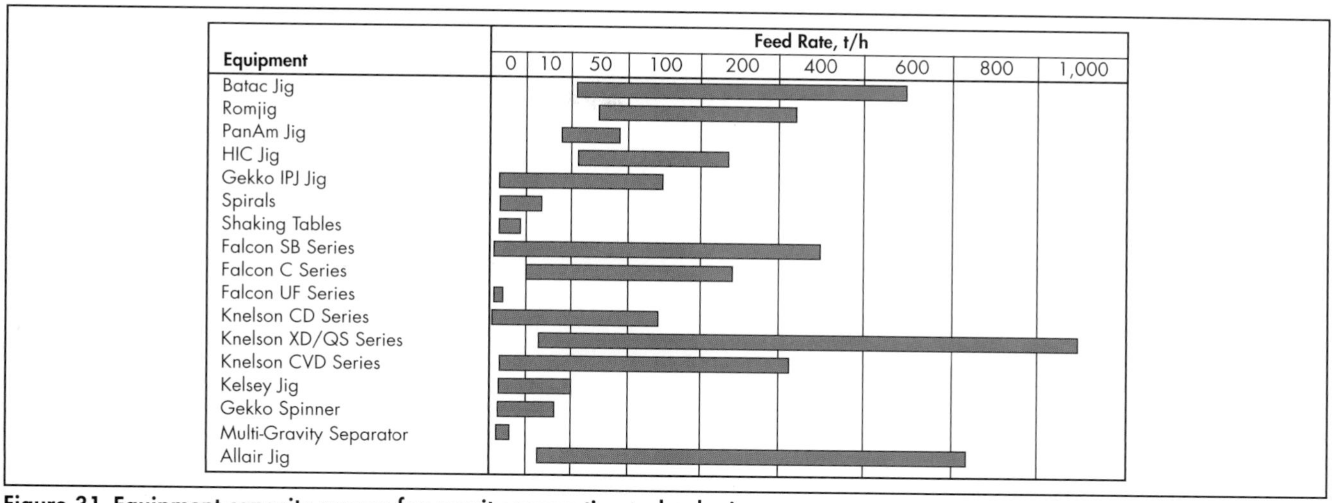

Figure 31 Equipment capacity ranges for gravity separation technologies

jig constantly maintains a reserve layer of high-density material over the outlet. This keeps the lower-density coal particles at a fixed distance from the screen bed, thereby minimizing misplacement.

Several industrial installations use air jigging technology worldwide, including operations located in Spain, India, Brazil, and the United States. Although air jigs are applicable for any lightweight material, most installations involve coal cleaning for the purposes of ash and/or sulfur reduction. Commercial air jigs are capable of treating up to 60 t/h of coarse (75 × 12 mm) coal and up to 40 t/h of fine (12 × 1 mm) coal (Snoby et al. 2009).

Air Tables

Riffled tables using air to create a fluidized bed of high-density fine particles were the most commonly used pneumatic separators in the early 1900s. Common technologies were the Birtley and Sutton-Steel tables. These technologies provided effective density separations for coal coarser than 6 mm and minerals having a particle size as small as 0.3 mm (Jarman 1942). Throughput capacity was around 5 t/h per table. Recent modifications to air tables include vertical suspension of the table, modern dust collection technology, and automation instrumentation, which increased throughput capacity to 120 t/h per table when treating coal in the particle size range of 75 × 6 mm (Lu et al. 2003).

The separation data obtained from coal studies conducted in China indicate that a riffled table has the potential to provide an effective separation for particles as coarse as 80 mm to a lower size limit of around 3 mm. The operational data also indicate that the process is relatively insensitive to surface moisture up to a value of around 7%–10% by weight. Performance evaluations conducted on full-scale units indicate the ability to provide a relatively high separation density (RD_{50}) of around 2.0 RD while achieving probable error values that range from 0.15 to 0.25 (Lu et al. 2003). A detailed evaluation conducted in the United States using a 5-t/h air table unit found that the table has the ability to provide significant upgrading for all ranks of coal (Honaker et al. 2008).

SUMMARY DATA FOR GRAVITY SEPARATORS

Equipment Treatment Rates

Capacities for commonly applied gravity concentration devices are shown in Figure 31, based on more detailed references, primarily in Cole et al. (2012) and Laplante and Gray (2005) as well as equipment specifications provided by manufacturers.

Feed Particle Size Range for Separators

The particle size ranges that can be treated in the more common gravity separators are shown in Figure 32.

GRAVITY SEPARATION TEST WORK

As with all test work, the samples need to be selected and subsampled correctly. Furthermore, samples should be properly characterized in respect to mineral composition, liberation, and chemical analysis. These topics are discussed elsewhere in the handbook.

Gekko-Wilfley Laboratory Table Test

The test procedure applied by Gekko Systems to establish yield–recovery relationships for in-line pressure jig (IPJ) applications is described by Cole et al. (2012). This test can be similarly applied for analysis of any continuous gravity concentration application. A quarter-size laboratory Wilfley shaking table is used to replicate the expected mass and metal recovery from an IPJ (Coward 2007). The IPJ can be configured in an open- or closed-circuit mode in industrial applications, and the laboratory procedure can simulate both modes of operation. The simulation of a closed-circuit process consists of four steps:

1. A sample preparation of the feed (usually 20 kg of –1-mm material) is fed as a slurry onto the shaking table and a table concentrate of ~10% of the feed mass is collected.
2. The table tailings are dewatered and the tailings are ground to a particle size P80 of 500 µm. The ground product is re-tabled and, again, a concentrate of ~10% of the feed is collected.

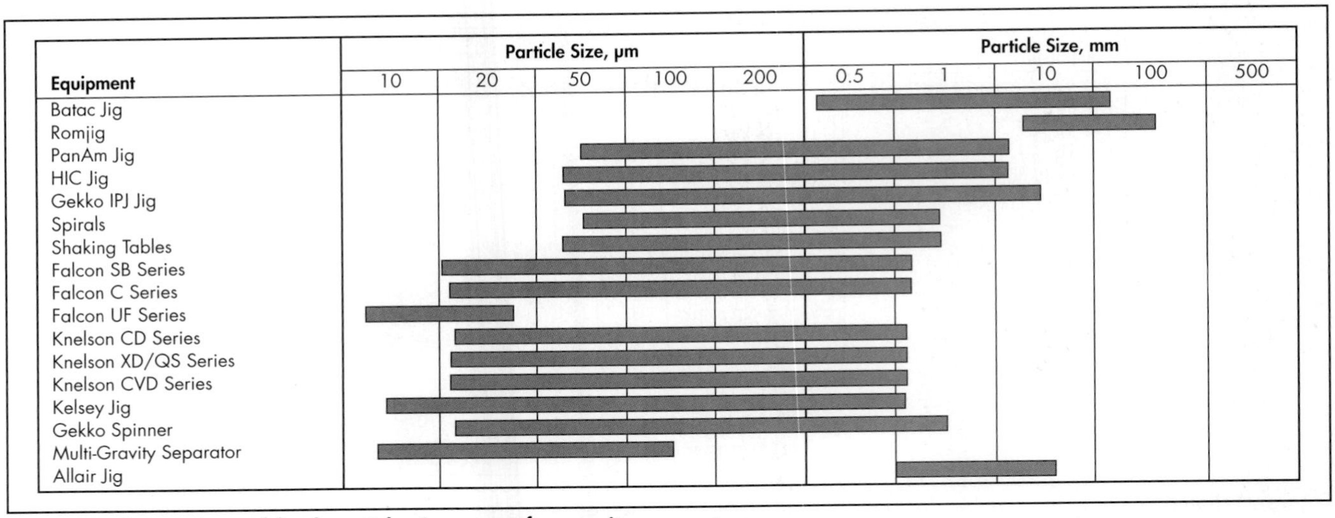

Figure 32 Comparison of feed particle size range for gravity separators

3. The table tailings are collected, dewatered, and reground, this time to a particle size P80 of 106 µm before tabling again. A concentrate of around 5% of the feed mass is collected for this test. The table tailings from this test are the final tailings product, and it is filtered, dried, and portions removed for chemical analysis.
4. The three table concentrates from steps 1, 2, and 3 are combined (i.e., some 25% of the original mass) and then re-tabled. Four samples are collected from this test (high-grade, medium-grade, low-grade, and low-grade tailings). The products are filtered, dried, and weighed, and assay portions are removed for assaying.

The data from the test-work program are used to develop a cumulative concentrate mass recovery versus metal recovery curve and allows for a decision to be made on the optimum mass of concentrate to be collected for the ore being evaluated. For an open-circuit IPJ simulation, a very simple shaking table procedure can be adopted whereby only a single test is performed and four concentrates and a tailing are collected.

Test-Work Methodology for Knelson and Falcon Centrifugal Separators

The laboratory test procedure for both the Knelson and Falcon laboratory batch machines are similar; however, operating parameters (water addition and agitation speed [g-force]) are particular to each machine as well as the ore being tested (Cole et al. 2012). The operating manuals provided with the laboratory machines provide guidance as to the operating parameters for setup. The design and operating parameters of the batch laboratory machines are similar to those of the larger industrial machines as described previously. The quantity of concentrate produced from both the Knelson and Falcon laboratory machines is similar, around 100 g (±20 g), depending on the relative density of the concentrate being collected. The volume of concentrate collected remains constant regardless of the amount of material fed to the machine. On this basis, the concentrate mass percent generated during a gravity test is mainly a function of the feed mass used for the test. For example, if 10 kg is fed to the batch machine, then the mass recovery to the concentrate (~100 g) will be 1%. For 1 kg of feed, the mass recovery to the concentrate will be around 10%, whereas 0.1% mass of concentrate is provided when 100 kg of feed are treated in the batch machine. The fluidization water flow rate is best determined by inspecting the concentrate retained in the riffles of the bowl after a setup test. The concentrate should not be packed hard but should just start to slump out of the lower riffle. Too high a water pressure will give a low concentrate mass, while too low a water pressure will have a higher-than-optimal mass recovery. Generally, the applied g-force is a function of the feed particle size. For coarse particle size, the g-force should be low (<50 g), whereas for very fine particles, it will be high (>200 g). The maximum particle size of the feed sample to the batch machines is usually around 1 mm.

Continuous Separations—Continuous Knelson Test

A procedure using a laboratory-scale Knelson concentrator is described by Sakuhuni et al. (2014) to establish mass (yield)–recovery relationships for continuous applications of centrifugal concentrators. This gravity release analysis consists of a series of rougher-scavenger-cleaner gravity tests based on the 3-in. laboratory Knelson concentrator (model MD3). The starting feed sample is a representative 5-kg (–1 mm) sample riffled from the original sample. The 5-kg sample is further riffle split into 1-kg subsamples. Each 1-kg subsample is treated twice in an MD3 concentrator, collecting the concentrate after each run. The tailings from each subsample are combined to constitute the final tailings. The concentrates are combined to constitute an ~1-kg rougher concentrate which is rerun in the MD3 four times, collecting the concentrates after each run. The concentrates are weighed and assayed for the target elements. The final tailing is weighed and a representative sample is assayed for the required elements.

Gravity Recovery Gold Test Procedure

The most commonly applied test procedure for precious metals gravity-circuit design and optimization is the gravity-recoverable gold (GRG) test procedure described by Laplante and Clarke (2006) and Giblett et al. (2013). The GRG test requires a large sample of ore (typically 80–100 kg) to be processed through a Knelson manual discharge MD3 concentrator at three successive liberation sizes, known as

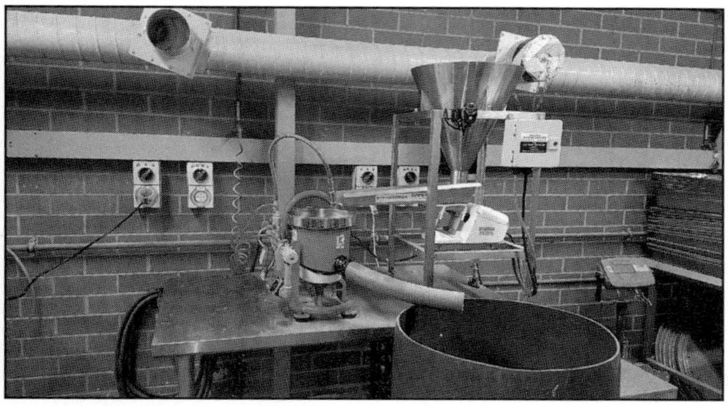

Courtesy of Gold Technology Group, Curtin University

Figure 33 MD3 treatment circuit for gravity test work

Stages 1–3. In Stage 1, the entire sample is crushed to 100% –850 µm and processed through the concentrator, generating a Stage 1 GRG concentrate. The Stage 1 tailings are then split to produce a smaller sample (25–35 kg), which is ground to 45%–55% passing 75 µm and then run through the Knelson MD3 in Stage 2 of the test, producing a Stage 2 GRG concentrate. Finally, the Stage 2 tailings are ground to 80% passing 75 µm and run through the MD3 again, producing a third (Stage 3) Knelson concentrate. A final mass balance determines the total amount of gold recovered as GRG and the percentage recovered in each individual stage of the test. In the absence of sufficient information to estimate the required sample mass, the guidance of Laplante and Doucet (1996) should be followed and a sample mass of between 40 kg and 70 kg selected for initial gravity testing, ideally following the Laplante GRG test procedure (referenced to avoid confusion with the single-stage test). However, for samples of lower gold grade (<1.5 g/t) and coarse gold particle size, up to 150 kg of sample may be required (Laplante 2000). Where the single-stage GRG test is to be used for gravity gold characterization, such as for variability testing, a smaller sample mass of nominally 20 kg can be applied (Laplante and Clarke 2006). Interpretations of test results are provided in Laplante (2000) and Giblett et al. (2012).

The Knelson MD3 concentrator is the recommended laboratory unit for the GRG test. Figure 33 shows an installed Knelson concentrator with dedicated feed bin and feeding system to allow consistent material flow to the concentrator. Access to water for feed dilution, concentrator fluidization, and cleanup is required. The ability to handle large volumes of dilute gravity unit tailings and to extract dust from transfer points when working with dry feed should be considered in the laboratory system design.

Gravity Circuits for Precious Metal Recovery

Gravity circuits in precious metal recovery applications are typically situated in the grinding circuit to allow the recovery of free gold before sufficient damage or coating of the gold particles occurs to make gravity separation, flotation, or cyanidation less effective. Where the grinding circuit involves semiautogenous grinding and ball milling, the gravity circuit will typically be located in the secondary grinding stage. Where flotation is applied downstream, it may also be feasible to apply gravity separation in the concentrate regrind circuit, where a recirculating load of finer free gold particles may occur as a natural result of classification by hydrocyclones (Fullam 2010).

Gravity concentration has also been successfully applied to the treatment of carbon-in-pulp tailings in gold flow sheets to recover gold-bearing sulfides (Delahey et al. 1992; Butcher and Laplante 2003).

The treatment of gravity concentrates by tabling or intensive cyanidation often requires a batch processing approach, which is facilitated by having suitable concentrate storage capacity to ensure the continuity of the process. Where batch-intensive cyanidation is employed, there is a need for sufficient concentrate storage volume to match the full capacity of the leach reactor. This allows concentrate to be accumulated in preparation for the next leach cycle and therefore metal production rates to be maximized without leach reactor downtime due to insufficient concentrate stocks. The efficiency of upgrading gravity concentrates by shaking tables is significantly improved by having a consistent solids feed rate and pulp density. Ideally, sufficient concentrate storage should be provided ahead of the table to ensure a constant feed rate during the operating cycle of the table.

ACKNOWLEDGMENTS

The chapter authors thank D.E Pickett, G.W. Riley, A.W Deurbrouck, W.W Agey, and F.B. Mitchell, authors of several chapters in the "Gravity Concentration" section of the 1985 edition of the *SME Mineral Processing Handbook*, who set a standard to be followed with their thorough and well-developed work. The authors also thank the following people for providing information that has been incorporated into the current chapter: Mike Fullam, Steve McAlister, Jonas Boehnke, Ajay Sihota, Shane Anderson, Hamid Sheriff, Brian Packer, John Christophersen, Sandy Gray, Anastasia Plischka, Randy Clarkson, Eric Spiller, and Andrew Jonkers.

REFERENCES

Allmineral. 2013. *Alljig, Allflux, Allair, Gaustec: Product Information.* Duisburg, Germany: Allmineral. www.allmineral.com/fileadmin/user_upload/allmineral/pdf/allmineral_productinfo_gb.pdf.

Arnold, B.J., Hervol, J.D., and Leonard, J.W. 1991. Dry particle concentration. In *Coal Preparation*, 5th ed. Edited by J.W. Leonard. Littleton, CO: SME.

Batchelor, G.K. 1967. *An Introduction to Fluid Dynamics.* Cambridge: Cambridge University Press.

Bradbury, W.F. 1912. Ore-separator. U.S. Patent 1,042,194.

Brown, A.S. 2000. Axsia acquires Richard Mozley. https://www.chemicalonline.com/doc/axsia-acquires-richard-mozley-0001. Accessed August 2018.

Burt, R.O. 1984. *Gravity Concentration Technology.* New York: Elsevier.

Burt, R.O. 1986. Selection of gravity concentration equipment. In *Design and Installation of Concentration and Dewatering Circuits.* Edited by A.L. Mular and M.A. Anderson. Littleton, CO: SME.

Burt, R.O. 1999. The role of gravity concentration in modern processing plants. *Miner. Eng.* 12(11):1291.

Butcher, G., and Laplante, A. 2003. Recovery of gold carriers at Granny Smith mine using Kelsey jigs. In *Advances in Gravity Separation.* Littleton, CO: SME. pp. 155–164.

Clarkson, R. 1990. *Placer Gold Recovery Research: Final Summary.* Whitehorse, Yukon: New Era Engineering.

Coggin, H.M. 1995. Jig recovery systems. Circular 52. Phoenix, AZ: Department of Mines and Mineral Resources. April. http://docs.azgs.az.gov/OnlineAccessMineFiles/Pubs/2013-02-0121.pdf.

Cole, J., Dunne, R., and Giblett, A. 2012. Review of current enhanced gravity separation technologies and applications. In *Separation Technologies for Minerals, Coal, and Earth Resources*. Edited by C.A. Young and G.H. Luttrell. Littleton, CO: SME.

Coward, M. 2007. Gravity concentration, progressive grind. In *Gekko Systems Laboratory Manual*. Gekko Systems.

Davies, E.D. 1902. *Machinery for Metalliferous Mines: A Practical Treatise for Mining Engineers*. New York: Van Nostrand.

Delahey, G., Martins, J.V., and Dunne, R. 1992. Plant practice at the New Celebration gold mine. In *Extractive Metallurgy of Gold and Base Metals*. Melbourne, Victoria: Australasian Institute of Mining and Metallurgy. pp. 95–101.

Deveau, C., and Young, S.R. 2005. Pushing the limits of gravity separation. SME Preprint No. 05-84. Littleton, CO: SME.

Doerner, K. 1997. Producing high-quality sands with sophisticated technology in the new sand beneficiation plant of Pinnow gravel works. *Aufbereit Tech. Miner Process.* 38(1):27.

Eccleston, C.W. 1923. Centrifugal separator. U.S. Patent 1,473,421.

Elder, F., Kow, W., Domenico, J., and Wyatt, D. 2001. Gravity concentration—A better way (or how to produce heavy mineral concentrate and not recirculating loads. In *Proceedings of the International Heavy Minerals Conference*. Melbourne, Victoria: Australasian Institute of Mining and Metallurgy. p. 115.

Falconer, A. 2003. Gravity separation: Old technique/new method. *Phys. Sep. Sci. Eng.* 12(1):31–48.

Ferree, T.J. 1973. An expanded role in minerals processing seen for Reichert Cone. *Min. Eng.* 25(3):29–31.

FLSmidth. 2014. Various Knelson concentrator specification sheets (CD, QS, XD series). www.flsmidth.com/en-US/Industries/Categories/Products/PreciousMetalExtraction/Gravity+Concentration/Batch+Concentrators. Accessed August 2018.

FLSmidth-Knelson. 2013. Knelson Concentrator Model KC-QS30 Specification Sheet. https://www.consep.com.au/assets/documents/Knelson%20KC-QS30%20rev1.pdf.

FLSmidth-Krebs. 2014a. *Krebs Products for the Coal Processing Industry: Krebs Hydrocyclones, millMAX Pumps, and Technequip Valves for Coal Processing.* www.flsmidth.com/~/media/PDF%20Files/Krebs/KrebsCyclonesPumpsandValvesforCOALIndustryweb11132015.ashx.

FLSmidth-Krebs. 2014b. *Krebs Spiral Concentrators for Mineral Applications.* Bulletin 4-202. https://www.progress-eq.com/wp-content/uploads/2014/10/Krebs-Spiral-for-Mineral.pdf.

Fullam, M. 2010. Predicting the benefit of gravity recovery prior to flotation. In *Proceedings of Gravity Gold.* Edited by S. Dominy. Melbourne, Victoria: Australasian Institute of Mining and Metallurgy.

Galvin, K.P. 2012. Application of the Reflux Classifier for measuring gravity recoverable product. In *Separation Technologies for Minerals, Coal, and Earth Resources.* Edited by C.A. Young and G.H. Luttrell. Littleton, CO: SME.

Galvin, K.P., Walton, K., and Zhou, J. 2009. How to elutriate particles according to their density. *Chem. Eng. Sci.* 64:2003.

Galvin, K.P., Walton, K., and Zhou, J. 2010. Application of closely spaced inclined channels in gravity separation of fine particles. *Miner. Eng.* 23:326.

Gaudin, A.M. 1939. *Principles of Mineral Dressing.* New York: McGraw-Hill.

Gekko Systems. 2010. InLine Spinner: Technical specifications. http://gekkos.com/technical-specifications-inline-spinner.

Giblett, A., Hillier, D., Parker, K., and Ramsell, V. 2012. The impact of gravity concentration at Kalgoorlie Consolidated Gold Mines. Presented at AusIMM Mill Operators Conference, Hobart, Tasmania.

Giblett, A., Bax, A., Wardell-Johnson, G., and Staunton, W. 2013. A review of best practice in gravity circuit design and operation. In *MetPlant 2013: Plant Design and Operating Strategies.* Melbourne, Victoria: Australasian Institute of Mining and Metallurgy.

Gray, A.H. 1997. Inline pressure jig—An exciting, low cost technology with significant operational benefits in gravity separation of minerals. In *Resourcing the 21st Century: The AusIMM 1997 Annual Conference.* Melbourne, Victoria: Australasian Institute of Mining and Metallurgy.

Harrison, H.L.H. 1962. Chapter 6. In *Alluvial Mining for Tin and Gold.* London: Mining Publications. pp. 119–141.

Honaker, R.Q., Saracoglu, M., Thompson, E., Bratton, R., Luttrell, G.H., and Richardson, V. 2008. Upgrading coal using a pneumatic density-based separator. *Inter. J. Coal Prep. Util.* 28(1):51.

Hoover, H.C., and Hoover, L.H. 1950. *De Re Metallica.* New York: Dover Publications.

Humphreys, I.B. 1975. Helical chute concentrator and method of concentrating. U.S. Patent 3,891,546.

Hutahaean, B.P., and Yudok, G. 2013. Analysis and proposed changes of tin ore processing system on cutter suction dredges into low grade to improve added value for the company. *Indonesian J. Bus. Admin.* 2(16).

IHC Technical Services. n.d. *Mineral Processing* [brochure]. The Netherlands: IHC Technical Services.

Iijima, Y., Shimoda, Y., Kubo, Y., Jinnouchi, Y., Sakata, S., and Konisha, M. 1998. Improvement of Baum jig into Vari-Wave type in Ikeshima colliery. In *Proceedings of the XII International Coal Preparation Congress.*

Indian Institute of Technology Kharagpur. n.d. Module 2: Solid Fossil Fuel (Coal); Lecture 9: Coal preparation and washing. NPTEL web course. https://nptel.ac.in/courses/103105110/m2l9.pdf.

Jarman, G.W. 1942. Special methods for concentrating and purifying industrial minerals. *Trans. AIME* 304.

Kelley, M., and Snoby, R. 2002. Performance and cost of air jigging in the 21st century. In *Proceedings of the 19th International Coal Preparation Conference.* Stamford, CT: Primedia Business Exhibitions. pp. 175–186.

Knelson. n.d.(a). *CVD: Continuous Variable Discharge Technology* [brochure]. http://knelsongravity.xplorex.com/sites/knelsongravity/files/CVD-single-p.pdf.

Knelson. n.d.(b). *Semi-Continuous Knelson Concentrators* [brochure]. https://www.multitechwa.com/file/Knelson%20Semi-Continuous%20Brochure.pdf.

Knelson, B. 1992. The Knelson concentrator metamorphosis from crude beginning to sophisticated world-wide acceptance. *Miner. Eng.* 5(10-12):1091–1097.

Kohmuench, J.N., Luttrell, G.H., and Mankosa, M.J. 2001. Coarse particle concentration using the hydrofloat separator. *Miner. Metall. Process.* 18(2):61.

Kohmuench, J.N., Mankosa, M.J., Kennedy, D.G., Yasalonis, J.L., Taylor, G.B., and Luttrell, G.H. 2007. Implementation of the HydroFloat separator at the South Fort Meade mine. *Miner. Metall. Process.* 24(4):264.

Laplante, A. 2000. Testing requirements and insight for gravity gold circuit design. Presented at the Randol Gold and Silver Forum, Vancouver, BC.

Laplante, A., and Clarke, J. 2006. A simple gravity-recoverable gold test. In *Proceedings of the 38th Annual Meeting of the Canadian Mineral Processors.* Montreal, QC: Canadian Institute of Mining, Metallurgy and Petroleum. pp. 23–38.

Laplante, A., and Doucet, R. 1996. A laboratory procedure to determine the amount of gravity recoverable gold. SME Preprint No. 96-174. Littleton, CO: SME.

Laplante, A.R., and Gray, A.H. 2005. Advances in gravity gold technology. In *Advances in Gold Ore Processing.* Amsterdam: Elsevier. pp. 280–306.

Leonard, J.W., and Mitchell, D.R., eds. 1968. *Coal Preparation*, 3rd ed. New York: AIME.

Lewis, R.M. 1990. The Lewis hydrosizer. In NC State Minerals Research Laboratory Publication No. 81-12-P.

Loveday, G., and Jonkers, A. 2002. The Apic jig and the Jigscan controller take the guesswork out of jigging. In *Proceedings of the XIV International Coal Preparation Congress.* Johannesburg: Southern African Institute of Mining and Metallurgy. p. 247.

Lu, M., Yang, Y., and Li, G. 2003. The application of compound dry separation technology in China. In *Proceedings of the 20th Annual International Coal Preparation Exhibition and Conference.* Greenwood Village, CO: Primedia Business Exhibitions.

Luttrell, G.H., Honaker, R.Q., and Phillips, D.I. 1995. Enhanced gravity separators: New alternatives for fine coal cleaning. In *Proceedings of the 12th International Coal Preparation Conference.* Deland, FL: Intertec Presentations.

Luttrell, G.H., Kohmuench, J.N., Stanley, F.L., and Trump, G.D. 1998. Improving spiral performance using circuit analysis. *Miner. Metall. Process.* 15(4):16.

Luttrell, G.H., Westerfield, T.C., Kohmuench, J.N., Mankosa, M.J., Mikkola, K.A., and Oswald, G. 2006. Development of high-efficiency hydraulic separators. *Miner. Metall. Process.* 23(1):33.

Mankosa, M.J., and Luttrell, G.H. 1999. Hindered-bed separator device and method. U.S. Patent 6,264,040.

Mankosa, M.J., and Luttrell, G.H. 2002. Air-assisted density separator device and method. U.S. Patent 6,425,485.

Mankosa, M., Stanley, F., and Honaker, R.Q. 1995. Combining hydraulic classification and spiral separation for improved efficiency in fine coal cleaning. In *High Efficiency Coal Preparation.* Edited by S.K. Kawatra. Littleton, CO: SME. pp. 99–108.

MBE Coal and Minerals Technology GmbH. 2011. *Batac and Romjig Jigging Technology.* Cologne, Germany: MBE.

McAlister, S.A., and Armstrong, K.C. 1998. Development of the Falcon concentrator. SME Preprint No. 98-172. Littleton, CO: SME.

McCracken, D. 2013. Setting the proper water velocity through a sluice box. Happy Camp, CA: The New 49'ers.

Michell, F.B. 1985. Flowing film concentrators. In *SME Mineral Processing Handbook.* Edited by N.L. Weiss. Littleton, CO: SME-AIME.

Miller, F.G., DeMull Jr., T.J., and Matoney, J.P. 1985. Centrifugal specific gravity separators. In *SME Mineral Processing Handbook.* Edited by N.L. Weiss. Littleton, CO: SME-AIME.

Min, I.Y.K. 2006. Gravel pump tin mining in Malaysia. *Jurutera* (May):6–11.

Mineral Technologies. n.d. Gemeni table. Queensland, Australia: MD Mineral Technologies. http://www.gravityrecovery.com/gemeni.pdf.

Mineral Technologies. 2017a. Technology: MD—Spiral separators [brochure]. Queensland, Australia: MD Mineral Technologies.

Mineral Technologies. 2017b. Technology: MD—Coal spiral separators [brochure]. Queensland, Australia: MD Mineral Technologies.

Mozley, R. 1990. Minerals separator. U.S. patent application 4,964,845.

Multotec. 2016. *Heavy Mineral Spiral Concentrators.* https://www.multotec.com/public/uploads/fileLibrary/73660%20Multotec%20Heavy%20Mineral%20Spiral%20Concentrators-Digital-df289.pdf.

Newton, I. 1726. Book II: Scholium to proposition XL [The motion of bodies]. In *Philosophiae Naturalis Principia Mathematica [The Mathematical Principles of Natural Philosophy]*, 3rd ed. London: Royal Society. Reprinted by University of Glasgow, 1871.

PhySep. n.d. *Spiral Concentrators.* Jacksonville, FL: PhySep.

Ponten, A. 1910. Centrifugal sluicing-machine. U.S. Patent 980,001.

Price, J.D., and Bertholf, W.M. 1959. Improvements in plant operations at Pueblo coal washery. *Min. Eng.* (December):1190–1195.

Reichert, E. 1965. Method and apparatus for the wet gravity concentration of ores. Australian Patent 628811965.

Richards, R.G., and Jones, T.A. 2004. The Kelsey centrifugal jig—An update on technology and application. SME Preprint No. 04-021. Littleton, CO: SME.

Richards, R.G., and Palmer, M.K. 1997. High capacity gravity separators—A review of current status. *Miner. Eng.* 10(9):973.

Richards, R.G., Hunter, J.L., and Holland-Batt, A.B. 1985. Spiral concentrators for fine coal treatment. *Coal Prep.* 1:207.

Richards, R.H., Locke, C.E., and Schuhmann, R. 1940. *Textbook of Ore Dressing*, 3rd ed. New York: McGraw-Hill.

Sakuhuni, G., Tong, L., and Klein, B. 2014. Gravity release analysis and its application on Myra Falls zinc flotation tailings. Presented at the XXVII International Mineral Processing Conference, Santiago, Chile, October 20–24.

Salter Cyclones. 2015. Multi-Gravity Separation Systems. https://www.saltercyclones.com/files/6814/4231/3314/multi-gravity_separation_systems_june_2015.pdf.

Sanders, G.J., and Ziaja, D. 2003. The role and performance of Romjig in modern coal preparation practice. In *Proceedings of the 20th International Coal Preparation Conference.* Greenwood Village, CO: Primedia Business Exhibitions. p. 207.

Sanders, G.J., Gnanaiah, E.U., and Ziaja, D. 2000. Application of the Humboldt de-stoning process in Australia. In *Proceedings of the Eighth Australian Coal Preparation Conference.* Newcastle, New South Wales: Australian Coal Preparation Society.

Sanders, G.J., Ziaja, D., and Kottman, J. 2002. Cost efficient beneficiation of coal by Romjigs and Batac jigs. In *Proceedings of XIV International Coal Preparation Congress*. Johannesburg: Southern African Institute of Mining and Metallurgy. p. 395.

Sepro Mineral Systems. 2018a. *Falcon Continuous (C) Concentrators.* Rev. 1. https://seprosystems.com/wp-content/uploads/2016/08/Falcon_C_Concentrator_2018.pdf.

Sepro Mineral Systems. 2018b. *Falcon Semi-Batch (SB) Concentrators.* Rev. 1. https://seprosystems.com/wp-content/uploads/2016/08/Falcon_SB_Concentrator_2018.pdf.

Sepro Mineral Systems. 2018c. *Falcon Ultra-Fine (UF) Concentrators.* Rev. 1. https://seprosystems.com/wp-content/uploads/2016/08/Falcon_UF_Concentrator_2018.pdf.

Seymour, C.E. 1893. Means for and process of separating metals from ores. U.S. Patent 489,101.

Short, M.A., Snoby, R.J., and Jungmann, A. 2001. Beneficiation of 3 mm × 0.15 mm fine coal with two-stage hindered settling equipment. SME Preprint No. 01-51. Littleton, CO: SME.

Snoby, R., Thompson, K., Mishra, S., and Snoby, B. 2009. Dry jigging coal: Case history performance. SME Preprint No. 09-052. Littleton, CO: SME.

Symonds, D.F. 1986. Selection and sizing of heavy media equipment. In *Design and Installation of Concentration and Dewatering Circuits.* Littleton, CO: SME.

Taggart, A.F. 1945. *Handbook of Mineral Dressing.* New York: John Wiley and Sons.

Taggart, A.F. 1951. *Elements of Ore Dressing.* New York: John Wiley and Sons.

Takakuwa, T., and Matsumura, M. 1954. A contribution toward the improvement of the air pulsated jig. In *Proceedings of the Second International Coal Preparation Congress.* West Germany.

Tanaka, M., Jinnouchi, Y., Sawata, Y., and Kawashima, S. 1990. Effect of the wave pattern of pulsation of an air-pulsated jig. In *Proceedings of 11th International Coal Preparation Congress.* Tokyo: Mining and Materials Processing Institute of Japan. p. 139.

Tucker, P. Chan, S.K., Mozley, R.H., and Childs, G.J. 1992. Modeling of the multi-gravity separator. *Les Techniques* 45:77–89.

Vance, W.H., and Moulton, R.W. 1965. A study of slip ratios for the flow of steam-water mixtures at high void fraction. *AIChE J.* 11(6):1114.

Wills, B.A. 2006. Gravity concentration. In *Wills' Mineral Processing Technology*, 7th ed. Oxford: Elsevier.

Wilson, K.C., Addie, G.R., Sellgren, A., and Clift, R. 2006. *Slurry Transport Using Centrifugal Pumps.* New York: Springer.

Ziaja, D., and Yannoulis, G. 2007. Is there anything new in coarse or intermediate coal cleaning? In *Designing the Coal Preparation Plant of the Future.* Edited by B. Arnold, M. Klima, and P. Bethell. Littleton, CO: SME.

CHAPTER 6.3

Dense Medium Separation

Robert C. Dunne, Rick Q. Honaker, and Graham Popplewell

Dense (or heavy) medium separation (DMS) is used for upgrading minerals to meet market specifications (e.g., coal and iron ore) or as a preconcentration step to remove gangue minerals before further downstream treatment (e.g., diamonds and metalliferous ores such as lead, zinc, nickel, and copper). The transition of the float–sink principle from a laboratory heavy-liquids method to continuous industrial practice offers the prospect of very precise mineral separations based on density differences alone.

The process offers some advantages over other gravity processes. It has the ability to make sharp separations at any required density (D), with a high degree of efficiency even in the presence of large amounts of "near-density" particles. The density of separation can be closely controlled, within a relative density (RD) of 0.005, and can be maintained under normal conditions for indefinite periods. The separating density can, however, be changed fairly quickly to meet varying requirements. The process is relatively expensive compared to other gravity separation processes, mainly because of the ancillary equipment required to clean the medium and the cost of the medium itself.

DMS is applicable to any ore where a suitable degree of mineral liberation by comminution exists. The process is most widely applied when the density difference occurs at a coarse particle size, as separation efficiency decreases with particle size. Particles larger than 0.5 mm can be treated by DMS. However, treatment rates are much lower at the finer particle sizes. Separation can be effectively carried out where RDs of 0.1 exist. Again, particle size affects separation efficiency, finer particles having a lower efficiency of separation.

Various solids can be used to make up a heavy liquid. The media used most widely are suspensions of finely ground magnetite or ferrosilicon in water. These solids are particularly suitable because they are relatively dense (with RDs of 5.1 and 6.8, respectively), are physically stable and inert, and have magnetic properties that enable simple and efficient recovery of the solids by use of a magnetic separator.

DMSs are employed most extensively in the beneficiation of coal and in the extraction of diamonds. The most important use of DMS is in coal preparation, where a relatively simple separation removes the low-ash coal from the heavier high-ash discard and associated shales and sandstones. In addition, DMS is used in the upgrading of iron ore and preconcentration of lead-zinc ores (Munro et al. 1982).

HEAVY MEDIUM SUSPENSIONS

Below a concentration of about 15% by volume, finely ground particle suspensions in water behave essentially as simple Newtonian fluids. Above this concentration, however, the suspension becomes non-Newtonian and a certain minimum stress, or yield stress, has to be applied before shear will occur and the movement of a particle can commence. Therefore, small particles, or those close to the medium density, are unable to overcome the rigidity offered by the medium before movement can be achieved. This can be overcome to some extent either by increasing the shearing forces on the particles or by decreasing the apparent viscosity of the suspension. The shearing force may be increased by substituting centrifugal force for gravity. The viscous effect may be decreased by agitating the medium, which causes elements of liquid to be sheared relative to each other. In practice, the medium is never static, as motion is imparted to it by paddles, by air, and also by the sinking material itself.

To produce a stable suspension of sufficiently high density, it is essential to use fine particles of high specific gravity with some means of agitation to maintain a homogeneous medium. The fine solid particles of the medium must be hard, with no tendency to slime, as degradation increases the apparent viscosity. The medium must be easily removed from the mineral surfaces by washing, be amenable to recovery from the washings by magnetic separation, and be corrosion resistant. The heavy medium for coal is magnetite, while ferrosilicon is used for all other DMSs (e.g., diamonds, iron ore, and base metals).

Robert C. Dunne, Principal, Robert Dunne Consulting, Gooseberry Hill, Western Australia, Australia
Rick Q. Honaker, Professor, Mining Engineering, University of Kentucky, Lexington, Kentucky, USA
Graham Popplewell, Technical Director & Senior Fellow, Diamond Processing, Fluor Corporation, Perth, Western Australia, Australia

DENSE MEDIUM SEPARATION VESSELS

DMS vessels can be categorized into two broad categories. Static, open-bath vessels (e.g., drums) are used for separations at coarser particle sizes, while dynamic separators are generally employed for finer particle size ranges. Open-bath vessels make use of the natural settling velocity of particles in a dense medium slurry (at standard gravity), whereas dynamic vessels, namely centrifugal devices, make use of centrifugal forces to enhance the settling forces acting on the particles, thereby effectively increasing the settling velocity of the particles (and in turn increasing the capacity of the process). Cyclones are the most common centrifugal DMS devices in most industries. However, multistage separators are gaining traction outside of North America. The trend is moving toward gravity-fed separators where only the medium is pumped into the separation vessel via one of the inlets. Feeding by gravity ensures a consistent head pressure, while feeding by variable-speed pumping allows for variation and control of the pressure to the separating vessel. This reduces power and maintenance costs.

Gravitational units comprise some form of vessel into which the feed and medium are introduced and the floats (light particles) are removed by lifters, or merely by overflow. Removal of the sinks (heavy particles) is the most difficult part of separator design. The aim is to discharge the sinks particles without removing enough of the medium to cause disturbing downward currents in the vessel.

IMPLEMENTATION OF DENSE MEDIUM SEPARATION

The operation of dense medium beneficiation circuits is fairly straightforward. A generic DMS circuit is shown in Figure 1. The medium, controlled to the density dictated by the separation desired, is mixed with feed ore prior to entering the separating vessel. The float and sink products, intimately mixed with medium, leave the bath and each proceeds to a separate medium recovery circuit. The system of recovery is identical in each case. The product is fed to a drain-and-rinse screen. Most of the medium is drained from the product at the head of the screen. This medium is undiluted, or is "at density," and can be recycled directly to the "correct medium" tank, ready for immediate reuse. After draining, the product is washed in two or more stages to remove any adhering medium; otherwise, such medium would be lost from the circuit. This washing is crucial, and inefficiencies at this stage lead to increases in the overall losses of medium and, hence, operating costs. The washings from the "dilute" medium are treated in a densifying circuit. This involves the extraction of the ferrosilicon from the medium entering the cyclone. This dilute medium is treated by low-intensity magnetic separators. The magnetic product is stored in the "over-dense" tank. This over-dense medium is then used for the making of more medium, on an intermittent basis, by the addition of appropriate quantities of water.

DENSE MEDIUM CIRCUITS

Although the separating vessel is the most important element of a DMS process, it is only one part of a relatively complex circuit. Other equipment is required to prepare the feed and to recover, clean, and recirculate the medium (Symonds and Malbon 2002). The feed to a dense medium circuit must be screened to remove fine ore. Slimes should be removed by washing to minimize any tendency of the slime content to create significant increases in medium viscosity.

Dense Medium Recovery

The medium recovery portion of a DMS plant consists of two circuits: the circulating medium and the dilute medium. The sinks and floats streams from the separation vessels each report to horizontal vibrating screens fitted with screen panel apertures finer than those of the feed preparation screen. The first portion of the screen is used to drain the dense medium.

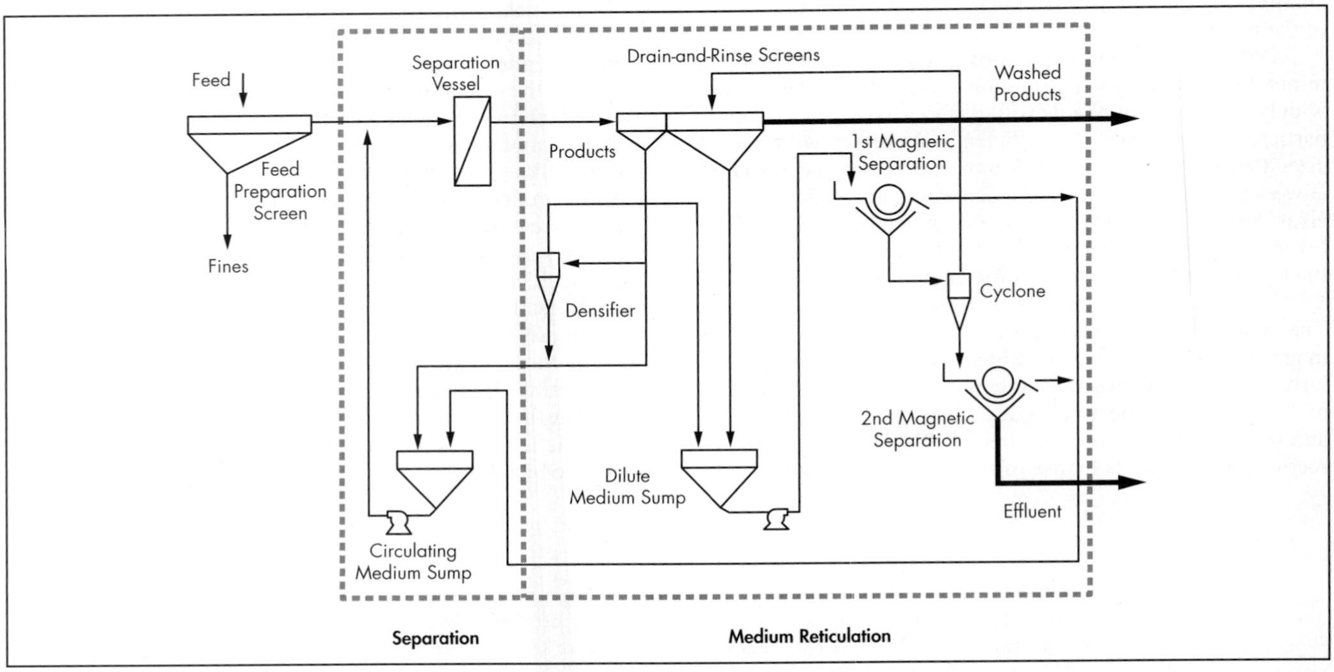

Adapted from Napier-Munn et al. 2013

Figure 1 Dense medium process flow sheet

The screen undersize then reports to the circulating medium tank. The second portion of the screen is fitted with water sprays to rinse any remaining ferromagnetic particles from the solid particles in the two streams. The cleaned solids are discharged from the screens for further treatment or disposal.

Densification of Washed Medium

The undersize products from the washing screens, consisting of medium, wash water, and fine solid particles, are too dilute and contaminated to be returned directly as medium to the separating vessel. They are treated individually as shown in Figure 1 or together, by magnetic separation, to recover the magnetic ferrosilicon or magnetite from the nonmagnetic fines. Reclaimed, cleaned medium is thickened to the required density by small cyclones or a pipe densifier, and continuously returned to the DMS vessel. In some circuits, the densified medium passes through a demagnetizing coil to ensure that a non-flocculated, uniform suspension is returned to the separating vessel.

Magnetic Separation

Dilute medium is pumped to a magnetic separation stage, which is the primary step in media recovery. Wet, low-intensity, permanent magnetic drum separators are almost exclusively used for this role, and typically in two stages to maximize medium recovery and mitigate the impact of magnetic separator feed fluctuations. The primary drum collects the majority of the medium, while the secondary drum is used as a scavenger. Criteria to consider in the design and operation of a magnetic separation circuit include the volumetric flow rate of the dilute medium, the pulp density of the dilute medium, and the percentage of magnetics in the dilute medium. These factors influence the performance of the magnetic separators, and pulp levels and magnet positions in the separators must be controlled to enable an acceptable balance between maximum medium recovery and minimum dilution of the magnetic concentrate.

Demagnetizing Coil

A demagnetizing coil is often used in the densification circuit to demagnetize ferrosilicon following recovery through a magnetic separator. Ferrosilicon has a tendency to form magnetic flocs when subjected to a magnetic field, and this can lead to medium instability in the DMS operating circuit. The demagnetizing coil essentially prevents flocculation of the magnetic particles and allows thorough mixing of a stable ferrosilicon medium.

Control

Most large DMS plants include automatic control of the feed medium density. The density of the densifying medium is usually measured with a gamma attenuation gauge. The signal is used to adjust the amount of water added to the medium to return it to the correct density. Control of the medium density is a major factor in the operation of the DMS process. The medium density must be kept close to the target or set-point density to avoid displacement of "near-density material," which will affect the yields and product grades. The density should be controlled to a precision of two decimal points, or ±0.005.

Medium Rheology

Medium rheology is critical to efficient operation of dense medium systems (Napier-Munn 1990; Bosman 2014), although the effects of viscosity are difficult to quantify (Reeves 1990; Dunglison et al. 1999). Management of viscosity includes selecting the correct medium specifications, minimizing operating density, and minimizing the content of clays and other fine contamination (Napier-Munn and Scott 1990). If the quantity of fines in the circuit reaches a high proportion, it is usual to increase the volume of medium diverted to the cleaning circuit. Many circuits have such a provision, allowing medium from the draining screen to be diverted into the washing screen undersize sump.

Water Balance

Control of the medium density requires removal of water added to and retained in the circulating or correct medium circuit. The main source of water entering the system is from the ore feed, which has been wet screened and has retained water because of the medium solids adhering to the float and sink products. Other sources of water entering the circulating medium circuit include rinse screen water sprays and medium pump gland water. Thus, to be able to control the medium density, the system must remove water from the circulating medium circuit. The most effective mode of control is to remove more water than enters the circuit, and control is established by adding water. This is referred to as positive water addition. The medium density is controlled by the operation of the densification and medium recovery circuits. In lower operating density separations (e.g., coal operating at 1.4–1.8 RD), the dilute medium circuit magnetic separator is sufficient to remove water from the circulating medium and to enable density control. For higher-density mineral separations operating at RDs of 2.5 and higher, centrifugal densifiers are usually required to maintain good density control. Furthermore, densifiers will reject lower-density contaminant minerals from the circulating medium (e.g., silicates and clays). However, in higher-density mineral separations, such as sulfides, it is recommended that a circulating medium bleed to the dilute medium circuit be incorporated to remove fine high-density nonmagnetic particles that can build up in the circuit and affect the medium density and viscosity.

Medium Losses

Despite the recovery systems of ferrosilicon through the float and sink screens and magnetic separator circuits, these recovery systems are not 100% efficient. Losses due to medium attached to the float and sink particles and effluent losses via the magnetic separators are unavoidable. Generally, ferrosilicon consumption is typically in the range of 0.15 to 0.25 kg/t of DMS feed. Losses are made up by mixing fresh ferrosilicon into the correct medium circuit. Magnetite losses of around 0.5 to 1.0 kg/t of coal treated are common.

The actual quantity of medium lost will depend on the size and porosity of the feed ore, the characteristics of the medium solids, and plant design (Napier-Munn et al. 1995). Losses increase for fine or porous ores, fine media, and high operating densities.

Effluent water always contains some entrained medium—the more that can be recycled back to the plant, the better

(Dardis 1987). Careful attention should also be paid to the quality of the medium used. Williams and Kelsall (1992) demonstrated that certain ferrosilicon powders are more prone to mechanical degradation and corrosion than others.

Operating Costs

The major costs in DMS are power (for pumping) and medium consumption. Medium losses can account for 10%–35% of total costs.

APPLICATION OF HEAVY LIQUIDS

Heavy liquids are dense fluids or solutions used to separate minerals of different density through their buoyancy. Minerals with a density greater than the heavy liquid will sink, while materials with a density less than the heavy liquid will float on the liquid surface. The aim of heavy liquids in the mineral industry is to separate the minerals in an ore sample into a series of fractions according to density, establishing the relationship between the high- and low-density minerals. Australian Standard AS 4350.2 (1999) and American standard D4371 (ASTM 1999) are used for heavy liquid separation.

Heavy liquids are used widely in metallurgical laboratories for the appraisal of gravity separation techniques on ores. Heavy liquid testing may be performed to determine the feasibility of DMS on a particular ore, and to determine the economic separating density, or it may be used to assess the efficiency of an existing dense medium circuit by carrying out tests on the sink and float products.

Given the cost and toxicity of heavy liquids, the range of available materials has been reduced considerably compared to previous decades. The liquids are mainly high-density organic compounds or aqueous salt solutions. Commonly used heavy liquids include the following:

- Tetrabromoethane (TBE), having a density of 2.96, is commonly used. It may be diluted with white spirit or carbon tetrachloride (1.58 SG [specific gravity]) to give a range of densities lower than 2.96.
- Bromoform (density 2.89) may be mixed with carbon tetrachloride to give densities ranging from 1.58 to 2.89. It is an expensive chemical but evaporates much quicker than other reagents, so sample processing is faster, in particular when dealing with large sample volumes.
- Perchloroethene (density 1.6).
- White spirits (density 0.78) is used as a diluent for bromoform and TBE.
- Diiodomethane or methylene iodide (density 3.325).
- Clerici solution (thallium formate–thallium malonate solution) allows separation at densities up to 4.2 at 20°C or 5.0 at 90°C. Clerici liquids are extremely poisonous and must be handled with extreme care.
- Sodium polytungstate, or SPT (density 3.100). Aqueous solutions of sodium polytungstate have certain advantages over organic liquids, such as being virtually nonvolatile, nontoxic, and of lower viscosity, and densities of up to 3.1 can be easily achieved (Anon. 1984).
- Lithium polytungstate (density 2.92) is an inorganic heavy liquid.

Two of the most commonly used organic heavy liquids are bromoform and TBE. Of these, bromoform has the lower viscosity (1.8 cP) but is considered more hazardous to work with because it has a higher vapor pressure (5.9 mm Hg at 25°C). TBE has a higher viscosity (9 cP) and a lower vapor pressure (0.02 mm Hg at 25°C). Another organic heavy liquid, used when higher densities are required, is diiodomethane, which has a density of 3.31 g/mL, a vapor pressure of 1.2 mm Hg at 25°C, and a low viscosity of 2.6 cP.

The highly soluble inorganic salt sodium polytungstate, when dissolved in water, can be used to replace the toxic organic liquids traditionally used for gravity separation work. Sodium polytungstate is mainly used for mineral separation and for density gradient centrifugation. The inorganic salt is available in powder form and results in an aqueous, colorless, transparent neutral solution. It has no obnoxious smell and no corrosive properties, is not flammable, and is ecologically safe and simple to handle. Losses are minimal (<5%) during treatment, washing, and reconcentration.

For the separation of fine particles, sufficient time is required for the particles to settle properly in the heavy liquid, especially because the viscosity of the liquid increases as the density increases. Fine materials are often centrifuged to reduce the settling time, but this should be done with care, as there is a tendency for the floats to become entrained in the sinks fraction. Unsatisfactory results are often obtained with porous materials, such as magnesite ores, because of the entrainment of liquid in the pores, which changes the apparent density of the particles.

Heavy suspensions can achieve float–sink separations at densities higher than 3.0. Cargille liquids, which are heavy metal particles dispersed in organic liquids, can provide densities ranging up to 7.5 (Browning 1961). However, the use of these liquids is limited to the separation of mineral particles coarser than 0.6 mm, given the physical properties of the suspension. Rhodes et al. (1993) showed that high-density heavy suspensions can be made from finely divided ferrosilicon added to solutions of SPT. Eroglu and Stallknecht (2000) took the same approach but added lithium heteropolytungstate. Recently, Koroznikova et al. (2007) demonstrated that individual mineral sand minerals in the particle size range –250 +150 μm can be separated by a mixture of ultrafine tungsten carbide (3.3 μm) and a solution of lithium heteropolytungstates.

Magnetohydrostatics separation utilizing the supplementary "weighting force" produced in a solution of a paramagnetic salt or ferrofluid placed in a magnetic field gradient is another technique to separate minerals on the basis of density (Parsonage 1980; Walker 1985). The separation is applicable primarily to nonmagnetic minerals with a lower limiting particle size of about 50 μm (Parsonage 1980; Domenico et al. 1994).

DETERMINING THE SUITABILITY OF DENSE MEDIUM SEPARATION

Laboratory testing may be performed on ores to assess the suitability of DMS and to determine the appropriate separating density. Liquids covering a range of densities in incremental steps are prepared, and the representative sample of crushed ore is introduced into the liquid of highest density. The floats product is removed and washed, then placed in the liquid of the next lower density. The float product from this liquid is then transferred to the next lower-density liquid, and so on. The sinks product is finally drained, washed, and dried, and then weighed, together with the final floats product, to give the density distribution of the sample by mass.

The next step is to submit the individual float and sink fractions for chemical analysis. The results from this assessment, in conjunction with the masses obtained from the heavy liquid test, provide a mineral/metal distribution by mass. Similarly, a distribution with density in place of mass can be generated for the sample. For coal, the technique is referred to as a washability curve (ASTM International 1999). Inspection of the distribution will provide the desired density of separation for the appropriate recovery of mineral into either the float or sink product.

Near-Gravity Material

The amount of near-gravity material present is sometimes regarded as being the mass of material in the range of ±0.1 or ±0.05 kg of the separating density, and separations involving feeds with less than ~7% of ±0.1 near-gravity material are regarded by coal preparation engineers as being fairly easy to control. Such separations are often performed in Baum jigs, as these are cheaper than dense medium plants, which entail expensive media-cleaning facilities, and also do not require feed preparation (i.e., removal of the fine particles by screening). However, the density of separation in jigs is not as easy to control to close limits as it is in dense medium baths; and for near-gravity material much above 7%, DMS is preferred.

Efficiency of Dense Medium Separation

While laboratory heavy liquid separation testing is done under near-ideal conditions, the dynamic conditions of a continuous DMS process introduce natural inefficiencies, with higher-density particles misplaced into the floats stream and lower-density particles misplaced into the sinks stream. The degree of inefficiency increases relative to the proportion of particles in the feed whose density is near the density of the separation (i.e., near-density material). The efficiency of the separation process can be determined through the generation of a partition curve (coal), also known as a Tromp curve (metalliferous ores and diamonds).

Other factors also play a role in determining the efficiency of separation. Fine particles generally separate less efficiently than coarse ones, again because of their slower settling rates. The properties of the medium, the design and condition of the separating vessel, and the feed conditions, particularly feed rate, will all influence the separation.

Partition Curve or Tromp Curve

The partition curve relates the partition coefficient or partition number—the percentage of the feed material of a particular specific gravity that reports to either the sinks product (generally used for minerals) or the floats product (generally used for coal)—to specific gravity. This is analogous to the classification efficiency curve, in which the partition coefficient is plotted against size rather than specific gravity.

The partition curve can be constructed after the separation of the sample has been completed by collecting representative samples from the floats and sinks streams during the separation and performing heavy liquid tests on each sample to assess the degree of misplaced material in each of the floats and sinks streams. The partition curve of the reconstituted feed can then be constructed.

Determining Efficiency Using a Tromp Curve

The efficiency of separation can be represented by the slope of a partition or Tromp curve, first introduced by K.F. Tromp (1937). It describes the separating efficiency for the separator, whatever the quality of the feed. The ideal partition curve reflects a perfect separation in which all particles having a density higher than the separating density report to sinks and those lighter report to floats (Figure 2). There is no misplaced material. The partition curve for a real separation shows that efficiency is highest for particles of density far from the operating density and decreases for particles approaching the operating density. The area between the two curves is called the "error area" and is a measure of the degree of misplacement of particles to the wrong product. Many partition curves give a reasonable straight-line relationship between distributions of 25% and 75%, and the slope of the line between these distributions is used to show the efficiency of the process (Terra 1938).

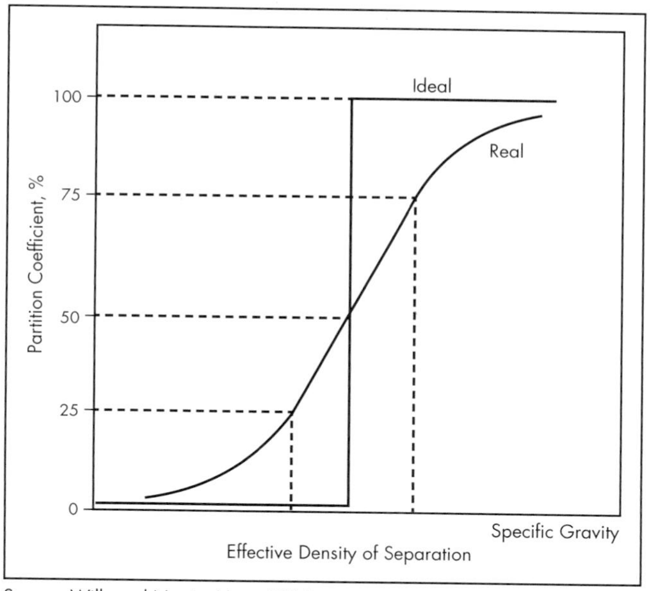

Source: Wills and Napier-Munn 2005

Figure 2 Partition or Tromp curve

The probable error of separation, or the Ecart probable (Ep), is defined as half the difference between the density where 75% is recovered to sinks and that at which 25% is recovered to sinks, that is, from Figure 2:

$$Ep = \frac{\rho_{75} - \rho_{25}}{2}$$

The density at which 50% of the particles report to sinks is shown as the effective density of separation, which may not be exactly the same as the medium density, particularly for centrifugal separators. The separating density is generally higher than the medium density. The lower the Ep, the nearer to vertical is the line between 25% and 75% and the more efficient is the separation. An ideal separation has a vertical line with an Ep = 0, whereas in practice, the Ep usually lies in the range of 0.01 to 0.10. The Ep is not commonly used as a method of assessing the efficiency of separation in units such as tables, spirals, cones, and so forth, due to the many operating variables (wash water, table slope, speed, etc.) that can affect the separation efficiency. It is, however, ideally suited to the relatively simple and reproducible DMS process (Jowett 1986). However, care should be taken in its application, as it does not reflect performance at the tails of the curve, which can be important.

Construction of Partition Curves for an Operating Plant

The partition curve for an operating dense medium vessel can be determined by sampling the sink and float products and performing heavy liquid tests to determine the amount of material in each density fraction. The range of liquid densities applied must cover the working density of the dense medium unit.

An alternative, rapid method of determining the partition curve of a separator is to utilize density tracers (Chironis 1987). Specially developed color-coded plastic tracers can be fed to the process, the partitioned products being collected and hand sorted by density (color). It is then a simple matter to construct the partition curve directly by noting the proportion of each density of tracer reporting to either the sink or the float product. Application of tracer methods has shown that considerable uncertainties can exist in experimentally determined Tromp curves unless an adequate number of tracers is used. Napier-Munn (1985) provides graphs that facilitate the selection of tracer sample size and the calculation of confidence limits.

Partition curves can be used to predict the products that would be obtained if the feed or separation gravity were changed. The curves are specific to the vessel for which they were established and are not affected by the type of material fed to it, provided the following are true:

- The feed particle size is the same—efficiency generally decreases with a decrease in size. Centrifugal separators are better than baths.
- The separating gravity is in approximately the same range—the higher the effective separating density, the greater the probable error, due to the increased medium viscosity. In fact, the Ep is directly proportional to the separating density, all other factors being the same (Gottfried 1978).
- The feed rate is the same.

The partition curve for a vessel can be used to determine the amount of misplaced material that will report to the products for any particular feed material.

Modeling of Tromp/Partition Curve

This method of evaluating the performance of a separator on a particular feed is tedious and is ideal for a spreadsheet, provided that the partition numbers for each density fraction are known. These can be represented by a suitable mathematical function (Tarjan 1974; Stratford and Napier-Munn 1986; Wood et al. 1987; Napier-Munn 1991; Scott and Napier-Munn 1992; Dunglison and Napier-Munn 1997). A large body of literature exists on the selection and application of such functions. Some are arbitrary, and others have some theoretical or heuristic justification. The key feature of the partition curve is its S-shaped character. In this it bears a passing resemblance to several probability distribution functions, and indeed the curve can be thought of as a statistical description of the DMS process, describing the probability with which a particle of given density (and other characteristics) reports to the sink product.

However, in practice, many real partition curves do not behave ideally. In particular, they are not asymptotic to 0% and 100% but exhibit evidence of short-circuit flow to one or both products. Stratford and Napier-Munn (1986) identified four attributes required of a suitable function to represent the partition curve:

1. It should have natural asymptotes, preferably described by separate parameters.
2. It should be capable of exhibiting asymmetry about the separating density; that is, the differentiated form of the function should be capable of describing skewed distributions.
3. It should be mathematically continuous.
4. Its parameters should be capable of estimation by accessible methods.

EQUIPMENT

Historical Development

The early DMS circuits employed a "static" bath-type separator. These machines treated feed particle sizes as coarse as 300 mm and as fine as +6 mm or +12 mm, depending on the application. The development of cyclone technology introduced the "dynamic" separators used to treat particles down to 0.5 mm. Recent large-cyclone developments allow the treating of particles with a top size as coarse as 100 mm.

The following selected chronology of the historical development of DMS is adapted from Burt and Mills (1984), Wills (1988), Leonard (1991), Hillman (2003), Holtham (2006), and Napier-Munn et al. (2013).

1858	Bessemer's patent covered the use of chlorides of iron, manganese, barium, and calcium as medium, but there was no serious industrial application until the 20th century.
1911	Du Pont developed the use of chlorinated hydrocarbons to achieve higher densities. The Chance process patent covered the use of a sand/water medium in a cone separator. This would probably be called a teetered-bed separator today.
1921	First Chance industrial process for cleaning anthracite.
1922	First experiments with magnetite medium for coal cleaning (Conklin process).
1930s	DMS plants for minerals using galena, in the United Kingdom and the United States.
1931	First general use of DMS plants.
1937	First use of FeSi in an ore concentrating plant in the United States, with medium recovery using cross-belt magnetic separators.
1938	Tromp process first to use magnetite medium commercially in Germany.
1940	First U.S. ferrosilicon patent. American Cyanamid Company introduced the DMS process for coal cleaning, including magnetic recovery of magnetite medium.
1942	Dense medium cyclone patented by Dutch State Mines, Holland.
1946	First use of dense medium cones in diamond processing.
1950s	Wemco (United States) developed bath (drum) using magnetite medium to treat –200/+6 mm feed and produce three products. Drewboy bath developed using magnetite medium. Teska (United States) implemented the bath method using magnetite medium.
1955	First use of dense medium cyclones in diamond processing.
1960s	Dyna Whirlpool developed in the United States for coal and later used for minerals. Vorsyl dynamic separators developed in the United Kingdom.

Table 1 Economies of scale in the coal industry for a 1,000-t/h plant

Equipment	1977	1987	1997	2007	2017
Number of modules	6	4	2	1	1
Number of deslime screens	6	4	2	1	1
Number and size of dense medium cyclones	12 × 500 mm	8 × 660 mm	2 × 1,200 mm	1 × 1,500 mm	1 × 1,500 mm
Number of drain-and-rinse screens	24	16	4	2	2
Total number of items in circuit	164	110	90	82	82
Capital expenditure, million $	26	23	20	18	18

Source: Osborne 2010

1970s Tri-Flo, a multistage separator similar to two Dyna Whirlpools bolted together in series.

1980 Larcodems, similar to the Dyna Whirlpool and developed by British Coal, as an efficient replacement for the Baum jig.

~1992 First use of cyclone diameters >1,000 mm in coal preparation (current maximum 1,500 mm).

~2005 First use of cyclone diameters >610 mm in mineral separations (current maximum 800 mm).

All the early dense medium processes were of the static bath type, in which the separation takes place under gravitational forces. Most of the development over the years was in the mechanical arrangement by which feed was introduced to the bath and the products removed. This led to many proprietary devices manufactured by different companies, including the Wemco cone, Wemco drum, Drewboy, Teska, Barvoy, and Norwalt bath trough-type separators, and many others (Osborne 1988; Wills and Napier-Munn 2005).The design innovations were developed principally in the United States and Europe (mostly in Germany and the United Kingdom).

Current Trends

The practice of treating coarse coal in static DMS vessels (Wemco and Teska drums) has changed over the last three decades to treating coarse material in large-diameter dense medium cyclones (DMCs). Finer coal (millimeter sized) has been processed in DMS cyclones for a long time. Spirals, water-only cyclones, and teetered-bed separators are used to process fine coal (+0.5–1 mm) while ultrafine coal is recovered by flotation. The trend toward using very large DMS cyclones over the last four decades is shown in Table 1.

The introduction of large-diameter DMCs into the coal industry and the belief that fine coal particles are not effectively beneficiated below the "breakaway" particle size led to a change in circuit design. Increasing the aperture of the desliming screens to 3 mm increased screen capacity and improved cyclone efficiency on the coarser product. The <3-mm particles are treated in teetered-bed separators and spirals (de Korte and Bosman 2006).

Static Gravitational Separators

Static gravitational units comprise some form of vessel into which the feed and medium are introduced, and the floats are removed by paddles or merely by overflow, and sinks by mechanical conveyors or lifters fixed to the inside of a rotating drum. The lifters empty into a sink launder when they pass the horizontal position. Removal of the sinks is the most difficult part of the design of these types of separators. The aim is to discharge the sinks particles without removing large quantities of medium that will cause disturbing downward currents in the vessel. Drum separators have very large sinks capacities. They are commonly used in the coal industry because of their simplicity, reliability, and relatively small maintenance needs. Drum, Wemco or Teska, and rectangular separators are still used in the minerals industry (Galvin and Iverson 2013) and available from engineering companies.

Wemco Dense Medium Drum

Wemco drum separators, patented in 1954 (Maust 1954), are built in several sizes, up to 4.3 m diameter by 6 m long, having a maximum capacity of 450 t/h, and able to treat feeds up to 300 mm in diameter (Wilkes 2006). Separation is accomplished by the continuous removal of the sink product through the action of lifters fixed to the inside of the rotating drum (Figure 3). The lifters collect solids into the sink launder when they pass the horizontal position. The float product overflows a weir and flows into the second compartment where the middlings and the true sinks are separated.

The Wemco drum consists of a steel shell equipped with two specially designed tire-and-collar assemblies (England et al. 2002). The shell is supported from and runs on four support roller assemblies. Removable and replaceable perforated-sinks lifter plates are attached to the inside of the shell. The drum shell is maintained in its operating position longitudinally on the support rollers by two thrust rollers that engage on both sides of one of the main tires as required. The support and thrust roller assemblies are located on a sub-frame that also carries the drive motor, V-belt drive, gearbox (and protective coupling where fitted), drive and idler sprocket assemblies, and drive chain. The drive chain engages with the drive sprockets bolted around the outside of the drum shell to provide a positive nonslipping drive to rotate the drum at the correct operating speed (Wilkes 2006).

The main internal components of the drum are the feed chute and feed box, sinks hopper, curtains, and media addition pipework. Media of the appropriate type and density are circulated through the Wemco drum. The media are normally introduced into the feed chute, behind one of the curtains, and into the sinks launder.

A twin-compartment (Figure 4), three-product Wemco drum is also available. In this design, two different media densities are used. The first compartment uses low-density medium and the second compartment uses high-density medium. In this way, three products can be made, namely, low-density floats, high-density floats (or middlings), and a final sinks.

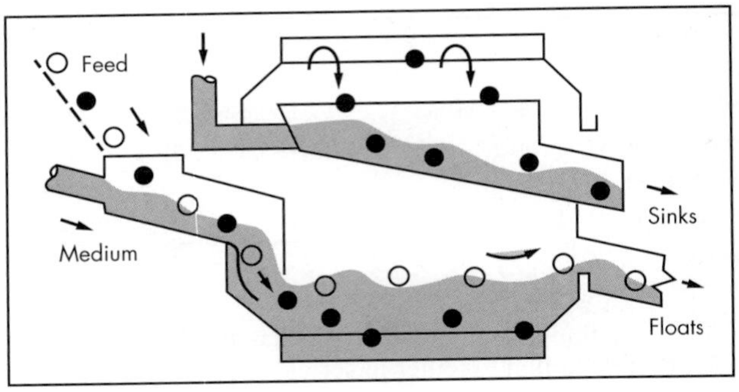

Courtesy of FLSmidth

Courtesy of Wemco

Figure 3 Wemco dense medium drum

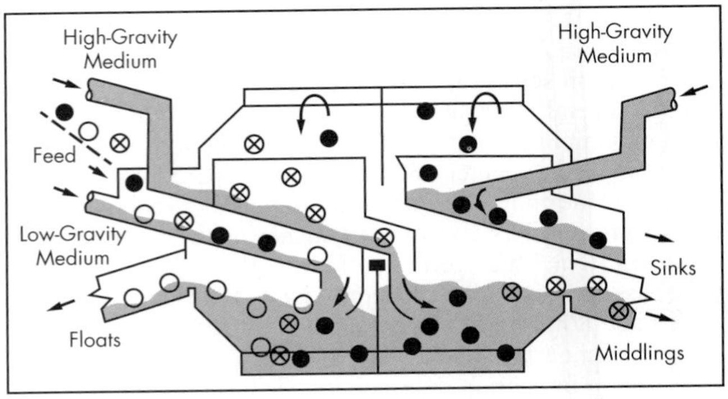

Courtesy of Wemco

Figure 4 Twin-compartment Wemco drum

This separator requires various treatments to the media addition points. Primary or low-density media are added to the feed chute and behind the primary curtain. Secondary or high-density media are added to the primary sinks launder (to introduce high-density media into the second compartment of the drum), behind the secondary curtain, and to the final secondary compartment sinks launder. The drained media from the screens are recirculated to the drum by a pump. For certain specific applications, it is possible to vary the twin-compartment drum design to treat two separate size fractions (one in each compartment) and to make a common sinks product.

The maximum size of coal that can be handled in the Wemco drum depends on the size of the inlet, outlet, and lifters. Material with a top size of 300–500 mm is not unusual if liberation of mineral matter from the coat is not a requirement. The Wemco drum can handle as much as 400 t/h feed, ~20–25 t/h/m^2 pool area, with very good separation efficiencies (Osborne 1988). High separation efficiency can be maintained through manipulation of the residence time in the vessel, either by supplying a large enough pool surface or by reducing the throughput of the raw coal. Ep values of 0.03 or better can be achieved (Osborne 1988).

Wemco drums are available in sizes from laboratory scale, 460 mm in diameter × 690 mm long, to industrial units up to 5.4 m in diameter and 8.1 m long and having a two- or three-product separation format. Maximum capacity on a Wemco drum for coal is around 900 t/h. Modeling information of drum dense medium separators is provided by Baguley and Napier-Munn (1996) and Boseman (2014).

Metals recycling. In metals recycling, the top size is usually set by the metal shredder and is around 100–125 mm. The bottom size is nominally 10 mm, with the small material going to jigs because DMCs are susceptible to blockage when treating metal/plastic wires present in most scrap metal feeds. Metals recycling plant design has its own challenges. For example, consideration needs to be given to removing fibrous materials, if present, from the media to facilitate suitable downstream separations. The entire feed material should be thoroughly prewashed, followed by a separation stage with water (normally upward current) to remove light materials such as plastics, light rubber, and so on. Metal separation is often achieved in a DMS Wemco drum. Excessive wear and high medium losses are usually the major problems. Medium tends to be lost on the underside of flat metal pieces and in crevasses of crushed and mangled metal fragments.

Metals recycling plants are smaller-tonnage operations, and because of their complexity and the highly variable size range of materials, it is common to use fairly large Wemco drum separators. However, the drums are modified internally to obtain the best results with the types of material encountered. Typical feeds can be the nonferrous components of shredded cars and airplane engines, and so forth. The simplified flow sheet in Figure 5 shows the basic process.

Teska Drum Separator

The Teska drum separator shown in Figures 6 and 7 has a slowly rotating bucket wheel that is joined to the separating compartment by mechanical seals (England et al. 2002). The bucket wheel for sinks is partitioned into compartments by means of perforated plates that enable removal of the sinks. The floats flow toward a discharge paddle wheel. The sealing system mounted between the rotary bucket wheel and stationary inlet and outlet section of the separating compartment prevents the escape of slurry to the tires and drive mechanism. The bath width is similar to the inside diameter of the bucket wheel. The chute for sinks can be positioned at the right or left (i.e., either concurrent with or opposite to the direction of feeding). Adjustable nozzles arranged over the bucket-wheel

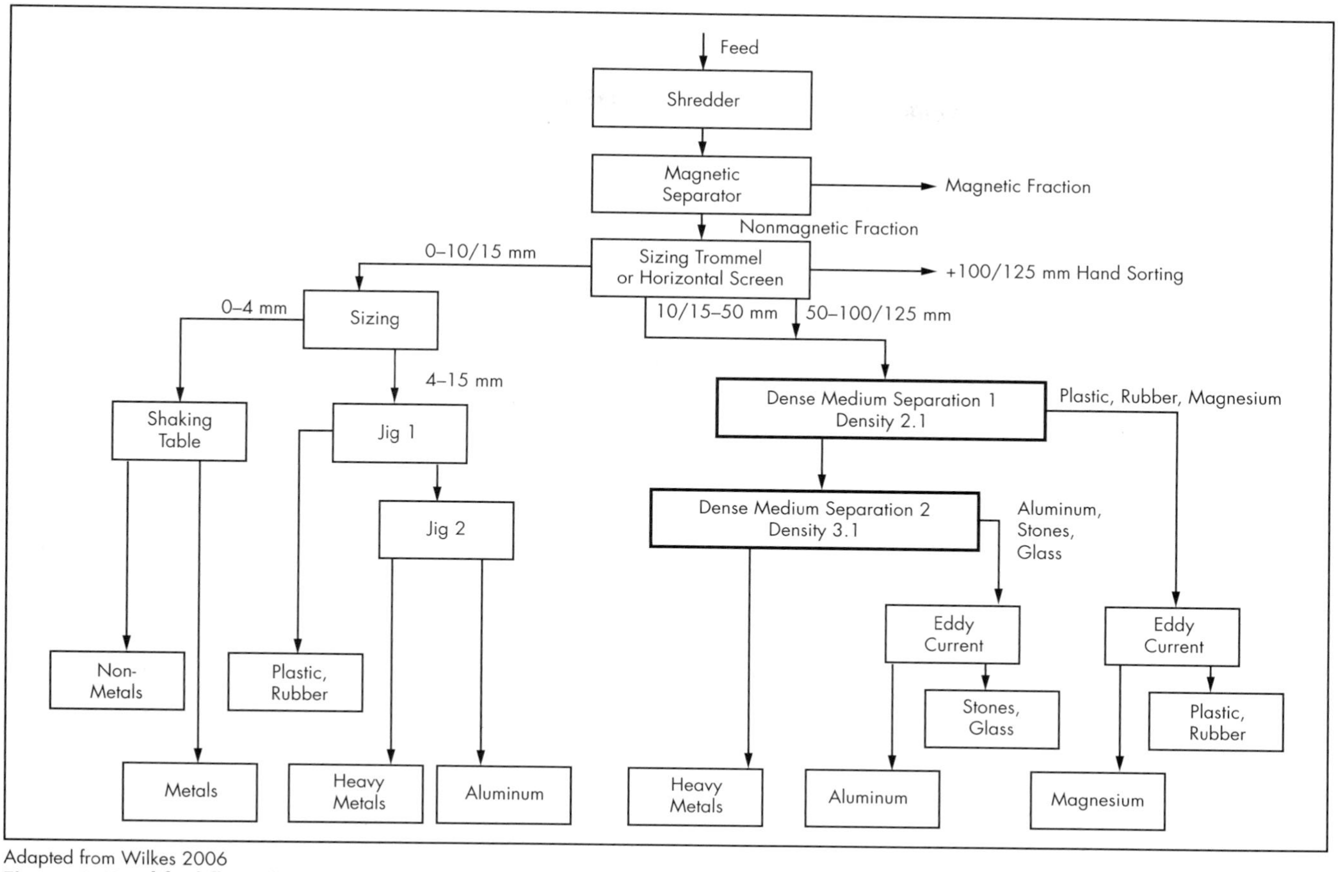

Adapted from Wilkes 2006

Figure 5 Simplified flow sheet for car scrap treatment

circumference induce a downward flow of dense medium and thus prevent undesirable density concentrations in the bath. At the same time, this feature avoids the formation of layers within the suspension of the bath. The discharge of dense medium is increased by the admission of additional working medium from above. A continuous flow is maintained via nozzles mounted over the complete bath width. One of the features of the Teska drum separator is optimal bath utilization relative to the separator's volume. To protect against wear, ceramic tiles are fitted to high-wear areas such as the inlet, the separating compartment, and the bucket wheel for sinks.

The Teska units can be supplied with bucket wheels up to 6.5 m in diameter, bucket widths of 1.5 m, and bath widths of 3.0 m. Feed particle sizes up to 1.2-m edge length can be processed. Treatment rates of 800 t/h on coal of +6 mm have been realized. Separation efficiency is controlled by residence time, or pool surface, and throughput.

Rectangular Heavy Medium Separators

The rectangular heavy medium separator shown in Figure 8 is manufactured by several engineering companies, such as The Daniels Company, Peters Equipment, and FLSmidth, among others (Symonds 1986; see also Figure 6 in Chapter 12.6, "Coal Preparation"). The separator can be manufactured in 3- to 8-m lengths with capacities for coal up to 650 t/h. The feed enters the vessel from one side and is immersed into the vessel medium bath by means of a submergence baffle to prevent rafting of heavy sink material. Medium enters just below the feed and carries the coal into the separation zone. Heavy particles settle to the bottom of the vessel and are removed by a flight conveyor system. Any floatable particles trapped with falling heavy particles are released by upflowing medium from the purge hoppers. Light particles float to the surface and are carried across the bath to the overflow weir. The medium stream exiting the bath flows by gravity to screens and into the return media sump before being pumped back to the washer. Any medium adhering to coal particles is rinsed with water that flows to a media recovery system.

The medium in the separation zone is controlled to maintain the appropriate flow without excess velocities that will cause poor separation. Upcurrent from the purge system prevents stratification of the media into layers of different densities. The washer requires minimal monitoring and adjustment. The gentle bath agitation minimizes the creation of fines. The separator is capable of treating coal with particles ranging between 6 and 200 mm.

The length of the unit is based on the overflow weir width needed to effectively carry the coarse material into the product overflow stream. Width is based on the amount of reject needed to be removed by the drag flight conveyor (see estimated capacities in Table 2). A general rule is that the length should be based on 75 t/h of feed material for every meter of overflow weir. This rule is based on a separation gravity below 1.55 and for a 50% yield to the product stream.

Prior to feeding coal to the vessel, the correct dense medium is pumped into the vessel through the feed washer

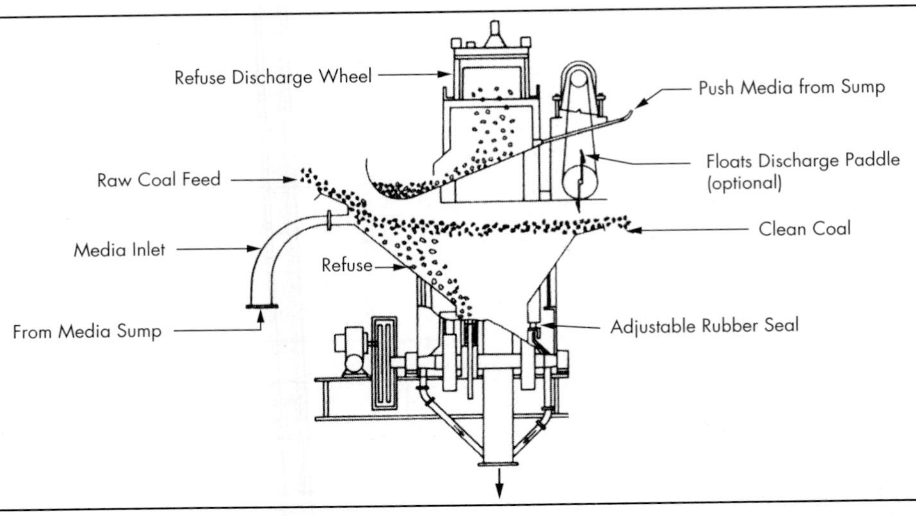

Source: England et al. 2002

Figure 6 Teska drum dense medium separator

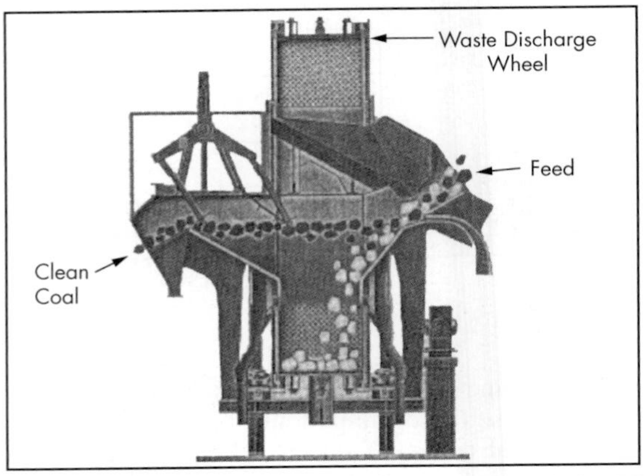

Source: Humboldt Wedag India 2008

Figure 7 Teska drum separator

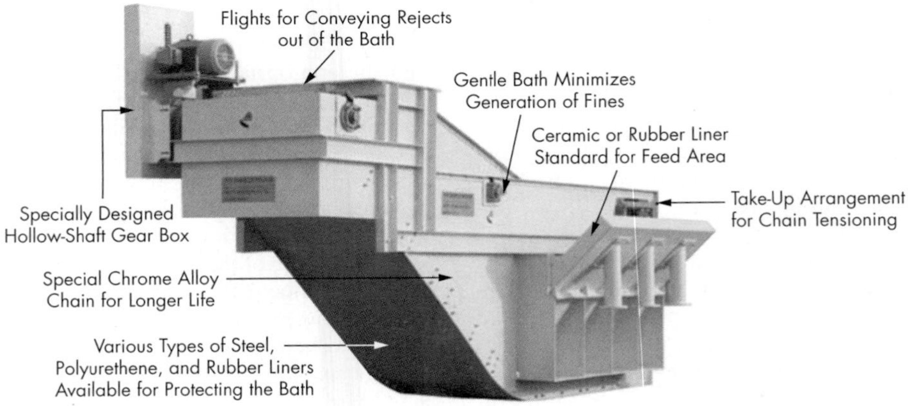

Courtesy of The Daniels Company

Figure 8 Rectangular heavy medium separator

Table 2 Flight conveyor sink capacities for rectangle dense medium separators

Height of Flights, mm	Approximate Sink Removal Rate, t/h		
	Width of Conveyor		
	1.21 m	1.37 m	1.52 m
203	260	300	330
229	300	330	336
254	330	336	363

Adapted from UAF 2018.
Note: Capacity values are based on a flight speed of 21 m/min. Conveyor capacity can be increased by either a change in speed or an increase in the number of flights, up to a critical value. The wear rates of the conveyor and vessel parts increase exponentially as the chain speed exceeds 23 m/min.

manifold and the purge hoppers. The vessel is filled until it is freely overflowing the weir. The proper inflow of medium is around 3.2 m^3/min for every meter of overflow weir length. Coal is fed to the vessel via the feed manifold. The depth of feed injected should be minimized to avoid any short-circuiting of light particles to the tailings stream. If refuse (gangue) particles adhere to coal particles, then the feeder plate needs further adjustment downward, where higher current flows are present. This plate is a high-wear item and should be checked periodically. Approximately 10% of the dense medium enters through the purge hoppers and provides a gentle up current that helps stabilize the medium. The purge hopper also serves as the drain when the unit is shut down.

The bypassing of low-density material in the separator may be attributed to

- Insufficient upward flow of medium,
- Plugged purge hoppers,
- Feed injection that is too deep,
- Unstable medium,
- Overloaded overflow weir, or
- False reading caused by laminated middlings.

Dense Medium Cyclone Systems

DMC separators are now widely used in the treatment of ores and coal. They provide a high centrifugal force and enable much finer separations than in gravitational (static) separators. Since the early 2000s, various design improvements have been made to DMCs to allow for ever-increasing throughput and sometimes better separation efficiency.

Pressure energy in the fluid (the mixture of medium and coal/ore particles) is converted into a centrifugal force in a rotational motion that causes the particles to separate from each other. The rotational motion is created by the tangential injection of the fluid into the vessel. At the entry point of the fluid, the vessel is cylindrical. Depending on the technology employed, the vessel can stay cylindrical in shape (Larcodems separator) or become conical (cyclones). The smaller the radius of the cylinder or cyclone, the higher the centrifugal forces on a particle. Hence, the centrifugal forces become larger than the gravitational forces acting on the particle. At relatively low tangential velocities in the cyclone, the gravitational forces are of no relevance to the separation. Indeed, a cyclone separator can work upside down without detrimental effects on separation efficiency.

Dyna Whirlpool and Tri-Flo Separators

The Dyna Whirlpool shown in Figure 9 is a cylindrical vessel that employs centrifugal forces to enhance the separation process (Hacioglu and Turner 1985). The dense medium is not introduced together with solids, but separately. Wills and Lewis (1980) provide information on the applications of the Dyna Whirlpool in the minerals industry.

The Tri-Flo separator can be regarded as two Dyna Whirlpool separators joined in series (Figure 10). Tri-Flo separators have been installed in numerous coal, metalliferous, and nonmetallic ore treatment plants (Burton et al. 1991; Kitsikopoulos et al. 1991; Ferrara et al. 1994). Involuted medium inlets and sink outlets are used. This produces less turbulence than tangential inlets.

The Tri-Flo separator can also be operated with two mediums of differing densities to provide sink products of individual controllable densities. Two-stage treatment using a single-medium density produces a float and two sinks products with only slightly different separation densities. With metalliferous ores, the second sink product can be regarded as a scavenging stage for the dense minerals, thus increasing the recovery. This second product may be subjected to a comminution step to increase liberation, and, after desliming, returned for retreatment in the separator. Where the separator is used for washing coal, the second stage cleans the float to produce a higher-grade product. Two stages of separation also increase the sharpness of separation. The Tri-Flo separator is capable of handling large-size coal particles of up to 100 mm.

Burton et al. (1991) and Ferrara et al. (1994) provide numerous examples of operations and pilot programs utilizing the Tri-Flo separator. Applications include ores containing fluorspar, chromite, lithium, lead-zinc-copper, tin, nickel, and iron. Sepro Minerals Systems markets the Tri-Flo separator under the brand name Condor dense medium separator (Sepro 2018). Available sizes are shown in Table 3. Performance

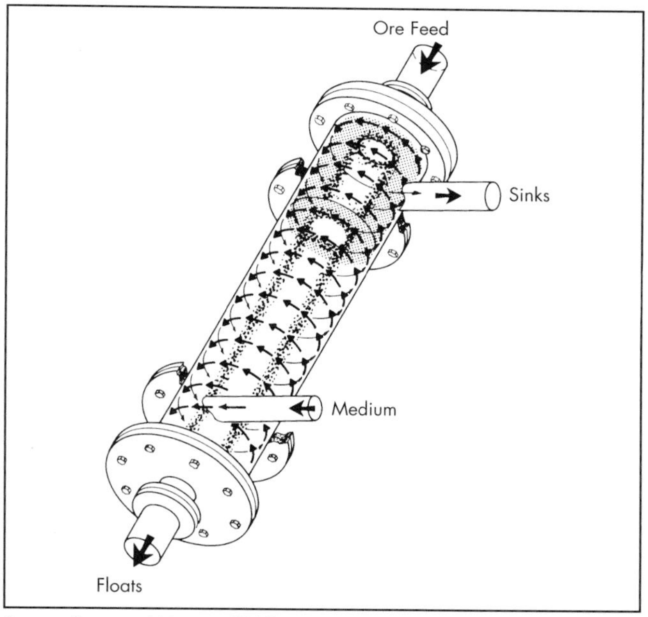

Source: Stortz and Houston 1963
Figure 9 Dyna Whirlpool

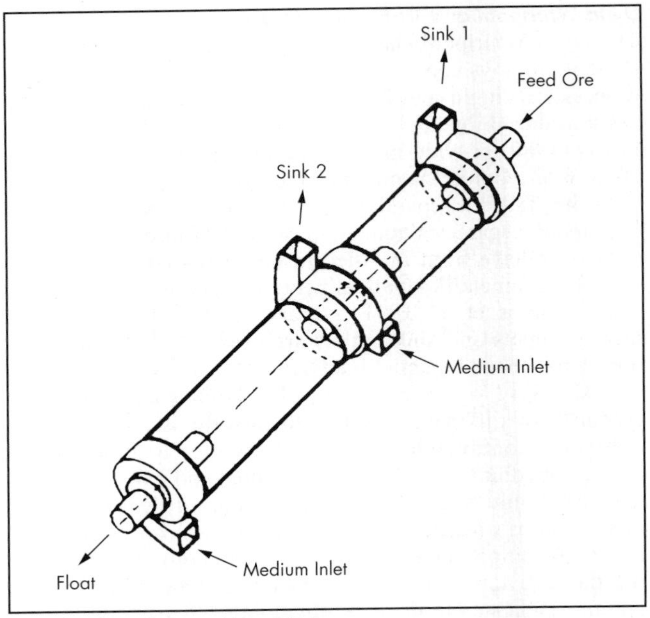

Source: Wills and Napier-Munn 2005
Figure 10 Tri-Flo separator

details of a Tri-Flo separator are provided by Dehghan and Aghaei (2014).

Larcodems Dense Medium Separator

The Larcodems separator in Figure 11 is cylindrical and mounted at an angle of approximately 30° to horizontal (Shah 1992; Rudman 2000). The medium and feed material flows are countercurrent, similar to the Dyna Whirlpool. The extractor is off-center, which gives improved control. The medium split between discard/rejects and product outlet can be varied from 60/40 to 40/60 to suit the yields of product and discard without detrimental effects on separation efficiency (Rudman 2000). Furthermore, the separation efficiency is not materially affected when handling a feed with a large percentage of discard material. This is an issue with DMCs. Applications include destoning of a 100 mm × 12 mm steaming coal feed to a power station at a feed rate of ~450 t/h feed. The relative density of separation was 1.81–1.88 with an Ep of 0.0095–0.0105. The available equipment sizes are shown in Table 4. A mathematical model for the separator is described by Álvarez et al. (2017). Suppliers include Mineral Processing Technologies and Don Valley Engineering.

Three-Product Dense Medium Cyclones

The three-product DMC consists of a cylindrical dense medium vessel, with a conventional DMC attached to the rejects outlet of the primary unit. The YTMC-series cyclones from Haiwang Technology Group (2016) are widely applied in raw coal separation processes and for pre-desliming washing in China. The cyclones can be pump fed or gravity fed. Figure 12 shows schematics for both a pump-fed and gravity-fed DMS cyclone, and Figure 13 shows an actual installation in China. The three-product DMS cyclone for coal beneficiation is the preferred option for new coal preparation plants in China (Jacobs and de Korte 2013). Available cyclone sizes and throughputs are shown in Table 5.

Table 3 Condor dense medium separator

Separator Dimensions		Feed Capacity, t/h	Maximum Feed Size, mm
Diameter, mm	Length, mm (approximate)		
250	2,450	10–30	20–25
300	2,850	30–50	30–38
400	3,250	50–80	40–50
500	3,300	80–120	60–75
600	3,940	120–170	70–88
700	4,580	170–230	80–100

Adapted from Sepro Mineral Systems 2018

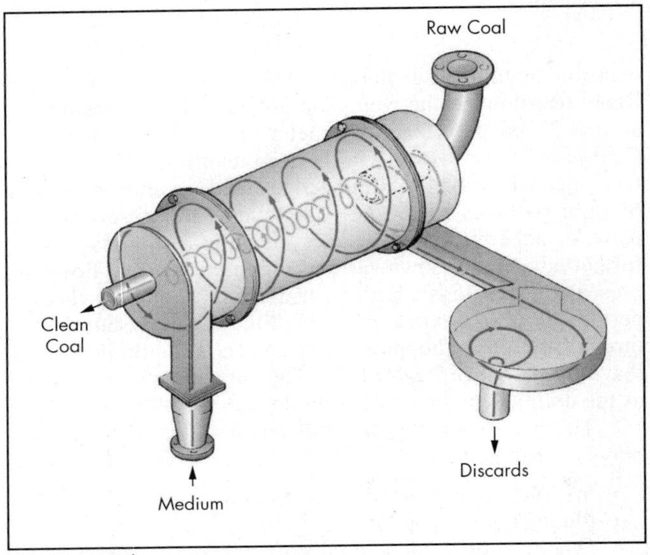

Adapted from Álvarez et al. 2017
Figure 11 Larcodems separator

Table 4 Sizes and capacities of the Larcodems separator

Separator Diameter, mm	Medium Flow Rate, m³/h	Solids Feed Rate, t/h	Maximum Feed Size, mm
850	500	200	75
1,000	700	250	85
1,200	850	350	100
1,350	1,250	450	120

Source: Rudman 2000

Pump-fed three-product DMCs. The largest size in this category has a primary cyclone with a diameter of 1,400 mm and secondary cyclone of 1,000 mm. The capacity is between 450 and 550 t/h, and it has an Ep of ≤0.03 in the primary cyclone and ≤0.05 in the second cyclone. The cyclone comes with an alumina ceramic lining.

Gravity-fed three-product DMCs. The largest size in this category has a primary cyclone with a diameter of 1,500 mm and secondary cyclone of 1,000 mm. The capacity is between 500 and 600 t/h, and it has an Ep of ≤0.03 in the primary cyclone and ≤0.05 in the second cyclone. The cyclone is supplied with an alumina ceramic lining.

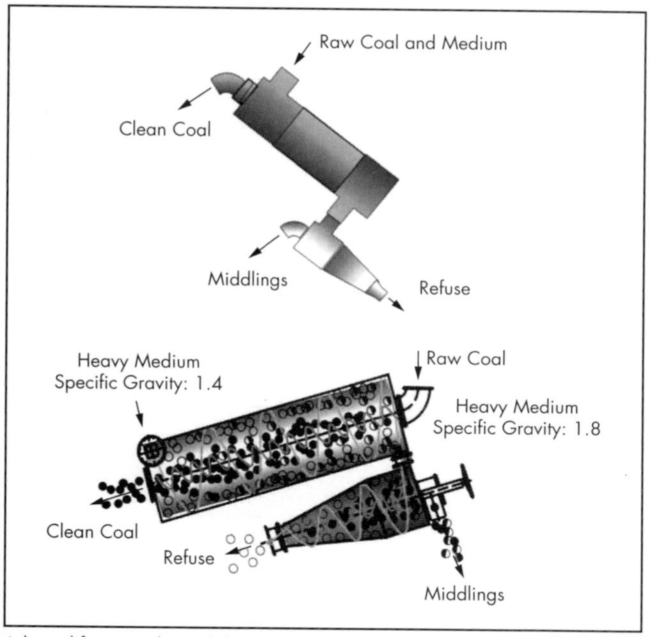

Adapted from Jacobs and de Korte 2013

Figure 12 Pump- and gravity-fed three-product DSM cyclones

Source: Jacobs and de Korte 2013

Figure 13 Installation of three-product DMS cyclone in China

Dense Medium Cyclone

Developed in the Netherlands by De Staatsmijnen, the DSM cyclone is the basis of all later cyclone developments (Driessen et al. 1951). The principle of operation of a DMC is similar to that of a conventional hydrocyclone. For pump-fed DMS cyclones, the ore is suspended in a dense medium and is introduced tangentially to the cyclone under pressure. Particles denser than the slurry will move to the wall of the cyclone and travel down to the apex and through the cyclone underflow, while less-dense particles migrate to the vortex and ultimately to the cyclone overflow. A typical schematic of a DSM cyclone is shown in Figure 14.

Cyclone feed pressures are low compared to classification hydrocyclones. In DMS, the cyclone feed pressure is generally held within 9–14 times the cyclone diameter. This pressure is required to maintain the stability of the medium. Excessive feed pressure will impart high gravitational forces (g-forces) on the particles of both the ore and the medium and will cause separation by size of the medium.

DMC separators are widely used in the treatment of ores and coal. They provide a centrifugal force and low viscosity in the medium, enabling much finer separations to be achieved than in static separators. Feed to these devices is typically deslimed at around 0.5 mm to avoid contamination of the medium with slimes and to minimize medium consumption. A finer medium is required than with gravitational vessels, to avoid medium instability. In recent years, work has been carried out in many parts of the world to extend the range of particle sizes treated by dense medium centrifugal separators to much finer particles, particularly those operating in coal preparation plants.

Cyclones are widely used for the beneficiation of coal. In general, the particle size of the raw coal ranges from ~50 mm to 0.5 mm. When the coal particle size becomes smaller, the separation becomes influenced by the medium stability and viscosity. Large-diameter cyclones (up to 1,500 mm in diameter) can treat particles as large as 100 mm and still have good efficiencies. The ore and medium are tangentially fed to the cyclone either by a pump or by gravity.

Gravity-fed DMCs are favored primarily because a static head can be maintained on the cyclone, which results in improved cyclone efficiencies. The effective cut point of the cyclone is higher than the density of the medium because of the centrifugal forces experienced. The various components

Table 5 Capacities of the pump-fed three-product heavy medium cyclones

Model YTMC	Diameter, mm		Feeding Size, mm	Inlet pressure, MPa	Capacity, t/h	Volume Capacity, m³/h
	1st Stage	2nd Stage				
500/350	500	350	≤20	0.06–0.10	40–60	200–300
600/400	600	400	≤30	0.08–0.12	50–80	300–400
710/500	710	500	≤35	0.09–0.14	70–120	400–550
780/550	780	550	≤40	0.10–0.15	100–160	550–650
850/600	850	600	≤45	0.13–0.16	120–180	650–750
900/650	900	650	≤50	0.15–0.18	140–200	750–950
1000/710	1,000	710	≤55	0.18–0.22	180–240	900–1,100
1100/780	1,100	780	≤60	0.20–0.24	220–300	1,100–1,400
1200/850	1,200	850	≤70	0.22–0.28	300–400	1,400–1,700
1300/920	1,300	920	≤80	0.26–0.32	350–450	1,600–1,900
1400/1000	1,400	1,000	≤90	0.30–0.40	450–550	1,900–2,300

Source: Haiwang Technology Group 2016

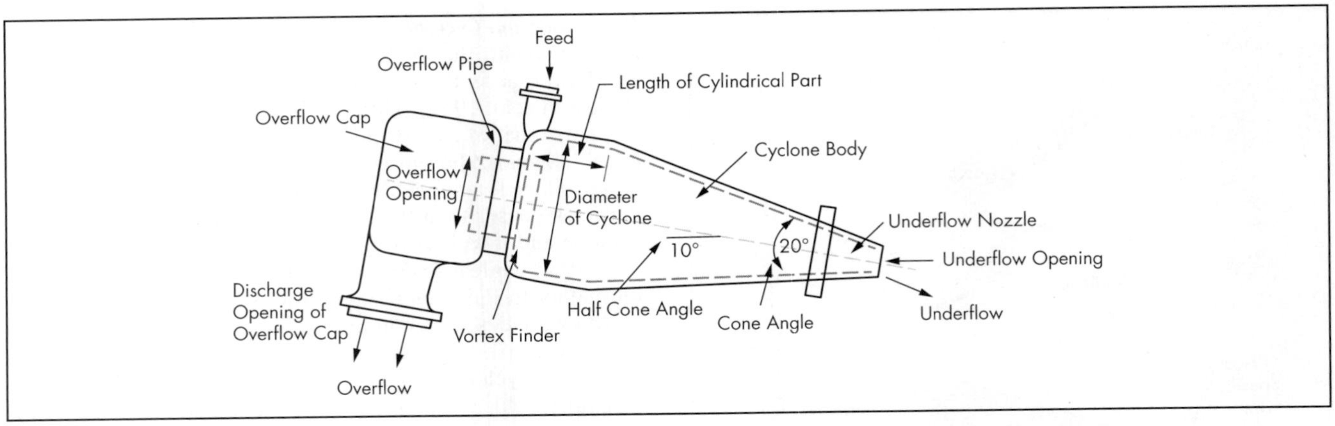

Source: Sanders 2007

Figure 14 Typical dense medium cyclone

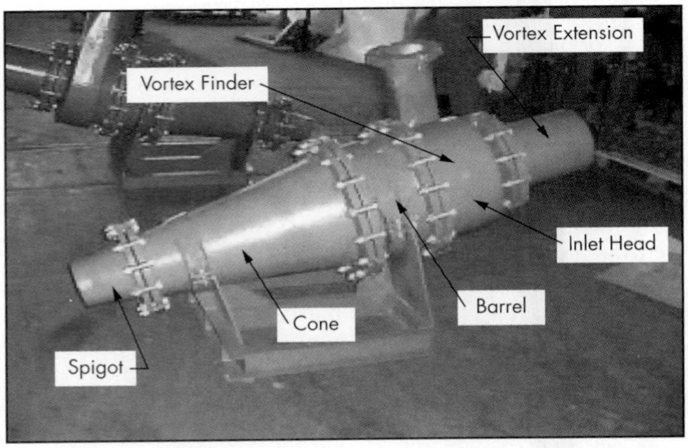

Source: Bornman 2014

Figure 15 Components of a dense medium cyclone

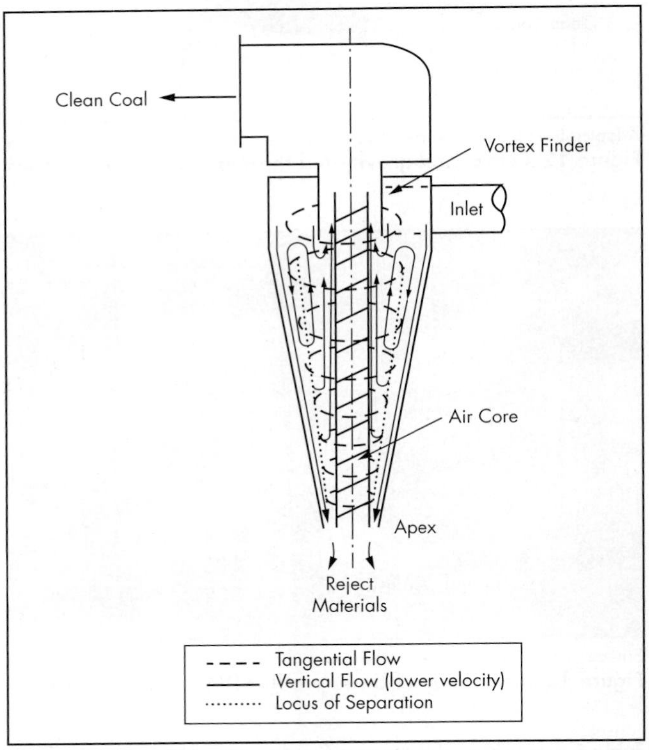

Source: Clarkson 1987

Figure 16 Flow pattern in a dense medium cyclone

of a DMC are shown in Figure 15. The flow patterns in the cyclone are shown schematically in Figure 16.

Constraints on DMC selection. The first step in choosing the appropriate cyclone dimensions and medium for a DMS separation is to identify the constraints and their sources that are applicable to the equipment selection process to prepare a flow sheet. These are given in Table 6. For most DMC applications, only one constraint will apply. In some cases, two will apply, and the second one is normally the breakaway size constraint. It is considered, but rarely is it imposed as a constraint.

Cyclone geometry. The standard DMC has a fixed geometry (Bosman 2008), as shown in Table 7.

DMC operating efficiency. The performance of a DMC can, as with other DMSs, be represented by a partition curve. In the day-to-day operation of an industrial DMS plant, the performance of a DMS cyclone is checked on a regular basis with density tracers. The tracers, which are essentially cubes, are manufactured in various sizes and densities. For more frequent (daily) cyclone performance testing, standard tracers of a single and small size at the designated separating density are normally tested. The efficiency or recovery of the cyclone is determined as the percentage of tracers recovered in the sink product from the cyclone feed. On a less frequent but regular basis, density tracers of varying sizes and colors are added to the DMS cyclone circuit, sometimes at varying feed densities. This allows partition curves or Tromp curves to be plotted and the efficiency of separation to be established.

Regular inspection and cyclone maintenance is also an essential part of ensuring high operating efficiencies. Measurement of the cyclone(s) vortex finder and apex diameter will monitor wear and allow replacement of parts before cyclone operation is adversely affected. Pressure gauges/transducers are often installed at the DMC inlet to monitor pressure, which can give a more immediate indication of potential problems within the cyclone. Gravity-fed DMCs

Table 6 Dense medium cyclone constraints and sources of constraint

Item	Constraint	Source of Constraint
Ore	Particle size distribution	Liberation
	Mass recovery/yield	Ore mineralogy
Cyclone	Cyclone diameter	Capacity (maximum flow rate)
		Breakaway size (reduced efficiency)
	Inlet dimensions	Top size (prevents blocking)
	Vortex finder diameter	Floats capacity (maximum)
	Spigot diameter	Sinks capacity (maximum)
	Spigot/vortex finder ratio	Maximum 0.85 (reduced efficiency)
Operational	Operating head	Minimum 9D (performance)
	Feed medium/ore ratio	Efficiency
	Medium type	Stability/viscosity (efficiency)

Adapted from Bosman 2008

Table 7 Fixed ratios for dense medium cyclones

Parameter	Ratio of Cyclone Diameter
Dc (cyclone diameter)	1
Di (cyclone inlet)	0.2
Do (cyclone overflow)	0.43
Barrel (length)	1.6–2.5
Cone angle	20°
Du—Standard (spigot)	70% of Do
Du—Maximum (spigot)	80% of Do

Adapted from Bosman 2008

are favored primarily because a static head can be maintained on the cyclone, which results in improved cyclone efficiency.

Operating parameters. The following information is for a tangential inlet design with a square opening (Bosman 2008). Other designs can result in significant differences, especially with regard to cyclone capacity.

- **Feed size distribution.** The smaller the particle, the less centrifugal force exerted, resulting in an increase in the fine sinks reporting to the floats. The separation efficiency (Ep value) is usually measured at selected particle size ranges. The smallest particle range will have the lowest efficiency of separation. Smaller-diameter cyclones are employed for finer feed materials to reduce the breakaway size. For a given cyclone diameter, there is a particle size, termed the *breakaway size*, below which Ep deteriorates rapidly. As cyclone diameter increases, the breakaway size becomes larger (Meyers et al. 2014).
- **Vortex-finder-to-spigot ratio.** The spigot size determines the mass/yield recovery to the underflow or overflow (coal). The spigot diameter also affects the differentials; the higher the spigot-to-vortex-finder ratio, the lower the differentials. The spigot diameter should not exceed 85% of the vortex finder diameter. The smaller the spigot, the higher the cut-point density and the lower the spigot capacity. Larger spigots reduce the hang-up of coarse particles and the excessive wear that takes place when hang-up occurs.
- **Density shift as a function of particle size.** The density separation of ore particles is time dependent; larger particles separate faster than small particles. The separation of large particles will take place at a higher level within the cyclone and hence generate lower density. As the feed solids get smaller, the d50 cut point for that particular size of solids increases. This is referred to as the density shift.
- **Head pressure.** At a high feed inlet pressure, the slurry flow rate to the cyclone increases. The tangential velocity of the particles also increases, which intensifies the centrifugal force. This, in turn, improves the separation process, especially of fine particles, despite the shorter residence time. Unfortunately, a high feed inlet pressure amplifies the classification of the media, and the density differential of the media increases, which is measurable by determining the underflow and overflow densities, respectively, and calculating the density differential. A fluctuating head affects the medium-to-ore ratio, especially when the solids feed rate is left unchanged.

  Generally, a fluctuation of 10% in cyclone pressure will result in a fluctuation of 15% in the inlet velocity to the cyclone. The intent should always be to keep the cyclone pressure as constant as possible. Cyclone operation should not be at too high a pressure (>20D [cyclone diameter]), as the pressure-to-wear-rate ratio is an exponential function. The differential over the cyclone can be calculated to determine whether the operation is in the acceptable range.

  Recommended feed heads are as follows (Bornman 2014):
  - Coal, 9 × D
  - Diamonds, 12 – 15 × D
  - Other minerals, 12 – 20 × D

  It is generally accepted that below 7D, the cyclone air core becomes unstable and the cyclone performance is likely to deteriorate (Bornman 2014).

  The operating pressure of a DMC is normally expressed in meters of head as a function of the cyclone diameter. For example, a cyclone of 500-mm diameter operating at a head of 9D will have an operating head of 4.5 m (9 × 0.5).
- **Medium-to-ore ratio.** The medium-to-ore ratio is important to prevent overcrowding of the ore. Typical ratios are as follows (Bornman 2014):
  - 3:1 for washing coal, as this a relatively easy separation
  - 5–7:1 for diamond washing
  - 5:1 for other minerals

  Higher ratios seem to have very little effect in terms of improved efficiency and typically result in increased equipment size, higher operating costs, and increased medium losses (Bornman 2014).
- **Medium differentials in feed, underflow, and overflow.** A DMC performs a separation within a separation. All the solids in the cyclone are classified (separated by size); hence, the medium solids will separate to a certain degree. Therefore the feed, overflow, and underflow will have different medium densities. The following factors affect differentials (Bornman 2014):
  - Medium viscosity/stability
  - Feed head (pressure)
  - Cyclone diameter
  - Inlet design
  - Barrel section
  - Spigot size
  - Medium density

The differential is calculated as follows (Bornman 2014):

$$\frac{RD_{feed} - RD_{overflow}}{RD_{feed}} = 3\% - 12\%$$

If the differentials are too large, hang-up of particles in the cyclone can occur. The amount of near-density material in the cyclone is usually the major factor leading to hang-ups. The acceptable medium densities during operation for diamonds are shown in Figure 17 (Bornman 2014).

- **Clays and slime.** *Slimes*, defined as <45-µm clay particles, cause viscosity problems in the medium, which will influence the separation of near-density particles. Therefore, slimes need to be removed (washing and desliming) ahead of treatment in the DMC.

Cyclone check and problem solving. Table 8 lists several observations and identifies possible problems for the observations as well as the recommended actions (Bosman 2014).

Multotec Dense Medium Cyclones

Multotec is one of the largest suppliers of DMS cyclones and provides a range of sizes having different capacities, as shown in Table 9. Multotec has developed several cyclones that have different inlet designs. The newest are the MAX range cyclones. The MAX DMS cyclones have a scrolled evolute inlet design with wear surfaces tiled in alumina.

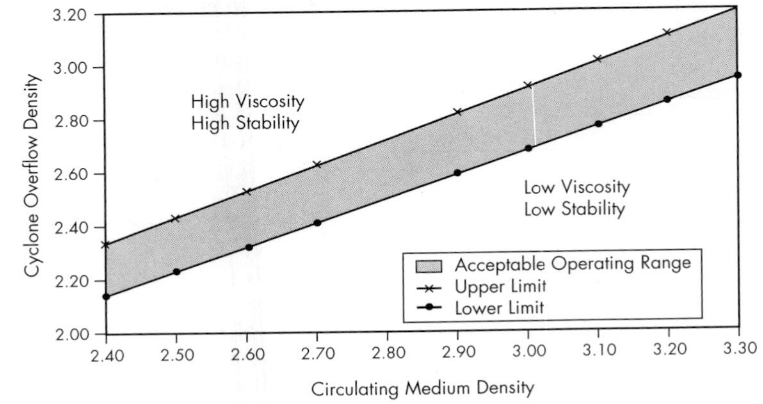

Source: Bornman 2014

Figure 17 Acceptable medium properties

Table 8 Dense medium cyclone separation fault finding

Observation	Possible Problem	Action
Feed to the preparation screen is fluctuating	Inconsistent feed rate from ore preparation section.	Investigate problem and rectify.
Not enough water on screens	Water pressure too low, water valves not open, spray nozzles blocked.	Verify water pressure, ensure that water valves are open, unblock spray nozzles, and add sufficient water to remove clay. Minimize clay balls and cyclone blockages by ensuring that the spray nozzles on the spray bars are open; better to have too much water than too little.
Screen blockage, no proper draining	Type of material treated, contamination with vegetation, requires different type of screen panel, screen motion.	De-blind screen where required, maintain aperture size, verify correct panels for the application, ensure screen stroke is sufficient.
	Contamination of medium.	Have a weir bar between the drain-and-rinse screens to prevent medium carryover to the rinse section; use enough water on the rinse section to minimize adhering medium losses.
Coarse material in effluent and medium	Holes in screen panels.	Check condition of screen panels and rectify.
Fluctuating cyclone pressure	Mixing box level not constant; insufficient medium in circuit.	Ensure header tank level. Check for correct medium sump level, and add medium if required.
	Medium to mixing box is fluctuating.	Maintain level in mixing box; open medium flow to mixing box. Check for obstructions in medium flow to mixing box.
	Poor cyclone efficiency.	Low pumping efficiency. Check condition of medium pump, and make sure ore feed rate is constant. Operate as constant as possible.
	Accelerated wear.	Ensure that cyclone pressure is in the specified operating range.
Cyclone pressure increase	Blocked cyclone inlet (product disappears from screens).	Stop plant and unblock inlet, and check for source of oversize.
	Increase in medium density.	Check density and rectify; add water.
	Mixing box level too high.	Adjust level by restricting medium flow to box.

Adapted from Bornman 2014

Table 9 Dimensions and capacities of Multotec dense medium cyclones

Standard-Capacity Cyclone			High-Capacity Cyclone		
Cyclone Diameter, mm	Maximum Particle Size, mm	Coal Feed, t/h	Cyclone Diameter, mm	Maximum Particle Size, mm	Coal Feed, t/h
510	34	54	510	51	99
610	41	81	610	61	145
660	44	97	660	66	175
710	47	114	710	71	207
800	53	149	800	80	270
900	60	196	900	94	355
1,000	67	249	1,000	100	454
1,150	77	351	1,150	115	638
1,300	87	468	1,300	130	854
1,450	97	608	1,450	145	1,108

Source: Bornman 2014

Table 10 Operating information for Krebs heavy medium cyclones

Model	Maximum Feed Size, mm	Feed Capacity* (dry), t/h	Pulp Flow Rate, m³/h	Head Equivalent, m
D20LSB	19	77	238	4.6
D26B	38	136	434	6.1
CoalMAX26	38	145	464	6.1
D263B	38	150	469	6.0
D30B	51	205	643	7.0
D33T154	63	264	829	7.6
D33T214	63	286	892	7.6
D40B	76	418	1,308	9.2
D44B-A	76	432	1,348	10.1
D44B-U	76	527	1,653	10.1
D48B	89	659	2,066	11.0
D55B	102	895	2,792	12.6

Source: FLSmidth-Krebs 2014

*Based on a 4:1 medium-to-coal ratio. Capacities represent maximums for units fitted with the largest inlet and vortex finder.

Krebs Dense Medium Separation Cyclones

Krebs heavy medium cyclones (HMCs) are designed specifically to clean coal. Krebs HMCs are usually operated in a near-horizontal orientation, allowing large apex sizes to be used for refuse/waste removal. The medium density primarily determines separating gravity. Pressure should be kept relatively low to reduce classification effects. The Krebs large-diameter cyclones allow the larger top size to be efficiently cleaned at higher feed rates. Cyclones of 760-mm diameter and larger utilize a proprietary ceramic "acceleration wedge" to modify capacity and g-forces. Inserting a smaller wedge will have an immediate effect on increasing capacity. Inserting a larger wedge will generate higher g-forces, which are required to clean fine coal more efficiently. The large cyclones can handle feeds with a particle top size of 100 mm. Capacities for the Krebs HMCs are shown in Table 10.

Minco Tech Dense Medium Cyclones

Minco Tech's design philosophy for their HMC separator is based on extensive research carried out by the Julius Kruttschnitt Mineral Research Centre in Queensland, Australia, during the 1990s. The research generated the JK cyclone simulation and modeling program. Further cyclone research was carried out by the Australian Coal Association Research Program. This work assisted with improving cyclone performance parameters, especially with cyclone feed inlet design. Minco Tech cyclones are modeled and designed with consideration given to the above-mentioned research and also supported by results from installations throughout the world. Available DMC units range in size from 50 to 1,500 mm with nominal feed rates of 10 to 1,000 t/h. Cyclones are available in various materials, from 27% chrome to pre-engineered high-alumina ceramic tiles. Wall thickness ranges from 25 to 38 mm. Cyclone capacities are shown in Figure 18 in terms of volumetric feed rates. Figure 19 shows a large-diameter Minco DMC (Minco Tech 2015).

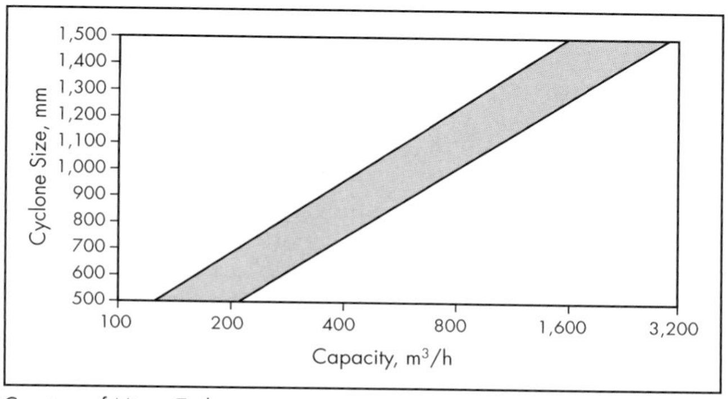

Courtesy of Minco Tech

Figure 18 Volumetric capacities of Minco dense medium cyclones

Haiwang DMS Cyclone

The two-product HMCs (FZJ series) marketed by Haiwang Technology Group are widely used in China for rougher and cleaner applications in primary and secondary coal separation plants. Cyclones range in size from 350 mm to 1,500 mm. The 1,500-mm-diameter cyclone has a capacity of 700–800 t/h and comes with an alumina ceramic lining (Haiwang Technology Group 2016).

Courtesy of Minco Tech
Figure 19 Minco large-diameter dense medium cyclone

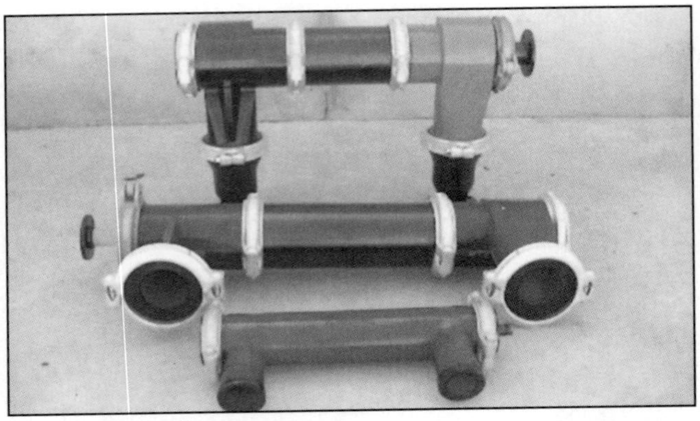

Courtesy of Concord Engineering
Figure 20 Concord densifier

Densifier Circuits and Equipment

Most dense medium plants use "circulating medium" densification methods, where a part of the medium is fed by gravity or pump to the densifier. Some operations use "dilute medium" densification, where the feed to the densifier is rinse medium/water from the product screens. This stream is pump-fed to the densifier to reduce height. Densifier overflows from both methods are treated by magnetic separators to recover contained medium. To allow for medium cleaning, some of the medium bypasses the densifier and goes directly to the magnetic separators. Loading on the magnetic separators is higher with circulating medium densification than with dilute medium densification. This is not much of an issue with modern-day magnetic separators.

A comparison of the two methods shows that the pipe densifier rejects significantly fewer magnetics to the overflow than the cyclone densifier. This greatly reduces the dilute medium circuit magnetics loading and, in turn, the size of the wet drum magnetic separator required. Additional advantages of correct medium densification using a pipe densifier are reduced magnetics classification and reduced fine magnetics losses from the dilute medium circuit wet drum magnetic separator.

Both types of densifier will reject lower-density contaminant minerals from the circulating medium (e.g., silicates and clays). However, in high-density mineral separations, such as sulfides, it is recommended that a circulating-medium bleed to the dilute medium circuit be incorporated to remove fine high-density nonmagnetics that can build up in the circuit and affect the medium density and viscosity.

Densifiers have become increasingly important in dense medium circuits for the following reasons:

- The production of an over-dense stream into the correct medium tank significantly reduces the start-up time required to achieve density, from typically 30 minutes using magnetic separators alone to less than 10 minutes using a combination of magnetic separator and densifier.
- Improved density control through the introduction of an over-dense stream into the correct medium tank.
- Cleaning nonmagnetics that continuously build up in the media. Analysis has shown that where 80% of the magnetics are recovered to the cyclone underflow, 65% of the nonmagnetics report to the cyclone overflow.

The Concord densifier (Figure 20) manufactured by Multotec is widely used in the DMS industry. It works over a broad range of operating pressures, from 300 to 700 kPa. It can also cope with much greater volumetric feed flow rates.

Work by Davidson et al. (1987) found that major medium losses were adherence losses and losses in the water recovery circuit. Operating density and medium blend had a significant effect on the medium consumption. Viscosity was thought to be the problem. The viscosity problem was resolved by adding dilution water to the cyclone spigot discharge to reduce viscosity. The result was a reduction in medium consumption of around 25%.

Densifying cyclones are available in diameters of 100, 165, 250, and 350 mm and vary in materials of construction:

- Long-lasting and high-elasticity polyurethane
- Twenty-seven percent high-chrome cast iron for highly abrasive applications
- Ceramic lined for pulp distribution and abrasion resistance

DESIGN OF DENSE MEDIUM CIRCUITS

Details of DMS circuit design that include equipment sizing are provided by Symonds (Symonds 1986; Symonds and Malbon 2002). Information on sizing and performance of dense media is provided by Chaston and Napier-Munn (1974), Dardis (1987), King and Juckes (1988), Reeves and Platt (1988), Lee et al. (1995), Rylatt and Poplewell (1999), England et al. (2002), Holtham (2006), and Napier-Munn et al. (2009). The economics of DMS is detailed by Burton et al. (1991) and Schena et al. (1990).

CYCLONE WEAR PERFORMANCE

The most common materials used for DMS cyclones are high-chrome cast iron and alumina ceramic tiles bonded to a mild steel body. Other materials include tungsten carbide and stainless steel. Table 11 provides the relative life of a DMC.

MEDIUM TYPE AND SELECTION

The most important factor that must be taken into account when selecting a medium grade for a particular application is the operating density. This is determined by the separation desired, the characteristics of the feed ore, and the nature of the separation vessel being used. Once the operating density has been selected, the stability and viscosity will then determine

Table 11 Relative life and costs of dense medium cyclones*

Materials of Construction	Relative Life	Relative Cost	Relative Cost per Unit of Duty
High-chrome cast iron	1.0	1.0	1.0
Alumina tile lined	2.0	1.25	1.6
Sintered silicon carbide	3.0	3.7	0.82
Reaction-bonded silicon carbide	3.0	3.7	0.82
Nitride-bonded silicon carbide	3.2	3.7	0.86
Zirconium oxide	No data	No data	No data

Source: Bookless 2008

*Based on plant data. Does not include labor cost of replacement, loss of production, or cost of storing the replacements.

the appropriate ferrosilicon grade. The ideal medium is one with high stability and low viscosity. Any selection involves a compromise between these two mutually exclusive properties.

Ferrosilicon Medium

The following general guidelines for the medium are recommended:

- For very high operating densities (say, higher than 3.2 RD), only atomized grades can be used, as the milled grades are too viscous at the high solids concentrations required.
- Atomized grades may also be appropriate where corrosive conditions or highly porous ores (leading to high medium losses and thus operating costs) are suspected. Otherwise, milled grades are preferred because of their lower cost.
- Dynamic separators require finer grades than bath separators, because of the greater tendency for instability (large differentials) in dynamic separators. High-pressure dynamic vessels (large static heads or high pump-delivery pressures) require finer grades than low-pressure systems, to preserve stability.
- All things being equal, coarser particle grades are preferred to finer particle grades because, under most conditions, medium losses occur preferentially in the finer sizes. In addition, the coarser grades are often less expensive.

The wide-scale acceptance of heavy media separation is due to its simplicity and efficiency. Conventional heavy medium separation of ores having a specific gravity in the range from 2.5 to 4.0 utilizes a mixture of either milled or atomized ferrosilicon powder containing 14%–16% Si and water.

One of the most important parameters of the heavy medium process is the rheology of the medium. Because rheological behavior is extremely complex, the medium grade—and, more specifically, the size distribution of the ferrosilicon required for a specific application—is still sometimes chosen arbitrarily or on the basis of subjective experience.

Media Type and Specification

The information in this section is based on information regarding Samancor ferrosilicon products (Blair 1983). The three grades represented in Table 12 are products produced by Samancor and represent 90% of ferrosilicon used in diamond heavy medium processing.

Table 12 Samancor ferrosilicon products

Specification	Medium Product		
	100D	150D	270D
Medium type	Milled	Milled	Milled
% Passing 20 µm	25–35	40–50	52–62
% Passing 45 µm	61–69	73–81	85–93
% Passing 75 µm	90–95	94–98	97–100
Silicon, %	14–16	14–16	14–16
Carbon, % maximum	1.3	1.3	1.3
Iron, % minimum	80	80	80
Sulfur, % maximum	0.05	0.05	0.05
Phosphorus, % maximum	0.15	0.15	0.15
Rust index, % maximum	1.2	1.2	1.2
% Nonmagnetics (Davis tube), maximum	0.75	0.75	0.75
Density, specific gravity	6.7–7.1	6.7–7.1	6.7–7.1

Source: Blair 1983

Table 13 Typical size analysis for milled ferrosilicon 14%–16% Si

Particle Size, µm	Medium Product, %				
	48D	65D	100D	150D	270D
+212	1.5	0.4	0.2	0.1	0.0
–212 +150	7.2	1.5	0.6	0.3	0.1
–150 +105	16.1	5.8	2.7	1.2	0.3
–105 +75	22.3	12.7	7.4	4.1	1.7
–75 +45	20.8	27.9	23.9	16.9	10.0
–45 +20	22.5	37.2	29.6	29.0	49.9
–20	8.8	14.5	35.7	48.4	38.0
–45 (typical)	3.1	52	65	77	88
–45 (limits)	20–38	38–58	58–70	70–85	85+

Source: Blair 1983

Some of the more important properties essential to a metal or alloy powder that is to be used as a heavy medium are as follows:

- Resistance to abrasion
- Resistance to corrosion
- High specific gravity
- Magnetism, which allows easy magnetic recovery with subsequent easy demagnetization
- Economical operating costs

The typical size analyses of available milled ferrosilicon 14%–16% Si are shown in Table 13, and the size analysis for atomized ferrosilicon is provided in Table 14. The typical chemical and physical properties of milled and atomized ferrosilicon are shown in Table 15. The respective chemical and physical data with sizing analyses given for both the milled and atomized ferrosilicon grades represent typical average values obtained and are in no way meant to constitute final specifications.

Particle-to-Particle Homogeneity

The "in circuit" performance of 14%–16% Si ferrosilicon medium is directly related to its chemistry and physical metallurgy. Particle-to-particle homogeneity is important, particularly in relation to the required properties of the medium. The

Table 14 Typical size analysis for atomized ferrosilicon 14%–16% Si

	Medium Product, %				
	Special		Cyclone		
Particle Size, μm	Coarse	Fine	60	40	20
+212	2.5	1.7	0.0	0.0	–
–212 +150	6.7	5.3	0.1	0.0	–
–150 +105	13.0	10.6	0.3	0.1	–
–105 +75	18.2	16.2	3.2	1.2	–
–75 +45	21.9	21.6	21.6	14.8	–
–45 (typical)	38	45	75	84	–
–45 (limits)	35–40	40–50	65–80	80–85	90+

Source: Blair 1983

Table 15 Chemical and physical properties for milled and atomized ferrosilicon

	Medium	
Specifications	Milled	Atomized
Si, %	14.7	14.9
Fe, %	80.7	82.9
C, %	1.0	0.3
P, %	0.04	0.06
Rust index, %	0.5	–
Nonmagnetics, %	0.5	0.2
Density	6.78	6.85
Satmagan, %	57	73
Spherical particles, no.	–	69

Source: Blair 1983

presence of a dual-phase structure (silica-enriched dendritic structure), which is characteristic of ferrosilicon alloys containing 15%–25% Si, affects the performance of atomized ferrosilicon to a greater extent than milled ferrosilicon.

Particle Shape

The manufacturing processes for milled and atomized ferrosilicon result in a marked difference in both the shape and nature of the particles of ferrosilicon powder produced. This difference in the individual particle shape has marked effects on the rheological properties of the aqueous suspensions of the two types of ferrosilicon.

Rheological Properties

The rheological properties of heavy medium suspensions play an important part both in the heavy medium separation itself and in the handling of the medium with regard to pumping and storage. The rheological properties of the medium that are regarded as process determining are its viscosity and stability. The details of medium viscosity measurement are provided by Ferrara and Schena (1986), Reeves (1990), and Myburgh (2002).

Viscosity

Because heavy media suspensions are unstable, they do not follow the same laws of fluid mechanics as ordinary liquids and are referred to as non-Newtonian. The main feature of the atomized grades is their ability to produce high pulp densities combined with low viscosities. This is due to the greater friction of collision of the irregularly shaped milled particles, which requires a larger energy input to achieve a given deformation of suspension.

Stability

The *stability* of a suspension can be defined as the reciprocal of the settling rate, because the consistency or apparent viscosity of the separation medium is a direct function of the settling rate of the media (i.e., the higher the consistency, the lower the settling rate). The settling rate or stability is a very valuable parameter for characterizing the condition of the medium. However, as in the case of viscosity, the stability increases with both the fineness of the ferrosilicon and the pulp density. The viscosity-effect grades of milled ferrosilicon show greater stability than grades of atomized ferrosilicon of equivalent size.

Modification of Rheology

The rheological properties of ferrosilicon suspensions can be modified in the following ways:

- Adding polymeric compounds or other reagents (positive impact)
- Extraneous contamination by ore slimes of clays (negative impact)
- Demagnetization of the circulating medium (positive impact)
- pH modification
- Increasing the proportion of spherical particles (for atomized applications only)

Magnetic Properties

Ferrosilicon is inherently magnetic, being an iron alloy, which enables easy recovery and cleaning of the medium in circuit. The magnetic properties are measured by a Satmagan balance, which effectively measures the magnetic moment of a sample when saturated in a strong magnetic field.

Corrosion Resistance

Corrosion resistance is important for the following reasons:

- Corrosion leads to loss of ferrosilicon.
- The finely divided products of corrosion increase the viscosity of the medium and affect separating efficiency.
- Corrosion in situ when stored wet can result in start-up difficulties after long shutdowns.
- Corrosion of ferrosilicon adversely affects the density of the medium.

Stewart and Guerney (1997) provide information on the detection and prevention of ferrosilicon corrosion in dense medium plants.

Density of Separation

In general, for actual separations above ~3.0 for static baths and 3.2 for cyclones, the use of atomized ferrosilicon is necessary, because above a true pulp density of ~3.0, the viscosity of the milled material can increase rapidly to unmanageable proportions. At higher densities, the question is one of selecting a grade of atomized ferrosilicon that exhibits adequate stability in suspension of the required density. The finer grades are used at the lower end of this range.

Particle Size of Ore

A finer medium is required in dynamic separators than for static separators because of the tendency of the medium to thicken and create high differentials as a result of the centrifugal forces involved in dynamic separators. Finer, more stable medium is required for smaller cyclones. As with most gravity separations, the narrower the size range of the feed, the greater the efficiency of the separation for small particles.

Sharpness of Separation

The sharpness of the separation achieved is influenced by the grade of ferrosilicon used, being higher for the finer, more-stable ferrosilicon and particularly in the finer particle sizes of ore. The flexibility of cyclone operation is greater with finer medium, enabling specific gravity of media feed, pressure, and cone ratio to be adjusted to give variations in quality and quantity of the apex discharge.

Selection of Medium for Industrial Application

The appropriate selection of a ferrosilicon medium for application in DMS equipment is discussed by Collins et al. (1974) and Hunt et al. (1986).

Magnetite Medium

The most popular manufactured media used in coal preparation are a fine magnetite suspension in water (1.25–2.2 RD), ferrosilicon suspension (2.9–3.4 RD), or a mixture of the two (2.2–2.9 RD), depending on the selected separating density. The effective density of such a slurry can be controlled by adjusting the solid concentration of the suspension (King 2001; Galvin and Iverson 2013; Gupta and Yan 2006). Various grades of magnetite, which are determined by fineness, are produced for coal cleaning processes (Robertson and Williams 1991). The grade of magnetite is typically marketed based on the percentage of ultrafine particles smaller than 44 µm. The most common grade of magnetite used in coal preparation is grade B, which is normally more than 95% below 44 µm. The physical and rheological properties are of considerable importance in determining the effectiveness of magnetite as a medium. Since the RD of magnetite is relatively high (about 5.0), the generation of low-viscosity dense media over a wide range of densities, even in the presence of slimes contamination, is possible given the low degree of solids concentration. Furthermore, highly effective medium recovery and regeneration are achievable because of the magnetic characteristics of magnetite (Osborne 1988; Sanders 2007).

Physical properties used to specify the quality and suitability of a particular source of magnetite must be assessable by standard techniques. Several countries have been actively involved in establishing magnetite specifications for heavy medium applications (Mikhail and Osborne 1990; Osborne 1988).

The general specifications for magnetite based on the British coal mining industry are

- Particle size distribution at a maximum of 5% by weight larger than 45 µm and 30% by weight smaller than 10 µm;
- RD of 4.9–5.2; and
- Magnetite content not less than 95% by weight.

Industrial experience and practices in several coal-producing countries have been used to develop standard procedures to test magnetite for coal preparation purposes. The international standard ISO 8813 specifies the properties for testing moisture content, particle size distribution, magnetic content, and relative density. The following properties of magnetite are important for coal washing (Robertson and Williams 1991):

- Ratio of ferrous to total iron content
- Hardness and resistance to degradation
- Coercive force of 38 Oe (oersteds) or less
- Miscibility with water
- Susceptibility (emu/g) at 30 Oe should be 0.035 or greater and at 800 Oe should be 0.05 or greater
- Saturation moment (emu/g) of 0.80 or greater.
- Purity:
  - Magnetics, ≥96%
  - Total iron, ~6%
  - Ferrous iron, ~22%

Magnetite for DMS is available in different particle sizes: fine (95% <45 µm), medium (85% <45 µm), and coarse (75% <45 µm) (Idwala Industrial Holdings 2018).

ACKNOWLEDGMENTS

The chapter authors thank F.F. Aplan, G. Miller, T.J. Demull, and J.P. Matoney, authors of several chapters in the "Gravity Concentration" section from the 1985 edition of the *SME Mineral Processing Handbook*, who set a standard to be followed with their thorough and well-developed work.

REFERENCES

Álvarez, M.M., Sierra, M.H., Lasheras, S.F., and de Cos Juez, J.F. 2017. A parametric model of the Larcodems heavy media separator by means of multivariate adaptive regression splines. *Materials* 10(7):729.

Anon. 1984. Sodium metatungstate, a new medium for binary and ternary density gradient centrifugation. *Makromol. Chem.* 185:1429.

AS 4350.2. 1999. *Heavy Mineral and Concentrates—Physical Testing; Part 2: Determination of Heavy Minerals and Free Quartz—Heavy Liquid Separation Method.* Strathfield, New South Wales: Standards Australia International.

ASTM International. 1999. Standard test method for determining the washability characteristics of coal. In *Annual Book of ASTM Standards*. Vol. 5.05, Standard No. D4371. West Conshohocken, PA: ASTM International.

Baguley, P.J., and Napier-Munn, T.J. 1996. Mathematical model of the dense medium drum. *Trans. Inst. Min. Metall.* 105:Cl–C8.

Blair, W.I. 1983. The properties and selection of ferrosilicon powders for heavy medium separation. Presented at 1st Samancor Symposium on Dense Media Separation.

Bookless, T. 2008. A review of cyclone geometry and materials of construction. Presented at the 10th International Dense Medium Symposium, South Africa, May.

Bornman, F. 2014. Practical operational aspects of dense medium cyclone separation. Presented at the 46th Annual Canadian Mineral Processors Operators Conference, Ottawa, Ontario, January 21–23.

Bosman, J. 2008. Dense medium cyclone selection—A size based approach. Presented at the 10th International Dense Medium Symposium, South Africa, May.

Bosman, J. 2014. The art and science of dense medium selection. *J. S. Afr. Inst. Min. Metall.* 114:529–536.

Browning, J.S. 1961. *Heavy Liquids and Procedures for Laboratory Separation of Minerals*. Information Circular No. IC 8007. Washington, DC: U.S. Bureau of Mines.

Burt, R.O., and Mills, C. 1984. *Gravity Concentration Technology*. Vol. 5. Amsterdam; New York: Elsevier.

Burton, M., Ferrara, G., Machiavelli, G., Porter, M., and Ruff, H. 1991. The economic impact of modern dense medium systems. *Miner. Eng.* 4(3-4): 225–243.

Chaston, I-R.M., and Napier-Munn, T.J. 1974. Design and operation of dense-medium cyclone plants for the recovery of diamonds in Africa. *J. S. Afr. Inst. Min. Metall.* 75(5):120–133.

Chironis, N.P. 1987. On-line coal-tracing system improves cleaning efficiencies. *Coal Age* 92(3):44.

Clarkson, C.J. 1987. Model of dense medium cyclone. In *Proceedings of the Dense Medium Operators' Conference.* Melbourne, Victoria: Australasian Institute of Mining and Metallurgy.

Collins, B., Napier-Munn, T.J., and Sciarone, M. 1974. The production, properties and selection of ferrosilicon powders for heavy medium separation. *J. S. Afr. Inst. Min. Metal.* 75(5):103–119.

Dardis, K.A. 1987. The design and operation of heavy medium recovery circuits for improved medium recovery. In *Proceedings of the Dense Medium Operators' Conference.* Melbourne, Victoria: Australasian Institute of Mining and Metallurgy. pp. 157–184.

Davidson, J.R., Napier-Munn, T.J., and Kojovic, T. 1987. Investigation into relationships between medium properties and medium loss in the dense medium cyclone circuit at Mt Newman. In *Proceedings of the Dense Medium Operators' Conference.* Melbourne, Victoria: Australasian Institute of Mining and Metallurgy.

de Korte, D., and Bosman, J. 2006. Optimal coal processing route for the 3 × 0.5 mm size fraction? DMS and Gravity Concentration Operations and Technology in South Africa. *J. S. Afr. Inst. Min. Metall.* 107(July):411–414.

Dehghan, R., and Aghaei, M. 2014. Evaluation of the performance of Tri-Flo separators in Tabas (Parvadeh) coal washing plant. *Res. J. Appl. Sci. Eng. Technol.* 7(3):510–514.

Domenico, J.A., Stouffer, N.J., and Faye, C. 1994. Magstream as a heavy liquid separation alternative for mineral sands exploration. SME Preprint No. 94-262. Littleton, CO: SME.

Driessen, A.G., Krijgsman, C., and Leeman, J.N.J. 1951. Process for the separation of solids of different specific gravity and grain size. U.S. Patent 2,543,689A.

Dunglison, M.E., and Napier-Munn, T.J. 1997. Development of a general model for the dense medium cyclone. Presented at the 6th Samancor Symposium on Dense Media Separation, Broome, Western Australia, May 6–7.

Dunglison, M.E., Napier-Munn, T.J., and Shi, F. 1999. The rheology of ferrosilicon dense medium suspensions. *Min. Proc. Ext. Rev.* 20:183–196.

England, T., Hand, P.E., Michael, D.C., Falcon, L.M., and Yell, A.D. 2002. *Coal Preparation in South Africa*, 4th ed. Pietermaritzburg, South Africa: Natal Witness Commercial Printers.

Eroglu, B., and Stallknecht, H. 2000. A laboratory density analysis developed using non-toxic heavy liquid. In *Mineral Processing on the Verge of the 21st Century: Proceedings of the Eighth International Mineral Processing Symposium.* Rotterdam: A.A. Balkema. pp. 111–116.

Ferrara, G., and Schena, G.D. 1986. Influence of contamination and type of ferrosilicon on viscosity and stability of dense media. *Trans. Inst. Min. Metall.* 95(December):C211.

Ferrara, G., Machiavelli, G., Bevilacqua, P., and Meloy, T.P. 1994. Tri-Flo: A multistage high-sharpness DMS process with new applications. *Miner. Metall. Process.* 296:63–73.

FLSmidth-Krebs. 2014. *Krebs Products for the Coal Industry: Krebs Hydrocyclones, millMAX Pumps, and Technequip Valves for Coal Processing.* www.flsmidth.com/~/media/PDF%20Files/Krebs/KrebsCyclonesPumpsandValvesforCOALIndustryweb11132015.ashx.

Galvin, K.P., and Iverson, K. 2013. Cleaning of coarse and small coal. In *The Coal Handbook, Vol. 1: Towards Cleaner Production—Coal Production.* Burlington: Elsevier. pp. 263–300.

Gottfried, B.S. 1978. A generalisation of distribution data for characterizing the performance of float-sink coal cleaning devices. *Int. J. Miner. Process.* 5:1.

Gupta, A., and Yan, D. 2006. *Mineral Processing Design and Operation: An Introduction*. Amsterdam: Elsevier Science.

Hacioglu, E., and Turner, J.F. 1985. A study of the Dyna Whirlpool. In *Proceedings of the XV International Mineral Processing Congress*. Saint-Etienne, France: GEDIM. pp. 244–257.

Haiwang Technology Group. 2016. Heavy medium cyclone. www.haiwangtec.com/hydrocyclones/Heavy-Medium-Cyclone.html. Accessed August 2018.

Hillman, J. 2003. *A History of British Coal Preparation*. Worksop, Notts: Minerals Engineering Society.

Holtham, P.N. 2006. Dense medium cyclones for coal washing—A review. *Trans. Indian Inst. Met.* 59(5):521–533.

Humboldt Wedag India. 2008. Company presentation. https://fossil.energy.gov/international/Publications/Coal_Beneficiation_Workshop/8th_Mustafhi_Article_on_Coal_Beneficiati.pdf. p. 14.

Hunt, M.S., Hansen, J.O., and Davy, A.T. 1986. The influence of ferrosilicon properties on dense medium separation plant consumption. *Bull. Proc. Australas. Inst. Min. Metall.* 291(October):73.

Idwala Industrial Holdings. 2018. Magnetite production information. www.idwala.co.za/products-listing/magnetite/. Accessed August 2018.

ISO 8813. 1992. *Earth-Moving Machinery—Lift Capacity of Pipelayers and Wheeled Tractors or Loaders Equipped with Side Boom.* Geneva: International Organization for Standardization.

Jacobs, J., and de Korte, G.J. 2013. The three-product cyclone: Adding value to South African coal processing. *J. S. Afr. Inst. Min. Metall.* 113:859–963.

Jowett, A. 1986. An appraisal of partition curves for coal-cleaning processes. *Int. J. Miner. Process.* 16(January):75.

King, R.P. 2001. *Modeling and Simulation of Mineral Processing Systems.* Burlington: Elsevier Science.

King, R.P., and Juckes, A.H. 1988. Performance of a dense medium cyclone when beneficiating fine coal. *Coal Prep.* 5:185–210.

Kitsikopoulos, H., et al. 1991. Industrial operation of the first two-density three-stage dense medium separator processing chromite ores. In *Proceedings of the XVII International Mineral Processing Congress.* Vol. 3. Freiberg, Saxony: Freiberg University of Mining and Technology.

Koroznikova, L., Klutke, C., McKnight, S., and Hall, S. 2007. The use of low-toxic heavy suspensions in mineral sands evaluation and zircon fractionation. In *Proceedings of the 6th International Heavy Minerals Conference: Back to Basics.* Johannesburg: Southern African Institute of Mining and Metallurgy. pp 21–29.

Lee, D., Holtham, P.N., Wood, C.J., and Hammond, R. 1995. Operating experience and performance evaluation of the 1150 mm primary dense medium cyclone at Warkworth Mining. In *Proceedings of the 7th Australian Coal Preparation Conference.* Broadmeadow, New South Wales: Australian Coal Preparation Society. pp. 71–85.

Leonard, J.W. 1991. *Coal Preparation.* Littleton, CO: SME.

Maust, E.J. 1954. Drum separator. U.S. Patent 2,696,300.

Meyers, A., Sherritt, G., Jones, A., and O'Keeffe, M. 2014. Large-diameter dense medium cyclone performance in low-density/high near-gravity environment. *Int. J. Coal Prep. Util.* 34:133–144.

Mikhail, N.W., and Osborne, D.G. 1990. Magnetite heavy media: Standards and testing procedures. *Int. J. Coal Prep. Util.* 8:111–112.

Minco Tech. 2015. Cyclones: Heavy medium. www.minco-tech.com/cyclones/heavy-medium-metric/. Accessed August 2018.

Munro, P.D., Schache, I.S., Park, W.G., and Watsford, R.M.S. 1982. The design, construction and commissioning of a heavy medium plant of silver-lead-zinc ore treatment—Mount Isa Mines Ltd. In *Proceedings of the XIV International Mineral Processing Congress.* Montreal, QC: Canadian Institute of Mining, Metallurgy and Petroleum.

Myburgh, H.A. 2002. The influence of the quality of ferrosilicon on the rheology of dense medium and the ability to reach higher densities. In *Proceedings of Iron Ore 2002.* Melbourne, Victoria: Australasian Institute of Mining and Metallurgy. pp. 313–317.

Napier-Munn, T.J. 1985. Use of density tracers for determination of the Tromp curve for gravity separation processes. *Trans. Inst. Min. Metall.* 94(March):C45.

Napier-Munn, T.J. 1990. The effect of dense medium viscosity on separation efficiency. *Coal Prep.* 8:145.

Napier-Munn, T.J. 1991. Modelling and simulating dense medium separation processes—A progress report. *Miner. Eng.* 4(3-4):329.

Napier-Munn, T.J., and Scott, I.A. 1990. The effect of demagnetisation and ore contamination on the viscosity of the medium in a dense medium cyclone plant. *Miner. Eng.* 3(6):607–613.

Napier-Munn, T.J., Kojovic, T., Scott, I.A., Shi, F., Masinja, J.H., and Baguley, P.J. 1995. Some causes of medium loss in dense medium plants. *Miner. Eng.* 8(6):659–678.

Napier-Munn, T.J., Gibson, G., and Bessen, B. 2009. Advances in dense medium cyclone plant design. In *Proceedings of the Tenth Mill Operators' Conference.* Melbourne, Victoria: Australasian Institute of Mining and Metallurgy. pp. 12–14.

Napier-Munn, T., Bosman, J., and Holtham, P. 2013. Innovations in dense medium separation technology. In *Mineral Processing and Extractive Metallurgy: 100 Years of Innovation.* Edited by C.G. Anderson, R.C. Dunne, and J.L. Uhrie. Englewood, CO: SME.

Osborne, D.G. 1988. *Coal Preparation Technology.* London: Graham and Trotman.

Osborne, D. 2010. Value of R&D in coal preparation development. In *Proceedings of the XVI International Coal Preparation Congress.* Englewood, CO: SME. pp. 845–858.

Parsonage, P. 1980. Factors that influence performance of pilot-plant paramagnetic liquid separator for dense particle fractionation. *Trans. Inst. Miner. Metall. C* 89(December):166.

Reeves, T.J. 1990. On-line viscosity measurement under industrial conditions. *Coal Prep.* 8:135.

Reeves, T.J., and Platt, B.I. 1988. Maximizing efficiency in dense medium plants. Presented at the Third Samancor Symposium on Dense Medium Separation, Vereeniging, South Africa, February 29–March 1.

Rhodes, D., Hall, S.T., and Miles, N.J. 1993. Density separation in heavy inorganic liquid suspensions. In *Proceedings of the XVIII International Mineral Processing Congress.* Melbourne, Victoria: Australasian Institute of Mining and Metallurgy. pp. 375–377.

Robertson, A.C., and Williams, J.D. 1991. Magnetite production for the coal industry. In *Proceedings of the Queensland Coal Symposium.* Melbourne, Victoria: Australasian Institute of Mining and Metallurgy. pp. 83–88.

Rudman, I. 2000. The coal plant of the 21st century—A contractors view. In *Colloquium: Coal Preparation.* Johannesburg: Southern African Institute of Mining and Metallurgy.

Rylatt, M.G., and Popplewell, G.M. 1999. Diamond processing at Ekati in Canada. *Min. Eng.* (February):19–25.

Sanders, G.J. 2007. *The Principles of Coal Preparation,* 4th ed. Broadmeadow, New South Wales: Australian Coal Preparation Society.

Schena, G.D., Gochin, R.J., and Ferrara, G. 1990. Preconcentration by dense-medium separation—An economic evaluation. *Trans. Inst. Min. Metall.* (January-April):C21.

Scott, I.A., and Napier-Munn, T.J. 1992. Dense medium cyclone model based on the pivot phenomenon. *Trans. Inst. Min. Metall.* 101:C61–C76.

Sepro Mineral Systems. 2018. *Sepro Condor Dense Medium Separators.* Rev. 2. Langley, BC: Sepro Mineral Systems. https://seprosystems.com/wp-content/uploads/2016/09/Sepro_Condor_DMS_2018.pdf.

Shah, C.L. 1992. Larcodems separator—Development of three-product unit. In *Hydrocyclones: Fluid Mechanics and Its Applications.* Vol. 12. Dordrecht: Springer.

Stewart, K.J., and Guerney, P.J. 1997. Detection and prevention of ferrosilicon corrosion in dense medium plants. In *Proceedings of the 6th Mill Operators Conference.* Melbourne, Victoria: Australasian Institute of Mining and Metallurgy. pp. 177–183.

Stortz, A.F., and Houston, D.W. 1963. Fine coal cleaning at Glen Alden. *Min. Congr. J.* 44(5):34–46.

Stratford, K.J., and Napier-Munn, T.J. 1986. Functions for the mathematical representation of the partition curve for dense medium cyclones. In *Proceedings of the 19th Application of Computers and Operations Research in the Mineral Industry.* Littleton, CO: SME. pp. 719–728.

Symonds, D.F. 1986. Selection and sizing of heavy media equipment. In *Design and Installation of Concentration and Dewatering Circuits.* Littleton, CO: SME.

Symonds, D.F., and Malbon, S. 2002. Sizing and selection of heavy media equipment: Design and layout. In *Mineral Processing Plant Design, Practice, and Control.* Edited by A.L. Mular, D.N. Halbe, and D.J Barratt. Littleton, CO: SME. pp. 1011–1032.

Tarjan, G. 1974. Application of distribution functions to partition curves. *Int. J. Miner. Process.* 1:261–265.

Terra, A. 1938. Theory of washing coal. *Revue Industrie Minerale* 425:383.

Tromp, K.F. 1937. New methods of computing the washability of coals. *Coll. Guard.* 154:955–959, 1009.

UAF (University of Alaska–Fairbanks). 2018. Mining Mill Operator Training: AMIT 145: Lesson 3, Dense Medium Separation. https://millops.community.uaf.edu/amit-145/amit-145-lesson-3/. Accessed September 2018.

Walker, M.S. 1985. A new method for the commercial separation of particles of differing densities using magnetic fluids. In *Proceedings of the XV International Mineral Processing Congress.* Saint-Étienne, France: GEDIM. pp. 307–312.

Wilkes, K.D. 2006. The practical application of the Wemco HMS drum separator in the mining and metals recycling industries. In *Proceedings of the SAIMM Present Status and Future for DMS Gravity Concentration in the South African Mining Industry Conference.* Johannesburg: Southern African Institute of Mining and Metallurgy.

Williams, R.A., and Kelsall, G.H. 1992. Degradation of ferrosilicon media in dense medium separation circuits. *Miner. Eng.* 5(1):57.

Wills, B.A. 1988. *Wills' Mineral Processing Technology*, 4th ed, Oxford: Pergamon Press.

Wills, B.A., and Lewis, P.A. 1980. Applications of the Dyna Whirlpool in the minerals industry. *Min. Mag.* (September):255.

Wills, B.A., and Napier-Munn, T.J. 2005. *Wills' Mineral Processing Technology*, 7th ed. Amsterdam; Boston, MA: Elsevier.

Wood, C.J., Davis, J.J., and Lyman, G.J. 1987. Towards a medium behaviour based performance model for coal-washing DM cyclones. In *Proceedings of the Dense Medium Operators' Conference.* Melbourne, Victoria: Australasian Institute of Mining and Metallurgy. pp. 247–256.

CHAPTER 6.4

Magnetic Separation

Daniel A. Norrgran and Michael J. Mankosa

The science of magnetic separation has experienced exceptional technological advancements in the last three decades. As a consequence, new applications and design concepts in magnetic separation have evolved. This has resulted in a wide variety of highly effective and highly efficient magnetic separator designs, ranging a wide spectrum of industrial applications.

Magnetic separation may be used as an integral part of the primary process or as a secondary or scavenger operation to produce a mineral concentrate. It may also be used in applications involving the removal of tramp metal prior to crushing and grinding systems.

The applications of magnetic separation are diverse and employed in many different industries. It can be used to remove tramp metal, such as shovel teeth or chain in a crusher line, or micrometer-sized iron of abrasion from a high-purity industrial minerals process stream. It may be used to selectively concentrate magnetic components such as magnetite or hematite minerals in an iron ore concentrator. It is also used in removing deleterious magnetic elements to purify nonmagnetic elements such as in the manufacture of ceramics or refractories. In addition, recycling and secondary recovery applications are being developed with increasing interest using magnetic separation as an explicit method for recovering residual values from various process streams and hazardous constituents from waste streams.

MINERAL CHARACTERISTICS

The concept of magnetic separation is based on the ability to magnetize a particular mineral and then physically collect it. The magnetic susceptibility of the mineral is an inherent characteristic directly proportional to the response of a magnetic field and is the single most important variable when addressing the characteristics of magnetic separation.

When subjected to a magnetic field, all particles respond in a particular manner and can be classified as one of three groups: ferromagnetic, paramagnetic, or diamagnetic. Minerals that have a very high magnetic susceptibility and are strongly induced by a magnetic field are ferromagnetic. Following are the few common ferromagnetic minerals that respond to magnetic separation:

- Magnetite
- Titaniferous magnetite
- Monoclinic pyrrhotite
- Martite

Minerals that have a low magnetic susceptibility and a weak response to a magnetic field are termed *paramagnetic*. Following are several common paramagnetic minerals that respond to magnetic separation:

- Biotite
- Chalcopyrite
- Chromite
- Columbite
- Garnet
- Geothite
- Hematite
- Hexagonal pyrrhotite
- Ilmenite
- Monazite
- Siderite
- Specularite
- Staurolite
- Tantalite

Minerals with a negative magnetic susceptibility are termed *diamagnetic* and for all practical purposes are nonmagnetic. Ferromagnetic and, to a lesser extent, paramagnetic materials become magnetized when placed in a magnetic field. The amount of magnetization induced on the particle depends on the mass and magnetic susceptibility of the particle and the intensity of the applied magnetic field. This can be expressed as

$$M = m\chi H$$

where M is the induced magnetization of the particle, m is the mass of the particle, χ is the specific magnetic susceptibility, and H is the magnetic field intensity.

Daniel A. Norrgran, Former Manager Minerals & Materials Processing Division, Eriez Manufacturing, Erie, Pennsylvania, USA
Michael J. Mankosa, Executive Vice President of Global Technology, Eriez Manufacturing, Erie, Pennsylvania, USA

MAGNETIC UNITS

The subject of magnetics is complicated by the lack of standardization throughout the scientific community on the units of magnetic quantities. The centimeter–gram–second (cgs) system was once prevalent and is still commonly encountered today. There has been, however, a movement to a single comprehensive system known as the International System of Units, abbreviated SI (from the French *Système International d'Unités*), which was officially adopted in 1960. Nearly all modem physics and electrical engineering textbooks as well as technical journals have adopted SI units.

Unfortunately, the conversion between SI units and cgs units is not straightforward. When a choice is to be made between SI units and cgs units, the choice is usually made on the basis of convenience. Because this chapter is for process engineers with a wide range of backgrounds, where appropriate, both systems of units are provided. Magnetic quantities and the associated system of units are a topic in itself. The intent is to provide the most basic understanding of the common units used in magnetic separation.

By convention, magnetic separators are rated on the magnetic induction or flux density. The magnetic induction or flux density is defined by the flux passing through a unit area normal to the direction of the flux. The unit of flux density in the SI system is Weber per square meter, but the term *tesla* (T) has now been adapted. The unit in cgs is measured in gauss (G), or the lines of flux passing through a square centimeter. The following equation can be used to convert the magnetic induction or flux density from SI units to centimeter–gram–second units:

$$H_{cgs} = H_{SI}(10^{-4})$$

or 1 G = 0.0001 T, and conversely, 1 T = 10,000 G.

MAGNETIC SEPARATOR CHARACTERISTICS

In the design of a magnetic separator, the magnetic field intensity and the magnetic field gradient are the two first-order variables that affect separation response. High-intensity magnetic separators typically operate in regions more than 5,000 G (0.5 T). Low-intensity separators are commonly referenced as those separators generating a magnetic field strength of less than 2,000 G (0.2 T), which is the practical limitation of conventional permanent ferrite magnets. The *magnetic field gradient* refers to the rate of change or the convergence of the magnetic field strength (Figure 1).

Case A has a very uniform pattern of flux lines without gradation. A magnetic particle entering this field is attracted to the lines of flux and remain stationary without migrating to either pole piece. Case B illustrates a converging pattern of flux lines displaying a high gradient. As these lines pass through a smaller area, there is a significant increase in the magnetic field intensity. A magnetic particle entering this field configuration is not only attracted to the lines of flux but also migrates to the region of highest flux density, which occurs at the tip of the bottom pole piece. This illustrates the methodology for magnetic separation. In simplified terms, the magnetic field intensity holds the particle while the magnetic field gradient moves the particle.

From the earlier equation for magnetization, the magnetic attractive force acting on a particle is the product of the particle magnetization and the magnetic field gradient and can be expressed as

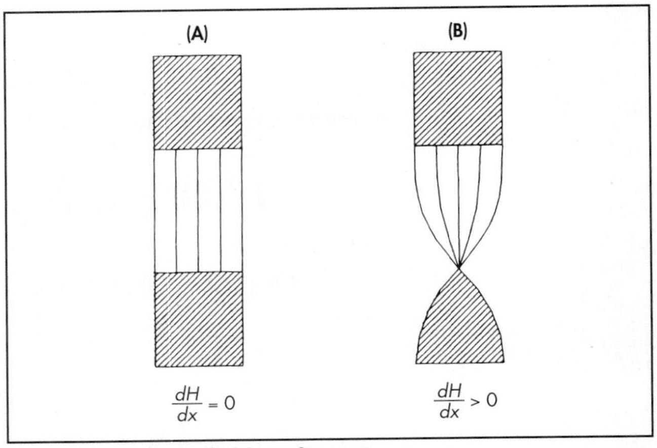

Figure 1 Magnetic field configurations

$$Fm = m\chi H\frac{dH}{dx} \text{ or } M\frac{dH}{dx}$$

where Fm is the magnetic attractive force, and dH/dx is the magnetic field gradient. Maximum magnetic dx force results only when both the magnetic field intensity and field gradient are maximized.

Two common methods for producing a magnetic gradient in a magnetic separator exist. The first method, which is typical of magnetic circuits utilizing permanent magnets, is to concentrate the lines of flux on a steel pole piece within the circuit. This can be simply accomplished by placing a steel pole piece between two magnets. The magnetic flux is concentrated in the steel pole piece, resulting in an area of extreme magnetic field intensity.

The second method involves positioning a steel matrix, such as a metal mesh, directly in a uniform magnetic field generated by an electromagnetic solenoid coil. This matrix consequently amplifies the magnetic field and converges the lines of flux to produce localized regions of extremely high magnetic field intensity.

MAGNETIC FIELD GENERATION

Either permanent magnets or an electromagnet generates the magnetic field on any given magnetic separator. When addressing the historical aspects of magnetic separation, a strong differentiation exists between the developments of those separators utilizing permanent magnets and those separators using electromagnets. The two separators were developed independently and oftentimes in competition.

Ferrite Permanent Magnets

Two distinct types of permanent magnets are utilized in magnetic separators. The first type of permanent magnet is a *ferrite* magnet and is used in low-intensity magnetic separators. The formulation is $SrFe_{12}O_{19}$. Ferrite is an inexpensive material with a moderate energy product ranging up to 32 kJ/m^3, or 4 MGOe (mega-gauss-oersted; Supermagnete, n.d.). The energy product of a permanent magnet is a relative measure of the intrinsic strength. Ferrite magnets are primarily used in drum-type separators that are used for collecting ferrous material as well as magnetite and pyrrhotite. These separators generate a magnetic field strength ranging up to 2,000 G (0.2 T).

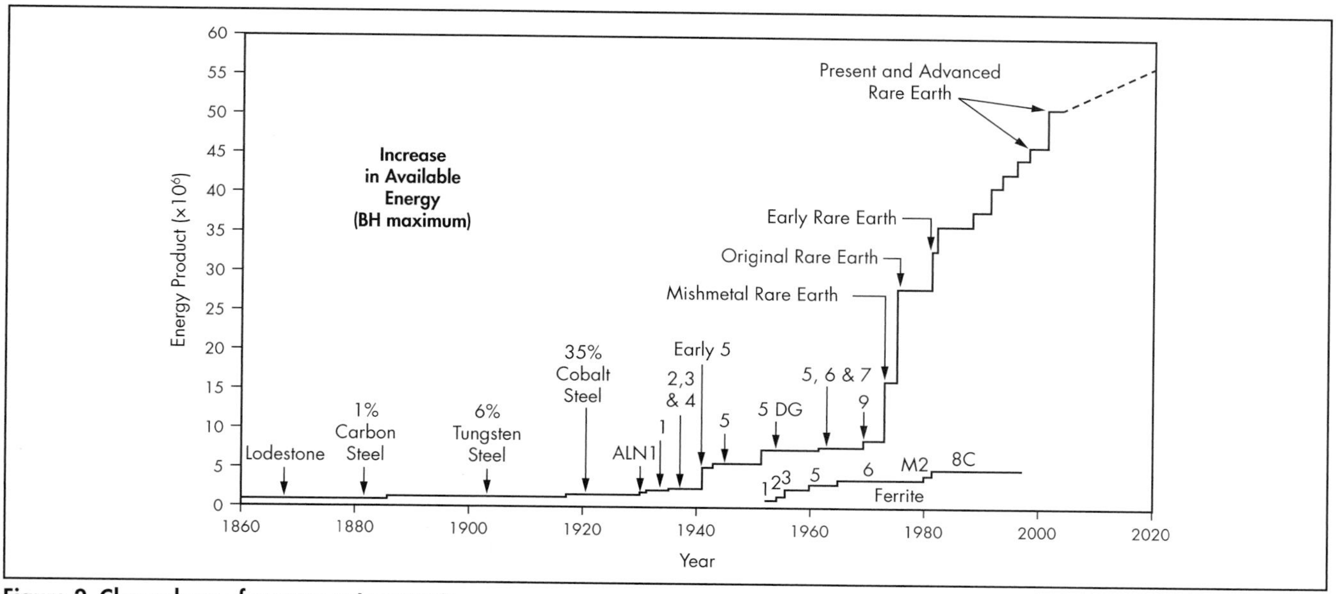

Figure 2 Chronology of permanent magnets

Rare Earth Permanent Magnets

The development of rare earth permanent magnets has revolutionized the field of magnetic separation. The advent of rare earth magnets has allowed for the design of high-intensity magnetic circuits operating energy free and surpassing the strength and effectiveness of electromagnets. Samarium and neodymium are the two most common elements used in the commercial manufacture of rare earth permanent magnets. The intermetallic compound $Nd_2Fe_{14}B$ is the third generation of rare earth magnets and is most prevalent today.

Neodymium magnets are experiencing tremendous growth. Much of the growth is strictly caused by the economics of using a much more abundant rare earth (neodymium is 10 times more abundant than samarium), coupled with inexpensive iron and boron. The energy product of the neodymium-based magnets now ranges up to 422 kJ/m^3, or 53 MGOe (Supermagnete, n.d.). Rare earth magnets are used in high-intensity drum- and roll-type separators that are effective for collecting paramagnetic minerals. Dependent on the magnetic circuit, these separators generate a magnetic field strength ranging up to 24,000 G (2.4 T). Presented in Figure 2 is a chronology of permanent magnets illustrating the increase in energy product.

Electromagnets

Industrial electromagnetic separators are typically designed utilizing a solenoid electromagnetic coil. Some separators use the bore of the solenoid coil as the separating zone. Other separators use the solenoid coil to convey the magnetic flux through a steel circuit or a C-frame circuit. The magnetic field in the gap, either the bore of the solenoid or the gap of the C frame, serves as the separation zone in a magnetic separator. Several operational factors have to be taken into consideration with a production magnetic separator designed for continuous use. There has to be a balance between the magnetic field strength, the associated steel circuit, the power requirements, and cooling factors.

Production-scale electromagnetic separators always employ a steel circuit to convey the magnetic flux and to pattern the magnetic field in the separating zone. The steel circuit tremendously enhances the magnetic field strength in the separating zone. The power input to the separator must also be designed to avoid saturating the steel circuit and overheating the electromagnet. Typically, electromagnetic separators operate up to approximately 20,000 G (2 T). At this magnetic field strength, the iron circuit begins to saturate.

MAGNETIC SEPARATION SYSTEMS

Both electromagnets and permanent magnets are used in separation systems. The following sections provide more detailed information.

Magnetic Field Generation

Two distinct methods can be used to generate a magnetic field. The first method, which has a long history, is the use of an electromagnet. Electromagnets employed in magnetic separators are almost exclusively solenoid coils. The second method is the use of a permanent magnet. Both methods of magnetic field generation have application in the advanced magnetic separation techniques of today. Each type of magnet, with its associated circuit design, has many parameters of distinction. In most instances, a specific type of magnetic circuit is selected to accommodate the application.

Mode of Separation

Two basic modes of magnetic separation systems exist. The first mode is the use of magnetic separation for the removal of tramp iron from a process stream. The second mode is the use of magnetic separation as an integral unit operation in the recovery process.

Tramp Metal Removal

The removal of tramp metal from a process stream is the most common application of magnetic separation. This application is straightforward and well documented. Following is an overview of this methodology. The basic premise is that tramp metal or ferrous contamination is present to some extent in all bulk materials. This contamination must be removed to (1) protect downstream equipment from potential damage, (2) ensure product quality, and (3) protect against fire or explosions in the process line.

Tramp metal may commonly originate in the following forms:

- Metal parts from earthmoving equipment
- Wire, cable, chain, reinforcing bar, or support structures from mining or quarrying operations
- Nails, clips, strapping, or other hardware from shipping containers
- Welding rod, nuts, bolts, and washers, or other hardware inadvertently introduced from maintenance and repair procedures
- Rust, chips, and fine iron of abrasion continually eroding from pipe lines, chutes, bins, and process equipment

When there is only an occasional piece of tramp metal or a very low level of ferrous contamination, a manual clean magnetic separator suffices. These separators are stationary systems utilizing either permanent or electromagnets to generate the magnetic field. The magnet effectively captures the tramp metal and holds it until it is manually removed.

A relatively large amount of ferrous contamination present in the process stream necessitates more frequent removal of the collected ferrous tramp. At this point, a self-cleaning separator such as a magnetic drum or a magnetic pulley is recommended.

Integral Separation

Magnetic separation as an integral unit process is used in many industries. The most notable applications are minerals, abrasives, ceramics, and secondary recovery processing. In these types of applications, the magnetic fraction and the nonmagnetic fraction may occur as a homogeneous blend. The magnetic separator is therefore subjected to continuous operation performing a continuous separation. In one case, the magnetic fraction can represent the product, and upgrading can take the form of the magnetic collection and recovery of the magnetic minerals. In another case, the nonmagnetic fraction can represent the product and upgrading can take the form of the magnetic removal of deleterious iron-bearing constituents from nonmagnetic minerals.

In some applications, the magnetic fraction represents the value. This is most prevalent in the concentration and recovery of iron-bearing magnetic minerals such as magnetite (Fe_3O_4) and hematite (Fe_2O_3). Iron ore concentrators typically treat thousands of tons per hour of milled ore, recovering the magnetic iron-bearing minerals from the nonmagnetic host rock.

Magnetite and ferrosilicon (FeSi) are commonly recovered from heavy media operations using magnetic separators. Other common magnetic minerals of value that are treated on large economies of scale are ilmenite ($FeTiO_3$) and chromite ($FeCr_2O_4$).

In other applications, the nonmagnetic fraction represents the value. The most notable application is the magnetic cleaning of industrial minerals such as silica, quartzite, feldspar, nepheline syenite, and fluorspar used as glass batch feedstock material. In these types of applications, the product has to essentially be free of iron-bearing particles. Increasingly, there is also a demand for higher-purity feedstock materials used in the manufacture of items such as specialty ceramic components, insulators, substrates, refractories, electrical components, specialty glasses, optical fibers, and grinding and polishing compounds. The brightness of kaolin clay used for paper coating is increased by removing paramagnetic contaminants.

The integral unit process necessitates a self-cleaning separator. Magnetic drums, pulleys, and roll-type separators are widely used as well as high-intensity and high-gradient matrix-type separators. For all practical purposes, magnetic separator selection can be reduced to the decision tree illustrated in Figure 3. The first node indicates that the separation is either tramp metal removal or an integral separation. The integral separation has subsequent nodes to indicate if the mineral is to be treated as either a slurry or as dry and free-flowing and the magnetic field strength necessary for effective separation.

MAGNETIC SEPARATORS FOR TRAMP METAL COLLECTION

Stationary, suspended, and trunnion permanent magnet separators are all used for tramp metal collection. Details of each of these magnets are provided in the following sections.

Stationary Permanent Magnetic Separators

The most enduring design incorporates simplicity. This is certainly the case when addressing stationary permanent magnetic separators. These separators, specifically plates, grates, and traps, represent the first industrial application of magnetic separation and perhaps the most prevalent type of separator in use today. These separators perform very well in removing tramp metal from a process and protecting downstream equipment as well as ensuring product quality.

Stationary permanent magnetic separators are configured with the common names of *plate*, *grate*, or *trap* and are illustrated in Figures 4, 5, and 6. These are simply enclosed barium ferrite or rare earth permanent magnets. The process stream flows over, around, or through the permanent magnets, and ferrous tramp metal is collected and held. For example, plate magnets are permanent magnets in a stainless-steel fixture and installed in a chute. Grate magnets are permanent magnets in stainless-steel tubes. Material flows around the tubes when placed in a dry process stream. Although trap magnets are also permanent magnets in stainless-steel tubes, this type of separator is placed in line with a slurry stream and the slurry flows around the tubes.

The prevalence of stationary permanent magnets can be attributed to several characteristics. They provide inexpensive insurance in the removal of tramp metal, which protects downstream equipment, such as mills, crushers, shredders, granulators, mixers, and fine screens, from potential damage. They are relatively low cost compared to most other types of magnetic separators. Permanent magnets are used to generate the magnetic field and require no utility for operation: There is no operating cost. Maintenance costs are virtually eliminated because there are no moving parts.

Suspended Magnetic Separators

When addressing the magnetic collection of tramp metal, the suspended electromagnet (SE) is undoubtedly the industry workhorse. Figure 7 shows a standard manual-cleaning SE; this separator has a deep magnetic field to effectively remove tramp ferrous iron. The SE is a widely used magnetic separator.

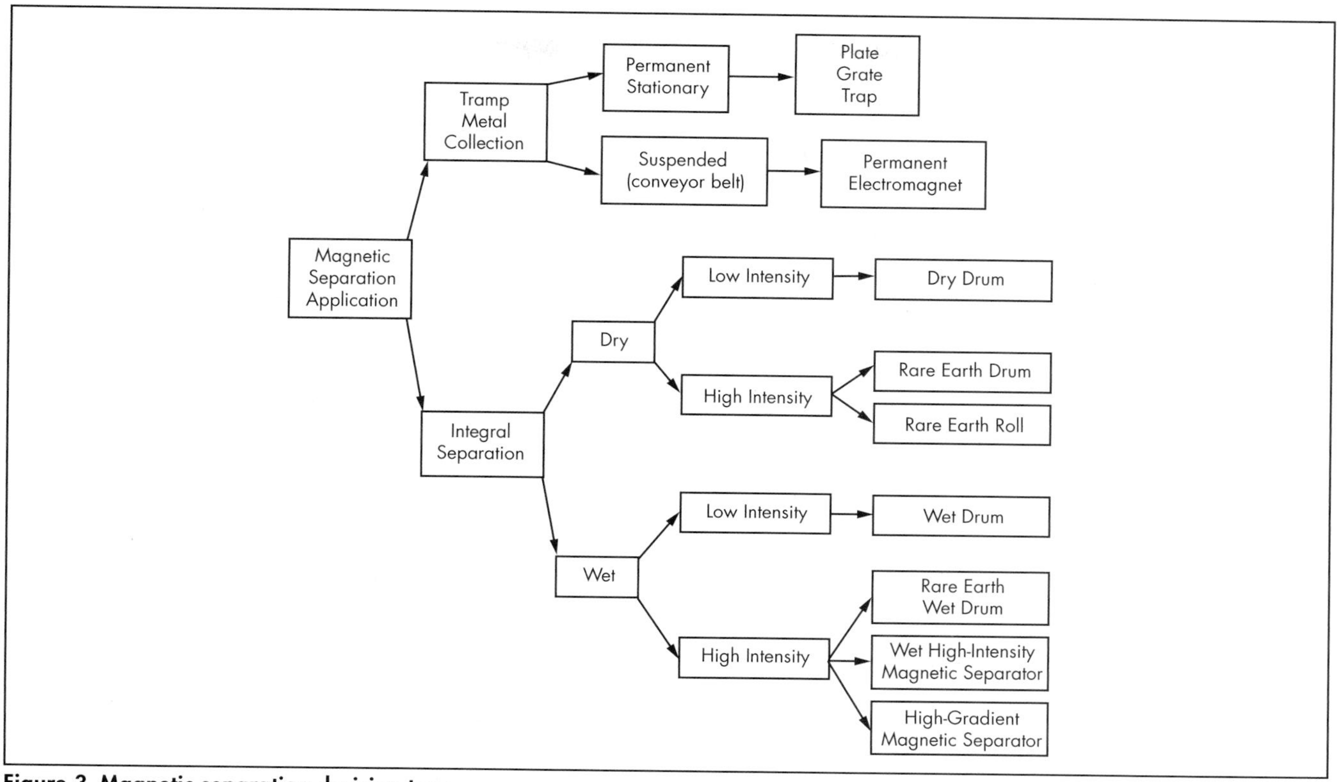

Figure 3 Magnetic separation decision tree

Figure 4 Plate magnet

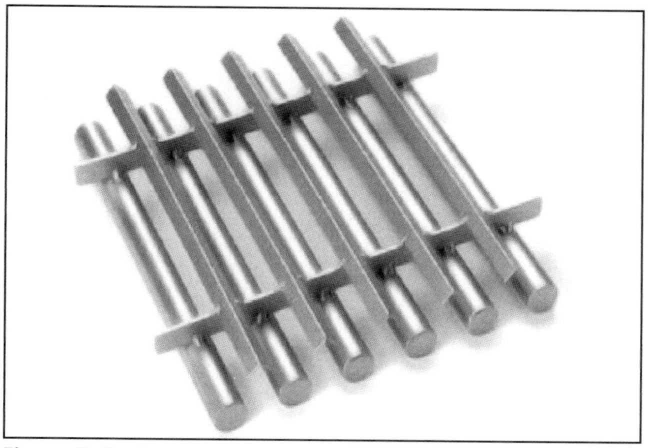

Figure 5 Grate magnet

The electromagnet is mounted or suspended over a conveyor belt to remove relatively large pieces of tramp metal that represent a potential hazard to downstream crushers, mills, and grinders. The SE consists of an aluminum conductor coil with a cylindrical steel core. The coil magnetically induces the steel core, which in turn configures the magnetic field for the collection of tramp metal from a moving conveyor belt. The coil is housed in a steel box that has the dual purpose of protecting the coil and completing the steel circuit. The coil is submerged in an oil bath within the housing to provide convection cooling. A rectifier is used to supply direct current to the magnet. Specifications of selected SEs are shown in Table 1.

Suspended Magnetic Separator Applications

The largest market for SEs is in coal mining, hard-rock mining, and aggregate products for removing shovel teeth, cable, tools, and bolting prior to crushing and grinding. The foremost factor in SE selection is the burden depth of the material on the conveyor belt. The burden depth accounts for the belt speed, belt width, capacity, and bulk density and determines the suspension height of the magnet and consequently the effective magnetic field strength at the belt surface. SEs are mounted in one of two positions over a conveyor belt, as shown in Figure 8. In position 1, the magnet is mounted directly over the conveyor belt prior to the head pulley and treats the entire

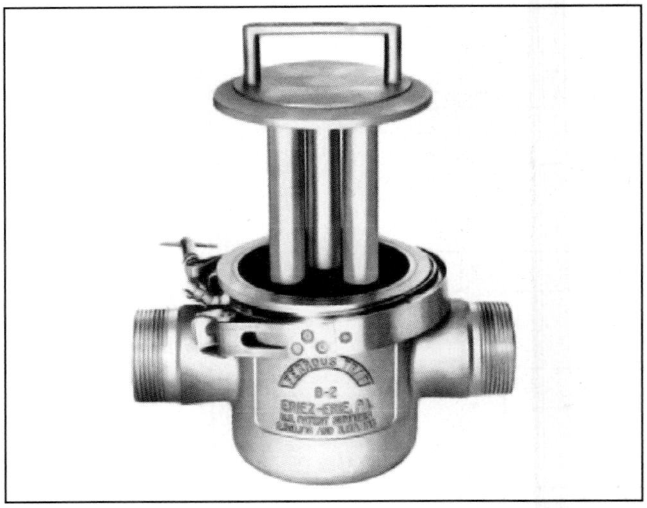

Figure 6 Trap magnet

Figure 7 Suspended electromagnet

material burden on the belt. This position requires higher magnetic field strengths to attract the ferrous component, shift the direction of momentum, and pull it through the bed of material.

In position 2, the magnet is mounted just over the stream of material leaving the head pulley and treats the material as it discharges from the belt. This position utilizes the full potential of the magnet as it reacts with the material in suspended trajectory. Tramp metal is easily pulled through the suspended burden. Further, the flow of material is directed toward the magnet face. Collection of the tramp metal from the material flow does not necessitate a change in direction. At conveyor belt speeds of less than 100 m/min, the suspended trajectory of the material is minimal and becomes near vertical. In this case, the magnet must be shifted to a position approaching directly over the head pulley.

SEs are available in either a manual cleaning style or a self-cleaning style. Manual-cleaning magnets are best suited where only small amounts or occasional pieces of tramp metal are encountered. The manual-cleaning magnets must be periodically turned off to remove tramp iron accumulation from the magnet face.

Self-cleaning magnets employ a cross-belt running around the magnet face to provide continuous removal of

Table 1 Suspended electromagnet specifications

Belt Width, m	Weight, kg	Power, W
0.9	823	3,800
1.2	1,916	6,300
1.5	3,731	9,940

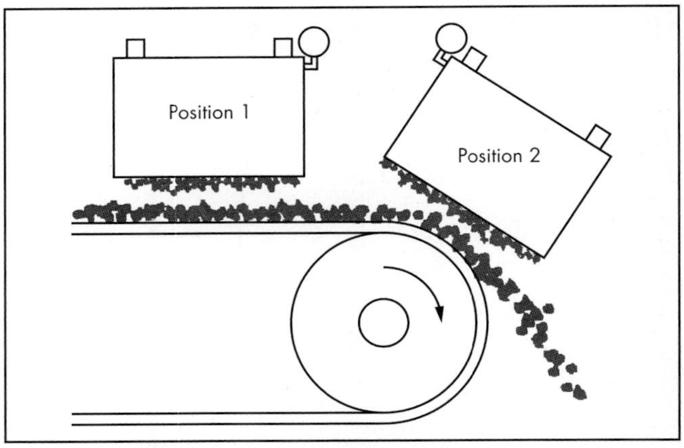

Figure 8 Positioning of suspended magnet

collected tramp iron. When tramp metal is attracted to the magnet face, the cross-belt intercepts it and discharges it to the side away from the conveyor belt. The self-cleaning magnet is best suited where a high level of tramp iron or large pieces of tramp iron are anticipated. The cross-belt is continuously driven around the magnet on a system of four pulleys driven with a small gearmotor.

In most applications, the magnet is suspended 5–8 cm over the top of the burden. Experience dictates that the operating magnetic field strength or the magnetic field strength at the belt surface should be 400–600 G (0.04–0.06 T) for adequate ferrous collection. As the magnet width increases to accommodate the belt width, the magnetic field strength increases at any given distance. A wider magnet simply has a larger core and coil.

Exceptions to the sizing of the SE exist. The response of the magnet is influenced by the following factors:

- **Belt speed.** As the belt speed increases, it becomes increasingly difficult to attract and collect ferrous components. This is especially difficult when the magnet is situated in position 1 where the momentum of the ferrous component has to change direction.
- **Burden depth.** As the burden depth increases, an increase in the magnetic field strength is required. A ferrous component situated on the surface of the belt, buried under a heavy burden, requires an increased magnetic attractive force for collection.
- **Size of ferrous component.** The suspension height of the magnet must be increased if relatively large ferrous components are anticipated. The suspension height of the magnet, with the ferrous component attracted to the face, must clear the burden on the conveyor belt and not impede the flow of material. The higher suspension height requires a stronger magnet to maintain the magnetic field strength at the working distance.

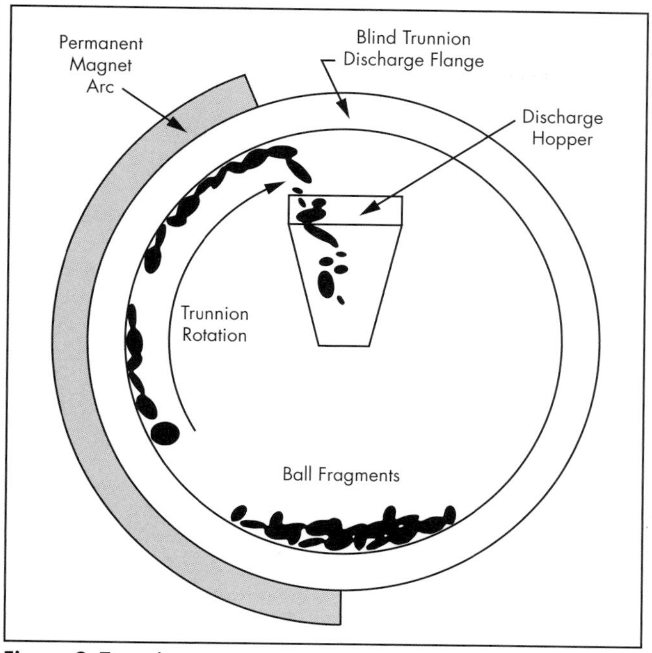

Figure 9 Trunnion magnet

Figure 10 Drum-type magnetic separator treating garnet concentrate

- **Shape of ferrous component.** The shape of the ferrous component is also a factor. Steel plate, for example, has a high surface area relative to its weight. This configuration reacts with a high magnetic attractive force when exposed to a magnetic field. In contrast, a sphere has the lowest surface area relative to its weight. This configuration reacts with a minimal magnetic attractive force.

As conveyor belts are increasing in width and speed to provide greater capacity, the magnets have had to increase in size and strength. Conventional suspended magnets weighing as much as 88 t (metric tons) are in service. For extremely demanding duty at a Chinese coal port, suspended magnets using superconducting coils have been installed. These generate a magnetic field of 4,000 G (0.4 T) at 500 mm from the magnet face. Protection of the pebble crusher commonly used in semiautogenous grinding mill comminution circuits is a typical application for self-cleaning SEs. The large grinding balls and chips can quickly damage the cone crusher.

Permanent suspended magnets are also available where operational conditions suffice. The advantage of a permanent suspended magnet is that it operates energy free without peripheral equipment. The drawback of a permanent magnet is that it cannot generate the depth of magnetic field required for treating high burden depths on the conveyor belt. The range of the burden depth is from 13 to 23 cm corresponding to a maximum suspension height of 20–30 cm. At these operating distances, the permanent magnet rivals the electromagnet.

Trunnion Magnets

A more recent development is the trunnion magnet, illustrated in Figure 9. This specific design removes balls and ball chips contained in the discharge of a ball mill. Removal of this unwanted steel not only reduces the damage to the pumps and hydrocyclones but can also increase the capacity of the mill by reducing the volume of steel in the mill. The trunnion magnet consists of a barrel that is attached to the rotating ball mill discharge trunnion and a 180-degree arc of stationary permanent magnets. The magnets attract and hold the steel to the inside of the rotating barrel where they are conveyed out of the mineral flow to the top of the trunnion where they are released and fall onto a chute that discharges them into a waste container. Typically, several tons of ball chips are removed from the mill during the initial days after installation, but then the system stabilizes and the discharge of chips is reduced.

INTEGRAL SEPARATION

The following sections discuss magnetic separators for the concentration and recovery of minerals, also known as *integral separation*.

Magnetic Drum Separators

The magnetic drum separator consists of a stationary, shaft-mounted magnetic circuit completely enclosed by a rotating drum. The magnetic circuit is typically comprised of several magnetic poles that span an arc of 120 degrees or more. When material is introduced to the revolving drum shell (concurrent at the 12 o'clock position), the nonmagnetic material discharges in a natural trajectory. The magnetic material is attracted to the drum shell by the magnetic circuit and is rotated out of the nonmagnetic particle stream. The magnetic material discharges from the drum shell when it is rotated out of the magnetic field. A magnetic drum separator is shown in Figure 10.

Separation Variables

The magnetic attractive force generated by a drum-type separator is opposed by centrifugal force. The primary variables affecting separation efficiency are the magnetic field strength,

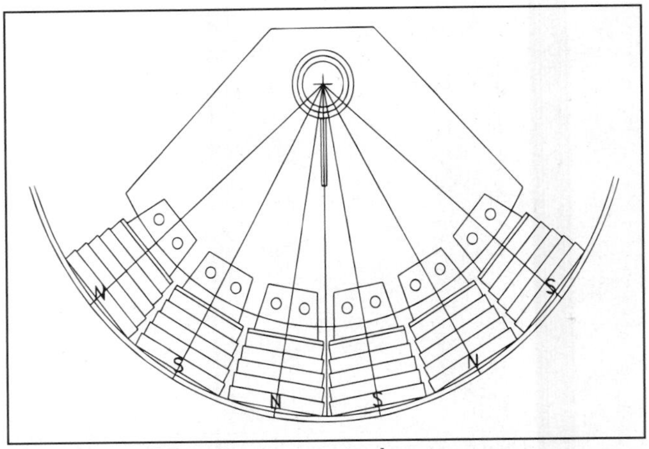

Figure 11 High-gradient magnetic element

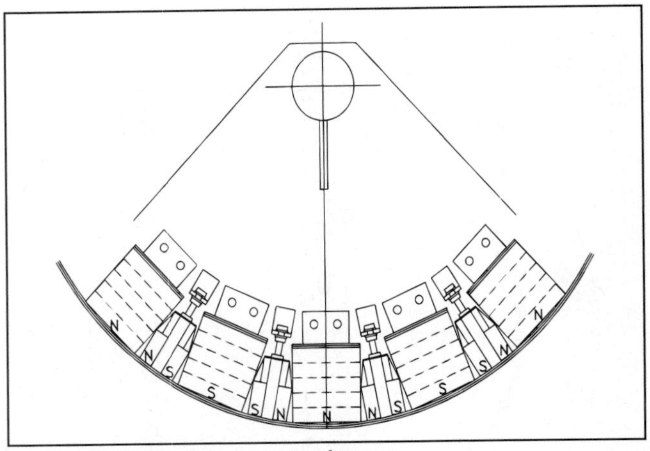

Figure 12 Interpole magnetic element

feed rate, linear speed of the separator surface, and particle size. An effective separation requires an equilibrium among the variables described as follows:

- A balance must be struck between an economic feed rate, product specifications, and recovery. As the feed rate increases, the layered particle bed on the separator surface increases in height, and the collection of magnetics decreases.
- The linear speed of the drum is also a primary variable related to the feed rate. As the linear speed is increased, the layered particle bed decreases in height, responding with an improved collection of the magnetic particles.
- The centrifugal force exerted by the drum or roll surface is the critical factor in providing separation. Beyond the critical speed, the centrifugal force overcomes the magnetic attractive force and the separation efficiency deteriorates.
- Particle size also affects separation efficiency independent of all other variables. Coarse particles provide a relatively high burden depth on the separator surface and respond with a relatively high magnetic attractive force.
- Coarse particles typically provide high unit capacities with high separation efficiencies. Fine particles with relatively low masses respond detrimentally to electrostatic forces. As a consequence, precise magnetic separations balancing magnetic forces against centrifugal forces deteriorate.

Dry Magnetic Separators

Dry magnetic separators include drum- and roll-type machines, both low and high intensity.

Low-Intensity Drum Type

Low-intensity drum magnetic separators are effective in collecting ferromagnetic materials and are used in various applications in the minerals processing industries. These separators combine the attributes of the permanent magnetic field with the self-cleaning feature. Magnetic drum-type separators are a staple in many mineral processing industries and are commonly used in two different modes of operation. The separator is equally effective in a treating a process stream for (1) concentrating and recovering a magnetic mineral and (2) rejecting magnetic minerals to produce a *clean* nonmagnetic mineral product.

An array of magnetic elements are available. Agitating magnetic elements are used in most integral applications where a clean magnetic product is desirable. The magnetic poles have alternating polarity to provide agitation. The agitating-type element minimizes the loss of nonmagnetics to the magnetic product (termed *misplacement*) because of physical entrapment.

In drum-type magnetic separators, two basic types of magnetic circuits are used. Figure 11 is a schematic of a low-intensity, high-gradient magnetic element consisting of alternating polarity magnetic poles. This element produces a high magnetic gradient near the surface of the drum, which measures 0.9 m in diameter.

The *high-gradient* element, as the name implies, is designed to produce very high magnetic field gradient and subsequently high attractive force. Several identical magnetic poles comprise the element. The poles are placed together, minimizing the air gap between them to produce the high surface gradient. Because of the high gradient, the attractive force is strongest closer to the drum, making it most effective when utilized with a relatively low material burden depth on the drum surface and lower unit capacity.

The interpole-style element uses a true ceramic *bucking* magnetic pole or *interpole* between each main pole. The magnetic field of the bucking elements are charged to oppose both of the adjacent main poles, resulting in increased magnetic field at depth. Therefore, the interpole elements allow relatively high material burden depths on the drum surface and thus higher unit capacity or improved separation efficiency in difficult applications. Figure 12 is a schematic of an interpole magnetic element. The interpoles oppose the polarity of the main poles to produce a deep magnetic field. Depending on the application, there are several variations of these two basic magnetic circuits. Table 2 presents general specifications of magnetic drum separators.

Magnetic drum separators incorporating either a radial or agitating magnetic element are commonly used to collect tramp metal from a mineral stream. The element typically contains large strontium ferrite poles to provide depth of field as well as a high surface gradient. This is necessary

Table 2 General specifications of magnetic drum separators

Drum Type	Attributes	Applications	Unit Capacity
Radial or agitating magnetic circuit for tramp metal collection	Radial or high-gradient agitating magnetic element using strontium ferrite magnets. Magnetic element designed to produce a deep magnetic field; 1,200 G (0.12 T) magnetic field strength on surface; 0.8 m/s drum speed; 0.4–0.8 m diameter.	Widely used drum throughout industries. Functional for tramp metal and iron of abrasion removal. Effective in removing large tramp metal such as nuts and bolts at high capacity. Effective in removing fine iron of abrasion at relatively low capacities.	Tramp metal applications can operate at high capacity; 0.4-m-diameter drum operates at 20–26 t/h/m width; 0.6-m-diameter drum operates at 30–40 t/h/m width. Reduce capacity by 50% for fine iron of abrasion.
Agitating magnetic circuit for coarse material up to 25 mm top size	High-gradient agitating magnetic element using strontium ferrite magnets; 1,200 G (0.12 T) magnetic field on drum surface; 10–15 magnetic poles with wide (8–10 cm wide) spacing. Magnetic element designed to produce a deep magnetic field. Up to 2.5 m/s drum speed; 0.9 m diameter.	Provides preconcentration of coarse ores up to 25 mm top size. Applications include magnetite recovery from iron ore and metallics recovery from slag.	Coarse minerals of 6–25 mm can generally be treated at high capacities. Unit capacities ranging up to 80 t/h/m width.
Agitating magnetic circuit for coarse material to 6 mm top size	High-gradient agitating magnetic element using strontium ferrite magnets. 1,100 G (0.11 T) magnetic field on drum surface; 20 magnetic poles with wide (5–8 cm wide) spacing. Magnetic element designed to produce a deep magnetic field. Up to 5.0 m/s drum speed; 0.9 m diameter.	Provides cleaning or concentration of minerals in the 100 mesh to 6 mm top size. Applications include cleaning industrial minerals such as alumina and silicate minerals; also magnetite recovery from iron ore and metallic recovery from slag.	Coarse minerals (6 mm) can be treated at unit capacities up to 65 t/h/m width. Minerals in the 100 mesh range can be treated at unit capacities up to 33 t/h/m width.
Agitating magnetic circuit for fine material (–100 mesh)	High-gradient agitating magnetic element using strontium ferrite magnets; 1,000 G (0.1 T) magnetic field on drum surface; 30–40 small magnetic poles with narrow (2–4 cm wide) spacing. Up to 7.6 m/s drum speed; 0.9 m diameter.	Provides cleaning or concentration of minerals in the –100 mesh size range. Applications include upgrading magnetite from milled preconcentrates, magnetite recovery from fly ash, and upgrading iron carbine.	High drum speed maintains a low burden depth, allowing unit capacities up to 33 t/h/m width.
Salient pole rare earth	Salient pole agitating magnetic element using rare earth magnets; 7,000 G (0.7 T) magnetic field on drum surface; 0.4, 0.6, 0.9, and 1.2 m in diameter. Magnetic element employs 5–12 magnetic poles spanning a 110-degree arc.	Effective in magnetically collecting an entire range of paramagnetic minerals such as hematite, specularite, and ilmenite. Effective in cleaning industrial minerals such as silica, quartzite, feldspar, nepheline syenite, alumina, and zircon.	Minerals in the 100 mesh range can be treated at unit capacities of 10–16 t/h/m width on a 0.4-m-diameter drum, 16–23 t/h/m width on a 0.6-m-diameter drum, and 26–33 t/h/m width on a 0.9-m-diameter drum.

to attract and hold ferromagnetic components when operating at a high burden depth. These separators can operate at high capacity when utilized for tramp metal such as nuts and bolts. A 0.6-m-diameter drum can operate at a unit capacity up to 40 t/h/m of drum width. The unit capacity is reduced by 50% for the collection of iron of abrasion.

Low-intensity drum separators for mineral separation applications combine a large-diameter drum of 0.9–1.2 m, with an agitating element and a high drum speed. This combination provides a high-capacity separation with a high level of precision. The separator typically incorporates an axial agitating (alternating polarity high-gradient style) magnetic element that spans 180–210 degrees. Strontium ferrite magnets are used for the collection and recovery of ferromagnetic materials. The high drum speed in combination with many agitating magnetic poles provides a high level of agitation as the ferromagnetics are transferred along the drum surface. This results in a near-complete rejection of nonmagnetic material, subsequently producing a very clean ferromagnetic concentrate.

The magnetic drum separators operate completely enclosed in housing. The housing incorporates a feed bin with an adjustable gate to deliver a controlled feed rate across the width of the drum. The housing also has magnetic and nonmagnetic product discharge chutes. The only utility necessary is the electrical input for the drive system. The parallel shaft gearmotor that drives the drum is mounted on the housing. The drive system uses a V-belt drive, which offers protection to both the motor and the drum shell in the event of a jam from oversize material. In many cases, the drive is controlled with a variable-frequency controller to allow variations in the drum speed. The 0.9- and 1.2-m-diameter drums typically have drum widths up to 3.0 m. The drive motor ranges from 3.7 to 18.6 kW, depending on the magnetic circuit configuration drum speed and width of the drum.

Coarse particles ranging in size from –25 mm to +6 mm respond well to a magnetic circuit employing large, wide magnetic poles. This magnetic circuit generates a *deep* magnetic field, effectively capturing coarse ferromagnetic or composite minerals. Coarse iron ore (magnetite) can be treated at a relatively high unit capacity. As an initial magnetic separation, the separator rejects a nonmagnetic product relatively free of iron while recovering the magnetite as a rougher concentrate. This concentrate represents an intermediate grade, containing locked particles, and requires milling to further liberate the iron minerals.

Crushed copper slag also responds well to this separator. The major component in this slag is iron-bearing fayalite, which constitutes approximately 90% of the feed. The fayalite is recovered as the magnetic fraction while the metallic copper is rejected as a nonmagnetic concentrate. Coarse slags and

Table 3 Applications and unit capacities of rare earth drums

Application	Feed Size	Drum Diameter, m	Unit Capacity, t/h/m of drum width
Preconcentrating hematite	–6 mm	0.6	66
Recovering specularite from a gravity concentrate	–35 mesh	0.6	13
Recovering final ilmenite product from heavy mineral concentrate	–35 mesh	0.4 0.6	10–13 13–16
Recovering high-iron ilmenite from low-iron ilmenite	–35 mesh	0.6	20
Recovering ilmenite from roasted concentrate	–4 mesh	0.9	33
Tabular alumina	–6 mesh +48 mesh	0.4	23
Cleaning tabular alumina	–48 mesh +325 mesh	0.4	16
Cleaning tabular alumina	–48 mesh	0.4	10
Cleaning silica and feldspar	–28 mesh	0.4 0.6	10–13 13–16
Cleaning garnet	–48 mesh	0.4	13–16
Collecting nickel bearing pyrrhotite from crushed nickel ore	–100 mm	0.9	66
Recovering metallic nickel from slag	–10 mm	0.9	40

iron ores are treated at a unit capacity of 80–115 t/h/m of drum width on a 0.9-m-diameter drum.

Fine particles in the size range of –100 mesh respond well to a magnetic circuit employing small, narrow magnetic poles. The magnetic circuit may consist of up to 40 separate magnetic poles providing a very high degree of agitation as the material is transferred through the magnetic field. Applications in these size ranges include iron ores as well as intermediate grade concentrates, slags, fly ash, ferrosilicon, and iron carbide. Iron ores and concentrates are treated at a unit capacity up to 33 t/h/m of drum width on a 0.9-m-diameter drum. Fly ash, which is very fine and has a low bulk density, is treated at a unit capacity of 16–25 t/h/m of drum width on a 0.9-m-diameter drum.

High-Intensity Drum Type

The high-intensity rare earth drum magnetic circuit is comprised of segments of alternating rare earth magnets and steel pole pieces. The steel poles are induced and project a high-intensity, high-gradient magnetic field. The peak magnetic field intensity on the drum is approximately 7,000 G (0.7 T) and is effective in removing many paramagnetic constituents.

The rare earth drum was developed to meet several operational objectives. Initially, the project focused on the dry treatment of ilmenite and hematite. It was anticipated that the separator could provide high capacity with relatively low capital and operating cost. Test work carried out on the prototype separator demonstrated that this design effectively recovers ilmenite from a heavy mineral concentrate and hematite from siliceous gangue. The unit capacities far exceed the typical capacities of cross-belts and induced magnetic roll separators. Currently, the applications utilizing the rare earth drum magnetic separator are diverse. Although many of the units are used primarily for the recovery and concentration of paramagnetic minerals, most are used for removing deleterious iron-bearing minerals from process streams such as alumina, silica, and feldspar. These various applications have confirmed that the rare earth drum is effective in treating a wide range of particle sizes and operates at relatively high capacities. Table 3 presents the unit capacities of specific applications using the rare earth drum magnetic separator.

High-Intensity Roll Type

The rare earth roll, generating peak magnetic field strengths in excess of 21,000 G (2.1 T), is very effective for concentrating or removing weakly magnetic minerals from a dry process stream. The rare earth roll magnetic separator is designed to provide peak separation efficiency and is typically used when a high-purity product is required. The roll is constructed of disks of neodymium–boron–iron permanent magnets sandwiched with steel pole pieces. The steel poles are magnetically induced to the saturation point of approximately 21,000 G (2.1 T). Magnetic roll diameters are typically 100 and 150 mm, although separators as large as 300 mm in diameter are available.

The machine is configured with a head pulley in the separator. A thin belt, usually measuring 0.125–0.50 mm (5–20 mil) thick, is used to convey the feed material through the magnetic field. When feed material enters the magnetic field, the nonmagnetic particles are discharged from the roll in their natural trajectory. The paramagnetic, or weakly magnetic, particles are attracted to the roll and are deflected out of the nonmagnetic particle stream. A splitter arrangement is used to segregate the two particle streams. The magnetic rolls are constructed up to 2 m wide. Typical feed rates to the separator range from 2 to 8 t/h/m of roll width. Coarse 6-mm material, such as chromite or hematite, can be treated at a very high rate, whereas fine material, such as silica, feldspar, nepheline syenite, or alumina, must be treated at lower rates. Sized silica sand can typically be treated at a rate of 4–6 t/h/m of roll width.

The separator can be configured with several magnetic rolls in series to provide a multiple-stage separation. Two to three magnetic rolls are typically placed in series (modular design) to provide optimum separation efficiency. When cleaning a nonmagnetic material, the rolls are usually arranged in a nonmagnetic repass configuration. The nonmagnetic material from the first stage of separation is delivered to the second roll to further remove residual paramagnetics. The rolls can also be arranged in a magnetic repass configuration to provide a cleaning stage of the magnetic fraction.

The rare earth roll is enclosed in a housing or framework. All product contact areas are fabricated from 304 stainless steel. The framework incorporates an electromagnetic feeder to provide a consistent feed rate across the entire width of the roll. A hopper is mounted over the feeder to contain and deliver incoming feed. Magnetic and nonmagnetic product discharge chutes are also used. A wide range splitter is used to segregate the magnetic fraction from the nonmagnetic product. A dust hood is mounted over each roll.

The only utility necessary is the electrical input for the electromagnetic vibratory feeder and the drive system. The vibratory feeder extends the entire width of the magnetic roll to provide a uniform distribution of feed. The vibratory feeder on a separator using a 1-m-wide magnetic roll draws ~0.8 kW. Each magnetic roll is driven independently. A 1-m-wide magnetic roll utilizes a 0.75-kW gearmotor. The roll speed is variable, controlled with a variable frequency drive. Belt speeds in a production mode typically range from 0.6 to 0.9 m/s. Treating silica sand on a 1-m-wide roll at a feed rate of 5 t/h and a belt speed of 0.8 m/s results in a burden depth of ~1 mm.

Wet Magnetic Separators

Wet magnetic separators include both low- and high-intensity drum-type machines.

Low-Intensity Drum Type

The wet drum magnetic separator is used extensively in iron ore and heavy media applications. It is also used for collecting magnetite or *black sands* from gold-bearing ores and tailings as well as pyrrhotite in sulfide mineral applications. The wet drum magnetic separator is well established with literally thousands of separators in current operation.

The wet drum magnetic separator consists of a rotating magnetic drum situated in a tank that receives the feed slurry. The magnetic drum consists of a stationary, shaft-mounted permanent magnetic circuit completely enclosed by a rotating drum. The magnetic circuit is comprised of segments of alternating permanent magnets that spans an arc of 120 degrees. The peak magnetic field intensity on the drum is approximately 2,000 G (0.2 T) and is effective in removing ferromagnetic minerals.

The magnetic drum is mounted in a tank. Slurry is fed to the tank and subsequently flows through the magnetic field generated by the drum. The magnetic minerals are attracted to the drum shell by the magnetic circuit and are rotated out of the slurry stream. The magnetic minerals discharge from the drum shell when rotated out of the magnetic field. Wet drum tanks are in three basic styles. Tank selection is based on the specifics of the application.

In a wet drum separator, the magnetic force acting on a ferromagnetic particle is predominately opposed by hydrodynamic drag force. This feature, when properly applied, provides the vehicle of separation, washing away the nonmagnetic particles while the ferromagnetic particles are collected in the magnetic field. The hydrodynamic drag force is also responsible for any losses of ferromagnetics.

Following are the variables that affect the collection of ferromagnetics in a wet drum magnetic separator:

- **Magnetic field strength.** The magnetic field strength must be sufficient to effectively collect ferromagnetic minerals.
- **Hydraulic capacity.** Ferromagnetic recovery is directly related to the flow rate through the separator. As the flow rate increases, the slurry velocity and, consequently, the fluid drag force increase, which tends to detach more magnetite particles from the opposing magnetic field.
- **Percent solids.** The percent solids of the feed directly affects the selectivity of the separation. As the percent solids increases, the slurry becomes more viscous, minimizing the effects of the fluid drag to assist in the separation of the silica.
- **Ferromagnetic content.** Any given wet drum magnetic separator has the characteristic of removing a limited amount of ferromagnetics based on the diameter of the drum, peripheral speed, and the magnetic field strength. This is referred to as the *magnetic loading*. Exceeding the limits of this magnetic loading results in increased magnetite losses.

Two distinct applications for wet drum magnetic separators exist. One application is the recovery of magnetite or ferrosilicon in a heavy media process. The other is the concentration and recovery of magnetite from iron ore. The use of wet drum magnetic separators has been well documented in each of these applications.

Drum Separators for Heavy Media Applications

Wet drum magnetic separators are extensively used for the recovery of fine magnetite in heavy media applications. The separator recovers the magnetite from a slurry stream toward the end of the gravity separation circuit and returns it to the heavy media separator.

Heavy media operations commonly employ wet drum magnetic separators, measuring 0.9 and 1.2 m in diameter, ranging up to 3.0 m in drum width. There can be two different tank styles—concurrent and counter-rotation. These tank styles are illustrated in Figures 13 and 14.

The drum rotates in the same direction as the slurry flow in the concurrent tank style. The slurry enters the feedbox and is channeled underneath the submerged drum. The slurry then flows into the magnetic field generated by the drum. The magnetite is attracted by the magnetic field, collected on the drum surface, and rotated out of the slurry flow. A shortcoming of this basic design is that any magnetite that is not immediately collected exits the tank with the nonmagnetics.

The counter-rotation wet drum tank style is preferred for heavy media applications. The drum rotates against the slurry flow in the counter-rotation tank style. The slurry enters the feedbox and flows directly into the magnetic field generated by the drum. The magnetite is attracted by the magnetic field, collected on the drum surface, and rotated out of the slurry flow. Any magnetite that is not immediately collected passes through to a magnetic scavenging zone. The short path that the magnetic material must be conveyed between the feed entry point and the magnetics discharge lip, combined with the magnetic scavenging zone, results in high magnetite recoveries.

Extensive sampling of wet drum magnetic separators in heavy media applications has indicated that the industry standard for magnetite losses is 0.25 g of magnetite per liter of nonmagnetic product. Figure 15 demonstrates the typical performance of a wet drum magnetic separator operating in a coal preparation plant employing a heavy media circuit. The wet drum magnetic separator is 0.9 m in diameter with a counter-rotation (Eriez Climaxx) style tank. This separator was isolated in the dilute heavy media circuit. The feed rate was incrementally increased with samples of the nonmagnetic

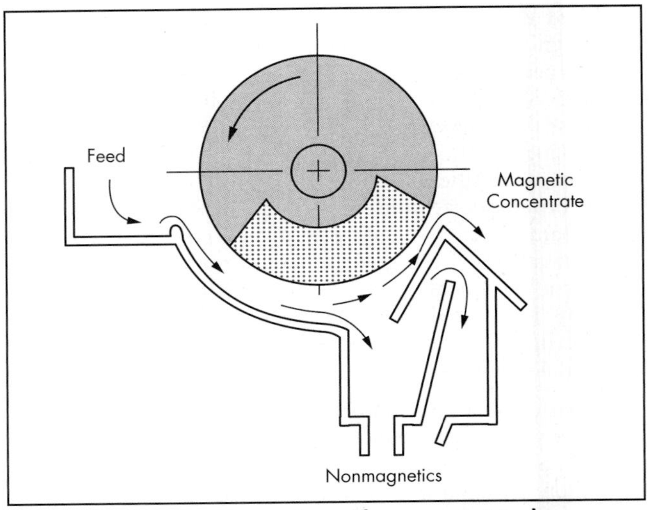

Figure 13 Wet drum separator with concurrent tank

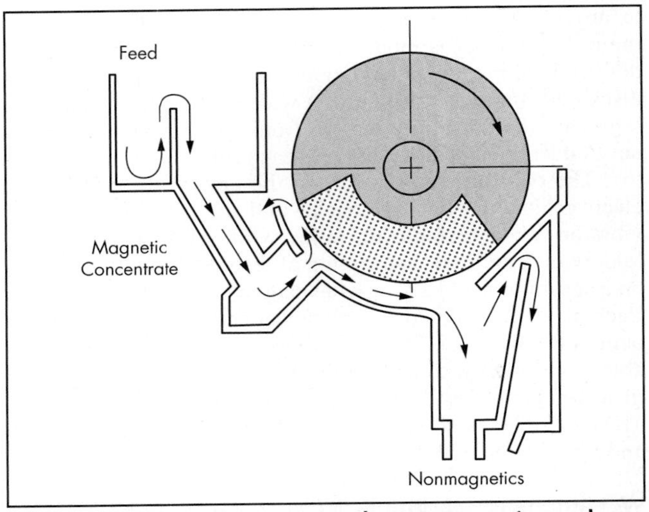

Figure 14 Wet drum separator with counter-rotation tank

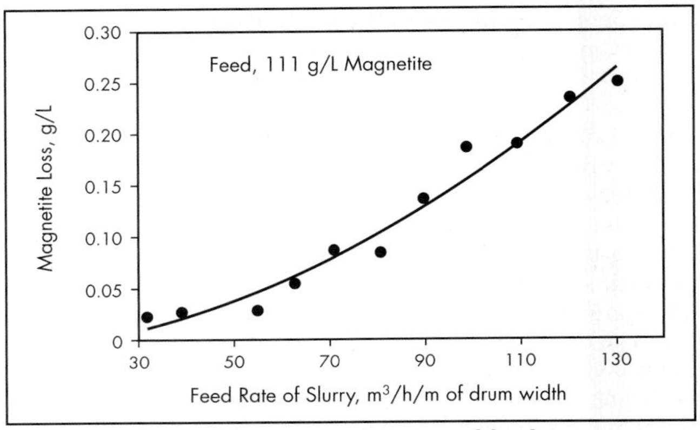

Figure 15 Magnetic losses as a function of feed rate

Table 4 Wet drum magnetic separator sizing parameters

Drum Diameter, m	Sizing Parameter	
	Hydraulic, $m^3/h/m$ of drum width	Magnetic Loading, t/h/m of drum width
0.9	90	16
1.2	110	24

product taken at each level. The magnetite level was measured in each nonmagnetic product sample.

The magnetite losses were then compared to the corresponding feed rate to the separator. The magnetite content of the feed was maintained at 111 g/L or 10.2% solids as magnetite. The graph demonstrates that the magnetite losses are relatively low throughout the range. The maximum magnetite losses of 0.25 g of magnetite per liter of nonmagnetic product occur at the maximum feed rate of 130 m^3 of slurry per meter of drum width.

Based on in-plant testing, general sizing parameters for heavy media wet drum magnetic separators have been established, as shown in Table 4. This is for a counter-rotation style tank and heavy media sized (grade E) magnetite feed slurry at less than 13% solids. The hydraulic capacity and the magnetic loading represent maximum sizing parameters. The separator should be sized on one of the parameters without exceeding the other parameter.

Drum Separators for Concentrating Applications

Wet drum magnetic separators are the most vital part of the upgrading process in magnetite concentration. The upgrading of primary magnetite is always accomplished with wet drum separators. Mill feed is typically upgraded to 65%–66% magnetic iron using a series of wet drum magnetic separators. The number of magnetic separation stages required to upgrade the ore is dependent on the magnetite content and the liberation characteristics of the ore.

Feed to the concentration process is crushed and milled in preparation for magnetic separation. The size distribution of the wet drum feed may have a top size of ~6 mm with the distribution extending down through 25 µm. The selected size distribution provides the minimum degree of liberation necessary for satisfactory magnetite recovery and silica rejection.

Upgrading the magnetite may require several stages of wet drum separation to achieve a final product while minimizing milling costs. To reject a substantial amount of weight prior to reprocessing, the initial magnetic separation or *cobber* stage is conducted on coarse milled ore. The concentrate is then reground to liberate the magnetite and subjected to additional cleaner and finisher stages of separation to produce a final concentrate.

Most cobber separators are double drums (concentrate re-treatment) to provide a two-stage separation and reject a maximum amount of weight. Counter-rotation style wet drum tanks are used in the cobber stage to provide a high level of recovery of the coarse ore. The magnetic element should generate a deep magnetic field for the collection of coarse locked or composite particles. High iron recoveries are achieved with the interpole-type magnetic element.

The cobber concentrate is reground to provide a high degree of liberation and subjected to additional wet drum separation stages. These subsequent stages are referred to as *finisher* stages. Most finisher separators are either double or triple drums (concentrate retreatment) to again provide a multiple-pass separation. The finisher stage produces a final magnetite concentrate. Countercurrent-style wet drum tanks are used in the finisher stage. A schematic of a countercurrent wet drum tank is illustrated in Figure 16.

The feed enters the separator at the bottom of the tank and flows concurrent to the drum rotation. Like the counter-rotation tank, this tank also has a magnetic scavenging zone. The nonmagnetics must migrate through the magnetic field to a full-width overflow. This design is most effective for producing a clean magnetite concentrate. The magnetic element incorporates several agitating magnetic poles (high-gradient design) to provide a high degree of cleaning.

Although most magnetite concentrators employ cobber and finisher magnetic separation stages, there is diversity among them. Specific features of the ore, variability in the ore body, and operating practices may require additional stages of magnetic separation. Many concentrators employ intermediate magnetic separation stages referred to as the *rougher* or *cleaner* stage. These intermediate stages typically treat a reground product to provide additional weight rejection prior to the final regrind stage.

Wet drum magnetic separators in iron ore concentrating operations are commonly 0.9 and 1.2 m in diameter, ranging up to 3.0 m in drum width. The vast majority of the wet drums supplied for magnetite concentration have been 1.2 × 3 m although drums as large as 1.5-m diameter × 5 m-wide have been built. Based on in-plant testing, general sizing parameters for iron ore (magnetite) concentrating wet drum (1.2 m in diameter) magnetic separators have been established, as shown in Table 5. The hydraulic capacity and the magnetic loading represent maximum sizing parameters. The separator should be sized on one of the parameters without exceeding the other parameter.

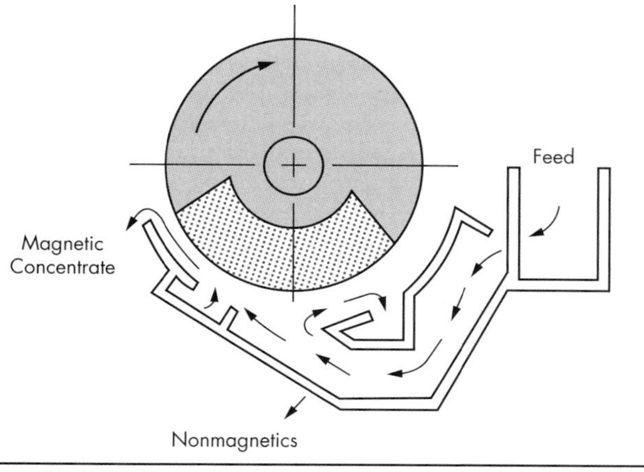

Figure 16 Wet drum with countercurrent tank

High-Intensity Drum Type

The wet drum magnetic separator can also employ a rare earth magnetic element to significantly increase the magnetic field strength. The salient pole rare earth magnetic element used in the dry drum is also used in the wet drum magnetic separator. The magnetic field strength on the drum surface is approximately 7,000 G (0.7 T).

The primary variables affecting separation efficiency are the same as described for the low-intensity magnetic drum separator. The increase in magnetic field strength extends the range of applications. Although the low-intensity magnetic drum is limited to the collection of ferromagnetic materials, the rare earth drum effectively collects many paramagnetic minerals. The drum is effective when used in a scalping operation prior to high-intensity or high-gradient matrix-type magnetic separators. Typical applications of the rare earth wet drum magnetic separator are the recovery of iron-bearing minerals, such as martite and hematite, as well as the collection of ilmenite from heavy mineral concentrates. It is also used in specialty minerals applications such as the magnetic cleaning of high-purity zinc and nickel for battery feedstock.

The rare earth wet drum is available in 0.6- and 0.75-m diameter with drum widths ranging up to 3.0 m. General sizing parameters for a counter-rotation style tank are provided in Table 6.

High-Intensity and High-Gradient Separators

Wet high-intensity magnetic separation (WHIMS) and high-gradient magnetic separation (HGMS) machines employ high-intensity electromagnets and a flux-converging matrix. These separators are utilized for the collection of fine paramagnetic minerals. Technically, WHIMS is distinguished from HGMS by the direction of slurry flow being perpendicular to the line of magnetic flux rather than parallel. In practical terms, WHIMS has come to refer to any continuous-feed carousel-type wet magnetic separator, whereas HGMS is associated with the canister-type filter that is batch operated.

Both WHIMS and HGMS use an electromagnet to generate a magnetic field. The magnetic field induces a matrix that creates a region of high magnetic gradient and provides the mineral collection zone. As slurry flows through the magnetized matrix, the paramagnetic particles are collected and held. The nonmagnetic particles flow through the matrix unaffected.

Table 6 High-intensity (rare earth) wet drum magnetic separator sizing parameters

Drum Diameter, m	Sizing Parameter	
	Hydraulic, $m^3/h/m$ of drum width	Magnetic Loading, t/h/m of drum width
0.6	34	10
0.75	45	13

Table 5 Wet drum magnetic separator sizing parameters for magnetite concentration

Application	Feed Size	Tank Style	% Solids	Sizing Parameter	
				Hydraulic, $m^3/h/m$ of drum width	Magnetic Loading, t/h/m of drum width
Cobber	–5 mm; –4 mesh	Counter-rotation	50	135	40
Rougher	–500 µm; –35 mesh	Counter-rotation	40	110	30–33
Cleaner or finisher	Nominal 75 µm; nominal 200 mesh	Countercurrent	35	105	26

Courtesy of Mineral Technologies

Figure 17 Wet high-intensity magnetic separator

The paramagnetic minerals are segregated as a separate product. This occurs when the magnetic field is de-energized and the paramagnetic minerals are flushed from the matrix.

The matrix is a ferromagnetic material that provides a high level of magnetic induction. It also has a high degree of open area to provide slurry flow without physical entrapment. The function of the matrix is actually threefold:

1. Amplifies the background magnetic field
2. Produces localized regions of extremely high gradient
3. Provides the collection sites for paramagnetic particle capture

Matrix material consists of screen cloth, expanded metal, wedge wire, steel balls, grooved plates, and steel wool. The matrix is selected based on the size distribution of the ore. The optimum matrix for a given application should provide a high magnetic loading capacity, high magnetic gradient, and have a relatively open structure for flow-through capacity. The matrix should be coarse enough to allow the largest particle to pass through unimpeded, yet fine enough to generate high magnetic field gradients and provide sufficient collection sites.

Wet high-intensity magnetic separation. WHIMS machines are built in two general configurations. They both consist of an annular ring that contains the matrix that rotates through the poles of an electromagnetic field. As the ring rotates through the electromagnet, the matrix is magnetically induced. Feed is continually delivered to the magnetized matrix. Paramagnetic minerals are captured while nonmagnetic minerals pass through unaffected. As this section of the matrix rotates out of the magnetic field, the paramagnetic minerals are rinsed out of the matrix and separately collected. The earliest-designed WHIMS machines all employed a horizontal carousel. These separators are widely used for concentrating hematite or for preconcentration of ilmenite from beach sands. Figure 17 show a particular WHIMS application that provides a rougher-cleaner magnetic separation.

Most recently, separators have been built with carousels each with magnetic feed zones. These 265-t separators have a capacity of 1,400 t/h of hematite tailings. The Chinese have developed a separator consisting of a vertical ring. The feed is introduced into the lower section of the ring, which is submerged. A jigging action is introduced in this area, which helps to rinse off any nonmagnetic particles. The Chinese have built thousands of these separators for concentrating paramagnetic ores, such as hematite, and removing paramagnetic contaminants from industrial minerals. Many have replaced earlier horizontal ring separators. Most of these separators are 2 m or less in diameter, but most recently, a 4-m-diameter separator has been installed. The rated capacity of this unit is 175–275 t/h. One manufacturer is providing two separate jigging mechanisms on each separator.

Permanent magnets are also used to generate the magnetic field in a WHIMS. The ferrous wheel is a matrix-type separator that generates a high-gradient magnetic field using a fine matrix and conventional barium ferrite or rare earth permanent magnets. The ferrous wheel utilizes a relatively fine matrix to produce high magnetic gradients. A 14-mesh screen cloth is used for the collection of fine hematite. A conceptual illustration of the ferrous wheel is shown in Figure 18. This particular application provides a rougher-cleaner magnetic separation.

The separating ring is vertical and rotates clockwise. The separating ring holds the matrix. The depth of the matrix is 0.15 m. The separating ring is open in the middle to allow the removal of the separated products. Magnets are mounted on each side of the separating ring and generate a magnetic field through the matrix. Two magnetic stations, each with ~60-degree magnetic arcs, are used on the separating ring.

The ferrous wheel magnetic separator provides a two-stage separation in the single vertical separating ring. The separator can be configured as either a rougher-scavenger or a rougher-cleaner. The feed enters the matrix in the magnetic field at the top of the separating ring. In this first separation stage, the feed enters from outside the separating ring and flows inward toward the center. The nonmagnetic particles pass through the matrix and are channeled out of the separator. This product represents the rougher tailings. The magnetic

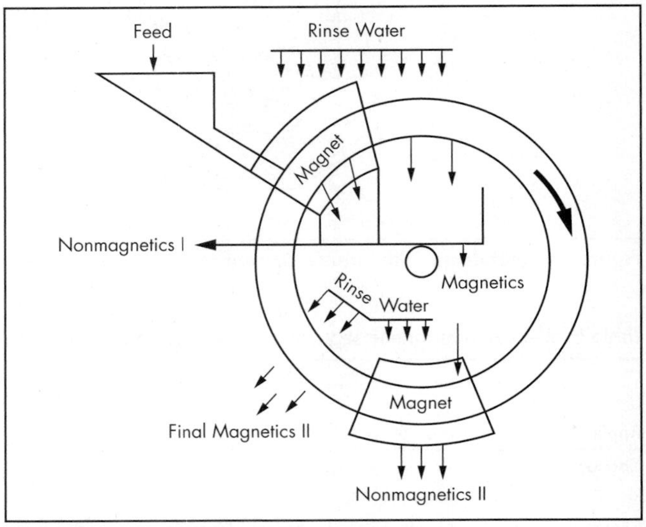

Figure 18 Ferrous wheel

(hematite) particles are collected and held in the magnetized matrix. As the matrix rotates out of the magnetic field, the magnetic particles are rinsed out of the matrix and collected in a hopper in the center of the separating ring. This magnetic product from the rougher separation stage is then repassed in a cleaner stage to wash out residual gangue minerals and further upgrade the hematite.

In the cleaner stage, the feed enters from inside the separating ring and flows outward. Again the nonmagnetic particles pass through the matrix and are channeled out of the system as a final tailing. The magnetic particles are collected and then rinsed out of the matrix. This cleaner magnetic product is channeled out of the separator and represents the final concentrate.

Several separating rings can be mounted together to form a single drum to provide the necessary production-capacity feed rates. Production separators have been fabricated with 15 separating rings in a single unit. The permanent magnets are placed between each ring to generate the magnetic field. The feed distributor consists of a box that runs the length of the separator. Each separating ring has a single feed pipe attached to the distributor box to accept new feed.

High-gradient magnetic separation. HGMS is associated with the canister-type filters that are batch operated. HGMS is also a common term for high-capacity magnetic filters used in the beneficiation of clay and other associated fine-grained earthy minerals. These separators were developed with specific features to provide peak separation effectiveness while operating at feed capacities ranging up to 227 m^3/h slurry. HGMS is operated with a controlled slurry flow rate through the matrix. The slurry is pumped into the bottom of the matrix canister and discharges out of the top. In this manner, the retention time in the matrix is controlled. The slurry must be well dispersed and the particles held in suspension to operate in this manner. Most applications are limited to fine-particle processing in the range of –10 μm, using a steel wool matrix.

The magnet in an HGMS system can either incorporate a hollow-core conductor coil that is water-cooled or a superconducting coil. In either system, the magnetic separation is performed in the exact manner; the only difference is the method of generating the magnetic field.

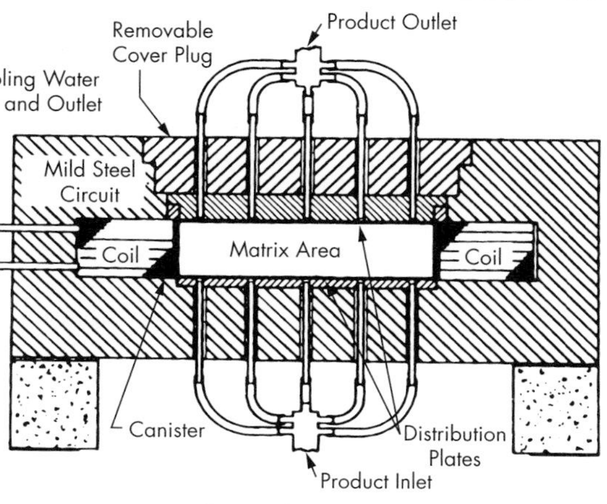

Figure 19 Production-scale high-gradient magnetic separator

Figure 20 Conventional water-cooled high-gradient magnetic separator

Conventional water-cooled high-gradient magnetic separation. The major components of a water-cooled hollow-core conductor system are illustrated in Figure 19, which depicts a production-scale high-gradient magnetic separator. The steel wool matrix is held in a canister in the center of the magnet. A solenoid coil surrounds the canister. The conductor is square with a hollow core to allow the passage of cooling water. Cooling water is pumped through the solenoid coil during operation. An external hydraulic system with heat exchangers provides continuous cooling of the water through the coils. The coil is enclosed in a steel circuit that provides magnetic field uniformity in the bore and minimizes stray magnetic fields. The feed slurry is pumped to the matrix from the bottom and collected at the top. This feed arrangement permits an even distribution of slurry through the matrix and allows control of the retention time through the matrix.

HGMS operates in a batch mode. Feed material is pumped through the separator for a predetermined amount of time. During this operating cycle, the matrix collects magnetics. Eventually the collection sites of the matrix are loaded with magnetics. At this stage, the magnet is de-energized and the matrix flushed of the magnetics. The *duty cycle* refers to the operating time of the HGMS in the separating mode.

Many hollow-core conductor or conventional HGMS systems are in operation. The standard canister size is 2.1 and 3.0 m in diameter. A conventional water-cooled HGMS system used for magnetically cleaning kaolin clay is shown in Figure 20. The power supply is shown on the left, and the magnet and feed distribution system are on the right.

The magnet incorporates a 2.1-m diameter canister with an operative height of 0.5 m. The nominal processing rate through the canister is 114 m^3/h feed slurry. The system consists of a power supply, iron enclosed magnet, and hydraulic cooling system. The magnet generates a field strength of 20,000 G (2 T) with a direct current input of 3,500 amperes, resulting in 400 kW. The solenoid coil is enclosed in an iron circuit to provide magnetic field uniformity within the canister. The magnet weighs 373 t. The open-loop cooling system requires ~1.14 m^3/min of water. The auxiliary system

Figure 21 Superconducting high-gradient magnetic separator

Courtesy of Outotec

Figure 22 Superconducting Cryofilter

also includes feed and discharge manifolds to provide even slurry distribution throughout the canister and remotely activated valves, timers, and controllers to properly activate the rinse cycle.

Superconducting high-gradient magnetic separation. HGMS systems incorporating superconducting electromagnetic coils were developed in the 1990s to significantly reduce power consumption. The first superconducting separators were similar in design to the conventional HGMS systems with the exception of the superconducting coil rather that a hollow conductor water-cooled coil.

A superconducting HGMS system with a 3.0-m diameter operating bore is shown in Figure 21. The system consists of an enclosed magnet, helium liquefier, power supply, compressor, and reservoirs for the liquid helium and liquid nitrogen. The magnet generates a field of 20,000 G (2 T). The niobium-titanium superconducting coil is cooled to –269°C with liquid helium. At this temperature, the conductor has no resistance to electrical current and consequently requires no power to generate the magnetic field. The helium is contained in a closed-loop refrigeration system that converts helium gas to liquid to supply the magnet. The refrigerator is precooled with liquid nitrogen. The refrigerator compressor requires approximately 60 kW for operation.

The power supply is used briefly at the start of each operating cycle to energize the magnet. Once the current in the coil reaches set point, the voltage input from the power supply shuts off. The magnet generates 20,000 G (2 T) with a direct current input of 840 amps at 0 V.

Recently, there have been advances in superconducting technology. The development of the Cryofilter by Carpco advanced the technology of magnetic field generation and the operating mode. The Cryofilter utilizes high-temperature superconducting leads and operates in a persistent mode (always energized), enabling the use of a compact cooler instead of a helium liquefier. The design of this system also results in magnetic field strengths up to 50,000 G (5 T).

The Cryofilter utilizes reciprocating canisters that allow a continuous-feed operation as opposed to batch mode operation. Two separate canisters are reciprocated in and out of the magnet. While one canister is in operation, the other canister is being flushed of magnetics. This essentially permit a continuous duty cycle resulting in an increase in capacity. It is reported that the Cryofilter operating on kaolin clay has reached feed rates of 100 t/h of feed slurry, consuming less than 15 kW (Figure 22).

EMPIRICAL BENCH-SCALE SEPARATORS

The initial step in assessing the magnetic separation of a mineral sample is to measure the response of the mineral to a magnetic field. The most common approach is to hold a hand magnet to the mineral sample. This is simply the first indicator to the presence of ferromagnetic material.

Two different bench-scale separators are commonly used for the quantitative analysis of magnetic minerals. These separators are specifically designed to provide a separation based on magnetic susceptibility. Both separators provide a high level of control and selectivity. The Davis tube tester provides a measure of ferromagnetic minerals, whereas the Frantz Isodynamic or barrier separators provide a measure of paramagnetic minerals.

Figure 23 illustrates a Davis tube tester consisting of a 25-mm-diameter glass tube that is agitated within an electromagnetic field. An electromagnetic coil on each side of the tube generates the magnetic field. The magnetic field strength in the separating zone can be varied.

The sample to be tested is slurried and introduced to the glass tube. The magnetic field holds the ferromagnetics within the tube during agitation. Wash water flowing through the tube removes the nonmagnetics. This apparatus essentially provides a perfect separation of the ferromagnetics. The variables controlling the selectivity of the separation are the magnetic field strength, wash-water volume, agitation, and slope of the glass tube. The Davis tube tester is extensively used in measuring the performance of wet drum magnetic separators in both the iron ore industry and in heavy media recovery applications. Samples of the nonmagnetic effluent are tested for magnetite content indicating recovery performance. The Davis tube also provides a measure of product quality for magnetite concentrates. Because the Davis tube essentially provides a perfect separation, any diluents (commonly silica and alumina) in the magnetite concentrate occur as locked particles.

The Frantz Isodynamic or barrier separators are used to separate paramagnetic minerals based on magnetic

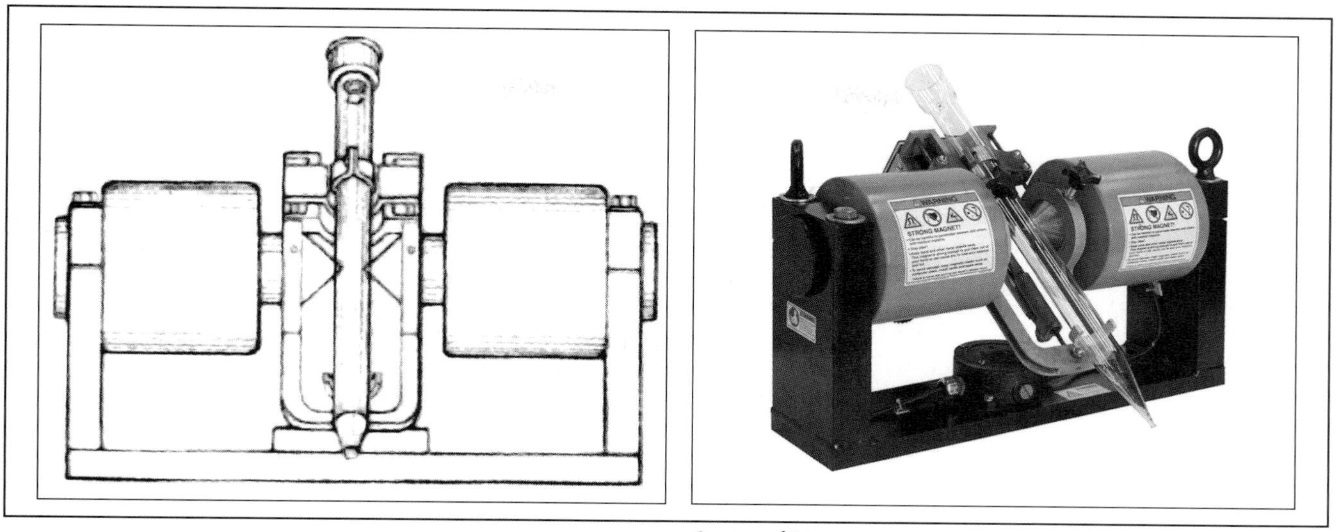

Courtesy of Eriez

Figure 23 Davis tube tester

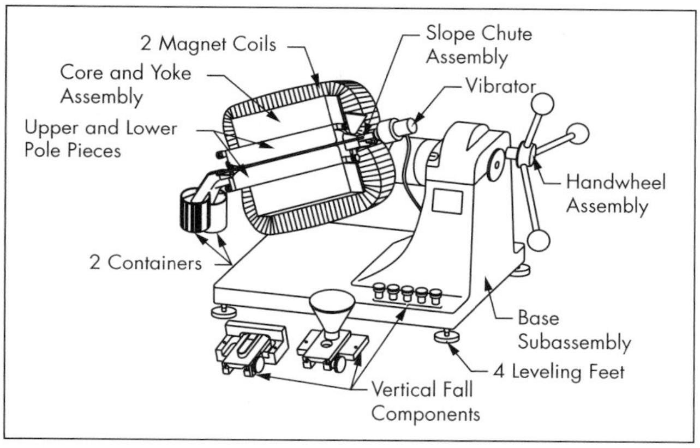

A. Frantz Isodynamic laboratory separator

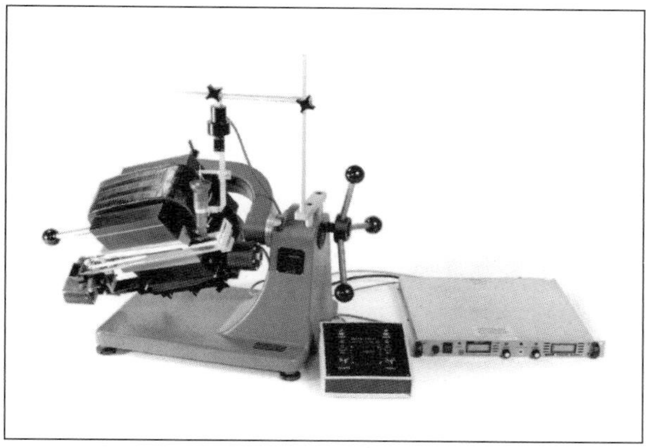

B. Frantz magnetic barrier laboratory separator

Courtesy of S.G. Frantz Company

Figure 24 Frantz Isodynamic and barrier separators

susceptibility. These instruments selectively separate minerals into products with a very narrow range of magnetic susceptibilities. Figure 24 illustrates the Frantz separators, which consist of a steel C-frame circuit coupled with electromagnetic coils. The magnetic field strength in the separating zone is variable.

The Frantz separator treats dry granular mineral samples. The gap or separating zone in the C-frame circuit is at a downward slope is shown in Figure 24. The mineral sample is fed through the magnetic field on an incline chute running the length of the electromagnetic circuit. As the mineral grains pass through regions of high magnetic gradient, those minerals with a selected threshold magnetic susceptibility are diverted from the mineral stream. The nonmagnetic minerals pass through the magnetic field unaffected. Several passes can be made through the separator at increasing magnetic field strength to perform a fractionation of the entire sample based on magnetic susceptibility. This particular series can yield several magnetic fractions with decreasing magnetic susceptibility. The variables controlling the selectivity of the separation are the magnetic field strength and the slope and pitch of the incline chute. This instrument is also used for assessing liberation. The various size fractions of a mineral sample are treated separately on the Frantz Isodynamic or barrier separator. A comparison of the chemical analysis of the magnetic products throughout the size range indicates the extent of liberation. The level of diluents contained with the magnetic fractions delineates the relative amount of locking.

PILOT-SCALE MAGNETIC SEPARATION

Pilot-scale magnetic separation tests provide the means to predict production plant performance. Once the bench-scale testing has defined the basic separation parameters, larger-scale test work provides other benefits to flow-sheet development.

Pilot-scale magnetic separation test work should closely resemble the production process and provide a higher degree of confidence for flow-sheet development. Several specific items can be identified from pilot-scale testing. These items can also be regarded as a particle characteristic or a separator characteristic. Several mineral characteristics can be assessed through pilot-scale test work:

- Magnetic susceptibility
- Concentration
- Recovery
- Variability of the ore
- Extent of liberation
- Temperature
- Moisture
- Interaction with different minerals (overlapping magnetic susceptibilities)

These characteristics are functions of the mineral. Some of these characteristics, such as magnetic susceptibility, are inherent to the mineral and can only be changed with some degree of difficulty. Other characteristics, such as extent of liberation, temperature, or moisture content, are specific to the sample preparation and can be altered in a straightforward manner. Conversely, there are a few separator characteristics that can be assessed through pilot-scale test work:

- Magnetic field strength
- Magnetic gradient
- Unit capacity

Several secondary variables related to unit capacity are separator characteristics. These relate to the design and operation of any specific separator. Many aspects of the feed parameters affect the capacity of the separator, such as percent solids, flow rate, drum speed, or belt speed.

SUMMARY

The field of magnetic separation is dynamic. The advent of rare earth permanent magnets and superconducting electromagnets coupled with new materials of construction have changed the scope of magnetic separation. New types of magnetic separators have been developed, making many older-style separators obsolete. Gone are the induced magnetic roll, cross-belt, and electromagnetic drum-type separators, to name a few. Many more new modifications and developments in magnetic separators will occur if the past is any indication of the future.

ACKNOWLEDGMENTS

This chapter has been adapted from the "Bench Scale and Pilot Plant Tests for Magnetic Concentration Circuit Design" and "Selection and Sizing of Magnetic Concentrating Equipment: Plant Design/Layout" chapters in *Mineral Processing Plant Design, Practice, and Control* (Norrgran and Mankosa 2002a, 2002b). All figures and tables are from Norrgran and Mankosa (2002a, 2002b) unless otherwise indicated.

REFERENCES

Norrgran, D.A., and Mankosa, M.J. 2002a. Bench scale and pilot plant tests for magnetic concentration circuit design. In *Mineral Processing Plant Design, Practice, and Control*. Edited by A.L. Mular, D.N. Halbe, and D.J. Barratt. Littleton, CO: SME. pp. 176–200.

Norrgran, D.A., and Mankosa, M.J. 2002b. Selection and sizing of magnetic concentrating equipment: Plant design/layout. In *Mineral Processing Plant Design, Practice, and Control*. Edited by A.L. Mular, D.N. Halbe, and D.J. Barratt. Littleton, CO: SME. pp. 1069–1093.

Supermagnete. n.d. Ferrite magnets. www.supermagnete.de/eng/ferrite-magnets. Accessed September 2018.

CHAPTER 6.5

Electrostatic Separation

Alex de Andrade and Mike Germain

Electrostatic separation is a generalized term describing processes that exploit differences in the electrical conductivity on the surfaces of various minerals. By nature, these are dry processes that rely on a free-flowing granular feed with particle size limits from approximately 40 μm to approximately 1 mm. There are principally three types of electrostatic separator in general industrial use: high-tension roll (HTR), electrostatic plate separator (EPS), and the tribostatic separator. The most commonly used separators are HTR and EPS units, which are extensively employed in the treatment of concentrates produced from mineral sand mining. These minerals are generally well liberated, which enables high product purity levels to be achieved. The effectiveness of separation of minerals using electrostatic and electrical technology depends on the degree of difference in the conductivity between the particles in the feed. As most minerals have some difference in conductivity, exploiting this feature could likely offer a means of separating them from each other. Moisture and surface contamination of the minerals are two of the factors that may reduce separation efficiency; these have to be considered by the process technology designers. Specialist reagents may also enhance conductivity and/or nonconductivity of targeted minerals for difficult feed types.

PRINCIPLES OF SEPARATION

High-Tension Roll Separation

HTR separation uses a rotating earth-grounded roll to carry feed particles under a thin electrode wire suspended above and parallel to the roll. The electrode typically carries between 20 and 30 kV, which causes partial ionization of the air between the wire and the roll. This allows movement of ions across the air gap and enables all the feed particles to accumulate a surface charge. This charge on the particles generates an attractive force against the surface of the earth-grounded roll, which is generally described as an *image force* (Figure 1). The image is effectively the result of the charged particle inducing an opposite charge in the surface of the roll such that there is attraction between the two opposing charges. This image force pulls the particles toward the rotating roll until they move past the wire electrode.

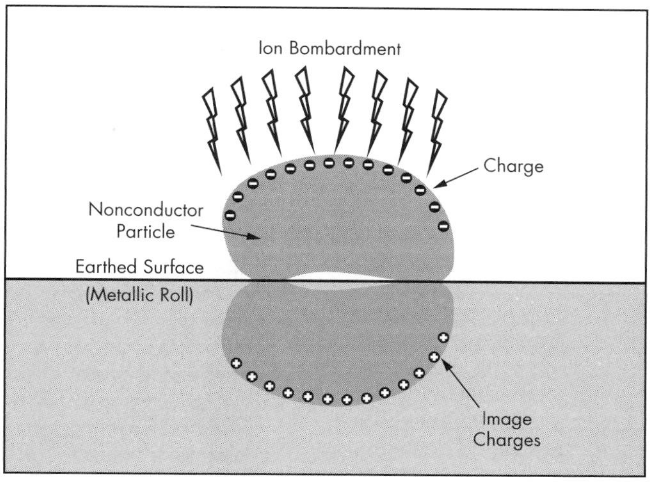

Figure 1 Image force and respective charge

As the particles are carried out of the charging zone beyond the wire electrode, the charge on the conducting particles will begin to discharge into the roll. When they have lost sufficient charge, the particles will follow a normal trajectory away from the roll, based on the speed imparted to them from the rotating roll. The most nonconductive particles will tend not to lose their charge and thus will be retained by the image force on the roll surface and be carried around to a brush that dislodges them from the roll surface. Mineral species exhibiting intermediate conductivity will discharge more quickly and fall from the roll at an earlier point into the middlings area. Hence, the different streams can be collected by the positioning of product splitters at appropriate locations beneath the roll (Figure 2).

Traditionally, the wire electrode has been supported by a metallic tube that can also act as a secondary nonionizing electrode, because the large diameter prevents ionizing of the air. When moved into a position sufficiently close to the roll, the nonionizing electrode provides a strong static field at the roll surface. This field induces the charge on a conductive

Alex de Andrade, General Manager Equipment & Technology, Mineral Technologies Pty Ltd., Gold Coast, Queensland, Australia
Mike Germain, Process Metallurgist, Mineral Technologies Pty Ltd., Gold Coast, Queensland, Australia

particle to cause separation. The charge carriers of a polarity opposite to the electrode are attracted toward it. The charge at the side closest to the roll flows to the earth-grounded roller, leaving the particle with a net charge of opposite polarity to the electrode. This process also accelerates the loss of corona charge by the conductive particles. The process is called *conductive induction*, and it is also the charging mechanism used by EPS units.

The effect of the secondary metallic bar electrode is limited in that it can only influence a small area of the roll surface, and hence some separators have been designed with a metal plate electrode instead of the bar. The limitation of the metallic electrode is that it cannot be brought too close to the roll without risk of arcing between the two conductive materials. Recent developments in HTR separators have incorporated secondary electrodes made from nonconductive materials, which have allowed them to be positioned much closer to the roll surface (Figure 3). This enables a stronger electric field to be generated, which significantly enhances the efficiency of separation. The nonconductive plate is able to impart additional induction charging of conductor minerals such that they are actively attracted toward the plate while coarse-sized nonconductors are held on the roll for a longer time. Thus, the resulting general trajectories of the conductor and nonconductor streams are farther apart, and the proportion of minerals misreporting to middlings is reduced.

HTR machine parameters that are normally adjusted to optimize the separation are the roll speed, electrode voltage, and product splitter positioning. Some separations may benefit from adjustments to the positioning of the electrodes. Introducing the feed material to the roll at an elevated temperature is also important for efficient separation. Increased roll speed will tend to improve the purity of the nonconductor stream, but reduce its weight yield. This may also cause increased levels of nonconductors to report with the conductor stream. Hence, there will be an optimum roll speed for any particular separation. HTR separators incorporating nonconductive secondary electrodes are generally most efficient when operated at the maximum electrode voltage that does not cause arcing between the wire and the roll (Figure 4). At a given roll speed, which has been set to give the most appropriate weight split between conductors and nonconductor species, the conductor quality can be adjusted by the positioning of the conductor splitter. Processing a narrow feed-size distribution with this technology will also improve the separation efficiency.

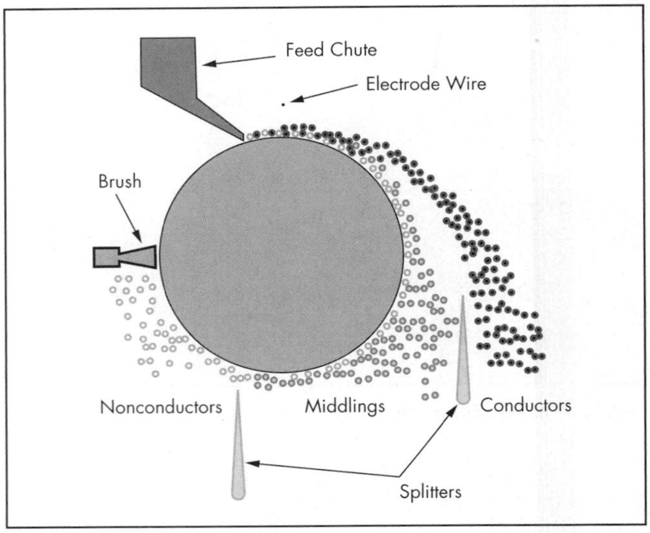

Figure 2 High-tension roll separator

Electrostatic Plate Separation

EPS allows feed particles to roll under gravity down an earth-grounded sheet metal plate. The particles pass under a nonionizing electrode in the form of an elliptical tube or plate, which is situated parallel to the length of the plate. The nonionizing electrode provides a strong static field at the plate surface. This field induces the charge on a conductive particle to separate. The charge carriers of polarity opposite to the electrode are attracted toward it. The charge at the side closest to the plate flows through the earth-grounded plate, leaving the particle with a net charge of opposite polarity to the electrode. The charging mechanism adopted is conductive induction.

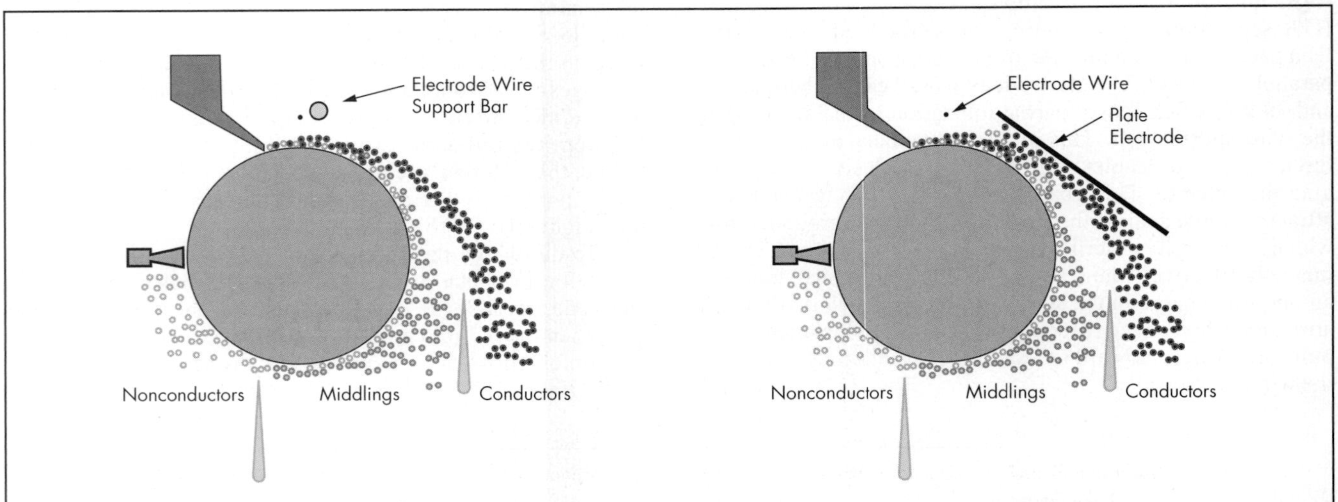

Figure 3 High-tension roll with bar- and plate-type secondary electrodes

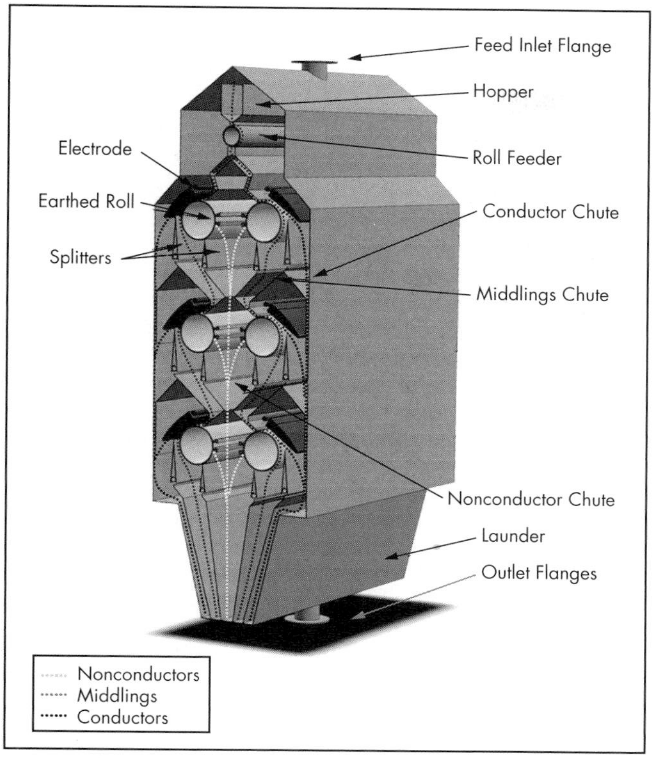

Figure 4 Front sectional view of a typical industrial multistage high-tension roll separator setup for middlings retreat

Conducting particles will easily collect charge from the plate and then be attracted up toward the oppositely charged electrode. This attraction, coupled with their momentum down the plate, will produce a trajectory that carries the particles over a suitably placed product splitter. Nonconducting particles will tend not to acquire charge from the plate and thus will be unaffected and continue on their natural trajectory. The charging mechanism is substantially weaker than that for the HTR separator and thus the separation forces are smaller. Consequently, the EPS unit is ideally suited to finer-sized conductor particles that are more easily deflected from their natural trajectory.

Two types of EPS units are *curved plate* and *screen plate* (Figure 5). The curved plate is the most commonly used and features an earth-grounded plate that is shaped to have an inflection point under the electrode, which assists the transfer of charge to the conductor particles. The screen plate has an earth-grounded plate with a simple concave curvature that directs the nonconducting particles to a coarse screen. The nonconductors fall through the screen while the conductors are lifted across it by the electrode and fall into a separate chute. Similar separations can be achieved with both types of EPS units; however, the curved plate type can be simpler to operate in a plant environment.

Tribostatic Separation

Tribostatic separation relies on the tendency for certain materials to exchange electrons during contact such that one component becomes negatively charged while the other becomes positively charged. The polarity of the charge retained is a function of the surface chemistry of the materials. The charging process depends on many factors, such as the contact velocity, temperature, and prevailing atmospheric conditions. A tribostatic separator will normally incorporate a means of charging the particles immediately ahead of the separation zone. This may simply consist of a cascade chute where the particles fall under gravity through a zigzag chute to provide opportunity for interparticle contact. Other charging mechanisms may include rotating tubes, brushes, or turbulent air currents.

The separation zone in a free-fall tribostatic unit consists of two vertically arranged electrodes of opposite polarity (Figure 6). The charged feed material is directed between the electrode plates where it falls under gravity. The particle trajectories are modified by the attraction toward the oppositely charged electrode, and thus the separated minerals can be collected at the bottom of the separation zone.

Triboelectrostatic Belt Separation

The triboelectrostatic belt separator uses triboelectric charging (Figure 7). Feed material is introduced into a narrow gap between two parallel planar electrodes. The action of a high-speed open-mesh belt moving between the electrodes causes interparticle collisions, resulting in triboelectrostatic charging of the particles. The different mineral particles are attracted toward the electrode plate of opposite charge. This causes them to engage with one side of the moving belt, which transports them to either end of the separator where they are captured in hoppers and chutes for product materials handling. The unique feature of this technology stems from the countercurrent flow of the separated particles making two distinct products.

This technology requires no additional chemicals, produces no emissions, and is suited for separation of particles with sizes from 1 to 500 µm. Throughput capability ranges from 1 to 50 t/h (metric tons per hour) per machine.

The ST Equipment and Technology triboelectrostatic belt separator (Figure 8) has demonstrated the capability to process fine particles from <0.001 to approximately 0.5 mm. These separators have been in operation since 1995, separating unburned carbon from fly ash minerals in coal-fired power plants in North America, Europe, and Asia to produce a concrete-grade pozzolan for use as a cement substitute (Bittner et al. 2013). Through pilot-plant testing, in-plant demonstration projects, and/or commercial operations, ST Equipment and Technology separators have demonstrated beneficiation of many minerals, including potash, barite, calcite, and talc (Bittner et al. 2014).

The primary interest in this technology has been its ability to process particles less than 0.1 mm; there are currently two sizes of the ST Equipment and Technology fine-particle triboelectrostatic belt separators available, with nominal capacities of 23 and 40 t/h.

The principles of operation of the ST Equipment and Technology separator are illustrated in Figures 9 and 10. The particles are charged by the triboelectric effect through particle-to-particle collisions in the air slide feed distributor and within the gap between the electrodes. The applied voltage on the electrodes is between ±4 and ±10 kV relative to ground, giving a total voltage difference of 8–20 kV. The belt, which is made of a nonconducting plastic, is a large mesh with approximately 60% open area. The particles can easily pass through the openings in the belt.

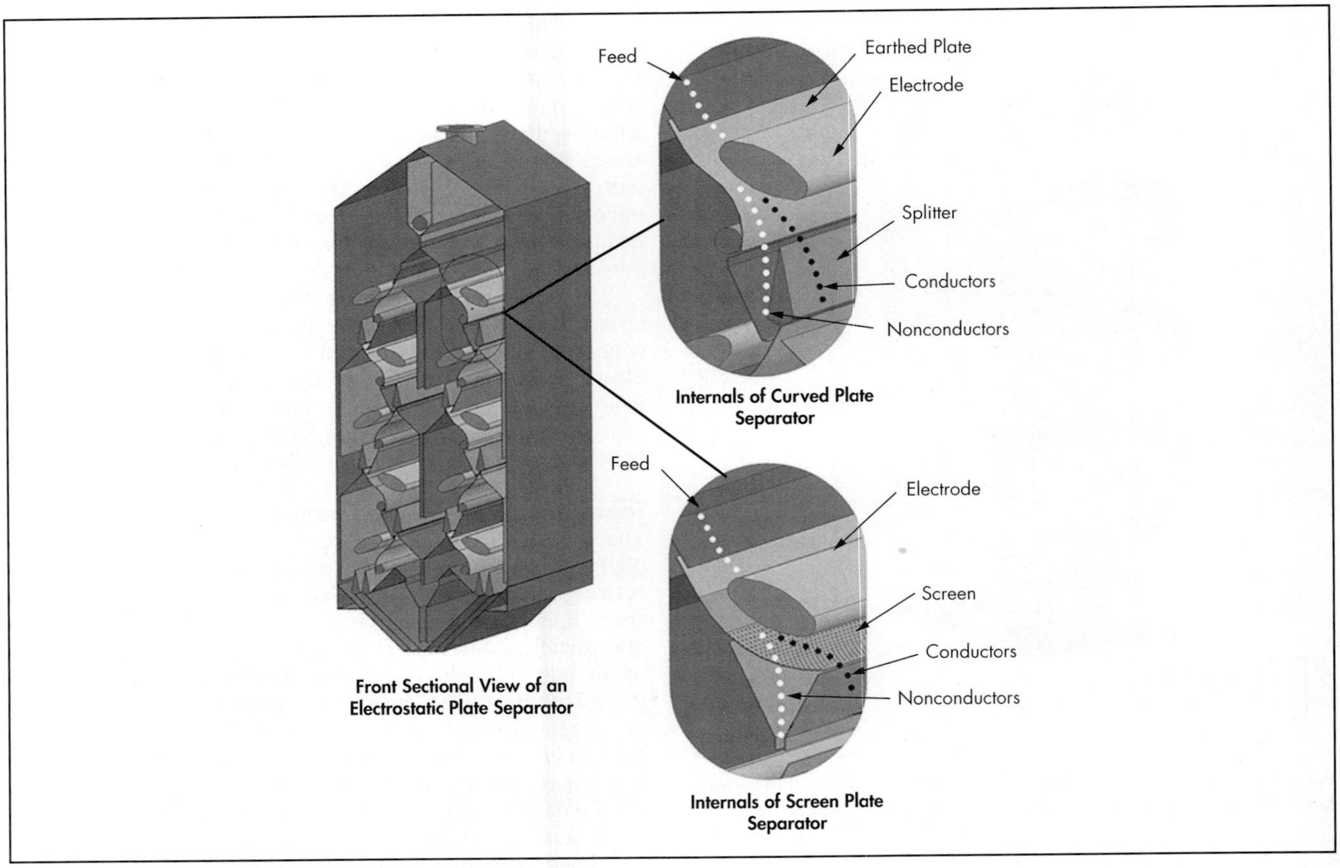

Figure 5 Curved plate and screen plate separators

The flow patterns and particle-to-particle contact within the electrode gap that are established by the moving belt are key to the effectiveness of the separator. On entry into the gap between the electrodes, the negatively charged particles are attracted by the electric field forces to the bottom positive electrode. The positively charged particles are attracted to the negatively charged top electrode. The speed of the continuous loop belt is variable from 4 to 20 m/s. The geometry of the belt cross-direction strands serves to sweep the particles off the electrodes, moving them toward the proper end of the separator and back into the high-shear zone between the oppositely moving sections of the belt. Because the particle number density is so high within the gap between the electrodes (approximately one-third the volume is occupied by particles) and the flow is vigorously agitated, there are many collisions between particles, and optimal charging occurs continuously throughout the separation zone. The countercurrent flow induced by the oppositely moving belt sections and the continual recharging and reseparation creates a countercurrent multistage separation within a single apparatus. This continuous charging and recharging of particles within the separator eliminates the need for any *charger* system prior to introducing material to the separator, thus removing a serious limitation on the capacity of electrostatic separation. The output of this separator is two streams, a concentrate and a residue, without a middlings stream. The efficiency of this separator is equivalent to approximately three stages of free-fall separation with middlings recycle.

The ST Equipment and Technology fine-particle triboelectrostatic belt separator has many process variables that enable optimization of the trade-off between product purity and recovery that is inherent in any beneficiation process. The coarse adjustment is the feed port through which the feed is introduced to the separation chamber. The port farthest from the discharge hopper of the desired product gives the best grade but at the expense of a lower recovery. A finer adjustment is the speed of the belt. The electrode gap, which is adjustable between 9 and 18 mm, and the applied voltage (± 4 to ± 10 kV) are also important variables. The polarity of the electrodes may be changed, which aids in the separation of some materials. Pretreatment of feed material by precise control of trace moisture content (as measured by feed relative humidity) is important to achieve optimum separation results. The addition of trace amounts of charge-modifying chemical agents can also aid in optimizing the process.

As stated earlier, the initial commercial application of the belt separator has been separation of unburned carbon from the glassy aluminosilicate particles in fly ash from coal-fired power plants. This technology is unique among electrostatic separators in its ability to separate fly ash, which typically has a mean particle size less than 0.02 mm. The fine-particle triboelectrostatic belt separator has also been proven to effectively separate magnesite from talc, halite from kieserite and sylvite, silicates from barite, and silicates from calcite. The mean particle size of all of these feed materials has ranged from 0.02 to 0.1 mm. Examples of separations for several materials

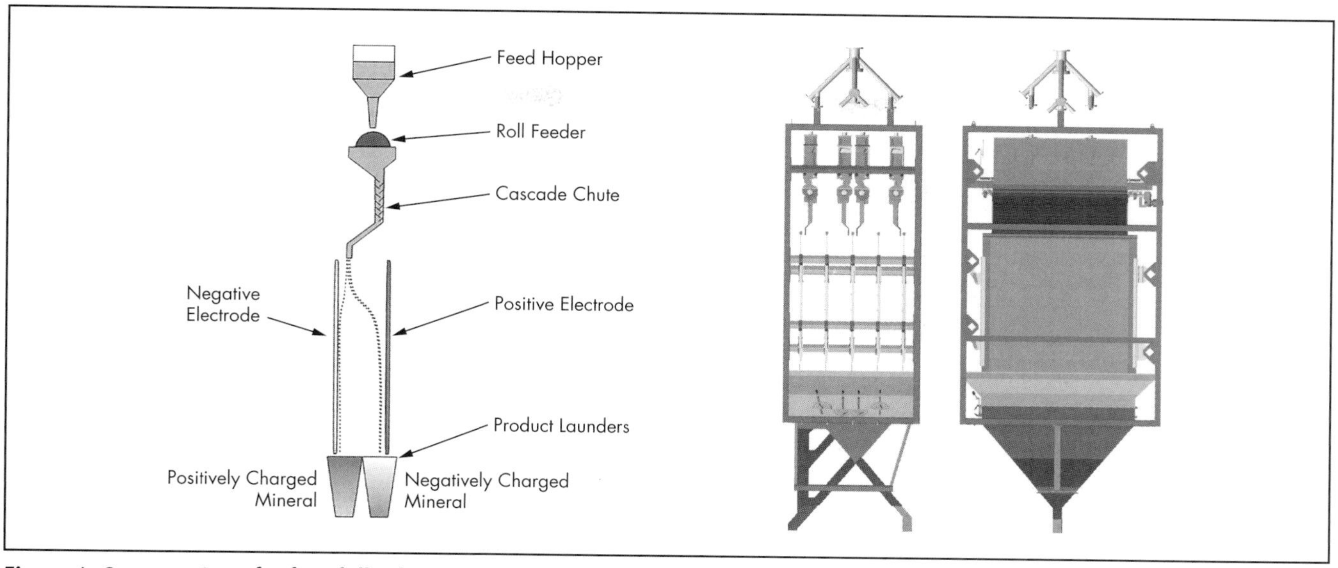

Figure 6 Cross section of a free-fall tribostatic separator

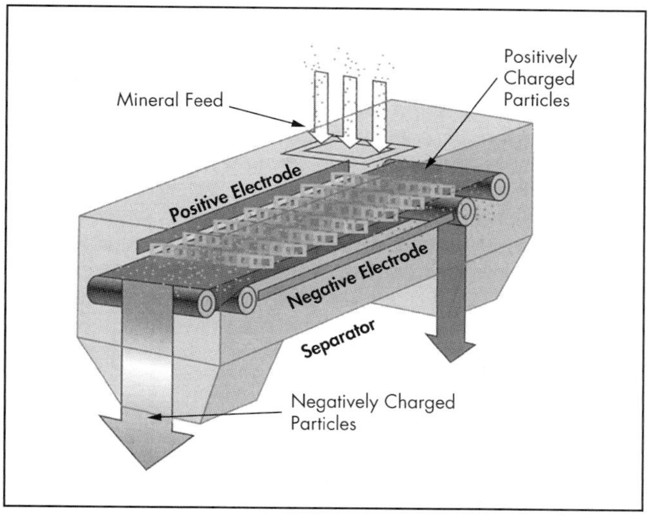

Courtesy of ST Equipment and Technology LLC
Figure 7 Operating principle of a triboelectrostatic belt separator

Courtesy of ST Equipment and Technology LLC
Figure 8 Industrial triboelectrostatic belt separator

are included in Table 1. The operating costs (Table 2) for the fine-particle triboelectrostatic belt separator are low, typically ranging from US$0.75 to US$2/ton feed (2015 prices, which include fine-particle triboelectrostatic belt separator consumables and electricity, excluding operating labor).

APPLICATIONS OF THE SEPARATION PRINCIPLES

HTR separators and EPS units are commonly used together in a plant flow sheet as their individual separation characteristics are complimentary. HTR separators will tend to generate a middlings product, enriched in finer-sized conductors and coarser nonconductors. This type of feedstock may be better suited to separation using EPS units as they are particularly effective at removing the fine-sized conductors from a coarse nonconductor stream.

The most common application for HTR separators and EPS units is in the treatment of mineral concentrates derived from mineral sand mining. After a mineral sand ore has been concentrated through wet gravity processes, a mixed mineral concentrate is produced, which generally contains up to 90% heavy minerals (minerals with specific gravity above 3.0). This concentrate would normally require additional upgrading to remove the remaining quartz and ensure that the mineral surfaces are clean and free of slime. This cleaned concentrate is then dried and treated through the *dry mill*, which consists of various stages of separation with HTR and EPS equipment along with magnetic separation as required. Depending on the constituents of the ore, varying proportions of the following minerals can be produced as final concentrates:

- Ilmenite
- Leucoxene

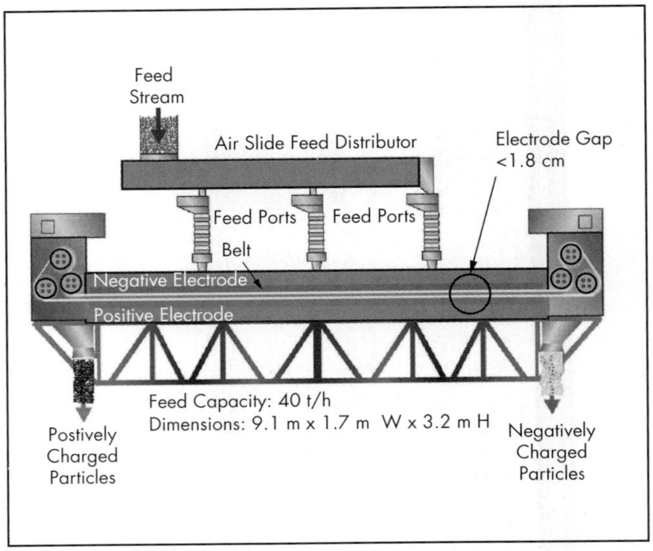

Courtesy of ST Equipment and Technology LLC

Figure 9 Schematic of fine-particle triboelectrostatic belt separator

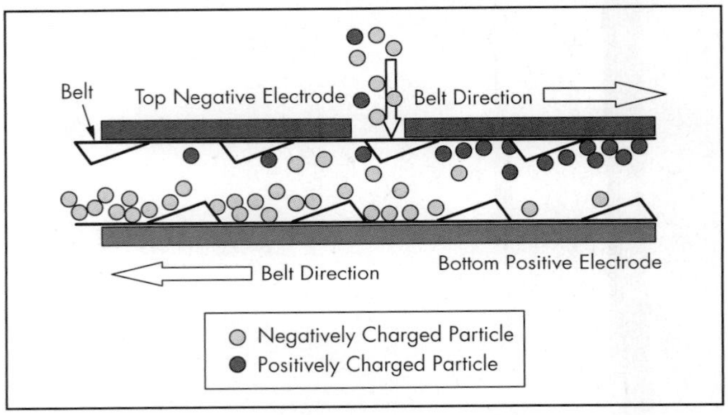

Courtesy of ST Equipment and Technology LLC

Figure 10 Electrode gap of fine-particle triboelectrostatic belt separator

- Rutile
- Zircon
- Garnet
- Sillimanite
- Monazite
- Xenotime

The conductor minerals in the preceding list are ilmenite, leucoxene, and rutile; all others are nonconductors. Hence, the final separation of the individual mineral species requires the use of magnetic separators and wet or dry gravity separators in conjunction with the electrostatic separators.

HTR separation is also commonly used in processing concentrates from alluvial tin mining where minerals such as cassiterite, ilmenite, and columbite are separated. It has also been applied to the rejection of nonconductive silicates from both iron ore and chromite concentrates. EPS units have also been used to reject weakly conductive contaminant grains from salt concentrates. Because most metal sulfides are more conductive than gangue minerals, there is also potential for dry separation of some sulfide ores.

Table 1 Separation examples

Separation	Feed	Product	Recovery
Calcium carbonate (silicates)	9.5% Acid insols	<1% A1	89% $CaCO_3$
Talc (magnesite)	58% Talc	95% Talc	77% Talc
		88% Talc	82% Talc
Kieserite + KCl – NaCl	11.5% K_2O	27.1% K_2O	90% K_2O
	12.2% Kieserite	31.8% Kieserite	94% Kieserite
	64.3% NaCl	14.3% NaCl	92% NaCl reject
Fly ash mineral (carbon)	6.3% Carbon	1.8% Carbon	88% Mineral
	11.2% Carbon	2.1% Carbon	84% Mineral
	19.3% Carbon	2.9% Carbon	78% Mineral

Courtesy of ST Equipment and Technology LLC

Table 2 Cost overview*

Overall Costs	Wet Beneficiation (flotation), %	Dry Beneficiation (fine-particle triboelectrostatic belt separator), %
Purchased major equipment	100	94.5
Total CAPEX	100	63.2
Annual OPEX	100	75.8
Unitary OPEX ($/ton conc.)	100	75.8
Total cost of ownership	100	70.0

Courtesy of ST Equipment and Technology LLC

*Capital expenditure (CAPEX) and operating expenditure (OPEX) of crushing are the same for both processes and are not included.

Free-fall tribostatic separation is used in the removal of quartz from lime sand or limestone and the removal of silica sand from feldspar. The tribostatic belt separator has proven application in the separation of the following mineral combinations:

- Potash–halite
- Talc–magnesite
- Limestone–quartz
- Brucite–quartz
- Iron oxide–silica
- Mica–feldspar–quartz
- Wollastonite–quartz
- Boron minerals and carbon–silica

Other potential applications being tested are

- Phosphate–calcite–silica,
- Barites–silicates,
- Zircon–rutile,
- Silver and gold slags,
- Beryl–quartz, and
- Fluorite–silica and fluorite–barite–calcite.

In theory, because particle charging depends on the triboelectric effect, any two minerals that are liberated from each other (conductor–conductor or nonconductor–conductor) can be separated by this method. Other potential applications include magnesite–quartz, feldspar–quartz, mineral sands, other potash mineral separations, and phosphate–calcite–silica separations.

FUNDAMENTALS OF EQUIPMENT SELECTION

Virtually every mineral suite will have its own peculiarities, and thus the quantity and arrangement of electrostatic separators required for beneficiation is likely to vary from site to site. While there are some general principles for machine selection that will give a first estimate of the quantity and type required, the only way to ensure the correct selection is through laboratory or pilot-scale testing.

The primary separation stages will generally employ HTR separators as these are able to treat material at a higher feed loading than EPS units. Indeed, the modern dual-electrode HTR separators are able to separate many mineral streams that once required EPS units. So the general principal in any test work program is to use HTR separators until it becomes apparent that they cannot perform a particular separation. Commonly, EPS units are used in the retreatment of difficult streams rejected from prior HTR separation stages.

EPS units are most efficient at separating finer-sized conductors from streams containing coarser-sized nonconductors. Such streams are commonly generated from the middlings product of HTR separators. Thus, microscopic inspection of the stream can give insight into when EPS units may be applicable. An EPS can sometimes differentiate between minerals exhibiting very similar conductivity where the HTR separator may be ineffective.

TYPICAL INDUSTRIAL APPLICATIONS

Various machine configurations exist to offer advanced techniques for electrostatic separation within a tribostatic panel. Some of the more common configurations are depicted in Table 3.

Particle Size Considerations

Most mineral separators operate more efficiently with a relatively narrow feed sizing. This is true of HTR separators in particular because the conditions ideally suited for separation of a fine particle are likely to require a higher roll speed than those for coarser particles of the same mineralogical composition. Thus, a mixed-size feed must adopt a compromise roll speed between the two ideals. That said, it is generally found that a ratio of 3:1 between the coarsest and finest particles allows a very reasonable separation to be achieved. Ratios higher than this can often still be handled adequately when the HTR separator is used in conjunction with the EPS unit or perhaps with screening of the final HTR products. For more difficult separations, reducing the size range to 2:1 can sometimes assist; however, this may be hard to justify in a plant situation unless the final product requires such narrow sizing.

Generally, the effectiveness of the HTR separation diminishes with separation of particles sized larger than 1 mm. This is caused by the difficulty of presenting the particles onto the rotating roll without having excessive bouncing of the particles. Also, the mass-to-surface-area ratio of large particles is increased such that there is insufficient charge accumulated to cause the particle to be retained on the roll surface in opposition to the centrifugal force and gravity.

An EPS unit can separate a wide particle size range, provided that the conductor minerals are preferentially at the fine

Table 3 Common equipment sizes

	High-Tension Roll	Electrostatic Plate Separator	Tribostatic Separator
Configurations	6 roll (2 × 3 configuration) 4 roll (2 × 2 configuration) 3 roll (1 × 3 configuration) Single-roll lab units Roll diameters: 6–16 in. (400 mm) Roll lengths: 300–2,000 mm	1 × 3 plates 2 × 3 plates 1 × 5 plates 2 × 5 plates 2 × 10 plates Single-plate lab unit	1 stage × 1 zone 1 stage × 2 zone 2 stage × 1 zone 2 stage × 2 zone
Mass range	250–2,500 kg	250–2,500 kg	350–1,500 kg
Size range	750–5,000 mm height 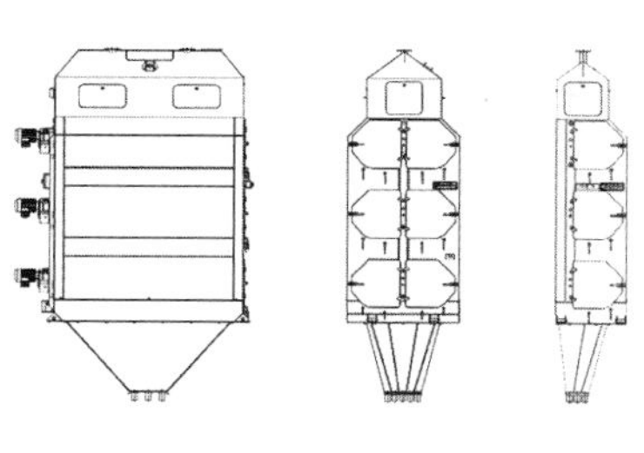	700–5,000 mm height 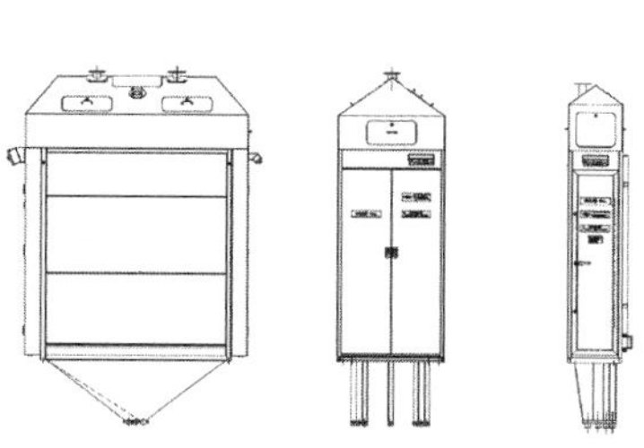	1,100–4,550 mm height 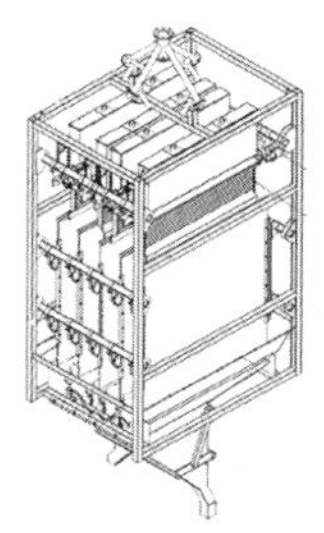
Throughput (subject to test work)	300 kg/h–10 t/h	100 kg/h–5 t/h	5 kg batch–40 t/h
Services required			
Power	0.08–5.5 kW Heating of feed to be considered Dust collection to be considered	0.08–0.5 kW Heating of feed to be considered Dust collection to be considered	0.25–5 kW
High-tension power supply	7.5–15 mA	7.5–15 mA	7.5–25 mA

end of the size range. If there is no size differentiation between the conductor and nonconductor minerals, then the same 3:1 sizing ratio applies as per the HTR separator.

The particle size also has a significant bearing on the capacity of the HTR and EPS units. As the feedstock becomes finer, the feed rate generally needs to be reduced to maintain a similar level of separation efficiency. For example, separation of a mineral sand suite, with a mean size of 130 μm, may typically be performed at around 1.8 t/h/m. A reduction in the mean size to 90 μm may require the loading to be reduced to around 1.2 t/h/m, while reduction to 50 μm may require loadings below 0.6 t/h/m. It is generally found that electrostatic separation is ineffective on particles below 30 μm as the interparticle attractions tend to prevent particles acting independently.

High-Tension Roll Size

Over the years, commercial HTR separators have been manufactured with various roll diameters ranging from 150 mm to approximately 400 mm. In general, a larger roll diameter will give an increase in capacity for a given mineral feed type, particularly in the primary separation stages. However, the degree of increase in capacity is subject to the feed composition and mineral characteristics. In the final cleaning stages of a circuit, there may be smaller differences between roll sizes larger than 250 mm as the majority of the feed must report to one or other product.

The use of larger-diameter rolls allows more area for charging of the mineral being fed onto the roll and permits more flexibility in positioning of the corona wire. The larger diameter also allows a greater surface area of the roll to be influenced by the plate electrode, which can significantly assist separation efficiency. Also, there is a longer retention time for the nonconductor particles that are adhered to the roll, which allows weakly conductive particles more time to lose their charge and be released from the roll into the middlings product. The effect of roll size on performance for a two-component feedstock is shown in Figure 11.

Feed Loading

To minimize the capital expense of a plant, the minimum number of separators is always sought. Hence, the unit feed loading to any stage of separation should be maximized within the constraints of the required grade and recovery targets. Increasing the feed loading on electrostatic separators will generally reduce the metallurgical efficiency of the separation, as shown in Figure 12.

With a complex feedstock containing many different mineral types, or where a very high product purity is required, it is expected that multiple stages of electrostatic separators will be required to achieve the final grade and recovery specifications. Thus, the primary HTR separation stage may employ a higher feed loading as the product streams generated will pass on to a second, and possibly third, cleaning stage, which would be able to remove the remaining contaminant minerals. For example, a feedstock with a mean particle size of around 130 μm may give an acceptable primary stage HTR separation at feed loadings between 1.7 and 1.9 t/h/m. To achieve the required product stream quality in the final stage of separation, the feed loading may need to be reduced to approximately 1.5 t/h/m. If the challenge is to make a sufficiently clean nonconductor product, a larger-diameter roll, which provides more time for the charge on the conductors to decay,

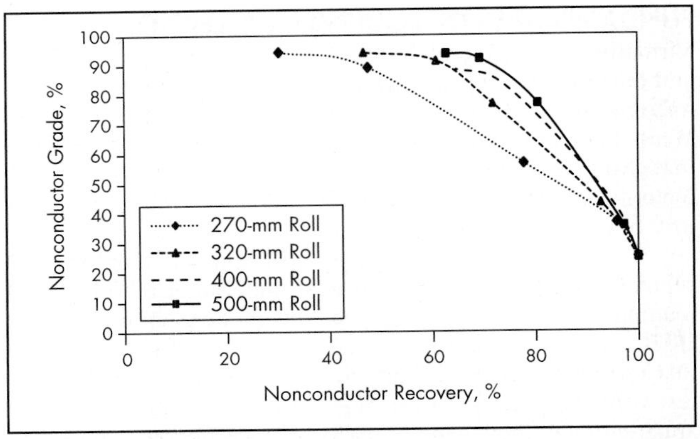

Courtesy of Mineral Technologies Pty Ltd.
Figure 11 Effect of high-tension roll diameter on separation of a two-component feedstock

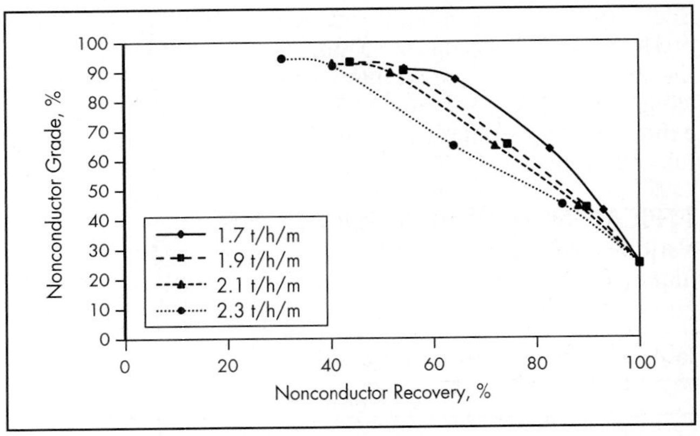

Courtesy of Mineral Technologies Pty Ltd.
Figure 12 Effect of feed loading on high-tension roll separation of a two-component feedstock

should be considered. Larger roll diameters can also be used for coarser feeds.

The use of higher feed rates will tend to reduce the quality of both the conductor and nonconductor streams as more entrainment and interparticle collisions occur. This can be mitigated to some degree by allowing a middlings stream to be taken. Thus, some improvement in product grades can be made, as feed rates increase the proportion of material reporting to middlings is also likely to increase.

Middlings Handling

In many applications, it is appropriate to take a middlings stream from the HTR separator and direct it to its own re-treatment stage. A reasonable proportion of the minerals in this middlings stream will simply be present because of misreporting, and historically, such streams have often been returned to the feed of the separator. Nevertheless, such streams tend to contain an elevated level of coarse nonconductors and fine conductors along with minerals having an intermediate conductivity. Returning such streams to the same separator does not provide the ideal separation conditions for such minerals and consequently results in a buildup of a circulating load, which can at times grow to represent a very significant proportion of the total feed to the separator.

It is almost impossible to predict the ultimate size of this circulating load from laboratory testing. Also, small changes in the feed characteristics can significantly affect the magnitude of the circulating load, making control of the plant very difficult. Consequently, most modern electrostatic flow sheets are designed to minimize the need for recirculation of middlings streams. When such a middlings stream represents a very small mass flow that does not justify an additional separation stage, a decision must be made as to whether it should be recirculated or directed to a stockpile for subsequent batch treatment. It is not uncommon for a very minor middlings stream, or rejects from a final HTR cleaning stage, to grow up to five times its original mass flow when recirculated.

Number of Separation Stages

Achieving a high-purity final product that meets the required specification normally requires multiple HTR separation stages to overcome the statistical chance of contaminant entrainment and to reject any minerals that have very similar conductivity to the product species. In doing this, a significant proportion of the product mineral will be lost to the various reject streams. Thus, additional HTR stages will be required to recover this lost mineral and maintain suitable recovery levels for the plant as a whole.

Historically, plant HTR separation machines have been supplied with single, dual, or triple tiers of separation rolls. Much of the space required for an electrostatic separation plant is involved with feed distribution and product laundering. A more compact overall design can generally be achieved using separators having multiple tiers of separation rolls. The modern dual-electrode HTR separators are able to generate a significant proportion of relatively clean conductor and nonconductor streams from a single roll while allowing the remaining unseparated minerals to report to a middlings stream. Rather than direct this middlings stream to a different separator, it can be immediately treated on the second-tier roll of a multiple-tier separator. The proportion of material reporting to such a middlings stream is typically around 50% of the original feed. It can be argued that the second-tier roll is underloaded and that fewer separation rolls would be required with a dedicated middlings treatment stage. However, as most HTR separations require operation at elevated temperature, this immediate re-treatment has the advantage that no heat is lost from the stream prior to the second separation roll. Also, as discussed earlier, the middlings will contain an increased proportion of minerals that are more difficult to separate, and the reduced feed rate enables a significant improvement in the separation efficiency. Indeed, this is so much so that the quality of the conductor and nonconductor streams from the second-tier roll can generally match that of the first tier and can be immediately combined with them. Inclusion of a third-tier roll, again treating the second-tier middlings, requires a minimal increase in plant space and allows the maximum recovery of both conductor and nonconductor minerals to their respective product streams while generating a very low mass flow to middlings, typically representing only 5%–10% of the original feed.

For some mineral separations where a high-purity product is sought, there may be the requirement to use the second- and/or third-tier roll for re-treating either the prior roll conductors or nonconductors rather than the middlings. Ideally, the separator design should be able to permit this flexibility by standard adjustments to the splitter configuration.

Temperature Considerations

Electrostatic separation exploits the difference in the surface conductivity of various minerals. The effect of moisture on the mineral surface has a significant effect on the resultant conductivity. Commonly, the feed to electrostatic separators is freshly dried after prior concentration in wet processes. While the minerals need to be fully dry to be free flowing, it is also required that the surfaces be free from moisture that condenses from the atmosphere as the mineral grains cool down. Hence, most electrostatic separations need to be performed at elevated temperature to minimize such condensation. This effect is more significant in humid climates, and it may be necessary to adjust the separation temperature and other separation parameters to compensate for local changes in atmospheric conditions. The absolute measure of the atmospheric moisture content is via the dew point rather than humidity. It is common for operating plants to install instruments for measurement of dew point to permit proactive adjustment of the separator parameters.

Typically, most HTR separations are performed at temperatures higher than 60°C, with optimum separation for many conductor minerals requiring temperatures around 100°C. Temperature loss between stages of HTR separators can be 10°–15°C, depending on the mass flow. Smaller mass flow streams will tend to lose more heat and thus may require reheating prior to the next stage of separation. The EPS units are generally operated with the feed temperature higher than 50°C, otherwise atmospheric moisture will cause significant amounts of nonconductor minerals to report to the conductor stream. With some minerals, there can be a maximum temperature for an EPS separation because excess heat can result in particles tending to adhere to the earth-grounded plate, which prevents the free flow required for separation.

Surface Preparation of Minerals

Similar to the effect of moisture, any contaminants on the mineral surfaces can affect conductivity. Thus, it is a requirement for efficient electrostatic separation that the feedstock be suitably prepared prior to entering the separators. This may simply involve washing in clean water to remove loose slime or dirt contaminants. If the contaminants are bonded to the mineral surfaces, then some form of mechanical scrubbing or attritioning will be required followed by desliming and clean water washing. If the minerals have been extracted from a saline groundwater situation, then thorough washing will be required to remove all traces of salt.

Some clays or organic surface contaminants may require the use of acidic or caustic reagents to clean the mineral surfaces. It is equally important that all traces of these reagents be removed from the mineral surfaces before separation. Hence, even though the process itself is dry, it can require water (and sometimes reagents) for surface cleaning and energy for drying.

Laboratory Test Work

Effective design of a flow sheet to suit a specific mineral suite can only be achieved by testing a representative sample with pilot-scale laboratory separators. Each stage of separation may require different parameter settings to optimize the separation, and thus a suitable quantity of feedstock should be prepared. It is essential to provide sufficient material so that the smallest mass flow stream can be properly tested, as it is often the case that these small middlings streams contain the most difficult

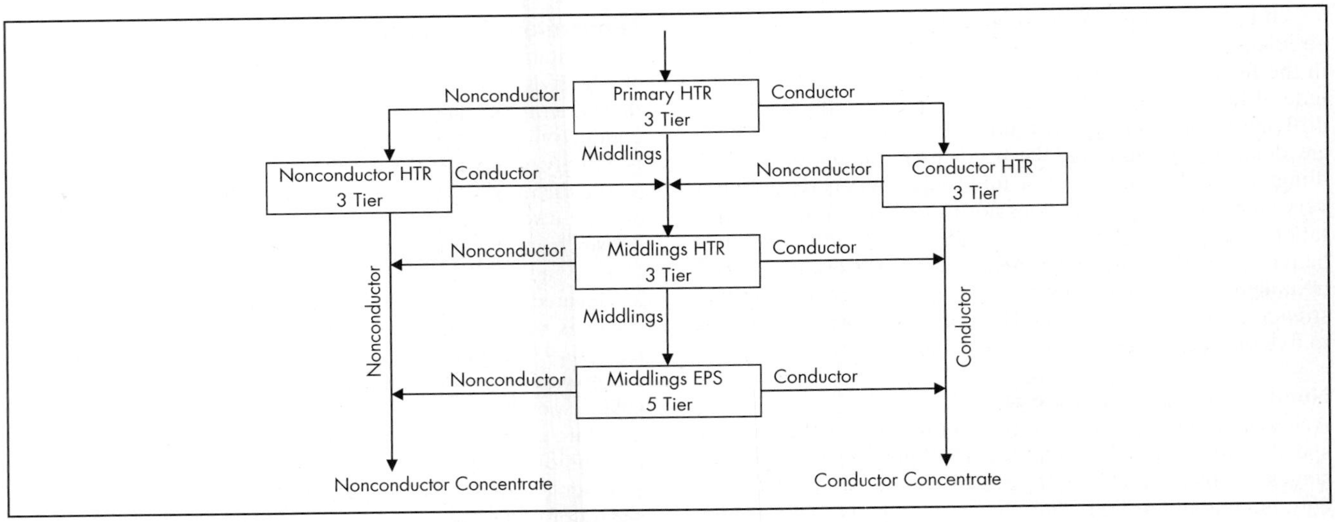

Courtesy of Mineral Technologies Pty Ltd.

Figure 13 Common flow-sheet structure for multistage electrostatic separation

minerals to separate. Without proper consideration, unwanted minerals contained in these streams can adversely affect the final product quality and jeopardize plant performance.

The tests must be performed with a view to what would be the most likely separation conditions found in an operating plant with regard to feed temperature and feed loading. An allowance must also be made for the likely variation in plant conditions caused by changes in ore type, upstream operational difficulties, and ongoing maintenance. It is relatively easy to precisely set up a single laboratory separator; however, such precision is unlikely to be achieved in a plant with multiple stages of separation and several separators in each stage. Thus, the typical metallurgical efficiency in a plant is likely to be below that achieved in laboratory tests performed under well-controlled conditions.

The materials handling of the mineral through an electrostatic separation plant will tend to generate some amount of dust as the particles rub on each other. The presence of this dust in the atmosphere inside the separators can affect the ability to maintain the high voltage required for separation. This can result in repetitive arcing from the corona wire to the roll, which is detrimental to the separation and can damage the high-voltage rectifiers. Thus, it is necessary to operate at a reduced voltage in the plant to prevent this arcing. This must be considered when performing laboratory test work as significantly higher voltages can be maintained in a laboratory separator where there is no dust and heat buildup. Typically, a maximum corona wire voltage for laboratory testing would be 0.43 kV/mm; however, with highly conductive and friable feedstocks, this may reduce to 0.38 kV/mm.

Common Flow-Sheet Structure

Each particular feedstock will require its own flow-sheet peculiarities. A common structure for a mixed feedstock containing reasonable proportions of both conductor and nonconductor minerals is shown in Figure 13. Additional stages of separation may be warranted at either of the conductor HTR or nonconductor HTR stages. Also, re-treatment of the middlings HTR and EPS products may be required for more difficult separations. Each of the conductor and nonconductor concentrates would commonly be directed to further separation processes, such as magnetic or gravity, so that the individual minerals contained therein can be separated. Following these processes, there may be a requirement for further stages of electrostatic separation to achieve specifications for very high-purity products.

CASE STUDY: LABORATORY-SCALE TEST WORK PROGRAM

High-Tension Roll Separation

Typically, a bench-scale laboratory unit will be fitted with a roll of the same diameter as the industrial plant-size unit (270–400 mm) but with a much shorter roll length. The dry test sample is fed onto the roll by a mechanized roll feeder with a feed-path length of normally around 100 mm.

Critical Test Work Variables

The following parameters are the most commonly investigated operating variables that influence the quality of the separation.

1. **Feed rate.** The feed rate must be set and should be a direct ratio of the feed-path width in the test work compared with the plant-scale unit for representative results.
2. **Feed temperature.** Electrostatic separation is much more effective at elevated temperatures where the conductivity of minerals, such as rutile and ilmenite, is enhanced. Typically, the feed is heated to between 80°C and 110°C for separation. Even higher temperatures may produce a better result, but the cost of heating has economic limitations.
3. **Electrode position.** The test unit may have a single ionizing electrode (corona wire), as in the older style "conventional" high-tension separators, or two electrodes (ionizing and plate electrodes), as in most separators built since 2000. The distances between the electrodes and the roll surface will influence separation. These distances should be set to a standard setting or a setting based on knowledge of the test sample or plant experience.
4. **Electrode potential.** The distance between the roll and the corona wire determines the absolute voltage. This is altered by changing the applied voltage on the electrode. Usually, increasing voltage will produce a higher-grade

conductor stream by increasing the mass yield reporting to the nonconductor fraction because more particles are *held* on the separating roll. Optimum voltage is achieved when increasing the applied voltage does not improve the quality of the conductor stream.

5. **Roll speed.** The rotational speed of the separating roll is one of the main operating variables. Roll speed influences the trajectory of the conductor fractions that are centrifugally thrown from the roll and can also influence how tightly nonconductive particles are held onto the roll. Normally, the interplay of electrode potential and roll speed are used to establish the optimum separation.
6. **Splitters.** Three or four adjustable splitters are used to divide the separated minerals and divert the flows into collection trays. The splitters are adjusted to achieve the desired mass yields or product quality. Typically, the conductor/middlings splitter is adjusted most frequently in the test work. Various forms of chute work exist to retreat different streams.
7. **Electrode polarity.** The polarity of the ionizing electrode may be changed. This variable is not commonly changed but may occasionally be necessary. An example of where it may be used is to assist in the adherence of fine quartz to the separating roll.
8. **Atmospheric conditions.** Electrostatic separation is significantly influenced by environmental conditions and, in particular, the water vapor content (measured as relative humidity) of the air. While this may be difficult to control or optimize in a laboratory environment, ideally the conditions should be as similar as possible to the plant environment.

Setup Considerations

The following procedure is typically adopted as best practice when setting up an HTR machine.

1. **Electrical isolation.** Ensure that the separator is electrically isolated and that the unit is properly earth-grounded.
2. **Cleaning.** Safely remove the covers and cautiously clean out the machine using compressed air and brushes, including inside the feed roller, any hangout points inside the main body, the product collection trays, and the brush at the rear of the separation roll (which removes nonconductive minerals from the roll).
3. **Set electrode positions.** Set the electrode positions to the desired gap settings; normally, this is initially a standard position. A spacing template is a useful tool for this setting.
4. **Power and heat up.** Turn on the isolation switch. Start the feed bin heater and set to the desired feed temperature. Turn on the separation roll and heater to the starting rotation rate and let the machine heat up for at least 30 minutes.
5. **Feed-rate measurement.** Using a measured weight of a small sample of feed, adjust the feed roller speed to produce the desired initial feed rate by taking a timed sample of products and calculating the rate. Once the desired feed rate is set, the machine should be cleaned again.
6. **Heat test sample.** The test sample should be kept at the desired temperature in an oven if possible. The temperature is confirmed using a digital thermometer prior to separation. For reheating feed during the test work, it is normally faster to use electric hot plates with continual turning of the sample to achieve the desired temperature consistently through the whole sample.

Test Work Methodology

The following methodology is typically adopted when performing metallurgical test work with an HTR machine.

1. **Initial test.** Switch on the power to the HTR separator and select a standard kilovolt setting. Ensure that there is no arcing. If arcing occurs, then the positioning of the electrodes must be incorrect. Set the roll speed and splitters to standard positions and pass the heated test sample. View the trajectory of the conductors (a spotlight or torch may be required) and adjust the conductor splitter if necessary (modern facilities use high-speed cameras for this evaluation). If arcing occurs while the sample is processing, then the feed temperature or voltage setting may be too high or the electrodes incorrectly positioned too close to the roll. Collect and examine the separated fractions.
2. **Optimization.** Reheat the test sample to the desired temperature and repass, using adjustments of splitters and voltage to improve the separation until it cannot be improved further. If the separation is unsatisfactory at this point, roll speed and feed-rate adjustments may be necessary, especially if the sample is unusually coarse or fine, or contains composite particles reducing the separation quality.
   - Typically, the range of parameters should be maintained between 22 and 30 kV.
   - The feed rate should be an equivalent of 3–6 t/h for a standard-sized plant-scale separator.
   - The feed temperature should be 85°–110°C.
3. **Quality.** If the separation quality is still not acceptable, it may be necessary to adjust the electrode positions. This is the most difficult parameter to adjust, so it should be done after other parameters have been evaluated and the separation is still unsatisfactory. Always ensure that the separator is electrically isolated before adjusting electrodes or reaching into the separation space at any stage.

Tribostatic Separation

The tribostatic separation is generally used in the separation of a two-component feedstock where both minerals are nonconductors. Thus, HTR separators would not be considered for this duty. For some applications, the EPS unit can be operated as a tribostatic separator where there is sufficient contact charging of the particles as they travel down the earth-grounded plates and through the various feed chutes. The EPS unit has been used commercially for the removal of residual quartz from zircon and sillimanite concentrates: however, feedstocks containing larger amounts of quartz would normally be directed to other separation processes.

Tribostatic charging generally only permits relatively weak charges to be accumulated on the mineral surfaces, and thus the condition of the mineral surfaces is more critical than for HTR and EPS separation. For effective tribostatic separation, it is a common requirement that the temperature of the feed material be higher than 100°C and temperatures around 120°C are typical. This ensures that there is absolutely no moisture on the mineral surfaces.

The selection of tribostatic separation can only be made following laboratory test work confirming its effectiveness.

Some feedstocks permit a degree of upgrading by tribostatic means. This, however, may be insufficient to permit a viable plant process to be developed.

A GUIDE TO OPERATOR TROUBLESHOOTING

Metallurgical Issues

In the event of unsatisfactory metallurgical performance of the HTR separator, the following actions may assist.

High-Tension Roll Separation

Problem 1: The nonconductor product contains too many conductor minerals.

- Move the nonconductor splitter farther under the roll.
- Increase the separation roll speed.
- Reduce the feed rate.
- Reduce the feed temperature.
- Decrease the voltage to the electrodes.
- Reduce the plate electrode gap.

Problem 2: The conductor product contains too many nonconductor minerals.

- Move the conductor splitter farther out from the roll.
- Decrease the separation roll speed.
- Reduce the feed rate.
- Increase the feed temperature.
- Increase the voltage to the electrodes.
- Increase the plate electrode gap.
- Clean the corona electrode wire.

Electrostatic Plate Separation

In the event of unsatisfactory metallurgical performance of the EPS unit, the following actions may assist.

Problem 1: The nonconductor product contains too many conductor minerals.

- Move the splitters closer to the earth-grounded plate.
- Increase the voltage to the electrodes.
- Reduce the feed rate.
- Reduce the feed temperature.
- Reduce the plate electrode gap.
- Clean the surface of the earth-grounded plate.

Problem 2: The conductor product contains too many nonconductor minerals.

- Move the splitters farther out from the earth-grounded plate.
- Decrease the voltage to the electrodes.
- Reduce the feed rate.
- Increase the feed temperature.
- Increase the plate electrode gap.
- Check for blocked screens on screen plate separator.

Tribostatic Separation

In the event of unsatisfactory metallurgical performance of the free-fall tribostatic unit, the following actions may assist:

- Adjust the splitters toward the flow of the off-specification stream.
- Reduce the feed rate.
- Vary the feed temperature.
- Vary the voltage to the electrodes.
- Adjust the position of the feed presentation.
- Clean the surfaces of the plate electrodes.

In a plant environment, there are many factors that can influence the overall performance of the various separators. If the preceding adjustments to the individual separators do not resolve an issue with the overall separation quality, then the following other external factors may affect the performance of all types of electrostatic and tribostatic separators:

- Atmospheric moisture (dew point)
- Mineralogical composition of feedstock
- Sizing of minerals to be separated
- Cleanliness of mineral surfaces

Mechanical Issues

The following points relate to mechanical issues that may influence the metallurgical performance of the equipment and should be considered. All types of electrostatic separators may suffer from feed material misreporting within the internal transfer chutes. This may be the result of blockages causing overflows or holed chutes. Regular monitoring is advised to avoid such occurrences.

High-Tension Roll Separators

Issues with the separation roll include

- Heavy scouring or grooved surfaces from over-tensioned brushes;
- Coating, slimes, or organic buildup not removed by the brushes;
- Poor or nonexistent earth-grounding via the brushes or other device; and/or
- Worn bearings causing vibration that prevents particles from properly adhering to the roll.

Issues with the high-voltage supply include

- Loose or broken high-voltage leads,
- Varying voltages between individual separation module electrodes,
- Poor reticulation to the machine and internally,
- Earth-ground leakage in the cable run from the high-voltage source,
- Non-corresponding indicated and actual voltages, and/or
- Corona wire with incorrect gauge.

Electrostatic Plate Separators

EPS machines have few moving parts; however, attention to the condition of fixed components is still advised, such as

- Deformed earth-ground plates, perhaps from overzealous cleaning;
- Deformed electrodes, also from overzealous cleaning;
- Damaged electrodes that have permitted feed material ingress; and/or
- Non-corresponding indicated and actual voltages.

High-voltage supply challenges include

- Loose or broken high-voltage leads,
- Varying voltages between individual separation module electrodes,
- Earth-ground leakage in the cable run from the high-voltage source, and/or
- Poor reticulation to the machine and internally.

Tribostatic Separators

Free-fall tribostatic separators have fewer moving parts than an EPS unit, and because of the separation principle, wear of the vertically positioned electrodes is found to be the majority of mechanical issues. High-voltage reticulation issues are also possible, and performing checks as per the EPS machine is advised.

The tribostatic separator uses high voltages (up to 70 kV). These voltages are potentially lethal. Safe procedures must be followed by certified fitters at all times during the operation of this machine. Earth-grounding is critical and the machine must be securely earth-grounded with zero electrical impedance between the machine frame and the plant electrical earth-ground. The machine must be operated with all doors and access panels closed. The access doors are fitted with safety cut-out switches to stop the machine and cut both standard-voltage and high-voltage circuits.

Larger industrial machines can weigh up to 10,000 kg and more than 15,000 kg if fully sanded, so lifting consideration and sturdy support structures for fixing the machine are essential. It is also recommended to allow around 30 minutes after stopping the machine to discharge all of the residual static charge on the electrodes before maintenance.

Wear on the moving belts of the tribostatic belt separator is to be expected, and hence a regular inspection and replacement schedule is warranted.

TROUBLESHOOTING CASE STUDY

A long-established mineral sands producer had noticed a gradual decline in the throughput of its dry separation plant, together with a constant struggle to achieve grade and recovery targets despite installation of new equipment in some parts of the circuit.

Through consultation with separator specialists, the decision was made to perform a comprehensive mechanical audit of each machine and the associated maintenance work practices. The audit identified areas where key components, either individually or when combined with a group of components, were negatively impacting the overall performance of the separator. This, together with a gradual decline in operational maintenance, contributed to the gradual worsening in overall plant performance.

Using the data compiled by the separator specialists, a program of refurbishing or replacing the components in the key areas was undertaken. The most common components identified were worn or out-of-specification separation rollers, alternative supply separation roller brushes that did not fit correctly in the brush holders, and ineffective earth-grounding. Also, the high-voltage leads were, in many cases, very old with deteriorated insulation causing voltage leakage, resulting in inconsistent voltage at the various electrodes within a single machine. All high-voltage leads were subsequently replaced.

Consequently, it became clear to senior operational personnel that, without revising work practices, the ability to maintain the operational integrity of the machines would be short-lived. Procedures for the day-to-day operation were implemented, together with structured personnel training for both new and existing employees. Also, the approach to separator maintenance was changed from a "fix it when it breaks" culture to a more structured program of planned maintenance.

The initial audit and refurbishment program took approximately 12 months to complete. Over that time, the previously noted decline in plant performance slowly reversed, and, ultimately, the performance returned to expected levels. The benefits of improved plant performance were not the only positive outcome. While the component replacements initially required additional expenditure, the maintenance costs decreased over the following two-year period and then stabilized.

REFERENCES

Bittner, J.D., Gasiorowski, S.A., Bush, T.W., and Hrach, F.J. 2013. Separation technologies' automated fly ash beneficiation process selected for new Korean power plant. In *Proceedings of the 2013 World of Coal Ash Conference*. www.flyash.info/2013/036-Gasiorowski-2013.pdf.

Bittner, J.D., Flynn K.P., and Hrach, F.J. 2014. Expanding applications in dry triboelectric separation of minerals. Presented at the 27th International Mineral Processing Congress, Santiago, Chile, October 20–24.

CHAPTER 6.6

Methods for Recycling Applications

Brajendra Mishra and Myungwon Jung

The recycling of metals from waste materials is important for conserving natural resources and reducing environmental issues. One of the benefits of metal recycling from scrap material is energy savings compared to the primary production of metals. For example, producing steel by remelting steel scraps saves about 74% more energy compared to producing steel from iron ores. In the case of aluminum, secondary aluminum production saves 95% more energy than primary production. For this reason, secondary aluminum production has outpaced primary production since 2001, and two nonoperating primary aluminum smelters were permanently shut down in 2015.

In general, primary metal production comprises exploration, mining, mineral processing, and extractive metallurgy as shown in Figure 1. Waste recycling follows the same general process steps, except the first two steps are (1) the identification of values in waste and (2) the collection and logistics of delivering these wastes to a central point for processing. For recycling applications, initial chemical/physical characterization of waste streams, including the use of automated mineralogy equipment such as the mineral liberalization analyzer (MLA) and Quantitative Evaluation of Minerals by Scanning Electron Microscopy (QEMSCAN), is required. Technologies that provide the basis of mineral processing and extractive metallurgy are fundamental to both recycling applications and primary metal production. Broadly, mineral processing utilizes physical techniques and methods, whereas extractive metallurgy processing is based on chemical separation principles. In metals and materials (waste) recycling, both mineral and extractive metallurgy processes are employed to separate and recover metals and other values, respectively.

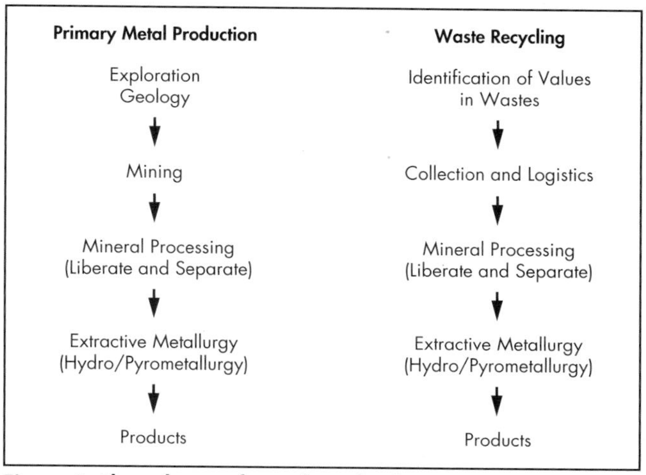

Figure 1 Flow sheets of metal production

MINERAL PROCESSING FOR ORES

The primary role of mineral processing is to concentrate (upgrade) the target material in an assembly of mixed materials (recyclable waste). When materials are intimately associated, comminution is often required to reduce the particle size for better liberation. Comminution is performed for the following reasons (Gupta 2003):

- Liberation of one or more economically important value components from the waste components in an ore or waste assemblage
- Exposure of a large surface area per unit mass of material to facilitate some specific chemical reaction, such as leaching
- Reduction of the raw material to the desired size for subsequent processing or handling
- Satisfaction of market requirements concerning particle size specifications

Basic methods of mechanical size reduction or comminution for recyclable wastes include crushing and grinding to break apart physically attached materials or phases. After the waste is reduced in size, separation techniques are used. Liberation and separation are the two main concepts in mineral processing. In general, liberation can be achieved by the comminution processes. After that, separation processes are required to yield concentrated products from the processed ore and other materials (Fuerstenau 2003; Hayes 2003; Kelly and

Brajendra Mishra, Director of Metal Processing Institute, Mechanical Engineering, Worcester Polytechnic Institute, Worcester, Massachusetts, USA
Myungwon Jung, Postdoctoral Researcher, Metal Processing Institute, Worcester Polytechnic Institute, Worcester, Massachusetts, USA

Table 1 Comparison chart for pyro- and hydrometallurgical processes

Terms of Reference	Pyrometallurgy	Hydrometallurgy
High-grade ores	More economical	Less economical
Low-grade ores	Unsuitable	Suitable
Complex ores	Unsuitable	Has flexibility
Secondary resources	Unsuitable in most cases	Suitable
Reaction rate	Rapid (high temperature)	Slow (low temperature)
Environmental pollution	Waste gas, noise, dust	Mainly wastewater

Adapted from Gupta and Mukherjee 1990

Spottiswood 1982; Pryor 2012). The major purposes of separation include the following (Gupta 2003):

- A crude product is purified by removing any contaminating impurities.
- A mixture of two or more products is separated into the individual pure products.
- The stream discharged from a process step may consist of a mixture of product and unconverted raw material, which must be separated and the unchanged raw material recycled to the reaction zone for further processing.
- A valuable substance, such as a metallic ore, dispersed in a mass of inert material must be liberated for recovery and the inert material discarded.

Prominent techniques for particulate separation are based on differences in specific gravity, surface chemistry (froth flotation), magnetic susceptibility, conductivity, and a whole host of identifiable characteristics (e.g., color, reflectance, chemistry, radioactivity) for sortation. These separation techniques are very useful for separating and concentrating materials of value in a mixed recyclable waste as it is often a mixture of metallic and nonmetallic materials, magnetic and nonmagnetic materials, and/or electrically conducting and insulating materials. Different densities of materials allow the use of gravity-based separation techniques (Anderson et al. 2014; Habashi 1997; Wills and Napier-Munn 2006).

EXTRACTIVE METALLURGY FOR ORES

Extractive metallurgical processes are divided into pyrometallurgy, hydrometallurgy, and electrometallurgy and are applied to the concentrate products produced by particulate separation in mineral processing (Rosenqvist 2004). In brief, pyrometallurgical processes rely on the application of heat, whereas hydrometallurgical processes rely on the use of solvents to obtain the desired chemical reaction. Electrometallurgy is generally considered a subcategory of hydrometallurgy, and it uses electric current to produce metals from their ions in solution. Typically, the same metal can be produced by different unit processes. The selection of the unit processes mainly depends on economic and environmental considerations. Based on the ore or waste types and their complexity, different metal extraction methods can be used, and a comparison chart for pyro- and hydrometallurgical processes is shown in Table 1.

Pyrometallurgical techniques provide the separation of valuable products as a liquid phase, separating the unwanted materials through the formation of a slag. On the other hand, hydrometallurgical methods generally begin with a leaching step to selectively dissolve or un-dissolve the target material. This separation is achieved by precipitation via pH control and/or oxidation, as well as by solvent extraction and ion-exchange principles (Habashi 1999). Metal recovery from leach solution is the final stage of hydrometallurgical processes. In general, metal recovery is done using several different methods, including cementation, ionic precipitation, crystallization, reduction with gas, and electrolysis of metal. Table 2 shows how specific metals can be recovered by one of these methods.

Table 2 Metal recovery methods and their applications

Recovery Methods	Solution Types	Applications	Products
Cementation	Dilute	Au, Ag, Cu, Cd	Metal (impure)
Ionic precipitation	Dilute/ concentrated	Fe, U, Th, Be, Cu, Co	Metal compounds
Crystallization	Concentrated	Al, Cu, Ni, Mo, W	Metal compounds
Reduction with gas	Concentrated	Cu, Ni, Ag, Mo, U	Metal and metal compounds
Electrolysis	Concentrated	Au, Ag, Cu, Ni, Co, Zn	Metal (pure)

Adapted from Gupta and Mukherjee 1990

APPLICATION OF MINERAL PROCESSING AND EXTRACTIVE METALLURGY FOR ORES TO WASTE RECYCLING

Mineral processing and extractive metallurgy techniques are used in the recycling of waste materials generated during manufacturing and use. Since the primary production involves ores that are chemically and compositionally different from the in-process or postconsumer waste forms, the process selection for treatment varies and is adjusted for the materials of interest. The environmental aspects are also different when recycling materials from industrial or postconsumer wastes. Recycling has requirements similar to those of low-energy, energy-efficient, environmentally compatible, and economically feasible processes for primary production. However, other challenges must be met to justify recycling, such as the following (Björklund and Finnveden 2005):

- The demand for materials must outstrip the supply.
- The developed recycling process must be energetically favorable.
- The process must not negatively affect the environment for air, water, and land.
- The process must be economically viable so there are economic gains for the recycler.
- Secondary resources must be available requiring high-volume collection systems.
- Separation and beneficiation (upgrading) must be technically possible.
- Feasible chemical/metallurgical processing must be available based on similar mineral processing and extraction techniques.
- A favorable market must exist for recycled products.
- The area of recycling needs to be supported politically and socially with proper policies and regulations for practical environment compatibility, economic incentives, and enforced disposal laws.
- The main sources of waste forms in metals and materials recycling come from Mine tailings and sludges;

- Materials rejected during physical concentration steps;
- Slags, baghouse dusts, and flue-dusts generated during pyrometallurgical operations;
- Fluid effluents during hydromet/electromet operations;
- Conversion of metals into alloys and engineering materials;
- Deformation and machining operations;
- Component fabrication rejects (out-of-specification components);
- Postconsumer wastes; and
- Landfill wastes.

Process Applications

Seven process application examples are presented in the following sections. They relate to (1) lithium-ion battery recycling, (2) recovery of metals and materials from red mud, (3) recycling phosphor dust for rare earth metals recovery, (4) recovery of aluminum from secondary aluminum production dust, (5) vanadium recovery from fly ash, (6) e-waste recycling, and (7) ferrous scrap recycling. All of these examples for recycling prominent waste materials exemplify the use of mineral processing and extractive metallurgy techniques.

Lithium-Ion Battery Recycling

Although many recycling processes are used industrially for lithium-ion batteries, they have some drawbacks (Al-Thyabat et al. 2013):

- Not all of them are lithium-ion battery dedicated processes.
- In most of the processes, not all of the materials are recycled.
- Usually the electrolyte, plastics, and organics are evaporated or are burned during pyrometallurgical steps and thus are not recovered.
- Hydrometallurgical processes can sometimes be questionable from an economical point of view because of the low solid-to-liquid ratio and the cost and handling of the reagents (especially if they are not recovered).
- Some battery components cannot be easily leached. An example is the strong bond between oxygen and cobalt, which makes leaching with commonly used reagents harder.

An efficient lithium-ion recycling process should adapt to the following requirements (Xu et al. 2008):

- Lithium-ion batteries can be composed of different materials, which are constantly being substituted by new ones, thus making it harder to find a process that would treat all the different compositions at the same time.
- The electrolyte ($LiPF_6$) should be recovered but is very hard to recover because of its high reactivity with water, moisture, and heating.
- The products produced from recycling must be pure enough to meet market needs and be competitive to new products.

Because of these difficulties, considerable research has been conducted over the last few years. The primary goal is to maximize recovery of the products and be environmentally friendly. A proposed approach to recycling lithium-ion batteries described by Marinos and Mishra (2015) is presented in Figure 2.

First, the material is shredded to liberate the constituents of the lithium-ion battery. This step requires a good exhaust system, such as a baghouse, to avoid dust and organic gas. Dust will be collected and all the fine material treated separately as shown in the flow sheet. Then the shredded material is passed through a low-intensity magnetic separator to remove the ferromagnetic material, which is primarily the steel battery shell.

The remaining material is leached at room temperature for one hour using distilled water as the leachant. Lithium is present in the cathode, usually as $LiCoO_2$ (insoluble in water), and in the electrolyte as $LiPF_6$. Thus, during the leaching process, lithium present in the electrolyte, as $LiPF_6$, goes into solution. This leaching step can be performed in a flotation cell at ambient temperature, pressure, and pH. After leaching is complete, the air valve is opened, forming bubbles and thus helping the low-specific-gravity particles (plastic and some fine particles that are stuck on the plastic) float. In this way, plastic and some fine particles are separated from the lithium-bearing leach solution. The main advantage of this system is that both leaching and separation are conducted in one step. The floated plastics and fine particles are then wet screened into coarse and fines to wash the plastics and separate the fine particles that are stuck on the plastic material. The coarse plastics can then be recycled and the fine material will be collected to treat separately.

The remaining pulp, pregnant solution, and solids of high specific gravity are wet screened into multiple sizes. The wet screening helps to clean the particles from the fine carbon and cathode powder. Each differently sized fraction will be treated separately. The very fine will be combined with the rest of the fines and go through the hydrocyclone and/or be leached to recover carbon and cathode materials. Carbon could also be recovered from fines using pine oil to float it selectively from the rest of the fines, and the cathode materials could be recovered in multiple ways that have been researched.

The pulp that passes through the smallest sieve gets filtered, and the pregnant solution is recovered. Fine particles that are filtered are combined with the fines from previous steps. Lithium concentration in solution is low, and to precipitate lithium carbonate from the pregnant solution, according to lithium carbonate solubility at 100°C, the concentration of lithium must be at least 1.4 g/L and at room temperature at least 2.4 g/L. The concentrations are based on the fact that lithium carbonate solubility decreases with increasing temperature. Thus, if the concentration of lithium in the pregnant solution is lower than 1.4 g/L at 100°C or 2.4 g/L at room temperature, the solution is recycled back to the leaching step to increase its concentration. If the concentration of lithium is higher than that, the lithium carbonate product is recovered using carbon dioxide or hydrochloric acid, to turn LiOH to LiCl, and sodium carbonate. This process is used industrially for the recovery of lithium.

The rest of the fractions that were obtained by sieving will be dried and sent separately for high-intensity magnetic separation and then eddy-current separation. High-intensity magnetic separation and, more specifically, rare earth roll magnetic separators operate at very high magnetic field intensities and thus could separate many paramagnetic materials from diamagnetic materials.

After the high-intensity magnetic separation, each fraction will be sent separately to an eddy-current separator. Eddy-current separation, depending on particle size, is very

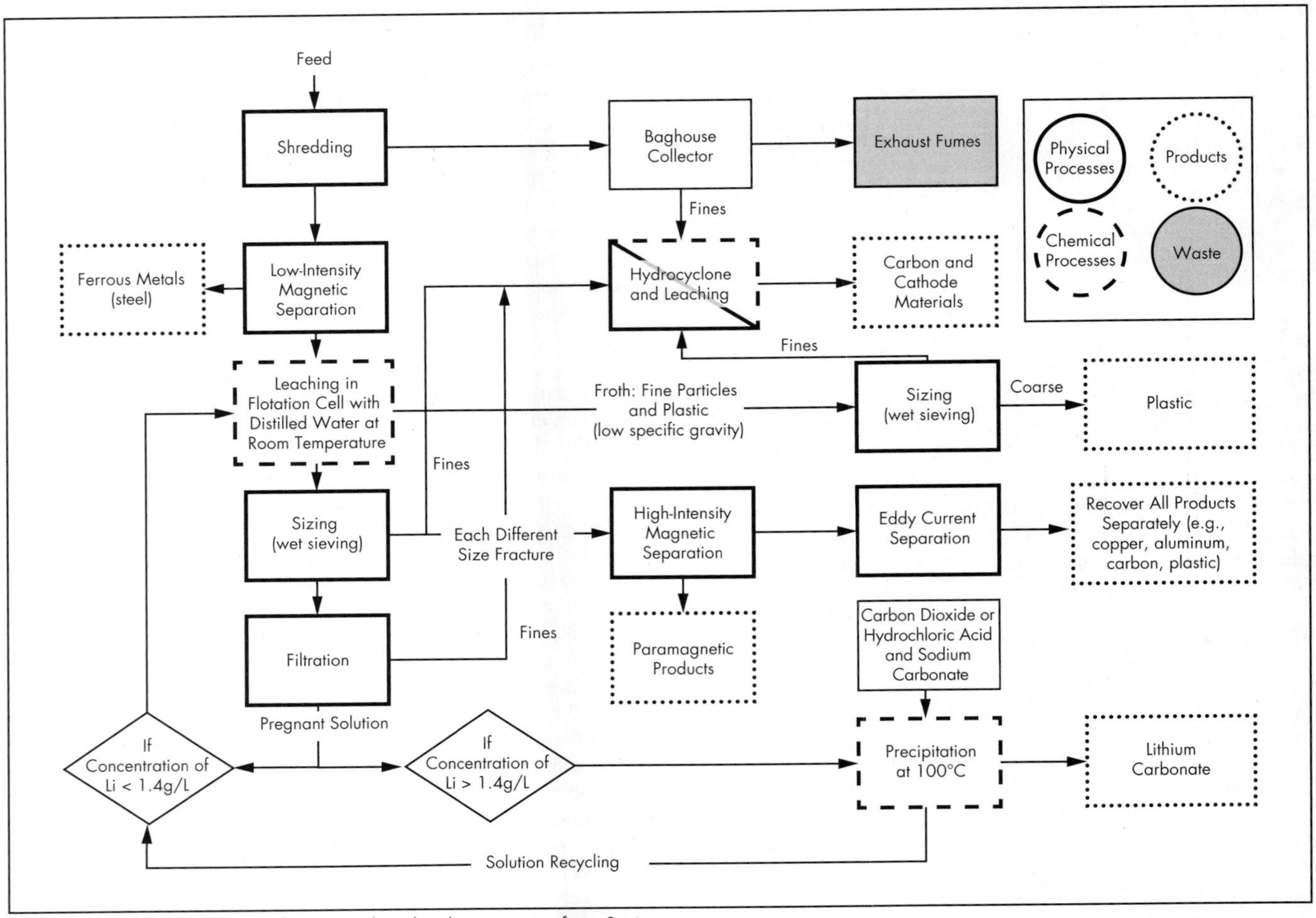

Source: Marinos and Mishra 2015, reproduced with permission from Springer

Figure 2 Experimentally investigated flow sheet for complete recycling of lithium-ion batteries

attractive for this process. Theoretically, if all the particles are liberated and the size of the particles is very similar, then a strong separation of multiple products can be achieved. In this case, all the material is completely liberated, and multiple sieves are used to achieve closely sized particles. In this step, all of the remaining products will be recovered separately (e.g., copper, any remaining carbon or plastic, or aluminum).

Certain steps in the process have been tested, such as leaching, flotation, and sieving. In research experiments, shredding was performed followed by magnetic separation (without optimizing the conditions) because these processes have already been researched and are successfully used by industries. The leaching conditions using distilled water were optimized. A flotation machine was used for leaching to remove the plastic. Sieving was performed to determine the weight and the chemical components present in each fraction. Then an indicative experiment was conducted to see if the rare earth rolls step and the eddy-current separation are feasible.

Recovery of Metals and Materials from Red Mud

Red mud, also known as bauxite residue, is a by-product of the Bayer process. During the Bayer process, crushed bauxite ores are leached in pressurized vessels to produce a concentrated sodium aluminate solution, and the pregnant leach solution is further treated through crystallization and calcination to produce aluminum oxide. After the leaching of bauxite ores, solid residue (red mud) is filtered and separated from the sodium aluminate solution. According to the International Aluminium Institute (2013), the global production of aluminum was about 101 Mt (million metric tons) in 2012, and it generated 120 Mt of red mud. Red mud contains the unreacted metallic oxides, such as iron oxide, aluminum oxide, titanium oxide, and silicon oxide, and their compositions are related to the quality of the bauxite ore. Because of its relatively high content of metal oxides, red mud is considered as a secondary raw material; however, red mud is discarded to tailings ponds and it causes environmental issues. For these reasons, much research has been conducted related to the recovery of metals and materials from red mud. For example, Figures 3 through 5 show various potential flow sheets using hydrometallurgical and pyrometallurgical techniques subsequent to mineral processing methodologies for the recovery of iron, aluminum, and titanium from red mud.

In addition to the flow sheets shown in Figures 3 through 5, recovery of metals from red mud is studied with different methods in the following examples.

Red mud contains about 2.2%–33% alumina. Alumina is recovered either by forming sodium aluminate (soda ash roast) or by incorporating organic reagents hydrometallurgically. Zhang et al. (2014) recovered alumina hydrochemically

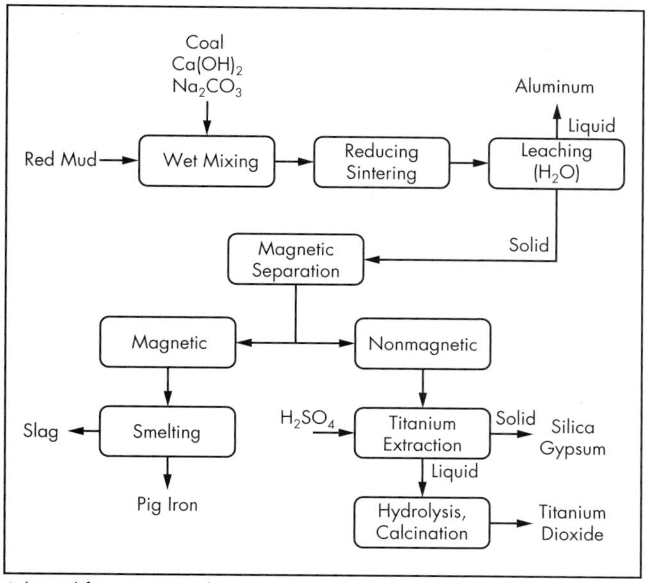

Adapted from Piga et al. 1993

Figure 3 Flow sheet for recovery and recycling of iron, aluminum, and titanium from red mud

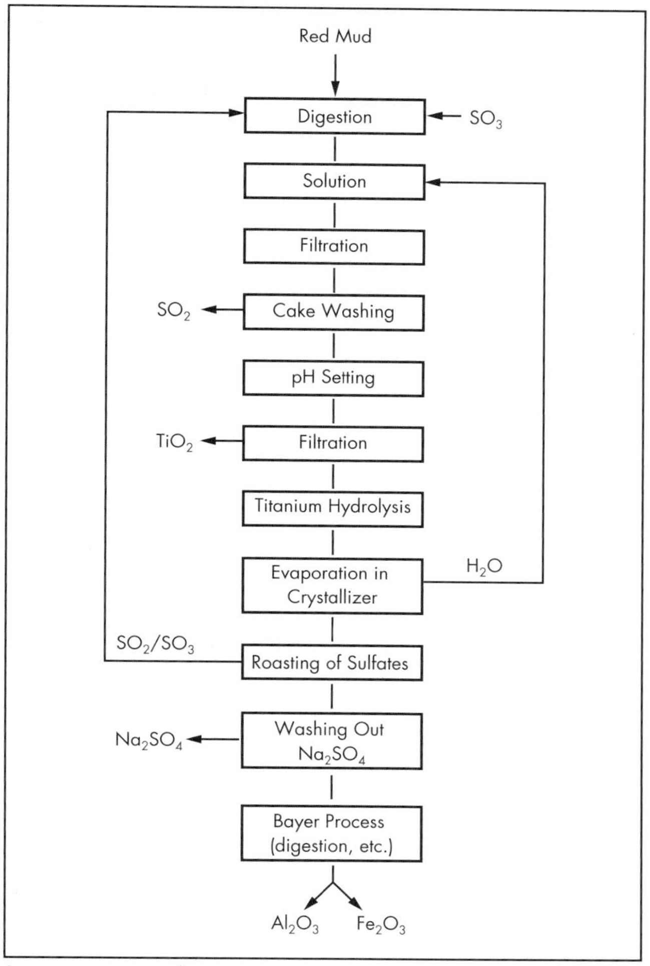

Source: Zimmer et al. 1978

Figure 4 Alternative flow sheet proposed for value recovery from red mud based on hydrometallurgical processing

by the formation of grossular hydrogarnet. In a study conducted by Zhong et al. (2009), about 87.8% of alumina was extracted from the red mud with mild hydrochemical process conditions utilizing NaOH pressure leaching, which is essentially a repeat of the Bayer process.

Bruckard et al. (2010) tried smelting on red mud, which was mixed with calcium carbonate at 1,400°C. The Al and Na could be easily recovered because of the phase transformation into $(Na,Ca)_{2-x}(Al,Fe^{3+})_{2-x}Si_xO_4$. The slag was subjected to water leaching at 60°C, and 55% Al and 90% Na could be recovered at 50% pulp density.

Li et al. (2009) have attempted to devise a simultaneous sintering process for alumina recovery and converting hematite to magnetite. They found that the sintering temperature and amount of carbon strongly influence alumina recovery. A high recovery of 75.7% Al and 80.7% Na can also be obtained by soda lime roasting followed by water roasting of red mud.

Iron is the major constituent in red mud. The concentration of Fe in iron oxides varies between 6.8 wt % and 71.9 wt %, depending on the grade of bauxite. A prior reduction of the hematite in red mud transforms it into a magnetic product that can be separated by magnetic separation. Li's team (2009) suggested that pulsating magnetic current utilization would provide better separations than the conventional magnetic current. A high-gradient magnetic separator can separate magnetic particles less than 100 µm in size. Li et al. observed that weakly magnetized particles are attracted.

Xiangong et al. (1996) found that adding calcium, sodium, or magnesium salt can improve the efficiency of carbon-based direct reduction of iron in red mud. A team led by Jayasankar (2012) has proposed that the addition of 10%–12% dolomite would increase iron recovery to 71%. A recovery of 94.5 wt % Fe can be obtained by employing 6% Na_2SO_4 and 6% Na_2CO_3 in the mixture.

The soda ash roasting and leaching process described by Li et al. (2009) generates a leach residue consisting of magnetite, hercynite, and $TiFe_2O_4$. The iron recovery decreases because of agglomeration of ferrous phases and other impurities.

Acid leaching for recovering iron was also investigated. A high recovery of 97.5% Fe can be obtained at a calcination temperature of 600°C and subsequent leaching in 6M H_2SO_4.

Recycling Phosphor Dust for Rare Earth Metals Recovery

Rare earth elements (REEs), which include 15 lanthanide series plus scandium and yttrium, are important components of modern technologies, such as electronics, clean energy, national defense, and others. The unique properties of the REEs, including luminescence, magnetism, and electronic properties, promote the performance of high-tech equipment by adding a small amount of REEs. Currently, more than 95% of world rare earth production comes from China, and the U.S. Department of Energy (DOE) addressed the challenges associated with the use of REEs in energy technologies. Thus, the DOE has supported the Critical Materials Institute and updated the criticality assessments of the materials since 2010. According to its *Critical Materials Strategy* report (DOE 2011), five REEs (dysprosium, neodymium, terbium, europium, and yttrium) and indium are most critical for clean

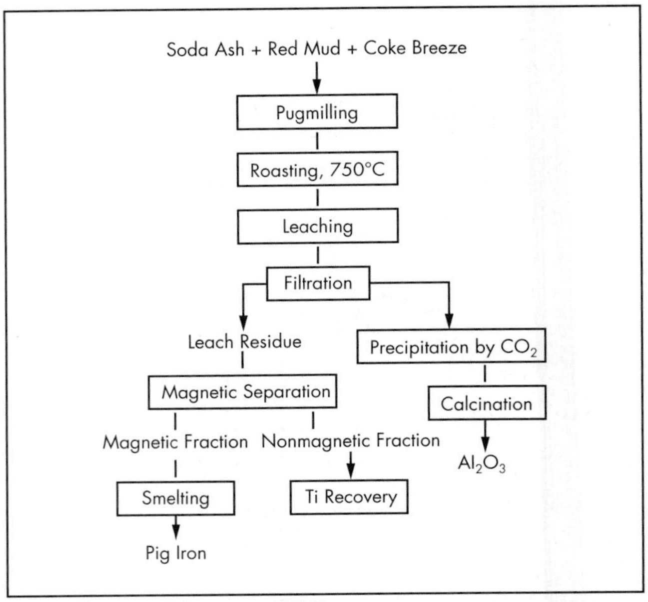

Source: Kumar and Srivastava 1998

Figure 5 Hydrometallurgical and pyrometallurgical processing of red mud for separation of iron, titanium, and aluminum

energy technologies. In general, fluorescent lighting phosphors contain europium, yttrium, and terbium, which are critical materials based on the DOE's materials strategy. Strauss (2016) and Eduafo (2016) conducted the recovery of REEs from phosphor dust, and the proposed flow sheet is shown in Figure 6.

Based on the flow sheet in Figure 6, a characterization of phosphor dust was completed for waste lamp phosphor dust. Various processes were used to quantify the dust. QEMSCAN found that the majority of the powder was quartz (54.30%), followed by apatite (16.64%), yttrium-bearing minerals (14.98%), and calcite (4.75%). Monazite was also present with trace amounts of xenotime, iron oxide, barite, celestine, and europium- and terbium-bearing minerals.

More than 60% of the rare-earth-containing minerals were smaller than 10 µm. In particular, more than 95% of the europium-containing minerals were less than 10 µm. In addition, more than 66% of the quartz was greater than 75 µm, indicating that size-based sorting could present a viable physical separation process for this aspproach. Future work should focus on physical separation methods, such as those using a cyclone, or magnetic separation unit operations.

Oxalic acid precipitation was optimized for the europium and yttrium leach solution. The optimized conditions were 25°C, 600 rpm agitation, pH = 0 (native pH). Neither

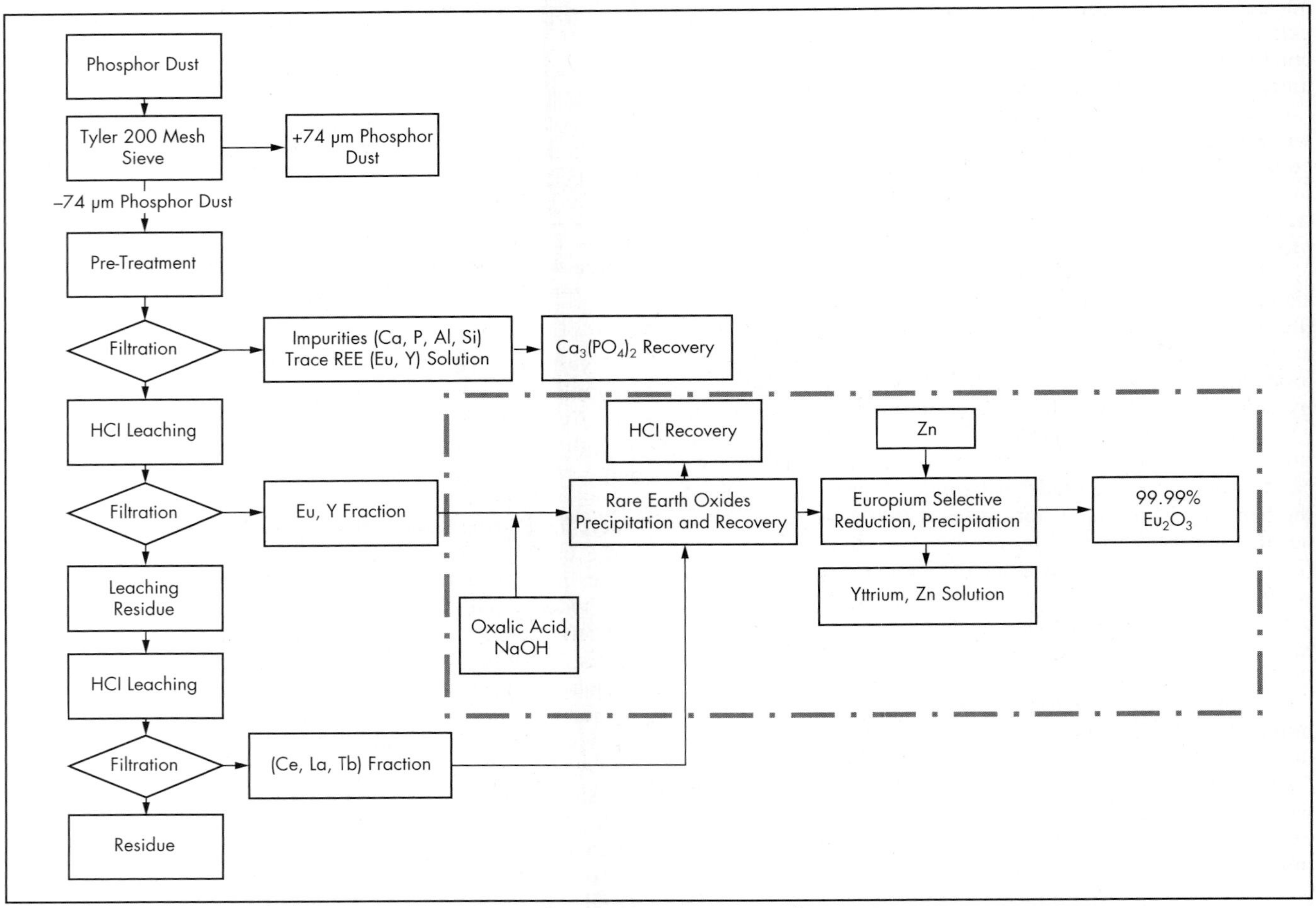

Source: Strauss 2016

Figure 6 Recycling of rare-earth metals from postconsumer products using low-energy hydrometallurgical processes

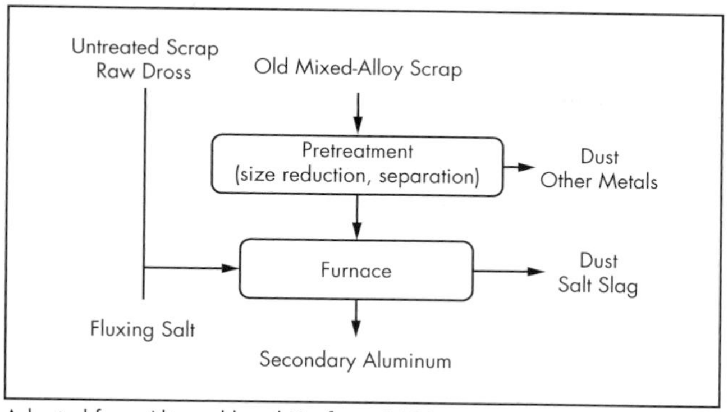

Adapted from Altenpohl and Kaufman 1998

Figure 7 Traditional processing of old scrap for secondary aluminum

temperature modification from room temperature nor pH modification with NaOH were necessary for optimization. More than 99% recovery and 99% purity were achieved for mixed yttrium and europium rare earth oxides (REOs). Future work should focus on conditions to decrease oxalic acid consumption to improve the economics of the process.

The selective reduction of europium(III) to europium(II) in solution along with precipitation of europium(II) sulfate is a successful method to separate europium from yttrium. More than 95% pure europium sulfate was created in the process at laboratory scale with more than 80% recovery. However, more research needs to be conducted to scale up this work to a larger apparatus and to verify that it can be economic. In addition, more simulations using HSC's Gibbs free energy minimization need to be conducted.

A very preliminary economic analysis of the recovery process was conducted to study creating a recycling facility for phosphor dust. Based on the REO six-year price average and REO two-year price average, the project is viable. However, at current prices, the project is highly uneconomic. More work needs to be conducted to study the cost of wastewater treatment as well as to add the value of acid recovery and terbium oxide recovery to the process.

Recovery of Aluminum from Secondary Aluminum Production Dust

Most metal production plants utilizing pyrometallurgical processes generate air pollutants including particulate matters and harmful gases. Thus, these plants often install several dust-collecting systems such as cyclones and baghouses to control the emission of air pollutants. In general, dust collected from a metal production plant also contains metal values. Therefore, metal recycling from the collected dust can be valuable by saving metal resources as well as reducing environmental issues. One example of industrial fine wastes is aluminum smelter baghouse dust, which is generated during shredding or smelting from secondary aluminum production facilities as shown in Figure 7. Secondary aluminum production requires about 5% of energy to produce aluminum compared to primary aluminum production. Since 2001, secondary aluminum production has outpaced primary aluminum production in the United States as shown in Figure 8. Moreover, global aluminum production will increase and is projected to reach around 100 Mt/yr by 2020 as shown in Figure 9.

Typically, dust from the aluminum industry that is larger than 20 mesh in size contains a large amount of aluminum. Thus, this particular size fraction is usually remelted. However, undersize fraction (smaller than 20 mesh) is landfilled. About 1 Mt of secondary aluminum production wastes, including

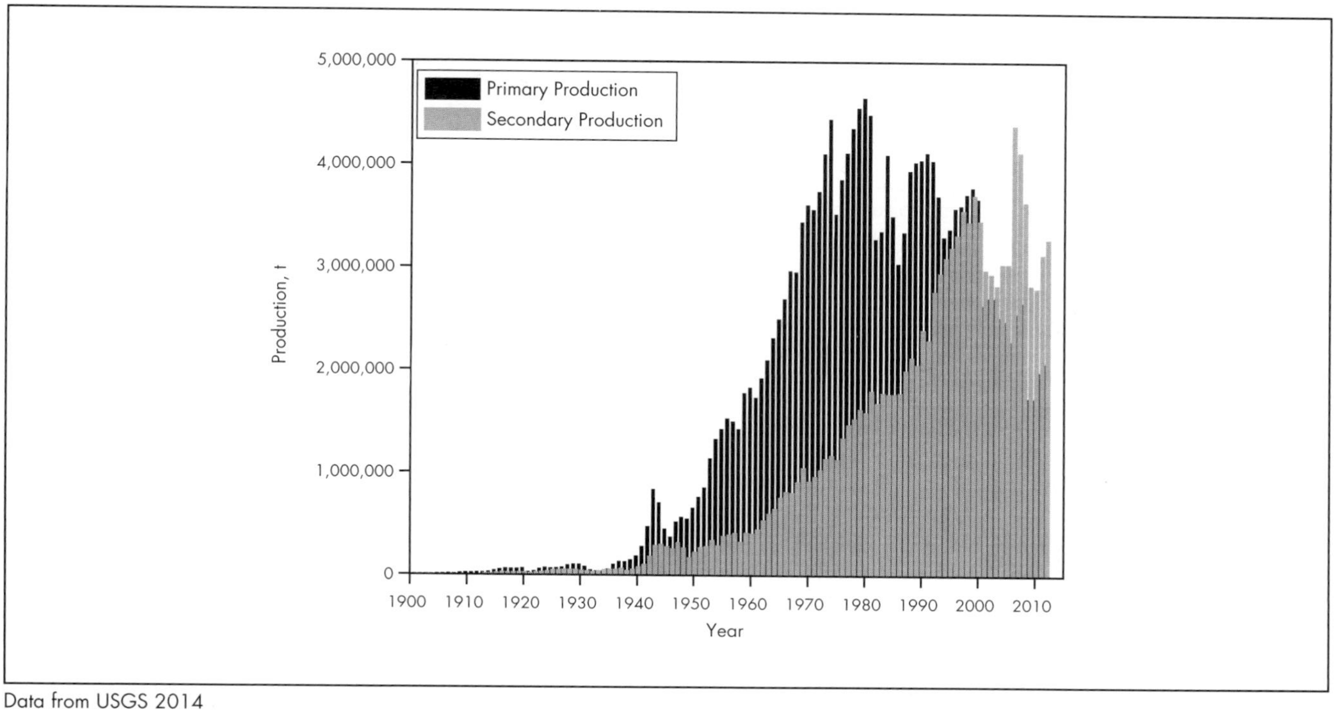

Data from USGS 2014

Figure 8 U.S. aluminum production by type

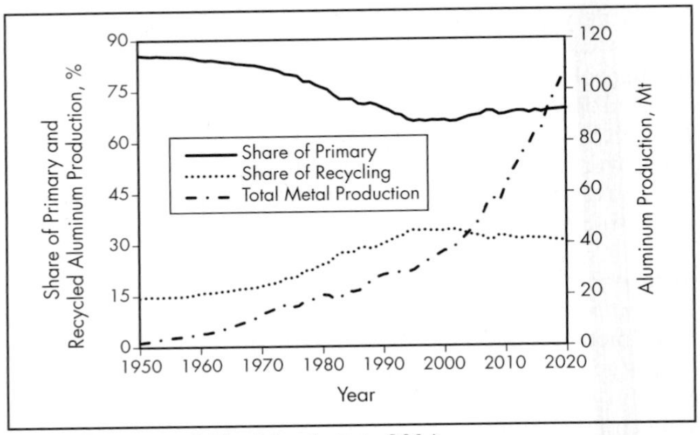

Source: International Aluminium Institute 2016

Figure 9 Forecast share of global primary and recycled metal production

salt cake and baghouse dust, are being generated annually and landfilled in the United States (DOE 1999). Therefore, disposal of aluminum industry dust will soon be an issue because of its flammable and irritant nature when the size of the dust is smaller than 420 µm (Aluminum Association, n.d.). In European Union countries, baghouse dust from aluminum production facilities is recognized as a hazardous waste.

Huang et al. (2015) investigated baghouse dust samples from 13 different secondary aluminum production facilities in the United States for mineral phases and metals in baghouse dust. Based on this study, the major aluminum phases are aluminum metal, aluminum oxide, aluminum nitride, and spinel. The average aluminum content is about 18% by weight and ranges from 2.6%–60%. Therefore, aluminum recovery from aluminum smelter baghouse dust could be valuable depending on the aluminum content of the waste streams.

Jung and Mishra (2016a, 2016b) have attempted to recover aluminum from aluminum smelter baghouse dust by hydrometallurgical processes. Based on inductively coupled plasma analysis, the aluminum contents in the sample ranged from 13%–20% by weight, depending on the particle size. Similarly, samples having a larger particle size showed a relatively large amount of aluminum content. In this type of sample, aluminum metal, quartz, and calcite are the major phases of the aluminum smelter baghouse dust based on X-ray powder diffraction (XRD) analysis as shown in Figure 10. One of the main advantages of hydrometallurgical processes is the selective leaching characteristic, depending on the reagent selection. Therefore, several leaching reagents, such as sulfuric acid, nitric acid, sodium carbonate, and sodium hydroxide, are used to extract aluminum. Based on the preliminary leaching tests, alkaline leaching reagents promote the selective leaching characteristic, and sodium hydroxide is a better reagent in terms of aluminum dissolution than is sodium carbonate. Because the major phase of the aluminum smelter baghouse dust is aluminum metal, the aluminum dissolution rate is relatively fast, and more than 75% of aluminum can be dissolved within 20 minutes. Based on XRD analysis, several reaction equations between aluminum and sodium hydroxide solution are proposed:

$$\mathrm{Al + 3H_2O(l) + NaOH(l) = Na[Al(OH)_4] + 1.5H_2(g)}\ \Delta G^\circ = -116.350 \text{ kcal at } 100°\mathrm{C} \quad \textbf{(EQ 1)}$$

$$\mathrm{2Al + 2H_2O(l) + 2NaOH(l) = 2NaAlO_2 + 3H_2(g)}\ \Delta G^\circ = -221.893 \text{ kcal at } 100°\mathrm{C} \quad \textbf{(EQ 2)}$$

$$\mathrm{2Al + 3H_2O(l) = Al_2O_3 + 3H_2(g)}\ \Delta G^\circ = -211.080 \text{ kcal at } 100°\mathrm{C} \quad \textbf{(EQ 3)}$$

$$\mathrm{Al + 3H_2O(l) = Al(OH)_3 + 1.5H_2(g)}\ \Delta G^\circ = -102.415 \text{ kcal at } 100°\mathrm{C} \quad \textbf{(EQ 4)}$$

For the same molarity of aluminum, the formation of $Na[Al(OH)_4]$ is thermodynamically favorable at 100°C since it shows the largest negative value of Gibbs free energy. An equilibrium composition diagram based on the HSC chemistry shows identical results based on the Gibbs free energy minimization.

In the primary production of aluminum, the Bayer process, which uses bauxite ores as an aluminum source, also uses sodium hydroxide to dissolve aluminum from the crushed and grounded bauxite ores. During pressure leaching, gibbsite, boehmite, and diaspore in bauxite ores also react with sodium hydroxide and produce $Na[Al(OH)_4]$ after leaching. The final product after leaching of bauxite ores and aluminum smelter baghouse dust is identical based on the thermodynamic modeling (Figure 11). Therefore, aluminum smelter baghouse dust can be used as an aluminum source to generate sodium aluminate and share the calcination plants to produce alumina. However, hydrogen gas generation during leaching, as shown in Equations 1 through 4, should be controlled properly because of its flammable and explosive nature. Additionally, hydrogen gas generated from aluminum recovery can be an alternative hydrogen production method, and several studies

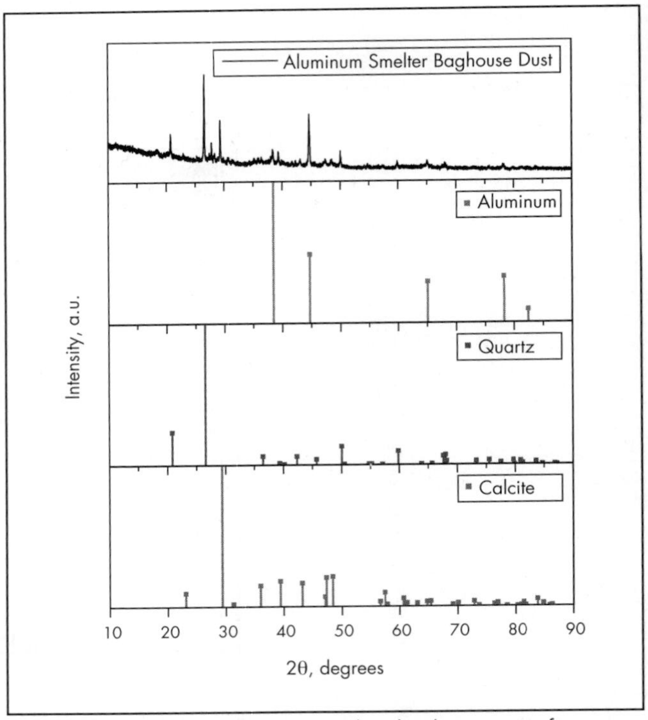

Source: Jung and Mishra 2016a, reproduced with permission from Springer

Figure 10 XRD analysis of the aluminum smelter baghouse dust

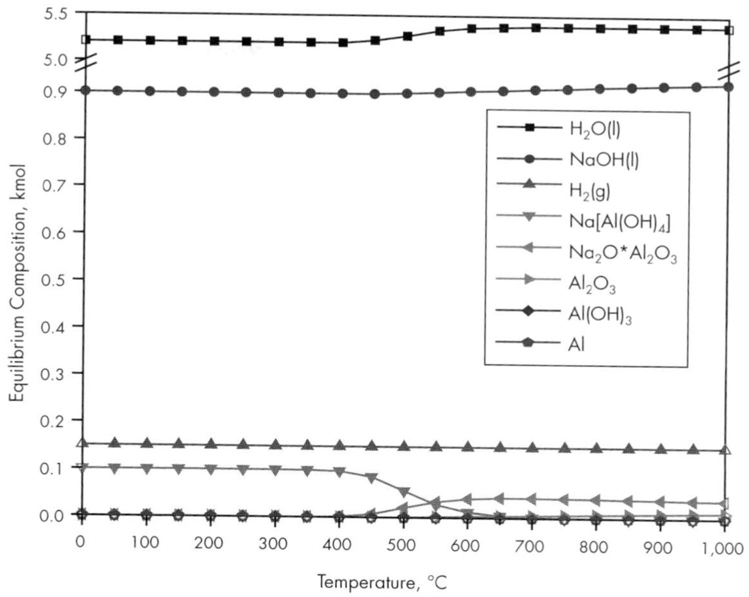

Source: Jung and Mishra 2016a, reproduced with permission from Springer

Figure 11 Equilibrium composition diagram of aluminum smelter baghouse dust leaching with NaOH

have been conducted based on the same or similar reaction equations between aluminum and sodium hydroxide.

Vanadium Recovery from Fly Ash

Several properties of vanadium, including its high melting point, low density, and relatively high strength, make it a valuable metal in applications in the steel, aerospace, chemical, and battery industries (Davis 2007). In 2012, about 90% of vanadium consumption came from steel applications (Figure 12) because adding less than 1% of vanadium to steel increases its tensile and high-temperature strength. In the aerospace industry, vanadium is used as an alloying element for titanium alloys because of its high strength-to-weight ratio. For example, the Ti-6Al-4V titanium alloy is widely used since it shows high strength, good toughness, and high-temperature stability. In the chemical industry, vanadium pentoxide is often used as a catalyst, such as for sulfuric acid production and selective catalytic reduction of nitrogen oxides in power plants. The vanadium redox battery is another industrial application of vanadium because vanadium can exist in solution as four oxidation states. The vanadium redox battery is a type of rechargeable battery that has an extremely large capacity and can be completely discharged without ill effects. Therefore, it is ideal for use in wind or solar energy storage.

In general, vanadium can be extracted from primary sources or secondary sources. Currently, titaniferous magnetite ores that contain about 1%–1.5% of vanadium are the most important source of the element. During the steelmaking process, vanadium-rich slag is co-produced from titaniferous magnetite and used as a source of vanadium. Additionally, certain energy-related resources containing vanadium, such as coal, oil (especially heavy fuel oil), petroleum coke, and the like, are burned and thus generate vanadium-bearing residues, such as ash, slag, spent catalysts, or residue (Moskalyk and Alfantazi 2003; Queneau et al. 1989; Tsai and Tsai 1998). These materials can be processed for vanadium recovery. A typical flow sheet of vanadium extraction is shown in Figure 13. First, vanadium in various sources is processed to generate vanadium concentrated product. Then, vanadium from concentrates is extracted by salt roasting followed by leaching. During the salt-roasting process, vanadium oxides are roasted with sodium carbonate or sodium chloride and converted to sodium meta-vanadate. Reaction equations between sodium chloride and vanadium oxide are as follows:

$$2NaCl + V_2O_5 + H_2O = 2NaVO_3 + 2HCl(g)$$

$$2NaCl + V_2O_3 + 1.5O_2(g) = 2NaVO_3 + Cl_2(g)$$

$$Na_2CO_3 + O_2(g) + V_2O_3 = 2NaVO_3 + CO_2(g)$$

$$Na_2CO_3 + V_2O_5 = 2NaVO_3 + CO_2(g)$$

Because sodium meta-vanadate is readily water soluble, it can be separated from gangue materials by leaching. After leaching, a solution purification step, such as solvent extraction or ion exchange, could be used to purify the vanadium-bearing solution before precipitation. Finally, vanadium pentoxide is precipitated from purified vanadium solution. During precipitation, ammonium sulfate is added to precipitate ammonium meta-vanadate, and the reaction equation is as follows:

$$2NaVO_3 + (NH_4)_2SO_4 = 2NH_4VO_3 + Na_2SO_4$$

After ammonium meta-vanadate is precipitated, it is filtered, dried, and calcined to vanadium pentoxide. There is a significant body of research on the subject covering methods other than salt roasting with water leaching.

Vitolo et al. (2001) investigated vanadium recovery from a previously burned heavy oil fly ash. Vanadium recovery is performed by preliminary burning followed by acid leaching and an oxidative precipitation of vanadium pentoxide,

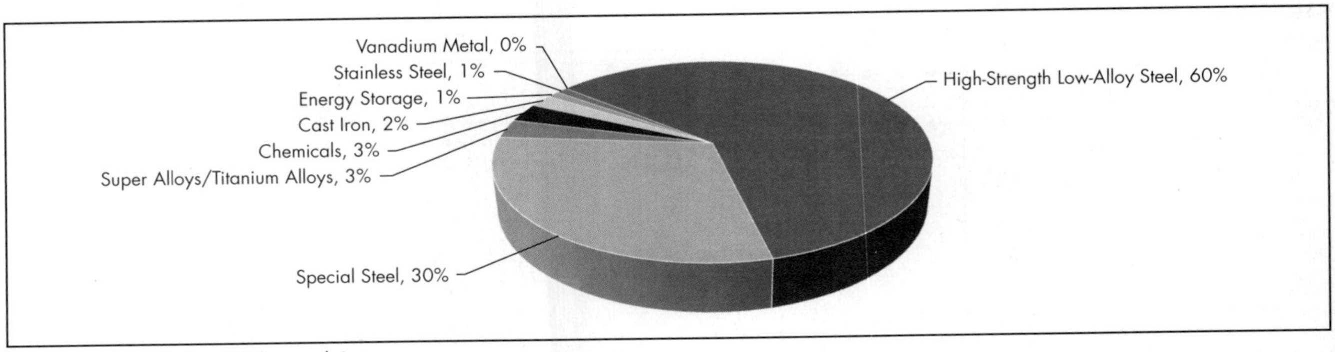

Courtesy of Terry Perles, TTP Squared, Inc.
Figure 12 Global vanadium consumption

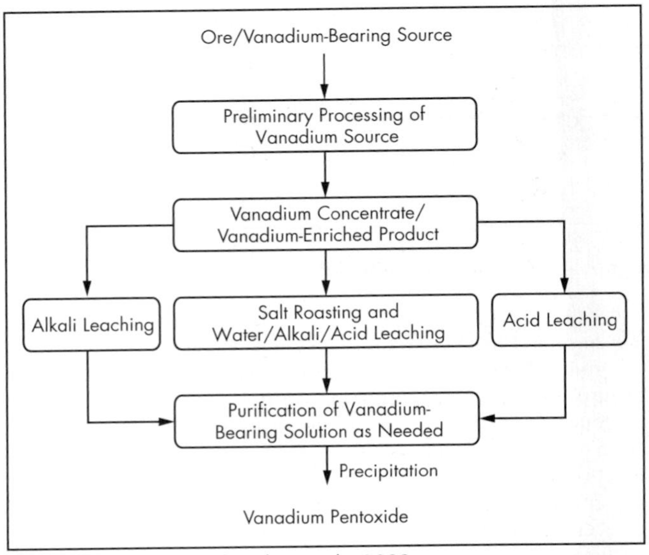

Adapted from Gupta and Krishnamurthy 1992
Figure 13 Flow sheet of vanadium extraction

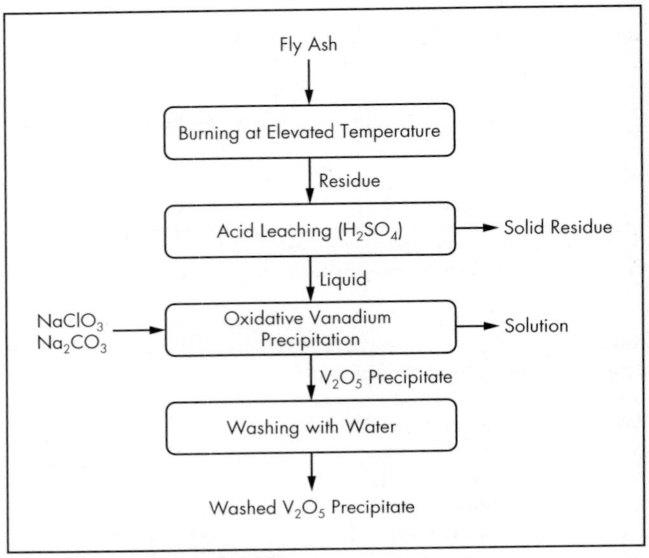

Adapted from Vitolo et al. 2001
Figure 14 Block diagram of the vanadium recovery process

as shown in Figure 14. In this method, a heavy oil fly ash containing 83.3 wt % carbon and 3.8 wt % vanadium is used for the vanadium recovery. During the preliminary burning step, the carbonaceous fraction is reduced and vanadium is concentrated from 3.8 wt % to 16~19 wt %. Then acid leaching is conducted using sulfuric acid to extract vanadium from the preliminary burned residue. The percentage of vanadium extraction is reported in the range of 91%–99% after burning at temperatures up to 950°C. However, when the burning temperature is higher than 1,050°C, formation of less leachable vanadium compounds hinders its extraction. After filtration, $NaClO_3$ is added to the leach solution to precipitate vanadium pentoxide at the boiling point of solution. During oxidative precipitation, typical red-brown color vanadium precipitates are observed, and the precipitation yield of vanadium is in the range of 72%–89%.

Navarro et al. (2007) also investigated vanadium recovery from oil fly ash from power plants. The researchers used a two-step process including leaching followed by solvent extraction or selective precipitation, as shown in Figure 15. The oil fly ashes collected from a thermal power plant contained 1.6% vanadium and 85% as a carbonaceous fraction. Alkaline leaching with sodium hydroxide shows better selective vanadium extraction compared to sulfuric acid leaching. During solvent extraction, vanadium is extracted with aluminum simultaneously, and stripping vanadium from the organic solvent is difficult. Therefore, using the solvent extraction for vanadium recovery has limited possibilities. However, a selective precipitation of aluminum followed by vanadium precipitation yielded 99% vanadium recovery from the leach solution.

Gomez-Bueno et al. (1981) investigated vanadium recovery from petroleum coke fly ash. The recovery processes include fly ash pretreatment, fly ash roasting, and leaching of roasted ash, as shown in Figure 16. During fly ash pretreatment, carbon in the as-received sample is burned off at 500°C. Then, the carbon-free fly ash is roasted in a quartz reactor in the presence of sodium chloride and water vapor. After that, the roasted sample is leached with sodium hydroxide. These recovery processes substantially improve vanadium dissolution compared to the sodium hydroxide leaching of nonroasted fly ash. The researchers also found that satisfactory roasting is achieved at temperatures below 920°C, and hydrogen chloride generation during the salt-roasting process could be recovered and used in pH adjustment of the leaching solution.

E-Waste Recycling

Recent developments in the world of information technology (IT) devices and electronic equipment makes life more convenient. On the other hand, rapid development of these devices also decreases their useful lifetime and creates a large amount of waste streams, or so-called e-waste (Kang and Schoenung 2005). As shown in Figure 17, waste streams from various types of electronic and electrical devices, such as temperature exchange equipment, screens, lamps, small or large equipment, and IT devices, are categorized as an e-waste. In general, these waste streams are landfilled, incinerated, exported to other countries, or recycled. Based on the U.S. Environmental Protection Agency's annual estimates, in 2013, the total generation of e-waste in the United States was about 3.1 Mt/yr, and the percentage of e-waste recycling has increased from 10%–40% since 2000 (Figure 18). Typically, e-wastes contain a relatively large amount of light/ferrous metals and plastics, as shown in Figure 19. Wires and printed circuit boards from e-waste also have a relatively large amount of gold and copper because they are very good conductors. However, e-waste includes a lot of toxic materials, such as chlorinated and brominated substances, plastic additives, and toxic metals, that must be properly separated and handled properly (Veit and Bernardes 2015). Thus, the European Union has introduced a directive to correctly manage e-waste. Today, e-waste is considered a good source of urban mining because of its high economic potential, as shown in Table 3, and many related studies, such as for dismantling, separation, and extraction, have been done.

Tuncuk et al. (2012) summarized metal recovery techniques from e-waste. As shown in Figure 20, e-waste recycling can be achieved by several different stages including dismantling, size reduction, separation, and extractive metallurgy. First, e-waste is dismantled and separated into various components, such as metals, plastics, batteries, and printed circuit boards. During these stages, valuable metals can be concentrated and hazardous materials can be removed from the recycling process. Therefore, the economic potential of e-waste recycling is closely related to the pretreatment steps. In general, mineral processing techniques are used to reduce the size of wastes and separate the materials. For example, shredders and hammer mills are often used to reduce particle size; the required particle size depends on the additional recycling processes. Physical separation utilizes different physical properties of the materials, such as specific gravity, magnetic

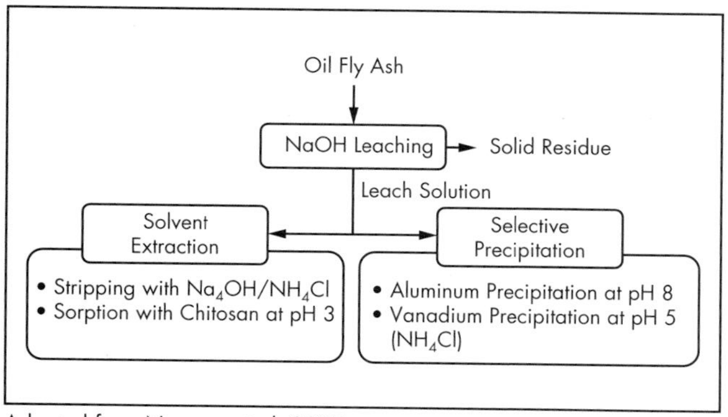

Adapted from Navarro et al. 2007

Figure 15 Integrated process for oil fly ash treatment

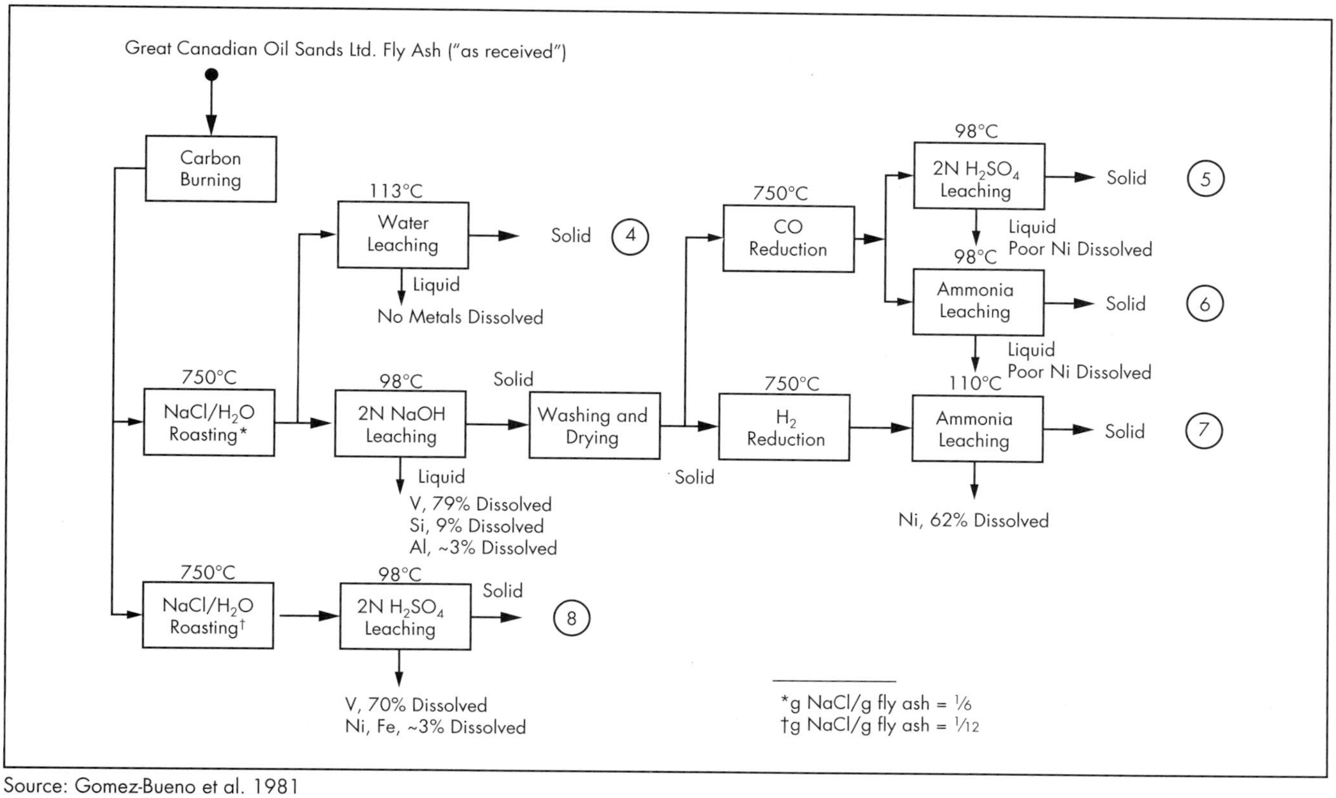

Source: Gomez-Bueno et al. 1981

Figure 16 Several treatments on carbon-free fly ash

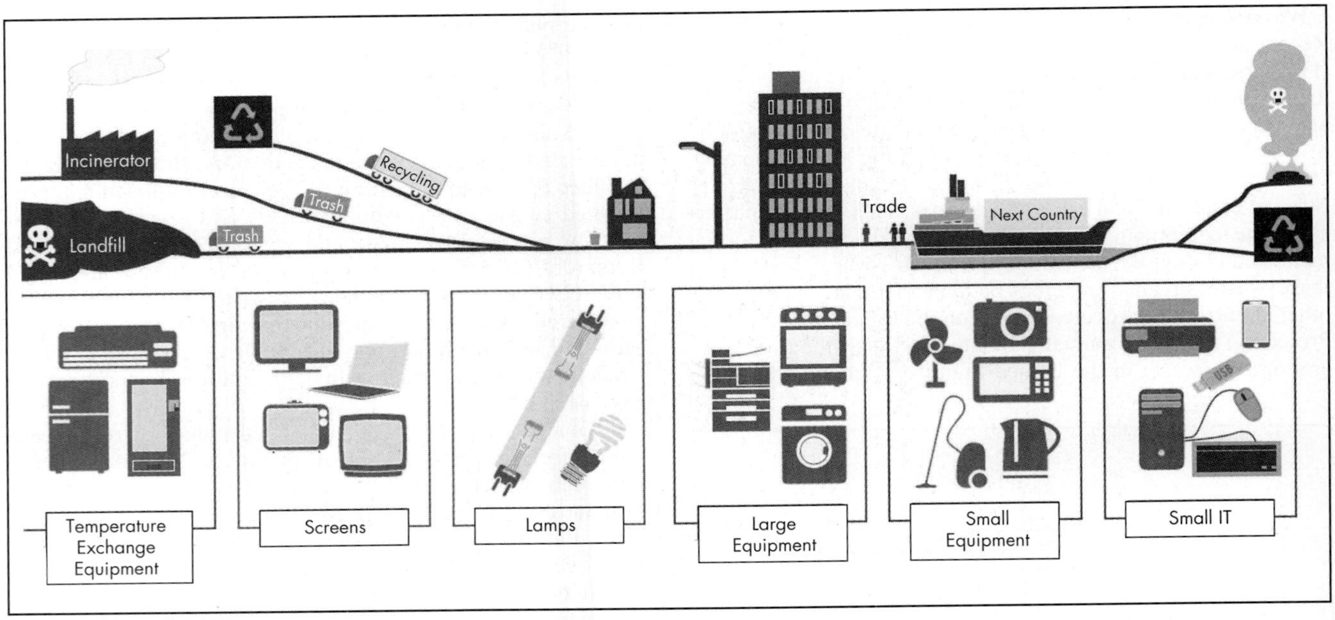

Source: Baldé et al. 2015
Figure 17 Various types of e-waste and their flows

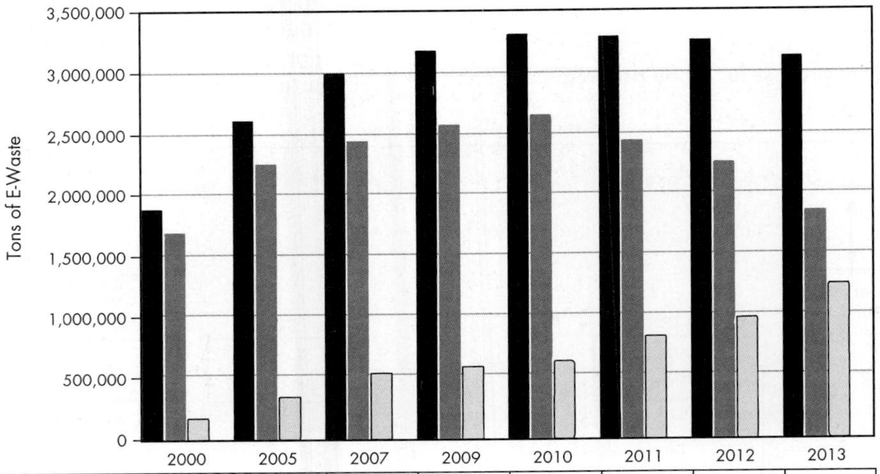

	2000	2005	2007	2009	2010	2011	2012	2013
■ Total e-waste generated	1,900,000	2,630,000	3,010,000	3,190,000	3,320,000	3,300,000	3,270,000	3,140,000
■ E-waste trashed	1,710,000	2,270,000	2,460,000	2,590,000	2,670,000	2,450,000	2,270,000	1,870,000
□ E-waste recycled	190,000	360,000	550,000	600,000	650,000	850,000	1,000,000	1,270,000
Percent recycled	10.0%	13.7%	18.3%	18.8%	19.6%	25.8%	30.60%	40.40%

Source: Electronics TakeBack Coalition 2015
Figure 18 E-waste generation and recycling 2000–2013

susceptibility, color, and electrical conductivity. The main purpose of physical separation in e-waste recycling is separating metals from nonmetal components. Then, hydrometallurgical or pyrometallurgical processes are used to extract metals from the preconcentrated materials, and valuable metals are recovered during reduction processes, as shown in Figure 20.

Bennett et al. (2009) demonstrated several separation techniques for e-waste recycling. The researchers mainly focused on plastics separation from e-waste based on 10 current techniques. Four different separation modes (sensor-based separation, shape/density separation, separation by impact milling, and separation by particle size distribution) were studied, and two—sensor-based and shape/density—generated acceptable plastic separation. Based on the report, near-infrared sortation can separate a mixture of acrylonitrile butadiene styrene and polystyrene into distinct polymer types; however, multiple

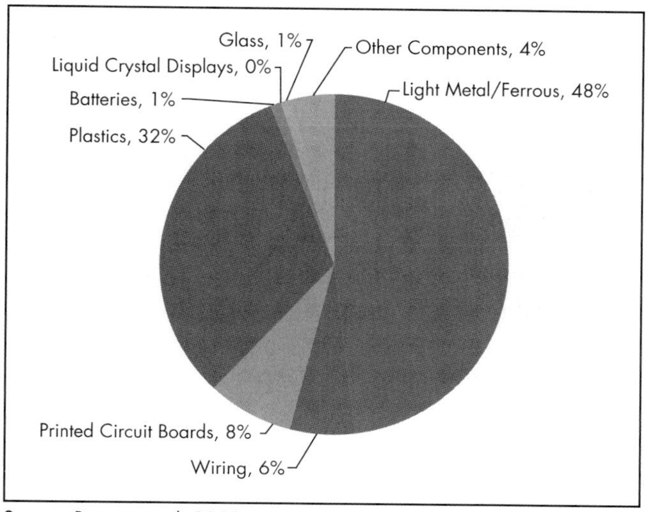

Source: Bennett et al. 2009

Figure 19 Typical composition of e-waste

Table 3 Opportunities for e-waste

Material	Kilotons	Million Euros
Fe	16,283	3,582
Cu	2,164	9,524
Al	2,472	3,585
Au	0.5	18,840
Ag	1.6	884
Pd	0.2	3,369
Plastics	12,230	15,043

Source: Baldé et al. 2017

passes are required for a better product purity. Wet shaking tables and upflow separators are also used to recover copper from copper-rich plastic. The wet shaking table can produce a very high-purity copper fraction (95% copper) when the sample size is reduced smaller than 2.3 mm. Upflow separators that utilize the principles of a fluidized bed can separate light, middle, and heavy fractions. When the sample size is reduced smaller than 2 mm by a hammer mill, the percentage of copper recovery is about 94% in the heavy fraction.

Van Kooy et al. (2004) invented a prototype kinetic gravity separator that can separate particles by differences in terminal velocity depending on the particle shape, size, and density. As shown in Figure 21A, the particles with different physical properties can be separated by a horizontal fluid motion. For example, particles that have a high vertical speed will be collected near the feeding point; however, particles settling at a lower speed will be collected far from the feeding point. The researchers reported that nonferrous scrap with particle sizes between 2 and 10 mm can be separated into light and heavy alloys by the kinetic gravity separator. Bennett et al. (2009) also demonstrated that the kinetic gravity separator can be used for copper recovery from plastics, obtaining good metal recovery from the e-waste plastics. The kinetic gravity separator has the advantage of being able to perform a separation using water. Therefore, this method is relatively simple and cheap compared to heavy medium gravity separation, and it is also suitable for separating metals from e-waste plastics.

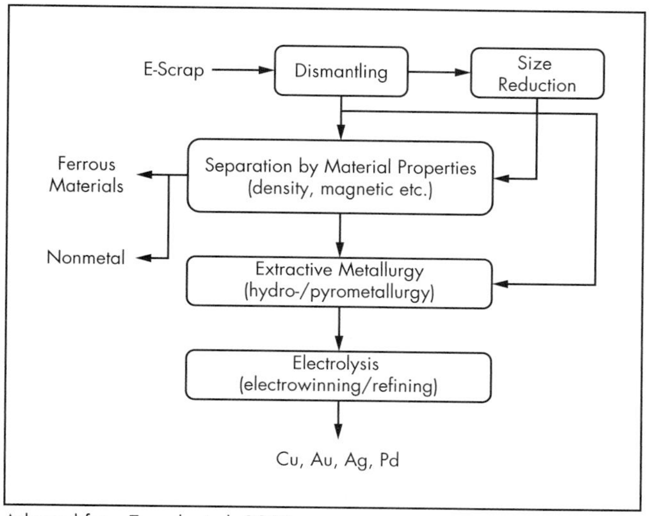

Adapted from Tuncuk et al. 2012

Figure 20 Flow sheet showing the potential process options for recovery of metals from e-scrap

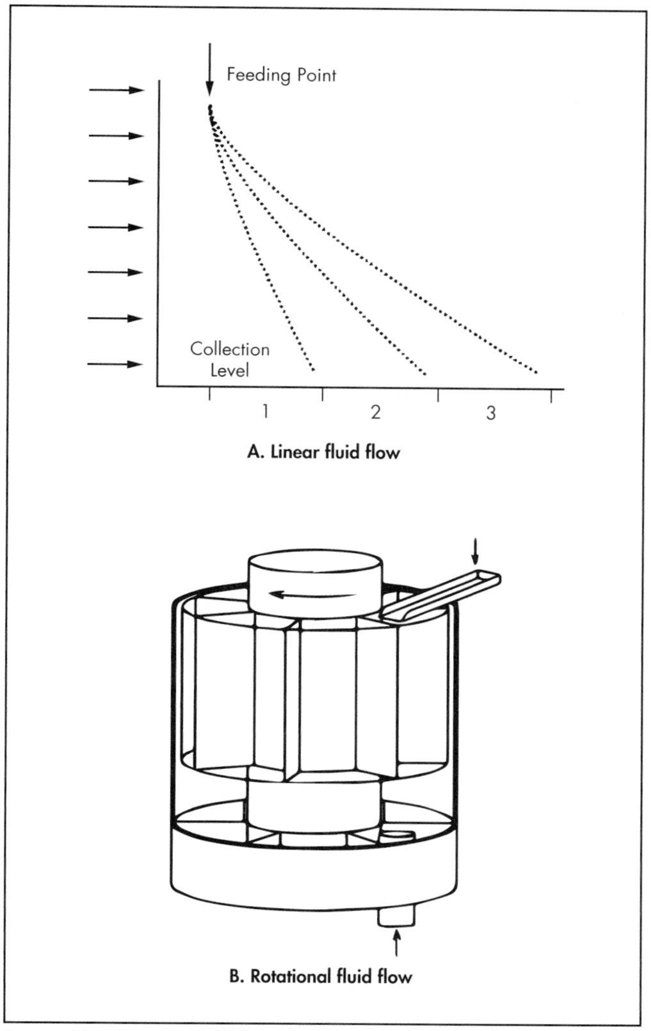

Source: Van Kooy et al. 2004

Figure 21 Principle of kinetic gravity separation

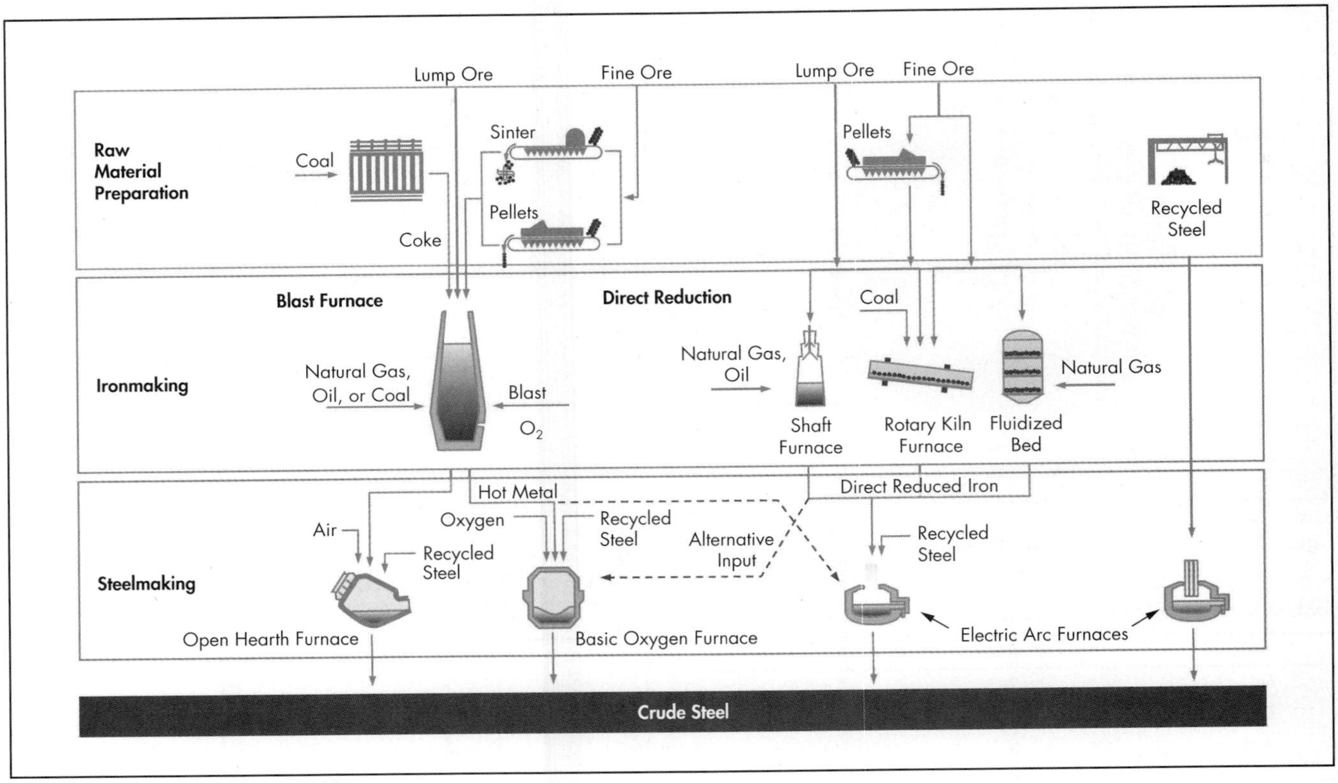

Source: World Steel Association 2015

Figure 22 Steel production routes

Ferrous Scrap Recycling

Properties such as formability, durability, and strength make steel the most important engineering and construction material. Additionally, its relatively low price attracts industries related to construction, transportation, energy, packaging, and appliances. Therefore, steel is widely used in many industrial applications. According to the World Steel Association (2015), since 1950, world crude steel production has increased from 189 to 1,621 Mt/yr to meet demand. Steel can be produced from various raw materials including lump ore, fine ore, and recycled steel. Based on the raw materials used, crude steel is produced in many ways. Energy-intensive processes are required for production of steel from ore resources; however, recycled steel is simply remelted in an electric arc furnace during secondary production of steel, as shown in Figure 22. The main advantages of using recycled steel are energy savings and resource conservation. In general, recycling steel saves 74% more energy compared to the primary production of steel and a large amount of iron ore, coal, and limestone. Moreover, steel does not lose its inherent physical properties during recycling processes. For these reasons, steel is the most recycled material worldwide. The overall recycling rate for steel in the United States was 88% in 2012 (World Steel Association 2015). As shown in Figure 23, preconsumer and postconsumer steel scraps are 100% recyclable, and about 650 Mt of steel are recycled annually. The reuse and remanufacturing of steel also contribute to its sustainability.

Bever (1976) summarized ferrous metals recycling, and Fenton (2003) investigated iron and steel recycling in the United States. Based on their work, ferrous scrap can be categorized as old, new, or home scrap, depending on its generation. In general, old scrap, also known as postconsumer scrap, is generated from automobiles, structures, household appliances, railroads, ships, and other sources at the end of their useful life. Because of the wide range of scrap sources, old scrap has a wide variety of physical and chemical properties. Therefore, old scrap processing often requires shredding and sorting, as shown in Figure 24. During these pretreatment steps, automobiles and other scraps are reduced to fist-size pieces by a shredder. The shredded scrap contains various metals, glass, and plastics. Next, these materials are separated by magnetic separation, eddy-current separation, heavy medium separation, or flotation. For scrap materials that are galvanized or tin coated, a de-zincing or de-tinning step is also required to separate zinc and tin from the steel.

New scrap, also known as prompt scrap, is generated from manufacturing steps such as cutting, drawing, and machining. Typically, the chemical and physical properties of new scrap are already known, and therefore the scrap is transported back to the steel plant for remelting, predominantly at electric arc furnaces, as shown in Figure 24. General Motors (GM) developed a similar process to recycle the clippings generated during automobile body production. GM simply remelts the clippings in an electric furnace and solidifies the molten steel by continuous casting to make gears.

Home scrap, also known as revert scrap, is generated from steel mills and foundries during the production of steel and cast iron. This type of scrap is always recycled by

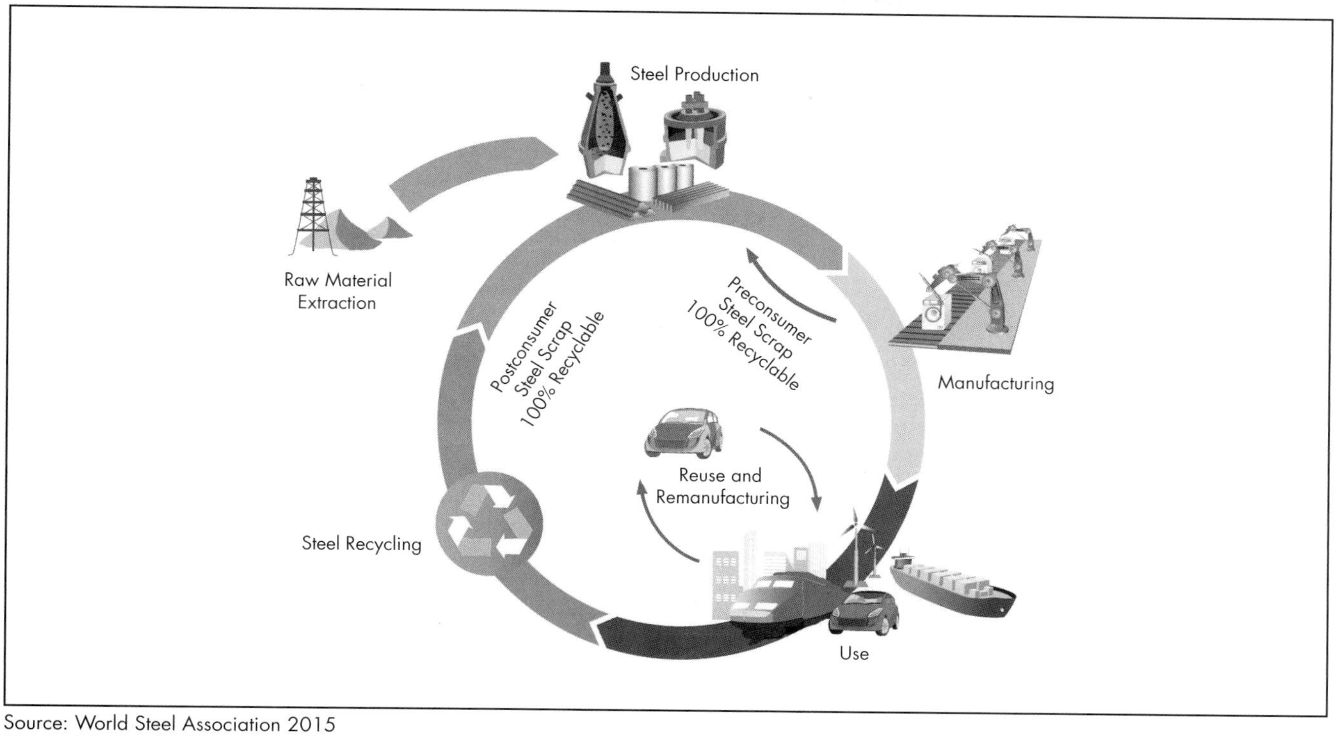

Source: World Steel Association 2015

Figure 23 Steel's life cycle

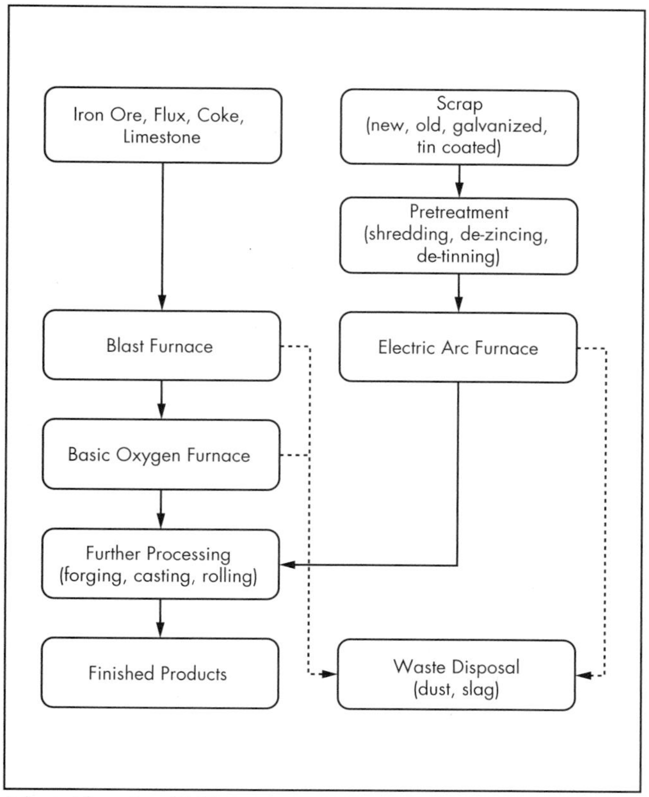

Adapted from Rao 2006

Figure 24 Schematic process flow sheet of steel recycling

remelting since the composition of the home scrap is already known. These recycling efforts are critically important from economic and environmental sustainability perspectives in industrial development.

REFERENCES

Altenpohl, D., and Kaufman, J.G. 1998. *Aluminum—Technology, Applications, and Environment: A Profile of a Modern Metal: Aluminum from Within*, 6th ed. Washington, DC: Aluminum Association.

Al-Thyabat, S., Nakamura, T., Shibata, E., and Iizuka, A. 2013. Adaptation of minerals processing operations for lithium-ion (LiBs) and nickel metal hydride (NiMH) batteries recycling: Critical review. *Miner. Eng.* 45:4–17.

Aluminum Association. n.d. *Guidelines for Handling Aluminum Fines Generated During Various Aluminum Fabricating Operations*. www.aluminum.org/sites/default/files/Safe%20Handling%20of%20Aluminum%20Fine%20Particles.pdf. Accessed September 2017.

Anderson, C.G., Dunne, R.C., and Uhrie, J.L., eds. 2014. *Mineral Processing and Extractive Metallurgy: 100 Years of Innovation*. Englewood, CO: SME.

Baldé, C.P., Wang, F., Kuehr, R., and Huisman, J. 2015. *The Global E-Waste Monitor 2014*. Bonn, Germany: United Nations University Institute for the Advanced Study of Sustainability.

Baldé, C.P., Forti, V., Gray, V., Kuehr, R., and Stegmann, P. 2017. *The Global E-Waste Monitor 2017: Quantities, Flows, and Resources*. Bonn, Germany: United Nations University.

Bennett, M., Edwards, L., Goyos-Ball, L., Hilder, R., Hall, P., Morrish, L., Morton, R., and Myles, N. 2009. *Separation of Mixed WEEE Plastics: Final Report*. Banbury, Oxon, U.K.: Waste and Resources Action Programme. www.wrap.org.uk/sites/files/wrap/Separation%20of%20mixed%20WEEE%20plastics%20-%20Final%20report.pdf. Accessed September 2017.

Bever, M.B. 1976. The recycling of metals—I. Ferrous metals. *Conserv. Recycl.* 1(1):55–69.

Björklund, A., and Finnveden, G. 2005. Recycling revisited—Life cycle comparisons of global warming impact and total energy use of waste management strategies. *Resour. Conserv. Recycl.* 44(4):309–317.

Bruckard, W.J., Calle, C.M., Davidson, R.H., Glenn, A.M., Jahanshahi, S., Somerville, M.A., Sparrow, G.J., and Zhang, L. 2010. Smelting of bauxite residue to form a soluble sodium aluminium silicate phase to recover alumina and soda. *Miner. Process. Extr. Metall.* 119(1):18–26.

Davis, T. 2007. Recovery of vanadium and titanium from iron slag. Ph.D. dissertation, Colorado School of Mines, Golden, CO.

DOE (U.S. Department of Energy). 1999. Recycling of aluminum dross/saltcake. Aluminum Project fact sheet. Washington, DC: DOE, Office of Industrial Technologies Energy Efficiency and Renewable Energy.

DOE (U.S. Department of Energy). 2011. *Critical Materials Strategy*. Washington, DC: DOE.

Eduafo, P.M. 2016. Experimental investigation of recycling rare earth elements from waste fluorescent lamp phosphors. Ph.D. dissertation, Colorado School of Mines, Golden, CO.

Electronics TakeBack Coalition. 2015. *E-Waste in Landfills*. www.electronicstakeback.com/designed-for-the-dump/e-waste-in-landfills/. Accessed September 2017.

EPA (U.S. Environmental Protection Agency). 2016. *Advancing Sustainable Materials Management: Facts and Figures 2014*. https://www.epa.gov/sites/production/files/2016-11/documents/2014_smm_tablesfigures_508.pdf. Accessed March 2018.

Fenton, M.D. 2003. *Iron and Steel Recycling in the United States in 1998*. Reston, VA: U.S. Geological Survey.

Fuerstenau, M.C. 2003. *Principles of Mineral Processing*. Littleton, CO: SME.

Gomez-Bueno, C.O., Spink, D.R., and Rempel, G.L. 1981. Extraction of vanadium from athabasca tar sands fly ash. *Metall. Trans. B.* 12(2):341–352.

Gupta, C.K. 2003. *Chemical Metallurgy: Principles and Practice*. Weinheim: Wiley-VCH.

Gupta, C.K., and Krishnamurthy, N. 1992. *Extractive Metallurgy of Vanadium*. Amsterdam: Elsevier Science.

Gupta, C.K., and Mukherjee, T.K. 1990. *Hydrometallurgy in Extraction Processes*. Vol. 1. Boca Raton, FL: CRC Press.

Habashi, F. 1997. *Handbook of Extractive Metallurgy*. Vol. 1. New York: Wiley-VCH.

Habashi, F. 1999. *A Textbook of Hydrometallurgy*. Quebec: Métallurgie Extractive Québec.

Hayes, P.C. 2003. *Process Principles in Minerals and Materials Production*. Queensland: Hayes Publishing.

Huang, X.L., El Badawy, A.M., Arambewela, M., Adkins, R., and Tolaymat, T. 2015. Mineral phases and metals in baghouse dust from secondary aluminum production. *Chemosphere* 134:25–30.

International Aluminium Institute. 2013. *Global Aluminum Recycling: A Cornerstone of Sustainable Development*. www.world-aluminium.org/media/filer_public/2013/01/15/fl0000181.pdf. Accessed September 2017.

International Aluminium Institute. 2016. Global aluminium flow model. 2016. www.world-aluminium.org/publications/tagged/mass%20flow/. Accessed December 2017.

Jayasankar, K., Ray, P.K., Chaubey, A.K., Padhi, A., Satapathy, B.K., and Mukherjee, P.S. 2012. Production of pig iron from red mud waste fines using thermal plasma technology. *Int. J. Miner. Metall. Mater.* 19(8):679–684.

Jung, M., and Mishra, B. 2016a. Kinetic and thermodynamic study of aluminum recovery from the aluminum smelter baghouse dust. *J. Sustainable Metall.* 2(3):257–264.

Jung, M., and Mishra, B. 2016b. Recovery of aluminum from the aluminum smelter baghouse Dust. In *Rewas 2016: Towards Materials Resource Sustainability*. pp. 255–260. Edited by R.E. Kirchain, B. Blanpain, C. Meskers, E. Olivetti, D. Apelian, J. Howarter, A. Kvithyld, B. Mishra, N.R. Neelameggham, and J. Spangenberger. Hoboken, NJ: John Wiley and Sons.

Kang, H.Y., and Schoenung, J.M. 2005. Electronic waste recycling: A review of U.S. infrastructure and technology options. *Resour. Conserv. Recycl.* 45(4):368–400.

Kelly, E.G., and Spottiswood, D.J. 1982. *Introduction to Mineral Processing*. New York: John Wiley and Sons.

Kumar, R., and Srivastava, J. 1998. Utilization of iron values of red mud for metallurgical applications. *Environ. Waste Manage.* 7:108–119.

Li, X.B., Wei, X.I.A.O., Wei, L.I.U., Liu, G.H., Peng, Z.H., Zhou, Q.S., and Qi, T.G. 2009. Recovery of alumina and ferric oxide from Bayer red mud rich in iron by reduction sintering. *Trans. Nonferr. Met. Soc. China* 19(5):1342–1347.

Marinos, D., and Mishra, B. 2015. An approach to processing of lithium-ion batteries for the zero-waste recovery of materials. *J. Sustainable Metall.* 1(4):263–274.

Moskalyk, R.R., and Alfantazi, A.M. 2003. Processing of vanadium: A review. *Miner. Eng.* 16(9):793–805.

Navarro, R., Guzman, J., Saucedo, I., Revilla, J., and Guibal, E. 2007. Vanadium recovery from oil fly ash by leaching, precipitation and solvent extraction processes. *Waste Manage.* 27(3):425–438.

Piga, L., Pochetti, F., and Stoppa, L. 1993. Recovering metals from red mud generated during alumina production. *J. Miner. Met. Mater. Soc.* 45(11):54–59.

Pryor, M.R. 2012. *Mineral Processing*. New York: Springer Science and Business Media.

Queneau, P.B., Hogsett, R.F., Beckstead, L.W., and Barchers, D.E. 1989. Processing of petroleum coke for recovery of vanadium and nickel. *Hydrometallurgy* 22(1):3–24.

Rao, S.R. 2006. *Resource Recovery and Recycling from Metallurgical Wastes*. Vol. 7. Oxford: Elsevier. p. 177.

Rosenqvist, T. 2004. *Principles of Extractive Metallurgy*. Trondheim: Tapir Academic Press.

Strauss, M.L. 2016. The recovery of rare earth oxides from waste fluorescent lamps. M.S. thesis, Colorado School of Mines, Golden, CO.

Tsai, S.L., and Tsai, M.S. 1998. A study of the extraction of vanadium and nickel in oil-fired fly ash. *Resour. Conserv. Recycl.* 22(3):163–176.

Tuncuk, A., Stazi, V., Akcil, A., Yazici, E.Y., and Deveci, H. 2012. Aqueous metal recovery techniques from e-scrap: Hydrometallurgy in recycling. *Miner. Eng.* 25(1):28–37.

USGS (U.S. Geological Survey). 2014. Aluminum statistics. In *Historical Statistics for Mineral and Material Commodities in the United States: U.S. Geological Survey Data Series 140*. Compiled by T.D. Kelly and G.R. Matos. http://minerals.usgs.gov/minerals/pubs/historical-statistics/. Accessed September 2017.

Van Kooy, L., Mooij, M., and Rem, P. 2004. Kinetic gravity separation. *Phys. Sep. Sci. Eng.* 13(1):25–32.

Veit, H.M., and Bernardes, A.M., eds. 2015. *Electronic Waste: Recycling Techniques*. Switzerland: Springer International.

Vitolo, S., Seggiani, M., and Falaschi, F. 2001. Recovery of vanadium from a previously burned heavy oil fly ash. *Hydrometallurgy* 62(3):145–150.

Wills, B.A., and Napier-Munn, T. 2006. *Wills' Mineral Processing Technology: An Introduction to the Practical Aspects of Ore Treatment and Mineral Recovery*. Oxford: Butterworth-Heinemann.

World Steel Association. 2015. *Steel in the Circular Economy*. www.worldsteel.org/en/dam/jcr:00892d89-551e-42d9-ae68-abdbd3b507a1/Steel+in+the+circular+economy+-+A+life+cycle+perspective.pdf. Accessed September 2017.

Xiangong, M., Mingliang, Y., Wenliang, Z., and Jin, C. 1996. Kinetics of the nucleation and grain growth of metallic phase in direct reduction of high-iron red mud with coal base. *J. Cent. South Univ. Nat. Sci.* 27(2):159–163.

Xu, J., Thomas, H.R., Francis, R.W., Lum, K.R., Wang, J., and Liang, B. 2008. A review of processes and technologies for the recycling of lithium-ion secondary batteries. *J. Power Sources* 177(2):512–527.

Zhang, T., He, Y., Wang, F., Ge, L., Zhu, X., and Li, H. 2014. Chemical and process mineralogical characterizations of spent lithium-ion batteries: An approach by multianalytical techniques. *Waste Manage.* 34(6):1051–1058.

Zhong, L., Zhang, Y., and Zhang, Y. 2009. Extraction of alumina and sodium oxide from red mud by a mild hydrochemical process. *J. Hazard. Mater. 172*(2-3):1629–1634.

Zimmer, E., Nafissi, A., and Winkhaus, G. 1978. Reclamation treatment of red mud. U.S. Patent 4,119,698.

PART 7

Flotation

CHAPTER 7.1

Mechanical Flotation

Michael G. Nelson and Dariusz Lelinski

Machines used for most mineral flotation applications may be designated as either pneumatic or mechanical. Pneumatic machines rely on fluid flow to mix solids, liquid, and air. They include flotation columns, Jameson cells, and others, and are not discussed in this chapter. Mechanical machines use a rotary mechanism to provide mixing and, in some cases, aeration. Mechanical machines are often called *cells*.

Mechanical machines may be further categorized based on the location of the mechanism and the method of introducing air into the process. *Multi-chamber* machines, supplied by Eriez and Woodgrove, use individual chambers for mixing and froth separation. In *single-chamber* machines, made by FLSmidth, Metso, and Outotec, one chamber serves for mixing and phase separation. The single-chamber machines include self-aerating machines, in which air is induced into the pulp by the action of the mechanism, and externally aerated machines, in which air is delivered into the pulp by a separately powered blower. A compressor may also be used, but this is seldom done in practice. Self-aerating mechanisms locate the rotor or impeller near the top of the tank, while externally aerated mechanisms are near the bottom. Figure 1 illustrates this taxonomy in more detail.

Mechanical flotation machines have been designed in a wide variety of configurations. Mechanical machines are designed for use in almost all flotation applications, with variations in size, mechanism power, aeration rate, and cell operation specified so that each machine functions as desired in the circuit. Many of these machines have been previously described (Nelson et al. 2002). This chapter provides current information on machine design and sizes available; describes new machines and new features on machines described previously; and provides detailed information on design, installation, start-up, and operation of mechanical flotation machines of all makes.

Specially designed machines are used for flash flotation, in which large, liberated particles of value are recovered into a concentrate immediately after grinding. All flash flotation machines are externally aerated, with mechanisms located near the bottom of the tank.

AVAILABLE EQUIPMENT

All the information in this section was provided by representatives of the respective manufacturers, or taken from manufacturers' websites and printed literature. Machine descriptions and benefits are given as provided by each manufacturer, and readers are left to form their own opinions regarding the material presented. The respective manufacturers are presented in alphabetical order, with no implication as to which (if any) the authors believe to be superior. The machines described in detail in this chapter are those most commonly used in North and South America, Europe, Africa, and Australia. Bateman and Delkor, both now owned by Takraf Tenova, have also supplied machines in these markets. Suppliers in China and Russia also provide mechanical flotation machines, but they are similar or identical in design to those described here (see Glembotskii et al. 1972 for examples).

As noted by Sherrell and Yoon (2005), flotation may be thought of in four subprocesses: bubble–particle contact, particle attachment, particle detachment, and froth recovery. (Some authors consider the first two subprocesses as a single subprocess and thus identify only three subprocesses.) For many years, all of the commercially available froth flotation machines have been designed to accomplish all four subprocesses in the same vessel, as shown in the image in Figure 2.

Innovations in machine design continue. Manufacturers continue to refine the respective designs of their mechanisms. Machines introduced by Eriez and Woodgrove each use nonconventional designs to accomplish the four subprocesses just mentioned. Both machines are based on a perceived need to execute the subprocesses in separate vessels, as described by Zhou (1996).

Eriez Manufacturing

The concept behind the dual-chamber design is to use one vessel for collection and one vessel for froth recovery, with the conditions in each vessel optimized for its designated function. The collection vessel is optimized for high-energy turbulent contacting, and the separation vessel can be optimized for

Michael G. Nelson, Professor & Chair, Mining Engineering Department, University of Utah, Salt Lake City, Utah, USA
Dariusz Lelinski, Global Product Manager of Flotation, FLSmidth Minerals, Salt Lake City, Utah, USA

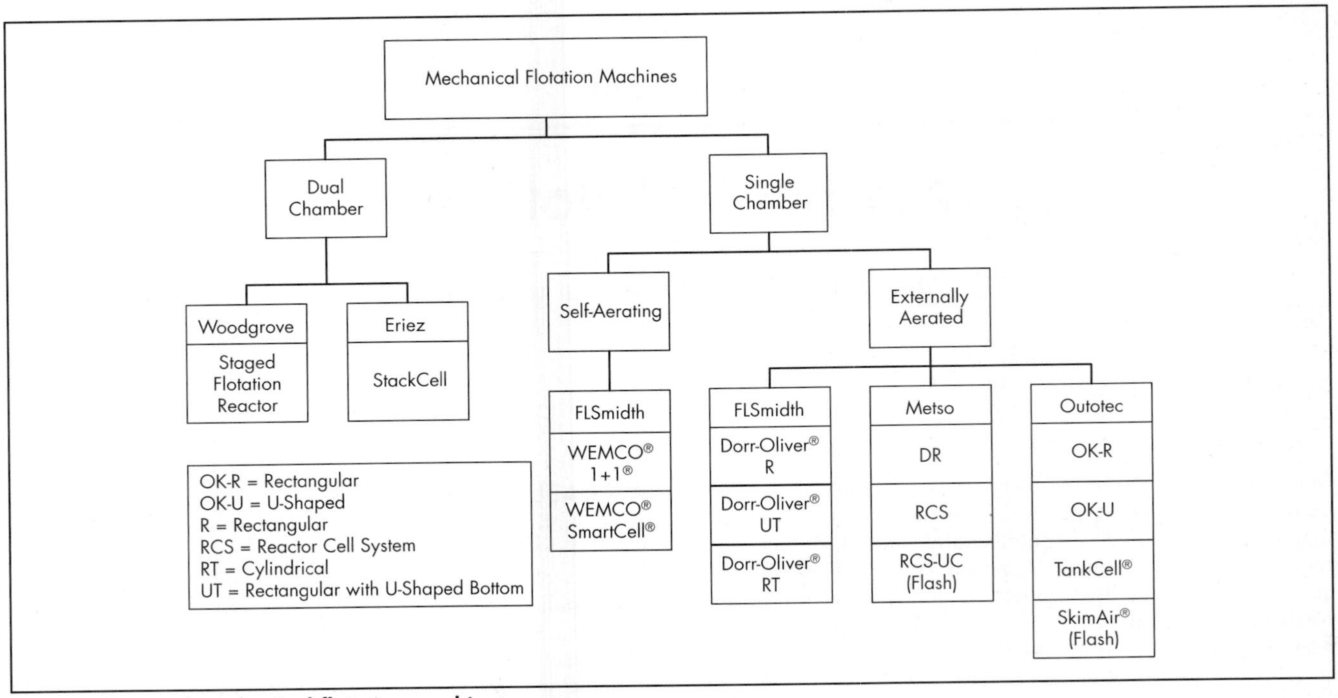

Figure 1 Types of mechanical flotation machines

gentle, nonturbulent bubble–particle separation from the pulp and froth recovery.

The Eriez StackCell® machine, shown in Figure 3, was developed to combine the metallurgical performance of a column with the footprint of a mechanical cell (Kohmuench et al. 2008).

Feed is introduced into an internal mixing and aeration canister. Intensely mixed, aerated slurry is discharged under pressure from the mixing canister into the larger cell, which operates under conditions designed to allow bubble–particle aggregates to rise into the froth and be recovered through an internal or external launder. The froth is washed by downflowing water to remove entrained hydrophilic particles. Because a rotor and stator are located inside the canister, mechanical energy is introduced into the system, as in a conventional cell. In comparison to conventional mechanical cells, the amount of mechanical energy introduced per volume into the canister is much higher, while the amount of mechanical energy introduced per volume into the overall system is much lower. This allows the StackCell machine to target and isolate the part of the system that benefits from mechanical energy.

In coal flotation, the StackCell machine is reported to deliver improved ash rejection and higher coal recovery. Its size and design also allows the installation of a bank of several cells in a vertical configuration. Such an installation increases retention time and minimizes the effects of short-circuiting, while the vertical configuration requires less floor space and decreases pumping costs by using gravity flow.

The StackCell machine has been applied in sulfide and gold flotation. Reportedly, equivalent metallurgical performance can be achieved compared to mechanical cells with a significant reduction in retention time or working volume of the flotation cell.

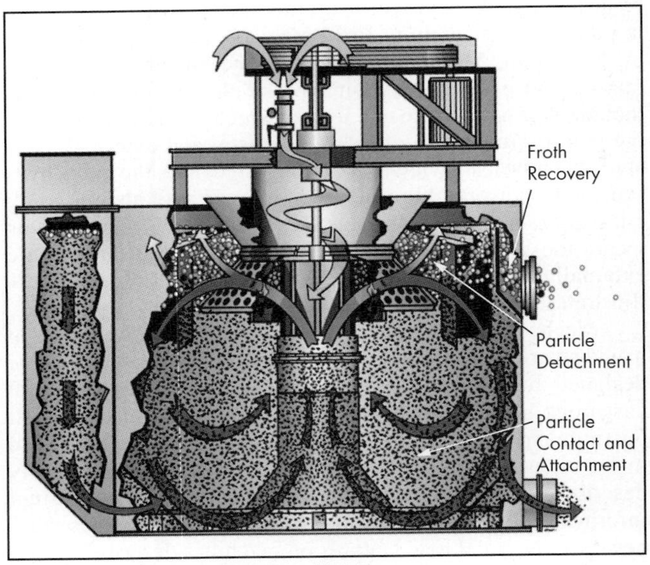

Reprinted with permission from FLSmidth A/S

Figure 2 WEMCO® SmartCell® machine diagram, showing subprocessng areas

FLSmidth

FLSmidth supplies two types of flotation machines: WEMCO® and Dorr-Oliver®. The WEMCO® machine is self-aerating, while the Dorr-Oliver® machine is externally aerated. FLSmidth has delivered more than 53,000 flotation cells to operations worldwide.

The principles of operation for self-aerating and externally aerated machines are similar in concept, but the

Courtesy of Eriez
Figure 3 Eriez StackCell machine

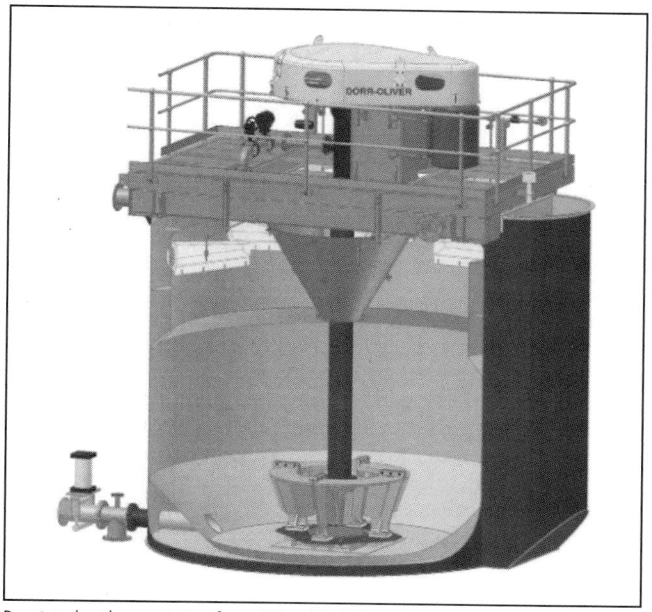

Reprinted with permission from FLSmidth A/S
Figure 4 FLSmidth Dorr-Oliver® machine

execution is different. The operating variables in the externally aerated machines are aeration rate, rotor speed, and froth depth. Aeration rate is controlled by adjusting valves in the air delivery line or by changing the blower speed. If this is varied independently of other variables, bubble size distribution in the cell will change. To a certain extent, aeration rate is independent of the site elevation and slurry density. However, the configuration of the blower and air delivery system may limit the ability to adjust the aeration rate. It is also important to realize that when the external aeration rate is changed, other parameters associated with aeration, such as superficial gas velocity, gas holdup, and bubble-surface-area flux, will also change.

Because WEMCO® machines are self-aerating, their flow is controlled differently than the flow in externally aerated machines, in which air is provided by blowers. The air intake for each WEMCO® machine can be fitted with manual or automatic control valves. Either of these may be used to decrease the airflow to the machine. The pumping of air and slurry by a WEMCO® machine is controlled by changing the position or speed of the rotor. In smaller machines, rotor position may be changed without removing the mechanism by raising or lowering the mechanism or changing the length of the draft tube. Rotor speed may be changed by changing belt position on a multi-sheave drive or through use of a variable-speed drive. Rotor position is described in terms of submergence and engagement. *Submergence* is the distance from the slurry surface to the top of the rotor blades; *engagement* is the depth to which the rotor engages the draft tube. Detailed description of these relationships can be found in Nelson and Lelinski (2000) and Nelson et al. (2002).

Figure 4 shows a typical Dorr-Oliver® machine. FLSmidth lists the advantages of these cells as follows: low power consumption, nonclogging, low-maintenance-design rotor, high air-dispersion capability, superior metallurgical performance, easier restarting mechanism, and low reagent costs.

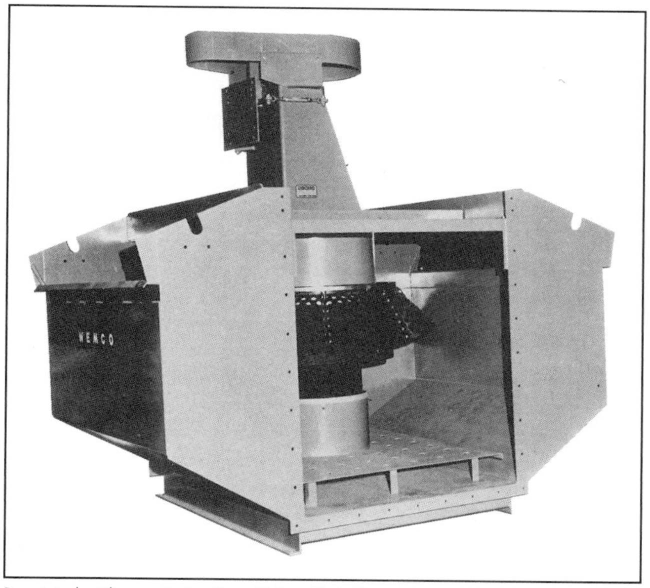

Reprinted with permission from FLSmidth A/S
Figure 5 FLSmidth WEMCO® 1+1® machine

The WEMCO® 1+1® machine is shown in Figure 5. FLSmidth literature states that the 1+1® provides the following advantages: higher recovery and grade with easier start-up, simpler operation, lower reagent consumption, longer mechanism life, and less required maintenance. The WEMCO® 1+1® cell uses a rotor-disperser design that delivers intense mixing and aeration. Ambient air is drawn into the cell and uniformly distributed throughout the pulp, providing optimum air-to-particle contact. In larger cells, a false bottom and draft tube channel slurry flow, ensuring high recirculation and eliminating sanding. The combination of efficient aeration and optimum solids suspension allow WEMCO® 1+1® cells to achieve high recovery and concentrate grade, with reduced reagent consumption.

The WEMCO® SmartCell® machine, shown in Figure 6, uses a redesigned 1+1® mechanism in a cylindrical tank. This cylindrical design improves mixing efficiency and air dispersion and also provides better surface stability, less pulp turbulence, lower capital costs, and reduced power consumption. It also features a hybrid draft tube and beveled tank bottom, which improve hydrodynamic mixing and coarse-particle recovery and increase solids suspension. Finally, the SmartCell® machine features radial launders and mixing baffles, which increase froth mobility, decrease froth residence time, increase recovery, and enhance froth stability.

FLSmidth has recently conducted extensive research into improving flotation efficiency in externally aerated machines by changing the design of the rotor and the stator. The rotor/stator (often called *the mechanism*) is responsible for solids suspension, gas dispersion, and bubble–particle attachment. Research on mechanism improvement started with laboratory testing of 200 different designs. The best designs were further tested in a 1.5-m^3 pilot cell using an industrial feed. This pilot work confirmed the results from the laboratory, and industrial prototypes were designed. The best three mechanism designs were then tested and compared at an industrial operation. After this extensive test work, the nextSTEP™ rotor and stator design was selected as FLSmidth's standard for flotation mechanisms. Figure 7 shows details of the nextSTEP mechanism.

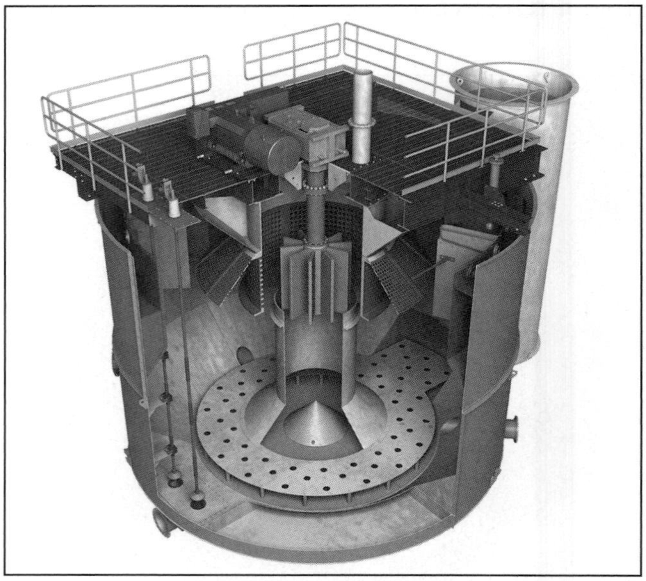

Reprinted with permission from FLSmidth A/S

Figure 6 WEMCO® SmartCell® machine

Hydrodynamic testing of the nextSTEP mechanism was conducted on slurry in a 660-m^3 SuperCell® machine. The influence of various operating variables on the absorbed specific power of the combined motor and blower was recorded, and statistical analysis was conducted to define the relationships among absorbed power, mechanism rotational speed (revolutions per minute), airflow, and solids suspension. These tests are described in detail by Lelinski et al. (2015). The key conclusions are as follows:

- The rpm–power relationship is very well defined and follows the generally accepted trend: the higher the rpm, the higher the absorbed power.
- The airflow–power relationship is different than that seen in smaller machines, where increased airflow *decreased* the absorbed power. In the 660-m^3 machine, this relationship was seen when just the motor power was considered, but when the blower power was included, the relationship was reversed in that increasing the airflow *increased* total absorbed power.
- With the nextSTEP mechanism installed, the specific power absorbed by the motor *decreased* to 0.2 kW/m^3.

Machines offered by FLSmidth are shown in Tables 1–3. Dorr-Oliver® machines are offered in three configurations: rectangular (R), rectangular with U-shaped bottom (UT), and cylindrical (RT).

Metso Group

Metso Group offers three types of externally aerated machines, the RCS™, the DR, and the RCS UC. The RCS UC is a flash flotation machine, and is discussed separately. The RCS machine is shown in Figure 8. The machine was designed to create the two classic zones within a flotation cell, an active lower zone for effective particle suspension and transportation and a relatively quiescent upper zone to minimize bubble–particle separation. A cylindrical tank was adopted to provide symmetrical hydraulic flow patterns with minimum upper-zone turbulence.

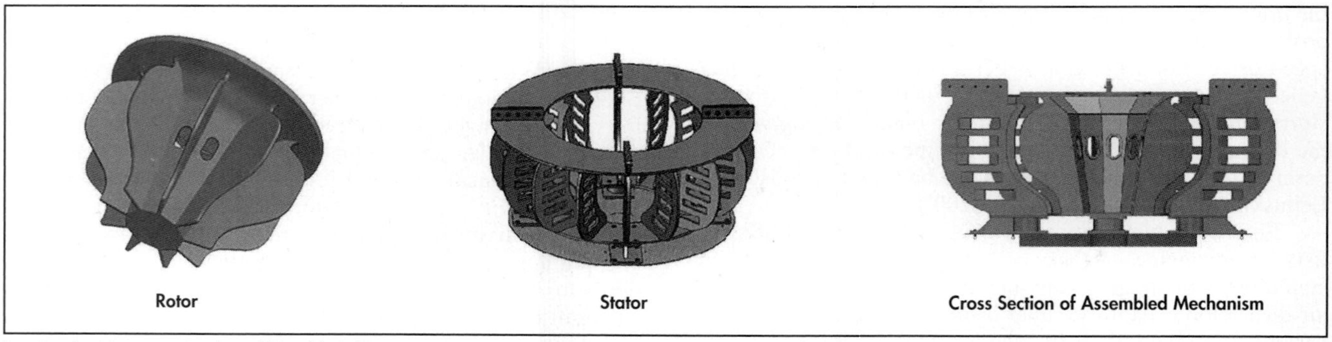

Reprinted with permission from FLSmidth A/S

Figure 7 FLSmidth nextSTEP mechanism

Table 1 Dorr-Oliver® machines

Cylindrical (RT) Tanks			Rectangular (R) and U-Bottom (UT) Tanks		
Model	Effective Cell Volume, m³	Motor, kW	Model	Effective Cell Volume, m³	Motor, kW
DO-1.5 RT (pilot)	1.5	7.5	DO-1 R	0.02	0.6
DO-5 RT	5	7.5	DO-10 R	0.24	1.1
DO-10 RT	10	14.9	DO-25 R	0.60	2.2
DO-20 RT	20	29.8	DO-50 R	1.2	3.7
DO-30 RT	30	37.3	DO-100 R	2.4	5.6
DO-40 RT	40	44.8	DO-300 UT	7.2	11.2
DO-50 RT	50	56	DO-600 UT	14.3	22.4
DO-70 RT	70	74.6	DO-1000 UT	23.8	29.8
DO-60 RT	60	74.6	DO-1350 UT	32.2	37.3
DO-100 RT	100	111.9	DO-1550 UT	36.9	44.8
DO-130 RT	130	149.2	DO-1550 UT	36.9	44.8
DO-160 RT	160	149.2			
DO-200 RT	200	186.5			
DO-330 RT	330	410			
DO-600 RT	600	500			
DO-660 RT	660	550			

Reprinted with permission from FLSmidth A/S

Table 2 WEMCO® 1+1® machines

Model	Effective Cell Volume, m³	Motor, kW
18	0.028	0.37
28	0.085	0.75
36	0.31	2.24
44	0.59	3.73
56	1.16	5.60
66	1.73	7.46
66D	2.83	11.2
84	4.25	11.2
120	8.50	18.7
144	14.16	22.4
164	28.32	44.8
190	42.48	74.6
225	84.96	149.2

Reprinted with permission from FLSmidth A/S

Features of the RCS machine include good particle–bubble contact, effective solids suspension during operation and resuspension after shutdown, effective air dispersion and distribution, flotation air provided by a separate air blower, manual or automatic control of aeration rate, V-belt drive standard up to 70-m³ cell volume, and gearbox drive with extended output shaft bearings and drywell construction for cell volumes larger than 70 m³.

The DR machine, shown in Figure 9, is based on the DR design developed by the former Denver Equipment Company. It features rectangular tanks, connected in the open, hog-trough configuration. It is especially suitable for small concentrators and for industrial minerals.

Table 3 WEMCO® SmartCell® machines

Model	Effective Cell Volume, m³	Motor, kW
1.5 (pilot)	1.5	7.5
5	5	30
10	10	37
20	20	50
30	30	75
40	40	90
50	90	90
60	60	150
70	70	150
100	100	150
130	130	185
160	160	185
200	200	250
250	250	315
300	300	373
350	350	373
600	600	800
660	660	875

Reprinted with permission from FLSmidth A/S

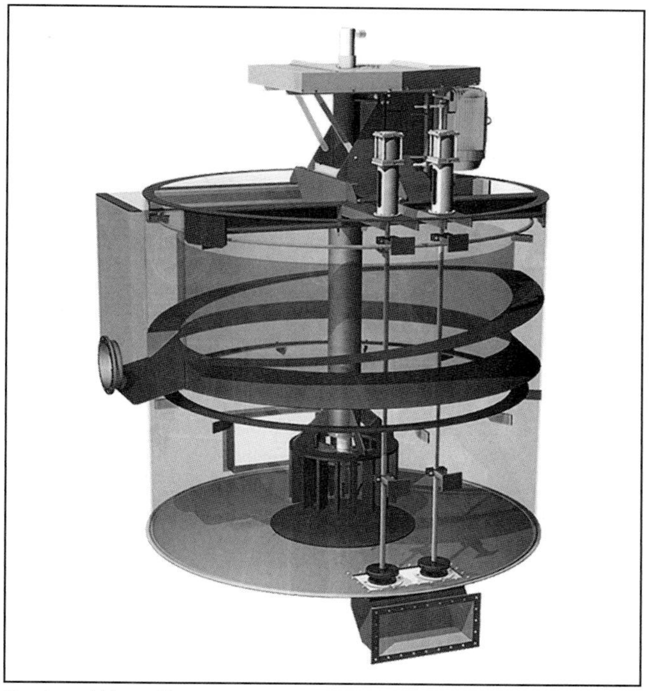

Courtesy of Metso Group

Figure 8 Metso RCS machine

Features of the DR machines include vertical circulation of pulp, achieved by combining a "recirculation well" with a reversible, top-feed impeller; external aeration; and minimized sanding, accomplished with the cell-to-cell design. Table 4 shows available models of the RCS and DR machines. Metso does not specify a motor size for each machine but selects the appropriate motor for each application.

Courtesy of Metso Group
Figure 9 Metso DR machines

Table 4 Current offerings of Metso RCS and DR machines

RCS Model	Effective Cell Volume, m³	DR Model	Effective Cell Volume, m³
RCS 0.8	0.8	DR 15	0.34
RCS 3	3	DR 18sp	0.71
RCS 5	5	DR 24	1.4
RCS 10	10	DR 100	2.8
RCS 15	15	DR 180	5.1
RCS 20	20	DR 300	8.5
RCS 30	30	DR 500	14.2
RCS 40	40		
RCS 50	50		
RCS 70	70		
RCS 100	100		
RCS 130	130		
RCS 160	160		
RCS 200	200		
RCS 300	300		
RCS 600	600–660		

Courtesy of Metso Group

Outotec

Outotec offers four types of machines, the OK-R, the OK-U, the TankCell®, and the SkimAir®. The SkimAir is a flash flotation machine and is discussed separately. All Outotec flotation cells are externally aerated. Outotec prioritizes maintaining maximum flexibility in terms of manipulating gas flow rate and number of bubbles. The first Outotec flotation cells were the U-shaped OK-16 and OK-38. The cylindrical cell, Outotec's TankCell, is now sold in sizes up to 630 m^3.

All Outotec cells are designed using hydrodynamic analysis and modeling with computational fluid dynamics (CFD). High-intensity microturbulence and macroscale laminar flow velocities must be correctly balanced to suit the particle sizes being floated.

The bottom part of the cell is considered the contact zone and is designed for suspending solids and placing air bubbles in contact with particles using turbulent conditions. Air is fed to the rotor through a hollow shaft. The flotation air is uniformly dispersed into the slurry through the slots of the rotor. Particles are collected by air bubbles forming aggregates in the contact zone, and particle–bubble aggregates rise by buoyancy toward the froth zone. The laminar flow field in the quiescent zone is dedicated for selectivity; it allows separation of valuable particles from unwanted particles. Particle–bubble aggregates form a froth phase, which acts as a cleaning zone, further rejecting unwanted solids and water, thereby upgrading the froth. The froth flows over the concentrate lip into the concentrate collection launder. Hydrophilic solid particles are carried out of the cell by the cell's flow fields through the valves at the bottom of the cell tank. Figure 10 shows a cutaway view of the Outotec TankCell machine.

Cell Mechanisms

The mechanisms in all Outotec cells are mounted near the bottom of the cell. For many years, Outotec offered two mechanisms, the FreeFlow and the MultiMix. Each was designed to maintain solids suspension, disperse the bulk gas flow from the shaft into small bubbles, and provide the acceleration required for the particles to attach to the bubbles. These mechanisms are described in detail in Nelson et al. (2002). The MultiMix design was recommended for fine and middlings particles (<100 µm), while the FreeFlow design was recommended for coarser particles. These mechanisms may still be found in older Outotec machines.

Outotec introduced the FloatForce® mechanism in 2009. This mechanism was designed using CFD simulation, with the intent of reaching an optimal combination of aeration and mixing, the two functions of a flotation mechanism. In the FloatForce mechanism, shown in Figure 11, the lower portion of the rotor functions primarily for slurry pumping, and air is introduced through the six air ports in the upper portion of the rotor that are connected through channels to the shaft. The air from these ports is discharged into six dispersion slots on the

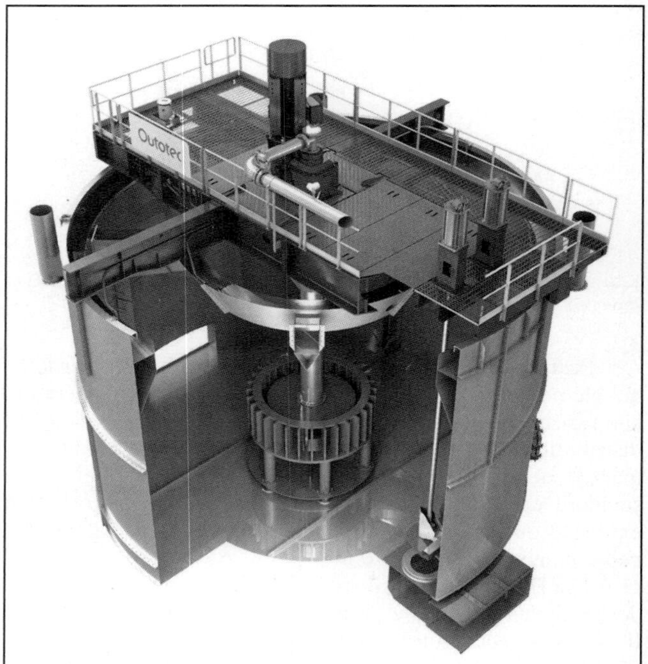

Courtesy of Outotec
Figure 10 Outotec TankCell with FloatForce® mechanism

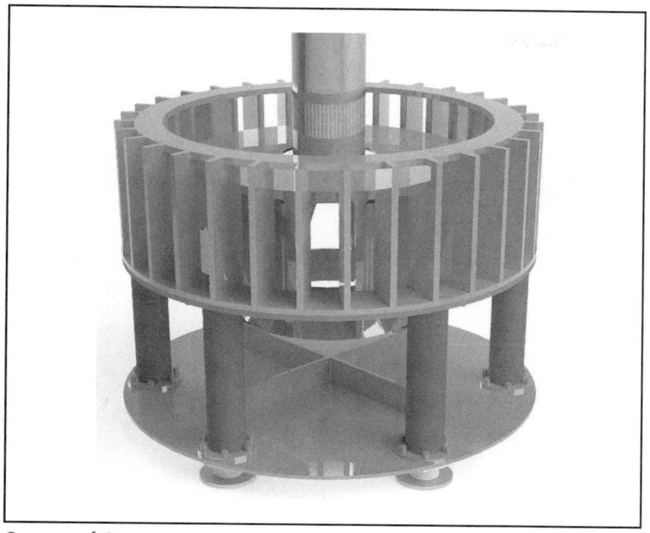

Courtesy of Outotec

Figure 11 Outotec FloatForce mechanism

periphery of the rotor so that air is dispersed closer to the stator. This minimizes the influence of the introduction of air on the pumping function of the rotor so that the mixing capacity remains high even when a high air feed rate is used.

The FloatForce mechanism was designed to provide improved metallurgical performance. It also provides decreased wear. Stator blades up to a certain size are individually bolted and light enough for one person to handle without a lifting device, which simplifies change-out. For larger sizes, blade sets are provided for one-quarter of the stator. This makes replacement of damaged blades quicker and less expensive than for conventional mechanisms, in which all stators were changed one-half at a time. Size, shape, and lining thickness of the stator blades have been optimized to provide maximum wear life.

The FloatForce mechanism can be retrofitted to existing Outotec flotation cells. It is available in the same sizes, configurations, and connecting dimensions as all previous mechanisms. For new projects, the FloatForce mechanism does not use different stator designs or bottom clearances, as was the case for the previous mechanisms. Instead, the adaptation for particle size is done by selecting the size and speed of the mechanism. For finer particles and easier duties, a mechanism one size smaller than normally used at higher speed is selected. For coarse particles, a larger mechanism at lower speed will provide sufficient mixing but minimize the high turbulence that breaks the connection between a large particle and a bubble.

Current Machines Offered

Table 5 shows rectangular (OK-R) and U-shaped (OK-U) machines; Table 6 shows cylindrical machines. The motor sizes shown are typical and may change as application conditions require.

Woodgrove Technologies

Woodgrove Technologies offers the staged flotation reactor (SFR) as an alternative to the large tank cells offered by other suppliers (Figure 12). Woodgrove notes that the trend in increasing tank cell sizes can result in "longer residence

Table 5 Outotec OK-R (rectangular) and OK-U (U-shaped) cells

Model	Effective Cell Volume, m^3	Installed Power, kW
OK-0.5-2R	0.5	7.5, dual drive
OK-1.5-2R	1.5	15, dual drive
OK-3-2R	4.3	11, single drive; 22, dual drive
OK-8-U	8	22, single drive
OK-16-U	18	30, single drive
OK-28-U	28	45, single drive
OK-38-U	38	55, single drive

Courtesy of Outotec

Table 6 Outotec TankCells

Model	Effective Cell Volume, m^3	Motor, kW
TankCell e5	5	11
TankCell e10	10	30
TankCell e20	20	30
TankCell e30	30	45
TankCell e50	50	110
TankCell e70	70	110
TankCell e100	100	110
TankCell e130	130	150
TankCell e160	160	175
TankCell e200	200	225
TankCell e300	300	315
TankCell e500	500	400
TankCell e630	630	500

Courtesy of Outotec

Courtesy of Woodgrove Technologies Inc.

Figure 12 SFR machine installation

times, massive infrastructure, and challenges in controlling froth drop-back" (Woodgrove Technologies 2015).

The SFR incorporates features of both conventional mechanical cells and flotation columns. Like the Eriez StackCell machine, it provides separate chambers for different subprocesses. The SFR, however, employs three chambers where the Eriez cell has only two. The three SFR chambers enable independent optimization of the subprocesses that make up the complete flotation process: bubble–particle contact, particle attachment, particle detachment, and froth recovery as shown in Figure 13.

Another notable difference from the Eriez cell is the location of the agitator. In the StackCell machine, the agitation chamber is *inside* the froth separation chamber, whereas the chambers of the SFR are side by side.

Froth drop-back is minimal and tightly controlled. The constrained froth recovery zone results in better recovery and upgrading. Benefits of the SFR compared with conventional mechanical cells are said to include the following:

- No power is used for solids suspension, so when combined with fewer stages, the SFR reduces circuit power consumption by 40%–50%.
- Required floor space is reduced to 40%–50% of that used by conventional mechanical cells, with corresponding savings in installation costs.
- Air consumption is significantly lower, typically 10%–20% of the air consumption of conventional tank cells.
- Fewer stages are needed to achieve equivalent concentrate grade.
- Underfroth water wash is introduced below the froth zone. This is effective in minimizing gangue dilution of the concentrate product by entrainment without limiting froth mobility.
- Operating with low air rate greatly enhances flotation selectivity. For example, this is very effective in the rejection of pyrite in copper circuits.
- Froth washing on roughers is an option in high-clay applications.
- High solids flux and low air consumption create high solids concentration in the froth as well as lowering downstream pump, piping, and even regrind sizes.
- High solids concentration in the froth creates a stable froth zone, minimizing froth drop-back and, in some cases, improving coarse-particle recovery.
- An isolated collection zone, in a separate vessel, allows addition of power for fine-particle collection without significantly affecting the overall footprint.
- Circuit control is improved.
- Wear and maintenance costs are reduced because of lower impeller tip speeds.

Five parameters are used in the design and sizing of SFRs: three hydrodynamic parameters, one slurry flux parameter, and one solids flux parameter. There are no standard sizes; rather, SFRs are custom designed to best match the variation of metal units over the life of the mine in each operation. It is this custom design for each application that allows the SFR to provide the stated benefits. To date, Woodgrove has carried out fabrication drawings for flow rates ranging from 20 to 2,850 m^3/h and has SFR designs to handle up to 4,600 m^3/h.

Courtesy of Woodgrove Technologies, Inc.

Figure 13 SFR machine, with agitation chamber on the left and separation chamber on the right

Flash Flotation Machines

The flotation of coarse, high-grade particles was investigated at North Broken Hill mine in 1932 (Garrett 1933) but not incorporated in routine operation (Lynch et al. 2010). The Denver Equipment Company introduced "unit cells" in the early 1940s (F. Seeton, personal communication), with the intent that "the mineral should be recovered as soon as it is liberated." A single, mechanical machine was often used in a ball mill–classifier circuit to recover the liberated, coarse mineral before further processing (Denver Equipment Company 1953). The concept reappeared many years later in a modified form, and was called *flash flotation* (Lynch et al. 2010). Flash flotation is designed to produce a high-grade concentrate by recovering liberated coarse particles from the circulating load of the grinding process. Flash flotation machines, shown in Figure 14, are offered by Metso and Outotec.

The Outotec SkimAir machine uses the MultiMix mechanism, which is designed to suspend coarse particles. It is designed with a deep froth crowder to accelerate froth removal and two outlets to provide options for separating middlings and tailings. The SkimAir is offered in six models, the SK-80, SK-240, SK-500, SK-1200, SK-1800, and SK-2400, where in each case the model number indicates the treatment capacity of the cell in tons per hour. The Metso RCS UC machine uses an RCS mechanism and integrates specific designs to handle coarse particles.

OPERATIONAL FUNCTIONS OF FLOTATION MACHINES

As discussed earlier, flotation machines are structurally classified by the design of their mechanisms. In operation, machines are classified by their functions. The simplest scheme classifies machines as roughers, cleaners, and scavengers. In some applications, additional categories are used, such as rougher-scavenger and cleaner-scavenger, unit cell, or flash flotation cell. In most cases, cells that have a given function are installed in groups or *banks*. The definitions and descriptions given here are those commonly used in the United States. In other locations, the nomenclature is different. For example, Chilean operators never use the terms *rougher-scavengers* or *cleaner-scavengers*. In Chile, the term *scavenger* is used only for cleaner-scavengers, while rougher-scavengers are understood to be the last few cells at the end of a row of roughers.

Roughers

Roughers are typically the first bank of cells in a flotation plant. They are designed and operated to achieve high recovery with acceptable selectivity, and to recover both coarse and fine particles of the valuable constituent. In complex ores, especially in a plant that produces more than one product, the roughers may be operated to achieve higher selectivity. Rougher operation is characterized by medium- to high-energy density, a medium to thick froth layer, and rapid froth pull. Rougher concentrate reports to the cleaner section; rougher tailings report either to the rougher-scavengers or to the final tailings.

Rougher-Scavengers

Rougher-scavengers are operated to attain high recovery of values from the rougher tailings. Depending on the mineralogy and the rougher operation, these values may be coarse, fine, or locked particles, or combinations of the three. Concentrate from rougher-scavengers reports either to a regrind circuit (for liberation of the value from locked particles) or to the cleaners. Scavengers operate with a shallower froth layer, and aeration may be changed to obtain the desired recoveries.

Rougher-scavengers may be a separate row or bank of cells, or many cells at the end of a row of roughers that are operated as scavengers. In the overall rougher section of a plant, the allocation of individual cells between the rougher and the rougher-scavenger function may be changed to accommodate changes in the feed.

Cleaners

Cleaners are operated to improve the concentrate grade to meet downstream specifications for content of product values and avoid penalties for unwanted constituents. They are much smaller than roughers and rougher-scavengers, because they are treating the combined products of the roughers and the regrind circuit from which most of the tailings have been eliminated. Cleaners are characterized by a thicker froth, which usually provides a higher-grade concentrate.

Cleaners also operate at a lower pulp density, with a finer particle-size distribution than that found in roughers and rougher-scavengers. Thus less energy is usually required from the cell mechanisms, which are operated at lower rotational speeds and, in self-aerating cells, with less submergence of the impeller in the draft tube. However, some applications, such as the flotation minerals bearing platinum group metals, require high energy input to cleaner cells to recover extremely fine, dense particles. In many applications, cleaning is accomplished in several stages. Final cleaners may be columns.

Cleaner-Scavengers

Cleaner-scavengers are similar to rougher-scavengers in that they are operated to attain high recovery values from cleaner tailings. They are different in that those values are almost always fine particles. Operation of cleaner-scavengers is similar to that of cleaners, with less intense agitation and sometimes a thicker froth layer.

Courtesy of Metso Group   Courtesy of Outotec

Figure 14 Flash flotation machines

Reverse Flotation

Reverse flotation refers to a process in which the gangue is floated away from the valuable constituent. It is most common in the concentration of iron ores, where silica is floated away from iron oxides (Lynch et al. 2010), and it is also used in coal processing where, in some cases, pyrite and ash are floated away from coal. In some reverse flotation processes, the concentrate volume will be larger than that of the tailings. Appropriate care should be exercised in the design of concentrate handling equipment—launders, pumps, and so forth.

Split-Feed Flotation

In some applications—notably the flotation of phosphate ores—feed to the flotation plant may be split into coarse and fine flowstreams, each of which is treated differently in the flotation process.

PROCESS DESIGN

The selection of flotation machines for a given application requires specification of the number and sizes of machines for each part of the flotation circuit. The residence time required to achieve the desired grade and recovery in the concentrate is determined in laboratory tests, and appropriate scale-up factors are applied to find the number and size of machines. In addition, the machine design must account for the physical requirements of froth removal to achieve adequate concentrate recovery. Cell sizes, numbers, and circuit configurations are determined by laboratory tests and the application of scale-up factors based on experience. The design of froth removal and froth handling systems cannot be based on laboratory tests because, in those tests, the froth is removed manually by scraping and thus does not exhibit the same behavior it will in operation. Furthermore, there is no reliable method known for the scale-up of froth transport distance.

Residence Time

The following discussion is summarized from Nelson et al. (2002). Residence time is the time that a unit volume of slurry takes to travel through a process or machine. If there is no short-circuiting or "dead volume" in the process, then residence time in minutes (min) is simply process volume in cubic meters (m^3) divided by the flow rate to the process in cubic meters per minute (m^3/min).

The required residence time for a given application is determined from laboratory or pilot-plant testing. The laboratory data are then used to calculate the parameters for a kinetic model, such as the one proposed by Klimpel (1980), where recovery, R, is given by

$$R = R_0 \left[1 - (1 - e^{-kT})/kT\right] \quad \textbf{(EQ 1)}$$

where

R_0 = recovery at time zero
k = flotation rate constant
T = time

When the required residence time is known, the flotation capacity is determined by a simple calculation (Poling 1980) and is summarized here:

$$C = Q_{\text{pulp}} \cdot R \quad \textbf{(EQ 2)}$$

where

C = total required flotation capacity, m^3
Q_{pulp} = volumetric flow rate of pulp, m^3/min
R = required residence time, min

In this calculation, the solids feed rate to the mill, the solids specific gravity, and the pulp density are used to calculate the volumetric flow rate of the pulp.

When the required total tank volume is determined, the number of machines of a given size is found from the following equation:

$$\tau = V_{\text{eff}}/Q_{\text{pulp}} \quad \textbf{(EQ 3)}$$

where

τ = theoretical residence time, min
V_{eff} = effective cell volume, m^3

The effective tank volume must subtract the volumes of the froth layer and the internal tank components. It must also account for the air holdup, which is the amount of air suspended in the pulp during machine operation.

In the past, when rectangular cells were used with no partition between several adjacent units (hog-trough configuration), there was extensive discussion of connection patterns for those cells. It was typical to connect two to four cells in "open" configuration, and connect one group to the next through a step-down junction box, with level control. There was much discussion of the best way to configure rows of flotation cells, to achieve the best mixing and minimize short-circuiting.

With the advent of large, cylindrical machines, the grouping of cells has become less common. Large machines are now often connected to one another individually, through connection boxes and level control valves. Extensive residence-time-distribution studies have shown this configuration poses no problems regarding retention time or short-circuiting, but it does limit the operator's options in dividing the froth for selective retreatment. Further, the minimum number of cells required in a row or section will typically be fewer than with smaller cells. For closed circuits (typically used in cleaning stages), where the tails are circulated back to other stages, only one cell per row may be needed, because the tailings may be further treated as required.

Froth Recovery

Froth removal is critical to the satisfactory function of any flotation circuit. Recent theoretical analyses have considered kinetics for *two* processes in flotation: (1) particle–bubble attachment and transport to the froth interface, and (2) froth recovery. Froth recovery is the fraction of particles reaching the froth interface that finally reports to the froth launder. In some cases, this has been reported to be as low as 10%.

The froth that develops at the top of a flotation cell must be removed to the next stage in the process. Because froth is continually being formed in the cell, the froth at the top will naturally overflow at the top. Froth recovery is characterized by the froth carry rate, measured in dry metric tons of concentrate per square meter of froth surface area per hour (t/m^2/h), and froth lip loading in dry metric tons of concentrate per meter of froth lip per hour (t/m/h). The two parameters are clearly related. If the lip length is too small, the thickness of the froth layer will increase until the froth layer begins to collapse. In addition, froth carry rate and froth lip loading are also related to the *pulling* rate at which the cell is being operated. The pulling rate affects the recovery and grade of the concentrate, which will be discussed further.

As mentioned earlier, designs of froth removal and froth handling systems cannot be determined by laboratory testing. Froth removal systems are usually sized by experienced personnel, who make calculations for each application and compare the results with their previous experience. Outotec recommends the following froth carry rates, all in (t/m^2/h): for roughers, 0.8–1.5, for scavengers, 0.3–0.8, and for cleaners, 1.0–2.0 (Heath 2013). Note that the *froth carry rate* must not be confused with the *froth factor*, which is discussed in detail later in the "Launders and Froth Removal" section.

SIZING AND DESIGN OF MACHINES AND CIRCUITS

A carefully planned laboratory test program is essential for the sizing of mechanical flotation cells and for proper circuit design in both rougher and cleaner applications. In addition, a good testing program requires properly selected samples. The importance of obtaining representative samples, along with careful preparation and preservation, cannot be overstated. The most careful and comprehensive test programs are worthless if the samples used are not representative. A successful sampling program requires detailed discussions with and cooperation from geology and mining groups. This is particularly important if the mineralogy of the ore body is highly varied. Diamond drill-core samples are preferred, but care must be taken to ensure that drilling or cutting fluids do not contaminate the core.

A laboratory testing program for flotation usually begins with a series of tests on several ore-type composites to evaluate appropriate reagent schemes and delineate optimum rougher flotation conditions such as primary grind size, flotation pulp density, flotation pulp pH, and retention time.

Once these parameters are established, a comprehensive ore variability program is undertaken to determine the response of various ore types to the optimum conditions established.

These ore variability programs often test hundreds of samples to ensure that the ore body is well understood.

Cleaner parameters, such as regrind size requirements, cleaner pH, pulp density, and number of stages, are defined. Semicontinuous locked-cycle tests are performed using the parameters established in the batch tests to determine the effect of the recirculation of middlings streams on flotation performance. The data generated and the experience gained in locked-cycle tests may be used to design and operate a continuous pilot plant if relatively large amounts of concentrate are required for further process development, such as optimization of by-product molybdenum production from bulk Cu-Mo concentrate.

The interpretation of data from the flotation test programs indicates the size of the rougher flotation circuit based on the single most important variable for sizing: retention time. Interpretation of laboratory test data is also the basis for cleaner circuit design and sizing. See Chapters 7.7 and 7.8, "Sulfide Flotation Testing" and "Non-Sulfide Flotation Testing," respectively, for detailed procedures.

SPECIFICATION OF NEW MACHINES

Drive

Large flotation machines use AC motors. The motors for machines larger than 250 m^3 may require medium-voltage power, which changes power-supply requirements and motor control configurations for the entire mill. The relative costs and advantages of medium-voltage power must be carefully considered for each site. Variable-frequency drives (VFDs) are often used for development and testing, but their high installed cost may preclude use in operations.

For machines with volumes up to 200 m^3, multiple V-belt connections are more reliable and easier to maintain. For larger sizes, gearbox drives are used by some suppliers. However, for externally aerated machines, additional cooling of the drive may be required to compensate for the heat added by the air being blown through the shaft. When belt drives are used for machines larger than 200 m^3, a redesign of the drive is required to transmit the large amount of torque required by the mechanism.

Power Density

Power density, usually measured in kW/m^3, characterizes the intensity of mixing or agitation in a flotation cell. Again, the important consideration is *absorbed* power density, as contrasted with *installed* power density. Installed power is based on the nameplate ratings of the drive motors, which in every case are oversized to account for atypical or upset conditions. Absorbed power uses the measurement of *actual kilowatts* (not just amperage and power factor) during routine operation and is thus determined by measuring voltage, current, and power factor, so the complex power relationships can be calculated. It is preferable to make these measurements for all three phases of a three-phase power system, because three-phase circuits may be unbalanced, especially after motors have operated for a long time.

In all cases, power density decreases as cell volume increases. In externally aerated cells, the tank diameter, and to some extent the tank's horizontal cross-sectional area, determine the required mixing intensity. Power is not consumed in moving pulp vertically because the buoyant gas bubbles carry the floatable material to the top of the cell. Sanding can be a problem in large cells when scale-up is not done properly and the power density is too low.

Experience has shown that higher density ores respond better to flotation in cells with higher power density. For example, in platinum flotation, 3 kW/m^3 is a standard value, while in copper flotation, the values range from 0.7 to 1.2 kW/m^3. This factor must be carefully analyzed for each application; the optimum power density is often not intuitively obvious. For example, in platinum cleaners, power density is greater than in platinum roughers; in copper cleaners, power density is lower than in copper roughers.

Mechanism

Mechanism designs vary widely among manufacturers. In all cases, it is important that the lower portion of the mechanism (shaft and rotor) be statically and dynamically balanced, *after* the installation of liners and *before* use. Careful balancing is especially important in the case of large cells, where small imbalances in the mechanism can cause large vibrations in the supporting structure. Vibration measurements should be made on each mechanism immediately after installation, and operating vibration should be checked against the baseline on at least a quarterly basis so that corrections can be made as needed.

Mechanical loading forces in the mechanism are easily calculated using first-order theoretical mechanics. Unfortunately, vibration characteristics are much more difficult to predict, and a systems approach is required. In mechanical flotation cells, besides mechanical properties (dimensions, masses, forces, etc.), the rheology of the slurry, the effects of air dispersion, generic and local flow patterns, and wear must be taken into account.

There are no standards for specifying allowable vibration levels in mechanical flotation cells. The closest applicable ISO standard is for rotating machines with excitation frequencies between 2 and 2,500 Hz, or 120–15,000 rpm (ISO 10816-3), but most large flotation cells operate below 2 Hz. Nonetheless, Outotec has found the guidelines for response to vibration given in ISO 10816-3, shown in Table 7, to be useful for flotation machines.

Table 7 ISO 10816-3 guidelines for response to vibrations

Vibration Level	0.0–1.4 mm/s	1.4–2.8 mm/s	2.8–4.5 mm/s	>4.5 mm/s
Condition and suggested response	Newly commissioned machine	Okay for normal, continuous operation	Maintenance needed on next shutdown	Maintenance required immediately

© ISO. Adapted from ISO 10816-3:2009 with permission from the American National Standards Institute (ANSI) on behalf of ISO. All rights reserved.

Dart Valves

Dart valves are the standard device for controlling flow between cells, and thus for regulating the pulp level in a cell. Dart valves typically use a conical closing mechanism, or plug, that fits against a circular seat. As the conical mechanism is moved by the valve rod, the opening gets larger, and flow through the valve increases. This characteristic is not necessarily linear and should be analyzed in combination with the characteristics of the controller to ensure that flow control can be maintained in the desired region. Plugs with a specially designed taper provide a more nearly linear response characteristic. On large cells, two valves are often used—one with high gain for large changes, and one with lower gain for fine control. A detailed description of dart valve design and operation may be found in Nelson et al. (2009).

Flotation cells cannot operate without correctly functioning dart valves. Dart valves must be carefully installed and correctly adjusted, calibrated, and tuned for level control after installation. Furthermore, dart valve systems must be diligently maintained. The valve rods and rod guides should be inspected and lubricated regularly. If the valves are pneumatically actuated, the air supply lines must be blown down regularly.

Tankage

The tankage associated with a machine includes feed boxes, the cell itself, connection boxes, and discharge boxes. Feed boxes provide a constant head for feed to a machine or row of machines. A feed box is usually designed to admit feed at the top and pass it out to the flotation process at the bottom through an open connection. The box should be designed to accommodate process upsets and should fit as closely as possible to the machine, minimizing required floor space. Each feed box typically has a sloping bottom to enhance flow into the flotation cell. The size of the opening that connects the feed box to the cell is based on the minimum linear velocity required to maintain slurry suspension. Of course, high velocities can cause high wear, so connections should be lined with wear-resistant materials. In concentrators with semiautogenous grinding (SAG) mills in the grinding circuit, large fluctuations in grinding output are common. In such installations, large feed boxes can provide valuable surge capacity in the flotation circuits.

Flotation cells come in two basic shapes, rectangular and cylindrical. Rectangular cells are usually of the hog-trough design to provide longer retention time and often have a U-shaped or horseshoe bottom to conform to hydrodynamic flow patterns and to minimize wear and sanding. The largest installed rectangular machine is the 85-m^3 WEMCO® 1+1®. Rectangular cells are still supplied for some portions of flotation circuits in smaller-volume plants. Cylindrical machines are symmetrical and thus provide better hydrodynamics. They eliminate stagnant areas found in the corners of rectangular cells and the turbulence resulting from corner effects. Plant layout using cylindrical cells requires careful consideration to minimize required floor space. However, some advantages can be achieved by configuring the layout so that connections can be modified using bypasses, allowing one cell to be taken out of service for maintenance without shutting down an entire section.

Connection and discharge boxes are practically the same. Both use dart valves to control the flow through the box. Connection boxes installed between adjacent cells have a lower hydrostatic head, usually just enough to maintain flow through the bank of cells and accommodate changes in throughput. Discharge boxes have a higher hydrostatic head to provide the required flow to the tailings circuit. The inlet to a feed box should be designed and positioned to avoid excess frothing and bubbling in the box, and every feedbox should be provided with an overflow to the froth launder system. Both boxes should be designed using the minimum linear velocity required to maintain slurry suspension. Again, high velocities can cause high wear, so connections should be lined with wear-resistant materials.

There are four designs for connection and discharge boxes used with cylindrical tanks. The first, most conventional design is based on the design still used with rectangular cells. These are rectangular boxes attached to the outside of the associated cell or cells. The second design, which saves floor space and provides easier access, is a small, cylindrical box that is again attached to the outside. The third design places the dart valves inside the cell. In this design, the discharge box is below the floor level and smaller. The advantages of this design are reduced floor space, lower equipment cost, easier access to the valve and seat wear parts, and no accumulation of air and froth in a separate box. In the three cases just described, the dart valve moves vertically. In the fourth design, the dart valve is inside the tank, but the tanks are connected directly on their adjoining walls. The valve plug is connected to the valve stem by a linkage that allows the valve stem's vertical motion to swing the valve plug in and out of an opening connecting the tanks. This fourth design saves both vertical and floor space but requires a more complex dart valve mechanism. It is available only from FLSmidth.

Launders and Froth Removal

Froth removal is critical to satisfactory function of any flotation circuit. *Froth recovery* is defined as the fraction of particles reaching the froth interface that finally report to the froth launder. As previously mentioned, this has been reported to be as low as 10%. In early flotation machines, froth transport occurred as a result of the natural mass flow through the cell. As cells became larger, and as new applications led to more voluminous froths, rotating paddles of various designs were used to accelerate removal of froth from the machine. Figure 15 shows froth flowing over the lip of a rectangular cell, aided by a froth paddle.

To further enhance froth recovery in a rectangular cell, a device called a *froth crowder* could be added to the far side of the cell, opposite the launder. As shown in Figure 16, the first froth crowders were plates welded to the back of the cell and inclined toward the mechanism, crowding the froth to the overflow lip on the opposite side of the cell. The presence of a crowder accelerated the flow of the froth as it rose to the surface of the cell, and decreased the distance on the surface

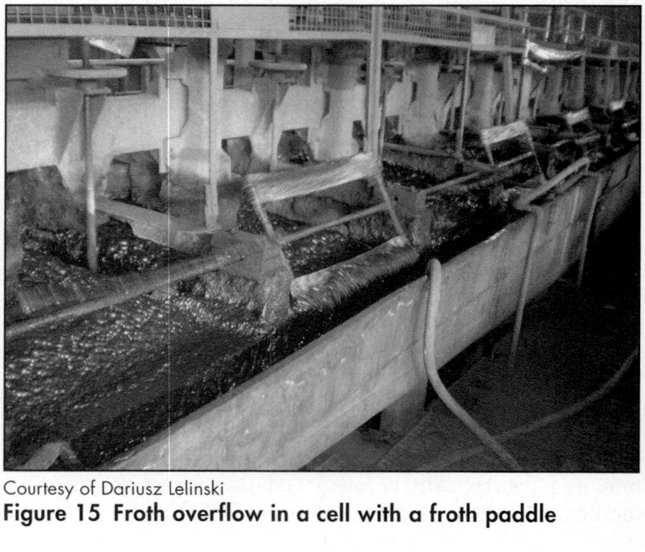

Courtesy of Dariusz Lelinski

Figure 15 Froth overflow in a cell with a froth paddle

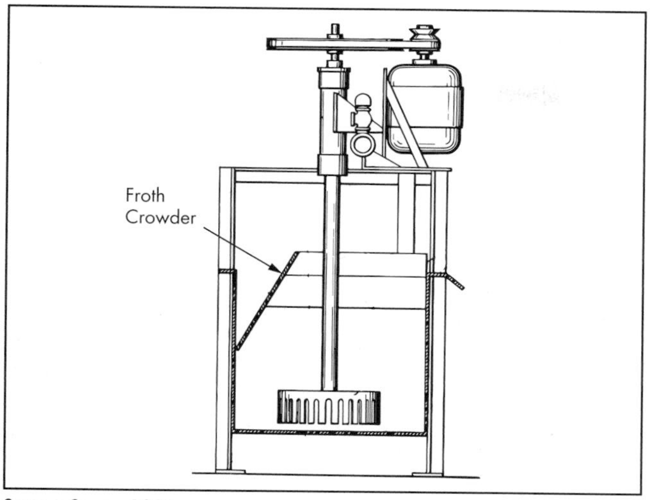

Source: Sayers 1956

Figure 16 Froth crowder in a rectangular cell

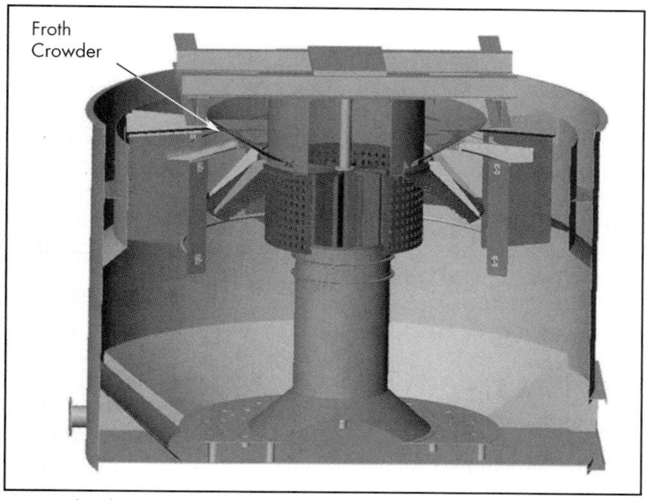

Reprinted with permission from FLSmidth A/S

Figure 17 Froth crowder in a WEMCO® SmartCell® machine

over which the froth had to travel to reach the launder. A modified crowder design, in which an inverted, truncated cone is installed around the mechanism, was developed in the 1990s (Degner 1997). Figure 17 shows a cylindrical cell with a froth crowder installed.

In large, cylindrical machines, froth recovery is more difficult for three reasons. First, the distance to a froth launder on the perimeter of the machine is larger. Second, in large cylindrical cells, the ratio of the surface area at the top of the cell, through which the froth flows, may not increase proportionally with increased volume, unless the aspect ratio of the cylinder is kept constant. Third, the circumference of the cell, over which the froth must flow for recovery, increases as the square root of the volume increases. This circumference is equivalent to the lip length in a rectangular cell. Flotation residence time is naturally related to volume, and gas flow rate through froth removal is related to surface area. Thus as cells become larger, concentrate flow rate does not increase proportionally to cell volume. Nonetheless, as lower-grade ore bodies are mined, more tonnage has to be treated to yield the same

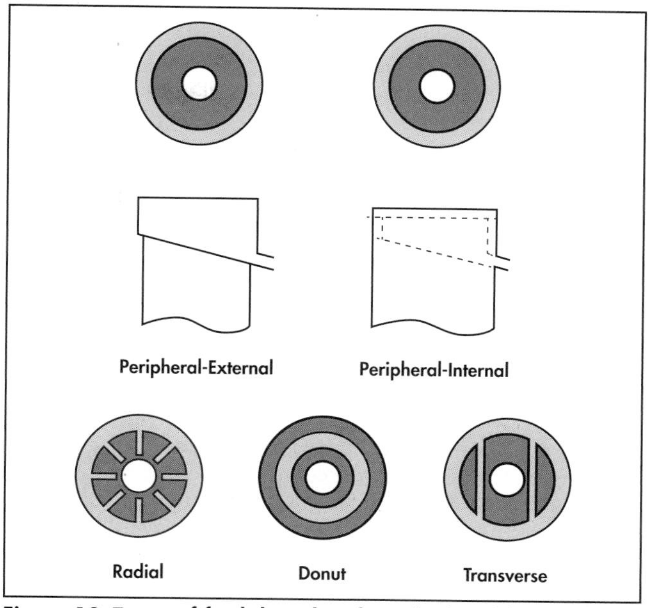

Figure 18 Types of froth launders for cylindrical flotation cells

amount of concentrate, and larger cells are better suited to this task. These constraints have led to the use of froth crowders on virtually all cylindrical machines.

The large quantity of froth produced in a big, cylindrical machine often requires additional launder capacity. Figure 18 shows five general types of launder design for cylindrical machines. In the figure, the dark shading indicates areas where froth is forming, the light shading indicates launders into which the froth moves, and the unshaded area indicates the froth crowder. Also, for the peripheral-internal launders, the lower drawing shows how the launder is mounted in the tank. Peripheral and radial launders are the most frequently used.

There are additional combinations of these five basic designs. All launders are sloped to direct the flow of concentrate to one or two discharge points. Peripheral-external and peripheral-internal launders are usually referred to as *external* and *internal* launders. They are simply attached to the outside or inside of the tank, respectively. Radial launders are always used in combination with a peripheral launder. Donut launders are often used alone but may also be installed in combination with an external or internal launder. The latter two systems are sometimes referred to, respectively, as a *double external launder* or a *double internal launder*.

The internal launder design is easy to install, requiring no additional structural support. The froth crowder directs the froth to the launder. Internal launders are widely used, but they reduce the effective volume of the flotation cell, which decreases residence time.

External launders are similar to internal launders in ease of design and installation; they are also widely used. They provide more lip length than peripheral-internal launders. Unlike internal launders, they do not decrease the effective cell volume but do require that cells be more widely spaced, thus increasing the overall footprint of the installed flotation equipment, and possibly requiring a larger building. External launders are often used on smaller cells with volumes less than 100 m^3, but may also be used on larger cells, where increased lip length and froth area are required. External launders should be designed to provide sufficient lip length for required froth flow, based on cell mass balance and froth surface area. Typical slope for froth launders is 12°–15° below horizontal, with steeper slopes used for sticky froths.

Radial launders are used where a larger lip length is required—for example, in the recovery of high-mass concentrates in iron ore processing. Like donut launders, they are more difficult to install and require additional internal supports. Figure 19 shows a 660-m^3 FLSmidth SuperCell® machine with radial launders installed. To provide adequate flow at a uniform rate, internal launders are often devised using previous external launder design. The use of internal-radial launders offers one distinct advantage in that launder capacity can be modified (usually increased) with relative ease by changing the number of radial launders.

Donut and transverse launders can also increase lip length, especially when used in combination with external or internal launders. Installation of both requires additional internal supports. Furthermore, both donut and transverse launders divide the upper portion of the froth layer into segments that tend to pull unevenly. This can lead to uneven loading of the launders.

Internal launders should be designed for relatively easy modification, for two reasons: first, to accommodate changes in froth extraction rates; second, to allow repairs after cell inspections and modification, when internal launders are often damaged. In addition, launder design should take into account the fact that that workers may stand on internal launders when conducting inspections and repairs.

The entire froth handling system must have adequate capacity to handle the highest anticipated volume of froth produced. This includes launders, piping, and pumps used to transport froth product to holding tanks or thickeners. The design of a froth handling system involves the use of the "froth factor," which is a semi-empirical quantity based on experience. It is essentially a multiplier for the volume of all equipment and parts that handle froth, as listed earlier. Froth handling equipment and parts are sized by volume. The volume of the solids in the concentrate is calculated based on the dry mass of the solids, the solids concentration, and the solids specific gravity. This volume is then multiplied by the froth factor to estimate the volume of the froth. For example, if the froth factor is estimated at a value of 3, and the calculated volumetric flow rate for the concentrate discharge pipe is 10 m^3/min, the pipe should be designed for a flow rate of 30 m^3/min. A froth factor of 3 is conservative and will almost always ensure that the sizing of the froth handling equipment (piping, valves, pumps, etc.) is adequate. Lower froth factors are sometimes used to save costs in the froth handling system, but this is not recommended.

Circuit design should include an estimation of the froth volume produced from each cell in a given time. Although only the dry mass concentrate flow for each section is typically provided for each section of the circuit, the dry volumetric flow rate can be calculated, and the froth volume estimated, by application of the froth factor. Dry mass concentrate flow is used to calculate the launder *lip length loading* in dry metric tons of concentrate per meter of froth lip per hour. To allow adequate removal of froth from the machine, lip length loading is typically below 1.0–1.5 t/m/h. As process flow moves down a row of cells, the amount of concentrate produced decreases dramatically from one cell to the next. The distribution of the concentrate produced in one row among the cells in that row varies among applications and operations. For example, in the row of 11 cells shown later in Figure 23, the distribution of the concentrate is apportioned thus: cell 1, 45%; cell 2, 26%; cell 3, 18%; cell 4, 5%; cell 5, 3%; cell 6, 2%; and cells 7–11, 1%.

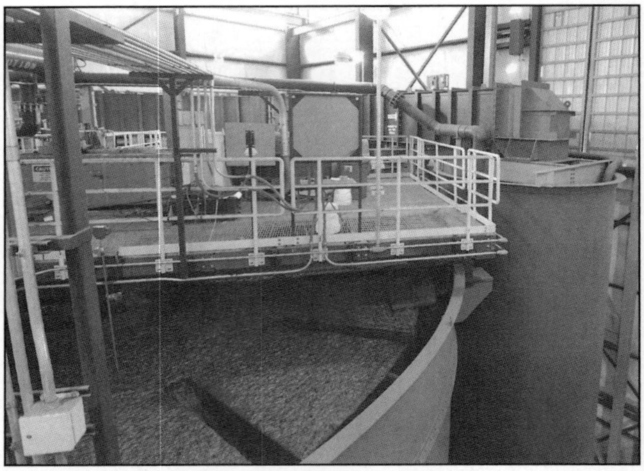
Reprinted with permission from FLSmidth A/S
Figure 19 FLSmidth 660-m^3 SuperCell® machine with radial launders

Froth crowders are frequently used to decrease the surface area at the top of the cell, forcing the froth to move more quickly to the launders. This can be especially important in scavenger cells, where the froths are relatively thin. Note however that in large cells, the decreasing ratio of surface area to cell volume can result in a condition of *overcrowding* in the froth, where particles that have attached to bubbles cannot travel through the froth for final recovery to the concentrate.

Liners

Liners are usually installed in two locations in flotation cells: the tank and launders and the mechanism. The tank and launders are lined to prevent corrosion and abrasion of the steel, so the liner should be robust, but not too thick. The most common liners used are hand-laid natural rubber and sprayed-on polyurethane, with a suitable, corrosion-inhibiting primer. Lining of larger tanks, and of tanks that are fabricated and shipped in segments for on-site assembly, requires special surface preparation and installation skills. These lining jobs should be done only by qualified contractors. In locations where large daily temperature fluctuations are expected, the lining system should be selected to withstand the *cold wall effect*, which is caused by a temperature differential across the interface between the liner and the substrate.

In all cases, metal surfaces must be properly prepared for liner installation. After cleaning and blasting the metal, the surface must be properly cleaned of grit and solvents. Liner application must be done in scrupulous conformance with the specifications given by the supplier of the liner material. In particular, some liner systems can only be applied in a narrow range of temperature and relative humidity. Finally, the chemistry and mineralogy of the flotation slurry must be considered in selection of liner material. This is particularly important with highly abrasive ores, unusual reagents, or process water with high or unusual ionic content. A detailed discussion of liners for mineral processing equipment is given by Nelson and Truss (2009).

Lifting and Transportation

Although often overlooked, lifting and transportation are key to smooth, successful installation and maintenance. With large cells, manufacturing location and methods for welding, applying liners, and other manufacturing functions must be analyzed very carefully in relation to the size of the tank or tank sections.

Lifting lugs should be installed before cell linings so that welding of lugs does not compromise lining integrity. Lugs or other lifting aids should be installed on all components that may need to be separately removed, including belt guards, mechanism components, launders, dart valves, and tank sections. Belt guards and other components should be designed so they can be separated into easily handled pieces. All bolt holes and connection points should be checked for alignment and compatibility before cell components are shipped to the plant site. For components that are lined with elastomer or other protective coatings, this check should be made before the lining is installed, and again afterwards.

Similarly, with large cells, method of shipment, shipment route, and shipping schedule must be carefully considered. Large cells must often be manufactured in sections to facilitate transport because they are usually shipped in pieces and assembled and leveled on-site. On-site assembly may include welding the tank components and installation of the cell linings. These operations must be carefully planned, with appropriate measures taken for working in confined spaces. In some cases, special permits for over-the-road shipping must be obtained, and the large sizes may make shipment by rail impossible. Where large cells are installed in existing plants, designers should carefully consider the sizes of doorways, overhead clearances, bridge crane capacities, and other factors, to ensure that the manufactured sections can be put into place as required.

INSTALLATION AND INFRASTRUCTURE

Safety

Of course, safety is the first consideration in the installation of all equipment. Particular mention is made here of the special precautions involved in the installation of large flotation cells. In such installations, large cranes and special handling techniques are required. All personnel should be trained in the lifting and movement of large, heavy loads, and all operations should be carried out by experienced, competent individuals.

Foundations

Any foundation for a large cell should be designed by a competent structural engineer. Large cell mechanisms generate high torque. For cells that have rotors near the bottom, this torque can cause flexing in the tank bottom if the support given by the foundation is inadequate. This can lead to failure of welds in the tank bottom or the parts of the mechanism attached to the bottom. The previously used mushroom design for cell foundations may not be suitable for larger cells.

The dynamic requirements for cell foundations are related to the excitation frequencies of the equipment installed and to the nominal or natural frequency of the entire system. Here, *the system* means the foundation structure supporting the cells, the cells *holding slurry*, and the cell mechanisms that provide the main excitation frequency. The system must be supported so that the nominal frequency is higher than the lowest excitation frequency by a safe margin. As previously mentioned, this margin is defined in ISO 10816-3. Support is considered stiff if the nominal frequency of the entire system is 25% higher than the main excitation frequency. Standard practice for most flotation cell installations falls into this category, but special care must be taken if the underlying earth is unstable because of moisture, permafrost, or some other anomalous condition. In any case, the cost of foundations for a given cell increases if the rotational speed of the flotation cell increases.

Flotation Buildings

Buildings and shelters should be designed with careful consideration of the local climate and possible extreme weather conditions, such as severe storms or wildfires. Building design should also consider both normal and upset operating conditions. Elevated floors should allow overflowing slurry to flow through to the lowest level, which should be designed for easy capture and recycling of overflowing slurry, with sloping floors, sumps, pumps, wash-down water, and so forth.

Buildings must have cranes large enough to lift all parts of the flotation cells and auxiliary equipment, including pumps, launders, and sampling devices. Roofs must be high enough to allow lifting and movement of cell components and auxiliary equipment. Maintenance and replacement of mechanisms, motors, dart valves, and dart valve boxes must be carefully considered, especially in the case of very large cells. If the flotation mechanisms are designed to be lifted out of the cells for maintenance, one or more maintenance bays should be included in the building design. Maintenance bays should provide for worker access on several levels, and for access by forklifts or other equipment on the lowest level. Each maintenance bay should have a stand on which mechanisms may be mounted for maintenance. One or more auxiliary cranes should be considered for use in maintenance. If the access to the mechanisms is through a port in the cell wall (as may be the case with large cells), the layout of the access level must allow travel by the maintenance equipment and workers.

The locations of control rooms should be carefully considered. Central control rooms, with live video feed from throughout the plant and video displays from a distributed control system, have great value. However, it is also beneficial for operators to regularly walk through the plant, observing the conditions in the cells. Visual observation of froth conditions and cell performance can be especially useful in plants where there are frequent changes in the feed. The installation of one or more satellite control rooms, strategically positioned among the flotation cells, can be useful when operators are making rounds.

Almost all concentrators eventually increase capacity above the original design level, so that sumps, launders, and pipelines can become undersized. It is preferable to oversize sumps and other utilitarian structures, in anticipation of the inevitable.

Walkways, Platforms, Stairways, and Railings

Walkways should be installed to provide access to all locations that may require attention from operating or maintenance personnel. This includes sampling ports; motor connections; drive belts; instrument mountings; manual and automatic control valves for slurry, air, or wash water; and drainage ports. Of course, all structures designed for human foot traffic must comply with applicable health and safety codes.

Access to the interior of the cells is frequently necessary for inspection and maintenance. In smaller cells, permanent ladders are often installed. In larger cells, access is through a maintenance hole near the bottom of the cell. In both cases, when operators enter the cell, the cell must be locked and tagged out, and operators must wear safety harnesses for recovery, as required by local health and safety codes.

Power Connections

As previously mentioned, the motors required for larger cells may require medium-voltage power and motor control centers. This possibility should be carefully considered in combination with the costs of power and installation at each site.

All machines should be provided with instruments to indicate real power consumption to the control operator. To a large degree, power consumption can indicate what is happening in the cell—whether it is sanded, the air intake is plugged, or some other issue—especially when power consumption is compared with feed rate.

VFDs are recommended by most suppliers for cells larger than 300 m^3. VFDs allow greater control of cell hydrodynamics, and when properly used, they can optimize recoveries and grades. They come at an additional cost, which must of course be considered.

Sampling Connections

Sampling connections have two purposes. First, they must provide a representative sample of a given flow stream; second, they must be capable of connecting to an online analysis system. Each connection should be configured so that even if it is connected to an online system, a separate sample can be manually removed at any time for independent laboratory analysis. Sampling connections are easily installed during construction, and connections for samplers should be installed so that every stream can be sampled if necessary. This is true even if there is no plan to sample a given stream when the operation is started. Sampling ports should be designed to be easily cleaned of blockages, and for easy attachment of connections to the sampling device or analyzer.

START-UP

Detailed, site-specific start-up procedures should be developed well in advance. All involved personnel must be trained with regard to the equipment, process, and procedures involved in the start-up. All procedures must follow local safety procedures for lockout and tagout, personal protective equipment, and so forth. Certain simple testing procedures during start-up can obviate serious problems during operation. These tests are first conducted with empty cells, then with water-filled cells, then with water-filled, fully aerated cells. All checkout procedures are made *before* slurry is introduced to the cells. (The checkout and start-up of the systems for reagent addition and process control are not covered in this chapter.)

On occasion, a project may be put on hold after the equipment is purchased. In this situation, it is important to maintain the full operability of the equipment while it is in storage. Electrical motors and equipment containing bearings must be especially considered. Both should be stored in a climate-controlled buildings at low humidity, and regularly rotated.

Empty Cell Tests

Electrical Systems

At the beginning of the start-up, a preliminary check of electrical components should be made. Bump the cell drive motors, blower motors (where used), and all pump motors, and check each for correct rotation direction. If motors start and rotate correctly when bumped, run each one long enough to confirm that its power draw is in the expected range. Finally, confirm correct operation of any VFDs.

Mechanical Systems

First, check each cell mechanism, by hand rotation on smaller machines and using appropriate tools on larger machines. Make sure each mechanism rotates freely and that the bearings are not binding. Make measurements to confirm that the rotor is concentric within the stator or draft tube following manufacturer's specifications. (This is done for two reasons: to make sure that the stator will not interfere with the rotor and to ensure that all the shaft flange connections are properly installed.) At the same time, check the height relationship between the rotor and the stator and draft tube, if present. Make a final inspection of welds in the structure and cell lining. Make sure that the bearing housing is greased according to the manufacturer's specification; if the cell is equipped with a gearbox, ensure that it is filled with the correct oil and that there is enough gear oil on-site for the first refill. (Manufacturers often require that the oil in a gearbox be changed shortly after start-up.) Be sure to install the breather valve or check valve on the gearbox—it is not usually installed when the gearbox is delivered.

Next, while the cells are still empty, power up each cell mechanism and check for vibration, visually and with a vibration sensor. After the mechanism has run for about 30 minutes, shut it down and again check the concentricity of the mechanism and the welds on the mechanism structure.

Control Systems

First, check the dart valves. Remember, dart valves are not designed to stop the flow out of the cell, only to control it. Typically, a procedure for seating the dart valve is provided by the supplier. It includes checking the stroke, concentricity, and calibration of all the dart valves. The dart valve must be very close to the seat or grommet when it is at the lowest possible position, but it should not touch the grommet. If the valve touches the grommet during operation, the shaft will be bent and the dart valve and the grommet can be damaged. While seating the dart valve, make sure that valve shafts are moving freely and not binding in the valve guides. Confirm that the valve plugs are concentric in the seats and that the plugs seat properly. Blow down the air lines for the valve control systems, then check the instrument air pressure and make sure that it functions over the correct pressure range.

Next, check control loop calibration and function, and confirm that control loop parameters are set as specified by the supplier. The final control loop calibration will take place after start-up and must be performed by a qualified person.

Where blowers are used, confirm the correct function of the blower control system. Cycle the blower through its prescribed range and make sure the blower and the flow sensors respond correctly. Also, confirm that any control valves or dampers in the flow control system function correctly.

Finally, if the cell mechanisms or pumps are fitted with VFDs, confirm that each VFD is functioning correctly. Cycle the motor through its design speed range and confirm that the VFD and all associated instruments are functioning correctly.

Water Test
In the water test, the electrical, mechanical, and control systems are tested simultaneously. Fill the cell with water. (Air is not introduced to the machine at this point. Self-aerated machines will draw in some air when the mechanism rotates, but this should be minimized by keeping the air control ports at the top of the draft tube closed.) Make a careful visual inspection to locate any static leaks in the feed box, cell, connection and discharge boxes, auxiliary tanks, and piping. Any leaks should be repaired before proceeding. After confirming that the water-filled system has no static leaks, start the mechanism and make another careful inspection for leaks throughout the system.

Now start the blower or open the air control ports at the top of the draft tube. When a blower is used, once again check the flow control function and confirm the correct operation of flow control valves or dampers in the air supply system. Run the fully aerated, water-filled machine for the time suggested by the manufacturer, and check the temperatures of the gearbox, shaft bearings, and air supply piping (in systems with blowers), and of all pump motors, couplings, and gearboxes. Finally, confirm the correct function of the level control system.

Slurry Test
After all the preceding checks have been made and everything is confirmed to be satisfactory, begin to introduce slurry to the feed box. Allow slurry to replace water in the system at its natural flow rate. As the slurry replaces the water, start the metallurgical operation and recheck the system functions: Vary the airflow, vary the level, and check the dart valve response. During this time, continue to monitor all temperatures and power draws, and frequently walk around each cell checking for leaks, vibration, or any other unusual circumstances.

The correct start-up sequence and timing are more important for the WEMCO® machines, where the mechanism is off the floor. This is because after the resuspension time guaranteed by the manufacturer for emergency shutdowns (usually 3 hours), it may not be possible to fluidize the settled solids below the bottom of the impeller. Similarly, before a routine shutdown, the slurry in these cells should be gradually diluted to only a few percent solids before stopping the mechanism.

CELL OPERATION
Successful and optimized flotation plant operation relies on human understanding, observation, and intervention. In particular, these control points should be monitored: pulp level, froth thickness, froth pull, airflow, and reagent addition. All are important in determining flotation response, but in every operation, *regardless of the machines being used*, they are closely interrelated, and no one variable can be changed without influencing the others.

Pulp Level
In self-aerating machines, pulp level is the variable most commonly changed by operators to influence the process outcome. If the pulp level is raised, the cell is said to be "pulling hard," and cell performance will move toward higher recovery and lower grade. The thickness of the froth layer will decrease, and the process will move toward increased recovery on the grade–recovery curve described in a following subsection. If the pulp level is lowered, the opposite occurs. If the pulp level is raised too high, the pulp will overflow into the froth launders, and the machine will not function properly. Operators should take special care not to pull the cells too hard if the froth/pulp interface is unstable.

Airflow
In externally aerated machines, it is more common to change airflow than pulp level. Increasing the airflow will make the machine pull harder, and move again toward increased recovery and lower grade. Decreasing airflow will have the opposite effect. Of course, there is a limit to the amount of air that can be distributed in the pulp. If this limit is exceeded, *burping* or *geysering* will occur on the surface of the cell, and stable operation will be disrupted.

In both types of machines, there is a combination of pulp level and airflow that will provide the best recovery for given feed conditions. Only experienced operators can consistently operate at the optimum point.

Reagent Addition
Reagent addition is not usually changed in routine operations. It is usually set based on experience and laboratory testing for the various ore types known to exist in the mine. Often it is impossible to know in advance when the ore type in the mill feed will change. Thus changes in the reagent addition scheme are usually made based on visual observations of the froth structure and the tailings, and on the online tailings analysis, after discussion among operations and engineering personnel.

Grade and Recovery
The grade–recovery relationship is one of the most important parameters in industrial flotation. It is unique for each operation and even for each ore type. The combination of grade and recovery is the most typical measure of separation performance. Both are expressed in terms of the valuable constituent in the ore. Recovery is the percentage of the valuable constituent, by mass, in the concentrate from the feed. Typical recoveries are between 60% and 95%. Grade represents the purity of the product. In the mill, the grade indicates the purity of the concentrate. The maximum grade of a concentrate is determined by the chemical composition of the mineral being recovered.

The grade–recovery relationship is associated with the mineralogy of the ore, the comminution process, and the flotation equipment and circuit design. Although the relationship can vary considerably in a given process, there is generally an inverse relationship between the two variables, as shown in Figure 20. In a higher-grade concentrate, the recovery will be lower, and the value left in the tailings will be higher. For example, in Figure 20, the concentrate grade at a 60% recovery is 22%, as shown by the first diamond-shaped point. When the recovery increases to 80%, the concentrate grade decreases to 13%.

It is crucial to monitor the grade–recovery curve of each part of the flotation plant on a regular basis. This curve is the most commonly used assessment of metallurgical performance, and knowledge of the grade–recovery curves for

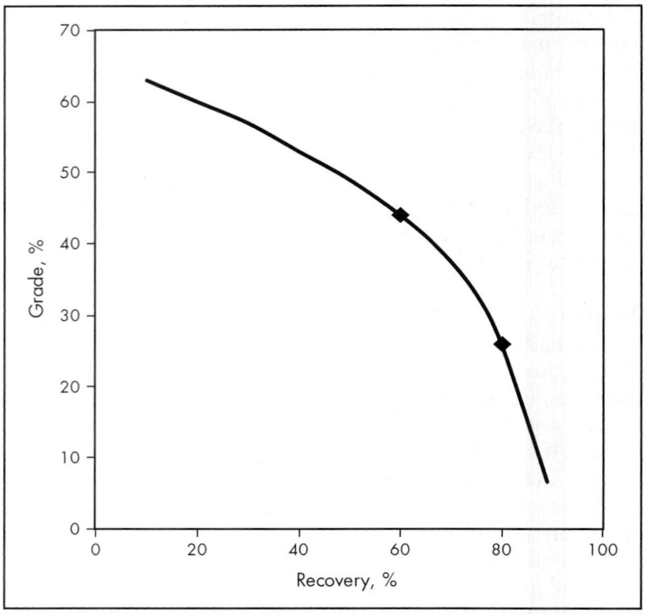

Figure 20 Example of a grade–recovery curve

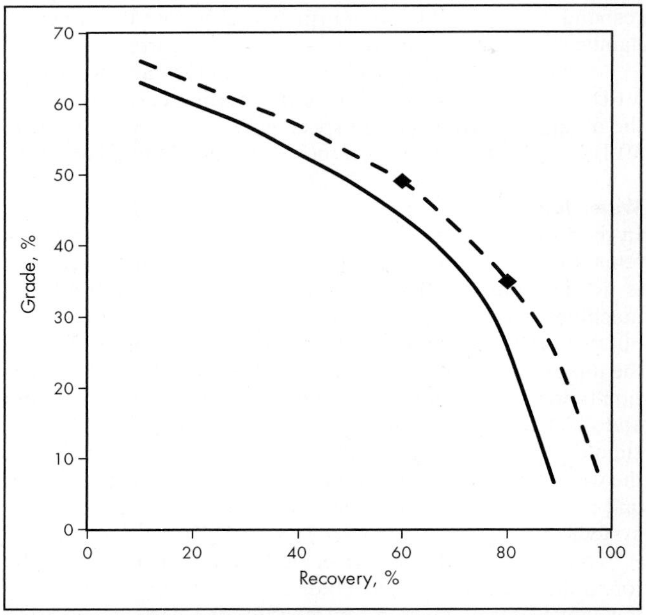

Figure 21 Example of a grade–recovery curve shifted by a change in practice

a given deposit allows the operator to know the achievable limits for different ore types and conditions. In most cases, changes in flotation operational parameters, such as froth depth and airflow (where applicable), will move the performance of the flotation process to different positions on the curve; if recovery is increased, then grade drops, and if recovery is decreased, then grade increases. Typically, shallower froth and/or higher airflow result in higher recovery and lower grade; deeper froth and/or lower airflow produce lower recovery and higher grade. However, if there is no change in ore mineralogy, comminution practice, or flotation circuit design, the flotation process will operate somewhere along the same curve. Higher recovery will be traded for lower grade, and vice versa, as shown in Figure 20.

Changes in ore mineralogy may often result in a completely different grade–recovery curve. However, certain changes in plant practice can cause a grade–recovery curve shift, resulting in higher recovery and higher grade at the same time. In Figure 21, the dashed line shows the new curve, shifted to the right of the earlier curve. With the new conditions, the concentrate grade at 60% recovery is 24%, and that at 80% recovery is 17%.

One way to shift the grade–recovery curve is to change comminution practice to achieve better liberation (smaller average particle size) of the valuable element. Improved liberation reduces the number of particles with grains of valuable mineral locked with gangue, which exposes hydrophobic material reporting to the concentrate, either naturally or after collector absorption.

A second way to shift the grade–recovery curve is to improve the flotation kinetics. This can be done by improving the reagent scheme, increasing the probability of bubble-to-particle contact, or increasing residence time. Reagent selection and testing is described in detail in Chapter 7.5, "Flotation Chemicals and Chemistry." Higher probability of contact can be achieved by improvement in the flotation mechanism. Residence time can be increased by adding more flotation

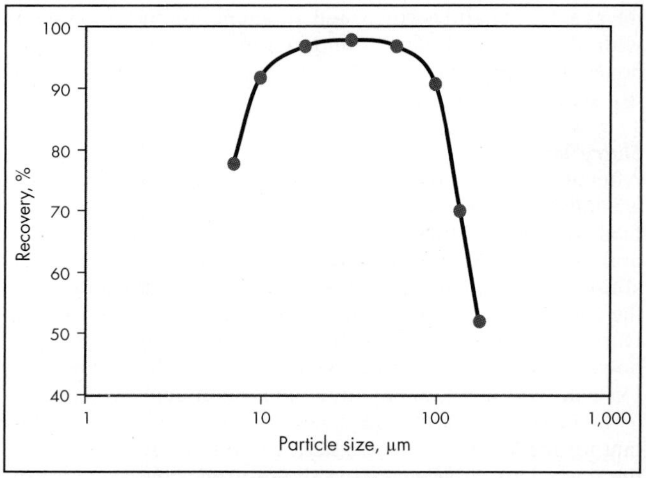

Figure 22 Recovery–particle size relationship

volume to the circuit, typically by installing supplementary flotation machines.

A third way to shift the grade–recovery curve is to improve the froth recovery. Froth recovery has been the focus of extensive research, and its importance cannot be overestimated. Recovery in the slurry phase has no effect if the resulting froth is not recovered, which means that if the froth recovery is zero, the overall recovery is also zero. Froth recovery has been observed to vary from 10% to 50%, depending on application. Typical improvements in froth recovery are mostly related to improvements in the froth carry rate and lip loading.

Of course, the goal is to have ultimate grade and recovery, but there are drawbacks associated with all approaches to the grade–recovery curve shift:

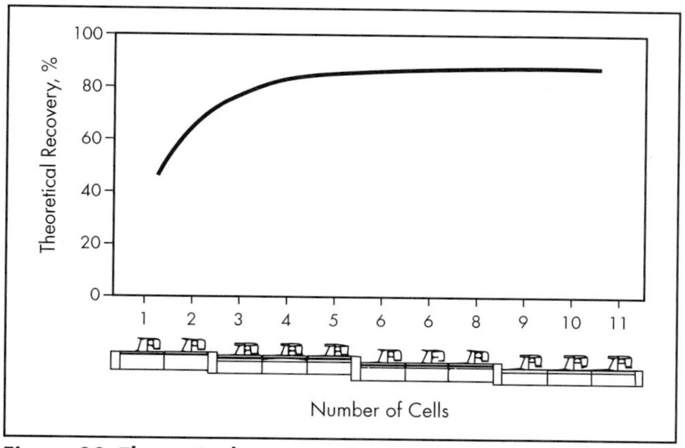

Figure 23 Theoretical recovery in a row of 11 perfectly mixed flotation cells

- Better liberation means smaller particle size entering the flotation process, resulting in a higher concentration of fines (particles smaller than 10 μm). It is well known that there is a particle-size range at which flotation gives the best results. Flotation response to particle size was described by Gaudin et al. (1942) and is shown conceptually in Figure 22. Note that particles between 10 and 120 μm float the most readily. Particles finer and coarser, outside the easy floating range, require specific conditions and higher residence time.
- Higher kinetics are often achieved with increased cost, associated with the use of more reagents, better flotation mechanisms, the cost of higher power required for increased contact frequency, or the purchase of additional flotation machines.
- Froth recovery is still the least understood part of the flotation process, in spite recent of research efforts by scientific institutions and flotation equipment suppliers. Even the measurement of froth recovery is associated with significant uncertainty. Improved understanding of froth recovery may increase the cost–benefit ratio of process changes that can shift the grade–recovery curve.

Recovery and Residence Time

The relation between recovery and residence time is usually determined during laboratory testing. Industrial flotation circuits are designed on the basis of this relationship, with the application of the appropriate scale-up factors. A theoretical recovery–residence-time relationship is shown in Figure 23; this example is based on a row of cells, as shown in the figure.

Total recovery in this example is 88%. In the first cell, 45% of this material is recovered; 94% is recovered in the first four cells, and the last 6% is recovered in the final seven cells. This example can be understood by referring to the size–recovery relationship shown in Figure 23. Material that floats easily is recovered quickly and is typically high grade. The last part of the flotation row is used to capture material that floats slowly because of liberation, particle size, or both.

For a given application, the detailed design of the flotation plant includes specifying the number of rows of cells, the number of cells in each row, and the allocation of the cell functions as roughers, cleaners, scavengers, and so forth. The design is almost always done with the assistance of a simulation software package. Such packages are available from several sources. In these models, the relationship shown in Figure 23 is accounted for by specifying several particle classes, ranging from fast- to slow-floating.

SAMPLING

The basis for evaluation of metallurgical performance of any process is a mass balance around the concentrate. Thus representative and regular samples, and accurate and timely analyses, are necessary. Each plant will have its own specific sampling schedule.

Routine Sampling

In large mills, it is typical to sample concentrates and tailings for online analysis two to four times per hour. Once or twice per hour, separate samples, taken by hand or with an automatic device, are also taken and composited into shift or daily samples. It is typical in many large operations to split the composite samples, with one split being analyzed for elemental composition and the other analyzed for mineralogy and mineral liberation, using a system that combines electron microscopy and X-ray spectroscopy.

Special Sampling

Routine sampling may not indicate long-term changes in the feed, the process, or the process equipment. Special sampling, or metallurgical auditing, should be undertaken on at least an annual basis to reestablish performance baselines for the plant. Special sampling includes down-the-row sampling, grind-release or size-by-size sampling, and residence-time-distribution analysis.

Down-the-Row Sampling

Down-the-row sampling is preferably conducted on two or three rows of each section (rougher, cleaner, scavenger, etc.) in a plant. Feed, concentrate, and tailings samples are taken from each bank in a given row. (A bank is a group of cells with common level control, and a row will include one or more banks.) The tailings from one bank are the feed to the next. In the evaluation, mass balances are calculated for each bank and each row. Down-the-row sampling is an arduous and time-consuming process. It requires preparation and careful forethought, especially in determining how and where to take representative samples. A carefully conducted, down-the-row sampling campaign can indicate long-term, gradual changes in the mechanical conditions of the cells, the properties of the ore, the performance of the grinding circuit, and other variables. By comparing results of down-the-row sampling with the results of grind-release studies in the laboratory, down-the-row sampling can give the metallurgist a good indication of where the plant is operating, in relation to the ideal grade–recovery curve. This, in turn, can provide worthwhile ideas for improving plant performance.

Grind-Release Sampling

Grind-release sampling can be conducted in the plant or in the laboratory. In-plant sampling, perhaps the more important method, includes sampling feed, concentrate, and tailings for the entire plant or any portion of the plant; screening the samples into appropriate size fractions; and analyzing each size fraction for content of the constituent of interest. For example, in a copper concentrator, a grind-release test may show that all the chalcopyrite particles larger than 200 mesh are reporting to the tailings. The results of grind-release sampling are especially valuable when used in conjunction with analysis by a mineral liberation analyzer.

Residence-Time-Distribution Analysis

Residence-time-distribution analysis has been previously described in detail (Nelson et al. 2002). Residence-time analyses are usually conducted by manufacturers on new flotation machine designs. It is important to realize that similar analyses can be conducted in the plant, using salt tracers, dyes, or radioactive tracers (where allowed). Ideally, the tracer is added *not* at the head of a flotation row, but in the feed distributor for a number of rows. Samples are then taken from the feed to each row to determine how well the feed distributor is working. Tailings samples are taken at the end of each row, and the residence time for each row calculated by the standard method. The tailings analyses can indicate gradual changes in the conditions of machines, connection and discharge boxes, and dart valves. These analyses also allow ready comparison of individual rows with one another, derived from baseline residence-time-distribution measurements performed on new machines.

All sampling programs should be statistically rigorous. Statistical design is especially important for in-plant sampling programs, which are almost always complex and expensive. Here it is important to ensure that samples are representative of the conditions in the plant at the time of sampling, and that the operating conditions at which samples are taken are characteristic of the range of conditions experienced in the plant. It is also important that there be a well-designed system for labeling, transport, preparation, and analysis of samples. A discussion of the design of flotation sampling programs is beyond the scope of this chapter. The general principles of statistical experimental design may be found in any one of many excellent texts available. For the design of laboratory flotation test programs, refer to Chapters 7.7 and 7.8, "Sulfide Flotation Testing" and "Non-Sulfide Flotation Testing," respectively. The design and execution of in-plant flotation tests has been thoroughly described in publications by Napier-Munn (1998, 2010, 2012), and Lelinski et al. (2014) offer a detailed description of one such test.

MAINTENANCE

The importance of good maintenance is obvious and cannot be overstated. Modern instruments and control systems make routine monitoring of important parameters, such as power draw, vibration, and bearing temperatures, relatively easy. However, regular and thorough visual inspections during shutdowns, performed by experienced operators and mechanics, remain an important component of a well-designed maintenance program.

Proper maintenance ensures optimal working conditions and increases the life span of the equipment, enabling high machine efficiency and ensuring high product quality. The most important part of regular and properly scheduled maintenance is ensuring that all equipment operates safely, with identified risks eliminated or minimized.

There are two conceptual approaches to maintenance: proactive and reactive. Proactive maintenance is planned maintenance and requires scheduled interruption of production; reactive maintenance is unplanned and takes place when machines or systems fail. Scheduled production interruption is preferable to the downtime caused by an unplanned breakdown and the required reactive maintenance. Because production interruptions are planned, they can occur at optimal times, and the workforce can almost always anticipate all issues that may be related to production interruption. In virtually all cases, planned maintenance takes less time and costs much less than emergency repairs and replacements. For example, mobilizing crew, equipment, and spare parts can be done ahead of time, reducing costs.

Proactive maintenance may be further divided into preventive, detective, and predictive approaches. Preventive maintenance is scheduled using parameters such as operating time or production quantity. For example, concentrators often have more than one "processing line," with each line comprising a SAG mill, one or more ball mills, and one row of rougher flotation cells. Such a mill might shut down one line for a few days each month, conducting routine inspection and maintenance on all the equipment in that line while the other lines continue to operate. Detective maintenance uses testing and measurement to locate equipment or components that have already failed but may not have yet caused shutdown of part or all of the plant. For example, instrumentation on a flotation cell mechanism may indicate that the motor for that cell has failed. The feed to the row can then be shut off before the cell in question sands in by filling with solids. Predictive maintenance uses condition monitoring to look for the indication of incipient failures. For example, high vibration on the drive for a compressor may indicate that a bearing is about to fail, so that the machine can be shut down and the bearing replaced before its failure results in further damage to the mechanism.

Clearly, predictive maintenance is the preferred approach, and even detective maintenance is superior to reactive maintenance. Predictive maintenance requires higher initial investments for condition monitoring equipment and for parts outfitted with the monitoring sensors, but it pays off very quickly.

Following are some specific recommendations for flotation cell maintenance based on the authors' experience.

Drive Assembly

Temperatures of the motor, bearings, and gear drive (if used) should be checked regularly. Statistical quality control methods may be used to detect important changes or long-term trends in readings.

Mechanism

Regular lubrication of shaft bearings is important, and bearing temperatures should be checked manually on a regular basis. The linings on the shaft, rotor, and stator should be inspected in each shutdown for wear, cracks, and chunk-type failure. Components with severe damage should be removed and replaced. Concentricity of the rotor inside the draft tube or stator should also be checked in each shutdown.

Mechanism rotation should be reversed every shutdown, to achieve even wear on both sides of the rotor blades. If rotation is not reversed regularly, symmetry will not be maintained, and then, when rotation is reversed, the motor may trip out because of excess current draw. Vibration should be checked against baseline values at least quarterly.

In self-aerating machines, the air openings should be checked regularly to ensure that they are clear. Similarly, in forced-air machines, the air intake openings in the rotor should be periodically cleared of any residual slurry or debris.

Liners
The integrity of tank liners should be inspected during every shutdown. Deterioration of liners can lead to liner failure over small or large areas, which in turn leads to increased corrosion and wear of the steel substrate, causing eventual failure of the tank. Liner integrity is especially important at the flow-through points between various tankage components—feed boxes and cells, or cells and connection or discharge boxes.

Dart Valves
Dart cones and seats should be carefully inspected for wear, proper alignment, and concentricity. Movement of the valve shaft in its guides should be checked to make sure that the shaft is not bent and that there is no binding or misalignment. The air lines and moisture traps on the dart valve controllers should be checked and blown down weekly. The flotation circuit need not be shut down to complete this operation.

Launders
Cell overflow lips and froth launders should be checked and washed down regularly to prevent accumulation of concentrate and ensure equal flow along the entire length of the lip. In large cells, connection points between the various components of interior launders may experience high wear and should be inspected on each shutdown.

Cleanout
Regular cleanout of the cell and its mechanism is important. Occasional upsets in the grinding circuit cannot be avoided, and oversize material will almost always accumulate in flotation cells. There is a limit to the amount of material that can be accumulated in any cell before performance is adversely affected. To ensure satisfactory operation, this material must be removed periodically.

The easiest way to remove oversize material is with water. In some cases, much of the oversize material can be removed by pumping water at a high flow rate through the circuit. In some plants, operators insert handheld, high-pressure water pipes into the cell to assist in moving oversize material into the flow. However, this practice should be used with caution, as it can damage the linings of the cell and mechanism. At points where the lining is damaged or detached, the high-pressure water tends to raise the lining and increase the damage and detachment. When oversize material cannot be removed with water, one alternative is the use of shovels and buckets. Large cells should be provided with access ports to allow easy access to the cell floor for cleanout.

ECONOMICS
Economic factors should be considered in the purchase and operation of flotation machines. Economics should be analyzed with a global perspective, considering capital costs, operating and maintenance costs, and metallurgical performance. Following are some examples.

First, in the design and construction of a new concentrator, the cost of the flotation equipment is usually a small part of the total capital cost for equipment. In some cases, a vendor may offer the flotation equipment for little or no additional cost if the customer agrees to purchase the comminution equipment from that vendor. While this option offers an attractive reduction in initial capital cost, it may not provide the best flotation equipment for the application.

Second, some flotation experts believe the only thing that matters is retention time, and the selection of flotation equipment can be made solely on the basis of cost per unit volume—literally, dollars per cubic meter. It is, however, important to ensure that all the performance parameters for the machines selected meet the requirements determined by laboratory tests, such as adequate power density to maintain slurry suspension, correct launder design to ensure rapid froth removal, and so forth.

Third, a given type of equipment may offer the lowest capital cost and best metallurgical performance, but its vendor may have limited presence in the operating location. Experienced service personnel may require several days to arrive on-site, and delivery of spare parts may be similarly slow. Such delays can outweigh other perceived advantages.

Finally, economies of scale often result in the design of very large mining and milling complexes, and in the use of the largest possible pieces of processing equipment. However, designers should carefully consider the challenges in fabricating, transporting, and installing large processing machines. They should also bear in mind that commodity prices are always subject to cycles, and that it is more difficult to reduce throughput in large plants that are fitted with large pieces of processing equipment.

ACKNOWLEDGMENTS
The authors of this chapter gratefully acknowledge the assistance of the following individuals: Eric Wasmund and Jaisen Kohmuench, Eriez Flotation Division; Asa Weber, FLSmidth; Kym Runge and Thierry Monredon, Metso; Rob Coleman, Outotec; and Glenn Dobby and Chris Bennett, Woodgrove Technologies.

REFERENCES
Degner, V.R. 1997. Flotation cell crowder device. U.S. Patent 5,611,917.

Denver Equipment Company. 1953. *Denver "Sub-A" Flotation*. Bulletin F10-B81. Denver, CO: Denver Equipment Company.

Garrett, J.E. 1933. Flotation of unclassified ball mill discharge for the recovery of the lead and zinc concentrates. *Proc. AusIMM* 9:475–499.

Gaudin, A.M., Schuhmann, R. Jr., and Schlechten, A.W. 1942. Flotation kinetics II: The effect of size on the behavior of galena particles. *J. Phys. Chem.* 46:902–910.

Glembotskii, V.A., Klassen, V.I., and Plaskin, I.N. 1972. *Flotation*. Translated by R.E. Hammond. New York: Primary Sources. First published 1963.

Heath, J. 2013. Frothing at the lip—stability in your flotation cell. *Outotec Output SEAP.* August. www.outotec.com/globalassets/newsletters/output/2013-2/04_frothcrowding-article_pg10-12.pdf.

ISO 10816-3. 2009. *Mechanical Vibration—Evaluation of Machine Vibration by Measurements on Non-Rotating Parts—Part 3: Industrial Machines with Nominal Power above 15 kW and Nominal Speeds between 120 r/min and 15,000 r/min When Measured In Situ*. Geneva: International Organization for Standardization.

Klimpel, R.R. 1980. Selection of chemical reagents for flotation. In *Mineral Processing Plant Design*. Edited by A.L. Mular and R.B. Bhappu. Littleton, CO: SME-AIME.

Kohmuench, J.N., Mankosa, M.J., and Yan, E.S. 2008. An alternative for fine coal flotation. *CPSA J.* 7(1):29–38.

Lelinski, D., Govender, D., Jespersen, M., Garcia, S., and Baker, T. 2014. The effect of rotor-stator treatments in a randomized trial at the Newmont Carlin concentrator (phase I). In *Proceedings of the XXVII International Mineral Processing Congress (IMPC 2014)*. Santiago, Chile: Gecamin.

Lelinski, D., Caldwell, K., Yang, Y., Olson, T., Traczyk, F., and Jespersen, M. 2015. Industrial application of the 660 m^3 SuperCell® equipped with the nextSTEP rotor and stator. Presented at the International Flotation Conference 2015, Cape Town, South Africa, November 15–19.

Lynch, A.J., Harbort, G.J., and Nelson, M.G. 2010. *History of Flotation*. Melbourne, Victoria: Australasian Institute of Mining and Metallurgy.

Napier-Munn, T.J. 1998. Analyzing plant trials by comparing recovery-grade regression lines. *Miner. Eng.* 11:949–958.

Napier-Munn, T.J. 2010. Designing and analyzing plant trials. In *Flotation Plant Optimisation: A Metallurgical Guide to Identifying and Solving Flotation Plants*. Edited by C.J. Greet. Melbourne, Victoria: Australasian Institute of Mining and Metallurgy.

Napier-Munn, T.J. 2012. Statistical methods to compare batch flotation grade–recovery curves and rate constants. *Miner. Eng.* 34:70–77.

Nelson, M.G., and Lelinski, D. 2000. Hydrodynamic design of self-aerating flotation machines. *Miner. Eng. Int.* 13:10–11.

Nelson, M.G., and Truss, R.W. 2009. Liners and coatings for mineral processing equipment. In *Recent Advances in Mineral Processing Plant Design*. Edited by D. Malhotra, P.R. Taylor, E. Spiller, and M. LeVier. Littleton, CO: SME. pp. 555–559.

Nelson, M.G., Traczyk, F.P., and Lelinski, D. 2002. Design of mechanical flotation machines. In *Mineral Processing Plant Design, Practice, and Control*. Vol. 1. Edited by A.L. Mular, D.N. Halbe, and D.J. Barratt. Littleton, CO: SME.

Nelson, M.G., Lelinski, D., and Gronstrand, S. 2009. Design and operation of mechanical flotation machines. In *Recent Advances in Mineral Processing Plant Design*. Edited by D. Malhotra, P.R. Taylor, E. Spiller, and M. LeVier. Littleton, CO: SME. pp. 168–189.

Poling, G.A. 1980. Selection and sizing of flotation machines. In *Mineral Processing Plant Design*. Edited by A.L. Mular and R.B. Bhappu. Littleton, CO: SME-AIME.

Sayers, M.J. 1956. Froth-crowding flotation machine and method. U.S. Patent 2,756,877.

Sherrell, I., and Yoon, R.-H. 2005. Development of a turbulent flotation model. In *Centenary of Flotation Symposium*. Melbourne, Victoria: Australasian Institute of Mining and Metallurgy.

Woodgrove Technologies. 2015. Staged flotation reactor. www.woodgrovetech.com/woodgrove-staged-flotation-reactor/. Accessed August 2017.

Zhou, Z-a. 1996. Gas nucleation and cavitation in flotation. Ph.D. dissertation, McGill University, Montreal, QC, Canada.

CHAPTER 7.2

Column Flotation

Michael J. Mankosa, Jaisen N. Kohmuench, and Gerald H. Luttrell

Froth flotation has been used for more than a century to efficiently and economically upgrade fine particles produced by the mining industry. The majority of industrial installations use conventional flotation machines that consist of a series of agitated tanks through which feed slurry is passed. Gas is introduced through a rotor–stator assembly that disperses air bubbles, suspends particles, and promotes bubble–particle collisions. Unfortunately, conventional flotation machines suffer from the contamination of froth products by the hydraulic entrainment of fine gangue particles. This problem is particularly severe when large amounts of ultrafine slimes are present in the feed stream. Nonselective entrainment has historically been minimized through the use of cleaner flotation circuits in which the froth product is repeatedly diluted with clarified water and refloated in multiple stages. This brute force approach of dealing with unwanted entrainment adds to both the cost and complexity of flotation plants.

In the early 1960s, an improved method of minimizing froth entrainment was patented by Canadian researchers Pierre Boutin and Remi Tremblay (1967). This invention consisted of a column-type flotation machine that introduced a countercurrent flow of wash water into a deep froth phase. The wash water practically eliminated the hydraulic entrainment of ultrafine slimes and substantially improved the purity of the froth product using only a single stage of flotation. Dobby (2002) has reported that the inventors conceived of column flotation as an analogy to solvent-in-pulp columns employed in the solvent extraction of uranium ore slurries. In solvent extraction, aqueous diluent is added to the top of the extraction column and allowed to flow countercurrently to rising droplets of solvent to reduce contamination of the solvent phase. The inventors first demonstrated column flotation technology for the removal of fine silica from iron ore using a 5-cm-diameter laboratory column and a 0.3-m-diameter pilot-scale column (Wheeler 1983). The technology was later demonstrated in Canada for copper sulfide flotation at the Opemisca Company concentrator using a 0.2-m^2 test column (Rubinstein 1995).

During the 1970s, numerous other laboratory and pilot-scale field trials were conducted by a newly formed company called the Column Flotation Company of Canada Ltd. (Wheeler 1983). This group, created and led by Don Wheeler, is credited with many of the early technical advancements associated with column technology. However, it was not until 1981 that the first industrially accepted production column was installed by Noranda Mine's Les Mines Gaspé concentrator within a molybdenum cleaner circuit (Coffin and Miszczak 1982). By 1987, the two-stage column circuit installed in this operation had replaced as many as 13 stages of conventional flotation machines (Finch and Dobby 1990). This success helped to spearhead the wider commercial acceptance of column technology as a standard unit operation in many flotation plants throughout the processing industry.

The rapid emergence of column flotation can be demonstrated from the number of scholarly works published on the topic since 1960. As shown in Figure 1, the number of publications associated with column technology increased rapidly during the late 1980s and early 1990s. This interest eventually peaked with the advent of the Canadian Institute of Mining, Metallurgy and Petroleum (CIM) symposium on column flotation in 1996. This was shortly after publication of the landmark book on the subject titled *Column Flotation* by James Finch and Glenn Dobby (1990). This text continues to be considered the most comprehensive source of both fundamental and applied information on column flotation technology. Scholarly interest in column technology waned after the CIM Column '96 symposium, which was largely in response to industry acceptance of the technology as a standard unit operation. Publication frequency started to rise again after 2009, largely because of a resurgence of new studies on column flotation in countries outside North and South America.

TECHNOLOGY

As the name implies, a column flotation machine (Figure 2) consists of a tubular tank with a relatively large height-to-diameter ratio. During operation, reagentized feed slurry

Michael J. Mankosa, Executive Vice President of Global Technology, Eriez Manufacturing, Erie, Pennsylvania, USA
Jaisen N. Kohmuench, Managing Director, Eriez Manufacturing Pty Ltd., Melbourne, Victoria, Australia
Gerald H. Luttrell, E. Morgan Massey Professor, Mining & Minerals Engineering, Virginia Tech, Blacksburg, Virginia, USA

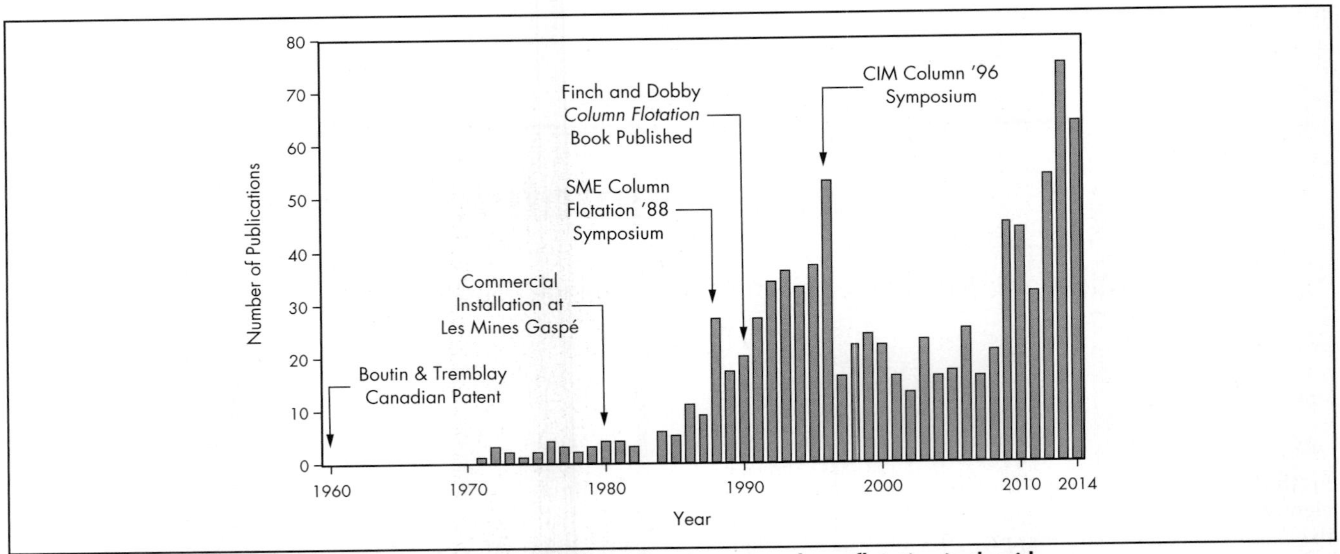

Figure 1 Annual number of publications that included *flotation column* or *column flotation* in the title

is introduced into the upper portion of the column though a header system that distributes the slurry over the column cross section. The slurry flows by gravity down through the column to a discharge port that regulates the underflow via an actuated control valve. The valve is normally configured to hold a pulp level sufficient to create 0.5–1.5 m of froth depth. Compressed air, mixtures of air–slurry or air–water, or other sources of gases are injected into the bottom of the column through various types of gas sparging systems. This arrangement ensures that the feed slurry moves in a countercurrent direction to rising gas bubbles. Hydrophobic particles collide and attach to the bubbles and are carried to the top of the column where a froth bed is formed. Wash water is then added to the froth bed through a distribution system that evenly spreads water over the entire cross section of the froth column. Sufficient wash water is added to prevent fine hydrophilic particles from entering the froth bed. The froth overflows into a collection launder where it reports as a high-grade hydrophobic product. Traditionally, columns do not make use of either an internal rotor–stator mechanism for gas dispersion/particle suspension or paddle mechanisms for froth removal. Gas compressors are normally required, however, to provide the gas flow and energy required for gas dispersion.

COLUMN ZONES

Up to three different aeration zones can be identified within a flotation column (Figure 3). The pulp zone is the region in the column in which bubbles and particles are brought together to form stable aggregates. This zone is located between the air injection point and the pulp level. The gas holdup (ε_p) in the pulp is typically 10%–20% by volume. Particles that do not become attached to bubbles in this zone are eventually discharged from the bottom of the column. The bubble–particle aggregates formed in the pulp zone eventually rise into the stabilized froth zone. This zone, which is located between the wash-water addition point and the pulp level, is largely responsible for the improved selectivity of columns over conventional flotation cells. In this zone, bubble coalescence is retarded by the downward flow of wash water that prevents the thinning of the froth lamella. Consequently, the gas holdup (ε_s) in the stabilized froth is relatively stable with height and is often in the range of 60%–70% by volume, which should be expected for packed spherical bubbles. The downward flow of water through the stabilized froth is commonly referred to as the bias flow. The bias flow stabilizes the froth and allows a deep, loosely packed bed of bubbles to be formed. The nonselective entrainment of gangue particles into this bed is prevented by the bias flow.

In addition, the circulation of floatable particles between the pulp and stabilized froth zones can be beneficial to selectivity. These features allow the performance of the column to closely approach that of the ideal separation by flotation predicted from release analysis (Davis et al. 1995). Although the stabilized froth allows high product grades to be achieved, it limits the particle recovery rate by restricting the amount of gas that can be passed through the column while still maintaining the proper bias flow. The stabilized froth places a constraint on throughput since only those particles that are attached to bubbles can enter the froth. This constraint, which is commonly referred to as the carrying capacity, places a limit on the mass rate of particles reporting to the froth product. Studies suggest that the carrying capacity limitation is particularly severe for the flotation of high-yield systems having fine particle size distributions. The carrying capacity limitation is almost always a major consideration in the design and operation of flotation columns.

The uppermost region in the column is the froth zone. This zone is located between the wash-water addition point and froth overflow lip and is similar to the drained froth phase present in conventional cells. As with conventional machines, the gas holdup (ε_f) in this zone is normally a strong function of height, with the surface reaching a gas fraction given by the gas rate, divided by the combined gas and slurry rate reporting to the froth overflow. This zone may be absent in some columns if the wash water is added above the froth bed using a drip pan or wash box. One purpose of this zone is to control the split of the wash water between the froth product and the stabilized froth zone. An adequate flow of water downward through the stabilized froth zone (i.e., bias flow) must be maintained to prevent the nonselective entrainment

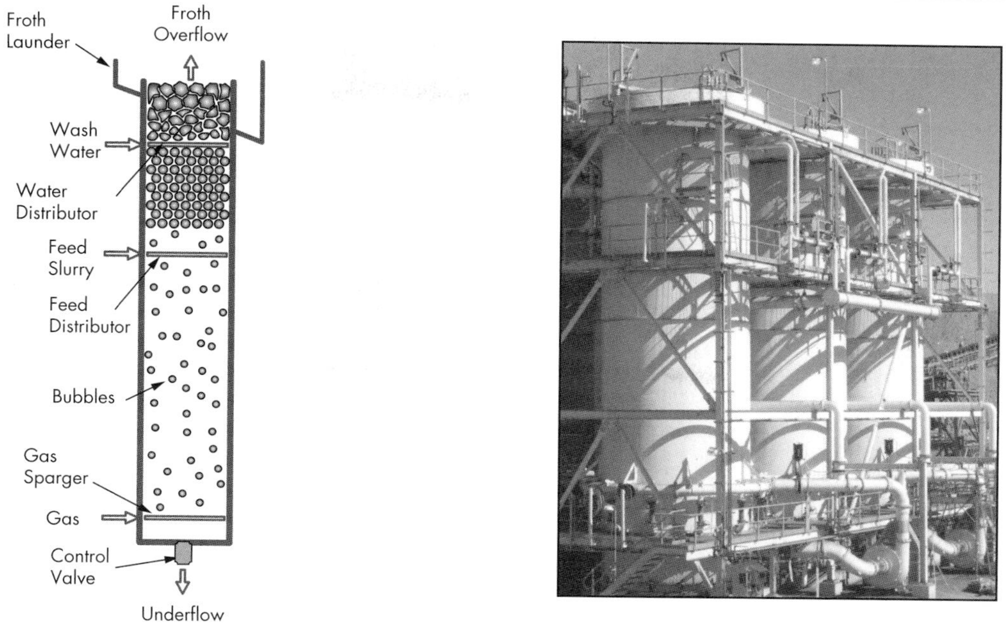

Figure 2 Basic features of a column flotation machine

of fine particles. A simple analysis can be used to show that the bias flow rate can be controlled to some extent by adjusting the aeration rate, wash-water addition point, and wash-water flow rate. Unfortunately, changes in these parameters may also affect the solids loading in the froth. If the froth becomes too heavily loaded with particles, it can become unstable and collapse. This phenomenon, which is referred to as froth overloading, places a limit on the solids content of the froth (Lynch et al. 1981). Potential problems associated with froth overloading must be recognized in the design and operation of columns.

SUPERFICIAL RATES

Because of the unique design of column cells, the flow rates passing through the machine are often reported as specific rates or superficial velocities. These values are calculated by dividing the volumetric flow rate or dry solids tonnage rate by the cross-sectional area of the column. In these calculations, the area may be reduced to correct for phenomena such as the holdup of gas within the column. The use of superficial velocities makes it easier to compare flow values for columns of different diameters and is generally more appropriate for describing the complex mechanisms that are the focus of column design calculations. Figures 4 and 5 show the superficial rates corresponding to different volumetric flow rates and dry solid tonnages, respectively, for columns of different diameters.

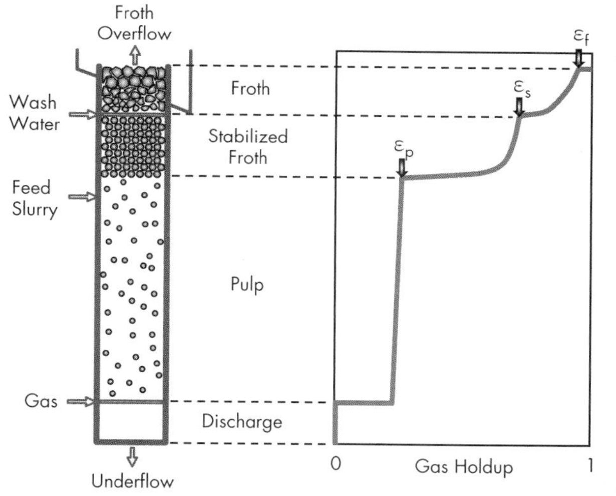

Figure 3 Gas holdup zones in a column flotation machine

DESIGN VARIATIONS

Standard Designs

Column flotation machines are designed, manufactured, and serviced by a variety of equipment vendors. While homemade columns were popular during the early years of column technology development, this trend diminished over the past two decades in response to well-documented failures associated with poorly designed units. An industry rule-of-thumb such as "column height = 0.9 × crane height" is obviously inadequate for designing a modern column flotation system. Today's leading manufacturers of standard column systems include Eriez, Metso, and Minnovex/SGS. The machines offered by these companies use a traditional tank design with a large height-to-diameter ratio. Tank designs are most commonly circular because of savings on construction costs, although square and rectangular columns are also in commercial use. Tanks are now offered with diameters up to 5 m and heights in excess of 15–20 m. Notable differences between column manufacturers primarily involve vendor-specific designs of systems for gas sparging, slurry/water/gas distribution, and process control. Some vendors also offer internal baffles that are designed to reduce unwanted internal mixing and short-circuiting of feed. Applications of standard columns include the flotation

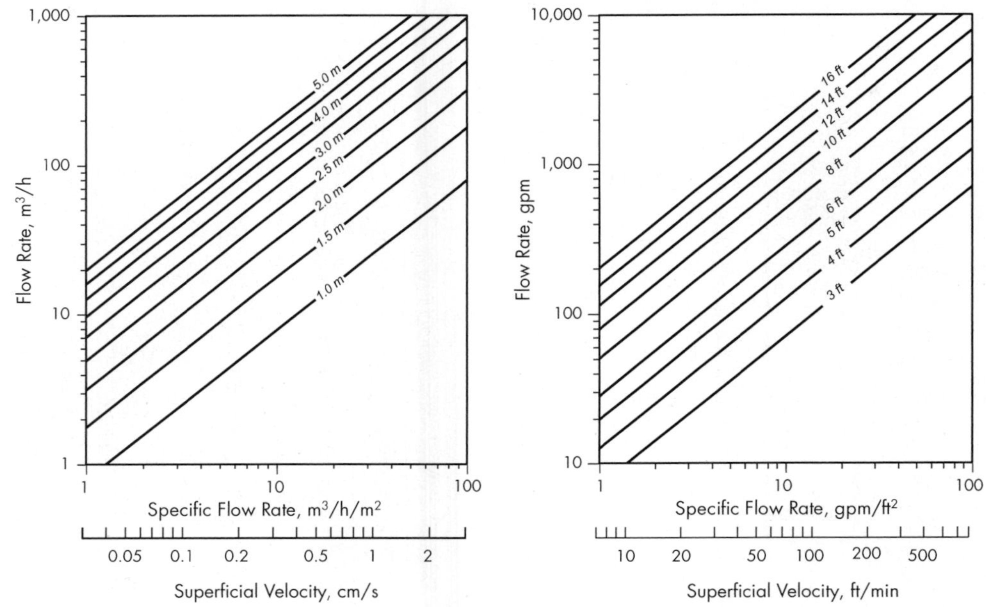

Figure 4 Volumetric flow rates for columns of different diameters, in metric and customary units

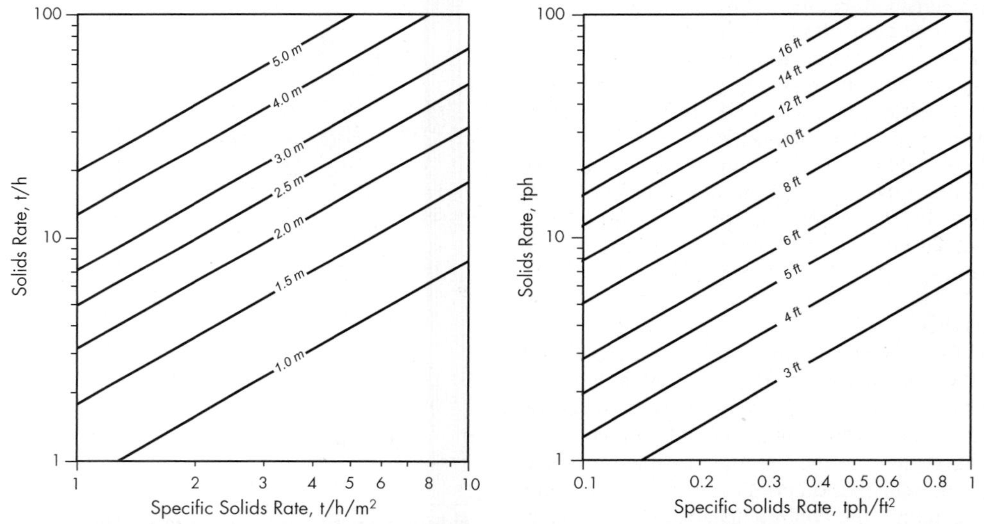

Figure 5 Dry solids rates for columns of different diameters, in metric and customary units

of particulates such as metallic and nonmetallic minerals, fine coal, solvent extraction organics, and bitumen.

Low-Profile Designs

Large-scale flotation columns can pose design challenges because of massive tank dimensions and associated large foundation loads. To combat this issue, several low-profile designs have been developed that incorporate column froth washing technology. Two popular examples of these machines include the Jameson Cell and StackCell technologies. The Jameson Cell (Figure 6) passes pressurized feed slurry though a series of downcomer tubes mounted vertically above the tank. The negative hydrostatic head within the downcomer draws gas into the feed slurry. Gas is dispersed into small bubbles by slurry injected through a wear-resistant nozzle in the top of the downcomer. The flow conditions create a slurry–gas mixture with a high gas holdup inside the downcomer that promotes efficient bubble–particle contacting. The mixture passes into a low-profile flotation tank where a deep froth bed is formed that is suitable for water washing.

Another low-profile column design is the StackCell technology (Figure 7). During operation, feed slurry enters the separator through either a bottom-fed or side-fed feed nozzle at which point low-pressure air is added. The slurry travels into an internal pre-aeration sparging device that provides significant shear and contacting prior to arrival into the separation

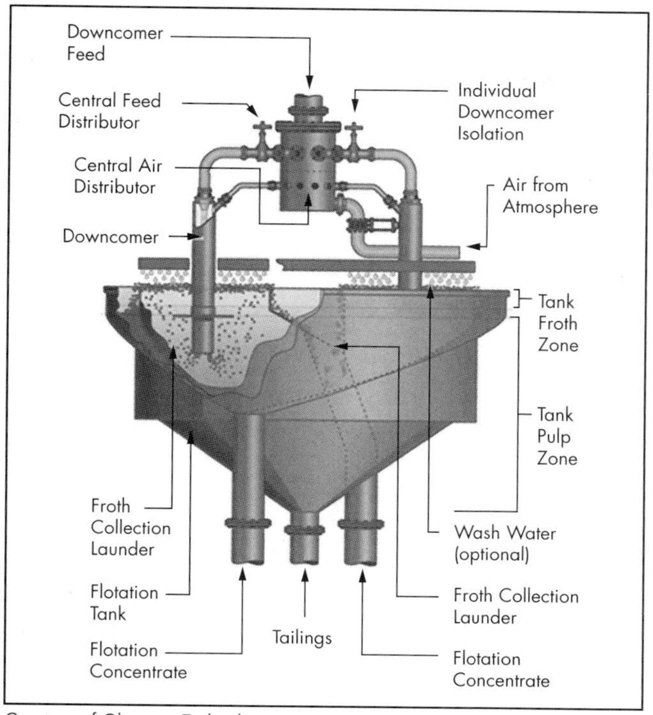

Courtesy of Glencore Technology
Figure 6 Low-profile Jameson Cell flotation

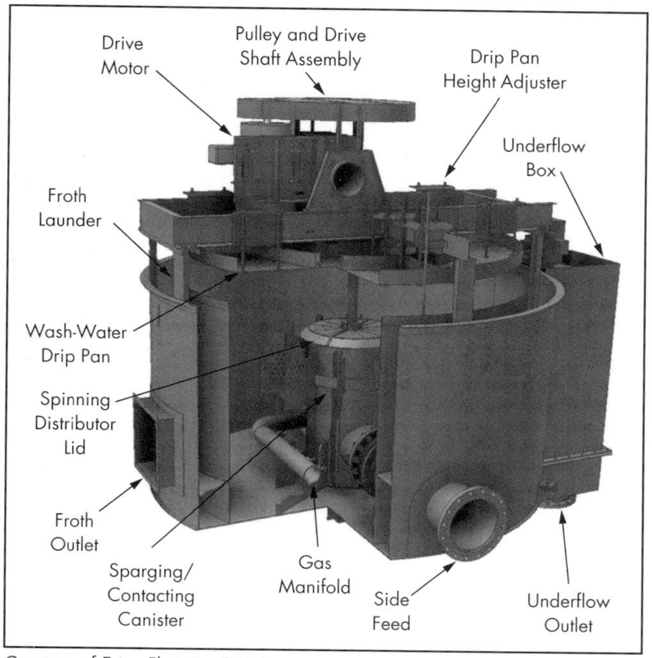

Courtesy of Eriez Flotation Division
Figure 7 Low-profile StackCell flotation

chamber. Intense bubble–particle contacting is conducted in the aeration chamber prior to injection into the primary tank where phase separation of the pulp and froth occurs. Studies suggest that the high-intensity agitation is necessary for ultra-fine particle flotation. A liquid slurry level is maintained inside the tank to provide a deep froth that can be washed, thereby providing a high-grade float product. The low-profile design allows the units to be easily stacked in series or placed ahead of existing conventional or column flotation cells.

FROTH WASHING

Bias Flow

The reduction in hydraulic entrainment by froth washing makes column flotation an attractive alternative to multistage cleaning. For the wash water to work effectively, a relatively deep froth (>1 m) is typically required and the flow of wash water should exceed the volumetric flow to the froth product. The difference between these two flow rates is normally referred to as the bias flow. The bias flow represents the amount of water that passes down through the stabilized froth to prevent fine hydrophilic slimes from entering the froth. As shown in Figure 8, a positive bias represents the desirable condition in which little or no water in the feed reports to the froth product, while a negative bias represents the undesirable condition in which a portion of the water in the feed reports to the froth product. Ideally, less than about 1% of the feed water will report to the froth product if the wash-water flow rate is properly controlled. In theory, a wash-water flow rate just above the flow rate of water demanded by the froth should provide a positive bias and prevent entrainment. In practice, however, mixing within the froth often requires that the wash-water flow rate be greater than the flow rate of water reporting with the froth phase. On the other hand, an excessive wash-water rate

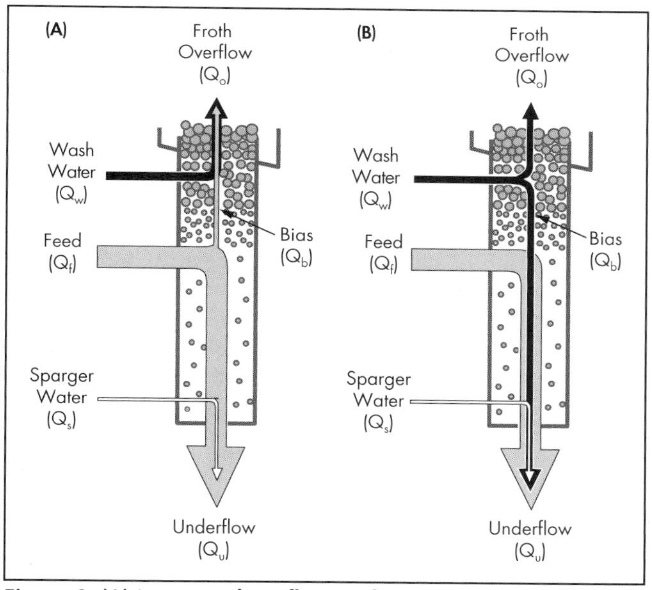

Figure 8 (A) Negative bias flow and (B) positive bias flow

must also be avoided since high water rates create an undesirable reduction in recovery in response to a reduction in slurry retention time. A very high wash-water flow rate can destabilize the froth by stripping/diluting the frother concentration. High wash-water flow rates can also have a detrimental impact on product grade by increasing axial froth mixing and reducing the wash-water effectiveness (Yianatos et al. 1988).

Design Criteria

Data collected from columns currently operating in the minerals industry suggest that a wash-water flow rate of

6.0–8.5 m^3/h (2.5–3.5 gpm/ft^2) of column cross-sectional area is normally adequate in a well-designed column to maintain a positive bias flow. However, the volumetric demand for wash water can be influenced by many interrelated factors that are associated with froth stability. A convenient method for evaluating the demand for wash water is to calculate the number of froth dilution washes (N_d), which is given by

$$N_d = W_w/W_o \qquad \text{(EQ 1)}$$

where W_w and W_o are the volumetric flow rates of wash water and froth overflow water, respectively. The effect of N_d on the quality of a froth product is shown in Figure 9. A value of $N_d = 1$ indicates that just enough wash water is being added to replace the water reporting to the froth product. However, for reasons described previously, the data show that a value in the range of $N_d = 1.5$ or higher is often required in practice to avoid deterioration of the froth product by nonselectively entrained solids. The data also indicate that the quality of the froth product is much less consistent when operating under negative bias conditions ($N_d < 1$) because fluctuations in froth stability change the amount of water and entrained solids entering the froth product.

The relationship between the volumetric flow of wash water (W_w), mass flow rate of dry solids reporting to the froth overflow (M_o), and the percent solids (S) of the froth product is given by

$$W_w = N_d (M_o) (100 - S)/S \qquad \text{(EQ 2)}$$

This relationship, which is plotted in Figure 10, effectively illustrates the rapid increase in wash-water demand as the froth solids content is reduced or dry solids rate of floatable particles is increased. A less stable froth will increase the froth solids content and will require less wash water. If the solids content is too high, however, an overloaded froth condition will occur that may limit froth mobility and capacity. Likewise, a too stable froth can lead to a negative bias condition as a result of the lower solids content. A highly stable froth can also require a water flow rate in excess of that which can penetrate the rising three-phase mixture of gas, solids, and water, thereby resulting in poor-quality froth even at high wash-water addition rates.

Water Distributors

The design of the wash-water distribution system can have a large impact on the effectiveness of the wash-water addition. Unfortunately, no standard design currently exists. A common layout for circular columns involves concentric rings of piping into which numerous small holes have been drilled for water distribution (Figure 11). For these systems, the distribution piping can be submerged below the cell lip so that a drained froth can form above the distributor and a stabilized froth below the distributor. This positioning allows the depth of the drained froth and the extent of froth drainage to be varied by raising or lowering the distributor; that is, the vertical position of the water distributor controls the split of water between the froth and bias streams. Multiple distributors placed at different depths may also be incorporated into the design to control froth stability and to balance the water flows (Yoon et al. 1992). For circular-shaped distribution rings, the inner rings are frequently placed above the outer rings to maintain the fluidity and mobility of the center of the froth. The multilevel distribution system can also be adjusted online to accommodate day-to-day variations in froth stability. More commonly, however, operators have found that it is necessary to locate the water distributor just above the top of the froth so that plugging problems can be visually monitored and corrected. This configuration, which eliminates the drained froth zone, also makes it possible to use drip pans instead of distribution pipes. A drip pan, which consists of an open-top perforated tray (Figure 12), is easier to routinely clean from above

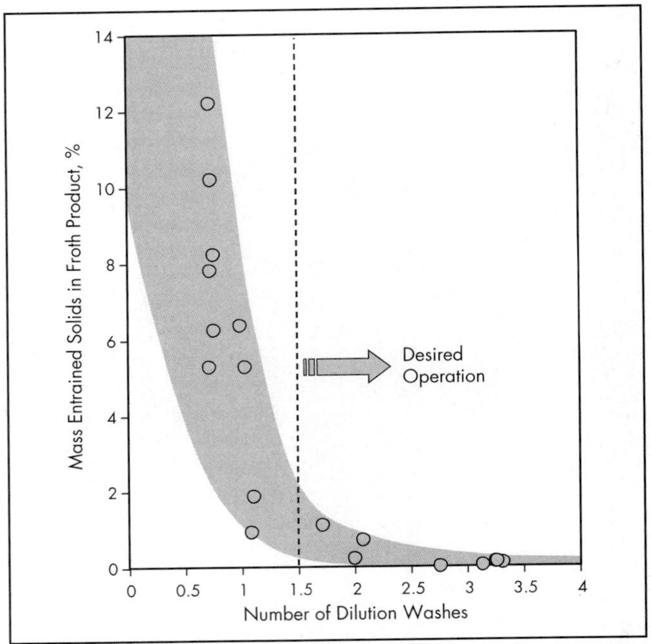

Figure 9 Effect of number of dilution washes on the amount of solids entrained into a froth

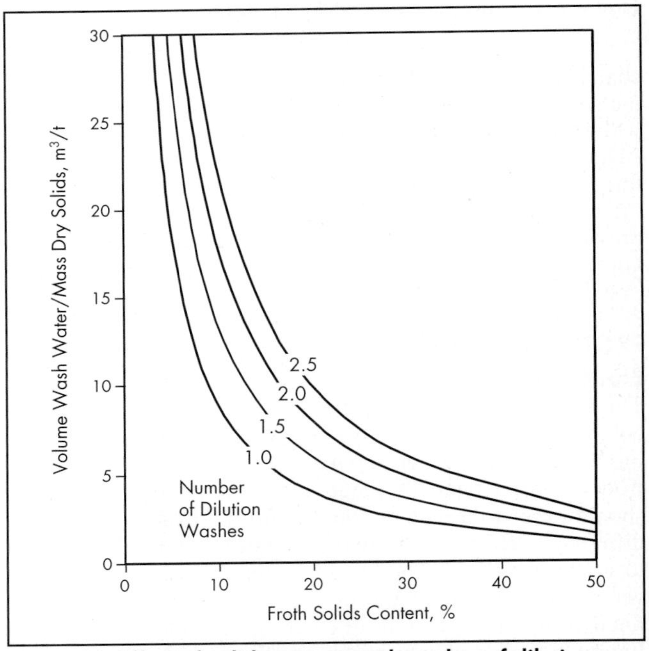

Figure 10 Effect of solids content and number of dilution washes on froth water demand

Figure 11 Pipe-ring wash-water distributor

Figure 12 Drip-pan wash-water distributor

using a simple water hose and brush. The major shortcoming of this design is that it does not allow the wash-water distributor to be raised or lowered to control the bias flow. Regardless of which design is used, the water should be gently introduced into the froth at low velocity to avoid jets that would disrupt the even upward flow of the froth (Finch and Dobby 1990).

The number, size, and spacing pattern of holes in the water distributor can vary considerably depending on wash-water demand, column area, and vendor-specific preferences. Typical hole sizes range from about 5–10 mm in diameter with center-to-center spacings ranging from 2–15 cm. The spacing pattern used should be chosen to provide an even water distribution over the entire froth area (Neethling et al. 2006). Failure to do so can create localized regions of positive and negative bias within the same froth that are detrimental to separation performance. On the other hand, work performed by Neethling and Cilliers (2001) suggests that washing performance can be enhanced by adding more water to the center of the column. Therefore, field adjustments by the column manufacturer are often required to fine-tune and optimize the performance of the wash-water distributor system for each industrial application.

SPARGING SYSTEMS

Superficial Bubble Surface Area Rate

The gas sparging system is the most important component in the design of a flotation column. An effective gas sparging system must be capable of producing small, uniformly sized bubbles at a desired aeration rate (Xu and Finch 1989; Groppo and Parekh 1992; Huls et al. 1991). Small bubbles can improve flotation kinetics by enhancing the capture efficiency of fine particles (Dobby and Finch 1986a; Yoon and Luttrell 1986). In addition, smaller bubbles may increase the ability of the froth to retain and carry solids by increasing the total interfacial flux of bubble surface area through the froth phase (Finch and Dobby 1990). An ideal sparging system should also be nonplugging and wear-resistant and allow for easy, online servicing.

A key measure of sparger performance is the superficial bubble surface area rate (S_b). This parameter, which has the units of 1/time, is defined as the total surface area of bubbles per unit of time passing through a given column cross-sectional area (Finch and Dobby 1990). This parameter can be calculated using the following equation:

$$\begin{aligned} S_b &= (Q_g/A_c) / (\pi D_b 3/6) \times \pi D_b^2 \\ &= 6Q_g/(A_c D_b) = 6\, J_g/D_b \end{aligned} \qquad \textbf{(EQ 3)}$$

where Q_g is the gas flow rate, A_c is the cross-sectional area of the column, D_b is the bubble diameter, and J_g is the superficial gas rate. This expression shows that amount of bubble surface area available for collecting hydrophobic particles is directly proportional to the ratio of superficial gas rate to bubble size. The impact of S_b on flotation recovery is illustrated by the test results given in Figure 13 (Kohmuench et al. 2004). The data for this case show that recovery increases sharply as S_b increases to about 80 s^{-1} and reaches a plateau at 120 s^{-1}. The data also indicate that the relationship between recovery and S_b is generally independent of sparger type. This finding agrees with other studies showing a linear correlation between S_b and flotation rate constant (Gorain et al. 1997, 1998). The problem is that many commercial sparging systems simply cannot produce high S_b values. Field tests also suggest that operation at S_b values above 120 s^{-1} can also produce a poor separation because of excessively wet froths that cannot be effectively washed. A proper combination of gas rate and bubble size will generally provide a gas holdup in the flotation pulp in the range of 12% to 18% by volume.

Sparger Types

Three primary types of column sparging systems have been used commercially: (1) porous bubblers fabricated from filter cloth, punctured rubber tubes, or sintered metals/plastics; (2) high-velocity gas or gas–water injectors; and (3) dynamic mixers that circulate gas–slurry mixtures through static mixing elements. Details of these systems have been described in the technical literature (Clingan and McGregor 1987; Xu and Finch 1989; Huls et al. 1991; Yoon et al. 1992; Groppo and Parekh 1992; Finch 1995; Rubinstein 1995).

Static Spargers

The first column flotation machines evaluated in industry used internal (static) porous bubblers fabricated from porous materials such as filter cloth, sintered metal/ceramic beads, or perforated rubber tubes. Porous metal and permeable ceramic tubes

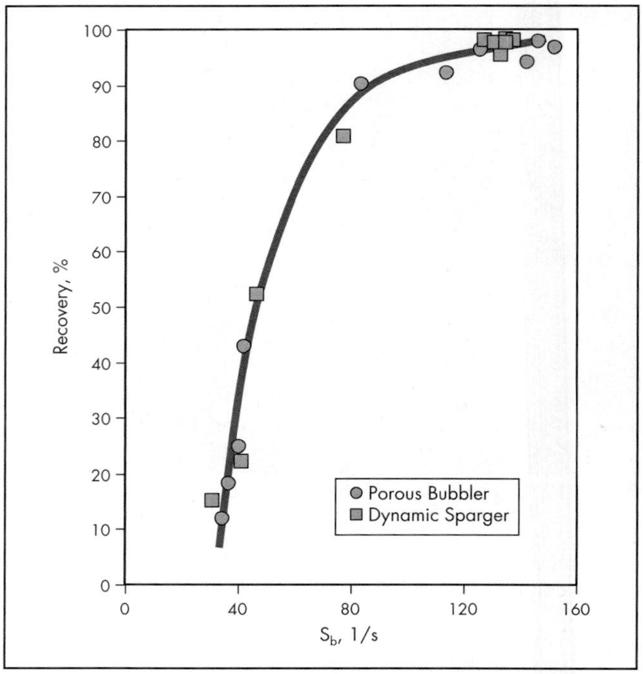

Figure 13 Effect of superficial bubble surface area rate on recovery for different spargers

can be used to produce small air bubbles, but they are prone to plugging and must be frequently removed and cleaned. This is a time-consuming and costly exercise that requires the column to be removed from service and drained. Plugging can also be a problem for static spargers that are designed to pass mixtures of compressed gas and pressurized water through small orifices since plant water and gas supplies are often contaminated with solids. Another issue with porous bubblers is scale-up capability. Xu and Finch (1989) and Clingan and McGregor (1987) showed that porous spargers should be scaled up by maintaining a constant ratio (R_s) between the column cross-sectional area and the sparger surface area. Unfortunately, this ratio can be difficult to maintain in industrial practice. For example, a 5-cm-diameter laboratory column equipped with a single 2-cm-diameter by 5-cm-long porous tube sparger would give $R_s = 0.625$. As such, a single 2-m-diameter full-scale column (3.14 m^2 of area) would require 5 m^2 of sparger surface area to perform similarly to the laboratory column. Installation of such a large area of porous spargers would be difficult in practice because of the space restrictions inside a column. This configuration would also present challenges in terms of plugging caused by slurry scaling and dirty gases.

High-Velocity Injectors

To overcome serious issues associated with static spargers, several high-velocity gas injection systems were developed for column sparging. One of the earliest systems, developed by the U.S. Bureau of Mines, passed gas through small (1–3 mm diameter) holes that were equally spaced along injector tubes that passed through the column wall and across the entire column diameter. A very small amount of pressurized clean water was added to the gas flow to improve bubble dispersion at the injector tips. This system was soon surpassed by a new generation of high-velocity gas injectors developed by groups such as Cominco and Minnovex/SGS. These new designs consisted of removable air lances with a single gas outlet at the end of the sparger tip. When operating properly, gas discharges from the tip at a sonic or near-sonic velocity. The intense turbulence created by the jet disperses the gas into fine bubbles. The new designs performed well but were still plagued with maintenance problems created by plugging during shutdown. Canadian Process Technologies solved this problem with the introduction of the SlamJet sparger (Figure 14). This technology incorporates a spring-loaded diaphragm that automatically drives a rod outward to close the end of the sparger tip when the gas pressure drops below a minimum set point (usually 200 kPa). The self-closing mechanism eliminates any backflow of slurry into the aeration system, thereby avoiding the plugging problems. Worldwide, more than 1,000 of these sparger systems are in use today for minerals processing applications. The recommended range of gas flow rates that can be generated per sparger tip is shown in Figure 15.

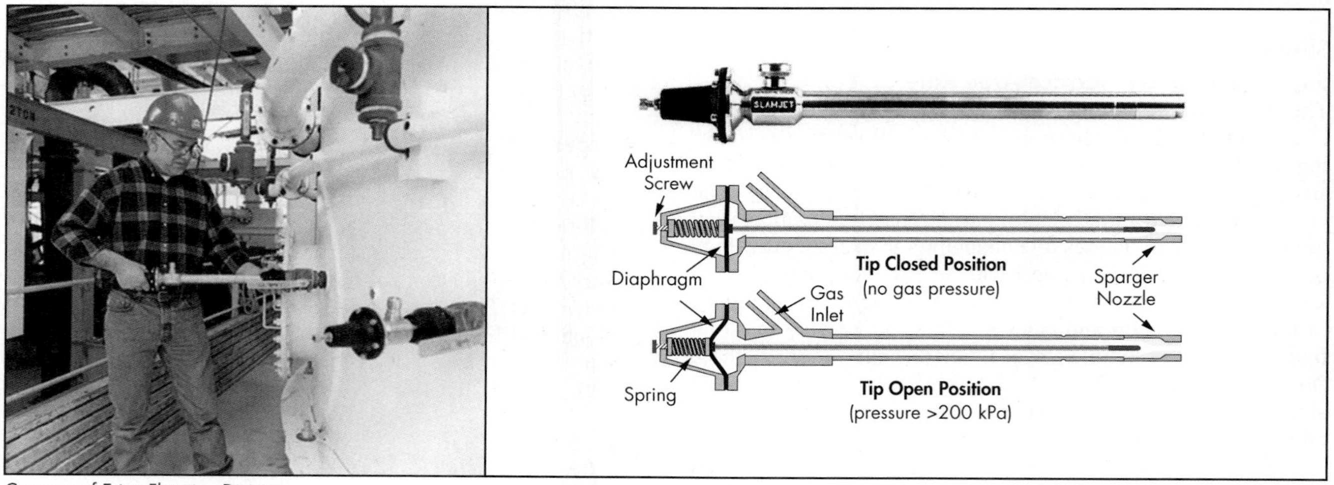

Courtesy of Eriez Flotation Division

Figure 14 Externally serviceable, self-closing, high-velocity gas sparger

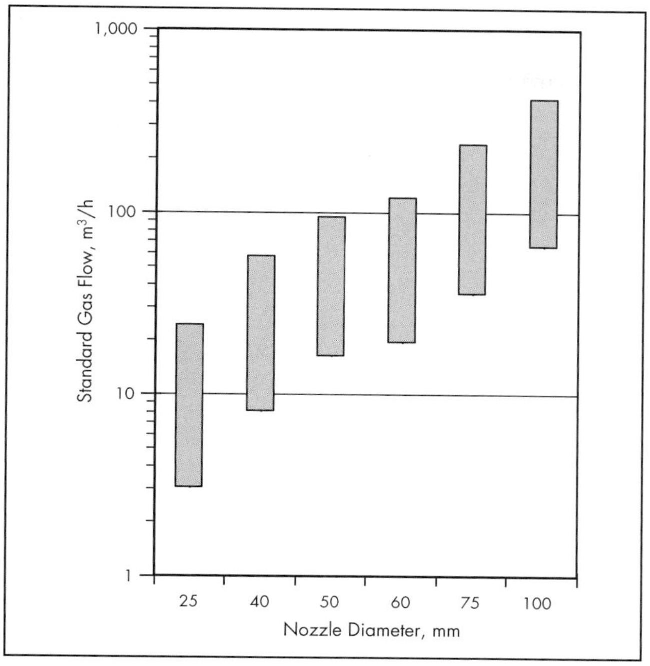

Figure 15 Recommended gas flow ranges for high-velocity gas injectors

Dynamic Spargers

Dynamic spargers are another popular design of gas aeration systems for column flotation machines. This type of sparger design uses a centrifugal pump to circulate slurry from the bottom section of the column through high-turbulence in-line contactors into which compressed gas is also injected. Depending on the manufacturer, the slurry–gas contactors may consist of static mixers, orifice plates, or venturi tubes. The intense mixing within the contactor produces small (<0.8 mm) bubbles and promotes bubble–particle contacting. Industrial studies indicate that these devices typically generate higher S_b values than other types of column spargers, thereby offering a higher capacity and/or recovery of fine particles (Pyecha et al. 2005).

Two popular designs of dynamic spargers are the Microcel and CavTube sparging systems (Figure 16). The Microcel technology uses in-line static mixers for gas dispersion and to scavenge any remaining floatable particles from the underflow prior to discharge. This configuration is designed to operate at relatively low back pressures, typically in the range of 100–250 kPa. The flow rate of circulated slurry is typically adjusted to provide a gas–slurry mixture with less than 40%–50% gas by volume. Figure 17 shows the recommended range of gas flow rates that can be effectively generated per sparger. The CavTube system is similar in function but is designed to operate at higher pressures sufficient to supersaturate the

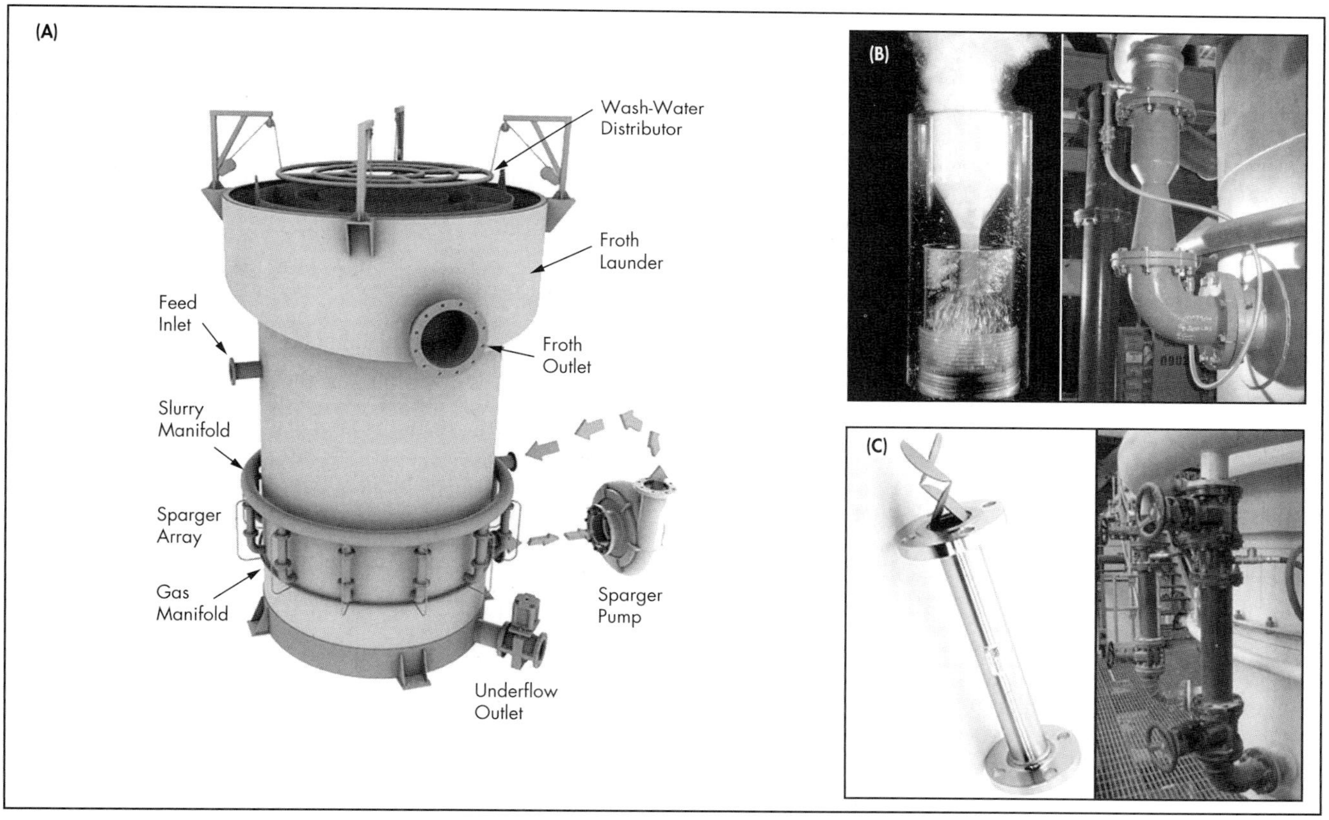

Figure 16 (A) Dynamic sparger system, including (B) CavTube and (C) Microcel systems

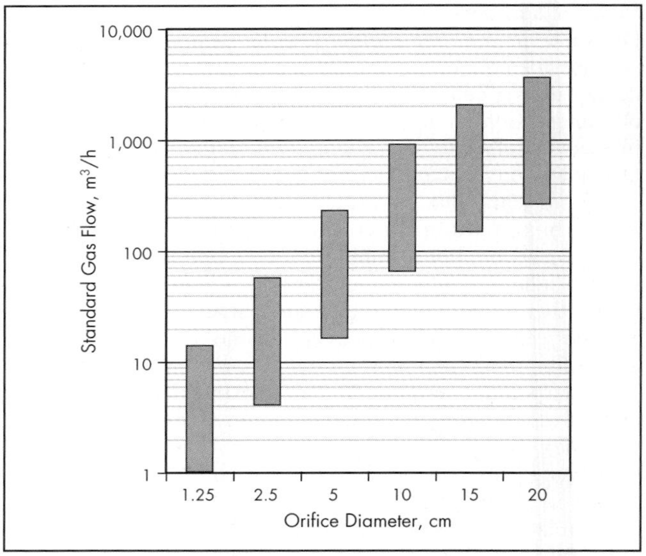

Figure 17 Recommended gas flow ranges for dynamic sparger

slurry with gas and to nucleate very tiny bubbles on the surfaces of hydrophobic particles. Although the nucleated gas bubbles represent less than 2% of the total introduced gas volume (Wasmund 2014), these tiny patches of gas enhance flotation kinetics by increasing the likelihood of successful bubble–particle attachments (Fan and Tao 2010). More than 200 of these systems are currently in operation in the minerals processing industry. CavTube spargers have also been used to pre-aerate the feed slurry to flotation columns and other types of flotation cells. Experimental studies have shown that CavTube pre-aeration can effectively improve flotation recovery, increase capacity, and reduce reagent consumption (Zhou et al. 1997; Honaker et al. 2011; Ahmed 2013).

Compressor Sizing

The majority of column cell installations require the use of compressed gas. Consequently, compressor sizing is an important task in the design of column flotation machines. Column manufacturers frequently report gas flow rates as a standard volumetric flow per time. This value is only valid for dry air at a pressure of 1 atm and temperature of 20°C. The actual flow rate specified by compressor manufacturers is typically reported in terms of inlet conditions (free air). Although this amount of air enters the compressor, it is not necessarily the amount of air delivered to the column because of compressor seal leakage. As a result, the actual flow may be only 95% of the inlet flow. Furthermore, corrections to the gas flow rate must be made to account for differences in elevation (atmospheric pressure) and humidity. Air temperature generally has little impact on the capacity of an oil-flooded screw compressor but may affect the performance of an air-cooled compressor (Sullair Corporation 1992). These complications generally require that professionals be consulted to ensure that the compressor is properly sized for the specified air requirements. However, as a sanity check, operators should expect to require about 0.1 kW of compressor power for each cubic meter per hour of gas flow demand. Caution should also be used during the metering of gas flow rates. A properly designed system should be equipped with a flowmeter that is calibrated to read correctly at a specified operating pressure. The operating pressure should be held constant by placing a pressure regulator ahead of the flowmeter. By placing the airflow control valve after the flowmeter, the flowmeter will always operate at its design pressure. If the flowmeter is placed after the control valve, then the operating pressure and true gas flow rate are both unknown. Improper metering of the gas flow rate can be a particularly serious problem when laboratory and pilot-scale tests are conducted to collect scale-up information.

SCALE-UP AND DESIGN

Flotation Kinetics

Froth flotation is commonly modeled as a first-order kinetic process. As such, the recovery (R) of floatable particles can be mathematically related to the residence time (τ), rate constant (k), and Peclet number (Pe) using Levenspiel's (1972) equation:

$$R = \frac{1 - 4A^{Pe/2}}{[(1+A)^2]^{APe/2} - [(1-A)^2]^{-APe/2}} \quad \textbf{(EQ 4)}$$

$$A = \sqrt{1 + 4k\tau Pe} \quad \textbf{(EQ 5)}$$

This relationship shows that recovery is a function of two dimensionless groups, k and Pe (Dobby and Finch 1985). For the recovery of particles from a flotation pulp, k has been shown (Finch and Dobby 1990) to be theoretically given by

$$k = \tfrac{1}{4} P S_b \quad \textbf{(EQ 6)}$$

which shows that the rate of flotation is proportional to the probability of attachment (P) and the superficial bubble surface area rate (S_b). Since k is normally assumed to remain unchanged during scale-up, both P and S_b must be held constant in both the small- and large-scale units. The mean residence time (τ_p) of small particles in the flotation column can be estimated from the active volume (V_p) of the flotation pulp and the volumetric flow rate of slurry to the underflow (Q_u) using

$$\tau_p = \frac{V_p}{Q_u} = H_p\left(\frac{1-\varepsilon_p}{J_u}\right) \quad \textbf{(EQ 7)}$$

in which H_p is the vertical height between the sparger inlet and base of the stabilized froth, ε_p is the gas holdup in the flotation pulp, and J_u is the underflow superficial velocity. This expression assumes that the residence time of slurry and particles is equivalent, and, hence, a correction is occasionally required if particle settling velocities become high. Finally, Pe is a dimensionless group that represents the degree of axial mixing within the flotation pulp. For column cells, Pe varies with both geometry and flow conditions (Dobby and Finch 1985) and may be estimated from an empirical expression derived by Mankosa et al. (1992):

$$Pe = 0.70\left(\frac{H_p}{D}\right)^{0.63}\left(\frac{J_u}{J_g}\right)^{0.30} \quad \textbf{(EQ 8)}$$

This expression suggests that columns become increasingly mixed in response to a reduction in either the height-to-diameter ratio or slurry-to-gas flow ratio.

Figure 18 shows an example of the relationship between recovery and residence time for a flotation column. In each test, the volumetric feed flow rate to the column was steadily increased to reduce the mean residence time. As shown, a good correlation was obtained between the values predicted

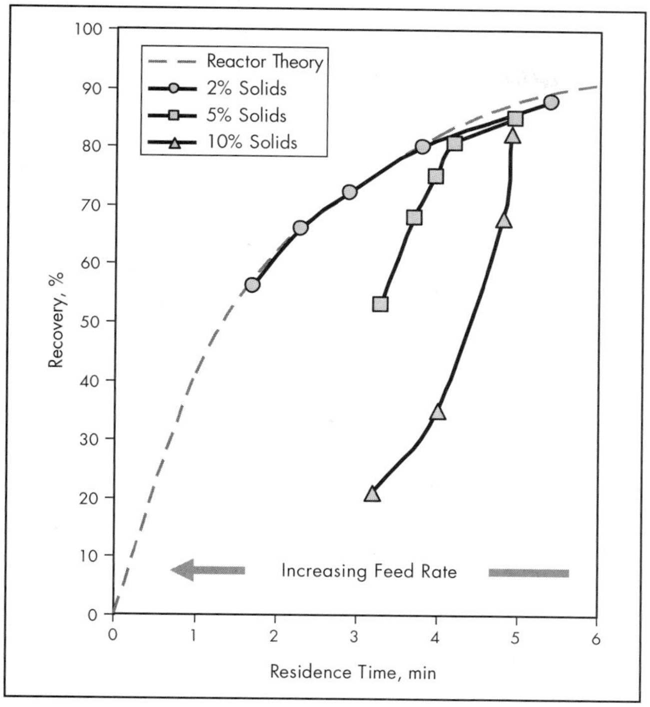

Figure 18 Froth overload condition created by increasing feed solids content

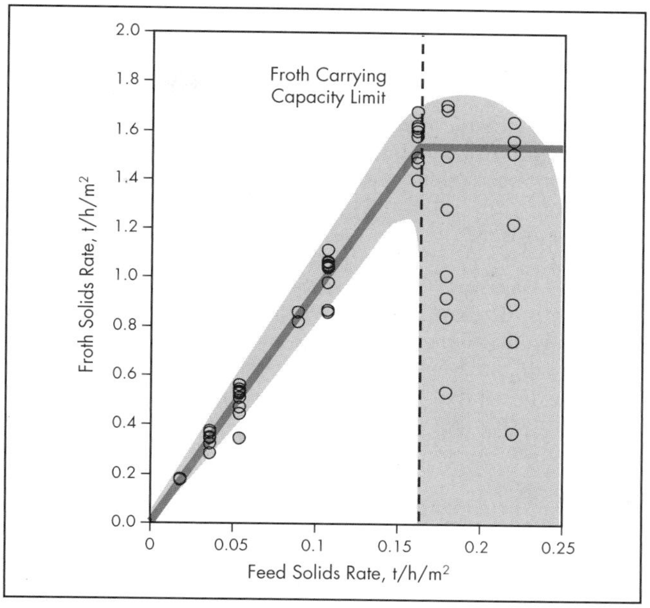

Figure 19 Graphical method for identifying the froth carrying capacity

from Equation 4 and the experimental data obtained using feed slurry with a very low solids content of 2%. The low solids content ensured that the column was limited by residence time and not by froth solids overloading. However, as the solids content of the feed stream was increased to 5% solids, the recovery dropped sharply and deviated substantially from the Levenspiel (1972) relationship. This deviation was even more pronounced when the solids content was doubled to 10% solids. The drop in recovery with increasing solids content can be attributed to limitations associated with froth carrying capacity. When this occurs, there is insufficient bubble surface area to carry all of the floatable particles collected in the pulp phase into the froth phase.

Froth Carrying Capacity

Although the carrying capacity restriction can often be ignored for conventional flotation circuits, it is important in the design of column cells, largely because the specific surface area of the cell (ratio of the cross-sectional area to the volume) is much higher for conventional cells. Column cells have specific surface areas in the range of 0.02–0.08 m^2/m^3, while those for conventional cells are in the 0.4–2.5 m^2/m^3 range. Theoretical studies indicate that carrying capacity, which is normally reported in terms of the rate of concentrate floated per unit cross-sectional area, is linearly related to the size and density of particles in the froth (Espinosa-Gomez et al. 1988; Sastri 1996). Froth carrying capacity may be estimated from laboratory and pilot-scale flotation tests by conducting experiments as a function of feed solids content (Finch and Dobby 1990). As shown in Figure 19, this procedure can be used to produce a plot of froth solids rate versus feed solids rate (Patwardhan and Honaker 2000). Prior to reaching the carrying capacity, an increase in feed solids rate produces a corresponding increase in the froth solids rate. Eventually, the froth solids rate reaches a plateau defined by the carrying capacity. The production rate of floatable solids becomes very sensitive to changes in operating parameters such as gas rate and frother dosage upon reaching this inflection point. For the case shown, a carrying capacity of 1.55 t/h/m^2 of column cross-sectional area should be expected. Table 1 shows the range of carrying capacities that have been observed for different mineral systems treated by column flotation. These values should be used with caution, however, because factors such as particle size and density can greatly alter these values (Espinosa-Gomez et al. 1988).

Scale-Up Projections

The methodology involved in the scale-up of laboratory and pilot-scale column data has been described in detail by Finch and Dobby (1990). Unfortunately, scale-up can be a challenging task for column designers. Many of the parameters used in these calculations may or may not remain constant in the full-scale column. For example, field observations suggest

Table 1 Examples of froth carrying capacities observed for different mineral systems

Mineral System	Carrying Capacity, t/h/m²
Coal	1.2
Copper	2.5
Lead	3.2
Molybdenum	3.7
Nickel	1.7
Oxides	4.4
Phosphates	5.0
Silica	5.0
Zinc	1.7

that the bubble size distributions in full-scale columns are often coarser than those used in the scale-up experiments. The buildup of recycle streams in the feed and contaminants in the process water can also impact the pulp chemistry and froth stability. Reputable manufacturers of column machines have long recognized these unforeseen issues and have developed proprietary corrections based on their historical experiences. Therefore, while a first-pass estimation procedure for column sizing is provided herein, column scale-up is generally best handled by experts working in the field.

The scale-up procedure for column cells primarily involves the determination of column diameter and height. For a given installation, the total cross-sectional area of the cells is normally dictated by the carrying capacity. This relationship may be expressed as

$$D = \sqrt{\frac{4M_f Y}{\pi C}} \quad \textbf{(EQ 9)}$$

where D is the column diameter, M_f is the mass feed rate of dry solids, Y is the expected mass yield of solids to the froth product, and C is the froth carrying capacity. For a first-pass estimate, it is generally more convenient to calculate the mass rate of froth overflow solids for a full-scale column ($M_{o,full}$) to that for a small laboratory or pilot-scale test column ($M_{o,lab}$) using

$$M_{o,full} = M_{o,lab}\left(\frac{D_{full}}{D_{lab}}\right)^n \quad \textbf{(EQ 10)}$$

where D_{full}/D_{lab} is the ratio of the diameters of the full-scale and laboratory test columns, respectively, and n is a scaling power factor. Historical field work indicates that an area-based scale-up (n~2) is valid for many industrial mineral flotation systems, provided that the column is well designed with adequate internal launder systems to minimize froth travel distances. According to this relationship, a 3-m-diameter column would have a theoretical froth solids capacity approaching 3,600 times that of a 5-cm-diameter test column. However, an area-based scale-up calculation appears to contradict data reported by Amelunxen (1990) suggesting that froth mobility problems make the column circumference the appropriate scale-up parameter (i.e., n~1). In practice, a value of $1 < n < 2$ may be used by column manufacturers based on specifications related to the launder design and information available from experimental testing and/or proprietary databases.

After determining the column diameter, the height of the column pulp zone must be estimated. This height, which is normally taken as the distance between the base of the froth and the air injection point, must be large enough to ensure that adequate residence time exists to achieve the target recovery dictated by the carrying capacity. Detailed procedures have been reported in the literature for determining this height (Dobby and Finch 1986b). However, as a first-pass estimate, the height of the collection zone for a full-scale column (H_{full}) relative to the height of a smaller test column (H_{lab}) can be determined from

$$H_{full} = H_{lab}\left(\frac{\tau_{full}}{\tau_{lab}}\right) = H_{lab}Z \quad \textbf{(EQ 11)}$$

where $Z = \tau_{full}/\tau_{lab}$, which is the ratio of the mean particle residence times for the full-scale plant column and laboratory-scale test column, respectively. It cannot be automatically assumed that Z = 1 since the full-scale column is typically more mixed and, hence, requires a longer residence time to achieve the same recovery. Therefore, for first-pass estimations, the plot provided in Figure 20 can be used to estimate the Z multiplier on residence time required for the full-scale column, provided that the flotation rate constant remains unchanged during scale-up. For example, consider a 5-cm-diameter laboratory column with a pulp height of 3.5 m. If the test column is near plug-flow (Pe = 0.05), then a target recovery of 86% in the test column would correspond to $k\tau$ = 2.2 for this set of operating conditions. Likewise, an 85% recovery in the full-scale column (Pe = 4) would require $k\tau$ = 5.2. Therefore, the height of the full-scale column pulp zone should be 5.2/2.2 = 2.4 times the laboratory pulp height to maintain the same recovery. This calculation dictates a full-scale column pulp height of 9.1 m (i.e., 3.5 × 2.4 = 8.4 m). An additional 1 m of column height should be added to the top of the pulp zone to account for the froth zones. Another half diameter (1.5 m) should also be added to the bottom of the pulp zone for underflow removal and sparger systems. (The height of the bottom section may vary significantly, however, depending on the mechanical design of the column.) Based on these estimates, the total estimated column height for this example would be 10.9 m (8.4 + 1 + 1.5).

Unfortunately, the Pe values for the small-scale test column and full-scale production column are usually not known in advance. In such cases, a conservative estimate for pulp height can be obtained by assuming that the test column is plug-flow (Pe = 0) and the full-scale column is perfectly mixed (Pe = ∞). In this case, the Z ratio can be mathematically derived as

$$Z = \frac{R}{(R-1)\ln(1-R)} \quad \textbf{(EQ 12)}$$

where R is the desired fractional recovery (D. Catarious, personal communication; Luttrell et al. 2000). For the current example, this expression gives a scale factor of $Z = H_{full}/H_{lab}$ = 3.1, which corresponds to a full-scale column pulp height of 10.9 m. While this pulp height is greater than the 8.4 m height predicted by the more rigorous approach, this shortcut method provides a quick means of estimating maximum column height for first-pass designs. For convenience, Equation 12 is plotted along with Equation 4 in Figure 20. Field data obtained from operating columns suggest that the shortcut method is reasonable provided that the height of the smaller test column is very close to that required to reach the carrying capacity limit.

INDUSTRIAL PRACTICE

Iron Ore Flotation

Columns have become widely accepted in the iron ore industry over the past two decades. Reverse flotation using columns has proven to be an economical and effective method for reducing the concentrate silica content to well below 2%. Additionally, the use of wash water has provided a means of obtaining low concentrate silica levels while keeping iron losses to a minimum. Furthermore, total energy per unit volume as well as capital and installation costs favor columns (Murdock et al. 1991). The Brazilian iron ore industry has led the world in adopting column technology for reducing the silica content of iron pellets (Sandvik et al. 1991). In fact, Samarco Mineração S.A., the first Brazilian producer to use columns, installed columns to increase capacity as part of a plant expansion program in the early 1990s. Since that time, additional columns have been installed for the recovery of fine iron from a desliming

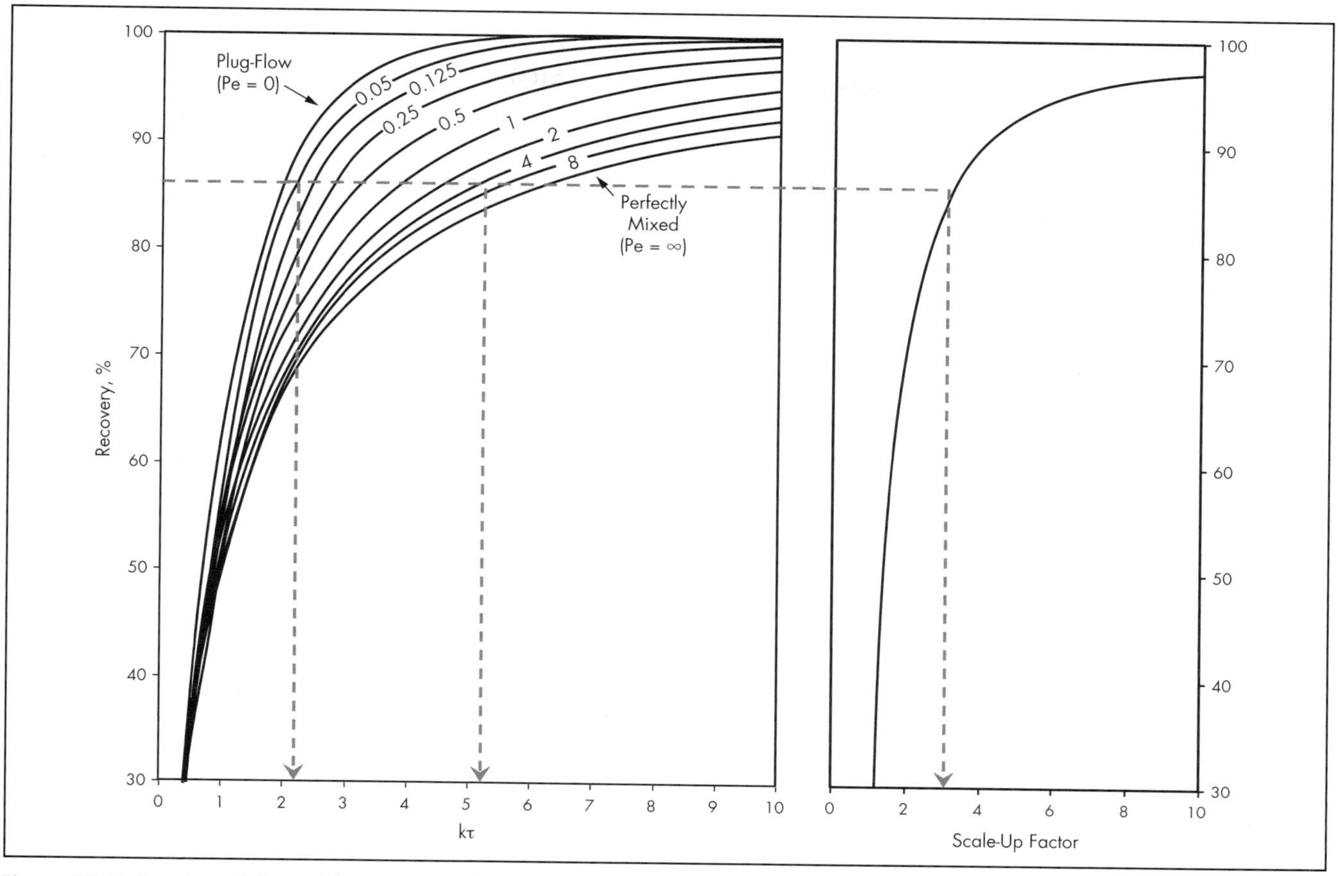

Figure 20 Estimation of the residence time scale-up factor for column flotation

circuit and for plant expansion programs. Many plants require the flexibility to produce more than one iron product, such as blast furnace (2.2% silicon dioxide [SiO_2]) and direct reduction (0.8% SiO_2) grades using the same equipment. When the higher silica blast furnace products are produced, the grinding requirements for liberation generally decrease and therefore the plants can operate at increased capacities. The higher feed rate affects the feed size distribution and must be taken into account in the circuit design. The following section includes examples of some of the flow-sheet configurations adopted by iron ore producers.

Cleaner–Scavenger Circuit

Columns are often used as cleaners in combination with mechanical cells. This is the case for a Brazilian flotation plant that originally consisted of four parallel mechanical flotation lines containing a rougher, cleaner, and two scavenging stages. The depletion of reserves at the existing mine and the subsequent development of a new ore zone necessitated the modification of the existing flow sheet because ore from the new mine is harder and slower floating (Sandvik et al. 1991). Several circuit alternatives were considered, including additional conventional cells, a parallel line of columns, and columns as re-cleaners. Pilot-plant test results revealed that the third option provided the best combination of plant throughput and iron recovery. The flow sheet for the modified circuit (Wyslouzil 2009a) is shown in Figure 21. In this approach, cyclone underflow from the desliming circuit feeds the flotation operations with a particle size range of 150 × 10 µm. One column was added as a re-cleaner to each processing line to preserve circuit flexibility. The re-cleaner column feed grade ranges between 1% and 6% silica; therefore, one column stage is sufficient to produce either blast-furnace or direct-reduction concentrate. A scavenger column was also added to the flow sheet to maximize fine iron recovery, particularly during periods when direct reducing-grade iron is being produced. Because these columns have been added to an existing silica flotation plant as re-cleaners, consideration was given to the effects of variations in the roughing and primary circuits. Generally, the fine, fast-floating silica is removed in the roughing stage and therefore the column feed tends to be enriched in coarse silica. Furthermore, any operational problems with the existing plant tend to flow through to the column, resulting in variable feed conditions.

Silica Reduction from Magnetite Concentrate

Columns can also be used in combination with magnetic separation circuits. The flow sheet in Figure 22 shows a 1,200 t/h flotation plant designed to lower the silica content of magnetite concentrate from 7%–9% to 2.2% SiO_2. New feed is ground to 80% passing 150 µm and is fed to primary magnetic separators. The magnetic concentrate is reground to approximately 75% passing 45 µm and is retreated in secondary magnetic separators that produce a concentrate containing 7%–9% SiO_2. The magnetic concentrate is conditioned with caustic starch solution and amine prior to being fed to a single-stage

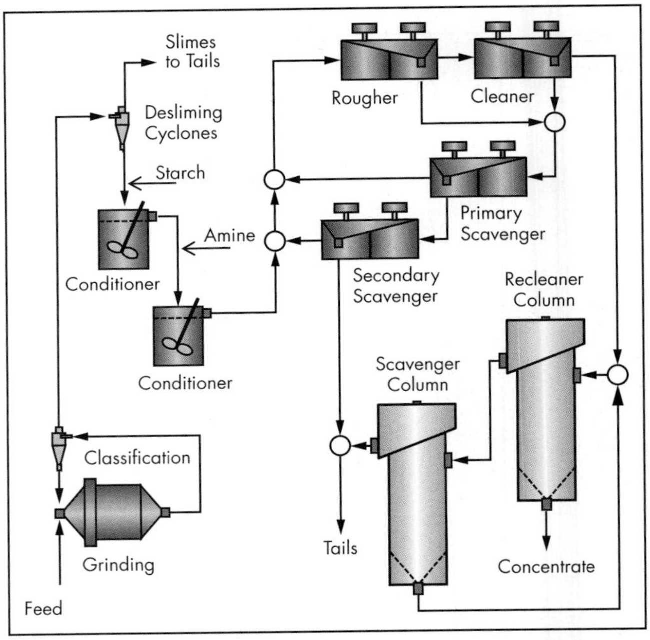

Figure 21 Columns added to an existing iron ore circuit

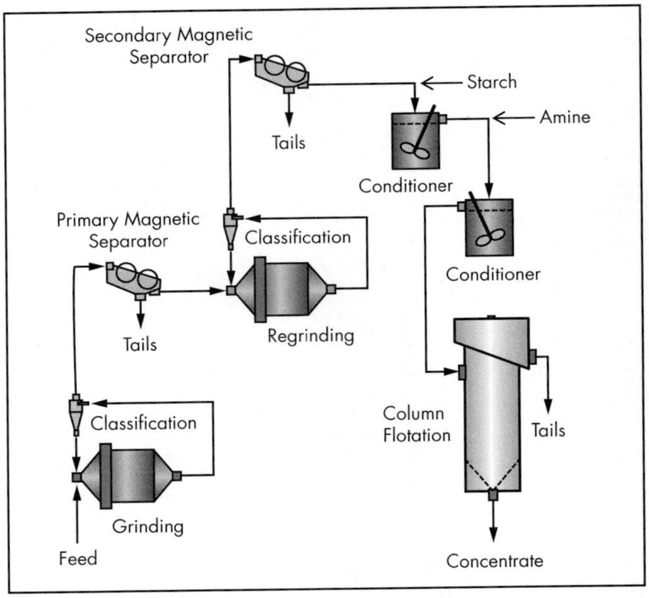

Figure 22 Column flotation circuit for magnetite treatment

flotation column where the silica content is reduced to below 2.2% SiO_2. In this case, the ore quality is highly variable, and therefore it was necessary to design the circuit to tolerate large fluctuations in feed rates. A high level of instrumentation was specified to provide online stream analysis, automatic reagent metering, and the ability to switch groups of columns into a standby mode. As a result, individual columns are automatically bypassed and idled with the reagent addition stopped during periods of decreased feed rates. When the feed increases to normal levels, the procedure is reversed and the columns resume normal operation.

Phosphate Flotation

Column flotation technology has proven to be an efficient method for improving the separation of valuable phosphate particles from the host matrix. As a result, many phosphate producers have installed column flotation systems as a means of boosting production while reducing operating costs. The approach has extended the size range of recoverable apatite particles from about 30 µm down to 5 µm (Wyslouzil 2009b). The high degree of selectivity achieved by this equipment has made it possible to treat material previously considered unrecoverable. A key characteristic of columns is the ability to operate with a deep froth, which is washed with fresh/clarified water, resulting in improved metallurgical performance. In some cases, the wash water can also partially offset the need for depressants, which lowers chemical dosage rates.

Conventional Versus Column

In Brazil, hard-rock phosphate deposits are volcanic in origin. These hard-rock ores contain apatite as the valuable component and, depending on the deposit, are associated with impurities such as barite, hematite, calcite, alumina, and silicates (e.g., quartz, micas, and clays). These ores are typically ground for liberation purposes and then subjected to coarse and fine flotation circuits. Mechanically agitated cells have been the standard choice for many years; however, this equipment often struggles to produce high-grade concentrates because of the fine particle sizes. Fine impurities such as magnetite and silica are easily carried into the froth by hydraulic entrainment, which lowers the quality of the final concentrate. Also, attrition because of the cell agitators results in the continuous generation and release of fresh fines into the circuit. Consequently, fine particle flotation circuits require chemical depressants and multiple cleaning stages to reach acceptable grades. As a result, in the mid-1990s producers began to install column cells based on a desire to improve separation performance, reduce operating costs, and simplify plant circuitry. Table 2 provides an example of the potential benefits that can be realized using column technology. The data compare the results obtained after installation of a column circuit at a major phosphate plant in Brazil. The upgrade reduced the number of

Table 2 Comparison of mechanical and column flotation cells

Circuit Data	Mechanical Cell	Column Cell	% Change
Number of cells	66	6	–91
Power, kW·h/t	108	100	–7
Collector, g/t			
Barite circuit	567	170	–70
Apatite circuit	2,803	404	–86
Barite recovery	95	97	2
Apatite recovery			
Natural fines	50	78	56
Generated fines	65	77	18
Coarse/regrind	71	72	1
Fine product			
P_2O_5	33.5	34.7	4
Fe_2O_3	6.1	5.8	–5
Coarse product			
P_2O_5	36.2	35.9	–1
Fe_2O_3	3.1	3.1	0

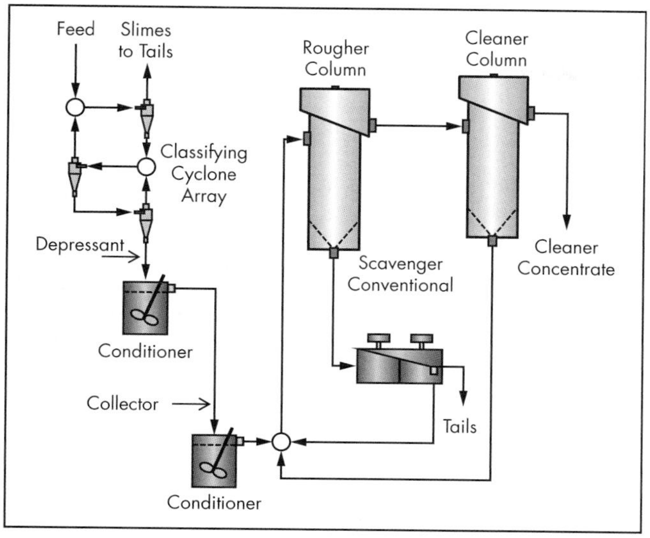

Figure 23 Circuit arrangement for ultrafine phosphate separation

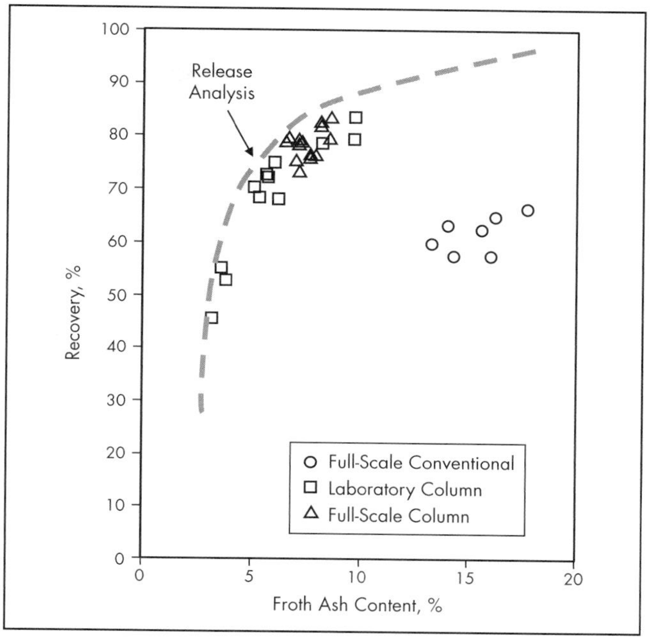

Source: Davis et al. 1995

Figure 24 Comparison of conventional and column flotation relative to release analysis

cells in the plant from 66 machines to six columns. Despite using fewer pieces of equipment, the recovery of coarse and fine apatite and barite was improved. The retrofit also reduced power costs and lowered collector demand.

Ultrafine Flow Sheet

Another column flow sheet for a phosphate operation in Brazil is shown in Figure 23 (Wyslouzil 2009b). In this case, the major source of phosphate is derived from apatite, which comprises approximately 30% of the minerals in the ore zones. The major impurities consist of iron oxides and silicate minerals. The ore is crushed, screened, and then fed to the concentrator, where it is subjected to grinding (rod and ball mills), classification (hydrocyclones), magnetic separation, desliming (hydrocyclones), flotation, and dewatering. Traditionally, the primary desliming stage was designed to remove particles finer than 25–30 μm, which were then discarded. The high surface area and high impurity content associated with this particle size class made it difficult to treat with conventional flotation equipment. The removal of these slimes constituted a major loss of the total phosphate values since they represented as much as 10%–15% of the total reserves. The application of column flotation made it possible to extend the lower size range of particles that can be treated by flotation to 5–10 μm. In this circuit, the primary slimes (–30 μm) are treated in a second stage of hydrocyclones with a size cut of 5–10 μm. The cyclone underflow is then fed to a series of conditioners where the pulp is treated with caustic soda, starch, and a collector prior to column flotation. The flotation cells are arranged in a rougher–scavenger–cleaner circuit with intermediate products recycled internally.

Results from this circuit modification indicate that an ultrafine phosphate product can be produced containing 33.5% P_2O_5. This represents a 72% recovery of P_2O_5 for the ultrafine circuit and an overall increase in plant yield of 3%–5%. By following this treatment scheme, it is possible to obtain an ultrafine concentrate ideally suited for the production of single superphosphate fertilizers. The fine particle size minimizes the costs of concentrate regrinding at the fertilizer plant, saving additional processing costs. Faced with difficult ores and low-grade deposits, Brazilian phosphate producers have been world leaders in adapting this technology to enhance fine phosphate recovery.

Coal Flotation

Over the past decade, columns have been used extensively in the coal industry as an efficient and economical means of recovering fine coal from preparation plant effluent streams. Coal flotation applications are particularly challenging because the flotation system must cope with a high mass recovery rate and, typically, an extensive amount of ultrafine clay. Under these conditions, a deep froth and high-volume flow of wash water are particularly beneficial. The benefit of wash water is illustrated by the data summarized in Figure 24, which compares column flotation technology with an existing bank of conventional cells. As shown, the product quality is significantly improved using wash water. In fact, column performance agrees with the separation curve predicted by release analysis (Dell 1963; Dell et al. 1972). A release analysis is an indication of the ultimate separation performance attainable for flotation. This figure suggests that columns provide a level of performance that would be difficult to achieve even after multiple stages of cleaning by conventional machines.

Two circuits are typically used for fine coal flotation (Figure 25). These include the traditional minus 0.150-mm by-zero circuit and the 0.150 × 0.045-mm deslime circuit. The by-zero approach incorporates raw coal cyclones that treat –1 mm feed. Generally, these cyclones are configured to cut at a nominal particle top size of 0.15 mm, with the finer fraction reporting directly to flotation. In some markets, the particle size cut can be as coarse as 0.5 mm, although it is recommended to not exceed 0.25 mm for columns.

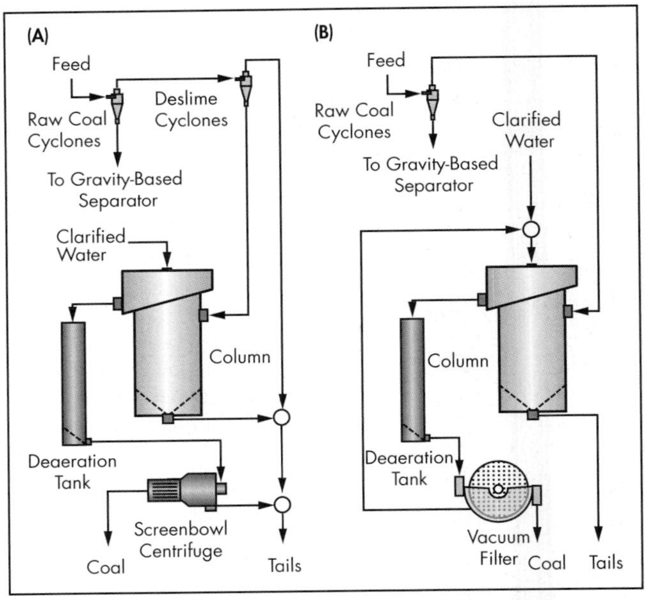

Figure 25 Coal flotation using (A) deslime column–screenbowl circuit and (B) by-zero column–filter circuit

By-Zero Column Circuits

The feed to flotation in a by-zero circuit is relatively dilute and typically ranges between 2.5% and 6.0% solids, by weight. As a result, volume flows in these circuits can be quite large. Consequently, significant cell volume must be available to achieve the desired retention time to ensure adequate combustible recovery. To attain the needed retention time, several large-diameter cells may be required, which can range in height from 8.5–16 m. The ultimate height is typically based on kinetic data generated from laboratory testing but is generally constrained by practical limits as dictated by both economic and site-specific engineering requirements. As a result, the height-to-diameter ratio for an industrial cell ranges from 3:1 to as low as 2:1. Also, because of the high flow volumes found in by-zero circuits, bypassing of feed can occur and is caused by the less-than-ideal mixing found within the large diameter and relatively short tanks commonly found in coal flotation installations. The high degree of mixing allows some material within the cell to report to tails without being influenced by the bubble swarm produced by the sparging system. As the residence time is increased, the probability of bypass is reduced. As a consequence of this relationship, the occurrence of bypass is more often found in traditional by-zero circuits where the volumetric feed rates are relatively high and retention times are typically low. More recent studies show that arranging column cells in series helps to minimize bypass (Stanley et al. 2006).

Deslime Column Circuits

The by-zero circuit works well in plants that serve the metallurgical coal market where ash is the primary quality specification; however, studies have shown that a by-zero approach is not the optimum circuitry for plants that are producing coal for the thermal market where other quality specifications such as moisture, heat content, and sulfur are also critical (Luttrell et al. 2009; Bethell and Luttrell 2005). These studies show that the popular deslime circuit is better suited to achieve a separation that maximizes profitability as dictated by a typical utility coal sales agreement. The deslime circuit achieves this rejection by incorporating a secondary bank of 150-mm-diameter cyclones, which is used to further classify the flotation feed at approximately 0.045 mm. This approach is superior for circuits that contain clay-rich flotation feeds where the benefit of wash water can be maximized.

Test results indicate that several benefits to the deslime flotation circuit are realized because of the rejection of high-surface-area particles. These include a lower-ash product, reduced froth moisture, an increase in flotation capacity, and an improvement in downstream froth handling. The improved froth handling is a result of two factors: a coarser float product and a significant reduction in total frother addition. The coarser froth product tends to be less stable and will readily collapse. The lower frother addition rate is a result of the volume split achieved in the smaller 150-mm-diameter cyclones, which substantially reduces the total volume of slurry that must be processed in flotation. Given that frother is added based on total flotation volume flow, this allows high concentrations of surfactant to be realized in the flotation circuit without causing upsets elsewhere in the plant. The obvious secondary benefit is the reduction in size and cost of the flotation and downstream dewatering equipment. Engineering estimates have shown that a deslime circuit may cost less than half of an equivalent by-zero circuit.

Low-Profile Column

More recently, there has been a move toward low-profile flotation cells designed specifically to overcome some of the challenges set forth by both end users and engineering firms with regard to traditional column cell installations. These challenges include cell size, foundation loads, and energy consumption. The benefits of this approach were demonstrated by Davis et al. (2011). In this case, StackCells were installed preceding two parallel column flotation cells to address a circuit-overloading condition. The net change in circuit performance is shown in Figure 26. As shown, the average coal recovery increased from 74.4%–83.7%. Likewise, the average tailings

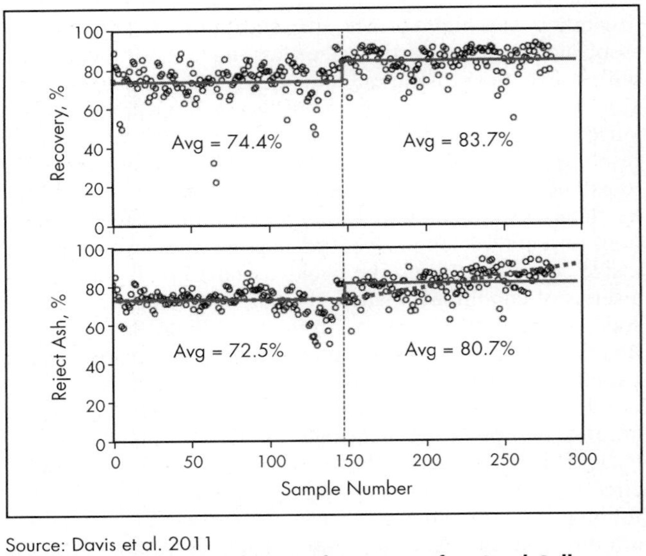

Source: Davis et al. 2011

Figure 26 Change in plant performance after StackCell installation

ash content increased from 72.5%–80.7%. More recent data indicate that the circuit regularly achieves a combustible recovery exceeding 90%. The improvement is attributed to the optimization of operating variables such as reagent dosage, froth depth, aeration rate, and wash-water addition rate. The increased recovery is significant considering that less than 10% more cell volume was added to the circuit via the installation of the StackCell technology.

Sulfide Flotation

The aforementioned case studies describe applications in which columns are used in complex circuits as well as in stand-alone applications (i.e., coal). However, because of the ability to provide a high upgrade ratio, columns are often used as cleaners after mechanical cells in many applications. Columns are ideally suited for this purpose, particularly in sulfide mineral applications. Because of the high tonnage rate and the low feed grade associated with finely disseminated sulfide ores, conventional flotation cells are typically used for roughing and scavenging operations with the objective of producing a final tailings stream. After this initial upgrading, column cells are then used as cleaners to produce a high-grade concentrate prior to secondary grinding. Likewise, columns can also be used after regrinding to produce final copper and molybdenum concentrates.

The ability to operate a column cell with a deep froth and froth washing typically provides a higher-grade final concentrate. One such example from a currently operating concentrator is shown in Figure 27. In this case, columns were selected to replace a bank of conventional flotation cells. Performance data collected before and after the upgrade show a substantial increase in copper grade while also achieving a slight increase in copper recovery. The latter is attributed to the need to run a lean froth in conventional machines when difficult feed conditions are encountered to achieve the required final concentrate grade. By contrast, the application of froth washing allows the columns to be operated under optimal recovery conditions while maintaining a consistent grade.

A recent report in the technical literature (Leonida 2015) describes the successful application of a cleaner column circuit installed in a copper–gold operation in Australia. According to the article, this operation, completed in 2013, is one of the Asia-Pacific region's premier high-grade copper mines and will produce up to 300,000 t of high-grade copper concentrate annually. The plant was initially designed with all conventional flotation cells. However, after extensive laboratory- and pilot-scale testing, columns were added to the cleaning circuit to improve overall plant metallurgy. Results show an improvement in both product quality and overall circuit copper recovery.

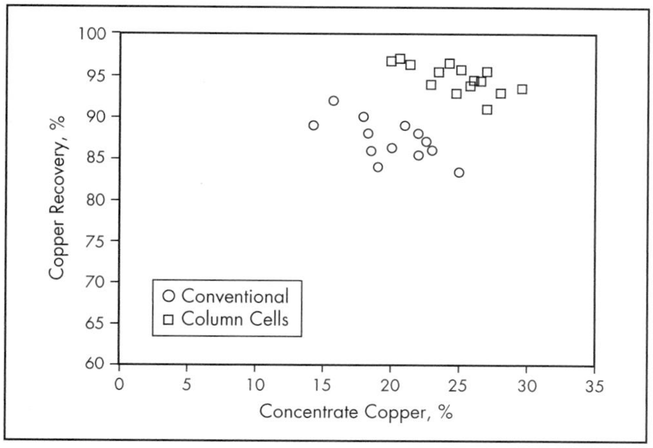

Figure 27 Column data for a copper cleaner application

In addition to copper sulfide circuits, columns are also advantageous in applications requiring a high upgrade ratio and/or a consistent final concentrate grade. They are routinely used for lead–zinc concentration, fine gold and silver recovery, and molybdenum concentration. Molybdenum typically occurs with copper sulfides and is recovered as a bulk sulfide flotation product. In this application, as discussed earlier, columns can be used for the bulk sulfide (copper and molybdenum) cleaner separation as well as the copper/molybdenum separation and subsequent cleaning. Columns with dynamic sparging systems are particularly beneficial in this latter application. The high superficial surface area rate (S_b) and the intense contacting of the sparging system are beneficial for ultrafine particle recovery situations.

REFERENCES

Ahmed S. 2013. Cavitation nanobubble enhanced flotation process for more efficient coal recovery. *Theses and Dissertations—Mining Engineering.* Paper 8. University of Kentucky.

Amelunxen, R.L. 1990. Column flotation: New carrying capacity considerations for scale-up, Presented at Expo Mineria, Santiago, Chile, May.

Bethell, P.J., and Luttrell, G.H. 2005. Effects of ultrafine desliming on coal flotation circuits. Presented at Centenary of Flotation Symposium, Brisbane, Australia.

Boutin, P., and Tremblay, R.J. 1967. Froth flotation method with counter-current separation. U.S. Patent 3,339,730.

Clingan, B.V., and McGregor, E.R. 1987. Column flotation experience at Magma Copper Company. *Miner. Metall. Process.* 3(3):121.

Coffin, V.L., and Miszczak, J. 1982. Column flotation at Mines Gaspé. Presented at 14th International Mineral Processing Congress, Paper IV-21, Toronto, Canada.

Davis, V.L., Bethell, P.J., Stanley, F.L., and Luttrell, G.H. 1995. Plant practices in fine coal column flotation. In *High Efficiency Coal Preparation: An International Symposium.* Edited by S.K. Kawatra. Littleton, CO: SME.

Davis, V., Stanley, F., Kiser, M., Bratton, R., Luttrell, G.H., Yan, E.S., Kohmuench, J.N., and Christodoulou, L. 2011. Industrial evaluation of the StackCell flotation technology. *Coal Prep. Soc. Am. J.* 10(3):22.

Dell, C.C. 1963. An improved release analysis procedure for determining coal washability. *J. Inst. Fuel* 37:149.

Dell, C.C., Bunyard, M.J., Rickelton, W.A., and Young, P.A. 1972. Release analysis: A comparison of techniques. *Trans. Inst. Min. Metall., Sect. C* 81(787):89.

Dobby, G. 2002. Column flotation. Technical Paper 2002-23. Switzerland: SGS Minerals Services.

Dobby, G.S., and Finch, J.A. 1985. Mixing characteristics of industrial flotation columns. *Chem. Eng. Sci.* 40(7):1061.

Dobby, G.S., and Finch, J.A. 1986a. Particle collection in columns—Gas rate and bubble size effects. *Can. Metall. Q.* 25(1):9.

Dobby, G.S., and Finch, J.A. 1986b. Flotation column scale-up and modelling. *CIM Bull.* 79:89.

Espinosa-Gomez, R., Yianatos, J.B., Finch, J.A., and Johnson, N.W. 1988. In *Column Flotation '88*. Edited by K.V.S. Sastry. Littleton, CO: SME.

Fan, M., and Tao, D. 2010. The role of picobubble enhanced coarse phosphate froth flotation. *Sep. Sci. Technol.* 43(1):1.

Finch, J.A. 1995. Column flotation: A selected review—Part IV: Novel flotation devices. *Miner. Eng.* 8(6):587.

Finch, J.A., and Dobby, G.S. 1990. *Column Flotation*. Elmsford, NY: Pergamon Press.

Gorain, B.K., Franzidis, J.P., and Manlapig, E.V. 1997. Studies on impeller type, impeller speed and air flow rate in an industrial-scale flotation cell. Part 4: Effect of bubble surface area flux on flotation kinetics. *Miner. Eng.* 10:367.

Gorain, B.K., Franzidis, J.P., and Manlapig, E.V. 1998. Studies on impeller type, impeller speed and air flow rate in an industrial scale flotation cell. Part 5: Validation of k-S_b relationship and effect of froth depth. *Miner. Eng.* 11:615.

Groppo, J.G., and Parekh, B.K. 1992. Comparison of bubble generation devices for column flotation of fine coal from refuse. SME Preprint No. 92-86. Littleton, CO: SME.

Honaker, R.Q., Saracoglu, M., Kohmuench, J.N., and Mankosa, M.J. 2011. Cavitation pretreatment of a flotation feedstock for enhanced coal recovery. Presented at the 28th International Coal Preparation Conference and Exhibit, Lexington, KY.

Huls, B.J., LaChance, C.D., and Dobby, G.S. 1991. Bubble generation assessment for an industrial flotation column. *Miner. Eng.* 4(1):37.

Kohmuench, J.N., Davy, M.S., Ingram, W.S., Brake, I.R., and Luttrell, G.H. 2004. Benefits of column flotation using the Eriez Microcel. Presented at the 10th Australian Coal Preparation Conference, New South Wales, Australia.

Leonida, C. 2015. Success for Eriez flotation column at Sandfire's DeGrussa mine. *Min. Mag.* March 16.

Levenspiel, O. 1972. *Chemical Reaction Engineering*. New York: Wiley.

Luttrell, G.H., Catarious, D.M., Miller, J.D., and Stanley, F.L. 2000. An evaluation of plantwide control strategies for coal preparation plants. In *Control 2000*. Edited by J.A. Herbst. Littleton, CO: SME.

Luttrell, G.H., Keles, S., and Honaker, R.Q. 2009. Implications of constant incremental quality on the design of fine coal dewatering circuitry. SME Preprint No. 09-093. Littleton, CO: SME.

Lynch, A.J., Johnson, N.W., Manlapig, E.V., and Thorne, C.G. 1981. *Mineral and Coal Flotation Circuits*. Atlanta: Elsevier.

Mankosa, M.J., Luttrell, G.H., Adel, G.T., and Yoon, R.H. 1992. A study of axial mixing in column flotation. *Int. J. Miner. Process.* 35:51.

Murdock, D.J., Tucker, R.J., and Jacobi, H.P. 1991. Column cells vs. conventional flotation—A cost comparison. Presented at Column 91, International Conference on Column Flotation, Sudbury, Canada.

Neethling, S.J., and Cilliers, J.J. 2001. Simulation of the effect of froth washing on flotation performance. *Chem. Eng. Sci.* 56(21–22):6303.

Neethling, S.J., Hadler, K., Cilliers, J.J., and Stradling, A.W. 2006. The use of FrothSim to optimize the water addition to a column flotation cell. *Miner. Eng.* 19(6–8):816.

Patwardhan, A., and Honaker, R.Q. 2000. Development of a carrying-capacity model for column froth flotation. *Int. J. Miner. Process.* 59(4):275.

Pyecha, J., Sims, S., Lacouture, B., Hope, G., and Stradl, A. 2005. Evaluation of a Microcel sparger in the Red Dog column flotation cells. In *Centenary of Flotation Symposium*. Melbourne, Victoria: Australasian Institute of Mining and Metallurgy.

Rubinstein, J.B. 1995. *Column Flotation: Processes, Designs and Practices*. Great Britain: Gordon and Breach Science Publishers.

Sandvik, K.L, Nybo, A.S., and Rushfeldt, O. 1991. Reverse flotation to low impurity levels by column flotation. Presented at Column 91, International Conference on Column Flotation, Sudbury, Canada.

Sastri, S.R.S. 1996. Technical note: Carrying capacity in flotation columns. *Miner. Eng.* 9(4):465.

Stanley, F., King, P., Horton, S., Kennedy, D., McGough, K., and Luttrell, G. 2006. Improvements in flotation column recovery using cell-to-cell circuitry. Presented at the 23rd International Coal Preparation Exhibition and Conference, Lexington, KY.

Sullair Corporation. 1992. *Sales Bulletin, Industrial Compressors—Compressor Capacity*. Bulletin E-78. Michigan City, IN: Sullair.

Wasmund, E.B. 2014. Flotation technology for coarse and fine particle recovery. Presented at Congreso Internacional de Flotacion de Minerals, Lima, Peru, August.

Wheeler, D.A. 1983. Column flotation. Seminar, McGill University, Montreal, Quebec, May.

Wyslouzil, H.E. 2009a. The production of high grade iron ore concentrates using flotation columns. Delta, BC: Eriez Flotation Division.

Wyslouzil, H.E. 2009b. The use of column flotation for the recovery of ultra-fine phosphates, Delta, BC: Eriez Flotation Division.

Xu, M., and Finch, J.A. 1989. Effect of sparger surface area on bubble diameter in flotation columns. *Can. Metall. Q.* 28(1):1.

Yianatos, J.B., Finch, J.A., and LaPlante, A.R. 1988. Selectivity in column flotation froths. *Int. J. Miner. Process.* 23:279.

Yoon, R.-H., and Luttrell, G.H. 1986. The effect of bubble size on fine coal flotation. *Coal Prep. Int. J.* 2:179.

Yoon, R-H., Adel, G.T., and Luttrell, G.H. 1992. Apparatus and process for the separation of hydrophobic and hydrophilic particles using microbubble column flotation together with a process and apparatus for generation of microbubbles. U.S. Patent 5,167,798.

Zhou, Z.A., Xu, Z., Finch, J.A., Hu, H., and Rao, S.R. 1997. Role of hydrodynamic cavitation in fine particle flotation. *Int. J. Miner. Process.* 51:139.

CHAPTER 7.3

Pneumatic Flotation

Greg Harbort

Pneumatic flotation cells are typically classified as high-intensity flotation machines with rapid kinetics, resulting in short residence times, often half that of mechanical flotation cells and a quarter of that of flotation columns. Wash water is often used to enhance concentrate grade. Typically used in cleaning duties and coal flotation, pneumatic flotation cells are also used in roughing and scavenging in a wide variety of mineral commodities and wastewater treatment. Numerous publications have provided regular updates on their development, including those by Finch (1995), Rubinstein (1995), Aksani (1998), Lynch et al. (2007, 2010), and Cheng and Liu (2015).

Pneumatic flotation machines were among the first machines used in flotation. However, with the advent of mechanical subaeration cells, the use of pneumatic flotation machines significantly declined. In 1928, five pneumatic cell designs were in industry use. Although this number increased to 11 machine types by 1945, the majority were derivatives of the earlier Callow, Forrester, and MacIntosh pneumatic cells.

The modern pneumatic flotation cell can most effectively be classified according to the theoretical and technical considerations of Bahr (1971). Although a variety of aerators, pulp feed arrangements, and separating vessel designs exist (Harbort and Clarke 2017), the applied principles and fundamental design remain unchanged. Because of the absence of an impeller, high fluid velocities are required to disperse the air into fine bubbles and to maintain particle suspension. Researchers realized that bubble–particle attachment and the separation of the phases could be considerably improved when carried out in separate units. Aerators are typically situated outside the separating vessel. The bubbles are produced by several means including sized pores, annular gaps, swirling annuli, and plunging jets. Air can either be supplied under high pressure or induced under vacuum.

The characteristics of pneumatic flotation can be summarized as follows:

- No moving parts exist for production of fine air bubbles.
- Particle adhesion to air bubbles occurs in the aerator (forced bubble adhesion).
- The aerator and separating vessel are physically separated.

Flotation columns have many features in common with pneumatic flotation, including the lack of moving parts and the use of wash water. The major differences compared with pneumatic flotation are

- Air-bubble/particle adhesion takes place within the separating column.
- There is a countercurrent flow of air bubbles and solid particles.
- Bubble–particle contact is of significantly lower intensity.

In general, the height/diameter ratio of column flotation cells is larger than that for pneumatic flotation. For column flotation, ratios of 2.2:10 are reported, while for pneumatic flotation plants, ratios of 1.5:3 are common.

DAVCRA CELL

The Davcra cell was developed in the 1960s in the Metallurgical Research Section of the Zinc Corporation. William Davis, who directed the cell's design and development, had proposed that the major proportion of mineral particle–bubble interaction occurred within the high-intensity zone of a flotation machine and was virtually time independent. Therefore, the body of the flotation cell was responsible for only a minor portion of recovery, and flotation residence time in the normally accepted sense had little meaning (Davis 1964, 1966).

The Davcra cell's operation is illustrated in Figure 1. Pulp under pressure enters the nozzle tangentially and issues from the nozzle orifice as a swirling annulus. Air under pressure passes through the orifice as a central air core. On entering the body of pulp in the tank, the interaction of the pulp annulus and air jet causes a shearing of fine bubbles with intimate mixing of these bubbles through the pulp annulus. The bubble–particle aggregates formed float into a mineralized froth. The feed pulp is prevented from flowing directly out of the tailing discharge by a baffle. This ensures that all tailings must first pass through the quiet deaeration zone where remaining bubble–particle aggregates rise into the froth. The level of pulp in the cell is maintained by hydraulic or mechanical restriction of flow in the tailing line (Cusack 1968).

Greg Harbort, Technical Director—Process, Wood plc, Brisbane, Queensland, Australia

Extensive pilot tests were conducted with Davcra cell prototypes at Broken Hill, New South Wales, Australia, progressively scaling up from 8 t/h (metric tons per hour) unit capacity to 60 t/h, and then to 200 t/h (Figure 2). By 1968, production units were in operation at Broken Hill in both lead roughing and scavenging duties (Cusack and Oley 1971). The largest cells had a nominal unit capacity of 363 t/h, in a tank 5.4 m long, 1.7 m wide, and 3.7 m deep. The aeration nozzle was the major wear component, with the use of titanium diboride ceramic providing an operating life >4,000 hours. The nozzle typically operated at a pulp pressure of 186 kPa, with compressed air supplied at 193 kPa.

The Davcra cell was conclusively proven to achieve higher recoveries at around half the residence time of mechanical flotation cells. By 1975, the cell was the most dominant of the pneumatic flotation cells installed in industry. Sites that operated successfully with it included Leinster Nickel in Australia; Bougainville Copper in Papua New Guinea; Anglo American Platinum in South Africa; ZCCM Mufulira copper concentrator in Zambia; Palabora Mining Company copper concentrator at Phalaborwa, South Africa; and Territory Enterprises copper mine and Coal Cliffs mine in Australia.

The Davcra cell was a robust, reliable flotation machine. However, it was developed by a company whose primary business was mining, and after a review of noncore business, the Zinc Corporation stopped manufacturing the Davcra cell.

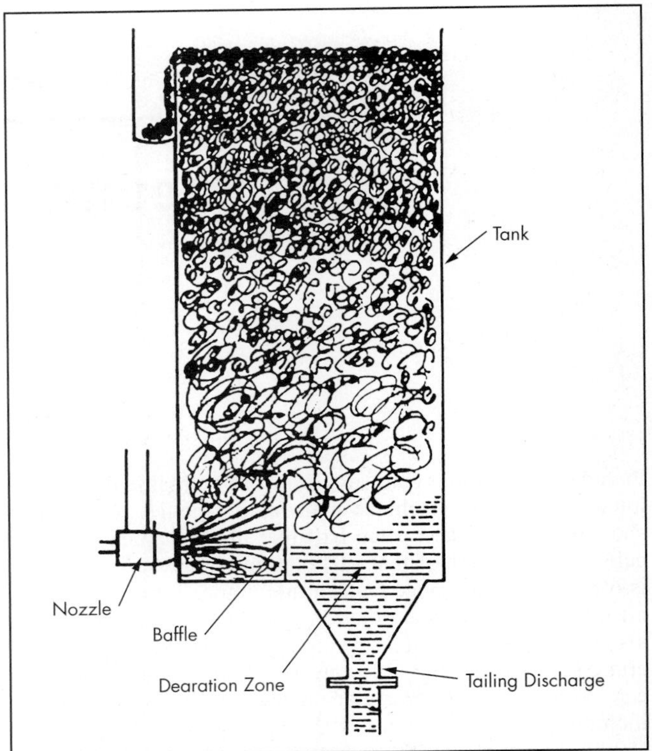

Source: Cusack 1968, reprinted with permission from the Australasian Institute of Mining and Metallurgy

Figure 1 Sectioned view of the Davcra flotation cell

Courtesy of Zinc Corporation

Figure 2 Davcra flotation cell (in background) at Broken Hill

BAHR CELL

Modern flotation machine developments in Germany commenced in the 1950s and centered on pneumatic flotation cells. In the ensuing decades, significant work was conducted by Bergbau-Forschung GmbH, the Clausthal University of Technology, the Technical University of Berlin, and KHD Humboldt Wedag. Individuals who had a preeminent role in this development included Albert Bahr, Wolfgang Simonis, and Rainer Imhof, among others.

Early work by KHD focused on "cyclone flotation," where compressed air was forced through porous walls of the flotation tank to generate bubbles. The pulp was fed tangentially into the cell for contacting with the air. After extensive laboratory and pilot tests, an apparently workable design was developed by 1961 (Salzmann and Koch 1964). A production unit was installed in 1965 at the Lüderich lead–zinc flotation plant. Its success was not overwhelming, given problems with blockages in the porous media, and this line of development was discontinued (Cordes 1997).

Ongoing investigation throughout the 1970s would eventually result in development of the Bahr cell in the early 1980s. This pneumatic flotation cell was developed jointly by Bergbau-Forschung GmbH, Clausthal University of Technology's Institute for Mineral Beneficiation, and Ruhrkohle AG's Westerholt mining company.

The design utilized external aerator units where compressed air flowed through annular rings via channels into the pulp. The aerator units were located beneath the main flotation tank and entered the tank vertically (Figure 3). Bahr confirmed that the optimum air-to-pulp ratio was 1.3:1.5 during coal flotation with the pneumatic cell. The optimum air-to-pulp ratio for the flotation of sulfide ore was much lower (0.15:0.48). Both high airflow rates and low pulp flow rates resulted in larger bubbles and lower recoveries (Changgen and Bahr 1992).

Production units were successfully installed at several plants, including the Dorfner kaolin plant in Germany, which secured the exclusive use of the Bahr cell in kaolin flotation, as well as at several coal operations in Germany. The success of the Bahr cell in Germany resulted in a large pilot-scale testing program in South Africa on coals from the Waterberg, Soutpansberg, and some of the Natal coalfields. This resulted in the successful installation of Bahr cells in the Iscor Durnacol coal plant (Gorzitzke and MacPhail 1989) and Grootegeluk coal mine (Ventert and Van Loggerenbergt 1992).

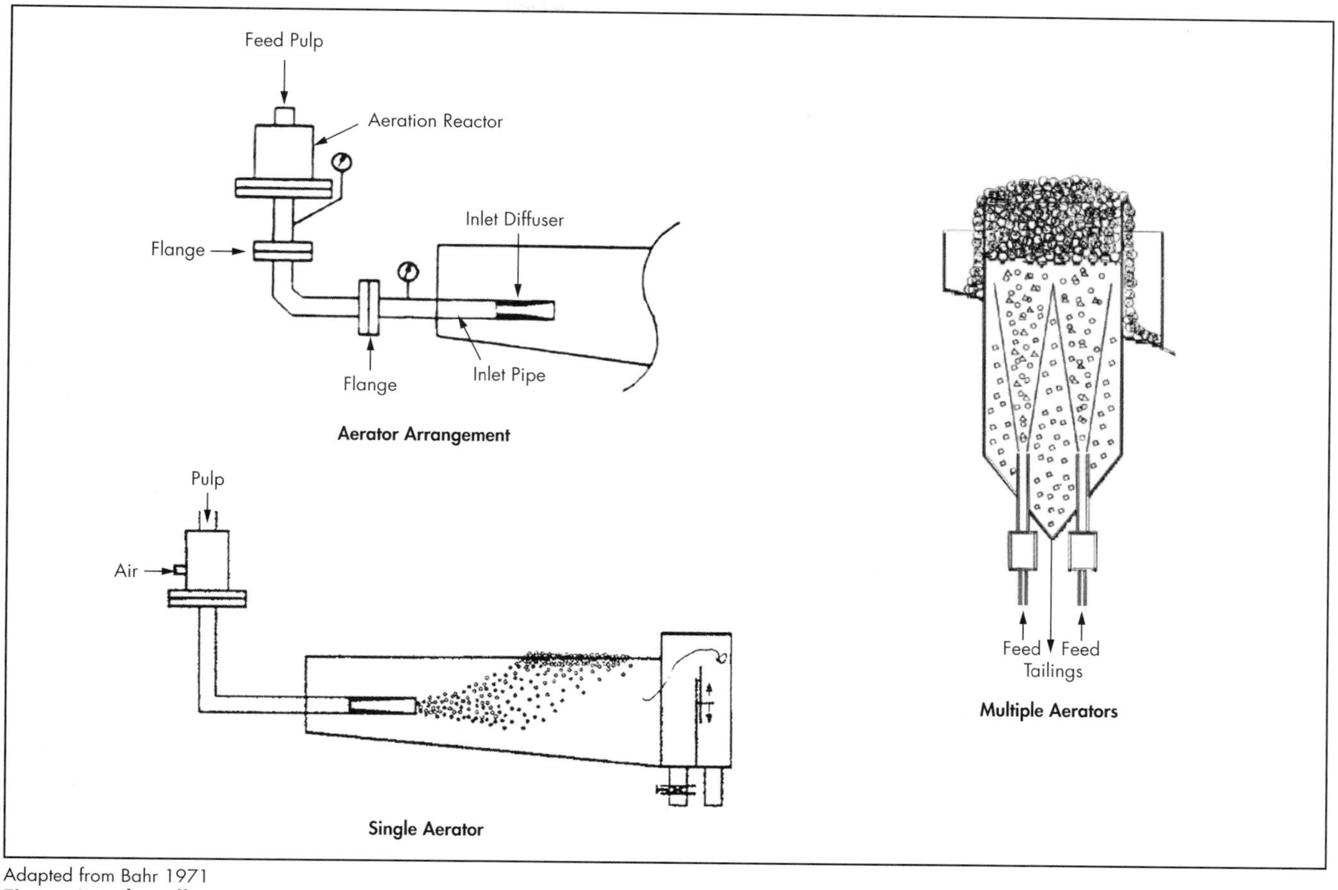

Adapted from Bahr 1971
Figure 3 Bahr cell

EKOF CELL

In the 1980s, KHD Humboldt Wedag commenced investigation into free jet flotation machines. The concept was developed from research conducted by Sigfried Heintges, Ali Alizadeh, and Wolfgang Simonis at the Technical University of Berlin (Heintges et al. 1984; Alizadeh and Simonis 1985). As shown in Figure 4, the conditioned pulp was pumped through a nozzle (4). Upon exiting the nozzle, air was entrained by the free jet (5) and broken into fine bubbles when it impinged upon the pulp. A downcomer (3) carried the aerated mixture into the main flotation tank (1), with the immersion depth of the downcomer used to vary bubble size distribution and the air entrainment rate adjusted by varying the nozzle feed pressure. Although initial tests on coal produced acceptable results, it would be another 10 years before the self-aspirating, free jet flotation machine was introduced as a commercial product.

From 1985 to 1987, KHD made a substantial effort to develop aeration reactors and flotation cell design. Extended pilot-scale tests were conducted at the Hugo and General Blumenthal coal plants of Ruhrkohle AG in Germany. The eventual pneumatic cell design would combine elements of both the earlier unsuccessful "cyclone flotation" cell and the Bahr cell. Marketed initially by KHD's subsidiary Erz- und Kohleflotation GmbH (Ekof), it was known as the Ekoflot, or Ekof flotation cell.

The Ekof cell used ring-shaped ceramic slot aerators housed within a polypropylene body. Compressed air passed through the discs with a gap of 50–75 μm before entering the pulp flow in a high-velocity region created by a displacement body. The highly aerated mixture was piped down to the flotation tank where its entry was tangential to the tank wall, creating a slow swirling motion in the pulp. The concentrate was removed through a central launder, with the circular motion enhancing transport and removal (Figure 5).

In May 1987, two 4.5-m-diameter Ekof cells with the tangential feed design were commissioned at the Pittston/Clinchfield coal operations in McClure, Virginia, United States. The units were operated in parallel, treating 1,600 m^3 of coal per hour. This installation was rapidly followed by one at the Fechner coal mine in Germany. The plant was commissioned in 1988 in a coal tailing retreatment duty using a 5.1-m-diameter cell, treating 1,600 m^3/h. At the time it was one of the highest capacity unit flotation cells installed.

PNEUFLOT CELL

In the mid-1990s, strategic restructuring within KHD saw the Ekof company return to a flotation reagent focus, with flotation machinery development directly handled by KHD. The Ekof cell became known as the Pneuflot cell. In addition to the tangentially fed tank, there was increased interest in a cell design using the aeration unit feeding into a vertical downcomer that discharged into the tank. The aerator is self-inducing and engineered to increase the velocity of the slurry as it runs over air inlets. The resulting Venturi effect causes air to be introduced as the slurry is pumped through the system. The aerator is also designed so that when the air enters the slurry, it sheers off in small bubbles and sufficient energy is generated to achieve particle–bubble attachment. This energy is generated in the aerator by the decreasing diameter into the aerator.

The single aerator feeds into a downcomer that passes through a central froth crowder and discharges into a circular perforated pipe that distributes the highly aerated pulp evenly across the cell cross section, as shown in Figure 6.

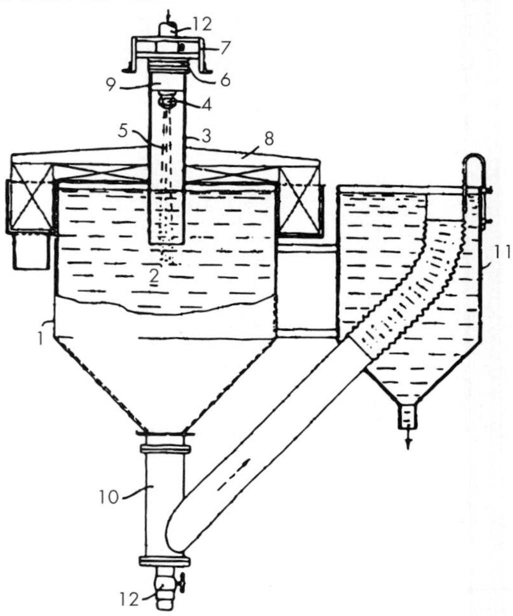

Note: Only relevant numbered items in the figure are discussed in the text.

Source: Heintges et al. 1984

Figure 4 Early-concept free jet pneumatic flotation cell

Initially, both designs were available from KHD (Sanders and Williamson 1996). At the Michilla copper operation in Chile, vertically fed, self-aerating Pneuflot cells were installed in one duty, with tangentially fed cells using compressed air in another duty (Sánchez-Pino et al. 2008). Within several years, the vertically fed cell would become the dominant Pneuflot design. Significant installations included the Magnesita magnesite operation, where six tangential Pneuflot cells were installed, and the Michilla copper operation, both in Chile (Ekof, n.d.; KHD 1997, 2006). A small but not insignificant number of Pneuflot cells were installed between 1992 and 1996; however, for eight years after that, the Pneuflot cell fell out of favor in the flotation industry. One reported difficulty was a shortage of pilot-plant equipment and technical support outside Europe. It was, however, to make a significant return in 2005–2006, with record numbers of cells installed (Markworth et al. 2007), including the Liuqiao and Zhujiadian coal mines in China and the Dartbrook coal preparation plant of Anglo Coal Australia. In 2009, McNally Bharat Engineering, a part of the Williamson Magor Group, took over the coal and mineral division of KHD. Under the name MBE-CMT, it continued to market the Pneuflot cell. Installations under the new owners have included phosphate roughing and cleaning in Kazakhstan and iron ore flotation in Brazil and Chile. In 2012, twelve Pneuflot cells ranging in diameter from 2.5 to 6.0 m were installed in copper roughing and scavenging duties in Kazakhstan.

ALLFLOT CELL

Another German company that was to achieve success with its pneumatic flotation machines was Allmineral Aufbereitungstechnik of Duisburg with its Allflot cell. The key to the device was the aeration unit developed in the mid-1980s by Andreas Jungmann and Ulrich Reilard (Figure 7). In the aerator, compressed air passed through a series of annular

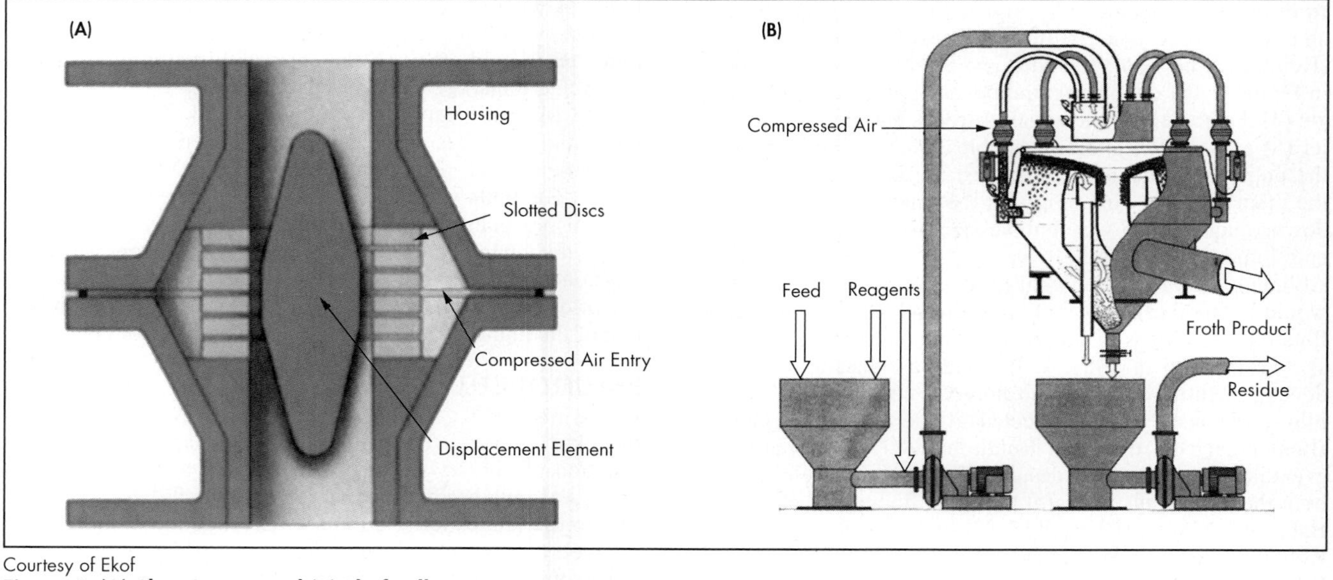

Courtesy of Ekof

Figure 5 (A) Slot aerator and (B) Ekof cell arrangement

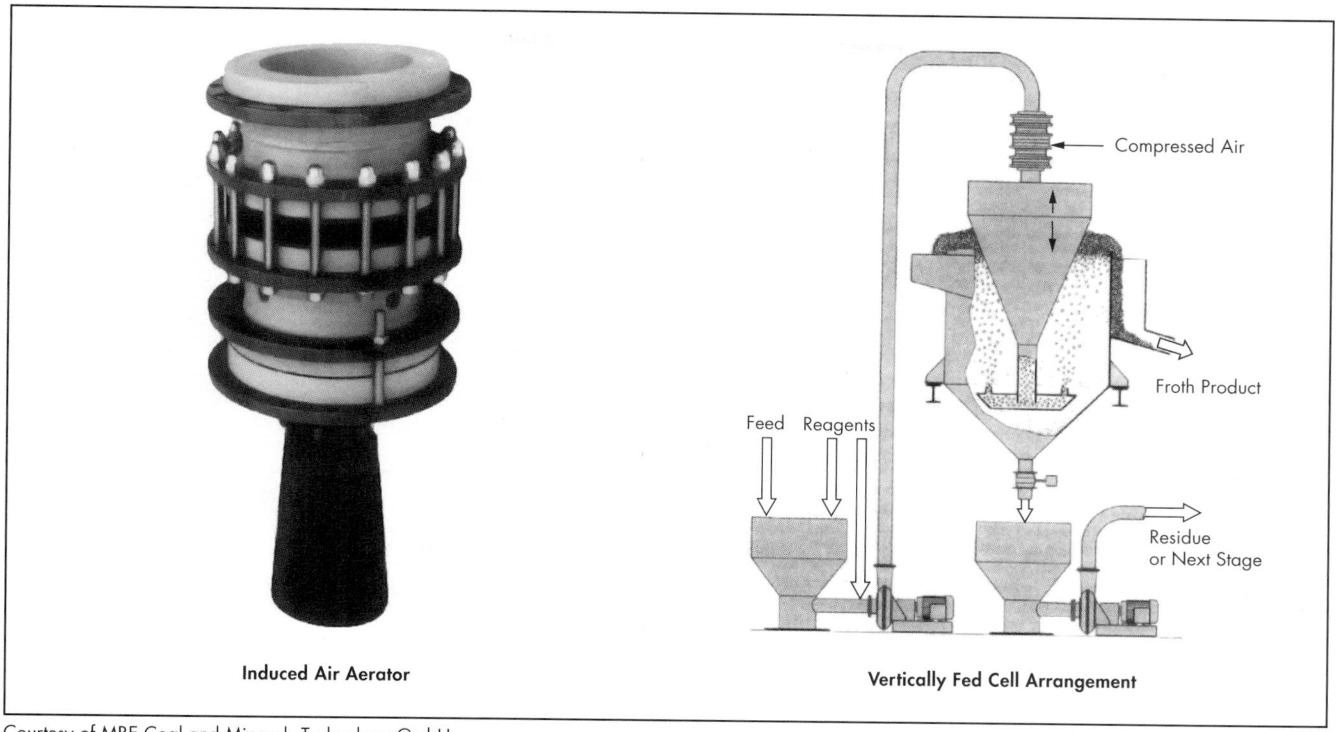

Courtesy of MBE Coal and Minerals Technology GmbH

Figure 6 Pneuflot cell

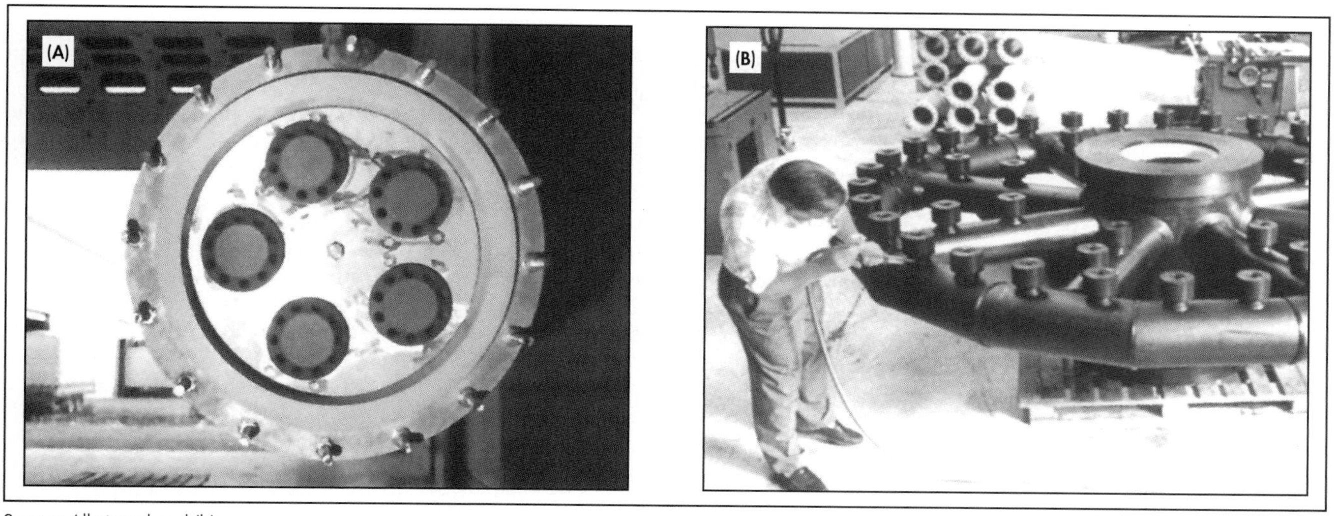

Source: Allmineral, n.d.(b)

Figure 7 (A) Feed side of 800 m^3/h Allflot aerator and (B) aerated pulp distribution manifold

slots surrounding the slurry stream. A high-velocity region created by a displacement body within the aerator produced intense contact between fine bubbles and pulp. A swirling motion was imparted to the aerated mixture by a series of screw-shaped strips within the mixing and dispersion sections. The highly aerated mixture entered a downcomer and passed to a dispersion ring to allow even dispersion of the mixture within the flotation tank (Jungmann and Reilard 1989; Allmineral, n.d.(a), n.d.(b)).

One major area of success for the Allflot cell was in coal flotation in Poland. Prior to 1990, a centrally managed economy operated there. Coal production costs were not an important issue, and losses on coal production were covered by government subsidies. After 1990, the transformation of Poland's centrally planned economy to an open-market economy saw significant changes in the country's coal industry. Nonprofitable mines were closed, profitable mines were modernized, and new coal preparation plants were constructed

Courtesy of Maelgwyn Mineral Services Limited
Figure 8 (A) Imhoflot V-cell nozzle arrangement and (B) ring distributor

(Blaschke and Gawlik 2004). There was also increased cooperation between German and Polish companies. Allmineral was one company that became a major supplier of equipment to Poland. Allflot cells were installed in the Jankowice, Bielszowice, and Knurów coal preparation plants treating –0.5-mm material (Blaschke 2000; Nycz and Zieleźny 2004). Tighter controls also saw increased quality requirements for coal-fired power plant feed stocks with the use of flotation in steaming coal preparation plants being introduced. This resulted in the installation of Allflot cells with a capacity of 600 m^3/h into the Jaworzno plant in 2002 and the Janina plant in 2004. Also of note was a copper slag flotation plant installed in Chile treating copper slag.

IMHOFLOT CELL

Also tracing their roots back to Clausthal University and the University of Berlin are Imhoflot pneumatic flotation cells. Rainer Imhof had been involved to some degree with the Germany pneumatic flotation machines for several decades. He continued their design and development, resulting in the Imhoflot pneumatic flotation process that has been successfully marketed by Maelgwyn Mineral Services of the United Kingdom.

The initial Imhoflot cell design was the V-cell. Two of its key components are shown in Figure 8. Pulp was pumped to a series of nozzles and the resultant energy aspirated air, which was sheared into fine bubbles. The aeration unit, or bubble generator, was constructed of fused silicon carbide components to protect it from wear (Imhof et al. 2005). The aerated mixture traveled through a downcomer tube and was introduced upwardly into the cell by a ring distributor system and nozzles.

An early installation of the Imhoflot V-cell was the Compañía Minera Huasco iron ore concentrator in Huasco, Chile. The twin-stream flotation plant incorporated six 4.5-m-diameter Imhoflot cells to treat 600 t/h of magnetic separator concentrates for removal of silicates. Extensive testing was conducted over several years with construction of the new flotation plant occurring in 2002 (MMS 2002a, 2003). Another significant installation took place in 2002 with another 4.5-m-diameter V-cell introduced to a coal tailings process plant in the Lugansk region of Ukraine. An Imhoflot plant was also used successfully for the El Romeral iron ore operation owned by Compañía Minera del Pacifico (CMP) of La Serena, Chile. The flotation plant incorporates two 5.2-m-diameter Imhoflot V-cells and has a design capacity of 965 m^3/h. The process is based on reverse flotation to remove silica impurities from the iron ore concentrate and is similar to that at CMP's Huasco pelletizing operation (MMS 2006a). A V-cell flotation plant was also installed at the Belaruskali potash plant in Turkmenistan in 2006. Operating in a saturated brine solution, the two 3.5-m flotation cells have a design throughput of 350 m^3/h treating 50–60 t/h solids (MMS 2006b). Figure 9 shows details of the V-cell arrangement and installation.

The second Imhoflot cell design was the G-cell, which incorporated centrifugal forces within the flotation tank. The aerator was self-aspirating and uses a high-shear ceramic multijet venturi system operating at around 250 kPa back pressure. Aerated feed is fed tangentially into the flotation tank. Unlike the Pneuflot tangentially fed cell, where rotational forces are not significant, the G-cell produces specific rotational speeds in the pulp, achieving centrifugal acceleration of 9.8 m/s^2. The resultant force on the pulp creates an angled pulp–froth interface. This enhances froth flow to the concentrate launder (Imhof et al. 2007). Figure 10 shows an assembled G-cell and the centrifugal flow of pulp within the tank.

In 2002, Ingwe Coal Corporation installed a 2.2-m-diameter G-cell with a feed capacity of 300 m^3/h in its Koornfontein operations in South Africa (MMS 2002b). During 2004, two 1.2-m G-cells were installed at the Barbrook mine of Caledonia Mining Corporation in South Africa to upgrade rougher flotation concentrates (MMS 2004). By 2005, another three 1.8-m-diameter Imhoflot cells had been installed in the Dorfner kaolin plant in Germany. The 2007 Imhoflot G-cell installation at ZapSib in Siberia, Russia, represented a quantum leap for the G-cell in terms of both scale-up and installation size, utilizing six 3.6-m-diameter cells to recover coal.

Courtesy of Maelgwyn Mineral Services Limited
Figure 9 Installation in Chile

JAMESON CELL

The Jameson cell is currently the most widely used pneumatic flotation cell, both in terms of installed capacity and number. The principles of Jameson cell operation have been discussed by numerous authors including Jameson (1988), Jameson et al. (1988), Atkinson (1994), Evans et al. (1995), and Harbort et al. (2002, 2003). The principles can be described with reference to Figures 11 and 12.

In its design, slurry is pumped to the Jameson cell slurry distributor and split between several downcomers. The downcomer is where primary contact of bubbles and particles occurs. The liquid feed is delivered to the nozzle under pressure, which results in a free jet of liquid being created as it passes through an orifice plate. As the free jet travels through the air-filled section of the downcomer, contact with the air results in a slight slowing of the jet, minor expansion of the jet diameter, and undulations on the jet surface, which in turn entrain small amounts of air. The free jet impinges on the liquid mixture within the downcomer, and the pressure of the impact creates a depression on the liquid surface, called the induction trumpet. Because of the fluted entry shape of the induction trumpet, air is channeled into the area at the base of the free jet. The free jet passes through the induction trumpet carrying with it this layer of entrained air. Additionally, the periodic collapse of the induction trumpet results in further air entrainment. Once the jet enters the main body of liquid in the downcomer through the bottom of the induction trumpet, it is referred to as a plunging jet. The high-shear rate of the plunging jet results in the entrained layer of air being broken down into a multitude of small bubbles, typically of 500-μm diameter, which are carried down the downcomer length by the inertia of the injected slurry and gravitational forces. The kinetic energy of the plunging jet is dispersed through the creation of a region of intense turbulence where momentum is transferred to the surrounding mixture and expands to occupy the downcomer cross section, creating recirculating eddies of aerated liquid. This region of jet dissipation is known as the *mixing zone* and is defined as the downcomer volume occupied by (1) the fluid inside the submerged jet below the induction trumpet

Courtesy of Maelgwyn Mineral Services Limited
Figure 10 (A) Assembled G-cell and (B) centrifugal flow of pulp

and (2) the body of recirculating fluid between the submerged jet and the downcomer wall.

Within the mixing zone, ongoing bubble coalescence and bubble creation continue to occur. Beneath the mixing zone is a region of uniform multiphase flow known as the pipe flow zone. Because the net downward slurry velocity counteracts the upward buoyancy of the bubbles, the bubbles pack together to form a moving expanded bubble bed of high void fraction. The pipe flow zone is characterized by a bubbly flow at lower air rates and moderately churn-turbulent flow at higher air rates.

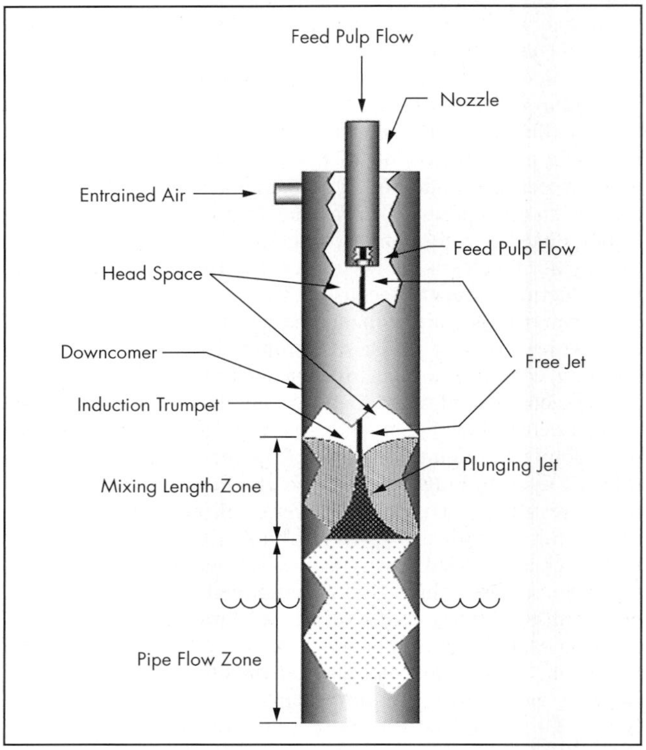

Source: Harbort 2006
Figure 11 Jameson cell downcomer

The bubbly mixture exits the downcomer into the tank pulp zone where secondary contacting of bubbles and particles occurs and bubbles disengage from the pulp. The velocity of the mixture and large density differential between it and the remainder of pulp in the tank results in recirculating fluid patterns, keeping particles in suspension without the need for mechanical agitation. Tank void fraction measurements and computational fluid dynamics modeling (Koh and Schwarz 2009) show that bubble patterns in general form a central, air-swept cone surrounding each downcomer (Figure 12) as described by Taggart (1945). The Jameson cell tank contains areas of high, localized air void throughout the pulp zone (Figure 13). The rising swarm of bubbles is governed by several factors including recirculating patterns within the tank, pulp flow volumes, and airflow volumes. The tank froth zone is where materials entrained in the froth are removed by froth drainage and/or froth washing.

Between June and December 1986, research and development staff at Mount Isa Mines in Queensland, Australia, evaluated column flotation on a range of streams in the lead–zinc concentrator and the copper concentrator for their upgrading properties in a pilot-plant-scale column operated correctly with a wash-water addition. As a result of the work, the company decided to install flotation columns. Before the column installations were completed, concerns were raised on possible future maintenance issues with the simple sparging systems that had to be used. Professor Graeme Jameson visited Mount Isa to describe in a presentation the options for bubble preparation in columns, including a newly developed laboratory-scale device developed at the University of Newcastle in Australia. The device was sent to Mount Isa for evaluation by research and development staff, based on the methodology created with the pilot-plant column. The device did not include a wash-water addition system. As a result of the benefits observed from the proper addition of wash water in the preceding tests of the pilot-plant column, this technology was

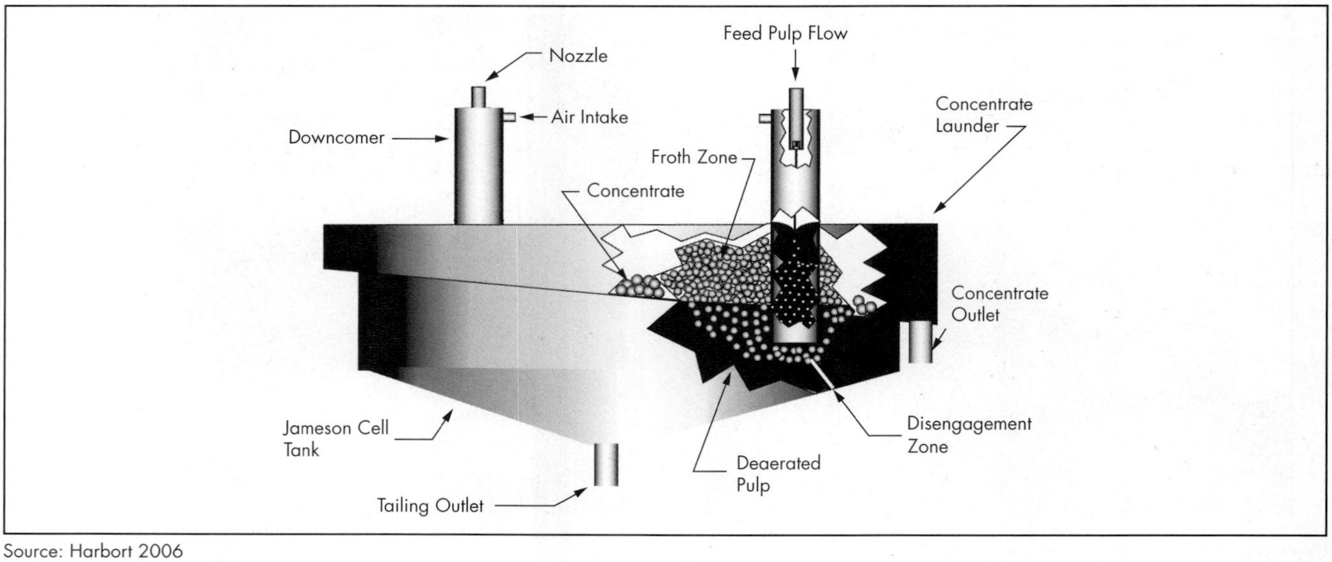

Source: Harbort 2006
Figure 12 Jameson cell

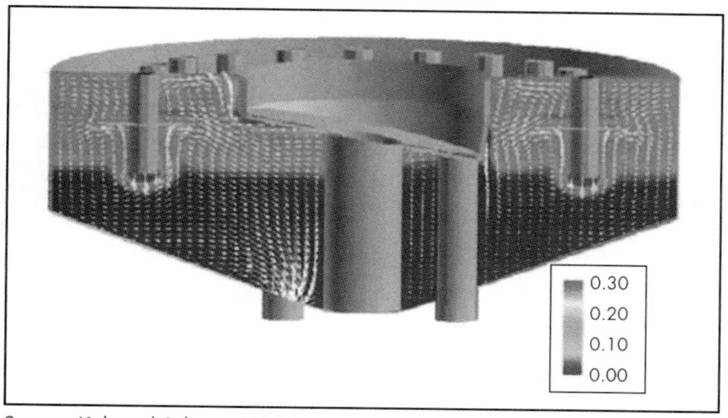

Source: Koh and Schwarz 2009

Figure 13 Air void fraction with Jameson cell

added to the device through the initiative of the research and development staff.

Research and development staff at Mount Isa Mines demonstrated the following:

1. The use of wash water in the device allowed the device to improve its performance (based on the position of its grade–recovery curve); further, this improved performance was equal to the pilot-plant column's performance for a given feed, again based on the position of the grade–recovery curve.
2. Successful scale-up of the device was demonstrated; the performance of a series of progressively enlarged versions of the device was evaluated during a period of collaboration between Mount Isa Mines and the University of Newcastle, where the enlarged versions were made at the University of Newcastle and sent to Mount Isa for evaluation.

In specific applications, equivalent performance could be achieved in the short tank design, now known as the Jameson cell (Jameson et al. 1988).

In 1989, two 1.9-m-diameter Jameson cells were installed in the Mount Isa lead–zinc concentrator cleaning lead slimes (Jameson and Manlapig 1991). This installation showed a vast difference in the flotation kinetics of the Jameson cell when compared to both mechanical cells and flotation columns. The enhanced Jameson cell kinetics were attributed to the short downcomer residence time allowing collection to occur prior to galena oxidation.

The initial success, however, was not replicated at other sites where the cell was tested. Conventional column flotation and the Jameson cell were both evaluated in the cleaning of Zn, Pb, Cu/Ag, and bulk rougher concentrates at the Luina concentrator in Australia in 1988. Both the column cell and the Jameson cell were suitable in cleaning Cu/Ag rougher concentrate. The very slow flotation rates and low rate differentials between pyritic gangue and sulfide values made the large volume, multistage conventional cleaning circuit more successful in other cleaning applications. Based on these results, a hybrid of Jameson technology and a conventional column was installed at the Hellyer mine in Australia for Cu/Ag cleaning (Lane et al. 1991). The hybrid system was chosen to maximize recovery in the recovery zone of the column. The feed was to be introduced to the column through a Jameson downcomer via a variable-speed pump. Because of pump problems, the Jameson downcomer column feed system was bypassed in the commissioning stages and the column operated with a standard sparging system. In 1990, Cominco tested the Jameson cell at the Sullivan concentrator in British Columbia, Canada, to evaluate the technology, potentially for the Red Dog concentrator in Alaska, United States. The Jameson cell was found to perform similarly to, but not better than, the existing mechanical cells at Sullivan. Size recovery analysis was evaluated on some of the Jameson product streams. The fine bubble size and high flotation rate suggested superior performance for the separation of fine (<20 µm) particles. No such superior performance, compared with the existing mechanical cell performance, was observed (Fairweather 2005). A Jameson cell was also tested as part of the Pasminco column flotation program in 1989. As discussed by Grimes (1991), it was unsuccessful because of the materials transportation difficulties presented by the coarse grind and high solids densities encountered in the company's South Concentrator zinc cleaning circuit. Several modifications to the pumping system failed to overcome the problems, and work was terminated in favor of devoting more time to the conventional column test program.

In 1989, two 3.0-m-diameter Jameson cells were installed at Mount Isa's new Hilton lead–zinc concentrator. The installation represented a major innovation. The lead rougher and rougher/scavenger concentrates were cleaned, with the Jameson cells baffled in such a way as to create four cells in series (Rohner 1993). Because of the larger diameter, an internal circular launder was installed. In recognition of earlier issues with unstable performance because of fluctuating feed pressures, a complex control logic was implemented to maintain pump box level and pressure. Incorrect downcomer positioning between the inner and outer concentrate launders resulted in low-grade concentrate being produced, and the inner launder was eventually closed. The pressure control logic did not work effectively, and the high percent solids of the feed resulted in low air entrainment rates. Although performance was generally below expectations, the installation did provide knowledge allowing the successful implementation of downcomer placement and internal launder design in subsequent Jameson cells.

The concept of control logic linking the feed pump level, dilution water, pump speed, and downcomer pressure was expanded for two Jameson cells installed in a zinc cleaning duty at the Kidd Creek mine in Canada. The logic followed similar operation for hydrocyclones, varying pump box water addition, varying pump speed, and automatically opening or closing downcomers depending on pressure. The system was inherently unstable, and the company eventually decided the Jameson cells could operate better on manual control.

In 1989, successful tests were conducted at the Peko Wallsend copper concentrator in Tennant Creek, Australia, which resulted in two 1.4-m-diameter cells being installed in December 1989 in a final cleaner role (Jameson et al. 1991). This represented the first installation in copper. The most significant Jameson cell installation of this period was at Newlands coal preparation plant in Australia where in 1988 and 1989, six 1.5-m × 3.5-m rectangular flotation cells were commissioned.

Over the next several years, the number of Jameson cells installed increased rapidly, such that by 1994 it had become the dominant nonmechanical flotation cell in use in mineral froth flotation. It remained in this position until 1999 and, during this five-year period, represented approximately 41% of installed

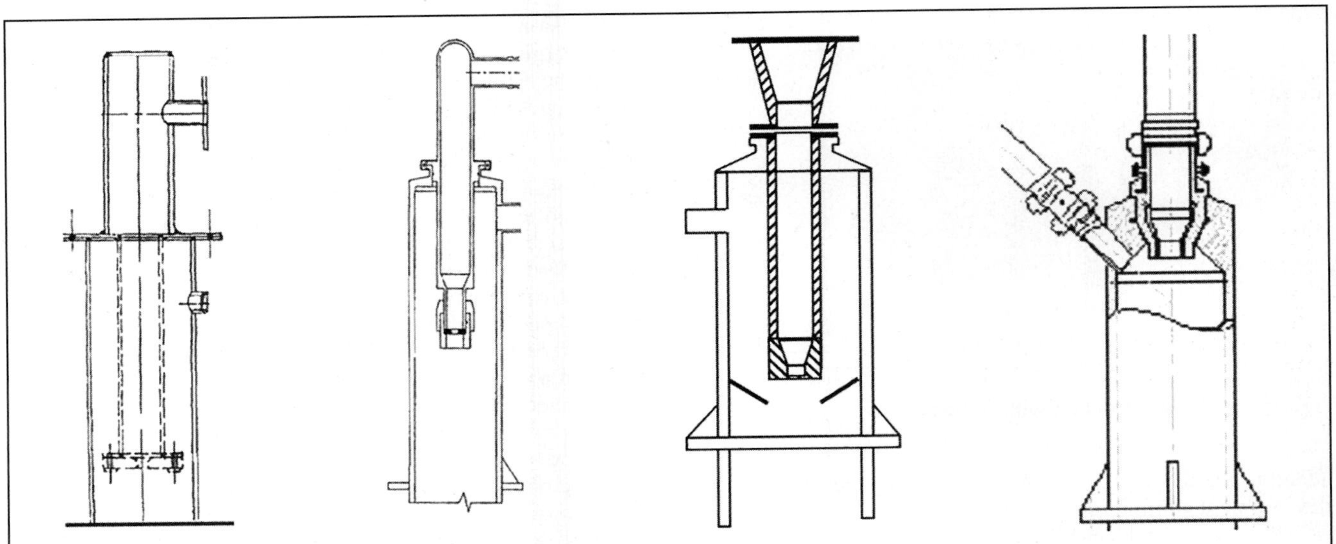

Source: Cowburn et al. 2005

Figure 14 Downcomer developments from 1989–2000

pneumatic plus column flotation capacity. Installations of note were Philex Mining copper cleaning and roughing circuits, treating 1,000 t/h; the Maricalum Mining copper cleaning and roughing circuits, treating 750 t/h; and the Alumbrera copper cleaner circuit (Harbort et al. 1994, 1997, 2000). During this time, the Jameson cell achieved almost total dominance in the Australian coal industry (Murphy et al. 2000), with installations at Blackwater, Goonyella, Riverside, North Goonyella, and Hail Creek, to name a few.

The period 1990–1999 was a time of extensive experimentation and development, with much conducted on operating sites. Initial work related to stabilizing the downcomer feed pressures to give stable flow and air entrainment. Typical feed pressures required are 150 kPa, to achieve a jet velocity of 15 m/s. The Jameson cell and feed systems were modified to operate at a higher volumetric throughput than the nominal fresh feed flow, with the difference composed of recycled tailing. Tailing was recycled to maintain a stable feed pump hopper level, either via hydraulic head equalization or via an instrumented control loop. Initially the hydraulic system was used in metalliferous operations, with the instrumented system used in higher flow coal operations. The instrumented system gradually became the standard. The amount of tailing recycled has varied from an initial 25% of downcomer feed to 40% as an enhancement to recovery and may now reach 80%.

In metalliferous operation, orifice wear was also an issue. Materials of construction of the orifice plate were investigated in 1991, including high-chromium hardened steel and various ceramics (Harbort et al. 1994). High-density alumina was deemed to have excellent wear properties and became the standard.

As discussed by Cowburn et al. (2005), a substantial effort was placed on improvement of performance of the nozzle and downcomer, resulting in the 1995 development of the "slurry lens." The major changes in the design are shown in Figure 14, which illustrates the initial steel downcomer and nozzle design, the use of high-density polyethylene, the first use of ceramic lining for roughing duties, and the current Jameson cell arrangement. The location of the slurry lens (instead of the previous orifice plate) allowed an increase in the mixing zone in the downcomer, improving residence time and allowing operation at higher air-to-pulp ratios, from ~0.6 to ~1.2, with superficial air rise velocities (Jg) up to 1.2 cm/s.

Substantial increases in downcomer and tank unit capacities also occurred. The cell diameter increased from 3.5 m in 1993 to 6.5 m in 2000, with downcomer capacities increasing from 30 m^3/h to 90 m^3/h. The larger flows increased the interaction of aerated slurry exiting neighboring downcomers, causing increased pulp turbulence. Significant testing was conducted to optimize the shape, location, and porosity of the bubble diffuser to minimize this turbulence. Diffusers were installed in an attempt to create uniform bubble rise velocities across the surface of the cell by slowing the superficial gas velocity in the high void faction area immediately around the downcomer.

The froth zone and carrying capacity limitations were also an area of investigation considering the Jameson cell's high unit capacity and frequent use in high mass recovery duties such as coal flotation and base metals cleaning. The 2000 review by Murphy et al. provides a detailed description of Jameson cell carrying capacity based on particle size distribution, aeration rate, and cross-sectional area. Both in-froth and above-froth wash-water addition systems have been used in the Jameson cell. In-froth washing produces a drier concentrate and, with washing occurring closer to the froth–pulp interface, allows increased time for bubble drainage in the froth phase, generally increasing washing efficiency. A reported difficulty of in-froth washing is blockage of water holes and uneven distribution of wash water. The current Jameson cell wash-water practice is to use above-froth wash-water addition systems in a shower arrangement. The increased use of recycle and higher Jg has resulted in greater dependency on wash water to achieve specified concentrate grade. Where earlier installations operated at a wash water to concentrate water ratio of 0.8:1.2, the modern norm is to operate with ratios of 1.2:1.5.

In 2000, Glencore Technology made a strategic decision to largely withdraw from metalliferous duties and focus on coal flotation. This resulted in the Jameson cell achieving wide acceptance in coal not only in Australia but also in the United States. During 2000–2012, 76 Jameson cells were installed in various coal plants. Jameson cells were also installed in the burgeoning Chinese coal industry, with varying success (Ren and Yang 2015).

By 2006, the Jameson cell had commenced a return to base metals flotation (Young et al. 2006). In addition to traditional cleaning roles, focus was placed on preflotation and scalping duties. Preflotation involves removing a fast-floating gangue prior to the main flotation circuit, rather than floating and then depressing the gangue material. Scalping involves recovering rapidly floating material to produce a final-grade concentrate at the head of either roughing or cleaning circuits, provided effective liberation and the correct electrochemical conditions exist. After scalping at the head of the rougher circuit, the valuable mineral remaining in the circuit is floated as a lower-grade concentrate that is then upgraded in the cleaner circuit. The removal of the liberated mineral prior to a regrinding stage allows the regrind and cleaning circuits to be designed and operated more appropriately to the middling material. This allows a greater efficiency of separation of composite particles. Minimizing the quantity of regrinding decreases slimes generation and reduces the losses that inevitably result from their presence (Gray et al. 1998; Pokrajcic et al. 2005). Huynh et al. (2014) also discuss the use of Jameson cells for the removal of penalty elements and their effect on circuit configuration. Jameson cell operation in prefloat duties include the Red Dog Pb-Zn installation (Smith et al. 2008), the Mount Isa copper concentrator (Carr et al. 2003), and the Century Pb-Zn mine (Rantucci et al. 2011). Operations in the scalper duty include the Phu Kham copper concentrator in Laos (Bennett et al. 2012), and the Telfer copper–gold concentrator (Seaman et al. 2012) and Prominent Hill copper concentrator, both in Australia (Barnes et al. 2009).

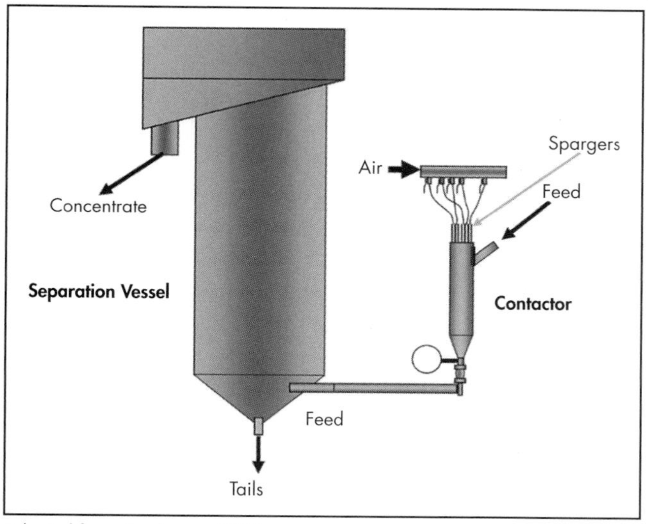

Adapted from Amelunxen 1993
Figure 15 Contact cell arrangement

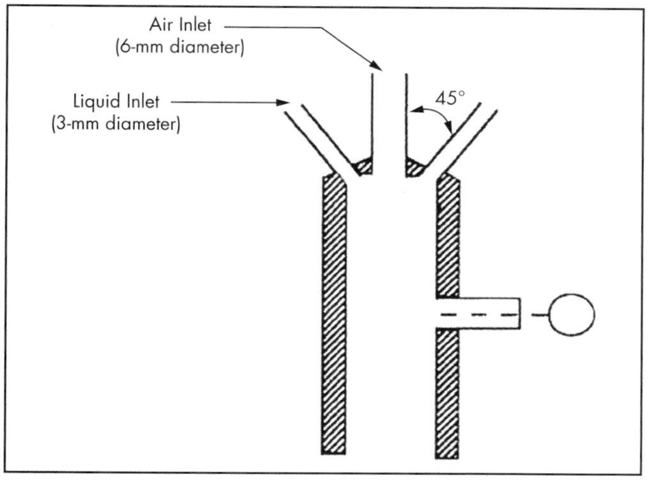

Source: Ityokumbul 1998
Figure 16 Early contact-cell aerator arrangement

CONTACT CELL

The contact cell was invented in 1992 by Roger Amelunxen as a high-capacity machine capable of producing high-purity concentrate grades (Amelunxen 1993). Feed slurry is placed in direct contact with air in a pressurized chamber outside the flotation tank (Figure 15). The resultant aerated pulp is released into the flotation tank where the mineral-loaded air bubbles are allowed to float and separate from the gangue. As with other pneumatic flotation cells, the sizing of the tank is not a function of residence time but rather mixing characteristics and carrying capacities of the froth. Figure 16 illustrates the arrangement of an external contactor tested by the Florida Industrial and Phosphate Research Institute (Ityokumbul 1998). The contactor is operated at a typical pressure of 140 kPa, with tank residence time reported being two-thirds to one-half of mechanical flotation cells.

Contact cells have been installed in several sites, treating various commodities. In Canada they have operated at the Agnico Eagle and Eskay Creek mines in copper cleaner duties. In the United States, they have operated at the Mountain Path rare earths operation in California and in Phoenix, Arizona, in copper cleaning. Four contact cells were installed in the Hustadmarmor mine in Norway for lime flotation.

At the Bulyanhulu concentrator in Tanzania, two contact cells were installed to treat gravity gold tailing. At the start of operations, frequent transitions between quartz-dominant ore and argillite-dominant ore impacted throughput and flotation performance. To mitigate these effects, the contact cells were added to recover gold from the grinding and flash flotation circuits. Figure 17 details a cutaway of the Bulyanhulu contact cell, mapping air holdup within the tank, which indicated a complex aeration pattern. A high air holdup and potentially turbulent region near the distributor outlet was evident, although this was rapidly dissipated. An area devoid of air exists at the lower extent of measurement at the cell wall. It is expected that this air-barren region would extend into the lower third of the tank below the distributor. The aerated pulp distributor is designed so the pulp exits at both ends, and this resulted in a reduction of the air content in the center of the tank. This was sufficient to cause air from high air holdup regions to move toward the center of the cell. There is also some distortion caused by the internal launder. The expected result is deterioration in the concentrate quality near the internal launder, with the problem increasing with higher aeration. At the time of the measurements, the internal launder was lidded and closed because of concentrate quality problems (Harbort and Schwarz 2007).

JET FLOTATION MACHINES

China has a long history of pneumatic flotation cells, as discussed by Hu and Liu (1988). One of the longest standing types of pneumatic cells is the jet flotation machine, under the models XPM, FJC, and FJCA.

The XPM series was first put into operation in 1967 at the Nanshan Colliery. Initially only a limited number of XPM jet flotation machines were installed in industry, although by the year 2000, 89 cells had been installed, overwhelmingly in coal flotation, with sizes ranging from 4 to 16 m^3. Reasons for the large number included the small unit capacity and the number of mechanical flotation cells that were retrofitted with the jet aerators.

In each cell, four aerators were arranged radially. A recirculating pump withdrew part of the pulp from the flotation tanks and pumped it through a jet chamber. Guide vanes within the jet created a circular motion, reportedly enhancing performance. The jet generated a vacuum, drawing in air. The aerated pulp passed through a downcomer prior to being dispersed via an umbrella into the flotation tank. The flotation tanks acted as a bank, with feed flowing down the bank to a regulated tailing outlet (Wu and Ma 1998). Figure 18 provides details of the FJC jet flotation machine, which has essentially the same arrangement as the XPM.

From 2000 to 2010, the number of jet flotation machines installed in China grew. During this period, nearly half of all pneumatic and column flotation machines installed worldwide were installed in China. Approximately 180 FJC jet flotation machines were installed in the Chinese coal industry during this period. However, the machines suffered from high wear in the downcomer and umbrella diffuser.

In 2006, the aerator mechanism was redesigned, and the new jet flotation machine was designated the FJCA. As shown in Figure 19, the FJCA design immersed the aerator, dramatically shortened the downcomer, and removed the umbrella diffuser. FJCA machines of 20-m^3 unit capacity were successfully installed in the high-throughput Linhuan and Wanfenanggang coal preparation plants with good aeration performance (Wu et al. 2010).

FLOKOB CELL

Major restructuring of the Polish coal mining industry commenced in 1990. The change from a centrally planned to a market economy enforced a different method for managing the coal mining industry. Unprofitable coal mines were gradually closed. In 1995, 65 coal mines were operating, but in 2004, only 41 coal mines existed, in which 43 coal preparation plants operated. There were also eight independent coal mines in Poland, which were closed coal mines purchased by new owners (Horsfall et al. 1994). Polish coals were difficult to float, and the FLOKOB pneumatic cell gained wide use in the country (Blaschke 2000; Blaschke and Gawlik 2004).

Research into the conditions and possibilities for the use of pneumatic flotation machines for the treatment of coal slurries commenced in the early 1980s in Poland (Brzezina 1982). By 1990, the Central Mining Institute in Katowice was at the leading edge of pneumatic cell flotation research, with the FLOKOB cell introduced to the Polish mining industry in 1991.

The publication by Brzezina and Sablik (1991) details the development and experimentation of the FLOKOB-3, a 3-m^3 flotation cell with a capacity of 60 m^3/h. Tests were conducted on various coals and investigated capacity, aeration, and immersion of the aerator downcomer. By 1996, the FLOKOB cell had successfully been scaled up to a 40-m^3 machine with a capacity of 520 m^3/h. Installations of significance included the Szczyglowice mine and the Rydultowy colliery preparation plant.

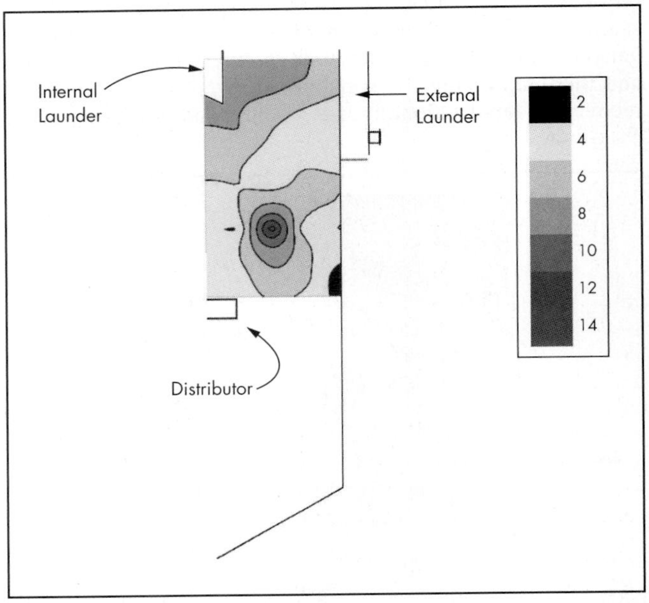

Source: Harbort and Schwarz 2007

Figure 17 Air holdup (%) within a contact cell

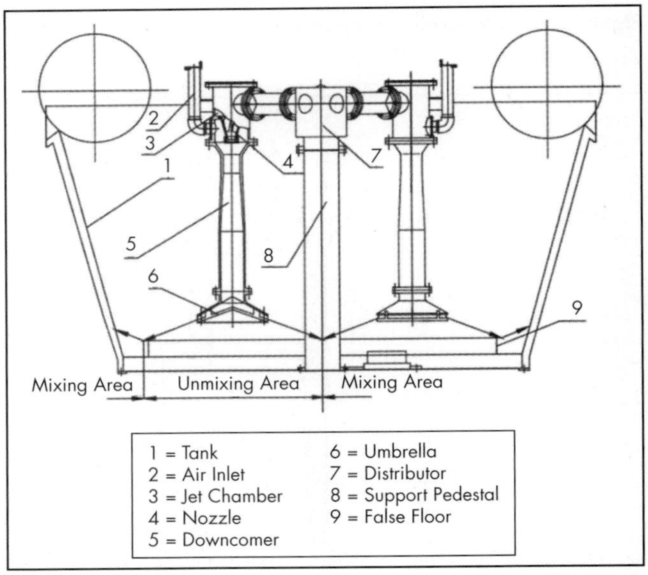

Source: Wu et al. 2010

Figure 18 FJC jet flotation machine

The FLOKOB cell operated with a dual aeration system, as shown in Figure 20. Feed was distributed to several aeration units. The slurry (at 120 kPa pressure) was mixed with air supplied at 30 kPa and entered a 1-m-long downcomer before entering the flotation tank. A portion of tailing was extracted from the cell and fed into a secondary aeration system submerged in the lower section of the cell (Brzezina and Sablik 1994).

AIR-SPARGED HYDROCYCLONE

One of the most extensively investigated pneumatic flotation machines is the air-sparged hydrocyclone (ASH). The ASH was initially developed at the University of Utah based on the concept of Miller (1981) with initial work supported by postgraduate researcher Van Camp (1981).

The basic features of an ASH are a porous wall through which air is sparged and a tangential flow of slurry orthogonal to the airflow. The ASH was envisioned to take advantage of a controlled high force field developed by cyclone fluid flow, together with a high density of bubbles, to achieve effective flotation of fine particles, with an increase in flotation kinetics. As discussed by Miller et al. (1982), many different designs incorporating these features were tested prior to a preferred design being selected. As shown in Figure 21 (Ye et al. 1988), the slurry is fed tangentially through a conventional cyclone header, passing through the separator as a thin layer in swirl flow, and travels downward countercurrent to the froth phase, which is moving upward in the center of the device. Hydrophilic particles are thrown to the porous cylinder wall and are discharged in the reject. Hydrophobic particles encounter the air bubbles sparged radially through the porous wall. The high-shear force at the wall generates small air bubbles and provides for intense particle–bubble interaction. Particle–bubble attachment occurs and the hydrophobic particles are transported into the froth phase; froth exits axially at the top of the cyclone through a vortex finder.

The effective recovery of fine particles is primarily because of the high centrifugal force field present in the ASH, with force fields of the order of 100 g increasing the inertia of the fine particles and facilitating the effective flotation of particles as small as 1 μm. The effectiveness of fine particle flotation is further enhanced by the presence of numerous, freshly formed, small air bubbles. An ASH bubble generation study by Lelinski et al. (1995) examined the influence of surfactant concentration, water flow rate, and porous tube pore size. Depending on the experimental conditions, the average bubble size ranged from about 100 to 300 μm in diameter.

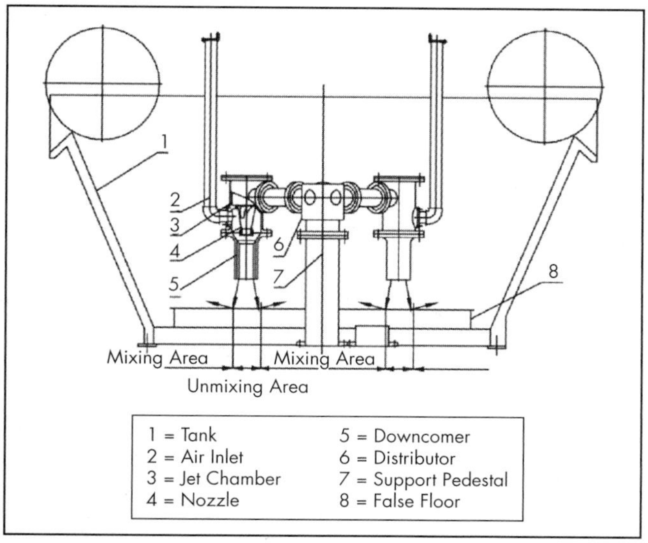

Source: Wu et al. 2010
Figure 19 FJCA jet flotation machine

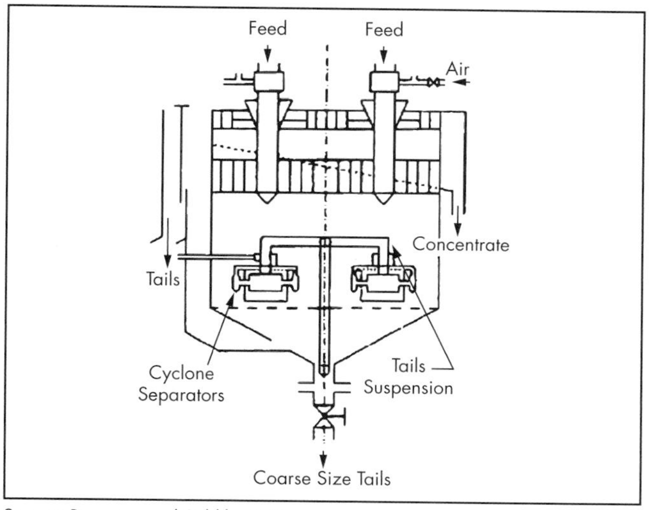

Source: Brzezina and Sablik 1994
Figure 20 FLOKOB flotation cell

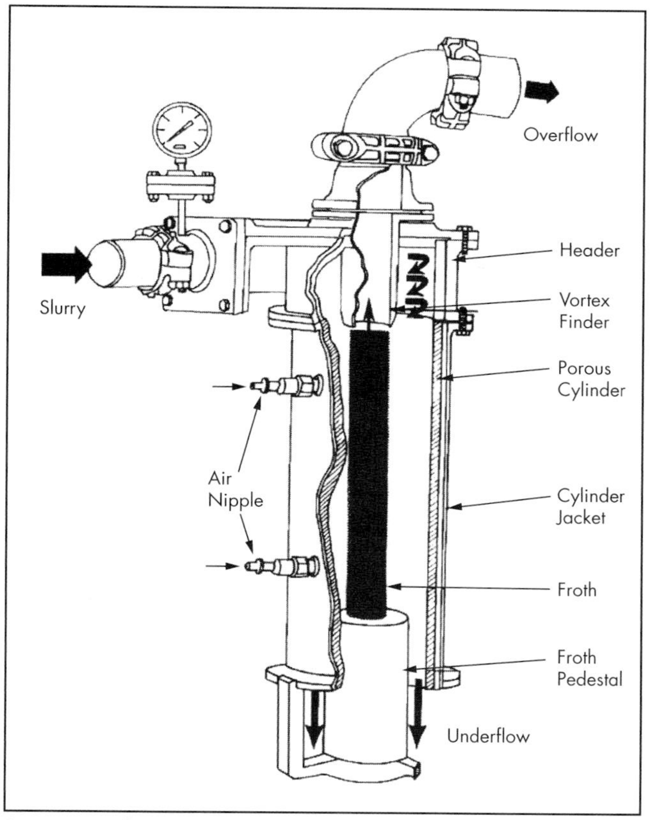

Source: Ye et al. 1988
Figure 21 Air-sparged hydrocyclone

The ASH is a machine of immense potential, although that potential has not resulted in significant installations in the minerals industry. In general, for most flotation testing with the unit, a relatively high reagent addition is required to achieve the same recovery as that achieved in mechanical bench-scale flotation. This high reagent demand was thought to arise because of the high centrifugal force generated by swirl flow during ASH flotation. Higher reagent addition is required to stabilize the bubble–particle attachment in the relatively high centrifugal field. With larger-diameter ASH units, the centrifugal force is smaller, and therefore the reagent addition was expected to be reduced significantly (Miller and Yu 1993). A significant decrease in the concentrate grade occurs if the length of the vortex finder exceeds a critical value. Some plant-site continuous operation of the ASH was limited to about six hours because of crud formation on the inner surface of the porous tube. This crud gradually built up, plugged the pores of the porous tube, and caused poor flotation separation apparently because of inadequate air distribution (Miller et al. 1999).

Many successful pilot tests have been reported. Miller et al. (1986) discuss gold flotation tests where a feed grade of 0.34 g/t Au was upgraded to 170 g/t, at 75% gold recovery. The results were achieved with a residence time of less than 1 second, compared to a mechanical bench-scale test that achieved a comparative performance in 10 minutes. Miller and Van Camp (1982) report coal flotation where comparable results were achieved with an ASH at <0.5-second residence time and with mechanical flotation having a 2-minute residence time. Pyrite was effectively recovered from a coarse Witwatersrand (South Africa) quartzitic ore with an ASH. Recoveries of pyrite from 85%–93% at grades of 35%–40% sulfur were obtained. The optimal recoveries and grades were obtained for particles between 38 μm and 106 μm. The results were achieved with a residence time of <1 second and were comparable with those attained in mechanical batch flotation tests after 5 minutes (Van Deventer et al. 1988). At the Cadjebut operation, Australia, the major reason for losses of galena during lead beneficiation was that the circuit was unable to recover the galena present in the finest size fractions. A pilot ASH test significantly enhanced performance in lead roughing recovery for particles smaller than 20 μm, as shown in Figure 22. Although the tests were not progressed, "ASH based flotation was considered to be viable and of considerable potential benefit to the Cadjebut concentrator … at high specific capacity without necessitating major alterations to the overall flotation reagent consumption" (Baker and Willey 1993).

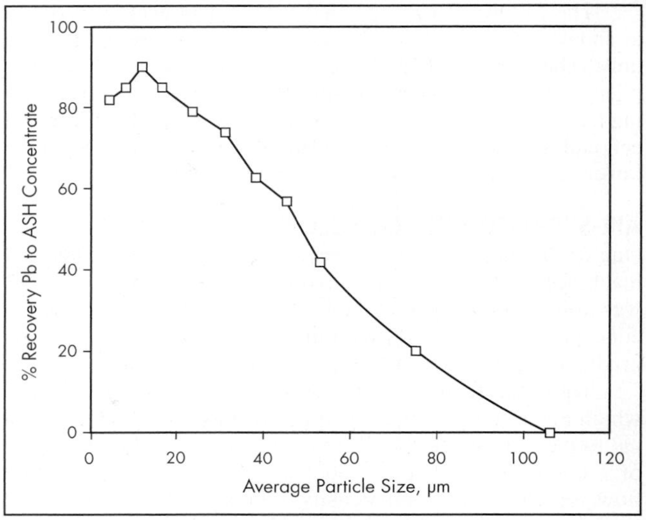

Source: Baker and Willey 1993, reprinted with permission from the Australasian Institute of Mining and Metallurgy

Figure 22 ASH size-by-size lead recovery at Cadjebut

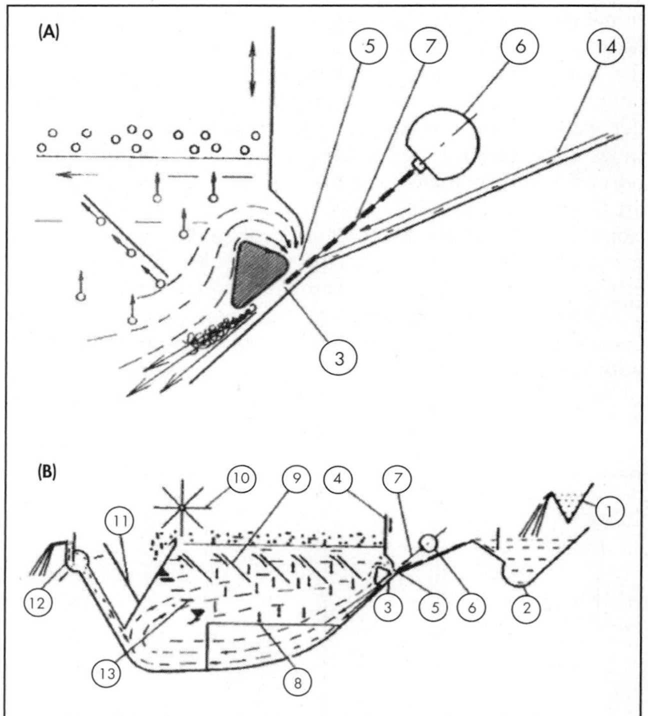

Note: Only relevant numbered items in the figure are discussed in the text.

Source: Chudacek et al. 1994, reprinted with permission from the Australasian Institute of Mining and Metallurgy

Figure 23 FASTFLOT (A) aerator and (B) cell arrangement

FASTFLOT CELL

The FASTFLOT pneumatic flotation cell underwent extensive plant testing at the Pasminco Broken Hill operations, Australia, during the 1990s (Chudacek et al. 1994, 1995).

In the machine, high-velocity water jets are used to entrain air and then enter the flotation tank generating froth. The FASTFLOT arrangement and operation are shown in Figure 23. Mixing intensity, shear rate, and acceleration levels were controlled by the velocity of the thin jets (7), controlled by the operating pressure in the manifold (6). The jets travelled at approximately 50 m/s, accelerating the surrounding air envelope before plunging between two layers of mineral slurry (5) and entering a diverging mixing nozzle (3). The shear rate, because of the velocity differential between these three streams, resulted in shredding of the gas envelope into very small bubbles and provided very intense agitation accompanied by high turbulence. The aerated flow was discharged into the cell bulk compartment where it flowed past longitudinal stabilizers. Froth was diverted by bubble guides (9) toward the froth discharge end of the cell. As the layer of froth travelled toward the froth weir, it drained liquid and

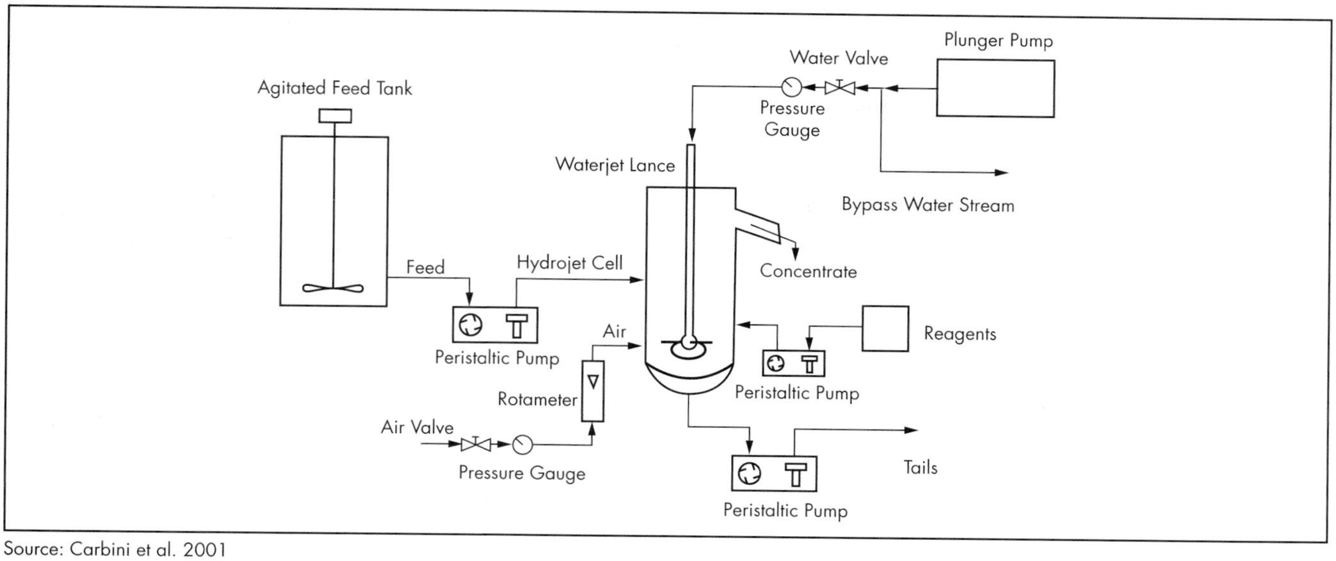

Source: Carbini et al. 2001

Figure 24 Hydrojet experimental arrangement

entrained gangue mineral particles and thickened because of crowding against the weir. It was subsequently discharged by a rotary paddle (10) into the froth launder (11). The slurry flow at the discharge end of the cell was partly released via flexible ducts into the discharge launder (12), which was adjustable vertically to control cell liquid level. The rest of the slurry flow was diverted upward by a curved bottom where, in turn, the slurry flow was diverted horizontally by a flow guide (13) toward the recycle gate (4) that controlled the recycle flow (5).

Operation at Broken Hill with a 1.5-m^3/h pilot cell achieved rougher zinc recoveries 10% higher than the operating plant at comparable concentrate grades. The FASTFLOT residence time was one-half to one-third that of the plant.

HYDROJET CELL

The Hydrojet cell also made use of high-pressure water jets for flotation. Developed at the Department of Geoengineering and Environmental Technologies, University of Cagliari, Italy, it made use of high-pressure (9 MPa) water lances immersed in a flotation tank. The lance exit was close to the curved base of the tank where compressed air also entered (Figure 24). The velocity of the water jet exiting the lance was 100 m/s, generating high turbulence and fine bubble formation. Successful tests were conducted on coal, barite, and zinc roughing. The comparative test on zinc roughing indicated that the Hydrojet cell in comparison to mechanical cell flotation would produce 10% higher recovery at equivalent concentrate grade and residence time (Carbini et al. 2001).

CYCLOJET CELL

The Cyclojet cell is a high-intensity jet, pneumatic flotation cell developed in Turkey in 2006. Figure 25 provides an arrangement of the cell. As described by Hacifazlioglu and Kursun (2012), feed pulp at a pressure of 10–60 kPa is pumped to the inlet of a hydrocyclone (1) in which the overflow has been closed (6). The tangential entry and conical shape generate centrifugal forces, resulting in a conic pulp jet (3) exiting the cyclone apex (2). The conic jet enters a conic tube (4), two-thirds of which is submerged in the flotation tank (5).

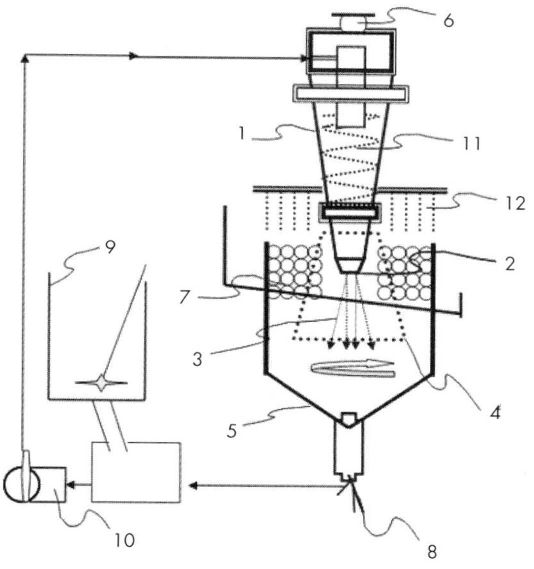

Note: Only relevant numbered items in the figure are discussed in the text.

Source: Hacifazlioglu and Kursun 2012

Figure 25 Cyclojet arrangement

An air gap is created by the effect of the cyclonic jet within the tube, generating a vacuum and allowing the suction of air from the atmosphere via a restricted air entry in the tube. The shear forces generate bubbles reportedly 0.3–0.5 mm in diameter. The aerated pulp exits the conic tube into the tank where disengagement of bubbles occurs to form a typical froth layer (7). Froth washing (12) can be utilized and a portion of the tailing (8) can be recirculated to the feed to maintain stable feed pressures.

The Cyclojet cell has been successfully tested in coal, barite, and feldspar flotation. Comparative tests between the

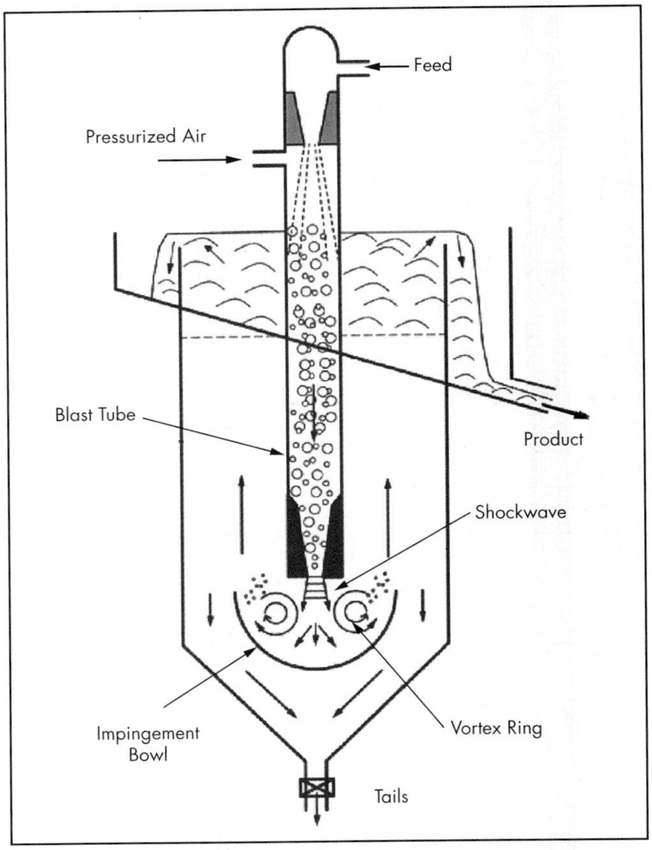

Source: Jameson 2010
Figure 26 Concorde cell

Cyclojet cell and the Jameson cell are reported to have produced similar recovery and concentrate grades (Demeekul et al. 2016).

CONCORDE CELL

In the Concorde cell pneumatic machine, preaerated feed is raised to supersonic velocities before passing into a high-shear zone in the flotation cell (Jameson 2010). A diagrammatic sketch of the Concorde cell is shown in Figure 26. It consists of two stages. In the first, small bubbles are formed in a blast tube under pressure. The feed enters as a vertical jet and mixes with air under pressure. In the second, the aerated mixture then passes through a choke. Downstream of the choke, a shock wave is created where there is a large change in pressure over a small spatial distance, and the bubbles reduce in size. Further particle–bubble contact occurs in the vortex ring that forms in the impingement bowl. Pilot-plant tests on platinum group metals in South Africa indicated that the rate of flotation in the Concorde cell was 100 times faster than the existing mechanical cell circuit.

SIMINE HYBRID FLOT CELL

The Simine Hybrid Flot cell was developed in 2007 by the Siemens Industrial Solutions and Services Group in the framework of a joint pilot project with the Minera Los Pelambres copper mine, Chile (Gaete 2007). As illustrated in Figure 27, feed pulp and gas are mixed inside the mixing chambers of an ejector system before being sprayed by high-pressure nozzles into the cell. A sparger system is installed in the lower section of the flotation tank to enhance recovery. The cell requires up to 70% less power when compared to systems of a similar capacity of 100–400 m^3/h (Krieglstein 2009).

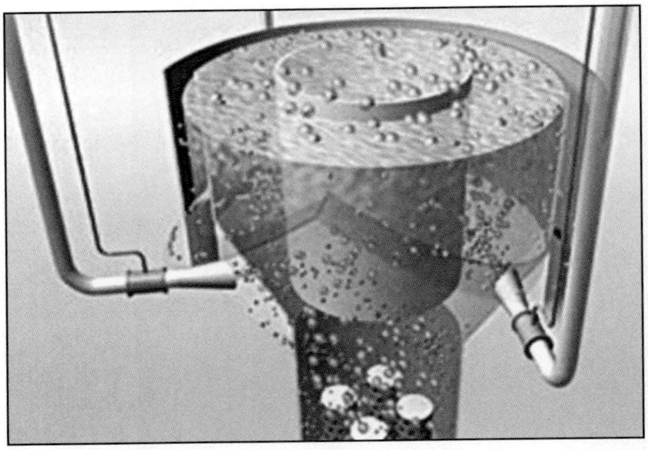

Courtesy of Siemens
Figure 27 Simine Hybrid Flot cell

Tests at the Minera Los Pelambres mine showed that the Simine Hybrid Flot cell in a selective copper–molybdenum process could achieve an increase in molybdenum recovery of more than 2% over the existing operation. The recovery increase was significantly in the –10 µm fraction (Siemens AG 2009). Two 16-m^3 Simine Hybrid Flot cells were installed as pre-roughers in the plant's copper–molybdenum selective process. A similar arrangement was planned for Codelco's Andina Division copper–molybdenum circuit. Successful piloting as an additional rougher line in the copper–molybdenum selective process was also conducted at Codelco's El Teniente Division (Carter 2011).

Ongoing development led to a new ejector working in a self-priming operation with a pulp pressure of 3.2–3.4 bar (Krieglstein et al. 2014). Bubble sizes of 500–750 µm have been reported (Hartmann et al. 2012), with a residence time approximately half that for installed mechanical cells.

CFM CELL

Another pneumatic machine that utilizes the dual aeration concept is the CFM cell. The CFM series was developed by the Institute of Material Studies and Metallurgy, Ural Federal University, Russia.

As detailed in Figure 28, pulp and air are mixed in ejectors and injected via a nozzle in the upper portion of the cell. Secondary aerators are positioned in the lower portion of the cell to enhance performance. An industrial CFM-1400VMG has a working chamber volume of 13.2 m^3 and a capacity of up to 500 m^3/h. Air consumption is up to 600 m^3/h, with a typical power consumption of 6 kW·h. The upper chamber diameter is 3.5 m, the lower chamber diameter is 1.4 m, and the height is 5.0 m. A residence time of 3.2 minutes has been reported to produce comparable performance to that of a mechanical cell circuit with a residence time of 10 minutes.

Four CFM-1400M cells have been installed at the JSC Svyatogor processing works (Russia) in various duties, including zinc and copper roughing. They have provided a 3.2% Cu

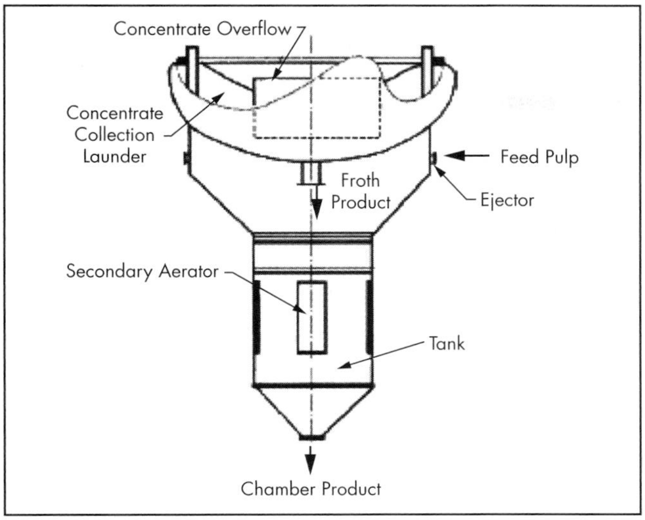

Source: Viduetsky et al. 2014

Figure 28 CFM pneumatic flotation machine (with ejectors and secondary aerators)

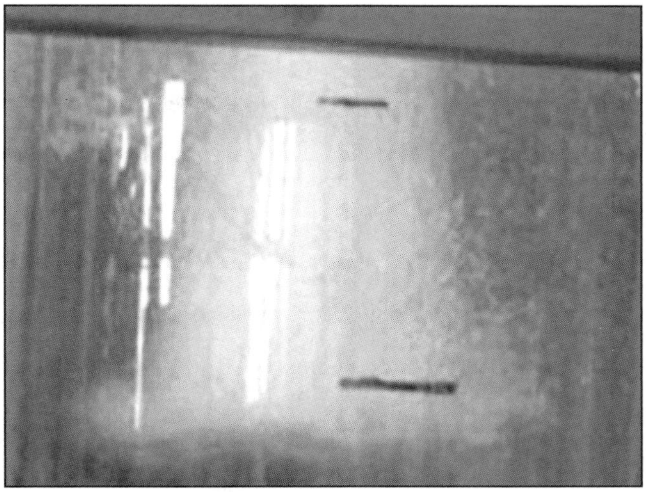

Source: Oteyaka et al. 2014a

Figure 29 Jet diffuser flotation cell modified downcomer exit

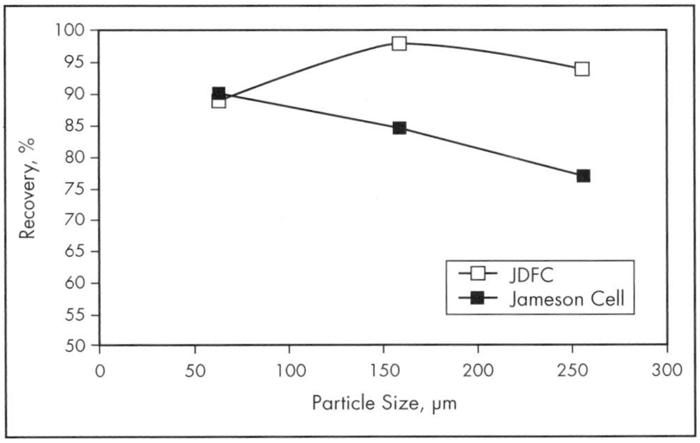

Source: Oteyaka et al. 2014a

Figure 30 Comparison of jet diffuser flotation cell and Jameson cell talc recovery by size

and a 2.1% Zn increase in plant concentrate grades, coupled with a 5.4% increase in Cu recovery and a 5.5% increase in Zn recovery (Viduetsky et al. 2014).

JET DIFFUSER FLOTATION CELL

The jet diffuser flotation cell (JDFC) was developed at Kütahya Dumlupinar University, Turkey, during a research project to increase coarse particle recovery in Jameson cells (Öteyaka et al. 2014a, 2014b). The main aim of the new design was to provide quiescent flow and more homogeneous dissipation of bubbles from the downcomer exit and decrease the turbulence occurring in the flotation tank. This was achieved by altering the shape of the downcomer exit to a fluted exit such that the diameter of the exit was approximately double the downcomer diameter (Figure 29). Pilot-scale operation of the JDFC was compared with an unmodified Jameson cell floating talc (Figure 30), indicating improved performance for particles >50 μm. In addition, the bubble plume exiting the downcomer decreased in length from two downcomer diameters for the unmodified downcomer to approximately half a downcomer diameter for the JDFC, reducing the risk of mineral-laden bubbles being carried into the tailing stream.

MODIFIED JET FLOTATION CELL

The modified jet flotation cell (MJC) was developed as part of a Brazilian project evaluating the use of Jameson cells for the recovery of oil emulsions from water in offshore oil platforms (Santander et al. 2011). Modifications are described with reference to Figure 31 and included placement of a concentric blind-end tube around the exit of the downcomer to allow floated material to enter immediately into the froth phase and a packed bed at the upper part of the concentric tube to stabilize the froth and facilitate the rise of the oil floc/bubble aggregates. Both laboratory- and pilot-scale tests were conducted with feed streams containing oil droplets of ~10 μm to 20 μm, and concentrations between 100 mg/L and 400 mg/L petroleum. The studies indicated that the MJC achieved an oil recovery of 85%, 5% higher than the Jameson cell.

FREE JET–TYPE FLOTATION SYSTEM

The free jet–type flotation system (FJFS) was evaluated by Istanbul Technical University, based on modifications to free jet flotation cells originally conceived at the Technical University of Berlin in the 1980s (Güney et al. 2016). The FJFS, as illustrated in Figure 32, makes use of a multiple orifice plate–single nozzle, operated at 100–120 kPa, to generate multiple free jets and entrain air from atmosphere via two air inlets. The aerated pulp enters a downcomer whose exit is close to the froth–pulp interface. Below the downcomer, a deflector plate is situated. The distance between the downcomer exit and the deflector creates a secondary turbulence zone. As the distance between the downcomer exit and deflector is increased, recovery increases with a deterioration in concentrate quality. Tests were conducted on asphaltite sourced from the Şırnak-Silopi region in southeast Anatolia, Turkey.

REFLUX FLOTATION CELL

The Reflux flotation cell allows the volumetric feed rate to be increased to nearly 10 times the typical conventional level, achieving low cell residence time, in the order of 25 seconds.

As described by Jiang et al. (2014) and Figure 33, bubbles approximately 500 μm in diameter are generated under a high shear rate within a rectangular, multichanneled, cuboidal downcomer. High liquid fluxes shear bubbles from a sintered surface sparger mounted flush to the channel wall before disengaging the downcomer flow into a vertical column. High feed fluxes, up to 15 cm/s, and high gas fluxes, up to 5.5 cm/s, ensure a high gas holdup beneath the downcomer. Bubble–liquid segregation is achieved using an arrangement of parallel inclined channels incorporated below the main vertical column.

A series of experiments was conducted to investigate flotation of fine coal particles from an extremely dilute feed of 0.35% solids using the Reflux flotation cell. The results demonstrated that the successful recovery of coal particles (92%) was possible (Dickinson et al. 2015).

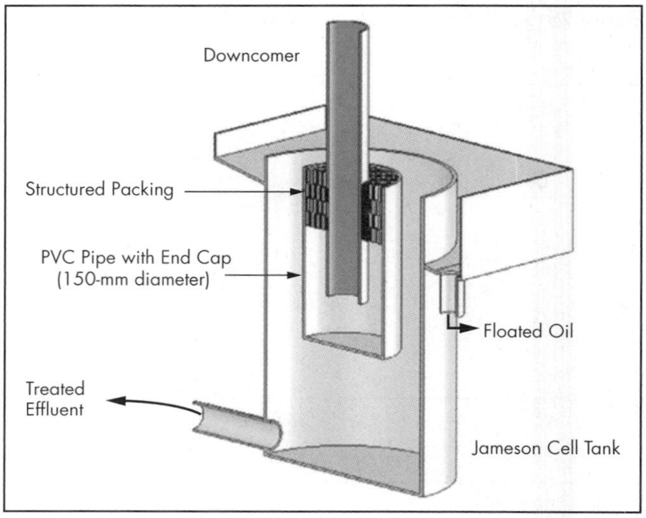

Source: Santander et al. 2011
Figure 31 Modified jet flotation cell

TURBOFROTH

The Turbofroth pneumatic cell was developed in South Africa as an alternative to tall flotation columns. The cell operated a double aeration system where primary aeration was accomplished either through downcomers as described by Steinmuller et al. 2000 (Figure 34) or via an external static mixer system in series with feed pipes that distributed the preaerated slurry into the top of the column. A second aeration step was provided at the base of the column via air spargers. The dual aeration system allowed reduction in machine height to between 3 and 5 m. During the 1990s, the Turbofroth was installed in numerous locations in Africa treating coal, copper, lead, zinc, and platinum group metals (Arnold and Terblanche 2001).

MULTICELL

The Multicell was developed by Anglo Coal during a flotation machine characterization program at Goedehoop Colliery, South Africa (Opperman et al. 2002). Throughout the program, researchers found that high mixing energy was required to achieve acceptable recoveries and optimal use of reagents. During Jameson cell tests, researchers found that the mixing power through the orifice plate in the downcomer was insufficient to achieve the recoveries expected on the Goedehoop fine coal.

Feed slurry is introduced into the chamber of a vertical spindle pump with its suction positioned at the base. Compressed air is blown into the suction of the pump at a controlled flow rate. Some of this air escapes from the pump alongside the shaft and is sheared by perforated paddles attached to the shaft. The slurry–air mixture exits the chamber through four perforated windows at the bottom of the pump shaft. This forms the first part of flotation in the Multicell and may be considered similar to the mixing and bubble generation zone within a mechanical cell. Pulp flows downward where it is drawn into the pump at the bottom of the tank. Inside the pump, the airstream is sheared into very small bubbles. The mixture of air and pulp is pumped to a second tank through a flow distributor. In this tank, mineral captured in the froth is allowed to separate from gangue material. This tank allows

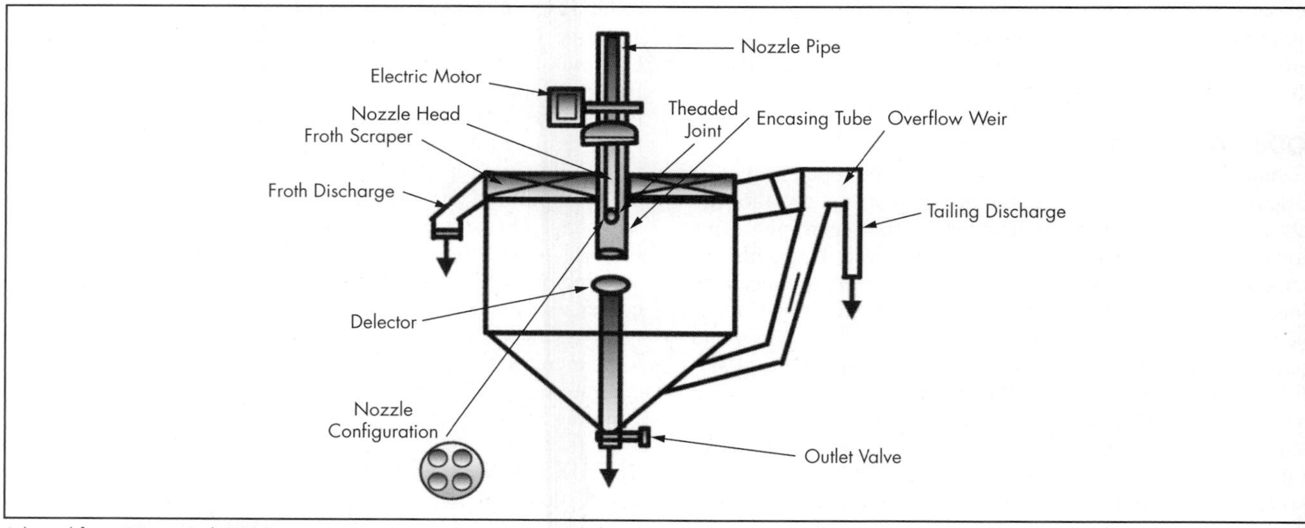

Adapted from Güney et al. 2002
Figure 32 Free jet type flotation system

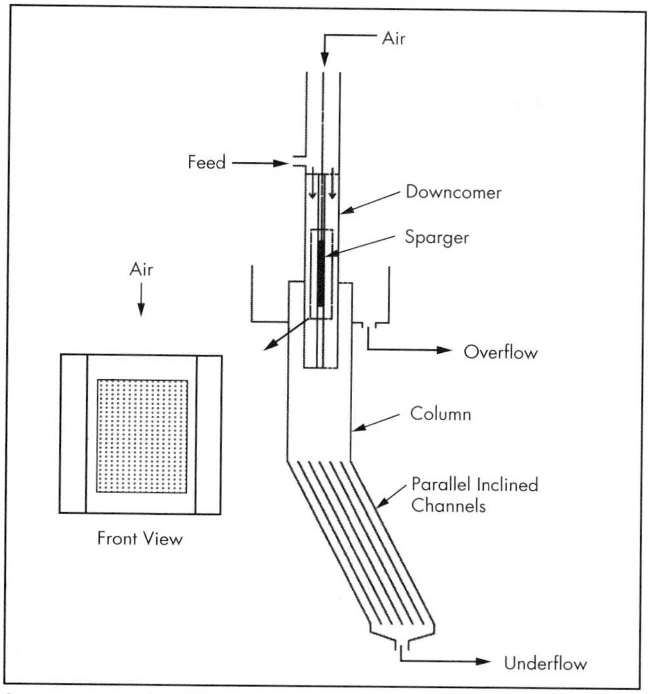

Source: Jiang et al. 2014
Figure 33 Reflux flotation cell

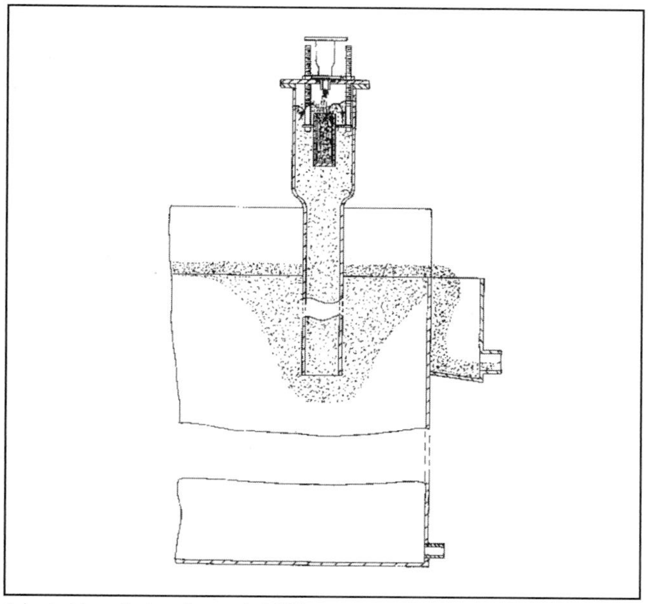

Adapted from Steinmuller et al. 2000
Figure 34 Turbofroth pneumatic cell

for an absolute quiescent zone, so little product is dislodged to tails and little mechanical entrainment of gangue in concentrate occurs. The Multicell was successfully scaled up to a feed capacity of 250 m^3/h.

CENTRIFUGAL FLOTATION CELL

Centrifugal flotation cells (CFCs) represent a series of pneumatic flotation machines where centrifugal forces are applied by various means to enhance flotation. Initially proposed in the 1980s, significant development was undertaken by Clean Earth Technologies in 1995 to separate oil from water. CFCs were subsequently applied to mineral froth flotation. The thesis by Jun-Xiang (2001) provides a detailed review of CFCs' early development and theory of operation.

The first CFC consisted of a rotating drum where separation occurred, a pulp feeding pipe, and an air injector. In practice, the pulp entered at the bottom of the rotating drum near the center. The centrifugal force caused the slurry to migrate outward across a screen of bubbles that were injected at high velocity through a jet at the bottom of the drum. Angular momentum was continuously being transferred to the fluid, and in a short time, the fluid inside the cell formed a rotating motion, creating a centrifugal force field in the range of 50–150g. The mineralized froth had a density smaller than that of the fluid, so it migrated inward, where it overflowed a weir and collected in the froth launder. The tailing was discharged through tailing ports (orifices) along the outer rim of the drum. One of the persistent problems with this CFC was the rapid blockage of the tailing discharge ports in the outer rim of the rotating drum, if the feed slurry had a pulp density higher than 30% solids.

To overcome this issue the CFC-1 was developed. This version is described by U.S. Patent 5,928,125 (Jian et al. 1999) and is illustrated in Figure 35. The conditioned pulp is pumped into the aerator (14) comprising a vertical sparger section (16) with a 2.0-µm compressed air tube sparger (22), a horizontal section (18), and a downcomer (20). The high superficial velocity slurry stream shears off the bubbles that have formed on the sparger wall. The aerated slurry continues downward through the downcomer into the bottom of a rotating bowl (36). An elongated shaft (24) extends vertically through a bearing housing and collar (26) mounted on the upper portions of the downcomer. The upper end of the shaft is connected to a variable-speed motor (28). The lower end of the shaft supports a rotating vessel (36). A froth column (42) with an upright annular wall (44) is placed between the downcomer (20) and the sidewalls of the rotating vessel. The upright annular wall provides a vertical weir that extends to a height above the sidewalls of the rotating vessel. As the vessel rotates, the air bubbles move toward the downcomer, rise to the pulp surface, and exit as overflow through the froth column in a concentrate launder. Tailing moves outward and upward, passes the annular passageway, and exits via the tailing outlet. The final prototype operated with a pulp feed rate of 0.3 m^3/h, an aeration rate up to 0.6 m^3/h, and a bowl rotating at between 100 and 400 rpm.

Further development led to the CFC-2, which had a substantially different structure. This version is described by U.S. Patent 6,126,836 (Jian et al. 2000) and is illustrated in Figure 36. The aeration unit was the same as for the CFC-1, but instead of rotating an internal tank, the downcomer was rotated. To prevent downward egress of froth-laden bubbles to tailing, a base plate was installed to close the bottom of the downcomer. Aerated pulp flowed horizontally through downcomer exit ports, with a plate parallel to the base plate enhancing lateral and radial discharge of the stream. Centrifugal force moved the rotating slurry to the wall of the flotation cell where mineral-laden bubbles rose to the froth layer and tailing descended to the tailing outlet.

Laboratory tests comparing the CFC-1 and CFC-2 performance against Denver flotation cell bench tests were

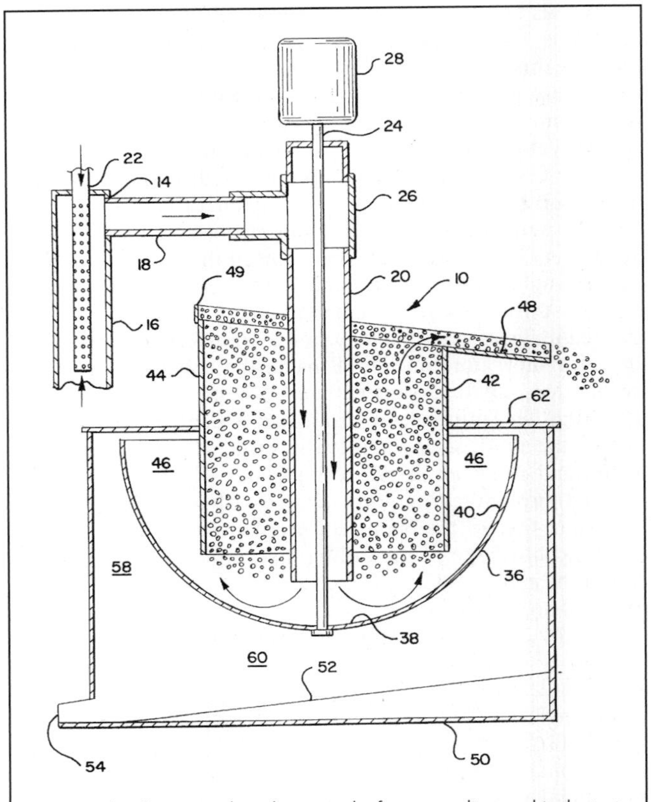

Note: Only relevant numbered items in the figure are discussed in the text.

Source: Jian et al. 1999
Figure 35 CFC-1

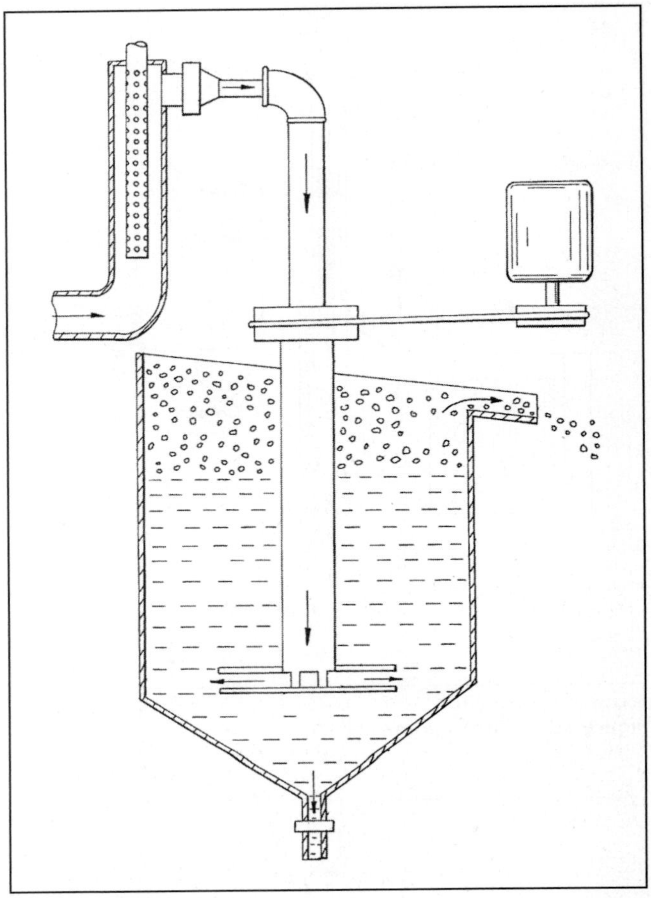

Adapted from Jian et al. 2000
Figure 36 CFC-2

conducted. These indicated that at comparable recovery and concentrate grade, the CFCs when floating galena achieved flotation kinetics 1.5–3.0 times greater than the Denver laboratory cell. When floating chalcopyrite, the CFCs achieved flotation kinetics 3.0–5.0 times greater than the Denver laboratory cell. Ultimate recoveries were comparable, with enhanced recovery of –10 µm mineral.

Research into CFCs has continued over the years, with more recent studies conducted by Yazd University in Iran (Ghaffari and Karimi 2012), where a significant computational fluid dynamics study was undertaken.

CYCLO-COLUMN CELL

The cyclo-column cell is another pneumatic flotation machine that utilizes centrifugal forces. As illustrated in Figure 37, the principal element is a circular, centrifugal column inside which a centrifugal force field is generated by pumping a pre-aerated flotation feed tangentially into its lower end. The column is closed at the bottom and open at the top. As a result of the tangential entry, the feed swirls inside the column as it moves upward. During the process, bubble–particle attachment takes place and the resulting bubble–particle aggregates collect around the central axis of the column. Inside the centrifugal column, there is a second column, called the bubble column, which captures the bubble–particle aggregates and transports them into the froth column above. The froth column is wider, enabling entrained gangue to drain from the froth,

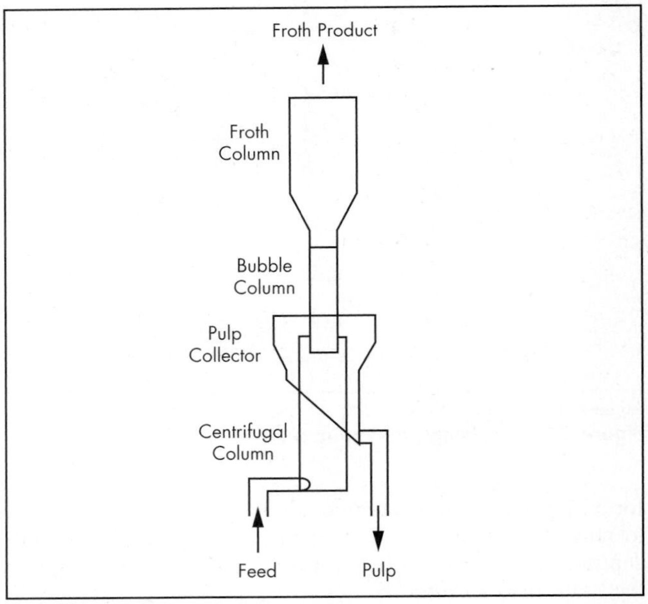

Source: Yalcin 1995
Figure 37 Cyclo-column cell

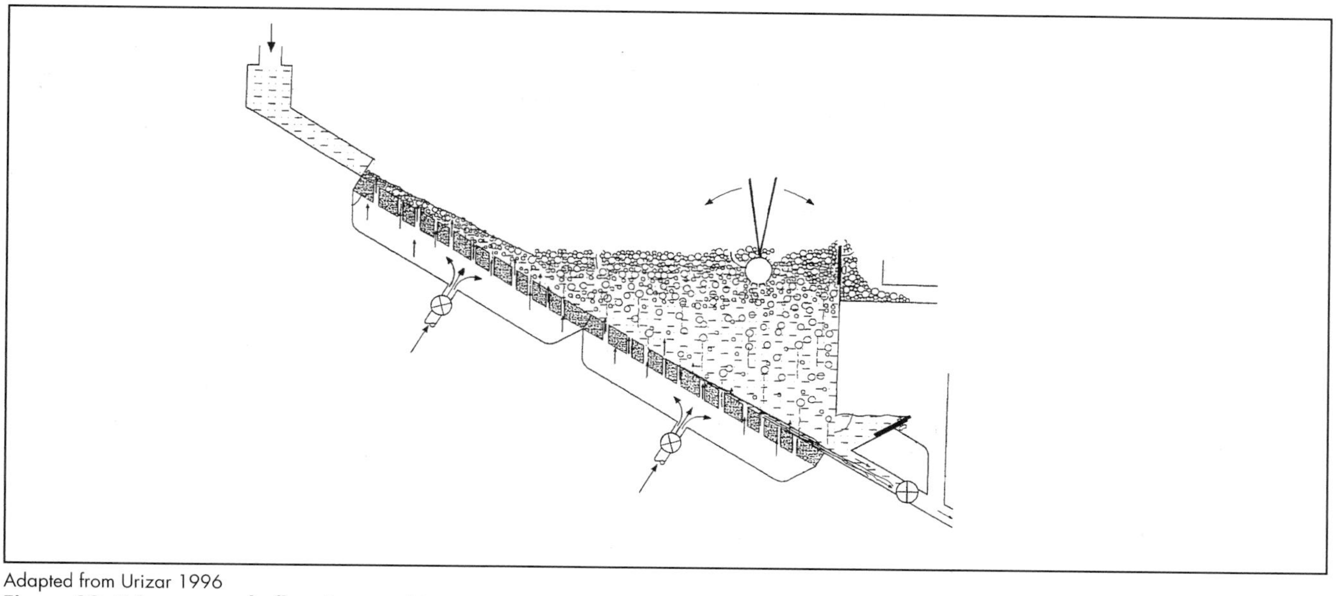

Adapted from Urizar 1996
Figure 38 Urizar cascade flotation machine

Courtesy of Amerigo Resources Ltd.
Figure 39 Cascade flotation at El Teniente

resulting in a clean concentrate. The material that does not enter the bubble column flows into a pulp collector and reports as tailing (Yalcin 1995).

CASCADE FLOTATION MACHINES

Cascade flotation machines historically have been simple and inexpensive. A modern design is shown in the development of Urizar (1996), where air is forced through a porous media into a thin film of flowing pulp (Figure 38).

Although rarely used in the modern flotation era, the size of the cascade flotation plant treating tailing from Codelco's El Teniente copper concentrators warrants mention. In 1992, Minera Valle Central S.A. started to recover copper from the fines fraction of El Teniente tailings using a simple cascade flotation system and a flotation cleaning circuit. Subsequent expansions introduced additional cascades, mechanical roughing cells, grinding and regrinding circuits, and flotation columns for cleaning. By 2003, the cascade flotation circuit had a capacity of 100,000 t/d and consisted of four lines of concrete cascade flotation cells, each 1.80 m × 0.70 m × 54 m in the first stage, and four lines in the second stage, each 1.80 m × 0.70 m × 126 m, providing 148 cascades in total (Figure 39). The cascade circuit reportedly produces approximately 4,500 t/d of low-grade concentrate at about 0.42% Cu. This concentrate is pumped directly to the rougher flotation (Maycock 2003).

PIPE FLOTATION

In 1996, Syncrude operated a pilot plant to evaluate the potential for addition of reagents and air to pipelines to allow initial bubble–droplet contacting. The pilot study showed that the bitumen recovery target of 95% was exceeded. The froth quality achieved was 59% bitumen and 20% solids (Mankowski et al. 1999). The new process operated at a temperature of 25°C, offering lower energy costs and lower capital costs. The arrangement is described by U.S. Patent 0249431 (Cymerman et al. 2006a).

The system was implemented in Syncrude's Aurora operations, which commenced in 2000. Oil sand slurry was transported from the mine site via pipelines (3–5 km) with a relatively high slurry velocity (4 m/s). Air was injected into the pipelines. The aerated oil sand slurry was then introduced into primary separation vessels through a tangential entry feed well. Bitumen droplets attached to air bubbles to form a primary froth for recovery. Wash water was added under the froth to reduce the amount of solids reporting to the bitumen froth. The implementation was partially successful with a bitumen recovery of 83% versus the target of 95% (Cymerman et al. 2006b).

RUBINSTEIN FPPM

Rubinstein's FPPM pneumatic cell was a Russian development in the 1970s (Rubinstein 2003). As shown in Figure 40, horizontal air spargers provided aeration at the base of the tank. The collection zone was divided into compartments of 0.04 m^2 cross-section by vertical partitions, allowing optimization of the slurry and airflow patterns in the cell. Two 40-m^3 cells (2.2 m × 4.4 m cross section by 6.0 m height) were installed in the Kuznetsk coal basin. Each cell had a unit capacity of approximately 400 m^3/h.

DE-INKING PNEUMATIC FLOTATION CELLS

Froth flotation is the most widely used separation process in modern paper mills, and the development of flotation de-inking cells has been pursued more aggressively than the technologies of any other segment of the pulp and paper industry. Flotation is traditionally the standard European de-inking system for old newspapers and is gaining increased use in other areas. The earliest work on utilization of froth flotation for de-inking of waste papers dates to the 1930s with the flotation de-inking patent by Hines (1933). The first commercial flotation de-inking system was installed in the United States in 1952; Europe installed its first one in 1959. As discussed by Jiang and Ma (2000), flotation de-inking has many similarities to mineral flotation with performance affected by factors such as pH, consistency, temperature, ink/fiber particle size, chemical types, water hardness, and air bubble size. Several types of de-inking pneumatic flotation machines are of interest to the minerals industry and are discussed in the sections that follow.

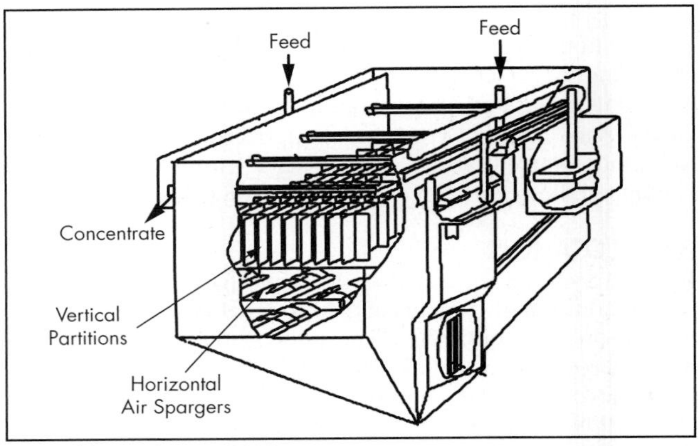

Adapted from Rubinstein 2003
Figure 40 Rubinstein FPPM pneumatic cell

Swemac Hellberg Cell

The Swemac Hellberg cell aerator provides independent control of bubble size and the air-to-liquid ratio, allowing improved selectivity. It features an annular flotation cell into which the liquid to be treated is introduced tangentially. As the liquid whirls around the cell, foam is blown radially inwardly by an airstream into a central region in which continued whirling of the foam facilitates its separation from the air. The foam drops into a recovery zone while the air is drawn off and recirculated (Hellberg 1980).

Escher Wyss Cells

The Escher Wyss FZ-U cell made use of a rectangular tank with an air distributor located at its base. Compressed air was forced through apertures in the distributor to form bubbles. One advantage of the cell was its ability to be stacked vertically (Holik and Mueller 1975). The CF (compact flotation) cell was designed for the removal of small ink particles, and the CFS (compact flotation–speck removal) cell was designed for the removal of large ink particles. Both cells are round with tangential stock feed through multiple lines, and foam removed by gravity overflow. The primary differences between these cells are in the aeration sections and in the tank construction. The CFC introduced in 1992 was a modification of the CF and the CFS cells. It is a totally enclosed modular design and may be installed in either a single cell or a stacked two-cell arrangement (Sauvé 1999).

Tubular Injector Cell, Elliptical Cell, and EcoCell

The tubular injector cell utilized a vertical venturi-type aerator with flotation conducted in a closed horizontal, cylindrical vessel. The tubular shape was subsequently compressed from cylindrical to elliptical to promote faster air removal. It was further refined into the EcoCell following replacement of the venturi with the Escher Wyss CFC design.

Shinhama Hi-Flo Flotator and Kamyr Gas-Sparged Cyclone

Both the Shinhama Hi-Flo flotator and the Kamyr gas-sparged cyclone (GSC) make successful use of the ASH for de-inking

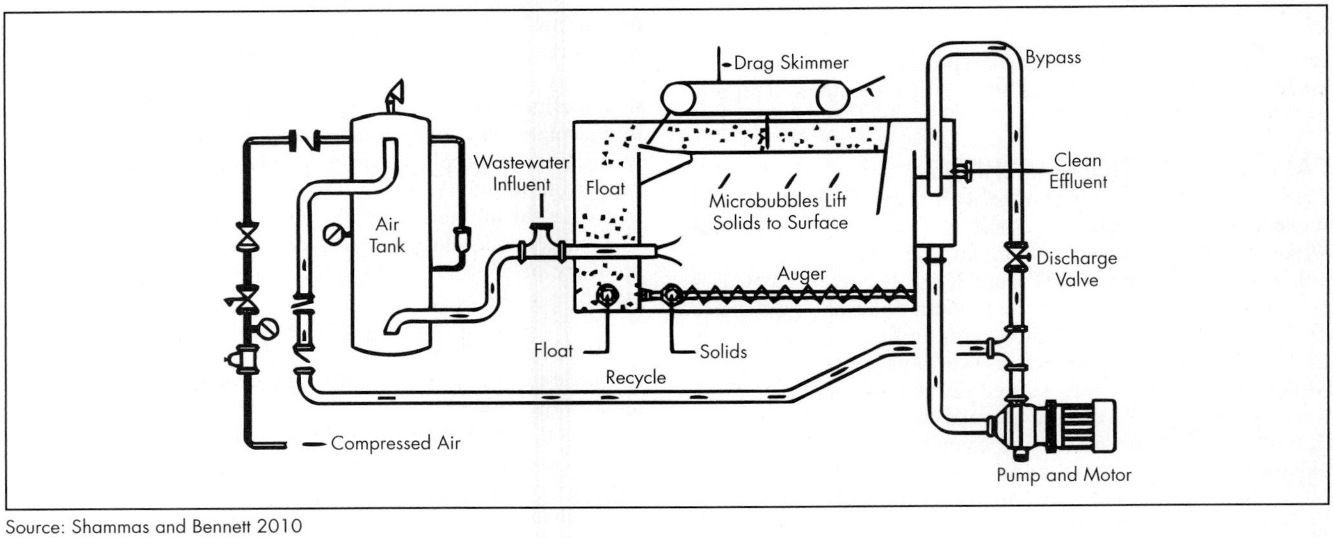

Source: Shammas and Bennett 2010
Figure 41 Dissolved air flotation arrangement

flotation. Where the GSC is a direct development of the ASH, the Hi-Flo is larger and does not use a porous medium for bubble generation.

DISSOLVED AIR FLOTATION

Dissolved air flotation is widely used to treat wastes from a wide variety of sources including paper making, refineries, ship's bilge and ballast waste, de-inking operations, metal plating, meat processing, laundries, iron and steel plants, soap manufacturing, chemical processing and manufacturing plants, barrel and drum cleaning, wash rack and equipment maintenance, glass plants, soybean processing, mill waste, and aluminum forming. The air is released from a supersaturated solution because of the reduction of pressure. As a result of this pressurization–depressurization, very small gas bubbles are formed and rise to the surface with oil and suspended solids attached. The amount of gas dissolved in solution and consequently the amount of gas released upon reduction of the pressure are both direct functions of the initial air pressure. Following pressurization, the water proceeds from the saturator, through a pressure-reducing valve, into the flotation basin; there the bubbles will first nucleate on any available low-energy sites on solid particles. If no sites are available, bubbles will nucleate homogeneously in the liquid phase. The bubbles will then grow until their growth is diffusion limited. Bubble sizes typically range from 45 to 115 μm, depending on operating conditions (Shammas and Bennett 2010). A typical dissolved air flotation arrangement is shown in Figure 41.

REFERENCES

Aksani, B. 1998. Flotation columns—Part 2: Comparative studies, application problems and alternative column designs. *Madencilik* 37(2):41–56.

Alizadeh, A., and Simonis, W. 1985. Flotation of finest and ultra-finest size coal particles *Aufbereitungs-Technik* 6:363–366.

Allmineral. n.d.(a). Pneumatic flotation Allflot MAF-45NS. Technical Drawing No. 20-4-006076, Project No. 5659. Duisburg, Germany: Allmineral Aufbereitungstechnik.

Allmineral. n.d.(b). Pneumatic flotation Allflot (promotional flyer). Duisburg, Germany: Allmineral Aufbereitungstechnik.

Amelunxen, R.L. 1993. The contact cell—A future generation of flotation machines. *Eng. Min. J.* 194:36–37.

Arnold, B.J., and Terblanche, N. 2001. Froth flotation with the Turbo flotation column. SME Preprint No. 01-163. Littleton, CO: SME.

Atkinson, B.W. 1994. Hydrodynamic characteristics of a plunging jet reactor. Ph.D. thesis, Department of Chemical Engineering, University of Newcastle, New South Wales, Australia.

Bahr, A. 1971. The mineralization of gas bubbles during the flotation process (Zur Mineralisation der Gasblasen beim Flotationsprozess). Habilitation thesis, Clausthal University of Technology, Clausthal-Zellerfeld, Germany.

Baker, M.W., and Willey, G.J. 1993. Potential Cadjebut plant performance enhancement by the air-sparged hydrocyclone. In *Proceedings of the XVIII International Mineral Processing Congress*. Melbourne, Victoria: Australasian Institute of Mining and Metallurgy. pp. 893–896.

Barnes, K.E., Colbert, P.J., and Munro, P.D. 2009. Designing the optimal flotation circuit—The Prominent Hill case. In *Proceedings of the 10th Mill Operators' Conference*. Melbourne, Victoria: Australasian Institute of Mining and Metallurgy. pp. 173–182.

Bennett, D., Crnkovic, I., and Walker, P. 2012. Recent process developments at the Phu Kham copper gold concentrator, Laos. In *Proceedings of the 11th Mill Operators' Conference*. Melbourne, Victoria: Australasian Institute of Mining and Metallurgy. pp. 257–272.

Blaschke, Z. 2000. Coal preparation in Poland: Present practice and perspectives. In *Mining in the New Millennium: Proceedings of the American–Polish Mining Symposium*. Edited by T.S. Golosinski. Oxfordshire, UK: Taylor and Francis. pp. 231–241.

Blaschke, W., and Gawlik L. 2004. Coal preparation in Poland—Current situation and development prospects. In *Proceedings of the 8th Conference on Environment and Mineral Processing, Part II*. Ostrau, Mähren, Czech Republic: Institut Environmentálního Inženýrství, pp. 305–312.

Brzezina, R. 1982. O warunkach i możliwościach stosowania maszyn flotacyjnych pneumomechanicznych do wzbogacania mułów węglowych. Zbiór referatów na czwarte naukowo-przemysłowe seminarium pt. Flotacja węgla. [The conditions and possibilities of the use of pneumatic flotation-mechanical machines enrichment of coal slurry. A collection of papers on the Fourth Scientific and Industrial Seminar, Flotation of Coal.] pp. 206–216.

Brzezina, R., and Sablik, J. 1991. The use of column flotation for coal slurry beneficiation. *Physicochem. Probl. Miner. Process.* 24:57–65.

Brzezina, R., and Sablik, J. 1994. The "FLOKOB" column flotation machine. In *Trends in Coal Preparation and Equipment—Proceedings of the 12th International Coal Preparation Congress*. Edited by W. Blaschke. London: Gordon and Breach. pp. 779–786.

Carbini, P., Ciccu, R., Ghiani, M., Tilocca, C., and Satta, F. 2001. Flotation of a sulphide ore using high velocity water jets. In *Proceedings of VI SHMMT/XVIII ENTMME*. Rio de Janeiro, Brazil: Centro de Tecnologia Mineral. pp. 365–369.

Carr, D., Harbort, G., and Lawson, V. 2003. Expansion of the Mount Isa Mines copper concentrator phase one cleaner circuit expansion. In *Proceedings of the Eighth Mill Operators' Conference*. Melbourne, Victoria: Australasian Institute of Mining and Metallurgy. pp. 53–62.

Carter, R.A. 2011. New flotation-cell designs and process-control tools keep bubbling to the surface. *Eng. Min. J. News*, August 11.

Changgen, L., and Bahr, A. 1992. Flotation of copper ore in a pneumatic flotation cell. *Miner. Metall. Process.* 9(1):7–12.

Cheng, G., and Liu, J.T. 2015. Development of column flotation technology. *J. Chem. Pharm. Res.* 7(2):540–549.

Chudacek, M.W., Marshall, S.H., Burgess, J., Burgess, F.L., and Stephenson, D.B. 1994. Initial testing of the Fastflot process at Pasminco Mining—Broken Hill. In *Proceedings of the AusIMM Annual Conference*. Melbourne, Victoria: Australasian Institute of Mining and Metallurgy. pp. 273–278.

Chudacek, M.W., Estrina, H., Marshall, S.H., Fichera, M.A., Burgess, J., and Burgess, F.L. 1995. Super-scavenging of zinc residue by the FASTFLOT process at Pasminco Mining, Broken Hill. In *Proceedings of the AusIMM Annual Conference*. Melbourne, Victoria: Australasian Institute of Mining and Metallurgy. pp. 207–212.

Cordes, H. 1997. Development of pneumatic flotation cells to their present day status. *Miner. Process. J.* 2(February):69–82.

Cowburn, J., Stone, R., Bourke, S., and Hill, B. 2005. Design developments of the Jameson Cell. In *Proceedings of the Centenary of Flotation 2005 Symposium*. Melbourne, Victoria: Australasian Institute of Mining and Metallurgy.

Cusack, B.L. 1968. The development of the Davcra flotation cell. In *Broken Hill Mines 1968*. AusIMM Monograph Series. Edited by M. Radmanovitch and J.T. Woodcock. Melbourne, Victoria: Australasian Institute of Mining and Metallurgy. pp. 481–487.

Cusack, B.L., and Oley, M.W. 1971. Operation of large Davcra cells in production and test circuits. Presented at the AusIMM Regional Conference, Adelaide, Australia.

Cymerman, G.J., Ng, S., Siy, R., and Spence, J. 2006a. Energy conservation measures at Syncrude oil sand processing operations. Paper 2191. Presented at the CIM Mining Conference and Exhibition, Vancouver, BC.

Cymerman, G.J., Spence, J.R., Ng, Y.M.S., and Siy, R.D. 2006b. Low energy process for extraction of bitumen from oil sands. U.S. Patent 20060249431 A1.

Davis, W.J.N. 1964. The development of a mathematical model of the lead flotation circuit at the Zinc Corporation Ltd. *Proc. Australas. Inst. Min. Metall.* 212(December):61–90.

Davis, W.J.N. 1966. The development of a mathematical model of the lead flotation circuit at the Zinc Corporation Ltd.—Discussion and contributions. *Proc. Australas. Inst. Min. Metall.* 220(December):79–85.

Demeekul, N., Sikong, L., and Masniyom, M. 2016. Influence of air flow rate and immersion depth of designed flotation cell on barite beneficiation. *Mater. Sci. Forum.* 867:66–70.

Dickinson, J.E., Jiang, K., and Galvin, K.P. 2015. Fast flotation of coal at low pulp density using the Reflux flotation cell. *Chem. Eng. Res. Des.* 101:74–81.

Ekof. n.d. Pneumatic flotation (promotional brochure). Bochum, Germany: Erz- und Kohleflotation GmbH.

Evans, G.M., Atkinson, B.W., and Jameson, G.J. 1995. The Jameson cell. In *Flotation Science and Engineering*. Edited by K.A. Matis. New York: Marcel Dekker. pp. 331–363.

Fairweather, M.J. 2005. The Sullivan concentrator—The last (and best) thirty years. In *Proceedings of the Centenary of Flotation Symposium*. Melbourne, Victoria: Australasian Institute of Mining and Metallurgy. pp. 835–841.

Finch, J.A. 1995. Column flotation: A selected review—Part IV: Novel flotation devices. *Miner. Eng.* 8(6):587–602.

Gaete, P. 2007. Siemens implements new Mo flotation cell at Los Pelambres. *Mining and Metals*. July 12. www.bnamericas.com/en/news/mining/Siemens_implements_new_Mo_flotation_cell_at_Los_Pelambres. Accessed February 2017.

Ghaffari, A., and Karimi, M. 2012. Numerical investigation on multiphase flow simulation in a Centrifugal Flotation Cell. *Int. J. Coal Prep. Util.* 32:3:120–129.

Gorzitzke, W.G., and MacPhail, N.G. 1989. Pneumatic flotation of some South African coals. Presented at the International Colloquium on Developments in Froth Flotation, Gordon's Bay, South Africa, August 3–4.

Gray, M.P., Harbort, G.J., and Murphy, A.S. 1998. Flotation circuit design utilising the Jameson cell. In *Mineral Processing and Hydrometallurgical Plant Design*. Edited by N. Jackson, G. Dunlop, and P. Cameron. Melbourne, Victoria: Australasian Institute of Mining and Metallurgy. pp. 217–225.

Grimes, A.W. 1991. Column flotation at Pasminco's South Mine, Broken Hill. In *Proceedings of the Fourth Mill Operators Conference*. Melbourne, Victoria: Australasian Institute of Mining and Metallurgy. pp. 129–132.

Güney, A., Güven, Ö, and Ergut, Ö. 2002. Beneficiation of fine coal by using the free jet flotation system. *Fuel Proces. Technol.* 75(2)141–150.

Güney, A., Burat, F., Kayaduman, M., and Kangal, O. 2016. Demineralization of asphaltite using free jet flotation. *Asia-Pac. J. Chem. Eng.* 12(1):42–49.

Hacifazlioglu, H., and Kursun, I. 2012. Beneficiation of low-grade feldspar ore using Cyclojet flotation cell, conventional cell and magnetic separator. *Physicochem. Probl. Miner. Process.* 48(2):381–392.

Harbort, G. 2006. A study of the hydrodynamics, bubble–particle interaction and flotation within the high intensity zone of a flotation machine and its implication for flotation machine design. Ph.D. thesis, University of Queensland, Australia.

Harbort, G., and Clarke, D. 2017. Fluctuations in the popularity and usage of flotation columns—An overview. *Miner. Eng.* 100:17–30.

Harbort, G.J., and Schwarz, S.E.E. 2007. Techniques for the optimisation of coal flotation. In *Proceedings of the Eleventh Australian Coal Preparation Conference*. Edited by P.N. Holtham. Melbourne, Victoria: Australasian Institute of Mining and Metallurgy. pp. 36–49.

Harbort, G.J., Jackson, B.R., and Manlapig, E.V. 1994. Recent advances in Jameson flotation technology. *Miner. Eng.* 7(2/3):319–332.

Harbort, G., Murphy, A., and Budod, A. 1997. Jameson Cell developments at Philex Mining Corporation. In *Proceedings of the Sixth Mill Operators' Conference*. Melbourne, Victoria: Australasian Institute of Mining and Metallurgy. pp. 105–113.

Harbort, G., Lauder, D., Miranda, J., and Murphy, A. 2000. Size by size analysis of operating characteristics of Jameson Cell cleaners at the Bajo De Alumbrera Copper/Gold Concentrator. In *Proceedings of the Seventh Mill Operators' Conference*. Melbourne, Victoria: Australasian Institute of Mining and Metallurgy. pp. 207–220.

Harbort, G.J., Manlapig, E.V., and DeBono, S.K. 2002. A discussion of particle collection within the Jameson Cell downcomer (*Trans. Inst. Min. Metall., Sect. C*: 111.) *Proc. Australas. Inst. Min. Metall.* 307(January/April):C1–C10.

Harbort, G.J., Manlapig, E.V., DeBono, S.K., and Monaghan, A.J. 2003. Air and fluid dynamics within a Jameson cell downcomer and its implications for bubble–particle contact in flotation. In *Proceedings of the XXII International Mineral Processing Congress*. Edited by L. Lorenzen and D.J. Bradshaw. Cape Town, South Africa: Southern African Institute of Mining and Metallurgy. pp. 715–724.

Hartmann, W., Wolfrum, S., Fleck, R., and Grossmann, L. 2012. Computational approach to bubble sizes in mixing nozzles working at high shear rates. Presented at PROCEMIN 2012, Santiago, Chile, November 20–23.

Heintges, S., Alizadeh, A., and Simonis, W. 1984. Process and flotation cell for flotation of coal and ore [translation from German]. German Patent 3140966 A1.

Hellberg, E.V. 1980. Apparatus for eliminating by flotation impurities in the form of solid particles contained in a liquid. U.S. Patent 4,186,094 A.

Hines, P.R. 1933. Process for deinking printed paper. U.S. Patent 2,005,742 A.

Holik, H., and Mueller, K. 1975. Flotation device for a fibrous suspension. U.S. Patent 3,865,719 A.

Horsfall, D.W., Btaszczyriski, S., Witold, Z., and Drogori, W.Z. 1994. The state of hard coal gravity cleaning process, Paper C1. Presented at the 12th International Coal Preparation Congress, May 23–27, Krakow, Poland.

Hu, W., and Liu, O. 1988. Design and operating experiences with flotation columns in China. In *Column Flotation '88—Proceedings of an International Symposium on Column Flotation*. Edited by K.V.S. Sastry. SME: Littleton, CO. pp. 35–42.

Huynh, L., Araya, R., Seaman, D.R., Harbort, G., and Munro, P.D. 2014. Improved cleaner circuit design for better performance using the Jameson Cell. In *Proceedings of the 12th AusIMM Mill Operators' Conference*. Melbourne, Victoria: Australasian Institute of Mining and Metallurgy. pp. 141–152.

Imhof, R., Battersby, M., Parra, F., and Sánchez-Pino, S. 2005. The successful application of pneumatic flotation technology for the removal of silica by reverse flotation at the iron ore pellet plant of Compañía Minera Huasco, Chile. In *Proceedings of the AusIMM Centenary of Flotation Symposium*. Melbourne, Victoria: Australasian Institute of Mining and Metallurgy. pp. 861–867.

Imhof, R., Fletcher, M., Vathavooran, A., and Singh, A. 2007. Application of IMHOFLOT G-cell centrifugal flotation technology. *J. South. Afr. Inst. Min. Metall.* 107(October):623–631.

Ityokumbul, M.T. 1998. *Evaluation of a Novel Contact Cell for Fine Phosphate Recovery*. Final report to the Pennsylvania State University. FIPR Publication No. 02-110-149. Bartow. FL: Florida Institute of Phosphate Research.

Jameson, G.J. 1988. A new concept in flotation column design. In *Column Flotation '88—Proceedings of an International Symposium on Column Flotation*. Edited by K.V.S. Sastry. SME: Littleton, CO. pp. 281–289.

Jameson, G. 2010. New directions in flotation machine design. *Miner. Eng.* 23:835–841.

Jameson, G.J., and Manlapig, E.V. 1991. Flotation cell design—Experiences with the Jameson Cell. Presented at the Fifth AusIMM Extractive Metallurgy Conference, Perth, Australia, October 2–4.

Jameson, G.J., Belk, M., Johnson, N.W., Espinosa-Gomez, R., and Andreaditis, J.P. 1988. Mineral flotation in a high intensity column. In *Proceedings of Chemeca 88, 16th Australian Conference on Chemical Engineering*, Sydney, Australia: Institution of Engineers. pp. 507–510.

Jameson, G.J., Harbort, G., and Riches, N. 1991. The development and application of the Jameson Cell. In *Proceedings of the Fourth Mill Operators' Conference*. Melbourne, Victoria: Australasian Institute of Mining and Metallurgy. pp. 45–50.

Jian, D., Yen, W-T., and Pindred, A.R. 1999. Centrifugal flotation cell with rotating drum. U.S. Patent 5,928,125.

Jian, D., Yen, W-T, and Pindred, A.R. 2000. Centrifugal flotation cell with rotating feed. U.S. Patent 6,126,836.

Jiang, C., and Ma, J. 2000. De-inking of waste paper: Flotation. In *Encyclopedia of Separation Science*. Cambridge, MA: Academic Press. pp. 2537–2544.

Jiang, K., Dickinson, J.E., and Galvin, K.P. 2014. Maximizing bubble segregation at high liquid fluxes. *Adv. Powder Technol.* 25:1205–1211.

Jungmann, A., and Reilard, U. 1989. Device for aerating fluids, in particular during flotation. U.S. Patent 4,840,753.

Jun-Xiang, G. 2001. Development and theory of centrifuge flotation cells. Ph.D. thesis, Queen's University, Canada.

KHD. 1997. *Pneumatic Flotation with Pneuflot Cells*. Promotional brochure 4-612e. Cologne, Germany: KHD Humboldt Wedag.

KHD. 2006. *Pneumatic Flotation with Pneuflot Cells*. Promotional brochure 4-613e. Cologne, Germany: KHD Humboldt Wedag.

Koh, P.T.L., and Schwarz, M.P. 2009. CFD models of Microcel and Jameson flotation cells. Presented at the Seventh International Conference on CFD in the Minerals and Process Industries, CSIRO, Melbourne, Australia, December 9–11.

Krieglstein, W. 2009. Higher yield at lower cost. *Met. Min.* 1:20–21.

Krieglstein, W., Grossmann, L., Fleck, R., and Hartmann, W. 2014. Flotation tests with porphyry copper ores using a laboratory-size hybrid flotation cell. In *Proceedings of XXVI IMPC 2012*. New Delhi, India: Indian Institute of Minerals. pp. 2540–2550.

Lane, G., Richmond, G., and Quilliam, B. 1991. Flotation cells at Hellyer—Horses for courses. In *Proceedings of the Fourth Mill Operator's Conference*. Melbourne, Victoria: Australasian Institute of Mining and Metallurgy. pp. 37–44.

Lelinski, D., Bokotko, R., Hupka, J., and Miller, J.D. 1995. Bubble generation in swirl flow during air-sparged hydrocyclone flotation. SME Preprint No. 95-173. Littleton, CO: SME.

Lynch, A.J., Watt, J.S., Finch, J.A., and Harbort, G.J. 2007. History of flotation technology. In *Froth Flotation: A Century of Innovation*. Edited by M.C. Fuerstenau, G.J. Jameson, and R.-H. Yoon. Littleton, CO: SME. pp. 65–91.

Lynch, A., Harbort, G., and Nelson, M. 2010. *History of Flotation*. Melbourne, Victoria: Australasian Institute of Mining and Metallurgy.

Mankowski, P., Ng, S., Siy, R., Spence, J., and Stapleton, P. 1999. Syncrude's low energy extraction process: Commercial implementation. In *Proceedings of the 31st Annual Meeting of the Canadian Mineral Processors*. Montreal, QC: Canadian Institute of Mining, Metallurgy and Petroleum. pp. 153–181.

Markworth, L, Jaspers, W., and Kottmann, J. 2007. PNEUFLOT®—Modern flotation technology in the 21st century. Presented at SAIMM Conference 2007, South Africa.

Maycock, A. 2003. Technical Review of *Operations at Minera Valle Central Rancagua Region VI, Chile for Amerigo Resources Ltd*. Chile: AMEC International (Chile) S.A.

Miller, J.D. 1981. Air-sparged hydrocyclone and method. U.S. Patent 4,279,743.

Miller, J.D., and Van Camp, M.C. 1982. Fine coal flotation in a centrifugal field with an air sparged hydrocyclone. *Min. Eng.* 34(11):1575–1580.

Miller, J.D., and Yu, Q. 1993. Preliminary evaluation of air-sparged hydrocyclone technology for scavenger flotation of western phosphates. In *Beneficiation of Phosphate: Theory and Practice*. Edited by H. El-Shall, B.M. Moudgil, and R. Wiegel. Littleton, CO: SME. pp. 349–360.

Miller, J.D., Kinneberg, D.J., and Van Camp, M.C. 1982. Principles of swirl flotation in a centrifugal field with an air sparged hydrocyclone. SME Preprint No. 82-167. Littleton, CO: SME.

Miller, J.D., Misra, M., and Gopalakrisbn, S. 1986. Gold flotation from Colorado River sand with the air-sparged hydrocyclone. *Miner. Metall. Process*. 3:145–148.

Miller, J.D., Wang, X., Lu, Y., Liu, N., and Yin, D. 1999. Plant-site evaluation of air-sparged hydrocyclone technology for phosphate flotation separation. In *Beneficiation of Phosphate: Theory and Practice*. Edited by H. El-Shall, B.M. Moudgil, and R. Wiegel. Littleton, CO: SME. pp. 195–209.

MMS (Maelwgyn Mineral Services). 2002a. *Delivery and Construction of IMF 45s in Chile*. General Information Release May 2002. Wales, UK: MMS.

MMS (Maelwgyn Mineral Services). 2002b. *Construction of G-Cell in South Africa*. General Information Release July 2002. Wales, UK: MMS.

MMS (Maelwgyn Mineral Services). 2003. *Commissioning of Iron Ore Flotation Plant*. General Information Release January 2003. Wales, UK: MMS.

MMS (Maelwgyn Mineral Services). 2004. *Caledonia Mining Installs MMS G-Cell and Aachen Aerator Technology at Barbrook*. General Information Release December 2004. Wales, UK: MMS.

MMS (Maelwgyn Mineral Services). 2006a. *MMS Supply Imhoflot V-Cell Plant for the Romeral Mine, Chile*. General Information Release February 2006. Wales, UK: MMS.

MMS (Maelwgyn Mineral Services). 2006b. *MMS to Supply Two Stage IMHOFLOT V-Cell Flotation Plant for Potash Recovery at Belaruskali*. General Information Release June 2006. Wales, UK: MMS.

Murphy, A.S., Honaker, R., Manlapig, E.S.V., Lee, D.J., and Harbort, G.J. 2000. Breaking the Boundaries of Jameson Cell Capacity. In *Proceedings of the Eighth Australian Coal Preparation Conference*. Edited by W.B. Membrey. Newcastle, New South Wales: Australian Coal Preparation Society.

Nycz, R., and Zieleńży, A. 2004. Kompania Węglowa S.A.—Coal beneficiation technology and quality of production. *J. Polish Miner. Eng. Soc*. July-December:2–19.

Opperman, S.N., Nebbe, D., and Power, D. 2002. Flotation at Goedehoop Colliery. *J. South. Afr. Inst. Min. Metall.* October:405–409.

Öteyaka, B., Şahbaz, O., Uçar, A., Bilir, K., and Gürsoy, H. 2014a. Design of jet diffuser flotation column in pilot scale. Presented at XXVII IMPC 2014, Santiago, Chile.

Öteyaka, B., Şahbaz, O., Uçar, A., Bilir, K., and Gürsoy, H. 2014b. Design of jet diffuser flotation column. In *Proceedings of the 14th International Mineral Processing Symposium*. Izmire, Turkey: Turkish Mining Development Foundation. pp. 255–259.

Pokrajcic, Z., Harbort, G.J., Lawson, V., and Reemeyer, L. 2005. Applications of the Jameson Cell at the head of base metal flotation circuits. In *Proceedings of the Centenary of Flotation Symposium*. Melbourne, Victoria: Australasian Institute of Mining and Metallurgy. pp. 165–170.

Rantucci, D.G., Akroyd, T.J., and Grattan, L.J. 2011. Carbon prefloat improvements at Century mine. In *Proceedings of the Metallurgical Plant Design and Operating Strategies (MetPlant 2011)*. Melbourne, Victoria: Australasian Institute of Mining and Metallurgy. pp. 430–441.

Ren, Z., and Yang, R. 2015. Reconstruction and practice of coarse slime system of Chengjiao Coal Preparation Plant. In *Proceedings of the 2015 AASRI International Conference on Industrial Electronics and Applications (IEA 2015)*. Paris: Atlantis Press. pp. 391–394.

Rohner, P. 1993. Lead–zinc–silver ore concentration practice at the Hilton Concentrator, Mount Isa Mines Limited, Mount Isa, Queensland. In *AusIMM Proceedings No. 1 1993*. Melbourne, Victoria: Australasian Institute of Mining and Metallurgy. pp. 23–26.

Rubinstein, J.B. 1995. *Column Flotation: Processes, Designs, and Practices*. Langhorne, PA: Gordon and Breach Science Publishers.

Rubinstein, J.B. 2003. Design, simulation and operation of column flotation. In *Proceedings of the 22nd International Mineral Processing Congress*. Edited by L. Lorenzen and D. Bradshaw. Cape Town, South Africa: International Mineral Processing Congress.

Salzmann, G., and Koch, W. 1964. Froth-flotation apparatus. Australian Patent Specification 275365.

Sánchez-Pino, S., Sanches-Baquedano, S., Rojos-Tapia, F., and Pereira-Cerda, O. 2008. G Cell pneumatic flotation technology: A high-performance Imhoflot technology for molybdenum applications. In *Proceedings of 40th Annual. Meeting of Canadian Mineral Processors*. Montreal, QC: Canadian Institute of Mining, Metallurgy and Petroleum. pp. 377–389.

Sanders, G.J., and Williamson, M.M. 1996. *Coal Flotation Technical Review*. ACARP Report C4047. Brisbane, Queensland: Australian Coal Industry's Research Program.

Santander, M., Rodrigues, R.T., and Rubio, J. 2011. Modified jet flotation in oil (petroleum) emulsion/water separations, Colloids and Surfaces A. *Physicochem. Eng. Aspects* 375(2011):237–244.

Sauvé, C.P. 1999. The effect of flotation deinking process parameters on air bubble size and deinking efficiency. ME thesis, Department of Chemical Engineering, McGill University, Montreal, QC.

Seaman, D.R., Burns, F., Adamson, B., Seaman, B.A., and Manton, P. 2012. Telfer processing plant upgrade—The implementation of additional cleaning capacity and the regrinding of copper and pyrite concentrates. In *Proceedings of the 11th Mill Operators' Conference 2012*. Melbourne, Victoria: Australasian Institute of Mining and Metallurgy. pp. 373–381.

Shammas, N.K., and Bennett, G.F. 2010. Principles of air flotation technology. In *Handbook of Environmental Engineering, Volume 12: Flotation Technology*. Edited by L.K. Wang, N.K. Shammas, W.A. Selke, and D.B. Aulenbach. New York: Springer Science. pp. 1–48.

Siemens AG. 2009. New flotation cell from Siemens: Higher material yield with lower costs. *Met. Min.* 2009:01.

Smith, T., Lin, D., Lacouture, B., and Anderson, G. 2008. Removal of organic carbon with a Jameson cell at Red Dog mine. In *Proceedings of the 40th Annual Meeting of the Canadian Mineral Processors*. Montreal, QC: Canadian Institute of Mining, Metallurgy and Petroleum. pp. 333–346.

Steinmuller, A., Moys, M.H., and Terblanche, A.N. 2000. Method and apparatus for aeration of liquids or slurries. U.S. Patent 6,092,667 A.

Taggart, A.F. 1945. *Handbook of Mineral Dressing*. New York: John Wiley and Sons. pp. 120–130.

Urizar, D. 1996. Procedure and apparatus for materials separation by pneumatic flotation. U.S. Patent 5,544,759 A.

Van Camp, M.C. 1981. Development of the air sparged hydrocyclone for flotation in a centrifugal field. M.S. thesis, University of Utah, Salt Lake City.

Van Deventer, J.S.J., Burger, A.J., and Cloete, F.L.D. 1988. Intensification of flotation with an air-sparged hydrocyclone. *J. South. Afr. Inst. Min. Metall.* 88(10):325–332.

Ventert, P.E., and Van Loggerenbergt, C. 1992. Modifications to the coal-preparation circuit at the Grootegeuk Coal Mine to improve its efficiency *J. South. Afr. Inst. Min. Metall.* 92(2):53–61.

Viduetsky, M., Maltsev, V., Garifulin, I., Purgin, A., Panshin, A., Kozlov, P., Tropnikov, D., Bondarev, A., and Sokolov, V. 2014. Application of pneumatic flotation machines CFM for dressing non-ferrous metals and zinc cakes. Presented at XXVII IMPC 2014.

Wu, D., and Ma, L. 1998. XPM-Series jet flotation machine. Presented at XIII International Coal Preparation Congress, Brisbane, Australia.

Wu, D., Yu, Y., Zhou, G., Zhu, J., and Jiang, M. 2010. A novel type of jet coal flotation machine. In *International Coal Preparation Congress 2010 Conference Proceedings*. Edited by R.Q. Honaker. Littleton, CO: SME. pp. 456–465.

Yalcin T. 1995. The effect of some design and operating parameters in the cyclo-column cell. *Miner. Eng.* 8(3):311–319.

Ye, Y., Gopalakrishnan, S., Pacquet, E., and Miller, J.D. 1988. Development of the air-sparged hydrocyclone—A swirl flow flotation column. In *Column Flotation '88—Proceedings of an International Symposium on Column Flotation*. Edited by K.V.S. Sastry. Littleton, CO: SME. pp. 325–332.

Young, M.F., Barns, K.E., Anderson, G.S., and Pease, J.D. 2006. Jameson cell: The "comeback" in base metals applications using improved design and flowsheets. In *Proceedings of the 38th Annual Canadian Minerals Processors Conference*. Montreal, QC: Canadian Institute of Mining, Metallurgy and Petroleum. pp. 311–332.

CHAPTER 7.4

Froth Management

Jason Heath and Kym Runge

ROLE OF FROTH IN FLOTATION CELLS

When air is released below the surface of a fluid, it will naturally want to move to the region of lowest pressure, usually the surface. An analogy is the rising of air in a freshly poured beer to form a frothy head. In the industrial process of froth flotation, air is added below the surface of a liquid–solid mixture and the air also rises to the surface of the vessel. However, in froth flotation the goal is for targeted hydrophobic mineral particles to be attached to and carried by the air bubbles to the surface. The buoyancy of the air–particle aggregates results in the accumulation of air, solid particles, and water in a low-density, three-phase mixture at the top of the cell. This layer is referred to as the froth phase or simply "the froth," and is the uppermost region in a typical flotation cell (Gorain et al. 2000). A flotation reagent known as frother is usually added to the process to stabilize bubbles formed in the flotation cell and also the froth phase itself.

The froth is usually visible from the top of a flotation cell and often used by flotation operators as a first indicator of cell performance. As hydrophobic particles tend to accumulate in the froth, they give it characteristics and properties that the operator can observe with the naked eye or determine with the use of a simple device such as a vanning plaque. Before the introduction of continuous online slurry analysis instruments, the ability to read the froth was a hallmark of a good flotation operator.

Several decades ago, experts acknowledged that there is a separation of both the slurry and froth regions in a flotation cell (Harris 1976), and the performance of each of these contributes to the overall cell performance. One of the key functions of the froth is to provide an environment for mineral-laden bubbles to accumulate and stabilize when they reach the pulp–air interface. Figure 1 shows a typical flotation cell froth structure, with new bubbles reaching the lower edge of the froth from the slurry below, interstitial water visible between the bubbles, and the uppermost bubbles being larger because of bubble coalescence.

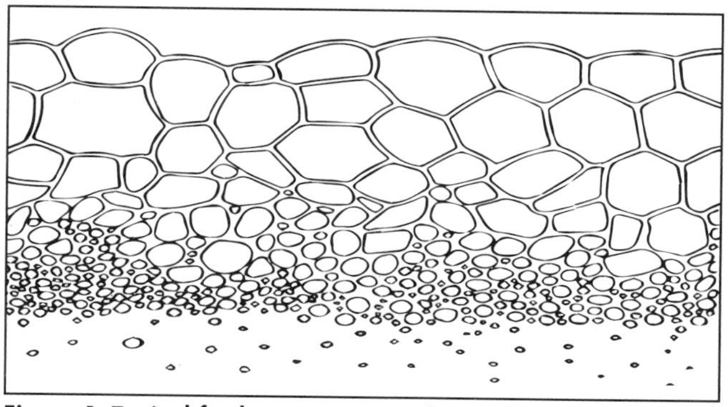

Figure 1 Typical froth structure in industrial froth flotation

To be effective in concentrating the targeted minerals, the froth must have sufficient stability to carry the floated particles until they can overflow the launder lip and be removed from the cell. Although not drawn in Figure 1, the launder lip would typically be configured to allow the uppermost portion of froth to overflow the lip, enabling continual removal of froth from the flotation cell.

The uppermost froth surface is normally where there is the highest concentration difference between the target hydrophobic minerals and other gangue minerals. This is due to the process of bubble coalescence and redistribution of the released particles. Bubble coalescence results in the release of the attached mineral particle load and interstitial water to the surrounding bubble surface, and the released particles must redistribute and compete for the remaining available bubble surface sites (Kaya and Laplante 1990; Barbian et al. 2005). This process of competition means that the most hydrophobically and physically suited (i.e., most energetically favorable) particles are then able to reestablish a bubble surface site, resulting in the rejection of less hydrophobic particles, which are typically lower in the target mineral and more hydrophilic. There is also a chance that, when no suitable bubble site is available, a hydrophobic particle will drop back into the bulk slurry phase. This is normally undesirable and is known as "particle drop-back."

Jason Heath, Senior Metallurgist, Talison Lithium, Greenbushes, Western Australia, Australia
Kym Runge, Program Leader, Separation, JKMRC, Sustainable Minerals Institute, University of Queensland, Brisbane, Queensland, Australia

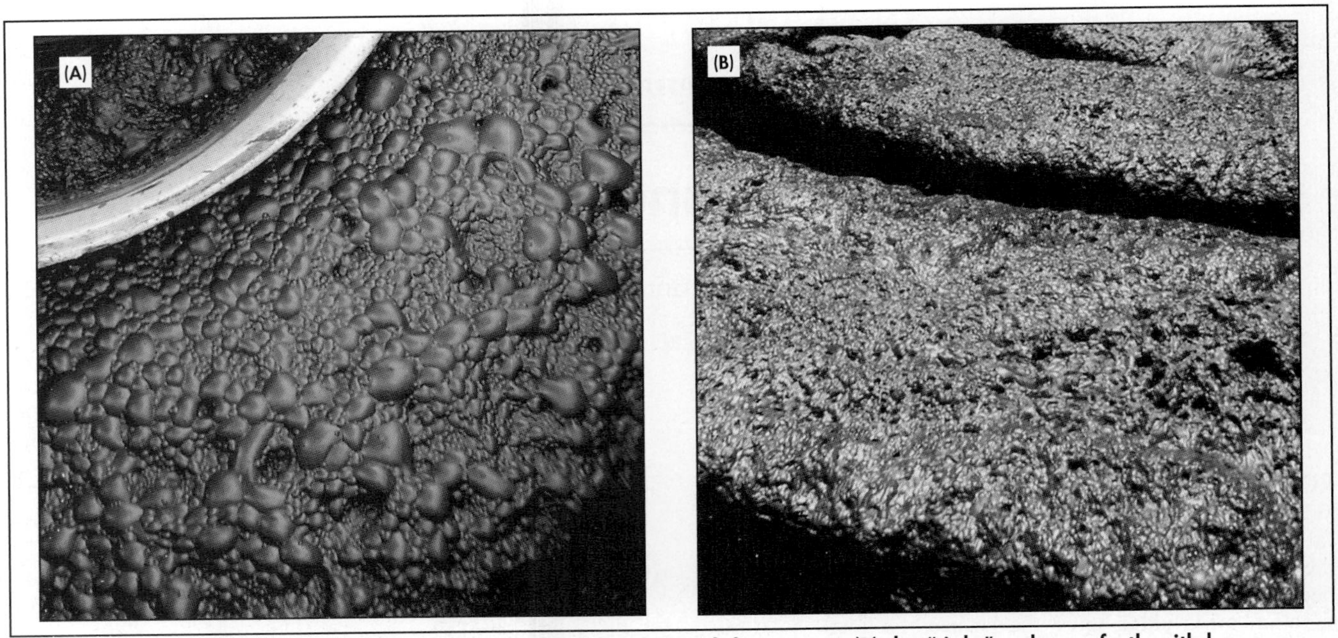

Figure 2 Compare (A) a well-structured froth with good mobility and drainage to (B) the "tight" or heavy froth with low mobility and static appearance

Aside from transportation of mineral particles, another important function of the froth phase is enabling selectivity of target minerals over entrained gangue. Suspended gangue particles in the upper regions of the pulp phase become trapped between the bubbles as they congest at the pulp–froth interface and are dragged upward into the froth. A well-stabilized and constructed froth provides drainage of interstitial water, taking with it any entrained hydrophilic gangue. The drainage of particles is strongly affected by the structure of the froth; for example, a "tight" froth composed of many small bubbles will have lower gangue drainage rates compared to froth composed of larger bubbles or of a mixture of bubble sizes including large bubbles (Ata et al. 2003). Additionally, the minimal bubble coalescence in a "tight" froth will further limit the difference in concentration between the lower and upper regions in the froth. An example of both of these froth types is shown in Figure 2.

The depth of the froth has a direct influence over the mineral concentration gradient seen from the bottom of the froth to the froth surface. Generally, with a deeper froth the average particle residence time in the froth is increased, which tends to increase the product grade. At the same time in a deep froth, particle recovery is usually decreased because of drop-back and other factors, negatively affecting recovery. These outcomes can be understood when considering the bubble coalescence and froth drainage principles as discussed previously in this chapter and in other publications (Ventura-Medina et al. 2003; Neethling and Cilliers 2003). Hence it is critical for a metallurgist to realize that froth depth is normally a very important parameter in optimizing flotation cell performance.

Consequently, without a suitably stable and well-constructed froth, no matter what froth depth is used, there will be poor froth recovery and transportation of attached mineral particles over the launder lip into the collection launder. This will generally result in overall poor performance of the flotation cell. Bubble coalescence in a froth is mainly a consequence of the rupture of thin liquid films, which occurs when the liquid content falls below a critical value. The particles in a flotation froth are largely responsible for producing the stable froth required for successful flotation, as they provide rigidity and impede water drainage. This property of particles in froth phases has been known for decades to play an important role (Lovell 1976) and was recently researched as part of the AMIRA P9 body of work (Achaye et al. 2015). The greater the amount of particles on the surface, the greater the froth stability, which is why flotation froths at the head of a flotation bank are significantly more stable than those farther down the bank where there are less hydrophobic particles reporting to the concentrate. Fine particles more densely pack on bubble surfaces than coarse particles, and this results in significantly higher bubble stability. Flotation performed with very little fines in the feed results in brittle froth and low froth recoveries (Rahman et al. 2012). Froth stability is optimum with moderately hydrophobic particles; highly hydrophobic particles were found to destabilize the froth in the work of Ata (2012).

Another important parameter of a froth is its mobility. Froth viscosity directly affects its mobility at the cell surface; a more viscous froth is less likely to move horizontally to the launder lip, resulting in a higher froth residence time and greater probability of froth collapse. Heavily laden dry froths with a small bubble size resist the flow of bubble over bubble that occurs as the froth moves toward the cell lip, resulting in a viscous immobile froth (Li et al. 2016). The froth mobility can be modified by changing the superficial gas rate and/or reagent concentrations, as these parameters affect the amount of water contained in the froth and modify the amount of particles present on bubble surfaces. For this reason, the use of depressants that minimize recovery of talcs or other gangue species can increase froth mobility and recoveries from a flotation cell.

Froth is also strongly affected by the type and concentration of frother, other reagents, and the process water quality, including the concentration of dissolved cations (Farrokhpay and Zanin 2012). These variables play a role in stabilizing the froth as well as affecting the amount of water drawn into the froth phase, which affects its mobility. The type of flotation cell, including the available froth surface area and transport distances, also play a role. For example, in larger flotation cells (i.e., greater than 100 m^3), it is widely accepted that a slightly stronger frother than normal will give better performance because of the typically larger transport distances involved with larger cells. Therefore, it is important for a metallurgist to adequately consider these points when looking to optimize flotation cell performance.

Particular attention to the froth-phase conditions is required when processing relatively coarse particles (>100 µm). Coarse particles exhibit much lower froth recoveries than fines because they readily drain through the froth when they detach from bubble surfaces, whereas detached fines tend to remain within the froth and still report to the concentrate. Coarse particle recovery tends to therefore be much more sensitive to froth-phase conditions, as demonstrated by Rahman et al. (2012) and Tabosa et al. (2013).

FROTH CARRY RATE, LIP LOADING, AND FLOTATION CIRCUITS

Froth is a three-phase mixture made up of air, solids, and water, with air contributing the largest component by volume. Froth stability is affected by many parameters. Regardless of the relative froth stability, as froth is largely made up of air, there are limits to how much mass froth can physically support and transport before it collapses upon itself.

Before 1995, lip loading was the only way metallurgists could measure a flotation cell's capacity to recover concentrate. The units used were metric tons per hour per meter (t/h/m) of lip length. By adopting column flotation parameters such as froth capacity in grams per square centimeter per minute (g/cm^2/min), Outotec came up with a new term for its TankCells using metric tons per square meter per hour (t/m^2/h). This is known as the *froth carry rate* (FCR) and is defined by how much material (metric tons) is transported by the froth over a specific area (square meters) in a given time frame (hours). Based on many years of experience together with metallurgical surveying of existing operating flotation operations, a design range of FCRs for different froth flotation duties has been established (Table 1).

Table 1 Typical froth carry rate ranges for banks of cells in base metal sulfide duties

Flotation Duty	Rougher	Scavenger	Cleaner
Froth carry rate, t/m^2/h	1.5–0.8	0.8–0.3	2.0–1.0

Source: Bourke 2002

The values given in Table 1 are design rules of thumb that refer to whole banks of flotation cells, and individual cells in the bank may have higher or lower values. Particle size and mineral type will also have an influence on these design values, as dry solid specific gravity will vary between different minerals; for example, galena = 7.5 t/m^3 is much denser than chalcopyrite = 4.2 t/m^3. The values given in Table 1 are representative of typical base metal sulfide operations with a feed particle size P80 of greater than 80 µm. For nonsulfide minerals, much higher FCRs have been reported. For example, in an iron ore reverse flotation application, the FCR varied from 10–40 t/m^2/h depending on which section of the flotation concentrator was surveyed (Miettinen et al. 2013). The FCR limits, both high and low, for a given duty are likely a function of particle concentration, particle size distribution, and particle specific gravity. However, no first-principal method of prediction of these limits exists, though attempts have been made by some researchers (Espinosa-Gomez et al. 1988).

A related concept to FCR and mass transport of the froth is *lip loading*. This is defined as the mass of solids passing over the launder lip per hour per unit length, and is expressed in units of metric tons per meter per hour. When considering if a given flotation cell froth surface area is suitable, it is useful to think first in terms of the overall "froth handling strategy." After the FCR has been confirmed in the correct range, the lip loading should also be calculated and kept at less than 1.5 t/m/h.

FCR and lip loading are important parameters to be aware of to avoid a situation where there is too much or too little solids in the froth phase. Both of these conditions have a negative effect on either the froth stability or mobility and hence performance of a flotation cell (Morar et al. 2014). With insufficient solid particles in the froth phase, the froth stability is significantly reduced and the froth is brittle and breaks down readily. This is often seen toward the end of a rougher bank of cells where very little mass is floating.

Conversely, when there are too many solids in the froth, the froth becomes immobile and the weight of the solids can overcome the carrying capacity of the froth, leading to froth collapse. This sometimes occurs at the beginning of a rougher bank if the flotation kinetics are extremely fast and a lot of mass is trying to float, or also in cleaning duties. Both of these situations result in a significant increase in particle drop-back to the bulk slurry phase, and poor transport of the mineral particles to the launder lip (froth recovery), which reduces the overall froth recovery and hence flotation cell performance.

The FCR must be taken into account when considering a typical base metal or coal flotation circuit. As eluded to earlier, the feed slurry will normally contain both fast- and slow-floating components, and this leads to a curved recovery response plot when performing a laboratory test, as shown in Figure 3.

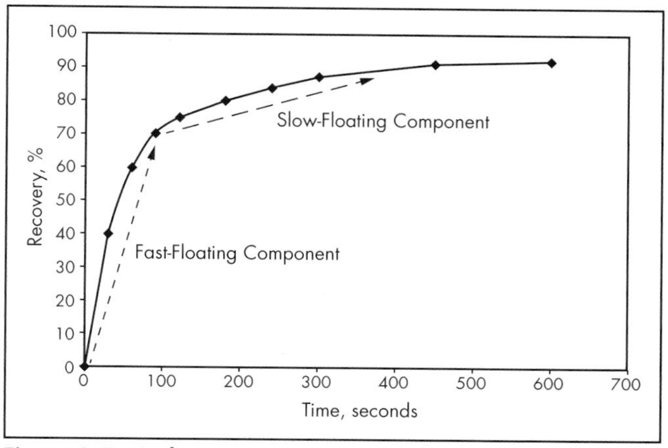

Figure 3 Typical recovery versus time plot for sample with fast- and slow-floating components

In the full-scale circuit, the recovery response shown in Figure 3 means that the majority of the mass will be recovered from the first few cells. So in a typical bank of six cells, the first two cells may be responsible for 70%–80% of the valuable mineral recovered from the entire bank, whereas the last four cells would recover the final 20%–30%. This is a significant difference and shows that a different approach is required in terms of how these cells are designed and operated.

The implications of this recovery distribution can often lead to problems with the FCR in a full-scale plant. If a flotation circuit was designed with all six cells having the same froth area, by simply calculating an average FCR over all cells using the aforementioned example, the first two cells would probably be overloaded (i.e., FCR too high) and the last few cells would be underloaded (i.e., FCR too low). Neither scenario is ideal, and both would lead to less-than-optimum flotation cell performance. The best solution is to tailor each flotation cell design to the expected mass recovery using detailed and robust laboratory or pilot test data. In practice, this usually means adjusting the flotation cell launder type and/or cell crowding configuration down the bank to achieve the best design. It is also beneficial to use an increasing gas rate profile as use of lower gas rates in the cells at the head of a bank reduces the risk of froth overloading and excessive entrainment recoveries. Selectivity and recovery have both been observed to increase in a flotation bank with an increasing gas rate profile to that achieved when the same gas rate was used in all cells (Cooper et al. 2004). Staged addition of collector down the bank of cells can also be used to better manage the distribution of the mass pulled to concentrate.

Having the FCR outside of the ideal range is also an issue at sites where the feed grade has changed significantly since the original process design was conducted. For example, if the feed grade is 50% lower than when the flotation circuit was designed, it can be expected that, all things being equal, the mass recovery would also decrease by a similar amount. The result will be that FCRs are probably lower than the ideal ranges, and froth stability and flotation cell recovery suboptimal. Fortunately, this situation can be rectified quite easily by retrofitting additional crowding to the flotation cells (to reduce the froth surface area and increase the FCR), and this is often conducted by equipment vendors.

Another positive aspect when retrofitting froth crowding or additional froth launders is that this usually results in a reduced average froth transport distance. Froth transport distance is the distance that a particle may have to travel from the froth surface to the closest launder lip, and it is heavily influenced by the selected launder configuration. Therefore, all other things being equal, the average residence time of a particle in the froth will be reduced and the probability of particle detachment and drop-back to the slurry phase is reduced. This can be especially important in improving recovery of leanly liberated coarse particles at the back of a rougher circuit.

WAYS TO MANIPULATE FROTH

FCR and lip loading in a flotation cell are important parameters when considering froth recovery and cell performance. These froth parameters are determined by the amount of mass recovered from a given flotation cell, together with the froth area or lip length available, which is in turn determined by the flotation cell design. For this reason, flotation cell vendors often have several launder types for a given cell size. Consider the cross-sectional froth areas of flotation tank cells with different launder types outlined in Figure 4. The flotation tank cells outlined in the figure all have the same tank diameter and similar central froth crowder size. The launder designs depicted represent some of the most commonly available launder configurations used in flotation cells, and it should be clear that, as drawn from left to right, the froth area decreases significantly because of the launder(s) placement and also the tank shape providing froth crowding.

The different launder configurations shown in Figure 4 do not significantly affect the tank volume and hence flotation residence time. Therefore, the different froth surface area and lip lengths can be used to optimize the flotation cell design for particular duties in a plant based on the expected FCR and lip loadings (e.g., roughing, scavenging, and cleaning). Some vendors have taken this concept a step further by offering multiple launder types within the same cell, as well as concentric designs.

Most of the strategies and comments mentioned so far in this chapter are for flotation cell design in terms of froth management for a greenfield project where new equipment will be designed to suit. However, there are also strategies for manipulating froth in an existing operation. In many ore bodies there is a very fast-floating component in the ore, which may overload the first cell resulting in very high FCRs. Operational parameters such as gas rate, froth depth, and the amount and type of reagent can be manipulated to change the mass recovery profile down a bank with the objective of improving concentrate grades and recoveries. Staged reagent addition can also be employed.

Another way to permanently adjust the amount of fast-floating material reporting to the first flotation cell is to split the new feed to both the first and second cells; that is, send a portion of the new feed straight to the second cell. This is known as split feeding. Split feeding takes advantage of the froth area of both the first and second cell to recover any

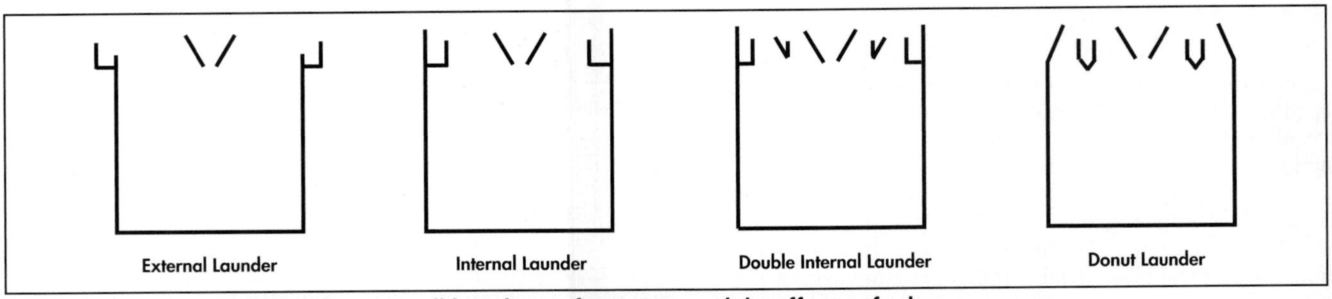

Figure 4 Commonly available flotation cell launder configurations and the effect on froth area

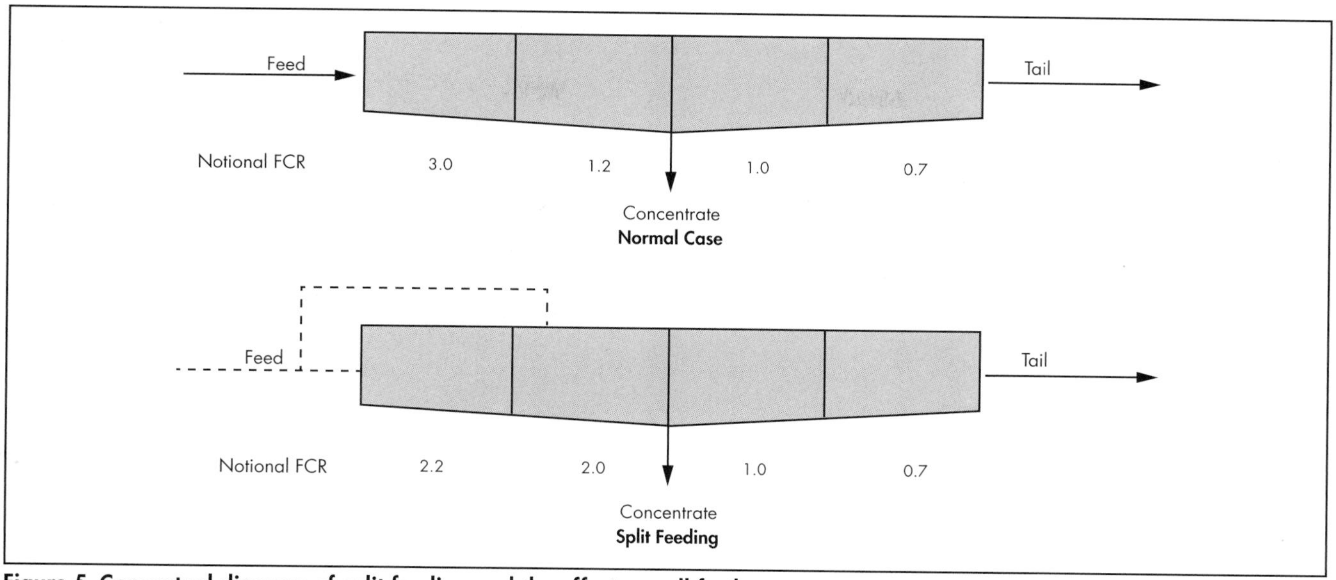

Figure 5 Conceptual diagram of split feeding and the effect on cell froth carry rate

fast-floating component in the ore instead of just the first cell. This thereby tends to average out the FCRs over both cells and produce an overall higher-grade concentrate (Figure 5).

The split feeding concept is particularly relevant for cleaner duties that have very fast kinetics. It is also relatively easy to implement as a trial in an operating flotation circuit using temporary piping and valves. With split feeding, sending a portion of the new feed directly to the second cell will reduce the number of flotation stages and residence time for that component of the feed. If the tailings from the bank of cells are open circuit, then measures should be taken to ensure that a suitable number of flotation stages and residence time remain after the second cell to minimize slurry short-circuiting and a potential loss in recovery.

In a typical base metal flotation concentrator scavenging circuit, it is commonplace to target the recovery of leanly liberated, coarse composite particles. Because of their lean liberation and coarse particle size, these particles are usually weakly attached to a bubble and hence easily detached in the froth phase, and therefore poorly recovered into the launder. One way to increase the recovery of these particles is to transport them as quickly as possible over the launder lip when they reach the froth phase, which can be done by minimizing the froth depth as well as froth transport distance.

Another method to optimize flotation cell design to suit the expected froth characteristics is to alter the overall tank shape. Two main mechanical flotation cell tank types are currently available: those based on a cylindrical tank (TankCell) and those based on a trough-type tank. An example of each is shown in Figure 6.

There are several differences between the tank types shown in Figure 6 in terms of launders, air control, reagent addition, and so forth. In terms of froth management, the trough-style tank generally has a much higher surface-area-to-volume ratio than a tank cell and hence is particularly suited to duties with high mass recoveries, such as coal flotation. Depending on the comparative cell sizes, the transport distance may also be significantly longer for a trough-type flotation cell. Hence there are several considerations when evaluating which tank shape is the best suited for a particular flotation application.

In terms of direct froth quality monitoring, most recent developments in this area have come from advances in camera technology and algorithms to continuously evaluate the camera output. Many vendors now offer small, easy-to-install video cameras designed to sit above the froth surface and continuously monitor the froth characteristics, which can then be used for process control. Froth properties that can be measured include froth velocity, bubble size, bubble stability (breakage), froth color, froth direction, and so on. To date, the most widely used parameter for process control is simply the froth velocity, which is normally used as a proxy for the actual mass recovery from the flotation cell.

The froth velocity (and froth recovery) can generally be raised by increasing the amount of air being added to the flotation cell. Too much air, however, has a detrimental effect on recovery, as it results in excessive turbulence at the pulp–froth interface, disturbing the froth phase and leading to excessive entrainment of unwanted mineral particles into the collection launder as well as more coarse particle drop-back (Hadler et al. 2010). The optimum air rate needs to be established for a particular flotation cell through experimental testing. The optimum air rate will have been exceeded if boiling is observed at the froth surface, and therefore this condition should be avoided.

Another way to increase froth velocity is to reduce the froth bed depth by increasing the flotation cell slurry height set point. With a shorter froth depth, the froth residence time is reduced and the solids concentration on the froth surface is also reduced, hence the froth has more mobility and tends to move to the launder faster (i.e., faster froth velocity). A shallower froth also means there is less bubble coalescence, and therefore, usually a lower-grade product is removed from the flotation cell but often at a higher overall recovery. To maximize recovery, Crosbie et al. (2009) recommend that the froth depth used should not exceed 70% of the maximum achievable froth height measured in a froth stability column. At this point, froth recovery was observed to decrease significantly

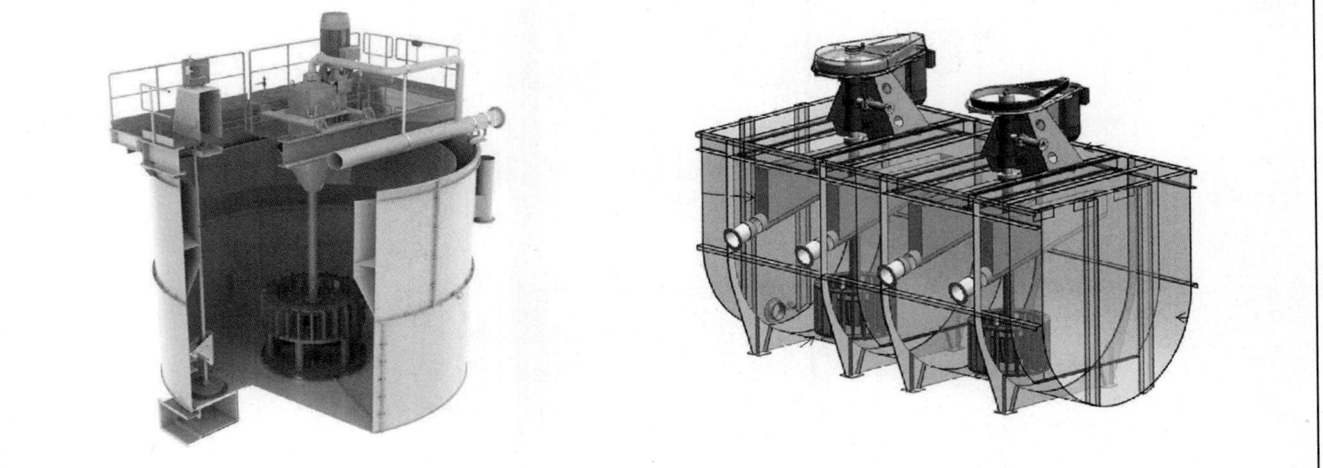

Courtesy of Outotec Pty Ltd.

Figure 6 Two main types of mechanical flotation cells used in the industry

because of significant froth coalescence. This point often cannot be maintained in a scavenger because the recommended froth depth is so low that it would result in pulping and therefore excessive gangue entrainment. Accurate froth depth measurement and control philosophies are therefore important in scavenging applications to enable low froth depths to be employed.

Air addition rate and froth depth are the two main variables to adjust froth velocity in a flotation cell fitted with a froth camera. It is also routine to change the froth properties by adjusting the amount of reagents added to a flotation cell. Generally, increasing the amount of frother or decreasing the amount of collector will increase the froth velocity. When using these variables to control the froth performance down a flotation bank, one also must consider the need to distribute the mass recovered by each cell to optimize froth stability and mobility and reduce gangue entrainment.

These reagents (and others, such as depressants), together with components in the ore itself, can often have synergistic effects. Consequently, the best strategy for reagent addition and control is usually determined on a plant-to-plant basis. To increase froth velocity, the cost of adjusting air addition rate or froth depth typically is significantly less than the cost of adding additional frother; therefore, one should adjust the air and froth depth first before adding additional reagents.

FROTH WASHING AND HANDLING

Two important aspects of froth management are the processes of froth washing and mechanical froth removal. Froth washing in the context of column flotation is discussed in Chapter 7.2, "Column Flotation." The discussions here focus mainly on mechanical flotation cells, such as Outotec high-grade cells, which have a deeper froth depth design and utilize froth washing.

The process of froth washing is carried out by introducing clean water into the froth, with that clean water displacing some or all of the water already in the froth. Froth washing improves metallurgical performance in flotation cells in two ways: first by increasing the flow of water draining between bubbles and washing out entrained gangue; and second by increasing bubble coalescence, which results in a crowding effect on the reduced bubble surface (Kaya and Laplante 1990). Froth washing is particularly useful in washing penalty elements present as hydrophilic minerals back into the slurry phase (e.g., some uranium and fluorine minerals). Several operating plants would incur concentrate smelting penalties if they had not implemented froth washing to remove these elements.

The increased bubble coalescence caused by froth washing also affects the flotation cell froth recovery. Therefore, although froth washing can technically be used anywhere in the flotation circuit, it is usually used in flotation duties where the product grade is of prime importance, such as the recleaning or final cleaning stages. Also, the additional water added to the froth during froth washing does result in the bulk slurry density decreasing, and this reduces the flotation bank residence time. In a greenfield project, this can be allowed for by increasing flotation capacity. However, an existing operation contemplating froth washing should take this extra volume of water into account when considering plant residence time. The overall effect of froth washing is usually a higher-grade product to be recovered to the launder at a lower cell stage recovery.

Typically, froth wash water is added to froth by sprays or streams of water delivered by a froth wash-water delivery system. The water can be introduced from a shallow trough suspended above the froth surface or directly from wash-water piping. An example of each of these systems is shown in Figure 7. Fine sprays are not normally used because of their propensity to become blocked with fine particles and scale from the water. In both of the examples in Figure 7, water is added above the froth surface.

Another way of introducing wash water to the froth is submersed delivery, that is, below the froth surface. This can be done by having the delivery system submersed below the froth surface as is sometimes seen in column cells. Another method of introducing wash water above or below the froth

Figure 7 Suspended (A) wash-water trough and (B) wash-water ring providing wash water

is via a somewhat novel combined froth scraping and washing system developed by Outotec (Figure 8). The combined froth scraping and washing system is typically fitted under the bridge in a tank-style flotation cell, and has an auxiliary drive and slewing ring arrangement to drive the blades continuously around the froth area. The blades sweep out an arc and continuously push the uppermost layers of froth toward and over the launder lip.

Fitted behind the scraper blades are froth wash-water channels. These channels fill with wash water and allow it to run out to the full length of the channel. Adjustable stems with exit holes are fitted, which allow the wash water to run down the stems and fan out. Both the blades and the stems are height adjustable, and the stems can be adjusted to above or below the scraped froth level.

The number of blades and wash channels are adjustable to suit the cell size and froth characteristics. To date, this combined froth scraping and washing system has found use in particularly tenacious froths (e.g., coal flotation, molybdenum cleaning), where mechanical removal of the froth has shown to improve froth recovery and cell performance.

Another mechanical method of froth removal that has been used in flotation machines for many decades is rotating froth paddles. These froth paddles are typically mounted adjacent and parallel to the launder lip (Figure 9). The paddles rotate at a fixed speed to assist with scooping froth over the lip into the collection launder.

The use of froth paddles is limited to trough-shaped flotation cells, as they are mounted adjacent to the launder lip. Froth mobility can often be an issue with trough-style cells because of the lack of internal froth crowding directing the froth flow. These paddles are still used today at processing plants with very sticky froth (e.g., rare earth oxide flotation, spodumene flotation, coal flotation) and are also seen in very coarse flotation duties (e.g., apatite flotation) where they help lift coarse particles of up to several millimeters in size over the launder lip.

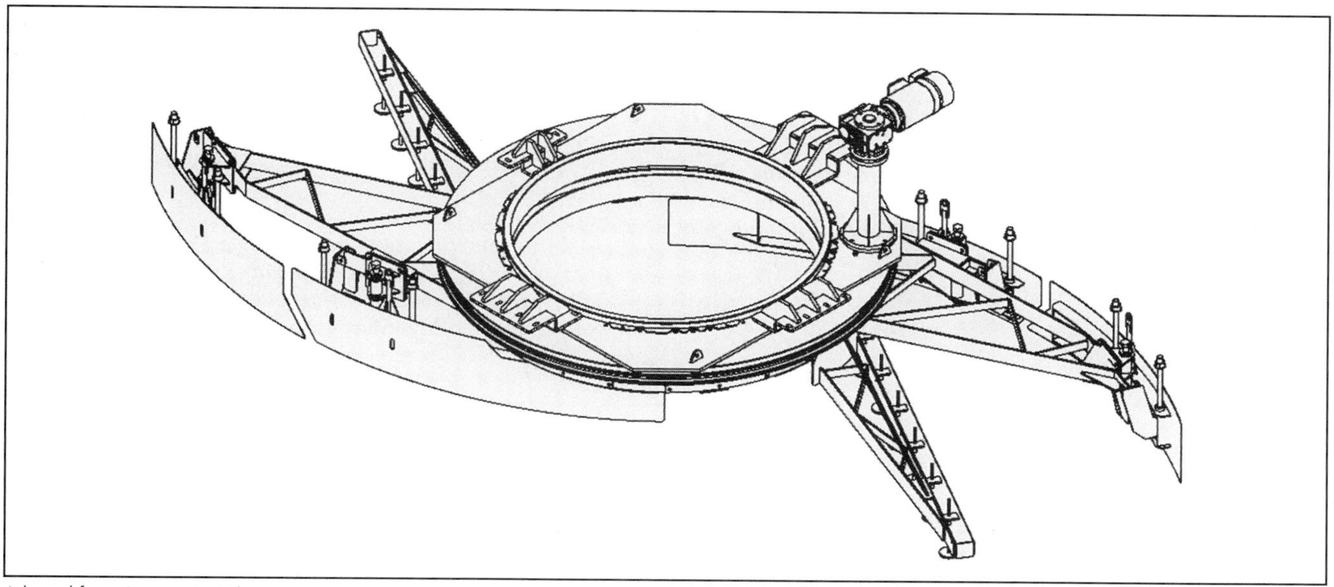

Adapted from Outotec Pty Ltd.

Figure 8 Combined froth scraping and washing system

Figure 9 Froth paddles together with trough-style flotation cells (partially empty)

REFERENCES

Achaye, I., Wiese, J., and McFanzean, B. 2015. Effect of particle size on froth stability. Presented at Flotation '15, Cape Town, South Africa, November 16–19.

Ata, S. 2012. Phenomena in the froth phase of flotation: A review. *Int. J. Miner. Process*. 102-103:1–12.

Ata, S., Ahmed, N., and Jameson., G.J. 2003. Gangue drainage in flotation froths. Presented at the XXII International Mineral Processing Congress, Cape Town, South Africa, September 28–October 3.

Barbian, N., Ventura-Medina, E., and Cilliers, J.J. 2005. Mineral attachment and bubble bursting in flotation froths. In *Centenary of Flotation Symposium*. Melbourne, Victoria: Australasian Institute of Mining and Metallurgy. pp. 321–327.

Bourke, P. 2002. Selecting flotation cells: How many and what size? *Output* 1:5.

Cooper, M., Dahlke, R., Scott, D., Gomez, C.O., and Finch, J.A. 2004. Impact of air distribution profile on banks in a Zn cleaning circuit. *CIM Bull.* 97(Oct.): Paper 13.

Crosbie, R., Runge, K., McMaster, J., Rivett, T., and Peaker, P. 2009. The impact of the froth zone on metallurgical performance in a 3m^3 RCS flotation cell. Presented at Flotation '09, Cape Town, South Africa, November 9–12.

Espinosa-Gomez, R., Yianatos, J., Finch, J., and Johnson, N.W. 1988. Carrying capacity limitations in flotation columns. In *Proceedings of the International Symposium on Column Flotation*. Edited by K.V.S. Sastry. Littleton, CO: SME. pp. 143–148.

Farrokhpay, S., and Zanin, M. 2012. Synergic effect of collector and frother on froth stability and flotation recovery: An industrial case. In *Proceedings of the 11th AusIMM Mill Operators Conference*. Melbourne, Victoria: Australasian Institute of Mining and Metallurgy. pp. 145–150.

Gorain, B.K., Franzidis, J.P., and Manlapig, E.V. 2000. Flotation cell design: Application of fundamental principles. In *Encyclopedia of Separation Science*. Vol. II. San Diego: Academic Press. pp. 1502–1512.

Hadler, K., Smith, C.D., and Cilliers, J.J. 2010. Recovery vs mass pull: The link to air recovery. *Miner. Eng.* 23(11-13): 994–1002.

Harris, C.C. 1976. Flotation machines. In *Flotation: A.M. Gaudin Memorial Volume 2*. Edited by M.C. Fuerstenau. New York: AIME. pp. 753–815.

Kaya, M., and Laplante, A.R. 1990. Froth washing and froth vibration in mechanical flotation machines. SME Preprint No. 90-30. Littleton, CO: SME.

Li, C., Runge, K., Shi, F., and Farrokhpay, S. 2016. Effect of flotation froth properties on froth rheology. *Powder Technol.* 294:55–65.

Lovell, V.M. 1976. Froth characteristics in phosphate flotation. In *Flotation: A.M. Gaudin Memorial Volume 2*. Edited by M.C. Fuerstenau. New York: AIME. pp. 597–621.

Miettinen, T., Uliana, A., Araujo de Souza Franco, G., Sergio de Oliveria, P., and Murphy, B. 2013. TankCell flotation at Samarco's Germano concentrators. In *Proceedings of Iron Ore 2013*. Melbourne, Victoria: Australasian Institute of Mining and Metallurgy. pp. 441–444.

Morar, S.H., Bradshaw, D.J., and Harris, M.C. 2014. Assessment of froth stability down a flotation bank. In *Proceedings of the XXVII International Mineral Processing Congress: IMPC 2014*. Edited by J. Yianatos. Santiago, Chile: IMPC. pp. 946–954.

Neethling, S.J., and Cilliers, J.J. 2003. The impact of operating variables on flotation performance: A simulation study. In *Proceedings of the XXII International Mineral Processing Congress*. Marshalltown, South Africa: Southern African Institute of Mining and Metallurgy. pp. 874–881.

Rahman, R.M., Ata, S., and Jameson, G.J. 2012. The effect of flotation variables on the recovery of different particle size fractions in the froth and the pulp. *Int. J. Miner. Process*. 106–109:70–77.

Tabosa, E., Runge, K., and Duffy, K. 2013. Strategies for increasing coarse particle flotation in conventional flotation cells. Presented at Flotation '13, Cape Town, South Africa, November 18–21.

Ventura-Medina, E., Barbian, N., and Cilliers, J.J. 2003. Froth stability and flotation performance. In *Proceedings of the XXII International Mineral Processing Congress*. Marshalltown, South Africa: Southern African Institute of Mining and Metallurgy. pp. 937–945.

CHAPTER 7.5

Flotation Chemicals and Chemistry

D.R. Nagaraj, Raymond S. Farinato, and Esau Arinaitwe

Successful practical mineral recovery by flotation requires careful manipulation and control of mineral surface and aquatic chemistry via flotation chemicals with concomitant attention paid to physical-mechanical and operational aspects of the integrated system. This chapter is a concise compilation and analysis of the vast, but somewhat fragmented, body of knowledge pertaining to flotation chemicals used to control mineral surface and aquatic chemistry, presenting guidelines that can be used to fulfill practical mineral flotation requirements. It is written for the flotation practitioner—novice to expert, including students and aspiring young professionals.

A considerable effort has been made to bring all the bits and pieces of information scattered in the published literature and in industrial practice into one complete compilation. This is particularly germane after a century of flotation practice and serves to (1) highlight and fill some of the knowledge gaps; (2) rectify misconceptions, misrepresentation, and myths; (3) and provide practical guidelines. Such a complete compilation is not readily available in the literature.

The historical and theoretical aspects, which are well documented in the literature, are highlighted here only when necessary to support the applied aspects of flotation chemicals and chemistry, reagent selection, evaluation, and optimization with major emphasis on relevance to operator needs and plant practice. Pitfalls and gaps arising from somewhat arbitrary and reductionist practice in reagent applications are identified, and the impact of environmental regulations on current and future reagent development is discussed. The importance of frothers and modifiers, which in the past has received less overall attention than collectors, is highlighted. The critical need for a rational, holistic ("total system") approach for reagent design, selection, and optimization is discussed with a view to provide remedies against pitfalls, thereby allowing development of robust reagent schemes.

Reagent selection for any given application is strongly dictated by the surface characteristics of minerals in the ore pulp and the particular set of chemical, physical-mechanical, and operational conditions that exist in the plant (reagent choice is not predictable based on surface properties of single minerals determined in simple model systems using a single reagent). Since one does not have real-time information on the in situ mineral surface compositions, mineralogy, and properties of minerals in the pulp or sufficiently sophisticated models that could incorporate such information, one must rely on (i) empirical ore flotation testing, (ii) information on bulk mineralogy and liberation, (iii) a holistic application of fundamental concepts and insights in reagent–mineral interactions and mineral aquatic chemistry generated over many decades, and (iv) experiential knowledge to screen, evaluate, and select reagents. The complexity of natural ore systems invariably requires more than one collector, modifier, or frother to achieve metallurgical performance targets (recovery, grade, economics, environmental, etc.).

A chronology of innovations and major periods of development of flotation chemicals are given in Nagaraj and Farinato 2013 and 2016 (Table 1 therein). The practitioner can learn more from several older publications (Fuerstenau 1962; Somasundaran and Moudgil 1987; Malhotra and Riggs 1986; Mulukutla 1994; Nagaraj and Ravishankar 2007).

TERMINOLOGY AND CLASSIFICATION

As is typical of any technology that has evolved over a long period of time, terminology in flotation is rather loose, somewhat arbitrary, and often highlights merely one function or property, among many, of the reagents in question. There is also considerable discrepancy in the terminology and classification used between industry and academe; this is perhaps a reflection of differences in point of view between practitioners and academicians. The purpose of this section is to describe the terminology (and classification) adopted in this chapter and at the same time to identify a common ground that is meaningful to both research and practice.

Classification of Flotation Chemicals

The term *flotation agent* is sometimes used to describe flotation chemicals or flotation reagents, but its usage is discouraged;

D.R. Nagaraj, Principal Research Fellow, Solvay Technology Solutions, Stamford, Connecticut, USA
Raymond S. Farinato, Senior Research Fellow, Solvay Technology Solutions, Stamford, Connecticut, USA
Esau Arinaitwe, Group Leader, Solvay Technology Solutions, Stamford, Connecticut, USA

the terms *flotation chemicals* or *flotation reagents* are more meaningful and accurate and are, therefore, preferred (see Holman 1930). *Chemicals* and *reagents* are used interchangeably in the industry, and thus this chapter follows this common usage. Throughout the chapter, *flotation reagent* refers to any of the three classes of reagents—collectors, frothers, or modifiers used in flotation (Leja 1982). As depicted in the flotation reagent triangle in Figure 1, each of these three classes of reagents is important for achieving successful mineral separation.

Deliberately added modifiers (sometimes referred to as modifying agents) are a broad category of compounds that include any chemicals that cannot be classified as a collector or frother; examples include depressants, dispersants, pH regulators, activators, gangue-control reagents, and slime binding reagents, to name a few. Collectors, frothers, and modifiers are of equally critical importance and they undergo complex interactions with each other and the various minerals in the pulp. There is a trade-off in terms of mineral recovery and selectivity depending on mineralogy and plant conditions (Nagaraj 2005). Table 1 shows examples of the function provided by each class of reagent. No matter how the flotation reagents are categorized, however, the ultimate goal is to choose the appropriate combination of chemicals that will most economically and selectively separate the desired mineral species with high recovery.

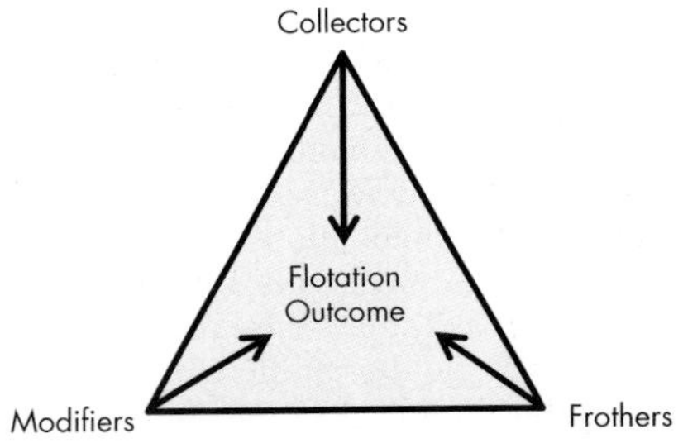

Source: Nagaraj 2005
Figure 1 Flotation reagent triangle

In the academic (research) community, collectors, for example, are variously classified into ionic (cationic and anionic) and nonionic surfactants, sulfhydryl, thio, thiol, non-thio, non-thiol, soluble, insoluble, oily, hydrolysable, ionizable, and so forth (Leja 1982; Rao 2004). In the case of sulfide minerals, these classifications are inadequate because many diverse collector families are used, collector adsorption is predominantly through surface chemical reactions (chemisorption and surface precipitation), and electrostatic effects are not very relevant. In the industry, collectors are invariably referred to and selected by their chemistry (e.g., dithiophosphates, thionocarbamates, and the associated chain lengths of hydrocarbon substituents), which immediately suggests a link between performance and a particular combination of functional group and substituents. In other words, collectors are not selected in practice based on the above-mentioned academic classifications, but more on their functionality. Thus, a classification by functional group and substituents is more appropriate and scientifically meaningful in both research and practice. Similarly in non-sulfide flotation, a classification based on functional group is very useful and revealing in terms of performance characteristics.

A functional group is a well-defined specific group of atoms containing important electron donor atoms (S, N, O), in a molecule with distinct chemical properties independent of other atoms in the molecule attached to the group. For example, —OH in alcohols, —C(=O)OH in fatty acids, —OC(=S)S in xanthate, and so on. The term *functional group* is used widely to characterize the chemistry of small organic, oligomeric, and polymeric molecules. However, in a broader sense, the term can be applied to characterize many inorganic compounds as well—for example, PO_4^{3-} in sodium phosphate or in apatite mineral or CO_3^{2-} in sodium carbonate or carbonate minerals. The functional group is sometimes referred to as a ligand. Sometimes the whole molecule carrying a particular functional group is referred to as a ligand. The term *ligand* is more commonly used in coordination chemistry. In this chapter, ligand is often used to refer to either a collector or

Table 1 Classification of reagents*

Reagent	Functional Category or Attribute	Practical Classification (in industry)	Research Classification (in academe)
Collectors	Sulfide collectors	Xanthate, dithiophosphate, thionocarbamate	Nonionic, anionic, cationic, hydrolysable, ionizable, soluble, insoluble, surfactants
	Non-sulfide collectors	Carboxylates, sulfonates, amines, hydroxamates	
Frothers	Froth stability, froth mobility, carrying capacity	Aliphatic alcohols, polyglycols	Surfactants, foamability, foam stability
Modifiers	pH and E_p† modifiers, activators, depressants, dispersants, slime or fines control reagents, coagulation, flocculation, metal hydroxy species formation and adsorption control, froth control	Acids (sulfuric), bases (NaOH [sodium hydroxide], lime, soda ash), redox reagents (NaSH [sodium hydrosulfide], Na_2S [sodium sulfide], Nokes), metal salts (Cu^{2+}), depressants and dispersants (silicate, phosphate, carboxylate, carbonate, sulfonates, polysaccharides), complexing reagents (cyanide, sulfite, sulfoxy species [S_xO_y]), alcohols, coagulants (high-charge metal ions, low-molecular-weight polymers), and flocculants (high-molecular-weight polymers) and many others	Reagents to control pH and E_h†; depressants, complexing reagents, dispersants, activators, flocculants, coagulants, surface charge density

*Although individual classification is given for collectors, frothers, and modifiers, formulations (blends) are used frequently for collectors and frothers (and sometimes modifiers) in industry, given the complex and routinely changing mineralogy/ore types in the plant and the wide size distribution of mineral particles that have to be treated without significant recovery losses in the coarse or fine size ranges and in concentrate grades.

†Ep is pulp potential, most often with respect to an Ag/AgCl reference electrode. E_h is solution or pulp potential with reference to a standard hydrogen electrode.

a functional group capable of forming metal complexes (via covalent or coordinate-covalent bonds).

The term *oily collectors* is used in several contexts, which can be confusing, because there is a distinction in the manner of their interaction with mineral surfaces. A useful categorization is based on whether or not there is an active functional group involved in adsorbing to a mineral surface. For example, thionocarbamates, thioureas, and xanthate esters all contain active sulfur functionalities involved in chemisorption to sulfide mineral surfaces. They are considered as oily collectors because they have very marginal solubility in water and carry no charge under typical use conditions, although they may ionize at extreme values of pH depending on their pK_a. Hydrocarbon oils (e.g., petroleum-based—mainly nonpolar; Seitz and Kawatra 1986), on the other hand, are also referred to as oily collectors (or extenders). Such chemicals, which are used in great quantities in sulfide and non-sulfide ore flotation, have very low solubility, do not contain active electron donor atoms (i.e., S, N, O), and work in a support fashion by co-adsorbing to mineral surfaces that are naturally hydrophobic (e.g., molybdenite) or have been made hydrophobic by the primary collector (e.g., fatty-acid-treated apatite; xanthate-treated sulfide minerals). Extenders in droplet form can act as a locus around which hydrophobic mineral particles can cluster, thereby improving collision efficiency with air bubbles; these clusters can also include air bubbles. Extenders co-adsorbed onto collector-coated particles can increase the effective area of the hydrophobic domain. Non-functionalized oils are also commonly used in froth modification and control.

Frothers are grouped into soluble and partially (or sparingly) soluble frothers (Booth and Freyberger 1962). The soluble frothers are polyglycols and their alkyl ethers. Partially soluble or sparingly soluble frothers comprise aliphatic and aromatic alcohols, terpineols, aliphatic aldehydes, ketones, and esters. In the industry, frothers are commonly referred to and selected by their chemistry, namely, alcohols and glycols and their mixtures; solubility is implicit when the chemistry is specified. Other classifications or selection criteria are being actively explored, for example, those based on hydrophilic-lipophilic balance, critical coalescence concentration, or by flotation kinetics and stability. However, these are not fully developed at this stage. Given that —OH is the most important functional group (other functional groups constitute a small part of the total frother usage in the industry), classification based on solubility differences is also quite acceptable. Although molecular weight (MW) and branching play a role, they can also be linked to solubility. In non-sulfide flotation, though a separate frother is not universally used, long-chain surfactants with a variety of functional groups and short-chain alcohols and glycols are sometimes used as froth modifiers. Even in these systems, a classification by functional group is meaningful.

The terminology used for modifiers is more contentious. Modifiers essentially enhance the activity and/or selectivity of collectors and frothers, and overall separation efficiency and selectivity. There is no single, generally accepted classification of modifiers. They are variously referred to as pH modifiers (or regulators), depressants, dispersants, flocculants, coagulants, activators, deactivators, promoters, froth modifiers, viscosity modifiers, and slime-blinding reagents. Often the terms *depressants* and *dispersants* are used interchangeably.

The term *promoter* has occasionally been mentioned in the literature to refer to certain modifiers; however, there is no category of modifier that should be referred to as a "promoter." *Promoter* usually refers to a chemical that is more appropriately termed a *collector*, but such usage is far less common. It is best to avoid the use of the term *promoter* altogether.

Slime-blinding reagents refer to chemicals such as dextrin or guar that adsorb onto fine particles (clays, amorphous carbon, etc.) and "blind" them to prevent these fine particles from interfering with flotation of value minerals.

The term *dispersant* has different connotations in sulfide and non-sulfide systems. It is used loosely in sulfide flotation systems where its primary function appears to be one of controlling slimes (i.e., to prevent them from interfering with recovery or grade of value mineral species) and pulp viscosity and not for providing colloidal stability to suspensions. In non-sulfide systems, a dispersant's multiple functions are widely exploited (e.g., to prevent aggregation to moderate pulp viscosity during conditioning, and/or improve selectivity of collector adsorption by preventing collector adsorption on unwanted mineral sites).

The term *modifier* is sometimes used in a very narrow sense to represent a specific function among many (e.g., a pH modifier). All such classification or categorization fails to recognize that in the complex flotation system, a single modifier invariably has multiple functions, which vary in number and intensity from system to system. However, it is difficult to establish a distinct classification that comprises the entire spectrum of applications. The best recommendation is to use the term *modifier* (distinct from *collectors* and *frothers*) in a broad sense and to recognize that it constitutes the third apex of the flotation reagent triangle (Figure 1) and it has important implications in flotation research and practice.

Qualitative descriptors are used routinely in research and practice to indicate reagent performance (e.g., use of the terms *strength* and *selectivity*). These are relative terms, though they are sometimes used as if they were absolute. They are used interchangeably to refer to strength or selectivity of adsorption of a reagent (thermodynamic or kinetic) on individual minerals or of flotation recovery of a value mineral at a given dosage or under a given set of plant conditions. Thus, descriptors such as "strong or weak collector (modifier, frother)" or "selective collector (modifier, frother)" are necessarily qualitative; implicit in the description is that some reference is used. Similarly, qualitative descriptors are routinely used for frothers (e.g., watery froth or dry froth). These terms will be used as such in this chapter, for they serve a useful purpose in plant practice in the absence of quantitative descriptors.

Ultimately, what matters the most in terms of plant practice is whether the selected reagents from the flotation reagent triangle are able to meet the selectivity and recovery targets in the plant for the desired value mineral species in the most economical manner.

Classification of Ores: Sulfides and Non-Sulfides

In the field of mineral processing, ores and minerals have been classified in many ways: metallic or nonmetallic; ferrous or nonferrous; industrial minerals; construction minerals; soluble or semi-soluble or salt-type; oxide ores; silicate ores; carbonate ores; base metal sulfide ores; precious metals ores; and so on. The myriad classifications are partly historical and partly based on differences in perspectives arising from different mineral industry segments, the academic community, ore/mineral product properties, their end use, and the particular value mineral/metal contained in the ores.

Whereas for sulfide minerals the classification is usually based on the dominant metallic element(s), for non-sulfide minerals the classification has been based on any of a variety of attributes, including crystal structure, solubility in water, or other physical properties such as metallic or nonmetallic character, and ferrous versus nonferrous and industrial uses. This leads to overlapping and possibly confusing categories in which one might misplace non-sulfide minerals. It would appear that there is really no need for any type of classification other than the distinction from sulfide minerals when one considers the types of collectors, modifiers, processing schemes, strategies, and equipment used in the industry for the flotation separation of all these types of non-sulfide minerals. Indeed in plant practice, the decisions on flow sheets and reagent schemes are not based on whether the ore is salt type or oxide or industrial mineral, and so forth. The various physical properties of non-sulfide minerals, and the mineralogical associations, merely suggest the conditions that might be useful in flotation separations. For example, both sylvite (KCl) and quartz (SiO_2) can be floated by the same amine at approximately the same pH, yet these minerals have very little in common. The most significant factor, which dictates the choice of a separation scheme, is that there are only small differences in surface properties between a value mineral and the gangue, both of which are non-sulfide minerals. Highly specific treatment conditions are frequently required to achieve adequate selectivity and satisfactory metallurgy, the most critical aspect of which is generally the quality and purity of the finished product.

Thus, in discussing the chemistry and application of flotation reagents, none of the mineral classification schemes noted previously are adequate or enlightening (or useful). Nagaraj (1987) provided arguments for a convenient and useful classification based on the predominant occurrence and source of elements in earth resources. *Thus from the perspective of flotation reagents, minerals/ores are classified into two distinct groups: sulfide minerals/ores, and non-sulfide minerals/ores.* The rationale/basis for this classification is found in Nagaraj and Ravishankar (2007).

CHEMICALS USED IN SULFIDE AND NON-SULFIDE FLOTATION—BASIC CONCEPTS AND REAGENT FUNDAMENTALS

Interfacial chemistry is at the heart of flotation separations. Interfacial chemistry can be modulated and controlled with the use of a wide variety of chemicals. These chemicals and the minerals in the ore participate in a wide range of chemical processes in the pulp and froth zones. That is, these processes occur both at the mineral–solution and air–solution interfaces as well as in the bulk aqueous phase. All these processes together constitute *mineral (or ore) aquatic chemistry*. Interfacial processes include hydration, dissolution, hydrolysis, reduction–oxidation (redox) reactions, adsorption (chemisorption, physisorption), surface reactions, and surface precipitation. These processes occur even in the absence of any deliberately added flotation chemicals. Bulk aqueous processes include hydrolysis, redox processes, complexation (Lewis acid–base interactions), and precipitation. All these processes must be considered together because they occur simultaneously (though at different rates). When chemicals are added, many of the interfacial and bulk aqueous phase processes are immediately altered via many pathways, and the added chemicals become a part of the system. The complexity increases substantially; however, many of the processes (or pathways) may become unimportant or even shut down. For example, addition of a collector may reduce solubility of a mineral on which it adsorbs, or passivate the mineral and thus shut down redox processes. Addition of lime may have similar effects. Ore aquatic chemistry begins when the ore is first exposed to the aqueous phase. An understanding of the basic concepts in ore aquatic chemistry is critical for the design, selection, and optimization of flotation reagents and for plant problem-solving. There is extensive literature (though fragmented) on ore or mineral aquatic chemistry. The topic of ore aquatic chemistry is very broad and beyond the scope of this chapter.

Although fundamental studies focus invariably on relatively pure, single reagents (justifiably so), in plant practice, invariably, combinations of reagents and single minerals (such as formulations of collectors or frothers) are used, and ores have multiple mineral phases often associated with each other. This creates numerous competitive processes in practical ore systems, which are absent in the simplified systems used for fundamental studies. Indeed, commercial-grade reagents are rarely pure reagents (i.e., they may have more than one active component as a result of the manufacturing process, or they may have impurities, which may influence reagent performance). Some of the frothers may be deliberate formulations of several raw materials (which may be a by-product or waste stream from a manufacturing process), which themselves may be mixtures. Thus, results of fundamental studies based on single-mineral, single pure-reagent systems are often difficult to translate to practical systems.

Mineral processing via flotation consists of a series of unit operations whose main physical success metric is metallurgical outcome (e.g., recovery and grade). The concepts and factors relevant to the chain of events leading to successful flotation are shown schematically in Figure 2. This provides context for understanding the parameters that can affect flotation performance and the factors that should be taken into account to optimize a chemical treatment program.

Successful separation via flotation is based on leveraging wettability differences of mineral particle surfaces in a mixture so that the more hydrophobic particle surfaces have a greater probability of attaching to bubbles, thereby selectively transporting them into the froth zone. Not only must the surfaces of the mineral particles to be floated be made hydrophobic, the surfaces of all the rest of the mineral particles must be maintained in a hydrophilic state. Thus *interfacial chemistry is at the heart of flotation*. The components of an effective flotation reagent package (collectors and modifiers), which are the principal means by which interfacial chemistry is affected, must operate in and take advantage of the environment of the pulp phase (water chemistry, pH, E_p, etc.). This allows maximizing the wettability difference between particles to be floated and gangue mineral particles.

FLOTATION REAGENTS—PLANT PRACTICE

Reagent design and selection for a given flotation separation is inevitably a compromise that must take into account not

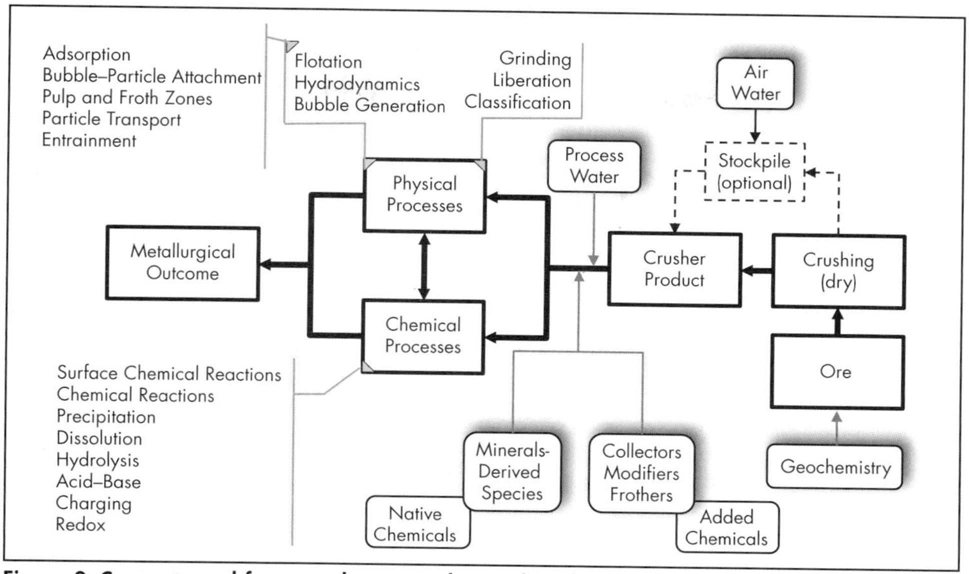

Figure 2 Concepts and factors relevant to chemical performance in mineral flotation

only the primary function of a reagent (e.g., collector, frother, or modifier) but any secondary functions (intended or inadvertent) and potential interactions with other components in the flotation cell (dissolved species, frother–collector interactions, etc.). As discussed in the previous section, most plants use combinations of collectors to solve their problems. This applies to modifiers and frothers as well.

Major factors that can affect reagent design and selection include mineralogy, toxicity, environmental impact, cost–performance, physical and chemical stability, ease of manufacture, raw material cost and availability, and performance attributes (e.g., strength, selectivity, froth properties, and breadth of applicability). Chemicals made from variable raw material feedstocks (e.g., animal or vegetable origin) tend to have a much greater variability in composition distribution than those made via synthetic or mineral-sourced feedstocks (petroleum-derived alcohols, soda ash, sulfuric acid from pyrite, etc.). Since flotation reagents for non-sulfide mineral recovery more typically fall in the first class (e.g., fatty acids, amines, petroleum sulfonates), more variability is encountered in product quality and therefore performance. Consequently, reagent selection is more dependent on the manufacturing process and raw material source. Impurities in the manufactured products may influence regulatory aspects and environmental impact. For naturally derived products, many of the regulations are grandfathered. Other custom-made reagents typically have tighter specifications for active components and impurities, making reagent selection less dependent on manufacturing process or raw material source. Reagents typically used in sulfide flotation fall into this second class, requiring no further discussion of raw material sources or manufacturing processes.

Many of these factors are not amenable to quantification, making it difficult to develop specific guidelines for use. Under such circumstances, one must resort to efficient empirical exploration of reagent combinations, doses, and operational conditions to arrive at optimized schemes (see the "Reagent Selection and Optimization" section later in this chapter).

Reagents for Base Metal Sulfides and Precious Metals

Classification

Sulfide ore flotation can be divided into the following four metal-value-based groups based on certain common features of ores and mineralogy in each group, flow sheets, and metallurgical objectives:

1. Cu, Cu-Mo, Cu-Co, and Cu-Au ores
2. Complex, polymetallic ores such as Pb-Cu-Zn-Au-Ag
3. Primary Au and platinum group metal (PGM) ores
4. Ni and Ni-Cu ores

Other minor, value elements (tellurium, rhenium, bismuth, antimony, cobalt, etc.) may be recovered from these ores, but these are invariably by-products that may or may not be economically feasible or sensible to recover. In groups 1 and 2, separation of sulfides from each other is necessary (Cu sulfides from pyrite, Cu sulfides from galena, etc.) in addition to separations of sulfides from non-sulfides, and selective reagents are invariably used judiciously. In group 3, the objective is to recover all recoverable value sulfides and precious metals at as high a grade as possible (lowest mass recovery), and strong collectors are used for this purpose. In group 4, both selective flotation and bulk flotation are practiced depending on the ore type (e.g., in massive sulfide ores, selective separation is required; in mafic and ultramafic ores, bulk flotation is practiced). There are significant differences in flow sheets, operating strategies, and operating philosophies among these groups. There are also differences in the particular reagent choices and their usage or application. In spite of this division and the differences, all four groups also have much in common, especially reagent chemistry (e.g., the same families of collectors used in all groups) and mechanisms of interaction with minerals (e.g., dominated by chemisorption and electrochemical processes), which is not surprising given that the important value minerals in these ores are based on metals that are all grouped together in the periodic table and exhibit similar chemical properties.

Collectors

It is well established that the main function of the collector is to impart sufficient hydrophobicity (i.e., reduced water wettability) to the mineral surface so that the probability of bubble–mineral attachment is increased and the probability of detachment is decreased. It is also well established that a collector molecule accomplishes this function by adsorbing on the mineral surface. A separate "collector" is not always necessary; a mineral can be floated without a collector if it is naturally hydrophobic (e.g., molybdenite) or made hydrophobic as a result of interfacial processes (e.g., chalcopyrite after mild oxidation resulting in a metal-deficient, sulfur-rich hydrophobic surface leading to collectorless flotation). Sometimes a modifier can remove hydrophilic coatings on a mineral, thus exposing a hydrophobic surface. However, in plant practice, a collector is almost always necessary to ensure maximum flotation rate and recovery of values or rejection of gangue; no plant relies on natural or induced hydrophobicity of one specific mineral. However, collector adsorption and the consequent development of hydrophobicity on the mineral surface do not guarantee the desired flotation outcome, since flotation is a probabilistic physicochemical process. Flotation is a delicate balance between buoyancy and gravitational forces; any perturbation (chemical, physical, or operational) can tip the balance. Flotation outcome cannot be predicted exactly, but probabilities of the various subprocesses can be estimated, some better than others. Collector, modifier, and frother work in concert. It is also necessary for the collector to *not* alter the wettability of the gangue minerals to provide adequate selectivity in the separation process. Collector adsorption onto gangue minerals (unless a reverse flotation process is in effect) may have to be actively suppressed through a judicious use of modifiers to achieve the best separation.

A perusal of flotation literature on chemical aspects over the past 100 years would indicate that collectors have been accorded paramount importance and more attention compared to modifiers and frothers. However, viewed from the perspective of a flotation plant's overall goals of maximizing recovery, grade, and profitability, such singular focus on collectors is extremely unbalanced. Although collectors do play a critical role in providing a part of the solution, their performance is intimately affected by the correct choice of modifiers and frothers. This is particularly important in non-sulfide mineral flotation. In other words, collectors are necessary but not sufficient to control a flotation operation. The poor availability of concise information on modifiers and frothers is an area for improvement. See the "Modifiers" section later in this chapter for general guidelines.

Currently used and recently developed collectors for separation of metal sulfides and precious metals are shown in Table 2. Additional details can be found in manufacturer's handbooks (e.g., *Mining Chemicals Handbook* [Cytec Industries 2010]). General guidelines for various ore types are provided later in this chapter.

Xanthates continue to account for the largest consumption of collectors in terms of volume used in sulfide mineral flotation, but not necessarily in terms of total value of the minerals recovered. Large quantities (or dosages) of xanthates are used for bulk flotation of all the sulfide minerals and precious metals in primary Au, PGMs, and certain Ni ores without regard for selectivity between floated species (selectivity against gangue is presumed). Even in these cases, other more selective collectors are also required to augment performance deficiencies of xanthates and maximize recovery of values. Where selectivity between sulfides is the key goal (as in Cu, Cu-Mo, Cu-Co, Cu-Au, Ni-Cu, complex sulfides, and polymetallic ores—all of which make up the bulk of sulfide flotation), xanthate is not the primary collector; other more selective collectors such as dithiophosphinates, thionocarbamates, and dithiophosphates serve as the primary collectors, with xanthate used as an auxiliary (or secondary) reagent. In fact, the use of specialty collectors has grown significantly over the past four decades, most significantly during the past two decades, in part due to the prevalence of difficult-to-process ores with complex mineralogy and the desire for safer, greener flotation reagents.

Those acquainted with plant practices would realize that a single collector cannot be universally effective for the large mineralogical diversity encountered in commercial flotation operations. Flotation practitioners have realized this fact and have adopted reagent combinations in response to the separation challenges faced in the plant.

Hydrocarbon oils are widely used as collectors for molybdenite, extenders for other primary collectors (to enhance particle hydrophobicity, particle-bubble-oil cluster formation, improved coarse particle recovery) and for froth modification and control to improve selectivity. Extenders play a role in froth modification and control. Hydrocarbon oils can produce a highly mineralized froth that drains more easily, thereby reducing entrainment of fine hydrophilic gangue and improving selectivity. Over-frothing due to the presence of slimes can be reduced or eliminated using these oils (Seitz and Kawatra 1986).

Significant advances have been made in the past three decades in reagent design, selection, and systematic structure–activity relationships. These are described in several notable articles and books (Jones and Oblatt 1984; Aplan and Chander 1987; Somasundaran and Moudgil 1987; Nagaraj et al. 1989; Pradip and Fuerstenau 1991; Mulukutla 1994; Nagaraj 1997; Pradip et al. 2002a, 2002b; Herrera-Urbina 2003; Nagaraj and Ravishankar 2007).

Frothers and Frothing

Frothers and the froth zone (also called the "froth phase") play a critical role in flotation, which will ultimately be judged based on the recovery of values and the grade of the flotation concentrate obtained for each ore type. Frothers aid in froth formation and in achieving the desirable (or optimum) characteristics of the froth to facilitate selective separation and transport of targeted particles. They reduce bubble size by reducing bubble coalescence in the pulp zone, which can affect water and solid transport into the froth zone. Frothers can also contribute to the transient stability of thin liquid films in the froth zone, thereby impacting bubble coalescence rates, water holdup, froth mobility, and carrying capacity. For example, all else being the same, a polyglycol frother usually results in a wetter froth than does methyl isobutyl carbinol (MIBC). Careful selection of the most appropriate frother, combined with practical application expertise, is essential for maximizing mineral separation. In the froth flotation process, as practiced in most operations using mechanical cells, columns, and even the emerging new designs (e.g., the Woodgrove devices;

Table 2 Collectors used in sulfide mineral flotation

No.	Chemical Class	Typical Compounds Used	Structure	Manufacture	Product Form and Aqueous Solubility	Plant Application	Hydrolytic Stability*	Resistance to Aqueous Oxidation*
1	Thiocarbamic acids (dithio derivatives)	Dithiocarbamate, dialkyl, sodium (R = R' = ethyl or butyl)	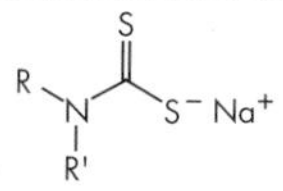	Reaction of dialkyl amines with carbon disulfide	Aqueous solutions of sodium salts	Limited use in PGMs; primary Au, Cu, and polymetallic ores	High	Medium
2	Thiocarbamic acids (thiono derivatives)	Thionocarbamate, alkoxycarbonyl alkyl (R' = ethyl or butyl; R = C_2, C_3, C_4 etc.). Note: Product where R' = ethyl has known toxicity†	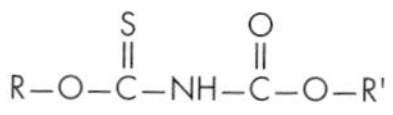	Reaction of isothiocyanate with alcohol	Oily; insoluble in water, but disperse well in water	Widely used for Cu, Cu-Mo, Cu-Au, and Cu-Co ores in a wide pH range (4-11); very selective against gangue iron sulfides	Medium	High
3	Thiocarbamic acids (allyl thiono derivatives)	Thionocarbamate, allyl alkyl (R = butyl)	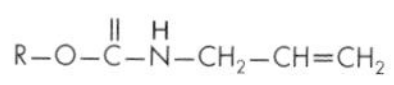	Reaction of allyl isothiocyanate with alcohol	Oily; insoluble in water, but disperse well in water	Widely used in Cu, Cu-Mo, Cu-Au, Zn, and Cu slag circuits; stronger than corresponding dialkyl thionocarbamates	High	High
4	Thiocarbamic acids (dialkyl thiono derivatives)	Thionocarbamate, dialkyl (R = isopropyl or butyl; R' = methyl or ethyl)	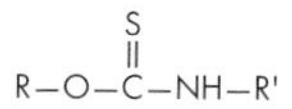	Reaction of alkyl xanthate with ethyl amine using catalyst	Oily; insoluble in water, but disperse well in water	Widely used for Cu, Cu-Mo ores; very selective for copper sulfides; selective against gangue sulfides	High	High
5	Thiophosphoric acid (dialkyl derivatives)	Dithiophosphate, dialkyl, sodium (R = ethyl, isopropyl, isobutyl, or isoamyl)	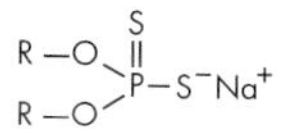	Reaction of alcohols with phosphorous pentasulfide	Aqueous solutions of sodium salts	Very widely used, next to xanthate; more selective than xanthate; used as primary or secondary collectors; general purpose collector for most sulfides and precious metals ores	High (at high pH)	Medium
6	Thiophosphoric acid (diaryl derivatives)	Dithiophosphate, dicresyl, sodium or ammonium, or in the acid form (R = methyl)	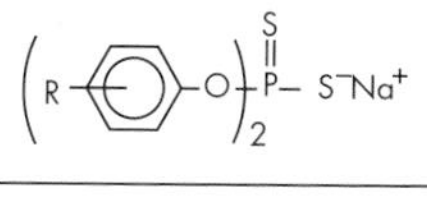	Reaction of cresols with phosphorous pentasulfide	Aqueous solutions of sodium and ammonium salts	Mostly for complex and polymetallic sulfide ores (Pb-Ag-Zn-Cu); sporadic use in Cu, Cu-Mo-Au ores	High (at high pH)	Medium
7	Thiophosphoric acid (dialkyl monothio derivative)	Monothiophosphate dialkyl, sodium (R = ethyl or isobutyl)	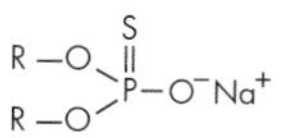	First hydrolysis of the corresponding dithiophosphoric acids	Aqueous solutions of sodium salts	Mostly for Cu-Au and Au ores; excellent collectors in acid circuits; selective precious metal flotation in alkaline circuits	High	High
8	Thiophosphoric acid (diaryl monothio derivatives)	Monothiophosphate dicresyl, sodium (R = methyl)	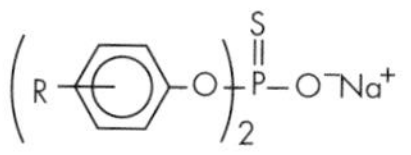	First hydrolysis of the corresponding dithiophosphoric acids	Aqueous solutions of sodium salts	Mostly for Cu-Au and Au ores; excellent collectors in acid circuits; selective precious metal flotation in alkaline circuits	High	High
9	Thiophosphinic acid	Dithiophosphinate, dialkyl, sodium (R = isobutyl)	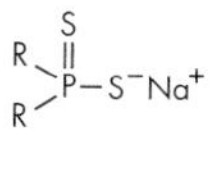	Phosphine reacted with alkene and sulfur	Aqueous solutions of sodium salts	Widely used for complex and polymetallic (Pb-Ag-Zn-Cu), Cu-Mo-Au, Au, and Cu-Ni ores; very selective against gangue sulfides; rapid flotation kinetics; low dosages	High	High

(continues)

Table 2 Collectors used in sulfide mineral flotation (continued)

No.	Chemical Class	Typical Compounds Used	Structure	Manufacture	Product Form and Aqueous Solubility	Plant Application	Hydrolytic Stability*	Resistance to Aqueous Oxidation*
10	Thiol	Mercaptan, alkyl (R = dodecyl)	R–SH	Reaction of alcohol with H_2S	Oily; insoluble in water	Very limited use as a smaller component of formulations	High	Low
11	Thiazole	Mercaptobenzothiazole	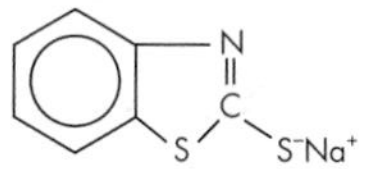	Reaction of aniline, carbon disulfide and sulfur under high pressure	Aqueous solutions of sodium salts	Mostly for complex sulfide, polymetallic, and Au ores, tarnished sulfides, and as auxiliary collectors	High	Medium
12	Thioethers	Sulfide, dialkyl (R = octyl and R' = ethyl)	R–S–R'		Oily; insoluble in water	Very limited use; often as a minor component in formulations	High	High
13	Thiourea	Thiourea, alcoxycarbonyl alkyl (R = C_2, C_3, C_4 etc.), R' = butyl or ethyl product where R' = ethyl has known toxicity	R–NH–C(=S)–NH–C(=O)–O–R'	Reaction of isothiocyanate with amine	Oily; insoluble in water, but disperse well in water	Similar to alkoxycarbonyl thionocarbamates, but with a few unique properties	High	High
14	Thioester (dithio derivatives)	Xanthate ester, *allyl* alkyl (R = amyl)	R–O–C(=S)–S–CH_2–CH=CH_2	Reaction of alkyl xanthate with allyl chloride	Oily; insoluble in water, but disperse well in water	Widely used for Cu, Cu-Mo, Cu-Au, and Au ores; very selective for copper sulfides and precious metals; excellent for molybdenite	Medium	High
15	Thioester (dithio derivatives)	Xanthogen formate, dialkyl (R' = ethyl; R = ethyl, isopropyl or isobutyl)	R–O–C(=S)–S–C(=O)–O–R'	Reaction of xanthates with alkyl chloroformate	Oily; insoluble in water, but disperse well in water	Very limited use; mostly for Cu, Cu-Mo; known to be acid circuit collectors	Low	Medium
16	Thiocarbonic acid (dithio derivatives)	Xanthate, alkyl (R = ethyl, isopropyl, n-butyl, isobutyl, amyl)	R–O–C(=S)–S^-Na^+	Reaction of alcohols with carbon disulfide	Solid form of sodium or potassium salts. High solubility in water. ~30% solutions are sold when practical.	Generic (nonselective) collectors for all sulfide ores; alloys and minerals carrying Au, Ag, PGMs; sulfidized oxide minerals of Cu, Pb, and Zn; primary collectors for Ni, Au, and PGM ores; secondary collectors for all other ores	Low	Low
17	Thiocarbonic acid (trithio derivatives)	Trithiocarbonate, alkyl, sodium salt (R = butyl)	R–S–C(=S)–S^-Na^+	Reaction of mercaptan with carbon disulfide	Aqueous solutions of sodium salts	Very limited use (stench); general purpose collector	Low	Low
18	Thiourea	Diphenyl thiourea (thiocarbanilide)	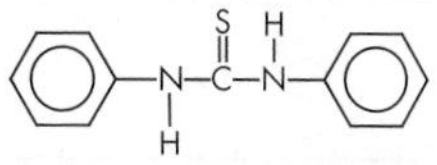	Reaction of aniline, carbon disulfide and sulfur	Solid; insoluble in water	Very limited use (complex sulfides); used only as a small component in a formulation	High	High
19	Guanidine	Diphenyl guanidine	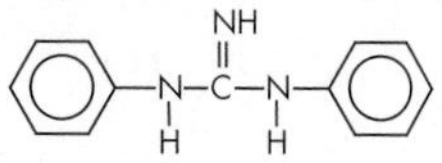	Reaction of cyanogen chloride with aniline	Solid; insoluble in water	Used for Cu-Ni matte separation in one plant	High	High
20	Hydrocarbon oils	Kerosene, diesel, burner oil, gas oil, fuel oil, pump oil, vapor oil, aromatic oils, vegetable and mineral oils	Complex mixtures of aliphatic and aromatic hydrocarbons, linear and branched.	Products of petroleum cracking and refining and natural oils	Liquids, oily; insoluble in water	Naturally hydrophobic minerals (MoS_2 [molybdenite], talc, sulfur, stibnite, graphite, and other forms of carbon); extender for other collectors; froth control	No hydrolysis except for vegetable oils	High

Adapted from Nagaraj and Ravishankar 2007
*Qualitative/relative ranking only.
†Residual ethyl carbamate in the derivative where R′ = ethyl has known toxicity (EPA 2000).

Bennett 2015; Dobby and Kosick 2013), the froth zone is the last separation stage through which all material that ends up in the launder must flow. In other words, froth zone characteristics directly impact metallurgical outcome. Empirical evidence from plant practice shows that many operational variables can influence froth characteristics to a varying degree. It would be naïve to think that the "froth" is controlled only by the "frother." Even statements such as "collector performance is strongly influenced by froth phase characteristics" begin to lose impact. The influence of chemical and physical-operational factors on froth characteristics (hence, flotation outcome) is easy to observe and demonstrate in the laboratory and plant and is, therefore, testable and self-evident. A change in any reagent or operating condition can change the appearance and properties of the froth zone. Indeed, operators run a circuit by managing the froth zone properties in the plant, though mostly qualitatively through visual observation and experience. For example, a fluctuation in pH, say, by 1 unit (from 10.5 to 9.5) in a copper circuit can have dramatic changes in the appearance and properties of the froth, hence, flotation outcome. Such changes can cover a wide range, from a highly destabilized froth (e.g., no froth layer, and bubbles bursting continually with fizzle) to a very stable froth (either with no mobility or a runaway froth), depending on the ore type and other plant conditions. It is almost impossible to predict this a priori or to generalize this across the industry given our current, limited understanding of the complexity of the froth zone. An extreme example of this complexity might be that a mere change in the addition point of a collector or modifier could have a large influence on froth characteristics, and similarly for other conditions in the plant.

Available frother types. Effective frothers available today fall into one of two categories of compounds: short chain aliphatic (mono) alcohols and polyglycols (see Table 3). The latter includes both unsubstituted and alkyl monoethers (Booth and Freyberger 1962; Klimpel and Hansen 1987; Tveter 1952). Of the thousands of other compounds developed as frothers, certain aldehydes, ketones, ethers, and esters do function as frothers; however, they are not used as primary frothers and are present most commonly as minor components in some frother feedstocks. MIBC is a more widely known alcohol frother and often serves as a benchmark, though it is considered a rather "weak" frother (its strong odor and flammability have also been raised as issues in a few plants). Polyglycols and their ethers, which are considered "strong" frothers, are also widely used. However, the industry is increasingly adopting tailored formulations of alcohol-glycol-glycol ether frothers to meet specific challenges (e.g., frothers for operations with a much coarser primary grind size, or for much finer grind size on finely disseminated ores, or frothers for processing poorer grade/quality ores). Such formulations are most often considered the best compromise to achieve optimum metallurgical outcome; many plants tend to evaluate such combinations on a frequent basis depending on their specific needs. Other notable frothers include cresylic acids (very limited use because of environmental concerns) and alkoxy-substituted paraffins, such as triethoxybutane (TEB; limited use). Pine oil used to be a popular frother until about three decades ago. Its usage has declined substantially. It is used to a limited extent in several plants, notably in China. Hydrocarbon oils are often used in conjunction with alcohols and cresylic acids to modify and stabilize the structure of the froth.

Table 3 lists the most important frothers currently used in flotation operations and their properties. Except for MIBC and 2-ethyl hexanol, commercial frothers are invariably mixtures either by design or by default (i.e., a result of the manufacturing process). Operators generally prefer to control frother independent of other reagents. However, frothers are also often used as a component in collector formulations to produce easy-to-handle products and to aid in physical compatibility of oily and aqueous components. One downside is potential loss in the operator's ability to control froth independently and possibly imposing an upper limit to usable dosage of the collector formulation, since an increase in collector dosage may inadvertently exceed the frother requirement threshold.

Alcohol frothers have very low solubility, whereas most of the glycols (and their ethers) used are completely miscible with water. At the low concentrations used in flotation, solubility and dispersibility are not a limiting factor. Frothing characteristics of alcohol frothers are significantly different from those of glycols, and these differences can be attributed to differences in solubilities, surface activity, critical coalescence concentration (CCC), and diffusion rates.

Guiding rules for frother selection. Currently, selection of frothers and controlling the froth zone in the plants are still based on general qualitative guidelines, empirical testing in the lab or plant to assess performance, visual observations in the plant (often aided by froth visualization cameras), and experience of the operators. An adequate level of practical knowledge of frother selection resides largely with flotation reagent manufacturers and a limited number of industry experts. Frother manufacturer brochures and product lists are a starting point for frother choice; however, invariably one has to rely on an in-house expert from one of the manufacturers, formulators, or consulting companies, and empirical testing, preferably in the plant. Booth and Freyberger (1962) provided an overview of frothers and frothing and summarized general guidelines, but this document is obviously quite old and reflects the lack of attention to the froth zone and frothers in the first 50 years of flotation. In the past three decades, however, there has been a gradual, but healthy, increase in research efforts to understand and model the physics and engineering aspects of froth zone and developing selection criteria for frothers based on quantifiable attributes. Several research groups around the world have contributed significantly and laid the foundation for continued research. A few notable examples of such groups (in alphabetical order) are Aalto University (Finland), Julius Kruttschnitt Mineral Research Centre (Australia), Ian Wark Research Institute (Australia), Imperial College group (United Kingdom), McGill University (Canada), University of British Columbia (Canada), University of Cape Town (South Africa), University of Newcastle (Australia), University of New South Wales (Australia), and University of Queensland (Australia). More recent information on frothers and froth zone may be found in works by Rao 2004; Khoshdast and Sam 2011; Ata 2012; Riggs 1986; Finch et al. 2008; Melo and Laskowski 2006, 2007; Laskowski et al. 2013; Cho and Laskowski 2002; Makhotla et al. 2015; Zhang et al. 2012; Zhou et al. 2016; Cappuccitti and Finch 2008; Zanin and Grano 2006. The vast majority of the studies have been in two-phase systems (i.e., water and air, in the absence of solids), though attempts to

Table 3 Important frothers used in mineral flotation applications

Chemical Class	Structure		Properties (values refer to the example structure only)				
			Example (structure)	Molecular Weight, g/mol	Density at 20°C, g/mL	Water Solubility, g/L	Critical Coalescence Concentration, mmol/L
Aliphatic alcohols	Linear; n = 4–7		$(CH_3)_2CHCH_2CH_2OH$	88.17	0.815	25	0.29
	Branched: 2-ethyl hexanol	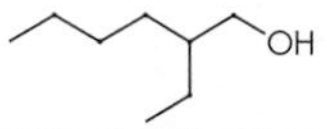	$CH_3(CH_2)_3CH(C_2H_5)CH_2OH$	130.2	0.833	4.5	0.122 (based on d50)
	MIBC (methyl isobutyl carbinol)	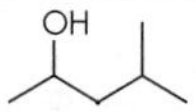	$CH_3CH(OH)CH_2CH(CH_3)_2$	102.2	0.807	17	0.11
Poly(ethylene) glycols (PEG), Poly(propylene) glycols (PPG), PPG monoalkyl ethers (MW range: 200–500)	PEG, R=H	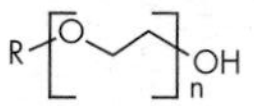	$C_4H_9(OC_2H_4)_3OH$	206.28	0.989	Miscible	0.11
	PPG	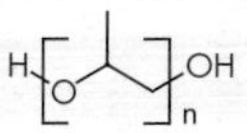	$H(OC_3H_6)_{6.5}OH$	395.6	1.009	Miscible	0.012
	PPGe ($R = C_1–C_4$)	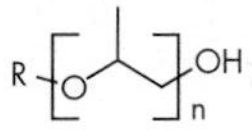	$CH_3(OC_3H_6)_3OH$	206.3	0.97	Miscible	0.089
Alkoxy compounds	Triethoxybutane	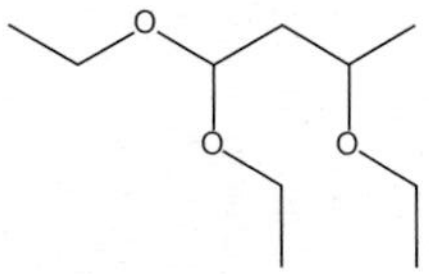					~0.11
Ketones, aldehydes, esters (not primary frothers; mostly minor components in feedstocks)	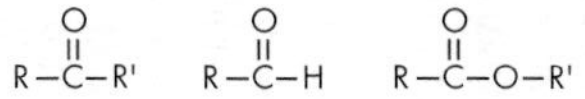						
Pine oil (α-terpineol) (limited use)	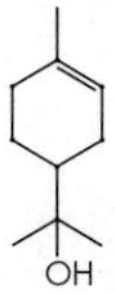						0.052
Cresols (isomers) (not used anymore)	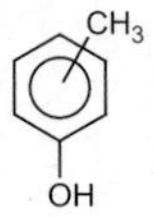						

Adapted from Nagaraj and Ravishankar 2007

Table 4 Typical qualitative attributes observed for alcohol and glycol* frothers in sulfide flotation*

Attribute	Alcohols	Polyglycols and Glycol Ethers†
Frothing power	Normal alcohols > other isomers; aliphatic alcohols > aromatic alcohols; saturated > unsaturated (opposite in terpene-based)	Strongest frothers; maintain frothing power without staged addition
Bubble size	Larger bubble size, less compact structure	Smaller bubble size, compact structure
Flotation kinetics	Faster kinetics	Slower kinetics
Sensitivity to other conditions	More sensitive to pH; tolerant to slimes	Less sensitive to pH; less tolerant to slimes
Stability	Brittle froth; less tenacity, less persistence; requires staged addition; may cause effervescence at high dosages	Persistent at low dosages; does not require staged addition; breaks down readily in launders; less prone to cause effervescence at high dosages
Selectivity	Lower water retention (greater froth drainage) resulting in greater selectivity; less downstream effects because of selectivity and low persistence	More water retention, lower selectivity, greater downstream effect because of persistence
Coarse particle recovery and load capacity	Less	More

Adapted from Nagaraj and Ravishankar 2007
*The froths resulting from fatty acids, sulfonates, and amine collectors in non-sulfide flotation are often closely textured, stable-bubble aggregates that may not break, even by the application of water jets and sprays. Flotation operators seek to avoid voluminous over-frothing and permanent, lathery froth structures, since these are not conducive to the production of high-grade concentrate and are difficult to handle. The addition of specific frothers along with the non-sulfide collectors are suggested to improve these froth conditions, particularly if large quantities of slimes are present in the ore pulp (Booth and Freyberger 1962; Klimpel 1994; Nagaraj and Ravishankar 2007); however, this may not be generalized and is best evaluated in the plant.
†Triethoxybutane (TEB) produces froths resembling those of pine oil (small-bubble, closely knit texture, lower drainage, lower grades, higher recoveries; breaks down readily in launders). However, like polyglycols, TEB does not flatten froth and produce effervescent froths at high dosages.

characterize the behavior of practical three-phase froths date back to 1930 (Taggart 1947). Results from two-phase studies do not translate well to practical flotation froths, which are dynamic three-phase systems. Several frother attributes have been developed or considered in two-phase systems (e.g., CCC and foam height); however, none of them has been linked to metallurgical outcomes from practical froths (see, for example, Melo and Laskowski 2007; Laskowski et al. 2013).

A system much closer to a plant froth zone is a simulated three-phase system (solid/liquid/gas) in the laboratory. In such systems, it has been clearly demonstrated that a small change in collector, either to a different class or to a different homologue in the same class, can have dramatic synergistic or antagonistic effects on froth characteristics. Such alterations can change an effervescent, loose, watery froth to become a tight, heavily mineralized froth that looks like (flocculated) curds. In extreme cases, heavily mineralized froths can become very dry and sluggish. Adding more or the "right" collector (or frother) may transport more particles to the froth zone, but conversely, adding too much or too strong a collector (or frother) might flatten a froth or destabilize it. Depressing some gangue minerals that were involved in stabilizing the froth might in fact destabilize it. Such effects can be linked to the wettability of solids in the froth zone; however, literature is scarce on this subject.

Importantly, much of the discussion in the literature is related to sulfide flotation where a distinct, designated frother is used, unlike the situation in non-sulfide flotation wherein a designated frother (such as alcohol or glycol) is not always used. In non-sulfide operations, collectors such as long-chain fatty acids and amines generate sufficient froth; however, a judicious use of a frother can often lead to significant metallurgical improvement. Although the overall froth zone properties are determined by collectors, modifiers, and other chemical and physical factors, a separate for-purpose chemical is still necessary in virtually all sulfide flotation systems to provide the basic froth. It is widely believed, though somewhat inaccurately, that most of the sulfide collectors, both water-soluble and water-insoluble, are (essentially) nonfrothing because of their shorter chain lengths and, therefore, require the use of specific (or designated) frothers. Some sulfide collectors, under certain specific conditions, are capable of producing froth that is sufficient to operate a circuit for a period of time—for example, the dithiophosphates, especially the longer-chain, water-soluble dialkyl and dicresyl dithiophosphate salts, and the oily dicresyl dithiophosphoric acid. Sometimes, the residual alcohol in these collectors contributes to froth formation as well. However, plants cannot rely on a collector alone to operate a circuit. More importantly, however, the use of frothers in these systems is predicated on the necessity to float coarse (100–250 μm), heavy particles (specific gravity 3.5–19)—both liberated and middlings—and to produce a froth zone that can support such particles (Klimpel and Hansen 1987). A judicious choice of frothers is also predicated on the need to target a certain mass pull (or selectivity).

Qualitative attributes of various alcohols and glycols are shown in Table 4. Impurities in frothers, even in small quantities, sometimes have a marked effect on frothing power and froth structure. Highly insoluble components are frequently observed to float off with the froth, to concentrate in localized areas at the froth surface, and to depress frothing (Booth and Freyberger 1962).

Parameters that affect and are used to describe froth zone characteristics are listed in Table 5. Operational variables, froth characteristics, and qualitative descriptors are listed in Table 6.

Operators manipulate chemical addition and process parameters to achieve multiple concomitant goals in an optimally performing flotation circuit. General goals are listed here, although the importance of each goal may vary from operation to operation:

Table 5 Froth zone parameter definitions

Parameter	Definition	Defining Equation	Units	Typical Range
Q_A	Volumetric air flow rate		L/min	
d_s	Bubble size (surface diameter)		μm	
d_v	Bubble size (volume diameter)		μm	
d_b	Bubble size (diameter); Sauter mean bubble diameter, d_{32}		μm	1,000–3,000
n	Number of bubbles per unit time		1/s	
S	Bubble surface area	$\pi \cdot d_b^2$	m^2	
J_g	Superficial gas rate or velocity [A = cross-sectional area of cell]	Q_A/A	m/s	0.001–0.03
S_b	Bubble surface area rate or flux	$[n \cdot S/A]$ or $[(Q_A)/(\pi/6 \cdot d_b^3)]$ or $[6J_g/d_b]$	1/s	30–100
C_a	Froth carrying capacity (rate of concentrate floated per unit cross-sectional area): d_p = particle diameter (m) ρ_p = particle density (kg/m^3)	$\pi \cdot d_p \cdot \rho_p \cdot J_g/d_b$	$t/h/m^2$	1–5
k	Overall flotation rate constant: R = recovery at time t R_∞ = ultimate recovery (t = ∞)	$-\ln[1-(R/R_\infty)]/t$	1/s	
K_c	Collection zone flotation rate constant (zero froth depth)	$-\ln[1-(R/R_\infty)]/t$	1/s	
R_f	Froth recovery	k/K_c	—	
h	Froth depth		m	0.05–0.5 (mechanical cells); 5–10 (column cells)
τ_f	Froth retention time index	h/J_g or $h \cdot A/Q_A$	s	
ε_g	Gas holdup (or froth factor); volume of bubbles/total pulp volume		%	10%–20% in mechanical cells
ε_w	Water holdup (volume fraction of water in the froth)		%	
ε_s	Solids holdup (volume fraction of solids in the froth)		%	
R_f	Froth recovery factor (efficiency with which particles arriving at the froth–pulp interface reach the concentrate). P is the mineral floatability factor.	$k/(P{*}S_b)$	%	

Table 6 Operational variables, froth characteristics, and qualitative descriptors

Operational Variables	Measurable Froth Characteristics	Qualitative Descriptors*
• Type, amount, and addition point of frother(s) • Type, amount, and addition point of collector(s) • Type, amount, and addition point of modifier(s) • Ore type and mineralogy • Particle size distribution • pH • Presence of slimes • Airflow rate • Hydrodynamics in the cell • Pulp level density • Water chemistry (dissolved inorganic salts, organic matter)	• Bubble size • Volume (height or depth) • Texture • Mobility • Stability • Persistence • Rheology • Carrying capacity	• Well-knit, close-knit • Watery, wet or dry • Lots of windows, clean windows • Stable • Evanescent (quickly fading) • Effervescent (unstable froth, creating a mist over froth surface) • Persistent • Sticky • Brittle • Free-flowing, mobile • Selective, nonselective • Loose • Lathery

Adapted from Nagaraj and Ravishankar 2007
*Common terms used by plant operators.

- Mass pull, mass flow, and grade targets
- Manageable and consistent froth characteristics down the entire bank of cells
  - Steady, uniform, and appropriate volumetric flow of froth from cell to launder.
  - Full mineralization, or, if partial mineralization, the windows on bubbles are clear (i.e., water film is not dirty with fine gangue).
  - Rougher cells exhibit good mass pull and high grade in the first few cells.
  - Cells at the end of the rougher bank are barren, yet there is still some manageable amount of froth.
  - No flooding occurs in the first or second cleaner; no or low insols in the cleaner circuit. (Insols are materials that are insoluble in acid digestion. They are a proxy [i.e., an incomplete accounting] for non-sulfide gangue and typically underestimate the total concentration of gangue in a sample.)
  - No froth occurs in the concentrate and tailings thickeners.

Operators typically employ combinations of frothers to manage and control their circuits. Those combinations and doses can vary across the flotation circuit. In addition to choice of frother, operators can alter the addition points of the frother. For example, frother is rarely added to the grind and more commonly to the head of the rougher circuit. Some frother may also be added to the scavenger circuit if the froth there is inadequate or if the tails assay (for value minerals) is too high. Frothers may even be used in the cleaner circuit; however, such use depends on the frother carried over from the rougher-scavenger circuit and frother coming from circulating loads.

The attribute of froth texture is a little more difficult to put in quantitative terms; however, operators have their own descriptions of froth texture, which become part of their experiential knowledge. These descriptions are nonetheless quite valuable in guiding the changes they make to the flotation system. Booth and Freyberger (1962) noted that froth structure or texture has a strong influence on recovery and grade: "Small-bubbled, closely knit froths will be conducive to high recovery; such froths support heavy mineral loads and prevent detachment of values from the froth. On the other hand, a loosely textured, larger-bubbled froth will tend to release gangue particles from the bubble aggregate and tend to produce concentrates of higher grade."

Another parameter that features importantly in managing and controlling a flotation circuit is pH. Many statements and generalizations about the effects of pH on froth appearance and behavior are prevalent in the industry. However, they are rarely linked to fundamental concepts and should be viewed with caution since they only provide information for specific systems and conditions and lack universality.

Klimpel and Hansen (1987) noted:

> It is sometimes assumed (especially in laboratory studies) that the choice of frother type and dosage is not as crucial as some of the other factors such as collector type and dosage. Plant practice does not support this assumption. Significant improvements can be achieved by managing froth phase, i.e., frother type and dosage. The scientific fundamentals involved in froth development and stability in the presence of particles and turbulence are extremely complex and not yet fully developed; consequently, a priori predictions of performance cannot be made for a given frother. Sometimes, a higher frother dosage, rather than higher collector dosage, is used to improve value recovery because frothers generally have lower unit price, in spite of the fact that this invariably produces poor selectivity.

Rao (2004) observed the following: "It is obvious that despite a considerable amount of attention being paid in the last 100 years to the characteristics of bubbles, of thin layers separating these bubbles in foams and froths, and of the lifetime of froths, there is no unanimity regarding parameters that govern the stability and the collapse of froths." Therefore, research is critically needed in the three-phase froths with direct relevance to plant systems.

Ultimately, frother performance is judged based on the metallurgical outcomes: value mineral recovery, by-product or co-value recovery, concentrate grade, rejection of penalty elements, reduction in mass pulls, ease of operation (in terms of managing froth stability, mobility), carrying capacity, rejection of gangue minerals (particularly those resulting from entrainment and entrapment), cost, and possibly some others specific to an operation.

Reagents for Non-Sulfide Minerals

With the exception of a few elements such as the base and precious metals, which are predominantly obtained from sulfides ores, most other elements or their minerals are obtained from non-sulfide ores. A chronology of reagent development of relevance to plant practice in non-sulfide flotation may be found in Nagaraj and Ravishankar (2007; Table 1 therein), and Nagaraj and Farinato (2014, 2016).

Aplan (1994) noted that the non-sulfide flotation literature is very fragmented. This is necessarily so if one considers the variety of non-sulfides that are beneficiated in the industry and the artisanry that surrounds these processes. However, as in sulfide flotation, a common theme emerges if one considers the types of reagents used (both collectors and modifiers) and the types of treatment methods employed. Redeker and Bentzen (1986) noted that "Flotation of nonmetallic minerals is sometimes viewed as alchemy. However, the art is, in reality, based on sound chemistry, experience and a little luck in the treatment of new ores. In the flotation of nonmetallic ores, as opposed to base metal sulfides, highly specific treatment conditions are required to accomplish separations." Aplan and Fuerstenau (1962) noted that "Selective flotation in non-sulfide systems is often difficult in that one class of compounds alone, the carboxylic acids, may float nearly the entire array of non-sulfide minerals under the appropriate conditions. The non-sulfide mineral flotation operator is confronted with a vast array of modifying reagents."

In the case of sulfide flotation, a wide variety of collectors, many of which exhibit a high degree of selectivity for a given sulfide, is available for the relatively small number of sulfide minerals floated. In contrast, only a small number of collectors, many of which are rather nonselective, are available for the large number of non-sulfide minerals floated. Indeed, modifiers sometimes assume a greater role than collectors in non-sulfide separations, and it is common to use several modifiers in a given separation. This obviously complicates the system considerably, puts it in the realm of art form, and makes it very challenging for a scientific study.

Pulp preparation prior to flotation separation is far more important in non-sulfide separations. This includes one or more of the processes such as washing and/or attrition scrubbing (usually with acid or alkali), blunging (typically for clays), desliming, hot pulp conditioning, and high solids conditioning. Preconditioning requirements are often less stringent in flotation with amines, in contrast to flotation with fatty acids. Conditioning in amine flotation is usually at the same pulp densities (~20–30 wt %) as flotation, unlike in fatty acid or sulfonate flotation wherein pulp densities could be as high as 70–75 wt %. A variety of modifiers is used in these operations to ensure that collectors added subsequently adsorb selectively onto the desired minerals. The critical factors are pH, contact time of reagents, agitation intensity, water quality, and frequently the temperature during scrubbing, conditioning time, and the number of conditioning and flotation stages.

Conditioning times with amines and petroleum sulfonates tend to be short (e.g., 30 seconds to 2 minutes), whereas for fatty acids and alkyl hydroxamates, conditioning times vary considerably, from 30 seconds to 30 minutes, depending on the particular separation.

Desliming is a very common and often critical step in non-sulfide flotation circuits, unlike in sulfide mineral flotation, because the slimes, being non-sulfide minerals (clays, iron oxides, etc.), can increase collector consumption, coat coarse particles, hinder particle attachment to bubbles, and seriously interfere with separations. The presence of slimes can reduce the ability of collectors to selectively adsorb onto targeted mineral surfaces. Since these slimes are usually discarded, they can account for a considerable loss in value minerals. The deslimed material is sometimes separated into coarse and fine fractions and treated separately to optimize flotation conditions for each size fraction. Some non-sulfide separations are virtually impossible if the feed to flotation is not properly deslimed. The term *slimes* has a loose meaning in non-sulfide separation and is system dependent. For example, –100 µm particles are considered as slimes in some phosphate operations, whereas in other systems, particles less than 1–10 mm are considered slimes.

Compared to sulfide flotation, the field of non-sulfide flotation is almost bewildering in terms of the diversity of separations and technologies employed. However, general guidelines are available in published literature and chemical supplier handbooks (Redeker and Bentzen 1986; Cytec Industries 2010; Arbiter and Williams 1980; Yang 1987; Holme 1986; Davis 1985). These are excellent starting points, and the practitioner must then conduct empirical testing to make the necessary changes to reagent schemes and processing schemes dictated largely by the mineralogy, particularly that of gangue minerals. The general approach to laboratory testing is provided in a separate chapter of this handbook.

Collectors

In stark contrast to sulfide flotation, two types of collectors, namely fatty acids and amines, cover approximately 90% of the separations in non-sulfide systems. The usage of collectors, excluding auxiliary collectors such as hydrocarbon oils, in non-sulfide flotation follows the order: fatty acids > amines > petroleum sulfonates > other surfactants. In recent years, there has been growing interest in the development of specialty collectors that have the potential to provide significantly improved performance. A great example is that of alkyl hydroxamates (discussed later in this chapter). The important active donor atom in almost all of the collectors is still the O. The important classes (or families) of collectors used in non-sulfide mineral flotation are listed in Table 7. Only generic v structures and typical chain lengths are given because, in practice, most of the reagents tend to be mixtures either resulting from the manufacturing process or deliberately formulated by chemical suppliers; exact compositions may not be available for these, whether for lack of analysis or proprietary reasons.

Fatty acids. Fatty acids can be considered to be almost universal collectors, because they can be made to float almost all minerals; consequently they tend to be nonselective. The use of appropriate modifiers alleviates this problem somewhat and allows even difficult mineral separations. Fatty acids can also be used to float sulfides, but they offer no advantages over the S-containing compounds that are much more selective for sulfides than are fatty acids. Fatty acids, however, have been used for the recovery of oxide copper minerals, though this is not widespread because of the poor selectivity and often the presence of several multivalent ions in the plant water (especially alkaline-earth ions, which can precipitate as fatty acid soaps).

Two main sources of fatty acids are oils and fats from vegetable and animal sources (tallow), and wood pulp (tall oil fatty acids or TOFA). Composition of fatty acids varies depending on the source. Fatty acids are best characterized by their iodine value (IV) and acid number (AN). IV indicates the amount of unsaturation in the molecule (i.e., the number of double bonds). AN is a measure of the amount of carboxylic acid groups in the product. TOFA and crude tall oil (less expensive than TOFA) are the most widely used. The composition of crude tall oil varies greatly, depending on the wood used (acid numbers can vary in the range of 125 to 165, depending on the wood used). By fractional distillation of crude tall oil (with rosin content of ~40% or higher), higher-grade fatty acids (with reduced rosin content) is obtained. A good quality TOFA is a distillation product with rosin content 10% or less. TOFA consists mostly of oleic acid. The choice between TOFA and crude tall oil is dependent on the price tolerance of a particular operation, though the performance of crude tall oil is typically less than that of TOFA. Suppliers may also market formulations of fatty acids with various additives; composition of such formulations is most often proprietary.

Amines. Five types of amine collectors can be recognized (see Table 7): (1) fatty acid condensate amine (and imidazoline), (2) alkyl ether amine (and diamine), (3) alkyl morpholine, (4) fatty amine, and (5) alkyl guanidine.

Typical dosages for amines are 0.05 to 0.5 kg/t. They generally require short conditioning time as they are known to have rapid adsorption. Amines are effective in a wide pH range (2.5–10.5), and the choice of the pH depends on the mineralogy and the required selectivity—for example, direct flotation of sylvite (KCl) in sylvite–halite (NaCl) separation; reverse flotation of halite in carnallite–halite separation; feldspar and mica flotation at pH 2.5–3.5; silica flotation from phosphate at pH ~ 6.5–7 (phosphate mineral depressed using sulfuric acid, starch, and others), and silica flotation from hematite at pH ~10.5 (hematite depressed using starch; silica activated and slimes dispersed using sodium silicate). Amines are often used in reverse flotation processes, that is, when the gangue (e.g., silica) is floated away from the value minerals.

Amines are characterized by amine number, which is a measure of the amount of amine groups and used to estimate the average MW. This is especially useful when the collector product is a mixture of amines and often with other formulation components. It is determined by titration of the amine acetate with hydrochloric acid (HCl).

Natural and synthetic petroleum sulfonates. The family of sulfonates and sulfates (all anionic surfactants; Table 7) is quite large. Several products from this family have been used extensively in flotation applications, and these constitute the third largest group (after fatty acids and amines) of collectors used in non-sulfide flotation. Natural petroleum sulfonates are manufactured by sulfonation of crude oil, crude distillates, or any portion of these distillates (Basu and Shravan 2008). Synthetic sulfonates are made by alkylation of an aromatic hydrocarbon with a selected olefin, followed by sulfonation

Table 7 Collectors used in non-sulfide mineral flotation*

No.	Chemical Class	Structure	Plant Application
1	Fatty acids	$R = C_{14}–C_{18}$  Oleic acid	Widely used for beneficiation of phosphates, fluorspar, kaolin clays; glass sands, heavy mineral sands, foundry/molding sands, kyanite, tungsten ores, and to a limited extent low-grade iron ores; common example is tall oil containing palmitic acid (C16:0), oleic acid (C18:1 cis-9), linoleic acid (C18:2 cis,cis-9,12). C_{18} unsaturated fatty acids (e.g., oleic, linoleic, and linolenic acids) are the most efficient collectors in tall oil mixtures. Long-chain saturated fatty acids ($>C_{12}$) are poor collectors (attributed to their much lower solubility and increasingly high melting points; Hsieh 1980) and, therefore, are not used as primary collectors (there may be small amounts of saturated fatty acids in commercial fatty acid products). Dosages vary from 250–3,000 g/t depending on the ore system and target mineral. Conditioning times vary from 30 seconds to 30 minutes. High solids (50–75 wt %) conditioning is typical, sometimes at high temperature (e.g., fluorite and kaolin).
2	Amines and their salts	Fatty amines: 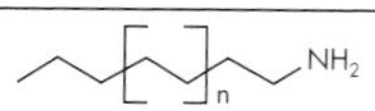 Amine salt: $R–NH_3^+\ Cl^-$ Condensate amines:  Alkyl ether amine: 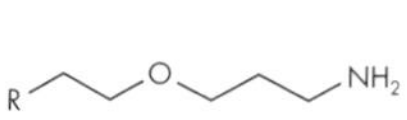 Alkyl ether diamine: 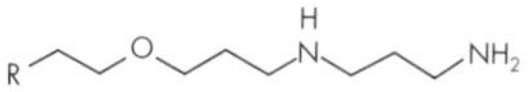 Alkyl imidazoline: 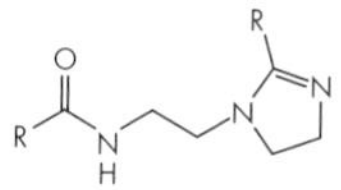 Alkyl morpholine: 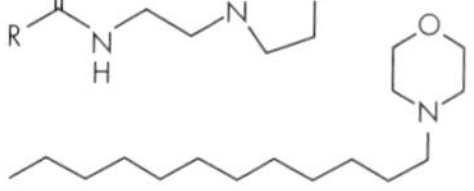 Dodecyl guanidine HCl: 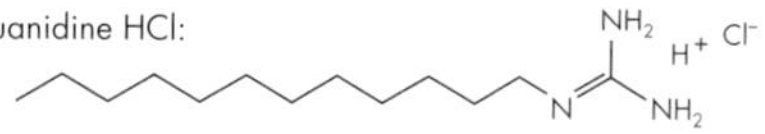	Typical $R = C_{12}–C_{18}$; widely used for the beneficiation of potash, iron ore, phosphate, calcium carbonate, feldspar and other silicates, and silica sand. Primary fatty amines, ether amines, and condensate amines constitute the bulk of cationic collector usage in the industry. Alkyl morpholine is used in the halite flotation from carnallite salts. Ether amines and ether diamines are used predominantly to float silica in Fe ore beneficiation. Ether amines are typically $C_{12–14}$ and are considered to have slightly weaker collecting property and therefore higher selectivity. Amines are typically pastes, difficult to handle, and have poor solubility and dispersibility. Consequently, they are used as custom blends of amine or its salt (e.g., full or half acetate) and frothers or other diluents (e.g., hydrocarbon oil, surfactants); these are used as dilute dispersions (1%–5%). Plants may prepare the salt on-site. Formulations provide more flexibility. Their composition is often proprietary. Ether amines are liquid products, more soluble and dispersible. Diamines tend to be stronger because they contain two N atoms (one secondary and the other primary), but they are also known to be more frothing.

(continues)

Table 7 Collectors used in non-sulfide mineral flotation (continued)

No.	Chemical Class	Structure	Plant Application
3	Petroleum sulfonates MW ~300–550	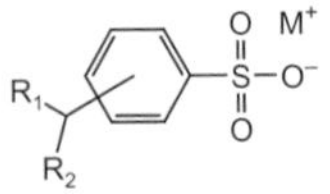	Widely used for removal of iron and other contaminants in the processing of glass sands, feldspar, and other ceramic raw materials. Also in formulations with fatty acids. Combinations of low- and high-molecular-weight compounds, or of synthetic and natural sulfonates, are also used commonly. Products tend to be viscous liquids or pastes that require specific reagent preparation steps such as tanks equipped with heaters and intense agitation to solubilize or disperse. Typically supplied as custom formulations that provide better handling characteristics and better metallurgical performance at reduced dosages.
4	Sulfated and sulfonated fatty acids	R = C_{14}–C_{18} 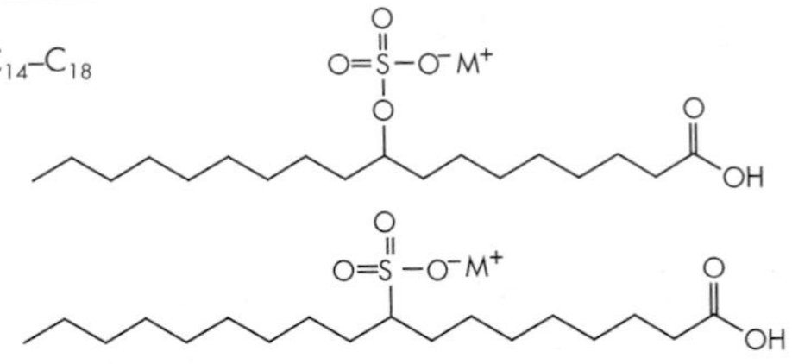	Used for reverse flotation of carbonates from phosphate ores in the pH range of 5–5.5 (Lawendy and McClellan 1993). These collectors are known to be more tolerant to Ca and Mg ions and, therefore, particularly useful in waters with high hardness.
5	Alkyl sulfosuccinamates	R = C_{12}–C_{18} 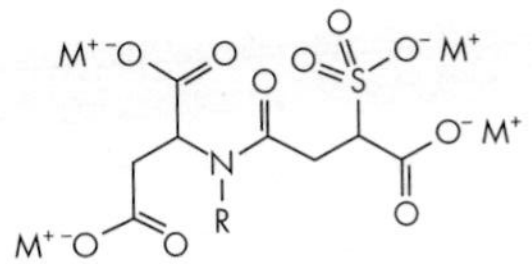	Mostly used for specialty applications, often in combination with fatty acids and petroleum sulfonates. They are used in the beneficiation of barite, fluorspar, scheelite, phosphates, cassiterite, wollastonite, celestite, and glass sands.
6	Dialkyl sulfosuccinates	R = C_6–C_8 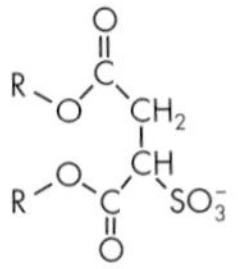	Mostly used for specialty applications, and in combination with fatty acids.
7	Alkyl sulfates Alkyl ether sulfates	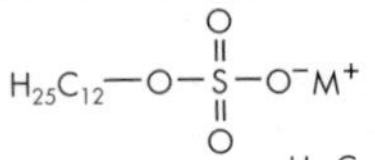 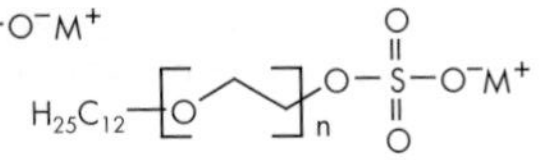	Limited use; frequently in combination with fatty acids (e.g., barite ore beneficiation).
8	Alkyl hydroxamic acids	R = C_8–C_{10} 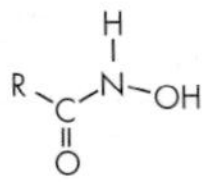	Mostly for specialty applications; sometimes combined with fatty acids. Important applications include kaolin clay (especially fine clay) beneficiation (colored mineral removal without activators or carrier minerals), rare earths, oxide copper minerals from oxide/sulfide/mixed ores.

(continues)

Table 7 Collectors used in non-sulfide mineral flotation (continued)

No.	Chemical Class	Structure	Plant Application
9	Aryl hydroxamic acids (e.g., salicyl-, hydroxynaphthyl-, and benzo-hydroxamic acid)	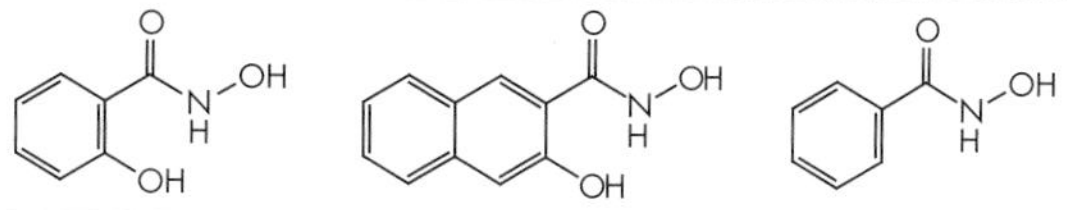	Limited use currently; rare earths and oxide Cu beneficiation.
10	Aryl and alkyl phosphonic acids	Styrene phosphonic acid (SPA) $R = C_8$ 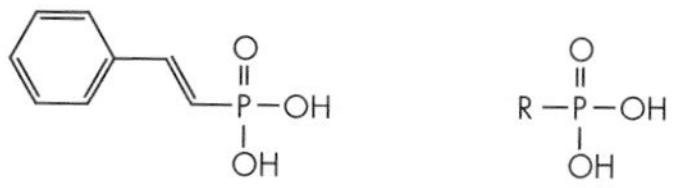	SPA used widely in cassiterite flotation; high unit cost of SPA is often offset by unique selectivity and efficiency. No other use.
11	Dialkyl phosphates	$R = C_8$ 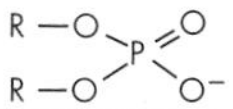	Very limited use (e.g., selective separation of phosphates from siliceous and carbonate gangue).
12	Amphoteric collectors	(low pH) R typically C_{12} (high pH) 	Very limited use (e.g., selective separations between phosphate, silica, and carbonate gangue); removal of phosphate impurities from iron ore.
13	Hydrocarbon oils (kerosene, diesel oil, burner oil, gas oil, fuel oil, pump oil, vapor oil, aromatic oils, vegetable oils, mineral oils)	Complex mixtures of aliphatic hydrocarbon, aromatic hydrocarbons, linear and branched. Carbon number range: C_8–C_{22}	Flotation of naturally hydrophobic minerals (talc, sulfur, graphite and other forms of carbon). Extender to fatty acids, amines and other collectors. Froth modification and control.

*Only generic idealized structures and typical chain lengths are given.

of this alkylate (Malmberg et al. 1982). In general, natural petroleum sulfonates are much more complex mixtures than synthetics (Basu and Shravan 2008). Petroleum sulfonates were originally developed as relatively inexpensive anionic collectors to selectively concentrate iron-bearing minerals in low-grade iron ores and to remove iron and other mineral contaminants by flotation in the processing of glass sands, feldspar, and other ceramic raw materials. Further development resulted in additional reagents with a broadened application to the selective flotation of a variety of non-sulfide minerals (Holme 1986; Falconer 1961).

Sulfated and sulfonated fatty acids. These date back to the mid-1800s, developed primarily as textile additives and soap. They are produced by reacting fats, oils, and tall oil with sulfuric acid or oleum or SO_3, sulfuric trioxide (current practice), followed by hydrolysis and saponification using sodium hydroxide. This results in addition of a sulfate or sulfonate group at the double bond in unsaturated fatty acids. The terms *sulfated* and *sulfonated* are used interchangeably, though they are chemically very different—sulfated species having the sulfur attached to the hydrocarbon chain via an oxygen atom (Table 7). The formation of a sulfate or sulfonate is determined largely by whether sulfuric acid or oleum or SO_3 is used and by the process conditions. Further complexity arises from the extent of hydrolysis and saponification. Sulfonated (sulfated) fatty acids are typically complex mixtures, and their characterization is often not readily available.

Sulfosuccinates and sulfosuccinamates. Sulfosuccinamates (Table 7) were introduced originally in 1945 for non-sulfide flotation and more formally in ~1965 as collectors for cassiterite flotation (Arbiter and Hinn 1968), and subsequently for the flotation of a wide variety of non-sulfide minerals, either alone or in combination with fatty acids and petroleum sulfonates (Day and Hartjens 1974, 1977). Industrial practice indicates that they have unique properties and provide synergistic effects with fatty acids and petroleum sulfonates. For example, when used in combination with fatty acids, at a level of ~10%–20% of the fatty acid dosage, they provide better froth structure and higher recovery of fine and coarse particles in phosphate flotation. They also serve to emulsify the fatty acid, thereby improving its performance and providing a stabilizing effect on the plant performance differences resulting from pH, fatty acid grade, and feed grade variations.

Alkyl hydroxamates. Alkyl hydroxamates (Table 7) were first suggested for flotation in 1941 (Pöpperle 1941), and their utility was expanded in the 1950s through 1970s, but plant usage was sporadic and minor, mostly in the old Soviet bloc. There is currently extensive literature on both fundamental studies and potential applications of alkyl hydroxamates for flotation of a wide variety of minerals (Fuerstenau and Pradip 1984; Assis et al. 1996; Hughes et al. 2007; Bhambhani et al. 2016; Jordens et al. 2016). Their usage prior to 1980s was rather limited. Two important innovations in the 1980s appear to have changed this situation. First was the use of alkyl hydroxamate for removing colored impurities from kaolin clays (Yoon and Hilderbrand 1986a, 1986b). The second was the development of a practical, safe, and economical process for the manufacture of alkyl hydroxamates (Wang and Nagaraj 1989), which resulted in a clean liquid product that was stable and easy to handle. Additional developments occurred in the manufacture of alkyl hydroxamates (Hughes 2004, 2006; Bhambhani et al. 2017). In addition to C_8–C_{10} alkyl hydroxamates, three other hydroxamic acids are now available: salicyl-, hydroxynaphthyl-, and benzo-hydroxamic acid (Wu and Zhu 2006; Jordens et al. 2016; Xia et al. 2015). However, their plant usage appears to be limited.

Although hydroxamates and hydroxamic acids are structurally closest to fatty acids (hydroxamates have the additional –NH in the functional group), their solution chemistry and interactions with minerals are vastly different from those of fatty acids (Nagaraj 1987; Fuerstenau et al. 1970). One of the most important differences is that hydroxamates are far more selective than fatty acids under most conditions.

Phosphonic acids, phosphoric acids, amphoterics, and hydrocarbon oils. The most well-known phosphonic acid is styrene phosphonic acid, or SPA (Table 7), which has shown some unique properties (Andrews 1989, 1990; Wottgen 1969). This is used in cassiterite flotation. One cheaper alternative, an amino phosphonic acid, has been successfully used in a tin operation (Codd 1984). There does not appear to be any other commercial use for SPA. Other phosphonic acids have been proposed (Wottgen and Lippman 1963), but none have found significant application.

Alkyl phosphoric acid esters have found niche applications, usually in conjunction with other collectors. Houot (1983) describes their use in selective separation of phosphates from siliceous and carbonate gangue. In one case, for example, the phosphates and carbonates can be floated into a bulk concentrate using alkyl phosphates, leaving the silica in the tails. The bulk concentrate is then scrubbed with sulfuric acid to selectively strip the collector off the phosphate surfaces and the carbonates are floated without any additional collector. In another case, the carbonates can be floated selectively with alkyl phosphates using sodium fluorosilicate or sulfuric acid as a depressant for phosphates. The silica remaining with the phosphates can then be floated selectively using an amine collector.

Amphoteric collectors have some unique properties and provide the potential for achieving selectivity over a wide range of pulp conditions.

Sodium N-n-dodecyl sarcosinate has been used commercially for selective separations between phosphate, silica, and carbonate gangue. Although reportedly used in Siilinjarvi, Finland (Chadwick 1981), there does not appear to be a widespread application of this collector. One of the important reasons could be the higher cost of the product compared to the fatty acids and amines used in phosphate beneficiation. Amphoteric collectors have also been used to remove phosphate impurity from iron ores. One advantage with an amphoteric collector is that the same reagent can be used at two different pH values to float different mineral species.

Large amounts (and high dosages) of hydrocarbon oils such as fuel oil, diesel oil, and kerosene are used in non-sulfide flotation as extenders or auxiliary collectors to enhance the flotation with fatty acids, amines, and petroleum sulfonates. Overall tonnage used is more than that of fatty acids.

Frothers

Because the collectors used in non-sulfide flotation provide adequate froth formation, the use of *separate* frothers (short chain alcohols and polyglycols) is generally not widespread. When used judiciously, frothers have been shown, however,

Table 8 Interaction of modifiers with minerals and the mechanism of modification of floatability

Adsorption Phenomena	
1	Adsorption of ions: electrostatic, specific adsorption, adsorption of potential determining ions (e.g., M^{n+} on oxides and sulfides; OH^- on sulfides or oxides, which then affects collector adsorption; HS^- on sulfides [driven by solubility product, chemical reaction])
2	Ion exchange and chemical reaction; Cu activation of sphalerite
3	Entropy-driven adsorption (due to bound counterion release; e.g., adsorption of polyacrylate on iron oxides)
4	Hydrogen bonding and hydrophobic bonding (e.g., adsorption of nonionic species on minerals, especially nonionic polymers)
5	Chemisorption and chemical reaction (e.g., NaCN [sodium cyanide] on sulfides, polyacrylate on calcite or apatite)
Mechanism by Which a Modifier Can Affect Mineral Floatability	
1	Selective blocking of all active sites to prevent collector adsorption (PO_4^{2-} on apatite preventing fatty acid adsorption)
2	Inadvertent activation and the consequent adsorption of collector (the active centers are not of the mineral but are of the impurity species causing inadvertent activation)
3	Concentration-dependent effect: At low concentrations sodium sulfide can activate certain oxides and tarnished sulfides, and at high concentrations it depresses these.
4	Adsorption of colloidal reaction products of modifiers on minerals (e.g., metal sols of sodium silicate, products of solution hydrolysis, sol formation and gel formation, or other colloidal precipitates)
5	Prevention of heterocoagulation of slimes to minerals (e.g., with dispersants such as polyacrylate, and polyphosphate)
6	Removal of active species (sites) from the surface by complexation. This is often concentration dependent. At low concentrations, certain complexing reagents can chemisorb on surfaces, but as the concentration increases, they can form soluble complexes that are then released into solution. NaCN is a good example; another example is polyphosphate.
7	Removal of hydrophobic species from surfaces. Chemical or physical methods. For example, Na_2S in Cu-Mo separation—removal of collectors from surfaces. Removal of oleate from apatite by sulfuric acid. Layers of mineral may be removed in this process. Physical means would be steaming, high-energy attrition.
8	Removal of hydrophilic patches to increase floatability, (e.g., sodium sulfide, dispersants). For example, a layer of limonite may be removed from the oxidized surface of chalcopyrite by treatment with oxalic acid and subsequent washing with water. Acid scrubbing of many sulfide and non-sulfide minerals and of glass sands to activate iron-bearing impurities and remove such impurities from silica surfaces. Treatment of beryl with hydrofluoric acid (HF) leads to removal of the silicate components such as silica and alumina, thereby increasing Be concentration on the surface.

Adapted from Nagaraj and Ravishankar 2007

to contribute to improved performance (e.g., better froth control, carrying capacity, and recovery of coarse and fine particles) in the beneficiation of phosphate, glass sand, feldspar, soluble salt, and calcium carbonate. Frothers are also often a component in collector formulations (e.g., those of fatty acids or amines) to produce easy-to-handle products and to aid in physical compatibility of oily and aqueous components. In some cases, however, when frother is already a part of the formulation, it reduces the ability to control froth independently and may impose an upper limit to dosage of the collector formulation (because an increase in collector dosage may inadvertently exceed the frother requirement threshold).

Kerosene and fuel oil are often used to effectively control the voluminous over-frothing caused on occasion by saponified fatty acids and tall oils.

Modifiers

This section discusses the broad collection of modifiers that can be used in both sulfide and non-sulfide mineral flotation. Unlike collectors, understanding modifier chemistry and complex interactions presents a formidable challenge; this is exacerbated by the limited coverage and discussion in the literature for modifiers that are of relevance to plant practice. The intent is to give relatively greater coverage to modifiers in this section.

Of the three major reagent categories forming the flotation reagent triangle, modifiers have perhaps the greatest effect on rate of flotation of particles; in the extreme, flotation is completely suppressed. This effect can be either positive or negative, and it can be temporary (i.e., the mineral can be refloated in another stage with relative ease, such as when using a polymeric modifier) or permanent (i.e., true depression, where the mineral can be refloated only after stronger measures have been taken, such as a change in chemical environment; for example cyanide depression of pyrite or sphalerite, reversed by addition of copper sulfate). Thus, one modifier may merely decrease the flotation rate of one mineral selectively and sufficiently so that selective flotation of another mineral is possible.

Modifier Function and Mode of Action

Modifiers are used to create and/or magnify the differences in the surface properties of minerals subject to separation. The overall action of modifiers on flotation is manifested in the activation or depression of individual minerals, allowing an increase in flotation selectivity. Modifiers can often participate in a variety of processes. The dominant forms of modifier action are those that proceed most rapidly and whose physical results have the greatest effect on the state and properties of the solid surfaces and air bubbles, and on pulp chemistry. The action of most modifiers on minerals is strongly influenced by the particular conditions in the reagent conditioning and flotation stages, and by the ore composition. Modifier action is best understood by determining their dominant effects on the individual subprocesses in a flotation system.

The chemistry of modifiers in flotation pulps is quite complex in comparison to that of collectors, and it is more complex in non-sulfide systems than it is in sulfide systems. The interactions of modifiers with minerals and the mechanism by which modifiers affect mineral floatability and selectivity are summarized in Table 8. Modifiers can affect multiple factors simultaneously in flotation pulps and, therefore, can have multiple functions (details of *predominant* functions of various modifiers are found in Eigeles 1977, Laskowski 1988, and Chander 1987). For example, addition of lime in a copper circuit can have the following effects:

- pH modification (and the resultant changes in interfacial chemistry, solubility of minerals, changes in water chemistry)
- Depression of pyrite
- Formation of colloidal calcium sulfate precipitate and its interaction with sulfide minerals

- Modification of froth characteristics and viscosity of pulp, thereby altering particle transport through froth, gas dispersion, and bubble–particle interactions
- Modification of collector adsorption
- Precipitation of multivalent metal ions as hydroxides
- Removal of slime coatings from sulfides
- Dispersion and depression of certain gangue minerals
- Reduction in grinding media wear

Thus, depending on the type, modifiers can change pH, pulp potential (E_p), composition of the aqueous phase, surface composition of the minerals and bubbles, and froth characteristics; in other words, they can affect properties of all three interfaces. Modifiers change mineral surfaces either directly by adsorbing at the solid–liquid interface (via physical or chemical adsorption) or indirectly by changing aqueous phase composition. Modifiers can also affect the liquid–gas interface (thus, froth characteristics) significantly, but this aspect has been inadequately studied. They are also known to affect dispersion and rise-velocity of bubbles. The same modifier (e.g., lime) in different circuits could produce very different flotation outcomes, or different modifiers used for pH control in a circuit can have very different results at the same nominal pH (e.g., lime versus NaOH).

The action of modifiers (alkalis, acids, and salts) on the dispersion of air bubbles and the intensity of frothing was described long ago (Gaudin 1957). The strong effect of pH on froth stability and mobility is all too well known in sulfide circuits, and the direction and magnitude of this effect is strongly influenced by gangue mineralogy and is quite variable. Generalizations are difficult to make. Salts of multivalent metals, such as copper and iron, even at low concentrations, reduce the stability (and/or mobility) of the froth formed by an alcohol; the magnitude of the effect is strongly influenced by pH. Thus, copper sulfate is often used to control excessive froth in certain sulfide circuits. In a zinc circuit, after the addition of copper sulfate and xanthate, a decrease in frothing is sometimes observed, and this can be restored after the addition of lime (Eigeles 1977). Slimes and colloidal precipitates that exist in flotation pulps strongly influence the froth zone and bubble–particle interactions. Their effect can be moderated by modifiers as well. Thus modifiers can affect froth flow, and particle transport from pulp into froth and from froth into launder.

In many cases, modifiers can have multiple modes of action. Some examples are shown in Table 9. The following general categories of modifier mode of action are illustrated with examples.

Classes of Modifiers

Modifiers fall into two groups: unintended and deliberately added chemicals. Deliberately added modifiers can be classified as inorganic, small organic, or polymeric reagents. Inorganic modifiers, which are the least expensive (unit price or price per kilogram) and have been around the longest, are used the most. Many small organic modifiers can provide good performance; however, their higher treatment cost relative to inorganic modifiers often limits their use. Polymeric modifiers include natural product-based (e.g., polysaccharides, tannins, and lignin) and synthetic low-MW polymers that have mineral-specific functional groups incorporated into the polymer (Nagaraj et al. 1987; Nagaraj 2000; Aimone and Booth 1956). Synthetic modifiers can be specifically tailored by incorporation of various functional groups to provide selective separations (Nagaraj 2000). With this ability, synthetic modifiers can help improve the sustainability of challenging separations and potentially serve as attractive alternatives to some hazardous chemicals currently in use. Polymeric modifiers typically require significantly lower doses compared to inorganic and small organic modifiers, making them more cost-effective.

In addition to modifiers added deliberately, significant amounts of unintended unavoidable ions and species released from the minerals are also present in flotation pulps, and they can act as unintended (or inadvertent) modifiers. Even minerals themselves can act as unintended modifiers. The unavoidable ions and species formed in this process react with water, with each other, with other reagents forming complex ions and molecules, and colloidal and coarsely dispersed suspensions and precipitates. These species, in turn, adsorb on or adhere to air bubbles, thereby hindering bubble–particle attachment. These species also adsorb on, or react with, the mineral surfaces and change their surface compositions. The products and kinetics of these processes depend on the concentration of the interacting components, temperature, and, to a large degree, hydrodynamics. Temperature is not an important factor in most of sulfide flotation circuits (only small tonnage separation circuits like Cu-Mo separation may operate at higher temperatures) because the tonnages processed are too large and it would be cost-prohibitive to attempt to control temperature. However, because of seasonal variations, temperature does become a factor in many plants. Temperature is an important factor in many non-sulfide flotation circuits (e.g., fluorite, kaolin clay, and rare earth ores). The effect of unintended species in flotation is further complicated in plant circuits by large recirculating loads (common in sulfide flotation circuits; not so common in non-sulfide circuits) and recycled water (also most common in sulfide circuits; not so in non-sulfide flotation). Unavoidable species generated from minerals in the ore can have a wide range of effects on both efficiency and selectivity of separation via (a) interfering with collector adsorption on desirable minerals; (b) reducing flotation of targeted minerals; (c) promoting collector adsorption on, and flotation of, undesirable mineral(s); (d) changing froth characteristics; (e) changing pulp viscosity; and so on. One major effect is inadvertent activation of certain value minerals (Nagaraj and Brinen 1995, 1997; Smart et al. 1998; 2003; Stowe et al. 1995; Rao 2004). For example, inadvertent activation of sphalerite by Cu or Pb causes significant amounts of Zn misreporting to Pb and Cu concentrates. Other important examples are inadvertent Cu activation of pyrite in sulfide circuits and Ca activation of non-sulfide minerals in non-sulfide flotation circuits. Inadvertent activation is often difficult to predict, because it is strongly influenced by mineralogy and pulp conditions that exhibit significant variability in a plant. Indeed, one of the main functions of modifiers added deliberately (e.g., lime, soda ash, sodium silicate, polyacrylates, sodium sulfide) is to control the effects of inadvertent and unavoidable species.

Inorganic Modifiers

Action of inorganic modifiers. Inorganic compounds constitute the bulk of modifiers used in flotation. The material presented in this chapter characterizes only the intended

Table 9 Multiple modes of modifier action

Modifier	Mode of Action						
	pH Regulator	E_p (pulp potential) Regulator	Depressant	Dispersant	Activator	Deactivator	Extender
Lime	✓		✓	✓	✓		
NaOH	✓		✓	✓			
NH_4OH	✓						
Na_2CO_3	✓		✓	✓	✓	✓	
H_2SO_4	✓		✓		✓	✓	
HF			✓		✓	✓	
NaCN		✓	✓		✓	✓	
$Zn(CN)_2$			✓				
SO_2	✓	✓	✓		✓	✓	
NaMBS ($Na_2S_2O_5$)		✓	✓				
Na_2SO_3		✓	✓				
$Na_2S_2O_3$		✓	✓				
$NaSH/Na_2S$		✓	✓		✓	✓	
Nokes (P_2S_5+NaOH)		✓	✓				
$Na_4Fe(CN)_6$ or $K_4Fe(CN)_6$			✓				
H_2O_2		✓	✓		✓		
$KMnO_4$		✓	✓		✓		
$K_2Cr_2O_7$		✓	✓				
O_3		✓	✓				
Guar			✓				
Other gums			✓				
Carboxymethylcellulose			✓	✓			
Starch			✓				
Dextrin			✓				
Lignin sulfonates			✓	✓			
Tannin (quebracho)			✓	✓			
Na_3PO_4			✓	✓			
Synthetic polymers			✓	✓		✓	
Metal ions			✓		✓		
$CuSO_4$					✓		
$ZnSO_4$			✓				
$NiSO_4$			✓				
$Pb(NO_3)_2$					✓		
Ca^{2+}			✓		✓	✓	
Organic complexing reagents			✓			✓	
Sodium silicate			✓	✓	✓	✓	
Metal sols			✓				
Polyphosphates			✓	✓		✓	
Hydrocarbon oils							✓

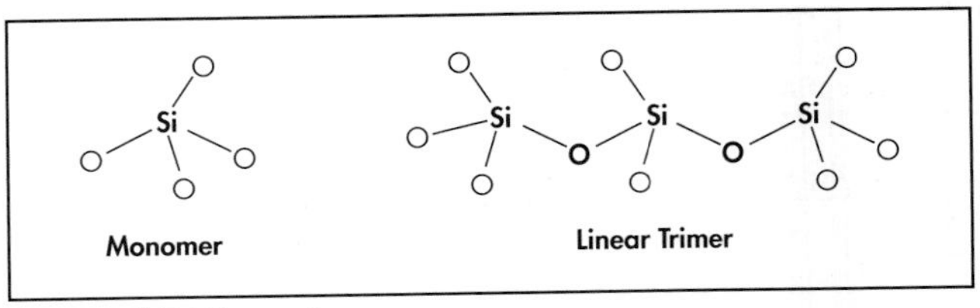

Courtesy of PQ Corporation
Figure 3 Silicate structures

function. Inorganic modifiers exhibit a wide range of effects on flotation performance depending on circuit conditions, pulp chemistry, modifier dosage, and mineral composition of the ore. Inorganic modifiers also find use in treating recycled water that may harbor deleterious species. Functions, mode of action, and usage of various inorganic modifiers are shown in Table 10.

Acid and bases. Control of the pH of the pulp using appropriate acids and bases is one of the most significant methods for controlling interfacial chemistry, bulk-phase chemistry, and overall flotation selectivity and efficiency. Acids and bases often have multiple effects in flotation systems. Table 10 lists examples of how the major acids and bases are used in commercial flotation operations. Notably, hydrofluoric acid is quite hazardous and it has come under increasing scrutiny and regulatory pressure.

Salts of weak, medium-strength, and strong acids. The action of metal salts of different acids on flotation systems is even more varied than that of alkalis and acids, and a large number of these compounds is used as modifiers. The various commercially important salts used to modify flotation performance are listed in Table 10 with application examples. Many of these chemicals (e.g., NaSH, Na_2S, chromates, dichromates, permanganates, Na fluoride and Na fluorosilicate) present significant hazards and have come under increasing regulatory scrutiny.

Complexing and colloidal modifiers. In the selective flotation of sulfide and non-sulfide ores, inorganic modifiers are used, whose action is related not only to ionic adsorption and to simple ionic reactions but also to the formation of complex compounds and to the formation and action of colloidal sols. The commercially important ones are listed in Table 10, along with examples of their use.

Sodium and other metal silicates.* These form a family of commonly used multifunctional modifiers in flotation applications. Sodium silicate was introduced in the early days of flotation (1910–1920). There is extensive literature on silicate chemistry. Iler's (1979) book, *The Chemistry of Silica*, is a classic; and while it does include sections on soluble silicates, most of the very extensive book is not relevant for practitioners in flotation. There are more appropriate references pertaining to flotation (Falcone 1982; Marinakis and Shergold 1985; Qi et al. 1993; Yang et al. 2008).

Soluble silicates ($Na_2O \cdot xSiO_2$) contain three main components: silica, alkali, and water. All soluble silicates can be differentiated by the ratio SiO_2/Na_2O, which determines the physical and chemical properties of the product. Commercial sodium silicates are available with ratios of SiO_2 to Na_2O of 1.6–3.25.

*Information in this section was largely provided by PQ Corporation.

Sodium silicate solutions contain numerous species of MW from 325–2,000 g/mole. Monomeric, oligomeric, and higher-MW linear polymeric arrangements of these structures exist in equilibrium in silicate solutions. At the molecular level, the fundamental building block of silicate species is SiO_4^{4-} anion, the silica tetrahedron consisting of the silicon atom at the center of an oxygen-cornered pyramid illustrated in Figure 3.

The mix of silicate anions in a soluble silicate solution is much more complex than a simple ratio may suggest, and this mix can substantially affect the properties and performance. In mineral pulps, soluble silicates undergo several simultaneous reactions to varying degrees depending on pH: (1) adsorption onto minerals, (2) complexation with metal ions, and (3) polymerization among silicates to form extended Si-O-Si networks in the form of sols, gels, and at very low pH as crystalline SiO_2.

Commercial sodium silicate with a ratio of 3.22 is most commonly used in mineral processing applications because of its high silica content and relatively low cost. It contains the highest amount of polymeric silica species. These polymeric species are highly charged and very reactive with polyvalent cations on mineral surfaces and in solution. Sodium silicates of lower ratios form a strongly alkaline pulp and will exhibit weaker depressing/dispersing abilities.

Various modes of action of Na silicate in mineral flotation are listed in Table 10.

Low-Molecular-Weight Organic Modifiers

Low-MW organic modifiers are used less frequently than inorganic modifiers, although their effect on the selectivity and efficacy in specific cases may be very significant. They often serve as complexing reagents or as depressants/dispersants. Applications of some selected modifiers are shown in Table 11.

Polymeric Modifiers

The types of polymeric modifiers in early use (pre-1950) were drawn from mostly naturally occurring water-soluble or water-dispersible polymers, such as starch, dextrin, glue, gelatin, gum arabic, and guar. Synthetic polymers were developed later.

These naturally derived polymers were used widely in depression of a variety of gangue minerals in sulfide and non-sulfide ore flotation, carbonaceous matter, molybdenite, and iron oxides (Nagaraj and Farinato 2016). The use of polysaccharides is reviewed in many articles (Mackenzie 1986; Pugh 1989). See Table 12 for more examples.

Naturally derived polymers tend to be nonspecific in their interaction with mineral surfaces (mainly via hydroxyl functional groups) and offer limited opportunities for incorporating mineral-specific functionalities. Synthetic polymers, which allow a wider range of functional groups and control of molecular architecture, can be designed to be more selective. Development of low-MW synthetic water-soluble polymers as modifiers dates back to the early 1950s. Aimone and Booth (1956) listed a large number of such polymers with various functional groups for flotation applications. The use of polyacrylates and hydrolyzed polyacrylamide as depressants and dispersants soon followed. While the use of polyacrylates for clays and slimes (Lin and Burdick 1987) increased, further

Table 10 Inorganic modifiers: Function, mode of action, effects, and uses

Modifier	Function/Mode of Action/Effects	Usage Examples*
Acid and bases	pH change leading to a change (beneficial or detrimental to flotation selectivity or efficiency) in (a) solubility of the mineral; (b) collector adsorption on value and/or gangue minerals; (c) ratio of individual ions and molecular forms of species (particularly weak acids, including ionic collectors) in the pulp; (d) formation of sparingly soluble colloidal precipitates (e.g., hydroxides, carbonates, and sulfates of multivalent metals, which can impact collector adsorption); (e) near-surface mineral composition, surface charge, and the attendant electrical double layer; (f) removal of mineral surface coatings that impede collector adsorption; (g) aggregation or de-aggregation of fine particles; (h) slime coating; (i) suspension stability and pulp viscosity, thereby altering gas dispersion and bubble–particle interactions; (j) water chemistry (bulk aqueous phase composition and dynamics thereof); and (k) froth characteristics, altering particle transport through froth. There is often a critical pH value above which flotation of many minerals is sharply depressed or promoted. Which result predominates depends on mineral composition in the pulp and the acid/base dosage.	**H_2SO_4 (sulfuric acid):** Surface scrubbing of sulfide minerals (e.g., pyrite, galena) to remove hydroxide and other hydrophilic coatings to improve flotation; increase floatability of sulfides (e.g., pyrite) that have been depressed with lime or cyanides; selective flotation of copper or bismuth sulfides (pH 2.5–4.5); depression of phosphate minerals in reverse flotation circuits; treatment of ilmenite to remove surface coatings; acid scrubbing of glass sands; sequential removal of impurities in feldspar and rare earth beneficiation.
		HF (hydrofluoric acid): Activation of beryl and selective depression of SiO_2 in flotation with anionic collectors; depression of spodumene, feldspar, and mica in flotation with anionic collectors; activation of feldspar and beryl flotation with cationic collectors; activation of ilmenite, chromite, spodumene, and zircon flotation with cationic collectors; depression of quartz, hematite, and pyrochlore flotation with cationic collectors; selective flotation of beryl from a bulk feldspar-beryl concentrate.
		H_3PO_4 (phosphoric acid): Depression of phosphates in the reverse flotation of carbonates from high-carbonate phosphate-dolomite-calcite ores.
		CaO, $Ca(OH)_2$, lime: Widely used in sulfide flotation (Cu, Cu-Mo, Cu-Au, Zn circuits, Ni circuits). Depressant for pyrite, marcasite and pyrrhotite; disperses many non-sulfide gangue minerals and slimes; precipitation of certain multivalent metal ions; pH control; pulp viscosity modifier; froth control; depression (at high lime dosages) of some copper minerals (e.g., enargite), galena, marmatite, and pentlandite; depression of silicates in flotation with cationic collectors; coagulation of fines.
		NaOH (sodium hydroxide): Preferred base when lime cannot be used (e.g., in fatty acid flotation); depression of certain sulfide minerals (e.g., pyrite, stibnite, galena, pyrrhotite); fluorite; niobium-bearing minerals; brookite; dispersion of slimes.
		Na_2CO_3 (sodium carbonate): Widely used in beneficiation of non-sulfide and polymetallic ores; dispersion and control of many problematic non-sulfide gangue minerals and slimes; sequestration of harmful multivalent metal ions; pH control; pulp viscosity modifier; depressant or activator in flotation with fatty acids; promotes spodumene flotation separation from beryl.
Salts of weak, medium-strength, and strong acids	Pulp potential regulation, sulfidization of surfaces of oxide or oxidized minerals (e.g., oxide copper minerals, tarnished sulfides); pH regulation; oxidation of mineral surface and sulfhydryl collectors; depression of sulfide minerals; sequestration and/or precipitation of harmful metal ions	**NaSH/Na_2S/Nokes (sodium thiophosphates; P_2S_5 + NaOH):** Activation of oxide copper minerals, tarnished sulfides at low concentrations; depression of collector-treated copper sulfide minerals, oxide copper minerals, iron sulfide minerals, cerussite, etc., at high concentrations; depression of oxides and silicates (quartz, feldspar, and micaceous shales) activated by iron cations; removal of non-sulfide slime coating and hydroxide coatings on sulfide minerals; enhancing flotation of molybdenite; precipitation of multivalent metal ions that may cause inadvertent activation (e.g., Cu^{2+} and Pb^{2+} ions); activation of cerussite, calamine, and smithsonite, especially at higher temperature.
		Na_2SO_3, SO_2, and other SxOy species: Used in combination with other modifiers (copper sulfate, zinc sulfate, NaCN, $Zn(CN)_2$); depress zinc and iron sulfides in polymetallic ore beneficiation; depression of galena in Cu-Pb separation; SO_2 with NaCN to depress copper sulfides; depression of pyrite in Cu circuits.
		$KCrO_4$ and $K_2Cr_2O_7$: Limited use due to environmental concerns. Depression of galena; selective flotation of copper sulfides from copper-lead bulk concentrate (in combination with Na sulfite); activates fluorite and selectively depresses barites to reduce fatty acid adsorption; calcite depression.
		$KMnO_4$: Limited use due to environmental concerns. Selectively depresses arsenopyrite (pH 5–7) and pyrrhotite (pH 7.5–9.0); depresses sphalerite and copper sulfides in the presence of galena; depresses bismuthinite in acid medium; depresses barites and, to a lesser extent, calcite and cassiterite when fatty acid collectors are used; activates fluorite; selective flotation of complex sulfide ores after activation with $CuSO_4$ and Na_2CO_3 ($KMnO_4$ depresses pyrrhotite and arsenopyrite); supplements Na silicate for depressing calcite in scheelite flotation in highly alkaline medium (pH ~11).

(continues)

Table 10 Inorganic modifiers: Function, mode of action, effects, and uses (continued)

Modifier	Function/Mode of Action/Effects	Usage Examples*
Salts of weak, medium-strength, and strong acids (continued)	Activation; depression	**NaF:** Activates ilmenite and fluorite flotation (by fatty acid); depresses apatite and some silicates; reacts with H_2SO_4 to produce HF in situ (e.g., activates ilmenite flotation from sulfide tailings with tall oil and kerosene).
		Na_2SiF_6: Depression of silicates and silica; depression of shale and sandstone; activation of tin and tungsten minerals and oxidized antimony ores; depression of zircon in the selective flotation of rutile-zircon concentrates by fatty acid and kerosene; depression of pyrochlore in the selective flotation from a pyrochlore-zircon concentrate by anionic collectors; direct flotation of martite in acid media (separation from iron-containing silicates, garnets, and others).
		$CuSO_4$ (copper sulfate): Activation of sphalerite, gold-bearing sulfide minerals; reverses the floatability of minerals depressed by cyanides (sphalerite, chalcopyrite, pyrite, and others).
		$CaCl_2$ (calcium chloride): Selective separation of fluorite from carbonate-fluorite ores; activates, at low concentration, oxide and silicate minerals in fatty acid flotation.
		$ZnSO_4$ (zinc sulfate): Depression of zinc sulfides in the selective flotation of Cu-Pb-Zn-Ag polymetallic ores (often in combination with cyanide and sulfoxy species).
		Iron salts ($FeSO_4$, $FeCl_3$) (not common): Depression of pyrite; slight depression of chalcopyrite and sphalerite and weak activation of galena in weakly acid media by $FeSO_4$; $FeSO_4$ depression of barite and wolframite; $FeCl_3$ depression of cassiterite and barite at low concentrations; forms Fe-Sols with sodium silicate (this is a more effective depressant).
		Aluminum salts ($Al_2(SO_4)_3$, $Al(NO_3)_3$ (not common): $Al_2(SO_4)_3$ depresses sphalerite (in combination with Na sulfite); depresses galena, pyrite, and sphalerite (in combination with Na thiosulfate); $Al_2(SO_4)_3$ depresses calcite and fluorite (in combination with Na silicate); activates barite; depresses fluorite and activates calcite at high pH; $Al(NO_3)_3$ aids in fluorite flotation (with fatty acid); stabilizes suspensions of barite.
Complexing and colloidal modifiers	Complexing compounds. Complexation results in soluble complexes, colloidal sols, gels, or precipitates. These can act as • Depressants by forming highly stable, hydrophilic complexes with metal sites on surfaces to prevent collector adsorption or remove adsorbed collector species, or by sequestering deleterious ions that might otherwise cause inadvertent activation; • Activators by removing hydrophilic coatings from mineral surfaces to enhance collector adsorption; and • Dispersants by increasing mineral surface charge in pulp, thereby increasing interparticle repulsions and reducing pulp viscosity, improving gangue drainage from froth (i.e., reduce entrainment) by modification of gangue surface charge and reducing froth viscosity.	**Cyanides (NaCN, $Na_4[Fe(CN)_6]$, $Na_2Zn(CN)_4$):** Depression of sulfides of zinc, iron, and nickel and, at high concentrations, also for copper sulfides (used only in alkaline media); Cu-Pb separation; ferrocyanide depression of secondary copper sulfides, chalcocite, bornite, and oxidized secondary copper sulfides in Cu-Mo separation.
		Na silicates ($Na_2O{\cdot}xSiO_2$): Widely used depressant and dispersant. Depressant and dispersant for quartz and silicate minerals (particularly clays) in phosphate, potash, tin, tungsten, iron ore (nonmagnetic taconite), and many other non-sulfide systems. Depressant for calcite, barite, and fluorite (activation at low Na silicate concentration); dispersant of slimes; in sulfide ore flotation, silicates are used in Cu-Mo separation circuits and in polymetallic flotation circuits. They are also used to reduce pulp viscosity when clays are present; improve selectivity by sequestering and complexing soluble metal cations (e.g., Ca^{2+}, Mg^{2+}, Fe^{3+}) in the mineral pulp; depression of Fe-activated quartz (in combination with anionic collectors); used together with primary depressants such as polyacrylates, hexametaphosphate, starch, and carboxymethylcellulose; hydrophilic metal silicate formation, including metal sols, results in enhanced depression.
		Phosphates and polyphosphates (Na_2HPO_4, Na_3PO_4, $(NaPO_3)_6$, $Na_5P_3O_{10}$, $Na_4P_2O_7$): Common depressant/dispersant. Depression of phosphate minerals in reverse flotation of carbonates; depression of fluorite and calcite by disodium phosphate (pH dependent); selective flotation of fluorite from a complex fluorite-rare earth ore with a high calcite content; dispersion of clays, slimes, and other gangue in sulfide flotation systems including oxide Cu/Pb/Zn ores; soften hard water; selective fatty acid flotation of hydrargillite from kaolinite-hydrargillite bauxites (with fatty acids).

*These are based on either results published in the literature or observed in plant practice and are not an exhaustive list. Specific effects are strongly influenced by mineral composition of the ore and pulp chemistry. Exceptions may be expected.

Table 11 Small and low-molecular-weight organic modifiers

Reagent	Usage Examples*
Polyhydric organic acids and their salts† (oxalic, citric, tartaric, lactic)	• Prevent formation and growth of colloidal precipitates in flotation pulps; remove hydrophilic patches from value minerals to improve floatability (e.g., Fe hydroxyoxides from sulfide mineral surfaces using oxalic acid). • Depressants for iron-containing minerals in the flotation of cassiterite and of pyrochlore at low pH. • Depressants for barites in flotation of fluorite; depressants for fluorite and apatite; depressants for mica in sulfide flotation. • Depressants for silicates in flotation of lithium-containing minerals such as lepidolite and spodumene.
Tannins (quebracho)	• Depressant for calcite, silicates, and metal sulfides (mostly Zn, Fe, and Pb sulfides) in fluorite flotation, and for calcite in scheelite flotation. Condensed tannins such as quebracho are preferred to hydrolysable tannins such as tannic acid since they are more selective in the separation of calcite from fluorite. • Dispersion of slimes. • Tannins have been tested in the lab and plant as pyrite, pyrrhotite, sphalerite, galena, and Cu sulfide depressants in sulfide flotation, but commercial use in the plant is minor.
Lignins and sulfonated lignins	Use of lignin and sulfonated lignins has been investigated at lab, pilot, and plant level as gangue dispersants and depressants in the following systems: • Dispersants for non-sulfide gangue minerals (e.g., clays in Cu-Au flotation). • Depressants for pyrite (both activated and un-activated). Large dosages are required for efficient pyrite depression. Loss of selectivity at such high dosages limits their application in the plant. Loss of efficiency at high pH (>10.5). • Depressants for calcite and Mg silicates during flotation of sulfide ores. • Depressants for hematite in the amine flotation of silica from Fe oxide ores. • Depressant for calcite in the flotation of fluorite ores. Despite extensive testing, commercial use in the plant in both sulfide and non-sulfide flotation appears to be limited.
Naphthalene formaldehyde sulfonic acid	Depressant/dispersant for carbonaceous matter (graphitic and amorphous) and pyrite.
Polyamines (DETA, TETA)‡	DETA and TETA are used as pyrrhotite and pyrite depressants in Cu/Ni flotation.

*These are based on either results published in the literature or observed in plant practice and are not an exhaustive list. Specific effects are strongly influenced by mineral composition of the ore and pulp chemistry. Exceptions may be expected.
†Small organic molecules typically require high dosages as compared to polymeric modifiers. They often change the aquatic chemistry of the system because of build-up of various organic species that can impact the flotation of value minerals.
‡DETA = diethylenetriamine; TETA = triethylenetetramine.

Table 12 Polymeric modifiers

Reagent	Usage Examples
Polysaccharides	• Starch, dextrin, and guar are the most commonly used polysaccharides as depressants of a variety of minerals. • Starch is used as a depressant for molybdenite; high concentrations of starch depresses sulfides of lead, silver, and other metals (often used in conjunction with other modifiers such as SO_2, sulfoxy species). • Starch and causticized (hydrolyzed) starch depress graphite, carbonaceous minerals, micas, quartz, silicates, hematite, calcite, apatite, pyrochlore, aegirine, fluorite, barite, and others. • Starch can depress even collector-coated minerals. • Guar is primarily used as a depressant for talc. • Dextrin depresses calcite and more strongly depresses barite. • Dextrin depresses molybdenite, carbonaceous matter, and sericite. • Dextrin may depress spodumene and iron oxides in the flotation of mica, feldspar, and quartz with a cationic collector.
Cellulosics	• Carboxymethylcellulose (CMC) is used as a depressant or dispersant for Mg and Mg-Fe silicates (serpentines, talc, amphiboles, chlorite) in Ni, Ni-Cu, PGM ore flotation; also used as a dispersant for other slime-forming minerals in both sulfide and non-sulfide flotation. Depresses clays and other slimes in potash flotation with amines. • CMC depresses molybdenite.
Generic synthetic water soluble polymers (anionic, nonionic)	• High-MW polymeric flocculants may be used in the selective flotation of finely dispersed ores (e.g., potash beneficiation). • Polyacrylates and hydrolyzed polyacrylamides are used as dispersants and depressants for a large number of non-sulfide gangue minerals (e.g., kaolinite beneficiation); also used as pulp viscosity modifiers. • Polymers containing both carboxyl and sulfonate groups are excellent dispersants for a variety of silicate minerals, oxides, hydroxyoxides, and carbonates in sulfide and non-sulfide systems; they can sequester deleterious multivalent metal ions, reduce pulp viscosity, and control or eliminate slime coating. • Low-MW nonionic polymers of ethylene oxide and propylene oxide are used for depression and dispersion of Mg silicates in Ni ore flotation (especially in saline waters); froth control; improved P_2O_5 recovery in phosphate flotation (MgO rejection) (personal communication from Huntsman Corp.).
Designed-for-purpose synthetic water-soluble polymers (low MW; <100,000)	• Polymers with thiourea functionality: Depression of copper sulfides and pyrite in Cu-Mo separation from final bulk copper concentrates; separation of talc/carbon from sulfide minerals; depression of iron sulfides in base metal sulfide flotation; depression of sphalerite in Pb or Cu circuits. • Polymers with hydroxyl, carboxyl, and amide functionality: Depression of Mg silicates, such as talc, serpentines, and pyroxene in Ni and PGM ores; depression of pyrrhotite in Ni and Zn circuits; depression of carbonaceous matter and sphalerite in Pb circuits. • Polymers with hydroxy-carboxyl functionality: Depression of phosphate minerals in the reverse flotation of silica from phosphate ores; depression of iron oxides in reverse flotation of silica from iron ores; depression of slime-forming minerals (e.g., limonite, sericite, Mg silicates) in sulfide and non-sulfide systems.

Table 13 Analytical methods for chemical analysis of flotation reagents

Method	Principle	Capability/Advantages	Limitations
Spectrophotometry	Absorption of light in the UV region	Simple and fast technique, applicable to many collectors	Method is prone to interference, making it difficult to distinguish between different components in the sample. Sample preparation is needed to remove UV-absorbing solids.
Potentiometry	Measurement of changes in potential during titration of silver or copper nitrate, or copper sulfate with a flotation reagent such as xanthate	Ion-selective electrodes are applicable to low concentrations of collectors. Technique is selective if the correct type of electrode is used. Can easily be adapted to plant conditions (e.g., Xanthoprobe).	Not suitable for strong alkaline solutions such as those commonly found in flotation plants. Method is prone to interferences from sulfides, thiosulfates, carbonates, and thiocarbonates; different electrodes needed for different collectors.
Chromatography (HPLC)	Normal/reverse-phase chromatography or ion-interaction chromatography followed by UV detection	Possible to analyze many compounds at a time	Requires sample preparation (removal of solids by filtration).
Chromatography (capillary electrophoresis)	Separation method based on different electrophoretic movements of ions in an electric field	Possible to separate and detect different flotation collectors in process water	Different methods have to be developed for different collectors; requires sample preparation (filtration to remove solids).
Voltammetry	Concentration of chemical in solution is obtained by measuring current as potential is varied	Method is applicable to many types of collectors	The technique is tedious because of the need for sample preparation (N_2 purging, pH adjustment). Method does not differentiate between species in the sample.

Adapted from Lam 1999; Sihvonen 2012

development of synthetic polymers was rather slow until the 1980s when advances of commercial significance were made (Nagaraj et al. 1987; Nagaraj 2000; Bhambhani et al. 2014). As depressants, they offer many advantages over traditional ones (many of which are hazardous; e.g., NaSH, NaCN), such as ease of structural modification to suit different applications and ore variability, lower toxicity, lower transportation costs and storage hazards, better performance at lower dosage and lower treatment cost (exploiting the efficiency of a polymer molecule), ease of handling, and batch-to-batch consistency. For polymeric modifiers, structure–activity relationships take on added dimensions in terms of complexity and possibilities. In addition to changes in functional group, factors such as MW, degree of substitution, segment sequence, and distribution of functional groups become important. The mechanism of modifier action of these polymers is different from that of the non-polymeric modifiers such as sodium cyanide or sodium hydrosulfide; for example, polymeric modifiers generally do not alter and do not require any adjustment or control of pulp redox potentials.

Table 12 shows major functions and applications of natural and synthetic polymeric modifiers. Polymer selectivity and efficacy depends on dosage, mineralogy, water quality, processing conditions, and interactions with other chemicals in the system (added or inadvertent). Exceptions may be expected. Empirical ore flotation testing, preferably with either process water or simulated process water, is necessary. In general, polymeric modifiers require significantly lower doses than do inorganic or small organic modifiers to be effective; hence, they can have superior cost–performance benefits.

Chemical Analysis of Flotation Reagents in Plant Practice

Residual reagents (collectors, modifiers, and frothers) in process waters can have an impact on the flotation process and the environment if present in significant *relative* amounts in recycled and tailings waters. Analysis of flotation pulps for residual reagents or their degradation products is, therefore, of some value since it could be used as a diagnostic tool for control and optimization of the flotation process (e.g., dosage control), or as a way to monitor the composition of water discharged into the environment.

Various techniques for characterizing flotation reagents are given in Table 13. Offline techniques such as ultraviolet (UV) spectroscopy, high-pressure liquid chromatography (HPLC), voltammetry, titrimetry, and capillary electrophoresis have successfully been used to analyze flotation reagents in plant waters. Most of them are tedious (e.g., sample preparation) and slow, and do not offer real-time value to the operational improvement of the flotation process. However, they are very valuable in assessing environmental impact. Ion-selective electrodes (Chan et al. 1988; Bugajski and Gamsjger 1987; Cabrera et al. 2001) are extremely rapid and can be used in situ, but they are not sufficiently specific and robust in the presence of different reagents and mineral-derived species present in flotation pulps.

The best option would be online analyses that offer close to real-time, continuous, rapid, and reliable monitoring of flotation reagents. Various techniques have been developed and tested for this purpose (Hao et al. 2008; Luukkanen et al. 2003; Knight and Knights 2011; Sullivan and Woodcock 1973), but few have had sustained practical application. The online techniques successfully tested for plant flotation pulps are based on a simple arrangement involving a filter inserted in the flotation pulp and analysis of the filtered sample using UV spectrophotometry or titrimetry. Most attention has been given to xanthate in view of its wide usage in the industry. Automated online instruments such as Mintek's Xanthoprobe (Knight and Knights 2011) for determination of free xanthate in plants have shown success in plants using only xanthate.

Other categories of collectors (e.g., dithiophosphates, dithiocarbamates) as well as modifiers and frothers have received limited attention. Considering that combinations of collectors are often used to improve flotation performance and that many other types of species are present in plant flotation pulps, the greatest challenge is in developing a method that can distinguish between all these types of compounds. Another complexity in achieving this objective is interference from different species such as frothers and metal ions (Bushell and Malnarich 1956). Some of the interfering species are unknown and therefore cannot be corrected for. Even in systems involving only xanthate, formation of collector oxidation species and other concurrent reactions resulting in formation of disulfides and trithiocarbonates at the highly alkaline conditions prevalent in flotation pulps complicate chemical analysis of free xanthate by spectrophotometric methods (Pomianowski and Leja 1963). In this regard, chromatographic techniques, particularly those employing ion-interaction methods, are very attractive, though they are not suitable for online measurements. In the laboratory, it has been shown that a combination of online monitoring of flotation pulps by UV spectrophotometry and manual measurements by HPLC can be used to determine individual xanthates and xanthate decomposition products in one sample (Hao et al. 2000, 2008). Ion-interaction chromatographic techniques have also been successfully used to simultaneously analyze different collectors such as mercaptans, xanthates, dithiophosphates, dithiophosphinates, and dithiocarbamates (Barnes and Pohlandt-Watson 1988, 1993). However, their utility in industrial settings may be limited by the long time it takes to complete the analysis and the expertise needed to analyze the data.

Interest in the analysis of different flotation reagents is growing. Recent research has shown that capillary electrophoresis, a separation method based on differences in electrophoretic motilities of ions in an electric field, offers promise in lab and plant settings. Sihvonen (2012) described two capillary zone electrophoresis methods for analysis of flotation collectors from process waters. One method was able to separate and detect sodium di-isobutyl dithiophosphate and sodium di-isobutyl dithiophosphinate in pure water and flotation process water at detection limits <6.7 mg/L in less than 10 minutes. This method, however, was not able to quantify ethyl and isobutyl xanthates at acceptable detection limits. A separate method was developed for xanthates. The technique is yet to be fully developed for reliable usage in the plant.

All the available techniques are quite useful in assessing environmental impact of residual reagents in plant waters, though not for real-time in situ monitoring in plant circuits.

ENVIRONMENTAL, TOXICITY, AND REGULATORY ISSUES*

This section provides a very brief overview of environmental, toxicity, and regulatory issues that must be considered in the selection and use of flotation chemicals. These topics are broad and complex. Flotation practitioners should obtain the most-up-to-date information from reputable chemical suppliers and regulatory agencies, and, if available, consult their own company specialists (safety, hygiene, environmental, and regulatory).

Many regulations govern the use of flotation chemicals in plants, and these regulations vary from country to country. The following countries have implemented chemical control legislation for *new substances* (i.e., a chemical that is in either a totally new product or an existing product that is newly being introduced into the country in question) and have established national inventory lists of chemical entities that have been approved for manufacture and/or import: Australia (Australian Inventory of Chemical Substances), Canada (Domestic Substances List), China (Inventory of Existing Chemical Substances Produced or Imported in China), the European Union (EU Registration, Evaluation, Authorisation and Restriction of Chemicals, or REACH), Japan (Existing and New Chemical Substances, also known as METI, or Ministry of Economy, Trade and Industry), Korea (Korea Existing Chemicals List), the Philippines (Philippine Inventory of Chemicals and Chemical Substances), Taiwan, Switzerland, and the United States (Toxic Substances Control Act, or TSCA). Several other countries are contemplating initializing a chemical inventory. The landscape is rapidly changing and becoming more complex.

A product containing a component not on a country's inventory list cannot be manufactured, imported, or sold in that country without some type of registration or exemption. For the most part, regulations are for distinct chemical species. Therefore, if a product contains several new chemicals, each one would require regulatory approval, and, in some cases, the complex mixture, including impurities, would require regulatory approval. The legislation differs from country to country in terms of the type of tests (e.g., chemical and physical properties, mammalian toxicity, aquatic toxicity, mutagenicity, and environmental fate) required to add a chemical to a country's inventory. The details of registration requirements may depend on the volume of the chemical brought into the country and it may be the case that each importer/manufacturer must do their own registration.

For new chemical substances in the United States, submission to the U.S. Environmental Protection Agency (EPA) of a Premanufacture Notification (PMN) is the sole requirement for inclusion in the TSCA inventory. Currently, there are no mandatory toxicology testing requirements set forth by the EPA, but the EPA can require data after their initial review process. With future TSCA reform, this could change. As part of product stewardship initiatives and new product introduction processes, the reagent manufacturer should conduct a preliminary set of toxicity tests to assess the hazards of products so they can accurately warn downstream users of potential hazards of their product. The EU under REACH legislation, however, requires extensive testing based on production volumes to register a new or existing chemical. This process of conducting toxicity testing and filing a dossier can take 9–18 months to complete. Australia, Canada, China, Japan, and Korea also require extensive testing as part of their registration processes. The higher the production/import volume, the greater is the potential testing requirement. Typical testing costs, in U.S. dollars, range from $60,000 to $100,000 in the United States, $65,000 to $400,000 in the EU, and $50,000 to $220,000 in Canada. A typical Japanese METI registration can range from $280,000 to $430,000.

Meeting the testing requirements for Europe will satisfy many requirements for other countries because of the mutual acceptance of studies conducted by Organisation for Economic Co-operation and Development test guidelines. China's testing requirements are evolving but currently appear

*Information in this section was produced with assistance from Amy Essenfeld of Solvay.

to be a combination of requirements for EU and Japan with some in-country testing requirements for aquatic toxicology studies. Efforts to harmonize registration requirements are continuing, but this effort will take many years to accomplish.

Flotation reagents cover a wide range of chemistries. Each chemical has its own set of hazards that should be reviewed prior to use. Chemical hazards are communicated in safety data sheets, or SDSs (formerly material safety data sheets, or MSDSs) supplied by the manufacturer. Plant health, safety and environment (HSE) specialists must (and indeed most do) thoroughly review the SDS of flotation reagents with consideration for use and assess their impact on (1) health and safety of operators, (2) surrounding communities, (3) air and water resources, (4) tailings ponds, and (5) the environment over a longer term. They are then responsible for ongoing training of plant personnel, especially operators who may be exposed to chemicals in the flotation plant. In cases where a facility does not have an HSE specialist, it is the responsibility of the practitioner to contact the chemical manufacturer to understand the hazards and then implement the appropriate safety, protection, and environmental measures. Although an SDS provides valuable information, it may not have information pertaining to the fate of a flotation reagent in use (i.e., from the arrival of the chemical in the plant to its fate during processing and afterward) including potential hazards of any degradation products and chemicals that may be present in the dispatched concentrates. Good sources for some of this additional information are reputable chemical manufacturers, many of whom have capabilities to conduct life-cycle analyses.

Attempts have been made to develop ranking schemes that more completely assess the HSE characteristics of a chemical, taking into account its manufacture, use, and ultimate fate through an industrial process. One example is the attempt to develop a Greenness Index by Shen et al. (2016a); however, this attempt is currently incomplete and was thwarted by the paucity of sufficient information on the fate of flotation reagents during use. This is an indication of the complexity of such a holistic analysis.

The importance of assessing a chemical through its use cycle in addition to its inherent chemical properties may be illustrated for the case of xanthates. Although the SDS for solid xanthate does not indicate any extreme hazard (the suggested U.S. Department of Transportation shipping label phrase is "spontaneously combustible"), it is known to hydrolyze when exposed to moisture or added to aqueous solutions and flotation pulps to continually generate toxic species such as carbon disulfide (CS_2) and carbonyl sulfide (COS). These species can accumulate in the flotation plants during use and can thereby present a significant HSE hazard. Potential health impacts can arise from these hazardous vapors accumulating in enclosed infrastructures and causing chronic operator exposure to these decomposition products. The permissible exposure limit (PEL) for CS_2 is 20 ppm (8-hour time-weighted average, or TWA) according to U.S. OSHA (Occupational Safety and Health Administration) as of this writing. The CS_2 PEL recommended by California OSHA is set even lower, at 1 ppm (OSHA 2017), and similarly in Canada.

Until recently, there was no study of xanthate decomposition in ore pulps under flotation conditions. Shen et al. (2016b) studied the fate of xanthate decomposition under laboratory conditions representative of flotation in ore pulps in a batch mode; however, this has not yet been extended to a continuous operation simulating plant practice, nor has it been extended to other flotation reagents.

In recent years, many countries have adopted a more harmonized way of classifying products and communicating their hazards. This Globally Harmonized System (GHS) has, unfortunately, not been adopted with the same requirements by all countries. Consequently, there are still some differences between countries, though there is now better alignment than in the past. The GHS uses standardized pictograms to alert people to the type and hazard level of a product.

These hazards are provided by the manufacturer in Section 2 of the GHS SDS. This section should be reviewed carefully prior to handling and storing the product. Special attention must be given to the following sections in the SDS:

- Section 3 (hazardous substances)
- Section 7 (handling instructions)
- Section 8 (personal protective equipment)
- Section 10 (stability and reactivity information)

The supplier should provide an overview of the product hazards and safe handling.

Sulfur-based reagents (mostly collectors and modifiers used in sulfide flotation) have the potential to present traces of hazardous sulfur-containing vapors such as carbon disulfide, carbonyl sulfide, and hydrogen sulfide in the head spaces of closed containers. These are typically either present in the reagent from manufacturing or are degradation products of the reagents after manufacture. It is important to monitor the levels of these vapors when the containers are opened to ensure that worker exposure is below safe levels. The PEL is often given as a TWA concentration of a chemical to which a person working an 8-hour shift can be exposed without suffering ill effects. The odor of many sulfur-containing compounds can often be detected at concentrations far below the PEL. The odor threshold of a compound is the lowest concentration that is perceivable by the human sense of smell. Smelling something offensive does not necessarily mean the PEL has been exceeded; however, it is always prudent to quantify potentially hazardous vapor concentrations and to evaluate exposure using direct measurement methods. It is important to note whether the odor threshold for a particular hazardous species is above or below its PEL value. Hazardous species concentrations must be kept below the established PELs in the workplace; adequate ventilation must be maintained. The PEL values for some common odiferous species are shown in Table 14.

REAGENT SELECTION AND OPTIMIZATION

Up to this point, a detailed account of flotation reagents and flotation chemistry for both sulfide and non-sulfide ore systems has been presented. This information coupled with the material later in Tables 17 and 18 provide the foundation for reagent selection and optimization. Additional details for specific systems can be found in the literature recommended throughout this chapter. In this section, the focus is on potential problems and pitfalls encountered in reagent selection and optimization, and presentation of a more holistic approach to dealing with these issues in general.

Reagent selection has two connotations: (1) selecting a single flotation reagent (e.g., a collector or frother or modifier)

Table 14 Odor thresholds and permissible exposure limits for some common odor compounds

Odor Compound (perceived odor)	Odor Threshold	Permissible Exposure Limit*
Carbon disulfide (foul sulfurous odor)	0.1–0.2 ppm†	OSHA PEL: 20 ppm TWA, 30 ppm ceiling ACGIH (TLV): skin, 1 ppm TWA
Hydrogen sulfide (rotten eggs)	0.005 ppm	OSHA PEL: 20 ppm ceiling ACGIH (TLV): 5 ppm STEL, 1 ppm TWA
Carbonyl sulfide (foul sulfurous odor)	0.05 ppm	ACGIH (TLV): 5 ppm TWA

*ACGIH = American Conference of Governmental Industrial Hygienists; OSHA = Occupational Safety and Health Administration; STEL = short-term exposure limit; TLV = threshold limit value; TWA = time-weighted average
†ppm = parts of vapor or gas per million parts of contaminated air by volume at 25°C and 760 torr.

or (2) a reagent scheme (i.e., type and dosage of collector, frother, modifier; pH; reagent preparation and conditioning; addition points and sequence). Each of these different meanings can be either for an existing operating plant (brownfield operation) for current ores being processed or for new/future ore types, or for a new mine/plant/project (greenfield project). The main difference is the level of uncertainty and unknowns—typically greenfield projects having more uncertainty and less constraints. Choice of a reagent or reagent scheme is intimately linked with mineralogy, plant metallurgical and economic constraints and objectives, mine plan, environmental factors, and flow sheet or circuit design. For some ore types, the choice of a reagent scheme may dictate a particular flow sheet, and vice versa. Examples include primary Au ores, polymetallic ores, Ni ores, and many non-sulfide ore systems. The presence of carbonaceous matter or certain penalty elements in primary gold ores may influence whether one changes the reagent choice or the flow sheet (e.g., preflotation or depression of carbonaceous matter). In polymetallic ores, reagent selection and flow-sheet design would depend on whether sequential or bulk flotation of Cu and Pb is chosen. In phosphate/carbonate ores, reagent selection and flow-sheet design depend on whether phosphate or carbonate is floated.

Reagent optimization has many connotations as well. To the plant metallurgists, for example, it may mean working with the reagent currently in use to find conditions (e.g., dosages, addition points) that provide *optimum* performance with respect to certain target metallurgical goals (e.g., improved recovery/grade, cost-effectiveness, improved HSE aspects). Or it may mean searching the market for an alternative reagent that provides performance better than the one currently in use. For a greenfield project, reagent optimization is typically not an important requirement given the uncertainties, limited number of drill-core samples available, and the absence of a plant to evaluate performance. To the chemical supplier, reagent optimization may mean modification of existing, or designing new, structures to provide reagents that perform better (for a given application, e.g., copper flotation) than currently marketed reagents, not just in any particular plant but across all plants in the industry. The ultimate goal is a robust solution to the needs in the dynamic plant system with all its variability. Although laboratory testing is an integral part of the process of optimization, its goal is merely to identify *potential* solutions, problems, and benefits. *True optimization occurs in the plant and not in the laboratory.*

The main driving factors in an operating plant include economics, value creation (maximizing cash flow within safety and environmental constraints), regulatory pressure, health, and safety. Reagent selection and optimization are intimately linked with each of these factors.

For a given separation problem (greenfield or brownfield), at first glance, the pool of reagent classes—and members within each class—available for selection may appear to be large, bewildering, complicated, and hopeless, especially for practitioners new to flotation. However, this is certainly not the case. This is partly the result of the evolutionary nature of reagent development over the decades. But more importantly, given the complexity and apparent idiosyncratic nature of the flotation system, a particular reagent class or a very specific member or homologue is required to produce the optimum results for the separation and ore type in question—the reasons for which are difficult to fully understand. General guidelines are available, including this chapter, for reagent selection in the form of an accumulated knowledge base: (1) in reagent developers' handbooks and product literature (e.g., the *Mining Chemicals Handbook* [Cytec Industries 2010]); (2) many useful articles describing plant practice in trade journals, conference proceedings, operators' forums; (3) expertise that exists in consulting/testing labs; and (4) experiential knowledge of metallurgists. There is also excellent literature on flotation chemistry. All these resources serve to establish a starting point and to narrow the selection for a given application; however, the metallurgist must still use additional criteria to refine the selection to a manageable level to move to the laboratory or plant testing phase where the selected reagents or schemes are properly evaluated and optimized to meet metallurgical and economic objectives. The major challenge in reagent selection and optimization is, therefore, not related to a lack of literature on the subject of reagent chemistry or selection but is one of incorporating the knowledge embodied in that literature to underpin an informed *approach* to reagent selection and optimization.

Problems and pitfalls are encountered when a reductionist and informal approach is applied to the two critical tasks—applying additional criteria (e.g., mineralogy, water chemistry, process variability) to refine the selection, and the design and execution of a laboratory or plant program. Flotation systems are complex, and problems in complex systems invariably have multiple solutions, each with its own degree of desirability. Even an arbitrary process of reagent (or reagent scheme) selection and evaluation will provide *a* solution to the separation problem in question (either in the lab or plant). However, such a solution may not be sufficiently robust or optimal (i.e., the selected reagent [or scheme] will have a narrow, unoptimized window of performance and may only be a temporary fix). The lack of reagent robustness with respect to system variability may not even be recognized if reagent performance is overwhelmed by the inherent variability in the system. A rigorous, statistical analysis of plant performance is needed to discern significant changes in plant performance outside of random noise or auto-correlated plant variations (Bruey 1999; Napier-Munn 2014; Napier-Munn and Meyer 1999). There is also the danger of a false positive (an inferior reagent or scheme yielding apparent improvement, leading to a mistaken conclusion that the reagent is superior) or a false negative (a superior reagent or scheme yielding apparent poorer performance, leading to a mistaken conclusion that there is deficiency in the reagent)—all due to improper design and execution of lab and/or plant testing. Even when the best reagent is developed for an unmet need in the industry, it is deemed a failure if we fail to apply it properly on the large scale. This issue becomes acute when one considers the high overall cost ($10 million or more from concept to commercialization) and the long time required (5–10 years) to develop and implement new reagents and technology. Compounding this problem is the exponential growth in environmental regulations and the high up-front cost of regulatory registration for new chemicals. Substantial improvements in operation effectiveness and value creation can be achieved at relatively low cost/performance ratios when optimal flotation chemicals are employed. A reductionist and informal approach is very costly in the long run and adds little to our knowledge base.

These issues are well documented in the literature (Nagaraj 1994, 2005; Cappuccitti 1994; Malhotra 1994; Klimpel 1994; Nagaraj and Bruey 2003; Nagaraj and Farinato 2016). These authors noted numerous reasons for the unfortunate prevalence of reductionist and informal approach to reagent selection and optimization over the past few decades.

Important factors that dictate reagent selection and trade-offs are listed in Table 15.

Factors grouped under constraints in Table 15 appear to be relatively straightforward; however, they can sometimes have an overriding impact on reagent selection. For example, a reagent selected on the basis of its superior metallurgical performance may be rejected for any of several possible reasons:

- Its odor is objectionable.
- It was not registered for use in the country where the plant is located.
- It might be perceived to present environmental problems.
- Its unit price is perceived to be unacceptable though the overall treatment cost would have been significantly lower.

Technical factors (see Table 15) present the greatest challenge in reagent selection, because many of these are difficult to fully assess in the preliminary reagent screening phase prior to laboratory evaluation. Consequently, a combination of available knowledge base, experience, and empirical diagnostic laboratory testing is used in an iterative process to assess the impact of the technical factors in reagent selection. The objective is to reduce the number of potential reagents and conditions that need to be evaluated in the laboratory. A holistic approach using highly efficient experimental designs and rigorous data analysis greatly facilitates the reagent selection and evaluation process and the chances of arriving at a robust solution rapidly.

The schematic shown in Figure 4 represents an expanded holistic view of the flotation system, especially of particular importance to reagent selection and optimization, depicting trade-offs as additional triangles. The expanded view of chemical factors reveals the flotation reagent triangle interacting with mineralogy (ore types). Only the chemical factors are expanded. The rationale is that much of the reagent selection process occurs in the laboratory stage where a plant's physical and operational factors are difficult to simulate. These factors are necessarily included when the reagent, identified in the laboratory, is taken to the plant. The resulting tetrahedron forms the basis for reagent selection and evaluation, in combination with triangles depicted for best practices, plant needs, and knowledge base—all emphasizing trade-offs.

A rational, holistic process comprising four critical phases is schematically represented in Figure 5. Although these four phases proceed in a logical sequence, they are indeed highly iterative. The discovery and definition phase provides all the necessary information for reagent selection and sets the objectives, goals, and success criteria. The reagent selection phase then begins with preliminary screening of available reagents using available knowledge base and expertise (e.g., of the reagent developer) to arrive at a subset of reagents that meet the requirements established in the discovery and definition phase. The selected reagents can then be screened in the laboratory by an iterative process. Historically, the preliminary screening phase has been rather simplistic and flawed as it involved merely selecting representative candidates from several different families, and essentially ignoring other important reagents in the circuit and operational variables. For example, if the objective is to screen collectors for a Cu ore, the candidate products might be xanthate, dithiophosphate, and thionocarbamates. These will then be evaluated one at a time while keeping all other reagents and conditions constant. This approach is used even for evaluating other variables such as pH, p80, ore type, and so forth. The erroneous assumption in this one-reagent-at-a-time or one-factor-at-a-time approach is that reagents (or variables) perform independently of one another.

A holistic approach for reagent selection is based on considering more than just the chemistry of one reagent in isolation or more than just one type of reagent for a given mineralogy. In fact, all the chemical and operational factors should be considered simultaneously. A combination of the technical factors, constraints, and issues delineated in Tables 15 and 16 must be considered to narrow down the candidates into a manageable subset for lab and plant testing, and optimization in plant use.

Rule-based expert systems for reagent selection have been developed more recently (developed by Cytec Industries in the early 2000s and mentioned by Franzidis 2005). These systems incorporate all the elements of the holistic approach, namely, definition of mineralogy and plant needs, reagent selection, and laboratory and plant best practices.

Collector selection is usually viewed as being easier than modifier and frother selection. However, even though there is

Table 15 Technical factors and constraints

Technical Factors	Constraints
Reagent chemistry	Reagent toxicology registrations
Ore types, mineralogy, and mineral chemistry	Handling
Frothing characteristics	Logistics
Water chemistry and gangue minerals	Toxicity and environmental issues
Reagent stability (both short term and long term)	Odor
Particle size distribution	Manufacture
Downstream effects	Cost
Co-value minerals recovery	Strategic fit
Plant circuit, process conditions, plant constraints	
Compatibility with other reagents	

Adapted from Nagaraj 2005

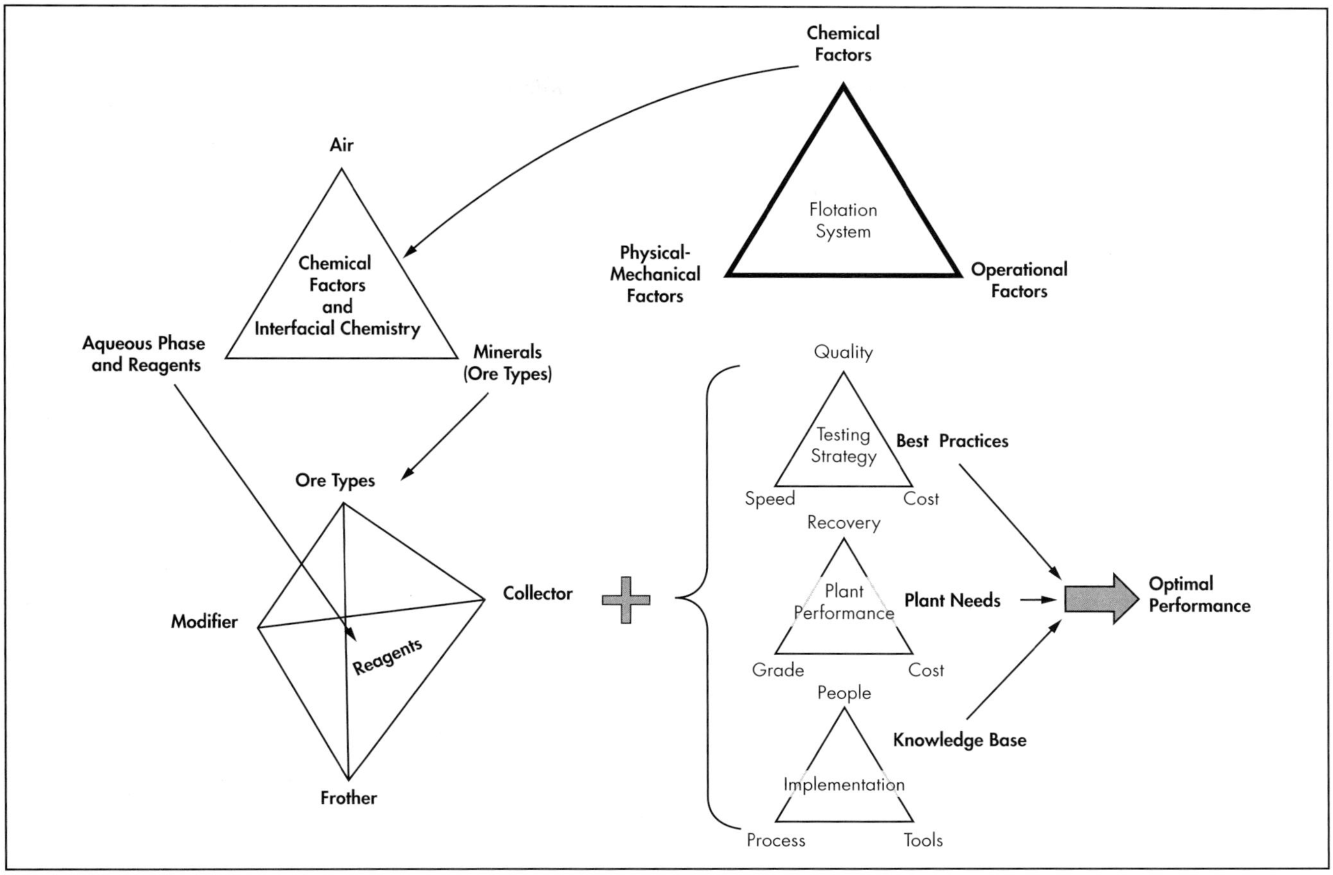

Adapted from Nagaraj 2005

Figure 4 Expanded holistic view of the flotation system in reagent selection and optimization

a greater body of literature on collectors relative to modifiers and frothers, such collector literature remains slanted toward a rather narrow set of conventional chemicals (e.g., xanthate/MIBC/lime for a sulfide ore; fatty acid/soda ash for a nonsulfide ore). This effectively results in a narrowed set of options, often leading to suboptimization with respect to chemical selection. Nevertheless, general guidelines are available for all classes of flotation reagents, and these can form the basis of selection for a given application (e.g., chemical supplier handbooks and technical data sheets).

The available pool of reagents is large for collectors and modifiers. In contrast, the choice of a frother is limited to two main classes of compounds in terms of fundamental chemistry. Frequently, formulations are used to fine-tune a frother for a given specific application. Consequently, frother selection is invariably based on empirical testing, most often in the plant. Laboratory evaluation of frothers is generally (though somewhat inaccurately) considered to be inadequate. One reason is that the froth zone in a plant operation is very dynamic, and its properties are variable and strongly influenced by mineralogy and physical, operational, and chemical factors. Based on an extensive survey of plant usage of frothers, Booth and Freyberger noted in 1962 that "the same frother may be used in different plants in a given geographical area, but treating quite dissimilar ore types; conversely, different frothers may be used by plants treating almost the same ore types and similar plant conditions." This finding is valid even today, and many examples can be found for frother selections that are not based solely on a single visible attribute or a specific property or a single metallurgical response. Nevertheless, laboratory testing, when conducted properly (using a multivariate approach), can still be useful for an empirical ranking of frothers and for assessing their potential benefits and problems in a plant circuit.

Initial selection of a frother for any particular flotation operation should be contingent upon how closely a product fulfills the basic requirements in terms of several "appearance attributes," such as froth volume, bubble size, texture, stability, persistence (down the bank of cells), gangue and water carryover, mineral loading (carrying capacity), and froth mobility. Numerous qualitative descriptors (listed previously) are used in the plants. In this initial selection stage, the same technical factors and constraints that were shown in Table 15 are used. In any case, frother selection must be adequate to meet or exceed metallurgical targets. Currently, there is no reliable (i.e., predictive) and robust frother evaluation protocol for *practical ore flotation* in the published literature, although many methods are described for two-phase foams.

Selection and evaluation (either in the lab or plant) of modifiers are far more challenging than for collectors and

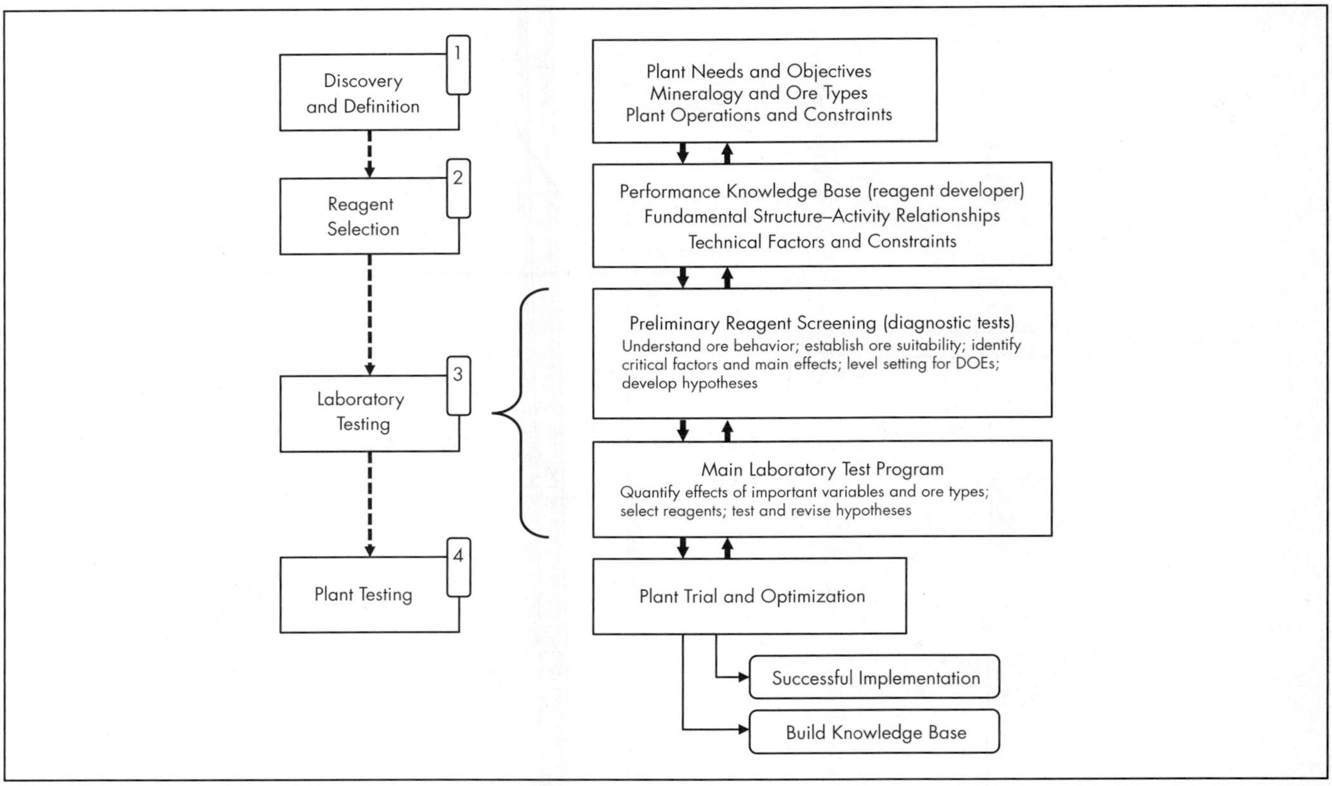

Adapted from Nagaraj 2005

Figure 5 Rational process for reagent selection and optimization

frothers. Many of the complex interactions and effects of modifiers are evident only in real flotation pulps and cannot be simulated in simple model systems such as single-mineral systems. Sometimes even laboratory ore flotation systems fall short of capturing effects of modifiers observed in plant circuits where the variability in mineralogy and operating conditions have a significant impact. Thus, effects observed in model systems used in laboratory investigation can be quite misleading. Single-mineral (or other model) systems do not capture the complexity and the dynamics of these interactions. Additionally, model system studies are conducted in the absence of other reagents (collectors and frothers). Consequently, results from such fundamental studies do not translate well to practical ore flotation. Thus, much of the published literature based on model systems is not relevant to modifier selection and optimization. They may, at best, provide some insights and concepts that could be tested in ore systems. For example, single-mineral studies may indicate that when lime is used as the modifier, adsorption of Ca^{2+} ions (or other metal ions) on sulfide particles, such as molybdenite or pyrite, may occur, resulting in depression of these minerals (Castro and Bobadilla 1995; Raghavan and Hsu 1984). However, ore flotation data and plant data do not seem to support results from single-mineral model system studies. Nevertheless, the belief that Ca ions depress MoS_2 and pyrite is quite prevalent.

General guidelines for modifier selection and evaluation were discussed earlier in this chapter, and numerous references are given there for additional information. As noted previously, mineral aquatic chemistry in flotation systems is complex and dynamic. Furthermore, flotation reagents work in concert, not independently of one another. In some systems, a change in collector type may impact selection of a modifier, or vice versa. Modifier function and the expected performance benefits are also highly dependent on pulp chemistry conditions and mineralogy—gangue minerals have a strong influence on modifier performance. In sulfide flotation systems, for example, certain non-sulfide gangue minerals—notably, clays, altered silicates (especially those of Mg-Fe), carbonates (e.g., calcite, dolomite, siderite), and hydroxyoxides (e.g., goethite, limonite)—can have severe adverse effects on both value minerals recovery and concentrate grade (Patra et al. 2012a, 2012b, 2014; Vasudevan et al. 2011, 2012; Farrokhpay 2012; Ndlovu et al. 2013; Zhang and Peng 2015; Cruz et al. 2013, 2015).

Prior to any selection and evaluation of modifiers, it is crucial to first diagnose the problems causing poor separation efficiency and selectivity. This is best done in ore flotation tests. For example, in a Cu-Mo-Au ore, the problem could be with concentrate grade, recovery of Cu/Au/Mo, or both. The root cause of the problem(s) could be (1) wrong collector, pH, or frother; (2) entrapment/entrainment of non-sulfide gangue (NSG) minerals; (3) hydrophobic NSG (e.g., C or talc); (4) NSG that increases pulp viscosity (e.g., high-aspect-ratio gangue); (5) sulfide gangue (e.g., pyrite); (6) slime (fine fraction) minerals (e.g., clays, hydroxyoxides); or (7) deleterious aqueous species derived from minerals, process water, or added chemicals (e.g., lime). Diagnosing the problem(s) can be accomplished via well-designed ore flotation testing, qualitative and quantitative mineralogy (especially the fine

Table 16 Factors in reagent selection and plant use

Factor	Issue	Solution
Cost	Perceived high cost of the proposed alternative reagent	Focus on delivered value, not the unit cost of the reagent.
Choice of reagent and application scheme	Not based on realistic needs, goals, commercial and technical viability, but based on personal experience, gut feeling, anecdotes.	Focus on "total system"; holistic selection criteria.
	Narrow focus on limited set of collectors and modifiers. Inadequate use of scientific method and existing knowledge base.	Options considered should be broadened to include the wider palette of reagents, and consideration of the inherent complexity of practical systems must be taken into account.
Application knowledge	Improper selection and use of reagents.	Use published and experiential knowledge and incorporate it into reagent selection and application.
Setting lab and plant test objectives	Testing objectives that do not meet or are inconsistent with plant needs and constraints (equipment, circuit configuration, etc.). Not all plant problems are solvable by chemical means.	Plant needs, constraints, and objectives must be clearly understood, defined, and agreed upon before a single test is run or a solution is recommended.
Lab and plant test execution	Inadequate or improper testing yielding misleading information regarding the efficacy of a flotation reagent, and leading to expensive failures in plant. The lack of standardization of laboratory flotation testing best practices and statistical tools (e.g., experimental design).	Use data-driven decision making. Employ well-designed testing plans based on sound hypotheses and a holistic approach, which account for interactions among the important factors.
Health, safety, and environmental (HSE) assessment	Failure to quantify HSE benefits of reagents leading to lopsided analyses and lost opportunities for value creation (e.g., focusing only on reagent unit cost and not accounting for HSE benefits).	Monetize the impact of environmental regulations and HSE aspects; then act to mitigate risk, and protect employees and the environment.
Societal	Hazardous chemical risk reduction; sustainable natural resource development, and better legacy.	Conduct proper evaluation and monetization of safer, alternative chemicals and processes.
Return on investment (ROI)	Faulty ROI calculations due to a lack of proper analysis (e.g., basing ROI only on reagent unit cost; not considering co-value recovery, penalty element rejection, logistics and handling issues, downstream effects).	Base ROI calculations on a complete, holistic analysis of costs, benefits, and risks.
Industry adoption and valuation of technology	Lack of plant's willingness to (1) test and adopt new flotation reagents, (2) consider alternative conditions to accommodate the new chemistry, (3) completely assess the value of new technology.	Valuation of a technology should include cost–performance metrics over the complete life and use cycle of the materials, and the overall (holistic) benefits to the entire plant operation.
Education and expertise	Shortage of qualified metallurgists; declining availability of education in mineral processing.	Industry, government, and academe should work closely to promote mineral industry education.
Economic pressures	Pressure on plants to find ways to further cut costs when ore grades decline in order to be competitive in the new global economy.	Proper reagent selection is paramount. This allows maximal leveraging of chemical solutions to hedge against declining ore quality, especially given the fixed nature of plant equipment.

Adapted from Nagaraj 1994; Nagaraj and Ravishankar 2007

size fraction), water chemistry analysis, pulp viscosity, batch settling behavior, and a thorough understanding of the plant flow sheet including plant data analysis. This yields a more rational basis for modifier selection.

Once modifiers are selected, they must be evaluated via empirical testing using a holistic approach in which mineralogy, flotation reagents, and conditions are all considered simultaneously and all the technical factors and constraints are included in the screening phase.

The ultimate choice of any flotation reagent is invariably based on empirical testing in the laboratory followed by rigorous testing in the plant, and to a large extent on the overall or global performance of the reagent in the plant (i.e., not just in one small part of the whole plant circuit). Details of the laboratory testing and plant testing phases in the process of reagent optimization are discussed elsewhere (Nagaraj and Bruey 2003; Nagaraj 2005, 2008).

MAJOR INDUSTRIAL SEPARATIONS

Base Metal Sulfide and Precious Metal Ores

Many of the industrial separations from sulfide and precious metal ores require the separation of value sulfides from each other or the separation of value sulfides and precious metals from NSG. In each case, a specific reagent scheme is necessary to effect the separation and produce a salable flotation concentrate. In some ores, the occurrence of penalty minerals, oxidized or tarnished mineral surfaces, or carbonaceous matter (i.e., graphite and amorphous carbon) greatly influences the choice of reagents employed. Table 17 is a compilation of features of the major industrial separations for base metal sulfide and precious metal ores. The material in this table is based on plant practice and many other sources (loc. cit.).

Major Industrial Separations—Non-Sulfide Ores

Table 18 is a compilation of features of the major industrial non-sulfide mineral flotation separations. The material in this table is based on plant practice and many other sources, including Manser 1975; Aplan 1994; Arbiter and Williams 1980; Aplan and Fuerstenau 1962; Cytec Industries 2010; Coghill and Clemmer 1935; Davis 1985; Redeker and Bentzen 1986; Falconer and Crawford 1947; Falconer 1961; Gaudin and Glover 1928; Holme 1986; Houot 1983; Jones and Oblatt 1984; Malhotra and Riggs 1986; Somasundaran and Moudgil 1987; Mulukutla 1994; Miller et al. 2002; Fuerstenau 1962; and Fuerstenau et al. 2007.

Table 17 Main industrial separations: Sulfide minerals

Ore	Main Minerals	Desired Separation	Typical Reagent Scheme(s)
Copper	Chalcopyrite, chalcocite, digenite, covellite, bornite; pyrite, minor amounts of acid-soluble Cu minerals (ASCu)	High selectivity against pyrite	Lime is used to maintain pH range ~9–11 in the roughers, and ~11–12 in the cleaners to depress pyrite. Alkyl dithiophosphates, functionalized thionocarbamates, dithiophosphinates, and dialkyl thionocarbamates, and formulations thereof, are the primary collectors to achieve both recovery and selectivity. A xanthate in small dosages is used if necessary as a secondary or scavenger collector. Dialkyl dithiophosphinates are finding increasing use as Cu sulfide collectors because of their strong collecting power, high selectivity against iron sulfides, low frothing contribution, and fast kinetics at much lower dosages than other primary collectors. Sulfoxy compounds (e.g., sulfite, metabisulfite) and NaCN are sometimes used as pyrite depressants. Functionalized polymeric depressants are also used; they are safer and easier to use. Alcohol, polyglycols, and combinations thereof, are used for desired selectivity or better kinetics
		Selectivity against As and Sb minerals (penalty minerals)	Functionalized thionocarbamates in combination with short-chain dithiophosphates are commonly used. Dialkyl dithiophosphinates may offer better overall performance because of their selectivity against certain penalty element minerals.
		Acid circuit flotation (pH <6)	Sulfuric or sulfurous acids are commonly used as pH modifiers. Alkyl monothiophosphates, mercaptobenzothiazole, and dialkyl dithiocarbamates (often with dithiophosphates) are primary collectors. Very limited use of dialkyl xanthogen formate.
		Flotation of oxidized or tarnished ores	Mercaptobenzothiazole and dithiophosphate collectors are used in conjunction with xanthate. Small dosages of alkyl hydroxamic acids may be used to improve recovery of ASCu.
Copper-gold	Chalcopyrite, chalcocite, covellite, bornite, native gold (alloys), pyrite	Selectivity against iron sulfides: when gold is present in copper sulfides	The objective is to float Cu sulfides and Au values selectively but not float pyrite. Lime is used to provide high pH required for depression of non-gold-containing pyrite. NaCN and/or functionalized polymeric depressants for pyrite control may be used. Primary collectors include dialkyl dithiophosphinates, aryl and alkyl monothiophosphates, oily dithiocarbamate, dithiophosphates (especially short chain), and formulations thereof. Functionalized (sulfonate and carboxyl) polymeric dispersants or depressants are often used for slime control. Frothers include alcohols, glycols, and combinations. Difficult-to-float coarse gold normally recovered by centrifugal concentration or flash flotation. Any unrecovered gold in the tailings can be recovered by cyanidation depending on economics and mineralogy.
		When significant gold is present in pyrite	A bulk copper-gold-pyrite concentrate using the preceding reagent scheme is typically obtained in the pH range of 7–10 to avoid depression of gold-containing pyrite. Copper and gold are then recovered from the bulk concentrate by first cleaning it and then making separate Cu-Au and pyrite-Au concentrates by depressing pyrite at high pH.
		When gold is associated with oxide Cu minerals	Improved copper recovery can be obtained by combining hydroxamate-based collectors with the copper-gold collectors listed previously, and by sulfidization-flotation.
Copper-molybdenum	Chalcopyrite, chalcocite, covellite, bornite, molybdenite, pyrite	Bulk Cu-Mo flotation followed by Cu-Mo separation	Cu-Mo bulk flotation: Primary collectors include alkyl dithiophosphates, functionalized thionocarbamates, dithiophosphinates, and dialkyl thionocarbamates, and formulations thereof. Hydrocarbon oil, oily xanthate ester, oily dithiocarbamate, and functionalized thionocarbamate to enhance recovery of molybdenite and any gold values. Cu-Mo separation: Depression of Cu sulfides and pyrite using NaSH, Na_2S, Nokes, $(NH_4)_2S$, and floating molybdenite using hydrocarbon oil. NaCN is sometimes used to depress residual Cu sulfides and pyrite in the Mo cleaners. New functionalized polymeric depressants can replace 50%–90% of the hazardous inorganic depressants mentioned previously while improving the efficiency of the Cu-Mo separation. Formulations of alcohol-glycols with hydrocarbons are widely used frothers.
Primary gold (refractory ores)	Native gold, electrum, tellurides, sylvanite, petzite, auricupride, maldonite, and many others; several sulfide minerals	Bulk flotation to recover Au values and associated minor amount of sulfides	Bulk gold flotation is achieved by using a strong xanthate collector such as potassium amyl xanthate in combination with secondary collectors such as oily dithiocarbamate, short-chain alkyl dithiophosphates or dithiophosphinates and mercaptobenzothiazole. Gold in bulk concentrate is recovered by cyanidation (after roasting or pressure oxidation); Au in tails is recovered by cyanidation depending on economics. Strong alcohol/glycol frother blends are used to enhance coarse Au recovery.

(continues)

Table 17 Main industrial separations: Sulfide minerals (continued)

Ore	Main Minerals	Desired Separation	Typical Reagent Scheme(s)
Primary gold (refractory ores) (continued)		Flotation of gold associated with sulfide minerals (pyrite, pyrrhotite, arsenopyrite, arsenian pyrite, marcasite, enargite, tennantite and tetrahedrite, realgar/orpiment)	The objective is to float all sulfide and Au values (pH 4–10 depending on mineralogy). Strong collectors such as amyl xanthate along with dithiophosphinate, oily dithiocarbamate, aryl or alkyl monothiophosphate, mercaptobenzothiazole, and dithiophosphate are employed. The concentrate is then treated by cyanidation after roasting or pressure oxidation. A significant amount of xanthate can often be fully replaced by a combination of the other collectors. Cu- or Pb-activation is practiced for many ores to enhance flotation of gold carriers such as pyrite, arsenopyrite, arsenian-pyrite, and other sulfide minerals. Success of this practice depends on particular pulp conditions such as pH, E_p. When the ore contains low amounts of acid-consuming minerals, flotation is conducted in acid-circuits (pH 3–7) to enhance flotation of sulfide minerals. Mercaptobenzothiazole with di- and mono-thiophosphates are quite effective collectors.
		Ores containing problematic non-sulfide gangue minerals such as Mg silicates, clay minerals, hydroxyoxides, carbonaceous matter	One or more modifiers (dispersants or depressants) are added to improve selectivity. The choice of modifier depends on the type of non-sulfide gangue mineral, pulp conditions, and flow sheet. Selective rejection of nuisance carbonaceous matter can be achieved by (1) implementing a carbon prefloat using collector/frother formulations based on hydrocarbon oils or more effective specialty formulations to enhance recovery of slow-floating carbon (the carbon prefloat concentrate may be reprocessed to recover Au values), or (2) carbon depression and "blinding" using dextrin (sometimes with controlled dosage of dyes for efficient C depression).
		Tarnished ores and ores containing significant quantities of arsenopyrite	Copper sulfate is typically used to activate sulfides. Mercaptobenzothiazole with mono and dithiophosphates and oily dithiocarbamates enhance arsenopyrite and gold recovery. Concentrate is pretreated by roasting, biooxidation, or pressure oxidation prior to cyanidation. Tailings are treated by cyanidation if they contain sufficient gold.
Primary molybdenum	Molybdenite, minor amounts of sulfides of Cu, Fe, and Pb	Flotation to produce a high-grade Mo concentrate	Hydrocarbon oils are used as collectors to enhance the natural floatability of molybdenite. Xanthate esters and functionalized thionocarbamates serve as secondary collectors. Alcohol-based frothers are used to enhance flotation kinetics in rougher flotation. A strong glycol based frother is often used to carry the rougher concentrate through multiple cleaning stages. Sodium silicate and/or soda ash are used for slime and pH control. Sulfide gangue minerals are depressed with Nokes reagent or polymeric modifiers.
Lead	Galena, cerrusite, anglesite, plumbo-jarosite	Recovery from high-grade galena ores	Galena is best floated at natural or slightly alkaline pH (~8.5) as it is depressed by high OH^- concentrations; Na_2CO_3 rather than lime is used for pH control (though lime is used in some cases). Primary collectors are dithiophosphinate, ammonium aryl dithiophosphate, and ethyl or isopropyl xanthate. Formulated collectors are often used as partial or full replacement of xanthate.
		High selectivity against zinc and iron sulfides	Stage addition of dithiophosphinate or ammonium aryl dithiophosphate offer better selectivity than xanthates and improve Ag recovery. Cyanide is sometimes used to provide better selectivity against zinc and iron sulfides.
		Recovery of Pb from oxidized lead ores	For slightly tarnished ores, mercaptobenzothiazole with dithiophosphate is very effective in addition to a sulfidizing agent such as sodium sulfide. For highly oxidized ores, lead sulfide is first recovered using collectors mentioned previously; oxide Pb minerals are then recovered by sulfidization followed by flotation using same collectors.
Lead-zinc	Galena, cerrusite, anglesite, sphalerite, marmatite, smithsonite	Recovery of high-grade Pb and Zn concentrates and rejection of iron sulfides	General practice is Pb flotation followed by Zn flotation. Soda ash is normally used for pH control, ~8–10. In the Pb circuit, zinc minerals are depressed with $ZnSO_4$, Na_2SO_3, $Na_2S_2O_5$, NaCN, $Zn(CN)_2$, or combinations. Primary collectors for Pb are same as those mentioned previously. Aryl dithiophosphate appears to be selective when Zn is somewhat activated. $CuSO_4$ is used as an activator of Zn minerals in the Zn circuit. pH is typically in the range 10–12 for pyrite and pyrrhotite depression. Primary collectors in the Zn circuit are xanthates, thionocarbamates, short-chain dithiophosphates, and combinations thereof. Alcohol frothers are more common, often with controlled amounts of glycols.
		Rejection of gangue carbonaceous minerals, dolomite, or magnesite	Organic depressants such as lignin sulfonates, quebracho, dextrin, and nigrosine are commonly used in the lead and zinc circuits for depression of gangue minerals.

(continues)

Table 17 Main industrial separations: Sulfide minerals (continued)

Ore	Main Minerals	Desired Separation	Typical Reagent Scheme(s)
Copper-zinc	Copper sulfides, sphalerite, Ag values, minor galena	Differential flotation of copper sulfides from sphalerite and rejection of pyrite and pyrrhotite	Most common plant practice is sequential flotation: Selective Cu flotation (depressing Zn minerals) followed by $CuSO_4$ activation of Zn (depressing any pyrite). Lime or soda ash to control pH ~9–10.5; lower pH (~6–8) used in some plants (especially when SO_2 is used in Cu circuit to depress Zn and Fe sulfides). $ZnSO_4$, Na_2SO_3, $Na_2S_2O_5$, SO_2, NaCN, $Zn(CN)_2$, or combinations used to depress Zn and Fe sulfides. Primary collectors are dithiophosphinate, dithiophosphate, thionocarbamates, and xanthate, and combinations to maximize Cu/Ag recovery and selectivity. Alcohol-type frothers are common; glycols and combinations are also used. Primary collectors in Zn circuit are the same as described previously.
Copper-lead-zinc	Copper sulfides, galena, sphalerite, Ag values	Bulk Cu-Pb flotation (depress Zn and Fe sulfides), Cu-Pb separation, and Zn flotation	Cu-Pb flotation: Primary collectors are dithiophosphinate, aryl dithiophosphate, and xanthate; depressants for Zn and Fe sulfides same as described previously (see copper-zinc minerals). Cu-Pb separation is achieved by either selectively depressing Pb or Cu while floating the other, depending on Cu/Pb ratio, feed grades, and mineralogy. Pb depressants: Na_2SO_3, $Na_2S_2O_5$, SO_2, starch, and $K_2Cr_2O_7$ or $Na_2Cr_2O_7$ (only if permitted). Cu depression (while floating Pb) is less common; achieved with NaCN; environmental issues with NaCN and potential risk of depressing Ag/Au values. $Zn(CN)_2$ is an alternative. Zn flotation after Cu-Pb bulk flotation is the same as listed previously.
		Sequential Cu, Pb, Zn flotation	This scheme requires the appropriate use of depressants ($ZnSO_4$, Na_2SO_3, $Na_2S_2O_5$, SO_2) for galena, sphalerite, and pyrite depressants during selective Cu flotation. Cu flotation: Very selective collectors, which are also poor galena collectors, such as the alkyl dithiophosphates or modified thionocarbamates are used to float Cu minerals. Pb flotation: Pulp is conditioned with cyanide to depress zinc minerals and activate Pb. Dithiophosphinates, ammonium salts of aryl dithiophosphates, and xanthate are used as collectors. Zinc minerals are floated next as described previously. Alcohol-type frothers are used in copper-lead-zinc separations for maximum selectivity.
Nickel (sulfide ores; lateritic ores are treated by hydrometallurgy)	Pentlandite, millerite, violarite, heazelwoodite, with varying amounts of pyrrhotite	Recovery of a high-grade nickel concentrate from nickel sulfide ores, rejection of mostly Mg silicates (talc and serpentines)	Xanthates are widely used collectors. Dithiocarbamate, dithiophosphinate, and dithiophosphate are used as secondary collectors. A significant amount of xanthate can be replaced by formulations based on alkyl dithiophosphates, functionalized thionocarbamates, dithiophosphinates, and dialkyl thionocarbamates while improving recoveries or selectivity. $CuSO_4$ activation may be used to enhance flotation of Ni minerals and pyrrhotite (if necessary). Rejection of Mg silicates is achieved with guar, carboxymethylcellulose (CMC), and functionalized polymers. Pyrrhotite depression (when necessary) is achieved with soda ash, DETA, and/or NaCN (if environmentally acceptable), sodium sulfite, and functionalized polymers (safer and relatively nonhazardous). Alcohol and alcohol/glycol blends are common.
Nickel-copper	Chalcopyrite, pentlandite, millerite, violarite, heazelwoodite, pyrrhotite	Sequential Cu, Ni flotation followed by Ni-Cu separation or bulk Cu-Ni flotation followed by Cu-Ni separation (choice dictated by Ni/Cu ratio)	Bulk Cu-Ni flotation: Primary collectors are xanthate in combination with dithiophosphinate, dithiocarbamate, and dithiophosphate. pH 7.5–9.5 (with soda ash or lime). Functionalized polymers may improve pyrrhotite rejection. Floated pyrrhotite is rejected after Cu-Ni separation. If pyrrhotite is a minor component, $CuSO_4$ activation may be used. Xanthates used in nickel-copper flotation can be replaced with certain formulations based on alkyl dithiophosphates, functionalized thionocarbamates, dithiophosphinates, and dialkyl thionocarbamates for increased recovery of metal values in the Ni-Cu bulk concentrate. Use of some formulations enables Cu-Ni flotation to be conducted under acidic conditions thereby improving recoveries. Talc and serpentines are depressed as described previously for nickel ores. Cu-Ni separation is achieved by depressing Ni minerals with lime (pH 11–12) and small amounts of cyanide (if permitted). Functionalized polymers, starch, dextrin may also be used.
Platinum group metals (PGMs)	PGM alloys, Au/Ag, PGM arsenides, tellurides, Cu, Ni, Fe sulfides	Bulk flotation of PGMs, copper and nickel sulfides	Xanthates (at high dosages) are primary collectors; dithiophosphates, dithiocarbamates, and small amounts of modified thionocarbamates and xanthate esters are secondary collectors. Guar, CMC, and functionalized polymers are depressants for talc and other Mg silicates. Weaker frothers are typical to ensure low mass pulls.

Table 18 Main industrial separations: Non-sulfide minerals

Ore	Main Value and Gangue Minerals	Desired Separation	Typical Reagent Scheme(s)
Barite	Barite and gangue minerals: fluorite, celestite, calcite, quartz, siderite, goethite, galena, sphalerite, and limonite	Recovery of high-grade barite	Sodium silicate is used as a pH modifier and slimes dispersant. Primary collectors are alkyl sulfates, petroleum sulfonates, or formulations. Collectors with high petroleum sulfonate content are particularly effective since they are tolerant to slimes. Partial replacement (~5%–20%) of petroleum sulfonate with alkyl sulfosuccinamate improves performance and circuit control.
		Selectivity against fluorite and calcite	Alkyl sulfosuccinamate as primary collector with moderate to high amounts of sodium silicate provides a high degree of selectivity against fluorite and calcite that cannot be obtained with petroleum sulfonates. Much lower dosages than with petroleum sulfonates.
Tin ores	Cassiterite with gangue minerals quartz, dolomite, siderite, Cu and Fe sulfides, tourmaline, wolframite, scheelite	Selectivity against gangue minerals	Flotation feed (gravity concentration tailings in most plants) is conditioned with sodium silicate, sodium fluoride, or sodium fluorosilicate as gangue depressants and/or dispersants. pH is then adjusted to 3–5 using H_2SO_4 to depress silicate minerals. If sulfides are present, they are removed using a xanthate, dithiophosphate, mercaptobenzothiazole. $CuSO_4$ may be used to activate sulfides. Tailings from the sulfide flotation stage are treated with cassiterite collectors. Styrene phosphonic acid (SPA), alkyl sulfosuccinamates, or combinations are the primary collectors. Use of arsonic acids is limited because of environmental restrictions. SPA is preferred for finely disseminated cassiterite ores. Phosphoric acid esters and alkyl sulfate may be used in minor amounts. Alkyl hydroxamates have shown promise as cassiterite collectors at the laboratory scale.
Coal (covered elsewhere in the handbook)	Coal, pyrite, silica, clays, and carbonates	Low-ash, low-sulfur coal concentrate (selectivity against pyrite and other mineral impurities)	Flotation is typically conducted on fine coal (–0.5 mm) fraction from coarse coal cleaning stage. Primary collectors are petroleum hydrocarbons such as diesel fuel, heavy fuel oils, and kerosene, often as formulations with alcohol and glycol frothers.
Feldspar	Orthoclase, microcline and albite, with gangue minerals silica sand, muscovite, biotite, tourmaline, garnets, ilmenite, and other iron oxides	Separation of feldspar from gangue minerals	Sequential flotation of mica, iron oxides (and other heavy minerals), and feldspar. The flotation feed is attrition scrubbed at ~70% solids and is *deslimed*. Mica flotation: Conditioning deslimed feed at 50%–60% solids at a pH of 3.0–3.5 using sulfuric acid. The feed is diluted to ~20%–30% solids and mica is floated using a cationic collector such as tallow amine with fuel oil (if necessary). Iron oxides and heavy minerals flotation: Any residual mica is also removed in this stage. Mica flotation tailings are dewatered and conditioned at 70%–75% solids at pH 2.5–3.0. The pulp is then diluted to 20%–30% solids, and flotation is conducted using strong petroleum sulfonates. Feldspar flotation: Tailings from the second stage are dewatered and conditioned at 50%–60% solids at pH 2–2.5 using sulfuric acid. HF is added to activate feldspar and depress silica. Feldspars are then floated at 20%–30% solids using a primary long-chain alkyl amine such as tallow amine with kerosene or No. 2 fuel oil.
Fluorspar	Fluorite and gangue minerals quartz, barite, calcite, and other carbonates	Rejection of gangue minerals to produce high-grade fluorspar concentrates (>97% CaF_2 for acid-grade concentrates, >80% CaF_2 for metallurgical- or ceramic-grade concentrates)	Selectivity in fluorite flotation is paramount because of the stringent specifications of acid-grade concentrates (>97% CaF_2, <1.0% SiO_2, 1.5% $CaCO_3$, 0.5% Fe_2O_3). The typical reagent scheme includes Na or NH_4 carbonate as a pH modifier and clay slimes dispersant. Sodium silicate, quebracho, starch, or tannin is used to depress calcite, silicates, and other carbonate minerals. High solids (~45%–75%) and high temperature (55°–85°C) conditioning is often needed to achieve desired selectivity. Fluorspar flotation is achieved using oleic acid or very-high-grade tall oil fatty acid. Partial replacement of fatty acid with the more selective alkyl sulfosuccinamate may improve performance. Alkyl sulfosuccinamate may also be used in conjunction with modifiers as the sole fluorspar collector. In ores with high phosphate content, the use of functionalized polymers to depress phosphate impurities prior to collector addition can improve fluorite concentrate grade.
Iron ore	Hematite, magnetite; gangue minerals: quartz, other silicates, goethite, apatite	Direct flotation of iron oxides and reject gangue minerals (not common)	High solids (50%–65%) conditioning at pH 3–5 using sulfuric acid and petroleum sulfonates. Depending on the type of gangue minerals present, formulated fatty acid can be used. Difficult ores that cannot be treated with petroleum sulfonates and fatty acids can be floated with hydroxamate collectors.
		Reverse flotation of quartz and other silicates (most common)	For high-iron/low-silica-grade ores, reverse flotation of silica is practiced using ether amines or diamines. Iron oxides are selectively flocculated and depressed using causticized starch. Sodium silicate is used as a dispersant. Desliming is often necessary to reject fines and slimes which interfere in flotation.

(continues)

Table 18 Main industrial separations: Non-sulfide minerals (continued)

Ore	Main Value and Gangue Minerals	Desired Separation	Typical Reagent Scheme(s)
Kaolin clay	Kaolinite, with impurities iron oxides, anatase, silica, feldspar, mica, sulfides, and organic matter	Rejection of colored impurities and other gangue minerals	Low- to medium-grade kaolin clays are typically produced using dry processes such as air flotation, sizing, and magnetic separation. It is common to incorporate froth flotation depending on the desired grade. High-grade kaolin clays, on the other hand, are mostly produced by the reverse flotation of heavy and colored impurities such as anatase and iron oxides using hydroxamates and/or fatty acids. Sodium silicate, polyacrylate, and hexametaphosphate are used as depressants and dispersants. High solids (50%–70%) and high-intensity conditioning is typical during blunging and after collector addition. pH is ~7.5–9. Column flotation is common. Also, the combined use of other processes such as high-gradient magnetic separation and selective flocculation enables production of kaolin clays that meet strict specifications such as high purity and brightness. Fatty acids require $CaCl_2$ as an activator. Hydroxamic acids do not require activator and provide far better performance (yield and brightness), even with "hard to process" clays.
Mineral sands	Rutile, anatase, zircon, tourmaline are the main heavy mineral sands with gangue quartz, feldspar, clays	Separation of heavy mineral sands from quartz	Desliming and conditioning at high pulp densities (68%–75%) with TOFA. For ores too coarse for flotation, grinding with the addition of a small amount of fuel oil may be required. pH ~8.5 adjusted with soda ash. Flotation concentrate is typically ~80% heavy minerals and ~20% quartz. Tailings are high-grade quartz product. In ores where feldspar occurs with quartz and the Al_2O_3 content of the final quartz product needs to be reduced, a quartz-rich concentrate is floated off from the tailings using an amine and a feldspar depressant. Liberated feldspar is rejected along with clays and other gangue minerals.
Potash	Sylvite and carnallite are the main value minerals. Halite is the main gangue mineral.	Direct flotation of sylvite and carnallite from halite and removal or depression of water-insoluble slimes	Flotation of sylvite (KCl) from halite (NaCl) in saturated brine at high temperature (30°–50°C); some plants try to operate at lower temperatures. Changes in temperature impact the solubility, and hence the performance of the amine collectors. Primary short-chain amines are preferred for low-temperature operations while long-chain amines are more effective at high temperatures. Recovery of coarse sylvite is an important goal. Therefore, hydrocarbon oils are used as "extenders." Ores containing high content of insoluble slimes, usually clays and carbonates, are difficult to process since these minerals compete with sylvite for the amine collector. Fully neutralized long-chain amines are preferred for these ores. Slime-blinding reagents such as carboxymethylcellulose, dextrin, guar gum, or functionalized polymeric depressants are added during the conditioning stage.
		Reverse flotation of halite from sylvite or carnallite	The reverse flotation of halite from sylvite/carnallite is preferred in some ores such as those found in the Dead Sea region in Israel and Jordan. In this process, alkyl morpholine type collectors are used.

(continues)

Table 18 Main industrial separations: Non-sulfide minerals (continued)

Ore	Main Value and Gangue Minerals	Desired Separation	Typical Reagent Scheme(s)
Phosphate	Phosphate minerals (fluorapatie, hydroxyapatite, francolite, phosphorite, collophase, etc.). Common gangue minerals are quartz, carbonates, mica and clays, and clay-type minerals.	Selective separation of phosphate minerals of igneous origin from gangue minerals	Apatite ores (with ~7% P_2O_5 or greater) with low carbonate content are relatively easier to process by flotation than ores with high mica and calcite content. Desliming is necessary. Feed is typically conditioned with sodium carbonate pH ~10. Sodium silicate, causticized starch, and guar are typically used as depressants for silicates, carbonates, and iron minerals. Primary collectors are TOFA, petroleum sulfonates, and alkyl sulfosuccinamate (which improves selectivity and recovery of fine apatite). If the ores contain magnetic or weakly magnetic iron oxides, the rougher concentrate is upgraded using magnetic separation to produce a high-grade phosphate concentrate (at least 36% P_2O_5).
		Selective separation of sedimentary phosphate minerals from gangue minerals	The "double flotation" (Crago) process is widely used for sedimentary ores. In the first stage, after desliming, high solids (~70%–72%) conditioning with soda ash or NaOH, a fatty acid formulation, and fuel oil (extender oil) at pH ~9.0–9.5, phosphate minerals are floated from silicates and clays. Short conditioning and flotation times are used. In the second stage, the phosphate concentrate is scrubbed with H_2SO_4 to remove collector and fuel oil coating from the solids (de-oiling), followed by washing and dewatering to remove reagents. The washed concentrate is conditioned with a fatty amine and diesel fuel oil at slightly acidic to neutral pH to float silica. The use of polymeric phosphate depressants improves selectivity. Short conditioning and flotation times are used. Some high-grade phosphate ores with low silica content can be processed without the need for the "double flotation" method. A high-grade concentrate is typically produced by floating the silica gangue into the concentrate using amine collectors and polymeric phosphate depressants. Ores with high carbonate content (dolomite and calcite), reverse flotation of carbonates is achieved at low pH (~5–5.5 with sulfuric acid) using formulated fatty acid collectors. Phosphoric acid and sodium polyphosphates are used as phosphate depressants. High solids conditioning with short conditioning time and flotation time are typical.
Tungsten	Scheelite and wolframite are the main value minerals	Separation of tungsten minerals from gangue minerals such as calcite, sulfides, and phosphate minerals	Tungsten ores are generally amenable to gravity concentration because of the large specific gravity difference between tungsten and gangue minerals. However, most tungsten minerals are finely disseminated in gangue minerals, which necessitates fine grinding and at least one flotation stage to recover the fine tungsten minerals. Flotation is usually conducted in alkaline pH (adjusted with soda ash, caustic, or sodium silicate) using TOFA. The use of formulated TOFAs provides better metallurgical performance at lower dosages than straight TOFAs. Presence of calcite necessitates the use of depressants such as quebracho or tannic acids. If the main value mineral is wolframite, flotation is typically performed in an acid circuit where petroleum sulfonates are used in conjunction with fuel oil. Alkyl hydroxamic acids may provide improved recovery and selectivity based on lab-scale experiments. For ores containing sulfide minerals, a preliminary sulfide flotation stage using xanthate is employed prior to gravity concentration (tabling) and flotation of tungsten minerals.

REFERENCES

Aimone, F.M., and Booth, R.B. 1956. Flotation of ores using addition polymers as depressants. U.S. Patent No. 2740522.

Andrews, P.R.A. 1989. *Review of Developments in Cassiterite Flotation in Respect of Physico-chemical Considerations.* CANMET report ISSN 0705-5196. Ottawa, ON: Canada Centre for Mineral and Energy Technology

Andrews, P.R.A. 1990. Review of developments in Cassiterite flotation. Presented at 14th Congress of the Council of Mining and Metallurical Institutions, Edinburgh, Scotland.

Aplan, F.F. 1994. Reagents for the flotation of non-sulfide ores In *Reagents for Better Metallurgy*. Edited by P.S. Mulukutla. Littleton, CO: SME. pp. 103–117.

Aplan, F.F., and Chander, S. 1987. Collectors for sulfide mineral flotation. In *Surfactant Science Series 27: Reagents in Mineral Technology*. Edited by P. Somasundaran and B.M. Moudgil. New York: Marcel Dekker. pp. 335–369.

Aplan, F.F., and Fuerstenau, D.W. 1962. Principles of nonmetallic mineral flotation. In *Froth Flotation: 50th Anniversary Volume.* Edited by D.W. Fuerstenau. New York: AIME.

Arbiter, N., and Hinn, H. 1968. Beneficiation of cassiterite ore by froth flotation. G.B. Patent No. 1110643.

Arbiter, N., and Williams, E.K.C. 1980. Conditioning in oleic acid flotation. In *Fine Particles Processing: Proceedings of the International Symposium on Fine Particles Processing*. Edited by P. Somasundaran. New York: AIME. pp. 802–831.

Assis, S.M., Montenegro, L.C.M., and Peres, A.E.C. 1996. Utilisation of hydroxamates in minerals froth flotation. *Miner. Eng.* 9(1):103–114.

Ata, S. 2012. Phenomena in the froth phase of flotation—A review. *Int. J. Miner. Process.* 102–103(0):1–12.

Barnes, D.E., and Pohlandt-Watson, C. 1988. *The Separation of Xanthanates by Ion-Interaction Chromatography*. South Africa: Council for Mineral Technology.

Barnes, D.E., and Pohlandt-Watson, C. 1993. Separation and determination of the sulph-hydryl flotation collectors using ion-interaction chromatography. *Fresenius J. Anal. Chem.* 345(1):36–42.

Basu, S., and Shravan, S. 2008. Preparation and characterization of petroleum sulfonate directly from crude. *Pet. Sci. Technol.* 26(13):1559–1570.

Bennett, C. 2015. The Woodgrove staged flotation reactor Procemin 2015. Presented at 11th International Mineral Processing Conference, Santiago, Chile.

Bhambhani, T., Nagaraj, D.R., Gupta, P., Lawrence, A., Peart, M., and Zarate, P. 2014. Practical aspects of Cu-Mo separations and alternatives to NaSH and Nokes reagent. Presented at IMPC XXVII, Santiago, Chile.

Bhambhani, T., Nagaraj, D.R., and Yavuzkan, O. 2016. Improving flotation recovery of oxide copper minerals. Presented at IMPC 2016: XXVIII International Mineral Processing Congress. Quebec.

Bhambhani, T., Freeman, J., and Nagaraj, D. 2017. Collector compositions and methods of using same in mineral flotation processes. WO Patent No. 2017091552 A1; U.S. Patent No. 20170144168.

Booth, R.B., and Freyberger, W.L. 1962. Froths and frothing agents. In *Froth Flotation: 50th Anniversary Volume*. Edited by D.W. Fuerstenau. New York: AIME. pp. 258–276.

Bruey, F. 1999. The refdist method for design and analysis of plant trials. SME Preprint No. 99-190. Littleton, CO: SME.

Bugajski, J., and Gamsjger, H. 1987. A new xanthate ion-specific electrode. *Fresenius Zeitschrift für Analytische Chemie* 327(3-4):362–363.

Bushell, C.H.G., and Malnarich, M. 1956. Reagent control in flotation. *[Trans. AIME 205] Min. Eng.* 8:734–737.

Cabrera, W.J., Maldonado, E.S., and Ríos, H.E. 2001. Effect of dodecyl alcohol on the potentiometric response of an isopropyl xanthate ion-selective electrode. *J. Colloid Interface Sci.* 237(1):76–79.

Cappuccitti, F.R. 1994. Current trends in the marketing of sulfide mineral collectors. In *Reagents for Better Metallurgy*. Edited by P.S. Mulukutla. Littleton, CO: SME.

Cappuccitti, F., and Finch, J.A. 2008. Development of new frothers through hydrodynamic characterization. *Miner. Eng.* 21(12-14):944–948.

Castro, S.H., and Bobadilla, C. 1995. The depressant effect of some inorganic ions on the flotation of molybdenite. In *Proceedings of the UBC-McGill Bi-Annual International Symposium on Fundamentals of Mineral Processing, 1st, Processing of Hydrophobic Minerals and Fine Coal.* Montreal, QC: Canadian Institute of Mining, Metallurgy and Petroleum. pp. 95–103.

Chadwick, J.R. 1981. How Siilinjarvi successfully floats low-grade apatite. *World Min.* 34(June):106–109.

Chan, W.H., Lee, A.W.M., and Fung, K.T. 1988. Ion-selective electrodes in organic analysis—determination of xanthate. *Talanta* 35(10):807–809.

Chander, S. 1987. Inorganic depressants for sulfide minerals. In *Reagents in Mineral Technology*. Edited by P. Somasundaran and B.M. Moudgil. New York: Marcel Dekker. pp. 429–469.

Cho, Y.S., and Laskowski, J.S. 2002. Effect of flotation frothers on bubble size and foam stability. *Int. J. Miner. Process.* 64(2-3):69–80.

Codd, D. 1984. Tin flotation, amino-phosphonic acid demostrates a high selectivity for cassiterite. *Eng. Min. J.* (October).

Coghill, W.H., and Clemmer, J.B. 1935. Soap flotation of the nonsulfides *Trans. AIME* 449–465.

Cruz, N., Peng, Y., Farrokhpay, S., and Bradshaw, D. 2013. Interactions of clay minerals in copper–gold flotation: Part 1—Rheological properties of clay mineral suspensions in the presence of flotation reagents. *Miner. Eng.* 50–51: 30–37.

Cruz, N., Peng, Y., and Wightman, E. 2015. Interactions of clay minerals in copper–gold flotation: Part 2—Influence of some calcium bearing gangue minerals on the rheological behaviour. *Int. J. Miner. Process.* 141:51–60.

Cytec Industries. 2010. *Mining Chemicals Handbook*. Woodland Park, NJ: Cytec Industries.

Davis, F.T. 1985. Nonmetallic industrial minerals. In *SME Mineral Processing Handbook*. Edited by N.L. Weiss. Littleton, CO: SME-AIME.

Day, A., and Hartjens, H. 1974. Mineral flotation with sulfosuccinamate and depressent. U.S. Patent No. 3830366A.

Day, A., and Hartjens, H. 1977. Tri-carboxylated and tetracarboxylated fatty acid aspartates as flotation collectors. U.S. Patent No. 4043902A.

Dobby, G.S., and Kosick, G.A. 2013. Froth flotation and apparatus for same. U.S. Patent No. 2013140218A.

Eigeles, M.A. 1977. *Modifiers in the Flotation Process*. Moscow: Nedra.

EPA (U.S. Environmental Protection Agency). 2000. Ethyl carbamate (urethane). 51-79-6. https://www.epa.gov/sites/production/files/2016-09/documents/ethyl-carbamate.pdf. Accessed January 2017.

Falcone, J.S. 1982. Recent advances in the chemistry of sodium silicates: Implications for ore beneficiation. *Min. Eng.* (October):1493–1494.

Falconer, S.A. 1961. Minerals beneficiation—Non-sulfide flotation with fatty acid and petroleum sulfonate promoters. *Trans. AIME* 217:207–215.

Falconer, S.A., and Crawford, B.D. 1947. Froth flotation of some nonsulphide minerals of strategic importance. *Trans. AIME* 527–542.

Farrokhpay, S. 2012. The importance of rheology in mineral flotation: A review. *Miner. Eng.* 36-38:272–278.

Finch, J.A., Nesset, J.E., and Acuña, C. 2008. Role of frother on bubble production and behaviour in flotation. *Miner. Eng.* 21(12-14):949–957.

Franzidis, J.P. 2005. Integration of science and practice in mineral flotation. *Min. Mag.* (May):12–17.

Fuerstenau, D.W., ed. 1962. *Froth Flotation: 50th Anniversary Volume*. New York: AIME.

Fuerstenau, D.W., and Pradip. 1984. Mineral flotation with hydroxamate collectors. In *Reagents in the Mineral Industry*. Edited by M.J. Jones and R. Oblatt. London: Institution of Mining and Metallurgy. pp. 161–168.

Fuerstenau, M.C., Harper, R.W., and Miller, J.D. 1970. Hydroxamate vs. fatty acid flotation of iron oxide. *Trans. AIME* 247:69–73.

Gaudin, A.M. 1957. *Flotation*. New York: McGraw-Hill.

Gaudin, A.M., and Glover, H. 1928. Flotation of some oxide and silicate minerals. Department Mining and Metallurgical Research, University of Utah Technical Papers 1:78–101.

Hao, F.P., Silvester, E., and Senior, G.D. 2000. Spectroscopic characterization of ethyl xanthate oxidation products and analysis by ion interaction chromatography. *Anal. Chem.* 72(20):4836–4845.

Hao, F., Davey, K.J., Bruckard, W.J., and Woodcock, J.T. 2008. Online analysis for xanthate in laboratory flotation pulps with a UV monitor. *Int. J. Miner. Process.* 89(1-4):71–75.

Herrera-Urbina, R. 2003. Recent developments and advances in formulations and applications of chemical reagents used in froth flotation. *Miner. Process. Extr. Metall. Rev.* 24(2):139–182.

Holman, B.W. 1930. Flotation reagents. *Inst. Min. Metall. Bull.* No. 314.

Holme, R.N. 1986. Sulfonate-type flotation reagents. In *Chemical Reagents in the Mineral Processing Industry*. Littleton, CO: SME. pp. 99–111.

Houot, R. 1983. Beneficiation of iron ore by flotation—Review of industrial and potential applications. *Int. J. Miner. Process.* 10(3):183–204.

Hsieh, S.S. 1980. Flotation studies on carboxylic acid components of tall oils. SME Preprint No. 80-300. Littleton, CO: SME.

Hughes, T.C. 2004. Preparation of fatty hydroxamate. U.S. Patent No. 2004059156A.

Hughes, T.C. 2006. Hydroxamate composition and method for froth flotation. U.S. Patent No. 7007805 (B2).

Hughes, T., Lee, K., Sheldon, G., Bygrave, J., and Mann, L. 2007. The application of Ausmelt's AM28 alkyl hydroxamate flotation reagent to Fox Resources' West Whundo copper ore at Radio Hill, Western Australia. In *Ninth Mill Operators' Conference*. Melbourne, Victoria: Australasian Institute of Mining and Metallurgy. pp. 51–54.

Iler, R.K. 1979. *The Chemistry of Silica: Solubility, Polymerization, Colloid and Surface Properties and Biochemistry of Silica*. New York: Wiley.

Jones, M.J., and Oblatt, R., eds. 1984. *Reagents in the Minerals Industry*. London: Institution of Mining and Metallurgy.

Jordens, A., Zhiyong, Y., and Cappuccitti, F. 2016. The application of Florrea 8920, a New hydroxamate-based flotation collector for the flotation of rare earth minerals. Presented at IMPC 2016: XXVIII International Mineral Processing Congress.

Khoshdast, H., and Sam, A. 2011. Flotation frothers: Review of their classifications, properties and preparation. *Open Miner. Process. J.* 4:25–44.

Klimpel, R.R. 1994. A discussion of traditional and new reagent chemistries for the flotation of sulfide minerals. In *Reagents for Better Metallurgy*. Edited by P.S. Mulukutla. Littleton, CO: SME. pp. 59–66.

Klimpel, R.R., and Hansen, R.D. 1987. Frothers. In *Reagents in Mineral Technology*. Edited by P. Somasundaran and B.M. Moudgil. New York: Marcel Dekker. pp. 385–409.

Knight, J.W., and Knights, B.D.H. 2011. Industrial trial of Mintek's xanthoprobe at the Eland platinum concentrator. In *Proceedings of the 6th Southern African Base Metals Conference 2011*. Johannesburg: Southern African Institute of Mining and Metallurgy. pp. 87–98.

Lam, K.S. 1999. *Biodegradation of Xanthate by Microbes Isolated from a Tailings Lagoon and a Potential Role for Biofilm and Plant/microbe Associations*. PhD dissertation, University of Western Sydney, Nepean.

Laskowski, J.S. 1988. Dispersing agents in mineral processing. In *Developments in Mineral Processing*. Vol. 9, *Froth Flotation*. Edited by S.H. Castro and J. Alvarez. Amsterdam: Elsevier. pp. 1–16.

Laskowski, J.S., Castro, S., and Ramos, O. 2013. Effect of seawater main components on frothability in the flotation of Cu-Mo sulfide ore. *Physicochem. Probl. Miner. Process.* 50(1):17–29.

Lawendy, T.A.B., and McClellan, G.H. 1993. Flotation of dolomitic and calcareous phosphate ores. In *Beneficiation of Phosphates: Theory and Practice*. Edited by H. El-Shall, B. Moudgil, and R. Wiegel. Littleton, CO: SME. pp. 231–244.

Leja, J. 1982. *Surface Chemistry of Froth Flotation*. New York: Plenum Press.

Lin, K.F., and Burdick, C.L. 1987. Polymeric depressants. In *Reagents in Mineral Technology*. Edited by P. Somasundaran and B.M. Moudgil. New York: Marcel Dekker. pp. 471–483.

Luukkanen, S., Parvinen, P., Miettinen, M., Stén, P., Lähteenmäki, S., and Tuikka, A. 2003. Monitoring the composition of water of flotation slurries with an on-line analyser. *Miner. Eng.* 16(11):1075–1079.

Mackenzie, M. 1986. Organic polymers as depressants. In *Chemical Reagents in the Mineral Processing Industry*. Littleton, CO: SME. pp. 139–145.

Makhotla, N., Hearn, S., and Boskovic, S. 2015. The Huntsman approach to flotation frothers. In *Proceedings of Copper Cobalt Africa, incorporating the 8th Southern African Base Metals Conference*. Johannesburg: Southern African Institute of Mining and Metallurgy.

Malhotra, D. 1994. Reagents in the mining industry: Commodities or specialty chemicals? In *Reagents for Better Metallurgy*. Edited by P.S. Mulukutla. Littleton, CO: SME. pp. 75–80.

Malhotra, D., and Riggs, W.F. 1986. *Chemical Reagents in the Mineral Processing Industry*. Littleton, CO: SME.

Malmberg, E.W., Gajderowicz, C.C., Martin, F.D., Ward, J.S., and Taber, J.J. 1982. Characterization and oil recovery observations on a series of synthetic petroleum sulfonates. *Soc. Pet. Eng. J.* (April):226–236.

Manser, R.M. 1975. *Handbook of Silica Flotation*. United Kingdom: DOB Services (Hitchin).

Marinakis, K.I., and Shergold, H.L. 1985. Influence of sodium silicate addition on the adsorption of oleic acid by fluorite, calcite and barite. *Int. J. Miner. Process.* 14(3):177–193.

Melo, F., and Laskowski, J.S. 2006. Fundamental properties of flotation frothers and their effect on flotation. *Miner. Eng.* 19(6-8):766–773.

Melo, F., and Laskowski, J.S. 2007. Effect of frothers and solid particles on the rate of water transfer to froth. *Int. J. Miner. Process.* 84(1-4):33–40.

Miller, J.D., Tippin, B., and Pruett, R. 2002. Nonsulfide flotation technology and plant practice. In *Mineral Processing Plant Design, Practice, and Control*. Edited by A.L. Mular, D.N. Halbe, and D.J. Barratt. Littleton, CO: SME. pp. 1159–1178.

Mulukutla, P.S., ed. 1994. *Reagents for Better Metallurgy*. Littleton, CO: SME.

Nagaraj, D.R. 1987. The chemistry and applications of chelating or complexing agents in mineral separations. In *Reagents in Mineral Technology*. Edited by P. Somasundaran and B.M. Moudgil. New York: Marcel Dekker. pp. 257–334.

Nagaraj, D.R. 1994. A critical assessment of flotation agents. In *Reagents for Better Metallurgy*. Littleton, CO: SME. pp. 81–90.

Nagaraj, D.R. 1997. Development of new flotation chemicals. *Trans. Indian Inst. Met.* 50(5):355–363.

Nagaraj, D.R. 2000. New synthetic polymeric depressants for sulfide and non-sulfide minerals. In *Proceedings of the XXI International Mineral Processing Congress*. Developments in Mineral Processing Series. New York: Elsevier. pp. 13:C8b-1–C8b-8.

Nagaraj, D.R. 2005. Reagent selection and optimization—The case for a holistic approach. *Miner. Eng.* 18(2):151–158.

Nagaraj, D.R. 2008. New approach to reagent development and applications in the processing of base and precious metals ores. In *Proceedings of the XXIV International Mineral Processing Congress*. Beijing: Science Press. pp. 1503–1512.

Nagaraj, D.R., and Brinen, J. 1995. SIMS study of metal ion activation in gangue flotation. In *Proceedings of XIX International Mineral Processing Congress—Flotation Operating Practices and Fundamentals*. Littleton, CO: SME. pp. 253–257.

Nagaraj, D.R., and Brinen, J.S. 1997. SIMS analysis of flotation collector adsorption and metal ion activation on minerals: Recent studies. *Trans. Indian Inst. Met.* 50(5):365–376.

Nagaraj, D.R., and Bruey, F. 2003. Reagent optimization in base metal sulfide flotation - pitfalls of "standard" practice. Flotation and flocculation—From fundamentals to applications. In *Proceedings of the Strategic Conference and Workshop, Kona Hawaii*. Adelaide, Australia: University of South Australia, Ian Wark Research Institute. p. 257.

Nagaraj, D.R., and Farinato, R.S. 2013. *Major Innovations in the Evolution of Flotation Reagents*. Englewood, CO: SME.

Nagaraj, D.R., and Farinato, R.S. 2014. Major innovations in the evolution of flotation reagents. In *Mineral Processing and Extractive Metallurgy: 100 Years of Innovation*. Englewood, CO: SME. pp. 159–175.

Nagaraj, D.R., and Farinato, R.S. 2016. Evolution of flotation chemistry and chemicals: A century of innovations and the lingering challenges. *Miner. Eng.* 96–97:2–14.

Nagaraj, D.R., and Ravishankar, S.A. 2007. Flotation reagents—A critical overview from an industry perspective. In *Froth Flotation: A Century of Innovation*. Edited by M.C. Fuerstenau, G.J. Jameson, and R.-H. Yoon. Littleton, CO: SME. pp. 375–424.

Nagaraj, D.R., Rothenberg, A.S., Lipp, D.W., and Panzer, H.P. 1987. Low molecular weight polyacrylamide-based polymers as modifiers in phosphate beneficiation. *Int. J. Miner. Process.* 20(3-4):291–308.

Nagaraj, D.R., Basilio, C., and Yoon, R.H. 1989. The chemistry and structure-activity relationships for new sulphide collectors. In *Processing of Complex Ores*. Edited by G.S. Dobby and S.R. Rao. New York: Pergamon Press. pp. 157–166.

Napier-Munn, T. 2014. *Statistical Methods for Mineral Engineers: How to Design Experiments and Analyse Data*. Indooroopilly, Queensland: Julius Kruttschnitt Mineral Research Centre.

Napier-Munn, T.J., and Meyer, D.H. 1999. A modified paired t-test for the analysis of plant trials with data autocorrelated in time. *Miner. Eng.* 12(9):1093–1100.

Ndlovu, B., Farrokhpay, S., and Bradshaw, D. 2013. The effect of phyllosilicate minerals on mineral processing industry. *Int. J. Miner. Process.* 125:149–156.

OSHA (Occupational Safety and Health Administration). 2017. Anotated OSHA Z-2 table. https://www.osha.gov/dsg/annotated-pels/tablez-2.html.

Patra, P., Bhambhani, T., Nagaraj, D.R., and Somasundaran, P. 2012a. Impact of pulp rheological behavior on selective separation of Ni minerals from fibrous serpentine ores. *Colloids Surf. A* 411:24–26.

Patra, P., Bhambhani, T., Vasudevan, M., Nagaraj, D.R., and Somasundaran, P. 2012b. Transport of fibrous gangue mineral networks to froth by bubbles in flotation separation. *Int. J. Miner. Process.* 104-105(0):45–48.

Patra, P., Bhambani, T., Nagaraj, D.R., and Somasundaran, P. 2014. Dissolution of serpentine fibers under acidic flotation conditions reduces inter-fiber friction and alleviates impact of pulp rheological behavior on Ni ore beneficiation. *Colloids Surf. A* 459:11–13.

Pomianowski, A., and Leja, J. 1963. Spectrophotometric study of xanthate and and dixanthogen solutions. *Can. J. Chem.* 41(9):2219–2230.

Pöpperle, J. 1941. Verfahren zur Schwimmaufbereitung von Erzen (Method for flotation of ores). German Patent No. 700735.

Pradip and Fuerstenau, D.W. 1991. The role of inorganic and organic reagents in the flotation separation of rare-earth ores. *Int. J. Miner. Process.* 32(1-2):1–22.

Pradip, Rai, B., Rao, T.K., Krishnamurthy, S., Vetrivel, R., Mielczarski, J., and Cases, J.M. 2002a. Molecular modeling of interactions of alkyl hydroxamates with calcium minerals. *J. Colloid Interface Sci.* 256:106–113.

Pradip, Rai, B., Rao, T. K., Krishnamurthy, S., Vetrivel, R., Mielczarski, J., and Cases, J.M. 2002b. Molecular modeling of interactions of diphosphonic acid based surfactants with calcium minerals. *Langmuir* 18:932–940.

Pugh, R.J. 1989. Macromolecular organic depressants in sulphide flotation—A review, 1. Principles, types and applications. *Int. J. Miner. Process.* 25(1-2):101–130.

Qi, G.W., Klauber, C., and Warren, L.J. 1993. Mechanism of action of sodium silicate in the flotation of apatite from hematite. *Int. J. Miner. Process.* 39(3):251–273.

Raghavan, S., and Hsu, L.L. 1984. Factors affecting the flotation recovery of molybdenite from porphyry copper ores. *Int. J. Miner. Process.* 12(1-3):145–162.

Rao, S.R. 2004. *Surface Chemistry of Froth Flotation, Volume 2: Reagents and Mechanisms*. New York: Kluwer Academic/Plenum Publishers.

Redeker, I.H., and Bentzen, E.H. 1986. Plant and laboratory practice in nonmetallic mineral flotation. In *Chemical Reagents in the Mineral Processing Industry*. Edited by D. Malhorta and W.F. Riggs. Littleton, CO: SME. pp. 3–20.

Riggs, W.F. 1986. Frothers—An operator's guide. In *Chemical Reagents in the Mineral Processing Industry*. Edited by D. Malhorta and W.F. Riggs. Littleton, CO: SME. pp. 113–116.

Seitz, R.A., and Kawatra, S.K. 1986. The role of nonpolar oils as flotation reagents. In *Chemical Reagents in the Mineral Processing Industry*. Edited by D. Malhotra and W.F. Riggs: Littleton, CO: SME. pp. 171–180.

Shen, Y., Lo, C., Nagaraj, D.R., Farinato, R., Essenfeld, A., and Somasundaran, P. 2016a. Development of Greenness Index as an evaluation tool to assess reagents: Evaluation based on SDS (safety data sheet) information. *Miner. Eng.* 94:1–9.

Shen, Y., Nagaraj, D.R., Farinato, R., and Somasundaran, P. 2016b. Study of xanthate decomposition in aqueous solutions. *Miner. Eng.* 93:10–15.

Sihvonen, T. 2012. Determination of collector chemicals from flotation process waters using capillary electrophoresis. M.S. thesis, Lappeenranta University of Technology, Lappeenranta, Finland.

Smart, R.S.C., Amarantidis, J., Skinner, W., Prestidge, C.A., LaVanier, L., and Grano, S. 1998. Surface analytical studies of oxidation and collector adsorption in sulfide mineral flotation. *Scan. Microsc.* 12(4):553–583.

Smart, R.S.C., Amarantidis, J., Skinner, W.M., Prestidge, C.A., La Vanier, L., and Grano, S.R. 2003. Surface analytical studies of oxidation and collector adsorption in sulfide mineral flotation. *Topics Appl. Phys.* 85:3–60.

Somasundaran, P., and Moudgil, B.M., eds. 1987. *Reagents in Mineral Technology*. Vol. 27 of Surfactant Science Series. New York: Taylor and Francis.

Stowe, K.G., Chryssoulis, S.L., and Kim, J.Y. 1995. Mapping of composition of mineral surfaces by TOF-SIMS. *Miner. Eng.* 8(4-5):421–430.

Sullivan, J.V., and Woodcock, J.T. 1973. A simple on-stream xanthate monitor. *Proc. Australas. Inst. Min. Metall.* 248:1–7.

Taggart, A.F. 1947. Flotation. In *Handbook of Mineral Dressing (Ores and Industrial Minerals)*. Edited by A.F. Taggart. New York: John Wiley and Sons. pp. 12-01–12-47.

Tveter, E.C. 1952. Frothing agents for flotation of ores. U.S. Patent No. 2611485.

Vasudevan, M., Patra, P., Bhambhani, T., Nagaraj, D.R., and Somasundaran, P. 2011. Processing of ultramafic Ni ores: Atypical grade-recovery curves. Presented at Conference of Metallurgists, Montreal, QC.

Vasudevan, M., Bhambhani, T., Nagaraj, D.R., and Farinato, R.S. 2012. Impact of dissolved gangue species and fine colloidal matter in process water on flotation performance. In *Water in Mineral Processing*. Englewood, CO: SME. pp. 247–259.

Wang, S.S., and Nagaraj, D.R. 1989. Novel collectors and processes for making and using same. U.S. Patent No. 4871466.

Wottgen, E. 1969. Adsorption of phosphonic acids on cassiterite. *Trans. Inst. Min. Metall.* 78:C91–97.

Wottgen, E., and Lippman, C. 1963. Phosphonic acids as collectors for cassiterite. *Bergakademie* 15(12):868–870.

Wu, X.Q., and Zhu, J.G. 2006. Selective flotation of cassiterite with benzohydroxamic acid. *Miner. Eng.* 19(14):1410–1417.

Xia, L., Hart, B., Douglas, K., and Zhong, H. 2015. Two new structures of hydroxamate collectors and their application to ilmenite and wolframite. In *Proceedings of the 47th Annual Canadian Mineral Processors Operators Conference*. Ottawa, ON: Canadian Mineral Processors Annual Conference.

Yang, D.C. 1987. Reagents in iron ore processing. In *Reagents in Mineral Technology*. Edited by P. Somasundaran and B.M. Moudgil. New York: Marcel Dekker. pp. 579–644.

Yang, X., Roonasi, P., and Holmgren, A. 2008. A study of sodium silicate in aqueous solution and sorbed by synthetic magnetite using in situ ATR-FTIR spectroscopy. *J. Colloid Interface Sci.* 328(1):41–47.

Yoon, R.-H., and Hilderbrand, T.M. 1986a. Purification of kaolin clay by froth flotation using hydroxamate collectors. G.B. Patent No. 2167687.

Yoon, R.-H., and Hilderbrand, T.M. 1986b. Purification of kaolin clay by froth flotation using hydroxamate collectors. U.S. Patent No. 4629556.

Zanin, M., and Grano, S. 2006. Selecting frothers for the flotation of specific ores by means of batch scale foaming tests. In *Proceedings of Metallurgical Plant Design and Operating Strategies 2006.* Melbourne, Victoria: Australasian Institute of Mining and Metallurgy. pp. 339–349.

Zhang, M., and Peng, Y. 2015. Effect of clay minerals on pulp rheology and the flotation of copper and gold minerals. *Miner. Eng.* 70:8–13.

Zhang, W., Nesset, J.E., Rao, R., and Finch, J.A. 2012. Characterizing frothers through critical coalescence concentration (CCC)95-hydrophile-lipophile balance (HLB) relationship. *Minerals* 2(3):208–227.

Zhou, X., Jordens, A., Cappuccitti, F., Finch, J.A., and Waters, K.E. 2016. Gas dispersion properties of collector/frother blends. *Miner. Eng.* 96–97:20–25.

CHAPTER 7.6

Entrainment in Flotation

Bill Johnson

The entrainment mechanism is a nonselective physical mechanism for transfer of particles, both hydrophilic and hydrophobic, from the pulp zone of a flotation cell to the concentrate launder. Some traditional flotation practices (e.g., the use of low percent solids by weight values in the pulp in cleaning circuits) result from the need to lower the contribution from the entrainment mechanism. Understanding the mechanism is therefore very important for industrial metallurgists who must ensure that in sizing flotation samples for the analysis of the entrainment mechanism, multiple narrow-size fractions are obtained in the subsieve range (e.g., –38 µm or –53 µm) by use of a Cyclosizer or a technique of similar capability.

Two broad categories can be recognized for the froth flotation process using standard mechanical flotation machines for reporting of the hydrophobic minerals to the froth product:

1. Conventional flotation process (hydrophobic valuable mineral to froth product)
2. Reverse flotation process (one or more hydrophobic gangue minerals to froth product)

Conventional flotation is the dominant category in industrial flotation. However, the much less common reverse flotation process has some important industrial applications. The entrainment mechanism causes inefficiencies in each flotation stage for the two types of industrial flotation processes as listed in Table 1.

For the finest size fractions (e.g., 0–11 µm) in conventional flotation, the entrainment mechanism is often the sole or dominant recovery mechanism for the hydrophilic minerals in the launder for the froth product. Similarly, for the finest size fractions (e.g., 0–11 µm) in the reverse flotation process, the entrainment mechanism is often the sole or dominant recovery mechanism for the valuable mineral in the launder for the froth product.

Table 1 Detrimental outcomes from the entrainment mechanism

Type of Flotation Process	Detrimental Outcome
Conventional flotation	Liberated hydrophilic gangue is entrained in the water derived from the pulp zone and reaches the launder for the froth product, *lowering the concentrate grade.*
Reverse flotation	Liberated hydrophilic valuable mineral is entrained in the water derived from the pulp zone and reaches the launder for the froth product, *lowering recovery of the valuable mineral.*

Before the entrainment mechanism was understood, it was assumed that the kinetic behavior displayed by gangue minerals (in particular, liberated gangue minerals) and the modeling of gangue minerals would be similar to the valuable minerals except that the values for the parameters or rate constants would be much lower. From laboratory-scale data, research results showed the reality: The behavior of gangue minerals in reaching the concentrate was connected to the behavior of the water reaching the concentrate (Jowett 1966). By obtaining sized data, the detailed entrainment behavior was reported in plant data for a porphyry copper ore and other ores by Johnson et al. (1974) who presented nonsulfide gangue recovery–size data and water–recovery data to improve understanding of the behavior of the nonsulfide gangue.

The improved understanding of the recovery of the liberated nonsulfide gangue also provided support for the benefits of a properly designed water addition system and a correct value for the water flow rate into the froth region of column flotation cells. In the preface of their book *Column Flotation*, Finch and Dobby (1990) referred to the importance of a positive bias (net flow of water from the froth region to the pulp region) in column flotation technology. Further, the authors indicated in later text that the column technology, operated with an appropriate water addition system and water flow rate, resulted in an important lowering of the number of cleaning stages needed in the production of a molybdenite concentrate, that is, an important simplification for a duty typically employing 6–10 or more cleaning stages with standard flotation cells in each stage.

The mechanism for recovery of liberated nonsulfide gangue in conjunction with the recovery of water is termed *entrainment*. Mass balance data for both water and nonsulfide gangue are required for examination of the contribution from the entrainment mechanism. Before 1970, it was not the

Bill Johnson, Principal Consultant & Adjunct Professor, Mineralis Consultants Pty Ltd. & University of Queensland, Brisbane, Queensland, Australia normal practice to obtain mass balance data for both water and nonsulfide gangue minerals from surveys of the froth flotation process.

For both laboratory batch tests and plant survey data, the water data must reflect the water that flowed over the cell lip and must not represent any contributions from water added beyond that point to assist transport of the froth product in plants or to assist transfer of particles to the concentrate tray in laboratory batch tests. The use of element-to-mineral conversions, a wider range of elemental assays, and other instrumental assay methods has allowed data to be obtained for nonsulfide gangue mineral(s) for the total solid and for size fractions.

There are three levels of detail at which mineral processing data can be obtained and analyzed for both conventional and reverse flotation processes:

1. Recoveries of minerals based on overall flows of minerals (i.e., non-sized data)
2. Mineral recovery–size data
3. Mineral recovery–size–liberation data

In the published literature on entrainment, the vast majority used data in levels 1 and 2, collected from a conventional flotation process; often, nonsulfide gangue minerals were the candidates for entrainment. Researchers and plant metallurgists sought to collect data from parts of circuits where there was likely to be the least interference in nonsulfide gangue data from, for example, nonsulfide gangue in composite particles with the hydrophobic valuable mineral. As one example, the pattern of behavior reflecting the entrainment of nonsulfide gangue was more likely to be evident at the commencement of a rougher section with a well-liberated feed than in the rougher-scavenger section at the end of the same bank where the intention was to recover mainly low-quality composite particles dominated by nonsulfide gangue with relatively low amounts of valuable mineral. In this case, there are two important mechanisms for recovery of nonsulfide gangue: the recovery of nonsulfide gangue in composites with a valuable mineral can become the dominant mechanism for recovery of nonsulfide gangue, and interpretation of the data is less straightforward.

The applications of the *reverse* flotation process are discussed in a later section. If the valuable mineral is well liberated, the entrainment mechanism may be dominant for losses of the valuable mineral in the fine-size fractions; if the valuable mineral is not well liberated, the contribution to losses from the valuable mineral in composites with the hydrophobic gangue mineral(s) will increase. Hence, the relevance of mineral recovery–size–liberation data for reverse flotation processes can also be envisioned.

Since approximately 1995, it has been possible to routinely obtain mineral recovery–size–liberation data for all sections (roughers, scavengers, cleaners, etc.) for both conventional and reverse flotation processes; the availability of automated electron microscopes for measurement of liberation levels has caused this change. Before 1995, such data were acquired by using optical microscopy to obtain point-count liberation data. Later, automated scanning electron microscopy and mineral liberation analyzers became relatively common in companies and commercial laboratories.

Liberated nonsulfide gangue recovery–size data (i.e., level 3) are now readily available, and metallurgists in the industry should be maximizing the use of this direct method for analysis of entrainment in conventional flotation processes as much as possible.

The first-order rate equation will be used in some algebraic manipulation of equations for entrainment. For the flotation process, the first-order rate equation can be applied to a range of species of interest including a mineral in a size fraction or water in general for calculation of the rate of recovery of a species of interest with units of mass/time:

$$\begin{matrix}\text{rate of recovery} \\ \text{of species}\end{matrix} = \begin{matrix}k \times \text{mass of species in} \\ \text{the pulp of the cell}\end{matrix} \qquad \textbf{(EQ 1)}$$

The first-order rate constant k has units of time^{-1}, and the form of the relationship in Equation 2 is used later in algebraic manipulations. Emphasis is placed on the first-order rate constant k for a mineral in Equation 1 being the "overall" rate constant for the cell as a result of the recovery processes for that mineral in both the pulp and froth regions for the cell:

$$k = \frac{(\text{rate of recovery of species})}{(\text{mass of species in the pulp of the cell})} \qquad \textbf{(EQ 2)}$$

This chapter has been prepared with an emphasis on analysis of industrial data. Entrainment is illustrated partly in this chapter with some graphs of industrial data from a review of entrainment (Johnson 2005) from the Centenary of Flotation conference. That review also provided some additional discussion and references on some aspects of entrainment and contained discussion of another mechanism—entrapment—observed in a low number of unusual systems for the recovery of dispersed, liberated, and hydrophilic particles in the froth product (Appendix A).

QUALITATIVE DESCRIPTION OF ENTRAINMENT

Figure 1 greatly assists in understanding the entrainment mechanism where a liberated nonsulfide gangue mineral in a conventional flotation system is used as a convenient example for the following discussion. It is assumed that the nonsulfide gangue in the figure is fully liberated, fully dispersed, and hydrophilic. However, in real systems, all particles and minerals, in both liberated and composite forms, are subject to the entrainment mechanism, noting that the significance of the mechanism for each mineral differs.

Two regions are depicted in the flotation cell: the pulp and the froth regions. Three cases are demonstrated:

- Case 1 is the simplest case where the pulp zone is perfectly mixed for all size fractions.
- Case 2 is a case where the pulp zone is perfectly mixed for the important finest size fraction and not perfectly mixed for the two coarser size fractions.
- Case 3 is a case where the pulp zone is not perfectly mixed for all size fractions.

In case 1, where the pulp region is perfectly mixed for all size fractions of solid present, the important consequence is that the mass of each mineral in each size fraction, including the nonsulfide gangue mineral, is known and constant in all parts of the pulp region. This can be calculated from the relative amounts of each mineral in each size fraction (i.e., from a sizing of a sample of the solid in the pulp and mineral assays for each size fraction of the solid) and the mass of water in the pulp region (i.e., from values for the percent solids by weight in the pulp).

For each size fraction, the mass of each mineral (defined as liberated in Figure 1) per unit mass of water at the top of

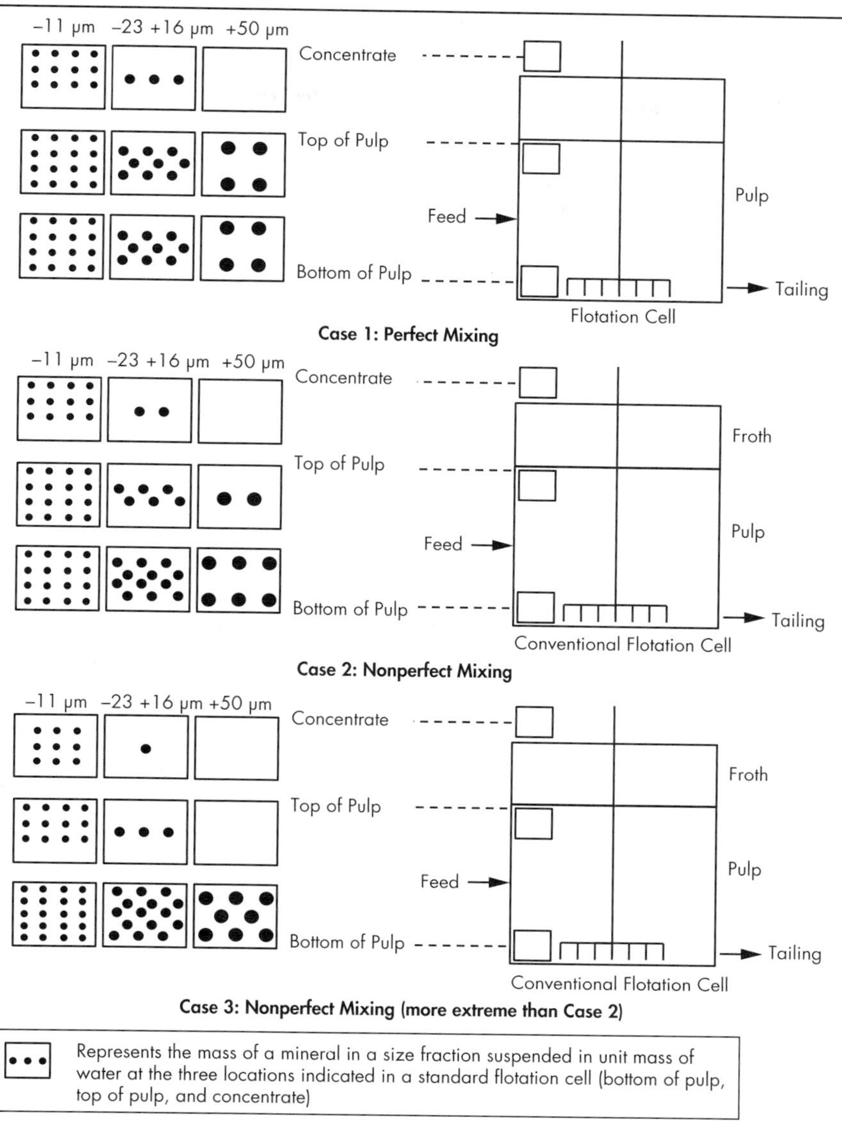

Note: Case 1 shows a perfect mixing example. Three size fractions (−11 μm, −23 +16 μm, and +50 μm) are given with entrainment efficiency values (ENT_i) of 0.75, 0.33, and 0.0, respectively. In the nonperfect mixing cases (cases 2 and 3), a schematic adjustment has been made to the mineral mass in unit mass of water depicted at the bottom of the pulp.

Figure 1 Entrainment mechanism in a conventional flotation system for three cases of mixing in the pulp zone

the pulp region (just below the interface with the froth region) is a very important commencement point for the entrainment mechanism. In case 1 with perfect mixing, the mass of each mineral per unit mass of water at the top of the pulp region is the same as in the bottom of the pulp zone.

Water is required to form the froth region; this water (complete with the nonsulfide gangue particles suspended in it) can only be supplied from the water at the top of the pulp region in a standard flotation cell. The transfer of unit weight of water from the top of the pulp region (along with its suspended nonsulfide gangue particles) represents the commencement of the entrainment mechanism for the transport of nonsulfide gangue particles through the froth region and into the concentrate launder. Notably, other categories of minerals (e.g., hydrophobic valuable particles) are also suspended in unit weight of water at the top of the pulp region, and the entrainment mechanism does contribute only a minor proportion to their total transfer into the froth region.

For case 1, the consequence of the entrainment mechanism is shown in Figure 1; for the concentrate, the mass of nonsulfide gangue per unit mass of water in each size fraction is displayed. By comparison of the mass of nonsulfide gangue per unit mass of water at the top of the pulp with the mass of nonsulfide gangue per unit mass of water in the concentrate for a given size fraction, the efficiency of the transportation of that size fraction can be seen.

It is typically observed that the transportation efficiency approaches unity for the −11 μm fraction, and it may approach zero in the region of 50 μm or coarser. Transportation efficiencies decline progressively for size fractions larger than −11 μm; that is, the entrainment mechanism becomes less important as larger size fractions are considered.

The values for the entrainment efficiency result from the structure of the froth and the subprocesses in the froth. Thin lamellae separating the bubbles (tens of micrometers in thickness) contain most of the attached hydrophobic particles; the Plateau border drainage channels (typically millimeters in diameter, i.e., much larger than the particles) contain most of the water in the froth region and provide the sites for drainage of particles from the froth region. There is settling of particles in the Plateau border drainage channels, and upward drag is exerted on particles in the froth region by water moving upward and ultimately to the concentrate launder. Cutting (1989) discussed additional mechanisms for the return of particles, and water, from the froth region to the pulp region.

The values for the efficiency of entrainment of liberated hydrophilic particles (liberated nonsulfide gangue in Figure 1) provide a practical indirect indicator of the conditions experienced in the froth region by the hydrophilic particles introduced to the base of the froth region in unit mass of water located at the top of the pulp region.

In case 2, where the pulp zone is perfectly mixed for some fine-size fractions (–11 μm) and not perfectly mixed for some other coarser size fractions (–23 +16 μm and +50 μm), the important consequence is that the mass of each mineral in each size fraction per unit mass of water at the top of the pulp zone, such as the nonsulfide gangue mineral, is not readily known for the size fractions that are not perfectly mixed. However, samples can be taken in relevant locations to obtain the information.

For case 2, the mass of nonsulfide gangue per unit mass of water at the top of the pulp region is shown schematically in Figure 1 for the two size fractions without perfect mixing. For these two size fractions, the unit mass of water is less heavily laden with nonsulfide gangue than for case 1, and this quantity of nonsulfide gangue per unit mass of water represents the starting condition for the entrainment mechanism. For case 2, the same efficiency for transportation through the froth region as for case 1 is expected for a given size fraction and has been used.

It can be seen in the figure that the altered extent of mixing for the two coarser size fractions (in case 2) had a small overall effect on the mass of nonsulfide gangue per unit mass of water in the concentrate for the –23 +16 μm size fraction and no overall effect for the +50 μm size fraction.

In case 3, an extreme scenario is represented for a flotation cell; alternatively, the scenario may arise from a cell designed to produce the type of mixing depicted in the pulp region. In case 3, where the pulp zone is not perfectly mixed for all size fractions in the figure, the important consequence is that the mass of each mineral in each size fraction per unit mass of water at the top of the pulp zone, such as the nonsulfide gangue mineral, is not readily known for all the size fractions. However, samples can be taken in relevant locations to obtain the information.

For case 3, the mass of nonsulfide gangue per unit mass of water at the top of the pulp region is shown schematically in Figure 1 for the three size fractions without perfect mixing. For these three size fractions, the unit mass of water is less heavily laden with nonsulfide gangue than for case 1, and this quantity of nonsulfide gangue per unit mass of water represents the starting condition for the entrainment mechanism. For case 3, the same efficiency for transportation through the froth region as for case 1 is expected for a given size fraction and has been used.

From Figure 1, it can be seen that the altered extent of mixing for all three size fractions (in case 3) had a moderate overall effect in comparison to case 2 on the mass of nonsulfide gangue per unit mass of water in the concentrate for the size fractions –11 μm and –23 +16 μm. In the progression from case 1 to cases 2 and 3, a decline in the mass of nonsulfide gangue per unit mass of water in the concentrate for the size fraction –23 +16 μm can be noted; for the size fraction –11 μm, a decline only occurred for case 3.

QUANTITATIVE DESCRIPTION OF ENTRAINMENT

In discussion of Figure 1, the possibility that the pulp region was not perfectly mixed for some or all size fractions was noted. Hence, in discussion of the quantitative evaluation of entrainment in this section, the following general equation is used for convenience, for size fraction i, noting that subscript con denotes concentrate:

$$\begin{array}{c}\text{entrainment}\\ \text{efficiency}\\ \text{value}\,(\text{ENT}_i)\end{array} = \frac{(\text{mass of liberated gangue per unit mass of water})_{\text{con},i}}{(\text{mass of liberated gangue per unit mass of water})_{\text{top of pulp},i}} \qquad \textbf{(EQ 3)}$$

For a conventional flotation system, typical values for ENT_i for a range of size fractions of siliceous nonsulfide gangue are provided in Table 2, where typical values decline from values approaching 1.0 for the finest size fractions to values approaching 0.0 for size fractions larger than 50 μm. The series of values in the table can be obtained for gangue minerals for which there is a reliable method for obtaining the necessary assays. For example, if a gangue mineral has an elevated specific gravity, the values for ENT_i in each size fraction can be determined to establish if there is any lowering of the values because of the elevated specific gravity. If a gangue mineral has an unusual shape—for example, platelike—the values for ENT_i in each size fraction will indicate if the unusual shape

Table 2 Example of ENT_i values for liberated siliceous nonsulfide gangue

Values for ENT_i, k_i, k_w, and k_i/k_w*	Cyclosizer Size Fraction					
	–C5 (–11†)	–C4 +C5 (–16 +11†)	–C3 +C4 (–23 +16†)	–C2 +C3 (–33 +23†)	–C1 +C2 (44 +33†)	–74 μm +C1 (–74 +44‡)
ENT_i values	0.77	0.46	0.30	0.24	0.13	0.03
k_i values (min^{-1})	0.00616	0.00368	0.00240	0.00192	0.00104	0.00024
k_w value (min^{-1})	0.00800	0.00800	0.00800	0.00800	0.00800	0.00800
k_i/k_w values‡	0.77	0.46	0.30	0.24	0.13	0.03

*k_i = overall first-order rate constant; k_w = first-order rate constant for water.
†Upper and lower values (μm) for size fraction.
‡The calculation for the values k_i/k_w (i.e., ENT_i) for each size fraction is given.

has caused a change in the entrainment efficiency in the froth region.

For a particular industrial flotation cell with a standard design, two samples will provide the necessary data for determination of the values for ENT_i in each size fraction, based on Equation 3:

1. A sample of the pulp at the top of the pulp region
2. A sample of the concentrate discharged from the froth region into the concentrate launder (noting that this sample must not contain any extraneous water from sprays in the launder or from other sources)

To obtain data on entrainment for the selected cell, flow rate data are not required; from Equation 3, only the ratios existing in the numerator and the denominator have to be determined. Of course, entrainment data can also be obtained as one of the outputs from a detailed bank or plant survey where flow rates exist. Before the detailed survey, if the industrial cells were perfectly mixed in the pulp region, then the sampling requirements for the survey are simplified, along with the extraction of the entrainment data. If the industrial cells were not perfectly mixed in the pulp region, then the sampling requirements for the survey will be increased. Additional samples will be required to obtain the term (mass of liberated gangue per unit mass of water)$_{\text{top of pulp},i}$, which is the denominator in Equation 3 for each industrial cell of interest. For flotation cells that are found not to be perfectly mixed, Zheng et al. (2005) proposed an exponential function to describe the measured solid suspension property of such cells.

To completely satisfy the requirements of Equation 3, the data from the samples have to be obtained on the appropriate size fractions and must cover in detail the subsieve region. Further, liberation data are required for the mineral(s) of interest including the nonsulfide gangue. This situation can be considered from two perspectives:

1. Analysis of the process weaknesses in mineral recovery–size–liberation data for an overall flotation bank, including the contribution from entrainment to these weaknesses (an activity of long standing in the industry by metallurgists or consultants)
2. Detailed modeling of the individual cells or groups of cells in a flotation bank, including the contribution from entrainment to the results from the bank (an activity most likely undertaken by research scholars, research metallurgists, or consultants)

For item 2, this activity has been described as the use of a mineral tracer for entrainment (Savassi 1998). This requires obtaining liberation data for a gangue mineral that is fully dispersed and hydrophilic. Hence, the data for the liberated portion of the selected mineral can be used to determine the contribution from entrainment.

In the past, an approach based on a compromise was often used in the initial investigations of the entrainment mechanism. In this compromise, the data from the samples were obtained on the appropriate size fractions in the subsieve region, and the samples were collected from sections of a flotation circuit where it was believed that the nonsulfide gangue reporting to the concentrate was predominantly liberated. This approach was taken because the measurement of liberation on size fractions was relatively rare compared to the present situation. The methods, which were based on traditional microscopy, existed but were not used often; however, automated instrumental methods have since become available and are now widely used.

For data derived from samples that have not been sized at all, it is clearly not possible to obtain the values in ENT_i; however, if water data (derived from percent solids values for the samples) have been obtained, it is possible to examine the data from a flotation section in a survey for an approximately linear relationship for one of the following graph types:

1. The relationship between nonsulfide gangue flow rate (y-axis) and water flow rate (x-axis) is shown in Figure 2A.
2. The relationship between nonsulfide gangue recovery (y-axis) and water recovery (x-axis) is shown in Figure 2B.

The data in the graphs in Figure 2 were obtained in the copper rougher-scavenger sections at the Philex Mining Corporation operation in the Philippines. The same original data set was used for each graphing option. When data of the type shown in this figure are collected over a very wide range of water recovery or flow rate values, the relationship can be slightly nonlinear (concave upward).

It is also necessary for the nonsulfide gangue mineral assay to be obtained by a reliable method. The nonsulfide gangue may refer to the combination of a group of nonsulfide

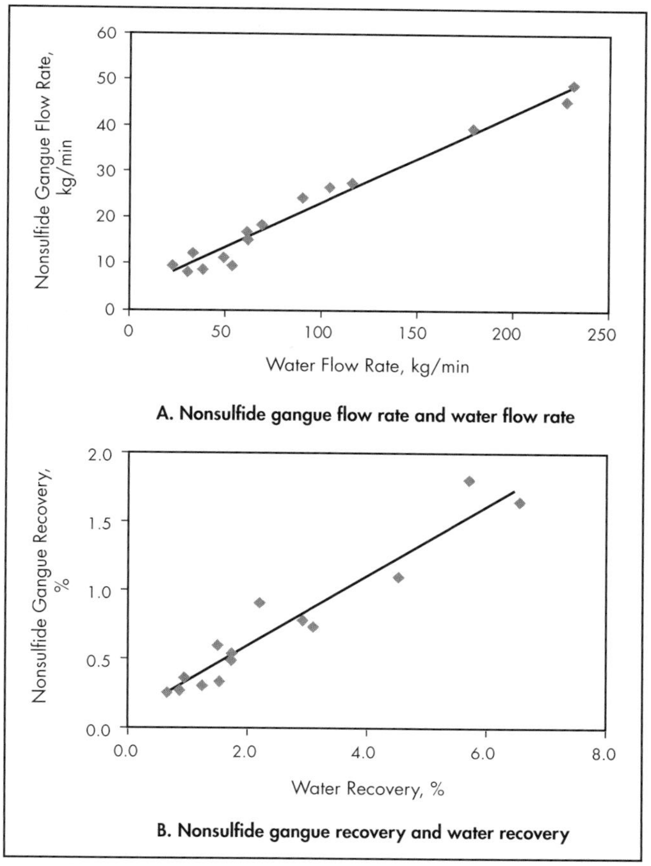

Source: Johnson 2005, reprinted with permission from the Australasian Institute of Mining and Metallurgy

Figure 2 Relationships of rougher-scavenger sections in the Philex Mining Corporation plant

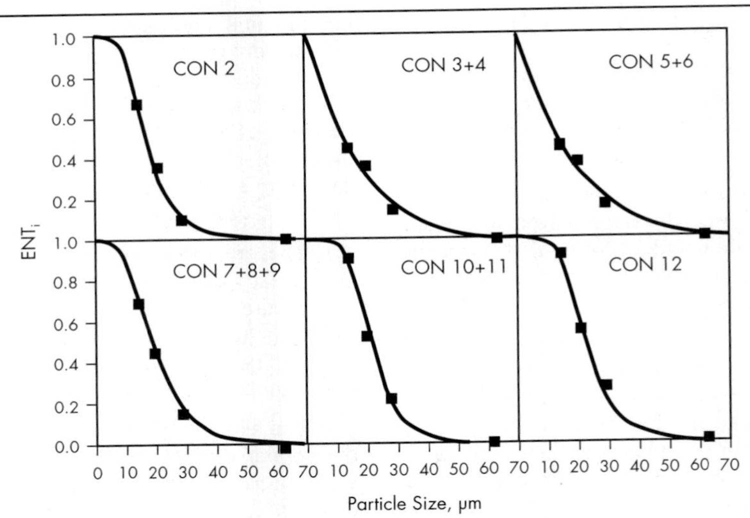

Source: Savassi 1998

Figure 3 Fitted relationship between ENT$_i$ values and size based on data from four size fractions for fully liberated nonsulfide gangue

gangue minerals or to a particular nonsulfide gangue mineral. The approximately linear relationship will only be observed convincingly for a flotation section where the nonsulfide gangue reporting to the concentrate is predominantly liberated and entrainment is the only or dominant mechanism. For plant survey data, the water data must reflect the water that flowed over the cell lip with the discharged concentrate and must not represent any contributions from launder, pump box, or other such water additions added beyond the cell lip.

The most powerful method for analysis of entrainment (mineral recovery–size–liberation level) is now considered. By having data for the liberated form of the relevant minerals in each size fraction, the ENT$_i$ values for each size fraction can be evaluated by inserting the required measured values as stated in Equation 3. The ENT$_i$ values can also be evaluated through a modified version of Equation 3 given as Equation 4 (Lynch et al. 1981). Equation 4 is derived from Equation 3 (see Appendix B). The values for the overall first-order rate constant (k_i) and the first-order rate constant for water (k_w) are able to be calculated from survey data (mineral recovery–size–liberation level of detail) for each cell, groups of cells, or for the whole bank, where the overall rate constant was defined earlier in this chapter.

$$ENT_i = k_i/k_w \quad \textbf{(EQ 4)}$$

Hence, for the *liberated* form of a particular mineral, the value for ENT$_i$ can be determined for each of the i-size fractions, using Equation 3 or Equation 4.

Some typical ENT$_i$ values for liberated siliceous nonsulfide gangue in various Cyclosizer fractions are given in Table 2. Further, for a liberated mineral displaying entrainment, the typical pattern of behavior in the size fractions for the rate constants for the liberated form of siliceous nonsulfide gangue relative to the water rate constant is also included in Table 2.

Some ENT$_i$ values for various flotation cells along a primary galena rougher bank at Mount Isa Mines Limited in Australia are given in Figure 3 (Savassi 1998). The ENT$_i$ values for four size fractions are provided for the following cells in the bank, noting that there was no production from cell 1: cell 2, cells 3+4, cells 5+6, cells 7+8+9, cells 10+11, and cell 12. The four size fractions were –75 +53 μm and Cyclosizer fractions C2, C3, and C4 with midpoints of 28.5, 20.2, and 14.0 μm, respectively, for nonsulfide gangue (Savassi 1998). Savassi also analyzed the entrainment data along the bank for any systematic changes using the methods summarized for this purpose in Savassi et al. (1998).

Equation 4 can be rewritten to a slightly different but useful form:

$$k_i = ENT_i\, k_w \quad \textbf{(EQ 5)}$$

Hence, in general metallurgical calculations of entrainment involving Equation 5, if a value for the water rate constant is known or assumed, and by use of a set of values for ENT$_i$, it is possible to calculate the value of k_i for each size fraction that results from entrainment of a liberated gangue mineral. To this point for discussion of sized data, rate constants or ENT$_i$ values have been employed to express the behavior of water and the entrained liberated mineral within the entrainment mechanism, noting that recovery values for size fractions have not been considered because recovery values do not appear in Equations 3 to 5, which are various forms of the fundamental equations for entrainment.

Figure 4 shows three presentations for the rougher portion of the Philex Mining Corporation data in Figure 2. For the three graphs, the samples were sized and the size fractions were assayed, but no detailed liberation data were obtained. The details of the three graphs are summarized in Table 3 and comments on each graph are also provided.

A selection can be made from the three presentation examples of the same data in Figure 4 depending on the purpose for displaying the data.

Equation 3 can also be rearranged, as demonstrated in Appendix B, to Equation 6 in which the recovery rate (mass/time) to the concentrate launder of liberated nonsulfide gangue in size fraction i by entrainment is on the left side of the equation.

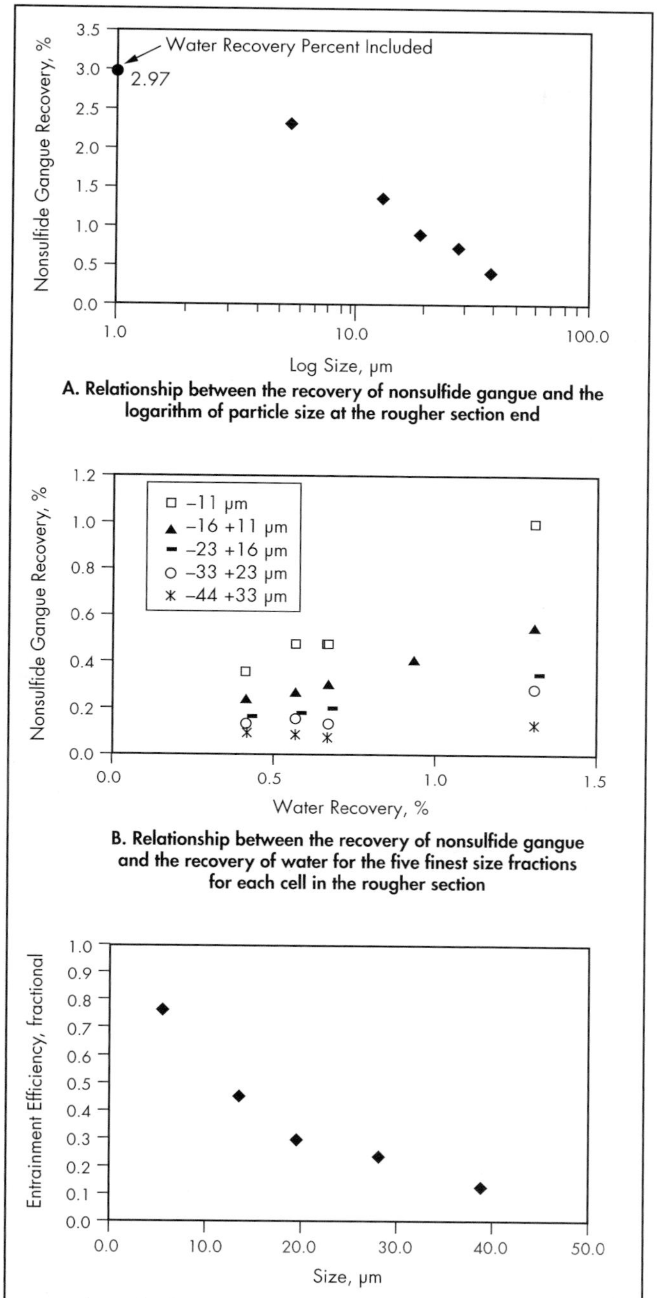

A. Relationship between the recovery of nonsulfide gangue and the logarithm of particle size at the rougher section end

B. Relationship between the recovery of nonsulfide gangue and the recovery of water for the five finest size fractions for each cell in the rougher section

C. Relationship between an entrainment efficiency factor for nonsulfide gangue and particle size at the rougher section end

Source: Johnson 2005, reprinted with permission from the Australasian Institute of Mining and Metallurgy

Figure 4 Recovery and entrainment efficiency relationships in the Philex Mining Corporation rougher

$$RRFG_{con,i} = ENT_i \frac{MFG_{top\ of\ pulp,i}}{MWAT_{top\ of\ pulp}} RRW_{con} \quad \textbf{(EQ 6)}$$

where

$RRFG_{con,i}$ = rate of recovery of free (liberated) gangue in the concentrate for size i

$MWAT_{top\ of\ pulp}$ = unit mass of water at the top of the pulp region

$MFG_{top\ of\ pulp,i}$ = mass of free (liberated) gangue in size i suspended in $MWAT_{top\ of\ pulp}$

RRW_{con} = rate of recovery of water in the concentrate

For each size fraction, Equation 6 contains the important term ($MFG_{top\ of\ pulp,i}$)/($MWAT_{top\ of\ pulp}$); for each size fraction, this term expresses the mass of a mineral suspended in unit mass of water at the top of the pulp region. This term is multiplied by another term (RRW_{con}), the number of water units derived from the pulp region reaching the concentrate launder; it is stressed that this term refers only to water originating from the top of the pulp zone and therefore carrying suspended particles. Finally, the transportation efficiency (ENT_i) for each size fraction is included in the equation. The term ($MFG_{top\ of\ pulp,i}$)/($MWAT_{top\ of\ pulp}$) has a connection with the percent solids in the pulp (lowering the percent solids value will lower the value for the term), but it is clearly not the same entity; the term describes the well-known benefits of the use of dilute pulps in industrial flotation machines of a normal type to lessen the consequences of the entrainment mechanism.

Various authors have proposed equations to describe the decay in values for ENT_i with increase in particle size, that is, the type of relationship demonstrated in Figures 3 and 4C. These equations have been developed by an empirical or a mechanistic approach by Bisshop and White (1976), Ross and Van Deventer (1988), Kirjavainen (1992), Maachar and Dobby (1992), Savassi et al. (1998), and Yianatos and Contreras (2010). Zheng et al. (2005) used the approach to entrainment described by Savassi et al. (1998) and complemented that approach with an exponential function to describe the measured solid suspension property of cells that were not perfectly mixed.

Figure 3 shows the fitted relationship between ENT_i and particle size. Savassi (1998) developed an equation (not shown in this chapter) that was suitable for fitting to ENT-size data and allowed the evaluation of convenient parameters in the equation. This is described in general by Savassi et al. (1998).

All of these equations, except those of Yianatos and Contreras (2010), were summarized in an appendix of a review on entrainment (Johnson 2005). The important step for many metallurgists in the industry is the evaluation of the numerical values of ENT_i from industrial results.

ANALYSIS FOR ENTRAINMENT IN INDUSTRIAL FLOTATION DATA

For industrial flotation data in general at both the mineral recovery–size–liberation level of detail and the less specific mineral recovery–size level, it is common for the data to be presented with a graph displaying mineral recovery (y-axis) and the logarithm of particle size (x-axis). The key point here is that mineral recovery values are displayed (and not rate constants, which were discussed in the previous section). Mineral recovery values are an output from normal metallurgical calculations; mineral recovery values have been calculated traditionally and hence have been employed commonly in the preparation of graphs summarizing the data.

A practice has evolved where metallurgists have sought to extend the use of these graphs to allow interpretation of the contribution from entrainment. The water recovery is displayed on the graph, and its value is compared with the

Table 3 Summary of various graphs* for the nonsulfide gangue in the rougher section

Graph	Y-Axis	X-Axis	Origin of Data†	Comments and Observations
A	Nonsulfide gangue recovery, %; water recovery noted	Log_{10} size, µm	Rougher section (to end)	This is a conventional mineral recovery–size relationship. Nonsulfide gangue recovery approached water recovery for the finest size fraction. This is a convenient and commonly used graphical depiction without integration of water–recovery data.
B	Nonsulfide gangue recovery, % for each of five size fractions	Water recovery, %	Cells in rougher section	Each size fraction displayed a linear relationship. The finest fraction displayed the steepest gradient. The gradient decreased in an orderly way from the finest to the other fractions. This is an example of integration of water–recovery data.
C	ENT_i (entrainment efficiency value)	Size, µm	Rougher section (to end)	The gradients for each of the linear relationships in Figure 4B are plotted on the y-axis with the corresponding size fraction on the x-axis. This is an example of integration of water data.

Adapted from Johnson 1972
*The three graphs are shown in Figure 4.
†Philex Mining Corporation rougher section.

recovery values for the liberated form of the gangue mineral(s) suspected of being principally recovered by entrainment.

The legitimacy of this approach is now considered with use of a commonly used equation that can be applied to the liberated form for each mineral (Equation 7) and for water (Equation 8). Equation 7 contains, for size i, terms for the mineral recovery (R_i), the first-order rate constant (k_i), and the cell residence time (λ) (Lynch et al. 1981) for a single, perfectly mixed flotation machine.

$$R_i = k_i \lambda /(1 + k_i \lambda) \quad \textbf{(EQ 7)}$$

$$R_w = k_w \lambda /(1 + k_w \lambda) \quad \textbf{(EQ 8)}$$

For a single, perfectly mixed flotation machine, Equation 8 contains terms for water recovery (R_w), the first-order rate constant for water (k_w), and the cell residence time (λ) (Lynch et al. 1981).

When the value for the first-order rate constant is very low (the typical situation for a gangue mineral reaching the concentrate launder by entrainment), Equations 7 and 8 can be approximated by Equations 9 and 10. The approximation is assisted when the value for residence time (λ) is also low.

$$R_i \cong k_i \lambda \quad \textbf{(EQ 9)}$$

$$R_w \cong k_w \lambda \quad \textbf{(EQ 10)}$$

In this situation, the conveniently observed ratio R_i/R_w from the described graphs (or calculated from the data used to construct the graphs) provides an approximate value for the desired ratio k_i/k_w, which is equal to ENT_i as shown in Equation 4 earlier.

Table 4 demonstrates where the approximations in Equations 9 and 10 can be used via some numerical examples. In Table 4, the effect of a range of values for k_i, k_w, and λ is demonstrated for the closeness of the calculated value for R_i/R_w to the corresponding value of k_i/k_w.

To summarize, for mineral recovery–size–liberation data, the use of the ratio k_i/k_w to obtain the value for ENT_i in each size fraction is the desirable method. When the absolute values for k_i and k_w are sufficiently low, as described earlier, the use of the ratio R_i/R_w to obtain the values for ENT_i in each size fraction is a reasonable approximation. From Table 4, the approximate method also becomes more reliable when the residence time (λ) is low in conjunction with low absolute values for the rate constants.

Table 4 Values for inputs substituted into Equations 7 and 8* and the resulting output

Values for Inputs to Calculation				Values for Output from Calculation
λ, min	k_i	k_w	k_i/k_w	R_i/R_w
1.0	0.08	0.1	0.80	0.82
1.0	0.008	0.01	0.80	0.80
1.0	0.0008	0.001	0.80	0.80
1.0	0.04	0.1	0.40	0.42
1.0	0.004	0.01	0.40	0.40
1.0	0.0004	0.001	0.40	0.40
5.0	0.08	0.1	0.80	0.86
5.0	0.008	0.01	0.80	0.81
5.0	0.0008	0.001	0.80	0.80
5.0	0.04	0.1	0.40	0.50
5.0	0.004	0.01	0.40	0.41
5.0	0.0004	0.001	0.40	0.40
10.0	0.08	0.1	0.80	0.89
10.0	0.008	0.01	0.80	0.82
10.0	0.0008	0.001	0.80	0.80
10.0	0.04	0.1	0.40	0.57
10.0	0.004	0.01	0.40	0.42
10.0	0.0004	0.001	0.40	0.40

*Calculations for residence times (λ) of 1.0, 5.0, and 10.0 minutes; and for k_w values (min^{-1}) of 0.10, 0.010, and 0.0010; and for k_i values set at 0.8 and 0.4 of the k_w values.

For the less detailed mineral recovery–size data, the behavior patterns associated with entrainment can be expected in banks for which the candidate minerals for entrainment are well liberated and entrainment is the dominant mechanism. In banks where the candidate minerals for entrainment have additional important recovery mechanisms, the information in the graphs of mineral recovery (y-axis) and the logarithm of particle size (x-axis) can be interrogated, and the described ratios can be examined; however, the patterns of behavior associated with entrainment may be less obvious. For the less

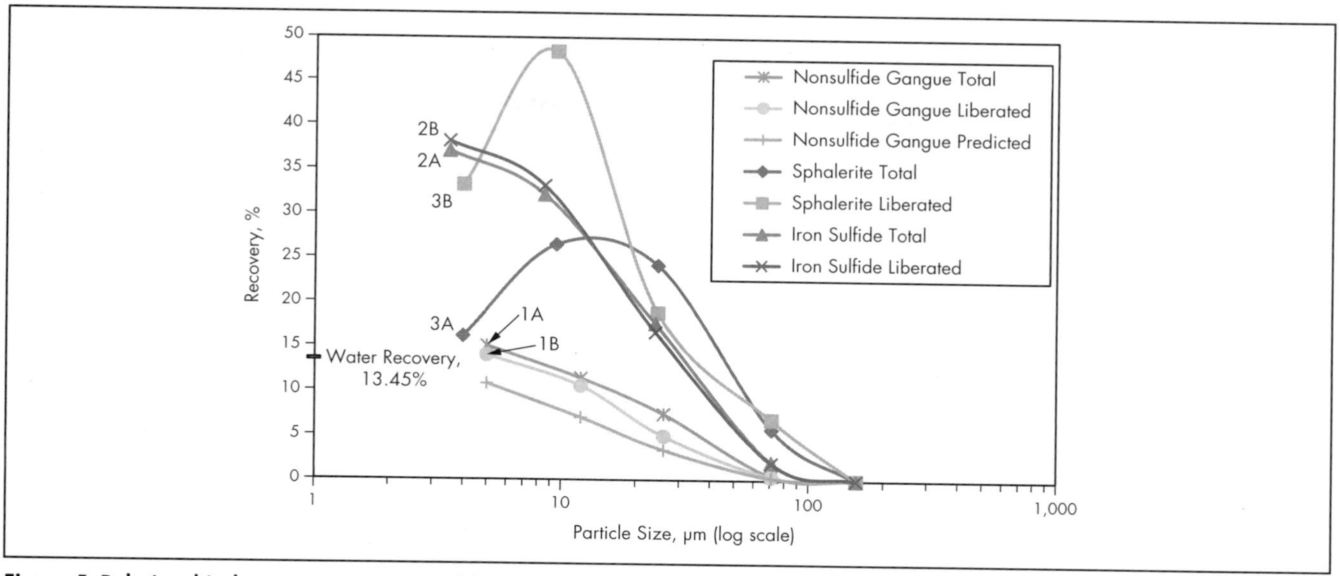

Figure 5 Relationship between recovery and log particle size for nonsulfide gangue, sphalerite, and iron sulfide in the galena primary rougher bank at Mount Isa Mines Limited

detailed mineral recovery–size data, the ratio k_i/k_W can only be calculated based on the total flow for the mineral in each size fraction (and not for the liberated portion). Calculation of the ratio k_i/k_W gives a value for ENT_i in each size fraction and remains the more reliable method. When the values for k_i and k_W are sufficiently low, as shown earlier in Table 4, the use of the ratio R_i/R_W to obtain the value for ENT_i in each size fraction is, again, a reasonable approximation.

For equivalent calculations to those in Table 4 with multiple perfectly mixed flotation machines in series in a bank, Equation 11 can be used for each size fraction:

$$R_i = 1 - (1 + k_i \lambda)^{-N} \qquad \textbf{(EQ 11)}$$

For size fraction i, Equation 11 contains terms for mineral recovery (R_i), the first-order rate constant (k_i) for the mineral, the cell residence time (λ), and the number (N) of perfectly mixed flotation machines in the bank each with residence time (λ) (Lynch et al. 1981); the same form of the equation can be used for water.

Conventional Flotation Process

From the perspective of minimal importance of the entrainment mechanism, the ideal scenario for application of conventional flotation (hydrophobic valuable mineral reporting to the froth region and the concentrate launder) is that the valuable mineral has a low head grade, with the following observations:

1. The solid recovery will be low (assuming a moderate or high grade for the concentrate).
2. The water recovery will also be low (assuming a typical percent solids by weight value for the concentrate e.g., 35%–50%).
3. The low recovery of water in the concentrate constrains the closely related recovery of the unwanted minerals by the entrainment mechanism alone to an even lower value.

The recovery of coal in a conventional flotation process is an example where the head grade of the hydrophobic mineral is high and is not an example of an ideal scenario for conventional flotation. However, it still represents an economic scenario.

In Figure 5, an example of industrial diagnostic data is given for the galena primary rougher bank at Mount Isa Mines Limited in 1981; this is the section from which the data in Figure 3 were obtained approximately 15 years later. An analysis is given for the contribution of entrainment to the recovery of minerals for which recovery was not sought in the galena roughing section: nonsulfide gangue, iron sulfide, and sphalerite. The purpose of the galena roughing stage was essentially to recover liberated galena; that is, collector additions were set at low values to recover liberated galena, not to recover substantial quantities of galena in composite particles with nonsulfide gangue and with other sulfide minerals. The recovery of galena was 63.3% in the initial galena rougher, and the mineral recovery–size (log scale) curves for the three minerals for which recovery was not sought in the section are shown. For each mineral in each size fraction, the recovery values for the total mineral (all liberated and all composite forms combined) and for the liberated form alone are given. For nonsulfide gangue, an additional curve is given with predicted values for the recovery of liberated nonsulfide gangue using a standard set of ENT_i values.

The implication of mineral recovery graphs at the mineral recovery–size level of data analysis is that all forms of the mineral, both liberated and in composite particles, are represented in the recovery values for each size fraction; the combined recovery of all forms (liberated and all types of composites) of the mineral is expressed.

For nonsulfide gangue, at the mineral recovery–size level of data analysis, it is common for the recovery–size (log scale) values to decline from elevated values (typically slightly lower than the water–recovery value) in the region 0–10 μm to lower values at slightly larger sizes and then reach, in the coarser fractions, a region of low values or to increase to a maximum before declining again. This effect is caused by the recovery of nonsulfide gangue in composites with the valuable mineral in the coarser fractions in scavenging. In this situation, because of the absence of scavenging conditions, the values decline from elevated values in the region below 10 µm to values approaching zero in the coarser fractions.

The water recovery was 13.5%, an important yardstick in determining the extent of entrainment for the unwanted minerals in the fine fractions. By using the approximation that $R_i = ENT_i R_w$ and the ENT_i values for nonsulfide gangue in Table 2 (obtained from a chalcopyrite ore), predicted recovery values of liberated nonsulfide gangue in the various size fractions in Figure 5 have been calculated and are shown in that figure. The predicted values for recovery of liberated nonsulfide gangue were 10.8% in the finest fraction and declined to 7.0%, 3.4%, 0.3%, and 0.0% for the larger fractions, in sequence. Because the size fractions are slightly different in Figure 5 than Table 2, the ENT_i values in the table were graphed, and the ENT_i values for the size fractions used in the figure were interpolated. The predicted recovery values for liberated nonsulfide gangue were calculated using the following interpolated ENT_i values commencing with the finest fraction: 0.80, 0.52, 0.25, 0.02, and 0.0. As a first approximation, the recovery values for liberated nonsulfide gangue in Figure 5 are assumed to also apply to the liberated iron sulfides and the liberated sphalerite in that figure in the following discussion.

The behavior of the minerals in Figure 5 is discussed on the basis of recovery values because the water rate constant (k_w) is sufficiently low (0.0136 min^{-1}), and, from Table 4, it is safe to use the convenient method based on mineral recovery values in that figure. In Figure 5, the *mineral recovery–size level of analysis* for the finest fraction will be discussed initially. This means that the recovery of all forms (liberated and in all types of composites) of each mineral is being considered for the finest size fraction.

- The recovery of nonsulfide gangue (15.0%, point 1A) is slightly greater than the recovery of water (13.5%) and the associated predicted value for recovery of liberated nonsulfide gangue (10.8%) by entrainment; that is, the entrainment mechanism explains most of the recovery of nonsulfide gangue.
- The recovery of iron sulfide (37.0%, point 2A) is very much greater than the recovery of water (13.5%) and the associated predicted value for its recovery by entrainment (10.8%); that is, the entrainment mechanism explains only a minority of the recovery of iron sulfide.
- The recovery of sphalerite (16.1%, point 3A) is slightly greater than the recovery of water (13.5%) and the associated predicted value for its recovery by entrainment (10.8%); that is, the entrainment mechanism explains most of the recovery of sphalerite.

For the finest size fraction, the *liberated* nonsulfide gangue, iron sulfide, and sphalerite in Figure 5 will now be interpreted. A graph of the liberated form of the mineral is one of the many graphs available for the mineral recovery–size–liberation level of analysis. Taking nonsulfide gangue as an example, the importance of the data for its liberated form is that the recovery of the nonsulfide gangue in composite particles with hydrophobic sulfide minerals in the feed is eliminated, making the resulting values more relevant for examination of the entrainment mechanism. The following comments can be made:

- The recovery of liberated nonsulfide gangue (14.0%, point 1B) is approximately the same as the recovery of water (13.5%) and slightly greater than the associated predicted value for recovery of liberated nonsulfide gangue by entrainment (10.8%); that is, the entrainment mechanism explains most of the recovery of nonsulfide gangue and the results are closer to the expected results for entrainment than in point 1A.
- The recovery of liberated iron sulfide (38.1%, point 2B) is very much greater than the recovery of water (13.5%) and very much greater than the associated predicted value for recovery of nonsulfide gangue by entrainment (10.8%); that is, the entrainment mechanism explains only a minority of the recovery of iron sulfide. This conclusion is the same as for point 2A because the iron sulfide in this size fraction was highly liberated in both the feed and the concentrate. It is also clear that the recovered liberated iron sulfide must have some level of hydrophobicity to allow its recovery. The cause of the hydrophobicity cannot be deduced from the results in Figure 5 alone.
- The recovery of liberated sphalerite (33.2%, point 3B) is much greater than the recovery of water (13.5%) and very much greater than the associated predicted value for recovery of nonsulfide gangue by entrainment (10.8%); that is, the entrainment mechanism explains only a minority of the recovery of sphalerite. This conclusion is quite different from point 3A because a significant portion of the recovered sphalerite was in composites with galena and also with iron sulfides. It is also clear that the recovered liberated sphalerite must have some level of hydrophobicity. The cause of the hydrophobicity cannot be deduced from the results in Figure 5 alone.

From interpretation of the graphs for nonsulfide gangue, iron sulfide, and sphalerite for the second finest fraction, similar conclusions can be drawn as for nonsulfide gangue (points 1A and 1B) and iron sulfide (points 2A and 2B) in the finest fraction; however, this is not the case for the sphalerite in the second finest fraction. Unlike the finest fraction and for total sphalerite, the recovery of sphalerite (26.5%) is much greater than the recovery of water (13.5%) and the associated predicted recovery of sphalerite by entrainment (7.0%); that is, the entrainment mechanism explains only a minor portion of the recovery of sphalerite. Correspondingly, the recovery of liberated sphalerite (48.3%) is very much greater than the recovery of water (13.5%) and its predicted entrainment value (7.0%); its recovery value (48.3%) is considerably higher than for the liberated sphalerite in the finest fraction (33.2%); that is, the entrainment mechanism explains only a small minority of the recovery of liberated sphalerite.

From the example in Figure 5, the following can be observed:

- The information on the level of entrainment in graphs of total mineral recovery–size is, in some cases, quite different from graphs of liberated mineral size. In the example, the information for sphalerite in the two types of graphs was quite different (Figure 5).
- The level of entrainment for nonsulfide gangue was slightly higher than is normally expected, but the entrainment mechanism explained most of the recovery of liberated nonsulfide gangue.

The ore in this case displays behavior that is observed for the various stratiform zinc–lead ores that are found in the Carpentaria Inlier in the Northern Territory and in northwestern Queensland in Australia. Principally, a portion of the iron sulfide minerals is affected by naturally occurring hydrocarbons, which existed at the time of ore formation and that exist principally on the surface of a portion of the iron sulfide (relatively fine-grained spheroidal pyrite). These hydrocarbons confer limited hydrophobicity on the iron sulfides but are sufficient for significant recovery of this unwanted sulfide (Croxford et al. 1961). The affected iron sulfides are sometimes referred to as *carbonaceous pyrite*. Supplementary types of flotation tests (Johnson and Jowett 1982), such as tests with frother as the only reagent, show that hydrophobicity of the iron sulfide causes the behavior observed for this mineral in the fine fractions in the example. A smaller proportion of the liberated sphalerite also acquires hydrophobicity from the hydrocarbons, and the liberated nonsulfide gangue can be affected slightly. Hydrocarbons conferring the hydrophobicity largely occur as very thin layers and are not able to be recorded in the measurement of liberation data.

In conventional flotation, including the example, it has been seen that entrainment can be important for the fine fractions of the gangue mineral(s); that is, the entrainment mechanism results in a lowering of the concentrate grade. In contrast, for reverse flotation processes, the entrainment mechanism is important for the valuable mineral; that is, the entrainment mechanism can result in a lowering of the recovery of the valuable mineral.

Reverse Flotation Process

In a reverse flotation process, the valuable mineral of interest is not in a hydrophobic condition; instead, at least one unwanted mineral is in a hydrophobic condition, arising either from the mineral being naturally hydrophobic (e.g., talc) or from deliberate collector adsorption on the unwanted mineral. Three cases for reverse flotation processes are listed:

1. Preflotation before conventional roughing to remove hydrophobic gangue minerals—for example, talc (or perhaps a related mineral with similar properties, pyrophyllite), carbonaceous pyrite, or native sulfur (valuable mineral in the rougher feed subject to loss by entrainment in the recovered water of the reverse flotation process).
2. An additional stage of upgrading by reverse flotation of a concentrate produced by conventional roughing and cleaning steps (valuable mineral in the original concentrate subject to loss by entrainment in the recovered water of the reverse flotation process).
3. A stage of reverse flotation upgrading of a concentrate prepared by another separation technique—for example, a concentrate prepared by magnetic concentration of valuable iron ore minerals (valuable iron-bearing minerals in the magnetic concentrate subject to loss by entrainment in the recovered water of the reverse flotation process).

In case 1 (preflotation), all the valuable mineral in the plant feed is exposed to potential loss in the preflotation concentrate removed to the launder. In cases 2 and 3, a concentrate containing a substantial to high proportion of the valuable mineral in the feed is exposed to potential loss in the reverse flotation concentrate containing unwanted gangue mineral(s) removed via the froth region to the launder. The important point is that the valuable mineral in a hydrophilic condition is exposed to entrainment in a reverse flotation process; there is a very strong economic incentive to minimize losses by entrainment, thereby maximizing recovery of the valuable mineral in the target product.

The same methods used for conventional flotation processes can be applied for the analysis of process weaknesses for reverse flotation; assaying for the valuable mineral subjected to entrainment is often direct and without the complications associated with gangue minerals. Because the valuable mineral is subjected to loss of recovery by the entrainment mechanism in a reverse flotation system, there may be a case to employ the most powerful method for lessening the importance of entrainment in a single stage of processing (cells designed with water addition to the froth region).

Sometimes, the reason for application of a reverse flotation process is that the selectivity (expressed as the ratio of the rate constants for the minerals of interest) is higher than for the related conventional flotation process:

Reverse Flotation Process		Conventional Flotation Process
Rate constant for liberated gangue mineral / Rate constant for liberated valuable mineral	>	Rate constant for liberated valuable mineral / Rate constant for liberated gangue mineral

Sometimes, flotation processes require the use of a particular gas instead of the use of air. This practice is used quite often in reverse flotation processes where gases such as nitrogen and carbon dioxide may be used to provide the required chemical conditions for the reverse flotation process and/or minimize consumption of reagents. The contact angles for a range of gases (air, oxygen, hydrogen, nitrogen, carbon dioxide, and sulfur dioxide) for a mineral of fixed hydrophobicity were shown to be quite similar (Wark and Cox 1932). Hence, the change in the type of gas does not appear to affect the principal recovery mechanism for the hydrophobic mineral based on particle–bubble collision and adhesion.

TECHNIQUES FOR REDUCTION OF ENTRAINMENT IN INDUSTRIAL FLOTATION

Several methods (based on Equation 6) are available to industry metallurgists to lower the importance of entrainment in the design of new plants and in the improvement of the efficiency of existing plants, for both conventional and reverse flotation processes:

1. The use of dilute pulps in standard flotation machines, particularly in cleaning
2. The use of more stages of cleaning with dilute pulps in standard flotation machines
3. In the absence of a water addition to the froth region, decreasing the flow of water in the concentrate discharged to the concentrate launder
4. The use of water additions to the froth region, with the correct flow rate of water and a suitable water distribution (typically used with column flotation machines, Jameson cells, and other cells designed to operate with relatively deep froth regions)
5. Use of coarser feed-sizing distributions where appropriate

From Equation 6, the important term $(MFG_{\text{top of pulp},i})/(MWAT_{\text{top of pulp}})$ (closely related to the percent solids by weight of the pulp) describes the well-known beneficial effect from the use of dilute pulps in conventional flotation cells to lower the importance of the entrainment mechanism. This beneficial effect is relevant for rougher sections, and it is particularly important in cleaning sections (often a series of cleaning stages arranged in closed circuit and with a major water addition to the final cleaning stage to provide the necessary dilute pulps in each of the cleaning stages in closed circuit).

With more dilute pulps, the mass of liberated nonsulfide gangue in size i per unit mass of water being transferred from the top of the pulp region into the froth region is lowered. In this reasoning, it is assumed that the flow rate of the water discharged to the concentrate launder and originating from the pulp zone (RRW_{con}), remains approximately fixed; that is, the percent solids value for the concentrate discharged to the concentrate launder remains similar. The objective remains to recover essentially the same flow rate of valuable mineral from the section (with some lowering of the flow rate of liberated nonsulfide gangue); hence, the total flow rate of solid discharged from the section will only be changed moderately, and, with a similar percent solids value for the concentrate discharged, the flow rate of the water discharged to the concentrate launder will remain largely unchanged.

In summary, the largely unchanged flow of water in the concentrate has been provided by water from the pulp region with a lower *load* of suspended gangue (lower mass of gangue per unit mass of water) because of the diluted pulp, leading to a lower load of gangue per unit mass of water in the concentrate, noting the largely unchanged flow of water in concentrate.

In an existing operation with dilute pulps in a series of cleaning stages arranged in closed circuit, one option for improved performance is the inclusion of one or more additional cleaning stages designed with the capacity to provide the necessary residence time for the valuable minerals, recognizing the large flow rates of water required to be processed by each additional cleaning stage.

In an existing operation with nondilute pulps in a series of cleaning stages arranged in closed circuit or in another format, an option is the inclusion of additional flotation-cell capacity in each cleaning stage to accommodate the increased water additions necessary for more dilute pulps. Such changes need to be designed with sufficient flotation-cell capacity to maintain the necessary residence time for the valuable minerals, recognizing the much larger flow rates of water required to be processed by each cleaning stage as a result of the higher water additions to achieve the necessary low percent solids values in the pulp. The percent solids by weight values in the various banks or stages in a cleaner section are commonly in the region 20%–30% by weight; the values can be set at a lower range for more effective lowering of the contribution of the entrainment mechanism.

Cleaning sections often rely on several mechanisms in parallel to achieve their objective, a large improvement in concentrate grade with low losses of the valuable mineral. In addition to the use of dilute pulps to minimize the detrimental entrainment mechanism for liberated hydrophilic gangue, elevated pH conditions and depressants lower the recovery of moderately hydrophobic, liberated gangue minerals. Such conditions and depressants can be expected to lower the flotation rate of the valuable mineral as well, but to a lesser extent than the gangue minerals with some hydrophobicity.

In Equation 6, the values for ENT_i are much higher for the finest fractions. Hence, lowering the proportion of nonsulfide gangue that exists in the finest size fractions in the pulp region will lower the flow rate of nonsulfide gangue recovered by entrainment. For example, if it is possible to increase the coarseness of the feed to a rougher section while maintaining the recovery of the valuable mineral in the rougher stage by recovery of more of the valuable mineral in the form of composites, a better result is obtained in terms of lowering entrainment.

The flow rate of water in the concentrate, derived from the pulp region (RRW_{con}), is another term in Equation 6, and its capability to alter the transfer of liberated gangue by the entrainment mechanism is now described. In a standard cell, there is sometimes scope for lowering the value of RRW_{con} by altering variables, such as lowering the air addition rate, increasing the vertical distance from the lip of the cell to the top of the pulp region, or lowering the frother addition. The *pulling rate* of the cell is lowered by such changes and can be expected to moderately lessen the flow rate of the hydrophobic mineral in the concentrate. The overall effect of such a change is a higher grade concentrate in conventional flotation and moderate lowering of the recovery of the hydrophobic mineral. The nature of the trade-off can be determined by sampling a selected cell at a range of pulling rates. When the residence time available in a bank is marginal, the scope for this option to be viable will be low.

For any type of cell designed to operate with a wash-water addition—for example, column cells, Jameson cells, and some other types of cells—the value for RRW_{con} (the water flow discharged in the concentrate originating from the pulp zone) can be greatly lowered to the extent that it approaches zero. In such cells designed for a wash-water addition, the water in the froth region, which is derived from the top of the pulp region (and its load of suspended particles), is replaced within the froth region by an externally supplied source of water containing no suspended particles. To obtain this effect, the wash water must be added in sufficient quantity to replace all the water that is required in the concentrate that discharges from the cell, with a small excess (Finch and Dobby 1990). Further, the wash water must be evenly distributed over the total area of the froth region to allow complete replacement of the original water from the pulp region containing suspended particles. Dissolved tracer can be used to check the transfer of water in the column feed (i.e., from the top of the pulp region) to the concentrate (Finch and Dobby 1990; Espinosa-Gomez et al. 1989) with appropriate operation of the wash-water system. The recovery of water from the feed was found by these authors to be very low, in the region of 1%, with most of the water in the concentrate being provided from the added wash water.

Because the replacement of the water from the pulp region with wash water occurs in cells designed to operate with wash water, the percent solids value in the pulp region (effectively the mass of liberated gangue in size i per unit mass of water at the top of the pulp region) is, in principle, not a factor for minimization of entrainment in such cells.

Two methods for the addition of water to lower entrainment have been discussed: water addition to the pulp region to dilute the pulp (standard cells) or water addition to the froth region (mainly column cells but occasionally standard cells). The most powerful method, when implemented and operated correctly, is water addition to the froth region, and there are many examples of this method in the industry. The degree of difficulty in its implementation and operation is not prohibitive, but it is more difficult to implement and operate correctly than pulp dilution. Where the economic incentive for lowering entrainment is large, it may be the appropriate approach. The traditional method, water addition to the pulp region in standard cells, is a less powerful method, requiring more stages to achieve an objective; however, it is easier to implement and operate correctly.

It has been reported that the shape of minerals also has an effect on mineral susceptibility to entrainment (Kirjavainen 1992) by demonstrating that the mica mineral phlogopite, with its platelike shape, has a higher propensity for entrainment. Kirjavainen (1992) proposed that this resulted from a change in the rheology in the froth region caused by the presence of fine phlogopite in sufficient concentration. From the findings of Kirjavainen (1992), a higher value for ENT_i for a mineral with a platelike shape would be expected than for a given size fraction and specific gravity of minerals with a normal shape. Silvester et al. (2011) noted that sericite, historically viewed as a troublesome mineral in some ores, is a mineral whose "recovery can be controlled by judicial use of dilution cleaning and/or froth washing" in recent work. This finding indicated that, for the systems investigated, the sericite was being recovered by the entrainment mechanism. Evaluation of the ENT_i values for size fractions for liberated sericite would demonstrate if this mineral displays higher values than other minerals of similar specific gravity; this is an example of the application of methods described in this section to improve understanding of the entrainment behavior of some gangue minerals.

Other possibilities for lessening of entrainment are now briefly mentioned. These proposals are from research papers; a small proportion of these proposals may eventually find economic application in industrial plants.

Deliberate design modifications may be taken in cell design to minimize the suspension of fine gangue at the top of the pulp zone, via altered hydrodynamic regimes and/or the use of physical devices (Harris et al. 1983; Guerra and Schubert 2003). This approach is the deliberate creation of the conditions depicted at the top of the pulp region for cases 2 and 3 in Figure 1. Sometimes, in existing standard cells, the general situation in case 2 may arise accidentally, particularly in large modern cells. The primary purpose of standard flotation cells in a conventional flotation process is to create conditions suitable for the recovery of the valuable minerals, and impairment of this capability would need to be minimized in any attempts to lessen entrainment.

Guerra and Schubert (2003) proposed that the effective particle size for gangue minerals in the entrainment mechanism could be increased by selective flocculation of the gangue, thereby shifting much of the gangue to a larger size fraction for which the ENT_i values are much lower. The intention in this approach is to achieve selective flocculation of all or the majority of the unwanted minerals. For such an approach to be viable, only minimal amounts of the hydrophobic valuable particles could be tolerated in the created floccules where the hydrophobic valuable mineral would not be accessible for recovery.

Examples of such experiments have been given by Liu et al. (2006) (use of high-molecular-weight polymers) and Gong et al. (2010) (use of polyethylene oxide). These workers used a combination of flocculants and collectors in a sequence and at appropriate addition rates.

A quite different situation was considered by Cao and Liu (2006). Sphalerite is a valuable sulfide mineral that is normally recovered after chalcopyrite and/or galena, and its depression is required during recovery of the other valuable mineral(s). Cao and Liu (2006) claimed that the conventional inorganic depressant, zinc sulfate, for sphalerite acts by making its surface hydrophilic and also by allowing floccules of sphalerite to form, thereby lowering its susceptibility to entrainment. In this situation, only a small portion of the feed would be selectively flocculated.

Kaya and Laplante (1990) reported that mechanical and ultrasonic vibration in the froth region of a conventional cell provided further rejection of entrained gangue beyond the effect of wash water in studies on sulfide flotation systems. In studies of coal flotation systems, Cheng et al. (1999; 2007) reported benefits in selectivity from mechanical vibration.

George et al. (2004) indicated that very small-diameter bubbles (150 μm) transported less gangue in their wake and that their use may lessen entrainment while also assisting recovery of the valuable mineral.

From flotation studies for remediation of soils and sediments involving fine particles (2–25 μm), Mulleneers et al. (2002) reported a modified Hallimond tube to prevent entrainment of very fine hydrophilic particles to the concentrate. The modified Hallimond tube contained an active countercurrent sedimentation region and an additional compartment for collection of the hydrophilic particles. Rubio (1996) had described earlier the addition of an extra compartment.

ANALYSIS OF ENTRAINMENT IN LABORATORY BATCH TESTS

For laboratory-batch flotation data at both the mineral recovery–size–liberation level of detail and the less detailed mineral recovery–size level, it is common for the data to be presented on a graph displaying mineral recovery (y-axis) and the logarithm of particle size (x-axis), that is, the same format as used earlier for plant data in Figures 4A and 5. The key point is that mineral recovery values are displayed and these are an output from normal metallurgical calculations.

As for the industrial data (described earlier), a practice has evolved where metallurgists have sought to extend the use of these types of graphs to allow interpretation of the contribution from entrainment. The water recovery can be displayed on the graph and its value compared with the recovery values for the liberated form of the mineral(s) suspected of being recovered principally by entrainment and not wanted in the concentrate. When liberation data are not available, that is, sized data are available, an equivalent procedure is adopted for the mineral of interest—for example, nonsulfide gangue but the recovery values are for all forms of that mineral and interpretation of the graph may be more difficult.

The legitimacy of this approach is now considered with use of a well-known equation, which can be applied to the liberated form for each mineral (Equation 12) and for water (Equation 13). Equation 12 contains, for size i, terms for the concentration of a mineral species remaining in the cell (C_i), the first-order rate constant (k_i) for the mineral, the flotation time (t), and the initial concentration of a mineral species in the cell (C_{oi}) (Lynch et al. 1981). For a particular size fraction i, a mineral species may be the liberated form or other forms of a particular mineral for recovery–size–liberation data or it may be all forms combined for the mineral for recovery–size data.

$$C_i = C_{oi} e^{-k_i t} \quad \textbf{(EQ 12)}$$

$$C_w = C_{ow}e^{-k_w t} \quad \textbf{(EQ 13)}$$

Equation 13 contains terms for the concentration of water remaining in the cell (C_w), the first-order rate constant (k_w) for water, the flotation time (t), and the initial concentration of the water in the cell (C_{ow}) (Lynch et al. 1981).

Equations 12 and 13 can be rearranged by bringing the rate constant to the left side to produce Equations 16 and 17; use of the following relationships for fractional mineral recovery (R_i) in Equation 14 and fractional water recovery (R_w) in Equation 15 was also required in the rearrangement.

$$R_i = 1 - (C_i/C_{oi}) \quad \textbf{(EQ 14)}$$

$$R_w = 1 - (C_w/C_{ow}) \quad \textbf{(EQ 15)}$$

$$k_i = -\log_e (1 - R_i)/t \quad \textbf{(EQ 16)}$$

$$k_w = -\log_e (1 - R_w)/t \quad \textbf{(EQ 17)}$$

Hence, it is possible to provide the following expression for calculating the ratio of interest (k_i/k_w) from Equations 16 and 17 in terms of R_i and R_w:

$$k_i/k_w = (\log_{10} (1 - R_i))/(\log_{10} (1 - R_w)) \quad \textbf{(EQ 18)}$$

The values for R_i and R_w are readily observed for a gangue mineral from a graph displaying mineral recovery (y-axis) and the logarithm of particle size (x-axis) on which the water recovery has been marked; alternatively, the values may be observed in tables of results. The similarity of the ratio R_i/R_w from the described graph to the desired ratio k_i/k_w (Equation 18) is examined in Table 5.

For laboratory batch tests, the water data must reflect the water that flowed over the cell lip with the concentrate and must not represent any contributions from water added for procedures associated with batch flotation tests, such as cleaning the extended cell discharge lip or concentrate scraper.

In this situation, the conveniently observed ratio R_i/R_w from the described graphs (or calculated from the data used to construct the graphs) provides an approximate value for the desired ratio k_i/k_w, which is equal to ENT_i as demonstrated in Equation 4. Use of the conveniently obtained ratio R_i/R_w is most reliable when the fractional water recovery is in the region from zero to 0.20 but remains tolerable for initial interpretation of the data up to fractional water recoveries in the region of 0.40.

To summarize, for mineral recovery–size–liberation data, the use of the ratio k_i/k_w to obtain the value for ENT_i in each size fraction is the desirable method. When the value for R_w is low (0.0–0.4), the conveniently obtained ratio R_i/R_w for calculation of the values for ENT_i in each size fraction is a reasonable approximation for the more desirable ratio, k_i/k_w.

For the less detailed mineral recovery–size data, the behavior patterns associated with entrainment can be expected from feeds in which the candidate minerals for entrainment are well liberated and entrainment is the dominant mechanism. For feeds where the candidate minerals for entrainment have additional important recovery mechanisms, the information in the graphs of mineral recovery (y-axis) and the logarithm of particle size (x-axis) can be interrogated, and the described ratios for R_i/R_w can be examined; however, the patterns of behavior associated with entrainment may be less obvious.

Further, for the less detailed mineral recovery–size data, the ratio k_i/k_w can be calculated based on the total flow for the mineral of interest for each size fraction (and not for the liberated portion). Calculation of the ratio k_i/k_w gives a value for ENT_i in each size fraction and remains the desirable method. When the value for R_w is sufficiently low (see Table 5), the ratio R_i/R_w is a reasonable and convenient approximation to obtain the value for ENT_i in each size fraction.

Table 5 Values for inputs* substituted into Equation 18 and the resulting output†

Values for Inputs to Calculation			Values for Output from Calculation
R_i	R_w	R_i/R_w	k_i/k_w
0.04	0.05	0.80	0.796
0.08	0.10	0.80	0.791
0.16	0.20	0.80	0.781
0.24	0.30	0.80	0.769
0.32	0.40	0.80	0.755
0.40	0.50	0.80	0.737
0.02	0.05	0.40	0.394
0.04	0.10	0.40	0.387
0.08	0.20	0.40	0.374
0.12	0.30	0.40	0.358
0.16	0.40	0.40	0.341
0.20	0.50	0.40	0.322

*R_i and R_w.
†k_i/k_w from the calculation with R_w values from 0.05 to 0.50 (i.e., water recoveries of 5% to 50%) and R_i/R_w ratios of 0.40 and 0.80.

Emphasis has been given in this chapter to the inadvertent recovery of hydrophilic particles by entrainment in concentrates. The proportion of the hydrophobic mineral that is recovered by entrainment is considered to be of lower importance for analysis because the hydrophobic mineral has reached the required product and a metallurgical inefficiency has not occurred.

The proportion of the hydrophobic mineral that is recovered by entrainment can be estimated as described for hydrophilic particles using the previously discussed approximation that $R_i = ENT_i\ R_w$. This method was used for estimating the underlying contribution from entrainment for iron sulfide and sphalerite in the text associated with Figure 5. With the related graphical methods, liberated mineral recovery–size graphs for the contribution from entrainment for the unwanted minerals in a given separation are preferable. In the absence of liberation data, benefit may be gained for many flotation systems by interpretation of the less detailed mineral recovery–size graphs. For both types of graphs, the water recovery should also be depicted.

There is much research based on batch flotation systems, which addresses the determination of the proportion of the hydrophobic mineral that is recovered by entrainment. These methods rely on the flexibilities associated with the laboratory batch testing method; the methods are not suitable for industrial cells. Some methods require more than one flotation test conducted on a given sample and batch flotation cell. Trahar's method (1981) requires a batch test with collector and frother, and a further test with frother alone, with comparison of the results for the mineral of interest at a fixed water recovery. Warren's method (1985) requires several tests with a range of settings for suitable physical variables to provide a range of water–recovery values. Warren's method can be described as employing a range of pulling rates in the various batch tests, allowing detection of the effect of pulling rate, expressed as water recovery, on the recovery of the valuable mineral.

The method of George et al. (2004) requires two batch tests, one with a hydrophobic mineral and another test with a hydrophilic mineral of the same size, allowing their comparison; it is suited to artificial feeds prepared for a research activity. George et al. employed silica (negative charge on surface) as the hydrophobic mineral (cationic collector added plus frother) in one batch test and alumina (positive charge on surface) as the hydrophilic mineral (same cationic collector added plus frother) in the other test. Ross's method (1990) has the advantage of requiring a single batch flotation test for a given sample and batch flotation cell.

ACKNOWLEDGMENTS

The many types of support from Mount Isa Mines Limited (now Glencore) over many decades for technical work in its plants are acknowledged fully and gratefully.

The assistance of Suzanne Munro (Mineralurgy Pty Ltd.) in preparation of various aspects of the section on entrainment is also acknowledged with gratitude.

REFERENCES

Bisshop, J.P., and White, M.E. 1976. Study of particle entrainment in flotation froths. *Trans. Inst. Min. Metall.* 85:C191–194.

Cao, M., and Liu, Q. 2006. Reexamining the functions of zinc sulfate as a selective depressant in differential sulfide flotation—The role of coagulation. *J. Colloid Interface Sci.* 301(2):523–531.

Cheng, H., Cai, C., Zhang, X., and Zhang, J. 1999. A highly selective flotation cell with oscillating separator. In *Mining Science and Technology 1999: Proceedings of the '99 International Symposium.* Edited by H. Xie and T. Golosinski. Boca Raton, FL: CRC Press. pp. 551–553.

Cheng, H., Lu, M., Shi, H., Zhang, X., and Shi, Y. 2007. Mechanism of oscillating flotation to improve selectivity. *J. China Coal Soc.* 32(5):531–534.

Croxford, N.J.W., Draper, N., and Harraway, D.H. 1961. Some aspects of the carbonaceous fraction of Mount Isa lead concentrate. *Proc. Australas. Inst. Min. Metall.* 197:149–161.

Cutting, G.W. 1989. Effect of froth structure and mobility on plant performance. *Miner. Process. Extr. Metall. Rev.* 5:169–201.

Espinosa-Gomez, R., Johnson, N.W., Pease, J.D., and Munro, P.D. 1989. The commissioning of the first three flotation columns at Mount Isa Mines Limited. In *Processing of Complex Ores.* Edited by G.S. Dobby and S.R. Rao. Oxford: Pergamon Press. pp. 293–302.

Finch, J.A., and Dobby, G.S. 1990. *Column Flotation.* Oxford: Pergamon Press. pp. ix, 5–7, 84–88.

Gaudin, A. 1957. *Flotation*, 2nd ed. New York: McGraw-Hill. pp. 362–366.

George, P., Nguyen, A.V., and Jameson, G.J. 2004. Assessment of true flotation and entrainment in the flotation of submicron particles by fine bubbles. *Miner. Eng.* 17:847–853.

Gong, J., Peng, Y., Bouajila, A., Ourriban, M., Yeung, A., and Liu, Q. 2010. Reducing quartz gangue entrainment in sulfide ore flotation by high molecular weight polyethylene oxide. *Int. J. Miner. Process.* 97:44–51.

Guerra, E.A., and Schubert, H. 2003. Effect of the suspension state on the entrainment in flotation machines. *Proceedings of the XXI International Mineral Processing Congress.* Vol. B. Edited by P. Massacci. Amsterdam: Elsevier Science. pp. B8a-8–B8a-15.

Harris, C.C., Arbiter, N., and Musa, M.J. 1983. Mixing and gangue dispersion in flotation machine pulps. *Int. J. Miner. Process.* 10:45–60.

Johnson, N.W. 1972. The flotation behavior of some chalcopyrite ores. Ph.D. dissertation, University of Queensland, Brisbane, Australia.

Johnson, N.W. 2005. A review of the entrainment mechanism and its modelling in industrial flotation processes. In *Centenary of Flotation Symposium.* Edited by G.J. Johnson. Melbourne, Victoria: Australasian Institute of Mining and Metallurgy. pp. 487–496.

Johnson, N.W., and Jowett, A. 1982. Comparison of lead primary rougher behavior of several Mount Isa lead zinc ores. In *Mill Operators' Conference North West Queensland Branch.* Melbourne, Victoria: Australasian Institute of Mining and Metallurgy. pp. 271–285.

Johnson, N.W., McKee, D.J., and Lynch, A.J. 1974. Flotation rates of non-sulfide minerals in chalcopyrite flotation processes. *Trans AIME* 256:204–209.

Jowett, A. 1966. Gangue mineral contamination of froth. *Brit. Chem. Eng.* 11:330–333.

Kaya, M., and Laplante, A.R. 1990. Froth washing and froth vibration in mechanical flotation machines. SME Preprint No. 90-30. Littleton, CO: SME.

Kirjavainen, V.M. 1992. Mathematical model for the entrainment of hydrophilic particles in froth flotation. *Int. J. Miner. Process.* 35:1–11.

Kirjavainen, V.M. 1996. Review and analysis of factors controlling the mechanical flotation of gangue minerals. *Int. J. Miner. Process.* 46:21–34.

Laplante, A.R., Kaya, M., and Smith, H.W. 1989. The effect of froth on flotation kinetics—A mass transfer approach. *Miner. Process. Extr. Metall. Rev.* 5:147–168.

Liu, Q., Wannas, D., and Peng, Y. 2006. Exploiting the dual functions of polymer depressants in fine particle flotation. *Int. J. Miner. Process.* 80:244–254.

Lynch, A.J., Johnson, N.W., Manlapig, E.V., and Thorne, C.G. 1981. *Mineral Coal and Flotation Circuits—Their Simulation and Control.* Amsterdam: Elsevier. pp. 64–78.

Maachar, A., and Dobby, C.S. 1992. Measurement of feed water recovery and entrainment solids recovery in flotation columns. *Can. Metall. Q.* 31:167—172.

Mulleneers, H.A.E., Koopal, L.K., Bruning, H., and Rulkens, W.H. 2002. Selective separation of fine particles by a new flotation approach. *Sep. Sci. Technol.* 37(9):2097–2112.

Ross, V.E. 1990. Flotation and entrainment of particles during batch flotation tests. *Miner. Eng.* 3(3-4):245–256.

Ross, V.E., and Van Deventer, J.S.J. 1988. Mass transport in flotation column froths. In *Column Flotation '88: Proceedings of an International Symposium on Column Flotation.* Edited by K. Sastry. Littleton, CO: SME. pp. 383–386.

Rubio, J. 1996. Modified column flotation of mineral particles. *Int. J. Miner. Process.* 48:183–196.

Savassi, O.N. 1998. Direct estimation of the degree of entrainment and the froth recovery of attached particles in industrial flotation cells. Ph.D. dissertation, University of Queensland, Brisbane, Australia.

Savassi, O.N., Alexander, D.J., Franzidis, J-P., and Manlapig, E.V. 1998. An empirical model for entrainment in industrial flotation plants. *Miner. Eng.* 11(3):243–256.

Silvester, E.J., Heyes, G.W., Bruckard, W.J., and Woodcock, J.T. 2011. The recovery of sericite in flotation concentrates. *Miner. Process. Extr. Metall. (Trans. Inst. Min. Metall. C)* 120:10–14.

Smith, P.G., and Warren, L.J. 1989. Entrainment of particles into flotation froths. *Miner. Process. Extr. Metall. Rev.* 5:123–145.

Trahar, W.J. 1981. A rational interpretation of the role of particle size in flotation. *Int. J. Miner. Process.* 8(289):327.

Vianna, S.M. 2004. The effect of particle size, collector coverage and liberation on the floatability of galena particles in an ore. Ph.D. dissertation, University of Queensland, Brisbane, Australia.

Wark, I.W., and Cox, A.B. 1932. Principles of flotation, I. An experimental study of the effect of xanthates on contact angles at mineral surfaces. *Trans. AIME* 112:189–244.

Warren, L.J. 1985. Determination of the contributions of true flotation and entrainment in batch flotation tests. *Int. J. Miner. Process.* 14:33–44.

Yianatos, J., and Contreras, F. 2010. Particle entrainment model for industrial flotation cells. *Powder Technol.* 197:260–267.

Zheng, X., Franzidis, J-P., Johnson, N.W., and Manlapig, E.V. 2005. Modelling of entrainment in industrial flotation cells: The effect of solids suspension. *Miner. Eng.* 18:51–58.

Zheng, X., Johnson, N.W., and Franzidis, J-P. 2006. Modelling of entrainment in industrial flotation cells: Water recovery and degree of entrainment. *Miner. Eng.* 19:1191–1203.

APPENDIX A
ENTRAPMENT

Another mechanism, entrapment, in the froth region has been reported for the recovery of dispersed, liberated, and hydrophilic gangue particles in a flotation concentrate (Gaudin 1957; Laplante et al. 1989; Smith and Warren 1989; Kirjavainen 1996). Entrainment is a mechanism that exists in all standard flotation machines, whereas entrapment is a mechanism that is *beyond entrainment* and may exist when unusual conditions occur in the process feed and the resulting froth region.

When the entrapment mechanism is important, the values for ENT_i for the liberated gangue in the various size fractions will exceed the values for entrainment mechanism alone, which will not exceed 1.0 for the finest fraction and decline in value for larger fractions.

Further, examples of sized data supporting the entrapment mechanism have been lacking. It can be speculated that this situation arises because systems are not common in which the recovery by entrainment has additional recovery overlain on it. Vianna (2004) studied an ore with unusual characteristics; he also demonstrated that there was no evidence of collector adsorbed on the liberated nonsulfide gangue particles. Hence, hydrophobicity of the nonsulfide gangue could not explain the observed behavior. The feed grade (23.5% Pb and 27.1% PbS) was very high and the percent solids values in the concentrate were also high (50%–64% solid); a wide range of particle sizes also existed in the system. In such a system with dry froths, steric hindrance may have severely restricted movement of the dispersed, liberated, and hydrophilic gangue particles in the water lamellae in the froth region. For a liberated gangue mineral in a given size fraction, the recovery by entrapment can be estimated by the difference between the total recovery of the liberated gangue mineral and the predicted contribution from entrainment of the liberated mineral by use of Equation 5.

For the system of Vianna (2004), the values for ENT_i for liberated gangue were in the region between 5.0 and 6.0 for size fractions below 30 µm and decreased to values slightly less than 1.0 at 90 µm (Johnson 2005). It was concluded that the entrapment mechanism was much more important than the entrainment mechanism in this unusual system.

Zheng et al. (2006) reported on a large number of tests with different values for physical variables in a 3-m^3 continuous cell, which was operated in parallel with a copper rougher bank at Mount Isa Mines Limited. The tests were analyzed at the mineral recovery–size level for the extent of entrainment of silica, the dominant nonsulfide gangue mineral. The expected values for ENT_i from entrainment were observed in the size fractions from 0 to 30 µm, but the values did not decay to zero in the region of 50 µm and in the coarser fractions. The authors attributed this to two mechanisms: recovery of silica in composites with the copper mineral and entrapment of the larger silica particles in the froth region. The absence of liberation data meant that the contribution from each could not be quantified.

APPENDIX B
DERIVATION OF EQUATIONS 4 AND 6 FROM EQUATION 3

Equation 3, the definition of ENT_i, is provided as the starting point for the derivation of Equation 4, which was discussed in the "Quantitative Description of Entrainment" section, noting in the derivation that $RRFG_{con,i}$ is the rate of recovery of free (liberated) gangue in the concentrate for size i and that RRW_{con} is the rate of recovery of water in the concentrate.

$$\begin{array}{c}\text{entrainment}\\ \text{efficiency}\\ \text{value}\,(ENT_i)\end{array} = \frac{(\text{mass of liberated gangue per unit mass of water})_{con,i}}{(\text{mass of liberated gangue per unit mass of water})_{top\ of\ pulp,i}} \quad \textbf{(EQ 3)}$$

$$= \frac{\left(\dfrac{MFG_{con,i}}{MWAT_{con}}\right)}{\left(\dfrac{MFG_{top\ of\ pulp,i}}{MWAT_{top\ of\ pulp}}\right)} \quad \textbf{(EQ 3a)}$$

$$= \frac{\left(\dfrac{RRFG_{con,i}}{RRW_{con}}\right)}{\left(\dfrac{MFG_{top\ of\ pulp,i}}{MWAT_{top\ of\ pulp}}\right)} \quad \textbf{(EQ 3b)}$$

$$= \frac{\left(\dfrac{RRFG_{con,i}}{MFG_{top\ of\ pulp,i}}\right)}{\left(\dfrac{RRW_{con}}{MWAT_{top\ of\ pulp}}\right)} \quad \textbf{(EQ 3c)}$$

With use of the definition of a first-order rate constant in Equation 2, the following equation can be written from Equation 3c:

$$ENT_i = k_i / k_w \quad \textbf{(EQ 4)}$$

Equation 6 can be derived by starting with Equation 3b and then bringing the term $RRFG_{con,i}$ to one side:

$$ENT_i = \frac{\left(\dfrac{RRFG_{con,i}}{MFG_{top\ of\ pulp,i}}\right)}{\left(\dfrac{RRW_{con}}{MWAT_{top\ of\ pulp}}\right)} \quad \textbf{(EQ 5)}$$

$$RRFG_{con,i} = ENT_i \frac{MFG_{top\ of\ pulp,i}}{MWAT_{top\ of\ pulp}} RRW_{con} \quad \textbf{(EQ 6)}$$

CHAPTER 7.7

Sulfide Flotation Testing

Philip Thompson, Kym Runge, and Robert C. Dunne

Laboratory batch flotation testing is a fundamental part of characterizing a sulfide ore to determine the merits of recovering valuable minerals from an ore. This chapter describes the laboratory procedures for characterizing the flotation response of sulfide ores and products from sulfide flotation plants.

LABORATORY TESTING FOR SULFIDE FLOTATION AND PROCESS DEVELOPMENT

A carefully planned laboratory test program is essential for the successful development of any sulfide flotation process. These programs usually begin with a series of batch tests on several ore-type composites to evaluate appropriate reagent schemes and delineate optimum rougher flotation conditions, such as primary grind size, flotation pulp density, flotation pulp pH, and retention time. Cleaner parameters such as regrind size requirements, cleaner pH, pulp density, and number of stages are then defined. Semicontinuous locked-cycle tests are performed using the parameters established in the batch tests to determine the effect of middling streams recirculation on flotation performance. The data generated and the experience gained in the locked-cycle tests may be used to design and operate a continuous pilot plant if relatively large amounts of concentrate are required for further process development.

A comprehensive ore variability program is undertaken to determine the response of various ore types to the optimum conditions established. These ore variability programs often test hundreds of samples in standardized batch rougher and cleaner flotation tests to ensure that the ore body is well understood (Thompson 2016).

The importance of obtaining representative samples, along with careful preparation and preservation, is essential to all metallurgical test-work programs. The most careful and comprehensive test programs are meaningless if the samples used are not representative. The block flow diagram in Figure 1 provides an overview of the development process.

Batch Laboratory Flotation Equipment

Several standard laboratory flotation machines are available. The most common are the Denver D12 (supplied by Metso) and the FLSmidth-Essa. The machines have impeller speed control, and air is controlled by a rotameter. Air pressure is typically set at ~50–75 kPa.

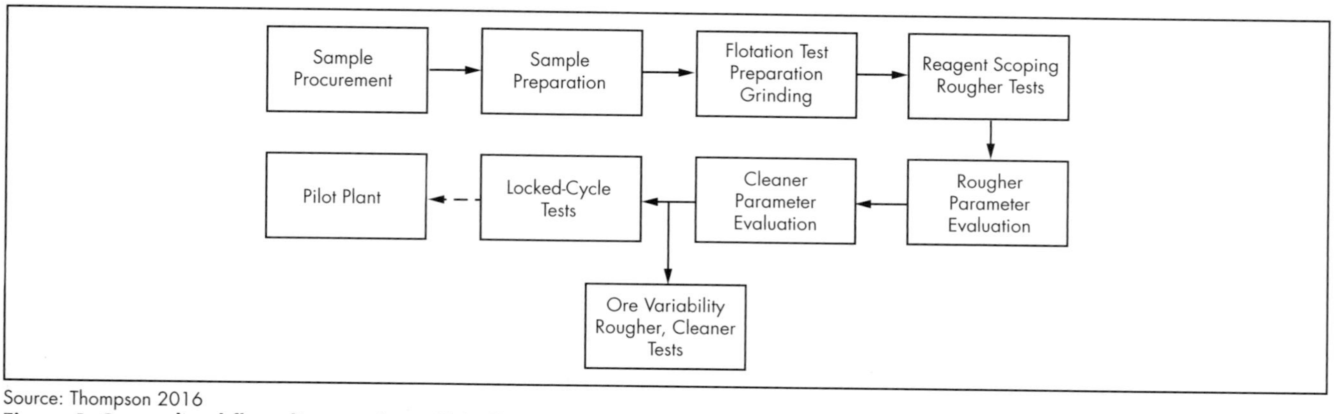

Source: Thompson 2016

Figure 1 Generalized flow diagram for sulfide flotation development

Philip Thompson, Senior Minerals Processing Advisor, FLSmidth, Salt Lake City, Utah, USA
Kym Runge, Program Leader, Separation, JKMRC, Sustainable Minerals Institute, University of Queensland, Brisbane, Queensland, Australia
Robert C. Dunne, Principal, Robert Dunne Consulting, Gooseberry Hill, Western Australia, Australia

Batch Laboratory Flotation Test Cell Sizes

The most common flotation test cell sizes used in sulfide flotation work range from 1.5 to 6 L and typically contain 500–2,000 g of ground solids. However, flotation cells as small as 1.25 L and as large as 60 L are available. The larger cells are normally employed when considerable amounts of concentrate are required for additional testing, such as selective flotation to recover by-products (i.e., molybdenite), leaching, or gravity separation. The smaller flotation cells, 1.25 and 1.5 L in size, are generally used for concentrate cleaning applications. The 2.2-L and 4.4-L Agitair cells are typically used in many laboratory flotation test-work programs. Sample masses are typically 1 kg for the 2.2-L cell and 2 kg for the 4.4-L cell.

Batch Flotation Operating Parameters

Pulp Density

Choosing an appropriate flotation pulp density is often dictated by the gangue constituents of the ore, particularly the clay or slimes content. These types of slimes can have a significant effect on pulp viscosity, particularly if a high lime pulp pH is required for pyrite rejection. Typical pulp densities range from 30% to 40% solids in roughing and 15% to 20% solids in cleaning. The higher pulp densities are normally employed when treating siliceous ores with very low slimes content.

Flotation Reagent Conditioning

Suggested impeller speed for reagent conditioning is typically between 750 and 950 rpm for a Denver batch flotation machine using a 2.5-L batch cell. For low-intensity conditioning, the impeller speed range can be decreased to 600–750 rpm, whereas for high-intensity conditioning, increased impeller speeds to 900–1,250 rpm are typical.

Rougher-Scavenger Flotation

Recommended operating conditions provided are to be used as guidelines and are not "cast in stone." Cell revolutions per minute are chosen, usually on a visual basis, to provide adequate agitation and suspension of solids but should not be too intense to cause spillage. The intensity of agitation reduces once air is introduced into the system. Typical impeller speeds for rougher-scavenger flotation range from 950 to 1,500 rpm, depending on the mixing characteristics of the ore. Aeration rates can range from 5 to 25 L/m and largely depend on the froth characteristics of the sample being tested.

Cleaner-Recleaner Flotation

Typical impeller speeds for cleaner-recleaner flotation range from 700 to 900 rpm, depending on the mixing characteristics of the ore. For concentrates with large quantities of coarse particles, the impeller speed may need to be increased to 800–1,100 rpm to keep particles in suspension. Aeration rates can range from 0.1 to 5 L/m and, again, depend largely on the froth characteristics of the concentrate being floated.

Pulp and Froth Levels

With no agitation and airflow, the pulp level is set at 50 mm below the froth overflow lip. The cell should be marked and calibrated for pulp level with agitation and no air so that with airflow, the pulp level is 15–20 mm below the froth overflow lip. Froth height should be level with the overflow lip. During the rougher test, water needs to be manually added to the flotation cell to maintain the "set" pulp level. For the scavenger part of flotation, the pulp level is usually increased by 5 mm below the froth lip. Depending on the pulp and froth characteristics of the ore being tested, the pulp levels will be altered by the operator (EMC 2012a).

Froth Scraping in Batch Flotation

Froth removal is usually by manual scraping by hand paddles, though mechanical scrapers can be incorporated into the system. The froth paddles are sized to extend 10 mm below the overflow lip (EMC 2012b). The typical scraping rate is one scrape (both sides of the cell) every 10–15 seconds. To increase the mass (yield) of concentrate, the scraping rate can be increased to 6–10 per minute.

Flotation Time

Rougher-scavenger flotation is typically carried out for 6–12 minutes. However, for ores where the sulfides are slow floating, the flotation time is extended to 20 minutes. For cleaning applications, floatation times are typically 70% of roughing times. Exceptions can occur, particularly when clayey ores are processed, where flotation times can be greater than the rougher retention time.

SCOPING FLOTATION TESTS FOR REAGENT SCHEME SELECTION

The first phase of most flotation test programs is the selection of a reagent scheme. This should be done in comparative rougher scoping tests before other important parameters, such as grind size, pulp pH, and pulp density, are optimized. A good starting point for sulfide systems is a primary grind P_{80} of ~106 µm, pulp pH of 8–9, and pulp density of 30% solids. Flotation reagents may be classified into the following categories: collectors, depressants, activators, frothers, and pulp dispersants. Pulp dispersants include sodium silicate, for slimey or clayey pulps, and activators include copper sulfate for pyrite activation and for sphalerite flotation. Both flotation systems are not discussed in this chapter but can be found elsewhere in the handbook. Frother evaluation in small bench-scale tests is not effective because of the small surface area in these machines. Most reagent scoping tests are therefore focused on the selection of appropriate collectors (Thompson 2002), which may be separated into the following classes:

- **Xanthates:** Powerful, nonselective, minimal frothing, and relatively inexpensive.
- **Dithiophosphates:** Moderately selective, with some frothing characteristics.
- **Thionocarbamates:** Selective, with some frothing characteristics.
- **Aerophines:** Very selective, with moderate frothing characteristics.
- **Mercaptobenzothiazole:** Selective, specialty reagent for tarnished sulfides.
- **Dual-action collectors:** Reagents tailored for dual mineral recovery (e.g., copper and gold, copper and molybdenum).

In sulfide flotation systems, pyrite is the most common sulfide gangue mineral, and most depressants are evaluated for rejection of this mineral. Notable exceptions include the use of zinc sulfate for sphalerite depression in galena flotation and sulfur dioxide gas to depress galena in copper sulfide flotation. The most common pyrite depressants are inorganic compounds such as lime (CaO), which is the most common depressant used; sulfites such as sodium sulfite, sodium metabisulfite, and sulfur dioxide gas; and sodium cyanide, which is the most powerful pyrite depressant but is seldom used because of environmental and social concerns. Several organic pyrite depressants are available, including calcium lignosulfonate, diethylenetriamine, sodium thioglycolate, 2-naphthelenesulfonic acid, modified polyacrylamide, and several starch/sodium hydroxide/sodium sulfide formulations.

For auriferous pyrite and arsenopyrire flotation, the standard collector is potassium xanthate in combination with a strong frother. To increase free gold recovery, if present, and improve overall performance, other flotation reagents are usually employed, as discussed by Kappes et al. (2011) and Dunne (2016).

The selection of a suitable reagent scheme is often made using an experience-based approach and begins with a review of existing operations that are treating ores with similar characteristics. Assistance from reagent vendors is recommended. Most screening flotation tests are performed in rougher kinetic format with at least three, and preferably four or five, timed concentrate samples taken.

ROUGHER FLOTATION PARAMETER EVALUATION

Rougher flotation parameters are normally evaluated in batch rougher flotation tests that are performed in a kinetic format after a suitable collector reagent scheme has been selected. These parameters include primary grind fineness, flotation time (kinetics), flotation pulp density, and flotation pulp pH. Primary grind fineness is the parameter that has the greatest impact on capital and operating expenses and, as such, is tested extensively. Rougher flotation testing focuses on valuable mineral recovery, but some selectivity against pyrite is often desired, and this is usually achieved by increasing the grinding and flotation pulp pH with lime. It is important to add lime to the grinding step prior to flotation to achieve maximum pyrite depression using this reagent.

Primary Grind Particle Size

The effect of the primary grind size of the primary grind on valuable mineral recovery is evaluated in a series of separate rougher kinetic tests at increasing grind fineness. Grind size P_{80} values varying from 75 to 250 μm are commonly used, but coarser and finer grinds are occasionally tested. Figure 2 shows a typical plot for recovery versus grind size for a copper sulfide ore.

The negative slope of the linear regression analysis indicates that recovery decreases as the grind size becomes coarser, which is from locking of valuable mineral, usually with non-sulfide gangue. Slopes ranging from –0.1200 to –0.0245 are common for copper-molybdenum porphyry ores.

Size-by-size analyses of the flotation feed, the overall concentrate, and the tailings at a single grind size are often performed to provide a size-by-size recovery profile and to confirm the results obtained in the grind variability study. In these tests, each size fraction of flotation concentrate and tailings is weighed and assayed to calculate recovery by size fraction. The size-by-size analysis of the feed is used as a check on the metallurgical balance. Typical results are presented in Figure 3. This size-by-size test is used in most ore variability studies.

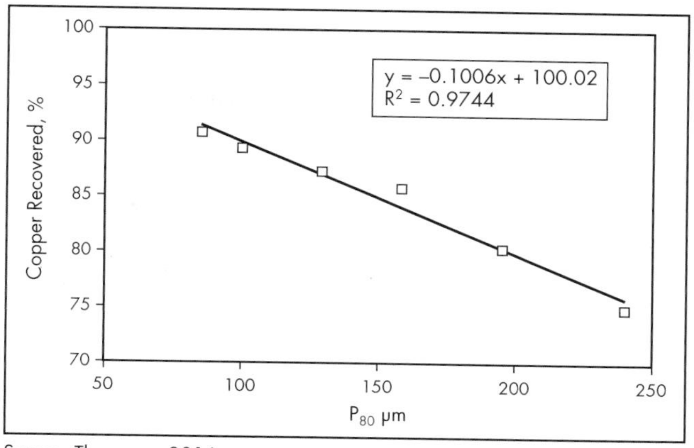

Source: Thompson 2016

Figure 2 Typical plot for copper recovery versus grind fineness

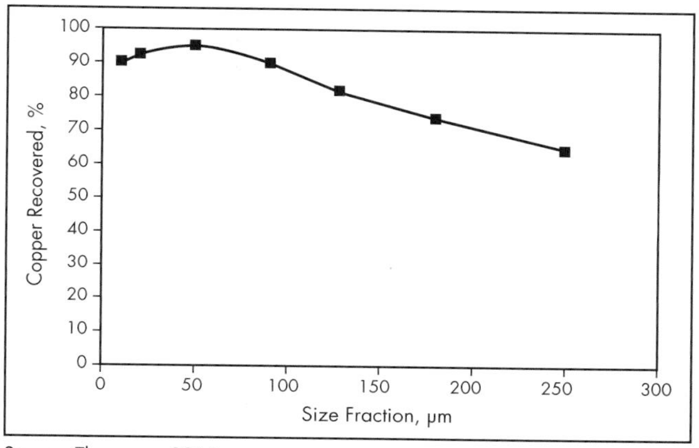

Source: Thompson 2016

Figure 3 Size-by-size recovery of copper for a typical ore

Occasionally, a more detailed size-by-size test is performed to evaluate both recovery and kinetics by size fraction. This type of test usually requires a 10- to 15-kg test charge because all the timed concentrate samples are screened, weighed, and assayed separately. The data obtained are usually worth the extra time, effort, and samples required. A schematic for this test and typical size-by-size kinetic curves are illustrated in Figures 4 and 5.

Size fractions >20 μm (635 Tyler mesh) are generated using standard wet screen sieving. Size fractions finer than 20 μm may be generated using a Marc Technologies Cyclosizer (formerly known as the Warman Cyclosizer). This device uses a series of either five or seven small, inverted cyclones in series to generate closely sized subsieve, or smaller than 20 μm, fractions. The actual size distributions of each cyclosizer product may be determined using an appropriate laser particle size analyzer. The size-by-size kinetics data presented in Figure 5 indicate much slower kinetics in the coarser size fractions. This is typical for many ores because of the partial locking of valuable mineral with non-sulfide gangue-inhibiting kinetics.

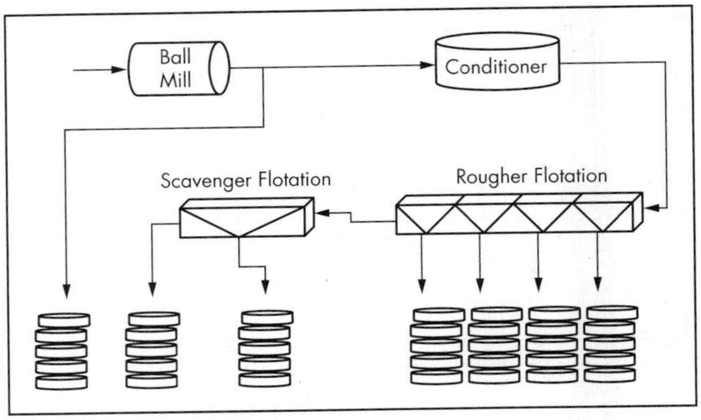

Adapted from Thompson 2016

Figure 4 Flow scheme for size-by-size kinetics

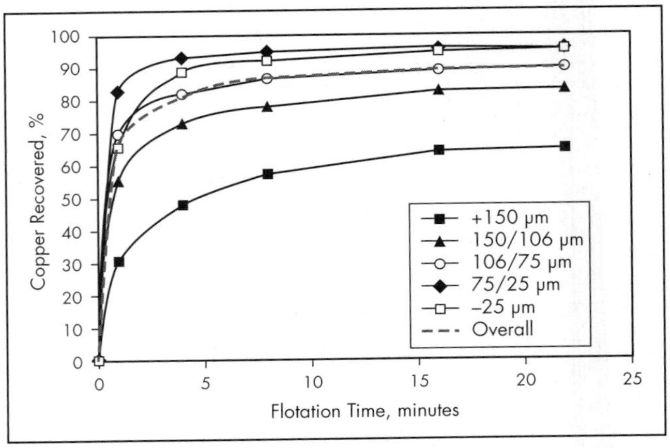

Adapted from Thompson 2016

Figure 5 Size-by-size kinetics for a typical copper ore

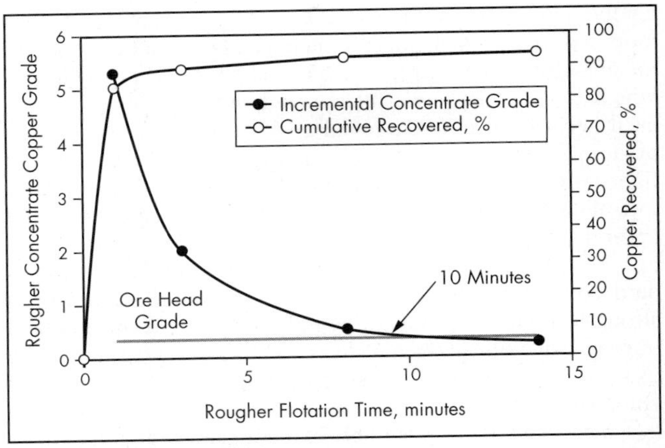

Adapted from Thompson 2016

Figure 6 Rougher retention time evaluation by incremental concentrate grade

Flotation Kinetics

Batch flotation tests are usually performed in kinetic format by taking flotation concentrates at several time increments until the froth is barren. The data are then presented in a kinetic graph similar to the one in Figure 5, using overall recovery because size-by-size kinetics are a special case.

The most widely used method for determining retention time in these batch tests is the cumulative recovery–incremental concentrate grade approach. Cumulative recovery and incremental grade of the individual time interval concentrates are plotted as a function of total flotation time. Flotation is considered complete when the incremental concentrate grade crosses the test feed head grade, drawn as a horizontal line. At this point, no additional concentration from the test feed material is occurring, and the test can be considered complete. Figure 6 illustrates this concept, which is valid for the evaluation of rougher, scavenger, and cleaner flotation.

Notably, flotation kinetics are generally influenced by two important factors, namely, the rate constant and the froth-carrying capacity. The second factor is often overlooked, but it can be an important one, particularly in cleaner circuits. Maximum froth-carrying capacity for industrial mechanical cells is provided in Chapter 7.4, "Froth Management," and froth-carrying capacity for column cells is discussed in Chapter 7.2, "Column Flotation." Typical values for sulfide ores tested in mechanical laboratory cells are 0.05 to 0.10 t/h/m^2 for roughing applications and 0.15 to 0.30 t/h/m^2 for cleaning applications. Kinetic limitations due to froth-carrying capacity in laboratory tests are generally limited to industrial minerals applications where high concentrate weight pulls are common. Column cleaning in some sulfide systems can also encounter this issue, particularly if the ore is high grade.

Flotation Pulp pH

In most sulfide flotation systems, the choice of flotation pulp pH is governed by the amount of pyrite rejection that is required. Ores containing high levels of pyrite often require rougher pulp pH levels of up to 12 for pyrite control, whereas ores containing more moderate or low amounts of pyrite may only require pH 8. Lime is the most common pH modifier used in the minerals industry. A series of rougher flotation tests at increasing pH values is recommended to determine the response of both pyrite and valuable mineral to this variable. Some collectors do not perform well at elevated pH values above 9. The sensitivity of cleaner flotation to pulp pH should also be investigated.

CLEANER CIRCUIT EVALUATION

Test programs for evaluating cleaner circuits should be undertaken only after rougher parameters are optimized. Although rougher testing focuses on maximizing recovery of valuable mineral, cleaner testing centers on producing a saleable concentrate of acceptable grade. In most sulfide systems, this is accomplished by separating the valuable mineral from nonsulfide gangue and sulfide gangue, usually pyrite. A delicate balance exists between grade and recovery in these tests, and the production of the highest possible grade concentrate may not always be the most economically viable, given the reduced recovery issues. Consequently, the most important product from cleaner flotation work is the grade–recovery curve.

Methods for Generating Grade–Recovery Curves

The standard approach for generating grade–recovery curves is a relatively simple method that is the most widely used in the minerals industry for most cleaning applications. Concentrates are refloated, and tailings from each stage are analyzed separately. Intermediate concentrates are back-calculated from

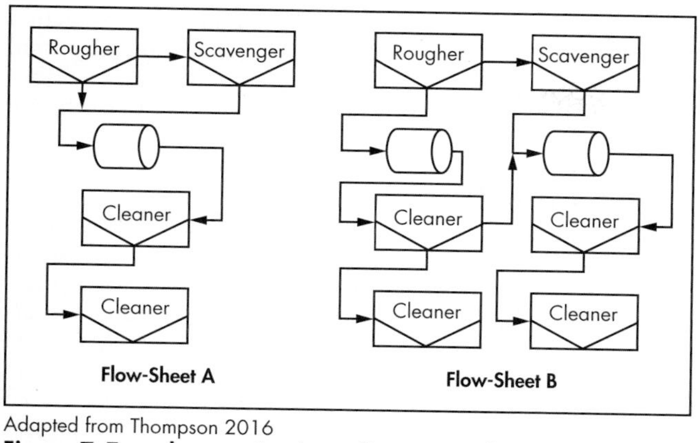

Adapted from Thompson 2016

Figure 7 Two cleaner circuit configurations for comparison of grade–recovery curves

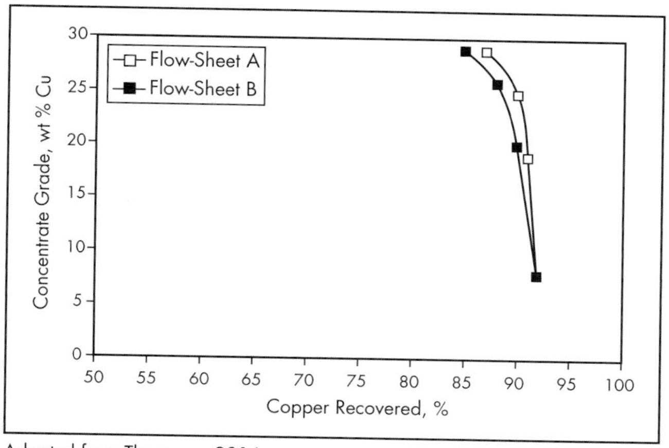

Adapted from Thompson 2016

Figure 8 Grade–recovery curves for two laboratory cleaner circuit configurations

concentrates and tailings. Two other methods (release analysis and three circuits) are discussed by Thompson (2016).

Cleaner Circuit Configurations

Several large copper-molybdenum-porphyry concentrators clean rougher and scavenger concentrates in separate cleaner circuits. These types of separate cleaning circuits are exceedingly difficult to test in the laboratory because of factors such as low weight pulls in the scavenger-cleaner section and middlings recirculation from the rougher-cleaner to scavenger-cleaner sections. Figure 7 shows a simple and relatively complex cleaning circuit (flow-sheet A and flow-sheet B, respectively), while Figure 8 provides the laboratory grade–recovery results for the two circuits. The curves are very similar, indicating that the relatively simple flow scheme A is acceptable for generation of laboratory grade–recovery data in most cases, and this example is used in the following sections that address cleaner circuit parameter optimization. These parameters include regrind fineness, flotation kinetics, number of cleaning stages required, circuit pH, and circuit pulp density.

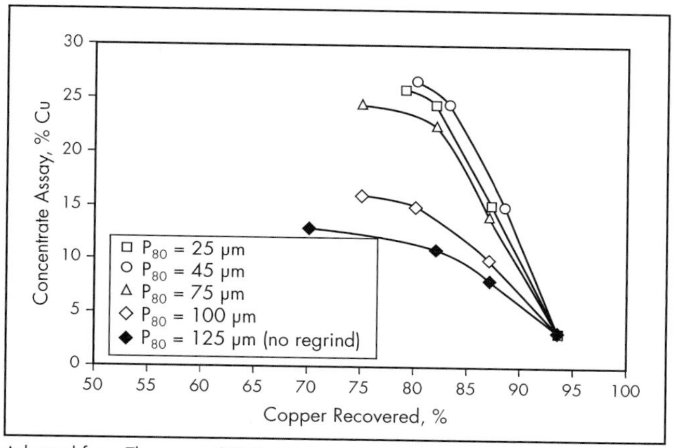

Adapted from Thompson 2016

Figure 9 Effect of regrind fineness on grade–recovery curves

Evaluation of Regrind Fineness

Most rougher and rougher-scavenger concentrates that are produced in the rougher evaluation stage of a test program contain locked particles of valuable mineral and either non-sulfide gangue, often silicates, or sulfide gangue, usually pyrite, or both. Regrind testing is performed to determine the grind fineness that is required to produce an acceptable final concentrate grade. This test work is usually complemented with liberation and locking analyses by an automated mineral analyzer program. Pyrite may also be present as liberated mineral may report to the rougher concentrate because of activation by collector, entrainment in the froth, or both. In many copper sulfide flotation circuits, rougher flotation is considered a "bulk float," and pyrite flotation is tolerated to maximize copper recovery. This liberated pyrite must also be rejected during cleaning. Lime is the most common depressant used in these cleaner circuits, and a high lime cleaner pH is normally used when testing regrind fineness for copper ores. However, auxiliary depressants such as sulfides or cyanide may be required, particularly when treating ores where the valuable mineral is sensitive to calcium ion, such as lead-zinc-silver ores.

A high-lime cleaner pH of 11.0–11.5 is normally used in the concentrate regrind test series for copper ores. The recommended procedure consists of performing a large-scale rougher test on 12–15 kg of ore under optimum conditions. The rougher concentrate is then split into four or five samples using a rotary slurry splitter or equivalent. Each rougher concentrate split is then subjected to regrinding for increasing time periods, then cleaned at least two times, often more, at pH 11.0–11.5. The rougher concentrate is screened at an appropriate size, usually 25 or 37 μm, and only the screen oversize is reground. This prevents excessive sliming of the cleaner feed. This procedure generates a family of grade–recovery curves similar to those in Figure 9 for a copper-porphyry ore. Steep grade–recovery curves indicate good upgrading, and the steeper the curve, the more selective the cleaning. Evaluation of these results suggests that a regrind fineness P_{80} of 45 μm is appropriate for this sample.

Number of Cleaning Stages

The shape of the grade–recovery curve and the desired final concentrate usually dictate the number of cleaner flotation stages required. Figure 10 shows two examples of different

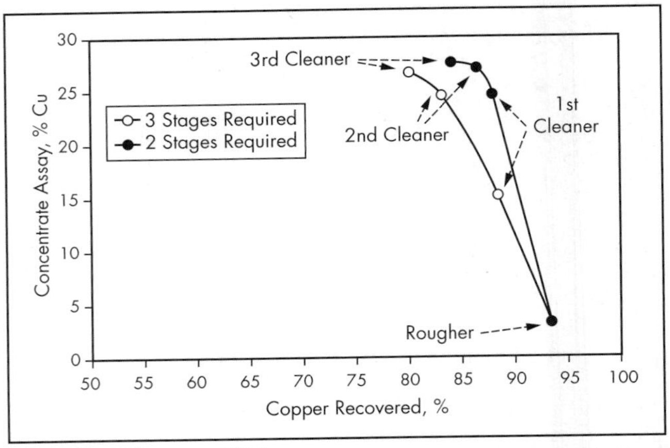

Adapted from Thompson 2016

Figure 10 Typical grade–recovery curves for two- and three-stage cleaning

cleaning requirements. Horizontal sections of the curve typically indicate no additional upgrading, and the cleaner stage at which the knee in the curve occurs may be considered the final stage. In some cases, the curve intersects the desired final concentrate grade before the knee in the curve—as in Figure 9 for the 45- and 25-μm regrinds if a 25% copper grade was targeted—and often there is no knee in the curve and the desired final concentrate grade may not be achieved. This is the case in Figure 9 for the P_{80} of 100-μm regrind and 125-μm (no regrind) conditions.

The evaluation of cleaner circuits in the laboratory with mechanical test cells can be complicated if column cells are being considered for the full-scale plant. Column flotation testing is difficult in the laboratory given the large amounts of flotation feed slurry required, particularly in cleaner flotation. It has been author Thompson's experience that approximately three stages of mechanical cleaning in the laboratory approximates one stage of column cleaning in the plant.

Most cleaner circuits include a scavenging step on the first cleaner tailings, and this is always the case when column cleaning is used. The cleaner-scavenger tailings are normally discarded as open-circuit tailings and combined with rougher tailings to produce general mill tailings in commercial plant operation. However, the rougher and cleaner-scavenger tailings are almost always kept separate in laboratory testing.

Cleaner Circuit pH and Pulp Density

The most common sulfide gangue in many sulfide flotation circuits is pyrite, and cleaner pulp pH values are usually maintained in a range from 11.0 to 11.5 using lime for pyrite rejection. These pH values may be too high for some ores, resulting in depression of valuable by-products. For these ores, a lower cleaner pH of 10.0 or 10.5 may be appropriate, as well as addition of a pyrite depressant such as sulfites or other polymer depressants. Sensitivity of the cleaner circuit to high pH is determined in a series of cleaner tests that are performed at an optimum regrind size and nominal pH values of 9, 10, 11, and 12. The evaluation of grade–recovery curves from these test series are similar to those described for regrind fineness testing.

The evaluation of cleaner pulp density in laboratory test programs is often difficult because of the limited amounts of concentrate generated and the limited number of flotation test cell sizes available. Cleaner pulp densities ranging from 15% to 20% solids are ideal, but laboratory tests may be conducted as low as 5% solids if low concentrate weight pulls are encountered.

LOCKED-CYCLE TESTING

Locked-cycle tests are performed using rougher and cleaner conditions that have been optimized in batch test work. Each test is a series of batch rougher-cleaner tests (cycles) wherein the middlings products—cleaner tailings and cleaner-scavenger concentrate—from one cycle are recirculated to the next cycle. This type of test is very powerful because it can be used to project continuous metallurgical performance from a series of batch test cycles. The data may be used to project mass and metallurgical balances for plant design or, in special cases, provide a baseline for pilot-plant design and operation.

Items of importance to discuss in greater detail are

- Locked-cycle test methodology;
- How to calculate test results, both overall and steady state;
- The importance of achieving steady state; and
- How to determine when steady state is achieved.

Figure 11 shows a flow sheet for a typical locked-cycle test, with dashed lines highlighting the recycle middlings products.

Locked-Cycle Test Methodology

A batch test is completed in the first cycle and the middlings streams are stored as slurries. Using Figure 11 as an example, the second cycle would incorporate the cleaner-scavenger concentrate and second cleaner tailings from the first cycle with the rougher-scavenger concentrate from the second cycle for regrinding and cleaning. The third cleaner tailings from the first cycle would be mixed with second cleaner concentrate from the second cycle as feed to the third cleaner. This procedure is repeated for several cycles. The minimum number of cycles required will vary with the ore type and flow-sheet

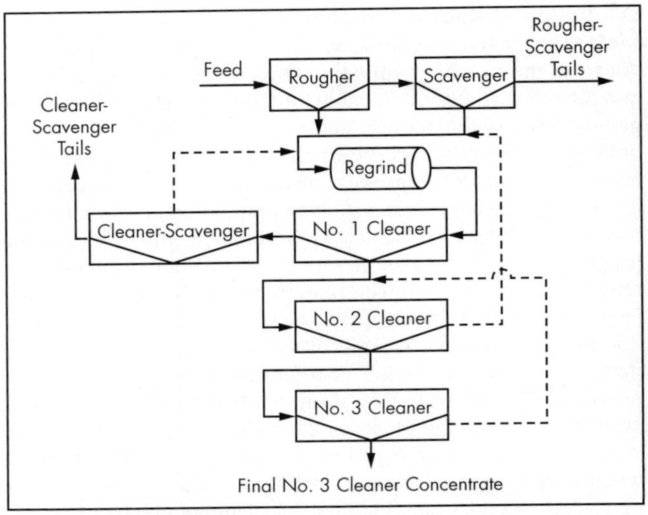

Adapted from Thompson 2016

Figure 11 Typical flow sheet for a locked-cycle test on a copper ore

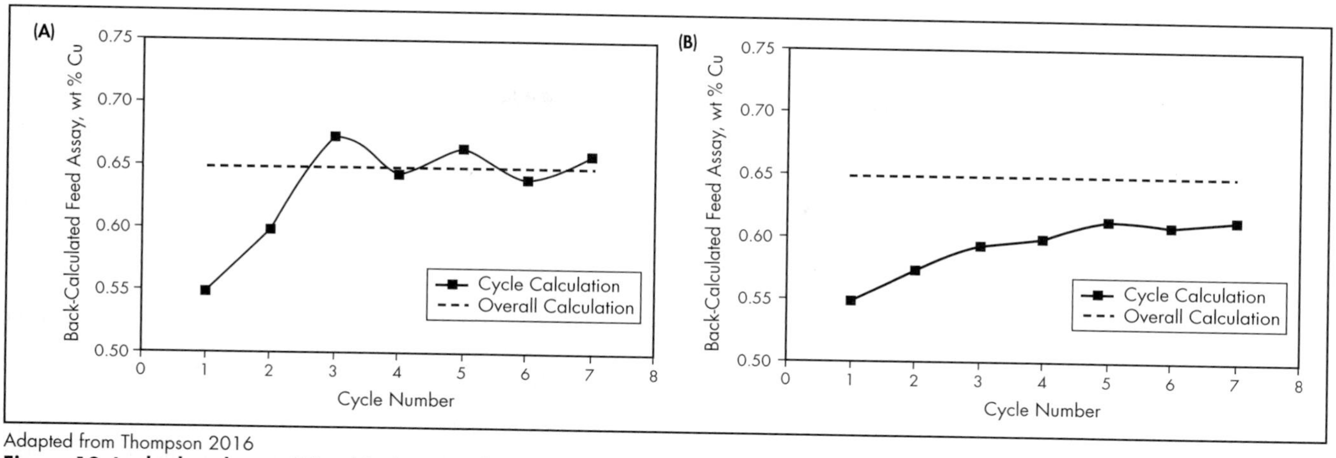

Adapted from Thompson 2016

Figure 12 Locked-cycle test (A) achieving steady state and (B) not in steady state

configuration. However, it has been author Thompson's experience that the minimum number of cycles should be equal to the number of cleaner stages plus three. In actual laboratory practice, most tests are performed using seven cycles. Wet filter cake weights of final products from each test cycle provide an indication of when, or if, the test achieves steady state because these weights should stabilize. The final concentrates are usually the most reliable indicator because these are typically the lowest-weight products from the test. The final cycle will include the recycle streams as separate test products in addition to the final products of concentrate and tailings.

The issue of water recycle is rather complicated. It is impossible to simulate plant process water in cycle testing for greenfield projects because these tests do not take into account recycle water aging in places such as thickeners and tailings ponds. If water is an issue (e.g., if seawater, brackish water, or sewage treatment plant effluent is used), then a pilot plant is probably warranted.

Locked-Cycle Calculations

Metallurgical balance calculations for locked-cycle tests are similar to those used in batch tests. Test product weights and assays are used to calculate metal units and distributions for the overall test. The overall metallurgical balance is examined to determine if the final concentrate weights and assays have stabilized in the later cycles, bearing in mind that variations should normally not exceed ±3%. If the test has stabilized, then the weights and assays of products from the last three cycles, excluding the recycle products, are used to calculate a "steady-state" metallurgical balance that should simulate continuous operation. The recycle stream weights and assays from the final cycle are then used to calculate the metal distributions in these streams as a percentage of new test feed. They do not enter into the steady-state balance.

Importance and Determination of Steady State

Meaningful data from locked-cycle testing can only be obtained if the test reaches steady state. This implies constancy and is usually achieved in the last three cycles of the test. Inputs equal outputs; for example, if a 2,000-g test feed charge per cycle is used, then the cycle products—concentrate and tailings—should sum to 2,000 g. This concept of input equals output must also hold true for metal balances in addition to weight (mass) balances. If this is not the case, then there is a buildup of recirculating middlings somewhere in the test flow circuit that must be addressed, usually in subsequent additional test work.

The achievement of steady state may be rigorously calculated by comparing the back-calculated head (cycle feed) assays of the individual test cycles with the overall test back-calculated head, including the recycle middlings from the last cycle. The individual cycle back-calculated heads should intersect the overall back-calculated head several times for the test to be in steady state. Figure 12 shows examples of a test in and out of steady state, respectively.

The achievement of steady-state conditions for ores containing valuable secondary minerals (i.e., molybdenite or gold) is often problematic. The tendency for copper minerals to crowd out the minor minerals can result in the building up of these minerals in middlings streams, particularly in the cleaner-scavenger concentrate.

ORE VARIABILITY AND MODELING

Ore variability and geometallurgy programs are extremely important parts of any flotation process development project. The variability testing is normally started after the rougher and cleaner parameters have been established. Selection of samples for testing may be based on ore types, locations in the ore body, or anticipated mining plan, such as when will a certain section of the ore body become flotation plant feed? These programs integrate mineralogical examination and characterization with comminution testing and flotation testing and distribute the test data across the mineral deposit. A typical ore variability program will involve mineralogical examination, comminution testing, and flotation testing of at least 100 samples, and programs of more than 300–400 samples are not uncommon, depending upon the size of the ore body.

Flotation tests in ore variability programs include batch rougher tests, batch cleaner tests, and locked-cycle tests. Rougher tests include size-by-size analyses of feed, concentrate, and tailings to obtain recovery by size data. A well-planned variability program should include at least 100 rougher tests and 25 cleaner tests. The cleaner tests may be performed on composite samples or selected individual samples and at least two cleaning stages are normally used, though the number of cleaner stages used will depend upon the results obtained in the batch cleaner optimization study. Locked-cycle tests should be performed on the same samples as the batch cleaners. This will provide a good comparison for upgrading in batch and semicontinuous cleaner tests.

BATCH FLOTATION TESTING FOR FLOTATION CIRCUIT OPTIMIZATION

Traditionally, batch laboratory flotation tests are performed using flotation feed samples and are used as a tool for circuit design or to optimize feed floatability properties. When performed on different streams of a flotation circuit during a circuit survey, they can also play an important role in diagnosing and optimizing a flotation circuit by providing a measure of ore floatability.

The properties that affect the propensity of an ore to float include particle size, mineralogy, degree of liberation, reagent coverage, and the presence of deleterious surface coatings (Trahar and Warren 1976; Sutherland 1989; Crawford and Ralston 1988). Because of system complexity, it is not yet possible to measure ore properties and "calculate" the ore floatability. Instead, it must be inferred from the results of a flotation process. This is not straightforward, as it cannot be determined directly from flotation circuit recoveries because these recoveries are also affected by the operating conditions used in the industrial flotation cells (airflow rate, froth depth, impeller speed, etc.). Batch laboratory flotation testing can provide a benchmarking tool, where operating conditions can be kept constant and differences in flotation response can be used as a measure of ore floatability.

These types of specially designed tests performed in conjunction with a flotation circuit survey can provide measures of ore floatability for use in flotation circuit modeling, as a diagnostic tool to assess flotation mechanisms, or to determine the effectiveness of ore floatability modification strategies (e.g., regrinding, reagent addition). In the following sections, the experimental procedure recommended to perform these tests are outlined and analysis options presented.

Experimental Considerations

Ore floatability characterization batch flotation tests are best performed using samples collected directly from the flotation circuit to be studied. In this way, the characteristics of the particles in the float test feed will truly reflect that in the process. The objective is for the batch flotation feed sample to have the same particle size distribution, mineralogy, degree of liberation, and surface chemistry in the test as that of the circuit stream being studied. Time between sample collection and flotation should be minimized to prevent particle surface oxidation.

The tests should be performed to minimize the impact of operating conditions on the result, using the same air rate, impeller speed, and froth depth in all experiments. Air rates should be high (>10 L/min) to minimize bubble overloading effects that occur where there is insufficient air to remove all particles that are "ready" to float. The impeller speed chosen should be that required to keep all particles in suspension but not so high that it results in particle surface modification by abrasion. The froth depth should be kept to a minimum and the entire froth phase should be removed by scraping at regular intervals during the experiment. Bottom-driven laboratory flotation cells where a fixed depth scraper can remove a constant depth of froth regularly throughout the experiment have proved ideal for doing these types of experiments (Runge 2010). Figure 13 shows examples of modified Agitair-style bottom-driven cells. These types of cells have the advantage that the froth flow is not impeded by the impeller and can be systematically removed using a fixed froth depth scraper to give very consistent results. However, top-driven laboratory machines with scrapers designed to move around the impeller can achieve similar consistencies in performance.

Tests should be performed over a time period sufficient to remove all of the floatable particles in the system. Concentrates should be collected at regular intervals such that a recovery-versus-time curve can be constructed and both the rate of flotation (i.e., how fast the target mineral is recovered)

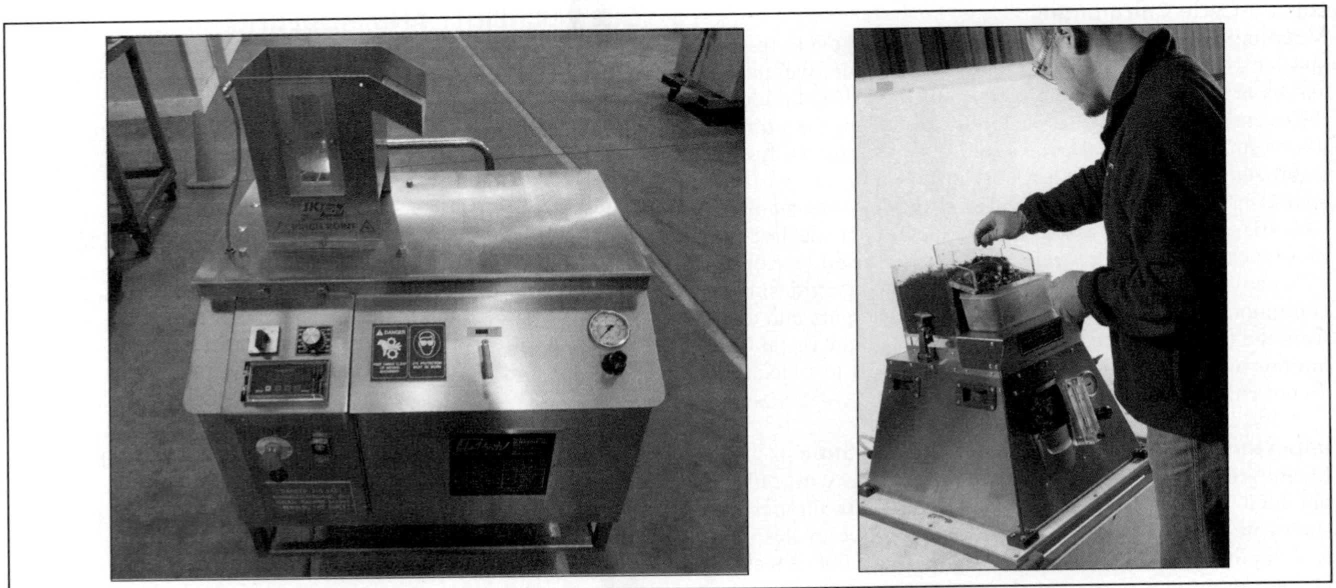

Courtesy of JKTech Courtesy of Magotteaux

Figure 13 Bottom-driven laboratory flotation machines

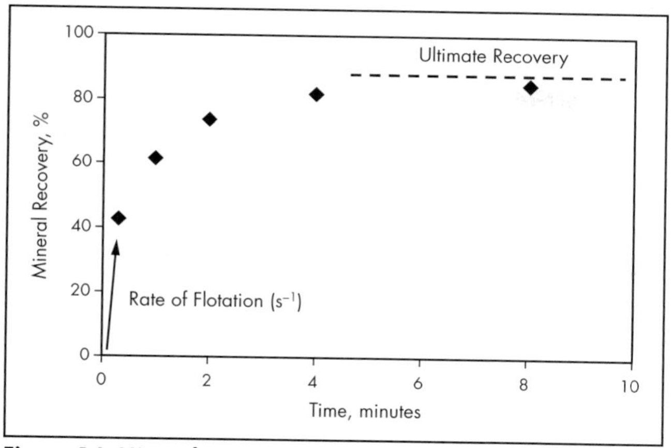

Figure 14 Mineral-recovery-versus-time curve depicting the mineral rate of flotation and the ultimate recovery achievable

and the ultimate recovery achievable can be established with confidence (Figure 14). The test begins when the air is switched on, and the end of the test is dictated by the need for the final concentrate collected to contain no floatable material. In this way, the point at which the recovery time curve asymptotes, which is used to determine the ultimate recovery, can be established with confidence.

As concentrate is removed from the batch flotation cell, it is important to add water to maintain a constant froth depth. This water should have the same chemical composition as that in the original sample to prevent changes in particle surface chemistry. This can be achieved by filtering another sample of the same stream and using the filtrate. Plant process water dosed with an appropriate amount of frother is usually a viable alternative.

Frother concentration needs to be high during the batch test to prevent bubble coalescence and to maintain a constant bubble size in the pulp and a stable froth to minimize particle drop-back. Excessive water recoveries, however, are an indication that the frother concentration is too high. For frothers normally used in flotation, a frother concentration of 30 µL/L is usually sufficient to achieve this aim. Frother should be added at the start of the test and also dosed at a similar concentration in the process water added to the cell throughout the experiment. This is because frother concentration tends to decrease with time during a flotation test as it is removed with the concentrate. No reagent, other than frother, should be added to these tests so that the floatability of the particles in the sample is not altered from that of the original stream.

Tests should be performed on all the major streams of the circuit—bank feeds, concentrates, and tailings, and not just the circuit feed. Since viscosity can play a complicating role at high percent solids, streams need to be diluted so that it is only the intrinsic particle floatability that affects the response. Recovery rates can also be lower than they should in concentrate streams performed at high percent solids because there is insufficient air to remove all the particles that would otherwise float. As a rule of thumb, flotation percent solids in the pulp when floating streams collected from the cleaners should be kept below 10% and rougher-scavenger feed and tailing streams kept below 30%.

Analysis Options

The information derived from the previously described types of ore characterization batch flotation tests can be used in a variety of ways, as outlined in the following sections.

Benchmarking

Most plants will be surveyed on a regular basis with a view to either benchmarking performance or testing the effectiveness of a circuit modification. Because of the large number of variables that can impact flotation, it is sometimes difficult to determine the reason for a change in result. Ore characterization batch tests performed on the feed during a survey can establish a measure of ore floatability that can be compared over time. The tests are performed using the same operating conditions so that any differences in a result can be confidently related to changes in grind size, liberation, mineralogy, or surface chemistry. Differences in performance of surveys where the ore floatability is the same can then be attributed with confidence to changes that have occurred within the circuit itself.

Assessment of Particle Modification Processes

Regrinding and reagent addition are employed within a flotation circuit to maximize the recovery and selectivity of the process. Batch flotation tests performed on streams collected before and after these types of processes can be compared to determine their effects on ore floatability (Runge et al. 2004; Runge 2007). This is a fairly simple process when there is only one input and output stream, as in a regrind. Batch flotation tests in this instance can directly be compared. To compare floatability across processes where there are multiple input and/or output streams (e.g., circuit or flotation bank), batch flotation data can be combined by weighting the data according to the relative flow of the mineral in the streams feeding or exiting the process (Equation 1). The aim is to produce a combined feed batch test result to compare to a combined product batch test result.

$$R^{\text{combined stream}, t \text{ minutes}}_{\text{mineral } j} = \frac{\sum_{s=1}^{n} F^{\text{stream } s}_{\text{mineral } j} \times R^{\text{stream } s, t \text{ minutes}}_{\text{mineral } j}}{\sum_{s=1}^{n} F^{\text{stream } s}_{\text{mineral } j}} \qquad \textbf{(EQ 1)}$$

where

$F^{\text{stream } s}_{\text{mineral } j}$ = flow rate of mineral j in stream s

$R^{\text{stream } s, t \text{ minutes}}_{\text{mineral } j}$ = cumulative recovery of mineral j in stream s after t minutes of flotation in the standard laboratory batch flotation test

By comparing batch flotation tests before and after a particle modification process such as reagent addition or regrinding, the metallurgist can deduce whether the strategy employed is achieving the desired result. The aim of the modification process is to increase the rate and amount of valuable mineral recoverable while improving selectivity of the valuable mineral with respect to the gangue recovery, as depicted in Figure 15.

These types of comparisons are not only useful to make before and after an individual process or unit, but they can also be used to assess the net effect of all process modification processes within a circuit by comparing the batch test recoveries in the circuit feed and combined circuit product streams.

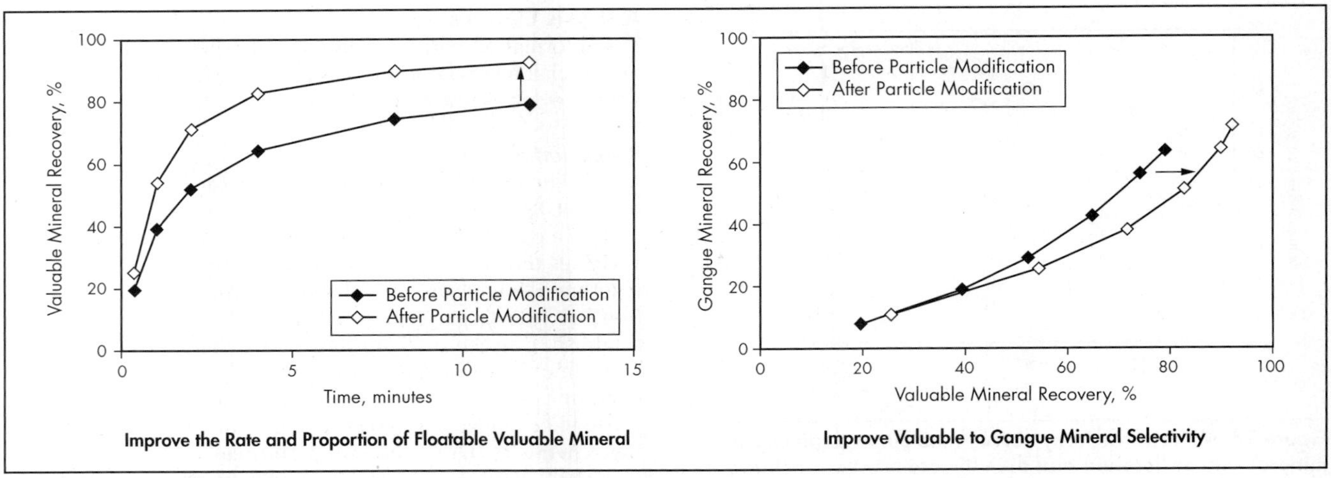

Source: Runge et al. 2004

Figure 15 Aim of particle modification processes

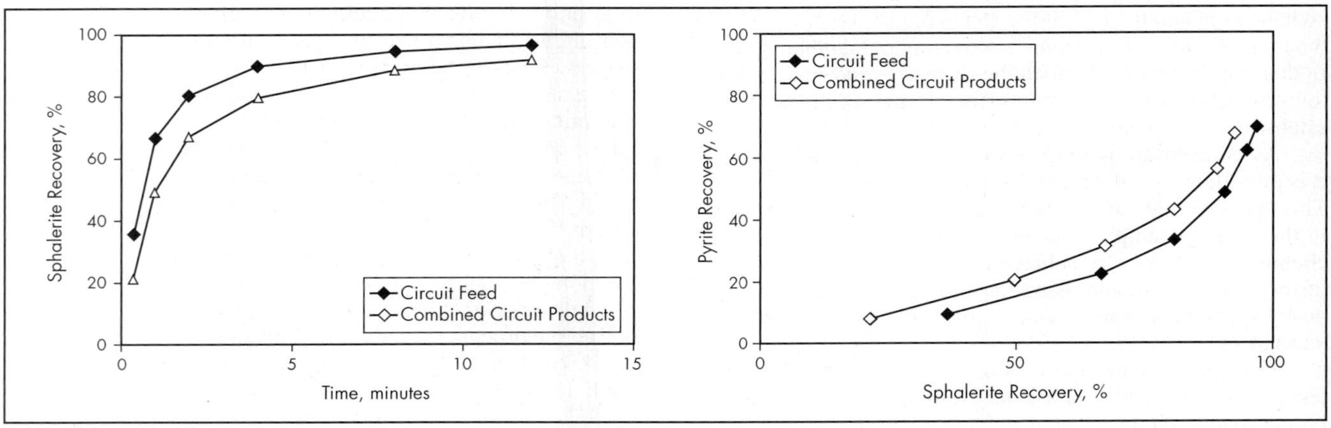

Adapted from Runge et al. 2004

Figure 16 Comparison of batch flotation test results in the feed and combined product streams of a sphalerite circuit within which lime addition and regrinding were employed, aiming to depress pyrite

In the example shown in Figure 16, batch test comparison was able to show that the lime addition and regrinding that had been employed in a sphalerite flotation circuit to depress pyrite was not successful. Pyrite recovery was depressed by the lime, but the sphalerite flotation was also depressed and to a greater extent than the pyrite, resulting in poorer sphalerite/pyrite selectivity.

The advantage with batch tests is that they enable the ore floatability to be analyzed in isolation to those effects related to circuit operation. What should also be kept in mind is that they are a measurement of pulp floatability only and are not an indicator of full-scale froth-phase performance. In the batch flotation tests, the froth-phase effects have been minimized by using shallow froth depths and by continually removing the froth. Ore properties will affect the froth phase, and this should also be kept in mind when considering the effectiveness of particle modification processes.

Determination of Flotation Mechanisms

Particles can be recovered to the concentrate of a flotation circuit by several mechanisms. They can attach to bubbles because of a hydrophobic surface. They can be locked in composite particles with a hydrophobic mineral and "come along for the ride" (e.g., gangue in unliberated particles, piggy-backing on particle surfaces). Alternatively, they can be entrained unselectively in the water recovered to concentrate.

To maximize the selectivity of the process, it is important to maximize the floatability of the valuable wanted mineral while minimizing the flotation recovery of the unwanted gangue minerals. It is important, therefore, to understand the mechanisms of recovery in a flotation circuit so that appropriate strategies for improvement can be developed.

Mineralogical analysis and surface chemistry measurements can be employed to judge whether gangue is reporting to concentrate because of composite particles or natural hydrophobicity. It is sometimes difficult, however, to judge the proportion of gangue recovery in a circuit due to entrainment. Batch flotation tests of concentrate streams in a circuit provide a means of making this judgment. Gangue not recovered in these tests is not floatable and can only be present in the concentrate stream because it has been entrained in the flotation bank from which it originates. Figure 17 shows recoveries

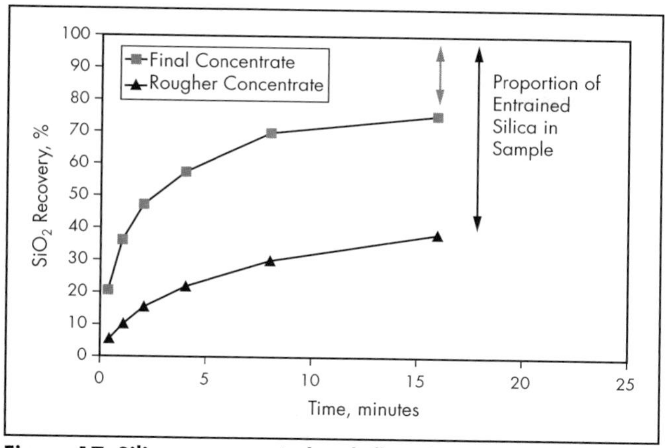

Figure 17 Silica recovery in batch flotation tests performed using a rougher and a final concentrate collected from an industrial copper concentrator

of silica measured in batch flotation tests performed on the rougher concentrate and final concentrate collected from a copper concentrator. It demonstrates how silica recovery in the rougher is largely due to entrainment, whereas in the final concentrate, after re-flotation, the proportion of gangue that does not float is significantly reduced. Excessive entrainment recovery to a final concentrate can be addressed by additional flotation stages or dilution of pulp density.

Batch flotation test data can also be used to assess valuable mineral recovery (or non-recovery) mechanisms. Fast flotation kinetics of the valuable mineral in a tailing stream is usually indicative of insufficient flotation capacity or poor flotation cell operation (e.g., use of very deep froth depths in scavengers). Slow kinetics for the floatable mineral, exhibiting poor selectivity with gangue, will be observed when the grind size is too fine (something that can occur after regrinding in a cleaning circuit). Low valuable mineral recovery in the tailing is an indication that improvements in circuit recovery will best be achieved by addressing ore floatability—either by improving flotation selectivity, improving the feed liberation, or addressing flotation chemistry issues.

Flotation Model Development

Flotation recovery and separability will be a strong function of the circuit arrangement, flotation bank residence times, ore floatability, and the cell operating parameters. Flotation modeling provides a tool for optimizing these various interactive effects and enables simulation of different strategies for circuit improvement.

Harris et al (2002) has summarized a methodology that uses ore floatability characterization batch tests performed on different streams of a flotation circuit as a way of deriving parameters of a floatability component model for use in circuit simulation. Many case studies have been published in the literature, highlighting the use of this methodology for flotation circuit optimization (Crosbie et al. 2009; Schwarz and Alexander 2007).

In floatability component models, the ore floatability of different minerals in different size classes is represented by fast, slow, and non-floating components. Because the batch flotation tests are all performed using the same operating conditions, any change in response in the different streams is due to a change in ore floatability. A model is fitted to the batch test data to determine the proportions and rates of the fast, slow, and non-floating components in the feed and all of the streams of the circuit. The data can also be used to develop models to modify ore floatability at points in the circuit where regrinding or reagent addition occurs.

Performing batch flotation tests during a circuit survey not only provides useful data for circuit diagnosis but also means that the information required to develop models of the circuit using the above techniques is collected. Without this type of ore characterization data, it is very difficult to isolate cell operating effects (e.g., poor froth recovery, insufficient air) from ore floatability effects (i.e., selectivity and recovery potential) from circuit survey data alone.

SUMMARY

The key concepts underlying a laboratory test program for the development of a copper sulfide flotation process are described. The testing program includes scoping flotation tests for reagent scheme selection, rougher flotation parameter evaluation, cleaner circuit evaluation, locked-cycle testing, and ore variability and modeling. The concepts explored can be applied or modified for most sulfide flotation systems.

Ore characterization batch tests for flotation plant optimization are relatively simple to perform and produce very repeatable results if performed using the systematic procedures outlined in this chapter. If performed routinely during circuit survey work, they provide important information about how the variables that affect ore floatability (size, mineralogy, chemistry) are impacting flotation circuit performance. They provide benchmarking of ore floatability between surveys and a diagnostic tool for offering information about recovery mechanisms or the effectiveness of particle modification processes. They can also be used to develop floatability component models that can simulate options for improving the grades and recoveries achievable in the circuit.

It is recommended that ore characterization batch flotation tests are performed in conjunction with other measurements. Although batch laboratory flotation tests indicate the overall result of a change, other information from sizing, mineralogical, and chemical analysis tools can be used to determine the reason for a particular result.

ACKNOWLEDGMENT

Thanks are extended to Peter Wilkinson, who kindly provided information on batch flotation cell operating conditions that have been incorporated into the chapter.

REFERENCES

Collins, D.A., Schwartz, S., and Alexander, D.J. 2009. Designing modern flotation circuits using JKFIT and JKSimFloat. In *Recent Advances in Mineral Processing Plant Design*. Edited by D. Malhotra, P.R. Taylor, E. Spiller, and M. LeVier. Littleton, CO: SME. pp. 197–201.

Crawford, R., and Ralston, J. 1988. The influence of particle size and contact angle in mineral flotation. *Int. J. Miner. Process.* 23:1–24.

Crosbie, R., Runge, K., Brent, C., Korte, M., and Gibbons, T. 2009. An integrated optimisation study of the Barrick Osborne concentrator: Part B–Flotation. In *Proceedings of the 10th Mill Operators' Conference*. Melbourne, Victoria: Australasian Institute of Mining and Metallurgy. pp. 189–197.

Dunne, R. 2016. Flotation of gold and gold bearing minerals. In *Gold Ore Processing. Project Development and Operations*, 2nd ed. Edited by M.D. Adams. San Diego, CA: Elsevier Science.

EMC (Eurus Mineral Consultants). 2012a. EMC flotation test procedure 2: Laboratory flotation test procedure for MF1 and MF2 tests and regrinding of cleaner tailings. Update 2 (January). www.eurusminerals.co.za/docs/EMC%20Lab%20Test%20Procedure%202%20Lab%20flotation%20MF1-MF2%20tests%20Update2%20Jan12.pdf. Accessed August 2018.

EMC (Eurus Mineral Consultants). 2012b. Laboratory flotation rate test procedure for PGM, base metal sulphide and oxide ores: For a MF1 rate test. Update 4 (January). https://www.911 metallurgist.com/blog/wp-content/uploads/2015/06/Laboratory-Flotation-Testing-Procedure.pdf. Accessed August 2018.

Harris, M.C., Runge, K.C., Whiten, W.J., and Morrison, R.D. 2002. JKSimFloat as a practical tool for flotation process design and optimisation. In *Mineral Processing Plant Design, Practice, and Control*. Edited by A.L. Mular, D.N. Halbe, and D.J. Barratt. Littleton, CO: SME. pp. 461–478.

Kappes, R., Fortin, C., and Dunne, R. 2011. The current status of the chemistry of gold flotation in industry. In *World Gold 2011*. Westmount, QC: Canadian Institute of Mining, Metallurgy and Petroleum. pp. 385–395.

Runge, K., 2007. Modelling of ore floatability in industrial flotation circuits. Ph.D. thesis, Julius Kruttschnitt Mineral Research Centre, The University of Queensland.

Runge, K. 2010. Laboratory flotation testing—An essential tool for ore characterisation. In *Flotation Plant Optimisation: A Metallurgical Guide to Identifying and Solving Problems in Flotation Plants*. Edited by C.J. Greet. Melbourne, Victoria: Australasian Institute of Mining and Metallurgy. pp. 155–173.

Runge, K.C., Manlapig, E., and Franzidis, J.-P. 2004. Optimisation of flotation pulp selectivity utilising nodal analysis. In *Proceedings of the 36th Annual Meeting of the Canadian Mineral Processors*. Montreal, QC: Canadian Institute of Mining, Metallurgy and Petroleum. pp. 575–598.

Schwarz, S., and Alexander, D. 2007. Optimisation of flotation circuits through simulation using JKSimFloat. In *APCOM 2007: Proceedings of the 33rd International Symposium on the Application of Computers and Operations Research in the Minerals Industry*. Edited by E.J. Magri. Santiago, Chile: Gecamin. pp. 461–466.

Sutherland, D.N. 1989. Batch flotation behaviour of composite particles. *Miner. Eng.* 2(3):351–367.

Thompson, P. 2002. The selection of flotation reagents via batch flotation tests. In *Mineral Processing Plant Design, Practice, and Control*. Edited by A.L. Mular, D.N. Halbe, and D.J. Barratt. Littleton, CO: SME. pp. 136–144.

Thompson, P. 2016. Laboratory testing for sulfide flotation process development. *Miner. Metall. Process.* 33(4):200–213.

Trahar, W.J., and Warren, L.J. 1976. The floatability of very fine particles—A review. *Int. J. Miner. Process.* 3(2):103–131.

CHAPTER 7.8

Non-Sulfide Flotation Testing

Howard Haselhuhn, Shashi Rao, and D.R. Nagaraj

With the exception of a few elements, such as the base and precious metals, which are predominantly obtained from sulfide minerals in sulfide ores, most other elements or their minerals are obtained from non-sulfide ores. Thus, non-sulfide mineral flotation from ores covers a wide variety of minerals of industrial importance. The term *non-sulfide minerals* (or *ores*) comprises minerals (or ores) belonging to the following classes: oxides, silicates, sulfates, phosphates, carbonates, and halides. These are variously described in the industry under terms such as *industrial minerals*, *nonmetallic minerals*, *construction materials*, *functional minerals*, and so on. Examples of important non-sulfide mineral ores include phosphate, iron oxides, kaolinite and bentonite, spodumene, potash, borates, trona, fluorite, calcite, dolomite, limestone, barite, mica, feldspar, quartz, silica sands, monazite, kyanite, magnesite, chromite, bauxite, ilmenite, rutile, manganese oxides, graphite, talc, and cassiterite. Strategies and approaches for laboratory and pilot-plant testing are discussed in this chapter using examples from flotation beneficiation of well-known commodities such as phosphate, iron ore, clays, potash, and feldspar. Details of the processing flow sheets are discussed elsewhere in separate chapters devoted to each mineral commodity. Historical aspects of the development of non-sulfide flotation can be found in Fuerstenau (2007), Nagaraj and Ravishankar (2007), and Nagaraj and Farinato (2016).

Flotation concentration of non-sulfide minerals is significantly more complicated than that of sulfide minerals. This complication arises primarily because of the similarities in surface chemical composition and properties in an aqueous medium between non-sulfide value minerals and non-sulfide gangue minerals. In sulfide mineral flotation, the primary separation task comprises selective flotation of sulfide minerals from the vastly different non-sulfide minerals. In non-sulfide flotation, the task is floating one non-sulfide mineral from many other non-sulfide minerals. This separation task requires significant know-how and art to manipulate and exploit small differences in surface properties between minerals and to develop specific process conditions. Redeker and Bentzen (1986) noted that "flotation of nonmetallic minerals is sometimes viewed as alchemy. However, the art is, in reality, based on sound chemistry, experience and a little luck in the treatment of new ores. In the flotation of nonmetallic ores, as opposed to base metal sulfides, highly specific treatment conditions are required to accomplish separations." A vast amount of published literature exists for both the fundamental and applied aspects of non-sulfide flotation, beginning with the pioneering work of U.S. Bureau of Mines investigators in the late 1920s (Coghill and Clemmer 1935); however, this literature is far more fragmented than that for sulfide flotation (Aplan 1994). This is necessarily so if one considers the variety of non-sulfides that are beneficiated in the industry and the artisanship that surrounds these processes. Separation schemes for non-sulfide minerals are distinctly different from those for base metal sulfide minerals. Such distinction can be readily understood by the fundamental differences that exist in physical and chemical properties between sulfide and non-sulfide minerals (see Nagaraj and Ravishankar 2007).

Plant practice is often consistent with the major differences between sulfides and non-sulfide minerals. Aplan and Fuerstenau (1962) noted that "selective flotation in non-sulfide systems is often difficult in that one class of compounds alone, the carboxylic acids, may float nearly the entire array of non-sulfide minerals under the appropriate conditions. The non-sulfide mineral flotation operator is confronted with a vast array of modifying reagents." This is in contrast to sulfide mineral flotation where a wide variety of collectors, many of which exhibit a high degree of selectivity for a given sulfide, is available (and used) for the relatively small number of sulfide minerals floated. Modifiers are thus far more critical to successful separations in non-sulfide systems than they are in sulfide systems.

Given the diversity of non-sulfide flotation plant practice and the rather fragmented literature information, a practitioner new to non-sulfide flotation may initially find the task of developing a laboratory program and flow sheet rather bewildering and daunting. However, this is certainly not the case. A

Howard Haselhuhn, Applications Engineer–Industrial Minerals, Solvay Technology Solutions, Stamford, Connecticut, USA
Shashi Rao, Metallurgical Engineer, Natural Resources Research Institute, Coleraine Labs, University of Minnesota, Coleraine, Minnesota, USA
D.R. Nagaraj, Principal Research Fellow, Solvay Technology Solutions, Stamford, Connecticut, USA

common theme and a set of guidelines emerge for non-sulfide flotation if one considers the types of reagents used (both collectors and modifiers) and the types of treatment methods and conditions employed. This chapter aims to provide basic guidelines for laboratory and pilot-plant flotation testing for a given non-sulfide ore flotation system to assist in the following goals:

1. Develop a new flow sheet or evaluate an existing one.
2. Optimize and improve performance of an existing plant process via a better reagent and/or engineering schemes.
3. Evaluate different ore types and ore variability.
4. Develop process models.
5. Produce sufficient samples for market utilization studies.

These guidelines are based on information gathered from published literature on both fundamental and applied aspects; articles describing plant practices in trade journals, conference proceedings, and operators' forums; and experiential knowledge of metallurgists. A compilation of features of the major industrial non-sulfide mineral flotation separations is provided in Table 18 of Chapter 7.5, "Flotation Chemicals and Chemistry," and in the following selected references: Aplan and Fuerstenau (1962); Arbiter and Williams (1980); Aplan (1994); Baarson et al. (1962); Coghill and Clemmer (1935); Cytec (2010); Davis (1985); Falconer and Crawford (1947); Falconer (1961); Fuerstenau (1962); Fuerstenau et al. (2007); Gaudin and Glover (1928); Holme (1986); Houot (1983); Jones and Oblatt (1984); Malhotra and Riggs (1986); Manser (1975); Miller et al. (2002, 2007); Mulukutla (1994); Peres et al. (2007); Redeker (1981); Redeker and Bentzen (1986); Somasundaran and Moudgil (1987). A few useful references for laboratory testing include Dunne et al. (2013); Macdonald and Brison (1962); Spedden (1985); and Williams et al. (2002). North Carolina State University Minerals Research Laboratory reports are also an excellent source for laboratory testing and development for a variety of non-sulfide mineral flotation systems (NC State University 2018).

The guidelines serve to establish a starting point and a basis for designing the laboratory program; the metallurgist still requires empirical testing to refine the reagent schemes, conditions, and flow sheets in an iterative manner to meet metallurgical and economic objectives. The pilot test methods presented in this chapter can be used to establish suitable flotation circuits. Analytical techniques employed to gather meaningful information from test work data are discussed in this chapter as well.

DISTINGUISHING FEATURES OF NON-SULFIDE FLOTATION

Several important distinguishing features of non-sulfide mineral systems, in comparison to sulfide flotation systems, dictate strategies and approach to laboratory flotation. These are listed in Table 1.

Table 1 Non-sulfide versus sulfide mineral flotation distinguishing factors*

Factor	Non-Sulfide Mineral System	Sulfide Mineral System
Ore type and mineralogy	Very diverse and broad classes (e.g., silicates). Value minerals and gangue minerals are both non-sulfides and have rather similar surface properties; thus, separation selectivity is a challenge (ores may contain gangue sulfide minerals such as pyrite, but these are easily removed using traditional sulfide collectors prior to non-sulfide mineral separation). Careful selection of collectors, modifiers, and pulp conditions is necessary. Many ores are the sedimentary type and contain a significant amount of slimes requiring desliming. Minerals can have a wide range of substitutions, compositions, and structures (e.g., the silicate minerals)—all of which can influence surface properties.	Value minerals are invariably sulfides and precious metal minerals. The majority of gangue minerals are non-sulfides; only a few gangue sulfide minerals are present. Collectors are invariably selective and specific for sulfide minerals; selection of modifiers and conditions are relatively more straightforward. Ores are predominantly igneous type.
Surface chemistry in the aqueous phase	Relatively small differences between a value mineral and the gangue. Surface chemistry is dominated by reactions involving H^+, hydroxyl (OH^-), and the lattice ions. All of the minerals in the system undergo strong surface charging, hydration, dissolution, and hydrolysis—all of which influence surface chemistry. Various species released from non-sulfide minerals can have complex interactions in the aqueous phase; the products of these processes, in turn, can interact with many minerals causing changes in flotation selectivity. Redox reactions, and complications arising from these, are absent in most systems (most minerals are insulators).	Very large differences between value and gangue minerals. Surface chemistry is dominated by complex multistep redox reactions (most sulfides are excellent conductors); these can have a significant adverse effect on flotation. Relatively smaller contribution from hydration, hydrolysis, and dissolution.
Water quality	Critical to achieving selectivity of separation. Imperative to conduct final testing in water from a source representative of local water to be used. Any change of water source during operations should be checked thoroughly before general use to determine its effect on selectivity and reagent consumption.	Not as critical; effects are a matter of degree rather than catastrophic.
Comminution	Grinding for liberation is not always necessary. In some systems, fine grinding is necessary to achieve adequate liberation; this results in substantial amounts of slimes. Many ores are from sedimentary deposits and can be processed without any comminution steps (e.g., phosphate, glass sands, clays, and potash ores). Liberation of value minerals from gangue is not a major limiting factor in these systems.	Grinding is critical to achieving liberation. Liberation is most often incomplete and/or inadequate.
Flotation conditions and strategy	Highly specific treatment conditions are frequently required to achieve adequate selectivity and satisfactory metallurgy, the most critical aspect of which is generally the quality and purity of the finished mineral product (which can go directly to use in many cases).	Treatment conditions are common to almost all of the systems. The mineral product is used for metal production.

(continues)

Table 1 Non-sulfide versus sulfide mineral flotation distinguishing factors* (continued)

Factor	Non-Sulfide Mineral System	Sulfide Mineral System
Flotation conditions and strategy (continued)	Both direct flotation (flotation of targeted value mineral away from gangue) and reverse flotation (flotation of gangue minerals away from value mineral) are plausible and frequently used options. The choice is dictated by the efficiency and selectivity of separation, composition, and mineralogy of the ore, choice of reagents, pulp conditions, and economics.	Direct flotation of sulfides is the most common practice.
	Flotation rates and particle sizes floated are large in some systems (e.g., phosphate ore, potash, and silica); floated particles must be collected quickly to prevent drop back. Some plants even refer to their process as *flash flotation*, likening it to a flash distillation process. For example, in phosphate flotation, retention times are a few minutes. (Of course, there are systems wherein the flotation rates and particles sizes are quite small, requiring long residence times [e.g., flotation of fine anatase from fine kaolin clays; residence time could be 30–60 minutes]; these systems are similar to the sulfide ore systems.)	Particle sizes are typically <212 μm. Slow flotation rates and long residence times are common.
Pulp preparation	Pulp preparation prior to flotation separation is critical in non-sulfide separations because of complex surface interactions. This includes processes such as washing (acid or caustic), scrubbing or attrition scrubbing, desliming, hot pulp conditioning, and high solids conditioning.	No special pulp preparation is required. High solids and high-temperature conditioning are rare. Most operations are at ambient temperature.
	Conditioning: A wide spectrum of conditioning requirements exists for reagents with respect to time, pulp density, agitation intensity, pH, order of reagent addition, types of modifiers, temperature, and water quality. Preconditioning requirements are often less stringent in amine flotation, in contrast to fatty acid flotation. Conditioning in amine flotation is usually at flotation pulp density, unlike in fatty acid flotation.	Short conditioning times are typical (from a few seconds to a few minutes). High intensity and high solids conditioning are rare.
	Desliming: Very common and often a critical step. Some non-sulfide separations are virtually impossible if the feed to flotation is not properly deslimed. The slimes† (clays, iron oxides, and hydroxyoxides, etc.) can consume collector, coat coarse particles, cause stable froths with reduced drainage, increase pulp viscosity, and seriously interfere in separations in other ways. The presence of slimes can reduce the ability of collectors to differentiate among various mineral surfaces. Because the slime fraction is usually discarded, it can account for a considerable loss in value minerals. The deslimed material is sometimes separated into coarse and fine fractions and treated separately to optimize flotation conditions for each size fraction.	Desliming is not a common practice (used only in a few niche systems).
Circuit complexity	Flotation circuits vary widely and can be rather complex, some involving several stages of sequential flotation separation, each stage having its own specific treatment conditions and preparation requirements. Plant flow sheets often employ other separation techniques prior to, within, and after flotation circuit to reach the target grade (e.g., gravity, magnetic, electrostatic, selective flocculation, air classification, and further comminution and sizing); specific examples are found in other chapters. Strategies employed in flotation may be influenced by separation units prior to and after flotation. High circulating loads and water reuse may not be possible in many non-sulfide ore systems, for example, the three stages of flotation in feldspar beneficiation, each stage requiring dewatering, washing, and use of fresh/clean water for the next stage.	Flotation circuits tend to be similar in most of the systems and relatively less complex. High circulating loads and water reuse are typical in all circuits. Occasionally, magnetic separation may be used. Gravity separation is used for ores containing gold and for platinum group metals.
Feed characteristics	Unlike sulfide ores, the concentration of valuable minerals in non-sulfide ores is typically high (e.g., kaolin clays, phosphate, iron ore, glass sands, feldspar, barite, potash, and others); however, tonnages processed are generally lower than those in sulfide ore systems.	Low concentration of valuable minerals in most ores. Very large tonnages (e.g., 100,000–300,000 t/d in copper ore operations).
Reagents	The large concentrations of valuable minerals, and their significant hydrophilicity (high degree of hydration and hydrogen bonding), typically require high dosages of long-chain collectors; for example, 500–1,000 g/t of C_{18} fatty acid in some systems (C_{12} considered minimum chain length). In some systems, collector adsorption is via metal complex formation and surface precipitation; in others, it is electrostatics. Mineral lattice anion has a greater role than metal ion (in contrast to sulfides). Modifiers are critical to achieving required selectivity and efficiency. Some separations are impossible without appropriate modifiers. Multiple modifiers may be required in some systems. Their dosages also tend to be high.	Very low dosages (2–50 g/t) and short-chain (C_2–C_5) collectors. Sulfide surfaces are generally far less hydrophilic (weaker hydration and hydrogen bonding). Role of modifiers is not as critical as in non-sulfide systems. pH modifiers are most common. Other modifiers are used as needed. Collector adsorption is dominated by coordination covalent bonding; metal ion of mineral is very important.
	Reagent solubility is often an issue requiring specialized preparation procedures prior to use. Examples of this include neutralization of amine collectors, emulsification of fatty acids, and high-temperature solubilizing of organic polymeric depressants.	No special equipment is necessary for reagent dosing.

*More specific information and details can be found in Aplan and Fuerstenau (1962); Aplan (1994); Baarson et al. (1962); Coghill and Clemmer (1935); Cytec (2010); Davis (1985); Falconer and Crawford (1947); Falconer (1961); Holme (1986); Malhotra and Riggs (1986); Manser (1975); Miller et al. (2002, 2007); Mulukutla (1994); Peres et al. (2007); Redeker (1981); Redeker and Bentzen (1986); Somasundaran and Moudgil (1987). Factors such as froth stability, pulp density, and others are not highlighted and distinguished because the ranges and observations appear to be common to both sulfide and non-sulfide flotation.

†The term *slimes* has a loose meaning in non-sulfide separation and is system dependent; for example, −100 μm particles are considered slimes in some phosphate operations, whereas in other systems, particles <1–10 μm are considered slimes.

STRATEGY AND APPROACH TO LABORATORY TESTING

The strategy and approach to laboratory testing are dictated by whether the goal is to improve efficiency and selectivity in an existing operating plant or to design a flow sheet for a new ore or project. The key aspects are given in Table 2.

Preparation Prior to Laboratory Testing

The following items need to be addressed prior to laboratory testing to ensure that the outcomes are meaningful:

- Appropriate selection of samples for project development
- Proper collection of samples for plant optimization testing
- Proper subsampling and preparation of samples for flotation testing
- Understanding of the mineralogy and liberation characteristics of the sample collected for testing
- Appropriated analytical procedures and incorporation of a suitable quality assurance and quality control program
- A robust design for laboratory tests that will provide statistically meaningful outcomes

Laboratory Test Work

If grinding is involved, a grinding study is performed to develop the appropriate procedure for grinding the feed to desired liberation (determined by mineralogy and economics) while maintaining as narrow a size distribution as possible. This is typically done using a laboratory rod or ball mill (ceramic grinding media used in systems where iron species create a problem). Size-by-size mineralogical analysis is then conducted to determine the mineral separation potential, valuable mineral content, the type of gangue minerals in the slime (or fine size) fractions, and the need for desliming to reject predominantly problematic gangue minerals with an acceptable loss in values. Desliming can be performed using hydroseparators (elutriators, desliming thickeners), cyclones, sieves, or simple sedimentation equipment (e.g., differential settling in a tall cylinder or a tank with a conical bottom). If aging of the pulp is an issue, then it is best to prepare the flotation feed charge prior to the flotation test.

Water Quality and Water Chemistry

Water quality and water chemistry in non-sulfide flotation are especially important because aqueous species—both anionic species and multivalent metal ions—interact with each other, with non-sulfide mineral surfaces, and with reagents via both chemical (e.g., calcium on phosphate mineral or carbonate mineral dictated by solubility products) and electrostatic pathways (calcium on quartz) to influence reagent adsorption and selectivity of separation. These aqueous species may be from deliberately added modifiers, from the water source available at the mine site, or from various minerals (both value minerals and gangue) in the system. Multivalent ions (both anionic and cationic) can be either beneficial or detrimental depending on the specific system in question and the type of collector considered. Species present in the water source and/or released from minerals (because of their finite and significant solubility) are called *unintended species* or *modifiers*. These can have significant adverse effects on selectivity and recovery, especially when water is recycled and circulating loads are present.

Although the effects of species from deliberately added (or intended) modifiers are reasonably well-known and targeted for a specific function, those of unintended species generated from minerals in the ore are less well-known and predictable. Unintended species can have a wide range of effects on both efficiency and selectivity of separation:

- They can interfere with collector adsorption on desirable minerals (e.g., Ca^{2+} adsorption on quartz can prevent amine collector adsorption; sulfate, SO_4^{2-}, ions in water can adsorb on barite and prevent anionic collector adsorption).
- Flotation of targeted minerals can be reduced.
- They can promote collector adsorption on, and flotation of, undesirable mineral(s) via inadvertent activation (inadvertent Ca^{2+} adsorption on gangue quartz to promote fatty acid adsorption and flotation). The pH of activation by metal ions is strongly influenced by the type of metal ion and the pH of formation of first hydroxide (or hydroxyl) species.
- Froth characteristics can change (highly variable and difficult to predict; froth can be influenced by the presence of organics and/or gelatinous/colloidal precipitates in the aqueous phase).
- Pulp viscosity can change (multivalent ions can stabilize/promote mineral network or form colloidal or gelatinous precipitates).

Inadvertent activation, for example, is often difficult to predict, because it is strongly influenced by mineralogy and pulp conditions that exhibit significant variability in a plant. Indeed, one of the main functions of modifiers deliberately added (e.g., lime, soda ash, sodium silicate, polyacrylates, sodium sulfide) is to control the effects of inadvertent and unavoidable species. Locked-cycle testing should be considered if water chemistry effects are anticipated (e.g., when the quality of the available water source is less than desirable, or when flotation conditions such as low pH cause excessive release of species from some minerals). A good example of the influence of water chemistry is the selective flocculation desliming of quartz from hematite. The presence of calcium and magnesium ions in iron ore pulps that are upgraded by selective desliming can result in indiscriminate flocculation of both iron oxide and siliceous gangue slimes and/or in interfering with amine adsorption on quartz. In other cases where fatty acids are used as collectors, cations such as Ca^{2+}, Mg^{2+}, and others (Fe^{3+}, Al^{3+}) will react with fatty acids to form insoluble compounds thereby consuming the collector (in some cases, metal soaps can precipitate on or near the mineral surface, and this can enhance hydrophobicity). In the case of phosphate ores containing apatite and calcite/dolomite, the surface of apatite can be converted to calcite and vice versa depending on pH and carbonate concentration. The dissolved mineral species can alter the mineral surface properties to such an extent that surface properties of various minerals become indistinguishable, leading to loss of selectivity in the adsorption of reagents.

In systems involving sequential flotation of gangue minerals with specific treatment conditions and collector types, the value product from each stage may require dewatering, thorough washing with fresh water, chemical treatment (sulfuric acid, bleach, etc.) to generate fresh surfaces, and use of fresh water in conditioning prior to addition of the collectors (e.g., beneficiation of beryl, spodumene, feldspar, phosphate).

Table 2 Flotation aspects for defining a non-sulfide mineral flotation strategy

Flotation Aspect	Description
Mineralogy	Mineralogical characterization is one of the most important aspects of determining the most effective approach to laboratory flotation of non-sulfide minerals. This includes mineral phase identification, mineral association, and liberation characteristics (if comminution is necessary). This can be accomplished by a combination of popular techniques, such as MLA (mineral liberation analyzer), QEMSCAN (Quantitative Evaluation of Minerals by Scanning Electron Microscopy), XRD (X-ray diffraction), SEM-EDX (scanning electron microscopy–energy dispersive X-ray), EPMA (electron probe microanalysis), optical microscopy, and elemental analysis. Information on impurity or gangue minerals, even if present in minor amounts (e.g., phosphate mineral in iron ores; Fe-Ti minerals in kaolin clays), will be necessary because their presence can have a large impact on the selectivity of separation and final product specification.
Literature review	A survey of plant practice or even past laboratory investigation on ores similar to the one under consideration provides a general framework. Very few ore types have not been explored by mineral separation technologists; strategies for successful separation can generally be gleaned from other operations with similar ores. Many sources for such information can be found, including Chapter 7.5, "Flotation Chemicals and Chemistry," and other chapters on individual mineral commodities in Part 12 of this handbook. These should serve as a starting point for working with a new ore at a bench scale.
Comminution	The need for, and extent of, comminution is dictated by rock or ore type, mineralogy and liberation characteristics, economics, and the grade specifications for the final mineral product. Some ores, such as sedimentary phosphate ores of central Florida (United States), do not require grinding for liberation, whereas hard taconite in northern Michigan (United States) can require extensive grinding with a liberation size of <25 µm (Kawatra and Carlson 2013; Siirak and Hancock 1988). Some ores may require the breaking of aggregates in the mined ore, which could be accomplished in high-shear, high solids conditioning; for example, kaolin clays of Georgia (United States).
Floated mineral(s)	The decision to consider either direct flotation or reverse flotation is largely dictated by the following: • The ratio of the amount of value mineral to gangue mineral. A small ratio favors direct flotation; a very large ratio favors reverse flotation; and for ratios in the intermediate region, either direct flotation or reverse flotation is an option. The decision will be based on economics, selectivity, overall recovery, and the level of complexity involved. • Gangue mineralogy. A sequential reverse flotation scheme to remove one or more impurities in each stage (e.g., feldspar flotation) can be considered when the gangue composition is complex, the ratio of value to gangue is large, and when the different gangue minerals cannot be floated selectively and effectively with the same reagent scheme. In some cases, even if the ratio of value to gangue is large and there are several gangue minerals to be depressed, direct flotation may be considered because a robust collector-modifier scheme is available for selective separation (e.g., direct flotation of diaspore from bauxite ores). Often, ores have gangue sulfide impurities such as pyrite; these can be easily removed by flotation with traditional sulfide collectors before non-sulfide mineral separation. Sometimes sulfides can also be removed by gravity separation. Any gangue talc can be removed easily by flotation with only a frother. • Economics and complexity. The decision to select direct or reverse flotation is dictated by overall economics and complexity of the flow sheet, number of unit operations and separation schemes (magnetic, flotation, gravity, etc., involving isolated water streams) required for each route. If a practical, economical and robust reagent scheme already exists for one route, then this may be preferred to developing a new scheme, which may be expensive and time-consuming unless the new scheme is potentially less complex and more economical. It is often preferable to remove gangue and impurity minerals by physical separations such as magnetic separation, size, and gravity separation prior to flotation if at all possible. This strategy reduces the complexity in flotation. • Product specifications (for either internal use or external sale). This is extremely important because many consumers will require a specific grade or absence of specific impurities. If this target cannot be reached, there is no need to further pursue the project.
Reagent treatment schemes	Reagent and treatment schemes are discussed in Chapter 7.5, "Flotation Chemicals and Chemistry," and in other chapters on individual mineral commodities in Part 12 of this handbook. The choice is dictated by whether direct or reverse flotation is used. Temperature is an important factor to consider when evaluating reagent schemes (e.g., fluorite, kaolin clay, and rare earth ores). Well-established and reputed reagent suppliers have expertise beyond the solutions suggested in the literature. Efforts must be made to simplify the reagent schemes as much as possible (removing gangue by other physical separation techniques can help).
Water availability	Water quality and chemistry have a critical influence (far more than in sulfide ore systems) on both selectivity of separation and recovery of value minerals. Knowledge of the type and composition of the water resources available at the mine site is necessary before any laboratory investigation. If only seawater is available, the laboratory strategy should be developed using seawater as a water source. If fresh water is available, extensive locked-cycle and piloting are necessary to ensure that the water circuit can be closed without interfering with plant performance. Water softening agents may be used for separations sensitive to water chemistry.

In these complex circuits, efforts must be made to isolate water from each circuit and prevent cross-contamination.

BENCH-SCALE FLOTATION EQUIPMENT

This section discusses the equipment necessary for a bench-scale flotation program.

Flotation Cells

The laboratory procedures and equipment designed for conventional flotation have remained substantially the same over the past 50 years, consisting primarily of subaeration-type flotation machines with induced air delivery such as the Metso Denver D12 and the Wemco Mineral-Master 600, or with forced air delivery such as the Essa FTM100 and the Agitair LA-500. Some improvements have been made to laboratory equipment design; however, these have mainly been for ease of use.

Conditioning

Pulp conditioning can be done either in the flotation cell (common when grinding is involved and/or attrition scrubbing is not necessary; e.g., for igneous ores) or in a separate vessel such as an attrition cell or attrition scrubber or attrition-conditioning cell (e.g., for sedimentary ores, no significant grinding is involved). Grinding achieves (1) size reduction, (2) liberation, and (3) generation of fresh surfaces. Attrition-conditioning provides the necessary high-shear scrubbing intensity, with minimal size reduction, to remove surface coatings and expose fresh mineral surfaces for the reagents. Attrition cells can also break up aggregates and disperse minerals effectively. They typically operate at high solids loading with sufficient water to fluidize the solids, typically 50%–80% depending on the system and the pulp viscosity. Small amounts of slimes generated in this operation are often discarded prior to reagent

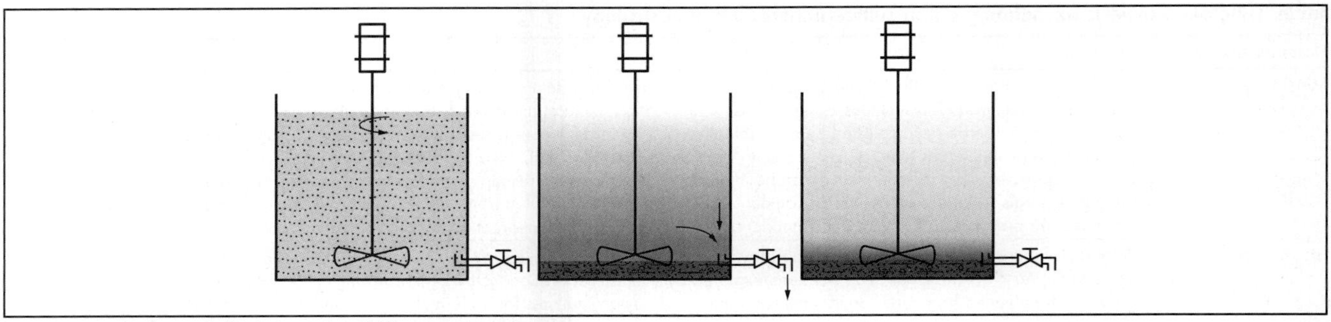

Figure 1 Laboratory desliming by differential settling

conditioning. This includes conditioning with a modifier and a collector. Modifiers could include a dispersant, a pH modifier, an activator or a depressant. Attrition cells can be used in both chemisorbing collector systems (such as fatty acids and hydroxamates) and physisorbing collector systems (such as amines and petroleum sulfonates). Some flotation machines are designed to be used as a separate conditioning device by changing the rotor, removing the stator(s), and using a cell of different geometry (e.g., cylindrical, hexagonal, or square cross-section). Separate laboratory conditioning machines can also be made from a standard drill press, a shaft impeller, and a cylindrical vessel (2 L is typical) mechanically held to the drill press table.

Desliming

Desliming by wet screening is a very common technique because it is easy and does not require special rigs. However, it is slow if material below ~38 μm (especially below 20 μm) has to be removed. The coarser fractions are first removed to protect fine screens and to reduce the amount of screening time. Consistency is hard to obtain with this method because of the varying techniques, water flow rates, and screening times that an operator might use. Differential settling is a straightforward and repeatable way to deslime a feed. It involves agitating a slurry containing slimes at approximately 5%–10% solids, allowing the coarse solids to settle for a set period, then siphoning or using a valve to drain the supernatant water containing slimes from the vessel (Figure 1). This can be repeated several times depending on the degree of desliming required. In-cell elutriation is often the quickest and easiest way of desliming a feed. It is performed by agitating the feed charge containing slimes in a float cell or attrition cell, then injecting water at a constant and consistent rate at the bottom of the float cell. Coarse particles settle and remain at the bottom, and the slimes leave the cell through the overflow launder. Dispersants, such as sodium silicate, sodium polyphosphate, or low-molecular-weight synthetic polymers, can be added to increase the effectiveness of all of these techniques.

Laboratory cycloning is an easy way of replicating many industrial desliming processes. An example of a bench-scale cyclone desliming unit is shown in Figure 2. The unit consists of a feed tank and a cyclone matched by flow characteristics to a feed pump. The slurry is fed into the feed tank, and the pump speed is adjusted to provide the optimum size cut. Once the optimum pump speed is established, the cyclone overflow is collected while water is added to the feed tank to maintain its level.

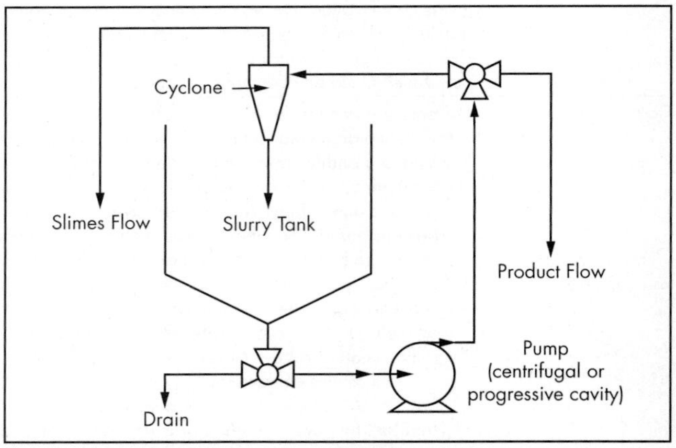

Figure 2 Bench-scale cyclone desliming unit

Bench-Scale Flotation Guidelines

Every ore presents its own challenges and requires a different series of steps and reagents to maximize metallurgical and economic performance. However, most bench-scale flotation experiments for non-sulfide minerals involve both conditioning and flotation. Basic considerations for laboratory testing are given in Table 3.

Flotation is typically carried out to completion. Most laboratory rougher flotation tests take less than 10 minutes for non-sulfide minerals. Scavenger and cleaner tests take approximately 60% of the time that a rougher flotation test takes. Table 4 shows optimum flotation times for laboratory rougher flotation of non-sulfide minerals with a comparison to plant scale.

LABORATORY FLOTATION TEST PROGRAMS

Design of Laboratory Tests

In designing a flotation test program, characteristics of the ore, empirical knowledge of related flotation separations, mineral product specifications, and economic considerations form the basis in the selection of reagent schemes and specific treatment conditions. As in any mineral processing operation, laboratory testing is also essential in overall risk assessment and in identifying potential problems that may be encountered when scaled up to the plant scale. This is well captured in Arthur Taggart's quote: "Make your mistakes on a small scale and make your profits on the large scale."

Rougher and cleaner or rougher and scavenger flotation tests are conducted to evaluate the major variables: grinding

Table 3 Cell size and flotation parameter considerations

Cell Volume, L	When to Use	Flotation Parameters	Examples of Use
0.5–1.5	• Finely liberated ores (<200 µm) • Limited availability of feed material (water and ore) • Valuable mineral constitutes more than 30% of the feed • Slurry solids 20%–25% typical but can range from 10% to 50% for certain applications (see literature for more details) • Scavenger and cleaner evaluation	• 200–500 g feed charge • 900–1,200 rpm • 3–5 L/min airflow	• Phosphate reverse flotation for carbonaceous ores • Finely liberated hematite • Scavengers and cleaners for most non-sulfide ores
2.5	• Most commonly used cell size • Feeds with top sizes <400 µm • Valuable mineral constitutes more than 10% of the feed • Slurry solids 20%–25% typical but can range from 10% to 50% for certain applications (see literature for more details) • Scavenger and cleaner evaluation	• 500–1,500 g feed charge • 1,000–1,400 rpm • 4–7 L/min airflow	• Rougher flotation of iron ores containing >30% Fe • Rougher flotation of phosphate ores containing carbonates and clays • Scavenger and cleaner flotation for coarse phosphate ores
3+	• When sufficient feed is available • Feeds with top sizes >400 µm • Valuable mineral constitutes <10% of the feed • Slurry solids 20%–25% typical but can range from 10% to 50% for certain applications (see literature for more details)	• 1–2 kg feed charge • 1,200–1,900 rpm • 5–10 L/min airflow	• Coarse sedimentary phosphate ores • On-site testing of any ore where ample supply of materials is available

Table 4 Laboratory rougher flotation times and plant-scale comparison

Mineral	Laboratory Flotation Time, minutes	Industrial Flotation Time, minutes
Barite	4–5	8–10
Calcite (from phosphate ore)	1–3	3–6
Coal	2–3	3–5
Feldspar	3–4	8–10
Fluorspar	4–5	8–10
Phosphate	1–3	4–6
Potash	2–3	4–6
Sand	3–4	7–9
Silica (from iron ore)	3–5	8–10
Silica (from phosphate ore)	2–3	4–6

Adapted from Kawatra 2011

stage factors such as fineness of grind, pulp density, type of grinding medium, and chemical additives; conditioning factors such as collectors, depressants, pH, time, intensity, and temperature; and flotation factors such as type of cell, collectors, frothers, depressants, pH, intensity of agitation, and time. Other variables may include desliming, magnetic separation, washing, and in continuous or locked-cycle tests, circulating middlings and water in grinding and flotation.

A laboratory test program can range from a simple flotation optimization for an existing operation to a full greenfield process development. Laboratory flotation testing should be considered in two stages: diagnostic and optimization.

Diagnostic Testing Stage

Diagnostic testing has three important objectives:

1. Determine the viability of flotation separation for the particular ore.
2. Determine the suitability of the ore sample for laboratory flotation testing.
3. Establish a benchmark laboratory flotation procedure.

Diagnostic testing should make extensive use of Six Sigma quality management principles such as "design of experiments." Because of the extensive number of factors that must be evaluated and the typical situation involving lack of enough material to perform a full factorial designed experiment, other types of two-level screening designs that can handle a large number of factors in a small number of tests are typically employed (Cytec 2010). Such designed tests provide valuable information on the main effects of variables (parameters) and serve to identify conditions for good performance and conditions to avoid. At least two replicates are recommended in these designed tests. The following are often the most commonly used parameters (there may be other factors that may be important in specific cases):

- Particle size distribution
- Desliming
- Ore type/mineralogy
- pH
- Collector type
- Collector dosage
- Collector addition point
- Frother type
- Frother dosage
- Frother addition point
- Modifier type
- Modifier dosage
- Modifier addition point
- Water chemistry
- Airflow rate
- Impeller speed
- Machine type
- Cell, impeller, and stator size
- Pulp level
- Pulp density in flotation

Many of these parameters may already be known, especially in the case of existing operation optimizations. Once the important parameters have been determined and a benchmark laboratory procedure has been established, the flotation conditions and reagents can be further optimized with a more focused laboratory test program. Appropriately designed experiments are required to generate statistically meaningful outcomes. During the design phase, the preliminary main and second-order interaction effects are determined with partial factorial experimental designs. Statistical analysis software packages (Minitab; JMP) are available to assist in the design and interpretation of results (Napier-Munn 2014; Cytec 2010).

Optimization Stage

The most impactful variables identified in the diagnostic-stage test work are the focus of the optimization stage. Further designed experiments using more extensive partial factorial or full factorial designed experiments with multiple levels are used to determine effects and nonlinear responses to process changes. This process is discussed further in other chapters of this handbook.

Rougher-Scavenger-Cleaner Laboratory Testing

Details of types of circuits investigated in the laboratory are given in Table 5. Rougher-scavenger, rougher-cleaner, and multistage flotation circuit designs are very common in non-sulfide systems. Their replication in the laboratory is not as simple as a rougher circuit alone, especially when the circuits involve multiple stages and presence of non-flotation separations such as magnetic or gravity separation. A guide on how to perform calculations commonly used in rougher-scavenger-cleaner flotation circuits can be found elsewhere in the literature (Kawatra 2011).

Bench-Scale Testing

Bench-scale testing methods (rougher, cleaner-recleaner, locked-cycle, kinetic) are described in Chapter 7.9, "Applied Flotation Modeling." Locked-cycle testing is a means of simulating plant conditions by recycling water and certain streams

Table 5 Laboratory investigated circuit types

Type of Circuit	Features and Use	Examples
Rougher-scavenger	• In direct flotation when rougher is capable of reaching the target grade, but the recovery is not optimal; the opposite in reverse flotation. • If the rougher float is a direct float, the tails (sinks) can require additional collector, frother, and modifier as well as further conditioning or grinding prior to scavenging. • If the rougher float is a reverse float, residual flotation reagents in the tails (floats) are, in many cases, sufficient for the scavenger float. • Maintain the pH between the rougher and scavenger. • Retaining the water from the rougher concentrate and using it in the scavenger float cell can give some indication to the effects of process water and residual reagents. • Size of the scavenger flotation cell can be reduced because of the low amount of solids in the rougher tails (feed to scavenger).	• Reverse flotation of silicate gangue from iron ores; rougher flotation with amine followed by single- or multistage scavenging
Rougher-cleaner	• Used when rougher is capable of reaching an acceptable recovery, but the grade is not optimal, because of complex ores or ores of wider size distributions. • If the rougher float is a reverse float, the concentrate (sinks) may require additional collector, frother, and modifier as well as further conditioning or grinding prior to scavenging. • If the rougher float is a direct float, residual flotation reagents in the concentrate (floats) are, in many cases, sufficient for the cleaner float. • Maintain the pH between the rougher and cleaner. • Retaining the water from the rougher tails and using it in the cleaner float cell can give some indication to the effects of process water and residual reagents.	• Barite flotation from oxide gangue; barite is floated using sulfate functionalized surfactants, and multiple cleaner floats are required to reach the desired grade.
Multistage flotation circuits	• Many flotation processes used for non-sulfide minerals are developed for complex ores in which a single flotation step may not be sufficient to provide the grade required. • Replication of a double-float or multistage flotation in the laboratory is more difficult. For example, de-oiling in the laboratory is typically performed by (1) conditioning the rougher concentrate at high solids in a separate cell with an acid, base, bleach, surfactant, or solvent; (2) allowing the solids a certain amount of time under agitation for the reagents to be scrubbed off the mineral surfaces; (3) washing over a wet sieve for a set amount of time with a set water flow rate to ensure reproducibility between tests; and (4) conditioning the scrubbed solids under new flotation conditions prior to reverse flotation. • Water in each stage will be segregated because different reagent schemes and pH are used.	• Double-float or Crago process (Crago 1942) for sedimentary phosphate ores; direct flotation of phosphate minerals with a fatty acid, followed by de-oiling of phosphate concentrate and reverse flotation of silica with an amine • Five-stage spodumene–mica–feldspar–quartz float previously used at Kings Mountain, North Carolina, United States (Redeker 1981; Kawatra and Carlson 2013)
Flow sheets with combination of flotation and non-flotation separation stages	• Most industrial flotation circuits have processing steps before or in between flotation stages. • Desliming is used with fine feeds to remove slime material prior to flotation. • Scrubbing is performed between flotation stages at some plants where multistage flotation is required. • Grinding can be performed between rougher and scavenger floats to liberate coarse middlings. • Size separation can be performed between flotation stages to remove coarse middlings particles via screens or cyclones.	• Magnetite beneficiation uses magnetic separators to reject liberated silicates prior to flotation. • Many carbonaceous phosphate ores are deslimed prior to flotation. • Gravity separation in many circuits (e.g., beryl, rare earths).

from test to test. For non-sulfide ores, this type of testing is especially important for ores containing minerals with appreciable solubility (e.g., carbonates, sulfates, phosphates). However, locked-cycle testing is cumbersome when the flow sheet involves multiple stages of flotation and/or non-flotation separations such as magnetic separation. Each stage may have to be isolated and subjected to locked-cycle testing. Pilot-plant testing is typically a far better option than locked-cycle testing for non-sulfide minerals, especially when complex flow sheets are involved.

PILOT-PLANT FLOTATION OF NON-SULFIDE MINERALS

Flow sheets for non-sulfide ore beneficiation tend to be more complex than those of sulfide ore systems, and simulating the whole flow sheet in batch laboratory flotation or locked-cycle testing is cumbersome. Pilot-plant testing is a far better option.

The sequence of testing is shown in Figure 3. If the ore sample or ore deposit under investigation is from, or similar to, an operating plant, then batch laboratory testing may be adequate, and pilot-plant testing can be bypassed. However, new flotation processes or those that have no close counterpart in current operating plants usually require piloting. A decision to have or not have a demonstration flotation plant should be based on the need to resolve potential process risks and/or unknowns in the proposed flow sheet, generate sufficient flotation concentrate to demonstrate downstream hydrometallurgical or pyrometallurgical processes, or evaluate mineral product end use.

Evaluation of Pilot-Plant Testing

The actual pilot-plant program is highly dependent on sample availability, mineralogy, metallurgical parameters, and flow-sheet considerations, as well as the budget.

Flotation Reagents for Pilot Testing

All reagents to be used during the pilot test should be prepared to the required strengths prior to commencement. Reagent preparation guidelines are available from the suppliers and in the published literature. Water-soluble reagents are added as aqueous solutions by metering pumps. For reagents that are insoluble or only sparingly soluble in water, special techniques are necessary, such as dissolving them in an organic solvent or emulsifying the reagent in water by mixing it with a suitable surfactant. In many instances, the insoluble reagent may exist as a liquid dispersion or a colloid. Accurate addition of reagents is necessary to quantify their effect on pilot metallurgical performance. With knowledge of feed rate of dry solids and the strength of the prepared reagents, the addition of the reagent can be calculated.

Equipment and Operations

Bench-scale parameters can be scaled up to the pilot scale, typically using a retention time required for the desired separation. The pilot-plant feed rates are usually guided by cell volumes and scaled retention times (Table 4). The slurry density guides the reagent addition rates. Typically, a flotation pilot plant is operated for several hours with minor reagent adjustments to obtain the steady-state conditions. The circuit is then operated for a few hours prior to sampling the circuit under these steady-state conditions. This usually lasts 8 hours to five days and is a continuous operation. Pilot flotation testing can be automated by using advanced control techniques to minimize experimental error.

Data Collection and Analysis

The pilot circuit is sampled over several hours to produce a composite sample of feed, concentrate, and tailings. Three to five composites are collected under steady-state conditions to

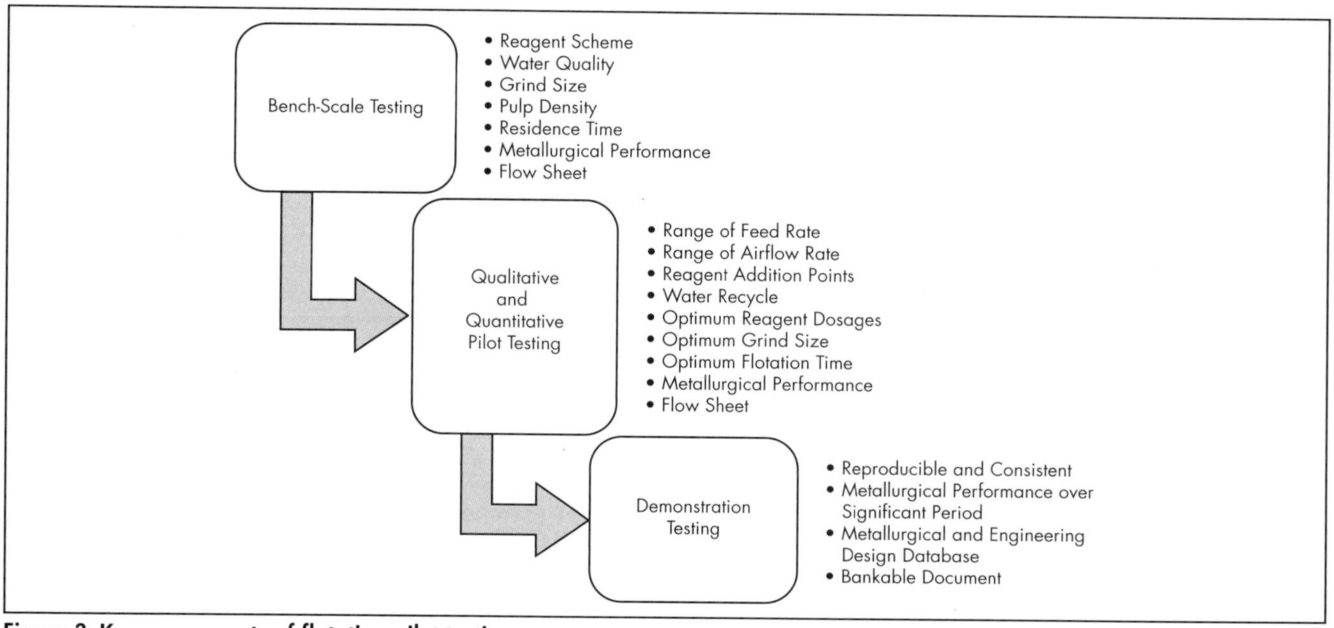

Figure 3 Key components of flotation pilot testing

Table 6 Summary of flotation development techniques

	Bench-Scale Testing	Locked-Cycle Testing	Pilot-Plant Testing
Objectives	• To evaluate major variables such as – Mineralogical variation – Head grade variation – Liberation and grind size requirements – Reagent types, quantities, conditioning time, and addition points – Concentrate regrinding requirements – Pulp density and viscosity effects – Water quality – Flow-sheet alternatives • To define the kinetics and separation response • To define valuable mineral recovery and grade of the saleable concentrate	• To determine the effect of recirculating material from cleaner circuit on recovery • To evaluate the variation in reagent dosages to compensate for the recycling of reagents, water, soluble metal species • To study the effect of the build-up of slimes or middlings in the cleaning stages of the test • To evaluate froth handling problems	• To confirm metallurgical response for continuous operation • To generate sufficient concentrate and tailings samples for downstream processes for end-use study • To obtain operating data for use in engineering design and scale-up purposes
Sample source	Existing operations, drill-core composite samples, variability samples	Existing operations, drill-core composite samples, variability samples	Bulk samples
Sample requirement	200–2,000 g per flotation test	10–50 kg per locked-cycle test	5–1,000 t
Flotation equipment	0.5–5 L mechanically agitated flotation machine	0.5–5 L mechanically agitated flotation machine	Forced air, self-aspirated, column, or shear contact pilot flotation cell banks
Limitations	• Does not allow metallurgical evaluation under conditions of continuous operation. • Does not allow evaluation of recycling of reagents or middlings. • The laboratory flotation feed preparation process (grinding and conditioning) produces different size distributions and chemical environments compared to an operating plant. • Difficult to simulate the whole flow sheet; difficult to simulate certain unit operations such as flash flotation, column flotation.	• Tests are time consuming, so it may not be possible to complete a test in one working day. • Requires rapid reporting of analytical results. • Aging of the intermediate products. • Cumbersome when the flow sheet is complex and contains non-flotation separations.	• Requires a large amount of representative and homogeneous feed sample. • High costs are involved in generating large quantities of sample for pilot-plant testing. • Requires storage and handling of tailings.

Note: Information in this table is from Barbery et al. 1986; Cameron and Dunlop 1990; Cytec 2010; Gochin and Smith 1987; Macdonald and Brison 1962; Williams et al. 2002; Wills and Napier-Munn 2006; and authors' experience.

estimate the standard deviation in the results. The data collected from a pilot-plant testing can be considerable, ranging from mass and water recoveries, to assay and mineralogical data, to pulp chemistry measurements. For pilot-plant control, pulp densities, plant feed rate, feed sizing, and reagent flows are regularly monitored. Concurrent with the pilot sampling, additional bench flotation tests may be performed on samples taken at the specified sample points in the flow sheet as a control; this is considered best practice.

Reporting

A typical pilot-plant report contains a comprehensive analysis and interpretation on circuits most applicable for treatment of the ore body, some design data and marketable grades, and recoveries possible. If required, the report can be produced to bankable document standards and include engineering design data (Cameron and Dunlop 1990; Williams et al. 2002).

SUMMARY

There are many factors to consider when developing flotation technology for non-sulfide minerals. A summary of the different techniques used for non-sulfide flotation development is shown in Table 6. Generally, the steps required to design a flotation circuit for non-sulfide minerals are as follows:

1. Sampling of the feed from drill cores or an existing operation
2. Mineralogical characterization of the feed
3. Liberation analysis of the feed
4. Grinding study if not adequately liberated
5. Bench-scale flotation testing
6. Locked-cycle testing
7. Pilot-plant testing
8. Full plant design implementation and evaluation

REFERENCES

Aplan, F.F. 1994. Reagents for the flotation of non-sulfide ores In *Reagents for Better Metallurgy*. Edited by P.S. Mulukutla. Littleton, CO: SME. pp. 103–117.

Aplan, F.F., and Fuerstenau, D.W. 1962. Principles of non-metallic mineral flotation. In *Froth Flotation: 50th Anniversary Volume*. Edited by D.W. Fuerstenau. New York: AIME. pp. 170–214.

Arbiter, N., and Williams, E.K.C. 1980. Conditioning in oleic acid flotation. In *Fine Particles Processing: Proceedings of the International Symposium on Fine Particles Processing*. Edited by P. Somasundaran. Littleton, CO: SME-AIME. pp. 802–831.

Baarson, R.E., Ray, C.L., and Treweek, H.B. 1962. Plant practice in nonmetallic mineral flotation. In *Froth Flotation: 50th Anniversary Volume*. Edited by D.W. Fuerstenau. New York: AIME. pp. 427–454.

Barbery, G., Bourassa, M., and Maachar, A. 1986. Laboratory testing for flotation circuit design. In *Design and Installation of Concentration and Dewatering Circuits*. Edited by A.L. Mular and M.A. Anderson. Littleton, CO: SME. pp. 419–432.

Cameron, P., and Dunlop, G. 1990. Pilot plant testing in mineral process plant design. In *Mining Industry Capital and Operating Cost Estimation Conference*. Melbourne, Victoria: Australasian Institute of Mining and Metallurgy. pp. 29–34.

Coghill, W.H., and Clemmer, J.B. 1935. Soap flotation of the nonsulfides. *Trans. AIME* 449–465.

Crago, A. 1942. Process of concentrating phosphate minerals. U.S. Patent 2,293,640.

Cytec. 2010. *Mining Chemicals Handbook*. Woodland Park, NJ: Cytec Industries.

Davis, F.T. 1985. Nonmetallic industrial minerals. In *SME Mineral Processing Handbook*. Edited by N.L. Weiss. Littleton, CO: SME-AIME.

Dunne, R.C., Lane, G.S., Richmond, G.D., and Dioses, J. 2013. Flotation data for the design of process plants: Part 1—testing and design procedures. *Miner. Process. Extr. Metall.* 199–204.

Falconer, S.A. 1961. Minerals beneficiation—Non-sulfide flotation with fatty acid and petroleum sulfonate promoters. *Trans. AIME* 217:207–215.

Falconer, S.A., and Crawford, B.D. 1947. Froth flotation of some nonsulphide minerals of strategic importance. *Trans. AIME* 527–542.

Fuerstenau, D.W., ed. 1962. *Froth Flotation: 50th Anniversary Volume*. New York: AIME.

Fuerstenau, D.W. 2007. A century of developments in the chemistry of flotation processing. In *Froth Flotation: A Century of Innovation*. Edited by M.C. Fuerstenau, G. Jameson, and R-H. Yoon. Littleton, CO: SME. pp. 3–64.

Fuerstenau, M.C., Jameson, G., and Yoon, R.-H. 2007. *Froth Flotation: A Century of Innovation*. Littleton, CO: SME.

Gaudin, A.M., and Glover, H. 1928. Flotation of some oxide and silicate minerals. Technical Paper No. 1. Salt Lake City: Department Mining and Metallurgical Research, University of Utah. pp. 78–101.

Gochin R.J., Smith M.R. 1987. The methodology of froth flotation testwork. In *NATO Science Series E: Applied Sciences*. Vol. 122, Mineral Processing Design. Edited by B. Yarar and Z.M. Dogan. Dordrecht, Germany: Springer.

Holme, R.N. 1986. Sulfonate-type flotation reagents. In *Chemical Reagents in the Mineral Processing Industry*. Edited by D. Malhotra and W. Riggs. Littleton, CO: SME. pp. 99–111.

Houot, R. 1983. Beneficiation of iron ore by flotation: Review of industrial and potential applications. *Int. J. Miner. Process.* 10:183–204.

Jones, M.J., and Oblatt, R., eds. 1984. *Reagents in the Minerals Industry*. London: Institution of Mining and Metallurgy.

Kawatra, S.K. 2011. Fundamental principles of froth flotation. In *SME Mining Engineering Handbook*, 3rd ed. Edited by P. Darling. Englewood, CO: SME. pp. 1517–1531.

Kawatra, S.K., and Carlson, J.T. 2013. *Beneficiation of Phosphate Ore*. Englewood, CO: SME.

Macdonald, R., and Brison, R. 1962. Applied research in flotation. In *Froth Flotation: 50th Anniversary Volume*. Edited by D.W. Fuerstenau. New York: AIME. pp. 298–327.

Malhotra, D., and Riggs, W.F. 1986. *Chemical Reagents in the Mineral Processing Industry*. Littleton, CO: SME.

Manser, R.M. 1975. *Handbook of Silica Flotation*. United Kingdom: DOB Services (Hitchin).

Miller, J.D., Tippin, B., and Pruett, R. 2002. Nonsulfide flotation technology and plant practice. In *Mineral Processing Plant Design, Practice, and Control*. Edited by A.L. Mular, D.N. Halbe, and D.J. Barratt. Littleton, CO: SME. pp. 1159–1178.

Miller, J.D, Abdel Khalek, N., Basilio, C., El-Shall, K., Fa, K., Forssberg, K.S.E., Fuerstenau, M.C., Mathur, S., Nalaskowski, J., Rao, K.H., Somasundaran, P., Wang, X., and Zhang, P. 2007. Flotation chemistry and technology of nonsulfide minerals. In *Froth Flotation: A Century of Innovation*. Edited by M.C. Fuerstenau, G. Jameson, and R-H. Yoon. Littleton, CO: SME. pp. 465–553.

Mulukutla, P.S., ed. 1994. *Reagents for Better Metallurgy*. Littleton, CO: SME.

Nagaraj, D., and Farinato, R. 2016. Evolution of flotation chemistry and chemicals: A century of innovations and the lingering challenges. *Miner. Eng.* 96-97:2–14.

Nagaraj, D., and Ravishankar, S. 2007. Flotation reagents—A critical overview from an industry perspective. In *Froth Flotation: A Century of Innovation*. Edited by M.C. Fuerstenau, G. Jameson, and R-H. Yoon. Littleton, CO: SME.

Napier-Munn, T. 2014. *Statistical Methods for Mineral Engineers: How to Design Experiments and Analyse Data*. Indooroopilly, Queensland: Julius Kruttschnitt Mineral Research Centre.

NC State University. 2018. Minerals Research Laboratory: Minerals research lab reports. https://mrl.ies.ncsu.edu/minerals-research-lab-reports/. Accessed August 2018.

Peres, A.E.C., Araujo, A.C., El-Shall, H., Abdel-Khalek, N.A. 2007. Plant practice: Nonsulfide minerals. In *Froth Flotation: A Century of Innovation*. Edited by M.C. Fuerstenau, G. Jameson, and R-H. Yoon. Littleton, CO: SME.

Redeker, I. 1981. Flotation of feldspar, spodumene, quartz and mica from pegmatites in North Carolina, USA. CMP Preprint No. 81-1-P. Ottawa, ON: Canadian Mineral Processors.

Redeker, I., and Bentzen, E. 1986. Plant and laboratory practice in nonmetallic mineral flotation. In *Chemical Reagents in the Mineral Processing Industry*. Edited by D. Malhotra and W. Riggs. Littleton, CO: SME. pp. 3–20.

Siirak, J., and Hancock, B.A. 1988. Progress in developing a flotation phosphorous reduction process at the Tilden iron ore mine. In *XVI International Mineral Processing Congress*. Amsterdam: Elsevier. pp. 1393–1404.

Somasundaran, P., and Moudgil, B.M., eds. 1987. *Reagents in Mineral Technology*. Surfactant Science Series 27. New York: Taylor and Francis.

Spedden, H.R. 1985. Sampling and testing. In *SME Mineral Processing Handbook*. Edited by N.L. Weiss. Littleton, CO: SME-AIME.

Williams, S., Ounpuu, M., and Sarbutt, K. 2002. *Bench and Pilot Plant Programs for Flotation Circuit Design*. Technical Paper No. 2002-04. Rutherford, NJ: SGS Mineral Services.

Wills, B.A., and Napier-Munn, T.J. 2006. Froth flotation. In *Wills' Mineral Processing Technology: An Introduction to the Practical Aspects of Ore Treatment*, 7th ed. Oxford: Elsevier Science and Technology Books. pp. 267–352.

CHAPTER 7.9

Applied Flotation Modeling

Peter Amelunxen and Richard LaDouceur

Froth flotation models are useful for many purposes, and each purpose often requires a unique set of model characteristics. As such, different modeling approaches have evolved to fit different purposes or objectives, and sometimes, they are not interchangeable. For example, a simulator developed to populate a block model with plant performance parameters will not be very useful for helping a cell manufacturer develop a more efficient cell mechanism. A dynamic flotation model, developed to predict flow surges and improve control loop response in real time, is not likely to be the primary tool for flow sheet development, which requires detailed ore characterization and some kind of economic optimization strategy. Given the diversity of modeling approaches, it was decided that this chapter, being part of an engineering handbook, should focus on the most common *applied* flotation models and their components, that is, those used for steady-state simulation, flotation circuit design, process optimization, and geometallurgy. Where appropriate, the reader is referred to sources covering alternative applications.

BACKGROUND

The practice of mathematical modeling far predates the development of the flotation process, and therefore, in many respects, the history of flotation modeling is a history of flotation itself. P.A. Amelunxen and Runge (2014) trace that history through the past 100 years or so, identifying several key developments in the field. Some were spawned by developments in our own field of extractive metallurgy; for example, the adoption of compartmental models was probably driven, in large part, by the commercial implementation of column flotation in the 1980s. Other milestones were achieved through advances outside mineral processing. Probably the single most important event in the history of flotation modeling is the widespread adoption of personal computers during the early 1990s, an event that gave researchers access to new tools to study flotation machines and gave plant designers and operators access to more advanced flotation simulation tools. Although it is true that many of the components of flotation models have been known for much longer (e.g., first-order kinetics models, hydraulic entrainment, and froth recovery have been recognized since at least the early 1960s), the packaging of these concepts into easily accessible and reliable simulation software is relatively recent.

When reading this chapter, we encourage the reader to remember that there is no magic involved in modeling froth flotation. Anybody can build a flotation model or simulator using a laptop, a spreadsheet, and a simple scripting or macro language (to handle iterations and circulating load calculations).

OBJECTIVES AND VALUE OF FLOTATION MODELS

There can be multiple objectives for flotation modeling and simulation. As with any engineering discipline, the objectives are focused on value creation:

- **Process design** reduces the cost and time requirements or decreases the risks associated with flotation process design, engineering, and/or construction projects.
- **Process optimization** increases the recovery of valuable minerals, improves the separation, reduces the costs, or increases the revenues of an existing flotation process or system.
- **Process control** improves overall circuit performance through online process control and optimization, usually as part of a broader advanced process control strategy and system (Shean and Cilliers 2011).
- **Geometallurgy** evaluates the susceptibility of a flotation circuit to changing ore characteristics, such as feed grades, mineralogy, liberation, and other parameters, and determines how this variability affects the economic return of a given flotation process or system.

There can be little doubt that flotation simulators over the past two decades have contributed significantly to the field of mineral processing (Lane et al. 2005). In the authors' experience, this is evidenced by the employment of simulators in the design and optimization of virtually every large, contemporary mineral flotation process.

Peter Amelunxen, President, Aminpro, Lima, Peru
Richard LaDouceur, Postdoctoral Research Scholar, Montana Tech, Butte, Montana, USA

THE FLOTATION SYSTEM

A froth flotation cell can be described as an open system because there is mass and energy exchange with the environment. The inputs include solids (feed), water, reagents, energy, and gas. The outputs include solids (concentrate and tails), water, energy (heat, noise), reagents, and gas. The system is also subject to atmospheric pressure, the magnitude of which depends on the design of the cell (pressurized, nonpressurized) and the elevation above sea level of the system. The system has several components, including a pulp zone, froth zone, and pulp–froth interface zone, and three phases (solids, water, and gas). For modeling purposes, de-inking, de-oiling, and other pseudo-liquid recovery applications are considered to be three-phase systems.

The system has several mass and energy transfer mechanisms as depicted in Figure 1, and some of these are interrelated and dynamic by nature. It is the purpose of a flotation model to provide a simplified mathematical framework to describe this complex system such that the objectives listed earlier can be met. The following "Flotation Machine Models" section describes the commonly used mathematical framework used to represent flotation systems.

Figure 1 clearly shows that the flotation system is complex, and even today, some of the underlying mass and energy transfer phenomena can be neither completely characterized nor fully understood. These gaps in our knowledge must be filled with either an assumption or a simplification, and, as a result, all flotation models and simulators contain some element of empiricism. This becomes extremely important when performing modeling and scale-up from laboratory flotation tests. It is precisely how these small elements of empiricism are handled that drives the confidence in the resulting model or scale-up calculation, and it is for this reason that one cannot discuss flotation modeling without also discussing laboratory test methodologies. This is presented later in the "Laboratory Testing" section of this chapter.

It is important to distinguish between a flotation model and a flotation simulator. A flotation model is a mathematical framework, while a flotation simulator is a software tool that contains models. Once the modeling framework and the laboratory test procedures have been determined, a flotation simulator can be chosen or constructed that has the necessary functionality for the job at hand. This functionality and some commonly available simulation approaches are described later in the "Simulators" section.

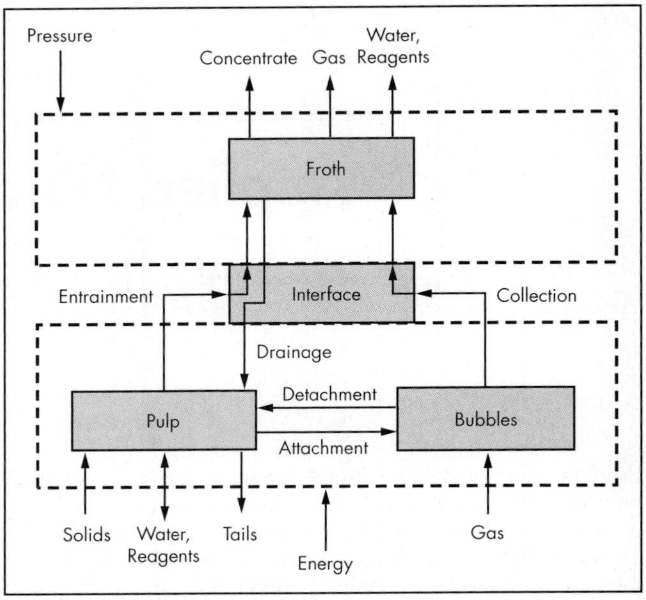

Adapted from P. Amelunxen et al. 2018

Figure 1 The flotation system

FLOTATION MACHINE MODELS

Mathematical models of continuous flotation machines can be divided into three classes: empirical models, phenomenological models, and physics-based models. Empirical models are relatively simple models that rely on an individual's experience or a set of industrial benchmark data to convert laboratory test results into industrial plant performance estimates. Phenomenological models seek to mathematically model the underlying rate processes or mass transfer using data gathered at the laboratory scale to predict industrial plant performance. Physics-based models seek to replicate the fundamental particle/fluid dynamics of the system by discretizing it into small individual elements, calculating the mass and energy balance for each element, and summing the elements to determine the performance of the whole. The method relies on a rigorous understanding of the interparticle and multi-fluid systems (a combination of discrete element methods and computational fluid dynamics), an appropriate programming shell to code that understanding, and sufficient computational processing power to simulate it. These models are often proprietary and are not used for designing circuits (yet). They are not discussed in this chapter.

Empirical Models

Empirical models originated during the 1950s, when test procedures focused on determining the flotation kinetics mainly for those elements that would generate value for the project; little was done to characterize the floatability of other minerals, such as pyrite and gangue, which provided no income. Because an accepted method for scaling up the results from a laboratory kinetics test to commercial level did not yet exist, many engineers developed their own methods from personal experience. One of the most common was the use of flotation time factors. For example, Lindgren and Broman (1976) wrote that the plant flotation time should be estimated by applying a factor of two to the laboratory time at the point where the cumulative recovery leveled off, while others recommend factors between 1.6 and 2.6 (Dunne et al. 2002). Others define the point at which the recovery levels off as the point at which the incremental concentrate grade drops below the feed grade.

There are several drawbacks to these kinds of approaches. One drawback is that extensive experience and benchmarking is required with the particular flotation application under design. As a result, different engineers, with different backgrounds and data, will arrive at different sets of design criteria. For smaller, simpler applications, this may not be a significant obstacle, because the engineer or project owner can correct for risk through engineering margins and safety factors—usually by adding some extra flotation time at the end of the bank, increasing pump and launder capacities, and so forth. However, for modern, low-grade, mass mining projects, the capital cost of these design margins and safety factors can be quite substantial, and the cost of under-designing the flotation circuit even more so.

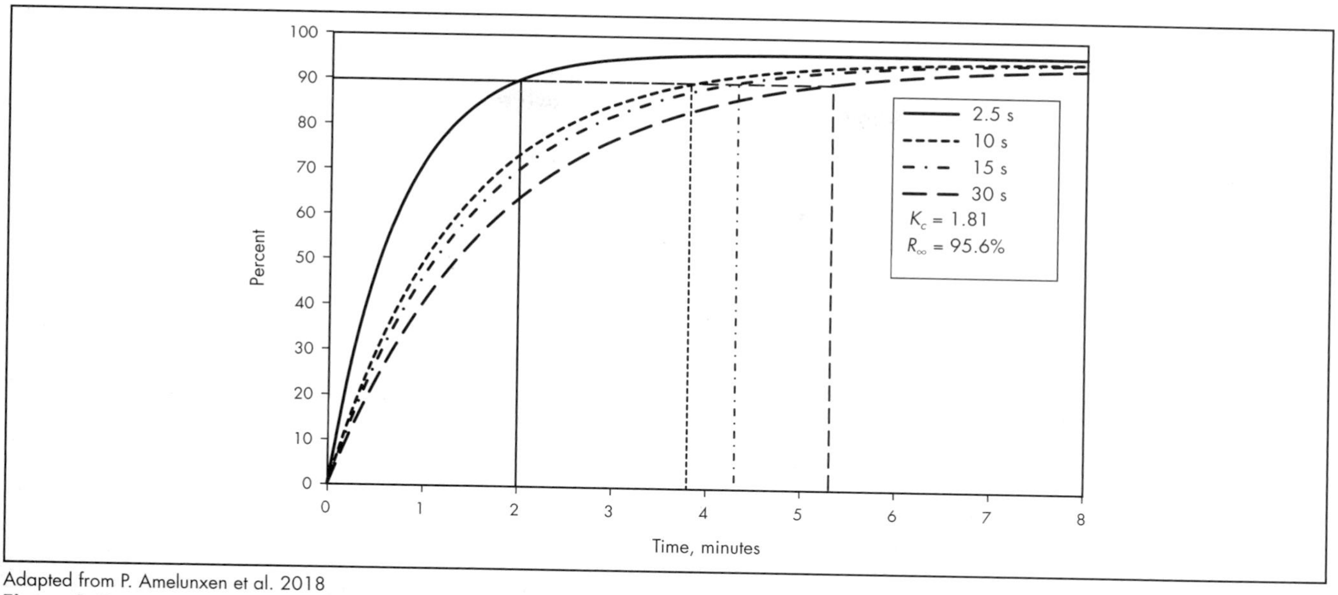

Adapted from P. Amelunxen et al. 2018

Figure 2 Recovery versus time for four scraping rates

Another drawback is that because each laboratory technician might scrape the froth at his or her characteristic speed, reproducibility of the test results is often relatively poor. Differences in water quality, froth and frother characteristics, and air flow rates also contribute to error. This can be seen in Figure 2, which shows the experimental recovery versus time curve for a copper mineral, for four scraping rates (2.5, 10, 15, and 30 seconds). Each curve is the average of triplicate tests (P.A. Amelunxen et al. 2014). Assuming a 2.5× scale-up factor for these conditions, the resulting continuous industrial bank retention times would be, respectively, 5, 9.5, 10.6, and 13 minutes.

It is mainly because of these drawbacks, combined with better understanding and accessibility of phenomenological models, that empirical models are usually only used for scoping or cursory level testing, or for applications that require low costs, such as geometallurgy.

Phenomenological Models

Because of the disadvantages of empirical models as discussed previously, large-scale flotation projects increasingly rely on phenomenological modeling methods for process design and optimization. The key components of phenomenological models are described in this section.

Mechanical Flotation Cells

Mechanical flotation machines are modeled using the so-called compartmental model of flotation, in which the various mass and energy transfer mechanisms shown in Figure 1 are simplified into two separate compartments or submodels; these are shown by the dashed line in Figure 1, and are

- The *collection zone*, where mineral is collected onto bubbles in the pulp, and
- The *froth zone*, where the concentrate is transported to the lip of the cell.

Figure 3 shows the graphical depiction of the compartmental model, including the relevant mass transfer vectors.

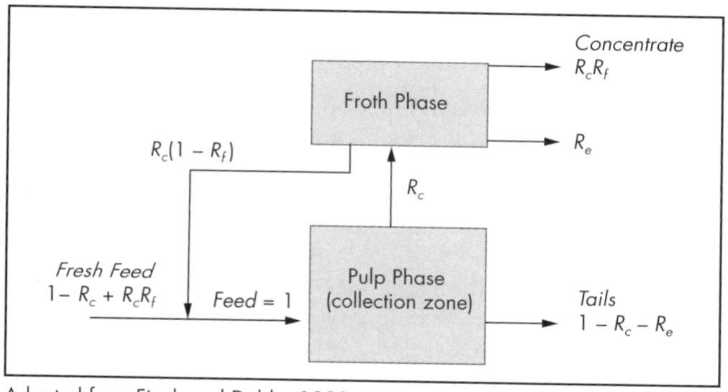

Adapted from Finch and Dobby 1990

Figure 3 Compartmental model (entrainment not shown)

The total recovery of a given mineral class, R, is the sum of the material recovered by entrainment and by collection, minus drainage.

If the fractional froth recovery, R_f, is defined as 1 minus the drainage, we can derive the so-called compartmental model of flotation (a better term is probably *dual-compartment model*). This is given by the following equation (Falutsu and Dobby 1989):

$$R = \frac{R_c R_f R_\infty}{R_c R_f + 1 - R_c} + R_e \quad \textbf{(EQ 1)}$$

where R_c is the collection recovery, R_∞ is the maximum recovery at infinite time (often defined by mineral size or liberation effects), and R_e is the entrainment recovery.

Collection zone. The collection recovery, R_c, of a given mineral particle class is usually expressed as a first-order rate process relative to time. For plug flow and continuous flow, the equations are, respectively,

$$R_c = 1 - e^{-kt} \quad \textbf{(EQ 2a)}$$

$$R_c = \frac{k\tau}{1 + k\tau} \quad \textbf{(EQ 2b)}$$

where t is the residence time in a plug flow reactor, τ is the mean residence time in a continuous perfectly mixed reactor, and k is the overall pseudo first-order rate constant of a given hydrophobic particle class, with respect to time (Garcia-Zuñiga 1935; Arbiter and Harris 1962; Dorenfeld 1962; Levenspiel 1962).

Once the cumulative recovery versus time profile is derived from the laboratory batch kinetics test, the first-order kinetic rate constant can be calculated independently for each mineral using curve-fitting techniques. There are three common methods for doing this. The most common approach is simply to apply Equation 2a directly to the time versus recovery profile, in which case, it is implied that entrainment in a laboratory test is zero while froth recovery is 100%. A second approach is to account for froth recovery and entrainment by combining Equation 2a with Equation 1 and solving for k, R_∞, R_f, and R_e for all minerals simultaneously using curve-fitting techniques (P.A. Amelunxen and Amelunxen 2009). A third approach involves calculating R_f and R_e from hindered particle settling theory and a water balance around the froth zone, and then solving for k and R_f independently for each mineral (P. Amelunxen et al. 2018).

Distributed rate constants. Equations 2a and 2b describe the classical approach, in which each mineral or element is assigned a single rate constant. It has been shown that the value of k for a given system is a function of the key parameters that define that system:

- Attachment and detachment rates (Schuhmann 1942)
- Feed characteristics, such as particle size (Gaudin et al. 1942; Morris 1952)
- Mineralogy (including particle oxidation), mineral association, and mineral surface expression (Imaizumi and Inoue 1965)
- Operating conditions such as reagent dosage or reagent surface coverage (Schuhmann 1942)
- Machine-dependent characteristics such as bubble saturation or impeller energy input (Sutherland 1948)

From a philosophical perspective, each individual particle would deserve its own rate constant and maximum recovery, depending on its size, shape, density, mineral constituents, hydrophobic surface expression, and hydrophobicity. This becomes mathematically unwieldy, and researchers have sought to reduce the complexity by expressing the distribution of particle species in a simplified form. The species distribution could be assumed to be discrete or continuous. Some proposed forms of discrete distributions include

- Single rate constant methods (classical) (Garcia-Zuñiga 1935; Thorne et al. 1965),
- Dual rate constant methods (Klimpel 1980), and
- Particle-size-based rate constants (Jowett 1974).

Some proposed forms of continuous distributions include

- Rectangular distribution (Huber-Panu et al. 1976; Klimpel 1980),
- Triangular distribution (Harris and Chakravarti 1970),
- Normal distribution (Chander and Polat 1994),
- Sinusoidal distribution (Diao et al. 1992),
- Weibull distribution (Dobby and Savassi 2005),
- Particle-size-based distribution (Morris 1952; Ofori et al. 2014; P.A. Amelunxen and Amelunxen 2009), and
- Gamma distribution (Imaizumi and Inoue 1965; Yianatos et al. 2010a).

The strengths and weaknesses of each of the preceding approaches have been discussed at length in the literature (Dowling et al. 1985; Polat and Chander 2000); this discussion is not repeated here. It is worth noting, however, that from the preceding list, there are three more common industrially and commercially applied methods:

1. The dual rate constant method, which forms the basis of JKTech's SimFloat simulator (Runge 2007) among others;
2. The Weibull distribution, which forms the basis of SGS's integrated geometallurgical simulation (iGS) flotation simulator (Dobby and Savassi 2005) among others; and
3. The particle size-based method, which forms the basis of Aminpro's AminFloat simulator (R.L. Amelunxen and Amelunxen 2009) among others.

Machine-dependent parameters. It has long been recognized that the characteristics of the flotation machine influence the value of the rate constant. Some practitioners vary the rate constant with bubble surface area flux (S_b). Bubble surface area flux is the superficial rise rate of bubble surface area per cross-sectional area. For a given flotation machine, this parameter can be directly related to flotation cell operating parameters and drives flotation rate (Gorain et al. 1998b):

$$k_c = PS_b \qquad \textbf{(EQ 3)}$$

where k_c is the collection zone rate constant with respect to time, and P is the proportionality constant (often termed the *ore floatability* or *floatability index* [Collins et al. 2009]). The bubble surface area flux is directly related to bubble size and the superficial gas velocity (Finch and Dobby 1990):

$$S_b = \frac{6J_g}{d_{32}} \qquad \textbf{(EQ 4)}$$

where J_g is the superficial gas velocity and d_{32} is the Sauter mean bubble size diameter. The importance of Equations 3 and 4 for interpreting laboratory kinetics tests and applying the results to simulating full-scale machine performance should not be underestimated. This can be seen in Figure 4, which shows the S_b plotted against gas velocity and bubble size for different flotation machines (laboratory, industrial mechanical cells, and industrial column cells). There can be a very large difference between the S_b of a laboratory cell and the S_b of an industrial cell, and applying Equation 3 may result in as much as a 400% adjustment to the observed collection rate scaling from laboratory to plant. For brownfield applications (plant optimization and some geometallurgy programs), the issue can be resolved by performing plant audits and validating the input parameters used for the scale-up procedure. For greenfields or other design applications, the plant S_b must be estimated from benchmark data and then fine-tuned (via impeller speed, impeller geometry, or frother strength) after the plant has been commissioned.

Although the preceding method is a common approach to incorporating machine influence, other methods exist. One alternative, reported by Yianatos et al. (2010b), is to multiply the laboratory rate constant by a factor, usually based on experience or benchmarking, to account for the machine-dependent parameters such as froth effects, mixing effects, and particle segregation effects.

Entrainment. Particles are also recovered through hydraulic entrainment, caused by viscous drag forces acting on the particle surface as water, froth, and collected solids

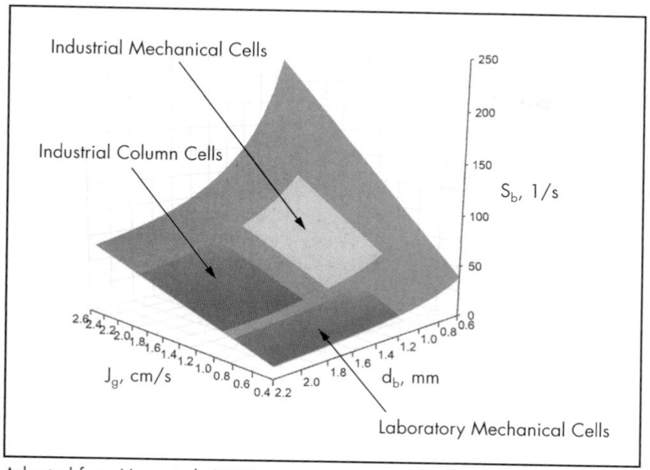

Adapted from Vera et al. 2002

Figure 4 Comparison of gas distribution parameters between laboratory, industrial mechanical, and column flotation cell

are recovered to the concentrate. When the particle settling velocity in the presence of these drag forces is lower than the net rise velocity of water, the particle will be entrained to the concentrate. Many factors affect the rate of hydraulic entrainment, including solids percentage; particle size, shape, and density; cell turbulence, particularly at the interface; froth height; gas flow rate; froth residence time; rheology; frother type and concentration; particle hydrophobicity; and concentration of hydrophobic particles (Wang 2016). The entrainment recovery of solids, R_e, is the percent of a given size class that is recovered by entrainment, and many mathematical models have been proposed to describe R_e as a function of the various system parameters (Bisshop 1974; Bisshop and White 1976; Moys 1978; Ross and Van Deventer 1988; Maachar and Dobby 1992; Savassi 1998; Neethling and Cilliers 2002, 2009; Yianatos and Contreras 2010). A common approach is to assume that for a given size class, R_e is proportional to water recovery (R_w) (Johnson et al. 1974; Smith and Warren 1989). Two commonly applied models are (Jowett 1966; Engelbrecht and Woodburn 1975; Laplante et al. 1989) shown in Equations 5a and 5b:

$$R_{e,i} = \frac{ENT_i R_w}{1 + R_w(ENT_i - 1)} \qquad \textbf{(EQ 5a)}$$

and its simplified form

$$R_{e,i} = ENT_i R_w \qquad \textbf{(EQ 5b)}$$

where ENT_i is a proportionality constant, also termed the *degree of entrainment*, and i is the particle size class.

These equations are simplifications and may break down near boundary conditions, such as at high R_w; very coarse particles; and high turbulence, particularly at the interface. Also, R_e is usually applied to gangue particles, but it should also be applied to uncollected hydrophobic particles. By convention, collection is calculated first and entrainment is applied to particles remaining after collection.

Water recovery. The water recovery of industrial cells is difficult to estimate from laboratory data without detailed knowledge of the machine and pulp characteristics—knowledge that is difficult to obtain for greenfield applications. For this reason, water recovery is one of the few parameters that is still often determined based on experience or empiricism (another is froth recovery). Commercially available models offer up to four methods for modeling water recovery in industrial cells (JKTech 2004):

1. **The concentrate percent solids set-point method.** The user specifies the mass fraction of solids in the concentrate slurry (not including air) and water recovery is back-calculated from the feed water flow rate.
2. **The first-order rate process method.** The user specifies a rate constant and water recovery is calculated from the corresponding first-order rate equation (plug flow or perfect mixer; in this case, froth recovery is ignored).
3. **The froth residence time method.** Water recovery is (usually linearly) proportional to the froth residence time, which is approximated from the froth height, air rate, and cell dimensions (the proportionality constant or constants are user-specified and must be derived from experience or plant surveys).
4. **The constant water recovery method.** The user simply specifies a fixed value for the water recovery.

Because of the relatively high water recovery that occurs in a laboratory kinetics test (relative to a continuous industrial cell), entrainment phenomena are often quite significant at the laboratory scale, and should, therefore, be accounted for when interpreting laboratory flotation kinetics tests. In this case, water recovery is known from a water balance derived from mass measurements on the feed and product stream. The effect of entrainment of laboratory rate constants is usually negligible for coarse, fast-floating minerals but can be very significant for fine, slow-floating minerals (examples include pyrite separation in reground intermediate concentrates and molybdenite separation from reground copper sulfides).

Froth recovery. Froth recovery should not be confused with air recovery or bubble recovery. It is specific to a given mineral class and is defined as the fraction of a mineral class recovered by collection that is subsequently recovered to the concentrate. If all of the collected mineral is recovered, the froth recovery is equal to unity ($R_f = 1$). If none of the collected mineral class is recovered, such as when a cell is not overflowing froth, then the $R_f = 0$. The froth recovery is driven by the operating and machine characteristics of the industrial cell. The following lists some of the recognized relationships between the froth recovery and the cell operating conditions:

- Higher air rates can lead to higher froth recovery, up to a peak (Hadler and Cilliers 2009).
- Deeper froth beds can lead to lower froth recovery (Feteris et al. 1987; Vera et al. 2002).
- Higher ore grades, higher fines concentration, and higher hydrophobicity can stabilize the froth film, and therefore lead to higher froth recovery (Morris et al. 2008).
- The type and concentration of frothers (Cho and Laskowski 2002) or electrolytes (Wang 2012; Biçak et al. 2012) can impact the froth stability, with higher concentrations leading to higher froth recovery.
- Shorter froth residence times, such as those created by internal launders or mechanical scrapers, lead to higher froth recoveries (Contreras et al. 2013).
- In some cases, extremely high carrying rates lead to lower froth recoveries, such as when the cell's carrying capacity per unit area is approached (Patwardhan and Honaker 2000).

It is difficult to accurately measure the froth recovery of a given system, and there are no predictive models that work well for greenfield design applications, although some froth models have been reported to be useful for process control (Neethling et al. 2006). For this reason, froth recovery is usually a user input and is based on experience. When flotation circuit operators vary the pulp level and air rate to a flotation cell, froth recovery and water recovery change, and for this reason, most commercial simulators link the froth recovery to the froth depth and superficial gas velocity (air rate may also be linked to collection recovery, per Equations 3 and 4). Some models relate the froth recovery directly to the froth depth and air rate; others to the froth residence time (Gorain et al. 1998a), which requires knowledge of the cross-sectional geometry of the cell, including froth crowder, concentrate launder, and lip length geometry.

Mechanical constraints. The two main mechanical constraints that are considered for process design are carrying capacity (*Ca*) and lip loading capacity. They apply to all machine types.

Carrying capacity is expressed in units of metric tons per hour of solids that may be recovered per square meter of cross-sectional area of the cell. It can be modeled empirically, using either the method of Finch and Dobby (1990),

$$C_a = K_1 \pi d_p \rho_p \left(J_g^{(1-q)}\right) \quad \textbf{(EQ 6)}$$

or the method of Patwardhan and Honaker (2000):

$$C_a = 522.54 d_{50}^{-0.542} \sigma^{0.2} \frac{n_p d_{50}^3 \rho_p}{d_b^3} J_g \quad \textbf{(EQ 7)}$$

where K_1 and q are fitted, ρ_p is the mineralized bubble density (in the aggregate), J_g is the superficial gas velocity in the cell, d_{50} is the 50% passing size, σ is the size modulus, and d_b is the bubble diameter. For detailed discussion of the usage and applicability of these empirical equations, the reader is referred to the cited reference.

Lip loading capacity is expressed in metric tons per hour of pulp (liquids and solids) that may be recovered per linear meter of lip length. It can be expressed as an empirical function of column diameter (R.L. Amelunxen 1991). The functions are mineral specific. Various methods are available to estimate pump, launder, and other slurry transport capacities; these are not discussed here.

Column flotation cells. The modeling of column flotation cells follows the same general framework presented for mechanical cells, with some key modifications. There are two more relevant modifications:

1. The collection zone recovery is modified to account for varying mean residence times between particle classes.
2. The entrainment recovery is modified to account for froth washing.

The first modification stems from recognition that the column cell may not be a perfectly mixed reactor, and therefore coarse particles may have higher internal settling velocities and shorter retention times than fine particles. There are several approaches; one commonly used is that described by Finch and Dobby (1990). The procedure involves several nonlinear interdependent formulas and requires a computational or iterative solution, as summarized in the following steps:

1. Start by assuming a value for slip velocity of a particle class relative to water, U_{sp}.
2. Calculate the particle Reynolds number, R_{ep}, for the mean particle diameter and the slip velocity using the following equation:

$$R_{ep} = \frac{d_p U_{sp} \rho_l (1-\varepsilon_g)}{\mu_f} \quad \textbf{(EQ 8)}$$

   where d_p is the mean particle diameter, ρ_l is the slurry specific gravity, ε_g is the fractional gas holdup, and μ_f is the fluid viscosity.
3. From the particle Reynolds number, recalculate the slip velocity from the following equation:

$$U_{sp} = \frac{g d_p^2 (\rho_p - \rho_{sl})(1-\varepsilon_g)^{2.7}}{18\mu_f (1 + 0.15 R_{ep}^{0.687})} \quad \textbf{(EQ 9)}$$

4. Repeat steps 2 and 3 until the slip velocity converges.
5. Calculate the dispersion number, N_d, from the slip velocity using the following equation:

$$N_d = \frac{0.63 d_c \left(\frac{J_g}{1.6}\right)^{0.3}}{\left[\left(\frac{J_{sl}}{1-\varepsilon_g}\right) + U_{sp}\right] H_c} \quad \textbf{(EQ 10)}$$

   where d_c is the column diameter (or the diameter of the internal partitions, if the column is portioned), J_{sl} is the superficial slurry velocity, and H_c is the effective column height (i.e., the vertical distance from the spargers to the pulp–froth interface). The effective column surface area also changes with the number of partitions, and this must be considered.
6. Calculate the collection zone recovery from the following equation (also called the *Wehner–Wilhelm equation*) (Wehner and Wilhelm 1956):

$$R_c = 100\left(1 - \frac{4ae^{\frac{1}{2N_d}}}{(1+a^2)e^{\left(\frac{a}{2N_d}\right)} - (1-a^2)e^{\left(\frac{-a}{2N_d}\right)}}\right) \quad \textbf{(EQ 11)}$$

   where a is defined as

$$a = \sqrt{1 + 4ktN_p} \quad \textbf{(EQ 12)}$$

7. Calculate the overall recovery from the compartmental model (Equation 1). Note that with wash water and positive bias (i.e., net flow of water downward), there is no recovery due to entrainment.

Equation 10 is the analytical solution of the mass transport differential equation that describes the concentration of reactants at a distance downstream from the point of injection in an axially mixed tubular reactor (i.e., no radial mixing) and ignoring the influence of nonuniform velocity profiles and short-circuiting. For plug flow conditions and continuous, perfectly mixed reactors, Equations 2a and 2b apply, respectively. For a more complete review of axially mixed reactor models applied to column flotation, refer to Tuteja et al. (1994).

Pneumatic Flotation Cells

There are currently no widely accepted phenomenological models for pneumatic cells, which is the class of flotation cells that includes Jameson cells, contact cells, and microcells. For this reason, most pneumatic flotation cells are scaled up from pilot plant testing. If models are required (i.e., for geometallurgy or variability purposes), practitioners tend to adopt one or more concepts from mechanical or column cell models, and then use empirical corrections to account for the differences in kinetics between the two classes of equipment.

LABORATORY TESTING

Flotation models and simulators are often only useful when the model inputs—including the maximum recovery and kinetics information—are derived from laboratory test data and not industrial plant data. As such, it can be overlooked that the efficacy or fidelity of a given flotation model is closely linked to the procedures and precision of the laboratory test that is used to generate the model inputs. In this section, flotation test procedures are reviewed and discussed with reference to their impact on the resulting flotation models.

Test Methods

Simple Batch Tests

Simple batch tests are nonkinetic, single time interval tests with chemical or mineralogical characterization of the feed, tails, and one concentrate stream. They do not provide kinetics information and are usually only conducted for scoping or exploratory purposes, that is, to determine optimum pulp chemistry or grind conditions. The flotation time can vary and usually depends on the experience of the technician, the standards of the laboratory, and other considerations.

Kinetics Tests

The flotation kinetics test is the most important test for flotation circuit design and scale-up. The test is conducted by putting a known mass of sample in an agitated flotation cell with water and reagents and adding air and agitation for fixed time intervals, over which concentrate samples are collected. A curve of cumulative mineral recovery as a function of time is determined from the resulting feed, concentrate, and tails characterization. Once the curve of cumulative recovery versus time is derived, usually from a combination of quantitative mineralogical and stoichiometric information, the kinetics are calculated for each mineral using curve-fitting techniques. Unlike open cycle and locked cycle tests, kinetics tests can be used for process modeling and design. In fact, they are the only tests that should be used for phenomenological scale-up, as discussed in the next subsections.

Open Cycle Tests

Open cycle tests are performed by conducting a rougher batch test and then re-treating the product streams as required. Usually, the rougher concentrate is reground and refloated in a series of cleaning circuits. The tails from each stage are assayed at the end of the test, and the overall recovery is calculated through a variety of methods. The method usually depends on the assumptions made regarding any mineral values remaining in the unprocessed intermediate streams (sometimes termed the *hanging streams*). Open cycle tests cannot be used to derive the flotation kinetics and hence are not useful for circuit simulation.

Locked Cycle Tests

Like open cycle tests, locked cycle tests are composed of a series of batch flotation tests with or without intermediate regrinding of the concentrate. The tails, concentrates, and water of the various cleaner or scavenger stages are returned to the prior stage of the subsequent cycle (depending on the particular configuration of the test). In this manner, the closed-circuit configuration of an industrial plant is mimicked, although it is constructed from a series of batch tests rather than from a series of continuously fed flotation banks. These tests have several drawbacks; chief among them is that they do not provide kinetics information. For this reason, they are rarely used for simulation purposes.

Pilot Plants

Pilot plants share some of the disadvantages of locked cycle tests. The higher water recovery resulting from the shallower froth heights typical of pilot scale cells often leads to excessive dilution in the cleaners. This can impact the concentrate quality and the separation kinetics and create bias when extrapolating the performance to plant conditions. For this reason, phenomenological models are usually developed with batch kinetics tests and then, if required, validated using pilot plant tests. If the model predicts the mineral distributions and concentrations for each stream in the pilot plant—given the water recoveries and froth recoveries of that system—then it is considered validated. The resulting validated model is then used to simulate the industrial plant, with corrections made for froth depth, air rate, and mechanical constraints.

Pilot plants are also useful for producing large samples of concentrate and tails, which is often necessary for downstream process test work and design.

Test Interpretation

The discussion on test interpretation is limited to kinetics tests, as these are the only tests that are capable of providing the first-order rate information required by phenomenological models. In general, there are three steps to interpreting a batch kinetics test and extracting the mineral-specific rate constants and maximum recoveries. They are listed as follows:

1. Ensure that the assays and the mass and water inputs and outputs are balanced using an accepted mass balancing method, of which there are many.
2. Convert the balanced assays into balanced minerals, calculating the assay of gangue by difference. The gangue assay can be checked by X-ray diffraction or other quantitative means if required.
3. Calculate the kinetics parameters for each mineral, including gangue, using a least-squares fitting method applied to Equations 1 and 2a.

In cases where one element can contribute to more than one mineral, it may be advantageous to model kinetics on a mineral basis rather than on a chemical assay basis, as different mineral phases may exhibit different behaviors regardless of their elemental constituents. A secondary advantage to modeling on a mineral basis is that by converting all elemental assays into their respective minerals, the mass fraction of gangue can be calculated and modeled as a separate mineral class with its own specific gravity, allowing for the calculation of specific gravity in the product streams. The disadvantages include higher costs associated with mineralogical determinations, and the additional complexity required to convert from assays to minerals and back again.

The fitting of Equation 1 to batch kinetic data is not straightforward, and the following considerations may be of assistance:

- It is helpful to treat each time interval as a separate *reactor*, with the tails from the first concentrate interval being the same as the feed to the second interval, and so on until the final test tails are reached. This allows for the use of separate froth recoveries for each stage (Vera et al. 2002; P.A. Amelunxen et al. 2014).
- Often it is preferable to fit the curve of entrainment versus particle size, rather than just the mean entrainment (for all particle size classes). This allows for modeling different size distributions for each concentrate period, such as when accounting for removal of fines because of entrainment in prior periods.
- It is important to always measure the percent solids of the concentrates; otherwise the cumulative water recovery cannot be determined, and therefore entrainment cannot be calculated or fitted.
- Unless using a phenomenological model for R_f and R_e (P. Amelunxen et al. 2018), the kinetics parameters for all minerals should be fitted simultaneously, as the test will not generate enough degrees of freedom to calculate them independently.
- Most engineers assume that the froth recovery in a batch test is 100%, but recent work has shown that this is decidedly not the case. The froth recovery is strongly dependent on the scraping rate, and even with continuous scraping (one complete scrape every 1–3 seconds), the R_f is significantly below 100% (P.A. Amelunxen et al. 2014). It is also important to validate the model through detailed plant surveys to correct for differences in froth recovery between the laboratory and the plant.

SIMULATORS

Steady-State Simulators

Most equations, such as the preceding, for industrial flotation circuit simulation are based on the assumption that the circuit being simulated is operating in dynamic equilibrium and that the mass, mineral, and energy flow into and out of the system are at steady state. This greatly simplifies the mathematical and computational handling of the calculation.

Steady-state flotation simulators are simply ensembles of flotation cell models. Each model is configured with the properties of the industrial cell (size, depth, wash water, etc.), which are then connected together in such a way that a circuit is formed. Several cells form a bank, several banks form a row, several rows form a stage, and several stages form a circuit. The actual method of connecting the various streams may require some ancillary models—splitters, combiners, and so forth. These can be subdivided as well. Splitters can be mass-based (header boxes) or size-based (hydrocyclone, screens, other classifiers). Most flotation circuit models involve circulating loads and iterative calculations, and, therefore, require some kind of simulation engine, which is the logic that drives the model convergence. This logic can be programmed in any language; some examples of platforms that have been used for flotation models include Fortran, Basic, Excel, Matlab, C++, C#, and R.

In cases where reagents are added midcircuit, it is necessary to account for this effect. Commonly, kinetics tests are performed using slurry collected from the plant at any point in which the surface chemistry or mineral liberation has been modified. For circuit design, it is often necessary to create an artificial sample for subsequent kinetics tests (P.A. Amelunxen et al. 2013).

To make it all work, two basic features are absolutely required and a few are recommended. The must-have features of a phenomenological steady-state flotation simulator are as follows:

- **The ability to track kinetics.** The R_∞ and k parameters that are measured in the laboratory, whether on an averaged or size-by-size basis, are only valid for the first cell in a rougher bank. The simple action of removing floatable material via the concentrate of the first cell reduces the kinetics of the feed to the second cell. The same must be true in reverse: When combining streams with different kinetics, the resultant kinetics must reflect that of the blend of streams. Models can account for this in various ways; some use the concept of differential kinetics (P.A. Amelunxen and Amelunxen 2009), and others track the floatability component of each mineral class separately (Runge 2007).
- **The ability to predict concentrate grades.** The phenomenological model must accurately predict stage recovery and concentrate grades for all minerals including gangue. This allows for the calculation of mass flow, bulk specific gravity, and slurry percent solids for each stream. These parameters are essential for calculating the residence times of subsequent cells and stages, and ultimately for circulating load and concentrate grade determinations.

In addition to the preceding, the following features should be given serious consideration. They are not absolutely critical, because they can be done manually (outside the framework of the simulation tool), even if they require more time to do so:

- **The ability to consider mechanical constraints.** Flotation machines have mechanical limits as well as metallurgical limits. The flotation model must consider such things as lip loading constraints, carrying rate limits, launder, pump and pipeline constraints, and froth crowding effects.
- **The ability to mimic some basic human logic.** When using a geometallurgical approach to plant design, it is common to change the process configuration assumptions based on the ore type being processed during any given period in the mine plan. For example, it is often desirable to change the regrind circuit specific energy based on the mineralogy and liberation characteristics of the ore type or to bypass a cleaner stage when processing higher grade ores. Modern flotation circuit design is an economic optimization exercise across a mine sequence plan. To avoid suboptimal outcomes (or time-consuming optimization tuning), the simulator should be flexible enough to incorporate some basic human logic to account for this.

Some other features are often found with flotation simulators:

- **A comminution circuit model** allows for simultaneous optimization of the comminution and flotation circuits as part of a more holistic approach to concentrator design.
- **An economic model** includes smelter treatment and refining charges, contaminant penalty charges, concentrate

freight costs, reagent consumptions and costs, grinding media consumption and costs, energy consumption and costs, capital costs, and other key economic parameters that are needed to evaluate a simulation based on economics, rather than mineral recoveries and product qualities.

- **An optimization engine** allows for searching through a range of possible operating conditions and configurations and identifying one or more economically optimal configurations that meet any required objectives. These can range from simple bisectional search routines to complex partial derivative and/or simplex algorithms.
- **Geometallurgical modeling functionality** includes the ability to simulate circuit performance across an array of ore types, head grades, grind sizes, and other conditions. This can be done on an individual sample basis, on a time period basis, or on a block model basis.
- **A Monte Carlo simulation** component allows for error propagation analysis, confidence estimation, risk evaluation, and other quality assurance and control studies.
- **A configurable process design output** supports rapid generation of key design documents, including process flow diagrams, process design criteria, equipment lists, and mass balance tables.

Dynamic Simulators

Up to this point, discussion on flotation models has focused on steady-state phenomenological models. Recently, there has been an increasing use of dynamic and physics-based flotation modeling software and methods. In practical applications, such as for the design and engineering of new concentrators, the use of these has been limited to the determination of required surge capacities, launder and pipe sizing, and process control requirements. In the future, it is possible that this class of model will become the more favored approach for process optimization and plant design (Herbst et al. 2005).

Limitations of Phenomenological Models and Simulators

All of the phenomenological models and simulators just discussed are merely mathematical approximations of a complex, dynamic process. Because they are only simplifications, they are neither capable of nor intended to completely and correctly describe all of the mass and energy transfer phenomena that occur in a flotation machine or circuit. Some of these simplifications are necessary because of inadequate scientific understanding, while others may stem from the inability to economically measure certain system parameters under the practical or economic constraints of industrial metallurgical laboratories or test programs. Regardless of the reason, it is the responsibility of the practitioner to understand, if not manage, the trade-off between fidelity and pragmatism. Some of the known limitations include the following:

- Equation 1 does not consider the energy input to the system, one of the most important variables and the key component of the operating cost of the equipment.
- The rate constant is the sum of attachment and detachment rates; most models do not allow for separate submodels of detachment and attachment rates. As the energy input changes, the attachment and detachment rates increase or decrease at different rates, changing the overall rate constant.
- Most phenomenological models do not include an interface zone.
- Most phenomenological models do not consider atmospheric pressure.

Researchers are working on addressing these shortcomings.

- Some have proposed amendments to Equation 4 to incorporate the role of energy intensity and turbulence (Sherrell 2004; Tabosa et al. 2011; Amini et al. 2015).
- In some works, the rate constant has indeed been expressed as the sum of the attachment and detachment rates, although quantifying them has been challenging. See, for example, Deglon et al. (1999) and Sherrell (2004).
- Some have proposed that a third compartment—the interface—be added to Equation 3, as proposed by Seaman et al. (2004).
- Work is progressing on characterizing the role of barometric pressure in froth recovery (Young 2007).

It should also be remembered that having a model is not the same thing as being able to scale up from the laboratory to the plant. Just because one has developed an equation or set of equations that can describe the flotation system to a desired level of fidelity does not mean that one can measure the required inputs at the laboratory scale, or use that equation to predict the behavior of a different system. By far, the most common commercially used laboratory flotation cell is the 1940s-era Denver benchtop flotation cell, and most laboratory tests can only characterize the inputs and outputs of the system—feed grades, particle size, water and reagent concentrations, air flow, and sometimes energy input. Many of the inputs and almost all of the intrinsic parameters remain unmeasured. These include entrainment, gas dispersion, attachment and detachment rate, drainage rate, and air recovery (or froth coalescence).

There is another complication associated with the flotation modeling methods described earlier. Many of the unmeasured or unmeasurable parameters change significantly from laboratory to plant. This introduces a very significant complexity into the scale-up process, because now not only must they be characterized (or, more likely, a value assumed for them) in the laboratory, but how they change between the laboratory scale and the plant scale must be understood. To illustrate, Table 1 lists some of the important system parameters, how they are thought to change between laboratory scale and plant scale, and what the effect on the flotation performance may be.

It is precisely the nature of, and differences between, these assumptions that require that any flotation model or simulator be validated. The process of model validation is straightforward; it involves performing a sampling campaign around a target flotation circuit to determine the mineral and mass flow rates of each stream. Kinetics tests are performed on the various nodes (as a rule, a flotation test should be performed on any node in which the surface chemistry has been modified, such as by regrinding, a change in pH or reagents, or a significant change in the slurry density).

The kinetics parameters are then extracted and input into the simulator. The simulator is run, and the froth recovery, water recovery, and degree of entrainment are adjusted

Table 1 Difference in flotation system parameters between plant and laboratory

Parameter	Laboratory	Plant	Effect
Feed solids			
P100	<425 µm	<1,180 µm	Lower R_{max} in the plant
P80	No change	No change	No change
P50/P80 ratio	0.4–0.6	0.25–0.5	Higher R_{max}, faster kinetics in laboratory
Time in grinding mill	15–70 minutes	3–15 minutes	Depends on media type; generally more pyrite passivation in plant
Feed flow	Batch (no flow)	Continuous	Plug-flow-reactor model in laboratory; continuous stirred-tank reactor model in plant
Feed and makeup water			
Water type	Usually tap or sea water	Reclaimed process water	Recycled process water contains residual frother, reagents. Very different frothing characteristics in plant
Water recovery	10%–65%	1%–10%	Higher entrainment rates in the laboratory
Water flow	Semi-continuous	Continuous	Higher water recovery in laboratory, higher oxidation potential in laboratory
Gas flow			
J_g	0.4–0.7	0.5–2.2	Slower kinetics in laboratory, lower R_f in laboratory
S_b	25–40	50–180	Slower kinetics in laboratory, lower R_f in laboratory
D_b distribution	Narrower	Wider	Faster kinetics in laboratory
Gas flow	Continuous	Continuous	No change
Froth			
Froth height	1–2 cm	4–150 cm	Faster kinetics, more entrainment in laboratory
Froth residence time	2–20 seconds	10–500 seconds	Faster kinetics, more entrainment in laboratory
Pulp–froth interface	Quiescent	Turbulent	Lower entrainment in laboratory
Pressure	Depends	Depends	Depends on elevation, cell type
Energy input	Higher	Lower	Higher attachment, detachment rates in laboratory
Degree of entrainment	Lower	Higher	Lower laboratory ENT for a given particle size, may be offset by higher R_w

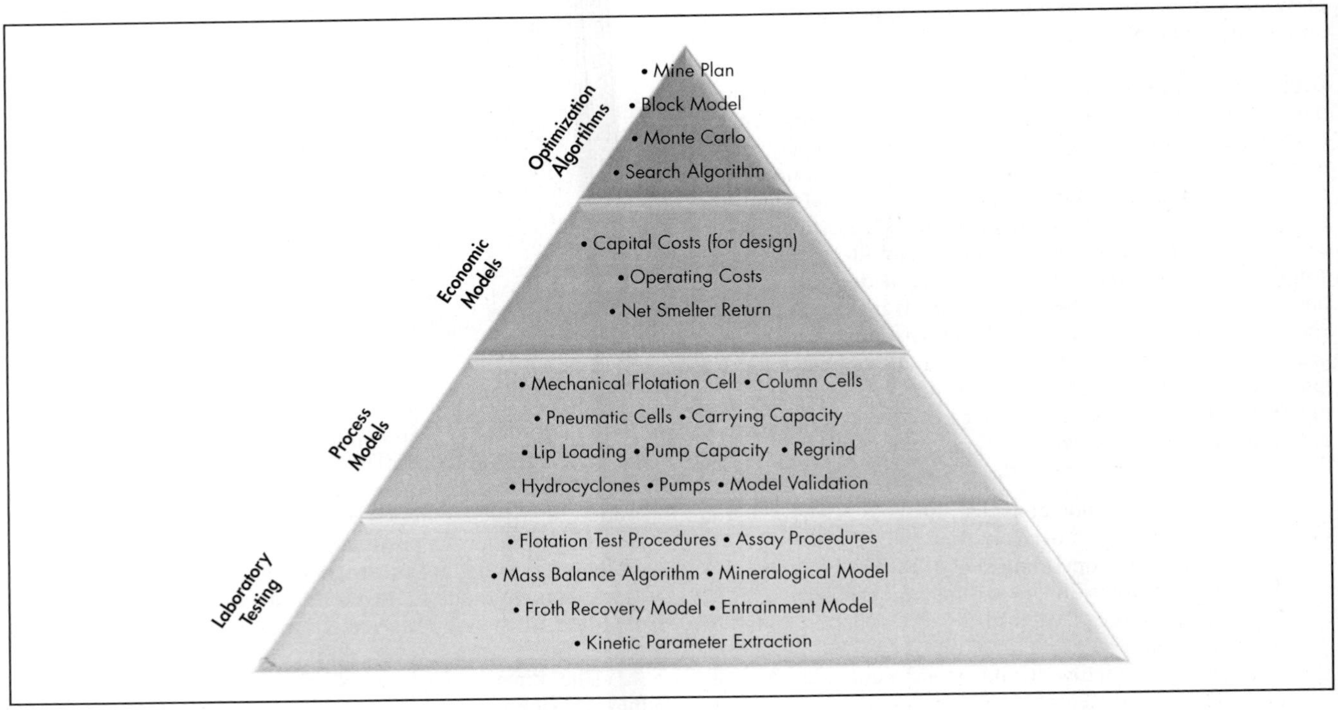

Courtesy of Aminpro

Figure 5 Simulation and scale-up methodology

until the model agrees with the plant balance. If the model is capable of predicting the solids flow rates for each stream, the mineral grades of each stream, and the mineral recoveries of each stage without the use of correction factors or tuning parameters, then the model is valid. Many examples of model validation studies can be found in the literature (Coleman et al. 2007; Dobby and Savassi 2005; P.A. Amelunxen and Amelunxen 2009).

This is a good place to comment that models do not always agree with the plant performance. More generally, flotation models are tools for *improving* our understanding of a complex, nonlinear process. They are not a *substitute* for understanding. In the authors' experience, it is precisely when the model fails to predict the plant performance that the most interesting lessons can be learned.

SUMMARY

The phenomenological modeling of flotation circuits can be thought of as a pyramid (Figure 5). At the base of the pyramid, forming the foundation, is the laboratory testing methodology, including the test procedures and equipment that are used for the work, the kind and quality of the data collected, and the treatment of that data to extract useful information for subsequent modeling and simulation.

The modeling and simulation step sits on top of the foundation of laboratory testing. This contains the basic mathematical and software framework of the model, including the dual-compartment model, lip loading and carrying capacity constraints, pumping constraints, and submodels for the various types of flotation cells—all linked together within the framework of a simulation platform.

Once the basic mathematical framework is complete, it is capable of basic simulation work. Whereas some simulators only contain these two levels, others are equipped with additional economic models and optimization algorithms, as shown in Figure 5, the pyramid diagram.

REFERENCES

Amelunxen, P.A., and Amelunxen, R.L. 2009. Methodology for using laboratory kinetic flotation parameters for plant design and optimization, Part 2. In *Procemin 2009: Proceedings of the VI International Mineral Processing Seminar*. Santiago, Chile: Gecamin.

Amelunxen, P.A., and Runge, K. 2014. Innovations in froth flotation modeling. In *Mineral Processing and Extractive Metallurgy: 100 Years of Innovation*. Edited by C.G. Anderson, R.C. Dunne, and J.L. Uhrie. Englewood, CO: SME. pp. 177–192.

Amelunxen, P.A., Amelunxen, R.L., Flores, L., McCord, T., and Lan, P. 2013. Integrated test programs for modern flotation flow sheet development. Presented at Flotation 2013, Cape Town, South Africa, November 17–21.

Amelunxen, P.A., Sandoval, G., Barriga, D., and Amelunxen, R.L. 2014. The implications of the froth recovery at the laboratory scale. *Miner. Eng.* 66-68(November):54–61.

Amelunxen, P., Ladouceur, R., Amelunxen, R., and Young, C. 2018. A phenomenological model of entrainment and froth recovery for interpreting laboratory flotation kinetics tests. *Miner. Eng.* 125:60–65.

Amelunxen, R.L. 1991. Lip loading considerations in flotation columns. In *Column '91: Proceedings of an International Conference on Column Flotation*, Vol. 2. Edited by G.E. Agar, B.J. Huls, and D.B. Hudyma. Montreal, QC: Canadian Institute of Mining, Metallurgy and Petroleum. pp. 661–672.

Amelunxen, R.L., and Amelunxen, P.A. 2009. Methodology for executing, interpreting, and applying kinetic flotation tests in scale-up, Part 1. In *Procemin 2009: Proceedings of the VI International Mineral Processing Seminar*. Santiago, Chile: Gecamin.

Amini, E., Xie, W., and Bradshaw, D.J. 2015. Enhancement of the precision and accuracy of the flotation scale up process by introducing turbulence parameters to the Amira P9 model. Presented at Flotation 2015, Cape Town, South Africa.

Arbiter, N., and Harris, C. 1962. Flotation kinetics. In *Froth Flotation*, 50th anniversary ed. Edited by D.W. Fuerstenau. New York: AIME. pp. 215–262.

Biçak, O., Ekmekçi, Z., Can, M., and Öztürk, Y. 2012. The effect of water chemistry on froth stability and surface chemistry of the flotation of a Cu-Zn sulfide ore. *Int. J. Miner. Process.* 102-103(January 25):32–37.

Bisshop, J. 1974. A study of particle entrainment in flotation froths. B.Sc. thesis, University of Queensland, Brisbane, Australia.

Bisshop, J.W.M., and White, M.E. 1976. Study of particle entrainment in flotation froths. *Trans. Inst. Min. Metall.* 85:C191–C194.

Chander, S., and Polat, M. 1994. In quest of a more realistic flotation kinetics model. In *Proceedings of the IV Meeting of the Southern Hemisphere on Mineral Technology and III Latin-American Congress on Froth Flotation*. Edited by S. Castro and J. Alvarez. Concepción, Chile: University of Concepción Department of Metallurgical Engineering.

Cho, Y.S., and Laskowski, J.S. 2002. Effect of flotation frothers on bubble size and foam stability. *Int. J. Miner. Process.* 64(2-3):69–80.

Coleman, R.G., Franzidis, J.-P., and Manlapig, E.V. 2007. Validation of the AMIRA P9 flotation model using the floatability characterisation test rig (FCTR). In *Ninth Mill Operators' Conference: Proceedings*. Melbourne, Victoria: Australasian Institute of Mining and Metallurgy.

Collins, D.A., Schwarz, S., and Alexander, D.J. 2009. Designing modern flotation circuits using JKFit and JKSimFloat. In *Recent Advances in Mineral Processing Plant Design*. Edited by D. Malhotra, P. Taylor, E. Spiller, and M. LeVier. Littleton, CO: SME. pp. 197–203.

Contreras, F., Yianatos, J., and Vinnett, L. 2013. On the froth transport modelling in industrial flotation cells. *Miner. Eng.* 41(February):17–24.

Deglon, D.A., Sawyer, F., and O'Connor, C.T. 1999. A model to relate flotation rate constant and bubble surface area flux in mechanical cells. *Miner. Eng.* 12(6):605.

Diao, J., Fuerstenau, D.W., and Hanson, J.S. 1992. Kinetics of coal flotation. SME Preprint No. 92-200. Littleton, CO: SME.

Dobby, G.S., and Savassi, O.N. 2005. An advanced modeling technique for scale-up of batch flotation results to plant metallurgical performance. In *Centenary of Flotation Symposium*. Melbourne, Victoria: Australasian Institute of Mining and Metallurgy.

Dorenfeld, A.C. 1962. Flotation circuit design. In *Froth Flotation*, 50th anniversary ed. Edited by D.W. Fuerstenau. New York: AIME. pp. 365–381.

Dowling, E.C., Klimpel, R.R., and Aplan, F.F. 1985. Model discrimination in the flotation of a porphyry copper ore. *Miner. Metall. Process.* 2(2):87–101.

Dunne, R.C., Lane, G.S., Richmond, G.D., and Dioses, J. 2002. Interpretation of flotation data for the design of process plants. In *Metallurgical Plant Design and Operating Strategies.* Melbourne, Victoria: Australasian Institute of Mining and Metallurgy.

Engelbrecht, J., and Woodburn, E. 1975. The effects of froth height, aeration rate and gas precipitation on flotation. *J. S. Afr. Inst. Min. Metall.* 76:125–132.

Falutsu, M., and Dobby, G. 1989. Direct measurement of froth performance in a laboratory column. In *Processing of Complex Ores*. Edited by G.S. Dobby and S.R. Rao. New York: Pergamon Press. p. 335.

Feteris, S.M., Frew, J.A., and Jowett, A. 1987. Modelling the effect of froth depth in flotation. *Int. J. Miner. Process.* 20(1-2): 121–135.

Finch, J.A., and Dobby, G.S. 1990. *Column Flotation*. New York: Pergamon Press.

Garcia-Zuñiga, H. 1935. The efficiency obtained by flotation is an exponential function of time. *Bol. Min. Soc. Nacional Min.* [Santiago, Chile]. 47:83–86.

Gaudin, A.M., Schuhmann, R. Jr., and Schlechten, A.W. 1942. Flotation kinetics. II: The effect of size on the behavior of galena particles. *J. Phys. Chem.* 46(8):902–910.

Gorain, B.K., Harris, M., Franzidis, J.-P., and Manlapig E. 1998a. The effect of froth residence time on the kinetics of flotation. *Miner. Eng.* 11(7):627–638.

Gorain, B.K., Napier-Munn, T.J., Franzidis, J.-P., and Manlapig, E.V. 1998b. Studies on impeller type, impeller speed, and air flow rate in an industrial scale flotation cell. Part 5: Validation of the k-Sb relationship. *Miner. Eng.* 16(6):739–744.

Hadler, K., and Cilliers, J.J. 2009. The relationship between the peak in air recovery and flotation bank performance. *Miner. Eng.* 22(5):451–455.

Harris, C.C., and Chakravarti, A. 1970. Semi-batch flotation kinetics: Species distribution analysis. *Trans. AIME* 247:162–172.

Herbst, J.A., Potapov, A., Pate, W., and Lichter, J. 2005. Advanced modelling for flotation process simulation. In *Centenary of Flotation Symposium*. Melbourne, Victoria: Australasian Institute of Mining and Metallurgy.

Huber-Panu, I., Ene-Danalache, E., and Cojocariu, D.G. 1976. Mathematical models of batch and continuous flotation. In *Flotation, A.M. Gaudin Memorial Volume.* Vol. 2. New York: AIME. pp. 675–724.

Imaizumi, T., and Inoue, T. 1965. Kinetic considerations of froth flotation. In *Mineral Processing: Proceedings of the Sixth International Congress*. Edited by A. Roberts. Oxford: Pergamon Press.

JKTech. 2004. *JKSimFloat Version 6.1 User Manual.* Indooroopilly, Queensland: JKTech.

Johnson, N.W., McKee, D.J., and Lynch, A.J. 1974. Flotation rates of nonsulfide minerals in chalcopyrite flotation processes. *Trans. AIME* 256:204–209.

Jowett, A. 1966. Gangue mineral contamination of froth. *Br. Chem. Eng.* 11:330–333.

Jowett, A. 1974. Refinements in methods of determining flotation rates. *Trans. Inst. Min. Metall.* 85:263–266.

Klimpel, R.R. 1980. Selection of chemical reagents for flotation. In *Mineral Processing Plant Design*, 2nd ed. Edited by A.L. Mular and R.B. Bhappu. New York: SME-AIME. pp. 907–934.

Lane, G., Brindely, S., Green, S., and McLeod, D. 2005. Design and engineering of flotation circuits (in Australia). In *Centenary of Flotation Symposium*. Melbourne, Victoria: Australasian Institute of Mining and Metallurgy.

Laplante, A.R., Kaya, M., and Smith, H.W. 1989. The effect of froth on flotation kinetics—a mass transfer approach. *Miner. Process. Extr. Metall. Rev.* 5(1-4):147–168.

Levenspiel, O. 1962. *Chemical Reaction Engineering*. New York: Wiley.

Lindgren, E., and Broman, P. 1976. Aspects of flotation circuit design. *Concentrates* 1:6–10.

Maachar, A., and Dobby, G.S. 1992. Measurement of feed water recovery and entrainment solids recovery in flotation columns. *Can. Metall. Q.* 31(3):167–172.

Morris, G., Pursell, M.R., Neethling, S.J., and Cilliers, J.J. 2008. The effect of particle hydrophobicity, separation distance and packing patters on the stability of a thin film. *J. Colloid Interface Sci.* 327(1):138–144.

Morris, T. 1952. Measurement and evaluation of the rate of flotation as a function of particle size. *Min. Eng.* 193:794.

Moys, M. 1978. A study of a plug-flow model for flotation froth behaviour. *Int. J. Miner. Process.* 5(1):21–38.

Neethling, S.J., and Cilliers, J.J. 2002. The entrainment of gangue into a flotation froth. *Int. J. Miner. Process.* 64(2-3):123–134.

Neethling, S.J., and Cilliers, J.J. 2009. The entrainment factor in froth flotation: Model for particle size and other operating parameter effects. *Int. J. Miner. Process.* 93(2):141–148.

Neethling, S.J., Hadler, K., Cilliers, J.J., and Stradling, A.W. 2006. The use of FrothSim to optimise the water addition to a column flotation cell. *Miner. Eng.* 19(6-8):816–823.

Ofori, P., O'Brian, G., Hapugoda, P., and Firth, B. 2014. Distributed flotation kinetics models—a new implementation approach for coal flotation. *Miner. Eng.* 66–68(November): 77–83.

Patwardhan, A., and Honaker, R.Q. 2000. Development of a carrying-capacity model for column froth flotation. *Int. J. Miner. Process.* 59(4):275–293.

Polat, M., and Chander, S. 2000. First-order flotation kinetics models and methods for estimation of the true distribution of flotation rate constants. *Int. J. Miner. Process.* 58(1-4):145–166.

Ross, V.E., and Van Deventer, J.S.J. 1988. Mass transport in flotation column froths. In *Column Flotation '88: Proceedings of an International Symposium on Column Flotation*. Edited by K.V.S. Sastry. Littleton, CO: SME.

Runge, K. 2007. Modelling of ore floatability in industrial flotation circuits. Ph.D. thesis, University of Queensland, Brisbane, Australia.

Savassi, O. 1998. Direct estimation of the degree of entrainment and the froth recovery of attached particles in industrial flotation cells. Ph.D. thesis, University of Queensland, Brisbane, Australia.

Schuhmann, R. 1942. Flotation kinetics. I: Methods for steady-state study of flotation problems. *J. Phys. Chem.* 46(8):891–902.

Seaman, D.R., Franzidis, J.-P., and Manlapig, E.V. 2004. Bubble load measurement in the pulp zone of industrial flotation machines—A new device for determining the froth recovery of attached particles. *Int. J. Miner. Process.* 74(1-4):1–13.

Shean, B.J., and Cilliers, J.J. 2011. A review of froth flotation control. *Int. J. Miner. Process.* 100(3-4):57–71.

Sherrell, I. 2004. Development of a flotation rate equation from first principles under turbulent flow conditions. Ph.D. dissertation, Virginia Polytechnic Institute and State University, Blacksburg, VA.

Smith, F.G., and Warren, L.J. 1989. Entrainment of particles into flotation froths. *Miner. Process. Extr. Metall. Rev.* 5(1-4):123–145.

Sutherland, K. 1948. Physical chemistry of flotation XI: Kinetics of the flotation process. *J. Phys. Chem.* 52(2):394–425.

Tabosa, E., Runge, K., Crosbie, R., McMaster, J., and Hotham, P. 2011. A study of the role of cell aspect ratio on flotation performance. Presented at Flotation 2011, Cape Town, South Africa.

Thorne, G.C., Manlapig, E.V., Hall, J.S., and Lynch, A.J. 1965. Modeling of industrial sulfide flotation circuits. In *Flotation, A.M. Gaudin Memorial Volume.* New York: AIME. pp. 725–752.

Tuteja, R.K., Spottiswood, D.J., and Misra, V.N. 1994. Mathematical models of the column flotation process—A review. *Miner. Eng.* 7(12):1459–1472.

Vera, M.A., Mathe, Z.T., Franzidis, J.P., Harris, M.C., Manlapig, E.V., and O'Connor, C.T. 2002. The modelling of froth zone recovery in batch and continuously operated laboratory flotation cells. *Int. J. Miner. Process.* 64(2-3):135–151.

Wang, L. 2012. Drainage and rupture of thin foam films in the presence of ionic and non-ionic surfactants. *Int. J. Miner. Process.* 102-103(January):58–68.

Wang, L. 2016. Entrainment of fine particles in froth flotation. Ph.D. thesis, University of Queensland, Brisbane, Australia.

Wehner, J.F., and Wilhelm, R.H. 1956. Boundary conditions of flow reactors. *Chem. Eng. Sci.* 6(89):89–93.

Yianatos, J., and Contreras, F. 2010. Particle entrainment model for industrial flotation cells. *Powder Technol.* 197(3):260–267.

Yianatos, J., Bergh, L., Vinnett, L., Contreras, F., and Diaz, F. 2010a. Flotation rate distribution in the collection zone of industrial cells. *Miner. Eng.* 23(11-13):1030–1035.

Yianatos, J., Contreras, F., Morales, P., Coddou, F., Elgueta, H., and Ortiz, J. 2010b. A novel scale-up approach for mechanical flotation cells. *Miner. Eng.* 23(11-13):877–884.

Young, C. 2007. A preliminary investigation into the effect of pressure on flotation performance. *JOM.* 59(10):48–52.

PART 8

Solid and Liquid Separation

CHAPTER 8.1

Sedimentation Equipment

Fred Schoenbrunn, Tim Laros, Brandt Henriksson, and Ian Arbuthnot

Sedimentation equipment is more commonly known as *thickeners* and *clarifiers* within the minerals industry, although other terms such as *subsiders* and *settlers* are also used. A wide variety of sedimentation equipment designs is available to the minerals industry. Selection and sizing of the proper piece of equipment depends on the process objectives. Details of the design of the equipment can vary not only with process objectives but also with site location and topography, material specific gravity and particle size, and availability of local building materials as well as plant operating philosophy.

The development of modern flocculants has led to high-rate and high-capacity designs that are optimized for their use. Modern flocculants have decreased the required sizing for sedimentation equipment by at least an order of magnitude. Flocculant is a significant operating cost and should be considered in an economic evaluation. Sometimes there is a trade-off between equipment size and required dosage, with higher throughput per unit area requiring a higher flocculant dosage, up to a point. It used to be that *conventional* meant that flocculant was not being used. Now it implies a mechanism that is not specifically designed for the optimal use of flocculant and is usually used as a reference point regarding the amount of risk in a design. In fact, past studies have shown that for this reason, flocculant dose in conventional thickeners can be at a similar rate to high-rate designs without real benefit. Therefore, a well-designed high-rate thickener should operate more efficiently and with lower risk than a larger well-designed conventional thickener using flocculant.

In the minerals industry, sedimentation equipment is ubiquitous. The most common applications in base metals involve thickening of tailings and concentrates. In alumina, nickel laterite, uranium, gold, and some of the other leach circuits, countercurrent decantation (CCD) is a fundamental part of the process, using as many as eight thickeners in series. Other common applications include clarification of plant discharge water, process water treatment, leach liquor clarification, removal of precipitates, pre-leach, and grind thickening.

There are three general types of thickener rake drive mechanism: bridge-mounted, center-column mounted, and traction drives. Bridge-mounted drives are centered on a bridge that spans the thickener, with a shaft attached to the rakes. Because of the bridge span, there is an upper limit on the size of machine this design can be economically applied to, generally around 40–50 m in diameter depending on the type of thickener.

Center-column drives are used on larger thickeners (or small ones without lifts) and are mounted on a center column that typically also supports an access bridge that spans one half of the tank. The shorter bridge is a relatively light structure, required for operating and maintenance access; the reduced weight compensates for the addition of the column and the use of a cage around the column to drive the rakes. These can be economical starting at about 35 m in diameter and are currently available up to a diameter of 130 m.

Traction thickeners use a peripheral drive mounted on one (or two) of the rake arms. These units can develop very high torques but do not have lift capabilities and are sensitive to environmental conditions regarding contact between the drive wheel and the rail or traction surface. In mineral processing plants, these are usually only considered for applications typically over 100 m in diameter.

Most thickener designs use center underflow outlets, with the floor slope determined by the type of thickener and, more correctly, the yield stress of the thickened slurry. In operation and as material is removed, thickened slurry flows by gravity toward the center of the tank, assisted by the rakes. In low slurry yield stress applications such as clarifiers, conventional, and many high-rate thickeners, floor slopes between 1:12 and 2:12 are common. Large-diameter machines often use a dual-slope design, with an inner slope of 2:12 and an outer slope of ½:12 or 1:12, to avoid making the machine excessively deep. As thickened slurry yield stress increases because of material behavior, as in the case of high-compression and paste thickeners, floor slope increases to aid slurry transport, and raking

Fred Schoenbrunn, Director for Thickeners, FLSmidth, Salt Lake City, Utah, USA
Tim Laros, Owner, Filtration Technologies LLC, Park City, Utah, USA
Brandt Henriksson, Vice President Thickeners & Clarifiers, Outotec OY, Espoo, Finland
Ian Arbuthnot, Retired, Outotec Pty Ltd., Perth, Western Australia, Australia

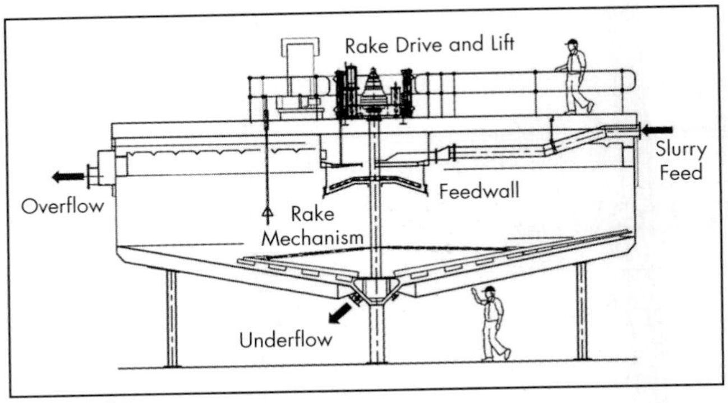

Courtesy of Outotec
Figure 1 Typical thickener with elevated tank

Courtesy of Outotec
Figure 2 Typical center-column thickener with on-ground concrete tank

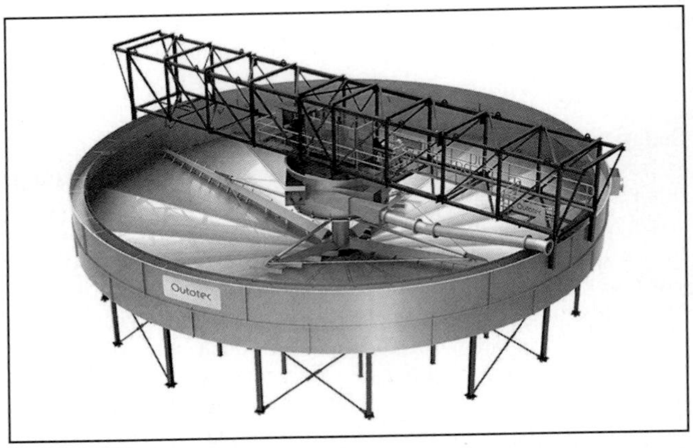

Courtesy of Outotec
Figure 3 Typical full-span bridge thickener with elevated steel tank

becomes critically important. Even some high-rate thickener applications, such as uranium yellowcake and magnetite, use steeper slopes such as 3:12. High-compression and paste thickeners typically use floor slopes of 3:12 and up to 1:1.

The tank sidewall depth varies between 1.8 m in small-diameter thickeners to about 3.6 m in larger-diameter conventional and high-rate thickeners. For high-compression and paste thickener sidewalls, between 4 and 12 m are utilized to achieve solids compression in the bed and increase the mud retention time. In determining sidewall depth, both process and mechanical considerations are factors in design. A freeboard of 150–300 mm is usually used on both the tank and feedwell. The center outlet is typically either a center discharge drum or a 45-degree cone on bridge-type thickeners or a trench on center-column thickeners. Either requires a scraper assembly attached to the rake structure to keep the material moving. A generalized thickener is shown in Figure 1 showing the major structures and process connections for feed slurry, overflow liquor, and underflow slurry. Figures 2 and 3 are models of typical center-column and bridge-mounted thickeners, respectively.

THICKENER COMPONENTS

In addition to overall tank design, thickener components include feed systems, flocculants, rakes, and launders.

Feed Systems

Thickeners and clarifiers are typically fed in the top center and use a feedwell to condition the feed stream and promote solids settling. The feedwell is also the most common point for flocculant addition. Properly designing the feed system, including the feedwell and flocculant-addition location to maximize the effectiveness of flocculation, can have great impact on thickener sizing and performance. It is not uncommon for poorly operating feed systems to exhibit lower underflow density, poor overflow clarity, high flocculant consumption, and even raking problems in the full-scale thickener operation.

When optimizing thickener performance, design of the feed system should consider slurry properties from upstream processes. Excessive aeration, high slurry velocity, or excessively low slurry velocity entering a thickener feed system will often create poor performance from the thickener. Design factors to consider include

- Reduction in slurry aeration through improved upstream piping design and plant layout;
- Inclusion of thickener feed tanks to allow slurry deaeration, especially in the case of flotation concentrates; and
- Inclusion of velocity break tanks to slow down high-velocity launder flow.

To summarize, the feed system in a modern thickener has the following functions:

- Promotes deaeration of the incoming slurry
- Dissipates excess incoming slurry energy
- Effectively mixes flocculant and slurry dilution water (if required)
- Prevents feed slurry short-circuiting
- Promotes aggregate growth and even distribution of aggregates into the thickener tank for sedimentation

The feed slurry concentration for optimum flocculation and most economic thickener size is not necessarily the slurry concentration reporting to the thickener from the upstream process. Quite often, the feed slurry requires dilution prior to addition of flocculant to achieve best flocculation and thickening performance. This effect is presented in Figure 4. This figure shows a maximum in the settling flux at a relatively low solids concentration. The settling flux is the amount of solids settling through a given area, which is the product of the solids settling rate and the concentration. The optimal concentration depends on the characteristics of the feed solids. Very fine solids may need to be diluted to <5 wt % whereas the maximum flux may occur at 15–25 wt % for a coarse grind tails. For clarification, the feed slurry may be too dilute and may benefit from increased concentration by recycling underflow slurry back to the feed slurry. This is discussed in the "Solids Contact Clarifiers" section.

Also apparent from Figure 4 is the increasing flocculant dose rate (in grams per metric ton of dry solids) required as

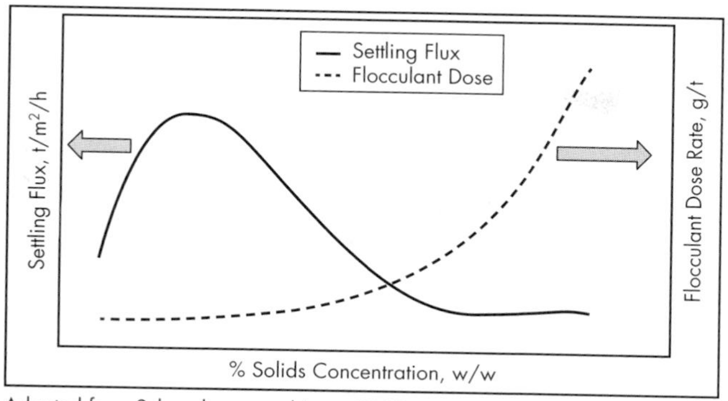

Adapted from Schoenbrunn and Laros 2002

Figure 4 Feed solids concentration, settling flux, and flocculent demand

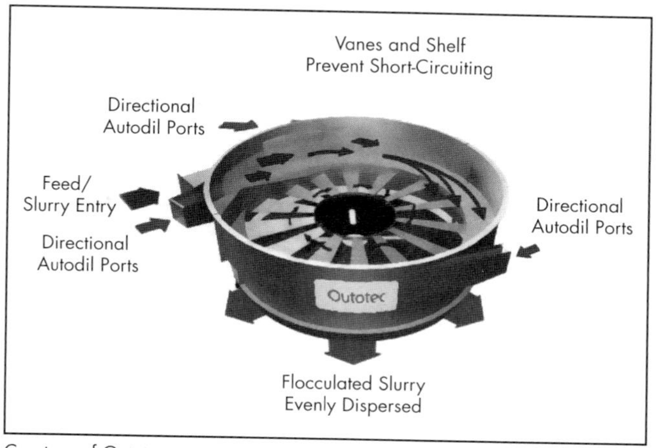

Courtesy of Outotec

Figure 5 Outotec Vane Feedwell with directional auto-dilution

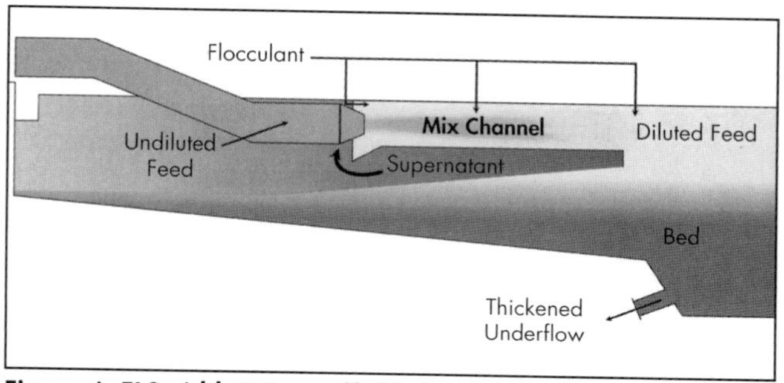

Figure 6 FLSmidth E-Duc self-diluting feedwell

solids density increases. Modern thickener feed systems use feed slurry dilution to both maximize solids settling rate and minimize flocculant dosage.

Many methods of feed dilution have been used since the advent of synthetic flocculants. Pumping of overflow liquor into the feed slurry from the overflow launder or directly from the top of the thickener with submersible pumps is quite common. This method requires a pump, which can be quite large and expensive in cases where large quantities of dilution are required. Early dilution methods without external pumping used slots or holes cut into the feedwell to draw dilution liquor into the feed slurry because of the density differential between the feed stream and the clarified liquor. These early types of feedwells are often referred to as Cross-type feedwells, as Harry Cross (1963) developed one of the first units.

Modern feedwells that include density-driven dilution systems and focus on energy dissipation include the Outotec Vane Feedwell shown in Figure 5. The Vane Feedwell uses dilution ports in the feedwell shell to direct dilution water from the thickener clarified zone into the feed slurry. The feedwell features a closed bottom design and is divided into upper and lower zones by a shelf and vanes. The upper zone mixes the feed stream, dilution water, and flocculant. The lower zone promotes aggregate growth and distributes the flocculated slurry evenly into the thickener for settling (Triglavcanin 2008).

An alternative method of dilution is to place an eductor or jet pump in the feed line to dilute, flocculate, and mix the slurry prior to entering the feedwell. The degree of dilution can be designed into the eductor through the geometry of the eductor and the driving head of the feed slurry through the eductor. This type of design is distinguished by the FLSmidth E-Volute feedwell that features the E-Duc dilution system. An example of this is shown in Figure 6.

Forced-feed dilution systems are also used in some applications. These typically consist of an electrically driven low-head/high-volume pump system in the top of the thickener to force dilution water into the feedwell. These may be required when operational process conditions do not allow effective use of the more simple density or flow-driven systems as described earlier. Two such systems include the Outotec Turbodil and FLSmidth P-Duc.

Flocculants

In a mineral suspension, there is usually a wide difference in particle size. Some particles may be large enough to settle out quickly, while very fine particles may not settle at all. The rate of settling of any given particle is dependent on its size, its density relative to that of the suspending medium, the viscosity of the medium, and the interactive forces between this and other suspended particles.

In mining applications, destabilization of the particle suspension is commonly achieved through polymer flocculation. Flocculation increases particle settling rate by agglomerating them into larger aggregates and is fundamental in the application of high-rate thickener design and operation. Typical polymer charges for mineral applications range from nonionic to highly anionic and can be applied over a wide range of pH and temperatures for most mineral slurries (Figure 7).

Flocculants are available in powder and liquid forms, both of which are made into solutions prior to dosing into the slurry entering a thickener. Solutions from powders are commonly made to 0.25–0.5% w/v concentration, whereas solutions from emulsion-type polymers are typically made to 1% w/v. This primary solution is made in dedicated polymer preparation plants, which can be fully or semiautomated.

Prior to dosing into the feed slurry, secondary dilution of the flocculant is recommended to a final concentration between 0.01% and 0.025% w/v. This increases mixing efficiency into the slurry and minimizes flocculant consumption. Care must be taken, however, with diluted flocculant solutions, as they are shear sensitive. Excess shear reduces polymer effectiveness because of molecule breakage, so secondary dilution of

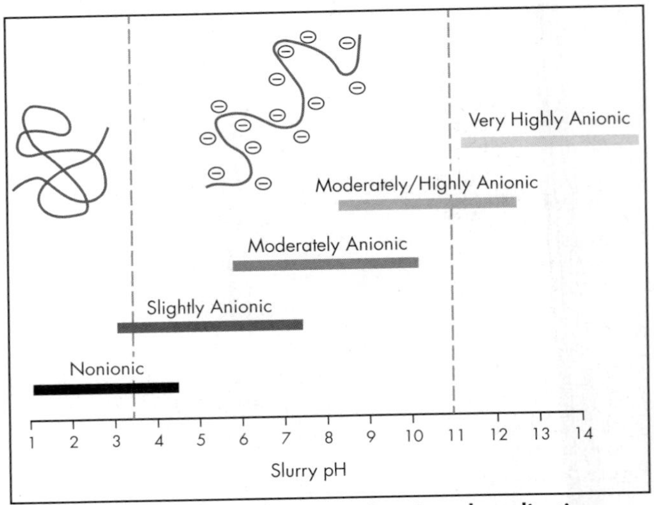

Figure 7 Common flocculent types in mineral applications

Courtesy of Outotec
Figure 8 Low-profile raking mechanism

primary solutions is best performed just prior to the dosing points in the thickener.

Flocculant is commonly dosed in multiple stages into dilute slurry. For most thickener applications, addition is performed in the feedwell of the thickener in multiple locations, ensuring good contact and mixing of the flocculant solution with the slurry particles. In some cases, a partial dose of the flocculant is also dosed upstream of the feed system or in the feed pipe. In all cases, flocculant addition is best done into dilute slurry to enhance mixing and to minimize flocculant consumption.

Thickener original equipment manufacturers include the flocculant dosing point design as part of the thickener scope of supply. Particle mixing and settling in feed systems makes the optimum placement and distribution of flocculant challenging, and some adjustment of dosing points is often required after initial start-up.

Rakes

Most thickeners use a set of arms that move through the pulp to help thicken the pulp and assist even movement of the thickened material to the underflow outlet. These typically have blades set at 30–45 degrees to the tangent of motion to push the solids toward the outlet. The rake arms must be strong enough to transmit the torque needed to push the solids toward the thickener discharge.

In most high-rate and conventional thickeners, most slurry flow toward the central discharge is by gravity. In this case, rakes assist and maintain even flow and solids distribution in the bottom of the thickener. In paste thickeners, slurry yield stress increases, and the slurry resistance to flow increases. In these applications, rake design and raking capacity become increasingly important to prevent thickener shutdown and to maintain performance targets.

Rake mechanisms are usually designed for specific process applications. Processes that produce a heavy scale buildup on the rakes, such as alumina refining, require a rake containing a minimum of steel surface area. Rakes for magnetite thickening may have spikes attached to the blades such that heavily thickened magnetite can be resuspended. Some rakes are of a streamlined low-profile design to help reduce the torque on the rake structure (Figure 8). Rakes may have pickets or thixo posts (posts to distance the blades from the rake arms) attached to them for processes in sticky or high yield stress applications. In general, rake design is ideally process dependent but must also meet the requirements of torque and deflection.

Rake Drives

The rake drive provides driving force (torque) to move the rake arms and blades against the resistance of the thickened solids. The drive also provides the vertical support for the rotating elements. Bridge-type thickeners can use spur gear, worm gear, planetary, or other commercial reducer drives. Column-mounted thickeners almost always use spur gear drives.

Rake design varies between manufacturers, with safety factors commonly between 1.0 and 1.5 times the 100% torque rating of the drive. It is therefore common for two or more levels of overload protection to be designed into the drive to prevent torque overloads and associated rake damage. Reliability is a critical issue with the drive, as failure frequently means digging out the thickener. Key drive elements are hardened steel gears, large precision bearings, oil bath lubrication, accurate torque measurement, and strong housings.

Torque descriptions used by thickener manufacturers include peak torque, design torque, maximum operating torque, normal operating torque, duty-rated torque, American Gear Manufacturers Association 20-year torque, cutout torque, and so forth (AGMA 2001-C95). This is partly due to the relationship of torque to gear life, where gears last a very long time at low torque, but exponentially shorter as the torque increases. It is important that equivalent design standards are used when comparing different thickeners, as drive life and operability can vary greatly.

There are two methods of motive power for drives: electric motors or hydraulic power packs. Hydraulic drives offer features such as soft starts, variable speed, torque indication by hydraulic pressure, low-speed hydraulic motors, excellent torque sharing on multiple pinion drives, and pressure relief as an overload protection. The downsides are low efficiency, higher cost, maintenance, and the added complexity of another system. Similar features are now available for electric drives using electric variable-frequency drives. Electric drive motors are relatively simple and can use mechanical, load cell type, or electronic load-sensing torque measurement and protection.

Table 1 Standard duties

Duty	Examples	K Factor
Light	River or lake water clarification, metal hydroxides, brine clarification	1–4
Standard	Magnesium oxide, lime softening, brine softening	5–9
Heavy	Flotation tails (copper, lead, zinc, nickel), iron tails, coal tails, flotation concentrates (coal, copper, zinc, nickel, or lead), clay, titanium oxide, gold tails, phosphate tails	20–40
Extra heavy	Uranium countercurrent decantation, iron ore concentrate, iron pellet feed, titanium ilmenite, coarse tailings	40–60

Adapted from Schoenbrunn and Laros 2002

Rake drive sizing is dependent on the application, with variables such as particle size distribution, specific gravity, flocculant use, solids loading, rake design, and design underflow concentration affecting the selection. Because the torque is related to the thickener diameter by a square power function, a K factor is usually used to refer to a drive size independent of diameter, where torque = K × D (diameter)^2. Depending on units used, the formula for rake torque is

$$\text{torque} = 14.6 \times K \times D^2$$

where

torque = maximum operating torque, N·m
D = diameter of thickener, m

Table 1 shows typical values for standard duties, covering clarifiers as well as conventional and high-rate thickeners. However, the selection of an appropriate K factor should also consider the preceding variables and may need to be significantly different from those shown in Table 1. For example, very high underflow densities may dictate a K factor to be an order of magnitude higher.

When selecting the drive K factor for high-compression and paste thickeners, the consolidated slurry yield stress becomes a dominating design criteria. For those applications, K factors between 150 and 400 are typically required.

Rake speed for thickeners is typically determined by tip speed—the velocity of the outer tip of the rake arm. For most thickeners, this may be 8–12 m/min and is driven by solids area loading. For clarifiers and low solid loading rates, this number is lower.

Rake Lifts

Rake lifts are used to protect the drive from high torque with a typical vertical lift between 300 and 600 mm. These can be used with either bridge or center-column designs and can be hydraulically or electrically actuated. A typical bridge-mounted thickener drive and rake lift is shown in Figure 9.

Traction-type drives are generally not compatible with rake lifts, although some have been supplied using the cable-type design described next. Lifts are typically powered by a separate motor from the drive and must be designed for precise control. Lifts can prevent shutdown of the machine in plant-upset conditions, for example, when large amounts of coarse material are encountered or, more generally, when the thickened material reaches a very high yield stress. Their use in minerals applications is fairly ubiquitous, although there are many applications that do not need a lift or where extra torque would make a lift unnecessary.

Courtesy of Outotec

Figure 9 Single planetary rake drive with hydraulic rake lift

Lifts have been used to aid in storing material, especially mineral concentrates, in a thickener. The rake is lifted allowing thickened concentrate to accumulate below the rake. The rake is then slowly driven into the dense concentrate to evacuate the thickener. This technique is not commonly used; however, it illustrates the need for the rakes and supporting structure, such as the bridge or column, to be designed to accommodate a downward force from the lift in addition to upward forces.

Cable-type lift designs use an arm hinged at the center connection and are supported and towed by cables, so that the arms can pivot upward when an obstruction or high torque is encountered. The advantages of this design include a streamlined arm design and low cost. The main disadvantages are the lack of positive control of the arm position and the inability of the rakes to lift significantly in the center, where the heaviest accumulations are usually found.

Drive Power

Thickener rake drive power is a function of torque and rake speed. In a mineral processing plant, thickener drive power consumption is a relatively low value compared to other unit operations in comminution, beneficiation, pumping, and filtration.

Rake drive power can be estimated by the formula

$$\text{power (kW)} = \frac{0.105 \times \text{speed} \times \text{torque}}{\text{efficiency}}$$

where

speed = rake speed, rpm
torque = rake torque, kN·m
efficiency = 0–1; mechanical efficiency of the drive system, which can be 0.9 for electric drives and 0.7 for hydraulic drives

From the preceding formula, the installed rake drive power can be determined from the maximum operating torque of the rake drive. Additional power is added for rake lift and is typically an incremental increase of rake drive power depending on rake lift capacity and may be between 20% and 30% of the drive power.

Actual operating power consumption can therefore be estimated as approximately 15%–20% of the total installed drive power in a correctly sized thickener drive.

Effluent Launders

Most thickeners use a peripheral effluent launder to collect the clarified overflow and bring it to a single or double discharge point. The effluent should uniformly flow into the launders around the periphery of the tank and should not be back-flooded into the thickener. V-notch weirs are often provided to assist in distributing the effluent around the periphery of the tank. The weirs can be built into the tank launder or can be a separate, adjustable element. Froth baffles can be located at the liquid level just inboard of the launder and are used to prevent floating material from getting directly to the launder. Flotation concentrate thickeners are almost always provided with froth baffles and often with some method of froth management such as water sprays and rotating booms.

Other methods of handling effluent include radial launders, single-point discharge, and bustle pipes. Radial launders are often used on solids contact clarifiers to help distribute the effluent removal evenly over the surface of the clarifier. Conversely, some applications can use a simple single-point discharge nozzle and still have acceptable overflow clarity. Bustle pipes are often used where liquor storage is desired at the top of the tank or the tank liquor level is variable, using submerged pipes with orifices to distribute the effluent discharge.

The size, slope, and number of discharge points of a launder are determined using hydraulic flow equations developed by the thickener manufacturers.

Tank Design

There are many possible tank styles that can be used, with attendant trade-offs. Most mineral applications require good access to the underflow piping, and it is usually good practice to locate underflow pumps near the thickener underflow outlet. This leads to the need for either elevated tanks or underflow access tunnels. Because of the complexity, cost, and safety issues with underflow tunnels, elevated tanks (Figure 10) are generally preferred for small- to medium-size thickeners, up to about 45 m in diameter, although larger elevated tanks to 65 m have been built. Other benefits of elevated tanks are storage space underneath, unhindered pump access, and ease of leak detection and repair. Elevated steel tanks can be of welded or bolted design. Bolted designs can offer significant installation savings by reducing site labor requirements for erection, although transport of resulting larger tank sections to remote sites can increase shipping costs. Steel tank materials offer various choices to suit process conditions from mild steel through stainless steel and duplex materials.

On-ground tanks do not require the structural steel needed for elevated tanks, and so they are generally preferred for applications where underflow pipe access is not critical, as well as for large-diameter thickeners and clarifiers, generally 50-m diameter and above (Figure 11). If the process allows the underflow pipe to be buried to the edge of the thickener without high risk of blockages, an on-ground tank can be significantly less expensive. As a result, on-ground tanks are fairly common in clarification and water treatment applications.

Courtesy of Outotec

Figure 10 Elevated thickener tank with bolted steel construction

Within the realm of on-ground tank design, there are many construction methods available using steel or concrete. *Anchor channel construction* refers to a steel channel embedded in a concrete footing at the wall, with the steel shell welded or bolted to the anchor channel. With this construction, the floor can be concrete, membrane, or fill. The other method of construction for a steel wall tank is to use a steel floor on a compacted foundation.

For concrete wall tanks, the floor can be concrete, membrane, or fill. Fill material used in construction should be engineered, properly sourced, placed, compacted, and inspected as part of the overall tank construction. Various linings, covers, and insulation can be applied to suit the process. The use of clay liners alone is limited because of groundwater contamination concerns. However, clay layers can be used effectively in combination with a variety of membranes in an engineered system.

SEDIMENTATION EQUIPMENT DESIGNS

Equipment designs for clarifiers and thickeners, including high-rate, high-compression, and paste thickeners.

Clarifiers

Standard clarifiers are often used for water and wastewater applications: These units usually have relatively low K factors and, hence, light mechanisms because the amount of raking, particle size, and solids density are usually on the light end of the spectrum. Rake lifts are frequently not needed. In general, clarifiers are very similar in appearance to thickeners. The feedwell is generally larger in volume and the tank depth greater, with tank walls typically 3 m high or more. This is done to provide more time for feed slurry flocculation in the feedwell, slower velocities exiting the feedwell, and a longer liquor detention time in the clarifier. These features are necessary to achieve optimum overflow liquor clarity.

Classic clarifier hydraulic loadings start at about 1 $m^3/h/m^2$ and typically go up as high as 6 $m^3/h/m^2$ for some solids contact clarifiers with optimal feed solids. Nominal sizing at 2.4 $m^3/h/m^2$ is a good starting point for many applications. These numbers consider using coagulants or flocculants,

Courtesy of Outotec

Figure 11 On-ground tank with concrete floor and bolted steel walls

which greatly aid the particle settling rate. Chemical addition is almost always used in clarification applications because of its effectiveness at flocculating very fine particles and increasing the effluent overflow clarity.

Solids Contact Clarifiers

This equipment is designed to internally or externally recirculate solids and to promote particle contact and flocculation in the reaction well or to enhance solids precipitation or hardness reduction. This can be done externally by pumping a portion of the underflow back to the feedwell. The same thing can be accomplished internally using a draft tube and a turbine to draw solids from the bottom of the clarifier tank.

The design philosophy for solids contact clarifiers is to maximize the size and concentration of aggregates in the reaction zone by having the pumping capability to suspend a high concentration of solids. Using the draft tube and turbine, solids are recirculated centrally and directly above the tank floor. The heavier particles necessary for improved settling velocities are mixed with the incoming feed to allow particle contact and aggregate growth. Flocculation and recirculation are accomplished symmetrically within the reaction well for the most efficient use of reactor volume and turbine energy.

A large feedwell provides the required detention time and allows precipitation to take place prior to the slurry entering the clarification zone. Chemical addition is introduced into the recirculation zone in the presence of previously formed precipitated solids prior to passing through the turbine for optimum mixing and flocculation.

Typical hydraulic loadings for solids contact clarifiers are 2.4–4.8 $m^3/h/m^2$, upward to 12–24 $m^3/h/m^2$ in some steel mill wastewater clarification applications.

Inclined Plate Clarifiers

Inclined plate clarifiers use plates to increase the effective settling area in a small tank. The plates are set at a 45–60-degree angle to allow settled solids to slough off. The plates are typically stacked with 50-mm spacing, although this can be varied for the application. The effective area is the sum of the horizontal projections of the plate areas. They can be supplied with either rakes or steep cone bottoms and are often available as packaged units in a square or rectangular configuration. These units are very effective at putting a lot of area in a small tank but are susceptible to solids buildup and plugging.

Conventional Thickeners

There are some applications for which polymer is either not needed or may adversely affect downstream processing, and hence it is not used. For these situations, a conventional thickener is all that is required. However, there are many conventional thickeners that use flocculant to improve overflow clarity, handle a higher tonnage, or aid in achieving the desired underflow density. These units are often fairly simple, although relatively sophisticated mechanisms are used for particular applications. Because of the relatively large size, they are somewhat forgiving in operation and may have the storage capacity to absorb some plant upsets without affecting downstream operations if sufficient torque and raking capacity is installed.

Typical features include a drive and rakes, a relatively open and shallow feedwell, and a bridge to support the feed pipe or launder and allow center access. The drive size is dependent on the application. Conventional thickeners can be either bridge, center-column, or traction design. Early thickeners (prior to 1970) mostly fall into this category.

High-Rate Thickeners

With the advent of synthetic flocculant, the terms *high rate* and *high capacity* emerged as a type of thickener, as the throughput rates for the now flocculated feed slurries were considerably higher than for unflocculated slurries. These terms now generally refer to designs optimized for use with flocculants, although further improvements on earlier optimized designs are evident as the technology has continued to evolve and improve. Thickener size or throughput is highly dependent on flocculant dose and feed slurry concentration, as presented in Figure 4. Consequently, most high-rate thickeners commonly use feed dilution systems when the incoming feed density is above the value for optimum settling. The size of these thickeners can be somewhat governed by capital and the primary operating cost of flocculant. A smaller thickener will have a lower installed cost but may be more costly in the long run because of a higher overall flocculant cost, and vice versa.

High-rate thickeners are commonly small- to medium-sized bridge-type thickeners up to 50 m in diameter, although large center-column thickeners processing very high tonnage can also fall into this category. Flocculation is required, and feed slurry dilution systems are often needed for optimal performance. This is the most common type of thickener in the minerals industry today. Grinding, tailings, leach, pre-leach, CCD, and concentrate thickeners often fall into this category. Because of the wide range of applications, solids loading rates vary considerably and in free settling applications can go to 1.5 $t/m^2/h$ or higher. Hydraulic capacity must also be considered in the design with values typically up to 6 $m^3/h/m^2$ and at the higher end to around 12 $m^3/h/m^2$. Table 2 contains typical sizing parameters for common high-rate thickener applications.

Common features include a self-diluting feedwell, heavy-duty drive, streamlined rake arms, and large effluent launders and underflow outlets. Because of relatively high solids loading rates, instrumentation and automated control systems are required on high-rate thickener applications to maintain steady-state operation and avoid downtime caused by rake *bogging*. Some examples of high-rate thickeners are shown in Figures 12 and 13.

High-Rate Rakeless Thickeners

This relatively new class of sedimentation equipment uses a deep tank and steep-bottom cone to achieve the underflow

Table 2 High-rate thickener sizing

Application	% Solids, w/w		Solids Loading Rate, t/m²/h†	Liquid Loading Rate, m³/h/m²‡	K Factor
	Feed*	Underflow			
Alumina, Bayer process					
Red mud clarifiers (settlers)	3–4	25–35	0.1–0.3	3.0–6.0	175–200§
Red mud washers	6–8	30–40	0.7–1.0	6.0–10.0	175–200§
Red mud final washers	6–8	35–45	0.7–1.0	6.0–10.0	175–200§
Hydrate	2–8	30–50	0.05–0.2	1.5–2.5	125–150§
Brine purification	–	–	0.01–0.02	0.5–1.2	10–25
Coal					
Coal tailings	0.5–6.0	25–40	0.1–0.4	1.0–6.0	20–30
Clean coal fines	4.0–7.0	30–40	0.2–0.4	2.0–5.0	75–150
Dense media (magnetite)	10–15	60–70	0.3–0.5	3.0–6.0	35–40
Flue dust					
Blast furnace	0.1–2.0	40–60	0.05–0.15	1.0–2.5	20–30
Base metal concentrates					
Copper concentrate	10–20	50–70	0.15–0.25	1.0–2.5	35–40
Zinc/lead concentrate	10–15	50–70	0.15–0.2	1.0–2.0	35–40
Nickel concentrate	10–15	50–65	0.15–0.25	1.0–2.5	35–40
Flotation tailings	10–15	45–65	0.15–1.00	1.5–6.5	30–45
Iron ore					
Concentrate	15–25	60–75	0.4–1.5	2.0–6.0	35–50
Tailings	1–10	50–65	0.2–0.6	2.0–6.0	20–35
Gold					
Countercurrent decantation (CCD)	10–15	40–55	0.3–0.8	3.0–6.0	25–35
Tailings	10–15	50–65	0.3–1.0	3.0–6.0	30–35
Copper CCD	10–15	55–65	0.5–0.8	3.0–6.0	35–40
Nickel laterite CCD	5–10	35–50	0.3–0.5	4.0–6.0	35–50
Molybdenum concentrate	8–10	35–55	0.05–0.15	0.5–1.5	30–40
Mineral sands slime	2–4	20–35	0.1–0.25	4.0–7.0	20–30
Platinum					
Concentrate	7–12	50–60	0.1–0.2	1.5–3.5	35–40
Tailings	8–15	45–55	0.15–0.5	3.0–6.0	30–35
Uranium					
Acid-leached ore	10–15	45–65	0.30–0.60	3.0–6.0	30–35
Uranium precipitate	1–2	10–30	0.01–0.05	0.5–2.5	20–25

Adapted from Schoenbrunn and Laros 2002
* Feed density for flocculation, which may include dilution.
† Solids loading rate = feed rate (t/h)/tank cross-sectional area (m^2).
‡ Liquid loading rate = feet rate (m^3/h)/tank cross-sectional area (m^2).
§ K factor based on scale load.

density while eliminating the rake and rake drive. The design is based on maximizing throughput rate in a small diameter while achieving good underflow density and overflow clarity. Flocculant is always used, and feed dilution is usually built into the design.

Because these units have a very low residence time, start-up and shutdown are quick, typically requiring only about 30 minutes to reach steady-state operation. They are operated as continuous process equipment and cannot be used for storage. The lack of a rake mechanism and smaller diameter makes these units a generally lower-cost alternative to rake thickeners. Operation is relatively simple; however in some applications, control can be unstable and plugging of the internals is sometimes a risk. It has also been demonstrated that in many applications, rakeless thickeners are unable to achieve the same consistently high density as properly configured raked thickeners. These limitations and the relatively small unit capacity has seen rakeless thickeners penetrate only niche applications in thickening. Current examples are the FLSmidth Eimco E-Cat clarifier-thickener and the Delkor Ultrasep ultra-high-rate thickener. A diagram of the internal flow pattern of an E-Cat clarifier-thickener is shown in Figure 14.

High-Compression or High-Density Thickeners

This technology is an extension of high-rate thickening, utilizing a deeper mud bed to augment the available compression forces in the bed, thus achieving higher underflow density compared to high-rate thickeners. High-compression thickeners typically add 1–3 m in sidewall depth to a high-rate design to aid in increasing the underflow density. Deeper mud beds increase the mud compressive force, which, combined with raking, increases underflow density.

As slurry yield stress increases with density, these machines use steeper floor slopes to aid slurry transport and require significantly more installed rake torque (two to five times) than high-rate machines. Mechanism design is also optimized for dewatering in deeper solids beds and may commonly feature dewatering pickets and thixo posts on the rake blades. Similar

Courtesy of Outotec

Figure 12 High-rate thickener, 65-m diameter

Courtesy of Outotec

Figure 13 High-rate thickener CCD circuit

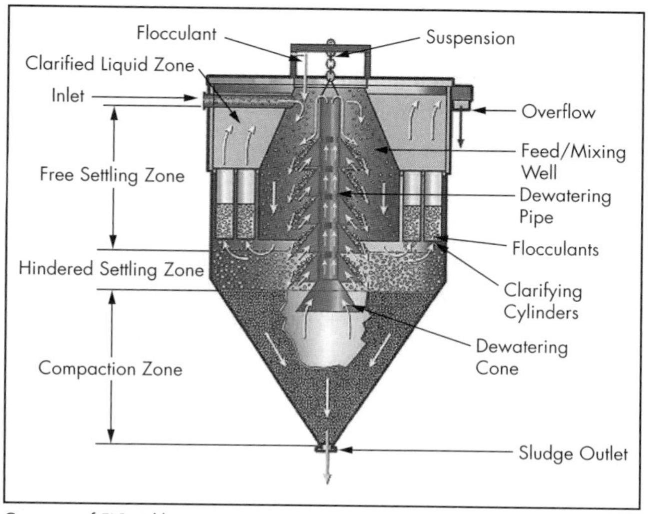

Courtesy of FLSmidth

Figure 14 E-Cat clarifier/thickener showing the internal flow patterns

to high-rate thickeners, high-compression or high-density thickeners require good instrumentation and process control systems to consistently achieve high-density underflow.

Paste and Deep Cone Thickeners

Today's paste thickeners originated from work done by the British Coal Board in the 1960s, utilizing steep-bottom tanks without rakes to produce underflows with very high solids contents. The goal of the work was to produce a material that could be put on a conveyor, and while this was not achieved consistently, very thick material could be produced (Klepper et al. 1998). Further development by Alcan in the alumina industry (Emmett et al. 1992) eventually led to commercialization of this technology outside alumina, with applications ranging from high-efficiency CCD washers to underground paste disposal and wet stacking of surface tailings. The FLSmidth Deep Cone paste thickener and Outotec paste thickener are examples of this design. In some applications, it is possible to produce material at the limits of pumpability with these units. An example of a paste thickener installation is shown in Figure 15.

Although underflow with the consistency of paste can be achieved by high-rate, high-rate rakeless, or high-compression machines, paste thickeners are currently the best technology for achieving maximum underflow densities utilizing sedimentation equipment alone. These units typically use very deep mud beds to take maximum advantage of mud compressive forces for dewatering and provide sufficient time for the mud to dewater to a paste consistency. The tank height-to-diameter ratio is frequently 1:1 or higher. Because of the high underflow yield stress, mechanism torques can be 5–10 times higher than high-rate machines on similar materials. Floor slope is also increased to between 30 and 45 degrees to aid in material transport. Rake mechanisms are specifically designed for dewatering in deep beds and to positively transport thickened slurry to the central discharge without causing rotation of the solids within the tank. Some suppliers may incorporate additional elements into the mechanism design to prevent rotation of agglomerated solids masses sometimes known as *donuts*.

Applications include surface tailings disposal by wet stacking, underground paste backfill, CCD, and pre-leach applications. In paste thickening, instrumentation and control

Courtesy of Yara

Figure 15 Paste thickener, 30-m diameter with 40-degree floor slope

systems are critical to achieve the required high-density underflow and to avoid equipment downtime.

INSTRUMENTATION AND PROCESS CONTROL IN RAKED THICKENERS

Although early and conventional thickeners used very little instrumentation and process control, the advent of high-rate designs and widespread use of flocculation to enhance settling has made process control essential for consistent thickener performance and to reduce costly machine downtime. To some degree, advances in thickener mechanical and process design have been outpacing the availability of reliable instrumentation and process control systems. Many thickeners are installed with traditional single-loop proportional integral controllers through plant distributed control systems. In many cases, these can be tuned to achieve control over steady-state operations. Process upsets coupled with slow and complex process dynamics often lead to poor control results. Many thickeners are then operated in a manual or semimanual operating mode.

Instrumentation and control strategies should be implemented based on several criteria: Application, thickener type, and process output priorities are key factors in deciding on a control strategy. The availability and capacity of resources to maintain sometimes complex control systems should also be considered.

New approaches to controlling thickeners using expert control systems or multivariable model predictive control (MPC) have shown advantages in overcoming the limitations of single-loop controllers (Kosonen et al. 2017). This application integrates highly interactive controls of solids inventory, underflow density, and overflow clarity into a single controller. The MPC is able to take changes in the incoming flow into account, deal with process constraints, and support prioritization between control variables. The results from this control system include increased stability and reduced variability around the desired process set points. An example of a thickener control schematic is shown in Figure 16.

TEST WORK AND SCALE-UP METHODS

A method widely used for conventional thickener sizing using static cylinder laboratory-scale tests was developed by Coe and Clevenger (1916). They identified four settling zones in a settling solid–liquid suspension, namely, clear water zone, initial or free settling zone, transition zone, and compression zone. By a series of batch settling tests, using a typical slurry sample, the limiting solids flux (solids settling rate per unit area) was established, which was then used for scale-up on a full-sized thickener. To allow for potential dynamic effects and material variations, it was usual to use a safety factor in scale-up (usually 20% or 25%). This method was simplified further by Kynch (1952), who applied a mathematical method to demonstrate that the limiting solids flux could be determined from a single batch test.

With the advent of high-rate thickener technology in the 1970s, a laboratory-scale dynamic test-work approach was used, both as a means of providing accurate sizing and a demonstration of the high-rate concept. Since the first thickeners of this type were put into service by Enviro-Clear, a dynamic test method has been adopted and further developed by many manufacturers including Supaflo (now Outotec), FLSmidth, Delkor, and others.

On the laboratory scale, a small thickener, typically 100 mm in diameter (Figure 17), is used, fitted out with all the elements of a full-scale thickener, that is, feed pipe, feedwell/flocculation tank, slurry deflector plate, a set of small rotating rakes, and an underflow offtake. This unit is fed from a mixing tank containing a suspension of a typical plant sample. The feed rate is calculated to present to the test unit a solids flux, measured in metric tons per square meter per hour, appropriate to the application. Flocculant is added prior to or at the feedwell, and the flocculated slurry is introduced into the thickener settling zone via the feedwell deflector plate

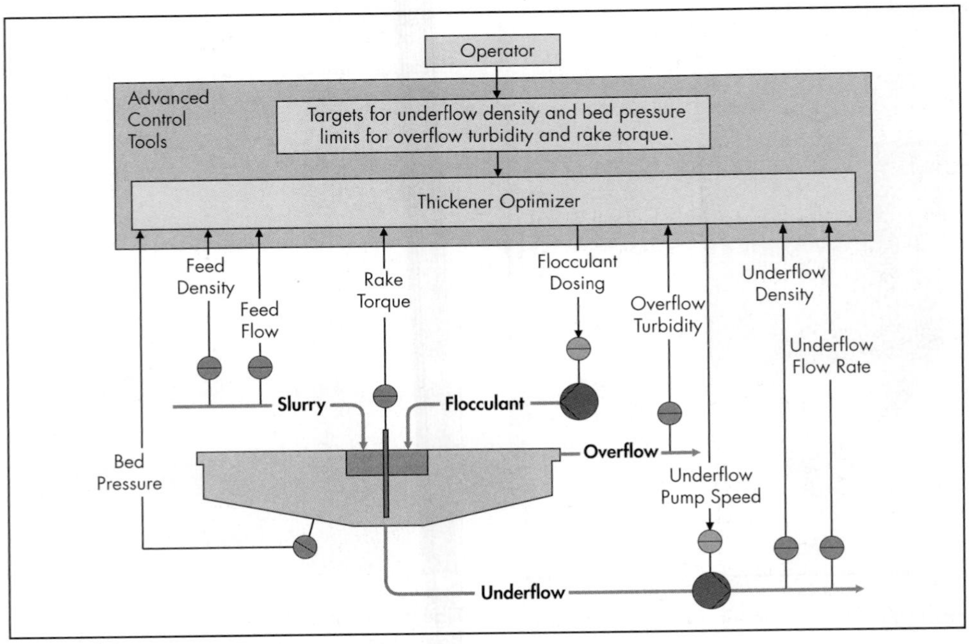

Courtesy of Outotec

Figure 16 Optimizing control for thickeners

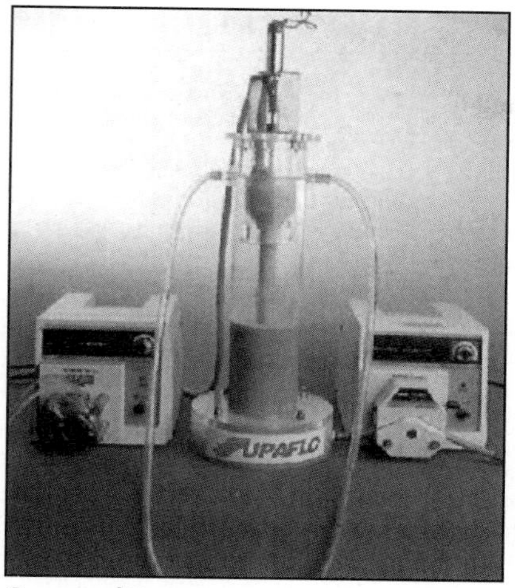

Courtesy of Outotec
Figure 17 Dynamic test thickener, 100 mm

Courtesy of Outotec
Figure 18 Pilot thickener, 1-m diameter

or similar. The rakes are rotated and underflow withdrawal started once a compression zone of settled solids is formed. This continues until steady-state operation is achieved, at which time the underflow and overflow are sampled and measured for density and clarity, respectively.

This test is repeated with a series of solids flux rates and flocculant addition rates until a data set is built up showing the relationship between the preceding variables. It is then possible to determine the appropriate flux rate and flocculant dosage to achieve the desired underflow density and overflow clarity directly from the data. Because the laboratory-scale thickener has a limited compression zone depth (usually no more than 250 mm), the flux rate selected on the basis of a required underflow density is regarded as conservative, and no safety factor is applied for scale-up to the full scale.

Test programs on pilot-scale thickeners have been used by some high-rate thickener manufacturers for situations in which large variations in feed conditions are anticipated, or where process conditions cannot be accurately simulated in the laboratory. Generally, pilot test work is carried out in the plant location and feed to the unit is taken directly from the appropriate plant process stream. The methods of data collection and scale-up are similar to the dynamic bench scale. However, because of the set-up time and need to cover a range of process conditions, pilot-scale test programs usually run to days or weeks and require significant personnel resources rather than one technician in a single day, as is the case with a typical laboratory test series. A typical 1-m-diameter pilot unit is shown in Figure 18.

TROUBLESHOOTING OPERATING THICKENERS AND CLARIFIERS

Table 3 is a list, by no means exhaustive, of common operating problems, together with some typical solutions.

FEED DEAERATION FOR FROTH CONCENTRATE THICKENERS

Flotation concentrates, which usually contain a high degree of air entrainment as a froth, present a particular challenge for thickener designers. The entrained air attaches as small bubbles to solids particles and causes solids to float to the surface of the thickener, resulting in dirty overflow and concentrate loss.

The conventional approach to this problem was to allow extra area in the thickener and include a large feed tank before the thickener to allow some of the entrained air to escape. These are still recommended design practices; however, at least one manufacturer uses a feed tank design that applies a centrifugal force to the incoming feed to accelerate the removal of entrained air. Even large amounts of froth in the feed have been shown to be removed by this type of feed tank (e.g., Outotec FrothBuster), and consequently, good thickener performance can be achieved without resorting to oversized feed tanks and thickeners. Alternatively, water sprays and froth booms can be effective for control.

MAJOR FACTORS INFLUENCING THICKENER DESIGN

The following issues can influence thickener design:

- Process requirements for the overflow liquor quality and underflow slurry density. These determine the thickener type, size, drive, and mechanism design.
- The quantity of solids to be handled. Usually expressed as weight of dry solids per unit area per hour, this typically determines the size and number of units.
- Hydraulic loading is the volume of feed divided by the cross-sectional area of the thickener and is typically expressed as rise rate in cubic meters per hour per square meter.
- The amount of material larger than 250 μm (+60 mesh) in the feed. This affects tank bottom slope, drive, and strength of mechanism. It may also require a rake-lifting device.
- Specific gravity of the solids. The greater the specific gravity, the more likely a stronger drive and mechanism will be required.
- Particle size, concentration, and settling characteristics will determine the feed system design, including the need for feed dilution.
- Feed, overflow, and underflow systems must be capable of handling additional material when other thickeners are out of service.

Table 3 Operating problems and solutions

Problem	Causes	Solutions
Low underflow density	• Can be caused by feeding a higher-than-design rate of solids to the thickener • Insufficient flocculant dosage • Inadequate flocculent mixing with the feed solids • Change in grind to a finer particle size • Low bed level caused by suboptimal thickener process control • Insufficient feed dilution	• Reduce solids feed rate • Increase flocculant dosage • Ensure correct flocculant secondary dilution • Increase solids inventory—bed level or bed mass • Move or optimize flocculant dosing locations • Change flocculant type • Ensure adequate feed dilution
Poor overflow clarity	• Sliming thickener: This occurs when the solid–liquid interface rises to the overflow weir, and slurry at a solids concentration similar to the thickener feed overflows the weir. • Overflow is murky, but not in a sliming condition: A portion of the feed solids is probably not being flocculated adequately.	• The usual solution to sliming is to increase flocculant dosage, increase underflow removal rate, or both. If necessary, the feed should be stopped until the interface drops back to normal. • Increasing feed dilution should be considered. • Increase flocculant dosage. • A different flocculant may be more effective, or a combination of inorganic settling aids (e.g., lime, iron chloride, alum) and synthetic flocculants (using laboratory-scale tests to identify the best flocculent combination is recommended). • If the problem persists, it could be a result of suboptimal mixing in the flocculation zone, which would imply a change to the flocculent/feed mixing chamber configuration.
High rake torque	• High torque with high-density underflow	• Mud load or solids inventory in the thickener needs to be reduced by increasing underflow pumping rate, reducing flocculant dosage, or both.
	• High torque with low-density underflow: This usually indicates short-circuiting of partially thickened material to the underflow and is commonly caused by a mass of agglomerated solids adhering to and rotating with the rake arms. The condition is variously referred to as a *donut, islands,* or *solids mass.*	• The usual cause is overflocculation of feed for a significant period. It is often difficult to remove without draining the thickener, although sometimes raising and lowering the rakes in a rapid sequence may move the mass. Flocculant dose rate should be reduced to a *correct* level for flocculation.

Adapted from Schoenbrunn and Laros 2002

- Underflow material characteristics may require special rake construction such as blades located a distance below the rake arms on posts (thixo posts).
- Scale buildup tendency of feed slurry may require special arms and drive (e.g., alumina applications).
- An operating requirement to accumulate solids for defined periods of time will require a special mechanism design, as it is not a normal operating procedure.
- Froth control or removal may require sprays, froth baffles, or skimmers.
- Slurry properties (pH, chlorides, etc.) will determine materials of construction for the wetted parts (tank and mechanism).
- Slurry temperature, vapors, gases, and so forth, may require covered and/or insulated tanks with attendant seals.
- Soil conditions and groundwater elevation affect foundation design and may determine tank and mechanism type.
- Climatic conditions may require special considerations, such as enclosures around the drive, tank insulation, and instrumentation.

CONCLUSIONS

A range of sedimentation equipment designs are available for various applications and process objectives. Utilizing the proper design for an application can make the difference between a smooth operating process and continual problems that prevent a plant from realizing its potential. Taking the time in the project planning stage to make sure the correct design is being used can make for a successful project.

REFERENCES

AGMA 2001-C95. 1995. *Fundamental Rating Factors and Calculation Methods for Involute Spur and Helical Geer Teeth*. Alexandria, VA: American Gear Manufacturers Association.

Coe, H.S., and Clevenger, G.H. 1916. Methods for determining the capacity of slime settling tanks. *Trans. AIME* 55:356–385.

Cross, H. 1963. A new approach to the design and operation of thickeners. *J. S. Afr. Inst. Min. Metall.* 63(7):271.

Emmett, R.C., Laros, T.J., and Paulson, K.A. 1992. Recent developments in solid/liquid separation technology in the alumina industry. In *Light Metals 1992*. Edited by E.R. Cutshall. Warrendale, PA: The Minerals, Metals & Materials Society. pp. 87–90.

Klepper, R., Laros, T., and Schoenbrunn, F. 1998. Deep paste thickening systems. In *Minefill '98: Proceedings of the Sixth International Symposium on Mining with Backfill.* Edited by M. Bloss. Melbourne, Victoria: Australasian Institute of Mining and Metallurgy.

Kosonen, M., Kauvosaari, S., Gao, S., and Henriksson, B. 2017. Performance optimisation of paste thickening. In *Proceedings of the 20th International Seminar on Paste and Thickened Tailings.* Edited by A. Wu and R. Jewell. Beijing, China: University of Science and Technology.

Kynch, G.J. 1952. A theory of sedimentation. *Trans. Faraday Soc.* 48:166–176.

Schoenbrunn, F., and Laros, T. 2002. Design features and types of sedimentation equipment. In *Mineral Processing Plant Design, Practice, and Control.* Edited by A.L. Mular, D.N. Halbe, and D.J. Barratt. Littleton, CO: SME.

Triglavcanin, R. 2008. The heart of thickener performance. In *Paste 2008: Proceedings of the 11th International Seminar on Paste and Thickened Tailings*. Edited by A. Fourie. Nedlands, Western Australia: Australian Centre for Geomechanics.

CHAPTER 8.2

Filtration

David G. Meadows

FILTRATION APPLIED THEORY AND BATCH FILTRATION

Filtration is used throughout the mining industry in a range of duties with some traditional steps and others more specific or unique to the process. Filtration in the minerals processing industry can be divided into several categories, and below highlights aspects of the batch filtration theory.

For batch filtration, two methods of operations are usually considered: filtration at a constant pressure and filtration at a constant volumetric rate. Typically, a combination of both methods is used, depending on its supporting (ancillary) equipment.

The process is described by Poiseuille's law using this basic expression as a starting point:

$$\frac{dV}{Adt} = \frac{\Delta P}{\frac{\mu(\alpha\omega V + r)}{A}}$$

where

V = volume of filtrate
A = area available for filtration
t = time
ΔP = pressure drop across filter
μ = liquid viscosity
α = specific cake resistance
ω = weight of dry cake solids per unit of volume filtrate
r = resistance of filter cloth and drainage network

Rearranging the equation:

$$\mu\alpha\omega V\frac{dV}{A^2} + r\frac{dV}{A} = \Delta P dt$$

If pressure is assumed as constant as well as temperature together with the feed conditions that means term α is supposedly a function of only ΔP and, therefore, considered as a constant too. By integrating the previous equation:

$$\frac{\mu\alpha\omega V}{2A^2} + \frac{\mu r V}{A} = \Delta P t$$

Dividing by V/A and ΔP:

$$\frac{t}{V/A} = \frac{\mu\alpha\omega V}{2\Delta P A} + \frac{\mu r}{\Delta P}$$

By plotting this equation $t/V/A$ versus V/A, it will produce a straight line with a slope of $\mu\alpha\omega/\Delta P$ in the following graph:

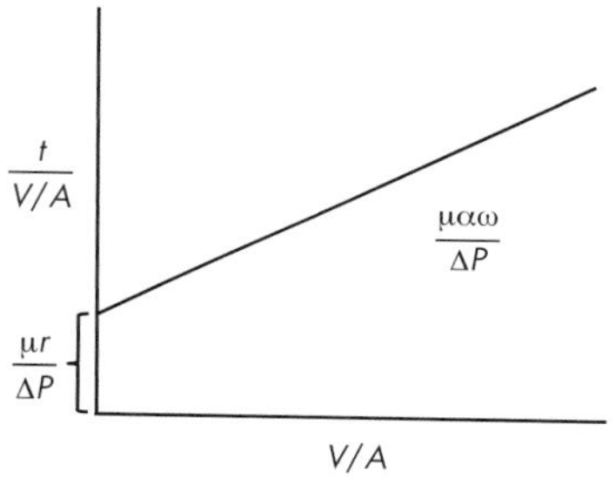

Because μ and ω can be measured, the terms of r and α can be readily solved in the preceding equation. Depending on the compressibility of the cake, the specific cake resistance can be measured as a function of the pressure. Results should be as shown in the following graph, where the slope (S) is the cake compressibility exponent ranging between 0 (incompressible) and 1 (highly compressible):

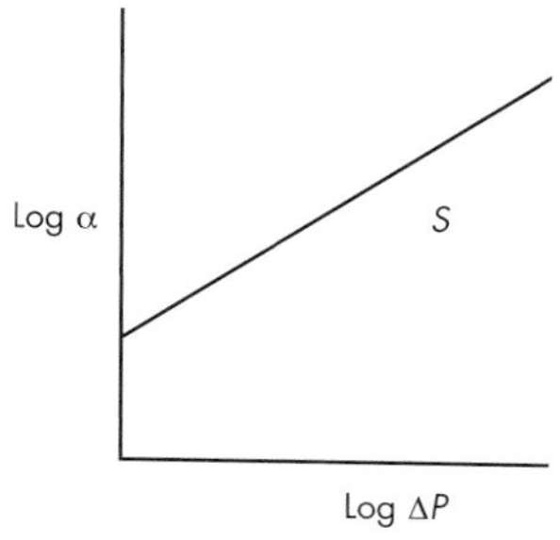

FILTER TESTING AND TESTING CAPABILITIES

Selection of the appropriate filtration equipment requires choosing the optimal piece of equipment for the specific application. For example, horizontal belt, rotary drum, and disc

David G. Meadows, Manager of Global Process Technology, Bechtel Mining and Metals, Phoenix, Arizona, USA types of vacuum filters, as well as horizontal or vertical filter presses, may all be candidates for filtration of many types of metallurgical streams and products. Each type, however, requires a different testing approach. Caution should be taken when comparing test method data that has used a different filtration mechanism to infer the filtration rate for the different units. Furthermore, process requirements must be understood and fully defined prior to initiating laboratory work. Some of the important questions include: Will the process require cake washing to reduce or remove the liquor solute content of the cake? What target moisture content is desired for the cake produced? These types of questions should be answered prior to starting a testing campaign and before including in a process design document. The process definition, operating parameters, and so forth, must be thoroughly considered so that all the data required to meet the process objectives are collected during the testing campaign. It is also important to take into account downstream requirements such as transportable moisture level or others predefined by a particular process.

Once objectives are understood, a test program can be conceived. Vacuum and/or pressure filtration tests can be planned as needed to determine unit production rates for each relevant piece of equipment. Planning should include the collection of all necessary data to describe various process options and their associated operational trade-offs. Cake washing rates and efficiency data should be collected if the filter is to be used for a washing application. Similarly, drying cycle data should adequately describe the trade-off between cake dryness and production rate. The effects of expected variations in feed conditions (such as feed slurry solids content, particle size [particularly in the finer sub-30-µm fractions], process temperature, applied vacuum, and ore mineralogy including gangue mineralogy such as clay content, precipitation conditions, etc.) should be considered in the testing scope as well.

Design Parameters

The following main parameters must be determined and established in the filter design:

- Feed slurry dewatering
- Cake washing
- Filtrate clarity requirements
- Cake moisture content
- Cake transportability requirements
- Filter production rate

Sample Characterization

Selection of representative samples is critical if test results are to apply to the actual processes. Samples should be matched as closely as possible to the actual expected stream and process conditions for which the filtration equipment is to be specified.

Sample characteristics that should be considered when determining whether a sample is representative include several physical and chemical parameters. Solids concentration, ore specific gravity, size distribution, slurry pH, redox potential, and chemical dosage (such as flotation reagents or flocculants) should be matched as closely as possible.

Variations in ore type or mineralogy in a mine can significantly affect filterability of flotation concentrates and should therefore be considered. Similarly, in industrial minerals applications, feed characteristic variation can influence the performance of the required filtration step. Supportive mineralogical characterization of samples will assist in interpreting filtration performance results.

Sample aging can also be extremely important. Many metallurgical solids are reactive and change significantly enough with time to alter the filtration characteristics. Samples for filtration testing should be as fresh as possible. Ideally, prolonged storage times should be avoided. Samples should not be allowed to freeze, dry, or be dewatered and then repulped prior to testing, since the resulting sample will probably not behave the same as it would have originally.

Expected process parameters should be closely matched when preparing the sample for filtration testing. Feed solids concentration, temperature, pH, and so forth, should be carefully matched to the process conditions. Temperature changes also influence filtrate viscosity and cloth blinding through salt deposition.

Feed solids concentration is especially important to consider, since any dewatering process will be greatly affected by the quantity of water present in the stream. Sometimes processes produce samples at solids concentrations that are not amenable to filtration. In these instances, it is generally recommended to pre-thicken the filtration feed with a standard style of thickener. If a flocculated underflow is feeding a filter, the shear conditions also need to be considered regarding the breakdown of the flocculant present. It may necessary to use a peristaltic pump to minimize the shear on the flocculated feed to a drum filter.

Another example would be flotation plant concentrates, which are normally always thickened prior to the final concentrate filtration. It is often advantageous to conduct all solid–liquid separation testing in concert, since (for example) the thickening underflow test material could be used for rheology and subsequent filtration testing. In general, it is better to perform the tests within the same laboratory facility, although this is sometimes not the case on more complex hydrometallurgical flow sheets involving multiple unit process steps and more complex solution chemistry considerations.

Equipment Needed

Vacuum Filtration Testing

The primary equipment required for vacuum filtration test work consists of a grid of known defined area covered with an appropriate filter cloth and surrounded by a metal or plastic shim to contain the pulp sample. This drainage grid or filter leaf is supported vertically on a vacuum flask. Alternately, a similar vacuum leaf can be attached to a flexible tube that connects to a vacuum flask. This alternate arrangement would allow for the leaf to be inverted and immersed in a container of filter feed slurry to simulate drum-type vacuum filter operation. The differential pressure is transmitted from the vacuum pump to the filter leaf surface through large-bore tubing and fittings. The vacuum pump should be equipped with an internal bypass system to control the vacuum level without the introduction of bleed air.

Pressure Filtration Testing

Bench-scale pressure filtration test work can be performed using a "pressure bomb device." The apparatus consists of a 250-mm section of nominal 50-mm pipe (to reduce the effect of any bypass of air or filtrate), capped with two flanges. The upper flange contains fittings for the air pressure connection and the sample feed inlet port. The lower flange contains an integral drainage grid, which supports the filter media. The filtrate port is centered in the bottom flange, below the filter media.

Data Collection

Vacuum filtration and washing tests should be conducted to collect a complete set of filtration and wash efficiency data to allow design and sizing of the vacuum filters. Testing should (as applicable) examine the effect of applied vacuum level, feed solids concentrations, cake thickness, dry time, filter aid (flocculant or diatomaceous earth, etc.), and the volume of applied wash solution on production rate, filter cake moisture, and wash efficiency. Similarly, pressure filtration and washing tests should be conducted to collect a general set of filtration and wash efficiency data sets to allow design and sizing of the pressure filters. Tests should examine the effect of applied pressure, feed solids concentration, cake thickness, air blow time, and the volume of applied wash solution on production rate, filter cake moisture, airflow rate, and wash efficiency.

Vacuum Filtration Testing

To produce a test filter cake from a slurry sample, a given weight of pulp at the appropriate process temperature and known to yield an approximate cake thickness should be poured onto the upturned test leaf while the ball valve connecting the leaf to the vacuum flask is opened to apply the differential pressure. As the remainder of the liquid phase disappears through the surface of the formed cake and the form time ends, a known volume of wash water is poured onto the surface of the newly formed cake (only applicable if a wash is to be tested). Once the last of the liquid wash solution disappears through the surface of the cake, the wash time is ended and noted and the subsequent dry time begins.

At the end of the dry time, the filter cake is discharged from the leaf, and the wet weight and cake thickness are measured and recorded. After drying, the final dry cake weight is determined and recorded for cake moisture calculations.

Other similar tests will allow additional cakes to be collected in the same manner, each at different test conditions, until the range of test variables selected has been adequately covered.

While performing vacuum filtration, the overall vacuum achieved in each cycle should also be considered. If there is loss of vacuum during the air-dry stage, this will dictate the vacuum that can be used for the form for most systems. Typically a single vacuum pump is used to drive both the form and the dry stage; therefore, the vacuum during the form must be adjusted to match the dry-vacuum requirement. This may be overcome by modifying the valves with two disc filters installed with two vacuum pumps. One pump is dedicated to the form and the other to the dry stage, and the pumps are connected to both filters.

Pressure Filtration Testing

To produce a test pressure filter cake from a slurry sample, a given weight of pulp at the desired process temperature and known to yield an approximate cake thickness is poured into the pressure chamber. The sample port is closed and air pressure applied above the feed slurry to facilitate initial cake formation and dewatering. As the last of the filtrate is produced, shown by a rapid airflow through the drainage grid, the form time is noted and recorded.

Immediately following the form time, the filter is depressurized and a known volume of wash solution is poured onto the surface of the newly formed cake (for washing cases only). The sample port is then closed and air pressure applied to facilitate cake washing and dewatering. Once the last of the liquid wash solution is produced, the wash time is noted and the subsequent dry time begins (only applicable if a wash was applied).

At the end of the timed air-blow step, the filter cake is discharged from the filter, and the wet weight and cake thickness are determined and recorded. After drying, the dry cake weight is determined and recorded for final cake moisture calculations. Additional cakes should be collected in the same manner, each at different test conditions, until the range of test variables selected has been adequately covered.

Wash Procedure Guideline

Experience has shown that the construction of a washing curve from data is often made more difficult than necessary by the choice of methods employed to collect these data sets. Hence, when washing effectiveness data are collected, the method of testing should be varied, as described in the following paragraphs.

A washing stage in vacuum and pressure filtration consists of cake liquor displacement by both air and the applied wash liquor. Remaining cake liquor content will be a function of time of air displacement, cake thickness, pressure drop across the cake, particle size characteristics, liquor viscosity, and other factors. The wash stage begins when the instant wash liquor is applied and continues as the fluid passes through the cake. The sequence is completed by a period of liquor displacement by air. The end of a wash stage may precede cake discharge from the filter, or the application of a subsequent wash (in cases where multiple wash stages are necessary).

A correlation of washing data should fall on a single curve and should illustrate the relationship between the volume of wash applied and the degree of solute removal without being affected by variations in the liquor concentration throughout the washing stage. It is possible to achieve this type of correlation using a parameter R, which is plotted as a function of a second parameter N, where N and R are defined as follows: N is the volume of wash applied, divided by the volume of liquor in the cake at the end of the wash stage, which includes some liquor displacement by air. If L_w is the volume of applied wash and L_2 is the volume of liquor in the cake at the end of the wash stage, then N is defined as

$$N = \frac{L_w}{L_2}$$

In situations where the wash fluid applied contains no soluble value, R is the concentration of solute in the cake liquor at the end of the wash stage, divided by the concentration of solute in the cake liquor at the beginning of the wash stage. In cases where the wash fluid contains soluble value, this concentration must be accounted for. Theoretically, with infinite wash applied, the concentration of the solute in the cake liquor would equal the concentration of solute in the wash liquor. Therefore, if C_2 is the concentration of solute in the cake liquor at the end of the wash stage, C_1 is the concentration of solute in the cake liquor at the beginning of the wash stage, and C_w is the concentration of solute in the applied wash fluid, then R is defined as

$$R = \frac{C_2 - C_w}{C_1 - C_w}$$

The method of determining cake liquor concentration at the end of a washing stage is critical in terms of the quality of the correlation achieved. Unsatisfactory results are obtained if either the wash filtrates or dry (washed) cake residues are analyzed for residual soluble value content.

The suggested and recommended approach consists of repulping each filter cake of a wash series in a known volume of pH-adjusted deionized water. After repulping, the resultant slurry is filtered using a new filter mat, a clean and dry glass Buchner funnel, and a vacuum flask for each cake. Filtrate reporting to the vacuum flask is submitted for assay while filter cake solids are dried to obtain dry suspended solids. Complete accountability for all cake solids must be maintained throughout this repulping procedure.

When the analytical results are obtained from the cake repulp liquor, the actual concentration of soluble value in the liquor associated with the cake upon discharge from the filter can be calculated by solving the following expression:

$$C_1 = \frac{\frac{W_1}{\rho} + V_r}{\frac{W_1}{\rho}} C_{r1}$$

where
C_1 = cake liquor concentration at discharge, g/mL
W_1 = weight of cake liquors, g
ρ = density of cake liquor, g/mL
V_r = volume of repulp fluid used, mL
C_{r1} = concentration of solute in repulp liquor (analytical result)

General Filtration Testing Data Interpretation and Correlation

Data collected during filtration testing can be correlated to show several general parameter relationships, which can subsequently be used to predict filter performance for a given set of process conditions (Choudhury and Dahlstrom 1957 and others). The key correlations noted include the following:

- Cake weight versus cake thickness
- Cake formation rate versus cake weight (or thickness)
- Cake moisture content versus relative dry cycle length
- Cake wash penetration versus cake thickness and applied wash volume
- Solute removal verses applied wash volume

These correlations can each be used to describe the filter operation during each portion of a complete filtration cycle, and when used together will describe the total filter performance and cycle time for the conditions tested. As an example, the following information is used to illustrate the correlations described in the preceding list. The specific correlations are shown in Figures 1–5.

During the testing campaign, several cakes of various thickness should be produced so that data for the first two correlations can be collected. The first correlation shown in Figure 1 demonstrates the relationship between wet filter cake thickness and dry filter cake weight, W. For design, for example, a possible cake thickness of 15 mm could be used for horizontal belt filters. According to Figure 1, the unit weight of a 15-mm cake would then be 2.9 dry kg/m^2 when flocculant is added (lower data points), while a 15-mm cake produced without flocculant would weigh 3.9 dry kg/m^2.

Figure 2 displays the logarithmic relationship of dry cake weight, W as a function of cake formation time. As predicted by theory, the slope of each curve is 0.5 (Silverblatt et al. 1974). The correlation shown in Figure 2 indicates that, for the case where flocculant is used, a 15-mm cake, which weighs 2.9 dry kg/m^2, will form in approximately 2.1 minutes, whereas when flocculant is not used, a 15-mm cake, which weighs 3.9 dry kg/m^2, will form in approximately 7.2 minutes.

The relationship between filter cake moisture at discharge and the dry time factor (θ_d/W) is shown in Figure 3. The dry time factor is the dry time (θ_d) divided by the dry cake weight per unit area (W). The dry time factor permits a correlation

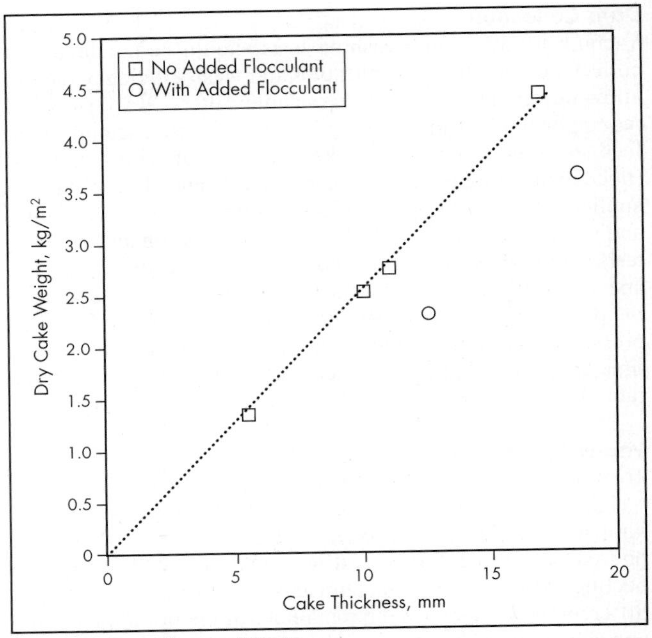

Source: Smith and Townsend 2002
Figure 1 Cake weight versus cake thickness

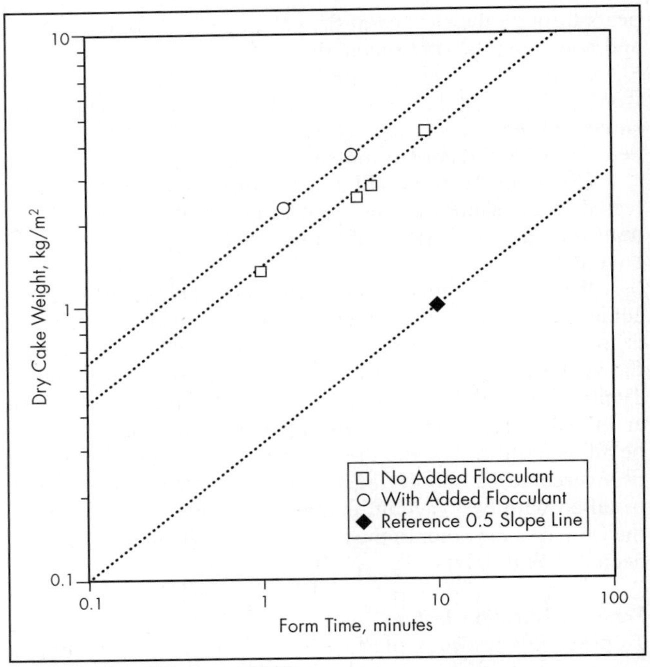

Source: Smith and Townsend 2002
Figure 2 Cake weight versus form time

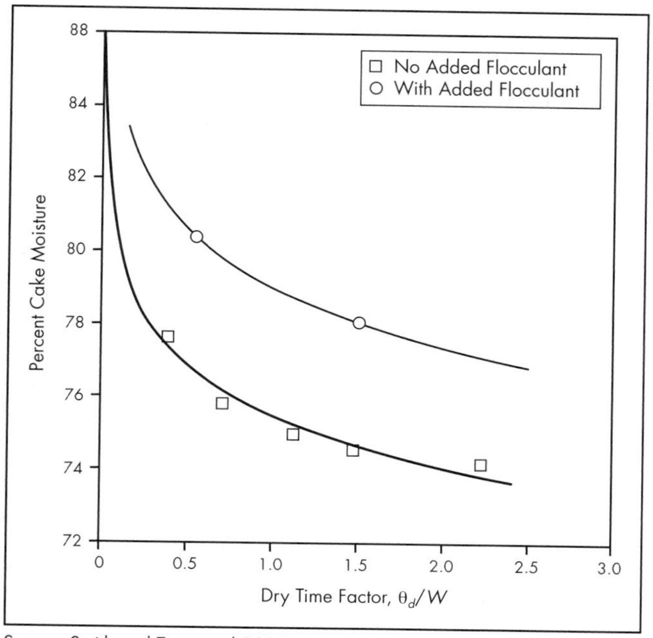

Source: Smith and Townsend 2002

Figure 3 Cake moisture versus dry time factor

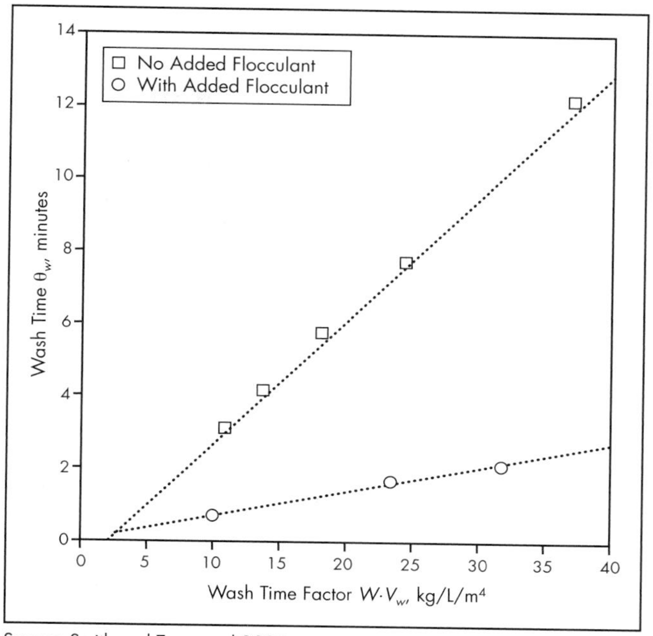

Source: Smith and Townsend 2002

Figure 4 Wash time versus wash time factor

between cake moisture and dry time for all variations in cake thickness, by normalizing the dry time for cake weight, which depends on cake thickness. The correlation indicates that when flocculant is employed, an ~2.9-minute dry time (θ_d/W = 1.0 min·mm²/kg and W = 2.90 dry kg/m²) following the cake wash will yield filter cake with 79% moisture. Similarly, when flocculant is not used, an ~4.9-minute dry time (θ_d/W = 1.27 min·mm²/kg and W = 3.9 dry kg/m²) following the cake wash will yield filter cake with 75% cake moisture.

Figure 4 shows the correlation of the wash time (θ_w) and the wash time factor ($W{\cdot}V_w$), plotted as θ_w versus $W{\cdot}V_w$. Similar to the dry time factor, the wash time factor permits a correlation between wash time and specific wash volume V_w for all variations of cake thickness. From this plot, it can be seen that for a $W{\cdot}V_w$ value of 27.82 kg·L/m⁴ (corresponding to a wash ratio of $N = 0.5$), a wash time of 8.98 minutes is required when no flocculant is added. Similarly, when flocculant is used, a $W{\cdot}V_w$ value of 19.89 kg·L/m⁴ (corresponding to a wash ratio of $N = 0.5$) a wash time of 1.37 minutes is required.

Figure 5 shows the relationship of fraction of solute remaining (R), determined from assay values on the repulped filter cakes, with respect to the wash ratio (N). The wash ratio N is the number of cake liquor displacements with wash solution when the filter cake is at the discharge moisture content.

The wash recovery data correlation displayed in Figure 5 (often referred to as a wash curve) allows a quick estimate of where increases in wash volume produce little effect, so that the optimal number of stages can be selected. The predicted "R" values also allow for construction of accurate material balances for a variety of washing filter types. An example of a material balance for a three-stage countercurrent washing belt filter is shown in Figure 6. Although this is a somewhat more complicated filter arrangement, data from the wash curve allows recovery prediction.

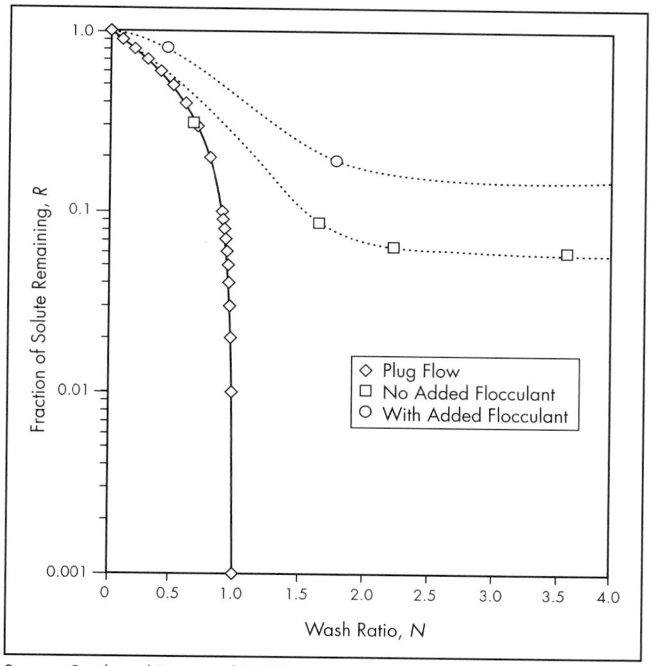

Source: Smith and Townsend 2002

Figure 5 Solute remaining versus wash ratio

Vacuum Filter Cycle Times and Production Rate Calculations

Horizontal Belt Vacuum Filter

The cycle time for horizontal belt filters is the summation of several steps. It includes the summation of the cake formation time, the cake washing time, and the final cycle cake dry time. The production rate (with units of weight per filter unit area

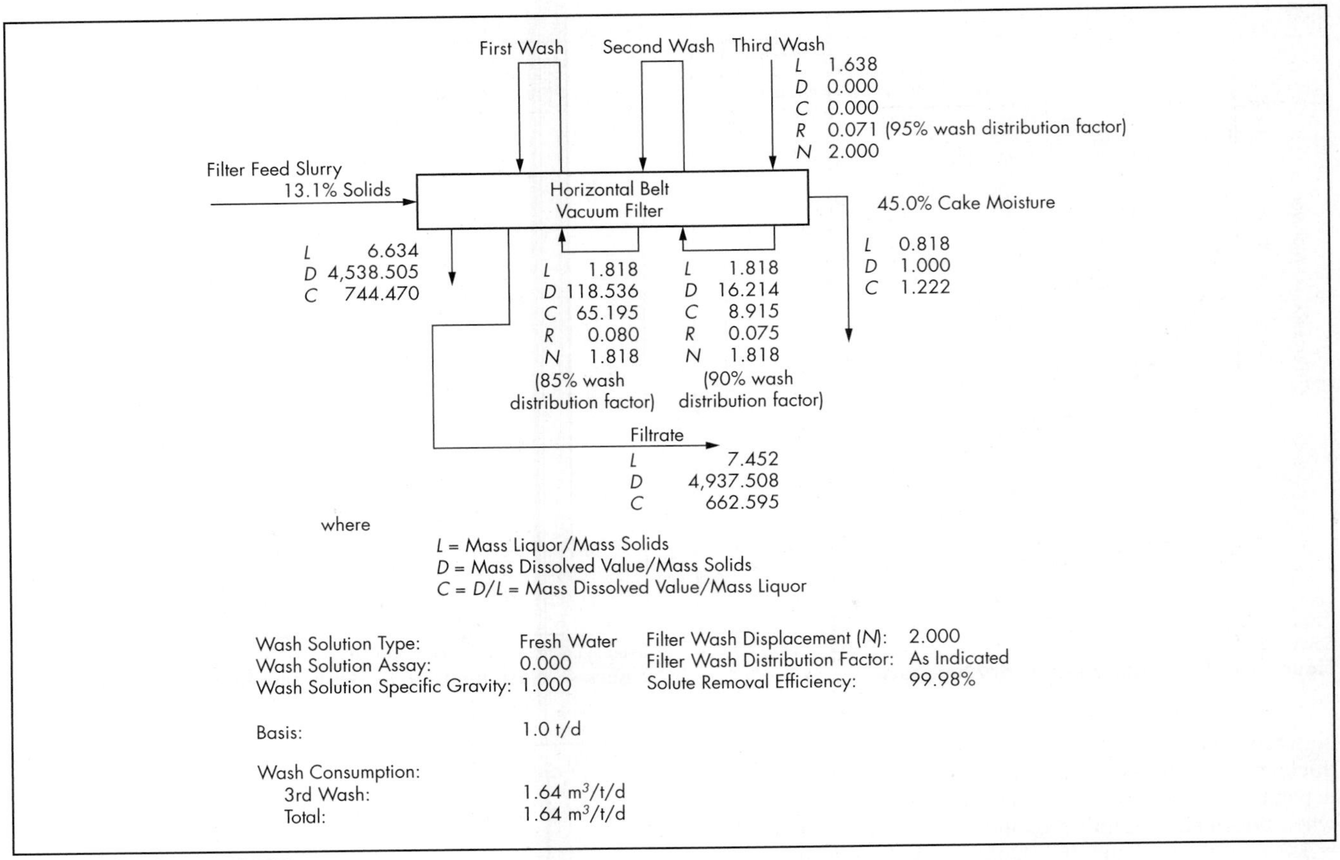

Source: Smith and Townsend 2002

Figure 6 Cake washing material balance

per time) for a horizontal belt vacuum filter is calculated as follows:

$$\text{production rate} = \frac{C \times W}{\text{cycle time}}$$

An empirical scale-up factor is included in the C term as well as the required constants for engineering unit conversions. Minimum cake thickness is considered during selection of cycle times used to calculate the production rates. The absolute minimum design cake thickness for horizontal belt vacuum filters ranges from 3.2 to 4.8 mm, depending upon the feed material characteristics.

Rotary Drum Vacuum Filter Configuration, Cycle Time, and Production Rate

A standard data and testing approach, such as that previously discussed for horizontal belt vacuum filters, is utilized in the same manner for rotary vacuum drum and disc filters with a few modifications that account for the configuration of a rotary vacuum filter. Form time, wash time (when applicable), and dry time, and thus total cycle time and production rate are generally based on vacuum drum filters described previously and are listed in Table 1. These fractions of the total cycle time result from the geometry of the drum filter itself. Figure 7 is an example of how a vacuum drum filter might be proportioned.

Hence, for the general case, cycle time is given by the following equation:

$$\text{cycle time} = \frac{\text{form time}}{\text{form zone fraction}} = \frac{\text{dry time}}{\text{dry zone fraction}}$$

Specifically, the cycle time for a rotary disc vacuum filter is given by

$$\text{cycle time} = \frac{\text{form time}}{0.30} = \frac{\text{dry time}}{0.375}$$

Filter operation can be form time limited or dry time limited based on which portion of the cycle is shown to be limiting during the testing. The cycle time used is always the largest part of the preceding sequence, since this slowest portion of the cycle limits the overall filter operation. A form- or dry-time-limited operation is possible within limits imposed by the maximum and minimum rotational speed of the drum. By determining the times needed for cake formation and drying for a given cake thickness and moisture target, it is possible to determine the total required cycle time.

The form cycle time can be limited by the minimum cake thickness required for discharging. The minimum design cake thickness for a rotary vacuum filter ranges from 9.5 to 12.7 mm, depending upon material characteristics. Some materials can be as low as 5–6 mm. If the thickness were lower than 6 mm, the cake would need to be a higher-specific-gravity material like a concentrate.

The limiting zone of the disc governs the cycle time for a complete circular rotation. The non-limiting zone may not

Table 1 Vacuum drum filter area apportionment

Drum Filter Zone	Drum Filter Without Wash		Drum Filter with Wash	
	Arc Length, degrees	% of Cycle	Arc Length, degrees	% of Cycle
Form zone	108	30	108	30
Wash zone	N/A*	N/A	90	25
Dry zone	180	50	65	18
Discharge/prewash/dead	72	20	97	27
Totals	360	100	360	100

Source: Smith and Townsend 2002
*N/A = not applicable.

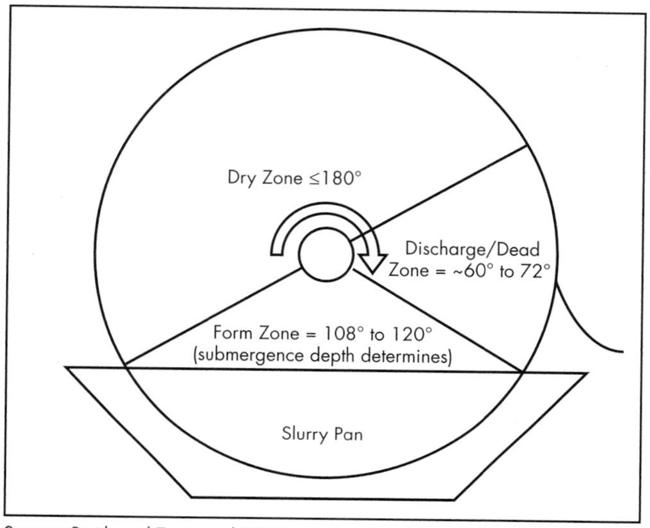

Source: Smith and Townsend 2002
Figure 7 Vacuum drum filter zone configuration without wash

require the full available zone to meet the design target. The form zone may be limited to the specific required form time by placing a rotary valve bridge block setting within this zone. There is also flexibility in the submergence of the disc or drum, which can be adjusted by the level in the feed trough. The drum can range from 5% to 30% submergence and the disc can be as low at 30% with larger units and up to 45%. This is influenced by the diameter of the barrel that supports the discs; consequently, the larger filters have more flexibility in the ratio of form to dry.

The production rate (with units of weight per filter unit area per time) for a rotary disc vacuum filter is calculated using the following equation:

$$\text{production rate} = \frac{C \times W}{\text{cycle time}}$$

where W is the dry cake weight per unit filter area, and C is a constant, including an empirical scale-up factor as well as required constants for engineering unit conversions.

Pressure Filter Cycle Time and Filter Press Sizing

Pressure filter cycle time will simply be the sum of the individual portions of the cycle time for each operation, that is, cake formation time, cake washing time, dry time (including diaphragm squeeze time, if applicable), and miscellaneous (or mechanical) time required to open and close the press, discharge the cake, and clean the cloth, and so on. Therefore, the total cycle time is given by the following equation:

$$\text{cycle time} = \text{form time} + \text{dry time} + \text{wash time} + \text{miscellaneous time}$$

Once a cake thickness is selected (based on filter press chamber thickness), the components of the cycle time can each be predicted by the correlations shown in Figures 1–4. Within a vertical recessed plate or plate-and-frame pressure filter chamber, two filtration surfaces exist; hence, a half-cake forms from each filtration surface simultaneously. This half-cake thickness must be used for form time prediction from the correlation in Figure 2. In pressure filters with horizontal plates, filtration is single-sided, so full-cake thickness is used for form time calculation. Total cake thickness (chamber thickness) should be used for all other calculations. The correlation in Figure 1 relates cake thickness to cake weight per unit area (W), and the correlation in Figure 2 relates form time to cake weight W. The correlations in Figures 3 and 4 similarly predict washing and dry times based on selected wash volume and cake moisture. The miscellaneous time will be equipment specific and should be supplied by the vendor of the selected equipment.

Once the pressure filter cycle time is known, the filter sizing can be undertaken. This is not as simple as determining the production rate based on test data. For example, if filtration rate alone determined pressure filter sizing, the production rate (with units of weight per filter unit area per time) would be given by

$$\text{theoretical production rate} = \frac{W}{\text{cycle time}}$$

This equation is generally only valid for hydraulically limited filters, for which the production rate is relatively small. Most pressure filters are actually limited by the volume capacity of the filter, which limits the amount of cake that can be produced in a single cycle.

Since pressure filters are closed-chambered devices, the fixed volume of the chambers limits the volume of cake that can be produced in a single cycle. Hence the dry bulk cake density (dry kilograms per cubic meter) is an extremely important experimental value. This value can be inferred from the correlation in Figure 1, which describes the dry cake weight per unit area per unit of cake thickness. With the dry bulk density known, the desired tonnage will set the total volume of cake to be produced in a given operating day. The cycle time can now be used to set filter sizing, since the total hours of filter operation per day will limit the number of cycles that can be run by the filter:

$$\text{cycles per day} = \frac{\text{hours per day worked}}{\text{cycle time (hours)}}$$

Therefore, the filter press required cake volume (in cubic meters) must be

$$\text{filter press volume} = \frac{C \times 1{,}000 \times \text{daily tonnage}}{\text{cake dry bulk density} \times \text{cycles per day}}$$

where C is an empirical scale-up factor, tonnage is in metric tons, and cake dry bulk density has units of kilograms per cubic meter. This volume specification can be considered to be the required filter sizing.

The effective production rate for a filter sized previously described would then be

$$\text{effective production rate} = \frac{\text{cake weight per cycle}}{\text{filter press area} \times \text{cycle time}}$$

If this effective production rate (for volume-limited filter presses) is less than the theoretical production rate previously described, the filter design can be considered to be volume limited. Since this happens to be the case in many instances, this shows the danger inherent in sizing a pressure filter based on production rate alone.

Pilot-Scale Testing

The laboratory testing described in the previous sections will generate filter sizing data using standard bench-scale test equipment. These types of test are particularly useful when only small quantities of representative sample are available and for earlier scoping-type work. The test data can be used to evaluate many vacuum and pressure filtration options as part of the flow sheet and equipment selection process. A short list of preferred filter types can then be established. While laboratory-scale testing focuses on the filtration process, the scope of pilot-scale testing is at a larger scale and includes data collection for full-scale plant design including some of the ancillary considerations. This also allows the generation of larger product samples that may be required for downstream uses.

If sufficient representative samples are available, for example at an existing production plant considering production upgrades, testing using pilot-scale equipment can be considered. Pilot-scale testing may not always be necessary but should be considered with respect to the following considerations:

- To evaluate a specific type of filter, and to demonstrate that it will work as predicted at laboratory scale
- To check filter operating characteristics that cannot be demonstrated at laboratory scale, such as cake discharge and cloth blinding
- To produce large samples of filter cake or filtrate to develop downstream unit operations, or for marketing purposes
- To collect engineering data for filter plant design and ancillary equipment selection
- To compare existing and proposed filters in parallel over time, and over a wide range of operating conditions
- To gain operating and maintenance experience on a filter as part of the selection process, or to train staff before the production unit is installed

Whereas standard laboratory equipment can be used to evaluate almost all types of vacuum and pressure filters, pilot-scale units tend to be more specific. Pilot-scale filters are often supplied by manufacturers and are designed to replicate the operation of that company's equipment. Many global laboratory testing centers also have a range of pilot-scale filters that they have collected over their many years of service (e.g., Hazen Research Inc. in Golden, Colorado, United States).

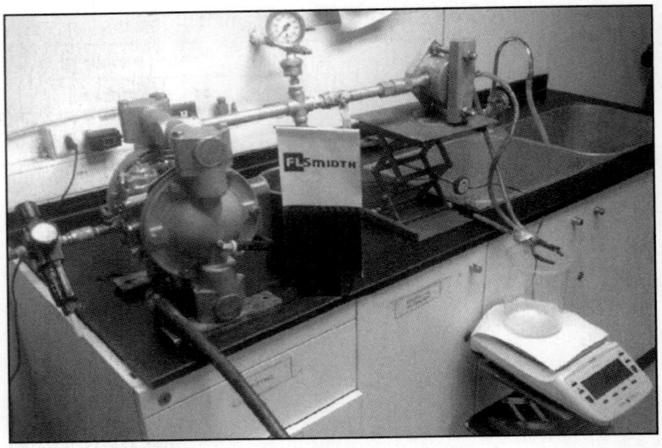

Courtesy of FLSmidth
Figure 8 Bench filter press

Pilot-scale testing is not rigidly defined, and test units can have filtration areas ranging from 0.1 to 5 m^2. Pilot plants may be located permanently, for example, at in a fixed location such as a laboratory research center. However, today, several of the equipment vendors such as FLSmidth and Outotec have mobile units located in multiple regions of the world. These machines can be transported and operated wherever needed once all the ancillary requirements are set up, such as power, water, compressed air, and so on. On-site testing gives greater testing options, and samples processed, in general, will always be fresh. Also, there is a greater level of attention and understanding to the sample collection at the site.

Smaller pilot-scale units are operated off-line in batch tests, whereas larger units can be integrated into an existing process for continuous operation. In both cases, preparations for pilot-scale testing must be made well in advance. Although some transportable pilot filters may be delivered with their own ancillaries, such as feed pumps and compressed air supply, others may need these to be provided at the test site. In all cases, plans must be made to provide power and water, analytical services, and assistance with installation and maintenance. A single technician can conduct laboratory-scale testing, but pilot-scale operation is significantly more demanding on resources.

A transportable filter press used for bench tests at project sites is shown in Figure 8. Slurry is fed by using a diaphragm pump powered by a compressor, and tests are run under different feed, pressure, and time conditions. The main purpose of this machine is to determine the specific filtration rate (kilograms per hour per square meter) for each particular feed and pressure condition. This equipment can either be used in a laboratory or sent to a site to compare its filtration rate results to those obtained in an industrial facility.

Similarly, Tenova's Delkor testing unit can perform test work. The Fast FP testing unit (Figure 9) is a 250 × 150 mm membrane plate with a working pressure up to 15 bar and an inflatable cushion filled with water for plate pack compression. Another Delkor testing unit is a 140 × 140 mm membrane

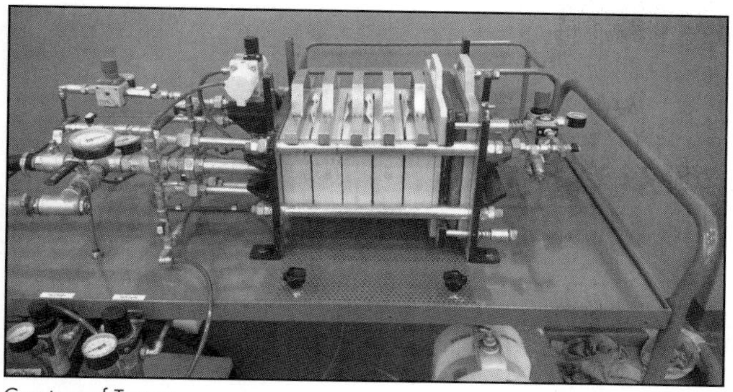

Courtesy of Tenova
Figure 9 Delkor Fast FP testing unit

plate that uses air for squeezing and has a cake-washing stage with up to 15 bar testing pressure. The compression with this unit is made manually.

Delkor also has another portable unit called the high-pressure portable FP test rig that is laboratory filter press of 140 × 140 mm with up to 25 bar testing pressure and a manually tightened screw for plate pack compression (Figure 10).

FILTRATION EQUIPMENT

With the development in technology and growth in vendor equipment sizes in the mineral processing industry, most filter applications employ mechanisms with continuous filtration. Types of continuous filters must first be considered because their characteristics will influence data interpretation and sizing. The percentage of the filter cycle that can be devoted to cake formation, cake washing, and cake dewatering will vary with basic units. Cake discharge methods can differ within a basic type, where cake washing can be practiced, to a more complex design where two or more countercurrent washes can be employed.

The filters forming the cake with the aid of gravity are generally applied to relatively fast filtering material and particularly if the solids cannot be left in suspension by a reasonable agitated filter feed tank system. The exception is the horizontal belt filter press where very thin cakes (as thin as 3 mm) can be discharged and countercurrent washing can be exploited while minimizing wash liquor addition.

Filter Selection

Mechanical dewatering means mechanical removal of liquids from a slurry to obtain the solids in a suitable form and/or recovery of valuable liquids for

- Further processing,
- Transportation,
- Agglomeration, or
- Disposal.

Several methods and products are available for mechanical dewatering, as depicted in Figure 11, depending on tonnage per hour (throughput) and final product quality (moisture):

- Gravimetric dewatering: spirals, screens, wheels
- Low-pressure dewatering: drum vacuum, belt drums, top feed
- Medium-pressure dewatering: air pressure filters (compression and through-blow)
- High-pressure dewatering: tube presses (compression and air purged)

Courtesy of Tenova
Figure 10 Delkor portable FP test rig

Filters that use pressure instead of a vacuum must use more complex mechanisms, especially for feed and discharge. The vacuum filter advantage has made them very popular in the mining industry and in general mineral processing flowsheet applications.

The main disadvantage of vacuum filters is the limitation on maximum pressure gradient for the process, which depends on local atmospheric pressure. Given that some of the copper and gold mining sites are located at elevations higher than 4,000 m, the pressure difference is low. This altitude limitation, together with technological breakthroughs, has made pressure filters the preferred equipment in the mining industry.

An interesting alternative is a combination of both filter types, pressure and vacuum filters. If a typical vacuum filter is set inside a pressurized vessel, the pressure gradient can be increased to optimal values needed for filtration. These kinds of filters are called hyperbaric and combine the construction and operation simplicity of a vacuum filter with the advantage of bigger pressures than pressure filters.

Vacuum Filters

The four types of vacuum filters are the drum, disc, tray (or pan), and horizontal belt. The first three are capable of producing cakes with about 12%–18% residual moisture while belt filters can reduce moisture to 8%–10%.

Drum Filters

A drum filter consists of a rotary drum covered with a filter media with its lower portion submerged in the slurry. The vacuum generated filtrates liquors through its interior, discharging it through proper pipelines while solids are retained in the filtering media for further dewatering and drying

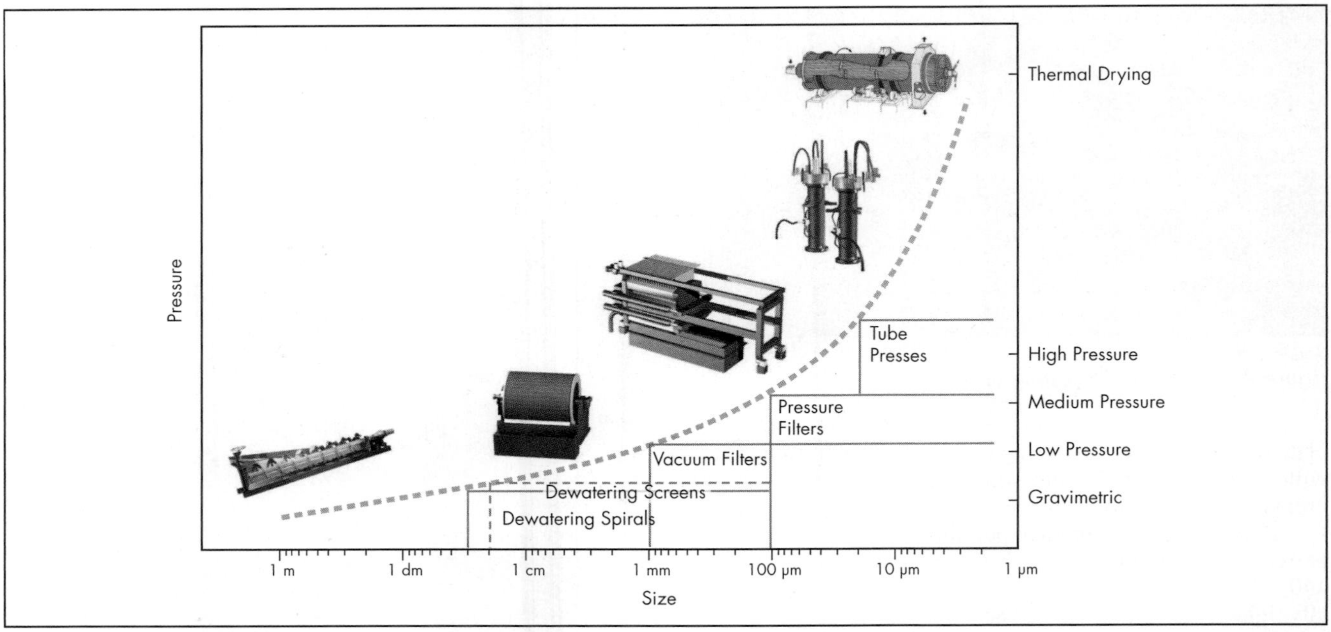

Courtesy of Metso

Figure 11 Dewatering equipment types, guidelines, and methods

during the drum rotation cycle, forming a cake that can also be washed using sprays. Once one full revolution is made, and before it contacts the slurry again, a scraping mechanism is used to remove the cake from the drum, unloading it to a bin. Next, a new filtration–drying–washing–drying–unloading cycle restarts.

Drum filters were very popular in the South African gold mining industry in the late 1970s and 1980s. They were used as the filtration step after the gold cyanidation step and ahead of the Merrill–Crowe zinc precipitation step. Typically for a 10,000–15,000-t/d (metric tons per day) gold plant, 15–20 filters were employed in the flow sheet with a treatment rate of ~30–40 t/h per drum filter.

The drum filtrate or pregnant solution would pass into the final zinc precipitation step and the filter cake would be discharged onto conveyor belts before transportation to the neutralization step and final tailings disposal. Dissolved gold losses on the filter cake were reported as 0.01–0.02 g/t Au (<1.5% soluble loss) and the filter cake moisture content was 22%–25%. Figure 12 shows a Dorr Oliver 3.65-m-diameter drum filter for gold processing.

Disc Filters

Disc filters consist of a central shaft supporting a limited number of discs made up of sectors covered with a filtering media, where each one is connected to a vacuum device (Figure 13). During rotation, each sector contacts the slurry at the lower part, working with a filtration–drying–washing–drying cycle. The system is equipped with a snap blow off that eases to release the cake.

One advantage of this machine compared to drum filter it is its available filtration area per layout area because the disc filter can filtrate using both disc faces and a single machine can handle several discs. Another advantage is that all discs are modular, making maintenance and cloth replacement very flexible.

Krauss-Maffei 84-m^2 disc filters were used in some of the uranium processing plants in South Africa in the 1980s. These filters were installed after the uranium leaching with a two-stage step ahead of the downstream clarification and solvent extraction plant. The filters were operated at a vacuum of 70 kPa with soluble losses of less than 0.005 g/L. Other applications of these filters include alumina, iron ore, and coal.

Vacuum disc filters are mainly used for dewatering coarse concentrates and in paste backfill plants (Figure 14). Vacuum disc filters "are a proven, industry-leading solution for dewatering iron, mineral concentrates, tailings, and industrial minerals. With all necessary ancillary equipment integrated into the equipment package, they are easy to install, operate, and maintain" (Outotec, n.d.[a]).

A variant for the disc filter is the ceramic filter (Figure 15). Ceramic filters are similar to disc filters in terms of operation; however, the panels are different. Panels on ceramic filters are

Courtesy of InfoMine

Figure 12 Dorr Oliver drum filter

made of a micropore ceramic material with a filtering media, based on alumina oxide. Ceramic plates use capillarity pressure to dewater the material.

Figure 16A shows material that has a 1.5-μm pore with a capillary point of 1.6 bar, and Figure 16B has a 2.0-μm pore with a capillary point of 1.2 bar.

Solid contents are accumulated at the disc surface and dewatering remains until no liquid is present. Capillarity filtration is a method that combines the advantages of a conventional vacuum filter because of its simple construction and capillary effect. Ceramic filters are used for copper concentrator filtration and industrial mineral applications.

Ceramic filters use a disc of microporous ceramic sectors instead of conventional filter cloths. Capillary action within the micropores creates a high suction that causes cake formation on the disc surface and ensures an extremely clear filtrate. The ceramic sectors may also require periodic washing, typically with nitric acid.

Four 45-m^2 ceramic filters have been successfully used at the Candelaria mine in Chile, producing a final copper concentrate of 8%–9.5% moisture.

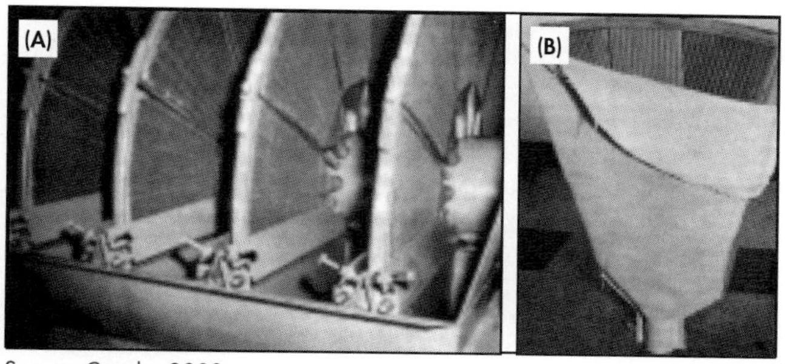

Source: Concha 2002

Figure 13 (A) Disc filters and (B) a single panel with filtering media (cloth)

Tray Filters

Tray filters consist of a set of open trays located horizontally on a plane, rotating around a single vertical shaft (Figure 17). Trays are made by numerous inclined trapezoid sectors point toward a central shaft connected to a common valve underneath the center of the equipment. Cake can be washed by water sprays once the filtration cycle has finished. Countercurrent washing is possible. Cake unloading is made by a spiral screw dragging it to the center or by tumbling trays. Filters can adapt very well to granular materials at high concentrations. Particularly in the sand mining industry, tray filters can employ washing and steam drying to lower the moisture further. The spiral screw also leaves a small amount of material on the surface to help protect the filter media. The main disadvantage of these units is that trays are used only on one side, needing a large footprint per unit of production.

Horizontal Belt Filters

This type of filter has some similarities to a conveyor belt, where the belt itself is made of a filtration cloth. The biggest advantage is its flexibility to set the filtration–washing–drying cycle time. Horizontal belt filters offer a wide variety of benefits to process, including good cake washing, continuous and visible processing, fully automatic and flexible operation, and relatively high speeds.

Modern horizontal belt filters have two moving belts. The main drainage belt is usually constructed of the appropriate elastomer and is driven by the drive pulley and slides over the vacuum pan with water lubrication from below. The drainage belt has grooves milled into the belt that facilitate the transport of the filtrate to the central vacuum box. The best design includes the South African style laydown curbs that are integral to the drainage belt. Laydown curbs are now available from many manufacturers. This feature eliminates the need for deckle seals that were originally used to contain the feed in the form zone prior to cake formation. This feature also allows for a simpler filter cloth design, as the cloth lays down

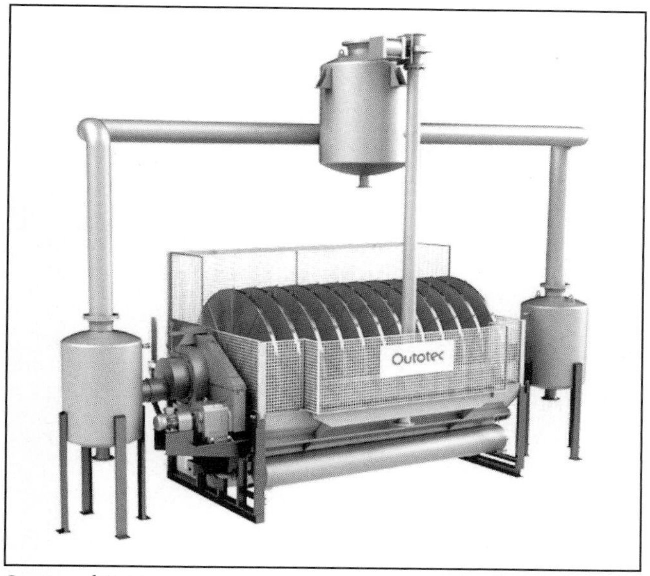

Courtesy of Outotec

Figure 14 Vacuum disc filter

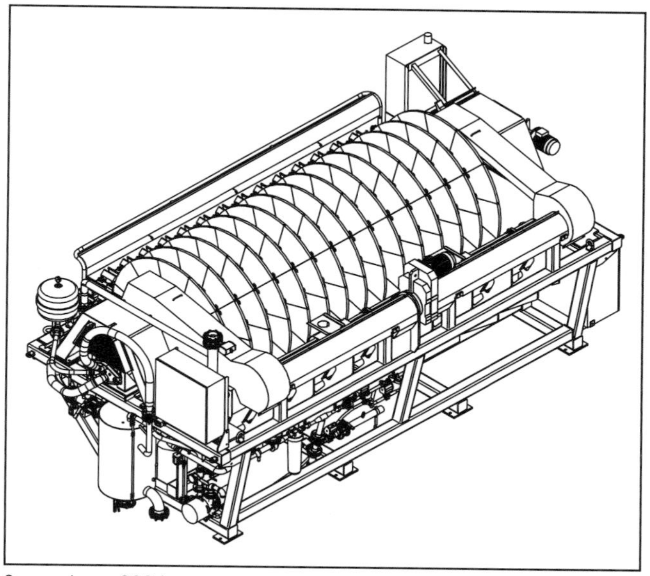

Source: Larox 2004

Figure 15 Ceramic filter

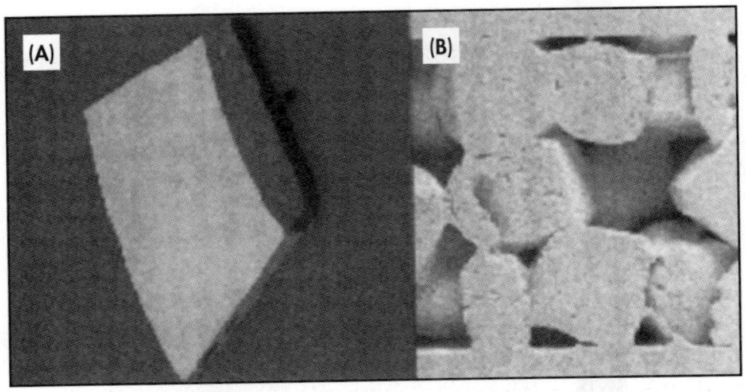

Source: Concha 2002

Figure 16 Ceramic filter plate cross section

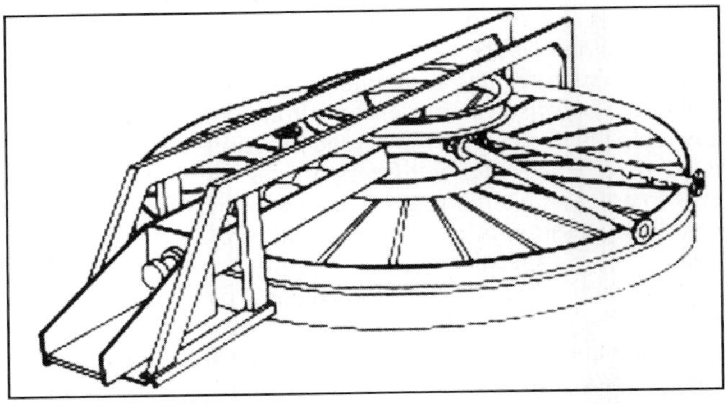

Source: Concha 2002

Figure 17 Tray filters

nicely inside the laydown curbs and facilitates cloth alignment without the need for side tensioning devices. At the discharge point, the filter cloth is separated from the drainage belt by passing over a discharge pulley, which then directs the cake to fall by gravity to the next process step. Depending on the cake release characteristics, a beater bar may be required to facilitate the cake release from the cloth. The beater bar is usually located below or after the discharge pulley while the cloth is returning and still over the discharge point so that any solids will fall with the main cake discharge.

Horizontal belt filters (Figure 18) provide high extraction efficiency, relatively low cake moisture, increased production, and reduced operating costs while achieving maximum filtration area in comparison to some of the other filter options. Horizontal belt filters maximize cake purity at minimum specific cost. These filters are especially suited for applications requiring low cake moisture and multi-stage washing with low energy consumption and a high filtration rate. Sizes are available up to a 257-m^2 filtration area.

Horizontal belt filters should be used when the following are required:

- Thorough cake washing
- Free settling solids
- Crystal breakage reduction of coarse products
- Full processing of feed materials
- Maximum product recovery
- Minimum operating cost
- A flexible range of operating conditions
- Simplicity of operation and maintenance
- Exotic construction materials for process contacts

Modern horizontal belt filter sizers have increased in overall area, and hence the width is now greater than in the past. In applications where there is a mixture of coarse and fine products that have initially been dewatered in separate operations (e.g., separate and parallel cycloning banks), it is extremely important that the final mixed feed to the filter feed point be of uniform particle size distribution and percent solids so that uniform cake forming, washing, and drying takes place. If there is not a uniform mixture, then portions of the cake will have much lower performance, resulting in nonuniform form times, washing and drying inefficiencies, and unacceptable levels of moisture in portions of the cake. For a downstream drying application, uniform cake moisture is extremely important for efficient drying operation from whatever drying equipment is employed.

A unique and interesting application of a horizontal belt filter is the filtration of the potash mineral langbeinite. Coarse langbeinite is cycloned in parallel with the cycloning of finer product fractions. The two separate underflows are then mixed in a ceramic-lined stationary mixing pot or chamber that is located directly under the underflow of the single coarse cyclone. The underflow from the fine cyclones then flows by gravity to the mixing pot where the two streams are mixed prior to feeding through their outlets down into an FLSmidth–Krebs-designed feedbox utilizing its "Varisieve" feedbox technology (Varisieve screening applications).

Rubber belt filters are tough enough for high capacities and offer outstanding performance in separation, washing, and cake dewatering (Figure 19). Some horizontal vacuum belt filters with reciprocating tray horizontal vacuum belt technology use a continuously moving filter cloth, which eliminates the need for a rubber support belt. The cloth is supported by a rigid vacuum tray and the filter is capable of cake washing in both cocurrent and countercurrent modes. The filter handles a wide range of feedstocks and produces high-purity products.

The horizontal belt filter is an efficient and reliable solid–liquid separation unit suitable for dewatering a wide range of materials, such as mineral sands, industrial minerals, iron ore,

Source: Concha 2002

Figure 18 Horizontal belt filter showing feed distribution box

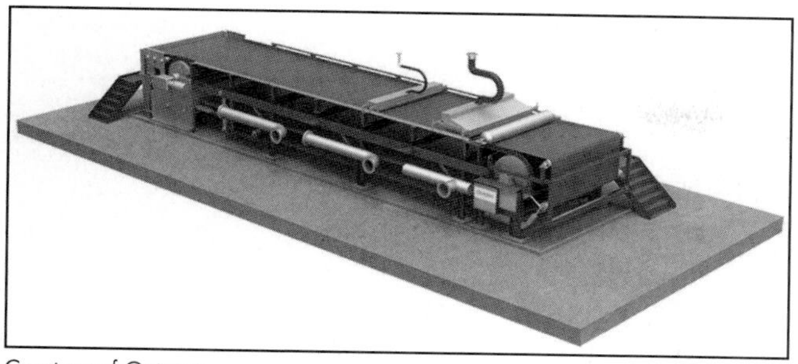

Courtesy of Outotec

Figure 19 Rubber belt filter

lead zinc concentrates, and fine coal flotation concentrates. During the 1970s, belt filters were used extensively in the Zambian copper belt. Furthermore, horizontal belt filters are being well accepted to dewater process plant tailings, to be used as mine backfill, or to be sent directly to the dry stack tailings impoundment.

Recent improvements include the large-diameter drive and tail pulleys, which have facilitated implementation of the South African–style drainage rubber belt that has the flexible laydown curbs on each side and has eliminated the need for the old deckle seals (Figure 20). This eliminates wear and tear on the filter media and prevents leakage of feed in the feed end of the filter.

At Kazphosphate LLP in Kazakhstan, three 85-m^2 horizontal vacuum belt filters were supplied. The filter had to be manufactured using corrosion-resistant material such as stainless-steel frames and air boxes and styrene-butadiene rubber for transportation belts, and glass-reinforced plastic fume hoods were used to contain acid fumes. Ancillary equipment such as vacuum pumps, air blowers, and the actual vessels were also delivered as part of supply. Similar examples have been supplied in India with eleven 110-m^2-belt vacuum filters (Figure 21).

Hyperbaric Filters

Vacuum filters have a pressure gradient limitation of 0.8 atm at the most favorable conditions, which means at sea level. If they are used at high geographic altitude, the pressure gradient drops drastically. During the 1980s, a revolutionary invention was made where a vacuum filter was placed inside a pressurized chamber, raising the pressure gradient (Figure 22). The machine consists of a vacuum filter, either disc, drum, or belt, inside a pressurized chamber. Like any pressurized filter, discharge might be a problem. With this machine, a final moisture content lower than 8% may be achieved.

Pressure Filters

Traditional filtering equipment, such as rotary and belt filters, were not accepted in the mining industry due to high residual moisture of the product passing to the dryers (before getting final product dispatch). Pressure filters are inherently a batch process. As with rotary filters, they work according to cycles, but unlike rotary filters, they must be stopped to load the slurry and unload the dry cake.

In pressure filtration, the three distinct equipment types are the vertical plate filter press, horizontal plate filter press, and belt filter press.

Vertical Plate Filter Press

In vertical plate filter presses, separation takes place between drainage surfaces of filtration plates that are kept together by using hydraulic pressure. These plates have holes to feed slurry and drain the filtrated product. Plates are mounted above and between two bars or suspended from beams. These beams or bars are connected to a fixed head end while the other end is connected to a mobile head for closure (Figure 23). Plates are compressed by using and hydraulic piston, which at one end is pushing every plate against the fixed head end, forming a single set of filtration plates or a pressing unit.

The chamber holding the cake is formed by two hollow plates forming a chamber or by two mid-level plates with a frame for the cake (similar to a picture frame). Both filtration plate faces have a drainage surface with a series of slots allowing filtrate to drain from filtering media behind, through an eye located at lower corners of the chambers. When the chambers fit together, a hole in the filter plate allows the slurry to pass through from one chamber to the next. At each chamber surface is a filtering media fitted by bolts and nuts to the plate. Figure 24 shows a 2 × 2-m plated filter press from FLSmidth.

Courtesy of Tenova

Figure 20 Horizontal vacuum belt filter showing a South Africa–style drainage belt

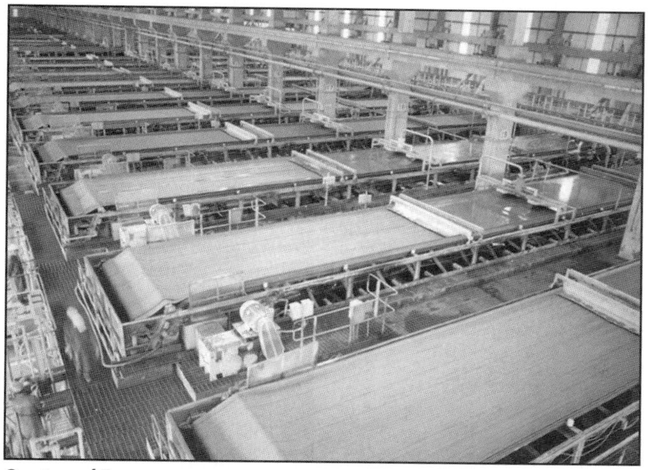

Courtesy of Tenova

Figure 21 110-m^2-belt vacuum filters at an installation in India

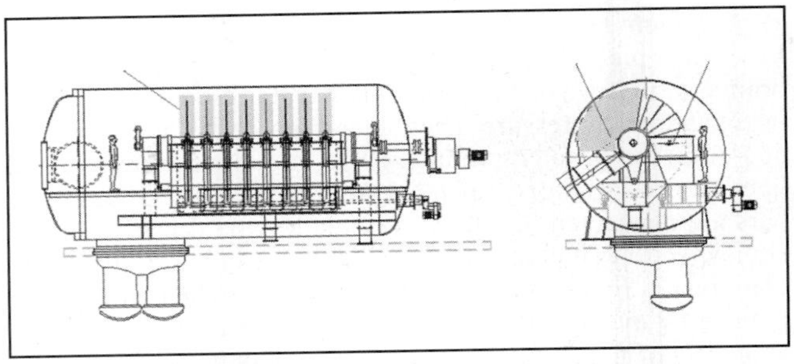

Source: Concha 2002
Figure 22 Hyperbaric filter

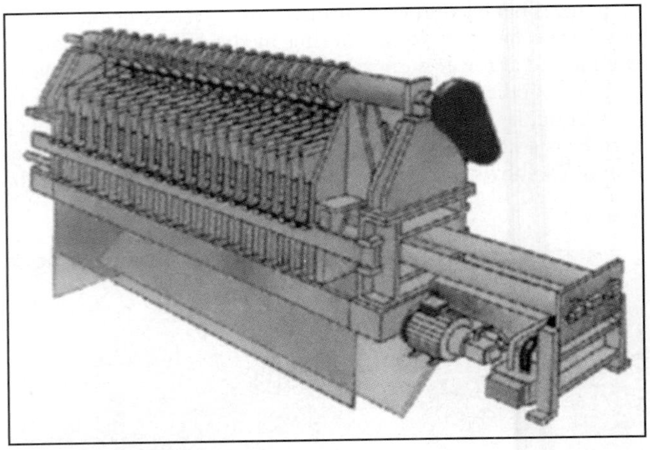

Source: Concha 2002
Figure 23 Vertical plate filter press

Consecutive filtration stages do not consider slurry or cake mechanical compression inside chambers (Figure 25). The cycle filtration process is as follows:

1. **Closed.** Plates are compressed at a high pressure by using a hydraulic piston/ram to avoid material leakage between plates and seal all the filtration plates.
2. **Feed.** Slurry feed is introduced through a core throughout the filter until all the chambers are full. As slurry begins to come under pressure, solid particles start to distribute between both faces of the filtering media, forming an initial cake layer or pre-coat. This pre-coat becomes the real filtration media, and, if filtration continues, the thickness of this layer continues to grow until cake from both surfaces touch each other. Pumping continues forming a cake compression until feed flow trends to zero. At this point, the feed pump stops.
3. **Cleaning (core removal).** Any material inside the core is washed with a pressurized countercurrent water jet. The remaining water inside the core is removed by using compressed air. This stage normally takes around 45 seconds.
4. **Air blowing.** High-pressurized air in introduced into the chambers, generating a displacement of the retained moisture inside the cake pores.
5. **Discharge.** Once filtration and compression stages are finished, reception bin trays open at the bottom section of the filter. Plates separate by retracting the hydraulic piston and the cake drops by gravity onto a conveyor belt.

Courtesy of FLSmidth
Figure 24 AFP 2020 filter press

6. **Cleaning (washing).** Before a new cycle starts, drip trays close and filtering media are washed with high-pressure water to keep media and plate surfaces free of particles and any debris (Figure 26). This avoids the cloth binding up and reduces plate wear.

Most pressure filters, and all filter presses, operate in batch cycles. Each cycle consists of a filling step, where the chambers are filled with slurry and the cake is formed; a dewatering step, where pressure is applied hydraulically (via the feed pump pressure) or mechanically (via rubber membranes) or pneumatically (compressed air is forced through the filter cakes); and the technical time required to actuate valves, shift the filter plates, discharge the cake, and regenerate the filter cloths. Figure 27 illustrates the pressure filter cycle showing how the filter cake moisture decreases during the cycle, while Figure 28 shows cake formation during the cycle.

For tailings dewatering, lower moisture content, although desired, is usually not as critical as with concentrate streams. The tailings target moisture content does have an impact on geotechnical aspects and also needs to be considered to make an engineered dry-stack configuration. For concentrates, the moisture above which the slurry will liquefy when vibrated or transported, referred to as the transportable moisture limit (TML), is critical for safe transportation. For concentrate transportation, the TML is normally referenced in the process design basis and on the material safety data sheet. This information for the concentrate must be specified on the material safety data sheet (MSDS). An example of the physical data that must be included on the MSDS (e.g., for copper concentrate) is as follows:

- **Particle size:** The 80% passing size of fresh copper concentrate is expected to be 30 µm based on pilot-plant results (ranging from 15 to 50 µm). The copper concentrate is expected to age and oxidize during storage, thus forming agglomerates and lumps.
- **Solubility in water:** Insoluble
- **Appearance in color:** Odorless, grayish, slurry or moist powder
- **Odor threshold (ppm):** Odorless
- **Corrosiveness to common metals:** Corrosive to aluminum and steel
- **Physical state:** Paste, typically as slurry or filter cake
- **pH:** 11.5–12.0

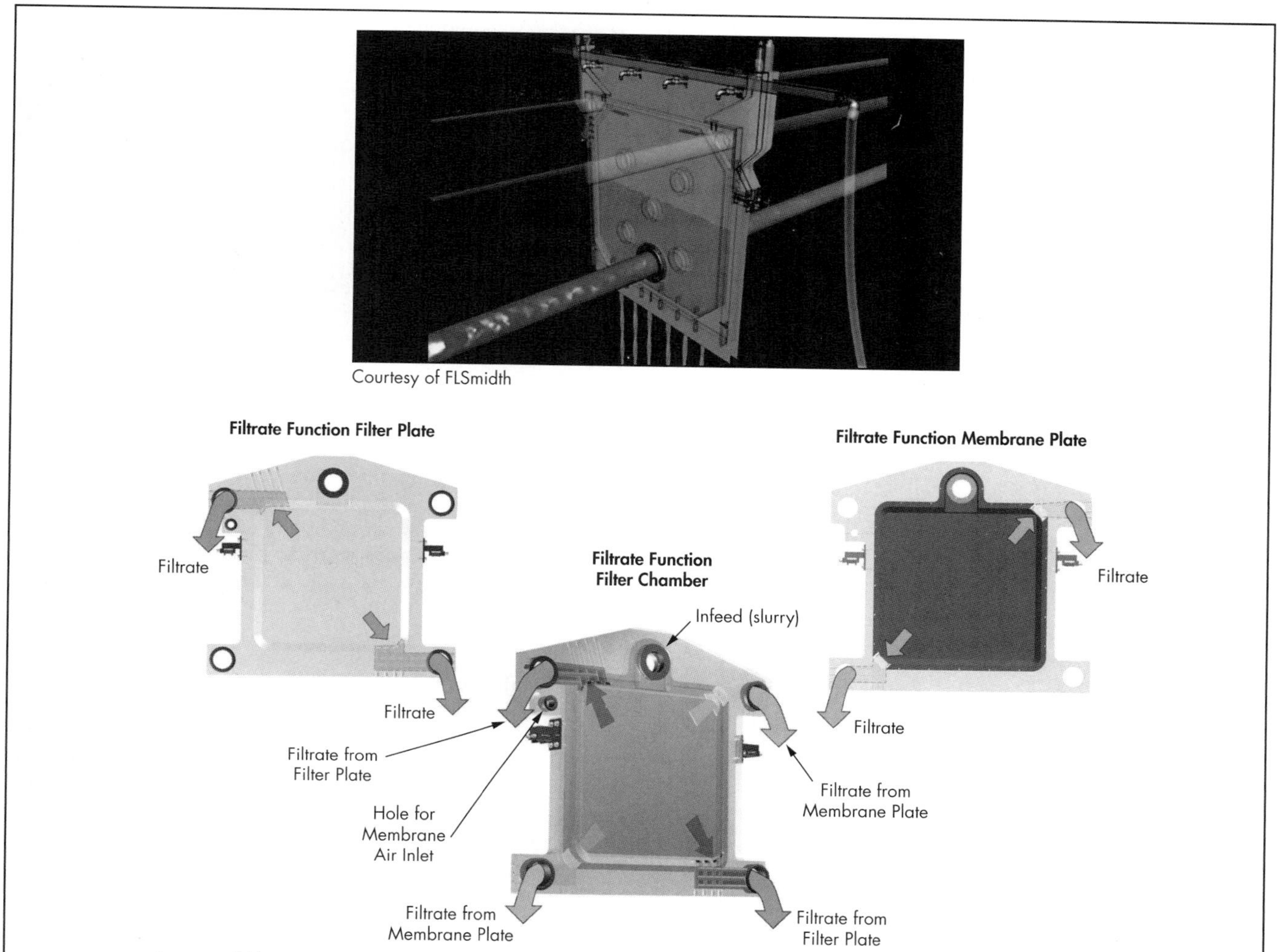

Courtesy of FLSmidth

Courtesy of Metso

Figure 25 Filter plate chamber being fed by slurry

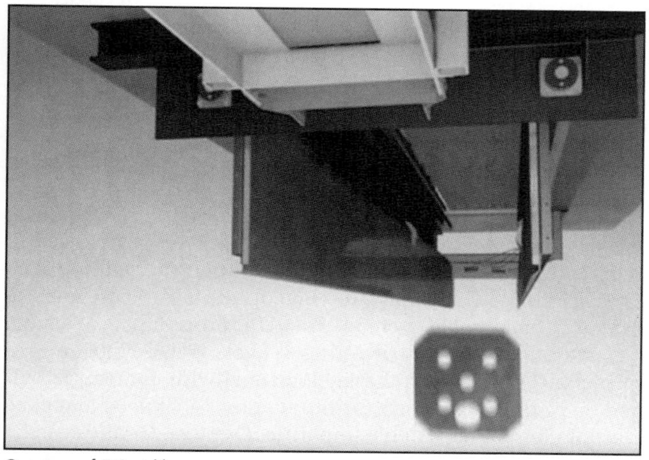

Courtesy of FLSmidth

Figure 26 Drip trays open under the filter. The discharge cake drops by gravity onto the conveyor below.

- **Specific gravity:** 4.0–4.5 g/cm³
- **Moisture:** Below bulk TML of 10.6%

Achieving filter cakes with moisture content below the TML does not always require air blowing or membrane squeezing. In those cases, the cycle is terminated at the end of the cake formation step; determining when the cycle can be terminated is critical to successful operation of the filter press. In cases where membrane squeezing or air blowing is required, determining the time in the cycle to apply these forces to achieve optimum operating conditions is still critical to successful operation.

Another filter press specifically developed for large-throughput, high-filterability products, such as mining concentrates and tailings, is the Diemme GHT F filter press (Figure 29). Diemme's models have an available area from 105 to 1,932 m^2 with 31 to 121 plates (Diemme Filtration 2017). The electrically driven movement of the mobile header and the resulting accordion-style opening of the plate pack considerably reduces the cake discharge time and the entire

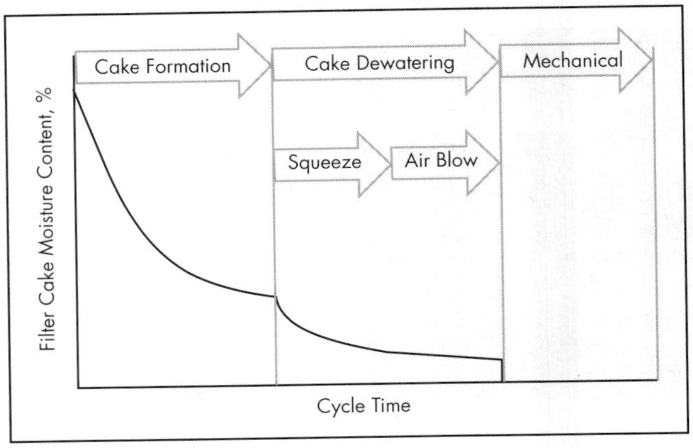

Courtesy of Metso

Figure 27 Pressure filtration cycle

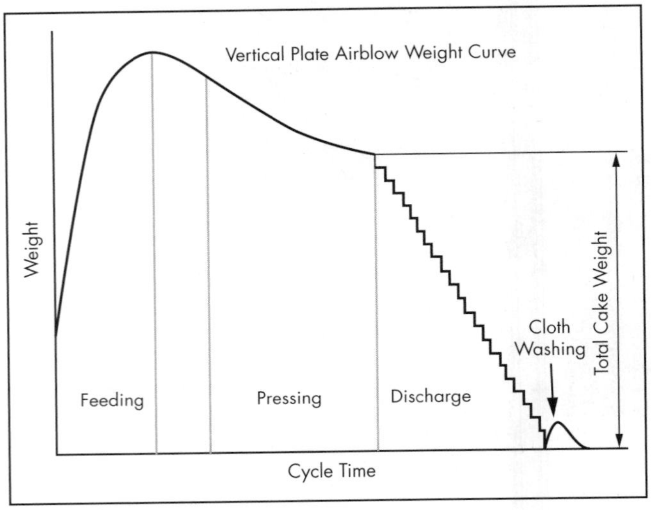

Courtesy of Metso

Figure 28 Cake formation during cycle

Source: Aqseptance Group, n.d.

Figure 29 Diemme GHT F filtration fast-cycling filter press

opening-and-closing sequence of the filter. The result is a fast-cycling filter press, which allows increased production per unit area of the filter.

The overhead side-beam structure of the filter facilitates the installation of different performance-improving devices, such as the cloth-rinsing system and the plate-shaking device. Additionally, routine maintenance operations are made easier because of an integrated platform that enables access from above the machine. The filter plate pack is sealed closed by means of four hydraulic cylinders. This system provides the correct functioning of the machine and compensates for possible misalignments of the plate pack. The plate construction material varies depending on the process temperature and chemical composition of the product to be filtered.

The automatic plate-shaking device, which is programmed to occur after the plate pack fully opens, ensures that the filter cake is properly discharged and ensures possible cake fragment removal from the cloths. This system, along with the cloth-rinsing system, allows the filter to run automatically and without operator assistance.

Like several of the other filters, the filter cloths are rinsed after each cycle to prevent the accumulation of cake fragments and fines buildup. Furthermore, the rinsing ensures that the sealing surfaces of the filter cloth are clean, providing a tight seal upon closing. The cloth-rinsing system, comprised of high-pressure wash-water pipes and directional sprays located at the top and lower portion of the plates on both sides of the filter, provide a thorough rinsing of the filter cloths while leaving the upper area of the machine free to perform routine maintenance operations. The automatic and programmable cloth-washing system, along with the cloth-rinse system, improves the machine efficiency and performance by preventing accumulation of cake fragments.

Placed under the supporting feet, load cells constantly monitor the machine weight to ensure that all cakes are discharged. The throughput of each load can also be recorded for accurate production tracking. The cake discharge time is significantly reduced compared to traditional plate opening/closing systems with the use of an accordion-style plate pack opening-and-closing system.

Most of these filter presses are provided with a programmable logic controller (PLC)-based automation system complete with a human machine interface, which simplifies monitoring of the filter operation and diagnosis of faults. Additionally, it allows for continuous adjustment of filtration parameters to optimize the variable requirements of the process.

The molded filter plates are formed with corrugated drainage surfaces in the chamber recesses and ports for slurry feed and filtrate drainage. Each face is covered with a filter cloth, and the plates are clamped together using a hydraulic ram. Slurry is pumped in under high pressure, filling the chambers with solids and pushing liquid out through the filter cloth. When no more solids can be forced into the chambers, the feed pumps are turned off and compressed air is used to remove interstitial water from pores in the filter cake. When the desired residual moisture content has been achieved, the filter is opened, cake is removed, and the procedure is repeated.

Additions to the basic filtration cycle include the removal of residual slurry from the feed channel with compressed air and/or water, an air purge of filtrate drain channels, and cake washing to remove soluble impurities. Very high solids can be achieved in cakes of low permeability through the use of filter plates lined with optional elastomer membranes. After the basic filtration cycle has been completed, these membranes are inflated with pressurized air or liquid, squeezing additional moisture from the filter cake.

An FLSmidth automated filter press, or AFP, provides a simple, rugged, and reliable design for a variety of solid–liquid separation processes. These filters not only provide high throughput and efficient solids capture but they also operate continuously, automatically, in relatively harsh environments, and can handle abrasive or corrosive slurries as well as more-difficult ultrafine higher-clay-content slurries. When coupled with simple systems and operating philosophies, the FLSmidth AFP yields competitive operating costs.

The heavy-duty design of these filter presses has been specially adapted for minerals service with several features designed to enhance their performance in these and other applications. Some units are designed with up to 2,000 m^2 of filtration area by using 2 × 4-m plates. Vibratory discharge assist is provided by a hydraulically actuated cloth shaker that transmits a vibration directly to the filter cloths when one or more chambers fail to empty completely. To minimize the effects of vibration on other system components, the shaker mechanism is supported on a separate frame system.

Various levels of automation can be achieved using the PLC within the filter control panel. Two wash headers run along the upper edges of the plate stack, with nozzles positioned to completely wash every plate face and filter cloth. To reduce the time required for plate shifting and cake discharge, filters utilize a long-stroke cylinder-driven crosshead to open and close the plate stack. Two short-stroke closing cylinders are mounted in a rolling carriage, which spans the filter sidebars. A hydraulically actuated chain drive or long-stroke cylinder moves the crosshead and linked plates quickly between open and closed positions. With the crosshead fully retracted, a uniform gap is opened between each plate, allowing all chambers to empty. The crosshead then returns the plates to filtration position where locking cylinders anchor the carriage. Closing pressure is applied to the plate stack and slurry feed resumed.

Hydraulically operated drip trays are located beneath the filter to prevent filtrate leakage and cloth wash water from rewetting previously dried discharge filter cake. Both single-slope and gable-type designs are available. Power for plate shifting, crosshead locking, filter closing, drip tray operation, and cloth shaking is supplied from a single electrically driven hydraulic power unit. A filter-mounted manifold contains all required valves, switches, and gauges for hydraulic control.

Filter plates and media support frames are designed for unobstructed lift out using a light-duty overhead crane or lifting frame. Cloths are then changed off-line while the filter continues to operate. Media cloths attach with a quick-release rod-and-clip arrangement for fast, simple exchange.

Metso vertical plate airblow (VPA) pressure filters are available with semi- or fully automatic control reaching over 250 t/h for its VPA 20 series with 60 chambers. They are designed to the latest Eurocode standards and ensure operator safety with perimeter protection and safety interlocks.

The Tenova Delkor filter press, with available filtration area from 8 to 720 m^2 available, allows shorter cycle times and higher throughput, availability, and safety, making the technology perfectly suitable for high-capacity and heavy-duty applications like concentrate and tailings dewatering (e.g., dry-stack tailings). It consists of a fabricated sidebar frame. As opposed to traditional filter presses, opening and closing of the plate pack is carried out by two synchronized nonreversible screws, while final compression of the plate pack is achieved using a pressurized cushion inflated with water at low pressure.

Some features of the Tenova Delkor Fast FP are as follows:

- Complete plate pack opened at once (up to 100 chambers), with the possibility to adjust the opening speed based on the actual process requirements
- Recessed mixed-membrane and full-membrane plate pack
- Filter plates with various feed and filtrate ports configured to suit process-specific requirements
- Automatic plate misalignment detection
- Double end feeding for fast filling
- Integrated bomb bay doors with no impact on surrounding items
- Filter cloth fixing system, barrel neck or single cloth with feed shoe
- Core blowing
- Cake air blowing/washing
- Fully automated plate/cloth shaker cake discharge
- High-pressure full face cloth washer and low-pressure wash down cloth washer
- Safety devices in compliance with process and local regulations
- Remote control of filter press to support operation and maintenance

Horizontal Plate Filter Press

Horizontal plate filter presses are designed for larger capacities and produce a low product moisture content. This kind of filter press consist of a horizontal filtration chamber set inside a main structural frame. By virtue of its design, it allows the addition of extra chambers mounted on top of each other, increasing the filtration area without increasing its footprint. This provides a very robust design that is flexible for operations with subtle feed changes over time.

Each filtration chamber has inflatable seals at both ends, which expand to seal the chamber. Chambers are fixed to the structure and have no movement during the filter operation. Each chamber has a filtrating conveyor mounted over a roller at each end, operating independently from other conveyors inside the equipment. Each roller has a hydraulic motor driving the conveyor during the discharge stage. At the top of each chamber is a flexible rubber diaphragm used to compress the cake and responsible for forming and reducing the cake final moisture.

Horizontal plate filter presses are fully automatic recessed-plate diaphragm filters with horizontally oriented chambers. The continuous filter cloth ensures thorough cloth washing and efficient cake discharge from each chamber at every cycle.

The filtration sequence for a horizontal plate filter press is as follows:

1. **Filtration.** Feed is pumped simultaneously into the filter chamber at a specific pressure. As the solids start to form, the filtrate starts to become displaced by more fresh slurry entering the chamber. As the solids continue to form, the pressure increases and filtrate release continues. Once the chamber is full, the feed is cut off.
2. **Diaphragm pressing 1.** High-pressure air or water automatically inflates the diaphragm located at the top of each chamber, reducing the chamber volume further and squeezing the solids to remove more filtrate. The solids filtration process and tightly woven filter cloth maximize

filtration efficiency. The diaphragm-pressing step allows for formation of homogeneous, dewatered solids of uniform thickness.

3. **Solids washing.** The filter design allows dewatered solids to be washed in situ to maximize solute removal or to recover the mother liquor with minimal dilution. The wash liquid is distributed evenly across the filter. The wash liquid flows through the solids, displacing the mother liquid with minimal mixing.
4. **Diaphragm pressing 2.** The diaphragms are reinflated, forcing the wash liquid uniformly through the bed of solids. This produces a very high washing efficiency with a consistently dry solids content.
5. **Air blowing.** The final dewatering is completed by an air-blow step. By setting of the pressure and duration of the air blow, the product's final moisture content can be adjusted accordingly.
6. **Cake release and cloth washing.** The plate pack opens and the dewatered solids are conveyed out of each of the chambers on the moving filter cloth. To reduce cloth blinding and filter performance deterioration, high-pressure water sprays wash both sides of the cloth. To complete the cycle, the belt is stopped when returning to its original position and the filtration cycle repeats automatically.

The FLSmidth Pneumapress horizontal plate filter press delivers a drier solids product; a cleaner, higher-quality filtrate; is easier to operate; and requires lower maintenance. It handles flow rates from 0.27 m^3/h to 2,271 m^3/h and utilized a filtering area from 0.14 m^2 to 92.9 m^2. It features isolatable filter plates, single-sided solids discharge, fully automatic operation with PLC, a touchscreen operator interface, individual filter cloth for each filter plate, automatic cloth wash system for each filter cloth, one hydraulic cylinder, and a cake thickness of 5–200 mm.

The main benefits of the FLSmidth Pneumapress filter are as follows:

- Fast cycles, typically less than 5 minutes
- High solids throughput per filter area
- Small equipment footprint
- No need for rubber membranes or diaphragms
- High equipment availability
- Versatility
- Can be highly automated and "hands off" operation
- High-temperature resistance capability (up to 200°C)

The Outotec Larox PF series filters have become one of the preferred choices for copper concentrate filtration applications in larger-capacity concentrators.

Belt Filter Press

The belt filter press is a mechanical pressure-filtration dewatering device used extensively for solid–liquid separation. Since the 1980s, belt filter presses have been used to for dewatering of coal tailings and in other industries such as chemical, food, pulp and paper, wastewater treatment, and sewage. The technology is designed to separate liquids from solids and create a compact product, referred to as cake, that is stable, easy to handle, and stackable. The key advantages of a belt filter press include continuous operation, relatively low capital and operating costs, fast start-up and shutdown times, less noise generation, and sufficient dewatering capabilities under variable conditions. They are also of high mechanical reliability. Belt filters operate at relatively low pressures. The belt filter press is typically used to process slurries of fine solid particles, with nearly all particles smaller than 600 µm.

The belt filter press is particularly effective compared to other dewatering technologies when slurries are ultrafine or have a high degree of colloidal solids or material that does not settle over time without polymer addition. It consists of a single chamber located over a filtrating media conveyor belt. A hydraulic piston supplies necessary force to produce filtration (Figure 30). The filter operation stages are as follows:

1. **Feed.** With the chamber closed, slurry is pumped into it, letting the liquid go through the filtering media. This is collected at the fixed end. All solid particles are retained on top of the media.
2. **Compression.** When no more slurry is admitted inside the chamber, this is compressed against the media by compressed air or gas action, forcing the liquid go through the solids and forming a dry cake.
3. **Cake discharge.** Finally, a hydraulic piston rises, freeing the dry cake, which is transported by the conveyor belt for later treatment, leaving the unloaded belt available for the next cycle.

Belt filter presses are sized for the throughput they can handle on a per-meter-of-belt-width-per-hour basis. For commercial tailings and waste sludges, available filter belt widths typically range from 1.2 to 3.0 m and the capacities range from 3 to 12 dt/h/m (dry tons per hour per meter) of belt width, depending on feed conditions. Along with the solids capacity or throughput of a machine, other critical performance measures are the discharge cake or solids content (weight percent), and the amount of material that has been retained (e.g., >90% yield).

As with any tailings dewatering technologies, many variables have a direct impact on belt filter press performance. Variables associated with feed conditions include particle size, specific gravity, volatiles content, ash content, feed density, homogeneity of slurry, and feed consistency.

Although the belt filter press is an established technology for coal and other tailings dewatering, alternative technologies have proven to be a viable solution in some cases and include the following:

- **Recessed-chamber filter press.** This is a high-pressure batch filtration process that typically yields lower cake-moisture results than belt filter presses. The capital cost is two to three times that of belt filter presses, but they do not always require the use of flocculants or preconditioning agents.
- **Membrane filter press.** Like recessed-chamber filter presses, membrane filter presses use high-pressure filtration, but with the additional step of further squeezing the filter cake by placing either air or water behind the membrane present in each chamber. Given that this method yields consistently lower cake-moisture results than belt filter presses, it is typically used in concentrate applications where low moisture is critical for a saleable product or downstream process.
- **Solid bowl decanter centrifuge.** This applies sedimentation and compression in a centrifugal field under continuous operation. It yields slightly higher cake-moisture results and has higher energy demands than belt filter presses.

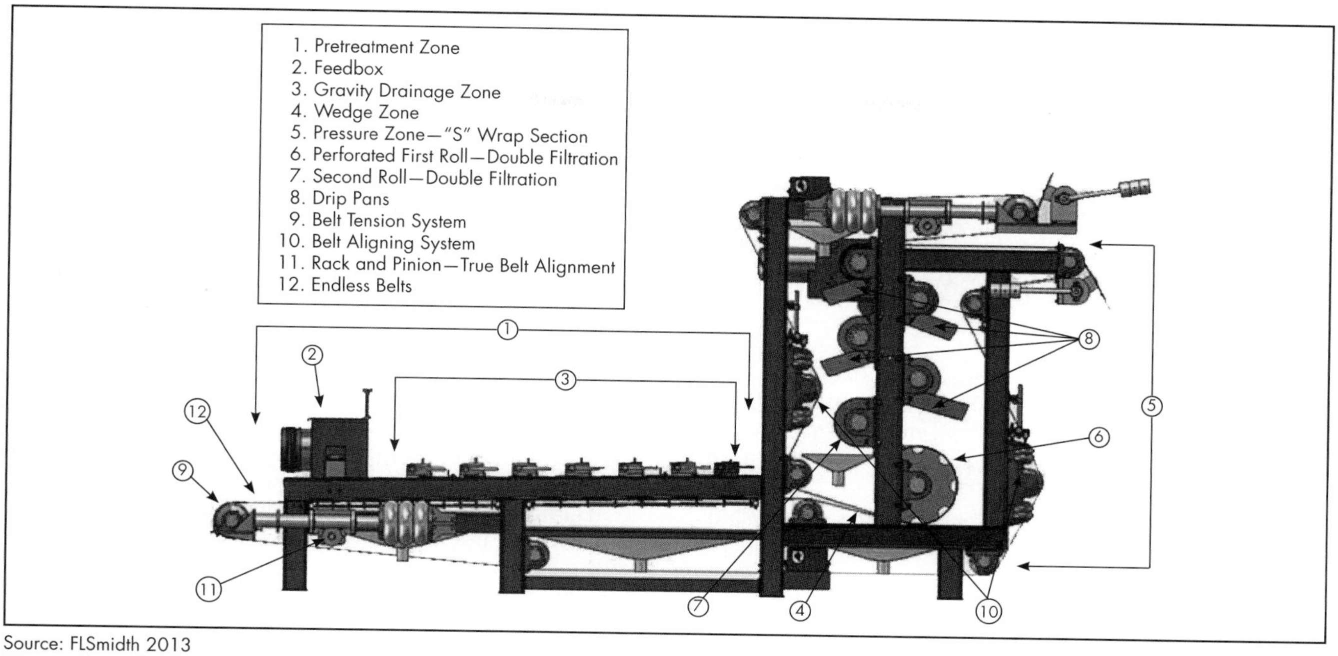

Source: FLSmidth 2013
Figure 30 Belt filter press

- **Rotary vacuum drum and disk filter.** In this process, slurry is collected through vacuum filtration and cake is discharged by scraper or the snap-blow process.

ANCILLARY SYSTEMS

Depending on the technology to be used for filtration, several ancillary systems must be considered during design, engineering, and ongoing operation and maintenance. In general, the following sections describe some of the common equipment pieces in a filtering system.

Feed Tank

It is always necessary to have a feed tank with enough capacity to hold between two and five cycles or a residence time equivalent to at least 1 hour of continuous operation (Figure 31). This will allow a steady filter operation and it can absorb some process variation and process upsets.

Feed tanks should have some valves before slurry discharge. Usually a manual on–off valve is followed by an automatic valve feeding a slurry pump or the filter system by gravity. The tank is normally level controlled by either a recycle stream from the slurry discharge pump or by the addition of liquid or additional feed slurry. The instrumentation used will depend on the nature of the slurry suspension. For most applications, an ultrasonic-level transmitter above the maximum liquid level provides the most trouble-free operation. The net positive suction head, or NPSH, is a function of the liquid vapor pressure, barometric pressure, intake friction losses, and the static intake head, and it must be greater than the minimum shown on the slurry pump curve for the design revolutions per minute and flow rate at the pump intake. Maintaining the intake head greater than the minimum NPSH requirement is essential to prevent pump cavitation. The tank is normally equipped with an agitator to keep the slurry in suspension at all times.

Feed Pumps

Feed pumps are an integral part of the filtering system. Pumps must have enough power to pressurize the system and deliver slurry in a short period to allow filling of the filter chambers. Filter presses usually use centrifugal feed pumps to fill filter chambers against filter pressure (Figure 32). The control and operation is first by flow, and then the operation is controlled by pressure. It is recommended to have standby and/or spare pumps for filtering feed applications.

Density Measurement

Given that filtration is usually a volumetric process, density plays an important role, because it is combined with flow to set the final filter mass feed rate. It is important to control slurry density at the feed, either manually or automatically. Feed samples can be obtained manually during operation every 2 or 3 hours, for control purposes. Using a simple Marcy Scale, the slurry feed density can be read off. The advantage of this method is simplicity, reliability, and punctual density recording. If more accurate data are needed, feed samples must be analyzed in a laboratory (calculating percent solids by drying samples). Automatic density measurement is achieved by a density gauge, which requires periodic calibration.

A density gauge uses a radiation source and detector on opposite sides of the pipe to measure the slurry density. Employment of this instrumentation requires the operator to have a radiation license and the proper training for nuclear gauges of this design. The units are periodically calibrated and are very reliable.

Gland Seal Water System

The gland seal water system should be located close to the filter-feed pumps. A robust gland seal water system, including dedicated gland seal water pumps for main distribution, are required. The gland seal system needs to be designed to minimize the water addition to reduce feed dilution effects, which

Courtesy of FLSmidth

Figure 31 Feed tank for a 2,000-m² filtration area filter press

Courtesy of FLSmidth–Krebs

Figure 32 Main slurry feed centrifugal pumps

will subsequently affect the filtration step. The system design must carefully analyze the pressure requirements.

For exterior or outdoor conditions, it is recommended to use a pressurized loop and a gland water surge tank. If weather conditions are extreme (extreme cold), heat tracing must be incorporated on the piping.

Filtrate (Recovered) Water Tank

Given that one of the main products produced by the filter is filtrate, all of this solution must be collected in a centralized storage tank, vessel, or clarifier. The tank must be large enough to cope with the instantaneous volume produced during the initial stages of filtration. The filtrate will periodically contain some suspended solids, so this storage vessel should include an agitator or, in the case of a clarifier-type storage, include a mechanism. Usually this water is reused by the filter itself (e.g., for washing purposes). Tank capacity must be sized to contain a few filter washing cycles.

Clarifier/Settler

A clarifier/settler might be needed if the water is of high turbidity and is reused as a "clean water source." If downstream processes need a high-water clarity, a proper settler must be considered. If so, a flocculant plant must be considered properly dimensioned and designed. It is advisable to run a flocculent screen test to choose a most efficient and economic flocculent reagent.

Air Compressor

A vital part of the filtration system is the auxiliary air supply. The air addition is associated with the cake formation and drying. The air supply system will include air receivers (accumulators) and various levels of process instruments such as valves and pressure gauges, pressure release valves, and so forth.

Hydraulic Unit

The hydraulic unit is one of the most important parts of the filtration unit (specifically for those using press filtration). The selected oil must be suitable for applications considering temperature variations (interior/exterior) and speed of all movable filter parts because thinner oils allow faster speeds.

During the construction stage, hydraulic system flushing must be done properly, leaving no material inside the pipework, as it may contaminate and affect the hydraulic system. Oil selected for flushing purposes will not be used during operation. All pistons and several of the valves are hydraulically operated. All hydraulics valves that fail close must have a backup system to operate individually in case of failure (Figure 33).

Hydraulic pressures can reach more than 8,274 kPa. It is highly advisable to have hydraulic pump units with correspondent backups and spares. Predictive maintenance and the proper oil type must be considered during operation. Also, in case of heavy weather conditions, the hydraulic unit must be covered and the oil reservoir must be temperature controlled by using cooling radiators and fans.

Equipment included in the hydraulic unit are as follows:

- Oil tank
- Directional valves
- Relief and safety valves
- Instrumentation
- Low-pressure/high-flow system (opening and closing)
- High-pressure/low-flow system
- Oil filters
- Safety breakers

Maintenance Crane

For most maintenance and operational activities, an overhead crane is needed in the filter area. The crane (portal, gantry, mobile, etc.) must be able to reach all parts of the filtering plant and plays an important role during construction and operational duties such as plate or filtering media change (Figure 34).

FILTERING MEDIA

Cloths

Vacuum and pressure filters employ filter cloth as the filtration medium. Experience has shown that the specification of cloth type is critical to the success of a filter application, in terms of process performance, operating cost and longevity, and filtrate clarity.

Media or cloth consumption is a major component of a filter's operating cost, expressed as cost per metric ton. Initial filter economics will include an assumption of the number of cycles or life expectancy of the filter cloth. Such assumptions

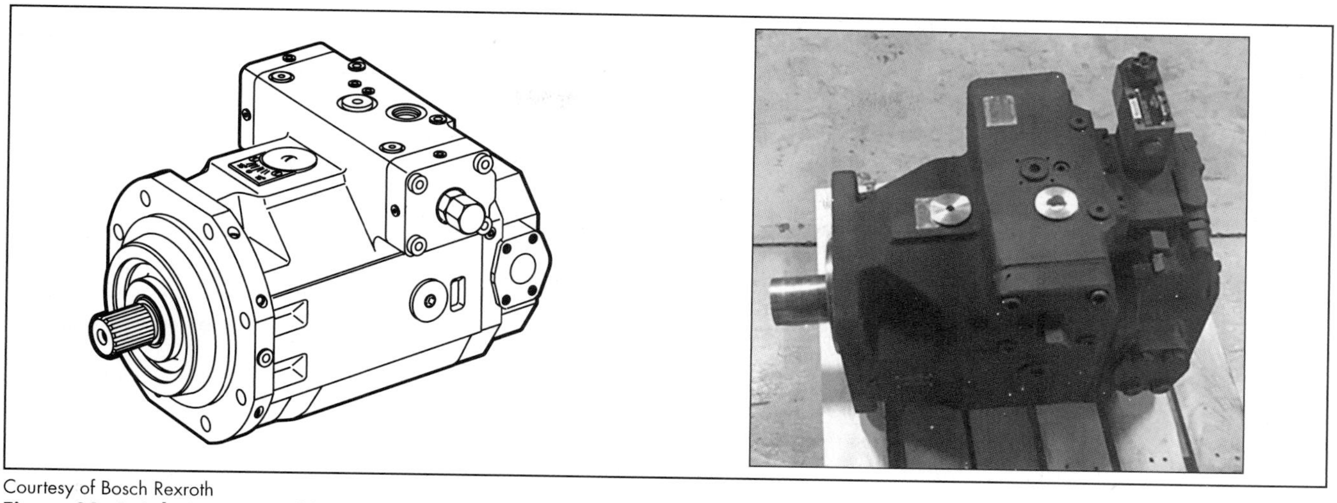

Courtesy of Bosch Rexroth

Figure 33 Axial piston variable hydraulic pump

will be based on experience with similar or like materials and industry benchmarking. However, sometimes the material encountered during early operation can have different characteristics regarding the degree of fines content, clay content, particle sizing, or abrasiveness. As such, the cloth preselected can blind or wear out prematurely. Assumed life cycles of, for example, 6,000 hours can degrade to less than 1,000 hours with a major combined impact on cloth consumption cost, maintenance labor hours, and filter availability. Therefore, it is important to define, as well as possible, the cloth specifications for a given application.

Following start-up, cloth performance will be monitored and reviewed throughout operations. Working closely with the filter supplier and their cloth suppliers is recommended to arrive at the best balance of cloth permeability, resistance to blinding and abrasion, chemical suitability, mechanical requirements, and overall economic impact. Cloth technology is constantly evolving, thus a good partnership with cloth suppliers will ensure that new innovations can be incorporated and performance can be continually improved upon.

Filter cloth basics are discussed in this section and a summary of filter cloth terms are presented in Table 2.

There are two types of cloth, woven and nonwoven. Generally, it has been determined that woven cloth offers the best combination of filtration efficiency and low blinding tendencies. Woven cloths have three basic yarn compositions: spun, multifilament, and monofilament. A cloth can be made from 100% of one of these yarns or 50% of two of them (with a different wrap and filling).

An all-spun fabric will have the highest particle capture rate for a given porosity, followed by multifilament and then monofilament, which has the lowest rating. As a trade-off, however, the reverse order is the case regarding the cloth's susceptibility to blinding. Thus, to meet a particular filter application requirement, today's cloths tend to be a blend of multi/spun or mono/multi or even mono/spun to take advantage of each yarn's strengths.

Yarns can be produced from several materials. The most popular is polypropylene, as it is the least expensive, most chemically resistant, and easiest with which to work. Nylon and polyester are also considered when temperature and/or chemical incompatibility issues preclude the use of polypropylene. Specialty fabrics like aramid, polyphenylene sulfide, and polytetrafluoroethylene, or PTFE (more commonly known as Nomex, Ryton, and Teflon), which are very expensive, are employed in the harshest of chemical and temperature environments.

The type of weave is also important in cloth performance. A plain weave is the tightest, and also the most susceptible to blinding. The twill weave, usually 2 × 2, was a standard way to offset blinding and give a thicker cloth for filtration. Current design favors the sateen weave, where a yam will travel over several perpendicular yarns, then go under one, then repeat the sequence. This produces a very smooth surface, which resists blinding and can be easily cleaned. It is a matter of determining the proper yarn or yarns and type of weave for optimum filtration.

Ceramic Media

In addition to conventional cloth technology, there have been developments and commercialized ceramic media used for vacuum disk application. A cross section of this ceramic media is presented in Figure 35. These ceramic discs are patented sintered alumina membranes with uniform micropores to create capillary action. This microporous filter media allows only liquid to flow through. Even at a perfect vacuum, there is

Figure 34 Gantry crane used for filter press plate change

Table 2 Filter cloth terms

Term	Detail
Blinding	The plugging up of a filter fabric, resulting in reduced flow rates and filtration efficiency.
Cake release	The ability of a filter cloth to completely discharge the cake from the cloth.
Monofilament	A single, long, continuous strand of a synthetic fiber extruded in fairly coarse diameter.
Multifilament	A smooth yam consisting of two or more monofilaments twisted tightly together.
Needled felt	A fabric (felt) made by the mechanical interlocking of individual fibers in a random orientation.
Nylon	A synthetic fiber manufactured from aliphatic or semi-aromatic polyamides. Has good resistance to alkalis, but is affected by strong acids and solutions in the pH range of 1 to 6. Maximum operating temperature is 107°C.
Oxford weave	A modified plain weave where both warp and weft yarns are multiple yarns but not of equal number.
Plain weave	The simplest and most common weave, produced by passing the weft thread over and under each successive warp thread. Offers low permeability and excellent particle retention; however, it is susceptible to blinding.
Polyester	A synthetic fiber manufactured from terephthalic acid and ethylene glycol, the resulting polymer has excellent resistance to most mineral acids and solutions in the pH range of 1 to 8. Maximum operating temperature is 140°C.
Polypropylene	A synthetic fiber manufactured from a petroleum industry by-product. Excellent resistance to full pH range and has maximum operating temperature of 32°C. Is affected by oxidizing agents.
Porosity (permeability)	The rate of flow of air under differential pressure through a cloth. Generally measured in liters per minute (at 12.7 mm water pressure).
Satin (or sateen) weave	A weave in which warp yarns are carried uninterruptedly over many weft yarns to produce a smooth-faced fabric. Offers superior cake release and excellent resistance to binding.
Spun (staple)	Yarns made from filaments that have been cut into short lengths, then twisted together. Staple fibers offer good particle retention; however, cake release may suffer due to the hairiness of the yarns.
3 × 1 Double weave	A special weave in which both sides of the cloth show a smooth 3 × 1 broken twill surface. Offers excellent stability, retention, and cake release properties.
Twill weave	A weave in which the weft threads pass over one and under two or more warp threads to give the look of diagonal lines. Offers medium retention and blinding properties with high abrasion resistance and good flow rates.
Warp	The yarns that run lengthwise in cloth as they are woven on a loom.
Weft	The yarns that run widthwise in cloth as they are woven on a loom. Also, known as filling yarns.

Source: Smith and Townsend 2002

insufficient pressure differential to force air through the media openings. The capillary phenomenon is based on the Young–Laplace law, which states that the pores of a certain diameter cause a capillary effect due to surface tension and the contact angle of the liquid. As the discs are immersed into the slurry basin, a pressure difference maintained with a small vacuum pump causes cake formation on the surface of the discs, and dewatering takes place as long as free liquid is present.

TRADE-OFF STUDIES

Currently, water recovery has become a very pertinent topic in mining projects. In countries like Chile and others with water scarcity, improving the overall water balance and using filtration steps remain very key and interesting considerations. Alternatives such as seawater supply alternatives are feasible but costly and can affect the actual flotation separation-step performance.

Tailings filtration can reduce water sent to the tailings facilities by up to 30%. As an example, in today's concentrator practice, the conventional tailings underflow percent solids can range from 52% to 64% solids, whereas with tailings filtration final solids can be deposited at 75%–80% solids, subject to the feed characteristics.

Although filtration technology is relatively new (at least regarding large-scale tailings filtration rates in the copper industry), it is recommended to evaluate the opportunities by means of a trade-off study. As an example, a trade-off study was made at a large-scale concentrator in Chile by selecting a base case study and three other options to evaluate them technical and economically. These case studies are discussed in the following sections. In this project, an incremental throughput increase of 50% above its original nameplate was one of the key objectives. The following information provides some comments on the advantages and disadvantages of the various options. This will typically be supported by accompanying capital and operating costs.

Case Studies

Base Case Study

For this base-case option, a fraction (35%) of the actual tailings production will be sent to a filtration plant incorporating three current overhauled thickeners and one new thickening unit. The new facility will send material to a new filtration plant with 11 filter presses (nine operating and two standby units). Filter cake with 18% final moisture will be transported via conveyor belts to a final stacking area located inside the existing tailings dam. Conventional and filter cake tailings will be disposed together with conventional tailings separated by a wall. Figure 36 shows the base case study schematic representation.

Case Study 1

This is a base-case optimization by partial filtrate tailings (Figure 37). A portion of tailings is sent to a tailings filtration plant but was previously thickened by using four high-density thickening units to increase water recovery. High-density thickening would be up to 61% w/w and is expected to increase the filtration rate (expected 18.5% more than base case) as feed percent solids increase. The other tailings portion (conventional) is sent to the tailings dam.

This new configuration would need additional recovered water ponds, including water pumping systems, a slurry pumping system to filtration plant, and ten tailings filtration units (eight operating and two standby). Filtrated material, together with conventional tailings, would be transported by conveyors and stacked by using front-end loaders.

Case Study 2

In this option, around 72% of the total tailings produced is sent to a tailings filtration plant with the remainder passing to a staged cycloning plant for coarse and fine particle separation (Figure 38). The cyclone fine fraction is then thickened to achieve 45% solids at discharge for final disposal. The cyclone coarse fraction is sent to filtration plant with 69% solids.

A new conventional thickener is required for this case, including a new cyclone cluster with its correspondent pumping system, and an extra thickener for the fine cyclone fraction and water recovery. Ten tailings filters (eight operating and two standby units) are required. All the filter cake material is transported via conveyor belts and disposed in the tailings dam area.

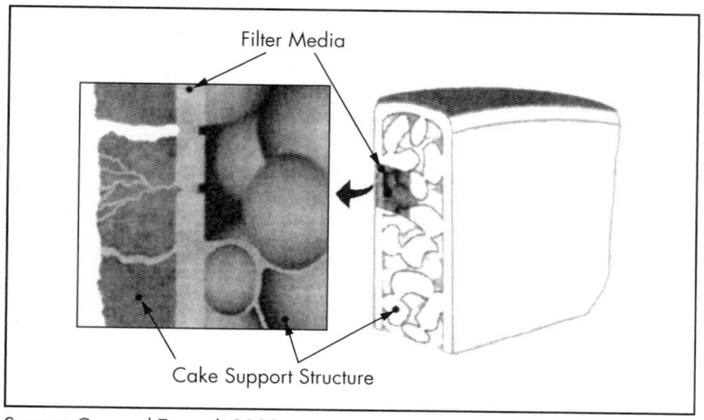

Source: Cox and Traczyk 2002
Figure 35 Ceramic disk media cross section

Case Study 3

In this case, all tailings produced are sent to a double-stage cyclone cluster where the coarse fraction (70%) is sent to dewatering screens with a screen oversize at 69% solids, while the fine fraction (30%) is sent to a thickener where its thickened underflow (at 45% solids) is sent directly to the tailings dam (Figure 39).

A new conventional thickener must be installed for this case as well as new cyclone clusters with 16 cyclones per cluster plus corresponding pumping stations. A new thickener for the fine fraction must also be installed. For final tailings dewatering, a set of 26 screens are considered with screen oversize at a 22% final moisture. The screen oversize material is transported by using conveyor belts.

Technical Analysis

Table 3 is a summary of the technical advantages and disadvantages found for each case study. According to this high-level technical evaluation, only the base case and case study 1 would be worthy of further economic evaluation. Further cost comparison data and information will allow a more detailed assessment of the options and proposed path forward.

Future Outlook

Tailings filtration is a feasible option to be considered during the conceptual and prefeasibility studies and the early engineering stages. Overall improvements in filter technology have provided opportunities to process larger throughputs with fewer machines. Presently, on a larger throughput scale, moisture contents of 20%–25% are achievable with larger filter presses. The gold industry has led the complete filtered

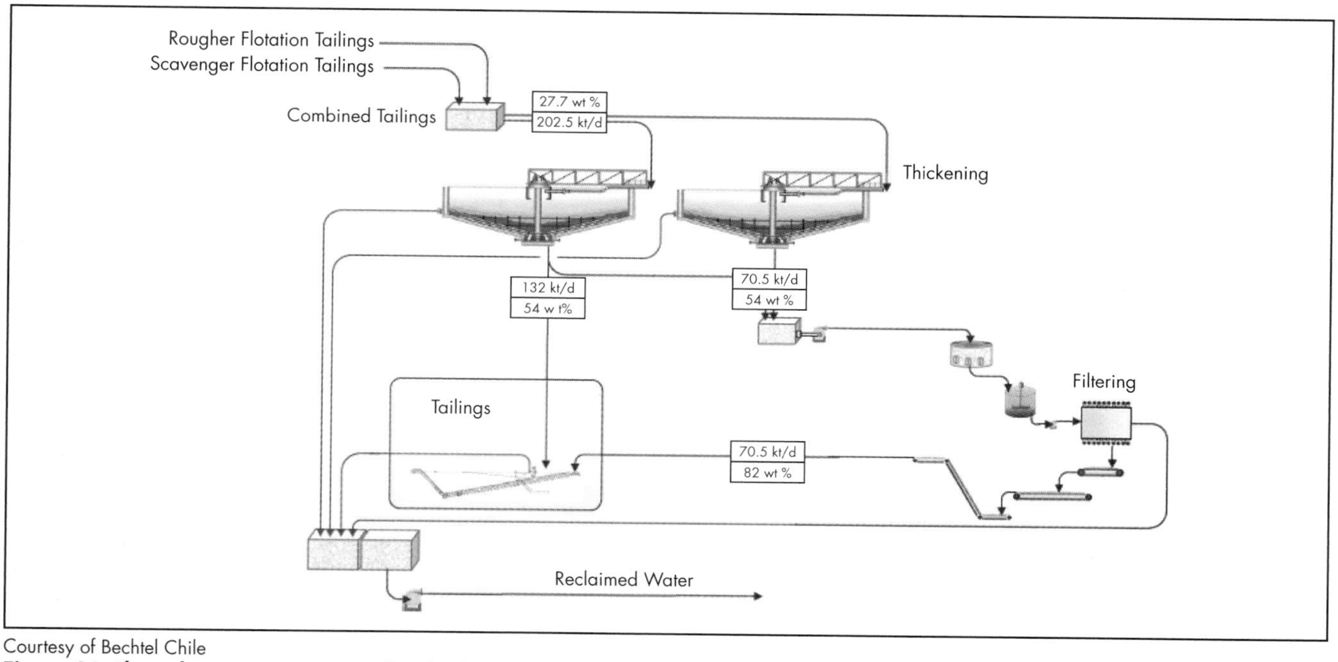

Courtesy of Bechtel Chile
Figure 36 Flow-sheet representation for the base case study

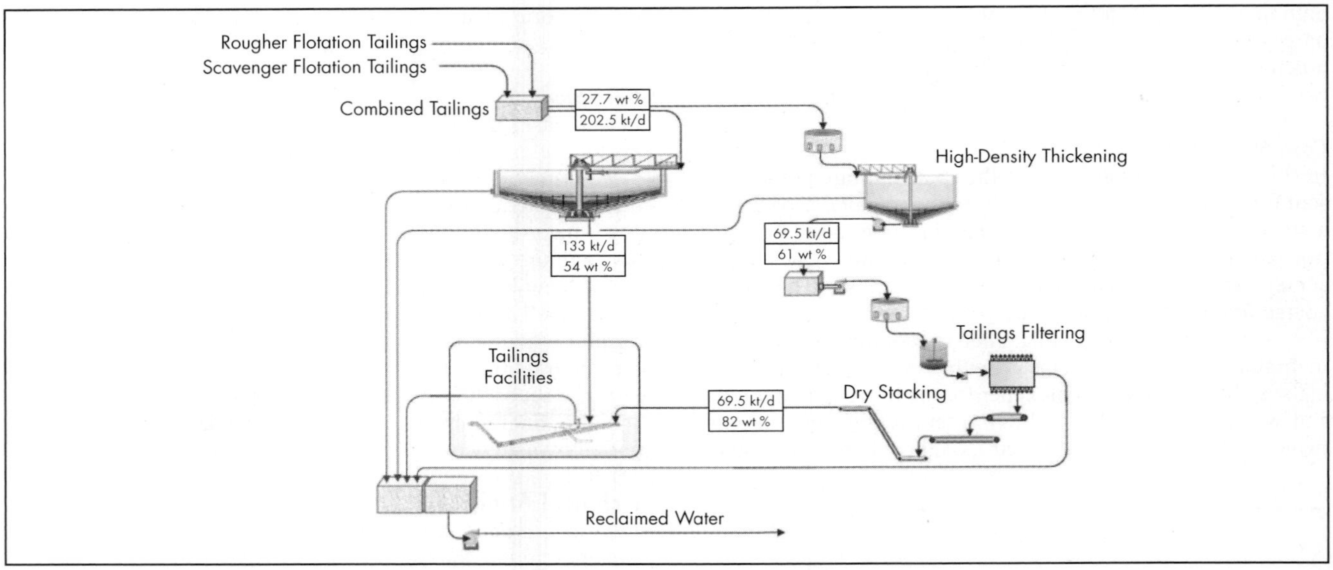

Courtesy of Bechtel Chile

Figure 37 Flow-sheet representation for case study 1

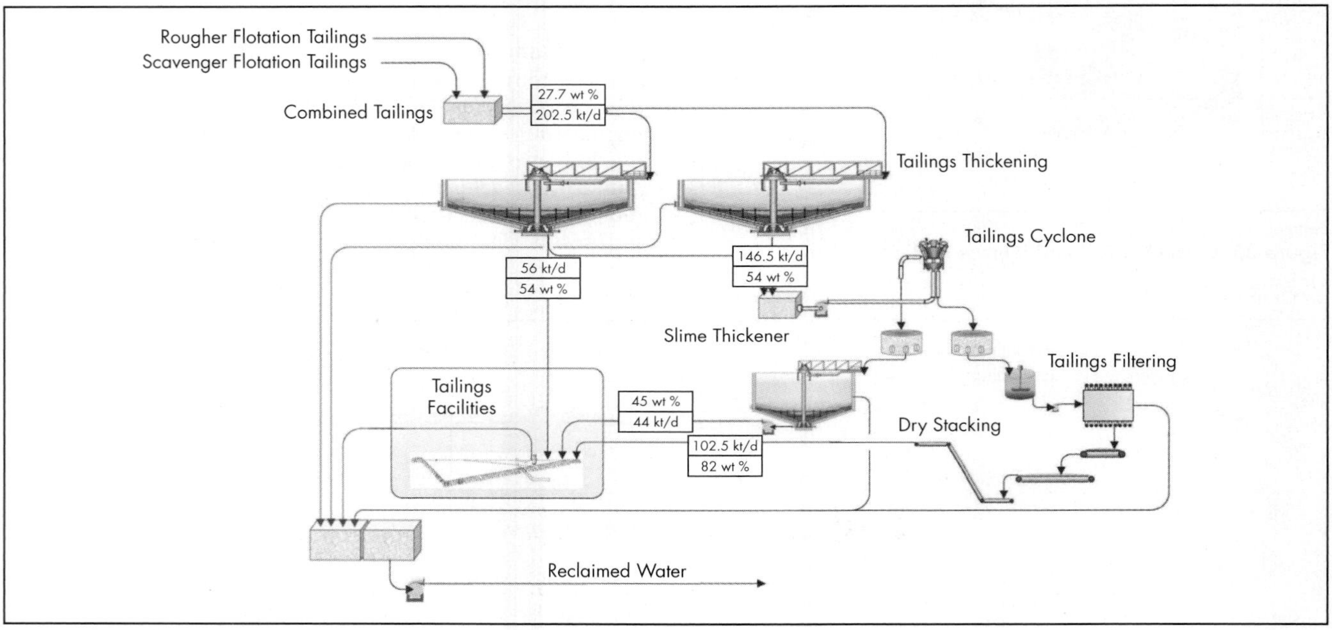

Courtesy of Bechtel Chile

Figure 38 Flow-sheet representation for case study 2

tailings applications with the seven high-pressure Diemme filters in Mexico, processing the entire tailings stream and the material subsequently being dry stacked.

Currently, the largest filter presses involve more than 2,000 m^2 of available filtration area and use 2 × 4-m plates. The filter vendors are continuing to explore these larger, robust machine designs with FLSmidth's AFP 2040 being tested in a tailings filtration application in South America during 2017. This test encompassed a single unit that was trialed over several months on plant tailings with a processing rate of up to 10,000 t/d. The filter cycle time for the trials was set to approximately 8 minutes. The overall filter cake moisture was in the range of 18% to 20% but also had performance fluctuations because of the influence of particle size and clay mineralogical content with the presence of minerals such as kaolinite and muscovite. The downstream impact of these variances still needs further geotechnical assessment to optimize the stacking process. One other important consideration in such trials is the sheer scale of the operation, the feed stream preparation, and all the important associated ancillary equipment around the filter itself.

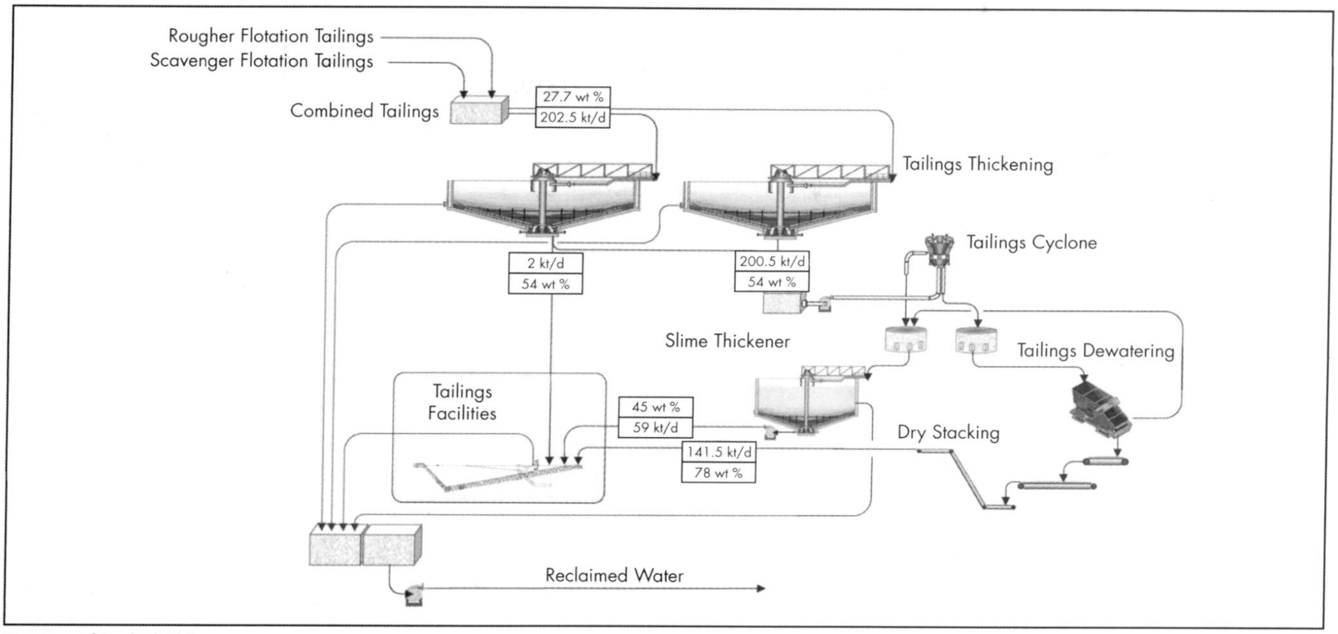

Courtesy of Bechtel Chile

Figure 39 Flow-sheet representation for case study 3

Table 3 Technical comparative table for the case studies

Case Study	Advantages	Disadvantages
Base	• Technical feasibility to reach water makeup. • By using tailings filters, there is an operational flexibility with a minimum impact on conventional tailings operations. • Tailings filtration produces larger amounts of water. • Well-known conventional thickening technology. • Less fine material recirculating. • Conveyor belts usage known by operations.	• High content of fine material might affect filtration. • Lower filter mechanical availability. Equipment maintenance is more complex. • Main operating expenditure (OPEX) is determined by filtration equipment. • A ramp-up curve is expected for new filtration systems. • Standby units are needed for operation. • Real filtering capacity to be proven.
1	• Technical feasibility to reach water makeup. • By using tailings filters, there is an operational flexibility with a minimum impact on conventional tailings operations. • Tailings filtration produces larger amounts of water. • Less impact over "conventional" tailings production. • Less number of filters as filtration rate increases. • Less fine material recirculating. • Conveyor belts usage is known by operations.	• High-density thickening needs a pumping system. • High-density thickening must be close to the filtering plant, so a new pumping system is needed to conduct material to it. • High content of fine material might affect filtration. • Lower filter mechanical availability. Equipment maintenance is more complex. • OPEX is determined mainly by filtration equipment. • A ramp-up curve is expected for new filtration systems.
2	• No fine material. Increase in filtration rate.	• New cyclone cluster must include pumping systems. • Cyclone underflow is too dilute. Not suitable for disposal at tailings dam. • Lower water recovery efficiency. • OPEX is determined mainly by cyclone cluster and pumping system.
3	• No fine material. Increase in filtration rate.	• New bank of cyclones must include pumping systems. • Cyclone underflow is too dilute. Not suitable for disposal at tailings dam. • Lower water recovery efficiency compared to filters. • OPEX is determined mainly by the bank of cyclones and pumping systems. • Dewatering screens have low treatment rates, increasing the large number of units.

Courtesy of Bechtel Chile

Potential opportunities, including 3 × 5-m plate filter presses, will allow lowering of the plant footprint and will improve the economic outlook.

Water recovery is an important issue to be addressed for capital expenditure (CAPEX) and operating expenditure (OPEX) calculations, particularly on the 100,000-t/d-plus projects). Decreasing final water content in disposal may alsohave advantages in terms of volume of the final disposal area.

Early geotechnical data have shown suitable stability for disposal and deposition of filter cake. This material must be sent by conveyors (or trucked) to a disposal area and stacked by using stacking equipment and/or heavy machinery (bulldozer) to compact the material.

In terms of the community's point of view, stable and dry tailings disposal may be better perceived by neighbors. Filtrate can be recirculated back to the plant and/or used by the filter itself, reducing general plant water requirements (makeup).

CAPITAL AND OPERATING COSTS

CAPEX and OPEX calculations are an integral part of evaluating the filter technology selection criteria as well as overall process circuit considerations (per trade-off-study evaluations on filtration versus other methodologies). Table 4 is a simplistic view comparing filter types and some relative relationship for CAPEX and OPEX.

Some general observations to be considered for each filter type include the following:

- Vacuum filters are relatively simple to operate compared to other filters but have high power consumption and deliver products with a residual moisture of 18%–25%.
- Horizontal vacuum belt filters are capable depending on the particle size distribution in the final cake, the level of vacuum in the form zone, the drying zone time, and airflow to produce moistures much lower than 18%. Moistures below 10% are possible with the proper design of the feed system to ensure good mixing of both the coarse and fine fractions prior to the form zone, resulting in a homogeneous cake with the maximum performance. Vacuum belt filters still have applications where they are the favored machine for a given flow sheet.
- Because ceramic filters use capillarity as their driving force, the material must be "fines free."
- Belt presses work better with compressible products, but it is not a good solution for hard rock applications.
- Filter presses can deliver product with 9%–11% final moisture on copper concentrates. In certain regions of the world, these are the preferred choice for this application.

In general terms, cycle times and cake weights must be considered, as they set the filtration area for vacuum filters. For filter press selection, cake density must also be considered. For more detailed analysis and filter selection/ranking, CAPEX estimation must be prepared to the appropriate level and use subtle information such as slurry yield stress and rheology, and so on. For the complete representation and impacts, all stages in a concentrator plant, including final tailings disposal (truck, conveyors, pumps, etc.) and geotechnical data must be included in any trade-off study.

The overall philosophy on plant availability, utilization, and redundancy must be carefully defined and agreed upon early in the evaluation. In the study, the following aspects must also be included in the complete evaluation:

- Overall capacity requirements
- Site-specific location and remoteness
- Availability of labor and skill sets
- Vendor support and infrastructure
- Immediate support equipment such as pumps, tanks, and mixers
- Plant infrastructure such as civil, piping, and buildings
- Instrumentation and control system

Other topics, including contingency and engineering, procurement, construction, and maintenance costs; owner's costs; and so forth; must also be included.

In terms of the OPEX calculations, the following items must be considered:

- Consumables such as filtering media cloth life and reagents as required
- General wear components
- Power demand, including all equipment and ancillaries
- Workforce
- Maintenance costs
- Contract maintenance and labor, if applicable
- Tailings area costs (earthmoving equipment and conveyors, etc.)

In all cases, power calculations and ancillary demands must be accounted for. With all the data acquired, the overall unitary cost ($/t) can be determined.

Table 4 CAPEX and OPEX comparison for different filter types

Filter Type	CAPEX	OPEX
Belt press	⇩	⇧
Vacuum filters	⇧	⇧
Ceramic filters	⇧	⇧
Filter press	⇧	⇩

Courtesy of Bechtel Chile

ACKNOWLEDGMENTS

The author gratefully acknowledges Dailey Jones of K&B Technical Services Inc. for his assistance regarding belt filter operation; and David Minson of MinTecProcess Consulting and Management Ltd. and Leonardo Soto of Bechtel Chile Limitada for their valuable contributions throughout this chapter.

REFERENCES

Aqseptence Group. n.d. *Fast Cycling Filter Press GHT F* [brochure]. Lugo, Italy: Aqseptence Group.

Choudhury, A.P.R., and Dahlstrom, D.A. 1957. Prediction of cake-washing results with continuous filtration equipment. *AIChE J.* 3:433.

Concha, F. 2002. *Manual de Filtración y Separación* [Manual of Filtration and Separation]. Chile: Universidad de Concepción Chile.

Cox, C., and Traczyk, F. 2002. Design features and types of filtration equipment. In *Mineral Processing Plant Design, Practice, and Control.* Edited by A.L. Mular, D.N. Halbe, and D.J. Barratt. Littleton, CO: SME.

Diemme Filtration. 2017. The Future of Mining Tailings Management [video]. September 21. https://www.youtube.com/watch?v=HAQRTLFmsFc.

FLSmidth. 2013. *Tailings Dewatering Press (TDP)* [brochure]. Midvale, UT: FLSmidth.

Larox. 2004. *Ceramec Technical Specification CC-15, -30, -45.* Document No. 121-1050-04D US. Larox Oyj.

Micronics. 2018. Filter cloth: The key ingredient to optimizing your filter press operations. Portsmouth, NH: Micronics.

Outotec. n.d.(a). Outotec Larox SC vacuum disc filter. www.outotec.com.

Silverblatt, C.E., Risbud, H., and Tiller, F.M. 1974. Batch, continuous processes for cake filtration. *Chem. Eng.* 4:127–136.

Smith, C.B., and Townsend, I.G. 2002. Testing, sizing and specifying of filtration equipment. In *Mineral Processing Plant Design, Practice, and Control.* Edited by A.L. Mular, D.N. Halbe, and D.J. Barratt. Littleton, CO: SME.

CHAPTER 8.3

Continuous Drying Technologies

Trond Muri

The drying operation of materials from particulate solids, through slurries and suspensions to continuous sheets, represents an important part of several processes in the chemical and mineral industries. The increasing demand of high productivity at low cost and minimal environmental impact makes it complex to evaluate and choose the best available drying technology for each industrial application.

A combination of heat, gas, and agitation are essential for process drying. The location and level of these key inputs are what distinguish direct from indirect drying systems (Kimball 2001). In the direct drying method (convection drying), the product is brought into direct contact with the drying medium, which may be hot gas or air. Large quantities of gas are necessary for transfer of the required energy. A special case of the direct dryer is the fluid bed dryer, in which the product floats on a cushion of air or gas. The process air is supplied to the bed through a perforated distributor plate and flows through the bed of solids at a velocity sufficient to support the weight of particles. Very high heat and mass transfer values are obtained as a result of the intimate contact with the solids and the relative velocities between individual particles and the fluidizing gas. In the indirect drying method (conduction or contact drying), the wet product is separated from the drying medium. The drying unit can consist of a rotor forming the indirect contact heating surface mounted in a stationary or rotating shell. The indirect method uses saturated or superheated steam, resulting in low energy consumption and minimal environmental impact.

The drying process typically consists of three periods. Figure 1 represents the drying process in an indirect dryer. In the first period, only heating of the wet product takes place. In the second period, the bulk of the water or liquid is evaporated at approximately constant product temperature. Finally, the removal of final moisture results in rising product temperature.

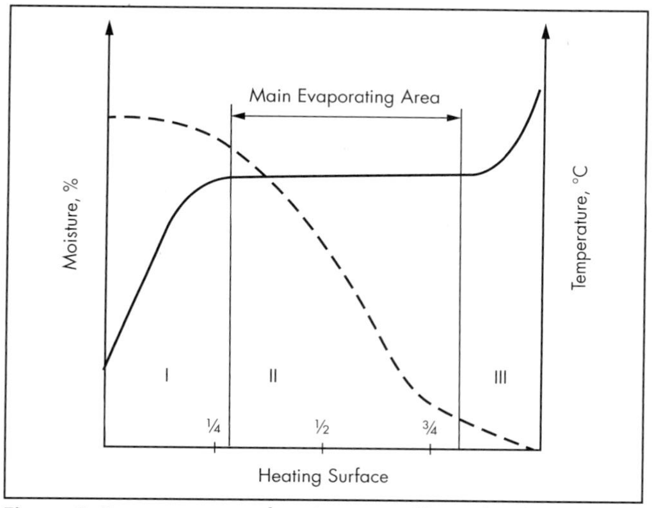

Figure 1 Temperature and moisture profile in the dryer

CLASSIFICATION OF DRYERS

The two most useful classifications of drying equipment are based on how the wet product is related to the drying medium or on the handling characteristics and physical properties of the wet material (Moyers and Baldwin 1999). A classification chart based on heat transfer (direct and indirect) with subclasses of continuous and batch-wise operation was defined by Marshall (1946). Batch operation is most useful for low product feed rates and when the residual time of the product must be controlled and uniform. Batch dryers are omitted in this study because mineral applications with high production rates are considered. Figure 2 shows a simplified chart for most of the dryers with continuous operation that are used in the mineral industry.

Trond Muri, Senior Project Engineer, PASSER, Vear, Vestfold, Norway

DIRECT DRYERS

Drying in a pneumatic conveying or flash dryer is often done in conjunction with grinding. The material is conveyed in high-temperature, high-velocity gases to a cyclone collector. This type can be used if the product is recirculated to make feed suitable for handling. Currently, the flash dryer performs the important role of effective utilization of off-gases from super heaters and the anode furnaces.

In direct rotary dryers, the material is conveyed and showered by flights inside a rotating cylinder where hot gases flow. The dryer is applicable to dry-product recirculation.

In a spray dryer, the wet feed must be capable of atomization by either a centrifugal disk or a nozzle. It is well suited for large capacities. High temperatures can be used with heat-sensitive materials. However, pressure-nozzle atomizers are subject to erosion.

In a fluid bed, the solids are fluidized in a stationary tank. The bed is either inert or mounted with a dry-solid recirculator. A fluid bed may also have indirect-heat coils to enhance the heat transfer.

INDIRECT DRYERS

An indirect drum dryer may be heated by steam or hot water under atmospheric or vacuum operation with a single, double, or twin drum. Maintenance costs can be high.

A screw-conveyer dryer is applicable to dry-product recirculation. Operation under vacuum is feasible to permit solvent recovery.

A steam-tube rotary dryer operates with rotating tubes and shell. It is also applicable to dry-product recirculation. Operation under slight negative pressure is feasible to permit solvent recovery. Dryer off-gas is directed to a bag filter for dust removal. All dust is returned to the dryer discharge drag conveyor.

In a steam dryer, the tubes or coils rotate, and the shell or vessel itself is stationary. Operation under vacuum is feasible to permit solvent recovery. Dust leakage is very low because the dryer body is stationary and no difficult sealings are needed. Another advantage with the stationary shell compared to a rotary shell is that the dust can be led directly back to the dryer and the dried product in the vessel. Additionally, the shell has inspection and service doors on the side opposite the product discharge.

To optimize the dryer, one seeks to increase the available heating surface per volume unit of the dryer. Coils offer a higher heating transfer surface compared to shovels, discs, or paddles. Additionally, the coil design is not subject to thermal expansion problems that are common in other types of steam dryers with longitudinal pipes. The configuration of parallel coils also permits an improved flow of product through the total volume of the dryer (Figure 3).

ADVANTAGES OF INDIRECT DRYERS

Indirect dryers offer several advantages over direct dryers (Raouzeos 2003):

- Cross-contamination is avoided because the product does not contact the heat transfer medium.
- Solvent recovery is easy due to the very small amount of non-condensable gas present.
- Extensive dust formation is generally avoided because of the small amount of vapor involved.

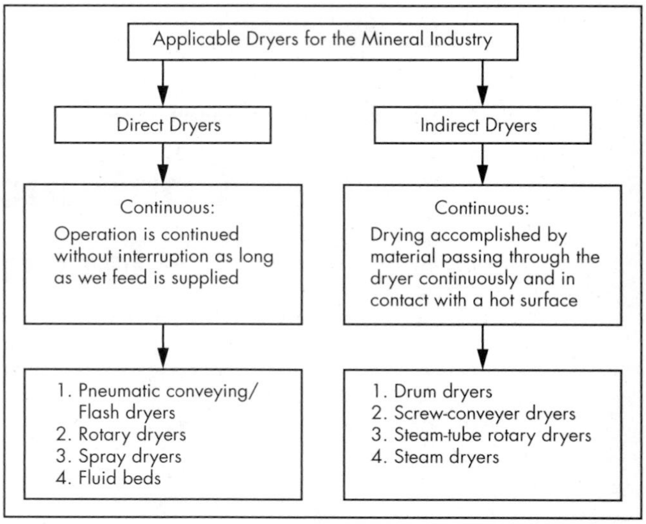

Source: Marshall 1946

Figure 2 Classification of dryers applicable to the mineral industry

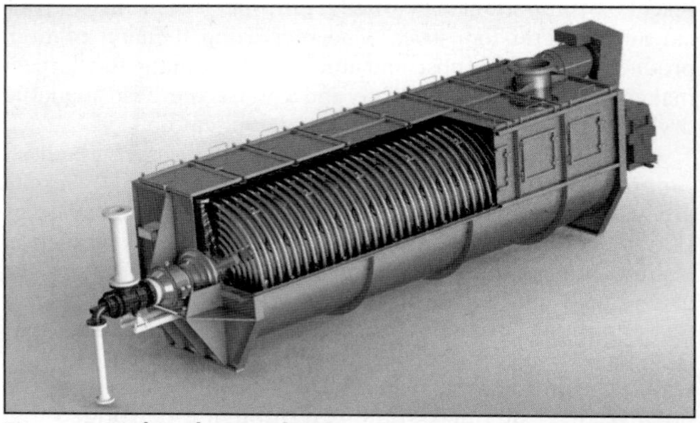

Figure 3 Multicoil, an indirect steam dryer with rotating coils and stationary vessel

- These dryers can be of closed design, thus containing toxic vapors and/or providing better control of explosive hazards.
- The *thermal efficiency*, defined as heat required per unit mass of evaporated liquid, is high.
- The final product has a higher bulk density than when the same product is dried in a spray dryer.
- Indirect dryers can be designed as pressure- and shock-resistant vessels.
- They usually require less erection space.

PRELIMINARY SELECTION OF THE DRYER

The preliminary selection of a drying system is determined by several factors (Moyers and Baldwin 1999):

- Product properties
  - Physical characteristics when wet and dry: pasty, sticky, or not sticky.
  - Particle size distribution: Small particle sizes result in better dryer efficiency and hence smaller dryer units (indirect dryers).

 - Corrosiveness: In material selection, high-grade steel is required.
 - Abrasiveness: In material selection, high-grade steel is required.
 - Toxicity: A closed drying system and recovery/capture of moisture/off-gas may be required.
 - Flammability: Fire and explosion protection devices should be considered. Indirect dryers are superior due to lower off-gas quantity.
- Drying characteristics of the material
 - Moisture type: Water or solvents.
 - Initial and final moisture content: High initial moisture may require drying in two steps, a combination of direct and indirect dryers. Indirect dryers are superior to achieve low final moisture content ≤0.1%.
 - Permissible drying temperature: Direct drying or indirect heating by hot water, for example.
 - Probable drying time: If a specific drying/product retention time is required, direct dryers (batch) are the most reliable.
- Flow of material
 - Capacity to be handled: Degree of material flowability is important.
 - Continuous or batch operation: Depends on product quantity and/or high-cost product.
 - Process before and after drying: Consider the need for dewatering/crushing of feed to smaller particles prior to feed inlet of the dryer unit, and cooling of the product below maximum operational temperature to the downstream conveying system.
- Product qualities
 - Gentle handling of product surface: Product testing is required.
 - Shrinkage: Dependent on product property; product testing is required.
 - Contamination
 - Uniformity of final moisture content: Dependent on product property and particle sizes.
 - Decomposition of product: Segregation of smaller/bigger particles is minimized by use of paddle dryers (direct dryers).
 - Overdrying: Indirect continuous dryer systems are not to be oversized due to the risk of overheating the product.
 - State of subdivision.
 - Product temperature: Heat sensitive or not.
 - Bulk density: Important due to dimensioning of installed power and drive systems.
- Recovery problems
 - Dust recovery: Heat tracing of off-gas system to avoid product buildup/blocking of filter bags.
 - Solvent recovery: Consider drying under vacuum.
 - Energy recovery: Indirect steam dryers are superior due to recovery of condensate energy to the steam boiler feed water tank.
- Facilities available at site
 - Space: Direct dryers require more space due to bigger dryer units and off-gas systems.
 - Temperature, humidity, and cleanliness of air: Site conditions are important in selecting an off-gas system and cooling/heating system on drive units.
 - Available fuels, drying medium (steam, cooling water, etc.).
 - Available electrical power.
 - Permissible noise, vibration, dust, or heat losses.
 - Source of wet feed: Stable and continuous feed is required for indirect dryers.
 - Exhaust-gas outlets: Direct dryer units have excessive off-gas quantities compared to indirect dryers.
- Investment cost: Consider spare parts needed over the lifetime of the dryer.
- Operational cost
 - Spare parts needed over the lifetime of the dryer.
 - Energy recovery of heating medium off-line. Indirect dryers are more energy-efficient due to recovery of condensate energy to the steam boiler feed water tank.
 - Availability of dryer: The dryer should be proven for at least 90% availability.
 - Maintenance: Easy access for maintenance and available service partner for dryer equipment.
- Environmental impact/emission: Less off-gas quantity on indirect dryers; selection of reliable off-gas system and material in filter bags.

These factors are helpful in the preliminary screening of the applicable dryer for a given application. The next step in the selection process is to obtain data from the equipment manufacturer to perform an adequate pre-selection. A comparison of drying systems is presented in Table 1 (Riekkola-Vanhanen 1999).

The main reasons for selecting a steam dryer instead of a conventional rotary dryer are the relatively simple off-gas system and its insensitivity to material variations (moisture, composition, and quantity) and the small space required (Gernerth and Willbrandt 1996). This means that the investment and operation costs are low.

Also, because the drying process is a high-energy-consuming stage in a production line, the obvious choice is an indirect dryer if a steam boiler is available. Using steam as heating medium allows recovery of the energy in the condensate from the dryer by return to the steam boilers' feed water tank. Compared to direct dryers, the heating medium off-gas around 60°–90°C is normally not usable for energy recovery.

Table 1 Comparison of drying systems

Criteria	Spray Dryer	Rotary Dryer	Steam Dryer	Flash Dryer	Fluid Bed Dryer
Investment cost	– –	+ +	+ + +	+	– –
Operating cost	– – –	+	+ + +	–	– –
Gas flow	– – –		+ + +	– –	–
Fugitive emission control	–	–	+ + +	–	–
Operational flexibility	–	+ +	+ + +	–	– –
Material handling	+	–	–	–	–
Future expansion	– –	+	+ + +	–	–
Environmental compliance	– –	–	+ + +	–	– –
Process sophistication	+	+	+ +	+	+
Process maturity	++	+ + +	+ + +	+ + +	+ +
Generality	+	+ + +	+ +	+ +	+

Source: Riekkola-Vanhanen 1999

+ Represents degree of advantage; – represents degree of disadvantage; blank represents neutral.

A steam boiler may not be available or feasible at the particular site, and therefore the option will be electric-heated air or air heated by excess energy from other processes on-site; hence direct dryers are a solution.

The selection of a dryer should be based on drying tests to determine the optimum operating conditions forming the basis for quotations from vendors. From the drying tests and quotations, the final selection can be made.

INDIRECT DRYING SYSTEM EXAMPLE

The following example is for a Multicoil indirect dryer unit with rotating coils and stationary vessel (Figure 3). Several of these dryer units are currently operating in copper smelters in Sweden, Germany, Spain, Canada, Mexico, Zambia, Australia, and India.

Typical Steam Dryer Plant Setup

Consider a Multicoil copper concentrate dryer plant with a drying capacity of 200 t/h (metric tons per hour) of wet product and 22 t/h of evaporated water at a steam pressure of 19 barg (bar gauge). Typical figures are 11% initial moisture content in the concentrate to 0.2% after drying, and the use of air as sweep gas at a rate of 1.5 kg air/kg evaporated water. Two dryers in parallel are used due to the high-capacity requirements. Each dryer is designed for a wet product capacity of 100 t/h and 11 t/h of evaporated water. This will result in a higher flexibility and reduce the risks associated with maintenance and eventual unexpected shutdowns. Two dryers in parallel offer the possibility to run the drying unit under partial load conditions when maintenance is carried out.

Heat Transfer Coefficient

An overall heat transfer coefficient K (W/m^2/°C) in a dryer is mandatory to determine scale-up of an industrial dryer size for a specific application.

An average heat transfer coefficient for the product could be determined by testing a product sample in a laboratory dryer. The test dryer should have a similar design to industrial units; however, a 3-m^2 heating surface is sufficient. A typical test is performed under medium operational steam pressure, 10 barg.

The dryer test starts with a product sample, for example, 150 kg, and with measured initial water content (WC). During the test, the following parameters are measured every 10 minutes:

- Product moisture, % WC
- Product temperature, °C (Cold product temperature is recorded at the start of the test.)
- Coil surface temperature, °C
- Off-gas temperature, °C

The energy consumed is calculated using the recorded temperatures for heating dry product and water, and including the total energy of the measured evaporated quantity of water at the end of the test.

Next, the logarithmic mean temperature difference (LMTD) is calculated. The LMTD is a logarithmic average of the temperature difference between the hot and cold feeds at each end of the heat exchanger:

$$\text{LMTD} = \frac{\Delta T_A - \Delta T_B}{\ln\left(\frac{\Delta T_A}{\Delta T_B}\right)}$$

The actual temperatures (T) are related to inlet (A) and outlet (B) ends of the dryer unit.

Finally, the heat transfer coefficient K (W/m^2/°C) is calculated based on known heat-exchanger surface, energy consumed, and a calculated logarithmic average of the temperature difference between hot and cold feeds. For minerals (copper concentrate), a typical K value is 160 W/m^2/°C in a Multicoil steam dryer.

Cost

The cost of a dryer depends on its size, production capacity and material, and current metal prices. A complete drying unit system includes the dryer, bag-filter and exhaust fan with drive motors, air-lock feeders for product inlet and outlet, condensate tank and pump, back-mixing system, local control panel, transport, engineering and project management, installation, commissioning and training, and so on, which define the total price of the package. The major players in the mineral market operate with prices within the same order of magnitude (Moyers and Baldwin 1999). The price of the drying equipment is small compared to the total investment cost of smelter construction; nevertheless, the functionality of the drying unit is essential to the production chain. If the dryer shuts down, the whole process comes to a complete stop.

The purchaser must also estimate the total cost during the life span of the dryer's operation by looking at the spare-parts cost and the approximately availability of the dryer (planned maintenance and unexpected shutdowns). This information can be difficult to find, unless the manufacturer provides reliable data from reference customers.

The availability of an indirect dryer for nonabrasive products is between 96% and 98%, with an average of only one to two days per year of unexpected stops for maintenance. The high efficiency, safety, and cleanliness characteristics of this technology are widely proven (Riekkola-Vanhanen 1999).

Materials of Construction

The dryer parts in contact with the concentrate are made of different materials, such as Super Duplex, depending on the corrosiveness of the concentrate. The shell or vessel is normally made of stainless-steel 316L and carbon steel. For all parts in contact with the concentrate and exhaust gas, the material of construction is a combination of Duplex (SAF2205) and 316L. Carbon steel is used for parts that are not in contact with the concentrate.

INDIRECT DRYER OPERATION AND PITFALLS

The following sections describe the most typical operational pitfalls and how to avoid dryer malfunctions by controls and proposed auxiliary equipment.

Product Overheating

Product overheating is due to improper management of steam pressure. Overheating causes an unnecessary increase in energy consumption and reduced capacity and may, in some cases, lead to a release of noxious gases in the vessel. Monitoring product temperature, which autoregulates the steam pressure, helps to avoid this phenomenon.

Overfilling

Overfilling the dryer could induce increased wear and erosion of the heating surfaces due to higher stresses. This is controlled by monitoring the amps on the rotor drive.

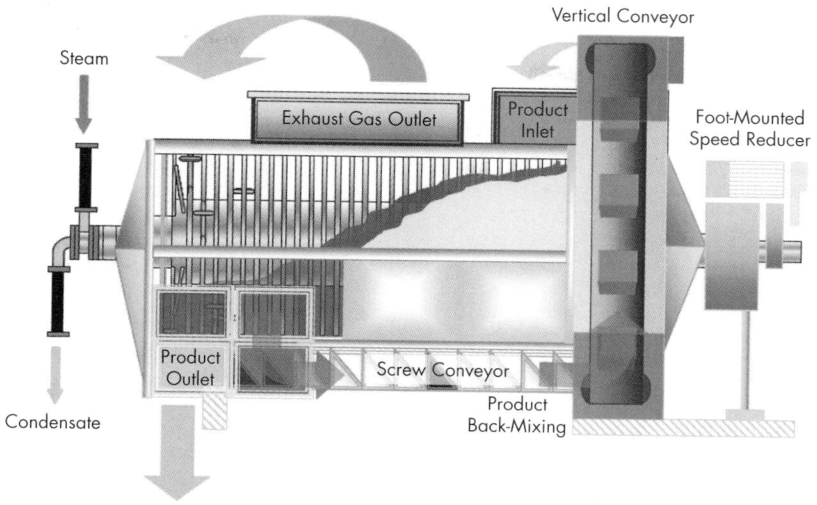

Figure 4 Drying unit with back-mixing system

Product Handling at Inlet Using a Back-Mixing System

The purpose of the back-mixing system is to mix wet and dried product in the inlet section of the dryer to avoid fouling the heating coils and to achieve better flowability of the wet feed (Figure 4). A pasty or sticky feed may plug between coils or stick to the heating surface in the inlet section. This will reduce the flow area and lower the heat transfer ratio in the dryer. When product sticks to the heated coils, the product layer will reduce the heat transfer and result in loss of production.

Because the product flow through the dryer is forced by slow rotation of the dryer rotor and gravity, it is critical that the product does not choke the inlet sections of the dryer or build up product layers on the heating surface. Acceptable flowability and no-stick performance will be achieved by reducing the average moisture level of the wet feed by introducing a parallel feed of dried product at the dryer inlet. The dryer rotor itself acts as a mixing device for wet and dried product.

A back-mixing system should be considered if the moisture content of the dryer feed is in the range of 11%–12% (wet basis) over longer periods. The back-mixing system is not necessary if the moisture content of the dryer feed is in the range of 8% to 9% (wet basis) during the main period of operation.

A back-mixing system may increase the interval between major maintenance of the rotor by 30%–50% and results in a longer rotor lifetime. An additional feature is that back-mixing enables proper operation at lower steam pressure, giving improved flexibility at partial load.

Maintenance

The maintenance of the dryer should be user-friendly. Operators should have easy access from the outside to all parts of the rotor or heat exchange unit. This allows a predictable maintenance routine that can be scheduled and conducted without unnecessarily long discontinuation of the operation. Some dryers have inspection doors available during operation. In this type of design, the rotor does not need to cool down before maintenance and/or repair work can take place. Steam leakages can be easily detected, and general maintenance and/or repair work requires a very short time.

CONCLUSION

This chapter provides an overview of the drying technologies available based on different drying methods. To facilitate the selection of a drying system for an application, the advantages and disadvantages are summarized with respect to issues such as investment, operating cost, maintenance, product capacity, product handling, and environmental impact. The final choice of dryer should be based on drying tests to determine the optimum operating conditions forming the basis for quotations from vendors. From the drying tests and vendors' quotations, the optimal selection can be made.

REFERENCES

Gernerth, S., and Willbrandt, P. 1996. Present changes at Norddeutsche affinerie's copper smelter. In *Proceedings of the Eighth International Flash Smelting Congress*. October 13–18. pp. 95–111.

Kimball, G. 2001. Direct vs. indirect drying: Optimizing the process. *Chem. Eng.* 108(5):74–81.

Marshall, W.R. 1946. *Heating, Piping and Air Conditioning*. Vol. 18. p. 71.

Moyers, C.G., and Baldwin, G.W. 1999. Psychrometry, evaporative cooling, and solids drying. In *Perry's Chemical Engineers' Handbook*, 7th ed. Edited by R.H. Perry, D.W. Green, and J.O. Maloney. New York: McGraw-Hill.

Raouzeos, G. 2003. The ins and outs of indirect drying. *Chem. Eng.* 110(13):30–37.

Riekkola-Vanhanen, M. 1999. *Finnish Expert Report on Best Available Techniques in Nickel Production*. Helsinki, Finland: Finnish Environment Institute.

PART 9

Disposal

CHAPTER 9.1

Tailings Disposal and Management

John Lupo and Vicki J. Scharnhorst

Tailings are the by-products generated from mechanical and chemical processes used to extract a desired product (such as gold and copper) from ore. After the ore is processed, the unrecoverable and uneconomic metals, minerals, chemicals, organics, and process water are discharged, normally as slurry, to a storage area commonly referred to as a tailings management facility, tailings storage facility (TSF), or residue disposal area.

The topics discussed in this chapter make up the basic elements of a tailings management framework. A tailings management framework describes the care and management of the facility through its entire life cycle (concept, design, construction, operation, and closure).

RECENT DEVELOPMENTS

Current trends in the mining industry are focused on minimizing water consumption, reducing risk associated with tailings management and storage, and reducing the tailings storage footprint. These trends include the following:

- An increased consideration of the use of thickened, paste, and filtered tailings to reduce water consumption and surface footprint
- Disposal of paste tailings into underground mines to provide ground support and to reduce the surface footprint
- Co-disposal of tailings within waste rock facilities to reduce surface footprint and to mitigate potential geochemical/drainage issues
- In-pit tailings disposal to reduce surface footprint
- Remote sensing and monitoring using satellite-based interferometric synthetic aperture radar (InSar) and drone technology. (These tools are being to monitor tailings facilities. They are used to inspect and measure deformation and seepage [via infrared] over time.)

As noted by Welch et al. (2012), failures and noncompliance of tailings facilities have serious safety, environmental, social, and economic impacts, and can result in loss of reputation and loss of investor confidence. In response, the mining industry continues to develop innovative methods to dispose of tailings to reduce risk and conserve water.

OBJECTIVES AND LIMITATIONS

It is not the intent of this chapter to fully describe a tailings management framework. Several guides and publications are available (MAC 2017; ANCOLD 2012; Martin and Davies 2000) that can be used to develop site-specific frameworks to manage tailings facilities. The primary objective is to provide an understanding of the tailings management framework and the role it plays overall in a mining project.

This chapter is limited to surface disposal of tailings and is not intended to be a comprehensive nor a step-by-step guide to the design, construction, and operation of tailings facilities. Although the basic function of a tailings facility is similar from operation to operation, the facility design, construction, operation, and closure are unique to each site. The discussions presented herein are intended to provide a general overview. Specific testing, procedures, operational controls, and so forth, must be developed specific to the site conditions, mineral processing method, and environmental and social constraints.

TAILINGS STORAGE FACILITY

The TSF is an engineered and constructed facility to contain tailings. The primary objective of a TSF is to provide a dedicated facility that will safely confine tailings solids and process water to a designated area, where they can be managed through life-of-mine, closure, and postclosure.

A TSF can take many forms and shapes depending on many factors, such as the following:

- Site topography
- Local and regional geotechnical and hydrological conditions
- Type of tailings to be contained
- Processing constraints
- Environmental constraints
- Seismicity
- Climate
- Facility construction constraints
- Country mining regulations

John Lupo, Senior Director Geotechnical and Hydrology, Newport Mining Corporation, Denver, Colorado, USA
Vicki J. Scharnhorst, Operations Manager, Tetra Tech Inc., Denver, Colorado, USA

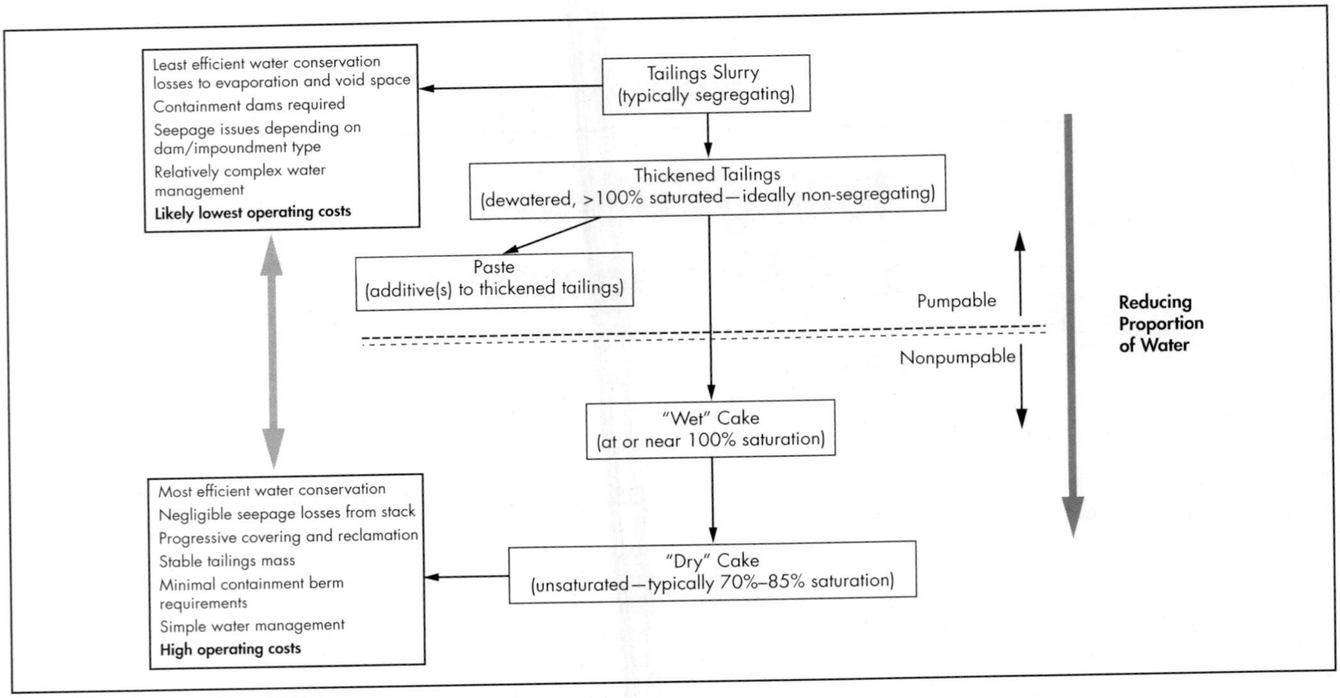

Adapted from Davies and Rice 2001

Figure 1 Tailings continuum

Tailings Continuum

Although a TSF contains tailings, not all tailings behave the same, which can affect the TSF design and management. The tailings continuum refers to the qualitative behavior of tailings at varying degrees of moisture content. Traditionally, tailings are deposited in a slurry form (e.g., a semiliquid mixture). However, as discussed earlier in this chapter, there is a trend to reduce the amount of water deposited with the tailings. As the proportion of water is reduced (by thickening or filtration), the tailings take on different behavior, which affects how tailings are deposited and managed. Additionally, changes in the proportion of water will affect the TSF water balance and water management for the site (discussed later in this chapter).

An illustration of the tailings continuum is shown in Figure 1. On one end of the continuum are slurry tailings, which behave as a semiliquid. On the other end of the continuum are cake or filtered tailings, which behave as a wet soil. In between these end points are thickened and paste tailings. Reducing the proportion of water in the tailings affects how they can be deposited (pumpable or non-pumpable), which can impact both capital and operating costs for tailings disposal. The benefits of reducing the water content of tailings include more effective use of water resources, lower water management costs (less water to handle), reduction in TSF footprint (from reducing water storage in the TSF), and potentially easier closure requirements (Davies et al. 2010).

The selection of the type of tailings to produce at an operation is largely driven by

- Capital and operating costs;
- Ore mineralogy and processing requirements;
- Availability of water for processing;
- Environmental constraints;
- Social acceptance;
- Land availability;
- Risk and risk tolerance; and
- Seismic, climate, geotechnical, and hydrological conditions at the TSF location.

At the start of a project, trade-off studies are recommended to evaluate which type of tailings may be better suited for a given site.

Tailings Storage Facility Configuration

For surface disposal of tailings (slurry, thickened, paste, or filtered), four main TSF storage configurations can be considered.

1. **Cross-valley.** A cross-valley TSF is formed by constructing containment embankments across a natural valley or drainage. The natural topography combined with the containment embankments forms the primary tailings storage basin. An example of a cross-valley TSF is shown in Figure 2.
2. **Paddock.** A paddock TSF (often referred to as a cell TSF, a ring dike TSF, or a turkey's nest TSF) is one where containment embankments are constructed around the entire facility, as shown in Figure 3. A paddock TSF is generally used in areas that are very flat and have little topographical relief.
3. **In-pit disposal.** As the name implies, an in-pit TSF is one where tailings are deposited into an inactive open pit, as illustrated in Figure 4.
4. **Comingling.** Comingling or co-disposal of tailings consists of mixing other materials, such as waste rock, in a single storage area. This approach allows the tailings to be "stacked" by filling in the voids within the waste rock or other media. Figure 5 presents an example of comingled tailings.

Courtesy of Teck Mining Company
Figure 2 Cross-valley tailings storage facility

Courtesy of Newmont Mining Corporation
Figure 3 Paddock tailings storage facility

Source: Hore and Luppnow 2015, reprinted with permission from the Australasian Institute of Mining and Metallurgy
Figure 4 In-pit disposal at Langer Heinrich mine, Africa

Figure 5 Comingled/co-disposal tailings

TSFs may incorporate a liner for environmental containment and/or to reduce loss of process water via seepage into the foundation. Some may have a compacted soil liner; compacted native or imported soils that provide a low-permeability layer. Other TSFs may have a composite liner (e.g., a geomembrane liner in combination with a compacted soil liner). Choosing the need for a liner and liner type depends on site conditions, quality of the tailings, geochemical characteristics, and environmental/social risks.

TSFs are seldom constructed for the life-of-mine tailings storage needs up front. Rather, they are periodically raised (i.e., the containment dams or embankments are constructed to a higher elevation) over time to provide additional tailings storage. This allows capital expenditures to be deferred until additional storage is needed. The first dam or embankment that is constructed in a TSF is often called the starter dam.

Three primary methods are used to raise TSF embankments (see Figure 6):

1. **Upstream raise.** The embankment is constructed over deposited tailings, upstream of the starter dam.
2. **Centerline raise.** The embankment is constructed so that the centerline remains more or less the same as the starter dam. In a modification of a centerline raise, the embankment can be constructed so that the centerline moves slightly upstream or downstream of the starter dam.
3. **Downstream raise.** The embankment is progressively constructed on the downstream end of the starter dam and TSF.

Strengths and weaknesses are associated with each of these TSF raise methods, as presented in Table 1.

The design and method of raising a TSF (for both existing and greenfield projects) requires input and discussion from many functions, including project execution, mine engineering, process engineering, environmental, geotechnical engineering, water management, land, and social. It is recommended to include these functions into the TSF planning and design stage so that the design and any associated risks are discussed and agreed on prior to finalizing the design.

Tailings Storage Facility Failures

Over the past several decades, the mining industry has experienced many TSF failures and incidents. The lessons learned from these events are improving safety and continue to support the development of enhanced practices in tailings management. International Committee on Large Dams (ICOLD) Bulletin 121 (ICOLD 2001) and Rico et al. (2008) provide a summary of TSF failure modes and lessons learned from the failures. Figure 7 presents a chart of the most common modes of failure for TSFs.

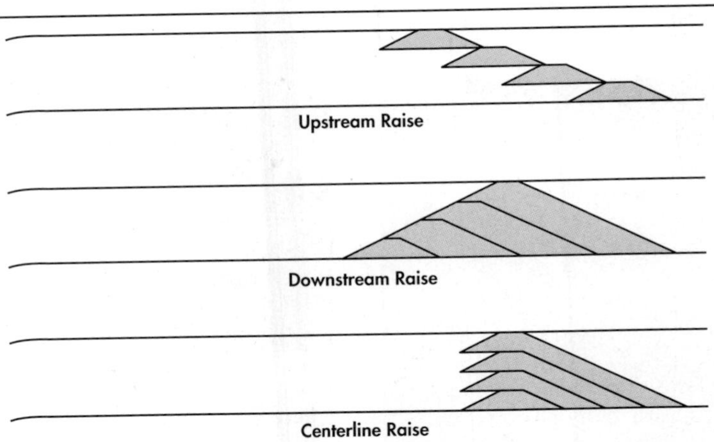

Source: Vick 1990

Figure 6 Tailings storage facility raise methods

Table 1 Tailings storage facility raise-method strengths and weaknesses

Upstream Raise	Centerline Raise	Downstream Raise
Strengths • Requires the least amount of earthwork, so often the lowest cost option. • Has the smallest area footprint compared to the other methods. • Allows concurrent reclamation.	**Strengths** • Has a smaller area footprint than a downstream raise. • Can be designed for water retention, if internal drainage is provided in containment embankment. • Generally, better seismic stability than upstream method.	**Strengths** • Each raise buttresses the previous raise, thereby providing a greater degree of stability. • Can be designed as a water retention structure. • Accommodates a composite liner, if required.
Weaknesses • Potentially susceptible to failure under seismic loading. Not recommended in highly seismic areas. • Cannot be used for water storage. Reclaim pond area must be minimized to maintain long, dry beaches. • Generally, limited to <3 m/yr rate of rise. • Must have sufficient underdrainage and/or internal drainage to maintain a drained condition in the containment embankment. • Very sensitive to deposition planning and water management. • Cannot have steep slopes on outer face of embankments (generally 3 horizontal to 1 vertical or flatter).	**Weaknesses** • Very difficult to implement a composite liner. • Can be susceptible to upstream slope failure, particularly in seismic events. • Requires internal drainage within containment embankment.	**Weaknesses** • Requires a greater volume of earthwork and fill. • Has a greater footprint compared to the other methods. • Requires internal drainage within containment embankment.

As noted in Figure 7, the top failure modes for TSFs are

- Slope stability issues resulting from poor construction or design, inadequate water management (excessive water in facility), and poor TSF management (e.g., incorrect beaching or reclaim pool location);
- Seismic events causing liquefaction of foundation and/or seismic instability because of inadequate design, construction, or lack of geotechnical investigation; and
- Overtopping because of storage of excessive water in the facility and/or poor reclaim pool management (e.g., location).

Failures of TSFs generally occur if there is a deficiency in the design, construction, and/or operation of the facility. Past TSF failures have shown that a well-designed and constructed TSF can still fail through poor operation. Therefore, it is important that all aspects (design, construction, and operation) of the facility be scrutinized to identify fatal flaws or weaknesses. Any flaws or weaknesses in the facility must be corrected to ensure safe operation. In addition, TSFs are designed for life-of-mine of the operation; however, changes in operation (e.g., throughput, process method, ore type, or mine plan) can occur during the life-of-mine that may be incompatible with the original TSF design. Care must be taken to assess the original intent of the TSF design with respect to anticipated operational changes so that, if needed, the TSF design can be modified. This process is referred to as *change management* and is key to successful TSF operation.

TSF DESIGN

A robust TSF design typically requires input from several functions, including process, environmental, social, operations,

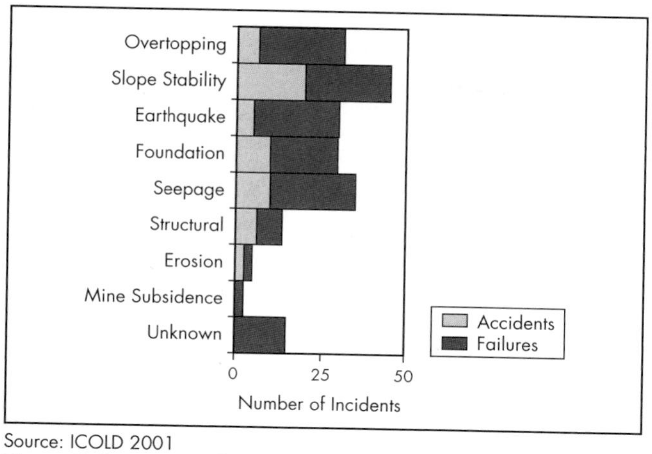

Source: ICOLD 2001

Figure 7 Common tailings storage facility failure modes

mine planning, geology, geotechnical, and mine water management. It is important that representatives from each of these functions be involved throughout the design stage so that the design and performance of the facility meets the desired intent.

Since the majority of TSFs are engineered aboveground structures, the design requires that sufficient knowledge be obtained regarding

- Subsurface geologic and geotechnical conditions,
- Subsurface hydrogeologic conditions,
- Surface water hydrology (rivers, streams, lakes, etc.),
- Geohazards (avalanches, landslides, sinkholes, etc.),
- Seismicity,
- Current and projected climatic conditions, and
- Tailings characteristics.

In addition, testing will be required on the tailings as well as foundation and construction materials to define the engineering parameters to support design.

Design Criteria

Design criteria define the basis for the TSF design. The criteria are generally established with input from the functions listed in the previous section and considered to be "fixed" throughout the design. If a change in criteria occurs (during the design stage or after the design is completed), the design must be reevaluated, and modified if needed, to ensure the facility meets the project objectives and can be safely operated.

Typically, TSF design criteria include the following (as applicable):

- Storage capacity
- Average tailings density
- Average tailings percent solids (by weight)
- Tailings delivery schedule
- Tailings deposition concept
- Design storm events (containment and emergency spillway)
- Stability (static and seismic) acceptance criteria (such as factor of safety and/or reliability)
- TSF lining system (if needed)
- Drainage elements (underdrains, embankment/dam internal drains, etc.)
- Average beach slope
- Seismic load
- Minimum embankment/dam crest width
- Regulatory and social requirements and constraints
- Closure considerations

The design criteria feed directly into engineering analyses that form the basis of the TSF (basin, embankments/dams, decant, spillway, etc.). Engineering analyses for the design of a TSF would generally include stability, seepage, liquefaction (static and seismic), settlement, and erosion. In addition to these engineering analyses, a water balance for the TSF must be developed. A TSF water balance is a critical element in *both* the design and operating phase of a TSF and is used to size the decant/reclaim system, establish operating thresholds, and estimate water demand from external sources. A TSF water balance also plays a critical role in the site-wide water management plan. The section on TSF operation and management later in this chapter presents a more thorough discussion on water management and water balances.

Operating Criteria

Operating criteria define the anticipated life-of-mine operating conditions for the facility. Operating criteria should be explicitly defined during the design stage so that the design is compatible with the anticipated operating conditions. Also, any changes in the operating criteria, before or after the design has been completed, should be evaluated against the design to identify potential conflicts or impacts to the desired facility performance.

Typically, the operating criteria for a TSF include the following (as applicable):

- Tailings delivery rate
- Tailings percent solids (by weight) operating range
- Tailings rheological parameters
- Maximum rate of rise (tailings)
- Maximum embankment/dam height
- Maximum reclaim pool extent
- Minimum freeboard
- Minimum beach widths
- Pore pressure or piezometric level thresholds
- Embankment/dam acceptable deformation threshold
- Seepage collection rate and threshold
- Decant operating criteria (maximum height, water level thresholds, structural thresholds)
- Hydraulic conveyance maximum pressures and flow rates

The operating criteria are also used to form the basis for the facility monitoring program. The TSF monitoring plan is discussed later in this chapter.

Geotechnical

The geotechnical components of a TSF design consist of field and laboratory work to characterize the TSF location, availability and characteristics of potential construction materials, and tailing characteristics.

Field Investigations

Geotechnical field investigations are required to gather information on the subsurface conditions beneath a TSF. The subsurface conditions will influence the type and design of a TSF that can be constructed. The most common geotechnical field investigation methods include the following.

Site reconnaissance and mapping. Site reconnaissance and field mapping consists of conducting a visual survey of the TSF area to identify surface features and geohazards that may affect the TSF design or performance.

Borehole investigation. A borehole investigation program consists of advancing a borehole to a specified depth, logging the borehole, conducting in situ testing, and collecting disturbed and undisturbed samples for geotechnical laboratory testing. Standard procedures, such as ASTM D1586 and AS 1289.6.3.1, are often used to guide the drilling, sampling, and testing procedures. Drilling methods may include auger, coring, and rotary methods. In situ testing typically consists of a standard penetration test, although other methods (e.g., large penetration tests or soil pressuremeter tests) may be used. Undisturbed samples are generally collected using a sampling tube, such as a split-spoon, Modified California Sampler, Shelby tube, and piston samplers.

Piezocone probes. Piezocone probes (often referred to as cone penetration tests, or CPTs) consist of pushing an instrumented cone into the ground at a controlled rate. CPTs are often used in active tailings impoundments to characterize the tailings deposit (e.g., density and drainage). Standard procedures, such as ASTM D5778 and AS 1289.6.5.1, are often used for CPT programs. The information from a CPT program can provide an estimate of shear strength, pore pressure gradients, pore pressure response to shear, susceptibility to liquefaction, and postliquefaction residual shear strength. The piezocone data also provide a discrete pore pressure profile (based on pore pressure dissipation tests during pauses in piezocone penetration) and a direct measurement of the in situ hydraulic conductivity (horizontal direction).

Test pits. Test pits are excavated holes or trenches dug into the ground to provide a detailed visual examination of the near-surface soil profile. Test pits can also be used to determine the depth to groundwater and bedrock. Both disturbed and undisturbed samples may be gathered from test pits.

Geophysical methods. Geophysical methods provide an inexpensive way to gather subsurface information (geologic profile, underground voids, depth to bedrock) over large areas. Commonly used geophysical methods include seismic reflection, seismic refraction, seismic tomography, resistivity, electromagnetics, and multichannel analysis of surface waves. A review of geophysical applications to geotechnical field work is presented in Anderson (2006).

Photogrammetry. Photogrammetric methods (aerial photographs or high-resolution satellite photographs) can be used to identify areas of existing or historic instability (e.g., landslides), surface subsidence, floodplains, and surface ruptures because of seismic events.

Geotechnical Laboratory Testing

Samples gathered from the geotechnical field investigation are typically submitted to a laboratory for characterization, strength, and hydraulic testing. The data from the laboratory testing program are used in basis for the TSF design. A typical laboratory test program would include the following.

Particle size analysis (ASTM C117/C136). A particle size analysis describes the distribution of particle sizes for a material (soil or tailings). This information can be used to characterize and qualitatively assess the behavior of a material.

Atterberg limits (ASTM D4318). Atterberg limits measure the critical water content of a soil at points where distinct changes occur in behavior or consistency. These tests are useful to qualitatively characterize the shear strength and permeability of a soil.

Moisture content (ASTM D2216). A moisture content test measures the amount of moisture or water in a soil or tailings sample. These tests are often correlated to the Atterberg limits to determine the state of the in-place soil. They can also be used to assess a potential source for construction materials.

Dry density (ASTM D7263). These tests are used to quantify the dry-state density of a soil or tailings sample. They are used to assess a potential source for construction materials or the density of the exposed tailings beach material.

Moisture–density relationship (ASTM D698/D1557). These tests correlate the maximum dry density of a soil to its moisture content. The moisture–density curve is a critical tool to guide earthworks and construction of earthen structures, such as dams or embankments.

Permeability (ASTM D5084). Permeability tests are conducted to measure the soil or tailings hydraulic conductivity. These tests are often conducted at varying pressures and are used in engineering calculations to estimate seepage through a foundation or dam.

Consolidation (ASTM D2435). A consolidation test measures the time-dependent settlement of a soil or tailings sample under an applied load. Consolidation test data is used to estimate settlement of the soil foundation beneath a TSF and the settlement of tailings after deposition.

Shear strength (ASTM D2850/D4767/D7181/D3080). Shear strength tests measure the strength of a soil sample under an applied load. The shear strength of a soil is used in the design of the dams or embankments. Shear strength tests typically include triaxial compression (drained and undrained) and direct shear testing methods.

Hydrological Field Investigations

In addition to geotechnical field investigations, hydrological field investigations should be conducted to support the TSF design work. Hydrological investigations should include both surface water and groundwater. The field investigations are conducted to evaluate

- The depth to groundwater,
- The permeability of the subsurface units,
- Groundwater gradient and/or flow paths,
- Surface water flows, and
- Surface water/groundwater interaction.

Groundwater field investigations typically consist of the installation of instrumented piezometers within and downgradient of the TSF location, installation of monitoring wells for sampling water quality and to measure changes in the water level, and installation of groundwater wells to conduct hydraulic testing to derive aquifer properties.

Surface water field investigations typically consist of stream/river gaging, water quality sampling, and establishing rainfall-runoff factors.

Tailings Characteristics Testing

Tailings characterization tests are typically conducted as part of the ore processing work effort. Typical tailings characterization test work includes

- Density (dry and slurry);
- Particle size analysis;
- Atterberg limits (including shrinkage limit);
- Percent solids (by weight) of the slurry, thickened, or paste tailings;

- Slurry consolidation testing;
- Tailings shear strength, from drained and undrained triaxial compression testing;
- Permeability at various loads;
- Slurry settling/flocculant tests; and
- Geochemistry.

If the tailings are to be discharged to the TSF via a pipeline or trench, then rheological tests are also conducted. Tailings rheology tests are used to quantify the viscosity of the tailings as a function of temperature, solids concentration, and applied shear. Tailings rheology tests typically consist of specific gravity, rotating viscometer tests, capillary tube viscometer tests, and loop tests.

Tailings rheological test results are used as input into pump and pipeline designs. Tailings rheology may also be correlated to beaching angles for thickened and pasted tailings. The details of tailings rheology cannot be covered in a single subsection. Several good references provide details of tailings rheological testing methods and data interpretation (Weir Minerals Division 2009; Laskowski 2001; Boger et al. 2006).

If filtration is being considered for tailings disposal, then bench-scale and pilot-scale tests are often conducted using vacuum belt and/or mechanical press technologies. The filter cake is then tested for

- Density (dry),
- Cake moisture content,
- Moisture–density relationship,
- Shear strength from drained and undrained triaxial compression testing,
- Settlement under load,
- Permeability at various loads (unsaturated hydraulic parameters via pressure-plate testing, may also be needed), and
- Geochemistry.

Other Considerations

Other considerations for the design of a TSF include the following.

Construction. An integral part of the design of a TSF is the development of construction technical specifications, issued-for-construction (IFC) drawings, and a construction quality assurance/construction quality control (CQA/CQC) plan. Technical specifications are developed as part of the design process to define which materials are to be used for construction of the TSF and how they are to be used. The intent of technical specifications is to ensure that the constructed facility will perform as per the design intent. IFC drawings are also developed as part of the design process. IFC drawings provide details on the layout, plans, sections, and survey control for construction of the TSF and all appurtenant structures. The CQA/CQC plan is developed to outline the testing, monitoring, and reporting requirements to ensure construction quality.

Water management. The TSF is just one component of a site-wide water management plan. In the effort to reduce overall water consumption, the TSF design and water balance should be integrated into the overall site-wide water management plan. This approach allows the development of a water management strategy that considers the efficient use of water resources.

Environmental and social impacts. Environmental and social impacts must also be considered so that the TSF design can be modified to be compatible with environmental and/or social constraints.

Operation, maintenance, and surveillance. All TSFs should have a prepared tailings operation, maintenance, and surveillance (OMS) manual that includes an emergency response plan (ERP). The ERP details roles, responsibilities, and contact information; monitoring thresholds and alarms that trigger a response; response details and protocols; and a clear communication plan.

Change management. Finally, there is the consideration for change management that may occur during the life-of-mine. Although it is not possible to predict change management, it is possible to add robustness to a design, through redundancies or additional capacity, so that the facility can accommodate potential operational changes in the future. Regardless, it is critical that the TSF design documentation be complete and clearly state the design basis, intent, and operational constraints so that any potential future changes are compatible with the original design.

TSF CONSTRUCTION QUALITY

Construction of a TSF often involves the management of several disciplines and contractors, such as earthworks, concrete, structural steel, geosynthetics, pumps, and pipelines, to execute the project. Successful project execution requires establishing clear battery limits, instituting lines of responsibility and accountability, and managing construction quality. Critical components to managing construction quality are the CQA and CQC plans that are developed as part of the TSF design. The primary objectives of the CQA and CQC plans are to ensure that the construction of the various TSF components complies with the project specifications and to demonstrate that the regulatory requirements for the construction are achieved.

A CQA program is independent of the CQC programs conducted by the manufacturers, installers, and contractors. A CQA program provides independent third-party verification and testing to demonstrate that the manufacturers, installers, and contractors have met their obligations in the supply and installation of TSF construction materials according to the design, project specifications, and contractual and regulatory requirements. A CQA program also identifies problems that may occur during construction and provides the means for resolution of these problems.

The CQA plan must address the following items:

- **Quality management structure.** Defines how the CQA plan fits within the construction project and how it will be managed.
- **Project definitions.** Clear and consistent definitions to be used throughout the project.
- **Roles, responsibilities, and inspecting authority.** Clear lines of authority, communication, and management.
- **Testing requirements.** Types of tests to be conducted, where the tests are to be conducted, and the names and qualifications of the testing personnel.
- **Testing frequencies.** Define the threshold (time, volume, weight) that triggers a test.
- **Surveying requirements.** Identify areas to be surveyed, acceptable survey tolerance, and qualifications of the surveyor.
- **Surveying frequencies.** Define the threshold (time, area, volume) that triggers surveying.

- **Nonconformance process.** A clear process on reporting nonconformance tests or survey and evidence of repair or resolution.
- **Reporting requirements and document control.** Documentation and reporting of testing results, survey results, approved changes or modifications to the design and/or technical specifications, and a summary of nonconformance resolution.

The objective of a CQC program is to ensure that proper materials, techniques, and procedures are used by manufacturers, installers, and contractors to meet the TSF design requirements. A CQC program is independent of the CQA program conducted by the third-party CQA organization. CQC testing may be performed at the point of manufacturing, processing, or stockpiles, or at the place of installation. This is in contrast with CQA testing that is mostly performed at the place of installation.

At a minimum, the CQC plan must address the following:

- **Quality management structure.** Defines how the CQC plan fits within the construction project and how it will be managed.
- **Project definitions.** Clear and consistent definitions to be used throughout the project.
- **Roles and responsibilities.** Clear lines of authority and communication.
- **Testing requirements.** Types of tests to be conducted, and where the tests are to be conducted and by whom.
- **Testing frequencies.** Define the threshold (time, volume, weight) that triggers a test.
- **Reporting requirement.** Documentation and reporting of testing results, approved changes or modifications to the design and/or technical specifications, and a summary of issues.

Upon completion of the TSF construction, an as-built report with drawings should be produced by the CQA organization. The as-built report represents the official record of what was constructed, how it was constructed, and any changes or modifications that occurred during the construction period. The as-built report should also include documentation from the CQA and CQC programs, reports on nonconformance and nonconformance resolution, and a discussion on site conditions that differed from the original design assumptions.

TSF OPERATION AND MANAGEMENT

An OMS manual must be completed for the TSF as part of the design stage. The manual should address the design intent, operating criteria, predicted behavior of tailings, daily operations and inspections, water management procedures, criteria for mechanical and electrical works (including pumps), surveillance, and maintenance and reporting requirements. Furthermore, the OMS manual should specify all requirements for operators and the minimum level of operator training. This section highlights the TSF operational aspects of the OMS. Monitoring and surveillance aspects of OMS are covered later in this chapter.

General Philosophy and Stewardship

Successful operation and management of a TSF focuses on four key areas, as applicable:

1. Controlled deposition of the tailings solids to form beaches that are consistent with the design geometry and design criteria
2. Management of the water within and around the TSF according to the conditions set forth in the design and operational criteria
3. Phased raising of the TSF dam(s)/embankment(s) to maintain sufficient storage capacity of storm events per the design
4. Frequent monitoring the performance of the TSF with respect to the design assumptions

These key areas are elements of a tailings management framework. The operation and management of a tailings facility should be governed by a tailings management framework that is site and operation specific.

The Mining Association of Canada (MAC) provides excellent guides for the management and monitoring of tailings facilities (MAC 2017, 2011a, 2011b, 2011c). The MAC guides provide information on safe and environmentally responsible management of tailings facilities; help companies develop tailings management systems that include environmental and safety criteria; and improve the consistency of application of sound engineering and management principles to tailings facilities.

In addition to the MAC guides, ICOLD has published several bulletins with respect to tailings dams. ICOLD publications pertaining to tailings dams include the following:

- Bulletin 74, *Tailings Dams Safety Guidelines* (ICOLD 1989)
- Bulletin 97, *Tailings Dams" Design of Drainage: Review and Recommendations* (ICOLD 1994)
- Bulletin 101, *Tailings Dams: Transport, Placement and Decantation* (ICOLD 1995)
- Bulletin 103, *Tailings Dams and Environment: Review and Recommendations* (ICOLD 1996a)
- Bulletin 104, *Monitoring of Tailings Dams: Review and Recommendations* (ICOLD 1996b)
- Bulletin 106, *A Guide to Tailings Dams and Impoundments: Design, Construction, Use and Rehabilitation* (ICOLD 1996c)
- Bulletin 139, *Improving Tailings Dam Safety: Critical Aspects of Management, Design, Operation, and Closure* (ICOLD 2011)
- Bulletin 153, *Sustainable Design and Post-Closure Performance of Tailings Dams* (ICOLD 2013)

The United Nations Environment Programme (UNEP) and the International Council on Metals and the Environment (ICME) have also published guidance related to tailings management (UNEP–ICME 1997, 1998). Key topics covered that relate to management of tailings dams include

- Corporate policies and procedures regarding stewardship of tailings facilities,
- Evolving regulatory climates and trends,
- Definitions of roles and responsibilities,
- Application of risk assessment techniques,
- Environmental management systems,
- Emergency preparedness and response, and
- Education and training.

These, and other guidelines, should be used to develop the tailings management framework for a specific TSF.

Tailings Deposition

The deposition of tailings into a TSF requires some forethought on the type of tailings to be deposited and the site conditions that may affect deposition. Tailings deposition considerations include but are not limited to the following:

- Variations in climate and climate change may have a significant impact on water management within the facility. For example, at start-up many operations run a negative water balance (e.g., the operation consumes more water than it has available, requiring makeup water from external sources). However, the water balance may become neutral or positive over time because of climatic conditions or changes in the mine plan (e.g., mining below the water table). Changes in the water balance need to be identified well in advance so that a water management strategy can be developed and implemented.
- Geometry of the tailings basin should be considered. For example, deposition in narrow valleys or rapid deposition in a confined area may result in a high rate of rise, impeding beach formation and beach drying. This can be an issue if the TSF was designed assuming beach formation and drying between deposition cycles.
- Paddock deposition using several separate cells is common in arid regions. This approach allows active deposition to be rotated between the cells, so that one cell is active while the others are allowed to "rest" and dry out. This approach has several benefits: (1) it provides a source of material that can be used for future raises; (2) by dividing deposition across several cells, seepage through the base of the facility is reduced; and (3) having multiple deposition cells distributes the "slimes" uniformly rather than from a central location, which may be difficult to reclaim at the end of the mine operation.
- Point of discharge flocculation may also be a consideration for slurry deposition. Dosing with flocculent at the point of discharge from a pipeline may result in a higher initial settled density and better water recovery.
- Many TSF designs are predicated on achieving a specific deposition geometry (e.g., beach length, slope, or slimes location). The inability to achieve the desired geometry may affect the performance and stability of the facility. Reconciling the actual geometry against the design is required so that potential impacts to performance or stability are identified early and appropriate risk mitigation efforts are implemented.

Three basic deposition methods are subaqueous, subaerial, and hydrocycloning. *Subaqueous deposition* is where the tailings surface is wholly or partially submerged below free water. In subaqueous deposition, the separation of the solids from the liquid phase occurs by sedimentation and submerged consolidation only. Subaqueous deposition is generally used where climatic conditions and/or operational controls ensure reliable year-round coverage by water, and where submerged tailings is determined to be essential for environmental or health and safety reasons dictated by the physical and/or chemical properties of the tailings. If subaqueous deposition is adopted, the TSF *must* be designed to account for water being in direct contact with the dam(s)/embankment(s). The TSF also must always have enough capacity to safely contain the design storm event to prevent discharging to the environment. Figure 8 presents a photograph of a subaqueous deposition facility.

Subaerial deposition is the most common method of deposition. Subaerial deposition occurs where placement of tailings forms an exposed beach over most of the storage area, with a reclaim pool at the end of the beach. With low solids content slurries, subaerial deposition promotes segregation of coarser particles from the finer particles (Blight 1994). This segregation occurs toward the top of the beach while finer particles (often referred to as "slimes") accumulate in the reclaim pool. The beach is exposed to evaporative drying, which increases the density of the beach tailings. The "slimes" generally remain submerged or partially submerged in the reclaim pool. Average beach angles for slurried tailings generally range from 0.2% to 1.0%. The coarser particles on the upper part of the beach may also be removed and used for dam/embankment construction. Thickened and paste tailings are also deposited subaerially; however, particle segregation does not occur in these materials.

Hydrocycloning is commonly used with low solids content slurry to produce sand for construction along the TSF dam/embankment (Figure 9). Kujawa (2011) provides practical guidance on potential sand production using hydrocyclones.

Figure 8 Subaqueous deposition facility

Figure 9 Hydrocycloning

Courtesy of Phibion

Figure 10 Mechanical drying of tailings

In higher solids content tailings (e.g., thickened and paste tailings), very little particle segregation occurs and only a minimal amount of free water may be present. With thickened and paste tailings, the achievable "beach" angle is a function of the tailings characteristics and deposition rate (McFayden 1998). The achievable beach angle for thickened and paste tailings may range from 1.0% to more than 5.0%. One of the many advantages of thickened and paste tailings (besides reduction in water consumption) is the ability to achieve a steeper beach as compared to low solids content slurries. The steeper beach angle can help reduce the TSF footprint and construction cost provided the topography and climate are suitable.

Some tailings, such as "mature fine tailings" in the oil sands industry, phosphate, and bauxite tailings, are composed mainly of clay-sized particles that remain in suspension in the process water. These materials can be very challenging as they retain moisture for long periods of time and are very difficult to consolidate, or densify (Caughill 1992; Kasperski 1992). These materials may require special flocculants or polymers (Li et al. 2008) and/or mechanical working (Munro and Smirk 2012), such as amphi-rollers (Figure 10), to achieve some level of densification.

Dry stacking involves dewatering the tailings (e.g., using vacuum or mechanical filters) so they can be stacked using earthmoving machinery or conveyors. Tailings that are dry stacked can be placed at slopes of up to 30% or more, if the material is compacted as placement occurs (Lupo and Hall 2010). The steep slopes that can be achieved by dry stacking can significantly reduce the TSF footprint and reduce construction costs, provided the topography and climate are suitable.

Deposition issues with dry stacking include material handling and scheduling, dust generation from the dry stack, erosion and water pooling on the surface of the stack, and static liquefaction (stacking at high rates).

Water Management

Water management within and around a TSF is a critical component to the overall management of the facility. As noted earlier in this chapter, poor water management has played a significant role in past TSF failures. Water management within a TSF is managing not only the size of the reclaim pool but also its location. Both the size and location of the pool can influence the stability of the TSF embankments/dams as well as seepage rates, beach geometry, tailings density, and the performance of the decant system.

The size and location of the decant pond can be affected by the method and direction of deposition, the geometry of the tailings facility, the climate, the decant recovery rate, and the settling characteristics of the tailings. For these reasons, the tailings deposition plan (discussed earlier in this chapter) must be integrated with the water management.

Two primary tools are used for effectively managing water in a TSF. These tools are (1) a water balance and (2) reconciliation/monitoring of the TSF performance.

Water Balance

A water balance is a methodology used to describe and quantify the flow of water into and out of a system. For a TSF, the primary components of a water balance include the following.

Inflows:

- Direct rainfall within the footprint of the facility
- Catchment runoff of precipitation into the facility
- Deposition water delivered within the tailings from the process plant
- Groundwater discharge into the facility (if present)

Outflows:

- Return/reclaim water from the TSF to the process plant
- Evaporation from the supernatant pond and beaches
- Water entrained in tailings
- Seepage capture from underdrains and leachate collection and recovery systems

A generalized schematic of a water balance for a TSF is presented in Figure 11.

It is common practice to develop a water balance during the design stage of a TSF. At this stage, the water balance is often based on limited climatic and site data, and assumed operational information; however, this is usually sufficient to provide an estimate to "bound" the water requirements for the operation. After the TSF goes into operation, the water balance should be updated annually and maintained using actual site information (daily precipitation, evaporation rates, production rates, percent solids, reclaim rate, etc.) as it becomes available.

A water balance is based on several simplifying assumptions. For example, the evaporation rates (which can be a significant outflow component) from the reclaim pool, wet beach, and dry beach are all different, yet the size of the pool, wet beach, and dry beach are constantly changing. So there is no practical means to estimate the loss to evaporation other than

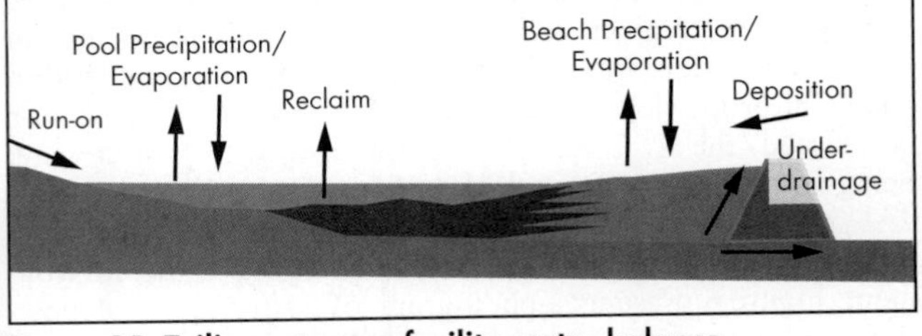

Figure 11 Tailings storage facility water balance

Figure 12 Tailings storage facility survey prism setup

Figure 13 Inclinometer (next to pickup truck)

using assumed areas and effective evaporation rates. Even though there are practical limitations to water balances, they are still important tools for water management. For example, water balances can be used to

- Assess potential changes in water management because of changes in processing (rate, percent solids, etc.),
- Estimate when changes in water demand and/or the buildup of excess water may occur,
- Estimate treatment requirements and/or discharge rates should the site have a positive (e.g., gaining) water balance, and
- Assist in managing external water resources.

TSF Reconciliation

A TSF is an engineered facility, and therefore, the performance of the facility should be reconciled against the original design intent and assumptions. TSF performance reconciliation is important to identify potential flaws in the design (e.g., when what was assumed conflicts with actual conditions) and to provide input should the TSF design need to be modified.

The reconciliation of the TSF performance is typically conducted via monitoring and surveillance programs. These programs should be developed as part of the TSF design stage and consider stability, settlement, seepage, beach configuration, reclaim pool location and size, and decant performance.

TSF MONITORING AND SURVEILLANCE PLAN

The objective of the TSF monitoring and surveillance plan is to define the requirements for monitoring the performance of the TSF through life-of-mine. The plan must cover all aspects of the TSF so that the "health" or performance of the facility can be evaluated.

Dam/Embankment Monitoring

Critical elements of a TSF are the containment dam(s) or embankment(s). These are the primary engineered structures that prevent uncontrolled discharge to the environment. TSF dams/embankments are typically constructed of earthen materials, tailings, rockfill, and/or concrete. A typical monitoring program for TSF dams/embankments consists of the following instruments.

Survey prisms. Prisms are reflective devices mounted on poles and are located along the dam/embankment to monitor surface deformation (vertical and lateral movements) of the structure. Prisms are useful for monitoring changes along the surface of the dam/embankment that may indicate a stability issue. Prisms can be surveyed by manual means or by a robotic total station. Figure 12 presents a typical prism survey setup for a TSF.

Inclinometers. Inclinometers are instruments that are placed within the dam/embankment and used to measure deformation within the structure. Inclinometers consist of a grooved casing and the inclinometer instrument that runs along the grooves (see Figure 13). The casing is installed in the dam/embankment by drilling a hole and grouting the casing within the hole. The inclinometer is lowered and raised in the casing and measures deflections along the casing. The deflections along the casing can be plotted with depth to show potential movement within the structure. Alternatively, shaped acceleration array inclinometers may also be considered. An example of inclinometer data is shown in Figure 14.

Piezometers. Piezometers are instruments that measure water pressure and are used to determine the hydraulic pressure (piezometric) heads within the dam/embankment, which is critical to assessing the stability. Piezometers are installed within the dam/embankment, either as direct placement within the fill during construction or grouted in borings after construction is complete (see Figure 15). It is recommended that multiple piezometers be installed with the dam/embankment at different elevations (including the foundation) and along the structure to monitor hydraulic conditions.

A recent development for TSF monitoring is the use of satellite-based interferometric synthetic aperture radar (InSar). InSar is a radar technique that can map surface deformations (millimeter scale). This remote sensing method is very effective for covering large areas and providing high-quality data. InSar is not meant to provide real-time monitoring but does provide an efficient means to gather periodic deformation information. Figure 16 provides an example of InSar output for a tailings facility.

Drones may also be employed for surveillance of TSFs to measure deformation. Infrared cameras may be used with drones to scan the facility for the presence of shallow seepage.

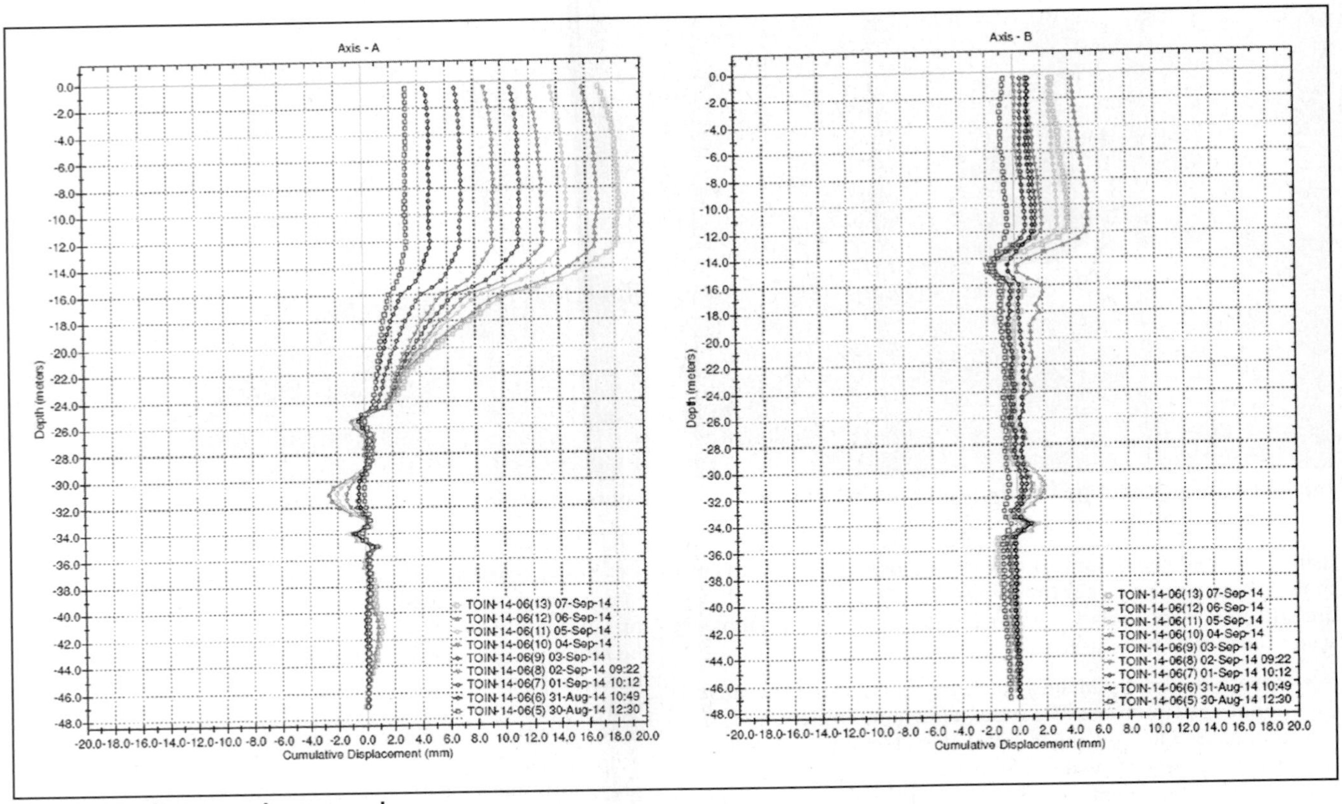

Figure 14 Inclinometer data example

Courtesy of Newmont Mining Corporation

Figure 15 Typical piezometer (fitted with a telemetry system)

Seepage Monitoring

Seepage monitoring from a TSF may consist of several elements, depending on the TSF design and site conditions, and generally consists of measuring both flow and water quality. Seepage monitoring provides critical information that relates to internal drainage of the facility, which can impact not only stability but also the effectiveness of the environmental containment.

Seepage monitoring of a TSF may include the following (as applicable):

- **Foundation underdrains.** Drainage systems within the foundation to control seepage pressures
- **Internal underdrains.** Drainage systems that may be placed on the surface of the storage basin to enhance consolidation of the tailings
- **Dam/embankment drains.** Drainage constructed within the dam/embankment to control seepage through the structure, such as blanket drains and chimney drains
- **Monitoring wells.** Located downgradient of the TSF to monitor water levels and water quality.

Seepage monitoring data should be analyzed routinely to look for trends (e.g., increasing flows or increasing constituent concentrations) that may affect how the TSF is managed.

Deposition Monitoring

Deposition monitoring is used to ensure that the performance of TSF operations meets the criteria set forth in the TSF design (see the discussion earlier in this chapter). At the TSF, deposition monitoring consists of visual inspections and surveying to quantify the following (as applicable):

- **Beach or stack quality (uniformity and condition), length, area and slope.** Used to reconcile with the design assumptions and available storage capacity.
- **Rate of rise.** Used to assess the rate of rise with respect to the design.
- **Reclaim pool size and location.** Used to estimate the pool volume and location to critical structures.

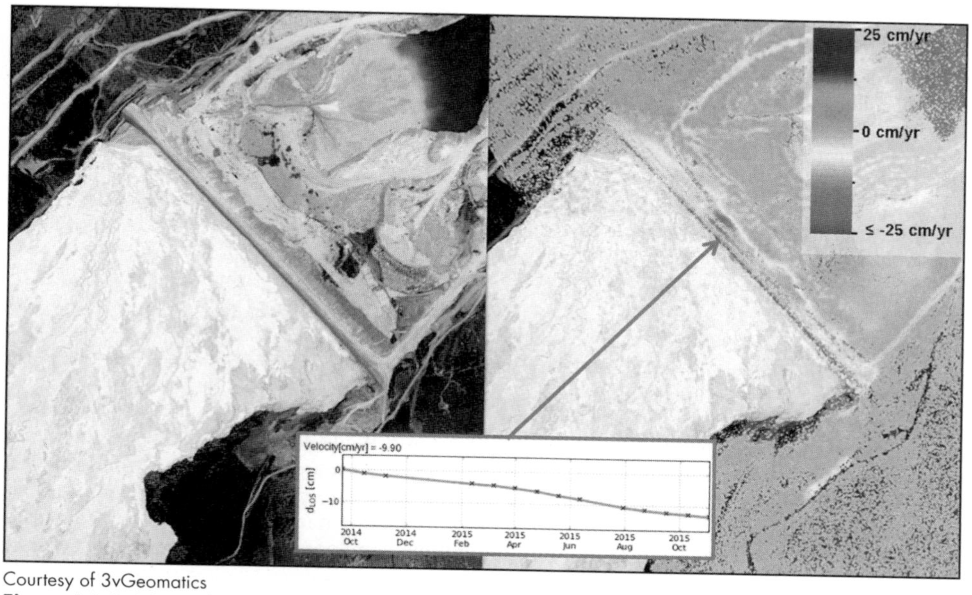

Courtesy of 3vGeomatics

Figure 16 InSAR tailings storage facility application

- **Available freeboard.** Used to quantify and verify the available dry freeboard to contain the design storm event and protect against overtopping because of wave runup.
- **Tailings deposit gradation.** Used to measure the effectiveness of cyclone operations. It can also be used to assess the effectiveness of spigot shapes and spacing.

At the process plant, deposition monitoring consists of measuring discharge rates to the TSF, percent solids, rheological parameters (primarily used for thickened and paste tailings), and water recovery rate.

Inspection and Review Program

A formal inspection and review program must be established for every TSF. This program should detail the level of inspections and reviews to be conducted, frequency of inspections, and documentation requirements. Inspection reports should be kept up to date and readily accessible so that changes in the TSF behavior can be identified.

The Australian National Committee on Large Dams recommends that an inspection program include the following (ANCOLD 2012):

- Groundwater monitoring with special emphasis on the environmental impacts of the TSF on groundwater (e.g., geochemical processes)
- Surface drainage and seepage monitoring, both visual observations and seepage measurement as a minimum, with chemical analysis also of value (e.g., acid drainage generation)
- Capacity monitoring (tailings, process water, water recovery, evaporation)
- Tailings monitoring (e.g., beach development, drainage, density, desiccation)
- Monitoring of instrumentation and instrumentation readings
- Monitoring of equipment and pipework
- Monitoring of dam movements, stresses, cracking and seepage
- Inspection reports (i.e., times, dates, observations)
- Incident reports (i.e., time date, nature, actions)
- Annual audit

The frequency of inspections is often dictated by regulatory or permit requirements. Table 2 presents suggestions from ANCOLD.

EMERGENCY RESPONSE PLANS

All TSFs where any persons, infrastructure, or environmental resources could be at risk if the facility collapses or fails should have an ERP. The ERP should be based on an appropriate dam break study with the conservative assumption of liquid tailings flow in the event of failure (unless a more sophisticated analysis of water and/or tailings flow can be justified).

ERPs are often developed following the guidance set forth in framework presented in *Awareness and Preparedness for Emergencies at Local Level* (UNEP 2015). ERPs should be tailored to suit the regulatory requirements, site conditions, operation, and local community involvement.

The basic elements of an ERP include clearly defined roles, responsibilities, and contact information during an emergency; emergency response guidelines and procedures (internal and external); detailed communications systems; and incident investigation procedures. ERPs should be periodically tested to evaluate effectiveness and potential gaps in the plan.

TSF CLOSURE AND RECLAMATION

This section discusses the closure and reclamation of TSFs, including closure objectives and constraints that may affect closure design and activities. According to the guidance set forth in ANCOLD (2012), the closure of a TSF should embrace the following objectives:

- Provide long-term physical, chemical, ecological and social stability.
- Minimize ongoing maintenance and expenditures other than normally required for similar land use.
- Not pose a risk to human health and safety.

Table 2 Recommended frequency of inspections

TSF Failure Consequence Category	Inspection Type			
	Comprehensive	Intermediate	Routine	Special
Extreme or high	After first year of operation, then every second year	Annual	Daily to three times per week	As needed
Significant	After first year of operation, then every fifth year	Annual	Twice a week to weekly	
Low	—	After first year of operation, then every fifth year	Monthly	

Source: ANCOLD 2012

- Not pose an unacceptable environmental risk.
- Provide an appropriate and sustainable land and water use, which meets stakeholder and community objectives and supports a sustainable ecology.

In addition, the TSF closure must also comply with local or regional regulatory requirements as well as corporate internal standards that may apply.

A preliminary TSF closure plan should be developed during the design stage. A closure plan sets forth the details on how the TSF will be closed and meets the overall closure objectives. This is done to ensure that the TSF design is compatible with the desired closure objectives. Over the life-of-mine, the closure plan should be revised to take into account any changes in operation or management that may impact closure. Robins (2004) provides an excellent review of current TSF closure practices and case histories. The basic design and control considerations for the closure of a TSF are summarized in Table 3. A cover for the TSF may be required as part of the closure. The details of cover design are beyond the scope of this chapter; however, the Mine Environment Neutral Drainage (MEND) program discusses the results from several case histories and provides insight on actual cover performance as well as lessons learned (MEND 2004b).

Physical and Climate Constraints

The closure of a TSF and its final landform is closely tied to the site physical and climate conditions. A closure plan for a TSF should be developed that incorporates the site physical and climate conditions, and the necessary design elements, to meet the objectives stated in the previous section.

Physical constraints that may be considered for a TSF closure include legal land boundaries, sloping ground, high seismicity, surface water bodies, public roads and land, national or protected forests, and nearby towns. These physical constraints can impact surface grading and resloping configuration, surface water diversion and controls, seepage collection, cover design, and the availability of suitable construction materials. It is important that during the TSF design phase, suitable attention is paid to the impact to closure as a result of the site physical constraints. Changes can often be made in the design stage that can simplify closure of the facility and reduce postclosure risk.

Site climate conditions will also play an important role in how a TSF will be closed. For example, a TSF in a wet climate will require more water management controls than one located in an arid region. Climate also plays a role in cover design and performance. Wide swings in temperature and precipitation are often difficult to accommodate in a cover design, resulting in poor performance (e.g., high infiltration rates or cover erosion) and requiring periodic maintenance to repair the cover. It is very common to conduct cover trials to assess the performance of different cover designs. Typically, cover trials are conducted several years prior to closure so cover performance under site conditions can be monitored and documented. This will reduce the uncertainty of the cover's suitability with site conditions.

Environmental and Social Constraints

Environmental and social concerns must be addressed in the TSF closure plan and design. Stakeholder engagement during the design phase is recommended so that the TSF design and postclosure use addresses concerns and desires for future land use, water quality, and visual/aesthetics impact.

The environmental and social constraints are unique to each site, so it is key that the driving environmental and social issues be clearly identified and discussed, preferably before or during the design work for the facility. These discussions should address any constraints or commitments to be considered in the design and operation of the facility.

Tailings and Cover Material Characteristics

The characteristics of the tailings will affect closure and cover design. MEND provides some guidance and the suite of testing

Table 3 Tailings storage facility closure considerations

Issues	Desired Outcome	Control
Physical stability • Dust • Erosion • Embankment/Dam • Drainage • Radon gas	• Dust control • Sediment control • Structural stability • Radon gas emission control	• Site selection • Facility design • Deposition method • Tailings characteristics • Cover characteristics and design • Water management and drainage control
Chemical stability • Metal leaching • Acidic drainage • Reagent drainage	Discharge of acceptable quality water (surface and groundwater)	• Cover design to control reactions • Effluent collection and treatment • Pretreatment of tailings • Facility liner design • Water management and drainage control
Land use • Loss of land use • Visual impacts	Restore to appropriate land use	• Landform design • Cover design • Revegetation

Adapted from MMSD 2002

that may be required to support cover design (MEND 2004a). The primary tailings characteristics that should be considered include the following.

Metal/reagent leaching potential. Often the leaching potential of tailings is assessed using the toxicity characteristic leaching procedure. Alternatively, the synthetic precipitation leaching procedure can be used. Both tests are used to simulate potential leaching and metals/reagent release under field conditions. The data from these tests can be used to understand what metals or reagents may be released and if the leachate would impact groundwater and vegetation growth.

Tailings and cover mineralogy. The mineralogy of the tailings and potential cover materials should be evaluated to identify minerals that may influence cover design. For example, the presence of expansive clays, such as smectite, can expand and shrink significantly, damaging a soil cover. Tailings that are radioactive from uranium processing require special cover designs to mitigate or attenuate radon gas emissions.

Hydraulic properties. The majority of TSF closure covers are designed to minimize infiltration and, in some cases, oxygen egress into the tailings. These types of covers require knowledge of the soil–water characteristic curve (SWCC) for both the tailings and the cover materials. The SWCC defines the relationship between soil suction and moisture content within a material. This relationship can then be used to derive the unsaturated hydraulic conductivity. The SWCC is used in engineering calculations to estimate the infiltration rate and saturation of the cover materials and tailings under site conditions. Often the SWCC is measured for several cover materials, so that a suitable cover design can be developed.

Soil organic matter. Most tailings do not contain any organic content nor nutritive value to support vegetative growth. Numerous studies have been conducted to evaluate amending various types of tailings with organic carbon and other materials to increase the ability of tailings to support vegetative growth (Lindsay et al. 2011; Fang 2015; Pitchel et al. 1994). Cover materials should also be evaluated for organic/carbon content and nutritional value to assess their suitability for use.

CLOSING COMMENTS

This chapter provides the reader with an appreciation for tailings disposal and management. Although tailings are a waste product from mining, the disposal of tailings carries significant cost and risks to a mining operation. These risks can be mitigated by completing a thorough design of the TSF; implementing a high-level of quality control during construction of the facility; and operating the facility fully within the design and operational criteria. TSFs are highly engineered structures and should be treated as such.

The material presented is meant to provide basic knowledge on the disposal and management of tailings, from design to closure. Readers are encouraged to read the references provided and conduct their own literature search of this topic as there are numerous pertinent and excellent resources available. Finally, the reader should understand that tailings disposal is unique for each operation around the world. The nuances for each operation must be understood so that a safe and reliable facility can be designed, constructed, and operated.

REFERENCES

ANCOLD (Australian National Committee on Large Dams). 2012. *Guidelines on Tailings Dams: Planning, Design, Construction, Operation and Closure*. Hobart, Tasmania: ANCOLD.

Anderson, N.L. 2006. Selection of appropriate geophysical techniques: A generalized protocol based on engineering objectives and site characteristics. In *Proceedings of the 2006 Highway Geophysics NDE Conference*, St. Louis, MO, December 4–7. pp. 29–47.

AS 1289.6.3.1-2004 (R2016). *Methods of Testing Soils for Engineering Purposes Soil Strength and Consolidation Tests—Determination of the Penetration Resistance of a Soil—Standard Penetration Test (SPT)*. Sydney, New South Wales: Standards Australia International.

AS 1289.6.5.1-1999 (R2013). *Methods of Testing Soils for Engineering Purposes Soil Strength and Consolidation Tests—Determination of the Static Cone Penetration Resistance of a Soil—Field Test Using a Mechanical and Electrical Cone or Friction-Cone Penetrometer.* Sydney, New South Wales: Standards Australia International.

ASTM C117-17. *Standard Test Method for Materials Finer than 75-µm (No. 200) Sieve in Mineral Aggregates by Washing*. West Conshohocken, PA: ASTM International.

ASTM C136/C136M-14. *Standard Test Method for Sieve Analysis of Fine and Coarse Aggregates*. West Conshohocken, PA: ASTM International.

ASTM D1557-12e1. *Standard Test Methods for Laboratory Compaction Characteristics of Soil Using Modified Effort (56,000 ft-lbf/ft³ (2,700 kN-m/m³))*. West Conshohocken, PA: ASTM International.

ASTM D1586-11. *Standard Test Method for Standard Penetration Test (SPT) and Split-Barrel Sampling of Soils*. West Conshohocken, PA: ASTM International.

ASTM D2216-10. *Standard Test Methods for Laboratory Determination of Water (Moisture) Content of Soil and Rock by Mass*. West Conshohocken, PA: ASTM International.

ASTM D2435/D2435M-11. *Standard Test Methods for One-Dimensional Consolidation Properties of Soils Using Incremental Loading*. West Conshohocken, PA: ASTM International.

ASTM D2850-15. *Standard Test Method for Unconsolidated-Undrained Triaxial Compression Test on Cohesive Soils*. West Conshohocken, PA: ASTM International.

ASTM D3080/D3080M-11. *Standard Test Method for Direct Shear Test of Soils Under Consolidated Drained Conditions*. West Conshohocken, PA: ASTM International.

ASTM D4318-17e1. *Standard Test Methods for Liquid Limit, Plastic Limit, and Plasticity Index of Soils*. West Conshohocken, PA: ASTM International.

ASTM D4767-11. *Standard Test Method for Consolidated Undrained Triaxial Compression Test for Cohesive Soils*. West Conshohocken, PA: ASTM International.

ASTM D5084-16a. *Standard Test Methods for Measurement of Hydraulic Conductivity of Saturated Porous Materials Using a Flexible Wall Permeameter*. West Conshohocken, PA: ASTM International.

ASTM D5778-12. *Standard Test Method for Electronic Friction Cone and Piezocone Penetration Testing of Soils*. West Conshohocken, PA: ASTM International.

ASTM D698-12e2. *Standard Test Methods for Laboratory Compaction Characteristics of Soil Using Standard Effort (12 400 ft-lbf/ft^3 (600 kN-m/m^3))*. West Conshohocken, PA: ASTM International.

ASTM D7181-11. *Method for Consolidated Drained Triaxial Compression Test for Soils*. West Conshohocken, PA: ASTM International.

ASTM D7263-09(2018)e1. *Standard Test Methods for Laboratory Determination of Density (Unit Weight) of Soil Specimens*. West Conshohocken, PA: ASTM International.

Blight, G.E. 1994. The master profile for hydraulic fill tailings beaches. *Proc. Inst. Civ. Eng. Geotech. Eng*. 107:27–40.

Boger, D.V., Scales, P.J., and Sofra, F. 2006. Rheological concepts. In *Paste and Thickened Tailings: A Guide*. 2nd ed. Edited by F. Jewell. Perth, Western Australia: Australian Centre for Geomechanics. pp. 25–37.

Caughill, D. 1992. Geotechnics of non-segregating oil sand tailings. M.Sc. thesis, University of Alberta, Civil and Environmental Engineering, Edmonton, AB.

Davies, M.P., and Rice, S. 2001. An alternative to conventional tailing management: "Dry stack" filtered tailings. In *Tailings and Mine Waste 2010*. Boca Raton, FL: CRC Press.

Davies, M.P., Lupo, J., Martin, T., McRoberts, E., Musse, M., and Ritchie, D. 2010. Dewatered tailings practice: Trends and observations. In *Tailings and Mine Waste 2010*. Boca Raton, FL: CRC Press.

Fang, Y. 2015. Rehabilitation of organic carbon and microbial community structure and functions in Cu-Pb-Zn mine tailings for in situ engineering technosols. Ph.D. thesis, University of Queensland, Queensland, Australia.

Hore, C., and Luppnow, D. 2015. In-pit tailings disposal at Langer Heinrich: Tailings storage facilities in a unique hydrogeological setting. In *Tailings and Mine Waste Management for the 21st Century*. Melbourne, Victoria: Australasian Institute of Mining and Metallurgy.

ICOLD (International Committee on Large Dams). 1989. *Tailings Dam Safety Guidelines*. Bulletin 74. Paris: ICOLD.

ICOLD (International Committee on Large Dams). 1994. *Tailings Dams: Design of Drainage: Review and Recommendations*. Bulletin 97. Paris: ICOLD.

ICOLD (International Committee on Large Dams). 1995. *Tailings Dams: Transport, Placement and Decantation*. Bulletin 101. Paris: ICOLD.

ICOLD (International Committee on Large Dams). 1996a. *Tailings Dams and Environment: Review and Recommendations*. Bulletin 103. Paris: ICOLD.

ICOLD (International Committee on Large Dams). 1996b. *Monitoring of Tailings Dams: Review and Recommendations*. Bulletin 104. Paris: ICOLD.

ICOLD (International Committee on Large Dams). 1996c. *A Guide to Tailings Dams and Impoundments: Design, Construction, Use and Rehabilitation*. Bulletin 106. Paris: ICOLD.

ICOLD (International Committee on Large Dams). 2001. *Tailings Dams: Risk of Dangerous Occurrences, Lessons Learnt*. Bulletin 121. Paris: ICOLD.

ICOLD (International Committee on Large Dams). 2011. *Improving Tailings Dam Safety: Critical Aspects of Management, Design, Operation, and Closure*. Bulletin 139. Paris: ICOLD.

ICOLD (International Committee on Large Dams). 2013. *Sustainable Design and Post-Closure Performance of Tailings Dams*. Bulletin 153. Paris: ICOLD.

Kasperski, K.L. 1992. A review of properties and treatment of oil sands tailings. *AOSTRA J. Res*. 8:11–53.

Kujawa, C. 2011. Cycloning of tailing for the production of sand as TSF construction material. *Tailings and Mine Waste 2011*. Vancouver, BC: Norman B. Keevil Institute of Mining Engineering.

Laskowski, J. 2001. Rheological measurements in mineral processing related research. Presented at the 9th Balkan Mineral Processing Congress, Turkey.

Li, H., Zhou, J., and Chow, R. 2008. Comparison of polymer applications to treatment of oil sands fine tailings. Presented at the First International Oil Sands Tailings Conference, Edmonton, AB.

Lindsay, B.J., Wakeman, K.D., Rowe, O.F., Grall, B.M., Ptacek, C.J., Blowes, D.W., and Barrie, D. 2011. Microbiology and geochemistry of mine tailings amended with organic carbon for passive treatment of pore water. *Geomicrobiology* 28(3):229–241.

Lupo, J.F., and Hall, J. 2010. Dry stack tailings design considerations. In *Tailings and Mine Waste 2010*. Boca Raton, FL: CRC Press.

MAC (Mining Association of Canada). 2011a. *A Guide to the Management of Tailings Facilities*. Ottawa, ON: MAC.

MAC (Mining Association of Canada). 2011b. *Developing an Operation, Maintenance and Surveillance Manual for Tailings and Water Management Facilities*. Ottawa, ON: MAC.

MAC (Mining Association of Canada). 2011c. *A Guide to Audit and Assessment of Tailings Facility Management*. Ottawa, ON: MAC.

MAC (Mining Association of Canada). 2017. *A Guide to the Management of Tailings Facilities*, 3rd ed. Ottawa, ON: MAC.

Martin, T.E., and Davies, M.P. 2000. Trends in the stewardship of tailings dams. In *Tailings and Mine Waste 2000*. Boca Raton, FL: CRC Press. pp. 393–407.

McFayden, P. 1998. Relationship of beach slope to yield stress. Thesis, University of Queensland, Department of Chemical Engineering, Queensland, Australia.

MEND (Mine Environment Neutral Drainage). 2004a. *Design, Construction and Performance Monitoring of Cover Systems for Waste Rock and Tailings, MEND 2.21.4*. Vol. 3: Site Characterization and Numerical Analysis of Cover Performance. Ottawa, ON: Natural Resources Canada.

MEND (Mine Environment Neutral Drainage). 2004b. *Design, Construction and Performance Monitoring of Cover Systems for Waste Rock and Tailings, MEND 2.21.4*. Vol. 5: Case Studies. Ottawa, ON: Natural Resources Canada.

MMSD (Mining, Minerals and Sustainable Development). 2002. *Breaking New Ground: Mining, Minerals, and Sustainable Development*. London: International Institute for Environment and Development.

Munro, L.D., and Smirk, D. 2012. Optimized bauxite residue deliquoring and consolidation. In *Proceedings of the 9th International Alumina Quality Workshop*, Perth, Western Australia, March 18–22. pp. 267–275.

Pitchel, J.R., Dick, W.A., and Sutton, P. 1994. Comparison of amendments and management practices for long-term reclamation of abandoned mine lands. *J. Environ. Qual.* 23:766–772

Rico, M., Benito, G., Salgueiro, A.R., Diez-Herrero, A., and Pereira, H.G. 2008. Reported tailings dam failures: A review of the European incidents in the worldwide context. *J. Hazard. Mat.* 152:846–852.

Robins, M. 2004. Closure of tailings facilities: Current practice review and guideline for success. M.S. thesis, University of the Witwatersand, Johannesburg, South Africa.

UNEP (United Nations Environment Programme). 2015. *Awareness and Preparedness for Emergencies at Local Level: A Process for Improving Community Awareness and Preparedness for Technological Hazards and Environmental Emergencies*, 2nd ed. Paris: UNEP.

UNEP–ICME (United Nations Environment Programme and International Council on Metals and the Environment). 1997. *Proceedings of the International Workshop on Managing the Risks of Tailings Disposal*. Stockholm: ICME.

UNEP–ICME (United Nations Environment Programme and International Council on Metals and the Environment). 1998. *Case Studies on Tailings Management*. Stockholm: ICME.

Vick, S. 1990. *Planning, Design and Analysis of Tailings Dams*. New York: John Wiley and Sons.

Weir Minerals Division. 2009. *Pumping Non-Newtonian Slurries*. Technical Bulletin No. 14, Version 2. Glasgow: Weir.

Welch, D., Kam, S., and Merry, P. 2012. Trends in tailings containment. In *Proceedings of the Canadian Dam Association, 2012 Annual Conference*. Toronto, ON: Canadian Dam Association.

CHAPTER 9.2

Water Balance

David Love and Leendert (Leon) Lorenzen

INTRODUCTION

The objective of this chapter is to explore the details of the water balance in the mining and mineral processing industry. This embraces the interaction between the supply of process water, the mining processes, and the environment with regard to effluent reuse or disposal of the resulting wastewater. Specifically, three main water streams are examined in some detail: process water, wastewater, and acid mine drainage. Examples of water treatment options are provided for the processing of gold, base metal, platinum, uranium ores, and coal.

Process Water

Recently, the water balance in many sites has been restricted and confined from all directions. The quantity and quality of primary water supplies have become limiting. This has resulted in an increasing need to use relatively impure primary water sources to use higher proportions of recycle from tailings dams, thickener overflows, and dewatering processes. Effluent discharge guidelines have also become more stringent. Some mining and mineral processing operations now operate under zero water discharge conditions or are required to control effluent release under much stricter quality control.

Primary water supplies from bores containing high levels of salinity (including calcium, magnesium, and iron salts as potential precipitates) are being used in some remote areas. Treated sewage effluent water with relatively high concentrations (>1,000 mg/L) of total organic carbon (TOC) is being used at some sites for make-up water supply. The return of cyclone underflow from primary grind and regrind operations in several circuits introduces high levels of dissolved species from soluble mineral phases (e.g., pyrrhotite, siderite, brucite, and other metal hydroxides, sulfates, and carbonates).

Variability of pH, dissolved oxygen, suspended solids, and dissolved ionic species in recycle water streams (thickener overflow, tailings dam return) contribute to a varying process water quality. Total dissolved solids (TDS) levels may also be raised by reagent addition for pH control (e.g., lime) and TOC levels increased by recycled organics (e.g., depressants, dispersants, flocculants, collector decomposition products). This variability adds another level of complexity to managing the process water and the ore extraction processes.

Wastewater Treatment

Figure 1 summarizes the groups of processes that are commonly used to treat wastewater on minerals industry sites, for the following main purposes:

- Recycle for specific end uses on-site
- Return to process water circuits
- Discharge off-site to a suitable environmental discharge point or to disposal

Neutralization

Lime is the most commonly used alkali for neutralization of the acidic wastewater streams that are often generated in mining sites. The net result of lime neutralization is the formation of metal precipitates such as hydroxides and carbonates. The high-density sludge (HDS) process has been commonly used for lime neutralization. The process features premixing of lime with recycled sludge and is operated at a pH of 10–11. The main attraction of the HDS process is the production of dense, high-solids-content waste sludge, which presents several benefits for waste disposal. Many variations of the HDS process have been developed over the years.

More recently, the use of limestone has become more common because it is lower in cost and overcomes some of the process constraints that are present with lime neutralization. Further innovation has been achieved by integration of neutralization and softening. In some cases, neutralization with caustic has been carried out, producing less sludge. This alternative is usually used for smaller facilities, because it is simpler to dose and has lower equipment costs.

Metals Removal

Mining wastewater often contains a range of dissolved and suspended metal contaminants. Some of these metals are dominant in the wastewater from the mining of certain ores, while other contaminants (e.g., iron) tend to be present in most wastewaters.

David Love, Principal Consultant & Director, Water Logics, Adelaide, Australia
Leendert (Leon) Lorenzen, Managing Director (Professor), Lorenzen Consultants, Mintrex (Stellenbosch Univ.), Perth, Australia (Stellenbosch, South Africa)

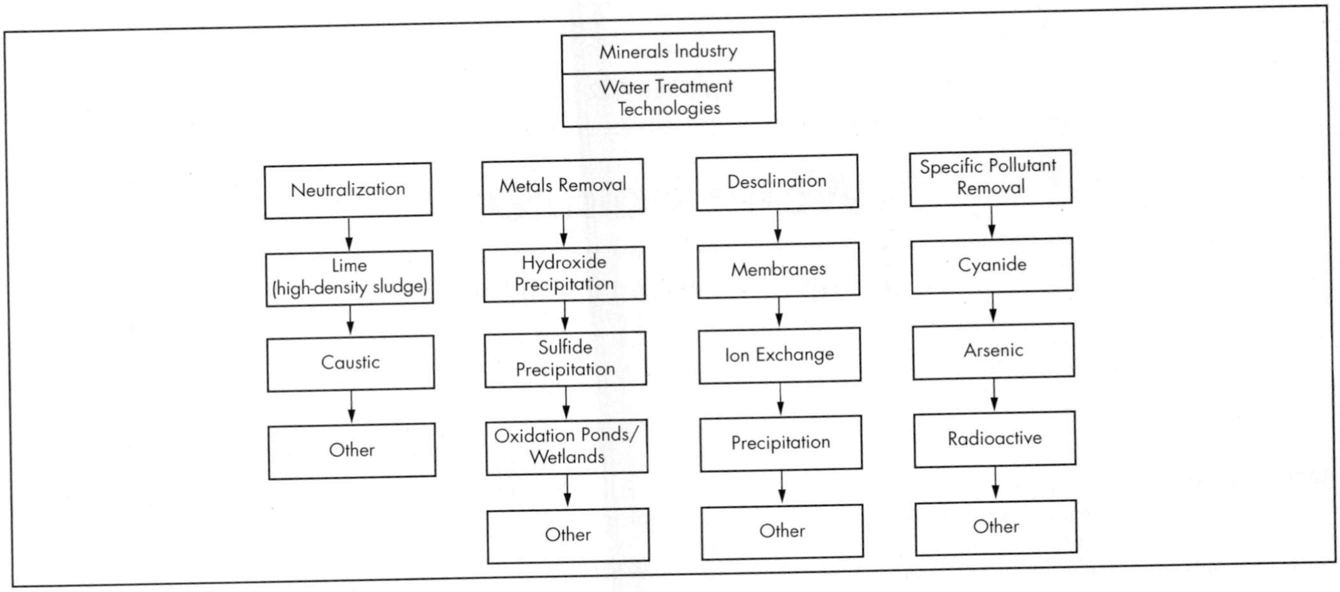

Source: Lorenzen 2015

Figure 1 Process groups used to treat wastewater

Most dissolved metal ions can be oxidized to their hydroxide forms. These hydroxides exhibit an inverted type of solubility curve that shows solubility decreasing as pH is increased until it reaches a minimum and then increasing again as the pH is raised further. By adjusting pH to that corresponding to the minimum solubility level, it is possible to achieve good removal of most metal ions by precipitation. However, the pH levels at which minimum hydroxide solubility takes place vary from metal ion to metal ion. This can often mean that several stages of neutralization are required to meet metal removal objectives.

In some cases, minimum hydroxide solubility levels are not low enough to meet target metal levels for specific uses. In this case, sulfide precipitation can be used to take advantage of the very low solubilities of metal sulfides. The solubility curves for metal sulfide do not exhibit inversion but continue to decrease as pH is increased. Because of costs and chemical handling safety issues, sulfide precipitation is usually used only when hydroxide precipitation will not meet the removal levels required.

Natural oxidation processes such as oxidation ponds and wetlands can also be used for dissolved and suspended metal reductions. This often requires large surface areas to give enough residence time for the slower oxidation rates that are involved.

Desalination

It is common in mining wastewaters to have significant levels of dissolved cations and anions, resulting in the TDS levels being quite high. This can often create performance issues with the processing of ores, as discussed later. The reduction of TDS levels can be achieved by using semipermeable membranes (e.g., nanofiltration, reverse osmosis) or by adsorption of the ion on resins (ion exchange). For these processes, one or more stages of pretreatment are usually required to reduce contaminants (such as suspended solids) down to low enough levels to avoid fouling or other damage to the membranes or resins. A generic desalination flow sheet is shown in Figure 2.

These processes can produce high-quality treated water that is suitable for high-quality uses on-site. However, they also produce brine waste streams that are very high in TDS. These wastes can be problematic for storage on-site or for transport off-site for disposal. Treatment of these brine wastes can involve using concentration processes such as natural evaporation, thermal evaporation, and crystallization. In the case of thermal evaporation and crystallization, up to 90% of the water can often be recovered for use on-site.

Specific Contaminant Removal

Where specific contaminants of concern are present, it may be necessary to target the treatment approach that is best to deal with removal of these contaminants. Examples include the removal of arsenic, cyanide, and uranium. Specific processes that are used will usually aim to remove significant levels of the contaminants of concern and to accept negligible removals of some other contaminants.

Acid Mine Drainage

Acid mine drainage (AMD) refers to the outflow of acidic water from metal mines and coal mines. AMD can originate from many locations within the site, such as

- Drainage from underground workings,
- Runoff from open pit workings,
- Seepage from waste rock dumps,
- Drainage from mill tailings,
- Drainage from ore stockpiles, and
- Drainage from heap leach operations.

The primary issues of concern to the environment from these AMD wastes are acidity, metals, and sulfate. Some of the wastewaters sourced from mines are highly acidic, with pH levels of ≤ 5, and laden with sulfate and metals. These are formed under natural conditions when the geologic strata-containing sulfide minerals (such as aluminum, iron, and manganese) are exposed to the atmosphere or to oxidizing environments.

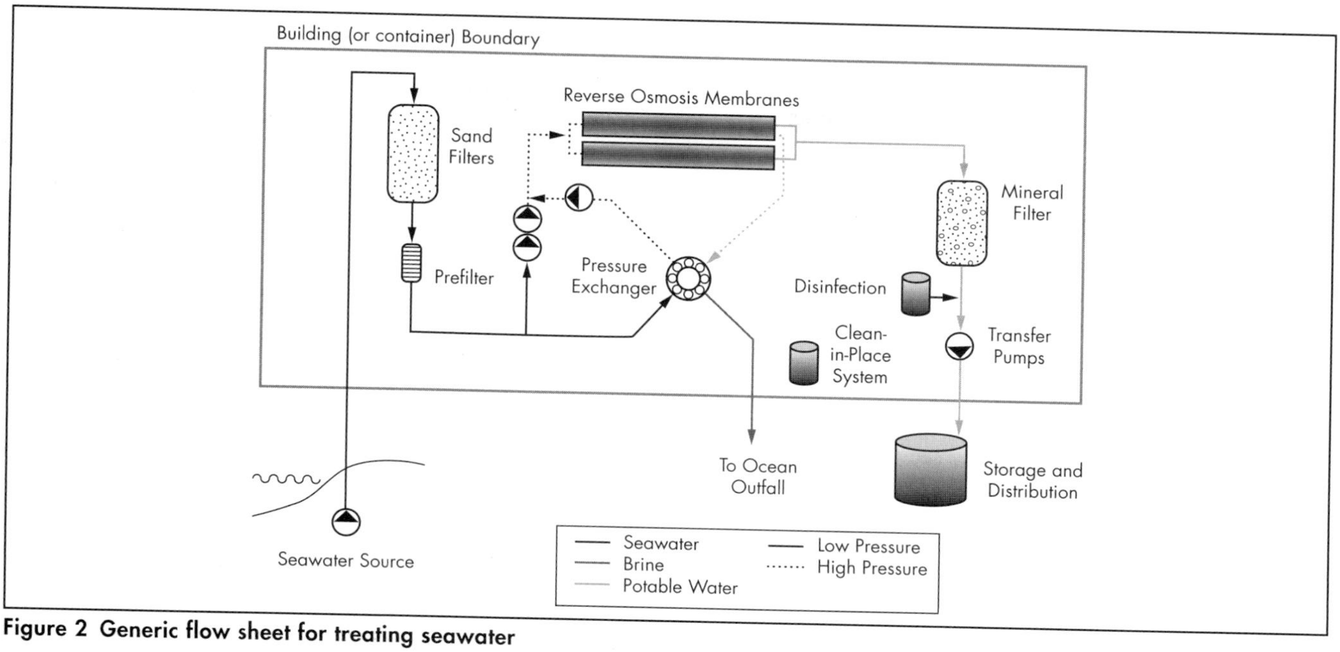

Figure 2 Generic flow sheet for treating seawater

PROCESS WATER

Raw water used for mining processes can be sourced from underground bores or from surface supplies. It is common for this raw water to be mixed with recycled water arising from treated process water. In many cases, the recycling of process water is desirable both for plant operating reasons and for environmental protection. The main advantages are better use of process chemicals, together with the reduction of polluting species discharged to the environment. However, there are many possible process problems with using recycled process water. Recycling can have a negative impact on the performance of the mining processes, sometimes only becoming evident some months after commencement of the recycling. The resulting quality of the process water, arising from the final water balance, is site specific and dependent on several factors, including the following:

1. **Quality of the incoming raw water.** Because of the remoteness of most mining sites, a main supply of potable quality water is usually not available as a raw water source. Groundwater from a bore system or surface water from a lake or reservoir will likely be used for potable water, cooling, steam generation, and other applications where higher quality is important. In some cases, groundwater can be high in dissolved salts, and surface water can be contaminated with suspended solids and organics.
2. **Extent of effluent treatment to render suitable for recycle.** The processes will often generate a combined effluent stream that varies in quality and quantity over time. Some effluent streams may be suitable for treatment, making them acceptable for recycle back into the process water circuits. This recycle stream may be either sent directly to specific uses or blended with raw water to achieve the required process water quality.
3. **Process water quality limitations set by the mining process.** In the case of many mining processes, process water is in direct contact with the ores or minerals being processed. It forms an integral part of the extraction or treatment processes, and its quality can have a significant impact on process performance.

This section focuses on the effects of item 3 in the preceding list by discussing the specific limitations that the process itself places on the quality of the process water being used. These limitations vary significantly and depend on the type of mining and extraction processes being used.

Gold Processing

For the extraction of gold, an alkaline cyanide solution (usually sodium cyanide or calcium cyanide) has commonly been used because of its high efficiency and relatively low cost. Dissolution of gold in dilute cyanide solutions (100–500 mg/L) is essentially an electrochemical process. The active cyanide ions react with the fine gold particles to produce a soluble gold cyanide complex $Au(CN)_2^-$.

Cyanide is very reactive, forming ionic complexes of varying strengths with numerous other metal and alkali earth cations. The stability of these salts is dependent on the cation and on the pH. The formation of these complexes can tie up cyanide that would otherwise be available to dissolve gold. The cyanide ion can also preferentially combine with free sulfur or sulfide minerals in the ore to form thiocyanate compounds (SCN^-).

The key factors requiring consideration for the process water that is used for gold ore processing are pH, magnesium buffering, oxygen compounds, and hydrogen cyanide.

pH

In water, cyanide ions (CN^-) hydrolyze in the following reversible reaction to form hydrogen cyanide (HCN). HCN is a colorless liquid that is highly soluble in water, but its solubility decreases with increased temperature and under highly saline conditions.

$$CN^- + H_2O \leftrightarrow HCN + OH^- \quad \text{(EQ 1)}$$

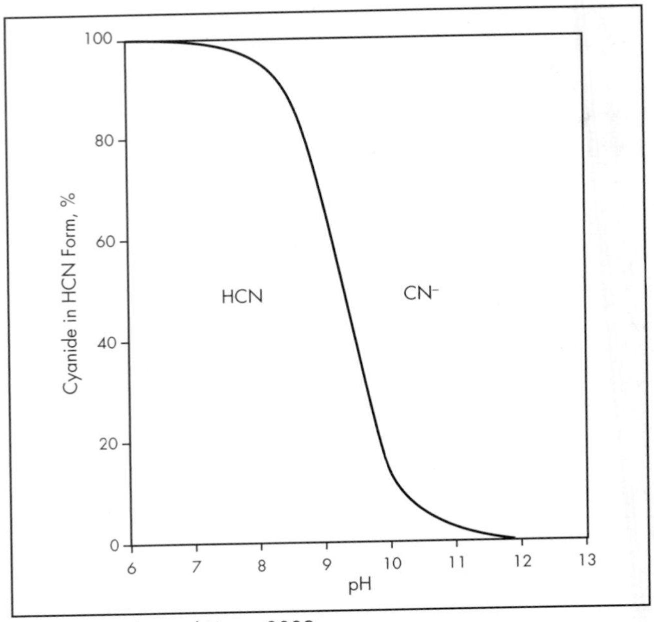

Source: Marsden and House 2009
Figure 3 HCN and CN^- as a function of pH

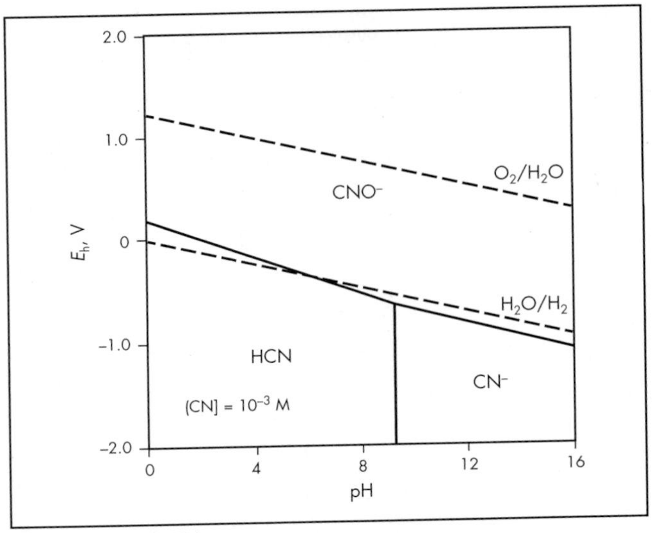

Source: Pourbaix 1974
Figure 4 Pourbaix (Eh–pH) diagram for cyanide

The extent of hydrolysis of the cyanide ion is pH dependent, with the relative proportions of hydrogen cyanide (HCN) and cyanide ion (CN^-) shown in Figure 3. At neutral or acidic pH levels, most of the cyanide ions will be converted to HCN. At a pH of 9.36, 50% of the total cyanide ions that are present will be converted to hydrogen cyanide. When the solution pH is further increased to 10.3 or more, the amount of dissolution will be negligible, and the cyanide ions concentration will be dominant. A good resource is *The Cyanide Compendium* (Mudder et al. 2001).

Equation 1 also shows that OH^- ions are produced, which can increase the solution pH above its starting pH. However, if the process water is high in hardness, the high levels of magnesium ions present (sea and hypersaline waters) will remove the OH^- ions by forming insoluble hydroxides. This results in an initial buffering of the solution pH in the range of about 8.8 to 9.5. This means that significantly higher dose rates of alkali will be required to overcome the buffering, as well as to raise the solution pH up to the desirable level of 10.3.

The leaching, or extraction capability, of a cyanide solution is determined by the free cyanide present rather than just the ionized species. The term *free cyanide* refers to both cyanide ions (CN^-) and hydrogen cyanide (HCN). Loss of free cyanide from the system can occur by the process of volatilization, where some of the HCN can be lost as a vapor.

The amount of cyanide that is lost by volatilization increases with

- Decreasing pH,
- Increased aeration of the cyanide solution,
- Increasing temperature,
- Ultraviolet (UV) radiation, and
- Increasing salinity.

Volatilization losses from leach tanks and tailings dams were investigated by Lotter (2005). Losses from the leach tanks were less than 10% of the cyanide losses when operating the leaching at normal pH levels. However, volatilization was much greater from tailings dams where climactic conditions on the surface of the dams played a significant role. Botz and Mudder (2000) developed a computer simulation to estimate the losses of free cyanide, weak acid dissociable (WAD) cyanide, and total cyanide due to dissociation, photolysis, and volatilization.

This relationship is specifically important because, in gold-cyanide extraction processes that are operated at or above a pH of 10.3, most of the free cyanide in the process slurry water will be present as CN^- ions. Because the ratio of HCN/CN^- is relatively low at this pH, the capacity for cyanide loss by volatilization is low. However, if the pH is decreased down toward 9.4 and less, the ratio of HCN/CN^- increases substantially. As a result, cyanide loss by volatilization becomes more and more problematic, both with regard to the HCN emissions themselves as well as leaving less cyanide available for leaching.

The electrical potential in the leaching circuit can also play an important role in leaching kinetics of gold as well as the HCN and CN^- formation (Figure 4). Thus, an operating control system that is capable of determining the total HCN, pH, and Eh is essential. The slurry may also be subject to other preconditioning such as preoxidation at the head of the circuit before cyanide is added, which will aid in keeping the pH higher.

Magnesium Buffering

Water quality can have a significant impact on slurry pH and HCN formation. If the water supply has high levels of dissolved magnesium salts, it can be very expensive to raise the process pH higher than 9.0 because of the natural buffering action of magnesium (Figure 5). This can force gold processing plants to operate at lower leach pH values, resulting in higher cyanide consumption rates. This leaves only a small margin for error on the pH process control.

The high sulfate levels in the process water can also promote the formation of gypsum, which leads to significant scale issues such as screen blockages, carbon fouling, and pipe blockages. Lower magnesium levels in the feed process water are advantageous from both a chemical cost and extraction

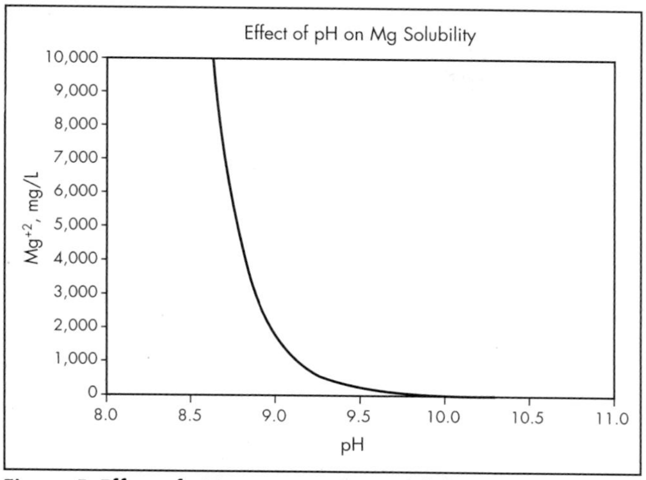

Figure 5 Effect of pH on magnesium solubility

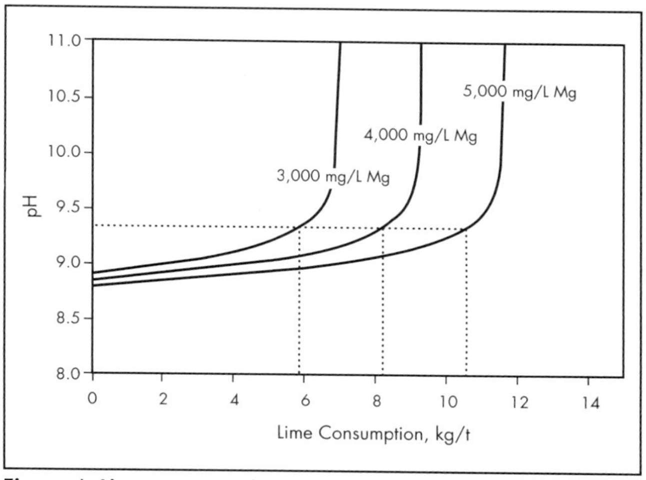

Figure 6 Lime consumption versus magnesium concentration

efficiency viewpoint. In certain cases, the removal of some of the magnesium by ion exchange treatment of the process feedwater may be justified. The addition of hydroxyl ions to water containing magnesium results in the precipitation of magnesium hydroxide, thereby removing hydroxyl ions from solution, as shown in the following equation:

$$Mg^{2+} + 2OH^- \rightarrow Mg(OH)_2\downarrow \quad \text{(EQ 2)}$$

When sufficient magnesium is precipitated at a pH of ~9.5, the pH increases, and this allows both the hydroxyl (lime) demand of the water and the original concentration of the magnesium to be calculated.

Curves indicating the quantity of lime required to obtain a pH value for various magnesium concentrations can be constructed as shown in Figure 6. These lime demand curves are for a system with a pulp density of 50% solids, 100% available lime, and low salinity. This illustrates a series of theoretical lime consumption versus pH curves for magnesium values within the range typical for raw water supply to Australian mines. In the presence of magnesium, the pH buffers at around 8.75–9.5, and larger quantities of lime are required to obtain a small change in the pH. Conversely, only small changes in the pH set point can dramatically reduce the lime consumption. Decreasing the magnesium content of the raw water, as well as increasing pulp densities, will result in a reduction in lime consumption.

Oxygen Compounds

The use of oxygen or per-oxygen compounds (instead of air) as an oxidant increases the leach rate and decreases the cyanide consumption. This is due to the inactivation of some of the cyanide-consuming species that are present in the slurry. If the pH of the slurry can be raised to about 10 (by dosing lime at the head of the leach circuit), this will ensure that when the cyanide is added, toxic hydrogen cyanide gas is not generated and the cyanide is kept in solution to dissolve the gold.

Cyanate (CNO^-) is the primary by-product of treatment using sulfur dioxide or hydrogen peroxide in the process. Cyanate is slowly hydrolyzed to ammonia and does exhibit its own toxicity. In this case, the primary approach in the elimination of cyanate is to select a process that does not form the compound as a by-product.

Hydrogen Cyanide

The presence of hydrogen cyanide (HCN) has far-reaching effects on refinery amine system performance. After hydrocarbon contamination, its presence is one of the primary reasons that amine systems suffer from accelerated corrosion and have operability and reliability problems. When HCN enters the amine system, its hydrolysis produces ammonia and formate, which is a heat-stable salt. Reaction of HCN with oxygen and hydrogen sulfide (H_2S) generates thiocyanate, another heat-stable salt. The heat-stable salts are known to chelate iron, which leads to accelerated corrosion. This also leads to faster formation of particulate iron sulfide, which can plug filter elements, foul equipment, produce stable foams, and lower plant capacity.

Base Metal Processing

Base metals such as copper, lead, and zinc are usually extracted from their sulfide ores by a flotation process. Many investigations have found that the quality of the process water that is used for flotation can have a significant impact on the flotation performance for extraction of each of these base metals.

Muzenda (2010) reported on a study to determine the effect of water quality on the flotation selectivity of complex sulfide minerals. Several water types were tested, from different thickener overflows and sewage effluent water (as process water) and also a potable water source. The water types differed in pH, TDS, total suspended solids (TSS), and conductivity. The process waters reduced both the concentrate recovery and mass pull but increased the concentrate grade. Conversely, the potable water increased the concentrate recovery and mass pull but decreased the concentrate grade. It is proposed that a combination of process water and potable water supply should be used in flotation circuits to balance the different effects that the different water types have on flotation efficiency.

The key factors requiring consideration for the process water that is used for based metal processing are pH and ion content.

pH

Ng'andu (2001) reported on the results of investigations in the Mufulira copper mine in Zambia. The investigation was initiated to find out the reason for the reduction of flotation recovery after changes in the site water balance. Traditionally, the mill water used for flotation consisted of 40% underground water (pH = 7–7.5) and 60% of recycled thickener overflow (pH >10.5).

The percentage of thickener overflow recycled was decreased to only 40%. The study found that the resulting pH in the primary grinding mill decreased from the previous range of 8.5 to 9.5 to less than 8.5. It was found that the lower mill water pH had a deleterious galvanic interaction between the sulfide minerals and the cast-iron grinding media inside the grinding mills. The consumption of oxygen created a strong reducing environment in the slurry fed to the flotation process, adversely affecting the proper adsorption and oxidation of xanthate. Further, the sulfide mineral surfaces, which tend to be cathodic, became prone to covering by hydrolytic oxide layers, hindering the flotation response. Moving the quicklime addition point from downstream of the milling process to the inlet of the primary ball mills had a positive effect and significantly reduced lime consumption.

Ion Content

Forssberg and Hallin (1989) carried out studies on the effect of recycle water on flotation process performance and concluded the following:

- The flotation process was stable in terms of sensitivity to different levels of anion and cation content of the recycled water, except the precipitation of gypsum. The calcium and sulfate ion products in excess of the solubility level of gypsum in a flotation system can affect the process.
- The use of recycled water always had an effect on frothing properties.
- There was not any accumulation of the analyzed ions in the recycled water.
- Flotation problems appear to relate to process chemical residues and different types of oils that accumulate in tailings ponds, when conditions for their degradation are poor.
- Naturally occurring organic surface-active substances like bacteria and humus acids can also interfere with the flotation process.

Coetzer et al. (2003) investigated the impact of ions present in various water sources (deionized, simulated circulation, lead concentrate thickener, lead tailings thickener, and borehole) on galena flotation to recover lead. The activation, depression, and selectivity effects were different for each water source. The borehole water had high salt content, which decreased the froth stability, requiring a stronger frother to be used. The water from the lead tailing thickener had the greatest effect (negative) on galena recovery. It was concluded that, to gain the full benefit from all of these water resources, it would be necessary to first remove some of the ions that have the greater negative impact on galena flotation performance.

Platinum Processing

Water chemistry influences the performance and selectivity of the flotation process because the milled ore is constantly in contact with water. Solution chemistry is extremely complex, and no one rule applies for all systems, or for a circuit as a total. Slatter et al. (2009) summarized several case histories where the process effects of water quality and impact of recycling were evaluated.

The key factors requiring consideration for the process water used for platinum ore processing are dissolved salts, reagent addition, ore dissolution, and organics.

Dissolved Salts

Slatter (2001) found that flotation reagents were more effective in process water with a higher TDS level (up to about 5,000 mg/L) rather than at lower TDS levels. In some sites, very little fresh water is available to the process, and so seawater (with TDS of 35,000–45,000 mg/L) or highly saline borehole waters are used in grinding (Dunstan 1999). Flotation in higher-salinity waters appears to cause little or no mineral dissolution, indicating little surface alteration (Shackleton et al. 2001).

At very high TDS levels, high levels of chlorides can have a negative effect on a smelter. To limit these negative effects, the concentrate is often washed with water that is low in chloride before it is sent to the smelter. The chlorides are also responsible for corrosion, and all materials from which the mills, flotation cells, launders, and so forth, are made need to be corrosion resistant, all of which adds to the capital costs of the process.

Research has shown that calcium and thiosulfate ions affect sulfide flotation differently:

- Calcium activates the adsorption of collector ions when the galvanic effect of mill iron is effective. An investigation into the effects of synthetic waters on synthetic minerals indicated that the adsorption of calcium onto pentlandite and pyrrhotite surfaces increased the minerals hydrophilicity. As a result, more xanthate was required to induce hydrophobicity of the minerals (Malysiak et al. 2003). In one process, substituting soda ash for lime decreased the amount of calcium in the process, which increases value recovery because calcium tends to precipitate on the mineral surface, causing depression (Smart et al. 1999).
- Thiosulfate, which also tends to be generated in flotation cells (Forssberg and Hallin 1989), decreases the adsorption of hydrophilic compounds such as metal hydroxides (Kirjavainen et al. 2002).

Reagent Addition

Flotation of minerals from gangue and from each other makes the use of their different surface properties (Arnold and Aplan 1986). These surface properties are affected by solution components (e.g., passivation of mineral surfaces may occur due to ion precipitation), and this has a negative effect on flotation because the surface chemistry of the mineral is changed. However, the effect of these components may be somewhat negated by using reagents to modify the surface properties (Smith and Hertzog 1985).

Reagent consumption decreases by about 50% when water from the process is reused compared with potable water (Forssberg and Hallin 1989). However, these recycled waters can cause very stable froths to form; consequently, measures need to be taken to ensure that water causing stable froth formation is not used. Cleaner circuits are the most sensitive to changes in froth properties, as the selectivity between sulfide particles and fine-sized gangue mainly depends on the water content in the froth. When the froth is stable and has a higher water content, selectivity tends to be reduced. MIBC (methyl isobutyl carbinol) has been used as the frother of choice in a closed water system because over-stable froths do not form (Johnson 2003).

Ore Dissolution

Ore dissolution can also cause various elements/compounds to accumulate in solution, which alter the chemistry of the system (Rey and Raffinot 1985). An example of ore dissolution effects is in a complex sulfide flotation plant where small amounts of copper may be dissolved during ore–water interaction. This effect can be further increased as a result of water recycling. Eventually, the level of dissolved copper may become high enough to cause the formation of sphalerite (which should be depressed at the start of the float) to be activated (Smith and Hertzog 1985).

Organics

Some natural organics (humic, fumic, tannic, and stearic acids) have a detrimental effect on flotation (Hoover 1980). High TOC levels also seem to cause frothing problems (Forssberg and Hallin 1989). In some cases, using recycled treated sewage effluent waters caused frothing problems across a flotation bank. Work has been performed showing that a foam fraction may improve the frothing problem but does decrease in metal recovery.

Uranium Processing

Uranium is extracted from the raw ore by oxidizing and leaching the ore material with an acid or alkaline leaching solution. As discussed in the NRC (2012) publication, the selection of either an acid or alkaline leaching solution depends on the nature of the ore. This requires extensive testing, economic studies, and environmental considerations to decide the final process choice.

Uranium is found in many deposits as uranium dioxide (as UO_2 in the +4 oxidation state). Given that uranium dioxide is insoluble, it needs to be converted first into a water-soluble form to enable extraction to take place. The first step in any uranium leaching operation is the oxidation of the uranium oxide from +4 oxidation state to +6. The second step is the stabilization of the uraniferous ions in the leaching solution. The uraniferous ions form stable, soluble complexes with sulfate or carbonate. Therefore, sulfuric acid is added as the source for sulfate ions when acid leaching is used. For alkaline leaching, sodium bicarbonate, sodium carbonate, or carbon dioxide are added to provide a carbonate ion source.

The key factors requiring consideration for the process water that is used for uranium ore processing are acid leaching and alkaline leaching.

Acid Leaching

Acid leaching is usually more effective with the difficult-to-treat uranium ores. Sulfuric acid is typically used because of the good solubility of uranyl sulfate complexes. The reaction is typically performed at slightly elevated temperatures (50°–60°C) and can often release H_2, H_2S, and carbon dioxide (CO_2) gases during the process. The uranium goes through a series of reactions, eventually leading to the formation of the desired uranyl sulfate complex. Oxidizers such as oxygen, hydrogen peroxide, sodium chlorate, or manganese dioxide are used to maintain the presence of the hexavalent U^{6+} cation.

Alkaline Leaching

Alkaline leaching solutions using sodium carbonate (30–60 m/L) and sodium bicarbonate (5–15 m/L) tend to be more selective to uranium minerals, so the leaching solution will contain fewer impurities. Consequently, the uranium oxide can be directly precipitated without further purification. The alkaline solutions are less corrosive and can be recycled without the annoyance of increasing impurity concentrations. However, because of the slower reactivity of the alkaline solutions, increased pressures and temperatures (90°–95°C) are often needed.

Because both acid and alkaline leaching require relatively high concentrations of reagents, it is likely that variations in the quality of the raw water and the process water have little impact on the uranium leaching performance. The ore slurry (with the uranium in solution) requires the separation of the solids from the uranium-containing liquid. This is commonly performed using filters (horizontal belt, pressure, or drum filters) or a series of thickeners or decanters. The slurry is washed with acidified water in the acid leach process, or water only in the case of the alkaline leach option, in what is termed *countercurrent decantation*. The washed solids (now referred to as tailings) are neutralized and then forwarded to a tailings impoundment facility for storage.

Coal Processing

Raw coal extracted from the mine needs to be cleaned using different physical processes, with or without chemical agents, to

- Remove inorganic material (ash), thus reducing ash handling in plants that are using coal for coke making or power generation;
- Increase the heating value of the coal; and
- Reduce the transportation cost per unit weight of coal.

The availability of good-quality process water in sufficient quantities is seen as an emerging issue in the coal industry in many countries. Potential approaches to meet this challenge are to

- Reduce water consumption in coal preparation,
- Increase recycling, and
- Use saline mine water (if available) on the mine site.

Most of the process water is supplied by thickening units, which settle out ultrafine suspended solids and enable the recycling of the clarified water back into the plant. A small amount of fresh makeup water from an external source is usually required to satisfy the balance between moisture contents of solids entering and exiting the plant. The clarification and recycling of process water provides an effective means of reducing freshwater demands and lowering environmental impacts. However, plant operators are faced with increasing difficulty in avoiding the buildup of contaminants in the process water. Evidence suggests that dissolved ions and suspended solids adversely impact the performance of dewatering processes.

Flotation is commonly used to treat fine coal (typically <500 μm in size). This is a complex process that is controlled by three facets: coal, chemistry, and machine. Flotation is most often used for treating metallurgical coal fines where the value of the product can justify the added treatment cost of cleaning and dewatering the product component. Because of improvements in flotation technology and the dewatering of both product and tailings, this avenue is becoming increasingly attractive for treating thermal coals. Deterioration of process water quality reduces sizing efficiency, lowers flotation recovery, and increases magnetite losses.

The key factors requiring consideration for the process water that is used for coal processing are salinity and suspended solids.

Salinity

Ofori et al. 2009 reported on a study to identify the impacts of using saline mine water as process water for coal preparation. The study examined the interaction between dissolved electrolytes, hydrophobic solids, hydrophilic solids, and air bubbles in terms of how they impacted coal preparation unit operations. The aim was to study the important physical forces that control the behavior of solids, liquid, and air bubbles in coal preparation processes. How the manipulation of these surface forces could mitigate negative effects of saline water use and thus maintain optimum process efficiencies was also studied.

In terms of the process, saline water had a positive impact, resulting in improved flotation recovery and/or reduced flotation reagent consumption. The impact was even greater when higher concentrations of divalent cations such as Mg^{2+} and Ca^{2+} and divalent anions such as SO_4^{2-} were present. The extent of this positive effect was determined to be dependent on the type of coal being processed. However, with regard to filtration and sedimentation rates, the impact of saline water was negative if no flocculent addition was being used. If flocculants were used, it had such an overwhelming positive impact on filtration and sedimentation rates that it overshadowed the negative effect due to salts. In this situation, the presence of the salts (in most cases) resulted in clearer sedimentation overflow, with less residual solids.

Suspended Solids

Physical separation processes such as dense medium cyclones are widely used for washing coal with particle sizes >0.5 mm. Process water used for these physical separation processes requires certain limits on suspended solids levels, depending on the individual equipment requirements. The water quality and size separation is specific to the usage and cannot be generalized. Even a specific duty such as filter cloth spray nozzles requires a drastically different particle cut size, depending on the type and aperture of nozzles used.

Bickert (2013) undertook a review of filtration technologies currently used with the potential to improve process water quality from a solids content perspective. Fully automatic backwash filters with nozzles, suction scanners, and/or mechanical screen cleaning were most suitable and cost-effective for solids removal from process water, where a separation size of 50–500 µm is sufficient. Where it is necessary to have essentially complete solids removal, or where the process contains mineral matter, media filters such as the continuous sand filter should be considered. However, these media filters should be fed with water with <200 mg/L solids, and so coarse screen filters may still be required for pre-filtration. Technologies such as flotation and gravity clarification are less suitable for solids removal from process water, because of their limited performance with low levels of solids in the water.

WASTEWATER TREATMENT

The main liquid waste streams that are generated from most mining sites arise from the various hydrometallurgical processes such as leaching, thickening, and flotation. The mining wastes include concentration process wastes, heap leaching products, and old slurry tailings from discontinued mining operations. The environmental threat imposed by these effluent streams can be attributed to various chemical compounds, toxic metal ions, and extreme pH variations. Some of these environmental threats exhibit the possibility of contaminating surface and groundwater through one or more of the following mechanisms:

- Acid mine drainage
- Release of residual cyanide and soluble metal species from heap leaching operations
- Mineral dissolution that results in the release of metal species and other anions

The extent to which these processes pose environmental threats highly depend on the process occurrence and containment techniques employed, as well as the nature and concentration of the species.

A detailed understanding of wastewater sources and their characteristics is essential for the purposes of selecting, as well as developing, efficient and reliable water management and wastewater treatment systems. Such knowledge, coupled with effluent design criteria, is utilized in establishing the level, type, and capacity of wastewater treatment that is required to meet the stringent environmental legislations, as prescribed by different environmental protection agencies globally.

The chemical characteristics are dependent on the complex interactions between the metallurgical processes and reagents, ore chemistry, and site hydrology. The fate of the effluent streams can be determined by the following scenarios:

- Suitability for treatment to recycle for specific uses onsite. This will be driven by the process performance factors.
- Treatment to meet stringent limitations for discharge to the environment. The actual treatment required will depend on the actual components that are in excess the discharge limits to be met.
- Containment on-site with no release to the environment. Effective containment can be achieved by using tailings impoundments, heap leach pads, or solution ponds that are constructed with impermeable barriers between the contained liquid and the surrounding ground. For smaller volumes, steel tanks or concrete sumps can also be used.

The specific issues with treatment of wastewater from gold, base metals, platinum, uranium, and coal ore processing facilities are discussed in the following sections.

Gold Processing

Effluent Characteristics

The significant variability of wastewater characteristics from gold mining processes, in terms of heavy metals and cyanide levels, is demonstrated in Table 1. The Yanacocha gold mine in Peru and the David Bell mine in Canada are good examples of technologies applied to reduce cyanide levels in the mine-site wastewater streams (Diaz et al. 1998; Meyer 1992).

Effluent for Recycle

The need for treatment of recycle streams is driven by the necessity to manage the quality of the process water. As discussed earlier, pH plays a significant role in the efficiency of gold extraction by cyanide. Recycle streams (untreated or treated) that contribute to an increased process water pH are desirable. These could increase alkalinity, or reduce magnesium that acts as a buffer on the cyanide solution.

Table 1 Chemical composition of barren, decant, and seepage solutions

Variable/Parameter	Concentration Range
Arsenic, mg/L	<0.020–10.0
Cadmium, mg/L	<0.005–0.02
Chromium, mg/L	<0.02–0.10
Copper, mg/L	0.1–400
Iron, mg/L	0.50–40
Lead, mg/L	<0.01–0.1
Manganese, mg/L	0.1–20
Mercury, mg/L	<0.0001–0.05
Nickel, mg/L	0.02–10
Selenium, mg/L	<0.02–6
Silver, mg/L	<0.005–2
Zinc, mg/L	0.05–100
Total cyanide, mg/L	0.5–1,000
WAD cyanide, mg/L	0.5–650
Free cyanide, mg/L	<0.01–200
Ammonia-N, mg/L	<0.1–50
Thiocynate, mg/L	<1–2,000
pH	2.0–11.5
Hardness (as $CaCO_3$), mg/L	200–1,500
Sulfate, mg/L	5–200,000
Temperature, °C	0–35

Adapted from Smith and Mudder 1991

In some cases, removal of magnesium by ion exchange or softening may be advantageous. Removal of sulfate ions may assist with reduction of gypsum formation and associated problems with scale deposition. Recycle of cyanide ion could contribute to lower cyanide dosing to maintain cyanide solution concentrations.

Effluent for Discharge

The U.S. Environmental Protection Agency (EPA) sets limits on discharge from sites that utilize processes for the beneficiation of gold ores. The limits shown in Table 2 are based on the degree of effluent reduction attainable by the application of best practical control technology (BPT). These limits are similar to those applied to copper, lead, and zinc processing.

Because cyanide solutions are commonly used for gold extraction, the effluent usually contains free cyanide concentrations that are well in excess of the safe level for release to the environment. There have been several different treatment processes used to treat effluent from gold mining/processing sites, and some of the more commonly used processes are discussed below. These processes have needed to take into account the local conditions such as ore chemistry and site hydrology as well as sludge treatment and disposal costs.

The most critical selection criterion for a certain treatment method is the stability and concentration of a given solution. This is because every facet of the mining operation affects the quantity and quality of wastewater produced. It is within this context that the process, or processes, selected must be reliable and flexible enough to maintain a consistently high-quality effluent throughout the life span of the mine, and even beyond mining.

Table 2 Guidelines for discharge from gold processing

Parameter	EPA* Guidelines for 1 Day	EPA† Guidelines for 30 Days
Cadmium, mg/L	0.1	0.05
Copper, mg/L	0.3	0.15
Lead, mg/L	0.6	0.3
Mercury, mg/L	0.002	0.001
pH	6.0–9.0	6.0–9.0
Total suspended solids, mg/L	30	20
Zinc, mg/L	1.5	0.75

Data from EPA 2018.
*Maximum for any one day.
†Average of daily values for 30 consecutive days.

Natural Degradation

The natural degradation of cyanide is achieved by the combination of the mechanisms of volatilization, UV degradation, the formation of strong complexes, adsorption, and biological oxidation (Rouse and Pyrih 1998). This process occurs primarily because of auto-oxidation. For example, the Homestake Mining site in New Mexico (United States) reported natural cyanide breakdown occurring in the tailings pond as well as in the mine (Halbe et al. 1979).

Research findings in Canada on natural degradation in the tailings ponds showed that natural degradation reduced the CN^- concentration from 68.7 to 0.08 mg/L during the warmer period (Schmidt et al. 1981). Most of the degradation was due to dissociation of cyanide complexes coupled with volatilization of molecular HCN. Oxidation of free CN^- to CNO^- only accounted for about 11% of the conversion, and photo- and biodegradation could not be detected. No degradation was detected during the colder winter months.

Though natural degradation is a simple method that reduces cyanide concentration to acceptable levels, its success depends upon the amount and species of cyanide present as well as the retention time the storage pond can provide. Even where the natural degradation process does not provide a final effluent with the desired level of CN^-, the process is useful in decreasing the amount of cyanide that may have to be treated further using chemical means (Ritcey 2005).

Natural degradation can be influenced by variables such as the cyanide species in solution and the relative concentrations, temperature, pH, aeration, sunlight, presence of bacteria, pond size, depth, and turbulence. Evaporation can be used effectively as a means of destroying cyanide in arid regions such as in Western Australia, South Africa, and the western United States. These arid areas are characterized by high evaporation rates, and therefore, the runoff into rivers and lakes is not a problem (Ritcey 2005). The key limitation of natural degradation is that it can take too long to detoxify a heap to meet regulatory limits; consequently, the operation and maintenance costs are higher (Marsden and House 2009).

Barren/Freshwater Rinse

Mosher and Figueroa (1996) described barren/freshwater rinse treatment as the washing of heaps with solution from the barren pond, only using fresh water to make up for the evaporative losses. No reagents were used and cyanide concentrations decreased, possibly due to a combination of native bacterial action, complexation, and volatilization processes. This treatment method suits mostly climate regions where there is a negative water balance as well as relatively cheap sources of water.

The method is characterized by a low requirement for additional capital equipment, no reagent additions, and it incurs no additional engineering costs. However, its shortcomings include high operation and maintenance costs as well as long waiting periods to rinse the heap down to the stipulated closure standards.

Artificial Wetlands

Greenway and Simpson (1996) presented results indicating that artificial wetlands can be used for treating gold mine leachate that contains cyanide. Their findings showed that the final effluent was of good quality and suitable for reuse purposes such as irrigation. Larsen et al. (2004) showed that cyanide can be removed using woody plants that have a high capacity to eliminate free cyanide.

The selection of suitable plants depends on the conditions at the mine (climate, soil, etc.) and their cyanide removal capacity. For applications in a humid climate zone, the willow hybrids have shown the most promising results so far (Rytter and Hansson 1996). It was found that a maximum of 1,100 kg of free cyanide could be removed by 1 ha of willows for a growth period of 200 days.

Bacterial Oxidation

Some bacteria species are able to metabolize cyanide and thiocyanate. Different species degrade cyanide into products such as ammonia and formate, based on the results of Raybuck (1992) and others. Free metals are either absorbed within the biofilm or precipitated from solution (Akcil 2003). The ease of degrading the metal cyanide complexes is generally in accordance with their chemical stability, with free cyanide being the most readily degradable and iron cyanide the least.

The first successful large-scale application of the biological oxidation method for the treatment of gold cyanide leaching effluent was carried out by Homestake Mining Company in South Dakota (United States) in 1984 and was operated successfully until the mine closed because of a lack of ore (Whitlock and Mudder 1986). The mechanism of the biological treatment of effluent-containing cyanide has been described in detail by Akcil (2003). The biological process is made up of two main stages:

1. Bacterial oxidation of CN^- and SCN^- to produce CO_3^{2-}, SO_4^{2-}, and NH_4^+, by using indigenous microorganisms that are acclimatized to cyanide and thiocyanate.
2. Nitrification of NH_3 into NO_3^{2-} by the Nitrosomonas and Nitrobacter anaerobic species.

Soda ash is added as a source of carbon to assist nitrification. Phosphorus is also required as a trace nutrient.

The bacteria that are attached to the discs in rotating contactors absorb the metals, once released from the cyanide. After the biodegradation and metal removal in the contactors, ferric chloride is added to aid clarification, followed by sand filtration. The key merits of this technique include

- Low reagent and operating costs;
- The potential to treat total cyanides without creating secondary waste streams; and
- The ability to remove thiocyanate, ammonia, and nitrate, unlike the costly chemical-based approaches (Given et al. 1998).

Although the initial investment cost is high, the operational costs are relatively low, so the net present worth is very low for a biological technique (Nelson et al. 1998). Biological processes, which are quite promising in terms of satisfying both the extraction and environmental control requirements, have been proven at large scale in well-understood engineering systems in countries such as the United States and Canada (Mudder et al. 1998).

Biotechnological treatment processes are under various stages of development; however, many proposed processes cannot currently compete with the conventional technologies and are unlikely to do so unless regulatory standards change. Processes that have been applied commercially are

- In-plant cyanide destruction,
- In situ cyanide destruction of spent heap leach piles,
- Metal and sulfate removal using active (in-plant) sulfate reduction, and
- Limited use of passive processes such as wetlands and ecological engineering for metals polishing.

Biotreatment of cyanide and associated species, such as ammonia and thiocyanate, in gold plant effluents can be readily adapted to handle large flows and the cyanide concentrations found in commercial gold operations. Commercial applications involving both in-plant treatment as an alternative to chemical processes and in situ cyanide destruction of spent heap leach piles are being used (Lawrence et al. 1998).

A full-scale combined biological and chemical treatment facility currently operating in Turkey consists of three steps (Akcil 2003):

1. Combined activated sludge treatment for the conversion of thiocyanate (SCN^-) to ammonia (NH_3) and for the oxidation of the ammonia (NH_3) formed to nitrate (NO_3^{2-});
2. Denitrification treatment to reduce NO_2 to nitrogen gas (N_2); and
3. HDS ferric sulfate treatment to precipitate arsenic and other metals as sulfate (SO_4^{2-}).

Trickling Filters

Trickling filters are the most frequently used fixed-film bioreactors for treating wastewaters containing free cyanide, carbon oxygen demand, thiocyanate, and copper and zinc ions (Evangelho et al. 2001). In these reactors, microorganisms are attached to a solid substratum in which they reach relatively high concentrations. The trickling filter is suitable for treating wastewaters with significant variations in organic and hydraulic loads.

The results indicate that approximately 90% of the free cyanide, thiocyanate, copper, and zinc in the influent are effectively removed. These removal efficiencies were obtained when the pilot bioreactor was operated without recirculation. Using recirculation brought about a decrease in pH and lowered the efficiency of zinc removal. The microbial activity was responsible for thiocyanate degradation and copper removal. The tests carried out in the reactor without biomass showed that the percentage of free cyanide removed by volatilization was low (22.6%) and was even lower when recirculation did not take place (7.7%). This confirms the important role that the biomass plays in the degradation of this pollutant.

Among the key merits of this process are that it is simple and easy to operate and has low energy requirements. It is possible to retain microorganisms with a slow growth rate, such as those responsible for cyanide degradation. However, its weakness lies in the inability to renew the biomass, as the metal accumulation may lead to biological activity inhibition.

Sulfur Dioxide–Inco Process

The SO_2–Inco process involves the oxidation of cyanide to cyanate (CNO^-) in the presence of Cu(II) ions as a catalyst, according to the following equation:

$$CN^- + SO_2 + H_2O + O_2 \rightarrow CNO^- + H_2SO_4 \quad \textbf{(EQ 3)}$$

The oxidation reaction appears specific to cyanide and cyanide–metal complexes. Hexacyanoferrates are not oxidized but are removed in the final treatment stage by precipitation. If present, some thiocyanate is degraded. If nickel is used as the catalyst, oxidation of thiocyanate proceeds at a slower rate.

The process has the added advantage of using a relatively inexpensive reagent. Also, the cyanate produced may be significantly less toxic than cyanide to receptor organisms such as fish, animals, and humans. However, the chemical-handling systems are rather more complex than for a hydrogen peroxide system, resulting in higher capital costs.

Ozonation

The oxidation reaction of ozone with cyanide is rapid. The rate of decomposition of complex cyanides with ozone varies with the stability of the metal complex. Nickel, zinc, and copper complexes are readily oxidized, whereas iron cyanides decompose with difficulty, unless performed at elevated temperatures or with UV light.

Reaction rates can be improved by metal ions such as Cu^{2+}, where the presence of 20 mg/L of copper doubles the rate of oxidation (Rowly and Otto 1980). The rate of decomposition is also increased by 50% at pH 12–13 compared to pH 9–11. A combination of both pH and copper addition can enhance the decomposition threefold.

Acidification, Volatilization, and Reneutralization

The AVR process (also known as the Merrill–Crowe process) exploits the fact that hydrogen cyanide is volatile at 26°C and 100 kPa. Early applications were aimed at reducing the cyanide consumption, as opposed to effluent control. In this process, the solution containing the cyanide is acidified using sulfuric acid (H_2SO_4), and the HCN gas is swept by air to an absorber tower and contacts lime slurry. The alkaline cyanide is returned to the process circuit. Ferrocyanide or thiocyanate are not destroyed by the treatment process, but they are precipitated completely by copper or zinc double-complex formation.

The unique processing steps in the AVR process result in the following benefits:

- Production of a metal salt by-product for recycling to the treatment process, if required.
- Availability of the metal salt by-product for recycling to a novel "polishing" step, which can reduce the metal cyanide content of the final effluent to ~0.1 mg/L.
- Rejection and stabilization of the iron content of the barren bleed for disposal to tailings.
- Cyanide is recoverable for reuse in the leaching process with a recovery potential better than 90%.
- Process efficiency is not sensitive to fluctuations in feed contaminant levels of CN^- or heavy metals concentrations.

However, the AVR process has several limitations:

- The hydrocyanic acid vapor is a hazard. Acidification and volatilization systems must be tightly sealed, and the plant area designated for this part of the operation must be well vented as precautionary safe practice. This increases the costs of operations.
- The process is more energy-intensive than other methods, such as ion exchange or chemical oxidation.
- It may not be suitable for applications where the effluent is discharged directly into any environmentally sensitive receiving watercourse without the utilization of on-property ponding or a polishing step.
- For very dilute barren bleed ($CN^- < 100$ mg/L), the capital equipment outlay is relatively high.

More recently, considerable research has been directed toward improving the AVR process because of the high cost of cyanide destruction. With various specific modifications, the AVR process has now been installed in several gold cyanidation plants throughout the world. It has also been successfully used in the application to silver cyanide leaching circuits for the recovery of cyanide for reuse (Botz and Mudder 1998). Prior to the mine closing in 1998, Kinross Gold Corporation's DeLamar mine in Idaho (United States), was able to achieve approximately 95% of the cyanide it recovered, leaving a residual of about 20 mg/L to be discharged to tailings and natural degradation. As an alternative to the AVR process, the SART (sulfidization, acidification, recycling, and thickening) process has been used in other mines, as detailed in McGrath et al. (2015).

Because the cyanide is usually complexed with base metals, there has been considerable work directed toward the breakdown of the complexes for the recovery of cyanide. With the modifications applied to liquors, similar processes have resulted in the presence of solids such as the AFR (acidification–filtration–reneutralization) process (Fleming and Trang 1998). In that process, as applied to tailings, it is necessary to achieve lower final-pH values to achieve optimum precipitation of Cu, Fe, CN^-, and SCN^-. The recovery of cyanide was reported to be in the order of 80%. Because most of the metal and cyanide contents can be recovered, the final waste solids contain no nonferrous base metals, and the precipitated metals can be returned to a mobile state in the tailings pond.

Alkaline Chlorination

Alkaline chlorination is a chemical process involving the oxidation and destruction of free and WAD forms of cyanide under alkaline conditions (pH range of 10.5 to 11.5). It is among the oldest and most widely recognized cyanide destruction technologies based on operational experience and engineering expertise (Smith and Mudder 1991). Cyanide is oxidized by the hypochlorite ion, which can originate from either chlorine gas or a hypochlorite salt (sodium or calcium). The hypochlorite ion oxidizes cyanide to cyanate according to the following simplified reaction:

$$OCl^- + CN^- \rightarrow CNO^- + Cl^- \quad \textbf{(EQ 4)}$$

WAD cyanide is also oxidized to cyanate, but hexacyanoferrates are not removed. Ferrocyanide complexes can be partially oxidized to ferric cyanides but are not removed with copper precipitation. Additionally, thiocyanate is oxidized preferentially to cyanide, increasing reagent consumption to meet a given WAD cyanide discharge limit.

This method has recently fallen out of favor because of the environmental implications of the process itself. Chemicals used in the process, whether sodium hypochlorite or chlorine gas, involve complex handling issues. Depending on the receiving water, it may also be unacceptable to add sodium. Notably, the excessive unreacted hypochlorite is toxic to fish. Finally, poor process control can cause the evolution of other toxics. These compounds, together with residual chlorine issues, have necessitated the development of alternatives of non-chlorine methods for the elimination of cyanides.

Hydrogen Peroxide

Hydrogen peroxide has replaced alkaline chlorination for many cyanide destruction applications, and it is often the method of choice for clear liquor detoxification, including heap leach decommissioning. Although relatively higher in cost, hydrogen peroxide presents some significant advantages over other chemical methods:

- Is a very strong oxidizing agent
- Does not introduce any other contaminants
- Simple process operating over a wide pH range
- Typically reduces cyanide levels down to below discharge limits
- Excess hydrogen peroxide decomposes to water and oxygen

Recent work by Kitis et al. (2005) has attempted to examine the effectiveness and kinetics of destroying cyanide using hydrogen peroxide in the tailings slurry from gold mines that are characterized by low sulfide and heavy metal content. The hydrogen peroxide oxidizes cyanide according to the following reaction:

$$CN^- + H_2O_2 \rightarrow CNO^- + H_2O \quad \textbf{(EQ 5)}$$

In the event that hydrogen peroxide is added in excess, nitrite and carbonate are formed, and eventually nitrate forms (Monteagudo et al. 2004). Also, copper or proprietary reagents may be used as catalysts. Hydrogen peroxide does not directly oxidize ferrocyanides, but these compounds can be removed by precipitation as copper ferrocyanide.

The primary drawback of hydrogen peroxide is its relatively high unit cost. Recently, the Degussa mine (part of Sandfire Resources) has carried out work with mixtures of hydrogen peroxide and Caro's acid, a promising approach that may reduce the overall reagent cost. The process is typically applied to achieve effluent cyanide levels suitable for discharge. This is a relatively simple process that is capable of operating over wide pH ranges without increasing total dissolved solids, as opposed to other chemical processes.

Base Metals Processing

Effluent Characteristics

The main streams contributing to the combined effluent from sites processing base metals (copper, zinc, and lead) usually originate from cyclone underflows, thickener overflows, and tailings dams. These sources contain varying amounts of anions, cations, dissolved heavy metals, suspended solids, and dissolved salts. Return of cyclone underflow from primary grind and regrind operations can introduce high levels of dissolved species from soluble mineral phases such as pyrrhotite, siderite, brucite, and other metal hydroxides, sulfates, and carbonates.

Effluent for Recycle

The opportunity for treatment of this effluent, either from the individual sources within the mine site or processing area, or as a combined stream, is driven by the need to maintain suitable quality of the water streams. For some of the individual effluents, their quality may be suitable for direct reuse for specific site uses such as dust control (with little or no treatment). In other cases, some treatment may be required before internal recycle to specific end uses. An example of mine-site water treatment is provided in Figure 7.

Effluent for Discharge

The EPA sets limits on discharge from sites that use froth flotation processes (alone, or in conjunction with other processes) for the beneficiation of copper ores, lead ores, and zinc ores. The limits shown in Table 2 for gold processing also apply to base metals processing and are based on the degree of effluent reduction attainable by the application of BPT. The limits shown in Table 2 are similar to those applied to discharges from silver ore processing (EPA 2018).

The effluent that is surplus to site requirements for recycling may need to be collected on-site before it can be disposed of by evaporation or discharged to a suitable water environment. Where one or more of the contaminants in the effluent exceeds the discharge guidelines, additional treatment over

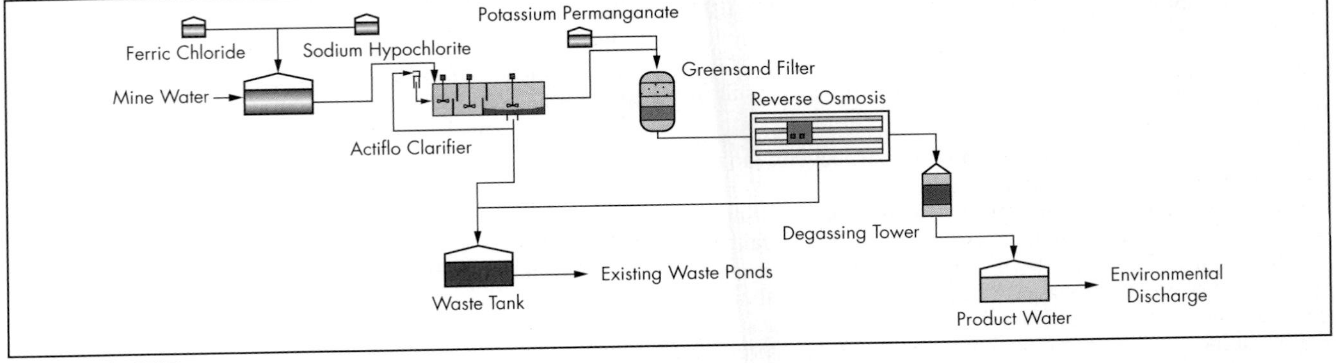

Source: Madin 2007

Figure 7 The Bendigo Mining contaminated mine water treatment process

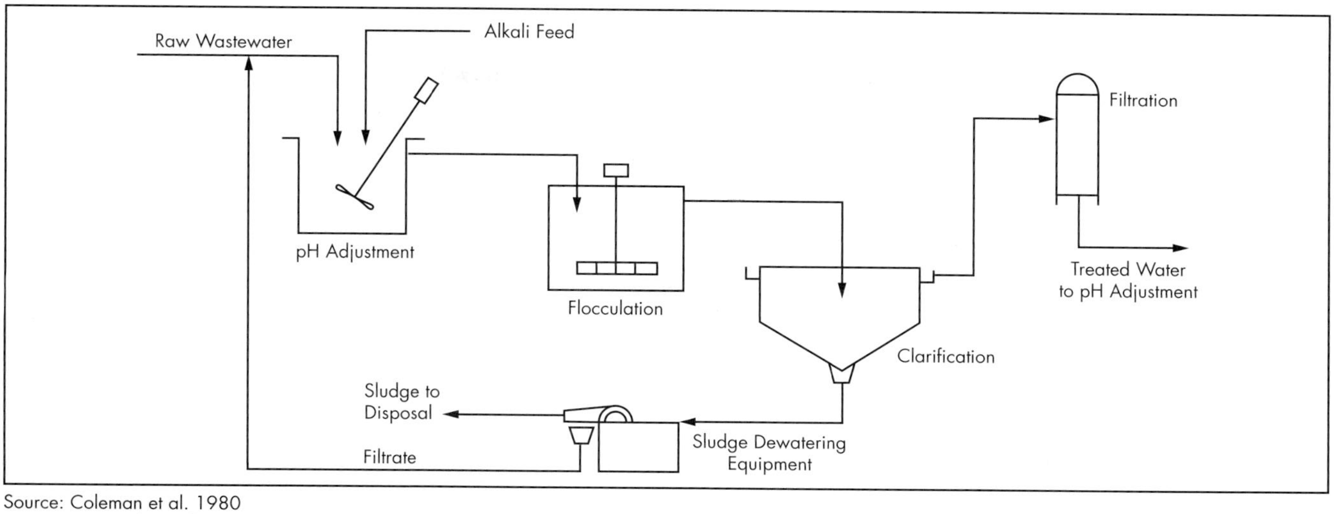

Source: Coleman et al. 1980
Figure 8 Hydroxide precipitation process

that necessary for internal recycle is likely to be necessary. Because effluents from these sites usually contain most of the metals listed, several stages of treatment may be required to address all of the limit exceedances.

Hydroxide Precipitation

Precipitation of metal hydroxides by adjustment of pH is commonly used for heavy metal removal. This process involves dosing an alkali to adjust pH before flocculation to promote the formation of hydroxide floc (Figure 8). This is followed by settling to separate the solids for dewatering and/or disposal. The alkali used is usually quicklime, hydrate lime, or caustic. The parameters affecting chemical dose rate include inlet pH, temperature, alkalinity, hardness, metals concentration, and the desired final pH. The actual determination of the required chemical dosage is best carried out using laboratory jar tests on samples of the actual wastewater to be treated.

Figure 9 shows the solubility curves for some heavy metals. Most of these exhibit inverted curves as pH increases and have different minimum concentrations. Selection of the desired pH to operate the hydroxide precipitation process will depend on the contaminant levels that are present and their relativity to the guideline limits for these contaminants. For example, copper, lead, and zinc have minimum solubilities within a relatively narrow pH range from 8.9 (copper) to 9.3 (lead). Selecting a pH control range of, say, 9.0 to 9.2 may be suitable for minimizing the concentrations of these three metals. However, the presence of one or more other heavy metals (e.g., nickel) at levels above the guideline limits may necessitate a two-stage pH-adjustment approach.

The theoretical minimum concentrations that can be reached for each heavy metal differ significantly. In the case of copper (0.001 mg/L) and zinc (0.1 mg/L), the minimum theoretical concentrations are well below the guideline limits. However, lead is more soluble and its minimum concentration of about 8 mg/L is well above its guideline limit.

In practice, however, the hydroxide precipitation process (Coleman et al. 1980) seems to yield residual concentrations that are different to the theoretical concentrations. For example, lead residual has been reported as being reduced to about 0.1 mg/L, which is much lower than the theoretical

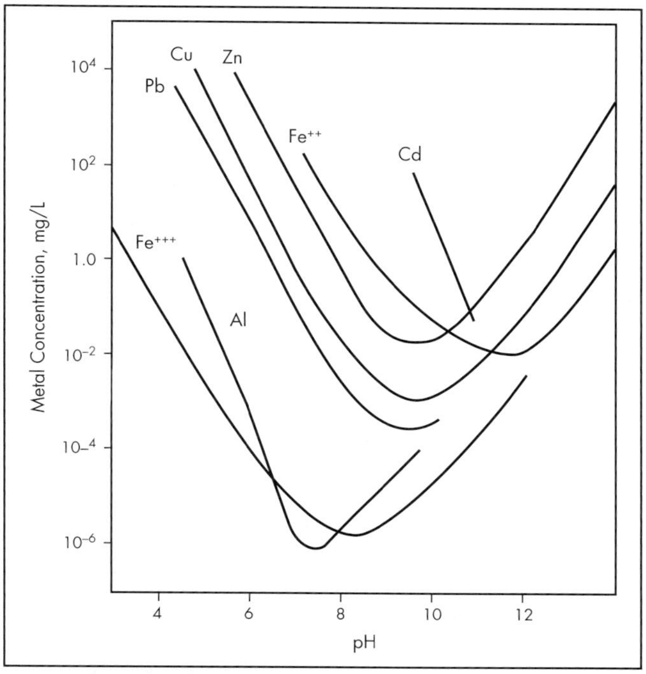

Source: Aubé and Zinck 2003
Figure 9 Solubilities of heavy metals as a function of pH

concentration. Several physical phenomena (ionic strength, co-precipitation, and adsorption) are believed to influence the effectiveness of precipitation, especially in solutions containing multiple metal ion species. Therefore, it is very important to carry out jar tests on the specific wastewater to determine likely residual concentrations under different pH conditions. In some wastewaters, hydroxide precipitation may not satisfy the treatment objectives for all the dissolved heavy metals, and further treatment may be required.

Sulfide Precipitation

In a similar process to hydroxide precipitation, dissolved metals can be removed by chemical formation of their sulfides that

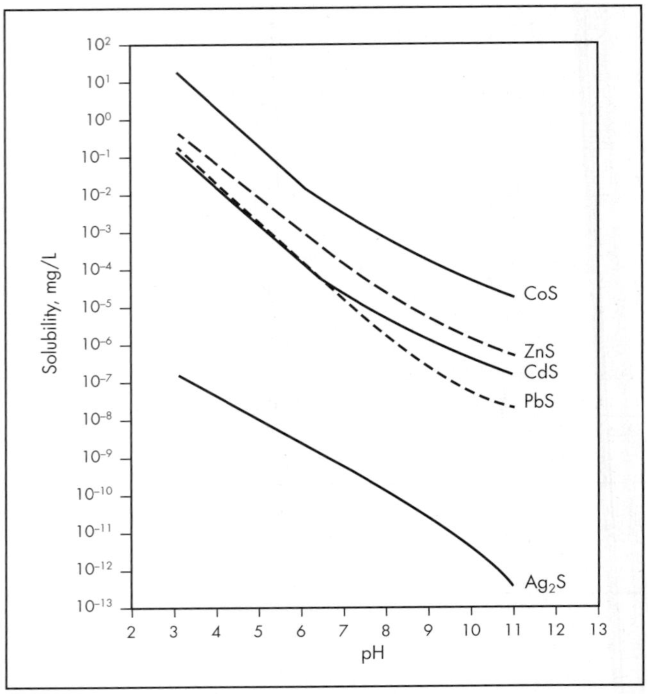

Source: Coleman et al. 1980

Figure 10 Solubilities of metal sulfides

are very insoluble. In general, sulfide solubilities are much lower than hydroxide solubilities and continue to decrease as pH increases, as shown in Figure 10. This has the advantage of not requiring adjustment of effluent pH levels to meet guideline limits. However, for a successful outcome, sulfide precipitation requires the following:

- Sufficient available sulfide to drive the precipitation reaction to completion.
- An efficient removal of the sulfide solids from the treated water. This usually requires a final filtration process to remove any unsettled solids from the effluent.

The most commonly used chemicals are the soluble sulfides, including sodium sulfide, sodium bisulfide, calcium sulfide, and barium sulfide. Excess sulfide must be avoided because of the potential formation of hydrogen sulfide gas. Therefore, it is very important that sulfide dosing be accurately controlled. One control method uses oxidation–reduction potential (ORP) measurement to monitor the electrical potential of the reaction solution. When the precipitation reactions are complete, there is a sharp decrease in the solution ORP.

Although the sulfide precipitation process usually produces less sludge than with hydroxide precipitation, sludge disposal is more problematic. If metal sulfide sludges are placed in a typical landfill, exposure to air can oxidize the sulfide and eventually create an acid environment. As the pH decreases, the heavy metals present can redissolve and be leached from the site via surface runoff or percolation into groundwater. Where a metal smelter is also on-site, the sulfide sludges can be recycled back to the smelter for metal recovery.

Ion Exchange

In general, ion exchange with strong acid cation resin works better at removing heavy metals when the wastewaters are in the pH range of 4 to 8, and when there are low suspended solids, iron, and aluminum concentrations. This process may be better suited to treatment of selected wastewater streams using prefiltration or as a polishing process. The more complex the mixture of contaminants, the more difficult it is to achieve the desired outcome. This is because of the following order of selection for strong acid cation resins to remove ions:

$$Pb^{2+} > Ca^{2+} > Ni^{2+} > Cd^{2+} > Cu^{2+} > Zn^{2+} > Mg^{2+} > K^+ > NH_4^+ > Na^{2+} > H^+$$

This means that the capability of the resin to remove ions that are lower down in the sequence (e.g., Zn^{2+}) can be retarded if there is significant presence of other dissolved metals that are higher in the selection sequence (e.g., Pb^{2+})

Ion exchange has an advantage over the chemical precipitation processes in that it produces much less waste for disposal. The disadvantages of ion exchange are that (1) it cannot handle concentrated metal solutions, particularly containing iron and aluminum, which foul the resin; and (2) ion exchange is nonselective and highly sensitive to the pH of the solution.

Adsorption

Recently, Barakat (2011) reported that adsorption has become one of the possible alternatives for treatment of wastewater laden with heavy metals. Basically, adsorption is a mass transfer process by which a substance is transferred from the liquid phase to the surface of a solid and becomes bound by the physical and/or chemical interactions.

Various low-cost adsorbents, derived from agricultural waste, industrial by-product, natural material, or modified biopolymers, have recently been developed and applied for the removal of heavy metals from metal-contaminated wastewater. Technical applicability and cost-effectiveness are the key factors that play major roles in the selection of the most suitable adsorbent to treat these wastewaters.

Platinum Processing

Effluent Characteristics

Different platinum ores have different gangue materials, and different grain sizes of precious metals and sulfides, and thus are processed in different concentrators under different conditions. In platinum processing plants, there are two main sources of effluent streams that can potentially be used for recycling on-site:

1. Thickener overflows, dewatering, and filtration units directly connected to the concentrator (called short recycle or internal recycle waters). These streams usually contain higher levels of suspended solids.
2. Tailings dams and clarification ponds (referred to as long recycle water). Typical contaminants are sulfate, chloride, fluoride, magnesium, calcium, sodium, potassium, sulfides, heavy metals, silicates, iron hydroxide, and natural organic matter. Other contaminant residuals that are present from the processing include frothers, collectors, activators, and depressants. TDS levels also tend to be higher than the raw water, as a consequence of evaporation from the open surfaces.

Treatment for Recycle

The main factors that influence the quality of the water that is available for recycle (Smith and Hertzog 1985) are

- Quality of available makeup water,
- Dissolution of contaminants from the gangue or mineral species,
- Flotation reagents and their degradation products,
- Tailings dam reactions,
- The extent to which recycle water is used and where it is used (this is specific to each metallurgical plant), and
- Biological processes.

In the case of short-recycle waters, the flotation reagents used have not usually had sufficient time in the water circuits to decompose. However, the suspended solids levels tend to be higher, and this can have a consequent negative impact on flotation (Rey and Raffinot 1985).

For long-recycle waters, time effects can become important, due to the delay in the water being returned to the plant (Forssberg and Hallin 1989). Tailings return water tends to have low redox potentials and low oxygen contents due to the oxidation processes taking place in the pond. Most problems connected to the use of recycled water tend to be caused by the process chemical residues and different types of oils that may accumulate in tailings return dams (Forssberg and Hallin 1989).

Because water recycling can cause an increase in total dissolved salts as well as an increase in the specific gravity, this can affect the process slurries. If the specific gravity of the slurry is required to be kept constant, this can result in lower solids percentage and throughput within a plant. Slurry viscosity may also escalate with increasing electrolyte concentrations because of particle aggregation, which can affect mineral floatability as well as classification and pumping.

When recycling water within a process, monitoring the composition of the process water is a necessity. This is needed to identify substances having a negative effect on the process and also to serve as a tool in acquiring information about the process. According to Forssberg and Hallin (1989), chemical consumption decreased when water from the process was reused. However, these recycled waters can cause very stable froths to form' consequently, measures need to be taken to ensure that recycled water causing stable froth formation is not used.

Generally, a flotation plant producing a single concentrate appears to be more amenable to recycling than one in which two or more concentrates are produced (Johnson 2003). Typically, if more than one concentrate is produced, it is essential that the recycle stream(s) be treated, or be matched to the concentrate, to ensure that deleterious reagents are not introduced into the other concentrate's water supply.

If treated sewage effluent is to be used as recycle water, it may need to be treated with activated carbon, or ion exchange, before it can be used in metallurgical processes. Different sources of activated carbon need to be tested to find the one that is most suitable for site-specific TOC removal. For example, sugar-based activated carbon was most suitable for the removal of organic sulfur and organic halides (Ng'andu 2001).

Interestingly, in one set of tests on molybdenite flotation using sewage effluent as a water source, the removal of anions with ion exchange was more beneficial than removal of TOC with activated carbon (Schnitzler et al. 1983). This finding was also observed for platinum flotation when treating process water with an anion exchange resin (Slatter 2001).

Table 3 Guidelines for discharge from platinum processing

Parameter	EPA* Guidelines for 1 Day	EPA† Guidelines for 30 Days
Cadmium, mg/L	0.1	0.05
Copper, mg/L	0.3	0.15
Lead, mg/L	0.6	0.3
Mercury, mg/L	0.002	0.001
Zinc, mg/L	1.0	0.5

Data from EPA 2018
*Maximum for any one day.
†Average of daily values for 30 consecutive days.

Treatment for Discharge

The EPA sets guidelines for discharges from mills using the froth flotation process alone, or in conjunction with other processes, for the beneficiation of platinum ores. The limits shown in Table 3 represent the degree of effluent reduction achievable by application of the BPT that is currently available (EPA 2018).

Effluent water that is excess to the site requirements for recycle will usually contain quantities of heavy metals, dissolved salts, and suspended solids. Some of these parameters will likely exceed the limits for discharge directly to the environment. Therefore, before discharge, site effluents usually require treatment to neutralize any chemicals and precipitate any dissolved metals.

Uranium Processing

Effluent Characteristics

Liquid effluents from uranium mining/processing sites typically contain both radioactive contaminants (uranium and radium) and nonradioactive contaminants (including nickel, arsenic, manganese, magnesium, molybdenum, selenium, fluorides, and sulfates). The characteristics of uranium liquid effluents depend on composition of the ore and the type of mining and processing used to extract the uranium. Local climate can also play a role in treatment and control of effluents.

Treatment for Recycle

Because of the diversity of operating environments and potential contaminating elements and compounds, a single treatment technology may not work for all potential contaminants. Therefore, different combinations of treatment strategies may be required, depending on the chemical composition of the effluents.

Water management within a mining project should start with the characterization of all potential water sources, possible usage, and possible contamination issues. This includes a site water balance analysis that assesses water flows and water quality as well as identifies water recycle options. This water balance analysis would account for seasonal variations and consider the use of cut-off berms, stormwater ponds, and possible evaporation ponds, all based on a probable-maximum-precipitation analysis with a suitable safety margin.

Excess water from mining operations and effluents from mineral processing can be internally recycled to minimize water usage and conserve process chemicals. Typical plant water usage will be on the order of 0.5–2.0 m^3 of water per ton of ore.

Table 4 Guidelines for discharges from uranium processing

Parameter	EPA* Guidelines for 1 Day	EPA† Guidelines for 30 Days
Arsenic, mg/L	1.0	0.5
Carbon oxygen demand, mg/L	—	500
NH_3, mg/L	—	100
pH	6.0–9.0	6.0–9.0
Ra-226 (dissolved), pCi/L	10	3
Ra-226 (total), pCi/L	30	10
Total suspended solids, mg/L	30	20
Zinc, mg/L	1.0	0.5

Data from EPA 2018
*Maximum for any one day.
†Average of daily values for 30 consecutive days.

Treatment for Discharge

The EPA sets guidelines for discharges from mills processing uranium by using acid leach, alkaline leach, or a combined acid leach and alkaline leach process, including sites where in situ leaching is used. The limits shown in Table 4 represent the degree of effluent reduction achievable by application of the BPT that is currently available (EPA 2018).

Effluent water that is in excess of the site requirements for recycle will usually contain quantities of heavy metals or suspended solids that will exceed the limits for discharge directly to the environment. Before discharge, site effluents usually require treatment to neutralize any chemicals, precipitate any dissolved metals, and to precipitate radium. A multistep process (Figure 11) is usually applied:

1. Heavy metals are removed by coagulation and sedimentation.
2. Ph is adjusted.
3. Radium is precipitated with barium chloride.

The selected treatment processes depend on the plant process, type of ore treated, and chemicals employed. After treatment, the water is discharged into holding ponds, where it is analyzed to ensure that the treated effluents meet environmental objectives before release.

Neutralization

Most conventional uranium mills use acid leach systems. Neutralization of the acidic slurries and liquors with lime before discharge is the preferred method for initial effluent treatment. With the exception of radium-226, the removal of heavy metals and radionuclides is typically greater than 99%. However, the resulting sludge contains concentrations of heavy metals, radionuclides, and dissolved salts that well exceed regulatory standards.

The sludge produced retains moisture and requires large storage volumes that limit the quantity of water that is available for recycle. It is generally known in Australia that recycling of the sludge and blending it with lime slurry can result in higher-density sludge with 50%–65% volume reduction. Recovery of up to 16% more neutralized water for reuse was also realized.

Radium Removal

Radium (as Ra-226) is usually still present in the effluent that overflows from the neutralization process, so this requires further specific treatment. Most operations dose barium chloride to remove the radium by precipitation. Also, special resins have now been developed that can efficiently remove radium from dilute streams such as pond water.

In the case of barium chloride dosing, the resulting sludge has high water content and needs large storage cells. In China, research has been conducted suggesting that recycling the sludge from barium chloride treatment reduces reagent consumption and improves settling properties so that less water is retained, reducing storage requirements and improving availability of water for recycling (Zhang et al. 2004a–d). Aerating the sludge produced during neutralization can also lead to formation of manganese hydroxide that absorbs radium, potentially eliminating the barium chloride step and saving on reagent costs (Zhang et al. 2004a–d). In effect, waste treatment is accomplished by using modified waste from the leaching process.

Biological Treatment

In Australia, two operating uranium mines have used wetlands to treat ore stockpile runoff and water from the open pit for 10 years. The principal solutes of concern are UO_2, Mn, NO_3, and SO_4 and, with the exception of sulfates, all are effectively removed in the wetland environment. Carbon limitation of bacterial activity has been identified as the cause of poor sulfate removal efficacy. Addition of biomass from a specific species of green algae has proven effective in promoting sulfate removal (Ring et al. 2004).

Other Treatments

Not all effluents require lime neutralization. For example, uranium extraction using alkaline leaching produces an effluent that mainly contains sodium sulfate and sodium bicarbonate. Physical treatment methods such as evaporation, electrodialysis, and reverse osmosis (RO) can be used. Some waste streams from acid leaching sites such as mine water, drainage from waste dumps, and contaminated surface waters may also be treated with physical methods without neutralization.

In Australia, nanofiltration is effective in removing ions from contaminated water. Nanofiltration is more cost effective than RO, where the removal of multivalent rather than monovalent ions is important. Research has been conducted in China on the use of macropore resins to remove uranium from mine water (Zhang et al. 2004a–d). The macropore resins have better adsorption properties than strong base ion exchange resins that are currently used in most mine water treatment systems.

Coal Processing

Effluent Characteristics

One of the main sources of effluent from coal processing is the coal preparation plant, where the run-of-mine coal undergoes a series of treatments, including washing, wet screening, sedimentation, and dewatering. This part of the site requires the circulation of large volumes of water from which large volumes of wastewater containing a variety of solid particles are generated.

Effluent from coal preparation plants can contain high percentages of ultrafine particles and inorganic impurities, which are composed of clay minerals such as kaolinite, illite, muscovite, quartz, and coal particles. Suspended solids levels in the effluent can range from 5,000 mg/L to as high as 30,000 mg/L. Wastewater can also be generated from other coal site sources such as

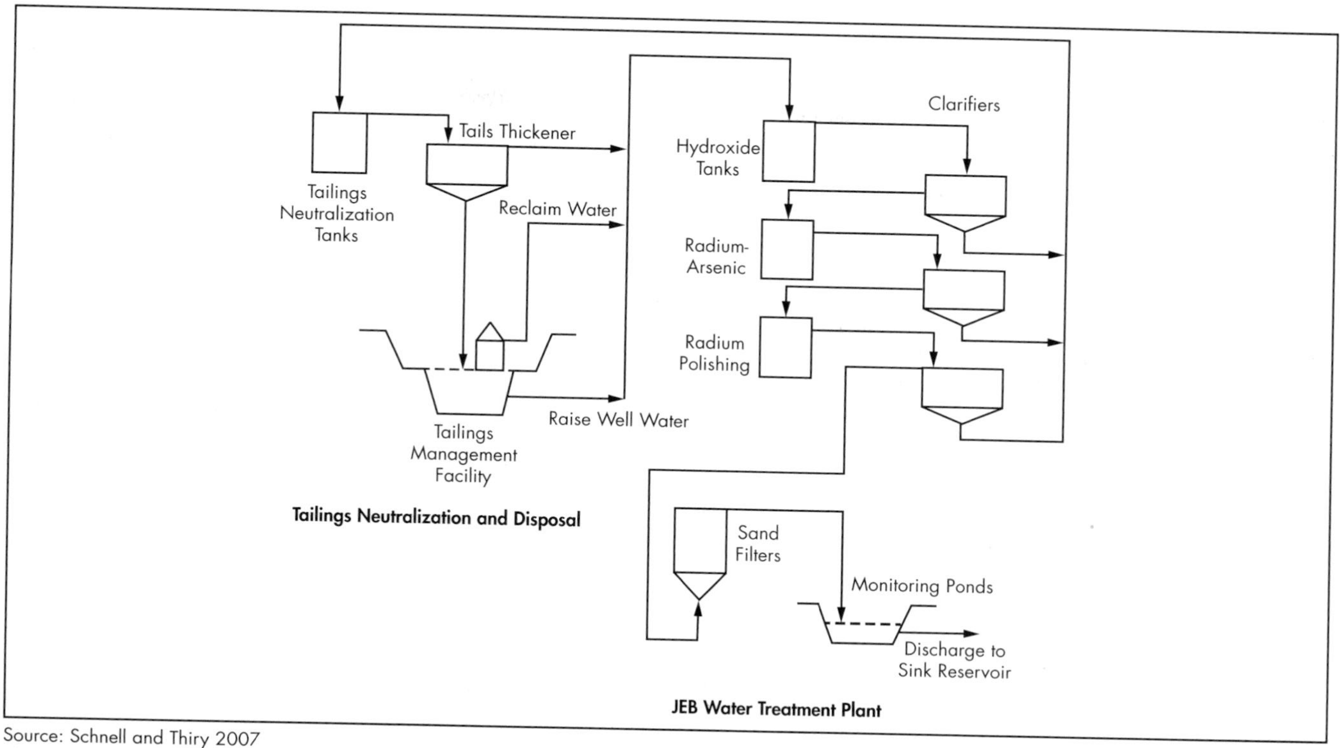

Source: Schnell and Thiry 2007

Figure 11 Typical uranium plant effluent treatment

- Groundwater produced during coal extraction,
- Water used by operators for equipment cooling and dust control,
- Precipitation entering mines, and
- Contaminated stormwater at coal storage facilities.

The combined effluent produced from these sources can be quite variable in quality. Some of these sources can be low pH, with high levels of dissolved salts and heavy metals. Other sources can produce near-neutral wastewater with low concentrations of heavy metals but with high hardness and silica levels.

Table 5 Guidelines for discharge from coal preparation

Parameter	EPA* Guidelines for 1 Day	EPA† Guidelines for 30 Days
Iron (total), mg/L	7.0	3.5
Manganese (total), mg/L	4.0	2.0
pH	6.0–9.0	6.0–9.0
Total suspended solids, mg/L	70	35

Data from EPA 2018

*Maximum for any one day.

†Average of daily values for 30 consecutive days.

Treatment for Recycle

Typical water treatment processes for coal-impacted water would include the following stages of treatment:

1. Removal of fine coal and suspended solids using media filtration or membrane filtration.
2. Removal of heavy metals, silica, phosphate, and hardness by using a chemical precipitation process.
3. Clarification followed by pH adjustment, ultrafiltration, and RO.

Treatment for Discharge

The EPA sets the guideline limits shown in Table 5 for the discharge of wastewater from coal preparation plant water circuits and coal storage, refuse storage, and ancillary areas related to the cleaning or beneficiation of coal of any rank including, but not limited to, bituminous, lignite, and anthracite (EPA 2018).

There are growing concerns that residual processing chemicals used in coal processing may be harmful to the environment. The vast amount of coal is upgraded without being in contact with any chemical additives using density-based separation processes. However, fine coal particles (typically <0.2 mm) are processed using froth flotation circuits, which require small dosages of reagents known as collectors and frothers. Collectors consist of oily hydrocarbons, such as diesel fuel, kerosene, and fuel oil, which are insoluble in water and coat fine coal particles. If required, dosages are typically in the range of 0.1 to 1.0 kg of collector per ton of fine coal processed. Likewise, frothers are added to all flotation systems to promote the formation of small air bubbles and to create a stable froth. Frothers are typically various types of alcohol and polyglycol surfactants.

Flocculation and Sedimentation

The treatment of tailings from the coal preparation plant requires special consideration of the nature of the ultrafine particles and inorganic impurities present. The natural sedimentation rate of these particles in their colloidal and finely divided suspended forms is very slow.

Flocculation technology is applied by using inorganic salts, polymeric flocculants, or both, depending on the chemical and physical characteristics of the pollutants that are present in suspended and dissolved states. Synthetic or natural polymeric reagents are generally used as flocculating agents. To achieve faster settling rates and water clarity, the optimization of process parameters such as suspension pH, polymer type, and dosage are very important. Various flocculent combinations such as cationic, anionic, and non-ionic have been used to achieve the highest settling rate with lowest turbidity.

Das et al. (2006) reported on an investigation into the use of magnetic carrier technology, which is an innovative way of selectively manipulating fine particles in suspension by coating them with a magnetic species. The addition of fine magnetite particles to the flocculation process achieved a significant reduction in final suspended solids after settling.

Reverse Osmosis

The potential for reuse of an effluent can often be limited by the presence of dissolved salts that arise from contact of the process water with the coal within the preparation processes. More recently, several coal mining operations in Australia have installed RO facilities to remove as much dissolved salts as possible, before recycling the treated water back into the site water balance.

Several stages of prefiltration are likely to be needed before RO can be used. Even after a good flocculation and sedimentation process performance, the treated water will still contain some colloidal suspended particles and have concentrations of TSS that are well in excess of that suitable for feeding direct to RO membranes.

Additional stages of filtration will be necessary:

- **Screen filtration.** Screen filters should hold screens with apertures of $\leq$150 µm and be of the automatic backwashable type to ensure good solids removal performance.
- **Membrane filtration such as ultrafiltration.** Hollow-fiber ultrafiltration membranes can accept particle sizes up to 150 µm and feed TSS levels up to 100 mg/L maximum. These membranes will produce a filtrate with a silt density index of <5 that is suitable for RO membrane feed.

ACID MINE DRAINAGE

AMD wastes are commonly formed in coal mines, both in aboveground and underground mines. The following reactions can occur under oxidizing conditions:

$$FeS_2 + 7/2O_2 + H_2O = Fe^{2+} + 2SO_4^{2-} + 2H^+ \quad \textbf{(EQ 6)}$$

$$Fe^{2+} + \tfrac{1}{4}O_2 + H^+ = Fe^{3+} + \tfrac{1}{2}H_2O \quad \textbf{(EQ 7)}$$

$$Fe^{3+} + 3H_2O = Fe(OH)_3 + 3H^+ \quad \textbf{(EQ 8)}$$

$$FeS_2 + 14Fe^{3+} + 8H_2O = 15Fe^{2+} + 2SO_4^{2-} + 16H^+ \quad \textbf{(EQ 9)}$$

The rate of pyrite (FeS_2) oxidation depends on several variables, such as reactive surface area of pyrite, form of pyritic sulfur, oxygen concentrations, solution pH, catalytic agents, and flushing frequencies. Reaction 6 produces hydrogen ions, which significantly reduce the pH levels in the waste. The oxidation reaction (7) is usually the rate-limiting step. This is due to the conversion of ferrous iron to ferric iron being relatively slow at the low pH (<5) levels that are commonly found in these wastes. Iron-oxidizing bacteria (principally *Thiobacillus*) can greatly accelerate this oxidation reaction, so the activities of bacteria are crucial for the generation of AMD.

If there is low porosity and permeability in the strata (e.g., soft shale), the availability of oxygen may be the rate-limiting step. In this case, oxidation will be limited to the upper few meters of the strata. In contrast, for porous and permeable strata (e.g., coarse sandstone), air convection driven by the heat generated from pyrite oxidation may provide high amounts of oxygen deep into the spoil.

The amount of acidity formed can be estimated empirically by using Equation 6:

$$\begin{aligned}\text{acidity (mg/L as } CaCO_3) = 50 \times [&(3 \times \text{Al}/27) \\ &+ (2 \times \text{Mn}/55) + (3 \times \text{Fe}/56) \\ &+ 1{,}000 \times 10 \text{ (power of } -\text{pH})]\end{aligned} \quad \textbf{(EQ 10)}$$

where Al, Mn, and Fe represent the concentrations of aluminum, manganese, and iron, respectively (in milligrams per liter).

The metals content of mine wastewater varies significantly depending on the following factors:

- Geology and geochemistry of the mine environment
- Specific ore being mined
- pH and ORP of the mine water, which governs the solubility of metals
- Climatic conditions

Many metals have an amphoteric property, with decreasing solubility up to a threshold pH, above which the metal solubility increases again because of the formation of soluble complexes. Dissolved metal levels in many sources of mine wastewater often exceed the limits for discharge to the environment. Although elevated sulfate levels have a lower toxicity impact than acidity or metals, sulfate in effluents is increasingly coming under the scrutiny of regulatory authorities.

Lime Neutralization Processes

Although there are many different biological and chemical technologies used for treatment of AMD, lime neutralization remains, by far, the most widely applied treatment method. This is largely due to the high efficiency in removal of dissolved heavy metals and because lime costs are low in comparison to alternatives.

Lime treatment involves bringing the pH of the drainage water to a point where the metals of concern are insoluble and precipitate out in the form of minuscule particles. Figure 9 shows the solubility of metal hydroxides as a function of pH. By controlling pH at 9.5, some of the metals such as iron, aluminum, and copper can be precipitated out to relatively low levels (<0.01 mg/L). However, some other metals may require a higher pH (10.5–11) to effectively precipitate out the hydroxides. Depending on the metal concentrations in the AMD, this may require two or more stages of treatment at different pH levels to achieve the overall limits in the treated water.

A separation process is then required to remove the precipitates and produce a clear effluent that meets regional discharge criteria. It is the method of separation of the precipitates that differs between some of the more recent and complex variations to this AMD treatment method. The older methods use lime less efficiently and do not allow for good control of the treatment system. The more recent processes require a greater capital investment but are considerably more efficient for lime usage and waste production.

The sludge formed depends on the applied process and can contain 1%–30% solids by weight. This sludge must be disposed of in an environmentally acceptable manner. Given that sludge disposal costs can be important, the most advanced processes minimize the volumes by creating an HDS. The sludge disposal and lime costs over the long term usually justify a more important capital investment due to significant savings in operating costs.

A common by-product of lime neutralization is gypsum. Gypsum precipitation occurs because the AMD is often rich is sulfate and the calcium added from lime will bring the solubility product well above saturation. This reaction is often responsible for scaling in treatment processes as well as increasing sludge production at sites where the feed sulfate concentrations can surpass 2,500 mg/L. This scale formation is particularly troublesome for plants where lime is added directly to water that contains significant concentrations of sulfate.

The advantages and disadvantages of several of the different methods of using lime neutralization are discussed in the following sections (Aubé 2004).

Pond Treatment

Pond treatment involves adding lime in a stream or mixing system and allowing the precipitates to settle out within a pond. The pond is often divided into primary and secondary sections. The primary pond serves to accumulate the precipitated sludge and can quickly be filled with solids. These ponds often require yearly dredging of sludge, which then requires a storage area. The secondary pond is usually larger and requires a longer retention time, with laminar flow conditions to allow for "polishing" of the effluent.

Pond treatment systems are often chosen for their simplicity and low capital costs when land is available. They can be used to treat very high flow rates and even important concentrations of metals but require a very large surface area when doing so. The greatest disadvantage of pond treatment systems is their low lime efficiency. A system that uses in-stream addition without any mechanical mixing may have less than 50% efficiency in lime dissolution. By using an agitator and pH control system, the lime usage efficiency can be increased significantly.

To ensure proper treatment in a pond system, the pH set point is often brought up much higher than is necessary for the targeted metals. For example, some pond systems for treating dissolved Zn often control the pH to more than 10.5, rather than the optimum of 9.3. This may necessitate the addition of carbon dioxide or sulfuric acid to reduce the final discharge pH level.

Sludge Recycle Process

In the sludge recycling process, the AMD is neutralized in a reactor tank with the controlled addition of lime to attain a desired pH set point. The slurry is then contacted to a flocculant and fed to a clarifier for solid–liquid separation. Sludge is collected from the bottom of the clarifier and some is recycled back to the reactor tank. The solids in the recycled sludge serve to increase the final sludge density due to precipitation on the surfaces of the existing particles.

The surplus sludge can either be pumped to a storage area or pressure filtered to increase its density prior to transport. The clarifier overflow can normally be released directly, but often a sand filtration system or polishing pond is used to reduce residual suspended solids.

High-Density Sludge Process

The HDS process is commonly used for AMD treatment. Instead of contacting the lime directly with the incoming AMD stream, this system contacts the recycled sludge with the lime slurry for neutralization. To do this, the sludge from the clarifier bottom is pumped to a lime/sludge mix tank, where sufficient lime is added to raise the pH to the desired pH set point. This step forces contact between the solids and promotes coagulation of lime particles onto the recycled precipitates.

The neutralized slurry feeds to the lime reactor where the precipitation reactions are completed. Aeration is often added to this reactor to oxidize ferrous iron to ferric. The slurry then overflows to a floc tank to contact the particles to a flocculant, properly agglomerate all precipitates, and promote efficient settling in the clarifier. The clarifier overflow can either be discharged or polished prior to discharge.

The key to this process lies in the mixing of lime and sludge prior to neutralization of the AMD. Precipitation reactions occur mostly on the surface of existing particles, thereby increasing their size and density. This process is capable of producing sludge of up to 20% solids concentration.

The HDS process minimizes scaling in reactors because the gypsum has other surfaces on which to precipitate. Gypsum is a crystal that preferentially forms on existing gypsum. When sludge is recycled, there are plenty of gypsum needles available in the slurry and these serve as preferred gypsum formation sites. If a plant operates without recycle and significant sulfate is present, some precipitation will occur on the reactor wall surfaces. Once these surfaces are coated, gypsum precipitation is facilitated and the rate of precipitation increases. HDS processes are therefore preferred for high-sulfate effluents, as they will minimize scaling in the reactors.

AMD with Arsenic

Treating AMD that contains arsenic with lime requires a very high pH level and tends to form an unstable sludge due to calcium arsenate. A more acceptable treatment is co-precipitation of arsenic with ferric and/or base metals such as copper and zinc. In all cases, pre-oxidation of As(III) to As(V) may be required, both for improved treatment efficiency and the production of a stable sludge. Many oxidants can be used with varying effectiveness and cost.

When treating with ferric iron, it is preferable to have a two-step process that precipitates most of the arsenic at a slightly acidic pH (4–6). This will form a more stable precipitate (ferric arsenate). In the second step, any other heavy metals present are precipitated and the remaining arsenic is essentially removed from solution.

Sludge recycle can be used to increase treatment efficiency. This can result in significant savings in reagent addition but can also cause the arsenic to be mostly adsorbed as opposed to co-precipitated. Adsorbed arsenic is slightly less stable than when it is co-precipitated as a ferric arsenate. The recycle will also help improve lime efficiency. An excess of iron can be applied with any process to improve sludge stability. The final sludge disposal system can weigh into the decision as to whether a recycle is applied and what ratio of iron to arsenic is used.

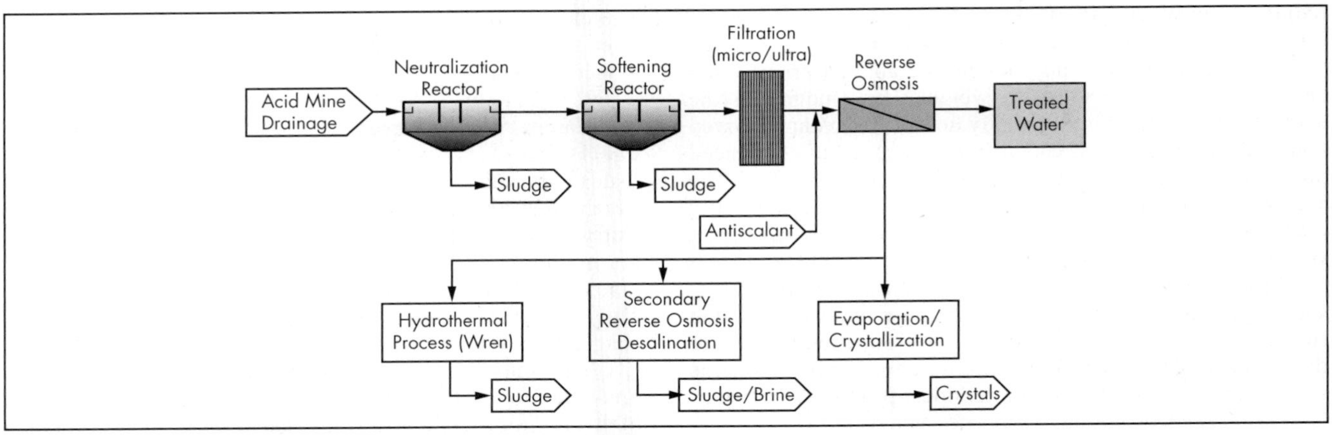

Source: INAP 2009

Figure 12 Typical flow sheet for treating AMD streams

AMD with Molybdenum

Molybdenum does not precipitate with lime addition as do the typical heavy metals. The most widely applied treatment process for AMD containing molybdenum is ferric co-precipitation. An example is given by Aubé and Stroiazzo (2000), where ferric sulfate is added to the alkaline AMD (about 3 mg/L of Mo) in the first step, and the pH is controlled to 4.5 in the second step. A flocculant is added to improve solid–liquid separation in the clarifier. To achieve a very low molybdenum level in the discharge (0.03 mg/L), a sand filter is also used to remove all residual molybdenum from the effluent.

AMD Treatment by Reverse Osmosis

RO is widely used in South America for treating AMD waters (Chesters et al. 2016). A typical flow sheet for treating AMD water is shown in Figure 12. Chesters et al. undertook a survey of 67 membrane plants worldwide and a summary of the information is provided in Tables 6 and 7. At the time, 25 plants were in Latin America, principally in Peru and Chile. The number of plants in Chile was boosted by 12 seawater RO units. Gold and copper account for 69% of the mines that have invested in RO plants. The largest application of RO plants is for AMD treatment. Only 15 of the RO plants were specifically designed for enhanced metal recovery.

Table 6 Membrane plans by geography and mine type

Location	Number of Plants	Type of Mine	Number of Plants
Latin America	25	Gold	30
Africa	18	Copper	16
North America	15	Coal	6
Australasia	7	Diamonds	3
Europe	1	Iron	3
Asia	1	Others	6

Source: Chesters et al. 2016

Table 7 Treatment by feed water type

Treated Water	Number of Plants
AMD cleanup	22
Seawater reverse osmosis	17
Metal recovery	15
Drinking water	8
Leachate	5

Source: Chesters et al. 2016

Sludge Disposal

The choice of method for sludge disposal depends on many factors, including regulatory considerations, sludge stability, sludge production rate, space availability, budget, and aesthetic considerations. Disposal in mine workings is often the least-expensive option but this choice requires more public and regulatory approval.

Other options for sludge disposal include disposal with tailings in engineered ponds, on waste rock piles, under a water cover, or in natural land depressions. Disposal with tailings can be done for operating mines. Another option is to dispose of sludge by placing it on the top of a closed tailings pile. Covering tailings with sludge may help reduce oxidation by providing a wet barrier, similar to that of engineered soil covers. Unfortunately, these options have not yet been sufficiently researched to confirm that this sludge would not eventually redissolve if tailings oxidation continues.

Engineered pond disposal is more expensive at first, but eventually results in the lowest volume of sludge generated. This is because these ponds can be designed to both drain water from the bottom and allow for evaporation from the top. Most sludges will densify further up to two or three times their original solid content. Therefore, some sites allow their sludge to densify in an engineered pond for a year or two before transporting it to a different location (on tailings or in mine workings).

Innovative Technologies

In recent years, several innovative technologies for treatment of AMD have been put into operation, such as permeable reactive barriers (PRBs), bioreactors, and constructed aerobic wetland technologies. Case studies on these innovative technologies are detailed by Costello (2003). The majority are in situ applications that manipulate natural processes to treat acidic and/or metals-contaminated water. The differences lie in their construction and their water source. Both bioreactors and wetlands almost always include collection and piping systems, whereas PRBs are simply placed in the flow path.

Permeable Reactive Barriers

PRBs have a subsurface reactive section through which groundwater flows as it follows its natural course. In some cases, there are impermeable walls to direct the flow of the water to the reactive section. The reactive media is usually compost that hosts sulfate-reducing bacteria, although there are a few other types of media used.

Bioreactors

In the case of bioreactor systems, the water (ground or surface) to be treated flows through a media bed, and natural biological reactions work to remove dissolved metals. Whether subsurface or exposed to the atmosphere, bioreactors are generally lined, filled with composted materials and/or alkaline agents, and, in some situations, include vegetation.

Aerobic Wetlands

Constructed aerobic wetlands are shallow (10–30 cm deep) and lined ponds that are filled with organic matter and/or alkaline agents and sometimes vegetation. Aerobic wetlands are most often used to treat alkaline waters that are high in dissolved iron and have low capacity to neutralize the acidity.

Oxygen infiltration is encouraged because it flows slowly through vegetation. Iron and other metals precipitate as oxyhydroxides, hydroxides, and carbonates by the mechanisms of absorption and adsorption. Aerobic wetlands can also remove manganese, but oxidation of manganese only starts when oxidation of iron is completed. To remove manganese, it is necessary to have a big pond area, which allows complete iron oxidation, or to add another wetland cell (Zipper et al. 2011).

REFERENCES

Akcil, A. 2003. Destruction of cyanide in gold mill effluents: biological versus chemical treatments. *Biotechnol. Adv.* 21:501–511.

Arnold, B.J., and Aplan, F.F. 1986. The effect of clay slimes on coal flotation, part II: The role of water quality. *Int. J. Miner. Process.* 17(3-4):243–260.

Aubé. B. 2004. The Science of Treating Acid Mine Drainage and Smelter Effluents. EnvirAube.

Aubé, B.C., and Stroiazzo, J. 2000. Molybdenum treatment at Brenda Mines. In *ICARD 2000: Proceedings of the 5th International Conference on Acid Rock Drainage.* Littleton, CO: SME.

Aubé, B., and Zinck, J. 2003. Comparison of AMD treatment processes and their impact on sludge characteristics. In *Proceedings of Sudbury '99: Mining and the Environment II.* Sudbury: ON: Laurentian University.

Barakat, M.A. 2011. New trends in removing heavy metals from industrial wastewater. *Arabian J. Chem.* 4:361–377.

Bickert, G. 2013. *Seal, Spray and Wash Water Filtration in Coal Preparation Plants.* Brisbane, Queensland: Australian Coal Industry Research Program.

Botz, M.M., and Mudder, T.I. 1998. Cyanide recovery for silver leaching operations; application of CCD-AVR circuits. In *Proceedings of the Randol Gold and Silver Forum.* Golden, CO: Randol International. pp. 295–297.

Botz, M.M., and Mudder, T.I. 2000. Modeling of natural cyanide attenuation in tailings impoundments. *Miner. Metall. Process.* 17(4).

Chesters, S.P., Morton, P., and Fazel, M. 2016. Membranes and minewater—Waste or revenue stream. In *Proceedings of IMWA 2016: Mining Meets Water—Conflicts and Solutions.* Freiberg: Technische Universität Bergakademie. pp. 1310–1322.

Coetzer, G., du Preez, H.S., and Bredenhann, R. 2003. Influence of water resources and metal ions on galena flotation of Rosh Pinah ore. *J. SAIMM* (April).

Coleman, R.T., Colley, J.D., Klausmeier, R.F., Meserole, N.P., Micheletti, W.C., and Schwitzgebel, K. 1980. *Sources and Treatment of Wastewater in the Nonferrous Metals Industry.* EPA-600/2-80-074. Cincinnati, OH: U.S. Environmental Protection Agency, Industrial Environmental Research Laboratory, Office of Research and Development.

Costello, C. 2003. *Acid Mine Drainage: Innovative Treatment Technologies.* Washington, DC: U.S. Environmental Protection Agency, Office of Solid Waste and Emergency Response.

Das, B., Prakash, S., Biswal, S.K., and Reddy, P.S.R. 2006. Settling characteristics of coal washery tailings using synthetic polyelectrolytes. *J. SAIMM* 106(October).

Diaz, M., Cavero, A., and Yataco, D. 1998. Water management at Minera Yanacocha. In *Latin American Perspectives: Exploration, Mining, and Processing.* Littleton, CO: SME. pp. 329–340.

Dunstan, A. 1999. The effects of water quality on mineral processing. In *Proceedings of WAMMO '99*, August 24–25. pp. 236–242.

EPA (U.S. Environmental Protection Agency). 2018. International cooperation. https://www.epa.gov/international-cooperation. Accessed July 2018.

Evangelho, M.R., Gonçalves, M.M.M., Sant'Anna Jr. G.L., and Villas Bôas, R.C. 2001. A trickling filter application for the treatment of a gold milling effluent. *Int. J. Miner. Process.* 62(1-4):279–292.

Fleming, C.A., and Trang, C.V. 1998. R eview of options for cyanide recovery at gold and silver mines. In *Proceedings of the Randol Gold and Silver Forum.* Golden, CO: Randol International. pp. 313–318.

Forssberg, K.S.E., and Hallin, M.I. 1989. Process Water Recirculation in a Lead-Zinc Plant and other sulphide flotation plants. In *Challenges in Mineral Processing.* Edited by K.V.S. Sastry and M.C. Fuerstenau. Littleton, CO: SME.

Given, B., Dixon, B., Douglas, G., Mihoc, R., and Mudder, T. 1998. Combined aerobic and anaerobic biological treatment of tailings solution at the Nickel Plate mine. In *The Cyanide Monograph.* London: Mining Journal Books.

Greenway, M., and Simpson, J.S. 1996. Artificial wetlands for wastewater treatment, water reuse and wildlife in Queensland, Australia. *Water Sci. Technol.* 33(10-11):221–229.

Halbe, D., Trautman, L., Neville, R.G., Hart, L., and Omnen, R. 1979. Destruction of cyanide in Homestake gold mill effluent—A status report. Presented at AIME Fall Meeting, October.

Hoover, M.R. 1980. Water chemistry effects in the flotation of sulphide ores—A review and discussion for molybdenite. In *Complex Sulphide Ores Conference.* Edited by M.J. Jones. London: Institution of Mining and Metallurgy. pp. 100–112.

INAP (International Network for Acid Prevention). 2009. *Global Acid Rock Drainage Guide*. Rev. 1. www.gardguide.com.

Johnson, N.W. 2003. Issues in maximisation of recycling of water in a mineral processing plant. In *Proceedings of Water in Mining 2003*. Melbourne, Victoria: Australasian Institute of Mining and Metallurgy. pp. 239–245.

Kirjavainen, V., Schreithofer, N., and Heiskanen, K. 2002. The effect of calcium and thiosulphate ions on flotation selectivity of nickel-copper ores. *Miner. Eng.* 15:1–5.

Kitis, M., Akcil, A., Karakaya, E., and Yigit, N.O. 2005. Destruction of cyanide by hydrogen dioxide in tailings slurries from low bearing sulphidic gold ores. *Miner. Eng.* 18(3):353–362.

Larsen, M., Trapp, S., and Pirandello, A. 2004. Removal of cyanide by woody plants. *Chemosphere* 54:325–333.

Lawrence, R.W., Poulin, R., Kalin, M., and Bechard, G. 1998. The potential of biotechnology in the mining industry. *Min. Process Extr. Metall. Rev.* 19:5–23.

Lorenzen, L. 2015. International trends on effluent treatment and management in the Gold Mining Industry. University of Pretoria, Republic of South Africa, November.

Lotter, N. 2005. Cyanide volatilisation from gold leaching operations and tailing storage facilities. University of Pretoria, Republic of South Africa, November.

Madin, C. 2007. Innovative treatment of gold mine water for sustainable benefit—Processes and case studies. In *World Gold 2013*. Melbourne, Victoria: Australasian Institute of Mining and Metallurgy.

Malysiak, V., Shackleton, N.J., and de Vaux, D. 2003. Effect of water quality on pentlandite-pyroxene floatability with an emphasis on calcium ions. In *Proceedings of the 22nd International Mineral Processing Congress*. Marshalltown, South Africa: Southern African Institute of Mining and Metallurgy.

Marsden, J.O., and House, C.I. 2009. *The Chemistry of Gold Extraction*. Littleton, CO: SME.

McGrath, T., Simons, A., Dunne, R., and Staunton, W.P. 2015. Cyanide recovery using SART—Current status. In *World Gold 2015*. Johannesburg: Southern African Institute of Mining and Metallurgy.

Meyer, K. 1992. The David Bell mine water quality management strategy. *CIM Bull.* (September):27–30.

Monteagudo, J.M., Rodriguez, L., and Villasenor, J. 2004. Advanced oxidation processes for destruction of cyanide from thermoelectric power station waste waters. *J. Chem. Technol. Biotechnol.* 79:117–125.

Mosher, J.B., and Figueroa, L. 1996. Biological oxidation of cyanide: A viable treatment option for the minerals processing industry? *Miner. Eng.* 9(5):573–581.

Mudder, T., Fox, F., Whitlock, J., Fero, T., Smith, G., Waterland, R., et al. 1998. The Homestake wastewater treatment process, part 2: Operation and performance. In *The Cyanide Monograph*. London: Mining Journal Books. pp. 365–388.

Mudder, T., Botz, M., and Smith, A. 2001. *The Cyanide Compendium* (CD-ROM). London: Mining Journal Books.

Muzenda, E. 2010. An investigation into the effect of water quality on flotation performance. *Int. J. Chem. Mol. Nucl. Mater. Metall. Eng.* 4(9).

Nelson, M.G., Kroeger, E.B., and Arps, P.J. 1998. Chemical and biological destruction of cyanide: comparative costs in a cold climate. *Min. Process Extr. Metall. Rev.* 19:217–26.

Ng'andu, D.E. 2001. The effect of underground mine water on the performance of the Mufulira flotation process. *J. SAIMM* (October).

NRC (National Research Council). 2012. *Uranium Mining in Virginia: Scientific, Technical, Environmental, Human Health and Safety, and Regulatory Aspects of Uranium Mining in Virginia*. Washington, DC: National Academies Press.

Ofori, P., Firth, B., McNally, C., and Nguyen, A. 2009. *Working Effectively with Saline Water in Coal Preparation—Report and Handbook*. Brisbane, Queensland: Australian Coal Industry Research Program.

Pourbaix, M. 1974. *Atlas of Electrochemical Equilibria in Aqueous Solutions*, 2nd English ed. Houston, TX: National Association of Corrosion Engineers.

Raybuck, S.A. 1992. Microbes and microbial enzymes for cyanide degradation. *Biodegradation* 3:3–18.

Rey, M., and Raffinot, P. 1985. Flotation of ores in seawater. In *Water Management and Treatment for Mining and Metallurgical Operations*. Vol. 6. Edited by H. von Michaelis. Golden, CO: Randol International. pp. 3167–3174.

Ring, R.J., Holden, P., McCulloch, J.K., Tapsell, G.J., Collier, D.E., Russell, R., MacNaughton, S., Marshall, K., and Stimson, D. 2004. Treatment of liquid effluent from uranium mines and mills during and after operation. In *Treatment of Liquid Effluent from Uranium Mines and Mills: Report of a Co-Ordinated Research Project 1996–2000*. IAEA-Tecdoc-1419. Vienna, Austria: International Atomic Energy Agency.

Ritcey, G.M. 2004. Prediction of weathering and contaminant migration in mine/mill tailings. In *Proceedings of the VI International Conference on Metallurgy, Refractories and Environment*. Edited by P. Palfy, M. Solc, and E. Vircikova. Slovakia: Kosice.

Ritcey, G.M. 2005. Tailings management in gold plants. *Hydrometallurgy* 78:3–30.

Rouse, J., and Pyrih, R. 1998. Natural geochemical attenuation of trace elements in migrating precious-metal process solutions. In *Proceedings of the Randol Perth International Gold Conference*. Golden, CO: Randol International.

Rowly, W.J., and Otto, F.D. 1980. Ozonation of cyanide with emphasis on gold mill waste waters. *Can. J. Chem. Eng.* 58(5):646–653.

Rytter, R.-M., and Hansson, A.-C. 1996. Seasonal amount, growth and depth distribution of fine roots in an irrigated and fertilized *Salix viminalis* L. plantation. *Biomass Bioenergy* 11:129–137.

Schmidt, J.W., Simovic, L., and Shannon, E. 1981. Natural degradation of cyanides in gold milling effluents. Seminar on Cyanide and Gold Mining Industry, Environment Canada, Ottawa, ON.

Schnell, H., and Thiry, J. 2007. Processing of high grade uranium ore. Presented at 2007 SME Annual Meeting, February 26, Denver, CO.

Schnitzler, M., Lévay, G., Kühn, W., and Sontheimer, H. 1983. Determination of organic halogen and sulphur compounds in water by group parameters. *Vom Wasser* 61:263–276.

Shackleton, N.J., Malysiak, V., and Slatter, K.A. 2001. Effect of water quality and surface passivation on flotation behaviour. Anglo American Metallurgical Symposium, November 1, Session 4, Paper 2.

Slatter, K.A. 2001. *Flotation of Waterval Mill Feed with Filtered Process Water and Process Water Treated by Ion Exchange.* Internal Report MP/09/01.

Slatter K.A., et.al. 2009. Water management in Anglo Platinum process operations: Effects of water quality on process operations. In *Proceedings of the International Mine Water Conference*, Pretoria, South Africa, October 19.

Smart, R.St.C., Skinner, W.M., and Levay, G. 1999. Process water quality and treatment: issues and methodology. In *Proceedings of WAMMO '99.* pp. 226–235.

Smith, A., and Mudder, T. 1991. *Chemistry and Treatment of Cyanidation Wastes.* London: Mining Journal Books.

Smith, M., and Hertzog, L.D. 1985. Seawater in flotation. In *Water Management and Treatment for Mining and Metallurgical Operations.* Vol. 6. Edited by H. von Michaelis. Golden, CO: Randol International. pp. 3163–3164.

Whitlock, J.L., and Mudder, T.I. 1986. The Homestake wastewater treatment process: biological removal of toxic parameters from cyanidation wastewaters and biossay effluent evaluation. In *Fundamental and Applied Biohydrometallurgy.* Edited by R.W. Lawrence, R.M.R. Branion, and H.G. Ebner. Amsterdam: Elsevier. pp. 327–339.

Zhang, J., Chen, S., Rao, S., and Qi, J. 2004a. Analysis and evaluation of water coming from several uranium processing areas. In *Treatment of Liquid Effluent from Uranium Mines and Mills: Report of a Co-Ordinated Research Project 1996–2000.* IAEA-Tecdoc-1419. Vienna, Austria: International Atomic Energy Agency.

Zhang, J., Chen, S., Rao, S., and Qi, J. 2004b. Barium chloride precipitation-sludge recycle to treat acidic uranium industrial effluent. In *Treatment of Liquid Effluent from Uranium Mines and Mills: Report of a Co-Ordinated Research Project 1996–2000.* IAEA-Tecdoc-1419. Vienna, Austria: International Atomic Energy Agency.

Zhang, J., Chen, S., and Qi, J. 2004c. Research on the removal of radium from uranium effluent by air-aeration hydrated manganese hydroxide adsorption. In *Treatment of Liquid Effluent from Uranium Mines and Mills: Report of a Co-Ordinated Research Project 1996–2000.* IAEA-Tecdoc-1419. Vienna, Austria: International Atomic Energy Agency.

Zhang, J., Chen, S., Qi, J., and Rao, S. 2004d. Study on the technology for the development of macro-porous resin adsorption for high purification of uranium effluent. In *Treatment of Liquid Effluent from Uranium Mines and Mills: Report of a Co-Ordinated Research Project 1996–2000.* IAEA-Tecdoc-1419. Vienna, Austria: International Atomic Energy Agency.

Zipper, C., Skousen, J., and Jage, C. 2011. Passive treatment of acid mine drainage. In *Powell River Project: Reclamation Guidelines for Surface Mined Land.* Virginia Cooperative Extension Publication 460-133. Blacksburg, VA: Virginia Tech.

CHAPTER 9.3

Treatment of Effluent Waste

Larry G. Twidwell

The *SME Mineral Processing Handbook* was published in 1985 (Weiss 1985). As predicted in that volume, the U.S. Environmental Protection Agency (EPA) has promulgated extensive additions and revisions to environmental regulations with respect to the mineral processing and extractive metallurgical industries; for example, Congress authorized the addition of the Hazardous and Solid Waste Amendments (EPA 1984) to the Resource Conservation and Recovery Act (EPA 2014c). These amendments directed the EPA to more aggressively manage some hazardous waste, and the RCRA restricted (banned) some hazardous waste from disposal without further stabilization. This initiated the land ban restrictions (LBR). Other environmental regulations have been formulated through the Clean Water Act (CWA) and its amendments, the Comprehensive Environmental Response, Compensation, and Liability Act (CERCLA, Superfund), and the Clean Air Act and its amendments that have significant consequences to the extractive metallurgical industries. The goal of this chapter is to supply a guide to the literature so that the user can evaluate and select appropriate technologies to treat effluent waste solutions and dispose of solid waste products in an environmentally safe manner.

The Society for Mining, Metallurgy & Exploration (SME) supports responsible environmental protection (Darling 2011):

> SME members recognize that there are environmental and social impacts that result from mining. Further, the industry strives to find innovative ways, based upon sound science and engineering principles, to explore the Earth, extract essential, critical and strategic minerals and to design and operate processing and manufacturing facilities that eliminate or minimize any adverse impacts. SME supports sustainable development principles that seek to return mined lands to useful and useable public and private lands for future generations. SME also recognizes the need for federal, state and local regulations to control and limit impacts of mining operations. There needs to be clear, concise and consistent interpretation and application of laws, regulations and guidance by the government that apply to the mining industry to assure environmental and economic justice for all citizens. In turn, the mining industry has a responsibility to participate in the public notice and comment process for new rules and then to comply with those rules and regulations that have been promulgated.

Hasen (2015) stated the following in an SME technical briefing:

> The mining, mineral processing and metallurgical industries support, and strictly follow, state and federal regulations to ensure protection of the environment and the health of industry workers.

This chapter focuses on the treatment of effluent wastewaters and solid waste, deemed in the literature as environmental hydrometallurgy. Doyle (2005) has characterized environmental hydrometallurgy as follows:

> *Environmentally compliant hydrometallurgy* refers to traditional hydrometallurgical processes (i.e., the production of metals and metal-bearing compounds using aqueous processing), but performed in such a way as to ensure little or no adverse impact on the environment.... *Hydrometallurgy for environmental compliance* covers hydrometallurgical processes used to assist in complying with waste discharge and management regulation, along with processes that lower the overall environmental impact of a particular process.

Both approaches are widely practiced in mineral and extractive metallurgical processing. Doyle makes an important point that hydrometallurgical unit operations are presently being used to treat waste solutions and/or waste solids to lower the impact of waste disposal by recycling treated solutions or solids to a process.

Larry G. Twidwell, (Retired) Emeritus Professor of Metallurgical Engineering, Metallurgy/Materials Eng. Dept., Montana Tech, Butte, Montana, USA

DEFINITIONS OF MINING, BENEFICIATION, AND MINERAL PROCESSING WASTE

The EPA, through the RCRA, regulates hazardous waste under Subtitle C, and individual states regulate nonhazardous waste under Subtitle D. Congress required that RCRA exclude "solid waste from the extraction, beneficiation, and processing of ores and minerals from regulation as a hazardous waste" (Housman 1999). In 1980, Congress passed the Bevill Amendment, which amended the RCRA to exempt these wastes from regulation under Subtitle C. For this reason, these wastes are commonly designated as Bevill wastes. See Housman (1999) and Luther (2013) for detailed histories of the selection of "special" excluded waste from regulation as a hazardous waste. EPA's excluded special wastes are presented in Table 1 (EPA 2009), and excluded mining and mineral processing wastes are listed in Table 2 (EPA 2009).

The 20 mineral processing wastes (designated Bevill wastes) listed in Table 2 are excluded from the EPA hazardous waste listing. In 40 CFR Section 261.4(b), the EPA (2009) states the following:

> Solid waste from the extraction, beneficiation, and processing of ores and minerals (including coal, phosphate rock, and overburden from the mining of uranium ore), except as provided by Sec. 266.112 of this chapter for facilities that burn or process hazardous waste.
>
> (i) For purposes of Sec. 261.4(b)(7) beneficiation of ores and minerals is restricted to the following activities; crushing; grinding; washing; dissolution; crystallization; filtration; sorting; sizing; drying; sintering; pelletizing; briquetting; calcining to remove water and/or carbon dioxide; roasting, autoclaving, and/or chlorination in preparation for leaching (except where the roasting (and/or autoclaving and/or chlorination)/leaching sequence produces a final or intermediate product that does not undergo further beneficiation or processing); gravity concentration; magnetic separation; electrostatic separation; flotation; ion exchange; solvent extraction; electrowinning; precipitation; amalgamation; and heap, dump, vat, tank, and in situ leaching.
>
> (ii) For the purposes of Sec. 261.4(b)(7), solid waste from the processing of ores and minerals includes only the following wastes as generated:
>
> (iii) A residue derived from co-processing mineral processing secondary materials with normal beneficiation raw materials or with normal mineral processing raw materials remains excluded under paragraph (b) of this section if the owner or operator:
>
> (A) Processes at least 50 percent by weight normal beneficiation raw materials or normal mineral processing raw materials; and,
>
> (B) Legitimately reclaims the secondary mineral processing materials.

The Bevill exclusion exempts the 20 wastes listed in Table 2 from Subtitle C hazardous waste regulation; however, it does not provide protection from liability under other EPA regulations, including CERCLA and the CWA. Luther (2013) states that

> Exemption from Subtitle C does not mean the waste is unregulated. As noted above, the waste is subject to other state or federal regulatory requirements. Those "other" requirements would include any established by individual states, including requirements established under their solid waste management programs. Potentially applicable federal regulatory requirements include those established under the

Table 1 EPA exclusions for "special" waste

§261.4(b)(1) Household Hazardous Waste
§261.4(b)(2) Agricultural Waste
§261.4(b)(3) Mining Overburden
§261.4(b)(4) Fossil Fuel Combustion Waste (Bevill)
§261.4(b)(5) Oil, Gas, and Geothermal Wastes (Bentsen Amendment)
§261.4(b)(6) Trivalent Chromium Wastes
§261.4(b)(7) Mining and Mineral Processing Wastes (Bevill)
§261.4(b)(8) Cement Kiln Dust (Bevill)
§261.4(b)(9) Arsenically Treated Wood
§261.4(b)(10) Petroleum Contaminated Media and Debris from Underground Storage Tanks
§261.4(b)(11) Injected Groundwater
§261.4(b)(12) Spent Chlorofluorocarbon Refrigerants
§261.4(b)(13) Used Oil Filters
§261.4(b)(14) Used Oil Distillation Bottoms
§261.4(b)(15) Landfill Leachate or Gas Condensate Derived from Certain Listed Wastes

Source: EPA 2009
Note: Emphasis added by chapter author.

Table 2 EPA exclusions for mining and mineral processing wastes (Bevill)

(A) Slag from primary copper processing
(B) Slag from primary lead processing
(C) Red and brown muds from bauxite refining
(D) Phosphogypsum from phosphoric acid production
(E) Slag from elemental phosphorus production
(F) Gasifier ash from coal gasification
(G) Process wastewater from coal gasification
(H) Calcium sulfate wastewater treatment plant sludge from primary copper processing
(I) Slag tailings from primary copper processing
(J) Fluorogypsum from hydrofluoric acid production
(K) Process wastewater from hydrofluoric acid production
(L) Air pollution control dust/sludge from iron blast furnaces
(M) Iron blast furnace slag
(N) Treated residue from roasting/leaching of chrome ore
(O) Process wastewater from primary magnesium processing by the anhydrous process
(P) Process wastewater from phosphoric acid production
(Q) Basic oxygen furnace and open hearth furnace air pollution control dust/sludge from carbon steel production
(R) Basic oxygen furnace and open hearth furnace slag from carbon steel production
(S) Chloride process waste solids from titanium tetrachloride production
(T) Slag from primary zinc processing

Source: EPA 2009
Note: Emphasis added by chapter author.

Clean Water Act (CWA) and Safe Drinking Water Act (SDWA). Commonly implemented by authorized states, both CWA and SDWA requirements apply to the management of some Bevill-Bentsen waste. For example, CWA requires that discharges of pollutants to surface waters (e.g., wastewater discharges to a river, bay, or ocean) must be authorized by a permit issued under the National Pollutant Discharge Elimination System (NPDES) program. Wastewater discharges to publicly owned treatment works (POTWs) are also subject to NPDES permitting requirements. Also, the SDWA regulates subsurface injection of fluids, including wastewater, pursuant to regulations established under the Underground Injection Control (UIC) program.

In general, the EPA views mining and beneficiation solids that are exempted from Subtitle C hazardous waste management to be those solids that have undergone physical changes but have not been transformed by chemical or temperature actions. The basic distinction between *beneficiation* and *mineral processing* wastes, as stated by Housman (1999), is as follows:

> Beneficiation operations typically serve to separate and concentrate the mineral values from waste material, remove impurities, or prepare the ore for further refinement. Beneficiation activities generally do not change the mineral values themselves other than by reducing (e.g., crushing or grinding), or enlarging (e.g., pelletizing or briquetting) particle size to facilitate processing. Where heat or chemicals, such as acid, are applied in a beneficiation operation, it is generally to drive off impurities (e.g., water), dissolve mineral values in a solution as a means of separation (leaching), or retrieve dissolved values from a solution (e.g., crystallization or solvent extraction). A chemical change in the mineral value does not typically occur in beneficiation.
>
> Mineral Processing operations, in contrast, generally follow beneficiation and serve to change the concentrated mineral into a more useful chemical form. This is often done by using heat or chemical reactions to change the chemical composition of the mineral. In contrast to beneficiation operations, processing activities often destroy the physical structure of the incoming ore or mineral feedstock such that the materials leaving the operation do not closely resemble those that entered the operation. Typically, beneficiation wastes are earthen in character, whereas mineral-processing wastes are derived from melting and chemical changes.

NON-EXCLUDED MINERAL PROCESSING WASTE

All mineral processing solid wastes not listed in Table 2 are not excluded as hazardous waste; and treatment, handling, storage, and disposal of those wastes must abide by the RCRA regulations for hazardous waste. This is also true for some extractive metallurgical wastewater and waste solids. Hazardous solid waste is categorized as "listed" waste and "characteristic" waste.

Listed Waste

The EPA has determined that some wastes are hazardous, and they must be handled in compliance with the RCRA established hazardous waste regulations. These wastes are designated as listed wastes and are categorized as *F* (wastes from nonspecific sources; EPA 1981a), *K* (wastes from specific sources; EPA 1981b), and *P* and *U* (discarded commercial chemical products; EPA 1981c). Handling and disposal of these wastes must follow Subtitle C regulations with respect to treatment, storage, and disposal requirements (EPA 2014a). The EPA has specified best demonstrated available technologies (BDATs) for all the listed wastes, both wastewater and non-wastewater (solids). The BDAT documents support the selection of the appropriate treatment technology to minimize mobility or toxicity of the hazardous constituents, and the EPA has established standards that all listed and new hazardous waste must meet. These standards must be met for the designated solid waste or wastewater; however, alternative treatments can be used if the resulting toxicity characterization leach procedure (TCLP) tests confirm that the hazardous constituent is controlled to levels less than the designated standard. BDAT documentation reports for all listed wastes can be found using the National Service Center for Environmental Publications (EPA 2015a).

Characteristic Waste

If a waste is not a listed waste, it may still be deemed a hazardous characteristic waste. A waste is defined as a characteristic waste by evaluating whether it shows any or all of four characteristics:

1. Ignitability (D001; EPA 2012a)
2. Corrosivity (D002; EPA 2012b)
3. Reactivity (D003; EPA 2012c)
4. Toxicity (D004, EPA 2012d)

The toxicity characteristic designation for wastewaters and waste solids created in some of the mineral processing and extractive metallurgical industries is of special importance. Toxicity of a waste solid is determined by a single test protocol, the TCLP. If a wastewater exceeds the TCLP listed concentrations, then it is deemed a hazardous water without being subjected to the TCLP protocol and it must be treated. The TCLP method is designated as EPA Method 1311. The TCLP test protocol consist of exposing a waste solid material to a buffered glacial acetic acid leach procedure (EPA 1992). Eight characteristic elements, four pesticides, two herbicides, and 26 organic compounds must be evaluated by the TCLP test. The results of applying the TCLP test to a solid waste determine whether the waste must be designated as hazardous or not; if the solution concentration in the TCLP filtrate contains greater than 100 times the maximum concentration level (MCL, drinking water standard), then the waste is deemed hazardous. The "characteristic" regulatory concentrations for the eight elements are presented in Table 3. Some states require a different or additional test procedure and additional element analyses; for example, California requires the waste extraction test (WET) and designates the soluble threshold limit concentrations as its regulatory concentrations. California's "characteristic" regulatory concentrations are set for 17 elements, 10 pesticides, and 10 organic compounds. The regulatory levels for the elements are presented in Table 4

(DTSC 2015a). A comparison of the differences between the TCLP test and WET is presented in Table 5 (DTSC 2015b).

Land Disposal Restrictions

According to the EPA (1998, 2005)

> The primary goal of the Resource Conservation and Recovery Act (RCRA) Subtitle C program is to protect human health and the environment from the dangers associated with generation, transportation, treatment, storage, and disposal of hazardous waste. Disposal of hazardous waste on the land is a practice of particular concern to the RCRA program. Land disposal units, such as landfills and surface impoundments, must comply with stringent requirements for liners, leak detection systems, and groundwater monitoring. The land disposal restrictions (LDR) provide a second measure of protection from threats posed by hazardous waste disposal. The LDR program ensures that hazardous waste cannot be placed on the land until the waste meets specific treatment standards to reduce the mobility or toxicity of the hazardous constituents in the waste.

Table 3 TCLP regulatory levels

Metals	TCLP Regulatory Level, mg/L	EPA Hazardous Waste Designation
As	5.0	D004
Ba	100.0	D005
Cd	1.0	D006
Cr	5.0	D007
Pb	5.0	D008
Hg	0.2	D009
Se	1.0	D010
Ag	5.0	D011

Source: EPA 1992

Table 4 Soluble threshold limit concentrations

Metals	TCLP Regulatory Level, mg/L	California Regulatory Level, mg/L
As	5.0	5.0
Ba	100.0	100.0
Cd	1.0	1.0
Cr	5.0	5.0
Pb	5.0	5.0
Hg	0.2	0.2
Se	1.0	1.0
Ag	5.0	5.0
Be		0.075
Co		80
Cu		25
Mo		350
Ni		20
Tl		7.0
V		24
Zn		250

Adapted from DTSC 2015a

All solid waste, including mineral processing and extractive metallurgical wastes that are not excluded but fail the TCLP test, must be treated to further stabilize the waste before it can be placed in an EPA-permitted treatment, storage, and disposal facility (TSDF). The stabilized waste must then pass the TCLP test to standards established under the land disposal restrictions (LDR; commonly called LBR) universal treatment standards before the disposal of the waste is allowed. The present universal treatment standard requirements for elements are presented in Table 6 with a comparison to the TCLP toxicity characterization concentrations.

Treatment, Storage, and Disposal Facilities

A defined hazardous waste must be placed in an EPA- or state-permitted TSDF. The regulations for different activities are cited in Figure 1.

According to federal regulations, a TSDF must perform one or more of the following functions (EPA 2014a):

Table 5 Comparison of TCLP and WET methods

Toxicity Characterization Leach Procedure	Waste Extraction Test
Involves 20-fold dilution of the solid portion of the waste to extraction fluid	Involves 10-fold dilution of the solid portion of the waste to extraction fluid
Acetic acid extractant	Citric acid extractant
18 hours extraction	48 hours extraction
7 inorganic (less aggressive than WET)	19 inorganic (in general, more aggressive than TCLP)
23 organic (volatiles use a zero-headspace extractor)	18 organic (zero-headspace extractor not used)

Source: DTSC 2015b

Table 6 Universal treatment standards as specified in the LDR regulations

Element	Toxicity Characterization Level, mg/L	LBR Universal Treatment Standard	
		Wastewater, mg/L	Non-Wastewater in TCLP, mg/L
Antimony	—	1.9	1.15
Arsenic	5.0	1.4	5.0
Barium	100	1.2	21
Beryllium	—	0.82	1.22
Cadmium	1.0	0.69	0.11
Chromium	5.0	2.77	0.6
Lead	5.0	0.69	0.75
Mercury (from retort)	0.2	—	0.2
Mercury—all others	0.2	0.15	0.025
Nickel	—	3.98	11
Selenium	1.0	0.82	5.7
Silver	5.0	0.43	0.14
Thallium	—	1.4	0.2
Vanadium	—	4.3	1.6
Zinc	—	2.61	4.3

Source: EPA 1998, 2015b

Summary Chart

Hazardous Waste Treatment, Storage, and Disposal Facilities (TSDF) Regulations

(Note: FDSys links are updated at the end of November; this version is 2014)

	Hazardous Waste Management/Permitting Activity	General Requirements	Permit Application Requirements	Closure/Post-Closure Requirements
1	General Facility Standards	264-B & 265-B	NA	NA
2	Preparedness and Prevention	264-C & 265-C	NA	NA
3	Contingency Plan and Emergency Procedures	264-D & 265-D	NA	NA
4	Manifest System, Recordkeeping and Reporting	264-E & 265-E	NA	NA
5	Releases from Solid Waste Management Units	264-F	NA	264.100
6	Groundwater Monitoring	265-F	NA	NA
7	Closure and Post-Closure	264-G & 265-G	NA	264-G & 265-G
8	Financial Assurance Requirements	264-H & 265-H	NA	NA
9	Use and Management of Containers	264-I & 265-I	270.15 & 270.27	264.178
10	Tank Systems	264-J & 265-J	270.16 & 270.27	264.197 & 265.197
11	Surface Impoundments	264-K & 265-K	270.17 & 270.27	264.228 & 265.228
12	Waste Piles	264-L & 265-L	270.18	264.258 & 265.258
13	Land Treatment	264-M & 265-M	270.20	264.280 & 265.280
14	Landfills	264-N & 265-N	270.21	264.310 & 265.310
15	Incinerators	264-O & 265-O & 63-EEE	270.19 & 270.62	264.351 & 265.351
16	Thermal Treatment	265-P	NA	265.381
17	Chemical, Physical, Biological Treatment	265-Q	NA	265.404
18	Underground Injection	265-R	NA	NA
19	Special Provisions for Cleanup	264-S	NA	NA
20	Drip Pads	264-W & 265-W	270.26	264.575 & 265.445
21	Miscellaneous Units	264-X	270.23	264.603
22	Air Emission Standards for Process Vents	264-AA & 265-AA	NA	NA
23	Air Emission Standards for Equipment Leaks	264-BB & 265-BB	NA	NA
24	Air Emission Standards for Tanks, Surface Impoundments and Containers	264-CC & 265-CC	NA	NA
25	Containment Buildings	264-DD & 265-DD	NA	264.1102 & 265.1102
26	Munitions and Explosives Storage	264-EE & 265-EE	NA	264.1202 & 265.1202
27	Solid Waste Management Units	264-F & 264-S	NA	264.101
28	Boilers/Industrial Furnaces	266-H & 63-EEE	270.66	NA
29	Land Disposal Restrictions	268	NA	NA
30	Hazardous Waste Permitting Program General Requirements	270	270	NA
31	Administrative Procedures Act Requirements for RCRA Permitting	124	124-B & 124-G	NA

Source: EPA 2014a

Figure 1 TSDF regulations by activity

> **Treatment**—Any method, technique, or process, including neutralization, designed to change the physical, chemical, or biological character or composition of any hazardous waste so as to neutralize such waste, or so as to recover energy or material resources from the waste, or so as to render such waste non-hazardous, or less hazardous; safer to transport, store, or dispose of; or amenable for recovery, amenable for storage, or reduced in volume.
>
> **Storage**—The holding of hazardous waste for a temporary period, at the end of which the hazardous waste is treated, disposed of, or stored elsewhere.
>
> **Disposal**—The discharge, deposit, injection, dumping, spilling, leaking, or placing of any solid waste or hazardous waste into or on any land or water so that such solid waste or hazardous waste or any constituent thereof may enter the environment or be emitted into the air or discharged into any waters, including ground waters.

Care must be exercised in selecting a TSDF for disposing of a mineral processing or extractive metallurgical waste. It is important that a qualified TSDF is chosen because, according to RCRA, the "generator" of a waste is responsible for that waste from "cradle to grave." A company will share in the cost of cleanup if the hazardous waste is mismanaged, even if the TSDF facility is permitted to operate.

National Pollutant Discharge Elimination System

The CWA (1972, with many amendments thereafter) authorized the establishment of the National Pollutant Discharge Elimination System (NPDES) program (EPA 2015e). The NPDES requires permits for any water discharges into the waters of the United States, including publicly (or privately) owned (wastewater) treatment works. The permitting program is, with some exceptions, administered by authorized states. The states provide the permits, and monitor and enforce the regulations for the water discharges. The EPA has oversight on the states' programs and offers guideline regulations to the states; the states can accept the guidelines or set their own regulations, which may be more stringent than the federal requirements. Section 518(e) and (h) of the CWA authorizes the EPA to treat tribes in a manner similar to states for the purposes of administering certain CWA programs for tribal waters.

Determining Whether Solid Waste Is Hazardous

The process of determining if a mineral processing or extractive metallurgy solid waste is deemed hazardous can be very confusing, and competent environmental and legal assistance are required. A simplified schematic of the procedure is presented in Figure 2. It shows a series of decision steps. Step 1 asks, Is the waste a solid waste? (A *solid waste* is defined as any material that is discarded by being either abandoned, inherently waste-like, or recycled.) Is it a sludge from a wastewater treatment plant? If the answer is yes, then the waste

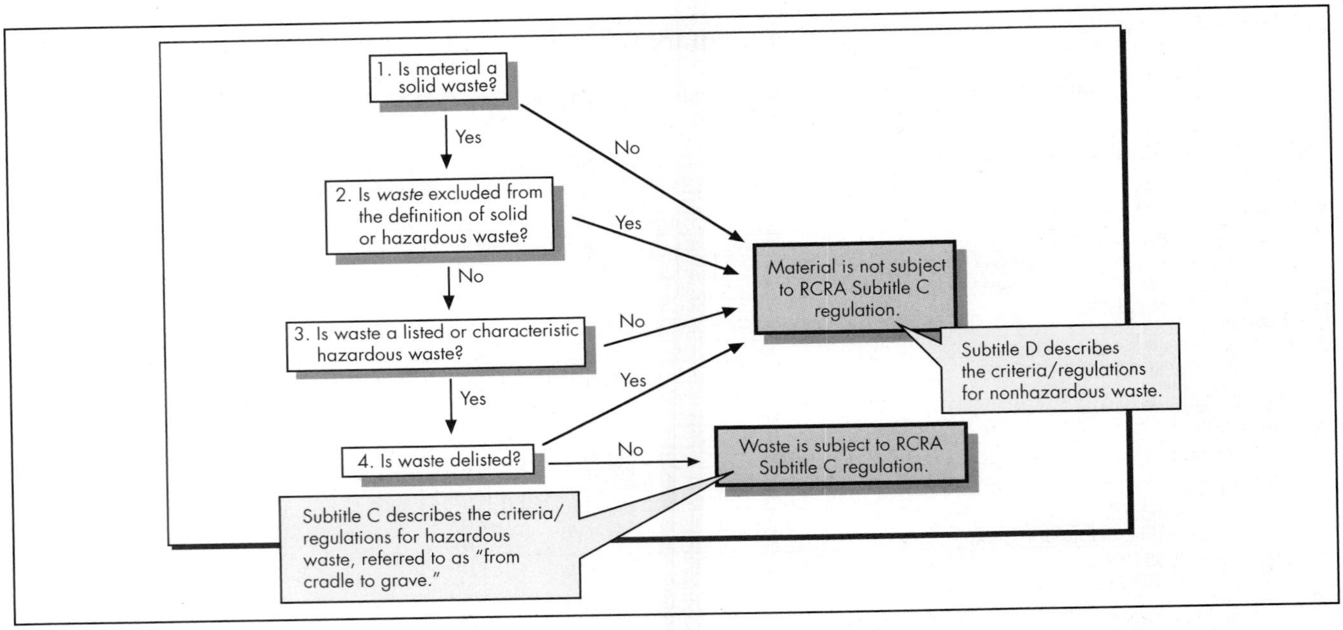

Source: EPA 1998

Figure 2 Waste treatment decision tree

consideration moves to step 2; if the answer is no, then the waste doses not need to be treated as an RCRA Subtitle C hazardous waste. Step 2 asks, Is the waste excluded from being required to follow hazardous waste protocols for handling and storage? If yes, then the waste does not have to follow hazardous protocols; if no, then the waste consideration moves to step 3. Step 3 asks, Is the waste a listed waste? If it is, it must follow the Subtitle C protocol; if it is a non-excluded waste, it must be evaluated to determine whether it is a characteristic waste, that is, whether the waste has one or more of the four characteristics (ignitability, corrosivity, reactivity, toxicity). Generally, mineral processing and extractive metallurgy wastes will need to be tested for toxicity. The determining test is the TCLP test. If the waste passes the TCLP test (where the hazardous constituent is less than 100 times the MCL limiting concentration), then it is deemed nonhazardous. If it fails the test, then the waste moves to step 4. Step 4 states, If the waste has been stabilized so that it passes the TCLP test, then it may be "delisted" and treated as a nonhazardous waste. If the waste is not delisted, then it must follow the Subtitle C regulation.

EFFLUENT SOLUTIONS

Mineral processing and extractive metallurgical operations create appreciable wastewater and waste solid products that have to be handled, treated for recycle, or treated for environmentally safe disposal. Complete coverage of all waste products is beyond the scope of this chapter. The topics covered here include acid mine drainage and hydrometallurgical effluent solutions, emphasizing arsenic, selenium, and metals.

Acid Mine Drainage

Acid mine drainage (AMD) or acid rock drainage (ARD) is often considered to be a major pollutant in countries that have historic or current mining activities. Its generation, mobility, and attenuation involve many complex processes that are based on combinations of physical, chemical, and biological interactions. Although AMD is an important topic for the mining, mineral processing, and extractive metallurgical industries, it is not the focus of this chapter. However, interested readers are referred to the following information sources.

Detailed discussions of ARD topics are presented by Verburg (2011) in the *SME Mining Engineering Handbook* (Darling 2011). The presented information is primarily based on the *Global Acid Rock Drainage Guide* (INAP 2014) and includes coverage of the following topics: formation of ARD and ARD management (characterization, prediction, prevention and mitigation, treatment, monitoring, and performance assessment). Nordstrom et al. (2015) discuss the hydrogeochemistry and microbiology of ARD:

> The extraction of mineral resources requires access through underground workings, or open pit operations, or through drillholes for solution mining. Additionally, mineral processing can generate large quantities of waste, including mill tailings, waste rock and refinery wastes, heap leach pads, and slag. Thus, through mining and mineral processing activities, large surface areas of sulfide minerals can be exposed to oxygen, water, and microbes, resulting in accelerated oxidation of sulfide and other minerals and the potential for the generation of low-quality drainage. The oxidation of sulfide minerals in mine wastes is accelerated by microbial catalysis of the oxidation of aqueous ferrous iron and sulfide. These reactions, particularly when combined with evaporation, can lead to extremely acidic drainage and very high concentrations of dissolved constituents. Although acid mine drainage is the most prevalent and damaging environmental concern associated with mining activities, generation of saline, basic and neutral drainage containing elevated concentrations of dissolved metals, non-metals, and metalloids has

> recently been recognized as a potential environmental concern. Acid neutralization reactions through the dissolution of carbonate, hydroxide, and silicate minerals and formation of secondary aluminum and ferric hydroxide phases can moderate the effects of acid generation and enhance the formation of secondary hydrated iron and aluminum minerals which may lessen the concentration of dissolved metals. Numerical models provide powerful tools for assessing impacts of these reactions on water quality.

Lapakko (2015) offers a preoperational assessment of mining operations:

> Environmental assessments are conducted prior to mineral development at proposed mining operations. Among the objectives of these assessments is prediction of solute release from mine wastes projected to be generated by the proposed mining and associated operations. This paper provides guidance to those engaged in these assessments and, in more detail, provides insights on solid-phase characterization and application of kinetic test results for predicting solute release from waste rock. The logic guiding the process is consistent with general model construction practices and recent publications. Baseline conditions at the proposed site are determined and a detailed operational plan is developed and imposed upon the site. Block modeling of the mine geology is conducted to identify the mineral assemblages present, their masses and compositional variations. This information is used to select samples, representative of waste rock to be generated, that will be analyzed and tested to describe characteristics influencing waste rock drainage quality. The characterization results are used to select samples for laboratory dissolution testing (kinetic tests). These tests provide empirical data on dissolution of the various mineral assemblages present as waste rock. The data generated are used, in conjunction with environmental conditions, the proposed method of mine waste storage, and scientific and technical principles, to estimate solute release rates for the operational scale waste rock Common concerns regarding waste rock are generation of acidic drainage and release of heavy metals and sulfate. Key solid phases in the assessments are those that dissolve to release acid and sulfate (iron sulfides, soluble iron sulfates, hydrated iron-sulfate minerals, minerals of the alunite–jarosite group), those that dissolve to neutralize acid (calcium and magnesium carbonates, silicate minerals), and those that release trace metals (trace metal sulfides, hydrated trace metal-sulfate minerals). Conventional mineralogic, petrographic, and geochemical analyses generally can be used to determine the quantities of these minerals present and to describe characteristics that influence their dissolution. A key solid-phase characteristic is the mineral surface area exposed for reaction, which is influenced by mode of occurrence (included, interstitial, and liberated) and the extent of mineral surface coating. Short-term dissolution tests can estimate the extent of hydrated sulfate minerals present. Longer term dissolution tests are necessary to describe the dependence of drainage pH and solute release rates on solid-phase variation. The extensive data compiled from baseline pre-development definition, the operational plan, solid-phase characterization, and dissolution testing are ultimately synthesized by means of a modeling exercise requiring considerable technical and scientific expertise. The predicted rates (model outputs) are expressed as probability distributions to allow assessment of risk. This exercise must be technically defensible and transparent so that regulators can confidently assess the results and evaluate the operational plan proposed. Technical and non-technical challenges involved in implementing such programs are identified to benefit management planning for both industry and government.

In addition, Parbhakar-Fox and Lottermoser (2015) provide a review of ARD prediction methods:

> Failure to accurately predict acid rock drainage (ARD) leads to long-term impacts on ecosystems and human health, in addition to substantial financial consequences and reputational damage to operators. Currently, a range of chemical static and kinetic tests are used to evaluate the acid producing nature of materials, from which risk assessments are prepared and waste classification schemes designed. However, these well-established tests and practices have inherent limitations, for example: (i) best-practice sampling is not pursued; (ii) risk assessments rely on limited static and kinetic test data, thus compromising the accuracy of resulting ARD block models; (iii) static tests are completed off-site and do not reflect actual field measurements; (iv) kinetic test data do not become available until later stages of mine development; (v) waste classification schemes generally categorise materials as only three types (i.e., PAF, NAF and UC) with other drainage forms (e.g., neutral metalliferous or saline) not considered; and (vi) conventional testing fails to consider that reactivity of waste is controlled by parameters other than chemistry (e.g., microbiology, type and occurrence of minerals, texture and hardness). Thus, accurate prediction is challenging because of the multifaceted processes leading to ARD. Hence, risk assessments need to consider mineralogical, textural and geometallurgical rock properties in addition to predictive geochemical test data. Instead, a new architecture of integrative, staged ARD testing should be pursued. Better ARD prediction must start with improving the definition of geoenvironmental models and waste units. Then, a range of low-cost and rapid tests for the screening of samples should be conducted on site prior to the performance of established tests and advanced analyses using state-of-the-art laboratories. Such an approach to ARD prediction would support more accurate and cost-effective waste management during operation, and ultimately less costly mine closure outcomes.

Included in the *SME Mining Engineering Handbook* is coverage of waste disposal and contamination management as related to mining, beneficiation, and mineral processing (Borden 2011). Topics covered include mineral waste, waste characterization and impact prediction, waste management strategy and facility environmental design, ongoing management and monitoring, nonmineral waste, hazardous materials and contamination control, contaminated water management, and contaminated site management. The discussion focuses on disposal of mining and beneficiation waste but does not cover wastewater and solid waste created during the hydrometallurgical processing of minerals to produce a metal product. Therefore, topics including in-plant processing unit operations are not covered, for example, heavily contaminated smelter blowdown wastewater and other unit operations applied to remove hazardous constituent as solid waste product. Recall from the discussion presented previously that EPA defines *mineral processing waste* differently than Borden does; that is, the EPA defines *mineral processing* as follows (Housman 1999):

> [Mineral] Processing operations, in contrast, generally follow beneficiation and serve to change the concentrated mineral into a more useful chemical form. This is often done by using heat or chemical reactions to change the chemical composition of the mineral. In contrast to beneficiation operations, processing activities often destroy the physical structure of the incoming ore or mineral feedstock such that the materials leaving the operation do not closely resemble those that entered the operation. Typically, beneficiation wastes are earthen in character, whereas mineral-processing wastes are derived from melting and chemical changes.

Other publications that address mining and beneficiation waste include Jamieson et al. (2015) who discuss the mineralogical characterization of mine waste:

> The application of mineralogical characterization to mine waste has the potential to improve risk assessment, guide appropriate mine planning for planned and active mines and optimize remediation design at closed or abandoned mines. Characterization of minerals, especially sulphide and carbonate phases, is particularly important for predicting the potential for acidic drainage and metal(loid) leaching. Another valuable outcome from mineralogical studies of mine waste is an understanding of the stability of reactive and metal(loid)-bearing minerals under various redox conditions. This paper reviews analytical methods that have been used to study mine waste mineralogy, including conventional methods such as X-ray diffraction and scanning electron microscopy, and advanced methods such as synchrotron-based microanalysis and automated mineralogy. We recommend direct collaboration between researchers and mining companies to choose the optimal mineralogical techniques to solve complex problems, to co-publish the results, and to ensure that mineralogical knowledge is used to inform mine waste management at all stages of the mining life cycle. A case study of arsenic-bearing gold mine tailings from Nova Scotia is presented to demonstrate the application of mineralogical techniques to improve human health risk assessment and the long-term management of historical mine wastes.

Kossoff et al. (2014) provide a review of mine tailings dams:

> On a global scale demand for the products of the extractive industries is ever increasing. Extraction of the targeted resource results in the concurrent production of a significant volume of waste material, including tailings, which are mixtures of crushed rock and processing fluids from mills, washeries or concentrators that remain after the extraction of economic metals, minerals, mineral fuels or coal. The volume of tailings is normally far in excess of the liberated resource, and the tailings often contain potentially hazardous contaminants. A priority for a reasonable and responsible mining organization must be to proactively isolate the tailings so as to forestall them from entering ground waters, rivers, lakes and the wind. There is ample evidence that, should such tailings enter these environments they may contaminate food chains and drinking water. Furthermore, the tailings undergo physical and chemical change after they have been deposited. The chemical changes are most often a function of exposure to atmospheric oxidation and tends to make previously, perhaps safely held contaminants mobile and available. If the tailings are stored under water, contact with the atmosphere is substantially reduced, thereby forestalling oxygen-mediated chemical change. It is therefore accepted practice for tailings to be stored in isolated impoundments under water and behind dams. However, these dams frequently fail, releasing enormous quantities of tailings into river catchments. These accidents pose a serious threat to animal and human health and are of concern for extractive industries and the wider community. It is therefore of importance to understand the nature of the material held within these dams, what best safety practice is for these structures and, should the worst happen, what adverse effects such accidents might have on the wider environment and how these might be mitigated. This paper reviews these factors, covering the characteristics, types and magnitudes, environmental impacts, and remediation of mine tailings dam failures.

The EPA's *Reference Guide to Treatment Technologies for Mining-Influenced Water* presents current information with respect to technology, constituents treated (metals, nitrate, arsenic, selenium), system operations, long-term maintenance, system limitations, costs, and effectiveness (EPA 2014b). An excerpt follows:

> This report provides an overview of select mining-influenced water (MIW) treatment technologies used or piloted as part of remediation efforts at mine sites. The report is intended to provide information on treatment technologies for MIW to federal, state and local regulators, site owners and operators,

consultants, and other stakeholders. The technologies described in this report are applicable to treatment of water from both coal and hard-rock mine operations. The report provides short descriptions of treatment technologies and presents information on the contaminants treated, pre-treatment requirements, long-term maintenance needs, performance and costs. Sample sites illustrate considerations associated with selecting a technology. Website links and sources for more information on each topic are also included.

MIW is defined as any water whose chemical composition has been affected by mining or mineral processing and includes acid rock drainage (ARD), neutral and alkaline waters, mineral processing waters and residual waters. MIW can contain metals, metalloids and other constituents in concentrations above regulatory standards. MIW affects over 10,000 miles of receiving waters in the United States, primarily by acidic drainage.

Hydrometallurgical Effluent Solutions

The EPA has established effluent limitation guideline standards for the treatment of industrial wastewater discharges to surface waters and municipal sewage treatment plants (EPA 2015c, 2015d). The guidelines are based on performance of present treatment and commonly used control technologies. The guidelines are included as a part of individual facility NPDES permits. The EPA has promulgated effluent guidelines and standards for the mineral, mining, and processing industries (40 CFR Part 436). The guidelines and regulations cover wastewater discharges from mine drainage, and mineral and extraction processing operations. Example effluent guidelines for smelter and refining operations are summarized in Table 7 (recommended global guidelines for all smelting and refining operations).

Table 7 Effluent levels for nickel, copper, lead, zinc, and aluminum smelting and refining

Pollutant	Smelting Type	Units	Guideline Value
Aluminum	Aluminum	mg/L	0.2
Arsenic	Copper, lead, and zinc	mg/L	0.05
Cadmium	Nickel, copper	mg/L	0.05
Chemical oxygen demand	All	mg/L	50
Copper	Copper	mg/L	0.1
Fluoride	Aluminum	mg/L	5
Hydrocarbons	Aluminum	mg/L	5
Lead	—	—	0.1
Mercury	All	mg/L	0.01
Nickel	Copper, lead, and zinc	mg/L	0.1
pH	All	mg/L	6–9
Temperature increase	All	°C	<3°
Total suspended solids	All	mg/L	20
Toxicity	—To be determined on a case-specific basis—		
Zinc	Copper, lead, and zinc	mg/L	0.2

Source: World Bank Group 2007

Arsenic

The removal of arsenic from hydrometallurgical solutions, wastewaters, and AMD waters has been and continues to be an important research and regulatory topic. The EPA has promulgated regulations based on its determination of the BDAT to be used for listed and characteristic waste-containing arsenic. The chosen BDAT is ferrihydrite adsorption of arsenic (Rosengrant and Fargo 1990). Therefore, a guide to the literature is presented here that is focused on the removal of arsenic from aqueous solutions using ferric precipitation and subsequent adsorption of arsenic and on the long-term outdoor storage of the arsenic-bearing products (Twidwell and McCloskey 2011). The following guide is arranged in the following order: summary of current industrial processes, a presentation for those who want further detailed information on ferrihydrite formation and properties, and ferrihydrite/arsenic formation and stability of arsenic-bearing waste products.

Summary of current industrial processes. Two ferric precipitation arsenic removal technologies are presently practiced by industry throughout the world: ambient temperature ferrihydrite arsenic adsorption/co-precipitation and elevated temperature, elevated pressure precipitation of scorodite/ferric arsenate. The ambient temperature technology is relatively simple, and the presence of commonly associated metals (copper, lead, zinc) and gypsum have a stabilizing effect on the long-term stability of the outdoor storage of the product. The disadvantages of the adsorption technology include the following:

- A relatively large amount of waste material is created (Fe/As mole ratio varies; it is usually approximately 3–4 but can be as high as 10), which may be difficult to filter.
- The arsenic must be present in the fully oxidized state (arsenate).
- The presence of competitive associated anionic species may negatively influence the adsorption of arsenate.
- The long-term stability of the product in the presence of reducing substances and bacterial activity must be determined.

The second technology practiced at several copper smelting facilities is arsenic removal by precipitation of scorodite in autoclaves. The advantage of the scorodite process over the ferrihydrite technology is that less waste is formed (Fe/As mole ratio is one, greater density, and better thermodynamic stability). The disadvantages of scorodite/ferric arsenate precipitation are that the treatment process is more capital and energy intensive; the compound may dissolve incongruently if the pH is greater than 3; and its long-term storage may not be stable under reducing and/or anaerobic bacterial conditions (Erbs et al. 2010; Kocar et al. 2010; Jones et al. 2000; Xie et al. 2009).

Currently, the ambient temperature/ambient pressure ferrihydrite/arsenate process is more commonly applied throughout the world. However, recent research has shown that scorodite can be formed under less intense non-autoclave operating conditions, that is, elevated temperature but ambient pressure processing. This technique will likely be adopted in the future; to date, one smelter, the Dowa smelter in Japan (Fujita et al. 2012), has adopted this process. The various scorodite formation processes are summarized as follows (Twidwell 2014a, 2014b):

1. Autoclave hydrothermal precipitation of scorodite from acidic solutions (pH ~1, ~150°C) containing Fe(III) and As(V) (Gomez et al. 2011).
2. Elevated temperature ambient pressure precipitation from acidic solutions (pH ~1, 90°–95°C) containing Fe(III) and As(V) or As(III) (Demopoulos 2008, 2005).
3. Intermediate temperature ambient pressure precipitation by in situ oxidation of Fe(II) in the presence of As(V) from acidic solutions (pH ~1, ~70°C, 95°C) (Fujita et al. 2012).
4. Intermediate temperature ambient pressure precipitation by biogenic in situ oxidation of Fe(II) in the presence of As(V) from acidic solutions (pH ~1, ~70°C) (Okibe et al. 2013).
5. Intermediate temperature ambient pressure precipitation by biogenic in situ oxidation of Fe(II) and As(III) from acidic solutions (pH ~1, ~70°C) (Okibe et al. 2014).

Riveros et al. (2001) concluded the following from their extensive review of the literature:

> For practical purposes, arsenical ferrihydrite can be considered stable provided the Fe/As molar ratio is greater than 3, the pH is slightly acidic and that it does not come in contact with reducing substances such as reactive sulfides or reducing conditions such as deep water, bacteria or algae.

Swash et al. (2000) concluded that from the point of view of safe disposal of arsenic, there is no clear experimental evidence favoring the low-temperature precipitates over the high-temperature precipitates. A detailed review of the stability of scorodite, ferrihydrite, and ferrihydrite/arsenate adsorption is presented by Welham et al. (2000). Their conclusions are as follows:

> There are significant problems with the use of jarosite and scorodite as phases for the disposal of iron and/or arsenic from metallurgical systems. Neither phase is stable under typical atmospheric weathering conditions with transformation to goethite predicted to occur. The currently permitted discharge level of arsenic is only achieved due to the slow kinetics of the transformation releasing arsenic over time. Crystalline scorodite is two orders of magnitude less soluble than amorphous iron(III) arsenate precipitate often formed in low temperature systems.

Ferrihydrite formation and properties (Twidwell and McCloskey 2011). Important reviews detailing conditions for formation and the stability of ferrihydrite are presented by Jambor and Dutrizac (1998) and Cornell and Schwertmann (1996, 2003). The reviews by Jambor and Dutrizac (314 references) and Cornell and Schwertmann (approximately 1,500 references) are indeed excellent sources of information on ferrihydrite occurrence, structure, chemical composition, adsorptive capacity for cations and anions, and transformation rate as well as a summary of the factors that influence ferrihydrite's transformation to hematite or goethite. Schwertmann and Cornell (2000) have published a "recipe" book that details how to prepare iron oxides in the laboratory, including ferrihydrite, hematite, and goethite. Many of the experimental studies reported in the literature reference this publication.

Literature-reported characteristic features of ferrihydrite and its conversion to more crystalline forms (goethite and hematite) are briefly summarized here. Ferrihydrite is characterized by X-ray diffraction as having a two-line or six-line structure, which relates to the number of broad peaks present. Two-line ferrihydrite is formed by rapid hydrolysis to pH 7 at ambient temperature. Six-line ferrihydrite is formed by rapid hydrolysis at elevated temperature and is generally more crystalline than two-line ferrihydrite. However, Schwertmann and Cornell (2000) have demonstrated that either can be formed at ambient temperature by controlling the rate of hydrolysis (i.e., less crystalline two-line forms at rapid hydrolysis rates, whereas six-line forms if the precipitation is conducted at lower rates, and lepidocrocite forms if the rate of addition of sodium hydroxide is slow enough). Crystallite sizes have been reported to be 2–4 nm and 5–6 nm for two-line and six-line ferrihydrite, respectively. The surface area of freshly precipitated two-line ferrihydrite is greater than 340 m^2/g (Paktunc et al. 2008). The particle size of aggregated ferrihydrite is usually quoted in the range of 3–10 µm. Hohn et al. (2006) determined the mean minimum particle size to be 1.5–2.0 nm at pH 7. Hohn et al. also demonstrated that two-line ferrihydrite and arsenic-loaded (7%) ferrihydrite prepared at pH 4 and 7 self-flocculate to a mean agglomerate size of 5–10 µm.

The rate of transformation of ferrihydrite to hematite or goethite when suspended in an aqueous medium has been discussed in great detail by Cornell and Schwertmann (2003) and by Jambor and Dutrizac (1998) and many others. The rate of transformation is a function of time, temperature, and pH. For example, conversion of pure two-line ferrihydrite to hematite at 25°C is half complete in 280 days at pH 4 but is completely converted at 100°C in 4 hours (Cornell and Schwertzman 2003). Transformation of ferrihydrite results in a relatively large change in surface area (e.g., freshly prepared two-line ferrihydrite showed a surface area of about 150 m^2/g that when converted to goethite at 25°C was reduced to 92 m^2/g; when converted to goethite at 90°C, the area was reduced to 9 m^2/g). Many investigators have pointed out that pure ferrihydrite converts rapidly and the conversion results in a significant decrease in surface area. However, in real industrial systems, the ferrihydrite conversion rate may be mitigated (changed from days to perhaps years) by the presence of other species and solution conditions during precipitation and subsequent storage. General factors that have been shown to decrease the rate of conversion of two-line ferrihydrite to more crystalline forms include lower pH; lower temperatures; and presence of silicate, aluminum, arsenic, manganese, metals, sulfate, and organics.

Ferrihydrite/arsenic formation and stability of arsenic-bearing waste products (Twidwell and McCloskey 2011). The EPA promulgated rules for BDATs to be used for the following listed and characteristic waste-containing As and Se: K031, K084, K101, K102, As wastes (D004), Se wastes (D010), and P and U wastes (Rosengrant and Fargo 1990). The specified BDAT for treatment of effluent solutions is arsenic adsorption on ferrihydrite. This technology has also been selected as one of the best available technologies for removing arsenic from drinking water. A brief discussion of the arsenical ferrihydrite literature follows.

Removal from solution. Studies investigating the removal of anions on ferrihydrite surfaces are not new. Municipal water treatment systems have used ferric, ferrous, and aluminum hydroxide precipitation for many years to cleanse heavy metals and phosphates from solution as a final polishing stage. In fact, doping manuals are available for ferric chloride, ferric sulfate, or aluminum sulfate (AWWA 1988a, 1988b). A wide range of arsenic removal results are reported in the literature for the adsorption or co-precipitation of arsenic-bearing iron products. The relatively wide range of results is to be expected because there are several experimental factors that influence the removal. These influencing factors include the ferrihydrite formation procedure; the iron/arsenic mole ratio; pH, time, and initial concentration of arsenic; the anionic environment; the method of agitation; the presence of co-precipitants; and the arsenic valence.

The presence of arsenite, As(III), and other dissolved aqueous species are important aspects that need to be considered when discussing ferrihydrite technology. The normal approach when considering ferrihydrite removal of arsenic is to consider ways to oxidize the As(III) to As(V). It is often stated that As(V) is much more effectively removed by ferrihydrite than As(III). However, the relative removal of As(V) and As(III) depends upon the Fe/As ratio, pH, and whether the arsenic species are present individually or as mixtures. As(III) is often found in appreciable quantities in ambient temperature metallurgical operations and in groundwater and surface waters (Wilson et al. 2001; Borho and Wilderer 1996). In fact, Borho and Wilderer state that approximately 30% of the arsenic present in ground and surface waters used for drinking waters is As(III). The presence of associated ions such as phosphate, sulfate, carbonate, and dissolved organic species can greatly influence the removal of arsenic and the relative long-term stability of ferrihydrite.

Stability of products. Excellent examples of synthesis and transformation rates for scorodite, ferric arsenate, and arsenical ferrihydrite are provided in publications by Paktunc et al. (2008) and Paktunc and Bruggeman (2010). Drahota and Filippi (2009) present a recent review of arsenic mineralization and a description of primary and secondary minerals found at mine and industrial sites. Riveros et al. (2001) have presented a review of the disposal practices and long-term stability of arsenic products in the Canadian metallurgical industry. Paktunc et al. (2008) have investigated the structure of scorodite, ferric arsenate, and arsenical ferrihydrite, and results are presented in their publications. The authors have clarified the distinction between the three forms. For example, scorodite, $FeAsO_4$, is a fully crystallized phase containing a Fe/As mole ratio of 1 ($FeAsO_4$:3.5H_2O); ferric arsenate ($FeAsO_4$:4–7H_2O) is less crystalline than scorodite; and arsenical ferrihydrite (5Fe_2O_3:9H_2O) is X-ray amorphous. Ferric arsenate forms at low pH, has a Fe/As mole ratio of one, and is rapidly transformed to scorodite at pH levels below approximately 1.7. Above that pH, ferric arsenate and arsenical ferrihydrite form up to approximately a pH of 4.5 for Fe/As ratios from 1–10. Ferric arsenate was not present at Fe/As ratios of 5 and greater; ferrihydrite and arsenical ferrihydrite predominate. Ferrihydrite converts to a six-line product at these Fe/As ratios and the conversion is a function of aging time. Paktunc et al. (2008) discuss the implications for arsenic releases and controls:

> Common industrial practice is to remove and stabilize arsenic using Fe/As ratios of 3–4. Therefore, the solids are likely mixtures of ferric arsenate and arsenical ferrihydrite. The thought is that ferric arsenate forms and that phase dissolves to form arsenical ferrihydrite which has abundant surface sites for adsorbing arsenate. The applications at higher Fe/As ratios ensure that the more soluble ferric arsenate is dissolved to form the more insoluble arsenical ferrihydrite. At Fe/As ratios above five the two-line ferrihydrite converts to six-line (more crystalline) ferrihydrite but the presence of arsenic in the 2-line structure retards this conversion. Quoting the authors "the kinetics of phase transformations and its dependence on pH would facilitate the development of new technologies for the stabilization of As in mine wastes. Identification of ferric arsenate in mine wastes destined for disposal should be considered critical for the prediction of the short-term and long-term behavior of arsenic in wastes. Disposal options would then need to consider the changes in pH and redox conditions that may occur with time and their bearings on potential As releases.

Generally, arsenical ferrihydrite and scorodite pass the EPA TCLP test and the waste products do not have to follow EPA Subtitle C disposal regulations. However, an important unknown at this time is whether the product from ferrihydrite adsorption of arsenic will be stable if storage conditions are anaerobic or may become anaerobic or contain microbial agents. Erbs et al. (2010) demonstrated that induced reduction conditions using hydroquinone resulted in arsenic and iron reduction and that co-precipitated ferrihydrite showed less arsenic release than adsorbed arsenic on previously precipitated ferrihydrite. Pederson et al. (2006) found that arsenate remained with ferrihydrite when exposed to ascorbic acid until the surface area was insufficient to retain the arsenate. Brannon and Patrick (1987) have demonstrated that anaerobic lake sediments convert As(V) to As(III) (pH 5–8.0). However, when the anaerobic conditions were shifted by aerobic leaching, the previously reduced As(III) was reconverted to more immobile As(V), which was associated with aluminum and iron oxyhydroxides. Chatain et al. (2005a) investigated the effect of controlling the solution redox potential (normal hydrogen electrode reference) and pH using sodium ascorbate (–7 to 345 mV) and sodium borohydride (–500 to 140 mV) to treat an arsenic-bearing gold mining soil (2.8% As, 1.8% on ferrihydrite). The release of arsenic from the soil under oxidizing conditions (410 mV) showed the normal ferrihydrite release of As(V) (i.e., ~300 mg/L), whereas the treatment with 0.046 mole/L sodium ascorbate at an E_H = –7 mV (pH ~6) released ~80,000 mg/L As(III). Although different results were obtained with sodium borohydride, the release of arsenic was significant. Also, it is known that the effect of bacterial reduction of ferrihydrite and arsenate can be extensive. Kocar et al. (2010) found that the effect of sulfate-reducing bacteria (that produces dissolved sulfide species) was to reduce ferrihydrite to other iron solids along with the reduction of arsenate to arsenite. Chatain et al. (2005b) investigated the influence of anaerobic conditions (at pH ~7) with indigenous bacterial activity on the release of arsenic (and other metals) from a contaminated mining soil (3% As, 0.3% on ferrihydrite). The results showed <4,000 µg/L arsenic release from baseline soil/water leaches (80 days) and ~100,000 µg/L As(III) for nutrient-fed indigenous bacteria. Langer and Inskeep (2000) have investigated the possible reduction of As(V) to As(III) on ferrihydrite. They adsorbed arsenate onto previously precipitated two-line ferrihydrite solids, added a reducing fulcose fermenting microorganism to a suspended slurry containing the precipitated arsenate and arsenate species in solution at pH 6.8, and aged for 24 days. The solution arsenate was reduced to arsenite in less than one day, but the precipitated As(V) and ferrihydrite were not reduced. Langmuir et al. (2006) reported on analytical tests conducted for pore waters collected from various depths at a tailings impoundment (approximately five years aging). The results showed the following: pH = 7.76 ± 0.24, total arsenic = 4.1 ± 3.6 mg/L (which decreased to 1–2 mg/L in the older tailings), and E_H ~369 – 450 mV. The presence of several milligrams per liter of arsenite was noted in pore waters at some depths. Many more relatively recent studies are available that are not quoted here.

Removal of arsenic from solution by ferric precipitation has been and is or has been practiced at numerous extractive metallurgical facilities, for example, the Noranda Horne Smelter (Rouyn-Noranda, Quebec, Canada), the Giant mine and the Con mine (Yellowknife, Northwest Territories, Canada), and the Teck-Corona mine (Marathon, Ontario, Canada); the Kennecott Utah Smelter (Magna, Utah, United States), Placer Dome Lonetree and Getchell mines (Nevada, United States) (on a periodic basis), and Barrick's gold mining operations in Nevada. Harris (2000) has tabulated worldwide industrial operating practice (as of 2000/2001) for removal and stabilization of arsenic by the ferrihydrite, autoclave, or lime neutralization processes, or by production of copper arsenate. Harris (2000) states the following:

> By far the most popular approach is arsenical ferrihydrite, although possibly not always with the requisite level of understanding. Inco's CRED plant in Sudbury has been operating for close to thirty years, with no sign of ferrihydrite breakdown, or of arsenic release. As noted earlier, it is, however, well known that the incorporation of small amounts of cations and anions into the ferrihydrite matrix appreciably slows down any crystallization process to the formation of goethite/and or hematite, and hence the consequent release of adsorbed ions[54]. To all intents and purposes, it appears that recrystallization in these ferrihydrite materials in these situations is virtually non-existent. Certainly, the EPA regards the arsenical ferrihydrite process as the BDAT, and operations applying it correctly (molar Fe(III)/As(V) ratio >4) have not reported any contamination of local groundwater.

Drahota and Filippi (2009) present a review of secondary arsenic minerals in the environment (150 references). Important to this discussion is the authors' detailed presentation of information concerning the identification and presence of arsenic-bearing secondary solids in mining and industrial sites. This review is extensive and also includes the identification of calcium–iron–arsenate minerals in a wide variety of products such as "waste mine dumps, cyanidation tailings, naturally contaminated soils, stream bed sediments and altered rocks from arsenopyrite weathering." The authors suggest that further work needs to be performed to "clarify the formation conditions and stability of these minerals, in particular to evaluate their suitability for As sequestering under natural and lime/ferric precipitation conditions." The authors note the work of Demopoulos and colleagues on the synthesis, characterization, and solubility of calcium-containing compounds (Gomez et al. 2010; Bluteau et al. 2009; Jia and Demopoulos 2008). Building evidence shows that calcium Fe(III) As(V) compounds form during or with aging in lime precipitated arsenate removal systems or in pore waters saturated with gypsum. Several investigators have proposed that the presence of gypsum formed during ferric arsenate and/or ferrihydrite assists in maintaining long-term stability of the solids. Jia and Demopoulos (2008) have demonstrated that stability is enhanced by conducting arsenate removal, by ferric additions, using lime instead of sodium hydroxide. Bluteau et al. (2009) have continued work to determine the reason for the stabilizing effect by studying the dissolution of scorodite in gypsum-saturated solutions. Bluteau et al. and Gomez et al. have proposed that the mineral yukonite [$Ca_2Fe_3(AsO_4)_4(OH){:}12H_2O$] likely forms from poorly crystalline scorodite with aging time, especially at pH levels of 7 and higher. Yukonite has been identified in mine tailings by other investigators as presented by Drahota and Filippi (2009). In their review, Johnston and Singer (2007) have evaluated the characteristics of ferrous arsenate (or symplesite) formation, and they suggest the following:

> Symplesite or other ferrous arsenate solid phases could be a significant sink for As(V) and Fe(II) in Bangladesh and other countries with similar geochemical environments. Symplesite could also control As(V) solubility in extreme environments such as alkaline lakes or in brines used for regeneration of anion exchange resins.

McCloskey et al. (2014) have developed and implemented an industrial application using ferrous arsenate formation for the removal of arsenate concentrations from 100 mg/L to <25 μg/L.

Paktunc and Bruggeman (2010) concluded the following from their extensive study of phase transformations of arsenic-bearing solids:

> Industrial practice to stabilize arsenic in metallurgical circuits is to form precipitates having Fe/As molar ratios than 3 or 4. Despite its important practical implication, the meaning of this ratio in terms of controlling arsenic releases has remained unknown. As described above, the precipitates with different Fe/As ratios, variably referred to as ferric arsenate or arsenical ferrihydrite, are not composed of a single phase. Instead, they are mixtures of ferric arsenate and ferrihydrite. Following the precipitation of ferric arsenate from arsenic-rich solutions, ferrihydrite forms at pH 2 and above. Its formation drives the solution composition to under saturation with respect to ferric arsenate and promotes dissolution of ferric arsenate. With this, the ferrihydrite content would increase at the expense of ferric arsenate. The increasing relative abundance of ferrihydrite would impose control on the As concentration in solution by providing additional sites for arsenate adsorption Accordingly, formation of ferrihydrite coupled with ferric arsenate dissolution would be considered as an efficient process in terms of minimizing As release.

Because of the extensive research performed during the last decade, important influencing factors for the removal of arsenic from solution are known and well characterized. Therefore, after the specific solution composition and conditions that exist at a treatment site are known, appropriate and effective removal systems can be designed. However, careful evaluation of the storage site must be performed with regard to the local conditions and what the local conditions will be in the future so that environmentally safe storage systems can be implemented and maintained.

Selenium

Selenium frequently occurs in natural and industrial waters throughout the world. Industrial sources include metals production activities, coal and mountaintop coal mining, coal-fired power plant effluents, oil refinery effluents, and agriculture activities. Examples of selenium sources are presented in Table 8. Selenium is often associated with metal sulfide mining and coal mining because of its occurrence with pyrite and other metal sulfides.

The EPA promulgated rules for the BDATs to be used for the following listed and characteristic waste containing As and Se: K031, K084, K101, K102, As wastes (D004), Se wastes (D010), and P and U wastes (Rosengrant and Fargo 1990). The specified BDAT for treatment of effluent solutions is selenium adsorption on ferrihydrite. The problem with this BDAT is that selenium removal is effective if it is present as selenite. However, selenium is usually present in natural and wastewaters as selenate; therefore, ferrihydrite adsorption is seldom used as the preferred treatment option because reduction of selenate to selenite is chemically and kinetically difficult. However, if the selenium is present as selenite, then the use of ferrihydrite is recommended. In general, the valence of selenium that occurs in some industry sectors is summarized in Table 9 (EPA 2016).

Recent reviews detailing process technologies for removing selenium from wastewaters include EPA (2000), Golder Associates (2009), North American Metals Council (NAMC) (Sandy and DiSante 2010), and Santos et al. (2015). Many of the referenced technologies have only been demonstrated in a laboratory or small pilot-scale application. Data are often lacking for large-scale treatments; costs are seldom presented. The most comprehensive recent review of selenium removal technologies was conducted for the NAMC. Pilot- and full-scale treatment technologies are discussed for four industrial sectors: mining, agriculture, power generation, and oil and gas industries (Sandy and DiSante 2010):

> While these physical, chemical, and biological treatment technologies have the potential to remove selenium, there are very few technologies that have successfully and/or consistently removed selenium in water to less than 5 µg/L at any scale. There are still fewer technologies that have been demonstrated at full-scale to remove selenium to less than 5 µg/L, or have been in full-scale for sufficient time to determine the long-term feasibility of selenium removal technology. No single technology has been demonstrated at full-scale to cost-effectively remove selenium to less than 5 µg/L for waters associated with all industry sectors. Therefore, performance of the technology must be demonstrated on a case-specific basis.

Technologies that have been demonstrated at a full-scale level are included in Table 10, which is based on NAMC Table ES-1 (Sandy and DiSante 2010). More detail is available in their summary table, including key design considerations and estimated capital and operating costs. Extensive discussions are presented in their text covering individual technologies such as physical (membranes, filtration), chemical (precipitation and adsorption), and biological (adsorption by biomasses and biological precipitation of selenium) treatments. For further information on laboratory and pilot studies, refer to the referenced publication.

Technologies for treating metals and nonmetal species in mine waters and hydrometallurgical effluent solutions. The EPA has recently published a report to provide a reference guide for the treatment of mining-influenced water (EPA 2014b).

> This report provides an overview of select mining-influenced water (MIW) treatment technologies used or piloted as part of remediation efforts at mine sites. The report is intended to provide information on treatment technologies for MIW to federal, state and local regulators, site owners and operators, consultants, and other stakeholders. The technologies described in this report are applicable to treatment of water from both coal and hard-rock mine operations. The report provides short descriptions of treatment technologies and presents information on the contaminants treated, pre-treatment requirements, long-term maintenance needs, performance and costs. Sample sites illustrate considerations associated with

Table 8 Total selenium concentrations in different contaminated waters and wastewaters

Water/Wastewater	Se Concentration	Reference*
Agricultural drainage water	140–1,400 µg/L	Y.Q. Zhang et al. 2005
Mining wastewaters	3 µg/L to >12 mg/L	Wasewar et al. 2009a
Coal mining pond water	8.8–389 µg/L	Mar Systems 2012
Uranium mine discharge	1,600 µg/L	Twidwell et al. 2005
Gold mines wastewater	0.2–33 mg/L	Twidwell et al. 2005
Petrochemical industries wastewater	7.5–55.9 µg/L	Cassella et al. 2009
Flue gas desulfurization wastewaters	1–10 mg/L	Vance et al. 2009
	0.2–1.8 mg/L	Sonstegard et al. 2010
Lead smelter scrubber water	3–7 mg/L	McCloskey et al. 2008

Source: Santos et al. 2015

*References as cited in Santos et al. 2015.

Table 9 Presence of selenium in various wastewater sources

Selenium Form	Sources
Selenate	Agricultural irrigation drainage Treated oil refinery effluent Mountaintop coal mining/valley fill leachate Copper mining discharge
Selenite	Oil refinery effluent Fly ash disposal effluent Phosphate mining overburden leachate
Organoselenium	Treated agricultural drainage (in ponds or lagoons)

Source: Presser and Ohlendorf 1987, Zhang and Moore 1996, and Cutter and Diego-McGlone 1990, as cited in EPA 2016

selecting a technology. Website links and sources for more information on each topic are also included.

This report focuses on passive treatment methods, but also includes recently developed or not widely utilized active treatment systems and passive-active hybrid systems. The report does not include all traditional active technologies, such as lime precipitation or high-density sludge systems.

The technologies discussed in the preceding EPA report for active treatments (the technologies applicable to mineral processing and extractive metallurgical processes) include

Table 10 Technology summary for full-scale demonstration projects*

Technology	Brief Description	Advantages	Disadvantages
ABMet	System is a bioreactor that is an attached growth (comprised of a biofilm, or a layer of microorganisms that grow on the surface of a solid phase media) down flow granular active carbon bed filter.	• Commercially available technology that has been demonstrated to remove selenium to low levels in pilot-scale and full-scale applications. *This process is capable of achieving EPA guideline concentration of <5 µg/L for both selenate and selenite.* • Uses naturally occurring microbes and molasses-based nutrient feed to maintain biomass; *removes selenite and selenate.* • Biologically reduced elemental selenium is in an insoluble form as nanoparticles integral to the biological solids. • *Also removes nitrate, metals, mercury, and arsenic.*	• Potential need for pretreatment to remove suspended solids. • Large footprint required. • External carbon source is required if soluble influent organic content or chemical oxygen demand is insufficient. • Wasted biomass residuals contain elemental selenium that may be hazardous depending on toxicity characterization leach procedure (TCLP) results. • Biological residuals will need to be thickened and dewatered for landfill disposal.
Ferrihydrite adsorption or iron co-precipitation	Two-step physical adsorption process in which a ferric salt is added to the water source at proper conditions such that a ferric hydroxide and ferrihydrite precipitate results in concurrent adsorption of selenium on the surface; also known as iron co-precipitation.	• Widely implemented at full scale throughout the industry. • Established by EPA as the best demonstrated available technology of selenium (e.g., selenite) removal. • Relatively simple and low-cost chemical adsorption technology. • *Ferrihydrite is an adsorbent that can concurrently remove selenium and other species present in the wastewater such as arsenic, cadmium, copper, lead, and zinc.*	• Selenium removal not proven to low µg/L (less than 5 µg/L). • Produces relatively large quantities of sludge that may need to be disposed as a hazardous waste depending upon outcome of TCLP testing. • pH dependent with optimal conditions in the range of pH 4 to 6; *Fe/As mole ratio 3–4.* • Not able to remove selenite effectively. • Potential release of selenium from ferrihydrite residuals. *Long-term outdoor storage stability unknown.*
Ferrous hydroxide	A two-step reduction oxidation and physical adsorption process where ferrous iron is added resulting in the reduction of selenite to selenite and subsequent physical adsorption of co-precipitation of selenite by ferrihydrite or ferric hydroxide.	• Widely implemented at full scale throughout the industry. • Relatively simple and low-cost chemical adsorption technology.	• Selenium removal not proven to low µg/L (less than 5 µg/L). • Large quantities of sludge that may need to be disposed as a hazardous waste. • Reduction and subsequent adsorption is pH dependent with optimal conditions in the range of pH 8 to 9. *Huge excess of ferrous may be required.* • Not as effective at the reduction of selenate to selenite as zero-valent iron. • *Green rust may form in the presence of high-sulfate solutions, which may not be stable for outdoor storage.* • *Potential release of selenium from ferrihydrite residuals. Long-term outdoor storage stability unknown.*
Reverse osmosis	A membrane separation process that uses high pressure to force a solution through a membrane that retains the soluble selenium (e.g., selenite and selenate) and other dissolved salts less than 0.001 µm on the reject side of the membrane and allows the purified water to pass to the permeate side.	• Demonstrated at full scale to remove selenium (selenite or selenate) to <5 µg/L. • Can remove high levels of total dissolved solids (TDS), approximately 90%–98% removal. • Produces a high water quality with relatively high recoveries as a function of scale treatment. • Small space requirements, modular-type construction, and easy expansion. • Concentrates the selenium reducing the volume for ultimate reduction treatment.	• Higher capital cost to purchase, install, and operate than other membrane separation processes. • Requirements for pretreatment and chemical addition to reduce scaling or fouling. • Pressure, temperature, and pH requirements to meet membrane tolerances. • Frequent membrane monitoring and maintenance. • Requires treatment and disposal of the brine. • Permeate stream will require treatment (pH and TDS buffering) prior to discharge to receiving waters to meet aquatic toxicity test. • Operating issues will result from viscosity changes at extreme low and high temperatures.

Adapted from Sandy and DiSante 2010
*Italic has been added by the chapter author.

fluidized bed reactors, reverse osmosis, zero-valent iron, ferrihydrite adsorption, electrocoagulation, ion exchange, biological reduction, and ceramic microfiltration. Sixteen applications are discussed that include the topics technology description, constituent treated, long-term maintenance, system limitations, costs, and effectiveness.

Technologies that have been demonstrated at a full-scale level are included in Table 11, which is based on Appendix A from EPA 2014b. Additional information is available in the EPA (2014b) summary table, including entries for pilot and laboratory research. Additional descriptive information is presented for the topics, including operations, long-term maintenance, and costs. The appendix presents extensive discussions on individual technologies such as physical (membranes, filtration), chemical (precipitation and adsorption) and biological (adsorption by biomasses and biological precipitation of selenium) treatments. For further information on laboratory and pilot studies, refer to the referenced publication.

SUMMARY

All effluent wastewaters have their own characteristics, such as hazardous constituent concentration, associated constituents, physical properties (pH, oxidation–reduction potential, temperature, turbidity) and chemical properties (element speciation, associated ions speciation, biological oxygen demand, dissolved oxygen, etc.). Therefore, each effluent wastewater must be fully characterized before a treatment process can be selected, and a treatment process must be tailored for that specific water. The implementation of a successful technology depends on several variables that can have a great influence on which technology is chosen. A few examples are presented in the following examples.

Example 1: Adsorption Technology

Many influencing factors should be considered when choosing to use an adsorption technology; for example, the removal of arsenic from an effluent solution is dependent on the following:

- pH: Less than neutral is better.
- Fe/As loading ratio: The higher the ratio, the more effective the removal, but 3–4 is usually selected to limit excessive sludge formation.
- Species of arsenic present: Hydrated oxyanions are better adsorbed than arsenate anions.
- Valence of the arsenic specie present: Arsenates are favored over arsenites.
- Method of exposure to the adsorbent: Methods include addition of ferric to the solution or oxidation of ferrous in situ within the solution.
- Presence of other dissolved species: Sulfate, phosphate, and carbonate all compete with arsenic for adsorption sites.
- Presence of cations: Ferrihydrite has an affinity for cupric, nickel, and lead species so they complete for adsorption sites.

Storage disposal of the adsorbent/adsorbed specie product must be considered. Careful consideration must be given to the long-term stability of the product within the chosen repository. Consideration also must be given to the oxidation–reduction potential (will the waste matrix retain the constituent or constituents of interest?), pH (will the stability be maintained at a pH different than that used to form the product?), bacterial reduction (is the microbial environment suitable?), and comingling with other waste products (such as mixing with sulfide-bearing tailings). Many adsorbents have been and are presently being investigated for removal of specific constituents, and essentially the same considerations apply.

Example 2: Microbiological Technology

Two widely used industrial microbial-based bioreactor technologies are available: microbial-induced reduction (ABMet and BioSolve; Sandy and Disante 2010) and microbial-induced sulfide precipitation processes (Blumenstein et al. 2008). The microbial-induced reduction is via the establishment of a biofilm on activated carbon or plastic sponges to produce an anoxic/anaerobic reducing environment that locally transforms an oxyanion specie to its elemental state that is embedded within the biofilm matrix. The microbial-induced sulfide precipitation is via the b iological reduction of sulfate to sulfide with subsequent reaction with metallic and nonmetallic species to form sulfide solids.

Microbial-Induced Reduction

This process is conducted in a bioreactor where the conditions must be conducive for the establishment of biofilms (on various substrates) to form and to remain stable. Influencing factors include the following:

- Pretreatment of effluent wastewater to remove suspended solids
- Establishment of anoxic/anaerobic conditions
- Elevated temperature (10°C and 38°C)
- Sufficient concentration of oxyanions (selenates, selenites, nitrates) to supply oxygen during bacterial respiration
- Addition of an organic material as a carbon source to supply energy to the bacteria
- Residence time
- Circumneutral pH

The resulting exiting effluent water must be treated aerobically to increase dissolved oxygen and solubilize organic species. The constituent loaded biomass must be periodically stripped, dried, and disposed of in a proper repository. In general, the recovered biomass is considered nonhazardous.

Microbial-Induced Sulfide Precipitation

This process is conducted in a bioreactor or in large-scale passive systems for treatment of acid mine waters. Bioreactors have been operated effectively to remove arsenic, chromium, copper, lead, selenium, thallium, zinc, and uranium (EPA 2014b). The conditions must be conducive for the establishment of anoxic-reducing conditions sufficient for sulfate-reducing bacteria to form and to remain stable. Influencing factors include the following:

- Pretreatment of effluent wastewater to remove suspended solids
- Establishment of anoxic/anaerobic conditions
- Sulfate concentration (1–3 g/L required; sulfate additions may be required)
- An organic source to promote organism growth (usually acetic acid or ethanol)
- pH (usually 5–7 where metal sulfides are stable)
- Residence time (usually requires a few to several hours).

Table 11 Full-scale treatment of active technologies as applied to the removal of various constituents

Technology	Brief Description	Treated Constituents	Limitations	Effectiveness
Bioreactors (active systems)	Biochemical reactors (BCRs), or bioreactors, treat mining-influenced water (MIW) by using microorganisms to transform contaminants and to increase pH in the treated water. BCRs can be designed as open ponds and buried ponds or within tanks or even in trenches between mine waste and a surface water body. The most commonly used BCRs for treating MIW are operated anaerobically. They are also called "sulfate-reducing" bioreactors.	Selenium, cadmium, copper, nickel, lead, zinc, arsenic, chromium	The primary target metals for sulfate-reducing BCRs are those that precipitate as a metal sulfide at pH values between 5 and 7. Additionally, some active metals and metalloids such as arsenic, chromium, selenium, thallium, and uranium may undergo a change in oxidation state, which is more amenable to precipitation in a neutral anaerobic environment. The size of the system is a function of metal loading rate, which includes factors such as flow, metals concentration, and required retention time. For example, a higher metal loading rate necessitates a larger system that requires more retention time and, in most cases, a lower flow rate. BCRs are most effective in conditions where relatively steady flow rates and loading rates can be maintained and pH levels balanced to optimize overall treatment efficiency. Thus, flow equalization or pH adjustment may be required as a pretreatment step.	The use of active BCRs to treat MIW would allow for the treatment of relatively high flows in a small footprint. The use of active BCRs, with low energy and chemical input but which recycle (semipassively) like the rock-based system at the Leviathan mine (California, United States), is an attractive alternative. The amount of human intervention varies with the type of reactor. The use of an active BCR allows various design configurations because the MIW may be piped into different chambers or the sulfate reduction occurs in one cell while the precipitation occurs in another chamber. When it comes to sludge or precipitate removal, the main cell is not affected, so the system maintains its sulfide-generating capacity with little or no interruption in operation.
Biological reduction of selenium (BSeR and GE's ABMet)	This process uses specially developed biofilms that contain specific proprietary microorganisms in anaerobic bioreactors to reduce selenium (in the form of selenite and selenate) to elemental selenium. The end product is a fine precipitate of elemental selenium that is removed from the bioreactor with back flushing.	Selenium, mercury, complete nitrate removal and removal of a variety of other metals	Requires pre- and post-treatment steps to remove suspended solids, backwashing to prevent plugging and short-circuiting of flow, and temperature dependence.	Process can treat flow rates as low as 5 gpm and as high as 1,400 gpm, while achieving up to 99% removal rates and discharging a treated effluent containing 5 µg/L of selenium. Nitrate (typical feed of 10–25 mg/L), possible effluent: not detectable <1 mg/L. Selenium (typical feed of 10,000 µg/L), possible effluent: <0.005 µg/L. Mercury (typical feed of 5 µg/L), possible effluent: <0.012 µg/L.
EcoBond	EcoBond forms a chemical chain that binds with metal ions forming insoluble metal complexes, reducing bioavailability. It produces a reaction that proceeds at ambient temperatures and does not produce secondary waste streams or gases.	Arsenic, aluminum, cadmium, chromium, lead, mercury, selenium, zinc, radionuclides	The EcoBond was not effective at reducing zinc and copper. Since the chemicals are applied with water, the reactions and subsequent effectiveness are limited to the surfaces that can be contacted. This makes treatment at depth or in large stockpiles difficult, since the flow paths in mine waste are complex.	Reduced Al by 7%, Fe by 26%, Mn by 55%, Ni by 64%, and sulfate by 31%. Limited inhibition of copper and zinc. The maximum percent reduction of total metals from the EcoBond-treated plot was less than 50%.
Electrodialysis reversal (EDR)	Electrodialysis (ED) is an electrochemical separation process in which ions are transferred through ion exchange membranes by means of a direct current voltage. EDR is a variation on the ED process, which uses electrode polarity reversal to automatically clean membrane surfaces.	Arsenic, radium, nitrate, dissolved solids	Total dissolved solids (TDS) economical up to 8,000 mg/L, but often run at waters of 1,200 mg/L • pH: 2.0 to 11.0 • Iron (Fe^{+2}): 0.3 ppm • Mn (+2): 0.1 ppm • H_2S: up to 1 ppm	When combined with reverse osmosis at a South African mine, the system treats MIW with 5,000 mg/L TDS with high calcium sulfate content down to <40 mg/L TDS.
Ferrihydrite adsorption (iron co-precipitation)	Ferrihydrite adsorption is a two-step physical adsorption process in which a ferric salt is added to the water source at proper conditions such that a ferric hydroxide and ferrihydrite precipitate results in concurrent adsorption of selenium on the surface. Also known as iron co-precipitation.	Selenium, arsenic	Consistent removal to regulatory levels of selenium has not been proven. Potential release of selenium from ferrihydrite residuals. Gravity sedimentation may be required to separate iron solids and adsorbed metals from the water matrix. High operational costs are typical of chemical treatment. Ion exchange capacity for selenium can be greatly reduced by competing ions (e.g., sulfates, nitrates).	At the Garfield Wetlands-Kessler Springs site Salt Lake City, Utah (United States), water contained 1,950 µg/L selenium, primarily as selenite. Using an iron concentration of 4,800 mg/L, the mean effluent selenium concentration was 90 µg/L. The minimum reported selenium concentration was 35 µg/L. Selenium removal is not proven to less than 5 µg/L.

(continues)

Table 11 Full-scale treatment of active technologies as applied to the removal of various constituents (continued)

Technology	Brief Description	Treated Constituents	Limitations	Effectiveness
Chemical precipitation	Hydroxide precipitation: Raising the pH with the use of alkaline agents causes certain dissolved metals (e.g., cadmium, copper, iron, lead, manganese, and zinc) to precipitate as hydroxides. A polymer may be added to enhance flocculation, and the solution may be transferred to a clarifier to separate the solids from the cleaned overflow effluent. The resultant metal-hydroxide sludge extracted from the bottom of the clarifier usually contains a large percentage of bound water, limiting the potential for reuse, and is disposed of as a solid waste. The amount of sludge generated can be reduced by employing a high-density sludge (HDS) treatment technique. In HDS processes, the precipitated hydroxide sludge is recycled to a conditioning tank where it is mixed with the alkali reagent. The sludge/alkali slurry is then metered into the MIW to raise the pH and cause additional metal precipitation. This reconditioning of the sludge provides for precipitation sites for the dissolved metals to bond, increasing the overall density of the sludge in the clarifier underflow. Sulfide precipitation: The addition of a sulfide induces precipitation of dissolved metals as metal sulfides. This treatment is very effective for many metals, including zinc and cadmium. It is less effective for some contaminants such as manganese. pH adjustment may be required prior to the sulfide precipitation. Benefits of sulfide precipitation include a reduction over hydroxide precipitation of the quantity of sludge generated. The sludge is more easily reprocessed to recover the metals and may offset the cost of treatment.	Cadmium, copper, iron, lead, manganese, and zinc	Limitations: • High cost • Not applicable for all cases • Requires operation and maintenance • Requires power • May generates a waste product Depending on the volume of water necessary for treatment, the cost of this technology can be high. Metal hydroxide sludge typically contains water bound in the chemical matrix and results in a large volume of waste requiring disposal. Generation of large volumes of sludge may be problematic. Lined impoundments must be considered if there is potentially hazardous waste that can migrate from the sludge disposal site. The waste product may contain free liquids or fail testing by the toxicity characterization leaching procedure and may necessitate subsequent disposal at an approved hazardous waste facility, incurring more costs.	Chemical precipitation is a proven, large-scale technology that offers permanent results. It can be used solo or in conjunction with other treatment technologies. Chemical precipitation has demonstrated achievement of stringent water quality limits. The performance and results are specific to initial water quality and site limitations.
Reverse osmosis	Reverse osmosis is the pressure-driven separation through a semipermeable membrane that allows water to pass through while rejecting contaminants.	Metals, sulfate, selenium, arsenic	Requires high operating pressure. Not practical above 10,000 mg/L TDS. Requirements for pretreatment and chemical addition to reduce scaling/fouling. Reverse osmosis permeate steam will require treatment before discharge to receiving waters to meet aquatic toxicity test. Frequent membrane monitoring and maintenance. May require temperature control at low and high temperatures to minimize viscosity effects.	Kennecott's Bingham Canyon Water Treatment Plant (Utah, United States) has consistently seen permeate production efficiencies in the range of 71% to 72%. Demonstrated at full scale to remove selenium to <5 µg/L. Can remove 90%–98% TDS. A TDS removal efficiency of 98.5% was observed during pilot testing of the membranes tested.

Source: EPA 2014b

In general, the sulfide product must be considered hazardous, and if the product cannot be recycled to a system that treats metal sulfides, then the waste must follow hazardous waste handling, transport, and disposal regulations.

REFERENCES

AWWA (American Water Works Association). 1988a. B403-88. *AWWA Standard for Al Sulfate—Liquid, Ground, or Lump*. Denver, CO: AWWA.

AWWA (American Water Works Association). 1988b. B407-88. *AWWA Standard for Fe Chloride—Liquid, Ground, or Lump*. Denver, CO: AWWA.

Blumenstein, E.P., Volberding, J., and Gusek, J.J. 2008. Designing a biochemical reactor for selenium and thallium removal, from bench scale testing though pilot construction. Presented at the 2008 National Meeting of the American Society of Mining and Reclamation, Richmond, VA, June 14–19.

Bluteau, M., Becze, L., and Demopoulos, G.P. 2009. The dissolution of scorodite in gypsum-saturated waters: Evidence of Ca-Fe-AsO_4 mineral formation and its impact on arsenic retention. *Hydrometallurgy* 97:221–227.

Borden, R.K. 2011. Waste disposal and contamination management. In *SME Mining Engineering Handbook*, 3rd ed., Vol. 2. Edited by P. Darling. Littleton, CO: SME. pp. 1733–1752.

Borho, M., and Wilderer, P. 1996. Optimized removal of As(III) by adaptation of oxidation and precipitation processes to the filtration step. *Water Sci. Technol.* 34:25–31.

Brannon, J., and Patrick, W. 1987. Fixation, transformation and mobilization of As in sediments. *Environ. Sci. Technol.* 21:450–459.

Chatain, V., Sanchez, F., Bayard, R., Moszkowicz, P., and Gourdon, R. 2005a. Effect of experimentally induced reducing conditions on the mobility of Arsenic from a mining soil. *J. Hazard. Mater*. B122(1-2):119–128.

Chatain, V., Sanchez, F., Bayard, R., Moszkowicz, P., and Gourdon, R. 2005b. Effect of indigenous bacterial activity on Arsenic mobilization under anaerobic conditions. *Environ. Int.* 31(2):221–226.

Cornell, R.M., and Schwertmann, U. 1996. *The Iron Oxides, Structure, Properties, Reactions, Occurrence and Uses*. New York: Wiley-VCH.

Cornell, R.M., and Schwertmann, U. 2003. *The Iron Oxides, Structure, Properties, Reactions, Occurrence and Uses*, 2nd ed. New York: Wiley-VCH.

Darling, P., ed. 2011. *SME Mining Engineering Handbook*, 3rd ed., Vol. 2. Littleton, CO: SME.

Demopoulos, G.P. 2005. On the preparation and stability of scorodite. In *Proceedings of Arsenic Metallurgy*. Edited by R.G. Reddy and V. Ramachandran. Warrendale, PA: The Minerals, Metals & Materials Society. pp. 25–50.

Demopoulos, G.P. 2008. Advanced hydrometallurgical technology for the immobilization of arsenic. In *Proceedings of the Arsenic Stabilization Workshop*, Santiago, Chile, May.

Doyle, F.M. 2005. Teaching and learning environmental hydrometallurgy. *Hydrometallurgy* 79:1–14.

Drahota, P., and Filippi, M. 2009. Secondary arsenic minerals in the environment: A review. *Environ. Int.* 35:1243–1255.

DTSC (Department of Toxic Substances Control). 2015a. California Hazardous Waste Classification: Persistent and Bioaccumulative Toxic Substances. http://ccelearn.csus.edu/wasteclass/mod6/mod6_04.html. Accessed February 2016.

DTSC (Department of Toxic Substances Control). 2015b. California Hazardous Waste Classification: TCLP and WET Test Methods. http://ccelearn.csus.edu/wasteclass/mod6/mod6_05.html. Accessed February 2016.

EPA (U.S. Environmental Protection Agency). 1981a. Title 40 CFR 261.31: Hazardous Wastes from Non-Specific Sources. www.gpo.gov/fdsys/pkg/CFR-2012-title40-vol27/xml/CFR-2012-title40-vol27-sec261-32.xml.

EPA (U.S. Environmental Protection Agency). 1981b. Title 40 CFR 261.32: Hazardous Wastes from Specific Sources. www.gpo.gov/fdsys/pkg/CFR-2012-title40-vol27/xml/CFR-2012-title40-vol27-sec261-32.xml.

EPA (U.S. Environmental Protection Agency). 1981c. Title 40 CFR 261.33: Discarded Commercial Chemical Products. www.gpo.gov/fdsys/pkg/CFR-2012-title40-vol27/xml/CFR-2012-title40-vol27-sec261-32.xml

EPA (U.S. Environmental Protection Agency). 1984. Hazardous waste; Codification rule for the 1984 RCRA amendments. *Fed. Reg.* 52(230):45788–45799.

EPA (U.S. Environmental Protection Agency). 1992. Method 1311: Toxicity Characteristic Leaching Procedure. https://www.epa.gov/sites/production/files/2015-12/documents/1311.pdf. Accessed March 2018.

EPA (U.S. Environmental Protection Agency). 1998. Land Disposal Restrictions Phase IV: Final Rule Promulgating Treatment Standards for Metal Waste and Mineral Processing Wastes; Mineral Processing Secondary Materials and Bevill Exclusion Issues; Treatment Standards for Hazardous Soils, and Exclusion of Recycled Wood Preserving Wastewaters; Final Rule. Title 40 CFR Part 26. *Fed. Reg.*63(100):28556–28753.

EPA (U.S. Environmental Protection Agency). 2000. Issues identification and technology prioritization report: Selenium. *U.S. EPA Mine Waste Technology Program*. Vol. VII, MWTP-106, MSE Technology Applications. Washington, DC: EPA.

EPA (U.S. Environmental Protection Agency). 2005. *Introduction to Land Disposal Restrictions (40 CFR Part 268)*. EPA530-K-05-013. Washington, DC: EPA, Solid Waste and Emergency Response.

EPA (U.S. Environmental Protection Agency). 2009. *Identification and Listing of Hazardous Waste: Exclusions: Solid Wastes Which Are Not Hazardous Wastes—A User-Friendly Reference Document*. Title 40 CFR 261.4(b) and (b)(7). www.epa.gov/osw/hazard/wastetypes/wasteid/pdfs/rcra2614b-ref.pdf.

EPA (U.S. Environmental Protection Agency). 2012a. Characteristic of ignitability. Title 40 CFR 261.21. www.gpo.gov/fdsys/pkg/CFR-2012-title40-vol27/xml/CFR-2012-title40-vol27-sec261-21.xml.

EPA (U.S. Environmental Protection Agency). 2012b. Characteristic of corrosivity. Title 40 CFR 261.22. www.gpo.gov/fdsys/pkg/CFR-2012-title40-vol27/xml/CFR-2012-title40-vol27-sec261-22.xml.

EPA (U.S. Environmental Protection Agency). 2012c. Characteristic of reactivity. Title 40 CFR 261.23. www.gpo.gov/fdsys/pkg/CFR-2012-title40-vol27/xml/CFR-2012-title40-vol27-sec261-23.xml.

EPA (U.S. Environmental Protection Agency). 2012d. Characteristic of toxicity. Title 40 CFR 261.24. www.gpo.gov/fdsys/pkg/CFR-2012-title40-vol27/xml/CFR-2012-title40-vol27-sec261-24.xml.

EPA (U.S. Environmental Protection Agency). 2014a. *Hazardous Waste Treatment, Storage, and Disposal Facilities (TSDF) Regulations—A User-Friendly Reference Document for RCRA Subtitle C Permit Writers and Permittees*, Version 4. EPA 530-R-11-006. Washington, DC: EPA.

EPA (U.S. Environmental Protection Agency). 2014b. *Reference Guide to Treatment Technologies for Mining-Influenced Water.* EPA 542-R-14-001. Washington, DC: EPA, Office of Superfund Remediation and Technology Innovation.

EPA (U.S. Environmental Protection Agency). 2014. *RCRA Orientation Manual 2014*. Solid Waste and Emergency Response. EPA 530-F-11-003. Washington, DC: U.S. Environmental Protection Agency.

EPA (U.S. Environmental Protection Agency). 2015a. National Service Center for Environmental Publications. https://www.epa.gov/nscep. Washington, DC: EPA.

EPA (U.S. Environmental Protection Agency). 2015b. Universal Treatment Standards table. Title 40 CFR Part 268.48. www.ecfr.gov/cgi-bin/text-idx?SID=0f3ab17628 9fcb15faeb11b01fa823f7&mc=true&node=se40.27.268_148&rgn=div8.

EPA (U.S. Environmental Protection Agency). 2015c. Effluent guidelines. http://water.epa.gov/scitech/wastetech/guide/index.cfm.

EPA (U.S. Environmental Protection Agency). 2015d. Mineral mining and processing effluent guidelines. http://water.epa.gov/scitech/wastetech/guide/mineral/index.cfm.

EPA (U.S. Environmental Protection Agency). 2015e. National Pollutant Discharge Elimination System. http://water.epa.gov/polwaste/npdes/basics/. Accessed February 2016.

EPA (U.S. Environmental Protection Agency). 2016. *Aquatic Life Ambient Water Quality Criterion for Selenium—Freshwater 2016*. EPA 822-R-16-006. Washington, DC: EPA, Office of Water, Office of Science and Technology.

Erbs, J.J., Berquo, T.S., Reinsch, B.C., Lowry, G.V., Banerjee, S.K., and Penn, R.L. 2010. Reductive dissolution of arsenic-bearing ferrihydrite. *Geochim. Cosmochim. Acta* 74:3382–3395.

Fujita, T., Fujieda, S., Shinoda, K., and Suzuki, S. 2012. Environmental leaching characteristics of scorodite synthesized with Fe(II) iron. *Hydrometallurgy* 111-112:87–102.

Golder Associates. 2009. *Literature Review of Treatment Technologies to Remove Selenium from Mining Influenced Water*. Golden, CO: Golder Associates.

Gomez, M.A., Becze, L., Blyth, R.I.R., Cutler, J.N., and Demopoulos, G.P. 2010. Molecular and structural investigation of Yukonite (synthetic and natural) and its relation to Arseniosiderite. *Geochim. Cosmochim. Acta* 74(20):10.

Gomez, M.A., Becze, L., Cutler, J.N., and Demopoulos, G.P. 2011. Hydrothermal reaction chemistry and characterization of ferric arsenate phases precipitated from $Fe_2(SO_4)_3$–As_2O_5–H_2SO_4 solutions. *Hydrometallurgy* 107:74–90.

Harris, B. 2000. The removal and stabilization of arsenic from aqueous process solutions: Past, present, future. In *Minor Elements 2000: Processing and Environmental Aspects of As, Sb, Se, Te, and Bi.* Edited by C. Young. Littleton, CO: SME.

Hasen, J. 2015. What is the role of arsenic in the mining industry? miningfacts.org. www.miningfacts.org/Environment/What-is-the-role-of-arsenic-in-the-mining-industry/?terms=arsenic. Accessed February 2016.

Hohn, J.W., Twidwell, L.G., and Robins, R.G. 2006. A comparison of ferrihydrite and aluminum modified ferrihydrite for removing arsenate by co-precipitation. In *Proceedings of the Third Symposium on Iron Control in Hydrometallurgy*. Edited by J. Dutrizac and P. Riveros. Montreal, QC: Canadian Institute of Mining, Metallurgy and Petroleum. pp. 911–926.

Housman, V.E. 1999. The scope of the Bevill exclusion for mining wastes. 24 ELR 10657. *Environmental Law Reporter*. https://elr.info/sites/default/files/articles/24.10657.htm. Accessed June 2018.

INAP (International Network for Acid Prevention). 2014. *Global Acid Rock Drainage Guide (GARD Guide)*. Melbourne, Victoria: INAP. www.gardguide.com/index.php?title=Main_Page. Accessed February 2016.

Jambor, J., and Dutrizac, J.E. 1998. Occurrence and constitution of natural and synthetic ferrihydrite, a widespread iron oxyhydroxide. *Chem. Rev.* 98:2549–2585.

Jamieson, H.E., Walker, S.R., and Parsons, M.B. 2015. Review: Mineralogical characterization of mine wastes. *Appl. Geochem*. 57(June):85–105.

Jia, Y., and Demopoulos, G.P. 2008. Coprecipitation of arsenate with iron(III) in aqueous sulfate media: Effect of time, lime as base and co-ions on arsenic retention. *Water Res.* 42(93):661–668.

Johnston, R.B., and Singer, P.C. 2007. Solubility of symplesite (ferrous arsenate): Implications for reduced groundwaters and other geochemical environments. *Soil Sci. Soc. Am. J* 71(1):101–107.

Jones, C., Langner, H., and Anderson, K. 2000. Rates of microbially mediated arsenate reduction and solubilization. *Soil Sci. Soc. Am. J.* 64:600–608.

Kocar, B., Borch, T., and Fendorf, S. 2010. Arsenic repartitioning during biogenic sulfidization and transformation of ferrihydrite. *Geochim. Cosmochim. Acta* 74:980–994.

Kossoff, D., Dubbin, W.E., Alfredsson, M., Edwards, S.J., Macklin, M.G., and Hudson-Edwards, K.A. 2014. Review: Mine tailings dams: characteristics, failure, environmental impacts, and remediation. *Appl. Geochem.* 51:229–245.

Langer, H.W., and Inskeep, W.P. 2000. Microbial reduction of arsenate in the presence of ferrihydrite. *Environ. Sci. Technol*. 34(15):3131–3136.

Langmuir, D., Mahoney, J., and Rowson, J. 2006. Solubility products of amorphous ferric arsenate and crystalline scorodite ($FeAsO_4.2H_2O$) and their application to arsenic behavior in buried mine tailings. *Geochim. Cosmochim. Acta* 70:2942–2956.

Lapakko, K.A. 2015. Review: Preoperational assessment of solute release from waste rock at proposed mining operations. *Appl. Geochem*. 57(June):106–124.

Luther, L. 2013. Background on the implementation of the Bevill and Bentsen exclusions in the Resource Conservation and Recovery Act: EPA authorities to regulate "Special Wastes." Congressional Research Service (CRS) 7-5700. R43149. www.crs.gov.

McCloskey, J., Twidwell, L.G., Huang, H.H., and Goldstien, L. 2014. Fundamental procedures to evaluate and design industrial waste water metals treatment systems, case study discussions. In *Hydrometallurgy 2014: Proceedings of the Seventh International Symposium on Hydrometallurgy*. Westmount, QC: Canadian Institute of Mining, Metallurgy and Petroleum.

Nordstrom, D.K., Blowes, D.W., and Ptacek, C.J. 2015. Hydrogeochemistry and microbiology of mine drainage: An update. *Appl. Geochem.* 57(February).

Okibe, N., Koga, M., Sasaki, K., Hirajima, T., Heguri, S., and Asano, S. 2013. Simultaneous oxidation and immobilization of arsenite from refinery waste water by thermoacidophillic iron-oxidizing archaeon, *Acidianus brierleyi*. *Miner. Eng.* 48:26–134.

Okibe, N., Koga, M., Morishita, S., Tanaka, M., Heguri, S., and Asano, S. 2014. Microbial formation of crystalline scorodite for treatment of As(III)-bearing copper refinery process solution using *Acidianus brierleyi*. *Hydrometallurgy* 143:34–41.

Paktunc, D., and Bruggeman, K. 2010. Solubility of nanocrystalline scorodite and amorphous ferric arsenate: Implications for stabilization of arsenic in mine wastes. *Appl. Geochem.* 25:674–683.

Paktunc, D., Dutrizac, J.E., and Gertsmann, V. 2008. Synthetic and phase transformations involving scorodite, ferric arsenate and arsenical ferrihydrite: Implications for arsenic mobility. *Geochim. Cosmochim. Acta* 72:249–272.

Parbhakar-Fox, A., and Lottermoser, B.G. 2015. A critical review of acid rock drainage prediction methods and practices. *Miner. Eng.* 82(March).

Pederson, H.D., Postma, D., and Jakobsen, R. 2006. Release of Arsenic associated with the reduction and transformation of iron oxides. *Geochim. Cosmochim. Acta* 70:4116–4129.

Riveros, P.A., Dutrizac, J.E., and Spenser, P. 2001. Arsenic disposal practices in the metallurgical industry. *Can. Metall. Q.* 40(4):395–420.

Rosengrant, L., and Fargo, L. 1990. *Final Best Demonstrated Available Technology (BDAT) Background Document for K031, K084, K101, K102, As Wastes (D004), Se Wastes (D010), and P and U Wastes Containing As and Se Listing Constituents.* EPA/530/SW-90/059A. Washington, DC: U.S. Environmental Protection Agency.

Sandy, T., and DiSante, C. 2010. *Review of Available Technologies for the Removal of Selenium from Water.* Final Report by CH2MHill for the North American Metals Council (NAMC). www.namc.org/docs/00062756.PDF.

Santos, S., Ungureanu, G., Boaventura, R., and Botelho, C. 2015. Selenium contaminated waters: An overview of analytical methods, treatment options and recent advances in sorption methods. *Sci. Total Environ.* 521-522: 246–260.

Schwertmann, U., and Cornell, R.M. 2000. *Iron Oxides in the Laboratory*, 2nd ed. New York: Wiley-VCH.

Swash, P., Monhemius, A., and Schaekers, J. 2000. Solubilities of process residues from biological oxidation pretreatments of refractory gold ores. In *Minor Elements 2000: Processing and Environmental Aspects of As, Sb, Se, Te, and Bi.* Edited by C. Young. Littleton, CO: SME. pp. 115–122.

Twidwell, L.G. 2014a. *Guide to Recent Literature Focused on Scorodite Properties 2014*. Project Report. Butte, MT: Montana Tech, Center for Advanced Mineral and Metallurgical Processing.

Twidwell, L.G. 2014b. *A Guide to Recent Literature Focused on Scorodite for Containment of Arsenic.* Project Report. Butte, MT: Montana Tech, Center for Advanced Mineral and Metallurgical Processing.

Twidwell, L.G., and McCloskey, J.W. 2011. Removing arsenic from aqueous solution long-term product storage. *J. Miner. Met. Mater. Soc.* 63(8):94–110.

Verburg, R. 2011. Mitigating acid rock drainage. In *SME Mining Engineering Handbook*, 3rd ed., Vol. 2. Edited by P. Darling. pp. 1721–1732.

Weiss, N.L., ed. 1985. *SME Mineral Processing Handbook*. Littleton, CO: SME-AIME.

Welham, N., Malati, K., and Vukcevic, S. 2000. The stability of iron phases presently used for disposal from metallurgical systems-a review. *Miner. Eng.* 13(8-9):911–931.

Wilson, S.D., Kelly, W.R., and Holm, T.R. 2001. *Arsenic Removal in Water Treatment Facilities: Survey of Geochemical Factors and Pilot Plant Experiments*. mtac.sws.uiuc.edu/mtacdocs/finalreports/arsenictreatment MTACFinal Report.pdf. Accessed February 2016.

World Bank Group. 2007. *Environmental, Health, and Safety Guidelines: Base Metal Smelting and Refining*. Washington, DC: World Bank Group. http://documents.worldbank.org/curated/en/605891489653831342/text/113559-WP-ENGLISH-Smelting-and-Refining-PUBLIC.txt.

Xie, X., Ellis, A., Wang, Y., Xie, Z., Duan, M., and Su, C. 2009. Geochemistry of redox-sensitive elements and sulfur isotopes in the high Arsenic groundwater system of Datong Basin, China. *Sci. Total Environ.* 407(12):3823–3835.

Index

Note: *f.* indicates figure; *t.* indicates table.
Pages 1–1176 are found in Volume 1; pages 1177–2204 are found in Volume 2.